Standard Handbook of Petroleum & Natural Gas Engineering

Second Edition

Standard Handbook of Petroleum & Natural Gas Engineering

Second Edition

Editors

William C. Lyons, Ph.D., P.E.

Gary J. Plisga, B.S.

ELSEVIER

AMSTERDAM • BOSTON • HEIDELBERG • LONDON • NEW YORK • OXFORD
PARIS • SAN DIEGO • SAN FRANCISCO • SINGAPORE • SYDNEY • TOKYO

Gulf Professional Publishing is an imprint of Elsevier

G|P
P|

Gulf Professional Publishing is an imprint of Elsevier
200 Wheeler Road, Burlington, MA 01803, USA
Linacre House, Jordan Hill, Oxford OX2 8DP, UK

Library of Congress Cataloging-in-Publication Data

Standard handbook of petroleum & natural gas engineering.—2nd ed./
editors, William C. Lyons, Gary J. Plisga.
     p. cm.
 Includes bibliographical references and index.
 ISBN-13: 978-0-7506-7785-1   ISBN-10: 0-7506-7785-6
 1. Petroleum engineering. 2. Natural gas. I. Title: Standard handbook of
petroleum and natural gas engineering. II. Lyons, William C. III. Plisga, Gary J.
 TN870.S6233 2005
 665.5–dc22

                                                                2004056285

British Library Cataloguing-in-Publication Data
A catalogue record for this book is available from the British Library.

ISBN-13: 978-0-7506-7785-1
ISBN-10: 0-7506-7785-6

For information on all Gulf Professional Publishing
publications visit our Web site at www.gulfpp.com

10   9   8   7   6

Printed in the United States of America

Contents

Contributing Authors

Egill Abrahamsen
Weatherford International Limited
Houston, Texas

Chip Abrant
Weatherford International Limited
Houston, Texas

Bo Anderson
Weatherford International Limited
Houston, Texas

Robert P. Badrak
Weatherford International Limited
Houston, Texas

Frederick Beck
Consultant
Denver, Colorado

Susan Beck
Weatherford International Limited
Houston, Texas

Joe Berry
Varco Incorporated
Houston, Texas

Daniel Boone
Consultant in Petroleum Engineering
Houston, Texas

Gordon Bopp
Environmental Technology and Educational
    Services Company
Richland, Washington

Ronald Brimhall
Consultant
College Station, Texas

Ernie Brown
Schlumberger
Sugarland, Texas

Tom Carlson
Halliburton Energy Services Group
Houston, Texas

William X. Chavez, Jr.
New Mexico Institute of Mining and Technology
Socorro, New Mexico

Francesco Ciulla
Weatherford International Limited
Houston, Texas

Vern Cobb
Consultant

Robert Colpitts
Consultant in Geology and Geophysics
Las Vegas, Nevada

Robert B. Coolidge
Weatherford International Limited
Houston, Texas

Heru Danardatu
Schlumberger
Balikpapan, Indonesia

Tracy Darr van Reet
Chevron—retired
El Paso, Texas

Robert Desbrandes
Louisiana State University
Baton Rouge, Louisiana

Aimee Dobbs
Global Santa Fe
Houston, Texas

Patricia Duettra
Consultant in Applied Mathematics and
    Computer Analysis
Albuquerque, New Mexico

Ernie Dunn
Weatherford International Limited
Houston, Texas

Michael Economides
University of Houston
Houston, Texas

Jason Fasnacht
Boart Longyear
Salt Lake City, Utah

Joel Ferguson
Weatherford International Limited
Houston, Texas

Jerry W. Fisher
Weatherford International Limited
Houston, Texas

Robert Ford
Smith Bits International
Houston, Texas

Kazimierz Glowacki
Consultant in Energy and Environmental Engineering
Krakow, Poland

Bill Grubb
Weatherford International Limited
Houston, Texas

Mark Heironimus
El Paso Production
El Paso, Texas

Matthew Hill
Unocal Indonesia Company
Jakarta, Indonesia

John Hosford
Chevron Texaco
El Paso, Texas

Phillip Johnson
University of Alabama
Tuscaloosa, Alabama

Harald Jordan
BP America, Inc.
Farmington, New Mexico

Mike Juenke
Weatherford International Limited
Houston, Texas

Reza Kashmiri
International Lubrication and Fuel, Incorporated
Rio Rancho, New Mexico

William Kersting, MS
New Mexico State University
Las Cruces, New Mexico

Murty Kuntamukkla
Westinghouse Savannah River Company
Aiken, South Carolina

Doug LaBombard
Weatherford International Limited
Houston, Texas

Julius Langlinais
Louisiana State University
Baton Rouge, Louisiana

William Lyons
New Mexico Institute of Mining and Technology
Socorro, New Mexico

James Martens
Weatherford International Limited
Houston, Texas

F. David Martin
Consultant
Albuquerque, New Mexico

George McKown
Smith Services
Houston, Texas

David Mildren
Dril Tech Mission
Fort Worth, Texas

Mark Miller
Pathfinder
Texas

Richard J. Miller
Richard J. Miller and Associates, Incorporated
Huntington Beach, California

Stefan Miska
University of Tulsa
Tulsa, Oklahoma

Tom Morrow
Global Santa Fe
Houston, Texas

Abdul Mujeeb
Henkels & McCoy, Incorporated
Blue Bell, Pennsylvania

Bob Murphy
Weatherford International Limited
Houston, Texas

Tim Parker
Weatherford International Limited
Houston, Texas

Pudji Permadi
Institut Teknologi Bandung
Bandung, Indonesia

Jim Pipes
Weatherford International Limited
Houston, Texas

Gary J. Plisga
Consultant in Hydrocarbon Properties
Albuquerque, New Mexico

Floyd Preston
University of Kansas
Lawrence, Kansas

Toby Pugh
Weatherford International Limited
Houston, Texas

Carroll Rambin
Weatherford International Limited
Houston, Texas

Bharath N. Rao
President, Bhavya Technologies, Inc.

Richard S. Reilly
New Mexico Institute of Mining and Technology
Socorro, New Mexico

Cheryl Rofer
Tammoak Enterprises, LLC
Los Alamos, New Mexico

Chris Russell
Consultant in Environmental Engineering
Grand Junction, Colorado

Jorge H.B. Sampaio, Jr.
New Mexico Institute of Mining and Technology
Socorro, New Mexico

Eddie Scales
National Oil Well
Houston, Texas

Ron Schmidt
Weatherford International Limited
Houston, Texas

Ardeshir Shahraki
Dwight's Energy Data, Inc.
Richardson, Texas

Paul Singer
New Mexico Institute of Mining
    and Technology
Socorro, New Mexico

Jack Smith
Weatherford International Limited
Houston, Texas

Mark Trevithick
T&T Engineering Services, Inc.
Houston, Texas

Adrian Vuyk, Jr.
Weatherford International Limited
Houston, Texas

Bill Wamsley
Smith Bits International
Houston, Texas

Sue Weber
Consultant in Computer and Mathematics

Jack Wise
Sandia National Labs
Albuquerque, New Mexico

Andrzej Wojtanowicz
Louisiana State University
Baton Rouge, Louisiana

Preface

Several objectives guided the preparation of this second edition of the *Standard Handbook of Petroleum and Natural Gas Engineering*. As in the first edition, the first objective in this edition was to continue the effort to create for the worldwide petroleum and natural gas exploration and production industries an engineering handbook written in the spirit of the classic handbooks of the other important engineering disciplines. This new edition reflects the importance of these industries to the modern world economies and the importance of the engineers and technicians that serve these industries.

The second objective of this edition was to utilize, nearly exclusively, practicing engineers in industry to carry out the reviews, revisions, and any re-writes of first edition material for the new second edition. The third objective was, of course, to update the information of the old edition and to make the new edition more SI friendly. The fourth objective was to unite the previous two volumes of the first edition into a single volume that could be available in both book and CD form. The fifth and final objective of the handbook was to maintain and enhance the first edition objective of having a publication that could be read and understood by any up-to-date engineer or technician, regardless of discipline.

The initial chapters of the handbook set the tone by informing the reader of the common language and notation all engineering disciplines utilize. This common language and notation is used throughout the handbook (in nearly all cases consistent with Society of Petroleum Engineers publication practices). The 75 contributing authors have tried to avoid the jargon that has crept into petroleum engineering literature over the past few decades.

The specific petroleum engineering discipline chapters cover drilling and well completions, reservoir engineering, production engineering, and economics (with valuation and risk analysis). These chapters contain information, data, and example calculations directed toward practical situations that petroleum engineers often encounter. Also, these chapters reflect the growing role of natural gas in the world economies by integrating natural gas topics and related subjects throughout the volume.

The preparation of this new edition has taken approximately two years. Throughout the entire effort the authors have been steadfastly cooperative and supportive of the editors. In the preparation of the handbook the authors have used published information from both the American Petroleum Institute and the Society of Petroleum Engineers.

The authors and editors thank these two institutions for their cooperation. The authors and editors would also like to thank all the petroleum production and service company employees that have assisted in this project. Specifically, editors would like to express their great appreciation to the management and employees of Weatherford International Limited for providing direct support of this revision. The editors would also like to specifically thank management and employees of Burlington Resources Incorporated for their long term support of the students and faculty at the New Mexico Institute of Mining and Technology, and for their assistance in this book. These two companies have exhibited throughout the long preparation period exemplary vision regarding the potential value of this new edition to the industry.

In the detailed preparation of this new edition, the authors and editors would like to specifically thank Raven Gary. She started as an undergraduate student at New Mexico Institute of Mining and Technology in the fall of 2000. She is now a new BS graduate in petroleum engineering and is happily working in the industry. Raven Gary spent her last two years in college reviewing the incoming material from all the authors, checking outline organization, figure and table organization, and references, and communicating with the authors and Elsevier editors. Our deepest thanks go to Raven Gary. The authors and editors would also like to thank Phil Carmical and Andrea Sherman at Elsevier for their very competent preparation of the final manuscript of this new edition. We also thank all those at Elsevier for their support of this project over the past three years. The authors would like to thank Malone Mitchell III of Riata Energy for he and his company's continued support of our efforts to develop new petroleum engineering text and professional books for the continuing education and training of the indutry's vital engineers.

All the authors and editors know that this work is not perfect. But we also know that this handbook has to be written. Our greatest hope is that we have given those that will follow us in future editions of this handbook sound basic material to work with.

William C. Lyons, Ph.D., P.E.
Socorro, New Mexico

and

Gary J. Plisga, B.S.
Albuquerque, New Mexico

1

Mathematics

Contributing Authors
Patricia Duettra
Sue Schrader

Contents

1.1 GENERAL

See Reference 1 for additional information.

1.1.1 Sets and Functions

A *set* is a collection of distinct objects or *elements*. The intersection of two sets S and T is the set of elements which belong to S *and* which also belong to T. The *union* (or inclusive) of S and T is the set of all elements that belong to S or to T (or to both).

A *function* can be defined as a set of ordered pairs, denoted as (x, y) such that no two such pairs have the same first element. The element x is referred to as the independent variable, and the element y is referred to as the dependent variable. A function is established when a condition exists that determines y for each x, the condition usually being defined by an equation such as $y = f(x)$ [2].

References

1. *Mark's Standard Handbook for Mechanical Engineers*, 8th Edition, Baumeister, T., Avallone, E. A., and Baumeister III, T. (Eds.), McGraw-Hill, New York, 1978.

1.2 GEOMETRY

See References 1 and 2 for additional information.

1.2.1 Angles

Angles can be measured using degrees or with radian measure. Using the degree system of measurement, a circle has 360°, a straight line has 180°, and a right angle has 90°. The radian system of measurement uses the arc length of a unit circle cut off by the angle as the measurement of the angle. In this system, a circle is measured as 2π radians, a straight line is π radians and a right angle is $\pi/2$ radians. An angle A is defined as *acute* if $0° < A < 90°$, *right* if $A = 90°$, and *obtuse* if $90° < A < 180°$. Two angles are *complementary* if their sum is 90° or are *supplementary* if their sum is 180°. Angles are *congruent* if they have the same measurement in degrees and line segments are congruent if they have the same length. A *dihedral angle* is formed by two half-planes having the same edge, but not lying in the same plane. A *plane angle* is the intersection of a perpendicular plane with a dihedral angle.

1.2.2 Polygons

A *polygon* is a closed figure with at least three line segments that lies within a plane. A *regular polygon* is a polygon in which all sides and angles are congruent. Two polygons are *similar* if their corresponding angles are congruent and corresponding sides are proportional. A segment whose end points are two nonconsecutive vertices of a polygon is a *diagonal*. The *perimeter* is the sum of the lengths of the sides.

1.2.3 Triangles

A *triangle* is a three-sided polygon. The sum of the angles of a triangle is equal to 180°. An *equilateral* triangle has three sides that are the same length, an *isosceles* triangle has two sides that are the same length, and a *scalene* triangle has three sides of different lengths.

A *median* of a triangle is a line segment whose end points are a vertex and the midpoint of the opposite side. An *angle bisector* of a triangle is a median that lies on the ray bisecting an angle of the triangle. The *altitude* of a triangle is a perpendicular segment from a vertex to the opposite side.

Two triangles are congruent if one of the following is given (where S = side length and A = angle measurement): SSS, SAS, AAS, or ASA.

1.2.4 Quadrilaterals

A *quadrilateral* is a four-sided polygon.

A *trapezoid* has one pair of opposite parallel sides. A *parallelogram* has both pairs of opposite sides congruent and parallel. The opposite angles are then congruent, and adjacent angles are supplementary. The diagonals bisect each other and are congruent. A *rhombus* is a parallelogram whose four sides are congruent and whose diagonals are perpendicular to each other.

A *rectangle* is a parallelogram having four right angles; therefore, both pairs of opposite sides are congruent. A rectangle whose sides are all congruent is a *square*.

1.2.5 Circles and Spheres

If P is a point on a given plane and r is a positive number, the *circle* with center P and radius r is the set of all points of the plane whose distance from P is equal to r. The *sphere* with center P and radius r is the set of all points in space whose distance from P is equal to r. Two or more circles (or spheres) with the same P but different values of r are *concentric*.

A *chord* of a circle (or sphere) is a line segment whose end points lie on the circle (or sphere). A line which intersects the circle (or sphere) in two points is a *secant* of the circle (or sphere). A *diameter* of a circle (or sphere) is a chord containing the center, and a *radius* is a line segment from the center to a point on the circle (or sphere).

The intersection of a sphere with a plane through its center is called a *great circle*.

A line that intersects a circle at only one point is a *tangent* to the circle at that point. Every tangent is perpendicular to the radius drawn to the point of intersection. Spheres may have tangent lines or tangent planes.

Pi (π) is the universal ratio of the circumference of any circle to its diameter and is approximately equal to 3.14159. Therefore, the circumference of a circle is πd or $2\pi r$.

1.2.6 Arcs of Circles

A *central angle* of a circle is an angle whose vertex is the center of the circle. If P is the center and A and B are points, not on the same diameter, which lie on C (the circle), the *minor arc* AB is the union of A, B, and all points on C in the interior of <APB. The *major arc* is the union of A, B, and all points on C on the exterior of <APB. A and B are the end points of the arc and P is the center. If A and B are the end points of a diameter, the arc is a semicircle. A *sector* of a circle is a region bounded by two radii and an arc of the circle.

1.2.7 Concurrency

Two or more lines are *concurrent* if there is a single point that lies on all of them. The three altitudes of a triangle (if taken as lines, not segments) are always concurrent, and their point of concurrency is called the *orthocenter*. The angle bisectors of a triangle are concurrent at a point equidistant from their sides, and the medians are concurrent two thirds of the way along each median from the vertex to the opposite side. The point of concurrency of the medians is the *centroid*.

1.2.8 Similarity

Two figures with straight sides are *similar* if corresponding angles are congruent and the lengths of corresponding sides are in the same ratio. A line parallel to one side of a triangle divides the other two sides in proportion, producing a second triangle similar to the original one.

1.2.9 Prisms and Pyramids

A *prism* is a three-dimensional figure whose bases are any congruent and parallel polygons and whose sides are parallelograms. A *pyramid* is a solid with one base consisting of any polygon and with triangular sides meeting at a point in a plane parallel to the base.

Prisms and pyramids are described by their bases: a *triangular prism* has a triangular base, a *parallelpiped* is a prism whose base is a parallelogram and a *rectangular parallelpiped* is a right rectangular prism. A *cube* is a rectangular parallelpiped all of whose edges are congruent. A *triangular pyramid* has a triangular base, etc. A *circular cylinder* is a prism whose base is a circle and a *circular cone* is a pyramid whose base is a circle.

1.2.10 Coordinate Systems

Each point on a plane may be defined by a pair of numbers. The coordinate system is represented by a line X in the plane (the *x-axis*) and by a line Y (the *y-axis*) perpendicular to line X in the plane, constructed so that their intersection, the *origin*, is denoted by zero. Any point P on the plane can be described by its two coordinates, which form an ordered pair, so that $P(x_1, y_1)$ is a point whose location corresponds to the real numbers x and y on the x-axis and the y-axis.

If the coordinate system is extended into space, a third axis, the *z-axis*, perpendicular to the plane of the x_1 and y_1 axes, is needed to represent the third dimension coordinate defining a point $P(x_1, y_1, z_1)$. The z-axis intersects the x and y axes at their origin, zero. More than three dimensions are frequently dealt with mathematically but are difficult to visualize.

The *slope* m of a line segment in a plane with end points $P_1(x_1, y_1)$ and $P_2(x_2, y_2)$ is determined by the ratio of the change in the vertical (y) coordinates to the change in the horizontal (x) coordinates or

$$m = (y_2 - y_1)/(x_2 - x_1)$$

except that a vertical line segment (the change in x coordinates equal to zero) has no slope (i.e., m is undefined). A horizontal segment has a slope of zero. Two lines with the same slope are parallel and two lines whose slopes are negative reciprocals are perpendicular to each other.

Because the distance between two points $P_1(x_1, y_1)$ and $P_2(x_2, y_2)$ is the hypotenuse of a right triangle, the length (L) of the line segment P_1P_2 is equal to

$$L = \sqrt{(x_2 - x_1)^2 + (y_2 - y_1)^2}$$

1.2.11 Graphs

A *graph* is a set of points lying in a coordinate system and a graph of a condition (such as x = y + 2) is the set of all points that satisfy the condition. The graph of the *slope-intercept equation*, y = mx + b, is a straight line which passes through the point (0, b), where b is the y-intercept (x = 0) and m is the slope. The graph of the equation

$$(x - a)^2 + (y - b)^2 = r^2$$

is a circle with center (a, b) and radius r.

1.2.12 Vectors

A *vector* is described on a coordinate plane by a *directed segment* from its initial point to its terminal point. The directed segment represents the fact that every vector determines a magnitude and a direction. A vector **v** is not changed when moved around the plane, if its magnitude and angular orientation with respect to the x-axis is kept constant. The initial point of **v** may therefore be placed at the origin of the coordinate system and $\vec{v}$ may be denoted by

$$\vec{v} = \langle a, b \rangle$$

where a is the x-component and b is the y-component of the terminal point. The magnitude may then be determined by

the Pythagorean theorem

$$\mathbf{v} = \sqrt{a^2 + b^2}$$

For every pair of vectors (x_1, y_1) and (x_2, y_2), the *vector sum* is given by $(x_1 + x_2, y_1 + y_2)$. The *scalar product* of the vector P = (x, y) and a real number (a *scalar*) r is rP = (rx, ry). Also see the discussion of polar coordinates in the Section "Trigonometry" and Chapter 2, "Basic Mechanics."

1.2.13 Lengths and Areas of Plane Figures

For definitions of trigonometric functions, see "Trigonometry."

- *Right triangle* (Figure 1.2.1)

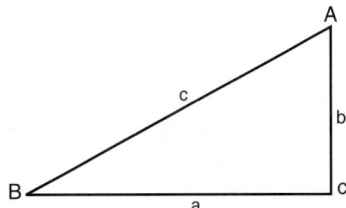

$$c^2 = a^2 + b^2 \text{(Pythagorean theorem)}$$
$$\text{area} = 1/2 \bullet ab = 1/2 \bullet a^2 \cot A$$
$$= 1/2 \bullet b^2 \tan A = 1/4 \bullet c^2 \sin 2A$$

- *Any triangle* (Figure 1.2.2)

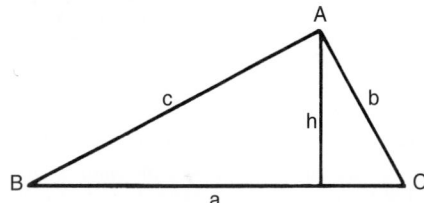

$$\text{area} = 1/2 \, \text{base} \bullet \text{altitude} = 1/2 \bullet ah = 1/2 \bullet ab \sin C$$
$$= \pm 1/2 \bullet \{(x_1y_2 - x_2y_1)$$
$$+ (x_2y_3 - x_3y_2)$$
$$+ (x_3y_1 - x_1y_3)\}$$

where (x_1, y_1), (x_2, y_2), (x_3, y_3) are coordinates of vertices.

- *Rectangle* (Figure 1.2.3)

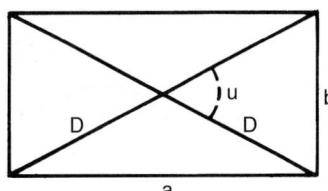

$$\text{area} = ab = 1/2 \bullet D^2 \sin u$$

where u = angle between diagonals D, D.

- *Parallelogram* (Figure 1.2.4)

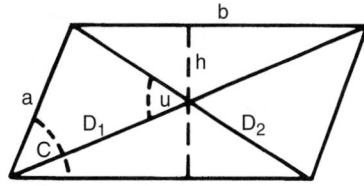

$$\text{area} = bh = ab \sin c = 1/2 \bullet D_1D_2 \sin u$$

where u = angle between diagonals D_1 and D_2.

- *Trapezoid* (Figure 1.2.5)

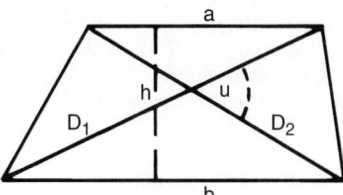

area $= 1/2 \bullet (a + b)h = 1/2 \bullet D_1 D_2 \sin u$
where u = angle between diagonals D_1 and D_2
and where bases a and b are parallel.

- *Any quadrilateral* (Figure 1.2.6)

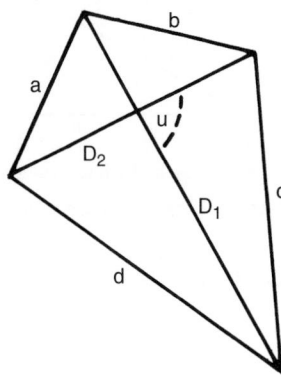

area $= 1/2 \bullet D_1 D_2 \sin u$
Note: $a^2 + b^2 + c^2 + d^2 = D_1^2 + D_2^2 + 4 m^2$
where m = distance between midpoints of D_1 and D_2.

- *Circles*
area $= \pi r^2 = 1/2 \bullet Cr = 1/4 \bullet Cd = 1/4 \bullet \pi d^2 = 0.785398 \ d^2$
where r = radius
    d = diameter
    C = circumference $= 2\pi r = \pi d$.

- *Annulus* (Figure 1.2.7)

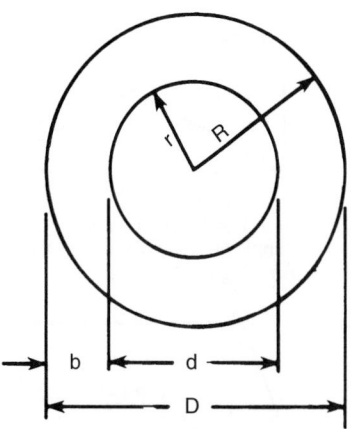

area $= \pi(R^2 - r^2) = \pi(D^2 - d^2)/4 = 2\pi R'b$
where R′ = mean radius $= 1/2 \bullet (R + r)$
b = R − r

- *Sector* (Figure 1.2.8)

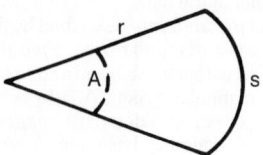

area $= 1/2 \bullet rs = \pi r^2 A/360° = 1/2 \bullet r^2 \text{ rad } A$
where rad A = radian measure of angle A
    s = length of arc = r rad A

- *Ellipse* (Figure 1.2.9)

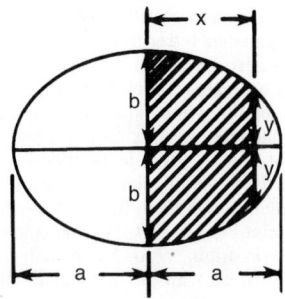

area of ellipse $= \pi ab$
area of shaded segment $= xy + ab \sin^{-1}(x/a)$
length of perimeter of ellipse $= \pi(a + b)K$,
where K $= (1 + 1/4 \bullet m^2 + 1/64 \bullet m^4 + 1/256 \bullet m^6 + \ldots)$
m $= (a - b)/(a + b)$

- *Hyperbola* (Figure 1.2.10)

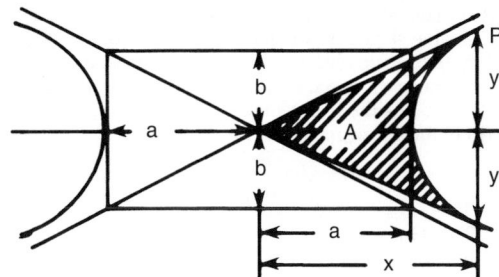

For any hyperbola,
shaded area A $= ab \bullet \ln[(x/a) + (y/b)]$
For an equilateral hyperbola (a = b),
area A $= a^2 \sinh^{-1}(y/a) = a^2 \cosh^{-1}(x/a)$
where x and y are coordinates of point P.

- *Parabola* (Figure 1.2.11)

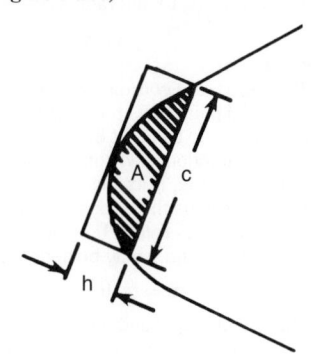

shaded area $A = 2/3 \bullet ch$

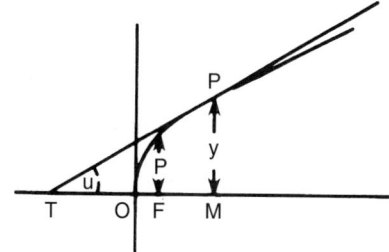

In Figure 1.2.12,
length of arc $OP = s = 1/2 \bullet PT + 1/2 \bullet p \bullet \ln [\cot(1/2 \bullet u)]$
Here c = any chord
$\quad p$ = semilatus rectum
$\quad PT$ = tangent at P
Note: $OT = OM = x$

1.2.14 Surfaces and Volumes of Solids
- *Regular prism* (Figure 1.2.13)

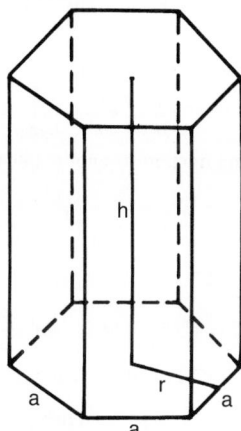

volume $= 1/2 \bullet nrah = Bh$
lateral area $= nah = Ph$
where n = number of sides
$\quad B$ = area of base
$\quad P$ = perimeter of base
- *Right circular cylinder* (Figure 1.2.14)

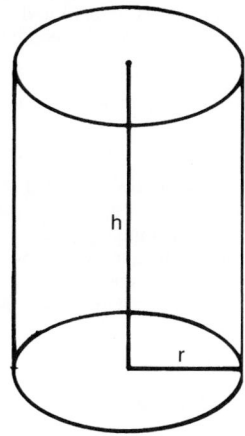

volume $= \pi r^2 h = Bh$
lateral area $= 2\pi rh = Ph$
where B = area of base
$\quad P$ = perimeter of base
- *Any prism or cylinder* (Figure 1.2.15)

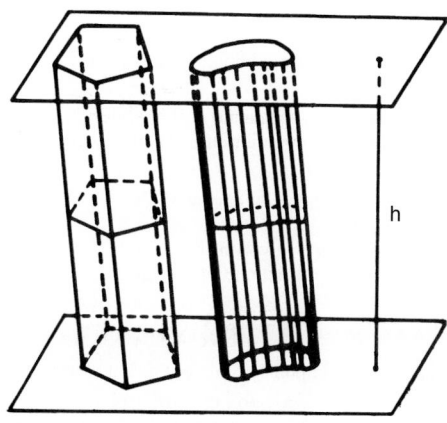

volume $= Bh = Nl$
lateral area $= Ql$
where l = length of an element or lateral edge
$\quad B$ = area of base
$\quad N$ = area of normal section
$\quad Q$ = perimeter of normal section
- *Hollow cylinder* (right and circular)
volume $= \pi h(R^2 - r^2) = \pi hb(D - b) = \pi hb(d + b) = \pi hbD' = \pi hb(R + r)$
where h = altitude
$r, R\ (d, D)$ = inner and outer radii (diameters)
b = thickness $= R - r$
D' = mean diam $= 1/2 \bullet (d + D) = D - b = d + b$
- *Sphere*
volume $= V = 4/3 \bullet \pi r^3 = 4.188790r^3 = 1/6 \bullet \pi d^3 = 0.523599d^3$
area $= A = 4\pi r^2 = \pi d^2$
where r = radius
$d = 2r$ = diameter $= \sqrt[3]{6V/\pi} = 1.24070\sqrt[3]{V}$
$\quad = \sqrt[3]{A/\pi} = 0.56419\sqrt{A}$
- *Hollow sphere,* or *spherical shell*
volume $= 4/3 \bullet \pi(R^3 - r^3) = 1/6 \bullet \pi(D^3 - d^3) = 4\pi R_1^2 t + 1/3 \bullet \pi t^3$
where R, r = outer and inner radii
D, d = outer and inner diameters
t = thickness $= R - r$
R_1 = mean radius $= 1/2 \bullet (R + r)$
- *Torus,* or *anchor ring* (Figure 1.2.16)

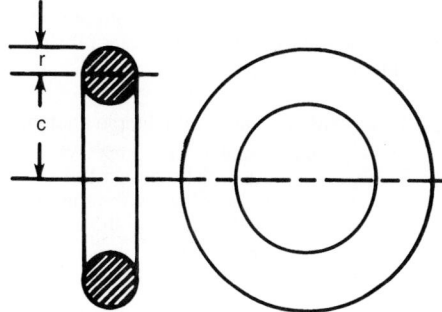

volume $= 2\pi^2 cr^2$
area $= 4\pi r^2 cr$ (proof by theorems of Pappus)

References
1. Moise, E. E., and Downs, Jr., F. L., *Geometry*, Addison Wesley, Melano Park, 1982.
2. Graening, J., *Geometry*, Charles E. Merrill, Columbus, 1980.

1.3 ALGEBRA

See Reference 1.3 for additional information.

1.3.1 Operator Precedence and Notation

Operations in an equation are performed in the following order of precedence:

1. Parenthesis and grouping symbols
2. Exponents
3. Multiplication or division (left to right)
4. Addition or subtraction (left to right)

For example:

$$a + b \bullet c - d^3/e$$

will be operated upon (calculated) as if it were written

$$a + (b \bullet c) - [(d^3)/e]$$

The symbol $|a|$ means "the absolute value of a," or the numerical value of a regardless of sign, so that

$$|-2| = |2| = 2$$

The n! means "n factorial" (where n is a whole number) and is the product of the whole numbers 1 to n inclusive, so that

$$4! = 1 \bullet 2 \bullet 3 \bullet 4 = 24$$

$$0! = 1 \text{ by definition}$$

The notation for the sum of any real numbers $a_1, a_2, \ldots, a_n$ is

$$\sum_{i=1}^{n} a_i$$

and for their product

$$\prod_{i=1}^{n} a_i$$

The notation "$x \infty y$" is read "x varies directly with y" or "x is directly proportional to y," meaning $x = ky$ where k is some constant. If $x \infty 1/y$, then x is inversely proportional to y and $x = k/y$.

1.3.2 Rules of Addition

$a + b = b + a$ (commutative property)
$(a + b) + c = a + (b + c)$ (associative property)
$a - (-b) = a + b$ and
$a - (x - y + z) = a - x + y - z$

(i.e., a minus sign preceding a pair of parentheses operates to reverse the signs of each term within if the parentheses are removed)

1.3.3 Rules of Multiplication and Simple Factoring

$a \bullet b = b \bullet a$ (commutative property)
$(ab)c = a(bc)$ (associative property)
$a(b + c) = ab + ac$ (distributive property)
$a(-b) = -ab$ and $-a(-b) = ab$
$(a + b)(a - b) = a^2 - b^2$
$(a + b)^2 = a^2 + 2ab + b^2$

and

$$(a - b)^2 = a^2 - 2ab + b^2$$
$$(a + b)^3 = a^3 + 3a^2 + 3ab^2 + b^3$$

and

$$(a - b)^3 = a^3 - 3a^2 + 3ab^2 - b^3$$

(For higher-order polynomials, see the "Binomial Theorem.") $a^n + b^n$ is factorable by $(a + b)$ if n is odd, and

$$a^3 + b^3 = (a + b)(a^2 - ab + b^2)$$

and $a^n - b^n$ is factorable by $(a - b)$, thus

$$a^n - b^n = (a - b)(a^{n-1} + a^{n-2}b + \ldots + ab^{n-2} + b^{n-1})$$

1.3.4 Fractions

The numerator and denominator of a fraction may be multiplied or divided by any quantity (other than zero) without altering the value of the fraction, so that, if $m \neq 0$,

$$\frac{ma + mb + mc}{mx + my} = \frac{a + b + c}{x + y}$$

To add fractions, transform each to a common denominator and add the numerators ($b, y \neq 0$):

$$\frac{a}{b} + \frac{x}{y} = \frac{ay}{by} + \frac{bx}{by} = \frac{ay + bx}{by}$$

To multiply fractions (denominators $\neq 0$):

$$\frac{a}{b} \bullet \frac{x}{y} = \frac{ax}{by}$$

$$\frac{a}{b} \bullet x = \frac{ax}{b}$$

$$\frac{a}{b} \bullet \frac{x}{y} \bullet \frac{c}{z} = \frac{axc}{byz}$$

To divide one fraction by another, invert the divisor and multiply:

$$\frac{a}{b} \div \frac{x}{y} = \frac{a}{b} \bullet \frac{y}{x} = \frac{ay}{bx}$$

1.3.5 Exponents

$$a^m \bullet a^n = a^{m+n} \quad \text{and} \quad a^m \div a^n = a^{m-n}$$

$$a^0 = 1 \ (a \neq 0) \quad \text{and} \quad a^1 = a$$

$$a^{-m} = 1/a^m$$

$$(a^m)^n = a^{mn}$$

$$a^{1/n} = \sqrt[n]{a} \quad \text{and} \quad a^{m/n} = \sqrt[n]{a^m}$$

$$(ab)^n = a^n b^n$$

$$(a/b)^n = a^n/b^n$$

Except in simple cases (square and cube roots), radical signs are replaced by fractional exponents. If n is odd,

$$\sqrt[n]{-a} = -\sqrt[n]{a}$$

but if n is even, the nth root of $-a$ is imaginary.

1.3.6 Logarithms

The logarithm of a positive number N is the power to which the base must be raised to produce N. So, $x = \log_b N$ means $b^x = N$. Logarithms to the base 10, frequently used in numerical computation, are called *common* or *denary logarithms*, and those to base e, used in theoretical work, are called *natural logarithms* and frequently notated as *ln*. In any case,

$$\log(ab) = \log a + \log b$$

$$\log(a/b) = \log a - \log b$$

$$\log(1/n) = -\log n$$

$$\log(a^n) = n \log a$$

$$\log_b(b) = 1, \text{ where b is either 10 or e}$$

$$\log 0 = -\infty$$

$$\log 1 = 0$$

$$\log_{10} e = M = 0.4342944819\ldots, \text{ so for conversion}$$

$$\log_{10} x = 0.4343 \log_e x$$

and since $1/M = 2.302585$, for conversion ($\ln = \log_e$)

$$\ln x = 2.3026 \log_{10} x$$

1.3.7 Binomial Theorem
Let

$$n_1 = n$$

$$n_2 = \frac{n(n-1)}{2!}$$

$$n_3 = \frac{n(n-1)(n-2)}{3!}$$

and so on. Then for any n, $|x| < 1$,

$$(1+x)^n = 1 + n_1 x + n_2 x^2 + n_3 x^3 + \ldots$$

If n is a positive integer, the system is valid without restriction on x and completes with the term $n_n x^n$.

Some of the more useful special cases follow [1]:

$$\sqrt{1+x} = (1+x)^{1/2} = 1 + \frac{1}{2}x - \frac{1}{8}x^2 + \frac{1}{16}x^3$$
$$- \frac{5}{128}x^4 + \ldots (|x| < 1)$$

$$\sqrt[3]{1+x} = (1+x)^{1/3} = 1 + \frac{1}{3}x - \frac{1}{9}x^2 - \frac{5}{81}x^3$$
$$- \frac{10}{243}x^4 + \ldots (|x| < 1)$$

$$\frac{1}{1+x} = (1+x)^{-1} = 1 - x + x^2 - x^3 + x^4 - \ldots (|x| < 1)$$

$$\frac{1}{\sqrt{1+x}} = (1+x)^{-1/2} = 1 - \frac{1}{2}x + \frac{3}{8}x^2 - \frac{5}{16}x^3$$
$$+ \frac{35}{128}x^4 - \ldots (|x| < 1)$$

$$\frac{1}{\sqrt[3]{1+x}} = (1+x)^{-1/3} = 1 - \frac{1}{3}x + \frac{2}{9}x^2 - \frac{14}{81}x^3$$
$$+ \frac{35}{243}x^4 - \ldots (|x| < 1)$$

$$\sqrt{(1+x)^3} = (1+x)^{3/2} = 1 - \frac{3}{2}x + \frac{3}{8}x^2 - \frac{1}{16}x^3$$
$$+ \frac{3}{128}x^4 - \ldots (|x| < 1)$$

$$\frac{1}{\sqrt{(1+x)^3}} = (1+x)^{-3/2} = 1 - \frac{3}{2}x + \frac{15}{8}x^2 - \frac{35}{16}x^3$$
$$+ \frac{315}{128}x^4 - \ldots (|x| < 1)$$

with corresponding formulas for $(1-x)^{1/2}$, etc., obtained by reversing the signs of the odd powers of x. Provided $|b| < |a|$:

$$(a+b)^n = a^n \left(1 + \frac{b}{a}\right)^n$$
$$= a^n + n_1 a^{n-1}b + n_2 a^{n-2}b^2 + n_3 a^{n-3}b^3 + \ldots$$

where n_1, n_2, etc., have the values given earlier.

1.3.8 Progressions
In an *arithmetic progression*, $(a, a+d, a+2d, a+3d, \ldots)$, each term is obtained from the preceding term by adding a constant difference, d. If n is the number of terms, the last term is $p = a + (n-1)d$, the "average" term is $1/2(a+p)$ and the sum of the terms is n times the average term or $s = n/2(a+p)$. The *arithmetic mean* between a and b is $(a+b)/2$.

In a *geometric progression*, $(a, ar, ar^2, ar^3, \ldots)$, each term is obtained from the preceding term by multiplying by a constant ratio, r. The nth term is ar^{n-1}, and the sum of the first n terms is $s = a(r^n - 1)/(r-1) = a(1 - r^n)/(1-r)$. If r is a fraction, r^n will approach zero as n increases and the sum of n terms will approach $a/(1-r)$ as a limit.

The *geometric mean*, also called the "mean proportional," between a and b is $\sqrt{ab}$. The *harmonic mean* between a and b is $2ab/(a+b)$.

1.3.9 Sums of the First n Natural Numbers
- To the first power:
$$1 + 2 + 3 + \ldots + (n-1) + n = n(n+1)/2$$
- To the second power (squared):
$$1^2 + 2^2 + \ldots + (n-1)^2 + n^2 = n(n+1)(2n+1)/6$$
- To the third power (cubed):
$$1^3 + 2^3 + \ldots + (n-1)^3 + n^3 = [n(n+1)/2]^2$$

1.3.10 Solution of Equations in One Unknown
Legitimate operations on equations include addition of any quantity to both sides, multiplication by any quantity of both sides (unless this would result in division by zero), raising both sides to any positive power (if $\pm$ is used for even roots) and taking the logarithm or the trigonometric functions of both sides.

Any *algebraic equation* may be written as a polynomial of nth degree in x of the form

$$a_0 x^n + a_1 x^{n-1} + a_2 x^{n-2} + \ldots + a_{n-1}x + a_n = 0$$

with, in general, n roots, some of which may be imaginary and some equal. If the polynomial can be factored in the form

$$(x-p)(x-q)(x-r)\ldots = 0$$

then p, q, r, ... are the roots of the equation. If $|x|$ is very large, the terms containing the lower powers of x are least important, while if $|x|$ is very small, the higher-order terms are least significant.

First-degree equations (*linear equations*) have the form

$$ax + b = c$$

with the solution $x = b - a$ and the root $b - a$.

Second-degree equations (*quadratic equations*) have the form

$$ax^2 + bx + c = 0$$

with the solution

$$x = \frac{-b \pm \sqrt{b^2 - 4ac}}{2a}$$

and the roots

$$\frac{-b + \sqrt{b^2 - 4ac}}{2a}$$

and

$$\frac{-b - \sqrt{b^2 - 4ac}}{2a}$$

The sum of the roots is $-b/a$ and their product is c/a.

Third-degree equations (*cubic equations*) have the form, after division by the coefficient of the highest-order term,

$$x^3 + ax^2 + bx + c = 0$$

with the solution

$$x_1^3 = Ax_1 + B$$

where $x_1 = x - a/3$
$A = 3(a/3)^2 - b$
$B = -2(a/3)^3 + b(a/3) - c$
Exponential equations are of the form

$$a^x = b$$

with the solution $x = (\log b)/(\log a)$ and the root $(\log b)/(\log a)$. The complete logarithm must be taken, not just the mantissa.

1.3.11 Solution of Systems of Simultaneous Equations

A set of *simultaneous equations* is a system of n equations in n unknowns. The solutions (if any) are the sets of values for the unknowns that satisfy all the equations in the system.

First-degree equations in 2 unknowns are of the form

$$a_1x_1 + b_1x_2 = c_1$$

$$a_2x_1 + b_2x_2 = c_2$$

The solution is found by multiplication of Equations 1.3.1 and 1.3.2 by some factors that will produce one term in each that will, upon addition of Equations 1.3.1 and 1.3.2, become zero. The resulting equation may then be rearranged to solve for the remaining unknown. For example, by multiplying Equation 1.3.1 by a_2 and Equation 1.3.2 by $-a_1$, adding Equation 1.3.1 and Equation 1.3.2 and rearranging their sum

$$x_2 = \frac{a_2c_1 - a_1c_2}{a_2b_1 - a_1b_2}$$

and by substitution in Equation 1.3.1:

$$x_1 = \frac{b_1c_2 - b_2c_1}{a_2b_1 - a_1b_2}$$

A set of *n first-degree equations in n unknowns* is solved in a similar fashion by multiplication and addition to eliminate $n - 1$ unknowns and then back substitution. *Second-degree equations in 2 unknowns* may be solved in the same way when two of the following are given: the product of the unknowns, their sum or difference, the sum of their squares. For further solutions, see "Numerical Methods."

1.3.12 Determinants

Determinants of the second order are of the following form and are evaluated as

$$\begin{vmatrix} a_1b_1 \\ a_2b_2 \end{vmatrix} = a_1b_2 - a_2b_1$$

and of the third order as

$$\begin{vmatrix} a_1b_1c_1 \\ a_2b_2c_2 \\ a_3b_3c_3 \end{vmatrix} = a_1 \begin{vmatrix} b_2c_2 \\ b_3c_3 \end{vmatrix} - a_2 \begin{vmatrix} b_1c_1 \\ b_3c_3 \end{vmatrix} + a_3 \begin{vmatrix} b_1c_1 \\ b_2c_2 \end{vmatrix}$$

and of higher orders, by the general rules as follows. To evaluate a determinant of the nth order, take the elements of the first column with alternate plus and minus signs and

form the sum of the products obtained by multiplying each of these elements by its corresponding *minor*. The minor corresponding to any element e_n is the determinant (of the next lowest order) obtained by striking out from the given determinant the row and column containing e_n.

Some of the general properties of determinants are

1. Columns may be changed to rows and rows to columns.
2. Interchanging two adjacent columns changes the sign of the result.
3. If two columns are equal or if one is a multiple of the other, the determinant is zero.
4. To multiply a determinant by any number m, multiply all elements of any one column by m.

Systems of simultaneous equations may be solved by the use of determinants using Cramer's rule. Although the example is a third-order system, larger systems may be solved by this method. If

$$a_1x + b_1y + c_1z = p_1$$
$$a_2x + b_2y + c_2z + p_2$$
$$a_3x + b_3y + c_3z = p_3$$

and if

$$D = \begin{vmatrix} a_1b_1c_1 \\ a_2b_2c_2 \\ a_3b_3c_3 \end{vmatrix} \neq 0$$

then

$$x = D_1/D$$
$$y = D_2/D$$
$$z = D_3/D$$

where

$$D_1 = \begin{vmatrix} p_1b_1c_1 \\ p_2b_2c_2 \\ p_3b_3c_3 \end{vmatrix}$$

$$D_2 = \begin{vmatrix} a_1p_1c_1 \\ a_2p_2c_2 \\ a_3p_3c_3 \end{vmatrix}$$

$$D_3 = \begin{vmatrix} a_1b_1p_1 \\ a_2b_2p_2 \\ a_3b_3p_3 \end{vmatrix}$$

References

1. Benice, D. D., *Precalculus Mathematics*, 2nd Edition, Prentice Hall, Englewood Cliffs, 1982.

1.4 TRIGONOMETRY

1.4.1 Directed Angles

If AB and AB' are any two rays with the same end point A, the directed angle $<BAB'$ is the ordered pair $(\overrightarrow{AB}, \overrightarrow{AB'})$. $\overrightarrow{AB}$ is the initial side of $<BAB'$ and $\overrightarrow{AB'}$ the terminal side. $<BAB' \neq <B'AB$ and any directed angle may be $\leq 0°$ or $\geq 180°$.

A directed angle may be thought of as an amount of rotation rather than a figure. If $\overrightarrow{AB}$ is considered the initial position of the ray, which is then rotated about its end point A to form $<BAB'$, $\overrightarrow{AB'}$ is its terminal position.

1.4.2 Basic Trigonometric Functions

A trigonometric function can be defined for an angle θ between $0°$ and $90°$ by using Figure 1.4.1.

1.4.3 Trigonometric Properties

$\sin \theta =$ opposite side/hypotenuse $= s_1/h$
$\cos \theta =$ adjacent side/hypotenuse $= s_2/h$
$\tan \theta =$ opposite side/adjacent side $= s_1/s_2 = \sin \theta/\cos \theta$

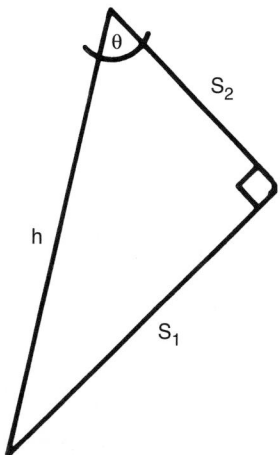

Figure 1.4.1 *Trigonometric functions of angles.*

and the reciprocals of the basic functions (where the function $\neq 0$)

$$\text{cotangent } \theta = \cot \theta = 1/\tan \theta = s_2/s_1$$
$$\text{secant } \theta = \sec \theta = 1/\cos \theta = h/s_2$$
$$\text{cosecant } \theta = \csc \theta = 1/\sin \theta = h/s_1$$

To reduce an angle to the first quadrant of the unit circle, that is, to a degree measure between $0°$ and $90°$, see Table 1.4.1. For function values at major angle values, see Tables 1.4.2 and 1.4.3. Relations between functions and the sum or difference of two functions are given in Table 1.4.4. Generally, there will be two angles between $0°$ and $360°$ that correspond to the value of a function.

The trigonometric functions sine and cosine can be defined for any real number by using the radian measure of the angle as described in the section on angles. The tangent function is defined on every real number except for places where cosine is zero.

1.4.4 Graphs of Trigonometric Functions
Graphs of the sine and cosine functions are identical in shape and periodic with a period of $360°$. The sine function graph

translated $\pm\ 90°$ along the x-axis produces the graph of the cosine function. The graph of the tangent function is discontinuous when the value of $\tan \theta$ is undefined, that is, at odd multiples of $90°$ $(\ldots, 90°, 270°, \ldots)$. For abbreviated graphs of the sine, cosine, and tangent functions, see Figure 1.4.2.

1.4.5 Inverse Trigonometric Functions
The inverse sine of x (also referred to as the arc sine of x), denoted by $\sin^{-1}x$, is the principal angle whose sine is x, that is,

$$y = \sin^{-1} x \text{ means } \sin y = x$$

Inverse functions $\cos^{-1}x$ and $\tan^{-1}x$ also exist for the cosine of y and the tangent of y. The principal angle for $\sin^{-1}x$ and $\tan^{-1}x$ is an angle a, where $-90° < a < 90°$, and for $\cos^{-1}x$, $0° < a < 180°$.

1.4.6 Solution of Plane Triangles
The solution of any part of a plane triangle is determined in general by any other three parts given by one of the following groups, where S is the length of a side and A is the degree measure of an angle:

- AAS
- SAS
- SSS

The fourth group, two sides and the angle opposite one of them, is ambiguous since it may give zero, one, or two solutions. Given an example triangle with sides a, b, and c and angles A, B, and C (A being opposite a, etc., and $A + B + C = 180°$), the fundamental laws relating to the solution of triangles are

1. Law of sines: $a/(\sin A) = b/(\sin B) = c/(\sin C)$
2. Law of cosines: $c^2 = a^2 + b^2 - 2ab \cos C$

1.4.7 Hyperbolic Functions
The *hyperbolic sine, hyperbolic cosine*, etc., of any number x are functions related to the exponential function e^x. Their definitions and properties are very similar to the trigonometric functions and are given in Table 1.4.5.

Table 1.4.1 *Angle Reduction to First Quadrant*

If	$90° < x < 180°$	$180° < x < 270°$	$270° < x < 360°$
$\sin x =$	$+\cos(x - 90°)$	$-\sin(x - 180°)$	$-\cos(x - 270°)$
$\cos x =$	$-\sin(x - 90°)$	$-\cos(x - 180°)$	$+\sin(x - 270°)$
$\tan x =$	$-\cot(x - 90°)$	$+\tan(x - 180°)$	$-\cot(x - 270°)$
$\csc x =$	$+\sec(x - 90°)$	$-\csc(x - 180°)$	$-\sec(x - 270°)$
$\sec x =$	$-\csc(x - 90°)$	$-\sec(x - 180°)$	$+\csc(x - 270°)$
$\cot x =$	$-\tan(x - 90°)$	$+\cot(x - 180°)$	$-\tan(x - 270°)$

Table 1.4.3 *Trigonometric Function Values at Major Angle Values*

Values at	$30°$	$45°$	$60°$
$\sin x$	$1/2$	$1/2\sqrt{2}$	$1/2\sqrt{3}$
$\cos x$	$1/2\sqrt{3}$	$1/2\sqrt{2}$	$1/2$
$\tan x$	$1/3\sqrt{3}$	1	$\sqrt{3}$
$\csc x$	2	$\sqrt{2}$	$2/3\sqrt{3}$
$\sec x$	$2/3\sqrt{3}$	$\sqrt{2}$	2
$\cot x$	$\sqrt{3}$	1	$1/3\sqrt{3}$

Table 1.4.2 *Trigonometric Function Values by Quadrant*

If	$0° < x < 90°$	$90° < x < 180°$	$180° < x < 270°$	$270° < x < 360°$
$\sin x$	$+0$ to $+1$	$+1$ to $+0$	-0 to -1	-1 to -0
$\cos x$	$+1$ to $+0$	-0 to -1	-1 to -0	$+0$ to $+1$
$\tan x$	$+0$ to $+\infty$	$-\infty$ to -0	$+0$ to $+\infty$	$-\infty$ to -0
$\csc x$	$+\infty$ to $+1$	$+1$ to $+\infty$	$-\infty$ to -1	-1 to $-\infty$
$\sec x$	$+1$ to $+\infty$	$-\infty$ to -1	-1 to $-\infty$	$+\infty$ to $+1$
$\cot x$	$+\infty$ to $+0$	-0 to $-\infty$	$+\infty$ to $+0$	-0 to $-\infty$

Table 1.4.4 *Relations Between Trigonometric Functions of Angles*

Single Angle

$\sin^2 x + \cos^2 x = 1$

$\tan x = (\sin x)/(\cos x)$

$\cot x = 1/(\tan x)$

$1 + \tan^2 x = \sec^2 x$

$1 + \cot^2 x = \csc^2 x$

$\sin(-x) = -\sin x, \cos(-x) = \cos x, \tan(-x) = -\tan x$

Two Angles

$\sin(x + y) = \sin x \cos y + \cos x \sin y$

$\sin(x - y) = \sin x \cos y - \cos x \sin y$

$\cos(x + y) = \cos x \cos y - \sin x \sin y$

$\cos(x - y) = \cos x \cos y + \sin x \sin y$

$\tan(x + y) = (\tan x + \tan y)/(1 - \tan x \tan y)$

$\tan(x - y) = (\tan x - \tan y)/(1 + \tan x \tan y)$

$\cot(x + y) = (\cot x \cot y - 1)/(\cot y + \cot x)$

$\cot(x - y) = (\cot x \cot y + 1)/(\cot y - \cot x)$

$\sin x + \sin y = 2 \sin[1/2(x + y)] \cos[1/2(x - y)]$

$\sin x - \sin y = 2 \cos[1/2(x + y)] \sin[1/2(x - y)]$

$\cos x + \cos y = 2 \cos[1/2(x + y)] \cos[1/2(x - y)]$

$\cos x - \cos y = -2 \sin[1/2(x + y)] \sin[1/2(x - y)]$

$\tan x + \tan y = [\sin(x + y)]/[\cos x \cos y]$

$\tan x - \tan y = [\sin(x - y)]/[\cos x \cos y]$

$\cot x + \cot y = [\sin(x + y)]/[\sin x \sin y]$

$\cot x - \cot y = [\sin(y - x)]/[\sin x \sin y]$

$\sin^2 x - \sin^2 y = \cos^2 y - \cos^2 x$

$\qquad = \sin(x + y) \sin(x - y)$

$\cos^2 x - \sin^2 y = \cos^2 y - \sin^2 x$

$\qquad = \cos(x + y) \cos(x - y)$

$\sin(45° + x) = \cos(45° - x), \tan(45° + x) = \cot(45° - x)$

$\sin(45° - x) = \cos(45° + x), \tan(45° - x) = \cot(45° + x)$

Multiple and Half Angles

$\tan 2x = (2 \tan x)/(1 - \tan^2 x)$

$\cot 2x = (\cot^2 x - 1)/(2 \cot x)$

$\sin(nx) = n \sin x \cos^{n-1} x - (n)_3 \sin^3 x \cos^{n-3} x$

$\qquad + (n)_5 \sin^5 x \cos^{n-5} x - \dots$

$\cos(nx) = \cos^n x - (n)_2 \sin^2 x \cos^{n-2} x$

$\qquad + (n)_4 \sin^4 x \cos^{n-4} x - \dots$

(Note: $(n)_2, \dots$ are the binomial coefficients)

$\sin(x/2) = \pm\sqrt{1/2(1 - \cos x)}$

$\cos(x/2) = \pm\sqrt{1/2(1 + \cos x)}$

$\tan(x/2) = (\sin x)/(1 + \cos x) = \pm\sqrt{(1 - \cos x)/(1 + \cos x)}$

Three Angles Whose Sum = 180°

$\sin A + \sin B + \sin C = 4 \cos(A/2) \cos(B/2) \cos(C/2)$

$\cos A + \cos B + \cos C = 4 \sin(A/2) \sin(B/2) \sin(C/2) + 1$

$\sin A + \sin B - \sin C = 4 \sin(A/2) \sin(B/2) \cos(C/2)$

$\cos A + \cos B - \cos C = 4 \cos(A/2) \cos(B/2) \sin(C/2) - 1$

$\sin^2 A + \sin^2 B + \sin^2 C = 2 \cos A \cos B \cos C + 2$

$\sin^2 A + \sin^2 B - \sin^2 C = 2 \sin A \sin B \cos C$

$\tan A + \tan B + \tan C = \tan A \tan B \tan C$

$\cot(A/2) + \cot(B/2) + \cot(C/2)$

$\qquad = \cot(A/2) \cot(B/2) \cot(C/2)$

$\sin 2A + \sin 2B + \sin 2C = 4 \sin A \sin B \sin C$

$\sin 2A + \sin 2B - \sin 2C = 4 \cos A \cos B \sin C$

The *inverse hyperbolic functions*, $\sinh^{-1} x$, etc., are related to the logarithmic functions and are particularly useful in integral calculus. These relationships may be defined for real numbers x and y as

$$\sinh^{-1}(x/y) = \ln(x + \sqrt{x^2 + y^2}) - \ln y$$
$$\cosh^{-1}(x/y) = \ln(x + \sqrt{x^2 - y^2}) - \ln y$$
$$\tanh^{-1}(x/y) = 1/2 \bullet \ln[(y + x)/(y - x)]$$
$$\coth^{-1}(x/y) = 1/2 \bullet \ln[(x + y)/(x - y)]$$

1.4.8 Polar Coordinate System

The *polar coordinate system* describes the location of a point (denoted as $[r, \theta]$) in a plane by specifying a distance r and an angle θ from the origin of the system. There are several relationships between polar and rectangular coordinates, diagrammed in Figure 1.4.3. From the Pythagorean theorem

$$r = \pm\sqrt{x^2 + y^2}$$

Also

$$\sin \theta = y/r \quad \text{or} \quad y = r \sin \theta$$
$$\cos \theta = x/r \quad \text{or} \quad x = r \cos \theta$$
$$\tan \theta = y/x \quad \text{or} \quad \theta = \tan^{-1}(y/x)$$

To convert rectangular coordinates to polar coordinates, given the point (x, y), using the Pythagorean theorem and the preceding equations.

$$[r, \theta] = \left[\sqrt{x^2 + y^2}, \tan^{-1}(y/x)\right]$$

To convert polar to rectangular coordinates, given the point $[r, \theta]$:

$$(x, y) = [r \cos \theta, r \sin \theta]$$

For graphic purposes, the polar plane is usually drawn as a series of concentric circles with the center at the origin and radii 1, 2, 3, $\dots$ Rays from the center are drawn at 0°, 15°, 30°, $\dots$, 360° or $0, \pi/12, \pi/6, \pi/4, \dots, 2\pi$ radians. The origin is called the *pole*, and points $[r, \theta]$ are plotted by moving a positive or negative distance r horizontally from the pole, and through an angle θ from the horizontal. See Figure 1.4.4 with θ given in radians as used in calculus. Also note that

$$[r, \theta] = [-r, \theta + \pi]$$

1.5 DIFFERENTIAL AND INTEGRAL CALCULUS

See References 1–4 for additional information.

1.5.1 Derivatives

Geometrically, the derivative of $y = f(x)$ at any value x_n is the slope of a tangent line T intersecting the curve at the point $P(x, y)$. Two conditions applying to differentiation (the process of determining the derivatives of a function) are

1. The primary (necessary and sufficient) condition is that

$$\lim_{\Delta x \to 0} \frac{\Delta y}{\Delta x}$$

exists and is independent of the way in which $\Delta x \to 0$

2. A secondary (necessary, not sufficient) condition is that

$$\lim_{\Delta x \to 0} f(x + \Delta x) = f(x)$$

A short table of derivatives will be found in Table 1.5.1.

1.5.2 Higher-Order Derivatives

The *second derivative* of a function $y = f(x)$, denoted $f''(x)$ or d^2y/dx^2 is the derivative of $f'(x)$ and the *third derivative*, $f'''(x)$ is the derivative of $f''(x)$. Geometrically, in terms of $f(x)$: if $f''(x) > 0$ then $f(x)$ is concave upwardly, if $f''(x) < 0$ then $f(x)$ is concave downwardly.

1.5.3 Partial Derivatives

If $u = f(x, y, \dots)$ is a function of two or more variables, the *partial derivative* of u with respect to x, $f_x(x, y, \dots)$ or $\partial u/\partial x$, may be formed by assuming x to be the independent variable and holding $(y, \dots)$ as constants. In a similar manner, $f_y(x, y, \dots)$ or $\partial u/\partial y$ may be formed by holding $(x, \dots)$ as constants. Second-order partial derivatives of $f(x, y)$ are denoted by the manner of their formation as f_{xx},

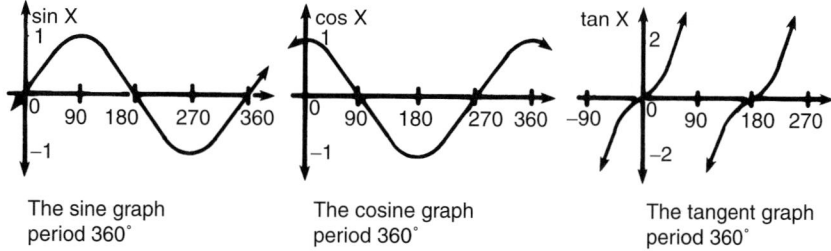

Figure 1.4.2 *Graphs of the trigonometric functions.*

Table 1.4.5 *Hyperbolic Functions*

$\sinh x = 1/2(e^x - e^{-x})$
$\cosh x = 1/2(e^x + e^{-x})$
$\tanh x = \sinh x/\cosh x$
$\operatorname{csch} x = 1/\sinh x$
$\operatorname{sech} x = 1/\cosh x$
$\coth x = 1/\tanh x$

$\sinh(-x) = -\sinh x$
$\cosh(-x) = \cosh x$
$\tanh(-x) = -\tanh x$

$\cosh^2 x - \sinh^2 x = 1$
$1 - \tanh^2 x = \operatorname{sech}^2 x$
$1 - \coth^2 x = -\operatorname{csch}^2 x$

$\sinh(x \pm y) = \sinh x \cosh y \pm \cosh x \sinh y$
$\cosh(x \pm y) = \cosh x \cosh y \pm \sinh x \sinh y$
$\tanh(x \pm y) = (\tanh x \pm \tanh y)/(1 \pm \tanh x \tanh y)$
$\sinh 2x = 2 \sinh x \cosh x$
$\cosh 2x = \cosh^2 x + \sinh^2 x$
$\tanh 2x = (2 \tanh x)/(1 + \tanh^2 x)$
$\sinh(x/2) = \sqrt{1/2(\cosh x - 1)}$
$\cosh(x/2) = \sqrt{1/2(\cosh x + 1)}$
$\tanh(x/2) = (\cosh x - 1)/(\sinh x) = (\sinh x)/(\cosh x + 1)$

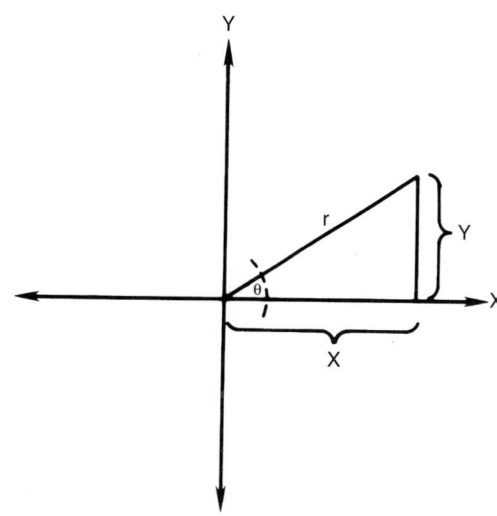

Figure 1.4.3 *Polar coordinates.*

f_{xy} (equal to f_{yx}), f_{yy} or as $\partial^2 u/\partial x^2$, $\partial^2 u/\partial x \partial y$, $\partial^2 u/\partial y^2$, and the higher-order partial derivatives are likewise formed.

Implicit functions (i.e., $f(x, y) = 0$) may be solved by the formula

$$\frac{dy}{dx} = -\frac{f_x}{f_y}$$

at the point in question.

1.5.4 Maxima and Minima

A *critical point* on a curve $y = f(x)$ is a point where $y' = 0$, that is, where the tangent to the curve is horizontal. A critical value of x is therefore a value such that $f'(x) = 0$. All roots of the equation $f'(x) = 0$ are critical values of x, and the corresponding values of y are the critical values of the function.

A function $f(x)$ has a *relative maximum* at $x = a$ if $f(x) < f(a)$ for all values of x (except a) in some open interval containing a and a *relative minimum* at $x = b$ if $f(x) > f(b)$ for all x (except b) in the interval containing b. At the relative maximum a of $f(x)$, $f'(a) = 0$, i.e., slope $= 0$, and $f''(a) < 0$, i.e., the curve is downwardly concave at this point, and at the relative minimum b, $f'(b) = 0$ and $f''(b) > 0$ (upward concavity). In Figure 1.5.1 A, B, C, and D are critical points and x_1, x_2, x_3, and x_4 are critical values of x. A and C are maxima, B is a minimum, and D is neither. D, F, G, and H are *points of inflection* where the slope is minimum or maximum. In special cases, such as E, maxima or minima may occur where $f'(x)$ is undefined or infinite.

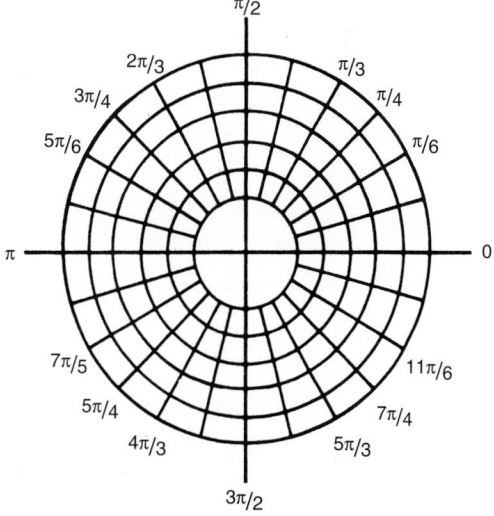

Figure 1.4.4 *The polar plane.*

The *absolute maximum* (or minimum) of $f(x)$ at $x = a$ exists if $f(x) \le f(a)$ (or $f(x) \ge f(a)$) for all x in the domain of the function and need not be a relative maximum or minimum. If a function is defined and continuous on a closed interval, it

Table 1.5.1 Table of Derivatives[a]

$\frac{d}{dx}(x) = 1$

$\frac{d}{dx}(a) = 0$

$\frac{d}{dx}(u + v \pm \dots) = \frac{du}{dx} \pm \frac{dv}{dx} \pm \dots$

$\frac{d}{dx}(au) = a\frac{du}{dx}$

$\frac{d}{dx}(uv) = u\frac{dv}{dx} + v\frac{du}{dx}$

$\frac{d}{dx}\frac{u}{v} = \frac{v\frac{du}{dx} - u\frac{dv}{dx}}{v^2}$

$\frac{d}{dx}(u^a) = nu^{a-1}\frac{du}{dx}$

$\frac{d}{dx}\log_a u = \frac{\log_a e}{u}\frac{du}{dx}$

$\frac{d}{dx}\log u = \frac{1}{u}\frac{du}{dx}$

$\frac{d}{dx}a^u = a^u \bullet \log_e a \bullet \frac{du}{dx}$

$\frac{d}{dx}e^u = e^u\frac{du}{dx}$

$\frac{d}{dx}u^v = vu^{v-1}\frac{du}{dx} + u^v\log_e u\frac{dv}{dx}$

$\frac{d}{dx}\sin u = \cos u\frac{du}{dx}$

$\frac{d}{dx}\cos u = -\sin u\frac{du}{dx}$

$\frac{d}{dx}\tan u = \sec^2 u\frac{du}{dx}$

$\frac{d}{dx}\cot u = -\csc^2 u\frac{du}{dx}$

$\frac{d}{dx}\sec u = \sec u \tan u\frac{du}{dx}$

$\frac{d}{dx}\csc u = -\csc u \cot u\frac{du}{dx}$

$\frac{d}{dx}\text{vers } u = \sin u\frac{du}{dx}$

$\frac{d}{dx}\sin^{-1}u = \frac{1}{\sqrt{1-u^2}}\frac{du}{dx}$

$\frac{d}{dx}\cos^{-1}u = -\frac{1}{\sqrt{1-u^2}}\frac{du}{dx}$

$\frac{d}{dx}\tan^{-1}u = \frac{1}{1+u^2}\frac{du}{dx}$

$\frac{d}{dx}\cot^{-1}u = -\frac{1}{1+u^2}\frac{du}{dx}$

$\frac{d}{dx}\sec^{-1}u = \frac{1}{u\sqrt{u^2-1}}\frac{du}{dx}$

$\frac{d}{dx}\csc^{-1}u = -\frac{1}{u\sqrt{u^2-1}}\frac{du}{dx}$

$\frac{d}{dx}\text{vers}^{-1}u = \frac{1}{\sqrt{2u-u^2}}\frac{du}{dx}$

$\frac{d}{dx}\sinh u = \cosh u\frac{du}{dx}$

$\frac{d}{dx}\cosh u = \sinh u\frac{du}{dx}$

$\frac{d}{dx}\tanh u = \sec h^2 u\frac{du}{dx}$

$\frac{d}{dx}\coth u = -\csc h^2 u\frac{du}{dx}$

$\frac{d}{dx}\text{sech} u = -\text{sech} u \tanh u\frac{du}{dx}$

$\frac{d}{dx}\text{csch} u = -\text{csch} u \coth u\frac{du}{dx}$

$\frac{d}{dx}\sinh^{-1}u = \frac{1}{\sqrt{u^2-1}}\frac{du}{dx}$

$\frac{d}{dx}\cosh^{-1}u = \frac{1}{\sqrt{u^2-1}}\frac{du}{dx}$

$\frac{d}{dx}\tanh^{-1}u = \frac{1}{1-u^2}\frac{du}{dx}$

$\frac{d}{dx}\coth^{-1}u = -\frac{1}{u^2-1}\frac{du}{dx}$

$\frac{d}{dx}\text{sech}^{-1}u = -\frac{1}{u\sqrt{1-u^2}}\frac{du}{dx}$

$\frac{d}{dx}\csc h^{-1}u = -\frac{1}{u\sqrt{u^2-1}}\frac{du}{dx}$

[a]The u and v represent functions of x. All angles are in radians.

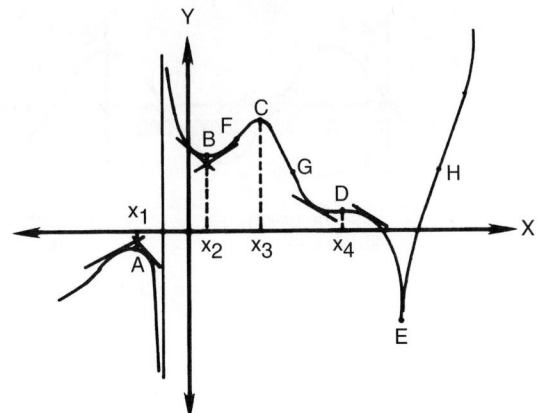

Figure 1.5.1 Maxima and minima.

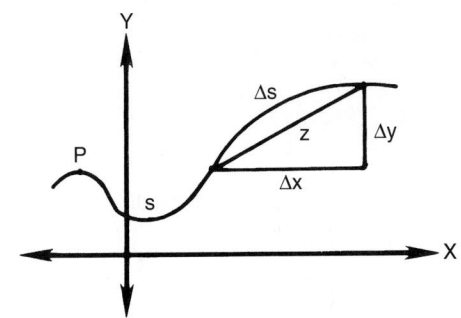

Figure 1.5.2 Radius of curvature in rectangular coordinates.

will always have an absolute minimum and an absolute maximum, and they will be found either at a relative minimum and a relative maximum or at the endpoints of the interval.

1.5.5 Differentials

If $y = f(x)$ and Δx and Δy are the increments of x and y, respectively, because $y + \Delta y = f(x + \Delta x)$, then

$$\Delta y = f(x + \Delta x) - f(x)$$

As Δx approaches its limit 0 and (since x is the independent variable) $dx = \Delta x$

$$\frac{dy}{dx} \cong \frac{f(x + \Delta x) - f(x)}{\Delta x}$$

and

$$dy \cong \Delta y$$

By defining dy and dx separately, it is now possible to write

$$\frac{dy}{dx} = f'(x)$$

as

$$dy = f' = (x)dx$$

In functions of two or more variables, where $f(x, y, \dots) = 0$, if dx, dy, ... are assigned to the independent variables x, y, ..., the differential du is given by differentiating term by term or by taking

$$du = f_x \bullet dx + f_y \bullet dy + \dots$$

If x, y, ... are functions of t, then

$$\frac{du}{dt} = (f_x)\frac{dx}{dt} + (f_y)\frac{dy}{dt} + \dots$$

expresses the rate of change of u with respect to t, in terms of the separate rates of change of x, y, ... with respect to t.

1.5.6 Radius of Curvature

The *radius of curvature* R of a plane curve at any point P is the distance along the normal (the perpendicular to the tangent to the curve at point P) on the concave side of the curve to the *center of curvature* (Figure 1.5.2). If the equation of the curve is $y = f(x)$

$$R = \frac{ds}{du} = \frac{[1 + f'(x)^2]^{3/2}}{f''(x)}$$

where the rate of change (ds/dx) and the differential of the arc (ds), s being the length of the arc, are defined as

$$\frac{ds}{dx} = \sqrt{1 + \left(\frac{dy}{dx}\right)^2}$$

and

$$ds = \sqrt{dx^2 + dy^2}$$

and $dx = ds \cos u$
  $dy = ds \sin u$
  $u = \tan^{-1}[f'(x)]$

with u being the angle of the tangent at P with respect to the x-axis. (Essentially, ds, dx, and y correspond to the sides of a right triangle.) The curvature K is the rate at which $<u$ is changing with respect to s, and

$$K = \frac{1}{R} = \frac{du}{ds}$$

If $f'(x)$ is small, $K \cong f''(x)$.

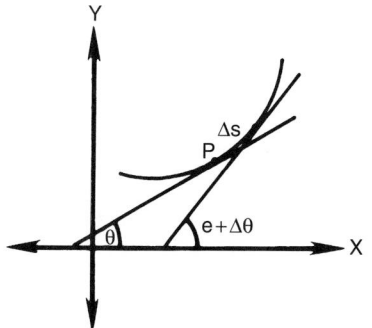

Figure 1.5.3 *Radius of curvature in polar coordinates.*

In polar coordinates (Figure 1.5.3), $r = f(\theta)$, where r is the radius vector and θ is the polar angle, and

$$ds = \sqrt{d\rho^2 + \rho^2 d\theta^2}$$

so that by $x = \rho \cos\theta$, $y = \rho \sin\theta$ and $K = 1/R = d\theta/ds$, then

$$R = \frac{ds}{d\theta} = \frac{[r^2 + (r')^2]^{3/2}}{r^2 - rr'' + 2(r')^2}$$

If the equation of the circle is

$$R^2 = (x - \alpha) + (y - \beta)^2$$

by differentiation and simplification

$$\alpha = x - \frac{y'[1 + (y')^2]}{y''}$$

and

$$\beta = y + \frac{1 + (y')^2}{y''}$$

The *evolute* is the locus of the centers of curvature, with variables α and β, and the parameter x (y, y', and y'' all being functions of x). If f(x) is the evolute of g(x), g(x) is the *involute* of f(x).

1.5.7 Indefinite Integrals
Definition: The indefinite integral (antiderivative) may be defined as follows:

$$\int f(x)dx = F(x) + C \text{ if } F(x) = f(x)$$

The constant C is called the constant of integration.
Integration by parts makes use of the differential of a product

$$d(uv) = udv + vdu$$

or

$$u\,dv = d(uv) - v\,du$$

and by integrating

$$\int u\,dv = uv - \int v\,du$$

where $\int v\,du$ may be recognizable as a standard form or may be more easily handled than $\int u\,dv$.
Integration by transformation may be useful when, in certain cases, particular transformations of a given integral to one of a recognizable form suggest themselves.

For example, a given integral involving such quantities as

$$\sqrt{u^2 - a^2}, \quad \sqrt{u^2 + a^2}, \quad \text{or} \quad \sqrt{a^2 - u^2}$$

may suggest appropriate trigonometric transformations such as, respectively,

$$u = a\csc\theta,$$
$$u = a\tan\theta,$$

or

$$u = a\sin\theta$$

Integration by partial fractions is of assistance in the integration of rational fractions. If

$$\frac{ax + b}{x^2 + px + q} = \frac{ax + b}{(x - \alpha)(x - \beta)} = \frac{A}{x - \alpha} + \frac{B}{x - \beta}$$

where $A + B = \alpha$

$$A\beta + B\alpha = -b$$

and A and B are found by use of determinants (see "Algebra"), then

$$\int \frac{(ax + b)dx}{(x - \alpha)(x - \beta)} = \int \frac{Adx}{x - \alpha} + \int \frac{Bdx}{x - \beta}$$

$$= A\log(x - \alpha) + B\log(x - \beta) + C$$

Integration by tables is possible if an integral may be put into a form that can be found in a table of integrals, such as the one given in Table 1.5.2. More complete tables may be found in Bois, "Table of Indefinite Integrals," Dover, and in others.

1.5.8 Definite Integrals
The *fundamental theorem of calculus* states that if f(x) is the derivative of F(x) and if f(x) is continuous in the interval [a, b], then

$$\int_a^b f(x)dx = F(b) - F(a)$$

Geometrically, the integral of f(x)dx over the interval [a, b] is the area bounded by the curve $y = f(x)$ from f(a) to f(b) and the x-axis from $x = a$ to $x = b$, or the "area under the curve from a to b."

1.5.9 Properties of Definite Integrals

$$\int_a^b = -\int_b^a$$

$$\int_a^c + \int_c^b = \int_a^b$$

The *mean value* of f(x), $\bar{f}$, between a and b is

$$\bar{f} = \frac{1}{b - a}\int_a^b f(x)dx$$

If the upper limit b is a variable, then $\int_a^b f(x)dx$ is a function of b and its derivative is

$$f(b) = \frac{d}{db}\int_a^b f(x)dx$$

To differentiate with respect to a parameter

$$\frac{\partial}{\partial c}\int_a^b f(x, c)dx = \int_a^b \frac{\partial f(x, c)}{\partial c}dx$$

Some methods of integration of definite integrals are covered in "Numerical Methods."

1.5.10 Improper Integrals
If one (or both) of the limits of integration is infinite, or if the integrand itself becomes infinite at or between the limits of integration, the integral is an *improper integral*. Depending on the function, the integral may be defined, may be equal to ∞, or may be undefined for all x or for certain values of x.

Table 1.5.2   *Table of Integrals^a*

1. $\int df(x) = f(x) + C$
2. $d \int f(x)dx = f(x)dx$
3. $\int 0 \bullet dx = C$
4. $\int af(x)dx = a \int f(x)dx$
5. $\int (u \pm v)dx = \int udx \pm \int vdx$
6. $\int udv = uv - \int vdu$
7. $\int \frac{udv}{dx}dx = uv - \int v\frac{du}{dx}dx$
8. $\int f(y)dx = \int \frac{f(y)dy}{\frac{dy}{dx}}$
9. $\int u^n du = \frac{u^{n+1}}{n+1} + C, n \neq -1$
10. $\int \frac{du}{u} = \log_e u + C$
11. $\int e^u du = e^u + C$
12. $\int b^u du = \frac{b^u}{\log_e b} + C$
13. $\int \sin u\, du = -\cos u + C$
14. $\int \cos u\, du = \sin u + C$
15. $\int \tan u\, du = \log_e \sec u + C = -\log_e \cos u + C$
16. $\int \cot u\, du = \log_e \sin u + C = -\log_e \csc u + C$
17. $\int \sec u\, du = \log_e(\sec u + \tan u) + C$
    $\qquad = \log_e \tan\left(\frac{u}{2} + \frac{\pi}{4}\right) + C$
18. $\int \csc u\, du = \log_e(\csc u - \cot u) + C = \log_e \tan \frac{u}{2} + C$
19. $\int \sin^2 u\, du = \frac{1}{2}u - \frac{1}{2}\sin u \cos u + C$
20. $\int \cos^2 u\, du = \frac{1}{2}u + \frac{1}{2}\sin u \cos u + C$
21. $\int \sec^2 u\, du = \tan u + C$
22. $\int \csc^2 u\, du = -\cot u + C$
23. $\int \tan^2 u\, du = \tan u - u + C$
24. $\int \text{ctn}^2 u\, du = -\cot u - u + C$
25. $\int \frac{du}{u^2+a^2} = \frac{1}{a}\tan^{-1}\frac{u}{a} + C$
26. $\int \frac{du}{u^2-a^2} = \frac{1}{2a}\log_e\left(\frac{u-a}{u+a}\right) + C$
    $\qquad = -\frac{1}{a}\coth^{-1}\left(\frac{u}{a}\right) + C, \text{ if } u^2 > a^2$
    $\qquad = \frac{1}{2a}\log_e\left(\frac{a-u}{a+u}\right) + C$
    $\qquad = -\frac{1}{a}\tanh^{-1}\left(\frac{u}{a}\right) + C, \text{ if } u^2 < a^2$
27. $\int \frac{du}{\sqrt{a^2-u^2}} = \sin^{-1}\left(\frac{u}{a}\right) + C$
28. $\int \frac{du}{\sqrt{u^2 \pm a^2}} \log_e(u + \sqrt{u^2 \pm a^2})^t + C$
29. $\int \frac{du}{\sqrt{au-u^2}} = \cos^{-1}\left(\frac{a-u}{a}\right) + C$
30. $\int \frac{du}{u\sqrt{u^2-a^2}} = \frac{1}{a}\sec^{-1}\left(\frac{u}{a}\right) + C = \frac{1}{a}\cos^{-1}\frac{a}{u} + C$
31. $\int \frac{du}{u\sqrt{a^2 \pm u^2}} = -\frac{1}{a}\log_e\left(\frac{a+\sqrt{a^2-u^2}}{u}\right)^t + C$
32. $\int \sqrt{a^2-u^2}\bullet du = \frac{1}{2}\left(u\sqrt{a^2-u^2} + a^2\sin^{-1}\frac{u}{a}\right) + C$
33. $\int \sqrt{u^2 \pm a^2}\bullet du$
    $\qquad = \frac{1}{2}\left[u\sqrt{u^2 \pm a^2} \pm a^2\log_e(u + \sqrt{u^2 \pm a^2})\right]^\dagger + C$
34. $\int \sinh u\, du = \cosh u + C$
35. $\int \cosh u\, du = \sinh u + C$
36. $\int \tanh u\, du = \log_e(\cosh u) + C$
37. $\int \coth u\, du = \log_e(\sinh u) + C$
38. $\int \text{sech}\, u\, du = \sin^{-1}(\tanh u) + C$
39. $\int \text{csch}\, u\, du = \log_e\left(\tanh \frac{u}{2}\right) + C$
40. $\int \text{sech}\bullet\tanh u\bullet du = -\text{sech}\, u + C$
41. $\int \text{csch}\, u\bullet \coth u\bullet du = -\text{csch}\, u + C$

^{a}The u and v represent functions of x.

$\dagger \log_e\left(\frac{u+\sqrt{u^2+a^2}}{a}\right) = \sinh^{-1}\left(\frac{u}{a}\right); \log_e\left(\frac{a+\sqrt{a^2-u^2}}{u}\right) = \text{sech}^{-1}\left(\frac{u}{a}\right);$

$\log_e\left(\frac{u+\sqrt{u^2-a^2}}{a}\right) = \cosh^{-1}\left(\frac{u}{a}\right); \log_e\left(\frac{a+\sqrt{a^2-u^2}}{u}\right) \text{csc h}^{-1}\left(\frac{u}{a}\right)$

1.5.11 Multiple Integrals

It is possible to integrate functions of several variables by using an *iterated integral*. An iterated integral is solved from the inner integral to the outer, and variables other than the variable of integration are held constant.

$$\iint f(x, y)\, dydx = \int \left(\int f(x, y)dy\right) dx$$

Definite multiple integrals may have variable inner limits of integration with respect to the outer variable of integration:

$$\int_c^d \int_{f(x)}^{g(x)} F(x, y)dydx$$

Uses for multiple integrals include finding areas, volumes, and the center of mass.

1.5.12 Differential Equations

An *ordinary differential equation* contains a single independent variable and a single unknown function of that variable, with its derivatives. A *partial differential equation* involves an unknown function of two or more independent variables, and its partial derivatives. The order of a differential equation is the order of the highest derivative in the equation. The general solution of a differential equation of order n is the set of all functions that possess at least n derivatives and satisfy the equation, as well as any auxiliary conditions.

1.5.13 Methods of Solving Ordinary Differential Equations

For *first-order equations*, if possible, separate the variables, integrate both sides, and add the constant of integration, C. If the equation is *homogeneous* in x and y, the value of dy/dx in terms of x and y is of the form $dy/dx = f(y/x)$ and the variables may be separated by introducing new independent variable $v = y/x$ and then

$$x\frac{dv}{dx} + v = f(v)$$

The expression $f(x, y)dx + F(x, y)dy$ is an *exact differential* if

$$\frac{\partial f(x, y)}{\partial y} = \frac{\partial F(x, y)}{\partial x} \qquad (= P, \text{ for example})$$

Then, the solution of $f(x, y)dx + F(x, y)dy = 0$ is

$$\int f(x, y)dx + \int \left[F(x, y) - \int P\, dx\right] dy = C$$

or

$$\int F(x, y)dy + \int \left[f(x, y) - \int P\, dy\right] dx = C$$

A *linear differential equation* of the first order such as

$$dy/dx + f(x)\bullet y = F(x)$$

has the solution

$$y = e^{-p}\left[\int e^p F(x)dx + C\right] \quad \text{where} \quad P = \int f(x)dx$$

In the class of nonlinear equations known as *Bernoulli's equations*, where

$$dy/dx + f(x)\bullet y = F(x)\bullet y^n$$

substituting $y^{1-n} = v$ gives

$$dv/dx + (1-n)f(x)\bullet v = (1-n)F(x) \quad [n \neq 0 \text{ or } 1]$$

which is linear in v and x. In *Clairaut's equations*

$$y = xp + f(p) \quad \text{where} \quad p = dy/dx,$$

the solution consists of the set of lines given by $y = Cx + f(C)$, where C is any constant, and the curve obtained by eliminating p between the original equation and $x + f'(p) = 0$ [1].

Some differential equation of the *second order* and their solutions follow:

For $d^2y/dx^2 = -n^2y$

$$y = C_1 \sin(nx + C_2)$$

$$= C_3 \sin nx + C_4 \cos nx$$

For $d^2y/dx^2 = +n^2y$

$$y = C_1 \sinh(nx + C_2)$$

$$= C_3 e^{nx} + C_4 e^{-nx}$$

For $d^2y/dx^2 = f(y)$

$$x = \int \frac{dy}{\sqrt{C_1 + 2P}} + C_2$$

where $P = \int f(y)dy$.

For $d^2y/dx^2 = f(x)$

$$y = \int P dx + C_1 x + C_2 \qquad \text{where } P = \int f(x)dx$$

$$= xP - \int xf(x)dx + C_1 x + C_2$$

For $d^2y/dx^2 = f(dy/dx)$, setting $dy/dx = z$ and $d^2/dx^2 = dz/dx$

$$x = \int dz/f(z) + C_1 \text{ and}$$

$$y = \int z dz/f(z) + C_2, \text{ then eliminating } z$$

For $d^2y/dx^2 + 2b(dy/dx) + a^2y = 0$ (the equation for damped vibration)

- If $a^2 - b^2 > 0$,

  then $m = \sqrt{a^2 - b^2}$

  $$y = C_1 e^{-bx} \sin(mx + C_2)$$

  $$= e^{-bx}[C_3 \sin(mx) + C_4 \cos(mx)]$$

- If $a^2 - b^2 = 0$,

  $$y = e^{-bx}(C_1 + C_2 x)$$

- If $a^2 - b^2 < 0$,

  then $n = \sqrt{b^2 - a^2}$ and

  $$y = C_1 e^{-bx} \sinh(nx + C_2)$$

  $$= C_3 e^{-(b+n)x} + C_4 e^{-(b-n)x}$$

For $d^2y/dx^2 + 2b(dy/dx) + a^2y = c$

$$y = c/a^2 + y_1$$

where y_1 is the solution of the previous equation with second term zero.

The preceding two equations are examples of linear differential equations with constant coefficients and their solutions are often found most simply by the use of Laplace transforms [1].

For the linear equation of the nth order

$$A_n(x)d^n y/dx^n + A_{n-1}(x)d^{n-1}y/dx^{n-1} + \ldots$$

$$+ A_1(x)dy/dx + A_0(x)y = E(x)$$

the general solution is

$$y = u + c_1 u_1 + c_2 u_2 + \ldots + c_n u_n,$$

where u is any solution of the given equation and u_1, $u_2, \ldots, u_n$ form a *fundamental system* of solutions to the homogeneous equation [E(x) ← zero]. A set of functions has linear independence if its Wronskian determinant,

$W(x)$, $\neq 0$, where

$$W(x) = \begin{vmatrix} u_1 & u_2 & \ldots & u_n \\ u_1 & u_2 & \ldots & u_n \\ \cdot & & \cdots & \cdot \\ u_1^m & u_2^m & \ldots & u_n^m \end{vmatrix}$$

and $m = n - 1$th derivative. (In certain cases, a set of functions may be linearly independent when $W(x) = 0$.)

1.5.14 The Laplace Transformation

The *Laplace transformation* is based on the Laplace integral which transforms a differential equation expressed in terms of time to an equation expressed in terms of a complex variable $\sigma + j\omega$. The new equation may be manipulated algebraically to solve for the desired quantity as an explicit function of the complex variable.

Essentially three reasons exist for the use of the Laplace transformation:

1. The ability to use algebraic manipulation to solve higher-order differential equations
2. Easy handling of boundary conditions
3. The method is suited to the complex-variable theory associated with the Nyquist stability criterion [1].

In Laplace-transformation mathematics, the following symbols and variables are used:

$f(t)$ = a function of time
s = a complex variable of the form $(\sigma + j\omega)$
$F(s)$ = the Laplace transform of f, expressed in s, resulting from operating on f(t) with the Laplace integral.
$\mathcal{L}$ = the Laplace operational symbol, i.e., $F(s) = \mathcal{L}[f(t)]$.

The Laplace integral is defined as

$$\mathcal{L} = \int_0^\infty e^{-st}dt \quad \text{and so}$$

$$\mathcal{L}[f(t)] = \int_0^\infty e^{-st}f(t)dt$$

Table 1.5.3 lists the transforms of some common time-variable expressions.

The transform of a first derivative of f(t) is

$$\mathcal{L}\left[\frac{d}{dt}f(t)\right] = sF(s) - f(0^+)$$

where $f(0^+)$ = initial value of f(t) as $t \to 0$ from positive values.

The transform of a second derivative of f(t) is

$$\mathcal{L}[f''(t)] = s^2 F(s) - sf(0^+) - f'(0^+)$$

and of $\int f(t)dt$ is

$$\mathcal{L}\left[\int f(t)dt\right] = \frac{f^{-1}(0^+)}{s} + \frac{F(s)}{s}$$

Solutions derived by Laplace transformation are in terms of the complex variable s. In some cases, it is necessary to retransform the solution in terms of time, performing an *inverse transformation*

$$\mathcal{L}^{-1}F(s) = f(t)$$

Just as there is only one direct transform F(s) for any f(t), there is only one inverse transform f(t) for any F(s) and inverse transforms are generally determined through use of tables.

References

1. Thompson, S. P., *Calculus Made Easy*, 3rd Edition, St. Martin's Press, New York, 1984.

Table 1.5.3 *Laplace Transforms*

f(t)	$F(s) = \mathcal{L}[f(t)]$
A	A/s
$1 = u(t)$	$1/s$
$e^{-\alpha t}$	$\frac{1}{s+\alpha}$
$1/re^{-t/\tau}$	$\frac{1}{\tau s+1}$
$Ae^{-\alpha t}$	$\frac{A}{s+\alpha}$
$\sin \beta t$	$\frac{\beta}{s^2+\beta^2}$
$\cos \beta t$	$\frac{s}{s^2+\beta^2}$
$\frac{1}{\beta}e^{-\alpha t}\sin \beta t$	$\frac{1}{s^2+2\alpha s+\alpha^2+\beta^2}$
$\frac{e^{-\alpha t}}{\beta-\alpha} - \frac{e^{-\beta t}}{\beta-\alpha}$	$\frac{1}{(s+\alpha)(s+\beta)}$
$\frac{Ae^{-\alpha t}-Be^{-\beta t}}{C}$	$\frac{s+a}{(s+\alpha)(s+\beta)}$
$A = a - \alpha$ where $B = a - \beta$ $c = \beta - \alpha$	
$\frac{e^{-\alpha t}}{A} + \frac{e^{-\beta t}}{B} + \frac{e^{-\delta t}}{C}$	$\frac{1}{(s+\alpha)(s+\beta)(s+\delta)}$
$A = (\beta - \alpha)(\delta - \alpha)$ where $B = (\alpha - \beta)(\delta - \beta)$ $C = (\alpha - \delta)(\beta - \delta)$	
t^n	$\frac{n!}{s^{n+1}}$
$d/dt[f(t)]$	$sF(s) - f(0^+)$
$d^2/dt^2[f(t)]$	$s^2F(s) - sf(0^+) - \frac{df}{dt}(0^+)$
$d^3/dt^3[f(t)]$	$s^3F(s) - s^2f(0^+)$ $-s\frac{df}{dt}(0^+) - \frac{d^2f}{dt^2}(0^+)$
$\int f(t)dt$	$\frac{1}{s}[F(s) + \int f(t)dt|o^+]$
$\frac{1}{\alpha}\sinh \alpha t$	$\frac{1}{s^2-\alpha^2}$
$\cosh \alpha t$	$\frac{s}{s^2-\alpha^2}$

2. Lial, M. L., and Miller, C. D., *Essential Calculus with Applications*, 2nd Edition, Scott, Foresman and Company, Glenview, 1980.
3. Oakley, C. O., *The Calculus*, Barnes and Noble, New York, 1957.
4. Thomas, G. B., and Finney, R. L., *Calculus and Analytic Geometry*, 9th Edition, Addison Wesley, Reading, 1995.

1.6 ANALYTIC GEOMETRY

1.6.1 Symmetry
Symmetry exists for the curve of a function about the y-axis if $F(x, y) = F(-x, y)$, about the x-axis if $F(x, y) = F(x, -y)$, about the origin if $F(x, y) = F(-x, -y)$, and about the 45° line if $F(x, y) = F(y, x)$.

1.6.2 Intercepts
Intercepts are points where the curve of a function crosses the axes. The x intercepts are found by setting $y = 0$ and the y intercepts by setting $x = 0$.

1.6.3 Asymptotes
As a point $P(x, y)$ on a curve moves away from the region of the origin (Figure 1.6.1a), the distance between P and some fixed line may tend to zero. If so, the line is called an asymptote of the curve. If $N(x)$ and $D(x)$ are polynomials with no common factor, and

$$y = N(x)/D(x)$$

where $x = c$ is a root of $D(x)$, then the line $x = c$ is an asymptote of the graph of y.

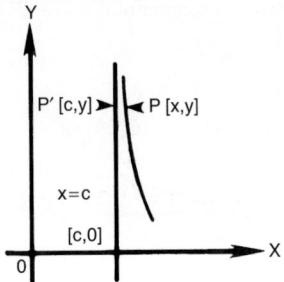

Figure 1.6.1a *Asymptote of a curve.*

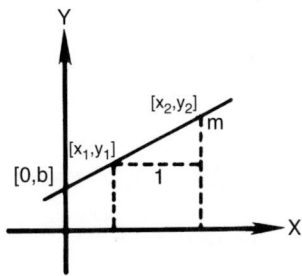

Figure 1.6.1b *Slope of a straight line.*

1.6.4 Equations of Slope and Straight Lines
1. Equation for slope of line connecting two points (x_1, y_1) and (x_2, y_2)

$$m = \frac{y_2 - y_1}{x_2 - x_1}$$

2. Two-point equation (Figure 1.25)

$$\frac{y - y_1}{x - x_1} = \frac{y_2 - y_1}{x_2 - x_1}$$

3. Point-slope equation (Figure 1.25)

$$y - y_1 = m(x - x_1)$$

4. Slope-intercept equation (Figure 1.6.1b)

$$y = mx + b$$

1.6.5 Tangents
In the slope m of the curve of $f(x)$ at (x_1, y_1) is given by (Figure 1.6.2)

$$m = \frac{dy}{dx(x_1, y_1)} = f'(x)$$

then the equation of the line tangent to the curve at this point is

$$y - y_1 = f'(x_1)(x - x_1)$$

and the normal to the curve is the line perpendicular to the tangent with slope m_2 where

$$m_2 = -1/m_1 = -1/f'(x)$$

or

$$y - y_1 = -(x - x_1)/f'(x_1)$$

1.6.6 Other Forms of the Equation of a Straight Line
• General equation

$$ax + by + c = 0$$

• Intercept equation

$$x/a + y/b = 1$$

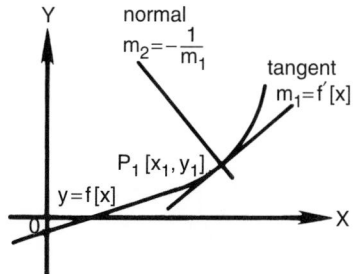

Figure 1.6.2 *Tangent and normal to a curve.*

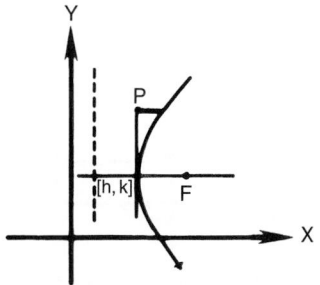

Figure 1.6.4 *Equation of a parabola.*

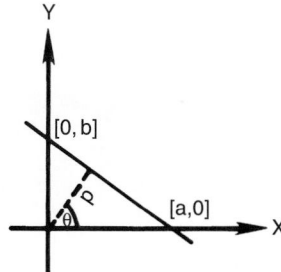

Figure 1.6.3 *Equation of a straight line (normal form).*

- Normal form (Figure 1.6.3)

$$x \cos \theta + y \sin \theta - p = 0$$

- Distance d from a straight line $(ax + by + c = 0)$ to a point $P(x_1, y_1)$

$$d = \frac{ax_1 + by_1 + c}{\pm \sqrt{a^2 + b^2}}$$

- If u is the angle between $ax + by + c = 0$ and $a'x + b'y + c' = 0$ then

$$\cos u = \frac{aa' + bb'}{\pm \sqrt{(a^2 + b^2)(a'^2 + b'^2)}}$$

1.6.7 Equations of a Circle (Center (h, k))
- $(x - h)^2 + (y - k)^2 = r^2$
- Origin at center

$$x^2 + y^2 = r^2$$

- General equation

$$x^2 + y^2 + Dx + Ey + F = 0$$

where center $= (-D/2, -E/2)$

$$\text{radius} = \sqrt{(D/2)^2 + (E/2)^2 - F}$$

- Tangent to circle at (x_1, y_1)

$$x_1 x + y_1 y + \frac{1}{2}D(x + x_1) + \frac{1}{2}E(y + y_1) + F = 0$$

- Parametric form, replacing x and y by

$$x = a \cos u$$

and

$$y = a \sin u$$

1.6.8 Equations of a Parabola (Figure 1.6.4)
A parabola is the set of points that are equidistant from a given fixed point (the focus) and from a given fixed line (the directrix) in the plane. The key feature of a parabola is that it is second degree in one of its coordinates and first degree in the other.

- $(y - k)^2 = 4p(x - h)$
- Coordinates of the vertex V(h, k) and of the focus $F(h + p, k)$
- Origin at vertex

$$y^2 = 4px$$

- Equation of the directrix

$$x = h - p$$

- Length of latus rectum

$$LL' = 4p$$

- Polar equation (focus as origin)

$$r = p/(1 - \cos \theta)$$

- Equation of the tangent to $y^2 = 2\,px$ at (x_1, y_1)

$$y_1 y = p(x + x_1)$$

1.6.9 Equations of an Ellipse of Eccentricity e (Figure 1.6.5)
- $\frac{(x-h)^2}{a^2} + \frac{(y-k)^2}{b^2} = 1$
- Coordinates of center C(h, k), of vertices V(h + a, k) and V'(h − a, k), and of foci F(h + ae, k) and F'(h − ae, k)
- Center at origin

$$x^2/a^2 + y^2/b^2 = 1$$

- Equation of the directrices

$$x = h \pm a/e$$

- Equation of the eccentricity

$$e = \frac{\sqrt{a^2 - b^2}}{a} < 1$$

- Length of the latus rectum

$$LL' = 2b^2/a$$

- Parametric form, replacing x and y by

$$x = a \cos u \qquad \text{and} \qquad y = b \sin u$$

- Polar equation (focus as origin)

$$r = p/(1 - e \cos \theta)$$

- Equation of the tangent at (x_1, y_1)

$$b^2 x_1 x + a^2 y_1 y = a^2 b^2$$

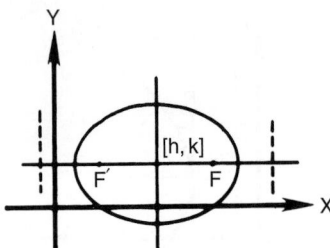

Figure 1.6.5 *Ellipse of eccentricity e.*

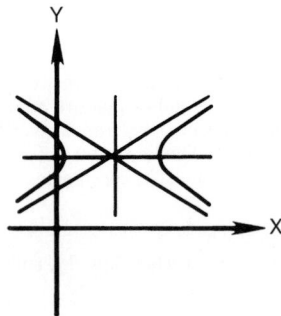

Figure 1.6.6 *Equation of a hyperbola.*

1.6.10 Equations of a Hyperbola (Figure 1.6.6)

- $\dfrac{(x-h)^2}{a^2} - \dfrac{(y-k)^2}{b^2} = 1$
- Coordinates of the center $C(h, k)$, of vertices $V(h + a, k)$ and $V'(h - a, k)$, and of the foci $F(h + ae, k)$ and $F'(h - ae, k)$
- Center at origin
$$x^2/a^2 - y^2/b^2 = 1$$
- Equation of the directrices
$$x = h \pm a/e$$
- Equation of the asymptotes
$$y - k = \pm b/a \bullet (x - k)$$
- Equation of the eccentricity
$$e = \frac{\sqrt{a^2 + b^2}}{a} > 1$$
- Length of the latus rectum
$$LL' = 2b^2/a$$
- Parametric form, replacing x and y
$$x = a \cosh u \quad \text{and} \quad y = b \sinh u$$
- Polar equation (focus as origin)
$$r = p/(1 - e \cos \theta)$$
- Equation of the tangent at (x, y)
$$b^2 x_1 x - a^2 y_1 y = a^2 b^2$$

1.6.11 Equations of Three-Dimensional Coordinate Systems (Figure 1.6.7)

- Distance d between two points
$$d = \sqrt{(x_2 - x_1)^2 + (y_2 - y_1)^2 + (z_2 - z_1)^2}$$
- Direction cosines of a line
$$\lambda = \cos \alpha, \qquad \mu = \cos \beta, \qquad v = \cos \gamma$$

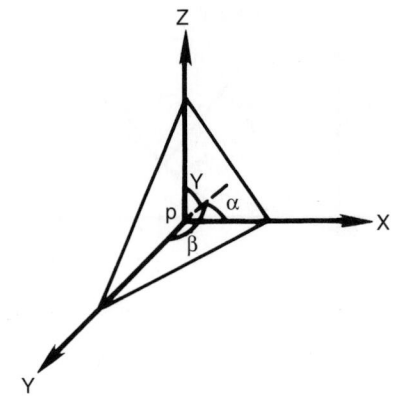

Figure 1.6.7 *Three-dimensional coordinate systems.*

- Direction numbers, proportional to the direction cosines with k
$$a = k\lambda, \qquad b = k\mu, \qquad c = kv$$

1.6.12 Equations of a Plane

- $ax + by + cz + d = 0$
- Intercept
$$x/a + y/b + z/c = 1$$
- Normal form
$$\lambda x + \mu y + vz - p = 0$$
- Distance from $ax + by + cz + d = 0$ to a point $P(x_1, y_1, z_1)$
$$D = \frac{ax_1 + by_1 + c_1 + d}{\pm\sqrt{a^2 + b^2 + c^2}}$$

1.6.13 Equations of a Line

- Intersection of two planes
$$\begin{cases} a_1 x + b_1 y + c_1 z + d_1 = 0 \\ a_2 x + b_2 y + c_2 z + x_2 = 0 \end{cases}$$

For this line
$$\lambda : \mu : v = \begin{vmatrix} b_1 & c_1 \\ b_2 & c_2 \end{vmatrix} : -\begin{vmatrix} a_1 & c_1 \\ a_2 & c_2 \end{vmatrix} : \begin{vmatrix} a_1 & b_1 \\ a_2 & b_2 \end{vmatrix}$$

- Symmetric form i.e., through (x_1, y_1, z_1) with direction numbers a, b, and c
$$(x - x_1)/a = (y - y_1)/b = (z - z_1)/c$$

Through two points
$$\frac{x - x_1}{x_2 - x_1} = \frac{y - y_1}{y_2 - y_1} = \frac{z - z_1}{z_2 - z_1}$$
where $\lambda : \mu : v = (x_2 - x_1) : (y_2 - y_1) : (z_2 - z_1)$

1.6.14 Equations of Angles

- Between two lines
$$\cos \theta = \lambda_1 \lambda_2 + \mu_1 \mu_2 + v_1 v_2$$

and the lines are parallel if $\cos \theta = 1$ or perpendicular if $\cos \theta = 0$

- Between two planes, given by the angle between the normals to the planes.

1.6.15 Equation (Standard Form) of a Sphere (Figure 1.6.8)

$$x^2 + y^2 + z^2 = r$$

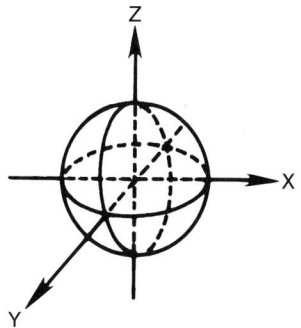

Figure 1.6.8   *Sphere.*

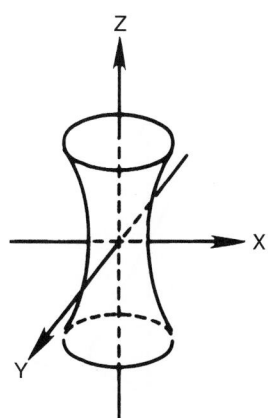

Figure 1.6.10   *Hyperboloid of one sheet.*

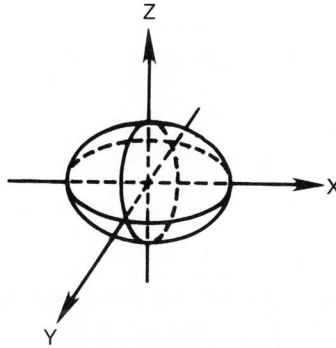

Figure 1.6.9   *Equation of an ellipsoid.*

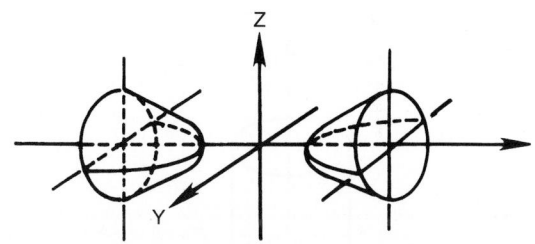

Figure 1.6.11   *Hyperboloid of two sheets.*

1.6.16 Equation (Standard Form) of an Ellipsoid
(Figure 1.6.9)

$$x^2/a^2 + y^2/b^2 + z^2/c^2 = 1$$

1.6.17 Equations (Standard Form) of Hyperboloids
- Of one sheet (Figure 1.6.10)

$$x^2/a^2 + y^2/b^2 - z^2/c^2 = 1$$

- Of two sheets (Figure 1.6.11)

$$x^2/a^2 - y^2/b^2 - z^2/c^2 = 1$$

1.6.18 Equations (Standard Form) of Paraboloids
- Of elliptic paraboloid (Figure 1.6.12)

$$x^2/a^2 + y^2/b^2 = cz$$

- Of hyperbolic paraboloid (Figure 1.6.13)

$$x^2/a^2 - y^2/b^2 = cz$$

1.6.19 Equation (Standard Form) of an Elliptic Cone
(Figure 1.6.14)

$$x^2/a^2 + y^2/b^2 - z^2/c^2 = 0$$

1.6.20 Equation (Standard Form) of an Elliptic Cylinder
(Figure 1.6.15)

$$x^2/a^2 + y^2/b^2 = 1$$

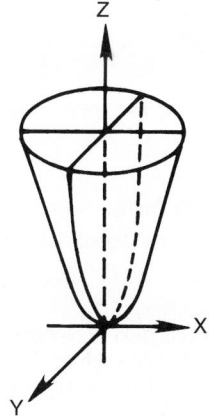

Figure 1.6.12   *Elliptic paraboloid.*

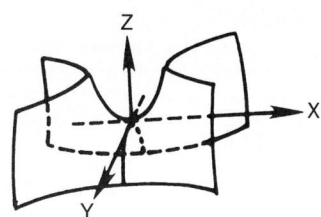

Figure 1.6.13   *Hyperbolic paraboloid.*

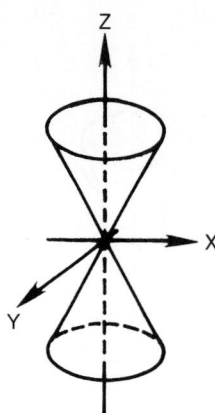

Figure 1.6.14 *Elliptic cone.*

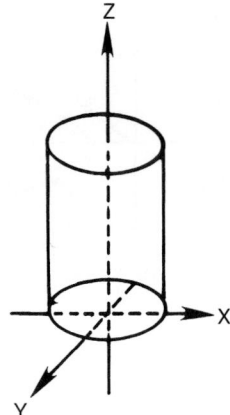

Figure 1.6.15 *Elliptic cylinder.*

1.7 NUMERICAL METHODS

See References 1–14 for additional information.

1.7.1 Expansion in Series

If the value of a function f(x) can be expressed in the region close to x = a, and if all derivatives of f(x) near a exist and are finite, then by the infinite power series

$$f(x) = f(a) + (x - a)f'(a) + \frac{(x - a)^2}{2!}f''(a) + \ldots$$

$$+ \frac{(x - a)^n}{n!}f^n(a) + \ldots$$

and f(x) is *analytic* near x = a. The preceding power series is called the *Taylor series expansion* of f(x) near x = a. If for some value of x as [x − a] is increased, the series is no longer convergent, then that value of x is outside the radius of convergence of the series.

The error due to truncation of the series is partially due to [x − a] and partially due to the number of terms (n) to which the series is taken. The quantities [x − a] and n can be controlled, and the truncation error is said to be of the order of $(x - a)^{n+1}$ or $\mathcal{O}(x - a)^{n+1}$.

1.7.2 Finite Difference Calculus

In the finite difference calculus, the fundamental rules of ordinary calculus are employed, but Δx is treated as a small quantity, rather than infinitesimal.

Given a function f(x) which is analytic (i.e., can be expanded in a Taylor series) in the region of a point x, where h = Δx, if f(x + h) is expanded about x, f′(x) can be defined at x = x_i as

$$f'(x_i) = f'_i = (f_{i+1} - f_i)/h + \mathcal{O}(h)$$

The first *forward difference* of f at x_i may be written as

$$\Delta f_i = f_{i+1} - f_i$$

and then

$$f'(x) = (\Delta f_i)/h + \mathcal{O}(h)$$

The first *backward difference* of f at x_i is

$$\nabla f_i = f_i - f_{i-1}$$

and f′(x) may also be written as

$$f'(x) = (\nabla f_i)/h + \mathcal{O}(h)$$

The second forward difference of f(x) at x_i is

$$\Delta^2 f_i = f_{i+2} - 2f_{i+1} + f_i$$

and the second derivative of f(x) is then given by

$$f''(x) = (\Delta^2 f_i)/h^2 + \mathcal{O}(h)$$

The second backward difference of f at x_i is

$$\nabla^2 f_i = f_i - 2f_{i-1} + f_{i-2}$$

and f″(x) may also be defined as

$$f''(x) = (\nabla^2 f_i)/h^2 + \mathcal{O}(h)$$

Approximate expressions for derivatives of any order are given in terms of forward and backward difference expressions as

$$f_i^{(n)} = (\Delta^n f_i)/h^n + \mathcal{O}(h) = (\nabla^n f_i)/h^n + \mathcal{O}(h)$$

Coefficients of forward difference expressions for derivatives of up to the fourth order are given in Figure 1.7.1 and of backward difference expressions in Figure 1.7.2.

More accurate difference expressions may be found by expanding the Taylor series. For example, f′(x) to ∇(h) is given by forward difference by

$$f'(x) = (-f_{i+2} + 4f_{i+1} - 3f_i)/(2h) + \mathcal{O}(h^2)$$

and a similar backward difference representation can also be easily obtained. These expressions are exact for a parabola. Forward and backward difference expressions of $\mathcal{O}(h^2)$ are contained in Figures 1.7.3 and 1.7.4.

A *central difference* expression may be derived by combining the equations for forward and backward differences.

$$\delta f_i = 1/2 \bullet (\Delta f_i + \nabla f_i) = 1/2 \bullet (f_{i+1} - f_{i-1})$$

The first derivative of f at x_i may then be given in terms of the central difference expression as

$$f'_i = (\delta f_i)/h + \mathcal{O}(h^2)$$

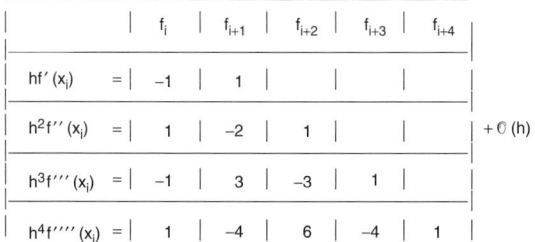

	f_i	f_{i+1}	f_{i+2}	f_{i+3}	f_{i+4}	
$hf'(x_i)$ =	−1	1				
$h^2f''(x_i)$ =	1	−2	1			+ $\mathcal{O}(h)$
$h^3f'''(x_i)$ =	−1	3	−3	1		
$h^4f''''(x_i)$ =	1	−4	6	−4	1	

Figure 1.7.1   *Forward difference coefficients of σ (h).*

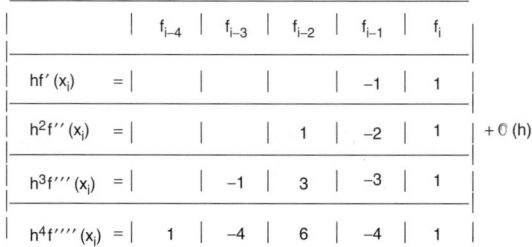

	f_{i-4}	f_{i-3}	f_{i-2}	f_{i-1}	f_i	
$hf'(x_i)$ =				−1	1	
$h^2f''(x_i)$ =			1	−2	1	+ $\mathcal{O}(h)$
$h^3f'''(x_i)$ =		−1	3	−3	1	
$h^4f''''(x_i)$ =	1	−4	6	−4	1	

Figure 1.7.2   *Backward difference coefficients of σ (h).*

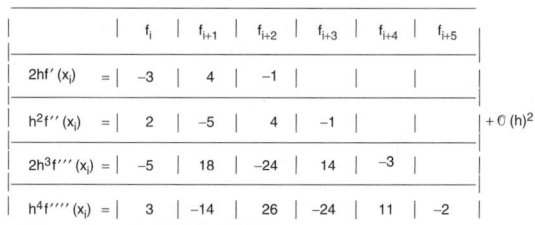

	f_i	f_{i+1}	f_{i+2}	f_{i+3}	f_{i+4}	f_{i+5}	
$2hf'(x_i)$ =	−3	4	−1				
$h^2f''(x_i)$ =	2	−5	4	−1			+ $\mathcal{O}(h)^2$
$2h^3f'''(x_i)$ =	−5	18	−24	14	−3		
$h^4f''''(x_i)$ =	3	−14	26	−24	11	−2	

Figure 1.7.3   *Forward difference coefficients of σ (h)².*

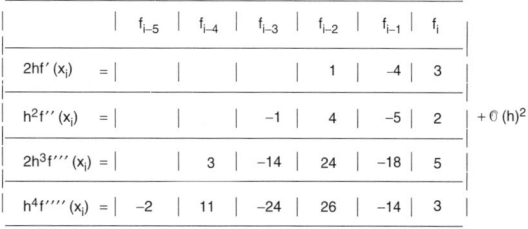

	f_{i-5}	f_{i-4}	f_{i-3}	f_{i-2}	f_{i-1}	f_i	
$2hf'(x_i)$ =				1	−4	3	
$h^2f''(x_i)$ =			−1	4	−5	2	+ $\mathcal{O}(h)^2$
$2h^3f'''(x_i)$ =		3	−14	24	−18	5	
$h^4f''''(x_i)$ =	−2	11	−24	26	−14	3	

Figure 1.7.4   *Backward difference coefficients of σ (h)².*

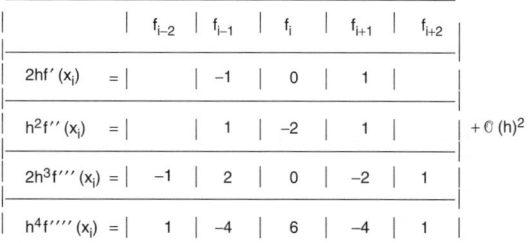

	f_{i-2}	f_{i-1}	f_i	f_{i+1}	f_{i+2}	
$2hf'(x_i)$ =	−1	0	1			
$h^2f''(x_i)$ =	1	−2	1			+ $\mathcal{O}(h)^2$
$2h^3f'''(x_i)$ =	−1	2	0	−2	1	
$h^4f''''(x_i)$ =	1	−4	6	−4	1	

Figure 1.7.5   *Central difference coefficients of σ (h)².*

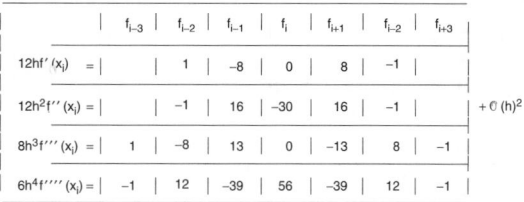

	f_{i-3}	f_{i-2}	f_{i-1}	f_i	f_{i+1}	f_{i+2}	f_{i+3}	
$12hf'(x_i)$ =		1	−8	0	8	−1		
$12h^2f''(x_i)$ =		−1	16	−30	16	−1		+ $\mathcal{O}(h)^2$
$8h^3f'''(x_i)$ =	1	−8	13	0	−13	8	−1	
$6h^4f''''(x_i)$ =	−1	12	−39	56	−39	12	−1	

Figure 1.7.6   *Backward difference coefficients of σ (h)⁴.*

and is accurate to a greater degree than the forward or backward expressions of f'. Central difference expressions for derivatives of any order in terms of forward and backward differences are given by

$$f_i^{(n)} = [\nabla^n f_{i+n/2} + \Delta^n f_{i-n/2}]/(2h^n) + \mathcal{O}(h^2), \ n \text{ even}$$

and

$$f_i^{(n)} = [\nabla^n f_{i+(n-1)/2} + \Delta^n f_{i-(n-1)/2}]/(2h^n) + \mathcal{O}(h^2), n \text{ odd}$$

Coefficients of central difference expressions for derivatives up to order four of $\mathcal{O}(h^2)$ are given in Figure 1.7.5 and of $\mathcal{O}(h^4)$ in Figure 1.7.6.

1.7.3 Interpolation

A *forward difference table* may be generated (see also "Algebra") using notation consistent with numerical methods as given in Table 1.7.1. In a similar manner, a *backward difference table* can be calculated as in Table 1.7.2. A *central difference table* is constructed in the same general manner, leaving a space between each line of original data, then taking the differences and entering them on alternate full lines and half lines (Table 1.7.3). The definition of the central difference δ is

$$\delta f_{i+1/2} = f_{i+1} - f_i$$

The quarter lines in the table are filled with the arithmetic mean of the values above and below (Table 1.7.4).

Given a data table with evenly spaced values of x, and rescaling x so that h = one unit, forward differences are usually used to find f(x) at x near the top of the table and backward differences at x near the bottom. Interpolation near the center of the set is best accomplished with central differences.

The *Gregory-Newton forward formula* is given as

$$f(x) = f(0) + x(\Delta f_0) + \frac{x(x-1)}{2!}\Delta^2 f_0$$

$$+ \frac{x(x-1)(x-2)}{3!}\Delta^3 f_0 + \ldots$$

and the *Gregory-Newton backward formula* as

$$f(x) = f(0) + x(\nabla f_0) + \frac{x(x+1)}{2!}\nabla^2 f_0$$

$$+ \frac{x(x+1)(x+2)}{3!}\nabla^3 f_0 + \ldots$$

To use central differences, the origin of x must be shifted to a base line (shaded area in Table 1.7.5) and x rescaled so one full (two half) line spacing = 1 unit. *Sterling's formula*

Table 1.7.1 *Forward Difference Table*

x	f(x)	Δf	$\Delta^2 f$	$\Delta^3 f$	$\Delta^4 f$	$\Delta^5 f$
0	0	2	-2	4	2	1
1	2	0	2	6	3	
2	2	2	8	9		
3	4	10	17			
4	14	27				
5	41					

Table 1.7.2 *Backward Difference Table*

x	f(x)	Δf	$\Delta^2 f$	$\Delta^3 f$	$\Delta^4 f$	$\Delta^5 f$
0	0					
1	2	2				
2	2	0	-2			
3	4	2	2	4		
4	14	10	8	6	2	
5	41	27	17	9	3	1

Table 1.7.3 *Central Difference Table (Original Data)*

x	f(x)	Δf	$\Delta^2 f$	$\Delta^3 f$	$\Delta^4 f$	$\Delta^5 f$
0	0					
		2				
1	2		-2			
		0		4		
2	2		2		2	
		2		6		1
3	4		8		3	
		10		9		
4	14		17			
		27				
5	41					

Table 1.7.4 *Central Difference Table (Filled)*

x	f(x)	Δf	$\Delta^2 f$	$\Delta^3 f$	$\Delta^4 f$	$\Delta^5 f$
0	0					
0.5	1	2				
1	2	1	-2			
1.5	2	0	0	4		
2	2	1	2	5	2	
2.5	3	2	5	6	2.5	1
3	4	6	8	7.5	3	
3.5	9	10	12.5	9		
4	14	18.5	17			
4.5	27.5	27				
5	41					

Table 1.7.5 *Central Difference Table with Base Line*

Old x	New x	f(x)	Δf	$\Delta^2 f$	$\Delta^3 f$	$\Delta^4 f$	$\Delta^5 f$
0	-2.5	0					
0.5	-2.0	1	2				
1	-1.5	2	1	-2			
1.5	-1.0	2	0	0	4		
2	-0.5	2	1	2	5	2	
2.5	0.0	3	2	5	6	2.5	1
3	+0.5	4	6	8	7.5	3	
3.5	+1.0	9	10	12.5	9		
4	+1.5	14	18.5	17			
4.5	+2.0	27.5	27				
5	+2.5	41					

(full lines as base) is defined as

$$f(x) = f(0) + x(\delta y_0) + \frac{x^2}{2!}(\delta^2 y_0) + \frac{x(x^2-1)}{3!}(\delta^3 y_0)$$

$$+ \frac{x^2(x^2-1)}{4!}(\delta^4 y_0) + \frac{x(x^2-1)(x^4-4)}{5!}(\delta^5 y_0) + \dots$$

and *Bessel's formula* (half line as base) as

$$f(x) = f(0) + x(\delta y_0) = \frac{\left(x^2 - \frac{1}{4}\right)}{2!}(\delta^2 y_0) + \frac{x\left(x^2 - \frac{1}{4}\right)}{3!}(\delta^3 y_0)$$

$$+ \frac{\left(x^2 - \frac{1}{4}\right)\left(x^2 - \frac{9}{4}\right)}{4!}(\delta^4 y_0)$$

$$+ \frac{x\left(x^2 - \frac{1}{4}\right)\left(x^2 - \frac{9}{4}\right)}{5!}(\delta^5 y_0) + \dots$$

Interpolation with nonequally spaced data may be accomplished by the use of *Lagrange Polynomials*, defined as a set of nth degree polynomials such that each one, $P_j(x)$ ($j = 0, 1, \dots, n$), passes through zero at each of the data points except one, x_k, where $k = j$. For each polynomial in the set

$$P_j(x) = A_j \prod_{\substack{i=0 \\ i \neq j}}^{n} (x - x_i)$$

where if

$$A_j = \frac{1}{\displaystyle\prod_{\substack{i=0 \\ i \neq j}}^{n} (x_j - x_i)}$$

then

$$P_j(x) = \begin{cases} 0, & k \neq j \\ 1, & k = j \end{cases}$$

and the linear combination of $P_j(x)$ may be formed

$$p_n(x) = \sum_{j=0}^{n} f(x_j) P_j(x)$$

It can be seen that for any x_i, $p_n(x_i) = f(x_i)$.

Interpolation of this type may be extremely unreliable toward the center of the region where the independent variable is widely spaced. If it is possible to select the values of x for which values of f(x) will be obtained, the maximum error can be minimized by the proper choices. In this particular case Chebyshev polynomials can be computed and interpolated [3].

Neville's algorithm constructs the same unique interpolating polynomial and improves the straightforward Lagrange implementation by the addition of an error estimate.

If P_i ($i = 1, \dots, n$) is defined as the value at x of the unique polynomial of degree zero passing through the point (x_i, y_i) and P_{ij} ($i = 1, \dots, n - 1, j = 2, \dots, n$) the polynomial of degree one passing through both (x_i, y_i) and (x_i, y_i), then the higher-order polynomials may likewise be defined up to $P_{123\dots n}$, which is the value of the unique interpolating polynomial passing through all n points. A table may be constructed (e.g., if $n = 3$):

$$\begin{array}{l} x_1 : y_1 = P_1 \\ x_2 : y_2 = P_2 \quad P_{12} \\ x_3 : y_3 = P_3 \quad P_{23} \end{array} \quad P_{123}$$

Neville's algorithm recursively calculates the preceding columns from left to right as

$$P_{i(i+1)\ldots(i+m)}$$

$$= \frac{(x - x_{i+m})P_{i(i+1)\ldots(i+m-1)} + (x_i - x)P_{(i+1)\ldots(i+m)}}{x_i - x_{i+m}}$$

In addition, the differences between the columns may be calculated as

$$D_{m+1,i} = \frac{(x_{i+m+1} - x)(C_{m,i+1} + D_{m,i})}{x_i - x_{i+m+1}}$$

$$C_{m+1,i} = \frac{(x_i - x)(C_{m,i+1} - D_{m,i})}{x_i - x_{i+m+1}}$$

and $P_{12\ldots n}$ is equal to the sum of any y_i plus a set of Cs and/or Ds that lead to the rightmost member of the table [14].

Functions with localized strong inflections or poles may be approximated by *rational functions* of the general form

$$R(x) = \frac{\sum_{i=0}^{n} a_i x^i}{\sum_{i=0}^{m} b_i x^i}$$

as long as there are sufficient powers of x in the denominator to cancel any nearby poles. Stoer and Bulirsch [8] give a Neville-type algorithm that performs rational function extrapolation on tabulated data

$$R_{i(i+1)\ldots(i+m)} = R_{(i+1)\ldots(i+m)} + \Big\{ \big(R_{(i+1)\ldots(i+m)}$$

$$- R_{i\ldots(i+m-1)}\big) \Big/ \big[(x - x_i)/(x - x_{i+m})\big]$$

$$\times \big[1 - \big(R_{(i+1)\ldots(i+m)} - R_{i\ldots(i+m-1)}\big)$$

$$\Big/\big(R_{(i+1)\ldots(i+m)} - R_{(i+1)\ldots(i+m-1)}\big)\big] - 1\Big\}$$

starting with $R_i = y_i$ and returning an estimate of error, calculated by C and D in a manner analogous with Neville's algorithm for polynomial approximation.

In a high-order polynomial, the highly inflected character of the function can more accurately be reported by the *cubic spline function*. Given a series of $x_i (i = 0, 1, \ldots, n)$ and corresponding $f(x_i)$, consider that for two arbitrary and adjacent points x_i and x_{i+j}, the cubic fitting these points is

$$F_i(x) = a_0 + a_1 x + a_2 x^2 + a_3 x^3$$

$$(x_i \leq x \leq x_{i+1})$$

The approximating cubic spline function $g(x)$ for the region $(x_0 \leq x \leq x_n)$ is constructed by matching the first and second derivatives (slope and curvature) of $F_i(x)$ to those of $F_{i-1}(x)$, with special treatment (outlined below) at the end points, so that $g(x)$ is the set of cubics $F_i(x)$, $i = 0, 1, 2, \ldots, n - 1$, and the second derivative $g''(x)$ is continuous over the region. The second derivative varies linearly over $[x_0, x_n]$ and at any $x(x_i \leq x \leq x_{i+1})$

$$g''(x) = g''(x_i) + \frac{x - x_i}{x_{i+1} - x_i}[g''(x_{i+1}) - g''(x_i)]$$

Integrating twice and setting $g(x_i) = f(x_i)$ and $g(x_{i+1}) = f(x_{i+1})$, then using the derivative matching conditions

$$F_i'(x_i) = F_{i+1}'(x_i) \quad \text{and} \quad F_i''(x_i) = F_{i-1}''(x_i)$$

and applying the condition for $i = [1, n - 1]$ finally yields a set of linear simultaneous equations of the form

$$[\Delta x_{i-1}]g''(x_{i-1}) + [2(x_{i+1} - x_{i-1})]g''(x_i) + [\Delta x_i]g''(x_{i+1})$$

$$= 6\left[\frac{f(x_{i+1}) - f(x_i)}{\Delta x_i} - \frac{f(x_i) - f(x_{i-1})}{\Delta x_{i-1}}\right]$$

where $i = 1, 2, \ldots, n - 1$

$$\Delta x_i = x_{i+1} - x_i$$

If the x_i are equally spaced by Δx, then the preceding equation becomes

$$[1]g''(x_{i-1}) + [4]g''(x_i) + [1]g''(x_{i+1})$$

$$= 6\left[\frac{f(x_{i+1}) - 2f(x_i) + f(x_{i-1})}{(\Delta x_i)^2}\right]$$

There are $n - 1$ equations in $n + 1$ unknowns and the two necessary additional equations are usually obtained by setting

$$g''(x_0) = 0 \quad \text{and} \quad g''(x_n) = 0$$

and $g(x)$ is now referred to as a *natural cubic spline*. $g''(x_0)$ or $g''(x_n)$ may alternatively be set to values calculated so as to make g' have a specified value on either or both boundaries. The cubic appropriate for the interval in which the x value lies may now be calculated (see "Solutions of Sets of Simultaneous Linear Equations").

Extrapolation is required if $f(x)$ is known on the interval [a, b], but values of $f(x)$ are needed for x values not in the interval. In addition to the uncertainties of interpolation, extrapolation is further complicated since the function is fixed only on one side. Gregory-Newton and Lagrange formulas may be used for extrapolation (depending on the spacing of the data points), but all results should be viewed with extreme skepticism.

1.7.4 Roots of Equations

Finding the root of an equation in x is the problem of determining the values of x for which $f(x) = 0$. *Bisection*, although rarely used now, is the basis of several more efficient methods. If a function $f(x)$ has one and only one root in [a, b], then the interval may be bisected at $x_m = (a + b)/2$. If $f(x_m) \bullet f(b) < 0$, the root is in $[x_m, b]$, while if $f(x_m) \bullet f(b) > 0$, the root is in $[a, x_m]$. Bisection of the appropriate intervals, where $x_m' = (a' + b')/2$, is repeated until the root is located $\pm \varepsilon$, ε being the maximum acceptable error and $\varepsilon \leq 1/2 \bullet$ size of interval.

The *Regula Falsa* method, or the method of false position, is a refinement of the bisection method, in which the new end point of a new interval is calculated from the old end points by

$$x_m = a - (b - a)\frac{f(a)}{f(b) - (a)}$$

Whether x_m replaces a or replaces b depends on the sign of a product, and

if $f(a) \bullet f(x_m) < 0$, then the new interval is $[a, x_m]$

or

if $f(x_m) \bullet f(b) < 0$, then the new interval is $[x_m, b]$.

Because of round off errors, the Regula Falsa method should include a check for excessive iterations. A *modified Regula Falsa method* is based on the use of a *relaxation factor*, i.e., a number used to alter the results of one iteration before inserting into the next. (See the section on relaxation methods and "Solution of Sets of Simultaneous Linear Equations.")

By iteration, the general expression for the *Newton-Raphson method* may be written (if f' can be evaluated and is continuous near the root):

$$x^{(n+1)} - x^{(n)} = \delta^{(n+1)} = -\frac{f(x^{(n)})}{f'(x^{(n)})}$$

where (n) denotes values obtained on the nth iteration and $(n+1)$ those obtained on the $(n+1)^{th}$ iteration. The iterations are terminated when the magnitude of $|\delta^{(n+1)} - \delta^{(n)}| < \varepsilon$, being the predetermined error factor and $\varepsilon \cong 0.1$ of the permissible error in the root.

The *modified Newton method* [4] offers one way of dealing with multiple roots. If a new function is defined

$$u(x) = \frac{f(x)}{f'(x)}$$

because $u(x) = 0$ when $f(x) = 0$ and if $f(x)$ has a multiple root at $x = c$ of multiplicity r, then Newton's method can be applied and

$$x^{(n+1)} - x^{(n)} = \delta^{(n+1)} = -\frac{u(x^{(n)})}{u'(x^{(n)})}$$

where

$$u'(x) = 1 - \frac{f(x)f''(x)}{[f'(x)]^2}$$

If multiple or closely spaced roots exist, both f and f' may vanish near a root and methods that depend on tangents will not work. Deflation of the polynomial $P(x)$ produces, by factoring,

$$P(x) = (x - r)Q(x)$$

where $Q(x)$ is a polynomial of one degree lower than $P(x)$ and the roots of Q are the remaining roots of P after factorization by synthetic division. Deflation avoids convergence to the same root more than one time. Although the calculated roots become progressively more inaccurate, errors may be minimized by using the results as initial guesses to iterate for the actual roots in P.

Methods such as Graeffe's root-squaring method, Muller's method, Laguerre's method, and others exist for finding all roots of polynomials with real coefficients [4, 7, 8].

1.7.5 Solution of Sets of Simultaneous Linear Equations

A matrix is a rectangular array of numbers, its size being determined by the number of rows and columns in the array. In this context, the primary concern is with square matrices, and matrices of column dimension 1 (column vectors) and row dimension 1 (row vectors).

Certain configurations of square matrices are of particular interest. If

$$C = \begin{vmatrix} c_{11} & \cdot & \cdot & c_{14} \\ \cdot & \cdot & \cdot & \cdot \\ \cdot & \cdot & \cdot & \cdot \\ c_{41} & \cdot & \cdot & c_{44} \end{vmatrix}$$

the diagonal consisting of c_{11}, c_{22}, c_{33} and c_{44} is the *main diagonal*. The matrix is *symmetric* if $c_{ij} = c_{ji}$. If all elements below the main diagonal are zero (blank), it is an *upper triangular matrix*, while if all elements above the main diagonal are zero, it is a *lower triangular matrix*. If all elements are zero except those on the main diagonal, the matrix is a *diagonal matrix* and if a diagonal matrix has all ones on the diagonal, it is the *unit, or identity, matrix*.

Matrix addition (or subtraction) is denoted as $S = A + B$ and defined as

$$s_{ij} = a_{ij} + b_{ij}$$

where A, B, and S have identical row and column dimensions. Also,

$$A + B = B + A$$
$$A - B = -B + A$$

Matrix multiplication, represented as $P = AB$, is defined as

$$P_{ij} = \sum_{k=1}^{n} a_{ik}b_{kj}$$

where n is the column dimension of A and the row dimension of B. P will have row dimension of A and column dimension of B. Also

$$AI = A$$

and

$$IA = A$$

while, in general,

$$AB \neq BA$$

Matrix division is not defined, although if C is a square matrix, C^{-1} (the *inverse* of C) can usually be defined so that

$$CC^{-1} = I$$

and

$$(C^{-1})^{-1} = C$$

The *transpose* of A if

$$A = \begin{vmatrix} a_{11} & a_{12} \\ a_{21} & a_{22} \\ a_{31} & a_{32} \end{vmatrix}$$

is

$$A^T = \begin{vmatrix} a_{11} & a_{21} & a_{31} \\ a_{12} & a_{22} & a_{32} \end{vmatrix}$$

A square matrix C is *orthogonal* if

$$C^T = C^{-1}$$

The *determinant* of a square matrix C (det C) is defined as the sum of all possible products found by taking one element from each row in order from the top and one element from each column, the sign of each product multiplied by $(-1)^r$, where r is the number of times the column index decreases in the product.

For a 2×2 matrix

$$C = \begin{vmatrix} c_{11} & c_{12} \\ c_{21} & c_{22} \end{vmatrix}$$

$$\det C = c_{11}c_{22} - c_{12}c_{21}$$

(Also see discussion of determinants in "Algebra.")

Given a set of simultaneous equations, for example, four equations in four unknowns:

$$c_{11}x_1 + c_{12}x_2 + c_{13}x_3 + c_{14}x_4 = r_1$$
$$c_{21}x_1 + c_{22}x_2 + c_{23}x_3 + c_{24}x_4 = r_2$$
$$c_{31}x_1 + c_{32}x_2 + c_{33}x_3 + c_{34}x_4 = r_3$$
$$c_{41}x_1 + c_{42}x_2 + c_{43}x_3 + c_{44}x_4 = r_4$$

and in matrix form

$$\begin{vmatrix} c_{11} & c_{12} & c_{13} & c_{14} \\ c_{21} & c_{22} & c_{23} & c_{24} \\ c_{31} & c_{32} & c_{33} & c_{34} \\ c_{41} & c_{42} & c_{43} & c_{44} \end{vmatrix} \begin{vmatrix} x_1 \\ x_2 \\ x_3 \\ x_4 \end{vmatrix} = \begin{vmatrix} r_1 \\ r_2 \\ r_3 \\ r_4 \end{vmatrix}$$

or

$$CX = R$$

The solution for x_k in a system of equations such as given in the matrix above is

$$x_k = (\det C_k)/(\det C)$$

where C_k is the matrix C, with its kth column replaced by R (Cramer's Rule). If det $C = 0$, C and its equations are singular and there is no solution.

Sets of simultaneous linear equations are frequently defined as [12]

- *Sparse* (many zero elements) and large
- *Dense* (few zero elements) and small. A *banded matrix* has all zero elements except for a band centered on the main diagonal, e.g.,

$$C = \begin{vmatrix} c_{11} & c_{12} & & \\ c_{21} & c_{22} & c_{23} & \\ & c_{32} & c_{33} & c_{34} \\ & & c_{43} & c_{44} \end{vmatrix}$$

then C is a banded matrix of bandwidth 3, also called a *tridiagonal matrix*.

Equation-solving techniques may be defined as *direct*, expected to yield results in a predictable number of operations, or *iterative*, yielding results of increasing accuracy with increasing numbers of iterations. Iterative techniques are in general preferable for very large sets and for large, sparse (not banded) sets. Direct methods are usually more suitable for small, dense sets and also for sets having banded coefficient matrices.

Gaussian elimination is the sequential application of the two operations:

1. Multiplication, or division, of any equation by a constant.
2. Replacement of an equation by the sum, or difference, of that equation and any other equation in the set, so that a set of equations

$$\begin{vmatrix} c_{11} & c_{12} & c_{13} & c_{14} \\ c_{21} & c_{22} & c_{23} & c_{24} \\ c_{31} & c_{32} & c_{33} & c_{34} \\ c_{41} & c_{42} & c_{43} & c_{44} \end{vmatrix} \begin{vmatrix} x_1 \\ x_2 \\ x_3 \\ x_4 \end{vmatrix} = \begin{vmatrix} r_1 \\ r_2 \\ r_3 \\ r_4 \end{vmatrix}$$

becomes, by division of the first equation by c_{11},

$$\begin{vmatrix} 1 & c'_{12} & c'_{13} & c'_{14} \\ c_{21} & c_{22} & c_{23} & c_{24} \\ c_{31} & c_{32} & c_{33} & c_{34} \\ c_{41} & c_{42} & c_{43} & c_{44} \end{vmatrix} \begin{vmatrix} x_1 \\ x_2 \\ x_3 \\ x_4 \end{vmatrix} = \begin{vmatrix} r'_1 \\ r_2 \\ r_3 \\ r_4 \end{vmatrix}$$

then, by replacement of the next three equations,

$$\begin{vmatrix} 1 & c'_{12} & c'_{13} & c'_{14} \\ 0 & c'_{22} & c'_{23} & c'_{24} \\ 0 & c'_{32} & c'_{33} & c'_{34} \\ 0 & c'_{42} & c'_{43} & c'_{44} \end{vmatrix} \begin{vmatrix} x_1 \\ x_2 \\ X_3 \\ x_4 \end{vmatrix} = \begin{vmatrix} r'_1 \\ r'_2 \\ r'_3 \\ r'_4 \end{vmatrix}$$

and finally

$$\begin{vmatrix} 1 & c'_{12} & c'_{13} & c'_{14} \\ & 1 & c'_{23} & c'_{24} \\ & & 1 & c'_{34} \\ & & & 1 \end{vmatrix} \begin{vmatrix} x_1 \\ x_2 \\ x_3 \\ x_4 \end{vmatrix} = \begin{vmatrix} r'_1 \\ r'_2 \\ r'_3 \\ r'_4 \end{vmatrix}$$

Gauss-Jordan elimination is a variation of the preceding method, which by continuation of the same procedures yields

$$\begin{vmatrix} 1 & & & \\ & 1 & & \\ & & 1 & \\ & & & 1 \end{vmatrix} \begin{vmatrix} x_1 \\ x_2 \\ x_3 \\ x_4 \end{vmatrix} = \begin{vmatrix} r''_1 \\ r''_2 \\ r''_3 \\ r''_4 \end{vmatrix}$$

Therefore, $x_1 = r''_1$, etc., i.e., the r vector is the solution vector. If the element in the current pivot position is zero or very small, switch the position of the entire pivot row with any row below it, including the x vector element, but not the r vector element.

If det $C \neq 0$, C^{-1} exists and can be found by *matrix inversion* (a modification of the Gauss-Jordan method), by writing C and I (the identity matrix) and then performing the same operations on each to transform C into I and, therefore, I into C^{-1}.

For any square matrix, a condition number can be defined as the product of the norm of the matrix and the norm of its inverse. If this number is large, the matrix is *ill conditioned*. For an ill-conditioned matrix, it can be difficult to compute the inverse. Two quick ways to recognize possible ill conditioning are

1. If there are elements of the inverse of the matrix that are larger than elements of the original matrix.
2. If the magnitude of the determinant is small, such as

$$\frac{\det C}{\sqrt{\sum_{i=1}^{n} \sum_{j=1}^{n} c_{ij}^2}} \leq 1$$

Gauss-Siedel method is an iterative technique for the solution of sets of equations. Given, for example, a set of three linear equations

$$c_{11}x_1 + c_{12}x_2 + c_{13}x_3 = r_1$$
$$c_{21}x_1 + c_{22}x_2 + c_{23}x_3 = r_2$$
$$c_{31}x_1 + c_{32}x_2 + c_{33}x_3 = r_3$$

solving for the unknowns yields

$$x_1 = \frac{r_1 - c_{12}x_2 - c_{13}x_3}{c_{11}}$$
$$x_2 = \frac{r_2 - c_{21}x_1 - c_{23}x_3}{c_{22}}$$
$$x_3 = \frac{r_3 - c_{31}x_1 - c_{32}x_2}{c_{33}}$$

By making an initial guess for x_1, x_2, and x_3, denoted as x_1^0, x_2^0, and x_3^0, the value of x_1 on the first iteration is

$$x_1^{(1)} = \frac{r_1 - c_{12}x_2^{(0)} - c_{13}x_3^{(0)}}{c_{11}}$$

Using the most recently obtained values for each unknown (as opposed to the fixed point or Jacobi method), then

$$x_2^{(1)} = \frac{r_2 - c_{21}x_1^{(1)} - c_{23}x_3^{(0)}}{c_{22}}$$
$$x_3^{(1)} = \frac{r_3 - c_{31}x_1^{(1)} - c_{32}x_2^{(1)}}{c_{33}}$$

If the equations have the proper characteristics, the iterative process will eventually converge. Commonly used convergence criteria are of two types:

1. *Absolute convergence* criteria of the form

$$\left| x_i^{(n+1)} - x_i^{(n)} \right| \leq \varepsilon$$

are most useful when approximate magnitudes of x_i are known beforehand so that ε may be chosen to be proportional to x_i.

2. *Relative convergence* criteria of the form

$$\left| \left(x_i^{(n+1)} - x_i^{(n)} \right) \Big/ x_i^{(n+1)} \right| \leq \varepsilon$$

is the choice if the magnitudes of x_i are uncertain.

Relaxation methods may also be used to modify the value of an unknown before it is used in the next calculation. The effect of the relaxation factor λ may be seen in the following equation, where $x_i^{(n+1)\bullet}$ is the value obtained at the present iteration.

$$x_i^{(n+1)} = \lambda x_i^{(n+1)\bullet} + (1 - \lambda)x_i^{(n)}$$

and $0 < \lambda < 2$. If $0 < \lambda < 1$, the effect is termed under relaxation, which is frequently employed to produce convergence in a nonconvergent process. If $1 < \lambda < 2$, the effect, overrelaxation, will be to accelerate an already convergent process.

1.7.6 Least Squares Curve Fitting

For a function f(x) given only as discrete points, the measure of accuracy of the fit is a function $d(x) = |f(x) - g(x)|$ where g(x) is the approximating function to f(x). If this is interpreted as minimizing $d(x)$ over all x in the interval, one point in error can cause a major shift in the approximating function towards that point. The better method is the least squares curve fit, where $d(x)$ is minimized if

$$E = \sum_{i=1}^{n} [g(x_i) - f(x_i)]^2$$

is minimized, and if $g(x_i)$ is a polynomial of order m

$$E = \sum_{i=1}^{n} \left[a_0 + a_1 x_i + a_2 x_i^2 + \ldots + a_m x_i^m - f(x_i)^2 \right]$$

Setting the partial derivatives of E with respect to each of the coefficients of g(x) equal to zero, differentiating and summing over $1, \ldots, n$ forms a set of $m + 1$ equations [9] so that

$$\begin{vmatrix} n & \sum x_i & \sum x_i^2 \ldots \\ \sum x_i & \sum x_i^2 & \sum x_i^3 \ldots \\ \ldots & & \end{vmatrix} \begin{vmatrix} a_0 \\ a_1 \\ \ldots \end{vmatrix} = \begin{vmatrix} \sum f(x_i) \\ \sum x_i f(x_i) \\ \ldots \end{vmatrix}$$

If the preceding solution is reduced to a linear approximation ($n = 1$), the matrix will be ($n = 1$)

$$\begin{vmatrix} n & \sum x_i \\ \sum x_i & \sum x_i^2 \end{vmatrix} \begin{vmatrix} a_0 \\ a_1 \end{vmatrix} = \begin{vmatrix} \sum f(x)_i \\ \sum x_i f(x_i) \end{vmatrix}$$

and for a parabola ($n = 2$), the first three rows and columns.

Another possible form is the exponential function

$$F(t) = \alpha e^{\beta t}$$

and although partial differentiation will produce two equations in two unknowns, they will be nonlinear and cannot be written in matrix form. However, a change in variable form may produce a model that is linear, for example, for the preceding equation

$$\ln(F) = \ln(\alpha) + \beta t$$

and if X is defined to be t, Y to be ln (F), $a_0 = \ln(\alpha)$, and $a_1 = \beta$, the equation becomes

$$Y(x) = a_0 + a_1 X$$

and linear least squares analysis may be applied.

In order to determine the quality (or the validity) of fit of a particular function to the data points given, a comparison of the deviation of the curve from the data to the size of the experimental error can be made. The deviations (i.e., the scatter off the curve) should be of the same order of magnitude as the experimental error, so that the quantity "chi-squared" is defined as

$$X^2 = \sum_{i=1}^{n} \frac{[y_i' - y_i]^2}{(\Delta y_i)^2}$$

where $y_i' =$ is the fitted function and y_i is the measured value of y at x_i, so that Δy_i is the magnitude of the error of y_i'. The sum is over n points and if the number of parameters in the model function is g_2, then if $\mathcal{O}(X^2) > \mathcal{O}(n - g)$, the approximating function is a poor fit, while if $\mathcal{O}(X^2) < \mathcal{O}(n - g)$, the function may be overfit, representing noise [10].

The simplest form of approximation to a *continuous* function is some polynomial. Continuous functions may be approximated in order to provide a "simpler form" than the original function. Truncated power series representations (such as the Taylor series) are one class of polynomial approximations.

Table 1.7.6   *Chebyshev Polynomials*

$T_0(x) = 1$
$T_1(x) = x$
$T_2(x) = 2x^2 - 1$
$T_3(x) = 4x^3 - 3x$
$T_4(x) = 8x^4 - 8x^2 + 1$
$T_5(x) = 16x^5 - 20x^3 + 5x$
$T_6(x) = 32x^6 - 38x^4 + 18x^2 - 1$
$T_7(x) = 64x^7 - 112x^5 + 56x^3 - 7x$
$T_8(x) = 128x^8 - 256x^6 + 160x^4 - 32x^2 + 1$

Table 1.7.7   *Inverted Chebyshev Polynomials*

$1 = T_0$
$x = T_1$
$x^2 = \frac{1}{2}(T_0 + T_2)$
$x^3 = \frac{1}{4}(3T_1 + T_3)$
$x^4 = \frac{1}{8}(3T_0 + 4T_2 + T_4)$
$x^5 = \frac{1}{16}(10T_1 + 5T_3 + T_5)$
$x^6 = \frac{1}{32}(10T_0 + 15T_2 + 6T_4 + T_6)$
$x^7 = \frac{1}{64}(35T_1 + 21T_3 + 7T_5 + T_7)$
$x^8 = \frac{1}{128}(35T_0 + 56T_2 + 28T_4 + 8T_6 + T_8)$

The Chebyshev polynomials $T_n(x)$ or T_n, which exist on the interval $-1 \leq x \leq 1$, form the series

$$T_0 = 1$$
$$T_1 = x$$
$$\cdot$$
$$\cdot$$
$$T_{n+1} = 2xT_n - T_{n-1}$$

(see Table 1.7.6 for a more complete list). If these polynomials are inverted, powers of x are given in terms of $T_n(x)$ (Table 1.7.7). Any finite interval $a \leq y \leq b$ can be mapped onto the interval $-1 \leq x \leq 1$ by the formula

$$x = (2y - b - a)/(b - a)$$

and the inverted Chebyshev polynomials can be substituted for powers of x in a power series representing any function f(x). Because the maximum magnitude for $T_n = 1$ because of the interval, the sum of the magnitudes of lower-order terms is relatively small. Therefore, even with truncation of the series after comparatively few terms, the altered series can provide sufficient accuracy.

See also the discussion on cubic splines in "Interpolation."

1.7.7 Numerical Integration

By assuming that a function can be replaced over a limited range by a simpler function and by first considering the simplest function, a straight line, the areas under a complicated curve may be approximated by the *trapezoidal rule*. The area is subdivided into n panels and

$$I = T_n = \frac{1}{2}\Delta x_n \left(f_a + 2\sum_{i=1}^{n-1} f_i + f_b \right)$$

where $\Delta x_n = (b - a)/n$ and f_i is the value of the function at each x_i. If the number of panels $n = 2^k$, an alternate form of the trapezoidal can be given, where

$$I = T_k = \frac{1}{2}T_{k-1} + \Delta x_k \sum_{\substack{i=1 \\ i \text{ odd}}}^{n-1} f(a + i\Delta x_k)$$

where $\Delta x_k = (b - a)/2^k$, $T_0 = \frac{1}{2}(f_a + f_b)(b - a)$, and the equation for T_k is repeatedly applied for $k = 1, 2, \ldots$ until sufficient accuracy has been obtained.

If the function $f(x)$ is approximated by parabolas, *Simpson's rule* is obtained, by which (the number of panels n being even)

$$I = S_n = \frac{1}{3}\Delta x \left[f_0 + 4\sum_{\substack{i=1 \\ i \text{ odd}}}^{n-1} f_i + 2\sum_{\substack{i=2 \\ i \text{ even}}}^{n-2} f_i + f_n \right] + E$$

where E is the dominant error term involving the fourth derivative of f, so that it is impractical to attempt to provide error correction by approximating this term. Instead, Simpson's rule *with end correction* (sixth order rather than fourth order) may be applied where

$$I = S_n = \frac{1}{15}\Delta x \left[14\left\{ \frac{1}{2}(f_0 + f_n) + \sum_{\substack{i=2 \\ i \text{ even}}}^{n-2} f_i \right\} \right.$$
$$\left. + 16\sum_{\substack{i=1 \\ i \text{ odd}}}^{n-1} f_i + \Delta x[f'(a) - f'(b)] \right]$$

The original Simpson's formula without end correction may be generalized in a similar way as the trapezoidal formula for $n = 2^k$ panels, using $\Delta x_k = (b - a)/2^k$ and increasing k until sufficient accuracy is achieved, where

$$S_k = \frac{1}{3}\Delta x_k \left[f_a + 4\sum_{\substack{i=1 \\ i \text{ odd}}}^{n-1} f(a + i\Delta x_k) + 2\sum_{\substack{i=2 \\ i \text{ even}}}^{n-2} f(a + i\Delta x_k) + f_b \right]$$

For the next higher level of integration algorithm, $f(x)$ over segments of [a, b] can be approximated by a cubic, and if this k^{th} order result is C_k, then *Cote's rule* can be given as

$$C_k = S_k + (S_k - S_{k-1})/15$$

and the next higher degree approximation as

$$D_k = C_k + (C_k - C_{k-1})/63$$

The limit suggested by the sequence

$$T_k \to S_k \to C_k \to D_k \to \ldots$$

is known as *Romberg integration*.

If a new notation $T_k^{(m)}$ is defined, where k is the order of the approximation ($n = 2^k$) and m is the level of the integration algorithm, then $m = 0$ (trapezoidal rule)

$$T_k^{(0)} = T_k$$

$m = 1$ (Simpson's rule)

$$T_k^{(1)} = S_k$$

$m = 2$ (Cote's rule)

$$T_k^{(2)} = C_k$$

$m = 3$

$$T_k^{(3)} = D_k$$

and so forth.

The generalization of the preceding definitions leads to the Romberg equation

$$T_k^{(m+1)} = T_k^{(m)} + \frac{T_k^{(m)} - T_{k-1}^{(m)}}{4^{(m+1)} - 1}$$

The procedure is to start with the one-panel trapezoidal rule

$$T_0 \to T_0^{(0)} = \frac{1}{2}(b - a)(f_a + f_b)$$

and then increase the order (k) of the calculation by

$$T_k = \frac{1}{2}T_{k-1} + \Delta x_k \bullet \sum_{\substack{i=1 \\ i \text{ odd}}}^{n-1} f(a + i\Delta k_k)$$

and next increase the level of the algorithm m by the equation for $T_k^{(m+1)}$ just shown. In terms of $T_k^{(m)}$ for the first few k and m [10].

m	0	1	2	3	4
k					
0	$T_0^{(0)}$				
1	$T_1^{(0)}$	$T_1^{(1)}$			
2	$T_2^{(0)}$	$T_2^{(1)}$	$T_2^{(2)}$		
3	$T_3^{(0)}$	$T_3^{(1)}$	$T_3^{(2)}$	$T_3^{(3)}$	
4	$T_4^{(0)}$	$T_4^{(1)}$	$T_4^{(2)}$	$T_4^{(3)}$	$T_4^{(4)}$

Increasing accuracy may be obtained by stepping down or across the table, while the most accurate approximation will be found on the lower vertex of the diagonal. The Romberg procedure is terminated when the values along the diagonal no longer change significantly, i.e., when the relative convergence criterion is less than some predetermined ε. In higher-level approximations, subtraction of like numbers occurs and the potential for round-off error increases. In order to provide a means of detecting this problem, a value is defined

$$R_k^{(m)} = \frac{1}{4^{(m+1)}} \frac{T_{k-1}^{(m)} - T_{k-2}^{(m)}}{T_k^{(m)} - T_{k-1}^{(m)}}$$

and since $R_k^{(m)}$ should approach 1 as a limit, a satisfactory criterion of error is if $R_k^{(m)}$ begins to differ significantly from 1.

An improper integral has one or more of the following qualities [38]:

1. Its integrand goes to finite limiting values at finite upper and lower limits, but cannot be integrated right on one or both of these limits.
2. Its upper limit equals ∞, or its lower limit equals $-\infty$.
3. It has an integrable singularity at (a) either limit, (b) a known place between its limits, or (c) an unknown place between its limits.

In the case of 3b, Gaussian quadrature can be used, choosing the weighting function to remove the singularities from the desired integral. A variable step size differential equation integration routine [Computer Applications, ref. 8] produces the only practicable solution to 3c.

Improper integrals of the other types whose problems involve both limits are handled by open formulas that do not require the integrand to be evaluated at its endpoints. One such formula, the extended midpoint rule, is accurate to the same order as the extended trapezoidal rule and is used when the limits of integration are located halfway between tabulated abscissas:

$$I = M_n = \Delta x(f_{3/2} + f_{5/2} + \ldots + f_{n-3/2} + f_{n-1/2})$$

Semi-open formulas are used when the problem exists at only one limit. At the closed end of the integration, the weights from the standard closed-type formulas are used and at the open end, the weights from open formulas are used. (Weights for closed and open formulas of various orders of error may be found in standard numerical methods texts.) Given a closed extended trapezoidal rule of one order higher than the preceding formula,

$$I = T_{2_c} = \Delta x \left[\frac{5}{12}(f_1 + f_n) + \frac{13}{12}(f_2 + f_{n-1}) + \sum_{i=2}^{n-2} f_i \right]$$

and the open extended formula of the same order of accuracy

$$I = T_{2_o} = \Delta x \left[\frac{23}{12}(f_1 + f_{n-1}) + \frac{7}{12}(f_3 + f_{n-2}) + \sum_{i=4}^{n-3} f_i \right]$$

a semi-open formula can be constructed that, in this example, is closed on the right and open on the left:

$$I = T_{2_s} = \Delta x \left[\frac{23}{12}f_2 + \frac{7}{12}f_3 + \left(\sum_{i=4}^{n-2} f_i \right) + \frac{13}{12}f_{n-1} + \frac{5}{12}f_n \right]$$

To eliminate the restriction of evenly spaced points, *Gauss quadrature* algorithms may be constructed. In these algorithms not only the function values are weighted, but the position of the function evaluations as well as the set of weight factors are left as parameters to be determined by optimizing the overall accuracy. If the function is evaluated at points $x_0, x_1, \ldots, x_n$, the procedure has $2n + 2$ parameters to be determined (the x_i, and the w_i for each x_i) and is required to be accurate for any polynomial of degree $N = 2n + 1$ or less.

These algorithms are frequently stated in terms of integrals over $[-1, 1]$, termed Gauss-Legendre quadrature, and the general formula then is

$$\int_{-1}^{1} f(x)dx \cong w_0 f_0 + w_1 f_1 + \ldots + w_n f_n$$

For example, for $n = 1$,

$$\int_{-1}^{1} f(x)dx \cong f\left(x = \frac{-1}{\sqrt{3}}\right) + f\left(x = \frac{1}{\sqrt{3}}\right)$$

For each choice of n (the number of points), the w_k and the n zeros (ξ_k) of the nth degree Legendre polynomial must be determined by requiring that the approximation be exact for polynomials of degree less than $2n + 1$. These have been determined for $n = 2$ through 95 and an abbreviated table for some n is given in Table 1.7.8. The interval $-1 \le \xi \le 1$ is transformed onto the interval $a \le x \le b$ by calculating for each $x_k (k = 1, \ldots, n)$

$$x_k = \frac{b+a}{2} + \frac{b-a}{2}\xi_k$$

and an approximation to the integral is then

$$I \cong \frac{b-a}{2} \sum_{k=1}^{m} w_k f(x_k)$$

Some other typical Gaussian quadrature formulas are

(a, b)	W(x)	Gauss-
$(-1, 1)$	$\frac{1}{\sqrt{1-x^2}}$	Chebyshev
$(0, \infty)$	$x^c e^{-x}$	Laguerre (c = 0, 1, . . .)
$(-\infty, \infty)$	e^{-x^2}	Hermite

Weights and zeros for these formulas (and for other Gaussian formulas) may be found in references such as Stroud (*Gaussian quadrature formulas*, Prentice-Hall, 1966).

Because the dominant error term in Gauss Quadrature involves very high-order derivatives, the best method for determining the accuracy of an integration is to compare

Table 1.7.8 *Sampling Points and Weight Factors for Gauss Quadratures*

n	1	x_1	ω_1
2	0	−0.5773502692	1.0000000000
	1	−0.5773502692	1.0000000000
3	0	−0.7745966692	0.5555555556
	1	0.0	0.8888888889
	2	0.7745966692	0.5555555556
5	0	−0.9061798459[a]	0.2369268850[a]
	1	−0.5384693101	0.4786286705
	2	0.0	0.5688888889
	3–4		
10	0	−0.9739065285[a]	0.0666713443[a]
	1	−0.8650633667	0.1494513492
	2	−0.6794095683	0.2190863625
	3	−0.4333953941	0.2692667193
	4	−0.1488743390	0.2955242247
	5–9		
20	0	−0.9931285992[a]	0.0176140071[a]
	1	−0.9639719273	0.0406014298
	2	−0.9122344283	0.0626720483
	3	−0.8391169718	0.0832767416
	4	−0.7463319065	0.1019301198
	5	−0.6360536807	0.1181945320
	6	−0.5108670020	0.1316886384
	7	−0.3737060887	0.1420961093
	8	−0.2277858511	0.1491729865
	9	−0.0765265211	0.1527533871
	10–19		

[a]Points and weight factors are symmetric with respect to zero.

the results for several different n. However, in certain cases, a comparison may result in a set of significantly different answers, due to the presence of one or more singularities in f(x) or to a highly oscillatory function. If very large values of n are employed, round-off error can cause a major deterioration in accuracy (see previous discussion of Romberg integration).

1.7.8 Numerical Solution of Differential Equations

The two major categories of *ordinary differential equations* are

1. *Initial value problems* where conditions are specified at some starting value of the independent variable.
2. *Boundary value problems* where conditions are specified at two (or, rarely, more) values of the independent variable.

(The solution of boundary value problems depends to a great degree on the ability to solve initial value problems.) Any n^{th}-order initial value problem can be represented as a system of n coupled first-order ordinary differential equations, each with an initial condition. In general

$$\frac{dy_1}{dt} = f_1(y_1, y_2, \ldots, y_n, t)$$

$$\frac{dy_2}{dt} = f_2(y_1, y_2, \ldots, y_n, t)$$

$$\frac{dy_n}{dt} = f_n(y_1, y_2, \ldots, y_n, t)$$

and

$$y_1(0) = y_{10}, y_2(0) = y_{20}, \ldots, y_n(0) = y_{n0}$$

The *Euler method*, while extremely inaccurate, is also extremely simple. This method is based on the definition of the derivative

$$\frac{dy}{dx} \cong \frac{y(x_i + \Delta x) - y(x_i)}{\Delta x}$$

or

$$y_{i+1} = y_i + f_i \Delta x$$

where $f_i = f(x_i, y_i)$ and $y(x = a) = y_0$ (initial condition).

Discretization error depends on the step size, i.e., if $\Delta x_i \to 0$, the algorithm would theoretically be exact. The error for Euler method at step N is $\mathcal{O}N(\Delta x)^2$ and total accumulated error is $\mathcal{O}(\Delta x)$, that is, it is a first-order method.

The *modified Euler method* needs two initial values y_0 and y_1 and is given by

$$y_n = y_{n-2} + f_{n-1}(2\Delta x) + \mathcal{O}(\Delta x)^2$$

If y_0 is given as the initial value, y_1 can be computed by Euler's method, or more accurately as

$$\Delta y_a = f(x_0, y_0)\Delta x$$

$$y_1 = y_0 + \Delta y_a$$

$$f_1 = f(x_1, y_1)$$

and

$$\Delta y_b = f_1 \Delta x$$

therefore

$$\Delta y = \frac{1}{2}(\Delta y_a + \Delta y_b)$$

and

$$y_1 = y_0 + \Delta y$$

Another improvement on the basic Euler method is to approximate the slope in the middle of the interval by the average of the slopes at the end points, or

$$y_{i+1} = y_i + \tfrac{1}{2}(f_i + f_{i+1})\Delta x$$

This form is a *closed-type* formula because it does not allow direct steps from x_i to x_{i+1} but uses the basic Euler's method to estimate y_{i+1}:

$$y_{i+1} = y_i + f_i \Delta x$$

$$f_{i+1} = f(x_{i+1}, y_{i+1})$$

$$y_{i+1} = y_i + \tfrac{1}{2}(f_i + f_{i+1})\Delta x$$

The *Runge-Kutta method* takes the weighted average of the slope at the left end point of the interval and at some intermediate point. This method can be extended to a fourth-order procedure with error $\mathcal{O}(\Delta x)^4$ and is given by

$$y_{i+1} = y_i + \frac{1}{6}[\Delta y_0 + 2\Delta y_1 + 2\Delta y_2 + \Delta y_3]$$

where $\Delta y_0 = f(x_i, y_i)\Delta x$

$$\Delta y_1 = f\left(x_i + \tfrac{1}{2}\Delta x, y_i + \tfrac{1}{2}\Delta y_0\right)\Delta x$$

$$\Delta y_2 = f\left(x_i + \tfrac{1}{2}\Delta x, y_i + \tfrac{1}{2}\Delta y_1\right)\Delta x$$

$$\Delta y_3 = f(x_{i+1}, y_i + \Delta y_2)\Delta x$$

Runge-Kutta formulas of the sixth and eighth orders are also available but are less commonly used.

If two values of y_{i+1} are calculated, y_{i+1} by using one step between x_i and x_{i+1} with Δx, and $\hat{y}_{i+1}$ by taking two steps with $\Delta x/2$, the estimate of the truncation error is

$$E_{i+1} \sim \frac{\hat{y}_{i+1} - y_{i+1}}{2^{-k} - 1}$$

where k is the order of the expression (e.g., $k = 4$ for the foregoing Runge-Kutta formula). The step size can be adjusted to keep the error E below some predetermined value.

The *Adams open formulas* are a class of multistep formulas such that the first-order formula reproduces the Euler formula. The second-order Adams open formula is given by

$$y_{i+1} = y_i + \Delta x \left[\tfrac{3}{2}f_i - \tfrac{1}{2}f_{i-1}\right] + \mathcal{O}(\Delta x)^2$$

This formula and the higher-order formulas are not self starting since they require f_{i-1}, f_{i-2}, etc. The common practice is to employ a Runge-Kutta formula of the same order to compute the first term(s) of y_i. The general Adams open formula may be written as

$$y_{i+1} = y_i + \Delta x \sum_{k=0}^{n} \beta_{nk} f_{i-k} + \mathcal{O}(\Delta x)^{n+2}$$

and the coefficients β are given in Table 1.7.9 for $n = 0, 1, \ldots, 5$.

Adams closed formulas require an iterative method to solve for y_{i+1}, because the right side of the expression requires a value of f_{i+1}. The iteration of estimating y, evaluating f, and obtaining a new estimate of y is repeated until it converges to the desired accuracy. The general formula is

$$y_{i+1} = y_i + \Delta x \sum_{k=0}^{n} \beta_{nk}^* f_{i+1-k} + \mathcal{O}(\Delta x)^{n+2}$$

and the coefficients β^* are given in Table 1.7.10.

Table 1.7.9 *Coefficients β_{nk} of the Open Adams Formulas*

n\k	0	1	2	3	4	5	$\mathcal{O}^a$
0	1						1
1	3/2	−1/2					2
2	23/12	−16/12	5/12				3
3	55/24	−59/24	37/24	−9/24			4
4	1901/720	−2774/720	2616/720	−1274/720	251/720		5
5	4277/1440	−7923/1440	9982/1440	−7298/1440	2877/1440	−475/1440	6

[a] $\mathcal{O}$ is the order of the method.

Table 1.7.10 *Coefficients β_{nk}^* of the Closed Adams Formulas*

n\k	0	1	2	3	4	5	$\mathcal{O}^a$
0	1						1
1	1/2	1/2					2
2	5/12	8/12	−1/12				3
3	9/24	19/24	−5/24	1/24			4
4	251/720	646/720	−264/720	106/720	−19/720		5
5	475/1440	1427/1440	−798/1440	482/1440	−173/1440	27/1440	6

[a] $\mathcal{O}$ is the order of the method.

A combination of open- and closed-type formulas is referred to as the *predictor-corrector method*. First the open equation (the predictor) is used to estimate a value of y_{i+1}, this value is then inserted into the right side of the corrector equation (the closed formula) and iterated to improve the accuracy of y. The predictor-corrector sets may be the low-order modified (open) and improved (closed) Euler equations, the Adams open and closed formulas, or the *Milne method*, which gives the following system

1. Predictor

$$y_{i+1} = y_{i-3} + \tfrac{4}{3}\Delta x(2f_i - f_{i-1} + 2f_{i-2})$$

2. Corrector

$$y_{i+1} = \tfrac{1}{3}\Delta x(f_{i+1} + 4f_i + f_{i-1})$$

although the Milne method, like the Adams formulas, is not self starting.

The *Hamming method* [4] applies a predictor y^0 and then a modifier $\hat{y}^0$, which provides a correction for the estimate of error in the predictor and corrector, and then iterates the corrector y^n as desired. The procedure is

1. Predictor

$$y_{i+1}^{(0)} = y_{i-3} + \tfrac{4}{3}\Delta x(2f_i - f_{i-1} + 2f_{i-2})$$

2. Modifier

$$\hat{y}_{i+1}^{(0)} = y_{i+1}^{(0)} + \tfrac{112}{121}\left(y_i - y_i^{(0)}\right)$$

3. Corrector

$$y_{i+1}^{(n+1)} = \tfrac{1}{8}(9y_i - y_{i-2}) + \tfrac{3}{8}\Delta x\left(f_{i+1}^{(n)} + 2f_i - f_{i-1}\right)$$

Truncation error estimates can be made to determine if the step size should be reduced or increased. For example, for the Hamming method,

$$E_{i+1} \sim \tfrac{9}{121}\left(y_{i+1} - y_{i+1}^{(0)}\right)$$

The Gear algorithm [15], based on the Adams formulas, adjusts both the order and mesh size to produce the desired local truncation error. *Bulirsch and Stoer method* [16, 22] is capable of producing accurate solutions using step sizes that are much smaller than conventional methods. Packaged Fortran subroutines for both methods are available.

One approach to second-order *boundary value problems* is a matrix formulation. Given

$$\frac{d^2y}{dx^2} + Ay = B, \quad y(0) = 0, \quad y(L) = 0$$

the function can be represented at i by

$$\frac{y_{i+1} - 2y_i + y_{i-1}}{(\Delta x)^2} + Ay_i = B$$

Because there are n equations of this form and n values of y, the set can be written in matrix form as

$$
\begin{vmatrix}
\alpha & 1 & & & & & \\
1 & \alpha & 1 & & & & \\
& 1 & \alpha & 1 & & & \\
& & - & - & - & & \\
& & & - & - & - & \\
& & & & 1 & \alpha & 1 \\
& & & & & 1 & \alpha \\
\end{vmatrix}
\begin{vmatrix}
y_1 \\ y_2 \\ y_3 \\ - \\ - \\ y_{n-1} \\ y_n
\end{vmatrix}
=
\begin{vmatrix}
B(\Delta x)^2 \\ B(\Delta x)^2 \\ B(\Delta x)^2 \\ - \\ - \\ B(\Delta x)^2 \\ B(\Delta x)^2
\end{vmatrix}
$$

where $\alpha = -2 + A(\Delta x)^2$ and the error is essentially second order. Row manipulation may be necessary if there are boundary conditions on the derivatives. Equations of higher order and sets of coupled ordinary differential equations may be solved this way if central difference representations of $\mathcal{O}(\Delta x)^2$ are used for the derivatives.

Shooting methods attempt to convert a boundary value problem into an initial value problem. For example, given

the preceding example restated as an initial value problem for which

$$y(0) = 0 \quad \text{and} \quad \frac{dy}{dx}(0) = U$$

U is unknown and must be chosen so that $y(L) = 0$. The equation may be solved as an initial value problem with predetermined step sizes so that x_n will equal L at the end point. Since $y(L)$ is a function of U, it will be denoted as $y_L(U)$ and an appropriate value of U sought so that

$$y_L(U) = y(L) = 0$$

Any standard root-seeking method that does not utilize explicitly the derivative of the function may be employed.

Given two estimates of the root U_{00} and U_0, two solutions of the initial value problem are calculated, $y_L(U_{00})$ and $y_L(U_0)$, a new estimate of U is obtained where

$$U_1 = U_0 - \frac{y_L(U_0)}{[y_L(U_0) - y_L(U_{00})]/(U_0 - U_{00})}$$

and the process is continued to convergence.

There are three basic classes of second-order *partial differential equations* involving two independent variables:

1. Parabolic

$$\frac{\partial^2 u}{\partial x^2} = \phi$$

2. Elliptic

$$\frac{\partial^2 u}{\partial x^2} + \frac{\partial^2 u}{\partial y^2} = \phi$$

3. Hyperbolic

$$\frac{\partial^2 u}{\partial x^2} - \frac{\partial^2 u}{\partial y^2} = \phi$$

where $\phi = \phi(x, y, u, \partial u/\partial x, \partial u/\partial y)$. Each class requires a different numerical approach. (For higher-order equations and equations in three or more variables, the extensions are usually straightforward.)

Given a parabolic equation of the form

$$\alpha\frac{\partial^2 u}{\partial x^2} = \frac{\partial u}{\partial y}$$

with boundary conditions

$$u(a, y) = u_a$$
$$u(b, y) = u_b$$
$$u(x, 0) = u_0$$

the equation can be written in a finite difference form, considered over the grid as shown in Figure 1.7.7. Using a central difference form for the derivative with respect to x and a forward difference form for the derivative with respect to

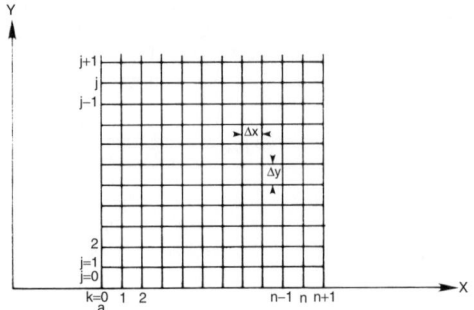

Figure 1.7.7 *Finite difference grid.*

y gives

$$\frac{u_{j,k+1} - 2u_{j,k} + u_{j,k-1}}{(\Delta x)^2} = \frac{1}{\alpha}\left[\frac{u_{j+1,k} - u_{j,k}}{\Delta y}\right]$$

and

$$u_{j+1,k} = \left[\frac{\alpha(\Delta y)}{(\Delta x)^2}\right]u_{j,k-1} + \left[1 - 2\frac{\alpha(\Delta y)}{(\Delta x)^2}\right]u_{j,k}$$
$$+ \left[\frac{\alpha(\Delta y)}{(\Delta x)^2}\right]u_{j,k+1}$$

If j is set to zero, the procedure can be used to take the first step after the initial conditions are set.

If $\partial^2 u/\partial x^2$ is represented by a central difference expression and $\partial u/\partial y$ by a backward difference expression an *implicit* solution may be obtained where

$$\frac{u_{j+1,k+1} - 2u_{j+1,k} + u_{j+1,k-1}}{(\Delta x)^2} = \frac{1}{\alpha}\left[\frac{u_{j+1,k} - u_{j,k}}{\Delta y}\right]$$

is written for all $k = 1, 2, \ldots, n$. A set of n linear algebraic equations in n unknowns is now defined, expressed in matrix form as

$$\begin{vmatrix} \beta & 1 & & & & \\ 1 & \beta & 1 & & & \\ & 1 & \beta & 1 & & \\ & - & - & - & & \\ & & - & - & - & \\ & & & 1 & \beta & 1 \\ & & & & 1 & \beta \end{vmatrix} \begin{vmatrix} u_{j+1,1} \\ u_{j+1,2} \\ u_{j+1,3} \\ - \\ - \\ u_{j+1,n-1} \\ u_{j+1,n} \end{vmatrix} = \begin{vmatrix} \Omega u_{j,1} - u_a \\ \Omega u_{j,2} \\ \Omega u_{j,3} \\ - \\ - \\ \Omega u_{j,n-1} \\ \Omega u_{j,n} - u_b \end{vmatrix}$$

where $\beta = -2 - [(\Delta x)^2]/[\alpha(\Delta y)]$
$\Omega = -[(\Delta x)^2]/[\alpha(\Delta y)]$

The *Crank-Nicholson method* is a special case of the formula

$$\alpha\left[\theta\frac{u_{j+1,k+1} - 2u_{j+1,k} + u_{j+1,k-1}}{(\Delta x)^2}\right.$$
$$\left. + (1-\theta)\frac{u_{j,k+1} - 2u_{j,k} + u_{j,k-1}}{(\Delta x)^2}\right] = \frac{u_{j+1,k} - u_{j,k}}{\Delta y}$$

where θ is the degree of implicitness, $\theta = 1$ yields implicit representation, $\theta = 1/2$ gives Crank-Nicholson method, and $\theta = 0$, the explicit representation. $\theta \geq 1/2$ is universally stable, while $\theta < 1/2$ is only conditionally stable.

Given a partial differential equation of the elliptic form

$$\frac{\partial^2 u}{\partial x^2} + \frac{\partial^2 u}{\partial y^2} = 0$$

and a grid as shown in Figure 1.46, then the equation may be written in central difference form at (j, k) as

$$\frac{u_{j,k+1} - 2u_{j,k} + u_{j,k-1}}{(\Delta x)^2} + \frac{u_{j+1,k} - 2u_{j,k} + u_{j-1,k}}{(\Delta y)^2} = 0$$

and there are mn simultaneous equations in mn unknowns $u_{j,k}$.

The most effective techniques for hyperbolic partial differential equations are based on the *method of characteristics* [11] and an extensive treatment of this method may be found in the literature of compressible fluid flow and plasticity fields.

Finite element methods [12, 13] have replaced finite difference methods in many fields, especially in the area of partial differential equations. With the finite element approach, the continuum is divided into a number of "finite elements" that are assumed to be joined by a discrete number of points along their boundaries. A function is chosen to represent the variation of the quantity over each element in terms of the value of the quantity at the boundary points. A set of simultaneous equations can be obtained that will produce a large, banded matrix.

The three primary advantages of the finite element approach over finite difference methods are [1]

1. Easy handling of irregularly shaped regions.
2. Variation in size of elements over a region, allowing smaller elements where strong variations occur.
3. Larger elements can produce comparable accuracy to smaller mesh elements of a finite difference grid, which is especially useful in handling elliptic partial differential matrices.

Other methods for solving partial differential matrices include Monte Carlo, spectral, and variational. Spectral methods in particular converge more rapidly than finite difference methods, but do not handle problems involving irregular geometries or discontinuities well.

References
1. Hornbeck, R. W., *Numerical Methods*, Prentice Hall, Englewood Cliffs, 1975.
2. Borse, G. J., *FORTRAN 77 and Numerical Methods for Engineers*, PWS Publishers, Boston, 1985.
3. Carnahan, B., Luther, H. A., and Wilkes, J. O., *Applied Numerical Methods*, John Wiley & Sons, New York, 1969.
4. Ralston, A., and Rabinowitz, P., *A First Course in Numerical Analysis*, 2nd Edition, McGraw Hill, New York, 1978.
5. Hamming, R. W., *Numerical Methods for Scientists and Engineers*, McGraw-Hill, New York, 1962.
6. Butcher, J. C., "On Runge-Kutta Processes of High Order," *Journal of the Australian Mathematical Society*, Vol. 4, pp. 179–194, 1964.
7. Gear, C. W., *Numerical Initial Value Problems in Ordinary Differential Equations*, Prentice Hall, Englewood Cliffs, 1971.
8. Stoer, J., and Bulirsh, R., *Introduction to Numerical Analysis*, Springer Verlag, New York, 1980.
9. Fox, L., *Numerical Solution of Ordinary and Partial Differential Equations*, Addison Wesley, Reading, 1962.
10. de G. Allen, D. N., *Relaxation Methods in Engineering and Science*, McGraw-Hill, New York, 1954.
11. Abbott, M. B., *An Introduction to the Method of Characteristics*, New York, American Elsevier Publishing, 1966.
12. Zienkiewicz, O. C., *The Finite Element in Engineering Science*, McGraw-Hill, London, 1971.
13. Strang, G., and Fix, G., *An Analysis of The Finite Element Method*, Prentice Hall, Englewood Cliffs, 1973.
14. Press, W. H., et al., *Numerical Recipes*, Cambridge University Press, New York, 1986.

1.8 APPLIED STATISTICS

See References 1–6 for additional information.

A *discrete random variable* is one that may take on only distinct, usually integer, values. A *continuous random variable* is one that may take on any value within a continuum of values.

1.8.1 Moments

The moments describe the characteristics of a sample or distribution function. The *mean*, which locates the average value on the measurement axis, is the first moment of values measured about the origin. The mean is denoted by μ for the population and $\overline{X}$ for the sample and is given for a continuous random variable by

$$\overline{X} \text{ or } \mu \int_{-\infty}^{\infty} Xf(X)dX$$

For a discrete random variable, the mean is given by

$$\overline{X} \text{ or } \mu = \sum_{i=1}^{n} X_i f(X_i)$$

and if each observation is given equal weight, $f(X) = 1/n$ and

$$\overline{X} \text{ or } \mu = \frac{1}{n} \sum_{i=1}^{n} X_i$$

The *variance* is the second moment about the mean and indicates the closeness of values to the mean. It is denoted by σ^2 (population) or S^2 (sample) and is given for a continuous random variable by

$$\sigma^2 \text{ or } S^2 = \int_{-\infty}^{\infty} (X - \mu)^2 f(X) dX$$

For a discrete random variable, the variance is

$$\sigma^2 \text{ or } S^2 = \sum_{i=1}^{n} (X_i - \mu)^2 f(X_i)$$

and if $f(X) = 1/n$

$$\sigma^2 \text{ or } S^2 = \frac{1}{n} \sum_{i=1}^{n} (X_i - \mu)^2 \text{(biased)}$$

or

$$\sigma^2 \text{ or } S^2 = \frac{1}{n-1} \sum_{i=1}^{n} (X_i - \mu)^2 \text{ (unbiased)}$$

The *standard deviation* is the square root of the variance and is denoted by σ (population) or S (sample).

The *skew*, the third moment about the mean, is a measure of symmetry of distribution and can be denoted by γ (population) or g (sample). It is given for a continuous random variable by

$$\gamma \text{ or } g = \int_{-\infty}^{\infty} (X - \mu)^3 f(X) dX$$

and for a discrete random variable by

$$\gamma \text{ or } g = \sum_{i=1}^{n} (X_i - \mu)^3 f(X_i)$$

A completely symmetrical distribution will have a skew = 0, and in a non-symmetric distribution the sign of the skew will depend upon the location of the tail of the distribution.

The *kurtosis*, the fourth moment, is a measure of peakedness or flatness of a distribution. It has no common notation (k is used here) and is given for a continuous random variable by

$$k = \int_{-\infty}^{\infty} (X - \mu)^4 f(X) dX$$

and for a discrete random variable by

$$k = \sum_{i=1}^{n} (X_i - \mu)^4 f(X_i)$$

1.8.2 Moment Ratios

The *moment ratios* are dimensionless coefficients used to compare characteristics of distributions measured on different scales.

The *coefficient of variation* is a measure of relative dispersion of a set of values and is given for the population by

$$C_v = \sigma/\mu$$

and for the sample by

$$C_v = S/\overline{X}$$

The *coefficient of skewness* is a measure of relative symmetry of a distribution and is given for the population by

$$\beta_1 = \gamma^2/\sigma^6$$

and for the sample by

$$\beta_1 = g^2/S^6$$

The *coefficient of kurtosis* is a measure of relative peakedness and is given by

$$\beta_2 = k/S^4$$

1.8.3 Common Probability Distributions for Continuous Random Variables

The *parameters* of a distribution control its geometric characteristics [1]:

1. A *location parameter* is the abscissa of a location point and may be a measure of central tendency, such as a mean.
2. A *scale parameter* determines the location of fractiles of the distribution relative to some specified point, often the value of the location parameter.
3. *Shape parameters* control the geometric configuration of a distribution. There may be zero, one, or multiple shape parameters.

A bounded continuous random variable with *uniform distribution* has the probability function

$$f(X) = \begin{vmatrix} 1/(\beta - \alpha) & \alpha \leq X \leq \beta \\ 0 & \text{otherwise} \end{vmatrix}$$

where α = location parameter, representing lower limit of the distribution

β = scale parameter, representing upper bound of the distribution

Probabilities are determined by integration over the necessary range:

$$P(X_1 < X < X_2) = \int_{x_1}^{x_2} 1/(\beta - \alpha) dX$$

The *normal (Gaussian) distribution* is the most frequently used probability function and is given by

$$f(X) = \frac{1}{\sqrt{2\pi}\sigma} \exp\left[-\frac{1}{2} \left(\frac{X - \mu}{\sigma} \right)^2 \right] \quad \text{for} -\infty < X < \infty$$

where μ = location parameter and σ = scale parameter. The cumulative function for this distribution is $\int f(X)$.

The *standard normal distribution* is determined by calculating a random variable z where

$$z = (X - \mu)/\sigma \quad \text{for the population}$$

$$z = \left(X - \overline{X}\right)/S \quad \text{for the sample}$$

The probability function for the standard normal distribution is then

$$f(z) = \frac{1}{\sqrt{2\pi}} e^{-.05z^2}$$

where z has mean of zero and a standard deviation of one. Probability estimates are evaluated by integrating f(z).

$$P(z_1 \leq z \leq z_2) = \int_{z_1}^{z_2} \frac{1}{\sqrt{2\pi}} e^{-.05z^2} dz$$

The *t (Student's t) distribution* is an unbounded distribution where the mean is zero and the variance is $v/(v - 2)$, v being the scale parameter (also called degrees of freedom). As $v \rightarrow \infty$, the variance $\rightarrow 1$ (standard normal distribution). A t table such as Table 1.8.1 is used to find values of the t statistic where v is located along the vertical margin

Table 1.8.1 *Critical Values for the t Distribution*

	Level of Significance for One-Tailed Test							
	.250	.100	.050	.025	.010	.005	.0025	.0005
	Level of Significance for Two-Tailed Test							
	.500	.200	.100	.050	.020	.010	.005	.001
v (degrees of freedom)								
1.	1.00	3.078	6.314	12.706	31.821	63.657	127.321	536.627
2.	.816	1.886	2.920	4.303	6.965	9.925	14.089	31.599
3.	.765	1.638	2.353	3.182	4.541	5.841	7.453	12.924
4.	.741	1.533	2.132	2.776	3.747	4.604	5.598	8.610
5.	.727	1.476	2.015	2.571	3.365	4.032	4.773	6.869
6.	.718	1.440	1.943	2.447	3.143	3.707	4.317	5.959
7.	.711	1.415	1.895	2.365	2.998	3.499	4.029	5.408
8.	.706	1.397	1.860	2.306	2.896	3.355	3.833	5.041
9.	.703	1.383	1.833	2.262	2.821	3.250	3.690	4.781
10.	.700	1.372	1.812	2.228	2.764	3.169	3.581	4.587
11.	.697	1.363	1.796	2.201	2.718	3.106	3.497	4.437
12.	.695	1.356	1.782	2.179	2.681	3.055	3.428	4.318
13.	.694	1.350	1.771	2.160	2.650	3.012	3.372	4.221
14	.692	1.345	1.761	2.145	2.624	2.977	3.326	4.140
15.	.691	1.341	1.753	2.131	2.602	2.947	3.286	4.073
16.	.690	1.337	1.746	2.120	2.583	2.921	3.252	4.015
17.	.689	1.333	1.740	2.110	2.567	2.898	3.222	3.965
18.	.688	1.330	1.734	2.101	2.552	2.878	3.197	3.922
19.	.688	1.328	1.729	2.093	2.539	2.861	3.174	3.883
20.	.687	1.325	1.725	2.086	2.528	2.845	3.153	3.850
21.	.686	1.323	1.721	2.080	2.518	2.831	3.135	3.819
22.	.686	1.321	1.717	2.074	2.508	2.819	3.119	3.792
23.	.685	1.319	1.714	2.069	2.500	2.807	3.104	3.768
24.	.685	1.318	1.711	2.064	2.492	2.797	3.091	3.745
25.	.684	1.316	1.708	2.060	2.485	2.787	3.078	3.725
26.	.684	1.315	1.706	2.056	2.479	2.779	3.067	3.707
27.	.684	1.314	1.703	2.052	2.473	2.771	3.057	3.690
28.	.683	1.313	1.701	2.048	2.467	2.763	3.047	3.674
29.	.683	1.311	1.699	2.045	2.462	2.756	3.038	3.659
30.	.683	1.310	1.697	2.042	2.457	2.750	3.030	3.646
40.	.681	1.303	1.684	2.021	2.423	2.704	2.971	3.551
50.	.679	1.299	1.676	2.009	2.403	2.678	2.937	3.496
60.	.679	1.296	1.671	2.000	2.390	2.660	2.915	3.460
70.	.678	1.294	1.667	1.994	2.381	2.648	2.899	3.435
80.	.678	1.292	1.664	1.990	2.374	2.639	2.887	3.416
90.	.677	1.291	1.662	1.987	2.368	2.632	2.878	3.402
100.	.677	1.290	1.660	1.984	2.364	2.626	2.871	3.390
150.	.676	1.287	1.655	1.976	2.351	2.609	2.849	3.357
200.	.676	1.286	1.653	1.972	2.345	2.601	2.839	3.340
∞	.6745	1.2816	1.6448	1.9600	2.3267	2.5758	2.8070	3.2905

and the probability is given on the horizontal margin. (For a one-tailed test, given the probability for the left tail, the t value must be preceded by a negative sign.)

The *chi-square distribution* gives the probability for a continuous random variable bounded on the left tail. The probability function has a shape parameter v (degrees of freedom), a mean of v, and a variance of 2v. Values of the χ^2 characteristic are obtained from a table such as Table 1.8.2, which is of similar construction as the t table (see Table 1.8.1).

The *F distribution* has two shape parameters, v_1 and v_2. Table 1.8.3 shows F values for 1% and 5% probabilities.

$$F\,(v_1, v_2) \neq F\,(v_2, v_1)$$

1.8.4 Probability Distributions for Discrete Random Variables

The *binomial distribution* applies to random variables where there are only two possible outcomes (A or B) for each trial and where the outcome probability is constant over all n trials. If the probability of A occurring on any one trial is denoted as p and the number of occurrences of A is denoted as x, then the *binomial coefficient* is given by

$$\binom{n}{x} = \frac{n!}{x!\,(n-x)!}$$

and the probability of getting x occurrences of A in n trials is

$$b\,(x; n, p) = \binom{n}{x}\,p^x\,(1-p)^{n-x} \quad \text{for } x = 0, 1, 2, \ldots, n$$

The cumulative probability of the binomial distribution is given by

$$B\,(x; n, p) = \sum_{i=0}^{x} b\,(i; n, p)$$

For the binomial distribution

$$\mu = np$$

$$\sigma = \sqrt{np\,(1-p)}$$

Table 1.8.2 *Critical Values for the Chi-Square Distribution*

Degrees of Freedom ν / α	.001	.005	.010	.020	.050	.100	.200	.500	.700	.900	.950	.975	.990	.995	.999
1.	10.809	7.905	6.637	5.405	3.842	2.706	1.642	.455	.148	.016	.004	.001	.000	.000	.000
2.	13.691	10.589	9.221	7.822	5.995	4.604	3.219	1.386	.713	.211	.102	.051	.020	.010	.002
3.	16.292	12.819	11.325	9.841	7.817	6.252	4.642	2.366	1.424	.584	.352	.216	.115	.071	.024
4.	18.432	14.824	13.280	11.660	9.492	7.782	5.989	3.357	2.195	1.064	.711	.484	.297	.205	.090
5.	20.751	16.762	15.088	13.385	11.073	9.237	7.291	4.352	3.000	1.610	1.145	.831	.553	.411	.209
6.	22.677	18.550	16.810	15.033	12.596	10.646	8.559	5.349	3.828	2.204	1.635	1.236	.871	.673	.377
7.	24.527	20.270	18.471	16.624	14.070	12.020	9.804	6.346	4.671	2.833	2.167	1.688	1.237	.988	.597
8.	26.318	21.938	20.082	18.171	15.512	13.363	11.031	7.344	5.527	3.489	2.732	2.179	1.642	1.341	.850
9.	28.061	23.563	21.654	19.683	16.925	14.686	12.243	8.343	6.393	4.168	3.324	2.699	2.086	1.728	1.135
10.	29.763	25.154	23.194	21.165	18.311	15.990	13.443	9.342	7.266	4.864	3.938	3.244	2.555	2.152	1.446
11.	31.431	26.714	24.755	22.623	19.681	17.278	14.633	10.341	8.148	5.576	4.574	3.815	3.047	2.597	1.819
12.	33.070	28.249	26.246	24.059	21.030	18.551	15.813	11.340	9.034	6.303	5.225	4.402	3.568	3.064	2.188
13.	34.683	29.878	27.717	25.477	22.367	19.814	16.986	12.340	9.926	7.041	5.890	5.006	4.102	3.560	2.577
14.	36.272	31.376	29.169	26.879	23.691	21.067	18.152	13.339	10.821	7.789	6.568	5.624	4.653	4.066	3.018
15.	37.842	32.857	30.605	28.266	25.000	22.310	19.312	14.339	11.721	8.546	7.260	6.260	5.226	4.588	3.449
16.	39.392	34.321	32.027	29.640	26.301	23.546	20.466	15.339	12.624	9.311	7.960	6.905	5.807	5.135	3.894
17.	40.926	35.771	33.435	31.002	27.593	24.771	21.616	16.338	13.530	10.083	8.670	7.560	6.400	5.687	4.350
18.	42.444	37.208	34.831	32.353	28.877	25.992	22.761	17.337	14.440	10.864	9.388	8.225	7.004	6.251	4.864
19.	43.949	38.633	36.216	33.694	30.148	27.206	23.902	18.338	15.351	11.650	10.114	8.904	7.627	6.825	5.351
20.	45.440	40.046	37.591	35.026	31.146	28.415	25.039	19.337	16.265	12.442	10.849	9.587	8.252	7.422	5.848
21.	46.919	41.449	38.957	36.350	32.678	29.619	26.173	20.337	17.182	13.238	11.590	10.278	8.886	8.018	6.398
22.	48.387	42.843	40.314	37.666	33.933	30.817	27.304	21.337	18.100	14.040	12.336	10.976	9.528	8.622	6.919
23.	49.845	44.228	41.662	38.975	35.178	32.012	28.431	22.337	19.020	14.846	13.088	11.685	10.187	9.247	7.447
24.	51.293	45.604	43.004	40.277	36.421	33.199	29.556	23.337	19.943	15.657	13.845	12.397	10.846	9.869	8.027
25.	52.732	46.973	44.338	41.573	37.660	34.384	30.678	24.337	20.866	16.471	14.607	13.115	11.510	10.498	8.576
26.	54.162	48.334	45.665	42.863	38.894	35.566	31.796	25.337	21.792	17.291	15.377	13.837	12.190	11.132	9.130
27.	55.584	49.688	46.986	44.147	40.119	36.745	32.913	26.336	22.718	18.113	16.149	14.565	12.868	11.789	9.735
28.	56.998	51.036	48.301	45.426	41.344	37.920	34.028	27.336	23.646	18.938	16.925	15.304	13.551	12.438	10.306
29.	58.405	52.378	49.610	46.699	42.565	39.092	35.140	28.336	24.576	19.766	17.705	16.042	14.240	13.092	10.882
30.	59.805	53.713	50.914	47.968	43.782	40.261	36.251	29.336	25.507	20.598	18.488	16.784	14.943	13.767	11.509
40.	73.490	66.802	63.710	60.443	55.753	51.796	47.261	39.337	34.879	29.055	26.508	24.423	22.139	20.699	17.846
50.	86.740	79.523	76.172	72.619	67.501	63.159	58.157	49.336	44.319	37.693	34.763	32.349	29.685	27.957	24.609
60.	99.679	91.982	88.396	84.586	79.078	74.390	68.966	59.336	53.814	46.463	43.187	40.474	37.465	35.503	31.678
70.	112.383	104.243	100.441	96.393	90.528	85.521	79.709	69.335	63.351	55.333	51.739	48.750	45.423	43.246	38.980
80.	124.901	116.348	112.344	108.075	101.876	96.572	90.400	79.335	72.920	64.282	60.391	57.147	53.523	51.145	46.466
90.	137.267	128.324	124.130	119.654	113.143	107.559	101.048	89.335	82.515	73.295	69.126	65.641	61.738	59.171	54.104
100.	149.505	140.193	135.820	131.147	124.340	118.493	111.662	99.335	92.133	82.362	77.929	74.216	70.049	67.303	61.869
150.	209.310	198.380	193.219	187.683	179.579	172.577	163.345	149.334	140.460	128.278	122.692	117.980	112.655	109.122	102.073
200.	267.579	255.281	249.455	243.191	233.993	226.017	216.605	199.334	189.051	174.838	168.279	162.724	156.421	152.224	143.807

Table 1.8.3 *Critical Values for the Cumulative F Distribution*

Degrees of Freedom

v2 \ v1	1	2	3	4	5	6	7	8	9	10	15	20	30	40	50	100	500	1000	10000
1.	161 / 4052.	200. / 5000.	216. / 5403.	225. / 5625.	230. / 5764.	234. / 5859.	237. / 5928.	239. / 5961.	241. / 6022.	242. / 6056.	246. / 6157.	248. / 6209.	250. / 6261.	251. / 6287.	252. / 6303.	253. / 6334.	254. / 6360.	254. / 6363.	254. / 6364.
2.	18.151 / 98.50	19.00 / 99.00	19.16 / 99.17	19.25 / 99.25	19.30 / 99.30	19.33 / 99.33	19.35 / 99.36	19.37 / 99.37	19.38 / 99.39	19.40 / 99.40	19.43 / 99.43	19.45 / 99.45	19.46 / 99.47	19.47 / 99.47	19.48 / 99.48	19.49 / 99.49	19.49 / 99.50	19.49 / 99.50	19.49 / 99.49
3.	10.13 / 34.12	9.55 / 30.82	9.28 / 29.46	9.12 / 28.71	9.01 / 28.24	8.94 / 27.91	8.89 / 27.67	8.85 / 27.49	8.81 / 27.34	8.79 / 27.23	8.70 / 26.87	8.66 / 26.69	8.62 / 26.50	8.59 / 26.41	8.58 / 26.35	8.55 / 26.24	8.53 / 26.15	8.53 / 26.14	8.52 / 26.12
4.	7.71 / 21.20	6.94 / 18.00	6.59 / 16.69	6.39 / 15.98	6.26 / 15.52	6.16 / 15.21	6.09 / 14.98	6.04 / 14.80	6.00 / 14.66	5.96 / 14.55	5.86 / 14.20	5.80 / 14.02	5.75 / 13.84	5.72 / 13.75	5.70 / 13.69	5.66 / 13.58	5.64 / 13.49	5.63 / 13.47	5.63 / 13.46
5.	6.61 / 16.26	5.79 / 13.27	5.41 / 12.06	5.19 / 11.39	5.05 / 10.97	4.95 / 10.67	4.88 / 10.46	4.82 / 10.29	4.77 / 10.16	4.74 / 10.05	4.62 / 9.72	4.56 / 9.55	4.50 / 9.38	4.46 / 9.29	4.44 / 9.24	4.41 / 9.13	4.37 / 9.04	4.37 / 9.03	4.37 / 9.02
6.	5.99 / 13.75	5.14 / 10.92	4.76 / 9.78	4.53 / 9.15	4.39 / 8.75	4.28 / 8.47	4.21 / 8.26	4.15 / 8.10	4.10 / 7.98	4.06 / 7.87	3.94 / 7.56	3.87 / 7.40	3.81 / 7.23	3.77 / 7.14	3.75 / 7.09	3.71 / 6.99	3.68 / 6.90	3.67 / 6.89	3.67 / 6.88
7.	5.59 / 12.25	4.74 / 9.55	4.35 / 8.45	4.12 / 7.85	3.97 / 7.46	3.87 / 7.19	3.79 / 6.99	3.73 / 6.84	3.68 / 6.72	3.64 / 6.62	3.51 / 6.31	3.44 / 6.16	3.38 / 5.99	3.34 / 5.91	3.32 / 5.86	3.27 / 5.75	3.24 / 5.67	3.23 / 5.66	3.23 / 5.65
8.	5.32 / 11.26	4.46 / 8.65	4.07 / 7.59	3.84 / 7.01	3.69 / 6.63	3.58 / 6.37	3.50 / 6.18	3.44 / 6.03	3.39 / 5.91	3.35 / 5.81	3.22 / 5.52	3.15 / 5.36	3.08 / 5.20	3.04 / 5.12	3.02 / 5.07	2.97 / 4.96	2.94 / 4.88	2.93 / 4.87	2.93 / 4.86
9.	5.12 / 10.56	4.26 / 8.02	3.86 / 6.99	3.63 / 6.42	3.48 / 6.06	3.37 / 5.80	3.29 / 5.61	3.23 / 5.47	3.18 / 5.35	3.14 / 5.26	3.01 / 4.96	2.94 / 4.81	2.86 / 4.65	2.83 / 4.57	2.80 / 4.52	2.76 / 4.41	2.72 / 4.33	2.71 / 4.32	2.71 / 4.31
10.	4.96 / 10.04	4.10 / 7.56	3.71 / 6.55	3.48 / 5.99	3.33 / 5.64	3.22 / 5.39	3.14 / 5.20	3.07 / 5.06	3.02 / 4.94	2.98 / 4.85	2.85 / 4.56	2.77 / 4.41	2.70 / 4.25	2.66 / 4.17	2.64 / 4.12	2.59 / 4.01	2.55 / 3.93	2.54 / 3.92	2.54 / 3.91
15.	4.54 / 8.68	3.68 / 6.36	3.29 / 5.42	3.06 / 4.89	2.90 / 4.56	2.79 / 4.32	2.71 / 4.14	2.64 / 4.00	2.59 / 3.89	2.54 / 3.80	2.40 / 3.52	2.33 / 3.37	2.25 / 3.21	2.20 / 3.13	2.18 / 3.08	2.12 / 2.98	2.08 / 2.89	2.07 / 2.88	2.07 / 2.87
20.	4.35 / 8.10	3.49 / 5.85	3.10 / 4.94	2.87 / 4.43	2.71 / 4.10	2.60 / 3.87	2.51 / 3.70	2.45 / 3.56	2.39 / 3.46	2.35 / 3.37	2.20 / 3.09	2.12 / 2.94	2.04 / 2.78	1.99 / 2.69	1.97 / 2.64	1.91 / 2.54	1.86 / 2.44	1.85 / 2.43	1.85 / 2.42
30.	4.17 / 7.56	3.32 / 5.39	2.92 / 4.51	2.69 / 4.02	2.53 / 3.70	2.42 / 3.47	2.33 / 3.30	2.27 / 3.17	2.21 / 3.07	2.16 / 2.98	2.01 / 2.70	1.93 / 2.55	1.84 / 2.39	1.79 / 2.30	1.76 / 2.25	1.70 / 2.13	1.64 / 2.03	1.63 / 2.02	1.62 / 2.01
40.	4.08 / 7.31	3.23 / 5.18	2.84 / 4.31	2.61 / 3.83	2.45 / 3.51	2.34 / 3.29	2.25 / 3.12	2.18 / 2.99	2.12 / 2.89	2.08 / 2.80	1.92 / 2.52	1.84 / 2.37	1.74 / 2.20	1.69 / 2.11	1.66 / 2.06	1.59 / 1.94	1.53 / 1.83	1.52 / 1.82	1.51 / 1.82
50.	4.03 / 7.17	3.18 / 5.06	2.79 / 4.20	2.56 / 3.72	2.40 / 3.41	2.29 / 3.19	2.20 / 3.02	2.13 / 2.89	2.07 / 2.78	2.03 / 2.70	1.87 / 2.42	1.78 / 2.27	1.69 / 2.10	1.63 / 2.01	1.60 / 1.95	1.52 / 1.82	1.46 / 1.71	1.45 / 1.70	1.44 / 1.68
100.	3.94 / 6.90	3.09 / 4.82	2.70 / 3.98	2.46 / 3.51	2.31 / 3.21	2.19 / 2.99	2.10 / 2.82	2.03 / 2.69	1.97 / 2.59	1.93 / 2.50	1.77 / 2.22	1.68 / 2.07	1.57 / 1.89	1.52 / 1.80	1.48 / 1.74	1.39 / 1.60	1.31 / 1.47	1.30 / 1.45	1.28 / 1.43
500.	3.86 / 6.69	3.01 / 4.65	2.62 / 3.82	2.39 / 3.36	2.23 / 3.05	2.12 / 2.84	2.03 / 2.68	1.96 / 2.55	1.90 / 2.44	1.85 / 2.36	1.69 / 2.07	1.59 / 1.92	1.48 / 1.74	1.42 / 1.63	1.38 / 1.57	1.28 / 1.41	1.16 / 1.23	1.14 / 1.20	1.12 / 1.17
1000.	3.85 / 6.66	3.00 / 4.63	2.61 / 3.80	2.38 / 3.34	2.22 / 3.04	2.11 / 2.82	2.02 / 2.66	1.95 / 2.53	1.89 / 2.43	1.84 / 2.34	1.68 / 2.06	1.58 / 1.90	1.47 / 1.72	1.41 / 1.61	1.36 / 1.54	1.26 / 1.38	1.13 / 1.19	1.11 / 1.16	1.08 / 1.12
10000.	3.84 / 6.64	3.00 / 4.61	2.61 / 3.78	2.37 / 3.32	2.21 / 3.02	2.10 / 2.80	2.01 / 2.64	1.94 / 2.51	1.88 / 2.41	1.83 / 2.32	1.67 / 2.04	1.57 / 1.88	1.46 / 1.70	1.40 / 1.59	1.35 / 1.53	1.25 / 1.36	1.11 / 1.16	1.08 / 1.11	1.03 / 1.05

a The upper and lower values in the table are for $F_{.05}$ and $F_{.01}$.

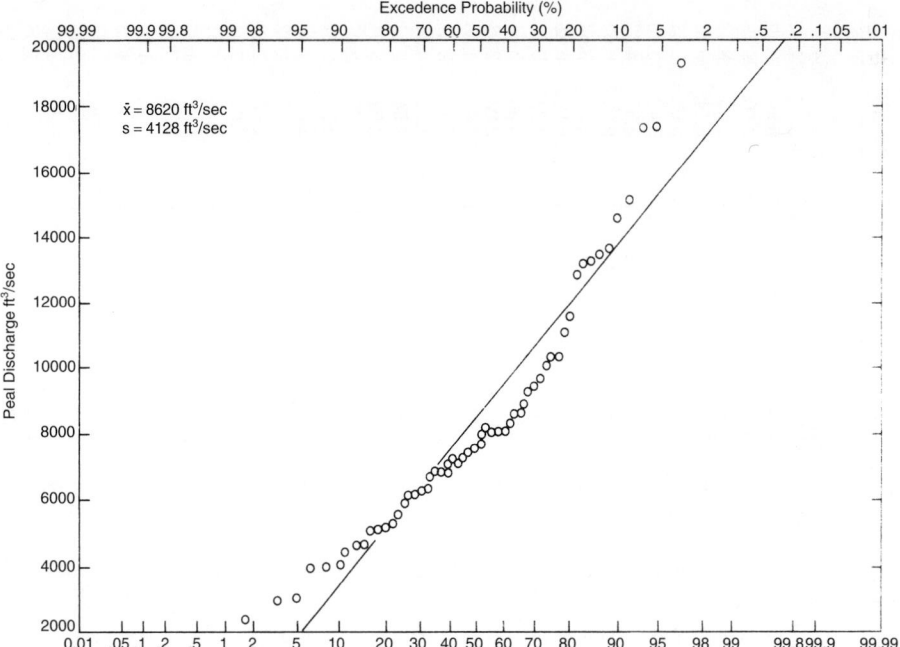

Figure 1.8.1 *Probability paper for frequency analysis.*

For $np \geq 5$ and $n(1-p) \geq 5$, an approximation of binomial probabilities is given by the standard normal distribution where z is a standard normal deviate and

$$z = \frac{x - np}{\sqrt{np(1-p)}}$$

The *negative binomial distribution* defines the probability of the kth occurrence of an outcome occurring on the xth trial as

$$b^-(x; k, p) = \binom{x-1}{k-1} p^k (1-p)^{x-k}$$

for $x = k, k+1, k+2, \ldots$

and

$$\mu = k(1-p)/p$$

$$\sigma^2 = k(1-p)/p^2$$

If the probabilities do not remain constant over the trials and if there are k (rather than two) possible outcomes of each trial, the *hypergeometric distribution* applies. For a sample of size N of a population of size T, where

$$t_1 + t_2 + \ldots + t_k = T, \text{ and}$$

$$n_1 + n_2 + \ldots + n_k = N$$

the probability is

$$h(n_i; N, t_i, T) = \frac{\binom{t_1}{n_1}\binom{t_2}{n_2}\cdots\binom{t_k}{n_k}}{\binom{T}{N}}$$

The *Poisson distribution* can be used to determine probabilities for discrete random variables where the random variable is the number of times that an event occurs in a single trial (unit of time, space, etc.). The probability function for a Poisson random variable is

$$P(x; \mu) = \frac{e^{-\mu}\mu^x}{x!} \quad \text{for } x = 0, 1, 2, \ldots$$

where μ=mean of the probability function (and also the variance)

The cumulative probability function is

$$F(X; \mu) = \sum_{i=0}^{x} P(i; \mu)$$

1.8.5 Univariate Analysis

For multivariate analysis, see McCuen, Reference 23, or other statistical texts.

The first step in data analysis is the selection of the best fitting probability function, often beginning with a *graphical analysis* of the frequency histogram. Moment ratios and *moment-ratio diagrams* (with β_1 as abscissa and β_2 as ordinate) are useful since probability functions of known distributions have characteristic values of β_1 and β_2.

Frequency analysis is an alternative to moment-ratio analysis in selecting a representative function. Probability paper (Figure 1.8.1) is available for each distribution, and the function is presented as a cumulative probability function. If the data sample has the same distribution function as the function used to scale the paper, the data will plot as a straight line.

The procedure is to fit the population frequency curve as a straight line using the sample moments and parameters of the proposed probability function. The data are then plotted by ordering the data from the largest event to the smallest and using the rank (i) of the event to obtain a probability plotting position. Two of the more common formulas are Weibull

$$pp_w = i/(n+1)$$

and Hazen

$$pp_h = (2i-1)/(2n)$$

where n is the sample size. If the data do not show a reasonable fit to the population curve, a different function should be investigated.

Estimation of model parameters is frequently accomplished by the *method of moments*. For example, for the uniform distribution, the mean is

$$\mu = \int_\alpha^\beta X\left(\frac{1}{\beta-\alpha}\right)dX = \frac{\beta+\alpha}{2} = \overline{X}$$

and the variance is

$$\sigma^2 = (\beta-\alpha)^2/12 = S^2$$

Solving for α and β gives

$$\alpha = \overline{X} - S\sqrt{3}$$
$$\beta = \overline{X} + S\sqrt{3}$$

1.8.6 Confidence Intervals
Confidence intervals provide a method of calculating the range of values that contains the true value of an estimate. A general equation for a two-sided confidence interval is

$$K_{est} = \pm FD$$

where K_{est} = estimated value of the K statistic
F = distribution factor
D = measure of dispersion

(For one-sided confidence intervals, the $\pm$ is replaced by + or by $-$.)
If σ is known, the *confidence interval on the mean* is

1. Two sided

$$\overline{X} - z_{\alpha/2}\frac{\sigma}{\sqrt{n}} \le \mu \le \overline{X} + z_{\alpha/2}\frac{\sigma}{\sqrt{n}}$$

2. One sided, lower limit

$$\overline{X} - z_\alpha\frac{\sigma}{\sqrt{n}} \le \mu \le \infty$$

3. One sided, upper limit

$$-\infty \le \mu \le \overline{X} + z_\alpha\frac{\sigma}{\sqrt{n}}$$

where $\overline{X}$ = sample mean = K_{est}
n = sample size
$z_\alpha, z_{\alpha/2}$ = values of random variables, with standard normal distribution, cutting off $(1-\gamma)$ and $(1-\gamma/2)$ respectively in the tail of the distribution, and $\alpha = 1 - \gamma$ (the level of significance) = F
$\sigma/\sqrt{n}$ = measure of dispersion

If σ is unknown, the equations are

1. Two sided

$$\overline{X} - t_{\alpha/2}\frac{s}{\sqrt{n}} \le \mu \le \overline{X} + t_{\alpha/2}\frac{s}{\sqrt{n}}$$

2. One sided, lower limit

$$\overline{X} - t_\alpha\frac{s}{\sqrt{n}} \le \mu \le \infty$$

3. One sided, upper limit

$$-\infty \le \mu \le \overline{X} + t_\alpha\frac{s}{\sqrt{n}}$$

where s = standard deviation of sample
$t_\alpha, t_{\alpha/2}$ = values of variables having a t distribution with $v = n - 1$, and $\alpha\%$ of distribution cut off in one tail and $\alpha/2\%$ in both tails = F
$s/\sqrt{n}$ = measure of dispersion

The *confidence interval on the variance* is computed by

1. $\frac{(n-1)S^2}{X_{\alpha/2}^2} \le \sigma^2 \le \frac{(n-1)S^2}{X_{(1-\alpha/2)}^2}$
2. $\frac{(n-1)S^2}{X_\alpha^2} \le \sigma^2 \le \infty$

3. $-\infty \le \sigma^2 \le \frac{(n-1)S^2}{X_{(1-\alpha)}^2}$

where $X_{\alpha/2}^2, X_\alpha^2$ = values of a random variable with a chi-square distribution cutting off $\alpha/2\%$ and $\alpha\%$, respectively, of the right tail
$X_{(1-\alpha/2)}^2, X_{(1-\alpha)}^2$ = values of a random variable with a chi-square distribution cutting off $(1-\alpha/2)\%$ and $(1-\alpha)\%$, respectively, of the left tail
$S^2 = K_{est}$
X^2 values = distribution factors

1.8.7 Correlation
Correlation analysis quantifies the degree to which the value of one variable can be used to predict the value of another. The most frequently used method is the *Pearson product-moment correlation coefficient*.

The *coefficient of determination* is the fraction of the variation that is explained by a linear relationship between two variables and is given by

$$R^2 = \frac{\sum_{i=1}^n \left(\widehat{Y}_i - \overline{Y}\right)^2}{\sum_{i=1}^n \left(Y_i - \overline{Y}\right)^2}$$

where Y = observation on the random variable
$\widehat{Y}$ = value of Y estimated from the best linear relationship with the second variable X
$\overline{Y}$ = mean of the observations on Y
and R is the *correlation coefficient*. A perfect association is indicated by R = 1 for a direct relationship and R = -1 for an inverse relationship. R = 0 indicates no linear association between X and Y.
A second definition of R is

$$R = \frac{1}{nS_xS_y}\sum_{i=1}^n \left(X_i - \overline{X}\right)\left(Y_i - \overline{Y}\right)$$

where n = number of observations on Y
S_x, S_y = biased (n degrees of freedom) estimates of the standard deviations of X and Y
Note: For small n, even high correlation may not indicate a significant relationship between the variables.

1.8.8 Regression
The relationship between a criterion variable and two or more predictor variables is given by a *linear multivariate model*:

$$\widehat{Y} = b_o + b_1x_1 + b_2x_2 + \ldots + b_px_p$$

where p = number of predictor variables
x_i = i^{th} predictor variable
b_i = i^{th} slope coefficient
b_o = intercept coefficient
$i = 1, 2, \ldots, p$
The coefficients b_i are the *partial regression coefficients*.
The *principle of least squares* is used to correlate Y with the X_i values. The error e (or residual) is defined as

$$e_i = \widehat{Y}_i - Y_i$$

where $\widehat{Y}_i$ = i^{th} predicted value of the criterion variable
Y_i = i^{th} measured value of the criterion variable
e_i = i^{th} error
The purpose of the principle of least squares is to minimize the sum of the squares of the errors so that

$$E = \min\sum_{i=1}^n \left(\widehat{Y}_i - Y\right)^2$$

where n = number of observations of the criterion variable (i.e., sample size)

$$E = \sum_{i=1}^{n} (b_o + b_1 X_i - Y_i)^2$$

By differentiating with respect to b_o and b_1 and setting the equations equal to zero, two equations in two unknowns are obtained (the summations are for $I = 1, \ldots, n$)

$$b_1 = \frac{\sum X_i Y_i - (\sum X_i \sum Y_i)/n}{\sum X_i^2 - (\sum X_i)^2/n}$$

and

$$b_o = \frac{\sum Y_i}{n} - \frac{b_1 \sum X_i}{n}$$

and by solutions of the set, b_o and b_1 can be obtained.

The standard deviation S_y gives the accuracy of prediction. If Y is related to one or more predictor variables, the error of prediction is reduced to the *standard error of estimate* S_e (the standard deviation of the errors), where

$$S_e = \left[\frac{1}{v} \sum_{i=1}^{n} \left(\widehat{Y}_i - Y_i\right)^2\right]^{0.5}$$

where v = degrees of freedom, or sample size − number of unknowns. For the general linear model with an intercept, there are $(p + 1)$ unknowns and $v = n - (p + 1)$.

If $S_e = 0$, then R = 1, and if $S_e = S_y$, then R = 0.

The *standardized partial regression coefficient* t is a measure of the relative importance of the corresponding predictor and is given by

$$t = (b_1 S_x)/S_y$$

where $-1 < t < 1$ for rational models

The two-sided confidence intervals for the coefficients b_o and b_1, when β_o and β_1 are random variables having t distributions with $(n - 2)$ degrees of freedom and error variances of

$$S_e^2(b_o) = \frac{S_e^2 \sum X_i^2}{n \sum (X_i - \overline{X})^2}$$

and

$$S_e^2(b_1) = \frac{S_e^2}{\sum \left(X_i - \overline{X}\right)^2}$$

are (if α is the level of significance)

$$b_o \pm t_{(\alpha/2, n-2)} S_e(b_o)$$

$$b_1 \pm t_{(\alpha/2, n-2)} S_e(b_1)$$

The confidence interval for a line of m points may be plotted by computing the confidence limits $\widehat{Y}_{ci}$ at each point, $\left(X_{xi}, \widehat{Y}_{ai}, i = 1, 2, \ldots, m\right)$ of the regression line when

$$\widehat{Y}_{ci} = \widehat{Y}_{ai} \pm S_e \sqrt{2F} \left[\frac{1}{n} + \frac{\left(X_{ai} - \overline{X}\right)^2}{\sum \left(X - \overline{X}\right)^2}\right]^{0.5}$$

where $\widehat{Y}_{ai} = \overline{Y} + b_1 \left(X_{ai} - \overline{X}\right)$

F = F statistic obtained for $(2, n - 2)$ degrees of freedom and a level of significance $\alpha = 1 - \gamma$, γ being the level of confidence.

The confidence interval for a single point, say X_o, can be computed using the interval

$$\widehat{Y} \pm t_{\alpha/2} S_e \left[\frac{1}{n} + \frac{\left(X_o - \overline{X}\right)^2}{\sum \left(X - \overline{X}\right)^2}\right]^{0.5}$$

where $t_{\alpha/2}$ = value of random variable having at t distribution with $(n - 2)$ degrees of freedom and a level of confidence $\alpha = 1 - \gamma$.

$$\widehat{Y} = \overline{Y} + b_1(X_o - \overline{X})$$

The confidence interval for a future value X_f is given by

$$\widehat{Y} \pm t_{\alpha/2} S_e \left[1 + \frac{1}{n} + \frac{\left(X_f - \overline{X}\right)^2}{\sum \left(X - \overline{X}\right)^2}\right]^{0.5}$$

where $\widehat{Y} = \overline{Y} + b_1 \left(X_f - \overline{X}\right)$.

References

1. McCuen, R. H., *Statistical Methods for Engineers*, Prentice Hall, London, 1985.
2. Gibra, I. N., *Probability and Statistical Inference for Scientists and Engineers*, Prentice Hall, Englewood Cliffs, 1973.
3. Kennedy, J. R., and Neville, A. M., *Basic Statistical Methods for Engineers and Scientists*, 2nd Edition, IEP, New York, 1976.
4. Robbins, H., and Van Ryzin, J., *Introduction to Statistics, Science Research Associates*, Inc., Chicago, 1975.
5. Arkin, H., and Colton, R. R., *Statistical Methods*, Barnes & Noble, New York, 1970.
6. Papoulis, A., *Probability, Random Variables, and Stochastic Processes*, McGraw-Hill, New York, 1965.

1.9 COMPUTER APPLICATIONS

See References 1–12 for additional information.

Some areas of digital computer use that are highly applicable to the engineering field include

- Numerical computations for design and modeling
- Information storage and retrieval
- Data sorting and reduction
- Computer-aided graphics for illustration and for design
- Word processing
- Communication networks and database access
- Artificial intelligence applications such as expert systems and neural networks (see Reference 40 for more information).

1.9.1 Problem Solving

The initial outline of a solution to a problem is the *algorithm* (i.e., a list of English-language instructions with the following properties):

1. The execution of the proposed algorithm must be completed after a finite number of operations, the number depending on the complexity of the problem and the degree of detail of the algorithm.
2. The representation of the solution must have a *unique interpretation,* so when executing the steps with the same input data, the same outputs are obtained.
3. The algorithm must present the computer with sufficient information and instructions to carry out the solution.
4. The scope of the algorithm may be predefined by the range of the inputs.

1.9.2 Programming Languages

A wide variety of programming languages are available ranging from *machine code,* composed of sequences of 0's and 1's representing either data or instructions, which is completely processor dependent and not transferable, through *assembly languages* consisting of mnemonics for instructions and usually hexadecimal representation of storage location addresses and of data to *high-level programming languages* such as FORTRAN, C, and JAVA. High-level languages may be divided into procedure-oriented languages and problem-oriented languages.

The term problem-oriented languages should be read as "special-purpose languages" or "applications-oriented languages," because in a more general context all high-level languages may be used to solve problems. Some of these languages have been designed for special applications such as electronic circuit analysis, while others are more general purpose, such as those written for simulation or statistical packages. An example is the programming language available in many computer algebra systems (CAS) such as Maple or Matlab.

Procedure-oriented languages, so called because they allow the programmer to concentrate on the process rather than on the machine architecture, include familiar languages, such as FORTRAN, as well as many more recently developed ones. Three of the high-level languages of common interest to engineers are:

1. Formula Translating Language (FORTRAN) the original, still most commonly used language for engineering computations. It is a compiled language (the entire program is translated to object code and saved before execution) that links to many libraries of subroutines and has a number of special purpose extensions. FORTRAN has been much improved by the addition of control structures which eliminate the necessity for many unstructured leaps of logic through the program. The basic structure of many programming languages is similar (see the FORTRAN language section for details on the structure and syntax of this language).
2. C is a programming language developed in the 1970s and originally associated with the Unix operating system. C++ is an update to C that adds *object-oriented* features to C. Object-oriented programming is a relatively new approach to programming that differs from traditional structured or procedural programming in that it is based on the use of *objects*, which are members of *classes* or categories of objects that have similar features [11].
3. Java is a fully object-oriented language that is based on the creation of *applets*, which are small programs that can be attached to Web pages and move across the Internet [11]. One important feature of Java is that, unlike other programs that require a specific operating system, a Java program runs on all operating systems and hardware platforms.

Packaged programs are available in many areas of general interest to engineers, including mathematics, statistics, and structural design. The popular Microsoft Office software contains a spreadsheet program, Excel, and a database management program, Access, that are used in many petroleum engineering applications. Excel has many built-in functions, including common statistical and mathematical functions, and it has the ability to easily perform operations on large amounts of data. A number of vendors also offer specialized petroleum engineering packages relating to such areas as EOR, drilling fluids, corrosion control, cementing, and well production histories. Some private vendors also maintain databases on specific subjects such as well production histories.

1.9.3 Common Data Types

Although the number of data types available varies with the programming language and a particular vendor's restrictions and extensions of the standard, the following types are particularly useful in scientific programming. (Some languages permit user definition of nonstandard data types, usually for the purpose of limiting the range of values accepted by a variable of that type.)

1. *Integer* — A signed number with no fractional part.
2. *Real* — A signed number with an integer part and a fractional part.
3. *Double precision* — Value stored as two words, rather than one, representing a real number, but allowing for approximately double the number of significant digits.
4. *Complex* — Value stored as two words, one representing the real part of the number and the other representing the imaginary part.
5. *Character* — Alphanumerical item (2, m, !, etc.) represented in memory as a binary code (see Table 1.9.1 for ASCII and Table 1.9.2 for EBCDIC).
6. *Logical* — Data type with only two possible values: True (represented as 1) and False (represented as 0), also referred to as Boolean.
7. *Pointer* — Identifies addresses of other data items; used to create linked data structures.

Table 1.9.1   ASCII (American Standard Code for Information Exchange)[a]

Left	Right Digits									
Digits	0	1	2	3	4	5	6	7	8	9
3				!	"	#	$	%	&	'
4	(	)	*	+	,	−	.	/	0	1
5	2	3	4	5	6	7	8	9	:	;
6	<	=	>	?	@	A	B	C	D	E
7	F	G	H	I	J	K	L	M	N	O
8	P	Q	R	S	T	U	V	W	X	Y
9	Z	[	\	]	^	_	`	a	b	c
10	d	e	f	g	h	i	j	k	l	m
11	n	o	p	q	r	s	t	u	v	w
12	x	y	z	{	|	}	~			

[a]Decimal codes 00 to 31 and 127 and higher represent nonprintable characters and special character codes.

Table 1.9.2   EBCDIC (Extended Binary Coded Decimal Interchange Code)[a]

Left	Right Digits									
Digits	0	1	2	3	4	5	6	7	8	9
6										
7					¢	.	<	(	+	|
8	&									
9	!	$	*	)	;	¬	−	/		
10							∧	,	%	−
11	>	?								
12			:	#	@	'	"			a
13	b	c	d	e	f	g	h	i		
14						j	k	l	m	n
15	o	p	q	r						
16		s	t	u	v	w	x	y	z	
17								\	{	}
18	[	]								
19				A	B	C	D	E	F	G
20	H	I								J
21	K	L	M	N	O	P	Q	R		
22							S	T	U	V
23	W	X	Y	Z						
24	0	1	2	3	4	5	6	7	8	9

[a] Decimal codes 00 to 63 and 250 to 255 represent nonprintable control characters.

1.9.4 Common Data Structures

The following data structures are available or can be constructed in most high-level languages.

1. *Variable* — Named data item of a specific type; may be assigned one or more values during the course of a program run (in some languages, a constant may be defined with a specified initial value that may not be changed).
2. *Array* — A collection of data items of the same type, referred to collectively by a single name. The individual items, the *array elements*, are ordered by their subscripts, the number of subscripts being determined by the dimensionality of the array. An element is referred to by the array name followed by its parenthesized subscripts (e.g., TEMP(I, J) might refer to the temperature at the Ith time increment and at the Jth pipe node).
3. *Record* — A collection of data items (fields) of various types, which may also be records themselves. If EMP1 is a record in the master file of employees, EMP1.NAME may be a character field, EMP1.ZIP an integer field, and EMP1.SAL a real field.
4. *File* — Collection of records that normally consist of matching types of fields. Records in a file may be accessed sequentially (the entire file must be read until the needed record is reached) or randomly (the record contains a key field, which determines its physical location in storage).
5. *Linked lists* — Data items linked by pointers. In the general form, each item, except the first, has one predecessor, and each item, except the last, has one successor, with pointers linking items to their successors. *Doubly linked lists* have pointers to both the predecessor and the successor of an item and a *circular list* has a pointer from the final item to the initial item (producing a predecessor to the initial item and a successor to the final item). Restricted lists also exist, such as *stacks*, where items may only be added (pushed) or deleted (popped) at one end (the top), and *queues*, where items must be inserted at one end and deleted from the other. *Trees* are linked lists in which each item (node) except the root node has one predecessor, but all nodes may have any finite number, or zero, successors; *graphs* contain both nodes and edges, which connect the nodes and define their relationships.
6. *Class* — A description of data and the procedures used to work with the data. Examples of classes common in *graphical user interfaces* (GUI) are Scrollbar and Button.
7. *Object* — An object is a specific member of a class.

1.9.5 Program Statements

The statements of which the program consists may be either executable or nonexecutable. *Nonexecutable statements* consist of comments, which explain the data and logic of the program, and declarations, which are orders to the translator or to other system programs and which usually serve to allocate memory space for data.

Executable statements, which are translated into machine code, are instructions by which operations are performed on data or by which the sequence of execution is changed. Statements producing operations on data are

1. *Assignment* — Assign a value, either a constant or a computed value, to a variable, an array element, a node, or a field.
2. *Input* — Transfer data from external devices, such as a keyboard or disk file, to the program.
3. *Output* — Transfer data from the program to an external device, such as a printed, screen, or a disk file.

Executable statements affecting the order in which the program instructions are executed include conditional (branching) statements, iterative (looping) statements, and statements which call subprogram units.

1. *Conditional statements* — Change the sequence in which instructions are executed depending upon the logical relationship(s) between variables and/or between variable(s) and set value(s).
2. *Iterative statements* — Force the repetition of instructions depending on preset conditions.
3. *Calling statements* — Transfer control to a subprogram unit.

1.9.6 Subprograms

The division of a program into a main program unit and one or more subprogram units allows logical organization of the program into sections of related operations and facilitates the coding, debugging, and replacement of units of the program. Data are passed from one unit (or module) to another through parameters (arguments) and/or through shared memory locations. There are two types of subprograms usually available:

1. *Function subprograms* — Return a single value as the value of the function name; these may be either extrinsic (user-defined) or intrinsic (provided as part of the system library).
2. *Subroutine or procedure subprograms* — Return values through parameters or global variables (see scope).

The *scope* of a data item in a program determines which program units may access (and change) the value of that data item. *Global* data may be accessed by all program units, whereas *local* data are visible only to the unit(s) in which they are defined. The method by which scope is determined depends on the type of language being used. In FORTRAN, a COMMON block defined by a COMMON statement in the main program allows data in that block to be accessed by any module in which the block is defined. Parameters allow two units to share data values; some types of parameters allow the passing of only the value and not the location of the item, so that the subunit can read, but not change, the value. In block structured languages, such as Pascal, the structure determines the scope of the variables; the scope of a variable is the block in which it was defined and all blocks contained therein. A variable in this case is global to a sub-unit if it was declared not in that sub-unit but in a higher-level unit which contains the sub-unit, e.g., all variables declared in the main program are global to all program units.

Recursion is available in languages having dynamic memory allocation. Direct recursion occurs when a program unit calls itself; indirect recursion occurs when a chain of subprogram calls results in the original calling unit being called again without returning to a higher-level unit, such as MAIN→SUBA→SUBB→SUBC→SUBA, where

MAIN
| | | |
A B C D

1.9.7 General Programming Principles

Two characteristics of well-designed programs are

1. *Generality* — To as great an extent as possible for a particular problem, a program should be able to operate on a wide variety of data sets with a minimum of program revision, and the necessary changes should be as simple to make as possible.
2. *Portability* — A program should adhere as closely as possible to a standard version of a language and avoid highly machine-dependent constants and constructions. Unidentified machine-dependent information should be localized and identified for simplification of transport.

With little extra effort, code can be written so as to minimize the difficulty a reader will encounter in comprehending the program logic. Several considerations in improving readability of a program are

1. Names should be as descriptive as possible (e.g., DEPTH rather than 12).
2. Comments should be liberally used to describe the data and logic of the program; they should be brief when placed in the body of a program unit, but may be longer and more descriptive at the beginning of a program unit.
3. Indentation, when possible, clarifies conditional and iterative constructions, and spacing improves the general readability of a program.
4. The use of subprogram units allows separation of the various operations of a program into modules, thereby clearly delineating the program logic. Specific types of calculations, input, and output may be done in distinct modules; in many cases, the main unit will consist primarily of calling statements to a few modules.

The proper handling of certain common errors can improve the run-time behavior of programs. In most cases, awareness of inherent problems in machine handling of data and attention to program details can avoid program crashes due to error. The following should be considered when the program code is being designed:

1. Input validation statements should be used to automatically check input data, to produce a clear message when an error is found, and to allow reentry of erroneous data. Data input should be echoed for user verification and an opportunity allowed for alteration of specific data items.
2. A method of exiting the program in case of a run-time error, which produces a message to the user as to the type and location of the error, should be provided. The possibility of certain errors occurring can be anticipated, and the use of flags and conditional constructions may provide a path to exit the program gracefully.
3. It is necessary to avoid predicting the exact value of a real variable, because after several operations, it may have been rounded off one or more times.
4. Side effects in subprogram units are unintentional changes in data values defined in other units. These frequently occur when the scope of a variable is mistakenly considered because of insufficient cross-checking to be local, when it is, in fact, global.

1.9.8 FORTRAN Language

FORTRAN names (e.g., unit, variable, array) consist of an initial letter (see defaults for real and integer types in the following) followed by letters or digits, the maximum length of which is 6 characters.

1.9.8.1 Data Types

See the section "Statements" for forms of declaration of type.

Integer — Variable names starting with I-N, unless otherwise declared.
Real — Variable names starting with A-H and O-Z, unless otherwise declared.
Double precision — Must be declared.
Complex — Must be declared.
Logical — Must be declared.
Character — Must be declared with length of string; default length is 1.

1.9.8.2 Data Structures

Variable — May be assigned value by a numerical or character constant, by input, or by an expression.
Array — May have up to seven dimensions; number and size (upper and lower boundaries) of dimensions are declared in DIMENSION or TYPE statements.
File — External (physical) only; may be sequential or random access.

1.9.8.3 Statements

Most statements, except where noted otherwise, begin in the seventh column of a page considered to be 80 columns wide; *continuation lines* are indicated by a "+" (symbol may vary with version of language) in the sixth column. The first five columns are reserved for *labels* (line numbers), which are only required if the line is referenced by another statement, and for *comment lines*, which are determined by a character in the first column. Columns 7 to 72 are reserved for statements; 73–80 are not read. (See Table 1.9.3 for required order of statements in FORTRAN.)

1.9.8.3.1 Nonexecutable Statements

Program unit heading — Program name, function name, or subroutine name.

Type declaration — Specifies data type to be represented by a variable name (overrides defaults):

```
REAL MSR
INTEGER COUNT, PNUM, AP
LOGICAL TEST1
CHARACTER*10 LNAME
```

Implicit declaration — Allows type specification for all names beginning with the given first letter(s)

```
IMPLICIT DOUBLE PRECISION (A-Z)
IMPLICIT COMPLEX (C)
```

Dimension statement — Specifies number and size of dimensions for each array (may be included in Type statement); lower bound default is 1

```
DIMENSION A1(5:10), A2(15)
DIMENSION A3(0:5, 0:10, 10:100), A4(10, 10, 10, 10)
```

Table 1.9.3 *Order of Statements in FORTRAN*

COMMENT Lines	PROGRAM, FUNCTION, SUBROUTINE or BLOCK DATA Statements		
	FORMAT and ENTRY Statements	IMPLICIT Statements	
		PARAMETER Statements	
		DIMENSION COMMON TYPE, and EQUIVALENCE Statements	
		DATA Statements	Statement Function Definitions
			Executable Statements
	END Statements		

or

INTEGER A1(5:10), A2(15)

Common statement — Defines a common block of global variables (one common block may have no specified name)

COMMON/blockname/varnames/blockname/varnames
COMMON X, Y, Z/C2BLK/A, B, C

Data may be entered into variables declared in a labeled COMMON block by an assignment or an input statement or through a BLOCK DATA subprogram as defined in subprogram statements.

Equivalence statement — Assigns two or more variable names to the same memory location

EQUIVALENCE(A, B, C)

Parameter statement — Declares the name of a constant whose value cannot be changed in the program

PARAMETER name

End statement — Compiler signal for end of unit

END

1.9.8.3.2 Executable Statements
Assignment statements may be DATA statements and are used mainly to assign initial values to variables

DATA A, B, C, D, E/1., 2., 3., 4., 5./X, Y, Z/3*10.

or values assigned by numerical or character constants or by expressions (see Table 1.9.4 for arithmetic operators and precedence)

PI = 3.1415927
C1 = 'This is a test'
$X(I, J) = X(I-1, J) + Y(J)*Z**3$

Input/output statements may be either list-directed (stream) or formatted. List-directed I/O statements may be

READ(device#, *)var1, var2, . . .
WRITE(device#, *)var1, var2, . . .
PRINT *var1, var2, . . . i

where device# refers to either a device such as a screen or printer or to a disk file. Usually, the default input device (referenced by an asterisk rather than a device#) is the keyboard and the default output device (also referenced by an asterisk) is the screen. PRINT connects only to the printer. Other devices and files may be assigned device numbers through fiie handling statements (see later). A statement requesting list-directed input from the default device (the keyboard)

might be, for example,

READ(*, *)TEMP, PRSSR, LENGTH

Formatted I/O statements require edit specifiers (Table 1.9.5) for each variable to be handled and may also include strings of characters enclosed in single quote marks. The format may be given in a separate format statement (referenced by a line number and labeled) or, in many systems, may be enclosed in quotes and parentheses in the I/O statement itself. The general form for formatted I/O statements is

READ(device#, label)var1, var2, . . .
WRITE(device#, label)var1, var2, . . .
PRINT label var1, var2, . . .
label FORMAT(list of specifications)

For example,

READ(*, 100)X(1), Y, I
100 FORMAT (1X, 2F12.4, I5)
    WRITE(6, 110)(TEMP(I), I = 1, 5)
110 FORMAT(1X, 5(F7.3, 2X))
    WRITE(*, '(''TEMP AT SURFACE IS'', F8.2)')TSURF

Table 1.9.5   *FORTRAN Edit Specifiers*

A	Character data fields
D	Double precision data fields
E	Real data fields-exponential (E) notation
F	Real data fields-decimal notation
G	General form
H	Character constants
I	Integer data fields
L	Logical data fields
P	Scale factors, used with D, E, F, and G specifiers to shift the decimal point or exponent size for output
S	Restores the optional + convention to the compiler
SP	Prints + with all subsequent positive data
SS	Suppresses + for all subsequent positive data
TL	Next character will be input/output the specified number of spaces left of the current position
TR	Next character will be input/output the specified number of spaces right of the current position
X	Skip the specified number of spaces before next character is input/output
:	Ends format control if no more data items in list
/	Skips a record

Specifiers for Input Operations Only

BN	Specifies that blank characters are to be ignored
BZ	Specifies blank characters are to be read as zeros

Specifications for Time: Four Most Common Data Types

Integer	Iw
Real	Fw.d
Character	Aw
Exponential	Ew.d

Table 1.9.4   *Precedence of FORTRAN Operators*

Class	Level	Symbol of Mnemonic
Exponential	First	**
	Second	−(negation) and +(identity)
Arithmetic	Third	*, /
	Fourth	+, −
Relational	Fifth	.GT., .GE., .LT., .LE., .EQ., .NE.
	Sixth	.NOT.
Logical	Seventh	.AND.
	Eighth	.OR.
	Ninth	.EQV. , .NEQV.

File handling statements allow the manipulation of sequential and random access files. (Because there is considerable variation from one system to the next, the following information is given in general terms only.) Devices are treated as sequential files, while disk files may be sequential or random access. The following statements are generally accepted forms.

OPEN (list of specifiers) — Connects an existing file to an I/O device or generates a new file, and specifies a device (unit) number. The specifiers (required and optional) are

UNIT =	file unit (device) number
IOSTAT	integer variable for I/O status
FILE	name of file
ERR	label for error transfer
STATUS	file status descriptor, may be OLD, NEW, SCRATCH, or UNKNOWN
ACCESS	may be SEQUENTIAL or DIRECT
FORM	may be FORMATTED or UNFORMATTED
RECL	record length, if file access is DIRECT
BLANK	specifies blank handling, either NULL or ZERO

CLOSE (list of specifiers) — Disconnects a file. The specifiers may be

UNIT =	file unit (device) number
IOSTAT	as in OPEN
ERR	as in OPEN
STATUS	may be KEEP or DELETE

INQUIRE (list of specifiers) — Returns information about the attributes of a file. Besides the UNIT= , IOSTAT, and ERR specifiers, the following specifiers may be included:

EXIST	returns .TRUE. if file exists, else .FALSE.
OPENED	returns .TRUE. if open, else .FALSE.
NUMBER	returns number of connected device
NAMED	returns .TRUE. or .FALSE.
NAME	returns name of file
FORM	returns FORMATTED or UNFORMATTED
RECL	returns record length in direct access file
NEXT REC	returns number of next record in direct access file
BLANK	returns whether blanks or zeros specified

The following statements must include the UNIT = and may include the IOSTAT and/or the ERR specifiers:

REWIND (list of specifiers)	— Causes a sequential file to be rewound to first record
BACKSPACE (list of specifiers)	— Causes a sequential file to rewind one record
ENDFILE (list of specifiers)	— Places an end-of-file mark on a sequential file

A few examples of file-handling statements are

OPEN (UNIT = 6, IOSTAT = FSTAT, FILE = 'PRINTER', STATUS = 'NEW')

OPEN (4, FILE = 'DATA1', STATUS = 'OLD')

CLOSE (6)
BACKSPACE (4, ERR = 500)

Control statements affect the flow of instructions within a program unit. (For control between units, see subprogram statements later.) These may be general, conditional, or iterative statements. *General control statements* follow.

PAUSE n — Interrupts program run, resumption on pressing "Enter"; n is an optional character constant or integer of less than 5 digits:

PAUSE 'VALUE INVALID'

STOP n — Halts program run (n as above)

STOP 10050

GOTO i — Transfers control to statement labeled i, where i is an integer constant or variable with value of label:

GOTO 700

Conditional statements are as follows.

IF(e)s1, s2, s3 — Arithmetic if, where e is an arithmetic expression and s1, s2, s3 are statement labels; transfers control to a labeled statement depending on whether e evaluates to a negative, zero, or positive value, respectively. For example,

IF(I – 5) 30, 40, 50

will transfer control to the statement labeled 40 if I = 5.

IF(e)st—Logical if, where e is a logical expression (see Table 1.9.5 for relational and logical operators) and st is any executable statement except DO, IF, ELSEIF, ENDIF, or END. For example,
IF(I.EQ.1) WRITE(*, *)'YES'

IF(e) THEN—Block if, where e is a logical expression, followed by a sequence of statements and completed by an ENDIF statement. The block may include sub-blocks introduced by one or more ELSEIF statements and/or one ELSE statement and all sub-blocks may contain nested IF-THEN-ELSE blocks within them. For example,

```
IF(I.EQ.J) THEN
    X = 4
    Y = 5
ENDIF
IF(I.EQ.J) THEN
    X = 4
ELSE
    X = 5
ENDIF
IF(I.EQ.J) THEN
    X = 4
ELSEIF(I.EQ.K) THEN
    X = 5
ELSE
    X = Y
ENDIF
```

The **iterative statement** in FORTRAN is the DO statement (although others may be constructed using conditional statements and GOTOs), where

DO st i = init, term, incr

introduces the repetitive section, and st is the label of the executable statement marking the end of the loop (usually, but not necessarily, a CONTINUE statement), i is the index or control integer variable and init, term, and incr (optional) are the initial value for i, the terminal value for i, and the increment of i to be used, respectively. These values must be integer constants, variables, or expressions in standard FORTRAN 77, but many extensions to the language allow real values to be used. DO loops may be nested to a level determined by a specific compiler. For example,

```
DO 100 I = 1, 10, 2
    PR(I) = PPE(I – 1)
```

```
    100 CONTINUE
       DO 100 I = 10, 1, - 1
          DO 100 J = 1, 10
    100 ATR(I, J) = P(J, I)*FRIC
          DO 100 I = −5, 25, 5
             PF(I) = TFP(I) − FFG
             IF(PF(I).GE.LMT) GOTO 110
    100 CONTINUE
    110 MAXC = I
```

Statement functions are defined before any other executable statements in the program and are called in the same way that subprogram or intrinsic functions are called (see later subprogram statements). They are one-line expressions that receive one or more parameters from the calling statement and return a single calculated value to the function name in the calling statement. For example, a statement function defined as

$$FDPT(X(I), Y(I), Z, I) = X(I)*Y(I) + Z**I$$

will calculate a value depending on the values of $X(I)$, $Y(I)$, Z, and I at the time of calling and return the calculated value to the calling statement through FDPT.

Subprogram statements are those used to transfer control between program units — the main program, functions, and subroutines. A *function call* is performed by invoking the name of the function module in an assignment statement, such as

$$X = FDPR(Z, Y(I))*PRF$$

which will transfer control to the function FDPR and pass the values Z, and Y(I) to that unit. An intrinsic function or a statement function may be called in the same way. (See Table 1.9.6 for a list of FORTRAN 77 intrinsic functions.) A *subroutine call* is performed by a statement such as

```
    CALL CALCSUB(MATFOR, I, J, PVAL)
```

which will transfer control to the subroutine CALCSUB and pass (and/or return) the values MATFOR, I, J, and PVAL. A subroutine may have an ENTRY name (parameter list) statement embedded within it, which when called in the same manner as the main subroutine call, will receive control transfer at that point. Control passes from the called unit back to the calling unit when a RETURN statement is encountered. Given a subroutine

```
    Subroutine CALCSUB(MAT, M, N, P1)
       REAL MAT(100)
       P1 = MAT(M) + MAT(N)
       RETURN
       ENTRY NEWCALC(MAT, M, N, P2)
       P2 = MAT(M) + MAT(N)
       RETURN
    END
```

a call to CALCSUB as before will return a value through P1 to PVAL and a call

```
    CALL NEWCALC(MAT, K, L, PVAL)
```

will transfer control in at the ENTRY statement and return a value through P2 to PVAL. A BLOCK DATA subprogram enters data into the variables declared in a labeled COMMON block and has the form

```
    BLOCK DATA
       (DATA, DIMENSION, IMPLICIT, TYPE, EQUIVA-
       LENCE, COMMON and PARAMETER statements)
    END
```

Table 1.9.6 *FORTRAN Intrinsic Functions*

Integer

IABS	Returns the absolute value of an argument
IDIM	Returns the positive difference between two arguments
IDINT	Converts a double-precision argument to integer by truncation
IFIX	Converts a real argument to integer by truncation
INT	Truncates the decimal part of an argument
ISIGN	Transfers the sign from one integer argument to the other
MAX0	Selects the largest value of several arguments
MAX1	Selects the largest value of several arguments, but converts any real result to integer
MIN0	Selects the smallest value of several arguments
MIN1	Selects the smallest value of several arguments, but converts any real result to integer
MOD	Returns the remainder from division of two arguments

Real

ABS	Returns the absolute value of an argument
ACOS	Returns the arc cosine of an argument
AIMAG	Returns the imaginary part of a complex number
AINT	Truncates the decimal part of an argument
ALOG	Returns the natural logarithm of an argument
ALOG10	Returns the common logarithm of an argument
AMAX1	Selects the largest value of several arguments
AMIN1	Selects the smallest value of several arguments
AMOD	Returns the remainder from division of two arguments
ANINT	Returns the whole number nearest in value to the argument
ASIN	Returns the arc sine of an argument
ATAN	Returns the arc tangent of an argument
ATAN2	Returns the arc tangent of two arguments $[\arctan(a_1/a_2)]$
COS	Returns the cosine of an argument
COSH	Returns the hyperbolic cosine of an argument
DIM	Returns the positive difference between two arguments
EXP	Returns the exponential e raised to the power of the argument
FLOAT	Converts an argument to a real number
NINT	Returns the nearest integer value
REAL	Converts a complex argument to a real value
SIGN	Transfers the sign from one argument to the other
SIN	Returns the sine of an argument
SINH	Returns the hyperbolic sine of an argument
SQRT	Returns the square root of an argument
SNGL	Converts a double precision argument to single precision
TAN	Returns the tangent of an argument
TANH	Returns the hyperbolic tangent of an argument

Double Precision

DABS	Returns the absolute value of an argument
DACOS	Returns the arc cosine of an argument
DASIN	Returns the arc sine of an argument
DATAN	Returns the arc tangent of one argument
DATAN2	Returns the arc tangent of two arguments $[\arctan(a_1/a_2)]$
DBLE	Converts an argument to double precision
DCOS	Returns the cosine of an argument

(continued)

Table 1.9.6 *Continued*

DCOSH	Returns the hyperbolic cosine of an argument
DDIM	Returns the positive difference between two arguments
DEXP	Returns the exponential e raised to the power of the argument
DINT	Truncates the decimal part of an argument
DLOG	Returns the natural logarithm of an argument
DLOG10	Returns the common logarithm of an argument
DMAX1	Selects the largest value of several arguments
DMIN1	Selects the smallest value of several arguments
DMOD	Returns the remainder from division of two arguments
DNINT	Returns the whole number closest in value to the argument
DPROD	Converts the product of two real arguments to double precision
DSIGN	Transfers the sign from one argument to the other
DSIN	Returns the sine of an argument
DSINH	Returns the hyperbolic sine of an argument
DSQRT	Returns the square root of an argument
DTAN	Returns the tangent of an argument
DTANH	Returns the hyperbolic tangent of an argument
IDINT	Converts the argument to the nearest integer value

Complex

CABS	Returns the absolute value of an argument
CCOS	Returns the cosine of an angle
CEXP	Returns the exponential e raised to the power of the argument
CLOG	Returns the natural logarithm of the argument
CMPLX	Converts the argument to a complex number
CONJ	Returns the conjugate of a complex function
CSQRT	Returns the square root of an argument
CSIN	Returns the sine of an argument

1.9.9 System Software

System software is the connection between the user and the machine. It provides management of the system resources and utilities which simplify development of applications programs. Essential system software includes the following:

1. **Translators** — Assemblers, interpreters, and/or compilers that translate symbolic language into machine code.
2. **Linkers and loaders** — Linkers resolve references between program units and allow access to system libraries; loaders place code into the main memory locations from which it will be executed.
3. **Operating systems** — Manage hardware resources of the computer system. Utilization may be of the *batch* method, in which program units, libraries, and data are submitted to the system along with the job control language statements needed to run the program. The operating system allocates the central processing unit to one batch job at a time, according to a hierarchical system. *Time-sharing* systems provide interactive sharing of resources by many users. The system must interweave allocation of resources to users and manage memory locations.
4. **Utility programs** — Simplify use by performing particular tasks for the programmer, such as editing, debugging, etc.
5. **File manager systems** — Maintain files and handle data input to and output from the files. *Database management*

systems (DBMS) contain integrated sets of files related by their use and provide uniform software interfaces for accessing data. The essential relationships between records in the files may be of several types, including sequential, associative, or hierarchical.
6. **Telecommunications monitors** — Supervise communications between remote terminals and the central computer.

1.9.10 System Hardware

System hardware consists of the central processor, the input devices (usually a keyboard), the output devices (probably both a video display terminal and a hardcopy printer), long-term storage devices, and perhaps communications components. In smaller systems, more than one of these components may be "built in" to one unit, while in larger systems there may be many units each of several components associated with the system.

The *central processing unit* (CPU) consists of the arithmetic-logic unit (ALU), the control unit, and the central storage (short-term memory) unit. The CPU is normally classified by size of the word (number of bits in one piece of information or address), size of memory (which is usually expressed in MB or megabytes, with a byte being a group of 8 bits and a bit being a single value, either 0 or 1) and by speed of operations (usually given in gigahertz or GHz). Input may also be from files stored on a disk or CD. There are many combinations of these factors, depending on the processor chip used and upon the architecture of the machine. Speed of operation and of data transfer is of major importance in large number-crunching programs. Memory size affects the size of the program and the amount of data that may be held at one time, while word size primarily affects the size of memory available. A useful addition to the system, if a need for large-scale number crunching is anticipated, is the arithmetic coprocessor chip, which performs high-speed numerical operations.

Keyboards are the most widely used *input devices*, but optical scanners and digital pads (for computer-aided design) are some additional input devices. Input may also be from files stored on a disk or tape.

Video display terminals for output are available in several sizes (measured diagonally in inches) and various resolutions. Laser printers are common forms of *hardcopy output devices*. Color and even photo-quality laser printers are readily available. For engineering and design use, plotters, either black-white or multicolored are frequently added to the system. Output may also be sent to a disk or CD for long-term storage.

Long-term memory storage devices include floppy disks, zip disks, and CDs. The choice between storage devices is based on the amount of available storage needed. Floppy disks can store up to 1.44 MB of data, and CDs can store up to 700 MB of data.

Systems may also include *modems*, which connect computers to other computers or servers. Computers may be connected to other computers over telephone lines, through cables, or using satellite transmission or other wireless methods. Modem speed is usually measured in bps or bits per second. Telephone modems usually have a rate of 56 Kbps.

References

1. *Petroleum Engineer International*, December 1985, 1985–1986 Engineering Software Directory.
2. *Access*, LEDS Publishing Company, Research Triangle.

3. *Simulation Journal, Transactions*, and other publications,
   The Society for Computer Simulation, San Diego.

4. SPE *Microcomp News*, SPE Microcomputer Users Group, Richardson.

5. Linger, R. C., Mills, H. D., and Witt, B. I., *Structural Programming: Theory and Practice*, Addison Wesley, Reading, 1979.

6. Wirth, Niklaus, *Algorithms + Data Structures = Programs*, Prentice Hall, Englewood Cliffs, 1976.

7. Hancock, L., and Krieger, M., *The C Primer*, McGraw-Hill, New York, 1982.

8. Nobles, M. A., *Using the Computer to Solve Petroleum Engineering Problems*, Gulf Publishing Company, Houston, 1974.

9. Zwass, V., *Introduction to Computer Science*, Barnes & Noble, New York, 1981.

10. Sedgewick, R., *Algorithms*, Addison Wesley, Reading, 1983.

11. Koosis, D., and Koosis, D., *JAVA*™ *Programming for Dummies*®, IDG Books, Foster City, 1988.

12. *Expert Systems in Engineering Applications*, SPE Reprint Series No. 41, Society of Petroleum Engineers, Richardson, 1995.

2

General Engineering and Science

Contributing Authors
Gordon R. Bopp, Ph.D
Ronald M. Brimhall, Ph.D., P.E.
William X. Chavez, Jr.
B.J. Gallaher, P.E.
Phillip W. Johnson, Ph.D., P.E.
Bill Kersting, MS
Murty Kuntamukkla, Ph.D.
William C. Lyons, Ph.D., P.E.
Richard S. Reilly
Cheryl Rofer
Jorge H.B. Sampaio, Jr.
Paul Singer

Contents

2.1 BASIC MECHANICS (STATICS AND DYNAMICS)

Mechanics is the physical science that deals with the effects of forces on the state of motion or rest of solid, liquid, or gaseous bodies. The field may be divided into the mechanics of rigid bodies, the mechanics of deformable bodies, and the mechanics of fluids.

A rigid body is one that does not deform. True rigid bodies do not exist in nature; however, the assumption of rigid body behavior is usually an acceptable accurate simplification for examining the state of motion or rest of structures and elements of structures. The rigid body assumption is not useful in the study of structural failure. Rigid body mechanics is further subdivided into the study of bodies at rest, *statics*, and the study of bodies in motion, *dynamics*.

2.1.1 Definitions, Laws, and Units
2.1.1.1 Fundamental Quantities
All of Newtonian mechanics is developed from the independent and absolute concepts of *space*, *time*, and *mass*. These quantities cannot be exactly defined, but they may be functionally defined as follows:

Space. Some fixed reference system in which the position of a body can be uniquely defined. The concept of space is generally handled by imposition of a coordinate system, such as the Cartesian system, in which the position of a body can be stated mathematically.

Time. Physical events generally occur in some causal sequence. Time is a measure of this sequence and is required in addition to position in space in order to fully specify an event.

Mass. A measure of the resistance of a body to changes in its state of motion.

2.1.1.2 Derived Quantities
The concept of space, time, and mass may be combined to produce additional useful measures and concepts.

Particle. An entity which has mass, but can be considered to occupy a point in space. Rigid bodies that are not subject to the action of an unbalanced couple often may be treated as particles.

Body. A collection of particles. A rigid body is a rigidly connected collection of particles.

Force. The action of one body on another. This action will cause a change in the motion of the first body unless counteracted by an additional force or forces. A force may be produced either by actual contact or remotely (gravitation, electrostatics, magnetism, etc.). Force is a vector quantity.

Couple. If two forces of equal magnitude, opposite direction, and different lines of action act on a body, they produce a tendency for rotation, but no tendency for translation. Such a pair of forces is called a couple. The magnitude of the moment produced by a couple is calculated by multiplying the magnitude of one of the two forces times the perpendicular distance between them. Moment is a vector quantity, and its sense of direction is considered to be outwardly perpendicular to the plane of counterclockwise rotation of the couple. The moment of a single force about some point A is the magnitude of the force times the perpendicular distance between A and the line of action of the force.

Velocity. A measure of the instantaneous rate of change of position in space with respect to time. Velocity is a vector quantity.

Acceleration. A measure of the instantaneous rate of change in velocity with respect to time. Acceleration is a vector quantity.

Gravitational acceleration. Every body falling in a vacuum at a given position above and near the surface of the earth will have the same acceleration, g. Although this acceleration varies slightly over the earth's surface due to local variations in its shape and density, it is sufficiently accurate for most engineering calculations to assume that $g = 32.2$ ft/s^2 or 9.81 m/s^2 at the surface of the earth.

Weight. A measure of the force exerted on a body of mass M by the gravitational attraction of the earth. The magnitude of this force is

$$\mathbf{W} = M\mathbf{g} \qquad [2.1.1]$$

where $\mathbf{W}$ is the weight of the body. Strictly speaking, weight is a vector quantity since it is a force acting in the direction of the gravitational acceleration.

2.1.1.3 General Laws
The foregoing defined quantities interact according to the following fundamental laws, which are based upon empirical evidence.

Conservation of mass. The mass of a system of particles remains unchanged during the course of ordinary physical events.

Parallelogram law for the addition of forces. Two forces, $\mathbf{F}_1$ and $\mathbf{F}_2$, acting on a particle may be replaced by a single force, $\mathbf{R}$, called their resultant. If the two forces are represented as the adjacent sides of a parallelogram, the diagonal of the parallelogram will represent the resultant (Figure 2.1.1).

Principle of transmissibility. A force acting at a point on a body can be replaced by a second force acting at a different point on the body without changing the state of equilibrium or motion of the body as long as the second force has the same magnitude and line of action as the first.

2.1.1.3.1 Newton's Laws of Motion
1. A particle at rest will remain at rest, and a particle in motion will remain in motion along a straight line with no acceleration unless acted upon by an unbalanced system of forces.
2. If an unbalanced system of forces acts upon a particle, it will accelerate in the direction of the resultant force at a rate proportional to the magnitude of the resultant force. This law expresses the relationship between force, mass, and acceleration and may be written as

$$\mathbf{F} = M\mathbf{a} \qquad [2.1.2]$$

where $\mathbf{F}$ is the resultant force, M is the mass of the particle, and $\mathbf{a}$ is the acceleration of the particle.
3. Contact forces between two bodies have the same magnitude, the same line of action, and opposite direction.

Gravitation. Two particles in space are attracted toward each other by a force that is proportional to the product of their masses and inversely proportional to the square of the distance between them. Mathematically this may be stated as

$$|\mathbf{F}| = \frac{Gm_1 m_2}{r^2} \qquad [2.1.3]$$

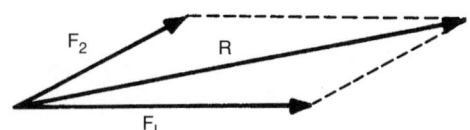

Figure 2.1.1 *Parallelogram law for addition of forces.*

where $|\mathbf{F}|$ is the magnitude of the force of gravitational attraction, G is the universal gravitational constant (6.673×10^{-11} m^3/kg $-$ s^2 or 3.44×10^{-8} ft^4/lb $-$ s^4), m$_1$ and m$_2$ are the masses of particles 1 and 2, and r is the distance between the two particles.

2.1.1.4 Systems of Units

Two systems of units are in common usage in mechanics. The first, the SI system, is an absolute system based on the fundamental quantities of space, time, and mass. All other quantities, including force, are derived. In the SI system the basic unit of mass is the kilogram (kg), the basic unit of length (space) is the meter (m), and the basic unit of time is the second (s). The derived unit of force is the Newton (N), which is defined as the force required to accelerate a mass of 1 kg at a rate of 1 m/s^2.

The U.S. customary or English system of units is a gravitational system based upon the quantities of space, time, and force (weight). All other quantities including mass are derived. The basic unit of length (space) is the foot (ft), the basic unit of time is the second (s), and the basic unit of force is the pound (lb). The derived unit of mass is the slug, which is the unit of mass that will be accelerated by a force of one pound at a rate of 1 ft/s^2. To apply the slug in practice, as in Equation 2.1.2, the weight in pounds mass must first be divided by g $=$ 32.2 ft/s^2, thus generating a working mass in units of lb $-$ s^2/ft, or slugs.

2.1.2 Statics

If there are no unbalanced forces acting on a *particle*, the particle is said to be in static equilibrium, and Newton's second law reduces to

$$\sum \mathbf{F} = 0 \qquad [2.1.4]$$

Solving a problem in particle statics reduces to finding the unknown force or forces such that the resultant force will be zero. To facilitate this process it is useful to draw a diagram showing the particle of interest and all the forces acting on it. This is called a *free-body diagram*. Next a coordinate system (usually Cartesian) is superimposed on the free-body diagram, and the forces are decomposed into their components along the coordinate axes. For the particle to be in equilibrium, the sum of the force components along each of the axes must be zero. This yields a set of algebraic equations to be solved for the forces in each coordinate direction.

Example 2.1.1

Block W, weighing 100 lb (see Figure 2.1.2) is attached at point A to a cable, which is, in turn, attached to vertical walls at points B and C. What are the tensions in segments AB and AC.

Breaking down the diagram into the various forces (Figure 2.1.2b):

- Force balance in the y direction:

$$\sum F_y = -100 + T_{AC} \sin 45° + T_{AB} \sin 15° = 0$$

$$0.707 T_{AC} + 0.259 T_{AB} = 100 \qquad [a]$$

- Force balance in the x direction:

$$\sum F_x = T_{AC} \cos 45° - T_{AB} \cos 15° = 0$$

$$0.707 T_{AC} - 0.966 T_{AB} = 0 \qquad [b]$$

and solving Example 2.1.1 Equations a and b simultaneously yields

$$T_{AB} = 81.6 \, \text{lb}$$

$$T_{AC} = 111.5 \, \text{lb}$$

If there are no unbalanced forces and no unbalanced moments acting on a *rigid body*, the rigid body is said to be in static equilibrium. That is, Equation 2.1.4 must be satisfied just as for particles, and furthermore

$$\sum \mathbf{M}_A = 0 \qquad [2.1.5]$$

where $\sum \mathbf{M}_A$ is the sum of the vector moments of all the forces acting on the body about any arbitrarily selected point A. In two dimensions this constitutes an algebraic equation because all moments must act about an axis perpendicular to the plane of the forces. In three dimensions the moments must be decomposed into components parallel to the principal, axes, and the components along each axis must sum algebraically to zero.

Example 2.1.2

A weightless beam 10 ft in length (see Figure 2.1.3a) supports a 10-lb weight, W, suspended by a cable at point C. The beam is inclined at an angle of 30° and rests against a step at point A and a frictionless fulcrum at point B, a distance of $L_1 = 6$ ft from point A. What are the reactions at points A and B?

Breaking the diagram down into the various forces (Figure 2.1.3b):

- Force balance in the x direction:

$$\sum F_x = R_{Ax} - 10 \cos 60° = 0$$

$$R_{Ax} = 5 \, \text{lb}$$

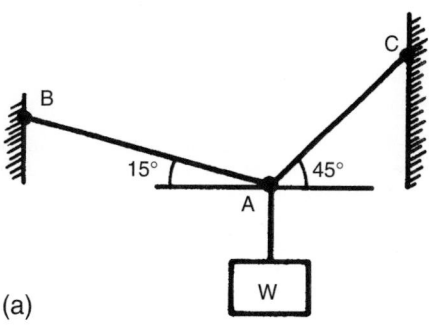

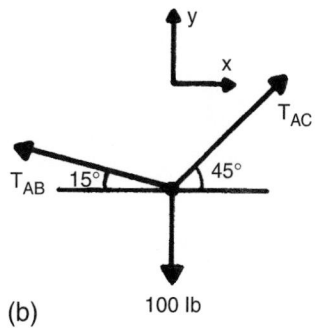

(a)   (b)   100 lb

Figure 2.1.2 *Diagram for Example 2.1.1.*

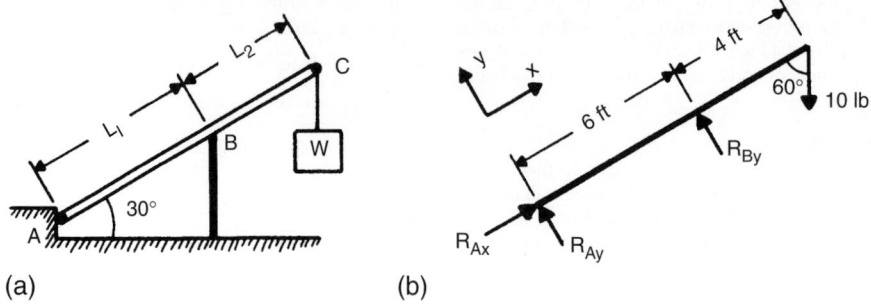

Figure 2.1.3 Diagram for Example 2.1.2.

Table 2.1.1 Centroids of Common Lines

Shape	Diagram	$\bar{x}$	$\bar{y}$	Length
Quarter-circular arc		$\dfrac{2r}{\pi}$	$\dfrac{2r}{\pi}$	$\dfrac{\pi r}{2}$
Semicircular arc		0	$\dfrac{2r}{\pi}$	πr
Arc of circle		$\dfrac{r\sin\alpha}{\alpha}$	0	$2\alpha r$

From Reference 2.

- Moment balance about point A:

$$+\sum M_A = (10)(10)\sin 60° - 6R_{By} = 0$$

$$R_{By} = 14.43 \text{ lb}$$

- Force balance in the y direction:

$$\sum F_y = R_{Ay} + R_{By} - 10\sin 60° = 0$$

$$R_{Ay} = 10\sin 60° - R_{By}$$

$$R_{Ay} = -4.43 \text{ lb}$$

Note that although the direction assumed for R_{Ay} was incorrect, the sign of the result indicates the correct direction.

Whenever the weight of a body is significant in comparison to the external forces, the weight, or body force, must be considered in both the force and moment balances.

The weight W of the body acts at the *center of gravity*, the Cartesian coordinates of which are found by

$$\bar{x} = \frac{1}{W}\int_v x\,dw \qquad [2.1.6]$$

$$\bar{y} = \frac{1}{W}\int_v y\,dw \qquad [2.1.7]$$

$$\bar{z} = \frac{1}{W}\int_v z\,dw \qquad [2.1.8]$$

The foregoing are volume integrals evaluated over the entire volume of the rigid body and dw is an infinitesimal element of weight. If the body is of uniform density, then the center of gravity is also called the *centroid*. Centroids of common lines, areas, and volumes are shown

in Tables 2.1.1, 2.1.2, and 2.1.3. For a composite body made up of elementary shapes with known centroids and known weights the center of gravity can be found from

$$\bar{x} = \frac{\Sigma_i \bar{x}_i W_i}{\Sigma_i W_i} \qquad [2.1.9]$$

$$\bar{y} = \frac{\Sigma_i \bar{y}_i W_i}{\Sigma_i W_i} \qquad [2.1.10]$$

$$\bar{z} = \frac{\Sigma_i \bar{z}_i W_i}{\Sigma_i W_i} \qquad [2.1.11]$$

Example 2.1.3

A mallet is composed of a section of a right circular cylinder welded to a cylindrical shaft, as shown in Figure 2.1.4a and b. Both components are steel, and the density is uniform throughout. Find the centroid of the mallet.

$$r = 2 \text{ in.} \qquad d = 1 \text{ in.}$$
$$L_1 = 6 \text{ in.} \qquad L_2 = 5 \text{ in.}$$
$$L_3 = 1.5 \text{ in.} \qquad \gamma = 0.283 \text{ lb/in.}^3$$

Letting the center of the bottom of the handle be the origin, the centroid of section 1, the handle, can be found by inspection as

$$\bar{x}_1 = 0$$

$$\bar{y}_1 = \frac{L_1}{2} = 3 \text{ in.}$$

$$w_1 = \frac{\pi}{4}d^2 L_1 \gamma = 1.33 \text{ lb}$$

Table 2.1.2 *Centroids of Common Areas*

Shape	Diagram	$\bar{x}$	$\bar{y}$	Area
Triangular area			$\dfrac{h}{3}$	$\dfrac{bh}{2}$
Quarter-circular area		$\dfrac{4r}{3\pi}$	$\dfrac{4r}{3\pi}$	$\dfrac{\pi r^2}{4}$
Semicircular area		0	$\dfrac{4r}{3\pi}$	$\dfrac{\pi r^2}{2}$
Quarter-elliptical area		$\dfrac{4a}{3\pi}$	$\dfrac{4b}{3\pi}$	$\dfrac{\pi ab}{4}$
Semielliptical area		0	$\dfrac{4b}{3\pi}$	$\dfrac{\pi ab}{2}$
Semiparabolic area		$\dfrac{3a}{8}$	$\dfrac{3h}{5}$	$\dfrac{2ah}{3}$
Parabolic area		0	$\dfrac{3h}{5}$	$\dfrac{4ah}{3}$
Parabolic spandrel		$\dfrac{3a}{4}$	$\dfrac{3h}{10}$	$\dfrac{ah}{3}$
General spandrel		$\dfrac{n+1}{n+2}a$	$\dfrac{n+1}{4n+2}h$	$\dfrac{ah}{n+1}$
Circular sector		$\dfrac{2r\sin\alpha}{3a}$	0	αr^2

From Reference 2.

In section 2, the integral formula, 2.1.7, is applied

$$\bar{x}_2 = 0$$

$$\bar{y}_2 = a + \frac{1}{W_2}\int_{L_1}^{b} y\,dW_2$$

where

$$a = L_1 - (r - L_3)$$
$$b = L_1 + L_3$$

Transforming the integral to polar coordinates as shown in Figure 2.1.4c:

$$dW_2 = L_2\gamma 2r\cos\theta\,dy$$

$$y = r\sin\theta$$

$$dy = r\cos\theta\,d\theta$$

$$\theta_1 = \sin^{-1}\left(\frac{r - L_3}{r}\right) = 14.5° = 0.253 \text{ rad}$$

$$\theta_2 = 90° = \frac{\pi}{2} \text{ rad}$$

$$\cos\theta_1 = 0.968 \text{ rad}$$

$$\bar{y}_2 = L_1 - (r - L_3) + \frac{2L_2\gamma r^3}{W_2}\int_{\theta_1}^{\theta_2}\sin\theta\cos\theta\,d\theta$$

$$\bar{y}_2 = 5.5 + \frac{6.85}{W_2}$$

$$W_2 = \int_{L_1}^{L_1+L_3} dW = 2r^2 L_2\gamma\int_{\theta_1}^{\theta_2}\cos^2\theta\,d\theta$$

$$W_2 = 6.09 \text{ lb}$$

Table 2.1.3 *Centroids of Common Volumes*

Shape		$\bar{x}$	Volume
Hemisphere		$\dfrac{3a}{8}$	$\dfrac{2}{3}\pi a^3$
Semiellipsoid of revolution		$\dfrac{3h}{8}$	$\dfrac{2}{3}\pi a^2 h$
Paraboloid of revolution		$\dfrac{h}{3}$	$\dfrac{1}{2}\pi a^2 h$
Cone		$\dfrac{h}{4}$	$\dfrac{1}{3}\pi a^2 h$
Pyramid		$\dfrac{h}{4}$	$\dfrac{1}{3}abh$

From Reference 2.

Substituting W_2 into the equation for $\bar{y}_2$

$$\bar{y}_2 = 6.62 \text{ in.}$$

For the entire body

$$\bar{x} = 0$$

$$\bar{y} = \frac{\bar{y}_1 W_1 + \bar{y}_2 W_2}{W_1 + W_2} = \frac{3 \times 1.33 + 6.62 \times 6.09}{1.33 + 6.09}$$

$$\bar{y} = 5.97 \text{ in.}$$

When two bodies are in contact and there is a tendency for them to slide with respect to each other, a tangential *friction* force is developed that opposes the motion. For dry surfaces this is called *dry friction* or *coulomb friction*. For lubricated surfaces the friction force is called viscous friction or lubricated, and it is treated in the study of fluid mechanics. Consider a block of weight **W** resting on a flat surface as shown in Figure 2.1.5. The weight of the block is balanced by a normal force **N** that is equal and opposite to the body weight. If some sufficiently small sidewise force **P** is

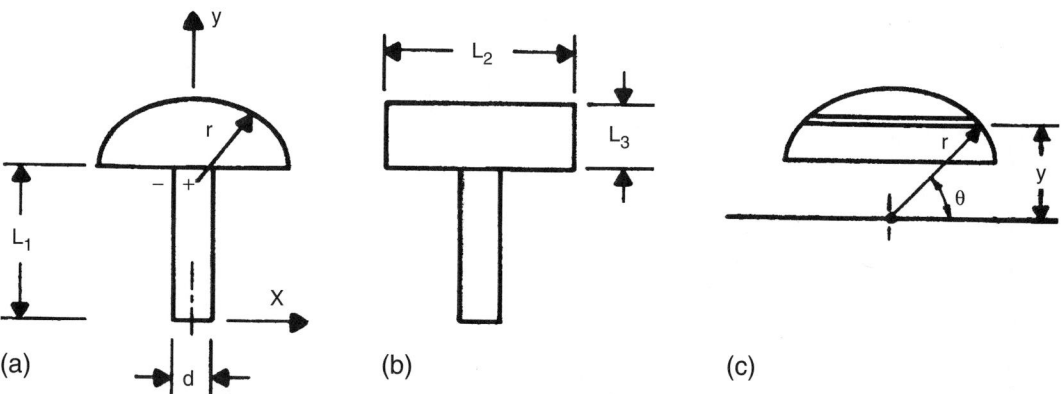

Figure 2.1.4 *Diagram for Example 2.1.3.*

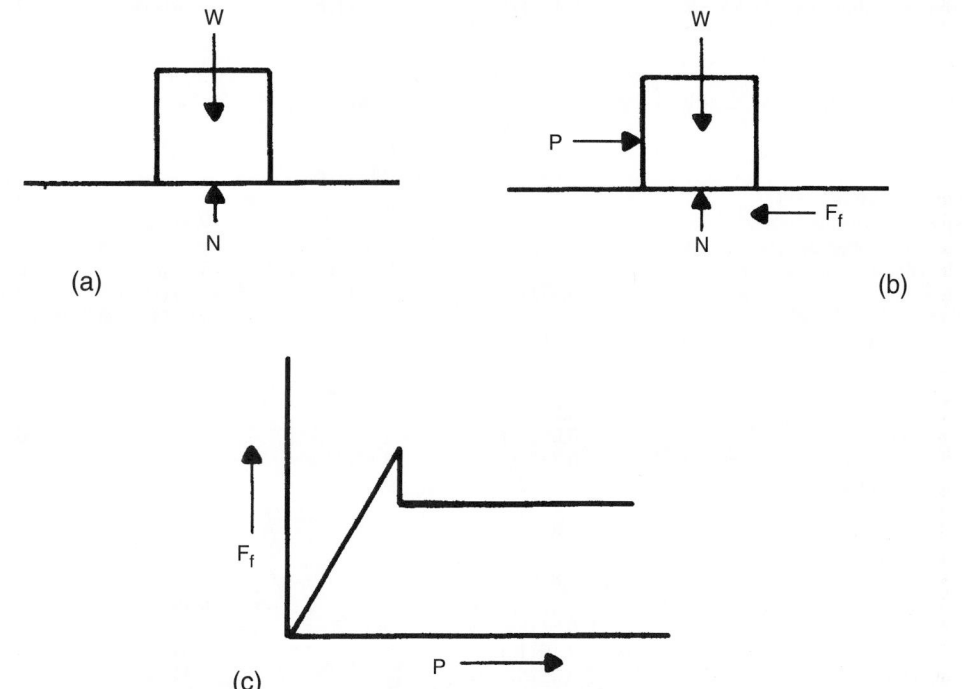

Figure 2.1.5 *Dry friction force.*

applied (Figure 2.1.5b), it will be opposed by a friction force **F** that is equal and opposite to **P**, and the block will remain fixed. If **P** is increased, **F** will simultaneously increase at the same rate until the maximum value of the static friction force is reached, at which point the block will begin to slide.

The maximum value of the static friction force is proportional to the normal force as

$$F_f = \mu_s N \qquad [2.1.12]$$

where μ_s is called the *coefficient of static friction*. Once the block begins to slide, the friction force decreases slightly and remains at a constant value defined by

$$F_f = \mu_k N \qquad [2.1.13]$$

where μ_K is the *coefficient of kinetic friction*. The magnitude of the friction force as a function of the applied force P is illustrated in Figure 2.1.5c, and typical values for μ_s and μ_k are give for both dry and lubricated surfaces in Table 2.1.4.

It is often necessary to compute the forces in *structures* made up of connected rigid bodies. A free-body diagram of the entire structure is used to develop an equation or equations of equilibrium based on the body weight of the structure and the external forces. Then the structure is decomposed into its elements and equilibrium equations are written for each element, taking advantage of the fact that by Newton's third law the forces between two members at a common frictionless joint are equal and opposite.

One of the simplest structures is the *truss*. A truss consists of straight members connected at their end points only (two force members). All loads, including the weight of the members themselves, are considered to be supported at the joints.

Table 2.1.4 *Typical Values for μ_s and μ_K for Dry and Lubricated Surfaces*

Materials	Static Dry	Static Greasy	Sliding Dry	Sliding Greasy
Hard steel on hard steel	0.78 (1)	0.11 (1, a)	0.42 (2)	0.029 (5, h)
		0.23 (1, b)		0.081 (5, c)
		0.15 (1, c)		0.080 (5, i)
		0.11 (1, d)		0.058 (5, j)
		0.0075 (18, p)		0.084 (5, d)
		0.0052 (18, h)		0.105 (5, k)
				0.108 (5, m)
				0.12 (5, a)
Mild steel on mild steel	0.74 (19)		0.57 (3)	0.09 (3, a)
				0.19 (3, u)
Hard steel on graphite	0.21 (1)	0.09 (1, a)		
Hard steel on babitt (ASTM No. 1)	0.70 (11)	0.23 (1, b)	0.33 (6)	0.16 (1, b)
		0.15 (1, c)		0.06 (1, c)
		0.08 (1, d)		0.11 (1, d)
		0.085 (1, e)		
Hard steel on Babbitt (ASTM No. 8)	0.42 (11)	0.17 (1, b)	0.35 (11)	0.14 (1, b)
		0.11 (1, c)		0.065 (1, d)
		0.09 (1, d)		0.07 (1, d)
		0.08 (1, e)		0.08 (11, h)
Hard steel on Babbitt (ASTM No. 10)		0.25 (1, b)		0.13 (1, b)
		0.12 (1, c)		0.06 (1, c)
		0.10 (1, d)		0.055 (1, d)
		0.11 (1, e)		
Mild steel on cadmium silver				0.173 (2, f)
Mild steel on phosphor bronze			0.34 (3)	0.173 (2, f)
Mild steel on copper lead				0.145 (2, f)
Mild steel on cast iron		0.183 (15, c)	0.23 (6)	0.133 (2, f)
Mild steel on lead	0.95 (11)	0.5 (1, f)	0.95 (11)	0.3 (11, f)
Nickel on mild steel			0.64 (3)	0.178 (3, x)
Aluminum on mild steel	0.61 (8)		0.47 (3)	
Magnesium on mild steel			0.42 (3)	
Magnesium on magnesium	0.6 (22)	0.08 (22, y)		
Teflon on Teflon	0.04 (22)			0.04 (22, f)
Teflon on steel	0.04 (22)			0.04 (22, f)
Tungsten carbide on tungsten carbide	0.2 (22)	0.12 (22, a)		
Tungsten carbide on steel	0.5 (22)	0.08 (22, a)		
Tungsten carbide on copper	0.35 (23)			
Tungsten on carbide on iron	0.8 (23)			
Bonded carbide on copper	0.35 (23)			
Bonded carbide on iron	0.8 (23)			
Cadmium on mild steel			0.46 (3)	
Copper on mild steel	0.53 (8)		0.36 (3)	0.18 (17, a)
Nickel on nickel	1.10 (16)		0.53 (3)	0.12 (3, w)
Brass on mild steel	0.51 (8)		0.44 (6)	
Brass on cast iron			0.30 (6)	
Zinc on cast iron	0.85 (16)		0.21 (7)	
Magnesium on cast iron			0.25 (7)	
Copper on cast iron	1.05 (16)		0.29 (7)	
Tin on cast iron			0.32 (7)	
Lead on cast iron			0.43 (7)	
Aluminum on aluminum	1.05 (16)		1.4 (3)	
Glass on glass	0.94 (8)	0.01 (10, p)	0.40 (3)	0.09 (3, a)
		0.005 (10, q)		0.116 (3, v)
Carbon on glass			0.18 (3)	
Garnet on mild steel			0.39 (3)	
Glass on nickel	0.78 (8)		0.56 (3)	
Copper on glass	0.68 (8)		0.53 (3)	
Cast iron on cast iron	1.10 (16)		0.15 (9)	0.070 (9, d)
				0.064 (9, n)
Bronze on cast iron			0.22 (9)	0.077 (9, n)
Oak on oak (parallel to grain)	0.62 (9)		0.48 (9)	0.164 (9, r)
				0.067 (9, s)

(continued)

Table 2.1.4   (continued)

Materials	Static		Sliding	
	Dry	Greasy	Dry	Greasy
Oak on oak (perpendicular)	0.54 (9)		0.32 (9)	0.072 (9, s)
Leather on oak (parallel)	0.61 (9)		0.52 (9)	
Cast iron on oak			0.49 (9)	0.075 (9, n)
Leather on cast iron			0.56 (9)	0.36 (9, t)
				0.13 (9, n)
Laminated plastic on steel			0.35 (12)	0.05 (12, t)
Fluted rubber bearing on steel				0.05 (13, t)

Reference letters indicate the lubricant used: numbers in parentheses give the sources.
(1) Campbell, Trans. *ASME*, 1939; (2) Clarke, Lincoln, and Sterrett, *Proc. API*, 1935; (3) Bears and Bowden, *Phil. Trans. Roy. Soc.*, 1935; (4) Dokos, Trans, ASME, 1946; (5) Boyd ani Robertson, Trans. *ASME*, 1945; (6) Sacha, Zed. *f. angeu. Math, und Mech.*, 1924; (7) Honda and Yania la . *Jour. I of M*. 1925; (8) Tomlinson, *Phil, Mag.,* 1929; (9) Morin, *Acad. Roy. des Sciences*, 1838; (10) Claypoole, *Trans. ASME*, 1943; (11) Tabor, Jour. Applied Phys., 1945; (12) Eyssen, General Discussion on Lubrication, *ASME*, 1937; (13) Brazier and Holland-Bowyer, General Discussion on Lubrication, ASME, 1937; (14) Burwell, *Jour. SAE*, 1942; (15) Stanton, "Friction," Longinans; (16) Ernst and Merchant, Conference on Friction and Surface Finish, M.I.T., 1940; (17) Gongwer, Conference on Friction and Surface Finish M.I.T., 1940; (18) Hardy and Bircumshaw, *Proc. Rey. Soc.*, 1925; (19) Hardy and Hardy, *Phil. Mag.*, 1919; M.I.T., 1940; (20) Bowden and Young, *Proc. Roy. Soc.*, 1951; (21) Hardy and Doubleday, *Proc. Roy. Soc.*, 1923; (22) Bowqden and Tabor, "The Friction and Lubrication of Solids," Oxford: (23) Shooter. *Research*, 4, 1951.
(a) Oleic acid; (b) Atlantic spindle oil (light mineral); (c) castor oil; (d) lard oil; (e) Atlantic spindle oil plus 2 percent oleic acid; (f) medium ineral oil; (g) medium mineral oil plus 1/2 percent oleic acid; (h) stearic acid; (i) grease (xinc oxide base); (f) graphite; (k) turbine oil plus 1 percent graphite; (1) turbine oil plus 1 percent stearic acid; (m) turbine oil (medium mineral); (n) olive oil; (p) palmitic acid; (q) ricinoleic acid; (r) dry soap; (s) lard; (t) water; (u rape oil); (v) 3-in-1 oil; (w) octyl alcohol; (x) triolein; (y) 1 percent lauric acid in paraffin oil.
From Reference 1.

Due to its construction and the assumption of loading at joints only, the members of a truss support only loads of axial tension or axial compression. A *rigid truss* or a rigid structure will not collapse and can only deform if its members deform. A *simple truss* is one that can be constructed, starting with three members arranged in a triangle, by adding new members in pairs, first connecting one end of each together to form a new joint, and then connecting the other ends at separate existing joints of the truss.

A *frame* is a structure with at least one member that supports more than two forces. Members of a frame may support lateral as well as axial forces. Connections in a frame need not be located at the ends of the members. Frames, like trusses, are designed to support loads, and are usually motionless. A *machine* also has multiforce members. It is designed to modify and transmit forces and, though it may sometimes be stationary, it always includes parts that move during some phase of operation.

Not all structures can be fully analyzed by the methods of statics. If the number of discrete equilibrium equations is equal to the number of unknown loads, then the structure is said to be *statically determinate* and rigid. If there are more

unknowns than equations, then the structure is *statically indeterminate*. If there are more equations than unknowns, then the structure is said to be *statically indeterminate and nonrigid*.

For further information on this subject, refer to References 1–5.

2.1.3 Dynamics
Dynamics is the study of the mechanics of rigid bodies in motion. It is usually subdivided into *kinematics*, the study of the motion of bodies without reference to the forces causing that motion or to the mass of bodies, and *kinetics*, the study of the relationship between the forces acting on a body, the mass and geometry of the body, and the resulting motion of the body.

2.1.3.1 Kinematics
Kinematics is based on one-dimensional differential equations of motion. Suppose a particle is moving along a straight line, and its distance from some reference point is S (see Figure 2.1.6a). Then its linear velocity and linear

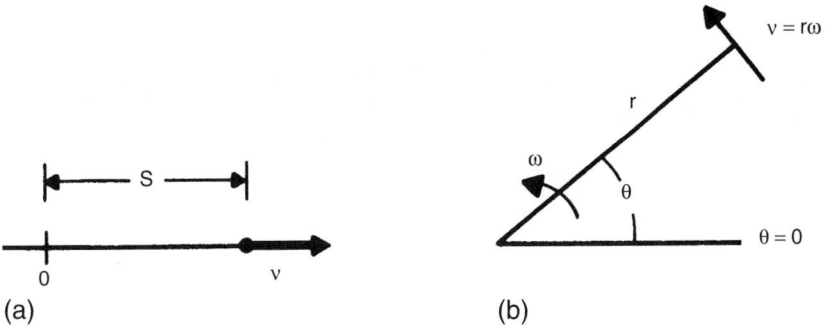

Figure 2.1.6   Diagrams of motion: (a) one-dimensional linear; (b) rotational.

Table 2.1.5 *One-dimensional Differential Equations of Motion and Their Solutions*

	Differential Equations	a/α = constant	$a=a(t)$; $\alpha = \alpha(t)$	$a = a(v)$; $\alpha = \alpha(\omega)$	$a = a(s)$; $\alpha = \alpha(\theta)$
Linear	$v = \dfrac{ds}{dt}$	$v_f = v_i = at$	$s_f = s_i + \displaystyle\int_0^t v(t)\,dt$		$t = \displaystyle\int_{s_i}^{s_f} \dfrac{ds}{v(s)}$
	$a = \dfrac{dv}{dt}$	$s_f = s_i + v_i t + \frac{1}{2}at^2$	$v_f = v_i + \displaystyle\int_0^t a(t)\,dt$	$t = \displaystyle\int_{v_i}^{v_f} \dfrac{dv}{a(v)}$	
	$v\,dv = a\,ds$	$v_f^2 = v_i^2 + 2a\Delta s$		$s_f = s_i + \displaystyle\int_{v_i}^{v_f} \dfrac{v\,dv}{a(v)}$	$v_f^2 = v_i^2 + 2\displaystyle\int_{s_i}^{s_f} a(s)\,ds$
Rotation	$\omega = \dfrac{d\theta}{dt}$	$\omega_f = \omega_i + \alpha t$	$\theta_r = \theta_i + \displaystyle\int_0^t w(t)\,dt$		$t = \displaystyle\int_{\theta_i}^{\theta_f} \dfrac{d\theta}{\omega(\theta)}$
	$\alpha = \dfrac{d\omega}{dt}$	$\theta_f = \theta_i + \omega_i t + \frac{1}{2}\alpha t^2$	$\omega_f = \omega_i + \displaystyle\int_0^t \alpha(t)\,dt$	$t = \displaystyle\int_{\omega_i}^{\omega_f} \dfrac{d\omega}{\alpha(\omega)}$	
	$\omega\,d\omega = \alpha\,d\theta$	$\omega_f^2 = \omega_i^2 + 2\alpha\Delta\theta$		$\theta_f = \theta_i + \displaystyle\int_{\omega_i}^{\omega_f} \dfrac{\omega\,d\omega}{\alpha(\omega)}$	$\omega_f^2 = \omega_i^2 + 2\displaystyle\int_{\theta_i}^{\theta_f} \alpha(\theta)\,d\theta$

acceleration are defined by the differential equations given in the top half of Column 1, Table 2.1.5. The solutions to these equations are in Columns 2–5, for the cases of constant acceleration, acceleration as a function of time, acceleration as a function of velocity, and acceleration as a function of position.

For rotational motion, as illustrated in Figure 2.1.6b, a completely analogous set of equations and solutions are given in the bottom half of Table 2.1.5. There ω is called the angular velocity and has units of radians/s, and α is called angular acceleration and has units of radians/s².

The equations of Table 2.1.5 are all scalar equations representing discrete components of motions along orthogonal axes. The axes along which the component ω or α acts is defined in the same fashion as for a couple. That is, the direction of ω is outwardly perpendicular to the plane of counterclockwise rotation (Figure 2.1.7).

The equations of Table 2.1.5 can be used define orthogonal components of motion in space, and these components are then combined vectorally to give the complete motion of the particle or point in question.

The calculation and combination of the components of particle motion requires imposition of a coordinate system. Perhaps the most common is the Cartesian system illustrated in Figure 2.1.8. Defining unit vectors $\hat{i}$, $\hat{j}$, and $\hat{k}$ along the coordinate axes x, y, and z, the position of some point in space, P, can be defined by a position vector, $\mathbf{r}_p$:

$$\mathbf{r}_p = x_p\hat{i} + y_p\hat{j} + z_p\hat{k} \qquad [2.1.14]$$

In Equation 2.1.14, x, y, and z represent the coordinates of point P. The velocity of P is the vector sum of the component velocities:

$$v_x = \frac{dx_p}{dt}, \qquad v_y = \frac{dy_p}{dt}, \qquad \text{and} \qquad v_z = \frac{dz_p}{dt}$$

$$v_p = v_x\hat{i} + v_y\hat{j} + v_z\hat{k} \qquad [2.1.15]$$

Likewise, the acceleration of P is the vector sum of the components of the accelerations where

$$a_x = \frac{dv_x}{dt}, \qquad a_y = \frac{dv_y}{dt}, \qquad \text{and} \qquad a_z = \frac{dv_z}{dt}$$

$$\mathbf{a}_p = a_x\hat{i} + a_y\hat{j} + a_z\hat{k} \qquad [2.1.16]$$

For any vector, the magnitude is the square root of the sum of the squares of the components. Thus the magnitude of the velocity of point P would be

$$|v_p| = (v_x^2 + v_y^2 + v_z^2)^{0.5} \qquad [2.1.17]$$

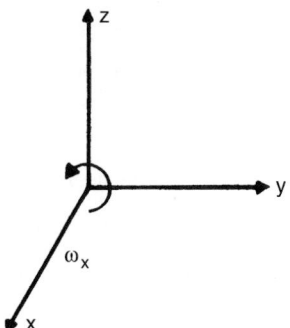

Figure 2.1.7 W_x is a vector of magnitude W_x acting along the x axis.

The angle between the total velocity (or any other vector) and any particular coordinate axis can be calculated from the scalar product of said vector and the unit vector along that axis. The scalar product is defined as

$$\mathbf{a}\cdot\mathbf{b} = a_x b_x + a_y b_y + a_z b_z = |a||b|\cos\theta$$

where θ is the angle between vectors **a** and **b**. Thus the angle between the velocity and the x axis is

$$\theta_x = \cos^{-1}\left(\frac{v\cdot\hat{i}}{|v|}\right)$$

$$= \cos^{-1}\left[\frac{v_x}{(v_x^2 + v_y^2 + v_z^2)^{0.5}}\right] \qquad [2.1.18]$$

If the magnitude and direction of a vector are known, its components are the products of the magnitude and the respective direction cosines. In the case of the velocity vector, for example, the components are

$$v_x = |v_p|\cos\theta_x$$
$$v_y = |v_p|\cos\theta_y \qquad [2.1.19]$$
$$v_z = |v_p|\cos\theta_z$$

Example 2.1.4
A projectile is fired at an angle of 30° to the surface of the earth with an initial velocity of 1000 ft/s (Figure 2.1.9).

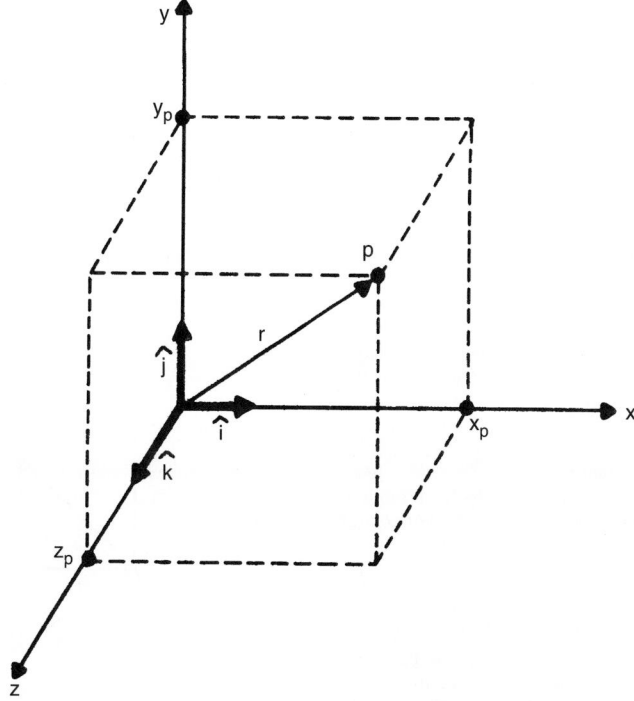

Figure 2.1.8 *Equations of motion in a Cartesian coordinate system.*

Figure 2.1.9 *Diagram for Example 2.1.4.*

What will be its velocity and the angle of its trajectory as a function of time?

x Component: The initial velocity in the x direction is

$$v_{xo} = |v| \cos \theta_x = 1000 \, \text{ft/s} \cos 30°$$

$$v_{xo} = 866 \, \text{ft/s}$$

Assuming no air friction, this velocity is constant; that is $v_x^{(t)} = v_{xo} = 866 \, \text{ft/s}$.

y Component: The initial velocity in the y direction is

$$v_{yo} = |v| \cos \theta_y = |v| \sin \theta_x = 1000 \, \text{ft/s} \sin 30°$$

$$v_{yo} = 500 \, \text{ft/s}$$

In the y direction, the projectile has a constant acceleration of $-g = -32.2 \, \text{ft/sec}^2$. By the first Equation, Column 2,

Table 2.1.5, its velocity as a function of time is

$$v_y(t) = v_{yo} + at = v_{yo} - gt$$

$$= 500 \, \text{ft/s} - \frac{32.2 \, \text{ft}}{s^2} t$$

and the total velocity vector is

$$v = 866\hat{i} + (500 - 32.2t)\hat{j}$$

The angle of the trajectory is found from Equation 2.1.18 as

$$\theta_x = \cos^{-1} \left\{ \frac{866}{[866^2 + (500 - 32.2t)^{0.5}} \right\}$$

It is often convenient to use some other coordinate system besides the Cartesian system. In the normal/tangential

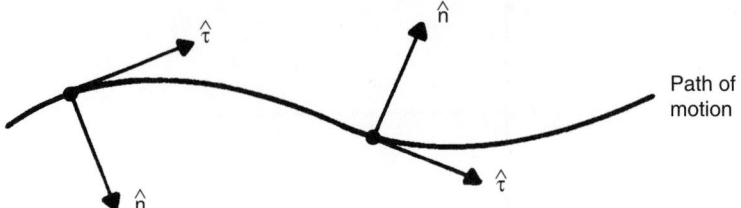

Figure 2.1.10 *Normal and tangential unit vectors at different points on a path.*

system (Figure 2.1.10), the point of reference is not fixed in space but is located on the particle and moves as the particle moves. There is no position vector and the velocity and acceleration vectors are written in terms of unit vectors $\hat{\tau}$, tangent to the path of motion, and $\hat{\eta}$, inwardly perpendicular to the path of motion.

The velocity is always tangent to the path of motion, and thus the velocity vector has only one component (Equation 2.1.20).

$$v = v\hat{\tau} \qquad [2.1.20]$$

The acceleration vector has a component tangent to the path $a_t = d|v|/d_t$, which the rate at which the magnitude of the velocity vector is changing, and a component perpendicular to the path $a_n = |v|^2/\rho$, which represents the rate at which the direction of motion is changing (Equation 2.1.21)

$$\mathbf{a} = a_t\hat{\tau} + a_n\hat{\mathbf{n}} = \frac{d|v|}{dt}\hat{\tau} + \frac{|v|^2}{\rho}\hat{\mathbf{n}} \qquad [2.1.21]$$

In Equation 2.1.21 ρ is the local radius of curvature of the path. The normal component of acceleration can also be expressed as $a_n = \rho|\omega|^2$ or $a_n = |v||\omega|$ where ω is the angular velocity of the particle.

Example 2.1.5
A car is increasing in speed at a rate of $10\,\text{ft/s}^2$ when it enters a curve with a radius of 50 ft at a speed of 30 ft/s. What is the magnitude of its total acceleration?

$$a_t = \frac{d|v|}{dt} = 10\,\text{ft/s}^2$$

$$a_n = \frac{v^2}{\rho} = \frac{(30\,\text{ft/s})^2}{50\,\text{ft}} = 18\,\text{ft/s}^2$$

$$\mathbf{a} = 10\hat{\tau} + 18\hat{\mathbf{n}}$$

$$|\mathbf{a}| = (10^2 + 18^2)^{0.5} = 20.6\,\text{ft/s}^2$$

In addition to the Cartesian and normal/tangential coordinate systems, the cylindrical (Figure 2.1.11) and spherical (Figure 2.1.12) coordinate systems are often used.

When dealing with the motions of rigid bodies or systems of rigid bodies, it is sometimes quite difficult to directly write out the equations of motion of the point in question as was done in Examples 2.1.4 and 2.1.5. It is sometimes more practical to analyze such a problem by *relative motion*. That is, first find the motion with respect to a nonaccelerating reference frame of some point on the body, typically the center of mass or axis or rotation, and vectorally add to this the motion of the point in question with respect to the reference point.

Example 2.1.6
Consider an arm 2 ft long rotating in the counter clockwise direction about a fixed axis at point A at a rate of 2 rpm

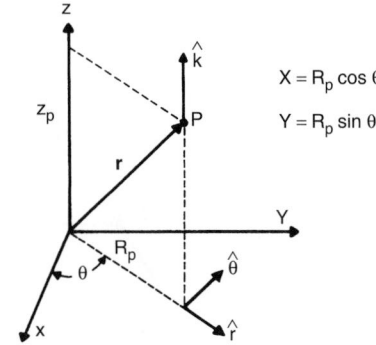

$$X = R_p \cos\theta$$
$$Y = R_p \sin\theta$$

$$\mathbf{r} = R_p\hat{r} + Z_p\hat{k}$$

$$\mathbf{v} = \left(\frac{dR_p}{dt}\hat{r}\right) + \left(R_p\frac{d\theta}{dt}\right)\hat{\theta} + \left(\frac{dz}{dt}\right)\hat{k}$$

$$\mathbf{a} = \left[\frac{d^2R_p}{dt^2} - R_p\left(\frac{d\theta}{dt}\right)^2\right]\hat{r} + \left[R_p\frac{d^2\theta}{dt^2} + 2\frac{dR_p}{dt}\frac{d\theta}{dt}\right]\hat{\theta} + \left[\frac{d^2z}{dt^2}\right]\hat{k}$$

Figure 2.1.11 *Equations of motion in a cylindrical coordinate system.*

(Figure 2.1.13a). Attached to the arm at point B is a disk with a radius of 1 ft, which is rotating in the clockwise direction about point B at a rate of 5 rpm. What is the velocity and acceleration of some arbitrary point C on the rim of the disk? (Consider that at $t_0 = 0$, the configuration is that of Figure 2.1.13d.)

By relative motion the position vector of C is the sum of the position vector of B and the position vector of C with respect to B

$$\mathbf{r}_C = \mathbf{r}_B + \mathbf{r}_{C/B} \qquad [2.1.22]$$

and likewise with the velocity and acceleration, which is

$$v_C = v_B + v_{C/B} \qquad [2.1.23]$$

$$\mathbf{a}_C = \mathbf{a}_B + \mathbf{a}_{C/B} \qquad [2.1.24]$$

Analyzing the motion of point B (Figure 2.1.13b),

$$\mathbf{r}_B = L\cos\theta_B\hat{i} + L\sin\theta_B\hat{j}$$

Assuming rotation starts at $\theta_B = 0$,

$$\theta_B = \omega_B t$$

$$v_B = \frac{d}{dt}(\mathbf{R}_B) = -L\omega_B\sin\omega_B t\hat{i} + L\omega_B\cos\omega_B t\hat{j}$$

$$\mathbf{a}_B = \frac{d}{dt}(v_B) = -L\omega_B^2\cos\omega_B t\hat{i} - L\omega_B^2\sin\omega_B t\hat{j}$$

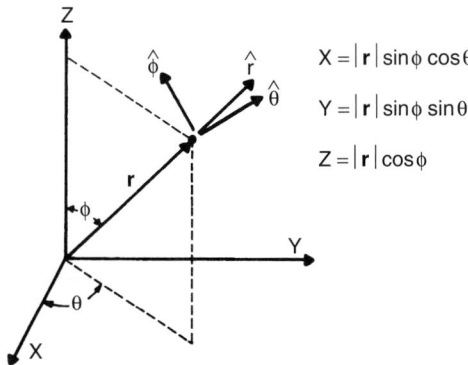

$$\mathbf{r} = |\mathbf{r}|\hat{r}$$

$$v = \left(\frac{d|\mathbf{r}|}{dt}\right)\hat{r} + \left(|\mathbf{r}|\frac{d\theta}{dt}\cos\phi\right)\hat{\theta} + \left(|\mathbf{r}|\frac{d\phi}{dt}\right)\hat{\phi}$$

$$\mathbf{a} = \left[\frac{d^2|\mathbf{r}|}{dt^2} - |\mathbf{r}|\left(\frac{d\phi}{dt}\right)^2 - |\mathbf{r}|\left(\frac{d\theta}{dt}\right)^2\cos^2\phi\right]\hat{r} + \left[\frac{\cos\phi}{|\mathbf{r}|}\frac{d}{dt}\left(|\mathbf{r}|^2\frac{d\theta}{dt}\right) - 2|\mathbf{r}|\frac{d\theta}{dt}\frac{d\phi}{dt}\sin\phi\right]\hat{\theta}$$

$$+ \left[\frac{1}{|\mathbf{r}|}\frac{d}{dt}\left(|\mathbf{r}|^2\frac{d\theta}{dt}\right) + |\mathbf{r}|\left(\frac{d\theta}{dt}\right)^2\sin\phi\cos\phi\right]\hat{\phi}$$

Figure 2.1.12 *Equations of motion in a spherical coordinate system.*

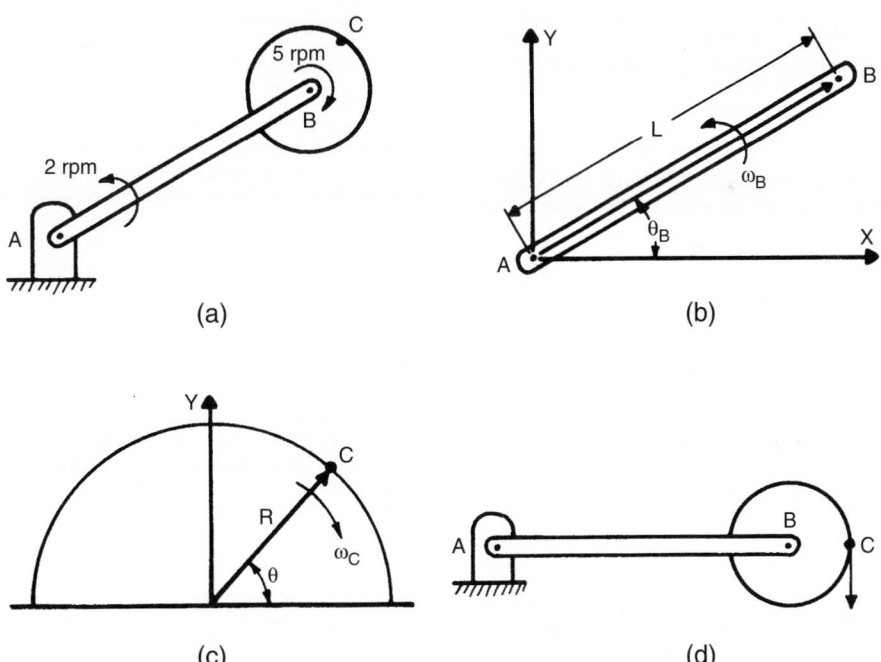

Figure 2.1.13 *Diagram for Example 2.1.6.*

Assuming point B is fixed and analyzing the motion of Point C with respect to B (see Figure 2.1.13c),

$$\mathbf{r}_{C/B} = R\cos\theta_C\hat{i} + R\sin\theta_C\hat{j}$$

$$\theta_c = \omega_c t$$

$$v_{C/B} = -R\omega_C\sin\omega_C t\hat{i} + R\omega_C\cos\omega_C t\hat{j}$$

$$\mathbf{a}_{C/B} = -R\omega_C^2\cos\omega_C t\hat{i} - R\omega_C^2\sin\omega_C t\hat{j}$$

The velocity of point C is

$$v_C = v_B + v_{C/B} = -(L\omega_B\sin\omega_B t + R\omega_C\sin\omega_C t)\hat{i}$$

$$+ (L\omega_B\cos\omega_B t + R\omega_C\cos\omega_C t)\hat{j}$$

$$\omega_B = \frac{2\text{rev}}{\text{min}} \left|\frac{1\,\text{min}}{60\text{s}}\right| \frac{2\pi\text{rad}}{\text{revolution}} = 0.209 \text{ rad/s}$$

$$L\omega_B = \frac{2\,\text{ft}}{}\left|\frac{0.209\,\text{rad/s}}{\text{s}}\right| = 0.418 \text{ ft/s}$$

$$\omega_c = -0.524 \text{ rad/s}$$

$$R\omega_c = -0.524 \text{ ft/s}$$

$$v_C = -[0.418\sin(0.209\,t) - 0.524\sin(-0.524\,t)]\hat{i}$$

$$+ [0.418\cos(0.209\,t) - 0.524\cos(-0.524\,t)]\hat{j}$$

At t = 0, for instance, the velocity is (Figure 2.1.13d)

$$v_C = -0.106\hat{j} \text{ ft/s}$$

Likewise, the general expression for the acceleration of point C is

$$a_C = a_B + a_{C/B} = -(L\omega_B^2\cos\omega_B t + R\omega_C^2\cos\omega_c t)\hat{i}$$

$$- (L\omega_B^2\sin\omega_B t + R\omega_C^2\sin\omega_c t)\hat{j}$$

At t = 0 this reduces to

$$a_C = -(L\omega_B^2 + R\omega_C^2)\hat{i} = -[2 \times 0.209^2 + 1 \times (-0.524)^2]\hat{i}$$

$$a_C = -0.362\hat{i} \text{ ft/s}^2$$

When looking for the velocities of points on a rigid body, the method of *instantaneous centers* can often be used. If the velocity of two points on the body are known, those points and all other points on the body can be considered to be rotating with the same angular velocity about some motionless central point. This central point is called the instantaneous center of zero velocity. The instantaneous center generally moves through space as a function of time and has acceleration. It does *not* represent a point about which acceleration may be determined.

Example 2.1.7

Link AB of length 2 ft (see Figure 2.1.14a) is sliding down a wall with point A moving downward at 4 ft/s when θ is 30°. What is the angular velocity of the link and linear velocity of point B?

Because v_A (Figure 2.1.14b) is parallel to the vertical wall, it is rotating about a point on a line through A perpendicular to the wall. Likewise B is rotating about a point on a line through B perpendicular to v_B. These two lines interest at C, the instantaneous center.

$$AC = L\cos\theta = Z \times 0.866 = 1.73 \text{ ft}$$

$$|v_A| = AC|\omega_{AB}|$$

$$\omega_{AB} = \frac{|v_A|}{AC} = \frac{4\,\text{ft}}{\text{s}}\left|\frac{}{1.73\,\text{ft}}\right|$$

where $\omega_{AB} = 2.31$ rad/s = the angular velocity of the link and of any line on the link

$$|v_B| = BC|\omega_{AB}| = L\sin\theta|\omega_{AB}| = \frac{2\,\text{ft}}{}\left|\frac{0.5}{}\right|\frac{2.31\,\text{rad}}{\text{s}}$$

$$|v_B| = 2.31 \text{ ft/s}$$

As an exercise, show that the locus of the instantaneous center of rotation represents a quarter circle. What is the radius of the circle?

2.1.3.2 Kinetics

In *kinetics*, Newton's second law, the principles of kinematics, conservation of momentum, and the laws of conservation of energy and mass are used to develop relationships between the forces acting on a body or system of bodies and the resulting motion.

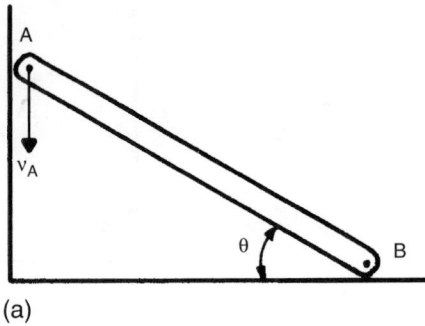

(a)

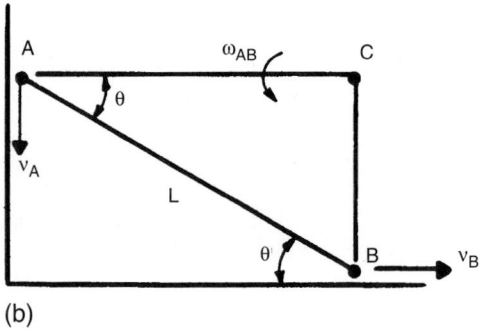

(b)

Figure 2.1.14 *Diagram for Example 2.1.7.*

2.1.3.3 Applications of Newton's Second Law

Problems involving no unbalanced couples can often be solved with the second law and the principles of kinematics. As in statics, it is appropriate to start with a free-body diagram showing all forces, decompose the forces into their components along a convenient set of orthogonal coordinate axes, and then solve a set of algebraic equations in each coordinate direction. If the accelerations are known, the solution will be for an unknown force or forces, and if the forces are known the solution will be for an unknown acceleration or accelerations.

Example 2.1.8

In Figure 2.1.15a, a 10-lb block slides down a ramp inclined at an angle of 30°. If the coefficient of kinetic friction between the block and the ramp is 0.1, what will be the acceleration of the block?

As shown in the free-body diagram of Figure 2.1.15b, all the motion of the block is parallel to the surface of the ramp, and there is a static force balance in the y direction.

$$\sum F_y = N - W\cos 30° = 0$$

$$N = W\cos 30°$$

$$\rightarrow F_f = \mu N = \mu W\cos 30°$$

By Newton's second law, the force in the x direction produces an acceleration a_x:

$$\sum F_x = W\sin 30° - F_f = ma_x = \frac{W}{g}a_x$$

$$a_x = g(\sin 30° - \mu\cos 30°)$$

$$a_x = 13.31 \text{ ft/s}^2$$

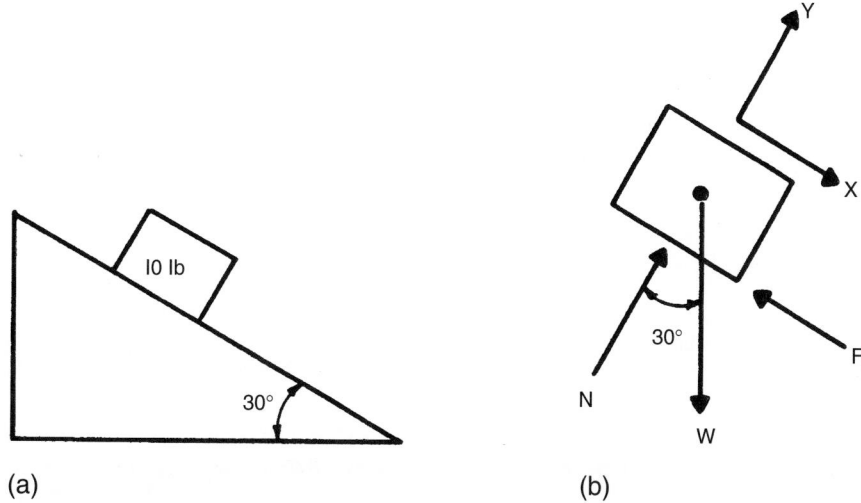

Figure 2.1.15 *Diagram for Example 2.1.8.*

When unbalanced couples are involved, a rotational analog to Newton's second law can be applied:

$$\sum \bar{\mathbf{M}} = \bar{\mathrm{I}}\alpha \qquad [2.1.25]$$

where $\sum \bar{\mathbf{M}}$ is the sum of all moments acting about the center of mass in the plane of rotation, $\bar{\mathrm{I}}$ is the mass moment of inertia about the center of mass, and α is the angular acceleration of the body. The mass moment of inertia is defined by

$$\mathrm{I} = \int \mathrm{r}^2 \mathrm{dm} = \mathrm{mk}^2 \qquad [2.1.26]$$

where r is the perpendicular distance from the axis of rotation to the differential element of mass, dm. I is sometimes expressed in terms of k, the radius of gyration, and m, the mass of the body. If the axis of rotation passes through the center of mass, then the mass moment of inertia is designated as $\bar{\mathrm{I}}$. Mass moments of inertia of common shapes are compiled in Tables 2.1.6 and 2.1.7.

It is often convenient to sum the moments about some arbitrary point 0, other than the mass center. In this case, Equation 2.1.25 becomes

$$\sum \mathbf{M}_0 = \bar{\mathrm{I}}\alpha + \mathrm{m}\bar{\mathrm{a}}\mathrm{d} \qquad [2.1.27]$$

where m is the mass of the body, $\bar{\mathrm{a}}$ is the linear acceleration of the mass center, and d is the perpendicular distance between the vector $\bar{\mathrm{a}}$ and point 0.

If 0 is a fixed axis or the instantaneous center of zero velocity, then Equation 2.1.27 reduces to

$$\sum \mathbf{M}_0 = \mathrm{I}_0 \alpha \qquad [2.1.28]$$

where I_0 is the mass moment of inertia about point 0, I_0 may be found from Equation 2.1.26, or it may be calculated from the parallel axis theorem

$$\mathrm{I}_0 = \bar{\mathrm{I}} + \mathrm{m}\bar{\mathrm{r}}^2 \qquad [2.1.29]$$

where $\bar{\mathrm{r}}$ is the distance from 0 to the center of mass. The parallel axis theorem may be used to find I_0 regardless of whether I_0 is a fixed axis of instantaneous center of zero velocity.

Example 2.1.9
In Figure 2.1.16a 10 lb cylinder with a 3-in. radius rolls down a 30° incline. What is its angular acceleration and the linear acceleration of its center of mass?

In the free-body diagram of Figure 2.1.16, the point of contact between the wheel and the ramp is the instantaneous center of zero velocity. Thus,

$$\sum \mathrm{M}_0 = \mathrm{rw}\sin 30° = \mathrm{I}_0 \alpha$$

$$\alpha = \frac{\mathrm{rw}\sin 30°}{\mathrm{I}_0}$$

From Tables 2.1.6 and 2.1.7 and the parallel axis theorem,

$$\mathrm{I}_0 = \frac{\mathrm{W}}{2\mathrm{g}}\mathrm{r}^2 + \frac{\mathrm{W}}{\mathrm{g}}\mathrm{r}^2 = \frac{3}{2}\frac{\mathrm{W}}{\mathrm{g}}\mathrm{r}^2$$

$$\alpha = \frac{2\mathrm{g}\sin 30°}{3\mathrm{r}} = 3.57 \ \mathrm{rad/s}$$

$$\mathrm{a}_\mathrm{x} = \mathrm{r}\alpha = 10.73 \ \mathrm{ft/s}^2$$

2.1.3.4 Conservation of Momentum
If the mass of a body or system of bodies remains constant, then Newton's second law can be interpreted as a balance between force and the time rate of change of momentum, *momentum* being a vector quantity defined as the product of the velocity of a body and its mass.

$$\mathbf{F} = \mathrm{m}\mathbf{a} = \frac{\mathrm{d}}{\mathrm{dt}}\mathrm{m}v = \frac{\mathrm{d}}{\mathrm{dt}}\mathbf{G} \qquad [2.1.30]$$

Integrating Equation 2.1.30 with respect to time yields the impulse/momentum equation

$$\int \mathbf{F}\mathrm{dt} = \Delta\mathbf{G} \qquad [2.1.31]$$

where $\mathbf{F}\mathrm{dt}$ is called the impulse, and $\Delta\mathbf{G}$ is the change in momentum. Equation 2.1.31 can be applied explicitly and is particularly useful when the force is known as a function of time.

In collisions between two bodies the contact force and the duration of contact are usually unknown. However, the duration of contact is the same for both bodies, and the force on the first body is the negative of the force on the second

Table 2.1.6   *Mass Centers and Moments of Inertia*

Body	Mass Center	Moments of Inertia
Circular Cylindrical Shell	—	$I_{xx} = \frac{1}{2}mr^2 + \frac{1}{12}ml^2$ $I_{x_1x_1} = \frac{1}{2}mr^2 + \frac{1}{2}ml^2$ $I_{zz} = mr^2$
Half Cylindrical Shell	$\bar{x} = \dfrac{2r}{\pi}$	$I_{xx} = I_{yy}$ $= \frac{1}{2}mr^2 + \frac{1}{2}ml^2$ $I_{x_1x_1} = I_{y_1y_1}$ $= \frac{1}{2}mr^2 + \frac{1}{3}ml^2$ $I_{zz} = mr^2$ $\bar{I}_{zz} = \left(1 - \dfrac{4}{\pi^2}\right)mr^2$
Circular Cylinder	—	$I_{xx} = \frac{1}{4}mr^2 + \frac{1}{12}ml^2$ $I_{x_1x_1} = \frac{1}{4}mr^2 + \frac{1}{3}ml^2$ $I_{zz} = \frac{1}{2}mr^2$
Semicylinder	$\bar{x} = \dfrac{4r}{3\pi}$	$I_{xx} = I_{yy}$ $= \frac{1}{4}mr^2 + \frac{1}{12}ml^2$ $I_{x_1x_1} = I_{v_1v_1}$ $= \frac{1}{4}mr^1 + \frac{1}{3}ml^2$ $I_{zz} = \frac{1}{2}mr^2$ $I_{zz} = \left(\dfrac{1}{2} - \dfrac{16}{9\pi^2}\right)mr^2$
Rectangular Parallelepiped	—	$I_{xx} = \frac{1}{12}m(a^2 + l^2)$ $I_{yy} = \frac{1}{12}m(b^2 + l^2)$ $I_{zz} = \frac{1}{12}m(a^2 + b^2)$ $I_{y_1y_1} = \frac{1}{12}mb^2 + \frac{1}{3}ml^2$

From Reference 3.

body. The net change in momentum is zero. This is called the principle of *conservation of momentum*.

If a collision is purely plastic, then the two colliding bodies will adhere to each other and move on as a single body. Knowing the initial velocities and masses allows calculation of the final velocity:

$$m_1v_1 + m_2v_2 = (m_1 + m_2)v \qquad [2.1.32]$$

If the collision is purely elastic or elasto-plastic, then the two bodies will depart the collision with different velocities.

$$m_1v_{11} + m_2v_{21} = m_1v_{12} + m_2v_{22} \qquad [2.1.33]$$

In this case, an additional equation is required before the final velocities may be found. Thus, the coefficient of restitution e is defined as the ratio of the velocity of separation to the velocity of approach:

$$e = \frac{v_{22x} - v_{12x}}{v_{11x} - v_{21x}} = \frac{v_{22y} - v_{12y}}{v_{11y} - v_{21y}} \qquad [2.1.34]$$

Note that e is defined in terms of the components of the velocities, not the vector velocities, whereas the momentum balance is defined in terms of the vector velocities. To solve Equations 2.1.32 and 2.1.33 when all the velocities are not colinear, one writes the momentum balances along the principal axes and solves the resulting equations simultaneously.

For purely elastic impacts, e = 1, and for purely plastic impacts, e = 0. For elastoplastic impacts, e lies between zero and one and is a function of both the material properties and the velocity of impact.

Example 2.1.10
Sphere 1 weights 1 lb and is traveling at 2 ft/s in the positive x direction when it strikes sphere 2, weighing 5 lb and traveling in the negative x direction at 1 ft/s. What will be the final velocity of the system if the collision is (a) plastic, or (b) Elastoplastic with e = 0.5?

(a) By Equation 2.1.32

$$v = \frac{m_1v_{11} - m_2v_{21}}{m_1 + m_2} = \frac{2 - 5}{6} = -0.5 \text{ ft/s}$$

(b) By Equation 2.1.34

$$e = \frac{v_{22} - v_{12}}{v_{11} - v_{21}}$$

$$v_{22} = v_{12} + 0.5[2 - (-1)] = v_{12} + 1.5$$

Table 2.1.7 *Area Centroids and Moment Areas*

Figure	Centroid	Area Moments of Inertia
Arc Segment	$\bar{r} = \dfrac{r \sin \alpha}{\alpha}$	—
Quarter and Semicircular Arcs	$\bar{y} = \dfrac{2r}{\pi}$	—
Triangular Area	$\bar{x} = \dfrac{a+b}{3}$ $\bar{y} = \dfrac{h}{3}$	$I_x = \dfrac{bh^3}{12}$ $\bar{I}_x = \dfrac{bh^3}{36}$ $I_{x_1} = \dfrac{bh^3}{4}$
Rectangular Area	—	$I_x = \dfrac{bh^3}{3}$ $\bar{I}_x = \dfrac{bh^3}{12}$ $\bar{J} = \dfrac{bh}{12}(b^2 + h^2)$
Area of Circular Sector	$\bar{x} = \dfrac{2}{3}\dfrac{r \sin \alpha}{\alpha}$	$I_x = \dfrac{r^4}{4}\left(\alpha - \dfrac{1}{2}\sin 2\alpha\right)$ $I_y = \dfrac{r^4}{4}\left(\alpha + \dfrac{1}{2}\sin 2\alpha\right)$ $J = \dfrac{1}{2}r^4\alpha$
Quarter Circular Area	$\bar{x} = \bar{y}, = \dfrac{4r}{3\pi}$	$I_x = I_y = \dfrac{\pi r^4}{16}$ $\bar{I}_x = \bar{I}_y = \left(\dfrac{\pi}{16} - \dfrac{4}{9\pi}\right)r^4$ $J = \dfrac{\pi r^4}{8}$
Area of Elliptical Quadrant Area $A = \dfrac{\pi ab}{4}$	$\bar{x} = \dfrac{4a}{3\pi}$ $\bar{y} = \dfrac{4b}{3\pi}$	$I_x = \dfrac{\pi ab^3}{16}, \; \bar{I}_x = \left(\dfrac{\pi}{16} - \dfrac{4}{9\pi}\right)ab^3$ $I_y = \dfrac{\pi a^3 b}{16}, \; \bar{I}_y = \left(\dfrac{\pi}{16} - \dfrac{4}{9\pi}\right)a^3 b$ $J = \dfrac{\pi ab}{16}(a^2 + b^2)$

From Reference 3, pp. 498–499.

By Equation 2.1.33

$$m_1 v_{11} + m_2 v_{21} = m_1 v_{12} + m_2 v_{22}$$

$$= m_1 v_{12} + m_2(v_{12} + 1.5)$$

$$= (m_1 + m_2)v_{12} + 1.5\, m_2$$

$$v_{12} = \frac{m_1 v_1 + m_2(v_2 - 1.5)}{m_1 + m_2} = \frac{1 \times 2 + 5(-1 - 1.5)}{1 + 5}$$

$$= -1.75 \text{ ft/s}$$

$$v_{22} = -0.25 \text{ ft/s}$$

The foregoing discussion of impulse and momentum applies only when no change in rotational motion is involved. There is an analogous set of equations for angular impulse and impulse momentum. The angular momentum about an axis through the center of mass is defined as

$$\bar{H} = \bar{I}\omega \qquad [2.1.35]$$

and the angular momentum about any arbitrary point 0 is defined as

$$H_o = \bar{I}\omega + m\bar{v}d \qquad [2.1.36]$$

where $\bar{v}$ is the velocity of the center of mass and d is the perpendicular distance between the vector $\bar{v}$ and the point 0.

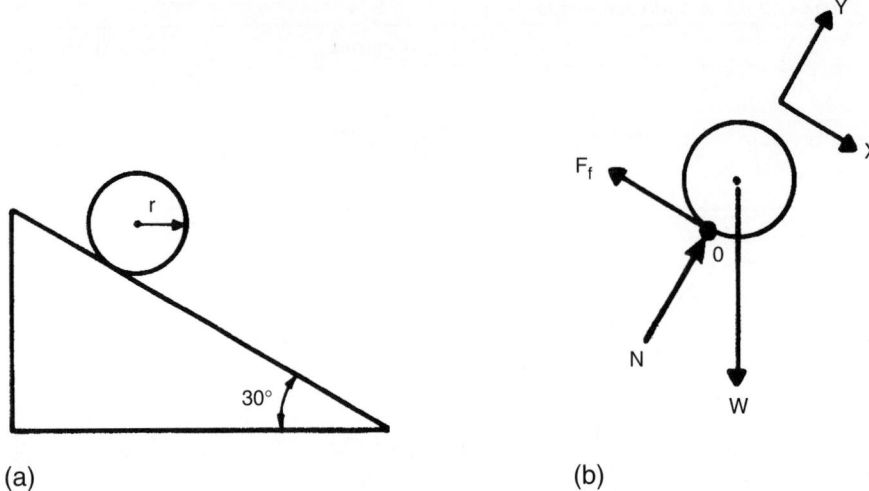

Figure 2.1.16   *Diagram for Example 2.1.9.*

And if 0 is a fixed axis or instantaneous center of zero velocity, then

$$\mathbf{H}_0 = I_0 \omega \qquad [2.1.37]$$

Likewise, the angular impulse is defined as

$$\int \mathbf{M}_0 dt = \Delta \mathbf{H}_0 \qquad [2.1.38]$$

In collisions, angular momentum, like linear momentum, is conserved.

2.1.3.5 Conservation of Energy

In a rigid-body system, energy is conserved in the sense that the net change in mechanical energy must be equal to the net work done on the system.

$$U = \Delta T + \Delta V_g + \Delta V_e \qquad [2.1.39]$$

U is the net work done on the system and is defined as the sum of the work done by external forces and external moments.

$$U = \int \mathbf{F} \cdot d\mathbf{s} + \int \mathbf{M} \cdot d\theta \qquad [2.1.40]$$

The work of the force **F** is positive if it acts in the direction of the displacement ds, and the work of the moment **M** is positive if it acts in the direction of rotation dθ.

ΔT is the change the kinetic energy, made up of a change in linear kinetic energy and rotational kinetic energy.

$$\Delta T = \frac{1}{2}m\left(\bar{v}_f^2 - \bar{v}_i^2\right) + \frac{1}{2}\bar{I}\left(\omega_f^2 - \omega_i^2\right) \qquad [2.1.41]$$

If the body in question has a fixed axis or an instantaneous center of zero velocity, then Equation 2.1.41 can be simplified to

$$\Delta T = \frac{1}{2}I_0\left(\omega_f^2 - \omega_i^2\right) \qquad [2.1.42]$$

ΔV_g is the net change in gravitational potential energy. This term is path independent and depends only on the initial and final heights, h_i and h_f, above some arbitrary reference height with respect to the surface of the earth.

$$\Delta V_g = mg(h_f - h_i) = W(h_f - h_i) \qquad [2.1.43]$$

ΔV_e is the net change in elastic energy stored in a massless spring, due to extension or compression (no spring is massless, but this assumption is reasonably accurate for most

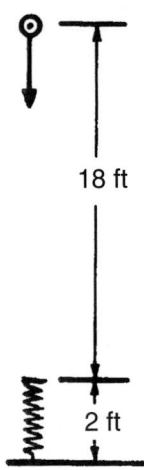

Figure 2.1.17   *Diagram for Example 2.1.11.*

engineering calculations).

$$\Delta V_e = \frac{1}{2}k\left(X_f^2 - X_i^2\right) \qquad [2.1.44]$$

The constant k, called the spring constant, represents the ratio of the force exerted by the spring to X, its net compression or extension from the rest length.

Example 2.1.11

A 1-lb sphere is dropped from a height of 20 ft to strike a 2-ft-long relaxed vertical spring with a constant of 100 lb/ft (Figure 2.1.17). What will be the velocity of the sphere at a height of 2 ft when it strikes the spring? What will be the maximum compression of the spring?

The sphere and the spring may be considered as a system in which no outside forces or moments are acting. The work term in Equation 2.1.39 is zero. Before the collision with the spring, $\Delta V_e = 0$ also, and Equation 2.1.39 reduces to

$$\Delta T + \Delta V_g = 0$$

$$\Delta T = \frac{1}{2}\frac{W}{g}\left(v_f^2 - v_i^2\right) = -\frac{1}{2}\frac{W}{g}v_f^2$$

$$\Delta V_g = W\,(h_f - h_i)$$

which can be solved for the impact velocity.

$$v_f = [2g(h_i - h_f)]^{0.5} = [2 \times 32.2(20 - 2)]^{0.5}$$

$$v_f = 34 \text{ ft/s}$$

At full compression the velocity of the sphere is zero. Equation 2.1.39 reduces to

$$\Delta V_e + \Delta V_g = 0$$

$$\Delta V_e = \frac{1}{2}k\left(X_f^2 - X_i^2\right) = \frac{1}{2}kX_f^2$$

$$\Delta V_g = W(h_f - h_i)$$

$$h_f = L - X_f$$

where L is the relaxed height of the spring which can be solved for X_f, the maximum compression of the spring.

$$\frac{1}{2}kX_f^2 + W(L - X_f - h_i) = 0$$

$$X_f^2 \frac{2W}{k}X_f + \frac{2W}{k}(L - h_i) = 0$$

$$X_f = \frac{2W/k \pm \left[(2W/k)^2 - 4\left((2W/k)(L - h_i)\right)\right]^{0.5}}{2}$$

$$= \frac{0.02 \pm [0.02^2 - 4[0.02(2 - 20)]]^{0.5}}{2}$$

$$X_f = 0.61 \text{ ft}$$

The negative root is ignored because it represents an extension of the spring rather than a compression.

For further information, refer to References 1–5.

References

1. Baumeister, T., et al., *Marks' Standard Handbook for Mechanical Engineers*, 8th Edition, McGraw-Hill, New York, 1979.
2. Beer, F. P., and Johnston, E. R., *Vector Mechanics for Engineers: Statics and Dynamics*, McGraw-Hill, New York, 1977.
3. Meriam, J. L., *Engineering Mechanics*, Vol. 2: *Dynamics*, John Wiley & Sons, New York, 1978.
4. Smith, Charles E., *Applied Mechanics: Dynamics*, John Wiley & Sons, New York, 1976.
5. Smith, Charles E., *Applied Mechanics: More Dynamics*, John Wiley & Sons, New York, 1976.

2.2 FLUID MECHANICS

In fluid mechanics, the principles of conservation of mass, conservation of momentum, the first and second laws of thermodynamics, and empirically developed correlations are used to predict the behavior of gases and liquids at rest or in motion. The field is generally divided into hydrostatics and hydrodynamics and further subdivided on the basis of compressibility. Liquids can usually be considered as incompressible, while gases are usually assumed to be compressible.

2.2.1 Fluid Statics

Pressure is the force per unit area exerted by or on a fluid. In a static fluid, the pressure increases with depth, but according to *Pascal's principle*, it is the same in all directions at any given depth. Pressure may be specified as either *absolute* or *gauge*, the relationship between the two being

$$P_g = P_a - P_{atm} \qquad [2.2.1]$$

where P_g is gauge pressure, P_a is absolute pressure, and P_{atm} is the *atmospheric pressure*. Fluid mechanics calculations are generally done in absolute pressure, and hereafter, P will represent absolute pressure.

The governing equation for the pressure within a fluid at any depth h is

$$dP = \rho g\,dh \qquad [2.2.2]$$

where ρ is the fluid density in mass per unit volume, and g is the acceleration due to gravity. In engineering calculations, it is often convenient to replace the quantity ρg with γ, the *specific weight*, which is a measure of the weight of the fluid per unit volume.

If γ can be considered to be constant, the fluid is said to be incompressible and Equation 2.2.2 can be solved to yield

$$P = P_0 + \gamma(h - h_0) \qquad [2.2.3]$$

where h_0 is some reference depth, h is depth increasing downward, and P_0 is the pressure at h_0. In a gas, the specific weight of the fluid is a function of pressure and temperature. The concept of an *ideal* or *perfect gas* as one in which the molecules occupy no volume and the only intermolecular forces are due to intermolecular collisions leads to the *ideal gas law:*

$$\gamma = \frac{PS}{RT} \qquad [2.2.4]$$

where P is the absolute pressure in pounds per square foot, T is the temperature in degrees Rankine, S is the specific gravity (the ratio of the density of the gas in question to the density of air at standard conditions), and R is Boltzman's constant (53.3 ft-lb/lb-°R). Under the assumption of an ideal gas at constant temperature, Equation 2.2.2 can be solved to yield

$$P = P_0 \exp\left[\frac{(h - h_0)S}{RT}\right] \qquad [2.2.5]$$

If the gas behavior deviates markedly from ideal, the real gas law can be written as

$$\gamma = \frac{Ps}{ZRT} \qquad [2.2.6]$$

where Z is an empirical compressibility factor that accounts for nonideal behavior (see Volume 2, Chapter 5).

Substituting the real gas law into Equation 2.2.3 yields

$$\frac{ZT}{P}dP = \frac{S}{R}dh \qquad [2.2.7]$$

Equation 2.2.7 can be integrated under the assumption that Z and T are constant to yield Equation 2.2.8, or, if extreme accuracy is required, it is necessary to account for variations in Z and T and a numerical integration may be required.

$$P = P_0 \exp\left[\frac{(h - h_0)s}{ZRT}\right] \qquad [2.2.8]$$

Example 2.2.1

Consider a 1,000-ft-deep hole. What will be the absolute pressure at the bottom if (a) it is filled with pure water or (b) it is filled with air at a constant temperature of 85°F?

(a)

$$P = P_0 + \gamma(h - h_0)$$

$$h_0 = 0$$

$$h = 1,000 \text{ ft}$$

$$P_0 = \frac{14.7 \text{ lb}}{\text{in.}^2} \left| \frac{144 \text{ in.}^2}{\text{ft}^2} \right. = 2,116.8 \text{ lb/ft}^2$$

$$\gamma = 62.4 \text{ lb/ft}^3$$

$$P = 2,116.8 + 62.4(1,000 - 0) = 64,516.8 \text{ lb/ft}^2$$

$$P = 448 \text{ psi}$$

(b)

$$P = P_0 \exp\left[\frac{(h - h_0)s}{ZRT}\right]$$

$$s = 1$$

Assume

$$Z = 1$$

$$T = 85 + 460 = 545°R$$

$$P = 2,116.8 \exp\left[\frac{(1,000 - 0)1}{1(53.3)(545)}\right] = 2,190.94 \text{ lb/ft}^2$$

$$P = 15.21 \text{ psi}$$

In a case where $Z \neq 1$, it is practical to assume $Z = 1$, perform Calculation (b), and then, based on the resultant, estimate for $P_{avg} = (P+P_0)/2$, find the value of Z, and repeat the calculation. Three iterations are generally sufficient. If T varies, it is usually sufficiently accurate to use an estimate of T_{avg} such as $T_{avg} = (T + T_0)/2$.

2.2.1.1 Fluid Dynamics

When fluids are in motion, the *pressure losses* may be determined through the principle of conservation of energy. For slightly compressible fluids, this leads to *Bernoulli's equation* (Equation 2.2.9), which accounts for static and *dynamic pressure losses* (due to changes in velocity), but does not account for frictional pressure losses, energy losses due to heat transfer, or work done in an engine.

$$\frac{P_1}{\gamma_1} + \frac{v_1^2}{2g} + h_1 = \frac{P_2}{\gamma_2} + \frac{v_2^2}{2g} + h_2 \qquad [2.2.9]$$

where $v_1, v_2 \equiv$ velocity at points 1 and 2
$\quad g \equiv$ the acceleration due to gravity
(See Figure 2.2.1.)

For flow in pipes and ducts, where frictional pressure losses are important, Equation 2.2.9 can be modified into

$$\frac{P_1}{\gamma_1} + \frac{v_1^2}{2g} + h_1 = \frac{P_2}{\gamma_2} + \frac{v_2^2}{2g} + h_2 + \frac{fLv^2}{2gD} \qquad [2.2.10]$$

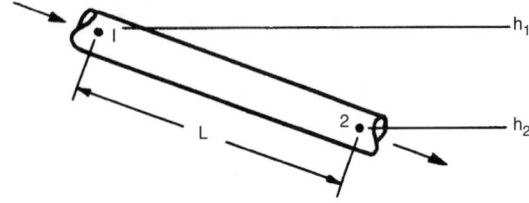

Figure 2.2.1 *Flow in an inclined pipe.*

where $f \equiv$ an empirical friction factor (Moody friction factor)
$\quad v \equiv$ the average velocity along the flow path
$\quad L \equiv$ the length of the flow path
$\quad D \equiv$ the hydraulic diameter, 2(flow area)/(wetted perimeter) (See Figure 2.2.1.)

If the fluid is highly compressible, Equation 2.2.9 must be further modified:

$$\frac{P_1}{\gamma_1}\left(\frac{k}{k-1}\right) + \frac{v_1^2}{2g} + h_1 = \frac{P_2}{\gamma_2}\left(\frac{k}{k-1}\right) + \frac{v_2^2}{2g} + h_2$$
$$[2.2.11]$$

where $k \equiv$ ratio of specific heats, c_p/c_v; see Table 2.2.1. (See Figure 2.2.1.)

The *Moody friction factor* in Equations 2.2.10 and 2.2.11 is a function of the *surface roughness* of the pipe and the *Reynolds number*. Typical surface roughnesses of new commercial pipes are shown in Table 2.2.2. Old or corroded pipes may have a significantly higher roughness.

The Reynolds number is the ratio of the inertia forces acting on the fluid to the viscous forces acting on the fluid. It is dimensionless and may be calculated as

$$R = \frac{\gamma Dv}{g\mu} = \frac{\rho Dv}{\mu} \qquad [2.2.12]$$

The term μ in Equation 2.2.12 is the *dynamic viscosity* of the fluid. The dynamic viscosity is the ratio of the shear stress to the shear rate. It has units of (force × time)/(area). The most common unit of viscosity is the centipoise

Table 2.2.1 *Critical Expansion Rates*

Gas	C_p Btu/ (lb mole) (°F)	$k = C_p/C_v$ at 1 atm, 60°F	P_e/P_1	Acoustical velocity at 60°F, ft/sec
Air	7.00	1.410	0.528	1,031
Helium	4.968	1.66	0.486	2,840
Methane	8.44	1.308	0.545	1,350
Ethane	12.30	1.193	0.565	967
Propane	17.10	1.133	0.577	793
Isobutane	22.4	1.097	0.585	681
n-Butane	23.0	1.094	0.585	680
0.6 gravity	8.84	1.299	0.546	1,309
0.7 gravity	9.77	1.279	0.550	1,035

From Reference 2.

Table 2.2.2 *Values of Absolute Roughness, New, Clean, Commercial Pipes*

Type of pipe or tubing	ε ft (0.3048 m) × 10^6 Range	ε ft (0.3048 m) × 10^6 Design	Probable max variation of f from design, %
Asphalted cast iron	400	400	−5 to +5
Brass and copper	5	5	−5 to +5
Concrete	1,000 10,000	4,000	−35 to 50
Cast iron	850	850	−10 to +15
Galvanized iron	500	500	0 to +10
Wrought iron	150	150	−5 to 10
Steel	150	150	−5 to 10
Riveted steel	3,000 30,000	6,000	−25 to 75
Wood stave	600 3,000	2,000	−35 to 20

Compiled from data given in "Pipe Friction Manual," Hydraulic Institute, 3d ed., 1961.

(1 centipoise = 0.01 g/cm-s). Dynamic viscosity may be a function of temperature, pressure, and shear rate.

For *Newtonian fluids* the dynamic viscosity is constant (Equation 2.2.13), for *power-law fluids* the dynamic viscosity varies with shear rate (Equation 2.2.14), and for *Bingham plastic fluids* flow occurs only after some minimum shear stress, called the yield stress, is imposed (Equation 2.2.15).

$$\tau = \mu \left(\frac{dv}{dy} \right) \quad \text{(Newtonian)} \qquad [2.2.13]$$

$$= \mu \left(\frac{dv}{dy} \right)^n \quad \text{(Power law)} \qquad [2.2.14]$$

$$= \tau_0 + \mu \left(\frac{dv}{dy} \right) \quad \text{(Bingham)} \qquad [2.2.15]$$

where τ is the shear stress in force per unit area, dv/dy is the shear rate (rate of change in velocity with respect to distance from measured perpendicular to the flow). The behavior of all three types of fluids is illustrated in Figure 2.2.2. Viscosities of common fluids that are normally Newtonian are given in Tables 2.2.3 and 2.2.4. Viscosities for hydrocarbon gases can be estimated from Figure 2.2.3.

If the calculated value of the Reynolds number is below 2,000, the flow will generally be laminar, that is, the fluid particles will follow parallel flow paths. For laminar flow the friction factor is

$$f = 64/R \qquad [2.2.16]$$

If the Reynolds number is greater than 4,000, the flow will generally be turbulent and the friction factor can be calculated from the Colebrook equation:

$$\frac{1}{\sqrt{f}} = -2\log_{10}\left[\frac{\varepsilon/D}{3.7} + \frac{2.51}{R\sqrt{f}} \right] \qquad [2.2.17]$$

where ε is the surface roughness. Equation 2.2.17 can be solved iteratively. If the Reynolds number falls between 2,000 and 4,000, the flow is said to be in the critical zone, and it may be either laminar or turbulent.

Equations 2.2.16 and 2.2.17 are illustrated graphically in Figure 2.2.4. This chart is called a Moody diagram, and it may be used to find the friction factor, given the Reynolds number and the surface roughness.

A better approach to determine the friction factor used in Equation 2.2.10 is to calculate f using Equation 2.2.16 (laminar flow) and Equation 2.2.17 (turbulent) and take the largest value.

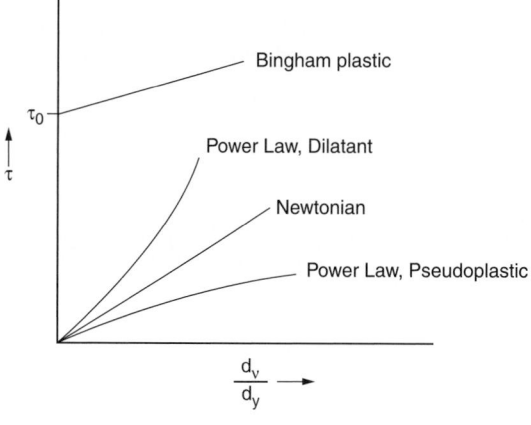

Figure 2.2.2 *Viscous behavior of fluids.*

The friction factor presented in this text is the called *Moody friction factor*. Other texts may use the term *Fanning friction factor*. Both use the Colebrook empirical implicit formula for calculation but differ in the definition of the friction factor. The Moody friction factor is defined by

$$f = \frac{(dp/dL)D}{e_k}$$

where dp/dL is the pressure drop gradient, D the diameter of the pipe, and e_k is the specific kinetic energy of the flowing fluid. The Fanning friction factor is defined as

$$f = \frac{\tau}{e_k}$$

where τ is the shear stress acting on the walls of the pipe. Working these two expressions shows that the Moody friction factor is four times the Fanning Friction factor. To use Equation 2.2.10, a friction factor obtained with the Fanning formula or read from the Stanton chart must be multiplied by four.

Example 2.2.2
Suppose 1,000 gal/min of light machine oil (see Table 2.2.3) flow through a 100-ft-long straight steel pipe with a square cross-section, 2 in. on a side. At the inlet of the pipe the pressure is 2,000 psi and the elevation is 150 ft. At the outlet the elevation is 100 ft. What will be the pressure at the outlet when the temperature is 32°F? When the temperature is 104°F?

$$\gamma = (0.907)(62.4 \text{ lb/ft}^3) = 56.60 \text{ lb/ft}^3$$

$$A = [(2 \text{ in.})/(12 \text{ in./ft})]^2 = 0.0278 \text{ ft}^2$$

$$C = 4(2 \text{ in.})/(12 \text{ in./ft}) = 0.667 \text{ ft}$$

$$D = 2A/C = 0.0833 \text{ ft}$$

$$v = Q/A = [(1,000 \text{ gal/min})(0.1337 \text{ ft}^3/\text{gal})/(60 \text{ s/min})]$$

$$= .0278 \text{ ft}^2 = 80.21 \text{ ft/s}$$

$$\mu = 7,380 \times 10^{-6} \text{ lb} - \text{s/ft}^2 @ 32°F$$

$$= 647 \times 10^{-6} \text{ lb} - \text{s/ft}^2 @ 104°F$$

$$g = 32.2 \text{ ft/s}^2$$

At 32°F

$$R = \frac{(56.60)(0.0833)(80.21)}{(32.2)(7,380 \times 10^{-6})} = 1,591$$

Because $R < 2000$ the flow is laminar and $f = 64/R = 0.04022$. Assuming that the fluid is incompressible implies that $\gamma_1 = \gamma_2$ and $v_1 = v_2$, and Equation 2.2.10 can be rewritten as

$$P_2 = P_1 + \gamma \left[(h_1 - h_2) - \frac{fLv^2}{2gD} \right]$$

$$= 2,000 + 56.60 \left[(150 - 100) - \frac{(0.04022)(100)(80.21)^2}{2(32.2)(0.0833)} \right] /$$

$$(144 \text{ in.}^2/\text{ft}^2)$$

$$= 124 \text{ psi}$$

At 104°F

$$R = \frac{(56.6)(0.0833)(80.21)}{(32.2)(647 \times 10^{-6})} = 18,150$$

Table 2.2.3 *Dynamic Viscosity of Liquids at Atmospheric Pressure*

Temp:						
°C	0	20	40	60	80	100
°F	32	68	104	140	176	212
Liquid	μ, (lbf·s)/(ft^2) [47.88 (N·s)/(m^2)] $\times 10^6$					
Alcohol, ethyl[a]	37.02	25.06	17.42	12.36	9.028	
Benzene[a]	19.05	13.62	10.51	8.187	6.871	
Carbon tetrachloride[a]	28.12	20.28	15.41	12.17	9.884	
Gasoline,[b] sp. gr. 0.68	7.28	5.98	4.93	4.28		
Glycerin[d]	252,000	29,500	5,931	1,695	666.2	309.1
Kerosene,[b] sp. gr. 0.81	61.8	38.1	26.8	20.3	16.3	
Mercury[a]	35.19	32.46	30.28	28.55	27.11	25.90
Oil, machine,[a] sp. gr. 0.907						
"Light"	7,380	1.810	647	299	164	102
"Heavy"	66,100	9,470	2,320	812	371	200
Water, fresh[e]	36.61	20.92	13.61	9.672	7.331	5.827
Water, salt[d]	39.40	22.61	18.20			

Computed from data given in
[a] *Handbook of Chemistry and Physics*, 52d Edition, Chemical Rubber Company, 1971–1972.
[b] *Smithsonian Physical Tables*, 9th Revised Edition, 1954.
[c] *Steam Tables*, ASME, 1967.
[d] *American Insitute of Physics Handbook*, 3d Edition, New York, McGraw Hill, 1972.
[e] *International Critical Tables*, McGraw-Hill.

Table 2.2.4 *Viscosity of Gases at 1 Atm*

Temp:									
°C	0	20	60	100	200	400	600	800	1000
°F	32	68	140	212	392	752	1112	1472	1832
Gas	μ, (lbf·s)/(ft^2) [47.88 (N·s)/(m^2)] $\times 10^8$								
Air[a]	35.67	39.16	41.79	45.95	53.15	70.42	80.72	91.75	100.8
Carbon dioxide[a]	29.03	30.91	35.00	38.99	47.77	62.92	74.96	87.56	97.71
Carbon monoxide[b]	34.60	36.97	41.57	45.96	52.39	66.92	79.68	91.49	102.2
Helium[a]	38.85	40.54	44.23	47.64	55.80	71.27	84.97	97.43	
Hydrogen[a,b]	17.43	18.27	20.95	21.57	25.29	32.02	38.17	43.92	49.20
Methane[a]	21.42	22.70	26.50	27.80	33.49	43.21			
Nitrogen[a,b]	34.67	36.51	40.14	43.55	51.47	65.02	76.47	86.38	95.40
Oxygen[b]	40.08	42.33	46.66	50.74	60.16	76.60	90.87	104.3	116.7
Steam[c]		18.49	21.89	25.29	33.79	50.79	67.79	84.79	

Computed from data given in
[a] *Handbook of Chemistry and Physics*, 52d Edition, Chemical Rubber Company, 1971–1972.
[b] *Tables of Thermal Properties of Gases*, NBS Circular 564, 1955.
[c] *Steam Tables*, ASME, 1967.

Because R > 4000 the flow is turbulent. From Table 2.2.2, $\varepsilon = 150 \times 10^{-6}$, and $\varepsilon/D = 0.0018$. Interpolating from Figure 2.2.4 yields f = 0.03. Now apply the Colebrook equation:

$$\frac{1}{\sqrt{f}} = -2\log_{10}\left[\frac{0.0018}{3.7} + \frac{2.51}{(18,150)(0.03)^{0.5}}\right]$$

$$f = 0.02991$$

$$P_2 = 2,000 + \left(\frac{56.60}{144}\right)$$

$$\times \left[(150 - 100) - \frac{(0.02991)(100)(80.21)^2}{2(32.2)(0.0833)}\right] = 610 \text{ psi}$$

In *piping systems* fittings, valves, bends, etc., all cause additional pressure drops. For such components the pressure drop can be estimated by modifying the frictional component of Equation 2.2.10 with a *resistance coefficient*, K = fL/D,

or an *equivalent length*, L/D. Typical resistance coefficients are given in Table 2.2.5 and typical equivalent lengths are given in Table 2.2.6. To correctly apply either the resistance coefficient or the equivalent length, the flow must be turbulent.

Example 2.2.3
Water flows from a horizontal 4-in. ID pipe into a horizontal 1-in. ID pipe at a rate of 1,000 gal/min (Figure 2.2.5). If the transition is abrupt, what will be the pressure change across the connection?

$$v_1 = \frac{Q}{A_1} = \frac{[(1,000 \text{ gal/min})(0.1337 \text{ ft}^3/\text{gal})(60 \text{ s/min})]}{\left[\pi/4 \, (4/12)^2 \text{ft}^2\right]}$$

$$= 25.53 \text{ ft/s}$$

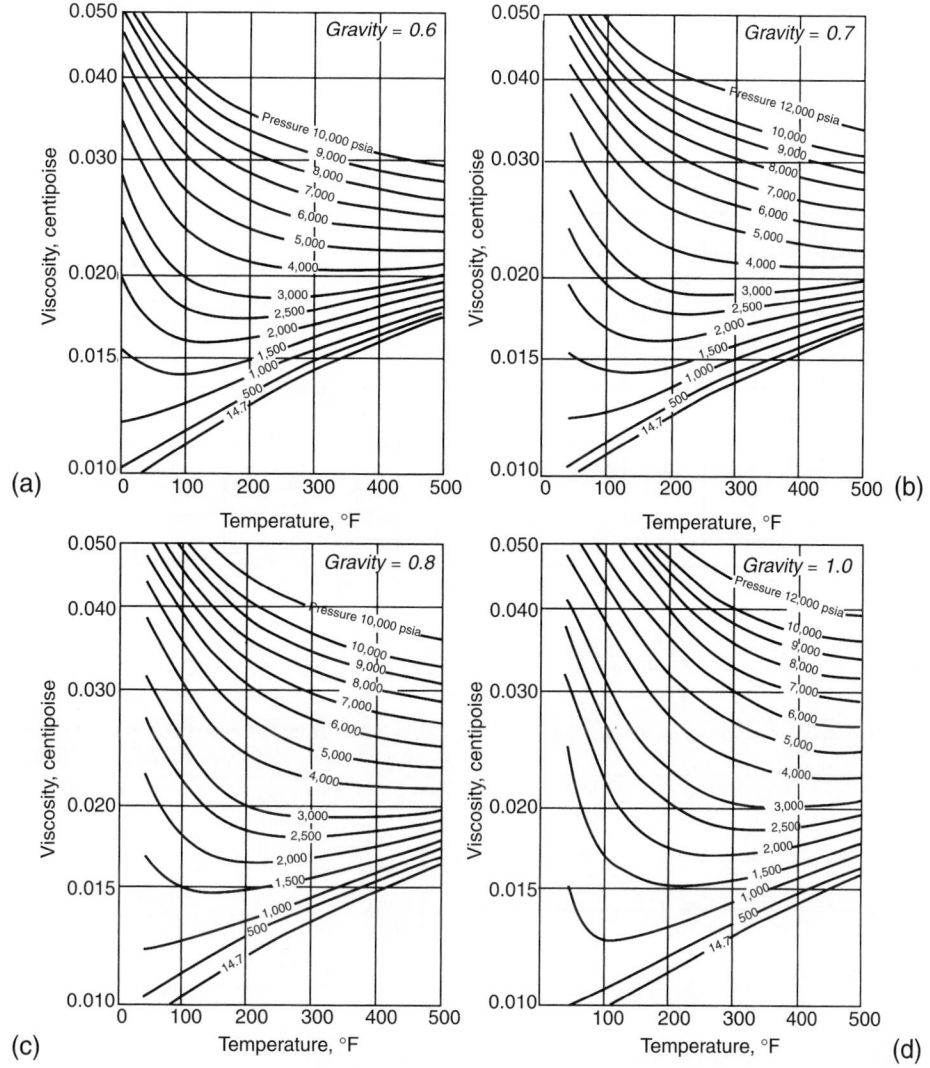

Figure 2.2.3 *Viscosity of natural gases: (a) 0.6 gravity; (b) 0.7 gravity; (c) 0.8 gravity; (d) 1.0 gravity.*

$$v_2 = \frac{[(1,000)(0.1337)/(60)]}{\left[\pi/4\,(1/12)^2\right]} = 408.6 \text{ ft/s}$$

$$\gamma_1 = \gamma_2 = 62.4 \text{ lb/ft}^3$$

From Table 2.2.1, $D_{in}/d_{out} = 4$; $k = 0.45$.

From Equation 2.2.10,

$$P_1 - P_2 = \Delta P = \frac{\gamma}{2g} \left[v_2^2 - v_1^2 + kv_2^2\right]$$

$$= \frac{\gamma}{2g} \left[v_2^2(1+k) - v_1^2\right]$$

$$= \frac{62.4}{2(32.2)} [408.6^2(1+0.45) - 25.53^2] \frac{1}{144}$$

$$= 1,625 \text{ psi}$$

The 1,120 psi of this pressure drop is a dynamic loss due to the change in velocity, and 505 psi is a frictional loss due to the fitting.

Components of a piping system that are connected in *series* produce additive pressure drops, while components that are connected in *parallel* must produce the same pressure drop.

While the modified energy equation provides for calculation of the flow rates and pressure drops in piping systems, the *impulse-momentum* equation is required in order to calculate the reaction forces on curved pipe sections. The impulse-momentum equation relates the force acting on the solid boundary to the change in fluid momentum. Because force and momentum are both vector quantities, it is most convenient to write the equations in terms of the scalar components in the three orthogonal directions.

$$\sum F_x = \dot{M}(V_{x_1} - V_{x_2})$$

$$\sum F_y = \dot{M}(V_{y_1} - V_{y_2}) \qquad [2.2.18]$$

$$\sum F_z = \dot{M}(V_{z_1} - V_{z_2})$$

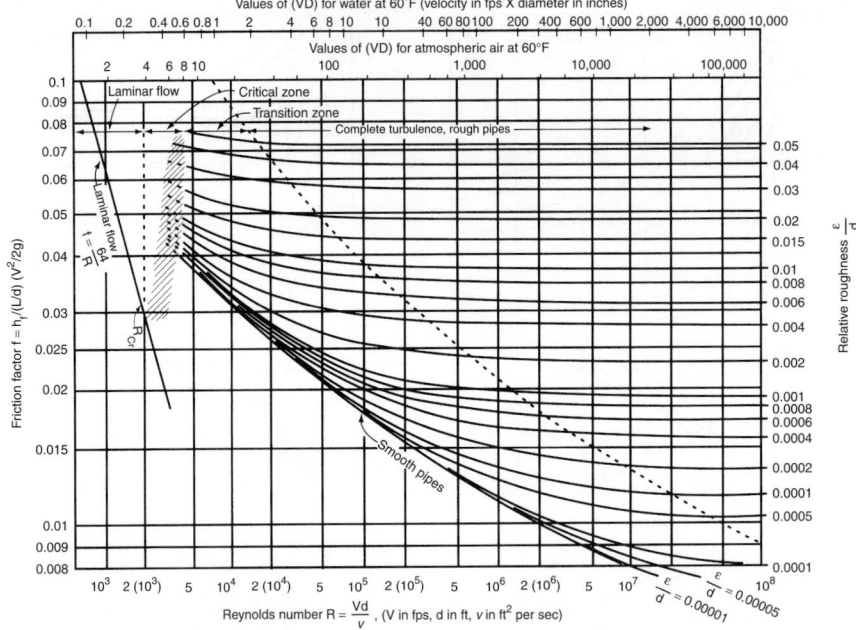

Figure 2.2.4 *Friction factor for flow in pipes [5].*

Table 2.2.5 *Representative values of Resistance Coefficient K*

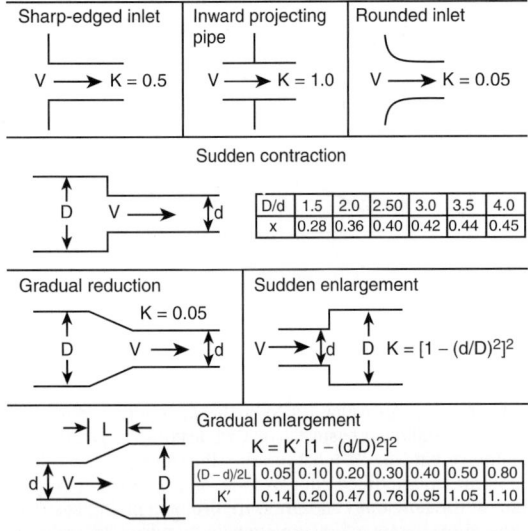

Sharp-edged inlet   Inward projecting pipe   Rounded inlet

V ⟶ K = 0.5   V ⟶ K = 1.0   V ⟶ K = 0.05

Sudden contraction

D/d	1.5	2.0	2.50	3.0	3.5	4.0
x	0.28	0.36	0.40	0.42	0.44	0.45

Gradual reduction   K = 0.05

Sudden enlargement $K = [1 - (d/D)^2]^2$

Gradual enlargement
$K = K' [1 - (d/D)^2]^2$

(D − d)/2L	0.05	0.10	0.20	0.30	0.40	0.50	0.80
K'	0.14	0.20	0.47	0.76	0.95	1.05	1.10

Exit loss = (sharp edged, projecting, Rounded), K = 1.0

Compiled from data given in "Pipe Friction Manual." 3d ed., Hydraulic Institute. 1961.

Table 2.2.6 *Representative Equivalent Length in Pipe Diameters $(L/D)_e$ of Various Valves and Fittings*

Globe valves, fully open	450
Angle valves, fully open	200
Gate valves, fully open	13
$\frac{3}{4}$ open	35
$\frac{1}{2}$ open	160
$\frac{1}{4}$ open	900
Swing check valves, fully open	135
In line, ball check valves, fully open	150
Butterfly valves, 6 in. and larger, fully open	20
90° standard elbow	30
45° standard elbow	16
90° long-radius elbow	20
90° street elbow	50
45° street elbow	26
Standard tee:	
Flow through run	20
Flow through branch	60

Data from Flow of Fluids, Crane Company Technical Paper 410, ASMF, 1971.

where $\dot{M}$ is the fluid mass flow rate, ΣF_x is the sum of forces in the x direction, V_{x_i} is the initial fluid velocity in the x direction, etc.

Example 2.2.4
Water flows through a 120°F reducing bend at a rate of 100 gpm. The inlet diameter of the bend is 2 in. and the outlet diameter is 1 in. (Figure 2.2.6). What is the reaction force on the bend?

Assuming that the flow is incompressible,

$$\dot{M} = \frac{(100 \text{ gal/min})(0.1337 \text{ ft}^3/\text{gal})(62.4 \text{ lb/ft}^3)}{[(32.2 \text{ ft/s}^2)(60 \text{ s/min})]}$$

$$\dot{M} = 0.4318 \frac{\text{lb} - \text{s}^4}{\text{ft}} = 0.4318 \text{ slugs/s}$$

$$A_1 = \frac{\pi}{4} \left(\frac{2}{12} \right)^2 = 0.02182 \text{ ft}^2$$

$$A_2 = \frac{\pi}{4} \left(\frac{1}{12} \right)^2 = 0.005454 \text{ ft}^2$$

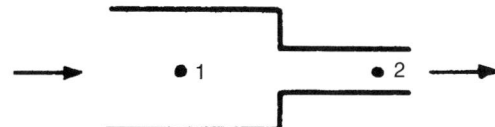

Figure 2.2.5 *Diagram for Example 2.2.3.*

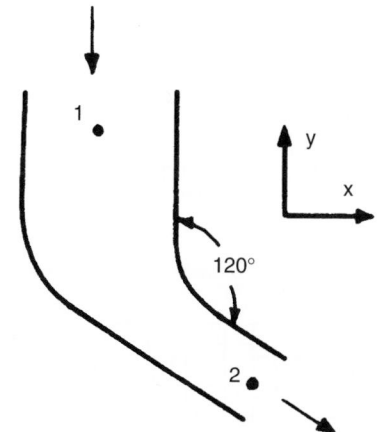

Figure 2.2.6 *Diagram for Example 2.2.4.*

$$v_1 = v_{y_1} = -\frac{Q}{A_1} = \frac{(100)(0.1337)(1/60)}{0.02182} = 10.21 \text{ ft/s}$$

$$v_2 = \frac{Q}{A_2} = \frac{(100)(0.1337)(1/60)}{0.005454} = 40.86 \text{ ft/s}$$

$$v_{y_2} = -v_2 \sin 30° = -20.43 \text{ ft/s}$$

$$v_{x_2} = v_2 \cos 30° = 35.38 \text{ ft/s}$$

$$F_x = \dot{M}(v_{x_1} - v_{x_2}) = 0.4318 \frac{\text{lb} \cdot \text{s}}{\text{ft}}(0 - 35.38 \text{ ft/s}) = -15.28 \text{ lb}$$

$$F_y = \dot{M}(v_{y_1} - v_{y_2}) = 0.4318(-10.21 + 20.43) = 4.413 \text{ lb}$$

Flow through *chokes and nozzles* is a special case of fluid dynamics. For *incompressible fluids* the problem can be handled by mass conservation and Bernoulli's equation. Bernoulli's equation is solved for the pressure drop across the choke, assuming that the velocity of approach and the vertical displacement are negligible. The velocity term is replaced by the volumetric flow rate times the area at the choke throat to yield

$$\Delta P = \frac{Q^2 \gamma}{2gC^2 A^2} \qquad [2.2.19]$$

C is a constant introduced to account for frictional effects. In general, $0.94 \le C \le 0.98$.

Example 2.2.5
Assume 100 ft³/min of water is to be pumped through a nozzle with a throat diameter of 3/4 in. What pressure drop should be expected?

$$Q = \frac{100}{60} = 1.67 \text{ ft}^3/\text{s}$$

$$A = \frac{\pi(0.75^2)}{4(144)} = 3.068 \times 10^{-3} \text{ ft}^2$$

$$\gamma = 62/4 \text{ lb/ft}^3$$

Assume $C = 0.95$; then

$$\Delta P = \frac{(1.67^2)(62.4)}{2(32.2)(0.95^2)(3.068 \times 10^{-3})^2(144)} = 2,210 \text{ psia}$$

To analyze *compressible flow through chokes* it is assumed that the entropy of the fluid remains constant. The equation of *isentropic* flow is

$$P_1 V_1^k = P_2 V_2^k \qquad [2.2.20]$$

where P_1 and V_1 are the pressure and specific volume of the fluid at point 1, immediately upstream of the choke, and P_2 and V_2 are the pressure and specific volume immediately downstream of the choke. Equation 2.2.20 can be combined with the ideal gas law to provide an estimate for the temperature drop across the choke

$$T_2 = T_1 \left(\frac{P_2}{P_1}\right)^{(k-1)/k} \qquad [2.2.21]$$

where T_2 and T_1 are temperatures in °R. Furthermore, the first law of thermodynamics can also be imposed, yielding the following equation for the volumetric flow rate:

$$Q = 864CA \frac{P_1}{(ST_1)^{0.5}} \left\{ \frac{k}{k-1} \left[\left(\frac{P_2}{P_1}\right)^{2/k} - \left(\frac{P_2}{P_1}\right)^{(k+1)/k} \right] \right\}^{0.5} \qquad [2.2.22]$$

where Q is the volumetric flow rate in scfm, C is a *discharge coefficient* that accounts for friction and velocity of approach (Figure 2.2.7). A is the choke area in square inches, P_1 is the inlet pressure in pounds per square inch absolute (psia), P_2 is the outlet pressure in psia, T_1 is the inlet temperature in °R, and S is the specific gravity of the gas.

Equations 2.2.21 and 2.2.22 apply only as long as the fluid velocity at the throat of the choke is subsonic. Sonic velocity is the speed of a pressure wave in a fluid. Once *sonic velocity* is achieved, the effects of the downstream pressure can no longer be transmitted to the upstream side of the choke.

There is a critical pressure ratio beyond which the flow at the throat is always sonic. This is termed *critical flow*.

$$\frac{P_2}{P_1} = \left(\frac{2}{k+1}\right)^{k/(k-1)} \qquad [2.2.23]$$

If the pressure ratio is less than or equal to that specified by Equation 2.2.23; the flow will be sonic at the choke throat and the temperature at the throat can be found from

$$T_2 = T_1 \left(\frac{2}{k+1}\right) \qquad [2.2.24]$$

The flow rate can be found from

$$Q = 610CA \frac{P_1}{(ST_1)^{0.5}} \left[k \left(\frac{2}{k+1}\right)^{(k+1)/(k-1)} \right]^{0.5} \qquad [2.2.25]$$

For critical flow the discharge coefficient is dependent upon the geometry of the choke and its diameter or the ratio β of its diameter to that of the upstream pipe (see Figure 2.2.7).

Example 2.2.6
A 0.6 gravity hydrocarbon gas flows from a 2-in. ID pipe through a 1-in. ID orifice plate. The upstream temperature and pressure are 75°F and 800 psia, respectively. The downstream pressure is 200 psia. Does heating need to be applied to assure that frost does not clog the orifice? What will be the flow rate?

Check for critical flow. From Table 2.2.1 it is determined that $k = 1.299$. Checking the pressure ratio, Equation 2.2.23, gives

$$\frac{P_2}{P_1} = 0.25 < \left(\frac{2}{k+1}\right)^{k/(k-1)} = 0.546$$

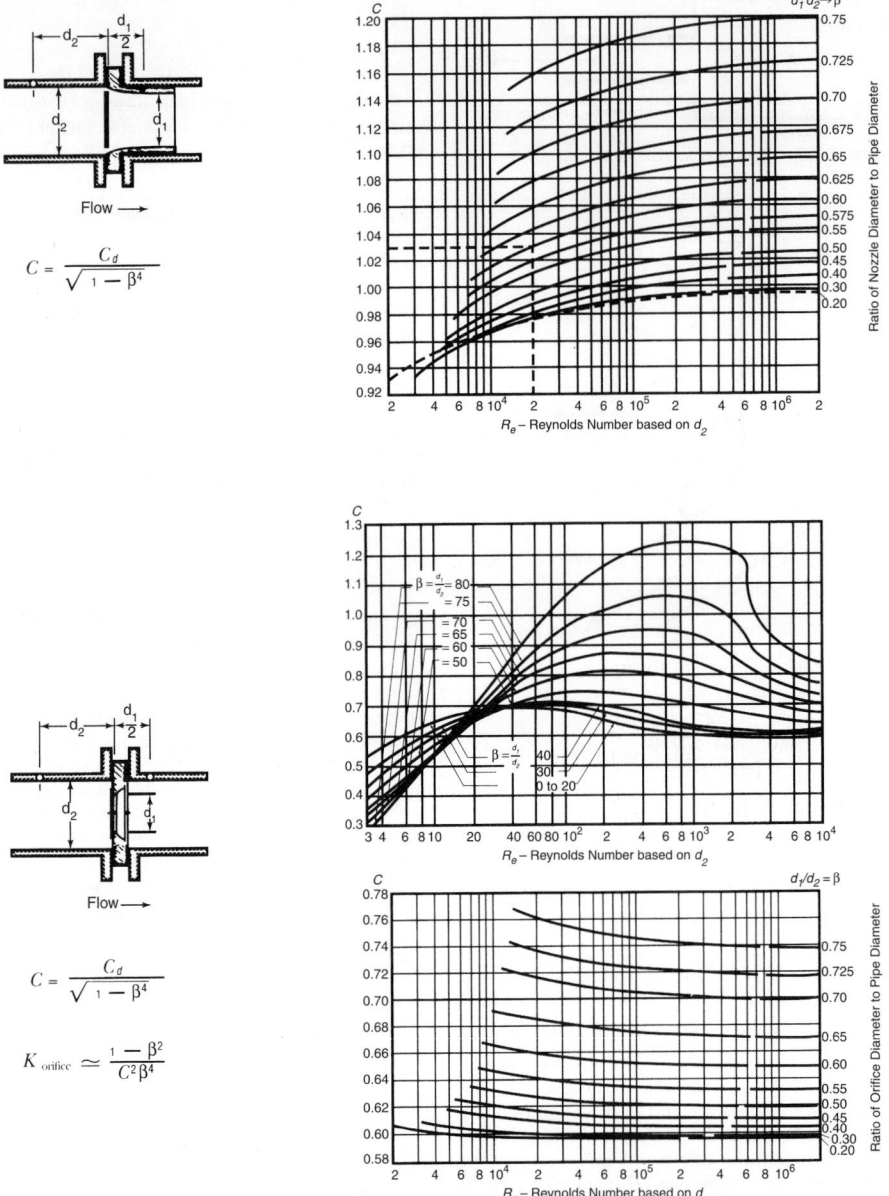

Figure 2.2.7   *Flow coefficient for nozzles and orifices (From Reference 1, p. 226).*

Therefore, critical flow conditions exist and the fluid velocity in the choke is sonic.

Temperature in the choke. From Equation 2.2.24, the temperature in the throat is calculated as

$$T_2 = T_1 \left(\frac{2}{k+1} \right) = 535 \left(\frac{2}{1+1.299} \right) = 465.4°R = 5.4°F$$

Therefore, if the gas contains water, icing or hydrate formation may occur, causing the throat to clog. A heating system may be needed.

Flow rate. Assume that the discharge coefficient is C = 1.0. The choke area is A = $\pi(1/2)^2 = 0.785$ in.^{2}. Thus, from

Equation 2.2.25,

$$Q = 610(1.0)(0.785)\frac{800}{[(0.6)(535)]^{0.5}}$$

$$\times \left[1.299 \left(\frac{2}{1.299+1} \right)^{(1.299+1)/(1.299-1)} \right]^{0.5}$$

$$= 14,021 \text{ scfm}$$

Now check on the discharge coefficient.

The Reynolds number for gases can be calculated directly in terms of flow rate and gas gravity as

$$R_e = \frac{28.8Qs}{\mu d} \qquad [2.2.26]$$

where Q is in scfm, s is the specific gravity of the gas, d is the pipe hydraulic diameter in inches, and μ is in centipoise. From Figure 2.2.3 the viscosity of the gas is

$$\mu = 0.0123 \text{ cp}$$

and

$$R_e = \frac{(28.8)(14021)(0.6)}{(0.0123)(2)} = 9,850,000$$

From Figure 2.2.7, using $\beta = 0.5$, the value of the discharge coefficient is read as C = 0.62, and a new estimate of Q is

$$Q = 0.62(14,021) = 8,693 \text{ scfm}$$

A further iteration produces no change in the estimated flow rate for this case.

In subcritical flow the discharge coefficient is affected by the velocity of approach as well as the type of choke and the ratio of choke diameter to pipe diameter. Discharge coefficients for subcritical flow are given in Figure 2.2.7 as a function of the diameter ratio and the upstream Reynolds number. Because the flow rate is not initially known, it is expedient to assume C = 1, calculate Q, use this Q to calculate the Reynolds number, and then use the charts to find a better value of C. This cycle should be repeated until the value of C no longer changes.

Example 2.2.7
A 0.65 gravity, natural gas (K = 1.25) flows from a 2-in. line through a 1.5-in. nozzle. The upstream temperature is 90°F. The upstream pressure is 100 psia, and the downstream pressure is 80 psia. Is icing a potential problem? What will be the flow rate?

Check for critical flow using Equation 2.2.23.

$$\frac{P_2}{P_1} = 0.8 > \left[\frac{2}{k+1}\right]^{k/(k-1)} = 0.549$$

The flow is clearly subcritical.

Check the outlet temperature using Equation 2.2.21.

$$T_2 = 550(0.8^{0.25/1.25}) = 506.86°R = 66°F$$

There will be no icing.

Calculate the flow rate.

$$A = \pi\left(\frac{1.5}{2}\right)^2 = 1.767 \text{ in.}^2$$

Assuming C = 1 and applying Equation 2.2.22 gives

$$Q = (864)(1)(1.767)\frac{100}{[(0.65)(530)]^{1/2}}$$

$$\times \left\{\frac{1.25}{1.25-1}\left[\left(\frac{80}{100}\right)^{2/1.25} - \left(\frac{80}{100}\right)^{(1.25+1)/1.25}\right]\right\}^{1/2}$$

$$= 3,214 \text{ scfm}$$

From Figure 2.2.3, the viscosity of the gas is

$$\mu = 0.0106 \text{ cp}$$

and from Equation 2.2.26

$$R_e = \frac{(28.8)(3,214)(0.65)}{(0.0106)(2)} = 2.84 \times 10^6$$

From Figure 2.2.7, using $\beta = 0.75$, the value of the discharge coefficient is read as c = 1.2. Now a new estimate of Q can be found as

$$Q = \left(\frac{1.2}{1}\right)3214 = 3,857 \text{ scfm}$$

Because further increases in the flow rate (see Figure 2.2.7) produce no increase in the discharge coefficient, it is unnecessary to do any further iterations.

For further information on this subject, refer to References 1–5.

References
1. Brown, Kermit E., and Beggs, H. Dale, *The Technology of Artificial Lift Methods*, Vol. 1, Penn Well Books, Tulsa, 1977.
2. Katz, Donald L., et al., *Handbook of Natural Gas Engineering*, McGraw-Hill, 1959.
3. Olsen, Reuben M., *Essentials of Engineering Fluid Mechanics*, Harper & Row, New York, 1980.
4. Perry, Robert H., and Chilton, Cecil H. (eds.), *Chemical Engineers' Handbook*, 5th Edition, McGraw-Hill, New York, 1973.

2.3 STRENGTH OF MATERIALS

The principles of strength of materials are applied to the design of structures to assure that the elements of the structures will operate reliably under a known set of loads. The field encompasses both the calculation of the strength and deformation of members and the measurement of the mechanical properties of engineering materials.

2.3.1 Stress and Strain

Consider a bar of length L and uniform cross-sectional area A to which an axial, uniformly distributed load with a magnitude, P, is applied at each end (Figure 2.3.1). Then within the bar there is said to be *uniaxial stress* σ, defined as the load, or force per unit area

$$\sigma = \frac{P}{A} \qquad [2.3.1]$$

If the load acts to elongate the bar, the stress is said to be *tensile* (+), and if the load acts to compress the bar, the stress is said to be *compressive* (−). For all real materials, an externally applied load will produce some *deformation*. The ratio of the deformation to the undeformed length of the body is called the *strain* ε. In the simple case illustrated in Figure 2.3.1, the strain is

$$\varepsilon = \delta/L \qquad [2.3.2]$$

where δ is the longitudinal deformation. The strain is tensile or compressive depending upon the sign of δ. The relationship between stress and strain in an axially loaded bar can be illustrated in a stress-strain curve (Figure 2.3.2). Such curves are experimentally generated through *tensile tests*.

In the region where the relationship between stress and strain is linear, the material is said to be *elastic*, and the constant of proportionality is E, *Young's modulus*, or the *elastic modulus*.

$$\sigma = E\varepsilon \qquad [2.3.3]$$

Equation 2.3.3 is called *Hooke's law*.

In the region where the relationship between stress and strain is nonlinear, the material is said to be *plastic*.

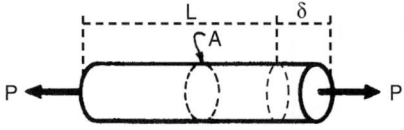

Figure 2.3.1 *Uniaxial loading of a bar.*

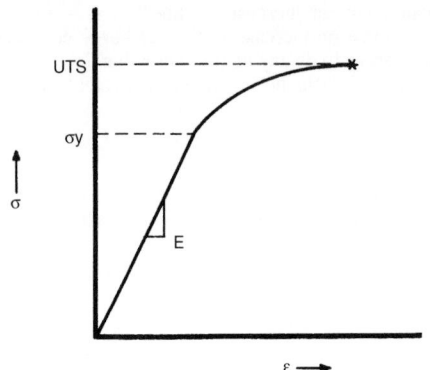

Figure 2.3.2 *Idealized stress–strain curve.*

Elastic deformation is recoverable upon removal of the load, whereas plastic deformation is permanent. The stress at which the transition occurs, σ_y, is called the *yield strength* or *yield point* of the material, and the maximum stress is called the *ultimate tensile strength*, UTS, of the material. Standard engineering practice is to define the yield point as 0.2% permanent strain.

When a bar is elongated axially, as in Figure 2.3.1, it will contract laterally. The negative ratio of the lateral strain to the axial strain is called *Poisson's ratio* ν. For *isotropic* materials, materials that have the same elastic properties in all directions, Poisson's ratio has a value of about 0.3.

Now consider a block to which a uniformly distributed load of magnitude P is applied parallel to opposed faces with area A (Figure 2.3.3). These loads produce a *shear stress* within the material τ.

$$\tau = P/A \qquad [2.3.4]$$

Note that in order for the block of Figure 2.3.3 to be in static equilibrium, there must also be a stress τ applied parallel to each of the faces B. Any given shear stress always implies a second shear stress of equal magnitude acting perpendicularly to the first so as to produce a state of static equilibrium. The shear stress will produce a deformation of the block, manifested as a change in the angle between the face perpendicular to the load and the face over which the load is applied. This change in angle is called the *shear strain* γ.

$$\gamma = \Delta\alpha \qquad [2.3.5]$$

For an elastic material the shear stress is related to the shear strain through a constant of proportionality G, called the *shear modulus*. The shear strain is dimensionless, and the shear modulus has units of force per unit area.

$$\tau = G\gamma \qquad [2.3.6]$$

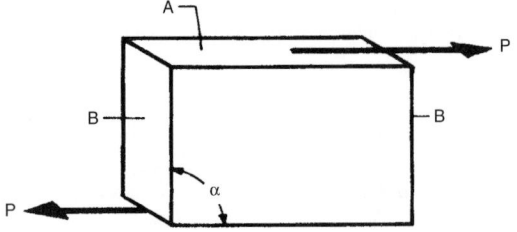

Figure 2.3.3 *Shear loading of a block.*

The shear modulus is related to Young's modulus and Poisson's ratio by

$$G = \frac{E}{2(1+\nu)} \qquad [2.3.7]$$

In practice, loads are not necessarily uniformly distributed nor uniaxial, and cross-sectional areas are often variable. It becomes necessary to define the stress at a point as the limiting value of the load per unit area as the area approaches zero. Furthermore, there may be tensile or compressive stresses (σ_x, σ_y, σ_z) in each of three orthogonal directions and as many as six shear stresses (τ_{xy}, τ_{yx}, τ_{xz}, τ_{zx}, τ_{yz}, τ_{zy}). The direction of the shear stress is indicated by two subscripts, the first of which indicates the direction normal to the plane in which the load is applied, and the second of which indicates the direction of the load. Note that for static equilibrium to exist, $\tau_{xy} = \tau_{yx}$, $\tau_{xz} = \tau_{zx}$, and $\tau_{yz} = \tau_{zy}$.

If a multidimensional state of stress exists, the effect of Poisson's ratio causes the tensile and compressive strains to be dependent on each of the components of stress.

$$\varepsilon_x = \frac{1}{E}[\sigma_x - \nu(\sigma_y + \sigma_z)] \qquad [2.3.8]$$

$$\varepsilon_y = \frac{1}{E}[\sigma_y - \nu(\sigma_x + \sigma_z)] \qquad [2.3.9]$$

$$\varepsilon_z = \frac{1}{E}[\sigma_z - \nu(\sigma_y - \sigma_z)] \qquad [2.3.10]$$

Likewise the stresses may be written in terms of the components of strain.

$$\sigma_x = \frac{E}{(1+\nu)(1-2\nu)}[(1-\nu)\varepsilon_x + \nu(\varepsilon_y + \varepsilon_z)] \qquad [2.3.11]$$

$$\sigma_y = \frac{E}{(1+\nu)(1-2\nu)}[(1-\nu)\varepsilon_y + \nu(\varepsilon_x + \varepsilon_z)] \qquad [2.3.12]$$

$$\sigma_z = \frac{E}{(1+\nu)(1-2\nu)}[(1-\nu)\varepsilon_z + \nu(\varepsilon_x + \varepsilon_y)] \qquad [2.3.13]$$

The components of the shear stress all obey Equation 2.3.6.

While the foregoing discussion of stress and strain is based on a Cartesian coordinate system, any orthogonal coordinate system may be used.

2.3.2 Elementary Loading Configurations
2.3.2.1 Torsion of a Cylinder
Consider a uniform cylindrical bar or tube to which some balanced torque T is applied (Figure 2.3.4). The bar will be subject to a *torsional stress*, or shear stress $\tau_{z\theta}$, which increases with the radial position within the bar.

$$\tau_{z\theta} = Tr/J \qquad [2.3.14]$$

where r is the radial distance from the z axis, and J is the polar moment of inertia. The polar moment of inertia for a hollow cylinder with an internal radius r_i and an external radius, r_o, is

$$J = \frac{\pi}{2}(r_o^4 - r_i^4) \qquad [2.3.15]$$

The strain due to the torque T is given by

$$\gamma_{z\theta} = \frac{2(1+\nu)}{E}\frac{Tr}{J} \qquad [2.3.16]$$

and the total angular deflection between any two surfaces perpendicular to the z axis is

$$\Theta = 2(1+\nu)\frac{TL}{EJ} \qquad [2.3.17]$$

where L is the distance between the two surfaces.

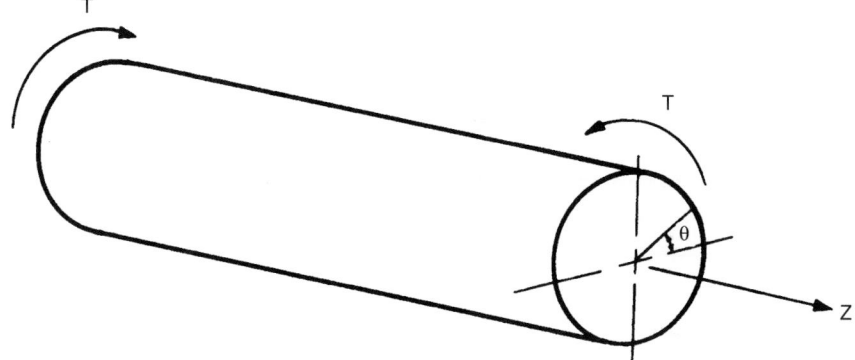

Figure 2.3.4 *Torsional loading of a right circular cylinder.*

Example 2.3.1
A uniform steel bar 3 ft long and 2.5 in. in diameter is subjected to a torque of 800 ft-lb. What will be the maximum
torsional stress and the angular deflection between the two
ends?

Assume $E = 30 \times 10^6$ psi and $\nu = 0.3$. Then

$$J = \frac{\pi}{2}(1.25^4 - 0^4) = 3.83 \text{ in.}^4$$

$$\tau_{z\theta} = \frac{800 \text{ ft-lb}}{} \left| \frac{12 \text{ in.}}{\text{ft}} \right| \frac{1.25 \text{ in.}}{3.83 \text{ in.}^4} = 3133 \text{ psi}$$

$$\theta = 2(1+0.3)\frac{800 \text{ ft-lb}}{} \left| \frac{12 \text{ in.}}{\text{ft}} \right| \frac{3 \text{ ft}}{} \left| \frac{12 \text{ in.}}{\text{ft}} \right| \frac{\text{in.}^2}{30 \times 10^6 \text{ lb}} \left| \frac{}{3.83 \text{ in.}^4} \right.$$

$$= 7.8 \times 10^{-3} \text{ rad} \frac{360°}{2\pi \text{ rad}} = 0.45°$$

2.3.2.2 Pure Bending

A beam is subjected to a pure bending moment when no
other loading (shear, torsion, or axial) exists. A typical situation is shown in Figure 2.3.5a. Because the moment $M = Fd$
is the same along the portion BC, this portion will bend in an
arc of a circle. An imaginary longitudinal fiber in the upper
side of the beam will be in compression (negative strain),
and an imaginary fiber in the lower side of the beam will
be in tension (positive strain). It is reasonable to conclude
that somewhere in between a longitudinal fiber will have no
strain, and therefore no axial stress. If the beam has a vertical
axis of symmetry, all zero strain lines will form a cylindrical
surface perpendicular to the plane of bending, called *neutral
plane of bending*. If ρ is the radius of the circle determined
by the neutral plane, the length L' of a fiber passing in a
point of the cross-section with coordinate y with respect to
the neutral plane is (Figure 2.3.5b).

$$L' = (\rho - y)\theta$$

Because the undeformed length of the fiber is $L = \rho\theta$, we
have for the strain of the deformed fiber

$$\varepsilon = \frac{L' - L}{L} = -\frac{y}{\rho}$$

Consequently, the stress in a point of coordinate y in any
cross-section of the beam is

$$\sigma = E\varepsilon = -\frac{E}{\rho}y \qquad [2.3.18]$$

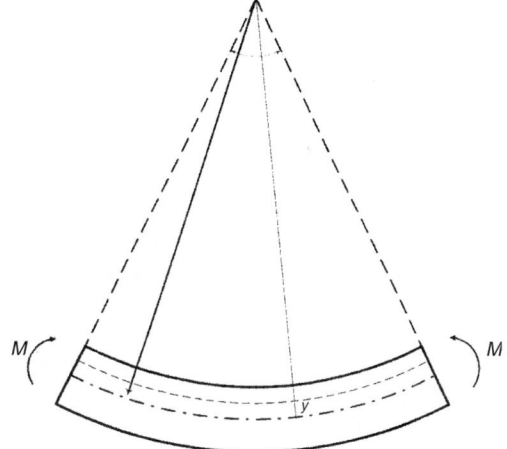

(a)

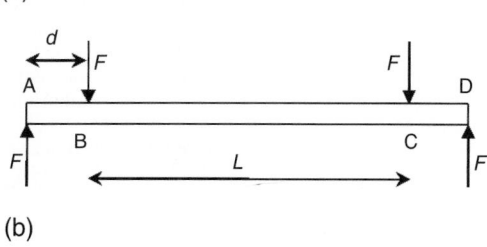

(b)

Figure 2.3.5 *Supported Beam.*

The equilibrium of forces acting on a cross-section is zero,
resulting in the following expression:

$$0 = \int_A \sigma dA = -\frac{E}{\rho}\int_A y dA$$

where A is the area of the cross-section of the beam. This
leads us to conclude that the neutral plane of bending passes
at the centroid of the cross section.

The equilibrium of moments acts on the cross-section
results in

$$M = -\int_A y\sigma A = \frac{E}{\rho}\int_A y^2 dA = \frac{EI_z}{\rho} \qquad [2.3.19]$$

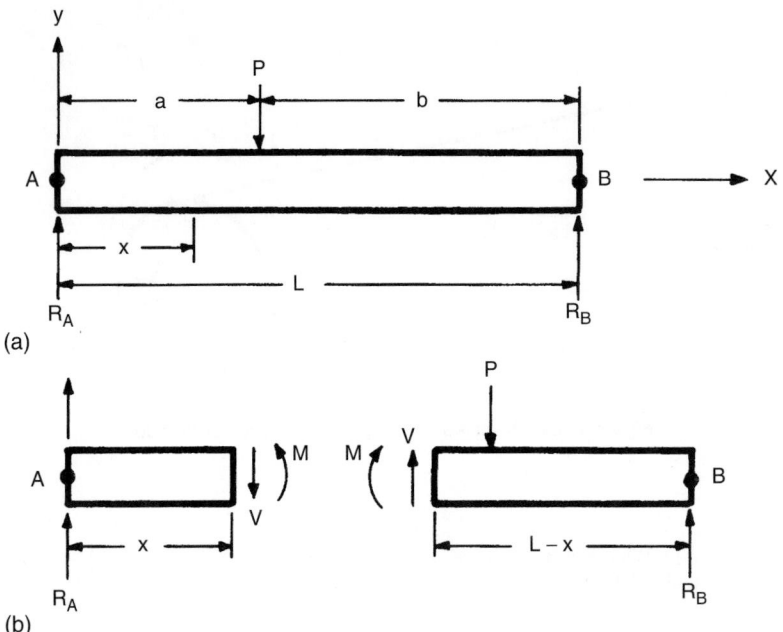

Figure 2.3.7 *Transverse loading of a beam.*

where I_z is the moment of inertia (or second moment) of the cross-section. Combining equation 2.3.21 with 2.3.22 results in

$$\sigma = -\frac{My}{I_z} \qquad [2.3.20]$$

2.3.2.3 Transverse Loading of Beams

A beam subjected to a simple transverse load (Figure 2.3.6a) will bend. Furthermore, if the beam is cut (Figure 2.3.6b) and free-body diagrams of the remaining sections are constructed, then a shear force V and a moment M must be applied to the curt ends to maintain static equilibrium.

The magnitude of the shearing force and the moment can be determined from the conditions of static equilibrium of the beam section. For the cut shown in Figure 2.3.6b, the shear force and the moment on the left-hand section are

$$V = -R_A = (-b/L)P \qquad [2.3.21]$$

$$M = R_A x = (b/L)Px \qquad [2.3.22]$$

If the cut had been to the right of the point of application of load P, then the shear force and the moment on the left-hand section would be

$$V = -R_A + P = (-b/L)P + P = (a/L)P \qquad [2.3.23]$$

$$M = R_A x - P(x-a) = P\left(\frac{bx}{L} - x + a\right) = \frac{a}{L}P(L-x) \qquad [2.3.24]$$

In the loading configuration of Figure 2.3.6, the beam will bend in the concave upward direction, thus putting the lowermost fiber in tension and the uppermost fiber in compression. The magnitude of this axial stress is

$$\sigma_x = -\frac{My}{I_z} \qquad [2.3.25]$$

The position of the centroidal axis of common areas is given in Table 2.3.1. The areal moments of inertia about any axis,

w for instance, is defined by

$$I_w = \int_A r^2 dA \qquad [2.3.26]$$

where A is the cross-sectional area of the body and r is the perpendicular distance from axis w to the differential element of area dA. Values for the areal moments of inertia of common cross-sections are given in Table 2.3.1.

The beam is also subject to a shear stress that varies over the beam cross-section.

$$\tau_{xy} = \frac{VQ}{I_z b} \qquad [2.3.27]$$

where b is the width of the beam. *The moment area about the z axis* Q is defined as

$$Q = \int_{y_0}^{y_{max}} y\, dA \qquad [2.3.28]$$

where y_0 is the location of the shear stress.

Example 2.3.2

Assuming that the beam in Figure 2.3.6 has a rectangular cross-section with a height of 1 ft and a width of 3 in. and given that L=10 ft, a=4 ft, and P=1,000 lb, what are the maximum values of the shear and tensile stresses within the beam?

Tensile stress. The maximum moment occurs at the point of application of the load P and has the value

$$M = \frac{ba}{L} = \frac{(4)(6)}{10}1,000 = 2,400 \text{ ft-lb} = 28,800 \text{ in.-lb}$$

The areal moment of inertia of the beam can be found from Table 2.3.1 as

$$I_z = \frac{1}{12}bh^3 = \frac{1}{12}(3)(12^3) = 432 \text{ in.}^4$$

Table 2.3.1 *Properties of Cross-Sections*

Section		Area	Centroid	Area Moments of Inertia
Rectangle		bh	$c_x = \dfrac{h}{2}$ $c_y = \dfrac{b}{2}$	$I_x = \dfrac{bh^3}{12}$ $I_y = \dfrac{hb^3}{12}$
Circle		$\dfrac{\pi d^2}{4}$	$c_x = \dfrac{d}{2}$ $c_y = \dfrac{d}{2}$	$I_x = \dfrac{\pi d^4}{64}$ $I_y = \dfrac{\pi d^4}{64}$ $I_z = J = \dfrac{\pi d^4}{32}$
Thick-walled tube		$\dfrac{\pi}{4}(d_e^2 - d_i^2)$	$c_x = \dfrac{d_o}{2}$ $c_y = \dfrac{d_o}{2}$	$I_x = \dfrac{\pi}{64}(d_e^4 - d_i^4)$ $I_y = \dfrac{\pi}{64}(d_e^4 - d_i^4)$ $I_z = J = \dfrac{\pi}{32}(d_o^4 - d_i^4)$
Thin-walled tube ($\bar{d} \gg t$)		$\pi \bar{d} t$	$c_x = \dfrac{\bar{d}}{2}$ $c_y = \dfrac{\bar{d}}{2}$	$I_x = \dfrac{\pi}{8} t \bar{d}^3$ $I_y = \dfrac{\pi}{8} t \bar{d}^3$ $I_x = J = \dfrac{\pi}{4} t \bar{d}^3$
Circular quadrant		$\dfrac{x}{4} r^2$	$c_x = \dfrac{4r}{3\pi}$ $c_y = \dfrac{4r}{3\pi}$	$I_x = \left(\dfrac{x}{16} - \dfrac{4}{9\pi}\right) r^4$ $I_y = \left(\dfrac{\pi}{16} - \dfrac{4}{9\pi}\right) r^4$ $P_{xy} = \left(\dfrac{1}{8} - \dfrac{4}{9\pi}\right) r^4$
Triangle		$\dfrac{1}{2} bh$	$c_x = \dfrac{h}{3}$ $c_y = \dfrac{b}{3}$	$I_x = \dfrac{bh^3}{36}$ $I_y = \dfrac{hb^3}{36}$ $P_{xy} = -\dfrac{b^2 h^2}{72}$

The maximum tensile stress occurs at the outer fiber of the beam (at $y = -6$ in.) and has the value

$$\sigma_{x_{max}} = -\frac{My}{I_z} = -\frac{(28,800)(-6)}{432} = 400 \text{ lb/in.}^2$$

Shear stress. For a rectangular cross-section, the maximum value of Q occurs at the neutral axis, and, because the width b of the beam is a constant 3 in., the maximum value of the shear stress occurs at the neutral axis.

$$Q = \int_0^6 yb \, dy = \frac{1}{2}by^2|_0^6 = \frac{1}{2}(3)(6)^2 = 54 \text{ in.}^3$$

$$V = -\frac{4}{10} \times 1,000 = -400 \text{ lb}$$

$$\tau_{xy} = \frac{(-400)(54)}{(3)(432)} = -16.67 \text{ lb/in.}^2$$

Thin-walled pressure vessels. A thin-walled pressure vessel is one in which the wall thickness t is small compared to the local radius of curvature r. At a point in the wall of the vessel where the radius of curvature varies with the direction, the wall stresses are

$$\frac{\sigma_\theta}{r_\theta} + \frac{\sigma_\alpha}{r_\alpha} = \frac{p}{t} \qquad [2.3.29]$$

where p is the net internal pressure and θ and α indicate any two orthogonal directions tangent to the vessel surface at the point in question. For a spherical vessel, $r_\theta = r_\alpha = r$, and Equation 2.3.29 reduces to

$$\sigma_\theta = \sigma_\alpha = \frac{Pr}{2t} \qquad [2.3.30]$$

For a cylindrical vessel, the radius of curvature in the axial direction is infinite, and the stress in the direction of the circumference, called the *hoop stress*, is

$$\sigma_\theta = \frac{Pr}{t} \qquad [2.3.31]$$

The stress in the axial direction in a cylindrical vessel is found by taking a cross-section perpendicular to the longitudinal axis and imposing the conditions of static equilibrium. This yields

$$\sigma_z = \frac{Pr}{2t} \qquad [2.3.32]$$

2.3.3 Prediction of Failure

For most practical purposes, the onset of plastic deformation constitutes failure. In an axially loaded part, the yield point is known from testing (see Tables 2.3.2 through 2.3.5), and failure prediction is no problem. However, it is often necessary to use uniaxial tensile data to predict yielding due to a multidimensional state of stress. Many failure theories have been developed for this purpose. For elastoplastic materials (steel, aluminum, brass, etc.), the *maximum distortion energy theory* or *von Mises theory* is in general application. With this theory the components of stress are combined into a single effective stress, denoted as σ_e, which can be compared to known data for uniaxial yielding. The ratio of the measure yield stress to the effective stress is known as the factor of safety.

$$\sigma_e = \left\{ \frac{1}{2}[(\sigma_x - \sigma_y)^2 + (\sigma_x - \sigma_z)^2 + (\sigma_y - \sigma_z)^2 \right.$$
$$\left. + 6(\tau_{xy}^2 + \tau_{yz}^2 + \tau_{xz}^2) \right\}^{1/2} \qquad [2.3.33]$$

For brittle materials such as glass or cast iron, the maximum shear-stress theory is usually applied.

Example 2.3.3
A cylindrical steel pressure vessel (AISI SAE 1035, cold rolled) with a wall thickness of 0.1 in. and an inside diameter of 1 ft is subject to an internal pressure of 1,000 psia and a torque of 10,000 ft-lb (see Figure 2.3.7). What is the effective stress at point A in the wall? What is the factor of safety in this design?

Hoop stress:

$$\sigma_\theta = \frac{(1,000 \text{ psi})(6 \text{ in.})}{(0.1 \text{ in.})} = 60,000 \text{ psi}$$

Axial stress:

$$\sigma_z = \frac{(1,000 \text{ psi})(6 \text{ in.})}{2(0.1 \text{ in.})} = 30,000 \text{ psi}$$

Torsion:

$$J = \pi/2(6.05^4 - 6.00^4) = 68.71 \text{ in.}^4$$

$$\tau_{z\theta} = \frac{10,000 \text{ ft-lb } 12 \text{ in. } 6.05 \text{ in.}}{|\text{ft}|68.71 \text{ in.}^4} = 10,566 \text{ psi}$$

Effective stress:

$$\sigma_e = \left\{ \frac{1}{2}[(60,000 - 30,000)^2 + (60,000 - 0)^2 \right.$$
$$\left. + (30,000 - 0)^2 + 6(10,566)^2] \right\}^{1/2}$$

$$= 55,090 \text{ psi}$$

From Table 2.3.2, the yield strength for AISI SAE 1035, cold rolled, is 67,000 psi. Thus the factor of safety is

$$SF = \frac{67,000}{55,090} = 1.22$$

For further information on this subject, see References 1–5.

References

1. Bolz, Ray E., et al., *Handbook of Tables for Applied Engineering Science*, CRC, Cleveland, 1970.
2. Budynas, Richard G., *Advanced Strength and Applied Stress Analysis*, McGraw-Hill, New York, 1977.
3. Rogers, Grover L., *Mechanics of Solids*, John Wiley & Sons, New York, 1964.
4. Timoshenko, S., and MacCullough, Gleason, H., *Elements of Strength of Materials*, Van Nostrand, Princeton, 1957.
5. Timoshenko, S., and Young, D. H., *Engineering Mechanics*, McGraw-Hill, 1940.

2.4 THERMODYNAMICS

2.4.1 Introduction

Historically, the word thermodynamics stems from the Greek words *therme* (heat) and *dynamis* (force), and the discipline grew out of considerations involving the motive power of heat (i.e., the capacity of hot bodies to produce *work*). Thermodynamics involves terms and concepts that often are used in a nontechnical setting. In thermodynamics, these terms are abstractions of the ordinary concepts and have precise meanings that may differ drastically from everyday usage. It is therefore necessary to begin the discussion of thermodynamics by defining some of these words and concepts.

System. A system is the particular part of the observable universe in which an investigator is interested. Typical thermodynamic systems are a quantity of a gas, a liquid and its vapor, a mixture of two liquids, a solution, and a crystalline solid.

Surroundings. Everything in the observable universe except the system is called the surroundings.

Table 2.3.2 Mechanical Properties of Metals and Alloys

Ferrous alloys comprise the largest volume of metal alloys used in engineering. The actual range of mechanical properties in any particular grade of alloy steel depends on the particular history and heat treatment. The steels listed in this table are intended to give some idea of the range of properties readily obtainable. Many hundreds of steels are available. Cost is frequently in important criterion in the choice of material; in general the greater the percentage of alloying elements present in the alloy, the greater will be the cost.

No.	Material	Nominal composition	Form and condition	Typical mechanical properties				Comments
				Yield strength (0.2% offset), 1000 lb/sq in.	Tensile strength, 1000 lb/sq in.	Elongation, in 2 in., %	Hardness, Brinell	
	FERROUS ALLOYS							
	IRON							
1	Ingot iron (Included for comparison)	Fe 99.9	Hot-rolled	29	45	26	90	
			Annealed	19	38	45	67	
	PLAIN CARBON STEELS							
2	AISI-SAE 1020	C 0.20 Mn 0.45 Si 0.25 Fe bal.	Hot-rolled	30	55	25	111	Bolts, crankshafts, gears, connecting rods; easily weldable
			Hardened (water-quenched, 1000°F-tempered)	62	90	25	179	
3	AISI 1025	C 0.25 Fe bal. Mn 0.45	Bar stock Hot-rolled	32	58	25	116	
			Cold-drawn	54	64	15	126	
4	AISE-SAE 1035	C 0.35 Mn 0.75	Hot-rolled	39	72	18	143	Medium-strength, engineering steel
			Cold-rolled	67	80	12	163	
5	AISI-SAE 1045	C 0.45 Fe bal. Mn 0.75	Bar stock Annealed	73	80	12	170	
			Hot-rolled	45	82	16	163	
			Cold-drawn	77	91	12	179	
6	AISI-SAE 1078	C 0.78 Fe bal. Mn 0.45	Bar stock Hot-rolled; spheroidized	55	100	12	207	
			Annealed	72	94	10	192	
7	AISI-SAE 1095	C 0.95 Mn 0.40						
8	AISI-SAE 1120	C 0.2 Mn 0.8 S 0.1	Cold-drawn	58	69	—	137	Free-cutting, leaded, resulphurized steel; high-speed, automatic machining
9	*ALLOY STEELS* ASTM A202/56	C 0.17 Mn 1.2 Cr 0.5 Si 0.75	Stress-relieved	45	75	18	—	Low alloy; boilers, pressure vessels

(continued)

Table 2.3.2 *(continued)*

No.	Material	Nominal composition	Form and condition	Yield strength (0.2% offset), 1000 lb/sq in.	Tensile strength, 1000 lb/sq in.	Elongation, in 2 in., %	Hardness, Brinell	Comments
10	AISI 4140	C 0.40, Cr 1.0, Mn 0.9, Si 0.3, Mo 0.2	Fully-tempered / Optimum properties	95 / 132	108 / 150	22 / 18	240 / —	High strength; gears, shafts
11	12% Manganese steel	12% Mn, C	Tempered 600°F / Rolled and heat-treated stock	200 / 44	220 / 160	10 / 40	— / 170	Machine tool parts; wear, abrasion-resistant
12	VASCO 300	Ni 18.5, Co 9.0, Mo 4.8, Ti 0.6, C 0.03	Solution treatment 1500°F; aged 900°F	110	150	18	—	Very high strength, maraging, good machining properties in annealed state
13	T1 (AISI)	W 18.0, Cr 4.0, V 1.0, C 0.7	Quenched; tempered				R(c)	High speed tool steel, cutting tools, punches, etc.
14	M2 (AISI)	W 6.5, Cr 4.0, V 2.0, Mo 5.0, C 0.85	Quenched; tempered				65–66	M-grade, cheaper, tougher
15	Stainless steel type 304	Ni 9.0, Cr 19.0, C 0.08 max	Annealed; cold-rolled	35 to 160	85 to 185	60 / 8	160 to 400	General purpose, weldable; nonmagnetic austenitic steel
16	Stainless steel type 316	Cr 18.0, Ni 11.0, Mo 2.5, C 0.10 max, Fe bal.	Annealed	30 to 120	90 to 150	50 / 8	165 / 275	For severe corrosive media, under stress; nonmagnetic austenitic steel
17	Stainless steel type 431	Cr 16.0, Ni 2.0, Mn 1.0, Si 1.0, C 0.20, Fe bal.	Annealed / Heat-treated	85 / 150	120 / 195	25 / 20	250 / 400	Heat-treated stainless steel, with good mechanical strength; magnetic
18	Stainless steel 17 4 PH	Cr 17.0, Ni 4.0, Cu 4.0, C 0.35, C 0.07, Fe bal.	Annealed	110	150	10	363	Precipitation hardening; heat-resisting type; retains strength up to approx. 600°F

CAST IRONS AND CAST STEELS

These alloys are used where large and or intricate-shaped articles are required or where over-all dimensional tolerances are not critical. Thus the article can be produced with the fabrication and machining costs held to a minimum. Except for a few heat-treatable cast steels, this class of alloys does not demonstrate high-strength qualities.

No.	Alloy	Composition, %	Condition					Applications
	CAST IRONS							
19	Cast gray iron ASTM A48 48, Class 25	C 3.4, Mn 0.5, Si 1.8	Cast (as cast)	—	25 min	0.5 max	180	Engine blocks, fly-wheels, gears, machine-tool bases
20	White	C 3.4, Mn 0.6, Si 0.7	Cast	—	25	0	450	
21	Malleable iron ASTM A47	C 2.5, Mn 0.55, Si 1.0	Cast (annealed)	33	52	12	130	Automotives, axle bearings, track wheels, crankshafts
22	Ductile or nodular iron (Mg-containing) ASTM A339	C 3.4, Mn 0.40, P 0.1 max, Mg 0.06, Ni 1%, Si 2.5, Fe bal.	Cast (as cast)	53	70	18	170	Heavy-duty machines, gears, cams, crankshafts
	ASTM A395		Cast (quenched tempered)	68	90	7	235	
23	Ni-hard type 2	C 2.7, Mn 0.5, Cr 2.0, Si 0.6, Ni 4.5, Fe bal.	Sand-cast	108	135	5	310	Strength, with heat- and corrosion-resistance
			Chill-cast (tempered)	—	55	—	550	
				—	75	—	625	
24	Ni-resist type 2	C 3.0, Mn 1.0, Cr 2.5, Si 2.0, Ni 20.0, Fe bal.	Cast (as cast)	—	27	2	140	
	CAST STEELS							
25	ASTM A27-62 (60-30)	C 0.3, Si 0.8, Cr 0.4, Mn 0.6, Ni 0.5, Mo 0.2		30	60	24	—	Low alloy, medium strength, general application
26	ASTM A148-60 (105-85)			85	105	17	—	High strength; structural application

(continued)

Table 2.3.2 (continued)

No.	Material	Nominal composition	Form and condition	Yield strength (0.2% offset), 1000 lb/sq in.	Tensile strength, 1000 lb/sq in.	Elongation, in 2 in., %	Hardness, Brinell	Comments
				Typical mechanical properties				
27	Cast 12 Cr alloy (CA-15)	C 0.15 max Mn 1.00 max Si 1.50 max Cr 11.5–14 Ni 1.00 max Fe bal.	Air-cooled from 1800°F; tempered at 600°F Air-cooled from 1800°F; tempered at 1400°F	150 75	200 100	7 30	390 185	Stainless, corrosion-resistant to mildly corrosive alkalis and acids
28	Cast 29 9 alloy (CE-30) ASTM A296 63T	C 0.30 max Mn 1.50 max Si 2.00 max Cr 26–30 Ni 8–11 Fe bal.	As cast	60	95	15	170	Greater corrosion resistance, especially for oxidizing condition
29	Cast 28 7 alloy (HD) ASTM A297-63T	C 0.50 max Mn 1.50 max Si 2.00 max Cr 26–30 Ni 4–7 Fe bal.	As cast	48	85	16	190	Heat resistant

SUPER ALLOYS

The advent of engineering applications requiring high temperature and high strength, as in jet engines and rocket motors, has led to the development of a range of alloys collectively called super alloys. These alloys require excellent resistance to oxidation together with strength at high temperatures, typically 1800°F in existing engines. These alloys are continually being modified to develop better specific properties, and therefore entries in this group of alloys should be considered "fluid." Both wrought and casting-type alloys are represented. As the high temperature properties of cast materials improve, these alloys become more attractive, since great dimensional precision is now attainable in investment castings.

No.	Material	Nominal composition	Form and condition	Yield strength (0.2% offset), 1000 lb/sq in.	Tensile strength, 1000 lb/sq in.	Elongation, in 2 in., %	Hardness, Brinell	Comments
	NICKEL BASE							
30	Hastelloy X	Co 1.5 max Fe 18.5 Cr 22.0 Mo 9.0 W 0.6 C 0.15 max (cast) C 0.20 max (wrought) Ni bal.	Wrought sheet Mill-annealed As investment cast	52 — 46.5	113.2 67 —	43 17 —	194 172 —	
31	Hastelloy C	Cr 16.0 Fe 6.0 W 4.0 C 0.15 max Mo 17.0 Ni bal.	Sand-cast (annealed) Rolled (annealed) Investment cast	50 71 50	78 130 80	5 45 10	199 204 215	

NICKEL BASE (Cont.)

No.	Alloy	Composition		Condition				
32	Inconel 713C	Ni (+Co) bal. Mo 4.5 Al 6.0	Cr 13.0 Cb 2.0 Ti 0.6	Investment cast	102	120	6	—
33	In 100	C 18.0 Mo 3.0 Al 55.0 V 1.0	Cr 10.0 Ti 4.7 Co 15.0	Cast				
34	Tax 8	C 125.0 Mo 4.0 W 4.0 Ta 8.0	Cr 6.0 Al 6.0 Zr 1.0 V 2.5	Cast				
35	Nimonic 90	Ni (+Co) 57.00 Mn 0.50 S 0.007 Cu 0.05 Al 1.65 Co 16.90	C 0.05 Fe 0.45 Si 0.20 Cr 20.55 Ti 2.60	Annealed: wrought	90	155	—	260	General elevated temperature applications
36	Inconel X	Ni (+Co) 72.85 Mn 0.65 S 0.007 Cu 0.05 Al 0.75 Cb (+Ta) 0.85	C 0.04 Fe 6.80 Si 0.30 Cr 15.0 Ti 2.50	Annealed Annealed; age-hardened	50 115	115 175	50 25	150 300
37	Waspaloy	C 0.08 Mo 4.3 Co 13.5	Cr 19.5 Ti 3.0	Cold-rolled	270	275	8	Rc 51
38	Rene 41	C 0.09 Mo 10.0 Al 1.5	Cr 19.0 Ti 3.1 Co 11.0	Wrought	100	145	—	—

(continued)

Table 2.3.2 *(continued)*

No.	Material	Nominal composition		Form and condition	Yield strength (0.2% offset), 1000 lb/sq in.	Tensile strength, 1000 lb/sq in.	Elongation, in 2 in., %	Hardness, Brinell	Comments
						Typical mechanical properties			
39	Udimet 700	C 0.08, Mo 5.0, Al 4.3, Ni 97.5	Cr 15.0, Ti 3.5, Co 18.5	Cold-rolled	280	285	6	Rc 53	
40	T. D. Nickel	Ni 97.5, ThO₂ 2.4		Extended and cold-worked	85	100	13	—	High temperature; jet engine parts
	COBALT BASE								
41	Haynes Stellite alloy 25 (L605)	C 0.15 max, Ni 10.0, Min 1.5	Cr 20.0, W 15.0, Co bal.	Wrought sheet; mill annealed	63	140	60	244	Wrought products
42	Haynes Stellite alloy 21 AMS 5385 (cast)	C 0.25, Ni 2.5, Cr 28.5	Mo 5.5, Co bal.	As investment cast	82	103	8	313 max	For castings

ALUMINUM ALLOYS

Although the strength of aluminum alloys is in general less than that attainable in ferrous alloys or copper-base alloys, their major advantage lies in their high strength-to-weight ratio due to the low density of aluminum. Aluminum alloys have good corrosion resistance for most applications except in alkaline solutions.

No.	Material	Nominal composition		Form and condition	Yield strength (0.2% offset), 1000 lb/sq in.	Tensile strength, 1000 lb/sq in.	Elongation, in 2 in., %	Hardness, Brinell	Comments
43	3003 ASTM B221	Cu 0.12, Mn 1.2	Al bal.	Annealed-O Cold-rolled-H14 Cold-rolled-H18	6 21 27	16 22 29	40 16 10	28 40 55	Good formability, weldable, medium strength; chemical equipment
44	2017 ASTM B221	Mn 0.5, Cu 4.0, 0.5	Al bal.	Annealed-O Heat-treated-T4	10 40	26 62	22 22	45 105	High strength; structural parts, aircraft, heavy forgings
45	2024 ASTM B211	Cu 4.5, Mg 1.5, Mn 0.6, 1.5	Al bal.	Heat-treated-T4	47	68	19	120	
46	5052 ASTM B211	Cr 0.25, Mg 2.5	Al bal.	Annealed-O Cold-rolled and stabilized-H34	13 31	28 38	30 14	47 68	Medium strength, good fatigue properties; street-light standards
47	ASTM B209			Cold-rolled and stabilized-H38	37	42	8	77	

No.	Alloy / ASTM	Composition	Condition					Applications
48	7075 ASTM B211	Cu 1.6, Cr 0.3, Zn 5.6, Mg 2.5, Al bal.	Annealed-O	15	33	17	60	High strength, good corrosion resistance
			Heat-treated and artificially aged-T6	73	83	11	150	
49	380 ASTM SC84B	Si 9.0, Cu 3.5, Al bal.	Die-cast	24	48	3	—	General purpose die-casting
50	195 ASTM C4A	Si 0.8, Cu 4.5, Al bal.	Sand-cast: heat-treated-T4	16	32	8.5	60	Structural elements, aircraft, and machines
			Sand-cast: heat-treated and artificially aged-T6	24	36	5	75	
51	214 ASTM G4A	Mg 3.8, Al bal.	Sand-cast-F	12	25	9	50	Chemical equipment, marine hardware, architectural
52	220 ASTM G10A	Mg 10.0, Al bal.	Sand-cast: heat-treated-T4	26	48	16	75	Strength with shock resistance: aircraft

COPPER ALLOYS

Because of their corrosion resistance and the fact that copper alloys have been used for many thousands of years, the number of copper alloys available is second only to the ferrous alloys. In general copper alloys do not have the high-strength qualities of the ferrous alloys, while their density is comparable. The cost per strength-weight ratio is high; however, they have the advantage of ease of joining by soldering, which is not shared by other metals that have reasonable corrosion resistance.

No.	Alloy / ASTM	Composition	Condition					Applications
53	Copper ASTM B152 ASTM B124, R133 ASTM B1, B2, B3	Cu 99.9 plus	Annealed	10	32	45	42	Bus-bars, switches, architectural, roofing, screens
			Cold-drawn	40	45	15	90	
			Cold-rolled	40	46	5	100	
54	Gilding metal ASTM B36	Cu 95.0, Zn 5.0	Cold-rolled	50	56	5	114	Coinage, ammunition
55	Cartridge 70-30 brass ASTM B14 ASTM B19 ASTM B36 ASTM B134 ASTM B135	Cu 70.0, Zn 30.0	Cold-rolled	63	76	8	155	Good cold-working properties: radiator covers, hardware, electrical
56	Phosphor bronze 10% ASTM B103 ASTM B139 ASTM B159	Cu 90.0, Sn 10.0, P 0.25	Spring temper	—	122	4	241	Good spring qualities, high-fatigue strength

(continued)

Table 2.3.2 (continued)

No.	Material	Nominal composition	Form and condition	Typical mechanical properties				Comments
				Yield strength (0.2% offset), 1000 lb/sq in.	Tensile strength, 1000 lb/sq in.	Elongation, in 2 in., %	Hardness, Brinell	
57	Yellow brass (high brass) ASTM B36 ASTM B134 ASTM B135	Cu 65.0, Zn 35.0	Annealed Cold-drawn Cold-rolled (HT)	18 55 60	48 70 74	60 15 10	55 115 180	Good corrosion resistance: plumbing, architectural
58	Manganese bronze ASTM B138	Cu 58.5, Zn 39.2, Fe 1.0, Sn 1.0, Mn 0.3	Annealed Cold-drawn	30 50	60 80	30 20	95 180	Forgings
59	Naval brass ASTM B21	Cu 60.0, Zn 39.25, Sn 0.75	Annealed Cold-drawn	22 40	56 65	40 35	90 150	Condensor tubing: high resistance to salt-water corrosion
60	Muntz metal ASTM B111	Cu 60.0, Zn 40.0	Annealed	20	54	45	80	Condensor tubes: valve stress
61	Aluminum bronze ASTM B169, alloy A ASTM B124 ASTM B150	Cu 92.0, Al 8.0	Annealed Hard	25 65	70 105	60 7	80 210	
62	Beryllium copper 25 ASTM B194 ASTM B197 ASTM B196	Be 1.9, Co or Ni 0.25, Cu bal.	Annealed, solution-treated Cold-rolled Cold-rolled	32 104 70	70 110 190	45 5 3	B60 (Rockwell) B81 C40	Bellows, fuse clips, electrical relay parts, valves, pumps
63	Free-cutting brass	Cu 62.0, Pb 2.5, Zn 35.5	Cold-drawn	44	70	18	B80 (Rockwell)	Screws, nuts, gears, keys
64	Nickel silver 18″ Alloy A (wrought) ASTM B112, No. 2	Cu 65.0, Ni 18.0, Zn 17.0	Annealed Cold-rolled Cold-drawn wire	25 70	58 85 105	40 4	70 (Rockwell) 170	Hardware, optical goods, camera parts
65	Nickel silver 13″ (cast) 10A ASTM B149, No. 10A	Ni 12.5, Sn 2.0, Zn 20.0, Pb 9.0, Cu bal.	Cast	18	35	15	55	Ornamental castings, plumbing: good machining qualities
66	Cupronickel 10″ ASTM B111 ASTM B171	Cu 88.35, Ni 10.0, Fe 1.25, Mn 0.4	Annealed Cold-drawn tube	22 57	44 60	45 15	— —	Condensor, salt-water piping

(*continued*)

No.	Alloy	Composition, %	Condition					Uses
67	Cupronickel	Cu 70.0, Ni 30.0	Wrought	17	35	25	60	Heat-exchange process equipment, valves
68	Red brass (cast) ASTM B30, No. 4A	Cu 85.0, Pb 5.0, Zn 5.0, Sn 5.0	As-cast					
69	Silicon bronze ASTM B30, alloy 12A	Si 4.0, Zn 4.0, Mn 1.0	Castings					Cheaper substitute for tin bronze
70	Tin bronze ASTM B30, alloy 1B	Sn 8, Zn 4.0	Castings					Bearings, high-pressure bushings, pump impellers
71	Navy bronze		Cast					

TIN AND LEAD-BASE ALLOYS

Major uses for these alloys are as "white" metal bearing alloys, extruded cable sheathing, and solders. Tin forms the basis of pewter used for culinary applications.

No.	Alloy	Composition, %	Condition					Uses
72	Lead-base Babbitt ASTM B23, alloy 19	Pb 85.0, Sb 10.0, Cu 0.5, Sn 5.0, As 0.6	Chill cast	—	10	5	19	Bearings, light loads and low speeds
73	Arsenical-lead Babbitt ASTM B23, alloy 15	Pb 83.0, Sb 16.0, Cu 0.6, Sn 1.0, As 1.1	Chill cast	—	10.3	2	20	Bearings, high loads and speeds, diesel engines, steel mills
74	Chemical lead	Pb 99.9, Bi 0.005 max, Cu 0.06	Rolled 95%	1.9	2.5	50	5	
75	Antimonial lead (hard lead)	Pb 94.0, Sb 6.0	Chill cast / Rolled 95%	—	6.8	22	(500 kg)	Good corrosion resistance and strength
76	Calcium lead	Pb 99.9, Cu 0.10, Ca 0.025	Extruded and aged	—	4.1 / 4.5	47 / 25	9	Cable sheathing, creep-resistant pipe
77	Tin Babbitt alloy ASTM B23-61, grade 1	Sb 4.5, Cu 4.5, Sn bal.	Chill cast	—	9.3	2	17	General bearings and die-casting
78	Tin die-casting alloy ASTM B102-52	Sb 13.0, Cu 5.0, Sn bal.	Die-cast	—	10	1	29	Die-casting alloy

Table 2.3.2 *(continued)*

No.	Material	Nominal composition	Form and condition	Yield strength (0.2% offset), 1000 lb/sq in.	Tensile strength, 1000 lb/sq in.	Elongation, in 2 in., %	Hardness, Brinell	Comments
79	Pewter	Sn 91.0, Cu 2.0, Sb 7.0	Rolled sheet, annealed	—	8.6	40	9.5	Ornamental and household items
80	Solder 50 50	Sn 50.0, Pb 50.0	Cast	4.8	6.1	60	14	General-purpose solder
81	Solder	Sn 20.0, Pb 80.0	Cast	3.6	5.8	16	11	Coating and joining, tilling seams on automobile bodies

MAGNESIUM ALLOYS

Because of their low density, these alloys are attractive for use where weight is at a premium. The major drawback to the use of these alloys is their ability to ignite in air (this can be a problem in machining); they are also costly. Magnesium alloys are used in both the wrought and die-cast forms, the latter being the most frequently used form.

No.	Material	Nominal composition	Form and condition	Yield strength (0.2% offset), 1000 lb/sq in.	Tensile strength, 1000 lb/sq in.	Elongation, in 2 in., %	Hardness, Brinell	Comments
82	Magnesium alloy AZ31B	Mn 0.20 min, Zn 1.0, AL 3.0	Rolled-plate (strain-hardened, then partially annealed)	24	37	18	—	Structural applications of medium strength
			Rolled-sheet (strain-hardened, then partially annealed)	32	42	15	73	
83	Magnesium alloy AZ80A	Mn 0.15 min, Zn 0.5, Al 8.5, Mg bal.	Annealed	22	37	21	56	General extruded and forged products
			Extruded	28	38	14	—	
			Extruded (age-hardened)	36	49	11	60	
			Forged (age-hardened)	39	53	6	82	
				34	50	6	72	
84	Magnesium alloy AZ92A	Mn 0.10 min, Zn 2.0, Al 9.0, Mg bal.	Sand-cast (as cast)	14	24	6	50	Pressure-tight sand and permanent mold castings: high UTS and good yield strength
			Sand-cast (solution heat-treated)	14	40	12	55	
			Sand-cast (solution heat-treated and aged)	19	40	5	83	
			Sand-cast (age-hardened)	16	30	18	—	
			Sand-cast and tempered	22	40	3	81	
85	Magnesium alloy ZK60A	Zn 5.7, Zr 0.55, Mg bal.	Extruded	43	52	12	82	

No.	Material	Composition (%)	Condition					Applications
86	Magnesium alloy AZ91A and AZ91B	Zn 0.6, Al 9.0, Mn 0.13 min, Mg bal.	Die-cast (as cast)	22	33	3	67	General die-casting applications
			BERYLLIUM					
87	Beryllium		Hot-pressed	27	33	1–3	—	Windows, X-ray tubes
				38	51			
			Cross-rolled	40	60	10–40	—	Moderator- and reflector-cladding nuclear reactors: heat-shield and structural-member missiles
				60	90			

Nickel and its alloys are expensive and used mainly either for their high-corrosion resistance in many environments or for high-temperature and strength applications. (See Super Alloys.)

No.	Material	Composition (%)	Condition					Applications
			NICKEL ALLOYS					
88	Nickel (cast)	Ni 95.6, Fe 0.5, Si 1.5, Cu 0.5, Mn 0.8, C 0.8	As cast	25	57	22	110	Good corrosion-resistance applications
89	K Monel	Ni (+Co) 65.25, Mn 0.60, S 0.005, Cu 29.60, Ti 0.45, C 0.15, Fe 1.00, Si 0.15, Al 2.75	Annealed	45	100	40	155	High strength and corrosion resistance; aircraft parts, valve stems, pumps
			Annealed, age-hardened	100	155	25	270	
			Spring	140	150	5	300	
			Spring, age-hardened	160	185	10	335	
90	A nickel	Ni (+Co) 99.40, Mn 0.25, S 0.005, Cu 0.05, C 0.06, Fe 0.15, Si 0.05	Annealed	20	70	40	100	Chemical industry for resistance to strong alkalis, plating nickel
	ASTM B160		Hot-rolled	25	75	40	110	
	ASTM B161		Cold-drawn	70	95	25	170	
	ASTM B162		Cold-rolled	95	105	5	210	
91	Duranickel	Ni (+Cu) 93.90, Mn 0.25, S 0.005, Cu 0.05, Ti 0.45, C 0.15, Fe 0.15, Si 0.55, Al 4.50	Annealed	45	100	40	160	High strength and corrosion resistance; pump rods, shafts, springs
			Annealed, age-hardened	125	170	25	330	
			Spring	—	175	5	320	
			Spring, age-hardened	—	205	10	370	
92	Cupronickel 55–45 (Constantan)	Cu 55.0, Ni 45.0	Annealed	30	60	45	—	Electrical-resistance wire; low temperature coefficient, high resistivity
			Cold-drawn	50	65	30	—	
			Cold-rolled	65	85	20	—	
93	Nichrome	Ni 80.0, Cr 20.0						Heating elements for furnaces

(continued)

Table 2.3.2 *(continued)*

No.	Material	Nominal composition		Form and condition	Yield strength (0.2% offset), 1000 lb/sq in.	Tensile strength, 1000 lb/sq in.	Elongation, in 2 in., %	Hardness, Brinell	Comments
94	"S" Monel	Ni 60.0 Fe 2.50 Si 4.0 max	Cu 29.0 Mn 1.5 Al 0.5 max	Sand-casting	80–115	110–145	2	270–350	High-strength casting alloy; good bearing properties for valve seats

TITANIUM ALLOYS

The main application for these alloys is in the aerospace industry. Because of the low density and high strength of titanium alloys, they present excellent strength-to-weight ratios.

No.	Material	Nominal composition		Form and condition	Yield strength (0.2% offset), 1000 lb/sq in.	Tensile strength, 1000 lb/sq in.	Elongation, in 2 in., %	Hardness, Brinell	Comments
95	Commercial titanium ASTM B265-58T	Ti 99.4		Annealed at 1100 to 1350°F (593 to 732°C)	70	80	20	—	Moderate strength, excellent fabricability; chemical industry pipes
96	Titanium alloy ASTM B265 58T 5 Ti 6 Al 4V			Water-quenched from 1750°F (954°C); aged at 1000 F (538 C) for 2 hr	160	170	13	—	High-temperature strength needed in gas-turbine compressor blades
97	Titanium alloy Ti 4 Al 4Mn			Water-quenched from 1450°F (788°C); aged at 900°F (482°C) for 8 hr	170	185	13	—	Aircraft forgings and compressor parts
98	Ti Mn alloy ASTM B265 58T 7	Fe 0.5 Mn 7.0	Ti bal. 8.0	Sheet	140	150	18	—	Good formability, moderate high-temperature strength; aircraft skin

ZINC ALLOYS

A major use for these alloys is for low-cost die-cast products such as household fixtures, automotive parts, and trim.

No.	Material	Nominal composition		Form and condition	Yield strength (0.2% offset), 1000 lb/sq in.	Tensile strength, 1000 lb/sq in.	Elongation, in 2 in., %	Hardness, Brinell	Comments
99	Zinc ASTM B69	Cd 0.35 Pd 0.08	Zn bal.	Hot-rolled	—	19.5	65	38	Battery cans, grommets lithographer's sheet
100	Zilloy-15	Cu 1.00 Mg 0.010	Zn bal.	Hot-rolled Cold-rolled	— —	29 36	20 25	61 80	Corrugated roofs, articles with maximum stiffness
101	Zilloy 40	Cu 1.00	Zn bal.	Hot-rolled Cold-rolled	— —	24 31	50 40	52 60	Weatherstrip, spun articles
102	Zamac-5 ASTM 25	Zn (99.99% pure remainder) Mg 0.03–0.08	Al 3.5–4.3 Cu 0.75 1.25	Die-cast	—	47.6	7	91	Die-casting for automobile parts, padlocks; used also for die material

ZIRCONIUM ALLOYS

These alloys have good corrosion resistance but are easily oxidized at elevated temperatures in air. The major application is for use in nuclear reactors.

No.	Material	Nominal composition		Form and condition	Yield strength (0.2% offset), 1000 lb/sq in.	Tensile strength, 1000 lb/sq in.	Elongation, in 2 in., %	Hardness, Brinell	Comments
103	Zirconium, commercial	O_2 0.07 Hf 1.90	C 0.15 Zn bal.	Annealed	40	65	27	B80 (Rockwell)	
104	Zircaloy 2	Hf 0.02 Fe 0.15 Su 1.46	Ni 0.05 Other 0.25 Zn bal.	Annealed	50	75	22	B90 (Rockwell)	Nuclear power-reactor cores at elevated temperatures

Table 2.3.3 *Typical Properties of Glass-Fiber–Reinforced Resins*

Property	Base resin				
	Polyester	Phenolic	Epoxy	Melamine	Polyurethane
Molding quality	Excellent	Good	Excellent	Good	Good
Compression molding					
Temperature, °F	170 to 320	280 to 350	300 to 330	280 to 340	300 to 400
Pressure, psi	250 to 2000	2000 to 4000	300 to 5000	2000 to 8000	100 to 5000
Mold shrinkage, in./in.	0.0 to 0.002	0.0001 to 0.001	0.001 to 0.002	0.001 to 0.004	0.009 to 0.03
Specific gravity	1.35 to 2.3	1.75 to 1.95	1.8 to 2.0	1.8 to 2.0	1.11 to 1.25
Tensile strength, 1000 psi	25 to 30	5 to 10	14 to 30	5 to 10	4.5 to 8
Elongation, %	0.5 to 5.0	0.02	4		10 to 650
Modulus of elasticity, 10^{-5} psi	8 to 20	33	30.4	24	
Compression strength, 1000 psi	15 to 30	17 to 26	30 to 38	20 to 35	20
Flexural strength, 1000 psi	10 to 40	10 to 60	20 to 26	15 to 23	7 to 9
Impact, Izod, ft-lb/in. or notch	2 to 10	10 to 50	8 to 15	4 to 6	No break
Hardness, Rockwell	M70 to M120	M95 to M100	M100 to M108		M28 to R60
Thermal expansion, per °C	2 to 5 $\times 10^{-5}$	1.6 $\times 10^{-5}$	1.1 to 3.0 $\times 10^{-5}$	1.5 $\times 10^{-5}$	10 to 20 $\times 10^{14}$
Volume resistivity at 50% RH					
23°C. ohm-cm	1 $\times 10^{14}$	7 $\times 10^{12}$	3.8 $\times 10^{15}$	2 $\times 10^{11}$	2 $\times 10^{11}$ to 10^{14}
Dielectric strength, $\frac{1}{2}$ in.					
Thickness, v/mil	350 to 500	140 to 370	360	170 to 300	330 to 900
Dielectric constant					
At 60 hz	3.8 to 6.0	7.1	5.5	9.7 to 11.1	5.4 to 7.6
At 1 khz	4.0 to 6.0	6.9			5.6 to 7.6
Dissipation factor					
At 60 hz	0.01 to 0.04	0.05	0.087	0.14 to 0.23	0.015 to 0.048
At 1 khz	0.01 to 0.05	0.02			0.043 to 0.060
Water absorption, %	0.01 to 1.0	0.1 to 1.2	0.05 to 0.095	0.09 to 0.21	0.7 to 0.9
Sunlight (change)	Slight	Darkens	Slight	Slight	None to slight
Chemical resistance	Fair**	Fair**	Excellent	Very good†	Fair
Machining qualities	Good		Good	Good	Good

Note: Filament-wound components with high glass content, highly oriented, have higher strengths. The decreasing order of tensile strength is: roving, glass cloth, continuous mat, and chopped-strand mat.
**Attacked by strong acids or alkalies.
†Attacked by strong acids.
*From "Reinforced Thermosets," C.A. Spang and G.I. Davis. *Machine Design* 40(29): 32 Dec 12, 1968.

Boundary. A boundary separates the system from the surroundings. Ideally, boundaries are mathematical surfaces that are endowed with various ideal properties, such as rigidity and impermeability. Real boundaries only approximate the properties or ideal thermodynamic boundaries. A system enclosed by a boundary impermeable to matter is called a *closed system*, and one enclosed by a permeable boundary is called an *open system*.

Thermodynamic Variables. Thermodynamic variables are quantities found from experimentation to be necessary or convenient to specify to give a macroscopic description of a system. Most such quantities are drawn from other branches of science or engineering; accordingly, no detailed, all-inclusive definitions of thermodynamic variables can be given.

State of a System. When all of the variables necessary to describe a system are specified, the *state of the system* is said to be specified. The specification of the state of a system gives no information on the process by which the system was brought to this state.

Processes and Cycles. When one or more properties of a system change, the system is said to have undergone a *change in state*. The path of the succession of states through which the system passes is called the *process*. When a system in a given initial state goes through a number of different changes of state or processes and finally returns to its initial state, the system has undergone a *cycle*.

Equilibrium. A fundamental concept in thermodynamics is that of equilibrium. The key idea is that the variables describing a system in equilibrium do not change with time. This idea does not form a sufficient basis for a definition of equilibrium, because it fails to exclude a number of steady-state processes (principally various types of flow processes) that cannot be handled by classic thermodynamic methods. To exclude these, a more restrictive definition is used: *A system is in equilibrium if and only if it is in a state from which no change is possible without net changes in the surroundings.* In steady-state processes, there must be continual changes in the surroundings to maintain the variables of the system at constant values. Classic thermodynamics deals only with systems in equilibrium.

Equilibrium is an abstraction, and real systems are never strictly in equilibrium, but so long as the variables do not change measurably during the time spent making a measurement on the system, the system can be considered to be in equilibrium, and thermodynamic reasoning can be applied to it. A system may be in equilibrium with respect to some variables but not with respect to others.

2.4.2 Fundamentals

Thermodynamics is the science that deals with "energy in transit" and is quantified in terms of heat and work and the properties of substances that bear a relation to heat and work. The science of thermodynamics deals very

Table 2.3.4 *Allowable Unit Stresses for Lumber*
SPECIES, SIZES, ALLOWABLE STRESSES, AND MODULUS OF ELASTICITY
Normal Loading Conditions: Moisture Content Not Over 19 Percent

Species and grades (visual grading)*	Sizes, nominal	Typical grading agency 1968*	Allowable unit stresses, psi*				Modulus of elasticity, psi
			Extreme fiber in bending	Tension parallel to grain	Compression perpendicular	Compression parallel	
Idaho white pine	2×4	W	850	500	240	1,050	1,120,000
	2×6 and wider		1,200	800	240	1,100	1,400,000
Ponderosa pine	2×4	W	850	500	280	1,000	950,000
	2×6 and wider		1,150	800	280	1,000	1,190,000
Lodgepole pine	2×4	W	1,400	850	250	1,100	1,030,000
	4×4		1,350	800	240	1,050	1,000,000
Southern pine	2×4	S	1,810	1,190	405	1,190	1,800,000
	4×4	S	1,810	1,190	405	1,300	1,700,000
Douglas fir	2×4	W	1,700	1,000	385	1,510	1,80,000
	2×6 and wider	W	1,900	1,250	385	1,800	1,810,000
Western hemlock	2×4	W	1,450	850	245	1,350	1,210,000
	2×6 and wider		1,650	1,100	245	1,450	1,520,000
Western spruce	2×4	W	1,150	650	220	950	920,000
	2×6 and wider		1,050	700	220	1,000	1,150,000
Wester cedar	2×4	W	850	500	295	1,150	860,000
	2×6 and wider		1,200	800	295	1,150	1,070,000
Redwood (unseasoned)	2″ and 4″ "Construction"	R	1,640	—	305	1,190	1,240,000

Note: Allowable unit stresses in horizontal sheat are in the range of 75 to 150 psi.

*There is no single grade designation that applies to all lumber. Values in the table apply approximately to "No. 1," although this designation is often modified by terms such as dense or dry. For grades better than No. 1, such terms as structural, heavy, select, dense, etc. are used. Lower grades are No. 2, No. 3 factory, light industrial etc., but there are seldom more than four grades of a single size in a given species. The allowable stresses are for "repetitive member" users.

*Most lumber is graded by the following agencies, although there are other grading organizations.

    W = Western Wood Products Association

    S = Southern Pine Inspection Bureau

    R = Redwood Inspection Service

*Load applied to joists or planks. For beam or stringer grades, stresses are for load applied to the narrow face.

*For engineered uses the allowable stresses are slightly lower; for kiln-dried lumber slightly higher. For short-term loads, such as wind, earthquake, or impact, higher unit stresses are allowed.

"Wood Handbook," U.S. Department of Agriculture Handbook No. 72, 1955.

"Timber Construction Manual," American Institute of Timber Construction, John Wiley & Sons, 1966.

"National Design Specification for Stress-Grade Lumber," National Forest Products Association, Washington, D.C. 1968.

broadly with concepts of how systems and processes work, why some systems and processes cannot work, and why some systems and processes do not work as intended. Like all sciences, the basis of thermodynamics is experimental observation. In thermodynamics, these findings have been formalized into certain basic laws, which are known as the first, second, and third law of thermodynamics. The zeroth law of thermodynamics, which is the definition of the thermodynamic temperature scale, precedes the first law. These laws of thermodynamics can be summarized as follows:

Zeroth Law of Thermodynamics	When two objects or bodies have equality of temperature with a third body, they have equality of temperature with each other.
First Law of Thermodynamics	The law of conservation of energy applies to all thermodynamic systems.
Second Law of Thermodynamics	This law deals with a quantity (entropy) that serves as a means of determining whether a process is possible. One way

of stating this law is as follows:

> *It is impossible to make any transformation whose only final result is the exchange of a non-zero amount of heat with less than two heat reservoirs and the appearance of a positive amount of work in the surroundings.*

Third Law of Thermodynamics	This law deals with the entropy of substances at the temperature of absolute zero. In essence, it states that the entropy of a perfect crystal is zero at absolute zero.

From a purely thermodynamic point of view, the third law is in a different class from the first and second laws. Its purely thermodynamic content is rather limited, and it is difficult to state the law so that there do not seem to be several experimentally observable exceptions. The second law is not believed to have any macroscopic exceptions; for example, it is not believed possible to construct a perpetual motion machine. However, there are experimental exceptions to many of the common statements of the third law.

Table 2.3.5 Physical, Mechanical, and Thermal Properties of Common Stones

Type of Stone	Density, lb per cu ft	Compressive strength $\times10^{-3}$, psi	Rupture modulus $\times10^{-3}$, psi (ASTM C99 52)	Shearing strength $\times10^{-3}$, psi	Young's modulus $\times10^{-6}$, psi	Modulus of rigidity $\times10^{-6}$, psi	Poisson's ratio	Abrasion-hardness index (ASTM C241-51)	Porosity, volume percent	48-hr water absorption (ASTM C97-47)	Thermal conductivity, Btu per ft per hr per deg F	Coefficient of thermal expansion $\times10^{-6}$ per deg F
Granite	160–190	13–55	1.4–5.5	3.5–6.5	4–16	2–6	0.05–0.2	37–88	0.6–3.8	0.02–0.38	20–35	3.6–4.6
Marble	165–179	8–27	0.6–4.0	1.3–6.5	5–11.5	2–4.5	0.1–0.2	8–42	0.4–2.1	0.02–0.45	8–36	3.0–8.5
Slate	168–180	9–10	6–15	2.0–3.6	6–16	25–6	0.1–0.3	6–12	0.1–1.7	0.01–0.6	12–26	3.3–5.6
Sandstone	119–168	5–20	0.7–2.3	0.3–3.0	0.7–10	0.3–4	0.1–0.3	2–26	1.9–27.3	2.0–12.0	4–40	3.9–6.7
Limestone	117–175	2.5–28	0.5–2.0	0.8–3.6	3–9	1–4	0.1–0.3	1–24	1.1–31.0	1.0–10.0	20–32	2.8–4.5

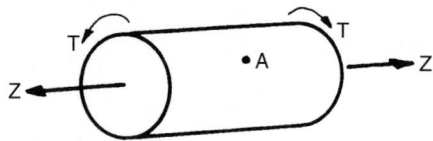

Figure 2.3.7 *Diagram to Example 2.3.3.*

Some care is required to frame the statement so that it is free of exceptions and does not claim anything that is not in principle susceptible of measurement on a macroscopic level. The following is the most satisfactory statement from a purely thermodynamic point of view:

The first and second laws of thermodynamics are applicable down to the limit of absolute zero, provided that at this limit entropy changes vanish for any reversible process.

This statement of the third law defines only entropy *differences* at absolute zero, not absolute values. Just as the absolute value of energy is undefined and only energy differences can be measured, so can only entropy differences be measured. However, if entropy values for all reversibly connected states are equal at absolute zero, this entropy value may as well be taken to be zero for convenience.

2.4.3 Units of Energy

A concept in the study of thermodynamics is the concept of energy. In classic thermodynamics, energy is a fundamental quantity, such as mass or force, and as is often the case with such concepts, it is difficult to define. A general definition that results in analytical predictions that can be verified by experimentation is energy is the property of a system to produce an effect. However, the absolute value of energy cannot be defined or calculated; only the changes in energy as a system transitions from one state to another state has any significance. Energy can be stored within a system in various macroscopic forms and can also be *transformed* from one form to another, such as potential energy and kinetic energy exchanges. It can be *transferred* between systems by various modes of heat transfer across the boundary or by the work associated with the motion of the boundary.

Work has units of force times distance, with the units of kinetic energy being the same as work. In the SI (Systeme International d'Unites) system, work and kinetic energy have units of newton (N)–meters (m), which is the definition of joules (J). In the English engineering system, work and kinetic energy have units of pound force (lbf)–foot (ft).

Heat transfer is a transit phenomenon and can be thought of as energy transfer across the boundary per unit time. In the SI system of units, heat transfer is measured in watts (W), which is the number of joules per unit time crossing the boundary. In the English engineering system of units, heat transfer is usually measured in British thermal units per hour or Btu/hr.

When working in the SI system and equating the heat transfer rate with the rate of doing work, the units are entirely consistent, because both are measured in joules/s, which are defined as watts. In the English engineering system, however, a conversion factor is required to maintain consistency of units because the heat transfer rate is specified in Btu/s but the rate of doing work is specified in ft-lbf/s. In addition, 550 ft-lbf/s is called 1 horsepower (hp). This conversion factor between Btu and ft-lbf is determined from the definition of Btu:

$$1 \text{ Btu} = 778.17 \text{ ft-lbf} \qquad [2.4.1]$$

This conversion factor is often referred to as the mechanical equivalent of heat. The following table expresses relationships between several other useful energy units:

$$1 \text{ J} = 0.3756 \text{ ft/lbf}$$

$$1 \text{ kJ} = 0.9478 \text{ Btu}$$

$$1 \text{ kJ/kg} = 0.42992 \text{ Btu/lbm}$$

$$1 \text{ ft-lbf} = 1.35562 \text{ J}$$

$$1 \text{ kcal} = 4.1868 \text{ kJ}$$

$$1 \text{ W} = 1 \text{ J/s} = 3.413 \text{ Btu/hr}$$

$$1 \text{ hp} = 550 \text{ ft-lbf/s}$$

$$1 \text{ hp} = 2545 \text{ Btu/h}$$

$$1 \text{ hp} = 0.7457 \text{ kW}$$

A = work function (Helmholtz free energy), Btu/lb_m or Btu
C = heat capacity, $\text{Btu/lb}_m\,^\circ\text{R}$
C_p = heat capacity at constant pressure
C_v = heat capacity at constant volume
F = (Gibbs) free energy, Btu/lb_m or Btu
g = acceleration due to gravity $= 32.174 \text{ ft/s}^2$
g_c = conversion factor between force and mass $= 32.174 \, (\text{lb}_m)(\text{ft/s}^2)/\text{lb}_f$
h, H = enthalpy or heat content, Btu/lb_m or Btu
κ = ratio C_P/C_V
Mw = molecular weight, $\text{lb}_m/\text{lb}_m\text{-mole}$
m, M = mass of fluid, lb_m
m, M = mass flow rate, lb_m/s
P = absolute pressure, lb_f/ft^2
P_s = entropy production rate, $\text{Btu}/^\circ\text{R}\cdot\text{s}$
Q = heat transferred to system across a system boundary, Btu/lb_m or Btu
Q = rate of heat transfer, Btu/s
R = universal gas constant, $\text{lb}_f - \text{ft}^3/\text{mole}\cdot\,^\circ\text{R}$
s, S = entropy, $\text{Btu/lb}_m\,^\circ\text{R}$ or $\text{Btu}/^\circ\text{R}$
T = absolute temperature, $^\circ\text{R}$
u, U = internal energy, Btu/lb_m or Btu
V = volume, ft^3/lb_m or ft^3
v = flow velocity, ft/s
W = work done by a system against its surroundings, Btu/lb_m or Btu
Z = height from center of gravity of a fluid mass to a fixed base level, ft

2.4.4 The First Law of Thermodynamics

The differential form of the first law as applied to a *closed* system, for which there is no exchange of matter between the system and its surroundings, is given by

$$dU = \delta Q - \delta W \qquad [2.4.2]$$

where dU represents an infinitesimal increase in the internal energy of the system, δQ is the heat absorbed by the system from its surroundings, and δW is the work done by the system on its surroundings. The state of a system is defined by its temperature, pressure, specific volume, and chemical composition. The change in internal energy expressed by Equation 2.4.2 depends only on the difference between the final and initial states and not upon the process or processes that occurred during the change. The heat and work terms depend on the process path. For a change from a state A to a state B, the first law becomes

$$\Delta U = U_B - U_A = Q - W \qquad [2.4.3]$$

Work interchange between a system and its surroundings can take on any of a variety of forms including mechanical shaft work, electrical work, magnetic work, surface tension, etc. For many applications, the only work involved is that of compression or expansion against the surroundings, in which case the work term in Equation 2.4.2 becomes

$$\delta W = PdV$$

or

$$W = \int_{V_A}^{V_B} PdV \qquad [2.4.4]$$

where V_B is the final volume and V_A the initial volume of the system, and P is the system pressure. Thus, for a *constant pressure* process,

$$W = P\Delta V = P(V_B - V_A) \quad \text{(constant pressure process)} \qquad [2.4.5]$$

or, combining Equations 2.4.3 and 2.4.4.

$$\Delta U = U_B - U_A = Q - PV_B + PV_A \qquad [2.4.6]$$

or

$$Q = (U_B + PV_B) - (U_A + PV_A) \qquad [2.4.7]$$

The combination of properties (U+PV) occurs so frequently in thermodynamics that it is given a special symbol, H, and termed the *enthalpy* or *heat content* of the system. Thus Equation 2.4.7 can be written as

$$Q = \Delta(U + PV) = H_B - H_A = \Delta H \text{ [constant pressure process]} \qquad [2.4.8]$$

Enthalpy is a *property* of the system independent of the path selected. Processes can be conveniently represented graphically. For example, a P–V diagram can be used to illustrate the work done when a system undergoes a change in state (see Figure 2.4.1) In each of the cases depicted in Figure 2.4.1, the work is equal to the shaded area under the P–V curve as shown.

Because the mass is fixed for a closed system, the equations in this discussion will be valid for the entire mass (M) or on a unit mass basis.

2.4.5 The First Law of Thermodynamics Applied to Open Systems

An open system is one which exchanges mass with its surroundings in addition to exchanging energy. For open systems, the first law is formulated from a consideration of the conservation of energy principle which can be stated as

follows:

$$\begin{pmatrix} \text{Net increase} \\ \text{of stored energy} \\ \text{of system} \end{pmatrix}$$

$$= \begin{pmatrix} \text{Stored} \\ \text{energy} \\ \text{of mass} \\ \text{entering} \end{pmatrix} - \begin{pmatrix} \text{Stored} \\ \text{energy} \\ \text{of mass} \\ \text{leaving} \end{pmatrix} + \begin{pmatrix} \text{Net energy} \\ \text{entering as} \\ \text{heat and all} \\ \text{forms of work} \end{pmatrix}$$

Consider the arbitrary open thermodynamic system illustrated in Figure 2.4.2. The foregoing statement of the first law for this open system can be written as

$$m_f U_f - m_i U_i + \frac{m_f v_f^2 - m_i v_i^2}{2g_c} + \frac{g}{g_c}(m_f Z_f - m_i Z_i)$$

$$= \int \left(H_1 + \frac{v_1^2}{2g_c} + \frac{g}{g_c} Z_1 \right) \delta m_1$$

$$- \int \left(H_2 + \frac{v_2^2}{2g_c} + \frac{g}{g_c} Z_2 \right) \delta m_2 + Q - W \qquad [2.4.9]$$

where δm refers to a differential mass of fluid, and the subscripts f and i refer to the entire system in its final state and initial state, respectively. Clearly, for a closed system defined as one which exchanges no mass with its surroundings, Equation 2.4.9 reduces to Equation 2.4.3.

For an open system at *steady state*, as in the case of turbines, compressors, pumps, etc., Equation 2.4.9 can be written (for unit mass flow rate) as

$$\begin{array}{ccccccccc} \Delta H & + & \frac{\Delta v^2}{2g_c} & + & \frac{g}{g_c}\Delta Z & = & Q & - & W_{net} \\ \uparrow & & \uparrow & & \uparrow & & \uparrow & & \uparrow \\ \text{Increase} & & \text{Increase} & & \text{Increase} & & \text{Net heat} & & \text{Net work} \\ \text{in enthalpy} & & \text{in kinetic} & & \text{in potential} & & \text{exchanged} & & \text{exchanged} \\ & & \text{energy} & & \text{energy} & & \text{with} & & \text{with} \\ & & & & & & \text{surroundings} & & \text{surroundings} \end{array}$$
$$[2.4.10]$$

where W_{net} is the net useful work (or shaft work) done by the fluid.

Example 2.4.1. Isobaric Compression of an Ideal Gas
One pound-mole of an ideal gas is compressed at a constant pressure of 1 atm in a piston-like device from an initial volume of 1.5 ft³ to a final volume of 0.5 ft³. The internal energy is known to decrease by 20 Btu. How much heat was transferred to or from the gas?

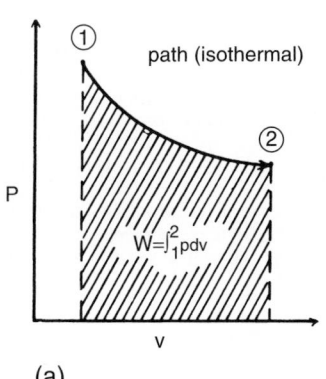

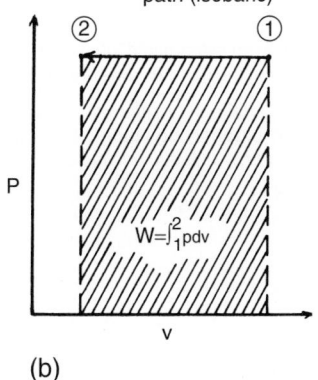

Figure 2.4.1 *P–V process diagrams: (a) isothermal expansion; (b) isobaric compression.*

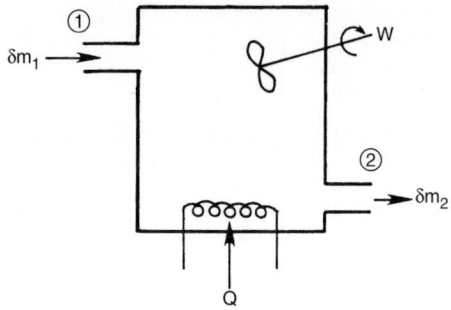

Figure 2.4.2   *An open thermodynamic system.*

2.4.5.0.1 Solution
This is an isobaric (constant pressure) process in a closed system. Equation 2.4.3 applies.

$$Q = \Delta U + W = \Delta U + \int_{v_1}^{v_2} P dV$$

$$= -20 \text{ Btu} + P(V_2 - V_1)$$

$$= -20 \text{ Btu} + (1 \text{ atm}) \left(\frac{14.7 \text{ lb}_f}{\text{in.}^2 \text{atm}} \right) \left(\frac{144 \text{ in.}^2}{\text{ft}^2} \right)$$

$$\times (0.5 \text{ ft}^3 - 1.5 \text{ ft}^3) \left(\frac{1 \text{ Btu}}{778 \text{ ft·lb}_f} \right)$$

$$= -20 \text{ Btu} - 2.72 \text{ Btu}$$

$$= -22.72 \text{ Btu}$$

Because Q is negative, 22.72 Btu of heat were transferred from the gas to its surroundings.

Example 2.4.2. Hydroelectric Power System
A hydroelectric power plant proposes to use 1500 ft³/s of river water to generate electricity. The water enters the system at 1 atm and 50°F and is discharged at 1 atm and 50.4°F after passing through a turbine generator. The discharge point is 600 ft below the inlet. The increase in enthalpy of the water is known to be 0.36 Btu/lb$_m$. Assuming 70% efficiency for the conversion, what power output can be expected from the power plant?

2.4.5.0.2 Solution
The following assumptions pertain to this *open* system:

1. Steady-state flow.
2. No heat transferred between system and surroundings.
3. Change in kinetic energy of the flow streams is negligible.

With these assumptions, we take as a reference point the discharge level of the water, and apply Equation 2.4.10. Thus, $Z_2 = 0$, $Z_1 = 600$ ft and the energy balance becomes

$$\Delta H + \frac{g}{g_c} \Delta Z = -W_{net}$$

where

$$\Delta H = \left(0.36 \frac{\text{Btu}}{\text{lb}_m} \right) \left(\frac{1,500 \text{ ft}^3}{\text{s}} \right) \left(\frac{62.4 \text{ lb}_m}{\text{ft}^3} \right) \left(\frac{3,600 \text{ s}}{\text{hr}} \right)$$

$$= 1.213 \times 10^8 \text{ Btu/hr}$$

$$= 35,553 \text{ kW}$$

and

$$\frac{g}{g_c} \Delta Z = \left(\frac{32.2 \text{ ft/s}^2}{32.2 \text{ ft·lb}_m/\text{lb}_f\text{·s}^2} \right) (0 - 600 \text{ ft}) \left(\frac{1,500 \text{ ft}^3}{\text{s}} \right)$$

$$\times \left(\frac{62.4 \text{ lb}_m}{\text{ft}^3} \right) \left(\frac{3,600 \text{ s}}{\text{hr}} \right)$$

$$= -2.599 \times 10^8 \text{ Btu/hr}$$

$$= -76,163 \text{ kW}$$

Therefore,

$$W_{net} = -35,553 \text{ kW} + 76,163 \text{ kW} = 40,610 \text{ kW}$$

At 70% efficiency, this would yield

$$W_{actual} = (0.70)(40,610) = 28,427 \text{ kW} = 28.427 \text{ MW}$$

2.4.6 Entropy and the Second Law
The second law of thermodynamics provides a basis for determining whether or not a process is possible. It is concerned with availability of the energy of a given system for doing work. All natural systems proceed towards a state of equilibrium and, during any change process, useful work can be extracted from the system. The property called *entropy*, and given the symbol S or s, serves as a quantitative measure of the extent to which the energy of a system is "degraded" or rendered unavailable for doing useful work.

For any *reversible process*, the sum of the changes in entropy for the system and its surroundings is zero. All natural or *real* processes are *irreversible* and are accompanied by a net *increase* in entropy.

Several useful statements have been formulated concerning the second law that are helpful in analyzing thermodynamic systems, such as:

- No thermodynamic cycle can be more efficient than a reversible cycle operating between the same temperature limits.
- The efficiency of all reversible cycles absorbing heat from a single-constant higher temperature and rejecting heat at a single-constant lower temperature must be the same.
- Every real system tends naturally towards a state of maximum probability.
- For any actual process, it is impossible to devise a means of restoring to its original state every system participating in the process.
- For any reversible process, the increase in entropy of any participating system is equal to the heat absorbed by that system divided by the absolute temperature at which the transfer occurred. That is, for a system, i,

$$dS_i = \frac{\delta Q_i}{T_i} \quad \text{(reversible processes)} \qquad [2.4.11]$$

Alternatively, for an ideal reversible process, the sum of all the changes in entropy must be zero or

$$\sum dS_i = \sum \frac{\delta Q_i}{T_i} = 0 \quad \text{(reversible processes)} \qquad [2.4.12]$$

Because all *real* processes are *irreversible* as a result of friction, electrical resistance, etc., any processes involving real systems experience an *increase in entropy*. For such systems

$$\sum dS_i > 0 \quad \text{(irreversible processes)} \qquad [2.4.13]$$

The entropy change of a system during any process depends only upon its initial and final states and not upon the path of the process by which it proceeds from its initial to its final state. One can devise a reversible idealized process to restore a system to its initial state following a change and thereby determine $\Delta S = S_{final} - S_{initial}$. This is one of the most useful aspects of the concepts of a reversible process.

2.4.7 Entropy Production: Flow Systems

In general, for all real processes, there is a net production of entropy and Equation 2.4.13 applies. Because many practical engineering processes involve open systems, it is useful to develop a generalized expression of the second law applied to such systems.

For the generalized control volume shown in Figure 2.4.3, and entropy balance can be stated as follows:

$$\begin{pmatrix} \text{Rate of entropy} \\ \text{in} \end{pmatrix} - \begin{pmatrix} \text{Rate of entropy} \\ \text{out} \end{pmatrix}$$

$$+ \begin{pmatrix} \text{Rate of entropy} \\ \text{production} \end{pmatrix} = \begin{pmatrix} \text{Rate of entropy} \\ \text{accumulation} \end{pmatrix}$$

$$\left[\sum_{in}(\dot{M}S) + \sum_{in} \frac{\delta \dot{Q}_i}{T_i} \right] - \left[\sum_{out} \dot{M}S + \sum_{out} \frac{\delta \dot{Q}_i}{T_i} \right] + \dot{P}_s = \frac{d(SM)}{dt}$$

[2.4.14]

In Equation 2.4.14, $\dot{P}_s$ is the rate of entropy production within the control volume; symbols with dots refer to the time rate of change of the quantity in question. The second law requires that the rate of entropy production be positive.

$$\dot{P}_s \geq 0 \qquad [2.4.15]$$

2.4.8 Heat Capacity

The heat capacity of a substance is extremely important in thermodynamic analysis involving both the first and second laws.

Heat capacity per unit mass is defined by the relationship

$$C \equiv \frac{\delta Q}{dT} \qquad [2.4.16]$$

where δQ is the heat absorbed by unit mass of the system over the infinitesimal temperature change, dT.

For *constant pressure* when only P–V work is involved, the first law yields

$$C_P = \left(\frac{\partial H}{\partial T} \right)_P \qquad [2.4.17]$$

For constant volume

$$C_V = \left(\frac{\partial U}{\partial T} \right)_V \qquad [2.4.18]$$

C_P is related to C_V by the following expression:

$$C_P - C_V = \left[P + \left(\frac{\partial U}{\partial V} \right)_T \right] \left(\frac{\partial V}{\partial T} \right)_P \qquad [2.4.19]$$

For an ideal gas for which $PV = RT/M_w$, with M_w representing the molecular weight, the enthalpy and internal energy are functions only of temperature and Equation 2.4.19 becomes

$$C_P - C_V = P \left(\frac{\partial V}{\partial T} \right)_P = \frac{R}{M_w} \qquad [2.4.20]$$

The ratio of the heat capacity at constant pressure to that at constant volume is

$$\frac{C_P}{C_V} = \kappa \qquad [2.4.21]$$

For an ideal gas, this ratio becomes

$$\kappa = 1 + \frac{R}{C_V M_w} \qquad [2.4.22]$$

2.4.9 Application of the Second Law
2.4.9.1 Heat Engines

The purpose of a heat engine is to remove heat Q_1 from a thermal reservoir at a higher absolute temperature T_1; extract useful work W; and reject heat Q_2 to a second thermal reservoir at a lower absolute temperature T_2. The device used to obtain the useful work is the heat engine.

Considering an ideal heat engine as the system, the first law as applied to the engine undergoing a series of reversible changes in a cyclical fashion becomes

$$\sum \Delta U_i = 0, \quad \text{or} \quad W = Q_1 - Q_2 \qquad [2.4.23]$$

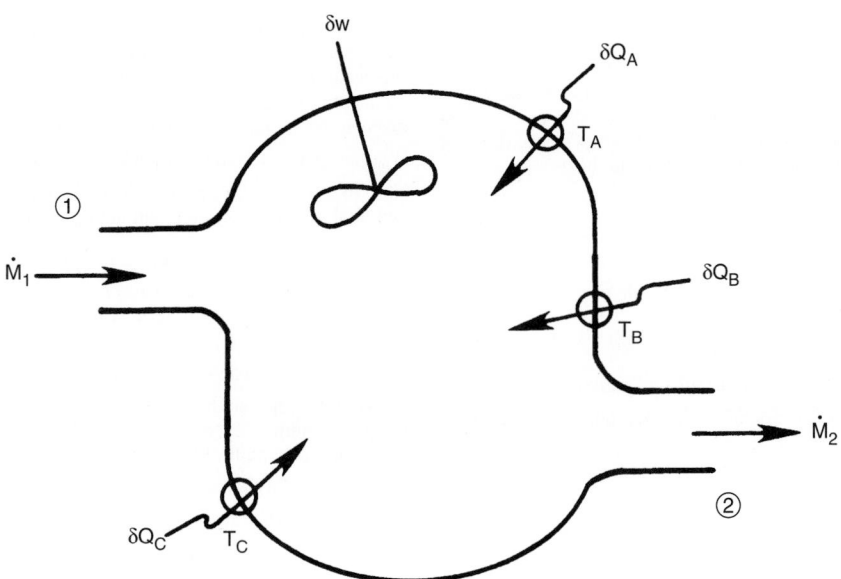

Figure 2.4.3 *Generalized control volume for a flow system.*

The second law yields

$$\sum \Delta S_i = 0 \quad \text{or} \quad \frac{Q_1}{T_1} - \frac{Q_2}{T_2} = 0 \quad \text{or} \quad \frac{Q_1}{Q_2} = \frac{T_1}{T_2} \qquad [2.4.24]$$

To obtain the fraction of the heat input Q_1 that is converted to useful work, Equations 2.4.23 and 2.4.24 are combined to give

$$\frac{W}{Q_1} = \frac{T_1 - T_2}{T_1} \qquad [2.4.25]$$

This important result is called the *Carnot engine efficiency* and yields the *maximum thermal efficiency* that can be achieved by *any* heat engine cycle operating between any two given temperature limits. Heat engines have been proposed to operate within the temperature gradients of the ocean as a means of harnessing the vast amounts of renewable energy available from that source.

2.4.9.2 Heat Pumps
A heat pump, which is the opposite of a heat engine, uses work energy to transfer heat from a cold reservoir to a "hot" reservoir. In households, the cold reservoir is often the surrounding air or the ground while the hot reservoir is the home. For an ideal heat pump system with Q_1 and T_1 referring to the hot reservoir and Q_2 and T_2 referring to the cold reservoir, the work required is, from the first and second laws,

$$\frac{W}{Q_1} = \frac{T_1 - T_2}{T_1} \qquad [2.4.26]$$

Application of this result shows that if 100 units of heat Q_1 are needed to maintain a household at 24°C (297°K) by "pumping" heat from the outside surroundings at 0°C (273°K), it would require a minimum of $(24 \times 100/297) = 8.08$ units of work energy.

2.4.9.3 Refrigeration Machines
Refrigerating machines absorb heat Q_2 from a cold reservoir at temperature T_2, and discharge heat Q_1, into a "hot" reservoir at T_1. To accomplish this, work energy must also be absorbed. The minimum required work is obtained as shown before, using the first and second laws:

$$\frac{W}{Q_2} = \frac{T_1 - T_2}{T_1} \qquad [2.4.27]$$

2.4.9.4 Reversible Work of Expansion or Compression
Many systems involve only work of expansion or compression of the system boundaries. For such systems the first law is written for unit mass of fluid as the basis:

$$dU = \delta Q - PdV \qquad [2.4.28]$$

where $\int_{v_1}^{v_2} PdV$ represents the reversible work of compression or expansion.

From the second law, for a reversible process,

$$dS = \frac{\delta Q}{T} \quad \text{or} \quad \delta Q = TdS \qquad [2.4.29]$$

Combining Equations 2.4.28 and 2.4.29 gives

$$dU = TdS - PdV \quad \text{(reversible, P–V work only)} \qquad [2.4.30]$$

2.4.9.5 Reversible Isobaric Processes
The second law is written as

$$\delta Q = TdS$$

but, since the heat capacity $C = \delta Q/dT$, the second law becomes

$$CdT = TdS \qquad [2.4.31]$$

For constant pressure processes, the entropy increase is written as

$$dS = C_p \frac{dT}{T} = C_p d(\ell n T) \quad \text{(reversible isobaric process)} \qquad [2.4.32]$$

2.4.9.6 Reversible Constant Volume Processes
A constant-volume process is called *isochoric*, and for such processes, the entropy increase is written as

$$dS = C_v \frac{dT}{T} = C_v d(\ell n T) \quad \text{(reversible isochoric process)} \qquad [2.4.33]$$

2.4.9.7 Reversible Isothermal Changes: Maximum Work
The variation of entropy with volume and pressure under conditions of constant temperature is determined by using Equation 2.4.30:

$$\left(\frac{\partial U}{\partial V} \right)_T = T \left(\frac{\partial S}{\partial V} \right)_T - P \qquad [2.4.34]$$

or rearranging:

$$\left(\frac{\partial S}{\partial V} \right)_T = \frac{1}{T} \left[\left(\frac{\partial U}{\partial V} \right)_T + P \right] \text{(constant T)} \qquad [2.4.35]$$

The variation of entropy with pressure is likewise written as

$$\left(\frac{\partial S}{\partial P} \right)_T = \frac{1}{T} \left[P \left(\frac{\partial V}{\partial P} \right)_T + \left(\frac{\partial U}{\partial P} \right)_T \right] \text{(constant T)} \qquad [2.4.36]$$

or

$$\left(\frac{\partial S}{\partial P} \right)_T = \frac{1}{T} \left[-V + \left(\frac{\partial H}{\partial P} \right)_T \right] \text{(constant T)} \qquad [2.4.37]$$

Combining the first law Equation 2.4.2 with the second law Equation 2.4.11 yields the expression

$$dU = TdS - \delta W = d(TS) - \delta W$$

or

$$\delta W = -d(U - TS)$$

which, upon integration between states 1 and 2, yields

$$W = -\Delta(U - TS) \text{(reversible isothermal processes)} \quad [2.4.38]$$

The combination of properties $U - TS$ occurs so frequently in thermodynamic analysis that it is given a special name and symbol, namely A, the *work function* or *maximum work* (because it represents the maximum work per unit mass, obtainable during any isothermal reversible change in any given system). Therefore, it is seen that

$$W_{max} = -\Delta A \text{(reversible isothermal process)} \qquad [2.4.39]$$

Note that the maximum work depends only upon the initial and final states of a system and not upon the path.

2.4.9.8 Maximum Useful Work: Free Energy
The first and second law expressions can be combined and written for constant temperature, constant pressure processes:

$$dU = TdS - \delta W = TdS - PdV - \delta W' \qquad [2.4.40]$$

where $\delta W'$ represents all work energy exchanged with the surroundings except P–V work that is written as PdV. Therefore, solving for $\delta W'$ gives

$$-\delta W' = dU + PdV - TdS \qquad [2.4.41]$$

or, because both T and P constant,

$$-\delta W' = dU + d(PV) - d(TS)$$

or

$$-\delta W' = -d(U + PV - TS) \qquad [2.4.42]$$

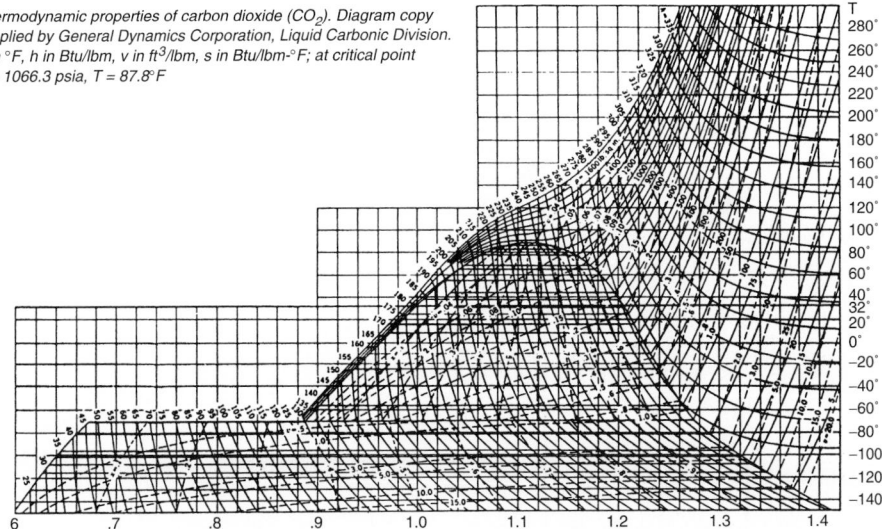

Thermodynamic properties of carbon dioxide (CO₂). Diagram copy supplied by General Dynamics Corporation, Liquid Carbonic Division. T in °F, h in Btu/lbm, v in ft³/lbm, s in Btu/lbm-°F; at critical point P = 1066.3 psia, T = 87.8°F

Figure 2.4.4 *Thermodynamic properties of carbon dioxide.*

By integration this becomes

$$W' = -\Delta(U + PV - TS)$$

$$= -\Delta(H - TS) \quad \text{(constant temperature and pressure)} \quad [2.4.43]$$

This expression shows that the maximum possible *useful work* (i.e., reversible work) that can be obtained from any process occurring at constant temperature and pressure is a function of the initial and final states only and is independent of the path. The combination of properties $U + PV - TS$ or $H - TS$ occurs so frequently in thermodynamic analysis that it is given a special name and symbol, F, the *free energy* (sometimes called the Gibbs free energy). Using this definition, Equation 2.4.44 is written

$$W'_{max} = -\Delta F \quad [2.4.44]$$

Because F is a function of temperature and pressure, its differential can be written as

$$dF = \left(\frac{\partial F}{\partial P}\right)_T dP + \left(\frac{\partial F}{\partial T}\right)_P dT \quad [2.4.45]$$

Because $F = U + PV - TS$, we can also write

$$dF = dU + PdV + VdP - TdS - SdT \quad [2.4.46]$$

Using Equation 2.4.30, this becomes

$$dF = VdP - SdT \quad [2.4.47]$$

Comparison with Equation 2.4.45 shows that

$$\left(\frac{\partial F}{\partial P}\right)_T = V \quad [2.4.48]$$

$$\left(\frac{\partial F}{\partial T}\right)_P = -S \quad [2.4.49]$$

Example 2.4.3
An inventor claims to have devised a CO_2 compressor that requires no shaft work. The device operates at steady state by transferring heat from a feed stream of 2 lb$_m$/s of CO_2 at 150 psia and 100°F. The CO_2 is compressed to a final pressure of 500 psia and a temperature of 40°F. Kinetic and potential energy effects are negligible. A cold source at −140°F "drives" the device at a heat transfer rate of 60 Btu/s. Check the validity of the inventor's claim.

2.4.9.8.1 Solution
The device will be impossible if it violates either the first or second law of thermodynamics. From Figure 2.4.4 the inlet and outlet properties are

State 1	State 2
$T_1 = 100°F$	$T_2 = 40°F$
$P_1 = 150$ psia	$P_2 = 500$ psia
$h_1 = 315$ Btu/lb$_m$	$h_2 = 285$ Btu/lb$_m$
$S_1 = 1.318$ Btu/lb$_m$°R	$S_2 = 1.215$ Btu/lb$_m$°R

Referring to Figure 2.4.5 (process diagram), the first law for this steady-state flow system becomes

$$\dot{M}(h_2 - h_1) + \dot{Q} = (2)(285 - 315) + 60 = 0 \text{ Btu/s}$$

Because energy is not created, the device does not violate the first law. Application of the second law (Equation 2.4.14) yields

$$\dot{P}_s = \dot{M}(S_2 - S_1) + \frac{\dot{Q}_c}{T_c} = (2)(1.215 - 1.318) + \frac{60}{(-140 + 460)}$$

$$= -0.206 + 0.1875 = -0.0185 \text{ Btu/s·°R}$$

Because the rate of entropy production is negative, the device violates the second law and is therefore impossible. Note that the device would be theoretically possible if the final pressure were specified as 400 psia or less by the inventor. That is, at $P_2 = 400$ psia, $T_2 = 40°F$, $h_2 = 290$ Btu/lb$_m$, and $S_2 = 1.25$ Btu/lb$_m$°R, the entropy production rate would be

$$\dot{P}_s = 2(1.25 - 1.318) + 0.1875 = +0.0515 \text{ Btu/s°·R}$$

Because entropy is produced in this case, the device is theoretically possible.

2.4.10 Summary of Thermodynamic Equations
The thermodynamic relations formulated earlier for a pure substance are summarized in Table 2.4.1 with unit mass of fluid as the basis. Several additional important relationships can be derived from them and these are shown in the third column of Table 2.4.1.

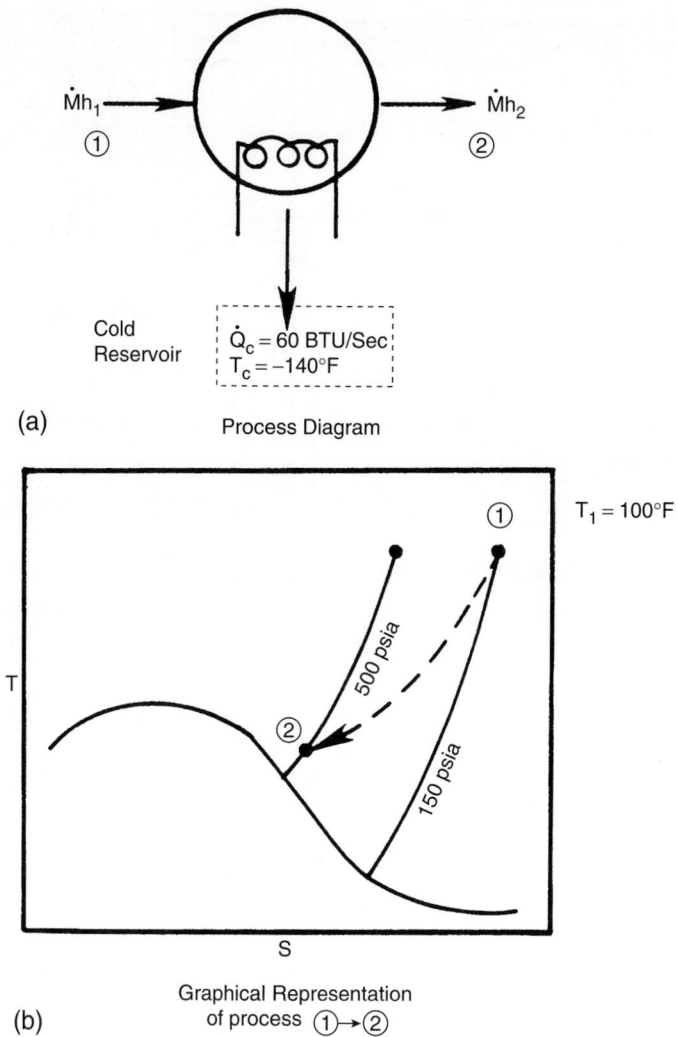

Figure 2.4.5 *(a) Process diagram. (b) Graphical representation of Processes 1 and 2.*

By mathematical manipulation, numerous additional relationships can be derived from those given in Table 2.4.1. Of particular significance are expressions that relate enthalpy H and internal energy U to the measurable variables, P, V, and T. Thus, choosing the basis as one pound mass,

$$\left(\frac{\partial H}{\partial P}\right)_T = V - T\left(\frac{\partial V}{\partial T}\right)_P \qquad [2.4.50]$$

and

$$\left(\frac{\partial U}{\partial V}\right)_T = -P + T\left(\frac{\partial P}{\partial T}\right)_V \qquad [2.4.51]$$

Equations 2.4.50 and 2.4.51 apply to any substance or system and are called *equations of state* because they completely determine the state of a system in terms of its thermodynamic properties.

2.4.11 Thermal Properties for Selected Systems

For practical applications of the numerous thermodynamic relationships, it is necessary to have available the properties of the system. In general, a given property of a pure substance can be expressed in terms of any other two properties to completely define the state of the substance. Thus, one can represent an equation of state by the functional relationship:

$$P = f(T, V) \qquad [2.4.52]$$

which indicates that the pressure is a function of the temperature and specific volume.

Plots of the properties of various substances as well as tables and charts are extremely useful in solving engineering thermodynamic problems. Two-dimensional representations of processes on P–V, T–S, or H–S diagrams are especially useful in analyzing cyclical processes. The use of the P–V diagram was illustrated earlier. A typical T–S diagram for a Rankine vapor power cycle is depicted in Figure 2.4.6.

For the Rankine cycle, the area enclosed by the line segments connecting points 1, 2, 3, 4, 1 on Figure 2.4.6 represents the net heat transferred into the system per unit

Table 2.4.1 *Summary of Thermodynamic Relations (Basis: Unit mass of Fluid)*

Function of Definition	Common Name for Terminology	Differential Equation	Derived Relationships Among Variables
$C_P = \left(\dfrac{\delta Q}{\partial T}\right)_P = \left(\dfrac{\partial H}{\partial T}\right)_P = T\left(\dfrac{\partial S}{\partial T_P}\right)_P$	Heat Capacity at Constant Pressure	$\delta Q_P = C_P dT$	$C_P = \left(\dfrac{\partial U}{\partial T}\right)_P + P\left(\dfrac{\partial V}{\partial T}\right)_P$
$C_V = \left(\dfrac{\partial Q}{\partial T}\right)_V = \left(\dfrac{\partial U}{\partial T}\right)_V = T\left(\dfrac{\partial S}{\partial T}\right)_V$	Heat Capacity at Constant Volume	$\delta Q_V = C_V dT$	$C_V = \left(\dfrac{\partial U}{\partial P}\right)_V \left(\dfrac{\partial P}{\partial T}\right)_V$
$\kappa = \dfrac{C_P}{C_V}$	Heat Capacity Ratio		$C_P - C_V = T\left(\dfrac{\partial P}{\partial T}\right)_V \left(\dfrac{\partial V}{\partial T}\right)_P$
			$C_P - C_V = \dfrac{R}{M_w}$ |Ideal Gas|
$\Delta U = Q - W$	Internal Energy (First Law of Thermodynamics)	$dU = TdS - PdV$	$\left(\dfrac{\partial T}{\partial V}\right)_S = -\left(\dfrac{\partial P}{\partial S}\right)_V$
			$\left(\dfrac{\partial U}{\partial S}\right)_V = T$
			$\left(\dfrac{\partial U}{\partial V}\right)_S = -P$
$H = U + PV$	Enthalpy	$dH = TdS + VdP$	$\left(\dfrac{\partial T}{\partial P}\right)_S = \left(\dfrac{\partial V}{\partial S}\right)_P$
			$\left(\dfrac{\partial H}{\partial S}\right)_P = T$
			$\left(\dfrac{\partial H}{\partial P}\right)_S = V$
$A = U - TS$	Work function or maximum work (Helmholtz free energy)	$dA = -SdT - PdV$	$\left(\dfrac{\partial S}{\partial V}\right)_T = \left(\dfrac{\partial P}{\partial T}\right)_V$
			$\left(\dfrac{\partial A}{\partial T}\right)_V = -S$
			$\left(\dfrac{\partial A}{\partial V}\right)_T = -P$
$F = H - TS$	Free energy or Gibbs Free Energy	$dF = -SdT + VdP$	$\left(\dfrac{\partial S}{\partial P}\right)_T = -\left(\dfrac{\partial V}{\partial T}\right)_P$
			$\left(\dfrac{\partial F}{\partial T}\right)_P = -S$
			$\left(\dfrac{\partial F}{\partial P}\right)_T = V$

mass, because

$$Q_{net} = \int_{s_4}^{s_1} TdS - \int_{s_2}^{s_3} TdS \qquad [2.4.53]$$

The efficiency of power cycles such as the Rankine cycle is given by the ratio of the net work out to the heat added. From Figure 2.4.6, the efficiency is

$$\eta = \frac{W_{net}}{Q_{in}} = \frac{Q_{in} - Q_{out}}{Q_{in}} \qquad [2.4.54]$$

The H–S plot is called a Mollier diagram and is particularly useful in analyzing throttling devices, steam turbines, and other fluid flow devices. A Mollier diagram for steam is presented in Figure 2.4.7 (standard engineering units) and in Figure 2.4.8 in SI units.

Thermodynamic properties may be presented in various ways, including:

• Equations of state (e.g., perfect gas laws, Van der Waals equation, etc.).
• Charts or graphs.
• tables.

Tables 2.4.2–2.4.7 present thermodynamic properties for several pure substances commonly encountered in petroleum engineering practice.

For further information on this subject, refer to References 1–8.

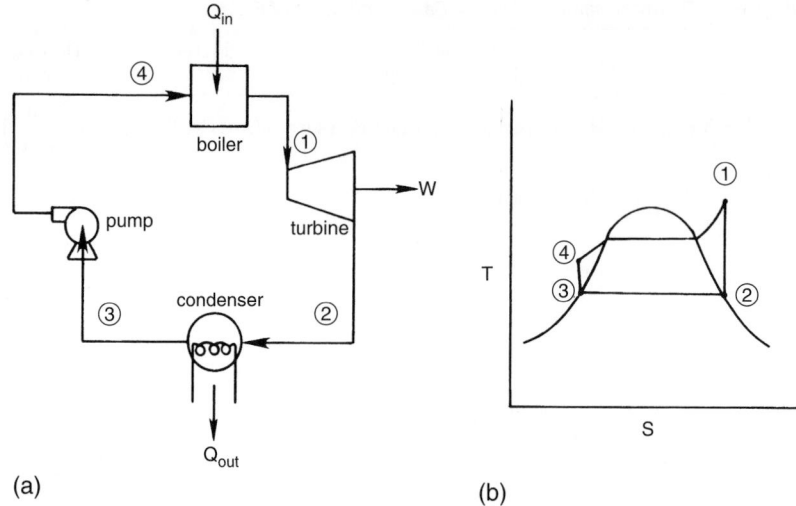

Figure 2.4.6 *(a) Schematic diagram of system. (b) T–S diagram of process.*

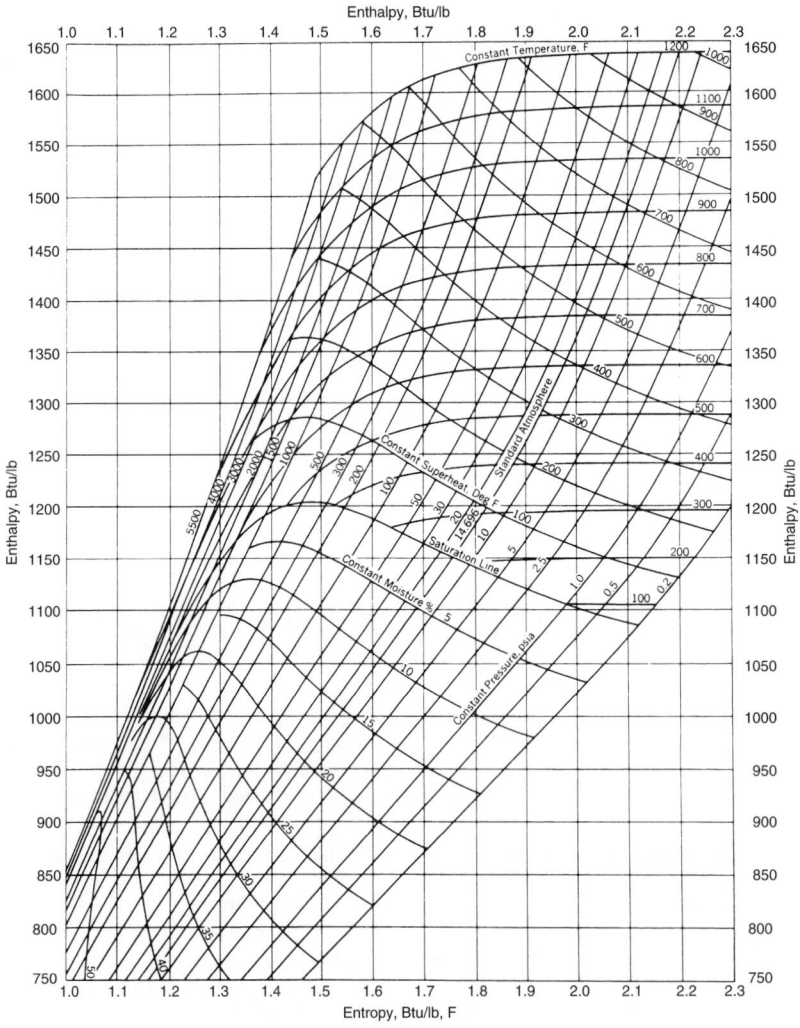

Figure 2.4.7 *Mollier diagram for steam.*
(SOURCE: *Steam Tables in SI-Units, Springer-Verlag, 1984.*)

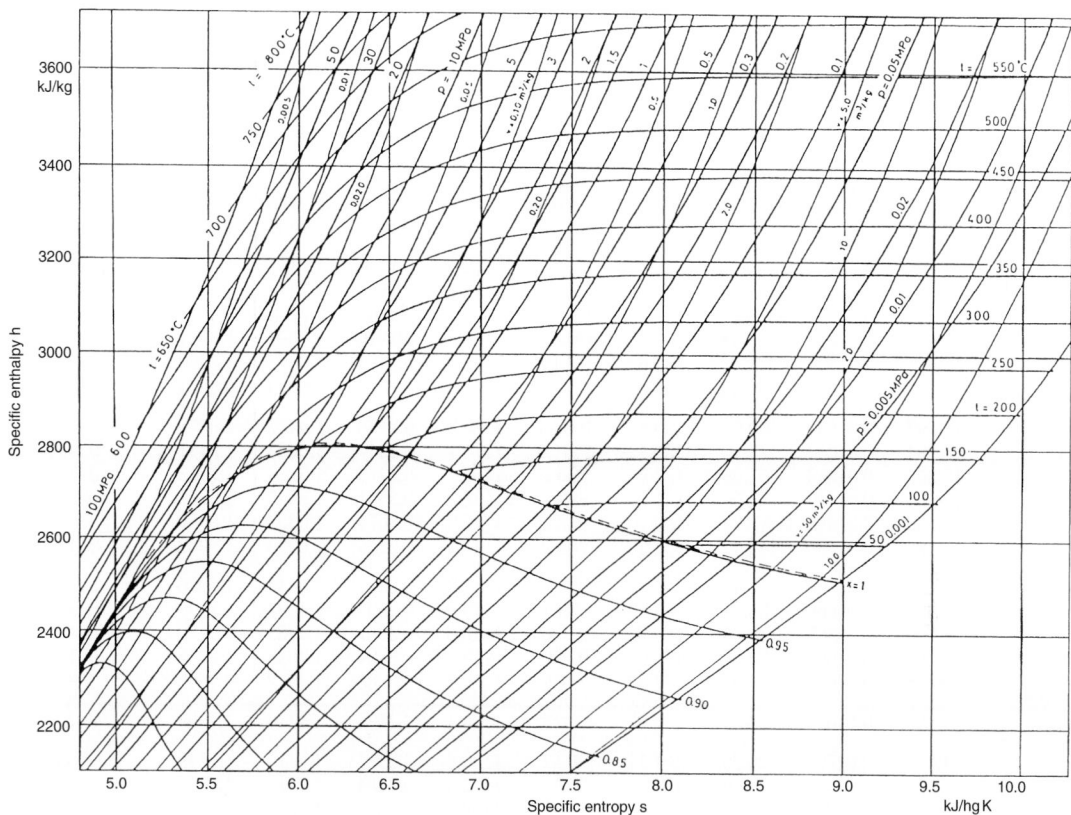

Figure 2.4.8 Mollier (enthalpy-entropy) diagram for steam.
(SOURCE: Steam, its generation and use, Babcock $ Wilcox Co., 1978.)

Table 2.4.2 *Properties of Saturated Steam (h_f and s_f are measured from 32°F)*

Abs press, psi	Temp, deg F	Specific volume		Enthalpy			Entropy			Internal energy
		Liquid	Vapor	Liquid	Evap	Vapor	Liquid	Evap	Vapor	Evap
1.0	101.74	0.01614	333.6	69.70	1036.3	1106.0	0.1326	1.8456	1.9782	974.6
1.2	107.92	0.01616	280.9	75.87	1032.7	1108.6	0.1435	1.8193	1.9628	970.3
1.4	113.26	0.01618	243.0	81.20	1029.6	1110.8	0.1528	1.7971	1.9498	966.7
1.6	117.99	0.01620	214.3	85.91	1026.9	1112.8	0.1610	1.7776	1.9386	963.5
1.8	122.23	0.01621	191.8	90.14	1024.5	1114.6	0.1683	1.7605	1.9288	960.6
2.0	126.08	0.01623	173.73	93.99	1022.2	1116.2	0.1749	1.7451	1.9200	957.9
2.2	129.62	0.01624	158.85	97.52	1020.2	1117.7	0.1809	1.7311	1.9120	955.5
2.4	132.89	0.01626	146.38	100.79	1018.3	1119.1	0.1864	1.7183	1.9047	953.3
2.6	135.94	0.01627	135.78	103.83	1016.5	1120.3	0.1916	1.7065	1.8981	951.2
2.8	138.79	0.01629	126.65	106.68	1014.8	1121.5	0.1963	1.6957	1.8920	949.2
3.0	141.48	0.01630	118.71	109.37	1013.2	1122.6	0.2008	1.6855	0.8863	947.3
4.0	152.97	0.01636	90.63	120.86	1006.4	1127.3	0.2198	1.6427	1.8625	939.3
5.0	162.24	0.01640	73.52	130.13	1001.0	1131.1	0.2347	1.6094	1.8441	933.0
6.0	170.06	0.01645	61.98	137.96	996.2	1134.2	0.2472	1.5820	1.8292	927.5
7.0	176.85	0.01649	53.64	144.76	992.1	1136.9	0.2581	1.5586	1.8167	922.7
8.0	182.86	0.01653	47.34	150.79	988.5	1139.3	0.2674	1.5383	1.8057	918.4
9.0	188.28	0.01656	42.40	156.22	985.2	1141.4	0.2759	1.5203	1.7962	914.6
10	193.21	0.01659	38.42	161.17	982.1	1143.3	0.2835	1.5041	1.7876	911.1
11	197.75	0.01662	35.14	165.73	979.3	1145.0	0.2903	1.4897	1.7800	907.8
12	201.96	0.01665	32.40	169.96	976.6	1146.6	0.2967	1.4763	1.7730	904.8
13	205.88	0.01667	30.06	173.91	974.2	1148.1	0.3027	1.4638	1.7665	901.9
14	209.56	0.01670	28.04	177.61	971.9	1149.5	0.3083	1.4522	1.7605	899.3
14.696	212.00	0.01672	26.80	180.07	970.3	1150.4	0.3120	1.4446	1.7566	897.5
15	213.03	0.01672	26.29	181.11	969.7	1150.8	0.3135	1.4415	1.7549	896.7
16	216.32	0.01674	24.75	184.42	967.6	1152.0	0.3184	1.4313	1.7497	894.3
17	219.44	0.01677	23.39	187.56	965.5	1153.1	0.3231	1.4218	1.7449	892.0
18	222.41	0.01679	22.17	190.56	963.6	1154.2	0.3275	1.4128	1.7403	889.9
19	225.24	0.01681	21.08	193.42	961.9	1155.3	0.3317	1.4043	1.7360	887.8
20	227.96	0.01683	20.089	196.16	960.1	1156.3	0.3356	1.3962	1.7319	885.8
21	230.57	0.01685	19.192	198.79	958.4	1157.2	0.3395	1.3885	1.7280	883.9
22	233.07	0.01687	18.375	201.33	956.8	1158.1	0.3431	1.3811	1.7242	882.0
23	235.49	0.01689	17.627	203.78	955.2	1159.0	0.3466	1.3740	1.7206	880.2
24	237.82	0.01691	16.938	206.14	953.7	1159.8	0.3500	1.3672	1.7172	878.5
25	240.07	0.01692	16.303	208.42	952.1	1160.6	0.3533	1.3606	1.7139	876.8
26	242.25	0.01694	15.715	210.62	950.7	1161.3	0.3564	1.3544	1.7108	875.2
27	244.36	0.01696	15.170	212.75	949.3	1162.0	0.3594	1.3484	1.7078	873.6
28	246.41	0.01698	14.663	214.83	947.9	1162.7	0.3623	1.3425	1.7048	872.1
29	248.40	0.01699	14.189	216.86	946.5	1163.4	0.3652	1.3368	1.7020	870.5
30	250.33	0.01701	13.746	218.82	945.3	1164.1	0.3680	1.3313	1.6993	869.1
31	252.22	0.01702	13.330	220.73	944.0	1164.7	0.3707	1.3260	1.6967	867.7
32	254.05	0.01704	12.940	222.59	942.8	1165.4	0.3733	1.3209	1.6941	866.3
33	255.84	0.01705	12.572	224.41	941.6	1166.0	0.3758	1.3159	1.6917	864.9
34	257.08	0.01707	12.226	226.18	940.3	1166.5	0.3783	1.3110	1.6893	863.5
35	259.28	0.01708	11.898	227.91	939.2	1167.1	0.3807	1.3063	1.6870	862.3
36	260.95	0.01709	11.588	229.60	938.0	1167.6	0.3831	1.3017	1.6848	861.0
37	262.57	0.01711	11.294	231.26	936.9	1168.2	0.3854	1.2972	1.6826	859.8
38	264.16	0.01712	11.015	232.89	935.8	1168.7	0.3876	1.2929	1.6805	858.5
39	265.72	0.01714	10.750	234.48	934.7	1169.2	0.3898	1.2886	1.6784	857.2

(continued)

Table 2.4.2 *(continued)*

Abs press, psi	Temp, deg F	Specific volume		Enthalpy			Entropy			Internal energy
		Liquid	Vapor	Liquid	Evap	Vapor	Liquid	Evap	Vapor	Evap
40	267.25	0.01715	10.498	236.03	933.7	1169.7	0.3919	1.2844	1.6763	856.1
41	268.74	0.01716	10.258	237.55	932.6	1170.2	0.3940	1.2803	1.6743	855.0
42	270.21	0.01717	10.029	239.04	931.6	1170.7	0.3960	1.2764	1.6724	853.8
43	271.64	0.01719	9.810	240.51	930.6	1171.1	0.3980	1.2726	1.6706	852.7
44	273.05	0.01720	9.601	241.95	929.6	1171.6	0.4000	1.2687	1.6687	851.6
45	274.44	0.01721	9.401	243.36	928.6	1172.0	0.4019	1.2650	1.6669	850.5
46	275.80	0.01722	9.209	244.75	927.7	1172.4	0.4038	1.2613	1.6652	849.5
47	277.13	0.01723	9.025	246.12	926.7	1172.9	0.4057	1.2577	1.6634	848.4
48	278.45	0.01725	8.848	247.47	925.8	1173.3	0.4075	1.2542	1.6617	847.4
49	279.74	0.01726	8.678	248.79	924.9	1173.7	0.4093	1.2508	1.6601	846.4
50	281.01	0.01727	8.515	250.09	924.0	1174.1	0.4110	1.2474	1.6585	845.4
51	282.26	0.01728	8.359	251.37	923.0	1174.4	0.4127	1.2432	1.6569	844.3
52	283.49	0.01729	8.208	252.63	922.2	1174.8	0.4144	1.2409	1.6553	843.3
53	284.70	0.01730	8.062	253.87	921.3	1175.2	0.4161	1.2377	1.6538	842.4
54	285.90	0.01731	7.922	255.09	920.5	1175.6	0.4177	1.2346	1.6523	841.5
55	287.07	0.01732	7.787	256.30	919.6	1175.9	0.4193	1.2316	1.6509	840.6
56	288.23	0.01733	7.656	257.50	918.8	1176.3	0.4209	1.2285	1.6494	839.7
57	289.37	0.01734	7.529	258.67	917.9	1176.6	0.4225	1.2255	1.6480	838.7
58	290.50	0.01736	7.407	259.82	917.1	1176.9	0.4240	1.2226	1.6466	837.8
59	291.61	0.01737	7.289	260.96	916.3	1177.3	0.4255	1.2197	1.6452	836.9
60	292.71	0.01738	7.175	262.09	915.5	1177.6	0.4270	1.2168	1.6438	836.0
61	293.79	0.01739	7.064	263.20	914.7	1177.9	0.4285	1.2140	1.6425	835.2
62	294.85	0.01740	6.957	264.30	913.9	1178.2	0.4300	1.2112	1.6412	834.3
63	295.90	0.01741	6.853	265.38	913.1	1178.5	0.4314	1.2085	1.6399	833.4
64	296.94	0.01742	6.752	266.45	912.3	1178.8	0.4328	1.2059	1.6387	832.6
65	297.97	0.01743	6.655	267.50	911.6	1179.1	0.4342	1.2032	1.6374	831.8
66	298.99	0.01744	6.560	268.55	910.8	1179.4	0.4356	1.2006	1.6362	831.0
67	299.99	0.01745	6.468	269.58	910.1	1179.7	0.4369	1.1981	1.6350	830.2
68	300.98	0.01746	6.378	270.60	909.4	1180.0	0.4383	1.1955	1.6338	829.4
69	301.96	0.01747	6.291	271.61	908.7	1180.3	0.4396	1.1930	1.6326	828.6
70	302.92	0.01748	6.206	272.61	907.9	1180.6	0.4409	1.1906	1.6315	827.8
71	303.88	0.01749	6.124	273.60	907.2	1180.8	0.4422	1.1881	1.6303	827.0
72	304.83	0.01750	6.044	274.57	906.5	1181.1	0.4435	1.1857	1.6292	826.3
73	305.76	0.01751	5.966	275.54	905.8	1181.3	0.4447	1.1834	1.6281	825.5
74	306.68	0.01752	5.890	276.49	905.1	1181.6	0.4460	1.1810	1.6270	824.7
75	307.60	0.01753	5.816	277.43	904.5	1181.9	0.4472	1.1787	1.6259	824.0
76	308.50	0.01754	5.743	278.37	903.7	1182.1	0.4484	1.1764	1.6248	823.3
77	309.40	0.01754	5.673	279.30	903.1	1182.4	0.4496	1.1742	1.6238	822.5
78	310.29	0.01755	5.604	280.21	902.4	1182.6	0.4508	1.1720	1.6228	821.7
79	311.16	0.01756	5.537	281.12	901.7	1182.8	0.4520	1.1698	1.6217	821.0
80	312.03	0.01757	5.472	282.02	901.1	1183.1	0.4531	1.1676	1.6207	820.3
81	312.89	0.01758	5.408	282.91	900.4	1183.3	0.4543	1.1654	1.6197	819.6
82	313.74	0.01759	5.346	283.79	899.7	1183.5	0.4554	1.1633	1.6187	818.9
83	314.59	0.01760	5.285	284.66	899.1	1183.8	0.4565	1.1612	1.6177	818.2
84	315.42	0.01761	5.226	285.53	898.5	1184.0	0.4576	1.1592	1.6168	817.5
85	316.25	0.01761	5.168	286.39	897.8	1184.2	0.4587	1.1571	1.6158	816.8
86	317.07	0.01762	5.111	287.24	897.2	1184.4	0.4598	1.1551	1.6149	816.1
87	317.88	0.01763	5.055	288.08	896.5	1184.6	0.4609	1.1530	1.6139	815.4
88	318.68	0.01764	5.001	288.91	895.9	1184.8	0.4620	1.1510	1.6130	814.8
89	319.48	0.01765	4.948	289.74	895.3	1185.1	0.4630	1.1491	1.6121	814.1

(continued)

Table 2.4.2   (continued)

Abs press, psi	Temp, deg F	Specific volume		Enthalpy			Entropy			Internal energy
		Liquid	Vapor	Liquid	Evap	Vapor	Liquid	Evap	Vapor	Evap
90	320.27	0.01766	4.896	290.56	894.7	1185.3	0.4641	1.1471	1.6112	813.4
91	321.06	0.01767	4.845	291.38	894.1	1185.5	0.4651	1.1452	1.6103	812.8
92	321.83	0.01768	4.796	292.18	893.5	1185.7	0.4661	1.1433	1.6094	812.2
93	322.60	0.01768	4.747	292.98	892.9	1185.9	0.4672	1.1413	1.6085	811.5
94	323.06	0.01769	4.699	293.78	892.3	1186.1	0.4682	1.1394	1.6076	810.9
95	324.12	0.01770	4.652	294.56	891.7	1186.2	0.4692	1.1376	1.6068	810.2
96	324.87	0.01771	4.606	295.34	891.1	1186.4	0.4702	1.1358	1.6060	809.6
97	325.61	0.01772	4.561	296.12	890.5	1186.6	0.4711	1.1340	1.6051	808.9
98	326.35	0.01772	4.517	296.89	889.9	1186.8	0.4721	1.1322	1.6043	808.3
99	327.08	0.01773	4.474	297.65	889.4	1187.0	0.4731	1.1304	1.6035	807.7
100	327.81	0.01774	4.432	298.40	888.8	1187.2	0.4740	1.1286	1.6026	807.1
102	329.25	0.01775	4.350	299.90	887.6	1187.5	0.4759	1.1251	1.6010	805.9
104	330.66	0.01777	4.271	301.37	886.5	1187.9	0.4778	1.1216	1.5994	804.7
106	332.05	0.01778	4.194	302.82	885.4	1188.2	0.4796	1.1182	1.5978	803.5
108	333.42	0.01780	4.120	304.26	884.3	1188.6	0.4814	1.1149	1.5963	802.4
110	334.77	0.01782	4.049	305.66	883.2	1188.9	0.4832	1.1117	1.5948	801.2
112	336.11	0.01783	3.981	307.06	882.1	1189.2	0.4849	1.1085	1.5934	800.0
114	337.42	0.01784	3.914	308.43	881.1	1189.5	0.4866	1.1053	1.5919	798.9
116	338.72	0.01786	3.850	309.79	880.0	1189.8	0.4883	1.1022	1.5905	797.8
118	339.99	0.01787	3.788	311.12	879.0	1190.1	0.4900	1.0992	1.5891	796.7
120	341.25	0.01789	3.728	312.44	877.9	1190.4	0.4916	1.0962	1.5878	795.6
122	342.50	0.01791	3.670	313.75	876.9	1190.7	0.4932	1.0933	1.5865	794.5
124	343.72	0.01792	3.614	315.04	875.9	1190.9	0.4948	1.0903	1.5851	793.4
126	344.94	0.01793	3.560	316.31	874.9	1191.2	0.4964	1.0874	1.5838	792.3
128	346.13	0.01794	3.507	317.57	873.9	1191.5	0.4980	1.0845	1.5825	791.3
130	347.32	0.01796	3.455	318.81	872.9	1191.7	0.4995	1.0817	1.5812	790.2
132	348.48	0.01797	3.405	320.04	872.0	1192.0	0.5010	1.0790	1.5800	789.2
134	349.64	0.01799	3.357	321.25	871.0	1192.2	0.5025	1.0762	1.5787	788.2
136	350.78	0.01800	3.310	322.45	870.1	1192.5	0.5040	1.0735	1.5775	787.2
138	351.91	0.01801	3.264	323.64	869.1	1192.7	0.5054	1.0709	1.5763	786.2
140	353.02	0.01802	3.220	324.82	868.2	1193.0	0.5069	1.0682	1.5751	785.2
142	354.12	0.01804	3.177	325.98	867.2	1193.2	0.5083	1.0657	1.5740	784.3
144	355.21	0.01805	3.134	327.13	866.3	1193.4	0.5097	1.0631	1.5728	783.3
146	356.29	0.01806	3.094	328.27	865.3	1193.6	0.5111	1.0605	1.5716	782.3
148	357.36	0.01808	3.054	329.39	864.5	1193.9	0.5124	1.0580	1.5705	781.4
150	358.42	0.01809	3.015	330.51	863.6	1194.1	0.5138	1.0556	1.5694	780.5
152	359.46	0.01810	2.977	331.61	862.7	1194.3	0.5151	1.0532	1.5683	779.5
154	360.49	0.01812	2.940	332.70	861.8	1194.5	0.5165	1.0507	1.5672	778.5
156	361.52	0.01813	2.904	333.79	860.9	1194.7	0.5178	1.0483	1.5661	777.6
158	362.03	0.01814	2.869	334.86	860.0	1194.9	0.5191	1.0459	1.5650	776.8
160	363.53	0.01815	2.834	335.93	859.2	1195.1	0.5204	1.0436	1.5640	775.8
162	364.53	0.01817	2.801	336.98	858.3	1195.3	0.5216	1.0414	1.5630	775.0
164	365.51	0.01818	2.768	338.02	857.5	1195.5	0.5229	1.0391	1.5620	774.1
166	366.48	0.01819	2.736	339.05	856.6	1195.7	0.5241	1.0369	1.5610	773.2
168	367.45	0.01820	2.705	340.07	855.7	1195.8	0.5254	1.0346	1.5600	772.3
170	368.41	0.01822	2.675	341.09	854.9	1196.0	0.5266	1.0324	1.5590	771.4
172	369.35	0.01823	2.645	342.10	854.1	1196.2	0.5278	1.0302	1.5580	770.5
174	370.29	0.01824	2.616	343.10	853.3	1196.4	0.5290	1.0280	1.5570	769.7
176	371.22	0.01825	2.587	344.09	852.4	1196.5	0.5302	1.0259	1.5561	768.8
178	372.14	001826	2.559	345.06	851.6	1196.7	0.5313	1.0238	1.5551	767.9

(continued)

Table 2.4.2   *(continued)*

Abs press, psi	Temp, deg F	Specific volume		Enthalpy			Entropy			Internal energy
		Liquid	Vapor	Liquid	Evap	Vapor	Liquid	Evap	Vapor	Evap
180	373.06	0.01827	2.532	346.03	850.8	1196.9	0.5325	1.0217	1.5542	767.1
182	373.96	0.01829	2.505	347.00	850.0	1197.0	0.5336	1.0196	1.5532	766.2
184	374.86	0.01830	2.479	347.96	849.2	1197.2	0.5348	1.0175	1.5523	765.4
186	375.75	0.01831	2.454	348.92	848.4	1197.3	0.5359	1.0155	1.5514	764.6
188	376.64	0.01832	2.429	349.86	847.6	1197.5	0.5370	1.0136	1.5506	763.8
190	377.51	0.01833	2.404	350.79	846.8	1197.6	0.5381	1.0116	1.5497	763.0
192	378.38	0.01834	2.380	351.72	846.1	1197.8	0.5392	1.0096	1.5488	762.1
194	379.24	0.01835	2.356	352.64	845.3	1197.9	0.5403	1.0076	1.5479	761.3
196	380.10	0.01836	2.333	353.55	844.5	1198.1	0.5414	1.0056	1.5470	760.6
198	380.95	0.01838	2.310	354.46	843.7	1198.2	0.5425	1.0037	1.5462	759.8
200	381.79	0.01839	2.288	355.36	843.0	1198.4	0.5435	1.0018	1.5453	759.0
205	383.86	0.01842	2.234	357.58	841.1	1198.7	0.5461	0.9971	1.5432	757.1
210	385.90	0.01844	2.183	359.77	839.2	1199.0	0.5487	0.9925	1.5412	755.2
215	387.89	0.01847	2.134	361.91	837.4	1199.3	0.5512	0.9880	1.5392	753.2
220	389.86	0.01850	2.087	364.02	835.6	1199.6	0.5537	0.9835	1.5372	751.3
225	391.79	0.01852	2.0422	366.09	833.8	1199.9	0.5561	0.9792	1.5353	749.5
230	393.68	0.01854	1.9992	368.13	832.0	1200.1	0.5585	0.9750	1.5334	747.7
235	395.54	0.01857	1.9579	370.14	830.3	1200.4	0.5608	0.9708	1.5316	745.9
240	397.37	0.01860	1.9183	372.12	828.5	1200.6	0.5631	0.9667	1.5298	744.1
245	399.18	0.01863	1.8803	374.08	826.8	1200.9	0.5653	0.9627	1.5280	742.4
250	400.95	0.01865	1.8438	376.00	825.1	1201.1	0.5675	0.9588	1.5263	740.7
260	404.42	0.01870	1.7748	379.76	821.8	1201.5	0.5719	0.9510	1.5229	737.3
270	407.78	0.01875	1.7107	383.42	818.5	1201.9	0.5760	0.9436	1.5196	733.9
280	411.05	0.01880	1.6511	386.98	815.3	1202.3	0.5801	0.9363	1.5164	730.7
290	414.23	0.01885	1.5954	390.46	812.1	1202.6	0.5841	0.9292	1.5133	727.5
300	417.33	0.01890	1.5433	393.84	809.0	1202.8	0.5879	0.9225	1.5104	724.3
320	423.29	0.01899	1.4485	400.39	803.0	1203.4	0.5952	0.9094	1.5046	718.3
340	428.97	0.01908	1.3645	406.66	797.1	1203.7	0.6022	0.8970	1.4992	712.4
360	434.140	0.01917	1.2895	412.67	791.4	1204.1	0.6090	0.8851	1.4941	706.8
380	439.60	0.01925	1.2222	418.45	785.8	1204.3	0.6153	0.8738	1.4891	701.3
400	444.59	0.0193	1.1613	424.0	780.5	1204.5	0.6214	0.8630	1.4844	695.9
420	449.39	0.0194	1.1061	429.4	775.2	1204.6	0.6272	0.8527	1.4799	690.8
440	454.02	0.0195	1.0556	434.6	770.0	1204.6	0.6329	0.8426	1.4755	685.7
460	458.50	0.0196	1.0094	439.7	764.9	1204.6	0.6383	0.8330	1.4713	680.7
480	462.82	0.0197	0.9670	444.6	759.9	1204.6	0.6436	0.8237	1.4673	675.7
500	467.01	0.0197	0.9278	449.4	755.0	1204.4	0.6487	0.8147	1.4634	1118.6
520	471.07	0.0198	0.8915	454.1	750.1	1204.2	0.6536	0.8060	1.4596	1118.4
540	475.01	0.0199	0.8578	458.6	745.4	1204.0	0.6584	0.7976	1.4560	1118.3
560	478.85	0.0200	0.8265	463.0	740.8	1203.8	0.6631	0.7893	1.4524	1118.2
580	482.58	0.0201	0.7973	467.4	736.1	1203.5	0.6676	0.7813	1.4489	1118.0
600	486.21	0.0201	0.7698	471.6	731.6	1203.2	0.6720	0.7734	1.4454	1117.7
620	489.75	0.0202	0.7440	475.7	727.2	1202.9	0.6763	0.7658	1.4421	1117.5
640	493.21	0.0203	0.7198	479.8	722.7	1202.5	0.6805	0.7584	1.4389	1117.3
660	496.58	0.0204	0.6971	483.8	718.3	1202.1	0.6846	0.7512	1.4358	1117.0
680	499.88	0.0204	0.6757	487.7	714.0	1201.7	0.6886	0.7441	1.4327	1116.7
700	503.10	0.0205	0.6554	491.5	709.7	1201.2	0.6925	0.7371	1.4296	1116.3
720	506.25	0.0206	0.6362	495.3	705.4	1200.7	0.6963	0.7303	1.4266	1116.0
740	509.34	0.0207	0.6180	499.0	701.2	1200.2	0.7001	0.7237	1.4237	1115.6
760	512.36	0.0207	0.6007	502.6	697.1	1199.7	0.7037	0.7172	1.4209	1115.2
780	515.33	0.0208	0.5843	506.2	692.9	1199.1	0.7073	0.7108	1.4181	1114.8
800	518.23	0.0209	0.5687	509.7	688.9	1198.6	0.7108	0.7045	1.4153	1114.4
820	521.08	0.0109	0.5538	513.2	684.8	1198.0	0.7143	0.6983	1.4126	1114.0
840	523.88	0.0210	0.5396	516.6	680.8	1197.4	0.7177	0.6922	1.4099	1113.6
860	526.63	0.0211	0.5260	520.0	676.8	1196.8	0.7210	0.6862	1.4072	1113.1
880	529.33	0.0212	0.5130	523.3	672.8	1196.1	0.7243	0.6803	1.4046	1112.6

(continued)

Table 2.4.2   (continued)

Abs press, psi	Temp, deg F	Specific volume		Enthalpy			Entropy			Internal energy
		Liquid	Vapor	Liquid	Evap	Vapor	Liquid	Evap	Vapor	Evap
900	531.98	0.0212	0.5006	526.6	668.8	1195.4	0.7275	0.6744	1.4020	1112.1
920	534.59	0.0213	0.4886	529.8	664.9	1194.7	0.7307	0.6687	1.3995	1111.5
940	537.16	0.0214	0.4772	533.0	661.0	1194.0	0.7339	0.6631	1.3970	1111.0
960	539.68	0.0214	0.4663	536.2	657.1	1193.3	0.7370	0.6576	1.3945	1110.5
980	542.17	0.0215	0.4557	539.3	653.3	1192.6	0.7400	0.6521	1.3921	1110.0
1,000	544.61	0.0216	0.4456	542.4	649.4	1191.8	0.7430	0.6467	1.3897	1109.4
1,050	550.57	0.0218	0.4218	550.0	639.9	1189.9	0.7504	0.6334	1.3838	1108.0
1,100	556.31	0.0220	0.4001	557.4	630.4	1187.8	0.7575	0.6205	1.3780	1106.4
1,150	561.86	0.0221	0.3802	564.6	621.0	1185.6	0.7644	0.6079	1.3723	1104.7
1,200	567.22	0.0223	0.3619	571.7	611.7	1183.4	0.7711	0.5956	1.3667	1103.0
1,250	572.42	0.0225	0.3450	578.6	602.4	1181.0	0.7776	0.5836	1.3612	1101.2
1,300	577.46	0.0227	0.3293	585.4	593.2	1178.6	0.7840	0.5719	1.3559	1099.4
1,350	582.35	0.0229	0.3148	592.1	584.0	1176.1	0.7902	0.5604	1.3506	1097.5
1,400	587.10	0.0231	0.3012	598.7	574.7	1173.4	0.7963	0.5491	1.3454	1095.4
1,450	591.73	0.0233	0.2884	605.2	565.5	1170.7	0.8023	0.5379	1.3402	1093.3
1,500	596.23	0.0235	0.2760	611.6	556.3	1167.9	0.8082	0.5269	1.3351	1091.0
1,600	604.90	0.0239	0.2548	624.1	538.0	1162.1	0.8196	0.5053	1.3249	1086.7
1,700	613.15	0.0243	0.2304	636.3	519.6	1155.9	0.8306	0.4843	1.3149	1081.8
1,800	621.03	0.0247	0.2179	648.3	501.1	1149.4	0.8412	0.4637	1.3049	1076.8
1,900	628.58	0.0252	0.2021	660.1	482.4	1142.4	0.8516	0.4433	1.2949	1071.4
2,000	635.82	0.0257	0.1878	671.7	463.4	1135.1	0.8619	0.4230	1.2849	1065.0
2,200	649.46	0.0268	0.1625	694.8	424.4	1119.2	0.8820	0.3826	1.2646	1053.1
2,400	662.12	0.0280	0.1407	718.4	382.7	1101.1	0.9023	0.3411	1.2434	1038.6
2,600	673.94	0.0295	0.1213	743.0	337.2	1080.2	0.9232	0.2973	1.2205	1021.9
2,800	684.99	0.0310	0.1035	770.1	284.7	1054.8	0.9459	0.2487	1.1946	1001.2
3,000	695.36	0.0346	0.0858	802.5	217.8	1020.3	0.9731	0.1885	1.1615	972.7
3,200	705.11	0.0444	0.0580	872.4	62.0	934.4	1.0320	0.0532	1.0852	898.4
3,206.2	705.40	0.0503	0.0503	902.7	0	902.7	1.0580	0	1.0580	872.9

Source: Baumeister T., and Marks, L. S., eds., *Standard Handbook for Mechanical Engineers*, Seventh edition, McGraw-Hill Book Co., New York, 1967.
Note: Specific volume in cu. ft. per lb_m, enthalphy and internal energy in Btu per lb_m, entropy in Btu per $lb_m °R$.

Table 2.4.3 *Superheated Steam Tables (v = specific volume, cu ft per lb; h = enthalpy, Btu per lb; s = entropy)*

Pressure, psia (Saturation temp. deg F)		Temperature of steam, deg F								
		340	380	420	460	500	550	600	650	700
20 (227.96)	v	23.60	24.82	26.04	27.25	28.46	29.97	31.47	32.97	34.47
	h	1210.8	1229.7	1248.7	1267.6	1286.6	1310.5	1334.4	1358.6	1382.9
	s	1.8053	1.8285	1.8505	1.8716	1.8918	1.9160	1.9392	1.9671	1.9829
40 (267.25)	v	11.684	12.315	12.938	13.555	14.168	14.930	15.688	16.444	17.198
	h	1207.0	1226.7	1246.2	1265.5	1284.8	1309.0	1333.1	1357.4	1361.9
	s	1.7252	1.7493	1.7719	1.7934	1.8140	1.8345	1.8619	1.8843	1.9058
60 (292.71)	v	7.708	8.143	8.569	8.988	9.403	9.917	10.427	10.935	11.441
	h	1203.0	1223.6	1243.6	1263.4	1283.0	1307.4	1331.8	1356.3	1380.9
	s	1.6766	1.7135	1.7250	1.7470	1.7678	1.7927	1.8162	1.8388	1.8605
80 (312.03)	v	5.718	6.055	6.383	6.704	7.020	7.410	7.797	8.180	8.562
	h	1198.8	1220.3	1240.9	1261.1	1281.1	1305.8	1330.5	1355.1	1379.9
	s	1.6407	1.6669	1.6909	1.7134	1.7346	1.7598	1.7836	1.8063	1.8281
100 (327.81)	v	4.521	4.801	5.071	5.333	5.589	5.906	6.218	6.527	6.835
	h	1194.3	1216.8	1238.1	1258.8	1279.1	1304.2	1329.1	1354.0	1378.9
	s	1.6117	1.6391	1.6639	1.6869	1.7085	1.7340	1.7581	1.7610	1.8029
120 (341.25)	v		3.964	4.195	4.418	4.636	4.902	5.165	5.426	5.683
	h		1213.2	1235.3	1256.5	1277.2	1302.6	1327.7	1352.8	1377.8
	s		1.6156	1.6413	1.6649	1.6869	1.7127	1.7370	1.7801	1.7822
140 (353.02)	v		3.365	3.569	3.764	3.954	4.186	4.413	4.638	4.861
	h		1209.4	1232.3	1254.1	1275.2	1300.9	1326.4	1351.6	1376.8
	s		1.5950	1.6217	1.6458	1.6683	1.6945	1.7190	1.7423	1.7645
160 (363.53)	v		2.914	3.098	3.273	3.443	3.648	3.849	4.048	4.244
	h		1205.5	1229.3	1251.6	1273.1	1299.3	1325.0	1350.4	1375.7
	s		1.5766	1.6042	1.6291	1.6519	1.6785	1.7033	1.7268	1.7491
180 (373.06)	v		2.563	2.732	2.891	3.044	3.230	3.411	3.588	3.764
	h		1201.4	1226.1	1249.1	1271.0	1297.6	1323.5	1349.2	1374.7
	s		1.5596	1.5884	1.6139	1.6373	1.6642	1.6894	1.7130	1.7355
200 (381.79)	v			2.438	2.585	2.726	2.895	3.060	3.221	3.380
	h			1222.9	1246.5	1268.9	1295.8	1322.1	1348.0	1373.6
	s			1.5738	1.6001	1.6240	1.6513	1.6767	1.7006	1.7232
220 (389.86)	v			2.198	2.335	2.465	2.621	2.772	2.920	3.066
	h			1219.5	1243.8	1266.7	1294.1	1320.7	1346.8	1372.6
	s			1.5603	1.5874	1.6117	1.6395	1.6652	1.6892	1.7120
260 (404.42)	v			1.8257	1.9483	2.063	2.199	2.330	2.457	2.582
	h			1212.4	1238.3	1262.3	1290.5	1317.7	1344.3	1370.4
	s			1.5354	1.5642	1.5897	1.6184	1.6447	1.6692	1.6922
300 (417.33)	v			1.5513	1.6638	1.7675	1.8891	2.005	2.118	2.227
	h			1204.8	1232.5	1257.6	1286.8	1314.7	1341.8	1368.3
	s			1.5126	1.5434	1.5701	1.5998	1.6268	1.6517	1.6751
350 (431.72)	v				1.3984	1.4923	1.6010	1.7036	1.8021	1.8980
	h				1224.8	1251.5	1282.1	1310.9	1338.5	1365.5
	s				1.5197	1.5481	1.5792	1.6070	1.6325	1.6563
400 (444.59)	v				1.1978	1.2851	1.3843	1.4770	1.5654	1.6508
	h				1216.5	1245.1	1277.2	1306.9	1335.2	1362.7
	s				1.4977	1.5281	1.5607	1.5894	1.6155	1.6398

(continued)

Table 2.4.3   (continued)

Pressure, psia (Saturation temp. deg F)		Temperature of steam, deg F								
		340	380	420	460	500	550	600	650	700
450 (456.28)	v	1.1231	1.2154	1.3005	1.3810	1.4584	1.5337	1.6074	1.7516	1.8926
	h	1238.4	1272.0	1302.8	1331.9	1359.9	1387.3	1414.3	1467.7	1521.0
	s	1.5095	1.5437	1.5735	1.6003	1.6250	1.6481	1.6699	1.7108	1.7486
500 (467.01)	v	0.9927	1.0798	1.1591	1.2333	1.3044	1.3732	1.4405	1.5715	1.6996
	h	1231.3	1266.7	1298.6	1328.4	1357.0	1384.8	1412.1	1466.0	1519.6
	s	1.4919	1.5279	1.5588	1.5663	1.6115	1.6350	1.6571	1.6982	1.7363
550 (476.94)	v	0.8852	0.9686	1.0431	1.1124	1.1783	1.2419	1.3038	1.4241	1.5414
	h	1223.7	1261.2	1294.3	1324.9	1354.0	1382.3	1409.9	1464.3	1518.2
	s	1.4751	1.5131	1.5451	1.5734	1.5991	1.6228	1.6452	1.6868	1.7250
600 (486.21)	v	0.7947	0.8753	0.9463	1.0115	1.0732	1.1324	1.1899	1.3013	1.4096
	h	1215.7	1255.5	1289.9	1321.3	1351.3	1379.7	1407.7	1462.5	1516.7
	s	1.4586	1.4990	1.5323	1.5613	1.5875	1.6117	1.6343	1.6762	1.7147
700 (503.10)	v		0.7277	0.7934	0.8525	0.9077	0.9601	1.0108	1.1082	1.2024
	h		1243.2	1280.6	1313.9	1345.0	1374.5	1403.2	1459.0	1513.9
	s		1.4722	1.5084	1.5391	1.5665	1.5914	1.6147	1.6573	1.6963
800 (518.23)	v		0.6154	0.6779	0.7328	0.7833	0.8308	0.8763	0.9633	1.0470
	h		1229.8	1270.7	1306.2	1338.6	1369.2	1398.6	1455.4	1511.0
	s		1.4467	1.4863	1.5190	1.5476	1.5734	1.5972	1.6407	1.6801
900 (531.98)	v		0.5264	0.5873	0.6393	0.6863	0.7300	0.7716	0.8506	0.9262
	h		1215.0	1260.1	1298.0	1332.1	1363.7	1393.9	1451.8	1508.1
	s		1.4216	1.4653	1.5002	1.5303	1.5570	1.5814	1.6257	1.6656
1,000 (544.61)	v		0.4533	0.5140	0.5640	0.6084	0.6492	0.6878	0.7604	0.8294
	h		1198.3	1248.8	1289.5	1325.3	1358.1	1389.2	1448.2	1505.1
	s		1.3961	1.4450	1.4825	1.5141	1.5418	1.5670	1.6121	1.6525
1,100 (556.30)	v			0.4632	0.5020	0.5445	0.5830	0.6191	0.6866	0.7503
	h			1236.7	1280.5	1318.3	1352.4	1384.3	1444.5	1502.2
	s			1.4251	1.4656	1.4989	1.5276	1.5535	1.5995	1.6405
1,200 (567.22)	v			0.4016	0.4498	0.4909	0.5277	0.5617	0.6250	0.6843
	h			1223.5	1271.0	1311.0	1364.4	1379.3	1440.7	1499.2
	s			1.4052	1.4491	1.4843	1.5142	1.5409	1.5879	1.6293
1,400 (587.10)	v			0.3174	0.3668	0.4062	0.4403	0.4714	0.5281	0.5805
	h			1193.0	1250.6	1295.5	1334.0	1369.1	1433.1	1493.2
	s			1.3639	1.4171	1.4567	1.4893	1.5177	1.5666	1.6093
1,600 (604.90)	v				0.3027	0.3417	0.3743	0.4034	0.4553	0.5027
	h				1227.3	1278.7	1320.9	1358.4	1425.3	1487.0
	s				1.3800	1.4303	1.4660	1.4964	1.5476	1.5914
1,800 (621.03)	v				0.2506	0.2907	0.3225	0.3502	0.3986	0.4421
	h				1200.3	1260.3	1307.0	1347.2	1417.4	1480.8
	s				1.3515	1.4044	1.4438	1.4765	1.5301	1.5752
2,000 (635.82)	v				0.2058	0.2489	0.2806	0.3074	0.3532	0.5935
	h				1167.0	1240.0	1292.0	1335.5	1409.2	1474.5
	s				1.3139	1.3783	1.4223	1.4576	0.5139	1.5380
2,200 (649.46)	v				0.1633	0.2135	0.2457	0.2721	0.3159	0.3538
	h				1121.0	1217.4	1276.0	1323.3	1400.8	1468.2
	s				1.2665	1.3515	1.4010	1.4393	1.4986	1.5465

Source: Baumeister T., and Marks, L. S., eds., *Standard Handbook for Mechanical Engineers*, Seventh edition, McGraw-Hill Book Co., New York, 1967.

Table 2.4.4 *Steam Table for Use in Condenser Calculations*

Temp deg F, t	Abs pressure		Specific volume	Enthalpy			Entropy	
	Psi	In. Hg	Sat vapor, V_g	Sat liquid, h_f	Evap. h_{fg}	Sat vapor, h_g	Sat liquid, s_f	Sat vapor, s_g
		p						
50	0.17811	0.3626	1703.2	18.07	1065.6	1083.7	0.0361	2.1264
52	0.19182	0.3906	1587.6	20.07	1064.4	1084.5	0.0400	2.1199
54	0.20642	0.4203	1481.0	22.07	1063.3	1085.4	0.0439	2.1136
56	0.2220	0.4520	1382.4	24.06	1062.2	1086.3	0.0478	2.1072
58	0.2386	0.4858	1291.1	26.06	1061.0	1087.1	0.0517	2.1010
60	0.2563	0.5218	1206.7	28.06	1059.9	1088.0	0.0555	2.0948
62	0.2751	0.5601	1128.4	30.05	1058.8	1088.9	0.0593	2.0886
64	0.2951	0.6009	1055.7	32.05	1057.6	1089.7	0.0632	2.0826
66	0.3164	0.6442	988.4	34.05	1056.5	1090.6	0.0670	2.0766
68	0.3390	0.6903	925.9	36.04	1055.5	1091.5	0.0708	2.0706
70	0.3631	0.7392	867.9	38.04	1054.3	1092.3	0.0745	2.0647
72	0.3886	0.7912	813.9	40.04	1053.2	1093.2	0.0783	2.0588
74	0.4156	0.8462	763.8	42.03	1052.1	1094.1	0.0820	2.0530
76	0.4443	0.9046	717.1	44.03	1050.9	1094.9	0.0858	2.0473
78	0.4747	0.9666	673.6	46.02	1049.8	1095.8	0.0895	2.0416
80	0.5069	1.3021	633.1	48.02	1048.6	1096.6	0.0932	0.0360
82	0.5410	1.1016	595.3	50.01	1047.5	1097.5	0.0969	2.0304
84	0.5771	1.1750	560.2	52.01	1046.4	1098.4	0.1005	2.0249
86	0.6152	1.2527	527.3	54.00	1045.2	1099.2	0.1042	2.0195
88	0.6556	1.3347	496.7	56.00	1044.1	1100.1	0.1079	2.0141
90	0.6982	1.4215	468.0	57.99	1042.9	1100.9	0.1115	2.0087
92	0.7432	1.5131	441.3	59.99	1041.8	1101.8	0.1151	2.0034
94	0.7906	1.6097	416.2	61.98	1040.7	1102.6	0.1187	1.9981
96	0.8407	1.7117	392.8	63.98	1039.5	1103.5	0.1223	1.9929
98	0.8935	1.8192	370.9	65.97	1038.4	1104.4	0.1259	1.9877
100	0.9492	1.9325	350.4	67.97	1037.2	1105.2	0.1295	1.9826
102	1.0078	2.0519	331.1	69.96	1036.1	1106.1	0.1330	1.9775
104	1.0695	2.1775	313.1	71.96	1034.9	1106.9	0.1366	1.9725
106	1.1345	2.3099	296.2	73.95	1033.8	1107.8	0.1401	1.9675
108	1.2029	2.4491	280.3	75.95	1032.7	1108.6	0.1436	1.9626
110	1.2748	2.5955	265.4	77.94	1031.6	1109.5	0.1471	1.9577
112	1.3504	2.7494	251.4	79.94	1030.4	1110.3	0.1506	1.9529
114	1.4298	2.9111	238.2	81.93	1029.2	1111.1	0.1541	1.9481
116	1.5130	3.0806	225.8	83.93	1028.1	1112.0	0.1576	1.9433
118	1.6006	3.2589	214.2	85.92	1026.9	1112.8	0.1610	1.9386
120	1.6924	3.4458	203.27	87.92	1025.8	1113.7	0.1645	1.9339
122	1.7888	3.6420	192.95	89.92	1024.6	1114.5	0.1679	1.9293
124	1.8897	3.8475	183.25	91.91	1023.4	1115.3	0.1714	1.9247
126	1.9955	4.0629	174.10	93.91	1022.3	1116.2	0.1748	1.9202
128	2.1064	4.2887	165.47	95.91	1021.1	1117.0	0.1782	1.9156
130	2.2225	4.5251	157.34	97.90	1020.0	1117.9	0.1816	1.9112
132	2.3440	4.7725	149.66	99.90	1018.8	1118.7	0.1849	1.9067
134	2.4712	5.0314	142.42	101.90	1017.6	1119.5	0.1883	1.9023
136	2.6042	5.3022	135.58	103.90	1016.4	1120.3	0.1917	1.8980
138	2.7432	5.5852	129.12	105.89	1015.3	1121.2	0.1950	1.8937
140	2.8886	5.8812	123.01	107.89	1014.1	1122.0	0.1984	1.8894
142	3.0440	6.1903	117.23	109.89	1012.9	1122.8	0.2016	1.8851
144	3.1990	6.5132	111.77	111.89	1011.7	1123.6	0.2049	1.8809
146	3.365	6.850	106.60	113.89	1010.6	1124.5	0.2083	1.8768
148	3.537	7.202	101.71	115.89	1009.4	1125.3	0.2116	1.8726
150	3.718	7.569	97.07	117.89	1008.2	1126.1	0.2149	1.8685

Source: Baumeister T., and Marks, L. S., eds., *Standard Handbook for Mechanical Engineers*, Seventh edition, McGraw-Hill Book Co., New York, 1967.

Table 2.4.5 Properties of Carbon Dioxide (h_f and s_f are measured form 32°F)

Temp. deg F, t	Pressure, psia, p	Density lb per cu ft		Enthalpy, Btu			Entropy	
		Sat liquid	Sat vapor	Sat liquid, h_f	Vaporization, h_{fg}	Sat vapor, h_g	Sat liquid s_f	Sat vapor, s_e
−40	145.87	69.8	1.64	−38.5	136.5	98.0	−0.0850	0.2400
−35	161.33	69.1	1.83	−35.8	134.3	98.5	−0.0793	0.2367
−30	177.97	68.3	2.02	−33.1	132.1	99.0	−0.0735	0.2336
−25	195.85	67.6	2.23	−30.4	129.8	99.4	−0.0676	0.2306
−20	215.02	66.9	2.44	−27.7	127.5	99.8	−0.0619	0.2277
−15	235.53	66.1	2.66	−24.9	125.0	100.1	−0.0560	0.2250
−10	257.46	65.3	2.91	−22.1	122.4	100.3	−0.0500	0.2220
−5	280.85	64.5	3.17	−19.4	120.0	100.6	−0.0440	0.2198
0	305.76	63.6	3.46	−16.7	117.5	100.8	−0.0381	0.2173
5	332.2	62.8	3.77	−14.0	115.0	101.0	−0.0322	0.2151
10	360.4	61.9	4.12	−11.2	112.2	101.0	−0.0264	0.2124
15	390.2	61.0	4.49	−8.4	109.4	101.0	−0.0204	0.2100
20	421.8	60.0	4.89	−5.5	106.3	100.8	−0.0144	0.2071
25	455.3	59.0	5.33	−2.5	103.1	100.6	−0.0083	0.2043
30	490.6	58.0	5.81	+ 0.4	99.7	100.1	−0.0021	0.2012
35	528.0	57.0	6.35	3.5	95.8	99.3	+ 0.0039	0.1975
40	567.3	55.9	6.91	6.6	91.8	98.4	0.0099	0.1934
45	608.9	54.7	7.60	9.8	87.5	97.3	0.0160	0.1892
50	652.7	53.4	8.37	12.9	83.2	96.1	0.0220	0.1852
55	698.8	52.1	9.27	16.1	78.7	94.8	0.0282	0.1809
60	747.4	50.7	10.2	19.4	74.0	93.4	0.0345	0.1767
65	798.6	49.1	11.3	22.9	68.9	91.8	0.0412	0.1724
70	852.4	47.3	12.6	26.6	62.7	89.3	0.0482	0.1665
75	909.3	45.1	14.2	30.9	54.8	85.7	0.0562	0.1587
80	969.3	42.4	16.2	35.6	44.0	79.6	0.0649	0.1464
85	1032.7	38.2	19.1	41.7	27.5	69.2	0.0761	0.1265
88	1072.1	32.9	25.4	Critical point at 88.43°F				

Source: Baumeister T., and Marks, L. S., eds., *Standard Handbook for Mechanical Engineers*, Seventh edition, McGraw-Hill Book Co., New York, 1967.

Table 2.4.6 Properties of Propane and Butane

	Propane (C₃H₄) (Heat measurement are from 0°F)						Butane (C₄H₁₀) (Heat measurements are from 0°F)					
			Enthalpy, Btu per lb		Entropy				Enthalpy, Btu per lb		Entropy	
Temp, deg F	Pressure, psia	Specific volume of vapor, cu ft per lb	Liquid h_f	Vapor h_g	Liquid s_f	Vapor S_g	Specific volume of vapor, cu ft per lb	Pressure, psia	Liquid h_f	Vapor h_g	Liquid S_f	Vapor S_g
−70	7.37	12.9	−37.0	152.5	−0.086	0.400						
−60	9.72	9.93	−32.0	155.0	−0.074	0.393						
−50	12.6	7.74	−26.5	158.0	−0.061	0.389						
−40	16.2	6.13	−21.5	160.0	−0.049	0.384						
−30	20.3	4.93	−16.0	163.0	−0.036	0.380						
−20	25.4	4.00	−11.0	165.0	−0.024	0.377						
−10	31.4	3.26	−5.5	168.0	−0.012	0.374						
0	38.2	2.71	0	170.5	0.000	0.371	11.10	7.3	0	170.5	0.000	0.371
+10	46.0	2.27	5.5	173.5	0.012	0.370	8.95	9.2	5.5	174.0	0.011	0.370
20	55.5	1.90	11.0	176.0	0.024	0.368	7.23	11.6	10.5	177.5	0.022	0.370
30	66.3	1.60	17.0	179.0	0.035	0.366	5.90	14.4	16.0	181.5	0.033	0.371
40	78.0	1.37	23.0	182.0	0.047	0.366	4.88	17.7	21.5	185.0	0.044	0.371
50	91.8	1.18	29.0	185.0	0.059	0.365	4.07	21.6	27.0	188.5	0.056	0.373
60	107.1	1.01	35.0	188.0	0.070	0.364	3.40	26.3	33.0	192.5	0.067	0.374
70	124.0	0.883	41.0	190.5	0.082	0.364	2.88	31.6	38.5	196.0	0.078	0.375
80	142.8	0.770	47.5	193.5	0.093	0.364	2.46	37.6	44.5	199.5	0.089	0.376
90	164.0	0.673	54.0	196.5	0.105	0.364	2.10	44.5	51.0	203.0	0.100	0.377
100	187.0	0.591	60.5	199.0	0.116	0.363	1.81	52.2	57.0	206.5	0.111	0.378
110	212.0	0.521	67.0	201.0	0.128	0.363	1.58	60.8	63.5	210.5	0.122	0.380
120	240.0	0.459	73.5	202.5	0.140	0.363	1.38	70.8	70.0	213.5	0.134	0.382
130							1.21	81.4	76.5	217.0	0.145	0.384
140							1.07	92.6	83.5	221.0	0.157	0.386

SOURCE: Baumeister T., and Marks, L. S., eds., *Standard Handbook for Mechanical Engineers*, Seventh edition, McGraw-Hill Book Co., New York, 1967.

Table 2.4.7 Properties of Freon 11 and Freon 12

| Temp, deg F | Freon 11 (CChF) (Heat measurement are from −40°F) | | | | | | Freon 12 (CCh F₂) (Heat measurements are from −40°F) | | | | | |
| | Pressure, psia | Specific volume of vapor, cu ft per lb | Enthalpy, Btu per lb | | Entropy | | Pressure, psia | Specific volume of vapor, cu ft per lb | Enthalpy, Btu per lb | | Entropy | |
			Liquid h_f	Vapor h_g	Liquid s_f	Vapor S_g			Liquid h_f	Vapor h_g	Liquid S_f	Vapor S_g
−40	0.739	44.2	0.00	87.48	0.0000	0.2085	9.3	3.91	0.00	73.50	0.0000	0.1752
−30	1.03	32.3	1.97	88.67	0.0046	0.2064	12.0	3.09	2.03	74.70	0.00471	0.1739
−20	1.42	24.1	3.94	89.87	0.0091	0.2046	15.3	2.47	4.07	75.87	0.00940	0.1727
−10	1.92	18.2	5.91	91.07	0.0136	0.2030	19.2	2.00	6.14	77.05	0.01403	0.1717
0	2.55	13.9	7.89	92.27	0.0179	0.2015	23.9	1.64	8.25	78.21	0.01869	0.1709
10	3.35	10.8	9.88	93.48	0.0222	0.2003	29.3	1.35	10.39	79.36	0.02328	0.1701
15	3.82	9.59	10.88	94.09	0.0244	0.1997	32.4	1.23	11.48	79.94	0.02556	0.1698
20	4.34	8.52	11.87	94.69	0.0264	0.1991	35.7	1.12	12.55	80.49	0.02783	0.1695
25	4.92	7.58	12.88	95.30	0.0285	0.1986	39.3	1.02	13.66	81.06	0.03008	0.1692
30	5.56	6.75	13.88	95.91	0.0306	0.1981	43.2	0.939	14.76	81.61	0.03233	0.1689
35	6.26	6.07	14.88	96.51	0.0326	0.1976	47.3	0.862	15.87	82.16	0.03458	0.1686
40	7.03	5.45	15.89	97.11	0.0346	0.1972	51.7	0.792	17.00	82.71	0.03680	0.1683
45	7.88	4.90	16.91	97.72	0.0366	0.1968	56.4	0.730	18.14	83.26	0.03903	0.1681
50	8.80	4.42	17.92	98.32	0.0386	0.1964	61.4	0.673	19.27	83.78	0.04126	0.1678
55	9.81	4.00	18.95	98.93	0.0406	0.1960	66.7	0.622	20.41	84.31	0.04348	0.1676
60	10.9	3.63	19.96	99.53	0.0426	0.1958	72.4	0.575	21.57	84.82	0.04568	0.1674
70	13.4	2.99	22.02	100.73	0.0465	0.1951	84.8	0.493	23.90	85.82	0.05009	0.1670
80	16.3	2.49	24.09	101.93	0.0504	0.1947	98.8	0.425	26.28	86.80	0.05446	0.1666
90	19.7	2.09	26.18	103.12	0.0542	0.1942	114.3	0.368	28.70	87.74	0.05882	0.0662
100	23.6	1.76	28.27	104.30	0.0580	0.1938	131.6	0.319	31.16	88.62	0.06316	0.1658
110	28.1	1.50	30.40	105.47	0.0617	0.1935	150.7	0.277	33.65	89.43	0.06749	0.1654
120	33.2	1.28	32.53	106.63	0.0654	0.1933	171.8	0.240	36.16	90.15	0.07180	0.1649
130	39.0	1.10	34.67	107.78	0.0691	0.1931	194.9	0.208	38.69	90.76	0.07607	0.1644

SOURCE: Baumeister T., and Marks, L. S., eds., *Standard Handbook for Mechanical Engineers*, Seventh edition, McGraw-Hill Book Co., New York, 1967.

References

1. Lewis, G. N., and Randall, M., *Thermodynamics*, 2nd Edition, revised by Pitzer and Brewer, McGraw-Hill, New York, 1961.
2. Zemansky, M. W., and Van Ness, H. C., *Basic Engineering Thermodynamics*, McGraw-Hill, New York, 1966.
3. Hougen, O. A., and Watson, K. M., *Chemical Process Principles, Part 2: Thermodynamics*, John Wiley & Sons, New York, 1947.
4. Doolittle, J. S., and Hale, F. J., *Thermodynamics For Engineers*, John Wiley & Sons, New York, 1983.
5. Burghardt, M. D., *Engineering Thermodynamics with Applications*, 2nd Edition, Harper and Row, New York, 1982.
6. Sonntag, R. E., Borgnakke, C., and Van Wylen, G. J., *Fundamentals of Thermodynamics*, Fifth Edition, John Wiley & Sons, New York, 1998.
7. Moran, M. J., and Shapiro, H. W., *Fundamentals of Engineering Thermodynamics*, Fourth Edition, John Wiley & Sons, New York, 2000.
8. Reynolds, W. C., and Perkins, A. C., *Engineering Thermodynamics*, 2nd Edition, McGraw-Hill, New York, 1977.

2.5 GEOLOGICAL ENGINEERING

Geology is the study of the Earth, its internal and surface composition, structure, and the processes that cause changes in its composition and structure. The Earth is constantly changing. The processes within the Earth and the history of these processes are important factors in determining how minerals deposits were formed, where they accumulated, and how they have been preserved. The composition and structure of the Earth and the history of the processes that resulted in the present geological settings of rocks are very important in the prediction of where accumulations of economically valuable hydrocarbons (oil and gas) may be found.

Studies of surface geological features and interpretations of past processes, coupled with surface geophysical investigation techniques such as seismic, gravity, magnetic, radioactive, electrical, and geochemical methods, are used to locate probable subsurface target regions that may contain economically valuable accumulations of hydrocarbons. However, only by drilling from the surface to these subsurface regions is it possible to definitely assess whether hydrocarbons exist and to determine quantitatively their distribution and composition. Drilling from the surface — whether on land or from floating platforms — allows drill holes to test subsurface rocks and provides a direct sampling of rocks and fluids from subsurface regions. These subsurface fluids, if present, can be assessed for their economic value.

Geology is important in exploring for hydrocarbons, and engineers must study the present composition and structure of the Earth to successfully drill boreholes. After hydrocarbons have been found and have proved to be economically recoverable, studies of the physical and chemical aspects of the Earth in such regions are important to production and reservoir engineering. These studies help to ensure that the accumulated hydrocarbons are recovered in an economic and a sustainable manner.

2.5.1 General Rock Types

The Earth is composed of three general rock types: igneous, sedimentary, and metamorphic. Reference 1 provides a general background of these three rock types and their relationships in the *rock cycle*, sometimes called the *rock recycle*.

Igneous rocks represent the solidified products of magmatism and volcanism — processes that bring molten materials close to the Earth's surface, allowing these materials to erupt on the surface (volcanic rocks) or freeze at shallow levels within (intrusive or plutonic rocks) the Earth. Such rocks comprise variable yet definable mineral assemblages, usually constituting relatively dense, compact, and weakly porous materials. Fracturing in igneous rocks may provide substantial permeability; however, because igneous rocks generally are not found in sedimentary basins that otherwise would preserve hydrocarbon accumulations, this rock type seldom hosts significant hydrocarbon concentrations.

Sedimentary rocks comprise materials derived from the weathering and erosion of preexisting rocks and soils. These materials have been variably consolidated and cemented (clastic sedimentary rocks) or precipitated directly from water (chemical sedimentary rocks). Particulate materials and dissolved rock components are transported by water, wind, and ice (as glaciers) to new locations, where they eventually assemble by means of the general process of lithification into a new rock mass. The eroded material from which sedimentary rocks are derived, called *protolith*, may be igneous, sedimentary, or metamorphic or some combination of these forms. During the process of weathering and erosion, especially by water, geochemical components of the original rock mass may be dissolved and may therefore be transported separately from the original rock fragments. This is an important characteristic of the genesis of sedimentary rocks because such dissolved components may reprecipitate — re-form from solution — and constitute the cementing agent in sedimentary rocks. These pore-filling cements are very important in determining the hydrocarbon storage capacity of sedimentary rock units and require assessment for any rock considered a potential hydrocarbon source.

Metamorphic rocks represent igneous, sedimentary, or other metamorphic rocks that have changed texturally and, in some cases, compositionally. Because the processes that engender these changes — called *metamorphism* — involve application of substantial heat and pressure to the original rock, changes may take place in the rock that preclude or inhibit completely the ability of the rock to produce or store hydrocarbons. These temperature-pressure effects may force rocks to endure conditions outside of the "petroleum window" defining suitable conditions for hydrocarbon generation and storage. These changes usually result in a reduction in pore volume by means of compaction and consolidation, including mineral deformation; therefore, metamorphic rocks, unless they are near the surface and well fractured, seldom make suitable storage media for hydrocarbons.

2.5.2 Historical Geology

Based on studies of radioactive isotopes derived from meteorites, the age of the Earth is estimated to be approximately 4,650 million years; the oldest known rocks, found in metamorphic rock assemblages in northern Canada and in Australia, are approximately 3,800 million years old [2]. Radiometric age dating, first devised in the early part of the 20th century, provides geologists with a means of establishing the absolute ages of rock units containing radioactive minerals that were formed at the same time as the enclosing rock. However, in the search for petroleum and gas resources and because most hydrocarbons occur in sedimentary rocks, petroleum geologists rely on the relative geological ages of rock units to unravel the sedimentary and structural history of potential hydrocarbon-bearing rocks.

Geologists generally use two characteristics of geological processes to determine the relative ages of rock units, especially of sedimentary rock sequences.

1. The *principle of uniformity* states that the internal and external processes affecting the Earth today have been operating unchanged and at the same rates throughout the developmental history of Earth. This means that an historical geological event preserved in the rock record can be identified and compared to similar events occurring at the present time in terms of the elapsed time and the geological processes necessary for that event to have occurred. Therefore, rates of deposition, erosion, igneous emplacement, and structural development are preserved in the geological column and can be compared with current similar processes [2,3].

2. *Relative time* is based on the occurrence of geological events relative to each other. Dating in this manner requires the development of a sequence of events that can be established on the basis of obvious consecutive criteria, as indicated later. This requires the geologist to identify a geological continuum such that the events within the sequence are sufficiently identifiable (events such as uplift, erosion, and deposition) and widespread (acting over geographically large areas on the scale of hundreds of square kilometers) to have practical significance. Such dating in relative time allows events to be identified throughout the world [2,3].

Examples of the types of observations used to establish relative time are given in the following list:

- *Superposition* is fundamental to the study of layered rocks. This means that in a normal layered sedimentary rock sequence, the oldest rocks were deposited first and are at the sedimentary base of the sequence. Younger rocks were deposited last and are at the top of the sequence.
- *Succession of flora and fauna* refers to the deposition of sedimentary material that includes the remains of contemporaneous plant and animal life. Fossils of these plants and animals may be preserved in the rock formations that result from the processes of deposition and lithification. The presence, absence, or change of the plant and animal life within a sequence of the geological column, as indicated by the presence, abundance, and diversity of fossil evidence, provides important information that allows geologists to correlate rock formations and thereby relative time from area to area. The fossil records within sedimentary sequences also provide important information regarding the evolution of life through geological time.
- *Inclusion* of one rock type within another indicates that the included rock is necessarily older than the rock in which it is enclosed. This is significant in interpretation of sedimentary sequences, especially those that have been structurally deformed by faulting or folding, because it allows a geologist to ascertain lateral and vertical continuity of rock units or structures that may comprise source regions or host rocks for hydrocarbons.
- *Cross-cutting relationships* are essential for the determination of the relative ages of rocks that show rock-to-rock contact. The idea is that any rock (especially igneous rocks) or structure (such as a fault or sedimentary bedding feature) that transects or "cuts" another must be younger than that feature. The feature that has been transected must have been present (older) before the cross-cutting feature (younger) existed.

- Physiographic development of the surface of the earth refers to the landforms and shapes of the landscape. These surface features are subject to continuous change from constructive (e.g., uplift, volcanic activity, deposition of sediments) and destructive (e.g., erosion) processes. Landform modifications are continuous and sequential. These modifications establish a predictable continuity that can be helpful in determining certain aspects of relative geologic ages.

The relationship between time units, time-rock units, and rock units is as follows:

Time Units	Time-Rock Units	Rock Units
Eon		
Era	Erathem	Group
Period	System	Formation
Epoch	Series	Member
Age	Stage	

This system for keeping track of these important units is used as the basis for the standard geologic time and the evolution of the animal life on earth. (See also Tables 2.5.1. and 2.5.2.) Table 2.5.3 gives the relationship between geologic time and important physical and evolutionary events that are used to aid in the identification of rock units in relative geologic time [3].

2.5.3 Petroleum Geology

Empirical field evidence has led geologists to conclude that nearly all economically important liquid and gas hydrocarbon accumulations are associated with sedimentary rocks. This evidence suggests that hydrocarbons represent thermally altered organic material derived from microscopic plant and animal life. This microscopic plant and animal life thrived in terrestrial and aquatic environments through geological time, but since the start of the Paleozoic era (see Table 2.5.1). Occasionally, these life forms were deposited in geochemically significant quantities along with fine-grain sediments, especially in marine environments along or adjacent to continental margins.

Organic material deposited with these marine sediments becomes entrained with the clastic debris and eventually forms sedimentary source rocks. Whether a sedimentary sequence becomes a substantial source of hydrocarbons is a function of a number of physical and geochemical parameters, including the origin and amount of original organic materials deposited, the burial and thermal history (see discussion of Equation 2.5.1) of the evolving sediment-to-rock mass, and the nature and continuity of permeability within the source rock. Conversion of organic matter to various liquid and gaseous hydrocarbons takes place as heat, pressure, and biological activity geochemically change the entrained organic materials in a reducing environment through a series of processes collectively called *maturation*. If geochemically significant oxygen is present during any stages of the hydrocarbon maturation process, at least some of the hydrocarbons will be destroyed by oxidation and will be converted to water plus carbon dioxide with or without sulfur.

Because trapping of organic materials with resultant prevention of organic material oxidation is essential to the generation of economically important hydrocarbon accumulations, rapid burial of organic-bearing sediments is geologically favorable; as such, sedimentary basins in which sediment input is high and basins in which reducing conditions are maintained over extended periods, along with

Table 2.5.1 *The Standard Geological Column [1]*

Relative Geologic Time				
Era	Period		Epoch	Atomic Time*
Cenozoic	Quaternary		Holocene	
			Pleistocene	
				—2–3—
	Tertiary		Pliocene	—12—
			Miocene	—26—
			Oligocene	—37–38—
			Eocene	—53–54—
			Paleocene	—65—
Mesozoic	Cretaceous		Late / Early	—136—
	Jurassic		Late / Middle / Early	
				—190–195—
	Triassic		Late / Middle / Early	—225—
Paleozoic	Permian		Late / Early	—280—
	Carboniferous	Pennsylvanian	Late / Middle / Early	
	Systems	Mississippian	Late / Early	
				—345—
	Devonian		Late / Middle / Early	—395—
	Silurian		Late / Middle / Early	—430–440—
	Ordovician		Late / Middle / Early	—500—
	Cambrian		Late / Middle / Early	—600—
Precambrian				3600

*Estimated ages of time boundaries (millions of years)

Table 2.5.2 *Geological Time and Evolution of Ancient Life [3]*

Era	Approx. Age in Millions of Years (Radioactivity)	Period or System — Period refers to a time measure; system refers to the rocks deposited during a period.
Cenozoic	7 / 26 / 37-38 / 53-54 / —65	Recent (Holocene) / Pleistocene / Pliocene / Miocene / Oligocene / Eocene / Paleocene — Neogene / Paleogene (Tertiary); Humans, Birds, Mammals
Mesozoic	136 / 190-195 / —225	Cretaceous / Jurassic / Triassic; Dinosaurs
Paleozoic	280 / 310 / 345 / 395 / 430-440 / 500 / —570	Permian / Pennsylvanian / Mississippian (Carboniferous) / Devonian / Silurian / Ordovician / Cambrian; Reptiles, Amphibians, Fish
Precambrian	700	First multi-celled organisms
	3,400	First one-celled organisms
	4,000	Approximate age of oldest rocks discovered
	4,500	Approximate age of meteorites

Table 2.5.3   *Geological Time and Important Events [3]*

Uniform Time Scale	Subdivisions Based on Strata/Time			Radiometric Dates (millions of years ago)	Outstanding Events	
		Systems/Periods	Series/Epochs		In Physical History	In Evolution of Living Things
Phanerozoic — 0 / 580	Cenozoic	Quaternary	Recent or Holocene Pleistocene	0 / —2?—	Several glacial ages	Homo sapiens
		Tertiary	Pliocene	—6—	Colorado River begins	Later hominids
			Miocene	—22—		Primitive hominids / Grasses; grazing mammals
			Oligocene	—36—	Mountains and basins in Nevada	
			Eocene	—58—	Yellowstone Park volcanism	Primitive horses
			Paleocene			
Phanerozoic / Precambrian	Mesozoic	Cretaceous		—63— / —145—	Rocky Mountains begin / Lower Mississippi River begins	Spreading of mammals / Dinosaurs extinct / Flowering plants / Climax of dinosaurs
		Jurassic		—210—		Birds
		Triassic		—255—	Atlantic Ocean begins	Conifers, cycads, primitive mammals / Dinosaurs
	Paleozoic	Permian		—280—	Appalachian Mountains climax	Mammal-like reptiles
		Pennsylvanian (Upper Carboniferous)		—320—		Coal forests, insects, amphibians, reptiles
		Mississippian (Lower Carboniferous)	(Many)	—360—		
		Devonian		—415—		Amphibians
		Silurian		—465—		Land plants and land animals
		Ordovician		—520—	Appalachian Mountains begin	Primitive fishes
		Cambrian		—580—		Marine animals abundant
	Precambrian (Mainly igneous and metamorphic rocks; no worldwide subdivisions.)			—1,000— / —2,000— / —3,000—	Oldest dated rocks	Primitive marine animals / Green algae / Bacteria, blue green algae
~4,650	Birth of Planet Earth			—4,650—		

input of elevated amounts of marine organic materials, are considered most prospective for hydrocarbon-source rocks. For this reason, geologists and geophysicists are employed to interpret the sedimentation and tectonic history and the lithologic settings of rock volumes that are considered favorable for petroleum and gas occurrence [3–6].

2.5.3.1 Source Rocks
Source regions for hydrocarbons are required to have a substantial input of organic material and, to preserve that material so that it may mature in to liquid and gaseous hydrocarbons, a geochemically reducing environment. Restricted sedimentary basins accumulating detrital continental debris and marine organic matter represent the most favorable environments for the storage and generation of hydrocarbons.

Generation of chemically precipitated sediments, notably limestones and related calcareous components of clastic rocks, usually results in at least initial trapping of organic matter. This matter may be released for migration after fracturing of the host carbonate or if the carbonate is subjected to thermal stress such that volatile organic materials are driven from the otherwise stable carbonate source rock. Other chemical precipitates such as gypsum (a sulfate), halite, and related halide minerals such as potash salts and anhydrite derived from gypsum may host organic matter as a burial

component and therefore may serve as source rocks under some sedimentary-geochemical conditions.

2.5.3.2 Migration
Organic matter buried in reducing environments (usually of marine nature) may accumulate in sufficient quantities to be considered geochemically and economically important source rocks. Because of thermal and load stresses applied to such sediments and the organic materials they host, this organic matter may change chemically and may be driven from the source rock along more permeable and porous pathways. These chemically and thermally changed hydrocarbons may, if structural and thermal conditions permit (see sections on "Structural Geology" and "Traps"), cease migration, accumulating in rock masses called *reservoir rock*.

This migration from source to reservoir is generated by hydrodynamic forces, thermal stress, or a combination of these factors. Water generally moves in conjunction with hydrocarbons and is usually segregated from gas and oil by buoyancy (density) contrast (Figure 2.5.1). Migration of hydrocarbons is attenuated or ceases entirely when applied stresses, the existing stress regime of enclosing rock, and hydrocarbon with or without water pore pressures reach transient equilibrium. Because such equilibrium is dynamic and therefore transient, fluids may continue to migrate if any of the factors controlling fluid movement are changed.

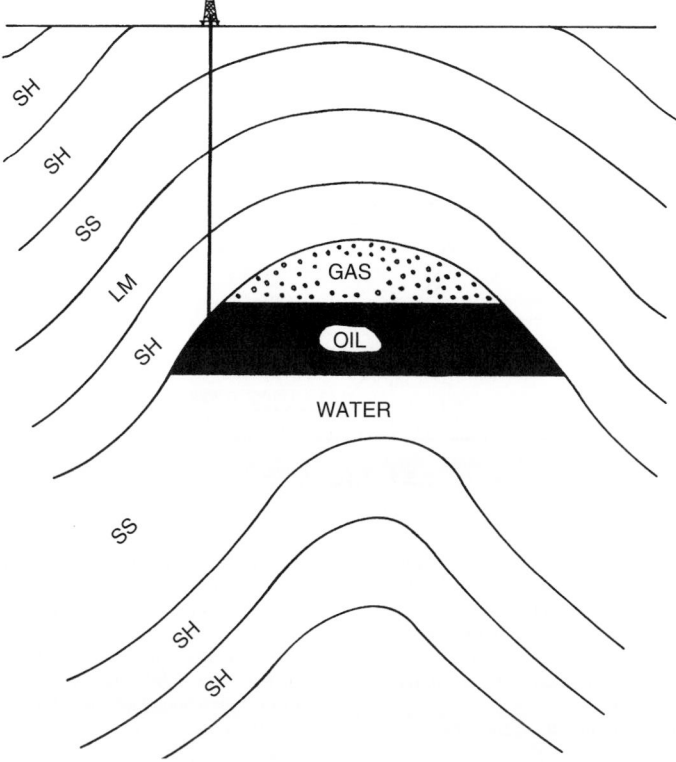

Figure 2.5.1 *Anticlinal hydrocarbon accumulation.*

2.5.3.3 Accumulation

Accumulation and storage of hydrocarbons occurs when structural settings appropriate to the preservation of oil and gas (see section on "Traps") receive hydrocarbons and, usually, water. The migration of hydrocarbons is strongly influenced by geological structures and the nature and lateral continuity of sedimentary rock bedding features (see Figure 2.5.10). Gases generally travel as dissolved components of liquid hydrocarbons or as a component of water. The amount of gas dissolved and therefore transported is a function of the temperature and pressure of the source region and adjacent rock masses through which the hydrocarbon-water fluids pass. After accumulation, hydrocarbons tend to stratify according to the relative densities of gas, oil, and water, with gases accumulating in the structurally highest portions of the reservoir, liquid hydrocarbons lower, and water lowest. Most reservoirs "leak" somewhat, with lighter components, such as gases and lighter liquid hydrocarbon fractions, capable of continuing migration, effectively leaving behind heavier and more viscous hydrocarbons.

After hydrocarbons accumulate in a rock mass serving as a reservoir, the hydrocarbons must be preserved from oxidation, from escape along other structural avenues, and from bacterial activity that might convert the hydrocarbons to sulfur, as in sulfur deposits of west Texas, the Gulf Coast region of the southern United States, the Misraq area of eastern Iraq, and southern Poland. Because the Earth is dynamic, the stability of any reservoir rock is transitory, and rocks that otherwise appear to be excellent storage media may no longer host hydrocarbons because of changes in local or regional structural, tectonic, or sedimentary regimes.

2.5.4 Structural Geology

The *law of original horizontality* states that all sediments are deposited essentially horizontally in response to gravity and the approximate Stoke's law of settling of particles. Sedimentary rocks, originally deposited as horizontal or near-horizontal sediment beds, undergo diagenetic changes that result in variable lithification. Because of the dynamic nature of Earth forces, sedimentary rocks are subject to myriad stresses and changes in stress regimes. Rock masses respond to such stress and stress changes by deforming, with deformation taking place on scales varying from microscopic to hundreds of kilometers.

Sedimentary rocks generally show deformation by the rearrangement of mineral grains (microscopic) and the larger-scale folding or fracturing of rock units; such folding and fracturing display variations in scale from microscopic offset of mineral grains to folds having wavelengths measured in tens to hundreds of kilometers.

These structural features of deformed sedimentary rock sequences comprise some of the most important reservoirs for hydrocarbons and water and are therefore of significant interest to petroleum geologists and geophysicists in the search for petroleum resources [3].

1. Faults are breaks in the Earth's crust along which there has been measurable movement, called *displacement*, of rock on one side of the fault relative to the other (Figure 2.5.2). Some definitions describing the most important structural features of faulted sedimentary rocks follow:

   - *Dip* — the angle the fault plane makes with the horizontal, measured from the horizontal to the fault plane.

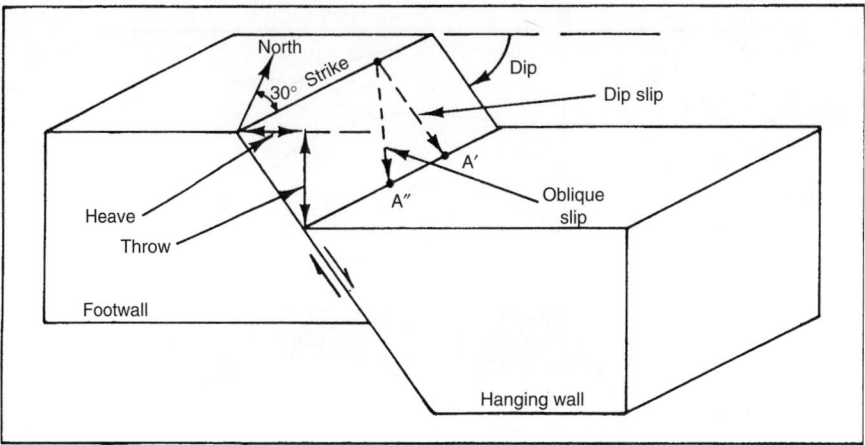

Figure 2.5.2 *Fault terminology [3].*

- *Strike* — a line on the horizontal surface represented by the intersection of the fault plane and the horizontal surface. The strike line is always horizontal, and since it has direction, it is measured either by azimuth or bearing. Strike is always perpendicular to the dip.
- *Heave* — the horizontal component of movement of the fault.
- *Throw* — the vertical component of movement of the fault.
- *Slip* — the actual linear movement along the fault plane.
- *Hanging wall* — the block located above and bearing down on the fault surface.
- *Footwall* — the block that occupies the position beneath the fault, regardless of whether the hanging wall has moved up or down.
- *Normal faulting* takes place in extensional regimes in which tension within rock masses produces slip such that the hanging wall block is displaced downward with respect to the footwall block Figure 2.5.2 is an example of a normal fault.
- *Reverse (thrust) faulting* takes place in generally compressional structural regimes, and produces displacement such that the hanging wall is thrust over footwall rocks.
- *Strike-slip faulting* occurs in structural environments in which more or less horizontally directed forces cause structural blocks to slide laterally along the fault surface. Transform faults are special cases of strike-slip faults, occurring as differential spreading takes place along mid-ocean ridges.

2. Displacement along faults varies from almost zero to tens of kilometers for normal and reverse faults and up to hundreds of kilometers for strike-slip structures. Faults seldom consist of only single surfaces; they typically are made of myriad faulted surfaces comprising a fault zone, along which each of the faults takes up — accommodates — a portion of the overall displacement. Because fault surfaces and the zones they comprise may represent relatively low permeability zones, such structural zones may serve as effective traps for hydrocarbons and water (see Figure 2.5.10).

3. Folds in sedimentary rocks are generated when rock mass strength and the rate at which pressure is applied to a sedimentary rock sequence are such that the rock is able to reaccommodate an applied stress, accumulating strain energy as deformed mineral grains. In some cases, because the rate of applied pressure is changed, folded rock sequences may rupture and fault, with stress being taken up by movement along the fault instead of as deformed mineral grains. Folds are generated on scales varying from millimeters to kilometers Figure 2.5.3.

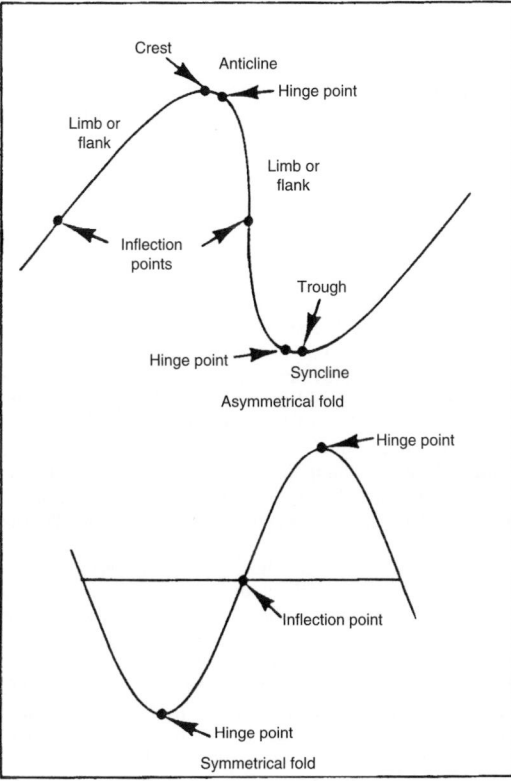

Figure 2.5.3 *Folding terminology [3].*

Fold terminology follows:

- *Anticline* — a fold with upward convexity.
- *Syncline* — a fold that is concave upward.
- *Hinge point* — the point of maximum curvature of a fold. The *hinge surface* is the locus of hinge lines within the fold.
- *Inflection point* — occurs when bed curvature in one direction changes to bed curvature in the opposite direction.
- *Limbs (or flanks) of a fold* — those portions adjacent to the inflection points of the fold.
- *Symmetrical fold* — a fold whose shape is a mirror image across the hinge point.
- *Asymmetrical fold* — a fold whose shape is not a mirror image across the huge point.
- *Recumbent fold* — characterized by a horizontal or nearly horizontal hinge surface (see Figure 2.5.4).
- *Overturned fold* — when the hinge surface is depressed below the horizontal (see Figure 2.5.4).
- *Concentric (parallel) folds* — rock formations parallel to each other such that their respective thicknesses remain constant (see Figure 2.5.5).
- *Nonparallel folds* — rock formations that do not have constant thickness along the fold (Figure 2.5.6).
- *Similar folds* — folds that have the same geometric form, but where shear flow in the plastic beds has occurred (Figure 2.5.7).
- *Disharmonic folds* — folds in layered rock that have variable thickness and competence and, thereby, fold in accordance to their ability (Figure 2.5.8).

2.5.5 Traps

Traps are structural features that represent permeability contrasts such that the migration of hydrocarbons or water is attenuated to an extent that permits temporary accumulation and storage. Because traps are essentially always of

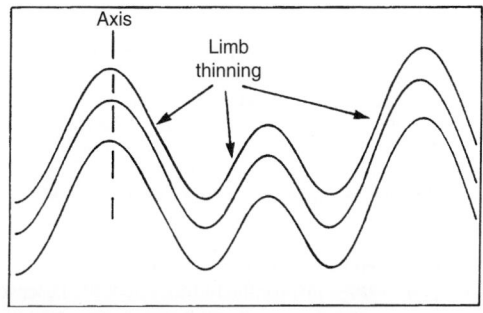

Figure 2.5.6   *Nonparallel folds [3].*

Figure 2.5.7   *Similar folds [3].*

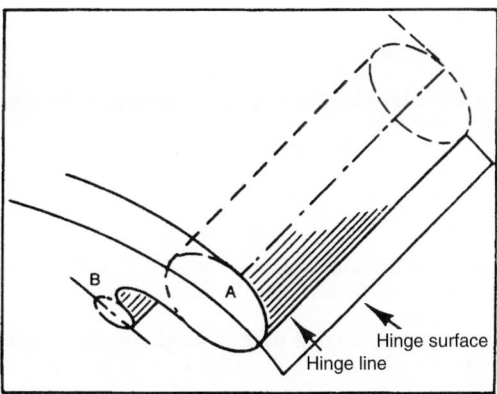

Figure 2.5.4   *Recumbent (A) and overturned (B) folds [3].*

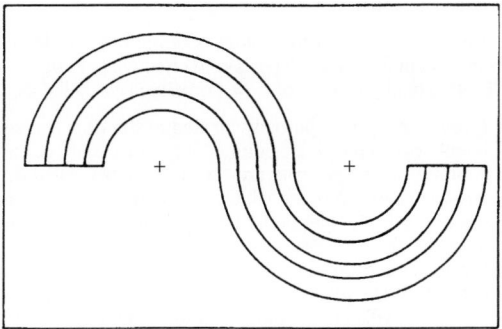

Figure 2.5.5   *Concentric folds [3].*

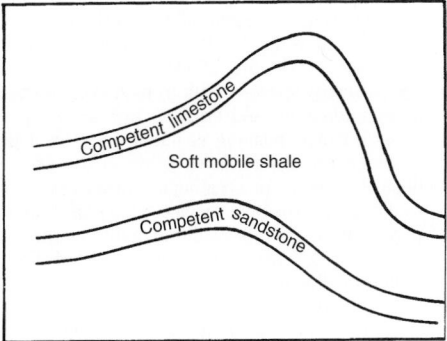

Figure 2.5.8   *Disharmonic folds [3].*

structural or stratigraphic origin, they may have finite life-times; if trap morphology, permeability, or size changes, the ability of a trap to store hydrocarbons and water may be altered substantially.

1. Structural traps result from deformation of rock masses, usually attributable to folding or faulting.

   - *Anticline trap* — because of the upwarping of sediments in this fold type (see Figures 2.5.1 and 2.5.8), hydrocarbons and water tend to accumulate in fold crests; as such, this type of fold represents one of the most important productive structural traps for hydrocarbons. In some cases, faulting, especially thrust faulting, may complicate or generate anticlinal structures and the traps they may form (Figure 2.5.9). The fold-and-thrust belt of the western United States (also called the overthrust belt) represents an excellent example of composite folding and faulting, and associated hydrocarbon production.
   - *Fault traps* — involve the movement of the reservoir rock formation to a position where the formation across the fault plane provides a seal preventing further migration of hydrocarbons (Figure 2.5.10).
   - *Salt-related traps* — formed when the plastic salt formations deform into domelike structures under the overburden forces of the beds above the salt beds. Such plastic flowing (and bulging) of the salt beds deforms the rock formations above producing anticline structures and faults in the rock formation astride the domelike structures (Figures 2.5.11 and 2.5.12).

2. *Stratigraphic traps* consist of permeability contrasts in sedimentary rocks. These contrasts might have been created by sedimentary processes associated with deposition of the original sediments; by post-depositional (diagenetic) processes, including lithification mechanisms; or by a combination of these processes. Biological activity also may form significant stratigraphy-related traps, such as reefs and fossiliferous carbonate banks.

   - *Sand body traps* — finite sand bodies such as channel sands, river delta sands, and sea or ocean beach or barrier bar sands. These sand body traps are deposited over well-defined regions. As deposition continues on a wider regional basis of shale-forming deposits, the sand body becomes enclosed by shale and becomes a trap for fluids, particularly hydrocarbons (Figure 2.5.13).
   - *Reef traps* — important hydrocarbon-producing geological features. The porosity and permeability in reefs can be excellent. As in sand body traps, reef traps are finite bodies that are deposited over well-defined regions. Continued deposition of silt and clay materials will eventually enclose such features in shale, allowing them to trap fluids, particularly hydrocarbons (Figure 2.5.14).
   - *Unconformity traps* — result from alternating periods of sedimentation and erosion; that is, the process of sediment deposition is interrupted by a period in which erosion removes some of the previously deposited sediment, creating a contrast in permeability along the erosional surface. This erosional period may occur in response to tilting of the sediment package (Figure 2.5.15) or as a consequence of climate or structural change.
   - *Combination traps* — sedimentary trap features that result from both stratigraphic and structural mechanisms. There can be many combinations for stratigraphic and structural traps. An example of such a

trap would be a reef feature overlaying a porous and permeable sandstone but in which the sequence has been faulted (Figure 2.5.16). Without the fault, which has provided an impregnable barrier, the hydrocarbons would have migrated further up dip within the sandstone.

2.5.6 Basic Engineering Properties of Rock

Because sedimentary rocks constitute the most important source and storage rocks for hydrocarbons and water, petroleum engineers emphasize the significance of understanding the nature of sedimentary rocks and the depositional environments that formed them.

Most liquid and gaseous hydrocarbons are generated from — and represent — the former remains of microscopic marine plant and animal life. Continental plant life, comprising grasses and woody materials, tends to generate coal and some low-molecular-weight gases rather than liquid hydrocarbons; this is attributable to the environment in which coal materials are accumulated, the size of the carbon molecules making up woody plants and grasses, and the nature of the sedimentary rocks with which the precursor coal organic materials are associated. Because marine life forms exist in the same aquatic environment that produces sediments and consequently sedimentary rocks, hydrocarbon genesis, migration, storage, and preservation are related closely to sedimentary processes. The two properties with which we will now be concerned are permeability and porosity; although these are different characteristics of hydrocarbon-host rocks, they are each significant in determining whether a rock unit may serve or might have served as a source or storage media for hydrocarbons.

2.5.6.1 Porosity

Porosity is a measure of the void space within a rock, which is expressed as a fraction (or percentage) of the bulk volume of that rock [8].

The general expression for porosity ϕ is

$$\phi = \frac{V_b - V_s}{V_b} = \frac{V_p}{V_b} \qquad [2.5.1]$$

where V_b is the bulk volume of the rock, V_s is the volume occupied by solids (also called grain volume), and V_p is the pore volume.

From an engineering point of view, porosity is classified as:

- *Absolute porosity* — total porosity of a rock, regardless of whether the individual voids are connected.
- *Effective porosity* — only that porosity due to voids that are interconnected.

It is the effective porosity that is of interest. All further discussion of porosity will pertain to effective porosity.

From a geologic point of view, porosity is classified as:

1. *Primary porosity* — porosity formed at the time the sediment was deposited. Sedimentary rocks that typically exhibit primary porosity are the *clastic* (also called *fragmental* or *detrital*) rocks, which are composed of erosional fragments from older beds. These particles are classified by grain size.
2. *Secondary porosity* — voids formed after the sediment was deposited. The magnitude, shape, size, and interconnection of the voids bears little or no relation to the form of the original sedimentary particles. Secondary porosity is subdivided into three classes.

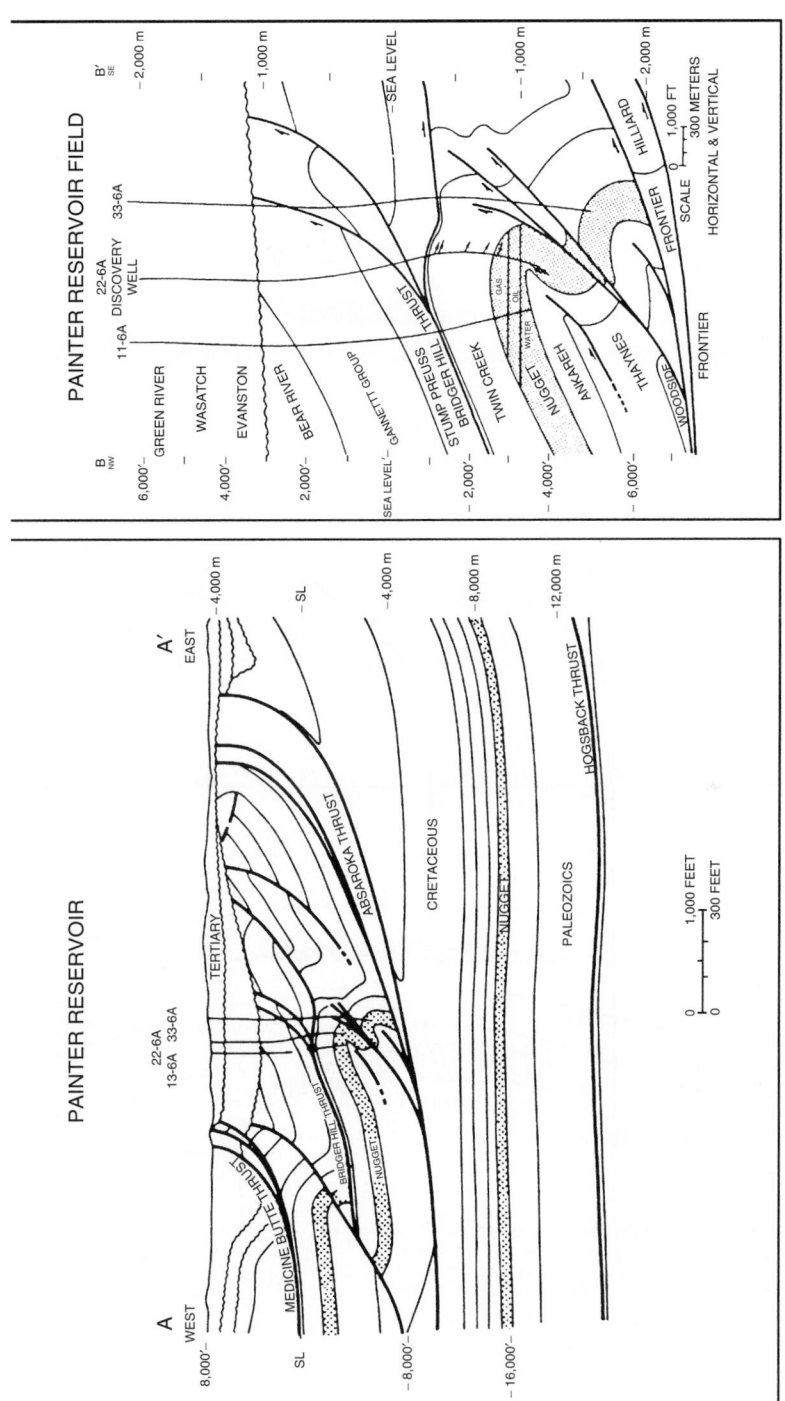

Figure 2.5.9 *Overthrust structural trap (Painter Reservoir, Wyoming) [3].*

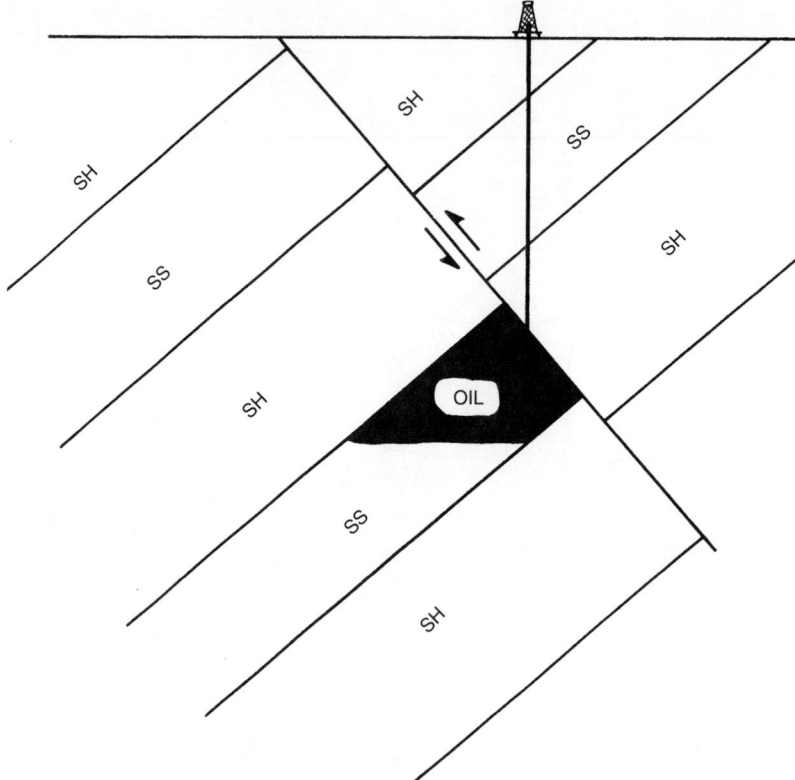

Figure 2.5.10   *Fault structural traps.*

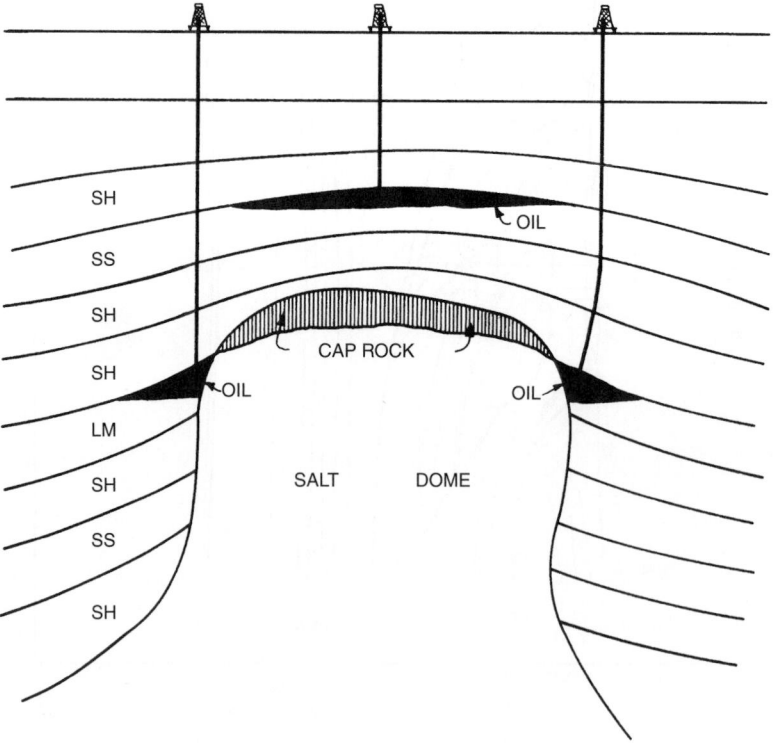

Figure 2.5.11   *Salt dome structural traps.*

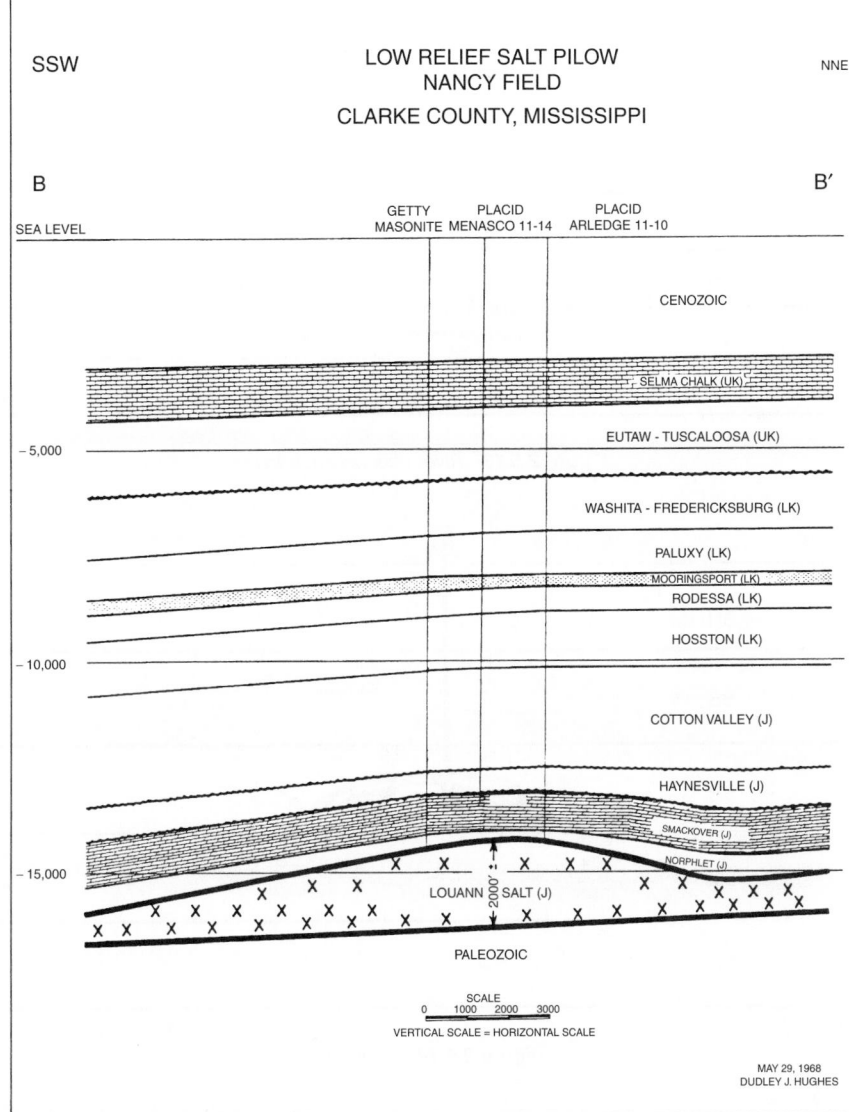

Figure 2.5.12 *Low-relief salt pillow trap (Nancy Field, Mississippi) (from Gulf Coast Association of Geological Societies).*

- *Solution porosity* refers to voids formed by the solution of the more soluble portions of the rock in the presence of subsurface migrating (or surface percolating) waters containing carbonic and other organic acids. Solution porosity is also called *vugular porosity* where individual holes are called vugs.
- *Fractures, fissures, and joints* represent planar surfaces formed in response to applied stresses that produce local rock failure. These structures may be of tensional or compressional nature, and they may individually contribute nominal to volumetrically substantial *permeability* to a rock mass by locally enhancing the connection between pores or between porous rock units.
- *Dolomitization* is the process by which limestones, dominated by the mineral calcite, are transformed into

a Ca-Mg carbonate, dolomite [Ca, Mg $(CO_3)_2$]. This process involves a significant change in rock density; because dolomite has a greater density than limestone, rocks undergoing dolomitization develop voids. These voids may result in very substantial increase in rock porosity (and permeability), making some dolomites excellent hydrocarbon reservoirs (e.g., the Ordovician-age Ellenberger dolomite of West Texas).

The typical value of porosity for a clean, consolidated, and reasonably uniform sand is 20%. The carbonate rocks (limestone and dolomite) normally exhibit lower values (e.g., 6–8%). These are approximate values and do not fit all situations. The principal factors that complicate intergranular porosity magnitudes are uniformity of grain size, degree of cementation, packing of the grains, and particle shape.

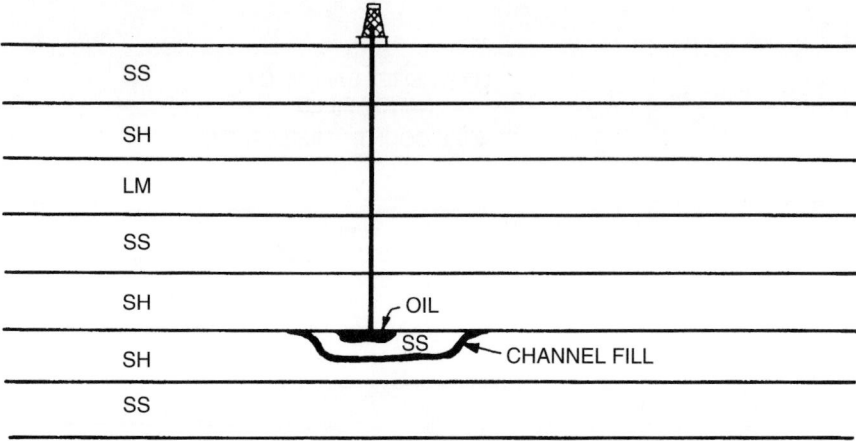

Figure 2.5.13 *River channel sand trap.*

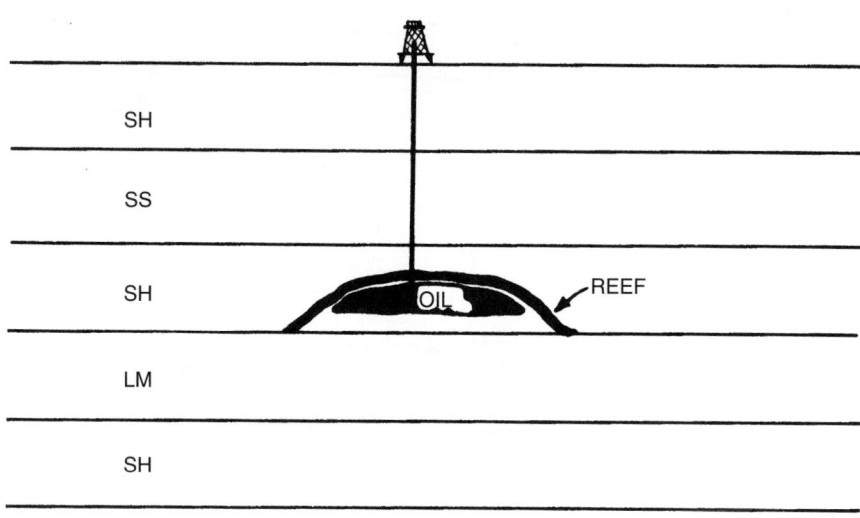

Figure 2.5.14 *Reef trap.*

2.5.6.2 Permeability

Permeability is defined as a measure of the ability of a rock to permit passage of fluids. In addition to rock porosity, the permeability of a rock unit is definitive in terms of allowing the migration accumulation, and storage of hydrocarbons and water in rock media. Fluids are transmitted through rock because of differential pressures created in the normal burial and tectonic environments of subsurface rocks. Permeability reflects the connected character of pores, fractures, joints, and faults within a rock unit; because each of these structures permits fluid flow to some extent, the overall nature of these structures to serve as interconnected passageways for fluids also represents the capacity of a rock mass to transmit fluid in response to pressure differences. The quantitative definition of permeability was first given in an empirical relationship developed by the French hydrologist Henry D'Arcy, who studied the flow of water through unconsolidated sands [8].

This law in differential form is

$$v = -\frac{k}{\mu}\frac{dp}{d\ell}$$

[2.5.2]

where v is the apparent flow velocity (cm/s), μ is the viscosity of the flowing fluid (centipoise), p is pressure (atmospheres), ℓ is the length (cm), k is permeability of the porous media (darcies).

Consider the linear flow system of Figure 2.5.17. The following assumptions are necessary to establish the basic flow equations:

- Steady-state flow conditions.
- Pore volume is 100% filled with flowing fluid; therefore, k is the absolute permeability.
- Viscosity of the flowing fluid is constant. In general, this is not true for most real fluids. However, the effect is negligible if μ at the average pressure is used.
- Isothermal conditions prevail.
- Flow is horizontal and linear.
- Flow is laminar.

Using the foregoing restrictions:

$$v = \frac{q}{A}$$

[2.5.3]

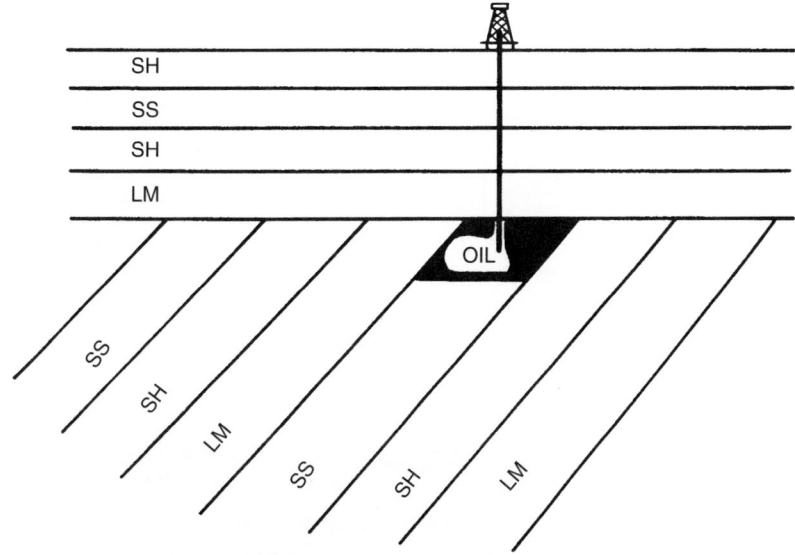

Figure 2.5.15   *Unconformity trap.*

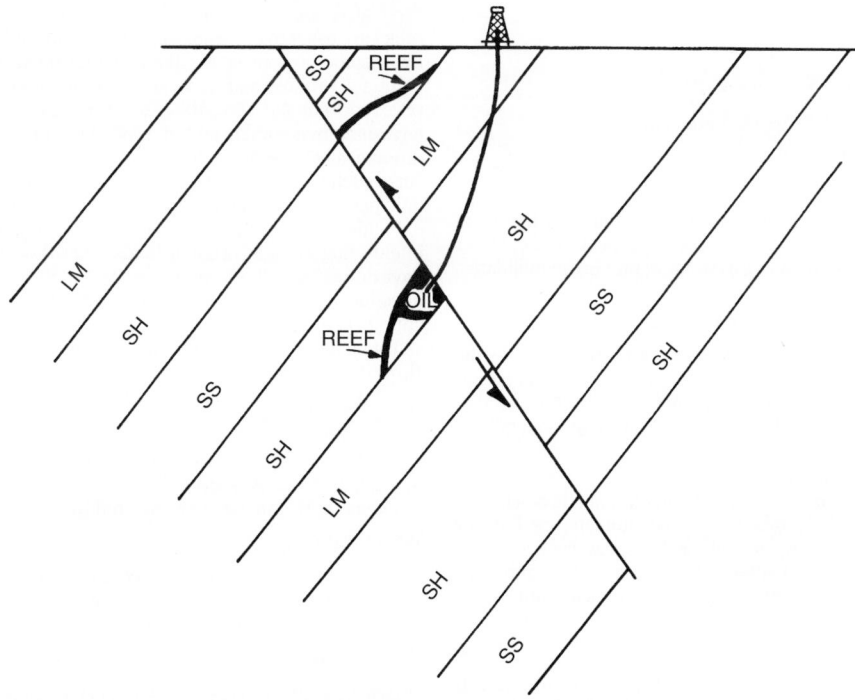

Figure 2.5.16   *Combination trap.*

where q is the volumetric rate of fluid flow (cm³/s); A is the total cross-sectional area perpendicular to flow direction (cm²).

This further assumption concerning velocity and the volumetric rate of flow restricts flow to the pores and not the full area. Therefore, v is an apparent velocity. The actual velocity, assuming a uniform medium, is

$$V_{actual} = \frac{V_{apparent}}{\phi} \qquad [2.5.4]$$

where ϕ is porosity defined in Equation 2.5.1.

Substituting Equation 2.5.3 into Equation 2.5.2 yields

$$\frac{q}{A} = -\frac{k}{\mu}\frac{dp}{d\ell} \qquad [2.5.5]$$

Separation of variables and using limits from Figure 2.5.17 gives

$$\frac{q}{A}\int_0^\ell d\ell = -\frac{k}{\mu}\int_{p_1}^{p_2} dp \qquad [2.5.6]$$

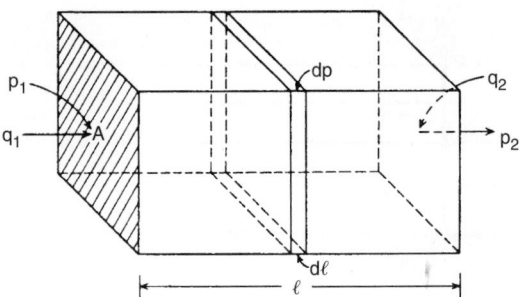

Figure 2.5.17 *Linear flow system.*

Integrating the preceding yields

$$q = \frac{kA(p_1 - p_2)}{\mu\ell} \qquad [2.5.7]$$

or

$$k = \frac{q\mu\ell}{A\Delta p} \qquad [2.5.8]$$

Equations 2.5.7 and 2.5.8 are the basic forms of the permeability relationship, and the following example serves to define the darcy unit:

If $q = 1 \text{ cm}^3/\text{s}$

$$A = 1 \text{ cm}^3$$

$$\mu = 1 \text{ cp}$$

$$\Delta p/\ell = 1 \text{ atm/cm}$$

then from Equation 2.5.8,

$$k = 1 \text{ darcy}$$

A permeability of 1 darcy is much higher than that commonly found in sedimentary rock, particularly reservoir rocks. Consequently, a more common unit is the millidarcy, where

$$1 \text{ darcy} = 1{,}000 \text{ millidarcies}$$

Typical values for sedimentary rock permeability for the flow of hydrocarbons and other fluids are 100 millidarcies (md) or greater. Rocks exhibiting permeabilities of 50 md or less are considered tight relative to the flow of most fluids.

2.5.6.3 Subsurface Temperature

Attributable to many causes, including lithostatic and hydrostatic pressures, frictional energy, accumulated and locally released strain energy, and radioactive decay, rock temperatures show a general tendency to increase with depth. This temperature depth relationship is commonly assumed to be a linear function.

$$t_d = t_s + \beta d \qquad [2.5.9]$$

where t_d is the temperature of the rock at depth, $d(^\circ F)$, t_s is the average surface temperature ($^\circ F$), β is the temperature gradient ($^\circ F/\text{ft}$), and d is the depth (ft).

In general, there is considerable variation in the geothermal gradient throughout the United States and the world. In many regions of the world where there is evidence of rather thin crust, the relationship between temperature at depth and depth may not be approximated by the linear function given in Equation 2.5.9. The increase in temperature with depth has important consequences for drilling and production equipment that is used in the petroleum industry. The viscosity of drilling and production fluids will, in general decrease with high temperatures. The sitting time for well cement will generally decrease with increased temperature,

as will the strength and fracture toughness of steels used in drilling, well completion, and production.

Temperature logs can be taken as wells are drilled and the temperature gradient determined for the particular region. These temperature logs taken at depth are used to determine the types of drilling fluids used as drilling progresses. The temperatures at depth will determine the cements used in well completion operations.

The average geothermal gradient used in most areas of the United States for initial predictions of subsurface temperatures is a value of $0.016^\circ F/\text{ft}$ [9].

Example 2.5.1

Determine the temperature at 20,000 ft using the approximate geothermal gradient $\beta \approx 0.016^\circ F/\text{ft}$.

Assume $t_s \approx 60^\circ F$.

$$t_d = t_s + 0.016d$$

$$t_d = 60 + 0.016 \,(20{,}000)$$

$$t_d = 380^\circ F$$

2.5.6.4 Subsurface Fluid Pressure (Pore Pressure Gradient)

The total pressure acting on a subsurface rock mass and the fluids it contains is a function of the weight of the rock (called *lithostatic pressure*) and fluid (called *hydrostatic pressure*) affecting the given subsurface rock mass. Because rocks are essentially "squeezed" by tectonic, lithostatic, and hydrostatic pressures, the fluids within rocks develop and exhibit a variable and generally substantial fluid pressure, referred to as *pore pressure*. Pore pressures represent the environmental equilibrium of a rock-fluid mass and change dramatically in response to local aberrations in applied pressure, such as the changes engendered when a surface drill hole penetrates a rock mass.

Fluids within rock pores may be inherited, such as sediments that contain original fluids from their depositional environments, called *connate waters*, or the fluids may be introduced, such as the gas-oil-water media that migrate into appropriate rock masses.

Total theoretical maximum overburden pressure, P (lb/ft^2), is

$$P_{ob} = \frac{W_r + W_{sw}}{A} \qquad [2.5.10]$$

where W_r is weight of rock particle grains (lb), W_{sw} is weight of water (lb), and A is area (ft^2).

The term W_r can be approximated by

$$W = (1 - \phi)Ad\gamma_m \qquad [2.5.11]$$

where ϕ is the fractional porosity, d is depth (ft), and γ_m is average mineral specific weight (lb/ft^3).

The term W can be approximated by

$$W_{sw} = \phi Ad\gamma_{sw} \qquad [2.5.12]$$

where γ_{sw} is the average saltwater specific weight (lb/ft^3).

Substitution of Equations 2.5.11 and 2.5.12 into Equation 2.5.10 yields

$$P_{ob} = (1 - \phi)d\gamma_m + \phi d\gamma_{sw} \qquad [2.5.13]$$

Equation 2.5.13 can be rewritten in terms of the specific gravities of average minerals S_m and salt water S_{sw}:

$$P = (1 - \phi)dS_m\gamma_w + \phi dS_{sw}\gamma_w \qquad [2.5.14]$$

where γ_w is the specific weight of fresh water (i.e., $62.4 \text{ lb}/\text{ft}^3$).

The average specific gravity of minerals in the earth's crust is taken to be 2.7. The average specific gravity of saltwater is taken to be 1.07. If the average sedimentary rock porosity is assumed to be 10%, then the total theoretical maximum overburden pressure gradient $(lb/ft^2)/ft$ becomes

$$\frac{P_{ob}}{d} = (1 - 0.10)(2.7)(62.4) + 0.10(1.07)(61.4)$$

[2.5.15]

$$\frac{P_{ob}}{d} = 158.3$$

Equation 2.5.15 can be expressed in normal gradient terms of psi/ft. Equation 2.5.15, which is the theoretical maximum overburden pressure gradient, becomes

$$\frac{p_{ob}}{d} = 1.10$$

[2.5.16]

where p_{ob} is pressure in (psi).

The foregoing theoretical overburden pressure gradient assumes that the sedimentary deposits together with the saline water are a mixture of materials and fluid. Such a mixture could be considered as a fluid with a new specific weight of

$$\gamma_f = 158.3 lb/ft$$

which in terms of drilling mud units would be

$$\bar{\gamma}_f = \frac{158.3}{7.48} = 21.1 ppg$$

where $1\ ft^3 = 7.48$ gal.

Immediately after and during the deposition of sediments in an aqueous environment, diagenetic processes may begin to cement (or dissolve) mineral grains, depositing by means of solution newly formed minerals within the interstices of the grains composing the original sediment. The effect of adding a cementing agent to otherwise unconsolidated sediment results in a substantial reduction in pore volume and concomitant reduction in permeability. If the pore space in a rock were occupied by cementing agents and water, and if the pores were connected, the calculated minimum pressure gradient applied to that rock would be that of the overlying column of water, given as follows:

$$\frac{P}{d} = 1.07(62.4) \quad \text{or}$$

$$\frac{P}{d} = 66.8$$

[2.5.17]

Equation 2.5.17 can be expressed in normal gradient terms of psi/ft. Equation 2.5.17 which is the minimum pressure gradient, becomes

$$\frac{p}{d} = 0.464$$

[2.5.18]

The foregoing minimum pressure gradient assumes also that the sedimentary column pores are completely filled with saline water and that there is communication from pore to pore within the rock column from surface to depth.

Figure 2.5.18 shows a plot of the theoretical maximum overburden pressure and the theoretical minimum pressure as a function of depth. Also plotted are various bottomhole fluid pressures from actual wells drilled in the Gulf Coast region [10]. These experimentally obtained pressures are the measurements of the pressures in the fluids that result from a combination of rock overburden

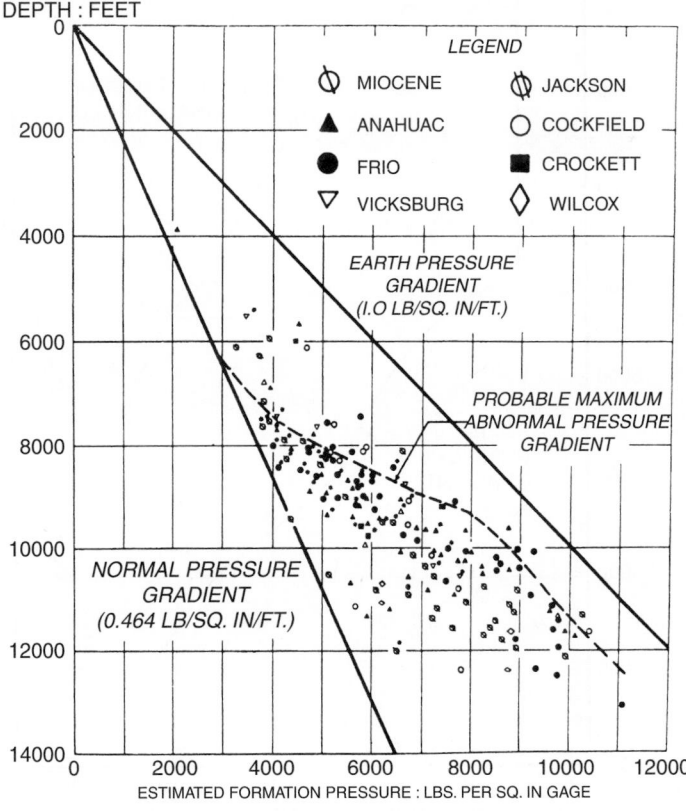

Figure 2.5.18 *Magnitude of abnormal pressure encountered in Gulf Coast region (from Cannon and Sullins, "Problems Encountered in Drilling Abnormal Pressure Formations," in API Drilling and Production Practices, 1946).*

and the fluid hydraulic column to the surface. These data show the bottomhole fluid pressure extremes. The abnormally high pressures can be explained by the fact that the sedimentary basins in the Gulf Coast region are immature basins and are therefore unconsolidated relative to older basins. In such basins, the cementing process is by no means complete, which results in pressures at depths approaching the maximum theoretical overburden pressures.

The fluid pressure in the rock at the bottom of a well is commonly defined as pore pressure (also called formation pressure, or reservoir pressure). Depending on the maturity of the sedimentary basin, the pore pressure will reflect geological column overburden that may include a portion of the rock particle weight (i.e., immature basins), or a simple hydrostatic column of fluid (i.e., mature basins). The pore pressure and therefore its gradient can be obtained from well log data as well are drilled. These pore pressure data are fundamental for the solution of engineering problems in drilling, well completions, production and reservoir engineering.

Because the geological column of sedimentary rock is usually filled with saline water, the pore pressure and pore pressure gradient can be obtained for nearly the entire column. Figure 2.5.19 shows a typical pore pressure gradient versus depth plot for a Gulf Coast region well.

2.5.6.5 Subsurface Rock Fracture Pressure (Fracture Pressure Gradient)

The Subsurface rock fracture pressure can be approximated by utilizing the known pore pressure at the same depth. The relationship between rock fracture pressure p_f (psi) and pore pressure p_p (psi) is [11]

$$p_f = (\sigma_{ob} - p_p)\left(\frac{v}{1-v}\right) + p_p \qquad [2.5.19]$$

where σ_{ob} is overburden stress (psi) and v is Poisson's ratio.

The subsurface rock fracture pressure gradient is

$$\frac{p_f}{d} = \left(\frac{\sigma_{ob}}{d} - \frac{p_p}{d}\right)\left(\frac{v}{1-v}\right) + \frac{p_p}{d} \qquad [2.5.20]$$

where d is the depth to the subsurface zone (ft).

Figure 2.5.20 shows the variation of Poisson's ratio versus depth for two general locations, the West Texas region and the Gulf Coast region.

The constant value of 0.25 for Poisson's ratio versus depth reflects the geology and the rock mechanics of the mature sedimentary basin in the West Texas region. Since mature basins are well cemented, the rock columns of West Texas will act a compressible, brittle, elastic materials.

The Cenozoic portions of the Gulf Coast sedimentary basins are immature; therefore, little cementing of the sediments has taken place. Poisson's ratio varies with depth for

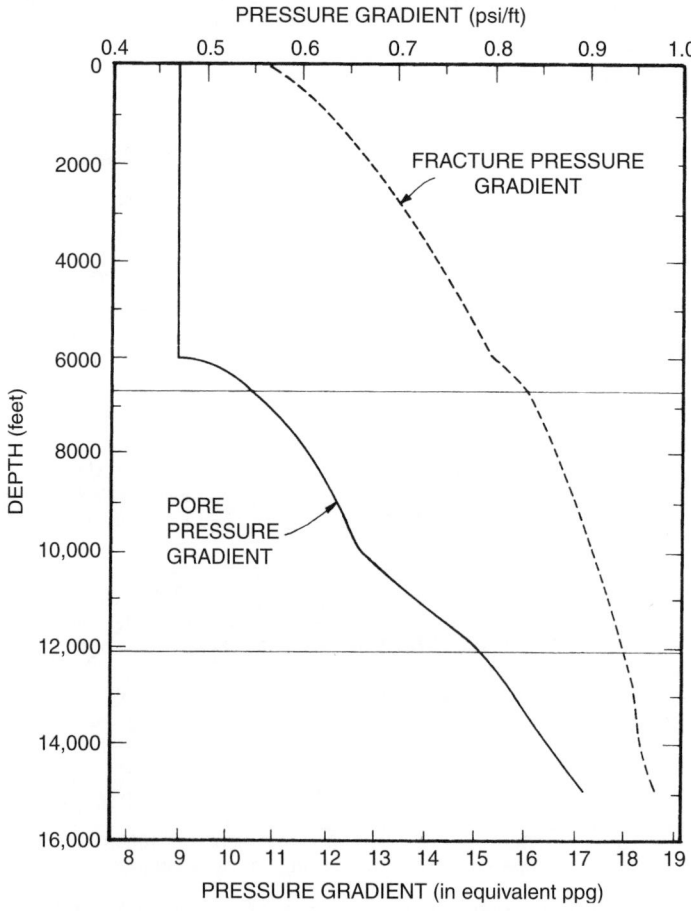

Figure 2.5.19 Pressure gradients vs. depth (Gulf Coast example).

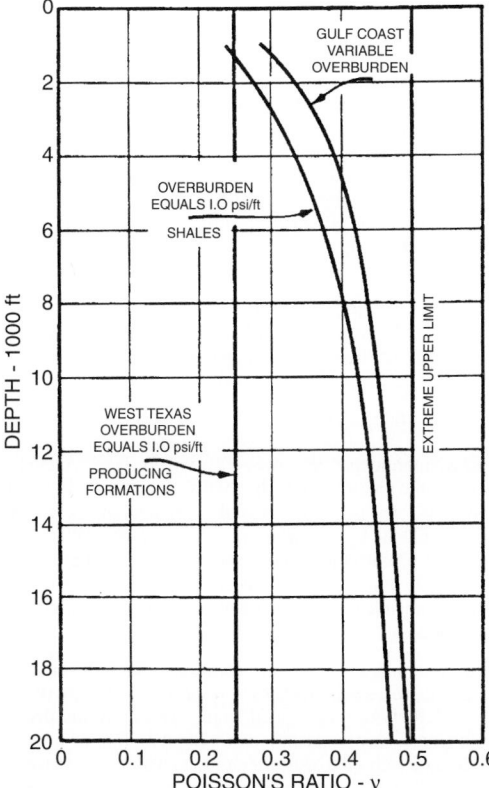

Figure 2.5.20 *Poisson's ratio vs. depth (from Eaton, "Fracture Gradient Prediction and Its Application in Oilfield Operations," Journal of Petroleum Technology, October 1969).*

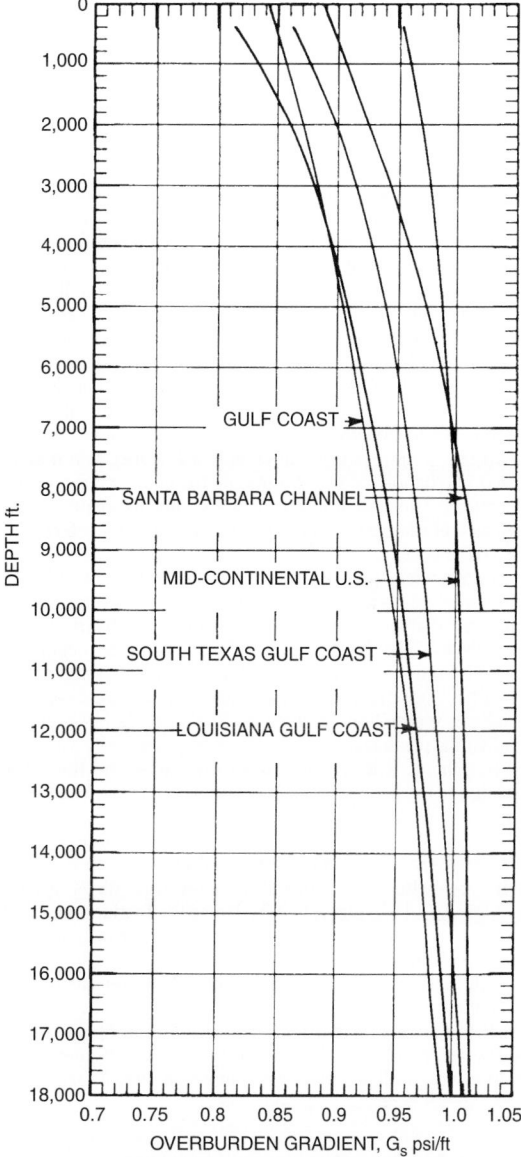

Figure 2.5.21 *Total overburden stress gradient vs. depth (from Engineering of Modern Drilling, Energy Publication Division of Harcourt Brace Jovanovich, New York, 1982, p. 82).*

such sedimentary columns, reflecting the variation of properties through the column. At great depth (i.e., approaching 20,000 ft), Poisson's ratio approaches that of incompressible, plastic materials (i.e., 0.5) [12].

Figure 2.5.21 gives typical total overburden stress gradients versus depths for several regions in North America [13].

The rock fracture pressure gradient at depth can be approximated by using Equation 2.5.20 and the variable Poisson's ratios versus depth data (Figure 2.5.20) and the variable total overburden stress gradients versus depth data (Figure 2.5.21).

Example 2.5.2

In Figure 2.5.19 the pore pressure gradient has been given as a function of depth for a typical Gulf Coast well. Determine the approximate fracture pressure gradient for a depth of 10,000 ft. From Figure 2.5.19, the pore pressure gradient at 10,000 ft is

$$\frac{p_p}{d} = 0.066 \text{ psi/ft}$$

From Figure 2.5.20, Poisson's ratio at 10,000 ft is (i.e., Gulf Coast curve)

$$v = 0.45$$

From Figure 2.5.21, the total overburden stress gradient is (i.e., Gulf Coast curve)

$$\frac{\sigma_{ob}}{d} = 0.95 \text{ psi/ft}$$

Substituting the foregoing values into Equation 2.5.20 yields

$$\frac{P_f}{d} = (0.95 - 0.66)\left(\frac{0.45}{1 - 0.45}\right) + 0.66 = 0.90 \text{ psi/ft}$$

This value of 0.90 psi/ft falls on the dashed line of Figure 2.5.19. The entire dashed line (fracture pressure gradient) in Figure 2.5.19 has been determined by using Equation 2.5.20.

In general, Equation 2.5.20 can be used to approximate fracture pressure gradients. To obtain an adequate approximation for fracture pressure gradients, the pore pressure gradient must be determined from well log data. Also, the overburden stress gradient and Poisson's ratio versus depth must be known for the region.

There is a field operation method by which the fracture pressure gradient can be experimentally verified. Such tests are known as leak-off tests. The leak-off test will be discussed in Chapter 4.

2.5.7 Basic Engineering Soil Properties

The surface rock formations of the earth are continually exposed to the weathering process of the atmosphere that surrounds the earth. The weathering process through time can change the rock exposed at the surface. Such changes are both mechanical and chemical. The altered surface rock is often used by lifeforms, particularly by plant life that directly draws nourishment from minerals at the surface. As the weathering process proceeds at the surface of the earth, rock at the surface disintegrates into small separate particles and is carried off and deposited at various locations around the original rock. Some particles are carried great distances to the sea to become source sediments for new rock formations. Some particles move only a few feet or a mile or two to be deposited with fragments from other nearby rock locations. These weathered particles, once deposited on the surface of the earth in land locations, are often referred to as soil [14].

To a farmer, soil is the substance that supports plant life. To a geologist soil is an ambiguous term that refers to the material that supports life plus the loose rock from which it was derived. To the engineer, soil has a broader meaning.

Soil, from the engineering point of view, is defined as any unconsolidated material composed of discrete solid particles that has either liquids or gases in the voids between the individual particles.

In general, soil overlays rock formations, and the soil is related to the rock since the rock was its source. Where the soil ends (in depth) and rock begins is not a well-defined interface. Basically, the depth to which soil is found is that depth where excavation by land methods can be employed. The area where the removal of material requires drilling, wedging, and blasting is believed to be the beginning of rock (in the engineering sense). The engineering properties of soil are of importance to petroleum engineering because it is soil that the drilling engineer first encounters as drilling is initiated. But, more important, it is soil that must support the loads of the drilling rig through an appropriately designed foundation. Further, the production engineer must support the well head surface equipment on soil through an appropriately designed foundation.

2.5.7.1 Soil Characteristics and Classification

The engineer visualizes a soil mass as an ideal, real, physical body incapable of resisting tensile stresses.

The ideal soil is defined as a loose, granular medium that is devoid of cohesion but possesses internal friction. In contrast, an ideal cohesive medium is one that is devoid of internal friction. Real soils generally fall between the foregoing two limiting definitions.

Soils can consist of rock, rock particles, mineral materials derived from rock formations, and organic matter.

- Bedrock is composed of competent, hard, rock formations that underlie soils. Bedrock is the foundation engineer's description of transition from soils to rock at depth. Such rock can be igneous, sedimentary, or metamorphic. Bedrock is very desirable for foundation placement.
- Weathered rock is bedrock that is deteriorating due to the weathering process. Usually, this is confined to the upper layers of the bedrock.
- Boulders are rock fragments over 10 in. in diameter found in soils.

- Cobbles are rock fragments from 2–4 in. in diameter found in soils.
- Pebbles are rock fragments from about 4 mm to 2 in. in diameter found in soils.
- Gravel denotes unconsolidated rock fragments from about 2 mm to 6 in. in size.
- Sand consists of rock particles from 0.05–2 mm in size.
- Silt and clay are fine-grained soils in which individual particle size cannot be readily distinguished with the unaided eye. Some classification systems distinguish these particles by size, other systems use plasticity to classify these particles.

Plasticity is defined as the ability of such particle groups to deform rapidly without cracking or crumbling. It also refers to the ability of such groups to change volume with relatively small rebound when the deforming force is removed.

- Silt, in one particle classification system, consists of rock particles from 0.005 to 0.05 mm in size.
- Clay, in one particle classification system, consists of inorganic particles less than 0.005 mm in size. In another system, clay is a fine-grained inorganic soil that can be made plastic by adjusting the water content. When dried, clay exhibits considerable strength (i.e., clay loses its plasticity when dried and its strength when wetted). Also, it will shrink when dried and expand when moisture is added.

Figure 2.5.22 shows a classification system developed by the Lower Mississippi Valley Division, U.S. Corps of Engineers. Percentages are based on dry weight. A mixture with 50% or more clay is classified as clay; with 80% or more silt, as silt; and with 80% or more sand, as sand. A mixture with 40% clay and 40% sand is a sandy clay. A mixture with 25% clay and 65% silt is a clay-silt (see intersection of dashed lines in Figure 2.5.22).

2.5.7.2 Index Properties of Soils

Easily observed physical properties of soils often are useful indexes of behavior. These index properties include texture and appearance, specific weight, moisture content, consistency, permeability, compressibility, and shearing strength [14,15].

- Soil texture, or appearance, depends on particle size, shape, and gradation. Therefore, using the classification in Figure 2.5.22 the soil texture can be specified as sandy clay or clay-sand.
- Soil specific weight is the measure of the concentration of packing of particles in a soil mass. It is also an index of compressibility. Less dense, or loosely packed, soils are much more compressible under loads. Soil specific weight may be expressed numerically as soil ratio and porosity (porosity for soils being basically the same definition as that for rocks discussed earlier in this section). Soil porosity e is

$$e = \frac{V_b - V_s}{V_b} \qquad [2.5.21]$$

where V_b is bulk volume of undisturbed soil (ft^3), and V_s is volume of solids in soil (ft).

The specific weight (unit weight) (lb/ft^3) of undisturbed soil is

$$\gamma = \frac{W_s}{W_b} \qquad [2.5.22]$$

where W_s is the weight of the soil solids relative to the undistributed soil bulk volume (lb).

Relative density D_d (a percent) is a measure of the compactness of a soil with void ratio e when the maximum void

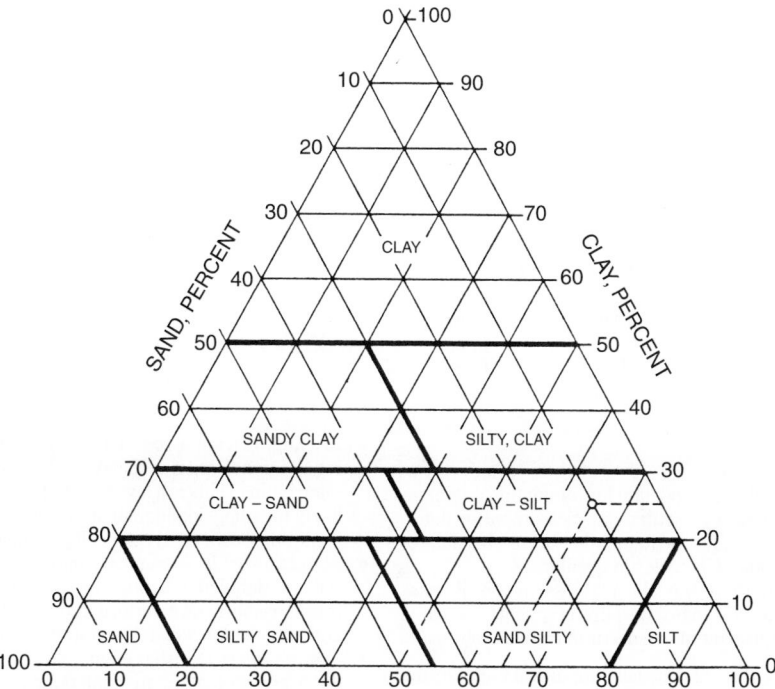

Figure 2.5.22   *Classification chart for mixed soils (from Lower Mississippi Valley Division, U.S. Corps of Engineers).*

ratio is e_{max} and the minimum e_{min}. Relative density is

$$D_d = \frac{e_{max} - e}{e_{max} - e_{min}}(100) \qquad [2.5.23]$$

Percentage compaction usually is used to measure soil density in a fill situation. Generally, the maximum Proctor specific weight (dry, lb/ft³), determined by a standard laboratory test, is set up as the standard for the soil. Normally, 90–100% compaction is specified.

- Moisture content, or water content, is an important influence on soil behavior. Water content w, dry-weight basis percent, is

$$W = \frac{W_w}{W_s} \qquad [2.5.24]$$

where W_w is weight of water in soil (lb) and W_s is weight of solids in soil (ld).

Total net weight of soils, W (lb), is

$$W_e = W_w + W_s \qquad [2.5.25]$$

Degree of saturation, S (percent) is

$$S = \frac{V_w}{V_b - V_s}(100) \qquad [2.5.26]$$

where V is volume of water in soil (ft³).

Saturation, porosity, and moisture content are related by

$$S_e = WG \qquad [2.5.27]$$

where G is the specific gravity of solids in the soil mass.

- Consistency describes the condition of fine-grained soils: soft, firm, or hard. Shearing strength and bearing capacity vary significantly with consistency. In consistency, there are four states: liquid, plastic, semisolid, and solid.

- Permeability is the ability of a soil to conduct or discharge water under a pressure, or hydraulic gradient (permeability for soils being basically the same definition as that for rocks discussed earlier in this section). For soils, the definition of coefficient of permeability is slightly different than that discussed earlier for petroleum reservoir rocks. Since civil engineers and hydrologists are always dealing with water, the coefficient of permeability, or more precisely, the hydraulic conductivity, k′ (cm/s)[1] is

$$k' = \frac{\gamma_w}{\mu_w}k \qquad [2.5.28]$$

where k is permeability as defined in Equation 2.5.8, γ_w is the specific weight of water, and μ_w is the viscosity of water.

Therefore, substituting Equation 2.5.8 into Equation 2.5.28 yields

$$k' = \frac{q}{iA} \qquad [2.5.29]$$

where i (cm/cm) is the hydraulic gradient and is expressed by

$$i = \frac{\Delta p}{\ell \gamma_w} \qquad [2.5.30]$$

and A (cm²) is the total cross-sectional area of soil through which flow occurs.

Table 2.5.4 gives some typical values for hydraulic conductivity and drainage characteristics for various soil types.

---

[1]Note: k′, hydraulic conductivity, is used for water flow in rocks as well as in soils.

Table 2.5.4 *Hydraulic Conductivity and Drainage Characteristics of Soils*

Soil Type	Approximate Coefficient of Permeability k (cm/s)	Drainage Characteristic
Clean gravel	5–10	Good
Clean coarse sand	0.4–3	Good
Clean medium sand	0.05–0.15	Good
Clean fine sand	0.004–0.02	Good
Silty sand and gravel	10^{-5}–0.01	Poor to good
Silty sand	10^{-5}–10^{-4}	Poor
Sandy clay	10^{-6}–10^{-5}	Poor
Silty clay	10^{-6}	Poor
Clay	10^{-7}	Poor
Colloidal clay	10^{-9}	Poor

From Reference 14.

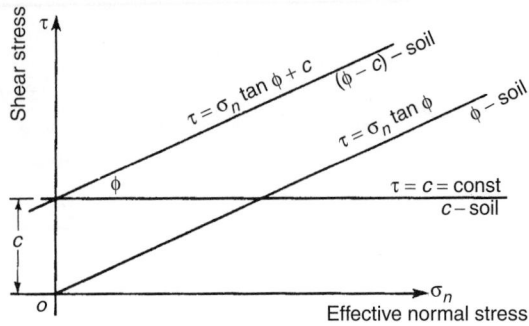

Figure 2.5.23 *Graphic shear strength of soils.*

- Soil compressibility is important for foundation engineering because it indicates settlement. Settlement or deformation of the soil under the foundation occurs because of change of position of particles in a soil mass.
- Shearing Strength is the shear stress in a soil mass at failure of the soil mass (usually cracking), or when continuous displacement can occur at constant stress.

Shearing strength of a soil is usually important in determining bearing capacity of the soil. Shearing stress, τ (lb/in.^{2}), of a soil is expressed as

$$\tau = c + \sigma_n \tan\theta \qquad [2.5.31]$$

where C is cohesion of the soil (lb/in.^{2}) σ_n is the normal, effective stress perpendicular to shear surface (lb/in.^{2}), and θ is the angle of internal friction of soil.

The shear strength of coarse particle soils like gravel and sand depend on the interlocking of their particles and, thereby, on intergranular friction. The shear strength of pure clay soils depends basically on cohesion. Figure 2.5.23 gives a graphical representation of the two limits of soil types, i.e., coarse particle soils are denoted as θ soils, pure clay soils are denoted as c soils. Most real soils are a combination (i.e., $\theta-c$) soils. Some approximate friction angles for typical cohesionless soil types are as follows [14]:

Soil Type	ϕ, deg	$\tan\phi$
Silt, or uniform fine to medium sand	26–30	0.5–0.6
Well-graded sand	30–34	0.6–0.7
Sand and gravel	32–36	0.6–0.7

2.5.8 Site Investigations and Laboratory Tests
Before the construction of a foundation, field investigation should be carried out to determine surface and subsurface conditions at the site.

2.5.8.1 Site Investigations
Numerous techniques are used for site investigations. The techniques vary in cost from relatively low-cost visual investigations to costly subsurface explorations and laboratory tests [16,17].

- Visual inspection is an essential primary step. Such inspections should provide data on surface soils, surface waters, and slopes.

- Probing, driving a rod or pipe into the soil and measuring the penetration resistance, obtains initial subsurface information. This is a low-cost method, but in general it is likely to supply inadequate information about subsurface conditions, especially on the depth and nature of bedrock.
- Augers provide subsurface data by bringing up material for detailed examination. Augers disturb the soil; therefore, little or no information can be obtained on the character of the soil in its natural undisturbed state.
- Test pits permit visual examination of the soil in place. Such pits also allow manual sampling of "undisturbed" soil samples. These samples can be taken from the side walls of the pit.
- "Dry" spoon sampling is a technique that is often used in conjunction with auger drilling. At certain depths in the augered borehole a spoon is driven into the undisturbed bottom of the borehole. The spoon sampler is a specified size (usually a 2-in. OD). The number of blows per foot to the spoon samples frequently are recorded, indicating the resistance of the soil. The spoon sampler is driven into the bottom of the hole with a free-falling weight. A 140-lb weight falling 30 in. onto the 2 in. O.D. spoon sampler is the standard method of driving the sampler into the borehole bottom. Table 2.5.5 shows a system of correlation of this technique of sampling. Figure 2.5.24 shows a typical subsurface soils log that describes the soils encountered at depth and number of blows per 6 in. (instead of per foot) on the spoon sampler.

Table 2.5.5 *Correlation of Standard-boring-spoon Penetration with Soil Consistency and Strength (2 in OD Spoon, 140 lb)*

Soil Consistency	Number of Blows per ft on Spoon	Unconfined Compressive Strength, tons/ft^2
Sand		
Loose	15	
Medium compact	16–30	
Compact	30–50	
Very compact	Over 50	
Clay		
Very soft	3 or less	0.3 or less
Soft	4–12	0.3–1.0
Stiff	12–35	1.0–4
Hard	Over 35	4 or more

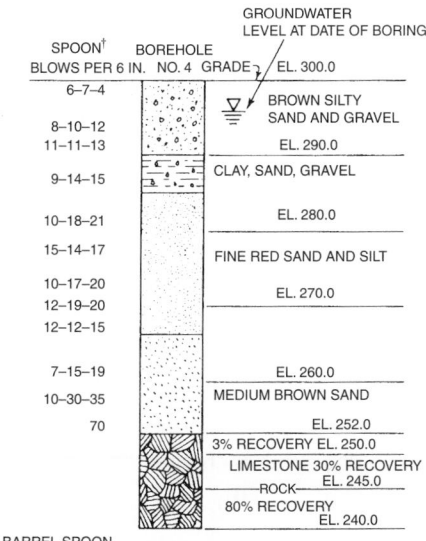

SPOON[†]
BLOWS PER 6 IN.

BOREHOLE
NO. 4

GROUNDWATER
LEVEL AT DATE OF BORING

GRADE EL. 300.0

6–7–4

8–10–12

11–11–13

BROWN SILTY
SAND AND GRAVEL

EL. 290.0

9–14–15

CLAY, SAND, GRAVEL

EL. 280.0

15–14–17

FINE RED SAND AND SILT

10–18–21

10–17–20
12–19–20
12–12–15

EL. 270.0

7–15–19

EL. 260.0

10–30–35

MEDIUM BROWN SAND

70

EL. 252.0
3% RECOVERY EL. 250.0
LIMESTONE 30% RECOVERY
EL. 245.0
ROCK
80% RECOVERY
EL. 240.0

[†]2-IN. SPLIT – BARREL SPOON
140 – LB HAMMER
30 – IN. DROP

Figure 2.5.24 *Typical soils boring log.*

2.5.8.2 Laboratory Tests

In addition to the site investigations just described and the on-site tests which can be carried out through test pits and augering, often specific laboratory tests are required to identify soils and determine their properties. Such laboratory tests are conducted on the soil samples that are recovered from various subsurface depths. These laboratory tests are used when there are questions as to the structural supporting capabilities of the soils at a particular site. Numerous types of laboratory tests are available that can aid the engineer in designing adequate foundation support for heavy or dynamic loads. In general, however, few tests are necessary for most foundation designs [14–17].

- Mechanical analyses determine the particle-size distribution in a soil sample. The distribution of coarse particles is determined by sieving, and particles finer than a 200 or 270-mesh sieve and found by sedimentation.
- Specific weight determinations measure the relative volumes of voids and solids in a soil.
- Compaction tests, such as the standard Proctor, determine the maximum specific weight or minimum void ratio that can be obtained for a soil, particularly a soil which is to be used for a fill. Specific weights of at least 95% of maximum are usually specified for compacted fills.
- In-place specific weight tests are used to correlate field compaction results with specified engineering requirements for specific weight.
- Moisture-content determinations provide data for estimating soil compaction and compressibility. If a soil is saturated, no volume change can occur without intake or discharge of water.
- Atterberg-limit tests determine the water content influence in defining liquid, plastic, semisolid and solid states of fine-grained soils. Permeability tests may be carried out in the laboratory or in the field. Such tests are used to determine the hydraulic conductivity coefficient k′.
- Confined compression tests are used to determine information pertaining to the behavior of foundations where

large volume changes of soil can occur under compression but in the vertical dimension only.

- Unconfined compression tests are used to estimate the shearing strength of cohesive soils.
- Consolidation tests are made on saturated silts and clays to determine the rate of volume change under constant load.
- Direct shear tests are made in the laboratory to obtain data for determining the bearing capacity of soils and the stability of embankments.
- Triaxial compression tests are another means of determining shearing strength of a soil. A complex device is used to apply pressure along the sides of a cylindrical specimen and axially down the axis of the cylindrical specimen. In general, triaxial tests are superior to direct shear tests since there is better control over intake and discharge of water from the specimen.
- California bearing ratio tests are used to evaluate subgrades for pavements. These tests may be carried out in the field or in the laboratory. Such tests determine the resistance to penetration of a subgrade soil relative to that of a standard crushed-rock base.
- Plate bearing tests are field tests that are also used to evaluate subgrades for pavements.

Foundation Loads and Pressures. Foundations should be designed to support the weight of the structure, the live load, and the load effect on the structure and its foundation due to such other loads as wind. In general, for foundation designs, a safety factor of 3 is used for dead loads or live loads independently. A safety factor of 2 is used for combination loads including transient loads [15,17].

In general, a foundation is designed for settlement and for pressure distribution. In designing for settlement the usual practice is to ignore transient loads. To keep differential settlements small, foundations are designed to apportion the pressure (between the foundation and the soil) equally over the soil. The assumption is that equal intensities of pressure will produce equal settlement. The accuracy of this assumption will vary with the soil uniformity beneath the foundation, the shape of the foundation, and the distribution of the load on the foundation. Pressures used in calculating bearing capacity or settlement are those in excess of the pressure due to the weight of the soil above. Thus, one should consider pressures composed by the adjacent foundation on the region below the foundation under design. Usual practice in foundation design is to assume that bearing pressure at the bottom of a foundation or on a parallel plane below the foundation is constant for concentrically loaded foundations. This assumption may not be entirely accurate, but more accurate and more complicated theories usually are not justified because of the lack of future knowledge of loading conditions or by the present knowledge of soil conditions. The assumption of constant pressure distribution has been the basis of design of many foundations that have performed satisfactorily for decades. This is especially true for rigid foundations on soils with allowable bearing pressures of 6,000 lb/ft^2 or more. Another commonly used assumption in foundation design is that the pressure spreads out with depth from the bottom of the foundation at an angle of 30° from the vertical or at a slope of 1 to 2 (Figure 2.5.25). Therefore, for a total load on a foundation of P, the pressure at the base of the footing would be assumed to be p = P/A, where A is the area of the foundation itself. As shown in Figure 2.5.25b, the pressure at a depth h would be taken to be P/A or, for a square foundation, as P/(b + h)^2. When a weak layer of soil underlies a stronger layer on which the foundation is founded, this method may be used to estimate the pressure on the lower layer. This method can be adopted

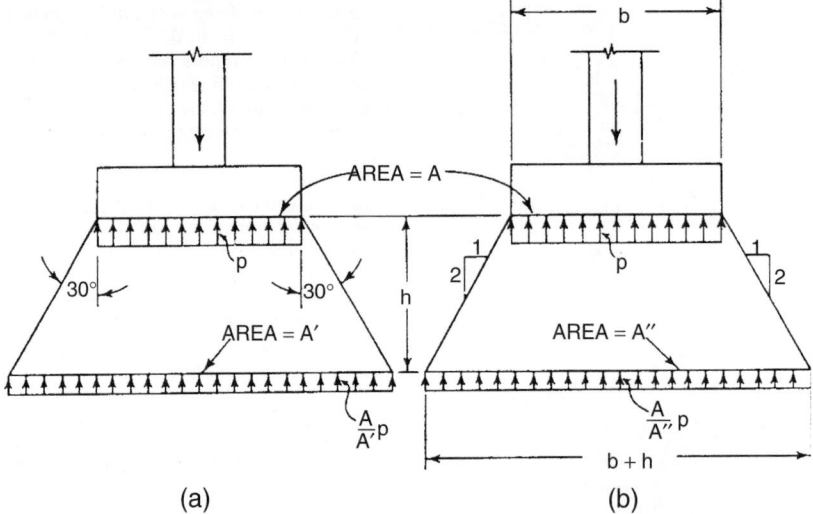

Figure 2.5.25   *Assumed pressure distribution under a foundation: (a) 30° spread; (b) 1×2 spread [14].*

to determine the total pressure resulting from overlapping pressure distribution from adjacent foundations.

Approximate allowable bearing pressures on sedimentary rock and soils may be taken from Table 2.5.6 [1,14]. Where questionable surface and subsurface soil conditions exist allowable bearing pressures can be determined with the aid of field sampling, field tests (both surface and subsurface through borings), and laboratory tests.

Spread Foundations. The purpose of the spread foundation is to distribute loads over a large enough area so that soil can support the loads safely and without excessive settlement. Such foundations are made of steel-reinforced concrete. When concrete is used for a foundation, it should be placed on undisturbed soil. All vegetation should be removed from the surface; therefore, the upper few inches of soil should be removed before concrete is laid down. The area of the foundation must be large enough to ensure that the bearing capacity of the soil is not exceeded or that maximum settlement is within acceptable limits. If details are known of the subsurface soil conditions, the foundation must be sized to that differential settlement will not be excessive. For uniform soil condition (both horizontal and vertical), this is accomplished by designing the foundation such that the unit pressure under the foundation is uniform for the working loads (usually dead load plus normal live loads) [15,17].

Example 2.5.3
It is intended that a spread foundation be designed for a concentric load of 300,000 lb (dead load plus live load). This foundation is to be placed on the surface (brown silty sand and gravel) of the soil and bedrock column shown in Figure 2.5.23. If a square foundation can be made to support the 300,000-lb load, what should be the dimensions of this foundation?

- The surface layer of soil (i.e., brown silty sand and gravel) has a minimum number of blows per foot on the 2-in. split-barrel spoon of 8 (see Figure 2.5.24). According to Table 2.5.5, this soil would have the classification of a loose silty sand and gravel layer. Table 2.5.6 indicates the allowable bearing capacity could be as low as 3,000 lb/ft². To improve the soil conditions where the foundation is to be laid, the initial few feet of soil are to be removed to

Table 2.5.6   *Allowable Bearing Capacities of Sedimentary Rock and Soils [14]*

Rock or Soil	Allowable Bearing Capacity lbs/ft²
Sedimentary rock: hard shales, sitstones, sandstones, requiring blasting to remove	20,000 to 30,000
Hardpan, cemented sand and gravel, difficult to remove by picking	16,000 to 20,000
Soft rock, disintegrated ledge; in natural ledge, difficult to remove by picking	10,000 to 20,000
Compact sand and gravel, requiring picking to remove	8,000 to 12,000
Hard clay, requiring picking for removal	8,000 to 10,000
Gravel, coarse sand, in natural thick beds	8,000 to 10,000
Loose, medium and coarse sand; fine compact sand	3,000 to 8,000
Medium clay, stiff but capable of being spaded	4,000 to 8,000
Fine loose sand	2,000 to 4,000
Soft clay	2,000

expose the subsurface layer which has a minimum number of blows per foot on the 2-in. split-barrel spoon of 16. This would require a removal of approximately 4 to 5 ft of soil. The brown silty sand and gravel in the layer at about 5 ft of depth can be classified as medium compact. Again referring to Table 2.5.6 this layer of soil should have an allowable bearing capacity above 3,000 lb/ft². The average between 3,000 and 8,000 lb/ft² is assumed, i.e., 5,500 lb/ft².

- The initial square foundation dimension that would be needed to meet the allowable bearing capacity of 5,500 lb/ft² is

$$A = \frac{300,000}{5,500} = 54.6 \text{ ft}^2$$

where A is the surface area of the foundation at the point where it contracts the soil. The dimensions of the square foundation are

$$b = (54.6)^{1/2} = 7.4 \text{ ft}$$

- There are two potentially weak subsurface soil layers in the boring log in Figure 2.5.24 that should be checked before the initial foundation design above is accepted.

  1. At the depth of approximately 10 ft below the surface, or 5 ft below the intended foundation soil interface, there is a layer of clay, sand, and gravel that has a minimum number of blows per foot on the 2-in. split-barrel spoon of 18. Referring to Table 2.5.5, the compressive strength of this layer could be as low as 2,000 lb/ft². Assuming a one-by-two spread of the subsurface pressure under the foundation, the pressure at that layer would be

$$P_1 = \frac{300,000}{(7.4+5)^2} = 1948 \text{ lb/ft}^2$$

  For further information on this subject, refer to References 15 through 17.

  The subsurface pressure calculated for this layer is below the assumed compressive strength of 2,000 lb/ft²; therefore, the initial foundation divisions are acceptable relative to the strength of this subsurface layer.

  2. At the depth of approximately 40 ft below the surface or 35 ft below the intended foundation-soil interface, there is a layer of medium brown sand that has a minimum number of blows per foot in the 2-in. split-barrel spoon of 14. Referring of Tables 2.5.5 and 2.5.6, this soil could be classified as a loose sand having a compressive strength as low as 3,000 lb/ft². Assuming a one-by-two spread of the subsurface pressure under the foundation, the pressure at that layer would be

$$P_2 = \frac{300,000}{(7.4+35)^2} = 167 \text{ lb/ft}^2$$

  The subsurface pressure calculated for this layer is well below the assumed compressive strength of 3000 lb/ft²; therefore, the initial foundation dimensions are acceptable relative to this subsurface layer also.

- Because the initial foundation dimensions result in acceptable foundation-soil interface bearing pressures and acceptable subsurface pressures on weaker underlying layers of soil, the square foundation of 7.4 ft is the final foundation design.

References

1. Plummer, C. C., McGreay, D., and Carlson, D. H., *Physical Geology*, 8th Edition, McGraw-Hill, New York, 1999, pp. 77–101, 121–189.
2. Tarbuck, Edward J., and Lutgens, F. K., 1999, *The Earth*, 6th Edition, Prentice Hall, Englewood Cliffs, 1999.
3. Link, P., *Basic Petroleum Geology*, OGCE Publications, 1982.
4. Lenorsen, A. I., *Geology of Petroleum*, 2nd Edition, W. H. Freeman, San Francisco, 1967.
5. Chapman, R. E., *Petroleum Geology, A Concise Study*, Elsevier, New York, 1973.
6. Dickey, P. A., *Petroleum Development Geology*, 2nd Edition, Penn Well, 1981.
7. Cox, A., et al., *Plate Techtonics and Geomagnetic Reversals*, W. H. Freeman, San Francisco, 1973.
8. Muskat, M., *Physical Principles of Oil Production*, Reprint Edition, IHRDC, 1981.
9. Nichols, E. A., "Geothermal Gradient in Mid-Continent and Gulf Coast Oil Fields," *Transactions ASME*, Vol. 170, 1947.
10. McGray, A. W., and Cole, F. W., *Oil Well Drilling Technology*, Oklahoma Press, 1959.
11. Hubbert, M., and Willis, D., "Mechanics of Hydraulic Fracturing," *Transactions AIME*, Vol. 210, 1957.
12. Eaton, B. A., "Fracture Gradient Prediction and Its Applications in Oilfield Operations," *Journal of Petroleum Technology*, October, 1969.
13. Schuh, F., *Engineering Essentials of Modern Drilling* (2nd Printing), Energy Publication Division of MBJ, 1982.
14. Merritt, F. S., *Standard Handbook for Civil Engineers*, 2nd Edition, McGraw-Hill, New York, 1976.
15. Sowers, G. B., and Sowers, G. F., *Introductory Soil Mechanics and Foundations*, 2nd Edition, MacMillan, 1961.
16. Jamibes, A. R., *Mechanics of Soils*, Van Nostrand, New York, 1964.
17. Tschebotarioff, G. P., *Soil Mechanics, Foundations, and Earth Structures*, McGraw-Hill, New York, 1951.

2.6 ELECTRICITY

2.6.1 Electrical Units

There are two principle unit systems used in electrical calculations, the centimeter-gram-second (cgs) and the meter-kilogram-second (mks) or International System (SI). Table 2.6.1 is a summary of common electrical quantities and their units. The magnitude of the units in Table 2.6.1 is often not conveniently sized for taking measurements or for expressing values, and Table 2.6.2 presents the prefixes that modify SI units to make them more convenient.

Energy or *work* W is defined by

$$W = \int_{t_1}^{t_2} p \, dt \qquad [2.6.1]$$

where W is work in joules, p is power in watts, and t is time in seconds.

Table 2.6.1 *Electrical Units*

Quantity	Symbol	MKS and SI Units
Current	I,i	Ampere
Charge (Quantity)	Q,q	Coulomb
Potential	V,v	Volt (V)
Electromotive Force	E,e	Volt (V)
Resistance	R,r	Ohm (Ω)
Resistivity	ℓ	Ohm-cm
Conductance	G,g	Mho, Siemens (Ω^{-1})
Conductivity	γ	Mho/cm
Capacitance	C	Farad (f)
Inductance	L	Henry (h)
Energy (work)	W	Joule (J), Watthour (Wh) Kilowatthour (KWh)
Power	P,p	Watt
Reactance, inductive	X_L	Ohm
Reactance, capacitive	X_C	Ohm
Impedance	Z	Ohm

Table 2.6.2 *SI Unit Dimensional Prefixes*

Symbol	Prefix	Multiple
T	tera	10^{12}
G	giga	10^{9}
M	mega	10^{6}
k*	kilo	10^{3}
h*	hecto	10^{2}
da*	deca	10^{1}
d	deci	10^{-1}
c	centi	10^{-2}
m	milli	10^{-3}
	micro	10^{-6}
n	nano	10^{-9}
p	pico	10^{-12}

*May also use capital letter for the symbol.

Relations between common units of energy are as follows:

watt-second = 1 joule
watt-second = 0.239 calorie
watt-second = 0.738 foot-pound
kilowatt-hour = 3413 Btu
kilowatt-hour = 1.34 horsepower-hours
kilowatt-hour = 3.6×10^{6} joules
electron-volt = 1.6×10^{-9} joule
erg = 10^{-7} joule

Power (P, p) is the time rate of doing work. For constant current I through an electrical load having a potential drop V, the power is given by

$$P = IV \qquad [2.6.2]$$

where P is power in watts, V is potential drop in volts, and I is current in amperes.

For a time-varying current i and potential drop v, the *average power* P_{av} is given by

$$P_{av} = \frac{1}{t} \int_{0}^{t} iv\,dt \qquad [2.6.3]$$

where P_{av} is average power in watts; i is current in amperes, v is potential drop in volts, and t is time in seconds.

Relations between common units of power are as follows:

$$1w = 1\,j/s$$

$$0.239\,c/sec$$

$$9.48 \times 10^{4}\,Btu/s$$

$$1/745\,hp$$

$$0.7375\,ft\text{-}lb/s$$

Electric *charge* or *quantity* Q, expressed in units of *coulombs*, is the amount of electricity that passes any section of an electric circuit in one s by a current of one *ampere*. A coulomb is the charge of 6.24×10^{18} electrons.

Current (I, i) is the flow of electrons through a conductor. Two principal classes of current are:

- *Direct* (dc) — the current always flows in the same direction.
- *Alternating* (ac) — the current changes direction periodically.

The unit of current is the *ampere* which is defined as one coulomb per second.

Electric *potential* (V, v), *potential difference*, or *electromotive force* (emf, E, e) have units of *volts* and refer to the energy change when a charge is moved from one point to another in an electric field.

Resistance (R, r) is an element of an electric circuit that reacts to impede the flow of current. The basic unit of resistance is the *ohm* (Ω), which is defined in terms of *Ohm's law* as the ratio of potential difference to current, i.e.,

$$R = \frac{V}{I} \qquad [2.6.4]$$

The resistance of a length of conductor of uniform cross-section is given by

$$R = \frac{\rho\ell}{A} \qquad [2.6.5]$$

where A is cross-section area of the conductor in square meters, ℓ is length of the conductor in meters, and ρ is resistivity of the material in ohm-meters; R is resistance in ohms.

Conductance (G, g) is the reciprocal of resistance and has units of reciprocal ohms or *mhos* (Ω^{-1}) or more properly in SI units, *seimens*.

Conductivity (γ) is the reciprocal of resistivity.

Capacitance (C) is the property that describes the quantity of electricity that can be stored when two conductors are separated by a dielectric material. The unit of capacitance is the *farad*. The capacitance of two equal-area, conducting parallel plates (see Figure 2.6.1) separated by a dielectric is given by

$$C = \frac{\varepsilon_r \varepsilon_v A}{d} \qquad [2.6.6]$$

where ε_r is dielectric constant of the material between the plates, ε_v is dielectric constant of free space or of a vacuum, A is the area of a plate in square meters, d is distance between the plates in meters, and C is capacitance in farads.

Inductance (L) is the property of an electric circuit that produces an emf in the circuit in response to a change in the rate of current, i.e.,

$$e = L\frac{di}{dt} \qquad [2.6.7a]$$

where e is emf induced in the circuit in volts, I is current in amperes, t is time in seconds, and L is coefficient of (self) inductance in henries.

For a coil (see Figure 2.6.2), the inductance is given by

$$L = KN^{2} \qquad [2.6.7b]$$

where N is number of turns, K is a constant that depends on the geometry and the materials of construction, and L is coefficient of inductance in henries.

Coils are described later in the section tilted "Magnetic Circuits."

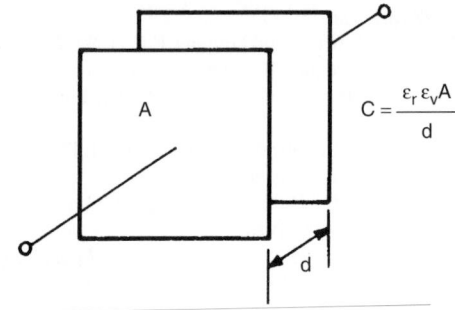

Figure 2.6.1 *Schematic of a parallel plate capacitor.*

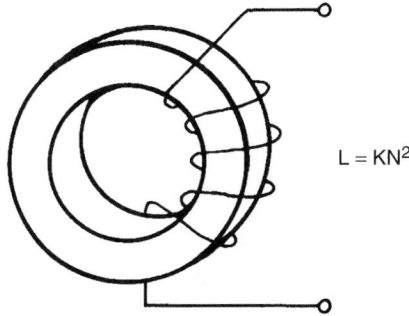

Figure 2.6.2 *Schematic of an induction coil.*

$$L = KN^2$$

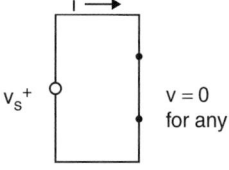

(a) SHORT CIRCUIT

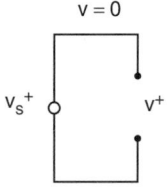

(b) OPEN CIRCUIT

Figure 2.6.3 *Schematic of the simplest circuit elements.*

2.6.2 Electrical Circuit Elements

Electrical phenomena may generally be classified as *static* or *dynamic*. In *static* phenomena, current or charges do not flow or the flow is only momentary. Most practical uses of electricity (at least in the conventional engineering sense) involve *dynamic* systems where current flows for useful periods of time. Systems that allow current to flow for extended periods of time are termed *circuits*. As the name implies, such systems are connected in a loop so that any beginning point in the circuit has electrical continuity with the ending point through *circuit elements*.

The simplest circuit element is the *short circuit*. Figure 2.6.3a illustrates the concept of a short circuit. A source of emf (labeled v_s) produces a current that flows relatively unimpeded through the conductor resulting in a nearly zero potential drop and an infinite current.

Figure 2.6.3b illustrates an *open circuit*. The conductor of the short circuit is interrupted and current cannot flow even though the emf produces a finite potential.

2.6.3 Passive Circuit Elements

Circuit elements may be classified as *passive* or *active*. *Passive elements* may store or transform electrical energy. *Active elements* may transform other forms of energy (including electrical energy from another system) into an increase in the electrical energy of the circuit or they may serve to dissipate circuit energy. In the latter case, the rate of increase or dissipation is controlled by conditions outside of the circuit. A simple example of an active element might be a generator that supplies power (emf) to the circuit. The rate of power generation depends on the mechanical shaft power input.

There are three, linear, *passive circuit elements:* resistors, capacitors, and inductors. *Resistors* dissipate energy in the circuit, i.e., electrical energy is transformed to heat energy and is lost from the circuit. *Capacitors* store electrical energy as charges on conductors separated by a dielectric material. *Inductors* store the electrical energy of current as magnetic potential in a manner analogous to the kinetic energy stored in a mass in motion. Table 2.6.3 summarizes the circuit element characteristics.

2.6.3.1 Series and Parallel Connection of Circuit Elements

Circuit elements may be connected in either a *series* or *parallel* configuration. In the *series* configuration, the same current flows through each and every element, and the circuit potential drop (or emf that is developed by the voltage source) is the algebraic sum of the potential drops of each individual element. For sources in series, the total emf developed is the algebraic sum of the emfs developed by each individual source.

In the *parallel* configuration, the same potential difference occurs across each and every element with the total current being the algebraic sum of the current flowing through each individual circuit element. Table 2.6.4 summarizes the equivalent resistance, conductance, capacitance, and inductance of series-parallel configurations of resistors, capacitors, and inductors.

2.6.3.2 Circuit Analysis

There are two fundamental laws used in circuit analysis, called Kirchhoff's laws:

1. At a branch point in an electric circuit, the sum of the currents flowing to the point equals the sum of the currents flowing from the point.
2. The electric potential measured between two points in an electric circuit is the same regardless of the path along which it is measured.

These laws apply to both DC and AC currents.

2.6.4 Transient and AC Circuits

In a transient or an AC circuit, we term the sum of resistance, inductance, and capacitance as *impedance*. Using complex notation, the energy storage properties of inductance and capacitance are represented as purely imaginary quantities, while the resistance is represented as a (+) real quantity. Capacitance is represented as the negative imaginary axis, and current through a pure capacitance is said to *lead* the potential by 90°. Inductance is represented as the positive imaginary axis, and the current through a pure inductance is said to *lag* the potential by 90°. These relationships may be expressed mathematically for an inductance and resistance in series as

$$Z = R + jX_L \qquad [2.6.8]$$

and for a capacitance and resistance in series as

$$Z = R - jX_C \qquad [2.6.9]$$

where Z is impedance in ohms, j is an imaginary unit, X_L is the impedance due to inductance or reactance in ohms $= 2\pi f L = \omega L$, X_C is impedance due to capacitance or *reactance* in ohms $= 1/2\pi f C = 1/\omega C$, R is resistance in *ohms*, L is coefficient of inductance in henries, C is capacitance in farads, f is frequency of the current (or voltage) in Hertz (cycles per second), and ω is angular velocity in radians per second.

Table 2.6.3 *Circuit Element Characteristics*

Element	Unit	Symbol	Characteristic
Resistance (Conductance)	ohm (mho)		$v = Ri$ $(i = Gv)$
Inductance	henry		$v = L\dfrac{di}{dt}$ $i = \dfrac{1}{L}\displaystyle\int_0^t v\,dt + I_0$
Capacitance	farad		$i = C\dfrac{dv}{dt}$ $v = \dfrac{1}{C}\displaystyle\int_0^t i\,dt + V_0$
Short circuit			$v = 0$ for any i
Open circuit			$i = 0$ for any v
Voltage source	volt		$v = v_s$ for any i
Current source	ampere		$i = i_s$ for any v

Table 2.6.4 *Series-Parallel Combinations*

Circuit Element	Series	Parallel
Resistor R_1, R_2	$R = R_1 + R_2$	$R = \dfrac{R_1 R_2}{R_1 + R_2}$ or $R = \left(\dfrac{1}{R_1} + \dfrac{1}{R_2}\right)^{-1}$
Inductor L_1, L_2	$L = L_1 + L_2$	$L = \dfrac{L_1 L_2}{L_1 + L_2}$ or $L = \left(\dfrac{1}{L_1} + \dfrac{1}{L_2}\right)^{-1}$
Capacitor C_1, C_2	$C = \dfrac{C_1 C_2}{C_1 + C_2}$ or $C = \left(\dfrac{1}{C_1} + \dfrac{1}{C_2}\right)^{-1}$	$C = C_1 + C_2$

For the impedance of the resistance and inductance in series, the current will lag the potential phase angle by

$$\theta = \arctan\left(\frac{X_L}{R}\right) \qquad\qquad [2.6.10a]$$

For the impedance of the resistance and capacitance in series, the current will lead the potential phase angle by

$$\theta = \arctan\left(\frac{-X_C}{R}\right) \qquad\qquad [2.6.10b]$$

When multiple circuit elements are involved, the resultant phase angle difference between the current and the potential will result from the contribution of each element.

2.6.5 AC Power

The power dissipated in an AC circuit with current of maximum amplitude I_m flowing through a resistance is less than the power produced by a constant DC current of magnitude I_m flowing through the same resistance. For a sinusoidal AC current, the *root mean square* (rms) value of current I is the magnitude of the DC current producing the same power as the AC current with maximum amplitude I_m. The rms value I is given by

$$I = \frac{1}{\sqrt{2}} I_m = 0.707 I_m \qquad [2.6.11]$$

The power dissipated in an AC circuit with only resistive elements is

$$P = I^2 R = \frac{V^2}{R} \qquad [2.6.12]$$

where V is rms value of the potential drop. This resistive power is termed *active power* with units of *watts*.

Reactive circuit elements (i.e., capacitors and inductors) store, not dissipate, energy. While the energy stored is periodically returned to the rest of the circuit, reactive elements do require increased potential or current to flow in the circuit. The power that must be supplied for the reactive elements is termed *reactive* power, and it is calculated as

$$Q_x = I^2 X = \frac{V^2}{X} \qquad [2.6.13]$$

where Q_x is reactive power in volt-amperes reactive (*VAR*) X is reactance in ohms, V is rms potential in volts, and I is current in amperes.

The reactive component of impedance is expressed as

$$X = Z \sin\theta \qquad [2.6.14]$$

where Z is impedance in *ohms*, and θ is leading or lagging phase difference between current and potential.

Note that $\sin\theta$ may be either positive or negative and lies between 0 and 1 for $|\theta| \leq 90°$.

The *apparent power* is the complex sum of the *active power* and the *reactive power*. By nothing that

$$R = Z \cos\theta \qquad [2.6.15]$$

we may calculate

$$P_A = VI\cos\theta + jVI\sin\theta \qquad [2.6.16]$$

where P_A is apparent power in volt-amperes (VA), V is rms potential, I is rms current, and θ is leading or lagging difference in phase angle between current and potential.

The *power factor* $\cos\theta$ is always a positive fraction between 0 and 1 (as long as $|\theta| \leq 90°$). The smaller the power factor, the greater the current that must be supplied to the circuit for a given active (useful) power output requirement. The increase in current associated with low power factors causes greater line losses or requires an increase in the capacity of the transmission equipment (wire size, transformers, etc.). As a result, for industrial applications there is often a power factor change in the rate structure for supplying electricity. The usual situation is for loads to be inductive, and the industrial consumer may add capacitance to their circuits to correct the lagging power factor.

2.6.6 Magnetism

Magnetic fields are created by the motion of electric charges. The charge motion may be a current in a conductor or, at the atomic level, the movements of orbital electrons. For certain materials, called *ferromagnetic* materials, the neighboring atoms align themselves so that the magnetic effects of their orbital electrons are additive. When the atoms of a piece of such a ferromagnetic material are aligned, the piece is called a *magnet*. Magnetic fields have north (N) and south (S) poles. When two magnets are brought together, like poles repel and unlike poles attract each other. In other (nonmagnetic) materials, the atoms are aligned randomly and the magnetic effects cancel.

Analogies exist between electric and magnetic fields. The *magnetic flux* (ϕ) is analogous to electric current and has SI units of webers (see Table 2.6.5). The *magnetic flux density* (B) is analogous to current density and has units of teslas. One tesla exists when the charge of one coulomb moving normal to the magnetic field with a velocity of one meter per second experiences a force of one newton. In vector notation this is expressed as

$$\bar{f} = q(\bar{u} \times \bar{B}) \qquad [2.6.17]$$

where $\bar{f}$ is the force vector in newtons, q is the charge in coulombs, $\bar{u}$ is velocity vector in meters per second, and $\bar{B}$ is magnetic flux density vector in teslas.

If a current flows through a conductor, the magnetic flux is oriented in a direction tangent to a circle whose plane is perpendicular to the conductor. For current flowing in an infinitely long, straight conductor, the magnetic flux density at a point in space outside the conductor is given as

$$B = \frac{\mu i}{2\pi D} \qquad [2.6.18]$$

where B is magnetic flux density in teslas, i is current in amperes, D is distance from conductors to the point in space in meters, and μ is permeability in webers per amp-meter or henries per meter.

The *permeability* (μ) is a property of the material surrounding the conductor. The *permeability of free space* (μ_0) is

$$\mu_0 = 4 \times 10^{-7} \text{ henries per meter} \qquad [2.6.19]$$

The ratio of the permeability of any material to the permeability of free space is termed the *relative permeability* (μ_r).

The magnetic field intensity (H) is given as

$$H = \frac{B}{\mu} \qquad [2.6.20]$$

or for the magnetic field induced by current in an infinite-length straight conductor,

$$H = \frac{i}{2\pi D} \qquad [2.6.21]$$

Table 2.6.5 *Magnetic Units*

Quantity	Symbol	Cgs Units	SI Units
Magnetomotive Force	$\mathcal{F}$	Gilberts	Amp-turns (NI)
Magnetic Field Intensity	H	Oersted (Oe)	Amp-turns/meter
Magnetic Flux	ϕ	Maxwell or line	Webers (Wb)
Magnetic Flux Density	B	Gauss	Teslas (Wb/m^2)

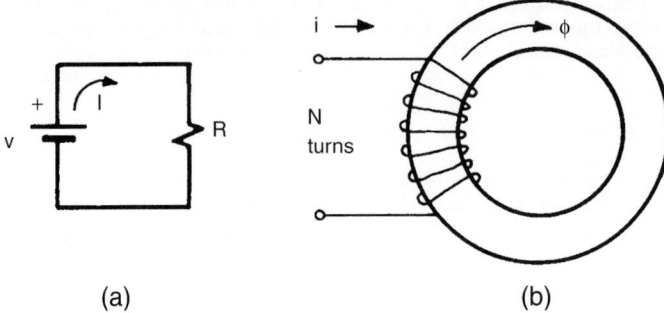

Figure 2.6.4 *Electric (a) and magnetic (b) circuits.*

2.6.7 Magnetic Circuits

Previously the analogy between electric fields and magnetic fields was introduced. Likewise, there are analogies between magnetic circuits and electric circuits. Figure 2.6.4 illustrates these analogies and allows us to define additional terms. In the electric circuit of Figure 2.6.4a Ohm's law applies, i.e.,

$$\frac{V}{I} = R = \frac{\rho\ell}{A} = \frac{\ell}{\gamma A} \qquad [2.6.22]$$

In Figure 2.6.4b, there is an analogous relationship, i.e.,

$$\mathcal{R} = \frac{\mathcal{F}}{\phi} = \frac{\ell}{\mu A} \qquad [2.6.23]$$

where $\mathcal{F}$ is magnetic magnetomotive force in ampere-turns, ϕ is magnetic flux in webers, $\mathcal{R}$ is reluctance — the resistance to magnetic flux, and μ is magnetic permeability. The magnetic magnetomotive force ($\mathcal{F}$) in Figure 2.6.4 is given by

$$\mathcal{F} = Ni \qquad [2.6.24]$$

where $\mathcal{F}$ is magnetomotive force in ampere-turns, N is number of turns, and i is current in amperes.

2.6.8 Transformers

Transformers are electromagnetic devices that allow electrical power supplied at one potential to be transformed into electrical power at another potential. The potential or voltage may be stepped up (increased) or stepped down (decreased). For instance, in the usual transmission of domestic power, the potential in the transmission lines is greater than the load requirements and a step down transformer is used to reduce the potential at the end use point.

Figure 2.6.5 illustrates the basics of a transformer. First there is a *core*, usually constructed of a material of high magnetic permeability to achieve a high magnetic flux density. The core has two *windings* of conductors, a *primary coil* (designated as N_1 in the figure) and a *secondary coil* (designated as N_2). Electric current through the primary coil causes a magnetic flux in the core and at the same time an impedance to the current and therefore an induced emf across the primary. The magnetic flux in the core in turn induces an emf across the secondary coil, causing a current to flow. The relation between the emf induced in the primary coil (note that this is not the source emf) and the emf induced in the secondary coil is given by

$$\frac{e_1}{e_2} = \frac{E_1}{E_2} \frac{N_1}{N_2} \qquad [2.6.25]$$

where e_1, e_2 is AC-induced emfs in volts, E_1, E_2 is rms values of e_1 and e_2, and N_1, N_2 is number of turns on the primary and secondary coils, respectively.

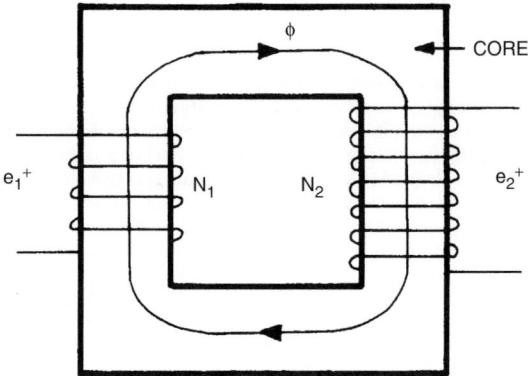

Figure 2.6.5 *Basic transformer operation.*

There will be inefficiencies within the transformer, and the voltages at the transformer terminals will vary a little from the previous relationships.

Likewise, the approximate relationship between the primary and secondary currents will be

$$\frac{I_1}{I_2} = \frac{N_2}{N_1} \qquad [2.6.26]$$

2.6.9 Rotating Machines

There is a class of electromechanical equipment in which mechanical energy is converted into electrical energy (or vice versa), all of which use either the response of conductors rotating through a magnetic field or a magnetic field rotating in the presence of stationary conductors. Because the machines rotate, power is transformed in a constant mode rather than the pulsating mode as occurs in similar translational devices.

Figure 2.6.6 is a schematic of perhaps the simplest rotating machine, the elementary *dynamo*. The elementary dynamo consists of a rectangular-shaped coil, which is free to rotate about an axis. In a practical device, the coil is physically attached to a shaft at the axis of rotation but is electrically insulated from this shaft. *Slip rings* connected to the coil are also attached but insulated from the shaft. The slip rings and brushes allow electrical contact with an external circuit while the shaft turns.

Consider the effect when the coil turns in the presence of an external magnetic field. An emf will be generated in the coil given by

$$e = NBA\omega\cos wt \qquad [2.6.27]$$

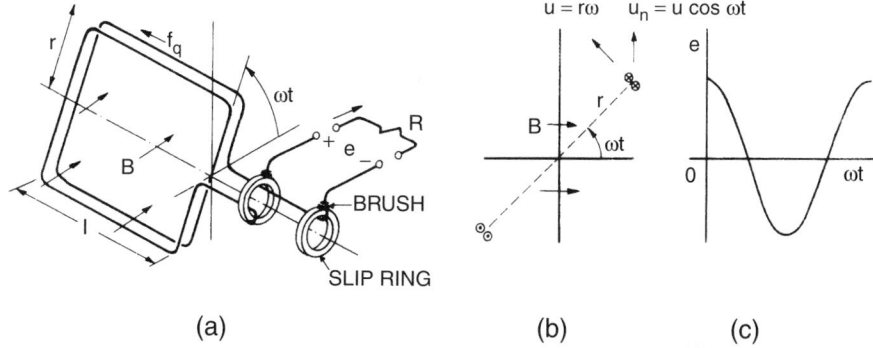

Figure 2.6.6 *Elementary dynamo construction and operation.*

where e is generated emf in volts, N is number of turns of wire in the coil, B is magnetic flux density in teslas, A is area surrounded by the coil $=2r\ell$, r is radius of rotation of the coil, ℓ is length of coil along the axis of rotation, ω is angular velocity in radians per second, and t = time in seconds.

A torque will have to be overcome to maintain the rotation of the coil. This torque is given as

$$\tau_d = NBAi\cos\omega t \qquad [2.6.28]$$

where N, B, A, ω, t are as previously defined, τ_d is developed torque in the elementary dynamo, and i is current.

Note that the direction of the applied torque will dictate the direction of the induced current in the elementary dynamo.

The "external" magnetic field could be due to a magnet or it could be due to the magnetomotive force induced by a current in a conductor (or another stationary coil). The relationship for torque developed when the fields of a stationary (*stator*) coil and a rotating (*rotor*) coil interact is given by

$$\tau_d = ki_s i_r \sin\delta \qquad [2.6.29]$$

where τ_d is torque developed in the dynamo, k is constant term, which includes the number of coil windings, dimensions of the dynamo, velocity of rotation, etc., i_s is current in the stator coil in amperes, i_r is current in the rotor coil in amperes, and δ is angle between the fields, called the *torque* or the *power angle*.

Because the current in each coil induces a magnetic field, the torque relationship may also be given as

$$\tau_d = k'B_s B_r \sin\delta \qquad [2.6.30]$$

where k' is constant, B_s is magnetic flux density associated with the stator in teslas, and B_r is magnetic flux density associated with the rotor in teslas.

Thus from Equation 2.6.30, we see that for a given dynamo geometry, the developed torque only depends on the interaction between two magnetic fields and their orientation with respect to each other. One or both of the magnetic fields may be induced by a current. If one of the fields is the field of a magnet, then it may be either in the rotor or the stator. If the rotation results from the imposition of mechanical power on the rotor, the device is called a *generator*. If the rotation is caused by the flow of current, the device is called a *motor*, i.e., converts electric power to mechanical power.

If the rotor of the elementary dynamo is turned in a uniform magnetic field, an AC emf and current are produced. If the speed of rotation is constant, the emf and current are sinusoidal as shown in Figure 2.6.6.

Figure 2.6.7 is a schematic of another AC generator called an *alternator*. A DC current is supplied to field windings on

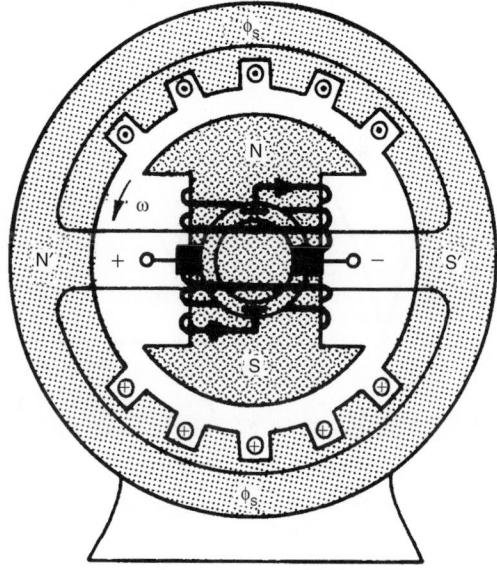

Figure 2.6.7 *An alternator.*

the rotor, which are then rotated inside the stator windings, producing the AC emf.

The device of Figure 2.6.7 can also operate as a motor if a DC current is applied to the rotor windings as in the alternator and an AC current is imposed on the stator windings. As the current to the stator flows in one direction, the torque developed on the rotor causes it to turn until the rotor and stator fields are aligned ($\delta = 0°$). If, at that instant, the stator current switches direction, then mechanical momentum will carry the rotor past the point of field alignment, and the opposite direction of the stator field will cause a torque in the same direction and continue the rotation.

If the slip rings of the elementary dynamo are replaced by a *split-ring commutator*, then a DC emf and current will be generated as shown in Figure 2.6.8. If the single coil of the elementary dynamo is replaced with multiple coils attached to opposing segments of a multisegment commutator, then the emf generated will be more nearly constant. There will, however, always be a momentary reduction in the emf at the times in the cycle when the spaces in the commutator pass

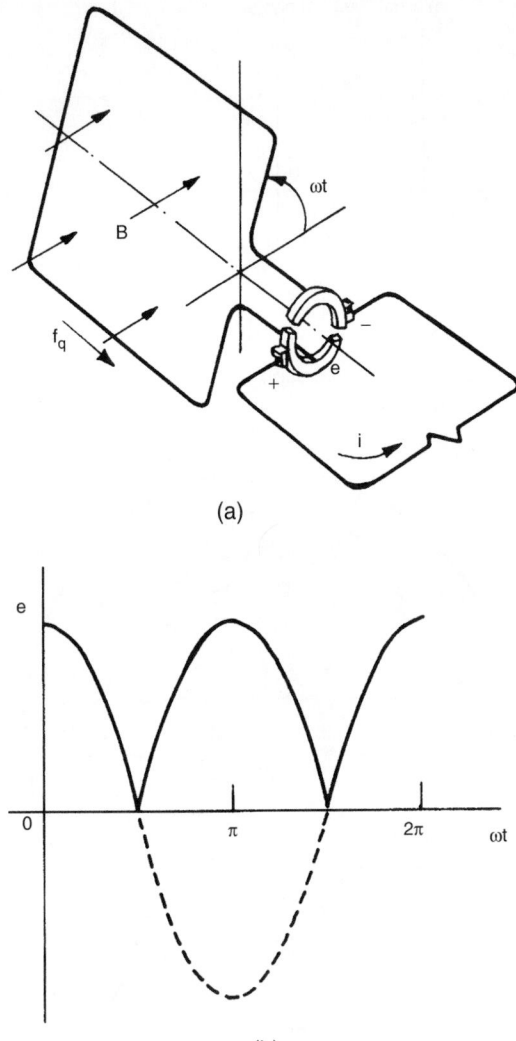

(a)

(b)

Figure 2.6.8 *Commutator operation (a) and generated voltage waveform (b).*

the brushes. By the proper orientation of the rotor (called the *armature*) windings in relation to the commutator segment in contact with the brushes, the torque angle ($\delta = 90°$) can be made to produce maximum torque.

The DC generator can be operated as a motor by imposing a dc current on the armature.

2.6.10 Polyphase Circuits

Circuits that carry AC current employing two, three, or more sinusoidal potentials are called *polyphase* circuits. Polyphase circuits provide for more efficient generation and transmission of power than single-phase circuits. Power in a three- (or more) phase circuit is constant rather than pulsating like the single-phase circuit. As a result, three-phase motors operate more efficiently than single-phase motors.

The usual situation is to have three phases, each generated by the same generator but with a difference of 120°

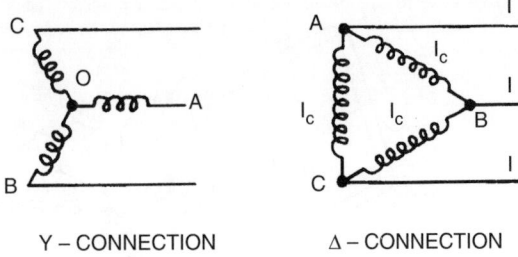

Y – CONNECTION   Δ – CONNECTION

Figure 2.6.9 *Three-phase connections.*

between each phase. Each phase of the generator could be operated independently of the other phases to supply single-phase loads, but to save wiring costs the phases are often run together. The three-phase generator may be connected in either *delta* (Δ) or Y configuration. Figure 2.6.9 illustrates the two types of generator connections. The coils in the figure are armature windings (see the section on "Rotating Machines"). For the *Y connection* with balanced (equal impedance and impedance phase angle on each phase) load, the *line and coil currents* are *equal*, but the line-to-line emfs are the square root of three times the coil emfs. In the Δ *configuration* with balanced loads, the line currents are the square root of three times the coil currents, and the *coil and line emfs are the same*.

The common connection of all three armature windings in the Y connection allows a fourth, or *neutral*, conductor to be used. This neutral point is often grounded in transmission and distribution circuits. Such a circuits is termed a *three-phase, four-wire* circuit.

In a balanced three-phase circuit, the total power is three times the power in each phase, or

$$P_{total} = 3P_p = 3V_p I_p \cos\theta \qquad [2.6.31]$$

where P_{total} is total circuit power, P_p is phase power, V_p is rms phase potential, I_p is rms phase current, and θ is phase angle difference between phase potential and phase current.

Regardless of whether the circuit is connected in Δ or Y, the total power is also

$$P_{total} = \sqrt{3}V_1 I_1 \cos\theta \qquad [2.6.32]$$

where P_{total} and θ are previously defined, V_1 is rms line-to-line potential, and I_1 is rms line current.

Note that θ is not the phase angle difference between line potential and line current.

Just as the armature coils of a three-phase generator may be connected in a Δ or Y configuration, the circuit loads may be connected in a Δ or Y configuration. The Δ-load configuration may be supplied from a source that is connected in either Δ or Y. The Y connection may include a neutral (fourth) wire, connected at the common connection of the circuit.

2.6.11 Power Transmission and Distribution Systems

Electric power is almost always transmitted as three-phase AC current. In domestic use, current is often distributed from a substation at voltages ranging from 4,160 to 34,500 volts, which are stepped down by a transformer close to the point of use to 600, 480, and 240 V for three-phase current for commercial power and 240 and 120 V for single-phase, three-wire current for household power and lights. If DC current is required, synchronous converters or rectifiers are used to convert the AC supply to DC.

Most devices are designed for a constant potential and, as a result, power is usually distributed to loads at constant potential. Two possible configurations for delivering a constant potential to multiple loads are illustrated in Figure 2.6.10. In the *parallel circuit*, the potential across the load decreases as the distance from the source increases.

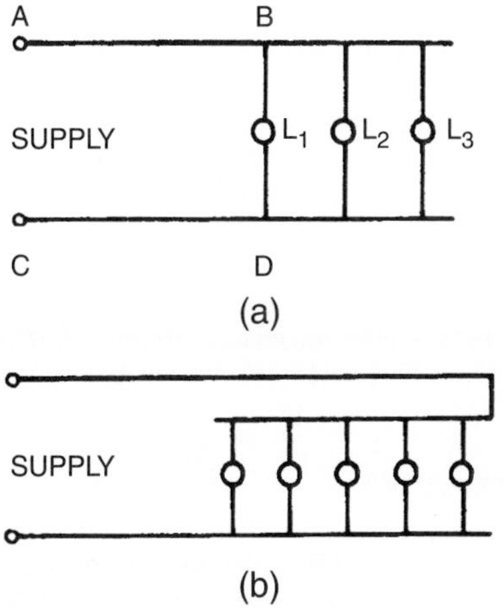

For the *loop circuit*, potentials are more nearly equal along the length of the circuit.

Transmission lines usually consist of two or more conductors separated by some form of insulation. Such a configuration exhibits a significant resistance, inductance, and capacitance. Figure 2.6.11 illustrates these effects. Solving the circuit of Figure 2.6.11 is quite involved, and transmission line electrical characteristics are often represented more simply as a lumped parameter model. Figure 2.6.12 depicts two common lumped models used as equivalent circuits to calculate line losses, changes in phase angle, etc.

The materials for transmission and distribution conductors are usually copper and aluminum. Copper is expensive, has a high conductivity, and has sufficient mechanical strength for many uses. Aluminum has the advantage that for a given weight of conductor, it has twice the conductance of copper. A disadvantage of aluminum is that its melting point is lower than copper, while its thermal expansion is greater and stability problems are sometimes encountered.

Conductor or wire sizes are expressed in terms of the *American Wire Gage* (AWG) system. In this system, the ratio of any wire diameter to the next smaller gage or diameter is 1.123. The AWG sizes range from 40 to 0000. Table 2.6.6 lists the AWG number, wire dimension, and resistance for solid copper wire. Wires larger than 0000 (as well as smaller wires) are stranded to maintain flexibility. Wire sizes greater than 0000 are expressed in circular mils. A circular mil is the square of the diameter of the conductor, where the diameter is expressed in mili-inches. (1000 mili-inch = 1 inch).

Figure 2.6.10   *Circuits: (a) parallel; (b) loop.*

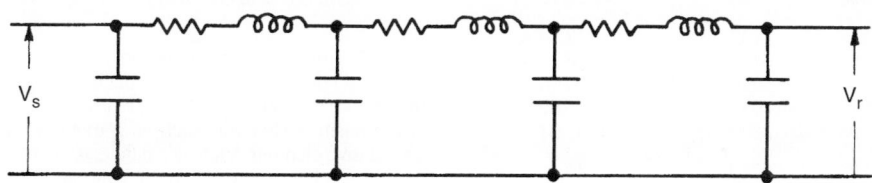

Figure 2.6.11   *Resistance, inductance, and capacitance of transmission lines.*

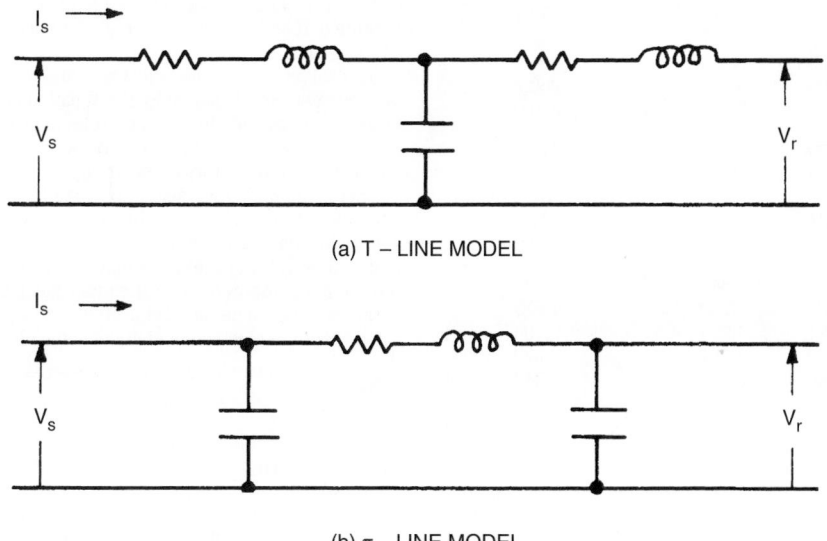

(a) T – LINE MODEL

(b) π – LINE MODEL

Figure 2.6.12   *Lumped element models of transmission line electrical characteristics.*

Table 2.6.6 *Wire Table for Copper*

AWG	Diameter, inches	Area, circular mils	Resistance at 20°C (68°F), ohms per 1000 feet of length
0000	0.4600	211,600	0.04901
000	0.4096	167,800	0.06180
00	0.3648	133,100	0.07793
0	0.3249	105,500	0.09827
1	0.2893	83,690	0.1239
2	0.2576	66,370	0.1563
3	0.2294	52,640	0.1970
4	0.2043	41,740	0.2485
5	0.1819	33,100	0.3133
6	0.1620	26,250	0.3951
7	0.1443	20,820	0.4982
8	0.1285	16,510	0.6282
9	0.1144	13,090	0.7921
10	0.1019	10,380	0.9989
11	0.09074	8,234	1.260
12	0.08081	6,530	1.588
13	0.07196	5,178	2.003
14	0.06408	4,107	2.525
15	0.05707	3,257	3.184
16	0.05082	2,583	4.016
17	0.04526	2,048	5.064
18	0.04030	1,624	6.385
19	0.03589	1,288	8.051
20	0.03196	1,022	10.15
21	0.02846	810.1	12.80
22	0.02535	642.4	16.14
23	0.02257	509.5	20.36
24	0.02010	404.0	25.67
25	0.01790	320.4	32.37
26	0.01594	254.1	40.81
27	0.01420	201.5	51.47
28	0.01264	159.8	64.90
29	0.01126	126.7	81.83
30	0.01003	100.5	103.2
31	0.008928	79.70	130.1
32	0.007950	63.21	164.1
33	0.007080	50.13	206.9
34	0.006305	39.75	260.9
35	0.005615	31.32	329.0
36	0.005000	25.00	414.8
37	0.004453	19.83	523.1
38	0.003965	15.72	659.6
39	0.003531	12.47	831.8
40	0.003145	9.888	1,049
41	0.002800	7.840	1,323
42	0.002494	6.200	1,673
43	0.002221	4.928	2,104
44	0.001978	3.881	2,672
45	0.001760	3.098	3,348

Interior wiring design and installation for most commercial and industrial uses should follow the *National Electrical Code* (NEC) which has been a national standard since 1970 with the passage of the Occupational Safety and Health Act (OSHA). Some localities, however, may not accept the NEC and require that their own (more stringent) standards be followed.

For further information on this subject, refer to References 1–5.

References

1. Bentley, James H., and Hess, Karen M., A *Programmed Review for Electrical Engineering*, Van Nostrand-Reinhold, New York, 1978.
2. Fitzgerald, A. E., Higginbotham, David E., and Grabel, Arvin, *Basic Electrical Engineering*, 5th Edition, McGraw-Hill, New York, 1981.
3. Kraybill, Edward K., *Electric Circuits for Engineers*, Macmillan, New York, 1951.
4. Smith, Ralph J., *Circuits, Devices, and Systems, A First Course in Electrical Engineering*, 2nd Edition, John Wiley & Sons, New York, 1971.
5. Summers, Wilford E., ed., *The National Electrical Code Handbook*, National Fire Protection Association, Boston, 1978.

2.7 CHEMISTRY

This section discusses basic chemistry, with emphasis on the compounds found in petroleum and selected physical chemistry topics likely to arise in petroleum engineering. Examples are provided to illustrate fundamental ideas and principles. More complete treatment of these topics and others can be found in standard textbooks of general, organic, and physical chemistry.

2.7.1 Chemical Bonds and Polarity

Chemical compounds may be considered *ionic* or *covalent*. In an ionic compound, electrons are transferred from a more electropositive element, such as a metal, to a more electronegative element, such as a halide, or to a more complex structure, such as nitrate (NO_3^-). Common salt, NaCl, is an example of an ionic compound. Ionic compounds are usually soluble in water because the *cation*, in this case Na^+, and the *anion*, Cl^-, can dissociate and become solvated by the water. Many inorganic compounds are at least partly ionic. Ionic compounds tend to be good conductors of electricity. Their melting and boiling points are higher than those of covalent compounds.

In covalent compounds, electrons are shared in the bond between two atoms, and anions and cations are not formed. Organic compounds found in petroleum are predominantly covalent. Covalent compounds are poorly soluble in water and are poor conductors of electricity.

Chemical bonds may also have mixed ionic and covalent character. As the difference in electronegativities of atoms sharing a chemical bond increases, the bond becomes more ionic. Although the molecule may be electroneutral, it can have a dipole moment, with the more electronegative atom carrying a partial negative charge.

The order of electronegativity of some elements [1] is

$$F > O > Cl > N > Br > I > S > C > H > B$$

$$> Si > Mg > Li > Na > K$$

Examples of nonpolar molecules in which both atoms are the same are H_2, N_2, O_2, I_2, and Cl_2. Compounds in which atoms are symmetrically arranged, such as methane (CH_4), also may be nonpolar.

Water is a covalent compound, with two bonds between oxygen and hydrogen in which the oxygen and hydrogen each contribute one electron. However, the electronegativity difference between oxygen and hydrogen is large enough that the bonds and therefore the molecule is polar:

$$\overset{\delta-}{O}$$
$$\underset{\delta+H}{\diagup}\,\,\,\,\underset{H\delta+}{\diagdown}\qquad\text{(bond angle = 105°)}$$

The oxygen in water also carries an electron pair, which is attracted to the hydrogens of other water molecules (i.e., hydrogen bonding) or to other electropositive species, such as cations or electropositive hydrogens in other compounds such as acids. The hydrogen bond is much stronger than most dipole-dipole interactions and accounts for many of the distinctive properties of water.

2.7.2 Hydrocarbons

Compounds formed from carbon and hydrogen only are called *hydrocarbons*, which make up the bulk of petroleum. Carbon atoms can bond to each other to form chains and rings, and they can share one, two, or three pairs of electrons in each bond. These characteristics allow the formation of a wide variety of compounds. Chemists group these compounds into *homologous series*, families of compounds that contain a common *functional group*. Each member of a homologous series differs from the next by a one-carbon unit.

Nonpolar molecules such as H_2, N_2, O_2, I_2, and Cl_2 have zero dipole moments, because $e=0$. On the other hand, hydrogen fluoride, HF, has a large dipole moment of 1.75 Debye and so is strongly polar. Simple carbon compounds with symmetric arrangement of like atoms (e.g., methane, CH_4, and carbon tetrachloride, CCl_4) have zero dipole moments and so are nonpolar.

In polar compounds, the operative intermolecular forces are dipole-dipole interactions, which refer to the attraction between the positive pole of one molecule and the negative pole of another. For this reason, polar compounds are relatively more stable than nonpolar substances. A particularly strong kind of dipole-dipole attraction is hydrogen bonding, in which a hydrogen atom acts as a bridge between tow electronegative atoms, holding one atom by a covalent linkage and the other by purely electrostatic forces, e.g.,

Hydrogen fluoride $H - F \cdots F - H$

Water $H - O \cdots O - H$
$\qquad\qquad\quad |\qquad\quad\ |$
$\qquad\qquad\quad H\qquad\quad H$

where the sequence of dots indicates the hydrogen bridge. This bond has a strength of 5 kcal/g-mole versus 50–100 kcal/g-mole for the covalent bond, but it is much stronger than other dipole–dipole interaction [21]. For hydrogen bonding to be important, both electronegative atoms must belong to the group: F, O, N.

The intermolecular forces operative in nonpolar compounds are also electrostatic in nature. These weak van der Waals forces involve attraction between nonbonded atoms and are effective over short ranges only.

The nomenclature of organic compounds is based on conventions adopted by the International Union of Pure and Applied Chemistry (IUPAC) [2]. A brief discussion of the various classes of hydrocarbons follows.

2.7.2.1 Alkanes

Alkanes are the simplest aliphatic compounds, containing only carbon and hydrogen held together by single bonds and not containing a ring. They have the general formula C_nH_{2n+2}, where n is the number of carbon atoms in the alkane molecule, and $n \geq 1$. These are also known as *paraffins* or *saturated aliphatic hydrocarbons*. Continuous or straight-chain alkanes are called normal paraffins or n-alkanes

(e.g., methane, CH_4; ethane, C_2H_6; propane, C_3H_8; n-butance, $n-C_4H_{10}$; etc.). The corresponding *alkyl* groups, methyl, $-CH_3$; ethyl, $-C_2H_5$; n-propyl, $-C_3H_7$; n-butyl, $-C_4H_9$; etc., are generally represented by the symbol R–.

Branched-chain alkanes, also known as "isoparaffins" or "isoalkanes," are possible when $n \geq 4$. The prefix "iso" is used when two methyl groups are attached to a terminal carbon atom of an otherwise straight chain and the prefix "neo" when three methyl groups are attached in that manner. Branched-chain alkanes are sometimes regarded as normal alkanes with attached substituent alkyl groups. An example is

$$CH_3 - CH - CH_2 - \underset{|}{\overset{CH_3}{\underset{CH_3}{C}}} - CH_3$$

(with CH_3 groups above the CH and C carbons)

2,2,4-trimethlypentane (isooctane)

Isomers are substances having the same molecular formula and molecular weight, but differing in physical and chemical properties. Branched and straight-chain alkanes with the same molecular formula can exist as distinct structures having different geometrical arrangement of the atoms, which are termed *structural isomers*. One example is C_4H_{10} (butane), which has two isomers:

$$H_3C - \overset{H}{\underset{H}{C}} - \overset{H}{\underset{H}{C}} - CH_3 \quad\text{and}\quad H - \overset{CH_3}{\underset{CH_3}{C}} - CH_3$$

$[CH_3CH_2CH_2CH_3\ (n\text{-butane})]$ $[(CH_3)_3CH\ (\text{isobutane})]$

As the carbon number increases, the number of possible structural isomers grows rapidly. For example, C_5H_{12} has only three isomers,

$$CH_3CH_2CH_2CH_2CH_3 \quad H - \overset{CH_3}{\underset{CH_3}{C}} - CH_2CH_3 \quad H_3C - \overset{CH_3}{\underset{CH_3}{C}} - CH_3$$

$[n-C_5H_{12}\,(n\text{-pentane})]$ $[(CH_3)_2CHC_2H_5$ (isopentane)] $[(CH_3)_4C\ (\text{neopentane})]$

while $C_{10}H_{22}$ (decane) has 75, and $C_{20}H_{42}$ (eicosane) has 366,319 possible isomers.

Although n-alkanes exhibit smooth and graded variations in physical properties (Table 2.7.1), the branched members do not [3]. The structural isomers of any alkane generally show dissimilar physical and chemical characteristics. A branched-chain isomer has a lower boiling point than a straight-chain isomer, and the more numerous the branches, the lower its boiling point. Alkanes are either nonpolar or weakly polar. They are soluble in nonpolar or weakly polar solvents (e.g., benzene, chloroform, ether), and are insoluble in water and other highly polar solvents. Alkanes can dissolve compounds of low polarity. Chemically, paraffins are unreactive at ordinary conditions. At high temperatures, they can burn completely in the presence of oxygen to yield CO_2 and H_2O as products. The combustion reaction is exothermic.

At ambient temperature and pressure, the first four members of the n-alkane series (methane to n-butane) are gases, the next thirteen (n-pentane through n-heptadecane) are liquids, and the higher members from $n=18$ on are solids.

Table 2.7.1 *Physical Properties of Normal Alkanes [3]*

Name	Molecular Formula	Melting Point (°C)	Boiling Point (°C)	Density[a]	State Under Atmospheric Conditions
Methane	CH_4	−183	−164	0.55	Gas
Ethane	C_2H_6	−183	−89	0.51	Gas
Propane	C_3H_8	−189	−42	0.50	Gas
n-Butane	C_4H_{10}	−138	0	0.58	Gas
n-Pentane	C_5H_{12}	−130	36	0.63	Liquid
n-Hexane	C_6H_{14}	−95	69	0.66	Liquid
n-Heptane	C_7H_{16}	−91	98	0.68	Liquid
n-Octane	C_8H_{18}	−57	126	0.70	Liquid
n-Nonane	C_9H_{20}	−51	151	0.72	Liquid
n-Decane	$C_{10}H_{22}$	−30	174	0.73	Liquid
n-Undecane	$C_{11}H_{24}$	−26	196	0.74	Liquid
n-Dodecane	$C_{12}H_{26}$	−10	216	0.75	Liquid
n-Tridecane	$C_{13}H_{28}$	−5	235	0.76	Liquid
n-Tetradecane	$C_{14}H_{30}$	6	254	0.76	Liquid
n-Pentadecane	$C_{15}H_{32}$	10	271	0.77	Liquid
n-Hexadecane	$C_{16}H_{34}$	18	287	0.77	Liquid
n-Heptadecane	$C_{17}H_{36}$	23	303	0.76	Liquid
n-Octadecane	$C_{18}H_{38}$	28	317	0.76	Solid
n-Nonadecane	$C_{19}H_{40}$	32	330	0.78	Solid
n-Eicosane	$C_{20}H_{42}$	37	343	0.79	Solid
n-Triacontane	$C_{30}H_{62}$	66	450	0.81	Solid

[a]Densities are given in g/mL at 20°C, except for methane and ethane, whose densities are given at their boiling points.

2.7.2.2 Alkenes

These are also called "olefins" and have the general formula C_nH_{2n}, with $n \geq 2$. They contain a C=C double bond (functional group) and are named in accordance with the IUPAC convention by specifying the location of the double bond from the terminal carbon atom nearest to it.

The first three members of the olefin series are ethene, propene, and butene. Structural isomers exist when $n \geq 4$, as a consequence of the positioning of the double bond in normal alkenes, or as a result of branching in branched alkenes. In addition, *geometric isomers* may be possible owing to restricted rotation of atoms about the C=C bond. For instance, C_4H_8 (butene) has four possible isomers:

$$CH_2CHCH_2CH_3 \qquad CH_3CHCHCH_3 \qquad (CH_3)_2CCH_2$$
$$\text{(1-butene)} \qquad\qquad \text{(2-butene)} \qquad\qquad \text{(isobutene)}$$

This occurs because 2-butene can exist in two different structures, the *cis* or the *trans* configurations, depending on whether the methyl groups are situated on the same side or on opposite sides of the main chain.

$$\text{(cis-2-butene)} \qquad\qquad \text{(trans-2-butene)}$$

With the inclusion of these two geometric isomers, butene has a total of four isomers.

As the carbon number increases, the number of possible isomeric structures for each member increases more rapidly than in the case of the alkane series.

The physical properties of alkenes [3] are not very different from the corresponding members of the alkane family. Alkenes are nonpolar or at most weakly polar. They are insoluble in water but are soluble in concentrated H_2SO_4 and liquid HF. Normal alkenes dissolve in nonpolar or weakly polar organic liquids such as ethers, CCl_4, and hydrocarbons. In general, the *cis* isomer has a slightly higher polarity, a higher boiling point, and a lower melting point than the *trans* isomer, but there are exceptions.

The double bond in olefins is reactive. Halogens can be added across the double bond:

$$CH_2=CH_2 + Br_2 \rightarrow CH_2BrCH_2Br$$

This reaction and its homologs are the basis for a common test for unsaturated compound, in which orange bromine water is shaken with a hydrocarbon. Disappearance of the orange color, caused by reaction of the bromine with double bonds, indicates the presence of unsaturated compounds.

More than one double bond may be present in a compound. The chemical behavior of compounds containing multiple double bonds is similar to that of compounds containing a single double bond.

Common *alkenyl* groups include

$$CH_2=CH- \quad CH_3=CHCH_2- \quad CH_3CH=CH- \quad CH_3CH=CHCH_2-$$
$$\text{(vinyl)} \qquad \text{(allyl)} \qquad \text{(propenyl)} \qquad \text{(crotyl)}$$

2.7.2.3 Alkynes

These contain a single triple bond and have the general formula C_nH_{2n-2}, with $n \geq 2$. Alkynes are also referred to as *acetylenic compounds*. The simple alkynes can be named as derivatives of acetylene:

$$HC{\equiv}CH \qquad \text{acetylene (or ethyne)}$$

$$CH_3C{\equiv}CH \qquad \text{methylacetylene or propyne}$$

Isomers are similar to the isomers of alkenes, with the double bond replaced by the triple bond, but *cis* and *trans* isomers are not possible because of the linear geometry of the triple bond.

Physical properties of alkynes [3] are similar to those of alkanes and alkenes. They are weakly polar and insoluble in water, but they are quite soluble in organic solvents of low polarity (e.g., ether, benzene, CCl_4). Chemically, the triple bond is less reactive than the double bond in some reagents and more reactive in others.

Diynes and *triynes* are alkynes containing two or three triple bonds; *poly-ynes* contain multiple triple bonds. A conjugated triyne is a straight-chain hydrocarbon with triple bonds alternating with single bonds. An examples is

$$CH_3-C{\equiv}C-C{\equiv}C-C{\equiv}C-CH_3 \quad \text{2,4,6-octatriyne}$$

When both double and triple bonds occur in the same molecule, the IUPAC system recommends the use of both endings -ene, and -yne, with the former always preceding the latter in the name. Common *alkynyl* groups are

$$HC{\equiv}C- \qquad \text{and} \qquad HC{\equiv}C-CH_2-$$
$$\text{(ethynyl)} \qquad\qquad\qquad \text{(propargyl)}$$

In cyclic hydrocarbons, the carbon atoms form a ring instead of an open chain. They are also called *carbocyclic* or *homocyclic* compounds. They are divided into two classes: alicyclic (or cycloaliphatic) and aromatic compounds.

Cyclic analogs of aliphatic hydrocarbons are also known as cycloalkanes, naphthenes, cycloparaffins, or saturated alicyclic hydrocarbons. They have the general formula C_nH_{2n} with $n \geq 3$. The first two members are

cyclopropane

cyclobutane

the next two being cyclopentane ⬠ and cyclohexane ⬡. For convenience, aliphatic rings are represented by polygons.

Their properties are similar to their open-chain aliphatic counterparts. Alicyclic hydrocarbons may have a single ring or multiple rings. Monocyclic aliphatic structures having more than 30 carbon atoms in the ring are known, but those containing 5 or 6 carbon atoms are more commonly found in nature. See, for example, the sections on steroids, prostaglandins, and terpenes in standard textbooks [3,4].

When substituent groups are present, they are identified and their positions indicated by numbers in naming the compound. As an example, dimethylcyclohexane has three structures:

(1,2-dimethylcyclohexane)   (1,3-dimethylcyclohexane)

(1,4-dimethylcyclohexane)

Cis-trans isomerism occurs in these disubstituted cycloalkanes because the substitutent groups may be located above or below the plane of the ring.

Physical properties of cycloalkanes [3] show reasonably gradual changes, but unlike most homologous series, different members may exhibit different degrees of chemical reactivity. For example, cyclohexane is the least reactive member in this family, whereas both cyclopropane and cyclobutane are more reactive than cyclopentane because of the strain involved in forming the ring.

Cycloalkenes may have one or more double bonds in the ring. Examples are

1-cyclopentene

1,4-cyclohexadiene

They are chemically as reactive as their straight-chain counterparts. Compounds are added across the double bond. In scission or cleavage reactions, the ring structure opens up into a straight chain.

Hydrocarbons containing both aliphatic and alicyclic parts may be named by considering either part as the parent structure and the other part as a substituent:

ethylcyclopropane

Cycloalkynes have a triple bond in the carbon ring, as shown in the following example:

cyclooctyne

Although it is possible to conceive of alicyclic hydrocarbons containing more than a triple bond or two double bonds in the carbocyclic ring, such ring structures are usually unstable.

Polycyclic aliphatic hydrocarbons contain two or more rings that share two or more carbon atoms. An example of a fused-ring system is

[decahydronaphthalene (or decalin)]

The aliphatic rings may be saturated or partially unsaturated.

2.7.2.4 Aromatic Hydrocarbons

Aromatic hydrocarbons are unsaturated cyclic compounds, usually with benzene or its derivatives as the common building block. Benzene, the simplest aromatic hydrocarbon, has the molecular formula C_6H_6. It is a flat and symmetrical molecule with six carbon atoms arranged in a hexagonal ring (bond angle $= 120°$) with a hydrogen atom attached to each carbon. Benzene is represented as having three double bonds and three single bonds between the carbon atoms, but the bonding electrons are delocalized around the ring, providing great stability. Reactions of benzene compounds usually involve substitutions of the atoms bonded to the carbon ring, rather than the ring itself. Severe conditions are required to break the ring. The molecule may be represented by

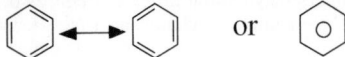

Nomenclature for benzene derivatives is a combination of the IUPAC system and traditional names. Many of the derivatives are named by the substituent group appearing as the prefix. Thus, alkylbenzenes are made up of a benzene ring and alkane units; alkenylbenzenes are composed of a benzene ring and alkene units; and alkynylbenzenes comprise a benzene ring and alkyne units. Examples of alkylbenzenes include

CH_3

or $C_6H_5CH_3$

[toluene (or methylbenzene)]

H_3C
CH
H_3C or $C_6H_5CH(CH_3)_2$

[cumene (or isopropylbenzene)]

If several groups are attached to the benzene ring, their names as well as their relative positions are indicated. For example, dimethylbenzene or xylene, $C_6H_4(CH_3)_2$, has three geometric isomers, with prefixes *ortho-*, *meta-*, and *para-*, indicating the relative positions of the two methyl groups.

CH_3
CH_3

[*ortho*-xylene (*o*-xylene or 1,2-dimethylbenzene)]

CH_3
CH_3

[*meta*-xylene (*m*-xylene or 1,3-dimethylbenzene)]

CH_3

CH_3

[*para*-xylene (*p*-xylene or 1,4-dimethylbenzene)]

Examples of an alkenylbenzene and an alkynylbenzene are given below:

$CH = CH_2$

and

$C \equiv CH$

[styrene (vinylbenzene)]

(phenylacetylene)

When the benzene ring is considered a substituent group, it is called the phenyl group and is represented by

or C_6H_5

An example is

$\overset{3}{CH_2} - \overset{2}{CH} = \overset{1}{CH_2}$ or $CH_2CHCH_2(C_6H_5)$

(3-phenylpropene)

More generally, aromatic substituent groups are called *aryl* groups.

Polynuclear aromatic hydrocarbons have complex structures made up of aromatic rings, or combinations of aromatic rings with aliphatic rings and chains. One such class of compounds is biphenyl and its derivatives, in which two benzene rings are connected by a single C–C bond. The structural formula of biphenyl (or phenylbenzene) is

m o   o m
p   p   or   or $C_6H_5C_6H_5$
β   α

in which the ortho-, meta-, and para-positions of the carbon atoms in the α and β rings are as marked.

Condensed-ring or fused-ring systems contain two or more aromatic rings that share a pair of carbon atoms. Examples include naphthalene, anthracene, and phenanthrene.

(naphthalene)   (anthracene)

or

(phenanthrene)

Other polynuclear hydrocarbons include bridged hydrocarbons, spiro hydrocarbons, mixed systems containing alicyclic and aromatic rings, and aliphatic chains, etc. Examples may be found in the *CRC Handbook* [3]. Physical properties of naphthalene are discussed in Reference 1.

Although many of the aromatic compounds have pleasant odors, they are usually toxic, and some are carcinogenic. Volatile aromatic hydrocarbons are highly flammable and burn with a luminous, sooty flame. The effects of molecular size (in simple arenes as well as in substituted aromatics) and of molecular symmetry (e.g., xylene isomers) are noticeable in physical properties [1,3].

2.7.3 Other Organic Compounds

Many other functional groups are possible. Other atoms, such as O, N, S, and Cl, combine with hydrocarbon chains to form homologous series. Table 2.7.2 lists common functional groups and examples of organic compounds containing them. Compounds in a homologous series show gradual variations in physical properties as the molecular size increases.

When compounds contain more than one functional group in their structures, they are referred to as *polyfunctional*

Table 2.7.2 *Selected Functional Groups and Representative Organic Compounds [3]*

Class	Functional Group	Example Molecular Formula	Example Compound Name
Alkene	>C=C<	$CH_3CH = CH_2$	Propylene
Alkyne	$-C\equiv C-$	$\begin{array}{c}CH_2 \\ | \diagdown \\ CHC \equiv CH \\ CH_2 \diagup \end{array}$	Cyclopropylacetylene
Alcohol	-OH	$CH_3CH_2\underset{\underset{OH}{|}}{C}HCH_3$	sec-Butyl alcohol
Aldehyde	$-\overset{\overset{O}{\|}}{C}-H$	C_6H_5CHO	Benzaldehyde
Ketone	$-\overset{\overset{O}{\|}}{C}-$	CH_3COCH_3	Acetone (Methyl ketone)
Carboxylic acid	$-\overset{\overset{O}{\|}}{C}-OH$	C_6H_5COOH	Benzoic acid
Ester	$-\overset{\overset{O}{\|}}{C}-O-$	$CH_3COOCH_2CH_3$	Ethyl acetate
Acid anhydride	$\begin{array}{c}-\overset{\overset{O}{\|\|}}{C}\diagdown \\ O \\ -\underset{\underset{O}{\|\|}}{C}\diagup \end{array}$	$(CH_3CO)_2O$	Acetic anhydride
Ether	-O-	$CH_3OC_6H_5$	Methylphenyl ether
Epoxide (Oxirane)	$\underset{>C - C<}{\overset{O}{\diagup\diagdown}}$	$\underset{O}{\overset{CH_2 - CH_2}{\diagdown\diagup}}$	Ethylene oxide
Peroxide	-O-O-	$(CH_3)_3COOC(CH_3)_3$	Di-tert-butyl peroxide
Halide	-X	$CHCl_3$	Trichloromethane (chloroform)
Acid halide	$-\overset{\overset{O}{\|}}{C}-X$	C_6H_5COCl	Benzoyl chloride
Amine	$-NH_2$	$(CH_3)_3N$	Trimethylamine

(continued)

Table 2.7.2   (continued)

Class	Functional Group	Example Molecular Formula	Example Compound Name
Nitrile	$-C\equiv N$	C_6H_5CN	Benzonitrile
Nitroso	$-N=O$	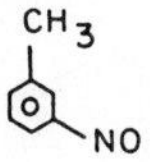 (structure with CH_3)	*meta*-Nitrosotoluene
Nitro	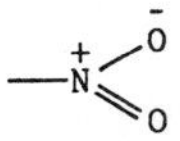 (structure $-N^+{}_{O^-}^{=O}$)	CH_2ONO_2 $CHONO_2$ CH_2ONO_2	Nitroglycerine
Amide	$-\overset{O}{\overset{\|}{C}}-NH_2$	$CONH_2$ $CONH_2$	Oxamide
Imide	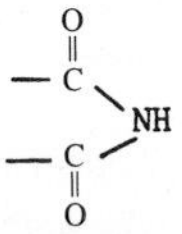 (structure)	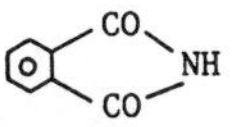 (structure)	Phthalimide
Isocyanate	$-N=C=O$	$CH_3(CH_2)_4NCO$	*n*-Amyl isocyanate
Oxime	$>C=N-OH$	C_6H_5CHNOH	Benzaldoxime
Mercaptan (Thiol)	$-SH$	C_6H_5SH	Phenyl mercaptan (Thiophenol)
Sulfide (Thioether)	$-S-$	$CH_3SCH_2CH_3$	Methylethyl sulfide
Disulfide	$-S-S-$	$CH_3CH_2SSCH_2CH_3$	Ethyldisulfide
Sulfonate (Salt of sulfonic acid)	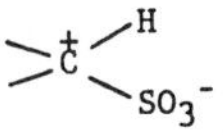 (structure $>C^+{}_{SO_3^-}^{H}$)	$C_{12}H_{25}SO_3Na$	Sodium dodecyl sulfonate (Sodium salt of dodecylsulfonic acid)
Organometallic	$-\overset{\|}{\underset{\|}{C}}-M$	$(CH_3)_4Sn$	Tetramethyltin
	R^-M^+	$\overset{-}{R}\ \overset{++}{Mg}\ \overset{-}{Br}$	Alkylmagnesium bromide

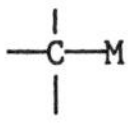

compounds. Examples include

$$\begin{array}{c} COOH \\ | \\ COOH \end{array}$$
(oxalic acid)

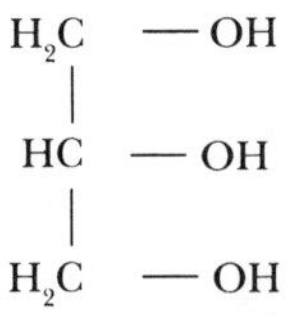

$$\begin{array}{c} H_2C-OH \\ | \\ HC-OH \\ | \\ H_2C-OH \end{array}$$
(glycerol)

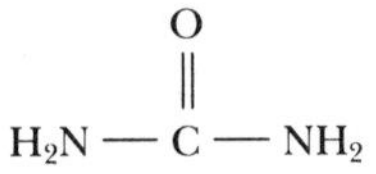

$$\begin{array}{c} O \\ \| \\ H_2N-C-NH_2 \end{array}$$
(urea)

are shown below:

$$\begin{array}{c} H_2C-CH_2 \\ \diagdown O \diagup \end{array}$$
(ethylene oxide)

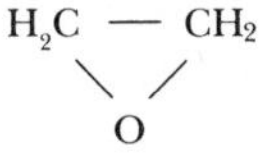

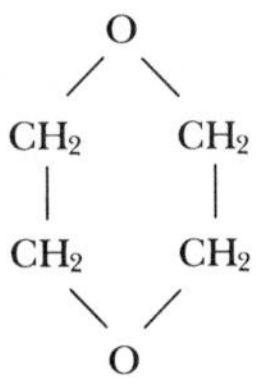

(1,4-dioxane)

(furan)

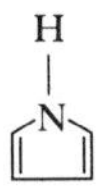

(pyrrole)

(thiophene)

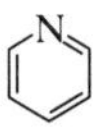

(pyridine)

Unlike the hydrocarbon rings (carbocyclic or homocyclic) discussed previously, *heterocyclic* compounds contain in their rings other atoms in addition to carbon. Some examples

Other structures may be found in Reference 1. Physical properties of heterocyclic compounds are given in Reference 5.

2.7.4 Stereoisomerism

Stereoisomers contain the same atoms bonded to each other in the same ways, but they differ with respect to the orientation of the atoms in space. *Enantiomers* are stereoisomers that are mirror reflections of each other. *Diastereomers* are stereoisomers that are not mirror reflections of each other.

Enantiomers come about because of the tetrahedral arrangement of singly-bonded atoms around carbon. If four different groups are bonded to a carbon atom, that atom is a *chiral center*, and the *chiral compound* will have two enantiomeric forms. Molecules may have more than one chiral center. Sugars contain multiple chiral centers.

The Cahn-Ingold-Prelog, or R-S, system is used to name enantiomers. Groups bonded to the chiral carbon are given priority numbers from 1 to 4. Priority numbers are assigned by atomic number of the atom bonded to the chiral carbon, starting with the lowest atomic number. Thus, H would always have the priority number of 1. If a distinction cannot be made between tow groups on this basis, the next atom is considered. For example, 2-butanol has a chiral center at the asterisked carbon:

$$CH_3 - C^*H - CH_2CH_3$$
$$|$$
$$OH$$

The groups attached to this carbon would be assigned priority numbers 1 for H, 2 for CH_3 3 for CH_2CH_3, and 4 for OH.

The formula or model is then rotated in space so that the lowest priority group is behind the chiral carbon. If the priority numbers of the other groups then increase counterclockwise, the enantiomer is designated R. If they increase clockwise, the enantiomer is designated S. The two enantiomers of 2-butanol are shown below.

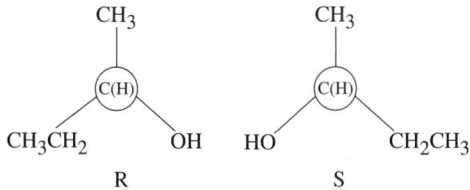

For further discussions of stereoisomerism, consult References 1, 2, and 4.

2.7.5 Petroleum Chemistry

Natural petroleum deposits are made up primarily of hydrocarbons in gaseous (*natural gas*), liquid (*crude oil* or *petroleum*), and solid (*tars* and *asphalts*) form. For general literature on the subject of petroleum chemistry consult References 6–9.

Methods of characterizing a crude oil include, among other techniques, the ultimate analysis for elemental composition and the classification, based on a standard distillation procedure, into various boiling fractions and residuum (Table 2.7.3). Selected examples of the types of compounds present in petroleum are illustrated in Tables 2.7.4 and 2.7.5.

The principal classes of hydrocarbons found in crude oils are paraffins, cycloparaffins (naphthenes), and aromatics. Compounds containing N, S, or O and metals and also found in crude oil. The molecular weight distribution of a crude oil ranges from 16 for methane to several thousands for asphaltenes. The lower boiling point fractions of most crude oils are dominated by saturated hydrocarbons (i.e., paraffins and napthenes), whereas the higher boiling fractions and residuum contain large proportions of aromatics and N/O compounds. The sulfur content tends to increase with the boiling point.

The ratio of hydrogen to carbon in typical crude oils is approximately 1.85:1; the other elements, chiefly sulfur, nitrogen, and oxygen account for less than 3% by weight in most light crudes. Table 2.7.6 illustrates the elemental compositions of typical crude oils and asphalts. Sour crudes contain larger amounts of sulfur-containing compounds. Traces of phosphorus and heavy metals such as vanadium, nickel, and iron are also present.

In refining operations, crude oils are fractionally distilled and separated into different fractions according to the boiling point range of the compounds and their end use or application (see Table 2.7.3).

A "base" designation is given in the refining industry to classify crude oils: (i) A *paraffin-base* crude contains predominantly paraffins and small amounts of naphthenes or asphalt. Upon distillation, it yields fine lubricating oils from the gas-oil fraction and paraffin wax from the solid residue. (ii) An *asphalt-base* crude contains mostly cyclic compounds (primarily naphthenes), which, upon distillation, produce high yields of black, pitchlike, solid residue, asphalt, and heavy fuel oil. (iii) A *mixed-base* crude has characteristics intermediate between the above two categories. (iv) An *aromatic-base* crude contains large amounts of low-molecular weight aromatics together with naphthenes, and small amounts of asphalt and paraffins.

Tars and Asphalts are semisolid or solid substances. They are also referred to as bitumens, waxes, and pitch. These materials consist of mixtures of complex organic molecules of high molecular weight. As with crude oils, an exact chemical analysis for identification and composition is impractical to perform on the solid deposits of petroleum.

Table 2.7.3 *Typical Crude Oil Fractions [1]*

Crude fraction	Boiling point, °C	Approximate chemical composition	Uses
Hydrocarbon gas	< 30	C_2–C_4	LP gas for heating
Gasoline	30–180	C_4–C_9	Motor fuel
Kerosene	160–230	C_8–C_{16}	Jet fuel, home heating
Diesel	200–320	C_{10}–C_{18}	Motor fuel
Heavy oil	300–450	C_{16}–C_{30}	Lubricating oil, bunker fuel
Petroleum	> 300 vacuum	C_{26}–C_{38}	Lubricating oil, home heating
"jelly" or "wax"		> C_{25}	Roofing compounds, paving asphalts
Residuum		> C_{35}	

Table 2.7.4 *Selected Examples of Cyclic Hydrocarbons Found in Crude Oils (MW = Molecular Weight, NBP = Boiling Point at 760 mm Hg, MP = Melting Point) [5]*

Naphthenes

	Methylcyclopentane (C_6H_{12})	Methylcyclohexane (C_7H_{14})	Cycloheptane (C_7H_{14})
MW =	84.2	98.2	98.2
NBP(°C) =	71.8	100.9	118.4
MP(°C) =	−142.5	−126.6	−8.0

Aromatics

	Benzene (C_6H_6)	Toluene (Methylbenzene) (C_7H_8)	m-xylene (1,3-dimethylbenzene) (C_8H_{10})	n-heptylbenzene $(C_{13}H_{20})$
MW =	78.1	92.1	106.2	176.3
NBP(°C) =	80.0	110.6	139.1	240
MP(°C) =	5.5	−94.9	−47.8	−48

	Indene (Indonaphthene) (C_9H_8)	Indan (2,3-dihydroindene) (C_9H_{10})	Naphthalene $(C_{10}H_8)$	Tetralin (1,2,3,4-Tetrahydronaphthalene) $(C_{10}H_{12})$
MW =	116.2	118.2	128.2	132.2
NBP(°C) =	182	177.9	217.9	207.6
MP(°C) =	−1.8	−51.4	80.2	−35.7

	Biphenyl (Diphenyl) (Phenylbenzene) $(C_{12}H_{10})$	Acenaphthylene Acenaphthalene $(C_{12}H_8)$	Acenaphthene 1,2-dihydroacenaphthylene $(C_{12}H_{10})$	Fluorene 2,2'-Methylenebiphene $(C_{13}H_{10})$
MW =	154.2	152.2	154.2	166.2
NBP(°C) =	256.1	280	279	295
MP(°C)	69	92.5	93.4	114.8

	Phenanthrene $(C_{14}H_{10})$	Fluoranthene 1,2-(1,8-Naphthylene)benezene $(C_{16}H_{10})$	Pyrene (Benzo [d e f] phenanthrene) $(C_{16}H_{10})$
MW =	178.2	202.3	202.3
NBP(°C) =	340	384	404
MP (°C) =	99.2	107.8	151.2

	Triphenylene (9,10-Benzophenanthrene) $(C_{18}H_{12})$	Chrysene (1,2-Benzophenanthrene) (Benzo [a] phenanthrene) $(C_{18}H_{12})$	Naphthacene 2,3-Benzanthracene) (Tetracene) $(C_{18}H_{12})$
MW =	228.3	228.3	228.3
NBP(°C) =	425	448	Sublimes
MP(°C) =	199	258.2	357

	Pentacene (Benzo [b] naphthacene) $(C_{22}H_{14})$	Perylene $(C_{20}H_{12})$
	278.4	252.3
	Sublimes	Sublimes
	257	274

Table 2.7.5 *Selected Examples of NSO Compound Types Found in Crude Oils (MW=Molecular Weight, NBP=Boiling Point at 760 mm Hg, MP=Melting Point) [5]*

	Phenol Hydroxybenzene (C_6H_6O)	Stearic acid Octadecanoic acid $(C_{18}H_{36}O_2)$	Furan Oxacyclopentadiene (C_4H_4O)	Benzofuran Coumarone (C_8H_6O)	Dibenzofuran 2,2'-Biphenylene oxide $(C_{12}H_8O)$
MW =	94.1	284.5	68.1	118.1	168.2
NBP(°C) =	181.8	350 (decomposes)	31.85	174	287
MP(°C) =	40.9	68.8	−85.76	< −18	86.5

CH₃(CH₂)₁₆COOH — Stearic acid

	Cyclopentanethiol Cyclopentylmercaptan $(C_5H_{10}S)$	Cyclohexanethiol Cyclohexylmercaptan $(C_6H_{12}S)$	Thiophene (Thiofuran) (C_4H_4S)	Benzothiophene (Thionaphthene) (C_8H_6S)	Dibenzothiophene $(C_{12}H_8S)$
MW =	102.2	116.2	84.1	134.2	184.3
NBP(°C) =	132.1	158.9	84.0	221	332.5
MP(°C) =	*	*	−39.4	32	99.5

	1H-Pyrrole (Azole) (C_4H_5N)	Pyridine (Azine) (C_5H_5N)	4-Hydroxypyridine 4-Pyridinol (C_5H_5NO)	Indole (2,3-Benzopyrrole) (C_8H_7N)	Quinoline 1-Azanaphthalne (C_9H_7N)
MW =	67.1	79.1	95.1	117.2	129.2
NBP(°C) =	129.8	115.2	> 350	253.6	237.1
MP(°C) =	−23.4	−41.6	149.8	52.5	−14.8

	Isoquinoline (Benzopyridine) (C_9H_7N)	6-Hydroxyquinoline 6-Quinolinol (C_9H_7NO)	1,2,3,4-Tetrahydroquinoline $(C_9H_{11}N)$	5,6,7,8-Tetrahydroquinoline $(C_9H_{11}N)$
MW =	129.2	145.2	133.2	133.2
NBP(°C) =	243.2	360	251	222
MP(°C) =	41.5	195	20	*

	Phenanthridine $(C_{13}H_9N)$	Acridine Dibenzo-[b, e] pyridinol $(C_{13}H_9N)$
MW =	179.2	179.2
NBP(°C) =	349	344.9
MP(°C) =	107.4	106,110 (different crystal forms)

	9H-Carbazole (Dibenzo [b, d] pyrrole) $(C_{12}H_9N)$	7H-Benzocarbazole $(C_{16}H_{11}N)$	1H-Imidazole 1,3-Diazole $(C_3H_4N_2)$	Pyridozine 1,2-Diazabenzene $(C_4H_4N_2)$
MW =	167.2	217.3	68.1	80.1
NBP(°C) =	354.7	448	257	208
MP(°C) =	246.2	134	90.5	−8

	Porphin (Tetramethenetetrapyrrole) $(C_{20}H_{14}N_4)$
MW =	310.4
NBP(°C) =	
MP(°C) =	darkens at 360

Table 2.7.6　*Elemental Composition of Natural Petroleum (Percentage by Weight) [7]*

Element	Crude Oils
Carbon	84–87
Hydrogen	11–14
Sulfur	< 0.1–8
Nitrogen	< 0.1–1.8
Oxygen	< 0.1–1.8
Metals (Ni, V, etc.)	trace – 1,000 ppm

Table 2.7.7　*Composition of Typical Petroleum Gases [6]*

Natural Gas	
Hydrocarbon	
Methane	70–98%
Ethane	1–10%
Propane	trace–5%
Butanes	trace–2%
Pentanes	trace–1%
Hexanes	trace–$\frac{1}{2}$%
Heptanes +	trace–$\frac{1}{2}$%
Nonhydrocarbon	
Nitrogen	trace–15%
Carbon dioxide*	trace–5%
Hydrogen sulfide*	trace–3%
Helium	up to 5%, usually trace or none

Gas from a Well that also produces Petroleum Liquid	
Hydrocarbon	
Methane	45–92%
Ethane	4–21%
Propane	1–15%
Butanes	$\frac{1}{2}$–7%
Pentanes	trace–3%
Hexanes	trace–2%
Heptanes +	none–$1\frac{1}{2}$%
Nonhydrocarbon	
Nitrogen	trace–up to 10%
Carbon dioxide	trace–4%
Hydrogen sulfide	none-trace–6%
Helium	none

*Occasionally natural gases are found which are predominately carbon dioxide or hydrogen sulfide.

Gas collected at the wellhead is mostly methane with decreasing amounts of heavier hydrocarbons. Typical compositions are given in Table 2.7.7. Natural gases are primarily mixtures of normal alkanes in the C_1 to C_4 range, although other paraffins and heavier hydrocarbons may also be present. Natural gases also contain water vapor, carbon dioxide (CO_2), hydrogen sulfide (H_2S), nitrogen and helium. At conditions of high pressure and low temperature, solid hydrates may form between H_2O and the hydrocarbons.

Natural gases are classified as sweet or sour (similar to crude oils), depending on the absence or presence, respectively, of significant amounts of hydrogen sulfide. Wet *gas* is capable of producing liquid hydrocarbons upon suitable treatment; dry gas does not have such ability. Processing of natural gas results in pure methane, liquefied petroleum gas (or LPG, which is mostly propane and some *n*-butane), and gasoline.

Saturated aliphatic hydrocarbons found in petroleum include normal alkanes as well as branched alkanes (isoalkanes). The paraffin content can vary widely from one crude oil to another. n-alkanes have been found throughout the boiling range of most crude oils, from n = 1 to 78. Some waxy crudes contain higher alkanes, up to n = 200. The pour point of a crude oil is strongly influenced by the amount and carbon-number distribution of n-alkanes present. For crude oils, it can range from −70 to 110°F (−57 to 43°C). Whereas n-alkanes tend to raise the pour point, other hydrocarbon types lower it.

All possible isoalkanes from C_4 through C_8 have been found in crude oils, along with several isomers of C_9 and some of C_{10}. Many isoprenoids (e.g., pristane and phytane) which serve as biomarkers to the genesis of petroleum have been detected in significant concentrations. The 2- and 3-methyl alkanes and pristane appear to be the dominant isoparaffins in crude oils.

Cycloalkanes occur in varying amounts in crude oils. The rings contain of five, six, or seven carbon atoms. Alkyl derivatives of cyclopentane and cyclohexane have been found, but not of cycloheptane. The most common naphthenes are methyl- and dimethyl-substituted cyclopentane and cyclohexane. The amount of naphthenes can be greater than 50 wt% of a crude oil, with the lighter boiling fractions containing less and the heavier fractions containing more. Fused polycyclic aliphatic structures such as decalin ($C_{10}H_{18}$) become prevalent in the heavier fractions. Some of these multiring compounds may contain up to seven rings.

Alkenes are found in petroleum in very low concentrations because of the reaction. Trace concentrations of multiring cycloalkenes such as hopenes and sterenes have been reported to be present in crude oils. Alkynes and cycloalkynes are not commonly found in natural petroleum.

Aromatic hydrocarbons are frequently found in crude oils. Benzene and alkylbenzenes, ranging from methyl through decyl groups, have been found in liquid petroleum, together with several C_{11}-alkyl isomers. Indane and tetrahydronaphthalenes as well as some of their methyl isomers have been identified. Biphenyl and its derivatives occur in lower concentrations than naphthalene and its derivatives. Polynuclear aromatics including phenanthrene, fluoranthene, pyrene, benz[a]anthracene, chrysene, triphenylene, benzopyrenes, perylene, etc., and some of their alkyl derivatives have been detected, but not anthracene. Crude oils also may contain aromatic compounds such as acenaphthene, acenaphthylene, fluorene, dibenzanthracenes, etc. Among the simpler aromatic molecules, toluene and meta-xylene are the most common; benzene, ethylbenzene, and other alkylbenzenes also occur in significant concentrations in distillates.

Aromatic content can vary considerably between crudes but rarely exceeds 15% of the total crude weight. The aromatic hydrocarbons appear throughout the boiling range but tend to be concentrated in the heavy fractions of petroleum, including the residuum. As a class, aromatics (e.g., toluene and xylenes) have the highest octane ratings among hydrocarbons and hence are used as additives to gasoline and other fuel oils. They show the largest viscosity changes with temperature and are therefore undesirable in the lubricating oil range. With rising boiling point, the heavy fractions contain increasing amounts of complex polycyclic aromatic compounds that are difficult to characterize. Some polynuclear aromatic molecules such as 3,4-benzopyrene and benz[a]anthracene are carcinogens.

Compounds containing nitrogen, sulfur, and oxygen are usually abbreviated *NSO compounds* and are sometimes referred to as *asphaltics*. The residuum contains a high percentage of NSO compounds. The strong odor of crude oil is imported by NSO compounds.

All crude oils contain sulfur in one of several forms including elemental sulfur, hydrogen sulfide, carbonyl sulfide (COS), and in aliphatic and aromatic compounds. The amount of sulfur-containing compounds increases with an increase in the boiling point of the fraction. Most of these compounds have one sulfur atom per molecule, but certain aromatic and polynuclear aromatic molecules found in low concentrations in crude oil contain two and three sulfur atoms.

Alkyl thiols (mercaptans) with normal or branched alkyl groups and with the thiol group in a primary, secondary or tertiary location have been found in petroleum, together with cycloalkyl thiols having rings of five or six carbon atoms. Continuous chain or branched alkyl sulfides and cyclic sulfides with four or five carbon atoms in their rings have been detected. Mixed alkyl cycloalkyl sulfides have also been found. Alkylpolycyclic sulfides containing one to eight cycloparaffin rings were identified in certain crudes. Aromatic compounds of sulfur include thiophenes, their benzo- and dibenzo-derivatives, and benzonaphthothiophenes. Thioindanes and alkylaryl sulfides are also present.

In general, mercaptans are more malodorous than sulfides and hydrogen sulfide. The presence of significant amounts of sulfur can poison catalyst for the refining of crude oil.

Most crude oils contain nitrogen; a large proportion of it occurs in the high boiling fractions and in the residuum. Examples of the nitrogen compounds present in petroleum include mono-, di-, and tri-alkylpyridines, quinoline and alkyl substituted quinolines, tetrahydroquinolines and dialkylbenz[h]quinolines. Carbazole and methyl- through decyl-substituted carbazoles have also been identified. The high boiling fractions from one crude oil contained a variety of nitrogen compound types (in excess of 0.1 wt% concentration) that included indoles, carbazoles, benzcarbazoles, pyridines, quinolines, and phenanthridines. Compounds containing both nitrogen and oxygen, such as amides, hydroxypyridines and hydroxyquinolines, as well as compounds containing two nitrogens such as azaindoles and azacarbazoles, were also found. Other molecular types including pyrroles, isoquinolines, benzoquinolines, and benzologues of acridine may be present in crude oil. Porphyrins are observed in the residuum, usually in association with metals. Certain aromatic nitrogen compounds (e.g., pyridines, quinolines) can cause coking on acid catalysts during petroleum processing.

Most crude oils contain only small amounts of oxygen. Oxygen compounds are mainly carboxylic acids, including straight-chain fatty acids, branched-chain acids, naphthenic acids, and dicarboxylic acids. Other molecular types observed in the higher boiling fractions include furans and their benzo-, dibenzo-, and benzonaphtho- derivatives. Oxygen may also be present in the form of phenols, alcohols, esters, and ketones and in combination with nitrogen.

Residuum is the undistilled fraction remaining at the end of distillation, which corresponds to an upper limit of $-565°C$ ($-1050°F$) at atmospheric pressure, or up to $-675°C$ ($-1250°F$) under vacuum. The residuum amounts to a small percentage of a very light crude oil and up to 30–40 wt% of a heavy crude. Its major constituents are resins, asphaltenes, and some high molecular weight oils and waxes. The residuum accounts for most of the total NSO content and the heavy metals. The resins and asphaltenes precipitate out when the residuum (or crude oil) is treated with liquid propane below 70°F. Additional treatment of this precipitate with n-pentane separates the soluble resins from the insoluble asphaltenes. The amount of resins always exceeds the asphaltene content of a crude oil. Resins are light to dark colored and range from thick viscous materials to amorphous solids. Asphaltenes are dark brown to black amorphous solids. Together, they may contain nearly 50% of the total nitrogen and sulfur in the crude oil, predominantly in the form of heterocyclic condensed rings structures. Asphaltenes may account for as much as 25 wt% of the residuum (up to 12% of the crude oil). Colorless oils are the most paraffinic, while asphaltenes are the most aromatic. Dark oils and resins show similar degrees of paraffinicity and aromaticity. Up to 40 wt% of saturated hydrocarbons may be present in the residuum; however, this comprises only 1–3 wt% of the total crude. The rest are aromatic and N/O-containing compounds. In the nonasphaltene fraction of the residuum, the typical aromatic structure is a highly substituted, condensed polynuclear aromatic molecule, with an average formula, $C_{100}H_{160}S$. The substituents are fused naphthenic rings, which in turn are substituted with long (C_{15}–C_{20}) alkyl side chains having intermittent methyl branches. The average structure for a N/O compound is similar in features excepting for slightly higher aromaticity and shorter (C_{10}–C_{15}) alkyl side chains. Other types of NSO compounds described previously may also be present.

In the asphaltene fraction, pure hydrocarbons become rare at molecular weights above 800, and polar functional groups become common. Asphaltenes exist as a dispersion of colloidal particles in an oily medium. Their molecular weight can be as high as 200,000. A typical asphaltene molecule has 10–20 condensed aromatic and naphthenic rings, with both alkyl and cycloalkyl substituted side chains. The observed polar functional groups in the high molecular weight compounds include carboxylic acids, amides, phenols, carbazoles and pyridines.

The *metals* present in crude oils usually exist in complexes of cyclic organic molecules—called porphyrins. The parent structure of the porphyrins is the tetramethene-tetrapyrrole ring ($C_{20}H_4N_4$). In a metalloporphyrin, the transition metal atom is held at the center of the porphyrin ring by coordination with the four pyrrole N atoms. Nickel and vanadium are present in petroleum in concentrations of less than 1 ppb to about 1000 ppm, in combined form with porphyrins. Some of the lighter metalloporphyrins are volatile, while the high molecular weight porphyrins appear in the nonvolatile residuum. When oil deposits occur together with saline water, the produced oil-water emulsions contain soluble salts of sodium, calcium, and magnesium. Other metals may be present in the inorganic state as suspended solids in the produced fluids along with any clays or mineral matter derived from the rock matrix and piping.

Figure 2.7.1 shows how the weight distributions of the different molecular types vary during the fractional distillation of a naphthenic crude oil. Saturated aliphatic hydrocarbons (i.e., paraffins and naphthenes) are the predominant constituents in the light gasoline fraction. As the boiling point is raised, the paraffin content decreases, and the NSO content increases. About 75 wt% of the residuum is composed of aromatics and NSO compounds.

Certain properties of a liquid fuel are measured routinely for characterization purposes. Besides density and viscosity, these properties include the *pour point*, the *cloud point*, and the *flash point*. Standard ASTM (American Society for Testing Materials) procedures are available for their determination.

The *pour point* represents the lowest temperature at which the liquid fuel will pour. This is a useful consideration in the transport of fuels through pipelines. To determine the pour point, an oil sample contained in a test tube is heated up to 115°F (46°C) until the paraffin waxes have melted. The tube is then cooled in a bath kept at about 20°F (11°C) below the estimated pour point. The temperature at which the oil does not flow when the tube is horizontally positioned is termed the pour point.

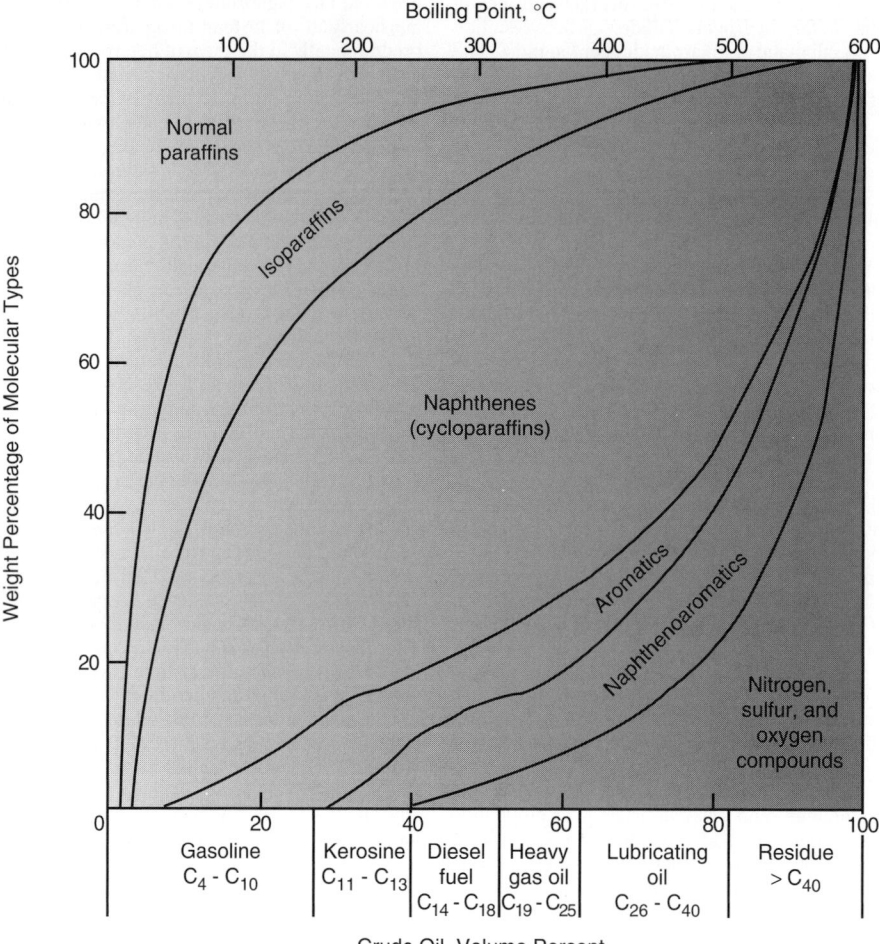

Figure 2.7.1 *Chemical composition of a naphthenic crude oil [9].*

The aniline *cloud point* is a measure of the paraffinicity of a fuel oil. A high value denotes a highly paraffinic oil, and a low value indicates an aromatic, a naphthenic, or a highly cracked oil. The *flash point* represents the temperature to which a liquid fuel can be heated before a flash appears on its surface upon exposure to a test flame under specified conditions. A knowledge of the flash point is needed to ensure safe handling and storage without fire hazards.

The specific gravity of petroleum or petroleum products is often expressed in terms of "degrees API" on a scale defined by

$$^\circ API = \frac{141.5}{G} - 131.5 \qquad [2.7.1]$$

where G is the specific gravity of the liquid at 60°F, with reference to water at 60°F. An *API gravity* of 10° corresponds to G=1.

U.O.P. characterization factor (K) is used as a qualitative index of the paraffinicity of an oil stock. By definition, $K = (T_B)^{1/3}/G$, where T_B is the average boiling point (°R) at one standard atmosphere, and G is the specific gravity at 60°F. The U.O.P. characterization factor has been found to vary within a homologous series; it is therefore not a true measure of the hydrocarbon type. Nevertheless, graphical correlations have been developed between K, API gravity,

molecular weights, T_B, and critical temperatures for a number of petroleum fractions [10]. Characterization factors have also been correlated with kinematic viscosities, API gravity, and T_B.

2.7.6 Physical Chemistry
2.7.6.1 Basic Definitions

The *formula weight* of an element (or a compound or a species) is the sum of the weights of the atoms making up its chemical formula. The formula weight of an element is its *atomic weight* and that of a compound is its *molecular weight*.

The *equivalent weight* of an ion is the ratio of its formula weight to its valence. Another definition of equivalent weight is the amount of a substance that reacts with one atomic weight of hydrogen or its chemical equivalent. Other definitions of equivalent weight may depend on the stoichiometry of a particular chemical reaction.

One *gram-atom* is the mass in grams of an element that is numerically equal to its atomic weight. Thus, the number of gram-atoms of an element is m/A, where m is its mass in grams and A, its atomic weight. Similarly, a pound-atom is the mass in grams of a given element that is numerically equal to its atomic weight. The following gram-quantities also have analogous pound-quantities. When the quantities are not specified as gram or pound, it may be assumed they refer to grams.

One *gram-mole* is the mass in grams of a compound that is numerically equal to its molecular weight (M). Thus, the number of moles of a compound contained in a mass of m grams is m/M.

One *gram-equivalent* is the mass in grams of a material that is numerically equal to its equivalent weight.

Avogadro's number is the number of atoms (or molecules) present in one gram-atom (or gram-mole) of any elementary substance (or compound). It has a value of 6.022142×10^{23} [5].

The *density* (ρ) of a substance is defined as its mass per unit volume, expressed as g/cm^3, lb/ft^3, etc. The *specific volume* ($\hat{V}$) of a substance is the reciprocal of its density and is expressed as cm^3/g, ft^3/lb, etc.

The *specific gravity* (G) of a substance is the ratio of the density (ρ) of the substance to the density (ρ_{ref}) of a reference substance at specified conditions. That is,

$$G = \frac{\rho}{\rho_{ref}}$$

For solids and liquids, density is a weak function of pressure and, therefore, only the temperatures T and T_{ref} are usually specified. The reference substance is commonly taken as water at $4°C$ at which $\rho_{ref} = 1.000 \, g/cm^3 = 62.43 \, lb/ft^3$. If a single temperature is specified, it implies that both densities have been measured at that temperature.

In the case of gases, it is imperative that both temperature and pressure be specified for specific gravity. Air is normally chosen as the reference fluid for gaseous substances.

Many resources for unit conversion and calculation of chemical and chemical engineering problems are available on the Internet. A good place to start is Martindale's The Reference Desk (http://www-sci.lib.uci.edu/HSG/RefCalculators3.html), which points to an enormous number of resources of this type. Specific references to these calculators will not be made in this handbook because of the changing nature of the Internet. However, for all the calculations in this section, relevant calculators and other help can be found in Martindale's.

2.7.6.2 Compositions of Multicomponent Systems

The relative amounts of individual components (or species) making up a mixture or solution can be expressed in a variety of ways: volumetric, mass, or molar.

Volumetric Basis. The volume fraction (v_i) of component i in a mixture is the fraction of the total volume (V) of the mixture that is attributable to that component at the stated temperature and pressure. If V_i denotes the actual volume of component i present in a total volume V of the mixture, then

$$v_i = \frac{V_i}{V} \qquad \sum v_i = 1 \qquad [2.7.2]$$

and the volume percent of $i = 100 \, v_i$. If V is the volume of one mole of the mixture, then V_i is the partial molar volume of component i.

For gases in which ideal gas behavior can be assumed, these relationships hold, and the total volume of the mixture equals the sum of the volumes (V_i^0) of the pure components, or

$$V = \sum V_i^0 \qquad \text{and} \qquad V_i = V_i^0 \qquad [2.7.3]$$

When a gas mixture contains water vapor, the volumetric analysis is stated either on a *wet* basis that includes the water vapor or on a *dry* basis (moisture-free basis) that excludes the water vapor.

For liquids and gases in which ideal gas behavior cannot be assumed, volume changes are observed on mixing of the components. For example, a mixture of ethyl alcohol and water has a lower volume than the sums of the individual volumes before mixing. In this case, the equalities above do not hold. The temperature and pressure dependences of these volume fractions can be complex.

Weight (or Mass) Basis. The terms *weight fraction* and *weight percent* are often used interchangeably with *mass fraction* and *mass percent*, respectively. If the total mass (m) of a mixture includes an amount m_i of component i, then the mass fraction of the component is given by

$$\omega_i = \frac{m_i}{m}$$

and the mass percent of $i = 100 \omega_i$. Note that $\sum m_i = m$ and $\sum \omega_i = 1$.

This representation is ordinarily used for solid and liquid systems and rarely for gases. In the absence of chemical reactions the mass composition of a system remains unchanged. Any consistent set of units for mass (or volume) may be selected in interpreting the compositions expressed on a mass (or volumetric) basis.

Molar Basis. The mole fraction (x_i) of a component i in a multicomponent mixture is defined as

$$x_i = \frac{\text{Number of moles of i}}{\text{Number of moles of mixture}} = \frac{n_i}{n} = \frac{n_i}{\sum n_i} \qquad [2.7.4]$$

The mole percent of $i = 100 x_i$ and $\sum x_i = 1$.

The volume fractions and mole fractions become identical in ideal gas mixtures at fixed conditions of pressure and temperature. In an isolated, nonreactive system, the molar composition does not vary with temperature.

Mixture Properties. The *average molecular weight* (M) of a homogeneous mixture is its total mass (m) divided by the total number of moles (n) of its components. If x_i represents the mole fraction of the ith component whose molecular weight is M_i, then

$$M = \frac{m}{n} = \sum x_i M_i \qquad [2.7.5]$$

In terms of mass fractions,

$$\frac{1}{M} = \sum \left(\frac{\omega_i}{M_i} \right) \qquad [2.7.6]$$

The *mass density* (ρ) of the mixture is the ratio of its total mass (m) to its total volume (V), whereas the *molar density* (c) of the mixture is defined as the ratio of the total number of moles (n) to the total volume of the mixture. Thus,

$$\rho = \frac{m}{V} \qquad \text{and} \qquad c = \frac{n}{V} \qquad [2.7.7]$$

and both are related by $\rho/c = m/n = M$.

Mass and Molar Concentrations of Components. The *mass concentration* (ρ_i) of the ith component (or species) in a homogeneous mixture is defined as the mass of that component present per unit volume of the mixture. The *molar concentration* (c_i) of component i is defined similarly but on a molar basis. Therefore,

$$\rho_i = \frac{m_i}{V} \qquad \text{and} \qquad c_i = \frac{n_i}{V} \qquad [2.7.8]$$

and both are related by

$$\frac{\rho_i}{c_i} = \frac{m_i}{n_i} = M_i \qquad [2.7.9]$$

Also,

$$\rho = \sum \rho_i \qquad \text{and} \qquad c = \sum c_i \qquad [2.7.10]$$

Compositions of Liquid Systems. A *solution* is a homogeneous mixture of two or more components. Petroleum is a solution, as are many of the liquids derived from it. Much of petroleum analysis and processing requires solutions of various kinds. Water is a common solvent. Because so many petroleum components are nonpolar, water is frequently not the best solvent, and organic solvents must be considered. A uniform solid mixture of two or more components is also called a solution.

There are a number of ways to quantify concentrations in solutions suited to different applications. On a weight basis, *grams of solute per gram of solvent* may be used. In the case of trace components, this may be expressed as *parts per million* (ppm), which is grams of solute per million grams of solvent. Also on a weight basis, the *molality* of a solution is the number of gram-moles of solute per 1000 grams of solvent and is usually designated by a small *m*.

Volumetric concentrations include

1. Grams of solute per unit volume of solution, or the *mass concentration* (e.g., grams of i/cm^3 of solution)
2. Gram-moles of solute per unit volume of solution, or the *molar concentration* (e.g., g-moles of i/cm^3 of solution), termed *molality*
3. Gram-equivalents of solute per liter of solution, termed *normality*.

Normality and *molarity* have widespread use in volumetric analysis. If a fixed amount of reagent is present in a solution, it can be diluted to any desired normality by application of the general dilution formula $V_1N_1 = V_2N_2$. Here, subscripts 1 and 2 refer to the initial solution and the final (diluted) solution, respectively; V denotes the solution volume (in milliliters) and N the solution normality. The product V_1N_1 expresses the amount of the reagent in gram-milliequivalents present in a volume V_1 ml of a solution of normality N_1. Numerically, it represents the volume of a one normal (1N) solution chemically equivalent to the original solution of volume V_1 and of normality N_1. The same equation $V_1N_1 = V_2N_2$ also applies in problems involving acid-base neutralization, oxidation-reduction, precipitation, or other types of titration reactions.

Example 2.7.1

The composition of an aqueous solution of H_2SO_4 is given as 40 mole % and its mass density as 1.681 g/cm^3 at 20°C. Find the mole density of the solution and also express its composition in the following ways: (i) weight percent, (ii) lb of solute/lb of solvent, (iii) lb-moles of solute/lb of solvent, (iv) g of solute/100 ml of solution, (v) molarity, (vi) normality, and (vii) molality.

2.7.6.2.1 Solution

Choose a *basis* of 100 g-moles of solution.

MW of H_2SO_4 (solute) = 98.07; MW of H_2O (solvent) = 18.02

Amount of solute = 40 g-moles = (40)(98.07) = 3922.8 g

Amount of solvent = 60 g-moles = (60)(18.02) = 1081.2 g

Total mass of solution = 5004.0 g

$$\text{Total volume of solution} = \frac{5004 g}{1.681 g/cm^3} = 2976.8 cm^3$$

Molar density c = n/V = 100/2976.8 = 3.36 × 10^{-2} g-moles/cm^3

(i) weight percent of solute = (3922.8)(100)/(5004)

$$= 78.39\%$$

weight percent of solvent = 100 − 78.39 = 21.61%

(ii) $\dfrac{\text{lb of solute}}{\text{lb of solvent}} = \dfrac{3922.8}{1081.2} = 3.628$

(iii) $\dfrac{\text{lb-moles of solute}}{\text{lb of solvent}} = \dfrac{40}{1081.2} = 3.70 \times 10^{-2}$

(iv) $\dfrac{\text{g of solute}}{100 \text{ ml of solution}} = \dfrac{(3922.8)}{(2976.8)}(100) = 131.78$

(v) Molarity = $\dfrac{\text{g-moles of solute}}{1000 \text{ ml of solution}}$

$$= \frac{(40)}{(2976.8)}(1000) = 13.44 M$$

(vi) Equivalent weight of solute = $\dfrac{MW}{valence}$

$$= \frac{98.07}{2} = 49.03 g/g\text{-eq.}$$

$$\text{Normality} = \frac{\text{g-eq of solute}}{1000 \text{ ml of solution}}$$

$$= 2(\text{Molarity}) = 26.87 N$$

(vii) Molality = $\dfrac{\text{g-moles of solute}}{1000 \text{ g of solvent}} = \dfrac{(40)(1000)}{(1081.2)}$

$$= 37.00 \text{ molal}$$

The pH scale and Ionic Strength. The *pH of a solution* is defined as the logarithm of the reciprocal of the hydrogen ion concentration (or activity) [H$^+$] of the solution:

$$pH = -\log[H^+] \qquad [2.7.11]$$

Because the ion product of water $K_w = [H^+][OH^-] = 1.04 \times 10^{-14}$ at 25°C, it follows that a neutral solution (e.g., pure water at 25°C) in which $[H^+] = [OH^-]$ has a pH = 7. Acids have a lower pH and bases a higher pH than this neutral value of 7. Hydrogen ion concentrations can cover a wide range, from ~1 g-ion/liter or more in acidic solutions to ~10^{-14} g-ion/liter or less in alkaline solutions [11]. *Buffer* action refers to the property of a solution in resisting a change of pH on addition of an acid or a base. Buffer solutions usually consist of a mixture of a weak acid and its salt (conjugate base) or of a weak base and its salt (conjugate acid).

The ionic strength (μ) of a solution is a measure of the intensity of the electrical field due to the ions present in the solution. It is defined by

$$\mu = \left(\frac{1}{2}\right) \sum c_i Z_1^2$$

where c_i is the molal concentration of the ionic species i (g-equivalent i/1000 g solvent), and z_i is the valence or charge on the ion i.

Further discussion of pH and other association constants can be found in References 1, 3, and 4.

Example 2.7.2

Find the ionic strength of (i) 0.05 molal sodium sulfate (Na_2SO_4) solution, and (ii) 0.25 molal nitric acid (HNO_3) and 0.4 molal barium nitrate ($Ba(NO_3)_2$) together in one solution.

2.7.6.2.2 Solution

(i) Ions Na^+ and SO_4^{2-} are present in concentrations of 0.10 and 0.05 molal, respectively.

$$\mu = \left(\frac{1}{2}\right)[0.10(1)^2 + (0.05)(2)^2] = 0.15$$

(ii) HNO_3 contributes the ions H^+ and NO_3^- in concentrations of 0.25 molal each, and $Ba(NO_3)_2$ contributes the ions Ba^{2+} and NO_3^- in concentrations of 0.4 and 0.8 molal, respectively. The NO_3^- contributions from both sources are summed.

$$\mu = \left(\frac{1}{2}\right)[0.25(1)^2 + (0.25 + 0.80)(1)^2 + (0.4)(2)^2] = 1.45$$

2.7.6.3 Material Balances

Chemical processes, such as petroleum refining or natural gas cleanup, transform raw materials into useful products. Matter and energy inputs and outputs must be tracked for process control. The principle of conservation of mass provides the basis for tracking matter.

A *system* must be defined before a materials balance can be calculated. The system is all the substances contained within defined boundaries, for example, a process vessel or a system of vessels and piping.

The physical and chemical changes occurring in this system must be enumerated. The definition of the system and the degree of detail necessary for the enumeration of these changes depends on the use intended for the material balance.

Processes may be classified as batch, continuous, or semibatch depending on how materials are transferred into and out of the system. Process operation may be characterized as unsteady state (i.e., transient) or steady state, depending on whether the process variables (e.g., pressure, temperature, compositions, flowrate, etc.) change with time, respectively. In a *batch* process, the entire feed material (charge) is added instantaneously to the system at the beginning of the process, and all the contents of the system including the products are removed later, at the end of the process. In a *continuous* process, the materials enter and leave the system as continuous streams. In a *semibatch* process, the feed may be added at once and the products removed continuously, or vice versa. Batch and semibatch processes are inherently unsteady state, whereas continuous processes may be operated in a steady or unsteady-state mode. Start-up and shut-down procedures of a continuous production process are examples of transient operation.

According to the law of conservation of mass, the total mass of an isolated system is invariant, even in the presence of chemical reactions. Thus, an *overall* material balance is performed on the entire material (or contents) of the system. If a mass balance is made on a component (chemical compound or atomic species) involved in the process, it is termed a *component* (or *species*) material balance. The general mass balance equation has the following form, and it can be applied to any material in any process.

$$\begin{pmatrix} \text{Mass} \\ \text{inflow} \\ \text{entering} \\ \text{the system} \end{pmatrix} - \begin{pmatrix} \text{Mass} \\ \text{outflow} \\ \text{leaving} \\ \text{the system} \end{pmatrix}$$

$$+ \begin{pmatrix} \text{Net internal} \\ \text{mass generation} \\ \text{due to chemical} \\ \text{reactions within} \\ \text{the system} \end{pmatrix} = \begin{pmatrix} \text{Net mass} \\ \text{accumulation} \\ \text{within the} \\ \text{system} \end{pmatrix}$$

or symbolically,

$$m_{in} - m_{out} + \mathcal{R} = \Delta m \qquad [2.7.12]$$

The third term on the left side of the equation $\mathcal{R}$, has significance for components of reactive systems only. It is positive when material is produced as a net result of all chemical reactions; it is negative if material is consumed by chemical reactions. The former situation corresponds to a source and the latter to a sink for the material under consideration. Because the total mass of reactants always equals the total mass of products in a chemical reaction, this term does not appear an overall material balance.

Two types of material balances, differential and integral, are applied in analyzing chemical processes. The differential mass balance is valid at any instant in time, with each term representing a rate (i.e., mass per unit time). A *general differential material balance* may be written on any material involved in any *transient* process, including semibatch and unsteady-state continuous flow processes:

$$\begin{pmatrix} \text{Net rate of} \\ \text{mass accumulation} \end{pmatrix} = \begin{pmatrix} \text{Rate of} \\ \text{mass inflow} \end{pmatrix}$$

$$- \begin{pmatrix} \text{Rate of} \\ \text{mass outflow} \end{pmatrix} + \begin{pmatrix} \text{Net rate of internal} \\ \text{mass generation by} \\ \text{chemical reactions} \end{pmatrix}$$

or symbolically

$$\frac{dm}{dt} = \dot{m}_{in}(t) - \dot{m}_{out}(t) + \dot{\mathcal{R}}(t) \qquad [2.7.13]$$

where each of the rate terms on the right side of the equations can be functions of time. The solution is obtained on integration over time between the initial and final states of the transient process.

A special case of the differential material balance applies to a *continuous steady-state flow process* when all of the rate terms are independent of time and the accumulation term is zero. Thus, the differential material balance for any component i is given by

$$\begin{pmatrix} \text{Rate of mass} \\ \text{outflow, } \dot{m}_{out} \end{pmatrix} - \begin{pmatrix} \text{Rate of mass} \\ \text{inflow, } \dot{m}_{in} \end{pmatrix}$$

$$= \begin{pmatrix} \text{Net rate of internal} \\ \text{mass generation, } \dot{\mathcal{R}} \end{pmatrix}_i$$

and the overall material balance is $\dot{m}_{in} = \dot{m}_{out}$. When chemical reactions are absent, we have

$$(\dot{m}_{in})_i = (\dot{m}_{out})_i \qquad [2.7.14]$$

for a continuous flow process at steady-state.

For a transient process that begins at time t_0 and is terminated at a later time t_f, the *general integral material balance* equation has the form

$$m(t_f) - m(t_0) = \int_{t_0}^{t_f} \dot{m}_{in} dt - \int_{t_0}^{t_f} \dot{m}_{out} dt + \int_{t_0}^{t_f} \dot{\mathcal{R}} dt \qquad [2.7.15]$$

Here, $m(t_f)$ is the mass of the system contents at final time t_f and $m(t_0)$ is the mass at initial time t_0. As before, both component and overall mass balances may be written in integral form.

A special case of Equation 2.7.15 is directly applicable to *batch* processes for which the mass flowrate terms are zero. The integral material balance for any component i in such a process is

$$\begin{pmatrix} \text{Net mass} \\ \text{accumulation of i} \end{pmatrix} = \Delta m_i = m_i(t_f) - m_i(t_0) = \dot{\mathcal{R}}_i \qquad [2.7.16]$$

and the overall mass balance is

$$\Delta m = m(t_f) - m(t_0) = 0 \qquad [2.7.17]$$

It must be kept in mind that the reaction term will not occur in the overall mass balance equations of reactive systems because $\sum \dot{\mathcal{R}}_i = 0$, i.e., there is no net mass gain or loss as a result of chemical reactions.

Stoichiometry in Reactive Systems. The use of molar units is preferred in chemical process calculations since the stoichiometry of a chemical reaction is always interpreted in terms of the number of molecules or number of moles. A stoichiometric equation indicates the relative proportions of the reactants and products in a given reaction. For example, the following stoichiometric equation represents the combustion of propane in oxygen:

$$C_3H_8(g) + 5O_2(g) \rightarrow 3CO_2(g) + 4H_2(\ell)$$
$$\nu_i = \quad 1 \qquad 5 \qquad\qquad 3 \qquad\quad 4$$

One molecule (or mole) of propane reacts with five molecules (or moles) of oxygen to produce three molecules (or moles) of carbon dioxide and four molecules (or moles) of water. These numbers are called *stoichiometric coefficients* (ν_i) of the reaction and are shown below each reactant and product in the equation. In a stoichiometrically balanced equation, the number of atoms of each element in the reactants must be the same as that in the products. Thus, there are three atoms of C, eight atoms of H, and ten atoms of O on either side of the equation. This is a consequence of the principle of conservation of mass. The combined mass of reactants is always equal to the combined mass of products in a chemical reaction, but the total number of moles may be different.

A *limiting reactant* is that reactant which is present in the smallest stoichiometric amount. In industrial reactions, the reactants may not be supplied in the proportions demanded by the stoichiometry of the equation. Under these circumstances, the reaction mixture at the conclusion of the process will include the products together with the unreacted reactants present in excess of their stoichiometric requirements. By identifying a limiting reactant, the *percent excess* is calculated for any *excess reactant* as

$$\% \text{ excess} = \frac{(n - n^*)}{n^*} 100 \qquad\qquad [2.7.18]$$

where n is number of available moles of excess reactant, and n* is the stoichiometric number of moles required to react with the limiting reactant.

Conversion (or degree of conversion) is the fraction of the feed or of some reactant in the feed that is converted into products.

The *degree of completion* of a reaction is the actual fraction of the limiting reactant that is converted into products.

The yield of a product is the actual amount of product divided by the stoichiometric amount of product. These quantities may be expressed in moles or in mass units.

For combustion of fuels, pure oxygen or air is supplied in amounts greater than the stoichiometric requirements for complete combustion. The terms "theoretical air" or "theoretical oxygen" are frequently encountered in combustion problems. The molar composition of dry air at atmospheric conditions [12]:

Gas	Mole%
Nitrogen	78.03
Oxygen	20.99
Argon	0.94
Carbon dioxide	0.03
H_2, He, Ne, Kr, Xe	0.01
	100.00
Average molecular weight =	28.97

In summary, the procedure for material balance calculations involving reactive systems is as follows:

1. Identify the system and physical and chemical changes.
2. Examine the stoichiometry of the chemical reaction, and identify the limiting reactant and excess reactants.
3. Perform an overall material balance and the necessary component material balances so as to provide the maximum number of independent equations. In the event the

balance is written in differential form, appropriate integration must be carried out over time, and the set of equations solved for the unknowns.

For a nonreactive system, the material balance may be done on either a mass or a molar basis.

Example 2.7.3
Consider the combustion of ethane (C_2H_6) in pure oxygen. If 100 lb of ethane are available and 10% excess oxygen is supplied to ensure complete combustion, calculate (1) the amount of oxygen supplied, and (2) compositions of the reactants and products on mass and molal bases.

2.7.6.3.1 Solution
Choose a *basis* of 100 lb of ethane (limiting reactant). The stoichiometric equation is

$$C_2H_6 + \left(\frac{7}{2}\right) O_2 \rightarrow 2CO_2 + 3H_2O$$

$$\nu_i = \quad 1 \qquad \frac{7}{2} \qquad\quad 2 \qquad\quad 3 \quad\text{ lb-moles}$$

$$M_i = 30.07 \quad 32.00 \qquad 44.01 \quad 18.02 \text{ lb/lb-moles}$$

$$\text{lb-moles of } C_2H_6 \text{ available} = 100/30.07$$
$$= 3.326$$

$$\text{Stoichiometric lb-moles of } O_2 \text{ required} = \left(\frac{7}{2}\right)(3.326)$$
$$= 11.641$$

With 10% excess, lb-moles of O_2 supplied $= 1.1(11.64)$
$$= 12.804$$

1. Amount of O_2 supplied $= 12.804(32.00) = 409.7$ lb
2. Calculation of the composition of the feed (reactant) mixture is shown below.

Reactant i	Mass m_i (lb)	Weight %	n_i (lb-moles)	Mole %
Ethane	100.0	19.62	3.326	20.66
Oxygen	409.7	80.38	12.804	79.34
Total	509.7	100.00	16.130	100.00

In the product mixture, amount of excess (unreacted) $O_2 = 0.1(11.64) = 1.164$ lb-moles $= 37.25$ lb.

Amount of CO_2 formed

$$= (2)(100)/30.07 = 6.651 \text{ lb-moles} = 292.7 \text{lb}$$

Amount of H_2O formed

$$= (3)(100)/30.07 = 9.977 \text{ lb-moles} = 179.8 \text{lb}$$

Hence, the composition of the product gas stream can be calculated as follows:

Product j	Mass m_i (lb)	Weight %	n_i lb-moles)	Mole %
CO_2	292.7	57.43 (88.72)	6.651	37.38 (85.11)
H_2O	179.8	35.28 (—)	9.978	56.08 (—)
O_2	37.2	7.29 (11.28)	1.164	6.54 (14.89)
Total	509.7	100.00	17.793	100.00

The values in parentheses show the product composition on a dry basis, excluding H_2O vapor.

2.7.6.4 Ideal Gas Laws

Ideal gas laws are derived from the approximation that gas molecules behave like hard spheres. This approximation is true for most gases at low pressures or elevated temperatures, for example, at atmospheric conditions. Real gas behavior at higher pressures and lower temperatures is frequently represented by modifications of the ideal gas laws. Thus, the ideal gas laws are a good starting place for calculations of gas properties.

Boyle's Law states that constant temperature (T), the volume (V) of a fixed mass of an ideal gas is inversely proportional to the absolute pressure (P). That is,

$$V \propto \frac{1}{P} \quad \text{or} \quad PV = \text{constant} \quad (T = \text{constant})$$

Charles' Law states that constant pressure (P), the volume (V) of a given mass of an ideal gas is directly proportional to the absolute temperature (T):

$$V \propto T \quad \text{or} \quad \frac{V}{T} = \text{constant} \quad (P = \text{constant})$$

These two laws may be combined to give the *ideal (or perfect) gas law*, an equation of state:

$$PV = nRT \quad \text{or} \quad P\hat{V} = \frac{RT}{M} \quad \text{or} \quad \rho = \frac{PM}{RT}$$

where n is m/M = number of moles of ideal gas, m is mass of gas present, M is molecular weight of gas in mass per mole, $\hat{V}$ is $V/M = 1/\rho$ = specific volume of ideal gas in volume per mass, and R is universal gas constant in energy per mole per absolute degree.

Values of R in different sets of units are given below:

8.314	joules/g-mole °K
8.314	m³ Pa/g-mole °K
82.057	cm³ atm/g-mole °K
83.14	cm³ bar/g-mole °K
62.36	liter mm Hg/g-mole °K
21.85	(ft)³(in.Hg)/lb-mole °R
0.7302	ft³ atm/lb-mole °R
10.73	ft³ psi/lb-mole °R
1545	(lb_f/ft²)(ft³)/lb-mole °R
1.987	cal/g-mole °K
1.987	Btu/lb-mole °R

When n moles of an ideal gas undergo a chemical reaction, then

$$P_1V_1 = nRT_1 \quad \text{and} \quad P_2V_2 = nRT_2$$

or

$$\frac{P_1V_1}{P_2V_2} = \frac{T_1}{T_2} \qquad [2.7.19]$$

where the subscripts 1 and 2 refer to the initial state and final state, respectively.

It is a common practice to evaluate the molal volume ($\tilde{V}$) of an ideal gas at a set of reference conditions known as the *standard state*. If the standard state is chosen to be

$$P = 1 \text{atm} = 760 \text{mmHg} = 14.696 \text{psia}$$

and

$$T = 0°C = 273.16°K = 32°F = 491.69°R$$

Then

$$\tilde{V} = \frac{V}{n} = \frac{RT}{P} \qquad [2.7.20]$$

represents the volume occupied by 1 mole of any perfect gas, and has the following values at the standard conditions of *t*emperature and *p*ressure (abbreviated as STP):

$$\tilde{V} = 22.414 \; \ell/\text{g-mole}$$

$$= 22,414.6 \text{ cm}^3/\text{g-mole}$$

$$= 359.046 \text{ ft}^3/\text{lb-mole}$$

The *specific gravity* (G) of an ideal gas relative to a reference gas (also assumed ideal) is given by

$$G = \frac{\rho}{\rho_{\text{ref}}} = \frac{M}{M_{\text{ref}}} \frac{P}{P_{\text{ref}}} \frac{T_{\text{ref}}}{T} \qquad [2.7.21]$$

or $G = M/M_{\text{ref}}$ when $T = T_{\text{ref}}$ and $P = P_{\text{ref}}$.

Although real gases deviate from ideal gas behavior and therefore require different equations of state, the deviations are relatively small under certain conditions. An error of 1% or less should result if the ideal gas law were used for diatomic gases when $\tilde{V} \geq 5\ell/\text{gm-mole}$ (80 ft³/lb-mole) and for other gases and light hydrocarbon vapors when $\tilde{V} \geq 20\ell/\text{gm-mole}$ (320 ft³/lb-mole) [19,20].

Example 2.7.4

Analysis of a volatile compound of chlorine gives 61.23% of oxygen and 38.77% of chlorine by weight. At 1 atm and 27 C, 1000 cm³ of its vapor weighs 7.44 g. Assuming ideal gas behavior for the vapor, estimate its molecular weight and deduce its molecular formula.

2.7.6.4.1 Solution

At $T = 27°C = 300.2°K$ and $P = 1$ atm,

$$\text{Mass density of the vapor} = \rho = \frac{7.44g}{1000 \text{cm}^3} = \frac{PM}{RT}$$

With $R = 82.057$ cm³-atm/g-mole °K, the molecular weight is

$$M = \frac{\rho RT}{P} = \frac{(7.44)(82.057)(300.2)}{(1000)(1)} = 183.3 \text{ g/g-mole}$$

If the unknown compound is made up of x atoms of chlorine and y atoms of oxygen, its molecular formula will be Cl_xO_y. Because the atomic weights of Cl and O are 35.45 and 16.00 g/g-mole, respectively, we can express the weight composition of the compound as

$$\frac{\text{weight of oxygen}}{\text{weight of chlorine}} = \frac{(y)16.00}{(x)(35.45)} = \frac{0.6123}{0.3877}$$

or $y = 3.5$ x. Making this substitution in the formula for the molecular weight of the substance, we have

$$183.3 = M = (x)35.45 + (y)16.00$$

or

$$183.3 = 35.45 + (16.00)(3.5)x = 91.45$$

$$\therefore x = (183.3)/(91.45) \approx 2$$

With $x = 2$ and $y = 7$, the compound has molecular formula Cl_2O_7 with $M = 182.9$.

Dalton's Law of Partial Pressures states that the total pressure (P) of a gaseous mixture equals the sum of the partial pressures of its components. By definition, the *partial pressure* of any component gas is the pressure it would exert if it occupied the entire volume (V) of the mixture at the same temperature (T). That is,

$$P = P_1 + P_2 \cdots + P_N = \sum P_i \qquad [2.7.22]$$

where the partial pressure of component gas i in a N-component gas mixture is

$$P_i = \frac{n_iRT}{V} \quad i = 1, 2, \ldots, N \qquad [2.7.23]$$

by the ideal gas law. If the mixture also behaves ideally, then

$$\frac{PV}{RT}=n=\sum n_i=\left(\sum P_i\right)\frac{V}{RT} \qquad [2.7.24]$$

Thus, the mole fraction of component gas i is

$$y_i=\frac{n_i}{n}=\frac{P_i}{P}=\frac{\text{Partial pressure of i}}{\text{Total pressure of mixture}} \qquad [2.7.25]$$

Amagat's Law states that the total volume of a gaseous mixture is the sum of the pure-component volumes. By definition, the *pure-component volume* of a component gas in a mixture is the volume that the component would occupy at the same temperature and total pressure of the mixture. By Amagat's law,

$$V=\sum V_i=V_1+V_2+\dots$$

If each component gas as well as the mixture obeys the ideal gas law, it follows that the pure-component volume of component is

$$V_i=\frac{n_iRT}{P}=y_i\frac{nRT}{P}=y_iV \qquad [2.7.26]$$

or

$$y_i=\frac{n_i}{n}=\frac{V_i}{V} \qquad [2.7.27]$$

In an ideal gas mixture, the mole fraction of each component is equal to its volume fraction (by Amagat's law) or the ratio of its partial pressure to the total pressure (by Dalton's law). For both laws to be applicable simultaneously, the mixture and its components must behave ideally.

Example 2.7.5
A natural gas has the following composition by volume at a temperature of 80°F and a gauge pressure of 40 psig— 87.2% methane, 4.5% ethane, 3.6% propane, 1.8% n-butane, 1.0% isobutene, and 1.9% nitrogen. Assuming the ideal-gas law is applicable, calculate (i) the average molecular weight of the mixture, (ii) density of the natural gas, (iii) specific gravity of the gas, (iv) volume occupied by 100 lb of gas at 1 atm and 60°F, (v) partial pressure of nitrogen, and (vi) pure-component volume of nitrogen per 1,000 ft^3 of gas.

2.7.6.4.2 Solution
Start with a *basis* of 1 lb-mole of the natural gas at $T=80°F=540°R$ and $P=40$ psig $=54.7$ psia. The volume percent and mole percent compositions are identical for a perfect-gas mixture.

(i) The molecular weight of the gas mixture is $M=\Sigma y_iM_i$ and its calculation is shown below in tabular form.

Component i	M_i (lb/lb-mole)	y_i	$m_i = y_iM_i$ (lb)
CH_4	16.04	0.872	13.987
C_2H_6	30.07	0.045	1.353
C_3H_8	44.10	0.036	1.588
$n-C_4H_{10}$	58.12	0.018	1.046
$iso-C_4H_{10}$	58.12	0.010	0.581
N_2	28.01	0.019	0.532
		1.000	19.087

$$\therefore M=\sum m_i=19.09 \text{ lb/lb-mole}$$

(ii) With $R=10.73$ ft^3-psia/lb-mole °R, mass density of the mixture

$$\rho=\frac{PM}{RT}=\frac{(54.7)(19.09)}{(10.73)(540)}=0.180 \text{ lb/ft}^3$$

Molar density $c=\rho/M=0.180/19.09=9.44 \times 10^{-3}$ lb-moles/ft^3

(iii) Assuming that air (reference substance) also obeys the ideal-gas law, the specific gravity of the mixture is given by

$$G(80°F, 54.7 \text{ psia})=\frac{\rho}{\rho_{air}}=\frac{M}{M_{air}}=\frac{19.09}{28.97}=0.659$$

(iv) At $T_1 = 60°F = 520°R$ and $P_1 = 1$ atm $= 14.696$ psia, the volume occupied by

$$n_1=\frac{100}{19.09} \text{ lb-moles of the gas}$$

is

$$V_1=\frac{n_1RT}{P_1}=\frac{(100)(10.73)(520)}{(19.09)(14.696)}=1,988.3 \text{ ft}^3$$

(v) At $P=40$ psig and $T=80°F$, by Dalton's law of partial pressures, $P_i=y_iP$ for any component i in the mixture. Therefore, partial pressure of nitrogen $= (0.019)(54.7)=1.039$ psia.

(vi) Applying Amagat's law at the same conditions, the pure component volume $V_i=y_iV$ for any i. For nitrogen, $V_i = (0.019)(1,000)=19.0$ ft^3.

2.7.6.5 Phase Rule and Phase Behavior
The phase rule derived by W.J. Gibbs applies to multiphase equilibria in multicomponent systems, in the absence of chemical reactions. It is written as

$$\mathcal{F}=\mathcal{C}-\mathcal{P}+2 \qquad [2.7.28]$$

where $\mathcal{F}$ is number of degrees of freedom or variance of the system, $\mathcal{C}$ is the number of independent chemical components, and $\mathcal{P}$ is number of phases.

The term $\mathcal{F}$ is the number of intensive variables that must be specified to describe the system. The intensive variables are temperature (T), pressure (P), and the composition of the mixtures (e.g., mole fractions, x_i). As an example, consider the triple point of water at which all three phases–ice, liquid water, and water vapor–coexist in equilibrium. According to the phase rule,

$$\mathcal{F}=\mathcal{C}-\mathcal{P}+2=1-3+2=0 \qquad [2.7.29]$$

The absence of any degrees of freedom implies that the triple point is a unique state that represents an *invariant* system, i.e., one in which any change in the state variables T or P will reduce the number of coexisting phases.

A pure substance can have at most three coexisting phases at equilibrium. At temperatures and pressures over than the triple point, a pure substance may exist as a single phase (solid, liquid, or vapor) or as a two-phase system. For the single-phase region ($\mathcal{P}=1$),

$$\mathcal{F}=\mathcal{C}-\mathcal{P}+2=1-1+2=2. \qquad [2.7.30]$$

indicating that P and T can vary simultaneously. In the two-phase region ($\mathcal{P}=2$), $\mathcal{F}=1-2+2=1$, so that either P to T can vary, but not both.

It is possible to graph the equilibrium phase behavior of a pure substance in three dimensions with pressure (P), specific volume* (V), and temperature (T) as coordinates. This is called a phase or P-V-T diagram (Figure 2.7.2). Orthogonal projections onto the P-V, V-T, and P-T planes give two-dimensional plots suited for determining phase properties for those combinations of variables. Isotherms (T=constant), isobars (P=constant), and isochors (V=constant) may be drawn as necessary to highlight aspects of phase behavior.

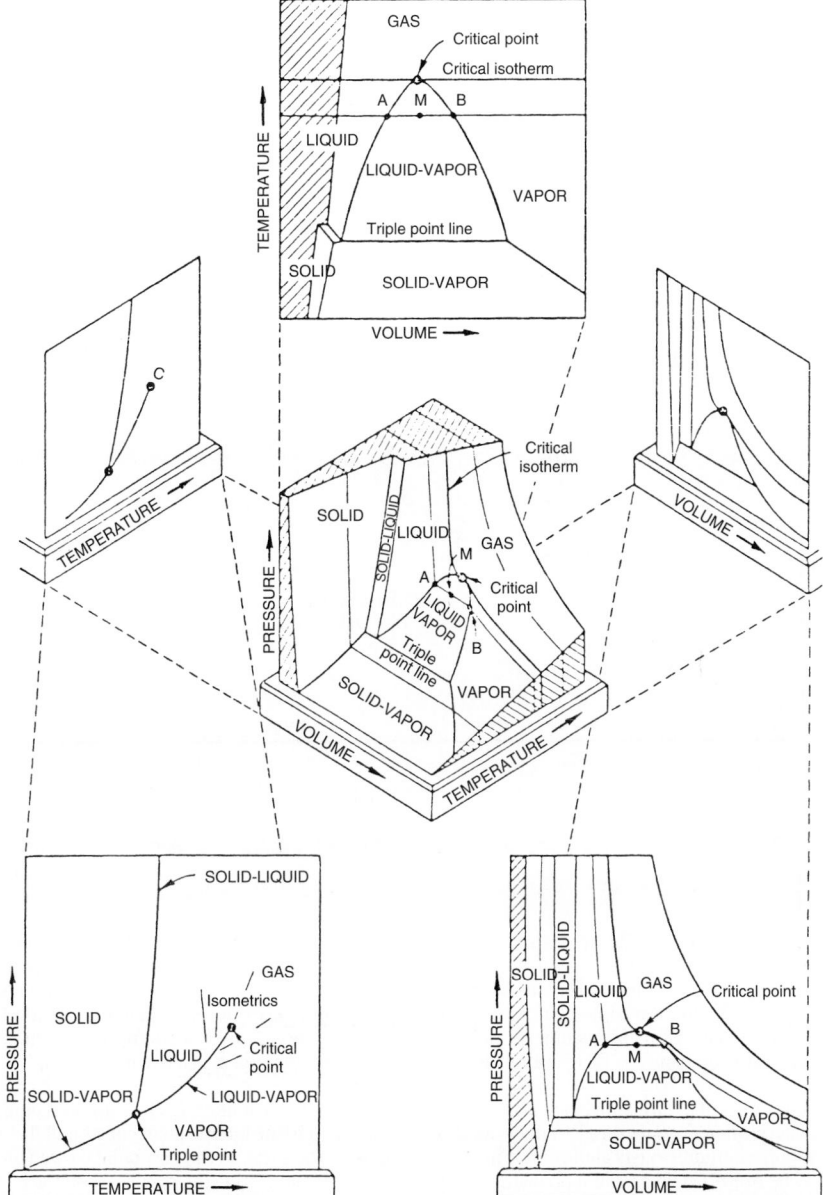

Figure 2.7.2 *Typical phase diagram for a pure substance showing P–V–T surface and its projections [18].*

The *vapor pressure* (P²) of a pure liquid at a given temperature (T) is the pressure exerted by its vapor in equilibrium with the liquid phase in a closed system. All liquids and solids exist in equilibrium with vapor. For instance, in Figure 2.7.3, lines BA and AC represent the equilibrium vapor pressure curves of the solid and liquid phases, respectively.

Phase transitions are equilibrium transformations from one phase to another. Examples are sublimation (solid to

vapor), boiling or vaporization (liquid to vapor), and freezing (liquid to solid). On a P–T diagram (see Figure 2.7.3) the phase transformations take place on the sublimation curve BA, vaporization curve AC, and melting curve AD, which separate the single-phase regions. Two phases coexist in equilibrium on each of these phase-boundary curves, with the exception of the triple point, A. The vaporization curve terminates at the critical point C, at which the distinction between liquid and vapor phases disappears and the latent heat of vaporization becomes zero. The single phase that exists above the critical point (P_c, T_c) is variously described as gas, dense fluid, or supercritical fluid. When T > T_c, it is impossible to liquefy a dense fluid by varying the pressure alone.

---

²The intensive variable for volume (V) can be either the specific volume (V̂, volume/mass) or the specific molal volume (V̄, volume/mole).

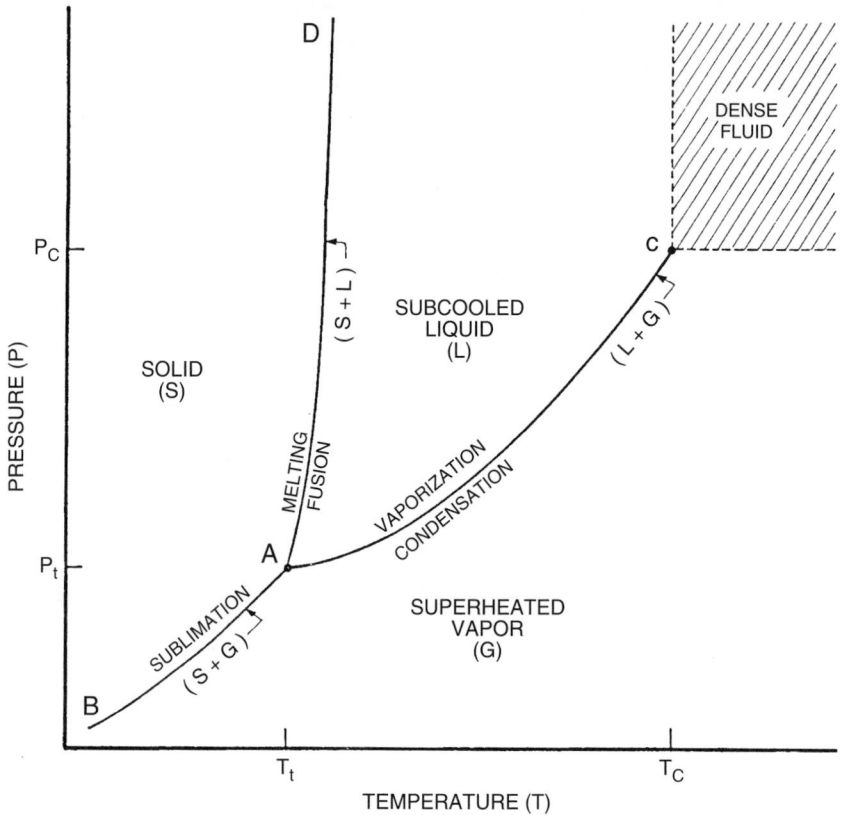

Figure 2.7.3 *Typical P–T diagram for a pure substance (A is the triple point and C is the critical point).*

Because a phase change is usually accompanied by a change in volume the two-phase systems of a pure substance appear on a P–V (or a T–V) diagram as regions with distinct boundaries. On a P–V plot, the triple point appears as a horizontal line, and the critical point becomes a point of inflection of the critical isotherm, $T = T_c$ (see Figure 2.7.2 and Figure 2.7.4).

The vapor–liquid region of a pure substance is contained within the *phase* or *saturation envelope* on a P–V diagram (see Figure 2.7.4). A vapor, whether it exists alone or in a mixture of gases, is said to be *saturated* if its partial pressure equals its equilibrium vapor pressure at the system temperature T. This temperature is called the *saturation temperature* or *dew point* T_{dp} of the vapor because any decrease in system temperature will result in condensation. For a pure substance, saturated vapor coexists in equilibrium with saturated liquid at every point on the vaporization curve AC (see Figure 2.7.3) of the P–T diagram. Its coordinates are given by P (system pressure) = P* (vapor pressure at temperature T), and T (system temperature) = T_b (boiling point of the substance at pressure P). The *normal boiling point* (nbp) refers to the value of T_b attained at a total pressure of 1 standard atmosphere.

A *superheated vapor* refers to a vapor that exists under conditions such that $P_i < P_i^*(T)$. The difference between its existing temperature (T) and its dew point (T_{dp}) is termed the *degrees of superheat* of the vapor. The region of superheated vapor for a pure substance is shown on the T–P diagram (see Figure 2.7.3) by the area lying to the right

of the vaporization curve AC. The term *bubble point* (T_{bp}) of a liquid is the temperature above which boiling will take place. For a pure component system, the bubble point and dew point temperatures are identical to the boiling point of the substance at that pressure. For binary and multicomponent systems of a given composition, $T_{dp} > T_{bp}$ at fixed P or, alternatively, P_{bp} (bubble point pressure) $> P_{dp}$ (dew-point pressure) at fixed T. The saturated liquid curve is thus equivalent to the bubble point curve, and the saturated vapor curve is the same as the dew point curve (see Figure 2.7.4). Although the isotherms on a P–V plot are horizontal lines in the vapor-liquid region of a pure substance, they become curves having negative slopes in system containing more than one component. A liquid is termed *subcooled* if its temperature (T) is less than the bubble-point temperature (T_{bp}) at given pressure (P). The *degrees of subcooling* are then given by the difference ($T_{bp}–T$). In the case of a pure substance, the area DAC lying to the left of the curve AC in the subcooled liquid region (see Figure 2.7.3).

In the vapor-liquid region of a pure substance, the composition of a two-phase system (at given T and P) varies from pure saturated liquid at the bubble point M to pure saturated vapor at the dew point N along the line MQN on the P–V diagram (Figure 2.7.4). For a *wet vapor* represented by an intermediate point Q, the *quality* (q) refers to the mass fraction of saturated vapor present in the two-phase mixture. If m_ℓ and m_g indicate the masses of saturated liquid and saturated vapor comprising the wet vapor at point Q, then its quality may be calculated by the application of the

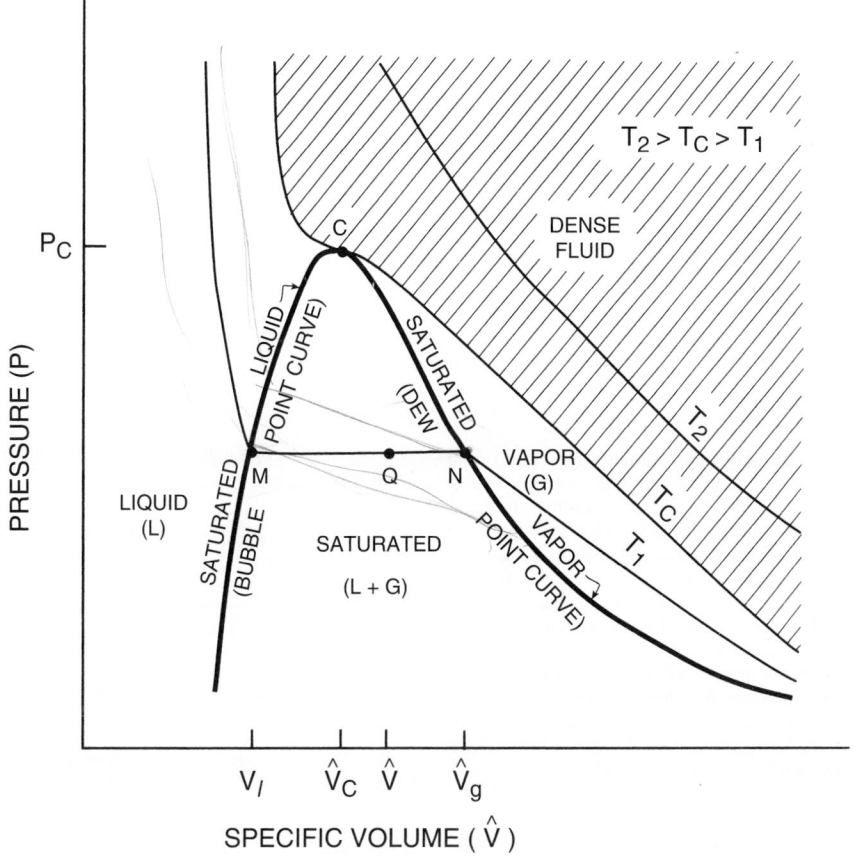

Figure 2.7.4 *Typical P–V diagram for a pure substance showing isotherms and saturation (phases) envelope.*

lever rule as

$$q \equiv \frac{m_g}{m_\ell + m_g} = \frac{\hat{V} - \hat{V}_\ell}{\hat{V}_g - \hat{V}_1} = \frac{\text{length MQ}}{\text{length MN}} \qquad [2.7.31]$$

The origin of this rule lies in the expression for the specific volume of the wet vapor, $\hat{V} = q\hat{V}_g + (1-q)\hat{V}_\ell$. The intensive properties of the wet vapor are obtained readily from the individual properties of its component phases by an analogous equation:

$$\mathcal{M} = q\mathcal{M}_g + (1-q)\mathcal{M}_\ell \qquad [2.7.32]$$

where the symbols $\mathcal{M}_g$, $\mathcal{M}_\ell$ and $\mathcal{M}$ refer to the specific properties of the saturated vapor, saturated liquid, and the two-phase mixture, respectively.

The *Clausius-Clapeyron Equation* was originally derived to describe the vaporization process of a pure liquid, but it can also be applied to other two-phase transitions of a pure substance. The Clausius-Clapeyron equation relates the variation of vapor pressure (P*) with absolute temperature (T) to the molar latent heat of vaporization, λ_v, i.e., the thermal energy required to vaporize one mole of the pure liquid:

$$\frac{dP^*}{dT} = \frac{\lambda_v}{T(\tilde{V}_g - \tilde{V}_\ell)} \approx \frac{\lambda_v}{T\tilde{V}_g} \qquad [2.7.33]$$

By neglecting the specific molar volume of the saturated liquid $\tilde{V}\ell$ in relation to that of the saturated vapor $\tilde{V}_g$, and by assuming the vapor phase to behave as an ideal gas, i.e.,

$P^*\tilde{V}_g = RT$, the above equation may be arranged into

$$\frac{d(\ell n P^*)}{d(1/T)} = -\frac{\lambda_v}{R} \qquad [2.7.34]$$

This suggests that a plot of $\ell n P^*$ against $1/T$ should yield a line having a local slope of $(-\lambda_v/R)$. A straight line is obtained only when λ_v is nearly constant, i.e., over a narrow range of temperatures. An integrated version of the Clausius-Clapeyron equation finds use in correlation of vapor pressure data:

$$P^*(T) \approx P_0^* \exp\left[-\frac{\bar{\lambda}_v}{R}\left(\frac{1}{T} - \frac{1}{T_0} \right) \right] \qquad [2.7.35]$$

where P_0^* is the known vapor pressure at reference temperature T_0, and $\bar{\lambda}_v$ is the average value of λ_v between T_0 and T. The above equation is reasonably valid when the range $|T_0 - T|$ is small and when the two-phase region of interest is far away from the critical point.

2.7.6.6 Vapor–Liquid Equilibria in Binary and Multicomponent Systems

Multicomponent systems do not exhibit a single boiling point at a given pressure as a pure substance does. Instead, at a constant pressure, a liquid mixture undergoes a change of phase to vapor over a range of temperatures lying between the bubble point and the dew point. Three different approaches — Raoult's law, Henry's law, and the concept of the equilibrium ratio or K factor—are available for computations involving vapor–liquid equilibria.

Raoult's Law. The molar composition of a liquid phase (ideal solution) in equilibrium with its vapor at any temperature T is given by

$$x_i = P_i/P_i^*(T) \quad i = 1, 2, \ldots, N \qquad [2.7.36]$$

If the vapor phase behaves as an ideal gas mixture, then by Dalton's law of partial pressures,

$$y_i = P_i/P = x_i P_i^*(T)/P \quad i = 1, 2, \ldots, N \qquad [2.7.37]$$

where x_i and y_i are the respective mole fractions of component i in the liquid and vapor phases, $P_i^*(T)$ is the equilibrium vapor pressure of pure liquid i at system temperature T, P_i is the partial pressure of i in the vapor phase, and P is the total pressure of the vapor phase. Each component i is distributed between the two phases to an extent dictated by the relative volatilities of the components at the system temperature. Note that $\sum x_i = 1, \sum y_i = 1$, and $P = \sum x_i P_i^*$.

Raoult's law is strictly applicable to ideal liquid solutions at all compositions, pressures, and temperatures. In an ideal of perfect solution, the components are mutually miscible in all proportions and there are no volume or thermal changes on mixing. Solutions that approach ideality include mixtures of nonpolar hydrocarbons belonging to a homologous family (e.g., paraffins). Whereas binary mixtures of propane-butane, n-hexane-n-heptane, benzene-toluene, etc., show ideal behavior over a range of compositions, nonideal solutions comprising a solvent (1) and a solute (2) obey Raoult's law only in the limit of $x_1 \rightarrow 1$ (i.e., in dilute solutions). Raoult's law finds use in molecular weight determination of nonvolatile solutes in dilute solutions, in estimation of equilibrium solubilities of noncondensable gases in nonpolar liquids, and in vapor-liquid equilibria calculations.

Molecular Weight Determination by Application of Raoult's Law. If a small amount (m_2 in grams) of a nonvolatile, nonionized substance (solute, 2) is dissolved in m_1 grams of a volatile liquid (solvent, 1), the vapor pressure is lower from the pure solvent value (P_1^*) to the solution value (P) at the system temperature. This is a consequence of Raoult's law because the total vapor pressure of the dilute solution ($x_2 \ll 1$) is given by $P^* = x_1 P_1^* + x_2 P_2^* \approx x_1 P_1^*$.

(a) The *relative lowering of vapor pressure* of the solvent is then

$$\frac{P_1^* - P^*}{P_1^*} = \frac{\Delta P^*}{P_1^*} \approx x_2 = \frac{(m_2/M_2)}{(m_1/M_1) + (m_2/M_2)} \approx \frac{m_2}{M_2} \frac{M_1}{m_1}$$

$$[2.7.38]$$

Measurement of P^* and P_1^* at one temperature allows determination of unknown M_2.

(b) The presence of the nonvolatile solute also causes an *elevation of the boiling point* of the liquid from pure solvent value T_b to solution value T. The above equation is coupled with Clausius-Clapeyron equation to yield

$$\Delta T_b = T - T_b = \frac{-RT_b^2}{\lambda_v} \ell n(1 - x_2) \approx \frac{RT_b^2}{\lambda_v} x_2 \approx \frac{RT_b^2}{\lambda_v} \frac{M_1}{M_2} \frac{m_2}{m_1}$$

where λ_v is the molal latent heat of vaporization of pure solvent at its boiling point (T_b) and R is the universal gas constant. By measuring ΔT_b and T_b at the same total pressure, x_2 can be computed and hence M_2.

(c) The presence of the solute also causes a *depression in the freezing point* of the solvent from the pure liquid value T_f to the solution value T. Application of Raoult's law together with Clausius-Clapeyron equation gives a similar result:

$$\Delta T_f = T_f - T \approx \frac{RT_f^2 x_2}{\lambda_f} \approx \frac{RT_f^2}{\lambda_f} \frac{M_1}{M_2} \frac{m_2}{m_1} \qquad [2.7.39]$$

where λ_f is the molal latent heat of fusion of the solvent at T_f. Techniques (b) and (c) provide more accurate data and therefore better estimates for M_2 than method (a).

Example 2.7.6

A solution is prepared by dissolving 0.911 g of carbon tetrachloride in 50.00 g of benzene. (i) Calculate the freezing point depression of the solution if pure benzene has a fp of 5.53°C and a latent heat of fusion of 30.45 cal/g. (ii) What will be the elevation in boiling point of the solution if pure benzene has a nbp of 80.1°C and a latent heat of vaporization of 7,700 cal/g-mole [5].

2.7.6.6.1 Solution

MW of solute (CCl_4) = M_2 = 153.82 m_2 = 0.911 g

MW of solvent (C_6H_6) = M_1 = 78.11 m_1 = 50.00 g

$$x_2 = \frac{(m_2/M_2)}{(m_1/M_1) + (m_2/M_2)} = 0.917 \times 10^{-2}$$

(i) Because $x_2 \ll 1$, the solution is dilute and Raoult's law may be applied

$$\Delta T_f = T_f \approx \frac{RT_f^2}{\lambda_f} x_2$$

With R = 1.987 cal/g-mole, T_f = 273.2 + 5.53 = 278.73°K, and λ_f = (30.45)(78.11) cal/g-mole, we obtain ΔT_f = 0.595°C, which compares well with an experimental observation of 0.603°C.

(ii) With T_b = 273.2 + 80.1 = 353.3°K, λ_v = 7,700 cal/g-mole, we get

$$\Delta T_b = T - T_b = \frac{RT_b^2 x_2}{\lambda_v} = 0.295°C$$

Henry's Law is an empirical formulation that describes equilibrium solubilities of noncondensable gases in a liquid when Raoult's law fails. It states that the mole fraction of a gas (solute i) dissolved in a liquid (solvent) is proportional to the partial pressure of the gas above the liquid surface at given temperature. That is,

$$x_i = P_i/H_i(T) \qquad [2.7.40]$$

where the constant of proportionality $H_i(T)$ is known as Henry's law constant, with units of pressure per mole fraction. It is a characteristic of the gas–liquid system and increases with T. Experimentally determined values of H_i are available in the standard references for various gas–liquid systems [13–16]. If the gas phase is assumed to be ideal, then the equilibrium mole fraction of component i in the gas phase is

$$y_i = P_i/P = x_i H_i(T)/P \qquad [2.7.41]$$

Henry's law is a reasonable approximation in the absence of gas–liquid reactions when total pressure is low to moderate. Deviations are usually manifested in the form of H_i dependence on P and phase compositions.

Example 2.7.7

A gas mixture has a molal composition of 20% H_2S, 30% CO_2 and 50% N_2 at 20°C and a total pressure of 1 atm. Use Henry's law to calculate the volumes of these gases that may be dissolved in 1,000 lb of water at equilibrium.

2.7.6.6.2 Solution

By Henry's law, the partial pressure of solute i in the gas phase is $P_i = H_i(T)x_i$, where x_i is the mole fraction of i in solution. Data on Henry's law constant are obtained from Chapter 14 of Perry and Chilton's *Chemical Engineers' Handbook* [14] for gas–water systems at 20°C.

Gas i:	H_2S	CO_2	N_2
H_i (atm/mole fraction):	4.83×10^2	1.42×10^3	8.04×10^4

Assuming ideal-gas law to hold, $P_i = y_i P$, where $P = 1$ atm and $y_i =$ mole fraction of i in the gas phase. The equilibrium mole fraction x_i of gas i in solution is then given by

$$x_i = P_i/H_i = y_i P/H_i$$

Gas i	y_i	P_i (atm)	x_i	n_i (lb-moles)	V_i (ft^3)
H_2S	0.20	0.20	4.14×10^{-4}	2.30×10^{-2}	8.86
CO_2	0.30	0.30	2.11×10^{-4}	1.17×10^{-2}	4.51
N_2	0.50	0.50	6.22×10^{-5}	3.45×10^{-3}	1.33

The table shows the rest of the required results for the number of moles n_i of each gas i present in the aqueous phase and the corresponding gas volume V_i dissolved in it. Since $x_i \ll 1$, the total moles of liquid is $n \approx (1000/18.01)$ lb-moles in 1000 lb of water and so $n_i = x_i n$ can be calculated. At $T = 20°C = 68°F = 528°R$ and $P = 1$ atm, the molal volume of the gas mixture is

$$\tilde{V} = \frac{RT}{P} = \frac{(0.7302)(528)}{(1)} = 385.55 \text{ ft}^3/\text{lb-mole}$$

The volume of each gas dissolved in 1,000 lb of water is then $V_i = n_i \tilde{V}$.

Equilibrium Distribution Ratio or K factor. This is also termed *distribution coefficient* in the literature; it is a widely accepted method of describing vapor–liquid equilibria in nonideal systems. For any component i distributed between the vapor phase and liquid phase at equilibrium, the distribution coefficient or K factor is defined by

$$K_i = y_i/x_i = K_i(T, P) \qquad [2.7.42]$$

The dimensionless K_i is regarded as a function of system T and P only and not of phase compositions. It must be experimentally determined. Reference 64 provides charts of K_i (T, P) for a number of paraffinic hydrocarbons. K_i increases with an increase in system T and decreases with an increase in P. Away from the critical point, it is assumed that the K_i values of component i are independent of the other components present in the system. In the absence of experimental data, caution must be exercised in the use of K-factor charts for a given application. The term *distribution coefficient* is also used in the context of a solute (solid or liquid) distributed between two immiscible liquid phases; y_i and x_i are then the equilibrium mole fractions of solute i in each liquid phase.

Example 2.7.8

The distribution coefficient for *n*-heptane (solute i) distributed between ethylene glycol (solvent 1) and benzene (solvent 2) at 25°C is given as the ratio of mass fractions $(\omega_{i,1}/\omega_{i,2}) = K_i = 0.30$. Suppose 60 g of *n*-heptane are added to a solvent mixture of 600 g of ethylene glycol and 400 g of benzene. Assuming that the solvents are immiscible, determine the amount of *n*-heptane dissolved in each liquid phase at equilibrium.

2.7.6.6.3 Solution

Consider the equilibrium state of the system assuming that m grams of *n*-heptane are dissolved in the benzene phase. The mass fraction of *n*-heptane in this phase is $\omega_{i,2} = m/(400 + m)$.

Because $(60 - m)$ g of i are present in the glycol phase, the equilibrium mass fraction of i in this phase is

$\omega_{i,1} = (60 - m)/(600 + 60 - m)$. Thus,

$$\frac{\omega_{i,1}}{\omega_{i,2}} = K_i = 0.30 = \frac{(60 - m)}{(660 - m)} \frac{(400 + m)}{m}$$

which must be solved for m. A quadratic equation in m is obtained by cross-multiplication and rearrangement of terms:

$$0.30m(660 - m) - (60 - m)(400 + m) = 0$$

or

$$0.70m^2 + 538m - 24,000 = 0$$

Its roots are

$$m = \frac{-538 \pm \sqrt{(538)^2 - 4(0.70)(-24,000)}}{2(0.70)}$$

Only the positive root is the physically meaningful value and so

$$m = (-538 + 597.197)/1.4 = 42.28 \text{g}$$

$$\therefore \omega_{i,2} = 42.28/442.28 = 0.0956 \text{ in benzene phase}$$

and

$$\omega_{i,1} = (60 - 42.28)/617.72 = 0.0287 \text{ in glycol phase}$$

Note that $w_{i,1}/w_{i,2} = 0.0287/0.0956 = 0.3$

2.7.7 Thermochemistry

Thermochemistry is the study of thermal effects associated with phase changes, formation of chemical compounds or solutions, and chemical reactions in general. The amount of heat (Q) liberated (or absorbed) is usually measured in a batch-type bomb calorimeter at fixed volume or in a steady-flow calorimeter at constant pressure. Under these operating conditions, $Q = Q_v = \Delta U$ (net change in the internal energy of the system) for the bomb calorimeter, while $Q = Q_p = \Delta H$ (net change in the enthalpy of the system) for the flow calorimeter. For a pure substance, the thermodynamic properties U and H are functions of the state variables, temperature (T) and pressure (P), and its state of aggregation [e.g., liquid (ℓ), gas or vapor (g), solid (s)]. We briefly review basic definitions and terminology employed in the area of thermochemistry and consider some applications pertaining to combustion of fuels.

Heat of Reaction (ΔH_r). The heat of a chemical reaction carried out at constant pressure (P) is given by the difference between the total enthalpies of the reactants and products.

$$\Delta H_r = H_{products} - H_{reactants} = \sum_P n_j H_j - \sum_R n_i H_i \qquad [2.7.43]$$

where the subscripts i and j refer to reactant i and product j, n is the number of moles, and symbols Σ_R and Σ_P imply summations over all reactants $(i = 1, 2, \dots)$ and all products $(j = 1, 2, \dots)$, respectively. ΔH_r has units of calories or BTUs; its value depends on the amounts and physical states of the reactants and products as well as on the reaction conditions. Note that ΔH_r is *negative* for an *exothermic* reaction in which heat is spontaneously liberated and is *positive* for an *endothermic* reaction in which heat is absorbed from the surroundings.

Standard-state enthalpy changes (ΔH^0). To expedite calculations, thermochemical data are ordinarily presented in the form of standard-state enthalpy changes of the system $\Delta H^0 (T, P)$, with the requirement that materials start and end at the same temperature (T) and pressure (P) and in their standard states of aggregation, i.e.,

$$\Delta H^0(T, P) = H^0_{final}(T, P) - H^0_{initial}(T, P) \qquad [2.7.44]$$

The reference state is usually chosen as $P = 1$ standard atmosphere and $T = 25°C$ (77°F). Examples include standard heats of reaction ΔH_r^0, heats of formation ΔH_f^0, heats of combusion ΔH_c^0, heats of vaporization ΔH_v^0 or λ_v, heats of solution ΔH_s^0, etc. To avoid confusion, the standard state of aggregation of each substance taking part in the thermochemical process must be specified by an appropriate letter symbol adjoining its chemical formula. The standard state for a gas is the ideal gas at 1 atm and specified T. The standard state for a solid is its stable crystalline form (e.g., rhombic sulfur) or amorphous form existing at the specified P and T. In the absence of such information, the normal state of aggregation of the material at given P and T is assumed. Tabulated values of standard-state enthalpy changes (ΔH^0) are readily available from a number of sources including handbooks and textbooks [13–18].

Standard Heat of Reaction. This is the standard enthalpy change accompanying a chemical reaction under the assumptions that the reactants and products exist in their standard states of aggregation at the same T and P, and stoichiometric amounts of reactants take part in the reaction to completion at constant P. With P = 1 atm and T = 25°C as the standard state, $\Delta H_r^0(T, P)$ can be written as

$$\Delta H_r^0(25°C, 1\,atm) = \sum_P v_i H_j^0(25°C, 1\,atm)$$

$$- \sum_R v_i H_i^0(25°C, 1\,atm) \qquad [2.7.45]$$

Because v represents the stoichiometric coefficient of a given species, the value of ΔH_r^0 clearly depends on the way the stoichiometric equation is written for the reaction. It is conventional to express the above equation in a simplified form as

$$\Delta H_r^0(25°C, 1\,atm) = \sum_P v_j H_j^0 - \sum_R v_i H_i^0 \qquad [2.7.46]$$

where it is understood that H_i^0 and H_j^0 are to be evaluated at the reaction conditions of 1 atm and 25°C. Another common practice is to use the number of moles n_i (or n_j) in place of the stoichiometric coefficient v_i (or v_j). Further discussion of heats of reaction can be found in Reference 11.

Relation between Q_p and Q_v. The heats of reaction measured at constant pressure and at constant volume are $Q_p = \Delta H_r(T, P)$ and $Q_v = \Delta U_r(T, P)$. Since $H = U + PV$, it follows that Q_p and Q_v are related by

$$\Delta H_r = \Delta U_r + P\Delta V \qquad [2.7.47]$$

where ΔV = total volume of products – total volume of reactants.

If all the species are gaseous and obey the ideal gas law, then

$$\Delta H_r = \Delta U_r + RT\Delta n$$

where $\Delta n = \sum_p n_j - \sum_r n_i$ = net change in the number of moles of the system.

The *standard heat of formation* (ΔH_f^0) of a chemical compound is the standard heat of reaction corresponding to the chemical combination of its constituent elements to form one mole of the compound, each existing in its standard state at 1 atm and 25°C. It has units of cal/g-mole.

$$\Delta H_f^0(1\,atm, 25°C) = H_{compound}^0 - H_{elements}^0 \qquad [2.7.48]$$

By convention, the standard-state enthalpies of the elements (H_i^0) are taken to be zero at 1 atm and 25°C so that $\Delta H_f^0 = H_{compound}^0$.

The *standard heat of combustion* (ΔH_c^0) of a chemical substance (usually an organic compound) is the standard heat of reaction for complete oxidation of 1 mole of the substance in pure oxygen to yield $CO_2(g)$ and $H_2O(\ell)$ as products. A reference state of 25°C and 1 atm is assumed in quoting

standard heats of combustion in cal/g-mole. The value of ΔH_c^0 is always negative because combustion is an exothermic reaction. Note that the standard heats of combustion for carbon and hydrogen are the same as the heats of formation for $CO_2(g)$ and $H_2O(\ell)$, respectively.

Laws of Thermochemistry. Lavoisier and Laplace (1780) found that the heat required to decompose a chemical compound into its elements was numerically equal to the heat generated in its formation under the same conditions of T and P. That is $\Delta H_d = -\Delta H_f$, where the subscript d refers to decomposition reaction [11,19].

An important corollary of this postulate is known as *Hess's law of constant heat summation* (1840): The overall heat of a chemical reaction is the same whether the reaction occurs in a single step or multiple steps.

The two basic principles permit the algebraic manipulation of chemical reactions for thermochemical calculations.

Applications. (1) Heats of formation of reactants and products can be used to calculate the standard heat of a chemical reaction by applying Hess's law:

$$\Delta H_r^0 = \sum_P v_j \Delta H_{f,j}^0 - \sum_R v_i \Delta H_{f,i}^0 \qquad [2.7.49]$$

where $\Delta H_{f,i}^0$ and $\Delta H_{f,j}^0$ are the standard heats of formation of reactant i and product j, respectively.

(2) For reactions involving only organic compounds as reactants, ΔH_r^0 can be determined using heats of combustion data.

$$\Delta H_r^0 = \sum_R v_i \Delta H_{c,i}^0 - \sum_P v_j \Delta H_{c,j}^0 \qquad [2.7.50]$$

These two approaches are useful when a direct measurement of ΔH_r^0 is not possible because of experimental difficulties.

Example 2.7.9
Calculate the heat of reaction at the standard reference state (1 atm, 25°C) for

$$(i) \quad n\text{-}C_4H_{10}(g) \rightarrow C_2H_4(g) + C_2H_6(g)$$

$$and \ (ii) \quad CO_2(g) + H_2(g) \rightarrow CO(g) + H_2O(\ell)$$

using heat of formation data. Also check results using heat of combustion data.

2.7.7.0.4 Solution
The required data are tabulated below in units of kJ/g-mole [5].

	$-\Delta H_f^0$	$-\Delta H_c^0$
n-C_4H_{10} (g)	125.7	2877.6
C_2H_4 (g)	−52.4	1411.2
C_2H_6 (g)	84.0	1560.7
CO_2 (g)	393.5	–
CO (g)	110.5	283.0
H_2 (g)	0	285.8
$H_2O(\ell)$	285.8	–

Using heat of formation data, by Hess' law

(i) $\Delta H_r^0 = \Delta H_f^0(C_2H_4) + \Delta H_f^0(C_2H_6) - \Delta H_f^0(C_4H_{10})$

$$= 52.4 - 84.0 - (-125.7)$$

$$= +94.1 \text{ kJ}$$

(ii) $\Delta H_r^0 = \Delta H_f^0(CO) + \Delta H_f^0(H_2O) - \Delta H_f^0(CO_2)$

$$- \Delta H_f^0(H_2)$$

$$= -110.5 - 285.8 - (-393.5)$$

$$= -2.8 \text{ kJ}$$

Using heat of combustion data,

(i) $\Delta H_r^0 = \Delta H_c^0(C_4H_{10}) - \Delta H_c^0(C_2H_4) - \Delta H_c^0(C_2H_6)$

$= -2877.6 - (-1411.2) - (-1560)$

$= +94.3 kJ$ for first reaction

(ii) $\Delta H_r^0 = \Delta H_c^0(CO_2) + \Delta H_c^0(H_2) - \Delta H_c^0(CO)$

$- \Delta H_c^0(H_2O)$

$= 0 - 285.8 - (-283.0) - 0$

$= -2.8 kJ$ for second reaction

Both methods yield identical results for the heats of reaction.

Very frequently ΔH_f^0 data are available for inorganic substances but not for organic compounds for which ΔH_c^0 values are more readily available. Because ΔH_f^0 of hydrocarbons are not easily measurable, they are often deduced by Hess's law from known ΔH_c^0 of the hydrocarbon and known ΔH_c^0 values of the products of combustion.

Example 2.7.10
Find the standard heat of formation of benzene (ℓ) given the following heats of combustion data (in kcal/g-mole) at 1 atm and 25°C:

(i) $C(s) + O_2(g) \rightarrow CO_2(g)$ $\Delta H_c^0 = -94.05$

(ii) $H_2(g) + 1/2 O_2(g)$
     $\rightarrow H_2O(\ell)$ $\Delta H_c^0 = -68.32$

(iii) $C_6H_6(\ell) + 15/2 O_2(g)$
      $\rightarrow 6CO_2(g) + 3H_2O(\ell)$ $\Delta H_c^0 = -780.98$

2.7.7.0.5 Solution
The desired formation reaction is

$6C(s) + 3H_2(g) \rightarrow C_6H_6(\ell)$ $\Delta H_f^0 = ?$

which is equivalent to [6 Equation (i) + 3 Equation (ii) – Equation (iii)]. By Hess' law,

$\Delta H_f^0 = \sum_R v_i \Delta H_{c,i}^0 - \sum_P v_j \Delta H_{c,j}^0$

$= 6(-94.05) + 3(-68.32) - 1(-780.98)$

$= -564.30 - 204.96 + 780.98 = 11.72$ kcal/g-mole

Effect of Temperature on the Heat of Reaction. It is possible to calculate the heat of a chemical reaction $\Delta H_r(T_1)$ at any temperature T_1 and pressure P, provided we know the standard heat of reaction $\Delta H_r^0(T_0)$ at reference conditions of T_0 and P (e.g., 25°C and 1 atm).

Temperature effects are incorporated into the calculation by a thermal cycle as shown in Figure 2.7.5. The terms ΔH_R and ΔH_P represent the sensible heats necessary to change the temperatures of reactants and products from T_0 to T_1. Thus,

$\Delta H_r^0(T_0) + \Delta H_P = \Delta H_R + \Delta H_r(T_1)$ [2.7.51]

or

$\Delta H_r(T_1) = \Delta H_r^0(T_0) + \Delta H_P - \Delta H_R$

$= \Delta H_r^0(T_0) + \sum_P \int_{T_0}^{T_1} v_j C_{p,j} dT - \sum_R \int_{T_0}^{T_1} v_i C_{p,i} dT$

The integral terms representing ΔH_P and ΔH_R can be computed from molal heat capacity data $C_p(T)$ for the reactants (i) and products (j). When phase transitions occur between T_0 and T_1 for any species, the appropriate latent heats of phase transformations for those pecies must be included in the evaluation of ΔH_R and ΔH_p. In the absence of phase

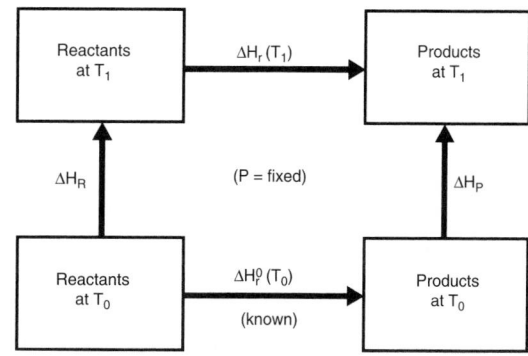

Figure 2.7.5 *Schematic representation to calculate the heat of reaction at temperature T_1.*

changes, let $C_p(T) = a + bT + cT^2$ describe the variation of C_p (cal/g-mole °K) with absolute temperature T (°K). Assuming that constants a, b, and c are known for each species involved in the reaction, we can write

$\Delta H_r(T_1) = \Delta H_r^0(T_0) + \alpha(T_1 - T_0)$

$+ \beta \frac{(T_1^2 - T_0^2)}{2} + \gamma \frac{(T_1^3 - T_0^3)}{3}$ [2.7.52]

where the coefficient, α, β, and γ are defined by

$\alpha \equiv \sum_P v_j a_j - \sum_R v_i a_i$

$\beta \equiv \sum_P v_j b_j - \sum_R v_i b_i$ [2.7.53]

$\gamma \equiv \sum_P v_j c_j - \sum_R v_i c_i$

In a more compact notation, the same result becomes

$\Delta H_r(T_1) = H_0 + \alpha T_1 + \frac{\beta T_1^2}{2} + \frac{\gamma T_1^3}{3}$ [2.7.54]

where

$H_0 \equiv \Delta H_r^0(T_0) - \left(\alpha T_0 + \frac{\beta T_0^2}{2} + \frac{\gamma T_0^3}{3} \right)$ [2.7.55]

is a function of T_0 only.

Sometimes tabulated values of the *mean molal heat capacities* $\bar{C}_p(T)$ are more easily accessible than $C_p(T)$ data, with respect to a reference temperature of $T_0 = 25°C$ (see Table 2.7.8). Since $\bar{C}_p$ is defined over the range T_0 and T_1 by

$\bar{C}_p = \frac{1}{(T_1 - T_0)} \int_{T_0}^{T_1} C_p(T) dT$ [2.7.56]

we can rewrite the expression for $\Delta H_r(T_1)$ as

$\Delta H_r(T_1) = \Delta H_r^0(T_0) + \left[\sum_P v_j \bar{C}_{p,j} - \sum_R v_i \bar{C}_{p,i} \right](T_1 - T_0)$ [2.7.57]

If mean molal heat capacity data are available with a reference temperature T_r other than $T_0 = 25°C$ (see Table 2.7.9 for data with $T_r = 0°C$), the following equation can be used to calculate $\Delta H_r(T_1)$:

$\Delta H_r(T_1) = \Delta H_r^0(T_0) + (T_1 - T_r)A - (T_0 - T_r)B$ [2.7.58]

where

$A, B = \sum_P v_j \bar{C}_{p,j} - \sum_R v_i \bar{C}_{p,i}$

Table 2.7.8 *Mean Molal Heat Capacities of Gases Between 25°C and T°C (Reference pressure = 0) (cal/(g-mole) (°C)) [13]*

T(°C)	H_2	N_2	CO	Air	O_2	NO	H_2O	CO_2	HCl	Cl_2	CH_4	SO_2	C_2H_4	SO_3	C_2H_6
25	6.894	6.961	6.965	6.972	7.017	7.134	8.024	8.884	6.96	8.12	8.55	9.54	10.45	12.11	12.63
100	6.924	6.972	6.983	6.996	7.083	7.144	8.084	9.251	6.97	8.24	8.98	9.85	11.35	12.84	13.76
200	6.957	6.996	7.017	7.021	7.181	7.224	8.177	9.701	6.98	8.37	9.62	10.25	12.53	13.74	15.27
300	6.970	7.036	7.070	7.073	7.293	7.252	8.215	10.108	7.00	8.48	10.29	10.62	13.65	14.54	16.72
400	6.982	7.080	7.136	7.152	7.406	7.301	8.409	10.462	7.02	8.55	10.97	10.94	14.67	15.22	18.11
500	6.995	7.159	7.210	7.225	7.513	7.389	8.539	10.776	7.06	8.61	11.65	11.22	15.60	15.82	19.39
600	7.011	7.229	7.289	7.299	7.616	7.470	8.678	11.053	7.10	8.66	12.27	11.45	16.45	16.33	20.58
700	7.032	7.298	7.365	7.374	7.706	7.549	8.816	11.303	7.15	8.70	12.90	11.66	17.22	16.77	21.68
800	7.060	7.369	7.443	7.447	7.792	7.630	8.963	11.53	7.21	8.73	13.48	11.84	17.95	17.17	22.72
900	7.076	7.443	7.521	7.520	7.874	7.708	9.109	11.74	7.27	8.77	14.04	12.01	18.63	17.52	23.69
1000	7.128	7.507	7.587	7.593	7.941	7.773	9.246	11.92	7.33	8.80	14.56	12.15	19.23	17.80	24.56
1100	7.169	7.574	7.653	7.660	8.009	7.839	9.389	12.10	7.39	8.82	15.04	12.28	19.81	18.17	25.40
1200	7.209	7.635	7.714	7.719	8.068	7.898	9.524	12.25	7.45	8.94	15.49	12.39	20.33	18.44	26.15
1300	7.252	7.692	7.772	7.778	8.123	7.952	9.66	12.39							
1400	7.288	7.738	7.815	7.824	8.166	7.994	9.77	12.50							
1500	7.326	7.786	7.866	7.373	8.203	8.039	9.89	12.69							
1600	7.380	7.844	7.922	7.929	8.260	8.092	9.95	12.75							
1700	7.421	7.879	7.958	7.965	8.305	8.124	10.13	12.70							
1800	7.467	7.924	8.001	8.010	8.349	8.164	10.24	12.93							
1900	7.505	7.957	8.033	8.043	8.383	8.192	10.34	13.01							
2000	7.548	7.994	8.069	8.081	8.423	8.225	10.43	13.10							
2100	7.588	8.028	8.101	8.115	8.460	3.255	10.52	13.17							
2200	7.624	8.054	8.127	3.144	8.491	8.277	10.61	13.24							

Table 2.7.9 *Mean Molal Heat Capacities of Combustion Gases Between 0°C and T°C (Reference temperature = 0°C; pressure = 1 atm) [12] ($\bar{C}_p$ in cal/(gm-mole) (°C) (To convert to J/kg mole) (°K), multiply by 4184.)*

T(°C)	N_2	O_2	Air	H_2	CO	CO_2	H_2O
0	6.959	6.989	6.946	6.838	6.960	8.595	8.001
18	6.960	6.998	6.949	6.858	6.961	8.706	8.009
25	6.960	7.002	6.949	6.864	6.962	8.716	8.012
100	6.965	7.057	6.965	6.926	6.973	9.122	8.061
200	6.985	7.154	7.001	6.955	7.050	9.590	8.150
300	7.023	7.275	7.054	6.967	7.057	10.003	8.256
400	7.075	7.380	7.118	6.983	7.120	10.360	8.377
500	7.138	7.489	7.190	6.998	7.196	10.680	8.507
600	7.207	7.591	7.266	7.015	7.273	10.965	8.644
700	7.277	7.684	7.340	7.036	7.351	11.221	8.785
800	7.350	7.768	7.414	7.062	7.428	11.451	8.928
900	7.420	7.845	7.485	7.093	7.501	11.68	9.070
1000	7.482	7.916	7.549	7.128	7.570	11.85	9.210
1100	7.551	7.980	7.616	7.165	7.635	12.02	9.348
1200	7.610	8.039	7.674	7.205	7.688	12.17	9.482
1300	7.665	8.094	7.729	7.227	7.752	12.32	9.613
1400	7.718	8.146	7.781	7.260	7.805	12.44	9.740
1500	7.769	8.192	7.830	7.296	7.855	12.56	9.86

excepting that the mean heat capacities are evaluated between T_r and T_1 for term A and between T_r and T_0 for term B.

Effects of Pressure on ΔH_r. Consider a reaction carried out at reference temperature $T = 25°C$ but at a constant pressure P_1 different from the initial standard-state pressure $P_0 = 1$ atm. The value of $\Delta H_r(P_1)$ can be found from an energy balance similar to the previous analysis:

$$\Delta H_r(P_1) = \Delta H_r^0(P_0) + \Delta H_P - \Delta H_R \qquad [2.7.59]$$

The terms ΔH_P and ΔH_R now denote the enthalpy changes associated with the change of pressure from P_0 to P_1. Thus

$$\Delta H_r(P_1) = \Delta H_r^0(P_0) + \sum_P \int_{P_0}^{P_1} \left(\frac{\partial H_j}{\partial P} \right)_T dP$$

$$- \sum_R \int_{P_0}^{P_1} \left(\frac{\partial H_i}{\partial P} \right)_T dP \qquad [2.7.60]$$

where $\partial/\partial P$ denotes partial differentiation with respect to pressure. For solids and liquids away from the critical point, the variation in enthalpy with pressure at constant T is quite small and, therefore, $\Delta H_r(P_1) \approx \Delta H_r^0(P_0)$ is assumed. For gaseous reactants and products that follow the ideal-gas law, $H = H(T)$ only, and the effect of pressure is zero, i.e., $\Delta H_r(P_1) = \Delta H_r^0(P_0)$. In nonideal gas systems, the enthalpy changes are nonzero, but the effect is usually small up to moderate pressures.

Heating Values of Combustion Fuels. The calorific value or heating value (HV) of a fuel (usually a hydrocarbon) is the negative of its standard heat of combustion at 1 atm and 25°C, expressed in cal/g or Btu/lb. It is termed *higher heating value* (HHV) if H_2O (ℓ) is a combustion product and is calculated as $HHV = (-\Delta H_c^0)/M$, where M is the molecular weight of the fuel. An appropriate ΔH_c^0 value must be used in referring to the *lower heating value* (LHV) based on H_2O (g) as a combustion product. Both are related by

$$HHV = LHV + (\nu_w \lambda_v)/M \qquad [2.7.61]$$

where ν_w is the stoichiometric coefficient for water in the combustion reaction of 1 mole of fuel, and λ_v is the molal latent heat of vaporization for water at 25°C amd 1 atm = 10,519 cal/g-mole = 18,934 Btu/lb-mole.

For a fuel mixture composed of combustible substances $i = 1, 2, \ldots$, the heating value is calculated as $HV = \sum \omega_i (HV)_i$, where ω_i is the mass fraction of the ith substance having a heating value of $(HV)_i$.

Adiabatic reaction temperature (T_{ad}). The concept of adiabatic or theoretical reaction temperature (T_{ad}), at which all of the heat liberated goes to heat the reaction products, plays an important role in the design of chemical reactors, gas furnaces, and other process equipment to handle highly exothermic reactions such as combustion. T_{ad} is defined as the final temperature attained by the reaction mixture at the completion of a chemical reaction carried out under adiabatic conditions in a closed system at constant pressure. Theoretically, this is the maximum temperature achieved by the products when stoichiometric quantities of reactants are completely converted into products in an adiabatic reactor. In general, T_{ad} is a function of the initial temperature (T_i) of the reactants and their relative amounts as well as the presence of any nonreactive (inert) materials. T_{ad} is also dependent on the extent of completion of the reaction. In actual experiments, it is very unlikely that the theoretical maximum values of T_{ad} will be realized, but the calculated results provide a basis for comparison of the thermal effects resulting from exothermic reactions. Lower feed temperatures (T_i), presence of inerts and excess reactants, and incomplete conversion tend to reduce the value of T_{ad}. The term *theoretical* or *adiabatic flame temperature* (T_{fl}) is preferred over T_{ad} in dealing with the combustion of fuels.

Calculation of T_{ad}. To calculate T_{ad} (or T_{fl} for a combustible fuel), we refer to Figure 2.7.6 and note that $Q = \Delta H = 0$ for the adiabatic reaction process. Taking 25°C ($= 298.2°K$) and 1 atm as the reference state, the energy balance can be expressed as

$$\Delta H = 0 = H_{products}(T_{ad}) - H_{reactants}(T_i)$$

$$= -\Delta H_R + \Delta H_r^0(25°C) + \Delta H_P \qquad [2.7.62]$$

or

$$\Delta H_P = -\Delta H_r^0 + \Delta H_R \qquad [2.7.63]$$

where

$$\Delta H_R = \sum_R \int_{298}^{T_{ad}} n_i C_{p,i} dT \qquad [2.7.64]$$

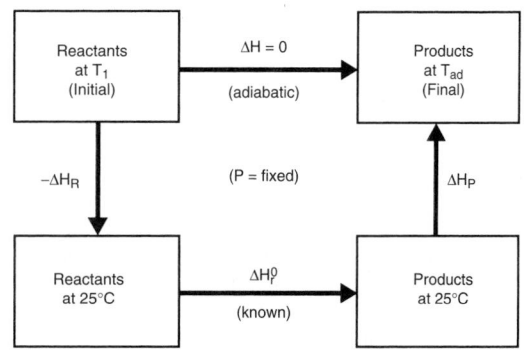

Figure 2.7.6 *Schematic representation to calculate the adiabatic reaction temperature (T_{ad}).*

and

$$\Delta H_P = \sum_P \int_{298}^{T_{ad}} n_j C_{p,j} dT + \sum_P n_j \lambda_{v,j} \qquad [2.7.65]$$

The second term on the right side of the expression for ΔH_p accounts for *any* phase changes that may occur between 25°C and T_{ad} for the final products; it should be deleted if not applicable. Using molal heat capacity data $C_p(T)$ for all the species present, the following equality is solved for T_{ad}.

$$\sum_P \int_{298}^{T_{ad}} n_j C_{p,j} dT = -\Delta H_r^0 - \sum_P n_j \lambda_{v,j} + \sum_R \int_{298}^{T_i} n_i C_{p,i} dT \qquad [2.7.66]$$

An alternative representation is useful when data on mean molal heat capacities $\bar{C}_p(T)$ are available with 25°C as the reference temperature (T_0). Then the equation

$$\sum_P n_j \bar{C}_{p,j}(T_{ad} - 298) = -\Delta H_r^0 - \sum_P n_j \lambda_{v,j} + \sum_R \bar{C}_{p,i}(T_i - 298) \qquad [2.7.67]$$

can be solved for T_{ad}, though it still requires a trial-and-error procedure.

Example 2.7.11

A natural gas having the volumetric composition 90% methane, 8% ethane, and 2% nitrogen at 1 atm and 25°C is used as fuel in a power plant. To ensure complete combustion 75% excess air is also supplied at 1 atm and 25°C. Calculate (i) the lower and higher heating values of the fuel at 25°C and (ii) the theoretical maximum temperature in the boiler assuming adiabatic operation and gaseous state for all products.

2.7.7.0.6 Solution

For ideal gas behavior, the volume composition is identical with the mole composition. Choose a *basis* of 1 g-mole of natural gas (at 1 atm and 25°C) which contains 0.90 g-mole of CH_4 (MW = 16.04), 0.08 g-mole of C_2H_6 (MW = 30.07), and 0.02 g-moles of N_2 (MW = 28.01). Then the molecular weight of the natural gas is

$$M = \sum y_i M_i = 0.90(16.04) + 0.08(30.07) + 0.02(28.01)$$

$$= 17.40 \, g/g\text{-mole}$$

(i) The combustion reactions of interest are

$$CH_4(g) + 2O_2(g) \rightarrow CO_2(g) + 2H_2O(\ell) \quad \Delta H_C^0 = -212.80$$

$$C_2H_6(g) + \frac{7}{2}O_2(g) \rightarrow 2CO_2(g) + 3H_2O(\ell) \quad \Delta H_C^0 = -372.82$$

where the standard heat of combustion ΔH_c^0 at 1 atm and 25°C are stated in kcal/g-mole.

With $H_2O(\ell)$ as a product, the higher heating value (HHV) of the fuel is calculated as

$$HHV = 0.9\left[-\Delta H_C^0(CH_4)\right] + 0.08\left[-\Delta H_C^0(C_2H_6)\right]$$

$$= 0.9(212.80) + 0.08(372.82) = 221.35 \text{ kcal/g-mole}$$

$$= \frac{221.35}{17.40} = 12.72 \text{ kcal/g}$$

Note that $H_2O(\ell) \to H_2O(g)$ at 1 atm and 25°C has $\lambda_v = \Delta H_v^0 = 10.519$ kcal/g-mole. With $H_2O(g)$ as a combustion product, the lower heating value (LHV) of the fuel is

$$LHV = HHV - n_w \lambda_v$$

$$= 221.35 - [0.9(2) + 0.08(3)](10.519) \text{ kcal/g-mole}$$

$$= 199.89 \text{ kcal/g-mole} = 11.49 \text{ kcal/g}$$

(ii) The adiabatic flame temperature T_{fl} must be calculated with all products in the gaseous state. So the appropriate standard heat of reaction at 1 atm and 25°C is the heat of combustion of the fuel with $H_2O(g)$ as a product, i.e., the negative of LHV.

$$\therefore \Delta H_r^0 = -(LHV) = -199.89 \text{kcal/g-mole}$$

Stoichiometric amount of O_2 required $= 2(0.9) + \dfrac{7}{2}(0.08)$

$$= 2.08 \text{ g-moles}$$

Because 75% excess air is supplied, amount of O_2 supplied $= 1.75(2.08) = 3.64$ g-moles.

Assuming dry air to contain 79 mole% N_2 and 21 mole% O_2, N_2 supplied $= (3.64)(0.79)/0.21 = 13.693$ g-moles.

The mole compositions of the feed gases entering (reactants i) at 25°C and of the products leaving the boiler at T_{fl} are tabulated:

Feed at 25°C	
Gas i	n_i (g-moles)
CH_4	0.90
C_2H_6	0.08
N_2	0.02 } 13.713
N_2 in air	13.693 }
Stoich. O_2	2.08 } 3.64
Excess O_2	1.56 }
Total	18.333

Products at T_{fl}	
Gas j	n_j (g-moles)
CO_2	$0.9(1) + 0.08(2) =$ 1.06
H_2O	$0.9(2) + 0.08(3) =$ 2.04
N_2	13.373
O_2	1.56
Total	18.373

Applying the heat balance equation, $\Delta H = 0 = \Delta H_r^0 - \Delta H_R + \Delta H_P$, we realize that $\Delta H_R = 0$ because the feed gases are supplied at the reference temperature of 25°C = 298°K.

$$\therefore \Delta H_P = \sum_P \int_{298}^{T_{fl}} n_j C_{p,j} dT = -\Delta H_r^0 = 199,890 \text{ cal} \qquad [2.7.68]$$

must be solved for unknown T_{fl}.

Method 1. Use $C_p(T)$ data for product gases from Tables 2.7.8 and 2.7.9. The equation is $C_p = a + bT + cT^2$, with T in °K and C_p in cal/g-mole °K.

Gas	a	$b \times 10^2$	$c \times 10^5$	Applicable range (°K)
$CO_2(g)$	6.393	1.0100	−0.3405	273–3700
$H_2O(g)$	6.970	0.3464	−0.0483	
$N_2(g)$	6.529	0.1488	−0.02271	273–3700
$O_2(g)$	6.732	0.1505	−0.01791	273–3700

The equation for T_{fl} now becomes

$$199,890 = \alpha(T_{fl} - 298) + \frac{\beta}{2}(T_{fl}^2 - 298^2) + \frac{\gamma}{3}(T_{fl}^3 - 298^3) \qquad [2.7.69]$$

where

$$\alpha = \sum_P n_j a_j$$

$$= 1.06(6.393) + 2.04(6.970) + 13.713(6.529) + 1.56(6.732)$$

$$= 6.777 + 14.219 + 89.532 + 10.502 = 121.029 \qquad [2.7.70]$$

Similar calculations yield

$$\beta = \sum_P n_j b_j = 4.053 \times 10^{-2} \qquad [2.7.71]$$

and

$$\gamma = \sum_P n_j c_j = -0.799 \times 10^{-5} \qquad [2.7.72]$$

A trial-and-error procedure is needed to solve the cubic equation in T_{fl}:

$$199,890 = 121.029(T_{fl} - 298) + 2.027 \times 10^{-2}(T_{fl}^2 - 298^2)$$

$$-0.266 \times 10^{-5}(T_{fl}^3 - 298^3)$$

As a first trial let $T_{fl} = 1,500°C = 1773°K$. Then

$$RHS = 178,518 + 61,909 - 14,774$$

$$= 225,652 \text{ cal (too high)}$$

Try $T_{fl} = 1,400°C = 1,673°K$ as a second trial value. Then

$$RHS = 166,415 + 54,925 - 12,402$$

$$= 206,566 \text{ (high)}$$

Similar calculations show that with $T_{fl} = 1345°C = 1618°K$, the RHS $= 159,759 + 51,257 - 11,212 = 199,804$ which is very close to LHS $= 199,890$.

$$\therefore \text{ The solution is } T_{fl} \approx 1345°C.$$

Method 2. Use mean heat capacity data from Table 2-45 with reference conditions of P=0 and T=25°C for the combustion gases. Then the equation for T_{fl} becomes

$$199,890 = \sum_P n_j \tilde{C}_{p,j}(T_n - 25)$$

$$= \Big[1.06\tilde{C}_p(CO_2) + 2.04\tilde{C}_p(H_2O)$$

$$+ 13.713\tilde{C}_p(N_2) + 1.56\tilde{C}_p(O_2)\Big]$$

$$\times (T_n - 25)$$

With a first trial value of $T_{fl} = 1400°C$, we have

$$RHS = [1.06(12.50) + 2.04(9.77) + 13.713(7.738)$$

$$+ 1.56(8.166)](1400 - 25)$$

$$= (152.031)(1375) = 209,043 \text{cal} \quad \text{(high)}$$

For the second trial, try $T_{fl} = 1350°C$. Then

$$RHS = [1.06(12.445) + 2.04(9.715) + 13.713(7.715)$$

$$+ 1.56(8.145)](1325) = 200,754 \text{ cal.}$$

which is nearer to the LHS. Thus, using interpolated values of $\bar{C}_p$ between the temperatures 1300°C and 1400°C, we find good agreement between RHS = 199,927 and LHS = 199,890 at $T_{fl} = 1345°C$.

Method 3. Use mean molal heat capacity data from Table 2.7.9 with reference conditions of 0°C and 1 atm. Then the equation for T_{fl} is

$$199,890 = \sum_P n_j \bar{C}_{p,j}(T_{fl} - 0) - \sum_P n_j \bar{C}_{p,j}(25 - 0)$$

or

$$\sum_P n_j \bar{C}_{p,j} T_{fl} = 199,890 + [1.06(8.716) + 2.04(8.012)$$

$$+ 13.713(6.960) + 1.56(7.002)](25)$$

$$= 199,890 + 3,299 = 203,189 \text{ cal}$$

which can be written as

$$203,189 = [1.06\bar{C}_p(CO_2) + 2.04\bar{C}_p(H_2O)$$

$$+ 13.713\bar{C}_p(N_2) + 1.56\bar{C}_p(O_2)]T_{fl}$$

Assume a trial value of $T_{fl} = 1400°C$ for which

$$RHS = [1.06(12.44) + 2.04(9.74) + 13.713(7.718)$$

$$+ 1.56(8.146)](1400)$$

$$= 212,240 \text{ (high)}$$

The results obtained with other trial values are shown below:

2nd trial: $T_{fl} = 1350°C$ RHS = 203,855

3rd trial: $T_{fl} = 1345°C$ RHS = 203,034

Because the last trial gives a close agreement with the LHS value of 203,189 cal, we take the solution to be $T_{fl} \approx 1,345°C$. It is concluded that all of the three methods of calculations provide the same final answer.

For further information on this subject, refer to References 1–2.

References

1. Loudon, G. M., *Organic Chemistry*, 4th Edition, Oxford, New York, 2002.

2. Panico, R., Powell, J. H., and Richter, J.-C., eds., *A Guide to IUPAC Nomenclature of Organic Compounds*, Blackwell Scientific Publications, Oxford, 1993.

3. Wade, L. G., *Organic Chemistry*, 5th Edition, Prentice-Hall, Upper Saddle River, NJ, 2003.

4. Bruice, P. Y., *Organic Chemistry*, Prentice-Hall, Upper Saddle River, NJ, 2001.

5. *CRC Handbook of Chemistry and Physics*, 83rd Edition, CRC Press, The Chemical Rubber Company, Boca Raton, FL, 2002–2003.

6. McChain, W. D., Jr., *The Properties of Petroleum Fluids*, 2nd Edition, PennWell Publishing Co., Tulsa, 1990.

7. Robbins, W. K., and Chang, S. H., in *Kirk-Othmer Encyclopedia of Chemical Technology*, Vol. 18, pp. 353–370, 4th Edition, John Wiley & Sons, New York, 1996.

8. Galli, A. F., in *Riegel's Handbook of Industrial Chemistry*, J. A. Kent, ed., 7th Edition, pp. 402–434, Van Nostrand Reinhold, New York, 1980.

9. Hunt, J. M., *Petroleum Geochemistry and Geology*, pp. 28–66, W. H. Freeman, San Francisco, 1979.

10. Watson, K. M., and Nelson, E. F., Ind. Eng. Chem., Vol. 25, p. 880, 1933.

11. Atkins, P., *Physical Chemistry*, 7th Edition, Freeman, New York, 2002.

12. Washburn, E. W. *International Critical Tables*, Vol. 1, p. 393, McGraw-Hill, 1926.

13. Hougen, O. A., Watson, K. M., and Ragatz, R. A., *Chemical Process Principles. Part I. Material and Energy Balances*, John Wiley & Sons, New York, 1954.

14. Perry, R. H., and Chilton, C. H., eds., *Chemical Engineer's Handbook*, 5th Edition, McGraw-Hill, New York, 1973.

15. Dean, J. A., ed., *Lange's Handbook of Chemistry*, 12th Edition, McGraw-Hill, New York, 1979.

16. Reid, R. C., Prausnitz, J. M., and Sherwood, T. K., *The Properties of Gases and Liquids*, 3rd Edition, McGraw-Hill, New York, 1977.

17. Felder, R. M., and Rousseau, R. W., *Elementary Principles of Chemical Processes*, John Wiley & Sons, New York, 1978.

18. Himmelblau, D. M., *Basic Principles and Calculations in Chemical Engineering*, 3rd Edition, Prentice-Hall, Englewood Cliffs, NJ, 1974.

19. Levine, I. N., *Physical Chemistry*, 5th Edition, McGraw-Hill, Boston, 2002.

20. McQuarrie, D. A., and Simon, J. D., *Physical Chemistry: A Molecular Approach*, University Science Books, Sausalito, CA, 1997.

21. Morrison, R. T., and Boyd, R. N., *Organic Chemistry*, 3rd Edition, Allyn and Bacon, Inc., Boston, 1973.

2.8 ENGINEERING DESIGN

2.8.1 Introduction

2.8.1.1 The Petroleum Engineer

Petroleum engineers are traditionally involved in activities known in the oil industry as the "front end" of the petroleum fuel cycle (petroleum is either liquid or gaseous hydrocarbons derived from natural deposits — reservoirs — in the earth). These front end activities are namely exploration (locating and proving out the new geological provinces with petroleum reservoirs that may be exploited in the future), and development (the systematic drilling, well completion, and production of economically producible reservoirs). Once the raw petroleum fluids (e.g., crude oil and natural gas) have been produced from the earth, the "back end" of the fuel cycle takes the produced raw petroleum fluids and refines these fluids into useful products.

Because of the complex interdisciplinary nature of the engineering activities of exploration and development, the petroleum engineer must be conversant with fundamentals of designing devices and systems particular to the petroleum industry [1,2]. The petroleum engineer must be competent in skills related to engineering design. Design involves planning, development, assembly and implementation of plans to achieve a prescribed and specified result.

2.8.1.2 Multidisciplinary Team

The exploration activities directed at locating new petroleum-producing provinces are cost-intensive operations. The capital expended in the search for new petroleum-producing provinces is always at risk because there are no guarantee that such searches will be economically successful. There is no way to actually "know" if crude oil or natural gas is present in a particular geologic formation except to drill into the formation and physically test it. Highly sophisticated geologic and geophysical methods can be used to identify the possible subsurface geological conditions that might contain crude oil and/or natural gas. However, the final test is to drill a well to the rock formation (reservoir) in question and physically ascertain whether it contains petroleum, and if it does, ascertain if the petroleum can be produced economically. Thus, in the early part of the exploration phase geologists, geochemists, geophysicists and petroleum engineers must form teams in order to carry out effective investigations of possible petroleum producing prospects [2]. In general, such teams are initially driven by the geologic, geochemical and geophysical sciences that are to be used to infer possible subsurface locations for new deposits of petroleum. However, once the subsurface locations have been identified, the process of discovery becomes driven by the necessity to drill and complete a test well to the prospective reservoir. The skills of the exploration geologist are important for successful drilling. These exploration (wildcat) wells are usually drilled in remote locations where little or no previous subsurface engineering experience is available, thus, they are inherently risky operations.

Development of new proven petroleum resources are also cost-intensive operations, but generally lack the high risk of nearly total capital loss that exists in exploration activities. Development operations begin after the exploration test wells have proven that crude oil and/or natural gas can be economically recovered from a new reservoir. These operations require that numerous development (or production) wells be drilled and completed into proven resources. Production equipment are placed in these wells and the crude oil and/or natural gas is recovered and transported to refineries (in the back of the fuel cycle). The placement of these wells and the selection of their respective production rates within a particular reservoir or field must be planned carefully. The pattern of well locations and production rates should be designed so that the largest fraction of the in-place petroleum can eventually be recovered even though the maximum recoverable fraction of the oil or natural gas in place will not reach 100 percent. As developmental wells are drilled and completed, the production geologists ensure that wells are drilled to correct subsurface targets.

The development effort is largely a engineering activity requiring a great deal of planning. This planning requires increased interaction of multiple engineering disciplines to assure that the wells are drilled, completed and produced in a safe, efficient and economical manner.

2.8.1.3 Engineering Arts and Sciences

There are three basic engineering fields of knowledge. These are called the engineering arts and sciences. They are the mechanical arts and sciences, the chemical arts and science, and the electrical arts and sciences. Every specific engineering discipline, such as petroleum engineering, utilizes one of the basic engineering fields of knowledge as their scholarly foundation.

The mechanical arts and sciences are based on classical physics (or Newtonian mechanics). The chemical arts and sciences are based on classical and modern chemistry. The electrical arts and sciences are based on modern particle physics. Table 2.8.1 gives a listing of the engineering disciplines that usually have the mechanical arts and sciences as their foundation. Table 2.8.2 gives a listing of the engineering disciplines that usually have the chemical arts and sciences as their foundation. And Table 2.8.3 gives a listing

Table 2.8.1   *Mechanical Arts and Sciences Engineering Disciplines*

• Mechanical engineering	• Environmental engineering
• Civil engineering	• Engineering mechanics
• Aeronautical and aerospace engineering	• Engineering science
• Materials and metallurgical engineering	• Industrial engineering
• Mining engineering	• Nuclear engineering
• Petroleum engineering	• Fire protection engineering
• Marine engineering	• Engineering physics
• Ocean engineering	

Table 2.8.2   *Chemical Arts and Sciences Engineering Disciplines*

• Chemical Engineering
• Petroleum Engineering
• Environmental Engineering
• Nuclear Engineering

Table 2.8.3   *Electrical Arts and Sciences Engineering Disciplines*

• Electrical Engineering
• Electronics Engineering
• Engineering Physics

of the engineering disciplines that usually have the electrical arts and sciences as their foundation.

Note that a few disciplines such as petroleum engineering, environmental engineering and nuclear engineering are listed under two of these general engineering fields. At one engineering university, for example, the petroleum engineering curriculum may have a mechanical arts and sciences foundation. At another university, the petroleum engineering curriculum could have a chemical arts and sciences foundation. However, most petroleum engineering curricula are based on the mechanical arts and sciences. Petroleum engineering curricula that are based on the mechanical arts and sciences usually have a balanced emphasis regarding drilling and completions, production and reservoir engineering subjects. Curricula that are based on the chemical arts and sciences generally concentrate on reservoir engineering subjects.

Similar situations occur with curricula in environmental engineering and in nuclear engineering.

2.8.2 Scientific and Engineering Philosophies
2.8.2.1 Scientific Method

The foundation philosophies of science and engineering are quite different. In the historical sense, it is obvious that scientific inquiry came first. In order for humans to "engineer," they had to be able to predict certain potentially useful physical phenomena. Thus, the scientific method evolved to aid humans in systematically discovering and understanding natural phenomena. The scientific method is a set of prescribed steps that are used to assist in understanding hitherto unknown natural phenomena. Table 2.8.4 gives a listing of the basic steps of the scientific method. The scientific method is an iterative procedure and in modern settings requires the development of a carefully devised plan of work. Such a plan requires both experimentation and analysis. It is necessary for the scientist to keep careful records in order to assure that little will be missed and that other investigators will be able to carry out any follow-on work. It is important to

Table 2.8.4 *Scientific Method*

- Observe a phenomenon
- Postulate a theory to explain the phenomenon
- Develop and conduct an experiment to test the validity of the theory
- Using the test results, draw conclusions as to the validity of the theory
- Re-postulate the theory in light of these conclusions
- Iterate the above steps and continue to refine theory

Table 2.8.5 *Engineering Method*

- A problem is recognized or a need is identified.
- Establish performance specifications (based on the details of the need stated above)
- Describe the preliminary characteristics of the life-cycle that the product design** should have (i.e., relative to production, distribution, consumption and retirement)
- Create first prototype design
- Evaluate feasibility of first prototype design from both the technical and economic point-of-view and modify first prototype design appropriately
- Evaluate performance characteristics of first prototype using models (usually both analytic and physical) and using the results, modify first prototype design appropriately
- Fabricate first prototype
- Test first prototype
- Iterate the above process until society is satisfied and accepts the product
- Plan for the maintenance and retirement of the product through its life-cycle

**The product in petroleum engineering may be as large and as inclusive as an entire oil field or it may be as small as a component of a machine or a process used in production.

understand that the act of seeking pure knowledge in mutually exclusive of any desire to apply the gained knowledge to anything useful. In essence, the scientist seeks a unique and comprehensive understanding (or solution) of phenomena.

2.8.2.2 Engineering Method

The scientific method has been successfully used throughout human history to enlighten us to our natural world and beyond. Since the earliest days there have been individuals who desired to use scientific knowledge for practical use within their respective societies. The modern engineer would have little trouble recognizing these individuals as our forerunners.

Modern engineering has its formal roots in France where, in 1675, Louis XIV established a school for a corps of military engineers. Shortly afterward, early in the reign of Louis XV, a similar school was created for civilian engineers [2]. These original engineering schools were created to provide trained individuals who could improve the well-being of French society by creating an economical infrastructure of water, sewer and transportation systems. By the time modern engineering was established in the United States, the idea of the dependence of engineering practice on scientific knowledge was well established. This is illustrated by the original charter of the Institution of Civil Engineers in 1828 (the first professional engineering society in the United States, know known as the American Society of Civil Engineers) by the following statement...

"... the art of directing the great sources of power in Nature for the use and convenience of man...."

In recent years there has been a greater appreciation of the word "use." As society and engineers have become more aware of the systems or process aspects of nature, it has become clear that the use of natural resources must stop short of abuse.

In early engineering practice, products (devices or systems) evolved slowly with step-by-step improvements made after studying the results of actual usage. In those early days, engineering judgments were made by intuition (i.e., parts are strong enough, etc.), and engineering design was the same as engineering drawing. As society's needs for new products expanded and as each product become more complicated, it also became necessary to develop an engineering methodology that would assure economic and systematic development of products In time, the engineering community developed the engineering method. Table 2.8.5 lists the basic steps of the engineering method.

The unique characteristics of the engineering method are:

1. The engineering method is initiated by the desire to create a product that will meet the needs of society, and
2. The created product will be economically affordable and "safe" for use by society.

However, like the scientific method, the engineering method is an interactive process and embedded in the engineering method is the systematic use of the scientific method

itself to predict behavior of a prototype device or process. This means both methods require the use of trial and error. The interdependency of trail and error was explained by someone who said. One cannot have trail and error without error.

Since the desire of society[3] to improve its condition will likely always exceed its complete scientific and economic knowledge to do so, the engineer must learn to deal with ambiguity — physical as well as societal. Ambiguities in engineering will always result in several useful engineering solutions (or alternative solutions) that can meet a perceived need. Thus, the engineering method, by its very nature, will yield alternative solutions. Society is then required to select one or more of those solutions to meet its need. However, society may decline all of the alternative solutions if the long term results of applying these solutions are unknown or if they have known detrimental outcomes. This was the situation in the United States and in other countries regarding the commercial application of nuclear energy.

2.8.3 The "Art" of Engineering Design

Engineering design is a creative act in much the same way that the creation of fine art is a creative act. This is why the three general fields of knowledge are described as "arts and sciences." This description acknowledges that engineering activity is a combination of creative "art" and knowledge in science. The act of design is directed at the creation of a product (for society's use) which can be, in broad terms, either a device or a system.

Probably the most difficult part of engineering design is the "definition of the problem." This part of engineering design is described by the first three steps of Table 2.8.5 (i.e., the engineering method). To define a problem for which the engineer may design several alternatives solutions requires extensive study of society's need and the analysis

[3]"Society" is used in a generic sense here. Depending on context, society may mean society at large. It may mean a local community, corporation, operating division of a company or corporation, or it may mean the immediate association of employees or personnel.

of applicable technologies that may be brought to bear. The most important aspect of engineering design is the assessment of whether society will accept or reject the engineer's proposed designs. Modern engineering practice is littered with engineered "solutions" to the wrong problem. Such situations emphasize the need for engineers to assure the correct problem is being solved.

2.8.3.1 Device Design

A device is a product having all major parts are essentially made by a single manufacture. Thus, the device design and its manufacturing operation are under the complete control of the designer. This definition does not dictate a particular product size or complexity, but does infer that such a product will have limited size and complexity.

The creation of these device designs is somewhat abstract with regards to providing products for the "needs of society." The device designs are in support of an industry which is vital to society's operation. Thus, the designer of a device for the petroleum industry designs, in a sense, for society's needs. In the petroleum industry, suppliers of hardware and services for operating companies function more in the area of designing devices than in the design and supply of systems (or processes) that produce natural petroleum resources.

2.8.3.2 System Design

A system design is a product that is made up of a combination of devices and components. As described above the devices within a system are under the control of the designer and are designed specifically for the system. Components, on the other hand, are other devices and/or subsystems which are not made to the specification of the system designer. Usually these components are manufactured for a number of applications in various systems. Thus, the system design and the fabrication of the system are under the control of the system designer. The definition of a system infer complexity in design and operation.

In the petroleum industry, the term system and process are often considered synonymous. This is basically because engineering designers in the industry create "systems" that will carry out needed operational "processes" (e.g., waterflood project).

2.8.3.3 Role of Models

Once the problem has been defined and the decision has been made to proceed with a design effort (device or system), the conceptual designs and/or parts of the device or system are modeled, both mathematically and physically. Engineers design by creating models that they can visualize, think about, and rearrange until they believe they have an adequate design concept for practical application. Therefore, modeling is fundamental to successful engineering design efforts and is used to predict the operating characteristics of a particular design concept in many operational situations.

Mathematical Models. In modern engineering practice it is usually possible to develop mathematical models that will allow engineers to understand the operational characteristics of a design concept without actually fabricating or testing the device or system itself. Thus, analysis (using mathematical models) usually allows for great economic savings during the development of a device or system. To carry out such analyses, it is necessary to separate a particular design concept into parts—each of which being tractable to mathematical modeling. Each of these parts can be analyzed to obtain needed data regarding the operating characteristics of a design concept. Although repetitive analyses of models are exceedingly powerful design tools, there is always the risk that some important physical aspect of a design will be overlooked by the designer. This could

be a hitherto unknown phenomenon, or an ignored phenomenon, or simply an interaction of physical phenomenon that the separate analysis models can not adequately treat.

Physical Models. At some point in the design process it is usually necessary to create physical models. Often these physical models are of separate parts of the design concept that have defined adequate mathematical modeling. But most often physical models are created to assist in understanding groups or parts of a design concept that are known to have important interactions. As in mathematical modeling, physical modeling usually requires that the design concept be separated into parts that are easily modeled. Physical modeling (like mathematical modeling) allows the engineer to carry out economic small scale experiments that give insight into the operational characteristics of the design concept. Physical modeling must be carried out in accordance to certain established rules. These rules are based on principals given in the technique known as dimensional analysis (also known as Lord Rayleigh's method) or in a related technique known as the Buckingham Π theorem [4–6].

Dimensional Analysis. In the design of rather simple devices or systems, dimensional analysis can be used in conjunction with physical model experimental investigations to gain insight into the performance of a particular design concept. It is usually possible to define the performance of a simple device or system with a certain number of well chosen geometric and performance related variables that describe the device or system. Once these variables have been selected, dimensional analysis can be used to:

1. Reduce the number of variables by combining the variables into a few dimensionless terms. These dimensionless terms can be used to assist in the design of a physical model that, in turn, can be used for experimentation (on an economic scale) that will yield insight into the prototype device or system.
2. Develop descriptive governing relationships (in equation or graphical form) utilizing the dimensionless terms (with the variables) and the model experimental results that describe the performance of the physical model.
3. Develop device or system designs utilizing the descriptive governing relationships between the dimensionless terms (and subsequently between the variables) that are similar to those of the tested physical model, but are the dimensional scale of the desired device or system.

Dimensional analysis techniques are especially useful for manufacturers that make families of products that vary in size and performance specifications. Often it is not economic to make full-scale prototypes of a final product (e.g., dams, bridges, communication antennas, etc.). Thus, the solution to many of these design problems is to create small scale physical models that can be tested in similar operational environments. The dimensional analysis terms combined with results of physical modeling form the basis for interpreting data and development of full-scale prototype devices or systems. Use of dimensional analysis in fluid mechanics is given in the following example.

Example 2.8.1

To describe laminar flow of a fluid, the unit shear stress τ is some function of the dynamic viscosity μ (lb/ft^2), and the velocity difference dV(ft/s) between adjacent laminae that are separated by the distance dy(ft). Develop a relationship for τ in terms of the variables μ, dV and dy.

The functional relationship between τ and m, dV and dy can be written as

$$\tau = K(\mu^a dV^b dy^c)$$

Table 2.8.6 *Dimensions and Units of Common Variables*

Symbol	Variable	Dimensions		Units	
		MLT	FLT	USCU*	SI
Geometric					
L	Length		L	ft	m
A	Area		L^2	ft^2	m^2
V	Volume		L^3	ft^3	m^3
Kinematic					
τ	Time		T	s	s
ω	Angular velocity		T^{-1}	s^{-1}	s^{-1}
f	Frequency		LT^{-1}	ft/s	m/s
V	Velocity		L^2T^{-1}	ft^2/s	m^2/s
ν	Kinematic viscosity		L^3T^{-1}	ft^3/s	m^3/s
Q	Volume flow rate		T^{-2}	s^{-2}	s^{-2}
α	Angular acceleration		LT^{-2}	ft/s^{-2}	m/s^2
a	Acceleration		LT^{-2}	ft/s^{-2}	m/s^2
Dynamic					
ρ	Density	ML^{-1}	$FL^{-4}T^2$	slug/ft^3	kg/m^3
M	Mass	M	$FL^{-1}T^2$	slugs	kg
I	Moment of inertia	ML^2	FLT^2	slug·ft^2	kg·m^2
μ	Dynamic viscosity	$ML^{-1}T^{-1}$	$FL^{-2}T$	slug/ft/s	kg/m/s
M	Mass flow rate	MT^{-1}	$FL^{-1}T^{-1}$	slug/s	kg/s
MV	Momentum				
Ft	Impulse	MLT^{-1}	FT	lbf·s	N·s
Mω	Angular momentum	ML^2T^{-1}	FLT	slug·ft^2/s	kg·m^2/s
γ	Specific weight	$ML^{-2}T^{-2}$	FL^{-3}	lbf/ft^3	N/m^3
p	Pressure				
τ	Unit shear stress	$ML^{-1}T^{-2}$	lbf/ft^3	N/m^2	
E	Modulus of elasticity				
σ	Surface tension	MT^{-2}	FL^{-1}	lbf/ft	N/m
F	Force	MLT^{-2}	F	lbf	N
E	Energy				
W	Work	ML^2T^{-2}	FL	lbf/ft	J
FL	Torque				
P	Power	ML^2T^{-3}	FLT^{-1}	lbf·ft/s	W
v	Specific volume	$M^{-1}L^4T^{-2}$	ft^3/lbm	m^3/kg	

*United States customary units.

Using Table 2.8.6, the dimensional form of the above is

$$(FL^{-2}) = K(FL^{-2}T)^a(LT^{-1})^b(L)^c$$

The equations for determining the dimensional exponents a, b and c are

Force F: $1 = a + 0 + 0$

Length L: $-2 = -2a + b + c$

Time T: $0 = a - b + c$

Solving the above equations for a, b and c yields

$$a = 1$$
$$b = 1$$
$$c = 1$$

Thus, inserting the exponent values gives

$$\tau = K(\mu^1 dV^1 dy^{-1})$$

which can be written as

$$\tau = K[\mu(dV/dy)]$$

where the constant K can be determined experimentally for the particular design application.

Buckingham Π. The Buckingham Π theorem is somewhat more sophisticated than the dimensional analysis technique. Although directly related to the dimensional analysis technique, Buckingham Π is usually used in design situations where there is less understanding of exactly which performance characteristics are ultimately important for prototype design. The Buckingham Π theorem is generally more applicable to the design of complex devices or systems. In particular, Buckingham Π is usually used when the number of dimensionless terms needed to describe a device or system exceeds four. The fundamental theorem is . . .

> "if an equation is dimensionally homogeneous (and contains all the essential geometry and performance variables) it can be reduced to a relationship among a complete set of dimensionless terms of the variables."

Applications of the Buckingham Π theorem results in the formulation of dimensionless terms called Π ratios. These Π ratios have no relation to the number 3.1416.

Example 2.8.2

Consider the rather complicated phenomenon of the helical buckling of drill pipe or production tubing inside a casing. This is a highly complicated situation that must be considered in many petroleum engineering drilling and production operations. In general, this involves the torsional post-buckling behavior of a thin walled elastic pipe. The mathematical model for this situation is very complicated and dependable closed-form solutions of this problem do not exist. But still the designer must obtain some insight into this phenomena before an engineering design can be completed. The problem can be approached by utilizing the principals of physical modeling based on the Buckingham Π theorem. What is sought is a set of dimensional relationships (the Π ratios) that are made up of the important variables of the problem. These Π ratios may then be used to created a small-scale experiment that will yield data that can be used to assist in the design the prototype (or full-scale pipe system). The important variables in this problem are EI (lb-in^2 or lb-ft^2) the flexural rigidity of the thin-walled elastic pipe (which is directly related to the torsional rigidity GJ through known material constants), the length of the pipe L(ft), the shortening of the pipe δ(ft) (as it helically twists in the casing), the annular distance between the outside of the pipe and the inside of the casing $(d_2 - d_1)$ (ft) (where d_1 is the outside diameter of the pipe and d_2 is the inside diameter of the casing), the applied torque to the pipe T(ft-lbs), and the applied tension in the pipe P(lbs).

Using Table 2.8.6 the variables are EI(FL^2), L(L), d(L), $(d_2 - d_1)$ (L), T(FL), and P(F). Note that this I is moment-area which is in the units of ft^4 (not to be confused with I given in Table 2.8.6 which is moment of inertia, see Chapter 2, Strength of Materials, for clarification). The number of Π ratios that will describe the problem is equal to the number of variables (6) minus the number of fundamental dimensions (F and L, or 2). Thus, there will be four Π ratios (i.e., $6 - 2 = 4$), Π_1, Π_2, Π_3, and Π_4. The selection of the combination of variables to be included in each n ratio must be carefully done in order not to create a complicated system of ratios. This is done by recognizing which variables will have the fundamental dimensions needed to cancel with the fundamental dimensions in the other included variables to have a truly dimensionless ratio. With this in mind, Π_1 is

$$\prod_1 = EI^a T^b L^c$$

and Π_2 is

$$\prod_2 = T^d P^e L^f$$

and Π_3 is

$$\prod_3 = L^g (d_2 - d_1)^h$$

and Π_4 is

$$\prod_4 = (d_2 - d_1)^i \delta^j$$

In dimensional form Π_1 and each variable of Π_1 becomes

$$(F^0 L^0) = (FL^2)^a (FL)^b (L)^c$$

and for Π_2

$$(F^0 L^0) = (FL)^d (F)^e (L)^f$$

and for Π_3

$$(F^0 L^0) = (L)^g (L)^h$$

and for Π_4

$$(F^0 L^0) = (L)^i (L)^j$$

For Π_1 the equations for a, b and c are

Force F: $0 = a + b$

Length L: $0 = 2a + b + c$

This yields b = $-$a and c = a, which is

$$\prod_1 = EI^a T^{-1} L^a$$

or

$$\prod_1 = [EI/TL]^a$$

and since Π_1 is a dimensionless ratio, then the actual value of a is not important. Therefore, the above becomes simply

$$\prod_1 = EI/TL.$$

Similarly it is found that

$$\prod_2 = T/PL$$

$$\prod_3 = L/(d_2 - d_1)$$

and

$$\prod_4 = (d_2 - d_1)/\delta$$

These Π ratios allow the following relationships to be made between the performance characteristics found for a small-scale experimental model and a full-scale prototype:

$$\left[\prod_1\right]_m = \left[\prod_1\right]_p$$

$$\left[\prod_2\right]_m = \left[\prod_2\right]_p$$

$$\left[\prod_3\right]_m = \left[\prod_3\right]_p$$

and

$$\left[\prod_4\right]_m = \left[\prod_4\right]_p$$

The Π ratios can be used as the basis for a small-scale experiment to obtain data that might allow an investigator to determine the number of helixes N, and the shortening δ, that will occur in a full-scale tubing string under particular torsional post-buckling conditions. The small-scale experiment will yield experimental data that can be related through the Π ratios to the full-scale prototype. The use of Π ratios and experimental data in engineering design requires extensive statistical and deterministic analyses.

2.8.3.4 Analysis, Synthesis and the Creative Art of Design

Evolution was an early engineering design methodology. This methodology (basically trial and error) is a primitive form of physical modeling. Prior to the development of more sophisticated engineering design methodologies, the only method to develop a useful device or system was to fabricate a conceptual design and put it into operation. If the design failed to meet the desired specifications, then the failure was analyzed and the appropriate changes made to the design and another product fabricated and placed into operation. However, design by evolution was not very economical and as engineering design methodologies were developed that would allow for design choices to be made without resorting to costly fabrication, design engineers were driven to accept these new methodologies. Modern engineering design methodologies are based on the design engineer's ability to appropriately utilize analysis and synthesis to create the desired device or system [7–10].

Nearly all modern engineering design situations require the use of models, both mathematical and physical. Once societal needs are defined, the designer specifies the performance characteristics required for a design. Also, at this point, the designer can often sketch or otherwise define the various initial alternative design concepts. The designer then separates each alternative design concept into parts that are tractable to either mathematical or physical modeling (or both). The designer utilizes a wide variety of analysis techniques (inherit in mathematical models but also part of the assessment of test data from physical models) to attempt to predict the performance characteristics of the various proposed alternative design concepts. This usually requires repeated analysis and the development of numerous sketches and fabrication drawings for the separate parts of each proposed alternative design concept. Thus, the design of a device or system is a lengthy iterative process.

Once the designer has developed confidence in the analysis techniques pertaining to the various parts of a design concept (whether derived from mathematical models or from physical models), the designer can begin the process of synthesis. Synthesis is basically the combining of the analyses (and any other pertinent information) to allow for the design concept to be viewed as a whole. In general, it is during the process of synthesis that the creative capabilities of the individual designer are utilized the most. The designer utilizes knowledge of the performance specifications, results from modeling, economics, experience with other similar projects, aesthetics (if needed), and intuition to create many easy-to-make choices concerning the design. Thus, the synthesis process often allows the designer to proceed to the final design of a device or system with little iteration.

Creativity in engineering is the art portion of engineering. It is a unique sum of an individual design engineer's experience, practice and intuition with a flair for the aesthetic. Creativity can be found in any part of the engineering design process, but it is usually in the synthesis portion of the process that a designer's aptitude for creativity can make the difference between a mediocre design and a truly successful design. Thus, creativity is an essential part of the engineering method.

2.8.3.5 Feasibility Studies and Economics

Once an engineering design has been identified (that will meet the performance specifications) it is necessary to ask a very important question, "Can and should the device or system be made?" In general, this question can be divided into three corollary questions [7]:

1. Can the design of the anticipated alternative devices or systems be physically realized?
2. Is the production of alternative devices or systems compatible with the capabilities and goals of the producing company?
3. Can the alternative devices or systems be economically produced and economically operated?

Only if all of the above three questions are answered in the affirmative can and should the device of system be made. The gathering of information to answer these questions is called the feasibility study.

Physical realizability is often the most difficult of the above three corollary questions to answer. In general, to answer this question it is necessary to know; 1) whether the materials and components required by the engineering design are available; and 2) whether the manufacturing (and/or fabrication) techniques and skilled craftsmen needed to fabricate the product are also available. These two assessments are difficult to make because they often involve the projection of future technological developments. Technological developments usually do not occur according to schedule.

Company compatibility requires the analysis of the engineering, manufacturing (with quality assurance), distribution, and sales capabilities of an organization to produce and realize a profit from a given engineering design or product. Often, an engineering department within a company may be capable of designing a particular device or system, but the production and sales departments are not capable of carrying out their respective tasks. Also, a new product line under consideration may be beyond the scope of the overall business goals and objectives of the company.

As in all business situations, the economics of the development of a newly designed device or system is of paramount importance. In order to fully understand the economic feasibility of a new design, the engineer must make the following economic assessments:

1. Evaluate two or more manufacturing techniques.
2. If components of a new system design are available from sources outside the company, then an evaluation must be made as to whether it is more economic to buy these components from outside or to manufacture them within.
3. Engineers will constantly seek ways of reducing manufacturing costs. As these design modifications are made, economic evaluations of changes must be made.
4. An overall economic evaluation must be made to ensure that the contemplated project for petroleum development will recover sufficient capital to pay for the total cost of development, installation and assembly effort of the company. Further, the capital return must yield a financial return consistent with the overall company risk involved. Thus, it is very important that the petroleum engineer designing a recovery system understand the company's evaluation criteria that will be used by management in judging whether a new project is to go forward. Chapter 7 (Petroleum Economics) has detailed discussions of general engineering economics and product (or project) evaluation criteria.

2.8.4 Design in the Petroleum Industry

2.8.4.1 Producing and Service Companies

The petroleum industry is a highly complex primary industry. Its function in society is to provide a reliable supply of liquid and gaseous hydrocarbon fuels and lubricants (and well as other related products) for both industrial and private customers. The industry is composed of two basic business elements 1) producing companies (often called operating companies 2) service companies.

Producing companies locate the subsurface petroleum resources; recover the oil, gas or condensate; and then market the recovered resource to customers. These activities involves planning and carrying out exploration, drilling and well completion, and the production activities.

The service companies provide a supporting role to the producing companies. They supply the products and services that allow the producing companies to explore, drill and complete wells, and produce the petroleum from the earth. Once the petroleum has been brought to the surface, the service companies also assist in providing safe transportation of the petroleum to refining facilities. There is a great variety of service companies. Many service companies are dedicated only to supporting the producing companies (e.g., geophysical surveying) while other service companies supply products and services to a number of primary industries.

2.8.4.2 Service Company Product Design

In general, engineering design of devices or systems takes place throughout the petroleum industry. However, service industries are responsible for supplying (thus, designing and manufacturing/fabricating) the vast majority of the devices and systems (and often the personnel to operate them) that are utilized by the producing companies. Therefore, the service companies do for more traditional device and system engineering design than the producing companies (e.g., geophysical survey tool design, drilling fluid design, drill bit design, cement design and placement system design, production pump design, etc.). Because of the complexity of the petroleum industry, these engineering design activities cover all three of the basic engineering fields (e.g., mechanical arts and sciences, the chemical arts and sciences, and the electrical arts and sciences).

2.8.4.3 Producing Company Project Design
The producing companies are responsible for carrying out the complex operations of exploration, drilling and completion, and production. These projects vary from the seeking oil and gas deposits at 15,000 ft of depth in the continental U.S. to seeking similar deposits at 15,000 ft below some remote ocean floor. The engineering planning and design of these complex projects is a special class of systems engineering design. Although dealing with quite different technologies, production operations have similar concerns related to overall system outcome as do large mining projects or large civil engineering projects (such as the construction of a dam or large transportation system). Thus, the engineering projects design and management aspects of systems engineering are of primary interest to the engineering staffs of the producing companies [12].

2.8.5 Engineering Ethics
2.8.5.1 The Profession of Engineering
An engineer is a professional in exactly the same context as a medical doctor, a lawyer, clergy, teacher, etc. The dictionary defines a "profession" as

"an occupation requiring advanced education."

In this context, an advanced education means an academic education beyond secondary education. The term 'professional' is defined as

"connected with, or engaged in a profession."

The professional should not be confused with the other occupations that make up the economic doers in modern society, namely craftsmen, tradesmen, and laborers. The difference between these other occupations and the professional is in the definition of the profession and professional. Very few occupations *require* an advanced education to be employed in that occupation. A plumber does not have to have an advanced academic education to be in the plumbing business. But the plumber does have to have certain manual skills in order to be successful in that occupation. Thus, the plumber is a craftsman. A banker does not have to have an advanced education to be in the banking business. Also, the banker does not have to have any particular manual skills to be a banker. Thus, the banker is considered to be a tradesman. It should be noted that the definition of the term "professional" does not depend upon whether the engineer (or medical doctor, etc.) is legally registered (with a state board) as a licensed professional engineer (or registered as a licensed medical doctor, etc.) [13].

There are certain unwritten accepted societal ideals that also separate professionals from the other occupations. The most important of these is that professionals, in carrying out the duties of their respective professions, usually are doing work that can affect society. This is particularly true of the engineering profession. The engineering profession designs many devices and systems that if not designed correctly and fail can have catastrophic effects on society as a whole. For example, when a civil engineer does not correctly design a community water supply and sewer system and the community becomes sick from contaminated drinking water, then society as a whole suffers from the civil engineer's incompetence. Likewise, when a petroleum engineer does not correctly design the reserve pits at the drilling location and drilling mud contaminated with crude oil is washed into a nearby stream, society as a whole suffers from the petroleum engineer's incompetence. Also, when a petroleum engineer plans and operates a drilling location that is quite near a high voltage power line and fails to get the local power company to either provide a by-pass line or rubber boot the line, and subsequently a crane operator on location is electrocuted, society as a whole suffers from the petroleum engineer's incompetence. In this case, "society" is represented by the craftsmen and laborers working at the location [14,15].

2.8.5.2 NSPE Code of Ethics
All professional groups in society have a code of ethics to guide their respective members as they carry out their professional duties for society. There is a general code of ethics of the National Society of Professional Engineers (NSPE) which basically covers all registered professional engineers regardless of specific discipline. The preamble and fundamental canons of the NSPE Code of Ethics is given below [16].

Preamble. Engineering is an important and learned profession. As members of this profession, engineers are expected to exhibit the highest standards of honesty and integrity. Engineering has a direct and vital impact on the quality of life for all people. Accordingly, the services provided by engineers require honesty, impartiality, fairness, and equity, and engineers must be dedicated to the protection of the public health, safety, and welfare. Engineers must perform under a standard of professional behavior that requires adherence to the highest principles of ethical conduct.

Fundamental canons. Engineers, in the fulfillment of their professional duties, shall

1. Hold paramount the safety, health and welfare of the public.
2. Perform services only in areas of their competence.
3. Issue public statements only in an objective and truthful manner.
4. Act for each employer or client as faithful agents or trustees.
5. Avoid deceptive acts.
6. Conduct themselves honorably, responsibly, ethically, and lawfully so as to enhance the honor, reputation, and usefulness of the profession.

2.8.5.3 SPE Code of Ethics
Each specific engineering discipline has developed its own code of ethics. These discipline specific codes of ethics generally reflect the working situations that are relevant that discipline. This can be seen in the Society of Petroleum Engineers (SPE), Guide for Professional Conduct. Because the devices and systems that petroleum engineers design are usually in operational locations that are remote from most of general public, the issue concerning the "safety, health and welfare of the public" is not as immediate to the petroleum engineer as it is to other disciplines (e.g., civil engineers). Thus, the "safety, health and welfare of the public" is found further down in this list of canons than in the NSPE Code of Ethics. The preamble, the fundamental principle, and

the canons of professional conduct of the SPE Guide for Professional Conduct is given below [17].

Preamble. Engineers recognize that the practice of engineering has a direct and vital influence on the quality of life of all people. Therefore, engineers should exhibit high standards of competency, honesty, and impartiality; be fair and equitable; and accept a personal responsibility for adherence to applicable laws, the protection of the public health, and maintenance of safety in their professional actions and behavior. These principles govern professional conduct in serving the interests of the public, clients, employers, colleagues, and the profession.

The Fundamental Principle. The engineer as a professional is dedicated to improving competence, service, fairness, and the exercise of well-founded judgment in the practice of engineering for the public, employers, and clients with fundamental concern for the public health and safety in the pursuit of this practice.

Canons of Professional Conduct.

1. Engineers offer services in the areas of their competence affording full disclosure of their qualifications.
2. Engineers consider the consequences of their work and societal issues pertinent to it and seek to extend public understanding of those relationships.
3. Engineers are honest, truthful, and fair in presenting information and in making public statements reflecting on professional matters and their professional role.
4. Engineers engage in professional relationships without bias because of race, religion, sex, age, national origin, or handicap.
5. Engineers act in professional matters for each employer or client as faithful agents or trustees disclosing nothing of a proprietary nature concerning the business affairs or technical processes of any present or former client or employer without specific consent.
6. Engineers disclose to affected parties known or potential conflicts of interest or other circumstances which might influence or appear to influence judgment or impair the fairness or quality of their performance.
7. Engineers are responsible for enhancing their professional competence throughout their careers and for encouraging similar actions by their colleagues.
8. Engineers accept responsibility for their actions; seek and acknowledge criticism of their work; offer honest criticism of the work of others; properly credit the contributions of others; and do not accept credit for work not theirs.
9. Engineers, perceiving a consequence of their professional duties to adversely affect the present or future public health and safety, shall formally advise their employers of clients and, if warranted, consider further disclosure.
10. Engineers act in accordance with all applicable laws and the canons of ethics as applicable to the practice of engineering as stated in the laws and regulations governing the practice of engineering in their country, territory, or state, and lend support to others who strive to do likewise.

2.8.6 Intellectual Property

Since one of the main focuses of the engineer is to provide technological solutions for the needs of society, then creative invention is a major part of the engineer's design activity. Therefore, the engineer must be familiar with the current laws (and any imminent future changes in the laws) for the protections of intellectual property. Intellectual property is composed of four separate legal entities. These are trade secrets, patents, trademarks, and copyrights.

2.8.6.1 Trade Secrets

Trade secrets are creative works (usually methods, processes, or designs) that are not covered by the strict definition of "an invention" required for a legal patent, or they are actual inventions that a company or individual does not want to expose to public scrutiny through the patent process (since the published patent must contain a written description of the invention).

Companies and individuals have the legal right to protect and maintain trade secrets. The key point, however, is that trade secrets must be something that can be kept secret. Anyone who discovers a trade secret via his or her own efforts may use it. The only legal protection for trade secrets is against persons who obtain these secrets by stealing them and then using them, or after having been entrusted with these secrets as a necessary part of their job and then using them [7,18].

The term trade secret is well understood by U.S. courts and is a legal term. In general, the U.S. courts have strongly upheld the protection of company- or individual-owned trade secrets. Cases against ex-employees or others who have obtained trade secrets by stealth are relatively easy to prosecute (compared to patent, trademark, or copyright infringement cases).

2.8.6.2 Patents

When the U.S. Constitution was framed in 1789 it contained this provision (Article 1, Section 8) [18–22]:

> "The Congress shall have power to promote the progress of science and useful arts by securing for limited times to authors and inventors the exclusive right to their respective writings and discoveries."

In 1790 the formal U.S. Patent system was founded. Since that time the U.S. Patent Office has undergone numerous changes that have generally improved legal protection for intellectual property owned by individual citizens.

A U.S. Patent is a legal document granted only to the individual inventor that allows the inventor to have the exclusive right for a limited time to exclude all others from making, using or selling his or her invention in the U.S. There are restrictions to this statement. An inventor does not have an absolute right to use his or her patented invention, if doing so would also involve making, using or selling another invention patented by someone else. The time limits for the exclusive use of an invention by the inventor are fixed by low and can only be extended in special circumstances. After this time limit expires, anyone may make and sell the invention without permission of the inventor. There are three legal categories of patents; utility patents, design patents, and plant patents.

Utility Patents. Utility patents are granted to individuals only who have invented or discovered any new and useful method, process, machine, manufacture, or matter composition. These patents must be useful to society (have utility). A utility patent has a total time limit for exclusive use by the inventor of 17 years.

Design Patents. Design patents are granted to individuals who have invented new, original and ornamental designs for an article of manufacture. A design patent has a total time limit for exclusive use by the inventor of 14 years.

Plant Patents. Plant patents are granted to individuals who have invented or discovered and a sexually reproduced a new and distinct variety of plant. This includes mutants, hybrids, and newly found seedlings. A plant patent has a total time limit for exclusive use by the inventor of 17 years.

Only the actual inventor may apply for a patent (utility, design, or plant patent). Once a patent has been obtained, the patent can be sold or mortgaged. Also, the owner of a patent may assign part or all interest in the patent to another individual or a company or other business entity. The owner of a patent may also grant licenses to others (either individuals or business entities), to make, use or sell the invention.

In addition to each of the above patent's unique characteristic (i.e., utility, ornamental or original design, be a plant), an invention, in order to be granted a U.S. Patent, must demonstrate two additional characteristics: the invention must be novel, and the invention must not be obvious. The steps to pursue to obtain a patent are as follows:

1. Make sure the invention is new and practical and is not already in use. This requires a preliminary search of the industrial literature to determine if the invention is practical and is not already in use in the public domain.
2. Keep records that document when and what was invented. It is important that accurate records are kept showing your original sketches with a disclosure statement describing what and how your invention works. It is useful to have someone witness this disclosure document and verify the date that this invention took place. It is often during this step that the invention concept is either modeled (mathematical or physical or both) and tested. Thus, accurate records of these analyzes or test results should also be kept. In the U.S. it is the first to invent that will obtain a patent in the event of two individuals inventing the same thing. Keep the disclosure document secret until the patent application is submitted to the patent office.
3. Search the existing patents relating to the invention. Carry out an extensive patent search of existing patents to find patents that are related to the new invention. Make sure the new invention is novel and unobvious and is not in conflict with existing patents. A search of U.S. patents can be made in the Search Room of the U.S. Patent and Trademark Office in Crystal Plaza, 2021 Jefferson Davis Highway, Arlington Virginia. An individual inventor or a patent attorney may hire agents that will carry out thorough patent searches. In addition, every state in the U.S. has a U.S. Patent Depository Library where patent searches may be conducted by private individuals.
4. Prepare and file the patent application documents. This document is called a "specification." The specification must be clear enough so that anyone skilled in the subject matter of the invention could recreate your invention and use it. In the written portion of the specifications the inventor must state the claims of the invention. These claims must show that the invention is novel and unobvious. Also, where applicable, illustrations must accompany the specification. There are filing fees for patent applications and if the application is successful there are maintenance fees to keep the patent in force. Literature describing patent applications and how they are to be submitted can be obtained from the Commissioner of Patents and Trademarks, Washington, D. C., 20231. This is also the address to submit properly prepared patent applications.

Infringement of a patent occurs when an individual or company makes, uses, or sells an invention without the permission of the inventor, assignee, owner, or licensee. The U.S. courts will award damages and place penalties on the infringer. But patent infringement cases are very costly and time consuming.

Any inventor may apply for a patent regardless of age, sex, or citizenship. Once a U.S. Patent has been granted most other countries allow the inventor up to one year to submit a patent application for a foreign patent. The exception is Japan which requires nearly immediate submittal of a patent application to the Japanese patent office at nearly the same time the inventor is submitting to the U.S. Patent office.

It should be noted that all other countries grant patents to inventors that are the first to submit patent applications to their respective patent offices. We are the only country the grants a patent to the first to invent (not the first to the patent office). However, in 2003 the U.S. may change to the "first to the patent office" method of granting patents.

2.8.6.3 Trademarks

A trademark is any work, name, symbol, or device or any combination of these adopted and used by manufacturers or merchants to identify their goods and distinguish them from those manufactured or sold by others. A service mark is a mark used in the sale or advertising of services to identify the services of one individual or company and distinguish them from the services of others [23,24].

Unlike a patent, but like a copyright, ownership of a trademark or service mark is acquired by use. An individual or company using the mark within a state may register the mark in that state. The Secretary of State within each state has the forms for the registration of a trademark or service mark. Once a mark has been used in interstate or foreign commerce, the mark may be registered in the U.S. Patent and Trademark Office. This is accomplished by filing an application to that office (see Reference 87). Prior to receiving federal registration of the mark, the symbols ^{TM} and ^{SM} may be used (these symbols give notice that the marks have been filed with the state). After the mark has received federal registration the symbol ® should be used.

There are two classes of registration for a mark. The Principal Register is for unique and distinctive marks that when applied to products and services are not likely to cause confusion or deception. To be registered on the Principal Register a mark must be in continuous, exclusive interstate use. Marks not registrable on the Principal Register may be registered on the Supplemental Register.

The term of a federal registered trademark and service mark is 20 years. In order to secure the mark for the full 20 years, an affidavit must be filed with the Commissioner of Patents and Trademarks in the sixth year showing the mark is still in use. It is advisable to conduct a search in the U.S. Patent and Trademark Office to determine whether a mark under consideration might conflict with existing registered marks. Trademarks and service mark registrations may be assigned after registration.

Once federal certification of a mark has been issued, the mark is protected nationwide. Any infringement by an individual or company of a mark after federal registration can be subject to damage claims by the mark owner in U.S. District Courts.

Foreign countries require registration of a mark in compliance with local laws. The local laws vary a great deal; thus, it is necessary to consult a local attorney or a U.S. attorney familiar with foreign patent and trademark law in order to register a mark in a foreign country.

Information concerning the registration of marks in the U.S. may be obtained from the Commissioner of Patents and Trademarks, Washington, D. C., 20231.

2.8.6.4 Copyrights

The copyright protects creative works such a literary works, musical works, dramatic works, pictorial, graphic, sculptural, motion pictures and other audio-visual works, and computer software. It is the latter that is of most concern to the engineer [25,26].

Registration for copyright protection may be obtained by application to the Register of Copyrights, Library of Congress, Washington, D. C., 20540. This registration must take place on or about the time of the first sale of the creative work. The forms for application are obtained from the address above (the application fee is $10).

The notice of copyright is given by "copyright" or the abbreviation "Copr" or by the symbol ©. The notice must be accompanied by the name of the copyright owner and the year of the first publication. Like a patent, only the author of the creative work can be the copyright owner. However, an assignee may file for the copyright on behalf of the author of a creative work. In general, the term of a copyright is for the life of the author (of the creative work) plus 50 years. Any infringement by an individual or company of a registered copyright can be subject to damage claims by the creative work owner or assignee in U.S. District Courts.

Under the Universal Copyright Convention, a U.S. citizen may obtain a copyright in most countries of the world by simply publishing within the U.S. using the © symbol and the appropriate author and date of publication notices (only the © symbol is recognized worldwide).

References

1. Graves, R. M., and Thompson, R. S., "The Ultimate Engineering Experience," *Proceedings of the ASEE Gulf-Southwest Section Meeting*, Austin, Texas, April 1–2, 1993.

2. Thompson, R. S., "Integrated Design by Design: Educating for Multidisciplinary Teamwork," *Proceedings of the 68th Annual Technical Conference and Exhibition of the SPE*, Houston, Texas, October 3–6, 1993.

3. Florman, S. C., "Engineering and the Concept of the Elite," *The Bridge*, Fall 1991.

4. Langhaar, H. L., *Dimensional Analysis and Theory of Models*, John Wiley and Sons, 1951.

5. Skoglund, V. J., *Similitude Theory and Applications*, International Textbook Company, 1967.

6. Taylor, E. S., *Dimensional Analysis for Engineers*, Clarendon Press, 1974.

7. Middendorf, W. H., *Design of Devices and Systems*, Second Edition, Marcel Dekker, 1990.

8. Wright, P. H., *Introduction to Engineering*, Second Edition, John Wiley, 1994.

9. Dym, C. L., *Engineering Design*, Cambridge, 1994.

10. Lewis, W. P., and Samuel, A. E., *Fundamentals of Engineering Design*, Prentice Hall, 1989.

11. Smith, A. A., Hinton, E., and Lewis, R. W., *Civil Engineering Systems Analysis and Design*, John Wiley, 1983.

12. Callahan, J., *Ethical Issues in Professional Life*, Oxford University Press, 1988.

13. McCuen, R. H. and Wallace, J. M., *Social Responsibilities in Engineering and Science*, Prentice Hall, 1987.

14. Martin, M. W., and Schinzinger, R., *Ethics in Engineering*, McGraw-Hill, 1989.

15. NSPE Publication as revised January, 2003.

16. SPE Approved by the Board of Directors February 25, 1985.

17. Rules of Practice of U.S. Patent and Trademark Office, 37 C. F. R., Title I.

18. Konold, W. G., et al, *What Every Engineer Should Know About Patents*, Marcel Dekker, 1989.

19. Robinson, W. C., *Law of Patents for Useful Inventions*, W. S. Hein, 1972.

20. Miller, A. R., and Davis, M. H., *Intellectual Property: Patents. Trademarks. and Copyrights*, West Publishing, 1983.

21. Amernick, B. A., *Patent Law for the Nonlawyer: A Guide for the Engineer, Technologist, and Manager*, Second Edition, Van Nos Reinhold, 1991.

22. Rules of Practice of U. S. Patent and Trademark Office, 37 C. F. R., Title IV.

23. Vandenburgh, E. C., *Trademark Law and Procedure*, Bobbs-Merrill Co., 1968.

24. Copyright Office Regulations, 37 C. F. R., Chapter 2.

25. Nimmer, M. B., *Nimmer on Copyright*, Vol. 1–5, Matthew Bender, 1978–1995.

3

Auxiliary Equipment

Contributing Authors
William C. Lyons, Ph.D., P.E.
Vern Cobb
Eddie Scales

Contents

This chapter describes the various auxiliary equipment that are important to the function of oil field activities. This equipment is used in drilling, well completion, production, and related operations. The discussions that follow will concentrate on the basic operation characteristics and specifications required by a user. The aim is to give the user the information needed to ascertain whether the equipment available for a particular field operation is adequate, or if the equipment available is not adequate, how to specify what equipment will be necessary.

3.1 PRIME MOVERS

The prime mover is the unit that first converts an energy source into a mechanical force. Typical prime movers are internal combustion motors, gas turbines, water turbines, steam engines and electrical motors. The discussion will be limited to the prime movers that are most used in modern well drilling and production operations. These are internal combustion motors, gas turbine motors and electric motors.

3.1.1 Internal Combustion Engines

The internal combustion engines considered in this section are piston-type engines. The combustion process in such engines are assumed to be constant volume, constant pressure, or some combination of both. Piston-type internal combustion engines are made up of a series of pistons that can move in enclosed chambers called cylinders. These engines can be designed in a variety of configurations. The most widely used configurations are the following [1]:

- An in-line type engine has all cylinders aligned and on one side of the crankshaft (Figure 3.1.1). These engines are found in sizes from 4 pistons (or cylinders) to as many as 16.
- A V-type engine has an equal number of cylinders aligned on two banks of the engine. These banks form a V shape (see Figure 3.1.2). These engines are found in sizes from 4 cylinders to as many as 12.
- An opposed-type engine has an equal number of cylinders aligned on two banks of the engine. These banks are horizontally opposed to one another on opposite sides of the crankshaft (see Figure 3.1.3). These engines are found in sizes from 4 cylinders to as many as 8.

One of the more important specifications of an internal combustion engine is its displacement volume.

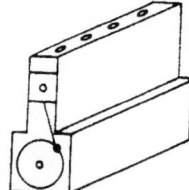

Figure 3.1.1   *In-line type engine [1].*

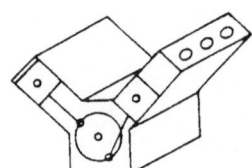

Figure 3.1.2   *V-type engine [1].*

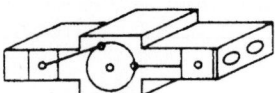

Figure 3.1.3   *Opposed type engine [1].*

The displacement volume, V_d (in^3), is (per revolution)

$$V_d \, (\text{in}^3) = \frac{\pi}{4} d^2 ln \qquad [3.1.1]$$

where d is the outside diameter of the piston (or bore of cylinder in in.), l is the stroke (or length of movement) of the piston within the cylinder in in., and n is the number of cylinders.

The nominal compression ratio (which is usually specified) is the displacement volume plus the clearance volume divided by the clearance volume. Because of the mechanics of intake value closing, the actual compression ratio r_a is less than the nominal. Thus, the compression pressure p (psia) may be estimated by

$$p = r_a \cdot p_m \qquad [3.1.2]$$

where p_m is the intake manifold pressure in psia.

There are two basic combustion cycles. These are

1. Spark-ignition, or otto cycle, engine which is fueled by a gas and air mixture.
2. Compression-ignition, or diesel cycle engine, which is fueled by a diesel oil and air mixture.

There can also be a combination of the two cycles in an engine. Such an engine is called a mixed, combination, or limited pressure cycle.

The spark-ignition cycle engine uses volatile liquids or gases as fuel. Such engines have compression ratios from 6:1 to 12:1. liquid gasoline is the most common fuel for this cycle, especially for automobile, truck, and airplane engines. Large stationery engines usually use commercial gases such as natural gas, produces gas, or coal gas. All engines operating on this cycle use carburetors, gas-mixing valves, or fuel injection systems to mix the volatile fuel with the appropriate volume of air for subsequent combustion. This mixing can be carried out prior to the mixture being placed in the cylinder of the engine, or mixed directly in the cylinder.

The compression-ignition cycle engine uses low volatile fuel. These are typically fuel oils. The compression ratios for these engine are from 11.5:1 to 22:1. These engines are used to power large trucks and buses. They are also used for large engines. The compression-ignition cycle engine usually is equipped with a fuel injection system which allows the fuel to be mixed with air directly in the cylinders of the engine.

The internal combustion engine can be operated as a two-stroke or a four-stroke engine.

The two-stroke engine requires two piston strokes (or one crackshaft revolution) for each cycle. In one upward (compression) stroke of the piston, the combustible mixture is brought into the cylinder through the intake valve and ignited near the top of the stroke as the piston is forced downward (to provide power); the exhaust valve is opened and the spent gases are allowed to escape. Near the bottom of their stroke the intake valve is opened again and the combustible mixture again is brought into the cylinder.

The four-stroke engine requires four piston strokes (or two crankshaft revolutions) for each cycle. In a downward stroke, the intake valve is opened and the combustible mixture is brought into the cylinder. In an upward stroke the fuel–air mixture is compressed and ignited near the top of the stroke. This forces the piston downward (to provide power).

In the next upward stroke the exhaust valve is opened and the spent gases are forced from the cylinder. In the following downward stroke, the cycle is repeated with the opening of the intake valve.

The standards for internal combustion engines have been established by the American Petroleum Institute (API), the Diesel Engine Manufacturers Association (DEMA), and the Internal Combustion Engine Institute (ICEI). In addition, some of the engine manufacturers have their own rating procedures. It is important to know which standards have been applied for the rating of an engine. A consistent set of standards should be used by the engineer when comparing the ratings of various engines for the purpose of selecting the appropriate design for field applications.

The API standards describe the method used in rating engines and the recommended practice for engine installation, maintenance, and operation [2,3]. In oil field operations the API rating standards are most frequently used.

The important definitions are

1. *Bare engine.* A bare engine shall be an engine less all accessories except those (built in or attached) absolutely required for running. All accessories normally required for operation of the engine, such as ignition, water pump, air cleaner, oil pump, governor, etc., shall be included.
2. *Power unit.* A power unit shall consist of a bare engine, plus other equipment such as a fan for air cooling, special water pumps, and so forth. When included, specific information must be given as to design factors such as ambient temperature and power consumption.
3. *Maximum standard brake horsepower.* At any rotational speed, maximum standard brake horsepower shall be the greatest horsepower, corrected to standard conditions, that can be sustained continuously under conditions as outlined under test procedure. The unit of horsepower is 33,000 ft-lb/min or 550 ft-lb/s. Standard conditions for the purpose of internal combustion engine testing and rating is 85°F (29.4°C) and 29.38 in. of mercury (99 kPa). Note these values are different from standard conditions for gas and air volume specifications.
4. *Maximum standard torque.* The maximum standard torque at any given rotational speed shall be that corresponding to the maximum standard brake horsepower at that speed.

Test engines shall be of exactly the same design and equipped with the same components and accessories as engines delivered to the purchaser [2]. The observed brake horsepower H_0 obtained during the testing of a bare engine or a power unit is converted to standard brake horsepower using

$$H_s = H_0 \frac{29.38}{P_0 - E_0} \frac{460 + t_0}{520} \qquad [3.1.3]$$

where P_0 is observed barometric pressure in in. Hg, E_0 is pressure of water vapor in air (from relative humidity data) in in. Hg, t_0 is observed air temperature in °F.

When an internal combustion engine is to be used at different operating conditions (altitude) other than the standard conditions that the engine was rated at, it is necessary to derate the engine specifications. The brake horsepower H at pressure and temperature conditions other that standard can be obtained from the following:

$$H = \frac{H_s (P - E)}{29.38} \frac{520}{t + 460} \qquad [3.1.4]$$

where P is the local barometric pressure in in. Hg, E is pressure of water vapor in air (from relative humidity data) in in. Hg, t is the local air temperature in °F.

At any given set of ambient conditions, the horsepower H is related to the torque the engine produces at a given speed from the following:

$$H = \frac{TN}{5,252} \qquad [3.1.5]$$

where T is the torque in ft-lb, N is the speed in rpms.

For the various transmission speed ratios, and the engine speed, the maximum available torque may be found using the maximum brake horsepower which can be produced by the engine.

Figures 3.1.4 and 3.1.5 are examples of a typical manufacturer's report of test results.

There is a reduction in the power available from an internal combustion engine when there is a decrease in ambient air pressure, an increase in ambient temperature, or an increase in relative vapor pressure. Such changes reduce the mass of oxygen available for combustion inside the engine cylinder. This reduction is standard brake horsepower as a function of altitude increase and temperature increase (or decrease) can be approximated by Table 3.1.1 [4]. This table gives the approximate percentage of power reduction for naturally aspirated and turbocharger internal combustion engines.

Internal combustion engines can be operated by a number of liquid and gaseous fuels. Table 3.1.2 gives the high value heat of combustion for various field available liquid and gaseous fuels [5].

The gas-powered drilling and production internal combustion engines can be set up at the manufacturer to operate on gasoline, natural gas, or liquified petroleum gas (LPG). The manufacturer can also set up the engine to operate on all three types of gas fuels. This is accomplished by providing engines with conversion kits that can be used to convert the engine in the field.

Power unit manufacture also produce diesel engines that can be converted to operate on a dual fuel carburation of about 10% diesel and natural gas. Such conversions are more difficult than converting spark-ignition engines to various gas fuels.

Actual fuel consumption data are available from the bare engine or power unit manufacturer. Conversion kits and alternate fuel consumption data are also usually available. However, often the field engineer does not have the time to obtain the needed data directly from the manufacturer regarding the various power units that are to be used for a particular field operation. Table 3.1.3 gives the approximate liquid fuel consumption for multicylinder engines for various fuels. Table 3.1.4 gives the approximate gaseous fuel consumption for multicylinder engines for various fuels.

Figure 3.1.6 shows average vacuum-load curves for several engine models with four-cycle engines of two or more cylinders constructed by six representative engine manufacturers [3]. These curves cannot be used for supercharged or turbocharger engines. Vacuum readings are obtained with a conventional vacuum gauge, containing a dial graduated in inches of mercury. The intake manifold vacuum reading is taken first with the engine running at normal speed with no load and then with the engine running at normal speed with normal load. The curve selected is the one which, at no load, most closely corresponds to the no-load intake manifold vacuum reading. The point on the curve is then located with the ordinate corresponding to the reading taken at normal loading. The abscissa of this point gives the percentage of full load at which the engine is operating. For instance, if the no-load reading is 17 in. Hg and the normal reading is 10 in. Hg, the curve selected in Figure 3.1.4 would be the one which shows 17 in. Hg at 0% of full load. This curve would be followed down to its intersection with the horizontal line

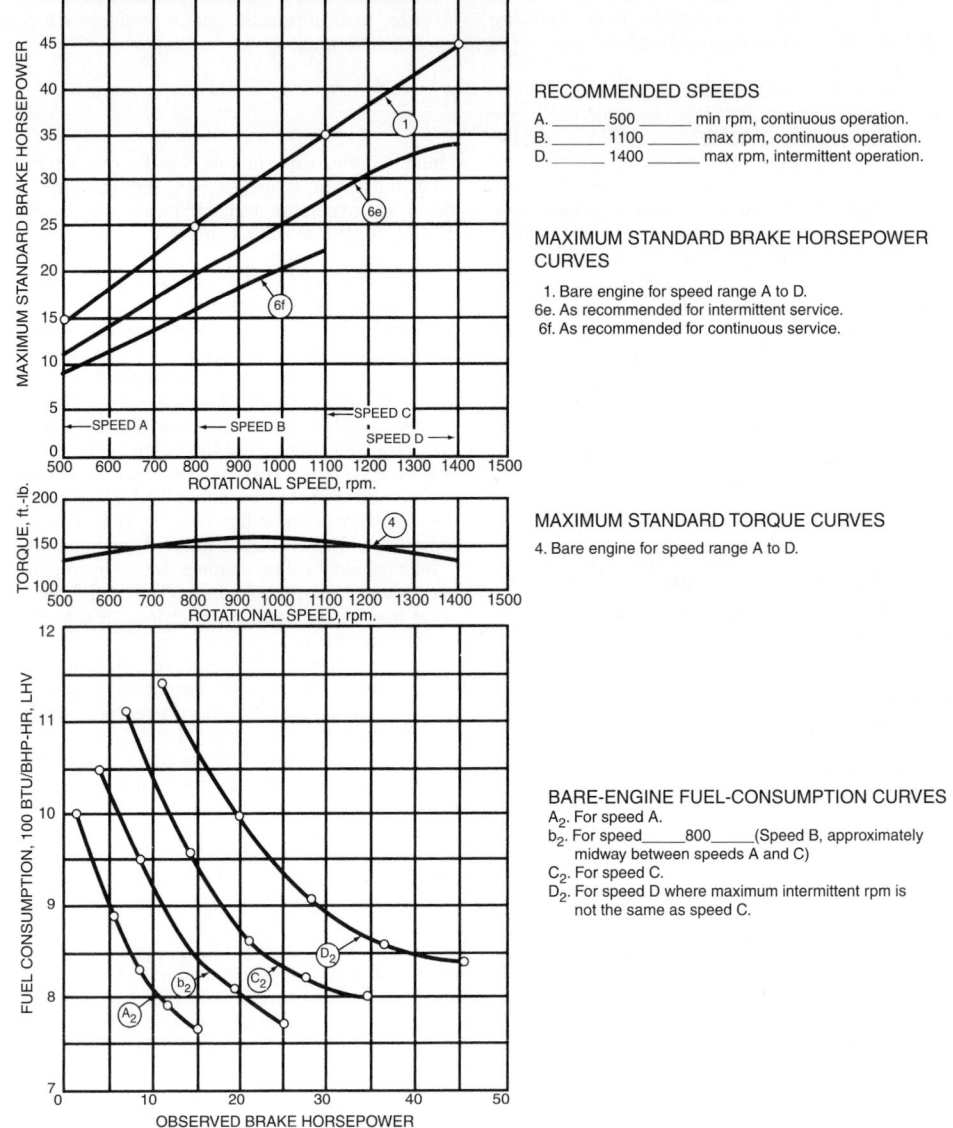

Figure 3.1.4 *Example of test data for internal-combustion bare engines [3].*

of 10 in. The abscissa of this point indicates that the engine is operating at 48% of full power.

The above method is useful in determining internal combustion engine loading conditions. Failure to duplicate former readings at no load and at normal speed indicates that the engine is in poor condition. Failure to duplicate former readings at normal load and normal speed indicates a change in either engine efficiency or load conditions.

3.1.2 Gas Turbine

Gas turbines are essentially constant speed machines and, therefore, generally not suited to be used for the mechanical driving of most equipment in the oil field. Gas turbines are used in combination electrical generating systems and

electrical motors to power drilling rigs and other auxiliary drilling and production equipment. In general, gas turbine are used where a constant power source is needed [1,5,6].

There are some rare situations where the gas turbine is used to power, through direct linkage, a mechanical unit, e.g., a hydraulic fracture pump.

The most common use of the gas turbine power system in the oil and gas industry is in combination with an electrical system (i.e., electric generators and electric motors). In 1965 such a system was used to power a rotary rig. This was a 3,000-hp rig developed by Continental-Ensco. The rig used three 1,100-hp Solar Saturn single-shaft gas turbines. These gas turbines operated at 22,300 rpm and were connected through double reduction gear transmissions to DC generators.

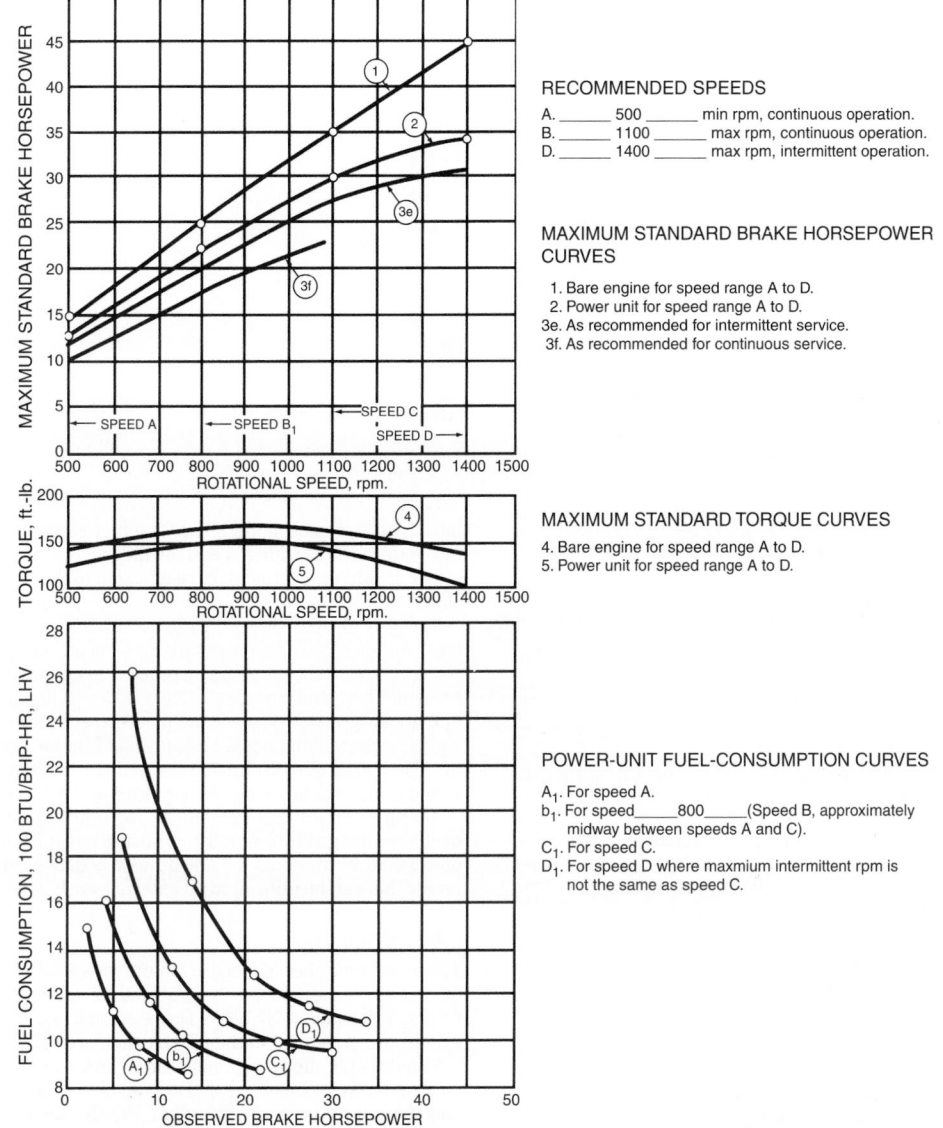

RECOMMENDED SPEEDS

A. _____ 500 _____ min rpm, continuous operation.
B. _____ 1100 _____ max rpm, continuous operation.
D. _____ 1400 _____ max rpm, intermittent operation.

MAXIMUM STANDARD BRAKE HORSEPOWER CURVES

1. Bare engine for speed range A to D.
2. Power unit for speed range A to D.
3e. As recommended for intermittent service.
3f. As recommended for continuous service.

MAXIMUM STANDARD TORQUE CURVES

4. Bare engine for speed range A to D.
5. Power unit for speed range A to D.

POWER-UNIT FUEL-CONSUMPTION CURVES

A_1. For speed A.
b_1. For speed_____800_____(Speed B, approximately midway between speeds A and C).
C_1. For speed C.
D_1. For speed D where maxmium intermittent rpm is not the same as speed C.

Figure 3.1.5 *Example of test data for internal combustion power units [3].*

Gas turbines can be fueled by either gas or liquid hydro-carbons. The gas turbine is often used as a means of utilizing what would otherwise be waste hydrocarbons gases and liquids to generate local electric power via the gas turbine power unit. This is called cogeneration.

The basic concept of the gas turbine is to inject the fuel into a steady flow of compressed air (see Figure 3.1.7). The compressed air is ignited and the expanding exhaust is forced to pass through a series of stationary and rotating turbine blades. These expanding gases force the turbine shaft to rotate. This shaft power is either directly connected or connected through speed reducers to an electric generator and the compressor used to initially compress the air entering the turbine combustor.

Table 3.1.1 *Internal Combustion Engine Power Reduction (Percent) [3]*

Engine	For Each 1000 ft Increase	For Each 10°F Increase or Decrease
Naturally Aspirated	3.66	0.72
Turbo-charged	2.44	1.08

As shown in Figure 3.1.7, a small electrical (or other type of motor) is needed to start the gas turbine. This is usually accomplished by disconnecting the generator load from the gas turbine (usually via a clutch mechanism on gear shaft

Table 3.1.2   *Net Heat of Combustion (High Heat Value) for Various Liquid and Gaseous Fuels [4]*

Fuel	BTU/lb	BTU/gal	BUT/SCF[a]	Specific Gravity	Specific Weight (lb/gal)
Methane	23890		995	0.54	
Natural Gas	26411		1100	0.65	
Propane	21670	91014			4.2
Butane	21316	104448			4.9
Motor Gasoline	20750	128650			6.2
Aviation Gasoline	21000	126000			6.0
Methanol	9758	42935			4.4
Ethanol	12770	80451			6.3
Kerosene	20000	134000			6.7
JP-4	18400	123280			6.7
JP-5	18300	128100			7.0
Diesel Grade 1-D (Save on No. 1 Fuel Oil and Grade 1-GT)	19940	137586			6.9
Diesel Grade 2-D (Save on No. 2 Fuel Oil and Grade 1-GT)	19570	140904			7.2

[a]SCF is at standard mid-latitude average atmospheric conditions of 14.696 psia and 59°F.

Table 3.1.3   *Approximate Liquid Fuel Consumption for Multicylinder Internal Combustion [4,5]*

Fuel	100% Load (lb/hp-hr)	75% Load (lb/hp-hr)	50% Load (lb/hp-hr)
Motor Gasoline	0.62	0.67	0.76
Propane/Butane	0.5	0.56	0.64
Diesel	0.5	0.53	0.63

Table 3.1.4   *Approximate Gaseous Fuel Consumption for Multicylinder Internal Combustion Engines [4,5]*

Fuel	100% Load (SCF/hp-hr)	75% Load (SCF/hp-hr)	50% Load (SCF/hp-hr)
Natural Gas	9.3	10.0	11.1

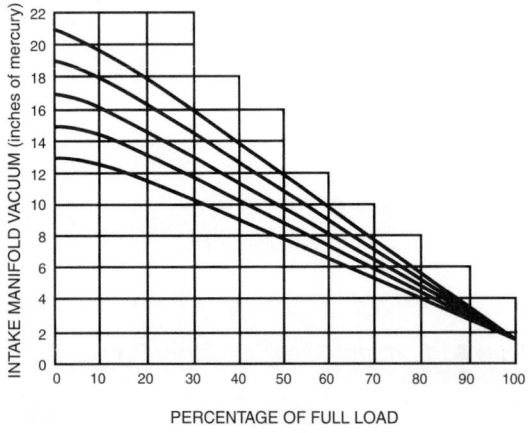

Figure 3.1.6   *Intake vacuum vs. load curves [3].*

in the transmission). The starter motor is actuated, which brings the gas turbine shaft up to a speed where the compressor is providing compressed air at the pressure and weight rate of flow that can sustain the gas turbine operation. Once the gas turbine is up to its critical start speed (usually about one-fourth to one-third of operating speed), fuel is injected

into the combustor and ignited. The fuel is then regulated to the desired operating speed for the system and the generator or other power take off re-engaged to the gas turbine shaft.

The gas turbine shown in Figure 3.1.7 is an open-cycle type. An open-cycle type gas turbine uses the same air that passes through the combustion process to operate the compressor. This is the type most often used for stationary power unit applications. A typical example of power requirements for an open-cycle type gas turbine would be for the unit to develop a total of 3,000 hp. However, about 2,000 hp of this would be needed to operate its compressor. This would leave 1,000 hp to operate the generator (or other systems connected to the gas turbine). Thus, such a gas turbine power unit would be rated as a 1,000-hp unit because this is the power that can be utilized to do external work.

3.1.3 Electric Motors
There are a number of electric motor types available. These motors are classified by the National Electrical Manufacturers Association (NEMA). These electric motor classifications are presented in the NEMA standards [7].

Figure 3.1.8 shows a graphic breakdown of the various electric motors available [8]. The outline in Figure 3.1.8 is based on an electrical classification. Besides their classification, NEMA also classified electric motors according to

Size
  Fractional horsepower
  Integral horsepower
Application
  General purpose
  Definite purpose (e.g., shell type)
  Special purpose
  Part-winding
Mechanical protection and cooling
  Open machine (with subclasses)
  Totally enclosed machine (with subclasses)
Variability of speed
  Constant speed
  Varying speed
  Adjustable speed
  Multispeed

In what follows, electric motors are described in accordance with their electric type as shown in Figure 3.1.8 [7,8].

Air inlet
Generator
Starting motor
Gas turbine
Axial flow compressor
Exhaust
Fuel
Combustor

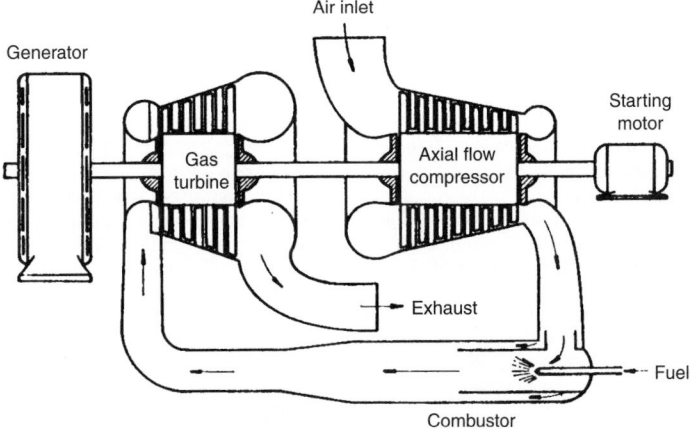

Figure 3.1.7 *Gas turbine (open-cycle) and electric generator.*

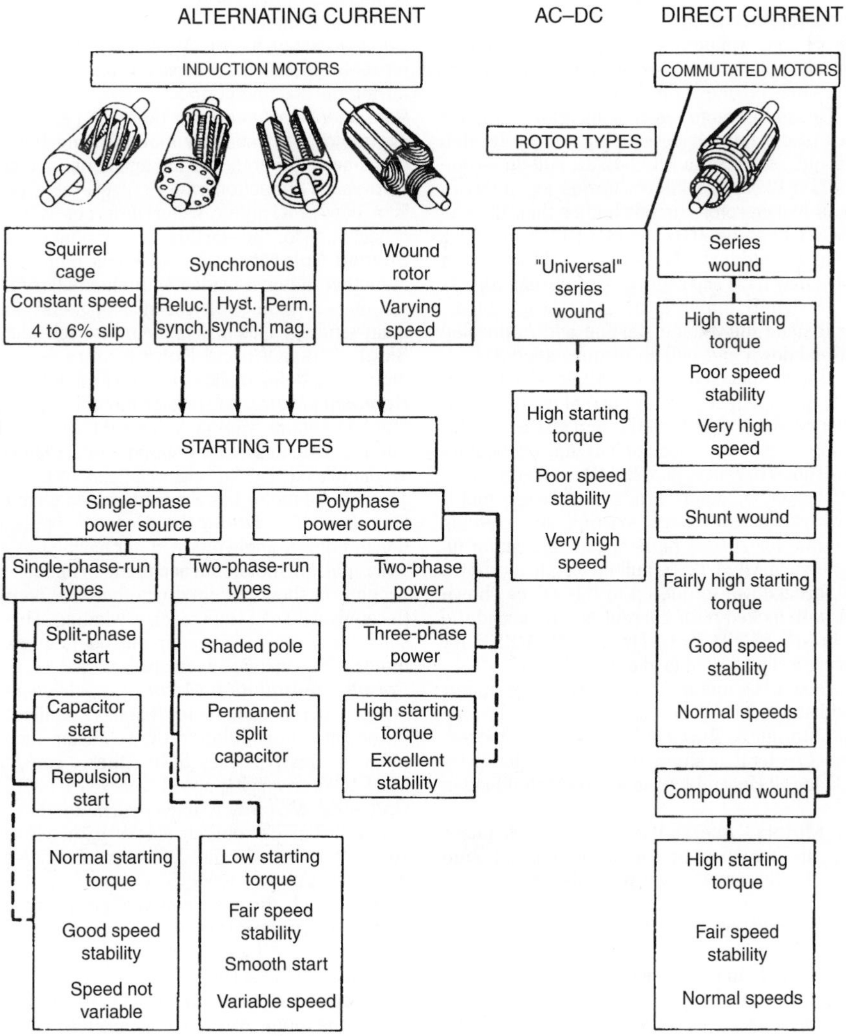

ALTERNATING CURRENT
AC–DC
DIRECT CURRENT
INDUCTION MOTORS
COMMUTATED MOTORS
ROTOR TYPES
Squirrel cage
Constant speed
4 to 6% slip
Synchronous
Reluc. synch.
Hyst. synch.
Perm. mag.
Wound rotor
Varying speed
"Universal" series wound
High starting torque
Poor speed stability
Very high speed
Series wound
High starting torque
Poor speed stability
Very high speed
STARTING TYPES
Single-phase power source
Polyphase power source
Single-phase-run types
Two-phase-run types
Two-phase power
Split-phase start
Shaded pole
Three-phase power
Capacitor start
Permanent split capacitor
High starting torque
Excellent stability
Repulsion start
Normal starting torque
Good speed stability
Speed not variable
Low starting torque
Fair speed stability
Smooth start
Variable speed
Shunt wound
Fairly high starting torque
Good speed stability
Normal speeds
Compound wound
High starting torque
Fair speed stability
Normal speeds

Figure 3.1.8 *Electric motor classification.*

3.1.3.1 Alternating-Current Motors

Alternating-current motors are of three general types, induction, synchronous, and series, and are defined as follows:

Induction Motors. An induction motor is an alternating-current motor in which a primary winding on one member (usually the stator) is connected to the power source and a polyphase secondary winding or a squirrel-cage secondary winding on the other member (usually the rotor) carries induced current. There are two types:

Squirrel-Cage Induction Motor. A squirrel-cage induction motor is one in which the secondary circuit consists of a squirrel-cage winding suitably disposed in slots in the secondary core.

Wound-Rotor Induction Motor. A wound-rotor induction motor is an induction motor in which the secondary circuit consists of a polyphase winding or coils whose terminals are either short circuited or closed through suitable circuits.

Synchronous Motor. A synchronous motor is a synchronous machine which transforms electrical power into mechanical power.

Series-Wound Motor. A series-wound motor is a motor in which the field circuit and armature circuit are connected in series.

Polyphase Motors. Alternating-current polyphase motors are of the squirrel-cage, wound-rotor, or synchronous types.

Design Letters. Polyphase squirrel-cage integral-horsepower induction motors may be one of the following:

Design A. A Design A motor is a squirrel-cage motor designed to withstand full-voltage starting and to develop locked-rotor torque as shown in MG 1-12.37, pull-up torque as shown in MG 1-12.39, breakdown torque as shown in MG 1-12.38 with locked-rotor current higher than the values shown in MG 1-12.34 for 60 Hz and MG 1-12.25 for 50 Hz and having a slip at rated load of less than 5%. Motors with 10 or more poles may have slip slightly greater than 5%.

Design B. A Design B motor is a squirrel-cage motor designed to withstand full-voltage starting and to develop locked-rotor, breakdown and pull-up torques adequate for general application as specified in MG 1-12.37, MG 1-12.38, and MG 1-12.39, drawing locked-rotor current not to exceed the values shown in MG 1-12.34 for 60 Hz and MG 1-12.35 for 50 Hz, and having a slip at rated load of less than 5%. Motors with 10 and more poles may have slip slightly greater than 5%.

Design C. A Design C motor is a squirrel-cage motor designed to withstand full-voltage starting and develop locked-rotor torque for special high-torque application up to the values shown in MG 1-12.37, pull-up torque as shown in MG 1-12.39, breakdown torque up to the values shown in MG 1-12.38, with locked-rotor current not to exceed the values shown in MG 1-12.34 for 60 Hz and MG 1-12.35 for 50 Hz, and having a slip at rated load of less than 5%.

Design D. A Design D motor is a squirrel-cage motor designed to withstand full-voltage starting and to develop high locked-rotor torque as shown in MG 1-1.37 with locked-rotor current not greater than shown in MG 1-12.34 for 60 Hz and MG 1-12.35 for 50 Hz, and having a slip at rated load of 5% or more.

Single-phase Motors. Alternating-current single-phase motors are usually induction or series motors, although single-phase synchronous motors are available in the smaller ratings.

Design Letters. Single-phase integral-horsepower motors may be one of the following:

Design L. A Design L motor is a single-phase integral-horsepower motor designed to withstand full-voltage starting and to develop a breakdown torque as shown in MG 1-10.33 with a locked-rotor current not to exceed the values shown in MG 1-12.33.

Design M. A Design M motor is a single-phase integral-horsepower motor designed to withstand full-voltage starting and to develop a breakdown torque as shown in MG 1-10.33 with a locked-rotor current not to exceed the values shown in MG 1-12.33.

Single-Phase Squirrel-cage Motors. Single-phase squirrel-cage induction motors are classified and defined as follows:

Split-Phase Motor. A split-phase motor is a single-phase induction motor equipped with an auxiliary winding, displaced in magnetic position from, and connected in parallel with, the main winding. *Note:* Unless otherwise specified, the auxiliary circuit is assumed to be opened when the motor has attained a predetermined speed. The term "split-phase motor," used without qualification, described a motor to be used without impedance other than that offered by the motor windings themselves, other types being separately defined.

Resistance-Start Motor. A resistance-start motor is a form of split-phase motor having a resistance connected in series with the auxiliary winding. The auxiliary circuit is opened when the motor has attained a predetermined speed.

Capacitor Motor. A capacitor motor is a single-phase induction motor with a main winding arranged for a direct connection to a source of power and an auxiliary winding connected in series with a capacitor. There are three types of capacitor motors, as follows:

Capacitor-start motor. A capacitor-start motor is a capacitor motor in which the capacitor phase is in the circuit only during the starting period.

Permanent-split capacitor motor. A permanent-split capacitor motor is a capacitor motor having the same value of capacitance for both starting and running conditions.

Two-value capacitor motor. A two-value capacitor motor is a capacitor motor using different values of effective capacitance for the starting and running conditions.

Shaded-Pole Motor. A shaded-pole motor is a single-phase induction motor provided with an auxiliary short-circuited winding or windings displaced in magnetic position from the main winding. This makes the motor self-starting.

Single-Phase Wound-Rotor Motors. Single wound-rotor motors are defined and classified as follows:

Repulsion Motor. A repulsion motor is a single-phase motor that has a stator winding arranged for connection to a commutator. Brushes on the commutator are short circuited and are so placed that the magnetic axis of the stator winding. This type of motor has a varying-speed characteristic.

Repulsion-Start Induction Motor. A repulsion-start induction motor is a single-phase motor having the same windings as a repulsion motor, but at a predetermined speed the rotor winding is short circuited or otherwise connected to give the equivalent of a squirrel-cage winding. This type of motor starts as a repulsion motor but operates as an induction motor with constant-speed characteristics.

Repulsion-Induction Motor. A repulsion-induction motor is a form of repulsion motor that has a squirrel-cage winding in the rotor in addition to the repulsion motor winding. A motor of this type may have either a constant-speed (see MG1-1.30) or varying-speed (see MG 1-1.31) characteristic.

Universal Motors. A universal motor is a series-wound motor designed to operate at approximately the same speed and output on either direct current or single-phase alternating current of a frequency not greater than 60 cycles/s and approximately the same rms voltage. There are two types:

Series-Wound motors. A series-wound motor is a commutator motor in which the field circuit and armature circuit are connected in series.

Compensated Series Motor. A compensated series motor is a series motor with a compensating field winding. (The compensating field winding and the series field winding may be combined into one field winding.)

3.1.3.2 Direct-Current Motors

Direct-current motors are of three general types, shunt wound, series wound, and compound wound, and are defined as follows:

Shunt-Wound Motor. A shunt-wound motor is a direct-current motor in which the field circuit and armature circuit are connected in parallel.

Straight Shunt-Wound Motor. A straight shunt-wound motor is a direct-current motor in which the field circuit is connected either in parallel with the armature circuit or to a separate source of excitation voltage. The shunt field is the only winding supplying field excitation.

Stabilized Shunt-Wound Motor. A stabilized shunt-wound motor is a direct-current motor in which the shunt field circuit is connected either in parallel with the armature circuit or to a separate source of excitation voltage, and which also has a light series winding added to prevent a rise in speed or to obtain a slight reduction in speed with increase in load.

Series-Wound Motor. A series-wound motor is a motor in which the field circuit and armature circuit are connected in series.

Compound-Wound Motor. A compound-wound motor is a direct-current motor which as two separate field windings. One, usually the predominating field, is connected in parallel with the armature circuit. The other is connected in series with the armature circuit.

Permanent Magnet Motor. A permanent magnet motor is a direct-current motor in which the field excitation is supplied by permanent magnets.

3.1.3.3 Rating, Performance, and Testing

The following defines and describes the commonly used terms of electric motor rating, performance, and testing [7]:

Rating of a Machine. The rating of a machine shall consist of the output power together with any other characteristics, such as speed, voltage, and current, assigned to it by the manufacturer. For machines that are designed for absorbing power, the rating shall be the input power.

Continuous Rating. The continuous rating defines the load that can be carried for an indefinitely long period of time.

Short-Time Rating. The short-time rating defines the load that can be carried for a short and definitely specified time.

Efficiency. The efficiency of a motor or generator is the ratio of its useful power output to its total power input and is usually expressed in percentage.

Power Factor. The power factor of an alternating-current motor or generator is the ratio of the kilowatt input (or output) to the kVA input (or output) to the kVA input (or output) and is usually expressed as a percentage.

Service Factor of Alternating-current Motors. The service factor of an alternating-current motor is a multiplier that, when applied to the rated horsepower, indicates a permissible horsepower loading that may be carried under the conditions specified for the service factor (see MG 1-14.35).

Speed Regulation of Direct-Current Motors. The speed regulation of a direct-current motor is the difference between the steady no-load speed and the steady rated-load speed, expressed in percent of rated-load speed.

Secondary Voltage of Wound-Rotor Motors. The secondary voltage of wound-rotor motors is the open-circuit voltage at standstill, measured across the slip rings, with rated voltage applied on the primary winding.

Full-Load Torque. The full-load torque of a motor is the torque necessary to produce its rated horsepower at full-load speed. In pounds at a 1-ft radius, it is equal to the horsepower multiplied by 5,252 divided by the full-load speed.

Locked-Rotor Torque (Static Torque). The locked-rotor torque of a motor is the minimum torque that it will develop at rest for all angular positions of the rotor, with rated voltage applied at rated frequency.

Pull-Up Torque. The pull-up torque of an alternating-current motor is the minimum torque developed by the motor during the period of acceleration from rest to the speed at which breakdown torque occurs. For motors that do not have a definite breakdown torque, the pull-up torque is the minimum torque developed up to rated speed.

Breakdown Torque. The breakdown torque of a motor is the maximum torque that it will develop with rated voltage applied at rated frequency, without an abrupt drop in speed.

Pull-Out Torque. The pull-out torque of a synchronous motor is the maximum sustained torque that the motor will develop at synchronous speed with rated voltage applied at rated frequency and with normal excitation.

Pull-In Torque. The pull-in torque of a synchronous motor is the maximum constant torque under which the motor will pull its connected inertia load into synchronism, at rated voltage and frequency, when its field excitation is applied. The speed to which a motor will bring its load depends on the power required to drive it. Whether the motor can pull the load into step from this speed depends on the inertia of the revolving parts, so that the pull-in torque cannot be determined without having the Wk^2 as well as the torque of the load.

Locked-Rotor Current. The locked-rotor current of a motor is the steady-state current taken from the line with the rotor locked and with rated voltage (and rated frequency in the case of alternating-current motors) applied to the motor.

Temperature Tests. Temperature tests are tests taken to determine the temperature rise of certain parts of the machine above the ambient temperature, when running under a specified load.

Ambient Temperature. Ambient temperature is the temperature of the surrounding cooling medium, such as gas of liquid, which comes into contact with the heated parts of the apparatus. *Note:* Ambient temperature is commonly known as "room temperature" in connection with air-cooled apparatus not provided with artificial ventilation.

High-Potential Test. High-potential tests are tests that consists of the application of a voltage higher than the rated voltage for a specified time for the purpose of determining the adequacy against breakdown of insulating materials and spacings under normal conditions (see MG 1, Part 3).

Starting Capacitance for a Capacitor Motor. The starting capacitance for a capacitor motor is the total effective capacitance in series with the starting winding under locked-rotor conditions.

Radial Magnetic Pull and Axial Centering Force

Radial Magnetic Pull. The radial magnetic pull of a motor or generator is the magnetic force on the rotor resulting from its radial (air gap) displacement from magnetic center.

Axial Centering Force. The axial centering force of a motor or generator is the magnetic force on the rotor resulting from its axial displacement from magnetic center. *Note:* Unless other conditions are specified, the value of radial magnetic pull and axial centering force will be for no load, with rated voltage, rated field current, and rated frequency applied, as applicable.

Induction Motor Time Constants. When a polyphase induction motor is open circuited or short circuited while running at rated speed, the rotor flux linkages generate a voltage in the stator winding. The decay of the rotor flux linkages, and the resultant open-circuit terminal voltage or short-circuit current, is determined by the

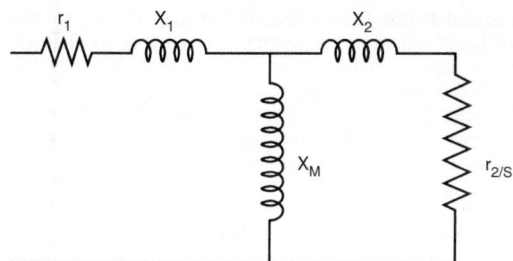

Figure 3.1.9 *Motor circuit [7].*

various motor time constants defined by the following equations:

Open-circuit AC time constant:

$$T_{do}'' = \frac{X_M + X_2}{2\pi f r_2} \quad \text{(s)} \qquad [3.1.6]$$

Short-circuit AC time constant:

$$T''d = \frac{X_S}{X_1 + X_M} T_{do}'' \quad \text{(s)} \qquad [3.1.7]$$

Short-circuit DC time constant:

$$T_a = \frac{X_S}{2\pi f r_1 \left(1 + \frac{LL_s}{kW_1}\right)} \quad \text{(s)} \qquad [3.1.8]$$

X/R ratio:

$$X/R = \frac{X_s}{r_1 \left(1 + \frac{LL_s}{kW_1}\right)} \quad \text{(rad)} \qquad [3.1.9]$$

Terms (Figure 3.1.9):

- r_1 = Stator DC resistance per phase corrected to operating temperature.
- r_2 = Rotor resistance per phase at rated speed and operating temperature referred to stator.
- X_1 = Stator leakage reactance per phase at rated current.
- X_2 = Rotor leakage reactance per phase at rated speed and rated current referred to stator.
- X_S = Total starting reactance (stator and rotor) per phase at zero speed and locked-rotor current.
- X_M = Magnetizing reactance per phase.
- LL_s = Fundamental-frequency component of stray-load loss in kilowatts at rated current.
- kW_1 = Stator I^2R loss in kilowatts at rated current and operating temperature.
- f = Rated frequency in hertz.
- s = Slip in per unit of synchronous speed.

3.1.3.4 AC Performance Examples

In general, the typical electric motor applications in the oil and gas industry are polyphase motors (either squirrel-cage or wound-rotor motors).

Squirrel-Cage Motor. This type of motor finds a broader application and a more extensive and general use than any other type of motor. This is because it is, inherently, the simplest type of electric motor and, also, has excellent characteristics and operates essentially at constant speed. It has greater reliability and low maintenance requirements and thus meets a broad range of applications.

Torque, horsepower, and speed requirements demanded in drives for most machines can be met with one of four designs of squirrel-cage polyphase induction motors. Each design offers a different combination of torque, speed, and current characteristics to meet the operating requirements of various industrial applications.

All four designs can withstand full-voltage starting directly across the power lines, that is, the motors are strong enough mechanically to withstand magnetic stresses and the locked-rotor torques developed at the time the switch is closed.

Design A produces exceptionally high breakdown torques but at the expense of high locket rotor currents that normally require provision for starting with reduced voltage. This motor is suitable for machines in which the friction and inertia loads are small.

Design B has normal starting torque adequate for a wide variety of industrial machine drives and a starting current usually acceptable on power systems. This design is suitable where slightly more than full load torque and low slip is required, also where relatively high breakdown torque is needed to sustain occasional emergency overloads, or where a low locked-rotor current is needed. These motors are for use in driving machine tools, blowers, centrifugal pumps, and textile machines.

Design C has high starting torque and a normal breakdown torque. Applications for this design are machines in which inertia loads are high at starting, but normally run at rated full load and are not subjected to high overload demands after running speed has been reached. Conveyors, plunger pumps, compressors that are not unloaded at starting, and over chain conveyors, also hoists, cranes, and machine tools where a quick start and reversal are required are typical examples of such machines.

Design D develops extremely high starting torque with moderate starting current. This design uses a high-resistance-type rotor to obtain variation of speed with load and has no sharply defined breakdown torque. This motor eases off in speed when surge loads are encountered and also develops high torque to recover speed rapidly. Typical applications for this motor are machines in which heavy loads are suddenly applied and removed at frequent intervals, such as hoists, machines with large flywheels, conventional punch presses, and centrifuges.

Slip ratings of the four designs are

Design	Slip, %
A	Less than 5
B	Less than 5
C	Less than 5
D	5 or more

Note that motors with 10 or more poles may have slip slightly greater than 5%.

Figure 3.1.10 shows the typical torque-speed performance curves for various designs of polyphase squirrel-cage induction motors [9].

Tables 3.1.5 and 3.1.6 give the typical locked-rotor torque developed by Designs A, B, and C motors [8].

Table 3.1.7 gives the typical breakdown torque for Design B and C motors with continuous ratings [8]. Breakdown torques for Design A motors are in excess of the values given for Design B motors.

It is usually good practice to apply motors at momentary loads at least 20% below the values given for maximum torque in order to offset that much torque drop caused by an allowable 10% voltage drop.

Wound Rotor. Characteristics of wound-rotor motors are such that the slip depends almost entirely upon the load on the motor. The speed returns practically to maximum when the load is removed. This characteristic limits the use of these motors on applications where reduced speeds at light loads are described.

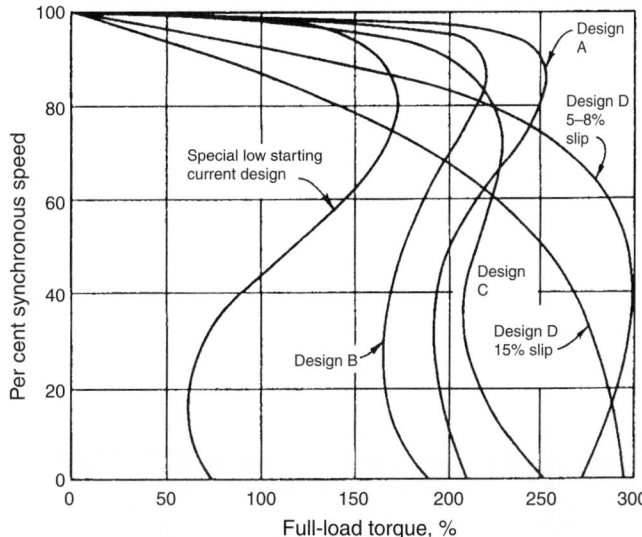

Figure 3.1.10 *Torque-speed curves for various designs of polyphase squirrel-cage induction motors [9].*

Table 3.1.5 *Locked-Rotor Torque of Design A and B Motors with Continuous Ratings [8]*

Hp	Synchronous speeds, rpm							
Frequency:								
60 cycles......	3600	1800	1200	900	720	600	514	450
50 cycles......	3000	1500	1000	750	600	500	428	375
1/2	...	...	...	150	150	115	110	105
3/4	...	...	175	150	150	115	110	105
1	...	275	175	150	150	115	110	105
1 1/2	175	265	175	150	150	115	110	105
2	175	250	175	150	145	115	110	105
3	175	250	175	150	135	115	110	105
5	150	185	160	130	130	115	110	105
7 1/2	150	175	150	125	120	115	110	105
10	150	175	150	125	120	115	110	105
15	150	165	140	125	120	115	110	105
20	150	150	135	125	120	115	110	105
25	150	150	135	125	120	115	110	105
30	150	150	135	125	120	115	110	105
40	135	150	135	125	120	115	110	105
50	125	150	135	125	120	115	110	105

Values are expressed in per cent of full-load torque and represent the upper limit of the range of application.
Values are based on rated voltage and frequency.

Table 3.1.6 *Locked Rotor Torque of Design C Motors with Continuous Ratings [8]*

Hp	Synchronous speeds, rpm		
Frequency:			
60 cycles......	1800	1200	900
50 cycles......	1500	1000	750
3	...	250	225
5	250	250	225
7 1/2	250	225	200
10	250	225	200
15	225	200	200
20	200	200	200
25 and larger	200	200	200

Values are expressed in per cent of full-load torque and represent the upper limit of the range of application.
Values are based on rated voltage and frequency.

Wound-rotor motors are suitable for constant-speed applications that require frequent starting or reversing under load or where starting duty is severe and exceptionally high starting torque is required.

A second major type of application is where speed adjustment is required. Controllers are used with these motors to obtain adjustable speed over a considerable range. But at any point of adjustment the speed will vary with a change in load. It is usually not practicable to operate at less than 50% of full speed by introducing external resistance in the secondary circuit of the motor. Horsepower output at 50% of normal speed is approximately 40% of rated horsepower.

Use of wound-rotor induction motors has been largely in continuous-duty constant-speed supplications where

particularly high starting torques and low starting currents are required simultaneously, such as in reciprocating pumps and compressors. These motors are also used where only alternating current is available to drive machines that require speed adjustment, such as types of fans and conveyors.

Typical torque-speed characteristics of wound-rotor motors are shown in the curves in Figure 3.1.11 for full voltage and for reduced voltage that are obtained with different values of secondary resistance.

3.1.3.5 DC Performance Examples
Important advantages of direct-current motors for machine drives are the adjustable speed over a wide range, the fact that speeds are not limited to synchronous speeds, and the variation of speed-torque characteristics [8].

Shunt-Wound Motors. These motors operate at approximately constant speed regardless of variations in load when connected to a constant supply voltage and with fixed field excitation. Maximum decrease in speed as load varies from no load to full load is about 10–12%.

Table 3.1.7 *Breakdown Torque of Design B and C, 60 and 50-cycle Motors with Continuous Ratings [8]*

Hp	Synchronous speed, rpm	Design B	Design C	Hp	Synchronous speed, rpm	Design B	Design C
1/2	900–750	250		3	1800–1500	275	
	Lower than 750	200			1200–1000	250	225
3/4	1200–1000	275			900–750	225	200
	900–750	250			Lower than 750	200	
	Lower than 750	200		5	3600–3000	225	
1	1800–1500	300			1800–1500	225	200
	1200–1000	275			1200–1000	225	200
	900–750	250			900–750	225	200
	Lower than 750	200			Lower than 750	200	
1 1/2	3600–300	300		7 1/2	3600–300	215	
	1800–1500	300			1800–1500	215	190
	1200–1000	275			1200–1000	215	190
	900–750	250			900–750	215	190
	Lower than 750	200			Lower than 750	200	
2	3600–3000	275		10	3600–3000	200	
	1800–1500	275			1800–1500	200	190
	1200–1000	250			1200–1000	200	190
	900–750	225			900–750	200	190
	Lower than 750	200			Lower than 750	200	
3	3600–3000	250		15 and larger	All speeds	200	190

Values are expressed in per cent of full-load torque and represent the upper limit of the range of application.
Values are based on rated voltage and frequency.

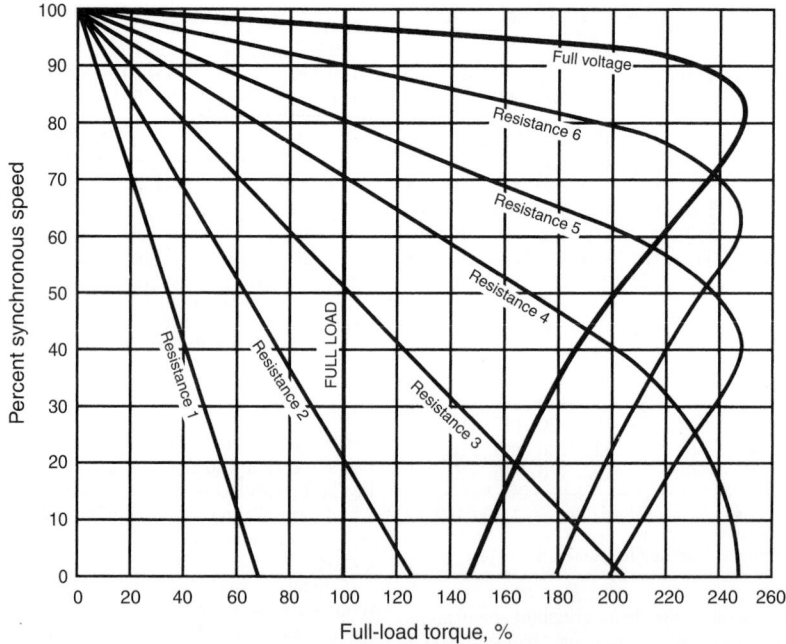

Figure 3.1.11 *Typical torque-speed characteristics of wound-rotor motors [8].*

Constant-speed motors are usually suited for a speed range of less than 3 to 1 by field control, but mechanical and electrical characteristics govern maximum safe speeds. With constant voltage on the armature, as the field is weakened the speed increases and the motor develops constant horsepower.

For speed ranges 3 to 1 or greater, adjustable-voltage operation is recommended. The adjustable-voltage drive usually includes a motor-generator set to supply an adjustable voltage supply to the armature of the drive motor. With this method of operation, a speed range of 10 to 1 readily obtained.

Regenerative braking can also be obtained with the variable-voltage drive. This system permits deceleration of the driven load by causing the motor to operate as a generator to drive the motor-generator set, thereby returning power to the alternating-current power supply.

With an adjustable voltage supply to the armature, at speeds below basic, the motor is suitable for a

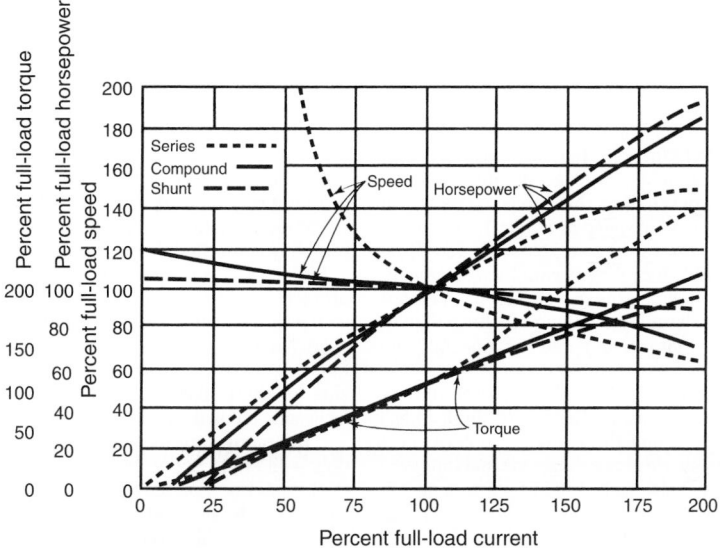

Figure 3.1.12 *Direct-current motor performance curves [8].*

constant-torque drive. Minimum speeds are limited by temperature rise, because the motor carries full-load current at the lower speeds and at low speeds the ventilation is reduced. Adjustable voltage drives are used for paper mills, rubber mill machinery, winders, machine-tool drives, and hoists.

Adjustable-speed shunt-wound motors with field control are designed for operation over speed ranges of 3 to 1 or greater. Standard adjustable-speed motors are rated in three ways: tapered horsepower, continuous, 40°C rise; constant horsepower, continuous, 40°C rise; and constant horsepower, 1 hr, 50°C rise, of the next larger horsepower rating than for continuous duty.

The 1 hr, 50°C open motors develop constant horsepower over the entire speed range. Semienclosing covers can be added without changing the rating.

The 40°C continuous-rated open motors develop the rated horsepower from 150% of minimum speed to the maximum speed. From minimum speed to 159% of minimum speed, the rated horsepower will be developed continuously without exceeding safe temperature limits.

Tapered horsepower motors develop the maximum rated horsepower at three times the minimum speed, the horsepower decreasing in direct proportion to the decrease in speed down to the horsepower rating at 150% of the minimum speed. Figure 3.1.12 plots characteristics of shunt, series, and compound-wound direct-current motors.

Series-Wound Motors. These motors are inherently varying-speed motors with charges in load. On light or no loads, the speed may become dangerously high. These motors should be employed only where the load is never entirely removed from the motor. They should never be connected to the driven machine by belt.

Series motors are used on loads that require very high starting torques or severe accelerating duty or where the high-speed characteristics may be advantageous, such as in hoists.

Compound-Wound Motors. These motors are used to drive machines that require high starting torque or in which the loads have a pulsating torque. Changes in load usually produce wide speed regulation. This motor is not suited for adjustable speed by field control.

The relative strength of the shunt and series fields of the motor determines to what extent the motor approaches the shunt or series characteristics.

From no load to full load, the drop in speed of compound-wound motors is approximately 25%. Compound-wound motors are used where reasonably constant speed is required and for loads where high starting torque is needed to accelerate the drive machine.

Compound-wound motors should be used on machines having flywheel or high inertia loads, wherein the dropping speed characteristics of the motor causes the flywheel to give up its energy as the shock load comes on, thereby cutting down the power peaks taken from the line and resulting in less heating of the motor.

3.1.3.6 Industry Applications

The oil and gas industry utilizes electric motors as prime movers in a number of industry operations. The motor applications and the electric motors typically utilized are now described [8,10].

Well Drilling and Completions. In the past, most drilling and completion rigs utilized direct drive diesel engines as prime movers. This was especially so of the large (deeper drilling) land rigs. The smaller land rigs utilize direct drive gas (including LPG and natural gas) engines. However, over the past decade, there has been an increasing trend to utilize diesel-engine-driven generators, which in turn operate direct-current motors for large land rigs and especially offshore drilling rigs. These direct-current electric motors are used for the drawworks and rotary table drives because of high starting-torque requirements for these operations.

Production. Torque requirements vary widely during the production pumping cycle, and peaks occur when the sucker-rod string and fluid load are lifted and when the counterweight is lifted. NEMA Design D motors, although relatively expensive, are well suited to this service, since they minimize current peaks and provide adequate torque under all service conditions, including automatic operation by time control. NEMA Design C motors may be used where operating conditions are less severe. NEMA Design B motors must be used with care in this service to avoid high cyclic current peaks, which may be objectionable on a small system, particularly if several wells should "get in step." The use of Design B motors can also lead to oversizing of motors in an attempt to obtain sufficient starting torque. This results in the operation of the motor at a relatively low load factor, with consequent low power factor.

Double or Triple-Rated Motors. There are special motors developed for oil-well pumping. They are totally enclosed, fan-cooled NEMA Design D motors that can be reconnected for 2–3-hp ratings at a common speed of 1,200r/min. Typical horsepower ratings are 20/15/10 and 50/40/30. They provide flexibility in the field since they permit the selection of the horsepower rating at which the motor may be operated most efficiently. They also permit changing the pumping speed by changing the motor pulley and reconnecting the motor.

Single-Phase Operation. If single-phase power only is available, it is advisable to consider the use of single-phase to three-phase converters and three-phases motors. This avoids the use of large single-phase capacitor start motors, which are relatively expensive and contain a starting switch that could be a source of trouble due to failure or to the presence of flammable gas in the vicinity of the well.

Oil-Well Control. A packaged control unit is available to control individual oil-well pumps. It contains, in a weatherproof enclosure, a combination magnetic starter, a time switch that can start and stop the motor according to a predetermined program, a timing relay that delays the start of the motor following a power failure, and lightning arresters. Pushbutton control is also provided.

Power-Factor Correction. The induction motors used for oil-well pumping have high starting torques with relatively low power factors. Also, the average load on these motors is fairly low. Therefore, it is advisable to consider the installation of capacitors to avoid paying the penalty imposed by most power companies for low-power factor. They will be installed at the individual motors and switched with them, if voltage drop in the distribution system is to be corrected as well as power factor. Otherwise they may be installed in large banks at the distribution center, if it is more economical to do so.

Oil Pipelines. The main pumps for an oil pipeline usually driven by 3,600 rpm induction electric motors having NEMA Design B characteristics. Full-voltage starting is used. Figure 3.1.13 shows a typical pumping station diagram. Such pumping stations are located at the beginning of the pipeline and at intervals along the line. Intermediate or booster stations must be capable of operating under varying conditions due to differences in liquid gravity, withdrawals at intermediate points, and the shutting down of other booster

stations. Pumping stations often contain two or three pumps connected in series, with bypass arrangements using check valves across each pump. The pumps may all be of the same capacity, or one of them may be half size. By operating the pumps singly or together, a range of pumping capacities can be achieved.

Throttling of pump discharge may also be used to provide finer control and to permit operation when pump suction pressure may be inadequate for full flow operation.

Pumping stations are often unattended and may be remotely controlled by radio or telephone circuits.

Motor enclosures for outdoor use are NEMA weather-protected Type II, totally enclosed, fan-cooled, or drip-proof with weather protection. Motors of the latter type are widely used. Not only are they less expensive than the other types, but they also have a service factor of 1.15. The above enclosure types are all suitable for the Class I, Group D, Division 2 classifications usually encountered.

If the pumps are located indoors, a Division 1 classification is likely to apply. Motors must be Class I, Group D, explosion-proof, or they may be separately ventilated with clean outside air brought to the motor by fans. Auxiliary devices such as alarm contacts on the motor must be suitable for the area classification. The installed costs, overall efficiencies, and service factors associated with the enclosures that are available will influence the selection.

Natural Gas Pipelines. Compressor drivers are usually reciprocating gas engines or gas turbines, to make use of the energy available in the pipeline. Electric motor drives use slow-speed synchronous motors for reciprocating compressors and four or six-pole induction motors with gear increases for high-speed centrifugal compressors. Motor voltages, types, and enclosures are selected as for oil pipeline pumps. Motors used with centrifugal compressors must develop sufficient torque at the voltage available under inrush conditions to accelerate the high inertia load. They must also have adequate thermal capacity for the long starting time required, which may be 20 or 30 s.

3.1.3.7 Motor Control
Efficient use of electric motors requires appropriate control systems. Typical control systems for AC and DC electric motor operations are now discussed [10].

Inverter-AC Motor Drives. An adjustable-frequency control of AC motors provide efficient operation with the use of brushless, high-performance induction, and synchronous motors. A typical system is shown in Figure 3.1.14. Such a system consists of a rectifier (which provides DC power from the AC line) and an inverter (which converts the DC power to adjustable-frequency AC power for the motor). Inverter cost

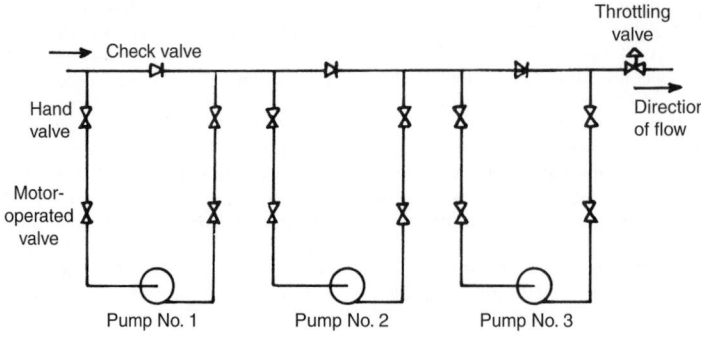

Figure 3.1.13 *Typical oil pipeline pump station [10].*

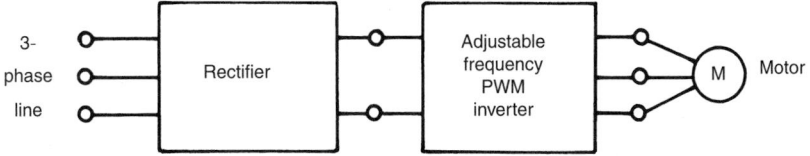

Figure 3.1.14 *Typical invester AC motor drive consisting of rectifier-DC link, adjustable-frequency inverter, and induction of synchronous motor [10].*

per kilowatt is about twice that of controller rectifiers; thus the power convertor for an AC drive can approach three times the cost of a DC drive.

These AC drive systems require the inverters to operate with either low-slip induction motors or reluctance-type synchronous-induction motors. Such systems are used where DC commutator motors are not acceptable. Examples of such applications are motor operations in hazardous atmospheres and high motor velocities.

The power convertor must provide the AC motor with low-harmonic voltage waveform and simultaneously allow the amplitude to be adjusted. This avoids magnetic saturation of the motor as the frequency is adjusted. For constant torque, from maximum speed to base speed, the voltage is adjusted proportional to frequency. Above base speed, the motor is usually operated at constant horsepower. In this region the voltage is held constant and the flux density declines. Also, the convertor must limit the starting current, ensure operation at favorable slip, and provide a path for reverse power flow during motor slowdown.

Inverters are designed with various power semiconductor arrangements. Power semiconductor elements of the inverter operate like switches by synthesizing the motor voltage waveform from segments of the DC bus voltage. For power ranges up to about 5 hp, convertors can use power transistors to synthesize six-step (per cycle), three-phase voltage for frequency ranges from 10 to 120 Hz for standard motors and from 240 to 1,200 Hz for high-frequency motors. For the conventional drive range from 5 to 500 hp, thyristor inverters are used to develop either six-step per cycle, twelve-step per cycle, or pulse-width modulated (pwm) voltages over typical frequency ranges from 10 to 120 Hz.

The voltage control of the six-step thyristor inverter can be accomplished in the rectifier, or the inverter. The output voltage of a six-step inverter is directly proportional to the DC bus voltage. Thus, control of the rectifier is usually sufficient provided that the inverter commutation capacitors are charged from a fixed-voltage bus. Another method that can be used in dealing with a fixed DC bus voltage is to operate two inverter bridges in series with adjustable relative phase shift to control the motor voltage. An additional method is to use an adjustable-ratio output transformer (Variac).

In general, the usage of the voltage control methods is being replaced by pulse-width-modulation (pwm) techniques. These techniques allow the DC bus voltage to remain fixed, and, thus, lower-cost systems with motor voltages of potentially lower harmonic content can be obtained. A variety of pwn techniques are now used in inverters. One technique requires that each half-cycle or motor voltage waveform be divided into a fixed number of pulses, (typically four or six) and the pulse width modulated to control the voltage. In another technique, the pulse number is increased with pulse period and the width modulated for voltage control. An additional technique requires that the pulse number be fixed, but the width is graded over the half-cycle as a sine function, and the relative widths are controlled.

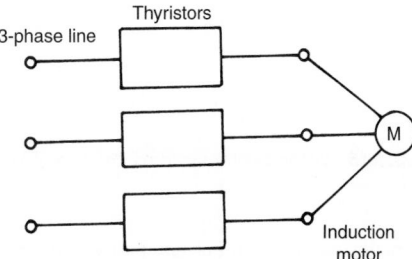

Figure 3.1.15 *Primary voltage central system [10].*

The inverter drive system that uses a current-controlled rectifier and parallel-capacitor commutation operates to both improve reliability and reduce cost. Such systems are built commercially for the ranges from 20 to 500 hp for the typical 20:1 constant-torque speed range.

Primary-Voltage-Control-AC Motor Driver. Induction motor torque at any slip s is proportional to primary V^2. Rotor-power dissipation is equal to s times the air-gap power. These two relationships define the boundary of operation of an induction motor with primary voltage control of speed. As the speed is reduced (s increased) at constant torque, the air-gap power remains fixed, but the power divides between rotor circuit dissipation and mechanical shaft power.

Solid-state primary voltage controllers are built with phase-controlled thyristors in each AC circuit and are used with high-slip NEMA Design D motors. The thyristors are either placed in the three supply lines (as shown in Figure 3.1.15) in the star ends of a wye motor winding or in the arms of a delta motor winding. Commercial systems are built to 100 hp or more for pumps and related applications. Such systems have the inherent limitation of thyristor rating and rotor heating.

Wound-Rotor Motor Drives. The wound-rotor induction motor, using adjustable rotor-circuit resistance, is rarely used. However, two versions that use solid-state auxiliary equipment are finding limited application.

Just as for the primary-voltage control system, the air-gap power at slip s divides between mechanical power and rotor-circuit power in the wound-rotor drive. Except for the losses in the rotor windings, the rotor-circuit power is extracted from the slip rings and is disposed of external to the machine. One type of solid-state system is shown in Figure 3.1.16a. In this system the power to a resistor is controlled by thyristors. Another type of solid-state system is shown in Figure 3.1.16b. In this system the slip-ring power is rectified and returned to the supply line through a line-commutated convertor. The first system is wasteful of power. The second system uses power in an economical manner.

The rating of the solid-state equipment depends upon the torque requirement as a function of speed. For a constant-torque drive, the auxiliary system at starting must handle

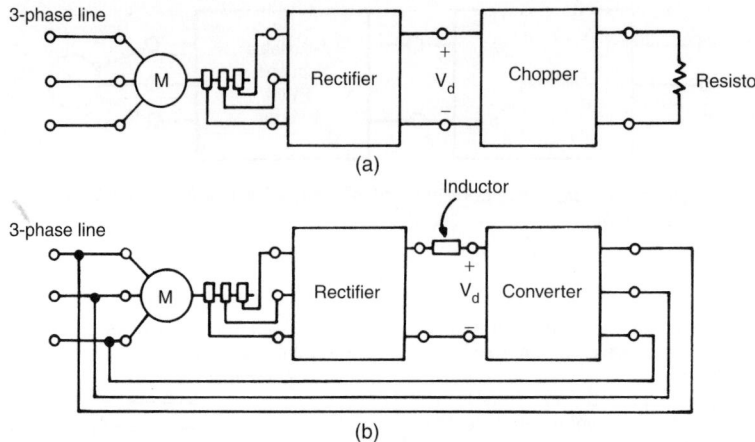

Figure 3.1.16 *Wound-rotor systems: (a) rotor power dissipated in resistance; (b) rotor power retirement to supply line [10].*

the full motor rated power. For pump drives, the auxiliary system ratings are reduced considerably. The convertor operates with phase-controlled firing signals from the supply line, such as for a DC drive rectifier and develops relatively constant-speed characteristics. The stator-rotor winding ratio is made slightly larger than unity to ensure that power flow is toward the supply line at the lowest speed (e.g., at stall).

Solid-State DC Drives. The controlled-thyristor rectifier and separate-field DC motor is the solid-state motor drive in greatest use. The combination provides control over at least a 10:1 speed range, plus an additional two to three times by field weakening. Depending upon the power level, the rectifier is operated directly from the AC supply lines, or via a transformer. Typical speed regulation of ±2% can be accomplished with a single control system. The horsepower and speed limitations are set by the DC motor, not by the semiconductor rectifiers. The DC motor and rectifier can be combined to any required power level.

Commercial solid-state DC motor drives fall into three general groups. Drives operating from single-phase lines are available in fractional horsepower sizes up to about 3 hp. Three-phase drives are available in horsepower sizes from 5 to 500 hp. Drivers above 500 hp are generally classed as special.

Speed adjustment from base speed downward is obtained by armature voltage control. The armature current and torque in this range is limited by the thyristor ratings or motor temperature rise. Control above base speed at constant horsepower is obtained by field weakening. An example system is shown in Figure 3.1.17a. In this example three or six thyristors are used in the rectifier. Reversible operation requires either the use of two rectifiers connected as shown in Figure 3.1.17b, a reversing armature contactor, or a field-reversing technique.

Speed adjustment, current limiting, and regulation against load, temperature, and line-voltage disturbances require a control system for the armature and field rectifier thyristors.

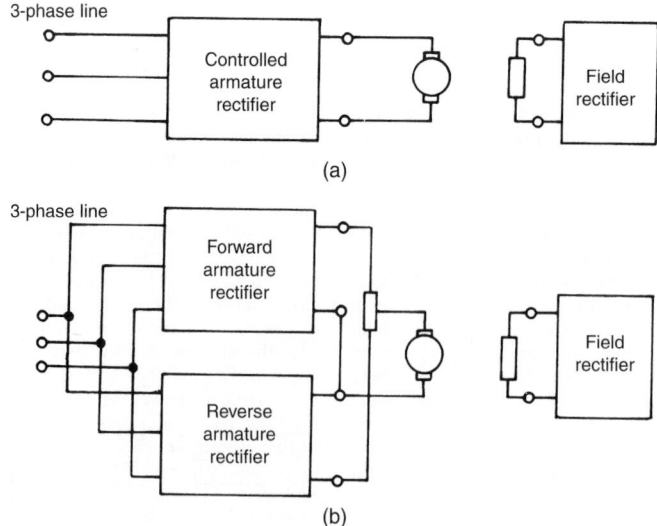

Figure 3.1.17 *Rectifier-DC motor driver: (a) unidirectional or field-reversing; (b) bidirectional operation with armature voltage reversing [10].*

For typical ±2% regulation, the speed signal is derived from the armature voltage, corrected for IR drop. For regulation to ±0.1%, the speed signal is obtained from analog or digital tachometers and processed in solid-state digital or analog circuits. Signals that override the constant-speed control system limit armature current during speed changes, acceleration, and reversal.

References
1. Kutz, M., *Mechanical Engineers' Handbook*, 12th Edition, John Wiley & Sons, New York, 1986.
2. "API Specification for Internal-Combustion Reciprocating Engines for Oil-Field Service," API STD 7B-11C, 9th Edition, 1994.
3. "API Recommended Practice for Installation, Maintenance, and Operation of Internal-Combustion Engines," API RP 7C-11F, 5th Edition, 1994.
4. *Atlas Copco Manual*, Fourth Edition, 1982.
5. Baumeister, T., *Marks' Standard Handbook for Mechanical Engineers*, 7th Edition, McGraw-Hill, New York, 1979.
6. Moore, W. W., *Fundamentals of Rotary Drilling*, Energy Publications, 1981.
7. "NEMA Standards, Motors and Generators," ANSI/NEMA Standards Publication, No. MG1-1978.
8. Greeenwood, D. G., *Mechanical Power Transmissions*, McGraw-Hill, New York, 1962.
9. Libby, C. C. *Motor Section and Application*, McGraw-Hill, New York, 1960.
10. Fink, D. G., and Beaty, H. W., *Standard Handbook for Electrical Engineers*, 11th Edition, McGraw-Hill, New York, 1983.

3.2 POWER TRANSMISSION

In nearly all mechanical power applications in the oil and gas industry, it is necessary to transmit the power generated by a prime mover to an operation (e.g., drawworks of a drilling rig, a production pumping system). The transmission of rotary power to such operation elements is carried out by a power transmission system. Mechanical power transmission is typically carried out by belting systems, chain systems, gear systems, hydraulic systems, or some combination of these four [1].

3.2.1 Power Belting
There are three basic types of power belting. They are round, flat, and V, as shown in Figure 3.2.1. In general, the choice of a power belt drive depends on factors such as speed, reduction ratio, positive-drive requirements, center distances, and shaft relationship (i.e., magnitude of skew and load).

3.2.1.1 Round-Belt Drives
Round belts, modern versions of the now obsolete rope drive are available in endless and spliced variations. Round belts can operate satisfactorily over pulleys in several different planes and are thus well suited for serpentine drives, reverse bends and 90° twists [2].

3.2.1.2 Flat-Belt Drives
Flat belts are used chiefly for conveyor belt systems rather than power transmission. Synchronous belt are generally used for control of critically timed rotating elements of a mechanical system and are not used in power transmission applications. V-ribbed belts are typically used in light-duty applications requiring high speed ratios.

3.2.1.3 V-Belt Drives
V belts were developed from the older rope drive systems. The grooved sheaves required for ropes became V shaped, and the belt itself was contoured to fit the groove. The reason V belts are able to transmit higher torque at smaller widths and tension than other types of belts is because of the wedging action of the V belts in the sheave groove. Figure 3.2.2 shows this wedging action. Figure 3.2.3 shows typical sheave designs for single and multiple V belts.

3.2.1.3.1 V Belt Types
There are several different configurations of V belts for power transmission applications. Cross-sections for standard or classic V belts and narrow V belts are shown in

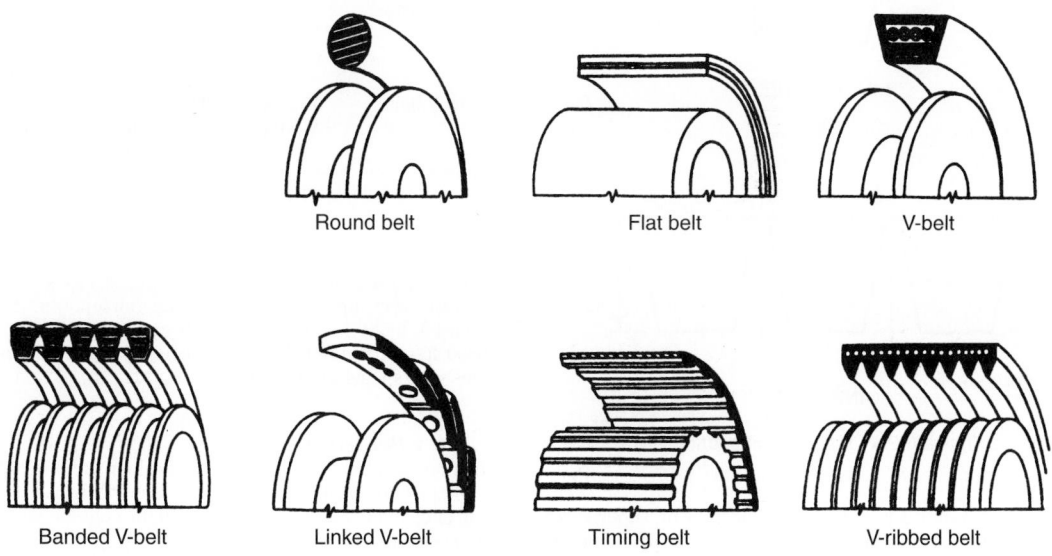

Figure 3.2.1 *Types of belts [2].*

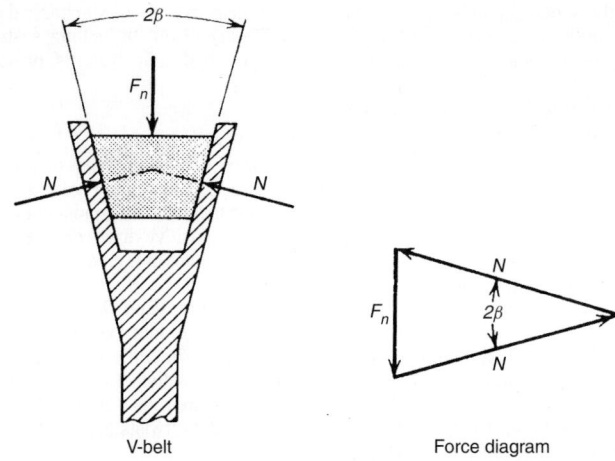

V-belt   Force diagram

Figure 3.2.2   *Element of a V belt in a sheave [2].*

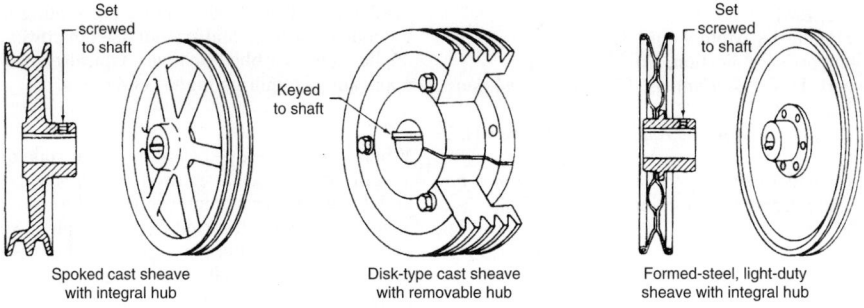

Figure 3.2.3   *Types of sheaves for V belts [2].*

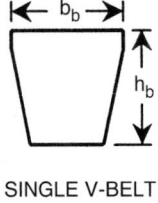

SINGLE V-BELT

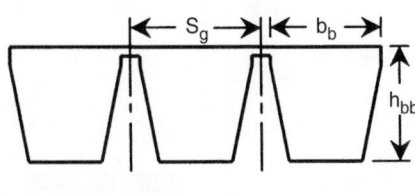

JOINED V-BELT

Figure 3.2.4   *V belt cross-sections [3].*

Figure 3.2.4 and Table 3.2.1. The narrow or superpower V belts were specifically designed to have a higher horsepower capacity than standard V belts. These superpower V belts allow shorter center distances and narrower sheaves

Table 3.2.1   *Nominal Dimensions of Cross-Sections, in Inches [3]*

Belt Type	Cross-section	b_b	h_b	h_{bb}	S_g^*
Classic V-belts	A, AX	0.50	0.31	0.41	0.625
	B, BX	0.66	0.41	0.50	0.750
	C, CX	0.88	0.53	0.66	1.000
	D	1.25	0.75	0.84	1.438
Narrow V-belts	3V, 3VX	0.38	0.31	0.38	0.406
	5V, 5VX	0.62	0.53	0.62	0.688
	8V	1.00	0.91	1.00	1.125

*S_g is sheave groove spacing.

without imposing any extra total bearing stresses. Such belts can carry up to three times the horsepower than standard V belts. In addition, speeds can be increased up to 6500 ft/min without dynamic balancing of the sheaves because the sheaves are smaller.

3.2.1.3.2 V Belt Terminology
- *Belt length (L):* V belts are of closed loop construction and come in standard lengths. V belt length is based on the pitch diameter of the sheaves and the center distance and is termed *pitch length*.
- *Center distance (C):* The distance between the center of the drive sheave to the center of the driven sheave.

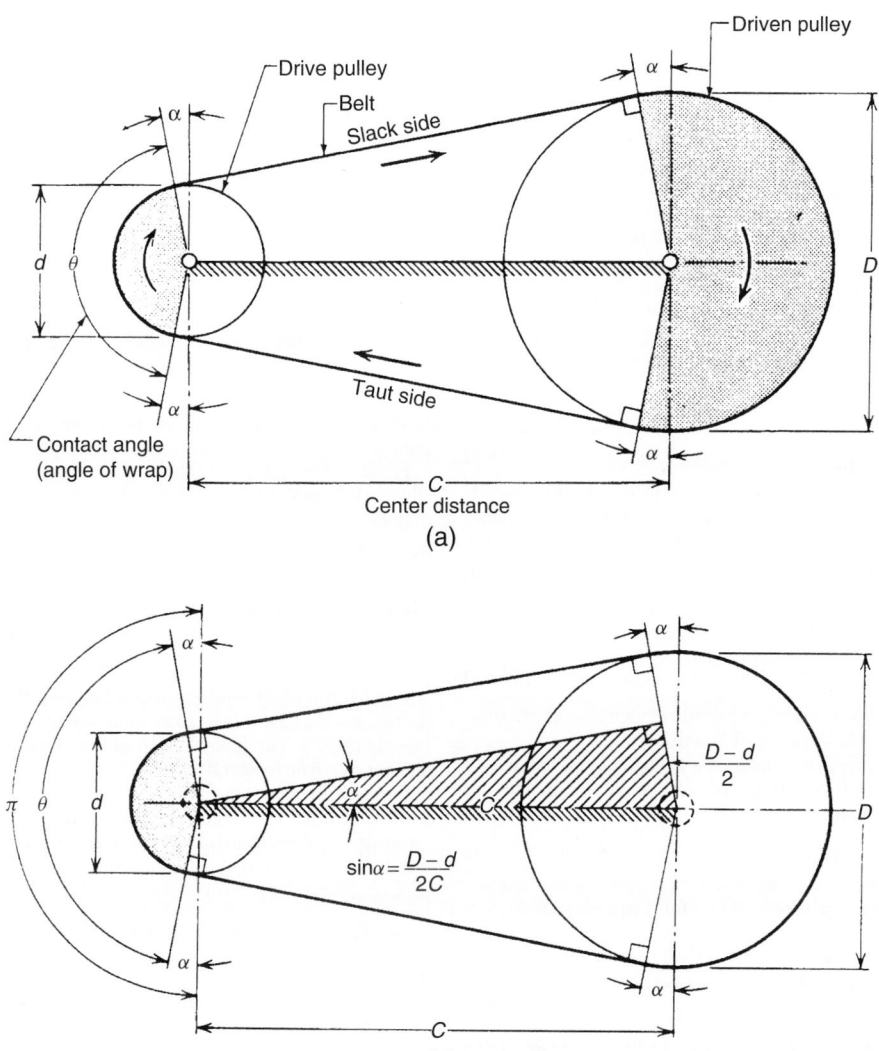

Figure 3.2.5 *V-belt geometry [2].*

- *Classic:* V belts with cross-sections including A, AX, B, BX, C, CX, and D.
- *Contact angle (θ):* The angle of belt wrap on the smaller sheave.
- *Creep:* Loss in speed of the driven sheave compared with the theoretical speed based on the speed ratio. It is caused by the alternate lengthening and shortening of the belt as it passes between the taut side and slack side of the drive.
- *Drive sheave:* This sheave is mounted on the motor or prime mover. It is usually the smaller of the two sheaves and has the higher rotational speed.
- *Driven sheave:* This sheave is located on the equipment being driven by the motor or prime mover. It is usually the larger of the two sheaves and thus has the lower rotational speed.
- *Narrow:* V belts with cross-sections including 3V, 3VX, 5V, 5VX, and 8V.
- *Pitch diameter (d, D):* The diameter at the point where the neutral axis of the belt contacts the sheave.

- *Service factor:* A value that is multiplied times the required horsepower to determine the design horsepower. This value compensates for those drives which intermittently operate above the required horsepower.
- *Speed ratio (i):* The ratio of the pitch diameter of the larger sheave to the pitch diameter of the smaller sheave. It can also be expressed as the speed of the faster sheave divided by the speed of the slower sheave.

3.2.1.3.3 V Belt Drive General Dimensions

Figure 3.2.5 shows the general dimensions of an open-belt drive. For V belt drives, the sheave diameters D and d are pitch diameters [2]. The quantity C is the center distance, θ is the contact angle for the smaller sheave, and 2x is the angular deviation from a 180° angle of contact.

From Figure 3.2.5, θ (radians) is approximately

$$\theta \approx \pi - \frac{D - d}{c} \qquad [3.2.1]$$

For V-belt applications, D and d are sheave pitch diameters. From the geometry of Figure 3.2.5, θ is always less than or equal to 180° or π radians. The lower guideline for θ is approximately 150°. Below this values, there will be increasing tension and slip, which will result in decreased life of the V belts. This limit on θ imposes a lower limit on the center distance and thus a practical limit on the speed ratio attainable is a given V-belt design.

3.2.1.3.4 Drive Center Distance and V-Belt Length

Shorter center distance is a very practical design objective. Such a design uses space economically and allows for a stable operation. In general, center distances are limited by the physical dimensions of the sheaves, or the minimum angle of θ (i.e., 150°). Maximum drive centers are limited only by available V-belt stocked lengths.

The center distance is

$$C = \frac{b + \sqrt{b^2 - 32(D-d)^2}}{16} \qquad [3.2.2]$$

where D is the pitch diameter of the larger sheave (usually the driven sheave) in in., d is the pitch diameter of the smaller sheave (usually the drive sheave) in in., and the quantity b is

$$b = 4L - 6.28(D+d) \qquad [3.2.3]$$

where L is the V-belt pitch length in inches.

From the above, L is

$$L = 2C + 1.57(D+d) + \frac{(D-d)^2}{4C} \qquad [3.2.4]$$

Modern V belts are nearly all of the closed-loop continuous type which come in standard lengths.

For V-belt drives with more than two sheaves, belt length is calculated from sheave coordinates and drive dimensions in layout drawings.

3.2.1.3.5 Speed Ratio

Most prime movers rotate at a higher speed than the operational driven equipment. Therefore, speed reduction is necessary for most belt drive systems. The speed ratio i between the prime mover drive shaft and the drive shaft is

$$i = \frac{\text{speed of drive shaft (rpm or ft/s)}}{\text{speed of driven shaft (rpm or ft/s)}} \qquad [3.2.5]$$

This can also be expressed in terms of sheave pitch diameter. This is

$$i = \frac{\text{pitch diam. of larger sheave (in.)}}{\text{pitch diam. of smaller sheave (in.)}} \qquad [3.2.6]$$

3.2.1.3.6 V-Belt Speed

Under normal operating conditions, speed reduction due to creep of the V-belt material is usually quite small. Neglecting these effects, the velocity v (in./s) of the V belt is

$$v = \frac{\pi D n_D}{60} = \frac{\pi d n_d}{60} \qquad [3.2.7]$$

where n_D is the large sheave speed and n_d is the small sheave speed, both in rpm.

In general, modern V-belt speeds are limited to about 6,500 ft/min (v = 108 ft/s), and the relationship between D, d, n_D, n_d, and i is

$$n_D = \frac{d}{D} n_d = \frac{n_d}{i} \qquad [3.2.8]$$

3.2.1.3.7 Creep and Slip

As a belt turns on a pulley, it tends to stretch on the contact arc of the driving pulley and shorten on the driven pulley. This local movement of the belt is known as *creep*. The loss in speed is about 0.5% and is neglected in sizing and speed calculations. Creep should not be confused with slip, which

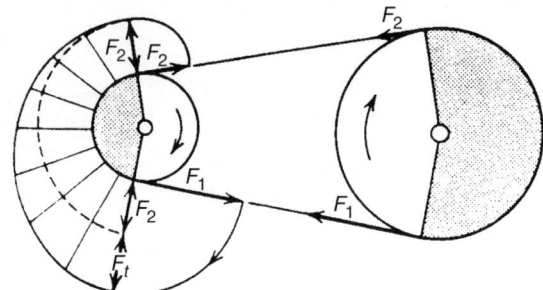

Figure 3.2.6 Equilibrium forces [2].

is an undesirable difference in speed between the belt and pulley.

From Figure 3.2.2, the relationship between the forces N (lb) normal to the V-belt sides and the force F_n (lb) can be obtained. This is

$$2N = \frac{F_n}{\sin \beta} \qquad [3.2.9]$$

The total friction force, F_f (lb) acting tangent to the sheave is

$$F_f = \frac{f F_n}{\sin \beta} \qquad [3.2.10]$$

where f is the coefficient of friction between the V belt and the sheave surface. The typical value for 2β is 32–40° (average value 36°). The typical value of f is about 0.30 (rubber against cast iron or steel).

Figure 3.2.6 shows the constant speed forces in the V belts with drive sheave transferring power via the V belt to the large driven sheave [2]. If centrifugal forces and slippage are neglected, then taut side tension force F_1 (lb) is

$$F_1 = F_2 e^{f\theta / \sin \beta} \qquad [3.2.11]$$

where F_2 (lb) is the slack side tension.

3.2.1.3.8 Power

The power P (hp) transmitted by the drive sheave to the drive sheave is

$$P = \frac{F_1(1 - e^{-(f\theta / \sin \beta)})v}{550} \qquad [3.2.12]$$

The force F_1 is the taut side of the V belt. In practice this force cannot exceed the allowable V-belt load F_{max} (lb). Equation 3.2.12 is valid only for rather slow speeds where centrifugal force can be neglected.

If centrifugal force is not neglected, then the power P transmitted is

$$P = \frac{(F_{max} - F_c)(1 - e^{-(f\theta / \sin \beta)})v}{550} \qquad [3.2.13]$$

where F_c (lb) is the centrifugal force created by the high speed of the V belt. The force F_c is

$$F_c = mv^2 \qquad [3.2.14]$$

where m is the mass per unit length of the V belt (lb-s²/ft²).

From Equations 3.2.13 and 3.2.14, an optimum velocity may be found for higher speed V-belt applications. This optimum velocity V_{op} (ft/s) is

$$V_{op} = \left(\frac{F_{max}}{3m}\right)^{1/2} \qquad [3.2.15]$$

and the optimum power P_{op} (hp) is

$$P_{op} = \frac{(F_{max} - mV_{op2})(1 - e^{-(f\theta / \sin \beta)})V_{op}}{550} \qquad [3.2.16]$$

Table 3.2.2 Belt Power Rating [3]

Section	K_1	K_2	K_3	K_4
A	1.004	1.652	15.547×10^{-4}	0.2126
AX	1.462	2.239	2.198×10^{-4}	0.4238
B	1.769	4.372	3.081×10^{-4}	0.3658
BX	2.051	3.532	3.097×10^{-4}	0.5735
C	3.325	12.070	5.828×10^{-4}	0.6886
CX	3.272	6.655	5.298×10^{-4}	0.8637
D	7.160	43.210	1.384×10^{-4}	1.4540
3V	1.204	1.904	2.069×10^{-4}	0.1763
3VX	1.169	1.530	1.523×10^{-4}	0.1596
5V	3.314	10.120	5.876×10^{-4}	0.4653
5VX	3.304	7.781	3.643×10^{-4}	0.4334
8V	8.663	49.320	15.810×10^{-4}	1.1670

Table 3.2.3 Speed Ratio Correction Factors [3]

Speed Ratio, D/d Range	K_{sr}
1.00 to and including 1.01	1.0000
Over 1.01 to and including 1.05	1.0096
Over 1.05 to and including 1.11	1.0266
Over 1.11 to and including 1.18	1.0473
Over 1.18 to and including 1.26	1.0655
Over 1.26 to and including 1.38	1.0805
Over 1.38 to and including 1.57	1.0956
Over 1.57 to and including 1.94	1.1089
Over 1.94 to and including 3.38	1.1198
Over 3.38	1.1278

3.2.1.3.9 Belt Power Rating

The horsepower rating for a belt of average length with 180° of contact is

$$P_r = d_p \cdot r \left[K_1 - \frac{K_2}{d_p} - K_3 \cdot (d_p \cdot r)^2 - K_4 \cdot LOG(d_p \cdot r) \right]$$

$$+ K_2 \cdot r \cdot \left(1 - \frac{1}{K_{sr}} \right) \qquad [3.2.17]$$

where d_p is the pitch diameter of the small sheave in inches, r is the RPM of the faster shaft divided by 1000, constants K_1 through K_4 are taken from Table 3.2.2, and the speed ratio correction factor K_{sr} is taken from Table 3.2.3.

For other lengths and arcs of contact, the power rating obtained from Equation 3.2.17 should be modified. The corrected horsepower rating per belt is

$$P_c = P_r \cdot K\theta \cdot K_l \qquad [3.2.18]$$

where $K\theta$ is the arc of contact correction factor from Table 3.2.4, and K_l is the length correction factor taken from Table 3.2.5 or 3.2.6.

3.2.1.3.10 Service Factors

Design horsepower equals the horsepower transmitted multiplied by the service factor. *Transmitted horsepower* is usually the nameplate rating of the motor. The service factor compensates for load fluctuations that cause the rating of the motor to be exceeded. Service factors for various types of oil field applications are given in Table 3.2.7.

3.2.1.3.11 Sizing a V Belt

API has developed specifications for V belting in oil-field power transmission applications [3]. Reference 3 is used to carry out detailed design calculations. Although this is an API publication, the specifications contained are consistent with specifications of other industrial groups. The basic

Table 3.2.4 Arc-of-Contact Correction Factors [3]

$\dfrac{D-d}{C}$	Arc-of-Contact on Small Sheave, Degrees	Correction Factor $K\theta$
0.00	180	1.00
0.10	174	0.99
0.20	169	0.97
0.30	163	0.96
0.40	157	0.94
0.50	151	0.93
0.60	145	0.91
0.70	139	0.89
0.80	133	0.87
0.90	127	0.85
1.00	120	0.82
1.10	113	0.80
1.20	106	0.77
1.30	99	0.73
1.40	91	0.70
1.50	83	0.65

C = center distance of drive; D = diameter of larger sheave; d = diameter of smaller sheave

Table 3.2.5 Length Correction Factors, Classic V Belts (K_l) [3]

Standard Length Designation	V-Belt Cross-Section			
	A, AX	B, BX	C, CX	D
26	0.78	—	—	—
31	0.82	—	—	—
35	0.85	0.80	—	—
38	0.87	0.82	—	—
42	0.89	0.84	—	—
46	0.91	0.86	—	—
51	0.93	0.88	0.80	—
55	0.95	0.89	—	—
60	0.97	0.91	0.83	—
68	1.00	0.94	0.85	—
75	1.02	0.96	0.87	—
80	1.04	—	—	—
81	—	0.98	0.89	—
85	1.05	0.99	0.90	—
90	1.07	1.00	0.91	—
96	1.08	—	0.92	—
97	—	1.02	—	—
105	1.10	1.03	0.94	—
112	1.12	1.05	0.95	—
120	1.13	1.06	0.96	0.88
128	1.15	1.08	0.98	0.89
144	—	1.10	1.00	0.91
158	—	1.12	1.02	0.93
173	—	1.14	1.04	0.94
180	—	1.15	1.05	0.95
195	—	1.17	1.06	0.96
210	—	1.18	1.07	0.98
240	—	1.22	1.10	1.00
270	—	1.24	1.13	1.02
300	—	1.27	1.15	1.04
330	—	—	1.17	1.06
360	—	—	1.18	1.07
390	—	—	1.20	1.09
420	—	—	1.21	1.10
480	—	—	—	1.13
540	—	—	—	1.15
600	—	—	—	1.17
660	—	—	—	1.18

Table 3.2.6 *Length Correction Factors, Narrow V Belts (K$_l$) [3]*

Standard Length Designation*	V-Belt Cross-Section		
	3V, 3VX	5V, 5VX	8V
250	0.83	—	—
265	0.84	—	—
280	0.85	—	—
300	0.86	—	—
315	0.87	—	—
335	0.88	—	—
355	0.89	—	—
375	0.90	—	—
400	0.92	—	—
425	0.93	—	—
450	0.94	—	—
475	0.95	—	—
500	0.96	0.85	—
530	0.97	0.86	—
560	0.98	0.87	—
600	0.99	0.88	—
630	1.00	0.89	—
670	1.01	0.90	—
710	1.02	0.91	—
750	1.03	0.92	—
800	1.04	0.93	—
850	1.06	0.94	—
900	1.07	0.95	—
950	1.08	0.96	—
1000	1.09	0.96	0.87
1060	1.10	0.97	0.88
1120	1.11	0.98	0.88
1180	1.12	0.99	0.89
1250	1.13	1.00	0.90
1320	1.14	1.01	0.91
1400	1.15	1.02	0.92
1500	—	1.03	0.93
1600	—	1.04	0.94
1700	—	1.05	0.94
1800	—	1.06	0.95
1900	—	1.07	0.96
2000	—	1.08	0.97
2120	—	1.09	0.98
2240	—	1.09	0.98
2360	—	1.10	0.99
2500	—	1.11	1.00
2650	—	1.12	1.00
2800	—	1.13	1.02
3000	—	1.14	1.03
3150	—	1.15	1.03
3350	—	1.16	1.04
3550	—	1.17	1.05
3750	—	—	1.06
4000	—	—	1.07
4250	—	—	1.08
4500	—	—	1.09
4750	—	—	1.09
5000	—	—	1.10

*Standard length designation is the effective length multiplied by 10.

Table 3.2.7 *V-Belt Service Factors [3]*

Types of Service	Service Factor
Compressors, reciprocating	1.6
Fans, propeller	1.5
Pumps, centrifugal	1.4
Pumps, rotary or vane	1.4
Pumps, duplex piston, except slush pumps	1.6
Pumps, duplex piston slush pumps	1.0*
Pumps, triplex plunger	1.5
Generators serving beam-pumping units	1.8
Generators with no beam-pumping load	1.5

*Consult manufacturer.

1. The following information is required to size a V-belt drive.
   a. Average horsepower transmitted
   b. The rpm of the faster shaft
   c. The rpm of the slower shaft
   d. Approximate desired center distance
2. Determine the design horsepower by multiplying the average horsepower transmitted by the service factor. Use Table 3.2.7 to select a service factor.
3. Select the proper V-belt cross-section from Figure 3.2.7 or 3.2.8 using the design horsepower (Step 2) and the speed of the faster shaft.
4. Calculate the speed ratio using Equation 3.2.5.
5. Select sheave diameters. Tables 3.2.8 and 3.2.9 show generally available sheave sizes for classic and narrow V belts. Select a set of sheaves that best match the desired speed ratio (Step 4) using Equation 3.2.6. Care should be taken such that the faster sheave is not smaller than the minimum recommended values given in Table 3.2.10.
6. Find the V-belt speed using Equation 3.2.7. The rim speed of the sheaves should not exceed 6,500 ft/min.
7. A desirable center distance should not be less than the diameter of the large sheave and preferably the sum of the diameters of the large and small sheaves. Using the approximate desired center distance, calculate the belt length using Equation 3.2.4, and then select a standard belt length that is the closest match to the calculated belt length. The actual center distance can be calculated using Equations 3.2.2 and 3.2.3.
8. Find the horsepower capability per V belt using Equations 3.2.17 and 3.2.18.
9. Determine the number of belts required by dividing the design horsepower (Step 2) by the horsepower capability per V belt (Step 7). If the number of V belts is a whole number and a fraction, the next higher whole number of belts should be used.
10. The V-belt drive can now be tabulated using
    a. V-belt type (Step 3)
    b. Number of V belts (Step 9)
    c. Drive sheave design (Steps 5 & 9)
    d. Driven sheave design (Steps 5 & 9)
    e. Center distance and V-belt length (Step 7)

calculation techniques above have been carried out for the stock V belts available. These data are tabulated in the API publication to simplify the design effect for V-belt power transmission systems.

Using the data contained in the API specifications for V belting the following steps can be used to find V-belt drive system dimensions.

3.2.2 Chains

Power transmission chains provide a positive drive even when operated under very adverse temperatures (−60 to 600°F) and other environmental conditions. These power transmission systems are very flexible with regards to their field applications. In general, chain drives are primarily selected for low-speed and medium-speed service. Some silent chain designs may be used in high-speed service [5,6].

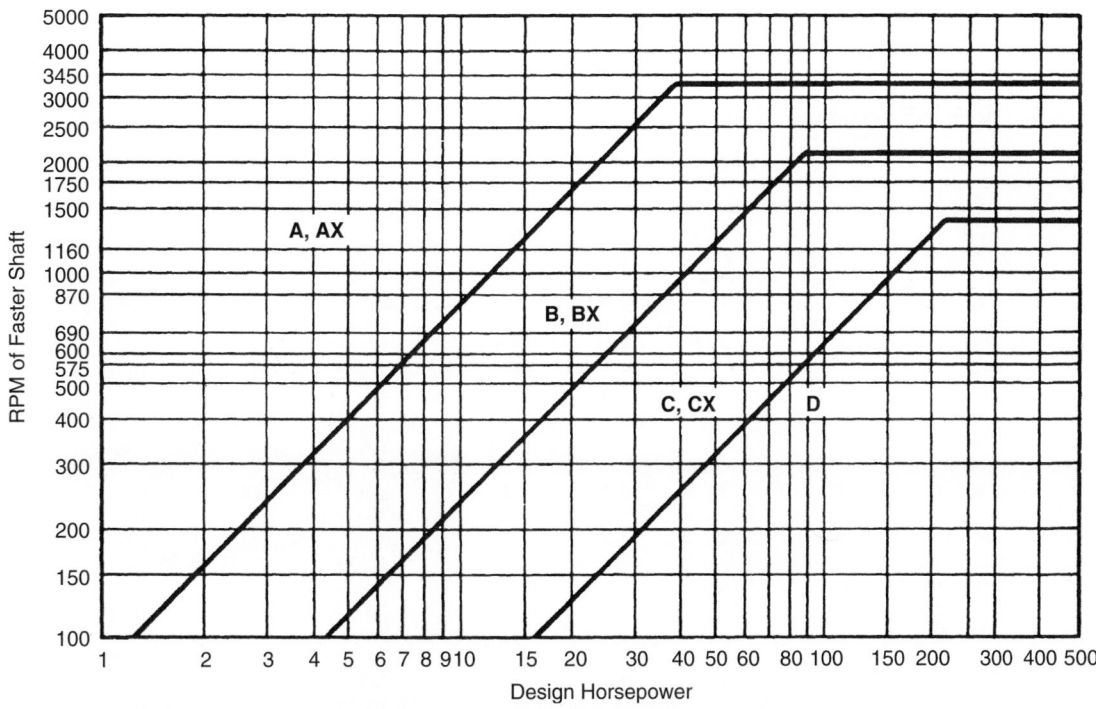

Figure 3.2.7   *Guide for selecting classic V-belt section [4].*

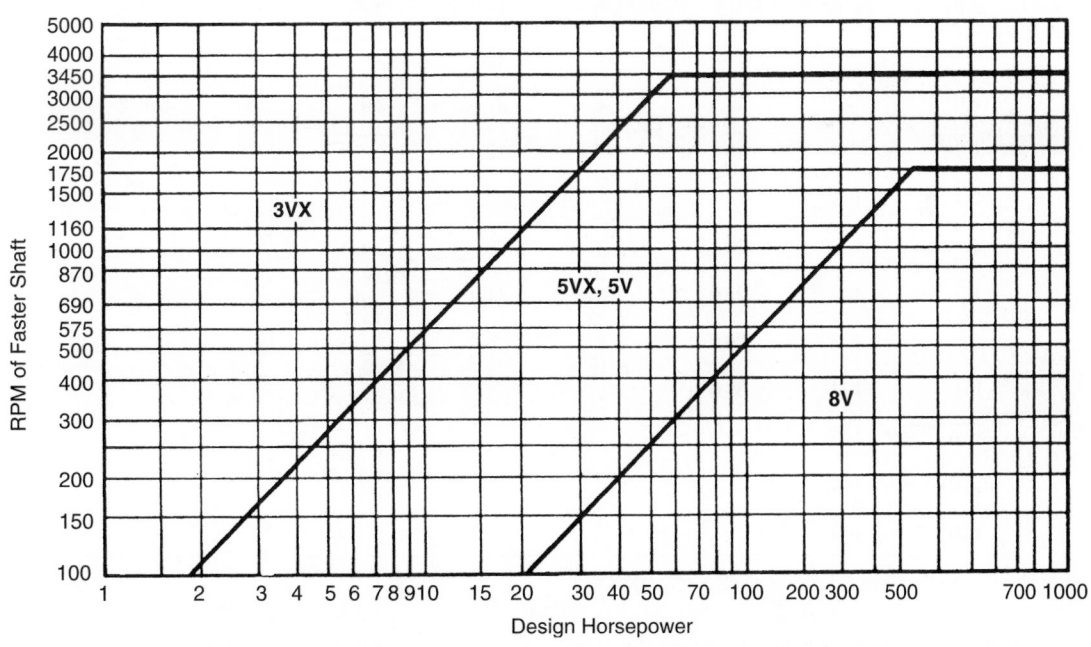

Figure 3.2.8   *Guide for selecting narrow V-belt section [4].*

There are six types of chains used for power transmission. These are roller, silent (inverted tooth), offset link (Ewart with bushing), detachable (open Ewart), pintle (closed Ewart), and bead.

Some of these are shown in Figure 3.2.9. Because most modern power transmissions use roller chains, silent chains or the offset link (Ewart) chains, these will be the only chain types discussed since they are quite important in oil field applications.

The term *chain* drive denotes a combination of chain and sprockets, with the sprockets mounted on rotating shafts.

Table 3.2.8 *Classic V-Belt Sheave Sizes Generally Available [3]*

A Section		B Section		C Section		D Section	
Diameter Datum	Number of Grooves	Diameter Datum	Number of Grooves	Diameter Datum	Number of Grooves	Diameter Datum	Number of Grooves
3.0	1–6	4.8	1–6	7.0	2–6	12.0	4–6, 8, 10, 12
3.2	1–6	5.0	1–6	7.5	2–6	13.0	4–6, 8, 10, 12
3.4	1–6	5.2	1–6	8.0	2–6, 8, 10	13.5	4–6, 8, 10, 12
3.6	1–6	5.4	1–6	8.5	2–6, 8, 10	14.0	4–6, 8, 10, 12
3.8	1–6	5.6	1–6	9.0	2–6, 8, 10, 12	14.5	4–6, 8, 10, 12
4.0	1–6	5.8	1–6	9.5	2–6, 8, 10, 12	15.0	4–6, 8, 10, 12
4.2	1–6	6.0	1–6	10.0	2–6, 8, 10, 12	15.5	4–6, 8, 10, 12
4.4	1–6	6.2	1–6	10.5	2–6, 8, 10, 12	16.0	4–6, 8, 10, 12
4.6	1–6	6.4	1–6	11.0	2–6, 8, 10, 12	18.0	4–6, 8, 10, 12
4.8	1–6	6.6	1–6	12.0	2–6, 8, 10, 12	20.0	4–6, 8, 10, 12
5.0	1–6	6.8	1–6	13.0	2–6, 8, 10, 12	22.0	4–6, 8, 10, 12
5.2	1–6	7.4	1–6	14.0	2–6, 8, 10, 12	27.0	4–6, 8, 10, 12
5.4	1–6	8.6	1–6	16.0	2–6, 8, 10, 12	33.0	4–6, 8, 10, 12
5.6	1–6	9.4	1–6	18.0	2–6, 8, 10, 12	40.0	4–6, 8, 10, 12
5.8	1–6	11.0	1–6	20.0	2–6, 8, 10, 12	48.0	5, 6, 8, 10
6.0	1–6	12.4	1–6	24.0	2–6, 8, 10, 12	58.0	5, 6, 8, 10
6.2	1–6	15.4	1–6	27.0	2–6, 8		
6.4	1–6	18.4	1–6	30.0	2–6, 8, 10, 12		
7.0	1–6	20.0	2–6, 8, 10	36.0	2–6, 8, 10, 12		
8.2	1–6	25.0	2–6, 8, 10	44.0	2–6, 8, 10, 12		
9.0	1–6	30.0	2–6, 8, 10	50.0	2–6, 8, 10, 12		
10.6	1–6	38.0	2–6, 8, 10				
12.0	1–6						
15.0	1–6						
18.0	1–6						

Standardization of chains is under the jurisdiction of the American National Standards Institute [7,8].

3.2.2.1 Chain Terminology
These are basic terminology terms that aid in the description of chains. These are (see Figure 3.2.10 [2]) the following:

- *Pitch* is the distance between any point on a link and the same point on the next link in a straight (unarticulated) chain.

- *Drive sprocket* is usually the sprocket that is provided with the shaft import power (usually the smaller diameter sprocket).
- *Driven sprocket* is the sprocket and shaft that the chain is transferring power to (usually the larger diameter sprocket).
- *Center distance* is the length between the centers of the drive and drive sprocket shafts.
- *Chain length* is the total length of the chain (usually measured in pitches).

Table 3.2.9 *Narrow V-Belt Sheave Sizes Generally Available [3]*

3V Section		5V Section		8V Section	
Diameter Datum	Number of Grooves	Diameter Datum	Number of Grooves	Diameter Datum	Number of Grooves
2.65	1–4	7.10	2–6, 8	12.50	4–6, 8, 10
2.80	1–4	7.50	2–6, 8	13.20	4–6, 8, 10
3.00	1–4	8.00	2–6, 8, 10	14.00	4–6, 8, 10
3.15	1–4	8.50	2–6, 8, 10	15.00	4–6, 8, 10
3.35	1–4	9.00	2–6, 8, 10	16.00	4–6, 8, 10
3.65	1–4	9.25	2–6, 8, 10	17.00	4–6, 8, 10
4.12	1–4	9.75	2–6, 8, 10	18.00	4–6, 8, 10
4.50	1–4	10.30	2–6, 8, 10	19.00	4–6, 8, 10
4.75	1–6, 8, 10	10.90	2–6, 8, 10	20.00	4–6, 8, 10
5.00	1–6, 8, 10	11.30	2–6, 8, 10	21.20	4–6, 8, 10
5.30	1–6, 8, 10	11.80	2–6, 8, 10	22.40	4–6, 8, 10
5.60	1–6, 8, 10	12.50	2–6, 8, 10	30.00	4–6, 8, 10
6.00	1–6, 8, 10	13.20	2–6, 8, 10	40.00	4–6, 8, 10
6.50	1–6, 8, 10	14.00	2–6, 8, 10	53.00	4–6, 8, 10
6.90	1–6, 8, 10	15.00	2–6, 8, 10		
8.00	1–6, 8, 10	16.00	2–6, 8, 10		
10.60	1–6, 8, 10	18.70	2–6, 8, 10		
14.00	1–6, 8, 10	21.20	2–6, 8, 10		
19.00	1–6, 8, 10	23.60	2–6, 8, 10		
25.00	2–6, 8, 10	28.00	3–6, 8, 10		
33.50	3–6, 8, 10	31.50	3–6, 8, 10		
		37.50	3–6, 8, 10		
		50.00	3–6, 8, 10		

Table 3.2.10 *Minimum Recommended Sheave Sizes [3]*

Classic		Narrow	
Section	Minimum Pitch Diameter, Inch	Section	Minimum Pitch Diameter, Inch
A	3.00	3V	2.65
AX	2.20	3VX	2.20
B	5.40	5V	7.10
BX	4.00	5VX	4.40
C	9.00	8V	12.50
CX	6.80		
D	13.0		

- *Pitch diameter* (of either the drive sprocket or driven sprocket) is generally on a theoretical circle described by the centerline of the chain as it passes over the sprocket.
- *Chain rating* is the load that a chain will satisfactorily handle over extended periods of time.
- *Angle of wrap* or *angle of contact* is the angular section of a sprocket that is in contact with the chain.

3.2.2.2 Roller Chain

Where accurate and higher-speed chain drives are required, roller chains are used. Roller chains are fundamentally a development of the block chain in which rollers have been inserted. The addition of the rollers increases the overall efficiency of the chain and permits it, in its ordinary form, to be operated at high speeds; a rate of 600–800 ft/min is usually recommended. The links can be furnished in a multitude of shapes, one common form being the "offset" shown in Figure 3.2.11. This type of chain is furnished in pitches (i.e., center distance between rollers) from $\frac{1}{8}$ up to $2\frac{1}{2}$ in. and in

breaking strengths, dependent upon the materials and the construction used, from about 10,000 to 20,000 lb [5].

Precision tools are used in the manufacture of roller chains, and a high degree of accuracy is maintained. The chain is made up of links, pins, and bushings, all fabricated from a high grade of steel, with the pins and the bushings ground to ensure accuracy of pitch.

Extreme care is used in the manufacture of some of the highest grades of roller chains, with the result that they can be operated at speeds as high as 4,000 ft/min. In general use, however, unless the chain is fully enclosed and well lubricated, the upper speed limit should be kept at about 1,400 ft/min. The chains operated at these high speeds must be fabricated from accurately ground rollers, bushings, and pins and be run over cast-iron or steel sprockets having accurately machined teeth. The general structure of a roller chain for this type of high-speed service is illustrated in Figures 3.2.12 and 3.2.13.

Roller chains can generally be furnished not only in single but in multiple widths as well. The use of the multiple roller chain makes possible a shorter pitch without sacrificing high-power capacity. Because the shorter pitch permits a greater number of teeth for an allowable sprocket diameter and a reduction in drive weight, quieter and smoother operation are obtained. The multiple chains may be operated at the same speed as a single chain of the same pitch. Theoretically, the power that can be transmitted by them is equal to the capacity of the single chain multiplied by the number of stands, but actually this quantity is usually reduced about 10%.

Sprockets. To secure full advantage of the modern roller chain, it should be operated on sprockets having accurately machined teeth, the profile of which has been specified or approved by the ANSI [7]. This profile, which is made up of circular arcs, is designed to compensate for the increase in pitch due to natural wear and thereby

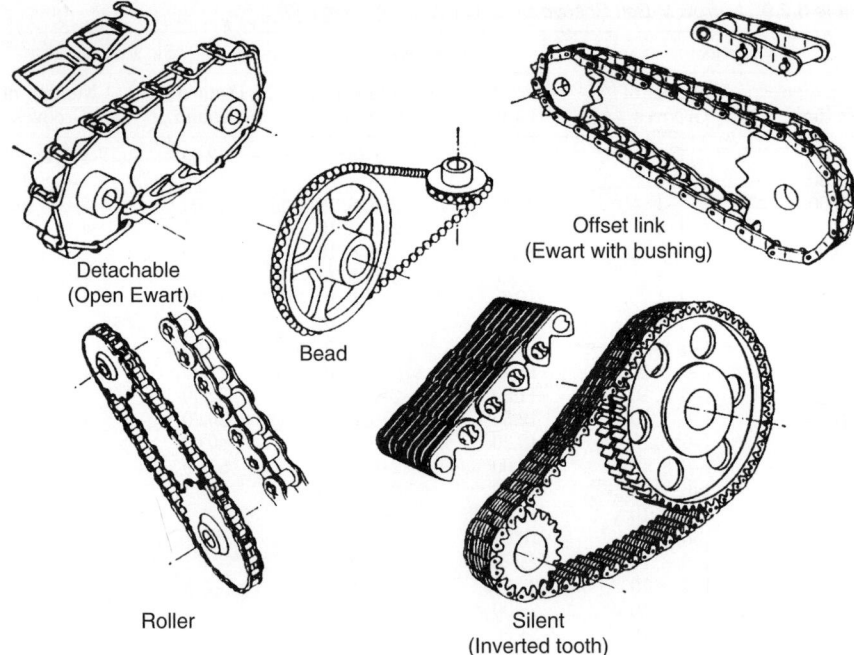

Figure 3.2.9   *Commonly used chain types [5].*

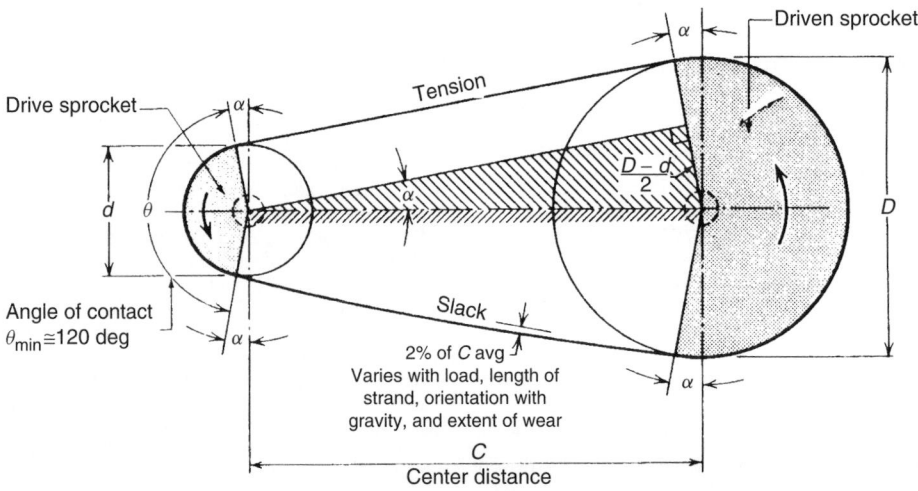

Figure 3.2.10   *Chain drive terminology and geometry [2].*

provides maximum efficiency throughout the life of the sprocket.

The shape (Figure 3.2.14) of the standard form of tooth used on the roller-chain sprockets permits the rollers to ride farther out on the teeth as the chain is stretched under load or the pitch is increased by wear. A further effect of the particular design is an apparent slight increase in pitch diameter as the chain and sprockets wear, thus distributing the load over a larger number of teeth and the wear over a larger portion of the tooth surfaces.

Although as few as five or six teeth and as many as 250 have been used on sprockets with roller chains, experience and research have proved that, if speed and efficiency are to be considered, the number of teeth should be neither too few nor too many. It has been generally conceded by all chain manufacturers that 17-tooth sprockets are the minimum to be used at high speeds and that 19 or 21-tooth sprockets are even better. The upper limit should be kept at about 125. It is considered good practice to use an odd number of teeth on the smaller sprocket. For a speed ratio of 7:1 or larger, a double-reduction drive often may be actually cheaper and undoubtedly will give longer life.

If the number of teeth is reduced below the recommended minimum, the increased shock or hammer action of the rollers engaging the teeth will increase wear and materially shorten the life of the chain. If more than 125 teeth

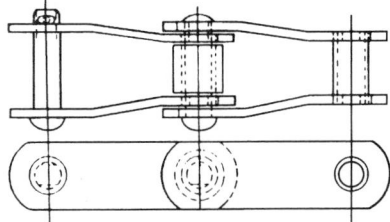

Figure 3.2.11   *Steel offset link [5].*

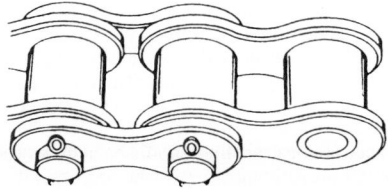

Figure 3.2.12   *Standard roller chain configuration [5].*

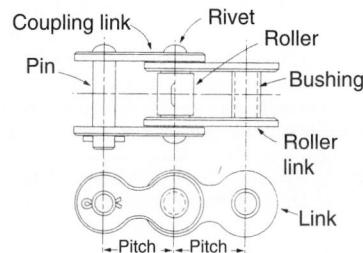

Figure 3.2.13   *Roller chain details [5].*

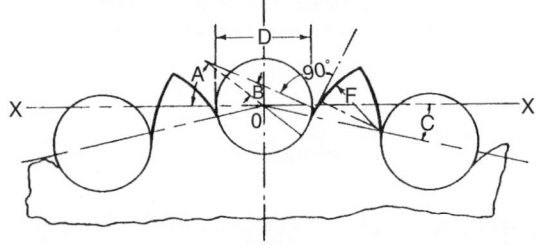

Figure 3.2.14   *ANSI roller chain sprocket tooth [5].*

are used, a small amount of pitch elongation will cause the chain to ride the sprocket long before it is actually worn out. Figure 3.2.15 shows a typical double-strand roller chain and its sprockets [2].

In general, four types of sprockets are available for roller chains. They are steel plate without hubs, cast iron or steel with hubs on one or both sides, split type, double-duty type.

The type of sprocket to be used depends entirely upon local or load conditions. The steel plate without hubs is the cheapest and is furnished for bolting to suitable hubs or flanges. The cast-iron or steel type fitted with hubs is made for direct mounting upon shaft and is fastened in place by either keys or setscrews or a combination of both. The split type is almost a necessity when the hub is mounted on a shaft with other pulleys or sheaves. Its construction facilitates installation and removal, but because of its extra cost

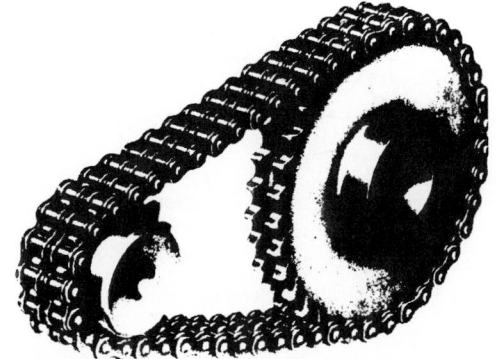

Figure 3.2.15   *Double-strand roller chain (courtesy of Borg Warner Corp.).*

it is usually not recommended except when solid hubs cannot be installed. The double-duty sprockets are made with steel rims or plates that may be removed or replaced without disturbing the hub, shaft, bearing, etc. Plates and hubs can be obtained either solid or split. They are particularly adapted for jobs requiring changing of drive ratios or where replacements must be made quickly.

Shaft Centers. It can be readily seen that on any chain drive the minimum center distances must be more than one-half the sum of the diameters of the two sprocket wheels. Experience has shown that best results are usually obtained when the center distance between shafts is 30 to 50 times the chain pitch. Forty times the chain pitch is about normal, and 80 times the pitch is maximum. In highly pulsating loads, 20 to 30 times the pitch is more nearly the correct center distance. Center distances of 10–12 ft are permissible with finished steel roller chains operated at moderate speeds without the use of idlers. On distances greater than this, an idler should be used to eliminate the possibility of swaying or flopping, which may cause the chain to jump the sprocket. When idlers are used, they should be placed on the slack strand of the chain. The number of teeth in the idler should be the largest possible and preferably not less than the number in the smaller sprocket of the drive.

3.2.2.3 Silent Chain

The expression "silent chain" may be somewhat misleading, for the type of chain which it is used to describe is not exactly silent, but it is much quieter in operation than the roller chain. The essential features of the silent chain are the straight-sided working jaws of the links, meshing with the straight-sided teeth of the sprocket, and the method of joining the links to form the chain. Figure 3.2.16 illustrates the principles of this chain, and shows that as the chain rides on the sprocket a rolling rather than a surface contact occurs. The construction tends to prevent undue vibration of the parts of the chain between sprockets [5].

In the silent chain the greatest wear occurs at the joint, and a number of developments and improvements have been made to reduce this to a minimum. In the first chain of this type (Figure 3.2.17a), the joint was made by means of a solid round pin that passed through circular holes punched in the links. The individual links bore directly upon the pins, and considerable wear resulted both in the links and the pins from the rubbing action produced by the wrapping of the chain around the sprockets. The first major improvement of this joint consisted of the placing of a bushing over the solid pin (Fig. 3.2.17b), thus producing a larger surface on

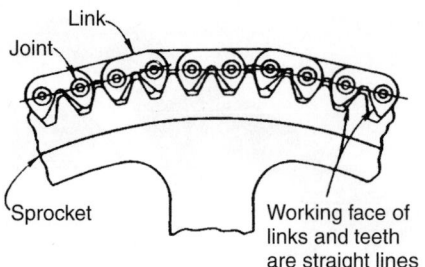

Figure 3.2.16 *Silent chain and differential-type sprocket teeth [5].*

which to distribute the load and a consequent reduction of the wear. This improvement was followed by still another one in which the solid bushing was replaced with a split or segmented one (Figure 3.2.17c). In still another design of joint (Figure 3.2.17d) the single round pin was replaced with two hardened alloy-steel specially shaped pins forming what is known as the "rocker-pin" joint. In this design the pin at one end of the link is the "seat pin" and that at the other is the "rocker," each securely held in their respective holes in the link and prevented from turning. Facing these pins in an individual link are the pins in the adjoining link, with the seat pin in contact with the rocker and the rocker turning on a seat pin. The rocker is shaped and located to provide a flat surface against the seat, while the chain is pulling in a straight line and yet rolls over its cylindrical section as the chain flexes and passes on the sprocket. In this way a rolling contact occurs as the joint parts turn; the wearing qualities are improved, and vibration is reduced to a minimum.

To guide silent chains on their sprockets, a number of arrangements are used. They can, however, all be grouped under one of two heads: flanged links as part of the chain, or flanged jaws of the sprockets.

In the flanged-link type of guide, there are again two general classes: the "middle guide" and the "side flange" (Figure 3.2.18a,b). This method of guiding is particularly

satisfactory when the drive is an integral part of the machine, and it is very reasonable at first cost. The choice between the two classes depends almost entirely upon the structural details of the sprockets. The middle guide is generally considered standard and is the one usually carried in stock by the chain manufacturer.

The "wire-flange" sprockets (Figure 3.2.18c) are assembled by simply inserting a preformed, crimped wire of the proper number of teeth and pitch along the edges of the teeth of the sprocket. This type of guide is recommended particularly in the case of independently mounted electric motors that are not likely to stay in accurate alignment.

The plate flange is used on the drive sprocket in those cases where heavy guiding action is expected such as is sometimes encountered in high-power devices.

Silent chains are furnished in pitches varying from $\frac{3}{16}$ to 3 in. and in weights capable of transmitting from a fraction of a horsepower to several thousand. They can be operated at speeds up to 3,500 ft/min, but where long life and low maintenance are desired, they should be held between 1,200 and 1,500 ft/min. Although they can be successfully operated at speeds below 1,200 ft/min, when extremely low speeds are required, it is usually more economical to use the roller chain.

Sprockets. Since the silent chain has not had the same wide usage as roller chains and because of the different types of construction that have been employed, there has been but little standardization of specializations. Each manufacturer has more or less carried out its own design and ratings, with the result that the sprockets to be used with this type of chain vary accordingly.

Practically all sprockets used with silent chain are made with cut teeth. Those with up to 25 teeth are usually cut from a steel forging and are furnished with a solid hub. For sprockets with more than 25 teeth, a semisteel is used, and the sprockets are available in either solid or split form.

Differential-Cut Wheels. When a wheel has 81 or more teeth, a patented differential-cut construction is available that assures maximum drive life. The object is to cut the teeth of a wheel having a relatively large number of teeth so that two

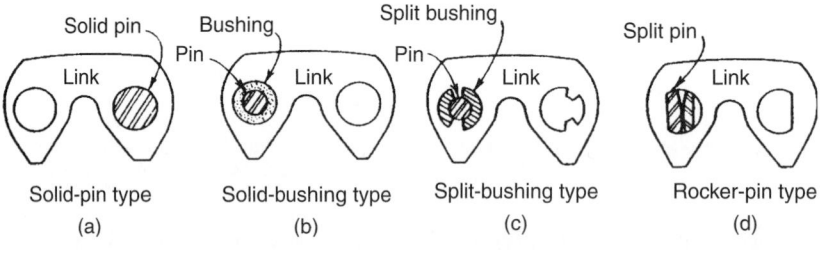

Figure 3.2.17 *Types of silent chain joints [5].*

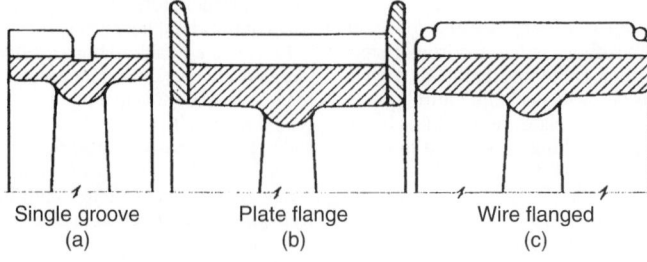

Figure 3.2.18 *Guides for silent chain sprockets [5].*

or more chain pitches, which are pivotally connected, act as a single unit to secure a greater angle of bend at the points of articulation and a corresponding larger component force acting toward the center of the wheel. For example, a 141-tooth, $\frac{1}{2}$-in. pitch wheel cut to a differential of 3 will have the same operating characteristics as a 47-tooth $1\frac{1}{2}$-in. pitch wheel and will provide better chain action and longer chain life than a 141-tooth single-cut wheel. Figure 3.2.16 shows this type of wheel.

As a protection for equipment against overload, sprockets are sometimes provided for silent chains with a "shearing pin" or a "break-pin hub." A pin of a known strength is used, and an overload of sufficient magnitude shears the pin and leaves the plate free to turn on the hub. After the overload condition is remedied, a new pin may be easily inserted and the drive again made ready to operate.

Shaft Centers. As in roller chains, the minimum distance between shaft centers for silent chains must, in order to provide tooth clearance, be greater than half the sum of the two sprocket diameters. On large-speed reductions, experience has shown that the center distance should not be less than eight-tenths the difference in diameters of the two sprockets.

3.2.2.4 Offset Link (Ewart with Bushing)

The offset link is used principally in conveying and elevating equipment, although it is also frequently employed to transmit very little loads at comparatively low speeds. This chain was originally patented over 60 years ago and has given a good account of itself ever since. It is furnished in either the open or closed type of link and is so designed as to permit ease of assembly and disassembly. The open-type link is illustrated in Figure 3.2.19 and the closed type in Figure 3.2.20. The use of the closed link results in greater strength and at the same time overcomes some objections by the exclusion of dust and grit.

The Ewart links are usually cast in one piece with no separate bushings or pins. The material used is generally malleable iron, although steel is also extensively employed. As a general rule the maximum speed at which this type of chain is operated is about 400 ft/min, and even at that speed it is apt to be quite noisy. Because the links used in this chain are not machined and the pitch is not very uniform, the teeth

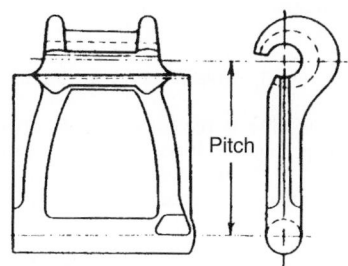

Figure 3.2.19 *Open Ewart link [5].*

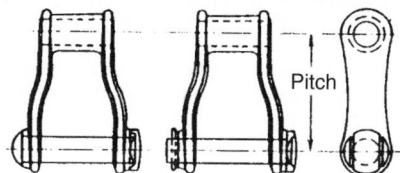

Figure 3.2.20 *Closed Ewart link or pintle link [5].*

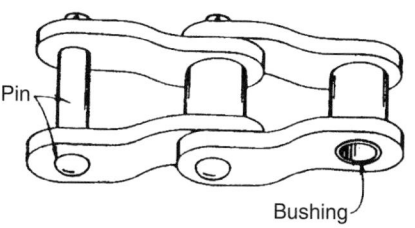

Figure 3.2.21 *Ewart link with bushing (API chain) [5].*

of the sprockets used with them are, in turn, generally not machined.

The offset link type of chain can be further improved by the use of a bushing, as shown in Figure 3.2.21. The improved or modified offset link can be used for more severe service conditions, but its safe maximum speed should, however, not exceed 400 ft/min. The offset link chain with bushing is known as the API chain. These chains are in only three nominal sizes, 3 in., $3\frac{1}{8}$ in., and 4 in. In general, the use of this older type of chain is not recommended.

3.2.2.5 Design Consideration

There are design considerations for the various chain power transmission applications that are common to nearly all chain types [5]. These are discussed in the following.

Roller Chains. These chains are primarily selected for low and medium-speed service and for conveyor work, although roller chains are sometimes used for higher speeds because they operate more smoothly and quietly.

The factors that will be discussed are often overlooked or misunderstood when applying chain. Other factors to be considered are found in chain manufacturers' catalogs along with chain sizes, sprocket types, and engineering examples.

Number of Teeth in Wheel. For roller chains, pinions should have 17 teeth or more for moderate-speed drives and 21 teeth or more for high-speed drives. Fewer teeth may be used for low-speed drives, with 12 teeth a recommended minimum.

For silent chain drives, pinions with 21 teeth or more are recommended for general applications and at least 25 teeth are recommended for high-speed applications. The recommended minimum is 17 teeth.

When space limits the diameter of the larger wheel, it may be necessary to select a wider chain with a smaller pitch to obtain a desirable number of teeth in the pinion.

Small roller-chain wheels should be hardened when used on moderate and high-speed drives, very low-speed heavily loaded drives, and when operating under abrasive conditions or when the drive ratio is greater than 4:1. Steel silent chain pinions should be hardened.

For 1:1 and 2:1 ratio drives, wheels of relatively large diameter should be selected. Large diameters assure that the distance between the two spans of chain is great enough to prevent them from striking after slack from normal joint wear has accumulated. This is of particular importance for drives operating on long fixed centers with the slack span on the chain on top.

Chordal Action. The chordal rise and fall of each chain pitch as it contacts a sprocket tooth is known as chordal action and results in repeated variations in linear chain speed. As shown in Figure 3.2.22, the amount of chordal movement and chain-speed variation becomes progressively smaller as the number of teeth in the pinion is increased. Smoother operation and longer chain life may be obtained by

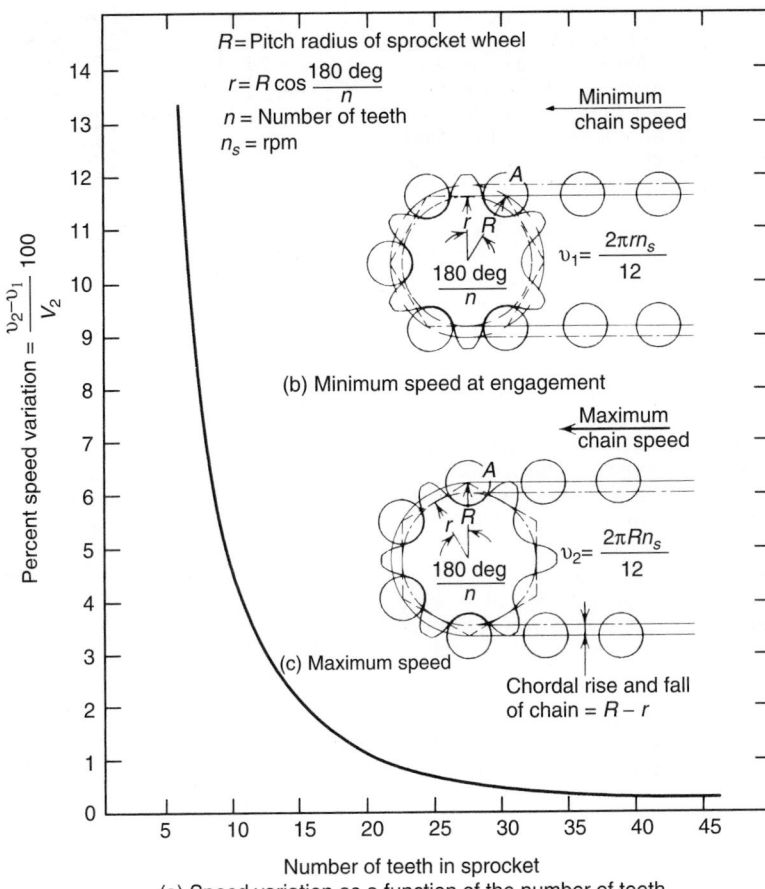

R = Pitch radius of sprocket wheel

$$r = R \cos \frac{180 \text{ deg}}{n}$$

n = Number of teeth

n_s = rpm

$$v_1 = \frac{2\pi r n_s}{12}$$

(b) Minimum speed at engagement

$$v_2 = \frac{2\pi R n_s}{12}$$

(c) Maximum speed

Chordal rise and fall of chain = $R - r$

(a) Speed variation as a function of the number of teeth

Figure 3.2.22   *Chordal action of a roller chain (courtesy of Link-Belt Co.).*

selecting pinions with 21 teeth or more because chain-joint articulation is reduced. Chordal action becomes negligible when a sprocket has 25 teeth or more.

Drive Ratio. Ratios in excess of 7:1 are generally not recommended for roller chains. If greater speed reduction is required, it is desirable and usually more economical to compound two or more drives.

Properly engineered silent chain drives having ratios as great as 12:1 will perform satisfactorily. However, it might be more economical to consider a compound drive where the ratio is 8:1 or larger.

Large reduction drives on minimum wheel centers are more economical if small-pitch, wide chains are being considered. Small reduction drives on long wheel centers are cheaper when larger pitch, narrow chains are used.

Wheel Centers. To avoid interference, wheel centers must be more than one-half the sum of the wheel outside diameters. Where ratios are 2:1 to 7:1, a center distance equal to the diameter of the large wheel plus one-half the diameter of the small wheel is recommended. Drives so proportioned will have ample chain wrap on the small wheel. A chain wrap of 90° is regarded as an absolute minimum for load-carrying sprocket wheels, and 120° or more of wrap is considered desirable.

Chain Tension. All chain drives should have some means of controlling the chain sag caused by normal joint wear. This is of utmost importance when the drive is subject to shock or pulsating loads or to reversals in direction of rotation. The most common methods taking up chain slack are (1) drive units mounted on adjustable base plat, slide rails, or similar units; these are used extensive in motor-driven applications; and (2) the use of adjustable idlers (Figure 3.2.23) and chain tensioners.

An adjustable idler is recommended for drives having fixed centers, particularly if the line of centers is vertical or near vertical. With such idlers, the required chain tension can be maintained for correct chain and sprocket-wheel contact.

An adjustable idler on relatively long-center drives that are subject to pulsating loads will eliminate whipping or thrashing of the chain. Such whipping action results in additional chain loading and joint wear, thereby reducing chain life. Also, an adjustable idler will provide sufficient chain wrap on the smaller sprocket wheel of large ratio, short-center drives.

Where maintenance service is infrequent, a counterweighted, spring-loaded, or automatic idler adjustment is best suited to promote long chain life. Manually adjusted idlers, if not periodically checked for proper chain tensioning, may become a destructive source in the drive system.

Adjustable-idler sprocket wheels must be securely mounted and should engage the slack or nonload carrying

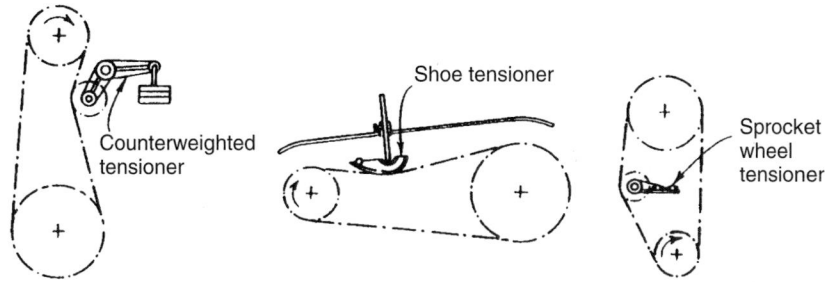

Figure 3.2.23 *Typical roller chain tensioning methods [5].*

side of the chain. At least three teeth of the idler must be in full engagement with the chain. To take up the slack of accumulative wear of the chain, an idler needs only to be adjustable for slightly more than two chain pitches. Finally, idler sprocket wheels should have at least 17 teeth.

A hardened-steel or hardwood shoe bearing against the back of the chain is another method of controlling chain tension. The method is satisfactory for small horsepower drives operating on fixed centers at slow or moderate speed with ample lubrication.

Offset couplers may be used to adjust chain tension or chain length when other methods are not feasible. This is done by removing a section of chain having one more pitch than the offset coupler and inserting the coupler in its place. **Drive Types.** Variable-speed drives may be selected on a chain-strength basis when operating with infrequent high chain loads such as from a torque converter, variable-speed motor, or multispeed transmission.

Speed-increasing drives should have at least 23 teeth in the smaller, faster running wheel. If possible, the taut span of the chain should be on top.

Vertical center drives require some form of chain tensioning or means for center adjustment to assure satisfactory operation and normal life. This is particularly important when the small wheel is in the lower position.

Fixed center drives should be selected on a conservative basis.

If an adjustable idler is not used, chain elongation may be retarded by using (1) a chain pitch larger than speed and horsepower indicate; (2) a larger service factor, thereby reducing the rate of wear in the chain joint; (3) as many teeth in the small sprocket wheel as the ratio will permit without exceeding 120 teeth in the larger wheel; and (4) good lubrication.

Lubrication. The primary purpose of chain lubrication is to maintain a film of oil between the bearing surfaces in the joints, thus assuring maximum operating efficiency. This clean oil film must be maintained at all load-carrying points where relative motion occurs, that is, between the pin and bushing on the chain, and the chain and wheel sprocket teeth. Table 3.2.11 gives some recommended methods for various speeds.

The method of application is primarily dictated by chain speed and the relative shaft positions. Some recommended methods are given below.

Manual Lubrication. This method is recommended for small horsepower drives with low chain speeds. Open running chains should not be exposed to abrasive dirt.

Drip-Cup Lubrication. This semiautomatic method is also suitable for small horsepower drives with low chain speeds. Cups should be located so that oil will drop to the center of the lower span at about 4 to 10 drops per minute.

Table 3.2.11 *Recommended Methods of Chain Lubrication [9]*

Chain Pitch, inch	Chain Speed, ft/min		
	Manual or Drip	Oil Bath or Slinger Disc	Oil Stream
0.500	up to 290	up to 2200	over 2200
0.625	240	1930	1930
0.750	210	1740	1740
1.000	170	1480	1480
1.250	145	1300	1300
1.500	125	1170	1170
1.750	110	1080	1080
2.000	100	1000	1000
2.250	90	930	930
2.500	85	880	880
3.000	75	790	790

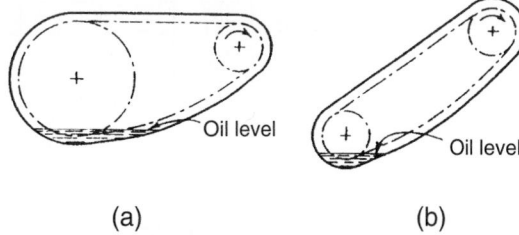

(a)   (b)

Figure 3.2.24 *Splash lubrication drive arrangements [5].*

Splash Lubrication. This is the simplest method of lubricating enclosed drives and is highly satisfactory for low and moderate-speed drives. In Figure 3.3.24 are shown a few arrangements for splash lubrication of different types of drives.

Oil-Disk Lubrication. This method is frequently used when a drive is not suitable for splash lubrication. It is highly satisfactory for moderate and semihigh-speed drives. The chain is kept above the oil level, and a circular disk, mounted to the lower wheel or shaft, dips into the oil about $\frac{1}{2}$ in. Figure 3.2.25 shows relative shaft positions best suited for oil-disk lubrication.

There must be sufficient disk speed to throw the oil. Figure 3.2.26 indicates the amount of oil delivered at various shaft speeds.

Forced lubrication is recommended for large-horsepower drives, heavily loaded drives, high-speed drives, or where

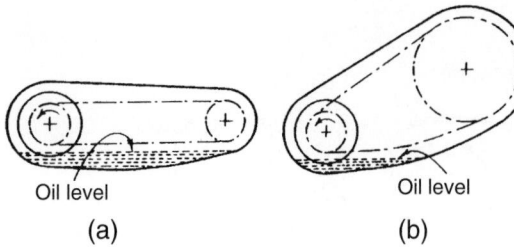

Oil level   Oil level

(a)   (b)

Figure 3.2.25 *Disk lubrication (courtesy of Link-Belt Co.).*

splash and disk lubrication cannot be used. An oil pump—capacity about 1 gal/min—supplies a continuous spray of oil to the inside of the lower span of chain. The circulation of the lubricant aids in the dissipation of heat and results in a well-lubricated, cooler operating device.

Trial Chain Sizes. To aid in the selection of an approximate chain size, Figure 3.2.27 gives chains of various pitches for horsepower from 1 to 300. More precise values for

chain capacity can be obtained in the various manufacturers catalogs.

Calculations. For a simple two-sprocket drive, the following guidelines for the initial design may be used [2].

Center Distance.

1. Minimum center distance is limited to the distance at which the two sprockets contact each other, or slightly more than half the sum of the outer diameters of the sprockets (Figure 3.2.28a).

2. Based on experience, a center distance equal to the diameter of the large sprocket plus half the diameter of the small sprocket is a good compromise with regard to angle of wrap, wear, and initial cost (see Figure 3.2.28b). The center distance C(n) is

$$C = D + 0.5d \quad \text{for} \quad D \gg d \qquad [3.2.19]$$

where d is the driven sprocket diameter in inches D is the driven sprocket diameter in inches.

3. With large ratios, the angle of contact becomes smaller and the number of teeth engaged with the chain decreases. For angles less than 120°, θ increasingly becomes a critical factor in the design of chain drives.

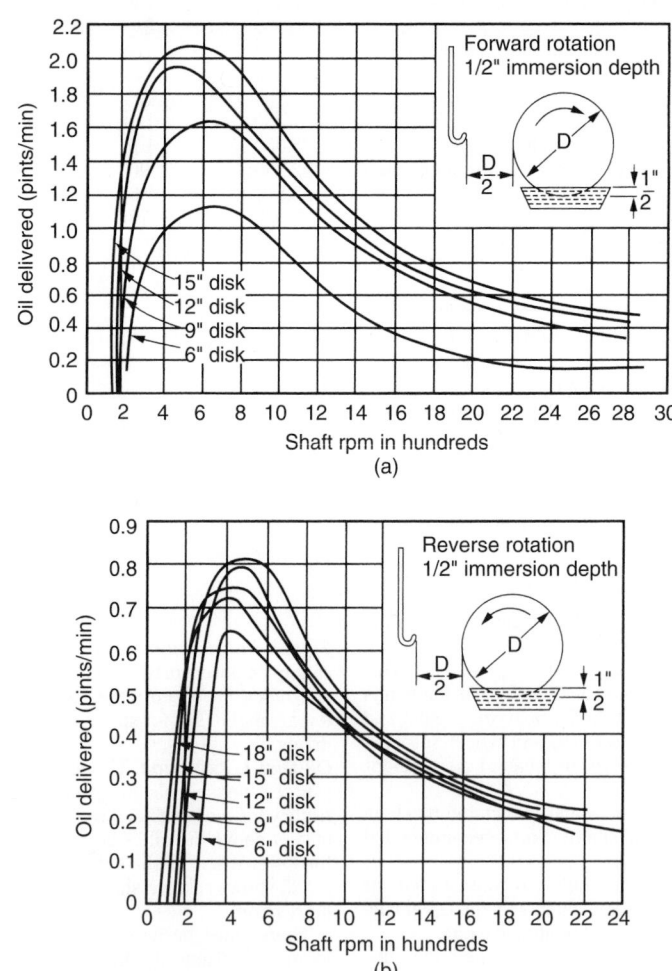

Figure 3.2.26 *Oil delivered to a chain by forward (a) and reverse (b) shafts [5].*

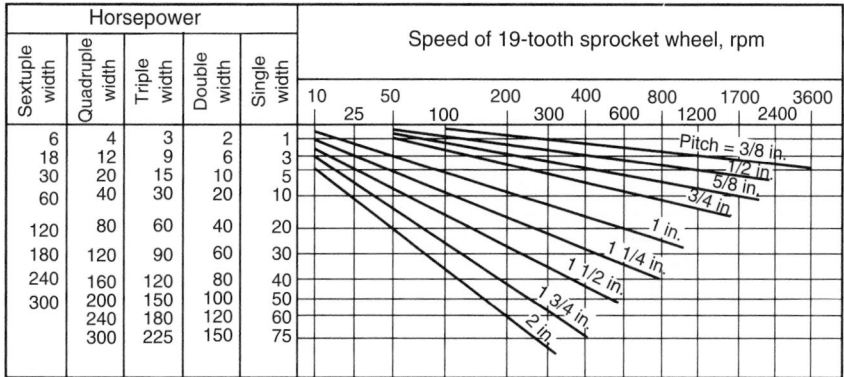

Figure 3.2.27 *Trial-chain chart enables an approximate to be chosen (courtesy of Link-Belt Co.).*

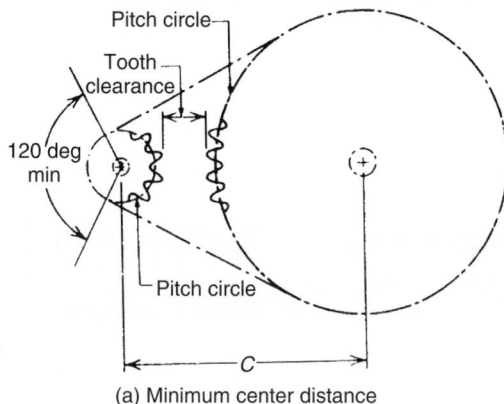

(a) Minimum center distance

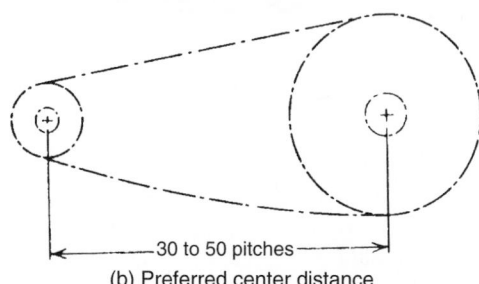

(b) Preferred center distance

Figure 3.2.28 *Shaft center distance (courtesy of American Chain Association).*

4. For maximum life and minimum wear, the center distance should be chosen so as to provide an even number of links in the chain. This arrangement, coupled with an odd number of teeth in each sprocket, will minimize wear.

5. A short center distance provides a compact design (desirable) and allows for a shorter, less expensive chain. But wear is more rapid on a drive with a short center distance because the chain has fewer links and each joint must therefore articulate more often.

6. When the center distance exceeds 60 pitches, a long chain will be needed and a manufacturer's representative should be consulted.

Angle of Wrap. The angle of wrap, θ (rad), is approximately

$$\theta = \pi - 2\alpha = \pi - \frac{D - d}{C} \qquad [3.2.20]$$

The chain length is generally measured in pitches because it consists of a whole number of links, each being of length p. Chain length is a function of the number of teeth in the two sprockets and the distance between sprocket centers. Thus, the chain length L (pitches) is

$$L = \frac{N + n}{2} + \frac{2C}{p} + \frac{p(N - n)^2}{39.5C} \qquad [3.2.21]$$

where n is the number of teeth in the drive sprocket, N is the number of teeth in the driven sprocket, C is the center distance in inches, p is pitch in inches.

Equation 3.2.21 does not often result in an even number of pitches. Thus, the results must be rounded off to a whole number, preferably an even whole number to avoid the specification of an offset link chain.

When several sprockets are used, center distances and chain lengths are best found by means of accurate engineering layouts and calculations.

Chain Speed. The chain speed v (ft/min) is

$$v = \pi n_s \frac{d}{12} \qquad [3.2.22]$$

where n_s is the drive shaft speed in rpm.

3.2.2.6 Design Data
In what follows, the pertinent design data are given for the roller chain, silent chain, and the offset link chain (API chain).

Roller Chain. Table 3.2.12 gives the typical service factor for roller chain drives [7]. Table 3.2.13 gives the basic roller chain design dimensions for ANSI standard roller chains by ANSI chain number [7].

Table 3.2.14a,b gives the horsepower ratings for single-strand roller chain drives. These data are given for each ANSI chain number [7].

Silent Chain. Table 3.2.15 gives the typical service factors for silent chain drives [10]. Tables 3.2.16 and 3.2.17 give the horsepower ratings per inch width for silent chain drives [10].

Drilling Applications. Drilling equipment utilizes chain drives in various applications on the drilling rig itself and its auxiliary equipment. The drives for this type of equipment are called compound drives. Such drive equipment are subject to very severe operational loads. Table 3.2.18 gives the typical service factors for these chain drives [9].

Table 3.2.12 *Service Factors for Roller Chain Drives [7]*

	Type of Input Power		
Type of Driven Load	Internal Combustion Engine with Hydraulic Drive	Electric Motor or Turbine	Internal Combustion Engine with Mechanical Drive
Smooth	1.0	1.0	1.2
Moderate shock	1.2	1.3	1.4
Heavy shock	1.4	1.5	1.7

Table 3.2.13 *Roller Chain Data and Dimensions, in Inches [10]*

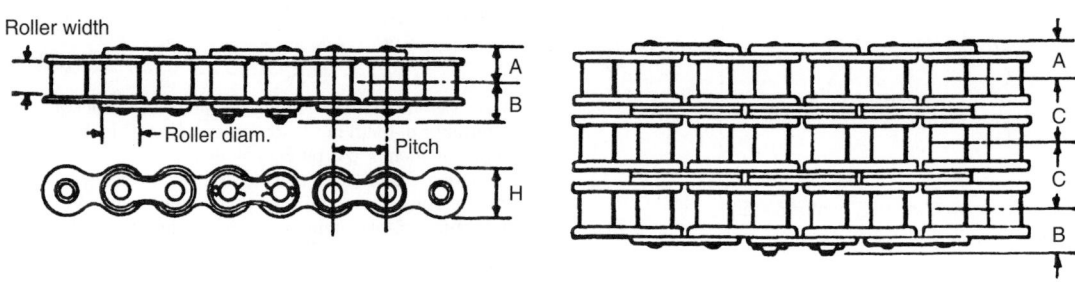

ANSI Chain no.	ISO Chain no.	*Roller*				*Roller Link Plate*		*Dimension*			Tensile Strength per Strand, lb	*Recommended Max Speed r/min*		
		Pitch	Width	Diam	Pin Diam	Thickness	Height H	A	B	C		12 Teeth	18 Teeth	24 Teeth
25	04C-1	1/4	1/8	0.130	0.091	0.030	0.230	0.150	0.190	0.252	925	5,000	7,000	7,000
35	06C-1	3/8	3/16	0.200	0.141	0.050	0.344	0.224	0.290	0.399	2,100	2,380	3,780	4,200
41	085	1/2	1/4	0.306	0.141	0.050	0.383	0.256	0.315		2,000	1,750	2,725	2,850
40	08A-1	1/2	5/16	0.312	0.156	0.060	0.452	0.313	0.358	0.566	3,700	1,800	2,830	3,000
50	10A-1	5/8	3/8	0.400	0.200	0.080	0.594	0.384	0.462	0.713	6,100	1,300	2,030	2,200
60	12A-1	3/4	1/2	0.469	0.234	0.094	0.679	0.493	0.567	0.897	8,500	1,025	1,615	1,700
80	16A-1	1	5/8	0.625	0.312	0.125	0.903	0.643	0.762	1.153	14,500	650	1,015	1,100
100	20A-1	1 1/4	3/4	0.750	0.375	0.156	1.128	0.780	0.910	1.408	24,000	450	730	850
120	24A-1	1 1/2	1	0.875	0.437	0.187	1.354	0.977	1.123	1.789	34,000	350	565	650
140	28A-1	1 3/4	1	1.000	0.500	0.218	1.647	1.054	1.219	1.924	46,000	260	415	500
160	32A-1	2	1 1/4	1.125	0.562	0.250	1.900	1.250	1.433	2.305	58,000	225	360	420
180		2 1/4	1 13/32	1.406	0.687	0.281	2.140	1.421	1.770	2.592	76,000	180	290	330
200	40A-1	2 1/2	1 1/2	1.562	0.781	0.312	2.275	1.533	1.850	2.817	95,000	170	260	300
240	48A-1	3	1 7/8	1.875	0.937	0.375	2.850	1.722	2.200	3.458	135,000	120	190	210

Table 3.2.14a *Horsepower Ratings for Single-Strand Roller Chain [10]*

ANSI no. and pitch, in	Number of teeth in small socket	Small sprocket, r/min										
		50	500	1,200	1,800	2,500	3,000	4,000	5,000	6,000	8,000	10,000
25 ¼	11	0.03	0.23	0.50	0.73	0.98	1.15	1.38	0.99	0.75	0.49	0.35
	15	0.04	0.32	0.70	1.01	1.36	1.61	2.08	1.57	1.20	0.78	0.56
	20	0.06	0.44	0.96	1.38	1.86	2.19	2.84	2.42	1.84	1.201	0.86
	25	0.07	0.56	1.22	1.76	2.37	2.79	3.61	3.38	2.57	1.67	1.20
	30	0.08	0.68	1.49	2.15	2.88	3.40	4.40	4.45	3.38	2.20	1.57
	40	0.12	0.92	2.03	2.93	3.93	4.64	6.00	6.85	5.21	3.38	2.42
35 ⅜	11	0.10	0.77	1.70	2.45	3.30	2.94	1.91	1.37	1.04	0.67	0.48
	15	0.14	1.08	2.38	3.43	4.61	4.68	3.04	2.17	1.65	1.07	0.77
	20	0.19	1.48	3.25	4.68	6.29	7.20	4.68	3.35	2.55	1.65	1.18
	25	0.24	1.88	4.13	5.95	8.00	9.43	6.54	4.68	3.56	2.31	1.65
	30	0.29	2.29	5.03	7.25	9.74	11.5	8.59	6.15	4.68	3.04	2.17
	40	0.39	3.12	6.87	9.89	13.3	15.7	13.2	9.47	7.20	4.68	—
41 ½	11	0.13	1.01	1.71	0.93	(0.58)	0.43	0.28	0.20	0.15	0.10	
	15	0.18	1.41	2.73	1.49	(0.76)	0.69	0.45	0.32	0.24	0.16	
	20	0.24	1.92	4.20	2.29	(1.41)	1.06	0.69	0.49	0.38		
	25	0.31	2.45	5.38	3.20	(1.97)	1.49	0.96	0.69	0.53		
	30	0.38	2.98	6.55	4.20	(2.58)	1.95	1.27	0.91	0.69		
	40	0.51	4.07	8.94	6.47	(3.97)	3.01	1.95	1.40			
40 ½	11	0.23	1.83	4.03	4.66	(3.56)	2.17	1.41	1.01	0.77	0.50	
	15	0.32	2.56	5.64	7.43	(4.56)	3.45	2.24	1.60	1.22	0.79	
	20	0.44	3.50	7.69	11.1	(7.03)	5.31	3.45	2.47	1.88		
	25	0.56	4.45	9.78	14.1	(9.83)	7.43	4.82	3.45	2.63		
	30	0.68	5.42	11.9	17.2	(12.9)	9.76	6.34	4.54	3.45		
	40	0.93	7.39	16.3	23.4	(19.9)	15.0	9.76	6.99			
50 ⅝	11	0.45	3.57	7.85	5.58	(3.43)	2.59	1.68	1.41	1.20	0.92	
	15	0.63	4.99	11.0	8.88	(5.46)	4.13	2.68	2.25	1.92		
	20	0.86	6.80	15.0	13.7	(8.40)	6.35	4.13	3.46	2.95		
	25	1.09	8.66	19.0	19.1	(11.7)	8.88	5.77	4.83			
	30	1.33	10.5	23.2	25.1	(15.4)	11.7	7.58				
	40	1.81	14.4	31.6	38.7	(23.7)	18.0					

The sections separated by heavy lines denote the method of lubrication as follows: type A (left section), manual; type B (middle section), bath or disk; type C (left section), oil stream.

Table 3.2.14b *Horsepower Ratings for Single-Strand Roller Chain [10]*

ANSI no. and pitch, in	Number of teeth in small sprocket	Small sprocket, r/min										
		10	50	100	200	500	700	1,000	1,400	2,000	2,700	4,000
60	11	0.18	0.77	1.44	2.69	6.13	8.30	11.4	9.41	5.51	(3.75)	1.95
3/4	15	0.25	1.08	2.01	3.76	8.57	11.6	16.0	15.0	8.77	(6.18)	3.10
	20	0.35	1.47	2.75	5.13	11.7	15.8	21.8	23.1	13.5	(9.20)	
	25	0.44	1.87	3.50	6.52	14.9	20.1	27.8	32.2	18.9	(12.9)	
	30	0.54	2.28	4.26	7.94	18.1	24.5	33.8	42.4	24.8	(16.9)	
	40	0.73	3.11	5.81	10.8	24.7	33.5	46.1	62.5	38.2		
80	11	0.42	1.80	3.36	6.28	14.3	19.4	19.6	11.8	6.93	4.42	
1	15	0.59	2.52	4.70	8.77	20.0	27.1	31.2	18.9	11.0	7.04	
	20	0.81	3.44	6.41	12.0	27.3	37.0	48.1	29.0	17.0		
	25	1.03	4.37	8.16	15.2	34.7	47.0	64.8	40.6	23.8		
	30	1.25	5.33	9.94	18.5	42.3	57.3	78.9	53.3	31.2		
	40	1.71	7.27	13.6	25.3	57.7	78.1	108	82.1	48.1		
100	11	0.81	3.45	6.44	12.0	27.4	37.1	23.4	14.2	8.29		
1 1/4	15	1.13	4.83	9.01	16.8	38.3	51.9	37.3	22.5	13.2		
	20	1.55	6.58	12.3	22.9	52.3	70.8	57.5	34.7	20.3		
	25	1.97	8.38	15.5	29.2	66.6	90.1	80.3	48.5	28.4		
	30	2.40	10.2	19.0	35.5	81.0	110	106	63.7	10.0		
	40	3.27	13.9	26.0	48.5	111	150	163	98.1			
120	11	1.37	5.83	10.9	20.3	46.3	46.3	27.1	16.4	9.59		
1 1/2	15	1.91	8.15	15.2	28.4	64.7	73.8	43.2	26.1			
	20	2.61	11.1	20.7	38.7	88.3	114	66.5	40.1			
	25	3.32	14.1	26.4	49.3	112	152	92.9	56.1			
	30	4.05	17.2	32.1	60.0	137	185	122	73.8			
	40	5.52	23.5	43.9	81.8	187	253	188	59.5			
140	11	2.12	9.02	16.8	31.4	71.6	52.4	30.7	18.5			
1 3/4	15	2.96	12.6	23.5	43.9	100	83.4	48.9	29.5			
	20	4.04	17.2	32.1	59.9	137	128	75.2	45.4			
	25	5.14	21.9	40.8	76.2	174	180	105	63.5			
	30	6.26	26.7	49.7	92.8	212	236	138				
	40	8.54	36.4	67.9	127	289	363	213				
160	11	3.07	13.1	24.4	45.6	96.6	58.3	34.1				
2	15	4.30	18.3	34.1	63.7	145	92.8	54.4				
	20	5.86	25.0	46.4	86.9	198	143	83.7				
	25	7.40	31.8	59.3	111	252	200	117				
	30	9.08	38.7	72.2	125	307	263	154				
	40	12.4	52.8	98.5	184	419	404					
180	11	4.24	18.1	33.7	62.9	106	64.1	37.5				
2 1/4	15	5.93	25.3	47.1	88.0	169	102	59.7				
	20	8.10	34.5	64.3	120	260	157	92.0				
	25	10.3	43.9	81.8	153	348	220					
	30	12.5	53.4	99.6	186	424	289					
	40	17.1	72.9	136	254	579	398					
200	11	5.64	24.0	44.8	83.5	115						
2 1/2	15	7.88	33.5	62.6	117	184						
	20	10.7	45.8	85.4	159	283						
	25	13.7	58.2	109	203	396						
240	11	9.08	38.6	72.1	135							
3	15	12.7	54.0	101	188							
	20	17.3	73.7	138	257							
	25	22.0	93.8	175	327							

The sections separated by heavy lines denote the method of lubrication as follows: type A (left section), manual; type B (middle section), bath or disk; type C (left section), oil stream.

Table 3.2.15 *Service Factors for Silent Chain Drives [10]*

Application	Fluid-Coupled Engine or Electric Motor		Engine with Straight Mechanical Drive		Torque Converter Drives	
	10 hr	24 hr	10 hr	24 hr	10 hr	24 hr
Agitators	1.1	1.4	1.3	1.6	1.5	1.8
Compressors	1.6	1.9	1.8	2.1	2.0	2.3
Cranes and hoists	1.4	1.7	1.6	1.9	1.8	2.1
Fans and blowers	1.5	1.8	1.7	2.0	1.9	2.2
Mixers	1.6	1.9	1.8	2.1	2.0	2.3
Oil field machinery	1.6	1.9	1.8	2.1	2.0	2.3
Pumps	1.6	1.9	1.8	2.1	2.0	2.3

Table 3.2.16 *Horsepower Ratings per Inch of Chain Width, Silent Chain Drive (Small Pitch) [10]*

Pitch, in	No. of teeth in small sprocket	Small sprocket, r/min											
		500	600	700	800	900	1,200	1,800	2,000	3,500	5,000	7,000	9,000
3/16	21	0.41	0.48	0.55	0.62	0.68	0.87	1.22	1.33	2.03	2.58	3.12	3.35
	25	0.49	0.58	0.66	0.74	0.82	1.05	1.47	1.60	2.45	3.13	3.80	4.10
	29	0.57	0.67	0.76	0.86	0.95	1.21	1.70	1.85	2.83	3.61	4.40	4.72
	33	0.64	0.75	0.86	0.97	1.07	1.37	1.90	2.08	3.17	4.02	4.85	—
	37	0.71	0.84	0.96	1.08	1.19	1.52	2.11	2.30	3.48	4.39	5.24	—
	45	0.86	1.02	1.15	1.30	1.43	1.83	2.53	2.75	4.15	5.21	—	—
Type*		I						II			III		

Pitch, in	No. of teeth in small sprocket	Small sprocket, r/min											
		100	500	1,000	1,200	1,500	1,800	2,000	2,500	3,000	3,500	4,000	5,000	6,000
3/8	21	0.58	2.8	5.1	6.0	7.3	8.3	9.0	10	11	12	12	12	10
	25	0.69	3.3	6.1	7.3	8.8	10	11	13	14	15	5	14
	29	0.80	3.8	7.3	8.5	10	12	13	15	16	18	19	19	18
	33	0.90	4.4	8.3	9.8	12	14	15	18	19	21	21	21	20
	37	1.0	4.9	9.1	11	14	15	16	20	21	24	24	24	—
	45	1.3	6.0	11	13	16	19	20	24	26	28	29	—	—
Type*		I			II			III					

Pitch, in	No. of teeth in small sprocket	Small sprocket, r/min										
		100	500	700	1,000	1,200	1,800	2,000	2,500	3,000	3,500	4,000
1/2	21	1.0	5.0	6.3	8.8	10	14	14	15	16	16	—
	25	1.2	5.0	7.5	10	13	16	18	20	21	21	20
	29	1.4	6.3	8.8	13	14	19	21	24	25	25	25
	33	1.6	7.5	10	14	16	23	24	28	29	30	29
	37	1.9	8.8	11	16	19	25	26	30	33	33	—
	45	2.5	10	14	19	23	30	30	36	39	—	—
Type*		I			II			III				

Pitch, in	No of teeth in small sprocket	Small sprocket, r/min									
		100	500	700	1,000	1,200	1,800	2,000	2,500	3,000	3,500
5/8	21	1.6	7.5	10	13	15	19	20	20	20	—
	25	1.9	8.8	11	16	19	24	25	26	26	24
	29	2.1	10	14	19	21	28	30	31	31	29
	33	2.5	11	16	21	25	33	34	36	36	34
	37	2.8	13	18	24	28	36	39	43	41	—
	45	3.4	16	21	29	34	44	46	—	—	—
Type*		I			II			III			

Pitch, in	No. of teeth in small sprocket	Small sprocket, r/min								
		100	500	700	1,000	1,200	1,500	1,800	2,000	2,500
3/4	21	2.3	10	14	18	20	23	24	25	24
	25	2.8	13	16	21	25	29	31	31	30
	29	3.1	15	20	26	30	34	36	38	38
	33	3.6	16	23	30	34	39	43	44	44
	37	4.0	19	25	34	39	44	48	49	49
	45	4.9	23	30	40	46	53	56	58	—
Type*		I			II			III		

*Lubrication types: Type I, manual, brush, or oil cup; Type II bath or disk; Type III, oil stream.

Table 3.2.17 *Horsepower Ratings per Inch of Chain Width, Silent Chain Drive (Large Pitch) [10]*

Pitch, in	No. of teeth in small sprocket	Small sprocket, r/min										
		100	200	300	400	500	700	1,000	1,200	1,500	1,800	2,000
1	21	3.8	7.5	11	15	18	23	29	31	33	33	—
	25	5.0	8.8	14	18	21	28	35	39	41	41	41
	29	5.0	11	16	20	25	33	11	46	50	51	50
	33	6.3	13	18	24	29	38	49	54	59	59	58
	37	6.8	14	20	26	33	43	54	60	65	66	—
	45	8.8	16	25	31	39	51	65	71	76	—	—
Type*		I		II				III				

Pitch, in	No. of teeth in small sprocket	Small sprocket, r/min										
		100	200	300	400	500	600	700	800	1,000	1,200	1,500
$1\frac{1}{4}$	21	6.3	11	18	23	26	30	33	36	40	41	—
	25	7.5	14	20	26	31	36	40	44	50	53	53
	29	8.6	16	24	31	38	43	48	53	59	63	64
	33	9.9	19	28	35	43	49	55	60	69	73	74
	37	11	21	30	40	48	55	63	68	76	81	—
	45	13	26	38	49	59	68	75	81	91	—	—
Type*		I		II				III				

Pitch, in	No. of teeth in small sprocket	Small sprocket, r/min										
		100	200	300	400	500	600	700	800	900	1,000	1,200
$1\frac{1}{2}$	21	8.8	16	24	30	36	40	44	46	49	49	—
	25	10	20	29	38	44	50	55	59	61	65	64
	29	13	24	34	44	51	59	65	70	74	75	76
	33	14	28	39	50	59	68	75	80	85	88	89
	37	16	30	44	59	66	76	84	90	96	99	—
	45	19	38	54	68	81	93	101	108	113	—	—
Type*		I		II				III				

Pitch, in	No. of teeth in small sprocket	Small sprocket, r/min								
		100	200	300	400	500	600	700	800	900
2	21	16	29	40	50	53	63	65	—	—
	25	18	35	49	61	70	78	83	85	85
	29	21	41	58	73	84	93	99	103	103
	33	25	46	66	83	96	106	114	118	118
	37	28	53	75	72	110	124	128	131	—
	45	34	64	90	113	131	144	151	—	—
Type*		I		II			III			

*Lubrication types: Type I, manual, brush, or oil cup; Type II, bath or disk; Type III, oil stream.

Table 3.2.18 *Drilling Equipment Service Factor [9]*

1	2	3
	Typical Service Factor	
Compound drives		
For hoisting service	12	16
For pump driving service	16	21
Pump final drive	20	25
Pump countershaft drive[1]	20	25
Drawworks input drive type A[2]	9	12
Drawworks input drive type B[2]	5	7
Drawworks transmission drives[3]	—	—
Drawworks low drum drive	2	3
Drawworks high drum drive	3.5	5
Rotary countershaft drive[4]	5	10
Rotary final drive[4]	5	10
Auxiliary brake drive[5]	—	—

[1] Pump countershaft drives are frequently very short centered, which reduces the heat liberating surface and for high horsepower they may require supplementary cooling.
[2] Drawworks input drives are of two types. The type A drive is located between the prime movers and drawworks transmission. The type B drive is located between the transmission and drawworks. Drawworks input drives have smaller service factors than compound drives, based on experience. Among the reasons for this are the fewer horsepower hours logged in a given period and the fact that they are somewhat removed from the influence of engine impulses.
[3] Service factors on drawworks transmission drives have been omitted from the Recommended Practices for two basic reasons. First, the recommended practices are basically intended for field design guides, and it is not expected that such designs will be made in the field. Second, transmission designs cover a wide range of conditions such as shaft speeds, number of ratios, and methods of obtaining ratios, so that no one family of service factors could be made to apply.
[4] Rotary countershaft and final drives have benefited only slightly from experience because too little has been known about rotary horsepowers through the years.
[5] Auxiliary brake chain drives have been omitted because of the indefinite nature of the load. Each manufacturer has established successful drives, but these are suited to specific conditions which cannot be reduced to simple terms. Most such drives violate the rules of chain speeds and loads and are successful only because of the short duration of use and deviation from calculated loads. It is recommended that the drawworks manufacturer be consulted.

References

1. Kutz, M., *Mechanical Engineers Handbook*, 12th Edition, John Wiley & Sons, New York, 1986.
2. Hindhede, U., et al., *Machine Design Fundamentals*, J. Wiley & Sons, New York, 1983.
3. API Specification 1B, 6th Edition, "Specification for Oil-Field V-Belt," American Petroleum Institute, Washington, D.C., January 1, 1995.
4. *Machinery's Handbook*, 25th Edition, Industrial Press, New York, 1996.
5. Greenwood, D. G., *Mechanical Power Transmissions*, McGraw-Hill, New York, 1962.
6. Faulkner, L. L., and Menkes, S. B., *Chains for Power Transmission and Materials Handling*, Marcel Dekker, New York, 1982.
7. "Precision Power Transmission Roller Chains, Attachments, and Sprockets," ANSI Standard B 29.1M-1993.
8. "Inverted Tooth (Silent) Chains and Sprockets," ANSI Standard B29.2M-1982 (R1999).
9. API Specification 7F, 6th Edition, "Specification for Oil Field Chain and Sprockets," American Petroleum Institute, Washington, D.C., August, 1999.
10. Avallone, E., and Baumeister, T., *Marks' Standard Handbook for Mechanical Engineers*, 10th Edition, McGraw-Hill, New York, 1996.

3.3 PUMPS

Pumps are a mechanical device that forces a fluid to move from one position to another. Usually a pump refers to the mechanical means to move incompressible (or nearly incompressible) fluid or liquid. Pumps are our earliest machine and are to this day one of our most numerous mechanical devices.

Pumps are a very essential part of the oil and gas industry. They are used throughout the industry, from drilling operations through to final delivery to the customer.

3.3.1 Classifications

Pumps are classified into two basic classes, displacement and dynamics. Figures 3.3.1/3.3.2 show the subclasses of pumps under each of these basic classes [1–3].

The most widely used pumps in the oil and gas industry are reciprocating displacement pumps (in particular piston plunger type), the rotary displacement pump, and the centrifugal dynamic pump. Only these pumps will be discussed in detail.

The reciprocating and rotary positive displacement pumps primary characteristic is that they have a nearly direct relationship between the motion of the pumping elements and the quantity of liquid pumped. Thus, in positive displacement pumps liquid displacement (or discharge from the device) is theoretically equal to the swept volume of the pumping element. Figure 3.3.3 shows the typical positive displacement plot of discharge rate Q (ft^3/s) versus pressure P (lbs/ft^2) [3]. The discharge rate remains the same (assuming a constant rate of rotation for the system) regardless of the pressure in the flow. The pressure in the flow is, of course, the result of resistance in the flow system the pump discharges to. If the resistance increases, rotation can be maintained and more force applied to each stroke of the pump (i.e., power). This is why the reciprocating piston plunger pump is also called a power pump. In practice, pressure does have some influence on the capacity of these pumps. This is because as the pressure increases, there is some leakage of the seals in the system. This leakage is

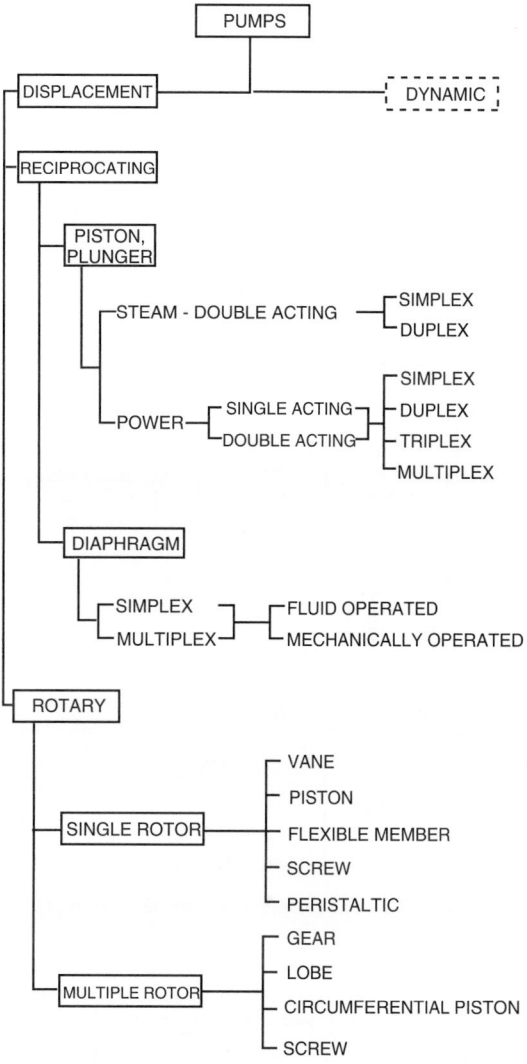

Figure 3.3.1 *Classification of displacement pumps.*

somewhat proportional to the pressure, particularly beyond some characteristic pressure related to the seals. The difference between theoretical flow and the actual flow of a pump is often referred to as slip. This slip is shown in Figure 3.3.3.

In the dynamic pump, in particular, the centrifugal pump, the discharge rate Q is determined by the resistance pressure P in the flow system the pump discharges to (assuming some given speed of the pump). This is illustrated in Figure 3.3.4.

3.3.2 General Calculations

There are several important calculations that are needed to properly evaluate and select the appropriate positive displacement pump [1,4–7].

The theoretical power **P** (hp), which is the power actually important to the fluid by the pump, is

$$\mathbf{P} = \frac{qp}{1,714} \qquad [3.3.1]$$

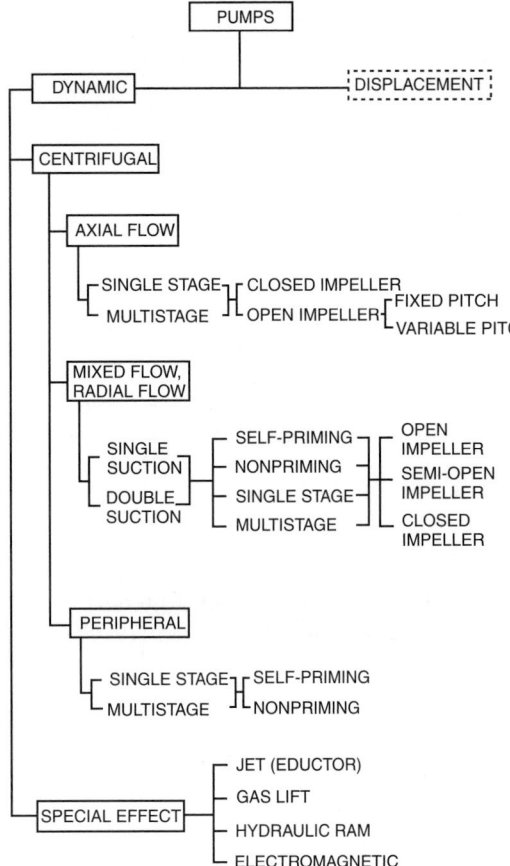

Figure 3.3.2 *Classification of dynamic pumps [1].*

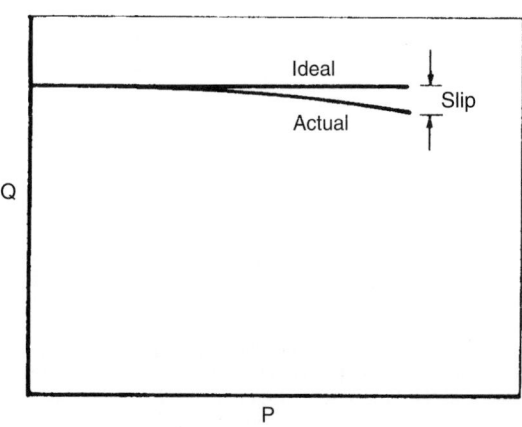

Figure 3.3.3 *Positive displacement pump [1].*

where q is the discharge rate in gpm and p is the differential pressure in psi (usually the actual gauge discharge pressure).

The total pump power input $\mathbf{P}pi$ is the net power delivered to the pump drive shaft by the prime mover system. This is

$$\mathbf{P}pi = \mathbf{P} + \mathbf{P}pl \qquad [3.3.2]$$

where $\mathbf{P}pl$ is the pump power loss.

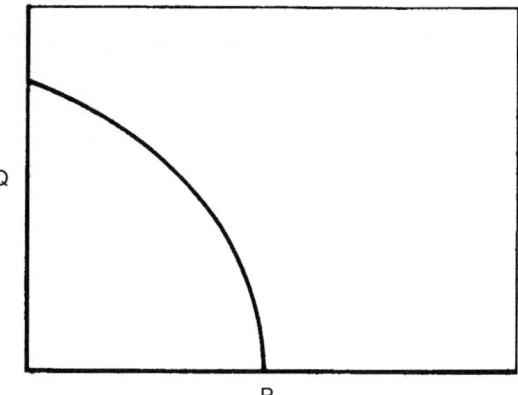

Figure 3.3.4 *Dynamic (centrifugal) pumps [1].*

The total power input $\mathbf{P}ti$ is the net power delivered by the prime mover. This is

$$\mathbf{P}ti = \mathbf{P}pi + \mathbf{P}pml \qquad [3.3.3]$$

where $\mathbf{P}pml$ is the prime mover loss.

Overall efficiency e_0 is

$$e_0 = \frac{\mathbf{P}}{\mathbf{P}ti} \qquad [3.3.4]$$

usually multiplied by 100 and expressed as a percentage.

Pump efficiency e_p is

$$e_p = \frac{\mathbf{P}}{\mathbf{P}pi} \qquad [3.3.5]$$

(usually expressed as a percentage).

Mechanical efficiency e_m (which usually refers to the efficiency of the prime mover to supply power to the pump) is

$$e_m = \frac{\mathbf{P}pi}{\mathbf{P}ti} \qquad [3.3.6]$$

(usually expressed as a percentage).

3.3.3 Reciprocating Pumps

The piston plunger pump is the simplest form of a positive displacement pump. These pumps can be powered by a variety of prime movers, internal combustion engines, and electric motors (and in some cases, powered by a gas turbine motor). In such applications, the separate pump unit is connected to the prime mover by a power transmission.

The capacity of a pump is determined by the number of plungers or pistons and the size of these elements (bore and stroke). A reciprocating pump is usually designed for a specific volumetric rate capacity and pressure capability. These factors are set by the application. Once the volumetric rate capacity and pressure capability are known, a designer can determine the plunger piston bore and stroke the rotation speed range and the power of the prime mover needed to complete the system.

Reciprocating pumps are fabricated in both horizontal and vertical configurations.

3.3.3.1 Single-Acting Pump

A single-acting pump has only one power (and discharge) stroke for its pistons. Such a pump brings fluid into its chamber through the inlet or suction value or the piston is drawn backward to open the chamber. To discharge the fluid, the inlet valve is closed and the outlet valve opened as the piston is forced forward to push the fluid from the chamber into

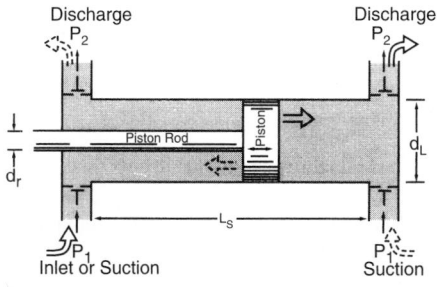

Figure 3.3.5 *Schematic of valve operation of single and double-acting pumps [7].*

the discharge line. The piston motion is accomplished by a rotating crankshaft that is connected to the piston by a piston rod much like an internal combustion piston engine. The rotating crankshaft of the pump is rotated by the rotational power of the prime mover (through a transmission) [7].

The single-action pump is usually available with three, five and even seven pistons. The odd number of pistons allows the pump to be rotationally balanced, and the use of at least three pistons reduces the discharge pulsation of these single-acting pumps. A three piston pump single-action pump is called a triplex pump. A five piston, or seven piston single-acting pump is called a multiplex pump.

3.3.3.2 Double-Acting Pump

Double-acting pumps have two power strokes. As a piston of the pump is pushed forward, the fluid is discharged from the forward chamber into the discharge line (much like a single-action piston). But during this same stroke, the chamber behind the piston (which contains the connecting rod) is being filled via that chamber's inlet valve (Figure 3.3.5). When the forward power stroke is complete and the fluid discharged from the chamber in front of the piston, the chamber behind the piston is filled. The crankshaft continues to rotate, requiring the piston to begin a rearward stroke. During this stroke the fluid behind the piston is forced from its chamber into the discharge line via the outlet valve and the chamber in front of the piston refills via its inlet valve [7].

Double-acting pumps are usually available with one or two pistons.

A one-piston double-action pump is called a *double-acting simplex* (since there are older single-action steam and pneumatic driven simplex pumps).

A two piston double-action pump is called a *duplex pump.*

3.3.3.3 Flow Characteristics

All reciprocating pumps have a pulsating discharge. This is the result of the piston motion as it stops and reverses. At this moment, the flow from that piston theoretically drops to zero. Thus, the discharge curves as a function of time are those illustrated in Figure 3.3.6. By having two or more pistons the pulsation of the discharge from the pump can be smoothed out and the magnitude of the pulsation reduced if the pistons motions are timed for proper dynamic balancing of the pump (Figure 3.3.7). For those pumps that have large pulsations, a cushion change (or accumulator) may be used in the discharge line to reduce or eliminate the pulsations (Figure 3.3.8).

3.3.3.4 Calculations

There are several important calculations that are needed in order to properly evaluate and select the appropriate

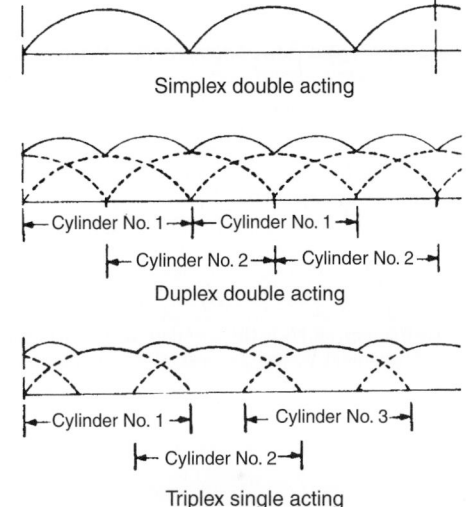

Figure 3.3.6 *Simplex (double-acting), duplex (double-acting), triplex (single-acting), pressure–time curves [7].*

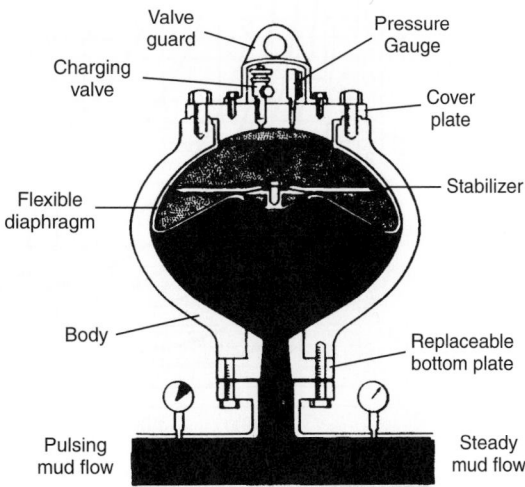

Figure 3.3.7 *Example pulsation dampener [7].*

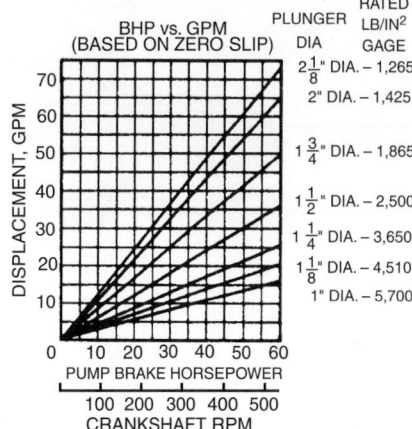

Figure 3.3.8 *Typical power-pump performance (courtesy of Ingersoll-Rand Co.).*

reciprocating piston pump [1]. These calculations are used in conjunction with Equations 3.3.1–3.3.6.

The capacity of a reciprocating piston pump, q (in^3/min), is

$$q = V_d(1 - S)N \qquad [3.3.7]$$

where V_d is the displacement of the pumps piston in in.^{3}, S is the slip fraction, and N is the speed of the pump in rpm. The displacement V_d (gals) of a single-acting pump is

$$V_d = \frac{nA_pL_s}{231} \qquad [3.3.8]$$

where n is the number of pistons, A_p is the cross-sectioned area of the piston in in.^{2}, and L_s is the length of the piston stroke in in.

The displacement V_d of a double-acting pump is

$$V_d = \frac{n(2A_p - a)L_s}{231} \qquad [3.3.9]$$

where a is the cross-sectioned area of the piston rod in in.^{2}.

The pressure p used in Equation 3.3.1 is the differential developed pressure (across the pump inlet and outlet). Since the inlet suction pressure is usually small compared to the discharge pressure, the discharge pressure is used. Thus, this is the application resistance pressure in most cases. Figure 3.3.8 shows a typical reciprocating pump performance.

The slip S is the fraction of suction capacity loss. It consists of the volumetric efficiency loss fraction l_e, the stuffing box loss fraction l_s, and the valve loss fraction l_v. The slip S is

$$S = l_e + l_b + l_v \qquad [3.3.10]$$

The volumetric efficiency loss l_e is

$$l_e = 1 - e_v \qquad [3.3.11]$$

where e_v is the volumetric efficiency. The volumetric efficiency is the ratio of discharge volume to suction volume and is expressed as a percentage. It is proportional to the ratio r and the developed pressure (Figure 3.3.9). The ratio r is the ratio of internal volume of fluid between valves when the piston is at the top of its back stroke to the piston displacement (Figure 3.3.10). Because volume cannot readily be measured at discharge pressure, it is taken at suction pressure. This will result in a higher e_v due to fluid compressibility which is neglected.

The stuffing box loss l_b is usually negligible.

The valve loss l_v is the loss due to the flow of fluid back through the valve during closing. This is of the order of 0.02 to 0.10 depending on the valve design.

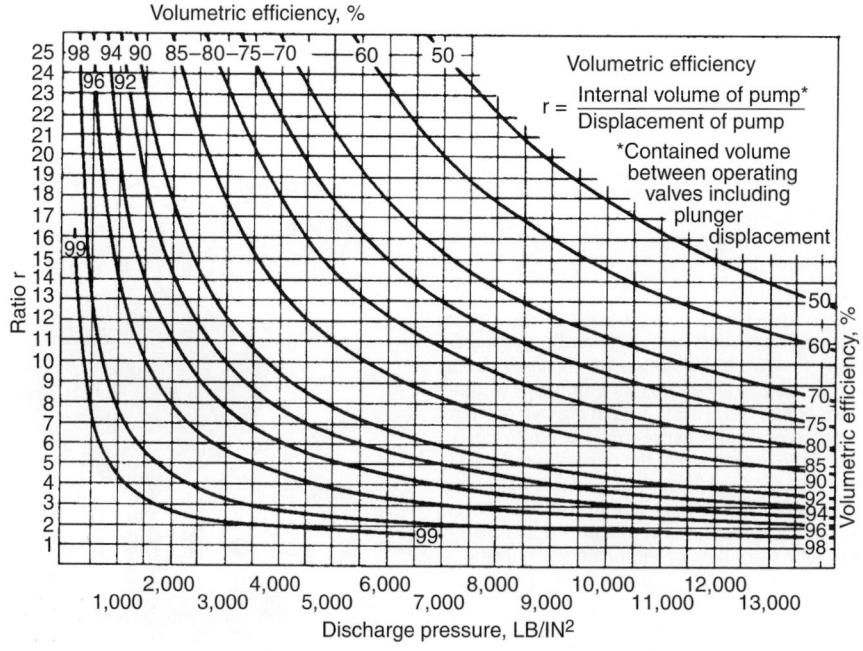

Figure 3.3.9 *Volumetric efficiency (courtesy of Ingersoll-Rand Co.).*

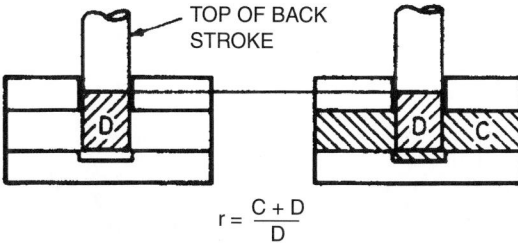

$$r = \frac{C + D}{D}$$

Figure 3.3.10 *Ratio r (courtesy Ingersoll-Rand Co.).*

Table 3.3.1 *Effects of Speed and Pressure on Mechanical Efficiency [1,7,8]*

Constant Speed	
% full-load developed pressure	Mechanical efficiency
20	82
40	88
60	90.5
80	92
100	92.5

Constant Developed Pressure	
% speed	Mechanical efficiency
44	93.3
50	92.5
73	92.5
100	92.5

The mechanical efficiency e_m of a power pump at full load of pressure and speed is 0.90 to 0.95. When the pump has a single built in gear as part of the power frame, the mechanical efficiency is 0.80 to 0.95. Table 3.3.1 shows the typical effects of speed and pressure on the mechanical efficiency of a power pump [1].

3.3.4 Rotary Pumps

Another important positive displacement pump is the rotary pump. This type of pump is usually of rather simple construction, having no valves and being lightweight. Such pumps can be constructed to handle small fluid capacities (i.e., less than a gal/min) to very large fluid capacities (i.e., 5,000 gal/min or greater). Rotary pumps are designed to operate at 1,000 psi discharge pressure, but the normal rotary pump design is for pressure of 25 to about 500 psi with mechanical efficiencies of 0.80 to 0.85.

Rotary pumps are classified in two basic groups and several subgroups:

Single-rotor pumps
  vane
  piston
  flexible vane
  screw
Multiple-rotor pumps
  lobe
  gear
  circumferential piston
  screw

These pumps require the maintenance of very close clearances between rubbing surfaces for continual volumetric efficiency. Some of the important pumps are discussed.

In general, rotary pumps with discharge pressure of up to 100 psi are considered low-pressure pumps. Rotary pumps with pressure between 100 and 500 psi are considered moderate-pressure pumps. Pumps with pressure beyond 500 psi are considered to be high-pressure pumps.

Rotary pumps with volume capacities up to 50 gal/min are considered to be small-volume capacity pumps. Pumps with volume capacities from 50 to 500 gal/min are moderate-volume capacity pumps. And pumps with volume capacities beyond 500 gal/min are large-volume capacity pumps.

3.3.4.1 Gear Pumps

Gear pumps are rotary pumps in which two or more gears mesh to provide the pumping action. Usually one of the gears is capable of driving the other(s). The simplest form of this type of rotary pump is the gerotor pump. Figure 3.3.11 shows a typical gerotor pump configuration. These pumps are often used for small-volume capacity applications where space is quite limited [4]. Such gear pumps can be used on a rotating shaft such as a crankshaft to provide oil lubrication of critical moving machine parts.

Figures 3.3.12 shows an external gear pump. This is an example of a double- (multiple) rotor pump. Such a gear pump has a drive gear and a driven gear that are encased in housing with minimum clearance between the housing and the tips of the gears. The simplest type of this pump uses spur gears. The large number of gear teeth in contact with the casing minimizes leakage around the periphery. The spur-gear type of pump is limited by its characteristics of trapping liquid. This occurs on the discharge side at the point of gear intermesh. This results in a noisy operation and low mechanical efficiency, particularly at high rotary speeds. Gear pumps can also be constructed with single or double-helical gears with 15–30° angles (0.26–0.52 rad). Such a helical (or even a herringbone) gear construction will nearly eliminate the problems of liquid trapping but increase the leakage.

One-Tooth Difference. These pumps are commonly known as gerotor pumps (see Figure 3.3.11). An outlet gear is mounted eccentrically with the outlet casing actuated by an internal gear rotating under the action of the central shaft (keyway connected). The internal gear has one less tooth than the outside ring gear. There is clearance between the outside impeller gear and its casing and the inner gear fixed to the shaft. As the shaft rotates, the inner gear forces the

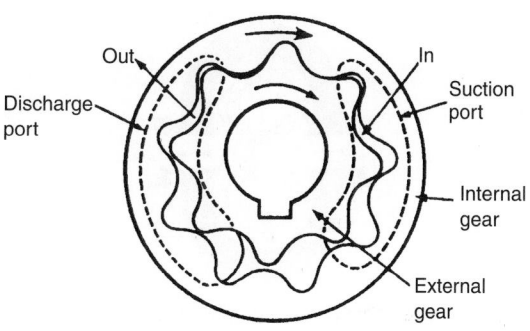

Figure 3.3.11 *Gerotor type pump [1].*

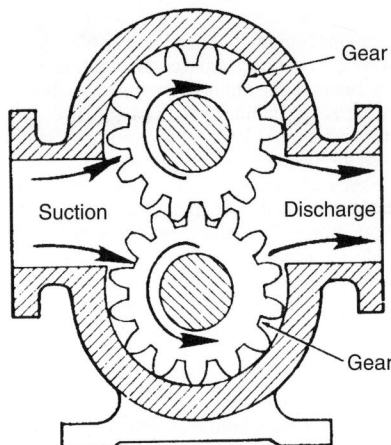

Figure 3.3.12   *Gear pump (double rotor type) [1].*

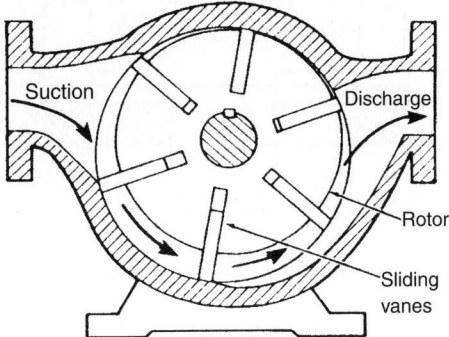

Figure 3.3.14   *Sliding vane pump (single rotor) [1].*

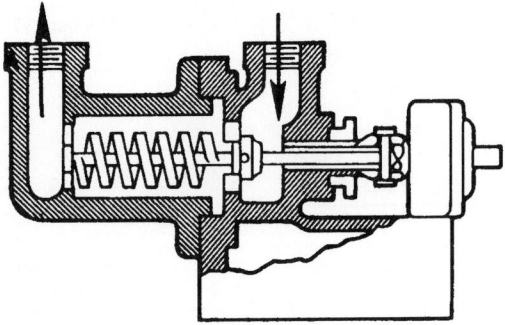

Figure 3.3.15   *Single screen pump (single rotor) [1].*

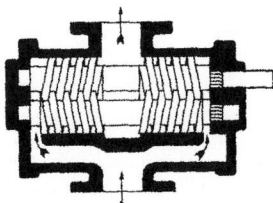

Figure 3.3.16   *Two-screw pump (double rotor) [1].*

outer gear to progress around the shaft at a rate slower than the shaft rotation. The liquid trapped between the two gears is forced from the space between the two gears as the rotation takes place.

Two-Teeth Difference. In this type of pump an abutment on one side plate is used to fill the clearance between the external and internal gear. This construction reduces leakage but involves the use of an overhung internal gear. Such a gear arrangement limits the application of these pumps to small and moderate-volume capacity pumps.

3.3.4.2 Circumferential Piston
Figure 3.3.13 shows the typical cross-section of this type of pump. In this pump liquid is pumped by the action of the rotation of the two eccentrically located piston surfaces. There is no contact between the piston surfaces.

3.3.4.3 Vane Pump
This example of a single-rotor type pump makes use of an eccentric shaft on which are mounted several sliding vanes (Figure 3.3.14). The vanes are forced against the bore of the housing by springs and the centrifugal force of the rotor rotation. The vanes are usually made of materials that will not damage the surface of the bore (e.g., bronze and bakelite). This type of pump is useful for small and moderate-volume capacities and low pressure. This is due to the rather low speeds such pumps must be operated at. High speeds result in rapid wear of the vanes.

Figure 3.3.13   *Circumferential-piston pump (double rotor) [1].*

3.3.4.4 Screw Pumps
Screw pumps are available in single or double-rotor (multiple) types.

Figure 3.3.15 shows the cross-section of the typical single-rotor screw pump. In this pump there is a single helical threaded rotating element. As the screw rotates the liquid progresses axially down the pump. This pump produces continuous flow with relatively little pulsation or agitation of the fluid. Screw pumps of this type are quiet and efficient. They are available with high-speed, high-pressure, and large-volume capacities.

Figure 3.3.16 shows the cross-section of the typical multiple-rotor screw pump. This screw pump incorporates right-hand and left-hand intermeshing helices on parallel shafts with timing gears. Such pumps are available with high-speed, high-pressure, and large-volume capacities.

For all screw pumps, flow is continuous.

3.3.4.5 Calculations
There are several important calculations which are needed to properly evaluate and select the appropriate rotary pump [1].

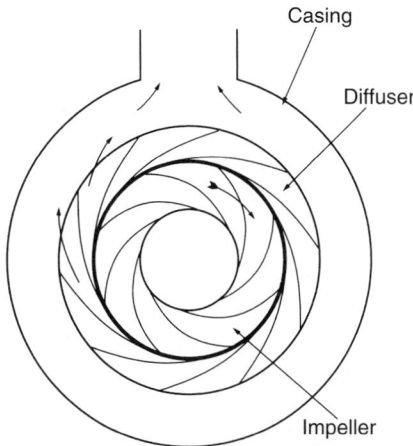

Figure 3.3.17 *Typical diffuser-type pump [1].*

These calculations are used in conjunction with Equations 3.3.1–3.3.6.

The capacity of a rotary pump operating with zero slip is called displacement capacity, q_d (gpm). Thus, the actual capacity of a rotary pump q (gpm) is

$$q = q_d - q_1 \qquad [3.3.12]$$

or

$$q = \frac{V_d N}{231} - q_1 \qquad [3.3.13]$$

where V_d is the pump displacement per revolution in in.^{3}, N is the pump speed in rpm, and q_1 is the pump loss in gpm.

Pump horsepower P is the power actually imparted to the fluid by the rotary pump in Equation 3.3.1.

The definitions for the various power terms and efficiencies for the rotary pumps are the same as those discussed above for the reciprocating pump; namely, Equations 3.3.2–3.3.6.

The volumetric efficiency e_v for the rotary pump is

$$e_v = \frac{q}{q_d} \qquad [3.3.14]$$

or

$$e_v = 1 - \frac{231 q_e}{V_d N} \qquad [3.3.15]$$

(usually expressed as a percentage).

3.3.5 Centrifugal Pumps

The centrifugal pump is the most important pump of the dynamic class of pumps for the oil and gas industry. Other pumps in this class are covered in other references [1,4].

In its simplest form, a centrifugal pump consists of a rotating impeller (with radical vanes) rotating at a rather high speed. The rotating impeller is encased in a rigid housing that directs the liquid within the pump (see Figure 3.3.17) [1]. Liquid is supplied to the inlet that feeds the liquid to the center section of the rotating impeller. The rotational motion of the impeller forces the liquid, via the centrifugal forces, to move radically outward with the aid of the stationary diffuser. The rigid housing around the impeller guides the

high-velocity fluid around the inside of the housing and out of the outlet of the pump.

The capacity of this type of pump depends on the pressure head against which the pump must act (see Figure 3.3.4).

When the liquid within the impeller is forced radially outward to the diffuser, a major portion of the velocity energy is converted into pressure energy by the stationary diffuser vanes (see Figure 3.3.17). This can also be accomplished by means of a volute, which is a part of the casing design (Figure 3.3.18) [1].

3.3.5.1 Centrifugal Pump Classifications

Centrifugal pumps with diffusion vanes are called *diffusion pumps* or, more recently, *vertical turbine pumps*. Those pumps with volute casings are called *volute pumps*.

Centrifugal pumps can also be classified by the design of the impeller. Centrifugal pumps may have radial-flow impellers, axial-flow impellers, and mixed-flow impellers (both radial-flow and axial-flow).

Pump impellers are further classified as to the inlet flow arrangement such as single suction (which has a single inlet on one side) and double suction (which has a double inlet on each side of the impeller).

Impellers can be further classified with regard to their physical design: a closed impeller has shrouds or sidewalls enclosing the fluid flow, an open impeller has no shrouds or sidewalls, and a semiopen impeller is a mix of the closed and open design.

Another centrifugal pump classification is whether the pump is a single-stage pump (the pressure head is developed by a single impeller) or a multistage pump (the pressure head is developed by two or more impellers).

Centrifugal pumps can be further classified by physical design or axially split, radially split and whether the axis of rotation of the impeller(s) is vertical or horizontal. Horizontal pumps can be classified according as end suction, side suction, bottom suction, and top suction.

In applications, centrifugal pumps can be supplied with liquid via piping, or the pump may be submerged. Vertical pumps are called dry-pit or wet-pit types. The wet-pit pump (submerged) discharges up through a pipe system to some point above the pump.

3.3.5.2 Calculations

There are important calculations that are needed to properly evaluate and select the appropriate centrifugal pump [6].

There are several similar relationships for centrifugal pumps that can be used if the effects of viscosity of the pumped fluid can be neglected. These relate the operating performance of any centrifugal pump for one set of operating conditions to the pump at another set of operating conditions, say conditions, and conditions 2.

The volumetric flow rate q (gpm) is related to rate q_2 (gpm) and the impeller speeds N_1 (rpm) and N_2 (rpm) by

$$\frac{q_1}{q_2} = \frac{N_1}{N_2} \qquad [3.3.16]$$

The pressure head H_1 (ft) related to the head H_2 (ft) and impeller speeds N_1 and N_2 by

$$\frac{H_1}{H_2} = \left(\frac{N_1}{N_2} \right)^{1/2} \qquad [3.3.17]$$

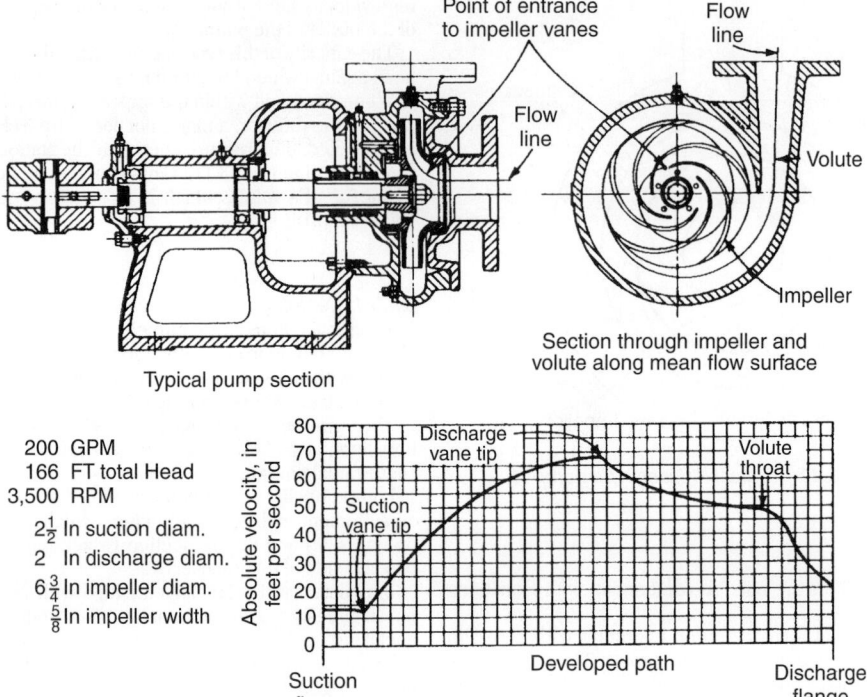

Figure 3.3.18 *Typical volute-type pump [1].*

The pump input power P_1 (Lp) is related to the pump input power P_2 (hp) and impeller speed N_1 and N_2 by

$$\frac{P_1}{P_2} = \left(\frac{N_1}{N_2}\right)^3 \qquad [3.3.18]$$

For a constant impeller speed, the relationship between q_1, q_2, and impeller diameter D_1 (ft) and D_2 (ft) is

$$\frac{q_1}{q_2} = \frac{D_1}{D_2} \qquad [3.3.19]$$

The heads are related by

$$\frac{H_1}{H_2} = \left(\frac{D_1}{D_2}\right)^2 \qquad [3.3.20]$$

and the power related by

$$\frac{P_1}{P_2} = \left(\frac{D_1}{D_2}\right)^3 \qquad [3.3.21]$$

The centrifugal pump specific speed N_{ds} (rpm) (or discharge specific speed) is

$$N_{ds} = \frac{Nq^{0.5}}{H^{0.75}} \qquad [3.3.22]$$

The centrifugal pump suction specific speed N_{ss} (rpm) is

$$N_{ss} = \frac{Nq^{0.5}}{H_{nps}^{0.75}} \qquad [3.3.23]$$

where H_{nps} is the net positive suction head (ft).

Figure 3.3.19 gives the upper limits of specific speeds of single-stage, single- and double-suction centrifugal pumps handling clear water at 85°F at sea level [8].

Figure 3.3.20 gives the upper limits of specific speeds of single-suction mixed-flow and axial-flow centrifugal pumps handling clear water at 85°F at sea level [8].

Figure 3.3.21 gives the approximate relative impeller shapes and efficiency variations for various specific speeds of centrifugal pumps.

Table 3.3.2 gives the specific speeds for various centrifugal pump types. Table 3.3.3 gives the suction specific speed ratings for single-suction and double-suction centrifugal pumps. These tables are for pumps handling clear water.

References

1. Karassik, I. J., et al., *Pump Handbook*, 3rd Edition, McGraw-Hill, New York, 2000.
2. Davidson, J. E., and Von Bertele, O., *Process Pumps Selection: A Systems Approach,* Mechanical Engineering Publications; 2nd Edition, London, 1999.
3. Lobanoff, V. S., and Ross, R. R., *Centrifugal Pumps: Design and Application*, Gulf Professional Publishing; 2nd Edition, 1992.
4. Matley, J., *Fluid Movers; Pump Compressors, Fans and Blowers*, McGraw-Hill, New York, 1979.
5. Gatlin, C., *Petroleum Engineering: Drilling and Well Completions*, Prentice-Hall, Englewood Cliffs, 1960.
6. Hicks, T. G., *Standard Handbook of Engineering Calculations*, 2nd Edition, McGraw-Hill, New York, 1985.
7. Bourgoyne, A. T., et al., *Applied Drilling Engineering*, SPE, 1986.
8. *Hydraulics Institute Standards for Centrifugal, Rotary, and Reciprocating Pumps*, 14th Edition, 1983.

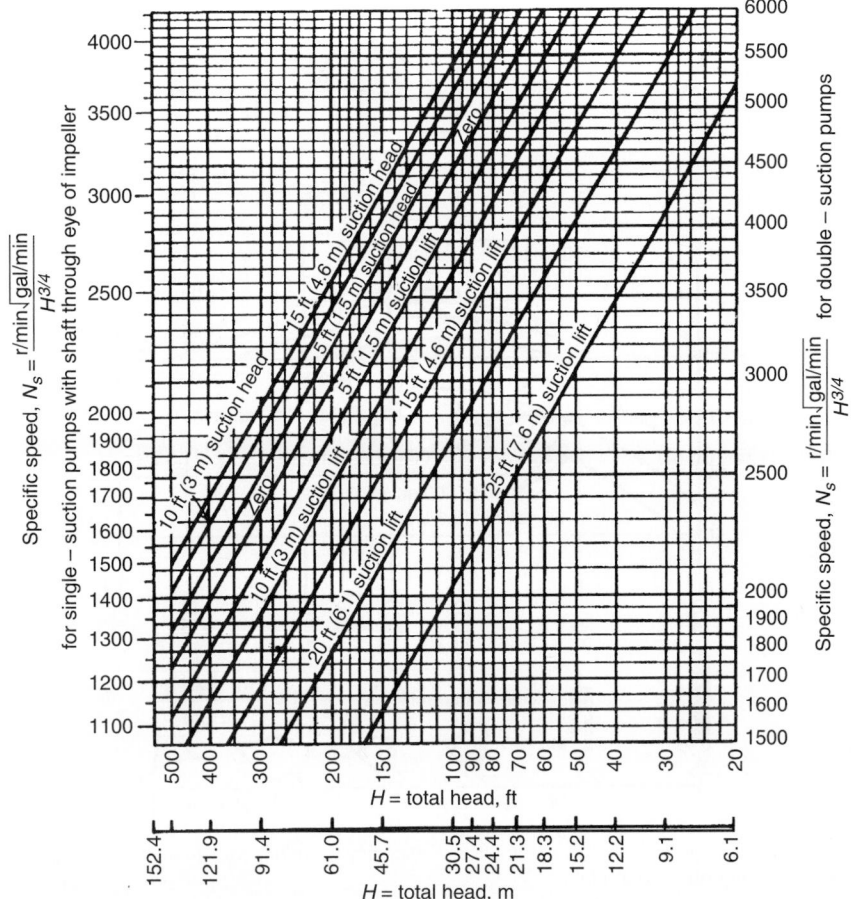

Figure 3.3.19 *Upper limits of specific speeds of single-stage, single and double-suction centrifugal pumps handling clear water at 85°F at sea level (courtesy of Hydraulic Institute).*

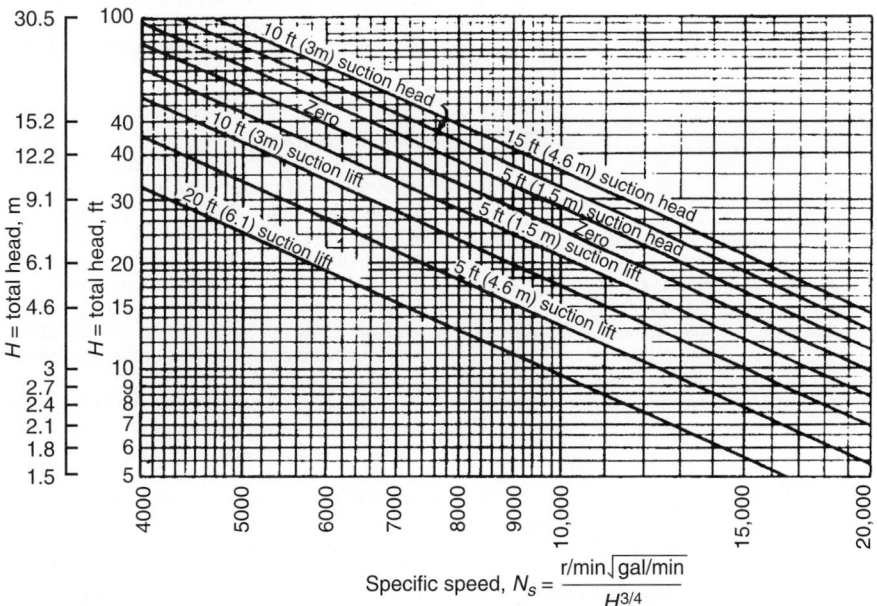

Figure 3.3.20 *Upper limits of specific speeds of single-suction, mixed flow and axial-flow pumps handling clear water at 85°F at sea level (courtesy Hydraulic Institute).*

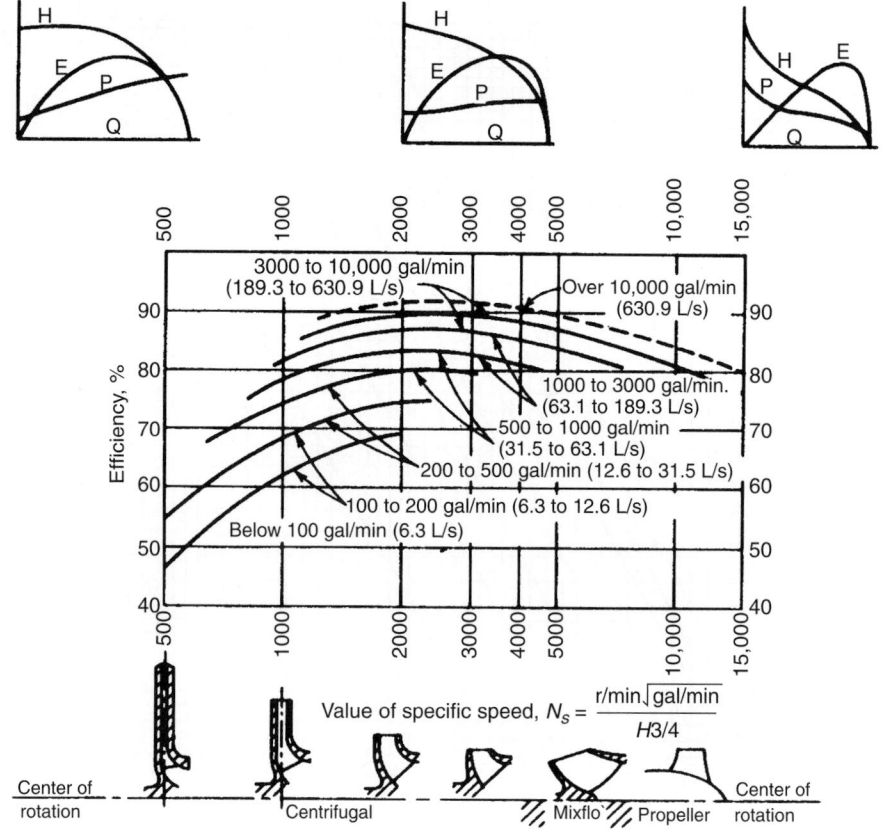

Figure 3.3.21 *Approximate relative impeller shapes and efficiency variations for various specific speeds of centrifugal pumps (courtesy Worthington Corporation).*

Table 3.3.2 *Pump Types Listed by Specific Speed*

Specific speed range	Type of pump
Below 2,000	Volute, diffuser
2,000–5,000	Turbine
4,000–10,000	Mixed-flow
9,000–15,000	Axial-flow

Table 3.3.3 *Suction Specific-Speed Ratings*

Single-suction pump	Double-suction pump	Rating
Below Above 11,000	Above 14,000	Excellent
9,000–11,000	11,000–14,000	Good
7,000–9,000	9,000–11,000	Average
5,000–7,000	7,000–9,000	Poor
Below 5,000	Below 7,000	Very poor

3.4 COMPRESSORS

Compressors of various designs and manufacturers are used in many operations throughout the oil and gas industry. Compressors are used in some drilling operations, in many production operations, and extensively used in surface transportation of oil and gas via pipelines.

Air or gas compressors are very similar in design and operation to liquid pumps discussed earlier. The air and gas compressor is a mover of compressed fluids; the pumps are movers of basically incompressible fluids (i.e., liquids).

3.4.1 Classifications

In much the same manner as pumps, compressors are classified as one of two general classes: positive displacement or dynamic (Figure 3.4.1) [1]. These two general classes of compressors are the same as that for pumps. The positive displacement class of compressors is an intermittent flow device, which is usually a reciprocating piston compressor or a rotary compressor (e.g., sliding vane, screws). The dynamic class of compressor is a continuous flow device, which is usually an axial-flow or centrifugal compressor (or mix of the two).

Each of the two general classes of compressors and their subclass types have certain advantages and disadvantages regarding their respective volumetric flow capabilities and the pressure ratios they can obtain. Figure 3.4.2 shows the typical application range in volumetric flow rate (actual cfm) and pressure ratio several of the most important compressor types can obtain [1].

In general, positive displacement compressors are best suited for handling high-pressure ratios (i.e., about 200), but with only moderate volumetric flow rates (i.e., up to about

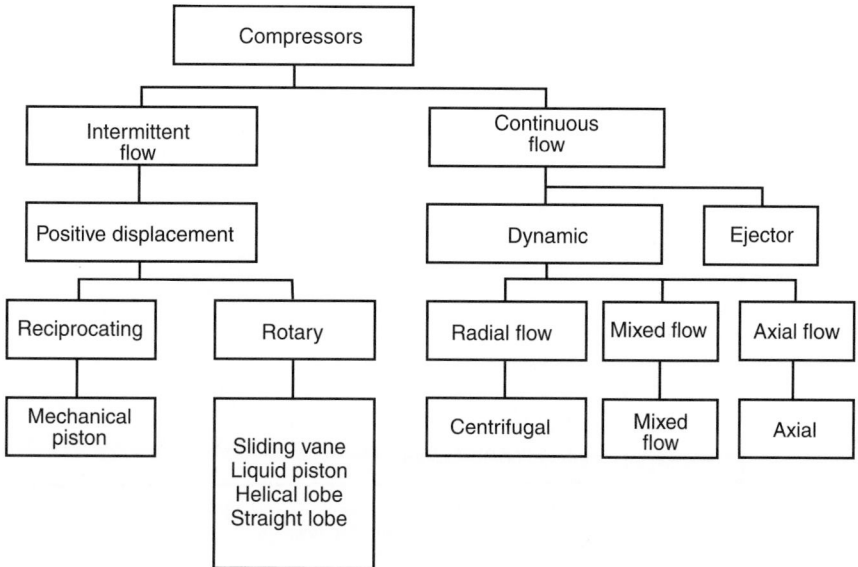

Figure 3.4.1 *Classification of compressor [1].*

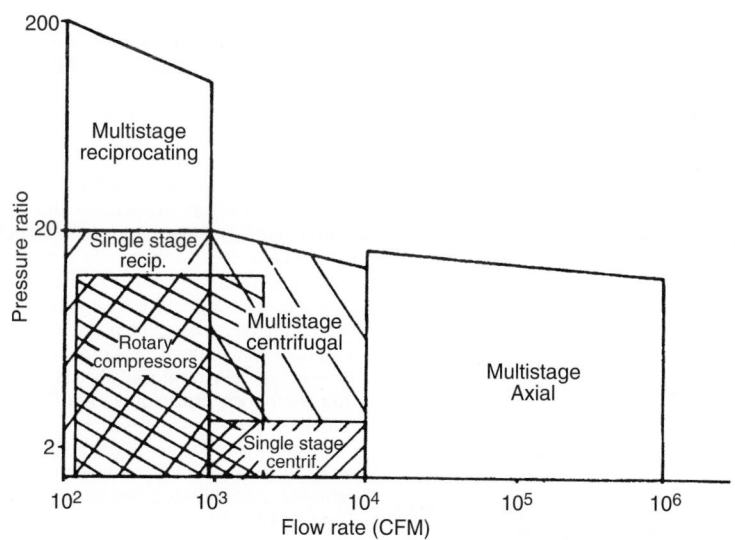

Figure 3.4.2 *Typical application ranges of various compressor types [1].*

10^3 actual cfm). Dynamic compressors are best suited for handling large volumetric flow rates (i.e., up to about 10^6 actual cfm), but with only moderate pressure ratios (i.e., about 20).

This is not the complete view of these two compressor classes. Figure 3.4.3 gives the general performance curves for various positive displacement and dynamic compressors. The positive displacement compressors, particularly the multistage reciprocating compressors, are very insensitive to pressure ratio changes. These compressors will produce their rated volumetric flow rate even when the pressure ratio approaches the design limit of the machine. This is less so for the rotary compressor. The dynamic compressors, however, are quite sensitive to pressure ratio changes. The

volumetric rate of flow will change drastically with changes in the pressure ratio around the pressure ratio the machine has been designed.

Thus, the positive displacement compressors are normally applied to industrial operations where the volumetric flow rate is critical and the pressure ratio is a variable. The dynamic compressors are generally applied to industrial operations where both the volumetric flow rate and pressure ratio requirements are relatively constant.

In general, only the reciprocating compressor allows for reliable flexibility in applying variable volumetric flow rate and variable pressure ratio in an operation. The rotary compressor does not allow for variation in either (except that of pressure through the decompression of the air or gas if the

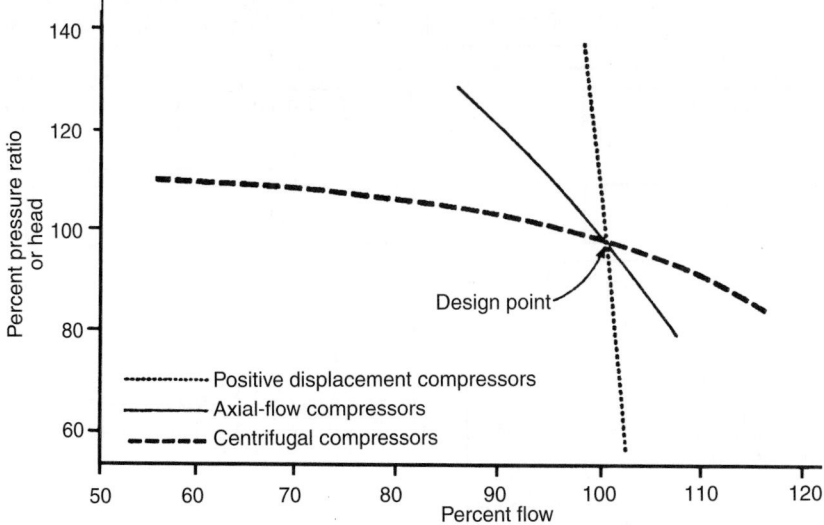

Figure 3.4.3   *General performance curve for various compressor types [1].*

system back pressure is below the design pressure of the machine). The dynamic compressors are designed for specific volumetric flow rates and pressure ratios and are not very useful when these design limits are altered. The positive displacement class of compressors is the most useful group for the oil and gas exploration, drilling and completion and production industry. This section will focus on this class of compressors.

3.4.2 Standard Units

In the United States, the unit of air or any gas if referenced to the standard cubic foot of dry air. The API Mechanical Equipment Standards standard atmospheric conditions for dry air is fixed at a temperature of 60°F (which is 459.67 + 60 = 519.67°R) and a pressure of 14.696 psia (760 mm, or 29.92 in. Hg) [2]. The equation of state for the perfect gas can be written as

$$\frac{P}{\gamma} = \frac{R_u T}{m_w} \qquad [3.4.1]$$

where P is the pressure in lb/ft² absolute, γ is the specific weight in lb/ft³, R_u is the universal gas constant in 1,545.4 ft-lb/lb-mole-°R, T is the temperature in °R, and m_w is the molecular weight of the gas in lb/lb-mole.

Thus, the specific weight γ or the weight of 1 ft³ of dry air will be

$$\gamma = \frac{14.696 \,(144)\,(28.96)}{1,545.4 \,(519.67)} = 0.0763 \text{ lb/ft}^3$$

where for dry air m_w = 28.96 lb/lb-mole. Thus, a dry cubic foot of air at the API Mechanical Equipment Standards standard atmosphere condition weighs 0.0763 lb and has a specific gravity of 1.000.

An alternate calculation method for obtaining the properties of gas is an expression commonly known as the gas engineering equation:

$$\frac{P}{\gamma} = \frac{R T}{S_g} \qquad [3.4.2]$$

where P is the pressure in lb/ft² absolute (or in absolute N/m²), γ is the specific weight in lb/ft³ (or in N/m³), **R** is

the engineering gas constant 53.36 lb-ft/lb-°R (or 29.28 N-m/N-K), T is the temperature in °R (or in °K), and S_g is the specific gravity of the gas (for API standard conditions air S_g = 1.0).

There are other organizations within the United States and regions around the world that have established different standards. The ASME standard atmosphere is at a temperature of 68°F, a pressure of 14.7 psia, and a relative humidity of 36%. The British use a standard atmosphere with a temperature of 60°F and a pressure of 30.00 in.Hg. The Europeans use a standard atmosphere with a temperature of 15°C (59°F) and pressure of 750 mmHg (14.5 psia) [3,4].

When selecting and sizing compressors, care should be taken in determining which standard has been used to rate a compressor under consideration, particularly if the compressor has been produced abroad. All further discussions in this section will utilize only the API Mechanical Equipment Standards standard atmosphere.

Compressors are rated as to their maximum input volumetric flow rate, and their maximum pressure output. These ratings are usually specified as *standard cubic feet per minute* (scfm) and psig (or standard cubic meters per minute and N/m² gage). The scfm of volumetric flow rate refers to the compressor intake. The pressure rating refers to the output pressure capability.

3.4.3 General Calculations

There are several important calculations needed to properly evaluate and select the appropriate compressors [1,3,4].

Compressors have nearly equal ratios of compression in each of the stages. Thus, the total pressure ratio r_t across the compressor (i.e., input pressure to output pressure prior to after cooling) is

$$r_t = \frac{p_{out}}{p_{in}} \qquad [3.4.3]$$

where p_{out} is the output pressure in psia, p_{in} is the input pressure in psia, and the pressure ratio for each stage r_s is

$$r_s = r_t^{1/n} \qquad [3.4.4]$$

where n is the number of equal compression stages in the compressor.

Thus, for a four-state compressor, if $p_1 = p_{in}$, then p_2 would be the pressure exiting the first stage compression. The pressure p_2 would be

$$p_2 = r_s p_1 \qquad [3.4.5]$$

The pressure entering the second stage would be p_2 and the pressure exiting would be p_3, which is

$$p_3 = r_s p_2 \qquad [3.4.6]$$

or

$$p_3 = r_s^2 p_1 \qquad [3.4.7]$$

The pressure entering the third stage would be p_3 and the pressure exiting would be p_4, which is

$$p_4 = r_s p_3 \qquad [3.4.8]$$

or

$$p_4 = r_s^3 p_1 \qquad [3.4.9]$$

The pressure entering the fourth stage would be p_4 and the pressure exiting would be p_5, which is

$$p_5 = r_s p_4 \qquad [3.4.10]$$

or

$$p_5 = r_s^4 p_1 \qquad [3.4.11]$$

The above pressure calculations assume intercooling between the first and second stages of compression, between the second and third stages of compression, and between the third and fourth stages of compression. The intercooling system in a multistage compressor ideally reduced the temperature of the gas leaving first stage (or the other stages) to ambient temperature, or at least the input temperature. This cooling of the gas moving from one stage to another is necessary for an efficient and economical design of the compressors. Further, if the gas moving to progressive stages were to get too hot, the machine could be severely damaged. This intercooling is normally accomplished using either a water-jacket, oil-jacket, or air-cooled finned pipes between the stages.

Assuming efficient intercooling between stages, then $T_1 = T_2 = T_3$. The temperature T_4 is

$$T_4 = T_1 \frac{P_4^{\frac{k-1}{k}}}{P_3} \qquad [3.4.12]$$

where k is ratio of specific heats.

Flow rate is often given in field units (e.g., g(cfm)). These field units are usually converted to consistent units $Q(\mathrm{ft}^3/\mathrm{s})$ (or in m^3/s).

The temperature T_4 at the exit of the compressor is cooled with an adjustable cooling system called an after-cooler. Such a system is useful in adjusting the output air for specific application purposes. The after cooling process is assumed to be a constant pressure process. The volumetric flow rate from the fourth stage is Q_4 $(\mathrm{ft}^3/\mathrm{s})$, then the volumetric flow rate after the flow or passed through the after cooler Q_4' $(\mathrm{ft}^3/\mathrm{s})$ will be

$$Q_4' = Q_4 \frac{T_4}{T_4'} \qquad [3.4.13]$$

where T_4' is the final temperature the after-cooler is to cool the output (°R).

If the compressor has a capability to compress a volumetric flow rate of Q_{in} $(\mathrm{ft}^3/\mathrm{s})$ then for the multistage compressor

$$Q_1 = Q_{in} \qquad [3.4.14]$$

This compressor design capability is usually stated in scfm, which means that this is the capability of the compressor

Table 3.4.1 *Atmosphere at Elevation (Mid-Latitudes) above Sea Level*

Surface Location Above Sea Level (ft)	Pressure (psi)	Temperature (°F)	Specific Weight (lb/ft³)
0	14.696	60.00	0.0763
2,000	13.662	51.87	0.0721
4,000	12.685	44.74	0.0679
6,000	11.769	37.60	0.0639
8,000	10.911	30.47	0.0601
10,000	10.108	23.36	0.0565

when at a sea level location. If the compressor is moved to a higher elevation, then the compressor will have to be derated and the scfm capability of the machine reduced, or the actual cfm term used to describe capability of the compressor. This derating can be carried using Table 3.4.1.

If a compressor is rated as having a volumetric flow capability of q_{in} (scfm) and is to be used at surface location of 6,000 ft, then the compressor still has the capability of its rated volumetric flow rate, but it is now written as q_{in} (actual cfm). Thus the actual weight rate of flow through the compressor w (lb/s), will be (see Table 3.4.1)

$$\dot{w} = \frac{0.0639 q_{in}}{60} \qquad [3.4.15]$$

The equivalent derated standard volumetric flow rate q_{indr} (scfm) is

$$q_{indr} = \frac{\dot{w}}{0.0763} (60) \qquad [3.4.16]$$

The theoretical power $\dot{W}_s$ (hp), which is actual power imparted to the gas by the multistage compressor shaft, is approximated as

$$\dot{W}_s = \frac{n_s k}{k-1} \frac{\dot{w}_g}{550} \left(\frac{R T_{in}}{S_g} \right) \left[r_t^{\frac{k-1}{n_s k}} - 1 \right] \qquad [3.4.17]$$

where T_{in} is the actual average temperature of the input air (or gas) (°R).

Mechanical efficiency e_m (which usually refers to the efficiency of the prime mover to supply power to the compressor) is

$$e_m = \frac{\dot{W}_s}{\dot{W}_{as}} \qquad [3.4.18]$$

where $\dot{W}_{as}$ (hp) is the actual shaft power required.

For the SI system, the theoretical power W_s (watts), is approximated by

$$\dot{W}_s = \frac{n_s k}{k-1} \dot{w}_g \left(\frac{R T_{in}}{S_g} \right) \left[r_t^{\frac{k-1}{n_s k}} - 1 \right] \qquad [3.4.19]$$

In the preceding equation,

$$r_t = \frac{P_{out}}{P_{in}}$$

and pressure is in the units of $\mathrm{N/m}^2$ absolute.

Mechanical efficiency, e_m, is

$$e_m = \frac{\dot{W}_s}{\dot{W}_{as}} \qquad [3.4.20]$$

where W_{as} (watts) is the actual shaft power required.

3.4.4 Reciprocating Compressors

The reciprocating compressor is the simplest example of the positive displacement class of compressors. This type of compressor is also the oldest. Like reciprocating pumps,

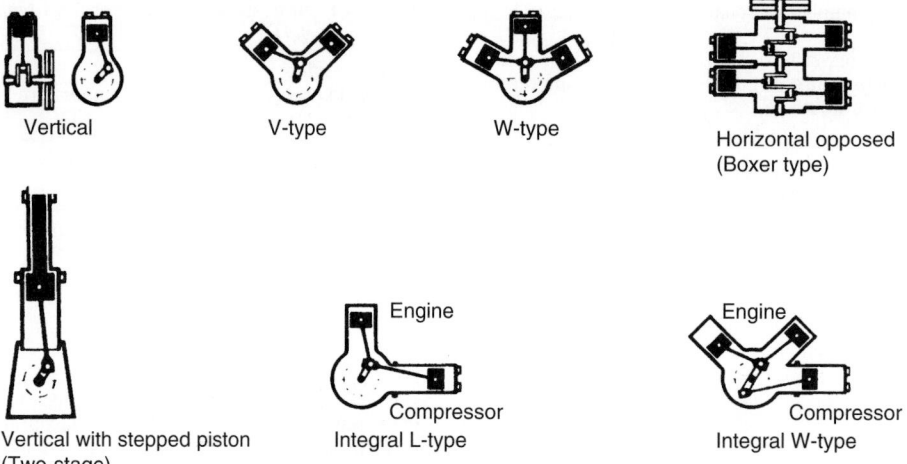

Figure 3.4.4   *Single-acting (trunk-type) reciprocating piston compressor [1].*

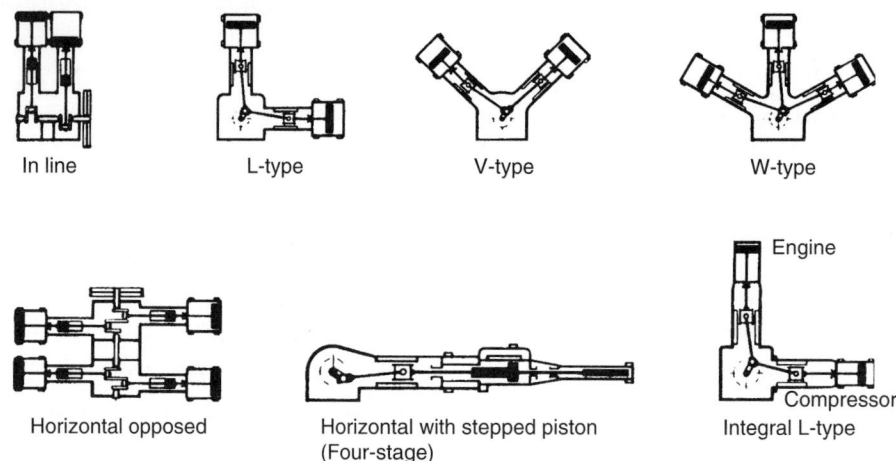

Figure 3.4.5   *Double-acting (crosshead-type) reciprocating piston compressor [5].*

reciprocating compressors can also be either single acting or double acting. Single-acting compressors are usually of the trunk type (Figure 3.4.4). Double-acting compressors are usually of the crosshead type (Figure 3.4.5) [4].

Reciprocating compressors are available in both lubricated and nonlubricated versions. The lubricated versions provide lubrication for the piston, moving pistons either through an oil lubricated intake air or gas stream, or via an oil pump and injection of oil to the piston sleeve. There are some applications where oil must be completely omitted from the compressed air or gas exiting the machine. For such applications where a reciprocating piston type of compressor is required, there are nonlubricated compressors. These compressors have piston rings and wear bands around the periphery of each piston. These wear bands are made of special wear-resistant dry lubricating materials such as polytetrafluorethylene. Trunk type nonlubricated compressors have dry crankcases with permanently lubricated bearings. Crosshead type compressors usually have lengthened piston

rods to ensure that no oil wet parts enter the compression space [4,5].

Most reciprocating compressors have inlet and outlet valves (on the piston heads) that are actuated by a pressure difference. These are called *self-acting valves*. There are some larger multistage reciprocating piston compressors that do have camshaft-controlled valves with rotary slide valves.

The main advantage of the multistage reciprocating piston compressor is that there is nearly total positive control of the volumetric flow rate which can be put through the machine and the pressure of the output. Many reciprocative piston compressors allow for the rotation to be adjusted, thus, changing the throughout of air or gas. Also, providing adequate power from the prime mover, reciprocating piston compressors will automatically adjust to back pressure changes and maintain proper rotation speed. These compressors are capable of extremely high output pressure (see Figure 3.4.1).

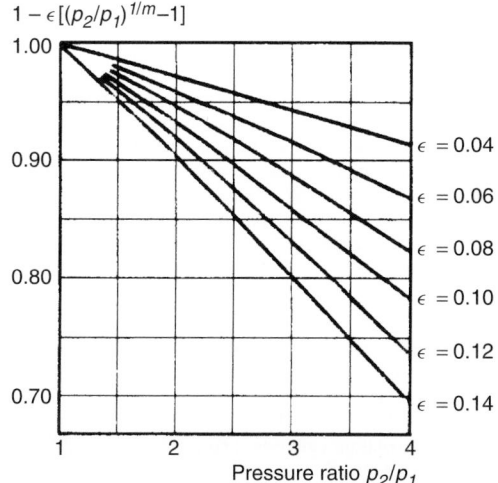

Figure 3.4.6 Volumetric efficiency for reciprocating piston compressors (with clearance) [4].

The main disadvantages to multistage reciprocating piston compressors is that they cannot be practically constructed in machines capable of volumetric flow rates much beyond 1,000 actual cfm. Also, the higher-capacity compressors are rather large and bulky and generally require more maintenance than similar capacity rotary compressors.

In a compressor, like a liquid pump, the real volume flow rate is smaller than the displacement volume. This is due to several factors:

- pressure drop on the suction side
- heating up of the intake air
- internal and external leakage
- expansion of the gas trapped in the clearance volume (reciprocating piston compressors only)

The first three factors are present in compressors, but they are small and on the whole can be neglected. The clearance volume problem, however, is unique to reciprocating piston compressors. The volumetric efficiency e_v estimates the effect of clearance. The volumetric efficiency can be approximated as

$$e_v = 0.96 \left[1 - \varepsilon \left(r_t^{1/k} - 1 \right) \right] \qquad [3.4.21]$$

where $\varepsilon = 0.04$–0.12. Figure 3.4.6 gives values of the term in the brackets for various values of ε and the r_t.

For a reciprocating piston compressor, Equation 3.4.18 becomes

$$\dot{W}_{as} = \frac{\dot{W}_s}{e_m e_v} \qquad [3.4.22]$$

3.4.5 Rotary Compressors

Another important positive displacement compressor is the rotary compressor. This type of compressor is usually of rather simple construction, having no valves and being lightweight. These compressors are constructed to handle volumetric flow rates up to around 2,000 actual cfm and pressure ratios up to around 15 (Figure 3.4.2). Rotary compressors are available in a variety of designs. The most widely used rotary compressors are sliding vane, rotary screw, rotary lobe, and liquid-piston.

The most important characteristic of this type of compressors is that all have a fixed built-in pressure compression

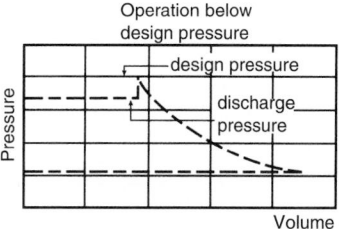

Figure 3.4.7 Rotary compressor with back pressure less than fixed pressure output [4].

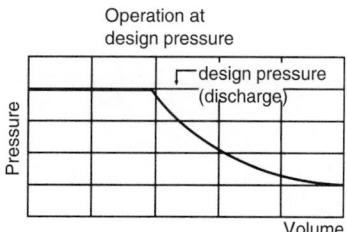

Figure 3.4.8 Rotary compressor with back pressure equal to fixed pressure output [4].

ratio for each stage of compression (as well as a fixed built-in volume displacement) [4]. Thus, at a given rate of rotational speed provided by the prime mover, there will be a predetermined volumetric flow rate through the compressor, and the pressure exiting the machine at the outlet will be equal to the design pressure ratio times the inlet pressure.

If the back pressure on the outlet side of the compressor is below the fixed output pressure, the compressed gas will simply expand in an expansion tank or in the initial portion of the pipeline attached to the outlet side of the compressor. Figure 3.4.7 shows the pressure versus volume plot for a typical rotary compressor operating against a back pressure below the design pressure of the compressor.

If the back pressure on the outlet side of the compressor is equal to the fixed output pressure, then there is no expansion of the output gas in the initial portion of the expansion tank or the initial portion of the pipeline.

Figure 3.4.8 shows the pressure versus volume plot for a typical rotary compressor operating against a back pressure equal to the design's pressure of the compressor.

If the back pressure in the outlet side of the compressor is above the fixed output pressure, then the compressor must match this higher pressure at the outlet. In so doing the compressor cannot expel the compressed volume within the compressor efficiently. Thus, the fixed volumetric flow rate (at a given rotation speed) will be reduced from what it would be if the back pressure were equal to or less than the fixed output pressure. Figure 3.4.9 shows the pressure versus volume plot for a typical rotary compressor operating against a back pressure greater than the design pressure of the compressor.

Nearly all rotary compressors can be designed with multiple stages. Such multistage compressors are designed with nearly equal compression ratios for each stage. Thus, since the volumetric flow rate (in actual cfm) is smaller from one stage to the next, the volume displacement of each stage is progressively smaller.

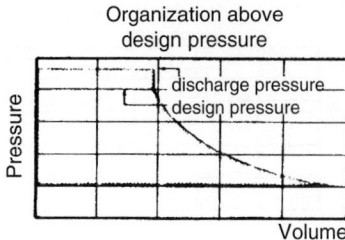

Figure 3.4.9 *Rotary compressor with back pressure greater than fixed pressure output.*

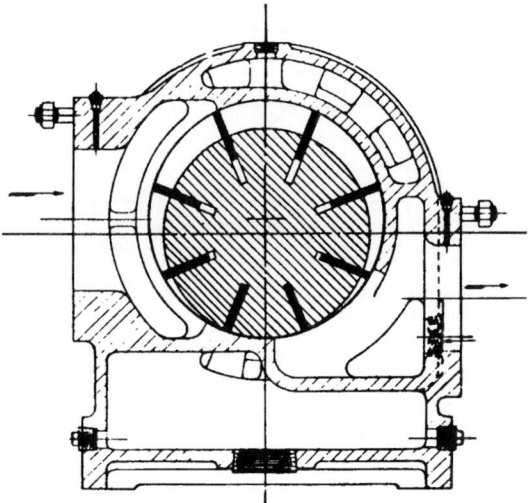

Figure 3.4.10 *Sliding vane compressor [4].*

3.4.5.1 Sliding Vane Compressor

The typical sliding vane compressor stage is a rotating cylinder located eccentrically in the bore of a cylindrical housing (Figure 3.4.10). The vanes are in slots in the rotating cylinder, and are allowed to move in and out of these slots to adjust to the changing clearance between the outside surface of the rotating cylinder and the inside bore surface of the housing. The vanes are always in contact with the inside bore due to either air pressure under the vane, or spring force under the vane. The top of the vanes slide over the inside surface of the bore of the housing as the inside cylinder rotates. Gas is brought into the compression stage through the inlet suction port. The gas is then trapped between the vanes, and as the inside cylinder rotates the gas is compressed to a smaller volume as the clearance is reduced. When the clearance is the smallest, the gas has rotated to the outlet port. The compressed gas is discharged to the pipeline system connected to the outlet side of the compressor. As each set of vanes reaches the outlet port, the gas trapped between the vanes is discharged. The clearance between the rotating cylinder and the housing is fixed, and thus the pressure ratio of compression for the stage is fixed, or built-in. The geometry (e.g., cylinder length, diameter) of the inside of each compressor stage determines the displacement volume and compression ratio of the compressor.

The principal seals within the sliding vane compressor are provided by the interface between the end of the vane and the inside surface of the cylindrical housing. The sliding vanes must be made of a material that will not damage the inside surface of the housing. Therefore, most vane material is phenolic resin-impregnated laminated fabrics (such as asbestos or cotton cloth). Also, some metals other than one that would gall with the housing can be used such as aluminum. Usually, vane compressors utilize oil lubricants in the compression cavity to allow for smooth action of the sliding vanes against the inside of the housing. There are, however, some sliding vane compressors that may be operated oil-free. These utilize bronze, or carbon/graphite vanes [4].

The volumetric flow rate for a sliding vane compression stage q_s (ft^3/min) is approximately

$$q_s = 2al(d_2 - mt)N \qquad [3.4.23]$$

where a is the eccentricity in ft, l is the length of the cylinder in ft, d_1 is the outer diameter of the rotary cylinder in ft, d_2 is the inside diameter of the cylindrical housing in ft, t is the vane thickness in ft, m is the number of vanes, and N is the speed of the rotating cylinder in rpm.

The eccentricity a is

$$a = \frac{d_2 - d_1}{2} \qquad [3.4.24]$$

Some typical values of a vane compressor stage geometry are $d_1/d_2 = 0.88$, $a = 0.06d_2$, $a = 0.06d_2$, and $l/d_2 = 2.00$ to 3.00. Typical vane up speed usually does not exceed 50 ft/s.

There is no clearance in a rotary compressor. However, there is leakage of air within the internal seal system and around the vanes. Thus, the typical volumetric efficiency for the sliding vane compression is of the order of 0.82 to 0.90. The heavier the gas, the greater the volumetric efficiency. The higher the pressure ratio through the stage, the lower the volumetric efficiency.

3.4.5.2 Rotary Screw Compressor

The typical rotary screw compressor stage is made up of two rotating shafts, or screws. One is a female rotor and the other a male rotor. These two rotating components turn counter to one another (counterrotating). The two rotating elements are designed so that as they rotate opposite to one another; their respective helix forms intermesh (Figure 3.4.11). As will all rotary compressors, there are no valves. The gas is sucked into the inlet post and is squeezed between the male and female portion of the rotating intermeshing screw elements and their housing. The compression ratio of the stage and its volumetric flow rate are determined by the geometry of the two rotating screw elements and the speed at which they are rotated.

Screw compressors operate at rather high speeds. Thus, they are rather high volumetric flow rate compressors with relatively small exterior dimensions.

Most rotary screw compressors use lubricating oil within the compression space. This oil is injected into the compression space and recovered, cooled, and recirculated. The lubricating oil has several functions

- seal the internal clearances
- cool the gas (usually air) during compression
- lubricate the rotors
- eliminate the need for timing gears

There are versions of the rotary screw compressor that utilize water injection (rather than oil). The water accomplishes the same purposes as the oil, but the air delivered in these machines is oil-free.

Some screw compressors have been designed to operate with an entirely oil-free compression space. Since the rotating elements of the compressor need not touch each other or the housing, lubrication can be eliminated. However, such rotary screw compressor designs require timing

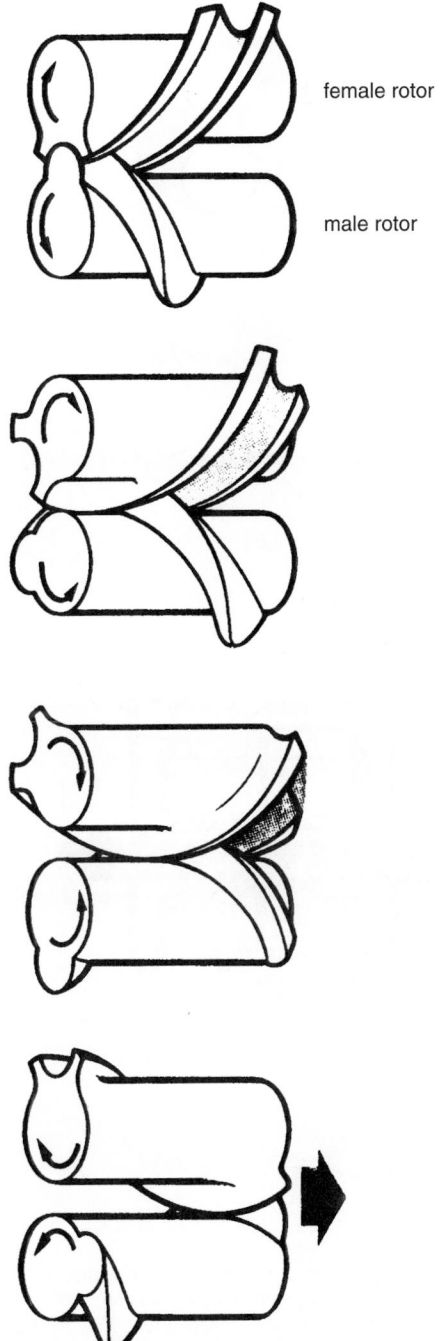

3.4.5.3 Rotary Lobe Compressor

The rotary lobe compressor stage is a rather low-pressure machine. These compressors do not compress gas internally in a fixed sealed volume as in other rotaries. The straight lobe compressor uses two rotors that intermesh as they rotate (Figure 3.4.12). The rotors are timed by a set of timing gears. The lobe shapes may be involute or cycloidal in form. The rotors may also have two or three lobes. As the rotors turn and pass the intake port, a volume of gas is trapped and carried between the lobes and the housing of the compressor. When the lobe pushes the gas toward the outlet port, the gas is compressed by the back pressure in the gas discharge line.

Volumetric efficiency is determined by the leakage at tips of the lobes. The leakage is referred to as slip. Slippage is a function of rotor diameter, differential pressure, and the gas being compressed.

For details concerning this low pressure compressor see other references [4–7].

3.4.5.4 Liquid Piston Compressor

The liquid piston compressor utilizes a liquid ring as a piston to perform gas compression within the compression space. The liquid piston compressor stage uses a single rotating element that is located eccentrically inside a housing (Figure 3.4.13). The rotor has a series of vanes extending radially from it with a slight curvature toward the direction of rotation. A liquid, such as oil, partially fills the compression space between the rotor and the housing walls. As rotation takes place, the liquid forms a ring as centrifugal forces and the vanes force the liquid to the outer boundary of the housing. Since the element is located eccentrically in the housing, the liquid ring (or piston) moves in an oscillatory manner. The compression space in the center of the stage communicates with the gas inlet and outlet parts and allows a gas pocket. The liquid ring alternately uncovers the inlet part and the outlet part. As the system rotates, gas is brought into the pocket, compressed, and released to the outlet port.

The liquid compressor has rather low efficiency, about 50%. The liquid piston compressor may be staged. The main advantage to this type of compressor is that it can be used to compress gases with significant liquid content in the stream.

3.4.6 Summary of Positive Displacement Compressors

The main advantages of reciprocating piston compressors are dependable, near constant volumetric flow rate, and variable pressure capability (up to the maximum pressure capacity of the compressor). The disadvantages are bulky construction, high initial capital costs (relative to rotary compressors of similar capabilities), and relatively high maintenance costs due to a great number of moving parts (relative to rotary compressors).

The advantages of rotary compressors are low initial capital costs, less bulky construction, and general ease of maintenance. The main disadvantages are an inability to adjust to flow line back pressure (i.e., fixed compression ratios), the need for frequent specific maintenance for rotating wear surfaces to prevent slippage, and the requirement of most rotary compressors must operate with some oil lubrication in the compression chambers [4,5].

Figure 3.4.11 *Screw compressor working principle [5].*

gears. These machines can deliver totally oil-free, water-free dry air (or gas).

The screw compressor can be staged. Often screw compressors are utilized in three- or four-stage versions.

Detailed calculations regarding the design of the rotary screw compressor are beyond the scope of this handbook. Additional details can be found in other references [4–7].

References

1. Brown, R. N., *Compressors: Selection and Sizing*, Gulf Publishing, 1986.
2. *API Specifications for Internal-Combustion Reciprocating Engines for Oil Field Service*, API Std. 7B-11C, 9th Edition, 1994.

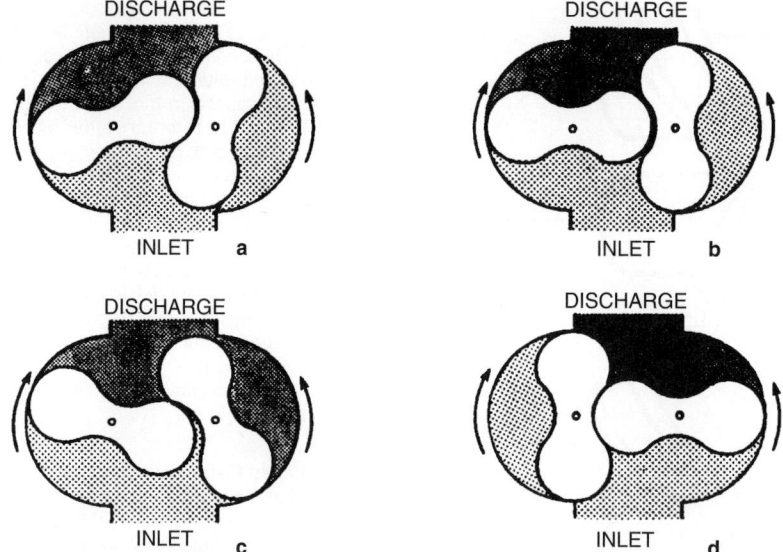

Figure 3.4.12   *Straight lobe rotary compressor operating cycle [4,23].*

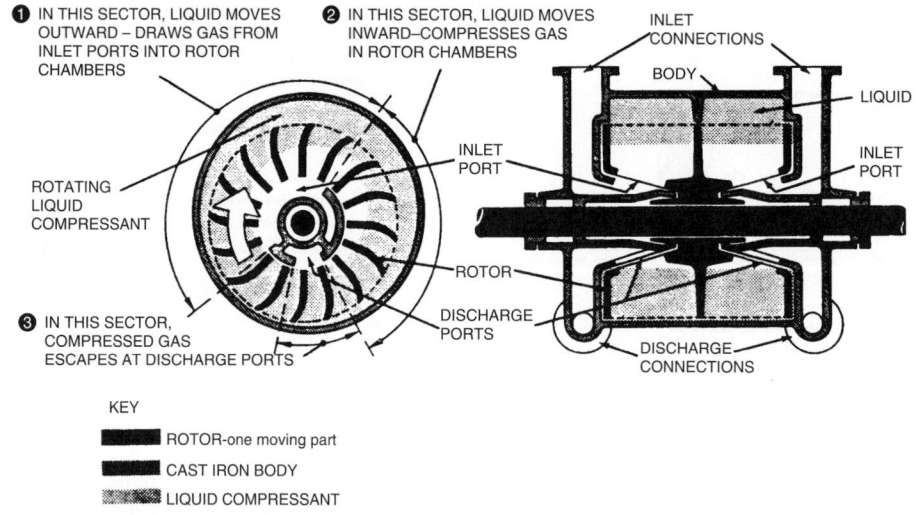

Figure 3.4.13   *Liquid piston compressor [1,5].*

3. Burghardt, M. D., *Engineering Thermodynamics with Applications*, Harper & Row, 2nd Edition, New York, 1982.
4. Loomis, A. W., *Compressed Air and Gas Data*, Ingersoll-Rand Company, 3rd Edition, 1980.
5. *Altas Copco Manual*, 4th Edition, 1982.
6. Pichot, P., *Compressor Application Engineering;* Vol. 1: *Compression Equipment*, Gulf publishing, 1986.
7. Pichot, P., *Compressor Application Engineering*, Vol. 2: *Drivers for Rotating Equipment*, Gulf Publishing, 1986.

4

Drilling and Well Completions

Contributing Authors

Egill Abrahamsen
Chip Abrant
Bo Anderson
Robert P. Badrak
Frederick E. Beck, Ph.D.
Joe Berry
Daniel E. Boone
Tom Carlson
Robert B. Coolidge
Tracy Darr van Reet
Aimee Dobbs
Robert DesBrandes, Ph.D.
Jason Fasnacht
Jerry W. Fisher
Robert Ford
Mark Heironimus, P.E.
John Hosford
Harald Jordan
William C. Lyons, Ph.D., P.E.
James Martens
George McKown
David Mildren
Mark Miller
Stefan Miska, Ph.D
Tom Morrow
Abdul Mujeeb
Bob Murphy
Charles Nathan, Ph.D., P.E.
Tim Parker
Carroll Rambin
Chris S. Russell, P.E.
Jorge H.B. Sampaio, Jr.
Ardeshir K. Shahraki, Ph.D.
Jack Smith
Mark Trevithick
Adrian Vuyk, Jr.
Bill Wamsley
Jack Wise
Andrzej K. Wojtanowicz, Ph.D.

Contents

4.1 DRILLING AND WELL SERVICING STRUCTURES

Structures for drilling and well servicing provide the clearance and structural support necessary for handling tubulars such as drill pipe and casing that are used in these operations and for handling and use of blowout preventors and well intervention equipment. Drilling structures are generally classified as derricks, masts, and substructures, but because they are used in a broad range of operational, environmental, and transportation conditions, there is a considerable blurring of additional characterizations.

Derricks are structural towers commonly used where disassembly is seldom or never required, such as on offshore floating vessels and fixed platforms, and they are rarely guyed. They are generally characterized by

- Having semi-permanent structures of square or rectangular cross-section
- Having members that are latticed or trussed on all four sides
- Using bolted connections
- Being designed to be assembled vertically in a member-by-member fashion

Masts are structural towers typically used in applications where the structure must be partially disassembled for transportation between well locations. They may be guyed or unguyed. They are more commonly used for onshore applications but are used offshore where occasional disassembly is required. They are characterized by

- Having a rectangular cross-section with at least one open face to facilitate handling of tubulars into well center
- Consisting of transportable sections that are assembled in a horizontal position near the ground
- Raising to the vertical operating position through an incorporated erection system

A *substructure* is a structural arrangement designed to support and elevate the working floor to allow the handling of blowout preventors below the floor and through which some or all drilling loads are transmitted.

Drilling and well servicing structures must be designed to safely carry all loads that are likely to be imposed during the structure's life [1]. For derricks and masts, the largest vertical dead load that will likely be applied to the structure is the heaviest casing string run into the borehole. The casing string may be floated in by plugging the lower end of the casing; the applied load at the hook is the weight of the casing string in air, reduced by the weight of displaced mud from the casing annulus. For substructures, casing loads can be applied simultaneously with the weight of the drill pipe string racked back on the drill floor.

These loads can be amplified by acceleration and impact while hoisting or setting the casing in the rotary slips. If the casing or drill string becomes stuck in the well bore, it may be necessary to apply an overpull to dislodge the string. For these reasons, it is customary for users to specify a required structural capacity somewhat higher than the maximum expected static drilling loads.

Drilling structures must also be designed to withstand environmental loads caused by wind, ice accumulation, vessel motion, and earthquake acting on the structures and racked tubulars. Because these loads can occur during operations, the designer and user must consider carefully combinations of loads that will limit the safe capacities of the structures. The designer must also consider dead weights and environmental loads on ancillary equipment that will be installed in the structures such as top drives, pipe handling equipment, BOP handling equipment, casing baskets, piping, cabling, and the like, because they will further reduce the capacity of the structures.

Improper design of foundations or failure to properly interface foundation design with design of the drilling structures is a leading cause of failures of drilling structures due to rig overturning. Freestanding drilling structures used onshore and some offshore structures during skidding are potentially subject to overturning under excessive wind and dynamic conditions. Overturning is also a concern for masts and substructures during erection. The designer must ensure that adequate stability against overturning is provided at all times, and such consideration should include the allowable bearing loading of the foundation. Weights of nonstructural equipment and other items used to stabilize the rig against overturning should be conservatively evaluated; caution should be used in considering weights of items that may be inadvertently left out or that cannot be in place at the time of erection. The rig operating instructions provided by the manufacturer must clearly state the requirements for such items, and the user must refer to and adhere to these instructions.

The API Standard 4F, Second Edition, June 1, 1995, "Specification for Drilling and Well Servicing Structures," was written to provide suitable steel structures for drilling and well servicing operations and to provide a uniform method of rating the structures for the petroleum industry. API Standard 4F supersedes API Standards 4A, 4D, and 4E; thus, many structures in service today may not satisfy all of the requirements of API Standard 4F [2–5].

For modern derrick and mast designs, API Standard 4F is the authoritative source of information* and much of this section is extracted directly from this standard. Drilling and well servicing structures that meet the requirements of API Standard 4F are identified by a nameplate securely affixed to the structure in a conspicuous place. The nameplate markings convey at least the following information:

4.1.1 Mast and Derrick Nameplate Information
a. Manufacturer's name
b. Manufacturer's address
c. Date of manufacture, including month and year
d. Serial number
e. Height in feet
f. Maximum rated static hook load in pounds, with guy lines if applicable, for stated number of lines to traveling block
g. Maximum rated wind velocity in knots, with guy lines if applicable, with rated capacity of pipe racked
h. The API specification and edition of the API specification under which the structure was designed and manufactured
i. Manufacturer's guying diagram—for structures as applicable
j. The following note: "Caution: Acceleration or impact, as well as setback and wind loads, will reduce the maximum rated static hook load capacity."
k. Manufacturer's load distribution diagram (which may be placed in mast instructions)
l. Graph of maximum allowable static hook load versus wind velocity
m. Mast setup distance for mast with guy lines

4.1.2 Substructure Nameplate Information
a. Manufacturer's name
b. Manufacturer's address

*The International Organization for Standardization (ISO), has distributed for approval what will be International Standard 13626 — Specification for drilling and well servicing structures; this document is based almost entirely on the API 4F specification. If approved, it is expected to be issued as an International Standard in 2004.

c. Date of manufacture, including month and year
d. Serial number
e. Maximum rated static rotary capacity
f. Maximum rated pipe setback capacity
g. Maximum combined rated static rotary and rated setback capacity
h. API specification and edition under which the structure was designed and manufactured

The manufacturer of structures that satisfy API Standard 4F must also furnish the purchaser with one set of instructions that covers operational features, block reeving diagram, and lubrication points for each drilling or well servicing structure. Instructions should include the raising and lowering of the mast and a facsimile of the API nameplate.

4.1.3 Definitions and Abbreviations
4.1.3.1 Definitions
The following terms are commonly used in discussing drilling structures:

Crown block assembly: The stationary sheave or block assembly installed at the top of a derrick or mast.
Design load: The force or combination of forces that a structure is designed to withstand without exceeding the allowable stress in any member.
Dynamic loading: The loading imposed upon a structure as a result of motion as opposed to static loading.
Dynamic stress: The varying or fluctuating stress occurring in a structural member as a result of dynamic loading.
Erection load: The load produced in the mast and its supporting structure during the raising and lowering operation.
Gin pole: An auxiliary structure typically mounted atop a derrick for use in maintenance and repair of crown components.
Guy line: A wire rope with one end attached to the derrick or mast assembly and the other end attached to a suitable anchor.
Guying pattern: A plane view showing the manufacturer's recommended locations and distance to the anchors with respect to the wellhead.
Height of derrick and mast without guy lines: The minimum clear vertical distance from the top of the working floor to the bottom of the crown block support beams.
Height of mast with guy lines: The minimum vertical distance from the ground to the bottom of the crown block support beams.
Impact loading: The loading resulting from sudden changes in the motion state of rig components.
Mast setup distance: The distance from the centerline of the well to a designated point on the mast structure defined by a manufacturer to assist in the setup of the rig.
Maximum rated static hook load: The sum of the weight applied at the hook and the traveling equipment for the designated location of the dead line anchor and the specified number of drilling lines without any pipe setback, sucker rod, or wind loadings.
Pipe lean: The angle between the vertical and a typical stand of pipe with the setback.
Racking platform: A platform located at a distance above the working floor for laterally supporting the upper end of racked pipe.
Rated static rotary load: The maximum weight being supported by the rotary table support beams.
Rated setback load: The maximum weight of tubular goods that the substructure can withstand in the setback area.
Rod board: A platform located at a distance above the working floor for supporting rods.
Static hook load: see *Maximum rated static hook load*.

4.1.3.2 Abbreviations
The following standard abbreviations are used throughout this section.

ABS — American Bureau of Shipping
AISC — American Institute of Steel Construction
AISI — American Iron and Steel Institute
ANSI — American National Standard Institute
API — American Petroleum Institute
ASA — American Standards Association
ASTM — American Society for Testing and Materials
AWS — American Welding Society
IADC — International Association of Drilling Contractors
SAE — Society of Automotive Engineers
USAS — United States of America Standard (ANSI)
RP — recommended practice

4.1.4 Load Capacities
All derricks and masts will fail under an excessively large load. Thus API makes it a practice to provide standard ratings for derricks and masts that meet its specifications. The method for specifying standard ratings has changed over the years; therefore, old structures may fail under one rating scheme and new structures may fail under another.

API Standard 4A (superseded by Standard 4F) provides rating of derrick capacities in terms of gross nominal capacity (GNC) as follows:

$$GNC = [(N + 4)/N] \times \text{Maximum Static Rated Hook Load} \quad [4.1.1]$$

where N is the number of lines strung between traveling block and crown block.

GNC served the purpose of providing a single capacity rating regardless of the number of drill lines strung to the block. Given a drill floor arrangement similar to that shown in Figure 4.1.1, statics gives the maximum leg load to be approximately equal to one fourth of the hook load plus the dead line load. Designing the legs for this load would make the derrick "capacity" four times the single leg capacity and thus numerically equal to GNC.

Most structures produced have different structural and floor arrangements for which the use of GNC could be highly

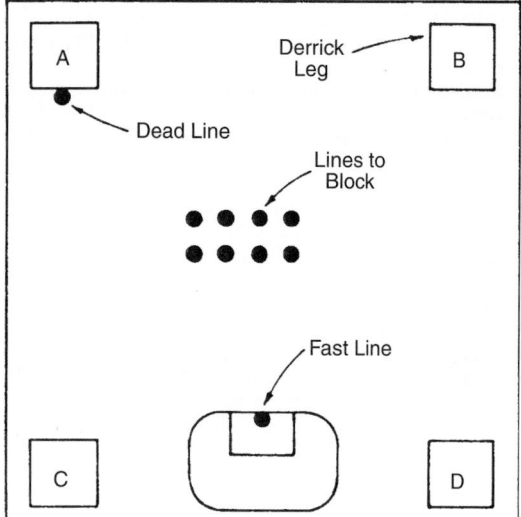

Figure 4.1.1 *Projection of fast-line and deadline locations on rig floor.*

inaccurate in establishing an allowable hook load for a given number of lines to the block, and the term was dropped in subsequent revisions of API Standard 4.

API Standard 4D (also superseded by Standard 4F) provides rating of portable masts as follows: For each mast, the manufacturer shall designate a maximum rated static hook load for each of the designated line reevings to the traveling block. Each load shall be the maximum static load that may be applied at the hook, for the designated location of deadline anchor and in the absence of any pipe-setback, sucker-rod, or wind loadings. The rated static hook load includes the weight of the traveling block and hook. The angle of mast lean and the specified minimum load guy line pattern shall be considered for guyed masts.

Under the rigging conditions given on the nameplate and in the absence of setback or wind loads, the static hook load under which failure may occur in masts conforming to this specification can be given as only approximately twice the maximum rated static hook load capacity.

The manufacturer shall establish the reduced rated static hook loads for the same conditions under which the maximum rated static hook loads apply, but with the addition of the pipe-setback and sucker-rod loadings. The reduced rated static hook loads shall be expressed as percentages of the maximum rated static hook loads.

The policy of Standard 4D, that the manufacturer specify the structure load capacity for various loading configurations, has been applied in detail in Standards 4E (superseded by Standard 4F) and 4F. Standard 4F calls for detailed capacity ratings that allow the user to look up the rating for a specific loading configuration. These required ratings are as follows.

4.1.4.1 Standard Ratings
Each structure shall be rated for the following applicable loading conditions. The structures shall be designed to meet or exceed these conditions in accordance with the applicable specifications set forth herein. The following ratings do not include any allowance for impact. Acceleration, impact, setback, and wind loads will reduce the rated static hook load capacity.

4.1.4.2 Derrick — Stationary Base
1. Maximum rated static hook load for a specified number of lines to the traveling block.
2. Maximum rated wind velocity (knots) without pipe setback.
3. Maximum rated wind velocity (knots) with full pipe setback.
4. Maximum number of stands and size of pipe in full setback.
5. Maximum rated gin pole capacity.
6. Rated static hook load for wind velocities varying from zero to maximum rated wind velocity with full rated setback and with maximum number of lines to the traveling block.

4.1.4.3 Mast with Guy Lines
1. Maximum rated static hook load capacity for a specified number of lines strung to the traveling block and the manufacturer's specified guying.
2. Maximum rated wind velocity (knots) without pipe setback.
3. Maximum rated wind velocity (knots) with full pipe setback.

*This is an API Standard 4A rating capacity and should not be confused with the actual derrick load that is discussed in the section titled "Derricks and Portable Masts."

4. Maximum number of stands and size of pipe in full setback.

4.1.4.4 Mast without Guy Lines
1. Maximum rated static hook load for a specified number of lines to the traveling block.
2. Maximum rated wind velocity (knots) without pipe setback.
3. Maximum rated wind velocity (knots) with full pipe setback.
4. Maximum number of stands and size of pipe in full setback.
5. Rated static hook load for wind velocities varying from zero to maximum rated wind velocity with full rated setback and with maximum number of lines to the traveling block.

4.1.4.5 Mast and Derricks under Dynamic Conditions
1. Maximum rated static hook load for a specified number of lines to the traveling block.
2. Hook load, wind load, vessel motions, and pipe setback in combination with each other for the following:
   a. Operating with partial setback.
   b. Running casing.
   c. Waiting on weather.
   d. Survival.
   e. Transit.

4.1.4.6 Substructures
1. Maximum rated static hook load, if applicable.
2. Maximum rated pipe setback load.
3. Maximum rated static load on rotary table beams.
4. Maximum rated combined load of setback and rotary table beams.

4.1.4.7 Substructure under Dynamic Conditions
1. Maximum rated static hook load.
2. Maximum rated pipe setback load.
3. Maximum rated load on rotary table beams.
4. Maximum rated combined load of setback and rotary table beams.
5. All ratings in the section titled "Mast and Derricks under Dynamic Conditions."

4.1.5 Design Loadings
Derricks and masts are designed to withstand some minimum loads or set of loads without failure. Each structure shall be designed for the following applicable loading conditions. The structure shall be designed to meet or exceed these conditions in accordance with the applicable specifications set forth herein.

4.1.5.1 Derrick — Stationary Base
1. Operating loads (no wind loads) composed of the following loads in combination:
   a. Maximum rated static hook load for each applicable string up condition.
   b. Dead load of derrick assembly.
2. Wind load without pipe setback composed of the following loads in combination:
   a. Wind load on derrick, derived from maximum rated wind velocity without setback (minimum wind velocity for API standard derrick sizes 10 through 18A is 93 knots, and for sizes 19 through 25 is 107 knots).
   b. Dead load of derrick assembly.

3. Wind load with rated pipe setback composed of the following loads in combination:
   a. Wind load on derrick derived from maximum rated wind velocity with setback of not less than 93 knots.
   b. Dead load of derrick assembly.
   c. Horizontal load at racking platform, derived from maximum rated wind velocity with setback of not less than 93 knots acting on full pipe setback.
   d. Horizontal load at racking platform from pipe lean.

4.1.5.2 Mast with Guy Lines

1. Operating loads (no wind load) composed of the following loads in combination:
   a. Maximum rated static hook load for each applicable string up condition.
   b. Dead load of mast assembly.
   c. Horizontal and vertical components of guy line loading.
2. Wind loads composed of the following loads in combination:
   a. Wind load on mast, derived from a maximum rated wind velocity with setback of not less than 60 knots.
   b. Dead load of mast assembly.
   c. Horizontal loading at racking board, derived from a maximum rated wind velocity with setback of not less than 60 knots, acting on full pipe setback.
   d. Horizontal and vertical components of guy line loading.
   e. Horizontal and vertical loading at rod board, derived from a maximum rated wind velocity with setback of not less than 60 knots, acting on rods in conjunction with dead weight of rods.
3. Wind loads composed of the following loads in combination:
   a. Wind load on mast, derived from a maximum rated wind velocity with setback of not less than 60 knots.
   b. Dead load of mast assembly.
   c. Horizontal loading at racking platform, derived from a maximum rated wind velocity with setback of not less than 60 knots, acting on full pipe setback.
   d. Horizontal and vertical components of guy line loading.
4. Wind loads composed of the following loads in combination:
   a. Wind load on mast derived from a maximum rated wind velocity without setback of not less than 60 knots.
   b. Dead load of mast assembly.
   c. Horizontal and vertical components of guy line loading.
5. Erection loads (zero wind load) composed of the following loads in combination:
   a. Forces applied to mast and supporting structure created by raising or lowering mast.
   b. Dead load of mast assembly.
6. Guy line loading (assume ground anchor pattern consistent with manufacturer's guying diagram shown on the nameplate).
   a. Maximum horizontal and vertical reactions from conditions of loading applied to guy line.
   b. Dead load of guy line.
   c. Initial tension in guy line specified by mast manufacturer.

4.1.5.3 Mast without Guy Lines

1. Operating loads composed of the following loads in combination:
   a. Maximum rated static hook load for each applicable string up condition.
   b. Dead load of mast assembly.

2. Wind load without pipe setback composed of the following loads in combination:
   a. Wind loading on mast, derived from a maximum rated wind velocity without setback of not less than 93 knots.*
   b. Dead load of mast assembly.
3. Wind load with pipe setback composed of the following loads in combination:
   a. Wind loading on mast, derived from a maximum rated wind velocity with setback of not less than 70 knots.*
   b. Dead load of mast assembly.
   c. Horizontal load at racking platform derived from a maximum rated wind velocity with setback of not less than 70 knots* acting on pipe setback.
   d. Horizontal load at racking platform from pipe lean.
4. Mast erection loads (zero wind load) composed of the following loads in combination:
   a. Forces applied to mast and supporting structure created by raising or lowering mast.
   b. Dead load of mast assembly.
5. Mast handling loads (mast assembly supported at its extreme ends).

When specifying masts for use offshore, it is recommended that the user specify the higher wind velocity requirements of derricks.

4.1.5.4 Derricks and Mast under Dynamic Conditions

All conditions listed in the section titled "Load Capacities," subsection titled "Mast and Derricks under Dynamic Conditions," are to be specified by the user. Forces resulting from wind and vessel motion are to be calculated in accordance with the formulas presented in the section titled "Design Specifications," paragraphs titled "Wind," "Dynamic Loading (Induced by Floating Hull Motion)."

4.1.5.5 Substructures

1. Erection of mast, if applicable.
2. Moving or skidding, if applicable.
3. Substructure shall be designed for the following conditions:
   a. Maximum rated static rotary load.
   b. Maximum rated setback load.
   c. Maximum rated static hook load (where applicable).
   d. Maximum combined rated static hook and rated setback loads (where applicable).
   e. Maximum combined rated static rotary and rated setback loads.
   f. Wind loads resulting from maximum rated wind velocity acting from any direction on all exposed elements. Wind pressures and resultant forces are to be calculated in accordance with the equations and tables in the section titled "Design Specifications," paragraph titled "Wind." When a substructure is utilized to react guy lines to the mast, these reactions from the guy lines must be designed into the substructure.
   g. Dead load of all components in combination with all of the above.

4.1.5.6 Substructure under Dynamic Conditions

All conditions listed in the section titled "Load Capacities," paragraph titled "Structure under Dynamic Conditions," are to be specified by the user. Forces resulting from wind and vessel motion are to be calculated in accordance with formulas from the section titled "Design Specifications," paragraphs titled "Wind" and "Dynamic Loading (Induced by Floating Hull Motion)."

4.1.6 Design Specifications
4.1.6.1 Allowable Stresses

AISC specifications for the design fabrication and erection of structural steel for buildings shall govern the design of these steel structures (for AISC specifications, see the current edition of Steel Construction Manual of the American Institute of Steel Construction). Only Part I of the AISC manual, the portion commonly referred to as elastic design, shall be used in determining allowable unit stresses; use of Part II, which is commonly referred to as plastic design, is not allowed. The AISC shall be the final authority for determination of allowable unit stresses, except that current practice and experience do not dictate the need to follow the AISC for members and connections subject to design, is not allowed. The AISC shall be the final authority for determination of allowable unit stresses, except that current practice and experience do not dictate the need to follow the AISC for members and connections subject to repeated variations of stress, and for the consideration of secondary stresses.

For purposes of this specification, stresses in the individual members of a latticed or trussed structure resulting from elastic deformation and rigidity of joints are defined as secondary stresses. These secondary stresses may be taken to be the difference between stresses from an analysis assuming fully rigid joints, with loads applied only at the joints, and stresses from a similar analysis with pinned joints. Stresses arising from eccentric joint connections, or from transverse loading of members between joints, or from applied moments, must be considered primary stresses.

Allowable unit stresses may be increased 20% from the basic allowable stress when secondary stresses are computed and added to the primary stresses in individual members. However, primary stresses shall not exceed the basic allowable stresses.

Wind and Dynamic Stresses (Induced by Floating Hull Motion)

Allowable unit stresses may be increased one-third over basic allowable stresses when produced by wind or dynamic loading, acting alone, or in combination with the design dead load and live loads, provided the required section computed on this basis is not less than required for the design dead and live loads and impact (if any), computed without the one-third increase.

Wire Rope

The size and type of wire rope shall be as specified in API Specification 9A and by API RP 9B (see section titled "Hoisting System").

1. A mast raised and lowered by wire rope shall have the wire rope sized to have a nominal strength of at least $2\frac{1}{2}$ times the maximum load on the line during erection.
2. A mast or derrick guyed by means of a wire rope shall have the wire rope sized so as to have a nominal strength of at least $2\frac{1}{2}$ times the maximum guy load resulting from a loading condition.

Wire ropes subjected to bending exhibit reduced breaking strengths compared with straight pull tests, and extensive testing has shown that the smaller a sheave is in relation to the diameter of the rope, the larger is the reduction in strength [6]. Figure 4.1.2 shows this relationship for 6×19 and 6×37 constructions, where D is the sheave tread diameter and d is the wire rope diameter. For example, a wire rope designed to the API 4F minimum design factor

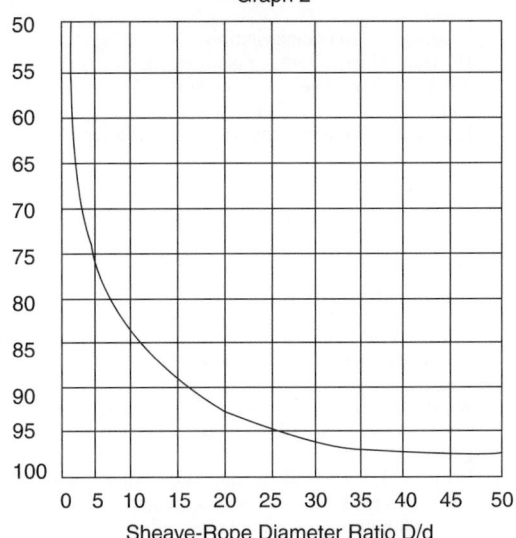

Efficiencies of Wire Ropes Bent Around
Stationary Sheaves
(Static Stresses Only)
Graph 2

Sheave-Rope Diameter Ratio D/d

Figure 4.1.2

of 2.5 and bent over a sheave with tread diameter less than 7.5 times the rope diameter will have an actual factor of safety to breaking of less than 2.0.

Ropes loaded to more than 50% of breaking strength suffer core damage, which can lead to premature and often catastrophic failure; the designer should carefully consider the effects of small-diameter erection and crown sheaves used with drilling structures. It is recommended that the minimum factors of safety required by API 4F (and API RP 9B for drilling lines) be increased by the inverse of the reduction factor from Figure 4.1.2.

Crown Shafting

Crown shafts, including fastline and deadline sheave support shafts, shall be designed to AISC specifications except that the safety factor in bending shall be a minimum of 1.67 to yield. Wire rope sheaves and bearings shall be designed in accordance with "API Specification 8A: Drilling and Production Hoisting Equipment."

4.1.6.2 Wind

Wind forces shall be applied to the entire structure.* The wind directions that result in the highest stresses for each component of the structure must be determined and considered. Wind forces for the various wind speeds shall be calculated according to

$$F = (P)(A) \tag{4.1.2}$$

where F = Force in lb
  P = Pressure in lb/ft^2
  A = Total area, in ft^2, projected on a plane, perpendicular to the direction of the wind, except that the exposed areas of two opposite sides of the mast or derrick shall be used.

Although the API 4F specification does not specifically address the wind areas of accessories and equipment, it seems obvious that calculations of wind loads must also include these areas.

When pipe or tubing is racked in more than one area, the minimum area of setback shall be no less than 120% of the area on one side; when rods are racked on more than one area, the minimum area of rods shall be no less than 150% of the area of one side to account for the effect of wind on the leeward area (Figure 4.1.3).

The pressure due to wind is

$$P = 0.00338\,(V_k^2)(C_h)(C_s) \qquad [4.1.3]$$

where P = pressure in lb/ft^2
$\quad$ V_k = wind velocity in knots
$\quad$ C_h = height coefficient (per Table 4.1.1).

For convenience, Table 4.1.2 tabulates wind pressures for various wind speeds with $C_s = 1.25$ and $C_h = 1.0$.

Height is the vertical distance from ground or water surface to the center of area. The shape coefficient C for a derrick is assumed as 1.25. C_s, and C_h were obtained from ABS, "Rules for Building and Classing Offshore Drilling Units, 1968."

4.1.6.3 Dynamic Loading
(Induced by Floating Hull Motion)
Forces shall be calculated according to the following [6]:

$$FP = \left(\frac{WL_1}{32.2}\right)\left(\frac{4\pi^2}{T_p^2}\right)\left(\frac{\pi\phi}{180}\right) + W\sin\phi \qquad [4.1.4]$$

$$FP = \left(\frac{W}{32.2}\right)\left(\frac{4\pi^2}{T_r^2}\right)\left(\frac{\pi\phi L}{180}\right) + W\sin\theta \qquad [4.1.5]$$

$$FH = W + \frac{W2\pi^2H}{T_h^2 g} \qquad [4.1.6]$$

Table 4.1.1 Height Coefficients, C_h

Height, feet (meters)				C_h
Over		Not Exceeding		
0	(0)	50	(15.2)	1.00
50	(15.2)	100	(30.5)	1.10
100	(30.5)	150	(45.7)	1.20
150	(45.7)	200	(61.0)	1.30
200	(61.0)	250	(76.2)	1.37
250	(76.2)	300	(91.4)	1.43
300	(91.4)	350	(106.7)	1.48
350	(106.7)	400	(121.9)	1.52
400	(121.9)	450	(137.2)	1.56
450	(137.2)	500	(152.4)	1.60
500	(152.4)	550	(167.6)	1.63
550	(167.6)	600	(182.9)	1.67
600	(182.9)	650	(198.1)	1.70
650	(198.1)	700	(213.4)	1.72
700	(213.4)	750	(228.6)	1.75
750	(228.6)	800	(243.8)	1.77
800	(243.8)	850	(259.1)	1.79
850	(259.1)			1.80

where W = dead weight of the point under consideration
$\quad$ L_1 = distance from pitch axis to the gravity center of the point under consideration in feet
$\quad$ L = distance from roll axis to the gravity center of the point under consideration in feet
$\quad$ H = heave (total displacement)
$\quad$ T_p = period of pitch in seconds
$\quad$ T_r = period of roll in seconds
$\quad$ T_h = period of heave in seconds
$\quad$ ϕ = angle of pitch in degrees
$\quad$ θ = angle of roll in degrees
$\quad$ g = gravity in 32.2 ft/s/s

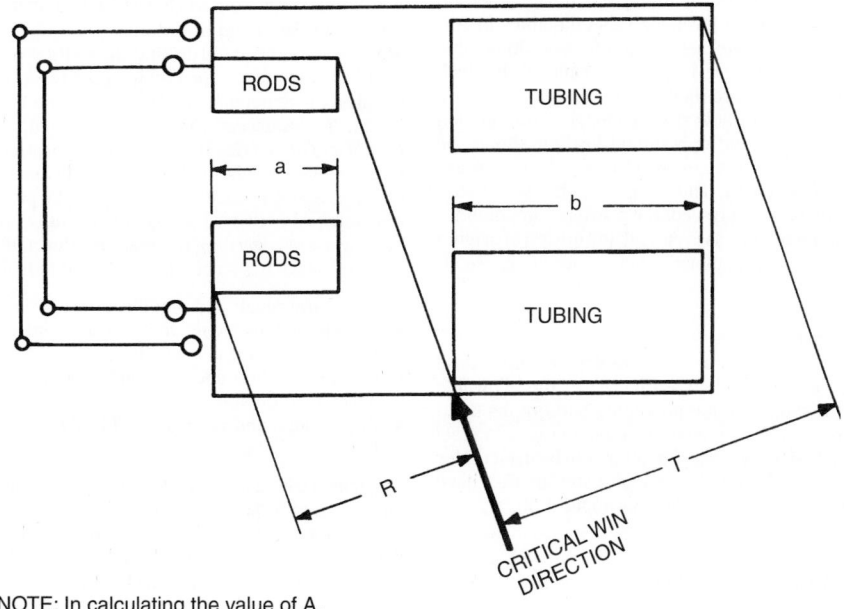

NOTE: In calculating the value of A,
If R is greater than 1.5a, use R. If not, use 1.5a.
If T is greater than 1.2b, use T. If not, use 1.2b.

Figure 4.1.3 Diagram of projected area [9].

Table 4.1.2 *Conversion Values (for 0–50 ft. height)*

Pressure		Wind Velocity		
Lb/Sq	Ft (kPa)	V_k Knots	Miles/ Hour	Meters/ Second
10	(68.9)	49	56	25.2
15	(103.4)	60	69	30.9
20	(137.9)	69	79	35.5
25	(172.4)	77	89	39.6
30	(206.8)	84	97	43.2
35	(241.3)	91	105	46.8
40	(275.8)	97	112	49.9
45	(310.3)	103	119	53.0
50	(344.7)	109	125	56.1
55	(379.2)	114	131	58.6

Unless specified, the force due to combined roll, pitch, and heave shall be considered to be the largest of the following:

1. Force due to roll plus force due to heave.
2. Force due to pitch plus force due to heave.
3. Force due to roll and pitch determined as the square root of the sum of squares plus force due to heave.

Angle of roll or pitch is the angle to one side from vertical. The period is for a complete cycle.

4.1.6.4 Earthquake

Earthquake is a special loading condition to be addressed when requested by the user. The user is responsible for furnishing the design criteria that includes design loading, design analysis method, and allowable response.

The design criteria for land units may be in accordance with local building codes using equivalent static design methods.

For fixed offshore platform units, the design method should follow the strength level analysis guidelines in API RP 2A. The drilling and well servicing units should be able to resist the deck movement, i.e., the response of the deck to the ground motion prescribed for the design of the offshore platform. The allowable stresses for the combination of earthquake, gravity and operational loading should be limited to those basic allowables with the one-third increase as specified in AISC Part I. The computed stresses should include the primary and the secondary stress components.

Guidance for the seismic assessment and design of drilling structures on offshore platforms can be found in Reference[7].

4.1.6.5 Extreme Temperature

Because of the effect of low temperatures on structural steel, it will be no use to change (decrease) the allowable unit stresses mentioned in the preceding paragraphs titled "Allowable Stresses." Low temperature phenomena in steel are well established in principle. Structures to be used under extreme conditions should use special materials that have been, and are being, developed for this application.

4.1.6.6 Other Loadings

The design of drilling structures often is governed by other loading conditions not specifically covered by the API 4F specification, such as lifting, handling, and transport before assembly and during rig moves. These loads can be highly variable, with significant impact loads. API RP 2A WSD is often used as a guide for the design of lifts made offshore.

Additional design loadings should be addressed by mutual agreement of the user and manufacturer and as required by good engineering practices.

4.1.7 Manufacturing, Inspection, and Delivery

In addition to meeting the design requirements of the specification, the materials and manufacture of drilling structures must meet design specifications, special process requirements for welding and heat treatment, and quality control and quality assurance requirements of API 4F and other applicable AWS, API, and AISC specifications.

The purchaser's inspector shall, with proper notice, have free entry at all times to all parts of the manufacturer's works concerned with the manufacture of the products ordered, so long as such inspections are conducted as to not interfere unnecessarily with the operations of the works. When requested by a prospective purchaser of the equipment for his use or by a user of the equipment, the manufacturer shall make available for examination details of computations, drawings, tests, or other supporting data as may be necessary to demonstrate compliance with the API 4F specification.

The manufacturer shall furnish to purchaser one set of instructions that covers operational features, block reeving diagram, and lubrication points for each drilling or well servicing structure. Instructions shall be included to cover erection and lowering of the mast and/or substructure. A facsimile of the nameplate shall be included in the instructions.

4.1.8 Standard Derricks

A standard derrick is a structure of square cross-section and set dimensions. At one time such standardization was desirable, but it is largely irrelevant today and standard derricks are seldom, if ever, specified. Appendix B in the API 4F specification lists standard derrick information.

4.1.9 Maintenance and Use of Drilling and Well Servicing Structures

Maintenance and use of drilling and well servicing structures should always be performed in accordance with the manufacturer's instructions and the most current edition of API RP 54, *Recommended Practices for Occupational Safety and Health for Oil and Gas Well Drilling and Servicing Operations.* In cases where this material is insufficient and the manufacturer is not available or is unable to provide sufficient guidance, API RP 4G [8] is an excellent source of information for users regarding the maintenance and use of drilling and well servicing structures. Covered topics include

- Structural repair and modification
- Raising line inspection and replacement
- Periodic structural inspections
- Guying, guyline anchors, and foundations for portable masts with guylines
- Precautions and procedures for low-temperature operations

A particularly useful presentation is made for reduced-footprint guying patterns for portable masts. While the use of such patterns requires higher guyline anchor capacities and derating of the masts maximum rated static hook load, they provide the user with solutions for avoidance of obstructions at the well-site such as rocks, pits, and electric lines.

Evaluation, repairs, and modifications of drilling and well servicing structures should be carried out by qualified personnel and should take into account the impact of such work on operating capacities, environmental loads, erection loads, transportation loads, lifting loads, structural overturning and foundation requirements, and other conditions as may be applicable. All work associated with such modifications should be done under the governance of a qualified quality assurance program in accordance with the instructions of the qualified individuals.

Both users and well-site owners and operators should exercise diligence in defining responsibilities of involved parties regarding the selection, installation, and maintenance of structures, foundations, cellars, and anchors for the intended job, as well as the selection and training of personnel for all work.

References

1. McCray, A. W., and F. W. Cole, *Oil Well Drilling Technology*, University of Oklahoma Press, Norman, Oklahoma, 1959.
2. American Petroleum Institute (API) Standard 4A, 16th Edition, "API Specification for Steel Derricks," API, Washington D.C., April 1967.
3. API Standard 4D, 6th Edition, "API Specification for Portable Masts," API, Washington D.C., March 1967.
4. API Standard 4E, 3rd Edition, "Specification for Drilling and Well Servicing Structures," API, Washington D.C., June 1, 1988.
5. API Standard 4F, 2nd Edition, "Specification for Drilling and Well Servicing Structures," API, Washington D.C., June 1, 1995.
6. Armco Steel Corporation, *Wire Rope Handbook*, 1973.
7. Turner, J. W., Effenberger, M., and Irick, J., "Seismic Assessment Procedure for Drilling Structures on Offshore Platforms," SPE, Richardson, Texas, 2002.
8. API Recommended Practice 4G, 2nd Edition, "Recommended Practice for Maintenance and Use of Drilling and Well Servicing Structures," API, Washington D.C., October, 1998. (The third edition of the recommended practice passed letter ballot within API in January 2004 and is scheduled for publication in early 2004.)

4.2 HOISTING SYSTEM

A hoisting system, as shown in Figure 4.2.1, is composed of the drawworks, traveling block, crown block, extra line storage spool, various clamps, hooks, and wire rope.

Normally, a hoisting system has an even number of working lines between the traveling block and the crown block. The fast line is spooled onto the drawworks' hoisting drum. The dead line is anchored to the rig floor across from the drawworks. The weight indicator is a load cell incorporated in the dead line anchor.

The mechanical advantage of the hoisting system is determined by the block and tackle and the number of working lines between the crown block and the traveling block [1].

Thus, for the static condition (i.e., no friction losses in the sheaves at the blocks), F_f (lb), the force in the fast line to hold the hook load, is

$$F_f = \frac{W_h}{j} \qquad [4.2.1]$$

where W_h is the weight of the traveling block plus the weight of the drill string suspended in the hole, corrected for buoyancy effects in pounds; and j is the number of working lines between the crown block and traveling block. Under these static conditions, F_d (lb), the force in the dead line, is

$$F_d = \frac{W_h}{j} \qquad [4.2.2]$$

The mechanical advantage (ma) under these static conditions is

$$ma(static) = \frac{W_h}{F_f} = j \qquad [4.2.3]$$

When the hook load is lifted, friction losses in crown block and traveling block sheaves occur. It is normally assumed that these losses are approximately 2% deduction per working line. Under dynamic conditions, there will be an efficiency factor for the block and tackle system to reflect these losses. The efficiency will be denoted as the hook-to-drawwork efficiency (e_h). The force in the fast line under dynamic conditions (i.e., hook is moving) will be

$$F_f = \frac{W_h}{e_h j} \qquad [4.2.4]$$

Equation 4.2.2 remains unchanged by the initiation of hook motion (i.e., the force in the dead line is the same under static or dynamic conditions). The mechanical advantage (ma) under dynamic conditions is

$$ma \text{ (dynamic)} = e_h j \qquad [4.2.5]$$

The total load on the derrick under dynamic conditions, F_t (lb), will be

$$F_t = W_h + \frac{W_h}{e_h j} + \frac{W_h}{j} + W_c + W_t \qquad [4.2.6]$$

where W_c is the weight of the crown block, and W_t is the weight of tools suspended in the derrick, both in pounds.

Example 4.2.1

For dynamic conditions, find the total load on a derrick that is capable of lifting a 600,000-lb drill string with an 8-working line block and tackle. The crown block weighs 9,000 lb and the traveling block weighs 4,500 lb. Assume that there are no other tools hanging in the derrick and that the deadline is attached to the rig floor across from the drawworks in its normal position (see Figure 4.2.1). Assume the standard

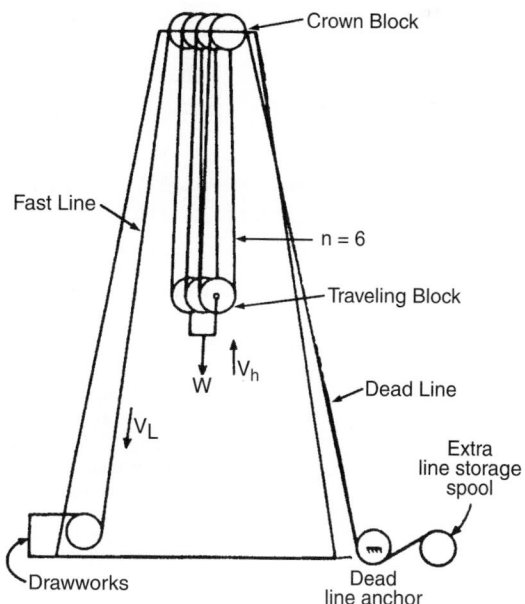

Figure 4.2.1 *Schematic of simplified hoisting system on rotary drilling rig [1].*

deduction of 2% per working line to calculate e_h.

$$e_h = 1.00 - 0.02(8) = 0.84$$

From Equation 4.2.6

$$F_t = 604,500 + \frac{600,000}{0.84(8)} + \frac{600,000}{8} + 9,000$$

$$= 604,500 + 89,286 + 75,000 + 9,000$$

$$= 786,786 \, \text{lb}$$

4.2.1 Drawworks

The drawworks is the key operating component of the hoisting system. On most modern rotary drilling rigs, the prime movers either operate the hoisting drum within the drawworks or operate the rotary table through the transmission within the drawworks. Thus the drawworks is a complicated mechanical system with many functions[1,7].

4.2.1.1 Functions

The drawworks does not carry out only hoisting functions on the rotary drilling rig. In general, the functions of the drawworks are as follows:

1. Transmit power from the prime movers (through the transmission) to its hoisting drum to lift drill string, casing string, or tubing string, or to pull in excess of these string loads to free stuck pipe.
2. Provide the braking systems on the hoist drum for lowering drill string, casing string, or tubing string into the borehole.

3. Transmit power from the prime movers (through the transmission) to the rotary drive sprocket to drive the rotary table.
4. Transmit power to the catheads for breaking out and making up drill string, casing string, and tubing string.

Figure 4.2.2 is a schematic of drawworks together with the prime mover power source.

4.2.1.2 Design

The drawworks basically contains the hoist drum, the transmissions, the brake systems, the cluth systems, rotary drive sprocket, and cathead. Figure 4.2.3 shows a schematic of the drawworks.

The power is provided to the drawworks by the prime movers at the master clutch (see Figure 4.2.2) and is transmitted to the master clutch shaft via sprockets and roller chain drives. The speed and the torque from the prime movers are controlled through the compound. The compound is a series of sprockets, roller chain drives, and clutches that allow the driller to control the power to the drawworks. The driller operates the compound and the drawworks (and other rig functions) from a driller's console (Figure 4.2.4).

With the compound, the driller can obtain as many as 12 gears working through the drawworks transmission.

In Figure 4.2.4, the driller's console is at the left of the drawworks. Also, the hoisting drum and sand reel can be seen. The driller's brake control is between the driller's console and the drawworks to control the brake systems of the hoisting drum.

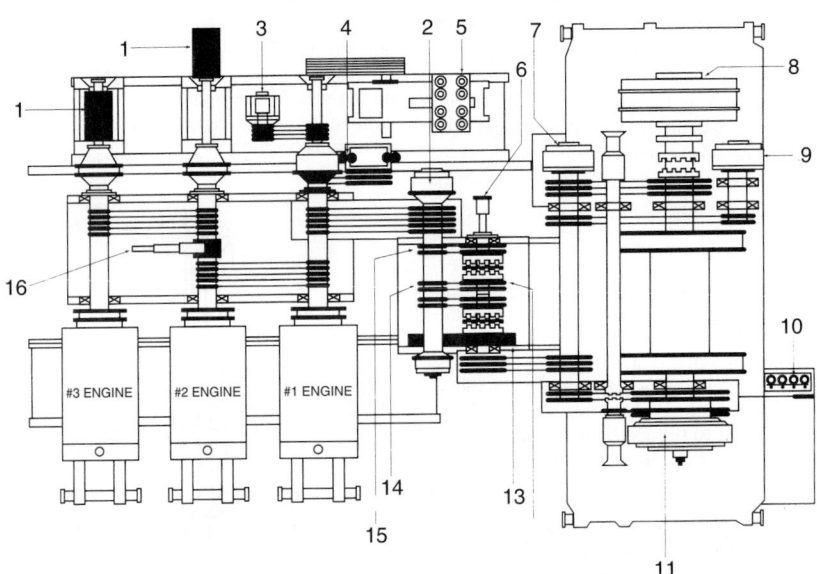

1. Drive to pump	9. Rotary drive air clutch countershaft
2. Master clutch	10. Driller's console
3. Generator	11. Drum low air clutch
4. Air compressor	12. High gear
5. Washdown pump	13. Reverse gear
6. Sand reel drive	14. Intermediate gear
7. Drum high air clutch	15. Low gear
8. Auxiliary brake	16. Power flow selector*

*Note: This item is shown as a manually operated clutch.
This, of course, on an actual rig would be air actuated.

Figure 4.2.2 *Power train on a drawworks with accessories. (Courtesy of Varco, Inc.)*

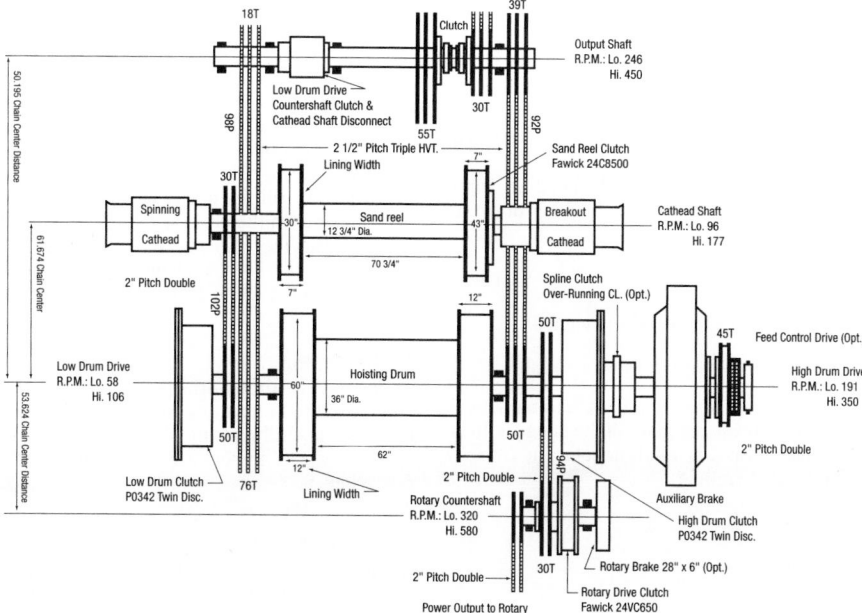

Figure 4.2.3 *The drive group of a large DC electric rig. Note that this rig may be equipped with either two or three traction motors. (Courtesy of Varco, Inc.)*

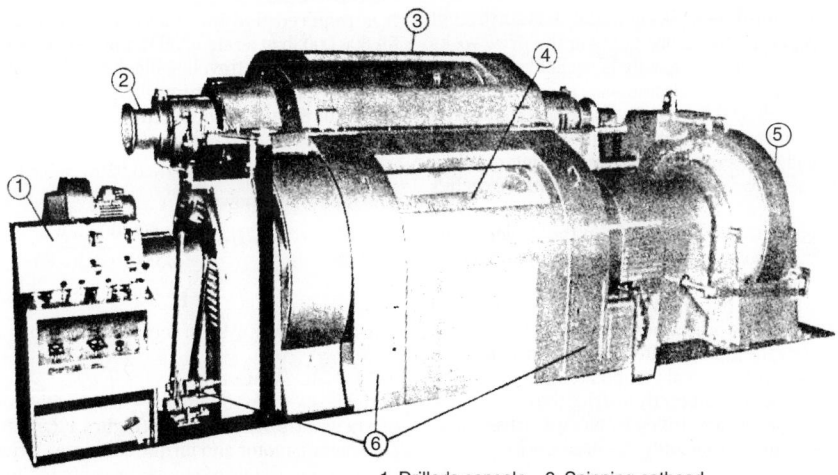

1. Driller's console.   2. Spinning cathead.
3. Sand reel.   4. Main drum (grooved).   5. Hydromatic brake.   6. Manual brakes (with inspection plates indicated).

Figure 4.2.4 *The hoist on the rig floor. (Courtesy of Varco, Inc.).*

Hoisting Drum

The hoisting drum (usually grooved) is probably the most important component on the drawworks. It is through the drum that power is transmitted to lift the drill string with the drilling line (wire rope) wound on the drum. From the standpoint of power requirements for hoisting, the ideal drum would have a diameter as small as possible and a width as great as possible. From the standpoint of drilling line wear and damage, the hoisting drum would have the largest drum diameter. Therefore, the design of the hoisting drum must be compromised to obtain an optimum design. Thus, the hoist drum is usually designed to be as small as practical,

but the drum is designed to be large enough to permit fast line speeds in consideration of operation and economy.

Often it is necessary to calculate the line-carrying capacity of the hoist drum. The capacity or length of drilling line in the first layer on the hoist drum L_1 (ft) is

$$L_1 = \frac{\pi}{12}(D + d)\frac{\ell}{d} \qquad [4.2.7]$$

where D is the drum diameter, d is the line diameter, and ℓ is the hoisting drum length, all in inches.

The length of the second layer, L_2 (ft) is

$$L_2 = \frac{\pi}{12}(D + 3d)\frac{\ell}{d}$$ [4.2.8]

The length of the nth layer L_n (f_t) is

$$L_n = \frac{\pi}{12}\left[D + (2n - 1)d\right]\frac{\ell}{d}$$ [4.2.9]

where n is the total number of layers on the hoisting drum.

The total length of drilling line on the hoisting drum, L_t (ft), will be the sum of all the layers;

$$L_t = \frac{\pi}{12}(D + h)\frac{\ell h}{d^2}$$ [4.2.10]

where h is the hoist drum flange height in inches.

Example 4.2.2

A hoist drum has an inside length of 48 in. and an outside diameter of 30 in. The outside diameter of the flange is 40 in. The drilling line diameter is 1 in. Find the total line capacity of the drum. The flange length is

$$h = \frac{40 - 30}{2} = 5 \text{ in.}$$

The total length (capacity) is

$$L_t = \frac{\pi}{12}(30 + 5)\frac{(48)(5)}{(1)^2}$$

$$= 2199 \text{ ft}$$

Transmission and Clutch

The transmission in the drawworks generally has six to eight spreads. Large rigs can have more gears in the drawworks transmissions. More gearing capacity is available when the compound is used. This transmission uses a combination of sprockets and roller chain drives and gears to accomplish the change of speeds and torque from the prime movers (via the compound). The clutches used in the transmitting of prime mover power to the drawworks are jaw-type positive clutches and friction-type clutches. In modern drawworks, nearly all clutches are pneumatically operated from the driller's console. The driller's console also controls the shifting of gears within the drawworks.

Torque converters used in most drawworks are designed to absorb shocks from the prime movers or the driven equipment and to multiply the input torque. Torque converters are used in conjunction with internal combustion prime movers when these engines are used directly to drive the drawworks. More modern drawworks are driven by electric drives since such prime movers usually simplify the drawworks.

Brakes

The brake systems of the drawworks are used to slow and stop the movement of the large weights that are being lowered into the borehole. The brake system will be in continuous use when a round trip is made. The principal brake of the drawworks is the friction-type mechanical brake system. But when this brake system is in continuous use, it would generate a great deal of heat. Therefore, an auxiliary brake system is used to slow the lowering speeds before the friction-type mechanical brake system is employed to stop the lowering motion. Hydraulic brake system and electromagnetic brake system are the basic types of auxiliary brake system in use. The hydraulic brake system uses fluid friction (much like a torque converter) to absorb power as equipment is lowered. The electromagnetic brake system uses two opposed magnetic fields supplied by external electrical current to control the speed of the hoisting drum. The auxiliary brake system can only control the speed of lowering and

cannot be used to stop the lowering as does the mechanical friction-type brake system.

Catheads

The catheads are small rotating spools located on the sides of the drawworks. The cathead is used as a power source to carry out routine operations on the rig floor and in the vicinity of the rig. These operations include making up and breaking out drill pipe and casing, pulling single joints of pipe and casing from the pipe rack to the rig floor. The sand reel is part of this mechanism. This small hoisting drum carries a light wire rope line (sand line) through the crown to carry out pulling operations on the rig floor or in the vicinity of the rig.

Power Rating

In general, the drawworks is rated by its input horsepower. But it used to be rated by depth capability along with a specific size of drill pipe to which the depth rating pertains. The drawworks horsepower input required HP_{in} for hoisting operations is

$$HP_{in} = \frac{Wv_h}{33,000e_h e_m}$$ [4.2.11]

where W is the hook load in lb, v_h is the hoisting velocity of the traveling block in ft/min, e_h is the hook-to-drawworks efficiency, and e_m is the mechanical efficiency within the drawworks and coupling between the prime movers and the drawworks (usually taken as about 0.85).

Example 4.2.3

It is required that the drawworks input power be able to lift 600,000 lb at a rate of 50 ft/min. There are eight working lines between the traveling block and the crown block. Three input power systems are available: 1,100, 1,400 and 1,800 hp. Which of the three will be the most appropriate? The value of e_h is

$$e_h = 1.00 - 0.02(8) = 0.84$$

The input horsepower is

$$HP = \frac{600,000(50)}{33,000(0.84)(0.85)}$$

$$= 1273.2$$

The input power system requires 1400 hp.

4.2.1.3 AC Drawworks

Variable-speed AC technology has enabled the development of a new class of drawworks (Figure 4.2.5). In the traditional DC traction motor and torque converter–driven drawworks, the speed of the prime mover was limited to approximately 1200 rpm. These low speeds required the drawworks to have shifting transmissions to get the correct speed and line pull combinations required for different drilling conditions. With AC technology, the motors operate at speeds up to 3000 rpm at continuous rated horsepower. This wider operating speed of the prime mover allow for drawworks designs without a multispeed transmission. Most of the new AC drawworks are gear driven, eliminating the roller chains used in DC drawworks.

The main advantage of an AC-powered, gear-driven drawworks is the ability to operated in hoist, lower, and stop modes without setting a parking brake or engaging and disengaging a drum clutch each time a change is made from one mode to another. The operating characteristics of the AC motors allow holding full torque at zero speed, and in the lowering mode, the motor absorbs energy in back EMF, which is then dissipated across a resistor bank or regenerated back into the total rig power grid. This four-quadrant

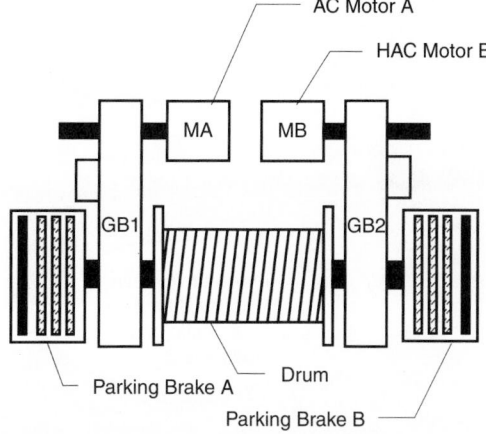

Figure 4.2.5 *Design safety factor and rating relationships. (Courtesy of Varco, Inc.)*

motor control allows frull operator control with a single joystick.

The AC drawworks, contains the hoist drum directly coupled to the hoist drum shaft, the spring applied, pneumatically released parking brakes, also directly coupled to the hoist drum shaft, and the gearboxes with the AC motors directly coupled. Some versions of AC drawworks use hydraulically released spring applied brakes acting directly on the flanges of the hoist drum. These machines generally do not have self-contained catheads or provisions for them.

Figure 4.2.6 shows the velocity profile for the operation of the AC drawworks in the hoist and lowering modes. The maximum allowable speed is a function of the load being lifted and the operator-commanded speed. The AC drawworks control system allows the maximum acceleration, maximum velocity, and maximum deceleration of the traveling assembly in both the hoisting and lowering directions while maintaining safe and reliable operation. By following the velocity profile, the drawworks control system optimizes the operator's ability to maximize the speed of trips between positions in the derrick (between the travel limits) while maintaining safe operation. Travel of the block assembly following the full-velocity profile (from upper travel limit to the lower travel limit and back again) is described below for operation under normal conditions.

Lower Deceleration Point
When lowering, the lower deceleration point is the lowest position at which mechanical braking can be utilized to safely stop the block without passing the lower travel limit. The lower deceleration point is calculated by adding the calculated stopping distance for downward travel to the lower

travel limit position. Stopping distance is calculated based on available braking torque and system inertia, which, in addition to constant system parameters, includes variables such as lines strung, hookload, and block position.

Lower Travel Limit
The lower travel limit is the driller-defined position of lowermost travel.

Hoisting Deceleration Point
When hoisting, the hoisting deceleration point is the uppermost position at which the block will begin deceleration to safely stop the blocks without passing the upper travel limit and keep tension in the drill line. The hoisting deceleration point is calculated by subtracting the stopping distance from the upper travel limit. Stopping distance is calculated based on available braking torque and system inertia, which in addition to constant system parameters, includes variables such as lines strung, hookload, and block position.

Upper Travel Limit
The upper travel limit is the operator-defined position of upper-most travel.

Velocity Control
The VFD system works as a velocity-control system. The position of the drawworks control joystick is a velocity command sent to the controller that commands the VFD to go to the desired block speed (as limited by the velocity profile). The central position of the joystick represents zero velocity (i.e., stationary drum). The difference between the desired velocity and the actual velocity is an error signal. The magnitude of this error signal determines the force to be applied to the load.

4.2.2 Drilling and Production Hoisting Equipment
Drilling and production hoisting equipment include the following [1,2]:

1. *Crown block sheaves and bearings:* The stationary pulley system at the top of the derrick or mast.
2. *Traveling blocks:* A heavy duty pulley system that hangs in the derrick and travels up and down with the hoisted tools. It is connected to the crown block with a wire rope that ultimately runs to the hoisting drum.
3. *Block-to-hook adapters:* A metal piece that attaches to the bottom of the traveling block and serves as the mount for the hook.
4. *Connectors and link adapters.*
5. *Drilling hooks:* The hook that attaches to the traveling block to connect the bail of te swivel.
6. *Tubing and sucker rod hooks:* Hooks connected to the traveling block for tubing and sucker-rod hoisting operations.
7. *Elevator links:* The elevator is a hinged clamp attached to the hook and is used to hoist drill pipe, tubing, and casing. The actual clamp is in a pair of links that in turn attaches to a bail supported on the hook.
8. *Casing, tubing and drill pipe elevators.*
9. *Sucker rod elevator.*
10. *Rotary swivel bail adaptors:* A bail adaptor that allows the bail of the swivel to be grasped and hoisted with elevators.
11. *Rotary swivels.* The swivel connecting the nonrotating hook and the rotating kelley while providing a nonrotating connection through which mud enters the kelley.
12. *Spiders:* The component of the elevator that latches onto the hoisted item.

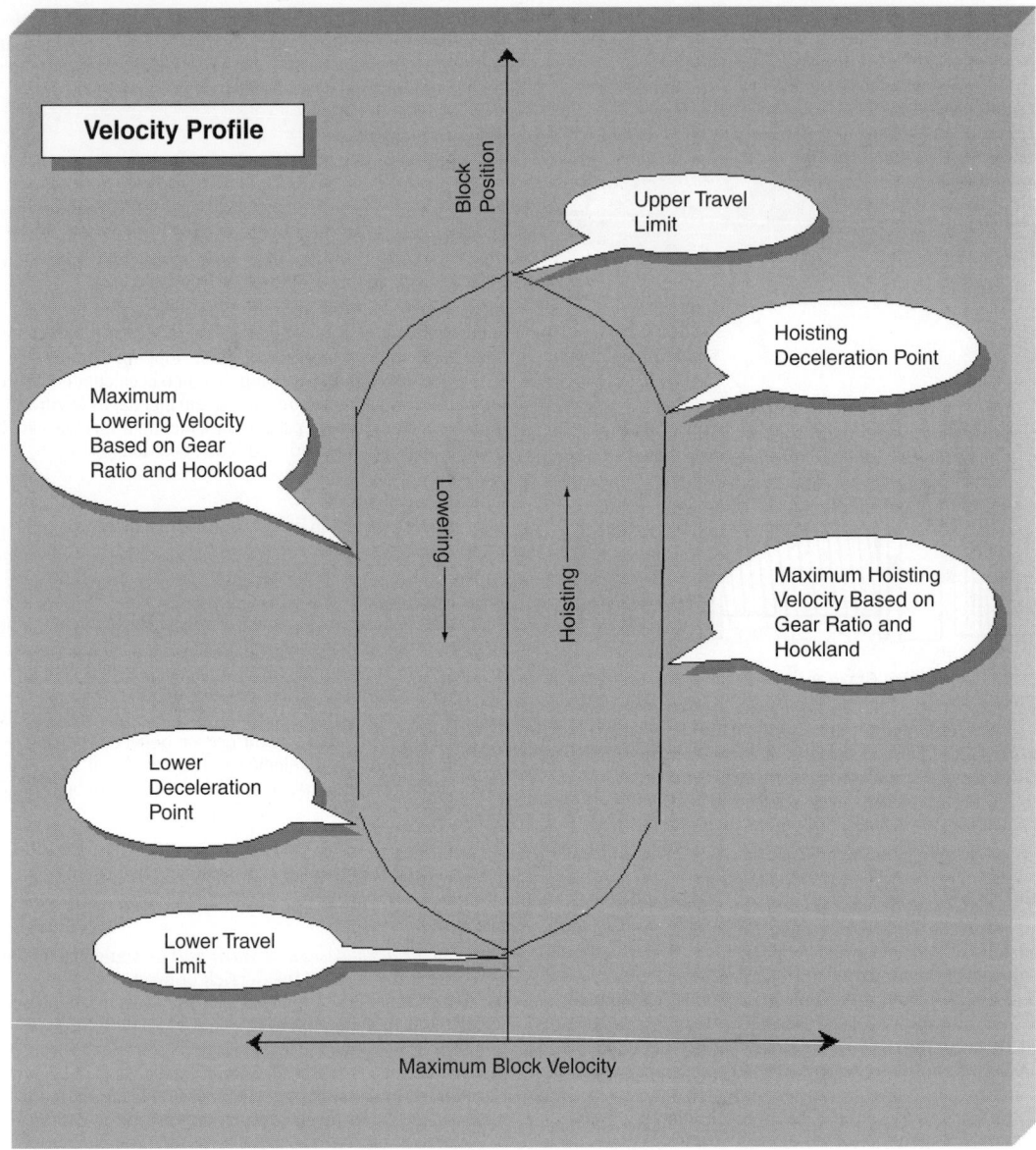

Figure 4.2.6 *Block velocity profile. (Courtesy of Varco, Inc.)*

13. *Deadline tiedowns:* The deadline is the nonmoving end of the wire rope from the hoisting down through the crown and traveling blocks. This end is anchored at ground level with a tiedown.
14. *Kelley spinners, when used as tension members:* An adapter between the swivel and the kelley that spins the Kelley for rapid attachment and disattachment to joints of drill pipe.
15. *Rotary tables, as structural members:* The rotary table rotates to turn the drill string. It is also used to support the drill string during some phases of opertion.
16. *Tension members of subsea handling equipment.*
17. *Rotary slips:* Wedging devices used to clamp the tool string into the rotary table. The wedging action is provided by friction.

4.2.2.1 Material Requirements

Castings

Steel castings used in the manufacture of the main load carrying components of the drilling and production hoisting equipment shall conform to ASTM A781: "Common Requirements for Steel and Alloy Castings for General Industrial Use," and either an individual material specification listed therein or a proprietary material specification that as a minimum conforms to ASTM A781.

Forgings

Steel forgings used in the manufacture of the main load carrying components of the equipment shall conform to ASTM A668: "Steel Forgings, Carbon and Alloy, for General Industrial Use" and ASTM A778: "Steel Forgings, General

Requirements." A material specification listed in ASTM A778 or a proprietary specification conforming to the minimum requirements of ASTM A788 may be used.

Plates, Shapes, and Bar Stock

Structural material used in the manufacture of main load carrying components of the equipment shall conform to applicable ASTM or API specifications covering steel shapes, plates, bars, or pipe, or a proprietary specification conforming to the minimum requirements of applicable ASTM or appropriate standard. Structural steel shapes having a specified minimum yield strength less than 33,000 psi, or steel pipe having a specified minimum yield strength less than 35,000 psi shall not be used.

4.2.2.2 Design Rating and Testing

All hoisting equipment shall be rated in accordance with the requirements specified herein. Such ratings shall consist of a maximum load rating for all items, and a main-bearing rating for crown blocks, traveling blocks, and swivels. *The traveling block and crown block ratings are independent of wire rope size and strength.* Such ratings shall be calculated as specified herein and in accordance with good engineering practices.

The ratings determined herein are intended to apply to new equipment only.

Maximum Load Rating

The maximum load ratings shall be given in tons (2,000-lb units). The size class designation shall represent the dimensional interchangeability and the maximum rated load of equipment specified herein. The recommended size classes are as follows (ton):

5	40	350
10	65	500
15	100	650
25	150	750
	250	1,000

For purpose of interchangeability contact radii shall comply with Table 4.2.1.

Maximum Load Rating Bases

The maximum load rating will be based on the design safety factor and the yield strength of the material. Crown block beams are an exception and shall be rated and tested in

Table 4.2.1 Recommended Hoisting Tool Contact Surface Radii (All dimensions in inches) [1]

1		2		3		4		5		6		7		8		9	
Rating		Traveling Block & Hook Bail See Fig. 4-16								Hook & Swivel Bail See Fig. 4-17							
Short tons	Metric tons	A_1 Max		A_2 Min		B_1 Min		B_2 Max		E_1 Min		E_2 Max		F_1 Max		F_2 Min	
		in.	mm	in.	mm	in.	mm	in.	mm	in.	mm	in.	mm	in.	mm	in.	mm
25-40	22.7-36.3	$2\frac{3}{4}$	69.85	$2\frac{3}{4}$	69.85	$3\frac{1}{4}$	82.55	3	76.20	2	50.80	$1\frac{1}{2}$	38.10	3	76.20	3	76.20
41-65	37.2-59	$2\frac{3}{4}$	69.85	$2\frac{3}{4}$	69.85	$3\frac{1}{4}$	82.55	3	76.20	2	50.80	$1\frac{3}{4}$	44.45	$3\frac{1}{2}$	88.90	$3\frac{1}{2}$	88.90
66-100	59.9-91	$2\frac{3}{4}$	69.85	$2\frac{3}{4}$	69.85	$3\frac{1}{4}$	82.55	3	76.20	$2\frac{1}{4}$	57.15	2	50.80	4	101.60	4	101.60
101-150	91.7-136	$2\frac{3}{4}$	69.85	$2\frac{3}{4}$	69.85	$3\frac{1}{4}$	82.55	3	76.20	$2\frac{1}{2}$	63.50	$2\frac{1}{4}$	57.15	$4\frac{1}{2}$	114.30	$4\frac{1}{2}$	114.30
151-250	137.1-227	4	101.60	4	101.60	$3\frac{1}{4}$	82.55	3	76.20	$2\frac{3}{4}$	69.85	$2\frac{1}{2}$	63.50	$4\frac{1}{2}$	114.30	$4\frac{1}{2}$	114.30
251-350	227.9-318	4	101.60	4	101.60	$3\frac{1}{4}$	82.55	3	76.20	3	76.20	$2\frac{3}{4}$	69.85	$4\frac{1}{2}$	114.30	$4\frac{1}{2}$	114.30
351-500	318.7-454	4	101.60	4	101.60	$3\frac{1}{2}$	88.90	$3\frac{1}{4}$	82.55	$3\frac{1}{2}$	88.90	$3\frac{1}{4}$	82.55	$4\frac{1}{2}$	114.30	$4\frac{1}{2}$	114.30
501-650	454.9-591	4	101.60	4	101.60	$3\frac{1}{2}$	88.90	$3\frac{1}{4}$	82.55	$3\frac{1}{2}$	88.90	$3\frac{1}{4}$	82.55	$4\frac{1}{2}$	114.30	$4\frac{1}{2}$	114.30
651-750	591.1-681	6	152.40	6	152.40	$3\frac{1}{4}$	88.90	$3\frac{1}{4}$	82.55	$4\frac{1}{4}$	107.95	4	101.60	$4\frac{1}{2}$	114.30	$4\frac{1}{2}$	114.30
751-1000	681.9-908	6	152.40	6	152.40	$6\frac{1}{4}$	158.75	6	152.40	$5\frac{1}{4}$	133.35	5	127.00	5	127.00	5	127.00

10		11		12		13		14		15		16		17		18	
Elevator Link & Hook Link Ear See Fig. 4-18								Elevator Link & Elevator Link Ear See Fig. 4-18								Rating	
C_1 Max		C_2 Min		D_1 Min		D_2 Max		G_1 Max		G_1 Min		H_1 Min		H_1 Max		Short tons	Metric tons
in.	mm	in.	mm	in.	mm	in.	mm	in.	mm	in.	mm	in.	mm	in.	mm		
$1\frac{1}{2}$	38.10	$1\frac{1}{4}$	38.10	$1\frac{1}{4}$	31.75	$\frac{7}{8}$	22.23			1	25.40			2	50.80	25-40	22.7-36.3
$2\frac{1}{2}$	63.50	$2\frac{1}{2}$	63.50	$1\frac{1}{4}$	31.75	$\frac{7}{8}$	22.23			1	25.40			2	50.80	41-65	37.2-59
$2\frac{1}{2}$	63.50	$2\frac{1}{2}$	63.50	$1\frac{1}{2}$	38.10	$1\frac{1}{8}$	28.58			1	25.40			2	50.80	66-100	59.9-91
$2\frac{1}{2}$	63.50	$2\frac{1}{2}$	63.50	$1\frac{1}{2}$	38.10	$1\frac{1}{8}$	28.58	$1\frac{5}{16}$	23.82	$1\frac{1}{2}$	38.10	2	50.80	2	50.80	101-150	91.7-136
4	101.60	4	101.60	$1\frac{3}{4}$	44.45	$1\frac{3}{8}$	34.93	$1\frac{7}{32}$	30.94	$1\frac{7}{8}$	47.63	$2\frac{3}{4}$	69.85	$2\frac{3}{4}$	69.85	151-250	137.1-227
4	101.60	4	101.60	$1\frac{3}{4}$	44.45	$1\frac{3}{8}$	34.93	$1\frac{5}{32}$	37.31	2	50.80	$2\frac{3}{4}$	69.85	$2\frac{3}{4}$	69.85	251-350	227.9-318
4	101.60	$4\frac{3}{4}$	120.65	$2\frac{1}{4}$	57.15	$1\frac{7}{8}$	47.63	$1\frac{7}{8}$	47.63	2	50.80	$3\frac{1}{4}$	82.55	$3\frac{1}{4}$	82.55	351-500	318.7-454
4	101.60	$4\frac{3}{4}$	120.65	$2\frac{1}{4}$	57.15	$1\frac{7}{8}$	47.63	$2\frac{1}{4}$	57.15	$2\frac{3}{8}$	60.32	5	127.00	5	127.00	501-650	454.9-591
4	101.60	5	127.00	$2\frac{1}{2}$	63.50	$2\frac{1}{2}$	63.50	$2\frac{1}{4}$	57.15	$2\frac{3}{8}$	60.32	5	127.00	5	127.00	651-750	591.1-681
$4\frac{1}{2}$	114.30	5	127.00	3	76.20	$2\frac{3}{4}$	69.85	$2\frac{3}{4}$	69.85	$2\frac{7}{8}$	73.03	$6\frac{1}{4}$	158.75	$6\frac{1}{4}$	158.75	751-1000	681.9-908

accordance with API Spec 4E, "Specification for Drilling and Well Servicing Structures."

Crown Block

For crown block design, see API Specification 4F. Crown block sheaves and bearings shall be designed in accordance with API Specification 8A.

Spacer Plates

Spacer plates of traveling blocks, not specifically designed to lend support to the sheave pin, shall not be considered in calculating the rated capacity of the block.

Sheave Pins

In calculations transferring the individual sheave loads to the pins of traveling blocks, these loads shall be considered as uniformly distributed over a length of pin equal to the length of the inner bearing race or over an equivalent length if an inner race is not provided.

Design Factor

The design safety factors shall be calculated as follows (Figure 4.2.7) for the relationship between the design safety factor and rating:

Calculated Rating (ton)	Yield Strength Design Safety Factor, SF_D
150 or less	3.00
Over 150 to 500	$3.00 - \dfrac{0.75(R^* - 150)}{350}$
Over 500	2.25

* R = rating in tons (2,000-lb units).

Mechanical Properties

The mechanical properties used for design shall be the minimum values allowed by the applicable material specification or shall be the minimum values determined by the manufacturer in accordance with the test procedures specified in ASTM A370: "Methods and Definitions for Mechanical Testing of Steel Products," or by mill certification for mill products. The yield point shall be used in lieu of yield strength for those materials exhibiting a yield point. Yield strength shall be determined at 0.2% offset.

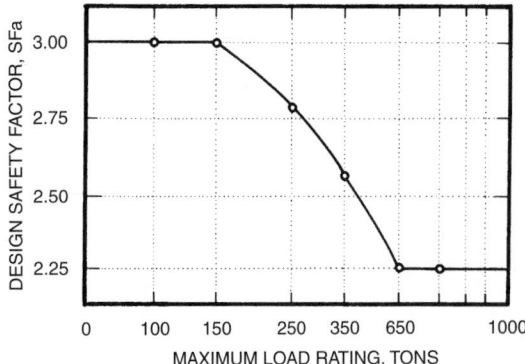

Figure 4.2.7 *Design safety factor and rating relationships [1].*

Shear Strength

For the purpose of calculations involving shear, the ratio of yield strength in shear-to-yield strength in tension shall be 0.58.

Extreme Low Temperature

Maximum load ratings shall be established at room temperature and shall be valid down to 0°F (-18°C). *The equipment at rated loads when temperature is less than 0°F is not recommended unless provided for by the supplement requirements. When the equipment is operating at lower temperatures, the lower impact absorbing characteristics of many steels must be considered.*

Test Unit

To ensure the integrity of design calculations, a test shall be made on one full size unit that in all respects represents the typical product. For a family of units of the same design concept but of varying sizes, or ratings, one test will be sufficient to verify the accuracy of the calculation method used, if the item tested is approximately midway of the size and rating range of the family, and the test results are applicable equally to all units in that family. Significant changes in design concept or the load rating will require supportive load testing.

Parts Testing

Individual parts of a unit may be tested separately if the holding fixtures simulate the load conditions applicable to the part in the assembled unit.

Test Fixtures

Test fixtures shall support the unit (or part) in essentially the same manner as in actual service, and with essentially the same areas of contact on the load-bearing surfaces.

Test Procedure

1. The test unit shall be loaded to the maximum rated load. After this load has been released, the unit shall be checked for useful functions. The useful function of all equipment parts shall not be impaired by this loading.
2. Strain gages may be applied to the test unit at all points where high stresses are anticipated, provided that the configuration of the units permits such techniques. The use of finite element analysis, models, brittle lacquer, etc., is recommended to confirm the proper location of strain gages. Three-element strain gages are recommended in critical areas to permit determination of the shear stresses and to eliminate the need for exact orientation of the gages.
3. The maximum test load to be applied to the test unit shall be $0.80 \times R \times SF_D$, but not less than 2R. Where R equals the calculated load rating in tons, SF_D is design safety factor.
4. The unit shall be loaded to the maximum test load carefully, reading strain gage values and observing for yielding. The test unit may be loaded as many times as necessary to obtain adequate test data.
5. Upon completion of the load test, the unit shall be disassembled and the dimensions of each part shall be checked carefully for evidence of yielding.

Determination of Load Rating

The maximum load rating may be determined from design and stress distribution calculations or from data acquired during a load test. Stress distribution calculations may be used to load rate the equipment only if the analysis has been shown to be within acceptable engineering allowances as verified by a load test on one member of the family of units of the same design. The stresses at that rating shall not exceed the allowed values. Localized yielding shall be permitted at areas of contact. In a unit that has been load tested, the critical permanent deformation determined by strain gages or other suitable means shall not exceed 0.002 in./in. If the stresses exceed the allowed values, the affected part or parts must be revised to obtain the desired rating. Stress distribution calculations may be used to load rate the equipment only if the analysis has been shown to be within acceptable engineering allowances as verified by a load test of one member of the family of units of the same design.

Alternate Test Procedure and Rating

Destructive testing may be used provided an accurate yield and tensile strength for the material used in the equipment has been determined. This may be accomplished by using tensile test specimens of the actual material and determining the yield strength to ultimate strength ratio. This ratio is then used to obtain the rating R (ton) of the equipment by the following equation:

$$R = \frac{\left(\dfrac{YS}{TS}\right) L_B}{SF_D} \qquad [4.2.12]$$

where SF_D = yield strength design safety factor
$\quad YS$ = yield strength in psi
$\quad TS$ = ultimate tensile strength in psi
$\quad L_B$ = breaking load in tons

Load Testing Apparatus

The load apparatus used to simulate the working load on the test unit shall be calibrated in accordance with ASTM E-4: "Standard Methods of Verification of Testing Machines," so as to assure that the prescribed test load is obtained.

Block Bearing Rating

The bearing rating of crown and traveling blocks shall be determined by

$$W_b = \frac{NW_r}{714} \qquad [4.2.13]$$

where W_b = calculated block bearing rating in tons
$\quad N$ = number of sheaves in the block
$\quad W_r$ = individual sheave bearing rating at 100 rpm for 3,000-hr minimum life for 90% of bearings in pounds.

Swivel Bearing Rating

The bearing rating of swivels shall be determined by

$$W_s = \frac{W_r}{1600} \qquad [4.2.14]$$

where W_s = calculated main thrust-bearing rating at 100 rpm in tons
$\quad W_r$ = main bearing thrust rating at 100 rpm for 3000-hr minimum life for 90% of bearings in pounds.

Traveling Block Hood Eye Opening Rating

The traveling block top handling member shall, for 500-ton size class and larger, have a static load rating based on safety factors given in the preceding paragraph titled "Design Factor."

Design Changes

When any change made in material, dimension, or construction might decrease the calculated load or bearing ratings, the unit changed shall be rerated, and retested if necessary. Parts of the modified unit that remain unchanged from the original design need not be retested, provided such omission does not alter the test results of the other components.

Records

The manufacturer shall keep records of all calculations and tests. When requested by prospective purchaser or by a user of the equipment, the manufacturer shall examine the details of computations, drawings, tests, or other supporting data necessary to demonstrate compliance with the specification. It shall be understood that such information is for the sole use of the user or prospective purchaser for checking the API rating, and the manufacturer shall not be required to release the information from his custody.

4.2.2.3 Elevators

Drill pipe elevators for taper shoulder and square shoulder weld-on tool joints shall have bore dimensions as specified in Table 4.2.2.

The permissible tolerance on the outside diameter immediately behind the tubing upset may cause problems with slip-type elevators.

4.2.2.4 Rotary Swivels

Rotary Swivel Pressure Testing

The assembled pilot model of rotary swivels shall be statically pressure tested. All cast members in the rotary swivel hydraulic circuit shall be pressure tested in production. This test pressure shall be shown on the cast member.

The test pressure shall be twice the working pressure up to 5,000 psi (incl.). For working pressures above 5,000 psi, the test pressure shall be one and one-half times the working pressure.

Swivel Gooseneck Connection

The angle between the gooseneck centerline and vertical shall be 15°. The swivel gooseneck connections shall be 2, $2\frac{1}{2}$, 3, $3\frac{1}{2}$, 4, or 5-in. nominal line pipe size as specified on the purchase order (see Figure 4.2.8). Threads on the gooseneck connection shall be internal line pipe threads conforming to API Standard 5B: "Threading, Gaging, and Thread Inspection of Casing, Tubing, and Line Pipe Threads." Rotary swivel gooseneck connections shall be marked with the size and type of thread, such as 3 API LP THD.

Rotary Hose Safety Chain Attachment

Swivels with gooseneck connections in 2 in. or larger shall have a suitable lug containing a $1\frac{1}{8}$-in. hole to accommodate the clevis of a chain having a breaking strength of 16,000 lb. The location of the lug is the choice of the manufacturer.

4.2.2.5 Sheaves for Hoisting Blocks

The sheave diameter shall be the overall diameter D as shown in Figure 4.2.9. Sheave diameters shall, wherever practicable, be determined in accordance with recommendations given in the section titled "Wire Rope."

Grooves for drilling and casing line sheaves shall be made for the rope size specified by the purchaser. The bottom of the groove shall have a radius R, Table 4.2.3, subtending an arc of 150°. The sides of the groove shall be tangent to the ends of the bottom arc. Total groove depth shall be a minimum of 1.33d and a maximum of 1.75d (d is the nominal rope diameter shown in Figure 4.2.9).

Table 4.2.2 *Drill Pipe Elevator Bores (All dimensions in inches) [1]*

1	2	3		4		5		6		7
		Weld-On Tool Joints								
		Taper Shoulder				Square Sholder				
Tool Joint Designation Reference	Drill Pipe Size and Style (All Weights and Grades)	Neck Diam. D_{TE} Max.[1] in.	mm	Elev. Bore in.	mm	Neck Diam. D_{SE} Max.[2] in.	mm	Elev. Bore in.	mm	Elev. Marking
NC 26 ($2\frac{3}{8}$ IF)	$2\frac{3}{8}$ EU	$2\frac{9}{16}$	65.09	$2\frac{21}{32}$	67.47	*		*		$2\frac{3}{8}$ EU
NC 31 ($2\frac{7}{8}$ IF)	$2\frac{7}{8}$ EU	$3\frac{3}{16}$	80.96	$3\frac{9}{32}$	83.34	$3\frac{3}{16}$	80.96	$3\frac{3}{8}$	87.73	$2\frac{7}{8}$ EU
NC 38 ($3\frac{1}{2}$ IF)	$3\frac{1}{2}$ EU	$3\frac{7}{8}$	98.43	$3\frac{31}{32}$	100.81	$3\frac{7}{8}$	98.43	$4\frac{1}{16}$	103.19	$3\frac{1}{2}$ EU
NC 40 (4 FH)	$3\frac{1}{2}$ EU	$3\frac{7}{8}$	98.43	$3\frac{31}{32}$	100.81	$3\frac{7}{8}$	98.43	$4\frac{1}{16}$	103.19	
NC 40 (4 FH)	4 IU	$4\frac{3}{16}$	106.36	$4\frac{9}{32}$	101.86	$4\frac{1}{8}$	104.78	$4\frac{5}{16}$	109.54	4 IU
NC 46 (4 IF)	4 EU	$4\frac{1}{2}$	114.30	$4\frac{23}{32}$	121.44	$4\frac{1}{2}$	114.30	$4\frac{13}{16}$	122.24	
	$4\frac{1}{2}$ IU	$4\frac{11}{16}$	119.06	$4\frac{25}{32}$	121.44	$4\frac{5}{8}$	117.48	$4\frac{13}{16}$	122.24	4 EU
	$4\frac{1}{2}$ IEU	$4\frac{11}{16}$	119.06	$4\frac{25}{32}$	121.44	$4\frac{5}{8}$	117.48	$4\frac{13}{16}$	122.24	$4\frac{1}{2}$ IU
$4\frac{1}{2}$ FH**	$4\frac{1}{2}$ IU	$4\frac{11}{16}$	119.06	$4\frac{25}{32}$	121.44	$4\frac{5}{8}$	117.48	$4\frac{13}{16}$	122.24	$4\frac{1}{2}$ IEU
	$4\frac{1}{2}$ IEU	$4\frac{11}{16}$	119.06	$4\frac{25}{32}$	121.44	$4\frac{5}{8}$	117.48	$4\frac{13}{16}$	122.24	
NC 50 ($4\frac{1}{2}$ IF)	$4\frac{1}{2}$ EU	5	127.00	$5\frac{1}{4}$	133.35	5	127.00	$5\frac{5}{16}$	134.94	$4\frac{1}{2}$ EU
	5 IEU	$5\frac{1}{4}$	130.18	$5\frac{1}{4}$	133.35	$5\frac{1}{8}$	130.18	$5\frac{5}{16}$	134.94	5 IEU
$5\frac{1}{2}$ FH**	5 IEU	$5\frac{1}{8}$	130.18	$5\frac{1}{4}$	133.35	$5\frac{1}{8}$	130.18	$5\frac{5}{16}$	134.94	
$5\frac{1}{2}$ FH**	5 IEU	$5\frac{11}{16}$	144.46	$5\frac{13}{16}$	147.64	$5\frac{11}{16}$	144.46	$5\frac{7}{8}$	149.23	$5\frac{1}{2}$ IEU
$6\frac{5}{8}$ FH	$6\frac{5}{8}$ IEU	$6\frac{57}{64}$	175.02	$7\frac{1}{32}$	178.66					$6\frac{5}{8}$

NOTE : *Elevators with the same bores are the same elevators.*
*Not manufactured.
**Obsolescent connection.

[1]Dimension D_{TE} from API Spec 7, Table 4.1.2.
[2]Dimension D_{SE} from API Spec 7. Appendix H.

Casing		Elevator Bores			
"D" Casing Dia.		"T_B" Top Bore ±1/64 ±.40 mm		"B_B" Bottom Bore +1/32 +.79 −1/64 −.40	
in.	*mm*	in.	*mm*	in.	*mm*
$4\frac{1}{2}$	*114.30*	4.594	*116.69*	4.504	*116.69*
5	*127.00*	5.125	*130.18*	5.125	*130.18*
$5\frac{1}{2}$	*139.70*	5.625	*142.88*	5.625	*142.88*
$6\frac{5}{8}$	*168.28*	6.750	*171.45*	6.750	*171.45*
7	*177.80*	7.125	*180.98*	7.125	*180.98*
$7\frac{5}{8}$	*193.68*	7.781	*197.64*	7.781	*197.64*
$7\frac{3}{4}$	*196.85*	7.906	*200.81*	7.906	*200.81*
$8\frac{5}{8}$	*219.08*	8.871	*223.04*	8.871	*223.04*
$9\frac{5}{8}$	*244.48*	9.871	*248.44*	9.871	*248.44*
$10\frac{3}{4}$	*273.05*	10.938	*277.83*	10.938	*277.83*
$11\frac{3}{4}$	*298.45*	11.938	*303.23*	11.938	*303.23*
$13\frac{3}{8}$	*339.73*	13.563	*344.50*	13.582	*344.50*
18	*406.40*	16.219	*411.96*	16.219	*411.96*
$18\frac{5}{8}$	*473.08*	18.875	*479.43*	18.875	*479.43*
20	*508.00*	20.281	*515.14*	20.281	*515.14*

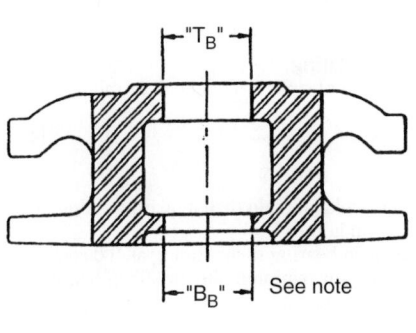

NOTE: *Bottom bore "B_B" is optional, some elevator designs do not have a bottom bore.*

Table 4.2.2 *continued*

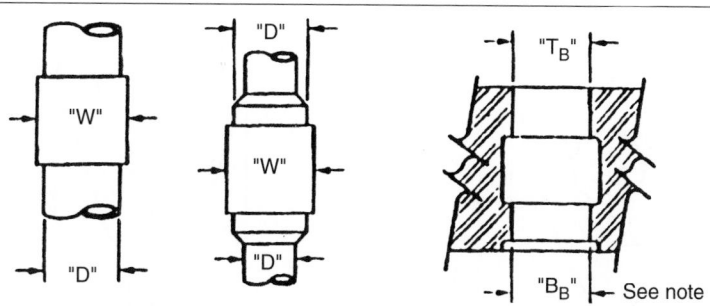

Tubing		Non-Upset Tubing						External Upset Tubing							
"D" Size O.D.		"W" Collar Dia.		"T_B" Top Bore		"B_B" Bottom Bore		"W" Collar Dia.		"D_4" Upset Dia.		"T_B" Top Bore		"B_B" Bottom Bore	
				+1.64 ± .40 mm		+1/32 +.79 − 1/64 −.40						+1.64 ± .40 mm		+1/32 +.79 − 1/64 −.40	
in.	mm	in.	mm	in.	mm	in.	mm	in.	mm	in.	mm	in.	mm	in.	mm
1.050	26.67	1.313	33.35	1.125	28.58	1.125	28.58	1.680	42.16	1.315	33.40	1.422	36.12	1.422	36.12
1.315	33.40	1.660	42.16	1.390	35.31	1.390	35.31	1.900	48.26	1.489	37.31	1.578	40.08	1.578	40.08
1.660	42.16	2.054	52.17	1.734	44.04	1.734	44.04	2.200	55.58	1.812	46.02	1.922	48.82	1.922	48.82
1.900	48.26	2.200	65.88	1.984	60.39	1.984	50.39	2.500	63.50	2.093	55.70	2.203	56.03	2.203	56.03
$2\frac{3}{8}$	60.32	2.875	73.03	2.453	62.31	2.453	62.31	3.063	77.80	2.593	65.89	2.703	68.58	2.703	68.58
$2\frac{7}{8}$	73.03	3.500	88.90	2.953	75.01	2.953	75.01	3.666	93.17	3.093	78.56	3.203	81.36	3.203	81.36
$3\frac{1}{2}$	88.90	4.250	107.95	3.678	90.88	3.578	90.88	4.500	114.30	3.750	96.25	3.859	98.02	3.859	98.02
4	101.60	4.750	120.65	4.078	103.58	4.078	109.58	5.000	127.00	4.250	107.95	4.359	110.74	4.359	110.74
$4\frac{1}{2}$	114.30	5.200	132.08	4.593	116.69	4.593	116.69	5.583	141.30	4.750	120.65	4.859	123.44	4.859	123.44

CAUTION: DO NOT USE EXTERNAL UPSET TUBING ELEVATORS ON NON-UPSET TUBING.

NOTE: *Bore "B_B" is optional, some elevator designs do not have a bottom bore.*

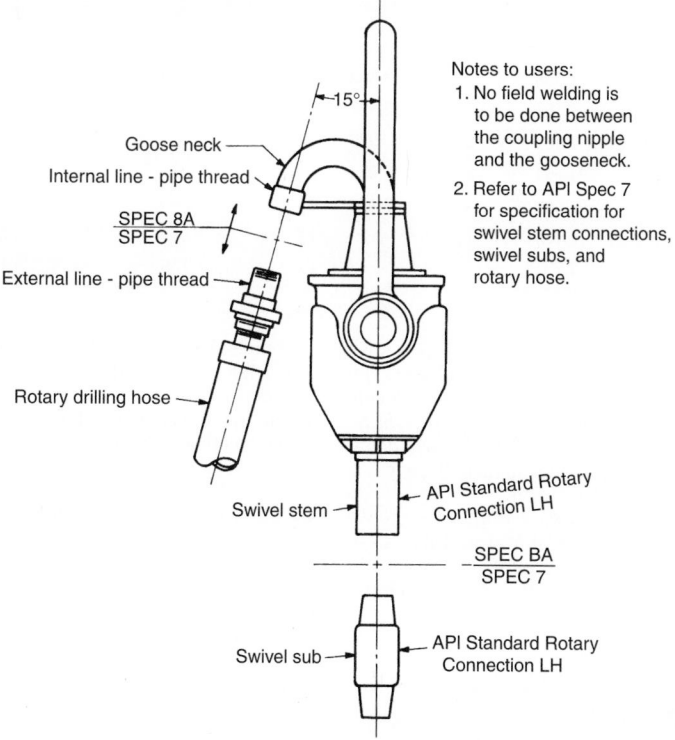

Notes to users:
1. No field welding is to be done between the coupling nipple and the gooseneck.
2. Refer to API Spec 7 for specification for swivel stem connections, swivel subs, and rotary hose.

Figure 4.2.8 *Rotary swivel connections [1].*

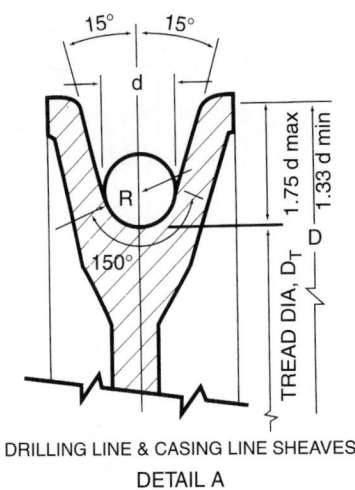
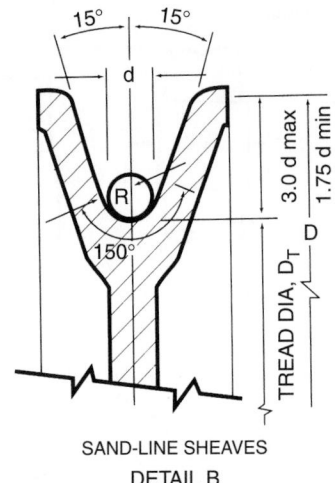

DRILLING LINE & CASING LINE SHEAVES
DETAIL A

SAND-LINE SHEAVES
DETAIL B

Figure 4.2.9 *Sheave grooves [1].*

Table 4.2.3 *Groove Radii for New and Reconditioned Sheaves and Drums (All dimensions in inches) [1]*

1	2	1	2	1	2
Wire Rope Nominal Size	Radii	Wire Rope Nominal Size	Radii	Wire Rope Nominal Size	Radii
$\frac{1}{4}$	.137	$1\frac{5}{8}$	.876	$3\frac{3}{8}$	1.807
$\frac{5}{16}$	.167	$1\frac{3}{4}$	.939	$3\frac{1}{2}$	1.869
$\frac{3}{8}$	.201	$1\frac{7}{8}$	1.003	$3\frac{3}{4}$	1.997
$\frac{7}{16}$	.234	2	1.070	4	2.139
$\frac{1}{2}$	.271	$2\frac{1}{8}$	1.137	$4\frac{1}{4}$	2.264
$\frac{9}{16}$	.303	$2\frac{1}{4}$	1.210	$4\frac{1}{2}$	2.396
$\frac{5}{8}$	.334	$2\frac{3}{8}$	1.273	$4\frac{3}{4}$	2.534
$\frac{3}{4}$	.401	$2\frac{1}{2}$	1.338	5	2.663
$\frac{7}{8}$	.468	$2\frac{5}{8}$	1.404	$5\frac{1}{4}$	2.804
1	.543	$2\frac{3}{4}$	1.481	$5\frac{1}{2}$	2.929
$1\frac{1}{8}$	.605	$2\frac{7}{8}$	1.544	$5\frac{3}{4}$	3.074
$1\frac{1}{4}$	.669	3	1.607	6	3.198
$1\frac{3}{8}$	.736	$3\frac{1}{8}$	1.664		
$1\frac{1}{2}$	.803	$3\frac{1}{4}$	1.731		

Table 4.2.4 *Groove Radii for Worn Sheaves and Drums (All dimensions in inches) [1]*

1	2	1	2	1	2
Wire Rope Nominal Size	Radii	Wire Rope Nominal Size	Radii	Wire Rope Nominal Size	Radii
$\frac{1}{4}$	.129	$1\frac{5}{8}$	.833	$3\frac{3}{8}$	1.730
$\frac{5}{16}$	.160	$1\frac{3}{4}$	.897	$3\frac{1}{2}$	1.794
$\frac{3}{8}$	.190	$1\frac{7}{8}$	.959	$3\frac{3}{4}$	1.918
$\frac{7}{16}$	.220	2	1.019	4	2.050
$\frac{1}{2}$	.256	$2\frac{1}{8}$	1.079	$4\frac{1}{4}$	2.178
$\frac{9}{16}$	.288	$2\frac{1}{4}$	1.153	$4\frac{1}{2}$	2.298
$\frac{5}{8}$	.320	$2\frac{3}{8}$	1.217	$4\frac{3}{4}$	2.434
$\frac{3}{4}$	.380	$2\frac{1}{2}$	1.279	5	2.557
$\frac{7}{8}$	.440	$2\frac{5}{8}$	1.339	$5\frac{1}{4}$	2.691
1	.513	$2\frac{3}{4}$	1.409	$5\frac{1}{2}$	2.817
$1\frac{1}{8}$	.577	$2\frac{7}{8}$	1.473	$5\frac{3}{4}$	2.947
$1\frac{1}{4}$	.639	3	1.538	6	3.075
$1\frac{3}{8}$	.699	$3\frac{1}{8}$	1.598		
$1\frac{1}{2}$	.759	$3\frac{1}{4}$	1.658		

Standard Machine Tolerance.

In the same manner, grooves for sand-line sheaves shall be made for the rope size specified by the purchaser. The bottom of the groove shall have a radius R, Table 4.2.2, subtending an arc of 150°. The sides of the groove shall be tangent to the ends of the bottom arc. Total groove depth shall be a minimum of 1.75d and a maximum of 3d, and d is nominal rope diameter (see Figure 4.2.9).

Sheaves should be replaced or reworked when the groove radius decreases below the values shown in Table 4.2.4. Use sheave gages as shown in Figure 4.2.10. Figure 4.2.10A shows a sheave with a aminimum groove radius, and Figure 4.2.10B shows a sheave with a tight groove.

4.2.2.6 Contact Surface Radii
Figures 4.2.11, 4.2.12, 4.2.13 and Table 4.2.1 show recommended radii of hoisting tool contact surfaces. These recommendations cover hoisting tools used in drilling, and tubing hooks, but all other workover tools. Contract radii are intended to cover *only points of contact* between two elements and are not intended to define other physical dimensions of the connecting parts.

4.2.2.7 Inspection, Nondestructive Examination, and Compliance
Inspection
While work on the contract of the purchaser is being performed, the purchaser's inspector, shall have reasonable access to the appropriate parts of the manufacturer's works concerning the manufacture of the equipment ordered hereunder. Inspection shall be made at the works prior to shipment, unless otherwise specified, and shall be conducted so

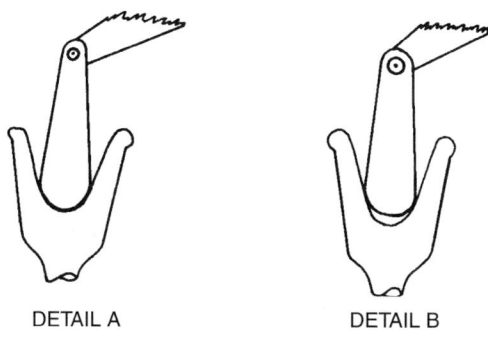

Figure 4.2.10 *Use of sheave gages [1].*

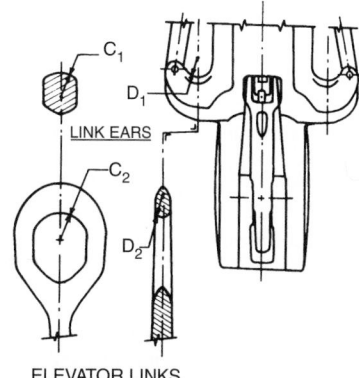

Figure 4.2.12 *Elevator link and link ear contact surface radii [1].*

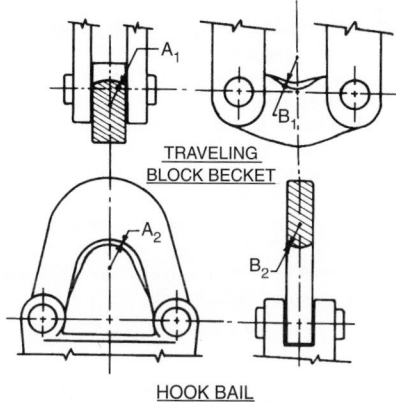

Figure 4.2.11 *Traveling block and hook bail contact surface radii [1].*

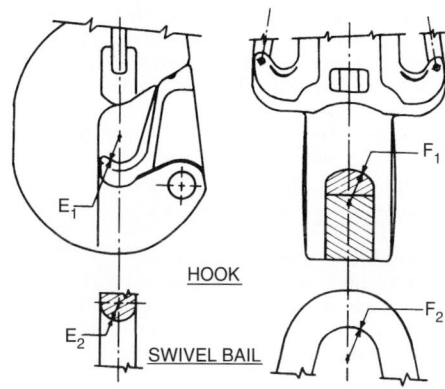

Figure 4.2.13 *Hook and swivel bail contact surface radii [1].*

as not to interfere unnecessarily with the works' operation or production schedules.

Nondestructive Examination
The manufacturer shall have a reasonable written nondestructive examination program to assure that the equipment manufactured is suitable for its intended use. If the purchaser's inspector desires to witness these operations, the manufacturer shall give reasonable notice of the time at which the examinations are to be performed.

Compliance
The manufacturer is responsible for complying with all of the provisions of the specification.

4.2.2.8 Supplementary Requirements
Magnetic Particle Examination
All accessible surfaces of the main load carrying components of the equipment shall be examined by a magnetic particle examination method or technique conforming to the requirements of ASTM E709: "Recommended Practice for Magnetic Particle Examination." Acceptance limits shall be as agreed upon by the manufacturer and the purchaser.

Liquid Penetrant Examination
All accessible surfaces of the main load carrying components of the equipment shall be examined by a liquid penetrant examination or technique conforming to the

requirements of ASTM E165: "Recommended Practice for the Liquid Penetrant Examination Method." Acceptance limits shall be as agreed upon by the manufacturer and the purchaser.

Ultrasonic Examination
Main load carrying components of the equipment shall be ultrasonically examined in accordance with applicable ASTM standards. The extent of examination, method of examination, and basis for acceptance shall be agreed upon by the manufacturer and purchaser.

Radiographic Examination
Main load carrying components of the equipment shall be examined by means of gamma rays or x-rays. The procedure used shall be in accordance with applicable ASTM standards. Types and degrees of discontinuities considered shall be compared to the reference radiographs of ASTM as applicable. The extent of examination and the basis for acceptance shall be agreed upon by the manufacturer and purchaser.

Traceability
The manufacturer shall have reports of chemical analysis, heat treatment, and mechanical property tests for the main load carrying components of the equipment.

Welding

Where welding is involved in the critical load path of main load carrying components, recognized standards shall be used to qualify welders and procedures.

Extreme Low Temperature

Equipment intended for operation at temperatures below 0°F may require special design and/or materials.

4.2.3 Hoisting Tool Inspection and Maintenance Procedures

4.2.3.1 Inspection

Frequency of Inspection

Field inspection of drilling, production, and workover hoisting equipment in an operating condition should be made on a regular basis. A thorough on-the-job shutdown inspection should be made on a periodic basis, typically at 90 to 120-day intervals, or as special circumstances may require.

Critical loads may be experienced; for example, severe loads, impact loads such as jarring, pulling on stuck pipe, and/or operating at low temperatures. If in the judgement of the supervisor a critical load has occurred, or may occur, an on-the-job shutdown inspection equivalent to the periodic field inspection should be conducted before and after the occurrence of such loading. If critical loads are unexpectedly encountered, the inspection should be conducted immediately after such an occurrence.

When necessary, disassembly inspection of hoisting equipment should be made in a suitably equipped facility.

Methods of Inspection

Hoisting equipment should be inspected on a regular basis for cracks, loose fits or connections, elongation of parts, and other signs of wear, corrosion, or overloading. Any equipment showing cracks, excessive wear, etc., should be removed from service.

The periodic or critical load inspection in the field should be conducted by the crew with the inspector. For the periodic or critical load inspection, all foreign matter should be removed from surfaces inspected. Total field disassembly is generally not practical, and is not recommended, except as may be indicated in the detailed procedure for each tool.

Equipment, if necessary, should be disassembled in a suitably equipped facility and inspected for excessive wear, cracks, flaws, or deformation. Corrections should be made in accordance with the recommendations of the manufacturer. Before inspection, all foreign material, such as dirt, paint, grease, oil, scale, etc., should be removed from the inspected areas by a suitable method. The equipment should be disassembled as much as necessary to permit inspection of all load bearing parts, and the inspection should be made by trained, competent personnel.

4.2.3.2 Maintenance and Repairs

A regular preventive maintenance program should be established for all hoisting tools. Written maintenance procedures should be given to the crew or maintenance personnel. Maintenance procedures should be specified for each tool, as well as the specific lubricants to be used, and should be based on the tool manufacturer's recommendation. This recommended practice includes generalized procedures that are considered a minimum program. Care should be taken that instruction plates, rating plates, and warning labels are not missing, damaged, or illegible.

If repairs are not performed by the manufacturer, such repairs should be made in accordance with methods or procedures approved by the manufacturer. Minor cracks or defects, which may be removed without influence on safety or operation of the equipment, can be removed by grinding or filing. Following repair, the part should again be inspected by an appropriate method to ensure that the defect has been completely removed.

Antifriction bearings play an important part in the safe performance of the tool. The most likely requirements for bearing placement are very loose or bent cages (retainers), corrosion, abrasion, inadequate (or improper) lubrication, and spalling from fatigue. Excessive clearance may indicate improper adjustment or assembly and should be corrected. Repair of antifriction bearings should not be attempted by field or shop personnel. Consultation with the equipment manufacturer is recommended in case of unexplained or repeated bearing failure.

If the tool or part is defective beyond repair, it should be destroyed immediately.

Welding should not be done on hoisting tools without consulting the manufacturer. Without full knowledge of the design criteria, the materials used and the proper procedures (stress relieving, normalizing, tempering, etc.), it is possible to reduce the capacity of a tool sufficiently to make its continued used dangerous.

Inspection and maintenance (lubrication) of wire rope used in hoisting should be carried out on a regular basis. Wire rope inspection and maintenance recommendations are included in API RP 9B, "Application, Care and Use of Wire Rope for Oil Field Service" (see "Wire Rope").

4.2.3.3 Inspection and Maintenance Illustrations

Figures 4.2.14 through 4.2.31 are self-explanatory illustrations of generalized inspection and maintenance recommendations for each of the hoisting tools.

4.2.4 Wire Rope

Wire rope includes (1) bright (uncoated), galvanized, and drawn-galvanized wire rope of various grades and construction, (2) mooring wire rope, (3) torpedo lines, (4) well-measuring wire, (5) well-measuring strand, (6) galvanized wire guy strand, and (7) galvanized structural rope and strand [3,4].

4.2.4.1 Material

Wire used in the manufacture of wire rope is made from (1) acid or basic open-hearth steel, (2) basic oxygen steel, or (3) electric furnace steel. Wire tested before and after fabrication shall meet different tensile and torsional requirements as specified in Tables 4.2.5 and 4.2.6.

Galvanized Wire Rope

Galvanized wire rope shall be made of wire having a tightly adherent, uniform and continuous coating of zinc applied after final cold drawing, by the electrodeposition process or by the hot-galvanizing process. The minimum weight of zinc coating shall be as specified in Table 4.2.7.

Drawn-Galvanized Wire Rope

Drawn-galvanized wire rope shall be made of wire having a tightly adherent, uniform, and continuous coating of zinc applied at an intermediate stage of the wire drawing operation, by the electrodeposition process or by the hot-galvanizing process. The minimum weight of zinc coating shall be as specified in Table 4.2.8.

4.2.4.2 Properties and Tests for Wire and Wire Rope

Selection of Test Specimens

For the test of individual wires and of rope, a 10-ft (3.05-m) section shall be cut from a finished piece of unusedand

(text continued on page 4-45)

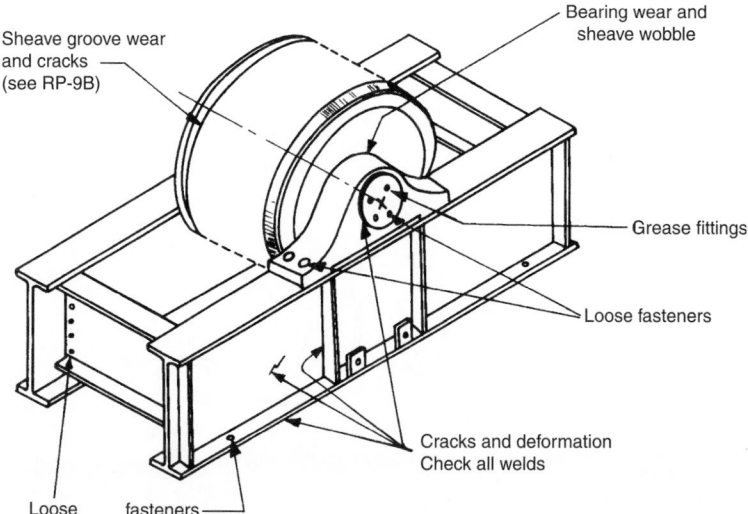

Sheave groove wear and cracks (see RP-9B)

Bearing wear and sheave wobble

Grease fittings

Loose fasteners

Cracks and deformation
Check all welds

Loose fasteners

MAINTENANCE:
1. Keep clean.
2. Lubricate bearings.
3. Remove and rust and weather protect as required.
4. Check and secure all fasteners.

Figure 4.2.14 *Crown block [2].*

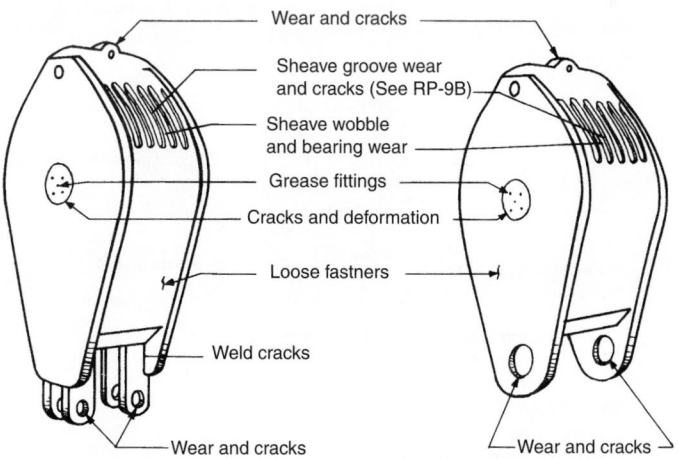

Wear and cracks

Sheave groove wear and cracks (See RP-9B)

Sheave wobble and bearing wear

Grease fittings

Cracks and deformation

Loose fastners

Weld cracks

Wear and cracks

Wear and cracks

MAINTENANCE:
1. Keep clean.
2. Lubricate bearings.
3. Remove any rust and weather protect as required.
4. Check and secure all fasteners.

Figure 4.2.15 *Traveling block [2].*

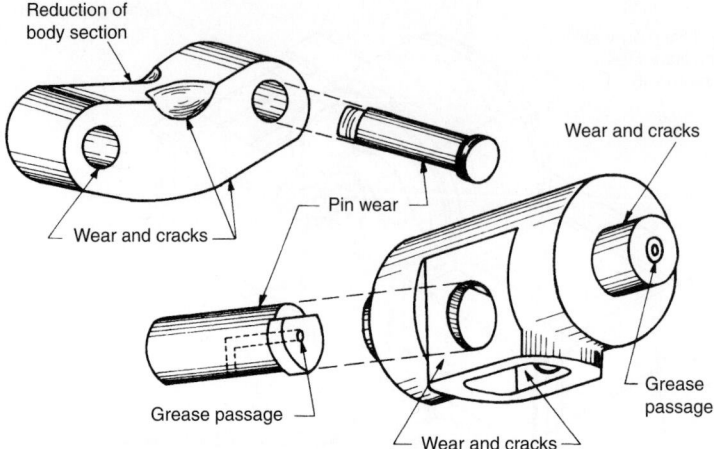

MAINTENANCE:
1. Keep clean.
2. Grease coat wear surface of clevis.
3. Remove any rust and weather protect as required.
4. Check and secure all pins.

Figure 4.2.16   *Block-to-hook adapter [2].*

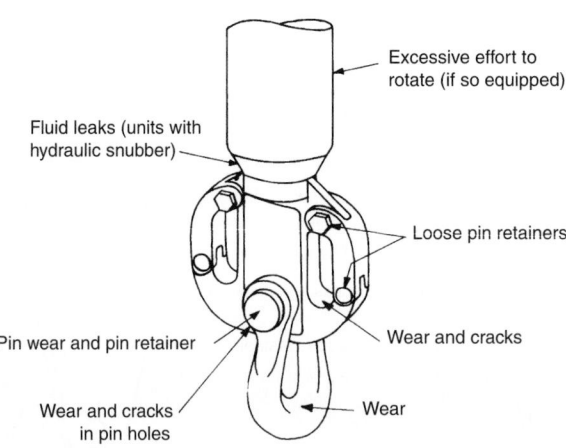

MAINTENANCE:
1. Keep clean.
2. Grease coat wear surfaces.
3. On units with hydraulic snubber check oil level and
   change oil at intervals recommended by manufacturer.
4. Oil pins not accessible to grease lubrication.
5. Remove any rust and weather protect as required.
6. Check and secure pins and fasteners.

Figure 4.2.17   *Link adapter [2].*

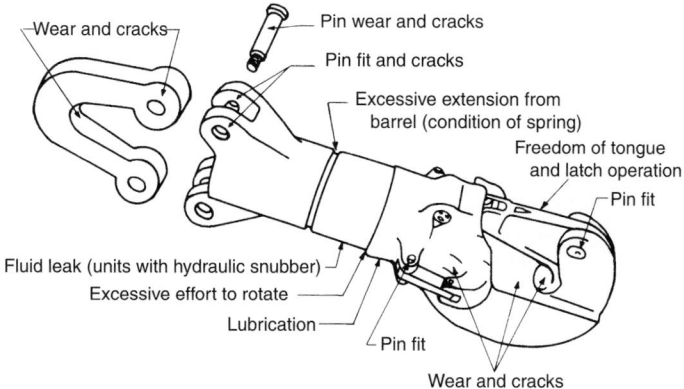

Wear and cracks

Pin wear and cracks

Pin fit and cracks

Excessive extension from barrel (condition of spring)

Freedom of tongue and latch operation

Pin fit

Fluid leak (units with hydraulic snubber)

Excessive effort to rotate

Lubrication

Pin fit

Wear and cracks

MAINTENANCE:
1. Keep clean.
2. Grease coat latching mechanism, link arms, and saddle.
3. Lube all grease fittings.
4. On units with hydraulic snubber check oil level and change oil at intervals recommended by manufacturer.
5. Oils pins not accessible to grease lubrication.
6. Remove any rust and weather protect as required.
7. Check and secure pins and fasteners.

Figure 4.2.18   *Drilling hook [2].*

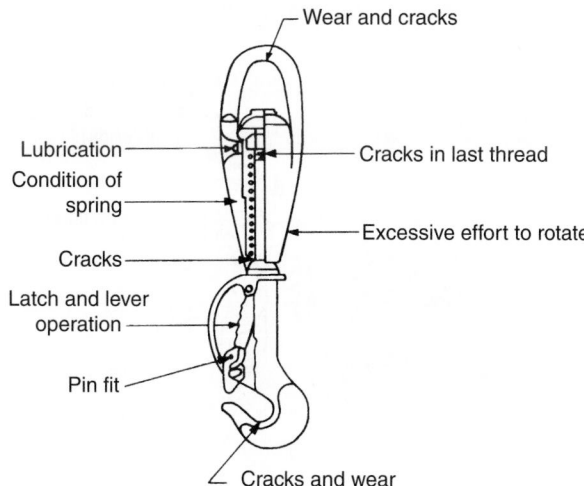

Wear and cracks

Lubrication

Condition of spring

Cracks in last thread

Cracks

Excessive effort to rotate

Latch and lever operation

Pin fit

Cracks and wear

MAINTENANCE:
1. Keep clean.
2. Grease coat latching mechanism, hook, and bail throat.
3. Grease main bearing.
4. Oil pins not accessible to grease lubrication.
5. Remove any rust and weather protect as required.
6. Check and secure pins and fasteners.

Figure 4.2.19   *Tubing and sucker rod hook [2].*

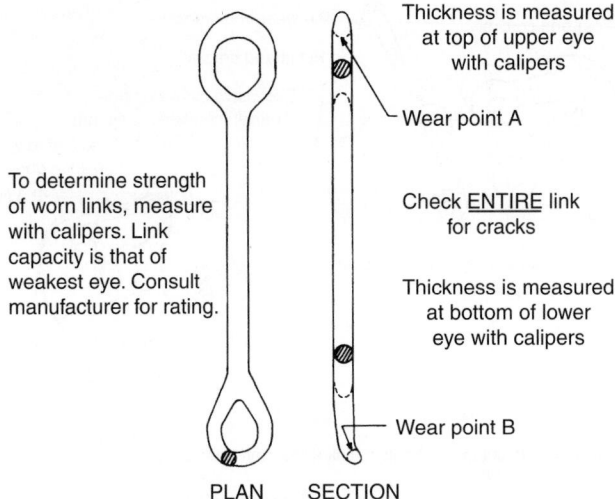

To determine strength of worn links, measure with calipers. Link capacity is that of weakest eye. Consult manufacturer for rating.

Thickness is measured at top of upper eye with calipers

Wear point A

Check <u>ENTIRE</u> link for cracks

Thickness is measured at bottom of lower eye with calipers

Wear point B

PLAN   SECTION

MAINTENANCE:
1. Keep clean.
2. Grease coat upper and lower eye wear surfaces.
3. Remove any rust and weather protect as required.

Figure 4.2.20   *Elevator link [2].*

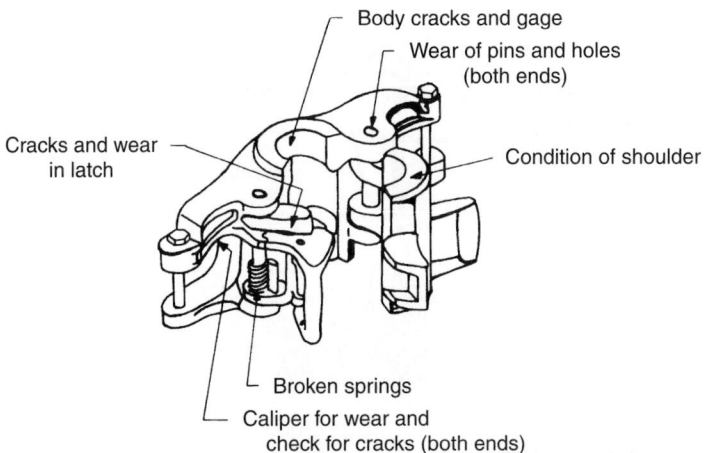

Body cracks and gage

Wear of pins and holes (both ends)

Cracks and wear in latch

Condition of shoulder

Broken springs

Caliper for wear and check for cracks (both ends)

MAINTENANCE:
1. Keep clean.
2. Grease coat link arm wear surfaces, latch lug and bore seat on bottleneck elevators.
3. Lubricate hinge pin.
4. Remove any rust and weather protect as required.
5. Check and secure pins and fasteners.

Figure 4.2.21a   *Casing, tubing, and drill pipe elevators and side door elevators [2].*

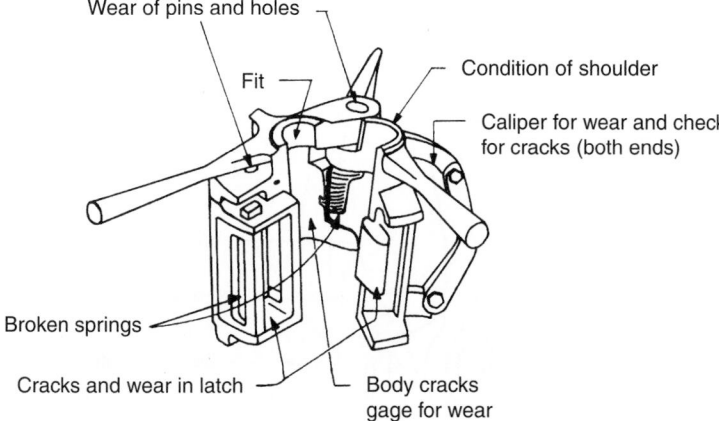

MAINTENANCE:
1. Keep clean.
2. Grease coat link arm wear surfaces, latch lug and bore seat on bottleneck elevators.
3. Lubricate hinge pin.
4. Remove any rust and weather protect as required.
5. Check and secure pins and fasteners.

Figure 4.2.21b   *Casing, tubing, and drill pipe elevators and center latch elevators [2].*

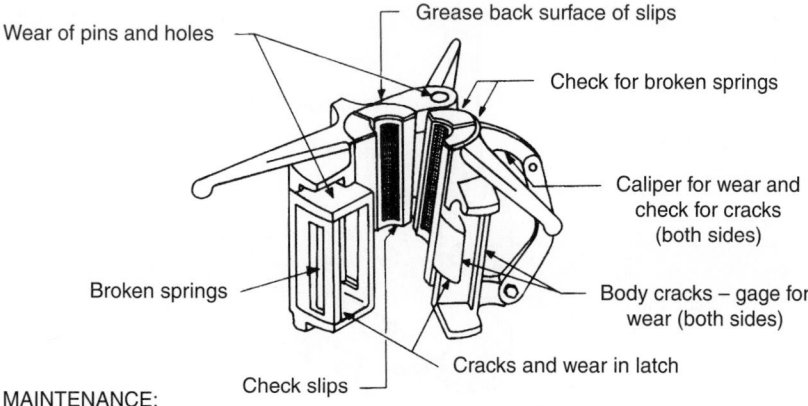

MAINTENANCE:
1. Keep clean.
2. Grease coat link arm wear surfaces and latch lug.
3. Lubricate hinge pin.
4. Remove any rust and weather protect as required.
5. Clean inserts. Replace when worn.
6. Tighten all loose fasteners.

Figure 4.2.21c   *Casing, tubing, and drill pipe elevators and slip type elevators [2].*

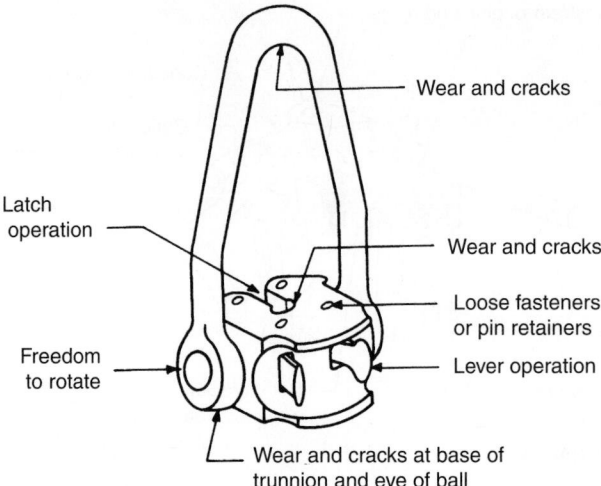

Wear and cracks

Latch operation

Wear and cracks

Loose fasteners or pin retainers

Freedom to rotate

Lever operation

Wear and cracks at base of trunnion and eye of ball

MAINTENANCE:
1. Keep clean.
2. Grease coat rod seating area, bail throat and latch mechanism.
3. Oil pins not accessible to grease lubrication.
4. Remove any rust and weather protect as required.
5. Check and secure pins and fasteners.

Figure 4.2.22   *Sucker rod elevators [2].*

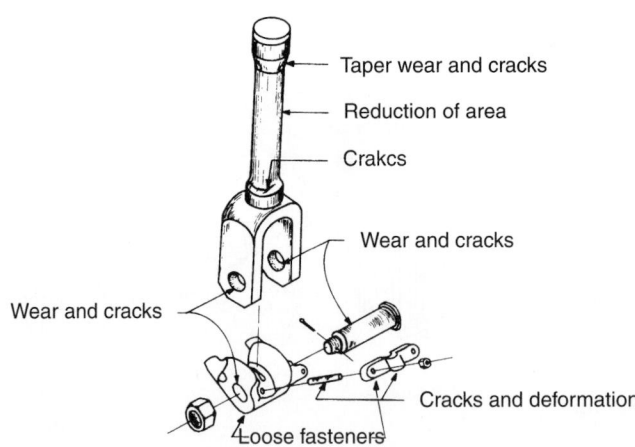

Taper wear and cracks

Reduction of area

Crakcs

Wear and cracks

Wear and cracks

Cracks and deformation

Loose fasteners

MAINTENANCE:
1. Keen clean.
2. Lubricate pivot and pin.
3. Remove any rust and weather protect as required.
4. Check and secure pins and fasteners.

Figure 4.2.23   *Swivel bail adapter [2].*

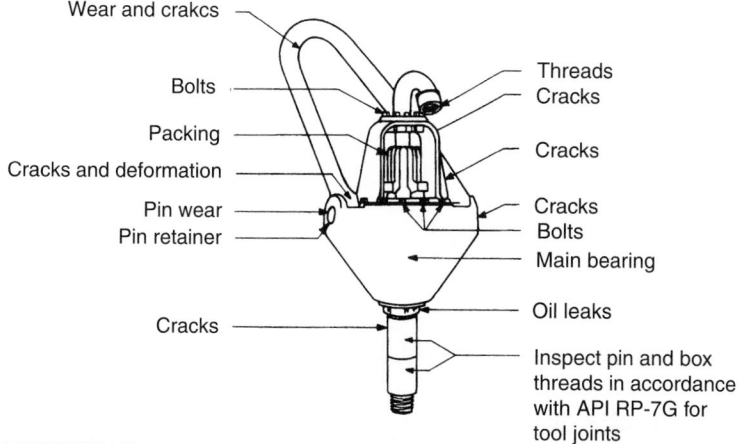

MAINTENANCE:
1. Keep clean.
2. Grease coat bail throat wear surface.
3. Lubricate bail pins, oil seals, upper bearing, and packing.
4. Check oil level as recommended by manufacturer.
5. Change oil at intervals recommended by manufacturer.
6. Remove any rust and weather protect as required.
7. Check and secure fasteners.
8. Protect threads at the gooseneck inlet and on the coupling
   nipple when not assembled during handling. Thread protection
   should be used.

Figure 4.2.24   *Rotary swivel [2].*

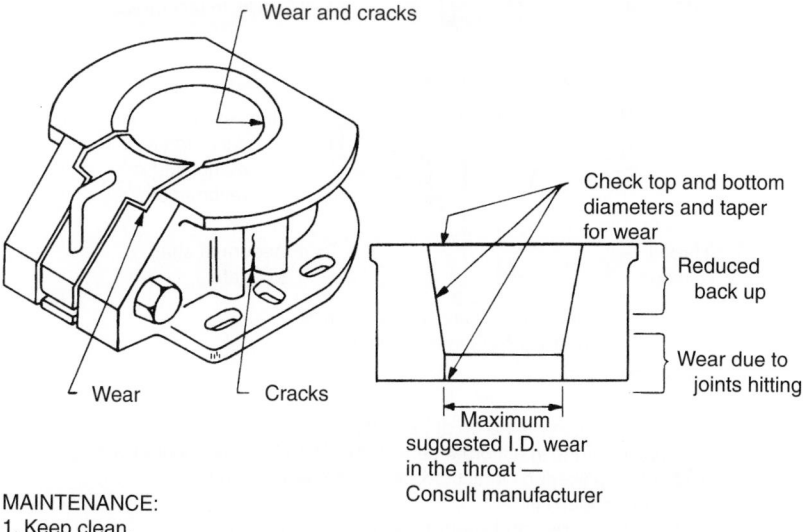

MAINTENANCE:
1. Keep clean.
2. Lubricate taper before each trip.
3. Remove any rust and weather protect as required.

Figure 4.2.25   *Spider [2].*

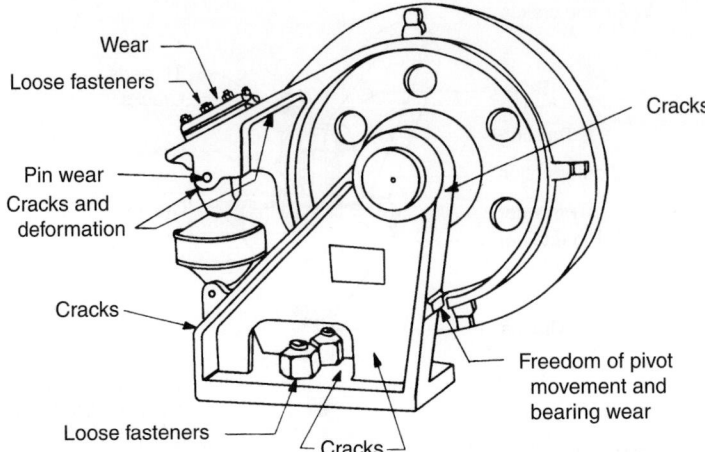

MAINTENANCE:
1. Keep clean.
2. Grease coat surface of wire line spool.
3. On units equipped with load cell for weight indicator, lubricate
   pivot bearing.
4. Remove any rust and weather protect as required.

Figure 4.2.26   *Deadline anchor [2].*

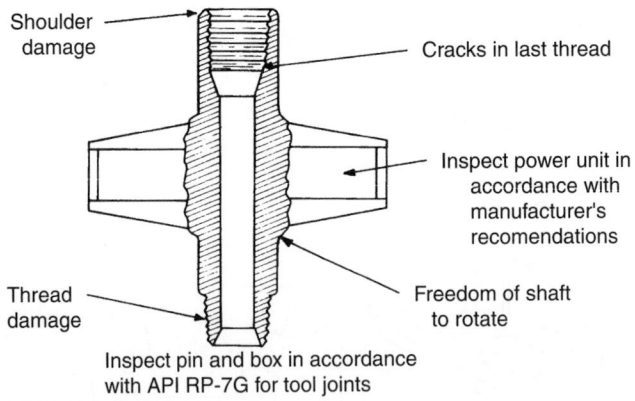

MAINTENANCE:
1. Keep clean.
2. Use thread compound on pin and box, and apply proper
   makeup torque in accordance with API RP-7G recommendations.
3. Maintain power unit in accordance with manufacturer's
   recommendations.
4. Remove any rust and weather protect as required.
5. Check and secure fasteners.

Figure 4.2.27   *Kelly spinner [2].*

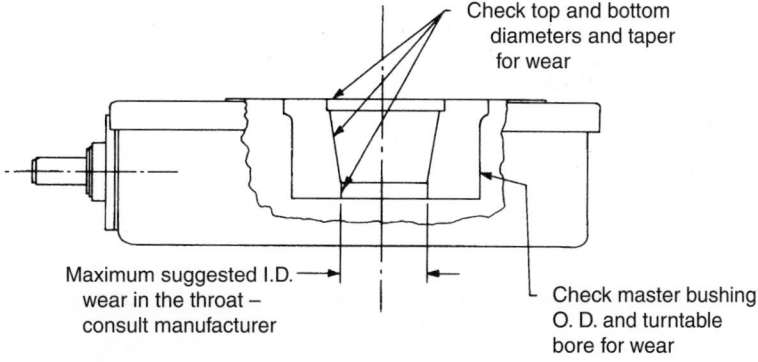

Check top and bottom diameters and taper for wear

Maximum suggested I.D. wear in the throat – consult manufacturer

Check master bushing O. D. and turntable bore for wear

MAINTENANCE:
1. Keep clean.
2. Remove any rust and weather protect as required.

Figure 4.2.28   *Rotary table [2].*

INSPECTION:
Heave compensator designs vary considerably among manufacturers, therefore, manufacturer's recommendations should be closely followed. In general, load carrying members shoud be checked for wear, cracks, flaws, and deformation. For compensators with integral traveling block and hook adaptor, the procedures defined in Figures. 4.2.11 and 4.2.12 apply to those parts of the assembly.

MAINTENANCE:
1. Keep clean.
2. Follow manufacturer's recommendations for specific unit.
3. For compensators with integral traveling block and hook adaptor, the procedures in Figures. 4.2.11 and 4.2.12 apply to those parts of the assembly.
4. Remove any rust and weather protect as required.

Figure 4.2.29   *Heave compensator [2].*

INSPECTION:
Tension members for sub sea handling equipment should be inspected according to the manufacturer's recommendations. In general, tension members should be checked for wear, cracks, reduction of area and elongation.

MAINTENANCE:
1. Tension members in sub sea handling equipment should be maintained in accordance with manufacturer's recommendations.

Figure 4.2.30   *Tension members of subsea handling equipment [2].*

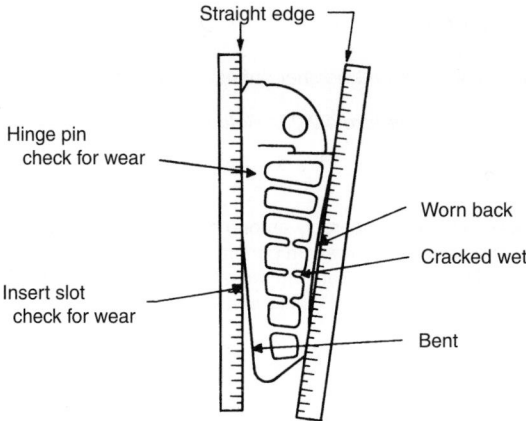

MAINTENANCE:
1. Keep clean.
2. Check the insert slots for wear and replace inserts as required.
3. Lubricate hinge pin.
4. Remove any rust and protect as required.
5. Use straight edge to detect uneven wear or damage.
6. *Caution:* Do not use wrong size slips—Match pipe and slip size.

Figure 4.2.31   *Rotary slips [2].*

Table 4.2.5 Mechanical Properties of Individual Rope Wires (After Fabrication) [4]

(1)	(2)	(3)	(4)	(5)	(6)	(7)	(8)	(9)	(10)	(11)	(12)	(13)	(14)	(15)	(16)	(17)	(18)	(19)	(20)	(21)	(22)
Wire Size Nominal Diameter		Level 2 Bright (Uncoated) or Drawn-Galvanized Breaking Strength					Level 3 Bright (Uncoated) or Drawn-Galvanized Breaking Strength					Level 4 Bright (Uncoated) or Drawn-Galvanized Breaking Strength					Level 5 Bright (Uncoated) or Drawn-Galvanized Breaking Strength				
		Individual Minimum		Average Minimum		Min. Tor.	Individual Minimum		Average Minimum		Min. Tor.	Individual Minimum		Average Minimum		Min. Tor.	Individual Minimum		Average Minimum		Min. Tor.
in.	mm	lb	N	lb	N		lb	N	lb	N		lb	N	lb	N		lb	N	lb	N	
0.010	0.25	17	76	17	76	241	20	89	21	93	222	21	93	23	102	202	23	102	25	111	176
0.011	0.28	20	89	22	98	219	23	102	25	111	202	26	116	28	125	183	28	125	30	133	160
0.012	0.30	24	107	26	116	201	28	125	30	133	185	31	138	33	147	168	33	147	35	156	146
0.013	0.33	28	125	30	133	185	33	147	35	156	171	36	160	38	169	155	39	173	41	182	135
0.014	0.36	33	147	35	156	172	38	169	40	178	159	42	187	44	196	144	45	200	47	209	126
0.015	0.38	38	169	40	178	161	44	196	46	205	148	48	214	50	222	134	52	231	54	240	117
0.016	0.41	43	191	45	200	150	50	222	52	231	139	55	245	57	254	126	59	262	62	276	109
0.017	0.43	49	218	51	227	142	56	249	58	258	130	61	271	65	289	118	66	294	70	311	103
0.018	0.46	55	245	57	254	134	62	276	66	294	124	69	307	73	325	112	74	329	78	347	97
0.019	0.48	60	267	64	285	126	70	311	74	329	117	77	342	81	360	106	83	369	87	387	93
0.020	0.51	67	298	71	316	120	77	342	81	360	110	85	378	89	396	100	92	409	96	427	87
0.021	0.53	74	329	78	347	114	85	378	89	396	105	94	418	98	436	95	100	445	106	471	83
0.022	0.56	81	360	85	378	109	94	418	98	436	101	102	454	108	480	90	110	489	116	516	79
0.023	0.58	89	396	93	414	105	102	454	108	480	96	112	498	118	525	86	121	538	127	565	75
0.024	0.61	97	431	101	449	100	111	494	117	520	92	122	543	128	569	82	132	587	138	614	71
0.025	0.64	104	463	110	489	96	120	534	126	560	88	133	592	139	618	78	142	632	150	667	68
0.026	0.66	113	503	119	529	92	130	578	136	605	85	143	636	151	672	75	154	685	162	721	65
0.027	0.69	122	543	128	569	88	140	623	148	658	82	154	685	162	721	72	166	738	174	774	63
0.028	0.71	131	583	137	609	86	151	672	159	707	79	166	738	174	774	69	178	792	188	836	60
0.029	0.74	140	623	148	658	83	162	721	170	756	76	177	787	187	832	66	191	850	201	894	58
0.030	0.76	150	667	158	703	80	173	770	181	805	73	190	845	200	890	64	205	912	215	956	55
0.031	0.79	160	712	168	747	77	184	818	194	863	71	203	903	213	947	62	218	970	230	1,023	54
0.032	0.81	171	761	179	796	74	196	872	206	916	68	215	956	227	1,010	59	232	1,032	244	1,085	51
0.033	0.84	181	805	191	850	72	209	930	219	974	67	229	1,019	241	1,072	57	247	1,099	259	1,150	50
0.034	0.86	192	854	202	898	70	221	983	233	1,036	65	244	1,085	256	1,139	56	261	1,161	275	1,223	49
0.035	0.89	204	907	214	952	68	234	1,041	246	1,094	63	257	1,143	271	1,205	54	277	1,232	291	1,294	47
0.036	0.91	215	956	227	1,010	67	248	1,103	260	1,156	61	273	1,214	287	1,277	52	293	1,303	309	1,374	45
0.037	0.94	227	1,010	239	1,063	65	261	1,161	275	1,223	59	288	1,281	302	1,343	50	309	1,374	325	1,446	43
0.038	0.97	240	1,068	252	1,121	63	276	1,228	290	1,290	58	303	1,348	319	1,419	49	326	1,450	342	1,521	43
0.039	0.99	253	1,125	265	1,179	61	291	1,294	305	1,357	56	319	1,419	335	1,490	47	343	1,526	361	1,606	41
0.040	1.02	265	1,179	279	1,241	59	305	1,357	321	1,428	54	335	1,490	353	1,570	46	361	1,606	379	1,686	40
0.041	1.04	279	1,241	293	1,303	58	321	1,428	337	1,499	53	352	1,566	370	1,646	45	378	1,681	398	1,770	39
0.042	1.07	293	1,303	308	1,370	56	336	1,495	354	1,575	52	370	1,646	388	1,726	43	397	1,766	417	1,855	37

(continued)

Table 4.2.5 continued

(1)	(2)	(3)	(4)	(5)	(6)	(7)	(8)	(9)	(10)	(11)	(12)	(13)	(14)	(15)	(16)	(17)	(18)	(19)	(20)	(21)	(22)
		Level 2 Bright (Uncoated) or Drawn-Galvanized Breaking Strength					Level 3 Bright (Uncoated) or Drawn-Galvanized Breaking Strength					Level 4 Bright (Uncoated) or Drawn-Galvanized Breaking Strength					Level 5 Bright (Uncoated) or Drawn-Galvanized Breaking Strength				
Wire Size Nominal Diameter		Individual Minimum		Average Minimum		Min. Tor.	Individual Minimum		Average Minimum		Min. Tor.	Individual Minimum		Average Minimum		Min. Tor.	Individual Minimum		Average Minimum		Min. Tor.
in.	mm	lb	N	lb	N		lb	N	lb	N		lb	N	lb	N		lb	N	lb	N	
0.043	1.09	306	1,361	322	1,432	55	352	1,566	370	1,646	50	387	1,721	407	1,810	42	416	1,850	438	1,948	36
0.044	1.12	320	1,423	336	1,495	54	369	1,641	387	1,721	49	405	1,801	425	1,890	41	436	1,939	458	2,037	36
0.045	1.14	334	1,486	352	1,566	52	385	1,712	405	1,801	48	423	1,882	445	1,979	40	455	2,024	479	2,131	35
0.046	1.17	349	1,552	367	1,632	51	402	1,788	422	1,877	48	442	1,966	464	2,064	39	475	2,113	499	2,220	34
0.047	1.19	365	1,624	383	1,704	50	419	1,864	441	1,962	47	461	2,051	485	2,157	38	495	2,202	521	2,317	33
0.048	1.22	380	1,690	400	1,779	49	437	1,944	459	2,042	46	481	2,139	505	2,246	37	517	2,300	543	2,415	32
0.049	1.24	396	1,761	416	1,850	48	455	2,024	479	2,131	45	500	2,224	526	2,340	36	538	2,393	566	2,518	32
0.050	1.27	411	1,828	433	1,926	48	474	2,108	498	2,215	44	521	2,317	547	2,433	35	560	2,491	588	2,615	31
0.051	1.30	428	1,904	450	2,002	47	492	2,188	518	2,304	43	541	2,406	569	2,531	34	582	2,589	612	2,722	30
0.052	1.32	445	1,979	467	2,077	46	512	2,277	538	2,393	42	563	2,504	591	2,629	34	605	2,691	636	2,829	29
0.053	1.35	462	2,055	486	2,162	45	531	2,362	559	2,486	41	584	2,598	614	2,731	33	628	2,793	660	2,936	29
0.054	1.37	479	2,131	503	2,237	44	551	2,451	579	2,575	40	605	2,691	637	2,833	32	651	2,896	685	3,047	28
0.055	1.40	496	2,206	522	2,322	43	571	2,540	601	2,673	39	628	2,793	660	2,936	31	676	3,007	710	3,158	27
0.056	1.42	515	2,291	541	2,406	42	592	2,633	622	2,767	39	650	2,891	684	3,042	31	700	3,114	736	3,274	27
0.057	1.45	532	2,366	560	2,491	41	612	2,722	644	2,865	38	674	2,998	708	3,149	30	724	3,220	762	3,389	26
0.058	1.47	551	2,451	579	2,575	41	634	2,820	666	2,962	37	697	3,100	733	3,260	29	750	3,336	788	3,505	26
0.059	1.50	569	2,531	599	2,664	40	655	2,913	689	3,065	36	721	3,207	757	3,367	29	775	3,447	815	3,625	25
0.060	1.52	589	2,620	619	2,753	39	678	3,016	712	3,167	36	745	3,314	783	3,483	28	800	3,558	842	3,745	24
0.061	1.55	608	2,704	640	2,847	38	700	3,114	736	3,274	35	769	3,421	809	3,598	27	827	3,678	869	3,865	23
0.062	1.57	628	2,793	660	2,936	38	722	3,211	760	3,380	35	795	3,536	835	3,714	27	854	3,799	898	3,994	23
0.063	1.60	648	2,882	682	3,034	37	745	3,314	783	3,483	34	820	3,647	862	3,834	26	881	3,919	927	4,123	23
0.064	1.63	668	2,971	702	3,122	36	768	3,416	808	3,594	33	845	3,759	889	3,954	26	909	4,043	955	4,248	22
0.065	1.65	689	3,065	725	3,225	36	793	3,527	833	3,705	33	872	3,879	916	4,074	25	937	4,168	985	4,381	22
0.066	1.68	710	3,158	746	3,318	35	816	3,630	858	3,816	32	898	3,994	944	4,199	25	965	4,292	1,015	4,515	22
0.067	1.70	731	3,251	769	3,421	35	840	3,736	884	3,932	32	924	4,110	972	4,323	24	994	4,421	1,044	4,644	21
0.068	1.73	753	3,349	791	3,518	34	865	3,848	909	4,043	31	952	4,234	1,000	4,448	24	1,023	4,550	1,075	4,782	21
0.069	1.75	774	3,443	814	3,621	34	890	3,959	936	4,163	31	979	4,355	1,029	4,577	23	1,053	4,684	1,107	4,924	20
0.070	1.78	797	3,545	837	3,723	33	916	4,074	962	4,279	30	1,007	4,479	1,059	4,710	23	1,083	4,817	1,139	5,066	20
0.071	1.80	819	3,643	861	3,830	33	942	4,190	990	4,404	30	1,035	4,604	1,089	4,844	22	1,113	4,951	1,171	5,209	20
0.072	1.83	841	3,741	885	3,936	32	967	4,301	1,017	4,524	29	1,064	4,733	1,118	4,973	22	1,144	5,089	1,202	5,346	19
0.073	1.85	864	3,843	908	4,039	32	994	4,421	1,044	4,644	29	1,093	4,862	1,149	5,111	22	1,175	5,226	1,235	5,493	19
0.074	1.88	887	3,945	933	4,150	31	1,021	4,541	1,073	4,773	29	1,122	4,991	1,180	5,249	21	1,207	5,369	1,269	5,645	18

0.075	1.91	911	4,052	957	4,257	31	1,047	4,657	1,101	4,897	29	1,152	5,124	1,212	5,391	21	1,239	5,511	1,303	5,796	18
0.076	1.93	935	4,159	983	4,372	30	1,075	4,782	1,131	5,031	29	1,183	5,262	1,243	5,529	20	1,271	5,653	1,337	5,947	18
0.077	1.96	958	4,261	1,008	4,484	30	1,103	4,906	1,159	5,155	28	1,213	5,395	1,275	5,671	20	1,304	5,800	1,370	6,094	17
0.078	1.98	983	4,372	1,033	4,595	29	1,131	5,031	1,189	5,289	28	1,244	5,533	1,308	5,818	20	1,337	5,947	1,405	6,249	17
0.079	2.01	1,008	4,484	1,060	4,715	29	1,159	5,155	1,219	5,422	27	1,275	5,671	1,341	5,965	19	1,371	6,098	1,441	6,410	17
0.080	2.03	1,033	4,595	1,085	4,826	29	1,188	5,284	1,248	5,551	27	1,307	5,814	1,374	6,112	19	1,405	6,249	1,477	6,570	16
0.081	2.06	1,058	4,706	1,112	4,946	29	1,217	5,413	1,279	5,689	27	1,339	5,956	1,407	6,258	19	1,439	6,401	1,513	6,730	16
0.082	2.08	1,083	4,817	1,139	5,066	29	1,246	5,542	1,310	5,827	26	1,371	6,098	1,441	6,410	18	1,473	6,552	1,549	6,890	16
0.083	2.11	1,110	4,937	1,166	5,186	28	1,276	5,676	1,342	5,969	26	1,404	6,245	1,476	6,565	18	1,509	6,712	1,587	7,059	16
0.084	2.13	1,136	5,053	1,194	5,311	28	1,306	5,809	1,372	6,103	26	1,436	6,387	1,510	6,716	18	1,544	6,868	1,624	7,224	15
0.085	2.16	1,162	5,169	1,222	5,435	28	1,337	5,947	1,405	6,249	25	1,470	6,539	1,546	6,877	18	1,580	7,028	1,662	7,393	15
0.086	2.18	1,189	5,289	1,249	5,556	27	1,367	6,080	1,437	6,392	25	1,503	6,685	1,581	7,032	18	1,617	7,192	1,699	7,557	15
0.087	2.21	1,216	5,409	1,278	5,685	27	1,398	6,218	1,470	6,539	25	1,538	6,841	1,616	7,188	18	1,654	7,357	1,738	7,731	15
0.088	2.24	1,243	5,529	1,307	5,814	27	1,429	6,356	1,503	6,685	24	1,573	6,997	1,653	7,353	17	1,691	7,522	1,777	7,904	15
0.089	2.26	1,270	5,649	1,336	5,943	26	1,462	6,503	1,536	6,832	24	1,607	7,148	1,689	7,513	17	1,728	7,686	1,816	8,078	15
0.090	2.29	1,299	5,778	1,365	6,072	26	1,493	6,641	1,569	6,979	24	1,642	7,304	1,726	7,677	17	1,766	7,855	1,856	8,225	15
0.091	2.31	1,326	5,898	1,394	6,201	26	1,525	6,783	1,603	7,130	23	1,678	7,464	1,764	7,846	17	1,804	8,024	1,896	8,433	15
0.092	2.34	1,355	6,027	1,425	6,338	25	1,558	6,930	1,638	7,286	23	1,714	7,624	1,802	8,015	16	1,843	8,198	1,937	8,616	14
0.093	2.36	1,384	6,156	1,454	6,467	25	1,591	7,077	1,673	7,442	23	1,750	7,784	1,840	8,184	16	1,882	8,371	1,978	8,798	14
0.094	2.39	1,413	6,285	1,485	6,605	25	1,624	7,224	1,708	7,597	23	1,786	7,944	1,878	8,353	16	1,921	8,545	2,019	8,981	14
0.095	2.41	1,442	6,414	1,516	6,743	24	1,658	7,375	1,743	7,753	22	1,823	8,109	1,917	8,527	16	1,961	8,723	2,061	9,167	14
0.096	2.44	1,471	6,543	1,547	6,881	24	1,692	7,526	1,778	7,909	22	1,861	8,278	1,957	8,705	15	2,001	8,900	2,103	9,354	13
0.097	2.46	1,501	6,676	1,577	7,014	24	1,726	7,677	1,814	8,069	22	1,898	8,442	1,996	8,878	15	2,041	9,078	2,145	9,541	13
0.098	2.49	1,531	6,810	1,609	7,157	24	1,761	7,833	1,851	8,233	22	1,936	8,611	2,036	9,056	15	2,082	9,261	2,188	9,732	13
0.099	2.51	1,561	6,943	1,641	7,299	23	1,795	7,984	1,887	8,393	21	1,975	8,785	2,077	9,238	15	2,123	9,443	2,231	9,923	13
0.100	2.54	1,592	7,081	1,674	7,446	23	1,830	8,140	1,924	8,558	21	2,013	8,954	2,117	9,416	15	2,165	9,630	2,276	10,124	13
0.101	2.57	1,622	7,215	1,706	7,588	23	1,866	8,300	1,962	8,727	21	2,052	9,127	2,158	9,599	15	2,206	9,812	2,320	10,319	13
0.102	2.59	1,654	7,357	1,738	7,731	23	1,902	8,460	2,000	8,896	21	2,092	9,305	2,200	9,786	15	2,249	10,004	2,365	10,520	12
0.103	2.62	1,685	7,495	1,771	7,877	22	1,938	8,620	2,038	9,065	20	2,131	9,479	2,241	9,968	15	2,291	10,190	2,409	10,715	12
0.104	2.64	1,717	7,637	1,805	8,029	22	1,974	8,780	2,076	9,234	20	2,172	9,661	2,284	10,159	15	2,335	10,386	2,455	10,920	12
0.105	2.67	1,749	7,780	1,839	8,180	22	2,011	8,945	2,115	9,408	20	2,212	9,839	2,326	10,346	14	2,378	10,577	2,500	11,120	12
0.106	2.69	1,781	7,922	1,873	8,331	22	2,048	9,110	2,154	9,581	20	2,253	10,021	2,369	10,537	14	2,422	10,773	2,546	11,325	12
0.107	2.72	1,814	8,069	1,907	8,482	21	2,086	9,279	2,192	9,750	20	2,294	10,204	2,412	10,729	14	2,466	10,969	2,592	11,529	12
0.108	2.74	1,847	8,215	1,941	8,634	21	2,124	9,448	2,232	9,928	19	2,336	10,391	2,456	10,924	14	2,511	11,169	2,639	11,738	12
0.109	2.77	1,880	8,362	1,976	8,789	21	2,162	9,617	2,272	10,106	19	2,377	10,573	2,499	11,116	14	2,555	11,365	2,687	11,952	12
0.110	2.79	1,913	8,509	2,011	8,945	21	2,200	9,786	2,312	10,284	19	2,420	10,764	2,544	11,316	13	2,601	11,569	2,735	12,165	12
0.111	2.82	1,946	8,656	2,046	9,101	21	2,239	9,959	2,353	10,466	19	2,462	10,951	2,588	11,511	13	2,647	11,774	2,783	12,379	12
0.112	2.84	1,980	8,807	2,082	9,261	20	2,278	10,133	2,394	10,649	19	2,505	11,142	2,633	11,712	13	2,693	11,978	2,831	12,592	12

(continued)

Table 4.2.5 continued

(1)	(2)	(3)	(4)	(5)	(6)	(7)	(8)	(9)	(10)	(11)	(12)	(13)	(14)	(15)	(16)	(17)	(18)	(19)	(20)	(21)	(22)
Wire Size Nominal Diameter		Level 2 Bright (Uncoated) or Drawn-Galvanized Breaking Strength					Level 3 Bright (Uncoated) or Drawn-Galvanized Breaking Strength					Level 4 Bright (Uncoated) or Drawn-Galvanized Breaking Strength					Level 5 Bright (Uncoated) or Drawn-Galvanized Breaking Strength				
		Individual Minimum		Average Minimum		Min. Tor.	Individual Minimum		Average Minimum		Min. Tor.	Individual Minimum		Average Minimum		Min. Tor.	Individual Minimum		Average Minimum		Min. Tor.
in.	mm	lb	N	lb	N		lb	N	lb	N		lb	N	lb	N		lb	N	lb	N	
0.113	2.87	2,014	8,959	2,118	9,421	20	2,317	10,306	2,435	10,831	18	2,548	11,334	2,678	11,912	13	2,739	12,183	2,879	12,806	11
0.114	2.90	2,048	9,110	2,154	9,581	20	2,356	10,479	2,476	11,013	18	2,592	11,529	2,724	12,116	13	2,786	12,392	2,928	13,024	11
0.115	2.92	2,084	9,270	2,190	9,741	20	2,396	10,657	2,518	11,200	18	2,635	11,720	2,771	12,325	13	2,833	12,601	2,979	13,251	11
0.116	2.95	2,118	9,421	2,226	9,901	20	2,436	10,835	2,560	11,387	18	2,679	11,916	2,817	12,530	12	2,880	12,810	3,028	13,469	11
0.117	2.97	2,154	9,581	2,264	10,070	19	2,477	11,018	2,604	11,583	18	2,724	12,116	2,864	12,739	12	2,928	13,024	3,078	13,691	11
0.118	3.00	2,189	9,737	2,301	10,235	19	2,517	11,196	2,647	11,774	17	2,769	12,317	2,911	12,948	12	2,977	13,242	3,129	13,918	11
0.119	3.02	2,224	9,892	2,338	10,399	19	2,558	11,378	2,690	11,965	17	2,814	12,517	2,958	13,157	12	3,024	13,451	3,180	14,145	11
0.120	3.05	2,260	10,052	2,376	10,568	19	2,599	11,560	2,733	12,156	17	2,860	12,721	3,006	13,371	12	3,074	13,673	3,232	14,376	10
0.121	3.07	2,296	10,213	2,414	10,737	19	2,641	11,747	2,777	12,352	17	2,906	12,926	3,055	13,589	12	3,123	13,891	3,283	14,603	10
0.122	3.10	2,333	10,377	2,453	10,911	18	2,683	11,934	2,821	12,548	17	2,951	13,126	3,103	13,802	11	3,173	14,114	3,335	14,834	10
0.123	3.12	2,370	10,542	2,492	11,084	18	2,725	12,121	2,865	12,744	17	2,998	13,335	3,152	14,020	11	3,222	14,331	3,388	15,070	10
0.124	3.15	2,406	10,702	2,530	11,253	18	2,768	12,312	2,910	12,944	17	3,045	13,544	3,201	14,238	11	3,273	14,558	3,441	15,306	9
0.125	3.18	2,444	10,871	2,570	11,431	18	2,811	12,503	2,955	13,144	16	3,092	13,753	3,250	14,456	11	3,324	14,785	3,494	15,541	9
0.126	3.20	2,481	11,035	2,609	11,605	18	2,854	12,695	3,000	13,344	16	3,140	13,967	3,301	14,683	11	3,374	15,008	3,548	15,782	9
0.127	3.23	2,519	11,205	2,649	11,783	18	2,897	12,886	3,045	13,544	16	3,187	14,176	3,351	14,905	11	3,426	15,239	3,602	16,022	9
0.128	3.25	2,557	11,374	2,689	11,961	17	2,941	13,082	3,091	13,749	16	3,235	14,389	3,401	15,128	10	3,478	15,470	3,656	16,262	9
0.129	3.28	2,595	11,543	2,729	12,139	17	2,984	13,273	3,138	13,958	16	3,284	14,607	3,452	15,354	10	3,530	15,701	3,711	16,507	9
0.130	3.30	2,634	11,716	2,770	12,321	17	3,029	13,473	3,185	14,167	16	3,333	14,825	3,503	15,581	10	3,582	15,933	3,766	16,751	9
0.131	3.33	2,672	11,885	2,810	12,499	17	3,074	13,673	3,232	14,376	16	3,381	15,039	3,555	15,813	10	3,635	16,168	3,821	16,996	8
0.132	3.35	2,711	12,059	2,851	12,681	17	3,119	13,873	3,279	14,585	15	3,431	15,261	3,607	16,044	10	3,687	16,400	3,877	17,245	8
0.133	3.38	2,751	12,236	2,893	12,868	17	3,164	14,073	3,326	14,794	15	3,481	15,483	3,659	16,275	10	3,741	16,640	3,933	17,494	8
0.134	3.40	2,790	12,410	2,934	13,050	17	3,210	14,278	3,374	15,008	15	3,530	15,701	3,712	16,511	9	3,795	16,880	3,989	17,743	8
0.135	3.43	2,830	12,588	2,976	13,237	16	3,256	14,483	3,422	15,221	15	3,580	15,924	3,764	16,742	9	3,849	17,120	4,047	18,001	8
0.136	3.45	2,870	12,766	3,018	13,424	16	3,301	14,683	3,471	15,439	15	3,631	16,151	3,817	16,978	9	3,904	17,365	4,104	18,255	8
0.137	3.48	2,911	12,948	3,061	13,615	16	3,347	14,887	3,519	15,653	15	3,683	16,382	3,871	17,218	9	3,959	17,610	4,162	18,513	8
0.138	3.51	2,951	13,126	3,103	13,802	16	3,394	15,097	3,568	15,870	15	3,733	16,604	3,925	17,458	9	4,014	17,854	4,220	18,771	8
0.139	3.53	2,992	13,308	3,146	13,993	16	3,441	15,306	3,617	16,088	14	3,785	16,836	3,979	17,699	9	4,069	18,099	4,277	19,024	8
0.140	3.56	3,033	13,491	3,189	14,185	16	3,489	15,519	3,667	16,311	14	3,837	17,067	4,033	17,939	8	4,125	18,348	4,337	19,291	7
0.141	3.58	3,074	13,673	3,232	14,376	16	3,535	15,724	3,717	16,533	14	3,889	17,298	4,089	18,188	8	4,181	18,597	4,395	19,549	7
0.142	3.61	3,116	13,860	3,276	14,572	16	3,583	15,937	3,767	16,756	14	3,942	17,534	4,144	18,433	8	4,237	18,846	4,455	19,816	7
0.143	3.63	3,158	14,047	3,320	14,767	15	3,632	16,155	3,818	16,982	14	3,995	17,770	4,199	18,677	8	4,294	19,100	4,514	20,078	7
0.144	3.66	3,200	14,234	3,364	14,963	15	3,680	16,369	3,868	17,205	14	4,048	18,006	4,256	18,931	8	4,351	19,353	4,575	20,350	7

0.145	3.68	3,242	14,420	15,159	3,408	15	3,728	16,582	3,920	17,436	14	4,102	18,246	4,312	19,180	8	4,409	19,611	4,635	20,616	7
0.146	3.71	3,285	14,612	15,359	3,453	15	3,777	16,800	3,971	17,663	14	4,155	18,481	4,369	19,433	8	4,466	19,865	4,696	20,888	7
0.147	3.73	3,328	14,803	15,559	3,498	15	3,827	17,022	4,023	17,894	14	4,209	18,722	4,425	19,682	8	4,525	20,127	4,757	21,159	7
0.148	3.76	3,371	14,994	15,759	3,543	15	3,876	17,240	4,074	18,121	13	4,264	18,966	4,482	19,936	8	4,583	20,385	4,819	21,435	7
0.149	3.78	3,413	15,181	15,964	3,589	15	3,925	17,458	4,127	18,357	13	4,318	19,206	4,540	20,194	7	4,642	20,648	4,880	21,706	6
0.150	3.81	3,457	15,377	16,168	3,635	15	3,976	17,685	4,180	18,593	13	4,374	19,456	4,598	20,452	7	4,701	20,910	4,943	21,986	6
0.151	3.84	3,501	15,572	16,373	3,681	14	4,026	17,908	4,232	18,824	13	4,428	19,696	4,656	20,710	7	4,761	21,177	5,005	22,262	6
0.152	3.86	3,545	15,768	16,578	3,727	14	4,076	18,130	4,286	19,064	13	4,484	19,945	4,714	20,968	7	4,820	21,439	5,068	22,542	6
0.153	3.89	3,589	15,964	16,782	3,773	14	4,127	18,357	4,339	19,300	13	4,541	20,198	4,773	21,230	7	4,881	21,711	5,131	22,823	6
0.154	3.91	3,634	16,164	16,991	3,820	14	4,179	18,588	4,393	19,540	13	4,596	20,443	4,832	21,493	7	4,941	21,978	5,195	23,107	6
0.155	3.94	3,679	16,364	17,200	3,867	14	4,230	18,815	4,446	19,776	13	4,653	20,697	4,891	21,755	7	5,002	22,249	5,258	23,388	6
0.156	3.96	3,724	16,564	17,409	3,914	14	4,281	19,042	4,501	20,020	13	4,710	20,950	4,952	22,026	7	5,063	22,520	5,323	23,677	6
0.157	3.99	3,768	16,760	17,626	3,962	14	4,334	19,278	4,556	20,265	13	4,767	21,204	5,011	22,289	7	5,125	22,796	5,387	23,961	6
0.158	4.01	3,814	16,965	17,836	4,010	14	4,386	19,509	4,610	20,505	12	4,824	21,457	5,072	22,560	7	5,186	23,067	5,452	24,250	6
0.159	4.04	3,859	17,165	18,046	4,057	14	4,438	19,740	4,666	20,754	12	4,882	21,715	5,132	22,827	7	5,248	23,343	5,518	24,544	6
0.160	4.06	3,905	17,369	18,259	4,105	13	4,491	19,976	4,721	20,999	12	4,940	21,973	5,194	23,103	6	5,311	23,623	5,583	24,833	5
0.161	4.09	3,952	17,578	18,477	4,154	13	4,544	20,212	4,778	21,253	12	4,999	22,236	5,255	23,374	6	5,373	23,899	5,649	25,127	5
0.162	4.11	3,998	17,783	18,695	4,203	13	4,597	20,447	4,833	21,497	12	5,057	22,494	5,317	23,650	6	5,437	24,184	5,715	25,420	5
0.163	4.14	4,044	17,988	18,913	4,252	13	4,651	20,688	4,889	21,746	12	5,116	22,756	5,378	23,921	6	5,500	24,464	5,782	25,718	5
0.164	4.17	4,091	18,197	19,131	4,301	13	4,704	20,923	4,946	22,000	12	5,175	23,018	5,441	24,202	6	5,563	24,744	5,849	26,016	5
0.165	4.19	4,138	18,406	19,349	4,350	13	4,759	21,168	5,003	22,253	12	5,235	23,285	5,503	24,477	6	5,628	25,033	5,916	26,314	5
0.166	4.22	4,186	18,619	19,571	4,400	13	4,814	21,412	5,060	22,507	12	5,294	23,548	5,566	24,758	6	5,692	25,318	5,984	26,617	5
0.167	4.24	4,232	18,824	19,794	4,450	13	4,868	21,643	5,118	22,765	12	5,355	23,819	5,629	25,038	6	5,756	25,603	6,052	26,919	5
0.168	4.27	4,280	19,037	20,016	4,500	13	4,923	21,898	5,175	23,108	12	5,415	24,086	5,693	25,322	6	5,821	25,892	6,119	27,217	5
0.169	4.29	4,329	19,255	20,243	4,551	13	4,977	22,138	5,233	23,276	11	5,476	24,357	5,756	25,603	6	5,886	26,181	6,188	27,524	5
0.170	4.32	4,377	19,469	20,465	4,601	13	5,033	22,387	5,291	23,534	11	5,537	24,629	5,821	25,892	6	5,951	26,470	6,257	27,831	5
0.171	4.34	4,426	19,687	20,692	4,652	12	5,089	22,636	5,349	23,792	11	5,597	24,895	5,885	26,176	6	6,018	26,768	6,326	28,138	5
0.172	4.37	4,474	19,900	20,923	4,704	12	5,145	22,885	5,409	24,059	11	5,659	25,171	5,949	26,461	6	6,084	27,062	6,396	28,449	5
0.173	4.39	4,523	20,118	21,150	4,755	12	5,201	23,134	5,467	24,317	11	5,721	25,447	6,015	26,755	6	6,150	27,355	6,466	28,761	5
0.174	4.42	4,572	20,336	21,377	4,806	12	5,257	23,383	5,527	25,584	11	5,784	25,727	6,080	27,044	6	6,217	27,653	6,535	29,068	5
0.175	4.45	4,622	20,559	21,613	4,859	12	5,314	23,637	5,586	24,847	11	5,846	26,003	6,146	27,337	6	6,284	27,951	6,606	29,383	5
0.176	4.47	4,670	20,772	21,840	4,910	12	5,371	23,890	5,647	25,118	11	5,909	26,283	6,212	27,631	6	6,351	28,249	6,677	29,699	5
0.177	4.50	4,720	20,995	22,071	4,962	12	5,429	24,148	5,707	25,385	11	5,971	26,559	6,277	27,920	6	6,419	28,552	6,749	30,020	5
0.178	4.52	4,771	21,221	22,307	5,015	12	5,486	24,402	5,768	25,656	11	6,034	26,839	6,344	28,218	6	6,487	28,854	6,819	30,331	5
0.179	4.55	4,820	21,439	22,542	5,068	12	5,544	24,660	5,828	25,923	11	6,098	27,124	6,410	28,512	6	6,555	29,157	6,891	30,651	5
0.180	4.57	4,871	21,666	22,778	5,121	12	5,601	24,913	5,889	26,194	11	6,162	27,409	6,478	28,814	6	6,624	29,464	6,964	30,976	5
0.181	4.60	4,922	21,893	23,014	5,174	12	5,660	25,176	5,950	26,466	11	6,226	27,693	6,546	29,117	6	6,692	29,766	7,036	31,296	5
0.182	4.62	4,973	22,120	23,254	5,228	12	5,718	25,434	6,012	26,741	10	6,291	27,982	6,613	29,415	6	6,762	30,077	7,108	31,616	5
0.183	4.65	5,023	22,342	23,490	5,281	11	5,777	25,696	6,073	27,013	10	6,355	28,267	6,681	29,717	6	6,832	30,389	7,182	31,946	5

(continued)

Table 4.2.5 continued

(1)	(2)	(3)	(4)	(5)	(6)	(7)	(8)	(9)	(10)	(11)	(12)	(13)	(14)	(15)	(16)	(17)	(18)	(19)	(20)	(21)	(22)
		Level 2 Bright (Uncoated) or Drawn-Galvanized Breaking Strength					Level 3 Bright (Uncoated) or Drawn-Galvanized Breaking Strength					Level 4 Bright (Uncoated) or Drawn-Galvanized Breaking Strength					Level 5 Bright (Uncoated) or Drawn-Galvanized Breaking Strength				
Wire Size Nominal Diameter		Individual Minimum		Average Minimum		Min. Tor.	Individual Minimum		Average Minimum		Min. Tor.	Individual Minimum		Average Minimum		Min. Tor.	Individual Minimum		Average Minimum		Min. Tor.
in.	mm	lb	N	lb	N		lb	N	lb	N		lb	N	lb	N		lb	N	lb	N	
0.184	4.67	5,075	22,574	5,335	23,730	11	5,836	25,959	6,136	27,293	10	6,419	28,552	6,749	30,020	5	6,901	30,696	7,255	32,270	5
0.185	4.70	5,127	22,805	5,389	23,970	11	5,896	26,225	6,198	27,569	10	6,485	28,845	6,187	30,322	5	6,971	31,007	7,329	32,599	5
0.186	4.72	5,178	23,032	5,444	24,125	11	5,955	26,488	6,261	27,849	10	6,550	29,134	6,886	30,629	5	7,041	31,318	7,403	32,929	5
0.187	4.75	5,230	23,263	5,498	24,455	11	6,015	26,755	6,323	28,125	10	6,616	29,428	6,956	30,940	5	7,113	31,639	7,477	33,258	5
0.188	4.78	5,283	23,499	5,553	24,700	11	6,074	27,017	6,386	28,405	10	6,683	29,726	7,025	31,247	5	7,184	31,954	7,552	33,591	5
0.189	4.80	5,335	23,730	5,609	24,949	11	6,135	27,288	6,449	28,685	10	6,748	30,015	7,094	31,554	5	7,255	32,270	7,627	33,925	5
0.190	4.83	5,387	23,961	5,663	25,189	11	6,195	27,555	6,513	28,970	10	6,815	30,313	7,165	31,870	5	7,326	32,586	7,702	34,258	5
0.191	4.85	5,441	24,202	5,720	25,443	11	6,257	27,831	6,577	29,254	10	6,882	30,611	7,234	32,177	5	7,398	32,906	7,778	34,597	5
0.192	4.88	5,493	24,433	5,775	25,687	11	6,317	28,098	6,641	29,539	10	6,949	30,909	7,305	32,493	5	7,470	33,227	7,854	34,935	4
0.193	4.90	5,547	24,673	5,831	25,936	11	6,378	28,369	6,706	29,828	10	7,016	31,207	7,376	32,808	5	7,543	33,551	7,929	35,268	4
0.194	4.93	5,600	24,909	5,888	26,190	11	6,440	28,645	6,770	30,113	10	7,084	31,510	7,448	33,129	5	7,615	33,872	8,005	35,606	4
0.195	4.95	5,654	25,149	5,944	26,439	11	6,501	28,916	6,835	30,402	10	7,152	31,812	7,518	33,440	5	7,688	34,196	8,082	35,949	4
0.196	4.98	5,708	25,389	6,000	26,688	11	6,564	29,197	6,900	30,691	10	7,220	32,115	7,590	33,760	5	7,762	34,525	8,160	36,296	4
0.197	5.00	5,761	25,625	6,057	26,942	10	6,626	29,472	6,966	30,985	10	7,288	32,417	7,662	34,081	5	7,835	34,850	8,237	36,638	4
0.198	5.03	5,816	25,870	6,114	27,195	10	6,689	29,753	7,032	31,278	10	7,357	32,724	7,735	34,405	5	7,909	35,179	8,315	36,985	4
0.199	5.05	5,870	26,110	6,172	27,453	10	6,751	30,028	7,097	31,567	10	7,427	33,035	7,807	34,726	5	7,983	35,508	8,393	37,332	4
0.200	5.08	5,925	26,354	6,229	27,707	10	6,814	30,309	7,164	31,865	10	7,496	33,342	7,880	35,050	5	8,057	35,838	8,471	37,679	4
0.201	5.11	5,980	26,599	6,286	27,960	10	6,877	30,589	7,229	32,155	10	7,565	33,649	7,953	35,375	5	8,132	36,171	8,550	38,030	4
0.202	5.13	6,035	26,844	6,345	28,223	10	6,940	30,869	7,296	32,453	10	7,634	33,956	8,026	35,700	5	8,208	36,509	8,628	38,377	4
0.203	5.16	6,091	27,093	6,403	28,481	10	7,004	31,154	7,364	32,755	10	7,704	34,267	8,100	36,029	5	8,283	36,843	8,707	38,729	4
0.204	5.18	6,146	27,337	6,462	28,743	10	7,068	31,438	7,430	33,049	10	7,775	34,583	8,173	36,354	5	8,358	37,176	8,786	39,080	4
0.205	5.21	6,202	27,586	6,520	29,001	10	7,132	31,723	7,498	33,351	10	7,846	34,899	8,248	36,687	5	8,434	37,514	8,866	39,436	4
0.206	5.23	6,258	27,836	6,578	29,259	10	7,196	32,008	7,566	33,654	10	7,916	35,210	8,322	37,016	5	8,510	37,852	8,946	39,792	4
0.207	5.26	6,314	28,085	6,638	29,526	10	7,261	32,297	7,633	33,952	10	7,987	35,526	8,397	37,350	5	8,586	38,191	9,026	40,148	4
0.208	5.28	6,371	28,338	6,697	29,788	10	7,326	32,586	7,702	34,258	10	8,058	35,842	8,472	37,683	5	8,663	38,533	9,107	40,508	4
0.209	5.31	6,427	28,587	6,757	30,055	10	7,391	32,875	7,771	34,565	10	8,131	36,167	8,547	38,017	5	8,740	38,876	9,188	40,868	4
0.210	5.33	6,484	28,841	6,816	30,318	10	7,457	33,169	7,839	34,868	10	8,202	36,482	8,622	38,351	5	8,817	39,218	9,269	41,229	4
0.211	5.36	6,540	29,090	6,876	30,584	10	7,522	33,458	7,908	35,175	10	8,274	36,803	8,698	38,689	5	8,895	39,565	9,351	41,593	4
0.212	5.38	6,598	29,348	6,936	30,851	10	7,587	33,747	7,977	35,482	10	8,346	37,123	8,774	39,027	5	8,972	39,907	9,432	41,954	4
0.213	5.41	6,655	29,601	6,997	31,123	10	7,654	34,045	8,046	35,789	10	8,419	37,448	8,851	39,369	5	9,050	40,254	9,514	42,318	4

0.214	5.44	6,713	29,859	7,057	31,390	10	7,720	34,339	8,116	36,100	10	8,492	37,772	8,928	39,712	5	9,129	40,606	9,597	42,687	4
0.215	5.46	6,770	30,113	7,118	31,661	10	7,786	34,632	8,186	36,411	10	8,564	38,093	9,004	40,050	5	9,207	40,953	9,679	43,052	4
0.216	5.49	6,829	30,375	7,179	31,932	10	7,853	34,930	8,255	36,718	10	8,639	38,426	9,082	40,392	5	9,286	41,304	9,762	43,421	4
0.217	5.51	6,886	30,629	7,240	32,204	10	7,920	35,228	8,326	37,034	10	8,712	38,751	9,158	40,735	5	9,365	41,656	9,845	43,791	4
0.218	5.54	6,945	30,891	7,301	32,475	10	7,987	35,526	8,397	37,350	10	8,786	39,080	9,236	41,082	5	9,445	42,011	9,929	44,164	4
0.219	5.56	7,003	31,149	7,363	32,751	10	8,054	35,824	8,468	37,666	10	8,860	39,409	9,314	41,429	5	9,524	42,363	10,012	44,533	4
0.220	5.59	7,063	31,416	7,425	33,026	10	8,122	36,127	8,538	37,977	9	8,934	39,738	9,392	41,776	4	9,604	42,719	10,096	44,907	4
0.221	5.61	7,121	31,674	7,487	33,302	10	8,190	36,429	8,610	38,297	9	9,009	40,072	9,471	42,127	4	9,685	43,079	10,181	45,285	4
0.222	5.64	7,181	31,941	7,549	33,578	10	8,257	36,727	8,681	38,613	9	9,083	40,401	9,549	42,474	4	9,765	43,435	10,265	45,659	4
0.223	5.66	7,240	32,204	7,612	33,858	10	8,326	37,034	8,752	38,929	9	9,158	40,735	9,628	42,825	4	9,846	43,795	10,350	46,037	4
0.224	5.69	7,300	32,470	7,674	34,134	10	8,395	37,341	8,825	39,254	9	9,234	41,073	9,708	43,181	4	9,926	44,151	10,436	46,419	4
0.225	5.72	7,359	32,733	7,737	34,414	10	8,463	37,643	8,897	39,574	9	9,309	41,406	9,787	43,533	4	10,007	44,511	10,521	46,797	4
0.226	5.74	7,419	33,000	7,799	34,690	10	8,532	37,950	8,970	39,899	9	9,385	41,744	9,867	43,888	4	10,089	44,876	10,607	47,180	4
0.227	5.77	7,479	33,267	7,863	34,975	10	8,601	38,257	9,043	40,223	9	9,461	42,083	9,947	44,244	4	10,171	45,241	10,693	47,562	4
0.228	5.79	7,540	33,538	7,926	35,255	10	8,671	38,569	9,115	40,544	9	9,537	42,421	10,027	44,600	4	10,253	45,605	10,779	47,945	4
0.229	5.82	7,600	33,805	7,990	35,540	10	8,740	38,876	9,188	40,868	9	9,614	42,763	10,108	44,960	4	10,335	45,970	10,865	48,328	4
0.230	5.84	7,661	34,076	8,053	35,820	10	8,810	39,187	9,262	41,197	9	9,691	43,106	10,187	45,312	4	10,418	46,339	10,952	48,714	4
0.231	5.87	7,722	34,347	8,118	36,109	10	8,879	39,494	9,335	41,522	9	9,768	43,448	10,268	45,672	4	10,501	46,708	11,039	49,101	4
0.232	5.89	7,782	34,614	8,182	36,394	10	8,950	39,810	9,408	41,847	9	9,845	43,791	10,349	46,032	4	10,584	47,078	11,126	49,488	4
0.233	5.92	7,844	34,890	8,246	36,678	9	9,021	40,125	9,483	42,180	9	9,923	44,138	10,431	46,397	4	10,667	47,447	11,214	49,880	4
0.234	5.94	7,905	35,161	8,311	36,967	9	9,091	40,437	9,557	42,510	9	10,001	44,484	10,513	46,762	4	10,750	47,816	11,302	50,271	4
0.235	5.97	7,967	35,437	8,375	37,252	9	9,162	40,753	9,632	42,843	9	10,078	44,827	10,594	47,122	4	10,834	48,190	11,390	50,663	4
0.236	5.99	8,029	35,713	8,441	37,546	9	9,233	41,068	9,707	43,177	8	10,157	45,178	10,677	47,491	4	10,918	48,563	11,478	51,054	4
0.237	6.02	8,091	35,989	8,505	37,830	9	9,304	41,384	9,782	43,510	8	10,235	45,525	10,759	47,856	4	11,002	48,937	11,566	51,446	4
0.238	6.05	8,153	36,265	8,571	38,124	9	9,376	41,704	9,856	43,839	8	10,314	45,877	10,842	48,225	4	11,087	49,315	11,655	51,841	4
0.239	6.07	8,215	36,540	8,637	38,417	9	9,448	42,025	9,932	44,178	8	10,393	46,288	10,925	48,594	4	11,172	49,693	11,744	52,237	4
0.240	6.10	8,278	36,821	8,702	38,706	9	9,519	42,341	10,007	44,511	8	10,472	46,579	11,009	48,968	4	11,256	50,067	11,834	52,638	3
0.241	6.12	8,340	37,096	8,768	39,000	9	9,591	42,661	10,083	44,849	8	10,550	46,926	11,092	49,337	4	11,342	50,449	11,924	53,038	3
0.242	6.15	8,404	37,381	8,834	39,294	9	9,664	42,985	10,160	45,192	8	10,630	47,282	11,176	49,711	4	11,427	50,827	12,013	53,434	3
0.243	6.17	8,466	37,657	8,900	39,587	9	9,736	43,306	10,236	45,530	8	10,709	47,634	11,259	50,080	4	11,513	51,210	12,103	53,834	3
0.244	6.20	8,529	37,937	8,967	39,885	9	9,809	43,630	10,313	45,872	8	10,790	47,994	11,344	50,458	4	11,600	51,597	12,194	54,239	3
0.245	6.22	8,593	38,222	9,033	40,179	9	9,882	43,955	10,388	46,206	8	10,870	48,350	11,428	50,832	4	11,685	51,975	12,285	54,644	3
0.246	6.25	8,657	38,506	9,101	40,481	9	9,955	44,280	10,465	46,548	8	10,950	48,706	11,512	51,205	4	11,772	52,362	12,376	55,048	3
0.247	6.27	8,720	38,787	9,168	40,779	9	10,029	44,609	10,543	46,895	8	11,031	49,066	11,597	51,583	4	11,859	52,749	12,467	55,453	3
0.248	6.30	8,785	39,076	9,235	41,077	9	10,102	44,934	10,620	47,238	8	11,112	49,426	11,682	51,962	4	11,946	53,136	12,558	55,858	3
0.249	6.32	8,848	39,356	9,302	41,375	9	10,176	45,263	10,698	47,585	8	11,193	49,786	11,767	52,340	4	12,032	53,518	12,650	56,267	3
0.250	6.35	8,912	39,641	9,370	41,678	9	10,249	45,588	10,775	47,927	8	11,275	50,151	11,853	52,722	4	12,120	53,910	12,742	56,676	3

Table 4.2.6   *Mechanical Properties of Individual Rope Wires (Before Fabrication) [4]*

(1)	(2)	(3)	(4)	(5)	(6)	(7)	(8)	(9)	(10)	(11)	(12)	(13)	(14)
		Level 2 Bright (Uncoated) or Drawn-Galvanized Breaking Strength			Level 3 Bright (Uncoated) or Drawn-Galvanized Breaking Strength			Level 4 Bright (Uncoated) or Drawn-Galvanized Breaking Strength			Level 5 Bright (Uncoated) or Drawn-Galvanized Breaking Strength		
Wire Size Nominal Diameter													
in.	mm	lb	N	Tor.	lb	N	Tor.	lb	N	Tor.	lb	N	Tor.
0.010	0.25	17	76	254	20	89	234	22	98	218	24	107	190
0.011	0.28	21	93	231	24	107	213	27	120	198	29	129	173
0.012	0.30	25	111	212	29	129	195	32	142	182	34	151	158
0.013	0.33	29	129	195	34	151	180	37	165	168	40	178	146
0.014	0.36	34	151	181	39	173	167	43	191	156	46	205	136
0.015	0.38	39	173	169	45	200	156	49	218	145	53	236	126
0.016	0.41	44	196	158	51	227	146	56	249	136	60	267	118
0.017	0.43	50	222	149	57	254	137	63	280	128	68	302	111
0.018	0.46	56	249	141	64	285	130	71	316	121	76	338	105
0.019	0.48	62	276	133	72	320	123	79	351	114	85	378	100
0.020	0.51	69	307	126	79	351	116	87	387	108	94	418	94
0.021	0.53	76	338	120	87	387	111	96	427	103	103	458	90
0.022	0.56	83	369	115	96	427	106	105	467	98	113	503	86
0.023	0.58	91	405	110	105	467	101	115	512	94	124	552	82
0.024	0.61	99	440	105	114	507	97	125	556	90	135	600	78
0.025	0.64	107	476	101	123	547	93	136	605	86	146	649	75
0.026	0.66	116	516	97	133	592	89	147	654	83	158	703	72
0.027	0.69	125	556	93	144	641	86	158	703	80	170	756	70
0.028	0.71	134	596	90	155	689	83	170	756	77	183	814	67
0.029	0.74	114	641	87	166	738	80	182	810	74	196	872	65
0.030	0.76	154	685	84	177	787	77	195	867	72	210	934	62
0.031	0.79	164	729	81	189	841	75	208	925	69	224	996	60
0.032	0.81	175	778	78	201	894	72	221	983	67	238	1,059	58
0.033	0.84	186	827	76	214	952	70	235	1,045	65	253	1,125	57
0.034	0.86	197	876	74	227	1,010	68	250	1,112	63	268	1,192	55
0.035	0.89	209	930	72	240	1,068	66	264	1,174	61	284	1,263	53
0.036	0.91	221	983	70	254	1,130	64	280	1,245	60	301	1,339	52
0.037	0.94	233	1,036	68	268	1,192	62	295	1,312	58	317	1,410	50
0.038	0.97	246	1,094	66	283	1,259	61	311	1,383	56	334	1,486	49
0.039	0.99	259	1,152	64	298	1,326	59	327	1,454	55	352	1,566	48
0.040	1.02	272	1,210	62	313	1,392	57	344	1,530	53	370	1,646	46
0.041	1.04	286	1,272	61	329	1,463	56	361	1,606	52	388	1,726	45
0.042	1.07	300	1,334	59	345	1,535	55	379	1,686	51	407	1,810	44
0.043	1.09	314	1,397	58	361	1,606	53	397	1,766	50	427	1,899	43
0.044	1.12	328	1,459	57	378	1,681	52	415	1,846	48	447	1,988	42
0.045	1.14	343	1,526	55	395	1,757	51	434	1,930	47	467	2,077	41
0.046	1.17	358	1,592	54	412	1,833	50	453	2,015	46	487	2,166	40
0.047	1.19	374	1,664	53	430	1,913	49	473	2,104	45	508	2,260	39
0.048	1.22	390	1,735	52	448	1,993	48	493	2,193	44	530	2,357	38
0.049	1.24	406	1,806	51	467	2,077	47	513	2,282	43	552	2,455	38
0.050	1.27	422	1,877	50	486	2,162	46	534	2,375	42	574	2,553	37
0.051	1.30	439	1,953	49	505	2,246	45	555	2,469	42	597	2,655	36
0.052	1.32	456	2,028	48	525	2,335	44	577	2,566	41	620	2,758	35
0.053	1.35	474	2,108	47	545	2,424	43	599	2,664	40	644	2,865	35
0.054	1.37	491	2,184	46	565	2,513	42	621	2,762	39	668	2,971	34
0.055	1.40	509	2,264	45	586	2,607	41	644	2,865	38	693	3,082	33
0.056	1.42	528	2,349	44	607	2,700	41	667	2,967	38	718	3,194	33
0.057	1.45	546	2,429	43	628	2,793	40	691	3,074	37	743	3,305	32
0.058	1.47	565	2,513	43	650	2,891	39	715	3,180	36	769	3,421	32

(continued)

Table 4.2.6 *continued*

(1)	(2)	(3)	(4)	(5)	(6)	(7)	(8)	(9)	(10)	(11)	(12)	(13)	(14)
Wire Size Nominal Diameter		Level 2 Bright (Uncoated) or Drawn-Galvanized Breaking Strength			Level 3 Bright (Uncoated) or Drawn-Galvanized Breaking Strength			Level 4 Bright (Uncoated) or Drawn-Galvanized Breaking Strength			Level 5 Bright (Uncoated) or Drawn-Galvanized Breaking Strength		
in.	mm	lb	N	Tor.	lb	N	Tor.	lb	N	Tor.	lb	N	Tor.
0.059	1.50	584	2,598	42	672	2,989	38	739	3,287	36	795	3,536	31
0.060	1.52	604	2,687	41	695	3,091	38	764	3,398	35	821	3,652	30
0.061	1.55	624	2,776	40	718	3,194	37	789	3,509	35	848	3,772	30
0.062	1.57	644	2,865	40	741	3,296	37	815	3,625	34	876	3,896	29
0.063	1.60	665	2,958	39	764	3,398	36	841	3,741	33	904	4,021	29
0.064	1.63	685	3,047	38	788	3,505	35	867	3,856	33	932	4,146	28
0.065	1.65	707	3,145	38	813	3,616	35	894	3,977	32	961	4,275	28
0.066	1.68	728	3,238	37	837	3,723	34	921	4,097	32	990	4,404	28
0.067	1.70	750	3,336	37	862	3,834	34	948	4,217	31	1,019	4,533	27
0.068	1.73	772	3,434	36	887	3,945	33	976	4,341	31	1,049	4,666	27
0.069	1.75	794	3,532	36	913	4,061	33	1,004	4,466	30	1,080	4,804	26
0.070	1.78	817	3,634	35	939	4,177	32	1,033	4,595	30	1,111	4,942	26
0.071	1.80	840	3,736	35	966	4,297	32	1,062	4,724	29	1,142	5,080	26
0.072	1.83	863	3,839	34	992	4,412	31	1,091	4,853	29	1,173	5,218	25
0.073	1.85	886	3,941	34	1,019	4,533	31	1,121	4,986	29	1,205	5,360	25
0.074	1.88	910	4,048	33	1,047	4,657	30	1,151	5,120	28	1,238	5,507	24
0.075	1.91	934	4,154	33	1,074	4,777	30	1,182	5,258	28	1,271	5,653	24
0.076	1.93	959	4,266	32	1,103	4,906	30	1,213	5,395	27	1,304	5,800	24
0.077	1.96	983	4,372	32	1,131	5,031	29	1,244	5,533	27	1,337	5,947	23
0.078	1.98	1,008	4,484	31	1,160	5,160	29	1,276	5,676	27	1,371	6,098	23
0.079	2.01	1,034	4,599	31	1,189	5,289	28	1,308	5,818	26	1,406	6,254	23
0.080	2.03	1,059	4,710	30	1,218	5,418	28	1,340	5,960	26	1,441	6,410	22
0.081	2.06	1,085	4,826	30	1,248	5,551	28	1,373	6,107	26	1,476	6,565	22
0.082	2.08	1,111	4,942	30	1,278	5,685	27	1,406	6,254	25	1,511	6,721	22
0.083	2.11	1,138	5,062	29	1,309	5,822	27	1,440	6,405	25	1,548	6,886	22
0.084	2.13	1,165	5,182	29	1,339	5,956	27	1,473	6,552	25	1,584	7,046	21
0.085	2.16	1,192	5,302	29	1,371	6,098	26	1,508	6,708	24	1,621	7,210	21
0.086	2.18	1,219	5,422	28	1,402	6,236	26	1,542	6,859	24	1,658	7,375	21
0.087	2.21	1,247	5,547	28	1,434	6,378	26	1,577	7,014	24	1,696	7,544	21
0.088	2.24	1,275	5,671	28	1,466	6,521	25	1,613	7,175	23	1,734	7,713	20
0.089	2.26	1,303	5,796	27	1,499	6,668	25	1,648	7,330	23	1,772	7,882	20
0.090	2.29	1,332	5,925	27	1,531	6,810	25	1,684	7,490	23	1,811	8,055	20
0.091	2.31	1,360	6,049	27	1,564	6,957	24	1,721	7,655	23	1,850	8,229	20
0.092	2.34	1,390	6,183	26	1,598	7,108	24	1,758	7,820	22	1,890	8,407	19
0.093	2.36	1,419	6,312	26	1,632	7,259	24	1,795	7,984	22	1,930	8,585	19
0.094	2.39	1,449	6,445	26	1,666	7,410	24	1,832	8,149	22	1,970	8,763	19
0.095	2.41	1,479	6,579	25	1,700	7,562	23	1,870	8,318	22	2,011	8,945	19
0.096	2.44	1,509	6,712	25	1,735	7,717	23	1,909	8,491	21	2,052	9,127	18
0.097	2.46	1,539	6,845	25	1,770	7,873	23	1,947	8,660	21	2,093	9,310	18
0.098	2.49	1,570	6,983	25	1,806	8,033	23	1,986	8,834	21	2,135	9,496	18
0.099	2.51	1,601	7,121	24	1,841	8,189	22	2,026	9,012	21	2,177	9,683	18
0.100	2.54	1,633	7,264	24	1,877	8,349	22	2,065	9,185	20	2,220	9,875	18
0.101	2.57	1,664	7,401	24	1,914	8,513	22	2,105	9,363	20	2,263	10,066	18
0.102	2.59	1,696	7,744	24	1,951	8,678	22	2,146	9,545	20	2,307	10,262	17
0.103	2.62	1,728	7,686	23	1,988	8,843	21	2,186	9,723	20	2,350	10,453	17
0.104	2.64	1,761	7,833	23	2,025	9,007	21	2,228	9,910	20	2,395	10,653	17
0.105	2.67	1,794	7,980	23	2,063	9,176	21	2,269	10,093	19	2,439	10,849	17
0.106	2.69	1,827	8,126	23	2,101	9,345	21	2,311	10,279	19	2,484	11,049	17
0.107	2.72	1,860	8,273	22	2,139	9,514	21	2,353	10,466	19	2,529	11,249	16
0.108	2.74	1,894	8,425	22	2,178	9,688	20	2,396	10,657	19	2,575	11,454	16
0.109	2.77	1,928	8,576	22	2,217	9,861	20	2,438	10,844	19	2,621	11,658	16

(continued)

Table 4.2.6 *continued*

(1)	(2)	(3)	(4)	(5)	(6)	(7)	(8)	(9)	(10)	(11)	(12)	(13)	(14)
		Level 2 Bright (Uncoated) or Drawn-Galvanized Breaking Strength			Level 3 Bright (Uncoated) or Drawn-Galvanized Breaking Strength			Level 4 Bright (Uncoated) or Drawn-Galvanized Breaking Strength			Level 5 Bright (Uncoated) or Drawn-Galvanized Breaking Strength		
Wire Size Nominal Diameter													
in.	mm	lb	N	Tor.	lb	N	Tor.	lb	N	Tor.	lb	N	Tor.
0.110	2.79	1,962	8,727	22	2,256	10,035	20	2,482	11,040	18	2,668	11,867	16
0.111	2.82	1,996	8,878	22	2,296	10,213	20	2,525	11,231	18	2,715	12,076	16
0.112	2.84	2,031	9,034	21	2,336	10,391	20	2,569	11,427	18	2,762	12,285	16
0.113	2.87	2,066	9,190	21	2,376	10,568	19	2,613	11,623	18	2,809	12,494	15
0.114	2.90	2,101	9,345	21	2,416	10,746	19	2,658	11,823	18	2,857	12,708	15
0.115	2.92	2,137	9,505	21	2,457	10,929	19	2,703	12,023	18	2,906	12,926	15
0.116	2.95	2,172	9,661	21	2,498	11,111	19	2,748	12,223	17	2,954	13,139	15
0.117	2.97	2,209	9,826	20	2,540	11,298	19	2,794	12,428	17	3,003	13,357	15
0.118	3.00	2,245	9,986	20	2,582	11,485	18	2,840	12,632	17	3,053	13,580	15
0.119	3.02	2,281	10,146	20	2,624	11,672	18	2,886	12,837	17	3,102	13,798	15
0.120	3.05	2,318	10,310	20	2,666	11,858	18	2,933	13,046	17	3,153	14,025	14
0.121	3.07	2,355	10,475	20	2,709	12,050	18	2,980	13,255	17	3,203	14,247	14
0.122	3.10	2,393	10,644	19	2,752	12,241	18	3,027	13,464	17	3,254	14,474	14
0.123	3.12	2,431	10,813	19	2,795	12,432	18	3,075	13,678	16	3,305	14,701	14
0.124	3.15	2,468	10,978	19	2,839	12,628	18	3,123	13,891	16	3,357	14,932	14
0.125	3.18	2,507	11,151	19	2,883	12,824	17	3,171	14,105	16	3,409	15,163	14
0.126	3.20	2,545	11,320	19	2,927	13,019	17	3,220	14,323	16	3,461	15,395	14
0.127	3.23	2,584	11,494	19	2,971	13,215	17	3,269	14,541	16	3,514	15,630	14
0.128	3.25	2,623	11,667	18	3,016	13,415	17	3,318	14,758	16	3,567	15,866	13
0.129	3.28	2,662	11,841	18	3,061	13,615	17	3,368	14,981	16	3,620	16,102	13
0.130	3.30	2,702	12,018	18	3,107	13,820	17	3,418	15,203	15	3,674	16,342	13
0.131	3.33	2,741	12,192	18	3,153	14,025	17	3,468	15,426	15	3,728	16,582	13
0.132	3.35	2,781	12,370	18	3,199	14,229	16	3,519	15,653	15	3,782	16,822	13
0.133	3.38	2,822	12,552	18	3,245	14,434	16	3,570	15,879	15	3,837	17,067	13
0.134	3.40	2,862	12,730	18	3,292	14,643	16	3,621	16,106	15	3,892	17,312	13
0.135	3.43	2,903	12,913	17	3,339	14,852	16	3,672	16,333	15	3,948	17,561	13
0.136	3.45	2,944	13,095	17	3,386	15,061	16	3,724	16,564	15	4,004	17,810	13
0.137	3.48	2,986	13,282	17	3,433	15,270	16	3,777	16,800	15	4,060	18,059	13
0.138	3.51	3,027	13,464	17	3,481	15,483	16	3,829	17,031	14	4,117	18,312	12
0.139	3.53	3,069	13,651	17	3,529	15,697	15	3,882	17,267	14	4,173	18,562	12
0.140	3.56	3,111	13,838	17	3,578	15,915	15	3,935	17,503	14	4,231	18,819	12
0.141	3.58	3,153	14,025	17	3,626	16,128	15	3,989	17,743	14	4,288	19,073	12
0.142	3.61	3,196	14,216	17	3,675	16,346	15	4,043	17,983	14	4,346	19,331	12
0.143	3.63	3,239	14,407	16	3,725	16,569	15	4,097	18,223	14	4,404	19,589	12
0.144	3.66	3,282	14,598	16	3,774	16,787	15	4,152	18,468	14	4,463	19,851	12
0.145	3.68	3,325	14,790	16	3,824	17,009	15	4,207	18,713	14	4,522	20,114	12
0.146	3.71	3,369	14,985	16	3,874	17,232	15	4,262	18,957	14	4,581	20,376	12
0.147	3.73	3,413	15,181	16	3,925	17,458	15	4,317	19,202	14	4,641	20,643	12
0.148	3.76	3,457	15,377	16	3,975	17,681	14	4,373	19,451	13	4,701	20,910	11
0.149	3.78	3,501	15,572	16	4,026	17,908	14	4,429	19,700	13	4,761	21,177	11
0.150	3.81	3,546	15,773	16	4,078	18,139	14	4,486	19,954	13	4,822	21,448	11
0.151	3.84	3,591	15,973	15	4,129	18,366	14	4,542	20,203	13	4,883	21,720	11
0.152	3.86	3,636	16,173	15	4,181	18,597	14	4,599	20,456	13	4,944	21,991	11
0.153	3.89	3,681	16,373	15	4,233	18,828	14	4,657	20,714	13	5,006	22,267	11
0.154	3.91	3,727	16,578	15	4,286	19,064	14	4,714	20,968	13	5,068	22,542	11
0.155	3.94	3,773	16,782	15	4,338	19,295	14	4,772	21,226	13	5,130	22,818	11
0.156	3.96	3,819	16,987	15	4,391	19,531	14	4,831	21,488	13	5,193	23,098	11
0.157	3.99	3,865	17,192	15	4,445	19,771	14	4,899	21,746	13	5,256	23,379	11
0.158	4.01	3,912	17,401	15	4,498	20,007	13	4,948	22,099	12	5,319	23,659	11
0.159	4.04	3,958	17,605	15	4,552	20,247	13	5,007	22,271	12	5,383	23,944	11

(continued)

Table 4.2.6 *continued*

(1)	(2)	(3)	(4)	(5)	(6)	(7)	(8)	(9)	(10)	(11)	(12)	(13)	(14)
		Level 2 Bright (Uncoated) or Drawn-Galvanized Breaking Strength			Level 3 Bright (Uncoated) or Drawn-Galvanized Breaking Strength			Level 4 Bright (Uncoated) or Drawn-Galvanized Breaking Strength			Level 5 Bright (Uncoated) or Drawn-Galvanized Breaking Strength		
Wire Size Nominal Diameter													
in.	mm	lb	N	Tor.	lb	N	Tor.	lb	N	Tor.	lb	N	Tor.
0.160	4.06	4,005	17,814	14	4,606	20,487	13	5,067	22,538	12	5,447	24,228	10
0.161	4.09	4,053	18,028	14	4,661	20,732	13	5,127	22,805	12	5,511	24,513	10
0.162	4.11	4,100	18,237	14	4,715	20,972	13	5,187	23,072	12	5,576	24,802	10
0.163	4.14	4,148	18,450	14	4,770	21,217	13	5,247	23,339	12	5,641	25,091	10
0.164	4.17	4,196	18,664	14	4,825	21,462	13	5,308	23,610	12	5,706	25,380	10
0.165	4.19	4,244	18,877	14	4,881	21,711	13	5,369	23,881	12	5,772	25,674	10
0.166	4.22	4,293	19,095	14	4,937	21,960	13	5,430	24,153	12	5,838	25,967	10
0.167	4.24	4,341	19,309	14	4,993	22,209	13	5,492	24,428	12	5,904	26,261	10
0.168	4.27	4,390	19,527	14	5,049	22,458	13	5,554	24,704	12	5,970	26,555	10
0.169	4.29	4,440	19,749	14	5,105	22,707	12	5,616	24,980	12	6,037	26,853	10
0.170	4.32	4,489	19,967	14	5,162	22,961	12	5,679	25,260	11	6,104	27,151	10
0.171	4.34	4,539	20,189	13	5,219	23,214	12	5,741	25,536	11	6,172	27,453	10
0.172	4.37	4,589	20,413	13	5,277	23,472	12	5,804	25,816	11	6,240	27,756	10
0.173	4.39	4,639	20,634	13	5,334	23,726	12	5,868	26,101	11	6,308	28,058	10
0.174	4.42	4,689	20,857	13	5,392	23,984	12	5,932	26,396	11	6,376	28,360	10
0.175	4.45	4,740	21,064	13	5,450	24,242	12	5,996	26,670	11	6,514	28,974	9
0.176	4.47	4,790	21,306	13	5,509	24,504	12	6,060	26,955	11	6,514	28,974	9
0.177	4.50	4,841	21,533	13	5,568	24,766	12	6,124	27,240	11	6,584	29,286	9
0.178	4.52	4,893	21,764	13	5,627	25,029	12	6,189	27,529	11	6,653	29,593	9
0.179	4.55	4,944	21,991	13	5,686	25,291	12	6,254	27,818	11	6,723	29,904	9
0.180	4.57	4,996	22,222	13	5,745	25,554	12	6,320	28,111	11	6,794	30,220	9
0.181	4.60	5,048	22,454	13	5,805	25,821	12	6,386	28,405	11	6,864	30,531	9
0.182	4.62	5,100	22,685	13	5,865	26,088	11	6,452	28,698	11	6,935	30,847	9
0.183	4.65	5,152	22,916	12	5,925	26,354	11	6,518	28,992	11	7,007	31,167	9
0.184	4.67	5,205	23,152	12	5,986	26,626	11	6,584	29,286	10	7,078	31,483	9
0.185	4.70	5,258	23,388	12	6,047	26,897	11	6,651	29,584	10	7,150	31,803	9
0.186	4.72	5,311	23,623	12	6,108	27,168	11	6,718	29,882	10	7,222	32,123	9
0.187	4.75	5,364	23,859	12	6,169	27,440	11	6,786	30,184	10	7,295	32,448	9
0.188	4.78	5,418	24,099	12	6,230	27,711	11	6,854	30,487	10	7,368	32,773	9
0.189	4.80	5,472	24,339	12	6,292	27,987	11	6,921	30,785	10	7,441	33,098	9
0.190	4.83	5,525	24,575	12	6,354	28,263	11	6,990	31,092	10	7,514	33,422	9
0.191	4.85	5,580	24,820	12	6,417	28,543	11	7,058	31,394	10	7,588	33,751	9
0.192	4.88	5,634	25,060	12	6,469	28,819	11	7,127	31,701	10	7,662	34,081	8
0.193	4.90	5,689	25,305	12	6,542	29,099	11	7,196	32,008	10	7,736	34,410	8
0.194	4.93	5,744	25,549	12	6,605	29,379	11	7,266	32,319	10	7,810	34,739	8
0.195	4.95	5,799	25,794	12	6,668	29,659	11	7,335	32,626	10	7,885	35,072	8
0.196	4.98	5,854	26,039	12	6,732	29,944	11	7,405	32,937	10	7,961	35,411	8
0.197	5.00	5,909	26,283	11	6,796	30,229	10	7,475	33,249	10	8,036	35,744	8
0.198	5.03	5,965	26,532	11	6,860	30,513	10	7,536	33,365	10	8,112	36,082	8
0.199	5.05	6,021	26,781	11	6,924	30,798	10	7,617	33,880	10	8,188	36,420	8
0.200	5.08	6,077	27,030	11	6,989	31,087	10	7,688	34,196	9	8,264	36,758	8
0.201	5.11	6,133	27,280	11	7,053	31,372	10	7,759	34,512	9	8,341	37,101	8
0.202	5.13	6,190	27,533	11	7,118	31,661	10	7,830	34,828	9	8,418	37,443	8
0.203	5.16	6,247	27,787	11	7,184	31,954	10	7,902	35,148	9	8,495	37,786	8
0.204	5.18	6,304	28,040	11	7,249	32,244	10	7,974	35,468	9	8,572	38,128	8
0.205	5.21	6,361	28,294	11	7,315	32,537	10	8,047	35,793	9	8,650	38,475	8
0.206	5.23	6,418	28,547	11	7,381	32,831	10	8,119	36,113	9	8,728	38,822	8
0.207	5.26	6,476	28,805	11	7,447	33,124	10	8,192	36,438	9	8,806	39,169	8
0.208	5.28	6,534	29,063	11	7,514	33,422	10	8,265	36,763	9	8,885	39,520	8
0.209	5.31	6,592	29,321	11	7,581	33,720	10	8,339	37,092	9	8,964	39,872	8

(continued)

Table 4.2.7 *continued*

(1)	(2)	(3)	(4)	(5)	(6)	(7)	(8)	(9)	(10)	(11)	(12)	(13)	(14)
		Level 2 Bright (Uncoated) or Drawn-Galvanized Breaking Strength			Level 3 Bright (Uncoated) or Drawn-Galvanized Breaking Strength			Level 4 Bright (Uncoated) or Drawn-Galvanized Breaking Strength			Level 5 Bright (Uncoated) or Drawn-Galvanized Breaking Strength		
Wire Size Nominal Diameter													
in.	mm	lb	N	Tor.	lb	N	Tor.	lb	N	Tor.	lb	N	Tor.
0.210	5.33	6,650	29,579	11	7,648	34,018	10	8,412	37,417	9	9,043	40,223	8
0.211	5.36	6,708	29,837	11	7,715	34,316	10	8,486	37,746	9	9,123	40,579	8
0.212	5.38	6,767	30,100	11	7,782	34,614	10	8,560	38,075	9	9,202	40,930	8
0.213	5.41	6,826	30,362	11	7,850	34,917	10	8,635	38,408	9	9,282	41,286	8
0.214	5.44	6,885	30,624	10	7,918	35,219	10	8,710	38,742	9	9,363	41,647	7
0.215	5.46	6,944	30,887	10	7,986	35,522	9	8,784	39,071	9	9,443	42,002	7
0.216	5.49	7,004	31,154	10	8,054	35,824	9	8,860	39,409	9	9,524	42,363	7
0.217	5.51	7,063	31,416	10	8,123	36,131	9	8,935	39,743	9	9,605	42,723	7
0.218	5.54	7,123	31,683	10	8,192	36,438	9	9,011	40,081	9	9,687	43,088	7
0.219	5.56	7,183	31,950	10	8,261	36,745	9	9,087	40,419	9	9,768	43,448	7
0.220	5.59	7,244	32,221	10	8,330	37,052	9	9,163	40,757	8	9,850	43,813	7
0.221	5.61	7,304	32,488	10	8,400	37,363	9	9,240	41,100	8	9,933	44,182	7
0.222	5.64	7,365	32,760	10	8,469	37,670	9	9,316	41,438	8	10,015	44,547	7
0.223	5.66	7,426	33,031	10	8,539	37,981	9	9,393	41,780	8	10,098	44,916	7
0.224	5.69	7,487	33,302	10	8,610	38,297	9	9,471	42,127	8	10,181	45,285	7
0.225	5.72	7,548	33,574	10	8,680	38,609	9	9,548	42,470	8	10,264	45,654	7
0.226	5.74	7,609	33,845	10	8,751	38,924	9	9,626	42,816	8	10,348	46,028	7
0.227	5.77	7,671	34,121	10	8,822	39,240	9	9,704	43,163	8	10,432	46,402	7
0.228	5.79	7,733	34,396	10	8,893	39,556	9	9,782	43,510	8	10,516	46,775	7
0.229	5.82	7,795	34,672	10	8,964	39,872	9	9,861	43,862	8	10,600	47,149	7
0.230	5.84	7,857	34,948	10	9,036	40,192	9	9,939	44,209	8	10,685	47,527	7
0.231	5.87	7,920	35,228	10	9,107	40,508	9	10,018	44,560	8	10,770	47,905	7
0.232	5.89	7,982	35,504	10	9,179	40,828	9	10,097	44,911	8	10,855	48,283	7
0.233	5.92	8,045	35,784	9	9,252	41,153	9	10,177	45,267	8	10,940	48,661	7
0.234	5.94	8,108	36,064	9	9,324	41,473	9	10,257	45,623	8	11,026	49,044	7
0.235	5.97	8,171	36,345	9	9,397	41,798	9	10,336	45,975	8	11,112	49,426	7
0.236	5.99	8,235	36,629	9	9,470	42,123	8	10,417	46,335	8	11,198	49,809	7
0.237	6.02	8,298	36,910	9	9,543	42,447	8	10,497	46,691	8	11,284	50,191	7
0.238	6.05	8,362	37,194	9	9,616	42,772	8	10,578	47,051	8	11,371	50,578	7
0.239	6.07	8,426	37,479	9	9,690	43,101	8	10,659	47,411	8	11,458	50,965	7
0.240	6.10	8,490	37,764	9	9,763	43,426	8	10,740	47,772	8	11,545	51,352	6
0.241	6.12	8,554	38,048	9	9,837	43,755	8	10,821	48,132	8	11,633	51,744	6
0.242	6.15	8,619	38,337	9	9,912	44,089	8	10,903	48,497	8	11,720	52,131	6
0.243	6.17	8,683	38,622	9	9,986	44,418	8	10,984	48,857	8	11,807	52,522	6
0.244	6.20	8,748	38,911	9	10,061	44,751	8	11,067	49,226	8	11,897	52,918	6
0.245	6.22	8,813	39,200	9	10,135	45,060	8	11,149	49,591	7	11,985	53,309	6
0.246	6.25	8,879	39,494	9	10,210	45,414	8	11,231	49,955	7	12,074	53,705	6
0.247	6.27	8,944	39,783	9	10,286	45,752	8	11,314	50,325	7	12,163	54,101	6
0.248	6.30	9,010	40,076	9	10,361	46,086	8	11,397	50,694	7	12,252	54,497	6
0.249	6.32	9,075	40,366	9	10,437	46,424	8	11,480	51,063	7	12,341	54,893	6
0.250	6.35	9,141	40,659	9	10,512	46,757	8	11,564	51,437	7	12,431	55,293	6

(*text continued from page 4-22*)

Table 4.2.7 *Weight of Zinc Coating for Galvanized Rope Wire [4]*

(1)	(2)	(3)	(4)
Diameter of Wire		Minimum Weight of Zinc Coating	
in.	mm	oz./ft²	kg/m²
0.028 to 0.047	0.71 to 1.19	0.20	0.06
0.048 to 0.054	1.22 to 1.37	0.40	0.12
0.055 to 0.063	1.40 to 1.60	0.50	0.15
0.064 to 0.079	1.63 to 2.01	0.60	0.18
0.080 to 0.092	2.03 to 2.34	0.70	0.21
0.093 and larger	2.36 and larger	0.80	0.24

Table 4.2.8 *Weight of Zinc Coating for Drawn-Galvanized Rope Wire [4]*

(1)	(2)	(3)	(4)
Diameter of Wire		Minimum Weight of Zinc Coating	
in.	mm	oz./ft²	kg/m²
0.018 to 0.028	0.46 to 0.71	0.10	0.03
0.029 to 0.060	0.74 to 1.52	0.20	0.06
0.061 to 0.090	1.55 to 2.29	0.30	0.09
0.091 to 0.140	2.31 to 3.56	0.40	0.12

undamaged wire rope; such sample must be new or in an unused condition. The total wire number to be tested shall be equal to the number of wires in any one strand, and the wires shall be selected from all strands of the rope. The specimens shall be selected from all locations or positions so that they would constitute a complete composite strand exactly similar to a regular strand in the rope. The specimen for all "like-positioned" (wires symmetrically placed in a strand) wires to be selected so as to use as nearly as possible an equal number from each strand. Any unsymmetrically placed wires, or marker wires, are to be disregarded entirely. Center wires are subject to the same stipulations that apply to symmetrical wires.

Selection and testing of wire prior to rope fabrication will be adequate to ensure the after-fabrication wire rope breaking strength and wire requirements can be met. Prior to fabrication, wire tests should meet the requirements of Table 4.2.6.

Conduct of Tests
The test results of each test on any one specimen should be associated and may be studied separately from other specimens.

If, when making any individual wire test on any wire, the first specimen fails, not more than two additional specimens from the same wire shall be tested. The average of any two tests showing failure or acceptance shall be used as the value to represent the wire. The test for the rope may be terminated at any time sufficient failures have occurred to be the cause for rejection.

The purchaser may at his or her expense test all of the wires if the results of the selected tests indicate that further checking is warranted.

Tensile Requirements of Individual Wire
Specimens shall not be less than 18 in. (457 mm) long, and the distance between the grips of the testing machine shall not be less than 12 in. (305 mm). The speed of the movable

head of the testing machine, under no load, shall not exceed 1 in./min (0.4 mm/s). Any specimen breaking within $\frac{1}{4}$ in. (6.35 mm) from the jaws shall be disregarded and a retest made.

Note: The diameter of wire can more easily and accurately be determined by placing the wire specimen in the test machine and applying a load not over 25% of the breaking strength of the wire.

The breaking strength of either bright (uncoated) or drawn-galvanized wires of the various grades shall meet the values shown in Table 4.2.5 or Table 4.2.6 for the size wire being tested. Wire tested after rope fabrication allows one wire in 6×7 classification, or three wires in 6×19 and 8×19 classifications and 18×7 and 19×7 constructions, or six wires in 6×37 classification or nine wires in 6×61 classification, or twelve wires in 6×91 classification wire rope to fall below but not more than 10% below, the tabular value for individual minimum. If, when making the specified test, any wires fall below, but not more than 10% below the individual minimum, additional wires from the same rope shall be tested until there is cause for rejection or until all of the wires in the rope have been tested. Tests of individual wires in galvanized wire rope and of individual wires in strand cores and in independent wire rope cores are not required.

Torsional Requirements of Individual Wire
The distance between the jaws of the testing machine shall be 8 in. ± $\frac{1}{16}$ in. (203 mm ± 1 mm). For small diameter wires, where the number of turns to cause failure is large, and in order to save testing time, the distance between the jaws of the testing machine may be less than 8 in. (203 mm). One end of the wire is to be rotated with respect to the other end at a uniform speed not to exceed sixty 360° (6.28 rad) twists per minute, until breakage occurs. The machine must be equipped with an automatic counter to record the number of twists causing breakage. One jaw shall be fixed axially and the other jaw movable axially and arranged for applying tension weights to wire under test. Tests in which breakage occurs within $\frac{1}{8}$ in. (3.18 mm) of the jaw shall be discarded.

In the torsion test, the wires tested must meet the values for the respective grades and sizes as covered by Table 4.2.8 or Table 4.2.9. In wire tested after rope fabrication, it will be permissible for two wires in 6×7 classification or five wires in 6×19 and 8×19 classifications and 18×7 and 19×7 constructions or ten wires in 6×37 classification or fifteen wires in 6×61 classification, or twenty wires in 6×91

Table 4.2.9 *Applied Tension for Torsional Tests [4]*

(1)	(2)	(3)	(4)
Wire Size Nominal Diameter		Minimum Applied Tension*	
(in)	(mm)	(lb)	(N)
0.011 to 0.016	0.28 to 0.42	1	4
0.017 to 0.020	0.43 to 0.52	2	9
0.021 to 0.030	0.53 to 0.77	4	18
0.031 to 0.040	0.78 to 1.02	6	27
0.041 to 0.050	1.03 to 1.28	8	36
0.051 to 0.060	1.29 to 1.53	9	40
0.061 to 0.070	1.54 to 1.79	11	49
0.071 to 0.080	1.80 to 2.04	13	58
0.081 to 0.090	2.05 to 2.30	16	71
0.091 to 0.100	2.31 to 2.55	19	85
0.101 to 0.110	2.56 to 2.80	21	93
0.111 to 0.120	2.81 to 3.06	23	102
0.121 to 0.130	3.07 to 3.31	25	111

*Weights shall not exceed twice the minimums listed.

classification rope to fall below, but not more than 30% below, the specified minimum number of twists for the individual wire being tested.

During the torsion test, tension weights as shown in Table 4.2.9 shall be applied to the wire tested.

The minimum torsions for individual bright (uncoated) or drawn-galvanized wire of the grades and sizes shown in Columns 7, 12, and 17 of Tables 4.2.5 and 4.2.6 shall be the number of 360° (6.28 rad) twists in an 8-in. (203 mm) length that the wire must withstand before breakage occurs. Torsion tests of individual wires in galvanized wire rope and of individual wires in strand cores and independent wire rope cores are not required.

When the distance between the jaws of the testing machine is less than 8 in. (203 mm), the minimum torsions shall be reduced in direct proportion to the change in jaw spacing, or determined by

$$T_s = \frac{(T_L)(L_s)}{(L_L)} \qquad [4.2.15]$$

where T_s = minimum torsions for short wire
T_L = minimum torsions for 8-in. (203-mm) length as given in Table 4.2.5 for size and grade of wire
L_s = distance between testing-machine jaws for short wire in in. (mm)
L_L = 8 in. (203 mm)

Breaking Strength Requirements for Wire Rope

The nominal strength of the various grades of finished wire rope with fiber core shall be as specified in Tables 4.2.10, 4.2.11, and 4.2.12. The nominal strength of the various grades of wire rope having a strand core or an independent wire-rope core shall be as specified in Tables 4.2.13 through 4.2.18. The nominal strength of the various types of flattened strand wire rope shall be specified in Table 4.2.19. The nominal strength of the various grades of drawn-galvanized wire rope shall be specified in Tables 4.2.10 through 4.2.19.

When testing finished wire-rope tensile test specimens to their breaking strength, suitable sockets shall be attached

(text continued on page 4-50)

Table 4.2.10   Classification Wire Rope, Bright (Uncoated) or Drawn-Galvanized Wire, Fiber Core [4]

(1)	(2)	(3)	(4)	(5)	(6)	(7)	(8)	(9)	(10)
						Nominal Strength			
Nominal Diameter		Approx. Mass		Plow Steel			Improved Plow Steel		
in.	mm	lb/ft	kg/m	lb	kN	Metric Tonnes	lb	kN	Metric Tonnes
$\frac{3}{8}$	9.5	0.21	0.31	10,200	45.4	4.63	11,720	52.1	5.32
$\frac{7}{16}$	11.5	0.29	0.43	13,800	61.4	6.26	15,860	70.5	7.20
$\frac{1}{2}$	13	0.38	0.57	17,920	79.7	8.13	20,600	91.6	9.35
$\frac{9}{16}$	14.5	0.48	0.71	22,600	101	10.3	26,000	116	11.8
$\frac{5}{8}$	16	0.59	0.88	27,800	124	12.6	31,800	141	14.4
$\frac{3}{4}$	19	0.84	1.25	39,600	176	18.0	45,400	202	20.6
$\frac{7}{8}$	22	1.15	1.71	53,400	238	24.2	61,400	273	27.9
1	26	1.50	2.23	69,000	307	31.3	79,400	353	36.0

Table 4.2.11   6×19 and 6×37 Classification Wire Rope, Bright (Uncoated) or Drawn-Galvanized Wire, Fiber Core [4]

(1)	(2)	(3)	(4)	(5)	(6)	(7)	(8)	(9)	(10)	(11)	(12)	(13)
							Nominal Strength					
Nominal Diameter		Approx. Mass		Plow Steel			Improved Plow Steel			Extra Improved Plow Steel		
in.	mm	lb/ft	kg/m	lb	kN	Metric Tonnes	lb	kN	Metric Tonnes	lb	kN	Metric Tonnes
$\frac{1}{2}$	13	0.42	0.63	18,700	83.2	8.48	21,400	95.2	9.71	23,600	105	10.7
$\frac{9}{16}$	14.5	0.53	0.79	23,600	106	10.7	27,000	120	12.2	29,800	132	13.5
$\frac{5}{8}$	16	0.66	0.98	29,000	129	13.2	33,400	149	15.1	36,600	163	16.6
$\frac{3}{4}$	19	0.95	1.41	41,400	184	18.8	47,600	212	21.6	52,400	233	23.8
$\frac{7}{8}$	22	1.29	1.92	56,000	249	25.4	64,400	286	29.2	70,800	315	32.1
1	26	1.68	2.50	72,800	324	33.0	83,600	372	37.9	92,000	409	41.7
$1\frac{1}{8}$	29	2.13	3.17	91,400	407	41.5	105,200	468	47.7	115,600	514	52.4
$1\frac{1}{4}$	32	2.63	3.91	112,400	500	51.0	129,200	575	58.5	142,200	632	64.5
$1\frac{3}{8}$	35	3.18	4.73				155,400	691	70.5	171,000	760	77.6
$1\frac{1}{2}$	38	3.78	5.63				184,000	818	83.5	202,000	898	91.6
$1\frac{5}{8}$	42	4.44	6.61				214,000	952	97.1	236,000	1050	107
$1\frac{3}{4}$	45	5.15	7.66				248,000	1100	112	274,000	1220	124
$1\frac{7}{8}$	48	5.91	8.80				282,000	1250	128	312,000	1390	142
2	52	6.72	10.0				320,000	1420	146	352,000	1560	160

Table 4.2.12 *18×7 Construction Wire Rope, Bright (Uncoated) or Drawn-Galvanized Wire, Fiber Core [4]*

(1)	(2)	(3)	(4)	(5)	(6)	(7)	(8)	(9)	(10)
				Nominal Strength					
Nominal Diameter		Approx. Mass		Improved Plow Steel			Extra Improved Plow Steel		
in.	mm	lb/ft	kg/m	lb	kN	Metric Tonnes	lb	kN	Metric Tonnes
$\frac{1}{3}$	13	0.43	0.64	19,700	87.6	8.94	21,600	96.1	9.80
$\frac{9}{16}$	14.5	0.55	0.82	24,800	110	11.2	27,200	121	12.3
$\frac{5}{8}$	16	0.68	1.01	30,600	136	13.9	33,600	149	15.2
$\frac{3}{4}$	19	0.97	1.44	43,600	194	19.8	48,000	214	21.8
$\frac{7}{8}$	22	1.32	1.96	59,000	262	26.8	65,000	289	29.5
1	26	1.73	2.57	76,600	341	34.7	84,400	375	38.3
$1\frac{1}{8}$	29	2.19	3.26	96,400	429	43.7	106,200	472	48.2
$1\frac{1}{4}$	32	2.70	4.02	118,400	527	53.7	130,200	579	59.1
$1\frac{3}{8}$	35	3.27	4.87	142,600	634	64.7	156,800	697	71.1
$1\frac{1}{2}$	38	3.89	5.79	168,800	751	76.6	185,600	826	84.2

*These strengths apply only when a test is conducted with both ends fixed. When in use, the strength of these ropes may be significantly reduced if one end is free to rotate.

Table 4.2.13 *6×19 Classification Wire Rope, Bright (Uncoated) or Drawn-Galvanized Wire, Independent Wire-Rope Core [4]*

(1)	(2)	(3)	(4)	(5)	(6)	(7)	(8)	(9)	(10)	(11)	(12)	(13)
				Nominal Strength								
Nominal Diameter		Approx. Mass		Improved Plow Steel			Extra Improved Plow Steel			Extra Extra Improved Plow Steel		
in.	mm	lb/ft	kg/m	lb	kN	Metric Tonnes	lb	kN	Metric Tonnes	lb	kN	Metric Tonnes
$\frac{1}{2}$	13	0.46	0.68	23,000	102	10.4	26,600	118	12.1	29,200	130	13.2
$\frac{9}{16}$	14.5	0.59	0.88	29,000	129	13.2	33,600	149	15.2	37,000	165	16.8
$\frac{5}{8}$	16	0.72	1.07	35,800	159	16.2	41,200	183	18.7	45,400	202	20.6
$\frac{3}{4}$	19	1.04	1.55	51,200	228	23.2	58,800	262	26.7	64,800	288	29.4
$\frac{7}{8}$	22	1.42	2.11	69,200	308	31.4	79,600	354	36.1	87,600	389	39.7
1	26	1.85	2.75	89,800	399	40.7	103,400	460	46.9	113,800	506	51.6
$1\frac{1}{8}$	29	2.34	3.48	113,000	503	51.3	130,000	678	59.0	143,000	636	64.9
$1\frac{1}{4}$	32	2.89	4.30	138,800	617	63.0	159,800	711	72.5	175,800	782	79.8
$1\frac{3}{8}$	35	3.50	5.21	167,000	743	75.7	192,000	854	87.1	212,000	943	96.2
$1\frac{1}{2}$	38	4.16	6.19	197,800	880	89.7	228,000	1010	103	250,000	1112	113
$1\frac{5}{8}$	42	4.88	7.26	230,000	1020	104	264,000	1170	120	292,000	1300	132
$1\frac{3}{4}$	45	5.67	8.44	266,000	1180	121	306,000	1360	139	338,000	1500	153
$1\frac{7}{8}$	48	6.50	9.67	304,000	1350	138	348,000	1550	158	384,000	1710	174
2	52	7.39	11.0	344,000	1630	156	396,000	1760	180	434,000	1930	197

Table 4.2.14 *6×37 Classification Wire Rope, Bright (Uncoated) or Drawn-Galvanized Wire, Independent Wire-Rope Core [4]*

(1)	(2)	(3)	(4)	(5)	(6)	(7)	(8)	(9)	(10)	(11)	(12)	(13)
				Nominal Strength								
Nominal Diameter		Approx. Mass		Improved Plow Steel			Extra Improved Plow Steel			Extra Extra Improved Plow Steel		
in.	mm	lb/ft	kg/m	lb	kN	Metric Tonnes	lb	kN	Metric Tonnes	lb	kN	Metric Tonnes
$\frac{1}{2}$	13	0.46	0.68	23,000	102	10.4	26,600	118	12.1	29,200	130	13.2
$\frac{9}{16}$	14.5	0.59	0.88	29,000	129	13.2	33,600	149	15.2	37,000	165	16.8
$\frac{5}{8}$	16	0.72	1.07	35,800	159	16.2	41,200	183	18.7	45,400	202	20.6
$\frac{3}{4}$	19	1.04	1.55	51,200	228	23.2	58,800	262	26.7	64,800	288	29.4
$\frac{7}{8}$	22	1.42	2.11	69,200	308	31.4	79,600	354	36.1	87,600	389	39.7
1	26	1.85	2.75	89,800	399	40.7	103,400	460	46.9	113,800	506	51.6
$1\frac{1}{8}$	29	2.34	3.48	113,000	503	51.3	130,000	578	59.0	143,000	636	64.9
$1\frac{1}{4}$	32	2.89	4.30	138,800	617	63.0	159,800	711	72.5	175,800	782	79.8
$1\frac{3}{8}$	35	3.50	5.21	167,000	743	75.7	192,000	854	87.1	212,000	943	96.2
$1\frac{1}{2}$	38	4.16	6.19	197,800	880	89.7	228,000	1010	103	250,000	1112	113
$1\frac{5}{8}$	42	4.88	7.26	230,000	1020	104	264,000	1170	120	292,000	1300	132
$1\frac{3}{4}$	45	5.67	8.44	266,000	1180	121	306,000	1360	139	338,000	1500	153
$1\frac{7}{8}$	48	6.50	9.67	304,000	1350	138	348,000	1550	158	384,000	1710	174
2	52	7.39	11.0	344,000	1530	156	396,000	1760	180	434,000	1930	197
$2\frac{1}{8}$	54	8.35	12.4	384,000	1710	174	442,000	1970	200	488,000	2170	221
$2\frac{1}{4}$	58	9.36	13.9	430,000	1910	195	494,000	2200	224	544,000	2420	247
$2\frac{3}{8}$	60	10.4	15.5	478,000	2130	217	548,000	2440	249	604,000	2690	274
$2\frac{1}{2}$	64	11.6	17.3	524,000	2330	238	604,000	2690	274	664,000	2950	301
$2\frac{5}{8}$	67	12.8	19.0	576,000	2560	261	658,000	2930	299	728,000	3240	330
$2\frac{3}{4}$	71	14.0	20.8	628,000	2790	285	736,000	3270	333	794,000	3530	360
$2\frac{7}{8}$	74	15.3	22.8	682,000	3030	309	796,000	3540	361	864,000	3840	392
3	77	16.6	24.7	740,000	3290	336	856,000	3810	389	936,000	4160	425
$3\frac{1}{8}$	80	18.0	26.8	798,000	3550	362	920,000	4090	417	1,010,000	4490	458
$3\frac{1}{4}$	83	19.5	29.0	858,000	3820	389	984,000	4380	447	1,086,000	4830	493
$3\frac{3}{8}$	87	21.0	31.3	918,000	4080	416	1,074,000	4780	487	1,164,000	5180	528
$3\frac{1}{2}$	90	22.7	33.8	982,000	4370	445	1,144,000	5090	519	1,242,000	5520	563
$3\frac{3}{4}$	96	26.0	38.7	1,114,000	4960	505	1,290,000	5740	585	1,410,000	6270	640
4	103	29.6	44.0	1,254,000	5580	569	1,466,000	6520	665	1,586,000	7050	720

Table 4.2.15 *6×61 Classification Wire Rope, Bright (Uncoated) or Drawn-Galvanized Wire, Independent Wire-Rope Core [4]*

(1)	(2)	(3)	(4)	(5)	(6)	(7)	(8)	(9)	(10)
				Nominal Strength					
Nominal Diameter		Approx. Mass		Improved Plow Steel			Extra Improved Plow Steel		
in.	mm	lb/ft	kg/m	lb	kN	Metric Tonnes	lb	kN	Metric Tonnes
$3\frac{1}{2}$	90	22.7	33.8	966,000	4300	438	1,110,000	4940	503
$3\frac{3}{4}$	96	26.0	38.7	1,098,000	4880	498	1,264,000	5620	573
4	103	29.6	44.0	1,240,000	5520	562	1,426,000	6340	647
$4\frac{1}{4}$	109	33.3	49.6	1,388,000	6170	630	1,598,000	7110	725
$4\frac{1}{2}$	115	37.4	55.7	1,544,000	6870	700	1,776,000	7900	806
$4\frac{3}{4}$	122	41.7	62.1	1,706,000	7590	774	1,962,000	8730	890
5	128	46.2	68.8	1,874,000	8340	850	2,156,000	9590	978

Table 4.2.16 *6×91 Classification Wire Rope, Bright (Uncoated) or Drawn-Galvanized Wire, Independent Wire-Rope Core [4]*

(1)	(2)	(3)	(4)	(5)	(6)	(7)	(8)	(9)	(10)
						Nominal Strength			
Nominal Diameter		Approx. Mass		Improved Plow Steel			Extra Improved Plow Steel		
in.	mm	lb/ft	kg/m	lb	kN	Metric Tonnes	lb	kN	Metric Tonnes
4	103	29.6	44.1	1,178,000	5240	534	1,354,000	6020	614
$4\frac{1}{4}$	109	33.3	49.6	1,320,000	5870	599	1,518,000	6750	689
$4\frac{1}{2}$	115	37.4	55.7	1,468,000	6530	666	1,688,000	7510	766
$4\frac{3}{4}$	122	41.7	62.1	1,620,000	7210	735	1,864,000	8290	846
5	128	46.2	68.7	1,782,000	7930	808	2,048,000	9110	929
$5\frac{1}{4}$	135	49.8	74.1	1,948,000	8670	884	2,240,000	9960	1016
$5\frac{1}{2}$	141	54.5	81.1	2,120,000	9430	962	2,438,000	10,800	1106
$5\frac{3}{4}$	148	59.6	88.7	2,296,000	10,200	1,049	2,640,000	11,700	1198
6	154	65.0	96.7	2,480,000	11,000	1,125	2,852,000	12,700	1294

Table 4.2.17 *8×19 Classification Wire Rope, Bright (Uncoated) or Drawn-Galvanized Wire, Independent Wire-Rope Core [4]*

(1)	(2)	(3)	(4)	(5)	(6)	(7)	(8)	(9)	(10)
						Nominal Strength			
Nominal Diameter		Approx. Mass		Improved Plow Steel			Extra Improved Plow Steel		
in.	mm	lb/ft	kg/m	lb	kN	Metric Tonnes	lb	kN	Metric Tonnes
$\frac{1}{2}$	13	0.47	0.70	20,200	89.9	9.16	23,400	104	10.5
$\frac{9}{16}$	14.5	0.60	0.89	25,600	114	11.6	29,400	131	13.3
$\frac{5}{8}$	16	0.73	1.09	31,400	140	14.2	36,200	161	16.4
$\frac{3}{4}$	19	1.06	1.58	45,000	200	20.4	51,800	230	23.5
$\frac{7}{8}$	22	1.44	2.14	61,000	271	27.7	70,000	311	31.8
1	26	1.88	2.80	79,200	352	35.9	91,000	405	41.3
$1\frac{1}{8}$	29	2.39	3.56	99,600	443	45.2	114,600	507	51.7

Table 4.2.18 *19×7 Classification Wire Rope, Bright (Uncoated) or Drawn-Galvanized Wire, Wire Strand Core [4]*

(1)	(2)	(3)	(4)	(5)	(6)	(7)	(8)	(9)	(10)
						Nominal Strength*			
Nominal Diameter		Approx. Mass		Improved Plow Steel			Extra Improved Plow Steel		
in.	mm	lb/ft	kg/m	lb	kN	Metric Tonnes	lb	kN	Metric Tonnes
$\frac{1}{2}$	13	0.45	0.67	19,700	87.6	8.94	21,600	96.1	9.80
$\frac{9}{16}$	14.5	0.58	0.86	24,800	110	11.2	27,200	121	12.3
$\frac{5}{8}$	16	0.71	1.06	30,600	136	13.9	33,600	149	15.2
$\frac{3}{4}$	19	1.02	1.52	43,600	194	19.8	48,000	214	21.8
$\frac{7}{8}$	22	1.39	2.07	59,000	262	26.8	65,000	289	29.5
1	26	1.82	2.71	76,600	341	34.7	84,400	375	38.3
$1\frac{1}{8}$	29	2.30	3.42	96,400	429	43.7	106,200	472	48.2
$1\frac{1}{4}$	32	2.84	4.23	118,400	527	53.7	130,200	579	59.1
$1\frac{3}{8}$	35	3.43	5.10	142,600	634	64.7	156,800	697	71.1
$1\frac{1}{2}$	38	40.8	6.07	168,800	751	76.6	185,600	826	84.2

*These strengths apply only when a test is conducted with both ends fixed. When in use, the strength of these ropes may be significantly reduced if one end is free to rotate.

Table 4.2.19 *6×25 "B," 6×27 "H," 6×30 "G," 6×31 "V" Flattened Strand Construction Wire Rope Bright (Uncoated) or Drawn-Galvanized Wire [4]*

(1)	(2)	(3)	(4)	(5)	(6)	(7)	(8)	(9)	(10)
						Typical Nominal Strength*			
Nominal Diameter		Approx. Mass		Improved Plow Steel			Extra Improved Plow Steel		
in.	mm	lb/ft	kg/m	lb	kN	Metric Tonnes	lb	kN	Metric Tonnes
$\frac{1}{2}$	13	0.47	0.70	25,400	113	11.5	28,000	125	12.7
$\frac{9}{16}$	14.5	0.60	0.89	32,000	142	14.5	35,200	157	16.0
$\frac{5}{8}$	16	0.74	1.10	39,400	175	17.9	43,400	193	19.7
$\frac{3}{4}$	19	1.06	1.58	56,400	251	25.6	62,000	276	28.1
$\frac{7}{8}$	22	1.46	2.17	76,000	330	34.5	83,800	373	38.0
1	26	1.89	2.81	98,800	439	44.8	108,800	484	49.3
$1\frac{1}{8}$	29	2.39	3.56	124,400	553	56.4	137,000	609	62.1
$1\frac{1}{4}$	32	2.95	4.39	152,600	679	69.2	168,000	747	76.2
$1\frac{3}{8}$	35	3.57	5.31	183,600	817	83.3	202,000	898	91.6
$1\frac{1}{2}$	38	4.25	6.32	216,000	961	98.0	238,000	1,060	108
$1\frac{5}{8}$	42	4.99	7.43	254,000	1,130	115	280,000	1,250	127
$1\frac{3}{4}$	45	5.74	8.62	292,000	1,300	132	322,000	1,430	146
$1\frac{7}{8}$	48	6.65	9.90	334,000	1,490	151	368,000	1,640	167
2	52	7.56	11.2	378,000	1,680	171	414,000	1,840	188

(text continued from page 4-46)

by the correct method. The length of test specimen shall not be less than 3 ft (0.91 m) between sockets for wire ropes up to 1-in. (25.4 mm) diameter and not less than 5 ft (1.52 m) between sockets for wire ropes $1\frac{1}{8}$-in. (28.6 mm) to 3-in. (77 mm) diameter. On wire ropes larger than 3 in. (77 mm), the clear length of the test specimen shall be at least 20 times the rope diameter. The test shall be valid if failure occurs 2 in. (50.8 mm) from the sockets or holding mechanism.

Due to the variables in sample preparation and testing procedures, it is difficult to determine the true strength. Thus, the actual breaking strength during test shall be at least 97.5% of the nominal strength as shown in the applicable table. If the first specimen fails at a value below the 97.5% nominal strength value, a second test shall be made, and if the second test meets the strength requirements, the wire rope shall be accepted.

4.2.4.3 Manufacture and Tolerances
Strand Construction

The 6×7 classification ropes shall contain six strands that are made up of 3 through 14 wires, of which no more than 9 are outside wires fabricated in one operations.* See Table 4.2.10 and Figure 4.2.32.

The 6×19 classification ropes shall contain 6 strands that are made up of 15 through 26 wires, of which no more than 12 are outside wires fabricated in one operation. See Tables 4.2.11 and 4.2.13 and Figures 4.2.33 to 4.2.38. The 6×37 classification ropes shall contain six strands that are made up of 27 through 49 wires, of which no more than 18 are outside wires fabricated in one operation. See Tables 4.2.11 and 4.2.14 and Figures 4.2.39 to 4.2.46.

The 6×61 classification ropes shall contain six strands that are made up of 50 through 74 wires, of which no more

than 24 are outside wires fabricated in one operation. See Table 4.2.15 and Figures 4.2.47. and 4.2.48.

The 6×91 classification wire rope shall have six strands that are made up of 75 through 109 wires, of which no more than 30 are outside wires fabricated in one operation. See Table 4.2.16 and Figures 4.2.49. and 4.2.50.

The 8×19 classification wire rope shall have eight strands that are made up of 15 through 26 wires, of which no more than 12 are outside wires fabricated in one operation. See Table 4.2.17 and Figures 4.2.51. and 4.2.52.

The 18×7 and 19×7 wire rope shall contain 18 or 19 strands, respectively. Each strand is made up of seven wires. It is manufactured counterhelically laying an outer 12-strand layer over an inner 6×7 or 7×7 wire rope. This produces a rotation-resistant characteristic. See Tables 4.2.12 and 4.2.18 and Figures 4.2.53 and 4.2.54.

The 6×25 "B," 6×27 "H," 6×30 "G," and 6×31 "V" flattened strand wire rope shall have six strands with 24 wires fabricated in two operations around a semitriangular shaped core. See Table 4.2.19 and Figures 4.2.55 to 4.2.58.

In the manufacture of uniform-diameter wire rope, wires shall be continuous. If joints are necessary in individual wires, they shall be made, prior to fabrication of the strand, by brazing or electric welding. Joints shall be spaced in accordance with the equation.

$$J = 24D \qquad [4.2.16]$$

where J = minimum distance between joints in main wires in any one strand in in. mm

D = nominal diameter of wire rope in in. mm

Wire rope is most often furnished preformed, but can be furnished non-preformed, upon special request by the purchaser. A preformed rope has the strands shaped to the helical form they assume in the finished rope before the strands have been fabricated into the rope. The strands of such preformed rope shall not spring from their normal position when the seizing bands are removed. Cable tool is one of the few application for which nonpreformed is still used.

---

*One operation strand–When the king wire of the strand becomes so large (manufacturer's discretion) that it is considered undesirable, it is allowed to be replaced with a seven-wire strand manufactured in a separate stranding operation. This does not constitute a two-operation strand.

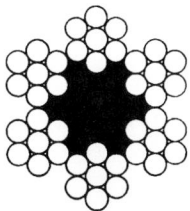

FIGURE 4.2.32
6 × 7 WITH FIBER CORE
6 × 7 CLASSIFICATION

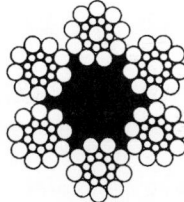

FIGURE 4.2.33
6 × 19 SEALE WITH
FIBER CORE

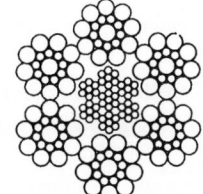

FIGURE 4.2.34
6 × 19 SEALE WITH INDE-
PENDENT WIRE-ROPE CORE

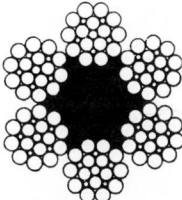

FIGURE 4.2.35
6 × 21 FILLER WIRE
WITH FIBER CORE

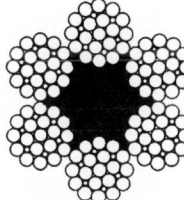

FIGURE 4.2.36
6 × 25 FILLER WIRE
WITH FIBER CORE

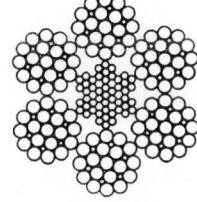

FIGURE 4.2.37
6 × 25 FILLER WIRE
WITH INDEPENDENT
WIRE-ROPE CORE
6 × 19 CLASSIFICATION

FIGURE 4.2.38
6 × 26 WARRINGTON SEALE
WITH INDEPENDENT
WIRE-ROPE CORE

TYPICAL WIRE-ROPE CONSTRUCTIONS WITH CORRECT ORDERING DESCRIPTIONS
(See the paragraph "Strand Construction," or construction which may be ordered with either fiber cores or independent wire rope cores.)

Figure 4.2.32   *6×7 with fiber core [4].*
Figure 4.2.33   *6×19 seale with fiber core [4].*
Figure 4.2.34   *6×19 seale with independent wire-rope core [4].*
Figure 4.2.35   *6×21 filler wire with fiber core [4].*
Figure 4.2.36   *6×25 filler wire with fiber core [4].*
Figure 4.2.37   *6×25 filler wire with independent wire-rope core [4].*
Figure 4.2.38   *6×26 Warrington seale with independent wire-rope core [4].*

The Lay of Finished Rope

Wire rope shall be furnished right lay or left lay and regular lay or Lang lay as specified by the purchaser (Figure 4.2.59). If not otherwise specified on the purchase order, right-lay, regular-lay rope shall be furnished. For 6×7 wire ropes, the lay of the finshed rope shall not exceed eight times the nominal diameter. For 6×19, 6×37, 6×61, 6×91 and 8×19 wire rope, the lay of the finished rope shall not exceed $7\frac{1}{4}$ times the nominal diameter. For flattened strand rope designations 6×25 "B," 6×27 "H," 6×30 "G," and 6×31 "V," the lay of the finished rope shall not exceed eight times the nominal diameter.

Diameter of Ropes and Tolerance Limits

The diameter of a wire rope shall be the diameter of a circumscribing circle and shall be measured at least 5 ft (1.52 m) from properly seized end with a suitable caliper (Figure 4.2.60). The diameter tolerance* of wire rope shall be

Nominal inch diameter: −0% to +5%

Nominal mm diameter: −1% to +4%

*A question may develop as to whether or not the wire rope complies with the oversize tolerance. In such cases, a tension of not less than 10% nor more than 20% of nominal required breaking strength is applied to the rope, and the rope is measured while under this tension.

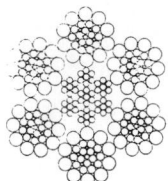

FIGURE 4.2.39
6 × 31 FILLER WIRE SEALE
WITH INDEPENDENT
WIRE-ROPE CORE

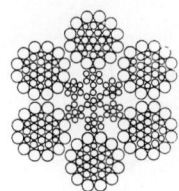

FIGURE 4.2.40
6 × 31 WARRINGTON SEARLE
WITH INDEPENDENT
WIRE-ROPE CORE

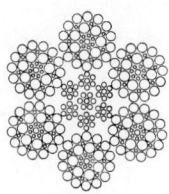

FIGURE 4.2.41
6 × 36 FILLER WIRE
WITH INDEPENDENT
WIRE-ROPE CORE

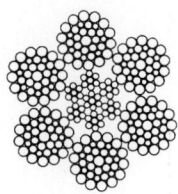

FIGURE 4.2.42
6 × 36 WARRINGTON SEALE
WITH INDEPENDENT
WIRE-ROPE CORE

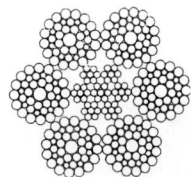

FIGURE 4.2.43
6 × 41 WARRINGTON SEALE
WITH INDEPENDENT
WIRE-ROPE CORE

FIGURE 4.2.44
6 × 41 FILLER WIRE
WITH INDEPENDENT
WIRE-ROPE CORE

FIGURE 4.2.45
6 × 46 FILLER WIRE
WITH INDEPENDENT
WIRE-ROPE CORE

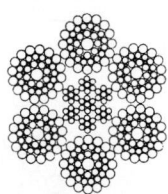

FIGURE 4.2.46
6 × 49 FILLER WIRE SEARLE
WITH INDEPENDENT
WIRE-ROPE CORE

6 × 37 CLASSIFICATION

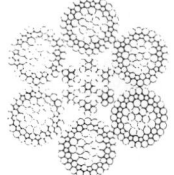

FIGURE 4.2.47
6 × 61 WARRINGTON SEALE
WITH INDEPENDENT
WIRE-ROPE CORE

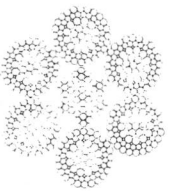

FIGURE 4.2.48
6 × 73 FILLER WIRE SEARLE
WITH INDEPENDENT
WIRE-ROPE CORE

6 × 61 CLASSIFICATION

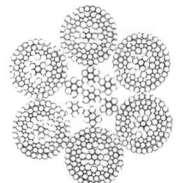

FIGURE 4.2.49
6 × 91 WITH INDEPENDENT
WIRE-ROPE CORE
(TWO-OPERATIONAL STRAND)

FIGURE 4.2.50
6 × 103 WITH INDEPENDENT
WIRE-ROPE CORE
(TWO-OPERATION STRAND)

6 × 91 CLASSIFICATION

TYPICAL WIRE-ROPE CONSTRUCTIONS WITH CORRECT ORDERING DESCRIPTIONS
(See the paragraph titled "Strand Construction," for construction which may be ordered with either fiber cores or independent wire rope cores.)

Figure 4.2.39   6×31 filler wire seale with independent wire-rope core [4].
Figure 4.2.40   6×31 Warrington seale with independent wire-rope core. [4].
Figure 4.2.41   6×36 filler wire seale with independent wire-rope core [4].
Figure 4.2.42   6×36 Warrington seale with independent wire-rope core [4].
Figure 4.2.43   6×41 Warrington seale with independent wire-rope core [4].
Figure 4.2.44   6×41 filler wire with independent wire-rope core [4].
Figure 4.2.45   6×46 filler wire with independent wire-rope core [4].
Figure 4.2.46   6×49 filler wire seale with independent wire-rope core [4].
Figure 4.2.47   6×61 Warrington seale with independent wire-rope core [4].
Figure 4.2.48   6×73 filler wire seale with independent wire-rope core [4].
Figure 4.2.49   6×91 with independent wire-rope core (two-operation strand) [4].
Figure 4.2.50   6×103 with independent wire-rope core (two-operation strand) [4].

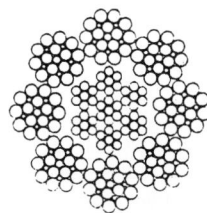 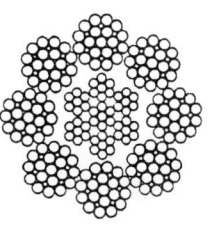

FIGURE 4.2.51
8 × 21 FILLER WIRE
WITH INDEPENDENT
WIRE-ROPE CORE

FIGURE 4.2.52
8 × 25 FILLER WIRE
WITH INDEPENDENT
WIRE-ROPE CORE

8 × 19 CLASSIFICATION

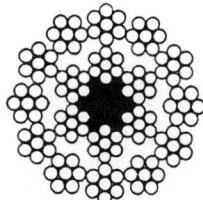 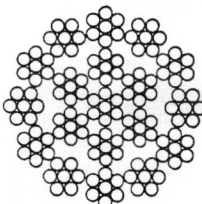 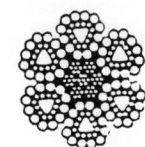

FIGURE 4.2.53
18 × 7 NON-ROTATING
WIRE-ROPE
FIBER CORE

FIGURE 4.2.54
19 × 7 NON-ROTATING
WIRE-ROPE

FIGURE 4.2.55
6 × 25 TYPE B
FLATTENED STRAND
WITH INDEPENDENT WIRE-ROPE CORE

18 × 7 AND 19 × 7 CONSTRUCTION

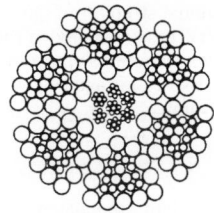 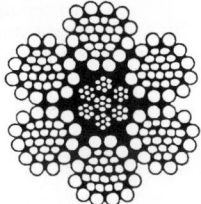 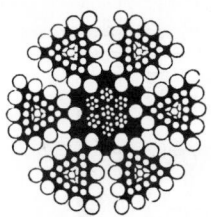

FIGURE 4.2.56
6 × 27 TYPE H
FLATTENED STRAND
WITH INDEPENDENT WIRE-ROPE CORE

FIGURE 4.2.57
6 × 30 STYLE G
FLATTENED STRAND
WITH INDEPENDENT WIRE-ROPE CORE

FIGURE 4.2.58
6 × 31 TYPE V.
FLATTENED STRAND
WITH INDEPENDENT WIRE-ROPE CORE

TYPICAL WIRE-ROPE CONSTRUCTIONS WITH CORRECT ORDERING DESCRIPTIONS
(See the paragraph titled "Strand Construction," for construction which may be ordered with either
fiber cores or independent wire rope cores.)

Figure 4.2.51   *8×21 filler wire with independent wire-rope core [4].*
Figure 4.2.52   *8×25 filler wire with independent wire-rope core [4].*
Figure 4.2.53   *18×7 nonrotating wire rope with fiber core [4].*
Figure 4.2.54   *19×7 nonrotating wire rope [4].*
Figure 4.2.55   *6×25 type B flattened strand with independent wire-rope core [4].*
Figure 4.2.56   *6×27 type H flattened strand with independent wire-rope core [4].*
Figure 4.2.57   *6×30 type G flattened strand with independent wire-rope core [4].*
Figure 4.2.58   *6×31 type V flattened strand with independent wire-rope core [4].*

Figure 4.2.59 *Right and left lay and regular and Lang lay [4].*

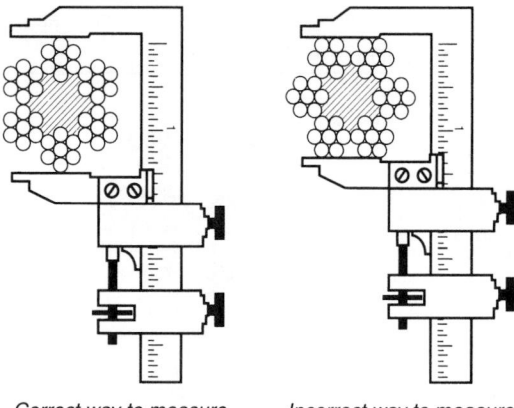

Correct way to measure the diameter of wire rope.   *Incorrect way to measure the diameter of wire rope.*

Figure 4.2.60 *Measurement of diameter [4].*

Diameter of Wire and Tolerance Limits

In seperating the wire rope for gaging of wire, care must be taken to separate the various sizes of wire composing the different layers of bright (uncoated), drawn-galvanized, or galvanized wires in the strand. In like-positioned wires total variations of wire diameters shall not exceed the values of Table 4.2.20.

Fiber Cores

For all wire ropes, all fiber cores shall be hard-twisted, best-quality, manila, sisal, polypropylene, or equivalent. For wire ropes of uniform diameter, the cores shall be of uniform diameter and hardness, effectively supporting the strands. Manila and sisal cores shall be thoroughly impregnated with a suitable lubricating compound free from acid. Jute cores shall not be used.

Lengths

Length of wire rope shall be specified by the purchaser. If minimum length is critical to the application, it shall be specified and conform to the following tolerances.

> 1300 ft (400 m): −0 to +5%
> >1300 ft (400 m) : Original tolerance
> +66 ft (20 m) per each additional 3280 ft (1000 m) or part thereof.

If minimum is not critical to the application, it shall conform to the following tolerances.

> 1300 ft (400 m): ±2.5%
> >1300 ft (400 m) : Original tolerance
> ±33 ft (10 m) per each additional 3280 ft (1000 m) or part thereof.

Lubrication

All wire rope, unless otherwise specified, shall be lubricated and impregnated in the manufacturing process with a suitable compound for the application in amounts best adapted to individual territories. This lubricant should thoroughly protect the ropes internally and externally to minimize rust or corrosion until the rope is put in service.

4.2.4.4 Mooring Wire Rope

Mooring wire rope is used as anchor lines in spread mooring systems, and shall comply with all the provisions of wire rope.

Wire rope for this use should be one operations, right lay, regular lay, independent wire rope core, preformed, galvanized or bright. The nominal strength of galvanized and bright mooring wire rope shall be as specified in Table 4.2.21. For bright mooring wire ropes, the wire grade shall comply with the requirements for Extra Improved Plow Steel, ISO Std 2232* value of 1770 N/mm².

4.2.4.5 Torpedo Lines

Torpedo lines shall be bright (uncoated) or drawn-galvanized, and shall be right, regular lay. The lay of the finished rope shall not exceed eight times the nominal diameter.

Torpedo lines shall be made of five strands of five wires each, or five strands of seven wires each. The strands of the 5×5 construction shall have one center wire and four outer wires of one diameter, fabricated in one operation. The five strands shall be laid around one fiber or cotton core (Figure 4.2.61). The strands of the 5×7 construction shall have one center wire and six outer wires of one diameter, fabricated in one operation. The strands shall be laid around one fiber or cotton core (Figure 4.2.62).

The four outer wires in each strand of the 5×5 construction [both bright (uncoated) and drawn-galvanized] and all the wires in each strand of the 5×7 construction [both bright (uncoated) and drawn-galvanized] shall have the breaking strengths as in Tables 4.2.11 and 4.2.12 for the specified grade and applicable wire size. The center wire of the 5×5 construction shall be hard drawn or annealed and shall not be required to meet the minimum breaking strength specified for the outer wires (the center wire represents about 5% of the total metallic area of the rope and is substantially a filler wire).

*International Organization for Standardization, Standard 2232-1973, "Drawn Wire for General Purpose Non-Alloy Steel Wire Ropes—Specifications," available from American National Standards Institute, 1430 Broadway, New York, New York 10018.

Table 4.2.20 *Wire Diameter Tolerance [4]*

(1)	(2)	(3)	(4)	(5)	(6)
			Total Variation		
		Uncoated (bright) and Drawn Galvanated Wires		Galvanized Wires	
Wire Diameters					
inches	mm	inches	mm	inches	mm
0.018-0.027	0.46-0.69	0.0015	0.038	—	—
0.028-0.059	0.70-1.50	0.0020	0.051	0.0035	0.089
0.060-0.092	1.51-2.34	0.0025	0.064	0.0045	0.114
0.093-0.141	2.35-3.58	0.0030	0.076	0.0055	0.140
0.142 and larger	3.59 and larger	0.0035	0.075	0.0075	0.190

Table 4.2.21 *6×19, 6×37, and 6×61 Construction Mooring Wire Rope, Independent Wire-Rope Core [4]*

1	2	3	4	5	6	7	8	9	10	11
					Nominal Strength					
Construction Classification	Nominal Diameter		Approximate Mass		Galvanized			Bright		
	in.	mm	lb/ft	kg/m	lb	kN	Metric Tonnes	lb	kN	Metric Tonnes
	1	26	1.85	2.75	93,060	414	42.2	95,800	426	43.5
	1⅛	29	2.34	3.48	117,000	520	53.1	119,000	530	54.1
	1¼	32	2.89	4.30	143,800	640	65.2	145,000	646	65.9
	1⅜	35	3.50	5.21	172,800	769	78.4	174,000	773	78.8
	1½	38	4.16	6.19	205,200	913	93.1	205,000	911	92.9
	1⅝	42	4.88	7.26	237,600	1,060	108	250,000	1,110	113
	1¾	45	5.67	8.44	275,400	1,230	125	287,000	1,280	130
	1⅞	48	6.50	9.67	313,200	1,390	142	327,000	1,450	148
	2	52	7.39	11.0	356,400	1,590	162	369,000	1,640	167
	2⅛	54	8.35	12.4	397,800	1,770	180	413,000	1,840	188
	2¼	58	9.36	13.9	444,600	1,980	202	461,000	2,050	209
	2⅜	60	10.4	15.5	493,200	2,190	224	528,000	2,350	239
	2½	64	11.6	17.0	543,600	2,420	247	604,000	2,690	274
	2⅝	67	12.8	18.6	595,800	2,650	270	658,000	2,930	299
	2¾	71	14.0	20.9	649,800	2,890	295	736,000	3,270	333
	2⅞	74	15.3	22.7	705,600	3,140	320	796,000	3,540	361
	3	77	16.6	24.6	765,000	3,400	347	856,000	3,810	389
	3⅛	80	18.0	26.6	824,400	3,670	374	920,000	4,090	417
	3¼	83	19.5	28.6	885,600	3,940	402	984,000	4,380	447
	3⅜	87	21.0	31.4	952,200	4,240	432	1,074,000	4,780	487
	3½	90	22.7	33.6	1,015,000	4,520	460	1,144,000	5,090	519
	3¾	96	26.0	38.2	1,138,000	5,060	516	1,290,000	5,740	585
	4	103	29.6	44.0	1,283,000	5,710	582	1,466,000	6,520	665
	4¼	109	33.3	49.3	1,438,000	6,400	652	1,606,000	7,140	728
	4½	115	37.4	54.9	1,598,000	7,110	725	1,774,000	7,890	805
	4¾	122	41.7	61.8	1,766,000	7,860	801	1,976,000	8,790	896

Construction classification brackets: 6 × 19, 6 × 37, 6 × 61

NOTE: For tests see Paragraph titled "Acceptance."

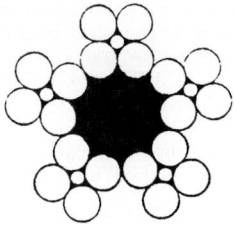

Figure 4.2.61 *5×5 construction torpedo line [4].*

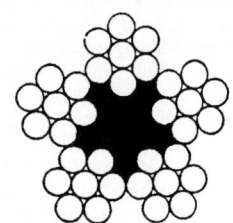

Figure 4.2.62 *5×7 construct torpedo line [4].*

Table 4.2.22 *5×5 Construction Torpedo Lines [4]*

(1)	(2)	(3)	(4)	(5)	(6)	(7)	(8)	(9)	(10)
						Nominal Strength			
Nominal Diameter		Approx. Mass		Plow Steel			Improved Plow Steel		
in.	mm	lb/100 ft	kg/100 m	lb	kN	Metric Tonnes	lb	kN	Metric Tonnes
$\frac{1}{8}$	3.18	2.21	3.29	1,120	4.98	0.51	1,290	5.74	0.59
$\frac{9}{64}$	3.57	2.80	4.16	1,410	6.27	0.64	1,620	7.21	0.74
$\frac{5}{32}$	3.97	3.46	5.15	1,740	7.74	0.79	2,000	8.90	0.91
$\frac{3}{16}$	4.76	4.98	7.41	2,490	11.80	1.13	2,860	12.72	1.30
$\frac{1}{4}$	6.35	8.86	13.91	4,380	19.48	1.99	5,030	22.37	2.28
$\frac{5}{16}$	7.94	13.80	20.54	6,780	30.16	3.08	7,790	34.65	3.53

Table 4.2.23 *5×7 Construction Torpedo Lines [4]*

(1)	(2)	(3)	(4)	(5)	(6)	(7)	(8)	(9)	(10)
						Nominal Strength			
Nominal Diameter		Approx. Mass		Plow Steel			Improved Plow Steel		
in.	mm	lb/100 ft	kg/100 m	lb	kN	Metric Tonnes	lb	kN	Metric Tonnes
$\frac{1}{8}$	3.18	2.39	3.56	1,210	5.38	0.55	1,400	6.23	0.64
$\frac{9}{64}$	3.57	3.02	4.49	1,530	6.81	0.69	1,760	7.83	0.80
$\frac{5}{32}$	3.97	3.73	5.55	1,890	8.41	0.86	2,170	9.65	0.98
$\frac{3}{16}$	4.76	5.38	8.01	2,700	12.01	1.23	3,110	13.83	1.41
$\frac{1}{4}$	6.35	9.55	14.21	4,760	21.17	2.16	5,470	24.33	2.48
$\frac{5}{16}$	7.94	14.90	22.17	7,380	32.83	3.35	8,490	37.77	3.85

The nominal strength of torpedo lines shall be as specified in Tables 4.2.22 and 4.2.23. When testing finished ropes to their breaking strength, suitable sockets or other acceptable means of holding small cords shall be used. The length of tension test specimen shall be not less than 1 ft (0.305 m) between attachments. If the first specimen fails at a value below the specified nominal strength, two additional specimens from the same rope shall be tested, one of which must comply with the nominal strength requirement.

The diameter of the ropes shall be not less than the specified diameter. Torpedo-line lengths shall vary in 500-ft (152.4 m) multiplies.

4.2.4.6 Well-Measuring Wire
Well-measuring wire shall be in accordance with Table 4.2.24 and shall consist of one continuous piece of wire without brazing of welding of the finished wire. The wire shall be made from the best quality of specified grade of material with good workmanship and shall be free from defects that might affect its appearance or serviceability. Coating on well-measuring wire shall be optional with the purchaser.

A specimen of 3-ft (0.91 m) wire shall be cut from each coil of well-measuring wire. One section of this specimen shall be tested for elongation simultaneously with the test for tensile strength. The ultimate elongation shall be measured on a 10-in. (254 mm) specimen at instant of rupture, which must occur within the 10-in. (254 mm) gage length. To determine elongation, a 100,000-psi (690-mPa) stress shall be imposed upon the wire at which the extensometer is applied. Directly to the extensometer reading shall be added 0.4% to allow for the initial elongation occurring before application of extensometer.

The remaining section of the 3-ft (0.91-m) test specimen shall be gaged for size and tested for torsional requirements.

If, in any individual test, the first specimen fails, not more than two additional specimens from the same wire shall be tested. The average of any two tests showing failure or acceptance shall be used as the value to represent the wire.

4.2.4.7 Well-Measuring Strand
Well-measuring strand shall be bright (uncoated) or drawn-galvanized.

Well-measuring strand shall be left lay. The lay of the finished strand shall not exceed 10 times the nominal diameter.

Well-measuring strands may be of various combinations of wires but are commonly furnished in 1×16 (1-6-9) and 1×19 (1-6-12) constructions.

Well-measuring strands shall conform to the properties listed in Table 4.2.25.

To test finished strands to their breaking strength, suitable sockets or other acceptable means of holding small cords shall be used.

4.2.4.8 Wire Guy Strand and Structural Rope and Strand
Galvanized wire guy strand shall conform to ASTM A-475: "Zinc-Coated Steel Wire Strand."* Aluminized wire guy strand can confirm to ASTM A-474: "Aluminum Coated Steel Wire Strand."* Galvanized structural strand shall conform to ASTM A-586: "Zinc-Coated Steel Structural Strand."* Galvanized structural rope shall conform to ASTM A-603: "Zinc-Coated Steel Structural Wire Rope:"*

*American Society for Testing and Materials, 1916 Race Street, Philadelphia, Pennsylvania 19103.

Table 4.2.24 *Requirements for Well-Measuring Wire, Bright or Drawn-Galvanized Carbon Steel* [4]*

		1	2	3	4	5	6	7
Nominal Diameter	in..................	0.066	0.072	0.082	0.092	0.105	0.108
	mm	1.68	1.83	2.08	2.34	2.67	2.74
Tolerance on diameter in.		±0.001	±0.001	±0.001	±0.001	±0.001	±0.001
mm		±0.03	±0.03	±0.03	±0.03	±0.03	±0.03
Breaking strength							
Minimum lb...........................		811	961	1239	1547	1966	2109
kN.................................		3.61	4.27	5.51	6.88	8.74	9.38
Maximum lb		984	1166	1504	1877	2421	2560
kN.................................		4.38	5.19	6.69	8.35	10.77	11.38
Elongation to 10 in. (254 mm), per cent							
Minimum		$1\frac{1}{2}$	$1\frac{1}{2}$	$1\frac{1}{2}$	$1\frac{1}{2}$	$1\frac{1}{2}$	$1\frac{1}{2}$
Torsions, minimum number of twists in							
8 in. (203 mm)		32	29	26	23	20	19

*For well-measuring wire of other materials or coatings, refer to supplier for physical properties.

Table 4.2.25 *Requirements for Well-Servicing Strand Bright or Drawn-Galvanized Carbon Steel [4]*

1		2	3
Nominal diameter..	Inches	$\frac{3}{16}''$	$\frac{1}{4}''$
	MM	4.8	6.4
Tolerances on diameter	Inches	$-0''$	$-0''$
		+.013"	+.015"
	MM	−.048	−.064
		+.288	+.320
Nominal breaking strength.............................	Lbs.	4700	8200
	KN	20.9	36.5
Approximate mass.......................................	Lbs./100'	7.3	12.7
	kg/100'	3.3	5.8

4.2.4.9 Packing and Marking

Finished wire rope, unless otherwise specified, shall be shipped on substantial round-head reels. Reels on which sand lines, drilling lines, or casing lines are shipped shall have round arbor holes of 5 in. (127 mm) to $5\frac{3}{4}$ in. (146 mm) diameter. When reel is full of rope, there shall be a clearance of not less than 2 in. (51 mm) between the full reel and the outside diameter of the flange.

The manufacturer shall protect the wire rope on reels from damage by moisture, dust, or dirt with a water-resistant covering of builtup material, such as tar paper and burlap, or similar material.

The following data shall be plainly marked on the face of the wire-rope reel:

1. Name of manufacturer.
2. Reel Number.
3. API monogram only by authorized manufacturers.
4. Grade (plow steel, improved plow steel, or extra improved plow steel).
5. Diameter of rope, in. (mm).
6. Length of rope, ft (m).
7. Type of construction (Warrington, Seale, or filler wire).
8. Type of core (fiber, wire, plastic, or fiber and plastic).

4.2.4.10 Inspection and Rejection

The manufacturer will, on request of the purchaser, conduct tests as called for in specifications on reasonable notice from the purchaser. During the tests, the manufacturer will afford opportunity to the purchaser's representative to present.

The manufacturer, when delivering wire rope with the API monogram and grade designation, should warrant that such material complies with the specification. The wire rope rejected under specifications should not be wound on reels bearing the API monogram, or sold as API wire rope. When the wire rope wound on reels bearing the API monogram is rejected, the monogram shall be removed.

It is recommended that whenever possible, the purchaser, upon receipt, shall test all new wire rope purchased in accordance with specifications. If a rope fails to render satisfactory service, it is impractical to retest such used rope. It is therefore required that the purchaser shall preserve at least one test specimen of all new rope purchased, length of specimen to be at least 10 ft (3.05 m), properly identified by reel number, etc. Care must be taken that no damage will result by storage of specimen.

If the purchaser is not satisfied with the wire rope service, he or she shall send the properly preserved sample or a sample of the rope from an unused section to any testing laboratory mutually agreed upon by the purchaser and the manufacturer, with instructions to make a complete API test, and notify the manufacturer to have a representative present. If the report indicates compliance with specifications, the purchaser shall assume cost of testing; otherwise, the manufacturer shall assume the expense and make satisfactory adjustments not exceeding full purchase price of the rope. If the report indicates noncompliance with specifications, the testing laboratory shall forward a copy of the test report to the manufacturer.

4.2.4.11 Wire-Rope Sizes and Constructions

Typical sizes and constructions of wire rope for oilfield service are shown in Table 4.2.26. Because of the variety of equipment designs, the selection of other constructions than those shown is justifiable.

In oil field service, wire rope is often referred to as wire line or cable. For clarity, these various expressions are incorporated in this recommended practice.

4.2.4.12 Field Care and Use of Wire Rope

Handling on Reel

When handling wire rope on a reel with a binding or lifting chain, wooden blocks should always be used between the rope and the chain to prevent damage to the wire or distortion of the strands in the rope. Bars for moving the reel should be used against the reel flange, and not against the rope. The reel should not be rolled over or dropped on any hard, sharp object to protect the rope, and should not be dropped from a truck or platform to avoid damage to the rope and the reel.

Rolling the reel in or allowing it to stand in any harmful medium such as mud, dirt or cinders should be avoided. Planking or cribbing will be of assistance in handling the reel as well as in protecting the rope against damage.

Handling during Installation

Blocks should be strung to give a minimum of wear against the sides of sheave grooves. It is also good practice in changing lines to suspend the traveling block from the crown on a single line. This tends to limit the amount of rubbing on guards or spacers, as well as chances for kinks. This is also very effective in pull-through and cutoff procedures.

The reel should be set up on a substantial horizontal axis, so that it is free to rotate as the rope is pulled off, and the rope will not rub against derrick members or other obstructions while being pulled over the crown. A snatch block with a suitable size sheave should be used to hold the rope away from such obstructions.

A suitable apparatus for jacking the reel off the floor and holding it so that it can turn on its axis is desirable. Tensions should be maintained on the wire rope as it leaves the reel by restricting the reel movement. A timber or plank provides satisfactory brake action. When winding the wire rope onto the drum, sufficient tension should be kept on the rope to assure tight winding.

To replace a worn rope with a new one, a swivel-type stringing grip for attaching the new rope to the old rope is recommended. This will prevent transferring the twist from one piece of rope to the other. Ensure that the grip is properly applied. The new rope should not be welded to the old rope to pull it through the system.

Care should be taken to avoid kinking a wire rope since a kink can be cause for removal of the wire rope or damaged section. Wire ropes should not be struck with any object, such as steel hammer, derrick hatchet, or crowbar, that may cause unnecessary nicks or bruises. Even a soft metal hammer can damage a rope. Therefore, when it is necessary to crowd wraps together, any such operation should be performed with the greatest care; and a block of wood should be interposed between the hammer and rope.

Solvent may be detrimental to a wire rope. If a rope becomes covered with dirt or grit, it should be cleaned with a brush.

After properly securing the wire rope in the drum socket, the number of excess or dead wraps or turns specified by the equipment manufacturer should be maintained. Whenever possible, a new wire rope should be run under controlled loads and speeds for a short period after installation. This will help to adjust the rope to working conditions. If a new coring or swabbing line is excessively wavy when first installed, two or four sinker bars may be added on the first few trips to straighten the line.

Care of Wire Rope in Service

The recommendations for handling a reel should be observed at all times during the life of the rope. The design factor should be determined by the following:

$$\text{Design factor} = \frac{B}{W} \qquad [4.2.17]$$

where B = nominal strength of the wire rope in pounds
W = fast line tension

When a wire rope is operated close to the minimum design factor, the rope and related equipment should be in a good operating condition. At all times, the operating personnel should minimize shock, impact, and acceleration or deceleration of loads. Successful field operations indicate that the following design factors should be regarded as minimum:

	Minimum Design Factor
Cable-tool line	3
Sand line	3
Rotary drilling line	3
Hoisting service other than drilling	3
Mast raising and lowering line	2.5
Rotary drilling line when setting casing	2
Pulling on stuck pipe and similar infrequent operations	2

Wire-rope life varies with the design factor; therefore, longer rope life can generally be expected when relatively high design factors are maintained.

To calculate the design factor for multipart string-ups, Figures 4.2.63 and 4.2.64 can be used to determine the value of W. W is the fast line tension and equals the fast line factor* times the hook load weight indicator reading. As an example,

drilling line	$1\frac{3}{8}$ in. (35 mm) EIPS
number of lines	10
hook load	400,000 lb (181.4 tons)

Sheaves are roller bearing type.

From Figure 4.2.63, Case A, the fast line factor is 0.123. The fast line tension is then 400,000 lb (181.41 t) ×0.123 = 49,200 lb (22.3 t) = W. Following the formula above, the design factor is then the nominal strength of $1\frac{3}{8}$ in. (35 mm) EIPS drilling line divided by the fast line tension, or 192,000 lb (87.1 tons) ÷ 49,200 lb (22.3 t) = 3.9.

When working near the minimum design factor, consideration should be given to the efficiencies of wire rope bent around sheaves, fittings, or drums. Figure 4.2.65 shows how rope can be affected by bending.

Rope should be kept tightly and evenly wound on the drum. Sudden, severe stresses are injurious to a wire rope, and should be reduced to a minimum. Experience has indicated that wear increases with speed; economy results from moderately increasing the load an diminishing the speed. Excessive speeds may injure wire rope. Care should be taken to see that the clamps used to fasten the rope for dead ending do not kink, flatten, or crush the rope.

---

*The fast line factor is calculated considering the tensions needed to overcome sheave bearing friction.

Table 4.2.26   *Typical Sizes and Constructions of Wire Rope for Oilfield Service [4]*

1	2	3	4
Service and Well Depth	**Wire Rope** in.	**Diameter** *(mm)*	**Wire Rope Description (Regular Lay)**
Rod and Tubing Pull Lines			
Shallow	½ to ¾ incl.	*(13 to 19)*	6x25 FW or 6x26 WS or 6x31 WS or 18x7′ or 19x7′.
Intermediate	¾, ⅞	*(19, 22)*	PF, LL′, IPS or EIPS, IWRC
Deep	⅞ to 1⅛ incl.	*(22 to 29)*	
Rod Hanger Lines	¼	*(6.5)*	6x19, PF, RL, IPS, FC
Sand Lines			
Shallow	¼ to ½ incl.	*(6.5 to 13)*	
Intermediate	½, 9⁄16	*(13, 14.5)*	6x7 Bright or Galv.[2], PF, RL, PS or IPS, FC
Deep	9⁄16, ⅝	*(14.5, 16)*	
Drilling Lines—Cable Tool (Drilling and Cleanout)			
Shallow	⅝, ¾	*(16, 19)*	
Intermediate	¾, ⅞	*(19, 22)*	6x21 FW, PF or NPF, RL or LL, PS or IPS, FC
Deep	⅞, 1	*(22, 26)*	
Casing Lines—Cable Tool			
Shallow	¾, ⅞	*(19, 22)*	
Intermediate	⅞, 1	*(22, 26)*	6x25 FW or 6x26 WS, PF, RL, IPS or EIPS, FC or IWRC
Deep	1, ⅛	*(26, 29)*	
Drilling Lines—Coring and Slim-Hole Rotary Rigs			
Shallow	⅞, 1	*(22, 26)*	6x26 WS, PF, RL, IPS or EIPS, IWRC
Intermediate	1, 1⅛	*(26, 29)*	6x19 S or 6x26 WS, PF, RL, IPS or EIPS, IWRC
Drillings Lines—Rotary Rigs			
Shallow	1, 1⅛	*(26, 29)*	
Intermediate	1⅛, 1¼	*(29, 32)*	6x19 S or 6x21 S or 6x25 FW or FS, PF, RL, IPS or
Deep	1¼ to 1¾ incl.	*(32, 45)*	EIPS, IWRC
Winch Lines—Heavy Duty	⅝ to ⅞ incl.	*(16 to 22)*	6x26 WS or 6x31 WS, PF, RL, IPS or EIPS, IWRC
	⅞ to 1⅛ incl.	*(22 to 29)*	6x36 WS, PF, RL, IPS or EIPS, IWRC
Horsehead Pumping-Unit Lines			
Shallow	½ to 1⅛ incl.[3]	*(13 to 29)*	6x19 Class or 6x37 Class or 19x7, PF, IPS, FC or IWRC
Intermediate	⅝ to 1⅛ incl.[4]	*(16 to 29)*	6x19 Class or 6x37 Class, PF, IPS, FC or IWRC
Offshore Anchorage Lines	⅞ to 2¾ incl.	*(22 to 70)*	6x19 Class, Bright or Galv., PF, RL, IPS or EIPS, IWRC
	1⅜ to 4¾ incl.	*(35 to 122)*	6x37 Class, Bright or Galv., PF, RL, IPS or EIPS, IWRC
	3¾ to 4¾ incl.	*(96 to 122)*	6x61 Class, Bright or Galv., PF, RL, IPS or EIPS, IWRC
Mast Raising Lines[5]	1⅜ and smaller	*(thru 35)*	6x19 Class, PF, RL, IPS or EIPS, IWRC
	1½ and larger	*(38 and up)*	6x37 Class, PF, RL, IPS or EIPS, IWRC
Guideline Tensioner Line	¾	*(19)*	6x25 FW, PF, RL, IPS or EIPS, IWRC
			Wire Rope Description (Lang Lay)
Riser Tensioner Line	1½, 2	*(38,51)*	6x37 Class or PF, RL, IPS or EIPS, IWRC

Abbreviations:

WS	— Warrington-Seale				
S	— Seale	IPS	— Improved plow steel	RL	— Right lay
FS	— Flattened strand	EIPS	— Extra improved plow steel	LL	— Left lay
FW	— Filler-Wire	PF	— Preformed	FC	— Fiber core
PS	— Plow steel	NPF	— Non-preformed	IWRC	— Independent wire rope core

[1]Single line pulling of rods and tubing requires left lay construction or 18 x 7 or 19 x 7 construction. Either left lay or right lay may be used for multiple line pulling.
[2]Bright wire sand lines are regularly furnished; galvanized finish is sometimes required.
[3]Applies to pumping units having one piece of wire rope looped over an ear on the horsehead and both ends fastened to a polished-rod equalizer yoke.
[4]Applies to pumping units having two vertical lines (parallel) with sockets at both ends of each line.
[5]See API Spec 4E: *Specification for Drilling and Well Servicing Structures.*

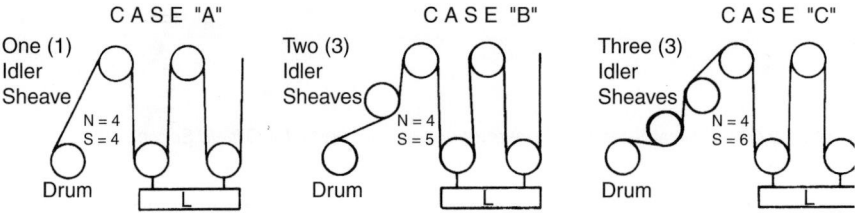

L = Load ;   S = No. of Sheaves ;   N = No. of Rope Parts Supporting Load

FAST LINE TENSION = FAST LINE FACTOR X LOAD

1	2	3	4	5	6	7	8	9	10	11	12	13
	Plain Bearing Sheaves						Roller Bearing Sheaves					
	K = 1.09*						K = 1.04*					
	Efficiency			Fast Line Factor			Efficiency			Fast Line Factor		
N	Case A	Case B	Case C	Case A	Case B	Case C	Case A	Case B	Case C	Case A	Case B	Case C
2	.880	.807	.740	.368	.620	.675	.943	.907	.872	.530	.551	.574
3	.844	.774	.710	.395	.431	.469	.925	.889	.855	.360	.375	.390
4	.810	.743	.682	.309	.336	.367	.908	.873	.839	.275	.286	.298
5	.778	.714	.655	.257	.280	.305	.890	.856	.823	.225	.234	.243
6	.748	.686	.629	.223	.243	.265	.874	.840	.808	.191	.198	.206
7	.719	.660	.605	.199	.216	.236	.857	.824	.793	.167	.173	.180
8	.692	.635	.582	.181	.197	.215	.842	.809	.778	.148	.154	.161
9	.666	.611	.561	.167	.182	.198	.826	.794	.764	.135	.140	.145
10	.642	.589	.540	.156	.170	.185	.811	.780	.750	.123	.128	.133
11	.619	.568	.521	.147	.160	.175	.796	.766	.736	.114	.119	.124
12	.597	.547	.502	.140	.152	.166	.782	.752	.723	.106	.111	.115
13	.576	.528	.485	.133	.145	.159	.768	.739	.710	.100	.104	.108
14	.556	.510	.468	.128	.140	.153	.755	.725	.698	.095	.099	.102
15	.537	.493	.452	.124	.135	.147	.741	.713	.685	.090	.094	.097

$$\text{EFFICIENCY} = \frac{(K^N - 1)}{K^S N (K-1)}$$

$$\text{Fast Line Factor} = \frac{1}{N \times \text{EFFICIENCY}}$$

NOTE: The above cases apply also where the rope is dead ended at the lower or traveling block or derrick floor after passing over a dead sheave in the crown.

*In these tables the K factor for sheave friction is 1.09 for plain bearings and 1.04 for roller bearings. Other K factors can be used if recommended by the equipment manufacturer.

Figure 4.2.63 *Efficiency of wire-rope reeving for multiple sheave blocks, Cases A, B, and C [3].*

Wire ropes are well lubricated when manufactured; however, this lubrication will not last throughout the entire service life of the rope. Periodically, therefore, the rope will need to be field lubricated. When necessary, lubricate the rope with a good grade of lubricant that will penetrate and adhere to the rope, and that is free from acid or alkali.

The clamps used to fasten lines for dead ending shall not kink, flatten, or crush the rope. The rotary line dead-end tie down is equal in importance to any other part of the system. The dead-line anchorage system shall be equipped with a drum and clamping device strong enough to withstand the loading, and designed to prevent wire line damage that would affect service over the sheaves in the system.

The following precautions should be observed to prevent premature wire breakage in drilling lines.

1. *Cable-tool drilling lines.* Movement of wire rope against metallic parts can accelerate wear. This can also create sufficient heat to form martensite, causing embrittlement of wire and early wire rope removal. Such also can be formed by friction against the casing or hard rock formation.

2. *Rotary drilling lines.* Care should be taken to maintain proper winding of rotary drilling lines on the drawworks drum to avoid excessive friction that may result in the formation of martensite. Martensite may also be formed by excessive friction in worn grooves of sheaves, slippage in sheaves, or excessive friction resulting from rubbing against a derrick member. A line guide should be employed between the drum and the fast line sheave to reduce vibration and to keep the drilling line from rubbing against the derrick.

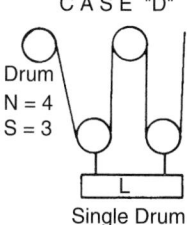
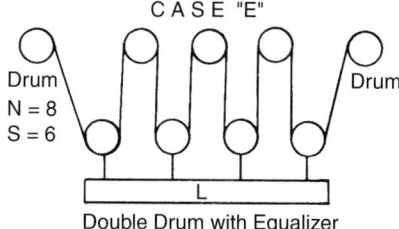

L = Load ;   S = No. of Sheaves ;   N = No. of Rope Parts Supporting Load
(Not counting equalizer)

FAST LINE TENSION = FAST LINE FACTOR X LOAD

1	2	3	4	5	6	7	8	9
	Plain Bearing Sheaves K = 1.09*				Roller Bearing Sheaves K = 1.04*			
	Efficiency		Fast Line Factor		Efficiency		Fast Line Factor	
N	Case D	Case E	Case D	Case E	Case D	Case E	Case D	Case E
2	.959	1.000	.522	.500	981	1.000	.510	.500
3	.920	. . . .	.362	. . . .	.962	. . . .	.346	. . . .
4	.883	.959	.283	.261	.944	.981	.265	.255
5	.848	. . . .	.236	. . . .	.926	. . . .	.216	. . . .
6	.815	.920	.204	.181	.909	.962	.183	.173
7	.784	. . . .	.182	. . . .	.892	. . . .	.160	. . . .
8	.754	.883	.166	.141	.875	.944	.143	.132
9	.726	. . . .	.153	. . . .	.859	. . . .	.130	. . . .
10	.700	.848	.143	.118	.844	.926	.119	.108
11	.674	. . . .	.135	. . . .	.828	. . . .	.110	. . . .
12	.650	.815	.128	.102	.813	.909	.101	.091
13	.628	. . . .	.122	. . . .	.799	. . . .	.096	. . . .
14	.606	.784	.118	.091	.785	.892	.091	.080
15	.586	. . . .	.114	. . . .	.771	. . . .	.086	. . . .

$$\text{CASE "D" EFFICIENCY} = \frac{(K^N - 1)}{K^S N (K-1)}$$

$$\text{FAST LINE FACTOR} = \frac{1}{N \times \text{EFFICIENCY}}$$

$$\text{CASE "E" EFFICIENCY} = \frac{2(K_T^N - 1)}{K_T^S N (K-1)}$$

$$\text{FAST LINE FACTOR} = \frac{1}{N \times \text{EFFICIENCY}}$$

NOTE: The above cases apply also where the rope is dead ended or the equalizer is located at the lower or traveling block or derrick floor after passing over a dead sheave in the crown.

*In these tables, the K factor for sheave friction is 1.09 for plain bearings and 1.04 for roller bearings. Other K factors can be used if recommended by the equipment manufacturer.

Figure 4.2.64 *Efficiency of wire-rope reeving for multiple sheave blocks, Cases D and E [3].*

Martensite is a hard, nonductile microconstituent formed when steel is heated above its critical temperature and cooled rapidly. In the case of steel of the composition conventionally used for rope wire, martensite can be formed if the wire surface is heated to a temperature near or somewhat in excess of 1400°F (760°C), and then cooled at a comparatively rapid rate. The presence of a martensite film at the surface of the outer wires of a rope that has been in service is evidence that sufficient frictional heat has been generated on the crown of the rope wires to momentarily raise the wire surface temperature to a point above the critical temperature range of the steel. The heated surface is then rapidly cooled by the adjacent cold metal within the wire and the rope structure, and an effective quenching results.

Figure 4.2.66A shows a rope that has developed fatigue fractures at the crown in the outer wires, and Figure 4.2.66B shows a photomicrograph (100× magnification) of a specimen cut from the crown of one of these outer wires. This photomicrograph clearly shows the depth of the martensite layer and the cracks produced by the inability of the martensite to withstand the normal flexing of the rope. The result is a disappointing service life for the rope. Most outer wire failures may be attributed to the presence of martensite.

Worn sheave and drum grooves cause excessive wear on the rope. All sheaves should be in proper alignment. The fast sheave should line up with the center of the hoisting drum. From the standpoint of wire-rope life, the condition and contour of sheave grooves are important and should

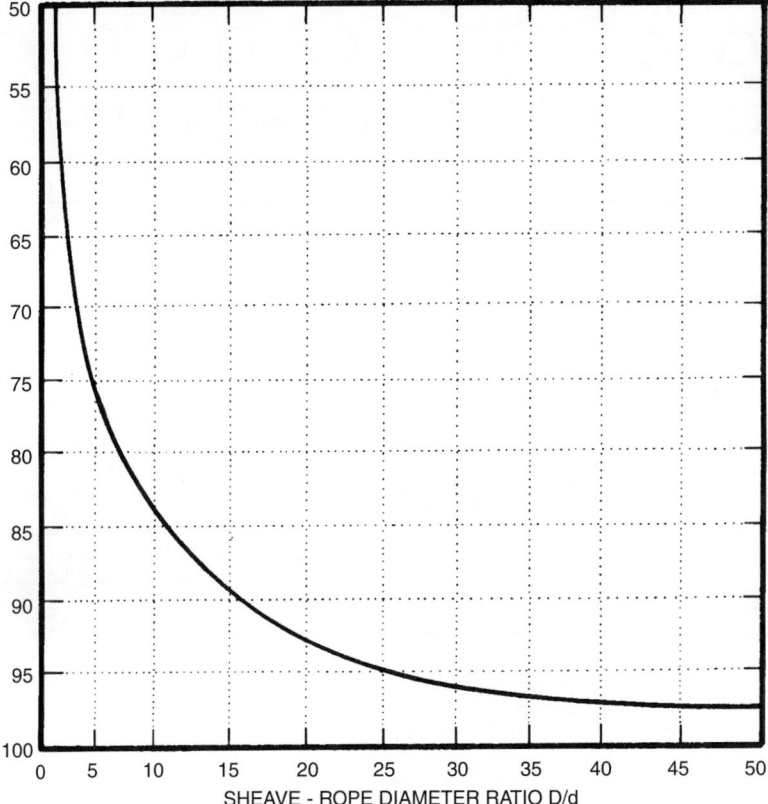

Figure 4.2.65 *Efficiencies of wire ropes bent around stationary sheaves (static stresses only) [3].*

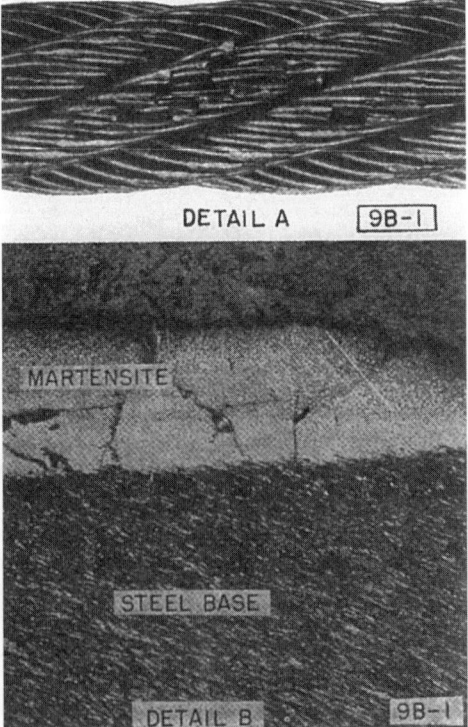

Figure 4.2.66 *Fatigue fractures in outer wires caused by the formation of martensite [3].*

Table 4.2.27 *Minimum Groove Radii for Worm Sheaves [3]*

1		2		1		2		1		2	
Wire Rope Nominal Size		Radii		Wire Rope Nominal Size		Radii		Wire Rope Nominal Size		Radii	
in.	(mm)	in.	(mm)	in.	(mm)	in.	(mm)	in.	(mm)	in.	(mm)
¼	(6.5)	.129	(3.28)	1⅝	(42)	.833	(21.16)	3⅜	(86)	1.730	(43.94)
5⁄16	(8)	.160	(4.06)	1¾	(45)	.897	(22.78)	3½	(90)	1.794	(45.57)
⅜	(9.5)	.190	(4.83)	1⅞	(48)	.959	(24.36)	3¾	(96)	1.918	(48.72)
7⁄16	(11)	.220	(5.59)	2	(52)	1.025	(26.04)	4	(103)	2.050	(52.07)
½	(13)	.256	(6.50)	2⅛	(54)	1.079	(27.41)	4¼	(109)	2.178	(55.32)
9⁄16	(14.5)	.288	(7.32)	2¼	(58)	1.153	(29.29)	4½	(115)	2.298	(58.37)
⅝	(16)	.320	(8.13)	2⅜	(60)	1.199	(30.45)	4¾	(122)	2.434	(61.82)
¾	(19)	.380	(9.65)	2½	(64)	1.279	(32.49)	5	(128)	2.557	(64.95)
⅞	(22)	.440	(11.18)	2⅝	(67)	1.339	(34.01)	5¼	(135)	2.691	(68.35)
1	(26)	.513	(13.03)	2¾	(71)	1.409	(35.79)	5½	(141)	2.817	(71.55)
1⅛	(29)	.577	(14.66)	2⅞	(74)	1.473	(37.41)	5¾	(148)	2.947	(74.85)
1¼	(32)	.639	(16.23)	3	(77)	1.538	(39.07)	6	(154)	3.075	(78.11)
1⅜	(35)	.699	(17.75)	3⅛	(80)	1.598	(40.59)				
1½	(38)	.759	(19.28)	3¼	(83)	1.658	(42.11)				

Sheaves worn to these sizes can be detrimental to wire rope service and should be regrooved or removed from service.

Table 4.2.28 *Minimum Groove Radii for New and Reconditioned Sheaves [3]*

1		2		1		2		1		2	
Wire Rope Nominal Size		Radii		Wire Rope Nominal Size		Radii		Wire Rope Nominal Size		Radii	
in.	(mm)	in.	(mm)	in.	(mm)	in.	(mm)	in.	(mm)	in.	(mm)
¼	(6.5)	.135	(3.43)	1⅝	(42)	.876	(22.25)	3⅜	(86)	1.807	(45.90)
5⁄16	(8)	.167	(4.24)	1¾	(45)	.939	(23.85)	3½	(90)	1.869	(47.47)
⅜	(9.5)	.201	(5.11)	1⅞	(48)	1.003	(25.48)	3¾	(96)	1.997	(50.72)
7⁄16	(11)	.234	(5.94)	2	(52)	1.085	(27.56)	4	(103)	2.139	(54.33)
½	(13)	.271	(6.88)	2⅛	(54)	1.137	(28.88)	4¼	(109)	2.264	(57.51)
9⁄16	(14.5)	.303	(7.70)	2¼	(58)	1.210	(30.73)	4½	(115)	2.396	(60.86)
⅝	(16)	.334	(8.48)	2⅜	(60)	1.271	(32.28)	4¾	(122)	2.534	(64.36)
¾	(19)	.401	(10.19)	2½	(64)	1.338	(33.99)	5	(128)	2.663	(67.64)
⅞	(22)	.468	(11.89)	2⅝	(67)	1.404	(35.66)	5¼	(135)	2.804	(71.22)
1	(26)	.543	(13.79)	2¾	(71)	1.481	(37.62)	5½	(141)	2.929	(74.40)
1⅛	(29)	.605	(15.37)	2⅞	(74)	1.544	(39.22)	5¾	(148)	3.074	(78.08)
1¼	(32)	.669	(16.99)	3	(77)	1.607	(40.82)	6	(154)	3.198	(81.23)
1⅜	(35)	.736	(18.69)	3⅛	(80)	1.664	(42.27)				
1½	(38)	.803	(20.40)	3¼	(83)	1.731	(43.97)				

Standard Machine Tolerance

be checked periodically. The sheave groove should have a radius not less than that in Table 4.2.27; otherwise, rope life can be reduced. Reconditioned sheave grooves should conform to the recommended radii for new and reconditioned sheaves as given in Table 4.2.28. Each operator should establish the most economical point at which sheaves should be regrooved by considering the loss in rope life that will result from worn sheaves as compared to the cost involved in regrooving. When a new rope is to be installed on used sheaves, it is particularly important that the sheave grooves be checked as recommended. To ensure a minimum turning effort, all sheaves should be kept properly lubricated.

Seizing

Before cutting, a wire rope should be securely seized on each side of the cut by serving with soft wire ties. For socketing, at least two additional seizings should be placed at a distance from the end equal to the basket length of the socket. The total length of the seizing should be at least two rope diameters and securely wrapped with a seizing iron.

This is very important, as it prevents the rope untwisting and ensures equal tension in the strands when the load is applied.

The recommended procedure for seizing a wire rope is as follows:

A. The seizing wire should be wound on the rope by hand as shown in Figure 4.2.67 (1). The coils should be kept together and considerable tension maintained on the wire.
B. After the seizing wire has been wound on the rope, the ends of the wire should be twisted together by hand in a counterclockwise direction so that the twisted portion of the wires is near the middle of the seizing (see Figure 4.2.67 (2)).
C. Using "Carew" cutters, the twist should be tightened just enough to take up the slack (Figure 4.2.67 (3)). Tightening the seizing by twisting should not be attempted.
D. The seizing should be tightened by prying the twist away from the axis of the rope with the cutters as shown in Figure. 4.2.67 (4).

Figure 4.2.67 *Putting a seizing on a wire rope [3].*

E. The tightening of the seizing should be repeated as often as necessary to make the seizing tight.
F. To complete the seizing operation, the ends of the wire should be cut off as shown in Figure 4.2.67 (5), and the twisted portion of the wire tapped flat against the rope. The appearance of the finished seizing is illustrated in Figure 4.2.67 (6).

4.2.4.13 Socketing (Zinc Poured or Spelter)
Wire Rope Preparation
The wire rope should be securely seized or clamped at the end before cutting. Measure from the end of the rope a length equal to approximately 90% of the length of the socket basket. Seize or clamp at this point. Use as many seizings as necessary to prevent the rope from unlaying.

After the rope is cut, the end seizing should be removed. Partial straightening of the strands and/or wire may be necessary. The wires should then be separated and broomed out and the cores treated as follows:

1. *Fiber core*—Cut back length of socket basket.
2. *Steel core*—Separate and broom out.
3. *Other*—Follow manufacturer's recommendations.

Cleaning
The wires should be carefully cleaned for the distance inserted in the socket by one of the following methods:

Acid cleaning
1. *Improved plow steel and extra improved plow steel, bright and galvanized.* Use a suitable solvent to remove lubricant. The wires then should be dipped in commercial muriatic acid until thoroughly cleaned. The depth of immersion in acid must not be more than the broomed length. The acid should be neutralized by rinsing in a bicarbonate of soda solution. Fresh acid should be prepared when satisfactory cleaning of the wires requires more than one minute. (Prepare new solution–do not merely add new acid to old.) Be sure acid surface is free of oil or scum. The wires should be dried and then dipped in a hot solution of zinc-ammonium chloride flux. Use a concentration of

1 lb (454 g) of zinc-ammonium chloride in 1 gal (3.8 L) of water and maintain the solution at a temperature of 180°F (82°C) to 200°F (93°C).
2. *Stainless steel.* Use a suitable solvent to remove lubricant. The wires then should be dipped in a hot caustic solution, such as oakite, then in a hot water rinse, and finally dipped in one of the following solutions until thoroughly cleaned:
   a. commercial muriatic acid.
   b. 1 part by weight of cupric chloride, 20 parts by weight of concentrated hydrochloric acid.
   c. 1 part by weight of ferric chloride, 10 parts by weight of concentrated nitric or hydrochloric acid, 20 parts by weight of water.
   Use the above solutions at room temperature. The wires should then be dipped in clean hot water. A suitable flux may be used.
   Fresh solution should be prepared when satisfactory cleaning of the wires requires more than a reasonable time. (Prepare new solutions–do not merely add new solution to old solution.) Be sure solution surface is free of oil and scum.
3. *Phosphor bronze.* Use a suitable solvent to remove lubricant. The wires should then be dipped in commercial muriatic acid until thoroughly cleaned.
4. *Monel Metal.* Use a suitable solvent to remove lubricant. The wires then should be dipped in the following solution until thoroughly cleaned: 1 part glacial acetic acid + 1 part concentrated nitric acid.

This solution is used at room temperature. The broom should be immersed from 30 to 90 s. The depth of immersion in the solution must not be more than broomed length. The wires should then be dipped in clean hot water.

Ultrasonic cleaning (all grades)
An ultrasonic cleaner suitable for cleaning wire rope is permitted in lieu of the acid cleaning methods described previously.

Other cleaning methods
Other cleaning methods of proven reliability are permitted.

Attaching Socket
Preheat the socket to approximately 200°F (93°C). Slip socket over ends of wire. Distribute all wires evenly in the basket and flush with top of basket. Be sure socket is in line with axis of rope.

Use only zinc not lower in quality than high grade per ASTM Specification B-6. Heat zinc to a range allowing pouring at 950°F (510°C) to 975°F (524°C). Skim off any dross accumulated on the surface of the zinc bath. Pour molten zinc into the socket basket in one continuous pour if possible. Tap socket basket while pouring.

Final Preparation
Remove all seizings. Apply lubricant to rope adjacent to socket to replace lubricant removed by socketing procedure. Socket is then ready for service.

Splicing
Splicing wire rope requires considerable skill, and the instructions for splicing wire rope will be found in the catalogues of most of the wire-rope manufacturers, where the operation sequence is carefully described, and many clear illustrations are presented. These illustrations give, in fact, most of the information needed.

Socketing (Thermo-Set Resin)
Before proceeding with thermo-set resin socketing, the manufacturer's instructions for using this product should

be carefully read. Particular attention should be given to sockets designed specifically for resin socketing. Other thermo-set resins used may have specifications that differ from those shown in this section.

Seizing and Cutting the Rope

The rope manufacturer's directions for a particular size or construction of rope are to be followed with regard to the number, position, and length of seizings, and the seizing wire size to be used. The seizing, which will be located at the base of the installed fitting, must be positioned so that the ends of the wires to be embedded will be slightly below the level of the top of the fitting's basket. Cutting the rope can best be accomplished by using an abrasive wheel.

Opening and Brooming the Rope End

Prior to opening the rope end, place a short temporary seizing directly above the seizing that represents the base of the broom. The temporary seizing is used to prevent brooming the wires to full length of the basket, and also to prevent the loss of lay in the strands and rope outside the socket. Remove all seizings between the end of the rope and the temporary seizing. Unlay the strands of the rope. Starting with the IWRC, or strand core, open each strand and each strand of the rope, and broom or unlay the individual wires. (A fiber core may be cut in the rope at the base of the seizing. Some prefer to leave the core in. Consult the manufacturer's instructions.) When the brooming is completed, the wires should be distributed evenly within a cone so that they form an included angle of approximately 60°. Some types of sockets require a different brooming procedure and the manufacturer's instructions should be followed.

Cleaning the Wires and Fittings

Different types of resin with different characteristics require varying degrees of cleanliness. The following cleaning procedure was used for one type of polyester resin with which over 800 tensile tests were made on ropes in sizes $\frac{1}{4}$-in. (6.5 mm) to $3\frac{1}{2}$-in. (90 mm) diameter without experiencing any failure in the resin socket attachment.

Thorough cleaning of the wires is required to obtain resin adhesion. Ultra-sonic cleaning in recommended solvents (such as trichloroethylene or 1,1,1-trichloroethane or other nonflammable grease-cutting solvents) is the preferred method in accordance with OSHA standards. If ultrasonic cleaning is not available, trichloroethane may be used in brush or dip-cleaning; but fresh solvent should be used for each rope end fitting and should be discarded after use. After cleaning, the broom should be dried with clean compressed air or in another suitable fashion before proceeding to the next step. Using acid to etch the wires before resin socketing is *unnecessary and not recommended.* Also, the use of flux on the wires before pouring the resin should be avoided since *this adversely affects bonding of the resin to the steel wires.* Since there is a variation in the properties of different resins, the manufacturer's instructions should be carefully followed.

Placement of the Fitting

The rope should be placed vertically with the broom up, and the broom should be closed and compacted to insert the broomed rope end into the fitting base. Slip on the fitting, removing any temporary bending or seizing as required. Make sure the broomed wires are uniformly spaced in the basket with the wire ends slightly below the top edge of the basket, and make sure the axis of the rope and the fitting are aligned. Seal the annular space between the base of the fitting and the exiting rope to prevent leakage of the resin from the basket. A nonhardening butyl rubber base sealant gives satisfactory performance. Make sure the sealant does not enter the socket base, so that the resin may fill the complete depth of the socket basket.

Pouring the Resin

Controlled heat-curing (no open flame) at a temperature range of 250 to 300° F (121 to 149°C) is recommended; and is required if ambient temperatures are less than 60°F (16°C) (which may vary with different resins). When controlled heat curing is not available and ambient temperatures are not less than 60°F (16°C), the attachment should not be disturbed and tension should not be applied to the socketed assembly for at least 24 hr.

Lubrication of Wire Rope after Socket Attachment

After the resin has cured, relubricate the wire rope at the base of the socket to replace the lubricant that was removed during the cleaning operation.

Resin Socketing Compositions

Manufacturer's directions should be followed in handling, mixing, and pouring the resin composition.

Performance of Cured Resin Sockets

Poured resin sockets may be moved when the resin has hardened. After ambient or elevated temperature cure recommended by the manufacturer, resin sockets should develop the nominal strength of the rope; and should also withstand, without cracking or breakage, shock loading sufficient to break the rope. Manufacturers of resin socketing material should be required to test to these criteria before resin materials are approved for this end use.

Attachment of Clips

The clip method of making wire-rope attachments is widely used. Drop-forged clips of either the U-bolt or the double-saddle type are recommended. When properly applied as described herein, the method develops about 80% of the rope strength in the case of six strand ropes.

When attaching clips, the rope length to be turned back when making a loop is dependent upon the rope size and the load to be handled. The recommended lengths, as measured from the thimble base, are given in Table 4.2.29. The thimble should first be wired to the rope at the desired point and the rope then bent around the thimble and temporarily secured by wiring the two rope members together.

The first clip should be attached at a point about one base width from the last seizing on the dead end of the rope and tightened securely. The saddle of the clip should rest upon the long or main rope and the U-bolt upon the dead end. All clips should be attached in this manner (Figure 4.2.68). The short end of the rope should rest squarely upon the main portion.

The second clip should be attached as near the loop as possible. The nuts for this clip should not be completely tightened when it is first installed. The recommended number of clips and the space between clips are given in Table 4.2.29. Additional clips should be attached with an equal spacing between clips. Before completely tightening the second and any of the additional clips, some stress should be placed upon the rope in order to take up the slack and equalize the tension on both sides of the rope.

When the clips are attached correctly, the saddle should be in contact with the long end of the wire rope and the U-bolt in contact with the short end of the loop in the rope as shown in Figure 4.2.68. The incorrect application of clips is illustrated in Figure.4.2.69.

The nuts on the second and additional clips should be tightened uniformly, by giving alternately a few turns to one

Table 4.2.29 *Attachment of Clips [3]*

1		2	3		4	
Diameter of Rope		Number of Clips	Length of Rope Turned Back		Torque	
in.	(mm)		in.	(mm)	ft-lb	(N · m)
$\frac{1}{8}$	(3)	2	$3\frac{1}{4}$	(83)	4.5	(6.1)
$\frac{3}{16}$	(5)	2	$3\frac{3}{4}$	(95)	7.5	(10)
$\frac{1}{4}$	(6.5)	2	$4\frac{3}{4}$	(121)	15	(20)
$\frac{5}{16}$	(8)	2	$5\frac{1}{4}$	(133)	30	(41)
$\frac{3}{8}$	(9.5)	2	$6\frac{1}{2}$	(165)	45	(61)
$\frac{7}{16}$	(11)	2	7	(178)	65	(88)
$\frac{1}{2}$	(13)	3	$11\frac{1}{2}$	(292)	65	(88)
$\frac{9}{16}$	(14.5)	3	12	(305)	95	(129)
$\frac{5}{8}$	(16)	3	12	(305)	95	(129)
$\frac{3}{4}$	(19)	4	18	(457)	130	(176)
$\frac{7}{8}$	(22)	4	19	(483)	225	(305)
1	(26)	5	26	(660)	225	(305)
$1\frac{1}{8}$	(29)	6	34	(864)	225	(305)
$1\frac{1}{4}$	(32)	7	44	(1117)	360	(488)
$1\frac{3}{8}$	(35)	7	44	(1120)	360	(488)
$1\frac{1}{2}$	(38)	8	54	(1372)	360	(488)
$1\frac{5}{8}$	(42)	8	58	(1473)	430	(583)
$1\frac{3}{4}$	(45)	8	61	(1549)	590	(800)
2	(51)	8	71	(1800)	750	(1020)
$2\frac{1}{4}$	(57)	8	73	(1850)	750	(1020)
$2\frac{1}{2}$	(64)	9	84	(2130)	750	(1020)
$2\frac{3}{4}$	(70)	10	100	(2540)	750	(1020)
3	(77)	10	106	(2690)	1200	(1630)

NOTE 1: If a pulley is used in place of a thimble for turning back the rope, add one additional clip.

side and then the other. It will be found that the application of a little oil to the threads will allow the nuts to be drawn tighter. After the rope has been in use a short time, the nuts on all clips should be retightened, as stress tends to stretch the rope, thereby reducing its diameter. The nuts should be tightened at all subsequent regular inspection periods. A half hitch, either with or without clips, is not desirable as it malforms and weakens wire rope.

4.2.4.14 Reeving of Wire Rope

Figure 4.2.70 illustrates, in a simplified form, the generally accepted methods of reeving (stringing up) in-line crown and traveling blocks, along with the location of the drawworks drum, monkey board, drill pipe fingers, and deadline anchor

in relation to the various sides of the derrick. Ordinarily, the only two variables in reeving systems, as illustrated, are the number of sheaves in the crown and traveling blocks or the number required for handling the load, and the location of the deadline anchor. Table 4.2.30 gives the right-hand string-ups. The reeving sequence for the left-hand reeving with 12 lines on a seven-sheave crown-block and six-sheave traveling block illustrated in Figure 4.2.70 is given in Arrangement No. 1 of Table 4.2.30. The predominant practice is to use left-hand reeving and locate the deadline anchor to the left of the derrick vee. In selecting the best of the various possible methods for reeving casing or drilling lines, the following basic factors should be considered:

1. Minimum fleet angle from the drawworks drum to the first sheave of the crown block, and from the crown block sheaves to the traveling block sheaves.
2. Proper balancing of crown and traveling blocks.
3. Convenience in changing from smaller to larger number of lines, or from larger to smaller number of lines.
4. Location of deadline on monkey board side for convenience and safety of derrickman.
5. Location of deadline anchor, and its influence upon the maximum rated static hook load of derrick.

4.2.4.15 Recommended Design Features

The proper design of sheaves, drums, and other equipment on which wire rope is used is very important to the service life of wire rope. It is strongly urged that the purchaser specify on his order that such material shall conform with recommendations set forth in this section.

The inside diameter of socket and swivel-socket baskets should be $\frac{5}{32}$ in. larger than the nominal diameter of the wire rope inserted. Alloy or carbon steel, heat treated, will best serve for sheave grooves. Antifriction bearings are recommended for all rotating sheaves.

Drums should be large enough to handle the rope with the smallest possible number of layers. Drums having a diameter of 20 times the nominal wire-rope diameter should be considered minimum for economical practice. Larger diameters than this are preferable. For well-measuring wire, the drum diameter should be as large as the design of the equipment will permit, but should not be less than 100 times the wire diameter. The recommended grooving for wire-rope drums is as follows:

A. On drums designed for multiple-layer winding, the distance between groove centerlines should be approximately equal to the nominal diameter of the wire rope plus one-half the specified oversize tolerance. For the best spooling condition, this dimension can vary according to the type of operation.
B. The curvature radius of the groove profile should be equal to the radii listed in Table 4.2.28.
C. The groove depth should be approximately 30% of the nominal diameter of the wire rope. The crests between grooves should be rounded off to provide the recommended groove depth.

Figure 4.2.68 *Correct method of attaching clips to wire rope [3].*

Figure 4.2.69 *Incorrect methods of attaching clips to wire rope [3].*

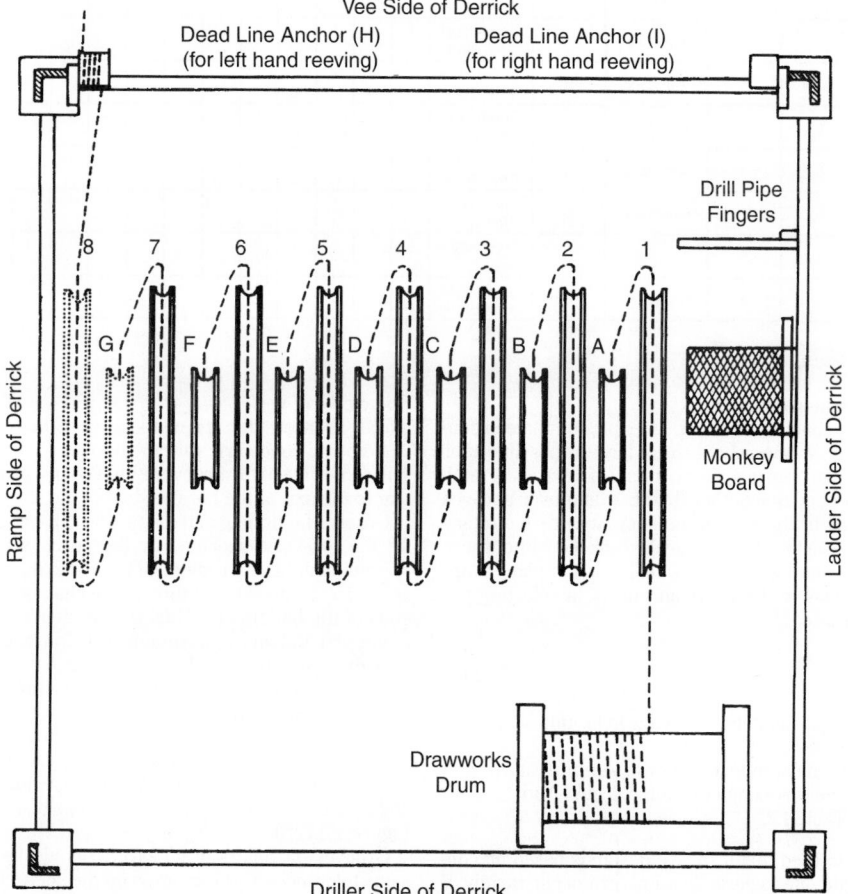

Figure 4.2.70 *Typical reeving diagram for 14-line string-up with eight-sheave crown block and seven-sheave traveling block: left-hand reeving [3].*

Table 4.2.30 *Recommended Reeving Arrangements for 12-, 10-, 8-, and 6-Line String-ups Using 7-Sheave Crown Blocks with 6-Sheave Traveling Blocks and 6-Sheave Crown Blocks with 5-Sheave Traveling Blocks [3]*

Reeving Sequence: Read From Left to Right Starting with Crown Block and Going Alternately From Crown to Traveling to Crown.

Arrange-ment No.	No. of Sheaves Crown Block	No. of Sheaves Trav. Block	Type of String-up	No. of Lines to	Block	r1	r2	r3	r4	r5	r6	r7	r8	r9	r10	r11	r12	r13	r14	r15
1	8	7	Left Hand	14	Crown Block	1		2		3		4		5		6		7		8
					Trav. Block		A		B		C		D		E		F		G	
2	8	7	Right Hand	14	Crown Block	8		7		6		5		4		3		2		1
					Trav. Block		G		F		E		D		C		B		A	
3	7	6	Left Hand	12	Crown Block	1		2		3		4		5		6		7		
					Trav. Block		A		B		C		D		E		F			
4	7	6	Right Hand	12	Crown Block	7		6		5		4		3		2		1		
					Trav. Block		F		E		D		C		B		A			
5	7	6	Left Hand	10	Crown Block	1		2		3				5		6		7		
					Trav. Block		A		B				D		E		F			
6	7	6	Right Hand	10	Crown Block	7		6		5				3		2		1		
					Trav. Block		F		E				C		B		A			
7	6	5	Left Hand	10	Crown Block	1		2		3		4		5		6				
					Trav. Block		A		B		C		D		E					
8	6	5	Right Hand	10	Crown Block	6		5		4		3		2		1				
					Trav. Block		E		D		C		B		A					
9	6	5	Left Hand	8	Crown Block	1		2		3				5		6				
					Trav. Block		A		B				D		E					
10	6	5	Right Hand	8	Crown Block	6		5		4				2		1				
					Trav. Block		E		D				B		A					
11	6	5	Left Hand	8	Crown Block	1		2		3		4		5						
					Trav. Block		A		B		C		D							G
12	6	5	Right Hand	8	Crown Block	6		5		4		3		2						
					Trav. Block		E		D		C		B							H
13	6	5	Left Hand	6	Crown Block			2		3		4		5						
					Trav. Block				B		C		D							G
14	6	5	Right Hand	6	Crown Block			5		4		3		2						
					Trav. Block				D		C		B							H
15	6	5	Left Hand	6	Crown Block	1				3		4				6				
					Trav. Block		A				C				E					
16	6	5	Right Hand	6	Crown Block	6				4		3				1				
					Trav. Block		E				C				A					

Diameter of Sheaves

When bending conditions over sheaves predominate in controlling rope life, sheaves should be as large as possible after consideration has been given to economy of design, portability, etc. When conditions other than bending over sheaves predominate as in the case of hoisting service for rotary drilling, the size of the sheaves may be reduced without seriously affecting rope life. The following recommendations are offered as a *guide* to designers and users in selecting the proper sheave size.

$$D_T = d \times F \qquad [4.2.18]$$

where D_T = tread diameter of sheave in in. (mm)
(see Figure 4.2.71)

d = nominal rope diameter in in. (mm), and

F = sheave-diameter factor, selected from Table 4.2.31.

It should be stressed that if sheave design is based on condition C, fatigue due to serve bending can occur rapidly. If other operation conditions are not present to cause the rope to be removed from service, this type of fatigue is apt to result in wires breaking where they are not readily visible to external examination. Any condition resulting in rope deterioration of a type that is difficult to judge by examination during service should certainly be avoided.

The diameter of sheaves for well-measuring wire should be as large as the design of the equipment will permit but not less than 100 times the diameter of the wire.

Sheave Grooves

On all sheaves, the arc of the groove bottom should be smooth and concentric with the bore or shaft of the sheave. The centerline of the groove should be in a plane perpendicular to the axis of the bore or shaft of the sheave.

Grooves for drilling and casing line sheaves shall be made for the rope size specified by the purchaser. The groove bottom shall have a radius R (Table 4.2.28) subtending an arc of 150°. The sides of the groove shall be tangent to the ends of the bottom arc. Total groove depth shall be a minimum of 1.33d and a maximum of 1.75d (d is the nominal rope diameter shown in Figure 4.2.71).

Grooves for sand-line sheaves shall be made for the rope size specified by the purchaser. The groove bottom shall have a radius R (Table 4.2.28) subtending an arc of 150°. The sides of the groove shall be tangent to the ends of the bottom arc. Total groove depth shall be a minimum of 1.75d and a maximum of 3d (d is nominal rope diameter shown in Figure 4.2.72B).

Grooves on rollers of oil savers should be made to the same tolerances as the grooves on the sheaves.

Sheaves conforming to the specifications (Specification 8A) shall be marked with the manufacturer's name or mark, the sheave groove size and the sheave OD. These markings shall be cast or stamped on the outer rim of the sheave groove and stamped on the nameplate of crown and traveling blocks. For example, a 36-in. sheave with $1\frac{1}{8}$ in. groove shall be marked

AB CO 1 1/8 SPEC 8A

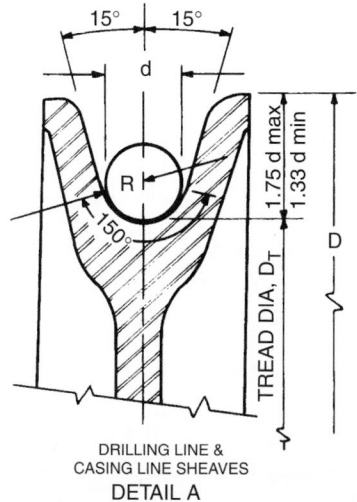
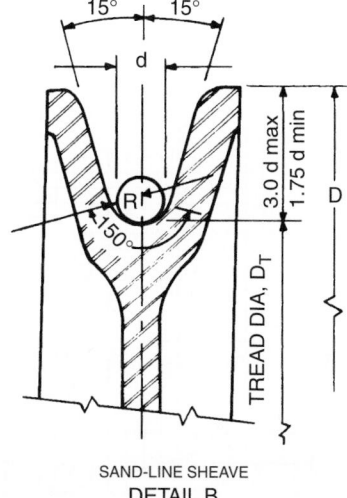

Figure 4.2.71 *Sheave grooves [3].*

Table 4.2.31 *Sheave-Diameter Factors [3]*

1	2	3	4
	Factor, F		
Rope Classification	Condition A	Condition B	Condition C
6×7	72	42	
6×17 Seale	56	33	
6×19 Seale	51	30	(See Fig. 3.1
6×21 Filler Wire	45	26	and
6×25 Filler Wire	41	24	Table 3.2)
6×31	38	22	
6×37	33	18	
8×19 Seale	36	21	
8×19 Warrington	31	18	
18×7 and 19×7	51	36	
Flattened Strand	51	45	*

* Follow manufacturer's recommendations.

Condition A—Where bending over sheaves is of major importance, sheaves at large as those determined by factors under condition A are recommended.

Conditon B—Where bending over sheaves is important, but some sacrifice in rope life is acceptable to achieve portability, reduction in weight, economy of design, etc., sheaves at least as large as those determined by factors under condition B are recommended.

Condition C—Some equipment is used under operating conditions which do not reflect the advantage of the selection of sheaves by factors under conditon A or B. In such cases, sheave-diameter factors may be selected from Figure 4.2.71 and Table 4.2.30. As smaller factors are selected, the bending life of the wire rope is reduced and it becomes an increasingly important condition of rope service. Some conception of relative rope service with different rope constructions and/or different sheave sizes may be obtained by multiplying the ordinate found in Figure 4.2.71 by the proper construction factor indicated in Table 4.2.30.

Sheaves should be replaced or reworked when the groove radius decreases below the values shown in Table 4.2.27. Use sheave gages as shown in Figure 4.2.72A shows a sheave with a minimum groove radius, and 4.2.72B shows a sheave with a tight groove.

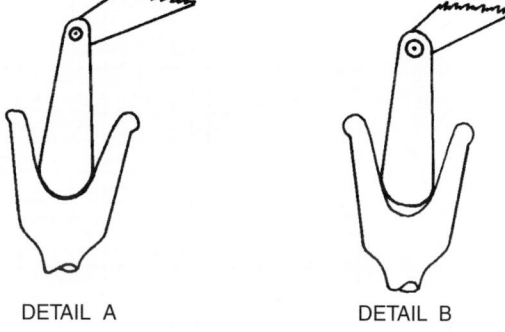

Figure 4.2.72 *Use Of Sheave Gage [3].*

4.2.4.16 Evaluation of Rotary Drilling Line.

The total service performed by a rotary drilling line can be evaluated by considering the amount of work done by the line in various drilling operations (drilling, coring, fishing, setting casing, etc.), and by evaluating such factors as the stresses imposed by acceleration and deceleration loadings, vibration stresses, stresses imposed by friction forces of the line in contact with drum and sheave surfaces, and other even more indeterminate loads. However, for comparative purposes, an approximate evaluation can be obtained by computing only the work done by the line in raising and lowering the applied loads in making round trips, and in the operations of drilling, coring, setting casing, and short trips.

Round-Trip Operations

Most of the work done by a drilling line is that performed in making round trips (or half-trips) involving running the string of drill pipe into the hole and pulling the string out of the hole. The amount of work performed per round trip should be determined by

$$T_r = \frac{D(L_s + D)W_m}{10,560,000} + \frac{D\left(M + \frac{1}{2}C\right)}{2,640,000} \qquad [4.2.19]$$

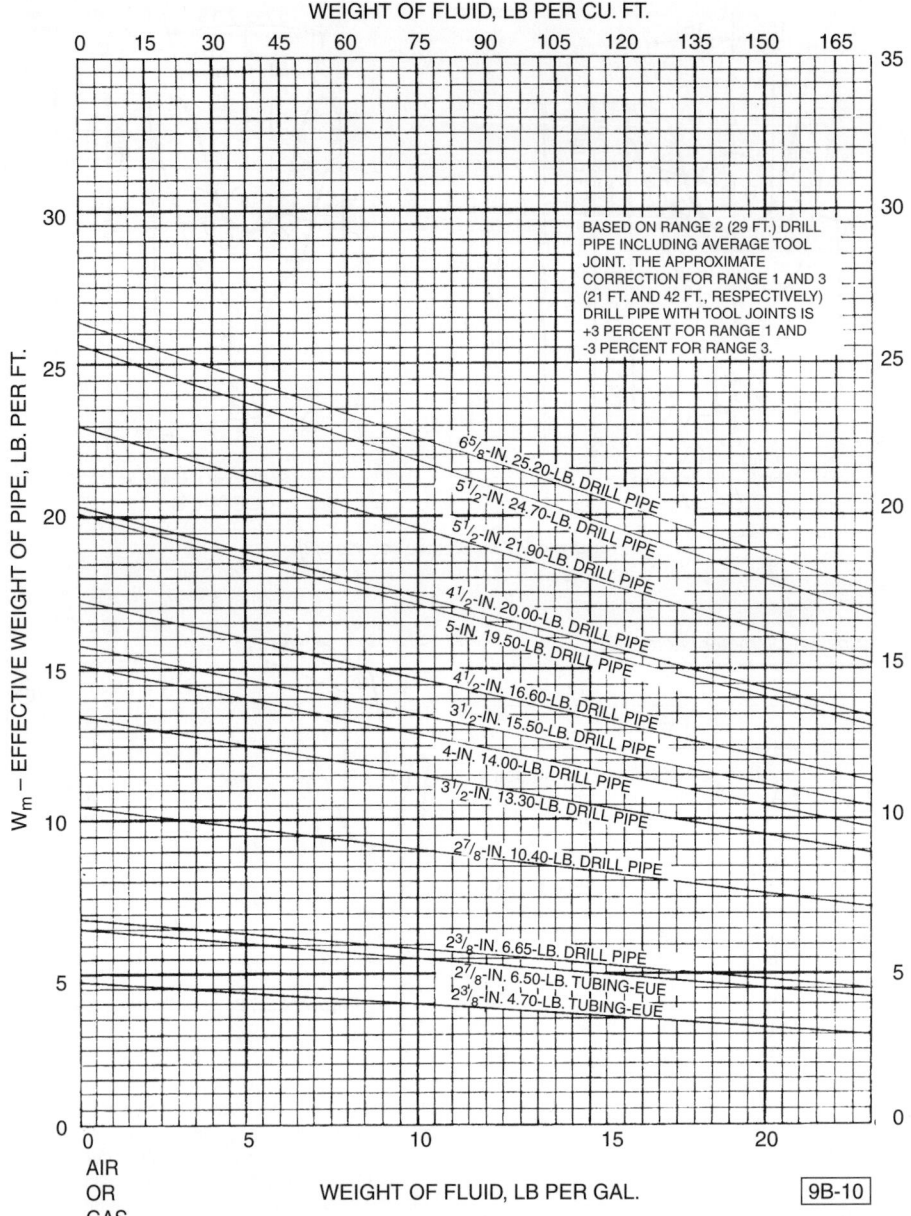

Figure 4.2.73 *Effective weight of pipe-in drilling fluid [3].*

where T_r = ton-miles (weight in tons times distance moved in miles)

D = depth of hole in feet

L_s = length of drill-pipe stand in feet

N = number of drill-pipe stands

W_m = effective weight per foot of drill-pipe from Figure 4.2.73.

M = total weight of traveling block-elevator assembly in pounds.

C = effective weight of drill-collar assembly from Figure 4.2.73 minus the effective weight of the same length of drill-pipe from Figure 4.2.73 in pounds

Drilling Operations

The ton-miles of work performed in drilling operations is expressed in terms of work performed in making round trips, since there is a direct relationship as illustrated in the following cycle of drilling oprations.

1. Drill ahead length of the kelly.
2. Pull up length of the kelly.
3. Ream ahead length of the kelly.
4. Pull up length of the kelly to add single or double.
5. Put kelly in rat hole.
6. Pick up single or double.
7. Lower drill stem in hole.
8. Pick up kelly.

Analysis of the cycle of operations shows that for any hole, the sum of operations 1 and 2 is equal to one round trip; the sum of operations 3 and 4 is equal to another round trip; the sum of operation 7 is equal to one-half a round trip; and the sum of operations 5, 6, and 8 may, and in this case does, equal another one-half round trip, thereby making the work of drilling the hole equivalent to three round trips to bottom, and the relationship can be expressed as

$$T_d = 3(T_2 - T_1) \qquad [4.2.20]$$

where T_d = ton-mile drilling
$\quad T_1$ = ton-miles for one round trip at depth D_1 (depth where drilling started after going in hole, in ft)
$\quad T_2$ = ton-miles for one round trip at depth D_2 (depth where drilling stopped before coming out of hole in ft)

If operations 3 and 4 are omitted, then formula 4.2.20 becomes

$$T_d = 2(T_2 - T_1) \qquad [4.2.21]$$

Coring Operations
The ton-miles of work performed in coring operations, as for drilling operations, is expressed in terms of work performed in making round trips, since there is a direct relationship illustrated in the following cycle of coring operations.

1. Core ahead length of core barrel.
2. Pull up length of kelly.
3. Put kelly in rat hole.
4. Pick up single.
5. Lower drill stem in hole.
6. Pick up kelly.

Analysis of the cycle of operation shows that for any one hole, the sum of operations 1 and 2 is equal to one round trip; the sum of operations 5 is equal to one-half a round trip; and the sum of operations 3, 4, and 6 may, and in this case does, equal another one-half round trip, thereby making the work of drilling the hole equivalent to two round trips to bottom, and the relationship can be expressed as

$$T_c = 2(T_4 - T_3) \qquad [4.2.22]$$

where T_c = ton-mile coring
$\quad T_3$ = ton-miles for one round trip at depth D_3 (depth where coring started after going in hole, in feet)
$\quad T_4$ = ton-miles for one round trip at depth D_4 (depth where coring stopped before coming out of hole, in feet)

Setting Casing Operations
The calculation of the ton-miles for the operation for setting casing should be determined as in round-trip operations as for drill pipe, but with the effective weight of the casing being used, and with the result being multiplied by one-half, since setting casing is a one-way (one-half round-trip) operation. Ton-miles for setting casing can be determined from

$$T_s = \frac{D \, (L_{cs} + D) \, (W_{cm})}{10,560,000} + \frac{D \left(M + \frac{1}{2}C \right)}{2,640,000} \left(\frac{1}{2} \right) \qquad [4.2.23]$$

Since no excess weight for drill collars need be considered, Equation 4.2.23 becomes

$$T_s = \frac{D \, (L_{cs} + D) \, (W_{cm})}{10,560,000} + \frac{DM}{2,640,000} \left(\frac{1}{2} \right) \qquad [4.2.24]$$

where T_s = ton-miles setting casing
$\quad L_{cs}$ = length of joint of casing in ft.
$\quad W_{cm}$ = effective weight per foot of casing in lb/ft

The effective weight per foot of casing W_{cm} may be estimated from data given on Figure 4.2.73 for drill pipe (using the approximate lb/ft), or calculated as

$$W_{cm} = W_{ca} \, (1 - 0.015B) \qquad [4.2.25]$$

where W_{ca} is weight per foot of casing in air in lb/ft
$\quad$ B is weight of drilling fluid from Figure 4.2.74 or Figure 4.2.75 in lb/gal

Short Trip Operations
The ton-miles of work performed in short trip operations, as for drilling and coring operations, is also expressed in terms of round trips. Analysis shows that the ton-miles of work done in making a short trip is equal to the difference in round trip ton-miles for the two depths in question.

$$T_{ST} = T_6 - T_5 \qquad [4.2.26]$$

where T_{ST} = ton-miles for short trip
$\quad T_5$ = ton-miles for one round trip at depth D_5 (shallower depth)
$\quad T_6$ = ton-miles for one round trip at depth D_6 (deeper depth)

For the comparative evaluation of service from rotary drilling lines, the grand total of ton-miles of work performed will be the sum of the ton-miles for all round-trip operations (Equation 4.2.19), the ton-miles for all drilling operations (Equation 4.2.20), the ton-miles for all coring operations (Equation 4.2.22), the ton-miles for all casing setting operations (Equation 4.2.23), and the ton-miles for all short trip operations (Equation 4.2.26). By dividing the grand total ton-miles for all wells by the original length of line in feet, the evaluation of rotary drilling lines in ton-miles per foot on initial length may be determined.

4.2.4.17 Rotary Drilling Line Service-Record Form
Figure 4.2.76 is a rotary drilling line service-record form. It can be filled out on the bases of Figure 4.2.77 and previous discussion.

4.2.4.18 Slipping and Cutoff Practice for Rotary Drilling Lines
Using a planned program of slipping and cutoff based upon increments of service can greatly increase the service life of drilling lines. Determining when to slip and cut depending only on visual inspection, will result in uneven wear, trouble with spooling (line "cutting in" on the drum), and long cutoffs, thus decreasing the service life. The general procedure in any program should be to supply an excess of drilling line over that required to string up, and to slip this excess through the system at such a rate that it is evenly worn and that the line removed by cutoff at the drum end has just reached the end of its useful life.

Initial Length of Line
The relationship between initial lengths of rotary lines and their normal service life expectancies is shown in Figure 4.2.78. Possible savings by the use of a longer line may be offset by an increased cost of handling for a longer line.

Service Goal
A goal for line service in terms of ton-miles between cutoffs should be selected. This value can initially be determined from Figures 4.2.79 and 4.2.80 and later adjusted in accordance with experience. Figure 4.2.81 shows a graphical method of determining optimum cutoff frequency.

(text continued on page 4-79)

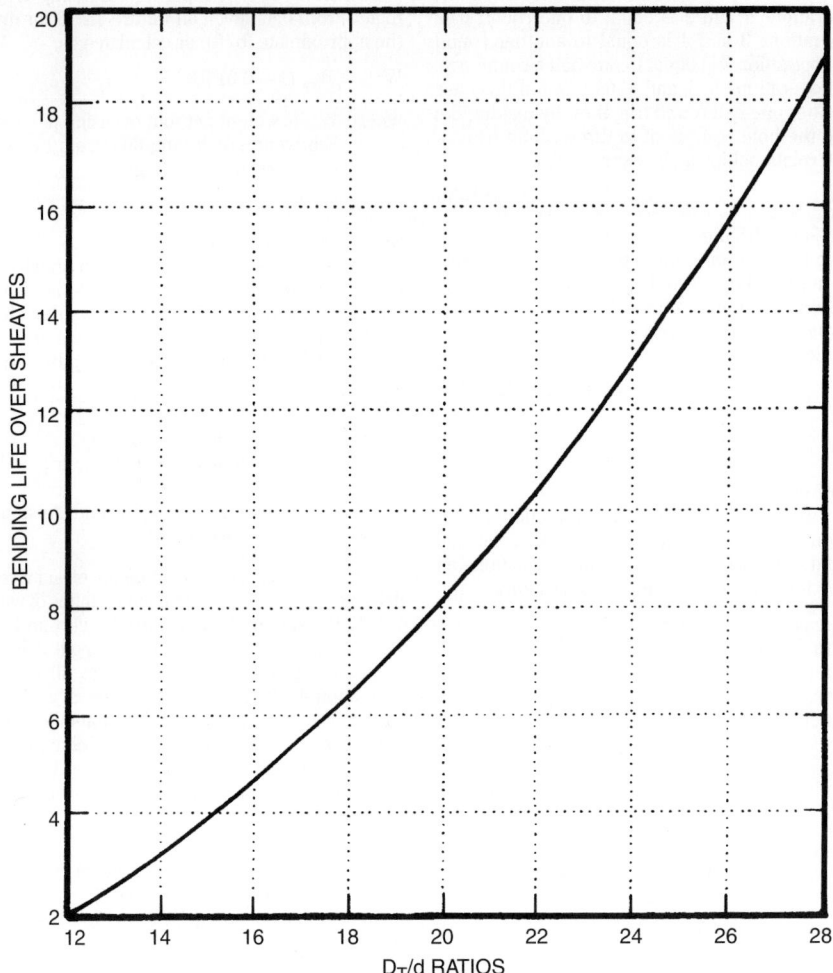

D_T = tread diameter of sheave, inches (*mm*). (See Fig. 4.2.67)
d = nominal rope diameter, inches (*mm*).

Figure 4.2.74   *Relative service for various D_T/d ratios for sheaves [3].*
See "Diameter of Sheaves," subparagraph title "Variation of Different Service Applications."
Based on laboratory tests involving systems consisting of sheaves only.

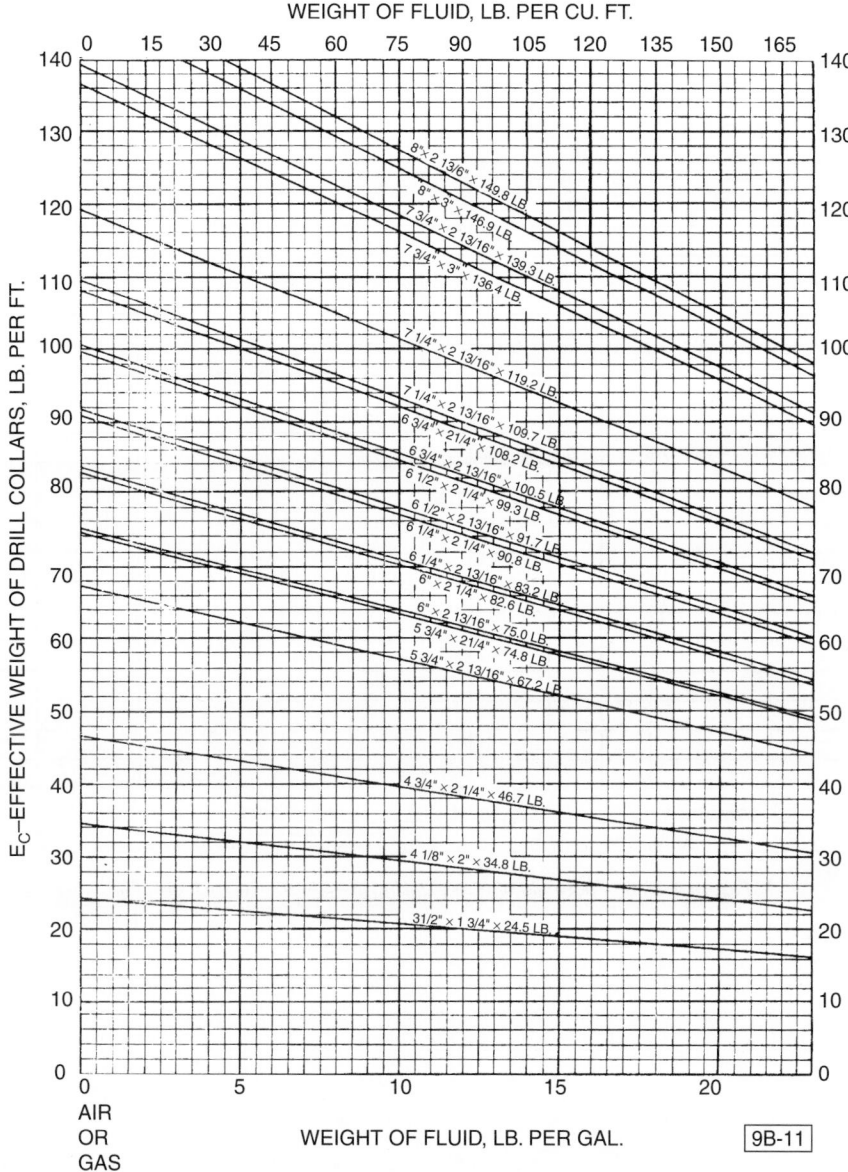

Figure 4.2.75 *Effective weight of drill collars in drilling fluid [3].*

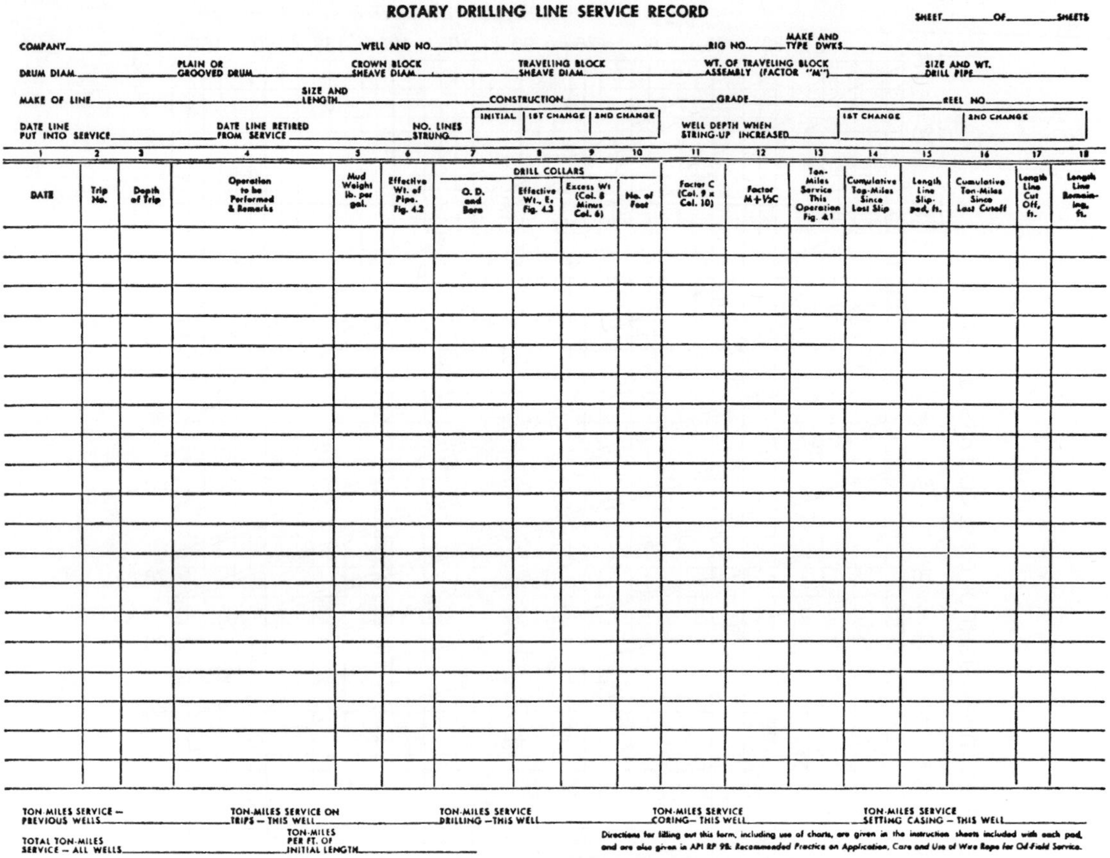

ROTARY DRILLING LINE SERVICE RECORD

SHEET_____OF_____SHEETS

COMPANY_____ WELL AND NO_____ RIG NO._____ MAKE AND TYPE DWKS_____

DRUM DIAM_____ PLAIN OR GROOVED DRUM_____ CROWN BLOCK SHEAVE DIAM_____ TRAVELING BLOCK SHEAVE DIAM_____ WT. OF TRAVELING BLOCK ASSEMBLY (FACTOR "M")_____ SIZE AND WT. DRILL PIPE_____

MAKE OF LINE_____ SIZE AND LENGTH_____ CONSTRUCTION_____ GRADE_____ REEL NO._____

DATE LINE PUT INTO SERVICE			DATE LINE RETIRED FROM SERVICE		NO. LINES STRUNG	INITIAL	1ST CHANGE	2ND CHANGE	WELL DEPTH WHEN STRING-UP INCREASED		1ST CHANGE		2ND CHANGE			
1	2	3	4	5	6	7	8	9	10	11	12	13	14	15	16	17	18
DATE	Trip No.	Depth of Trip	Operation to be Performed & Remarks	Mud Weight lb. per gal.	Effective Wt. of Pipe. Fig. 4.2	DRILL COLLARS O.D. and Bore	Effective Wt., E, Fig. 4.3	Excess Wt (Col. 8 Minus Col. 6)	No. of Feet	Factor C (Col. 9 x Col. 10)	Factor M + ½C	Ton-Miles Service This Operation Fig. 4.1	Cumulative Ton-Miles Since Last Slip	Length Line Slip-ped, ft.	Cumulative Ton-Miles Since Last Cutoff	Length Line Cut Off, ft.	Length Line Remain-ing, ft.

TON-MILES SERVICE — PREVIOUS WELLS_____

TON-MILES SERVICE ON TRIPS — THIS WELL_____

TON-MILES SERVICE DRILLING — THIS WELL_____

TON-MILES SERVICE CORING— THIS WELL_____

TON-MILES SERVICE SETTING CASING — THIS WELL_____

TOTAL TON-MILES SERVICE — ALL WELLS_____

TON-MILES PER FT. OF INITIAL LENGTH_____

Directions for filling out this form, including use of charts, are given in the instruction sheets included with each pad, and are also given in API RP 9B. Recommended Practice on Application, Care and Use of Wire Rope for Oil-Field Service.

Figure 4.2.76　*Facsimile of rotary drilling line service-record form [3].*

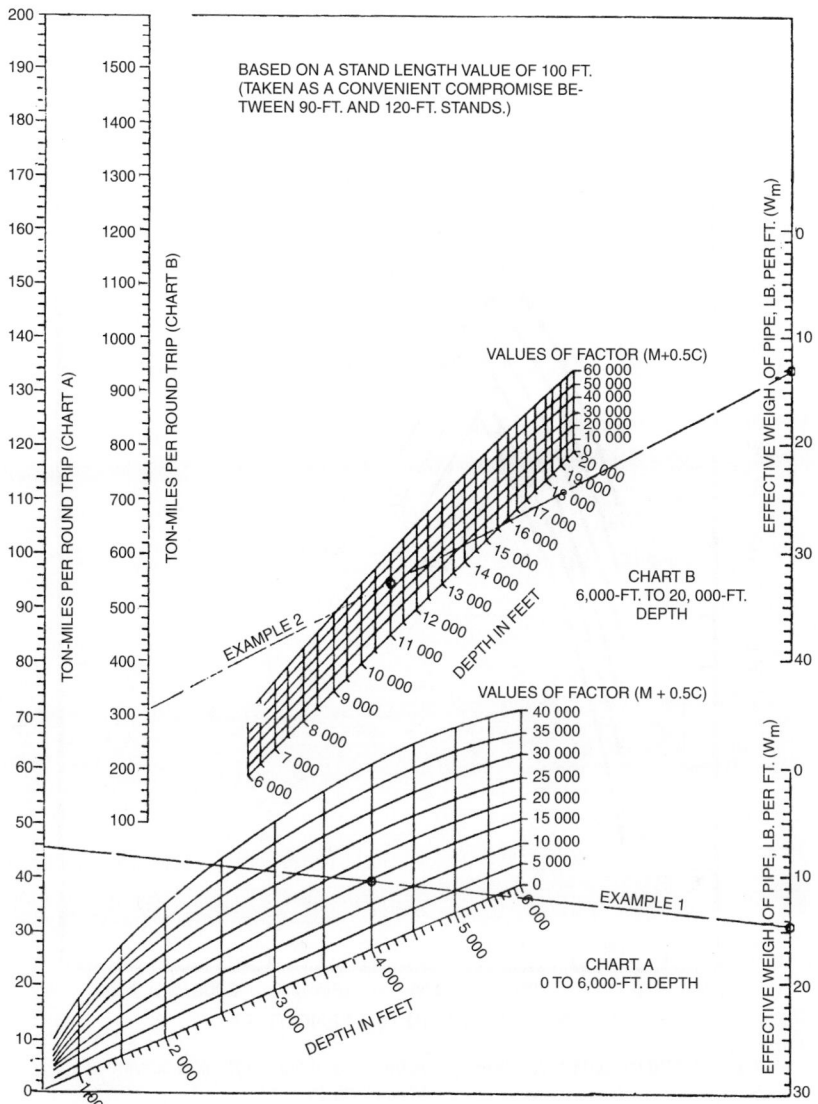

Figure 4.2.77 *Rotary-drilling ton-mile charts [3].*

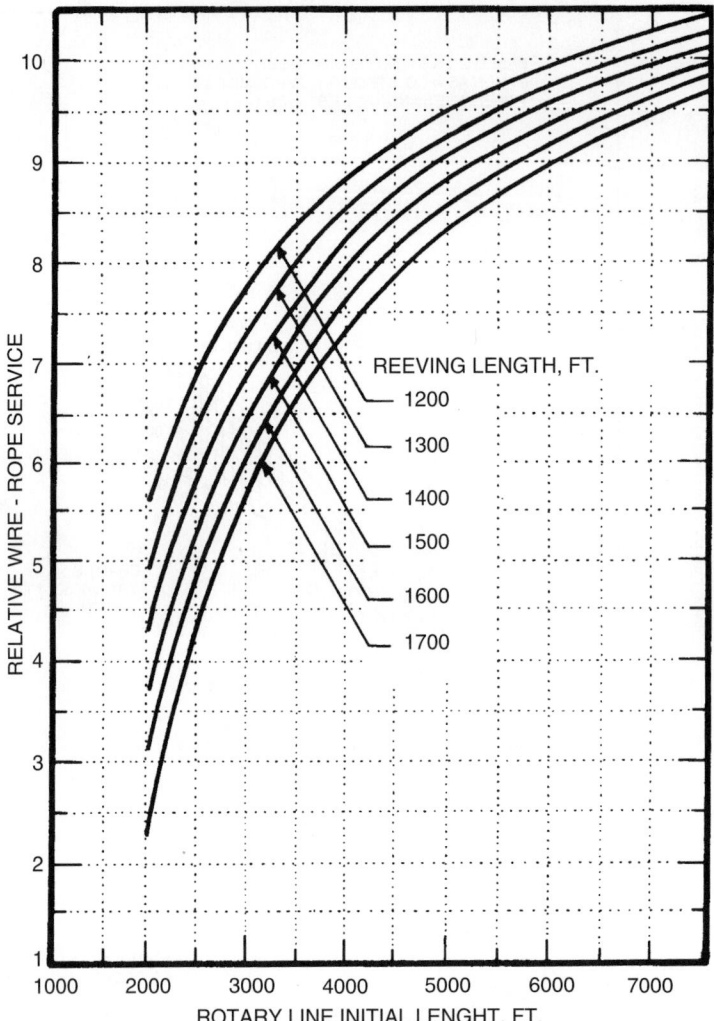

Figure 4.2.78   *Relationship between rotary-line initial length and service life [3].*
Empirical curves were developed from general field experience.

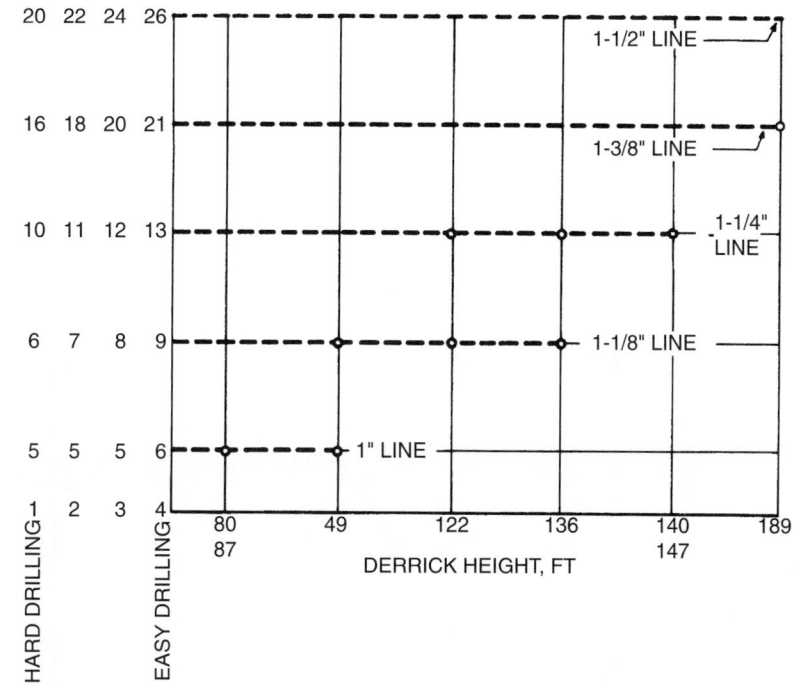

Explanation:

To determine (approximately) the desirable ton-miles before the first cutoff on a new line, draw a vertical line from the derrick height to the wireline size used. Project this line horizontally to the ton-mile figure given for the type of drilling encountered in the area. Subsequent cutoffs should be made at 100 ton-miles less than those indicated for $1^1/_8$-in. and smaller lines, and at 200 ton-miles less than $1^1/_4$-in. and $1^3/_8$-in. lines.

Figure 4.2.79 *Ton-mile derrick height and line-size relationships [3].*

The values for ton-miles before cutoff, as given in Figure 4.2.79 were calculated for improved plow steel with an independent wire-rope core and operating at a design factor of 5. When a design factor other than 5 is used, these values should be modified in accordance with Figure 4.2.80. The values given in Figure 4.2.79 are intended to serve as a guide for the selection of initial ton-mile values are explained in Par. "Service Goal." These values are conservative, and are applicable to all typical constructions of wire rope as recommended for the rotary drilling lines shown in Table 4.2.5.

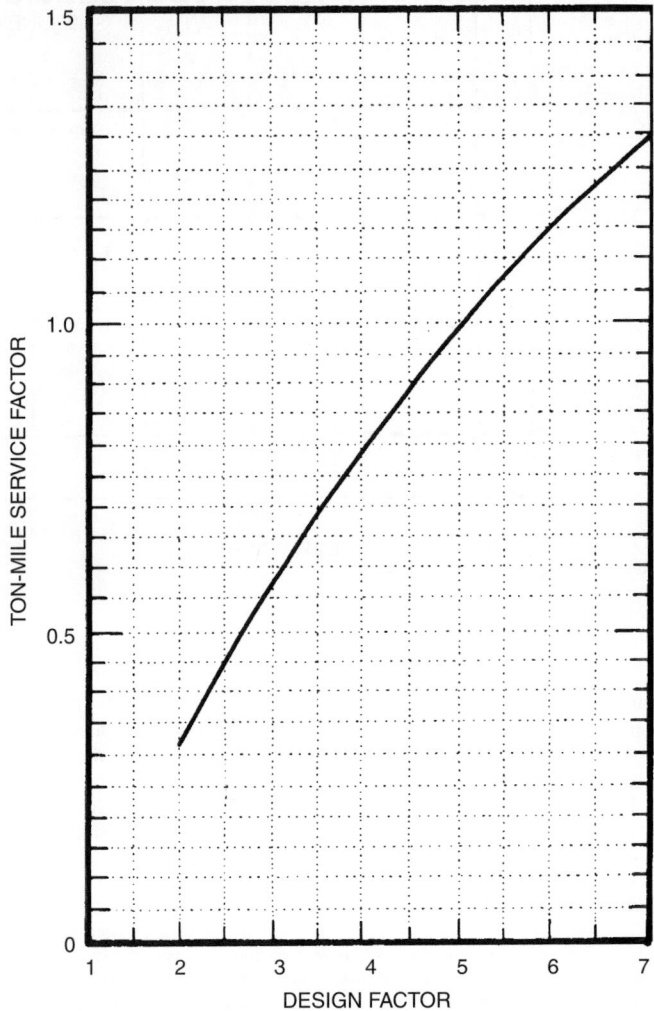

Figure 4.2.80 *Relationship between design factors and ton-mile service factors [3].*
NOTE: Light loads can cause rope to wear out from fatigue prior to accumulation of anticipated ton-miles.
Based on laboratory tests on bending over sheaves.

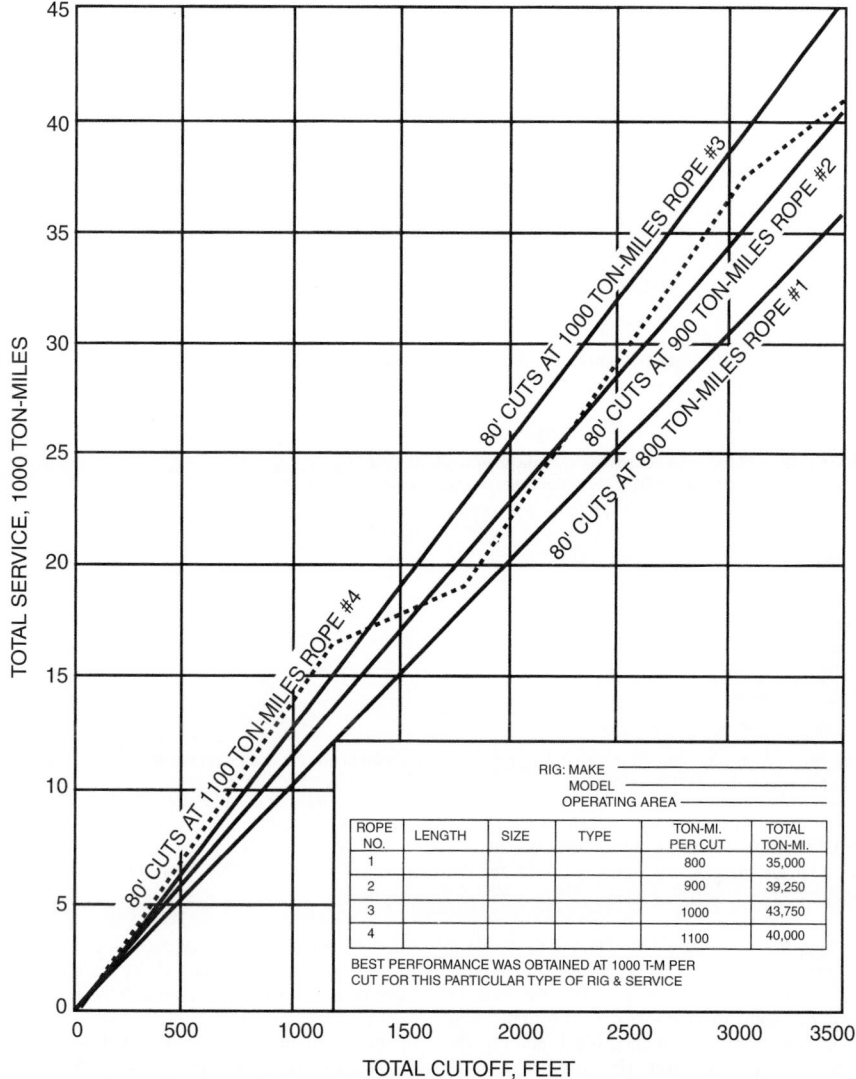

Figure 4.2.81 *Graphic method of determining optimum frequency of cutoff to give maximum total ton-miles for a particular rig operating under certain drilling conditions [3].*

(text continued from page 4-71)

Variations in Line Service

Ton-miles of service will vary with the type and condition of equipment used, drilling conditions encountered, and the skill used in the operation. A program should be "tailored" to the individual rig. The condition of the line as moved through the reeving system and the condition of the cutoff portions will indicate whether the proper goal was selected. In all cases, visual inspection of the wire rope by the operator should take precedence over any predetermined procedures. (See Figure 4.2.81 for a graphical comparison of rope services.)

Cutoff Length

The following factors should be considered in determining a cutoff length:

1. The excess length of line that can conveniently be carried on the drum.

2. Load-pickup points from reeving diagram.
3. Drum diameter and crossover points on the drum.

The crossover and pickup points should not repeat. This is done by avoiding cutoff lengths that are multiples of either drum circumference, or lengths between pickup points. Successful programs have been based on cutoff lengths ranging from 30 to 150 ft. Table 4.2.32 shows a recommended length of cutoff (number of drum laps) for each height derrick and drum diameter.

Slipping Program

The number of slips between cutoffs can vary considerably depending upon drilling conditions and the length and frequency of cutoffs. Slips should be increased if the digging is rough, if jarring jobs occur, etc. Slipping that causes too much line piles up on the drum, particularly an extra layer on the drum, before cutoff should be avoided.

Table 4.2.32 *Recommended Cutoff Lengths in Terms of Drum Laps* See Paragraph Titled "Cut off Length" [3]*

1	2	3	4	5	6	7	8	9	10	11	12	13	14	15
	Drum Diameter, in.													
Derrick or Mast Height. ft	11	13	14	16	18	20	22	24	26	28	30	32	34	36
	Number of Drum Laps per Cutoff													
151 Up										$15\frac{1}{2}$	$14\frac{1}{2}$	$13\frac{1}{2}$	$12\frac{1}{2}$	$11\frac{1}{2}$
141 to 150							$13\frac{1}{2}$	$12\frac{1}{2}$	$11\frac{1}{2}$	$11\frac{1}{2}$	$10\frac{1}{2}$			
133 to 140						$15\frac{1}{2}$	$14\frac{1}{2}$	$12\frac{1}{2}$	$11\frac{1}{2}$	$11\frac{1}{2}$	$10\frac{1}{2}$	$9\frac{1}{2}$		
120 to 132				$17\frac{1}{2}$	$15\frac{1}{2}$	$14\frac{1}{2}$	$12\frac{1}{2}$	$12\frac{1}{2}$	$11\frac{1}{2}$	$10\frac{1}{2}$	$9\frac{1}{2}$	$9\frac{1}{2}$		
91 to 119		$19\frac{1}{2}$	$17\frac{1}{2}$	$14\frac{1}{2}$	$12\frac{1}{2}$	$11\frac{1}{2}$	$10\frac{1}{2}$	$9\frac{1}{2}$	$9\frac{1}{2}$	$8\frac{1}{2}$				
73 to 90		$17\frac{1}{2}$	$14\frac{1}{2}$	$12\frac{1}{2}$	$11\frac{1}{2}$									
Up through 72	$12\frac{1}{2}$	$11\frac{1}{2}$												

*In order to insure a change of the point of crossover on the drum, where wear and crushing are most severe, the laps to be cut off are given in multiples of one-half lap or one quarter lap dependent upon the type of drum grooving.

Example:

Assumed conditions:

  A. Derrick height: 138 ft
  B. Wire-line size: $1\frac{1}{4}$ in.
  C. Type drilling: #3
  D. Drum diameter: 28 in.
  E. Design factor: 3.

Solution:

  1. From Fig. 4.2.79 determine that (for a line with a design factor of 5) the first cutoff would be made after 120 ton-miles and additional cut-offs after each successive 1000 ton-miles.
  2. Since a design factor of 3 applies, Fig. 4.2.80 indicates that these values should be multiplied by a factor of 0.58. Hence the first cutoff should be made after 696 ton-miles and additional cutoffs after each successive 580 ton-miles.
  3. From Table 4.2.32 determine that $11\frac{1}{2}$ drum laps (84 ft) should be removed at each cutoff.
  4. Slip 21 ft every 174 ton-miles for four times and cut off after the fourth slip. Thereafter, slip 21 ft every 145 ton-miles and cut off on the fourth slip.

In slipping the line, the rope should be slipped an amount such that no part of the rope will be located for a second time in a position of severe wear. The positions of severe wear are the point of crossover on the drum and the sections in contact with the traveling-block and crown-block sheaves at the pickup position. The cumulative number of feet slipped between cutoffs should be equal to the recommended feet for ton-mile cutoff. For example, if cutting off 80 ft every 800 ton-miles, 20 ft should be slipped every 200 ton-miles, and the line cut off on the fourth slip.

4.2.4.19 Field Troubles and Their Causes

All wire rope will eventually deteriorate in operation or have to be removed simply by virtue of the loads and reversals of load applied in normal service. However, many conditions of service or inadvertent abuse will materially shorten the normal life of a wire rope of proper construction although it is properly applied. The following field troubles and their causes give some of the field conditions and practices that result in the premature replacement of wire rope. It should be borne in mind that in all cases the contributory cause of removal may be one or more of these practices or conditions.

Wire-Rope Trouble	**Possible Cause**
Rope broken (all strands)	Overload resulting from severe impact, kinking, damage, localized water, weakening of one or more strands, or rust-bound condition and loss of elasticity. Loss of metallic area due to broken wires caused by severe bending.
One or more whole strands parted.	Overloading, kinking, divider interference, localized wear, or rust-bound condition. Fatigue, excessive speed, slipping, or running too loosely. Concentration of vibration at dead sheave or dead-end anchor.
Excessive corrosion.	Lack of lubrication. Exposure to salt spray, corrosion gases, alkaline water, acid water, mud, or dirt. Period of inactivity without adequate protection.
Rope damage by careless handling in hauling to the well or location	Rolling reel over obstructions or dropping from car, truck, or platform. The use of chains for lashing, or the use of lever against rope instead of flange. Nailing through rope to flange.

Wire-Rope Trouble	Possible Cause
Damage by improper socketing.	Improper seizing that allows slack from one or more strands to work back into rope; improper method of socketing or poor workmanship in socketing, frequently shown by rope being untwisted at socket, loose or drawn.
Kinks, dog legs, and other distorted places.	Kinking the rope and pulling out the loops such as improper coiling or unreeling. Improper winding on the drum. Improper tie-down. Open-drum reels having longitudinal spokes too widely spaced. Divider interference. The addition of improperly spaced cleats increase the drum diameter. Stressing while rope is over small sheave or obstacle.
Damage by hooking back slack too tightly to girt.	Operation of walking beam causing a bending action on wires at clamp and resulting in fatigue and cracking of wires, frequently before rope goes down into hole.
Damage or failure on a fishing job.	Rope improperly used on a fishing job, resulting in damage or failure as a result of the nature of the work.
Lengthening of lay and reduction of diameter.	Frequently produced by some type of overloading, such as an overload resulting in a collapse of the fiber core in swabbing lines. This may also occur in cable-tool lines as a result of concentrated pulsating or surging forces that may contribute to fiber-core collapse.
Premature breakage of wires.	Caused by frictional heat developed by pressure and slippage, regardless of drilling depth.
Excessive wear in spots.	Kinks or bends in rope due to improper handling during installation or service. Divider interference; also, wear against casing or hard shells or abrasive formations in a crooked hole. Too infrequent cutoffs on working end.
Spliced rope.	A splice is never as good as a continuous piece of rope, and slack is liable to work back and cause irregular wear.
Abrasion and broken wires in a straight line. Drawn or loosened strands. Rapid fatigue breaks.	Injury due to slipping rope through clamps.
Reduction in tensile strength or damage to rope.	Excessive heat due to careless exposure to fire or torch.
Distortion of wire rope.	Damage due to improperly attached clamps or wire-rope clips.
High strands.	Slipping through clamps, improper seizing, improper socketing or splicing, kinks, dog legs, and core popping.
Wear by abrasion.	Lack of lubrication. Slipping clamp unduly. Sandy or gritty working conditions. Rubbing against stationary object or abrasive surface. Faulty alignment. Undersized grooves and sheaves.
Fatigue breaks in wires.	Excessive vibration due to poor drilling conditions, i.e., high speed, rope, slipping, concentration of vibration of at dead sheave or dead-end anchor, undersized grooves and sheaves, and improper selection of rope construction. Prolonged bending action over spudder sheaves, such as that due to hard drilling.
Spiraling or curling.	Allowing rope to drag or rub over pipe, sill, or any object during installation or operation. It is recommended that a block with sheave diameter 16 times the nominal wire-rope diameter, or larger, be used during installation of the line.
Excessive flattening or crushing.	Heavy overload, loose winding on drum, or cross winding. Too infrequent cutoffs on working end of cable-tool lines. Improper cutoff and moving program for cable-tool lines.
Bird-caging or core-popping.	Sudden unloading of line such as hitting fluid with excessive speed. Improper drilling motion or jar action. Use of sheave or too small diameter or passing line around sharp bend.
Whipping off of rope	Running too loose.
Cutting in on drum.	Loose winding on drum. Improper cutoff and moving program for rotary drilling lines. Improper or worn drum grooving or line turnback plate.

References

1. API Specification 8A, 11th Edition: "API Specification for Drilling and Production Hoisting Equipment," API, Dallas, May 1, 1985.

2. API Recommended Practice 8B, 4th Edition: "Recommended Practice for Hoisting Tool Inspection and Maintenance Procedures," API, Dallas, April 1979.

3. API Recommended Practice 8B, 9th Edition: "Recommended Practice on Application, Care, and Use of Wire Rope for Oilfield Service," API, Dallas, May 30, 1986.

4. API Specification 9A, "API Specification for Wire Rope," API, Dallas, May 28, 1984.

4.3 ROTARY EQUIPMENT

Rotary equipment refers to all pieces of surface equipment in drilling operations that actually rotate or introduce rotational force to the drill string. This equipment includes the swivel and the rotary hose, the upper and lower Kelly valves, the Kelly, the Kelly bushings, the master bushing, and the rotary table.

4.3.1 Swivel and Rotary Hose
4.3.1.1 Swivel
The swivel (Figure 4.3.1) is suspended from the hook and traveling block and forms the top part of the drill stem. It allows for the rotation of the Kelly and provides a pressure tight connection from the rotary hose to the drill string to allow the circulation of drilling fluids.

The swivel is made up of a stationary and a rotating part and has to support the entire string weight during drilling operations. The rotating part turns over the main axial bearing, which is mounted in an oil filled housing. The fluid-tight connection between the two parts consists of a wash pipe assembly. The seal unit is the most vulnerable part of the swivel (Figure 4.3.2). Daily greasing of the seals and ensuring sufficient hydraulic fluid level will prevent leakage and decrease wear on the wash pipe assembly.

4.3.1.2 Rotary Hose
Although the rotary hose is not a rotating element, it is included here because of its connection to the swivel. It serves as a flexible high-pressure conduit between the stand pipe and the swivel and allows vertical travel of the swivel and block (Figure 4.3.3).

Usual length of the rotary hose is 45 ft. The minimum burst pressure rating should be $2\frac{1}{2}$ times the allowable working pressure. A pressure test should be conducted before drilling a new hole section. Further rotary hose specifications are provided in API RP7G.

4.3.2 Drill-Stem Subs
Different types of drill-stem subs are depicted in Figures 4.3.4 and 4.3.5. The classification of drill-stem subs is presented in Table 4.3.1. Their location within the string can be seen at Figure 4.3.6.

4.3.2.1 Swivel Sub
The outside diameter of the swivel sub is at least equal to the diameter of the outside of the upper Kelly box. The swivel sub should have a minimum of 8 in. of tong space length. The minimum diameter is equal to the Kelly diameter. The swivel sub is machined with a pin-up and pin-down left-handed rotary-shouldered connection.

4.3.2.2 Upper Kelly Valve
The upper Kelly valve or Kelly cock (Figure 4.3.7) is located between the swivel sub and the Kelly. The Kelly cock is a flapper-type valve. This valve must be manually closed to isolate the Kelly and drill string from the swivel and rotary hose. The valve can be partially opened by starting up the mud pumps, but it must be fully opened manually to prevent mud cutting.

4.3.2.3 Lower Kelly Valve
The lower Kelly valve or Kelly cock is located between the Kelly and the lower Kelly saver sub. This valve is a full-opening, ball-type valve typically operated with a hex-type wrench (Figure 4.3.8). By closing this valve before breaking a joint, this valve will prevent mud spilling on to the drill floor.

If working space is restricted and length is an issue, a shorter, more compact single-piece Kelly valve can be used instead (Figure 4.3.9).

The spring-loaded mud check valve (Figure 4.3.10) can be used as an alternative to the manually operated valves. When the mud pumps are shut off, this valve closes automatically to prevent mud spilling onto the rig floor. This tool is designed as full opening or full closing to avoid a pressure loss across the valve during circulation.

4.3.3 Kelly
The Kelly is a square or hexagonally shaped pipe (Figure 4.3.11). It transmits the power from the rotary table to the drill string. At the same time it serves as a conduit for the drilling fluid, it absorbs torque from the drill string and carries the entire axial load. It is the most heavily loaded component of the drill string. Refer to API RP7G for dimensions and further specifications.

Square Kellies are normally forged over the drive section, whereas hexagonal Kellies have been machined from bar stock. Typically, the machined flats have a greater resistance to wear than the forged surfaces.

Kellies are manufactured from AISI 4145-modified, fully heat-treated alloy steel with a Brinell hardness range of 285-341 and a minimum average Charpy impact value of 40 ft-lb.

Upper and lower upsets are kept long so that the Kelly can be rethreaded in case of thread damage.

For Kellies with the same outside diameter, a hexagonal Kelly will have a stronger drive section than a square Kelly (Figure 4.3.12). Under a given load the stress levels in the hexagonal Kelly will be lower, leading to a longer service life.

Service life as mentioned previously is greatly influenced by the maintenance of the drive bushing. Frequent adjustments of the rollers should be conducted to minimize play between the Kelly and the drive bushing.

Frequent inspections of the following items will increase the service life of all components:

- The corners of the drive section (surface wear)
- Transition between the upset and the drive section (cracking)
- Straightness of the Kelly

Operating with a bent Kelly will cause

- Severe vibrations at surface
- Additional wear on the drill pipe tools joints and the BOP and well head
- Additional bending stresses in the drill string
- Rapid wear on the Kelly and master bushings.

4.3.4 Rotary Table and Bushings
4.3.4.1 Rotary Table
The rotary table (Figure 4.3.13) serves a dual function: It provides the rotary movement to the Kelly and applies torque to the drill string. As the rotary table runs, the master bushing, the Kelly, the drill pipe, and the bit also turn. The rotary table is driven by an independent electric motor or on diesel mechanical drives by a compound drive by the drawworks engine.

When tripping pipe in and out of the well, the rotary table supports the drill string when the load is not suspended from the hook. At those times, it carries the entire weight of the drill string.

The rotary table (Figure 4.3.14) rests on reinforced beams in the rig floor that transfer the loads to the sub structure of the rig. The rotary has a cast or reinforced steel base that acts as a foundation to provide the strength and stability for this piece of equipment. The turntable rotates in the base

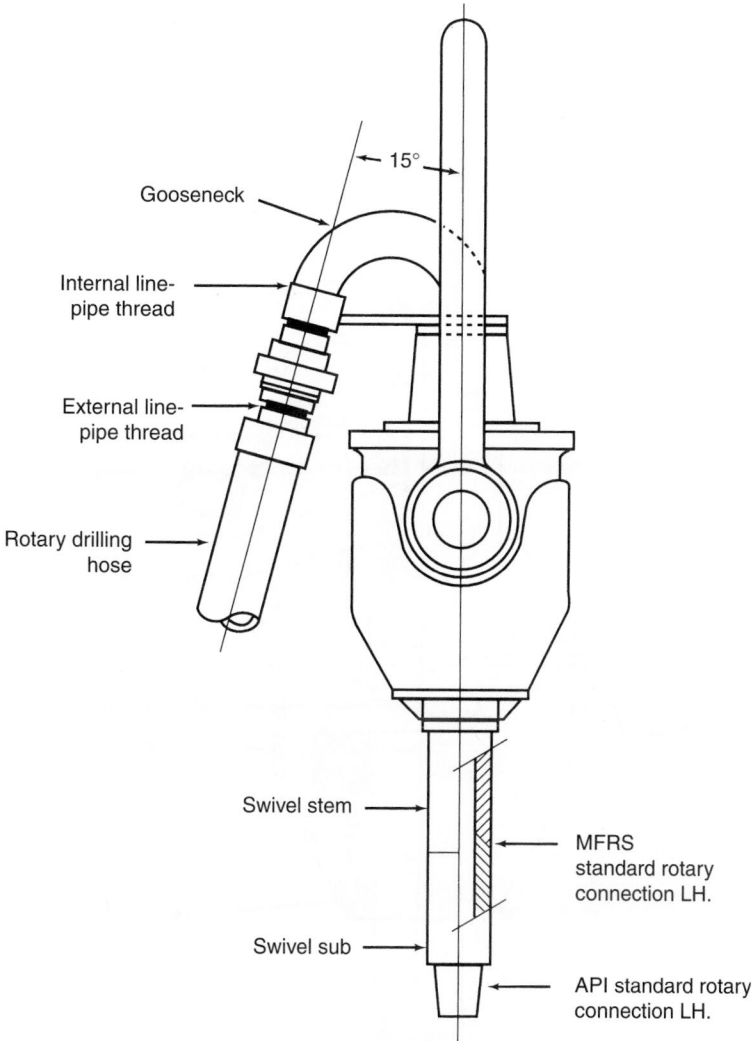

Figure 4.3.1 *Swivel nomenclature.*

using a heavy ball bearing and a ring gear shrunk over the turntable.

The ring gear is driven by a pinion shaft with roller bearings and seals. A detachable sprocket is connected to the pinion shaft on diesel mechanical drives or a coupling connector when driven by an electric motor. The rotary table is equipped with a locking device that is used when making up or breaking out the drill bit. It is also used to lock the drill string in place when directionally drilling with a motor. This lock should not be used when making up or breaking out pipe during a round trip. The combination of torque and bending moment can lead to premature pipe failure. The turntable has an opening in the center that houses the master bushing.

4.3.4.2 Master Bushing
The master bushing fits into the turntable and makes it possible to transfer the rotational movement of the rotary table to the Kelly. It also accommodates the slips when pipe is to be suspended in the rotary table. Because the bushings are

a removable insert, they also function as a wear sleeve for the rotary table.

There are two types of master bushings:

- Square drive master bushings (Figure 4.3.15)
- Pin drive master bushings (Figure 4.3.16)

Master bushings are constructed as a solid single piece (Figure 4.3.17), as a split bushing in two halves (Figure 4.3.18), or as a hinged type (Figure 4.3.19). All are machined with a taper to accommodate a range of pipe and slip sizes.

The API requirements for rotary table openings for square drive master bushings, and the sizes of the square drive and pindrive master bushings are specified in API RP7G.

4.3.4.3 Kelly Bushing
The Kelly bushing engages with the master bushing. It locks into the master bushing and transfers the rotary torque to

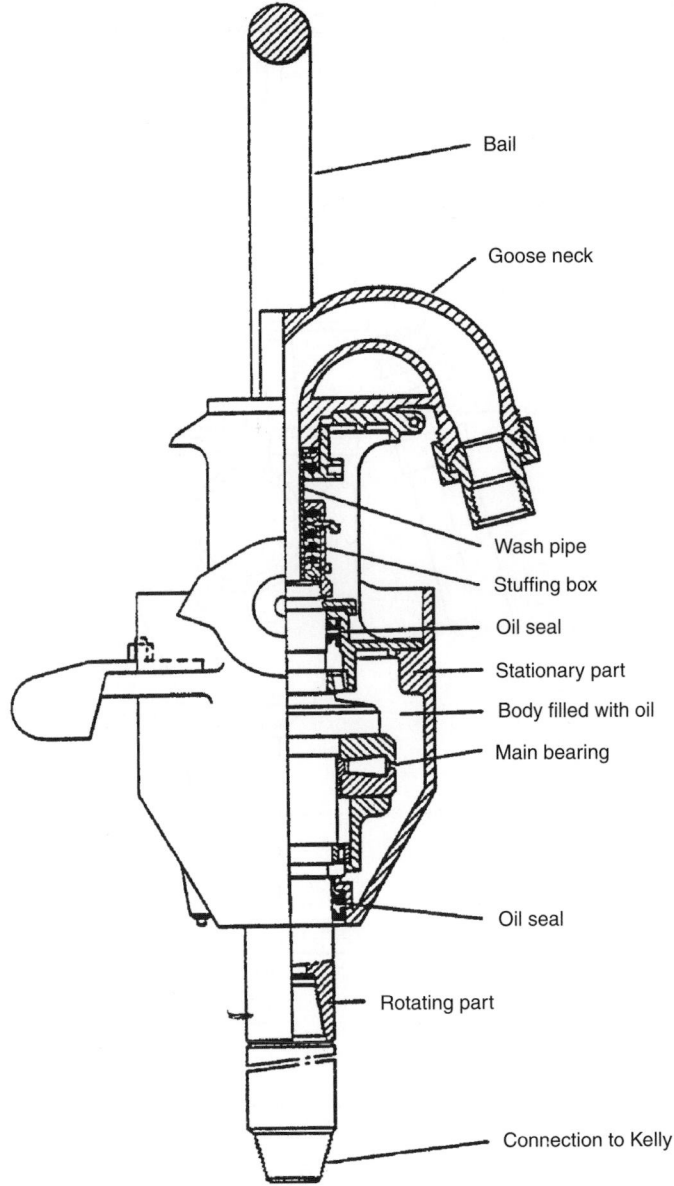

Figure 4.3.2 *National (ideal) swivel.*

the Kelly. There are two types of Kelly bushings:

- Square drive Kelly bushing that engages in a square master bushing (Figure 4.3.20).
- Pin drive Kelly bushing that engages in the pin drive holes of the master bushing (Figure 4.3.21).

The Kelly bushing is equipped with rollers that allow axial movement of the Kelly while the rotary is stationary or rotating.

For a square Kelly, the Kelly bushing is fitted with four plane rollers. For a hexagonal Kelly, the Kelly bushing is fitted with two plane rollers and two 120° V-rollers (Figure 4.3.22) or with six flat rollers.

Another version of the Kelly bushing is the double roller–type Kelly bushing (Figure 4.3.23). This Kelly drive has two rollers per Kelly flat stacked above each other in a roller assembly block. This Kelly bushing is used in high-torque applications. It reduces wear to the Kelly and itself through a wider force distribution.

For installation and maintenance every Kelly bushing is equipped with a cover that can be removed to allow access and removal of the rollers.

All rollers are adjustable to minimize the available play of the Kelly in the bushing. This will reduce vibration on the rotary and the drill string and reduce wear on the Kelly.

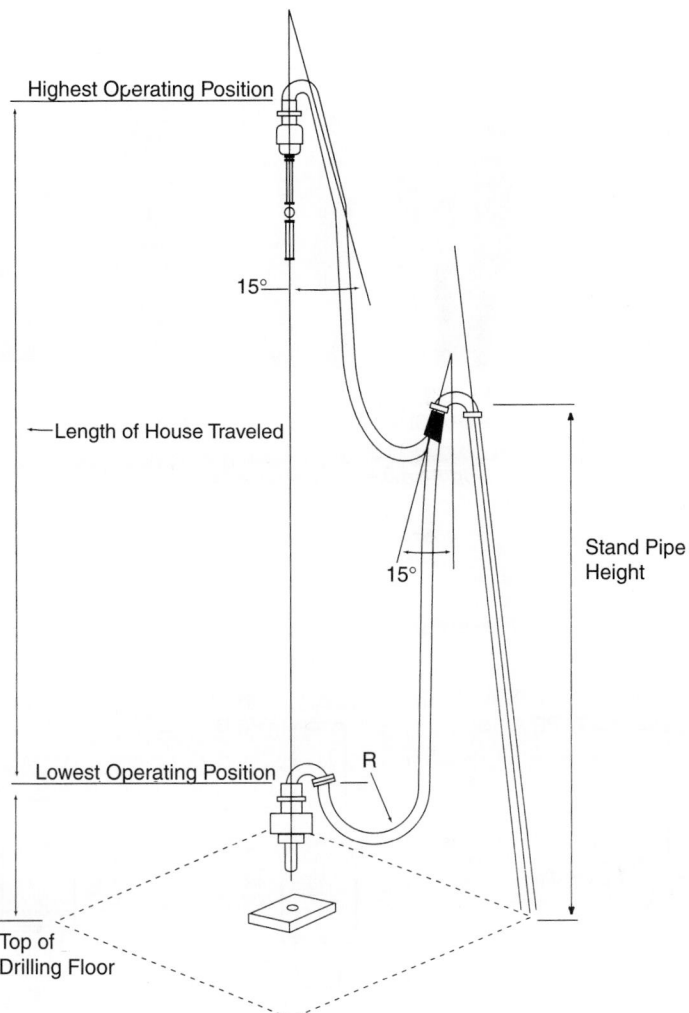

Highest Operating Position

15°

Length of House Traveled

15°

Stand Pipe Height

Lowest Operating Position

R

Top of Drilling Floor

Figure 4.3.3 *Rotary hose.*

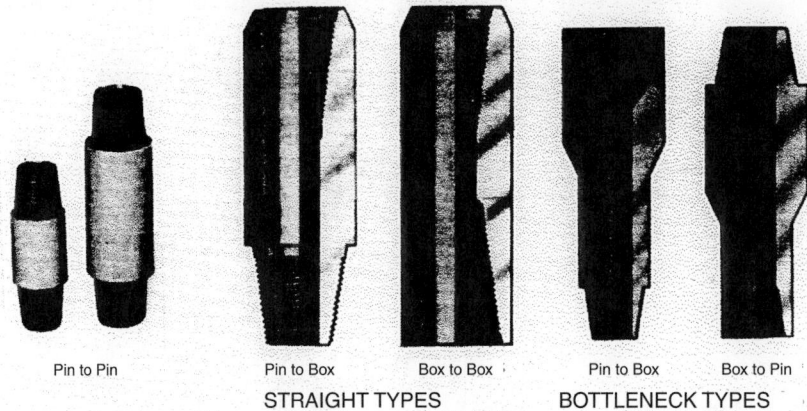

Figure 4.3.4   *Drill-stem subs.*

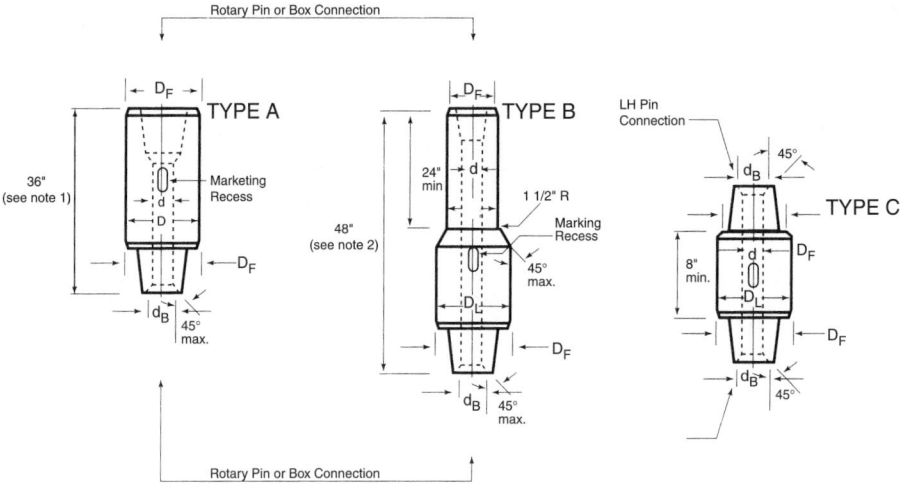

Figure 4.3.5   *Types of drill-stem subs.*

Table 4.3.1   *Drill-Stem Subs*

1	2	3	4
Type	Class	Upper Connection to Assemble w/	Lower Connection to Assemble w/
A or B	Kelly sub	Kelly	Tool joint
"	Tool joint sub	Tool joint	Tool joint
"	Crossover sub	Tool joint	Drill collar
"	Drill collar sub	Drill collar	Drill collar
"	Bit sub	Drill collar	Bit
C	Swivel sub	Swivel sub	Kelly

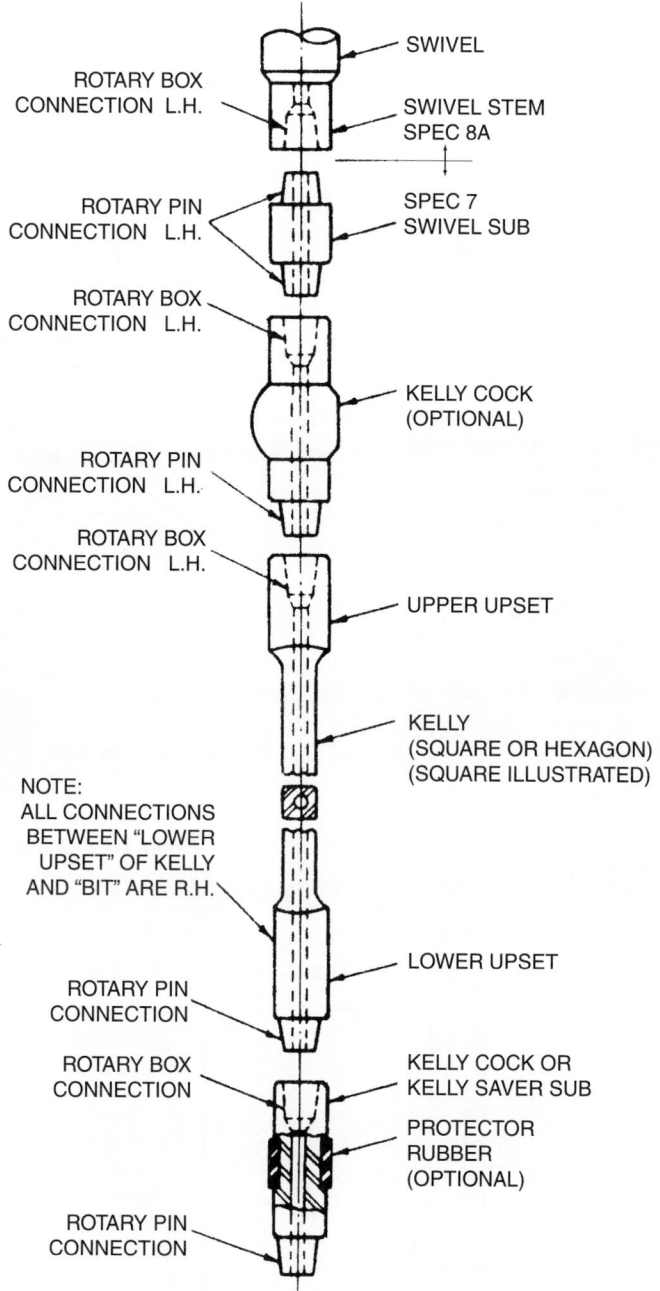

ROTARY BOX
CONNECTION L.H.

SWIVEL

SWIVEL STEM
SPEC 8A

ROTARY PIN
CONNECTION L.H.

SPEC 7
SWIVEL SUB

ROTARY BOX
CONNECTION L.H.

KELLY COCK
(OPTIONAL)

ROTARY PIN
CONNECTION L.H.

ROTARY BOX
CONNECTION L.H.

UPPER UPSET

KELLY
(SQUARE OR HEXAGON)
(SQUARE ILLUSTRATED)

NOTE:
ALL CONNECTIONS
BETWEEN "LOWER
UPSET" OF KELLY
AND "BIT" ARE R.H.

LOWER UPSET

ROTARY PIN
CONNECTION

ROTARY BOX
CONNECTION

KELLY COCK OR
KELLY SAVER SUB

PROTECTOR
RUBBER
(OPTIONAL)

ROTARY PIN
CONNECTION

Figure 4.3.6 *Placement of drill-stem subs.*

E
A
D
Wrench
Ram Shaft
Blind Nut
Shaft Washer
Shaft Capscr.
Friction Ring
Shaft Spacer
Ram Body
Ram Rubber
Ram key
Shaft Packing
Junk Ring-Female
Blind Nut Key
Junk Ring-Male
Retainer Pl. Scr.
Valve Body
Retainer Plate
Lower Sub
C
B

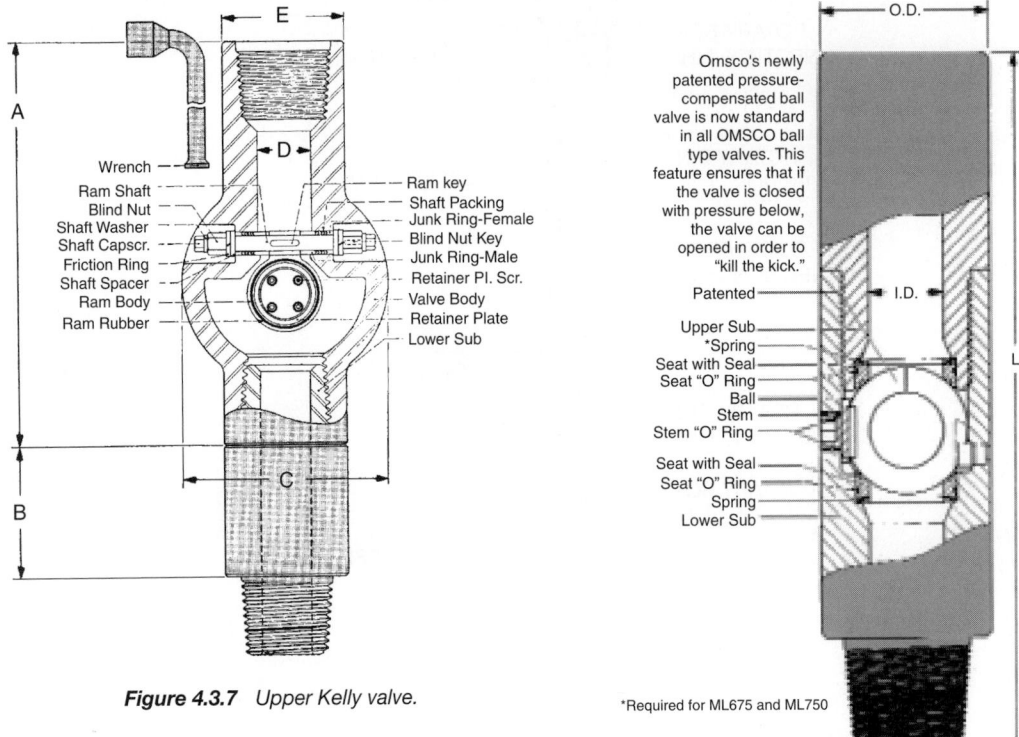

Figure 4.3.7 *Upper Kelly valve.*

O.D.
Omsco's newly patented pressure-compensated ball valve is now standard in all OMSCO ball type valves. This feature ensures that if the valve is closed with pressure below, the valve can be opened in order to "kill the kick."
Patented
I.D.
Upper Sub
*Spring
Seat with Seal
Seat "O" Ring
Ball
Stem
Stem "O" Ring
Seat with Seal
Seat "O" Ring
Spring
Lower Sub
L
*Required for ML675 and ML750

Figure 4.3.8 *Lower Kelly valve.*

Stem "O" Rings
(1 Set)
O.D
Snap Ring
Retainer Funnel
Funnel "O" Ring B
Split Retainer
Funnel "O" Ring A
Upper Seat
Lower Seat
Spring
Length
Wrench
Stem
Ball
Lower Seat "O" Ring
Valve Body
I.D.
Upper Seat "O" Ring

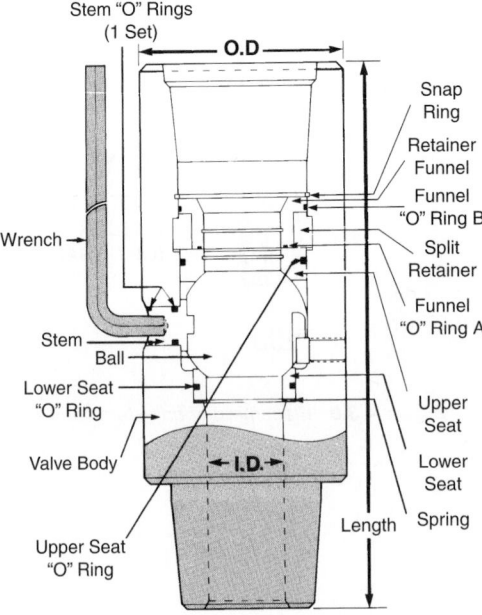

Figure 4.3.9 *Single-piece lower Kelly valve.*

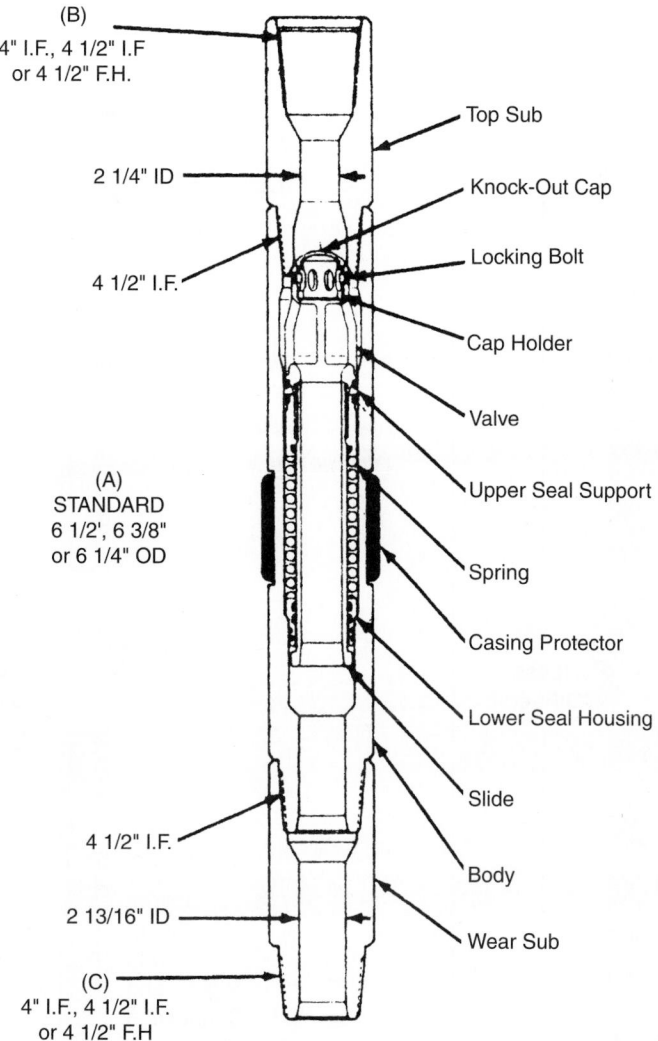

(B)
4" I.F., 4 1/2" I.F
or 4 1/2" F.H.

2 1/4" ID

4 1/2" I.F.

(A)
STANDARD
6 1/2', 6 3/8"
or 6 1/4" OD

4 1/2" I.F.

2 13/16" ID

(C)
4" I.F., 4 1/2" I.F.
or 4 1/2" F.H

Top Sub

Knock-Out Cap

Locking Bolt

Cap Holder

Valve

Upper Seal Support

Spring

Casing Protector

Lower Seal Housing

Slide

Body

Wear Sub

Figure 4.3.10 *Spring-loaded mind saver valve.*

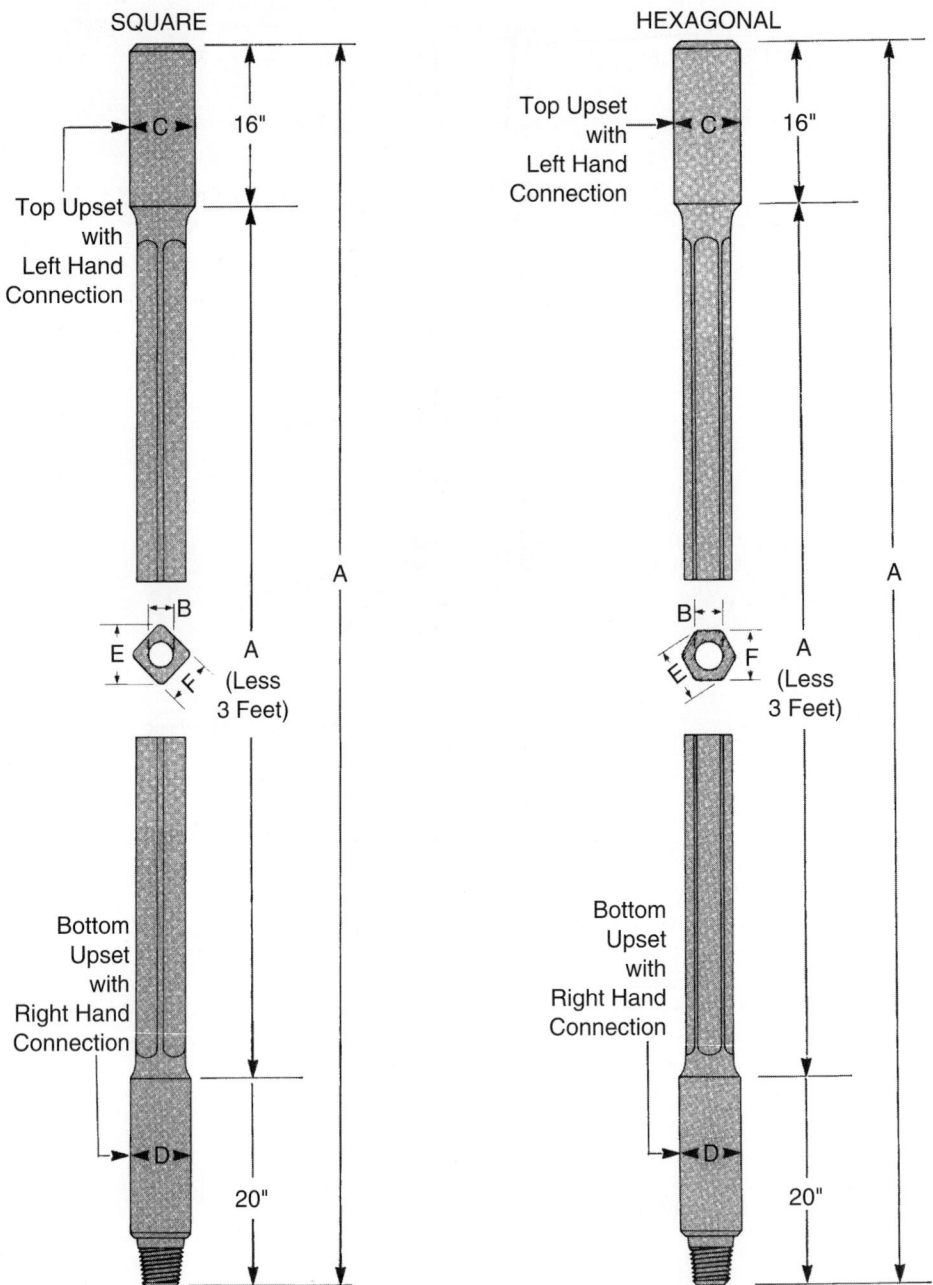

Figure 4.3.11   *Square and hexagonal Kelly.*

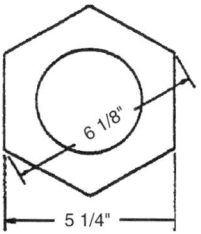
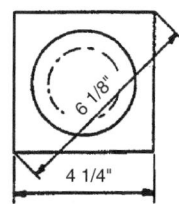

Figure 4.3.12 *Hexagonal and square Kelly cross-section.*

Figure 4.3.13 *Rotary table with pin drive master bushings.*

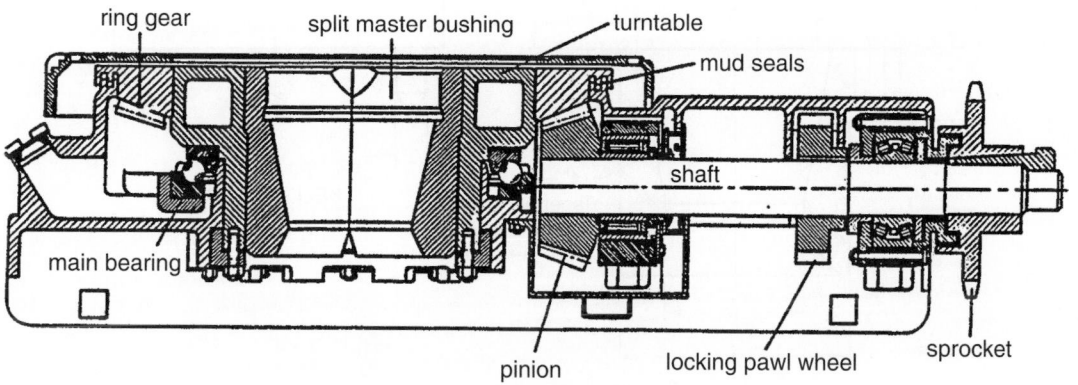

Figure 4.3.14 *Rotary table with split master bushing.*

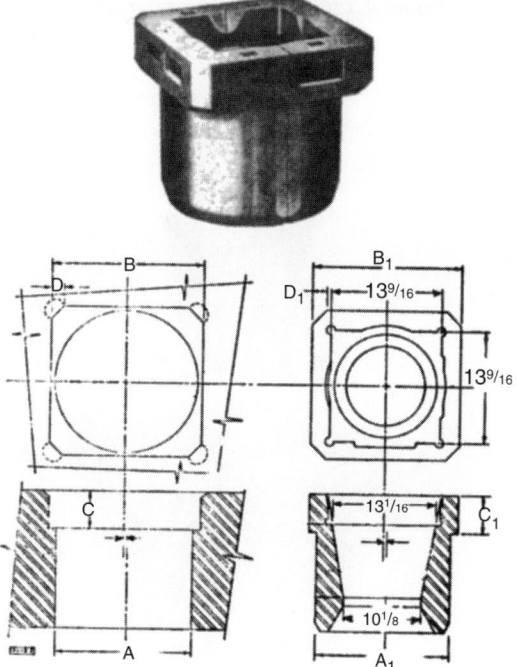

Figure 4.3.15 *Rotary table opening and square drive master bushing.*

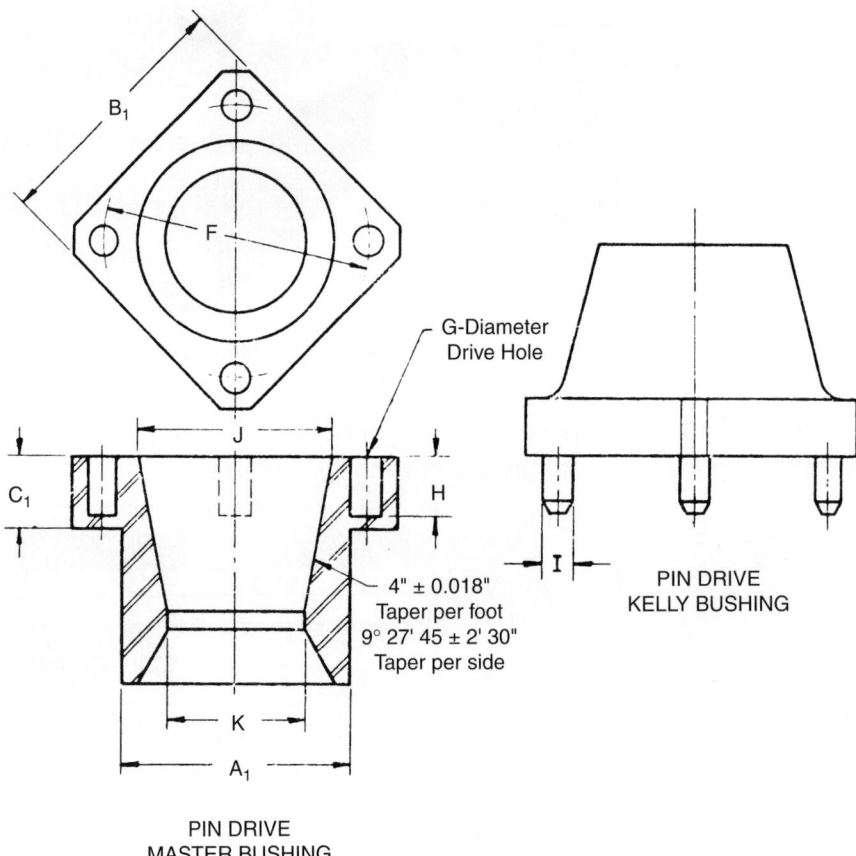

G-Diameter
Drive Hole

4" ± 0.018"
Taper per foot
9° 27' 45 ± 2' 30"
Taper per side

PIN DRIVE
MASTER BUSHING

PIN DRIVE
KELLY BUSHING

Figure 4.3.16 *Pin-drive master bushing.*

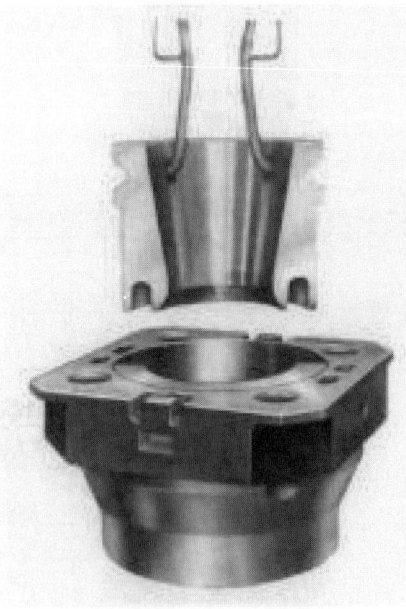

Figure 4.3.17 *Solid-body master bushing.*

Figure 4.3.18 *Split-body pin drive master bushing.*

Figure 4.3.19 *Hinged-body pin drive master bushing.*

Figure 4.3.20 *Square drive Kelly bushing for square Kelly.*

Figure 4.3.21   *Pin drive Kelly bushing for hexagonal Kelly.*

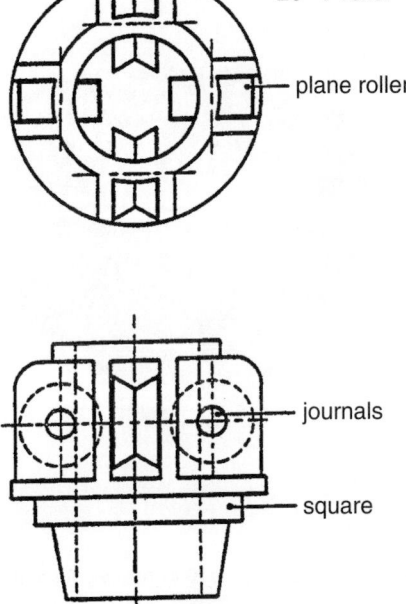

20° V-roller

plane roller

journals

square

E463

Figure 4.3.22   *Kelly bushing for hexagonal Kelly.*

Figure 4.3.23   *Double roller pin drive Kelly bushing.*

4.4 MUD PUMPS

Mud pumps consume more than 60% of all the horsepower used in rotary drilling. Mud pumps are used to circulate drilling fluid through the mud circulation system while drilling. A pump with two fluid cylinders, as shown in Figure 4.4.1, is called a duplex pump. A three-fluid-cylinder pump, as shown in Figure 4.4.2, is called a triplex pump. Duplex pumps are usually double action, and triplex pumps are usually single action. Pumps with six chambers are commercially available as well (Figure 4.4.3).

Mud pumps consists of a power input end and a fluid output end. The power input end, shown in Figure 4.4.4 transfers power from the driving engine (usually diesel or electric) to the pump crankshaft. The fluid end does the actual work of pumping the fluid. A cross-section of the fluid end is shown in Figure.4.4.5.

4.4.1 Pump Installation
4.4.1.1 Suction Manifold
The hydraulic horsepower produced by mud pumps depends mainly on the geometric and mechanical arrangement of the suction piping. If suction-charging centrifugal pumps (e.g., auxiliary pumps that help move the mud to the mud pump) are not used, the pump cylinders have to be filled by the hydrostatic head.

Incomplete filling of the cylinders can result in hammering, which produces destructive pressure peaks and shortens the pump life. Filling problems become more important with higher piston velocities. The suction pressure loss through the suction valve and seat is from 5 to 10 psi. Approximately 1.5 psi of pressure is required for each foot of suction lift. Since the maximum available atmospheric pressure is 14.7 psi (sea level), suction pits placed below the pump should be eliminated. Instead, suction tanks placed level with or higher than the pump should be used to ensure a positive suction head. Figure 4.4.6 shows an ideal suction arrangement with the least amount of friction and low inertia.

A poorly designed suction entrance to the pump can produce friction equivalent to 30 ft of pipe. Factors contributing to excessive suction pipe friction are an intake connection with sharp ends, a suction strainer, suction pipe with a small diameter, long runs of suction pipe, and numerous fittings along the suction pipe. Minimizing the effect of inertia requires a reduction of the suction velocity and mud weight. It is generally practical to use a short suction pipe with a large diameter.

When a desirable suction condition cannot be attained, a charging pump becomes necessary. This is a common solution used on many modern rigs.

4.4.1.2 Cooling Mud
Mud temperatures of 150° can present critical suction problems. Under low pressure or vacuum existing in the cylinder on the suction stroke, the mud can boil, hence decreasing the suction effectiveness. Furthermore, hot mud accelerates the deterioration of rubber parts, particularly when oil is present. Large mud tanks with cooling surfaces usually solve the problem.

4.4.1.3 Gas and Air Separation
Entrained gas and air expands under the reduced pressure of the suction stroke, lowering the suction efficiency. Gas in water-base mud may also deteriorate the natural rubber parts used. Gases are usually separated with baffles or by changing mud composition.

4.4.1.4 Settling Pits
The normally good lubricating qualities of mud can be lost if cuttings, particularly fine sand, are not effectively separated from the mud. Adequate settling pits and shale shakers usually eliminate this trouble. Desanders are used occasionally.

4.4.1.5 Discharge Manifold
A poorly designed discharge manifold can cause shock waves and excessive pressure peaks. This manifold should be as short and direct as possible, avoiding any sharps turns. The conventional small atmospheric air chamber, often furnished with pumps, supplies only a moderate cushioning effect. For best results, this air chamber should be supplemented by a large atmospheric air chamber or by a precharged pulsation dampener.

4.4.2 Pump Operation
4.4.2.1 Priming
A few strokes of the piston in a dry liner may ruin the liner. When the pump does not fill by gravity or when the cylinders have been emptied by standing too long or by replacement of the piston and liner, it is essential to prime the pump through the suction valve cap openings.

4.4.2.2 Cleaning the Suction Manifold
Suction lines are often partly filled by settled sand and by debris from the pits, causing the pump to hammer at abnormally low speeds. Frequent inspection and cleaning of the suction manifold is required. The suction strainer can also be a liability if it is not cleaned frequently.

4.4.2.3 Cleaning the Discharge Strainer
The discharge strainer often becomes clogged with pieces of piston and valve rubber. This may increase the pump pressure that is not shown by the pressure gauge beyond the strainer. The strainer should be inspected and cleaned frequently to prevent a pressure buildup.

4.4.2.4 Lost Circulation Materials
Usually special solids, such as nut shells, limestone, expanded perlite, etc., are added to the drilling muds to fill or clog rock fractures in the open hold of a well. Most of these lost circulation materials can shorten the life of pump parts. They are especially hard on valves and seats when they accumulate on the seats or between the valve body and the valve disc.

4.4.2.5 Parts Storage
Pump parts for high-pressure service are made of precisely manufactured materials and should be treated accordingly. In storage at the rig, metal parts should be protected from rusting and physical damage, and rubber parts should be protected from distortion and from exposure to heat, light, and oil. In general, parts should remain in their original packages where they are usually protected with rust-inhibiting coatings and wrappings and are properly supported to avoid damage. Careless stacking of pistons may distort or cut the sealing lips and result in early failures. Hanging lip-type or O-ring packings on a hook or throwing them carelessly into a bin may ruin them. Metal parts temporarily removed from pumps should be thoroughly cleaned, greased, and stored like new parts.

4.4.3 Pump Performance Charts
The charts indicating the pump output per stroke are shown in Table 4.4.1 and 4.4.2 for duplex and triplex pumps, respectively [2,3].

Figure 4.4.1   *Duplex slush (mud) pump. (Courtesy of National Oilwell.)*

Figure 4.4.2   *Triplex mud pumps. (Courtesy of Weatherford International, Inc.)*

Figure 4.4.3 *Six chamber mud pump. (Courtesy of National Oilwell, Inc.)*

Figure 4.4.4 *Power end of mud pump. (Courtesy of LTV Energy Products Company.)*

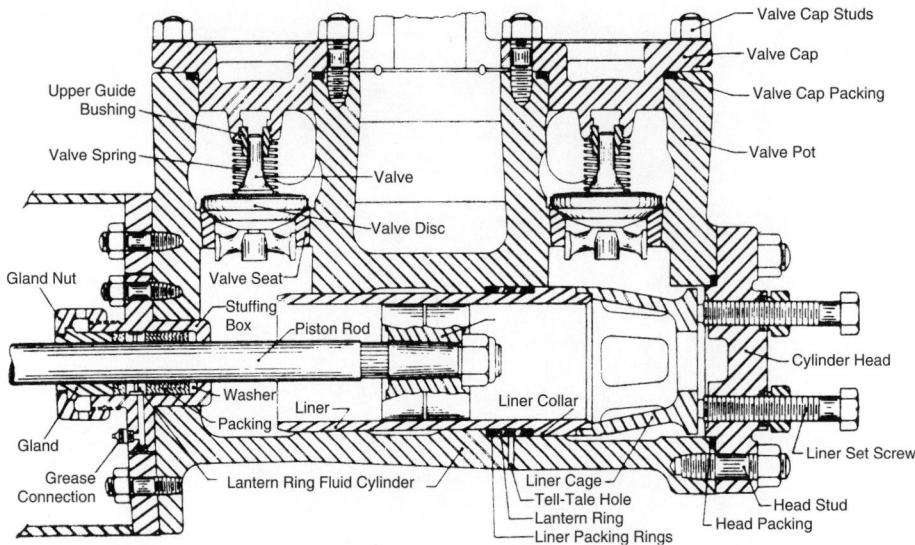

Figure 4.4.5 *Cross-section of fluid end of mud pump. (Courtesy of International Association of Drilling Contractors.)*

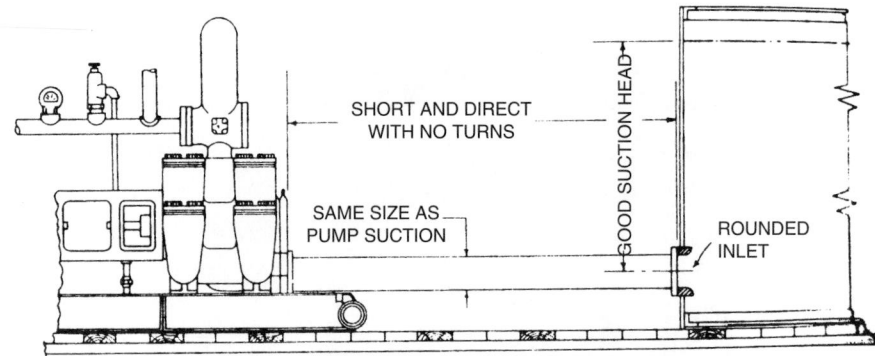

Figure 4.4.6 *Installation of mud pump suction piping. (Courtesy of International Association of Drilling Contractors.)*

4.4.4 Mud Pump Hydraulics

The required pump output can be approximated as follows [4–6]: Minimum Q (gal/min):

$$Q_{min} = (30 \text{ to } 50) \, D_h \qquad [4.4.1]$$

or

$$Q_{min} = 481 \frac{D_h^2 - D_p^2}{\overline{\sigma} D_h} \qquad [4.4.2]$$

where D_h = hole diameter in in.
D_p = pipe diameter in in.
$\overline{\sigma}$ = mud specific weight in lb/gal

The required pump working pressure PWP (psi) can be calculated as

$$PWP = \Delta P_s + \Delta P_d + \Delta P_a + \Delta P_b \qquad [4.4.3]$$

where ΔP_s = pressure loss through surface equipment in psi
ΔP_d = pressure loss through the inside of the drill string in psi
ΔP_a = pressure loss in annulus in psi
ΔP_b = pressure drop through bit nozzles in psi

Table 4.4.3 shows the diameters and areas of various nozzle sizes.

The required pump hydraulic horsepower (PHHP) can be calculated as

$$PHHP = HHP_{circ} + HHP_{bit} \qquad [4.4.4]$$

where HHP_{circ} = total HHP loss due to pressure losses in the ciruclating system
HHP_{bit} = hydraulic horsepower required at the bit

Table 4.4.1 *Duplex Pump Output**

Bore (in.)	Stroke (in.)	100% Efficiency		90% Efficiency	
		Cu. Ft.	Bbls.	Cu. Ft.	Bbls.
4	8	0.2328	0.0413	0.2095	0.0372
$4\frac{1}{2}$	8	0.2944	0.0526	0.2650	0.0473
5	8	0.3637	0.0648	0.3274	0.0603
4	10	0.2910	0.0517	0.2619	0.0465
$4\frac{1}{2}$	10	0.3680	0.0660	0.3312	0.0594
5	10	0.4547	0.0810	0.4092	0.0729
5	12	0.5456	0.0972	0.4909	0.0875
6	12	0.7854	0.1396	0.7069	0.1256
$6\frac{1}{2}$	12	0.9217	0.1580	0.8295	0.1422
7	12	1.069	0.1904	0.9621	0.1714
$7\frac{1}{2}$	12	1.227	0.2184	1.104	1.1966
5	16	0.7272	0.1296	0.6545	0.1206
6	16	1.047	0.1861	0.9423	0.1675
$6\frac{1}{2}$	16	1.229	0.2187	1.106	0.1968
7	16	1.425	0.2539	1.283	0.2285
$7\frac{1}{2}$	16	1.636	0.2912	1.572	0.2621
5	18	0.8181	0.1458	0.7363	0.1312
6	18	1.178	0.2094	1.060	0.1885
$6\frac{1}{2}$	18	1.383	0.2460	1.245	0.2214
7	18	1.604	0.2856	1.444	0.2570
$7\frac{1}{2}$	18	1.841	0.3276	1.657	0.2948
7	20	1.782	0.3173	1.604	0.2856
$7\frac{1}{2}$	20	2.046	0.3640	1.841	0.3276
8	20	2.330	0.4147	2.095	0.3732
$8\frac{1}{2}$	20	2.630	0.4680	2.361	0.4212
7	24	2.138	0.3808	1.924	0.3427
$7\frac{1}{2}$	24	2.455	0.4368	2.210	0.3931
8	24	2.792	0.4976	2.513	0.4478
$8\frac{1}{2}$	24	3.153	0.5616	2.838	0.5054
9	24	3.534	0.6296	3.181	0.5666

*Volume indicated are one complete cycle or revolution and represent four individual pump strokes. Multiply output/cycle by pump rpm to get vol./min. Courtesy of Weatherford International, Inc.

Table 4.4.2 *Triplex Pump Output**

Bore (in.)	Stroke (in.)	100% Efficiency		90% Efficiency	
		Cu. Ft.	Bbls.	Cu. Ft.	Bbls.
3	4	0.0491	0.0087	0.0442	0.0078
$3\frac{1}{4}$	4	0.0576	0.0103	0.0518	0.0093
$3\frac{1}{2}$	4	0.0668	0.0119	0.0601	0.0107
$3\frac{3}{4}$	4	0.0767	0.0137	0.0690	0.0123
4	4	0.0873	0.0155	0.0786	0.0140
$4\frac{1}{2}$	4	0.1104	0.0197	0.0994	0.0177
5	4	0.1364	0.0243	0.1228	0.0219
6	4	0.1963	0.0350	0.1767	0.0315
8	4	0.3491	0.0622	0.3142	0.0560
3	6	0.0737	0.0131	0.0663	0.0117
$3\frac{1}{4}$	6	0.0864	0.0155	0.0777	0.0140
$3\frac{1}{2}$	6	0.1002	0.0179	0.0902	0.0161
$3\frac{3}{4}$	6	0.1151	0.0206	0.1035	0.0185
4	6	0.1310	0.0233	0.1179	0.0210
$4\frac{1}{2}$	6	0.1656	0.0291	0.1491	0.0266
5	6	0.2046	0.0365	0.1842	0.0329
6	6	0.2945	0.0525	0.2651	0.0473
8	6	0.5237	0.0933	0.4713	0.0840
3	8	0.0982	0.0174	0.0884	0.0156
$3\frac{1}{4}$	8	0.1152	0.0206	0.1036	0.0186
$3\frac{1}{2}$	8	0.1336	0.0238	0.1202	0.0214
$3\frac{3}{4}$	8	0.1534	0.0274	0.1380	0.0246
4	8	0.1746	0.0310	0.1572	0.0280
$4\frac{1}{2}$	8	0.2208	0.0384	0.1988	0.0354
5	8	0.2728	0.0586	0.2456	0.0438
6	8	0.3926	0.0700	0.3534	0.0630
8	8	0.6982	0.1244	0.6284	0.1120
3	10	0.1228	0.0218	0.1105	0.0195
$3\frac{1}{4}$	10	0.1440	0.0258	0.1295	0.0233
$3\frac{1}{2}$	10	0.1670	0.0298	0.1503	0.0268
$3\frac{3}{4}$	10	0.1918	0.0343	0.1725	0.0308
4	10	0.2183	0.0388	0.1965	0.0350
$4\frac{1}{2}$	10	0.2760	0.0478	0.2485	0.0443
5	10	0.3410	0.0708	0.3070	0.0548
6	10	0.4908	0.0875	0.4418	0.0788
8	10	0.8728	0.1555	0.7855	0.1400

*Volumes indicated are for one complete cycle or revolution. Multiply output/cycle by pump rpm to get vol./min. For quintuplex pump, multiply output by 1.67. Courtesy of Weatherford International, Inc.

The general hydraulic horsepower is

$$HHP = \frac{Q\,\Delta P}{1714} \qquad [4.4.5]$$

where Q = flow rate in gal/min
ΔP = pressure difference in psi

The minimum bit HHP is shown in Figure 4.4.7. The maximum useful bit HHP is shown in Figure 4.4.8 and Figure 4.4.9 [2].

4.4.5 Useful Formulas

Theoretical output Q_t (gal/min) for a double action duplex pump is

$$Q_t = .0136NS\left(Dl^2 - \frac{d^2}{2}\right) \qquad [4.4.6]$$

where N = strokes per minute
S = stroke length, in.

Theoretical output Q_t (gal/min) for a single action triplex pump is

$$Q_t = 0.0102\ NS\ Dl^2 \qquad [4.4.7]$$

The volumetric efficiency η_v for duplex pumps or triplex pumps is

$$\eta_v = \frac{Q_a}{Q_t} \qquad [4.4.8]$$

where Q_a = actual volumetric flow rate in gal

Table 4.4.3 *Common Flow Diameters and Areas*

Nozzle size	Fractional Diameter, in.	Decimal Diameter, in.	Area, in²	Fractional Diameter, in.	Decimal Diameter, in.	Area, in²	Fractional Diameter, in.	Decimal Diameter, in.	Area, in²
6	$\frac{3}{16}$	0.1875	0.03	$7\frac{7}{8}$	7.875	48.707	17	**17.000**	226.98
7	$\frac{7}{32}$	0.2188	0.03758	8	**8.000**	50.265	$17\frac{1}{8}$	17.125	230.33
8	$\frac{1}{4}$	0.2500	0.04909	$8\frac{1}{8}$	8.125	51.849	$17\frac{1}{4}$	17.250	233.71
9	$\frac{9}{32}$	0.2813	0.06213	$8\frac{1}{4}$	8.250	53.456	$17\frac{3}{8}$	17.375	237.10
10	$\frac{5}{16}$	0.3125	0.07670	$8\frac{3}{8}$	8.375	55.088	$17\frac{1}{2}$	17.500	240.53
11	$\frac{11}{32}$	0.3438	0.09281	$8\frac{1}{2}$	8.500	56.745	$17\frac{5}{8}$	17.625	243.98
12	$\frac{3}{8}$	0.3750	0.1104	$8\frac{5}{8}$	8.625	58.426	$17\frac{3}{4}$	17.750	247.45
13	$\frac{13}{32}$	0.4063	0.1296	$8\frac{3}{4}$	8.750	60.132	$17\frac{7}{8}$	17.875	250.95
14	$\frac{7}{16}$	0.4375	0.1503	$8\frac{7}{8}$	8.875	61.862	18	**18.000**	254.47
15	$\frac{15}{32}$	0.4688	0.1726	9	**9.000**	63.617	$18\frac{1}{8}$	18.125	258.02
16	$\frac{1}{2}$	0.5000	0.1963	$9\frac{1}{8}$	9.125	65.397	$18\frac{1}{4}$	18.250	261.59
18	$\frac{9}{16}$	0.5625	0.2485	$9\frac{1}{4}$	9.250	67.201	$18\frac{3}{8}$	18.375	265.18
20	$\frac{5}{8}$	0.6250	0.3068	$9\frac{3}{8}$	9.375	69.029	$18\frac{1}{2}$	18.500	268.80
22	$\frac{11}{16}$	0.6875	0.3712	$9\frac{1}{2}$	9.500	70.882	$18\frac{5}{8}$	18.625	272.45
24	$\frac{3}{4}$	0.7500	0.4418	$9\frac{5}{8}$	9.625	72.760	$18\frac{3}{4}$	18.750	276.12
26	$\frac{13}{16}$	0.8125	0.5185	$9\frac{3}{4}$	9.750	74.662	$18\frac{7}{8}$	18.875	279.81
28	$\frac{7}{8}$	0.8750	0.6013	$9\frac{7}{8}$	9.875	76.589	19	**19.000**	283.53
30	$\frac{15}{16}$	0.9375	0.6903	10	**10.000**	78.540	$19\frac{1}{8}$	19.125	287.27
32	1	**1.000**	0.7854	$10\frac{1}{8}$	10.125	80.516	$19\frac{1}{4}$	19.250	291.04
	$1\frac{1}{8}$	1.125	0.9940	$10\frac{1}{4}$	10.250	82.516	$19\frac{3}{8}$	19.375	294.83
	$1\frac{1}{4}$	1.250	1.227	$10\frac{3}{8}$	10.375	84.541	$19\frac{1}{2}$	19.500	298.65
	$1\frac{3}{8}$	1.375	1.485	$10\frac{1}{2}$	10.500	86.590	$19\frac{5}{8}$	19.625	302.49
	$1\frac{1}{2}$	1.500	1.767	$10\frac{5}{8}$	10.625	88.664	$19\frac{3}{4}$	19.750	306.35
	$1\frac{5}{8}$	1.625	2.074	$10\frac{3}{4}$	10.750	90.763	$19\frac{7}{8}$	19.875	310.24
	$1\frac{3}{4}$	1.750	2.405	$10\frac{7}{8}$	10.875	92.886	20	**20.000**	314.16
	$1\frac{7}{8}$	1.875	2.761	11	**11.000**	95.033	$20\frac{1}{8}$	20.125	318.10
	2	**2.000**	3.142	$11\frac{1}{8}$	11.125	97.205	$20\frac{1}{4}$	20.250	322.06
	$2\frac{1}{8}$	2.125	3.547	$11\frac{1}{4}$	11.250	99.402	$20\frac{3}{8}$	20.375	326.05
	$2\frac{1}{4}$	2.250	3.976	$11\frac{3}{8}$	11.375	101.62	$20\frac{1}{2}$	20.500	330.06
	$2\frac{3}{8}$	2.375	4.430	$11\frac{1}{2}$	11.500	103.87	$20\frac{5}{8}$	20.625	334.10
	$2\frac{1}{2}$	2.500	4.909	$11\frac{5}{8}$	11.625	106.14	$20\frac{3}{4}$	20.750	338.16
	$2\frac{5}{8}$	2.625	5.412	$11\frac{3}{4}$	11.750	108.43	$20\frac{7}{8}$	20.875	342.25
	$2\frac{3}{4}$	2.750	5.940	$11\frac{7}{8}$	11.875	110.75	21	**21.000**	346.36
	$2\frac{7}{8}$	2.875	6.492	12	**12.000**	113.10	$21\frac{1}{8}$	21.125	350.50
	3	**3.000**	7.069	$12\frac{1}{8}$	12.125	115.47	$21\frac{1}{4}$	21.250	354.66
	$3\frac{1}{8}$	3.125	7.670	$12\frac{1}{4}$	12.250	117.86	$21\frac{3}{8}$	21.375	358.84
	$3\frac{1}{4}$	3.250	8.296	$12\frac{3}{8}$	12.375	120.28	$21\frac{1}{2}$	21.500	363.05
	$3\frac{3}{8}$	3.375	8.946	$12\frac{1}{2}$	12.500	122.72	$21\frac{5}{8}$	21.625	367.28
	$3\frac{1}{2}$	3.500	9.621	$12\frac{5}{8}$	12.625	125.19	$21\frac{3}{4}$	21.750	371.54
	$3\frac{5}{8}$	3.625	10.321	$12\frac{3}{4}$	12.750	127.68	$21\frac{7}{8}$	21.875	375.83
	$3\frac{3}{4}$	3.750	11.045	$12\frac{7}{8}$	12.875	130.19	22	**22.000**	380.13
	$3\frac{7}{8}$	3.875	11.793	13	**13.000**	132.73	$22\frac{1}{8}$	22.125	384.46
	4	**4.000**	12.566	$13\frac{1}{8}$	13.125	135.30	$22\frac{1}{4}$	22.250	388.82
	$4\frac{1}{8}$	4.125	13.364	$13\frac{1}{4}$	13.250	137.89	$22\frac{3}{8}$	22.375	393.20
	$4\frac{1}{4}$	4.250	14.186	$13\frac{3}{8}$	13.375	140.50	$22\frac{1}{2}$	22.500	397.61
	$4\frac{3}{8}$	4.375	15.033	$13\frac{1}{2}$	13.500	143.14	$22\frac{5}{8}$	22.625	402.04
	$4\frac{1}{2}$	4.500	15.904	$13\frac{5}{8}$	13.625	145.80	$22\frac{3}{4}$	22.750	406.49
	$4\frac{5}{8}$	4.625	16.800	$13\frac{3}{4}$	13.750	148.49	$22\frac{7}{8}$	22.875	410.97
	$4\frac{3}{4}$	4.750	17.721	$13\frac{7}{8}$	13.875	151.20	23	**23.000**	415.48
	$4\frac{7}{8}$	4.875	18.665	14	**14.000**	153.94	$23\frac{1}{8}$	23.125	420.00

(continued)

Table 4.4.3 *continued*

Nozzle size	Fractional Diameter, in.	Decimal Diameter, in.	Area, in²	Fractional Diameter, in.	Decimal Diameter, in.	Area, in²	Fractional Diameter, in.	Decimal Diameter, in.	Area, in²
5	**5.000**	19.635	$14\frac{1}{8}$	14.125	156.70	$23\frac{1}{4}$	23.250	424.56
$5\frac{1}{8}$	5.125	20.629	$14\frac{1}{4}$	14.250	159.48	$23\frac{3}{8}$	23.375	429.13
$5\frac{1}{4}$	5.250	21.648	$14\frac{3}{8}$	14.375	162.30	$23\frac{1}{2}$	23.500	433.74
$5\frac{3}{8}$	5.375	22.691	$14\frac{1}{2}$	14.500	165.13	$23\frac{5}{8}$	23.625	438.36
$5\frac{1}{2}$	5.500	23.758	$14\frac{5}{8}$	14.625	167.99	$23\frac{3}{4}$	23.750	443.01
$5\frac{5}{8}$	5.625	24.850	$14\frac{3}{4}$	14.750	170.87	$23\frac{7}{8}$	23.875	447.69
$5\frac{3}{4}$	5.750	25.967	$14\frac{7}{8}$	14.875	173.78	24	**24.000**	452.39
$5\frac{7}{8}$	5.875	27.109	15	**15.000**	176.71	$24\frac{1}{8}$	24.125	457.11
6	**6.000**	28.274	$15\frac{1}{8}$	15.125	179.67	$24\frac{1}{4}$	24.250	461.86
$6\frac{1}{8}$	6.125	29.465	$15\frac{1}{4}$	15.250	182.65	$24\frac{3}{8}$	24.375	466.64
$6\frac{1}{4}$	6.250	30.680	$15\frac{3}{8}$	15.375	185.66	$24\frac{1}{2}$	24.500	471.44
$6\frac{3}{8}$	6.375	31.919	$15\frac{1}{2}$	15.500	188.69	$24\frac{5}{8}$	24.625	476.26
$6\frac{1}{2}$	6.500	33.183	$15\frac{5}{8}$	15.625	191.75	$24\frac{3}{4}$	24.750	481.11
$6\frac{5}{8}$	6.625	34.472	$15\frac{3}{4}$	15.750	194.83	$24\frac{7}{8}$	24.875	485.98
$6\frac{3}{4}$	6.750	35.785	$15\frac{7}{8}$	15.875	197.93	25	**25.000**	490.87
$6\frac{7}{8}$	6.875	37.122	16	**16.000**	201.06	$25\frac{1}{8}$	25.125	495.79
7	**7.000**	38.485	$16\frac{1}{8}$	16.125	204.22	$25\frac{1}{4}$	25.250	500.74
$7\frac{1}{8}$	7.125	39.871	$16\frac{1}{4}$	16.250	207.39	$25\frac{3}{8}$	25.375	505.71
$7\frac{1}{4}$	7.250	41.282	$16\frac{3}{8}$	16.375	210.60	$25\frac{1}{2}$	25.500	510.71
$7\frac{3}{8}$	7.375	42.718	$16\frac{1}{2}$	16.500	213.82	$25\frac{5}{8}$	25.625	515.72
$7\frac{1}{2}$	7.500	44.179	$16\frac{5}{8}$	16.625	217.08	$25\frac{3}{4}$	25.750	520.77
$7\frac{5}{8}$	7.625	45.664	$16\frac{3}{4}$	16.750	220.35	$25\frac{7}{8}$	25.875	525.84
$7\frac{3}{4}$	7.750	47.173	16 7/8	16.875	223.65	26	**26.000**	530.93

Note: Area = 0.7854*(diameter)². Courtesy of Weatherfold International, Inc.

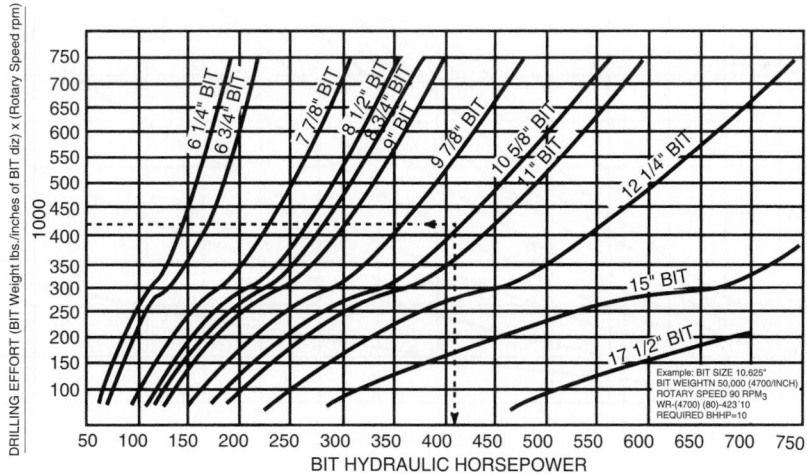

Figure 4.4.7 *Minimum bit HHP to prevent hydraulic flounder [4]. (Courtesy of Smith International, Inc.)*

Input engine power IHP (hp) required for a given pump theoretical output Q_t and pump working pressure PWP is

$$\text{IHP} = \frac{\text{PWP}(Q_t)}{1714\,\eta_m} \qquad [4.4.9]$$

where η_m = mechanical efficiency of the pump

Pressure loss correction for mud weight change is

$$\Delta P_2 = \Delta P_1 \frac{\overline{\gamma}_2}{\overline{\gamma}_1} \qquad [4.4.10]$$

where ΔP_1 = pressure loss in system calculated using mud weight $\overline{\gamma}_1$ in psi
ΔP_2 = Pressure loss in system calculated using mud weight $\overline{\gamma}_2$ in psi
$\overline{\gamma}_1, \overline{\gamma}_2$ = mud weight in lb/gal

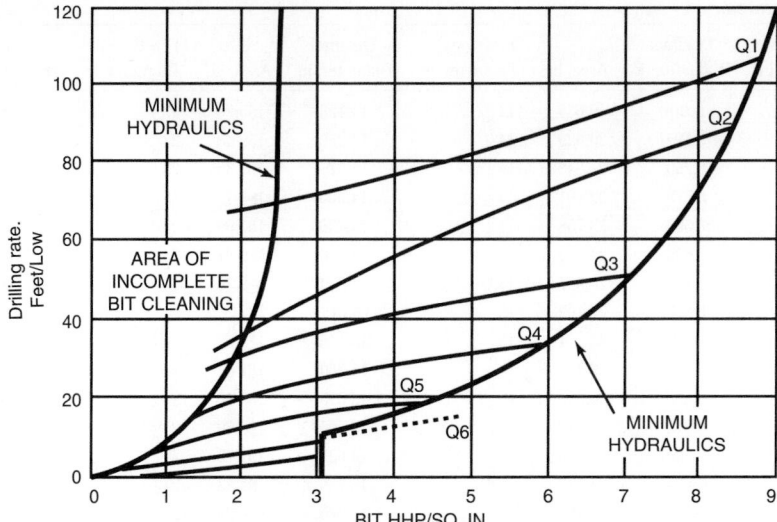

NOTE: PLACE POINT WHERE CONTROL BIT RUN IS LOCATED; USING SHIP
CURVE, FIT Q-LINE BELOW POINT; MOVE SHIP CURVE *UP* TO POINT.
EFFECT OF A CHANGE IN HYDERAULICS CAN THEN BE TRACED. POINTS
OF COMPLETE BIT CLEANING ARE ON THE "MAXIMUM HYDRAULICS"
CURE.

Figure 4.4.8 *Required bit hydraulic horsepower [5]. (Courtesy of Hant Publications. All rights reserved.)*

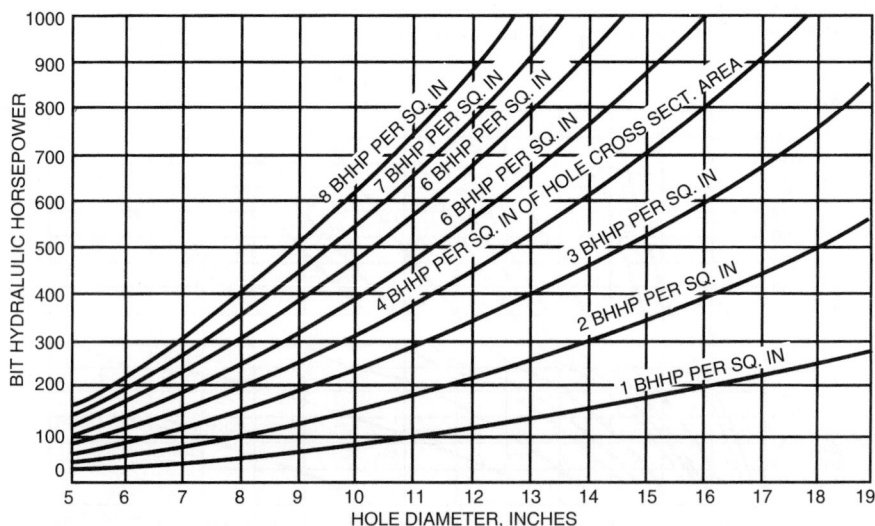

Figure 4.4.9 *Bottomhole hydraulic horsepower chart [2]. (Used by permission of the American Petroleum Institute, Production Department.)*

References

1. Hughes Tool Company, *Hughes Practical Hydraulics*, Houston, 1976.
2. API Bulletin D10, 2nd Edition: "Procedure for Selecting Rotary Drilling Equipment," API, Dallas, January 1982.
3. American Association of Drilling Contractors, *Tool Pusher's Manual*, API, Dallas, 1955.
4. Fullerton, H. B., *Constant Energy Drilling System Well Programming*, Sii Smith Tool, p. 6, Irvine (adapted from original work by Hal B. Fullerton, Jr., United Drilling Services).
5. Lummus, James L., "Drilling in the Seventies, Part II: Analysis of Mud-Hydraulics Interaction," *Petroleum Engineer*, February 1974.
6. Randal, B. V., "Optimum Hydraulics in the Oil Patch," *Petroleum Engineer*, September, 1975.
7. *Weatherford Technical Data Handbook*, Weatherford US, LP, Houston, TX, 2001.

4.5 DRILLING MUDS AND COMPLETION FLUIDS

4.5.1 Functions of Drilling Muds

4.5.1.1 Drilling Fluid Definitions and General Functions

Results of research has shown that penetration rate and its response to weight on bit and rotary speed is highly dependent on the hydraulic horsepower reaching the formation at the bit. Because the drilling fluid flow rate sets the system pressure losses and these pressure losses set the hydraulic horsepower across the bit, it can be concluded that the drilling fluid is as important in determining drilling costs as all other "controllable" variables combined. Considering these factors, an optimum drilling fluid is properly formulated so that the flow rate necessary to clean the hole results in the proper hydraulic horsepower to clean the bit for the weight and rotary speed imposed to give the lowest cost, provided that this combination of variables results in a stable borehole which penetrates the desired target. This definition incorporates and places in perspective the five major functions of a drilling fluid.

4.5.1.2 Cool and Lubricate the Bit and Drill String

Considerable heat and friction is generated at the bit and between the drill string and wellbore during drilling operations. Contact between the drill string and wellbore can also create considerable torque during rotation and drag during trips. Circulating drilling fluid transports heat away from these frictional sites, reducing the chance of premature bit failure and pipe damage. The drilling fluid also lubricates the bit tooth penetration through the bottom hole debris into the rock and serves as a lubricant between the wellbore and drill string, reducing torque and drag.

4.5.1.3 Clean the Bit and the Bottom of the Hole

If the cuttings generated at the bit face are not immediately removed and started toward the surface, they will be ground very fine, stick to the bit, and in general retard effective penetration into uncut rock.

4.5.1.4 Suspend Solids and Transport Cuttings and Sloughings to the Surface

Drilling fluids must have the capacity to suspend weight materials and drilled solids during connections, bit trips, and logging runs, or they will settle to the low side or bottom of the hole. Failure to suspend weight materials can result in a reduction in the drilling fluids density, which can lead to kicks and potential of a blowout.

The drilling fluid must be capable of transporting cuttings out of the hole at a reasonable velocity that minimizes their disintegration and incorporation as drilled solids into the drilling fluid system and able to release the cuttings at the surface for efficient removal. Failure to adequately clean the hole or to suspend drilled solids can contribute to hole problems such as fill on bottom after a trip, hole pack-off, lost returns, differentially stuck pipe, and inability to reach bottom with logging tools.

Factors influencing removal of cuttings and formation sloughings and solids suspension include

- Density of the solids
- Density of the drilling fluid
- Rheological properties of the drilling fluid
- Annular velocity
- Hole angle
- Slip velocity of the cuttings or sloughings

4.5.1.5 Stabilize the Wellbore and Control Subsurface Pressures

Borehole instability is a natural function of the unequal mechanical stresses and physical-chemical interactions and pressures created when supporting material and surfaces are exposed in the process of drilling a well. The drilling fluid must overcome the tendency for the hole to collapse from mechanical failure or from chemical interaction of the formation with the drilling fluid. The Earth's pressure gradient at sea level is 0.465 psi/ft, which is equivalent to the height of a column of salt water with a density (1.07 SG) of 8.94 ppg.

In most drilling areas, the fresh water plus the solids incorporated into the water from drilling subsurface formations is sufficient to balance the formation pressures. However, it is common to experience abnormally pressured formations that require high-density drilling fluids to control the formation pressures. Failure to control downhole pressures can result in an influx of formation fluids, resulting in a kick or blowout. Borehole stability is also maintained or enhanced by controlling the loss of filtrate to permeable formations and by careful control of the chemical composition of the drilling fluid.

Most permeable formations have pore space openings too small to allow the passage of whole mud into the formation, but filtrate from the drilling fluid can enter the pore spaces. The rate at which the filtrate enters the formation depends on the pressure differential between the formation and the column of drilling fluid and the quality of the filter cake deposited on the formation face.

Large volumes of drilling fluid filtrate and filtrates that are incompatible with the formation or formation fluids may destabilize the formation through hydration of shale and/or chemical interactions between components of the drilling fluid and the wellbore.

Drilling fluids that produce low-quality or thick filter cakes may also cause tight hole conditions, including stuck pipe, difficulty in running casing, and poor cement jobs.

4.5.1.6 Assist in the Gathering of Subsurface Geological Data and Formation Evaluation

Interpretation of surface geological data gathered through drilled cuttings, cores, and electrical logs is used to determine the commercial value of the zones penetrated. Invasion of these zones by the drilling fluid, its filtrate (oil or water) may mask or interfere with interpretation of data retrieved or prevent full commercial recovery of hydrocarbon.

4.5.1.7 Other Functions

In addition to the functions previously listed, the drilling fluid should be environmentally acceptable to the area in which it is used. It should be noncorrosive to tubulars being used in the drilling and completion operations. Most importantly, the drilling fluid should not damage the productive formations that are penetrated.

The functions described here are a few of the most obvious functions of a drilling fluid. Proper application of drilling fluids is the key to successfully drilling in various environments.

4.5.2 Classifications

A generalized classification of drilling fluids can be based on their fluid phase, alkalinity, dispersion, and type of chemicals used in the formulation and degrees of inhibition. In a broad sense, drilling fluids can be broken into five major categories.

4.5.2.1 Freshwater Muds—Dispersed Systems

The pH value of low-pH muds may range from 7.0 to 9.5. Low-pH muds include spud muds, bentonite-treated muds, natural muds, phosphate-treated muds, organic thinned muds (e.g., red muds, lignite muds, lignosulfonate muds), and organic colloid–treated muds. In this case, the lack of salinity of the water phase and the addition of chemical dispersants dictate the inclusion of these fluids in this broad category.

4.5.2.2 Inhibited Muds—Dispersed Systems
These are water-base drilling muds that repress the hydration and dispersion of clays through the inclusion of inhibiting ions such as calcium and salt. There are essentially four types of inhibited muds: lime muds (high pH), gypsum muds (low pH), seawater muds (unsaturated saltwater muds, low pH), and saturated saltwater muds (low pH). Newer-generation inhibited-dispersed fluids offer enhanced inhibitive performance and formation stabilization; these fluids include sodium silicate muds, formate brine-based fluids, and cationic polymer fluids.

4.5.2.3 Low Solids Muds—Nondispersed Systems
These muds contain less than 3–6% solids by volume, weight less than 9.5 lb/gal, and may be fresh or saltwater based. The typical low-solid systems are selective flocculent, minimum-solids muds, beneficiated clay muds, and low-solids polymer muds. Most low-solids drilling fluids are composed of water with varying quantities of bentonite and a polymer. The difference among low-solid systems lies in the various actions of different polymers.

4.5.2.4 Nonaqueous Fluids
Invert Emulsions
Invert emulsions are formed when one liquid is dispersed as small droplets in another liquid with which the dispersed liquid is immiscible. Mutually immiscible fluids, such as water and oil, can be emulsified by shear and the addition of surfactants. The suspending liquid is called the *continuous phase*, and the droplets are called the *dispersed* or *discontinuous phase*. There are two types of emulsions used in drilling fluids: oil-in-water emulsions that have water as the continuous phase and oil as the dispersed phase and water-in-oil emulsions that have oil as the continuous phase and water as the dispersed phase (i.e., invert emulsions).

Oil-Base Muds (nonaqueous fluid [NAF])
Oil-base muds contain oil (refined from crude such as diesel or synthetic-base oil) as the continuous phase and trace amounts of water as the dispersed phase. Oil-base muds generally contain less than 5% (by volume) water (which acts as a polar activator for organophilic clay), whereas invert emulsion fluids generally have more than 5% water in mud. Oil-base muds are usually a mixture of base oil, organophilic clay, and lignite or asphalt, and the filtrate is all oil.

4.5.3 Testing of Drilling Fluids
To properly control the hole cleaning, suspension, and filtration properties of a drilling fluid, testing of the fluid properties is done on a daily basis. Most tests are conducted at the rig site, and procedures are set forth in the API RPB13B. Testing of water-based fluids and nonaqueous fluids can be similar, but variations of procedures occur due to the nature of the fluid being tested.

4.5.3.1 Water-Base Muds Testing
To accurately determine the physical properties of water-based drilling fluids, examination of the fluid is required in a field laboratory setting. In many cases, this consists of a few simple tests conducted by the derrickman or mud Engineer at the rigsite. The procedures for conducting all routine drilling fluid testing can be found in the American Petroleum Institute's API RPB13B.

Density
Often referred to as the mud weight, density may be expressed as pounds per gallon (lb/gal), pounds per cubic foot (lb/ft^3), specific gravity (SG) or pressure gradient (psi/ft). Any instrument of sufficient accuracy within ± 0.1 lb/gal or ± 0.5 lb/ft^3 may be used. The mud balance is the instrument most commonly used. The weight of a mud cup attached to one end of the beam is balanced on the other end by a fixed counterweight and a rider free to move along a graduated scale. The density of the fluid is a direct reading from the scales located on both sides of the mud balance (Figure 4.5.1).

Marsh Funnel Viscosity
Mud viscosity is a measure of the mud's resistance to flow. The primary function of drilling fluid viscosity is a to transport cuttings to the surface and suspend weighing materials. Viscosity must be high enough that the weighting material will remain suspended but low enough to permit sand and cuttings to settle out and entrained gas to escape at the surface. Excessive viscosity can create high pump pressure, which magnifies the swab or surge effect during tripping operations. The control of equivalent circulating density (ECD) is always a prime concern when managing the viscosity of a drilling fluid. The Marsh funnel is a rig site instrument used to measure funnel viscosity. The funnel is dimensioned so that by following standard procedures, the outflow time of 1 qt (946 ml) of freshwater at a temperature of $70 \pm 5°$F is 26 ± 0.5 seconds (Figure 4.5.2). A graduated cup is used as a receiver.

Direct Indicating Viscometer
This is a rotational type instrument powered by an electric motor or by a hand crank (Figure 4.5.3). Mud is contained in the annular space between two cylinders. The outer cylinder or rotor sleeve is driven at a constant rotational velocity; its rotation in the mud produces a torque on the inner cylinder or bob. A torsion spring restrains the movement of the bob. A dial attached to the bob indicates its displacement on a direct reading scale. Instrument constraints have been adjusted so that plastic viscosity, apparent viscosity, and yield point are obtained by using readings from rotor sleeve speeds of 300 and 600 rpm.

Figure 4.5.1　*API mud balance.*

Figure 4.5.2　*Marsh funnel.*

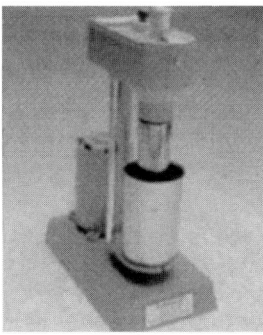

Figure 4.5.3 *Variable speed viscometer.*

Plastic viscosity (PV) in centipoise is equal to the 600 rpm dial reading minus the 300 rpm dial reading. Yield point (YP), in pounds per 100 ft^2, is equal to the 300-rpm dial reading minus the plastic viscosity. Apparent viscosity in centipoise is equal to the 600-rpm reading, divided by two.

Gel Strength

Gel strength is a measure of the inter-particle forces and indicates the gelling that will occur when circulation is stopped. This property prevents the cuttings from setting in the hole. High pump pressure is generally required to "break" circulation in a high-gel mud. Gel strength is measured in units of lbf/100 ft^2. This reading is obtained by noting the maximum dial deflection when the rotational viscometer is turned at a low rotor speed (3 rpm) after the mud has remained static for some period of time (10 seconds, 10 minutes, or 30 minutes). If the mud is allowed to remain static in the viscometer for a period of 10 seconds, the maximum dial deflection obtained when the viscometer is turned on is reported as the *initial gel* on the API mud report form. If the mud is allowed to remain static for 10 minutes, the maximum dial deflection is reported as the *10-min gel*. The same device is used to determine gel strength that is used to determine the plastic viscosity and yield point, the Variable Speed Rheometer/Viscometer.

API Filtration

A standard API filter press is used to determine the filter cake building characteristics and filtration of a drilling fluid (Figure 4.5.4). The API filter press consists of a cylindrical mud chamber made of materials resistant to strongly alkaline solutions. A filter paper is placed on the

Figure 4.5.5 *Sand content kit.*

bottom of the chamber just above a suitable support. The total filtration area is 7.1 ($\pm$0.1) in.^{2}. Below the support is a drain tube for discharging the filtrate into a graduated cylinder. The entire assembly is supported by a stand so 100-psi pressure can be applied to the mud sample in the chamber. At the end of the 30-minute filtration time, the volume of filtrate is reported as API filtration in milliliters. To obtain correlative results, one thickness of the proper 9-cm filter paper—Whatman No. 50, S&S No. 5765, or the equivalent—must be used. Thickness of the filter cake is measured and reported in 32nd of an inch. The cake is visually examined, and its consistency is reported using such notations as "hard," "soft," tough," "rubbery," or "firm."

Sand Content

The sand content in drilling fluids is determined using a 200-mesh sand sieve screen 2 inches in diameter, a funnel to fit the screen, and a glass-sand graduated measuring tube (Figure 4.5.5). The measuring tube is marked to indicate the volume of "mud to be added," water to be added and to directly read the volume of sand on the bottom of the tube.

Sand content of the mud is reported in percent by volume. Also reported is the point of sampling (e.g., flowline, shale shaker, suction pit). Solids other than sand may be retained on the screen (e.g., lost circulation material), and the presence of such solids should be noted.

Liquids and Solids Content

A mud retort is used to determine the liquids and solids content of a drilling fluid. Mud is placed in a steel container and heated at high temperature until the liquid components have been distilled off and vaporized (Figure 4.5.6). The vapors are passed through a condenser and collected in a graduated cylinder. The volume of liquids (water and oil) is

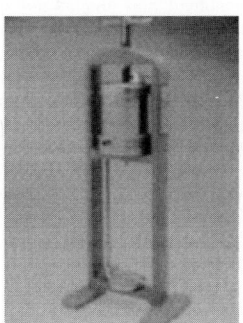

Figure 4.5.4 *API style filter press.*

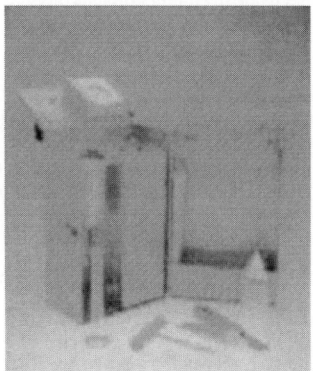

Figure 4.5.6 *Retort kit (10 ml).*

Table 4.5.1 *High- and Low-Gravity Solids in Drilling Fluids*

Specific Gravity of Solids	Barite, Percent by Weight	Clay, Percent by Weight
2.6	0	100
2.8	18	82
3.0	34	66
3.2	48	52
3.4	60	40
3.6	71	29
3.8	81	19
4.0	89	11
4.3	100	0

Figure 4.5.7 *pH Meter.*

then measured. Solids, both suspended and dissolved, are determined by volume as a difference between the mud in container and the distillate in graduated cylinder. Drilling fluid retorts are generally designed to distill 10-, 20-, or 50-ml sample volumes.

For freshwater muds, a rough measure of the relative amounts of barite and clay in the solids can be made (Table 4.5.1). Because both suspended and dissolved solids are retained in the retort for muds containing substantial quantities of salt, corrections must be made for the salt. Relative amounts of high- and low-gravity solids contained in drilling fluids can be found in Table 4.5.1.

pH
Two methods for measuring the pH of drilling fluid are commonly used: (1) a modified colorimetric method using pH paper or strips and (2) the electrometric method using a glass electrode (Figure 4.5.7). The paper strip test may not be reliable if the salt concentration of the sample is high. The electrometric method is subject to error in solutions containing high concentrations of sodium ions unless a special glass electrode is used or unless suitable correction factors are applied if an ordinary electrode is used. In addition, a temperature correction is required for the electrometric method of measuring pH.

The paper strips used in the colorimetric method are impregnated with dyes so that the color of the test paper depends on the pH of the medium in which the paper is placed. A standard color chart is supplied for comparison with the test strip. Test papers are available in a wide range, which permits estimating pH to 0.5 units, and in narrow range papers, with which the pH can be estimated to 0.2 units.

The glass electrode pH meter consists of a glass electrode, an electronic amplifier, and a meter calibrated in pH units. The electrode is composed of (1) the glass electrode, a thin-walled bulb made of special glass within which is sealed a suitable electrolyte and an electrode, and (2) the reference electrode, which is a saturated calomel cell. Electrical connection with the mud is established through a saturated solution of potassium chloride contained in a tube surrounding the calomel cell. The electrical potential generated in the glass electrode system by the hydrogen ions in the drilling mud is amplified and operates the calibrated pH meter.

Resistivity
Control of the resistivity of the mud and mud filtrate while drilling may be desirable to permit enhanced evaluation of the formation characteristics from electric logs. The determination of resistivity is essentially the measurement of the resistance to electrical current flow through a known sample

configuration. Measured resistance is converted to resistivity by use of a cell constant. The cell constant is fixed by the configuration of the sample in the cell and id determined by calibration with standard solutions of known resistivity. The resistivity is expressed in ohm-meters.

Filtrate Chemical Analysis
Standard chemical analyses have been developed for determining the concentration of various ions present in the mud. Tests for the concentration of chloride, hydroxyl, and calcium ions are required to fill out the API drilling mud report. The tests are based on filtration (i.e., reaction of a known volume of mud filtrate sample with a standard solution of known volume and concentration). The end of chemical reaction is usually indicated by the change of color. The concentration of the ion being tested can be determined from a knowledge of the chemical reaction taking place.

Chloride
The chloride concentration is determined by titration with silver nitrate solution. This causes the chloride to be removed from the solution as $AgCl^-$, a white precipitate. The endpoint of the titration is detected using a potassium chromate indicator. The excess Ag present after all Cl^- has been removed from solution reacts with the chromate to form Ag_9CrO_4, an orange-red precipitate. Contamination with chlorides generally results from drilling salt or from a saltwater flow. Salt can enter and contaminate the mud system when salt formations are drilled and when saline formation water enters the wellbore.

Alkalinity and Lime Content
Alkalinity is the ability of a solution or mixture to react with an acid. The *phenolphthalein alkalinity* refers to the amount of acid required to reduce the pH of the filtrate to 8.3, the phenolphthalein end point. The phenolphthalein alkalinity of the mud and mud filtrate is called the P_m and P_f, respectively. The P_f test includes the effect of only dissolved bases and salts, whereas the P_m test includes the effect of both dissolved and suspended bases and salts. The $_m$ and $_f$ indicate if the test was conducted on the whole mud or mud filtrate. The M_f alkalinity refers to the amount of acid required to reduce the pH to 4.3, the methyl orange end point. The methyl orange alkalinity of the mud and mud filtrate is called the M_m and M_f, respectively. The API diagnostic tests include the determination of P_m, P_f, and M_f. All values are reported in cubic centimeters of 0.02 N (normality = 0.02) sulfuric acid per cubic centimeter of sample. The lime content of the mud is calculated by subtracting the P_f from the P_m and dividing the result by 4.

The P_f and M_f tests are designed to establish the concentration of hydroxyl, bicarbonate, and carbonate ions in the aqueous phase of the mud. At a pH of 8.3, the conversion of hydroxides to water and carbonates to bicarbonates

Table 4.5.2 *Alkalinity*

Criteria	OH^- (mg/L)	CO_3^{2-} (mg/L)	HCO_3^- (mg/L)
$P_f = 0$	0	0	$1,220\,M_f$
$2P_f < M_f$	0	$1,200\,P_f$	$1,220\,(M_f - 2P_f)$
$2P_f = M_f$	0	$1,200\,P_f$	0
$2P_f < M_f$	$340\,(2P_f - M_f)$	$1,200\,(M_f - P_f)$	0
$P_f = M_f$	$340\,M_f$	0	0

is essentially complete. The bicarbonates originally present in solution do not enter the reactions. As the pH is further reduced to 4.3, the acid reacts with the bicarbonate ions to form carbon dioxide and water.

$$ml\ N/50\ H_2SO_4\ to\ reach\ pH = 8.3$$

$$CO_3^{2-} + H_2SO_4 \rightarrow HCO_3^- + HSO_4$$

carbonate + acid → bicarbonate + bisulfate

$$OH^- + H_2SO_4 \rightarrow HOH + SO_4 =$$

hydroxyl + acid → water + sulfate salt

The P_f and P_m test results indicate the reserve alkalinity of the suspended solids. As the $[OH^-]$ in solution is reduced, the lime and limestone suspended in the mud will go into solution and tend to stabilize the pH (Table 4.5.2). This reserve alkalinity generally is expressed as an excess lime concentration, in lb/bbl of mud. The accurate testing of P_f, M_f, and P_m are needed to determine the quality and quantity of alkaline material present in the drilling fluid. The chart below shows how to determine the hydroxyl, carbonate, and bicarbonate ion concentrations based on these titrations.

Total Hardness
The total combined concentration of calcium and magnesium in the mud-water phase is defined as total hardness. These contaminants are often present in the water available for use in the drilling fluid makeup. In addition, calcium can enter the mud when anhydrite ($CaSO_4$) or gypsum ($CaSO_4 \cdot 2H_2O$) formations are drilled. Cement also contains calcium and can contaminate the mud. The total hardness is determined by titration with a standard (0.02 N) versenate hardness titrating solution (EDTA). The standard versenate solution contains sodium versenate, an organic compound capable of forming a chelate when combined with Ca^2 and Mg^2.

The hardness test sometimes is performed on the whole mud as well as the mud filtrate. The mud hardness indicates the amount of calcium suspended in the mud and the amount of calcium in solution. This test usually is made on gypsum-treated muds to indicate the amount of excess $CaSO_4$ present in suspension. To perform the hardness test on mud, a small sample of mud is first diluted to 50 times its original volume with distilled water so that any undissolved calcium or magnesium compounds can go into solution. The mixture then is filtered through hardened filter paper to obtain a clear filtrate. The total hardness of this filtrate then is obtained using the same procedure used for the filtrate from the low-temperature, low-pressure API filter press apparatus.

Methylene Blue Capacity (CEC or MBT)
It is desirable to know the cation exchange capacity (CEC) of the drilling fluid. To some extent, this value can be correlated to the bentonite content of the mud. The test is only qualitative because organic material and other clays present in the mud also absorb methylene blue dye. The mud sample is treated with hydrogen peroxide to oxidize most of the

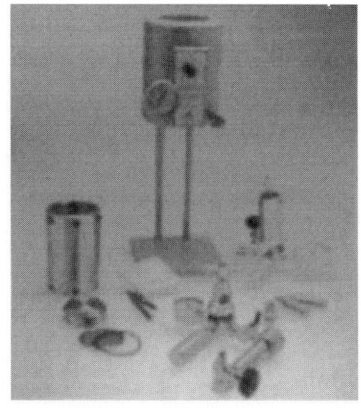

Figure 4.5.8 *HPHT fluid loss testing device.*

organic material. The cation exchange capacity is reported in milliequivalent weights (mEq) of methylene blue dye per 100 ml of mud. The methylene blue solution used for titration is usually 0.01 N, so that the cation exchange capacity is numerically equal to the cubic centimeters of methylene blue solution per cubic centimeter of sample required to reach an end point. If other adsorptive materials are not present in significant quantities, the montmorillonite content of the mud in pounds per barrel is calculated to be five times the cation exchange capacity. The methylene blue test can also be used to determine cation exchange capacity of clays and shales. In the test, a weighed amount of clay is dispersed into water by a high-speed stirrer or mixer. Tiration is carried out as for drilling muds, except that hydrogen peroxide is not added. The cation exchange capacity of clays is expressed as milliequivalents of methylene blue per 100 g of clay.

4.5.3.2 Oil-Base and Synthetic-Base Muds
(Nonaqueous Fluids Testing)
The field tests for rheology, mud density, and gel strength are accomplished in the same manner as outlined for water-based muds. The main difference is that rheology is tested at a specific temperature, usually 120°F or 150°F. Because oils tend to thin with temperature, heating fluid is required and should be reported on the API Mud Report.

Sand Content
Sand content measurement is the same as for water-base muds except that the mud's base oil instead of water should be used for dilution. The sand content of oil-base mud is not generally tested.

HPHT Filtration
The API filtration test result for oil-base muds is usually zero. In relaxed filtrate oil-based muds, the API filtrate should be all oil. The API test does not indicate downhole filtration rates. The alternative high-temperature–high pressure (HTHP) filtration test will generally give a better indication of the fluid loss characteristics of a fluid under downhole temperatures (Figure 4.5.8).

The instruments for the HTHP filtration test consists essentially of a controlled pressure source, a cell designed to withstand a working pressure of at least 1,000 psi, a system for heating the cell, and a suitable frame to hold the cell and the heating system. For filtration tests at temperatures above 200°F, a pressurized collection cell is attached to the delivery tube. The filter cell is equipped with a thermometer well,

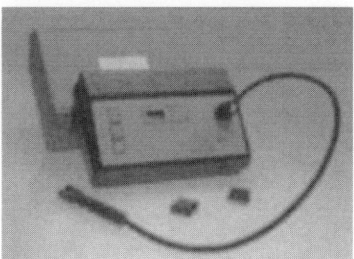

Figure 4.5.9 *Electrical stability meter.*

oil-resistant gaskets, and a support for the filter paper (Whatman no. 50 or the equivalent). A valve on the filtrate delivery tube controls flow from the cell. A nonhazardous gas such as nitrogen or carbon dioxide should be used as the pressure source. The test is usually performed at a temperature of 220 – 350°F and a pressure of 500 psi (differential) over a 30-minute period. When other temperatures, pressures, or times are used, their values should be reported together with test results. If the cake compressibility is desired, the test should be repeated with pressures of 200 psi on the filter cell and 100 psi back pressure on the collection cell. The volume of oil collected at the end of the test should be doubled to correct to a surface area of 7.1 inches.

Electrical Stability

The electrical stability test indicates the stability of emulsions of water in oil mixtures. The emulsion tester consists of a reliable circuit using a source of variable AC current (or DC current in portable units) connected to strip electrodes (Figure 4.5.9). The voltage imposed across the electrodes can be increased until a predetermined amount of current flows through the mud emulsion-breakdown point. Relative stability is indicated as the voltage at the breakdown point

and is reported as the electric stability of the fluid on the daily API test report.

Liquids and Solids Content

Oil, water, and solids volume percent is determined by retort analysis as in a water-base mud. More time is required to get a complete distillation of an oil mud than for a water mud. The corrected water phase volume, the volume percent of low-gravity solids, and the oil-to-water ratio can then be calculated.

The volume oil-to-water ratio can be found from the procedure below:

Oil fraction 100

$$\times \frac{\% \text{ by volume oil or synthetic oil}}{\% \text{ by volume oil or synthetic oil} - \% \text{ by volume water}}$$

Chemical analysis procedures for nonaqueous fluids can be found in the API 13B bulletin [1].

Alkalinity and Lime Content (NAF)

The whole mud alkalinity test procedure is a titration method that measures the volume of standard acid required to react with the alkaline (basic) materials in an oil mud sample. The alkalinity value is used to calculate the pounds per barrel of unreacted, "excess" lime in an oil mud. Excess alkaline materials, such as lime, help to stabilize the emulsion and neutralize carbon dioxide or hydrogen sulfide acidic gases.

Total Salinity (Water-Phase Salinity [WAF] for NAF)

The salinity control of NAF fluids is very important for stabilizing water-sensitive shales and clays. Depending on the ionic concentration of the shale waters and of the mud water phase, an osmotic flow of pure water from the weaker salt concentration (in shale) to the stronger salt concentration (in mud) will occur. This may cause dehydration of the shale and, consequently, affect its stabilization (Figure 4.5.10).

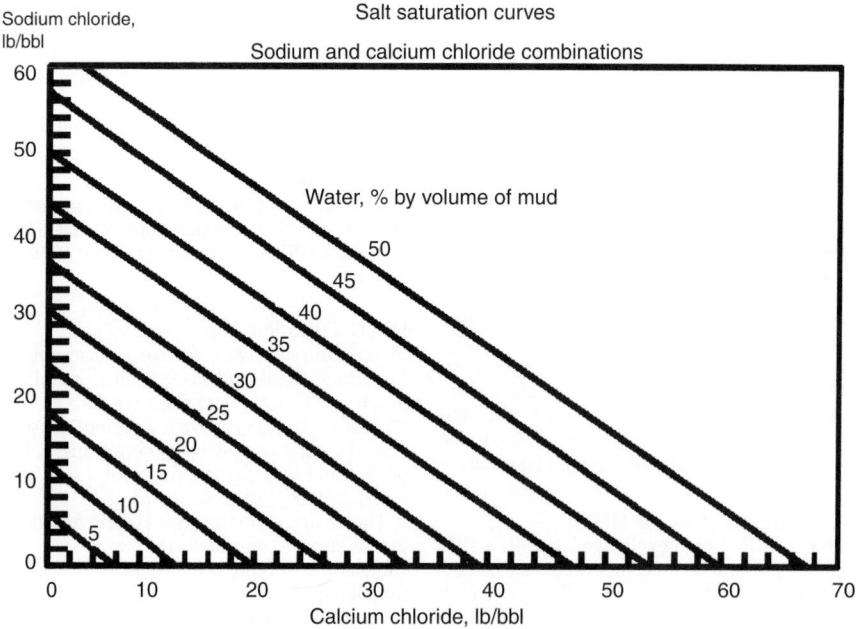

Figure 4.5.10 *Salt saturation curves.*

4.5.3.3 Specialized Tests

Other, more advanced laboratory-based testing is commonly carried out on drilling fluids to determine treatments or to define contaminants. Some of the more advanced analytical tests routinely conducted on drilling fluids include:

Advanced Rheology and Suspension Analysis

FANN 50 — A laboratory test for rheology under temperature and moderate pressure (up to 1,000 psi and 500°F).

FANN 70 — Laboratory test for rheology under high temperature and high pressure (up to 20,000 psi and 500°F).

FANN 75 — A more advanced computer-controlled version of the FANN 70 (up to 20,000 psi and 500°F).

High-Angle Sag Test (HAST)

A laboratory test device to determine the suspension properties of a fluid in high-angle wellbores. This test is designed to evaluate particle setting characteristics of a fluid in deviated wells.

Dynamic HAST

Laboratory test device to determine the suspension properties of a drilling fluid under high angle and dynamic conditions.

4.5.3.4 Specialized Filtration Testing

FANN 90

Dynamic filtration testing of a drilling fluid under pressure and temperature. This test determines if the fluid is properly conditioned to drill through highly permeable formations. The test results include two numbers: the dynamic filtration rate and the cake deposition index (CDI). The dynamic filtration rate is calculated from the slope of the curve of volume versus time. The CDI, which reflects the erodability of the wall cake, is calculated from the slope of the curve of volume/time versus time. CDI and dynamic filtration rates are calculated using data collected after twenty minutes. The filtration media for the FAN 90 is a synthetic core. The core size can be sized for each application to optimize the filtration rate.

Particle-Plugging Test (PPT)

The PPT test is accomplished with a modified HPHT cell to examine sealing characteristics of a drilling fluid. The PPT, sometimes known as the PPA (particle-plugging apparatus), is key when drilling in high-differential-pressure environments.

Aniline Point Test

Determine the aniline point of an oil-based fluid base oil. This test is critical to ensure elastomer compatibility when using nonaqueous fluids.

Particle-Size Distribution (PSD) Test

The PSD examines the volume and particle size distribution of solid sin a fluid. This test is valuable in determining the type and size of solids control equipment that will be needed to properly clean a fluid of undesirable solids.

Luminescence Fingerprinting

This test is used to determine if contamination of a synthetic-based mud has occurred with crude oil during drilling operations.

Lubricity Testing

Various lubricity meters and devices are available to the industry to determine how lubricous a fluid is when exposed to steel or shale. In high-angle drilling applications, a highly lubricious fluid is desirable to allow proper transmission of weight to the bit and reduce side wall sticking tendencies.

4.5.3.5 Shale Characterization Testing

Capillary Suction Time (CST)

Inhibition testing looks at the inhibitive nature of a drilling fluid filtrate when exposed to formation shale samples. The CST is one of many tests that are run routinely on shale samples to optimize the mud chemistry of a water-base fluid.

Linear-Swell Meter (LSM)

Another diagnostic test to determine the inhibitive nature of a drilling fluid on field shale samples. The LSM looks at long-term exposure of a fluid filtrate to a formation shale sample. Test times for LSM can run up to 14 days.

Shale Erosion

Shale inhibition testing looks at the inhibitive nature of a drilling fluid and examines the erodability of a shale when exposed to a drilling fluid. Various tests procedures for this analytical tool.

Return Permeability

Formation damage characterization of a fluid through an actual or simulated core is accomplished with the return permeability test. This test is a must when designing specialized reservoir drilling fluids to minimize formation impairment.

Bacteria Testing

Tests for the presence of bacteria in water-base muds; this is especially important in low-pH fluids because bacterial growth is high in these types of fluids.

Static Aging

The aging test is used to determine how bottom-hole conditions affect mud properties. Aging cells were developed to aid in predicting the performance of drilling mud under static, high-temperature conditions. If the bottom-hole temperature is greater than 212°F, the aging cells can be pressurized with nitrogen, carbon dioxide, or air to a desired pressure to prevent boiling and vaporization of the mud.

After the aging period, three properties of the aged mud are determined before the mud is agitated or stirred: shear strength, free oil (top oil separation in NAF), and solids setting. Shear strength indicates the gelling tendencies of fluid in the borehole. Second, the sample should be observed to determine if free oil is present. Separation of free oil is a measure of emulsion instability in the borehole and is expressed in 32nd of an inch. Setting of mud solids indicates the formation of a hard or soft layer or sediment in the borehole. After the unagitated sample has been examined, the sample is sheared, and the usual tests for determining rheological and filtration properties are performed.

4.5.3.6 Drilling Fluid Additives

Each drilling fluid vendor provides a wide array of basic and specialty chemicals to meet the needs of the drilling industry. The general classification of drilling fluid additives below is based on the definitions of the International Association of Drilling Contractors (IADC):

A. Alkalinity or pH control additives are products designed to control the degree of acidity or alkalinity of a drilling fluid. These additives include lime, caustic soda, and bicarbonate of soda.

B. Bactericides reduce the bacteria count of a drilling fluid. Para-formaldehyde, caustic soda, lime, and starch are commonly used as preservatives.

C. Calcium removers are chemicals used to prevent and to overcome the contaminating effects of anhydride and gypsum, both forms of calcium sulfate, which can wreck the effectiveness of nearly any chemically treated mud. The most common calcium removers are caustic soda, soda ash, bicarbonate of soda, and certain polyphosphates.

D. Corrosion inhibitors such as hydrated lime and amine salts are often added to mud and to air-gas systems. Mud containing an adequate percentage of colloids, certain emulsion muds, and oil muds exhibit, in themselves, excellent corrosion-inhibiting properties.

E. Defoamers are products designed to reduce foaming action, particularly that occurring in brackish water and saturated saltwater muds.

F. Emulsifiers are used for creating a heterogeneous mixture of two liquids. These include modified lignosulfonates, certain surface-active agents, anionic and nonionic (negatively charged and noncharged) products.

G. Filtrate, or fluid loss, reducers such as bentonite clays, sodium carboxymethyl cellulose (CMC), and pregelatinized starch serve to cut filter loss, a measure of the tendency of the liquid phase of a drilling fluid to pass into the formation.

H. Flocculants are used sometimes to increase gel strength. Salt (or brine), hydrated lime, gypsum, and sodium tetraphosphates may be used to cause the colloidal particles of a suspension to group into bunches of "flocks," causing solids to settle out.

I. Foaming agents are most often chemicals that also act as surfactants (surface-active agents) to foam in the presence of water. These foamers permit air or gas drilling through water-production formations.

J. Lost circulation materials (LCM) include nearly every possible product used to stop or slow the loss of circulating fluids into the formation. This loss must be differentiated from the normal loss of filtration liquid and from the loss of drilling mud solids to the filter cake (which is a continuous process in an open hole).

K. Extreme-pressure lubricants are designed to reduce torque by reducing the coefficient of friction and thereby increase horsepower at the bit. Certain oils, graphite powder, and soaps are used for this purpose.

L. Shale control inhibitors such as gypsum, sodium silicate, chrome lignosulfonates, as well as lime and salt are used to control caving by swelling or hydrous disintegration of shales.

M. Surface-active agents (surfactants) reduce the interfacial tension between contacting surfaces (e.g., water—oil, water—solid, water—air); these may be emulsifiers, de-emulsifiers, flocculants, or deflocculents, depending upon the surfaces involved.

N. Thinners and dispersants modify the relationship between the viscosity and the percentage of solids in a drilling mud and may further be used to vary the gel strength and improve "pumpability." Tannins (quebracho), various polyphosphates, and lignitic materials are chosen as thinners or as dispersants, because most of these chemicals also remove solids by precipitation or sequestering, and by deflocculaton reactions.

O. Viscosifiers such as bentonite, CMC, Attapulgite clays, sub-bentonites, and asbestos fibers are employed in drilling fluids to ensure a high viscosity–solids ratio.

P. Weighting materials, including barite, lead compounds, iron oxides, and similar products possessing extraordinarily high specific gravities, are used to control formation pressures, check caving, facilitate pulling dry drill pipe on round trips, and aid in combating some types of circulation loss.

The most common commercially available drilling mud additives are published annually by *World Oil*. The listing includes names and descriptions of more than 2,000 mud additives.

4.5.3.7 Clay Chemistry

Water-base drilling fluids normally contain a number of different types of clays. Most of the clays are added to attain certain physical properties (e.g., fluid loss, viscosity, yield point) and eliminate hole problems.

The most common clays incorporated into the drilling fluid from the formation (in the form of drill solids) are calcium montmorillonite, illites, and kaolinites. The most used commercial clay is sodium montmorillonite.

Bentonite is added to water-base drilling fluids to increase the viscosity and gel strength of the fluid. This results in quality suspension properties for weight materials and increases the carrying capacity for removal of solids from the well. The most important function of bentonite is to improve the filtration and filter cake properties of the water-base drilling fluid.

Clay particles are usually referred to as clay platelets or sheets. The structure of the sodium montmorillonite platelet has sheets consisting of three layers. The platelet, if looked at under an electron microscope, reveals that the sections are honeycombed inside the three layers. The three-layered (sandwich-type) sheet is composed of two silica tetrahedral layers with an octahedral aluminum center core layer between them. The section layers are bonded together in a very intricate lattice-type structure.

Cations are absorbed on the basal surface of the clay crystals to form a natural forming structure. This occurred in the earth over a period of 100 million years. The positive sodium or calcium cations compensate for the atomic substitution in the crystal structure (the isomorphic substitution that took place in forming of the clay). This is the primary way that sodium clays are differentiated from calcium clays.

Sodium montmorillonite absorbs water through expansion of the lattice structure. There are two mechanisms by which hydration can occur:

1. Between the layers (osmotic). The exposure of the clay to water vapor causes the water to condense between the layers, expanding them. The lower the concentration of sodium and chloride in the water, the greater the amount of water that can be absorbed into the clay lattice structure.

2. Around layers (crystalline). There is a layer of water that surrounds the clay particles (a cloud of Na^+ with water molecules held to the platelet by hydrogen bonding to the lattice network by the oxygen on the face of the platelet). The structure of water and clay is commonly called an envelope. (It must be remembered that the water envelope has viscosity.)

Aggregation

Clays are said to be in the aggregated state when the platelets are stacked loosely in bundles. When clay is collapsed and its layers are parallel, the formation is like a deck of cards stacked in a box. This is the state of sodium bentonite in the sack having a moisture content of 10%. When added to freshwater (does not contain a high concentration of chlorides), diffuses of water into the layers occurs, and swelling or dispersion results.

In solutions with high chloride concentrations, the double layer is compressed still further, and aggregation occurs.

Consequently, the size of the particle is reduced, and the total particle area per unit volume decreases. This occurs because the chloride ion has a strong bond with the H_2O, and free water is not available to enter the clay and hydrate effectively. In muds in which the clay is aggregated, the viscosity is low.

The relationship between the type and concentration of the salt in the water determines the point at which aggregation (inhibition) will occur:

- Sodium chloride (NaCl) 400 mEq/L
- Calcium chloride ($CaCl_2$) 20 mEq/L
- Aluminum chloride ($AlCl_2$) 20 mEq/L

It may be inferred that the higher the chloride content and the higher the valence of the cation salts in solution, the more the clay will be inhibited from swelling. It is also true that the tendency of the dispersed clays to revert to an aggregated (inhibited) state is measurable.

Dispersion

The subdivision of particles from the aggregated state in a fluid (water) to a hydrated colloid particle is the dispersing of that particle. In freshwater dispersion, the clay platelets drift about in an independent manor or in very small clusters. There are times when the platelets configure in random patterns. This usually occurs in a static condition and is termed *gel strength of dispersed day*. The random movement and drifting of a positively charged edge toward a negatively changed face happens slowly in a dispersed state. When bentonite is in a dispersed state, the positive ion cloud presents an effective "shield" around the clay and sometimes slows this effect. The ionized Na^+ surrounds the clay to form a weak crystalline barrier.

Dispersed clay state is characterized by

- High viscosity
- High gel strength
- Low filtrate

Flocculation (NaCl)

The most common cause of flocculation of clays in the field is the incorporation of NaCl in to a fresh water mud. When the Na^+ content is raised toward 1%, the water becomes more positively charged. The ionized envelope cloud that "protected" the platelet is of a lower charge than the bulk water. The positive Al^{3+} edge joins with the oxygen face, and the drift of edge to face is accelerated.

The viscosity rises, and water loss is uncontrollable when the clay flocculates edge to face in a "House of Cards structure," and the increase in viscosity and water loss is dramatic. As the NaCl content increases to 5%, the free water is tied up by the chloride ion, and the ion and the clays collapse and revert to the aggregated state. The water is removed from the clay platelet body.

When the NaCl content increases to 15% to 30% by weight, the agglomerates flocculate into large edge to face groups. This leads to extreme viscosities and very poor fluid loss control. This also depends on the solids content. In diluted suspensions, the viscosity usually is reduced by increasing salt concentrations, and clay platelets are in the aggregated state. Viscosity will go through a "hump."

Deflocculation (Chemical Dispersion)

One way to deflocculate, or chemically disperse, a clay platelet is with a large molecule having many carboxyl and sulfonate anions at scattered intervals on the cellulose chain. In deflocculated or chemically dispersed muds, the viscosity will be lower than it was in the flocculated state.

Lignosulfonate works to deflocculate by the anionic charges that latch onto the positive edges of the clay platelet.

The remainder of this huge (flat) cellulosic molecule is repelled from the negative clay face and rolls out from the edges.

The edge-to-face flocculation that occurred becomes virtually impossible. The polyanionic encapsulator can be rendered neutral if the pH drops below 9.5. The NaCl flocculant is still present in the solution, but its flocculating effects are rendered ineffective if the pH is maintained above 9.5.

Flocculation (Calcium)

When calcium is induced into a drilling fluid, its solubility depends on the pH of the water in the fluid. The double-positive charge on the calcium ion will attract itself to the face of the bentonite platelet at an accelerated rate, because this attraction is far superior to the sodium's ability to retain its place on the clay face. The divalent calcium ions will still partially hydrate, but the amount of water is less around the clay platelet. This will allow flocculation to occur much faster, because there is little water structure around the clay in this situation.

Calcium can cause flocculation in the same manner as salt (NaCl) in that edge-to-face groupings are formed. Calcium is a divalent cation, so it holds onto two platelet faces, which causes large groups to form, and then the edge-to-face grouping to take hold. Because calcium (Ca^{2+}) has a valence of 2, it can hold two clay platelets tightly together, and the flocculation reaction starts to happen at very low concentrations. To achieve flocculation with salt (NaCl), it takes 10 times the concentration for the edge-to-face groupings to form.

In the flocculated state, a dispersant (thinner) will work to separate the flocculating ions and encapsulate the platelets by mechanical shear. This is a short-term answer to the problem, however, because the contaminating ion is still active in the system, and it must be reduced to a normal active level for drilling to continue.

Deflocculation (Calcium Precipitation)

The most effective way to remove the flocculating calcium ion from the system is to chemically precipitate it. Two common chemicals can be used to accomplish removal of the calcium ion. They are Na_2Co_3 (soda ash) and $NaHCO_3$ (bicarbonate of soda). Because calcium is lodged between two platelets and holding them together, the two chemicals will, with mechanical help, bond together with the flocculant calcium as shown in the formula below:

$$Ca^{2+}(OH)_2 + Na_2HCO_3 \rightarrow CaCO_3 + NaOH + H_2O$$

Lime + Sodium Bicarbonate
$\rightarrow$ Calcium Carbonate + Caustic + Water

$$CaSO_4 + Na_2CO_3 \rightarrow CaCO_3 + Na_2SO_4$$

Calcium sulfate + Sodium carbonate
$\rightarrow$ Calcium carbonate + Sodium sulfate

In the previous chemical equation, calcium is precipitated and rendered inert. There is no longer a possible flocculating calcium ion to deal with.

Inhibition (NaCl)

When a water solution contains more than 12,000 mg/l of NaCl, it can inhibit clays from swelling or hydrating. This happens because the sodium ion content is high in the water, and the sodium ions on the clay face cannot leave to allow space for the water to enter the clay platelet. The chloride ion has an ability to tightly hold onto water molecules, which leaves few free ions to envelope or surround the clay. When the clay (aggregation) platelet does not hydrate, the state is the same as it is in the sack. In this instance, the ion is controlling the swelling of clays and is referred to as *inhibition*.

Controlling these various clay states in water-base drilling fluids is important for the success of any well using this chemistry. It can be said that flocculation causes and increases viscosity and that aggregation and deflocculation decrease viscosity.

4.5.3.8 WATER-BASE MUDS

A water-base drilling fluid is one that has water as its continuous or liquid phase. The types of drilling fluids are briefly described in the following sections.

Freshwater muds are generally lighly treated or untreated muds having a liquid phase of water, containing small concentrations of salt, and having a pH ranging from 8.0 to 10.5. Freshwater muds include the following types.

Spud Muds

These muds are prepared with available water and appropriate concentrations of bentonite and/or premium commercial clays. They are generally untreated chemically, although lime, cement, or caustic soda is occasionally added to increase viscosity and give the mud a fluff to seal possible lost return zones in unconsolidated upper hole surface formations. Spud muds are used for drilling the surface hole. Their tolerance for drilled solids and contaminants is very limited.

Natural Mud

Natural or native muds use native drilled solids incorporated into the mud for viscosity, weight, and fluid loss control. They are often supplemented with bentonite for added stability and water loss control. Surfactants can be used to aid in controlling mud weight and solids buildup. Natural muds are generally used in top hole drilling to mud-up or to conversion depth. They have a low tolerance for solids and contamination.

Saltwater Muds

Muds ordinarily are classified as saltwater muds when they contain more than 10,000 mg/L of chloride. They may be further classified according to the amount of salt present and/or the source of makeup water (see Table 4.5.3):

Amount of chloride in mg/L

1. Saturated salt muds (315,000 ppm as sodium chloride)
2. Salt muds (over 10,000 mg/L chloride but not saturated)

Source of make-up water

A. Brackish waer
B. Sea Water

Saltwater muds may be purposely prepared, or they may result from the use of salty makeup water, from drilling into salt domes or stringers, or when saltwater flows are encountered. Saltwater muds include the following types.

Table 4.5.3 *Seawater Composition*

Constituent	Parts per Million	Equivalent Parts per Million
Sodium	10440	454.0
Potassium	375	9.6
Magnesium	1270	104.6
Calcium	410	20.4
Chloride	18970	535.0
Sulfate	2720	57.8
Carbon dioxide	90	4.1
Other constituents	80	n/a

Seawater or Brackish Water Muds

These muds are prepared with available makeup water, both commercial and formation clay solids, caustic soda, and lignite and/or a lignosulfonate. CMC is usually used for fluid loss control, although concentration of lignites and lignosulfonates are also often used for this purpose. Viscosity and gel strength are controlled with caustic soda, lignosulfonate, and/or lignites. Soda ash is frequently used to lower the calcium concentration. CMC or lignosulfonates are used for water loss control, and pH is controlled between 8.5 to 11.0 with caustic soda. Seawater muds and brackish or hard water muds are used primarily because of the convenience of makeup water, usually open sea or bays. The degree of inhibitive properties varies with the salt and calcium concentration in the formulated fluid.

Saturated Salt Muds

Saturated salt water (natural or prepared) is used as makeup water in these fluids. Prehydrated bentonite (hydrated in freshwater) is added to give viscosity, and starch is commonly used to control fluid loss. Caustic soda is added to adjust the pH, and lignosulfonates are used for gel strength control. Occasionally, soda ash may be used to lower filtrate calcium and adjust the pH. Saturated salt muds are used to drill massive salt sections (composed mainly of NaCl) to prevent washouts and as a work-over or completion fluid. Freshwater bentonite suspensions are converted by adding NaCl to reach saturation. Conversion is carried out by diluting the freshwater mud to reduce the viscosity "hump" seen in breakovers. Saturated salt muds usually are used at mud weights below 14.0 lb/gal.

Composition of NaCl mud

- Brine NaCl
- Density — salt, barite, calcium carbonate or hematite
- Viscosity — CMC HV, Prehydrated bentonite, XC-polymer (xanthan gum)
- Rheology — lignosulfonate
- Fluid Loss — CMC LV or PAC (polyanionic cellulose)
- pH – P_f (alkalinity) — caustic potash or caustic soda

Chemically Treated Mud (No Calcium Compounds)

This type of mud is made up of a natural mud that has been conditioned with bentonite and treated with caustic soda and lignite or lignosulfonate (organic thinner). No inhibiting ions are found in this type of fluid.

Lignite/Lignosulfonate Mud

This fluid is prepared from freshwater and conditioned with bentonite. Lignosulfonate is added as a thinner and lignite for filtration control and increased temperature stability. CMC or PAC may be used for additional filtration control when the bottom-hole temperature does not exceed 121°C (250°F). This type of mud is applied at all mud weights and provides a relatively low pH system (pH values for calcium lignosulfonates will be 10.0–11.0). This type of fluid is stable at reasonably high temperatures (325°F) and has good resistance to contamination.

Calcium Treated Muds

Calcium-treated fluids are prepared from any low or high pH mud by the addition of appropriate amounts of lime or gypsum, caustic soda, and thinner (lignite or lignosulfonate). Calcium-treated muds include lime and gypsum muds.

Lime Muds

Lime muds include low- and high-lime muds. They are prepared from available muds by adding calcium lignosulfonate, lignite, caustic soda or KOH, lime, and a filtration-control

Table 4.5.4 *Gaseous Drilling Mud Systems*

Type of Mud	Density, ppg	PH	Temp. Limit °F	Application Characteristics
Air/gas	0	—	500	High-energy system. Fastest drilling rate in dry, hard formations. Limited by water influx and hole size.
Mist	0.13–0.8	7–11	300	High-energy system. Fast penetration rates. Can handle water intrusions. Stabilizes unstable holes (mud misting).
Foam	0.4–0.8	4–10	400	Very-low-energy system. Good penetration rates. Excellent cleaning ability regardless of hole size. Tolerates large water influx.

material, PAC or starch. Caustic soda is used to maintain the filtrate alkalinity (P_f values) and lime to control the mud alkalinity (P_m values) and excess lime. Lime muds offer resistance to salt, cement, or anhydrite contamination even at high mud weights.

Gypsum Mud
Commonly called "gyp muds," they are prepared from freshwater and conditioned with bentonite or from available gel and water mud. Caustic soda is added for pH control. Gypsum, lignosulfonate, and additional caustic soda are added simultaneously to the mud. CMC may be added for filtration control. This fluid is used for drilling in mildly reactive shale or where gypsum or anhydrite must be drilled. It resists contamination from cement or salt. Use is limited by the temperature stability of the filtration control materials, CMC (250°F ±)

4.5.3.9 Special Muds
In addition to the most common mud systems discussed previously, there are other muds that do not fall neatly into one category or another in the classification scheme.

Low-Density Fluids and Gaseous Drilling Mud (Air-Gas Drilling Fluids)
The basic gaseous drilling fluids and their characteristics are presented in Table 4.5.4.

This system involves injecting air or gas downhole at the rates sufficient to attain annular velocity of 2,000 to 3,000 ft/min. Hard formations that are relatively free from water are most desirable for drilling with air-gas drilling fluids. Small quantities of water usually can be dried up or sealed off by various techniques.

Air-gas drilling usually increases drilling rate by three or four times over that when drilling with mud, as well as one-half to one-fourth the number of bits are required. In some areas, drilling with air is the only solution; these are (1) severe lost circulation, (2) sensitive producing formation that can be blocked by drilling fluid (skin effect), and (3) hard formations near the surface that require the use of an air hammer to drill.

There are two major limitations with using air as a drilling fluid: large volumes of free water and size of the hole. Large water flows generally necessitate converting to another type of drilling fluid (mist or foam). Size of the hole determines a volume of air required for good cleaning. Lift ability of air depends annular velocity entirely (no viscosity or gel strength). Therefore, large holes require an enormous volume of air, which is not economical.

Mist Drilling Fluids
Misting involves the injection of air and mud or water and foam-making material. In the case of "water mist," only enough water and foam is injected into the air stream to clear the hole of produced fluids and cuttings. This unthickened water can cause problems due to the wetting of the exposed formation, which can result in sloughing and caving of water-sensitive shale into the wellbore. Mud misting, on the other hand, coats the walls of the hole with a thin film and has a stabilizing effect on water-sensitive formations. A mud slurry that has proved adequate for most purposes consists of 10 ppb of bentonite, 1 ppb of soda ash, and less than 0.5 ppb of foam-stabilizing polymer such as high-viscosity CMC. If additional foam stability is needed, additional foamer is used.

Nondispersed (Low-Solids) Muds
The term *low-solids mud* covers a wide variety of mud types, including clear water (fresh, salt, or brine), oil-in-water emulsions, and polymer or biopolymer fluids (muds with polymer and no other additives).

Extended Bentonite Muds
Low-solids nondispersed mud is generally prepared from freshwater with little or no drilled solids and bentonite, along with a dual-action polymer for extending the bentonite and flocculating drilled solids. This type of mud is designed for low-solids content and to have low viscosity at the bit for for high drilling rates. The polymers used greatly increases the viscosity contributed by the bentonite and serve as flocculants for native clay solids, making them easier to remove by solids-control equipment. These polymers or bentonite extenders permit the desired viscosity to be maintained with about half of the amount of bentonite normally required. No deflocculant is used, so a flocculated system is maintained. The flocculation and lower solids content permit the mud to have a relatively low viscosity at the bit and at the bottom of the hole, where shear rates are high, and a relatively high viscosicy at the lower shear rates in the annulus for good hole cleaning. One problem with this type of fluid is that filtration rates are fairly high, because the solids are flocculated and their quantity is low. This means that they do not pack tightly in the filter cake. Sodium polyacrylates or small amounts of CMC may be added for filtration control.

The temperature limitation of extended bentonite fluids is 200–275F°. Other benefits include improved hydraulics and less wear on bits and pump parts.

Inhibitive Salt/Polymer Muds
An *inhibitive mud* is one that does not appreciably alter a formation once it has been cut by the bit. The term covers a

large number of mud systems, among them saltwater muds with more than 10,000 mg/L of sodium chloride, calcium-treated muds (lime and gyp), and surfactant-treated muds. Under the category of inhibiting salt/polymer muds, however, we are speaking specifically about muds containing inhibitive salts such as KCl, NaCl, or diammonium phosphate along with complex, high-molecular-weight polymers. In thse muds, prehydrated bentonite and polymer are added for viscosity and gel strength, polyanionic cellulose (PAC) or CMC are added for fluid loss control, and corrosion inhibitors and oxygen scavengers often are used to protect tubular goods. These muds are used for drilling and protecting water-sensitive formations and are good for minimizing formation damage due to filtrate invasion when the formation contains hydratable clay solids. Good hole cleaning and shear thinning are characteristics of these fluids. High-solids concentrations cannot be tolerated, however, making good solids control very important. Temperature limitations of 200–250°F are also characteristic. Among the muds of this type is KCl/lime mud. This mud system uses pre-hydrated bentonite or KCl for inhibition, lignosulfonate and/or lignite as a thinner, KOH (caustic potash) or caustic soda for alkalinity, lime for alkalinity and inhibition, and polymers such as CMC or PAC for filtration control.

Surfactant Muds

Surfactant muds were developed primarily to replace calcium-treated muds when high temperature becomes a problem. The term *surfactant* means surface-acting agent, or a material that is capable of acting on the surface of a material. In drilling muds, surfactants are additives that function by altering the surface properties of the liquid and solid phases of the mud or by imparting certain wetting characteristics to the mud. The composition of the surfactant mud system tends to retard hydration or dispersion of formation clays and shales. The pH of these muds is kept from 8.5 to 10.0 to give a more stable mud at higher temperatures.

The surfactant mud usually encountered is a lignite surfactant mud system. This mud is made up from freshwater using bentonite, lignite, and the surfactant. Small amounts of defoamer may be required with the addition of the lignite. The pH of this mud is maintained within closely fixed limited (8.5 – 10.0) for maximum solubility of the thinner (lignite). Tolerance to salt, gyp, and cement contamination is limited. To retain satisfactory flow properties at high temperatures, the clay content of the mud must be kept low (1–1.6 CEC capacity) through the use of dilution and solids-control equipment. The combination of lignite with surfactant in this mud enables its use at extremely high bottom-hole temperatures. This is due to the temperature stability of lignite and the effect of the surfactant in providing viscosity control and minimizing gel strength development at higher temperatures.

High-Temperature Polymer Muds

Development of a high-temperature polymer system evolved from a need for a mud system with low solids and nondispersive performance at higher temperatures.
System capabilities:

- Good high-temperature stability
- Good contaminant tolerance
- Can formulate temperature stable nondispersed polymer mud system
- Can be used in wide variety of systems for good shale stability

- Minimum dispersion of cuttings and clays
- Flexibility of general application

Application of the high-temperature polymer system primarily consists of five products: (1) polymeric deflocculant, (2) acrylamide copolymer, (3) bentonite, (4) caustic soda or potassium hydroxide, and (5) oxygen scavenger. Barite, calcium carbonate, or hematite is then used as a weighting agent.

The polymeric deflocculant is a low-molecular-weight, modified polyacrylate deflocculant used to reduce rheological properties of the system. If differs from lignosulfonates in that it does not require caustic soda or an alkaline environment to perform. Limited amounts of the polymer may be used in low-mud-weight systems, but larger additions will be needed at higher mud weights and when adding barite to increase the fluid density.

The backbone of the system is an acrylamide copolymer used to control fluid loss. In freshwater systems, 1 to 2 lb/bb will be the range required to control the API fluid loss. In seawater systems, the concentration will range from 4 to 5 lb/bbl. HPHT fluid loss can also be controlled with the polymer. It is not affected by salinity or moderate levels of calcium. At higher concentrations of contaminants, some increase in viscosity will result.

Caustic soda and/or potassium hydroxide are alkaline agents used to control the pH of the system. Either is used to maintain the system pH between 8.3 and 9.0.

Oxygen scavengers serve two purposes in this system. First, because of the low pH characteristic of the system, it should be added to protect the drill pipe. (Run corrosion rings in the drill pipe to determine treatment rates for the corrosion that may be occurring.) Second, as the temperature of the mud exceeds 300°F, any oxygen present will react with the polymers and reduce their efficiency. Additional treatment will be required to replace affected or degraded polymers.

New-Generation Water-Based Chemistry

Several companies have developed water-base fluids that provide the inhibition formerly seen only when using oil-base fluids. Novel chemistry such as sodium silicates, membrane-efficient water-base muds, and highly inhibiting encapsulating polymers make these new systems unique and high in performance. Product development in the area of highly inhibitive polymers will no doubt result in the total replacement of invert emulsions. The need to provide more environmentally acceptable products drive the research and development of many drilling fluids by vendors around the world.

Oil-Base Mud Systems and Nonaqueous Fluids (NAF)

Oil-base muds are composed of oil as the continuous phase, water as the dispersed phase, emulsifiers, wetting agents, and gellants. Other chemicals are used for oil-base mud treatment, such as degellants, filtrate reducers, and weighting agents.

The oil for an oil-base mud can be diesel oil, kerosene, fuel oil, selected crude oil, mineral oil, vegetable esters, linear paraffins, olefins, or blends of various oils. There are several desired performance requirements for any oil:

- API gravity = 36° – 37°
- Flash point = 180°F or above
- Fire point = 200°F or above
- Aniline point = 140°F or above

Emulsifiers are very important in oil-base mud because water contamination on the drilling rig is very likely and can be deterimental to oil mud. Thinners, on the other hand, are far more important in waterbase mud than in oil-base mud; oil is dielectric, so there are no interparticle electric forces to be nullified.

The water phase of oil-base mud can be freshwater or various solutions of calcium chloride ($CaCl_2$), soidum chloride (NaCl), or formates. The concentration and composition of the water phase in oil-base mud determines its ability to solve the hydratable shale problem.

The external phase of oil-base mud is oil and does not allow the water to contact the formation; the shales are thereby prevented from becoming wet with water and dispersing into the mud or caving into the hole.

The stability of an emuslion mud is an important factor that has to be closely monitored while drilling. Poor stability results in coalescence of the dispersed phase, and the emulsion will separate into two distinct layers. Presence of any water in the HPHT filtrate is an indication of emulsion instability.

The advantages of drilling with emulsion muds rather than with water-base muds are

- High penetration rates
- Reduction in drill pipe torque and drag
- Less bit balling
- Reduction in differential sticking

Oil-base muds are generally expensive and should be used when conditions justify their application. As in any situation, a cost-benefit analysis should be done to ensure that the proper mud system is selected. Oil-based fluids are well suited for the following applications:

- Drilling troublesome shales that swell (hydrate) and disperse (slough)
- Drilling deep, high-temperature holes in which water-base muds solidify
- Drilling water-soluble formations such as salt, anhydride, camallite, and potash zones
- Drilling the producing zones

For additional applications, oil muds can be used

- As a completion and workover fluid
- As a spotting fluid to relieve stuck pipe
- As a packer fluid or a casing pack fluid

Drilling in younger formations such as "gumbo," a controlled salinity invert fluid is ideally suited. Gumbo, or plastic, flowing shale encountered in offshore Gulf of Mexico, the Oregon coast, Wyoming, West Africa, Venezuela, the Middle East, Western Asia, and the Sahara desert, benefits from a properly designed salinity program. Drilling gumbo with water-base mud shale disperses into the mud rapidly, which reduces the drilling rate and causes massive dilution of the mud system to be required. In some cases, the ROP must be controlled to prevent plugging of the flowline with hydrated "gumbo balls." Solids problems also are encountered with water-based fluid drilling gumbo such as bit balling, collar balling, stuck pipe, and shaker screens plugging.

Properly designed water-phase salinity invert fluids will pull water from the shale (through osmosis), which hardens the shale and stabilizes it for long-term integrity.

Generally, oil-base mud is to delivered to the rig mixed to the desired specifications. In some cases, the oil-base mud can be mixed on location, but this process can cost expensive rig time. In the latter case, the most important principles are (1) to ensure that ample energy in the form of shear is applied to the fluid and (20 to strictly follow a prescribed order of mixing. The following mixing procedure is usually recommended:

1. Pump the required amount of oil into the tank.
2. Add the calculated amounts of emulsifiers and wetting agent. Stir, agitate, and shear these components until adequate dispersion is obtained.
3. Mix in all of the water or the $CaCl_2$-water solution that has been premixed in the other mud tank. This requires shear energy. Add water slowly through the submerged guns; operation of a gun nozzle at 500 psi is considered satisfactory. After emulsifying all the water into the mud, the system should have a smooth, glossy, and shiny appearance. On close examination, there should be no visible droplets of water.
4. Add all the other oil-base mud products specified.
5. Add the weighting material last; make sure that there are no water additions while mixing in the weighting material (the barite could become water wet and be removed by the shale shakers).

When using an oil-base mud, certain rig equipment should be provided to control drilled solids in the mud and to reduce the loss of mud at the surfaces:

- Kelly valve—a valve installed between the Kelly and the drill pipe will save about one barrel per connection.
- Mud box—to prevent loss of mud while pulling a wet string on trips and connections; it should have a drain to the bell nipple and flow line.
- Wiper rubber—to keep the surface of the pipe dry and save mud.

Oil-base mud maintenance involves close monitoring of the mud properties, the mud temperature, and the chemical treatment (in which the order of additions must be strictly followed). The following general guidelines should be considered:

A. The mud weight of an oil mud can be controlled from 7 lb/gal (aerated) to 22 lb/gal. A mud weight up to 10.5 lb/gal can be achieved with sodium chloride or with calcium chloride. For densities above 10.5 lb/gal, barite, hematite, or ground limestone can be used. Calcium carbonate can be used to weight the mud up to 14 lb/gal; it is used when an acid-soluble solids fraction is desired, such as in drill-in fluids or in completion/workover fluids. Iron carbonate may be used to obtain weights up to 19.0 lb/gal when acid solubility is necessary (Table 4.5.5).
B. Mud rheology of oil-base mud is strongly affected by temperature. API procedure recommends that the mud

Table 4.5.5 *Estimated Requirements for Oil Mud Properties*

Mud Weight, ppg	Plastic Viscosity, cP	Yield Point, lbs/sq ft^2	Oil-Water Ratio	Electrical Stability
8–10	15–30	5–10	65/35–75/25	200–300
10–12	20–40	6–14	75/25–80/20	300–400
12–14	25–50	7–16	80/20–85/15	400–500
14–16	30–60	10–19	85/15–88/12	500–600
16–18	40–80	12–22	88/15–92/8	Above 600

temperature be reported along with the funnel viscosity. The general rule for maintenance of the rheological properties of oil-base muds is that the API funnel viscosity, the plastic viscosity, and the yield point should be maintained in a range similar to that of comparable-weight water muds. Excessive mud viscosity can be reduced by dilution with a base oil or with specialized thinners. Insufficient viscosity can be corrected by adding water (pilot testing required) or by treatment with a gallant, usually an organophilic clay or surfactant.

C. Low-gravity solids contents of oil-base muds should be kept at less than 6% v/v. Although oil muds are more tolerant for solids contamination, care must be taken to ensure that solids loading does not exceed the recommended guidelines. A daily log of solids content enables the engineer to quickly determine a solids level at which the mud system performs properly.

D. Water-wet solids is a very serious problem; in severe cases, uncontrollable barite setting may result. If there are any positive signs of water-wet solids, a wetting agent should be added immediately. Tests for water-wet solids should be run daily.

E. Temperature stability and emulsion stability depend on the proper alkalinity maintenance and emulsifier concentration. If the concentration of lime is too low, the solubility of the emulsifier changes, and the emulsion loses its stability. Lime maintenance has to be established and controlled by alkalinity testing. The recommended range of lime content for oil-base muds is 0.1 to 4 lb/bbl, depending on base oil being used. Some of the newer ester-base muds have a low tolerance for hydroxyl ions; in this case, lime additions should be closely controlled.

F. $CaCl_2$ content should be checked daily to ensure the desired levels of inhibition are maintained.

G. The oil-to-water ratio influences funnel viscosity, plastic viscosity, and HTHP filtration of the oil-base mud. Retort analysis is used to detect any change in the oil-water ratio, because changes to the oil-water ration can indicate an intrusion of water.

H. Electrical Stability is a measure of how well the water is emulsified in the continuous oil phase. Because many factors affect the electrical stability of oil-base muds, the test does not necessarily indicate that a particular oil-base muds, the test does not necessarily indicate that a particular oil-base mud is in good or in poor condition. For this reason, values are relative to the system for which they are being recorded. Stability measurement should be made routinely and the values recorded and plotted so that trends may be noted. Any change in electrical stability indicates a change in the system.

I. HTHP filtration should exhibit a low filtrate volume (< 6 ml). The filtrate should be water free; water in the filtrate indicates a poor emulsion, probably caused by water wetting of solids.

4.5.3.10 Environmental Aspects of Drilling Fluids

Much attention has been given to the environmental aspects of the drilling operation and the drilling fluid components. Well-deserved concern about the possibility of polluting underground water supplies and of damaging marine organisms, as well as effects on soil productivity and surface water quality, has stimulated widespread studies on this subject.

Drilling Fluid Toxicity

There are three contributing mechanisms of toxicity in drilling fluids: chemistry of mud mixing and treatment, storage and disposal practices, and drilled rock. The first group conventionally has been known the best because

it includes products deliberately added to the system to build and maintain the rheology and stability of drilling fluids.

Petroleum, whether crude or refined products, needs no longer to be added to water-base muds. Adequate substitutes exist and are economically viable for most situations. Levels of 1% or more of crude oil may be present in drilled rock cuttings, some of which will be in the mud.

Common salt, or sodium chloride, is also present in dissolved form in drilling fluids. Levels up to 3,000 mg/L of chloride and sometimes higher are naturally present in freshwater muds as a consequence of the salinity of subterranean brines in drilled formations. Seawater is the natural source of water for offshore drilling muds. Saturated-brine drilling fluids become a necessity when drilling with water-base muds through salt zones to get to oil and gas reservoirs below the salt. In onshore drilling, there is no need for chlorides above their background levels. Potassium chlordie has been added to some drilling fluids as an aid to controlling problem shale formations. Potassium acetate or potassium carbonate are acceptable substitutes in most of these situations.

Heavy metals are present in drilled formation solids and in naturally occurring materials used as mud additives. The latter include barite, bentonite, lignite, and mica (sometimes used to control mud lossess downhole). There are background levels of heavy metals in trees that carry through into lignosulfonate made from them.

Attention has focused on heavy metal impurities found in sources of barite. Proposed U.S. regulations would exclude many sources of barite ore based on levels of contamination. European and other countries are contemplating regulations of their own.

Chromium lignosulfonates are the biggest contributions to heavy metals in drilling fluids. Although studies have shown minimal environmental impact, substitutes exist that can result in lower chromium levels in muds. The less-used chromium lignites (trivalent chromium complexes) are similar in character and performance, with less chormium. Nonchromium substitutes are effective in many situations. Typical total chromium levels in muds are 100–1000 mg/L.

Zinc compounds such as zinc oxide and basic zinc carbonate are used in some drilling fluids. Their function is to react out swiftly sulfide and bisulfide ions originating with hydrogen sulfide in drilled formations. Because human safety is at stake, there can be no compromising effectiveness, and substitutes for zinc have not seemed to be effective. Fortunately, most drilling situations do not require the addition of sulfide scavengers.

Indiscriminate storage and disposal practices using drilling mud reserve pits can contribute toxicity to the spent drilling fluid. The data in Table 4.5.6 is from the EPA survey of the most important toxicants in spent drilling fluids. The survey included sampling active drilling mud (in circulating

Table 4.5.6 Toxicity Difference between Active and Waste Drilling Fluids

Toxicant	Active Mud	Detection Rate	Reserve Pit	Detection Rate, %
Benzene	No	—	Yes	39
Lead	No	—	Yes	100
Barium	Yes	100	Yes	100
Arsenic	No	—	Yes	62
Fluoride	No	—	Yes	100

system) and spent drilling mud (in the reserve pit). The data show that the storage disposal practices became a source of the benzene, lead, arsenic, and fluoride toxicities in the reserve pits because these components had not been detected in the active mud systems.

The third source of toxicity in drilling discharges are the cuttings from drilled rocks. A study of 36 cores collected from three areas (Gulf of Mexico, California, and Oklahoma) at various drilling depths (300 to 18,000 ft) revealed that the total concentration of cadmium in drilled rocks was more than five times greater than the cadmium concentration in commercial barites. It was also estimated, using a 10,000-ft model well discharge volumes, that 74.9% of all cadmium in drilling waste may be contributed by cuttings, but only 25.1% originate from the barite and the pipe dope.

Mud Toxicity Test for Water-Base Fluids
The only toxicity test for water-base drilling fluids having an EPA approval is the Mysid shrimp bioassay. The test was developed in the mid-1970s as a joint effort of the EPA and the oil industry. The bioassay is a test designed to measure the effect of a chemical on a test population of marine organisms. The test is designed to determine the water-leachable toxicity of a drilling fluid or mud-coated cutting. The effect may be a physiological or biochemical parameter, such as growth rate, respiration, or enzyme activity. In the case of drilling fluids, lethality is the measured effect. For the Mysid test, all fluids must exceed a 30,000 concentration of whole mud mixed in a 9:1 ratio of synthetic seawater.

Nonaqueous Fluid (NAF) and Drilling Fluid Toxicity
Until the advent of synthetic-based invert emulsion fluids in the early 1990s, the discharge of NAF was prohibited due to the poor biodegradability of the base oils. In 1985, a major mud supplier embarked on a research program aimed at developing the first fully biodegradable base fluid. The base fluid would need to fulfill a number of criteria, regarded as critical to sustain drilling fluids performance while eliminating long-term impact on the environment:

- Technical performance — the fluid must behave like traditional oil-base muds and offer all of their technical advantages
- The fluid must be nontoxic, must not cause tainting of marine life, not have potential to bioaccumulate, and be readily biodegradable.

Research into alternative biodegradable base fluids began with common vegetable oils, including peanut, rapeseed, and soy bean oils. Fish oils such as herring oil were also examined. However, the technical performance of such oils was poor as a result of high viscosity, hydrolysis, and low temperature stability. Such performance could only be gained from a derivative of such sources, so these were then examined.

Esters were found to be the most suitable naturally derived base fluids in terms of potential for use in drilling fluids. Esters are exceptional lubricants, show low toxicity, and have a high degree of aerobic and anaerobic biodegradability. However, there are a vast number of fatty acids and alcohols from which to synthesize esters, each of which would have unique physical and chemical properties.

After 5 years of intensive research, an ester-based mud that fulfilled all of the design criteria was ready for field testing. This fluid provided the same shale stabilization and superior lubricity as mineral oil-based mud but also satisfied environmental parameters. The first trial, in February 1990, took place in Norwegian waters and was a technical and economic success. Since then, over 400 wells have been drilled world wide using this ester-base system, with full approval based on its environmental performance. This history of field use is unrivalled for any synthetic drilling fluid on a global basis, and no other drilling fluid has been researched in such depth. The research program included

- Technical performance testing using oil-base mud as a baseline
- Toxicity to six marine species, including water column and sediment reworker species
- Seabed surveys
- Fish taint testing
- Aerobic and anaerobic biodegradability testing
- Human health and safety factors

The release of ester-base fluids onto the market marked the beginning of the era of synthetic-base invert drilling fluids. Following the success of esters, other drilling fluids were formulated that were classed as synthetics, these fluids included base oils derived from ethylene gas and included linear alpha olefins, internal olefins, and poly-alpha olefins.

Summary of Flashpoint and Aromatic Data
With the introduction of synthetic-base muds into the market, the EPA moved to provide guidelines on the quality and quantity of the synthetic oils being discharged into the Gulf of Mexico. In addition to the water column aquatic testing done for water-base fluids, the EPA set forth guidelines for examining toxicity to organisms living in the sediments of the seafloor. A Leptocheirus sedimentary reworker test was instituted in February of 2002 for all wells being drilled with synthetic-base muds to examine how oil-coated cuttings being discharged into the Gulf of Mexico would impact the organisms living on the seafloor.

Two standards were set forth to govern the discharge of synthetic-base muds. A *stock standard* test is required for the base oil looking at the biodegradability of a synthetic base oil and as well as a new test for the Leptocheirus sedimentary reworker. This stock standard is done once per year to certify that the base oils being used are in compliance with the regulation.

When a well is drilled with a synthetic-base mud, monthly and end-of-well tests are required for the *Mysid* and *Leptocheirus*, organisms to ensure that the synthetic-base oil being used meets a certain standard of environmental performance. There are two standards that can be used for these annual and well-to-well tests: an ester standard and a C 1618 Internal Olefin standard.

The test used as a standard is based on the type of base oil being tested against a similarly approved standard. Base oils that are less toxic and highly biodegradable would be compared with esters, while all others would have to meet or exceed the C1618 IO standard (Figure 4.5.11 and Table 4.5.7).

With synthetic-based muds being widely used in the Gulf of Mexico, especially in deep water, controlling the quality of these materials is extremely important to the environment.

From US-EPA 2001 NPDES General Permit for New and Existing Sources in the Offshore Subcategory of the Oil and Gas Extraction Category for the Western Portion of the Outer Continental Shelf of the Gulf of Mexico (GMC290000) 66 Fed. Reg. No. 243, p. 65209, December 18, 2001.

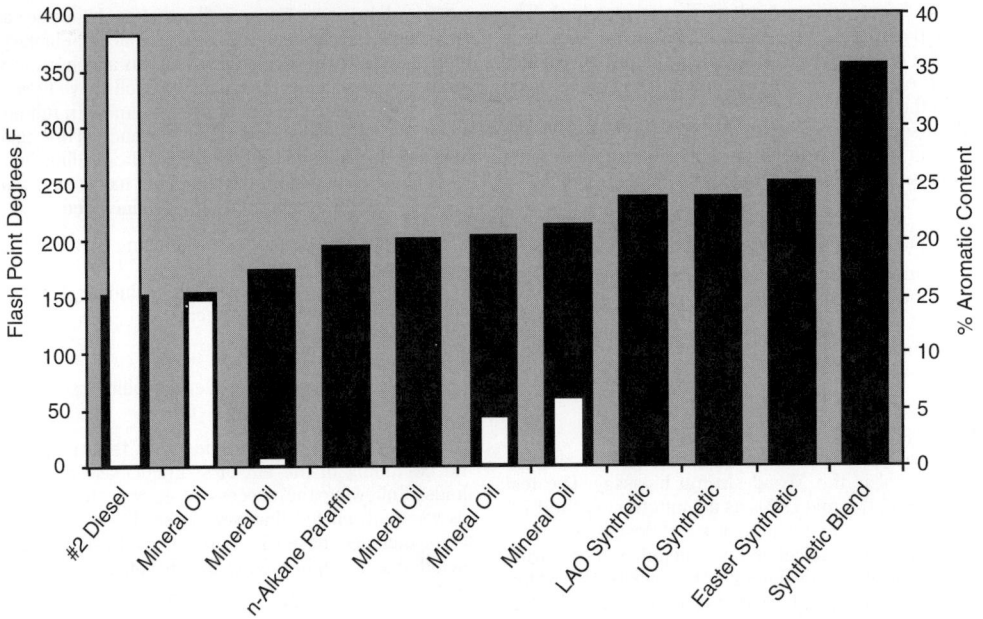

Figure 4.5.11 *Base fluid type.*

Table 4.5.7 *Aromatic Content Standards*

	Fluids Meeting Internal Olefin Standard	Performance of New Blended-Base Fluid
Base fluid biodegradation	Equal to or better than a 65:35 blend of C16 C18 internal olefin in a 275-day test. Tested once per year. Ratio of IO result compared with base fluid must be calculated at 1.0 or less.	Ratio = 0.8
Leptocheirus base fluid toxicity	Ten-day Leptocheirus LC50 must be equal to or less toxic than a 65:35 blend of C16 C18 internal olefin tested at least annually. Ratio of IO result compared with base fluid must be calculated at 1.0 or less	Ratio = 0.8
PAH content of base fluid	High-performance liquid chromatography/ UV-EPA method 1654 must give a PAH (as phenanthrene) content of less than 10 ppm	Below 1 ppm limit of detection

Summary of permit requirements and performance of blended drilling fluid, base fluid against those permit mandated C16–18 Internal olefin standards.
From US-EPA (2001) NPDES General Permit for New and Existing Sources in the Offshore Subcategory of the Oil and Gas Extraction Category for the Western Portion of the Outer Continental Shelf of the Gulf of Mexico (GMG290000) 66 Fed. Reg. No. 243, p. 65209, December 18, 2001.

4.5.4 Completion and Workover Fluids

Completion and workover fluids are any fluids used in the completion of a well or in a workover operation. These fluids range from low-density gases such as nitrogen to high-density muds and packer fluids. The application and requirements vary for each fluid.

Workover fluids are fluids used during the reworking of a well after its initial completion. They may be gases such as nitrogen or natural gas, brine waters, or muds. Workover fluids are used during operations such as well killing, cleaning out a well, drilling into a new production interval, and plugging back to complete a shallower interval.

Completion fluids are used during the process of establishing final contact between the productive formation and the wellbore. They may be a water-base mud, nitrogen, oil mud, solids-free brine, or acid-soluble system. The most significant requirement is that the fluid does not damage the producing formation and does not impair production performance.

Packer fluids are fluids placed in the annulus between the production tubing and casing. Packer fluids must provide the required hydrostatic pressure, must be nontoxic and non-corrosive, must not solidify or settle out of suspension over long periods of time, and must allow for minimal formation damage.

Various types of fluids may be used for completion and workover operations:

1. Oil fluids

   a. Crude
   b. Diesel
   c. Mineral oil

2. Clear water fluids

    a. Formation salt water

    b. Seawater

    c. Prepared salt water such as calcium chloride, potassium chloride or sodium chloride salt and zinc, calcium, or sodium-based bromides

3. Conventional water-base mud

4. Oil-base or invert emulsion muds

Completion or workover fluids may be categorized as

1. Water-base fluids containing oil-soluble organic particles

2. Acid-soluble and biodegradable

3. Water base with water-soluble solids

4. Oil-in-water emulsions

5. Oil-base fluids

Three types of completion or workover fluids are

1. Clear liquids (dense salt solutions)

2. Weighted suspensions containing calcium carbonate weighting material, a bridging agent to increase the density above that of saturated solutions

3. Water-in-oil emulsions made with emulsifiers for oil muds

Clear liquids have no suspended solids and can be referred to as solids-free fluids. Weighted suspensions are fluids with suspended solids for bridging or added density. These fluids can be referred to as solids-laden fluids.

For solids-free fluids, water may be used in conjunction with a defoamer, viscosifier, stabilized organic colloid, and usually a corrosion inhibitor. Solids-free completion and workover fluids have densities ranging from 7.0 to 19.2 pounds per gallon (ppg) (0.84 – 2.3 SG).

Solids-laden fluids may be composed of water, salt, a defoamer, suspension agent, stabilized organic colloid, pH stabilizer, and a weighting material/bridging agent.

4.5.4.1 Solids-Free Fluids

Brines used in completion and workover applications may be single-salt brines, two-salt brines, or brines containing three different salt compounds.

4.5.4.2 Single-Salt Brines

Single-salt brines are made with freshwater and one salt such as potassium chloride, sodium chloride, or calcium chloride. They are the simplest brines used in completion and workover fluids. Because they contain only one salt, their initial composition is easily understood. Their density is adjusted by adding either salt or water. Single-salt brines are available in densities of up to 11.6 ppg and are the least expensive brines used in completions.

Potassium chloride (KCl) brines are excellent completion fluids for water-sensitive formations when densities over (9.7 ppg) (1.16 SG) are not required. Corrosion rates are reasonably low and can be reduced even more by keeping the pH of the system between 7 and 10 and using corrosion inhibitors (1% by volume). Sodium chloride is one of the most used single-salt brines. Advantages of sodium chloride brines are low cost and wide availability. Densities up to 10.0 ppg are achievable for this single-salt brine. Calcium chloride ($CaCl_2$) brines are easily mxied at densities up to 11.6 ppg 1.39 SG. Sodium bromide brines can be used when the density of a calcium chloride brine is desired, but the presence of acid gas is possible. Sodium bromide has low corrosion rates even without the use of corrosion inhibitors. Although these brines are more expensive than $CaCl_2$ brines, they are useful in CO_2 environments.

4.5.4.3 Two-Salt Brines

The basic ingredient of calcium chloride/calcium bromide brines

($CaCl_2$/$CaBr_2$) is a calcium bromide solution that ranges in density from 14.1 to 14.3 ppg (1.72 SG); the pH range is 7.0 to 7.5. The density of $CaBr_2$ brine can be increased by adding calcium chloride pellets or flakes. However, a 1.81 S.G. $CaCl_2$/$CaBr_2$ solution crystallizes at approximately 65°F (18°C). $CaCl_2$/$CaBr_2$ brine can be diluted by adding a $CaCl_2$ brine weighing 11.6 ppg (1.39 SG). The corrosion rate for $CaCl_2$/$CaBr_2$ is no more than 5mm per year on N-80 steel coupons at 300°F (149°C). If a corrosion inhibitor is desired, a corrosion inhibitor microbiostat is recommended.

The viscosity of $CaCl_2$/$CaBr_2$ brine can be increased by adding liquefied HEC viscosifier. Reduction in filtration may be obtained by the addition of $CaCO_3$ weighting material/bridging agent or by increasing the viscosity with polymeric materials.

There is not much of a crystallization problem with calcium chloride/calcium bromide brines at densities between 11.7 and 13.5 ppg 1.40 and 1.62 SG. However, the heavier $CaCl_2$/$CaBr_2$ brines require special formulation in cold weather applications.

4.5.4.4 Three-Salt Brines

Three-salt brines such as calcium chloride/calcium bromide/zinc bromide brines are composed of $CaCl_2$, $CaBr_2$, and $ZnBr_2$. At high temperatures, corrosion reates in brines containing $ZnBr_2$ are very high and can result in severe damage to equipment. For use at high temperatures, the brine should be treated with corrosion inhibitors. The corrosion rate of the treated brine is usually less than 3 mm per year.

4.5.4.5 Classification of Heavy Brines

Properties and Characteristics of Completion and Workover Fluids

Although the properties required of a completion or workover fluid vary depending on the operation, formation protection should always be the primary concern.

Density

The first function of a completion of workover fluid is to control formation pressure.(Table 4.5.8). The density should be no higher than necessary to accomplish that function. Increased density can be obtained by using weighting materials such as calcium carbonate ($CaCO_3$), iron carbonate ($FeCO_3$), barite ($BaSO4$), or by using soluble salts such as NaCl, KCl, NaBr, $CaCl_2$, $CaBr_2$, or $ZnBr_2$. The Table 4.5.9 below shows the specific weight range and acid solubility of each type of solids-laden fluid.

Solids – Laden Fluids

The density of a brine solution is a function of temperature. When measured at atmospheric pressure, brine densities decrease as temperature increases.

Viscosity

In many cases, the viscosity of the fluid must be increased to provide lifting capacity required to bring sand or cuttings to the surface at reasonable circulating rates. A popular viscosifier for completion and workover fluids is hydroxyethylcellulose (HEC). It is a highly refined, partially water soluble, and acid-soluble polymer with very little residue when acidized. Other materials used as viscosifiers include guar gums and biogums (xanthan). Although these materials are applicable in certain instances, they do not meet the acid or water-solubility standards of HEC. HEC is the most common viscosifier for all types of brines.

Table 4.5.8 *Expansibility of Heavy Brine at 12,000 psi from 24° to 92°C or 76° to 198°F*

Compressibility of Brine	Heavy Brine at 198°F Density SG – lb/gal	198°F at 2,000–12,000 psi (lb/gal/1,000 psi)
Nacl	1.14–9.49	0.019
$CaCl_2$	1.37–11.46	0.017
NaBr	1.50–12.48	0.021
$CaBr_2$	1.72–14.30	0.022
$ZnBr_2/CaBr_2/CaCl_2$	1.92–16.01	0.022
$ZnBr_2/CaBr_2$	2.31–19.27	0.031

Table 4.5.9 *Solids-Laden Fluids*

Weight Materials	Pounds/gallon	Acid Solubility
$CaCO_3$	10–14	98%
$FeCO_3$	10–18	90%
Barite	10–21	0%

In some cases, solids must be suspended at low shear or static conditions. Available alternatives include clays and polymers. The most widely used suspension agent in completion and workover fluids is xanthan gum.

Suspension or Filtration Control: Filtration

In most applications, some measure of filtration control is desirable. The standard approach to filtration control in completion and workover fluids is the use of property sized calcium carbonate particles for bridging in conjunction with colloidal size materials such as starch or CMC. The reason for the popularity of calcium carbonate is that it is acid soluble and can be removed. In some cases, oil-soluble resins are used as bridging agents, as are sized salts when used in saturated salt brines.

The seasonal ambient temperature must be considered when selecting a completion or workover fluid. If the temperature drops too low for the selected fluid, the fluid will crystallize or freeze. Each brine solution has a point at which crystallization or freezing occurs. Two definitions are important: FCTA, or fist crystal to appear, is the temperature at which the first crystal appears as a brine is cooled. LCTD, or last crystal to dissolve, is temperature at which the last salt crystal disappears as the brine is allowed to warm. Although this type of visual check may be somewhat inexact, it is an important part of the analysis of brines. Once the crystallization point of a fluid is determined, you can be reasonably sure that the fluid is safe at a temperature equal to or higher than the crystallization point. The FTCA and LCTD are run under normal-pressure conditions; pressure can greatly alter the formation of crystals in a brine, and more sophisticated tests are required to determine this value.

Special brine formulations are used to accommodate seasonal changes in temperature. Summer blends are fluids appropriate for use in warmer weather. Their crystallization points range from approximately 7–20°C_2 (45°F–68°F). Winter blends are used in colder weather or colder climates and have crystallization points ranging from approximately 20° F (− 6°C) to below 0°F (− 18°C). At times, a crystallization point between those of summer and winter blends is desirable. Special formulations are then used to prepare fluids that can be called fall, spring, or intermediate blends.

At first, it may seem practical to consistently formulate fluids having lower than necessary, and therefore safe, crystallization points. Although this approach may be easier, it is likely to be much more expensive. Generally, the lower the crystallization point, the more the fluid costs. If you provide a fluid having a crystallization point much lower than necessary, you are likely to be providing a fluid with a considerably higher cost than necessary. This is just one of numerous factors to consider when selecting a fluid that is both effective and economical.

Preparing Brines

The typical blending procedure for NaCl and KCl brines is to begin with the required volume of water and then add sacked salt. Calcium chloride/calcium bromide brines and calcium chloride/calcium bromide/zinc bromide brines require special blending procedures.

Calcium Chloride/Calcium Bromide Solutions

The ingredients in $CaCl_2/CaBr_2$ solutions must be added in a specific order. The necessary order of addition is as follows:

1. Start with the $CaBr_2$ brine.
2. Add sacked $CaCl_2$.
3. Allow approximately 45 minutes for most of the sacked $CaCl_2$ to dissolve.

Calcium Chloride/Calcium Bromide/Zinc Bromide Solutions

For $CaCl_2/CaBr_2/ZnBr_2$ solutions using 15.0 ppg (1.80 SG) $CaCl_2/CaBr_2$ brine and 19.2 ppg (2.28 SG) $CaBr_2ZnBr_2$, the proper order of addition is as follows:

1. Start with the 15.0 ppg (1.80 SG) $CaCl_2/CaBr_2$ brine.
2. Add the 19.2 ppg (2.28 SG). CaBr/$ZnBr_2$ brine.

Rules of thumb for blending of brines

1. DO NOT CUT SACKS. Exception: $CaBr_2$ brines from 11.6 to 15.1 ppg.
2. An increase of $CaBr_2$ decreases the crystallization temperature for $CaBr_2$ brines.
3. An increase of $ZnBr_2$ decreases the crystallization temperature in any blend.
4. A decrease in crystallization temperature increases the cost of the fluid.
5. Do not mix fluids containing divalent ions (Ca^{2+}, Zn^{2+}) with fluids containing monovalent ions (Na^+, K^+), Precipitation may occur.
6. Do not increase the pH of $CaBr_2$ or $ZnBr_2$ fluids or precipitation may occur.
7. Do not add large volumes of water to $CaBr_2$ or $ZnBr_2$ brines or precipitation may occur.

Fluid Selection

A good approach to selecting a fluid is to decide what functions the fluid is to perform and then select a base fluid and additives that will most effectively do the job. The first decision in selecting a fluid is identification of the required functions or properties. The next step is the selection of the type of fluid to be used. The properties or functions of the fluid dictate the type of fluid to be used. If the decision is made in reverse order, a poor performance is likely to result.

Completion and Workover Fluids Weighting Materials

Calcium Carbonate

Calcium carbonate is available in five grades: 5, 50, 150, 600, and 2,300. At least 50% of the particles in each grade are

Table 4.5.10 *Specified, Gravity and Specific Weight of Common Materials*

Material	Specific gravity	lb/gal	lb/bbl
Barite	4.2–4.3	35.0–35.8	1470–1504
Calcium carbonate	2.7	22.5	945
Cement	3.1–3.2	25.8–26.7	1085–1120
Clays and/or drileld solids	2.4–2.7	20.0–22.5	840–945
Diesel oil	0.84	7.0	294
Dolomite	2.8–3.0	23.3–25.0	980–1050
Feldspar	2.4–2.7	20.0–22.5	840–945
Fresh water	1.0	8.33	350
Galena	6.5	54.1	2275
Gypsum	2.3	19.2	805
Halite (rock salt)	2.2	18.3	770
Iron	7.8	65.0	2730
Iron oxide (hematite)	5.1	42.5	1785
Lead	11.4	95.0	3990
Limestone	2.7–2.9	22.5–24.2	945–1015
Slate	2.7–2.8	22.5–23.3	945–980
Steel	7.0–8.0	58.3–66.6	2450–2800

large than the size (μm) indicated by the number. Other grind sizes can be made available.

Typical Physical Constants
Hardness (Mohr's scale)	3
Specific gravity	2.7
Bulk density, lb/ft^3, or ppg	168.3 or 22.5

Typical Chemical Composition
Total carbonates (Ca^{2+}, Mg^{2+})	98.0% (min)
Total impurities (Al_2O_3, Fe_2O_3, SiO_2, Mn)	2.0% (max)

$CaCO_3$ (5) (93% will pass through a 325 mesh) can be used alone or with ($FeCO_3$) to increase the densities of freshwater or brine fluids beyond their saturation limits. It may also be used to increase the density of oil base fluids (Table 4.5.10).

$CaCO_3$ (5) can be used instead of clays to provide wall cake buildup for acid-soluble fluids.

It may also be used as an acid-soluble bridging agent for formations having pore sizes up to 15 μm.

$CaCO_3$ 50, 150, 600, and 2,300 grades are recommended for use as bridging agents for lost circulation problems, in squeeze mixtures, and in other similar applications. The particle size distribution is maintained in the slurry to provide effective bridging at the surface of the pay zone.

Acid solubility
$CaCO_3$ is 98% soluble in 15% HCl solution. One gallon of 15% HCl dissolves 0.83 kg or 1.84 lb of $CaCO_3$.

Bridging agent
Normal treatment is 2.27 to 5 kg (5 to 10 ppb) of the appropriate grade(s). From 5% to 10% of the material added should have particle size at least one third of the formation pore diameter.

Iron Carbonate
Iron carbonate is used to achieve densities in excess of 14.0 ppg in (1.68 SG) solids-laden systems. The maximum density of a $CaCO_3$ fluid is approximately 14.0 ppg

(1.68 SG) and the maximum density of iron carbonate fluids is 17.5–18.0 ppg (2.10–2.16 SG). For weighting fluids in the 13.0–16.5 ppg (1.56–1.98 SG) range, a blend is recommended.

The following precautions should be considered when using iron carbonate:

1. Iron carbonate is only 87% acid soluble, and after acidizing, 13% of the solids added may be left to plug the formation or may be flushed out, depending on the size and distribution of the formation flow channels.
2. *Mud acid, a combination of hydrofluoric acid and hydrochloric acid, should not be used with iron carbonate.* The hydrofluoric acid reacts with iron carbonate to produce insoluble salts of acidic and basic nature (iron fluoride and iron hydroxide). When using iron carbonate, use only hydrochloric acid.

4.5.5 Safety Aspects of Handling Brines
4.5.5.1 Potassium Chloride
Toxicity
No published data indicate that potassium chloride is a hazardous material to handle. It is toxic only if ingested in very large amounts. It is considered a mild irritant to the eyes and skin. Inhalation of potassium chloride dust leaves a taste and causes mild irritation to mucous membranes in the nose and throat. Potassium is toxic to the Mysid shrimp used for aquatic toxicity testing in U.S. Federal waters. Potassium levels over 4% will likely fail the 30,000 ppm minimum required for discharge.

Safety precautions
Prolonged contact with skin and eyes should be avoided. Inhalation of potassium chloride dust should be avoided as much as possible. Eye protection should be worn according to the degree of exposure, and dust masks should be used in severe dusting conditions. Personal protective equipment (PPE) should always be used when mixing or handling brines and all fluids used in drilling and completions operations.

First Aid Measures
The following first aid measures should be used:

1. For contact with eyes, flush promptly with plenty of water for 15 minutes.
2. For contact with skin, flush with plenty of water to avoid irritation.
3. For ingestion, induce vomiting and get medical attention.
4. For inhalation, if illness occurs, remove the person to fresh air, keep him or her warm, and quiet, and get medical attention.

4.5.5.2 Sodium Chloride
Toxicity
There are no published data indicating that salt is a hazardous material to handle. Sodium chloride is considered a mild irritant to the eyes and skin. Inhalation of dust leaves a taste and causes mild irritation to mucous membranes in the nose and throat.

Safety Precautions
Prolonged contact with skin and eyes should be avoided. The inhalation of sodium chloride dust should be avoided as much as possible. Eye protection should be worn according to the degree of exposure, and dust masks may be needed in severe exposure.

First Aid Measures

The following first aid measures should be used:

1. For contact with eyes, flush promptly with plenty of water for 15 minutes.
2. For contact with skin, flush with plenty of water to avoid irritation.

4.5.5.3 Calcium Chloride

Toxicity

Three to five ounces of calcium chloride may be a lethal dose for a 45-kg (100-pound) person. However, calcium chloride is not likely to be absorbed through the skin in toxic amounts. Strong solutions are capable of causing severe irritation, superficial skin burns, and permanent eye damage. Normal solutions cause mild irritation to eyes and skin, and dust may be irritating.

Safety Precautions

Contact with eyes and prolonged skin contact should be avoided. Clean, long-legged clothing must be worn. Hand and eye covering may be required, depending on the severity of possible exposure. For severe exposure, chemical goggles and a dust respirator should be worn. Cool water (27°C, 80°F or cooler) should always be used when dissolving calcium chloride. Because of an exothermic reaction, calcium chloride can burn bare hands if solids have been added. Barrier creams should always be used when handling brines such as calcium chloride, calcium bromide, and zinc bromide.

First Aid Measures

The following first aid measures should be used:

1. For contact with eyes and skin, flush promptly with plenty of water for 15 minutes. Get medical attention in the event of contact with eyes. Remove contaminated clothing, and wash before reuse.
2. For inhalation, if illness occurs, remove the person to fresh air, keep him or her warm and quiet, and get medical attention.
3. For ingestion, induce vomiting, and get medical attention.

4.5.5.4 Calcium Bromide

Toxicity

There are no published data indicating that calcium bromide is a hazardous material to handle. However, it is considered toxic when ingested in large amounts. It is also a mild irritant to the skin and eyes. Inhalation results in irritation of the mucous membranes in the nose and throat. Because of an exothermic reaction, calcium bromide fluid can burn bare hands when sacked $CaBr_2$ is added to the solution. Burns caused by these fluids are the result of a chemical reaction with moisture on the skin.

Safety Precautions

Prolonged contact with the skin and eyes should be avoided. Clean, long-legged clothing and rubber boots should be worn. Eye protection should be worn and a dust respirator used for severe exposure. Contaminated clothing should be changed. Barrier creams should always be used when handling these brines.

First Aid Measures

The following first aid measures should be used:

1. For contact with eyes and skin, flush promptly with plenty of water.

2. For inhalation, if illness occurs, remove the victim to fresh air, keep him or her warm and quiet, and get medical attention.
3. For ingestion, induce vomiting, and get medical attention.

Environmental Considerations

Local regulations should be observed. Care should be taken to ensure that streams, ponds, lakes, or oceans are not polluted with calcium bromide.

4.5.5.5 Zinc Bromide

Toxicity

There are no published data indicating that zinc bromide is a hazardous material to handle. It is considered toxic when ingested in large amounts. Zinc bromide is also a severe irritant to the skin and eyes. Inhalation results in irritation of mucous membranes in the nose and throat. Because of an exothermic reaction, zinc bromide fluid can burn bare hands if sacked materials have been added. Never expose eyes to zinc bromide; blindness can occur.

Safety Precautions

Contact with skin and eyes should be avoided. Long-legged clothing and proper eye protection should be worn. Barrier creams should always be used when handling zinc bromide brines. Rubber boots and rubber protective clothing also is suggested. Contaminated clothing should be washed off or changed, because contact with the skin can cause burns.

First Aid Measures

1. For contact with eyes and skin, flush promptly with plenty of water. Wash skin with mild soap, and consider seeking medical attention.
2. For inhalation, if illness occurs, remove victim to fresh air, kept him or her warm and quiet, and get medical attention.
3. For ingestion, induce vomiting, and get medical attention.

Environmental Considerations

Local regulations should be observed, and care should be taken to avoid polluting streams, lakes, ponds, or oceans. Regulations in the United States prohibit the discharge of zinc into federal waters. Zinc bromide fluids should be disposed of in the same matter as oil fluids.

Safety Rules of Thumb

1. Do not wear leather boots.
2. Wear eye goggles for $CaCl_2$, $CaBr_2$, and $ZnBr_2$.
3. Wear rubber gloves underneath regular gloves while tripping.
4. Wear slicker suits while tripping pipe in most brines.
5. Wash off $CaCl_2$ or $ZnBr_2$ spills within 15 minutes; reapply barrier creams.
6. Change clothes within 30 minutes for $CaCl_2$ and within 15 minutes for $CaBr_2$ or $ZnBr_2$ if a spill occurs.
7. Do not wear shoes or boots for more than 15 minutes if they have $CaCl_2$, $CaBr_2$, or $ZnBr_2$ spilled in them.
8. Use pipe wipers when tripping.

4.5.6 Preventing Contamination

4.5.6.1 Brine Filtration

Filtration is a critical step if a well is to produce at its full potential and remain on line for a longer period. Although filtering can be expensive and time consuming, the net production can be enough to pay the difference in only a matter of days.

Filtration can be defined as the removal of solids particles from a fluid. Because these particles are not uniform in size, various methods of removal must be used (Table 4.5.11).

Table 4.5.11 *Drilling Fluids Contaminants Removal*

Contaminant To Be Removed	Chemical Used To Remove Contaminant	Conversion Factor mg/L (Contaminant) × Factor = lb/bbl Chemical to Add
Ca^{2+}	Soda ash	0.000925
Ca^{2+}	Sodium bicarbonate	0.000734
Mg^{2+}	Caustic soda	0.00115
CO_3^{-2}	Lime	0.00043
HCO_3^{-1}	Lime	0.00043
H_2S	Lime	0.00076
H_2S	Zinc carbonate	0.00128
H_2S	Zinc oxide	0.000836

Filtration has evolved from the surface filtering systems with low-flow volumes to highly sophisticated systems. Regardless of which system is used, a case for filtering fluid can be made for every well completed, every workover, and every secondary recovery project.

The purpose of filtering any fluid is to prevent the downhole contamination of the formation with undesirable solids present in the completion fluid. Contamination can impact production and shorten the productive life of the well. Contamination can occur during perforating, fracturing, acidizing, workover, water flooding, and gravel packing of a well. Any time a fluid containing solids is put into the wellbore, a chance of damaging the well exists.

Contaminants in fluids come in many sizes and forms. Cuttings from drilling operations, drilling mud, rust, scale, pipe dope, paraffin, undissolved polymer, and any other material on the casing or tubing string contributes to the solids in the fluid. At times, it is virtually impossible, because of particle size, to remove all of the solids from the fluid, but by filtering, the chance of success can be increased almost to 100%.

How clean does the fluid need to be? What size particle do we need to remove? Typically, the diameter of the grains of sand is three times the size of the pore throat, assuming the sand is perfectly round. Particles greater than one-third the diameter of the pore throat bridge instantly on the throat and do not penetrate the formation. These particles represent a problem, but one that can be remedied by hydraulic fracturing of the well and blowing the particles from the perforation tunnels, by perforation washing tools, or by acid. Particles less than one-tenth the diameter pass through the throat and through the formation without bridging or plugging. However, particles between one-third and one-tenth the pore throat diameter invade the formation and bridge in the pore throat deeper in the formation. These particles cause the serious problems because, with the pore throats plugged and no permeability, acid cannot be injected into the formation to clean the pore throats. Suggested guidelines for the degree of filtration are

Formation sand size (mesh)	Filtration level (μm)
11.84	40
5.41	80
2.49	2.09

In various stages of the completion process, we are faced with fluids contaminated by a high concentration of particles over a wide range of sizes. To maintain production, the best filtering process should use a number of steps to remove

contaminants, starting with the largest and working down to the smallest. This includes, in order of use, shale shaker or linear motion shaker, desilter, centrifuge, and cartridge filters.

In summary, successful completions primarily depend on following a set procedure without taking shortcuts and on good housekeeping practices. A key element in the entire process is using clean fluids, which is made possible in large part through filtration techniques.

4.5.6.2 Cartridge Filters

Each field, formation and well site has unique characteristics and conditions. These include reservoir rock permeabilities, pore sizes, connate fluid composition, downhole pressures, and so on. These conditions dictate the brine composition and level of clarity needed for a proper completion, which determine the level of filtration needed to achieve the fluid clarity level required.

Disposable cartridge filters are widely available around the world. They can be used alone, in combination (series), or in tandem with other types of filtration equipment. When very large particles or high solids concentrations are present, conventional solids control equipment should be used as prefilters if they are thoroughly washed and cleaned before use. After the filtration requirement is established, the goal becomes one of optimizing a filtration system design. This involves putting together a properly sized and operated system of prefilters and final filters to meet the filtration efficiency objective at the lowest operating cost. Cartridge filters are available in different configurations and various materials of construction. Filter media include yarns, felts, papers, resin-bonded fibers, woven-wire cloths, and sintered metallic and ceramic structures.

The cartridge is made of a perforated metal or plastic tube, layered with permeable material or wrapped with filament to form a permeable matrix around the tube. Coarser particles are stopped at or near the surface of the filter, and the finer particles are trapped within the matrix. Pleated outer surfaces are used to provide larger surface areas. Cartridge filters are rated by pore sizes such as 1, 2, 4, 10, 25, and 50 μm, which relate to the size of particles that the filter can remove. This rating is nominal or absolute, depending on how the cartridges are constructed. A nominally rated filter can be expected to remove approximately 90% of the particles that are larger than its nominal rating. Actually, solids larger than the rating pass through these filters, but the concentration of the larger particles is reduced. High flow rates and pressures cause their efficiencies to fall. They must be constantly monitored and changed when they begin to plug, or the fluid will begin to bypass the filters.

Absolute rated filters achieve a sharp cutoff at its rated size. They should remove all the particles larger than their rating and generally become plugged much faster than nominally rated cartridges. Cartridge filters are most often used downstream of other filters for final clarification.

4.5.6.3 Tubular Filters

Tubular filters consist of a fabric screen surrounding a perforated stainless steel tube. Dirty fluid flows from the outside, through the fabric, where solids are stopped, and the filtrate passes into the center tube. The fabric can remove particles down to 1 to 3 μm. Because the solids are trapped on the outside surface of fabric, the element is easy to backwash and clean. Backwashing is accomplished by changing the valving and forcing clean brine back through the filter in the opposite direction. In 8–15 seconds, the element can be filtering again.

4.6 DRILL STRING: COMPOSITION AND DESIGN

The drill string is defined here as drill pipe with tool joints and drill collars. The stem consists of the drill string and other components of the drilling assembly that may include the following items:

- Kelly
- Subs
- Stabilizers
- Reamers
- Shock absorbing tools
- Drilling jars
- Junk baskets
- Directional tools
- Information gathering tools
- Nonmagnetic tools
- Nonmetallic tools
- Mud motors

The drill stem (1) transmits power by rotary motion from the surface to a rock bit, (2) conveys drilling fluid to the rock bit, (3) produces the weight on bit for efficient rock destruction by the bit, and (4) provides control of borehole direction. The drill pipe itself can be used for formation evaluation (drill stem testing [DST]), well stimulation (e.g., fracturing, acidizing), and fishing operations. The drill pipe is used to set tools in place that will remain in the hole. Therefore, the drill string is a fundamental part, perhaps on of the most important parts, of any drilling activity.

The schematic, typical arrangement of a drill stem is shown in Figure 4.6.1. The illustration includes a Kelly. Kellys are used on rigs with a rotary table to turn the pipe. Many rigs use top drives to supply rotational power to the pipe. Top drive rigs do not have Kellys.

4.6.1 Drill Collar

The term *drill collar* derived from the short sub originally used to connect the bit to the drill pipe. A modern drill collar is about 30 ft long, and the total length of the string of drill collars may range from about 100 to 700 ft or longer. The purpose of drill collars is to furnish weight to the bit. However, the size and length of drill collars have an effect on bit performance, hold deviation, and drill pipe service life. Drill collars may be classified according to the shape of their cross-section as round drill collars (i.e., conventional drill

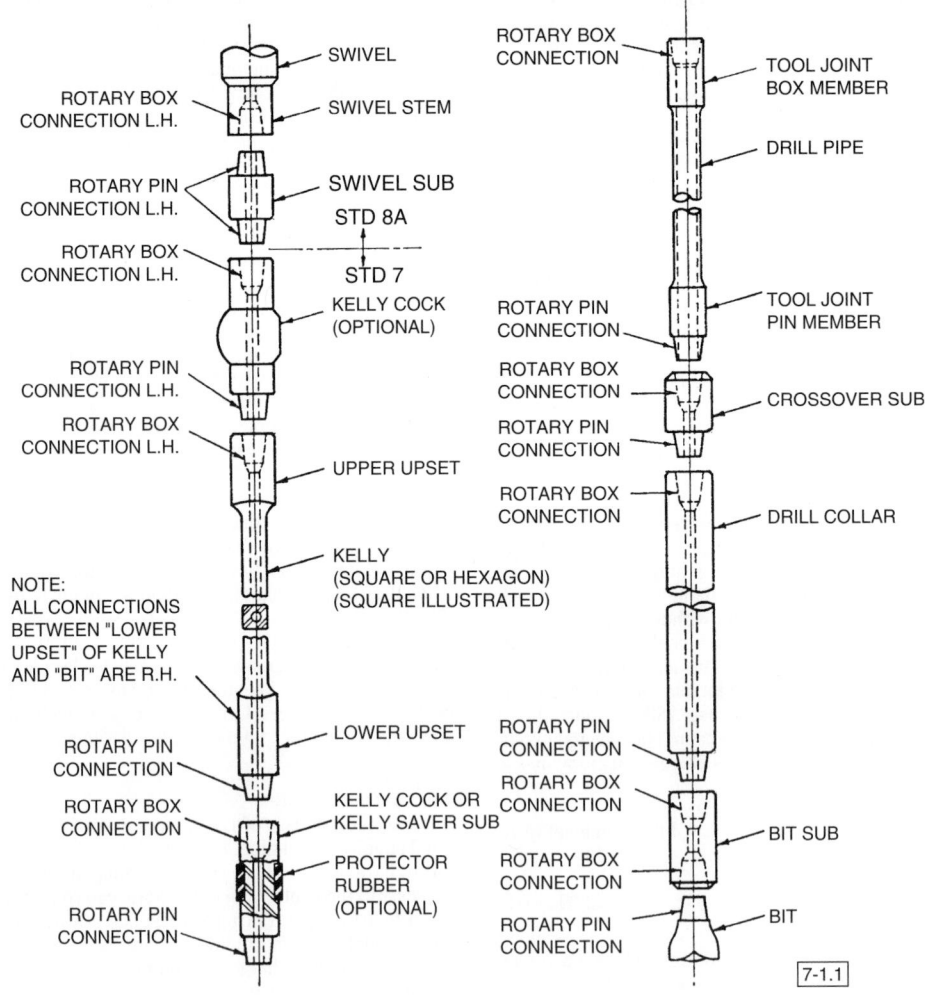

Figure 4.6.1 *Typical drill-stem assembly [1].*

Table 4.6.1 *Drill Collars [1]*

Drill Collar Number	Outside Diameter (in.)	Bore (in.)	Length (ft)	Bevel Diameter (in.)	Ref. Bending Strength Ratio
NC23-31	$3\frac{1}{8}$	$1\frac{1}{4}$	30	3	2.57:1
NC26-35	$3\frac{1}{2}$	$1\frac{1}{2}$	30	$3\frac{17}{64}$	2.42:1
NC31-41	$4\frac{1}{8}$	2	30 or 31	$3\frac{61}{64}$	2.43:1
NC35-47	$4\frac{3}{4}$	2	30 or 31	$4\frac{33}{64}$	2.58:1
NC38-50	5	$2\frac{1}{4}$	30 or 31	$4\frac{49}{64}$	2.38:1
NC44-60	6	$2\frac{1}{4}$	30 or 31	$5\frac{11}{16}$	2.49:1
NC44-60	6	$2\frac{13}{16}$	30 or 31	$5\frac{11}{16}$	2.84:1
NC44-62	$6\frac{1}{4}$	$2\frac{1}{4}$	30 or 31	$5\frac{7}{8}$	2.91:1
NC46-62	$6\frac{1}{4}$	$2\frac{13}{16}$	30 or 31	$52\frac{9}{32}$	2.63:1
NC46-65	$6\frac{1}{2}$	$2\frac{1}{4}$	30 or 31	$6\frac{3}{32}$	2.76:1
NC46-65	$6\frac{1}{2}$	$2\frac{13}{16}$	30 or 31	$6\frac{3}{32}$	3.05:1
NC46-67	$6\frac{3}{4}$	$2\frac{1}{4}$	30 or 31	$6\frac{9}{32}$	3.18:1
NC50-70	7	$2\frac{1}{4}$	30 or 31	$6\frac{31}{64}$	2.54:1
NC50-70	7	$2\frac{13}{16}$	30 or 31	$6\frac{31}{64}$	2.73:1
NC50-72	$7\frac{1}{4}$	$2\frac{13}{16}$	30 or 31	$6\frac{43}{64}$	3.12:1
NC56-77	$7\frac{3}{4}$	$2\frac{13}{16}$	30 or 31	$7\frac{19}{64}$	2.70:1
NC56-80	8	$2\frac{13}{16}$	30 or 31	$7\frac{31}{64}$	3.02:1
$6\frac{5}{8}$ Reg	$8\frac{1}{4}$	$2\frac{13}{16}$	30 or 31	$7\frac{45}{64}$	2.93:1
NC61-90	9	$2\frac{13}{16}$	30 or 31	$8\frac{3}{8}$	3.17:1
$7\frac{5}{8}$ Reg	$9\frac{1}{2}$	3	30 or 31	$8\frac{13}{16}$	2.81:1
NC70-97	$9\frac{3}{4}$	3	30 or 31	$9\frac{5}{32}$	2.57:1
NC70-100	10	3	30 or 31	$9\frac{11}{32}$	2.81:1
$8\frac{5}{8}$ Reg	11	3	30 or 31	$10\frac{1}{2}$	2.84:1

collars), square drill collars, or spiral drill collars (i.e., drill collars with spiral grooves).

Square drill collars are used to increase the stiffness of the drill string and are used mainly for drilling in crooked hole areas. Square drill collars are rarely used. Spiral drill collars are used for drilling formations in which the differential pressure can cause sticking of drill collars. The spiral grooves on the drill collar surface reduce the area of contact between the drill collar and wall of the hole, which reduces the sticking force. Conventional drill collars and spiral drill collars are made with a uniform outside diameter (except for the spiral grooves) and with slip and elevator recesses. Slip and elevator recesses are grooves ranging from about $\frac{1}{4}$ in. deep on smaller collars to about $\frac{1}{2}$ in. deep on larger sizes. The grooves are about 18 in. long and placed on the box end of the collar. They provide a shoulder for the elevators and slips to engage. They reduce drill collar handling time while tripping by eliminating the need for lift subs and safety clamps. A lift sub is a tool about 2 ft long that makes up into the box of the drill collar, which has a groove for the elevator to engage. A safety clamp is a device clamped around the drill collar that provides a shoulder for the slips to engage.

The risk of drill collar failure is increased with slip and elevator grooves because they create stress risers where fatigue cracks can form. The slip and elevator grooves may be used together or separately.

Dimensions, physical properties, and unit weight of new, conventional drill collars are specified in Tables 4.6.1, 4.6.2, and 4.6.3, respectively. Technical data on square and spiral drill collars are available from manufacturers.

Table 4.6.2 *Physical Properties of New Drill Collars [1]*

Drill Collar OD Range (in.)	Minimum Yield Strength (psi)	Minimum Tensile Strength (psi)	Elongation, Minimum
$3\frac{1}{8}$–$6\frac{7}{8}$	110,000	140,000	13%
7–10	100,000	135,000	13%

4.6.1.1 Selecting Drill Collar Size

Selection of the proper outside and inside diameter of drill collars is usually a difficult task. Perhaps the best way to select drill collar size is to study results obtained from offset wells previously drilled under similar conditions.

The most important factors in selecting drill collar size are

1. Bit size
2. Coupling diameter of the casing to be set in a hole
3. Formation's tendency to produce sharp changes in hole deviation and direction
4. Hydraulic program
5. Possibility of washing over if the drill collar fails and is lost in the hole

To avoid an abrupt change in hole deviation (which may make it difficult or impossible to run casing) when drilling in crooked hole areas with an unstabilized bit and drill collars, the required outside diameter of the drill collar placed right above the bit can be found from the following formula [5]:

$$D_{dc} = 2 \cdot OD_{cc} - OD_{bit} \qquad [4.6.1]$$

Table 4.6.3 *Drill Collar Weight (Steel) (Pounds per Foot) [4]*

(1)	(2)	(3)	(4)	(5)	(6)	(7)	(8)	(9)	(10)	(11)	(12)	(13)	(14)
Drill Collar OD,							Drill Collar ID, in.						
in.*	1	$1\frac{1}{4}$	$1\frac{1}{2}$	$1\frac{3}{4}$	2	$2\frac{1}{4}$	$2\frac{1}{2}$	$2\frac{13}{16}$	3	$3\frac{1}{4}$	$3\frac{1}{2}$	$3\frac{3}{4}$	4
$2\frac{7}{8}$	19	18	16										
3	21	20	18										
$3\frac{1}{8}$	22	22	20										
$3\frac{1}{4}$	26	24	22										
$3\frac{1}{2}$	30	29	27										
$3\frac{3}{4}$	35	33	32										
4	40	39	37	35	32	29							
$4\frac{1}{8}$	43	41	39	37	35	32							
$4\frac{1}{4}$	46	44	42	40	38	35							
$4\frac{1}{2}$	51	50	48	46	43	41							
$4\frac{3}{4}$			54	52	50	47	44						
5			61	69	56	53	50						
$5\frac{1}{4}$			68	65	63	60	57						
$5\frac{1}{2}$			75	73	70	67	64	60					
$5\frac{3}{4}$			82	80	78	75	72	67	64	60			
6			90	88	85	83	79	75	72	68			
$6\frac{1}{4}$			98	96	94	91	88	83	80	76	72		
$6\frac{1}{2}$			107	105	102	99	96	91	89	85	80		
$6\frac{3}{4}$			116	114	111	108	105	100	98	93	89		
7			125	123	120	117	114	110	107	103	98	93	84
$7\frac{1}{4}$			134	132	130	127	124	119	116	112	108	103	93
$7\frac{1}{2}$			144	142	139	137	133	129	126	122	117	113	102
$7\frac{3}{4}$			154	152	150	147	144	139	136	132	128	123	112
8			165	163	160	157	154	150	147	143	138	133	122
$8\frac{1}{4}$			176	174	171	168	165	160	158	154	149	144	133
$8\frac{1}{2}$			187	185	182	179	176	172	169	165	160	155	150
9			210	208	206	203	200	195	192	188	184	179	174
$9\frac{1}{2}$			234	232	230	227	224	220	216	212	209	206	198
$9\frac{3}{4}$			248	245	243	240	237	232	229	225	221	216	211
10			261	259	257	254	251	246	243	239	235	230	225
11			317	315	313	310	307	302	299	295	291	286	281
12			379	377	374	371	368	364	361	357	352	347	342

*See API Specification 7, Table 13 for API standard drill collar dimensions. For special configurations of drill collars, consult manufacture for reduction in weight.

where OD_{bit} = outside diameter of bit size
OD_{cc} = outside diameter of casing coupling
OD_{dc} = outside diameter of drill collar

Example 4.6.1
The casing string for a certain well is to consist of $13\frac{3}{8}$ casing with a coupling outside diameter of 14.375 in. Determine the required outside diameter of the drill collar to avoid possible problems with running casing if the borehole diameter is assumed to be $17\frac{1}{2}$ in.

Solution

$$D_{dc} = 2 \cdot 14.375 - 17.5 = 11.15$$

Being aware of standard drill collar sizes (see Table 4.6.1), an 11-in. or 12-in. drill collar should be selected. To avoid such a large drill collar OD, a stabilizer or a proper-sized square drill collar (or a combination of the two) should be

Table 4.6.4 *Popular Hole and Drill Collar Sizes [3]*

$4\frac{3}{4}$	$2\frac{1}{8} \times 1\frac{1}{4}$ with $2\frac{7}{8}$ PAC or $2\frac{3}{8}$ API Reg	$3\frac{1}{2} \times 1\frac{1}{2}$ with $2\frac{7}{8}$ PAC or $2\frac{3}{8}$ API Reg
$5\frac{7}{8}-6\frac{1}{8}$	$4\frac{1}{8} \times 2$ with NC31	$4\frac{3}{4} \times 2$ with NC31 or $3\frac{1}{2} \times$ H
$6\frac{1}{2}-6\frac{3}{4}$	$4\frac{3}{4} \times 2\frac{1}{4}$ with NC38	5 or $5\frac{1}{4} \times 2$ with NC38
$7\frac{5}{8}-7\frac{7}{8}$	$6 \times 2\frac{13}{16}$ with NC46	$6\frac{1}{4}$ or $6\frac{1}{2}$ OD $\times 2$ or $2\frac{1}{4}$ with NC46
$8\frac{1}{2}-8\frac{3}{4}$	$6\frac{1}{4} \times 2\frac{13}{16}$ with NC46 or $6\frac{1}{2} \times 2\frac{13}{16}$ with NC46 or NC50	$6\frac{3}{4}$ or $7 \times 2\frac{1}{4}$ with NC50
$9\frac{1}{2}-9\frac{7}{8}$	$7 \times 2\frac{13}{16}$ with NC50 $8 \times 2\frac{13}{16}$ with NC50	$7 \times 2\frac{1}{4}$ with NC50 $8 \times 2\frac{13}{16}$ with $6\frac{5}{8}$ API Reg
$10\frac{5}{8}-11$	$7 \times 2\frac{13}{16}$ with NC 50 or $8 \times 2\frac{13}{16}$ with $6\frac{5}{8}$ API Reg	$8 \times 2\frac{13}{16}$ with $6\frac{5}{8}$ API Reg $9 \times 2\frac{13}{16}$ $7\frac{5}{8}$ API Reg
$12\frac{1}{4}$	$8 \times 2\frac{13}{16}$ with $6\frac{5}{8}$ API Reg	$8 \times 2\frac{13}{16}$ with $6\frac{5}{8}$ API Reg $9 \times 2\frac{13}{16}$ $7\frac{5}{8}$ API Reg $10 \times 2\frac{13}{16}$ or 3 with $7\frac{5}{8}$ API Reg
$17\frac{1}{2}$	$8 \times 2\frac{13}{16}$ with $6\frac{5}{8}$ API Reg	$8 \times 2\frac{13}{16}$ with $6\frac{5}{8}$ API Reg $9 \times 2\frac{13}{16}$ $7\frac{5}{8}$ API Reg $10 \times 2\frac{13}{16}$ or 3 with $7\frac{5}{8}$ API Reg 11×3 with $8\frac{5}{8}$ API Reg
$18\frac{1}{2}-26$	Drill collar programs are the same as for the next reduced hole size.	

placed above the rock bit. If there is no tendency to cause an undersized hole, the largest drill collars that can be washed over are usually selected.

In general, if the optimal drilling programs require large drill collars, the operator should not hesitate to use them. Typical hole and drill collar sizes used in soft and hard formations are listed in Table 4.6.4.

4.6.1.2 Length of Drill Collars

The length of the drill collar string should be as short as possible but adequate to create the desired weight on bit. In vertical holes, ordinary drill pipe must never be used for exerting bit weight. In deviated holes, where the axial component of drill collar weight is sufficient for bit weight, heavy-weight pipe is often used in lieu of drill collars to reduce rotating torque. In highly deviated holes, where the axial component of the drill collar weight is below the needed bit weight, drill collars are not used, and heavy weight pipe is put high in the string in the vertical part of the hole. When this is done, bit weight is transmitted through the drill pipe. Extra care and planning must be done when apply bit weight through the pipe because buckling of the pipe can occur, which can lead to fatigue failures and accelerated wear of the pipe or tool joints.

The required length of drill collars can be obtained from the following formula:

$$L_{dc} = \frac{DF \cdot W}{W_{dc} \cdot K_b \cdot \cos \alpha} \qquad [4.6.2]$$

where DF = is the design factor (DF = 1.2–1.3), and the weight on bit (lb) is determined by the following:

W_{dc} = unit weight of drill collar in air (lb/ft)
K_b = buoyancy factor
$K_b = 1 - \frac{\lambda_m}{\lambda_{st}}$

where λ_m is the drilling fluid density (lb/gal), λ_{st} is the drill collar density (lb/gal) (for steel, $\lambda_{st} = 65.5$ lb/gal), and α is the hole inclination from vertical (degrees).

When possible, the drill pipe should be in tension. The lower most collar has the maximum compressive load, which is transmitted to the bit, and the upper most collar has a tensile load. This means there is some point in the drill collars between the bit and the drill pipe that has a zero axial load. This is called the *neutral point*. The design factor (DF) is needed to place the neutral point below the top of the drill collar string. This will ensure that the pipe is not in compression because of axial vibration or bouncing of the bit and because of inaccurate handling of the brake by the driller.

The excess of drill collars also helps to prevent transverse movement of drill pipe due to the effect of centrifugal force. While the drill string rotates, a centrifugal force is generated that may produce a lateral movement of drill pipe, which causes bending stress and excessive torque. The centrifugal force also contributes to vibration of the drill pipe. Hence, some excess of drill collars is suggested. The magnitude of the design factor to control vibration can be determined by field experiments in any particular set of drilling conditions. Experimental determination of the design factor for preventing compressive loading on the pipe is more difficult. The result of running the pipe in compression can be a fatigue crack leading to a washout or parted pipe.

The pressure area method (PAM), occasionally used for evaluation of drill collar string length, is wrong because it does not consider the triaxial state of stresses that actually occur. Hydrostatic forces cannot cause any buckling of the drill string as long as the density of the string is greater than the density of the drilling fluid.

Example 4.6.2

Determine the required length of 7 in. $\times 2\frac{1}{4}$ in. drill collars with the following conditions:

Desired weight on bit: W = 40,000 lb

Drilling fluid density: $\lambda_m = 10$ lb/gal

Hole deviation from vertical: $\alpha = 20°$

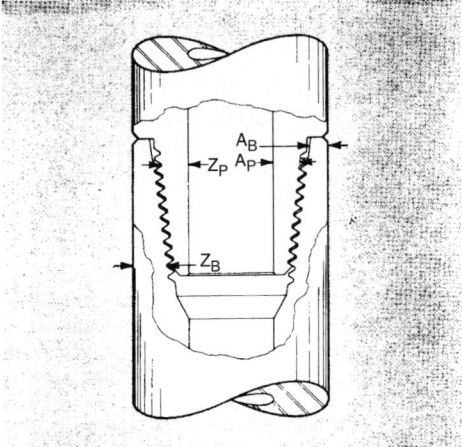

The section modulus, Z_b, of the box should be $2\frac{1}{2}$ times greater than the section modulus, Z_p, of the pin in a drill collar connection. On the right side of the connection are the spots at which the critical area of both the pin (A_p) and box (A_b) should be measured for calculating torsional strength.

Figure 4.6.2 *The drill collar connection. (After C. E. Wilson and W. R. Garrett).*

Solution
From Table 4.6.3, the unit weight of a $7 \times 2\frac{1}{4}$ drill collar is 117 lb/ft. The buoyancy factor is

$$K_b = 1 - \frac{10}{65.5} = 0.847$$

Applying Equation 4.6.2 gives

$$L_{dc} = \frac{(1.3)(40,000)}{(147)(0.847)(\cos 20)} = 444 \text{ ft}$$

The closest length, based on 30-ft collars, is 450 ft, which is 15 joints of drill collars.

4.6.1.3 Drill Collar Connections
It is current practice to select the rotary shoulder connection that provides a balanced bending fatigue resistance for the pin and the box. It has been determined empirically that the pin and boxes have approximately equally bending fatigue resistance if the section modulus of the box at its critical zone is 2.5 times the section modulus of the pin at its critical zone. This number is called the *bending strength ratio* (BSR). These critical zones are shown in Figure 4.6.2. Section modulus ratios from 2.25 to 2.75 are considered to be very good and satisfactory performance has been experienced with ratios from 2.0 to 2.3 [6].

The previous statements are valid if the connection is made up with the recommended makeup torque. A set of charts is available from the Drilco Division of Smith International, Inc. Some of these charts are presented in Figures 4.6.3 to 4.6.10. The charts are used as follows:

- The best group of connections are those that appear in the **shaded** sections of the charts.
- The second best group of connections are those that lie in the unshaded section to the left of the shaded section.
- The third best group of connections are those that lie in the unshaded section to right of the shaded section.
- The nearer the connection lies to the reference line, either to the right or the left, the more desirable is its selection.

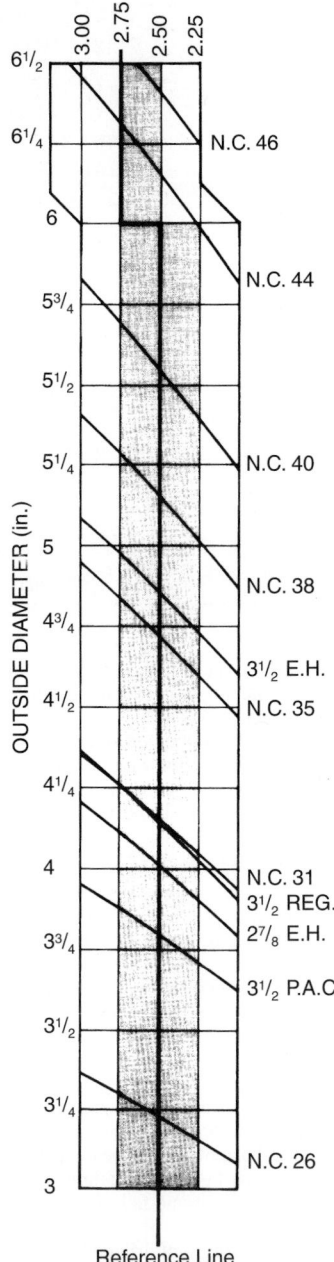

Figure 4.6.3 *Practical chart for drill collar selection—2-in. ID. (From Drilco, Division of Smith International, Inc.).*

Example 4.6.3
Select the best connection for $9\frac{3}{4}$ in. $\times 2\frac{9}{16}$ in. drill collars.

Solution
For average conditions, select in the following order:

- Best: NC70; shaded area and nearest reference line
- Second best: $7\frac{5}{8}$ REG low torque face; light area to left of reference line

2¹/₄" ID

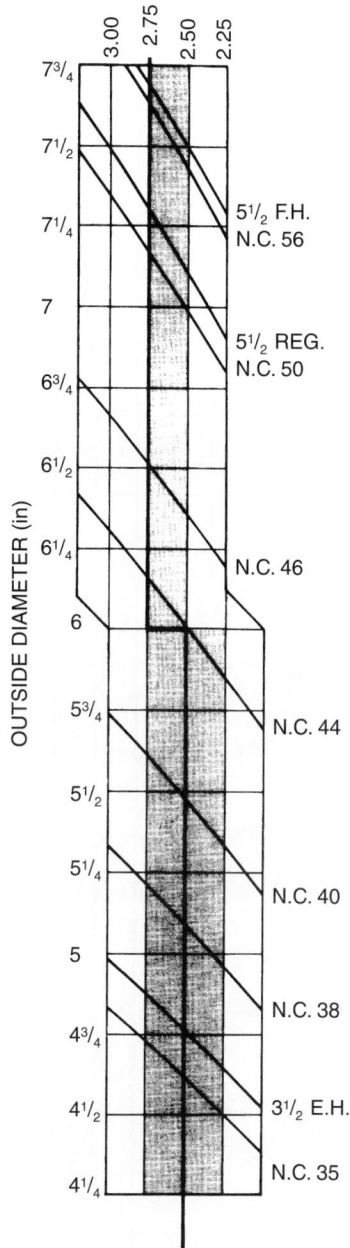

2¹/₂" ID

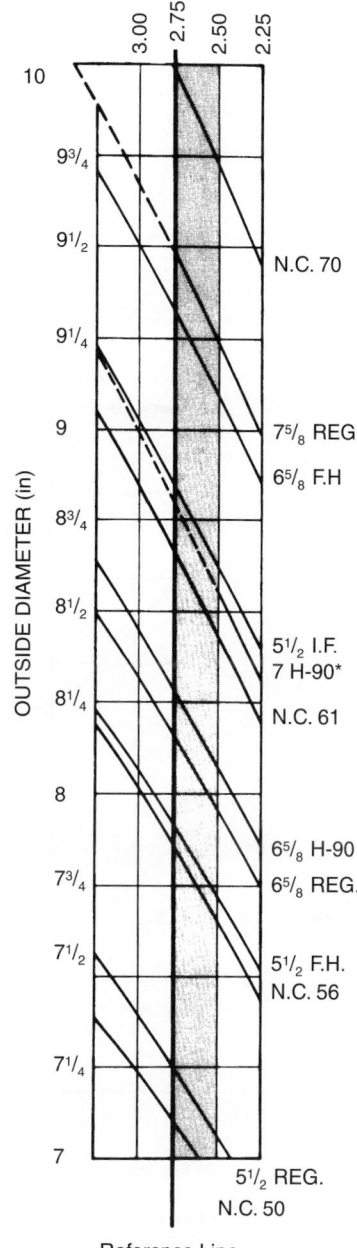

Figure 4.6.4 *Practical chart for drill collar selection—2-¹/₄-in. ID. (From Drilco, Division of Smith International, Inc.).*

Figure 4.6.5 *Practical chart for drill collar selection—2-¹/₂-in. ID. (From Drilco, Division of Smith International, Inc.).*

- Third best: $7\frac{5}{8}$ H-90 low torque face; light area to right of reference line

In extremely abrasive or corrosive conditions, the following selections may be made. In this environment, box wear could be a factor, which means as the drill collars are used, the box diameter decreases and the BSR decreases. Corrosion decreased the fatigue resistance of material. The box is exposed to the corrosive environment, but the pin is not. A larger OD of the box increases its stiffness and decreases the bending stress. More bending is taken by the pin.

- Best: $7\frac{5}{8}$ REG low troque face; which has the strongest box; light area to left of reference line
- Second best: NC70, second strongest box
- Third best: $7\frac{5}{8}$ H-90, weakest box

2¹/₂" ID

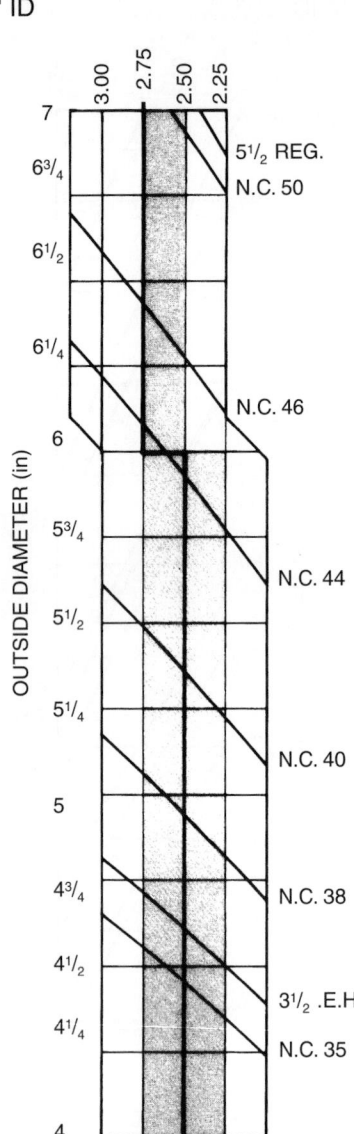

Figure 4.6.6 *Practical chart for drill collar selection—2-$\frac{1}{4}$-in. ID. (From Drilco, Division of Smith International, Inc.).*

2¹³/₁₆" ID

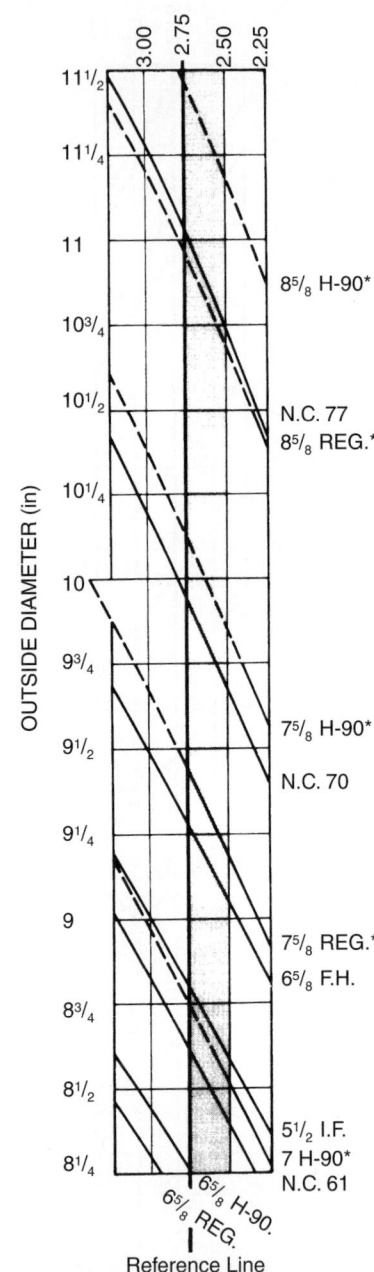

Figure 4.6.7 *Practical chart for drill collar selection—2-$\frac{13}{16}$-in. ID. (From Drilco, Division of Smith International, Inc.).*

4.6.1.4 Recommended Makeup Torque for Drill Collars

The rotary shouldered connections must be made up with a torque high enough to prevent the makeup shoulders from separating downhole. Shoulder separation can occur when rotating in doglegs, when drill collars are buckled by compressive loads, or when tensile loads, are applied to the drill collars. This is important because the makeup shoulder is the only seal in the connection. The threads themselves do not seal because they are designed with clearance between the crest of one thread and the root of its mating thread. The clearance acts as a channel for thread compound and

small solid particles that could be in the drilling fluid. A second reason to prevent shoulder separation is because when the shoulders separate, the box no longer takes any bending load, and the total bending load is felt by the pin.

To keep the shoulders together, the shoulder load must be high enough to create a compressive stress at the shoulder face, capable of offsetting the bending that occurs due to drill collar buckling or rotating in doglegs. This compressive

$2^{13}/_{16}$" ID

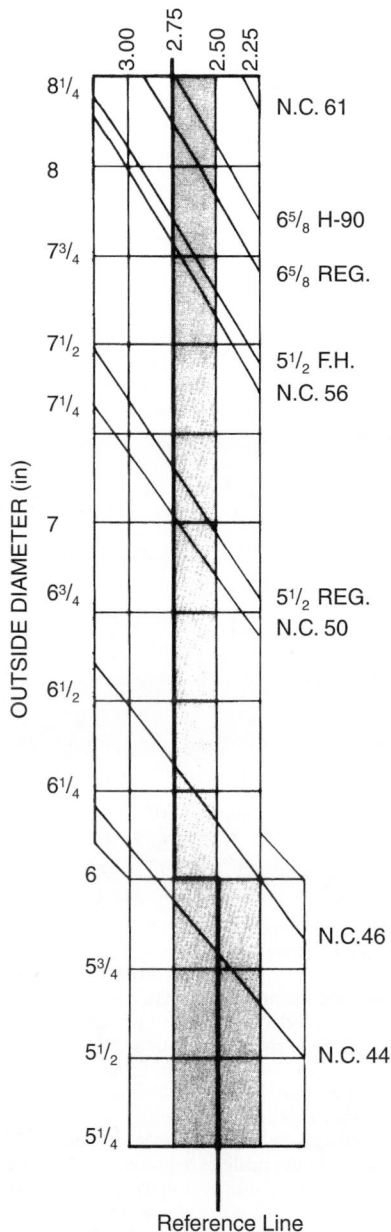

Figure 4.6.8 *Practical chart for drill collar selection—2-$\frac{13}{16}$-in. ID. (From Drilco, Division of Smith International, Inc.).*

3" ID

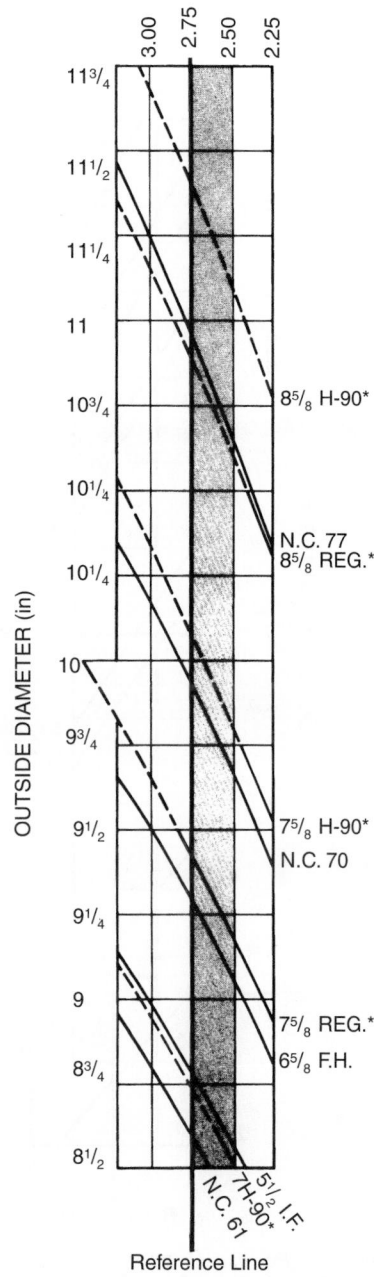

Figure 4.6.9 *Practical chart for drill collar selection—3-in. ID. (From Drilco, Division of Smith International, Inc.).*

shoulder load is generated by makeup torque. Field observations indicate that the makeup torque should create an average axial stress of 62,500 psi in the pin or box, whichever is has the lower cross-sectional area. The cross-section of the pin is taken at $\frac{3}{4}$ in. from the makeup shoulder. The cross-sectional area of the box is taken at $\frac{3}{8}$ in. from the makeup shoulder. The formulas for calculating the cross-sectional areas and the makeup torque can be found in API Recommended Practice for Drill Stem Design and Operating Limits (API RP7G). Makeup torque creates a tensile stress in the pin and can reduce the fatigue life if makeup torque is too high.

The recommended makeup torque for drill collars is given in Table 4.6.5.

4.6.1.5 Drill Collar Buckling

In a straight vertical hole with no weight on the bit, a string of drill collars remains straight. As the weight on the bit

3" ID

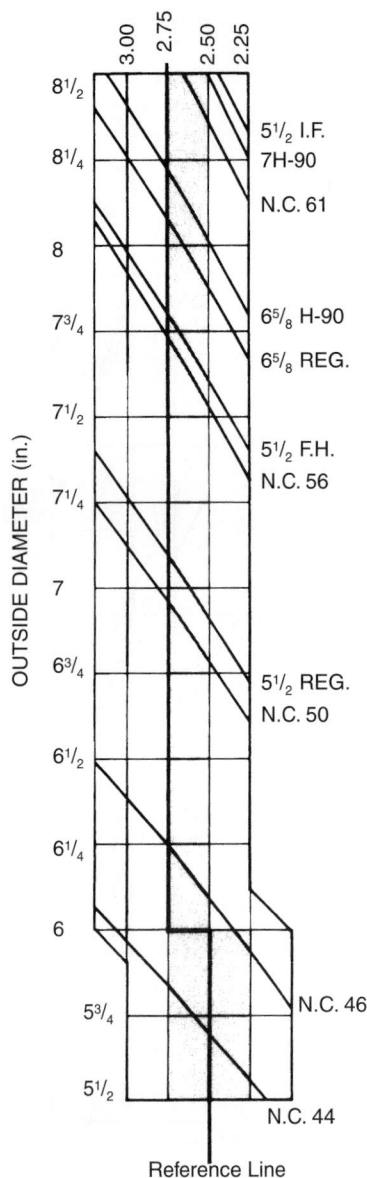

Figure 4.6.10 *Practical chart for drill collar selection—3-in. ID. (From Drilco, Division of Smith International, Inc.).*

is increased, compressive loads are induced into the collars, with the highest compressive load just above the bit and decreasing to zero at the neutral point. As weight is increased, the drill collars or drill pipe buckles and contacts the wall of the hole. If weight is increased further, the string buckles a second time and contacts the borehole at two points. With still further increased weight on the bit, the third and higher order of buckling occurs. The problem of drill collars buckling in vertical holes has been studied by A. Lubinski [8], and the weight on the bit that results in

first- and second-order buckling can be calculated follows:

$$W_{crI} = 1.94(EIp^2)^{1/3} \qquad [4.6.3]$$

$$W_{crII} = 3.75(EIp^2)^{1/3} \qquad [4.6.4]$$

where E is the modulus of elasticity for drill collars (lb/in.^{2}). For steel, $E = 4320 \times 10^6$ lb/ft^2, p is the unit weight of drill collar in drilling fluid (lb/ft), and I is the moment of inertial of drill collar cross-section with respect to its diameter (ft).

$$I = \frac{\pi}{64}(D^4 - d^4) \qquad [4.6.5]$$

where D is the outside diameter of drill collar (ft) and d is the inside diameter of drill collar (in.).

Example 4.6.4
Find the magnitude of the weight on bit and corresponding length of $6\frac{3}{4}$ in. $\times 2\frac{1}{4}$ in. drill collars that result in second-order buckling. Mud weight = 12 lb/gal.   • ▲

Solution
Moment of inertia:

$$I = \frac{\pi}{64}\left[\left(\frac{6.75}{12}\right)^4 - \left(\frac{2.25}{12}\right)^4\right] = 4.853 \times 10^{-3} \text{ ft}^4$$

Unit weight of drill collar in air = 108 lb/ft.
Unit weight of drill collar in drilling fluid:

$$p = 108\left(1 - \frac{12}{65.4}\right) = 88.18 \text{ lb/ft}$$

For weight on the bit that results in second-order buckling, use Equation 4.6.4:

$$W_{crII} = 3.75\,[(4320 \times 10^6)(48.53 \times 10^{-3})(88.18)^2]^{\frac{1}{3}}$$

$$= 20,485 \text{ lb}$$

Corresponding length of drill collars:

$$L_{dc} = \frac{20,485}{88.18} = 232 \text{ ft}$$

L_{dc} is the distance in which the drill collars are in compression or subject to buckling. It is the distance from the bit to the neutral point.

Lubinski also found [8] that do drill a vertical hole in homogeneous formations, it is best to carry less weight on the bit than the critical value of the first order at which the drill string buckles. However, if such weight is not sufficient, it is advisable to avoid the weight that falls between the first- and second-order buckling and to carry a weight close to the critical value of the third order.

In many instances, the previous statement holds true if formations being drilled are horizontal. When drilling in dipping formations, a proper drill collar stabilization is required for vertical or nearly vertical hole drilling. In an inclined hole, a critical value of weight on the bit that produces buckling may be calculated from the formula given by R. Dawson and P. R. Palsy [7].

$$W_{crit} = 2\left(\frac{EIp\sin\alpha}{r}\right) \qquad [4.6.6]$$

where α is the hole inclination measured from vertical (degrees), r is the radial clearance between drill collar and borehole wall (ft), and E, I, and p are as defined in Equations 4.6.3, 4.6.4, and 4.6.5.

It can be seen by this equation that the critical weight is very high in highly deviated holes. As the hole angle increases, the resistant to buckling increases the drill collar is supported by the borehole wall and lateral component of the buckling forces must overcome the gravity forces.

(text continued on page 4-137)

Table 4.6.5 *Recommended Makeup Torque[1] for Rotary Shouldered Drill Collar Connections (See footnotes for use of this table.) [4]*

(1)	(2)	(3)	(4)	(5)	(6)	(7)	(8)	(9)	(10)	(11)	(12)	(13)	(14)	(15)
Connection						Minimum Makeup Torque ft-lb [2]								
		OD, in.				Bore of Drill Collar, inches								
Size	Type		1	$1\frac{1}{4}$	$1\frac{1}{2}$	$1\frac{3}{4}$	2	$2\frac{1}{4}$	$2\frac{1}{2}$	$2\frac{13}{16}$	3	$3\frac{1}{4}$	$3\frac{1}{2}$	$3\frac{3}{4}$
API NC23		3	∗2,508	∗2,508	∗2,508									
		$3\frac{1}{8}$	∗3,330	∗3,330	2,647									
		$3\frac{1}{4}$	4,000	3,387	2,647									
$2\frac{3}{8}$	Regular	3		∗2,241	∗2,241	1,749								
		$3\frac{1}{8}$		∗3,028	2,574	1,749								
		$3\frac{1}{4}$		3,285	2,574	1,749								
$2\frac{3}{8}$	PAC[3]	3		∗3,797	∗3,797	2,926								
		$3\frac{3}{8}$		∗4,966	4,151	2,926								
		$3\frac{1}{4}$		5,206	4,151	2,926								
$2\frac{3}{8}$	API IF	$3\frac{1}{2}$		∗4,606	∗4,606	3,697								
API NC 26		$3\frac{3}{4}$		5,501	4,668	3,697								
$2\frac{7}{8}$	Regular	$3\frac{1}{2}$		∗3,838	∗3,838	∗3,838								
		$3\frac{3}{4}$		5,766	4,951	4,002								
		$3\frac{7}{8}$		5,766	4,951	4,002								
$2\frac{7}{8}$	Slim hole													
$2\frac{7}{8}$	Extra hole	$3\frac{3}{4}$		∗4,089	∗4,089	∗4,089								
$3\frac{1}{2}$	Double streamline	$3\frac{7}{8}$		∗5,352	∗5,352	∗5,352								
$2\frac{7}{8}$	Mod. open	$4\frac{1}{8}$		∗8,059	∗8,059	7,433								
$2\frac{7}{8}$	API IF	$3\frac{7}{8}$		∗4,640	∗4,640	∗4,640	∗4,640							
API NC 31		$4\frac{1}{8}$		∗7,390	∗7,390	∗7,390	6,853							
$3\frac{1}{2}$	Regular	$4\frac{1}{8}$		∗6,466	∗6,466	∗6,466	∗6,466	5,685						
		$4\frac{1}{4}$		∗7,886	∗7,886	∗7,886	7,115	5,685						
		$4\frac{1}{2}$		10,471	9,514	8,394	7,115	5,685						
$3\frac{1}{2}$	Slim hole	$4\frac{1}{4}$		∗8,858	∗8,858	8,161	6,853	5,391						
		$4\frac{1}{2}$		10,286	9,307	8,161	6,853	5,391						
API NC 35		$4\frac{1}{2}$				∗9,038	∗9,038	∗9,038	7,411					
		$4\frac{3}{4}$				12,273	10,826	9,202	7,411					
		5				12,273	10,826	9,202	7,411					
$3\frac{1}{2}$	Extra hole	$4\frac{1}{4}$				∗5,161	∗5,161	∗5,161	∗5,161					
4	Slim hole	$4\frac{1}{2}$				∗8,479	∗8,479	∗8,479	8,311					
$3\frac{1}{2}$	Mod. open	$4\frac{3}{4}$				∗12,074	11,803	10,144	8,311					
		5				13,283	11,803	10,144	8,311					
		$5\frac{1}{4}$				13,283	11,803	10,144	8,311					
$3\frac{1}{2}$	API IF	$4\frac{3}{4}$				∗9,986	∗9,986	∗9,986	∗9,986	8,315				
API NC 38		5				∗13,949	∗13,949	12,907	10,977	8,315				
$4\frac{1}{2}$	Slim hole	$5\frac{1}{4}$				16,207	14,643	12,907	10,977	8,315				
		$5\frac{1}{2}$				16,207	14,643	12,907	10,977	8,315				
$3\frac{1}{2}$	H-90[4]	$4\frac{3}{4}$				∗8,786	∗8,786	∗8,786	∗8,786	∗8,786				
		5				∗12,794	∗12,794	∗12,794	∗12,794	10,408				
		$5\frac{1}{4}$				∗17,094	16,929	15,137	13,151	10,408				
		$5\frac{1}{2}$				18,522	16,929	15,137	13,151	10,408				
4	Full hole	5				∗10,910	∗10,910	∗10,910	∗10,910	*10,910				
API NC 40		$5\frac{1}{4}$				∗15,290	∗15,290	∗15,290	14,969	12,125				
4	Mod. open	$5\frac{1}{2}$				∗19,985	18,886	17,028	14,969	12,125				

(continued)

Table 4.6.5 (continued)

(1)	(2)	(3)	(4)	(5)	(6)	(7)	(8)	(9)	(10)	(11)	(12)	(13)	(14)	(15)
Connection			Minimum Makeup Torque ft-lb [2]											
		OD,				Bore of Drill Collar, inches								
Size	Type	in.	1	$1\frac{1}{4}$	$1\frac{1}{2}$	$1\frac{3}{4}$	2	$2\frac{1}{4}$	$2\frac{1}{2}$	$2\frac{13}{16}$	3	$3\frac{1}{4}$	$3\frac{1}{2}$	$3\frac{3}{4}$
$4\frac{1}{2}$	Double streamline	$5\frac{3}{4}$				20,539	18,886	17,028	14,969	12,125				
		6				20,539	18,886	17,028	14,969	12,125				
4	H-90[4]	$5\frac{1}{4}$				*12,590	*12,590	*12,590	*12,590	*12,590				
		$5\frac{1}{2}$				*17,401	*17,401	*17,401	*17,401	16,536				
		$5\frac{3}{4}$				*22,531	*22,531	21,714	19,543	16,536				
		6				25,408	23,671	21,714	19,543	16,536				
		$6\frac{1}{4}$				25,408	23,671	21,714	19,543	16,536				
$4\frac{1}{2}$	API regular	$5\frac{1}{2}$					*15,576	*15,576	*15,576	*15,576	*15,576			
		$5\frac{3}{4}$					*20,609	*20,609	*20,609	19,601	16,629			
		6					25,407	23,686	21,749	19,601	16,629			
		$6\frac{1}{4}$					25,407	23,686	21,749	19,601	16,629			
API	NC 44	$5\frac{3}{4}$					*20,895	*20,895	*20,895	*20,895	18,161			
		6					*26,453	25,510	23,493	21,157	18,161			
		$6\frac{1}{4}$					27,300	25,510	23,493	21,157	18,161			
		$6\frac{1}{2}$					27,300	25,510	23,493	21,157	18,161			
$4\frac{1}{2}$	API full hole	$5\frac{1}{2}$						*12,973	*12,973	*12,973	*12,973	*12,973		
		$5\frac{3}{4}$						*18,119	*18,119	*18,119	*18,119	17,900		
		6						*23,605	*23,605	23,028	19,921	17,900		
		$6\frac{1}{4}$						27,294	25,772	23,028	19,921	17,900		
		$6\frac{1}{2}$						27,294	25,772	23,028	19,921	17,900		
$4\frac{1}{2}$	Extra hole	$5\frac{3}{4}$							*17,738	*17,738	*17,738	*17,738		
API	NC 46	6							*23,422	*23,422	22,426	20,311		
4	API IF	$6\frac{1}{4}$							28,021	25,676	22,426	20,311		
$4\frac{1}{2}$	Semi IF	$6\frac{1}{2}$							28,021	25,676	22,426	20,311		
5	Double Streamline	$6\frac{3}{4}$							28,021	25,676	22,426	20,311		
$4\frac{1}{2}$	Mod. open													
$4\frac{1}{2}$	H-90[4]	$5\frac{3}{4}$							*18,019	*18,019	*18,019	*18,019		
		6							*23,681	*23,681	23,159	21,051		
		$6\frac{1}{4}$							28,732	26,397	23,159	21,051		
		$6\frac{1}{2}$							28,732	26,397	23,159	21,051		
		$6\frac{3}{4}$							28,732	26,397	23,159	21,051		
5	H-90[4]	$6\frac{1}{4}$							*25,360	*25,360	*25,360	*25,360	23,988	
		$6\frac{1}{2}$							*31,895	*31,895	29,400	27,167	23,988	
		$6\frac{3}{4}$							35,292	32,825	29,400	27,167	23,988	
		7							35,292	32,825	29,400	27,167	23,988	
$4\frac{1}{2}$	API IF	$6\frac{1}{4}$							*23,004	*23,004	*23,004	*23,004	*23,004	
API	NC 50	$6\frac{1}{2}$							*29,679	*29,679	*29,679	*29,679	26,675	
5	Extra hole	$6\frac{3}{4}$							*36,742	35,824	32,277	29,966	26,675	
5	Mod. open	7							38,379	35,824	32,277	29,966	26,675	
$5\frac{1}{2}$	Double streamline	$7\frac{1}{4}$							38,379	35,824	32,277	29,966	26,675	
5	Semi-IF	$7\frac{1}{2}$							38,379	35,824	32,277	29,966	26,675	
$5\frac{1}{2}$	H-90[4]	$6\frac{3}{4}$							*34,508	*34,508	*34,508	34,142	30,781	
		7							*41,993	40,117	36,501	34,142	30,781	
		$7\frac{1}{4}$							42,719	40,117	36,501	34,142	30,781	
		$7\frac{1}{2}$							42,719	40,117	36,501	34,142	30,781	
$5\frac{1}{2}$	API regular	$6\frac{3}{4}$							*31,941	*31,941	*31,941	*31,941	30,495	
		7							*39,419	*39,419	36,235	33,868	30,495	

(continued)

Table 4.6.5 *(continued)*

(1)	(2)	(3)	(4)	(5)	(6)	(7)	(8)	(9)	(10)	(11)	(12)	(13)	(14)	(15)
Connection			Minimum Makeup Torque ft-lb [2]											
		OD,						Bore of Drill Collar, inches						
Size	Type	in.	1	$1\frac{1}{4}$	$1\frac{1}{2}$	$1\frac{3}{4}$	2	$2\frac{1}{4}$	$2\frac{1}{2}$	$2\frac{13}{16}$	3	$3\frac{1}{4}$	$3\frac{1}{2}$	$3\frac{3}{4}$
		$7\frac{1}{4}$						42,481	39,866	36,235	33,868	30,495		
		$7\frac{1}{2}$						42,481	39,866	36,235	33,868	30,495		
$5\frac{1}{2}$	API full hole	7						*32,762	*32,762	*32,762	*32,762	*32,762		
		$7\frac{1}{4}$						*40,998	*40,998	*40,998	*40,998	*40,998		
		$7\frac{1}{2}$						*49,661	*49,661	47,756	45,190	41,533		
		$7\frac{3}{4}$						54,515	51,687	47,756	45,190	41,533		
API	NC 56	$7\frac{1}{4}$							*40,498	*40,498	*40,498	*40,498		
		$7\frac{1}{2}$							*49,060	48,221	45,680	42,058		
		$7\frac{3}{4}$							52,115	48,221	45,680	42,058		
		8							52,115	48,221	45,680	42,058		
$6\frac{5}{8}$	API regular	$7\frac{1}{2}$							*46,399	*46,399	*46,399	*46,399		
		$7\frac{3}{4}$							*55,627	53,346	50,704	46,936		
		8							57,393	53,346	50,704	46,936		
		$8\frac{1}{4}$							57,393	53,346	50,704	46,936		
$6\frac{5}{8}$	H-90[4]	$7\frac{1}{2}$							*46,509	*46,509	*46,509	*46,509		
		$7\frac{3}{4}$							*55,708	*55,708	53,629	49,855		
		8							60,321	56,273	53,629	49,855		
		$8\frac{1}{4}$							60,321	56,273	53,629	49,855		
API	NC 61	8							*55,131	*55,131	*55,131	*55,131		
		$8\frac{1}{4}$							*65,438	*65,438	*65,438	61,624		
		$8\frac{1}{2}$							72,670	68,398	65,607	61,624		
		$8\frac{3}{4}$							72,670	68,398	65,607	61,624		
		9							72,670	68,398	65,607	61,624		
$5\frac{1}{2}$	API IF	8							*56,641	*56,641	*56,641	*56,641	*56,641	
		$8\frac{1}{4}$							*67,133	*67,133	*67,133	63,381	59,027	
		$8\frac{1}{2}$							74,626	70,277	67,436	63,381	59,027	
		$8\frac{3}{4}$							74,626	70,277	67,436	63,381	59,027	
		9							74,626	70,277	67,436	63,381	59,027	
		$9\frac{1}{4}$							74,626	70,277	67,436	63,381	59,027	
$6\frac{5}{8}$	API full hole	$8\frac{1}{2}$							*67,789	*67,789	*67,789	*67,789	*67,789	67,184
		$8\frac{3}{4}$							*79,554	*79,554	*79,554	76,706	72,102	67,184
		9							88,582	83,992	80,991	76,706	72,102	67,184
		$9\frac{1}{4}$							88,582	83,992	80,991	76,706	72,102	67,184
		$9\frac{1}{2}$							88,582	83,992	80,991	76,706	72,102	67,184
API	NC70	9							*75,781	*75,781	*75,781	*75,781	*75,781	*75,781
		$9\frac{1}{4}$							*88,802	*88,802	*88,802	*88,802	*88,802	*88,802
		$9\frac{1}{2}$							*102,354	*102,354	*102,354	101,107	96,214	90,984
		$9\frac{3}{4}$							113,710	108,841	105,657	101,107	96,214	90,984
		10							113,710	108,841	105,657	101,107	96,214	90,984
		$10\frac{1}{4}$							113,710	108,841	105,657	101,107	96,214	90,984
API	NC77	10							*108,194	*108,194	*108,194	*108,194	*108,194	*108,194
		$10\frac{1}{4}$							*124,051	*124,051	*124,051	*124,051	*124,051	*124,051
		$10\frac{1}{2}$							*140,491	*140,491	*140,491	140,488	135,119	129,375
		$10\frac{3}{4}$							154,297	148,965	145,476	140,488	135,119	129,375
		11							154,297	148,965	145,476	140,488	135,119	129,375
7	H-90[4]	8							*53,454	*53,454	*53,454	*53,454	*53,454	*53,454
		$8\frac{1}{4}$							*63,738	*63,738	*63,738	*63,738	60,971	56,382

(continued)

Table 4.6.5 (continued)

(1)	(2)	(3)	(4)	(5)	(6)	(7)	(8)	(9)	(10)	(11)	(12)	(13)	(14)	(15)
Connection									Minimum Makeup Torque ft-lb [2]					
		OD,							Bore of Drill Collar, inches					
Size	Type	in.	1	$1\frac{1}{4}$	$1\frac{1}{2}$	$1\frac{3}{4}$	2	$2\frac{1}{4}$	$2\frac{1}{2}$	$2\frac{13}{16}$	3	$3\frac{1}{4}$	$3\frac{1}{2}$	$3\frac{3}{4}$
$7\frac{5}{8}$ API regular		$8\frac{1}{2}$							*74,478	72,066	69,265	65,267	60,971	56,382
		$8\frac{1}{2}$							*60,402	*60,402	*60,402	*60,402	*60,402	*60,402
		$8\frac{3}{4}$							*72,169	*72,169	*72,169	*72,169	*72,169	*72,169
$7\frac{5}{8}$ API regular		9							*84,442	*84,442	*84,442	84,221	79,536	74,529
		$9\frac{1}{4}$							96,301	91,633	88,580	84,221	79,536	74,529
		$9\frac{1}{2}$							96,301	91,633	88,580	84,221	79,536	74,529
$7\frac{5}{8}$ H-90[4]		9							*73,017	*73,017	*73,017	*73,017	*73,017	*73,017
		$9\frac{1}{4}$							*86,006	*86,006	*86,006	*86,006	*86,006	*86,006
		$9\frac{1}{2}$							*99,508	*99,508	*99,508	*99,508	*99,508	96,285
$8\frac{5}{8}$ API regular		10							*109,345	*109,345	*109,345	*109,345	*109,345	*109,345
		$10\frac{1}{4}$							*125,263	*125,263	*125,263	*125,263	*125,263	125,034
		$10\frac{1}{2}$							*141,767	*141,767	141,134	136,146	130,077	125,034
$8\frac{5}{8}$ H-90[4]		$10\frac{1}{4}$							*113,482	*113,482	*113,482	*113,482	*113,482	*113,482
		$10\frac{1}{2}$							*130,063	*130,063	*130,063	*130,063	*130,063	*130,063
7 H-90[4] (with low torque face)		$8\frac{3}{4}$								*68,061	*68,061	67,257	62,845	58,131
		9								74,235	71,361	67,257	62,845	58,131
$7\frac{5}{8}$ API regular (with low torque face)		$9\frac{1}{4}$									*73,099	*73,099	*73,099	*73,099
		$9\frac{1}{2}$									*86,463	*86,463	82,457	77,289
		$9\frac{3}{4}$									91,789	87,292	82,457	77,289
		10									91,789	87,292	82,457	77,289
$7\frac{5}{8}$ H-90[4] (with low torque face)		$9\frac{3}{4}$								*91,667	*91,667	*91,667	*91,667	*91,667
		10								*106,260	*106,260	*106,260	104,171	98,804
		$10\frac{1}{4}$								117,112	113,851	109,188	104,171	98,804
		$10\frac{1}{2}$								117,112	113,851	109,188	104,171	98,804
$8\frac{5}{8}$ API regular (with low torque face)		$10\frac{3}{4}$									*112,883	*112,883	*112,883	*112,883
		11									*130,672	*130,672	*130,672	*130,672
		$11\frac{1}{4}$									147,616	142,430	136,846	130,871
$8\frac{5}{8}$ H-90[4] (with low torque face)		$10\frac{3}{4}$									*92,960	*92,960	*92,960	*92,960
		11									*110,781	*110,781	*110,781	*110,781
		$11\frac{1}{4}$									*129,203	*129,203	*129,203	*129,203

[1] Torque figures preceded by an asterisk (*) indicate that the weaker member for the corresponding outside diameter (OD) and bore is the BOX; for all other torque values, the weaker member is the PIN.

[2] In each connection size and type group, torque values apply to all connection types in the group, when used with the same drill collar outside diameter and bone, i.e., $2\frac{3}{8}$ API IF, API NC 26, and $2\frac{7}{8}$ Slim Hole connections used with $3\frac{1}{2} \times 1\frac{1}{4}$ drill collars all have the same minimum make-up torque of 4600 ft. lb., and the BOX is the weaker member.

[3] Stress-relief features are disregarded for make-up torque.

[1] Basis of calculations for recommended make-up torque assumed the use of a thread compound containing 40-60% by weight of finely powdered metallic zinc or 60% by weight of finely powdered metallic lead, with not more than 0.3% total active sulfur (reference the caution regarding the use of hazardous materials in Appendix G of Specification 7) applied thoroughly to all threads and shoulders and using the modified Screw Jack formula in A.8, and a unit stress of 62,500 psi in the box or pin, whichever is weaker.

[2] Normal torque range is tabulated value plus 10 percent. Higher torque values may be used under extreme conditions.

[3] Makeup torque for 2? PAC connection is based on 87,500 psi stress and other factors listed in footnote 1.

[4] Makeup torque for H-90 connection is based on 56,200 psi stress and other factors listed in footnote 1.

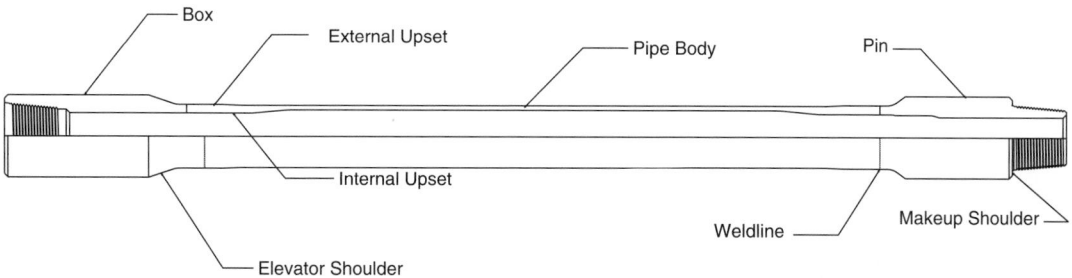

Figure 4.6.11

(text continued from page 4-132)

This explains why heavy-weight drill pipe is successfully used for creating weight on the bit in highly deviated holes. However, in drilling a vertical or nearly vertical hole, a drill pipe must never be run in effective compression; the neutral point must always reside in the drill collar string.

4.6.1.6 Rig Maintenance of Drill Collars
It is recommended practice to break a different joint on each trip, giving the crew an opportunity to look at each pin and box every third trip. Inspect the shoulders for signs of loose connections, galls, and possible washouts.

Thread protectors should be used on pins and boxes when picking up or laying down the drill collars.

Periodically, based on drilling conditions and experience, a magnetic particle inspection should be performed using a wet fluorescent and black light method. Before storing, the drill collars should be cleaned. If necessary, reface the shoulders with a shoulder refacing tool, and remove the fins on the shoulders by beveling. A good rust-preventative or drill collar compound should be applied to the connections liberally, and thread protectors should be installed.

4.6.2 Drill Pipe
The major portion of drill string is composed of drill pipe. Drill pipe consists of three components: a tube with a pin tool joint welded to one end and a box tool joint welded to the other. Figure 4.6.11 shows a sectioned view of a drill pipe assembly. Before the tool joints are welded to the tube, the tube is upset, or forged, on each end to increase the wall thickness. After upsetting, the tube is heat treated to the proper grade strength. All tool joints are heat treated to the same material yield strength (120,000 psi), regardless of the grade of pipe to which they are attached. Most drill pipe is made from material similar to AISI 4125/30 steel seamless tube. Most tool joints are made from material similar to AISI 4140 steel forgings, tubing, or bars stock.

Drill pipe dimensional and metallurgical specifications are defined by the American Petroleum Institute (API) and published in *API Spec 7 Specifications for Drill Stem Elements* and *API Spec 5D Specifications for Drill Pipe*. Performance characteristics, guidelines for drill pipe use and inspection standards are in *API RP7G Recommended Practice for Drill Stem Design and Operating Limits*. Drill pipe specifications and performance characteristics can also be found in ISO 10407-1, ISO 10407-2, ISO 10424-1, ISO 10424-2, and ISO 11961.

A list of drill pipe assemblies is shown in Table 4.6.6. This list is not all inclusive; it contains drill pipe assemblies with various tool joints but not all OD and ID combinations of tool joints.

The following items are required to completely identify a length of drill pipe:

Pipe size is the pipe OD (in., mm). Drill pipe sizes and wall thicknesses are shown in Table 4.6.8. The API specification for the drill pipe tube is API 5D.

Pipe weight (lb/ft, kg/m) is for the tube only exclusive of tool joints and upset ends and is used to specify wall thickness. Except in a few cases, the tabulated pipe weight is not the calculated pipe weight.

Pipe grade is the pipe yield strength. The API grades are listed in Tables 4.6.7 and 4.6.8. Drill pipe manufacturers may offer higher strength grades or grades designed for specific applications such as drilling in H_2S environments.

Pipe upset The drill pipe tubes are upset on each end to increase the wall thickness. The thicker wall is needed to compensate for loss of material strength during the welding process. There are three configurations of upsets:

- Internal upset (IU), in which the wall thickness is increased by decreasing the ID. This allows smaller OD tool joints to be welded to the pipe. This pipe is sometimes referred to as a *slim hole pipe* and is used in smaller-diameter holes.
- External upset (EU), in which the wall thickness is increased by increasing the OD. This allows larger tool joints to be welded to the pipe. The larger tool joints provide more torsional strength and create a lower pressure drop than those used on IU pipe.
- Internal external upset (IEU), in which the wall thickness is increased by increasing the OD and decreasing the ID. This is the most common upset type on pipe larger than 4 in.

Tool joint type. Table 4.6.9 is a tool joint interchangability chart which shows a number of tool joint types. The API tool joint types are printed in bold. These include API Reg (used mostly for drill collars, bits, subs and other bottomhole assembly components), NC numbered connections, and $5\frac{1}{2}$ and $6\frac{5}{8}$ full hole (FH). Before the NC-type connections were established, tool joint manufacturers often produced interchangeable tool joints with different names. The NC

(text continued on page 4-177)

Table 4.6.6 *Drill Pipe Assembly Properties*

									Pipe Data
Size OD	Nominal Weight	Grade and Upset Type	Torsional Yield Strength	Tensile Yield Strength	Wall Thickness	Nominal ID	Pipe Body Section Area	Pipe Body Moment of Inertia	Pipe Body Polar Moment of Inertia
in.	lb/ft		ft-lb	lb	in.	in.	sq in.	cu in.	cu in.
$2\frac{3}{8}$	4.85	E-75 EU	4,800	97,800	0.190	1.995	1.304	0.784	1.568
	4.85	E-75 EU	4,800	97,800	0.190	1.995	1.304	0.784	1.568
	4.85	E-75 EU	4,800	97,800	0.190	1.995	1.304	0.784	1.568
	4.85	E-75 EU	4,800	97,800	0.190	1.995	1.304	0.784	1.568
$2\frac{3}{8}$	4.85	X-95 EU	6,000	123,900	0.190	1.995	1.304	0.784	1.568
	4.85	X-95 EU	6,000	123,900	0.190	1.995	1.304	0.784	1.568
	4.85	X-95 EU	6,000	123,900	0.190	1.995	1.304	0.784	1.568
	4.85	X-95 EU	6,000	123,900	0.190	1.995	1.304	0.784	1.568
$2\frac{3}{8}$	4.85	G-105 EU	6,700	136,900	0.190	1.995	1.304	0.784	1.568
	4.85	G-105 EU	6,700	136,900	0.190	1.995	1.304	0.784	1.568
	4.85	G-105 EU	6,700	136,900	0.190	1.995	1.304	0.784	1.568
	4.85	G-105 EU	6,700	136,900	0.190	1.995	1.304	0.784	1.568
$2\frac{3}{8}$	4.85	S-135 EU	8,600	176,100	0.190	1.995	1.304	0.784	1.568
	4.85	S-135 EU	8,600	176,100	0.190	1.995	1.304	0.784	1.568
	4.85	S-135 EU	8,600	176,100	0.190	1.995	1.304	0.784	1.568
	4.85	S-135 EU	8,600	176,100	0.190	1.995	1.304	0.784	1.568
$2\frac{3}{8}$	4.85	Z-140 EU	8,900	182,600	0.190	1.995	1.304	0.784	1.568
	4.85	Z-140 EU	8,900	182,600	0.190	1.995	1.304	0.784	1.568
	4.85	Z-140 EU	8,900	182,600	0.190	1.995	1.304	0.784	1.568
$2\frac{3}{8}$	4.85	V-150 EU	9,500	195,600	0.190	1.995	1.304	0.784	1.568
	4.85	V-150 EU	9,500	195,600	0.190	1.995	1.304	0.784	1.568
	4.85	V-150 EU	9,500	195,600	0.190	1.995	1.304	0.784	1.568
$2\frac{3}{8}$	6.65	E-75 EU	6,300	138,200	0.280	1.815	1.843	1.029	2.058
	6.65	E-75 EU	6,300	138,200	0.280	1.815	1.843	1.029	2.058
	6.65	E-75 EU	6,300	138,200	0.280	1.815	1.843	1.029	2.058
	6.65	E-75 EU	6,300	138,200	0.280	1.815	1.843	1.029	2.058
$2\frac{3}{8}$	6.65	X-95 EU	7,900	175,100	0.280	1.815	1.843	1.029	2.058
	6.65	X-95 EU	7,900	175,100	0.280	1.815	1.843	1.029	2.058
	6.65	X-95 EU	7,900	175,100	0.280	1.815	1.843	1.029	2.058
	6.65	X-95 EU	7,900	175,100	0.280	1.815	1.843	1.029	2.058
$2\frac{3}{8}$	6.65	G-105 EU	8,800	193,500	0.280	1.815	1.843	1.029	2.058
	6.65	G-105 EU	8,800	193,500	0.280	1.815	1.843	1.029	2.058
	6.65	G-105 EU	8,800	193,500	0.280	1.815	1.843	1.029	2.058
	6.65	G-105 EU	8,800	193,500	0.280	1.815	1.843	1.029	2.058
$2\frac{3}{8}$	6.65	S-135 EU	11,300	248,800	0.280	1.815	1.843	1.029	2.058
	6.65	S-135 EU	11,300	248,800	0.280	1.815	1.843	1.029	2.058
	6.65	S-135 EU	11,300	248,800	0.280	1.815	1.843	1.029	2.058
	6.65	S-135 EU	11,300	248,800	0.280	1.815	1.843	1.029	2.058
	6.65	S-135 EU	11,300	248,800	0.280	1.815	1.843	1.029	2.058
$2\frac{3}{8}$	6.65	Z-140 EU	11,700	258,000	0.280	1.815	1.843	1.029	2.058
	6.65	Z-140 EU	11,700	258,000	0.280	1.815	1.843	1.029	2.058
	6.65	Z-140 EU	11,700	258,000	0.280	1.815	1.843	1.029	2.058
	6.65	Z-140 EU	11,700	258,000	0.280	1.815	1.843	1.029	2.058
$2\frac{3}{8}$	6.65	V-150 EU	12,500	276,400	0.280	1.815	1.843	1.029	2.058
	6.65	V-150 EU	12,500	276,400	0.280	1.815	1.843	1.029	2.058
	6.65	V-150 EU	12,500	276,400	0.280	1.815	1.843	1.029	2.058
	6.65	V-150 EU	12,500	276,400	0.280	1.815	1.843	1.029	2.058
$2\frac{7}{8}$	6.85	E-75 IU	8,100	135,900	0.217	2.441	1.812	1.611	3.222
	6.85	E-75 IU	8,100	135,900	0.217	2.441	1.812	1.611	3.222
	6.85	E-75 EU	8,100	135,900	0.217	2.441	1.812	1.611	3.222
	6.85	E-75 IU	8,100	135,900	0.217	2.441	1.812	1.611	3.222

(continued)

Table 4.6.6 (continued)

					Pipe Data				
Size OD	Nominal Weight	Grade and Upset Type	Torsional Yield Strength	Tensile Yield Strength	Wall Thickness	Nominal ID	Pipe Body Section Area	Pipe Body Moment of Inertia	Pipe Body Polar Moment of Inertia
in.	lb/ft		ft-lb	lb	in.	in.	sq in.	cu in.	cu in.
	6.85	E-75 EU	8,100	135,900	0.217	2.441	1.812	1.611	3.222
	6.65	E-75 IU	8,100	135,900	0.217	2.441	1.812	1.611	3.222
	6.85	E-75 EU	8,100	135,900	0.217	2.441	1.812	1.611	3.222
	6.85	E-75 EU	8,100	135,900	0.217	2.441	1.812	1.611	3.222
$2\frac{7}{8}$	6.85	X-95 IU	10,200	172,100	0.217	2.441	1.812	1.611	3.222
	6.85	X-95 IU	10,200	172,100	0.217	2.441	1.812	1.611	3.222
	6.85	X-95 EU	10,200	172,100	0.217	2.441	1.812	1.611	3.222
	6.85	X-95 IU	10,200	172,100	0.217	2.441	1.812	1.611	3.222
	6.85	X-95 EU	10,200	172,100	0.217	2.441	1.812	1.611	3.222
	6.65	X-95 IU	10,200	172,100	0.217	2.441	1.812	1.611	3.222
	6.85	X-95 EU	10,200	172,100	0.217	2.441	1.812	1.611	3.222
	6.85	X-95 EU	10,200	172,100	0.217	2.441	1.812	1.611	3.222
$2\frac{7}{8}$	6.85	G-105 IU	11,300	190,300	0.217	2.441	1.812	1.611	3.222
	6.85	G-105 IU	11,300	190,300	0.217	2.441	1.812	1.611	3.222
	6.85	G-105 EU	11,300	190,300	0.217	2.441	1.812	1.611	3.222
	6.85	G-105 IU	11,300	190,300	0.217	2.441	1.812	1.611	3.222
	6.85	G-105 EU	11,300	190,300	0.217	2.441	1.812	1.611	3.222
	6.65	G-105 IU	11,300	190,300	0.217	2.441	1.812	1.611	3.222
	6.85	G-105 EU	11,300	190,300	0.217	2.441	1.812	1.611	3.222
	6.85	G-105 EU	11,300	190,300	0.217	2.441	1.812	1.611	3.222
$2\frac{7}{8}$	6.85	S-135 IU	14,500	244,600	0.217	2.441	1.812	1.611	3.222
	6.85	S-135 IU	14,500	244,600	0.217	2.441	1.812	1.611	3.222
	6.85	S-135 EU	14,500	244,600	0.217	2.441	1.812	1.611	3.222
	6.85	S-135 IU	14,500	244,600	0.217	2.441	1.812	1.611	3.222
	6.85	S-135 EU	14,500	244,600	0.217	2.441	1.812	1.611	3.222
	6.65	S-135 IU	14,500	244,600	0.217	2.441	1.812	1.611	3.222
	6.85	S-135 EU	14,500	244,600	0.217	2.441	1.812	1.611	3.222
	6.85	S-135 EU	14,500	244,600	0.217	2.441	1.812	1.611	3.222
$2\frac{7}{8}$	6.85	Z-140 IU	15,100	253,700	0.217	2.441	1.812	1.611	3.222
	6.65	Z-140 IU	15,100	253,700	0.217	2.441	1.812	1.611	3.222
	6.85	Z-140 EU	15,100	253,700	0.217	2.441	1.812	1.611	3.222
	6.85	Z-140 EU	15,100	253,700	0.217	2.441	1.812	1.611	3.222
$2\frac{7}{8}$	6.85	V-150 IU	16,200	271,800	0.217	2.441	1.812	1.611	3.222
	6.65	V-150 IU	16,200	271,800	0.217	2.441	1.812	1.611	3.222
	6.85	V-150 EU	16,200	271,800	0.217	2.441	1.812	1.611	3.222
	6.85	V-150 EU	16,200	271,800	0.217	2.441	1.812	1.611	3.222
$2\frac{7}{8}$	10.4	E-75 EU	11,600	214,300	0.362	2.151	2.858	2.303	4.606
	10.4	E-75 EU	11,600	214,300	0.362	2.151	2.858	2.303	4.606
	10.4	E-75 IU	11,600	214,300	0.362	2.151	2.858	2.303	4.606
	10.4	E-75 EU	11,600	214,300	0.362	2.151	2.858	2.303	4.606
	10.4	E-75 EU	11,600	214,300	0.362	2.151	2.858	2.303	4.606
	10.4	E-75 IU	11,600	214,300	0.362	2.151	2.858	2.303	4.606
	10.4	E-75 IU	11,600	214,300	0.362	2.151	2.858	2.303	4.606
	10.4	E-75 EU	11,600	214,300	0.362	2.151	2.858	2.303	4.606
	10.4	E-75 IU	11,600	214,300	0.362	2.151	2.858	2.303	4.606
	10.4	E-75 EU	11,600	214,300	0.362	2.151	2.858	2.303	4.606
$2\frac{7}{8}$	10.4	X-95 EU	14,600	271,500	0.362	2.151	2.858	2.303	4.606
	10.4	X-95 IU	14,600	271,500	0.362	2.151	2.858	2.303	4.606
	10.4	X-95 IU	14,600	271,500	0.362	2.151	2.858	2.303	4.606
	10.4	X-95 EU	14,600	271,500	0.362	2.151	2.858	2.303	4.606
	10.4	X-95 EU	14,600	271,500	0.362	2.151	2.858	2.303	4.606
	10.4	X-95 IU	14,600	271,500	0.362	2.151	2.858	2.303	4.606
	10.4	X-95 IU	14,600	271,500	0.362	2.151	2.858	2.303	4.606
	10.4	X-95 EU	14,600	271,500	0.362	2.151	2.858	2.303	4.606

(continued)

Table 4.6.6   (continued)

					Pipe Data				
Size OD	Nominal Weight	Grade and Upset Type	Torsional Yield Strength	Tensile Yield Strength	Wall Thickness	Nominal ID	Pipe Body Section Area	Pipe Body Moment of Inertia	Pipe Body Polar Moment of Inertia
in.	lb/ft		ft-lb	lb	in.	in.	sq in.	cu in.	cu in.
	10.4	X-95 IU	14,600	271,500	0.362	2.151	2.858	2.303	4.606
	10.4	X-95 EU	14,600	271,500	0.362	2.151	2.858	2.303	4.606
$2\frac{7}{8}$	10.4	G-105 EU	16,200	300,100	0.362	2.151	2.858	2.303	4.606
	10.4	G-105 IU	16,200	300,100	0.362	2.151	2.858	2.303	4.606
	10.4	G-105 IU	16,200	300,100	0.362	2.151	2.858	2.303	4.606
	10.4	G-105 EU	16,200	300,100	0.362	2.151	2.858	2.303	4.606
	10.4	G-105 EU	16,200	300,100	0.362	2.151	2.858	2.303	4.606
	10.4	G-105 IU	16,200	300,100	0.362	2.151	2.858	2.303	4.606
	10.4	G-105 IU	16,200	300,100	0.362	2.151	2.858	2.303	4.606
	10.4	G-105 EU	16,200	300,100	0.362	2.151	2.858	2.303	4.606
	10.4	G-105 IU	16,200	300,100	0.362	2.151	2.858	2.303	4.606
	10.4	G-150 EU	16,200	300,100	0.362	2.151	2.858	2.303	4.606
$2\frac{7}{8}$	10.4	S-135 EU	20,800	385,800	0.362	2.151	2.858	2.303	4.606
	10.4	S-135 IU	20,800	385,800	0.362	2.151	2.858	2.303	4.606
	10.4	S-135 IU	20,800	385,800	0.362	2.151	2.858	2.303	4.606
	10.4	S-135 EU	20,800	385,800	0.362	2.151	2.858	2.303	4.606
	10.4	S-135 EU	20,800	385,800	0.362	2.151	2.858	2.303	4.606
	10.4	S-135 IU	20,800	385,800	0.362	2.151	2.858	2.303	4.606
	10.4	S-135 IU	20,800	385,800	0.362	2.151	2.858	2.303	4.606
	10.4	S-135 EU	20,800	385,800	0.362	2.151	2.858	2.303	4.606
	10.4	S-135 IU	20,800	385,800	0.362	2.151	2.858	2.303	4.606
	10.4	S-135 EU	20,800	385,800	0.362	2.151	2.858	2.303	4.606
	10.4	S-135 EU	20,800	385,800	0.362	2.151	2.858	2.303	4.606
$2\frac{7}{8}$	10.4	Z-140 IU	21,600	400,100	0.362	2.151	2.858	2.303	4.606
	10.4	Z-140 EU	21,600	400,100	0.362	2.151	2.858	2.303	4.606
	10.4	Z-140 IU	21,600	400,100	0.362	2.151	2.858	2.303	4.606
	10.4	Z-140 EU	21,600	400,100	0.362	2.151	2.858	2.303	4.606
	10.4	Z-140 EU	21,600	400,100	0.362	2.151	2.858	2.303	4.606
$2\frac{7}{8}$	10.4	V-150 IU	23,100	428,700	0.362	2.151	2.858	2.303	4.606
	10.4	V-150 EU	23,100	428,700	0.362	2.151	2.858	2.303	4.606
	10.4	V-150 IU	23,100	428,700	0.362	2.151	2.858	2.303	4.606
	10.4	V-150 EU	23,100	428,700	0.362	2.151	2.858	2.303	4.606
	10.4	V-150 EU	23,100	428,700	0.362	2.151	2.858	2.303	4.606
$3\frac{1}{2}$	9.50	E-75 EU	14,100	194,300	0.254	2.992	2.590	3.432	6.865
	9.50	E-75 IU	14,100	194,300	0.254	2.992	2.590	3.432	6.865
	9.50	E-75 IU	14,100	194,300	0.254	2.992	2.590	3.432	6.865
	9.50	E-75 EU	14,100	194,300	0.254	2.992	2.590	3.432	6.865
	9.50	E-75 EU	14,100	194,300	0.254	2.992	2.590	3.432	6.865
	9.50	E-75 IU	14,100	194,300	0.254	2.992	2.590	3.432	6.865
	9.50	E-75 EU	14,100	194,300	0.254	2.992	2.590	3.432	6.865
$3\frac{1}{2}$	9.50	X-95 EU	17,900	246,100	0.254	2.992	2.590	3.432	6.865
	9.50	X-95 IU	17,900	246,100	0.254	2.992	2.590	3.432	6.865
	9.50	X-95 IU	17,900	246,100	0.254	2.992	2.590	3.432	6.865
	9.50	X-95 EU	17,900	246,100	0.254	2.992	2.590	3.432	6.865
	9.50	X-95 EU	17,900	246,100	0.254	2.992	2.590	3.432	6.865
	9.50	X-95 IU	17,900	246,100	0.254	2.992	2.590	3.432	6.865
	9.50	X-95 EU	17,900	246,100	0.254	2.992	2.590	3.432	6.865
$3\frac{1}{2}$	9.50	G-105 EU	19,800	272,000	0.254	2.992	2.590	3.432	6.865
	9.50	G-105 IU	19,800	272,000	0.254	2.992	2.590	3.432	6.865
	9.50	G-105 IU	19,800	272,000	0.254	2.992	2.590	3.432	6.865
	9.50	G-105 EU	19,800	272,000	0.254	2.992	2.590	3.432	6.865
	9.50	G-105 EU	19,800	272,000	0.254	2.992	2.590	3.432	6.865
	9.50	G-105 IU	19,800	272,000	0.254	2.992	2.590	3.432	6.865
	9.50	G-105 EU	19,800	272,000	0.254	2.992	2.590	3.432	6.865

(continued)

Table 4.6.6 *(continued)*

							Pipe Data		
Size OD	Nominal Weight	Grade and Upset Type	Torsional Yield Strength	Tensile Yield Strength	Wall Thickness	Nominal ID	Pipe Body Section Area	Pipe Body Moment of Inertia	Pipe Body Polar Moment of Inertia
in.	lb/ft		ft-lb	lb	in.	in.	sq in.	cu in.	cu in.
$3\frac{1}{2}$	9.50	S-135 EU	25,500	349,700	0.254	2.992	2.590	3.432	6.865
	9.50	S-135 IU	25,500	349,700	0.254	2.992	2.590	3.432	6.865
	9.50	S-135 IU	25,500	349,700	0.254	2.992	2.590	3.432	6.865
	9.50	S-135 EU	25,500	349,700	0.254	2.992	2.590	3.432	6.865
	9.50	S-135 EU	25,500	349,700	0.254	2.992	2.590	3.432	6.865
	9.50	S-135 IU	25,500	349,700	0.254	2.992	2.590	3.432	6.865
	9.50	S-135 EU	25,500	349,700	0.254	2.992	2.590	3.432	6.865
$3\frac{1}{2}$	9.50	Z-140 IU	26,400	362,600	0.254	2.992	2.590	3.432	6.865
	9.50	Z-140 EU	26,400	362,600	0.254	2.992	2.590	3.432	6.865
	9.50	Z-140 IU	26,400	362,600	0.254	2.992	2.590	3.432	6.865
	9.50	Z-140 EU	26,400	362,600	0.254	2.992	2.590	3.432	6.865
$3\frac{1}{2}$	9.50	V-150 IU	28,300	388,500	0.254	2.992	2.590	3.432	6.865
	9.50	V-150 EU	28,300	388,500	0.254	2.992	2.590	3.432	6.865
	9.50	V-150 IU	28,300	388,500	0.254	2.992	2.590	3.432	6.865
	9.50	V-150 EU	28,300	388,500	0.254	2.992	2.590	3.432	6.865
$3\frac{1}{2}$	13.30	E-75 EU	18,600	271,600	0.368	2.764	3.621	4.501	9.002
	13.30	E-75 IU	18,600	271,600	0.368	2.764	3.621	4.501	9.002
	13.30	E-75 IU	18,600	271,600	0.368	2.764	3.621	4.501	9.002
	13.30	E-75 EU	18,600	271,600	0.368	2.764	3.621	4.501	9.002
	13.30	E-75 EU	18,600	271,600	0.368	2.764	3.621	4.501	9.002
	13.30	E-75 IU	18,600	271,600	0.368	2.764	3.621	4.501	9.002
	13.30	E-75 EU	18,600	271,600	0.368	2.764	3.621	4.501	9.002
$3\frac{1}{2}$	13.30	X-95 EU	23,500	344,000	0.368	2.764	3.621	4.501	9.002
	13.30	X-95 IU	23,500	344,000	0.368	2.764	3.621	4.501	9.002
	13.30	X-95 IU	23,500	344,000	0.368	2.764	3.621	4.501	9.002
	13.30	X-95 EU	23,500	344,000	0.368	2.764	3.621	4.501	9.002
	13.30	X-95 EU	23,500	344,000	0.368	2.764	3.621	4.501	9.002
	13.30	X-95 IU	23,500	344,000	0.368	2.764	3.621	4.501	9.002
	13.30	X-95 EU	23,500	344,000	0.368	2.764	3.621	4.501	9.002
$3\frac{1}{2}$	13.30	G-105 EU	26,000	380,200	0.368	2.764	3.621	4.501	9.002
	13.30	G-105 IU	26,000	380,200	0.368	2.764	3.621	4.501	9.002
	13.30	G-105 IU	26,000	380,200	0.368	2.764	3.621	4.501	9.002
	13.30	G-105 EU	26,000	380,200	0.368	2.764	3.621	4.501	9.002
	13.30	G-105 EU	26,000	380,200	0.368	2.764	3.621	4.501	9.002
	13.30	G-105 IU	26,000	380,200	0.368	2.764	3.621	4.501	9.002
	13.30	G-105 EU	26,000	380,200	0.368	2.764	3.621	4.501	9.002
$3\frac{1}{2}$	13.30	S-135 EU	33,400	488,800	0.368	2.764	3.621	4.501	9.002
	13.30	S-135 IU	33,400	488,800	0.368	2.764	3.621	4.501	9.002
	13.30	S-135 IU	33,400	488,800	0.368	2.764	3.621	4.501	9.002
	13.30	S-135 EU	33,400	488,800	0.368	2.764	3.621	4.501	9.002
	13.30	S-135 EU	33,400	488,800	0.368	2.764	3.621	4.501	9.002
	13.30	S-135 IU	33,400	488,800	0.368	2.764	3.621	4.501	9.002
	13.30	S-135 EU	33,400	488,800	0.368	2.764	3.621	4.501	9.002
	13.30	S-135 EU	33,400	488,800	0.368	2.764	3.621	4.501	9.002
$3\frac{1}{2}$	13.30	Z-140 IU	34,600	506,900	0.368	2.764	3.621	4.501	9.002
	13.30	Z-140 EU	34,600	506,900	0.368	2.764	3.621	4.501	9.002
	13.30	Z-140 IU	34,600	506,900	0.368	2.764	3.621	4.501	9.002
	13.30	Z-140 EU	34,600	506,900	0.368	2.764	3.621	4.501	9.002
	13.30	Z-140 EU	34,600	506,900	0.368	2.764	3.621	4.501	9.002
$3\frac{1}{2}$	13.30	V-150 IU	37,100	543,100	0.368	2.764	3.621	4.501	9.002
	13.30	V-150 EU	37,100	543,100	0.368	2.764	3.621	4.501	9.002
	13.30	V-150 IU	37,100	543,100	0.368	2.764	3.621	4.501	9.002
	13.30	V-150 EU	37,100	543,100	0.368	2.764	3.621	4.501	9.002
	13.30	V-150 EU	37,100	543,100	0.368	2.764	3.621	4.501	9.002

(continued)

Table 4.6.6 *(continued)*

					Pipe Data				
Size OD	Nominal Weight	Grade and Upset Type	Torsional Yield Strength	Tensile Yield Strength	Wall Thickness	Nominal ID	Pipe Body Section Area	Pipe Body Moment of Inertia	Pipe Body Polar Moment of Inertia
in.	lb/ft		ft-lb	lb	in.	in.	sq in.	cu in.	cu in.
$3\frac{1}{2}$	15.50	E-75 EU	21,100	322,800	0.499	2.602	4.304	5.116	10.232
	15.50	E-75 EU	21,100	322,800	0.499	2.602	4.304	5.116	10.232
	15.50	E-75 EU	21,100	322,800	0.499	2.602	4.304	5.116	10.232
$3\frac{1}{2}$	15.50	X-95 EU	26,700	408,800	0.499	2.602	4.304	5.116	10.232
	15.50	X-95 EU	26,700	408,800	0.499	2.602	4.304	5.116	10.232
	15.50	X-95 EU	26,700	408,800	0.499	2.602	4.304	5.116	10.232
$3\frac{1}{2}$	15.50	G-105 EU	29,500	451,900	0.499	2.602	4.304	5.116	10.232
	15.50	G-105 EU	29,500	451,900	0.499	2.602	4.304	5.116	10.232
	15.50	G-105 EU	29,500	451,900	0.499	2.602	4.304	5.116	10.232
	15.50	G-105 EU	29,500	451,900	0.499	2.602	4.304	5.116	10.232
$3\frac{1}{2}$	15.50	S-135 EU	38,000	581,000	0.499	2.602	4.304	5.116	10.232
	15.50	S-135 EU	38,000	581,000	0.499	2.602	4.304	5.116	10.232
	15.50	S-135 EU	38,000	581,000	0.499	2.602	4.304	5.116	10.232
	15.50	S-135 EU	38,000	581,000	0.499	2.602	4.304	5.116	10.232
	15.50	S-135 EU	38,000	581,000	0.499	2.602	4.304	5.116	10.232
	15.50	S-135 EU	38,000	581,000	0.499	2.602	4.304	5.116	10.232
$3\frac{1}{2}$	15.50	Z-140 EU	39,400	602,500	0.499	2.602	4.304	5.116	10.232
	15.50	Z-140 EU	39,400	602,500	0.499	2.602	4.304	5.116	10.232
	15.50	Z-140 EU	39,400	602,500	0.499	2.602	4.304	5.116	10.232
	15.50	Z-140 EU	39,400	602,500	0.499	2.602	4.304	5.116	10.232
$3\frac{1}{2}$	15.50	V-150 EU	42,200	645,500	0.499	2.602	4.304	5.116	10.232
	15.50	V-150 EU	42,200	645,500	0.499	2.602	4.304	5.116	10.232
	15.50	V-150 EU	42,200	645,500	0.499	2.602	4.304	5.116	10.232
	15.50	V-150 EU	42,200	645,500	0.499	2.602	4.304	5.116	10.232
4	11.85	E-75 IU	19,500	230,800	0.262	3.476	0.3077	5.400	10.800
	11.85	E-75 IU	19,500	230,800	0.262	3.476	0.3077	5.400	10.800
	11.85	E-75 IU	19,500	230,800	0.262	3.476	0.3077	5.400	10.800
	11.85	E-75 IU	19,500	230,800	0.262	3.476	0.3077	5.400	10.800
	11.85	E-75 IU	19,500	230,800	0.262	3.476	0.3077	5.400	10.800
4	11.85	X-95 IU	24,700	292,300	0.262	3.476	0.3077	5.400	10.800
	11.85	X-95 IU	24,700	292,300	0.262	3.476	0.3077	5.400	10.800
	11.85	X-95 IU	24,700	292,300	0.262	3.476	0.3077	5.400	10.800
	11.85	X-95 IU	24,700	292,300	0.262	3.476	0.3077	5.400	10.800
	11.85	X-95 IU	24,700	292,300	0.262	3.476	0.3077	5.400	10.800
4	11.85	G-105 IU	27,300	323,100	0.262	3.476	0.3077	5.400	10.800
	11.85	G-105 IU	27,300	323,100	0.262	3.476	0.3077	5.400	10.800
	11.85	G-105 IU	27,300	323,100	0.262	3.476	0.3077	5.400	10.800
	11.85	G-105 IU	27,300	323,100	0.262	3.476	0.3077	5.400	10.800
	11.85	G-105 IU	27,300	323,100	0.262	3.476	0.3077	5.400	10.800
4	11.85	S-135 IU	35,100	415,400	0.262	3.476	0.3077	5.400	10.800
	11.85	S-135 IU	35,100	415,400	0.262	3.476	0.3077	5.400	10.800
	11.85	S-135 IU	35,100	415,400	0.262	3.476	0.3077	5.400	10.800
	11.85	S-135 IU	35,100	415,400	0.262	3.476	0.3077	5.400	10.800
	11.85	S-135 IU	35,100	415,400	0.262	3.476	0.3077	5.400	10.800
4	11.85	Z-140 IU	36,400	430,700	0.262	3.476	0.3077	5.400	10.800
	11.85	Z-140 IU	36,400	430,700	0.262	3.476	0.3077	5.400	10.800
	11.85	Z-140 IU	36,400	430,700	0.262	3.476	0.3077	5.400	10.800
4	11.85	V-150 IU	38,900	461,500	0.262	3.476	0.3077	5.400	10.800
	11.85	V-150 IU	38,900	461,500	0.262	3.476	0.3077	5.400	10.800
	11.85	V-150 IU	38,900	461,500	0.262	3.476	0.3077	5.400	10.800
4	14.00	E-75 IU	23,300	285,400	0.330	3.340	3.805	6.458	12.915
	14.00	E-75 IU	23,300	285,400	0.330	3.340	3.805	6.458	12.915
	14.00	E-75 IU	23,300	285,400	0.330	3.340	3.805	6.458	12.915

(continued)

Table 4.6.6 *(continued)*

				Pipe Data					
Size OD	Nominal Weight	Grade and Upset Type	Torsional Yield Strength	Tensile Yield Strength	Wall Thickness	Nominal ID	Pipe Body Section Area	Pipe Body Moment of Inertia	Pipe Body Polar Moment of Inertia
in.	lb/ft		ft-lb	lb	in.	in.	sq in.	cu in.	cu in.
	14.00	E-75 IU	23,300	285,400	0.330	3.340	3.805	6.458	12.915
	14.00	E-75 EU	23,300	285,400	0.330	3.340	3.805	6.458	12.915
	14.00	E-75 IU	23,300	285,400	0.330	3.340	3.805	6.458	12.915
	14.00	E-75 IU	23,300	285,400	0.330	3.340	3.805	6.458	12.915
4	14.00	X-95 IU	29,500	361,500	0.330	3.340	3.805	6.458	12.915
	14.00	X-95 IU	29,500	361,500	0.330	3.340	3.805	6.458	12.915
	14.00	X-95 IU	29,500	361,500	0.330	3.340	3.805	6.458	12.915
	14.00	X-95 IU	29,500	361,500	0.330	3.340	3.805	6.458	12.915
	14.00	X-95 EU	29,500	361,500	0.330	3.340	3.805	6.458	12.915
	14.00	X-95 IU	29,500	361,500	0.330	3.340	3.805	6.458	12.915
	14.00	X-95 IU	29,500	361,500	0.330	3.340	3.805	6.458	12.915
4	14.00	G-105 IU	32,600	399,500	0.330	3.340	3.805	6.458	12.915
	14.00	G-105 IU	32,600	399,500	0.330	3.340	3.805	6.458	12.915
	14.00	G-105 IU	32,600	399,500	0.330	3.340	3.805	6.458	12.915
	14.00	G-105 IU	32,600	399,500	0.330	3.340	3.805	6.458	12.915
	14.00	G-105 EU	32,600	399,500	0.330	3.340	3.805	6.458	12.915
	14.00	G-105 IU	32,600	399,500	0.330	3.340	3.805	6.458	12.915
	14.00	G-105 IU	32,600	399,500	0.330	3.340	3.805	6.458	12.915
4	14.00	S-135 IU	41,900	513,600	0.330	3.340	3.805	6.458	12.915
	14.00	S-135 IU	41,900	513,600	0.330	3.340	3.805	6.458	12.915
	14.00	S-135 IU	41,900	513,600	0.330	3.340	3.805	6.458	12.915
	14.00	S-135 IU	41,900	513,600	0.330	3.340	3.805	6.458	12.915
	14.00	S-135 EU	41,900	513,600	0.330	3.340	3.805	6.458	12.915
	14.00	S-135 IU	41,900	513,600	0.330	3.340	3.805	6.458	12.915
	14.00	S-135 IU	41,900	513,600	0.330	3.340	3.805	6.458	12.915
4	14.00	Z-140 IU	43,500	532,700	0.330	3.340	3.805	6.458	12.915
	14.00	Z-140 IU	43,500	532,700	0.330	3.340	3.805	6.458	12.915
	14.00	Z-140 IU	43,500	532,700	0.330	3.340	3.805	6.458	12.915
	14.00	Z-140 IU	43,500	532,700	0.330	3.340	3.805	6.458	12.915
	14.00	Z-140 IU	43,500	532,700	0.330	3.340	3.805	6.458	12.915
4	14.00	V-150 IU	46,600	570,700	0.330	3.340	3.805	6.458	12.915
	14.00	V-150 IU	46,600	570,700	0.330	3.340	3.805	6.458	12.915
	14.00	V-150 IU	46,600	570,700	0.330	3.340	3.805	6.458	12.915
	14.00	V-150 IU	46,600	570,700	0.330	3.340	3.805	6.458	12.915
	14.00	V-150 IU	46,600	570,700	0.330	3.340	3.805	6.458	12.915
4	15.70	E-75 IU	25,800	324,100	0.380	3.240	4.322	7.157	14.314
	15.70	E-75 IU	25,800	324,100	0.380	3.240	4.322	7.157	14.314
	15.70	E-75 IU	25,800	324,100	0.380	3.240	4.322	7.157	14.314
	15.70	E-75 EU	25,800	324,100	0.380	3.240	4.322	7.157	14.314
	15.70	E-75 IU	25,800	324,100	0.380	3.240	4.322	7.157	14.314
	15.70	E-75 IU	25,800	324,100	0.380	3.240	4.322	7.157	14.314
4	15.70	X-95 IU	32,700	410,500	0.380	3.240	4.322	7.157	14.314
	15.70	X-95 IU	32,700	410,500	0.380	3.240	4.322	7.157	14.314
	15.70	X-95 IU	32,700	410,500	0.380	3.240	4.322	7.157	14.314
	15.70	X-95 EU	32,700	410,500	0.380	3.240	4.322	7.157	14.314
	15.70	X-95 IU	32,700	410,500	0.380	3.240	4.322	7.157	14.314
	15.70	X-95 IU	32,700	410,500	0.380	3.240	4.322	7.157	14.314
4	15.70	G-105 IU	36,100	453,800	0.380	3.240	4.322	7.157	14.314
	15.70	G-105 IU	36,100	453,800	0.380	3.240	4.322	7.157	14.314
	15.70	G-105 IU	36,100	453,800	0.380	3.240	4.322	7.157	14.314
	15.70	G-105 EU	36,100	453,800	0.380	3.240	4.322	7.157	14.314
	15.70	G-105 IU	36,100	453,800	0.380	3.240	4.322	7.157	14.314
	15.70	G-105 IU	36,100	453,800	0.380	3.240	4.322	7.157	14.314

(continued)

Table 4.6.6 *(continued)*

					Pipe Data				
Size OD	Nominal Weight	Grade and Upset Type	Torsional Yield Strength	Tensile Yield Strength	Wall Thickness	Nominal ID	Pipe Body Section Area	Pipe Body Moment of Inertia	Pipe Body Polar Moment of Inertia
in.	lb/ft		ft-lb	lb	in.	in.	sq in.	cu in.	cu in.
4	15.70	S-135 IU	46,500	583,400	0.380	3.240	4.322	7.157	14.314
	15.70	S-135 IU	46,500	583,400	0.380	3.240	4.322	7.157	14.314
	15.70	S-135 IU	46,500	583,400	0.380	3.240	4.322	7.157	14.314
	15.70	S-135 EU	46,500	583,400	0.380	3.240	4.322	7.157	14.314
	15.70	S-135 IU	46,500	583,400	0.380	3.240	4.322	7.157	14.314
	15.70	S-135 IU	46,500	583,400	0.380	3.240	4.322	7.157	14.314
	15.70	S-135 IU	46,500	583,400	0.380	3.240	4.322	7.157	14.314
4	15.70	Z-140 IU	48,200	605,000	0.380	3.240	4.322	7.157	14.314
	15.70	Z-140 IU	48,200	605,000	0.380	3.240	4.322	7.157	14.314
	15.70	Z-140 IU	48,200	605,000	0.380	3.240	4.322	7.157	14.314
	15.70	Z-140 IU	48,200	605,000	0.380	3.240	4.322	7.157	14.314
4	15.70	V-150 IU	51,600	648,200	0.380	3.240	4.322	7.157	14.314
	15.70	V-150 IU	51,600	648,200	0.380	3.240	4.322	7.157	14.314
	15.70	V-150 IU	51,600	648,200	0.380	3.240	4.322	7.157	14.314
	15.70	V-150 IU	51,600	648,200	0.380	3.240	4.322	7.157	14.314
$4\frac{1}{2}$	16.60	E-75 IEU	30,800	330,600	0.337	3.826	4.407	9.610	19.221
	16.60	E-75 EU	30,800	330,600	0.337	3.826	4.407	9.610	19.221
	16.60	E-75 IEU	30,800	330,600	0.337	3.826	4.407	9.610	19.221
	16.60	E-75 IEU	30,800	330,600	0.337	3.826	4.407	9.610	19.221
	16.60	E-75 IEU	30,800	330,600	0.337	3.826	4.407	9.610	19.221
	16.60	E-75 EU	30,800	330,600	0.337	3.826	4.407	9.610	19.221
	16.60	E-75 EU	30,800	330,600	0.337	3.826	4.407	9.610	19.221
	16.60	E-75 IEU	30,800	330,600	0.337	3.826	4.407	9.610	19.221
	16.60	E-75 IEU	30,800	330,600	0.337	3.826	4.407	9.610	19.221
	16.60	E-75 EU	30,800	330,600	0.337	3.826	4.407	9.610	19.221
$4\frac{1}{2}$	16.60	X-95 IEU	39,000	418,700	0.337	3.826	4.407	9.610	19.221
	16.60	X-95 EU	39,000	418,700	0.337	3.826	4.407	9.610	19.221
	16.60	X-95 IEU	39,000	418,700	0.337	3.826	4.407	9.610	19.221
	16.60	X-95 IEU	39,000	418,700	0.337	3.826	4.407	9.610	19.221
	16.60	X-95 IEU	39,000	418,700	0.337	3.826	4.407	9.610	19.221
	16.60	X-95 EU	39,000	418,700	0.337	3.826	4.407	9.610	19.221
	16.60	X-95 EU	39,000	418,700	0.337	3.826	4.407	9.610	19.221
	16.60	X-95 IEU	39,000	418,700	0.337	3.826	4.407	9.610	19.221
	16.60	X-95 IEU	39,000	418,700	0.337	3.826	4.407	9.610	19.221
	16.60	X-95 EU	39,000	418,700	0.337	3.826	4.407	9.610	19.221
$4\frac{1}{2}$	16.60	G-105 IEU	43,100	462,800	0.337	3.826	4.407	9.610	19.221
	16.60	G-105 EU	43,100	462,800	0.337	3.826	4.407	9.610	19.221
	16.60	G-105 IEU	43,100	462,800	0.337	3.826	4.407	9.610	19.221
	16.60	G-105 IEU	43,100	462,800	0.337	3.826	4.407	9.610	19.221
	16.60	G-105 IEU	43,100	462,800	0.337	3.826	4.407	9.610	19.221
	16.60	G-105 EU	43,100	462,800	0.337	3.826	4.407	9.610	19.221
	16.60	G-105 EU	43,100	462,800	0.337	3.826	4.407	9.610	19.221
	16.60	G-105 IEU	43,100	462,800	0.337	3.826	4.407	9.610	19.221
	16.60	G-105 IEU	43,100	462,800	0.337	3.826	4.407	9.610	19.221
	16.60	G-105 EU	43,100	462,800	0.337	3.826	4.407	9.610	19.221
$4\frac{1}{2}$	16.60	S-135 IEU	55,500	595,000	0.337	3.826	4.407	9.610	19.221
	16.60	S-135 EU	55,500	595,000	0.337	3.826	4.407	9.610	19.221
	16.60	S-135 IEU	55,500	595,000	0.337	3.826	4.407	9.610	19.221
	16.60	S-135 IEU	55,500	595,000	0.337	3.826	4.407	9.610	19.221
	16.60	S-135 IEU	55,500	595,000	0.337	3.826	4.407	9.610	19.221
	16.60	S-135 EU	55,500	595,000	0.337	3.826	4.407	9.610	19.221
	16.60	S-135 EU	55,500	595,000	0.337	3.826	4.407	9.610	19.221
	16.60	S-135 IEU	55,500	595,000	0.337	3.826	4.407	9.610	19.221
	16.60	S-135 IEU	55,500	595,000	0.337	3.826	4.407	9.610	19.221

(continued)

Table 4.6.6 (continued)

									Pipe Data
Size OD	Nominal Weight	Grade and Upset Type	Torsional Yield Strength	Tensile Yield Strength	Wall Thickness	Nominal ID	Pipe Body Section Area	Pipe Body Moment of Inertia	Pipe Body Polar Moment of Inertia
in.	lb/ft		ft-lb	lb	in.	in.	sq in.	cu in.	cu in.
	16.60	S-135 EU	55,500	595,000	0.337	3.826	4.407	9.610	19.221
	16.60	S-135 IEU	55,500	595,000	0.337	3.826	4.407	9.610	19.221
$4\frac{1}{2}$	16.60	Z-140 IEU	57,500	617,000	0.337	3.826	4.407	9.610	19.221
	16.60	Z-140 EU	57,500	617,000	0.337	3.826	4.407	9.610	19.221
	16.60	Z-140 IEU	57,500	617,000	0.337	3.826	4.407	9.610	19.221
	16.60	Z-140 IEU	57,500	617,000	0.337	3.826	4.407	9.610	19.221
	16.60	Z-140 EU	57,500	617,000	0.337	3.826	4.407	9.610	19.221
	16.60	Z-140 IEU	57,500	617,000	0.337	3.826	4.407	9.610	19.221
$4\frac{1}{2}$	16.60	V-150 IEU	61,600	661,100	0.337	3.826	4.407	9.610	19.221
	16.60	V-150 EU	61,600	661,100	0.337	3.826	4.407	9.610	19.221
	16.60	V-150 IEU	61,600	661,100	0.337	3.826	4.407	9.610	19.221
	16.60	V-150 IEU	61,600	661,100	0.337	3.826	4.407	9.610	19.221
	16.60	V-150 EU	61,600	661,100	0.337	3.826	4.407	9.610	19.221
	16.60	V-150 IEU	61,600	661,100	0.337	3.826	4.407	9.610	19.221
$4\frac{1}{2}$	20.00	E-75 IEU	36,900	412,400	0.430	3.640	5.498	11.512	23.023
	20.00	E-75 EU	36,900	412,400	0.430	3.640	5.498	11.512	23.023
	20.00	E-75 IEU	36,900	412,400	0.430	3.640	5.498	11.512	23.023
	20.00	E-75 IEU	36,900	412,400	0.430	3.640	5.498	11.512	23.023
	20.00	E-75 EU	36,900	412,400	0.430	3.640	5.498	11.512	23.023
	20.00	E-75 EU	36,900	412,400	0.430	3.640	5.498	11.512	23.023
	20.00	E-75 IEU	36,900	412,400	0.430	3.640	5.498	11.512	23.023
	20.00	E-75 EU	36,900	412,400	0.430	3.640	5.498	11.512	23.023
$4\frac{1}{2}$	20.00	X-95 IEU	46,700	522,300	0.430	3.640	5.498	11.512	23.023
	20.00	X-95 EU	46,700	522,300	0.430	3.640	5.498	11.512	23.023
	20.00	X-95 EU	46,700	522,300	0.430	3.640	5.498	11.512	23.023
	20.00	X-95 IEU	46,700	522,300	0.430	3.640	5.498	11.512	23.023
	20.00	X-95 EU	46,700	522,300	0.430	3.640	5.498	11.512	23.023
	20.00	X-95 EU	46,700	522,300	0.430	3.640	5.498	11.512	23.023
	20.00	X-95 IEU	46,700	522,300	0.430	3.640	5.498	11.512	23.023
	20.00	X-95 EU	46,700	522,300	0.430	3.640	5.498	11.512	23.023
$4\frac{1}{2}$	20.00	G-105 IEU	51,700	577,300	0.430	3.640	5.498	11.512	23.023
	20.00	G-105 EU	51,700	577,300	0.430	3.640	5.498	11.512	23.023
	20.00	G-105 IEU	51,700	577,300	0.430	3.640	5.498	11.512	23.023
	20.00	G-105 IEU	51,700	577,300	0.430	3.640	5.498	11.512	23.023
	20.00	G-105 EU	51,700	577,300	0.430	3.640	5.498	11.512	23.023
	20.00	G-105 EU	51,700	577,300	0.430	3.640	5.498	11.512	23.023
	20.00	G-105 IEU	51,700	577,300	0.430	3.640	5.498	11.512	23.023
	20.00	G-105 EU	51,700	577,300	0.430	3.640	5.498	11.512	23.023
$4\frac{1}{2}$	20.00	S-135 IEU	66,400	742,200	0.430	3.640	5.498	11.512	23.023
	20.00	S-135 EU	66,400	742,200	0.430	3.640	5.498	11.512	23.023
	20.00	S-135 IEU	66,400	742,200	0.430	3.640	5.498	11.512	23.023
	20.00	S-135 IEU	66,400	742,200	0.430	3.640	5.498	11.512	23.023
	20.00	S-135 EU	66,400	742,200	0.430	3.640	5.498	11.512	23.023
	20.00	S-135 EU	66,400	742,200	0.430	3.640	5.498	11.512	23.023
	20.00	S-135 IEU	66,400	742,200	0.430	3.640	5.498	11.512	23.023
	20.00	S-135 EU	66,400	742,200	0.430	3.640	5.498	11.512	23.023
	20.00	S-135 IEU	66,400	742,200	0.430	3.640	5.498	11.512	23.023
$4\frac{1}{2}$	20.00	Z-140 IEU	68,900	769,700	0.430	3.640	5.498	11.512	23.023
	20.00	Z-140 EU	68,900	769,700	0.430	3.640	5.498	11.512	23.023
	20.00	Z-140 IEU	68,900	769,700	0.430	3.640	5.498	11.512	23.023
	20.00	Z-140 EU	68,900	769,700	0.430	3.640	5.498	11.512	23.023
	20.00	Z-140 IEU	68,900	769,700	0.430	3.640	5.498	11.512	23.023
$4\frac{1}{2}$	20.00	V-150 IEU	73,800	824,700	0.430	3.640	5.498	11.512	23.023
	20.00	V-150 EU	73,800	824,700	0.430	3.640	5.498	11.512	23.023

(continued)

Table 4.6.6 *(continued)*

									Pipe Data
		Grade and Upset Type	Torsional Yield Strength	Tensile Yield Strength	Wall Thickness	Nominal ID	Pipe Body Section Area	Pipe Body Moment of Inertia	Pipe Body Polar Moment of Inertia
Size OD	Nominal Weight								
in.	lb/ft		ft-lb	lb	in.	in.	sq in.	cu in.	cu in.
	20.00	V-150 IEU	73,800	824,700	0.430	3.640	5.498	11.512	23.023
	20.00	V-150 EU	73,800	824,700	0.430	3.640	5.498	11.512	23.023
	20.00	V-150 IEU	73,800	824,700	0.430	3.640	5.498	11.512	23.023
5	19.50	E-75 IEU	41,200	395,600	0.362	4.276	5.275	14.269	28.538
	19.50	E-75 IEU	41,200	395,600	0.362	4.276	5.275	14.269	28.538
	19.50	E-75 IEU	41,200	395,600	0.362	4.276	5.275	14.269	28.538
	19.50	E-75 IEU	41,200	395,600	0.362	4.276	5.275	14.269	28.538
	19.50	E-75 IEU	41,200	395,600	0.362	4.276	5.275	14.269	28.538
5	19.50	X-95 IEU	52,100	501,100	0.362	4.276	5.275	14.269	28.538
	19.50	X-95 IEU	52,100	501,100	0.362	4.276	5.275	14.269	28.538
	19.50	X-95 IEU	52,100	501,100	0.362	4.276	5.275	14.269	28.538
	19.50	X-95 IEU	52,100	501,100	0.362	4.276	5.275	14.269	28.538
	19.50	X-95 IEU	52,100	501,100	0.362	4.276	5.275	14.269	28.538
5	19.50	G-105 IEU	57,600	553,800	0.362	4.276	5.275	14.269	28.538
	19.50	G-105 IEU	57,600	553,800	0.362	4.276	5.275	14.269	28.538
	19.50	G-105 IEU	57,600	553,800	0.362	4.276	5.275	14.269	28.538
	19.50	G-105 IEU	57,600	553,800	0.362	4.276	5.275	14.269	28.538
	19.50	G-105 IEU	57,600	553,800	0.362	4.276	5.275	14.269	28.538
	19.50	G-105 IEU	57,600	553,800	0.362	4.276	5.275	14.269	28.538
5	19.50	S-135 IEU	74,100	712,100	0.362	4.276	5.275	14.269	28.538
	19.50	S-135 IEU	74,100	712,100	0.362	4.276	5.275	14.269	28.538
	19.50	S-135 IEU	74,100	712,100	0.362	4.276	5.275	14.269	28.538
	19.50	S-135 IEU	74,100	712,100	0.362	4.276	5.275	14.269	28.538
	19.50	S-135 IEU	74,100	712,100	0.362	4.276	5.275	14.269	28.538
	19.50	S-135 IEU	74,100	712,100	0.362	4.276	5.275	14.269	28.538
5	19.50	Z-140 IEU	76,800	738,400	0.362	4.276	5.275	14.269	28.538
	19.50	Z-140 IEU	76,800	738,400	0.362	4.276	5.275	14.269	28.538
	19.50	Z-140 IEU	76,800	738,400	0.362	4.276	5.275	14.269	28.538
	19.50	Z-140 IEU	76,800	738,400	0.362	4.276	5.275	14.269	28.538
5	19.50	V-150 IEU	82,300	791,200	0.362	4.276	5.275	14.269	28.538
	19.50	V-150 IEU	82,300	791,200	0.362	4.276	5.275	14.269	28.538
	19.50	V-150 IEU	82,300	791,200	0.362	4.276	5.275	14.269	28.538
	19.50	V-150 IEU	82,300	791,200	0.362	4.276	5.275	14.269	28.538
5	25.60	E-75 IEU	52,300	530,100	0.500	4.000	7.069	18.113	36.226
	25.60	E-75 IEU	52,300	530,100	0.500	4.000	7.069	18.113	36.226
	25.60	E-75 IEU	52,300	530,100	0.500	4.000	7.069	18.113	36.226
	25.60	E-75 IEU	52,300	530,100	0.500	4.000	7.069	18.113	36.226
5	25.60	X-95 IEU	66,200	671,500	0.500	4.000	7.069	18.113	36.226
	25.60	X-95 IEU	66,200	671,500	0.500	4.000	7.069	18.113	36.226
	25.60	X-95 IEU	66,200	671,500	0.500	4.000	7.069	18.113	36.226
	25.60	X-95 IEU	66,200	671,500	0.500	4.000	7.069	18.113	36.226
5	25.60	G-105 IEU	73,200	742,200	0.500	4.000	7.069	18.113	36.226
	25.60	G-105 IEU	73,200	742,200	0.500	4.000	7.069	18.113	36.226
	25.60	G-105 IEU	73,200	742,200	0.500	4.000	7.069	18.113	36.226
	25.60	G-105 IEU	73,200	742,200	0.500	4.000	7.069	18.113	36.226
	25.60	G-105 IEU	73,200	742,200	0.500	4.000	7.069	18.113	36.226
5	25.60	S-135 IEU	94,100	954,300	0.500	4.000	7.069	18.113	36.226
	25.60	S-135 IEU	94,100	954,300	0.500	4.000	7.069	18.113	36.226
	25.60	S-135 IEU	94,100	954,300	0.500	4.000	7.069	18.113	36.226
	25.60	S-135 IEU	94,100	954,300	0.500	4.000	7.069	18.113	36.226
	25.60	S-135 IEU	94,100	954,300	0.500	4.000	7.069	18.113	36.226
5	25.60	Z-140 IEU	97,500	989,600	0.500	4.000	7.069	18.113	36.226
	25.60	Z-140 IEU	97,500	989,600	0.500	4.000	7.069	18.113	36.226
	25.60	Z-140 IEU	97,500	989,600	0.500	4.000	7.069	18.113	36.226

(continued)

Table 4.6.6   *(continued)*

					Pipe Data				
Size OD	Nominal Weight	Grade and Upset Type	Torsional Yield Strength	Tensile Yield Strength	Wall Thickness	Nominal ID	Pipe Body Section Area	Pipe Body Moment of Inertia	Pipe Body Polar Moment of Inertia
in.	lb/ft		ft-lb	lb	in.	in.	sq in.	cu in.	cu in.
5	25.60	V-150 IEU	104,500	1,060,300	0.500	4.000	7.069	18.113	36.226
	25.60	V-150 IEU	104,500	1,060,300	0.500	4.000	7.069	18.113	36.226
	25.60	V-150 IEU	104,500	1,060,300	0.500	4.000	7.069	18.113	36.226
$5\frac{1}{2}$	21.90	E-75 IEU	50,700	437,100	0.361	4.778	5.828	19.335	38.670
	21.90	E-75 IEU	50,700	437,100	0.361	4.778	5.828	19.335	38.670
	21.90	E-75 IEU	50,700	437,100	0.361	4.778	5.828	19.335	38.670
	21.90	E-75 IEU	50,700	437,100	0.361	4.778	5.828	19.335	38.670
$5\frac{1}{2}$	21.90	X-95 IEU	64,200	553,700	0.361	4.778	5.828	19.335	38.670
	21.90	X-95 IEU	64,200	553,700	0.361	4.778	5.828	19.335	38.670
	21.90	X-95 IEU	64,200	553,700	0.361	4.778	5.828	19.335	38.670
	21.90	X-95 IEU	64,200	553,700	0.361	4.778	5.828	19.335	38.670
$5\frac{1}{2}$	21.90	G-105 IEU	71,000	612,000	0.361	4.778	5.828	19.335	38.670
	21.90	G-105 IEU	71,000	612,000	0.361	4.778	5.828	19.335	38.670
	21.90	G-105 IEU	71,000	612,000	0.361	4.778	5.828	19.335	38.670
	21.90	G-105 IEU	71,000	612,000	0.361	4.778	5.828	19.335	38.670
	21.90	G-105 IEU	71,000	612,000	0.361	4.778	5.828	19.335	38.670
$5\frac{1}{2}$	21.90	S-135 IEU	91,300	786,800	0.361	4.778	5.828	19.335	38.670
	21.90	S-135 IEU	91,300	786,800	0.361	4.778	5.828	19.335	38.670
	21.90	S-135 IEU	91,300	786,800	0.361	4.778	5.828	19.335	38.670
	21.90	S-135 IEU	91,300	786,800	0.361	4.778	5.828	19.335	38.670
	21.90	S-135 IEU	91,300	786,800	0.361	4.778	5.828	19.335	38.670
$5\frac{1}{2}$	21.90	Z-140 IEU	94,700	816,000	0.361	4.778	5.828	19.335	38.670
	21.90	Z-140 IEU	94,700	816,000	0.361	4.778	5.828	19.335	38.670
	21.90	Z-140 IEU	94,700	816,000	0.361	4.778	5.828	19.335	38.670
	21.90	Z-140 IEU	94,700	816,000	0.361	4.778	5.828	19.335	38.670
	21.90	Z-140 IEU	94,700	816,000	0.361	4.778	5.828	19.335	38.670
$5\frac{1}{2}$	21.90	V-150 IEU	101,400	874,200	0.361	4.778	5.828	19.335	38.670
	21.90	V-150 IEU	101,400	874,200	0.361	4.778	5.828	19.335	38.670
	21.90	V-150 IEU	101,400	874,200	0.361	4.778	5.828	19.335	38.670
	21.90	V-150 IEU	101,400	874,200	0.361	4.778	5.828	19.335	38.670
	21.90	V-150 IEU	101,400	874,200	0.361	4.778	5.828	19.335	38.670
$5\frac{1}{2}$	24.70	E-75 IEU	56,600	497,200	0.415	4.670	6.630	21.571	43.141
	24.70	E-75 IEU	56,600	497,200	0.415	4.670	6.630	21.571	43.141
	24.70	E-75 IEU	56,600	497,200	0.415	4.670	6.630	21.571	43.141
	24.70	E-75 IEU	56,600	497,200	0.415	4.670	6.630	21.571	43.141
	24.70	E-75 IEU	56,600	497,200	0.415	4.670	6.630	21.571	43.141
$5\frac{1}{2}$	24.70	X-95 IEU	71,700	629,800	0.415	4.670	6.630	21.571	43.141
	24.70	X-95 IEU	71,700	629,800	0.415	4.670	6.630	21.571	43.141
	24.70	X-95 IEU	71,700	629,800	0.415	4.670	6.630	21.571	43.141
	24.70	X-95 IEU	71,700	629,800	0.415	4.670	6.630	21.571	43.141
$5\frac{1}{2}$	24.70	G-105 IEU	79,200	696,100	0.415	4.670	6.630	21.571	43.141
	24.70	G-105 IEU	79,200	696,100	0.415	4.670	6.630	21.571	43.141
	24.70	G-105 IEU	79,200	696,100	0.415	4.670	6.630	21.571	43.141
	24.70	G-105 IEU	79,200	696,100	0.415	4.670	6.630	21.571	43.141
	24.70	G-105 IEU	79,200	696,100	0.415	4.670	6.630	21.571	43.141
$5\frac{1}{2}$	24.70	S-135 IEU	101,800	895,000	0.415	4.670	6.630	21.571	43.141
	24.70	S-135 IEU	101,800	895,000	0.415	4.670	6.630	21.571	43.141
	24.70	S-135 IEU	101,800	895,000	0.415	4.670	6.630	21.571	43.141
	24.70	S-135 IEU	101,800	895,000	0.415	4.670	6.630	21.571	43.141
	24.70	S-135 IEU	101,800	895,000	0.415	4.670	6.630	21.571	43.141
$5\frac{1}{2}$	24.70	Z-140 IEU	105,600	928,100	0.415	4.670	6.630	21.571	43.141
	24.70	Z-140 IEU	105,600	928,100	0.415	4.670	6.630	21.571	43.141
	24.70	Z-140 IEU	105,600	928,100	0.415	4.670	6.630	21.571	43.141

(continued)

Table 4.6.6 (continued)

									Pipe Body
		Grade							Polar
		and	Torsional	Tensile			Pipe Body	Pipe Body	Moment of
	Nominal	Upset	Yield	Yield	Wall	Nominal	Section	Moment of	Inertia
Size OD	Weight	Type	Strength	Strength	Thickness	ID	Area	Inertia	
in.	lb/ft		ft-lb	lb	in.	in.	sq in.	cu in.	cu in.
	24.70	Z-140 IEU	105,600	928,100	0.415	4.670	6.630	21.571	43.141
	24.70	Z-140 IEU	105,600	928,100	0.415	4.670	6.630	21.571	43.141
$5\frac{1}{2}$	24.70	V-150 IEU	113,100	994,400	0.415	4.670	6.630	21.571	43.141
	24.70	V-150 IEU	113,100	994,400	0.415	4.670	6.630	21.571	43.141
	24.70	V-150 IEU	113,100	994,400	0.415	4.670	6.630	21.571	43.141
	24.70	V-150 IEU	113,100	994,400	0.415	4.670	6.630	21.571	43.141
	24.70	V-150 IEU	113,100	994,400	0.415	4.670	6.630	21.571	43.141
$5\frac{7}{8}$	23.40	E-75 IEU	58,600	469,000	0.361	5.153	6.254	23.868	47.737
$5\frac{7}{8}$	23.40	X-95 IEU	74,200	594,100	0.361	5.153	6.254	23.868	47.737
$5\frac{7}{8}$	23.40	G-105 IEU	82,000	656,600	0.361	5.153	6.254	23.868	47.737
$5\frac{7}{8}$	23.40	S-135 IEU	105,500	844,200	0.361	5.153	6.254	23.868	47.737
$5\frac{7}{8}$	23.40	Z-140 IEU	109,400	875,500	0.361	5.153	6.254	23.868	47.737
$5\frac{7}{8}$	23.40	V-150 IEU	117,200	938,000	0.361	5.153	6.254	23.868	47.737
$5\frac{7}{8}$	26.40	E-75 IEU	65,500	533,900	0.415	5.045	7.119	26.680	53.360
$5\frac{7}{8}$	26.40	X-95 IEU	83,000	676,300	0.415	5.045	7.119	26.680	53.360
$5\frac{7}{8}$	26.40	G-105 IEU	91,700	747,400	0.415	5.045	7.119	26.680	53.360
$5\frac{7}{8}$	26.40	S-135 IEU	117,900	961,000	0.415	5.045	7.119	26.680	53.360
$5\frac{7}{8}$	26.40	Z-140 IEU	122,300	996,600	0.415	5.045	7.119	26.680	53.360
$5\frac{7}{8}$	26.40	V-150 IEU	131,000	1,067,800	0.415	5.045	7.119	26.680	53.360
$6\frac{5}{8}$	25.20	E-75 IEU	70,600	489,500	0.330	5.965	6.526	32.416	64.831
	25.20	E-75 IEU	70,600	489,500	0.330	5.965	6.526	32.416	64.831
	25.20	E-75 IEU	70,600	489,500	0.330	5.965	6.526	32.416	64.831
$6\frac{5}{8}$	25.20	X-95 IEU	89,400	620,000	0.330	5.965	6.526	32.416	64.831
	25.20	X-95 IEU	89,400	620,000	0.330	5.965	6.526	32.416	64.831
	25.20	X-95 IEU	89,400	620,000	0.330	5.965	6.526	32.416	64.831
$6\frac{5}{8}$	25.20	G-105 IEU	98,800	685,200	0.330	5.965	6.526	32.416	64.831
	25.20	G-105 IEU	98,800	685,200	0.330	5.965	6.526	32.416	64.831
	25.20	G-105 IEU	98,800	685,200	0.330	5.965	6.526	32.416	64.831
$6\frac{5}{8}$	25.20	S-135 IEU	127,000	881,000	0.330	5.965	6.526	32.416	64.831
	25.20	S-135 IEU	127,000	881,000	0.330	5.965	6.526	32.416	64.831
	25.20	S-135 IEU	127,000	881,000	0.330	5.965	6.526	32.416	64.831
	25.20	S-135 IEU	127,000	881,000	0.330	5.965	6.526	32.416	64.831
$6\frac{5}{8}$	25.20	Z-140 IEU	131,700	913,700	0.330	5.965	6.526	32.416	64.831
	25.20	Z-140 IEU	131,700	913,700	0.330	5.965	6.526	32.416	64.831
	25.20	Z-140 IEU	131,700	913,700	0.330	5.965	6.526	32.416	64.831
	25.20	Z-140 IEU	131,700	913,700	0.330	5.965	6.526	32.416	64.831
$6\frac{5}{8}$	25.20	V-150 IEU	141,200	978,900	0.330	5.965	6.526	32.416	64.831
	25.20	V-150 IEU	141,200	978,900	0.330	5.965	6.526	32.416	64.831
	25.20	V-150 IEU	141,200	978,900	0.330	5.965	6.526	32.416	64.831
	25.20	V-150 IEU	141,200	978,900	0.330	5.965	6.526	32.416	64.831
$6\frac{5}{8}$	27.70	E-75 IEU	76,300	534,200	0.362	5.901	7.123	35.040	70.080
	27.70	E-75 IEU	76,300	534,200	0.362	5.901	7.123	35.040	70.080
	27.70	E-75 IEU	76,300	534,200	0.362	5.901	7.123	35.040	70.080
$6\frac{5}{8}$	27.70	X-95 IEU	96,600	676,700	0.362	5.901	7.123	35.040	70.080
	27.70	X-95 IEU	96,600	676,700	0.362	5.901	7.123	35.040	70.080
	27.70	X-95 IEU	96,600	676,700	0.362	5.901	7.123	35.040	70.080
$6\frac{5}{8}$	27.70	G-105 IEU	106,800	747,900	0.362	5.901	7.123	35.040	70.080
	27.70	G-105 IEU	106,800	747,900	0.362	5.901	7.123	35.040	70.080
	27.70	G-105 IEU	106,800	747,900	0.362	5.901	7.123	35.040	70.080
$6\frac{5}{8}$	27.70	S-135 IEU	137,300	961,600	0.362	5.901	7.123	35.040	70.080
	27.70	S-135 IEU	137,300	961,600	0.362	5.901	7.123	35.040	70.080

(continued)

Table 4.6.6   (continued)

					Pipe Data				
Size OD	Nominal Weight	Grade and Upset Type	Torsional Yield Strength	Tensile Yield Strength	Wall Thickness	Nominal ID	Pipe Body Section Area	Pipe Body Moment of Inertia	Pipe Body Polar Moment of Inertia
in.	lb/ft		ft-lb	lb	in.	in.	sq in.	cu in.	cu in.
	27.70	S-135 IEU	137,300	961,600	0.362	5.901	7.123	35.040	70.080
	27.70	S-135 IEU	137,300	961,600	0.362	5.901	7.123	35.040	70.080
$6\frac{5}{8}$	27.70	Z-140 IEU	142,400	997,200	0.362	5.901	7.123	35.040	70.080
	27.70	Z-140 IEU	142,400	997,200	0.362	5.901	7.123	35.040	70.080
	27.70	Z-140 IEU	142,400	997,200	0.362	5.901	7.123	35.040	70.080
	27.70	Z-140 IEU	142,400	997,200	0.362	5.901	7.123	35.040	70.080
$6\frac{5}{8}$	27.70	V-150 IEU	152,600	1,068,400	0.362	5.901	7.123	35.040	70.080
	27.70	V-150 IEU	152,600	1,068,400	0.362	5.901	7.123	35.040	70.080
	27.70	V-150 IEU	152,600	1,068,400	0.362	5.901	7.123	35.040	70.080
	27.70	V-150 IEU	152,600	1,068,400	0.362	5.901	7.123	35.040	70.080

					Tool Joint Data				
Internal Pressure	Collapse Pressure	Connection Type	Outside Diameter	Inside Diameter	Torsional Yield Strength	Tensile Yield Strength	Make-up Torque	Torsional Ratio Tool Joint to Pipe	* Pin Tong Space
psi	psi		in.	in.	ft-lb	lb	ft-lb		in.
10,500	11,040	NC26	$3\frac{3}{8}$	$1\frac{3}{4}$	6,900	313,700	3,900	1.44	9
10,500	11,040	$2\frac{3}{8}$ OH	$3\frac{3}{8}$	$1\frac{3}{4}$	6,600	294,600	3,700	1.38	9
10,500	11,040	$2\frac{3}{8}$ WO	$3\frac{3}{8}$	$1\frac{7}{8}$	5,400	241,300	3,000	1.13	9
10,500	11,040	$2\frac{3}{8}$ SLH90	$3\frac{3}{8}$	2	5,200	202,900	2,700	1.08	9
13,300	13,984	NC26	$3\frac{3}{8}$	$1\frac{3}{4}$	6,900	313,700	3,900	1.15	9
13,300	13,984	$2\frac{3}{8}$ OH	$3\frac{3}{8}$	$1\frac{3}{4}$	6,600	294,600	3,700	1.10	9
13,300	13,984	$2\frac{3}{8}$ WO	$3\frac{3}{8}$	$1\frac{7}{8}$	5,400	241,300	3,000	0.90	9
13,300	13,984	$2\frac{3}{8}$ SLH90	$3\frac{3}{8}$	$1\frac{7}{8}$	6,400	248,500	3,400	1.07	9
14,700	15,456	NC26	$3\frac{3}{8}$	$1\frac{3}{4}$	6,900	313,700	3,900	1.03	9
14,700	15,456	$2\frac{3}{8}$ OH	$3\frac{3}{8}$	$1\frac{3}{4}$	6,600	294,600	3,700	0.99	9
14,700	15,456	$2\frac{3}{8}$ WO	$3\frac{3}{8}$	$1\frac{7}{8}$	5,400	241,300	3,000	0.81	9
14,700	15,456	$2\frac{3}{8}$ SLH90	$3\frac{3}{8}$	$1\frac{7}{8}$	6,400	248,500	3,400	0.96	9
18,900	19,035	NC26	$3\frac{3}{8}$	$1\frac{3}{4}$	6,900	313,700	3,900	0.80	9
18,900	19,035	$2\frac{3}{8}$ OH	$3\frac{3}{8}$	$1\frac{1}{2}$	8,300	371,200	4,700	0.97	9
18,900	19,035	$2\frac{3}{8}$ WO	$3\frac{3}{8}$	$1\frac{5}{8}$	7,300	323,800	4,000	0.85	9
18,900	19,035	$2\frac{3}{8}$ SLH90	$3\frac{3}{8}$	$1\frac{3}{4}$	7,500	291,200	4,000	0.87	9
19,600	19,588	XT24	$3\frac{1}{8}$	$1\frac{1}{2}$	10,000	261,500	6,000	1.12	10
19,600	19,588	XT26	$3\frac{3}{8}$	$1\frac{3}{4}$	11,600	290,900	7,000	1.30	10
19,600	19,588	HT26	$3\frac{3}{8}$	$1\frac{3}{4}$	8,700	313,700	5,200	0.98	9
21,000	20,661	XT24	$3\frac{1}{8}$	$1\frac{1}{2}$	10,000	261,500	6,000	1.05	10
21,000	20,661	XT26	$3\frac{3}{8}$	$1\frac{3}{4}$	11,600	290,900	7,000	1.22	10
21,000	20,661	HT26	$3\frac{3}{8}$	$1\frac{3}{4}$	8,700	313,700	5,200	0.92	9
15,474	15,599	NC26	$3\frac{3}{8}$	$1\frac{3}{4}$	6,900	313,700	3,900	1.10	9
15,474	15,599	HT26	$3\frac{3}{8}$	$1\frac{3}{4}$	8,700	313,700	5,200	1.38	9
15,474	15,599	$2\frac{3}{8}$ OH	$3\frac{1}{4}$	$1\frac{3}{4}$	6,500	294,600	3,700	1.03	9
15,474	15,599	$2\frac{3}{8}$ SLH90	$3\frac{1}{4}$	$1\frac{13}{16}$	6,900	270,200	3,700	1.10	9

(continued)

Table 4.6.6   *(continued)*

<table>
<thead>
<tr><th colspan="10" align="center">Tool Joint Data</th></tr>
<tr>
<th>Internal
Pressure</th>
<th>Collapse
Pressure</th>
<th>Connection
Type</th>
<th>Outside
Diameter</th>
<th>Inside
Diameter</th>
<th>Torsional
Yield
Strength</th>
<th>Tensile
Yield
Strength</th>
<th>Make-up
Torque</th>
<th>Torsional
Ratio Tool
Joint to
Pipe</th>
<th>*
Pin
Tong
Space</th>
</tr>
<tr>
<th>psi</th><th>psi</th><th></th><th>in.</th><th>in.</th><th>ft-lb</th><th>lb</th><th>ft-lb</th><th></th><th>in.</th>
</tr>
</thead>
<tbody>
<tr><td>19,600</td><td>19,759</td><td>NC26</td><td>$3\frac{3}{8}$</td><td>$3\frac{1}{4}$</td><td>6,900</td><td>313,700</td><td>3,900</td><td>0.87</td><td>9</td></tr>
<tr><td>19,600</td><td>19,759</td><td>HT26</td><td>$3\frac{3}{8}$</td><td>$3\frac{1}{4}$</td><td>8,700</td><td>313,700</td><td>5,200</td><td>1.10</td><td>9</td></tr>
<tr><td>19,600</td><td>19,759</td><td>$2\frac{3}{8}$ OH</td><td>$3\frac{1}{4}$</td><td>$3\frac{1}{4}$</td><td>6,500</td><td>294,600</td><td>3,700</td><td>0.82</td><td>9</td></tr>
<tr><td>19,600</td><td>19,759</td><td>$2\frac{3}{8}$ SLH90</td><td>$3\frac{1}{4}$</td><td>$1\frac{13}{16}$</td><td>6,900</td><td>270,200</td><td>3,700</td><td>0.87</td><td>9</td></tr>
<tr><td>21,663</td><td>21,839</td><td>NC26</td><td>$3\frac{3}{8}$</td><td>$1\frac{3}{4}$</td><td>6,900</td><td>313,700</td><td>3,900</td><td>0.78</td><td>9</td></tr>
<tr><td>21,663</td><td>21,839</td><td>HT26</td><td>$3\frac{3}{8}$</td><td>$1\frac{3}{4}$</td><td>8,700</td><td>313,700</td><td>5,200</td><td>0.99</td><td>9</td></tr>
<tr><td>21,663</td><td>21,839</td><td>$2\frac{3}{8}$ OH</td><td>$3\frac{1}{4}$</td><td>$1\frac{3}{4}$</td><td>6,500</td><td>294,600</td><td>3,700</td><td>0.74</td><td>9</td></tr>
<tr><td>21,663</td><td>21,839</td><td>$2\frac{3}{8}$ SLH90</td><td>$3\frac{1}{4}$</td><td>$1\frac{13}{16}$</td><td>6,900</td><td>270,200</td><td>3,700</td><td>0.78</td><td>9</td></tr>
<tr><td>27,853</td><td>28,079</td><td>NC26</td><td>$3\frac{5}{8}$</td><td>$1\frac{1}{2}$</td><td>9,000</td><td>390,300</td><td>4,900</td><td>0.80</td><td>9</td></tr>
<tr><td>27,853</td><td>28,079</td><td>HT26</td><td>$3\frac{3}{8}$</td><td>$1\frac{3}{4}$</td><td>8,700</td><td>313,700</td><td>5,200</td><td>0.77</td><td>9</td></tr>
<tr><td>27,853</td><td>28,079</td><td>$2\frac{3}{8}$ OH</td><td>$3\frac{1}{4}$</td><td>$1\frac{3}{4}$</td><td>6,500</td><td>294,600</td><td>3,700</td><td>0.58</td><td>9</td></tr>
<tr><td>27,853</td><td>28,079</td><td>$2\frac{3}{8}$ SLH90</td><td>$3\frac{1}{4}$</td><td>$1\frac{13}{16}$</td><td>6,900</td><td>270,200</td><td>3,700</td><td>0.61</td><td>9</td></tr>
<tr><td>27,853</td><td>28,079</td><td>GPDS26</td><td>$3\frac{1}{2}$</td><td>$1\frac{3}{4}$</td><td>8,800</td><td>313,700</td><td>5,300</td><td>0.78</td><td>9</td></tr>
<tr><td>28,884</td><td>29,119</td><td>XT24</td><td>$3\frac{1}{8}$</td><td>$1\frac{1}{2}$</td><td>10,000</td><td>261,500</td><td>6,000</td><td>0.85</td><td>10</td></tr>
<tr><td>28,884</td><td>29,119</td><td>XT26</td><td>$3\frac{3}{8}$</td><td>$1\frac{3}{4}$</td><td>11,600</td><td>290,900</td><td>7,000</td><td>0.99</td><td>10</td></tr>
<tr><td>28,884</td><td>29,119</td><td>HT26</td><td>$3\frac{3}{8}$</td><td>$1\frac{3}{4}$</td><td>8,700</td><td>313,700</td><td>5,200</td><td>0.74</td><td>9</td></tr>
<tr><td>28,884</td><td>29,119</td><td>GPDS26</td><td>$3\frac{1}{2}$</td><td>$1\frac{11}{16}$</td><td>9,700</td><td>333,900</td><td>5,800</td><td>0.83</td><td>9</td></tr>
<tr><td>30,947</td><td>31,199</td><td>XT24</td><td>$3\frac{1}{8}$</td><td>$1\frac{3}{8}$</td><td>10,900</td><td>295,400</td><td>6,500</td><td>0.87</td><td>10</td></tr>
<tr><td>30,947</td><td>31,199</td><td>XT26</td><td>$3\frac{3}{8}$</td><td>$1\frac{3}{4}$</td><td>11,600</td><td>290,900</td><td>7,000</td><td>0.93</td><td>10</td></tr>
<tr><td>30,947</td><td>31,199</td><td>HT26</td><td>$3\frac{3}{8}$</td><td>$1\frac{3}{4}$</td><td>8,700</td><td>313,700</td><td>5,200</td><td>0.70</td><td>9</td></tr>
<tr><td>30,947</td><td>31,199</td><td>GPDS26</td><td>$3\frac{1}{2}$</td><td>$1\frac{5}{8}$</td><td>10,500</td><td>353,400</td><td>6,300</td><td>0.84</td><td>9</td></tr>
<tr><td>9,907</td><td>10,467</td><td>NC26</td><td>$3\frac{3}{8}$</td><td>$1\frac{3}{4}$</td><td>6,900</td><td>313,700</td><td>3,900</td><td>0.85</td><td>9</td></tr>
<tr><td>9,907</td><td>10,467</td><td>HT26</td><td>$3\frac{3}{8}$</td><td>$1\frac{3}{4}$</td><td>8,700</td><td>313,700</td><td>5,200</td><td>1.07</td><td>9</td></tr>
<tr><td>9,907</td><td>10,467</td><td>NC31</td><td>$4\frac{1}{8}$</td><td>$2\frac{5}{32}$</td><td>11,500</td><td>434,500</td><td>6,200</td><td>1.42</td><td>9</td></tr>
<tr><td>9,907</td><td>10,467</td><td>$2\frac{7}{8}$ PAC</td><td>$3\frac{1}{8}$</td><td>$1\frac{1}{2}$</td><td>5,700</td><td>273,000</td><td>3,200</td><td>0.70</td><td>9</td></tr>
<tr><td>9,907</td><td>10,467</td><td>$2\frac{7}{8}$ OH</td><td>$3\frac{3}{4}$</td><td>$2\frac{3}{8}$</td><td>6,300</td><td>252,100</td><td>3,500</td><td>0.78</td><td>9</td></tr>
<tr><td>9,907</td><td>10,467</td><td>XT 26</td><td>$3\frac{3}{8}$</td><td>$1\frac{3}{4}$</td><td>11,600</td><td>290,900</td><td>7,000</td><td>1.43</td><td>10</td></tr>
<tr><td>9,907</td><td>10,467</td><td>HT31</td><td>4</td><td>$2\frac{5}{32}$</td><td>14,900</td><td>434,600</td><td>8,900</td><td>1.84</td><td>9</td></tr>
<tr><td>9,907</td><td>10,467</td><td>XT31</td><td>4</td><td>$2\frac{3}{8}$</td><td>12,900</td><td>309,100</td><td>7,700</td><td>1.59</td><td>10</td></tr>
<tr><td>12,548</td><td>12,940</td><td>NC26</td><td>$3\frac{1}{2}$</td><td>$1\frac{1}{2}$</td><td>8,800</td><td>390,300</td><td>4,900</td><td>0.86</td><td>9</td></tr>
<tr><td>12,548</td><td>12,940</td><td>HT26</td><td>$3\frac{3}{8}$</td><td>$1\frac{3}{4}$</td><td>8,700</td><td>313,700</td><td>5,200</td><td>0.85</td><td>9</td></tr>
<tr><td>12,548</td><td>12,940</td><td>NC31</td><td>$4\frac{1}{8}$</td><td>$2\frac{5}{32}$</td><td>11,500</td><td>434,500</td><td>6,200</td><td>1.13</td><td>9</td></tr>
<tr><td>12,548</td><td>12,940</td><td>$2\frac{7}{8}$ PAC</td><td>$3\frac{1}{8}$</td><td>$1\frac{1}{2}$</td><td>5,700</td><td>273,000</td><td>3,200</td><td>0.56</td><td>9</td></tr>
<tr><td>12,548</td><td>12,940</td><td>$2\frac{7}{8}$ OH</td><td>$3\frac{3}{4}$</td><td>$2\frac{3}{8}$</td><td>6,300</td><td>252,100</td><td>3,500</td><td>0.62</td><td>9</td></tr>
<tr><td>12,548</td><td>12,940</td><td>XT26</td><td>$3\frac{3}{8}$</td><td>$1\frac{3}{4}$</td><td>11,600</td><td>290,900</td><td>7,000</td><td>1.14</td><td>10</td></tr>
<tr><td>12,548</td><td>12,940</td><td>HT31</td><td>4</td><td>$2\frac{5}{32}$</td><td>14,900</td><td>434,500</td><td>8,900</td><td>1.46</td><td>9</td></tr>
<tr><td>12,548</td><td>12,940</td><td>XT31</td><td>4</td><td>$2\frac{3}{8}$</td><td>12,900</td><td>309,100</td><td>7,700</td><td>1.26</td><td>10</td></tr>
<tr><td>13,869</td><td>14,020</td><td>NC26</td><td>$3\frac{5}{8}$</td><td>$1\frac{3}{4}$</td><td>7,200</td><td>313,700</td><td>3,900</td><td>0.64</td><td>9</td></tr>
<tr><td>13,869</td><td>14,020</td><td>HT26</td><td>$3\frac{3}{8}$</td><td>$1\frac{3}{4}$</td><td>8,700</td><td>313,700</td><td>5,200</td><td>0.77</td><td>9</td></tr>
<tr><td>13,869</td><td>14,020</td><td>NC31</td><td>$4\frac{1}{8}$</td><td>$2\frac{5}{32}$</td><td>11,500</td><td>434,500</td><td>6,200</td><td>1.02</td><td>9</td></tr>
<tr><td>13,869</td><td>14,020</td><td>$2\frac{7}{8}$ PAC</td><td>$3\frac{1}{8}$</td><td>$1\frac{7}{8}$</td><td>5,700</td><td>273,000</td><td>3,200</td><td>0.50</td><td>9</td></tr>
<tr><td>13,869</td><td>14,020</td><td>$2\frac{7}{8}$ OH</td><td>$3\frac{7}{8}$</td><td>$2\frac{5}{32}$</td><td>8,800</td><td>345,500</td><td>4,800</td><td>0.78</td><td>9</td></tr>
<tr><td>13,869</td><td>14,020</td><td>XT26</td><td>$3\frac{3}{8}$</td><td>$1\frac{3}{4}$</td><td>11,600</td><td>290,900</td><td>7,000</td><td>1.03</td><td>10</td></tr>
</tbody>
</table>

(continued)

Table 4.6.6   (continued)

Tool Joint Data

Internal Pressure	Collapse Pressure	Connection Type	Outside Diameter	Inside Diameter	Torsional Yield Strength	Tensile Yield Strength	Make-up Torque	Torsional Ratio Tool Joint to Pipe	* Pin Tong Space
psi	psi		in.	in.	ft-lb	lb	ft-lb		in.
13,869	14,020	HT31	4	$2\frac{5}{32}$	14,900	434,500	8,900	1.32	9
13,869	14,020	XT31	4	$2\frac{3}{8}$	12,900	309,100	7,700	1.14	10
17,832	17,034	NC26	$3\frac{5}{8}$	$1\frac{1}{2}$	9,000	390,300	4,900	0.62	9
17,832	17,034	HT26	$3\frac{1}{2}$	$1\frac{1}{2}$	12,100	390,300	7,300	0.83	9
17,832	17,034	NC31	$4\frac{1}{8}$	$2\frac{1}{8}$	11,900	447,100	6,400	0.82	9
17,832	17,034	$2\frac{7}{8}$ PAC	$3\frac{1}{8}$	$1\frac{1}{2}$	5,700	273,000	3,200	0.39	9
17,832	17,034	$2\frac{7}{8}$ OH	$3\frac{7}{8}$	$2\frac{5}{32}$	8,800	345,500	4,800	0.61	9
17,832	17,034	XT26	$3\frac{3}{8}$	$1\frac{3}{4}$	11,600	290,900	7,000	0.80	10
17,832	17,034	HT31	4	$2\frac{5}{32}$	14,900	434,500	8,900	1.03	9
17,832	17,034	XT31	4	$2\frac{3}{8}$	12,900	309,100	7,700	0.89	10
18,492	17,500	HT26	$3\frac{1}{2}$	$1\frac{1}{2}$	12,100	390,300	7,300	0.80	9
18,492	17,500	XT26	$3\frac{3}{8}$	$1\frac{3}{4}$	11,600	290,900	7,000	0.77	10
18,492	17,500	HT31	4	$2\frac{5}{32}$	14,900	434,500	8,900	0.99	9
18,492	17,500	XT31	4	$2\frac{3}{8}$	12,900	309,100	7,700	0.85	10
19,813	18,398	HT26	$3\frac{1}{2}$	$1\frac{1}{2}$	12,100	390,300	7,300	0.75	9
19,813	18,398	XT26	$3\frac{3}{8}$	$1\frac{3}{4}$	11,600	290,900	7,000	0.72	10
19,813	18,398	HT31	4	$2\frac{5}{32}$	14,900	434,500	8,900	0.92	9
19,813	18,398	XT31	4	$2\frac{3}{8}$	12,900	309,100	7,700	0.80	10
16,526	16,509	NC31	$4\frac{1}{8}$	$2\frac{5}{32}$	11,500	434,500	6,200	0.99	9
16,526	16,509	NC26	$3\frac{1}{2}$	$1\frac{1}{2}$	8,800	390,300	4,900	0.76	9
16,526	16,509	$2\frac{7}{8}$ PAC	$3\frac{1}{8}$	$1\frac{1}{2}$	5,700	273,000	3,200	0.49	9
16,526	16,509	$2\frac{7}{8}$ OH	$3\frac{7}{8}$	$2\frac{5}{32}$	8,800	345,500	4,800	0.76	9
16,526	16,509	$2\frac{7}{8}$ SLH90	$4\frac{1}{8}$	$2\frac{5}{32}$	11,500	382,800	5,900	0.99	9
16,526	16,509	2-78 HTPAC	$3\frac{1}{8}$	$1\frac{1}{2}$	8,500	273,000	5,100	0.73	9
16,526	16,509	HT26	$3\frac{1}{2}$	$1\frac{1}{2}$	12,100	390,300	7,300	1.04	9
16,526	16,509	HT31	$4\frac{1}{8}$	$2\frac{5}{32}$	16,000	434,500	9,600	1.38	9
16,526	16,509	XT26	$3\frac{1}{2}$	$1\frac{1}{2}$	14,900	367,400	8,900	1.28	10
16,526	16,509	XT31	4	$2\frac{5}{32}$	17,700	402,500	10,600	1.53	10
20,933	20,911	NC31	$4\frac{1}{8}$	2	13,200	495,700	7,100	0.90	9
20,933	20,911	NC26	$3\frac{1}{2}$	$1\frac{1}{2}$	8,800	390,300	4,900	0.60	9
20,933	20,911	$2\frac{7}{8}$ PAC	$3\frac{1}{8}$	$1\frac{1}{2}$	5,700	273,000	3,200	0.39	9
20,933	20,911	$2\frac{7}{8}$ OH	$3\frac{7}{8}$	$2\frac{5}{32}$	8,800	345,500	4,800	0.60	9
20,933	20,911	$2\frac{7}{8}$ SLH90	$4\frac{1}{8}$	$2\frac{5}{32}$	11,500	382,800	5,900	0.79	9
20,933	20,911	2-78 HTPAC	$3\frac{1}{8}$	$1\frac{1}{2}$	8,500	273,000	5,100	0.58	9
20,933	20,911	HT26	$3\frac{1}{2}$	$1\frac{1}{2}$	12,100	390,300	7,300	0.83	9
20,933	20,911	HT31	$4\frac{1}{8}$	$2\frac{5}{32}$	16,000	434,500	9,600	1.10	9
20,933	20,911	XT26	$3\frac{1}{2}$	$1\frac{1}{2}$	14,900	367,400	8,900	1.02	10
20,933	20,911	XT31	4	$2\frac{5}{32}$	17,700	402,500	10,600	1.21	10
23,137	23,112	NC31	$4\frac{1}{8}$	2	13,200	495,700	7,100	0.81	9
23,137	23,112	NC26	$3\frac{1}{2}$	$1\frac{1}{2}$	8,800	390,300	4,900	0.54	9
23,137	23,112	$2\frac{7}{8}$ PAC	$3\frac{1}{8}$	$1\frac{3}{8}$	6,500	306,900	3,600	0.40	9
23,137	23,112	$2\frac{7}{8}$ OH	$3\frac{7}{8}$	$2\frac{5}{32}$	8,800	345,500	4,800	0.54	9
23,137	23,112	$2\frac{7}{8}$ SLH90	$4\frac{1}{8}$	2	13,300	444,000	6,900	0.82	9

(*continued*)

Table 4.6.6   *(continued)*

					Tool Joint Data				
Internal Pressure	Collapse Pressure	Connection Type	Outside Diameter	Inside Diameter	Torsional Yield Strength	Tensile Yield Strength	Make-up Torque	Torsional Ratio Tool Joint to Pipe	*Pin Tong Space
psi	psi		in.	in.	ft-lb	lb	ft-lb		in.
23,137	23,112	$2\frac{7}{8}$ HTPAC	$3\frac{1}{8}$	$1\frac{3}{8}$	9,800	306,900	5,900	0.60	9
23,137	23,112	HT26	$3\frac{5}{8}$	$1\frac{1}{2}$	13,100	390,300	7,900	0.81	9
23,137	23,112	HT31	$4\frac{1}{8}$	$2\frac{5}{32}$	16,300	434,500	9,600	0.99	9
23,137	23,112	XT26	$3\frac{1}{2}$	$1\frac{1}{2}$	14,900	367,400	8,900	0.92	10
23,137	23,112	XT31	4	$2\frac{5}{32}$	17,700	402,500	10,600	1.09	10
29,747	29,716	NC31	$4\frac{3}{8}$	$1\frac{5}{8}$	16,900	623,800	9,000	0.81	9
29,747	29,716	NC26	$3\frac{5}{8}$	$1\frac{1}{2}$	9,000	390,300	4,900	0.43	9
29,747	29,716	$2\frac{7}{8}$ PAC	$3\frac{1}{8}$	$1\frac{1}{2}$	5,700	273,000	3,200	0.27	9
29,747	29,716	$2\frac{7}{8}$ OH	$3\frac{7}{8}$	2	10,400	406,700	5,700	0.50	9
29,747	29,716	$2\frac{7}{8}$ SLH90	$4\frac{1}{8}$	2	13,300	444,000	6,900	0.64	9
29,747	29,716	$2\frac{7}{8}$ HTPAC	$3\frac{1}{8}$	$1\frac{1}{2}$	8,500	273,000	5,100	0.41	9
29,747	29,716	HT26	$3\frac{5}{8}$	$1\frac{1}{2}$	13,100	390,300	7,900	0.63	9
29,747	29,716	HT31	$4\frac{1}{8}$	2	18,900	495,700	11,300	0.91	9
29,747	29,716	XT26	$3\frac{1}{2}$	$1\frac{3}{8}$	16,000	401,300	9,600	0.77	10
29,747	29,716	XT31	4	$2\frac{5}{32}$	17,700	402,500	10,600	0.85	10
29,747	29,716	GPDS31	$4\frac{1}{8}$	2	17,200	495,700	10,300	0.83	9
30,849	30,817	HT26	$3\frac{5}{8}$	$1\frac{3}{8}$	14,500	424,100	8,700	0.67	9
30,849	30,817	HT31	$4\frac{1}{8}$	2	18,900	495,700	11,300	0.88	9
30,849	30,817	XT26	$3\frac{1}{2}$	$1\frac{1}{4}$	16,500	432,200	9,900	0.76	10
30,849	30,817	XT31	4	$2\frac{5}{32}$	17,700	402,500	10,600	0.82	10
30,849	30,817	GPDS31	$4\frac{1}{8}$	2	17,200	495,700	10,300	0.80	9
33,052	33,018	HT26	$3\frac{5}{8}$	$1\frac{1}{4}$	15,300	455,100	9,200	0.66	9
33,052	33,018	HT31	$4\frac{1}{8}$	2	18,900	495,700	11,300	0.82	9
33,052	33,018	XT26	$3\frac{1}{2}$	$1\frac{1}{4}$	16,500	432,200	9,900	0.71	10
33,052	33,018	XT31	4	2	20,100	463,700	12,100	0.87	10
33,052	33,018	GPDS31	$4\frac{1}{8}$	$1\frac{7}{8}$	18,200	541,400	10,900	0.79	9
9,525	10,001	NC38	$4\frac{3}{4}$	$2\frac{11}{16}$	18,100	587,300	9,700	1.28	10
9,525	10,001	NC31	$4\frac{1}{8}$	$2\frac{1}{8}$	11,900	447,100	6,400	0.84	9
9,525	10,001	HT31	$4\frac{1}{8}$	$2\frac{1}{8}$	16,600	447,100	10,000	1.18	9
9,525	10,001	HT38	$4\frac{3}{4}$	$2\frac{11}{16}$	25,300	587,300	15,200	1.79	10
9,525	10,001	$3\frac{1}{2}$ SLH90	$4\frac{3}{4}$	$2\frac{11}{16}$	18,688	534,200	11,100	1.33	10
9,525	10,001	XT31	4	$2\frac{1}{8}$	18,300	415,100	11,000	1.30	10
9,525	10,001	XT38	$4\frac{5}{8}$	$2\frac{13}{16}$	24,000	473,000	14,400	1.70	10
12,065	12,077	NC38	$4\frac{3}{4}$	$2\frac{11}{16}$	18,100	587,300	9,700	1.01	10
12,065	12,077	NC31	$4\frac{1}{8}$	2	13,200	495,700	7,100	0.74	9
12,065	12,077	HT31	$4\frac{1}{8}$	$2\frac{1}{8}$	16,600	447,100	10,000	0.93	9
12,065	12,077	HT38	$4\frac{3}{4}$	$2\frac{11}{16}$	25,300	587,300	15,200	1.41	10
12,065	12,077	$3\frac{1}{2}$ SLH90	$4\frac{3}{4}$	$2\frac{11}{16}$	18,700	534,200	11,100	1.04	10
12,065	12,077	XT31	4	$2\frac{1}{8}$	18,300	415,100	11,000	1.02	10
12,065	12,077	XT38	$4\frac{5}{8}$	$2\frac{13}{16}$	24,000	473,000	14,400	1.34	10
13,335	13,055	NC38	$4\frac{3}{4}$	$2\frac{11}{16}$	18,100	587,300	9,700	0.91	10
13,335	13,055	NC31	$4\frac{1}{8}$	2	13,200	495,700	7,100	0.67	9

(continued)

Table 4.6.6 *(continued)*

					Tool Joint Data				
Internal Pressure	Collapse Pressure	Connection Type	Outside Diameter	Inside Diameter	Torsional Yield Strength	Tensile Yield Strength	Make-up Torque	Torsional Ratio Tool Joint to Pipe	* Pin Tong Space
psi	psi		in.	in.	ft-lb	lb	ft-lb		in.
13,335	13,055	HT31	$4\frac{1}{8}$	2	18,900	495,700	11,300	0.95	9
13,335	13,055	HT38	$4\frac{3}{4}$	$2\frac{11}{16}$	25,300	587,300	15,200	1.28	10
13,335	13,055	$3\frac{1}{2}$ SLH90	$4\frac{3}{4}$	$2\frac{11}{16}$	18,700	534,200	11,100	0.94	10
13,335	13,055	XT31	4	$2\frac{1}{8}$	18,300	415,100	11,000	0.92	10
13,335	13,055	XT38	$4\frac{5}{8}$	$2\frac{13}{16}$	24,000	473,000	14,400	1.21	10
17,145	15,748	NC38	5	$2\frac{9}{16}$	20,300	649,200	10,700	0.80	10
17,145	15,748	NC31	$4\frac{1}{8}$	2	13,200	495,700	7,100	0.52	9
17,145	15,748	HT31	$4\frac{1}{8}$	2	18,900	495,700	11,300	0.74	9
17,145	15,748	HT38	$4\frac{3}{4}$	$2\frac{11}{16}$	25,300	587,300	15,200	0.99	10
17,145	15,748	$3\frac{1}{2}$ SLH90	$4\frac{3}{4}$	$2\frac{9}{16}$	20,900	596,100	12,400	0.82	10
17,145	15,748	XT31	4	2	20,100	463,700	12,100	0.79	10
17,145	15,748	XT38	$4\frac{5}{8}$	$2\frac{13}{16}$	24,000	473,000	14,400	0.94	10
17,780	16,158	HT31	$4\frac{1}{8}$	2	18,900	495,700	11,300	0.72	9
17,780	16,158	HT38	$4\frac{3}{4}$	$2\frac{11}{16}$	25,300	587,300	15,200	0.96	10
17,780	16,158	XT31	4	2	20,100	463,700	12,100	0.79	10
17,780	16,158	XT38	$4\frac{5}{8}$	$2\frac{13}{16}$	24,000	473,000	14,400	0.91	10
19,050	16,943	HT31	$4\frac{1}{4}$	$1\frac{3}{4}$	23,400	584,100	14,000	0.83	9
19,050	16,943	HT38	$4\frac{3}{4}$	$2\frac{11}{16}$	25,300	587,300	15,200	0.89	10
19,050	16,943	XT31	4	2	20,100	463,700	12,100	0.71	10
19,050	16,943	XT38	$4\frac{5}{8}$	$2\frac{13}{16}$	24,000	473,000	14,400	0.85	10
13,800	14,113	NC38	$4\frac{3}{4}$	$2\frac{11}{16}$	18,100	587,300	9,700	0.97	10
13,800	14,113	NC31	$4\frac{1}{8}$	2	13,200	495,700	7,100	0.71	9
13,800	14,113	HT31	$4\frac{1}{8}$	$2\frac{1}{8}$	16,600	447,100	10,000	0.89	9
13,800	14,113	HT38	$4\frac{3}{4}$	$2\frac{11}{16}$	25,300	587,300	15,200	1.36	10
13,800	14,113	$3\frac{1}{2}$ SLH90	$4\frac{3}{4}$	$2\frac{13}{16}$	16,400	469,400	9,800	0.88	10
13,800	14,113	XT31	4	$2\frac{1}{8}$	18,300	415,100	11,000	0.98	10
13,800	14,113	XT38	$4\frac{3}{4}$	$2\frac{11}{16}$	27,900	537,800	16,700	1.50	10
17,480	17,877	NC38	5	$2\frac{9}{16}$	20,300	649,200	10,700	0.86	10
17,480	17,877	NC31	$4\frac{1}{8}$	2	13,200	495,700	7,100	0.56	9
17,480	17,877	HT31	$4\frac{1}{8}$	2	18,900	495,700	11,300	0.80	9
17,480	17,877	HT38	$4\frac{3}{4}$	$2\frac{11}{16}$	25,300	587,300	15,200	1.08	10
17,480	17,877	$3\frac{1}{2}$ SLH90	$4\frac{3}{4}$	$2\frac{11}{16}$	18,700	534,200	11,100	0.80	10
17,480	17,877	XT31	4	$2\frac{1}{8}$	18,300	415,100	11,000	0.78	10
17,480	17,877	XT38	$4\frac{3}{4}$	$2\frac{11}{16}$	27,900	537,800	16,700	1.19	10
19,320	19,758	NC38	5	$2\frac{7}{16}$	22,200	708,100	11,700	0.85	10
19,320	19,758	NC31	$4\frac{1}{8}$	2	13,200	495,700	7,100	0.51	9
19,320	19,758	HT31	$4\frac{1}{8}$	2	18,900	495,700	11,300	0.73	9
19,320	19,758	HT38	$4\frac{3}{4}$	$2\frac{11}{16}$	25,300	587,300	15,200	0.97	10
19,320	19,758	$3\frac{1}{2}$ SLH90	$4\frac{3}{4}$	$2\frac{9}{16}$	20,900	596,100	12,400	0.80	10
19,320	19,758	XT31	$4\frac{1}{8}$	2	20,900	463,700	12,500	0.80	10
19,320	19,758	XT38	$4\frac{3}{4}$	$2\frac{11}{16}$	27,900	537,800	16,700	1.07	10
24,840	25,404	NC38	5	$2\frac{1}{8}$	26,500	842,400	14,000	0.79	10
24,840	25,404	NC31	$4\frac{1}{8}$	2	13,200	495,700	7,100	0.40	9

(continued)

Table 4.6.6 (continued)

									Tool Joint Data

Internal Pressure	Collapse Pressure	Connection Type	Outside Diameter	Inside Diameter	Torsional Yield Strength	Tensile Yield Strength	Make-up Torque	Torsional Ratio Tool Joint to Pipe	* Pin Tong Space
psi	psi		in.	in.	ft-lb	lb	ft-lb		in.
24,840	25,404	HT31	$4\frac{1}{8}$	2	18,900	495,700	11,300	0.57	9
24,840	25,404	HT38	$4\frac{3}{4}$	$2\frac{9}{16}$	26,900	649,200	16,100	0.81	10
24,840	25,404	$3\frac{1}{2}$ SLH90	$4\frac{3}{4}$	$2\frac{9}{16}$	20,900	596,100	12,400	0.63	10
24,840	25,404	XT31	$4\frac{1}{8}$	$1\frac{7}{8}$	23,100	509,400	13,900	0.69	10
24,840	25,404	XT38	$4\frac{3}{4}$	$2\frac{11}{16}$	27,900	537,800	16,700	0.84	10
24,840	25,404	GPDS38	$4\frac{7}{8}$	$2\frac{9}{16}$	25,700	649,200	15,400	0.77	10
25,760	26,345	HT31	$4\frac{1}{8}$	$1\frac{7}{8}$	19,900	541,400	11,900	0.58	9
25,760	26,345	HT38	$4\frac{3}{4}$	$2\frac{11}{16}$	25,300	587,300	15,200	0.73	10
25,760	26,345	HT31	$4\frac{1}{8}$	$1\frac{3}{4}$	24,800	552,100	14,900	0.72	10
25,760	26,345	XT38	$4\frac{5}{8}$	$2\frac{11}{16}$	27,500	537,800	16,500	0.79	10
25,760	26,345	GPDS38	$4\frac{7}{8}$	$2\frac{7}{8}$	27,400	679,000	16,400	0.79	10
27,600	28,226	HT31	$4\frac{1}{8}$	$1\frac{3}{4}$	23,400	584,100	14,000	0.63	9
27,600	28,226	HT38	$4\frac{3}{4}$	$2\frac{11}{16}$	25,300	587,300	15,200	0.68	10
27,600	28,226	XT31	$4\frac{1}{8}$	$1\frac{3}{4}$	24,800	552,100	14,900	0.67	10
27,600	28,226	XT38	$4\frac{3}{4}$	$2\frac{9}{16}$	31,500	599,600	18,900	0.85	10
27,600	28,226	GPDS38	5	$2\frac{7}{16}$	29,200	708,100	17,500	0.79	10
16,838	16,774	NC38	$4\frac{3}{4}$	$2\frac{9}{16}$	19,200	649,200	10,700	0.91	10
16,838	16,774	HT38	$4\frac{3}{4}$	$2\frac{11}{16}$	25,300	587,300	15,200	1.20	10
16,838	16,774	XT38	$4\frac{3}{4}$	$2\frac{9}{16}$	31,500	599,600	18,900	1.49	10
21,328	21,247	NC38	5	$2\frac{7}{16}$	22,200	708,100	11,700	0.83	10
21,328	21,247	HT38	$4\frac{3}{4}$	$2\frac{11}{16}$	25,300	587,300	15,200	0.95	10
21,328	21,247	XT38	$4\frac{3}{4}$	$2\frac{9}{16}$	31,500	599,600	18,900	1.18	10
23,573	23,484	NC38	5	$2\frac{1}{8}$	26,500	842,400	14,000	0.90	10
23,573	23,484	HT38	$4\frac{3}{4}$	$2\frac{9}{16}$	26,900	649,200	16,100	0.91	10
23,573	23,484	NC40	$5\frac{1}{4}$	$2\frac{9}{16}$	27,800	838,300	14,600	0.94	9
23,573	23,484	XT38	$4\frac{3}{4}$	$2\frac{9}{16}$	31,500	599,600	18,900	1.07	10
30,308	30,194	NC38	5	$2\frac{1}{8}$	26,500	842,400	14,000	0.70	10
30,308	30,194	HT38	$4\frac{3}{4}$	$2\frac{7}{16}$	28,400	708,100	17,000	0.75	10
30,308	30,194	NC40	$5\frac{1}{2}$	$2\frac{1}{4}$	32,900	980,000	17,100	0.87	10
30,308	30,194	XT38	$4\frac{3}{4}$	$2\frac{9}{16}$	31,500	599,600	18,900	0.83	10
30,308	30,194	XT39	5	$2\frac{9}{16}$	40,800	729,700	24,500	1.07	10
30,308	30,194	GPDS38	5	$2\frac{3}{8}$	30,800	736,400	18,500	0.81	10
31,430	31,312	HT38	$4\frac{3}{4}$	$2\frac{7}{16}$	28,400	708,100	17,000	0.72	10
31,430	31,312	XT38	$4\frac{3}{4}$	$2\frac{7}{16}$	34,400	658,500	20,600	0.87	10
31,430	31,312	XT39	5	$2\frac{9}{16}$	40,800	729,700	24,500	1.04	10
31,430	31,312	GPDS38	5	$2\frac{3}{8}$	30,800	736,400	18,500	0.78	10
33,675	33,549	HT38	5	$2\frac{7}{16}$	33,000	708,100	19,800	0.78	10
33,675	33,549	XT38	$4\frac{3}{4}$	$2\frac{7}{16}$	34,400	658,500	20,600	0.82	10
33,675	33,549	XT39	5	$2\frac{9}{16}$	40,800	729,700	24,500	0.97	10
33,675	33,549	GPDS38	5	$2\frac{1}{4}$	33,900	790,900	20,300	0.80	10
8,597	8,381	NC40	$5\frac{1}{4}$	$2\frac{13}{16}$	23,500	711,600	12,400	1.21	9
8,597	8,381	4 SH	$4\frac{3}{4}$	$2\frac{9}{16}$	15,300	512,000	8,100	0.78	9
8,597	8,381	HT38	$4\frac{3}{4}$	$2\frac{11}{16}$	25,300	587,300	15,200	1.30	10

(*continued*)

Table 4.6.6   *(continued)*

					Tool Joint Data				
Internal Pressure	Collapse Pressure	Connection Type	Outside Diameter	Inside Diameter	Torsional Yield Strength	Tensile Yield Strength	Make-up Torque	Torsional Ratio Tool Joint to Pipe	* Pin Tong Space
psi	psi		in.	in.	ft-lb	lb	ft-lb		in.
8,597	8,381	XT38	$4\frac{3}{4}$	$2\frac{11}{16}$	27,900	537,800	16,700	1.43	10
8,597	8,381	XT39	5	$2\frac{7}{8}$	31,000	569,500	18,600	1.59	10
10,889	9,978	NC40	$5\frac{1}{4}$	$2\frac{13}{16}$	23,500	711,600	12,400	0.95	9
10,889	9,978	4 SH	$4\frac{3}{4}$	$2\frac{9}{16}$	15,300	512,000	8,100	0.62	9
10,889	9,978	HT38	$4\frac{3}{4}$	$2\frac{11}{16}$	25,300	587,300	15,200	1.02	10
10,889	9,978	XT38	$4\frac{3}{4}$	$2\frac{11}{16}$	27,900	537,800	16,700	1.13	10
10,889	9,978	XT39	5	$2\frac{7}{8}$	31,000	569,500	18,600	1.26	10
12,036	10,708	NC40	$5\frac{1}{4}$	$2\frac{13}{16}$	23,500	711,600	12,400	0.86	9
12,036	10,708	4 SH	$4\frac{3}{4}$	$2\frac{9}{16}$	15,300	512,000	8,100	0.56	9
12,036	10,708	HT38	$4\frac{3}{4}$	$2\frac{9}{16}$	26,900	649,200	16,100	0.99	10
12,036	10,708	XT38	$4\frac{3}{4}$	$2\frac{11}{16}$	27,900	537,800	16,700	1.02	10
12,036	10,708	XT39	5	$2\frac{7}{8}$	31,000	569,500	18,600	1.14	10
15,474	12,618	NC40	$5\frac{1}{2}$	$2\frac{9}{16}$	28,100	838,300	14,600	0.80	9
15,474	12,618	4 SH	$4\frac{3}{4}$	$2\frac{9}{16}$	15,300	512,000	8,100	0.44	9
15,474	12,618	HT38	$4\frac{3}{4}$	$2\frac{7}{16}$	28,400	708,100	17,000	0.81	10
15,474	12,618	XT38	$4\frac{3}{4}$	$2\frac{11}{16}$	27,900	537,800	16,700	0.79	10
15,474	12,618	XT39	5	$2\frac{7}{8}$	31,000	569,500	18,600	0.88	10
16,048	12,894	HT38	$4\frac{3}{4}$	$2\frac{7}{16}$	28,400	708,100	17,000	0.78	10
16,048	12,894	XT38	$4\frac{3}{4}$	$2\frac{11}{16}$	27,000	537,800	16,700	0.77	10
16,048	12,894	XT39	5	$2\frac{7}{8}$	31,000	569,500	18,600	0.85	10
17,194	13,404	HT38	5	$2\frac{7}{16}$	33,000	708,100	19,800	0.85	10
17,194	13,404	XT38	$4\frac{3}{4}$	$2\frac{9}{16}$	31,500	599,600	18,900	0.81	10
17,194	13,404	XT39	5	$2\frac{7}{8}$	31,000	569,500	18,600	0.80	10
10,828	11,354	NC40	$5\frac{1}{4}$	$2\frac{13}{16}$	23,500	711,600	12,400	1.01	9
10,828	11,354	HT38	$4\frac{3}{4}$	$2\frac{11}{16}$	25,300	587,300	15,200	1.09	10
10,828	11,354	4 SH	$4\frac{3}{4}$	$2\frac{7}{16}$	17,100	570,900	9,100	0.73	9
10,828	11,354	HT40	$5\frac{1}{4}$	$2\frac{13}{16}$	31,900	711,600	19,100	1.37	9
10,828	11,354	NC46	6	$3\frac{1}{4}$	33,600	901,200	17,600	1.44	9
10,828	11,354	XT38	$4\frac{3}{4}$	$2\frac{11}{16}$	27,900	537,800	16,700	1.20	10
10,828	11,354	XT39	5	$2\frac{13}{16}$	33,100	603,000	19,900	1.42	10
13,716	14,382	NC40	$5\frac{1}{4}$	$2\frac{13}{16}$	23,500	711,600	12,400	0.80	9
13,716	14,382	HT38	$4\frac{3}{4}$	$2\frac{11}{16}$	25,300	587,300	15,200	0.86	10
13,716	14,382	4 SH	$4\frac{3}{4}$	$2\frac{7}{16}$	17,100	570,900	9,100	0.58	9
13,716	14,382	HT40	$5\frac{1}{4}$	$2\frac{13}{16}$	31,900	711,600	19,100	1.08	9
13,716	14,382	NC46	6	$3\frac{1}{4}$	33,600	901,200	17,600	1.14	9
13,716	14,382	XT38	$4\frac{3}{4}$	$2\frac{11}{16}$	27,900	537,800	16,700	0.95	10
13,716	14,382	XT39	5	$2\frac{13}{16}$	33,100	603,000	19,900	1.12	10
15,159	15,896	NC40	$5\frac{1}{4}$	$2\frac{13}{16}$	23,500	711,600	12,400	0.72	9
15,159	15,896	HT38	5	$2\frac{9}{16}$	29,600	649,200	17,800	0.91	10
15,159	15,896	4 SH	$4\frac{3}{4}$	$2\frac{7}{16}$	17,100	570,900	9,100	0.52	9
15,159	15,896	HT40	$5\frac{1}{4}$	$2\frac{13}{16}$	31,900	711,600	19,100	0.98	9
15,159	15,896	NC46	6	$3\frac{1}{4}$	33,600	901,200	17,600	1.03	9
15,159	15,896	XT38	$4\frac{3}{4}$	$2\frac{11}{16}$	27,900	537,800	16,700	0.86	10
15,159	15,896	XT39	5	$2\frac{13}{16}$	33,100	603,000	19,900	1.02	10

(continued)

Table 4.6.6   (continued)

				Tool Joint Data					
Internal Pressure	Collapse Pressure	Connection Type	Outside Diameter	Inside Diameter	Torsional Yield Strength	Tensile Yield Strength	Make-up Torque	Torsional Ratio Tool Joint to Pipe	* Pin Tong Space
psi	psi		in.	in.	ft-lb	lb	ft-lb		in.
19,491	20,141	NC40	$5\frac{1}{2}$	$2\frac{9}{16}$	28,100	838,300	14,600	0.67	9
19,491	20,141	HT38	5	$2\frac{7}{16}$	33,000	708,100	19,800	0.79	10
19,491	20,141	4 SH	$4\frac{3}{4}$	$2\frac{7}{16}$	17,100	570,900	9,100	0.41	9
19,491	20,141	HT40	$5\frac{1}{4}$	$2\frac{11}{16}$	35,900	776,400	21,500	0.86	9
19,491	20,141	NC46	6	3	39,200	1,048,400	20,500	0.94	9
19,491	20,141	XT38	5	$2\frac{9}{16}$	31,800	599,600	19,100	0.76	10
19,491	20,141	XT39	5	$2\frac{13}{16}$	33,100	603,000	19,900	0.79	10
19,491	20,141	GPDS40	$5\frac{1}{4}$	$2\frac{11}{16}$	32,700	776,400	19,600	0.78	9
20,213	20,742	HT38	5	$2\frac{7}{16}$	33,000	708,100	19,800	0.76	10
20,213	20,742	HT40	$5\frac{1}{4}$	$2\frac{11}{16}$	35,900	776,400	21,500	0.83	9
20,213	20,742	XT38	5	$2\frac{9}{16}$	31,800	599,600	19,100	0.73	10
20,213	20,742	XT39	5	$2\frac{13}{16}$	33,100	603,000	19,900	0.76	10
20,213	20,742	GPDS40	$5\frac{1}{4}$	$2\frac{5}{8}$	34,600	807,700	20,800	0.80	9
21,656	21,912	HT38	5	$2\frac{7}{16}$	33,000	708,100	19,800	0.71	10
21,656	21,912	HT40	$5\frac{1}{4}$	$2\frac{11}{16}$	35,900	776,400	21,500	0.77	9
21,656	21,912	XT38	5	$2\frac{7}{16}$	35,200	658,500	21,100	0.76	10
21,656	21,912	XT39	5	$2\frac{13}{16}$	33,100	603,000	19,900	0.71	10
21,656	21,912	GPDS40	$5\frac{1}{4}$	$2\frac{1}{2}$	37,300	868,100	22,400	0.80	9
12,469	12,896	NC40	$5\frac{1}{4}$	$2\frac{13}{16}$	23,500	711,600	12,400	0.91	9
12,469	12,896	HT40	$5\frac{1}{4}$	$2\frac{13}{16}$	31,900	711,600	19,100	1.24	9
12,469	12,896	4 H90	$5\frac{1}{2}$	$2\frac{13}{16}$	35,400	913,700	20,400	1.37	9
12,469	12,896	NC46	6	$3\frac{1}{4}$	33,600	901,200	17,600	1.30	9
12,469	12,896	XT39	5	$2\frac{13}{16}$	33,100	603,000	19,900	1.28	10
12,469	12,896	XT40	$5\frac{1}{4}$	$2\frac{13}{16}$	44,100	751,600	26,500	1.71	10
15,794	16,335	NC40	$5\frac{1}{4}$	$2\frac{9}{16}$	27,800	838,300	14,600	0.85	9
15,794	16,335	HT40	$5\frac{1}{4}$	$2\frac{13}{16}$	31,900	711,600	19,100	0.98	9
15,794	16,335	4 H90	$5\frac{1}{2}$	$2\frac{13}{16}$	35,400	913,700	20,400	1.08	9
15,794	16,335	NC46	6	$3\frac{1}{4}$	33,600	901,200	17,600	1.03	9
15,794	16,335	XT39	5	$2\frac{13}{16}$	33,100	603,000	19,900	1.01	10
15,794	16,335	XT40	$5\frac{1}{4}$	$2\frac{13}{16}$	44,100	751,600	26,500	1.35	10
17,456	18,055	NC40	$5\frac{1}{2}$	$2\frac{7}{16}$	30,100	897,200	15,600	0.83	9
17,456	18,055	HT40	$5\frac{1}{4}$	$2\frac{13}{16}$	31,900	711,600	19,100	0.88	9
17,456	18,055	4 H90	$5\frac{1}{2}$	$2\frac{13}{16}$	35,400	913,700	20,400	0.98	9
17,456	18,055	NC46	6	$3\frac{1}{4}$	33,600	901,200	17,600	0.93	9
17,456	18,055	XT39	5	$2\frac{13}{16}$	33,100	603,000	19,900	0.92	10
17,456	18,055	XT40	$5\frac{1}{4}$	$2\frac{13}{16}$	44,100	751,600	26,500	1.22	10
22,444	23,213	NC40	$5\frac{1}{2}$	2	36,400	1,080,100	18,900	0.78	9
22,444	23,213	HT40	$5\frac{1}{2}$	$2\frac{9}{16}$	39,900	838,300	23,900	0.86	9
22,444	23,213	4 H90	$5\frac{3}{4}$	$2\frac{11}{16}$	38,400	978,500	21,800	0.83	9
22,444	23,213	NC46	6	3	39,200	1,048,400	20,500	0.84	9
22,444	23,213	XT39	5	$2\frac{11}{16}$	37,000	667,800	22,200	0.80	10
22,444	23,213	XT40	$5\frac{1}{4}$	$2\frac{13}{16}$	44,100	751,600	26,500	0.95	10
22,444	23,213	GPDS40	$5\frac{1}{4}$	$2\frac{9}{16}$	36,400	838,300	21,800	0.78	9
23,275	24,073	HT40	$5\frac{1}{2}$	$2\frac{9}{16}$	39,900	838,300	23,900	0.83	9

(continued)

Table 4.6.6 *(continued)*

					Torsional Yield Strength	Tensile Yield Strength	Make-up Torque	Torsional Ratio Tool Joint to Pipe	* Pin Tong Space
Internal Pressure	Collapse Pressure	Connection Type	Outside Diameter	Inside Diameter					
psi	psi		in.	in.	ft-lb	lb	ft-lb		in.
23,275	24,073	XT39	5	$2\frac{11}{16}$	37,000	667,800	22,200	0.77	10
23,275	24,073	XT40	$5\frac{1}{4}$	$2\frac{13}{16}$	44,100	751,600	26,500	0.91	10
23,275	24,073	GPDS40	$5\frac{3}{8}$	$2\frac{1}{2}$	38,400	868,100	23,000	0.80	9
24,938	25,793	HT40	$5\frac{1}{2}$	$2\frac{9}{16}$	39,900	838,300	23,900	0.77	9
24,938	25,793	XT39	5	$2\frac{9}{16}$	40,800	729,700	24,500	0.79	10
24,938	25,793	XT40	$5\frac{1}{4}$	$2\frac{13}{16}$	44,100	751,600	26,500	0.85	10
24,938	25,793	GPDS40	$5\frac{1}{2}$	$2\frac{7}{16}$	40,300	897,200	24,200	0.78	9
9,829	10,392	NC46	$6\frac{1}{4}$	$3\frac{1}{4}$	34,000	901,200	17,600	1.10	9
9,829	10,392	$4\frac{1}{2}$ OH	$5\frac{7}{8}$	$3\frac{3}{4}$	27,300	714,000	14,600	0.89	9
9,829	10,392	$4\frac{1}{2}$ FH	6	3	34,800	976,200	17,600	1.13	9
9,829	10,392	$4\frac{1}{2}$ H90	6	$3\frac{1}{4}$	39,000	938,400	18,800	1.27	9
9,829	10,392	HT46	$6\frac{1}{4}$	$3\frac{1}{4}$	47,600	901,200	28,600	1.55	9
9,829	10,392	NC50	$6\frac{3}{8}$	$3\frac{3}{4}$	37,700	939,100	19,800	1.22	9
9,829	10,392	HT50	$6\frac{1}{4}$	$3\frac{3}{4}$	52,700	939,100	31,600	1.71	9
9,829	10,392	XT40	$5\frac{1}{4}$	3	37,500	648,900	22,500	1.22	10
9,829	10,392	XT46	6	$3\frac{1}{2}$	58,100	910,300	34,900	1.89	10
9,829	10,392	XT50	$6\frac{3}{8}$	$3\frac{3}{4}$	75,200	1,085,500	45,100	2.44	10
12,450	12,765	NC46	$6\frac{1}{4}$	$3\frac{1}{4}$	34,000	901,200	17,600	0.87	9
12,450	12,765	$4\frac{1}{2}$ OH	$5\frac{7}{8}$	$3\frac{1}{2}$	33,900	884,800	18,200	0.87	9
12,450	12,765	$4\frac{1}{2}$ FH	6	3	34,800	976,200	17,600	0.89	9
12,450	12,765	$4\frac{1}{2}$ H90	6	$3\frac{1}{4}$	39,000	938,400	18,800	1.00	9
12,450	12,765	HT46	$6\frac{1}{4}$	$3\frac{1}{4}$	47,600	901,200	28,600	1.22	9
12,450	12,765	NC50	$6\frac{3}{8}$	$3\frac{3}{4}$	37,700	939,100	19,800	0.97	9
12,450	12,765	HT50	$6\frac{1}{4}$	$3\frac{3}{4}$	52,700	939,100	31,600	1.35	9
12,450	12,765	XT40	$5\frac{1}{4}$	3	37,500	648,900	22,500	0.96	10
12,450	12,765	XT46	6	$3\frac{1}{2}$	58,100	910,300	34,900	1.49	10
12,450	12,765	XT50	$6\frac{3}{8}$	$3\frac{3}{4}$	75,200	1,085,500	45,100	1.93	10
13,761	13,825	NC46	$6\frac{1}{4}$	3	39,700	1,048,400	20,500	0.92	9
13,761	13,825	$4\frac{1}{2}$ OH	6	$3\frac{1}{4}$	40,300	1,043,800	21,500	0.94	9
13,761	13,825	$4\frac{1}{2}$ FH	$6\frac{1}{4}$	$2\frac{3}{4}$	40,200	1,111,600	20,100	0.93	9
13,761	13,825	$4\frac{1}{2}$ H90	6	$3\frac{1}{4}$	39,000	938,400	18,800	0.90	9
13,761	13,825	HT46	$6\frac{1}{4}$	$3\frac{1}{4}$	47,600	901,200	28,600	1.10	9
13,761	13,825	NC50	$6\frac{3}{8}$	$3\frac{3}{4}$	37,700	939,100	19,800	0.87	9
13,761	13,825	HT50	$6\frac{1}{4}$	$3\frac{3}{4}$	52,700	939,100	31,600	1.22	9
13,761	13,825	XT40	$5\frac{1}{4}$	3	37,500	648,900	22,500	0.87	10
13,761	13,825	XT46	6	$3\frac{1}{2}$	58,100	910,300	34,900	1.35	10
13,761	13,825	XT50	$6\frac{3}{8}$	$3\frac{3}{4}$	75,200	1,085,500	45,100	1.74	10
17,693	16,773	NC46	$6\frac{1}{4}$	$2\frac{3}{4}$	44,900	1,183,900	23,200	0.81	9
17,693	16,773	$4\frac{1}{2}$ OH	6	2	43,400	1,191,100	24,600	0.78	9
17,693	16,773	$4\frac{1}{2}$ FH	$6\frac{1}{4}$	$2\frac{3}{4}$	40,200	1,111,600	20,100	0.72	9
17,693	16,773	$4\frac{1}{2}$ H90	$6\frac{1}{4}$	$2\frac{3}{4}$	51,500	1,221,100	24,600	0.93	9
17,693	16,773	HT46	$6\frac{1}{4}$	$3\frac{1}{4}$	47,600	901,200	28,600	0.86	9
17,693	16,773	NC50	$6\frac{3}{8}$	$3\frac{1}{2}$	44,700	1,109,900	23,400	0.81	9

(continued)

Table 4.6.6 (continued)

					Tool Joint Data				
Internal Pressure	Collapse Pressure	Connection Type	Outside Diameter	Inside Diameter	Torsional Yield Strength	Tensile Yield Strength	Make-up Torque	Torsional Ratio Tool Joint to Pipe	* Pin Tong Space
psi	psi		in.	in.	ft-lb	lb	ft-lb		in.
17,693	16,773	HT50	$6\frac{3}{8}$	$3\frac{1}{2}$	65,700	1,109,900	39,400	1.18	9
17,693	16,773	XT40	$5\frac{1}{4}$	$2\frac{13}{16}$	44,100	751,600	26,500	0.79	10
17,693	16,773	XT46	6	$3\frac{1}{2}$	58,100	910,300	34,900	1.05	10
17,693	16,773	XT50	$6\frac{3}{8}$	$3\frac{3}{4}$	75,200	1,085,500	45,100	1.35	10
17,693	16,773	GPDS46	6	$3\frac{1}{4}$	42,900	901,200	25,700	0.77	9
18,348	17,228	HT46	$6\frac{1}{4}$	$3\frac{1}{4}$	47,600	901,200	28,600	0.83	9
18,348	17,228	HT50	$6\frac{3}{8}$	$3\frac{1}{2}$	65,700	1,109,900	39,400	1.14	9
18,348	17,228	XT40	$5\frac{1}{4}$	$2\frac{13}{16}$	44,100	751,600	26,500	0.77	10
18,348	17,228	XT46	6	$3\frac{1}{2}$	58,100	910,300	34,900	1.01	10
18,348	17,228	XT50	$6\frac{3}{8}$	$3\frac{3}{4}$	75,200	1,085,500	45,100	1.31	10
18,348	17,228	GPDS46	6	$3\frac{3}{16}$	45,500	939,100	27,300	0.79	9
19,658	18,103	HT46	$6\frac{1}{4}$	$3\frac{1}{4}$	47,600	901,200	28,600	0.77	9
19,658	18,103	HT50	$6\frac{3}{8}$	$3\frac{1}{2}$	65,700	1,109,900	39,400	1.07	9
19,658	18,103	XT40	$5\frac{1}{4}$	$2\frac{13}{16}$	44,100	751,600	26,500	0.72	10
19,658	18,103	XT46	$6\frac{1}{4}$	$3\frac{1}{4}$	70,200	1,069,300	42,100	1.14	10
19,658	18,103	XT50	$6\frac{3}{8}$	$3\frac{1}{2}$	81,200	1,256,300	48,700	1.32	10
19,658	18,103	GPDS46	6	$3\frac{1}{8}$	48,000	976,300	28,800	0.78	9
12,542	12,964	NC46	$6\frac{1}{4}$	$3\frac{1}{4}$	34,000	901,200	17,600	0.92	9
12,542	12,964	$4\frac{1}{2}$ OH	6	$3\frac{1}{2}$	34,100	884,800	18,200	0.92	9
12,542	12,964	$4\frac{1}{2}$ H90	6	$3\frac{1}{4}$	39,000	938,400	18,800	1.06	9
12,542	12,964	HT46	$6\frac{1}{4}$	$3\frac{1}{4}$	47,600	901,200	28,600	1.29	9
12,542	12,964	NC50	$6\frac{3}{8}$	$3\frac{5}{8}$	41,200	1,026,000	21,600	1.12	9
12,542	12,964	HT50	$6\frac{1}{4}$	$3\frac{3}{4}$	52,700	939,100	31,600	1.43	9
12,542	12,964	XT46	6	$3\frac{1}{2}$	58,100	910,300	34,900	1.57	10
12,542	12,964	XT50	$6\frac{3}{8}$	$3\frac{1}{2}$	81,200	1,256,300	48,700	2.20	10
15,886	16,421	NC46	$6\frac{1}{4}$	3	39,700	1,048,400	20,500	0.85	9
15,886	16,421	$4\frac{1}{2}$ OH	$6\frac{1}{4}$	$3\frac{1}{4}$	40,700	1,043,800	21,500	0.87	9
15,886	16,421	$4\frac{1}{2}$ H90	6	$3\frac{1}{4}$	39,000	938,400	18,800	0.84	9
15,886	16,421	HT46	$6\frac{1}{4}$	$3\frac{1}{4}$	47,600	901,200	28,600	1.02	9
15,886	16,421	NC50	$6\frac{3}{8}$	$3\frac{5}{8}$	41,200	1,026,000	21,600	0.88	9
15,886	16,421	HT50	$6\frac{1}{4}$	$3\frac{3}{4}$	52,700	939,100	31,600	1.13	9
15,886	16,421	XT46	6	$3\frac{1}{2}$	58,100	910,300	34,900	1.24	10
15,886	16,421	XT50	$6\frac{3}{8}$	$3\frac{1}{2}$	81,200	1,256,300	48,700	1.74	10
17,558	18,149	NC46	$6\frac{1}{4}$	$2\frac{3}{4}$	44,900	1,183,900	23,200	0.87	9
17,558	18,149	$4\frac{1}{2}$ OH	$6\frac{1}{4}$	3	46,600	1,191,100	24,600	0.90	9
17,558	18,149	$4\frac{1}{2}$ H90	$6\frac{1}{4}$	3	45,700	1,085,700	21,800	0.88	9
17,558	18,149	HT46	$6\frac{1}{4}$	$3\frac{1}{4}$	47,600	901,200	28,600	0.92	9
17,558	18,149	NC50	$6\frac{3}{8}$	$3\frac{1}{2}$	44,700	1,109,900	23,400	0.86	9
17,558	18,149	HT50	$6\frac{1}{4}$	$3\frac{3}{4}$	52,700	939,100	31,600	1.02	9
17,558	18,149	XT46	6	$3\frac{1}{2}$	58,100	910,300	34,900	1.12	10
17,558	18,149	XT50	$6\frac{3}{8}$	$3\frac{1}{2}$	81,200	1,256,300	48,700	1.57	10
22,575	23,335	NC46	$6\frac{3}{8}$	$2\frac{1}{2}$	49,900	1,307,600	25,600	0.75	9
22,575	23,335	$4\frac{1}{2}$ OH	$6\frac{3}{8}$	$2\frac{3}{4}$	52,200	1,326,600	27,400	0.79	9
22,575	23,335	$4\frac{1}{2}$ H90	$6\frac{3}{8}$	$2\frac{3}{4}$	51,700	1,221,100	24,600	0.78	9

(continued)

Table 4.6.6 *(continued)*

								Tool Joint Data	

Internal Pressure	Collapse Pressure	Connection Type	Outside Diameter	Inside Diameter	Torsional Yield Strength	Tensile Yield Strength	Make-up Torque	Torsional Ratio Tool Joint to Pipe	* Pin Tong Space
psi	psi		in.	in.	ft-lb	lb	ft-lb		in.
22,575	23,335	HT46	$6\frac{1}{4}$	3	57,700	1,048,400	34,600	0.87	9
22,575	23,335	NC50	$6\frac{1}{2}$	$3\frac{1}{4}$	51,400	1,269,000	26,800	0.77	9
22,575	23,335	HT50	$6\frac{3}{8}$	$3\frac{1}{2}$	65,700	1,109,900	39,400	0.99	9
22,575	23,335	XT46	6	$3\frac{1}{2}$	58,100	910,300	34,900	0.88	10
22,575	23,335	XT50	$6\frac{3}{8}$	$3\frac{1}{2}$	81,200	1,256,300	48,700	1.22	10
22,575	23,335	GPDS46	6	3	52,900	1,048,400	31,700	0.80	9
23,411	24,199	HT46	$6\frac{1}{4}$	3	57,700	1,048,400	34,600	0.84	9
23,411	24,199	HT50	$6\frac{3}{8}$	$3\frac{1}{2}$	65,700	1,109,900	39,400	0.95	9
23,411	24,199	XT46	6	$3\frac{1}{2}$	58,100	910,300	34,900	0.84	10
23,411	24,199	XT50	$6\frac{3}{8}$	$3\frac{1}{2}$	81,200	1,256,300	48,700	1.18	10
23,411	24,199	GPDS46	6	$2\frac{15}{16}$	55,300	1,083,400	33,200	0.80	9
25,083	25,927	HT46	$6\frac{1}{4}$	3	57,700	1,048,400	34,600	0.78	9
25,083	25,927	HT50	$6\frac{3}{8}$	$3\frac{1}{2}$	65,700	1,109,900	39,400	0.89	9
25,083	25,927	XT46	$6\frac{1}{4}$	$3\frac{1}{4}$	70,200	1,069,300	42,100	0.95	10
25,083	25,927	XT50	$6\frac{3}{8}$	$3\frac{1}{2}$	81,200	1,256,300	48,700	1.10	10
25,083	25,927	GPDS46	6	$2\frac{3}{4}$	60,700	1,183,900	36,400	0.82	9
9,503	9,962	NC50	$6\frac{5}{8}$	$3\frac{3}{4}$	38,100	939,100	19,800	0.92	9
9,503	9,962	HT50	$6\frac{5}{8}$	$3\frac{3}{4}$	53,300	939,100	32,000	1.29	9
9,503	9,962	$5\frac{1}{2}$ FH	7	$3\frac{3}{4}$	62,900	1,448,400	33,400	1.53	10
9,503	9,962	XT46	6	$3\frac{1}{2}$	36,500	910,300	21,900	0.89	10
9,503	9,962	XT50	$6\frac{1}{2}$	4	38,700	902,900	23,200	0.94	10
12,037	12,026	NC50	$6\frac{5}{8}$	$3\frac{1}{2}$	45,100	1,109,900	23,400	0.87	9
12,037	12,026	HT50	$6\frac{5}{8}$	$3\frac{3}{4}$	53,300	939,100	32,000	1.02	9
12,037	12,026	$5\frac{1}{2}$ FH	7	$3\frac{3}{4}$	62,900	1,448,400	33,400	1.21	10
12,037	12,026	XT46	6	$3\frac{1}{2}$	58,100	910,300	34,900	1.12	10
12,037	12,026	XT50	$6\frac{1}{2}$	4	62,500	902,900	37,500	1.20	10
13,304	12,999	NC50	$6\frac{5}{8}$	$3\frac{1}{4}$	51,700	1,269,000	26,800	0.90	9
13,304	12,999	HT50	$6\frac{5}{8}$	$3\frac{3}{4}$	53,300	939,100	32,000	0.93	9
13,304	12,999	$5\frac{1}{2}$ FH	7	$3\frac{3}{4}$	62,900	1,448,400	33,400	1.09	10
13,304	12,999	XT46	6	$3\frac{1}{2}$	58,100	910,300	34,900	1.01	10
13,304	12,999	XT50	$6\frac{1}{2}$	4	62,500	902,900	37,500	1.09	10
13,304	12,999	GPDS50	$6\frac{1}{2}$	$3\frac{3}{4}$	47,500	939,400	28,500	0.82	9
17,105	15,672	NC50	$6\frac{5}{8}$	$2\frac{3}{4}$	63,400	1,551,700	32,900	0.86	9
17,105	15,672	HT50	$6\frac{5}{8}$	$3\frac{1}{2}$	66,200	1,109,900	39,700	0.89	9
17,105	15,672	$5\frac{1}{2}$ FH	$7\frac{1}{4}$	$3\frac{1}{2}$	72,500	1,619,200	37,400	0.98	10
17,105	15,672	XT46	6	$3\frac{1}{2}$	58,100	910,300	34,900	0.78	10
17,105	15,672	XT50	$6\frac{1}{2}$	$3\frac{3}{4}$	77,000	1,085,500	46,200	1.04	10
17,105	15,672	GPDS50	$6\frac{1}{2}$	$3\frac{1}{2}$	60,200	1,110,200	36,100	0.81	9
17,738	16,079	HT50	$6\frac{5}{8}$	$3\frac{1}{2}$	66,200	1,109,900	39,700	0.86	9
17,738	16,079	XT46	6	$3\frac{1}{2}$	58,100	910,300	34,900	0.76	10
17,738	16,079	XT50	$6\frac{1}{2}$	$3\frac{3}{4}$	77,000	1,085,500	46,200	1.00	10
17,738	16,079	GPDS50	$6\frac{1}{2}$	$3\frac{1}{2}$	60,200	1,110,200	36,100	0.78	9
19,005	16,858	HT50	$6\frac{5}{8}$	$3\frac{1}{2}$	66,200	1,109,900	39,700	0.80	9
19,005	16,858	XT46	$6\frac{1}{4}$	$3\frac{1}{4}$	70,200	1,069,300	42,100	0.85	10

(continued)

Table 4.6.6   *(continued)*

									Tool Joint Data

Internal Pressure	Collapse Pressure	Connection Type	Outside Diameter	Inside Diameter	Torsional Yield Strength	Tensile Yield Strength	Make-up Torque	Torsional Ratio Tool Joint to Pipe	* Pin Tong Space
psi	psi		in.	in.	ft-lb	lb	ft-lb		in.
19,005	16,858	XT50	$6\frac{1}{2}$	$3\frac{3}{4}$	77,000	1,085,500	46,200	0.94	10
19,005	16,858	GPDS50	$6\frac{1}{2}$	$3\frac{3}{8}$	66,200	1,191,200	39,700	0.80	9
13,125	13,500	NC50	$6\frac{5}{8}$	$3\frac{1}{2}$	45,100	1,109,900	23,400	0.86	9
13,125	13,500	HT50	$6\frac{5}{8}$	$3\frac{3}{4}$	53,300	939,100	32,000	1.02	9
13,125	13,500	$5\frac{1}{2}$ FH	7	$3\frac{1}{2}$	62,900	1,619,200	37,400	1.20	10
13,125	13,500	XT50	$6\frac{5}{8}$	$3\frac{3}{4}$	77,300	1,085,500	46,400	1.48	10
16,625	17,100	NC50	$6\frac{5}{8}$	3	57,800	1,416,200	30,000	0.87	9
16,625	17,100	HT50	$6\frac{5}{8}$	$3\frac{1}{2}$	66,200	1,109,900	39,700	1.00	9
16,625	17,100	$5\frac{1}{2}$ FH	7	$3\frac{1}{2}$	62,900	1,619,200	37,400	0.95	10
16,625	17,100	XT50	$6\frac{5}{8}$	$3\frac{3}{4}$	77,300	1,085,500	46,400	1.17	10
18,375	18,900	NC50	$6\frac{5}{8}$	$2\frac{3}{4}$	63,400	1,551,700	32,900	0.87	9
18,375	18,900	HT50	$6\frac{5}{8}$	$3\frac{1}{4}$	78,000	1,269,000	46,800	1.07	9
18,375	18,900	$5\frac{1}{2}$ FH	$7\frac{1}{4}$	$3\frac{1}{2}$	72,500	1,619,200	37,400	0.99	10
18,375	18,900	XT50	$6\frac{5}{8}$	$3\frac{3}{4}$	77,300	1,085,500	46,400	1.06	10
18,375	18,900	GPDS50	$6\frac{1}{2}$	$3\frac{1}{2}$	60,200	1,110,200	36,100	0.82	9
23,625	24,300	NC50	$6\frac{5}{8}$	$2\frac{3}{4}$	63,400	1,551,700	32,900	0.67	9
23,625	24,300	HT50	$6\frac{5}{8}$	3	88,800	1,416,200	53,300	0.94	9
23,625	24,300	$5\frac{1}{2}$ FH	$7\frac{1}{4}$	$3\frac{1}{4}$	78,700	1,778,300	41,200	0.84	10
23,625	24,300	XT50	$6\frac{5}{8}$	$3\frac{3}{4}$	77,300	1,085,500	46,400	0.82	10
23,625	24,300	GPDS50	$6\frac{1}{2}$	$3\frac{3}{16}$	74,700	1,307,200	44,800	0.79	9
24,500	25,200	HT50	$6\frac{5}{8}$	3	88,800	1,416,200	53,300	0.91	9
24,500	25,200	XT50	$6\frac{5}{8}$	$3\frac{3}{4}$	77,300	1,085,500	46,400	0.79	10
24,500	25,200	GPDS50	$6\frac{1}{2}$	$3\frac{1}{8}$	77,400	1,344,300	46,400	0.79	9
26,250	27,000	HT50	$6\frac{5}{8}$	3	88,800	1,416,200	53,300	0.85	9
26,250	27,000	XT50	$6\frac{5}{8}$	$3\frac{1}{2}$	90,700	1,256,300	54,400	0.87	10
26,250	27,000	GPDS50	$6\frac{5}{8}$	3	82,900	1,416,500	49,700	0.79	9
8,413	8,615	$5\frac{1}{2}$ FH	7	4	57,900	1,265,800	31,200	1.14	10
8,413	8,615	HT55	7	4	77,200	1,265,800	46,300	1.52	10
8,413	8,615	XT54	$6\frac{3}{4}$	$4\frac{1}{4}$	70,400	960,700	42,200	1.39	10
8,413	8,615	XT57	7	$4\frac{3}{8}$	85,600	1,107,100	51,400	1.69	10
10,019	10,912	$5\frac{1}{2}$ FH	7	$3\frac{3}{4}$	65,100	1,448,400	35,700	1.01	10
10,019	10,912	HT55	7	4	77,200	1,265,800	46,300	1.20	10
10,019	10,912	XT54	$6\frac{3}{4}$	$4\frac{1}{4}$	70,400	960,700	42,200	1.10	10
10,019	10,912	XT57	7	$4\frac{3}{8}$	85,600	1,107,100	51,400	1.33	10
10,753	12,061	$5\frac{1}{2}$ FH	$7\frac{1}{4}$	$3\frac{1}{2}$	75,000	1,691,200	40,000	1.06	10
10,753	12,061	HT55	7	4	77,200	1,265,800	46,300	1.09	10
10,753	12,061	XT54	$6\frac{3}{4}$	$4\frac{1}{4}$	70,400	960,700	42,200	0.99	10
10,753	12,061	XT57	7	$4\frac{3}{8}$	85,600	1,107,100	51,400	1.21	10
10,753	12,061	GPDS55	7	4	74,200	1,292,500	44,500	1.05	10
12,679	15,507	$5\frac{1}{2}$ FH	$7\frac{1}{2}$	3	90,200	1,925,500	47,700	0.99	10
12,679	15,507	HT55	7	4	77,200	1,265,800	46,300	0.85	10
12,679	15,507	XT54	$6\frac{3}{4}$	$4\frac{1}{4}$	70,400	960,700	42,200	0.77	10
12,679	15,507	XT57	7	$4\frac{3}{8}$	85,600	1,107,100	51,400	0.94	10
12,679	15,507	GPDS55	7	4	74,200	1,292,500	44,500	0.81	10

(continued)

Table 4.6.6 (continued)

					Tool Joint Data				
Internal Pressure	Collapse Pressure	Connection Type	Outside Diameter	Inside Diameter	Torsional Yield Strength	Tensile Yield Strength	Make-up Torque	Torsional Ratio Tool Joint to Pipe	* Pin Tong Space
psi	psi		in.	in.	ft-lb	lb	ft-lb		in.
12,957	16,081	$5\frac{1}{2}$ FH	$7\frac{1}{2}$	3	90,200	1,925,500	47,700	0.95	10
12,957	16,081	HT55	7	4	77,200	1,265,800	46,300	0.82	10
12,957	16,081	XT54	$6\frac{3}{4}$	$4\frac{1}{4}$	70,400	960,700	42,200	0.74	10
12,957	16,081	XT57	7	$4\frac{3}{8}$	85,600	1,107,100	51,400	0.90	10
12,957	16,081	GPDS55	7	4	74,200	1,292,500	44,500	0.78	10
13,473	17,230	$5\frac{1}{2}$ FH	$7\frac{1}{2}$	3	90,200	1,925,500	47,700	0.89	10
13,473	17,230	HT55	7	4	77,200	1,265,800	46,300	0.76	10
13,473	17,230	XT54	$6\frac{3}{4}$	4	86,600	1,155,100	52,000	0.85	10
13,473	17,230	XT57	7	$4\frac{1}{4}$	94,300	1,208,700	56,600	0.93	10
13,473	17,230	GPDS55	$7\frac{1}{8}$	$3\frac{7}{8}$	82,000	1,385,200	49,200	0.81	10
10,464	9,903	$5\frac{1}{2}$ FH	7	4	57,900	1,265,800	31,200	1.02	10
10,464	9,903	HT55	7	4	77,200	1,265,800	46,300	1.36	10
10,464	9,903	XT54	$6\frac{3}{4}$	$4\frac{1}{4}$	70,400	960,700	42,200	1.24	10
10,464	9,903	XT57	7	$4\frac{3}{8}$	85,600	1,107,100	51,400	1.51	10
12,933	12,544	$5\frac{1}{2}$ FH	$7\frac{1}{4}$	$3\frac{1}{2}$	75,000	1,619,200	40,000	1.05	10
12,933	12,544	HT55	7	4	77,200	1,265,800	46,300	1.08	10
12,933	12,544	XT54	$6\frac{3}{4}$	$4\frac{1}{4}$	70,400	960,700	42,200	0.98	10
12,933	12,544	XT57	7	$4\frac{3}{8}$	85,600	1,107,100	51,400	1.19	10
14,013	13,865	$5\frac{1}{2}$ FH	$7\frac{1}{4}$	$3\frac{1}{2}$	75,000	1,619,200	40,000	0.95	10
14,013	13,865	HT55	7	4	77,200	1,265,800	46,300	0.97	10
14,013	13,865	XT54	$6\frac{3}{4}$	$4\frac{1}{4}$	70,400	960,700	42,200	0.89	10
14,013	13,865	XT57	7	$4\frac{3}{8}$	85,600	1,107,100	51,400	1.08	10
14,013	13,865	GPDS55	7	4	74,200	1,292,500	44,500	0.94	10
17,023	17,826	$5\frac{1}{2}$ FH	$7\frac{1}{2}$	3	90,200	1,925,500	47,700	0.89	10
17,023	17,826	HT55	7	$3\frac{3}{4}$	87,700	1,448,400	52,600	0.86	10
17,023	17,826	XT54	$6\frac{3}{4}$	4	86,600	1,155,100	52,000	0.85	10
17,023	17,826	XT57	7	$4\frac{1}{4}$	94,300	1,208,700	56,600	0.93	10
17,023	17,826	GPDS55	$7\frac{1}{8}$	$3\frac{7}{8}$	82,000	1,385,200	49,200	0.81	10
17,489	18,486	$5\frac{1}{2}$ FH	$7\frac{1}{4}$	$3\frac{1}{2}$	75,000	1,691,200	40,000	0.71	10
17,489	18,486	HT55	7	4	77,200	1,265,800	46,300	0.73	10
17,489	18,486	XT54	$6\frac{3}{4}$	4	86,600	1,155,100	52,000	0.82	10
17,489	18,486	XT57	7	$4\frac{1}{4}$	94,300	1,208,700	56,600	0.89	10
17,489	18,486	GPDS55	$7\frac{1}{8}$	$3\frac{7}{8}$	82,000	1,385,200	49,200	0.78	10
18,386	19,807	$5\frac{1}{2}$ FH	$7\frac{1}{2}$	3	90,200	1,925,500	47,700	0.80	10
18,386	19,807	HT55	7	$3\frac{3}{4}$	87,700	1,448,400	52,600	0.78	10
18,386	19,807	XT54	$6\frac{3}{4}$	4	86,600	1,155,100	52,000	0.77	10
18,386	19,807	XT57	7	$4\frac{1}{4}$	94,300	1,208,700	56,600	0.83	10
18,386	19,807	GPDS55	$7\frac{1}{8}$	$3\frac{3}{4}$	89,300	1,475,100	53,600	0.79	10
7,453	8,065	XT57	7	$4\frac{1}{4}$	94,300	1,208,700	56,600	1.61	10
8,775	10,216	XT57	7	$4\frac{1}{4}$	94,300	1,208,700	56,600	1.27	10
9,362	11,291	XT57	7	$4\frac{1}{4}$	94,300	1,208,700	56,600	1.15	10
10,825	14,517	XT57	7	$4\frac{1}{4}$	94,300	1,208,700	56,600	0.89	10
11,023	15,054	XT57	7	$4\frac{1}{4}$	94,300	1,208,700	56,600	0.86	10
11,376	16,130	XT57	7	$4\frac{1}{4}$	94,300	1,208,700	56,600	0.80	10
9,558	9,271	XT57	7	$4\frac{1}{4}$	94,300	1,208,700	56,600	1.44	10

(continued)

Table 4.6.6 (continued)

					Tool Joint Data				
Internal Pressure	Collapse Pressure	Connection Type	Outside Diameter	Inside Diameter	Torsional Yield Strength	Tensile Yield Strength	Make-up Torque	Torsional Ratio Tool Joint to Pipe	* Pin Tong Space
psi	psi		in.	in.	ft-lb	lb	ft-lb		in.
11,503	11,744	XT57	7	$4\frac{1}{4}$	94,300	1,208,700	56,600	1.14	10
12,414	12,980	XT57	7	$4\frac{1}{4}$	94,300	1,208,700	56,600	1.03	10
14,892	16,688	XT57	7	$4\frac{1}{4}$	94,300	1,208,700	56,600	0.80	10
15,266	17,306	XT57	7	$4\frac{1}{4}$	94,300	1,208,700	56,600	0.77	10
15,976	18,543	XT57	7	4	106,200	1,403,100	63,700	0.81	10
4,788	6,538	$6\frac{5}{8}$ FH	8	5	73,700	1,448,400	38,400	1.04	10
4,788	6,538	HT65	8	5	99,700	1,448,400	59,800	1.41	10
4,788	6,538	XT65	8	5	135,300	1,543,700	81,200	1.92	10
5,321	8,281	$6\frac{5}{8}$ FH	8	5	73,700	1,448,400	38,400	0.82	10
5,321	8,281	HT65	8	5	99,700	1,448,400	59,800	1.12	10
5,321	8,281	XT65	8	5	135,300	1,543,700	81,200	1.51	10
5,500	9,153	$6\frac{5}{8}$ FH	$8\frac{1}{4}$	$4\frac{3}{4}$	86,200	1,678,100	44,600	0.87	10
5,500	9,153	HT65	8	5	99,700	1,448,400	59,800	1.01	10
5,500	9,153	XT65	8	5	135,300	1,543,700	81,200	1.37	10
6,036	11,768	$6\frac{5}{8}$ FH	$8\frac{1}{2}$	$4\frac{1}{4}$	109,200	2,102,300	56,100	0.86	10
6,036	11,768	HT65	8	5	99,700	1,448,400	59,800	0.79	10
6,036	11,768	XT65	8	5	135,300	1,543,700	81,200	1.07	10
6,036	11,768	GPDS65	8	$4\frac{15}{16}$	102,000	1,538,600	61,200	0.80	10
6,121	12,204	$6\frac{5}{8}$ FH	$8\frac{1}{2}$	$4\frac{1}{4}$	109,200	2,102,300	56,100	0.83	10
6,121	12,204	HT65	8	5	99,700	1,448,400	59,800	0.76	10
6,121	12,204	XT65	8	5	135,300	1,543,700	81,200	1.03	10
6,121	12,204	GPDS65	8	$4\frac{7}{8}$	107,500	1,596,400	64,500	0.82	10
6,260	13,075	$6\frac{5}{8}$ FH	$8\frac{1}{2}$	$4\frac{1}{4}$	109,200	2,102,300	56,100	0.77	10
6,260	13,075	HT65	8	5	99,700	1,448,400	59,800	0.71	10
6,260	13,075	XT65	8	5	135,300	1,543,700	81,200	0.96	10
6,260	13,075	GPDS65	$8\frac{1}{4}$	$4\frac{3}{4}$	119,000	1,709,800	71,400	0.84	10
5,894	7,172	$6\frac{5}{8}$ FH	8	5	73,700	1,448,400	38,400	0.97	10
5,894	7,172	HT65	8	5	99,700	1,448,400	59,800	1.31	10
5,894	7,172	XT65	8	5	135,300	1,543,700	81,200	1.77	10
6,755	9,084	$6\frac{5}{8}$ FH	$8\frac{1}{4}$	$4\frac{3}{4}$	86,200	1,678,100	44,600	0.89	10
6,755	9,084	HT65	8	5	99,700	1,448,400	59,800	1.03	10
6,755	9,084	XT65	8	5	135,300	1,543,700	81,200	1.40	10
7,103	10,040	$6\frac{5}{8}$ FH	$8\frac{1}{4}$	$4\frac{3}{4}$	86,200	1,678,100	44,600	0.81	10
7,103	10,040	HT65	8	5	99,700	1,448,400	59,800	0.93	10
7,103	10,040	XT65	8	5	135,300	1,543,700	81,200	1.27	10
7,813	12,909	$6\frac{5}{8}$ FH	$8\frac{1}{2}$	$4\frac{1}{4}$	109,200	2,102,300	56,100	0.80	10
7,813	12,909	HT65	8	5	99,700	1,448,400	59,800	0.73	10
7,813	12,909	XT65	8	5	135,300	1,543,700	81,200	0.99	10
7,813	12,909	GPDS65	8	$4\frac{7}{8}$	107,500	1,596,400	64,500	0.78	10
7,881	13,387	$6\frac{5}{8}$ FH	$8\frac{1}{2}$	$4\frac{1}{4}$	109,200	2,102,300	56,100	0.77	10
7,881	13,387	HT65	8	5	99,700	1,448,400	59,800	0.70	10
7,881	13,387	XT65	8	5	135,300	1,543,700	81,200	0.95	10
7,881	13,387	GPDS65	$8\frac{1}{4}$	$4\frac{3}{4}$	119,000	1,709,800	71,400	0.84	10
7,970	14,343	$6\frac{5}{8}$ FH	$8\frac{1}{2}$	$4\frac{1}{4}$	109,200	2,102,300	56,100	0.72	10
7,970	14,343	HT65	8	5	99,700	1,448,400	59,800	0.65	10
7,970	14,343	XT65	8	5	135,300	1,543,700	81,200	0.89	10
7,970	14,343	GPDS65	$8\frac{1}{4}$	$4\frac{3}{4}$	119,000	1,709,800	71,400	0.78	10

(continued)

Table 4.6.6 *(continued)*

			Assembly Data			
* Box Tong Space	Adjusted Weight	Minimum Tool Joint O.D. for Prem. Class	Drift Diameter	Capacity	Displacement	Size O.D.
in.	lb/ft	in.	in.	US gal/ft	US gal/ft	in.
10	5.52	$3\frac{1}{8}$	$1\frac{5}{8}$	0.160	0.085	$2\frac{3}{8}$
10	5.52	3	$1\frac{5}{8}$	0.160	0.085	
10	5.42	$3\frac{1}{16}$	$1\frac{3}{4}$	0.161	0.083	
10	5.30	$2\frac{31}{32}$	$1\frac{7}{8}$	0.162	0.081	
10	5.52	$3\frac{3}{16}$	$1\frac{5}{8}$	0.160	0.085	$2\frac{3}{8}$
10	5.52	$3\frac{1}{16}$	$1\frac{5}{8}$	0.160	0.085	
10	5.42	$3\frac{1}{8}$	$1\frac{3}{4}$	0.161	0.083	
10	5.42	3	$1\frac{3}{4}$	0.161	0.083	
10	5.52	$3\frac{7}{32}$	$1\frac{5}{8}$	0.160	0.085	$2\frac{3}{8}$
10	5.52	$3\frac{3}{32}$	$1\frac{5}{8}$	0.160	0.085	
10	5.42	$3\frac{5}{32}$	$1\frac{3}{4}$	0.161	0.083	
10	5.42	$3\frac{1}{32}$	$1\frac{3}{4}$	0.161	0.083	
10	5.52	$3\frac{9}{32}$	$1\frac{5}{8}$	0.160	0.085	$2\frac{3}{8}$
10	5.72	$3\frac{3}{16}$	$1\frac{3}{8}$	0.158	0.087	
10	5.63	$3\frac{7}{32}$	$1\frac{1}{2}$	0.159	0.086	
10	5.52	$3\frac{1}{8}$	$1\frac{5}{8}$	0.160	0.085	
15	5.70	$2\frac{3}{4}$	$1\frac{3}{8}$	0.157	0.087	$2\frac{3}{8}$
15	5.79	$2\frac{15}{16}$	$1\frac{5}{8}$	0.160	0.089	
12	5.61	NA	$1\frac{5}{8}$	0.160	0.086	
15	5.70	$2\frac{25}{32}$	$1\frac{3}{8}$	0.157	0.087	$2\frac{3}{8}$
15	5.79	$2\frac{31}{32}$	$1\frac{5}{8}$	0.160	0.089	
12	5.61	NA	$1\frac{5}{8}$	0.160	0.086	
10	7.19	$3\frac{3}{16}$	$1\frac{5}{8}$	0.134	0.110	$2\frac{3}{8}$
12	7.27	N/A	$1\frac{5}{8}$	0.134	0.111	
10	7.07	$3\frac{1}{16}$	$1\frac{5}{8}$	0.134	0.108	
10	7.02	$3\frac{1}{32}$	$1\frac{11}{16}$	0.134	0.107	
10	7.19	$3\frac{1}{4}$	$1\frac{5}{8}$	0.134	0.110	$2\frac{3}{8}$
12	7.27	NA	$1\frac{5}{8}$	0.134	0.111	
10	7.07	$3\frac{5}{32}$	$1\frac{5}{8}$	0.134	0.108	
10	7.02	$3\frac{3}{32}$	$1\frac{11}{16}$	0.134	0.107	
10	7.19	$3\frac{9}{32}$	$1\frac{5}{8}$	0.134	0.110	$2\frac{3}{8}$
12	7.27	NA	$1\frac{5}{8}$	0.134	0.111	
10	7.07	$3\frac{3}{16}$	$1\frac{5}{8}$	0.134	0.108	
10	7.02	$3\frac{1}{8}$	$1\frac{11}{16}$	0.134	0.107	
10	7.65	$3\frac{13}{32}$	$1\frac{3}{8}$	0.132	0.117	$2\frac{3}{8}$
12	7.27	NA	$1\frac{5}{8}$	0.134	0.111	
10	7.07	NA	$1\frac{5}{8}$	0.134	0.108	
10	7.02	$3\frac{7}{32}$	$1\frac{11}{16}$	0.134	0.107	
10	7.32	$3\frac{11}{32}$	$1\frac{5}{8}$	0.134	0.112	
15	7.35	$2\frac{29}{32}$	$1\frac{3}{8}$	0.131	0.112	$2\frac{3}{8}$
15	7.43	$3\frac{1}{16}$	$1\frac{5}{8}$	0.134	0.114	

(continued)

Table 4.6.6 (continued)

			Assembly Data			
* Box Tong Space	Adjusted Weight	Minimum Tool Joint O.D. for Prem. Class	Drift Diameter	Capacity	Displace- ment	Size O.D.
in.	lb/ft	in.	in.	US gal/ft	US gal/ft	in.
12	7.27	NA	$1\frac{5}{8}$	0.134	0.111	
10	7.37	$3\frac{5}{16}$	$1\frac{9}{16}$	0.133	0.113	
15	7.44	$2\frac{29}{32}$	$1\frac{1}{4}$	0.130	0.114	$2\frac{3}{8}$
15	7.43	$3\frac{3}{32}$	$1\frac{5}{8}$	0.134	0.114	
12	7.27	NA	$1\frac{5}{8}$	0.134	0.111	
10	7.42	$3\frac{11}{32}$	$1\frac{1}{2}$	0.133	0.114	
10	7.24	$3\frac{9}{32}$	$1\frac{5}{8}$	0.237	0.111	$2\frac{7}{8}$
12	7.32	NA	$1\frac{5}{8}$	0.236	0.112	
11	7.93	$3\frac{11}{16}$	$2\frac{1}{32}$	0.240	0.121	
10	7.18	3	$1\frac{3}{8}$	0.235	0.110	
11	7.23	$3\frac{1}{2}$	$2\frac{1}{4}$	0.242	0.111	
15	7.48	$2\frac{19}{32}$	$1\frac{5}{8}$	0.235	0.114	
13	7.89	$3\frac{1}{2}$	$2\frac{3}{97}$	0.240	0.121	
15	7.79	$3\frac{7}{16}$	$2\frac{1}{4}$	0.242	0.119	
10	7.55	$3\frac{3}{8}$	$1\frac{3}{8}$	0.235	0.116	$2\frac{7}{8}$
12	7.32	NA	$1\frac{5}{8}$	0.236	0.112	
11	7.93	$3\frac{3}{4}$	$2\frac{1}{32}$	0.240	0.121	
10	7.18	$3\frac{1}{8}$	$1\frac{3}{8}$	0.235	0.110	
11	7.23	$3\frac{9}{16}$	$2\frac{1}{4}$	0.242	0.111	
15	7.48	3	$1\frac{5}{8}$	0.235	0.114	
13	7.89	$3\frac{19}{32}$	$2\frac{1}{32}$	0.240	0.121	
15	7.79	$3\frac{1}{2}$	$2\frac{1}{4}$	0.242	0.119	
10	7.50	$3\frac{13}{32}$	$1\frac{5}{8}$	0.237	0.115	$2\frac{7}{8}$
12	7.32	NA	$1\frac{5}{8}$	0.236	0.112	
11	7.93	$3\frac{13}{16}$	$2\frac{1}{32}$	0.240	0.121	
10	7.18	NA	$1\frac{3}{8}$	0.235	0.110	
11	7.61	$3\frac{19}{32}$	$2\frac{1}{32}$	0.240	0.116	
15	7.48	$3\frac{1}{16}$	$1\frac{5}{8}$	0.235	0.114	
13	7.89	$3\frac{5}{8}$	$2\frac{1}{32}$	0.240	0.121	
15	7.79	$3\frac{17}{32}$	$2\frac{1}{4}$	0.242	0.119	
10	7.69	$3\frac{17}{32}$	$1\frac{3}{8}$	0.235	0.118	$2\frac{7}{8}$
12	7.66	$3\frac{5}{16}$	$1\frac{3}{8}$	0.234	0.117	
11	7.97	$3\frac{29}{32}$	2	0.240	0.122	
10	7.18	NA	$1\frac{3}{8}$	0.235	0.110	
11	7.61	$3\frac{23}{32}$	$2\frac{1}{32}$	0.240	0.116	
15	7.48	$3\frac{7}{32}$	$1\frac{5}{8}$	0.235	0.114	
13	7.89	$3\frac{23}{32}$	$2\frac{1}{32}$	0.240	0.121	
15	7.79	$3\frac{21}{32}$	$2\frac{1}{4}$	0.242	0.119	
12	7.66	$3\frac{11}{32}$	$1\frac{3}{8}$	0.234	0.117	$2\frac{7}{8}$
15	7.48	$3\frac{7}{32}$	$1\frac{5}{8}$	0.235	0.114	
13	7.89	$3\frac{3}{4}$	$2\frac{1}{32}$	0.240	0.121	
15	7.79	$3\frac{21}{32}$	$2\frac{1}{4}$	0.242	0.119	

(*continued*)

Table 4.6.6   *(continued)*

				Assembly Data		
* Box Tong Space	Adjusted Weight	Minimum Tool Joint O.D. for Prem. Class	Drift Diameter	Capacity	Displacement	Size O.D.
in.	lb/ft	in.	in.	US gal/ft	US gal/ft	in.
12	7.66	$3\frac{3}{8}$	$1\frac{3}{8}$	0.234	0.117	$2\frac{7}{8}$
15	7.48	$3\frac{9}{32}$	$1\frac{5}{8}$	0.235	0.114	
13	7.89	$3\frac{25}{32}$	$2\frac{1}{32}$	0.240	0.121	
15	7.79	$3\frac{11}{16}$	$2\frac{1}{4}$	0.242	0.119	
11	11.16	$3\frac{13}{16}$	$2\frac{1}{32}$	0.189	0.171	$2\frac{7}{8}$
10	10.80	$3\frac{13}{32}$	$1\frac{3}{8}$	0.183	0.165	
11	10.46	NA	$1\frac{3}{8}$	0.183	0.160	
11	10.84	$3\frac{19}{32}$	$2\frac{1}{32}$	0.189	0.166	
11	11.16	$3\frac{19}{32}$	$2\frac{1}{32}$	0.189	0.171	
13	10.51	$2\frac{31}{32}$	$1\frac{3}{8}$	0.183	0.161	
12	10.89	$3\frac{3}{16}$	$1\frac{3}{8}$	0.183	0.167	
13	11.27	$3\frac{5}{8}$	$2\frac{1}{32}$	0.189	0.172	
15	11.05	$2\frac{31}{32}$	$1\frac{3}{8}$	0.182	0.169	
15	11.25	$3\frac{13}{32}$	$2\frac{1}{32}$	0.189	0.172	
11	11.32	$3\frac{29}{32}$	$1\frac{7}{8}$	0.187	0.173	$2\frac{7}{8}$
10	10.80	NA	$1\frac{3}{8}$	0.183	0.165	
11	10.46	NA	$2\frac{1}{32}$	0.183	0.160	
11	10.84	$3\frac{23}{32}$	$2\frac{1}{32}$	0.189	0.166	
11	11.16	$3\frac{11}{16}$	$1\frac{3}{8}$	0.189	0.171	
13	10.51	$3\frac{1}{8}$	$1\frac{3}{8}$	0.183	0.161	
12	10.89	$3\frac{5}{16}$	$\frac{3}{8}$	0.183	0.167	
13	11.27	$3\frac{23}{32}$	$\frac{3}{8}$	0.189	0.172	
15	11.05	$3\frac{3}{32}$		0.182	0.169	
15	11.25	$3\frac{17}{32}$	$\frac{1}{32}$	0.189	0.172	
11	11.32	$3\frac{15}{16}$	$1\frac{7}{8}$	0.187	0.173	$2\frac{7}{8}$
10	10.80	NA	$1\frac{3}{8}$	0.183	0.165	
11	10.54	NA	$1\frac{1}{4}$	0.183	0.161	
11	10.84	$3\frac{3}{4}$	$2\frac{1}{32}$	0.189	0.166	
11	11.32	$3\frac{23}{32}$	$1\frac{7}{8}$	0.187	0.173	
13	10.60	NA	$1\frac{1}{4}$	0.182	0.162	
12	11.03	$3\frac{3}{8}$	$1\frac{3}{8}$	0.183	0.169	
13	11.27	$3\frac{25}{32}$	$2\frac{1}{32}$	0.189	0.172	
15	11.05	$3\frac{5}{32}$	$1\frac{3}{8}$	0.182	0.169	
15	11.25	$3\frac{9}{16}$	$2\frac{1}{32}$	0.189	0.172	
11	11.98	$4\frac{1}{16}$	$1\frac{1}{2}$	0.184	0.183	$2\frac{7}{8}$
10	10.94	NA	$1\frac{3}{8}$	0.183	0.167	
11	10.46	NA	$1\frac{3}{8}$	0.183	0.160	
11	11.00	NA	$1\frac{7}{8}$	0.187	0.168	
11	11.32	$3\frac{27}{32}$	$1\frac{7}{8}$	0.187	0.173	
13	10.51	NA	$1\frac{3}{8}$	0.183	0.161	
12	11.03	$3\frac{9}{16}$	$1\frac{3}{8}$	0.183	0.169	
13	11.44	$3\frac{27}{32}$	$1\frac{7}{8}$	0.187	0.175	

(continued)

Table 4.6.6   (continued)

			Assembly Data			
* Box Tong Space	Adjusted Weight	Minimum Tool Joint O.D. for Prem. Class	Drift Diameter	Capacity	Displace- ment	Size O.D.
in.	lb/ft	in.	in.	US gal/ft	US gal/ft	in.
15	11.15	$3\frac{5}{16}$	$1\frac{1}{4}$	0.181	0.171	$2\frac{7}{8}$
15	11.25	$3\frac{23}{32}$	$2\frac{1}{32}$	0.189	0.172	
11	11.32	$3\frac{15}{16}$	$1\frac{7}{8}$	0.187	0.173	
12	11.12	$3\frac{9}{16}$	$1\frac{1}{4}$	0.182	0.170	$2\frac{7}{8}$
13	11.44	$3\frac{7}{8}$	$1\frac{7}{8}$	0.187	0.175	
15	11.24	$3\frac{5}{16}$	$1\frac{1}{8}$	0.180	0.172	
15	11.25	$3\frac{3}{4}$	$2\frac{1}{32}$	0.189	0.172	
11	11.32	$3\frac{15}{16}$	$1\frac{7}{8}$	0.187	0.173	
12	11.20	$3\frac{9}{16}$	$1\frac{1}{8}$	0.181	0.171	$2\frac{7}{8}$
13	11.44	$3\frac{29}{32}$	$1\frac{7}{8}$	0.187	0.175	
15	11.24	$3\frac{3}{8}$	$1\frac{1}{8}$	0.180	0.172	
15	11.43	$3\frac{23}{32}$	$1\frac{7}{8}$	0.187	0.175	
11	11.44	$3\frac{15}{16}$	$1\frac{3}{4}$	0.186	0.175	
12.5	11.14	$4\frac{13}{32}$	$2\frac{9}{16}$	0.361	0.170	$3\frac{1}{2}$
11	10.58	$3\frac{7}{8}$	2	0.355	0.162	
13	10.70	$3\frac{11}{16}$	2	0.354	0.164	
15.5	11.38	$4\frac{5}{32}$	$2\frac{9}{16}$	0.360	0.174	
12.5	11.14	$4\frac{3}{16}$	$2\frac{9}{16}$	0.361	0.170	
15	10.68	$3\frac{1}{2}$	2	0.353	0.163	
15	10.92	4	$2\frac{11}{16}$	0.362	0.167	
12.5	11.14	$4\frac{15}{32}$	$2\frac{9}{26}$	0.361	0.170	$3\frac{1}{2}$
11	10.70	4	$1\frac{7}{8}$	0.354	0.164	
13	10.70	$3\frac{13}{16}$	2	0.354	0.164	
15.5	11.38	$4\frac{1}{4}$	$2\frac{9}{16}$	0.360	0.174	
12.5	11.14	$4\frac{9}{32}$	$2\frac{9}{16}$	0.361	0.170	
15	10.68	$3\frac{5}{8}$	2	0.353	0.163	
15	10.92	$4\frac{3}{32}$	$2\frac{11}{16}$	0.362	0.167	
12.5	11.14	$4\frac{17}{32}$	$2\frac{9}{16}$	0.361	0.170	$3\frac{1}{2}$
11	10.70	$4\frac{1}{16}$	$1\frac{7}{8}$	0.354	0.164	
13	10.83	$3\frac{27}{32}$	$1\frac{7}{8}$	0.353	0.166	
15.5	11.38	$4\frac{9}{32}$	$2\frac{9}{16}$	0.360	0.174	
12.5	11.14	$4\frac{5}{16}$	$1\frac{9}{16}$	0.361	0.170	
15	10.68	$3\frac{11}{16}$	1	0.353	0.163	
15	10.92	$4\frac{5}{32}$	$1\frac{11}{16}$	0.362	0.167	
12.5	11.75	$4\frac{21}{32}$	$2\frac{7}{16}$	0.359	0.180	$3\frac{1}{2}$
11	10.70	NA	$1\frac{7}{8}$	0.354	0.164	
13	10.83	4	$1\frac{7}{8}$	0.353	0.166	
15.5	11.38	$4\frac{7}{16}$	$2\frac{9}{16}$	0.360	0.174	
12.5	11.31	$4\frac{7}{16}$	$2\frac{7}{16}$	0.359	0.173	
15	10.82	$3\frac{13}{16}$	$1\frac{7}{8}$	0.351	0.166	
15	10.92	$4\frac{9}{32}$	$2\frac{11}{16}$	0.362	0.167	

(continued)

Table 4.6.6   (continued)

			Assembly Data			
* Box Tong Space	Adjusted Weight	Minimum Tool Joint O.D. for Prem. Class	Drift Diameter	Capacity	Displace- ment	Size O.D.
in.	lb/ft	in.	in.	US gal/ft	US gal/ft	in.
13	10.83	$4\frac{1}{32}$	$1\frac{7}{8}$	0.353	0.166	$3\frac{1}{2}$
15.5	11.38	$4\frac{15}{32}$	$2\frac{9}{16}$	0.360	0.174	
15	10.82	$3\frac{27}{32}$	$1\frac{7}{8}$	0.351	0.166	
15	10.92	$4\frac{5}{16}$	$2\frac{11}{16}$	0.362	0.167	
13	11.26	4	$1\frac{5}{5}$	0.350	0.172	$3\frac{1}{2}$
15.5	11.38	$4\frac{1}{2}$	$2\frac{9}{16}$	0.360	0.174	
15	10.82	$3\frac{29}{32}$	$1\frac{7}{8}$	0.351	0.166	
15	10.92	$4\frac{11}{32}$	$2\frac{11}{16}$	0.362	0.167	
12.5	14.30	$4\frac{1}{2}$	$2\frac{9}{16}$	0.311	0.219	$3\frac{1}{2}$
11	13.88	$4\frac{1}{32}$	$1\frac{7}{8}$	0.303	0.212	
13	13.86	$3\frac{27}{32}$	2	0.304	0.212	
15.5	14.51	$4\frac{1}{4}$	$2\frac{9}{16}$	0.310	0.222	
12.5	14.13	$4\frac{9}{32}$	$2\frac{11}{16}$	0.312	0.216	
15	13.82	$3\frac{21}{32}$	2	0.303	0.211	
15	14.47	$4\frac{1}{32}$	$2\frac{9}{16}7$	0.310	0.221	
12.5	14.90	$4\frac{19}{32}$	$2\frac{7}{16}$	0.309	0.228	$3\frac{1}{2}$
11	13.88	NA	$1\frac{7}{8}$	0.303	0.212	
13	13.99	$3\frac{15}{16}$	$1\frac{7}{8}$	0.302	0.214	
15.5	14.51	$4\frac{3}{8}$	$2\frac{9}{16}$	0.310	0.222	
12.5	14.30	$4\frac{3}{8}$	$2\frac{9}{16}$	0.311	0.219	
15	13.82	$3\frac{13}{16}$	2	0.303	0.211	
15	14.47	$4\frac{5}{32}$	$2\frac{9}{16}$	0.310	0.221	
12.5	15.07	$4\frac{21}{32}$	$2\frac{5}{16}$	0.307	0.231	$3\frac{1}{2}$
11	13.88	NA	$1\frac{7}{8}$	0.303	0.212	
13	13.99	4	$1\frac{7}{8}$	0.302	0.214	
15.5	14.51	$4\frac{7}{16}$	$2\frac{9}{16}$	0.310	0.222	
12.5	14.47	$4\frac{7}{16}$	$2\frac{7}{16}$	0.309	0.221	
15	14.16	$3\frac{13}{16}$	$1\frac{7}{8}$	0.301	0.217	
15	14.47	$4\frac{7}{32}$	$2\frac{9}{16}$	0.310	0.221	
12.5	15.45	$4\frac{13}{16}$	2	0.303	0.236	$3\frac{1}{2}$
11	13.88	NA	$1\frac{7}{8}$	0.303	0.212	
13	13.99	NA	$1\frac{7}{8}$	0.302	0.214	
15.5	14.69	$4\frac{9}{16}$	$2\frac{7}{16}$	0.309	0.225	
12.5	14.47	$4\frac{19}{32}$	$2\frac{7}{16}$	0.309	0.221	
15	14.29	4	$1\frac{3}{4}$	0.300	0.219	
15	14.47	$4\frac{13}{32}$	$2\frac{9}{16}$	0.310	0.221	
12.5	14.68	$4\frac{11}{16}$	$2\frac{7}{16}$	0.309	0.225	
13	14.12	NA	$1\frac{3}{4}$	0.301	0.216	$3\frac{1}{2}$
15.5	14.51	$4\frac{5}{8}$	$2\frac{9}{16}$	0.310	0.222	
15	14.42	4	$1\frac{5}{8}$	0.299	0.221	
15	14.25	$4\frac{7}{16}$	$2\frac{9}{16}$	0.311	0.218	
12.5	14.77	$4\frac{11}{16}$	$2\frac{3}{8}$	0.308	0.226	

(continued)

Table 4.6.6   (continued)

			Assembly Data			
* Box Tong Space	Adjusted Weight	Minimum Tool Joint O.D. for Prem. Class	Drift Diameter	Capacity	Displace- ment	Size O.D.
in.	lb/ft	in.	in.	US gal/ft	US gal/ft	in.
13	14.42	$4\frac{1}{4}$	$1\frac{5}{8}$	0.300	0.221	$3\frac{1}{2}$
15.5	14.51	$4\frac{11}{16}$	$2\frac{9}{16}$	0.310	0.222	
15	14.42	$4\frac{1}{16}$	$1\frac{5}{8}$	0.299	0.221	
15	14.65	$4\frac{13}{32}$	$2\frac{7}{16}$	0.309	0.224	
12.5	15.07	$4\frac{11}{16}$	$2\frac{5}{16}$	0.307	0.231	
12.5	16.56	$4\frac{17}{32}$	$2\frac{7}{16}$	0.276	0.253	$3\frac{1}{2}$
15.5	16.60	$4\frac{5}{16}$	$2\frac{9}{16}$	0.278	0.254	
15	16.73	$4\frac{1}{32}$	$2\frac{7}{16}$	0.276	0.256	
12.5	17.16	$4\frac{21}{32}$	$2\frac{5}{16}$	0.274	0.262	$3\frac{1}{2}$
15.5	16.60	$4\frac{7}{16}$	$2\frac{9}{16}$	0.278	0.254	
15	16.73	$4\frac{5}{32}$	$2\frac{7}{16}$	0.276	0.256	
12.5	17.54	$4\frac{23}{32}$	2	0.270	0.268	$3\frac{1}{2}$
15.5	16.77	$4\frac{7}{16}$	$2\frac{7}{16}$	0.276	0.256	
12	17.30	$4\frac{15}{16}$	$2\frac{7}{16}$	0.276	0.265	
15	16.73	$4\frac{7}{32}$	$2\frac{7}{16}$	0.276	0.256	
12.5	17.54	$4\frac{29}{32}$	2	0.270	0.268	$3\frac{1}{2}$
12.5	16.72	$4\frac{19}{32}$	$2\frac{5}{16}$	0.274	0.256	
12.5	18.35	$5\frac{3}{32}$	$2\frac{1}{8}$	0.271	0.281	
15	16.73	$4\frac{7}{16}$	$2\frac{7}{16}$	0.276	0.256	
15	17.20	$4\frac{7}{16}$	$2\frac{7}{16}$	0.276	0.263 ·	
12.5	17.24	$4\frac{11}{16}$	$2\frac{1}{4}$	0.273	0.264	
12.5	16.72	$4\frac{5}{8}$	$2\frac{5}{16}$	0.274	0.256	$3\frac{1}{2}$
15	16.91	$4\frac{13}{32}$	$2\frac{5}{16}$	0.274	0.259	
15	17.20	$4\frac{15}{32}$	$2\frac{7}{16}$	0.276	0.263	
12.5	17.24	$4\frac{23}{32}$	$2\frac{1}{4}$	0.273	0.264	
12.5	17.16	$4\frac{11}{16}$	$2\frac{5}{16}$	0.274	0.262	$3\frac{1}{2}$
15	16.91	$4\frac{15}{32}$	$2\frac{5}{16}$	0.274	0.259	
15	17.20	$4\frac{17}{32}$	$2\frac{7}{16}$	0.276	0.263	
12.5	17.39	$4\frac{23}{32}$	$2\frac{1}{8}$	0.272	0.266	
12	13.35	$4\frac{3}{4}$	$2\frac{11}{16}$	0.482	0.204	4
12	12.85	$4\frac{3}{8}$	$2\frac{7}{16}$	0.479	0.197	
15.5	13.02	$4\frac{9}{32}$	$2\frac{9}{16}$	0.479	0.199	
15	12.93	$4\frac{1}{16}$	$2\frac{9}{16}$	0.479	0.198	
15	13.11	$4\frac{3}{16}$	$2\frac{3}{4}$	0.482	0.200	
12	13.35	$4\frac{27}{32}$	$2\frac{11}{16}$	0.482	0.204	4
12	12.85	$4\frac{1}{2}$	$2\frac{7}{16}$	0.479	0.197	
15.5	13.02	$4\frac{13}{32}$	$2\frac{7}{16}$	0.479	0.199	
15	12.93	$4\frac{3}{16}$	$2\frac{9}{16}$	0.479	0.198	
15	13.11	$4\frac{5}{16}$	$2\frac{3}{4}$	0.482	0.200	
12	13.35	$4\frac{29}{32}$	$2\frac{11}{16}$	0.482	0.204	4
12	12.85	$4\frac{9}{16}$	$2\frac{7}{16}$	0.479	0.197	

(continued)

Table 4.6.6 *(continued)*

			Assembly Data			
* Box Tong Space	Adjusted Weight	Minimum Tool Joint O.D. for Prem. Class	Drift Diameter	Capacity	Displace- ment	Size O.D.
in.	lb/ft	in.	in.	US gal/ft	US gal/ft	in.
15.5	13.20	$4\frac{13}{32}$	$2\frac{7}{16}$	0.477	0.202	
15	12.93	$4\frac{1}{4}$	$2\frac{9}{16}$	0.479	0.198	
15	13.11	$4\frac{3}{8}$	$2\frac{3}{4}$	0.482	0.200	
12	14.17	$5\frac{1}{16}$	$2\frac{7}{16}$	0.478	0.217	4
12	12.85	$4\frac{23}{32}$	$2\frac{7}{16}$	0.479	0.197	
15.5	13.38	$4\frac{17}{32}$	$2\frac{5}{16}$	0.475	0.202	
15	12.93	$4\frac{7}{16}$	$2\frac{9}{16}$	0.479	0.198	
15	13.11	$4\frac{17}{32}$	$2\frac{3}{4}$	0.482	0.200	
15.5	13.38	$4\frac{9}{16}$	$2\frac{5}{16}$	0.475	0.205	4
15	12.93	$4\frac{15}{32}$	$2\frac{9}{16}$	0.479	0.198	
15	13.11	$4\frac{9}{16}$	$2\frac{3}{4}$	0.482	0.200	
15.5	13.86	$4\frac{5}{8}$	$2\frac{5}{16}$	0.475	0.212	4
15	13.11	$4\frac{15}{32}$	$2\frac{7}{16}$	0.477	0.201	
15	13.11	$4\frac{5}{8}$	$2\frac{3}{4}$	0.482	0.200	
12	15.58	$4\frac{13}{16}$	$2\frac{11}{16}$	0.447	0.238	4
15.5	15.23	$4\frac{3}{8}$	$2\frac{9}{16}$	0.444	0.233	
12	15.25	$4\frac{7}{16}$	$2\frac{5}{16}$	0.442	0.233	
15	15.88	$4\frac{19}{32}$	$2\frac{11}{16}$	0.446	0.243	
12	16.64	$4\frac{9}{32}$	$3\frac{1}{8}$	0.454	0.254	
15	15.20	$5\frac{5}{32}$	$2\frac{9}{16}$	0.444	0.232	
15	15.47	$4\frac{1}{4}$	$2\frac{11}{16}$	0.446	0.237	
12	15.58	$4\frac{15}{16}$	$2\frac{11}{16}$	0.447	0.238	4
15.5	15.23	$4\frac{17}{32}$	$2\frac{9}{16}$	0.444	0.233	
12	15.25	$4\frac{19}{32}$	$2\frac{5}{16}$	0.442	0.233	
15	15.88	$4\frac{23}{32}$	$2\frac{11}{16}$	0.446	0.243	
12	16.64	$5\frac{3}{8}$	$3\frac{1}{8}$	0.454	0.254	
15	15.20	$4\frac{5}{16}$	$2\frac{9}{16}$	0.444	0.232	
15	15.47	$4\frac{3}{8}$	$2\frac{11}{16}$	0.446	0.237	
12	15.58	5	$2\frac{11}{16}$	0.447	0.238	4
15.5	15.89	$4\frac{17}{32}$	$2\frac{7}{16}$	0.441	0.243	
12	15.25	$4\frac{21}{32}$	$2\frac{5}{16}$	0.442	0.233	
15	15.88	$4\frac{25}{32}$	$2\frac{11}{16}$	0.446	0.243	
12	16.64	$5\frac{7}{16}$	$3\frac{1}{8}$	0.454	0.254	
15	15.20	$4\frac{3}{8}$	$2\frac{9}{16}$	0.444	0.232	
15	15.47	$4\frac{7}{16}$	$2\frac{11}{16}$	0.446	0.237	
12	16.40	$5\frac{3}{16}$	$2\frac{7}{16}$	0.443	0.251	4
15.5	16.07	$4\frac{11}{16}$	$2\frac{5}{16}$	0.440	0.246	
12	15.25	NA	$2\frac{5}{16}$	0.442	0.233	
15	16.07	$4\frac{29}{32}$	$2\frac{9}{16}$	0.444	0.246	
12	17.04	$5\frac{9}{16}$	$3\frac{7}{8}$	0.450	0.261	
15	15.85	$4\frac{17}{32}$	$2\frac{7}{16}$	0.442	0.242	

(continued)

Table 4.6.6 (continued)

			Assembly Data			
* Box Tong Space	Adjusted Weight	Minimum Tool Joint O.D. for Prem. Class	Drift Diameter	Capacity	Displace- ment	Size O.D.
in.	lb/ft	in.	in.	US gal/ft	US gal/ft	in.
15	15.47	$4\frac{21}{32}$	$2\frac{11}{16}$	0.446	0.237	
12	15.76	5	$2\frac{9}{16}$	0.445	0.241	
15.5	16.07	$4\frac{23}{32}$	$2\frac{5}{16}$	0.440	0.246	4
15	16.07	$4\frac{15}{16}$	$2\frac{9}{16}$	0.444	0.246	
15	15.85	$4\frac{9}{16}$	$2\frac{7}{16}$	0.442	0.242	
15	15.47	$4\frac{11}{16}$	$2\frac{11}{16}$	0.446	0.237	
12	15.85	5	$2\frac{1}{2}$	0.444	0.242	
15.5	16.07	$4\frac{25}{32}$	$2\frac{5}{16}$	0.440	0.246	4
15	16.07	5	$2\frac{9}{16}$	0.444	0.246	
15	16.03	$4\frac{19}{32}$	$2\frac{5}{16}$	0.440	0.245	
15	15.47	$4\frac{3}{4}$	$2\frac{11}{16}$	0.446	0.237	
12	16.02	$5\frac{1}{32}$	$2\frac{3}{8}$	0.442	0.245	
12	17.17	$4\frac{7}{8}$	$2\frac{11}{16}$	0.422	0.263	4
15	17.45	$4\frac{5}{8}$	$2\frac{11}{16}$	0.421	0.267	
12	17.63	$4\frac{32}{32}$	$2\frac{11}{16}$	0.421	0.270	
12	17.90	$5\frac{5}{16}$	$3\frac{1}{8}$	0.428	0.274	
15	17.04	$4\frac{9}{32}$	$2\frac{11}{16}$	0.421	0.261	
15	17.54	$4\frac{5}{16}$	$2\frac{11}{16}$	0.421	0.268	
12	17.52	5	$2\frac{7}{16}$	0.418	0.268	4
15	17.45	$4\frac{25}{32}$	$2\frac{11}{16}$	0.421	0.267	
12	17.63	$5\frac{3}{32}$	$2\frac{11}{16}$	0.421	0.270	
12	17.90	$5\frac{7}{16}$	$3\frac{1}{8}$	0.428	0.274	
15	17.04	$4\frac{7}{16}$	$2\frac{11}{16}$	0.421	0.261	
15	17.54	$4\frac{15}{32}$	$2\frac{11}{16}$	0.421	0.268	
12	18.14	$5\frac{1}{16}$	$2\frac{5}{16}$	0.416	0.278	4
15	17.45	$4\frac{27}{32}$	$2\frac{11}{16}$	0.421	0.267	
15	17.96	$5\frac{5}{32}$	$2\frac{11}{16}$	0.421	0.275	
12	17.90	$5\frac{15}{32}$	$3\frac{1}{8}7$	0.428	0.274	
15	17.04	$4\frac{17}{32}$	$2\frac{11}{16}$	0.421	0.261	
15	17.54	$4\frac{17}{32}$	$2\frac{11}{16}$	0.421	0.268	
12	18.66	$5\frac{1}{4}$	$1\frac{7}{8}$	0.411	0.285	4
15	18.34	$4\frac{15}{16}$	$2\frac{7}{16}$	0.417	0.280	
15	18.69	$5\frac{5}{16}$	$2\frac{9}{16}$	0.418	0.286	
12	18.30	$5\frac{21}{32}$	$2\frac{7}{8}$	0.424	0.280	
15	17.24	$4\frac{11}{16}$	$2\frac{9}{16}$	0.419	0.264	
15	17.54	$4\frac{3}{4}$	$2\frac{11}{16}$	0.421	0.268	
12	17.52	$5\frac{1}{32}$	$2\frac{7}{16}$	0.418	0.268	
15	18.34	$4\frac{31}{32}$	$2\frac{7}{16}$	0.417	0.280	4
15	17.24	$4\frac{23}{32}$	$2\frac{9}{16}$	0.419	0.264	
15	17.54	$4\frac{25}{32}$	$2\frac{11}{16}$	0.421	0.268	
12	17.83	$5\frac{1}{16}$	$2\frac{3}{8}$	0.417	0.273	

(continued)

Table 4.6.6 (continued)

				Assembly Data		
* Box Tong Space	Adjusted Weight	Minimum Tool Joint O.D. for Prem. Class	Drift Diameter	Capacity	Displace- ment	Size O.D.
in.	lb/ft	in.	in.	US gal/ft	US gal/ft	in.
15	18.34	$5\frac{1}{16}$	$2\frac{7}{16}$	0.417	0.280	4
15	17.42	$4\frac{23}{32}$	$2\frac{7}{16}$	0.417	0.266	
15	17.54	$4\frac{27}{32}$	$2\frac{11}{16}$	0.421	0.268	
12	18.14	$5\frac{3}{32}$	$2\frac{5}{16}$	0.416	0.278	
12	19.17	$5\frac{13}{32}$	$3\frac{1}{8}$	0.586	0.293	$4\frac{1}{2}$
12	17.75	$5\frac{15}{32}$	$3\frac{5}{8}$	0.596	0.272	
12	19.05	$5\frac{3}{8}$	$2\frac{7}{8}$	0.583	0.291	
12	18.64	$5\frac{11}{32}$	$3\frac{1}{8}$	0.587	0.285	
15	19.62	$5\frac{13}{32}$	$3\frac{1}{8}$	0.585	0.300	
12	18.80	$5\frac{23}{32}$	$3\frac{5}{8}$	0.596	0.288	
15	18.90	$5\frac{13}{16}$	$3\frac{5}{8}$	0.596	0.289	
15	17.94	$4\frac{7}{8}$	$2\frac{7}{8}$	0.581	0.274	
15	18.68	$5\frac{5}{8}$	$3\frac{3}{8}$	0.590	0.286	
15	19.34	$5\frac{31}{32}$	$3\frac{5}{8}$	0.596	0.296	
12	19.17	$5\frac{17}{32}$	$3\frac{1}{8}$	0.586	0.293	$4\frac{1}{2}$
12	18.20	$5\frac{19}{32}$	$3\frac{3}{8}$	0.591	0.278	
12	19.05	$5\frac{1}{2}$	$2\frac{7}{8}$	0.583	0.291	
12	18.64	$5\frac{15}{32}$	$3\frac{1}{8}$	0.587	0.285	
15	19.62	$5\frac{13}{32}$	$3\frac{1}{8}$	0.585	0.300	
12	18.80	$5\frac{27}{32}$	$3\frac{5}{8}$	0.596	0.288	
15	18.90	$5\frac{13}{16}$	$3\frac{5}{8}$	0.596	0.289	
15	17.94	$4\frac{7}{8}$	$2\frac{7}{8}$	0.581	0.274	
15	18.68	$5\frac{5}{8}$	$3\frac{3}{8}$	0.590	0.286	
15	19.34	$5\frac{31}{32}$	$3\frac{5}{8}$	0.596	0.296	
12	19.59	$5\frac{19}{32}$	$2\frac{7}{8}$	0.582	0.300	$4\frac{1}{2}$
12	18.88	$5\frac{21}{32}$	$3\frac{1}{8}$	0.587	0.289	
12	19.97	$5\frac{9}{16}$	$2\frac{5}{8}$	0.578	0.305	
12	18.64	$5\frac{17}{32}$	$3\frac{1}{8}$	0.587	0.285	
15	19.62	$5\frac{13}{32}$	$3\frac{1}{8}$	0.585	0.300	
12	18.80	$5\frac{29}{32}$	$3\frac{5}{8}$	0.596	0.288	
15	18.90	$5\frac{13}{16}$	$3\frac{5}{8}$	0.596	0.289	
15	17.94	$4\frac{7}{8}$	$2\frac{7}{8}$	0.581	0.274	
15	18.68	$5\frac{5}{8}$	$3\frac{3}{8}$	0.590	0.286	
15	19.34	$5\frac{31}{32}$	$3\frac{5}{8}$	0.596	0.296	
12	19.97	$5\frac{25}{32}$	$2\frac{5}{8}$	0.578	0.305	$4\frac{1}{2}$
12	19.28	$5\frac{13}{16}$	$2\frac{7}{8}$	0.583	0.295	
12	19.97	$5\frac{3}{4}$	$2\frac{5}{8}$	0.578	0.305	
12	18.97	$5\frac{11}{16}$	$2\frac{5}{8}$	0.578	0.305	
15	19.62	$5\frac{1}{2}$	$3\frac{1}{8}$	0.585	0.300	
12	19.26	$6\frac{1}{16}$	$3\frac{3}{8}$	0.591	0.295	
15	19.71	$5\frac{13}{16}$	$3\frac{3}{8}$	0.590	0.301	

(*continued*)

Table 4.6.6   (continued)

			Assembly Data			
* Box Tong Space	Adjusted Weight	Minimum Tool Joint O.D. for Prem. Class	Drift Diameter	Capacity	Displace- ment	Size O.D.
in.	lb/ft	in.	in.	US gal/ft	US gal/ft	in.
15	18.24	$4\frac{15}{16}$	$2\frac{11}{16}$	0.578	0.279	
15	18.68	$5\frac{5}{8}$	$3\frac{3}{8}$	0.590	0.286	
15	19.34	$5\frac{31}{32}$	$3\frac{5}{8}$	0.596	0.296	
12	18.64	$5\frac{19}{32}$	$3\frac{1}{8}$	0.587	0.285	
15	19.62	$5\frac{17}{32}$	$3\frac{1}{8}$	0.585	0.300	$4\frac{1}{2}$
15	19.71	$5\frac{13}{16}$	$3\frac{3}{8}$	0.590	0.301	
15	18.24	$4\frac{21}{32}$	$2\frac{11}{16}$	0.578	0.279	
15	18.68	$5\frac{5}{8}$	$3\frac{3}{8}$	0.590	0.286	
15	19.34	$5\frac{31}{32}$	$3\frac{5}{8}$	0.596	0.296	
12	18.75	$5\frac{19}{32}$	$3\frac{1}{16}$	0.586	0.287	
15	19.62	$5\frac{19}{32}$	$3\frac{1}{8}$	0.585	0.300	$4\frac{1}{2}$
15	19.71	$5\frac{13}{16}$	$3\frac{3}{8}$	0.590	0.301	
15	18.24	$5\frac{1}{16}$	$2\frac{11}{16}$	0.578	0.279	
15	19.77	$5\frac{5}{8}$	$3\frac{1}{8}$	0.585	0.302	
15	19.85	$5\frac{31}{32}$	$3\frac{3}{8}$	0.590	0.304	
12	18.85	$5\frac{21}{32}$	3	0.585	0.288	
12	22.51	$5\frac{1}{2}$	$3\frac{1}{8}$	0.533	0.344	$4\frac{1}{2}$
12	21.81	$5\frac{17}{32}$	$3\frac{3}{8}$	0.538	0.334	
12	21.98	$5\frac{7}{16}$	$3\frac{1}{8}$	0.534	0.336	
15	22.93	$5\frac{13}{32}$	$3\frac{1}{8}$	0.533	0.351	
12	22.38	$5\frac{13}{16}$	$3\frac{1}{2}$	0.540	0.342	
15	22.25	$5\frac{13}{16}$	$3\frac{5}{8}$	0.543	0.340	
15	21.99	$5\frac{5}{8}$	$3\frac{3}{8}$	0.538	0.336	
15	23.17	$5\frac{31}{32}$	$3\frac{3}{8}$	0.538	0.354	
12	22.92	$5\frac{21}{32}$	$3\frac{7}{8}$	0.529	0.351	$4\frac{1}{2}$
12	22.77	$5\frac{11}{16}$	$3\frac{1}{8}$	0.534	0.348	
12	21.98	$5\frac{9}{16}$	$3\frac{1}{8}$	0.534	0.336	
15	22.93	$5\frac{13}{32}$	$3\frac{1}{8}$	0.533	0.351	
12	22.38	$5\frac{15}{16}$	$3\frac{1}{2}$	0.540	0.342	
15	22.25	$5\frac{13}{16}$	$3\frac{5}{8}$	0.543	0.340	
15	21.99	$5\frac{5}{8}$	$3\frac{3}{8}$	0.538	0.336	
15	23.17	$5\frac{31}{32}$	$3\frac{3}{8}$	0.538	0.354	
12	23.30	$5\frac{23}{32}$	$2\frac{5}{8}$	0.526	0.356	$4\frac{1}{2}$
12	23.17	$5\frac{3}{4}$	$2\frac{7}{8}$	0.530	0.354	
12	22.92	$5\frac{5}{8}$	$2\frac{7}{8}$	0.529	0.351	
15	22.93	$5\frac{7}{16}$	$3\frac{1}{8}$	0.533	0.351	
12	22.61	$6\frac{1}{32}$	$3\frac{3}{8}$	0.538	0.346	
15	22.25	$5\frac{13}{16}$	$3\frac{5}{8}$	0.543	0.340	
15	21.99	$5\frac{5}{8}$	$3\frac{3}{8}$	0.538	0.336	
15	23.17	$5\frac{31}{32}$	$3\frac{3}{8}$	0.538	0.354	

(continued)

Table 4.6.6 (continued)

		Assembly Data				
* Box Tong Space	Adjusted Weight	Minimum Tool Joint O.D. for Prem. Class	Drift Diameter	Capacity	Displace- ment	Size O.D.
in.	lb/ft	in.	in.	US gal/ft	US gal/ft	in.
12	23.94	$5\frac{15}{16}$	$2\frac{3}{8}$	0.522	0.366	$4\frac{1}{2}$
12	23.82	$5\frac{31}{32}$	$2\frac{5}{8}$	0.526	0.364	
12	23.58	$5\frac{27}{32}$	$2\frac{5}{8}$	0.525	0.361	
15	23.38	$5\frac{9}{16}$	$2\frac{7}{8}$	0.528	0.358	
12	23.32	$6\frac{7}{32}$	$3\frac{1}{8}$	0.534	0.357	
15	23.03	$5\frac{13}{16}$	$3\frac{3}{8}$	0.538	0.352	
15	21.99	$5\frac{5}{8}$	$3\frac{3}{8}$	0.538	0.336	
15	23.17	$5\frac{31}{32}$	$3\frac{3}{8}$	0.538	0.354	
12	22.39	$5\frac{21}{32}$	$2\frac{7}{8}$	0.530	0.342	
15	23.38	$5\frac{19}{32}$	$2\frac{7}{8}$	0.528	0.358	$4\frac{1}{2}$
15	23.03	$5\frac{27}{32}$	$3\frac{3}{8}$	0.538	0.352	
15	21.99	$5\frac{21}{32}$	$3\frac{3}{8}$	0.538	0.336	
15	23.17	$5\frac{31}{32}$	$3\frac{3}{8}$	0.538	0.354	
12	22.49	$5\frac{21}{32}$	$2\frac{13}{16}$	0.529	0.344	
15	23.38	$5\frac{21}{32}$	$2\frac{7}{8}$	0.528	0.358	$4\frac{1}{2}$
15	23.03	$5\frac{29}{32}$	$3\frac{3}{8}$	0.538	0.352	
15	23.07	$5\frac{5}{8}$	$3\frac{1}{8}$	0.532	0.353	
15	23.17	$5\frac{31}{32}$	$3\frac{3}{8}$	0.538	0.354	
12	22.76	$5\frac{21}{32}$	$2\frac{5}{8}$	0.526	0.348	
12	22.16	$5\frac{7}{8}$	$3\frac{5}{8}$	0.735	0.339	5
15	22.61	$5\frac{13}{16}$	$3\frac{5}{8}$	0.734	0.346	
12	23.24	$6\frac{3}{8}$	$3\frac{5}{8}$	0.734	0.356	
15	21.71	$5\frac{5}{8}$	$3\frac{3}{8}$	0.729	0.332	
15	21.88	$5\frac{31}{32}$	$3\frac{7}{8}$	0.739	0.335	
12	22.63	$6\frac{1}{32}$	$3\frac{3}{8}$	0.730	0.346	5
15	22.61	$5\frac{13}{16}$	$3\frac{5}{8}$	0.734	0.346	
12	23.24	$6\frac{1}{2}$	$3\frac{5}{8}$	0.734	0.356	
15	21.71	$5\frac{5}{8}$	$3\frac{3}{8}$	0.729	0.332	
15	21.88	$5\frac{31}{32}$	$3\frac{7}{8}$	0.739	0.335	
12	23.08	$6\frac{3}{32}$	$3\frac{1}{8}$	0.726	0.353	5
15	22.61	$5\frac{27}{32}$	$3\frac{5}{8}$	0.734	0.346	
12	23.24	$6\frac{9}{16}$	$3\frac{5}{8}$	0.734	0.356	
15	21.71	$5\frac{5}{8}$	$3\frac{3}{8}$	0.729	0.332	
15	21.88	$5\frac{31}{32}$	$3\frac{7}{8}$	0.739	0.335	
12	21.87	$5\frac{15}{16}$	$3\frac{5}{8}$	0.735	0.335	
12	23.87	$6\frac{5}{16}$	$2\frac{5}{8}$	0.718	0.365	5
15	23.12	$5\frac{15}{16}$	$3\frac{3}{8}$	0.728	0.354	
12	24.41	$6\frac{3}{4}$	$3\frac{3}{8}$	0.729	0.373	
15	21.71	$5\frac{23}{32}$	$3\frac{3}{8}$	0.729	0.332	
15	22.43	$5\frac{31}{32}$	$3\frac{5}{8}$	0.733	0.343	
12	22.35	$6\frac{1}{32}$	$3\frac{3}{8}$	0.730	0.342	

(*continued*)

Table 4.6.6 (continued)

			Assembly Data			
* Box Tong Space	Adjusted Weight	Minimum Tool Joint O.D. for Prem. Class	Drift Diameter	Capacity	Displacement	Size O.D.
in.	lb/ft	in.	in.	US gal/ft	US gal/ft	in.
15	23.12	$5\frac{31}{32}$	$3\frac{3}{8}$	0.728	0.354	5
15	21.71	$5\frac{25}{32}$	$3\frac{3}{8}$	0.729	0.332	
15	22.43	$5\frac{31}{32}$	$3\frac{5}{8}$	0.733	0.343	
12	22.35	$6\frac{3}{32}$	$3\frac{3}{8}$	0.730	0.342	
15	23.12	$6\frac{1}{32}$	$3\frac{3}{8}$	0.728	0.354	5
15	22.79	$5\frac{23}{32}$	$3\frac{1}{8}$	0.723	0.349	
15	22.43	$5\frac{31}{32}$	$3\frac{5}{8}$	0.733	0.343	
12	22.57	$6\frac{3}{32}$	$3\frac{1}{4}$	0.728	0.345	
12	28.13	$6\frac{1}{32}$	$3\frac{3}{8}$	0.643	0.430	5
15	28.06	$5\frac{13}{16}$	$3\frac{5}{8}$	0.647	0.429	
12	29.21	$6\frac{1}{2}$	$3\frac{3}{8}$	0.642	0.447	
15	28.20	$5\frac{31}{32}$	$3\frac{5}{8}$	0.647	0.431	
12	28.98	$6\frac{7}{32}$	$2\frac{7}{8}$	0.634	0.443	5
15	28.57	$5\frac{13}{16}$	$3\frac{3}{8}$	0.642	0.437	
12	29.21	$6\frac{21}{32}$	$3\frac{3}{8}$	0.642	0.447	
15	28.20	$5\frac{31}{32}$	$3\frac{5}{8}$	0.647	0.431	
12	29.36	$6\frac{9}{32}$	$2\frac{5}{8}$	0.631	0.449	5
15	29.05	$5\frac{13}{16}$	$3\frac{1}{8}$	0.637	0.444	
12	29.87	$6\frac{23}{32}$	$3\frac{3}{8}$	0.642	0.457	
15	28.20	$5\frac{31}{32}$	$3\frac{5}{8}$	0.647	0.431	
12	27.84	$6\frac{1}{32}$	$3\frac{3}{8}$	0.643	0.426	
12	29.36	$6\frac{17}{32}$	$2\frac{5}{8}$	0.631	0.449	5
15	29.49	$5\frac{31}{32}$	$2\frac{7}{8}$	0.632	0.451	
12	30.33	$6\frac{15}{16}$	$3\frac{1}{8}$	0.637	0.464	
15	28.20	$6\frac{1}{16}$	$3\frac{5}{8}$	0.647	0.431	
12	28.39	$6\frac{5}{32}$	$3\frac{1}{16}$	0.638	0.434	
15	29.49	$6\frac{1}{32}$	$2\frac{7}{8}$	0.632	0.451	5
15	28.20	$6\frac{3}{32}$	$3\frac{5}{8}$	0.647	0.431	
12	28.49	$6\frac{3}{16}$	3	0.637	0.436	
15	29.49	$6\frac{3}{32}$	$2\frac{7}{8}$	0.632	0.451	5
15	28.72	$6\frac{1}{16}$	$3\frac{3}{8}$	0.641	0.439	
12	28.98	$6\frac{7}{32}$	$2\frac{7}{8}$	0.634	0.443	
12	24.85	$6\frac{15}{32}$	$3\frac{7}{8}$	0.913	0.380	$5\frac{1}{2}$
15	25.35	$6\frac{13}{32}$	$3\frac{7}{8}$	0.911	0.388	
15	24.08	$6\frac{7}{32}$	$4\frac{1}{8}$	0.917	0.368	
15	24.45	$6\frac{15}{32}$	$4\frac{1}{4}$	0.920	0.374	
12	25.46	$6\frac{5}{8}$	$3\frac{5}{8}$	0.908	0.389	$5\frac{1}{2}$
15	25.44	$6\frac{13}{32}$	$3\frac{7}{8}$	0.911	0.389	
15	24.08	$6\frac{7}{32}$	$4\frac{1}{8}$	0.917	0.368	
15	24.45	$6\frac{15}{32}$	$4\frac{1}{4}$	0.920	0.374	
12	26.61	$6\frac{11}{16}$	$3\frac{3}{8}$	0.902	0.407	$5\frac{1}{2}$
15	25.44	$6\frac{13}{32}$	$3\frac{7}{8}$	0.911	0.389	

(*continued*)

Table 4.6.6 (continued)

Assembly Data						
* Box Tong Space	Adjusted Weight	Minimum Tool Joint O.D. for Prem. Class	Drift Diameter	Capacity	Displace- ment	Size O.D.
in.	lb/ft	in.	in.	US gal/ft	US gal/ft	in.
15	24.08	$6\frac{7}{32}$	$4\frac{1}{8}$	0.917	0.368	
15	24.45	$6\frac{51}{32}$	$4\frac{1}{4}$	0.920	0.374	
12	24.85	$6\frac{7}{16}$	$3\frac{7}{8}$	0.913	0.380	
12	28.21	$6\frac{29}{32}$	$2\frac{7}{8}$	0.893	0.431	$5\frac{1}{2}$
15	25.44	$6\frac{5}{8}$	$3\frac{7}{8}$	0.911	0.389	
15	24.08	$6\frac{5}{16}$	$4\frac{1}{8}$	0.917	0.368	
15	24.45	$6\frac{15}{32}$	$4\frac{1}{4}$	0.920	0.374	
12	24.85	$6\frac{11}{16}$	$3\frac{7}{8}$	0.913	0.380	
12	28.21	$6\frac{15}{16}$	$2\frac{7}{8}$	0.893	0.431	$5\frac{1}{2}$
15	25.44	$6\frac{21}{32}$	$3\frac{7}{8}$	0.911	0.389	
15	24.08	$6\frac{11}{32}$	$4\frac{1}{8}$	0.917	0.368	
15	24.45	$6\frac{15}{32}$	$4\frac{1}{4}$	0.920	0.374	
12	24.85	$6\frac{23}{32}$	$3\frac{7}{8}$	0.913	0.380	$5\frac{1}{2}$
12	28.21	7	$2\frac{7}{8}$	0.893	0.431	
15	25.44	$6\frac{23}{32}$	$3\frac{7}{8}$	0.911	0.389	
15	24.66	$6\frac{9}{32}$	$3\frac{7}{8}$	0.911	0.377	
15	24.76	$6\frac{15}{32}$	$4\frac{1}{8}$	0.917	0.379	
12	25.44	$6\frac{23}{32}$	$3\frac{3}{4}$	0.910	0.389	$5\frac{1}{2}$
12	27.41	$6\frac{17}{32}$	$3\frac{7}{8}$	0.874	0.419	
15	27.88	$6\frac{13}{32}$	$3\frac{7}{8}$	0.872	0.427	
15	26.51	$6\frac{7}{32}$	$4\frac{1}{8}$	0.879	0.406	
15	26.88	$6\frac{15}{32}$	$4\frac{1}{4}$	0.882	0.411	$5\frac{1}{2}$
12	29.07	$6\frac{11}{16}$	$3\frac{3}{8}$	0.863	0.445	
15	27.88	$6\frac{13}{32}$	$3\frac{7}{8}$	0.872	0.427	
15	26.62	$6\frac{7}{32}$	$4\frac{1}{8}$	0.879	0.407	
15	26.99	$6\frac{15}{32}$	$4\frac{1}{4}$	0.882	0.413	$5\frac{1}{2}$
12	29.07	$6\frac{25}{32}$	$3\frac{3}{8}$	0.863	0.445	
15	27.88	$6\frac{15}{32}$	$3\frac{7}{8}$	0.872	0.427	
15	26.62	$6\frac{7}{32}$	$4\frac{1}{8}$	0.879	0.407	
15	26.99	$6\frac{15}{32}$	$4\frac{1}{4}$	0.882	0.413	
12	27.30	$6\frac{17}{32}$	$3\frac{7}{8}$	0.874	0.418	$5\frac{1}{2}$
12	30.66	7	$2\frac{7}{8}$	0.854	0.469	
15	28.44	$6\frac{5}{8}$	$3\frac{5}{8}$	0.867	0.435	
15	27.20	$6\frac{9}{32}$	$3\frac{7}{8}$	0.873	0.416	
15	27.30	$6\frac{15}{32}$	$4\frac{1}{8}$	0.879	0.418	
12	27.88	$6\frac{23}{32}$	$3\frac{3}{4}$	0.871	0.427	$5\frac{1}{2}$
12	29.07	$7\frac{1}{32}$	$3\frac{3}{8}$	0.863	0.445	
15	27.88	$6\frac{25}{32}$	$3\frac{7}{8}$	0.872	0.427	
15	27.20	$6\frac{11}{32}$	$3\frac{7}{8}$	0.873	0.416	
15	27.30	$6\frac{15}{32}$	$4\frac{1}{8}$	0.879	0.418	
12	27.88	$6\frac{25}{32}$	$3\frac{3}{4}$	0.871	0.427	$5\frac{1}{2}$
12	30.66	$7\frac{3}{32}$	$2\frac{7}{8}$	0.854	0.469	

(continued)

Table 4.6.6 *(continued)*

			Assembly Data			
* Box Tong Space	Adjusted Weight	Minimum Tool Joint O.D. for Prem. Class	Drift Diameter	Capacity	Displacement	Size O.D.
in.	lb/ft	in.	in.	US gal/ft	US gal/ft	in.
15	28.44	$6\frac{23}{32}$	$3\frac{5}{8}$	0.867	0.435	
15	27.20	$6\frac{7}{16}$	$3\frac{7}{8}$	0.873	0.416	
15	27.30	$6\frac{9}{16}$	$4\frac{1}{8}$	0.879	0.418	
12	28.14	$6\frac{13}{16}$	$3\frac{5}{8}$	0.869	0.430	$5\frac{7}{8}$
15	26.46	$6\frac{15}{32}$	$4\frac{1}{8}$	1.059	0.405	$5\frac{7}{8}$
15	26.46	$6\frac{15}{32}$	$4\frac{1}{8}$	1.059	0.405	$5\frac{7}{8}$
15	26.46	$6\frac{15}{32}$	$4\frac{1}{8}$	1.059	0.405	$5\frac{7}{8}$
15	26.46	$6\frac{15}{32}$	$4\frac{1}{8}$	1.059	0.405	$5\frac{7}{8}$
15	26.46	$6\frac{17}{32}$	$4\frac{1}{8}$	1.059	0.405	$5\frac{7}{8}$
15	26.46	$6\frac{5}{8}$	$4\frac{1}{8}$	1.059	0.405	$5\frac{7}{8}$
15	29.10	$6\frac{15}{32}$	$4\frac{1}{8}$	1.017	0.445	$5\frac{7}{8}$
15	29.10	$6\frac{15}{32}$	$4\frac{1}{8}$	1.017	0.445	$5\frac{7}{8}$
15	29.10	$6\frac{15}{32}$	$4\frac{1}{8}$	1.017	0.445	$5\frac{7}{8}$
15	29.10	$6\frac{5}{8}$	$4\frac{1}{8}$	1.017	0.445	$5\frac{7}{8}$
15	29.10	$6\frac{21}{32}$	$4\frac{1}{8}$	1.017	0.445	$5\frac{7}{8}$
15	29.68	$6\frac{5}{8}$	$3\frac{7}{8}$	1.011	0.454	$6\frac{5}{8}$
13	28.81	$7\frac{7}{16}$	$4\frac{7}{8}$	1.423	0.441	
16	29.40	$7\frac{11}{32}$	$4\frac{7}{8}$	1.420	0.450	
15	29.20	$7\frac{11}{32}$	$4\frac{7}{8}$	1.421	0.447	$6\frac{5}{8}$
13	28.81	$7\frac{5}{8}$	$4\frac{7}{8}$	1.423	0.441	
16	29.40	$7\frac{11}{32}$	$4\frac{7}{8}$	1.420	0.450	
15	29.20	$7\frac{11}{32}$	$4\frac{7}{8}$	1.421	0.447	$6\frac{5}{8}$
13	30.25	$7\frac{11}{16}$	$4\frac{5}{8}$	1.415	0.463	
16	29.40	$7\frac{13}{32}$	$4\frac{7}{8}$	1.420	0.450	
15	29.20	$7\frac{11}{32}$	$4\frac{7}{8}$	1.421	0.447	$6\frac{5}{8}$
13	32.32	$7\frac{29}{32}$	$4\frac{1}{8}$	1.402	0.494	
16	29.40	$7\frac{5}{8}$	$4\frac{7}{8}$	1.420	0.450	
15	29.20	$7\frac{11}{32}$	$4\frac{7}{8}$	1.421	0.447	
13	28.98	$7\frac{21}{32}$	$4\frac{13}{16}$	1.421	0.443	$6\frac{5}{8}$
13	32.32	$7\frac{31}{32}$	$4\frac{1}{8}$	1.402	0.494	
16	29.40	$7\frac{11}{16}$	$4\frac{7}{8}$	1.420	0.450	
15	29.20	$7\frac{11}{32}$	$4\frac{7}{8}$	1.421	0.447	
13	29.14	$7\frac{21}{32}$	$4\frac{3}{4}$	1.419	0.446	$6\frac{5}{8}$
13	32.32	$8\frac{1}{32}$	$4\frac{1}{8}$	1.402	0.494	
16	29.40	$7\frac{3}{4}$	$4\frac{7}{8}$	1.420	0.450	
15	29.20	$7\frac{11}{32}$	$4\frac{7}{8}$	1.421	0.447	
13	30.25	$7\frac{11}{16}$	$4\frac{5}{8}$	1.415	0.463	$6\frac{5}{8}$
13	30.64	$7\frac{1}{2}$	$4\frac{7}{8}$	1.394	0.469	
16	31.22	$7\frac{11}{32}$	$4\frac{7}{8}$	1.391	0.478	
15	31.03	$7\frac{11}{32}$	$4\frac{7}{8}$	1.392	0.475	$6\frac{5}{8}$
13	32.08	$7\frac{11}{16}$	$4\frac{5}{8}$	1.386	0.491	
16	31.22	$7\frac{3}{8}$	$4\frac{7}{8}$	1.391	0.478	

(continued)

Table 4.6.6 (continued)

Box Tong Space *	Adjusted Weight	Minimum Tool Joint O.D. for Prem. Class	Drift Diameter	Capacity	Displacement	Size O.D.
in.	lb/ft	in.	in.	US gal/ft	US gal/ft	in.
15	31.03	$7\frac{11}{32}$	$4\frac{7}{8}$	1.392	0.475	$6\frac{5}{8}$
13	32.08	$7\frac{3}{4}$	$4\frac{5}{8}$	1.386	0.491	
16	31.22	$7\frac{15}{32}$	$4\frac{7}{8}$	1.391	0.478	
15	31.03	$7\frac{11}{32}$	$4\frac{7}{8}$	1.392	0.475	$6\frac{5}{8}$
13	34.14	8	$4\frac{1}{8}$	1.373	0.522	
16	31.22	$7\frac{23}{32}$	$4\frac{7}{8}$	1.391	0.478	
15	31.03	$7\frac{11}{32}$	$4\frac{7}{8}$	1.392	0.475	
13	30.97	$7\frac{23}{32}$	$4\frac{3}{4}$	1.390	0.474	$6\frac{5}{8}$
13	34.14	$8\frac{1}{32}$	$4\frac{1}{8}$	1.373	0.522	
16	31.22	$7\frac{3}{4}$	$4\frac{7}{8}$	1.391	0.478	
15	31.03	$7\frac{11}{32}$	$4\frac{7}{8}$	1.392	0.475	
13	32.08	$7\frac{11}{16}$	$4\frac{5}{8}$	1.386	0.491	$6\frac{5}{8}$
13	34.14	$8\frac{1}{8}$	$4\frac{1}{8}$	1.373	0.522	
16	31.22	$7\frac{27}{32}$	$4\frac{7}{8}$	1.391	0.478	
15	31.03	$7\frac{7}{16}$	$4\frac{7}{8}$	1.392	0.475	
13	32.08	$7\frac{25}{32}$	$4\frac{5}{8}$	1.386		

*Box Tong Space

Table 4.6.7 Drill Pipe Material Properties [1]

Material	Yield Strength (psi)		Tensile Strength (psi)	Elongation
	Min	Max	Min	Min
Pipe grade				
E	75,000	105,000	100,000	See 7.2.4
X	95,000	125,000	105,000	See 7.2.4
G	105,000	135,000	115,000	See 7.2.4
S	135,000	165,000	145,000	See 7.2.4
Tool joint	120,000		140,000	13%

(text continued from page 4-137)

connections were established to reduce the number of tool joint types. Tool joints with the same thread form and pitch diameter at the gage point are usually interchangeable. Drill pipe manufacturers may also offer proprietary tool joints, such as different variations of double shoulder tool joints for increased torsional strength and tool joints with special threads.

Tool joint OD and ID. The OD and ID (in., mm) of the tool joint dictates its strength. Generally, the torsional strength of the box is dictated by the tool joint OD and that of the pin is dictated by the ID. Because the box strength does not depend on its ID, the box ID of most drill pipe assemblies is the maximum available regardless of the pin ID. The tool joint OD affects the fish ability of the length and

the equivalent circulation density (ECD). The tool joint ID affects the drilling fluid pressure losses in the string.

Tool joint tong length. The tong length (in, mm) is the length of the cylindrical portion of the tool joint where the tongs grip. API specifies tong lengths for API connections. Drill pipe is often produced with tong lengths greater than the API-specified tong length, usually in 1-inch increments, to allow more thread recuts. Each time a damaged thread is recut, about $\frac{3}{4}$ in. of tong space is lost.

Drill pipe length. API-defined drill pipe is available in three standard lengths: range 1, range 2, and range 3, which are approximately 22, $31\frac{1}{2}$, and 45 ft long, respectively.

Table 4.6.8 *Dimensional Properties of API Drill Pipe Tubes [6]*

Drill Pipe Size	Nominal Wt (lb/ft)	Wall Thickness	Plain End Wt (lb/ft)	ID (in.)	Section Area Body of Pipe A (in.^{2})	Polar Section Modulus Z (in.^{3})
$2\frac{3}{8}$	4.85	0.109	4.43	1.995	1.304	1.321
$2\frac{3}{8}$	6.65	0.280	6.26	1.815	1.843	1.733
$2\frac{7}{8}$	6.85	0.217	6.16	2.441	1.812	2.241
$2\frac{7}{8}$	10.40	0.362	9.72	2.151	2.858	3.204
$3\frac{1}{2}$	9.50	0.254	8.81	2.992	2.590	3.923
$3\frac{1}{2}$	13.30	0.368	12.31	2.764	3.621	5.144
$3\frac{1}{2}$	15.50	0.449	14.63	2.602	4.304	5.847
4	11.85	0.262	10.46	3.476	3.077	5.400
4	14.00	0.330	12.93	3.340	3.805	6.458
4	15.70	0.380	14.69	3.240	4.322	7.157
$4\frac{1}{2}$	13.75	0.271	12.24	3.958	3.600	7.184
$4\frac{1}{2}$	16.60	0.337	14.98	3.826	4.407	8.543
$4\frac{1}{2}$	20.00	0.430	18.69	3.640	5.498	10.232
$4\frac{1}{2}$	22.82	0.500	21.36	3.500	6.283	11.345
5	16.25	0.296	14.87	4.408	4.374	9.718
5	19.50	0.362	17.93	4.276	5.275	11.415
5	25.60	0.500	24.03	4.000	7.069	14.491
$5\frac{1}{2}$	19.20	0.304	16.87	4.892	4.962	12.221
$5\frac{1}{2}$	21.90	0.361	19.81	4.778	5.828	14.062
$5\frac{1}{2}$	24.70	0.415	22.54	4.670	6.630	15.688
$5\frac{7}{8}$	23.40	0.361	23.40	5.153	6.254	16.251
$5\frac{7}{8}$	24.17	0.415	24.17	5.045	7.119	18.165
$6\frac{5}{8}$	25.20	0.330	22.19	5.965	6.526	19.572
$6\frac{5}{8}$	27.70	0.362	24.19	5.901	7.123	21.156

Hardbanding. Drill pipe is often produced with hardbanding on the box. The hardbanding reduces the wear rate of the tool joints, reduces casing wear, and reduces the frictional drag of the pipe rotating and sliding in the hole. Sometimes, hardbanding is applied to the pin as well as the box.

4.6.2.1 Classification of Drill Pipe
Torsional, tensile, collapse, and internal pressure data for new drill pipe are given in Tables 4.6.10, 4.6.11, and 4.6.12. Dimensional data is given in Table 4.6.13.

Drill pipe is classified based on wear, mechanical damage and corrosion of the tool joint, and tube. The classifications are premium, class 2, and class 3. Descriptions and specifications for the classes are found in Table 4.6.14 and API RP7G. The pipe is marked with colored bands as shown in Figure 4.6.12.

The torsion, tension, collapse and internal pressure resistance for new, premium class, and class 2 drill pipe are shown in Tables 4.6.10, 4.6.11, and 4.6.12, respectively.

Calculations for the minimum performance properties of drill pipe are based on formulas given in Appendix A of API RP7G. The numbers in Tables 4.6.10, 4.6.11, and 4.6.12 have been calculated for uniaxial sterss, (e.g., torsion only, tension only). The tensile strength is decreased when the drill string is subjected to both axial tension and torque; the collapse pressure rating is also decreased when the drill pipe is simultaneously affected by collapse pressure and tensile loads.

4.6.2.2 Load Capacity of Drill Pipe
In normal drilling operations and in operations such as DST or washover, drill pipe is subjected to combined effects of stresses.

To evaluate the load capacity of drill pipe (e.g., allowable tensile load while simultaneously applying torque), the maximum distortion energy theory is usually applied. This theory is in good agreement with experiments on ductile materials such as steel. According to this theory, the equivalent stress may be calculated from the following formula [10]:

$$2\sigma_e^2 = (\sigma_z - \sigma_t)^2 + (\sigma_t - \sigma_r)^2 + (\sigma_r - \sigma_z)^2 + 6\tau^2 \qquad [4.6.7]$$

where σ_e = equivalent stress in psi
σ_z = axial stress in psi ($\sigma_z > 0$ for tension, $\sigma_z < 0$ for compression)
σ_t = tangential stress in psi ($\sigma_t > 0$ for tension, $\sigma_t < 0$ for compression)
σ_r = radial stress in psi (usually neglected for the drill pipe strength analysis)
τ = shear stress in psi

Yielding of pipe does not occur provided that the equivalent stress is less than the yield strength of the drill pipe. For practical calculations, the equivalent stress is taken to be equal to the minimum yield strength of the pipe as specified by API. The stresses being considered in Equation 4.6.7 are effective stresses that exist beyond any isotropic stresses caused by hydrostatic pressure of the drilling fluid.

Consider a case in which the drill pipe is exposed to an axial load (P) and a torque (T). The axial stress (σ_z) and the

(text continued on page 4-185)

Table 4.6.9 *Interchangeability Chart for Tool Joints*

Style	Common Name	Size	Threads per inch	Taper (in./ft)	Thread Form	Same As or Interchangeable With
Internal flush (IF)		$2\frac{3}{8}$	4	2	V-0.065 (V-0.038R)	$2\frac{7}{8}$ slim hole NC26
		$2\frac{7}{8}$	4	2	V-0.065 (V-0.038R)	$3\frac{1}{2}$ slim hole NC31
		$3\frac{1}{2}$	4	2	V-0.065 (V-0.038R)	$4\frac{1}{2}$ slim hole NC38
		4	4	2	V-0.065 (V-0.038R)	$4\frac{1}{2}$ XH NC46
		$4\frac{1}{2}$	4	2	V-0.065 (V-0.038R)	5 XH $5\frac{1}{2}$ double streamline NC50
Full hole (FH)		4	4	2	V-0.065 (V-0.038R)	$4\frac{1}{2}$ double streamline NC40
Extra hole (XH, EH)		$2\frac{7}{8}$	4	2	V-0.065 (V-0.038R)	$3\frac{1}{2}$ double streamline
		$3\frac{1}{2}$	4	2	V-0.065 (V-0.038R)	4 slim hole $4\frac{1}{2}$ external flush
		$4\frac{1}{2}$	4	2	V-0.065 (V-0.038R)	4 internal flush NC46
		5	4	2	V-0.065 (V-0.038R)	$4\frac{1}{2}$ internal flush NC50 $5\frac{1}{2}$ double streamline
Slim hole (SH)		$2\frac{7}{8}$	4	2	V-0.065 (V-0.038R)	$2\frac{3}{8}$ IF NC26
		$3\frac{1}{2}$	4	2	V-0.065 (V-0.038R)	$2\frac{7}{8}$ IF NC31
		4	4	2	V-0.065 (V-0.038R)	$3\frac{1}{2}$ XH $4\frac{1}{2}$ IF
		$4\frac{1}{2}$	4	2	V-0.065 (V-0.038R)	$3\frac{1}{2}$ IF NC38
Double sreamline (DSL)		$3\frac{1}{2}$	4	2	V-0.065 (V-0.038R)	$2\frac{7}{8}$ XH
		$4\frac{1}{2}$	4	2	V-0.065 (V-0.038R)	4 FH NC40
		$5\frac{1}{2}$	4	2	V-0.065 (V-0.038R)	$4\frac{1}{2}$ IF 5 XH NC50
Numbered conn (NC)		26	4	2	V-0.38R	$2\frac{3}{8}$ IF $2\frac{7}{8}$ SH
		31	4	2	V-0.38R	$2\frac{7}{8}$ IF $3\frac{1}{2}$ SH
		38	4	2	V-0.38R	$3\frac{1}{2}$ IF $4\frac{1}{2}$ SH
		40	4	2	V-0.38R	4 FH $4\frac{1}{2}$ DSL
		46	4	2	V-0.38R	4 IF $4\frac{1}{2}$ XH
		50	4	2	V-0.38R	$4\frac{1}{2}$ IF 5 XH $5\frac{1}{2}$ DSL
External flush (EF)		$4\frac{1}{2}$	4	2	V-0.065 (V-0.038R)	4SH $3\frac{1}{2}$ XH

Table 4.6.10 New Drill Pipe: Torsional, Tensile, Collapse, and Internal Pressure Data

OD	Nominal Wt (lbs/ft)	Wall Thickness (in.)	Section Area (in.²)	Polar Section Modulus (in.³)	Torsional Yield Strength (ft-lb)				Tensile Yield Strength (lb)				Collapse Pressure (psi)				Internal Pressure (psi)			
					E-75	X-95	G-105	S-135	E-75	X-95	G-105	S-135	E-75	X-95	G-105	S-135	E-75	X-95	G-105	S-135
2 3/8	4.85	0.190	1.304	1.321	4,763	6,033	6,668	8,574	97,817	123,902	136,944	176,071	11,040	13,984	15,456	19,035	10,500	13,300	14,700	18,900
2 3/8	6.65	0.280	1.843	1.733	6,250	7,917	8,751	11,251	138,214	175,072	193,500	248,786	15,599	19,759	21,839	28,079	15,474	19,600	21,663	27,853
2 7/8	6.85	0.217	1.812	2.241	8,083	10,238	11,316	14,549	135,902	172,143	190,263	244,624	1,503	2,189	2,580	3,941	9,907	12,548	13,869	17,832
2 7/8	10.4	0.362	2.858	3.204	11,554	14,635	16,176	20,798	214,344	271,503	300,082	385,820	16,509	20,911	23,112	29,716	16,526	20,933	23,137	29,747
3 1/2	9.5	0.254	2.590	3.923	14,146	17,918	19,805	25,463	194,264	246,068	271,970	349,676	10,001	12,077	13,055	15,748	9,525	12,065	13,335	17,145
3 1/2	13.3	0.368	3.621	5.144	18,551	23,498	25,972	33,392	271,569	343,988	380,197	488,825	14,113	17,877	19,758	25,404	13,800	17,480	19,320	28,840
3 1/2	15.5	0.449	4.304	5.847	21,086	26,708	29,520	37,954	322,775	408,848	451,885	580,995	16,774	21,247	23,484	30,194	16,838	21,328	23,573	30,308
4	11.85	0.262	3.077	5.400	19,474	24,668	27,264	35,054	230,755	292,290	323,057	415,360	8,381	9,978	10,708	12,618	8,597	10,889	12,036	15,474
4	14.00	0.330	3.805	6.458	23,288	29,498	32,603	41,918	285,359	361,454	399,502	513,646	11,354	14,382	15,896	20,141	10,828	13,716	15,159	19,491
4	15.70	0.380	4.322	7.157	25,810	32,692	36,134	46,458	324,118	410,550	453,765	583,413	12,896	16,335	18,055	23,213	12,469	15,794	17,456	22,444
4 1/2	13.75	0.271	3.600	7.184	25,907	32,816	36,270	46,633	270,034	342,043	378,047	486,061	7,173	8,412	8,956	10,283	7,904	10,012	11,066	14,228
4 1/2	16.60	0.337	4.407	8.543	30,807	39,022	43,130	55,453	330,558	418,707	462,781	595,004	10,392	12,765	13,825	16,773	9,829	12,450	13,761	17,693
4 1/2	20.00	0.430	5.498	10.232	36,901	46,741	51,661	66,421	412,358	522,320	577,301	742,244	12,964	16,421	18,149	23,335	12,542	15,886	17,558	22,575
4 1/2	22.82	0.500	6.283	11.345	40,912	51,821	57,276	73,641	471,239	596,903	659,734	848,230	14,815	18,765	20,741	26,667	14,583	18,472	20,417	26,250
5	16.25	0.296	4.374	9.718	35,044	44,389	49,062	63,079	328,073	415,559	459,302	590,531	6,938	8,108	8,616	9,831	7,770	9,842	10,878	13,986
5	19.50	0.362	5.275	11.415	41,167	52,144	57,633	74,100	395,595	501,087	553,833	712,070	9,962	12,026	12,999	15,672	9,503	12,037	13,304	17,105
5	25.60	0.250	3.731	8.441	30,439	38,556	42,614	54,790	279,798	354,411	391,717	503,637	4,831	5,377	5,562	6,090	6,563	8,313	9,188	11,813
5 1/2	19.20	0.304	4.962	12.221	44,074	55,827	61,703	79,332	372,181	471,429	521,053	669,925	6,039	6,942	7,313	8,093	7,255	9,189	10,156	13,058
5 1/2	21.90	0.361	5.828	14.062	50,710	64,233	70,994	91,278	437,116	553,681	611,963	786,809	8,413	10,019	10,753	12,679	8,615	10,912	12,061	15,507
5 1/2	24.70	0.415	6.630	15.688	56,574	71,660	79,204	101,833	497,222	629,814	696,111	894,999	10,464	12,933	14,013	17,023	9,903	12,544	13,865	17,826
5 7/8	23.40	0.361	6.254	16.251	58,605	74,233	82,047	105,489	469,013	594,083	656,619	844,224	7,453	8,775	9,362	10,825	8,065	10,216	11,291	14,517
5 7/8	24.17	0.415	7.119	18.165	65,508	82,977	91,711	117,915	533,890	676,261	747,446	961,002	9,558	11,503	12,414	14,892	9,271	11,744	12,980	16,688
6 5/8	25.20	0.330	6.526	19.572	70,580	89,402	98,812	127,044	489,464	619,988	685,250	881,035	4,788	5,321	5,500	6,036	6,538	8,281	9,153	11,768
6 5/8	27.70	0.362	7.123	21.156	76,295	96,640	106,813	137,330	534,198	676,651	747,877	961,556	5,894	6,755	7,103	7,813	7,172	9,084	10,040	12,909

Table 4.6.11 Premium Drill Pipe: Torsional, Tensile, Collapse, and Internal Pressure Data

OD	Nominal Wt (lb/ft)	New Wall Thickness (in.)	Premium Wall Thickness (in.)	Premium OD (in.)	ID (in.)	Section Area (in.²)	Polar Section Modulus (in.³)	Torsional Yield Strength (ft-lb)				Tensile Yield Strength (lb)				Collapse Pressure (psi)				Internal Pressure (psi)			
								E-75	X-95	G-105	S-135	E-75	X-95	G-105	S-135	E-75	X-95	G-105	S-135	E-75	X-95	G-105	S-135
2 3/8	4.85	0.190	0.152	2.299	1.995	1.025	1.033	3,725	4,719	5,215	6,705	76,875	97,375	107,625	138,375	8,522	10,161	10,912	12,891	9,600	13,400	14,700	17,280
2 3/8	6.65	0.280	0.224	2.263	1.815	1.434	1.334	4,811	6,093	6,735	8,659	107,550	136,313	150,570	193,590	13,378	16,945	18,729	24,080	14,147	19,806	21,663	25,465
2 7/8	6.85	0.217	0.174	2.788	2.441	1.425	1.756	6,332	8,020	8,865	11,397	106,875	135,465	149,625	192,375	7,640	9,017	9,633	11,186	9,057	12,680	13,869	16,303
2 7/8	10.4	0.362	0.290	2.730	2.151	2.220	2.456	8,858	11,220	12,401	15,945	166,500	210,945	233,100	299,700	14,223	18,016	19,912	25,602	15,110	21,153	23,137	27,197
3 1/2	9.5	0.254	0.203	3.398	2.992	2.039	3.076	11,094	14,052	15,531	19,969	152,925	193,774	214,095	275,265	7,074	8,284	8,813	10,093	8,709	12,192	13,335	15,675
3 1/2	13.3	0.368	0.294	3.353	2.764	2.828	3.982	14,361	18,191	20,106	25,850	212,100	268,723	296,940	381,780	12,015	15,218	16,820	21,626	12,617	17,664	19,320	22,711
3 1/2	15.5	0.449	0.359	3.320	2.602	3.341	4.477	16,146	20,452	22,605	29,063	250,575	317,452	350,805	451,035	14,472	18,331	20,260	26,049	15,394	21,552	23,573	27,710
4	11.85	0.262	0.210	3.895	3.476	2.426	4.245	15,310	19,392	21,434	27,557	181,950	230,554	254,730	327,510	5,704	6,508	6,827	7,445	7,860	11,004	12,036	14,148
4	14.00	0.330	0.264	3.868	3.340	2.989	5.046	18,196	23,048	25,474	32,753	224,175	283,963	313,845	403,515	9,012	10,795	11,622	13,836	9,900	13,860	15,159	17,820
4	15.70	0.380	0.304	3.848	3.240	3.384	5.564	20,067	25,418	28,094	36,120	253,800	321,544	355,320	456,840	10,914	13,825	15,190	18,593	11,400	15,960	17,456	20,520
4 1/2	13.75	0.271	0.217	4.392	3.958	2.843	5.658	20,403	25,844	28,564	36,725	213,225	270,127	298,515	383,805	4,686	5,190	5,352	5,908	7,227	10,117	11,066	13,008
4 1/2	16.60	0.337	0.270	4.365	3.826	3.468	6.694	24,139	30,576	33,795	43,451	260,100	329,460	364,140	468,180	7,525	8,868	9,467	10,964	8,987	12,581	13,761	16,176
4 1/2	20.00	0.430	0.344	4.328	3.640	4.305	7.954	28,683	36,332	40,157	51,630	322,875	409,026	452,025	581,175	10,975	13,901	15,350	18,806	11,467	16,053	17,558	20,640
4 1/2	22.82	0.500	0.400	4.300	3.500	4.900	8.759	31,587	40,010	44,222	56,856	367,500	465,584	514,500	661,500	12,655	16,030	17,718	22,780	13,333	18,667	20,417	24,000
5	16.25	0.296	0.237	4.882	4.408	3.455	7.655	27,607	34,969	38,650	49,693	259,125	328,263	362,775	466,425	4,490	4,935	5,067	5,661	7,104	9,946	10,878	12,787
5	19.50	0.362	0.290	4.855	4.276	4.153	8.953	32,285	40,895	45,200	58,114	311,475	394,612	436,065	560,655	7,041	8,241	8,765	10,029	8,688	12,163	13,304	15,638
5	25.60	0.250	0.200	4.900	4.500	2.953	6.669	24,049	30,462	33,668	43,287	221,475	280,544	310,065	398,655	2,954	3,282	3,387	3,470	6,000	8,400	9,188	10,800
5 1/2	19.20	0.304	0.243	5.378	4.892	3.923	9.640	34,764	44,035	48,670	62,576	294,225	372,730	411,915	529,605	3,736	4,130	4,336	4,714	6,633	9,286	10,156	11,939
5 1/2	21.90	0.361	0.289	5.356	4.778	4.597	11.054	39,864	50,494	55,809	71,754	344,775	436,721	482,685	620,595	5,730	6,542	6,865	7,496	7,876	11,027	12,061	14,177
5 1/2	24.70	0.415	0.332	5.334	4.670	5.217	12.290	44,320	56,139	62,048	79,776	391,275	495,627	547,785	704,295	7,635	9,011	9,626	11,177	9,055	12,676	13,865	16,298
5 7/8	23.40	0.361	0.289	5.731	5.153	4.937	12.793	46,134	58,437	64,588	83,042	370,275	469,044	518,385	666,495	4,922	5,495	5,694	6,204	7,374	10,323	11,291	13,273
5 7/8	24.17	0.415	0.332	5.709	5.045	5.608	14.255	51,408	65,116	71,971	92,534	420,600	532,785	588,840	757,080	6,699	7,798	8,269	9,368	8,477	11,867	12,980	15,258
6 5/8	25.20	0.330	0.264	6.493	5.965	5.166	15.464	55,766	70,637	78,072	100,379	387,450	490,790	542,430	697,410	2,931	3,252	3,353	3,429	5,977	8,368	9,153	10,759
6 5/8	27.70	0.362	0.290	6.480	5.901	5.632	16.691	60,191	76,242	84,268	108,345	422,400	535,063	591,360	760,320	3,615	4,029	4,222	4,562	6,557	9,180	10,040	11,803

Table 4.6.12 Class 2 Drill Pipe: Torsional, Tensile, Collapse, and Internal Pressure Data

OD	Nominal Wt (lb/ft)	New Wall Thickness (in.)	Class 2 Wall Thickness (in.)	Class 2 OD (in.)	ID (in.)	Section Area (in.²)	Polar Section Modulus (in.³)	Torsional Yield Strength (ft-lb)				Tensile Yield Strength (lb)				Collapse Pressure (psi)				Internal Pressure (psi)			
								E-75	X-95	G-105	S-135	E-75	X-95	G-105	S-135	E-75	X-95	G-105	S-135	E-75	X-95	G-105	S-135
2 3/8	4.85	0.190	0.133	2.261	1.995	0.889	0.894	3,224	4,083	4,513	5,802	66,686	84,469	93,360	120,035	6,852	7,996	8,491	9,664	8,400	10,640	11,760	15,120
2 3/8	6.65	0.280	0.196	2.207	1.815	1.238	1.145	4,130	5,232	5,782	7,434	92,871	117,636	130,019	167,167	12,138	15,375	16,993	21,849	12,379	15,680	17,331	22,282
2 7/8	6.85	0.217	0.152	2.745	2.441	1.237	1.521	5,484	6,946	7,677	9,871	92,801	117,548	129,922	167,043	6,055	6,963	7,335	8,123	7,925	10,039	11,095	14,265
2 7/8	10.4	0.362	0.253	2.658	2.151	1.914	2.105	7,591	9,615	10,627	13,663	143,557	181,839	200,980	258,403	12,938	16,388	18,113	23,288	13,221	16,746	18,509	23,798
3 1/2	9.5	0.254	0.178	3.348	2.992	1.770	2.666	9,612	12,176	13,457	17,302	132,793	168,204	185,910	239,027	5,544	6,301	6,596	7,137	7,620	9,652	10,668	13,716
3 1/2	13.3	0.368	0.258	3.279	2.764	2.445	3.429	12,365	15,663	17,312	22,258	183,398	232,304	256,757	330,116	10,858	13,753	15,042	18,396	11,040	13,984	15,456	19,872
3 1/2	15.5	0.449	0.314	3.231	2.602	2.879	3.834	13,828	17,515	19,359	24,890	215,967	273,558	302,354	388,741	13,174	16,686	18,443	23,712	13,470	17,062	18,858	24,246
4	11.85	0.262	0.183	3.843	3.476	2.108	3.683	13,282	16,823	18,594	23,907	158,132	200,301	221,385	284,638	4,310	4,702	4,876	5,436	6,878	8,712	9,629	12,380
4	14.00	0.330	0.231	3.802	3.340	2.591	4.364	15,738	19,935	22,034	28,329	194,363	246,193	272,108	349,853	7,295	8,570	9,134	10,520	8,663	10,973	12,128	15,593
4	15.70	0.380	0.266	3.772	3.240	2.929	4.801	17,315	21,932	24,241	31,166	219,738	278,334	307,633	395,528	9,531	11,468	12,374	14,840	9,975	12,635	13,965	17,955
4 1/2	13.75	0.271	0.190	4.337	3.958	2.471	4.912	17,715	22,439	24,801	31,887	185,390	234,827	259,546	333,702	3,397	3,845	4,016	4,287	6,323	8,010	8,853	11,382
4 1/2	16.60	0.337	0.236	4.298	3.826	3.010	5.798	20,908	26,483	29,271	37,634	225,771	285,977	316,080	406,388	5,951	6,828	7,185	7,923	7,863	9,960	11,009	14,154
4 1/2	20.00	0.430	0.301	4.242	3.640	3.726	6.862	24,747	31,346	34,645	44,544	279,501	354,035	391,302	503,103	9,631	11,598	12,520	15,033	10,033	12,709	14,047	18,060
4 1/2	22.82	0.500	0.350	4.200	3.500	4.233	7.532	27,161	34,404	38,026	48,891	317,497	402,163	444,496	571,495	11,458	14,514	16,042	20,510	11,667	14,778	16,333	21,000
5	16.25	0.296	0.207	4.822	4.408	3.004	6.648	23,974	30,368	33,564	43,154	225,316	285,400	315,442	405,568	3,275	3,696	3,850	4,065	6,216	7,874	8,702	11,189
5	19.50	0.362	0.253	4.783	4.276	3.605	7.758	27,976	35,436	39,166	50,356	270,432	342,547	378,605	486,778	5,514	6,262	6,552	7,079	7,602	9,629	10,643	13,684
5	25.60	0.250	0.175	4.850	4.500	2.570	5.799	20,913	26,490	29,279	37,644	192,766	244,170	269,873	346,979	2,248	2,370	2,374	2,374	5,250	6,650	7,350	9,450
5 1/2	19.20	0.304	0.213	5.318	4.892	3.412	8.377	30,208	38,264	42,291	54,375	255,954	324,208	358,335	460,717	2,835	3,128	3,215	3,265	5,804	7,351	8,125	10,447
5 1/2	21.90	0.361	0.253	5.283	4.778	3.993	9.589	34,582	43,804	48,415	62,247	299,533	379,409	419,346	539,160	4,334	4,733	4,899	5,465	6,892	8,730	9,649	12,405
5 1/2	24.70	0.415	0.291	5.251	4.670	4.527	10.644	38,383	48,619	53,737	69,090	339,534	430,076	475,347	611,160	6,050	6,957	7,329	8,115	7,923	10,035	11,092	14,261
5 7/8	23.40	0.361	0.253	5.658	5.153	4.291	11.105	40,049	50,729	56,069	72,088	321,861	407,691	450,605	579,350	3,608	4,023	4,215	4,553	6,452	8,172	9,033	11,613
5 7/8	24.17	0.415	0.291	5.626	5.045	4.869	12.356	44,559	56,441	62,382	80,206	365,201	462,588	511,282	657,362	5,206	5,863	6,105	6,561	7,417	9,395	10,384	13,351
6 5/8	25.20	0.330	0.231	6.427	5.965	4.496	13.448	48,497	61,430	67,896	87,295	337,236	427,166	472,131	607,026	2,227	2,343	2,346	2,346	5,230	6,625	7,322	9,414
6 5/8	27.70	0.362	0.253	6.408	5.901	4.899	14.505	52,308	66,257	73,231	94,154	367,454	465,442	514,436	661,418	2,765	3,037	3,113	3,148	5,737	7,267	8,032	10,327

Table 4.6.13 *Dimensional Data and Mechanical Properties Heavy-Weight Drill Pipe [Smith International]*

		Tube Dimensions				Mechanical Properties Tube Section		Tool Joint			Mechanical Properties			Approximate Weight Including Tube and Tool Joints (lb)	
Size (in.)	ID (in.)	Wall Thickness	Area (in.²)	Center Upset (in.²)	Elevator Upset (In.)	Tensile Yield (lb)	Torsional Yield (ft-lb)	OD (in.)	ID (in.)	Connection Size	Tensile Yield (lb)	Torsional Yield (ft-lb)	Makeup Torque (ft-lb)	Wt/ft	Wt/jt
3½	2¼	0.625	5.645	4	3⅝	310,475	18,460	4¾	2⅜	NC38	675,045	17,575	10,000	23.4	721
4	2 9/16	0.719	7.41	4½	4⅛	407,550	27,635	5¼	2 11/16	NC40	711,475	23,525	13,300	29.9	920
4½	2¾	0.875	9.965	5	4⅝	548,075	40,715	6¼	2⅞	NC60	1,024,500	38,800	21,800	41.1	1,265
5	3	1	12.566	5 5/12	5⅛	691,185	56,495	6⅜	3 1/16	NC50	1,266,000	51,375	29,200	50.1	1,543
5½	3⅜	1.063	14.812	6	5 11/16	814,660	74,140	7	3½	5-½ FH	1,349,365	53,080	32,800	57.8	1,770
6⅝	4½	1.063	18.567	7⅛	6¾	1,021,185	118,845	8	4⅝	6-⅝ FH	1,490,495	73,215	45,800	71.3	2,193
Range 3															
4½	2¾	0.875	9.965	5	4⅝	548,075	40,715	6¼	2⅞	NC 46	1,024,500	38,800	21,800	39.9	1,750
5	3	1	12.566	5½	5⅛	691,185	56,495	6⅜	3 1/16	NC 50	1,266,000	51,375	29,200	48.5	2,130

Capacity – Volume of Fluid Necessary to Fill the ID

Size (in.)	Gal per Joint*	BBL per Joint*	Gal per 100 ft	BBL per 100 ft
3½	6.29	0.150	21.0	0.500
4	8.13	0.194	27.1	0.645
4½	9.37	0.223	31.2	0.743
5	11.14	0.265	37.1	0.883
5½	14.08	0.335	46.9	1.117
6⅝	24.79	0.590	82.6	1.967

Displacement – Open Ended (Metal Displacement Only)

Size (in.)	Gal per Joint*	BBL per Joint*	Gal per 100 ft	BBL per 100 ft
3½	10.62	0.253	35.4	0.843
4	12.62	0.300	42.1	1.002
4½	18.82	0.448	62.7	1.493
5	22.62	0.539	75.4	1.796
5½	26.48	0.630	88.3	2.102
6⅝	32.77	0.780	109.2	2.600

Table 4.6.14 *Classification of Used Drill Pipe [4]*

Classification Condition	Premium Class, Two White Bands	Class 2, One Yellow Band
Exterior Conditions		
OD wear	Remaining wall not less than 80%	Remaining wall not less than 70%
Dents and mashes	OD not less than 97%	OD not less than 96%
Crushing and necking	OD not less than 97%	OD not less than 96%
Slip area: cuts and gouges	Depth not more than 10% of average adjacent wall, and remaining wall not less than 80%	Depth not more than 20% of average adjacent wall, and remaining wall not less than 80% for transverse (70% for longitudinal)
Stretching	OD not less than 97%	OD not less than 96%
String shot	OD not more than 103%	OD not more than 104%
External corrosion	Remaining wall not less than 80%	Remaining wall not less than 70%
Longitudinal cuts and gouges	Remaining wall not less than 80%	Remaining wall not less than 70%
Cracks	None	None
Internal Conditions		
Corrosion pitting	Remaining wall not less than 80%	Remaining wall not less than 70%
Erosion and internal wall wear	Remaining wall not less than 80%	Remaining wall not less than 70%
Cracks	None	None

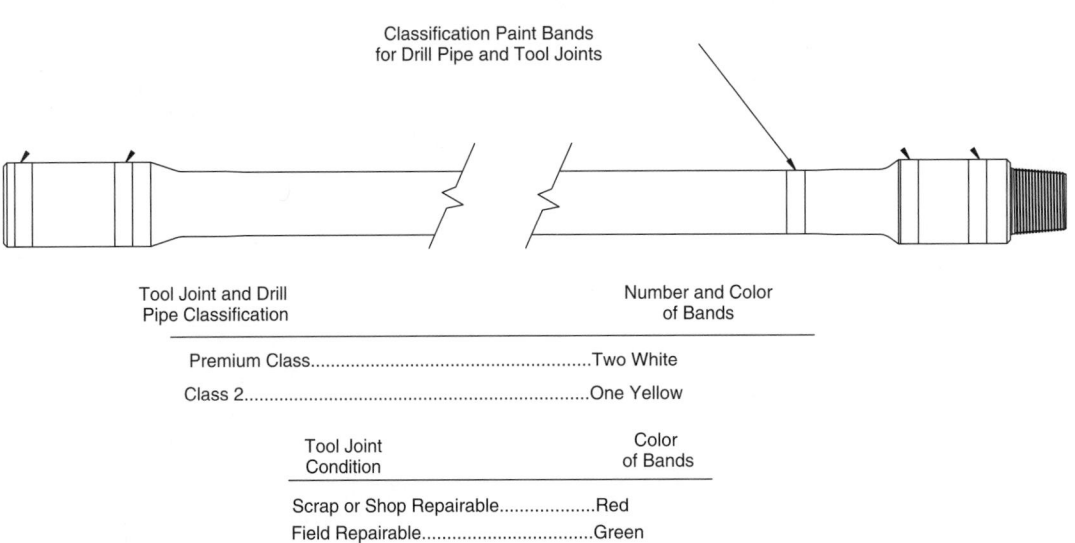

Tool Joint
Condition Bands

Classification Paint Bands
for Drill Pipe and Tool Joints

Tool Joint and Drill Pipe Classification	Number and Color of Bands
Premium Class	Two White
Class 2	One Yellow

Tool Joint Condition	Color of Bands
Scrap or Shop Repairable	Red
Field Repairable	Green

Figure 4.6.12

(text continued from page 4-178)
shear stress (τ) are given by the following formulas:

$$\sigma_z = \frac{P}{A} \qquad [4.6.8]$$

$$\tau = \frac{T}{Z} \qquad [4.6.9]$$

where P = axial load (lb)
  A = cross-sectional area of drill pipe (in.^{2})
  T = torque (in.-lb)
  Z = polar section modulus of drill pipe (in.^{3})
  $Z = 2J/D_{dp}$
  $J = (\pi/32)(D_{dp}^4 - d_{dp}^4)$
  D_{dp} = outside diameter of drill pipe (in.)
  d_{dp} = inside diameter of drill pipe (in.)

Substituting Equation 4.6.8 and Equation 4.6.9 into Equation 4.6.7 and putting $\sigma_z = Y_m$, $\sigma_t = 0$, (tangential stress equals zero in this case), the following formulas are obtained:

$$P = A\left[Y_m^2 - 3\left(\frac{T}{Z}\right)^2\right]^{\frac{1}{2}} \qquad [4.6.10]$$

$$P = \left[P_t^2 - 3\left(\frac{A \cdot T}{Z}\right)^2\right]^{\frac{1}{2}} \qquad [4.6.11]$$

where $P_t = Y_m$, A = tensile load capacity of drill pipe in uniaxial tensile stress (lb).

Equation 4.6.11 permits calculation of the tensile load capacity when the pipe is subjected to rotary torque (T).

Example 4.6.5
Determine the tensile load capacity of a $4\frac{1}{2}$ in., 16.60 lb/ft grade X-95 premium drill pipe subjected to a rotary torque of 12,000 ft-lb if the required safety factor is 2.0.

Solution
From Table 4.6.11, the following data are obtained:

  A = 3.468 in.^{2}
  P_t = 329, 460 lb
  Z = 6.694 in.^{4}

Using Equation 4.6.11,

$$P = \left[(329,460)^2 - 3\left(\frac{3.468 \times 144,000}{6.694}\right)^2\right]^{\frac{1}{2}} = 303,063$$

Because of the safety factor of 1.2, the tensile capacity of the drill pipe is 303,063/1.2 = 252,552 lb.

Example 4.6.6
Calculate the maximum value of a rotary torque that may be applied to the drill pipe as specified in Example 4.6.5 if the actual working tension load P = 275,000 lb. (e.g., pulling and trying to rotate a differentially stuck drill string).

Solution
From Equation 4.6.11, the magnitude of rotary torque is

$$T = \frac{Z}{A}\left(\frac{P_T^2 - P^2}{3}\right)^{\frac{1}{2}}$$

so

$$T = \frac{6.694}{3.468}\left[\frac{329,460^2 - 275,000^2}{3}\right]^{\frac{1}{2}}$$

$$= 202,194 \text{ in./lb or } 16,850 \text{ ft/lb}$$

Caution: No safety factor is included in this example calculation. Additional checking must be done if the obtained

value of the torque is not greater than the recommended makeup torque for tool joints.

During normal rotary drilling processes, because of frictional pressure losses and the pressure drop across the bit nozzles, the pressure inside the drill string is greater than that in the annulus outside drill string. The greatest difference between these pressures is at the surface.

If the drill string is thought to be a thin-walled cylinder with closed ends, the drill pipe pressure produces the axial stress and tangential stress given by the following formulas (for stress calculations, the pressure in the annulus may be ignored):

$$\sigma_a = \frac{P_{dp}D_{dp}}{4t} \qquad [4.6.12]$$

$$\sigma_t = \frac{P_{dp}D_{dp}}{2t} \qquad [4.6.13]$$

where σ_a = axial stress (psi)
  σ_t = tangential stress (psi)
  P_{dp} = internal drill pipe pressure (psi)
  D_{dp} = outsdie diameter of drill pipe (in.)
  t = wall thickness of drill pipe (in.)

Substituting Equations 4.6.12, 4.6.13, and 4.6.8 into Equation 4.6.7 and solving for the tensile load capacity of drill pipe yields

$$P = \left[P_t^2 - \frac{3}{16}\left(\frac{P_{dp}D_{dp}A}{t}\right)^2 - 3\left(\frac{AT}{Z}\right)^2\right]^{\frac{1}{2}} \qquad [4.6.14]$$

Example 4.6.7
Find the tensile load capacity of a 5-in., 19.50 lb/ft S-135 premium class drill pipe with an internal pressure of 3000 psi and an applied torque of 15,000 ft-lb.

Solution
From Table 4.6.11, 5-in. premium class pipe has a wall thickness of 0.290 in. and an OD of 4.855 in. The cross-sectional area is 4.153 in.^{2} The polar section modulus is 8.953 in.^{3} The tensile capacity is 560,655 lb. Using Equation 4.6.14,

$$P = \left\{560,655^2 - \frac{3}{16}\left[\frac{(3000)(4.855)(4.153)}{0.290}\right]^2 \right.$$

$$\left. - 3\left[\frac{(4.513)(15,000)(12)}{8.953}\right]^2\right\}^{\frac{1}{2}}$$

$$= 534,099 \text{ lb}$$

The reduction in tensile capacity of the drill pipe is $560,655 - 534,099 = 26,556$ lb. For practical purpose, depending on drilling conditions, a reasonable safety factor should be applied.

During DST operations, the drill pipe may be affected by a combined effect of collapse pressure and tensile load. For such a case,

$$\frac{\sigma_t}{Y_m} = \frac{P_{cc}}{P_c} \qquad [4.6.15]$$

or

$$\sigma_t = Y_m\frac{P_{cc}}{P_t} \qquad [4.6.16]$$

where P_c = minimum collapse pressure resistance. (psi) (see Table 4.6.11)
  P_{cc} = corrected collapse pressure resistance for effect of tension (psi)
  Y_m = minimum yield strength of pipe (psi)

Substituting Equation 4.6.16 in Equation 4.6.8 and solving for P_{cc} yields (note: $\sigma_r = 0$, $\tau = 0$, $\sigma_c = Y_m$)

$$P_{cc} = P_c \left\{ \left[-\left(\frac{P}{2AY_m} \right)^2 \right]^{\frac{1}{2}} - \frac{P}{2AYm} \right\} \qquad [4.6.17]$$

or

$$P_{cc} = P_c \left\{ \left[1 - 0.75 \left(\frac{\sigma_z}{Y_m} \right)^2 \right]^{\frac{1}{2}} - 0.5 \frac{\sigma_z}{Y_m} \right\} \qquad [4.6.18]$$

Equation 4.6.18 indicates that increased tensile load results in decreased collapse pressure resistance. The decrement of collapse pressure resistance during normal DST operations is relatively small; nevertheless, under certain conditions, it may be considerable.

Example 4.6.8
Determine if the drill pipe is strong enough to satisfy the safety factor on collapse of 1.1 for the DST conditions as below:

- Drill pipe: $4\frac{1}{2}$ in., 16.60 lb/ft, G-105, premium class
- Drilling fluid with a density of 12 lb/gal and drill pipe empty inside
- Packer set at a depth of 8,500 ft
- Tension load of 45,000 lb applied the drill pipe

Solution
From Table 4.6.11, the collapse pressure resistance in uniaxial state of stress (P_c) is 9,467 psi. The cross-sectional area is 3.468 in^2. The axial tensile stress at the packer level is

$$\sigma_z = \frac{45,000}{3.468} = 12,946 \cdot \text{psi}$$

The corrected collapse pressure resistance from Equation 4.6.18 is

$$P_c = 9,467 \left\{ \left[1 - 0.75 \left(\frac{12,946}{105,000} \right)^2 \right] - 0.5 \frac{12,946}{105,000} \right\}$$

$$= 8,775 \, \text{psi}$$

Hydrostatic pressure of the drilling fluid behind the drill string at the packer level is

$$P_h = (0.52)(12)(8,500) = 5,304 \, \text{psi}$$

$$\text{Safety factor} = 8,775/5,304 = 1.65$$

The required safety factor is 1.1; therefore, the pipe may be run empty.

4.6.2.3 Tool Joints
Individual lengths of drill pipe are connected to each other with threaded rotary shouldered connections called *tool joints* (Figure 4.6.13). Four types of tool joints (or thread forms) are used commonly used.

API tool joints. Tool joints with specifications defined by the API include all NC connections, $5\frac{1}{2}$ FH, $6\frac{5}{8}$ FH, and API regular. connections. Detailed information on the dimensional properties of these tool joints can be found in API Spec 7 and Section B of the IADC Drilling Manual.

Single shoulder non-API tool joints. For tool joints similar to (and sometimes interchangeable with) API tool joints, detailed information on the dimensional properties can be found in Section B of the IADC Drilling Manual. Examples of non-API tool joints are extra hole (EH or XH), open hole (OH), and many others. Table 4.6.9 shows the interchangability of single-shoulder tool joints with API tool joints.

Double shoulder tool joints. Double-shoulder tool joints have a shoulder on the nose of the pin that contacts an internal shoulder in the box. These additional shoulders limit the amount of axial deformation in the pin and provide more torsional strength.

Other proprietary threads. There are other tool joints that do not fall into the previous categories.

Table 4.6.6 contains the dimensions and performance characteristics of many possible tool joint, pipe size, pipe weight, and pipe grade combinations.

4.6.2.4 Makeup Torque
Makeup toque provides a preload in the connection that prevents downhole makeup, shoulder separation, wobble, or relative movement between the pin and box when subjected to the varied service loads the pipe sees and when rotating in doglegs. The preload provides the bearing stress in the shoulders to provide the seal in the tool joint. Tapered threads are used on tool joints to make it easier to put the pin and the box when adding lengths of pipe to the string, the taper does not seal.

Tool joints are generally made up to about 60% of their yield torque. Makeup torque values are shown in Table 4.6.6 and in API RP7G. The makeup torque for tool joints is calculated to induce a tensile stress at the last engaged thread of the pin or counterbore section of the box of 72,000 psi. A tolerance of $\pm 10\%$ is often used to speed up adding additional lengths of pipe to the drill string. As a deterrent to tool joint fatigue failures, Grant Prideco, who produced Table 4.6.6, includes this footnote in the drill pipe table of their product catalogue: "The makeup torque of the tool joint is based on the lower of 60% of the tool joint torsional yield strength or the 'T3' value calculated per the equation paragraph A.8.3 of API RP7G. Minimum makeup torques of 50% of the tool joint torsional strength may also be used."

The T3 value is the torsional load required to produce additional makeup of the tool joint when the shoulders are separated by an external tensile load that produces yield stress in the tool joint pin. This unlikely load case would occur when the pipe is stuck.

An explanation of these formulas and dimensional values for tool joints needed to solve them can be found in API RP7G.

4.6.2.5 Heavy-Weight Drill Pipe
Heavy-weight drill pipe with wall thicknesses of approximately 1 in. is frequently used for drilling vertical and directional holes (see Figure 4.6.14). It is often placed directly above the drill collars in vertical holes because it has been found to reduce the rate of fatigue failures in drill pipe just above the collars.

The best performance of individual members of the drill string is achieved when the bending stress ratio (BSR) of adjoining lengths of pipe is less than 5.5 [2]. BSR in this case is defined as a ratio of the bending section moduli of two connecting members (e.g., between the top most drill collar and the pipe just above it). For severe drilling conditions such as hole enlargement, corrosive environment, or hard formations, reduction of the BSR to 3.5 helps to reduce the frequency of drill pipe failure. To maintain the BSR of less than 5.5 between adjacent members of the drill string, the drill collar string must sometimes be made up of collars with different sizes. Heavy-weight pipe can be used at the transition from drill collars to drill pipe to bring the BSR into the desired range.

Sometimes, heavy-weight pipe is used instead of drill collars to provide weight to the bit. Pipe can be used because

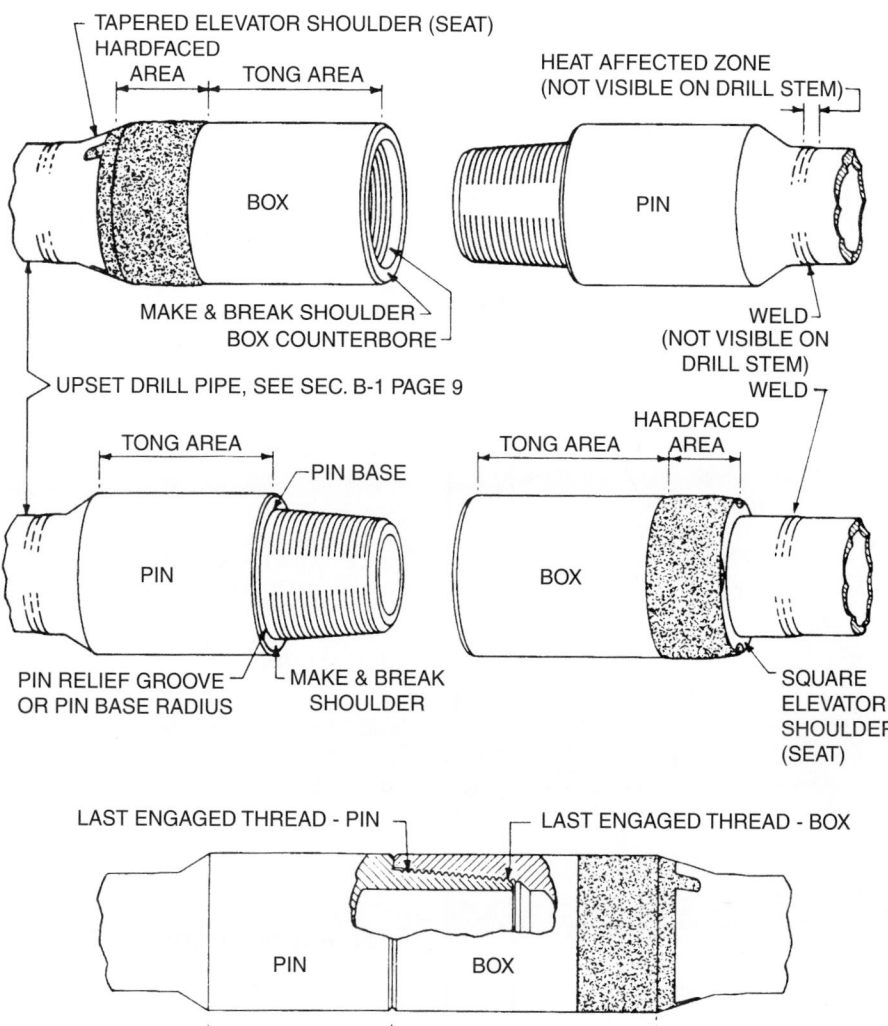

Figure 4.6.13 *Tool joint nomenclature [30].*

drill collars create excessive torque and drag or because there may be a deferential sticking problem. In directional holes, heavy-weight pipe is often used to provide weight to the bit. In many directional holes in which the deviation from vertical is too much for the pipe to slide, heavy-weight pipe is placed above the build zone.

The dimensional and mechanical properties of heavy-weight drill pipe manufactured by the Drilco division of Smith International are given in Table 4.6.13.

Example 4.6.9
Calculate the required length of $4\frac{1}{2}$ in. heavy-weight drill pipe for the following conditions:

- Hole size: $9\frac{1}{2}$ in.
- Hole angle: 40deg
- Desired weight on bit: 40,000 lb

- Drill collars: $7 \times 2\frac{13}{6}$ in.
- Length of drill collars: 330 ft.
- Drilling fluid specific gravity: 1.2
- Desired safety factor for neutral point: 1.15

Solution
Check to see if the BSR of drill collar and heavy-weight drill pipe is less that 5.5. The bending section modulus of heavy-weight drill pipe is

$$\frac{\pi}{16}\left[\frac{(4.5)^4 - (2.75)^4}{4.5}\right] = 15.397 \text{ in.}^3$$

$$\text{BSR} = \frac{65.592}{15.397} = 4.26 < 5.5$$

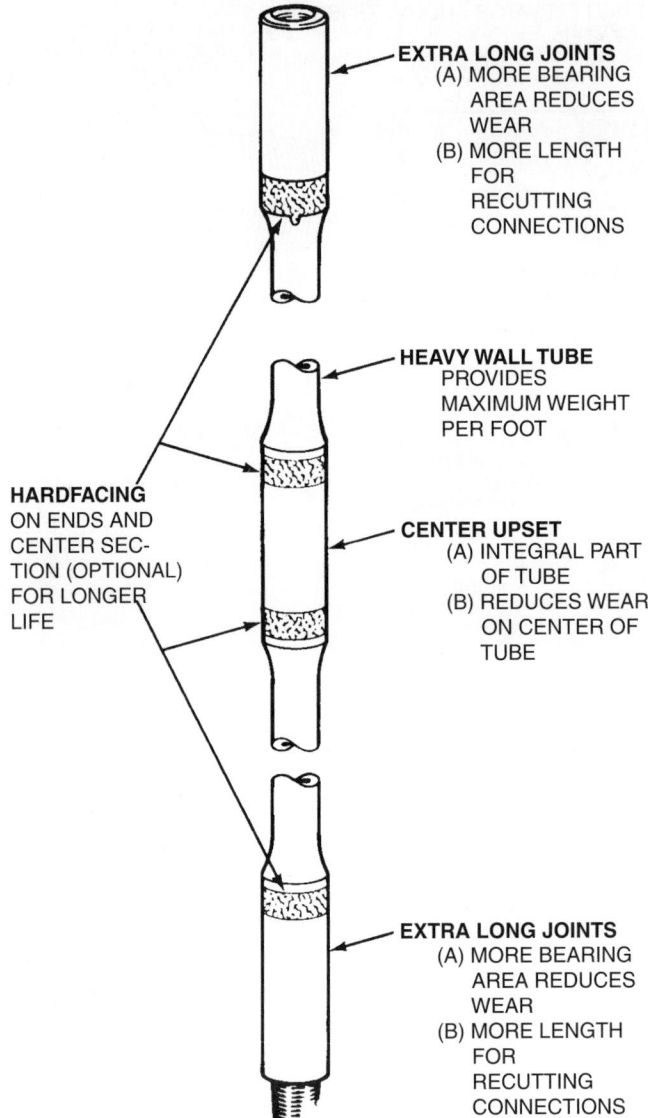

EXTRA LONG JOINTS
(A) MORE BEARING
     AREA REDUCES
     WEAR
(B) MORE LENGTH
     FOR
     RECUTTING
     CONNECTIONS

HEAVY WALL TUBE
PROVIDES
MAXIMUM WEIGHT
PER FOOT

HARDFACING
ON ENDS AND
CENTER SEC-
TION (OPTIONAL)
FOR LONGER
LIFE

CENTER UPSET
(A) INTEGRAL PART
     OF TUBE
(B) REDUCES WEAR
     ON CENTER OF
     TUBE

EXTRA LONG JOINTS
(A) MORE BEARING
     AREA REDUCES
     WEAR
(B) MORE LENGTH
     FOR
     RECUTTING
     CONNECTIONS

Figure 4.6.14   *Drilco's Hevi-Wate® drill pipe [42].*

The unit weight of drill collar in drilling fluid is

$$110 \left(1 - \frac{1.2}{7.85} \right) = 93.18 \, \text{lb/ft}.$$

The unit weight of heavy-weight drill pipe in drilling fluid is $(41)(0.847) = 34.72$ lb/ft. Part of the weight on bit may be created by using drill collars $= (93.18)(330)(\cos 40) = 23,555$ lb. The required length of heavy-weight drill pipe is

$$\frac{(40,000 - 23,555)(1.5)}{(34.72)(\cos \cdot 40)} = 711 \, \text{ft}$$

Assuming an average length of one joint of heavy-weight drill pipe to be 30 ft, 24 joints are required.

4.6.2.6 Fatigue Damage to Drill Pipe
Most drill pipe and tool joint failures occur as a result of fatigue. Fatigue damage is caused by the cyclic bending loads induced in the drill pipe during service. Cyclic stress results in a crack, which is the first stage of a fatigue failure. The crack grows to the point where the remaining cross-section is not great enough to support the service loads, and the drill pipe separates, this is the second stage. A washout often precedes a catastrophic failure. If the driller is able to detect a drop in standpipe pressure or a change in the sound of the pumps, the pipe can be pulled out of the hole intact.

Fatigue cracks most often occur on the outside surface because that is where bending stresses are highest and because material defects such as corrosion pitting are on the surface. Fatigue cracks can also begin on the inside surface of stress raisers such as pitting exist.

The resistance of a material to fatigue, or its endurance limit, is proportional to its ultimate strength. The endurance limit of a material is the cyclic stress below which a fatigue failure will not occur or the pipe will run under those conditions "forever." Forever in terms of drill pipe is usually between 2 and 3 million cycles in the laboratory. Endurance limits of particular materials are determined by conducting cyclic bending tests on a small specimen. These tests show that the endurance limit of the small, carefully machined samples is about 50% of the material's ultimate strength. The endurance limit of a full-size joint of drill pipe tested in the laboratory free from a corrosive environment is between 20,000 and 30,000 psi.

In general terms, the endurance limit is an indicator of how long, if ever, it will take a crack to form under cyclic loading. The stronger the pipe, the greater the endurance limit. The impact strength of the material is an indicator of how long it will take the crack to propagate to the point of separation. Generally, the stronger the pipe, the quicker the crack will propagate.

Based on work done by A. Lubinsky, J. E. Hansford, and R. W. Nicholson, API RP7G contains the formulas for the maximum permissible hole curvature to avoid fatigue damage to drill pipe.

$$c = \frac{432,000}{\pi} \frac{\sigma_b}{ED_{dp}} \frac{\tanh(KL)}{KL} \qquad [4.6.19]$$

$$K = \left(\frac{T}{EI}\right)^{\frac{1}{2}} \qquad [4.6.20]$$

For grade E drill pipe

$$\sigma_b = 19,500 - (0.149)\sigma_t \qquad [4.6.21]$$

For grade S drill pipe

$$\sigma_b = 20,000 \left(1 - \frac{\sigma_t}{145,000}\right) \qquad [4.6.22]$$

$$\sigma_t = \frac{T}{A} \qquad [4.6.23]$$

where

    c = maximum permissible dogleg severity in (degrees/100 ft)
    σ_b = maximum permissible bending stress (psi)
    σ_t = tensile stress from weight of suspended drill string (psi)
    E = Young's modulus, $E = 30 \times 10^6$ (psi)
    D_{dp} = outside diameter of drill pipe (in.)
    L = half the distance between tool joints (in.), L = 180 in. for range 2 drill pipe.
    T = weight of drill pipe suspended below the dogleg (lb)
    I = drill pipe moment of inertia with respect to its diameter (in.^{4})
    A = cross-sectional area of drill pipe (in.^{2})

By intelligent application of these formulas, several practical questions can be answered at the borehole design state and while drilling.

Example 4.6.10
Calculate the maximum permissible hole curvature for the following drill string:

- 5-in., 19.50 lb/ft S-135 range 2 premium class pipe with $6\frac{5}{8}$ OD $\times 2\frac{3}{4}$ ID tool joints
- Drill collars: $7 \times 2\frac{1}{4}$ in., unit weight of 117 lb/ft
- Length of drill collars: 550 ft
- Drilling fluid density: 12 lb/gal
- Anticipated length of hole below the dogleg: 8,000 ft
- Assume hole is vertical below dogleg

Solution
From Table 4.6.11, $D_{dp} = 4.855$ in., $d_{dp} = 4.276$ in., and $A = 4.153$ in.^{2}, and from Table 4.6.6, the unit weight of drill pipe adjusted for tool joint (W_{dp}) is 23.87 lb/ft. Even though this example is for premium class pipe, the weight for new pipe is used because it cannot be known for certain that the wall thickness of all the pipe is uniformly worn to 80% of its new thickness.

The weight of the drill collar string is

$$(550(117) \left(1 - \frac{12}{65.4}\right) = 52,543 \text{ lb}$$

The weight of the drill pipe is

$$(8,000 - 550)(23.87) \left(1 - \frac{12}{65.4}\right) = 145,202 \text{ lb}$$

Weight suspended below the dogleg (T = 197,745 lb) is

$$\text{Tensile stress } \sigma_t = \frac{197,745}{4.153} = 47,615 \text{ psi}$$

Maximum permissible bending stress is

$$\sigma_b = 20,000 \left(1 - \frac{47,615}{145,000}\right) = 13,432 \text{ lb}$$

Drill pipe moment of inertia is

$$I = \frac{\pi}{64} \left[(4.855)^4 - (4.276)^4\right] = 10.862 \text{ in.}^4$$

$$K = \sqrt{\frac{197,745}{(30)(10)^6 (10.862)}} = 2.463 \times 10^{-2} \text{in.}^{-1}$$

Maximum permissible hole curvature is

$$c = \frac{432,000}{\pi} \frac{13,432}{(30)(10)^6 (5)} \frac{\tanh[(2.406 \times 10^{-2})(180)]}{(2.406 \times 10^{-2})(180)}$$

$$= 2.842° / 100 \text{ ft}$$

The results of these calculations indicate that if a length of drill pipe were rotated in a dogleg under these conditions, a fatigue crack eventually would develop.

4.6.3 Drill String Inspection Procedure
Drill pipe should be periodically inspected to determine the following:

- Reduction of pipe body wall thickness that reduces pipe strength
- Reduction of tool joint OD that reduces tool joint strength
- Cracks in pipe body or tool joint threads that will lead to washouts or a parted string
- Damage to tool joint threads and makeup shoulders that could cause galling or washouts

- Extent of corrosion damage that could reduce pipe strength or lead to washouts
- Excessive slip damage that could lead to washouts
- Condition of internal plastic coating. Look for cracking, specifically under the pin threads (indicates stretched pin) and near the internal upset area where corrosion could lead to washouts.
- Condition of hardbanding
- Changes in dimensions caused by excessive service loads such as stretched pipe, stretched pin threads, and swollen boxes.

The frequency of inspection for the above varies. Some are visual inspections that can be done when the pipe is tripped out of the hole or while tripping back in. Generally, pipe is inspected after each job. More detailed information on types of inspections, procedures and frequencies can be found in API RP7G and ISO 10407-2.

Table 4.6.14, taken from API RP7G, shows the allowable wear and damage limits for Premium and Class 2 pipe.

4.6.3.1 Drill String Design

The drill string design should determine the optimum combination of drill pipe sizes, weights, and grades for the lowest cost or incorporate the performance characteristics to successfully accomplish the expected goals. The drill string is subjected to many service loads that may exist as static loads, cyclic loads, or dynamic loads. These loads include tensile loads, torsional loads, bending loads, internal pressure, and compressive loads. In certain circumstances, the drill string experiences external pressure. In addition to the varied service loads the pipe will see, the designer must consider hole drag and the risk of becoming stuck.

The designer should simultaneously consider the following conditions:

1. The working load at any part of the string must be less than or equal to the load capacity of the drill string member under consideration divided by the safety factor.
2. Ratio of section moduli of individual string members should be less than 5.5.
3. Pressure losses through the pipe going downhole and around the pipe going back to surface should be calculated so that the combination of pipe size and tool joint size are right for the diameter of the hole.

Selecting the size, weight, and grade of drill pipe and the tool joint OD and ID may be an iterative process, but the pipe size usually can be determined early on based on past experience and pipe availability. After calculating the strength requirements of the pipe and the lengths of each section of tapered strings, it may be necessary to chose a different pipe size from the one initially picked.

In a vertical hole, the maximum length of the pipe above the heavy-weight pipe or the drill collars is determined by the following equation:

$$(L_{dc}W_{dc} + L_{hw}W_{hw} + L_{dp1}W_{dp1})K_b = \frac{P_1}{SF} \quad [4.6.24]$$

where
L_{dc} = length of drill collar string (ft)
W_{dc} = unit weight of drill collar in air (lb/ft)
L_{hw} = length of heavy-weight drill pipe if used in the string (ft)
W_{hw} = unit weight of heavy-weight drill pipe (lb/ft)

L_{dp1} = length of drill pipe under consideration above the heavy-weight drill pipe (ft)
W_{dp1} = unit weight of drill pipe in section 1 (lb/ft)
K_b = buoyancy factor
P_1 = tensile load capacity of drill pipe in section 1 (lb)
SF = safety factor

Solving Equation 4.6.24 for L_{dp1} yields

$$L_{dp1} = \frac{P_1}{SF \cdot K_b \cdot W_{dp1}} - \frac{L_{dc}W_{dc}}{W_{dp1}} - \frac{L_{hw}W_{hw}}{W_{dp1}} \quad [4.6.25]$$

If the sum of $L_{dc} + L_{hw} + L_{dp1}$ is less than the planned borehole depth, stronger pipe must selected; that is, P_1 must be increased, or a stronger section of pipe must be placed above section 1.

The maximum length of the upper part in a tapered string in a vertical hole may be calculated from Equation 4.6.26:

$$L_{dp2} = \frac{P_2}{SF \cdot K_b \cdot W_{dp2}} - \frac{L_{dc}W_{dc}}{W_{dp2}}$$
$$- \frac{L_{hw}W_{hw}}{W_{dp2}} - \frac{L_{dp1}W_{dp1}}{W_{dp2}} \quad [4.6.26]$$

where P_2 = tensile load capacity of next (upper) section of drill pipe (lb)
L_{dp2} = length of drill pipe section 2 (ft)
W_{dp2} = unit weight of drill pipe section 2 (lb/ft)

Normally, not more that two sections are designed, but if necessary, three or more sections can be used. To calculate the tensile load capacity of drill pipe, it is suggested to apply Equation 4.6.11 and use the recommended makeup torque of the weakest tool joint for the rotary torque. The rotary torque should not exceed the makeup torque of any tool joint in the string.

The magnitude of the safety factor is very important and usually ranges from 1.4 to 2.8, depending on downhole conditions, drill pipe condition, and acceptable degree of risk. It is recommended that a value of safety factor be selected to produce a margin of overpull of at least about 70,000 lb.

$$MOP = P_n - K_b WT_{string} \quad [4.6.27]$$

where MOP = margin of overpull (lb)
P_n = tensile load capacity of top section of pipe (lb)
WT_{string} = combined air weight of entire drill string

Horizontal or deviated holes. Drill string design is more difficult on horizontal and deviated holes. Calculating frictional drag, the effects of compressive loading, and other factors by hand can be cumbersome and inaccurate. For these type holes, it is better to use torque and drag software for drill string design.

Slip loading. An additional check, especially in deep drilling, should be done to avoid drill pipe crushing in the slip area. The maximum load that can be suspended in the slips can be found from Equation 4.6.28:

$$W_{max} = \frac{P_1}{SF\left[1 + \frac{D_{dp}}{2}\frac{K}{L_s} + \left(\frac{D_{dp}}{2}\frac{K}{L_s}\right)^2\right]^{1/2}} \quad [4.6.28]$$

$$K = \frac{(1 - f \cdot \tan\alpha)}{f + \tan\alpha} \quad [4.6.29]$$

where

W_{max} = maximum allowable drill string load that can be suspended in the slips (lb)

P_t = tensile load capacity of drill pipe (lb)

D_{dp} = outside diameter of drill pipe (in)

L_s = length of slips (in)

K = lateral load factor of slip

F = coefficient of friction between slips and slip bowl, often taken as 0.08

α = slip taper (9° 27'45'')

SF = safety factor to account for dynamic loads when slips are set on moving drill pipe (SF = 1.1)

Example 4.6.11

Design a drill string for the conditions specified:

- Hole depth: 10,000 ft
- Hole size: $9\frac{7}{8}$ in.
- Mud weight: 12 lb/gal
- Maximum weight on bit: 60,000 lb
- Neutral point design factor: 1.15
- No crooked hole tendency
- Safety factor for tension: 1.4
- Required margin of overpull: 100,000 lb
- From offset wells, it is known that six joints of heavy-weight drill pipe are desirable
- Assume a vertical hole

Solution

Select the drill collar size from Table 4.6.3: $7\frac{3}{4} \times 2\frac{13}{16}$ in., 139 lb/ft. Such drill collars can be caught with overshot or washed over with washpipe.

$$\text{Length of drill collars} = \frac{(60,000)(1.15)}{(139)(0.816)} = 608 \text{ ft}$$

$$\text{Buoyancy factor } K_b = \left(1 - \frac{12}{65.4}\right) = 0.816$$

Select 21 joints of $7\frac{3}{4} \times 2\frac{13}{16}$ in. drill collars. This is a drill collar length of 630 ft. The calculated section modulus of the drill collars is 89.6 in.[3]

Example 4.6.12

To maintain a BSR of less than 5.5, select 5-in. heavy-weight drill pipe with a section modulus of 21.4 in[3] and a unit weight of 49.3 lb/ft (see Table 4.6.13).

$$\text{Length of heavy-weight drill pipe } L_{hw} = (6)(30) = 180 \text{ ft}$$

Select 5 in., 19.50 lb/ft grade G-105 premium class pipe with $6\frac{5}{8} \times 3\frac{1}{2}$ tool joints (see Table 4.6.6).

The unit weight of drill pipe with tool joints is 22.63 lb/ft (weight calculation based on new pipe with no wear). The section modulus of this pipe can be calculated to be 4.5 in.[3] From Table 4.6.6, the minimum tensile load capacity of the selected pipe is 436,065 lb. From Table 4.6.6, the recommended makeup torque for this tool joint is 23,400 ft-lb.

The tensile load capacity of the drill pipe, corrected for the effect of the maximum allowable torque, according to

Equation 4.6.11 is

$$P = \left\{ (436,065)^2 - 3\left[\frac{(4.153)(23,400)(12)}{8.953}\right]^2 \right\}^{\frac{1}{2}}$$

$$= 373,168 \text{ lb}$$

Determine the maximum allowable length of the selected drill pipe from Equation 4.6.25:

$$L_{dp} = \frac{373,168}{(1.4)(0.816)(22.63)} - \frac{(630)(139)}{22.63} - \frac{(180)(49.3)}{22.63}$$

$$= 10,172 \text{ ft}$$

Required length of drill pipe $L_{dp} = 10,000 - (630 + 180) = 9,190$ ft

Because the required length of drill pipe (9,190 ft) is less than the maximum allowable length (10,172 ft), it is apparent that the selected drill pipe satisfies the tensile load requirements.

Margin of overpull can be calculated in this way:

$$MOP = (0.90)(436,065) - [(630)(139)$$
$$+ (180)(49.3) + (9,190)(22.63)](0.816)$$
$$= 144,057 \text{ lb (greater than required 100,000 lb)}$$

In this example, the cost of the drill string is not considered. From a practical standpoint, the calculations outlined should be performed for various drill pipe unit weights and steel grades, and the design that produces the lowest cost should be selected.

The maximum load that can be suspended in the slips, from Equation 4.6.28 (assume K = 2.36, L_s = 12) is

$$W_{max} = \frac{436,065}{1.1\left[1 + \frac{5}{2}\frac{2.36}{12} + \left(\frac{5}{2}\frac{2.36}{12}\right)^2\right]^{\frac{1}{2}}} = 301,099 \text{ lb}$$

Total weight of string = 248,401 lb

The drill pipe will not be crushed in the slips. The drill string design satisfies the specified criteria.

References

1. API Specification 7, 40th Edition, "Specification for Rotary Drill Stem Elements," November 2001.

2. *Drilco Drilling Assembly Handbook*, Drilco Division of Smith International, Inc., 1977.

3. Wilson, G. E., and W. R. Garrett, "How To Drill A Usable Hole," Part 3, *World Oil*, October 1976.

4. API Recommended Practice 7G, 16th Edition, "Recommended Practice for Drill Stem Design and Operating Limits." December 1998.

5. Grant Prideco Product Catalog, December 2002.

6. API Specification 5D, 4th Edition, "Specification for Drill Pipe." January 2000.

7. Dawson R, and P. R. Paslay, "Drilling Pipe Buckling in Included Holes," SPE Paper 11167 (presented at the 57th Annual Fall Technical Conference and Exhibition of the SPE of AIME held in New Orleans, September 26–29, 1983).

8. Lubinski, A., "Maximum Permissible Dog-legs in Rotary Boreholes," *Journal of Petroleum Technology*, February 1961.

9. Rowe, M. E., "Heavy Wall Drip Pipe, a Key Member of the Drill Stem," Publ. No.45, Drilco, Division of Smith International, Inc., Houston, 19XX.

10. Timoshenko, S., and D. H. Young, *Elements of Strength of Materials,* Fifth Edition, D. Van Nostrand Co., New York, 1982.

4.7 BITS AND DOWNHOLE TOOLS

4.7.1 Roller Cone Drill Bits

A wide variety of rotary bit designs is available from a number of manufacturers. Each of these is intended to provide optimal performance in specific formations and in particular drilling environments. Bit designers must anticipate formational and environmental influences into their designs. To accomplish this, top manufacturers meticulously collect information on the operation of all bits with the purpose of enhancing the productive efficiency of their respective designs. Modern drill bits incorporate significantly different cutting structures and use vastly improved materials from those that were used just a few years ago. As a result, the productive efficiency of bit designs has improved systematically through the years. Variations in operating practices, types of equipment used, and hole conditions can require design adjustments, and manufacturers usually work closely with drilling companies to ensure that opportunities for design revision are expeditiously identified and implemented.

4.7.1.1 Roller Cone Bit Design

Rotary drill bits are designed to

> Accurately cut gage throughout the life of the bit
> Provide stable and vibration free operation at the intended rotational speed and weight on bit (WOB)
> Function at a low cost per foot drilled
> Have a long downhole life that minimizes requirements for tripping

To achieve these goals, bit designers must consider many factors. Among these are the formation to be drilled and the drilling environment, expected rotary speed, expected WOB, hydraulic arrangements, and anticipated wear rates from both abrasion and impact. The bit body, cone configurations, and cutting structures are design focal points, and metallurgical, tribological, and hydraulic considerations also have significant influence on design solutions.

4.7.1.1.1 Basic Design Principles

The performance of a drill bit is influenced by the environment in which it operates. Operating choices such as applied WOB, rotary speed, and hydraulic arrangements all have important implications on the way in which bits are designed and on their operating performance.

Environmental factors such as the nature of the formation to be drilled, hole depth and direction, characteristics of drilling fluids, and the way in which a drill rig is operated are also of critical importance to bit performance and design. Engineers consider these factors for all designs, and every design begins with close cooperation between the designer and the drilling company to ensure all applicable inputs contribute to the bit's design.

Design activities are principally focused in four general areas: material selection for the bit body and cones, geometry and type of cutting structure to be employed, mechanical operating requirements, and hydraulic requirements. The dimensions of a bit at the gage (outside diameter) and pin (arrangement for attachement to a drill stem) are fixed, usually by industry standards, and resultant design dimensions always accommodate them (Figure 4.7.1).

Structural Materials. Steels are selected that have appropriate yield strength, ability to harden, impact resistance, machineability, heat-treated properties, and ability to accept hardfacing without damage. Materials are described in more detail later.

Cutting Structure Arrangements. Cutting structures must be designed to provide efficient penetration of the formations to be drilled and accurately cut gage. The cutting structure must also prevent the bottom hole from contacting bit body structures (Figure 4.7.2). Cutting structures are described in more detail later.

Bearing, Seals, and Lubrication Design. The importance of roller cone bit bearing reliability cannot be understated. In an operational sense, bearings, seals, and

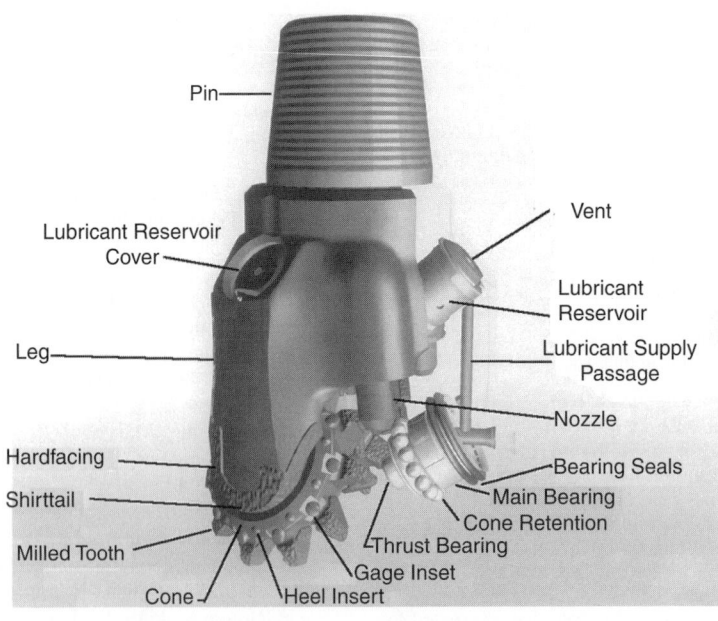

Figure 4.7.1 Roller cone bit general nomenclature.

Figure 4.7.2 *Typical roller cone bit cutting structure.*

lubrication arrangements function as a unit and their designs are closely interrelated (Figure 4.7.3). Bearing systems must function normally under high loads from WOB, in conditions of large impact loads, while immersed in abrasive and chemical laden drilling fluids, and in some cases in relatively high temperature environments (bearings, seals, and lubrication are described in more detail later).

Hydraulic Arrangements. Hydraulic configurations are designed so as to efficiently remove cuttings from the cutting structure and bottom hole and evacuate all cuttings efficiently (Figure 4.7.4).

4.7.1.1.2 Design Methods and Tools
How Teeth and Inserts "Drill." To understand design parameters for roller cone bits, it is first necessary to understand how roller cone bits drill. Two quite different types of drilling action take place. The first of these is a crushing action that takes place when weight applied to the bit forces its inserts (or teeth), into the formation. The second is a skidding, gouging type of action that partly results because the axis of cone rotation is slightly angled to the axis of bit rotation. Skidding and gouging also takes place because rotary motion of the bit does not permit a penetrated insert to rotate out of the crushed zone it has created without a certain amount of lateral force. The axis of cone rotation more dramatically produces the skidding-plowing action, but both effects contribute to cutting action (Figure 4.7.5).

Bentson Bit Design Method. The bit geometry and cutting structure engineering method of H. G. Bentson, reported in his paper entitled "Roller Cone Bit Design," has since 1956 been the root from which most roller cone bit design methods have been based [1]. Although modern engineering techniques and tools have advanced dramatically from those used in 1956, Benson's method is still the basis from which the design process has been derived. It continues to be useful as a background explanation of elements of bit design and is outlined later.

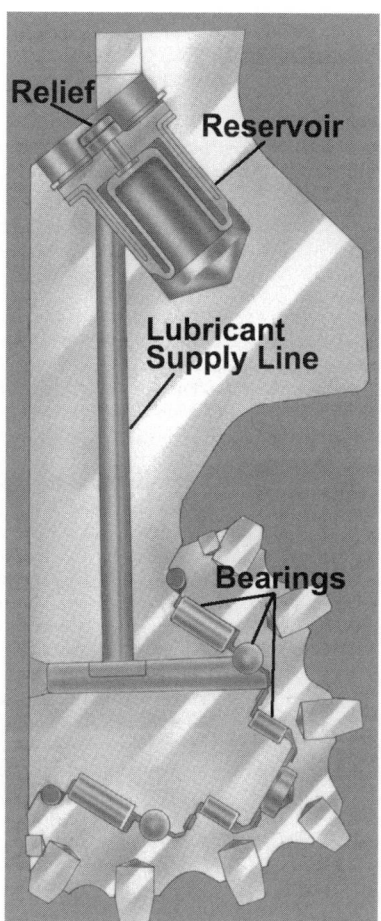

Figure 4.7.3 *Bearing, seal, and lubrication system.*

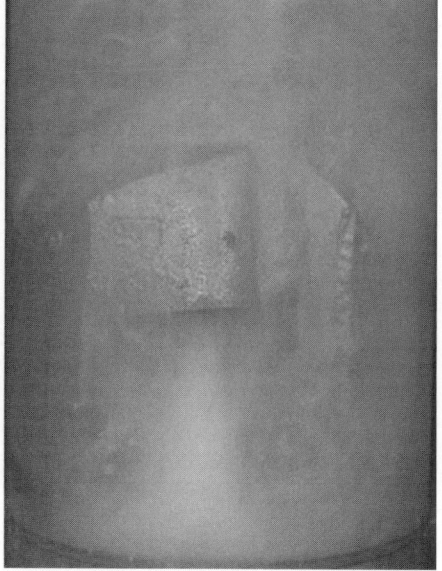

Figure 4.7.4 *Hydraulic flow around bit body.*

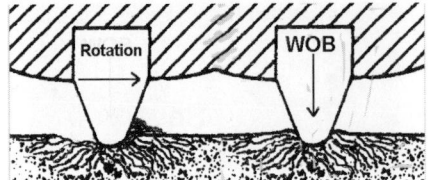

Figure 4.7.5 *Cutting actions for roller cone bits.*

Bit Diameter and Available Space. Well diameter and the required bit diameter to achieve it influence every design feature incorporated into any useful bit. The first consideration in the physical design of a roller cone bit is one of bit diameter, or to use other words, of available space. Every element of a rock bit must be confined within a circle representative of the hole to be drilled (Figure 4.7.6). The American Petroleum Institute has issued a specification establishing permissible tolerances for standard bit diameters [2]. In plan view, these prescribed diameters each inscribe a circle that is the maximum permissible diameter for a roller cone bit of a given size. Designers then use all available space within the circle or diameter when designing a bit. That is what specifications require and the largest possible structure within specification will normally be the strongest.

Most geometric and mechanical features of a bit are tied to the bit's circular cross-section. For example, the size of journals, bearings, cones, and hydraulic and lubrication features are collectively governed by circular cross-section. Individually, the sizing of the various elements can, to an extent, be varied. An increase in the physical size of one element will necessitate a corresponding decrease in one or more of the others, however. Proportions between component parts

such as a cone and a bearing are a compromise between bearing dimensions, in which larger usually has advantages, and cone shell thickness, in which increased thickness is, again, generally advantageous. The design of the final assembly of mating journals, bearings, and cones involves an even more complex array of compromises. Repositioning or altering size or shape of a single component nearly always requires subsequent additional changes in one or more of the other features. In smaller bits, finding good compromises can be difficult because of their shortage of available space.

Journal Angle. The term *journal angle* describes the angle formed by a line perpendicular to the axis of the bit and the axis of the bit leg journal (Figure 4.7.7). Journal angle is usually the first element in a roller cone bit design. It is introduced into design to optimize bit insert (or tooth), penetration into the formation being drilled; generally, bits with relatively small journal angles are best suited for drilling in softer formations while those with larger angles perform best in harder formations.

Cone Geometry. Cone shape is another critical design element and the relationship between cones is such that a change in shape or size of one cone always affects adjacent cones (Figure 4.7.7).

Cone Offset Angle. To increase the skidding-gouging action, bit designers generate additional working force by skewing (or offsetting) the centerlines of the cones so they do not intersect at a common point on the bit. The angle of skew is called the *cone offset angle.* Cone offset forces a cone to turn within the limits of the hole, rather than on its own axis.

Cone offset angle is the horizontal distance between the axis of a bit and the vertical plane through the axis of its journal. Offset is a dimension measured from the journal

Figure 4.7.6 *Confinement of three-cone bit elements within bit diameter.*

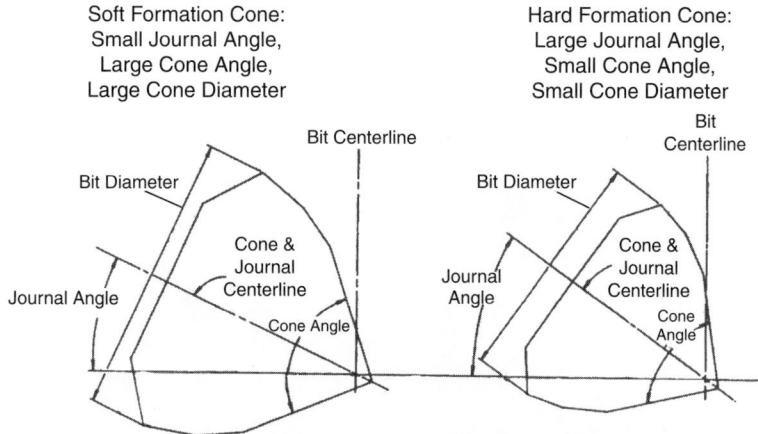

Figure 4.7.7 *Effect of journal angle on bit cones.*

centerline to the bit centerline. The line segment representing the amount of offset is perpendicular to both journal and bit centerlines. Offset is established by moving the centerline of a cone away from the centerline of the bit in such a way that a vertical plane through the cone centerline is parallel to the vertical centerline of the bit.

Basic cone geometry is directly affected by increasing or decreasing either the journal angle or the offset angle and a change in one of the two requires a compensating change in the other. For example, an increase in the journal angle requires an offsetting decrease in the cone offset angle.

Skidding-gouging improves penetration in soft and medium formations at the expense of increased insert or tooth wear. Bit designers limit the use of offset so results just meet requirements for formation penetration. In abrasive formations, offset can reduce cutting structure service life to an impractical level. As a result, bits designed for abrasive applications typically use small offsets that minimize skidding-gouging action.

Teeth and Inserts. Tooth and insert design is primarily governed by

Structural requirements for the insert or tooth
Formation requirements such as penetration, impact, abrasion, and stickiness

With borehole diameter and knowledge of formation requirements, the designer selects (or designs new), structurally satisfactory inserts that provide the optimal insert or tooth pattern necessary to efficiently drill the formation.

Factors that must be considered to design an efficient insert or tooth and establish an advantageous bottom-hole pattern include

Bearing assembly arrangement
Cone offset angle
Journal angle
Cone profile angles
Insert or tooth count
Insert or tooth spacing

When these requirements have been satisfied, remaining space can be allocated to insert or tooth contour and a cutting structure geometry that is best suited for the formation.

In general, the physical appearance of cutting structures designed for soft, medium, and hard formations can be

readily recognized by the length and geometric arrangement of their inserts.

Additional Design Criteria. With the exception of journal angle and offset values, the previous discussion of fundamental design has described roller cone bit design factors imposed by geometric limitations. Basic design criteria relate to more arbitrary factors that are primarily a function of previous experience. These include design criteria such as

Clearance between inserts or teeth and features on adjacent cones
Bottom-hole coverage
Cone-shell thickness
Bearing configuration

A typical insert or tooth design layout for a roller cone bit intended for use in medium-hard formations, to which these design criteria have been applied is shown in Figure 4.7.8. Because the design incorporates an offset, cone sections lie in parallel planes, one above the other, by an amount that is a calculated function of offset value. Design work of this sort is usually accomplished through three-dimensional modeling.

Insert or tooth clearances relative to adjacent cones must be closely controlled. Within limits, these clearances are maintained to the smallest possible dimension to gain maximum use of space. Excessive clearance results in inadequate bottom-hole coverage and accelerated wear on inserts or teeth. Excessive dimensions sacrifice insert or tooth depth, cone-shell thickness, and permissible bearing size. Insufficient clearance can cause interference between adjacent cones.

The composite view of the cones shown later the cone cluster is required to determine the bottom-hole coverage. This layout presents an overall check on the general insert or tooth spacing and is required for proper engineering of insert or tooth balance and insert or tooth arrangement. Spaces representing uncut bottom are kept to a minimum and, within limits, are balanced. If these spaces are kept uniform and within the dimensional limits shown, teeth on an offset bit will wipe away any uncut bottom.

Cone-Shell Thickness. Cone-shell thickness is of crucial importance from a strength standpoint. Insufficient shell thickness can result in cone failure and loss of downhole components. Varying shell thicknesses are used for different

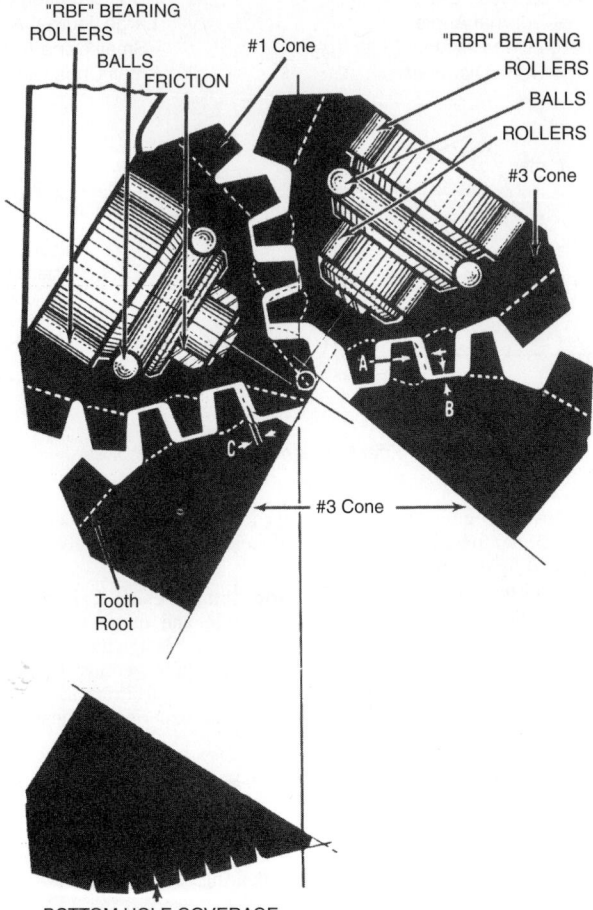

Figure 4.7.8 *Insert or milled tooth bit design showing clearances, bottom-hole coverage, cone-shell thickness, and bearing configuration.*

bit types, but design engineers always adhere to established minimum values (Figure 4.7.9).

Bearings Factors. In operation, radial loads are supported by the outer and inner bearings. The primary function of the ball bearings in roller cone bits is to secure cones to leg journals; under certain loadings, however, ball bearings can also help resist thrust loading although this is not their purpose (Figure 4.7.10).

Design criteria for roller cone bit bearing structures are critical. Bearings must routinely withstand extremely high values of unit and impact loading. Under the most adverse loading and lubrication conditions, bearing life is expected to exceed that of the inserts. Correct physical proportions for a cutting structure must be established to achieve this. Within the limited space available, overly large roller and ball bearings necessitate reduced journal diameters. This condition decreases wear life of the journal and increases its potential for fatigue failure. Conversely, excessive journal diameters require the use of smaller rollers and balls that will fail more readily under compression loads.

Length-to-diameter ratios of bearing rollers must fall within engineered limits. Rollers that do not have proper proportions tend to skew in their race or fail from beam loading. Roller and ball race flange thicknesses must be adequate to eliminate failure under the severest conditions. Balance between these elements must be maintained to prevent excessive wear or premature failure by one of them.

Conversely, the number of inserts or teeth must be few enough so that, when worn, sufficient unit force can still be applied to cause bit penetration.

Relatively larger bearings can be installed in hard formation bits because their shorter inserts or teeth consume less available cone space. Increased bearing size enables bearings to withstand the larger weights on bit required to drill hard formations.

4.7.1.1.3 Design as Applied to Cutting Structure

To this point, fundamentals of roller cone bit design common to all bit types have been discussed. The design of a bit for use in a specific category of formations requires the application of additional design factors, including

Journal angle and offset values
Rolling characteristics of the cones
The effect of insert or tooth depth on bearing-structure
  size

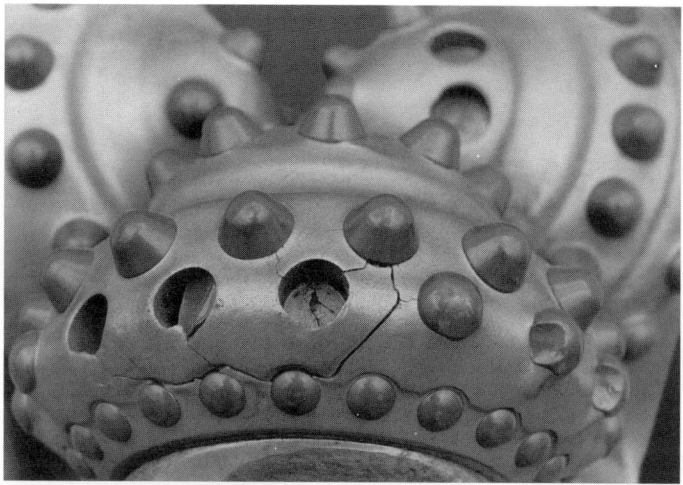

Figure 4.7.9 *Failed cone shell.*

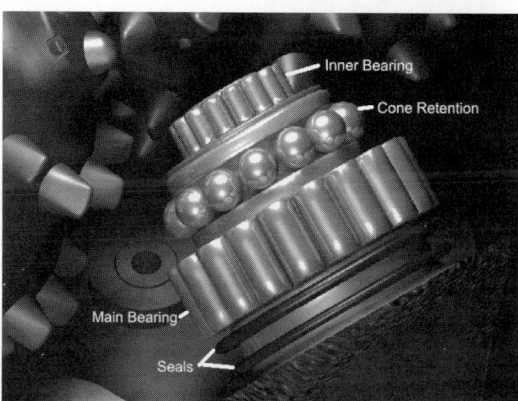

Figure 4.7.10 *Typical roller bearing system.*

The application of these design factors is best exemplified by comparison of two illustrations (Figure 4.7.11). The cutting structure on the left is designed for the softest formation types, that on the right for formations that are harder.

The action of bit cones on a formation is of prime importance to achieving a desirable penetration rate. Soft-formation bits require a gouging-scraping action. Hard-formation bits require a chipping-crushing action. These actions are primarily governed by the degree to which the cones roll and the degree to which they skid. Maximum gouging-scraping (soft-formation) actions require a significant amount of skid. Conversely, a chipping-crushing (hard-formation) action requires that cone roll approaches the "true roll" condition and skids very little.

For soft formations, a combination of the smallest journal angle, largest offset angle, and greatest variation in cone-profile angles. These develop a cone action that skids more than it rolls. Conversely, hard formations require a combination of the largest journal angle, no offset, and least variation in the cone profile. These will result in a cone action that closely approaches true roll with no skidding (Figure 4.7.12).

Relationships of Bearing Size, Insert or Teeth Dimensions, and Cone Shell. Cone insert or tooth length (the two are not exactly the same but have the same effect) affects bearing-structure size. Insert length describes the required shell thickness value for the drilled hole in which an insert is supported. Within the confine of available space, longer teeth are achieved by reducing cone shell thickness.

Figure 4.7.11 *Cutting structure for soft (left) and hard (right) formations.*

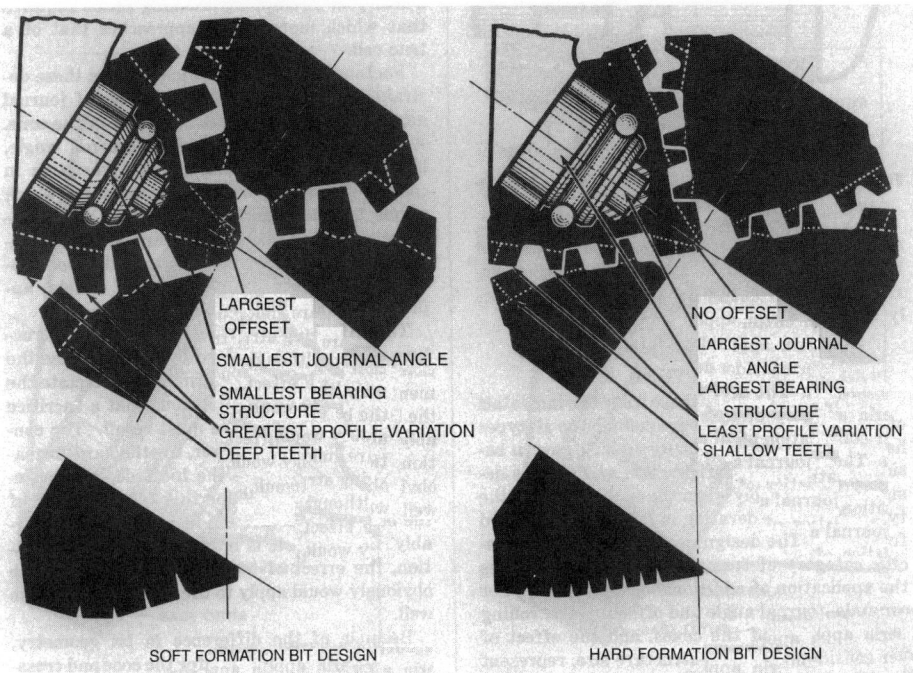

LARGEST
OFFSET

SMALLEST JOURNAL ANGLE

SMALLEST BEARING
STRUCTURE
GREATEST PROFILE VARIATION
DEEP TEETH

NO OFFSET

LARGEST JOURNAL
ANGLE
LARGEST BEARING
STRUCTURE
LEAST PROFILE VARIATION
SHALLOW TEETH

SOFT FORMATION BIT DESIGN

HARD FORMATION BIT DESIGN

Figure 4.7.12 *Cutting structure for soft (left) and hard (right) formations.*

Table 4.7.1 *Interrelationship between Inserts & Teeth, Hydraulic Requirements and the Formation*

Formation Characteristics	Insert/Tooth Spacing	Insert/Tooth Properties	Penetration & Cuttings Production	Cleaning/Hydraulic Flow Rate requirement
Soft	Wide	Long & Sharp	High	High
Medium	Relatively Wide	Shorter & Stubbier	Relatively High	Relatively High
Hard	Close	Short & Rounded	Relatively Low	Relatively Lower

Soft-formation bits require the use of longer inserts or teeth. This usually requires an offsetting reduction in bearing-structure size. Viewed from the opposite perspective, the size of required cone and bearing structures can be a limiting factor for insert or tooth length. Hard-formation bits require shorter inserts or teeth. In this case, space available will permit incorporation of relatively larger bearing structures.

Inserts or Teeth and the Cutting Structure. Because formations are not homogeneous, sizable variations exist in their drillability and have a large impact on cutting structure geometry. For a given WOB, wide spacing between inserts or teeth results in improved penetration and in relatively higher lateral loading on the inserts or teeth. Closely spacing inserts or teeth reduces loading at the expense of reduced penetration. The design of inserts and teeth, themselves, depends largely on the hardness and drillability of the formation. Penetration of inserts and teeth, cuttings production rate, and hydraulic requirements are interrelated as shown in Table 4.7.1.

Formation and cuttings removal impact cutting structure design. Soft, low compressive strength formations require long, sharp, and widely spaced inserts or teeth. Penetration

rate in this type of formation is partially a function of insert or tooth length and maximum insert or tooth depth must be used. Limits on the magnitude of maximum insert or tooth length are dictated by minimum requirements for cone-shell thickness and for bearing-structure size. Insert or tooth spacing must be sufficiently large to ensure efficient cleaning and cutting removal.

Requirements for hard, high-compressive-strength formation bits are usually the direct opposite of those for soft-formation types. Inserts are shallow, heavy, and closely spaced. Due in part to the abrasiveness of most hard formations and, in part, to the chipping action associated with drilling hard formations, the teeth must be closely spaced (Figure 4.7.13). This distributes loading widely to minimize insert or tooth wear rates and limit lateral loading on individual teeth. At the same time, inserts are stubby and milled tooth angles large to withstand heavy WOB loadings required to overcome high formation compressive strength. Close spacing often limits the size of inserts or teeth.

In softer and, to some extent, medium hardness formations, formation characteristics are such that provisions for efficient cleaning require careful attention from designers. If cutting structure geometry does not promote cuttings removal, bit penetration will be impeded and force the rate

Figure 4.7.13 *Comparison of softer IADC 437X (left) and harder 837Y (right) cutting structures.*

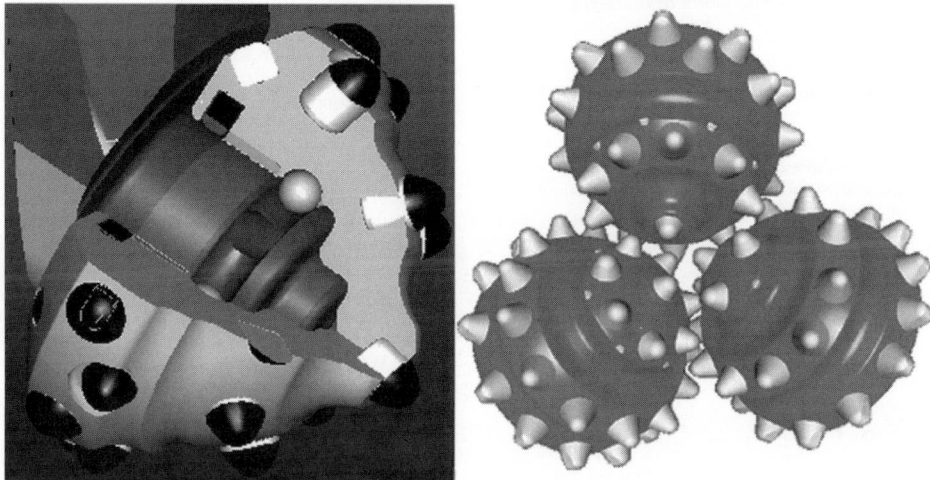

Figure 4.7.14 *Three dimensional engineering modeling.*

of penetration to decrease. Conversely, successful cutting structure engineering encourages both cone shell cleaning and cuttings removal.

Variations in downhole conditions and operating limitations often require use of different bit types even though drillability of the formations may be similar. Crooked-hole problems, for example, generally necessitate use of reduced WOB and higher rotary speeds. A different bit type may be required from that for the same formation under conditions that permit heavier weight and lower rotary speed. Weight on bit or rotary-speed limitations imposed by other factors would have a similar effect on the bit-type requirements. Knowledge of these design factors and their application to specific bit types is important to making productive bit choices.

The Mechanics of Roller Cone Bit Design. Modern roller cone bit design requires sophisticated engineering software tools. Such tools integrate many separate design and manufacturing related functions including CAD/CAM, preparation of bills of material, and error checking. These rely extensively on three dimensional modeling that allows the design engineer to flexibly and efficiently construct very complex assemblies. Software also has the capability

to evaluate models against existing standards and to add new features to existing standards. Models can be structurally evaluated using advanced simulation methods such as Finite Element Modeling and software verifies that parts, assemblies, and drawings comply with established modeling standards and design practices. All of this takes place in a single, seamless environment (Figure 4.7.14).

Structural Evaluations. The design tools described previously make it possible for engineers to evaluate, understand, and optimize the static and dynamic structural performance of their designs using parametric modeling that simulates the operating environment. Precise representations of CAD geometry, available for design models, together with adaptive solution technologies automatically provides accurate solutions that improve product quality and decrease development time requirements for prototyping.

Structural response simulation interrogates static and dynamic structural performance of a design in conditions specified by the engineer. These simulations employ advanced vibration tools to ascertain natural frequencies and vibration modes in a design and often eliminate the need for hardware prototypes.

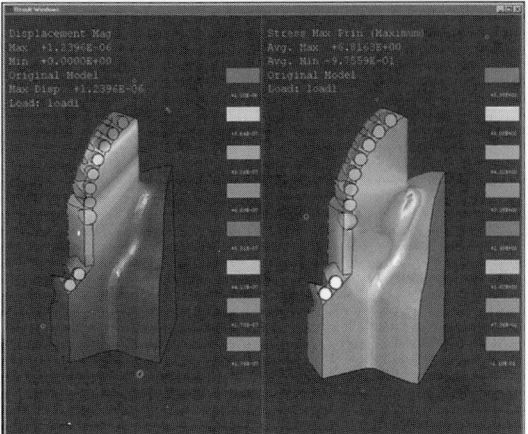

Figure 4.7.15 *Finite element modeling.*

Separately, finite element solutions for solid models automatically mesh with design environments or according to specifications (Figure 4.7.15). These allow interactive refinement that enables designers to quickly evaluate different configurations under a variety of environmental and geometric conditions.

Computational Fluid Dynamics. Computational Fluid Dynamics is also closely integrated with modern engineering software tools to provide simulation of drilling fluid flows around the bit. Some software incorporates fluid flow simulation into the design process (Figure 4.7.16). As parts or

assemblies are being designed, engineers can perform simulations directly into the geometry created in their CAD system. The need for physical prototype testing is greatly reduced by these hydraulic tools. Physical testing produces only "go or no-go" answers, whereas software provides analytical insight. Modern tools increase flow efficiencies for drilling fluids and lower the cost of hydraulic engineering relative to empirical methods used in the relatively recent past.

Advanced Engineering Methods. Bit design is undergoing very rapid evolution. The refined products of modern engineering tools, are being linked to practical expertise and formal scientific research regarding materials, geometries, hydraulic performance, and operating conditions in the environment of real formations. This real world approach is suggesting a great number of exceptionally efficient, innovative roller cone bit design solutions such as that shown in Figure 4.7.17.

An example of advanced engineering methods used by a leading bit manufacturer has several steps:

1. Identify and obtain samples of rock in which drilling will take place.
2. Under conditions that duplicate drilling by simulation (downhole confining pressure, WOB, and rotation speed), measure penetration and abrasive characteristics of the rock.
3. Under conditions that duplicate drilling in a hydraulics simulator (Flow and nozzle arrangements) measure hydraulic characteristics of cutting removal and bit cleaning properties.
4. Process mechanical testing data with proprietary software to deliver bit design outputs. This research method predicts rate of penetration at a given rotary speed, optimal WOB, and rock formation properties. By its use, bit

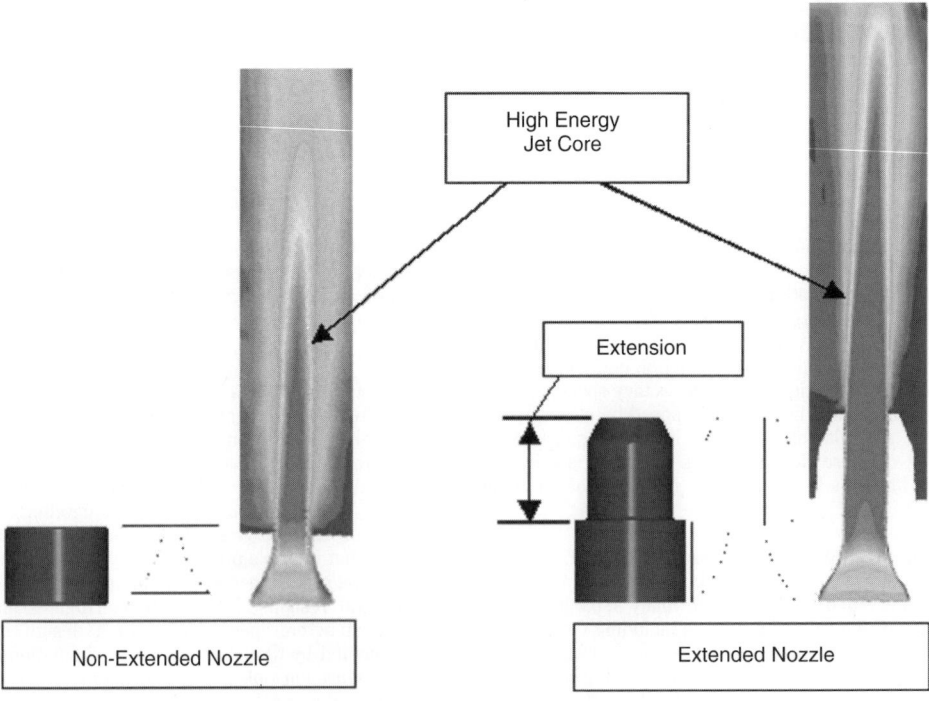

Figure 4.7.16 *Computational fluid dynamics.*

Figure 4.7.17 *Bit designs.*

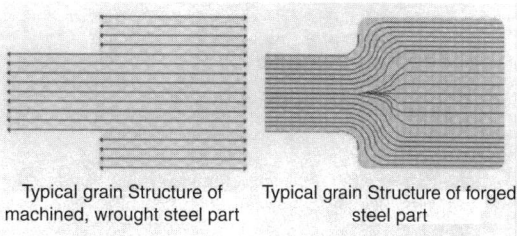

Typical grain Structure of machined, wrought steel part   Typical grain Structure of forged steel part

Figure 4.7.18 *Wrought-vs-forged grain structure.*

profile design; bottom-hole pattern; insert or tooth sharpness, shape, and counts; journal angle; and, required cone offset can all be optimized.

4.7.1.1.4 Materials Design

Materials properties are a crucial aspect of roller cone bit performance. Components must be resistant to abrasive wear, erosion, and impact loading. Metallurgical characteristics such as heat treatment properties, weldability, capability to accept hardfacing without damage, and machineability all figure into the eventual performance and longevity result for a bit. Physical properties for bit components are contingent on the raw material from which the component is made, the way that material has been processed, and of the type of heat treatment that has been applied to it. Steels employed in roller cone bit components are all melted to exacting chemistries, cleanliness, and interior properties. All are wrought because of grain structure refinements obtained by the rolling process. Most manufacturers begin

with forged blanks for both cones and legs because of further refinement and orientation of microstructure that results from the forging process.

Material specifications for legs and cones are different. Structural requirements and the need for abrasion or erosion resistance are different for legs and cones. Further, different sections of a component often require different physical properties. Leg journal sections, for example, require high hardenabilities that resist wear from bearing loads whereas the upper portion of legs are configured to provide high tensile strengths that can support large structural loads.

Legs and cones are commonly made for forged blanks. Forgings have superior metallurgical structures compared with those for components machined from wrought stock. The forging process increases metal density and orients metal grain structure to result in optimal load-bearing capabilities (Figure 4.7.18). Forgings are manufactured from wrought materials, whose properties were previously described.

Roller cone bit legs and cones are manufactured from low alloy steels. Legs are made of a material that is easily machinable before heat treatment, is weldable, has high tensile strength, and is capable of being hardened to a relatively high degree. Cones are made from materials that can be easily machined when soft, are weldable when soft, and can be case hardened to provide resistance to abrasion, and erosion (Figure 4.7.19).

4.7.1.1.5 Inserts and Wear-Resistant Hardfacing Materials

Tungsten carbide is one of the hardest materials known to man. This hardness makes it extremely useful as a cutting

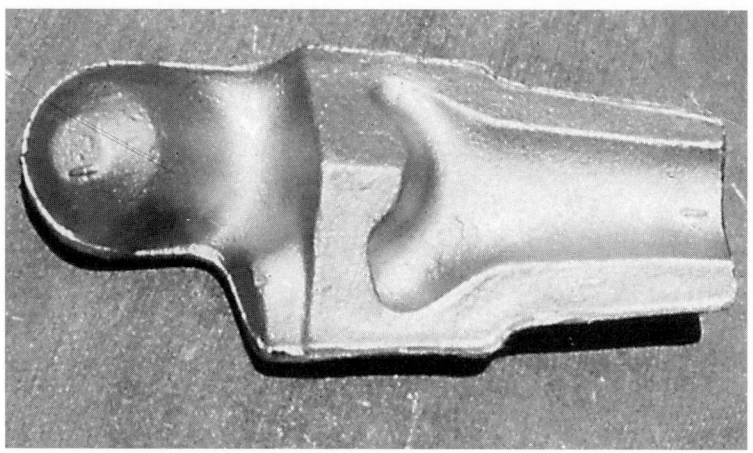

Figure 4.7.19 *Roller cone bit leg forging (three required).*

and abrasion resisting material for roller cone bits. The compressive strength of tungsten carbide is much greater than its tensile strength. It is a material whose usefulness is to be gained only within design arrangement that ensure compressive loading while protecting against shear and tension. Tungsten carbide bearing inserts are the most popular material for drill bit cutting elements. Hardfacing materials containing tungsten carbide grains are the standard for protection against abrasive wear on bit surfaces.

When most say "tungsten carbide," they do not refer to the chemical compound (WC), but rather to a sintered composite of tungsten carbide grains imbedded in and metallurgically bonded to a ductile matrix, or binder phase. Such materials are included in a family of materials called *cermets* or *ceramic-metal*. Binders support tungsten carbide grains and provide tensile strength. Because of binder, cutters can be formed into useful shapes that orient tungsten carbide grains so they will be loaded under compression. Tungsten carbide cermets can also be polished to a very smooth finish that reduces sliding friction. By control of grain size and binder content, the hardness and strength of these materials can be tailored to yield optimal properties for individual applications.

The most common binder metals for use with tungsten carbide are iron, nickel, and cobalt. These materials are related on the periodic table of elements and have an affinity for tungsten carbide (cobalt has the greatest affinity). Tungsten carbide cermets normally have binder contents in the range 6% to 16% by weight. Because tungsten carbide grains are metallurgically bonded to the binder, there is no porosity at boundaries between the binder and grains of tungsten carbide and the cermet is less subject to damage by shear and shock.

4.7.1.1.6 Properties of Tungsten Carbide Composites
The process of designing cermet properties makes it possible to exactly match a material to the requirements for a given drilling application. Tungsten carbide particle size (normally 2 to 6 μm), shape, and distribution and binder content (as a weight percent) affect composite material hardness, toughness, and strength. Generally, increasing binder content for a given tungsten carbide grain size causes the hardness to decrease and fracture toughness to increase. Conversely, increasing tungsten carbide grain size affects hardness and "toughness." Smaller tungsten carbide particle size and reducing binder content produces higher hardness,

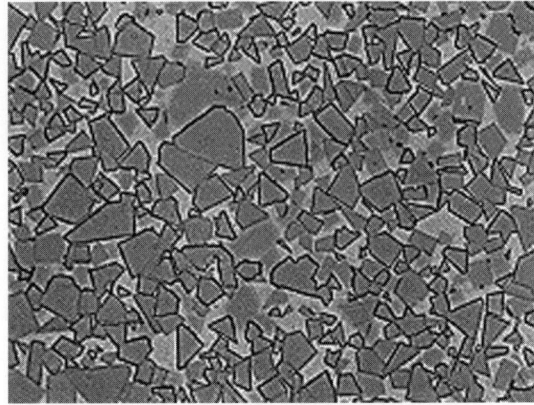

Figure 4.7.21   *Photomicrograph of a WC–Co composite.*

higher compressive strength and better wear resistance. In general, cermet grades are developed in a range in which hardness and toughness vary oppositely with changes in either particle size or binder content. In any case, subtle variations in tungsten carbide content, size, distribution, and porosity can markedly affect material performance (Figure 4.7.20).

The photomicrograph (Figure 4.7.21) shows a magnified, polished section of a tungsten carbide, with cobalt binder. The tungsten carbide grains in the photomicrograph appear as trapezoids and triangles. Average grain size is nominally 5 μm with a binder content of 10% by weight.

Dual-Composite (DC) Tungsten Carbide Inserts. DC tungsten carbide is a new class of carbide that overcomes fracture toughness limitations associated with conventional tungsten carbide grades. The material is characterized by large tungsten carbide particles that are manufactured in a multi-stage process and by a microstructure that is more homogeneous than conventional cemented tungsten carbides. DC tungsten carbide increases fracture toughness by as much as 50% over tough, cemented tungsten carbide grades without affecting wear performance and, in some cases, have improved both toughness and wear resistance.

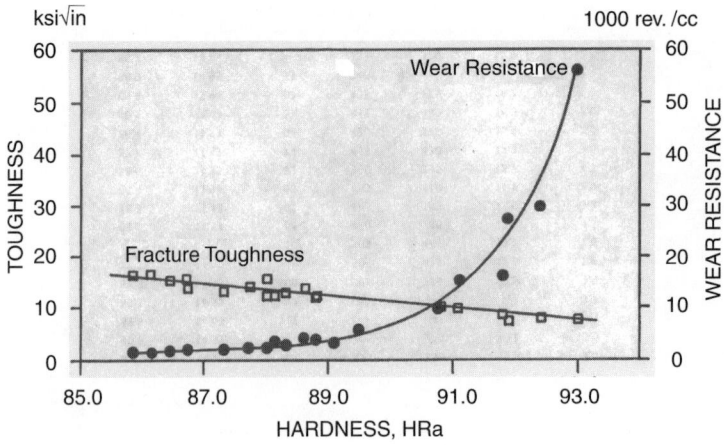

Figure 4.7.20   *Hardness, toughness, and wear resistance of cemented tungsten carbide.*

Comparison of Diamond, Polycrystalline Diamond, and Tungsten Carbide Materials					
	Knoop Hardness (Kg/mm^2)	Coefficient of Friction	Coefficient of Thermal Expansion ($10^{-4}/°C$)	Density (g/cm^3)	Fracture Toughness (Kic)
Tungsten Carbide (6% Co)	1475	0.2	4.3–5.6	14.95	10.8
Polycrystalline Diamond, (PCD)	5000–8000	0.08–0.15	1.5–3.8	3.80–4.10	6.1–8.9
Natural Diamond	6000–9000	0.05–0.10	0.8–4.8	3.52	3.4

Figure 4.7.22   Comparison of DEI and WC+Co Properties.

DC tungsten carbide maximizes fracture toughness by increasing mean separation between tungsten carbide particles. This results in a more homogenous distribution of tungsten carbide particles and a larger amount of the toughness providing cobalt binder around and between adjacent tungsten carbide composite granules. Wear improvement results from the large tungsten carbide grains. Because of the toughness and thermal conductivity of binder content in DC carbides, they have superior resistance to thermal damage. This combination of properties is well suited to high ROP drilling in soft to medium formations [5–7].

Diamond-Enhanced Tungsten Carbide Inserts. Diamond-enhanced inserts (DEI) are frequently used to prevent wear in the highly loaded, highly abraded gage area of bits or in all insert positions if drilling conditions are difficult. They are composed of polycrystalline diamond compact (usually called PDC), which is chemically bonded, synthetic diamond grit supported in a matrix of cemented tungsten carbide. PDC has both higher compressive strength and higher hardness than cemented tungsten carbide. Diamond materials are largely unaffected by chemical interactions with rock and are less sensitive to heat than cemented tungsten carbides. These properties make it possible for diamond-enhanced materials to function normally in drilling environments in which cemented tungsten carbide grades deliver disappointing or unsatisfactory results (Figure 4.7.22) [6, 7, 9].

When designing diamond-enhanced inserts, incorporation of higher diamond densities increases impact resistance and ability to economically penetrate abrasive formations. As would be expected, increasing diamond density increases insert cost. In the past, diamond-enhanced inserts have been available only in symmetric shapes. The first of these was the semi-round top insert. Today, some manufacturers have developed processes that make it possible to produce complex, diamond-enhanced insert shapes.

Cellular Diamond Inserts. Cellular diamond is an innovative new insert material that places exceptionally high diamond densities at the work point of an insert. Cellular diamond is a polycrystalline diamond compact organized in long strand-like cells (Figure 4.7.23). It has a honeycomb structure when viewed from the end that can be seen without magnification. The interior of the cells is composed of diamond grit in a tungsten carbide cermet matrix. Boundaries between cells are cemented tungsten carbide. The columnar nature of cellular diamond inhibits the propagation of cracking.

A particular advantage of cellular diamond material is that its geometry and structure can be engineered and scaled for particular applications. For example, complex insert shapes that previously could not have been diamond enhanced now are protected by cellular diamond [6].

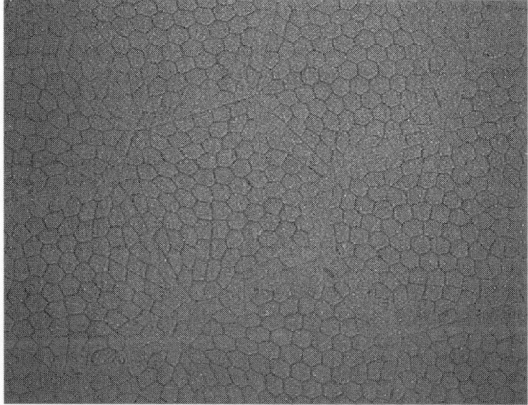

Figure 4.7.23   Cellular diamond microstructure.

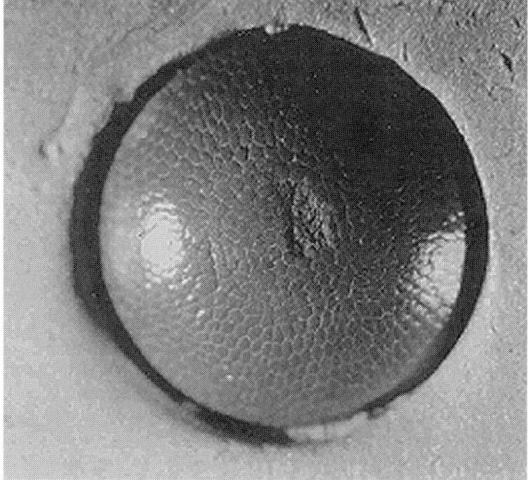

Figure 4.7.24   Pitted cellular diamond insert showing crack inhibition.

If there is local damage to a cellular diamond insert, the nature of the cellular structure inhibits formation of cracks, resulting only in localized pitting. Similar damage to tungsten carbide inserts would eventually lead to their failure (Figure 4.7.24). Pitting does not have significant adverse effects on insert performance and does not cause insert failure.

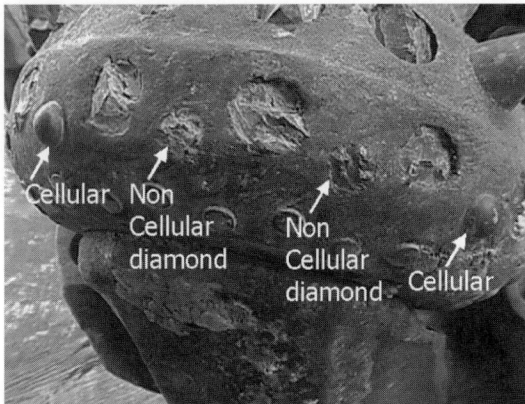

Figure 4.7.25 *Comparative cellular diamond performance on a failed experimental dull.*

As expected of a diamond-enhanced insert, resistance to abrasion is far superior to that for tungsten carbide. Figure 4.7.25 shows a comparative dull test bit that employed both diamond-enhanced tungsten carbide and cellular diamond gage inserts in a difficult environment.

Tungsten Carbide Hardfacing. Hardfacing materials are designed to provide wear resistance (abrasion, erosion and impact), for the bit. To be effective, they must be resistant to loss of material by flaking, chipping, and bond failure with the bit and are designed with these structural requirements in mind, as well. Hardfacing is used to provide wear protection on the lower (shirttail), area of all roller cone bit legs, and as a cutting structure material on milled tooth bits (Figure 4.7.26).

Hardfacing is most commonly, manually deposited by oxyacetylene welding. A hollow steel tube containing appropriately sized grains of tungsten carbide is held in a flame until the steel rod melts and bonds through surface melting with the bit feature being hardfaced. In the process, tungsten carbide grains flow as a solid, with melted steel from the rod, onto the bit. The steel solidifies around the tungsten carbide particles, firmly attaching them to the bit.

4.7.1.1.7 Cutting Structure Design
Principals. Roller cone bits drill soft formations differently than hard formations. Soft formations are drilled by penetration of the insert or tooth followed by a prying action as the cutting element is rotated during cone revolution. Cuttings from deeply penetrated, soft formations are large and the volume of cuttings produced is high, and rate of penetration is high. Hard formations are difficult to penetrate with available weights on bit and insert or tooth structures appropriately sized for use in roller cone bits. To overcome these problems, hard structures are drilled with blunt inserts or teeth that crush rather than penetrate the formation. Chips are small, volume of cuttings is small, and rate of penetration is low.

Inserts or teeth for soft and hard formation drilling are as different as their means of drilling suggest. Soft formation inserts or teeth are as long as available space permits. The limit for external length is established later the point at which interferences with adjacent cones or cutting structures would occur. Soft formation inserts are desirably sharp. There are limits on sharpness as well, of course. There must be sufficient structure to ensure reasonable expectations for

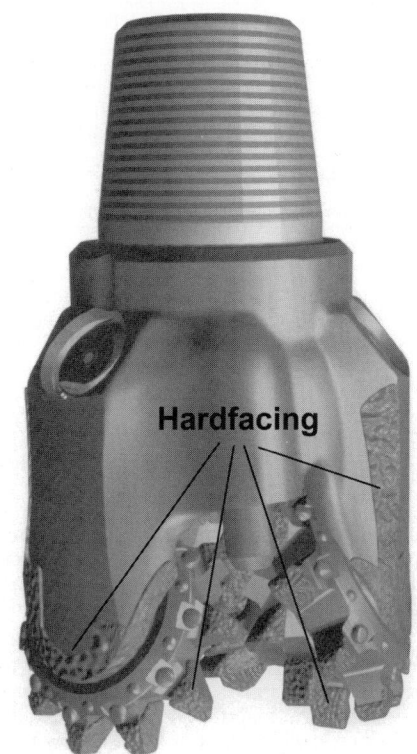

Figure 4.7.26 *Typical hardfacing applications.*

insert or tooth life. In this regard, use of inserts can result in a relatively sharper cutting structure than milled teeth. In any event, the extent of sharpness and the rate of penetration it promises are a compromise with requirements for bit life, limitations on tripping, and other operational factors.

Hard formation inserts or teeth are primarily designed with only structural considerations in mind. They have blunt profiles that can withstand large, crushing loads. These loads are decreased through use of a relatively higher number of inserts or teeth to focus WOB on the largest possible number of inserts. Dome-shaped inserts, often of compression- and abrasion-resistant diamond-enhanced materials, withstand these punishing conditions and are the preferred choice for extreme conditions. Figure 4.7.27 compares soft and hard formation cutting structures.

Tungsten Carbide Insert Design. The design of tungsten carbide inserts takes the properties of both tungsten carbide materials and the geometric efficiency for drilling of a particular rock formation into account. Softer materials require geometries that are long and sharp to encourage rapid penetration. Impact loads are low but abrasive wear can be high. Hard materials are drilled more by a crushing and grinding action than by penetration. Impact loads and abrasion can be very high. Tough materials, such as carbonates, are drilled by a gouging action and can sustain high impact loads and high operating temperatures. Variations in the way that drilling is accomplished, and rock formation properties govern the shape and grade of the correct tungsten carbide inserts to be selected.

The shape and grade of tungsten carbide inserts is influenced by their respective location on a cone. Inner rows

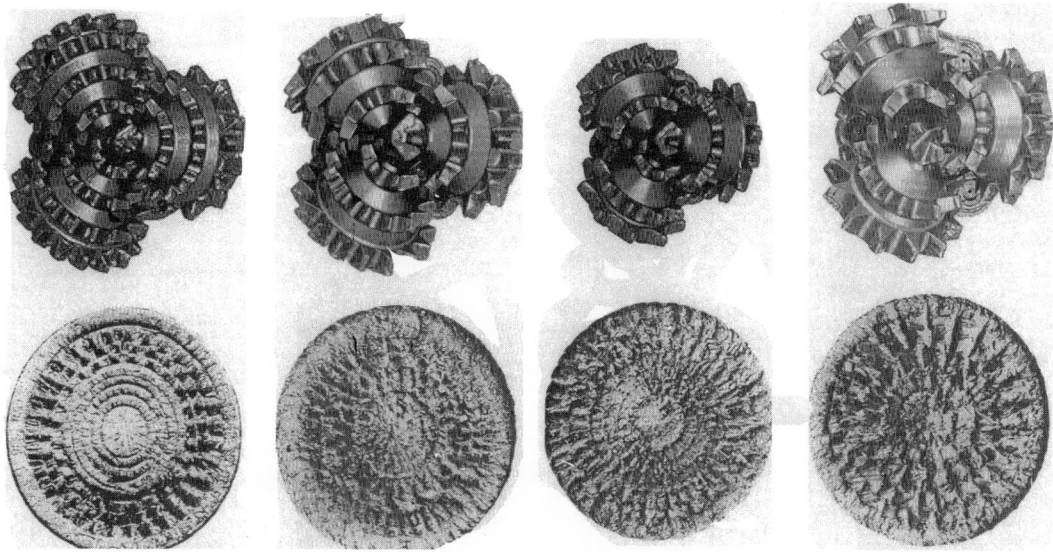

Figure 4.7.27   Soft and hard formation cutting structures.

Figure 4.7.28   Insert types.

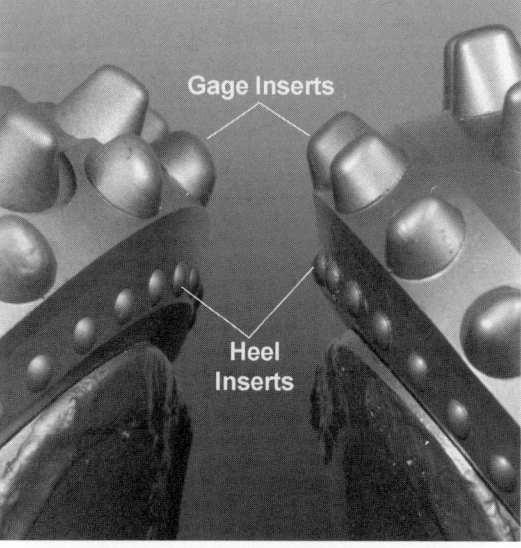

Figure 4.7.29   Typical heal row inserts.

of inserts function differently from outer rows. Inner rows have relatively lower rotational velocities about both the cone and bit axes. As a result, they have a natural tendency to gouge and scrape rather than roll. Inner insert rows generally use softer, tougher insert grades that best withstand crushing, gouging, and scraping actions. Gage inserts are commonly constructed of harder, more wear resistant grades that best withstand severe abrasive wear. It can be seen that as requirements at a specific bit location dictate, the designer must be prepared to provide different solutions. Fortunately, a large range of insert geometries, sizes, and grades through which bit performance can be optimized are available (Figure 4.7.28) [11].

Gage Cutting Structure. The most critical cutting structure feature is the gage row. Gage cutting structures must cut both the bottom of the hole and its outside diameter. Because of the severity of gage demands on a bit, both milled tooth and insert type bits can employ tungsten carbide or diamond-enhanced inserts on the gage. Under abrasive conditions, severe wear or gage-rounding is common, and at high rotary speeds, the gage row can experience temperatures that lead to heat checking, chipping, and eventually breakage.

Special Cutting Structure Features. Special cutting features include heel row cutters and ridge cutter inserts.

Heel Row Cutters. The high rotary speeds and side-loading forces generated by steerable systems can cause bits to prematurely become under gage. To counteract this, specially shaped tungsten carbide side-cutting chisel elements keep the bit in-gage longer without interfering with the steerability of the bit. In unusually abrasive conditions, use of semi-round top diamond-enhanced inserts on the heel minimize wear and ensure full gage holes. All insert bits and many milled tooth bits employ tungsten carbide or diamond-enhanced inserts on the heel (Figure 4.7.29).

Ridge Cutter Inserts. When drilling soft formations with insert type bits, it is fairly common to have uncut ridges develop between cutter rows. Ridges can become sufficiently

Ridge Cutters

Figure 4.7.30   *Ridge cutter application.*

high that they contact the bit cone impeding penetration and causing abrasive wear. Cutting structures intended for use in soft drilling environments commonly employ ridge cutter inserts to break up ridges that might develop. Ridge cutters are small inserts located between the principal cutter rows (Figure 4.7.30).

4.7.1.1.8 Hydraulic Features

Nozzles, Bosses, and Flow Tubes. Drilling fluids circulate through a drill string to nozzles at the bit and back to a reserve pit via the system annulus. They provide three crucial functions to drilling: cleaning the cutting structure, cuttings removal from the hole bottom and efficient cuttings evacuation. The hydraulic energy that causes fluid circulation is one of only three variable energy inputs (WOB, rotary speed, and hydraulic flow), available on a drill rig for optimization of drilling performance.

Many roller cone bit options such as nozzle selection, flow tubes, vectored flow tubes, and center nozzle ports help optimize hydraulic performance. These hydraulic features give alternatives to drilling operators for precise placement of hydraulic energy according to needs at the well bottom.

Generating cuttings is the first step needed to achieve high penetration rates; cleaning cuttings from the cone and from the hole bottom and lifting them through the annulus to the rig reserve pit is the remaining part of a hydraulic solution. Computer modeling supported by laboratory testing is the most common approach to development and verification of hydraulic designs. Efficient velocity profiles deliver hydraulic energy where it is most needed, particularly in cases in which drilling flow rates are compromised.

Normally, several different nozzles can be interchangeably used with the provided boss size on a particular bit. Nozzles are commonly classified into standard, extended, and diverging categories (Figure 4.7.31). Extended nozzles release the flow at a point closer than standard to the hole bottom. Diverging nozzles release flows in a wider than normal, lower velocity stream. They are primarily designed for use in center jet installations [10].

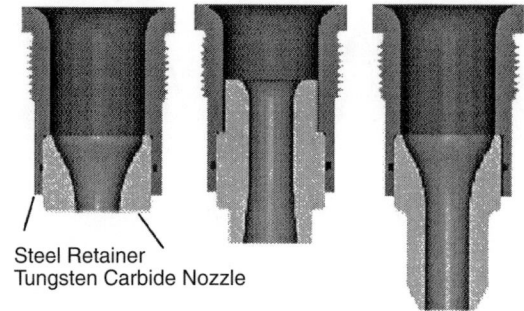

Steel Retainer
Tungsten Carbide Nozzle

Figure 4.7.31   *Typical standard, diffused, and extended nozzles showing a steel retainer and tungsten carbide wear insert.*

Extended Nozzles. Extended nozzles reduce flow dissipation, increase flow velocity at the impingement point, and increase hydraulic energy delivered to the well bottom. Higher impingement pressures help dislodge cuttings from the bottom and move them away from the bit. Higher velocity fluids are less likely to entrain surrounding fluids and to pull them back into the cones. This minimizes cuttings regrind and improves rates of penetration. Extended nozzles are particularly useful in marginal hydraulic environments and high overbalance applications.

Flow Tubes. Flow Tubes are extended nozzle tubes designed for use in high flow applications of up to 300GPM per nozzle. They are either welded attachments or are directly incorporated into bit leg forgings. Flow tubes accept both standard and extended nozzles and improve overall hydraulic efficiency by placing flow discharge energy nearer to the hole bottom. Flow tubes are particularly useful for directional wells in which hydraulic performance is critical. Flow tubes generally increase bit life by reducing abrasive wear from

Flow Tube
with extended Nozzle

Figure 4.7.32 Flow tube.

cuttings regrind and can be used in conjunction with most nozzles (Figure 4.7.32).

Vectored Flow Tubes. Vectored flow tubes are a generally available option. Vectored flows discharge, at an angle to bit centerline, onto the preceding cone face to enhance cone cleaning. They are particularly beneficial in intervals

with sticky formations or in motor runs in which available hydraulic energy at the bit is limited (Figure 4.7.33).

Diverging Nozzles. Diverging nozzles spread a flow as it leaves the nozzle. Cross-sectional area at the discharge orifice is increased to cause the flow rate to decelerate into a wide, low velocity spray. Dispersed, low pressure center nozzle flows improve cone cleaning in the cone apex areas and reduce tendencies toward bit balling without increasing potential for cone shell erosion. Dispersed center flows can also contribute to reductions in the loss of nose and middle row inserts by improving insert cleaning in the cone apex area.

For maximum benefit of a diverged center flow, approximately 20% of total hydraulic flow should be programmed through the center jet; 10% or less is generally too little to be effective while 30% or more erodes the cone area excessively.

Center Nozzles. Center nozzles are installed on the vertical centerline of a bit to provide a discharge onto the apex area of the three roller cones and enhance cone cleaning. Center nozzles are typically standard equipment on bit larger than 16 in.

Blank Nozzles. A plug that is used to close or block a fluid discharge port on a roller cone bit. Blank nozzles are threaded and can be field installed. They are typically used to create cross-flow configurations in bit nozzling.

Asymmetric Nozzle Configurations and Cross-flow. A bit has a symmetric nozzle configuration when three nozzles, at the same level on the periphery of a bit, of the same size and type are installed 120° to each other. A bit with an asymmetric nozzle configuration has at least two different nozzle sizes or types installed around the periphery of the bit.

When the fluid from a nozzle impinges on the well bottom, it moves away from the point of impingement in a 360°, fanlike spray. A boundary is formed where the fluid from two different jets meet. The fluid at these boundaries creates stagnant zones known as *dead zones*. In the case of a symmetric nozzle configuration, dead zones occur under the middle part of the cones (Figure 4.7.34). In an asymmetric nozzle configuration, the dead zones move away from the impingement zone of the larger jet and toward that of the smaller jet. This has the benefit of moving a stagnation zone away from the middle of the cone. Asymmetric flows ensure

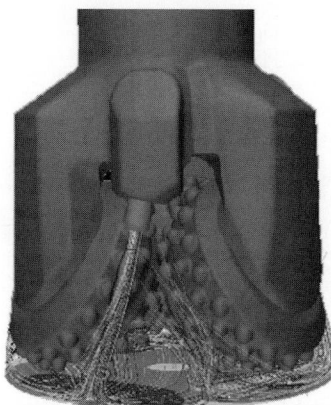

Figure 4.7.33 Vectored flow tube.

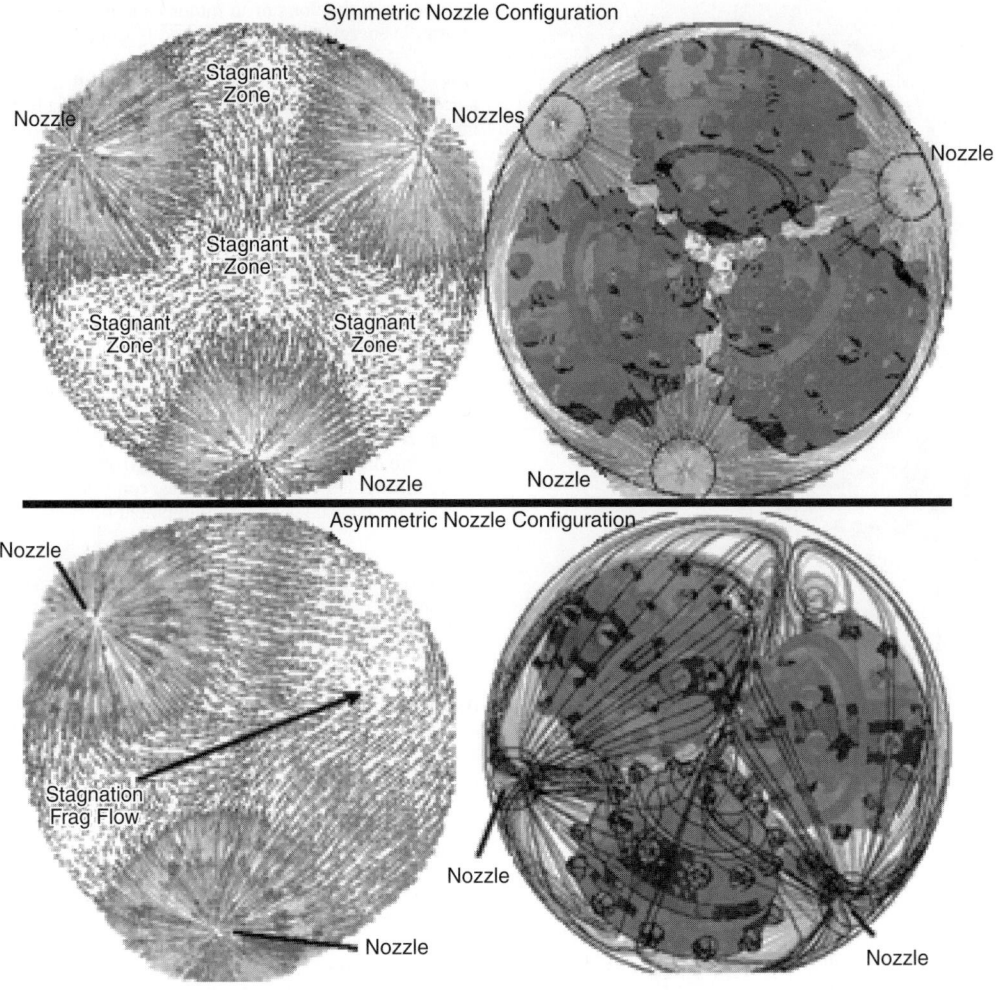

Figure 4.7.34 *Symmetric and asymmetric flow.*

that cuttings do not become impounded under a bit causing the inefficiencies of regrind, lower rates of penetration, and erosion wear to the bit.

Cross-flow is a subset of asymmetric nozzle sizing in which one jet is installed with a blank. The blanked side of the bit leaves a natural exit path for the fluid from the opposing two jets. The flow from the remaining two jets sweeps under two of the cones to improve bottom-hole cleaning and chip removal (Figure 4.7.35).

Hydraulic Computation. Hydraulic computation requires a knowledge of hydraulic nomenclature and energy, fluid velocity, and system pressure loss.

Hydraulic Nomenclature:

A_n = Area of a nozzle (sq in)
BHCP = Bottom hole circulating pressure (psi)
C_r = Chip rate (fpm)
ECD = Effective circulating density (lb/gal)
d_c = Chip diameter (in) (use 0.25)
D_h = Hole diameter (in)
D_j = Diameter of a nozzle (in)
HSI = Bit hydraulic horsepower (sq in.)

ID_P = Pipe inside diameter (in.)
I_f = Jet impact force (lb)
L = Pipe or hole depth (ft)
OD_P = Pipe outside diameter (in)
P_c = Density of a rock chip (lb/gal) use 21
P = Fluid density (lb/gal)
PV = Plastic viscosity (cp)
Q = Flow rate (gpm)
V_a = Annulus fluid velocity (fpm)
V_{ca} = Annulus critical velocity (fps)
V_{cp} = Pipe bore critical velocity (fps)
V_p = Pipe bore fluid velocity (fpm)
V_n = Jet velocity (fps)
V_s = Slip rate (fpm)
YP = Yield Point (lb/100 ft.)
ΔP_a = Annulus pressure loss (psi)
ΔP_b = Jet nozzle pressure loss (psi)
ΔP_p = Bore pressure loss (psi)
μ = Bingham plastic effective viscosity (cp)

Hydraulic Energy. Energy is the rate of doing work. A practical aspect of energy is that it can be transmitted or transformed from one form to another; for example from

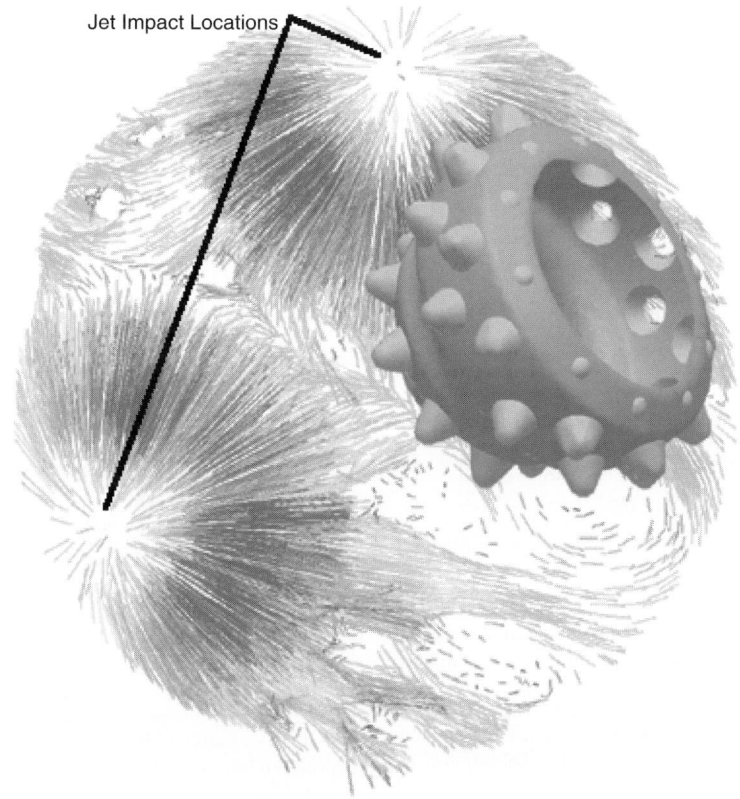

Figure 4.7.35 *Cross flow test.*

an electrical form to a mechanical form by a motor. A loss of energy always occurs during transformation or transmission. In drilling fluids, energy is called hydraulic energy or, commonly, hydraulic horsepower.

The basic equation for hydraulic energy is

$$HHP = \frac{(P)(Q)}{1714} \qquad [4.7.1]$$

in which

$$HHP = \text{Hydraulic horsepower}$$

$$P = \text{Pressure (psi) (or kP}_a)$$

$$Q = \text{Flow rate (gal/min) (or L/min)}$$

$$1714 = \text{Conversion of (psi-gal/min) to HHP}$$

$$\text{(or kP}_a - \text{L/min)} = 44{,}750$$

Rig pumps are the source of hydraulic energy carried in the drilling fluids. This hydraulic energy is commonly called the total hydraulic horsepower or pump hydraulic horsepower.

$$HHP_t = \frac{(P_t)(Q)}{1714} \qquad [4.7.2]$$

in which

$$HHP_1 = \text{Total hydraulic energy (hydraulic horsepower)}$$

$$P_1 = \text{Actual or theoretical rig pump pressure (psi)}$$

See prior equation for metric conversion.

The rig pump pressure (P_1) is the same as the total pressure loss or the system pressure loss. HHP_1 is the total hydraulic energy (rig pump) required to counteract all the friction energy (loss) starting from the Kelly hose (surface line) and Kelly, down the drill string, through the bit nozzles, and up the annulus, at a given flow rate (Q).

Bit hydraulic energy is the energy needed to counteract frictional energy (loss) at the bit, or can be expressed as the energy expended at the bit.

$$HHP_b = \frac{(P_b)(Q)}{1714} \qquad [4.7.3]$$

See prior equation for metric conversion.

Fluid Velocity. The general formula for fluid velocity is

$$V = \frac{Q}{A} \qquad [4.7.4]$$

in which

$$V = \text{Velocity (ft/min) or (m/min)}$$

$$Q = \text{Flow rate (gal/min) or (L/min)}$$

$$A = \text{Area of flow (ft}^2) \text{ (or m}^2)$$

The average velocity of a drilling fluid passing through a bit's jet nozzles is derived from the fluid velocity equation:

$$V_j = \frac{(0.32086)(Q)}{A_n} \qquad [4.7.5]$$

in which

V_j = Average jet velocity of bit nozzles (ft/sec) (or m/sec)

A_n = Total bit nozzle area (in^2) (or cm^2)

Nozzle sizes are expressed in 32nd of an inch (inside diameter) increments. Examples are 9/32 in. and 12/32 in. The "32nd" is usually not mentioned, but understood to be in 32nd of an inch. For example, 9/32-in. and 12/32-in. nozzle sizes are, respectively, expressed as 9 and 12.

The impact force of the drilling fluid at velocity (V_{j1}) can be derived using Newton's Second Law of Motion, force equals mass times acceleration. Assuming all the fluid momentum is transferred to the bottom of the hole:

$$I_j = (0.000518)(MW)(V_j) \qquad [4.7.6]$$

in which

I_j = Impact force of nozzle jets (lb of force) (or k)

MW = Mud weight (lb/gal) (or kg/L)

Q = Flow rate (gal/min) (or L/min)

V_j = Average jet velocity from bit nozzles (ft/sec)

(or m/sec)

System Pressure Loss. Pressure losses inside the drill string basically result from turbulent conditions during drilling. Viscosity has very little effect on pressure losses in turbulent flow. At higher Reynolds numbers, a larger variation results only in a small variation in the friction factor. The calculated pressure loss equations are based on turbulent flow and are corrected for mud weight instead of viscosity.

$$\Delta P = P_b = \frac{(MW)(Q)^2}{(10858)(A_n)^2} \qquad [4.7.7]$$

in which

A_n = Total combined area of the bit nozzles (in.^{2})

(or cm^2)

MW = Mud weight (lb/gal) (or kg/L)

P_b = Bit nozzle jets pressure loss (psi) (or kPa)

Q = Flow rate (gal/min) (or L/min)

Practical Hydraulic Guidelines The following table is a summary of accepted, starting hydraulics configurations (Table 4.7.2).

Table 4.7.2 *Rules of Thumb for Optimization of Roller Cone Bit Hydraulic Performance*

1. Use mini-extended nozzles except in severe balling conditions.
2. Use asymmetric nozzle configurations.
3. Configure nozzles for cross flow.
4. Re-evaluate and optimize hydraulics as applications and equipment change.
5. Design the hydraulic program to address requirements of key lithologies in the bit run.
6. Run center jets in sediments that have bit balling tendencies.
7. Use diffusing center-jets to minimize cone erosion and fluid wash.

4.7.1.2 Roller Cone Bit Components
4.7.1.2.1 Bearing, Seal, and Lubrication Systems
Roller cone bearing systems are designed to be fully operational when the cutting structure of the bit is worn out. To achieve this standard of bearing performance, modern goals for seal and bearing system life are for a million or more revolutions of a bit without failure, compared with 500,000 or fewer revolutions in the recent past. To achieve these advances, there is a major concentration of research into bearing, seal, and lubricant designs and into materials that can improve seal and bearing life.

Roller cone bits primarily use two types of bearings: roller bearings, and, journal bearings (sometimes called friction bearings). Each type is normally composed of a number of separate components, including: primary bearings, secondary bearings, seal system, features that resist thrust loading, cone retention balls, and the lubrication system (Figure 4.7.36).

Primary bearings, the main load-bearing component, are normally as large as possible within the limits of available space. Secondary bearings are smaller, reduced diameter, bearings located adjacent to the apex area of a cone. Secondary bearings provide additional load-bearing capability in an area too small in diameter for the primary bearings. Primary and secondary bearings can be roller bearings or journal bearings. It is common for a bearing system to be composed of combinations of the two, such as a primary roller bearing and secondary journal bearing.

Seals prevent the entrance of cuttings and drilling fluids into the bearing system and prevent lubricant from escaping the bearing system. Thrust washers are located on the end of leg journals and between the primary and secondary bearing surfaces to resist axial loading. Most roller cone bits incorporate what appears to be a ball type bearing. This is the cone retention locking system rather than a bearing. The bearing balls are the bit feature that retains cones on their mating journals. The lubrication system contains the stock of lubricant that will, throughout the life of the bearing system provide lubrication to bearings and seals. Each of these features is described in more detail later.

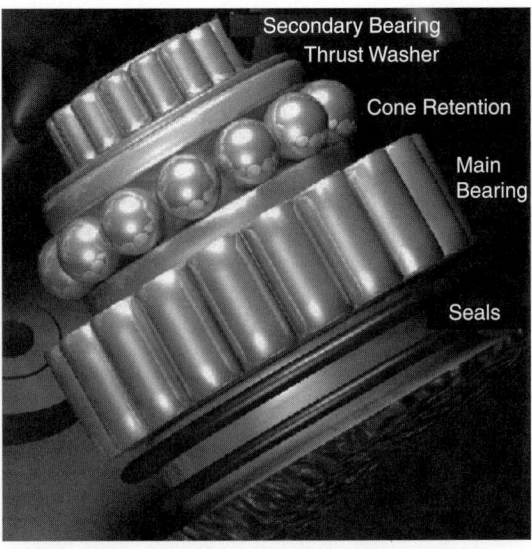

Figure 4.7.36 *Roller bearing features.*

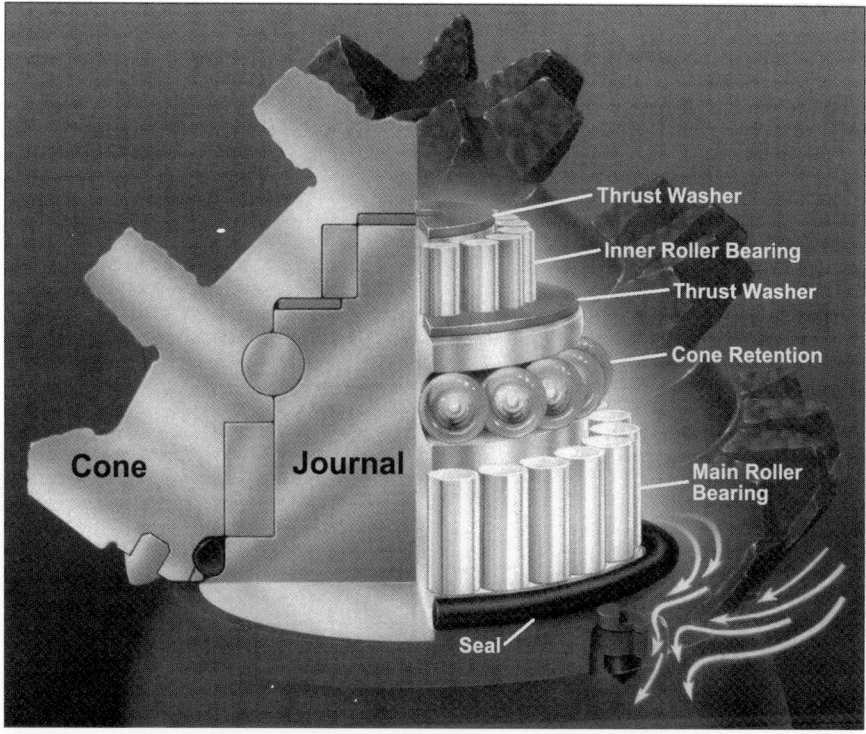

Figure 4.7.37 *Typical roller bearing arrangement.*

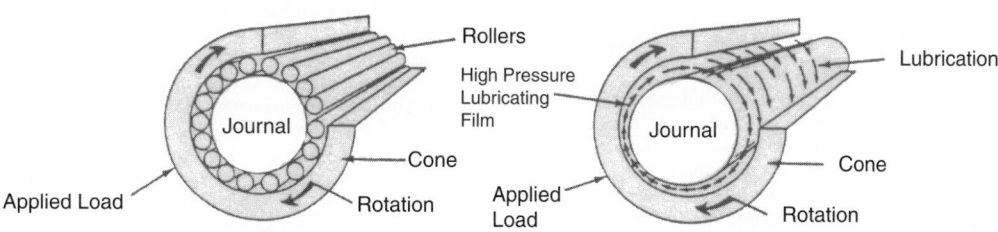

Figure 4.7.38 *Comparison of journal and roller bearing arrangements.*

Roller Bearing Systems. Roller bearings are a common bit bearing system because they can reliably support large loads and generally perform will in the drilling environment (Figure 4.7.37).

To enhance bearing life, leading manufacturers continually research bearing materials, sizing and shape. Bearing rollers tend to be loaded more at the ends than in the middle. This tends to cause spalling at the static contact point. To avoid damage from end-loads, rollers are sometimes contoured so the central portion of the roller has a slightly larger diameter than the ends forcing uniform load distribution across the entire roller length. Rollers of this type are called contoured rollers. They lower bearing operating temperatures and reduce bearing fatigue.

Bearings are commonly manufactured to precise roller tolerances (0.0005 in.) ensure uniform load distribution and are in some cases individually matched to increase bearing life. These improve bearing efficiency and help control temperature. Precise tolerances, in conjunction with contoured rollers, significantly increase bearing capacity compared with previous bearing systems, making it possible for bits

to function normally under higher weights on the bit and rotary speeds and still provide longer bearing life.

Journal Bearing Systems. Journal bearings consist of at least one rotating surface separated by a film of lubricant. The surfaces are specially designed so that the lubricant film keeps them separated; were they to touch, mating bearing components would gall or possibly even fuse. Journal bearings are generally inexpensive to manufacture and, when functioning properly, are very efficient. A comparison of roller bearing and journal bearing assemblies is presented in Figure 4.7.38.

Hydrodynamic Lubrication. Journal bearings are able to function efficiently and avoid wear because of a hydrodynamically generated film of lubricant that separates the rotating surfaces. The thickness of this lubricant film depends on the following variables:

Radial clearance between the leg journal and cone
Radius of the leg journal

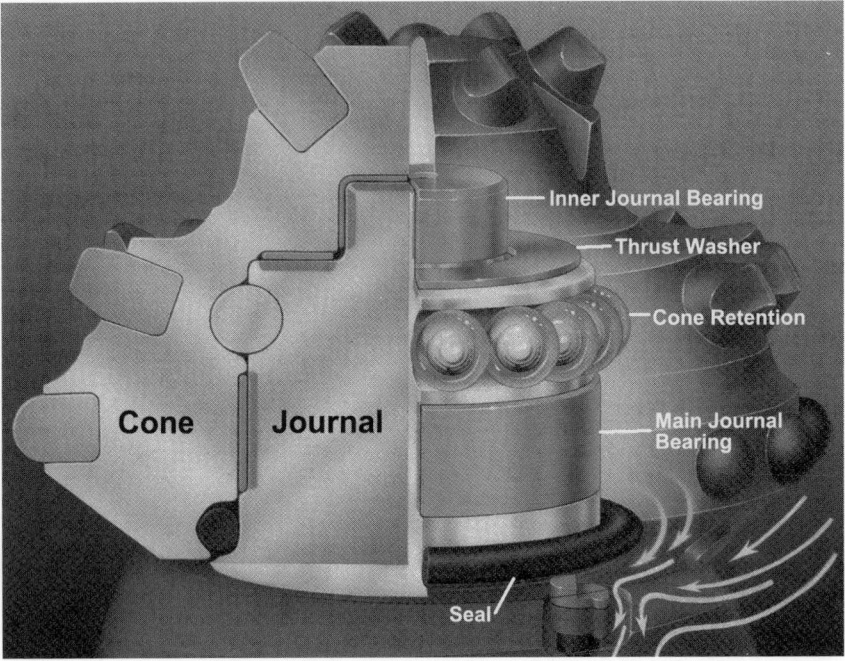

Inner Journal Bearing

Thrust Washer

Cone Retention

Main Journal Bearing

Cone   Journal

Seal

Figure 4.7.39 *Typical journal bearing system.*

Lubricant viscosity
Rotational velocity of the cone
Applied load

So long as the hydrodynamic conditions exist to create a lubricating film between the two rotating surfaces, the life of a journal bearing is virtually limitless. If the film breaks down for any reason (e.g., high loads, low speeds, low viscosity), the bearing surfaces will come into contact with each other, and a phenomenon called *boundary lubrication* takes place. Film thickness decreases as rotational speed decreases or as bearing loading increases.

In hydrodynamic lubrication, heat generated is a function of lubricant viscosity, applied load, and relative speed between cone and journal. As bearing temperature increases, lubricant viscosity and film thickness decrease creating a potential for journal bearing malfunction. Heat generated varies with the square of the speed. A twofold increase in bearing rotational speed produces a fourfold increase in the generation of heat in the bearings. In larger bits, heat generation can be significant and large bearing design must include provisions for removal of heat from the bearing area.

Boundary Lubrication. Boundary lubrication takes place in a transition between hydrodynamic lubrication and metal-to-metal contact in which there is no lubricant film at all. This transition exists because of surface roughness existing on the journal and cone bearing surfaces.

Peaks in surface roughness are called *asperities.* Asperities exist on all machined surfaces; even mirror finishes do not eliminate microscopic asperities on a bearing surface.

As a hydrodynamic film thins due to speed, load or viscosity, asperities touch and boundary lubrication occurs. The

degree to which they can be expected to touch varies. Bearing materials are carefully selected so that damage that will be done to a bearing by asperity interaction is minimized. The surfaces are designed so that a polishing effect is produced and improved, "worn-in" performance results.

In boundary lubrication, heat generated is a function of lubricant film strength, applied load, and the metallurgical properties of the cone and journal surfaces. Rough surfaces generate more heat than those that are smooth, and hard surfaces generate more heat than the soft. As a high pressure lubricating film becomes thinner, asperity interaction increases. Heat generated by boundary lubrication is greater than that generated by hydrodynamic lubrication and will eventually cause bearing failure.

Design of Journal Bearings. Journal bearings must be designed to provide a balanced bearing geometry, adequate journal strength, and maximize the thickness of the high pressure lubricating film developed during hydrodynamic lubrication. Surface areas, journal and cone diameters, and clearances between journal and cone are factors that affect the thickness of lubricating films (Figure 4.7.39).

The metallurgy of the bearing must be balanced to minimize heat generation during boundary lubrication. Cone-bearing surfaces are steel. Soft, silver-plated, sleeves are installed on the journal. Silver polishes easily and minor surface irregularities from machining are quickly smoothed. This ensures low friction operation, and uniform lubricant flow over the bearing surface.

Manufacturing tolerances must be precise so that surfaces run true to their design condition. Roundness of journal and cone surfaces is important. If any part of a bearing is out of round, the effectiveness of the lubrication regime will be adversely affected.

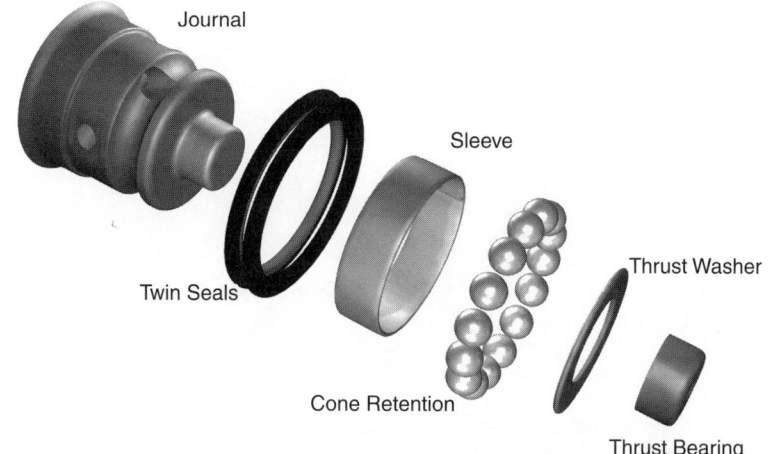

Figure 4.7.40 *Typical journal bearing sleeve.*

Journal Bearing Materials. Specifically formulated, high tensile strength and ductile metal alloys are used for journal bearing sleeves and thrust washers (Figure 4.7.40). Tensile strength provides load-bearing and high-rotational-speed capabilities; ductility increases impact resistance. These alloys are specifically designed to minimize the effect of asperity deformation on bearing performance. Surfaces are commonly silver plated because the softness of silver lends itself to smoothing under asperity deformation. Bearing smoothness encourages hydrodynamic lubrication, resulting low-friction, low-temperature operation and long bearing life.

Open Bearing Systems. Air-cooled, nonsealed roller bearings are commonly used for air drilling applications. When air, gas, or mist is used as a drilling fluid, such as blast hole drilling in open pit mines, nonsealed roller bearing bits are used. Open bearing designs rely on drilling fluid for cooling, cleaning, and lubrication of the bearings. If water is injected into the air or gas, bits are equipped with a water separator to prevent water in contact with loaded bearing surfaces and thereby reduce their life. Backflow valves prevent cuttings suspended in water from backing up through the bit into the drill pipe when the flow of air or gas is interrupted.

Seal Systems. In general, seal systems are classified as either static or dynamic. Roller cone bits use both types of seals. Dynamic seals involve sealing across surfaces that are moving in relationship to one another, as would be the case for a bearing seal. Seal parts or surfaces that do not move in relationship to one another during bit operation, such as the seal between a hydraulic nozzle and the bit to prevent leakage around the joint, are static seals.

Roller cone bearing seals operate in an exceptionally harsh environment. Drilling mud and most cuttings are extremely abrasive. Drilling fluids often contain chemicals, and operating temperatures can be sufficiently high to break down the elastomers from which seals are made. Pulses often occur in downhole drilling fluids that apply lateral loading on seals that must be resisted.

Bearing Seals. On a purely practical level, bearing seals have two functions: to prevent foreign materials such as

mud, cuttings, chemicals, and water from entering the bearings and to prevent bearing lubricant stocks from escaping the bit.

Visualize the difference in the nature of these two duties. On the interior side, the seal is excluding clean, functional lubricant from escaping the bit, whereas on the exterior side, the seal is excluding dirt and chemicals from penetrating the bit. The separation of these two extremely different functions takes place at a small point between the two sides of a seal. If either of the functions breaks down, the bearings and the bit could be destined for damage or failure. These facts cause most bit manufacturers to aggressively engage in sealing research.

Seal Definitions. In a rotating bearing, two working sides of a seal are respectively called the energizer, and, the dynamic wear face. These two parts are directly opposite, one to the other, with the energizing portion bearing on the gland and the dynamic wear face bearing against the rotating unit. For the energizing portion of the seal to function properly, it must have a surface against which to react. This is provided by a channel called a seal gland.

The wearing portion of the seal must have the capability to withstand the heat and abrasion generated as the rotating surface passes over it. The energizer, when functioning correctly, is not a high wear area. Ideally, it simply bears against the gland and provides the pushing energy that keeps the wear surface firmly in contact with the rotating unit.

O-Rings. Donut-shaped O-rings are used in many roller cone bit applications. O-rings are manufactured from elastomers (i.e., synthetic rubbers) that withstand the temperatures, pressures, and chemicals encountered in the drilling environment. They are a traditional but still consistently reliable seal system.

An O-ring is installed in a seal gland to form a seal system. The gland holds the O-ring in place and is sized so the O-ring is compressed between the gland and the bearing hub at which sealing is required. This compression, typically 9% to 12% in rock bits, is the sealing force between the gland and hub. It firmly presses the interior wall of the O-ring against the hub and the exterior diameter of the O-ring against the gland. This latter force tends to prevent the seal from turning

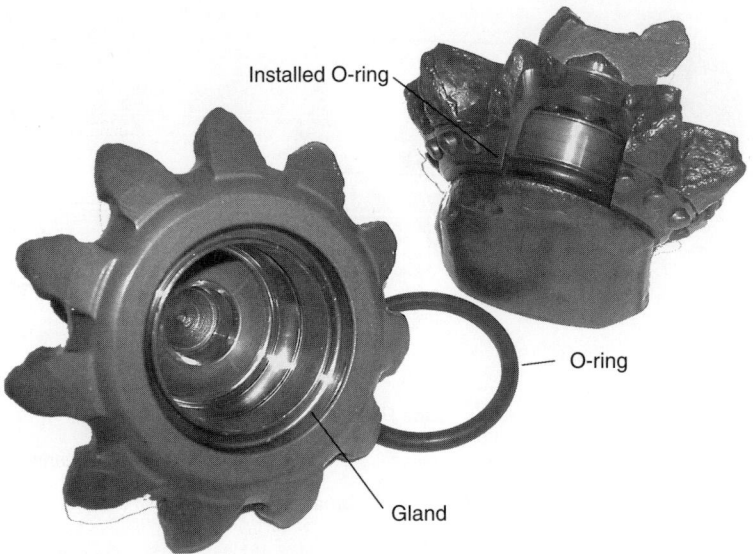

Figure 4.7.41   *O-Ring and gland.*

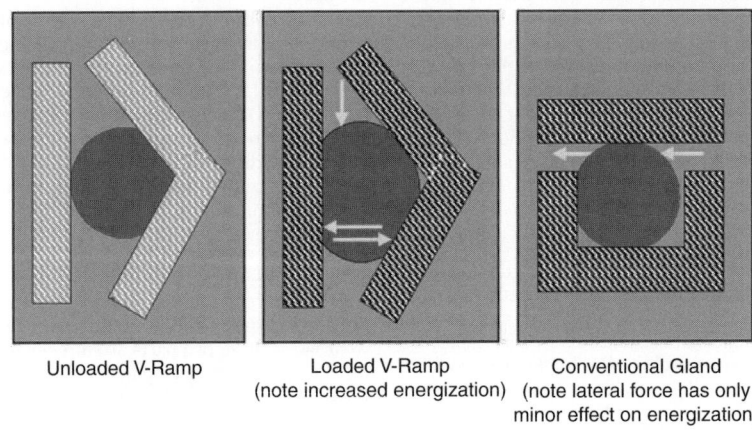

Unloaded V-Ramp

Loaded V-Ramp
(note increased energization)

Conventional Gland
(note lateral force has only
minor effect on energization)

Figure 4.7.42   *O-Ring with modern "V" Type gland.*

in the gland and experiencing wear on the outer surfaces from rotating contact with the gland (Figure 4.7.41).

Many roller cone bits use traditional O-ring seal systems. In some cases, O-ring seal efficiency is enhanced by innovative glands, such as a V shape, designed to increase energization entropy as load increases (Figure 4.7.42). The two-point contact between the V-surface and O-ring effectively prevents seal rotation. This arrangement also responds particularly well to the periodic pulse loadings that are often propagated through drilling fluids. If a pulse occurs, in either direction, it moves the O-ring in the direction in which the pulse is moving. As this movement develops, geometry of the V surface causes a wedge-like action against the O-ring that increases seal energization.

Shaped, Bi-material Seals. As with all parts of a modern roller cone bit, new materials and engineered shapes play a large role in seal performance. Different materials are combined to optimize energization and resist seal wear. These would include a highly entopic, soft elastomer to provide energization and a very hard elastomer or polymer fabric in the dynamic wear area. Seal shape is engineered to geometrically optimize energization and wear resistance. The modern seal solution shown later is shaped so a small contact area on the energization side increases pressure at the point of contact with the gland and prevents undesirable seal rotation. The dynamic wear portion of the seal is a relatively large surface resulting in low pressure on the wear area. This encourages low running temperatures and wear rates (Figure 4.7.43).

Dual-seal systems optimize the two sealing functions, excluding entry of contaminants into the bearing system and preventing loss of lubricant from the system (Figure 4.7.44). The individual seal that prevents contaminant entry is called the secondary seal. As one looks at the seal and bearing system, the secondary seal is the farther of the two seals from the bearings. The seal that prevents system lubricant loss is called the primary seal; it is located closest to the bearing.

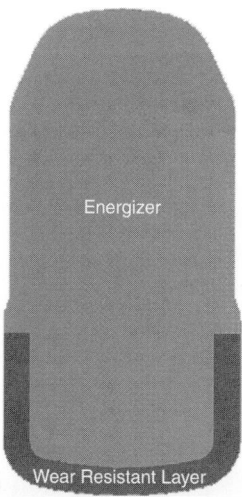

Figure 4.7.43 Cross-section of modern bi-material seal.

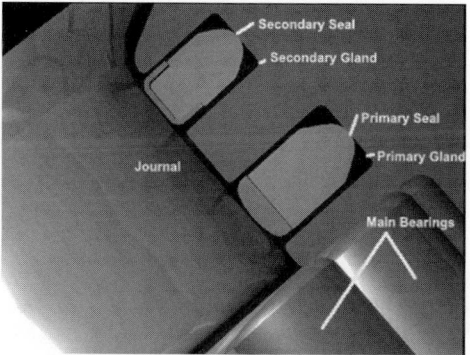

Figure 4.7.44 Typical dual-seal system.

The design, both shape and composition, of primary and secondary seals is very different. Dynamic wear surfaces for the pair are individually optimized to the different environments in which they work. Because the cavity between the primary and secondary seals is filled with lubricant under pressure, an added impedance to the entry of foreign solids and fluids into the bearing is introduced. Single seal systems must provide both sealing functions with a single seal.

Twin seals stand up well under cycling and uneven bit loading. Seal integrity is maintained in harsh drilling environments such as high rotational speeds, high weights on bit, high dogleg severity, high mud weights, and deep, hot environments.

Lubrication of Seals. Seals must be lubricated to prevent high wear rates and excessive temperatures that could lead to seal material failures. Lubricant for the bearings also serves to lubricate the seals.

4.7.1.2.2 Lubrication Systems and Lubricants
Lubricants play a vital role in bearing performance. They provide lubrication for both bearings and seals, and they provide a medium for heat transfer away from the bearings. In popular dual-seal bearing systems, lubricant between primary and secondary seals further serves as an additional

sealing element by impeding the migration of foreign matter that may penetrate the primary seal. To achieve these functions, lubricants are specially engineered and continually improved. Lubrication systems are engineered to provide reserve storage, positive delivery to the bearing system, capacity for thermal expansion, and pressure equalization with fluids on the bit exterior.

Lubrication systems include a resupply reservoir sufficiently large to ensure availability of lubricant for all lubrication functions throughout the life of the bit. A small positive pressure differential in the system ensures flow from reservoir to bearings. The system is vented to equalize internal and external reservoir pressures. Without equalization, a pressure differential between bit exterior and interior could be sufficient to cause seal damage and bearing failure.

Lubricants. High drilling temperatures and high pressures in the lubrication system, together with the potential of exposure to water and chemicals, require high performance from lubricants. Most bit lubricants are specially formulated. Leading bit manufacturers employ scientists to develop and test lubricants. Superior lubricants are stable to temperatures above 300°F, and many also function normally at temperatures down to about 0°F. They are also hydrophobic (i.e., repel water) and retain their stability if water succeeds in penetrating the bit, and are resistant to chemicals normally found in drilling fluids. Superior lubricants are environmentally safe and do not contain lead additives that have traditionally been included to help resist high pressures.

Lubricant Supply. Roller cone bits typically contain one lubricant reservoir in each leg (Figure 4.7.45). For a three-cone bit there are three reservoirs. Each must have the capacity for sufficient reserves of lubricant for operation of the bearing assembly it serves throughout the bit's life.

Reservoirs are more complex than most would expect. As temperatures in a bit rise, lubricants expand thermally. Some lubrication systems vent over-volumes resulting from thermal expansion to the bit exterior. Better designs, however, include extra capacity in the lubricant reservoir to accept expansion volumes without necessity for loss of lubricant by venting. Reservoirs are also a point at which heat

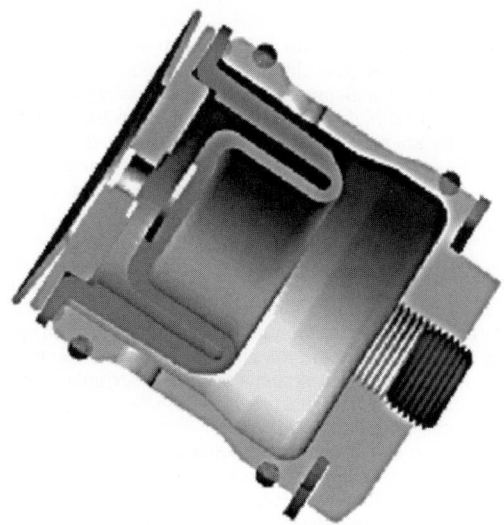

Figure 4.7.45 Typical roller cone bit lubricant reservoir.

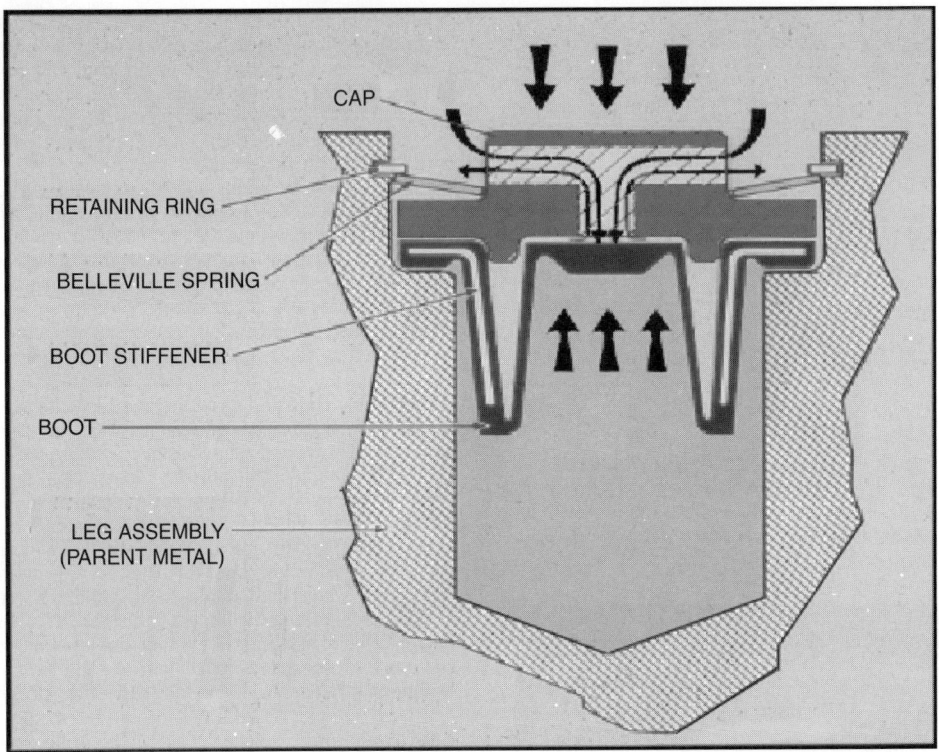

Figure 4.7.46 Pressure equalization.

removed from the bearings by means of the lubricant can be dissipated.

Well-designed reservoirs always include pressure relief capacity. Pressure beyond a set point is relieved. An over pressure condition in the reservoir system will cause a pressure relief valve to lift and release grease until the undesirable differential pressure is equalized. Such releases are rarely required.

Pressurization, Pressure Equalization, and Relief. The column of drilling fluids and cuttings, contained in a well, exert very high pressures on a bit operating at the well bottom. These high pressures are applied to the seal system. They are resisted by lubricant in the seal and bearing system. At installation, lubricant is at atmospheric pressure and it cannot provide significant resistance to well bottom pressures. Internal lubrication system pressures in well-designed systems accordingly equalize themselves with external bit pressures to prevent seal failure. Equalization is accomplished by two small valves installed in the bit. Seal gap compensation valves equalize any pressure differential that may occur between the primary and secondary seals in dual-seal systems (Figure 4.7.46), and lubricant reservoir dome vent valves. Twin seal systems require a secondary pressure equalization valve that prevents formation of a pressure differential between the seals.

Typical, high-performance lubrication systems establish positive lubricant flows to the bearings. The reservoir is also equipped with a rubber diaphragm open to contact with drilling fluids. When the bit is tripped toward the well bottom, drilling fluid is forced through a small opening in the dome of the bit onto the diaphragm at the hydrostatic pressure of the fluid column. This causes the diaphragm to deflect inward into the reservoir thereby pressurizing the lubricant. As lubricant is consumed or escapes through the seal system, this pressure causes lubricant to flow to the bearings.

4.7.1.2.3 Protective Features
Design engineers give a good deal of consideration to means by which roller cone bit legs can be protected from abrasive wear. Quality bits are shipped with hardfacing protection in the shirttail area. A number of additional protective features are commonly added, depending on abrasive severity (such as the extremes of horizontal or high angle wells, bent motor subs and housings, formations prone to caving, and inadequate hole cleaning). These include hardfacing on the upper leg, use of flush mounted tungsten carbide inserts in the leg, use of exposed tungsten carbide or semi-round top diamond-enhanced inserts in a variety of patterns, and lug pads equipped with carbide of diamond-enhanced inserts (Figures 4.7.47 and 4.7.48).

4.7.1.3 Special Purpose Designs
4.7.1.3.1 Mono-cone Bits
Mono-cone bits were first used in the 1930s because they have a number of theoretical advantages. In practice, the design has not been broadly used. Bit research, encouraged by advances in cutting structure materials, continues to keep this concept in mind because it has the room for extremely large bearings and has a very low cone rotation velocities that suggest potential for long bit life. Although they are of obvious general interest, these are particularly advantageous in smaller diameter bits in which bearing sizing presents significant engineering problems. Mono-cone bits drill efficiently at lower torques than those required by three-cone bits. This

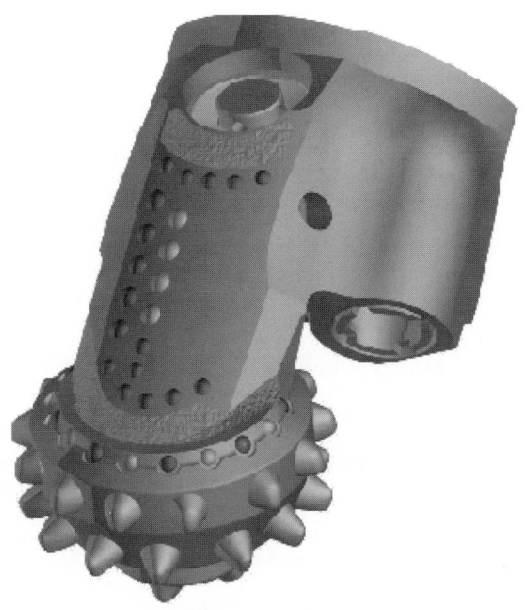

Figure 4.7.47 *Leg protection.*

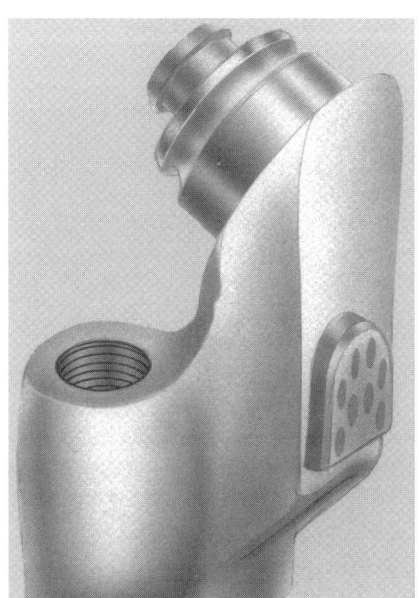

Figure 4.7.48 *Typical lug pad with tungsten carbide inserts.*

combination of characteristics are promising for small diameter bits and for motor driven directional applications (Figure 4.7.49).

Mono-cone bits drill differently from three-cone bits. Drilling properties can be similar to both beneficial crushing properties of roller cone bits and shearing action of PDC bits. Cutting structure research focuses partly on exploitation of both mechanisms encouraged by the promise of efficient shoe drill-outs and drilling in formations with hard stingers interrupting otherwise "soft" conditions.

The design provides ample space to enable nozzle placements for efficient bottom hole and cutting structure cleaning. Ample space also permits the use of large inserts.

Figure 4.7.49 *Mono-cone bit.*

Figure 4.7.50 *Two-cone bit.*

Unlike three-cone bits, most mono-cone inserts cut gage. This renders the bit under gage as inserts wear. Modern, ultra-hard cutter materials properties can almost certainly extend insert life and expand the range of applications in which this design could be profitable.

4.7.1.3.2 Two-Cone Bits

Two cone bit designs are another product whose origin lies well into the past. The first roller cone patent, issued in August 1909, covered a two-cone bit. Two cone bits have a reputation for drilling straight wells. As with mono-cone bits, two cone bits have available space for larger bearings and rotate at lower speeds than three-cone bits. Bearing and seal life for a particular bit diameter is greater than for a comparable three-cone bit. Two cone bits, while not common, are available and used in special applications (Figure 4.7.50). More importantly, their advantages always keep them in

mind for research and incorporation of modern materials into the design could renew interest in them.

The cutting action of two-cone bits similar to that of three-cone bits but the nature of cutting structure geometry is such that fewer inserts simultaneously contact the hole bottom. Penetration per insert is, accordingly, increased providing a particularly beneficial result in applications in which capabilities to place WOB are limited.

The additional space available in two cone designs has several advantages. It is possible to have large cone offset angles that produce increased scraping action at the gage. Space also enables excellent hydraulic characteristics by room for placement of nozzles close to the bottom. It also allows for use of large inserts or teeth that can extend bit life or efficiency.

Two cone bits do have a tendency to bounce and vibrate. This characteristic is especially a concern for directional drilling. This and advances in three-cone bearing life and cutting structures, cause use of two cone bits to be presently uncommon. Good designers have not forgotten them, however; the concept has advantages that cannot be ignored. As with many roller cone bit designs, modern materials and engineering capabilities frequently resolve problems that have been troublesome in the past.

4.7.1.4 Roller Cone Bit Nomenclature

Roller cone bits are generally classified as either tungsten carbide insert bits or milled tooth bits. To assist in comparison of similar products from various manufacturers, the International Association of Drilling Contractors (IADC) has established a unified, bit classification system for the naming of drill bits. This system is described later.

4.7.1.4.1 IADC Roller Cone Bit Classification Method

The IADC Roller Cone Bit Classification is an industry wide descriptive standard for description of milled tooth and insert type roller cone bits. It is a coding system based on key design and application related criteria. The presently used version was introduced in 1992 using criteria that were cooperatively developed by drill bit manufacturers under the auspices of the Society of Petroleum Engineers [3].

IADC Classification. The IADC classification system is a four-character design and application related code. The first three characters are always numeric; the last character is always alphabetic. The first digit refers to bit series, the second to bit type, the third to bearings and gage arrangement, and the fourth (alpha) character to bit features.

Examples of IADC 135M, 447X, and 637Y bits can be seen in the following photograph (Figure 4.7.51). These IADC codes represent the following bits and environments:

135M: Soft formation (1), milled tooth bit (3); roller bearings with gage protection (5); motor application (M)

135M   447X   627Y

Figure 4.7.51 *IADC codes.*

Table 4.7.3 *Typical Compressive Strengths of Common Rock Types*

Hardness	UCS (psi)	Examples
Ultra-Soft	<1000	gumbo, clay
Very Soft	1000–4000	unconsolidated sands, chalk, salt, claystone
Soft	4000–8000	coal, siltstone, schist, sands
Medium	8000–17,000	sandstone, slate, shale, limestone, dolomite
Hard	17,000–27,000	quartzite, basalt, gabbro, limestone, dolomite
Very Hard	>27,000	marble, granite, gneiss

UCS = Uniaxial Unconfined Compressive Strength.

447X: Soft formation (4), insert bit (4); friction bearings with gage protection (7); chisel type inserts (X)
637Y: Medium-hard (6), insert bit (3); friction bearing with gage protection (7); conical type inserts (Y)

Series. Series, the first character in the IADC System, defines general formation characteristics and divides milled tooth and insert type bits. Eight Series (or categories) are used to describe roller cone rock bits.

Series 1, 2, and 3 apply to milled tooth bits. Series 4, 5, 6, 7, and 8 apply to insert type bits. The higher the series number, the harder or more abrasive is the rock type. Series 1 represents the softest (easiest drilling applications) for milled-tooth bits. Series 3 represents the hardest and most abrasive applications for milled tooth bits. Series 4 represents the softest (easiest drilling applications) for insert type bits. Series 8 represents very hard and abrasive applications for insert-type bits.

Unfortunately, rock hardness is not clearly defined by the IADC System. The meaning of "hard" sandstone or "medium-soft" shale, for example are subjective and open to a degree of interpretation. Information should, accordingly, be used only in a descriptive sense; actual rock hardness will vary considerably depending on factors such as depth, overbalance and hydrostatic pressure, porosity and other factors that are difficult to quantify (Table 4.7.3).

Type. The second character in the IADC Categorization System represents bit type, insert, or milled tooth and describes a degree of formation hardness. Types range from 1 through 4.

Bearing Design and Gage Protection. The third IADC character defines both bearing design and gage protection. IADC defined seven categories of bearing design and gage protection. These are

1. Nonsealed roller bearing (also known as open bearing bits)
2. Air-cooled roller bearing (designed for air or foam or mist drilling applications)
3. Nonsealed roller bearing, gage protected
4. Sealed roller bearing
5. Sealed roller bearing, gage protected
6. Sealed friction bearing
7. Sealed friction bearing, gage protected

Gage protected indicates only that a bit has some feature that protects or enhances the gage of the bit. It does not specify the nature of the feature. For example, increased tooth count, modified insert shape or material, and diamond-enhanced inserts can all be categorized as gage protection.

Table 4.7.4 *Roller Cone Bit Major Dull Characteristics Codes*

Code	Description	Code	Description
BC	Broken Cone	LN	Lost Nozzle
BT	Broken Teeth/Cutters	LT	Lost Teeth/Cutters
BU	Balled-Up Bit	NO	No Dull Characteristics
CC	Cracked Cone	OC	Off Center Wear
CD	Cone Dragged	PB	Pinched Bit
CI	Cone Interference	PN	Plugged Nozzle or Flow Passage
CR	Cored	RG	Rounded Gage
CT	Chipped Teeth/Cutters	SD	Shirttail Damage
ER	Erosion	SS	Self-Sharpening Wear
FC	Flat Crested Wear	TR	Tracking
HC	Heat Checking	WO	Washed Out Bit
JD	Junk Damage	WT	Worn Teeth/Cutters
LC	Lost Cone		

Included Features. The fourth character used in the system defines the features available. IADC considers this category to be optional. This alpha character is not always recorded on bit records but is commonly used within bit manufacturer's catalogs and brochures. IADC categorization assigns and defines 16 identifying features (Table 4.7.4). Only one alphabetic feature character can be used under IADC rules. As bit designs commonly combine several of these features, the most significant feature is normally listed.

Limitations of IADC Classification. When using IADC coding, keep in mind that it is the bit manufacturer's responsibility to create IADC codes for their bit designs. There is no independent authority monitoring coding. Given the subjectivity of the system, this can put into question whether a bit has been properly positioned.

Bit designs can vary substantially within a single IADC code; three $7\frac{7}{8}$-in. insert type bits using a manufacturer's model designations of A, B, and C, are categorized within the IADC Classification as 517 despite large differences in the bits. Insert count for the inner row cutting structure increases by 30% from A to C (50 versus 65 inserts) and by 18% for the outer rows. This variation in insert counts could have a substantial impact on performance in a particular application and points out a weakness in the IADC Coding System (Figure 4.7.52).

It is very important to understand that an IADC coding does not limit the drilling application for a particular bit. The following paragraph from the IADC/SPE clearly highlights this point: "The fact that each bit has a distinct code does not mean that it is limited to drilling only the narrow range of formations defined by a single box on the chart. All bits, within reason, drill effectively in both softer and harder formations than specified by the IADC code. Competitive products with the same IADC code are built for similar applications but they may be quite different in design detail, quality and performance" [4].

4.7.1.5 IADC Roller Cone Bit Dull Grading

The International Association of Drilling Contractors (IADC), in conjunction with the Society of Petroleum Engineers (SPE), has established a systematic method by which the cause for bit failures can be clearly communicated. The intent of the system is to facilitate and accelerate product and operating development based on accurately recorded experience. This system is referred to as Dull Grading. The Dull

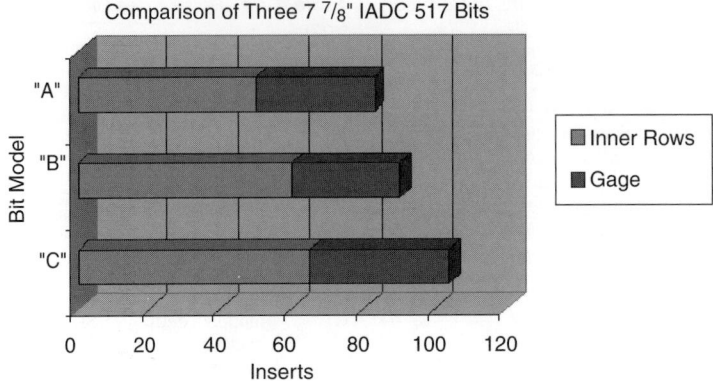

Figure 4.7.52 *Insert counts for 7 7/8" IADC 517 F14, F16, and F17 bits.*

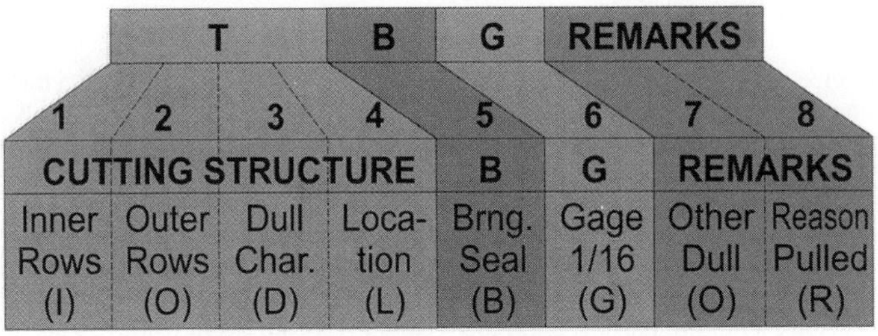

T				B	G	REMARKS	
1	2	3	4	5	6	7	8
CUTTING STRUCTURE				B	G	REMARKS	
Inner Rows (I)	Outer Rows (O)	Dull Char. (D)	Location (L)	Brng. Seal (B)	Gage 1/16 (G)	Other Dull (O)	Reason Pulled (R)

Figure 4.7.53 *IADC dull grading system.*

Grading Protocol requires a systematic evaluation of specific bit areas, reports the reason why the bit was removed from service, and provides a standard method for reporting.

Partly as a result of dull analyses, bit design processes and product operating efficiencies evolve quite rapidly. Engineers identify successful design features that can be reapplied and unsuccessful features that must be corrected or abandoned; manufacturing units receive feedback on product quality; sales personnel migrate performance gains and avoid duplication of mistakes between similar applications, and so forth. All bit manufacturers require the collection of dull information for every bit run.

4.7.1.5.1 IADC Eight-Point Dull Grading System

The IADC Eight-Point Dull Grading System reviews bit wear in four general categories: Cutting Structure, Bearings and Seals, Gage, and Remarks. These, and their subcategories, are described by the following graphic (Figure 4.7.53). Dull Grading for diamond and PDC matrix bits differs from that for roller cone bits in the way that cutting structures are measured and graded. When grading a diamond or PDC bit, specific measurement rules appropriate to that group must be used.

Cutting Structure Grading (T). For dull grading purposes, cutting structures are subdivided into four subcategories: inner rows, outer rows, major dull characteristic of the cutting structure, and location on the face of the bit where the dull characteristic occurs. Figure 4.7.54 provides information on insert or tooth row identification and the convention for reporting their wear.

Insert or Tooth Height Measurement. Measurements of cutting structure condition requires an evaluation of bit tooth or insert wear status. Wear is reported using an eight-increment wear scale in which minimal wear is represented by the notation T1, and 100% worn is represented by T8 (Figure 4.7.55).

General Rule for Cone Identification. Typically, the #1 cone contains the centermost cutting element. On some bits, the #1 and #3 cones both appear to have the centermost cutting element; in general, the cone with the greatest distance between the A and B rows is the #1 cone. Cones #2 and #3 follow the #1 cone in a clockwise direction. This rule is does not correctly define the first cone for some modern bits, however, and it may be necessary to verify with the manufacturer.

Inner Rows Cutting Structure (I). Dull grading begins with evaluation of wear on the inner rows of inserts or teeth, that is, with the cutting elements not touching the wall of the hole bore. Grading involves measurement of combined inner rows structure reduction due to loss, wear, and breakage using the measurement method described previously.

Outer Rows Cutting Structure (O). The second step is the outer rows of inserts or teeth, that is, those that touch the wall of the hole bore. Grading involves measurement of combined outer rows teeth or insert structure reduction due to loss, wear, and breakage using the measurement method described previously.

Cutting Structure Dull Characteristic (D). The cutting structure dull characteristic is the observed characteristic that is

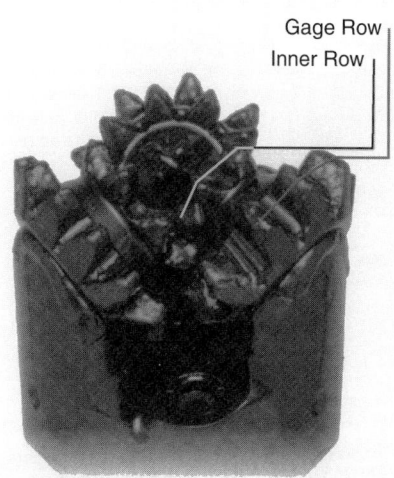

Figure 4.7.54   *Identifying insert and milled tooth rows.*

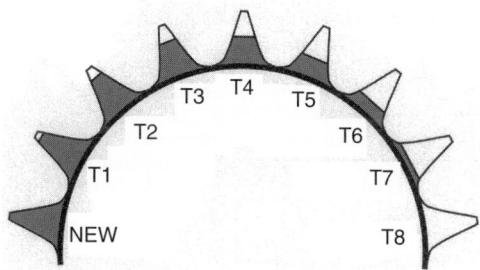

Figure 4.7.55   *Tooth height measurement.*

most likely to limit further usage of the bit in the intended application. A two-letter code is used to indicate the major dull characteristics of the cutting structure. These codes are listed in Table 4.7.4.

Location (L). A two-letter code is also used to indicate the location of the wear or failure that necessitated removal of the bit from service. These codes are listed in Table 4.7.5.

Bearings and Seals Grading (B). IADC provides separate protocols for estimation of bearing and seal wear in nonsealed and sealed bearing assemblies.

Estimating Wear on Nonsealed Bearings. For nonsealed bearings, wear is estimated on a 0 to 8 linear scale (0 is new, 8 is 100% expended.)

Table 4.7.5   *Roller Cone Bit Dull Location Codes*

Code	Description	Code	Description
N	Nose Row	1	Cone #1
M	Middle Row	2	Cone #2
G	Gage Row	3	Cone #3
A	All Rows		

Table 4.7.6   *Roller Cone Bit Bearing/Seal Evaluation Check List*

Condition	Acceptable Condition
Ability to rotate cone	Rotates normally
Cone spring-back	Spring-back exists
Seal squeak	Seal squeak exists
Internal sounds	No internal noises exist
Weeping grease	No lubricant leaks exist
Shale burn	No shale burn exists
Shale packing	If packing exists, remove before measuring
Gaps at the back-face or throat	No bearing gaps exist
Inner or outer bearing letdown	No bearing letdown exists

Estimating Wear on Sealed Bearings. A checklist for seal and bearing system condition is provided in Table 4.7.6. The grading protocol is

If no seal problems are encountered, use the grading code E.

If either any component in the assembly has failed, use the grading code F.

If any portion of the bearing is exposed or missing, it is considered an ineffective assembly; again use the grading code F.

Use grading code N if it is not possible to determine the condition of both the seal and the bearing.

Grade each seal and bearing assembly separately, by cone number (If grading all assemblies as one, report the worst case.)

Gage Grading (G). The Gage Category of the Dull Bit Grading System is used to report an under gage condition for cutting elements intended to touch the wall of the hole bore. For diamond and PDC bits only, gage is measured with a nominal ring gage (API specifications for nominal ring gages for roller cone bits have not been issued).

Mesured Distance

2 Cone Bits

AMOUNT OUT OF GAGE =
MEASURED DISTANCE

Measured Distance

3 Cone Bits

AMOUNT OUT OF GAGE =
MEASURED DISTANCE X 2/3

Figure 4.7.56 *Measuring out of gage.*

Undergage increments of $\frac{1}{16}$ in. are reported. If a bit is $\frac{1}{16}$ in. under gage, the gage report is "1." If a bit is $\frac{1}{8}$ in. ($\frac{2}{16}$ in.) undergage, the gage report is "2." If a bit is $\frac{3}{16}$ in. undergage, the gage report is "3," and so forth. Round to the nearest $\frac{1}{16}$ in. Gage rules apply to cutting structure elements only. Measurement are taken at the closer to gage of either the gage or heel cutting elements (Figure 4.7.56).

For two-cone bits, gage is the measured distance from the closer to gage of the gage or heel elements. For bits with bearing or seal failures, out of gage amount can be measured provided that cones do not have axial or radial movement. If nominal ring gages are not available, the true gage condition cannot be determined.

Remarks. The Remarks category is more important than its title implies. It enables explanation of dull characteristics that do not correctly fit into other categories, and it is the category in which the reason why a bit was removed from service is confirmed.

Other Dull Characteristics (O). Other dull characteristics can be used to report dull characteristics other than those reported under "Cutting Structure Dull Characteristics (D)." Codes for "Other Dull Characteristics" are shown in Table 4.7.7.

Table 4.7.7 *Other Dull Characteristics Codes*

Code	Description	Code	Description
BC	Broken Cone	LT	Lost Teeth/Cutters
BT	Broken Teeth/Cutters	NO	No Dull Characteristics
BU	Balled UP	OC	Off Center Wear
CC	Cracked Cone	PB*	Pinched Bit
CD	Cone Dragged	PN*	Plugged Nozzle or Flow Passage
CI	Cone Interference	RG	Rounded Gage
CT	Chipped Teeth Cutters	SD*	Shirttail Damage
ER	Erosion	SS	Self-Sharpening Wear
FC	Flat Crested Wear	TR	Tracking
HC	Heat Checking	WO*	Washed Out Bit
JD	Junk Damage	WT	Worn Teeth/Cutters
LN*	Lost Nozzle		

*Used only in the Other Dull Characteristics column.

Table 4.7.8 *Reason Pulled Codes*

Code	Description	Code	Description
BHA	Change bottom hole assembly	LOG	Run logs
CM	Condition mud	PP	Pump pressure
DMF	Downhole motor failure	PR	Penetration rate
DP	Drill plug	RIG	Rig repair
DSF	Drill string failure	RS	Retrieve survey
DST	Drill stem testing	TD	Total depth
DTF	Downhole tool failure	TQ	Torque
FM	Formation change	TW	Twist off
HP	Hole problems	WC	Weather conditions
HR	Hours on bit	WO	Washout in drill string
LIH	Left in Hole		

Reason Pulled (R). Codes for "Reason Pulled" are shown in Table 4.7.8.

4.7.1.6 Roller Cone Bit Economics

No matter what or how good a new product or method may be to a drilling operation, the final result is always measured in terms of cost per foot or meter. Lowest cost per foot indicates to drilling engineers and supervisors which products to use most advantageously in each situation. Reduced costs lead directly to higher profits or, in some cases, to the difference between profit and loss.

For those in administration, engineering, manufacturing, and sales, cost calculations are used to evaluate the effectiveness of any product or method, new or old. Because drilling costs are so important, everyone involved should know how to make a few simple cost calculations.

For example, the cost of a PDC bit can be up to twenty times the cost of a mill tooth bit and up to four times the cost of a tungsten carbide insert bit. The choice of either, the PDC bit, a milled tooth bit, or an insert roller cone bit must be economically justified by its performance. Occasionally, this performance justification is accomplished by simply staying in the hole longer. In such cases, the benefits of using it are intangible.

The main reason for using a bit, however, is that it saves money on a cost-per-foot basis. A PDC bit, to be economical, must make up for its additional cost by either drilling faster or staying in the hole longer. Because the bottom line on drilling costs is dollars and cents, a bit's performance is based on the cost of drilling each foot of hole.

Break-even analysis on a bit is the most important aspect of an economic evaluation. A break-even analysis is necessary to determine if the added bit cost can be justified for a particular application.

The break-even point for a bit is simply the footage and hours needed to equal the cost per foot that would be obtained on a particular well if the bit were not used. To get the break-even, a good offset well must be used for comparative purposes.

If the following bit record were used, we could determine if a bit would be economical (Table 4.7.9).

4.7.1.7 Example of Offset Well Performance

Total rotating time = 212.5 hr

Total trip time = 54.3 hr (Tripping rate is computed at 1000 ft/hr on average. This rate depends on rig type and operation.)

Table 4.7.9 *Bit Record*

Bit Size (in.)	Type	Bit Cost	Depth Out (ft.)	Footage Drilled (ft.)	Hours	ROP
$8\frac{1}{2}$	FDS+	$2,820	9,618	756	26	29.1
$8\frac{1}{2}$	FDT	$1,848	10,271	653	24	27.2
$8\frac{1}{2}$	FDG	$1,848	10,699	428	22	19.5
$8\frac{1}{2}$	F2	$4,816	11,614	915	71.5	12.8
$8\frac{1}{2}$	F2	$4,816	12,242	628	69	9.1
Total:		$16,148		3,380	212.5	Avg: 15.9

Rig operating cost = $300/hr

Total bit cost = $16,148

Total footage = 3,380 ft

Therefore, the offset cost per foot for this interval (8,862 to 12,242 ft) is calculated using the standard cost-per-foot equation:

$$C = \frac{R * (T + D) + B}{F} \qquad [4.7.8]$$

in which

C = drilling cost per foot ($/ft)

R = rig operating cost ($/hr)

   (Plus add-on equipment, such as downhole motor.)

T = trip time (hr)

D = drilling time (hr)

B = bit cost ($)

F = footage drilled (ft)

Using the data provided in the earlier example, the cost per foot is

$$C = \frac{(212.5 + 54.3)\$300 + \$16,148}{3,380} = 28.46 \; \$/ft.$$

When determining if an application can be suitable for a bit, the offset performances are given but the performance of the bit must be estimated. We must assume the footage the bit will drill or the penetration rate it will obtain. If the footage is assumed, then we use the following formula to calculate the break-even ROP:

$$ROP = \frac{R}{\dfrac{C - (RT + B)}{F}} \qquad [4.7.9]$$

in which

R = rig operating cost ($/hr)

C = offset cost per foot ($)

T = trip time of bit (hr)

B = bit cost ($)

F = assumed bit footage (ft)

In the previous example

R = $300

C = $28.46

T = 12 hr

B = $18,300

F = 3,380 ft

Therefore

$$ROP = \frac{\$300}{\dfrac{\$28.46 - (300 \times 12 + \$18,300)}{3,380 \; ft}} = \$13.7 \; ft/hr$$

The bit must drill the 3,380 ft at a penetration rate of 13.7 ft/hr to equal the offset cost per foot of $528.46 for the same 3,380 ft.

If a penetration rate is assumed, use the following equation to calculate the break-even footage:

$$F_{BE} = \frac{RT + B}{C - (R/ROP)} \qquad [4.7.10]$$

In the previous example, if we assume a penetration rate of 20 ft per hour, we have

$$F_{BE} = \frac{\$300 \times 12 + \$18,300}{\$28.46 - 300/20}$$

In this case, the bit must drill 1,627 ft to attain the break-even point.

4.7.1.8 Roller Cone Bit Glossary

Abrasive wear: wearing away by contact friction between two surfaces in relative motion to one another.

Air bit: a type of roller cone bit that uses compressed air or natural gas for cooling and cuttings removal.

Air drilling: a method of rotary drilling that uses compressed air, instead of water or mud, as the drilling fluid.

Bit gauge: a circular ring used to determine whether a bit has the correct outside diameter.

Bit record: a report that lists each bit used during a drilling operation, and reporting on wear using the IADC Dull Grading protocol.

Bit run: placing a bit on the bottom of a hole, drilling the prescribed amount of hole, or until it wears out, and pulling it from the hole.

Cellular diamond: extruded strands of Polycrystalline Diamond Compact encased within a tungsten carbide with binder that are bundled into groups, shaped, and used as a high performance cutting element.

Cone: usually conically shaped, rotating components of a roller cone bit that support the teeth or insert type cutting elements that penetrate the formation during drilling.

Cone alignment: the way in which roller cones are aligned with a bit's center axis.

Cone angle: the face angle of a bit cone.

Cone speed: the rotational velocity at which a roller cone bit cone rotates about its centerline. Three-cone bit cone rotation speed is normally higher than bit rotation speed; one cone bit rotation speed is normally lower than bit rotational speed.

Drilling fluid: fluids circulated in a well to clean and cool the bit and remove cuttings.

Erosion wear: surface wear caused by fluid born abrasives. For example, drilling fluids used in conjunction with cuttings removal from roller cone bits can cause severe erosion of exposed cone surfaces.

Gage or Gauge: the diameter of a bit or a hole drilled by a bit.

Gage area: the outside diameter of a bit that drills the wall of a well.

Gage row: the inserts or teeth in the outermost row on the cones of a bit. The gage row determines a hole's gage or diameter.

IADC: International Association of Drilling Contractors.

Impingement angle: with respect to roller cone bits, the angle at which a drilling fluid flows from a bit nozzle strikes the well bottom.

Inner bearings: bearings that lie near the apex of a roller cone (also called secondary or thrust bearing).

Insert: a tungsten carbide shape used as a cutting element on roller cone bits.

Journal: that portion of a roller cone bit leg that supports the bearing system, lubrication system, and cone retention system for one of the bits cones.

Journal bearing: a type of roller cone bit bearing designed to rotate on a film of lubricant rather than on roller type bearings.

Leg: a three-cone bit component comprising one third of the bit (120° of arc) when viewed in plan view. Three legs are assembled to complete a bit. Each leg contains a journal that supports the bearing system, seals, lubrication system, and cone retention system for one of the bit's three cones.

Milled-tooth bit: a roller cone bit in which the cutting elements are rows of hardfaced steel teeth.

Mono-cone bit: a type of roller cone bit in which a single semi-hemispherical cutting device rotates as the bit drills.

Offset: a roller cone bit term describing an engineered orientation of the cone axes so they do not intersect at the bit centerline. Offset prevents free rolling of the cone and sets up a skidding action that improves drilling efficiency.

Outer bearings (or primary bearings): the main bearings of a roller cone bit that are located near the open end of a cone.

Out-of-gage bit: a bit whose diameter is more or less than that specified for the hole to be drilled.

Out-of-gage hole: a hole that is smaller or larger than the specified diameter of the hole to be drilled.

Rate of penetration (ROP): a commonly used measure of drilling rate, normally expressed in terms of feet or meters per hour.

Rotary speed: the rotational velocity at which a bit turns about its centerline.

Sealed bearing: a type of bearing that is not exposed to the drilling fluids. Seals prevent drilling fluid from entering the bearings.

Shirttail: the rounded bottom portion of a roller cone bit leg.

Three-cone bit: a type of roller cone bit composed of three, rotating, intermeshing cones.

Tungsten carbide: a compound of tungsten with carbon. This term often refers to a composite material composed of tungsten carbide grains cemented with a metallic binder such as cobalt, nickel, or iron that is used for cutting and resisting abrasion.

TCI bit: a type of roller cone bit using tungsten carbide inserts for cutting elements.

Two-cone bit: a type of roller cone bit composed of two, cone-shaped cutting devices that rotate as the bit drills.

Undergage bit: a bit whose outside diameter is less than the specified minimum for a well.

Weight on bit (WOB): force, normally expressed in pounds or kilograms, applied to bit in the direction of the bit centerline.

References
1. Bentson, H. G., Roller Cone Bit Design, Pacific Coast District, API Division of Production, Smith International, Inc., Los Angeles, May 1956.
2. American Petroleum Institute, Specification for Rotary Drilling Equipment (API Specification 7), 37th Edition, Section 7, Washington, DC, August 1990.
3. McGehee et al., Roller Cone Bit Classification System, International Association of Drilling Contractors/Society of Petroleum Engineers (IADC/SPE) Drilling Conference, SPE Paper 23937, 1992.
4. McGehee et al., The IADC Roller Bit Dull Grading System, International Association of Drilling Contractors/Society of Petroleum Engineers (IADC/SPE) Drilling Conference, SPE Paper 23938, 1992.
5. Fang, Zhigang et al., Chipping Resistant Polycrystalline Diamond and Carbide Composite Materials for Roller Cone Bits, Society of Petroleum Engineers (SPE) Annual Technical Conference, SPE Paper 71394, 2001.
6. Keshavan, M. K. et al., Diamond-Enhanced Insert: New Compositions and Shapes for Drilling Soft-to-Hard Formations, Society of Petroleum Engineers (SPE) Drilling Conference, SPE Paper 25737, 1993.
7. Fang, Zhigang, et al., A Dual Composite of WC-Co, *Metallurgical and Materials Transactions A*, Vol 30A, December 1999.
8. Salesky, W.J., and Payne, B.R., Preliminary Field Test Results of Diamond-Enhanced Inserts for Three Cone Rock Bits, Society of Petroleum Engineers (SPE) Drilling Conference, SPE Paper 16115, 1987.
9. Salesky, W.J., et al., Offshore Tests of Diamond-Enhanced Rock Bits, Society of Petroleum Engineers (SPE) 63rd Annual Technical Conference, SPE Paper 18039, 1988.
10. Chia, R., and Smith, R., A New Nozzle System to Achieve High ROP Drilling, Society of Petroleum Engineers (SPE) 61st Annual Technical Conference, SPE Paper 15518, 1986.

11. Portwood, G., et al., Improved Performance Roller Cone Bits for Middle Eastern Carbonates, Society of Petroleum Engineers, Middle East Drilling Technology Conference, SPE Paper 72298, 2001.

4.7.2 Diamond Matrix Bits
4.7.2.1 Diamond-Impregnated Core Bits
Diamond matrix bits are being employed to a greater extent because of the advancements in mud motors. High rpms can destroy roller rock bits very quickly. On the other hand, diamond bits rotating at high rpm usually have longer life since there are no moving parts.

4.7.2.2 Diamond-Impregnated Core Bit Selection
The industry is moving to a new selection method for diamond-impregnated core bits based on the Mohs hardness scale. Advancements in metallurgy and diamond coating technologies have improved impregnated bits in relation to different rock types. On the low end of the Mohs scale, selection of a low-series bit is ideal, and the opposite is true at the upper end (Figure 4.7.57). In addition to the scale, a bit may be made specifically to cut abrasive or competent rock formations. The Mohs scale appears to be the new standard, because the selection criteria have scientific methodology for determining the right bit for the job, as opposed to the older color coded system [11].

4.7.2.3 Diamond Selection
Diamonds used as the cutting elements in the bit metal matrix have the following advantages:

1. Diamonds are the hardest material.
2. Diamonds are the most abrasion-resistant material.
3. Diamonds have the highest compressive strength.
4. Diamonds have a high thermal conductivity.

Diamonds also have some disadvantages as cutting elements: they are very weak in shear strength, have a very low shock impact resistance, and can damage or crack under extremely high temperatures.

When choosing diamonds for a particular drilling situation, there are basically three things to know. First, the quality of the diamond chosen should depend on the formations being drilled. Second, the size of the diamond and its shape will be determined by the formation and anticipated penetration rate. Third, the number of diamonds employed is dictated by the formation and anticipated penetration rate.

There are two types of diamonds, synthetic and natural. Synthetic diamonds are man made and are used in PDC STRATAPAX type bit designs. STRATAPAX PDC bits are best suited for extremely soft formations. The cutting edge of synthetic diamonds are round, half-moon shaped or pointed.

Natural diamonds are divided into three categories. First are the carbonate or black diamonds. These are the hardest and most expensive diamonds. They are used primarily as gage reinforcement at the shockpoint. Second are the West African diamonds. These are used in abrasive formations and usually are of gemstone quality. About 80% of the West African diamonds are pointed in shape and, therefore, 20% are the desirable spherical shape. Third are the Congo or coated diamonds. These are the most common category. Over 98% of these diamonds are spherical by nature. They are extremely effective in soft formations. The other 2% are usually cubed shaped, which is the weakest of the shapes available.

By studying specific formations, diamonds application can be generalized as follows:

Soft, gummy formations—Congo, cubed shaped
Soft formation—large Congo, spherically shaped
Abrasive formation—premium West Africa
Hard and abrasive—Special premium West Africa

Diamond Bit Design. Diamond drill bit geometry and descriptions are given in Figure 4.7.58 [18]. Diamond core bit geometry and descriptions are given in Figure 4.7.59 [9].

There are two main design variables of diamond bits, the crown profile and face layout (fluid course configuration).

The crown profile dictates the type of formation for which the bit is best suited. Properties include the round, parabolic, tapered, and flat crowns used, respectively, in hard to extremely hard formations, medium to hard formations, soft formations, and for fracturing formations, or sidetracks and for kick-offs.

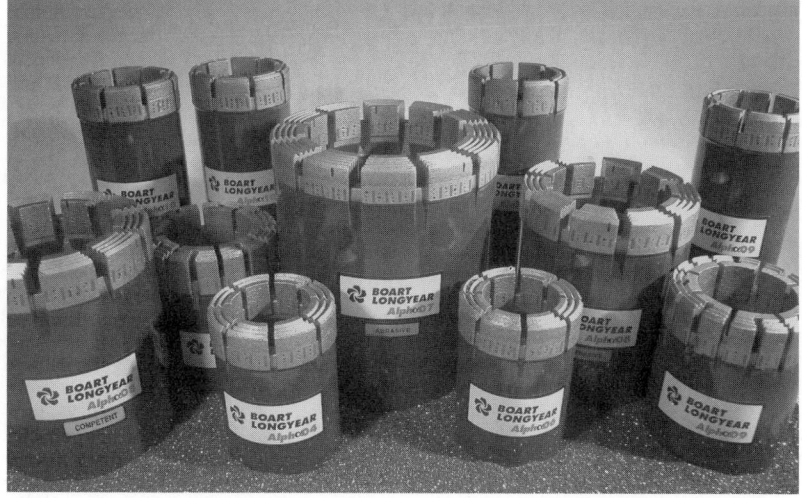

Figure 4.7.57 *Sample Alpha α bit product line as available in series 1 through 10 (courtesy of the Boart Longyear Company).*

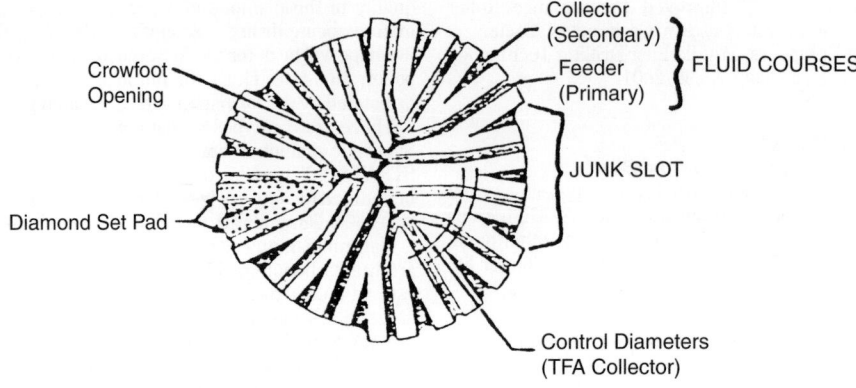

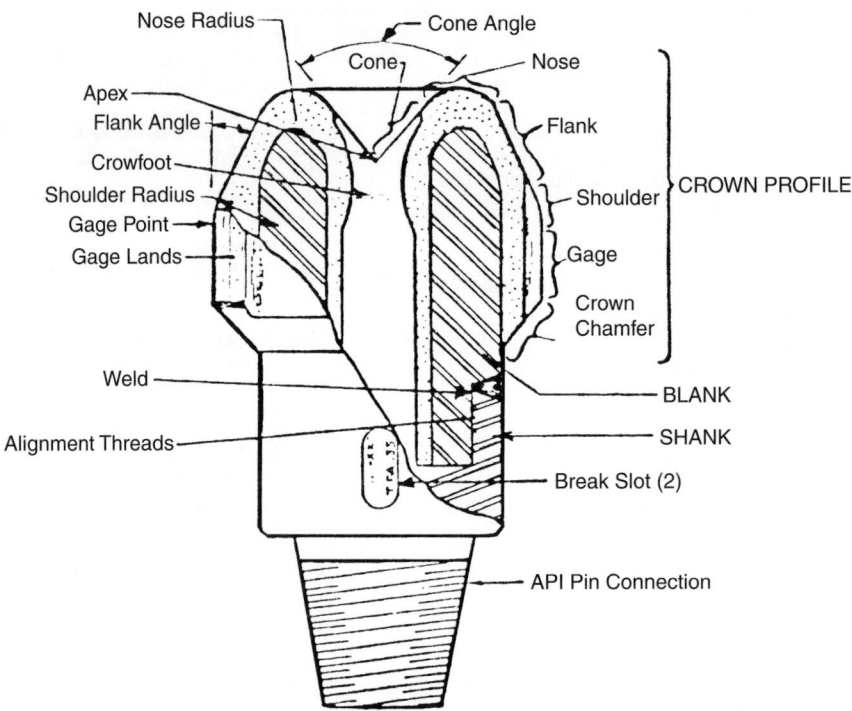

Figure 4.7.58   *Diamond drill bit nomenclature (courtesy of Hughes Christensen).*

Cone angles and throat depth dictate the bit best suited for stabilization. Cone angles that are steep (60° to 70°), medium (80° to 90°), or flat (100° to 120°) are best suited for highly stable, stable, and fracturing formation, respectively.

Diamond drill bits with special designs and features include

1. Long gage bits, used on downhole motors for drilling ahead in vertical boreholes.
2. Flat-bottom, shallow-cone bit designs, used on sidetracking jobs or in sidetracking jobs with downhole motors.
3. Deep cones having a 70° apex angle are normally used in drill bits to give built-in stability and to obtain greater diamond concentration in the bit-cone apex.

Diamond Bit Hydraulics. The hydraulics for diamond bits should accomplish rapid removal of the cuttings, and cooling and lubrication of the diamonds in the bit metal matrix.

Bit Hydraulic Horsepower. An effective level of hydraulic energy (hydraulic horsepower per square inch) is the key to optimum bit performance. The rule-of-thumb estimate of diamond bit hydraulic horsepower HP_h and penetration rates is shown in Table 4.7.10. The bit hydraulic horsepower is dependent upon the pressure drop across the bit and the flowrate.

Bit Pressure Drop. The pressure drop across the bit is determined on the rig as the difference in standpipe pressure when the bit is on bottom, and when the bit is off bottom, while maintaining constant flowrate.

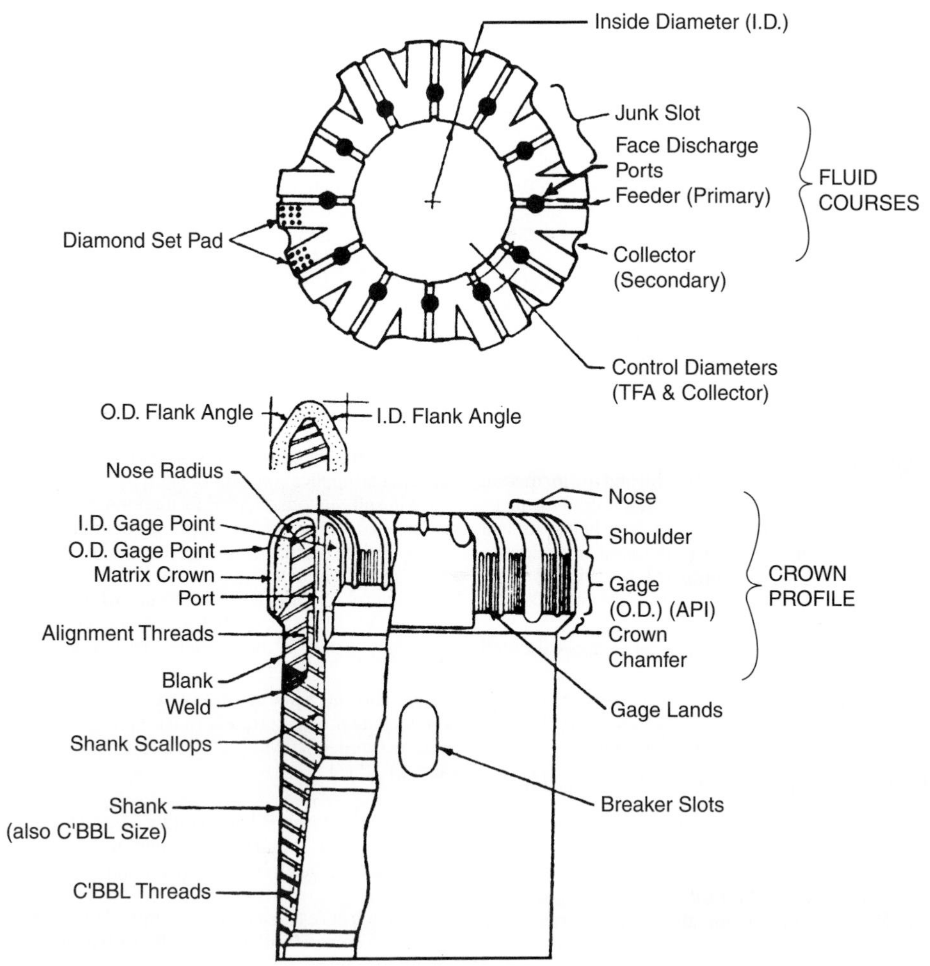

Figure 4.7.59 *Diamond core bit nomenclature (courtesy of Hughes Christensen).*

Table 4.7.10 *Bottomhole Hydraulic Horsepower Required for Diamond Drilling [2]*

Penetration Rate, ft/hr	1–2	2–4	4–6	6–10	over 10
Hydraulic Horsepower Required, HP_h;/sq. inch	1–1.5	1.5–2	2–2.5	2.5–3	3–3.5

Maximum Drilling Rate. In fast drilling operations (soft formations), the maximum penetration rate is limited by the maximum pressure available at the bit. This is the maximum allowable standpipe pressure minus the total losses in the circulating system.

Optimum Pump Output. In harder formations where drilling rates are limited by maximum available bit weight and rotary speed, the optimum value of flow rate should be adjusted to achieve the bit hydraulic horsepower required. The minimum pump discharge required to maintain annular velocity and bit cooling is shown in Figure 4.7.60.

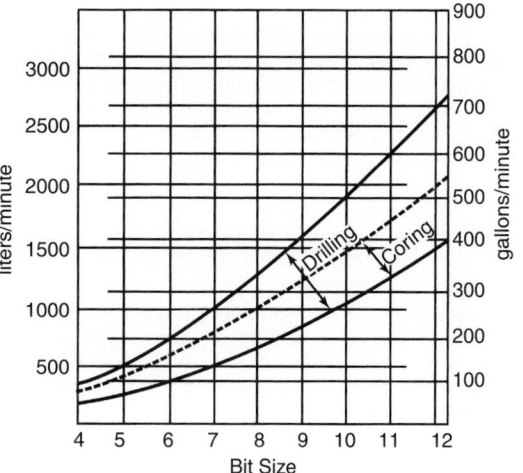

Figure 4.7.60 *Pump discharge for diamond bits [9]. (courtesy of Hughes Christensen).*

Table 4.7.11   *Core Barrels: Recommended Makeup Values [9]*

Core Barrel Size		3.5×1.75	4.12×2.12	4.50×2.12	4.75×2.62	5.75×3.50	6.25×3	6.25×4	6.75×4	8.0×5.25
Recommended Make up Torque foot-pounds		1,700 to 2,050	3,000 to 3,600	5,000 to 6,000	4,050 to 4,850	7,400 to 8,800	14,900 to 17,800	8,150 to 9,800	9,900 to 12,000	19,000 to 22,700
Pounds Line Pull for Different Length Tong Levers	60″	375	660	1,100	890	1,620	3,270	1,795	2,190	4,170
	55″	410	720	1,200	970	1,770	3,570	1,960	2,390	4,550
	53″	425	750	1,250	1,010	1,830	3,700	2,030	2,480	4,720
	47″	480	840	1,400	1,140	2,070	4,170	2,290	2,800	5,320
	44″	510	900	1,500	1,210	2,200	4,460	2,450	2,990	5,690
	42″	540	940	1,570	1,270	2,310	4,670	2,560	3,130	5,960
	36″	625	1,100	1,830	1,480	2,700	5,450	2,990	3,650	6,950

Courtesy of Hughes Christensen.

Hydraulic Pumpoff. The bit pressure drop acts over the bit face area between the cutting face of the bit and the formation and tends to lift the bit off the bottom of the hole. This force is large at the higher bit hydraulic horsepower being utilized today and in some cases may require additional bit weight to compensate. For example, the pumpoff force on an $8\frac{7}{16}$ in. diamond bit having a pressure drop across the bit of 900 psi would be about 6,000 lb.

The hydraulic pumpoff force F_{po} (lb) can be approximated by [4]:

$$F_{po} = 1.29\,(\Delta P_b)\,(d_h - 1) \qquad [4.7.11]$$

for the radial flow watercourse design bits, and

$$F_{po} = 0.32\,(\Delta P_b)\,(d_h - 1) \qquad [4.7.12]$$

for cross-flow watercourse system (refer to IADC classification of fixed-cutter bits).

4.7.2.4 Diamond Bit Weight on Bit and Rotary Speed

Weight on Bit. Drilling weight should be increased in increments of 2,000 lb as the penetration rate increases. As long as no problems are encountered with the hydraulics and torque, weight can be added. However, when additional weight is added and the penetration rate does not increase, the bit may be balling up, and the weight on the bit should be decreased.

Rotary Speed. Diamond bits can usually be rotated at up to 150 rpm without any problem when hole conditions and drill string design permit. Rotary speeds of 200 and 300 rpm can be used with stabilized drill strings in selected areas. Diamond bits have also operated very successfully with downhole motors at 600 to 900 rpm. The actual rotary speed limits are usually imposed by safety.

4.7.2.5 Core Bits

Most core barrels utilize diamonds as the rock cutting tool. There are three types of core barrels.

Wireline Core Barrel Systems. The wireline system can be used for continuous drilling or coring operations. The inner barrel or the drill plug center of the core bit can be dropped from the surface and retrieved without pulling the entire drill string.

Marine Core Barrels. Marine barrels were developed for offshore coring where a stronger core barrel is required. They are similar to the conventional core barrels except that they have heavier outer tube walls.

Rubber Sleeve Core Barrels. Rubber sleeve core barrels are special application tools designed to recover undisturbed core in soft, unconsolidated formations. As the core is cut, it is encased in the rubber sleeve that contains and supports it. Using face discharge ports in the bit, the contamination of the core by circulating fluid is reduced. The rubber sleeve core barrel has proven to be a very effective tool, in spite of the fact that the rubber sleeve becomes weak with a tendency to split as the temperature increases about 175°F.

Core Barrel Specifications. Core barrel sizes, recommended make-up torques, maximum recommended pulls and recommended fluid capacities are shown in Tables 4.7.11 and 4.7.12 [9].

4.7.2.6 Weight on Bit and Rotary Speed for Core Bits

Weight on Bit. Figure 4.7.61 shows the drilling weights for diamond core bits in various formations. These are average values determined in field tests [9]. The proper weight on the bit for each core run can be determined by increasing the bit weight in steps of 1,000 to 2,000 lb, with an average speed of 100 rpm. Coring should be continued at each interval while carefully observing the penetration rate. Optimum weight on the bit has been reached when additional weight does not provide any further increase in penetration rate or require excessive torque to rotate the bit. Using too much weight can cause the diamonds to penetrate too deeply into a soft formation with an insufficient amount of mud flow able to pass between the diamonds and the formation, resulting in poor removal of the cuttings. The core bit could clog or even

Table 4.7.12   *Core Barrels Characteristics [9]*

Size	(ft) Standard Length	# of Turns in Safety Joint	(GPM) Fluid Capacity	Recommended Maximum Pull
3-1/2×1-3/4	30	7	118	74,000
4-1/8×2-1/8	60	7	141	101,400
4-1/2×2-1/8	60	7	141	194,700
4-3/4×2-5/8	60	13	164	137,400
5-3/4×3-1/2	60	6	204	200,000
6-1/4×3	60	7	245	290,000
6-1/4×4	60	6	227	193,500
6-3/4×4	60	6	387	275,000
8×5-1/4	60	7	295	310,000

The Maximum Pull is based upon the ultimate tensile strength in the pin thread area with a safety factor of three.
Courtesy of Hughes Christensen.

burn, and penetration rate and bit life will be reduced. In harder formations, excessive weight will cause burning on the tips of the diamonds or shearing with a resulting loss in salvage.

Rotary Speed. The best rotational speed for coring is usually established by the limitations of the borehole and drill string. The size and number of drill collars in the string and the formation being cored must be considered when establishing the rotational speed. Figure 4.7.62 shows the recommended rotating speed range for optimal core recovery in different formations [8]. Concern should also be given to the harmonic vibrations of the drill string. Figure 4.7.63 gives critical rotary speeds that generate harmonic vibrations.

4.7.2.7 Polycrystalline Diamond Compacts (PDC) Bits

PDC bits get their name from the polycrystalline diamond compacts used for their cutting structure. The technology that led to the production of STRATAPAX drill blanks grew from the General Electric Co. work with polycrystalline manufactured diamond materials for abrasives and metal working tools. General Electric Co. researched and developed the STRATAPAX (trade name) drill blank in 1973 and Christensen, Inc. used these in PDC bit field tests. The bits

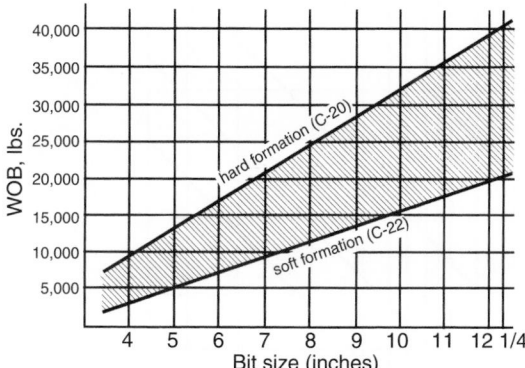

Figure 4.7.61 *Bit weight for core bits (courtesy of Hughes Christensen).*

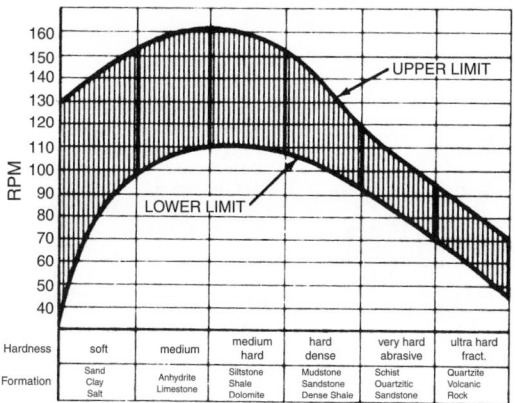

Figure 4.7.62 *Recommended rotary speed for core bits [8] (courtesy of Hughes Christensen).*

were successfully applied in offshore drilling in the North Sea area in the late 1970s and in on-shore areas in the United States in the early 1980s. In some areas, the PDC bits have out-drilled roller rock bits, reducing overall cost per foot by 30% to 50% and achieving four times the footage per bit at higher penetration rates. [11, 12].

On a worldwide basis, drag bits operated in soft and medium-hardness formations have set performance records for single-run footage (>22,000 ft), cumulative footage (>18,000 ft in 26 runs), and single-run penetration rate (>2,200 ft/hr) [13]. This success has motivated continuing efforts to expand the range of applications for drag bits to increasingly harder formations. Laboratory studies indicate that PDC drag bits can outperform conventional rollerbits in hard rock when they are operated under conditions that minimize frictional heat generation and bit chatter [14]. Field test have confirmed that a properly controlled standard PDC drag-bit design is capable of drilling an extended hard-rock interval at a high rate of penetration. [15].

Figure 4.7.64 shows the major components and design of the PDC bit. Typical polycrystalline diamond compacts are shown in Figure 4.7.65. The polycrystalline diamond compacts consist of a thin layer of synthetic diamonds on a tungsten carbide disk. These compacts are produced as an integral blank by a high pressure, high temperature process. The diamond layer consists of many tiny crystals bonded together at random orientations for maximum strength and wear resistance.

The tungsten carbide backing provides mechanical strength and further reinforces the diamond compact wear-resistant properties. During drilling, the polycrystalline diamond cutter wears down slowly with a self-sharpening effect. This helps maintain sharp cutters for high penetration-rate drilling throughout the life of the bit.

Variations in cutter design and processing parameters can strongly affect in-service cutter loads, wear, and durability. Geometric design variables include the diamond-table thickness, the cutting-edge chamfer design, and the diamond-table/tungsten-carbide interface configuration. Three non-planar interface configurations are illustrated in Figure 4.7.66. Material design variables for the diamond table itself include the binder composition and the distribution of diamond grain sizes. Manufacturing process variations can involve changes in pressure/temperature peaks and cycle history during high-pressure/high-temperature compaction. Appropriate adjustments in cutter design and processing can yield dramatic improvements in wear and/or impact performance [4.7.3].

PDC Bit Design. Figures 4.7.67 and 4.7.68 show typical PDC bits. Figure 4.7.67 is for soft formation. Figure 4.7.68 is for hard and abrasive formation [2].

Bit Body Material (Matrix). There are two common body materials for PDC bits, steel and tungsten carbide. Heat-treated steel body bits are normally a "stud" bit design, incorporating diamond compacts on tungsten carbide posts. These stud cutters are typically secured to the bit body by interference fitting and shrink fitting and with brazing. Steel body bits also generally incorporate three or more carbide nozzles (often interchangeable) and carbide buttons on gauge. Steel body bits have limitations of erosion of the bit face by the drilling mud and wear of the gauge section. Some steel body bits are offered with wear-resistant coatings applied to the bit face to limit mud erosion.

Greater bit design freedom is generally available with matrix body bits because they are "cast" in a mold like natural diamond bits. Thus, matrix body bits typically have more complex profiles and incorporate cast nozzles and

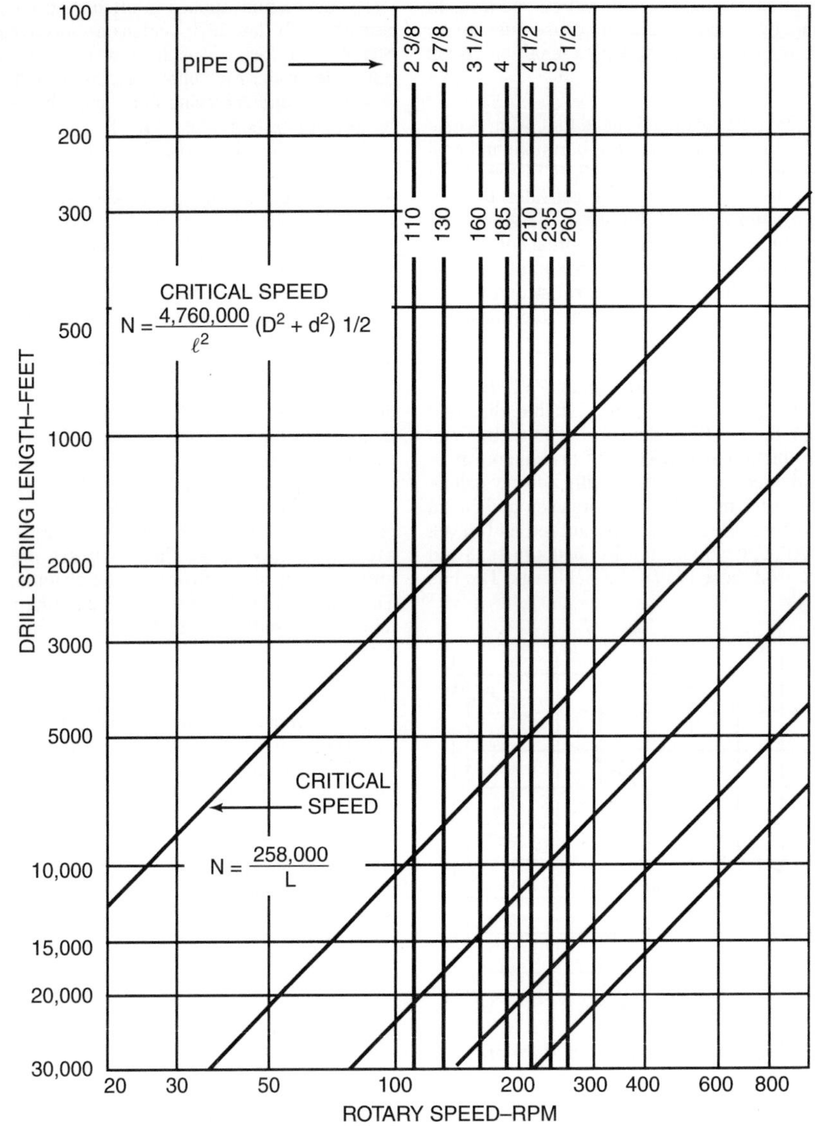

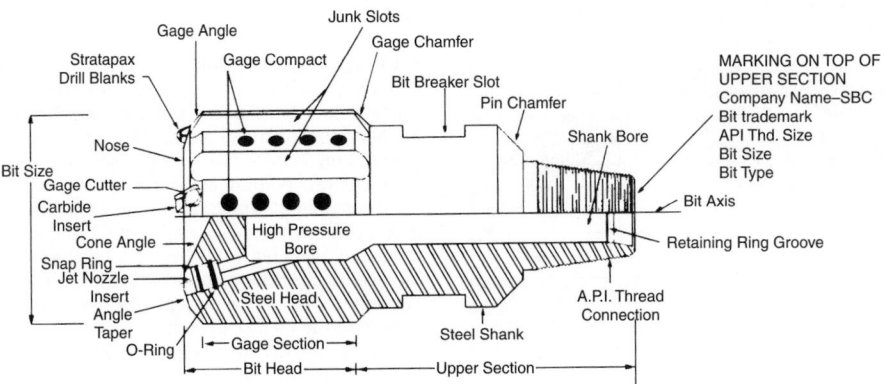

Figure 4.7.63 *Critical rotary speed for core bits [10] (courtesy of API).*

Figure 4.7.64 *PDC bit nomenclature (courtesy of Strata Bit Corp). Source: Strata Bit Corporation, 600 Kenrick, Suite A-1, Houston, TX 77060 ph (713) 999-4530 unknown booklet of the company.*

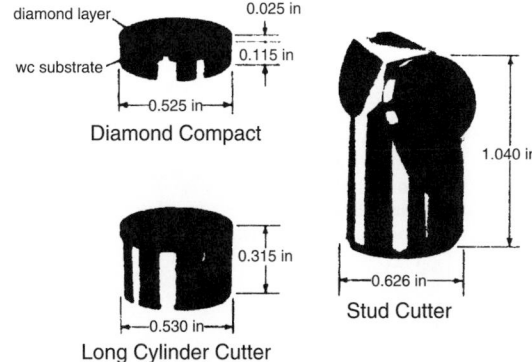

Figure 4.7.65 *Polycrystalline diamond compacts [2] (courtesy of Hughes Christensen).*

waterways. In addition to the advantages of bit face configuration and erosion resistance with matrix body bits diamond compact matrix bits often utilize natural diamonds to maintain full gage hole. Matrix body bits generally utilize long, cylinder-shaped cutters secured to the bit by brazing.

Bit Profile. Bit profile can significantly affect bit performance based upon the influence it has on bit cleaning, stability, and hole deviation control. The "double-cone" profile will help maintain a straight hole even in crooked hole country. The sharp nose will attack and drill the formation aggressively while the apex and reaming flank stabilize the bit. This sharp profile may be more vulnerable to damage when a hard stringer is encountered as only the cutters on the sharp nose will support the impact loading. A shallow cone profile appears to be the easiest to clean due to the concentration of hydraulics on the reduced surface area of the bit face. This profile relies heavily upon the gage section for directional stability. The shallow cone profile will hold direction and angle with sufficient gauge length and proper stabilization of the bit.

Cutter Exposure. Figure 4.7.69 shows the types of cutter exposure on PDC bits [1]. Cutter exposure is the distance between the cutting edge and the bit face. Stud bits typically have full exposure that proves very aggressive in soft formation. In harder formations, less than full exposure may be preferred for added cutter durability and enhanced cleaning.

Matrix body bits are designed with full or partial exposure depending on formation and operating parameters.

Cutter Orientation. Figure 4.7.70 shows the cutter orientation for typical PDC bits. The displacement of cuttings can be affected by side and back rake orientation of the cutters. Back rake angle typically varies from 0 to −25°. The greater the degree of back rake, generally the lower the rate of penetration, but the greater the resistance to cutting edge damage when encountering a hard section. Side rake has been found to be effective in assisting bit cleaning in some formations by mechanically directing cuttings toward the annulus. Matrix body bits allow greater flexibility in adjusting cutter orientation for best drilling performance in each formation.

References
1. Garrett, W. R., "The effect of a downhole shock absorber on drill bit stem performance," Publ. No. 41. Drilco, Division of Smith International, Inc., Houston, 19XX.
2. *Christensen Diamond Compact Bit Manual,* Christensen, Inc., March 1983.
3. Winters, W. J., et al., "Application of the 1987 IADC/roller bit classification system, SPE/DC 16143, Proc. SPE/IADC Drilling Conference, New Orleans, March 15–18, 1987, pp. 819–837.
4. *Drilling Manual,* Canadian Association of Drilling Contractors, Calgary, 1979.
5. Hampton, S. D., et al., "Application of the 1987 roller bit dull grading system," SPE/IADC 16146, Proc. of 1987 SPE/IADC Drilling Conference, New Orleans, March 15–18, 1987, pp. 859–867.
6. "Engineering essentials of modern drilling," Energy Publication Division of HBJ, Dallas, 1982.
7. McLean, R. H., "Crossflow and impact under jet bits," *Journal of Petroleum Technology,* November 1964.
8. *Diamond Drill Bit Technology,* student manual, Christensen, Inc.
9. *Diamond Coring Technology,* student manual, Christensen, Inc.
10. APR Recommended Practice 7G, Fourteenth Edition: "API Recommended Practice for Drill Stem Design and Operation Limits," August 1996.
11. Offenbacher, L. A., J. D. McDermain, and C. R. Paterson, "PDC bits find applications in Oklahoma drilling," paper presented at the IADC/SPE 1983 Drilling Conference, New Orleans.

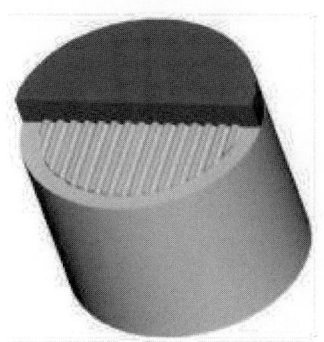
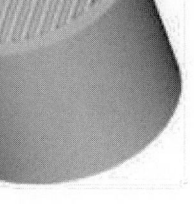

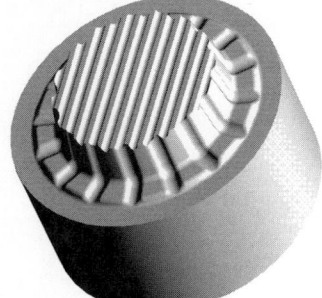

Modified   Honeycomb   HM160

Figure 4.7.66 *Non planar interface configurations.*

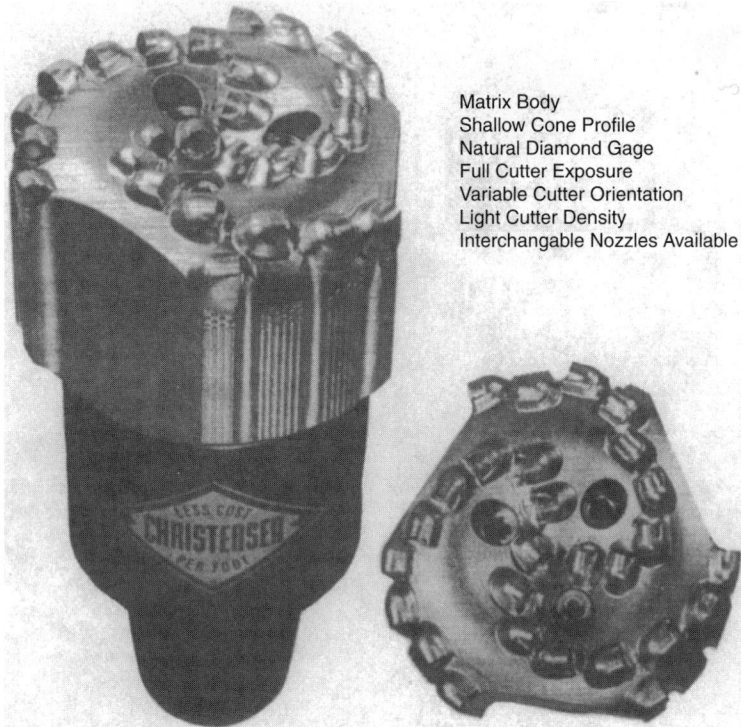

Matrix Body
Shallow Cone Profile
Natural Diamond Gage
Full Cutter Exposure
Variable Cutter Orientation
Light Cutter Density
Interchangable Nozzles Available

Figure 4.7.67   *PDC bit designed for soft formations [2] (courtesy of Hughes Christensen).*

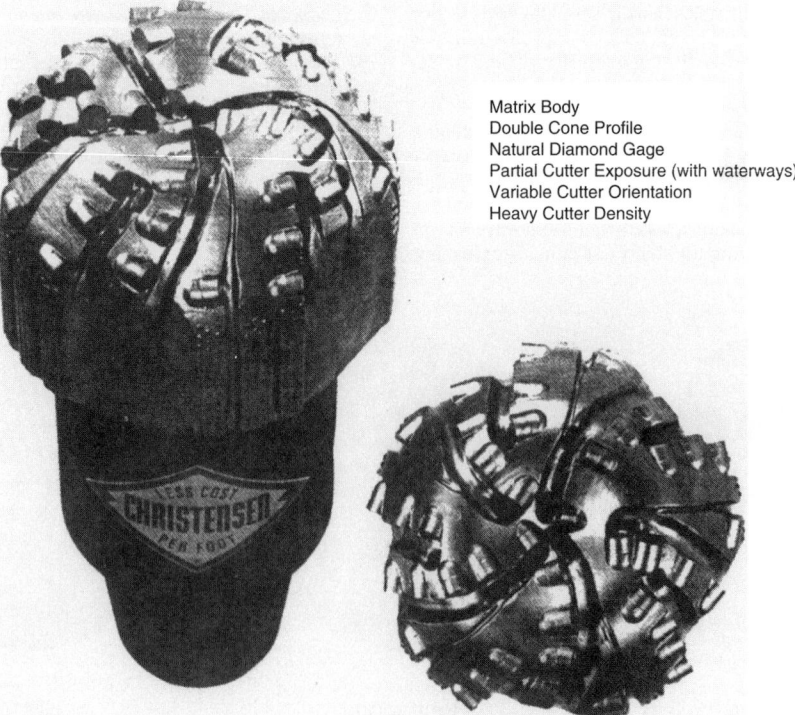

Matrix Body
Double Cone Profile
Natural Diamond Gage
Partial Cutter Exposure (with waterways)
Variable Cutter Orientation
Heavy Cutter Density

Figure 4.7.68   *PDC bit designed for hard and abrasive formations [2] (courtesy of Hughes Christensen).*

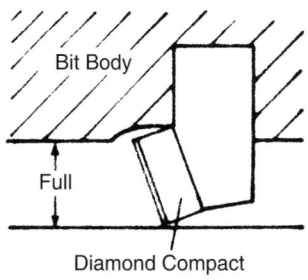 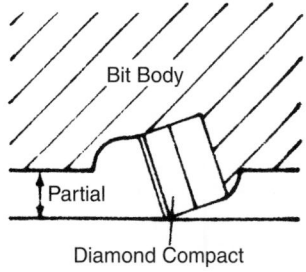

Figure 4.7.69 *Types of cutter exposure [2] (courtesy of Hughes Christensen).*

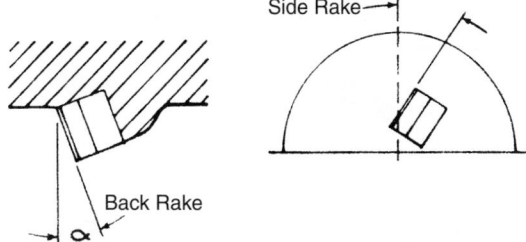

Figure 4.7.70 *Cutter orientation [2] (courtesy of Hughes Christensen).*

12. Smith, R., G. Hollingsworth, C. Joiner, and J. Bernard: "Chevron optimizes drilling procedures in the Tuscaloosa trends," *Petroleum Engineering International*, March 1982.
13. Perdue, J. M., "1999 Drilling and Production Yearbook," Hart's *Petroleum Engineer International*, March 1999, pp. 43–55.
14. Raymond, D. W., "PDC Bits Demonstrate Benefit over Conventional Hard-Rock Drill Bits," *Geothermal Resources Council Transactions*, Vol. 25, 2001, pp. 125–132.
15. Wise, J. L., J. T. Finger, A. J. Mansure, S. D. Knudsen, R. D. Jacobson, J. W. Grossman, W. A. Pritchard, and O. Matthews, "Hard-Rock Drilling Performance of a Conventional PDC Drag Bit Operated With, and Without, Benefit of Real-Time Downhole Diagnostics," *Geothermal Resources Council Transactions*, Vol. 27, 2003, pp. 197–205.
16. Wise, J. L., D. W. Raymond, C. H. Cooley, and K. Bertagno "Effects of Design and Processing Parameters on Performance of PDC Drag Cutters for Hard-Rock Drilling," *Geothermal Resources Council Transactions*, Vol. 26, 2002, pp. 201–206.

4.7.3 IADC Fixed Cutter Bit Classification System

The term fixed cutter is used as the most correct description for the broad category of non–roller-cone rock bits. The cutting elements may be comprised of any suitable material. To date, several types of diamond materials are used almost exclusively for fixed cutter petroleum drilling applications. This leads to the widespread use of the term "diamond" bits and PDC bits in reference to fixed cutter designs.

The IADC Drill Bits Subcommittee began work on a new classification method in 1985. It was determined from the outset that (1) a completely new approach was required, (2) the method must be simple enough to gain widespread acceptance and uniform application, yet provide sufficient detail to be useful, (3) emphasis should be placed on describing the form of the bit, i.e., "paint a mental picture of the design", (4) no attempt should be made to describe the function of the bit, i.e., do not link the bit to a particular formation type or drilling technique since relatively little is certain yet about such factors for fixed cutter bits, (5) every bit should have a unique IADC code, and (6) the classification system should be so versatile that it will not be readily obsolete.

The resultant four-character diamond bit classification code was formally presented to the IADC Drilling Technology Committee at the 1986 SPE/IADC Drilling Conference. It was subsequently approved by the IADC Board of Directors and designated to take effect concurrent with the 1987 SPE/IADC Drilling Conference. A description of the 1987 IADC Fixed Cutter Bit Classification Standard follows [1]

Four characters are utilized in a prescribed order (Figure 4.7.71) to indicate seven fixed cutter bit design features : cutter type, body material, bit profile, fluid discharge, flow distribution, cutter size, and cutter density. These design traits were selected as being most descriptive of fixed cutter bit characteristics.

The four-character bit code is entered on an IADC-API Daily Drilling Report Form as shown in Figure 4.7.72. The space requirements are consistent with the four-character IADC roller bit classification code. The two codes are readily distinguished from one another by the convention that diamond bit codes begin with a letter, while roller bit codes begin with a number.

Each of the four characters in the IADC fixed cutter bit classification code are further described as follows:

Cutter Type an Body Material. The first character of the fixed cutter classification code describes the primary cutter type and body material (see Figure 4.7.71). Five letters are presently defined: D—natural diamond/matrix body, M—PDC/matrix body, S—PDC/steel body, T—TSP/matrix body/O—other.

The term PDC is defined as "polycrystalline diamond compact." The term TSP is defined as "thermally stable polycrystalline" diamond TSP materials are composed of manufactured polycrystalline diamond which has the thermal stability of natural diamond. This is accomplished through the removal of cobalt binder material and trace impurities and, in some cases, the filling of lattice structure pore spaces with a material of compatible thermal expansion coefficient.

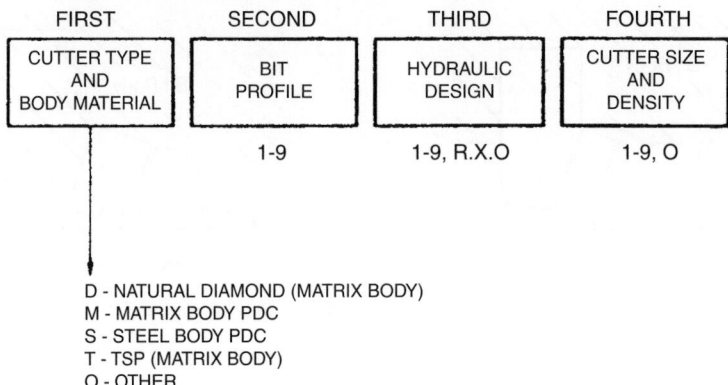

Figure 4.7.71 *Four-character classification code for fixed-cutter bits [1] (courtesy of SPE).*

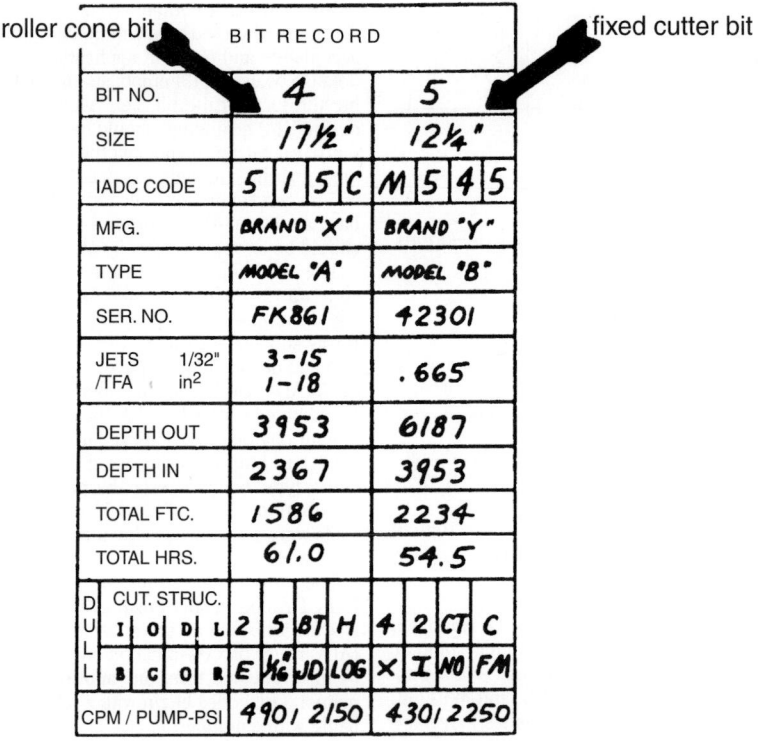

Figure 4.7.72 *Fixed-cutter bit code entry in IADC-API Daily Report [1] (courtesy of SPE).*

The distinction of *primary* cutter types is made because fixed cutter bits often contain a variety of diamond materials. Typically one type of diamond is used as the primary cutting element while another type is used as backup material.

Profile. The number 1 through 9 in the second character of the fixed cutter classification code refer to a bit's cross-sectional profile (Figure 4.7.73). The term profile is used here to describe the cross-section of the cutter/bottomhole pattern. This distinction is made because the cutter/bottomhole profile is not necessarily identical to the bit body profile.

Nine basic bit profiles are defined by arranging two profile parameters–outer taper(gage height) and inner concavity (cone height)–in a 3 × 3 matrix (Figure 4.7.74). The rows and columns of the matrix are assigned high, medium and low values for each parameter. Gage height systematically decreases from top to bottom. Cone height systematically decreases form left to right. Each profile is assigned a number.

Two versions of the profile matrix are presented. One version (see Figure 4.7.73) is primarily for the use of manufacturers in classifying their bit profiles. Precise ranges of high, medium, and low values are given. In Figure 4.7.73

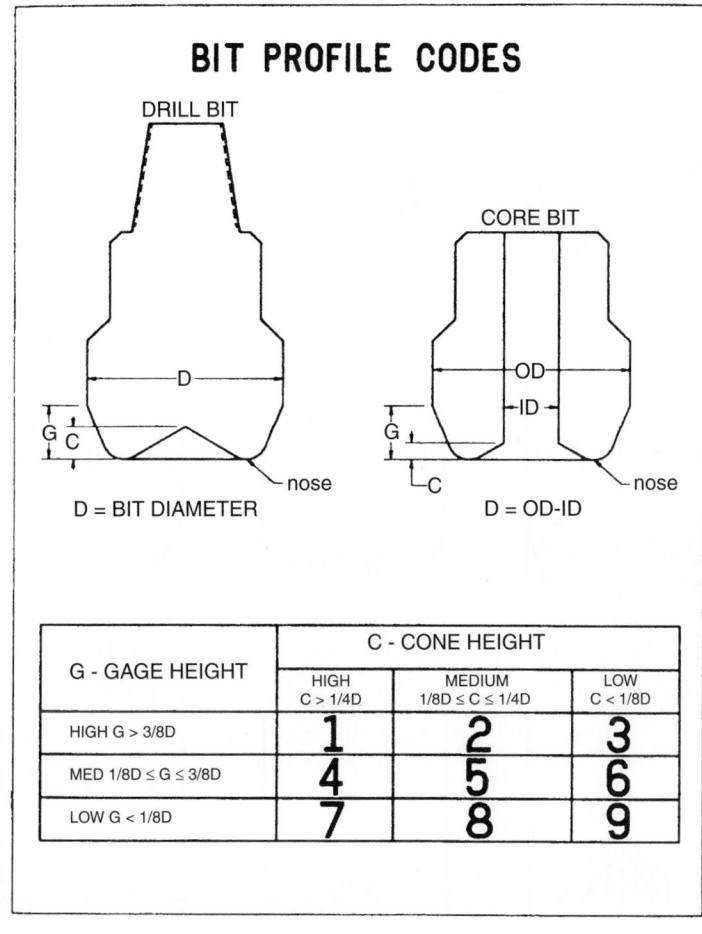

Figure 4.7.73 *Bit profile codes for fixed cutter bits [1] (courtesy of SPE).*

gage height and cone height dimensions are normalized to a reference dimension which is taken to be the bit diameter for drill bits and the (OD-ID) for core bits. Figure 4.7.74 provides a visual reference which is better suited for user by field personnel. Bold lines are drawn as examples of typical bit profiles in each category. Cross-hatched areas represent the range of variation for each category. Each of the nine profiles is given a name. For example, "double cone" is the term used to describe the profile in the center of the matrix (code 5). The double-cone profile is typical of many natural diamond and TSP bits.

The number 0 is used for unusual bit profiles which cannot be described by the 3×3 matrix of Figure 4.7.73. For example, a "bi-center" bit which has an asymmetrical profile with respect to the bit pin centerline should be classified with the numeral 0.

Hydraulic Design. The numbers 1 through 9 in the third character of the fixed cutter classification code refer to the hydraulic design of the bit (Figure 4.7.75). The hydraulic design is described by two components: the type of fluid outlet and the flow distribution. A 3×3 matrix of orifice types and flow distributions defines 9 numeric hydraulic design codes. The orifice type varies from changeable jets to fixed ports to open throat from left to right in the matrix. The flow

distribution varies from bladed to ribbed to open face from top to bottom. There is usually a close correlation between the flow distribution and the cutter arrangement.

The term *bladed* refers to raised, continuous flow restrictors with a standoff distance from the bit body of more than 1.0 in. In most cases cutters are affixed to the blades so that the cutter arrangement may also be described as bladed. The term *ribbed* refers to raised continuous flow restrictors with a standoff distance from the bit body of 1.0 in. or less. Cutters are usually affixed to most of the ribs so that the cutter arrangement may also be described as ribbed. The term *open face* refers to nonrestricted flow arrangements. Open face flow designs generally have a more even distribution of cutters over the bit face than with bladed or ribbed designs.

A special case is defined: the numbers 6 and 9 describe the crowfoot/water course design of most natural diamond and many TSP bits. Such designs are further described as having either radial flow, crossflow (feeder/collector), or other hydraulics. Thus, the letters R (radial flow). X (cross flow), or O (other) are used as the hydraulic design code for such bits.

Cutter Size and Placement Density. The numbers 1 through 9 and 0 in the fourth character of the fixed cutter classification code refer to the cutter size and placement density on the bit

BIT PROFILES

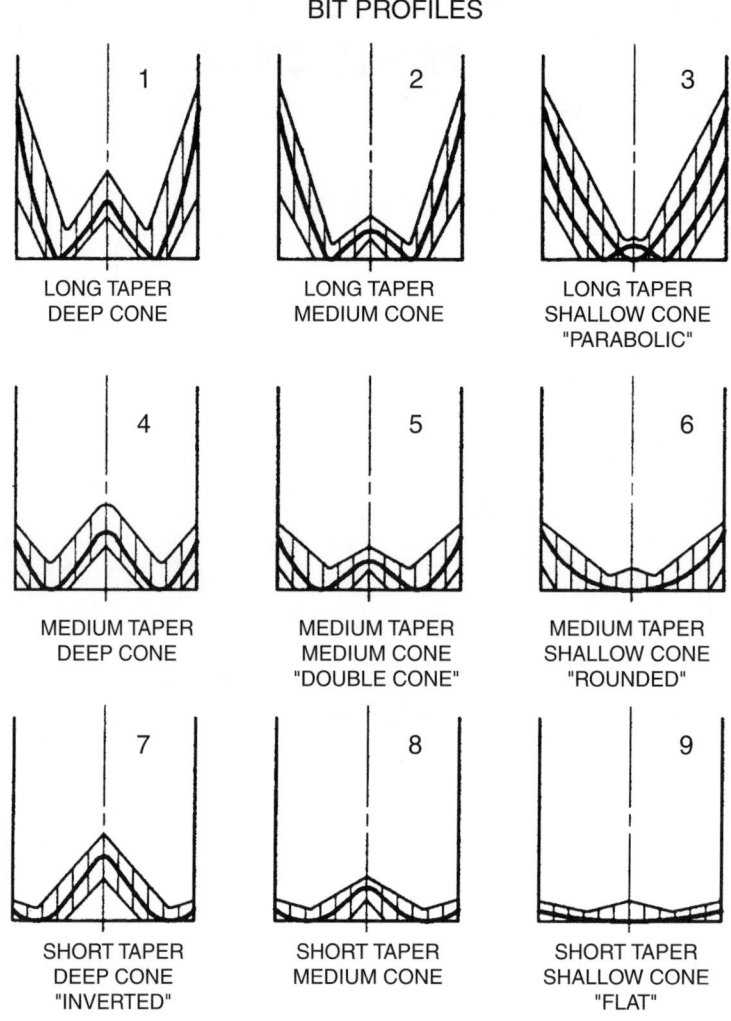

Figure 4.7.74   *Nine basic profiles of fixed-cutter bits [1] (courtesy of SPE).*

(Figure 4.7.76). A 3 × 3 matrix of cutter sizes and placement densities defines 9 numeric codes. The placement density varies from light to medium to heavy from left to right in the matrix. The cutter size varies from large to medium to small from top to bottom. The ultimate combination of small cutters set in a high density pattern is the impregnated bit, designated by the number 0.

Cutter size ranges are defined for natural diamonds based on the number of stones per carat. PDC and TSP cutter sizes are defined based on the amount of usable cutter height. Usable cutter height rather than total cutter height is the functional measure since various anchoring and attachment methods affect the "exposure" of the cutting structure. The most common type of PDC cutters, which have a diameter that is slightly more than $\frac{1}{2}$ in., were taken as the basis for defining medium size synthetic diamond cutters.

Cutter density ranges are not explicitly defined. The appropriate designation is left to the judgement of the manufacturer. In many cases manufacturers build "light-set" and "heavy-set" versions of a standard product. These can be distinguished by use of the light, medium, or heavy designation which is encoded in the fourth character of the IADC fixed cutter bit code. As a general guide, bits with minimal

cutter redundancy are classified as having light placement density and those with high cutter redundancy are classified as having heavy placement density.

4.7.3.1 Examples of Fixed-Cutter Bit Classification
Figure 4.7.77 shows a natural diamond drill bit which has a long outer taper and medium inner cone, radial flow fluid courses, and five to six stones per carat (spc) diamonds set with a medium placement density. Using the definitions in Figures 4.7.71, 4.7.73, 4.7.74 and 4.7.75, the characteristics of this bit are coded D 2 R 5 as follows:

Cutter/body type	D—natural diamond, matrix body
Bit profile	2—long taper, medium cone
Hydraulic design	R—open throat/open face radial flow
Cutter size/density	5—med. cutter size, med. placement density

Figure 4.7.78 shows a steel body PDC bit with standard-size cutters lightly set on a deep inner cone profile. This bit has changeable nozzles and is best described as having

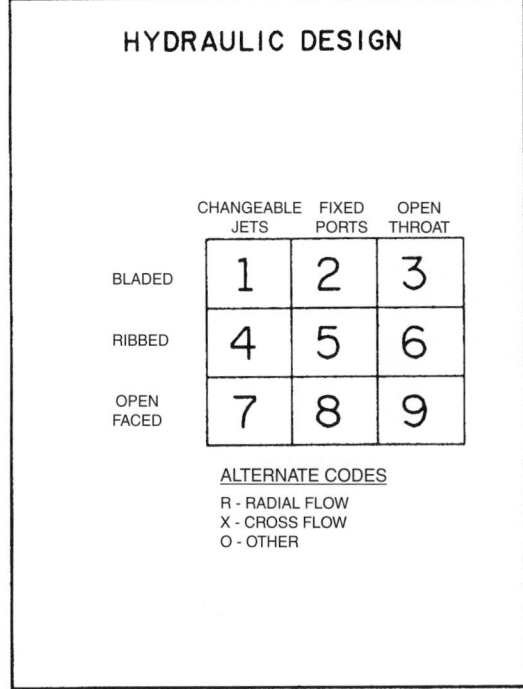

Figure 4.7.75 *Hydraulic design code for fixed-cutter bits [1] (courtesy of SPE).*

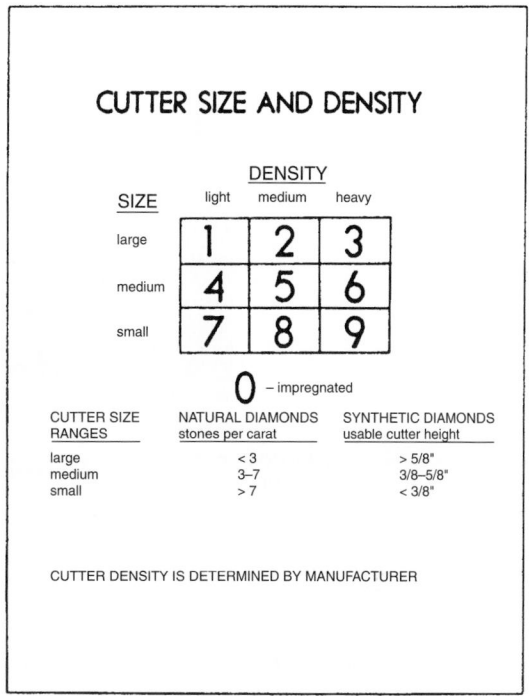

Figure 4.7.76 *Code of cutter size and placement [1] (courtesy of SPE).*

a ribbed flow pattern although there are open face characteristics near the center and bladed characteristics near the gage. The IADC classification code in this case is S 7 4 4.

Figure 4.7.79 shows a steel body core bit with a long-taper, stepped profile fitted with impregnated natural diamond blocks as the primary cutting elements. The bit has no inner cone. Since there is no specific code for the natural diamond/steel body combination, the letter O (other) is used as the cutter type/body material code. The profile code 3 is used to describe the long outer taper with little or no inner cone

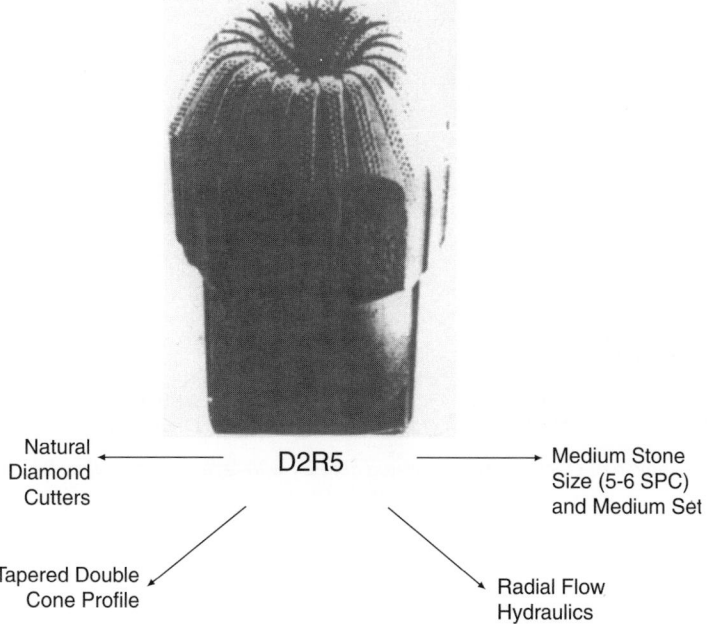

Figure 4.7.77 *Example of natural diamond bit with radial flow hydraulic design [1] (courtesy of SPE).*

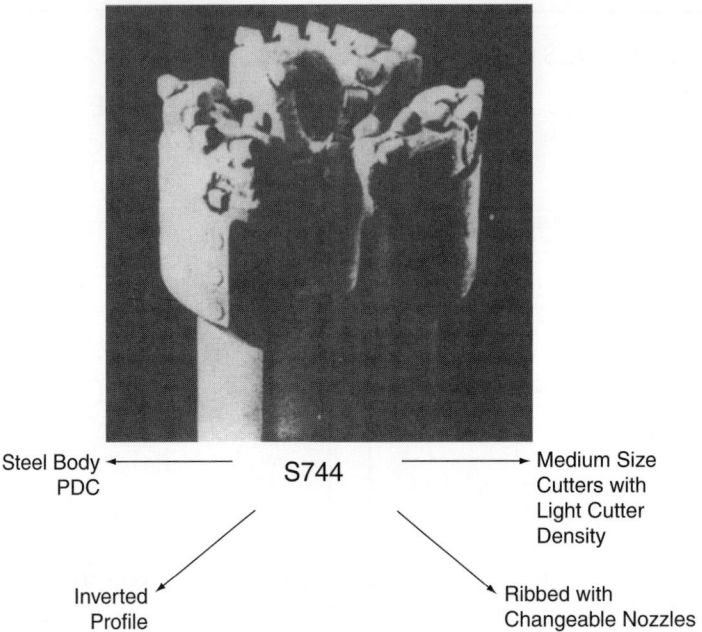

Steel Body ←——————— **S744** ——————→ Medium Size
PDC   Cutters with
                                           Light Cutter
                                           Density

Inverted   Ribbed with
Profile   Changeable Nozzles

Figure 4.7.78   *Example of steel body PDC bit with inverted profile [1] (courtesy of SPE).*

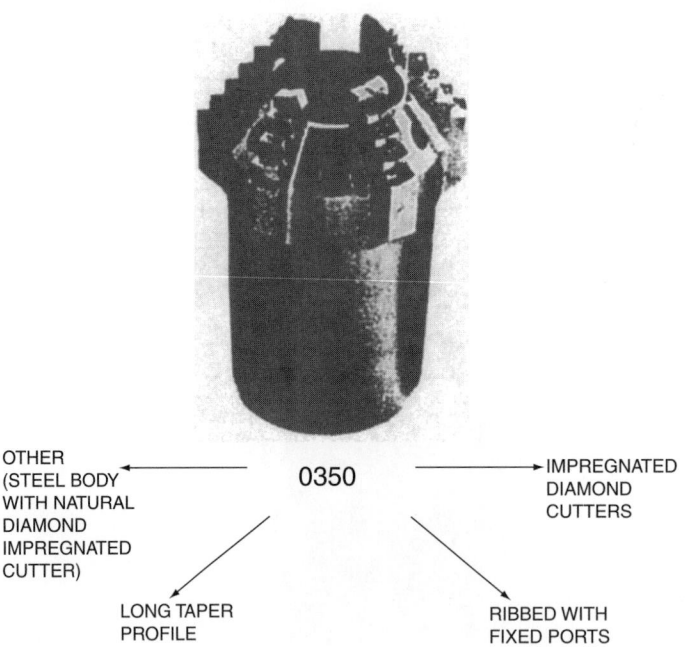

OTHER   IMPREGNATED
(STEEL BODY ←——————— **0350** ——————→ DIAMOND
WITH NATURAL   CUTTERS
DIAMOND
IMPREGNATED
CUTTER)

LONG TAPER   RIBBED WITH
PROFILE   FIXED PORTS

Figure 4.7.79   *Example of steel body impregnated core bit with face discharge flow [1] (courtesy of SPE).*

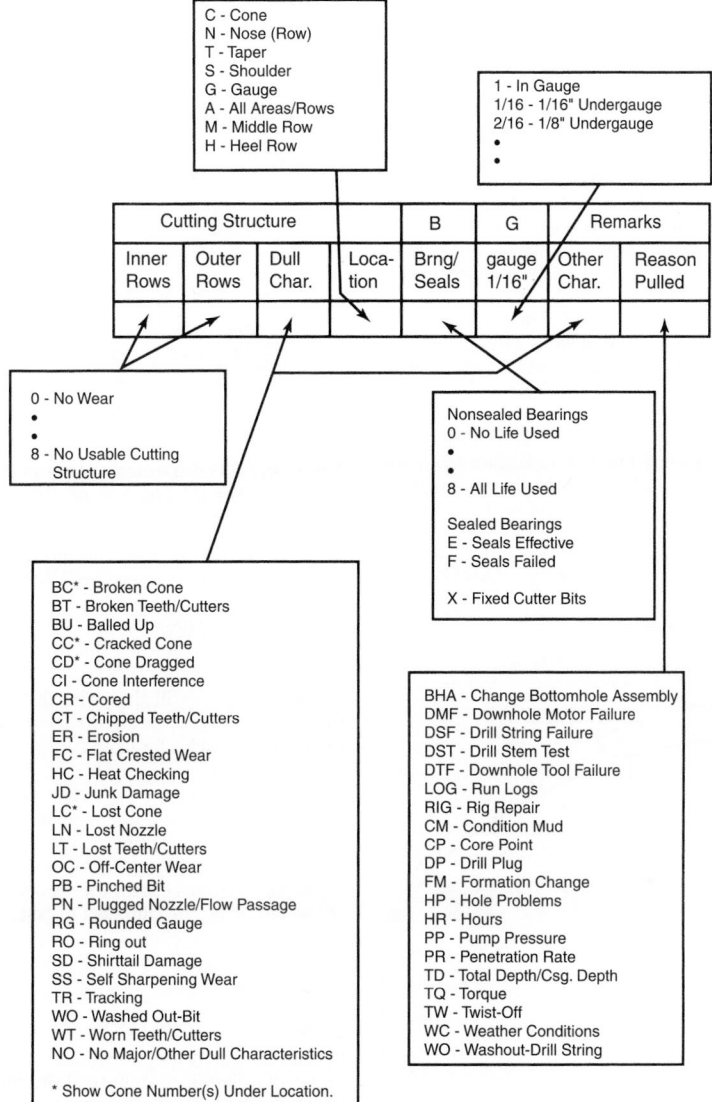

Figure 4.7.80 IADC bit dull grading codes—bold characters for fixed-cutter bits [18] (courtesy of SPE).

depth. The hydraulic design code 5 indicates a fixed port, ribbed design. Finally, the number 0 is used for impregnated natural diamond bits. Therefore the complete IADC classification code for this fixed cutter bit is 0 3 5 0. Although the classification code for this bit does not explicitly indicate the cutter type and body material, it can be referred from the rest of the code that this is an impregnated natural diamond, nonmatrix body bit, in which case steel is the most likely body material.

4.7.3.2 Dull Grading for Fixed Cutter Bits
This section describes the first IADC standardized system for dull grading natural diamond, PDC, and TSP (thermally stable polycrystalline diamond) bits, otherwise known as fixed cutter bits [2]. The new system is consistent with the recently revised dull grading system for roller bits. It describes the condition of the cutting structure, the primary

(with location) and secondary dull characteristics, the gage condition, and the reason the bit was pulled.

The format of the dull grading system is shown in Figure 4.7.80. For completeness, Figure 4.7.80 contains all of the codes needed to dull grade fixed cutter bits and roller bits. Those codes which apply to fixed cutter bits are in boldface.

Eight factors about a worn fixed cutter bit can be recorded. The first four spaces are used to describe the cutting structure. In the first two spaces, the amount of cutting structure wear is recorded using the linear scale 0 to 8, based on the initial useable cutter height. This is consistent with grading tooth wear on roller bits. The amount of cutter wear represented by 0 through 7 is shown schematically in Figure 4.7.81. An 8 means there is no cutter left. This same scale is to be used for TSP and natural diamond bits, with 0 meaning no wear, 4 meaning 50% wear, and so forth.

The first space of the dull grading format is used for the cutting structure in the inner two-thirds of the bit radius, and the second space applies to the cutting structure in the outer one-third of the bit radius, as shown schematically in Figure 4.7.81. When grading a dull bit, the average amount of wear in each area should be recorded. For example, in Figure 4.7.81 the five cutters in the inner area would be graded a 2. This is calculated by averaging the grades of the individual cutters in the inner area as follows: $(4+3+2+1+0)/5 = 2$. Similarly, the grade of the outer area would be a 6. On an actual bit the same procedure would be used. Note that for a core bit, the centerline in Figure 4.7.81 would be the core bit ID.

The third space is used to describe the primary dull characteristic of the worn bit; i.e., the obvious physical change from its new condition. The dull characteristics which apply to fixed cutter bits are listed in Figure 4.7.80.

The location of the primary dull characteristic is described in the fourth space. There are six choices: cone, nose, taper, shoulder, gauge, and all areas. Figure 4.7.82. shows four possible fixed cutter bit profiles with the different areas labeled.

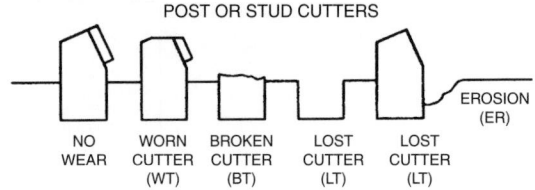

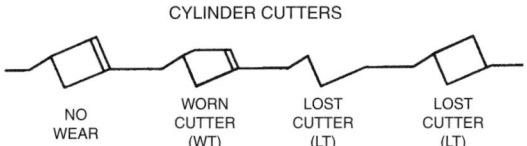

Figure 4.7.83 *Schematic of common dull characteristics [2] (courtesy of SPE).*

It is recognized that there are profiles for which the exact boundaries between areas are debatable and for which certain areas may not even exist. Notice that in the last profile there is no taper area shown. However, using Figure 4.7.83 as a guide, it should be possible to clearly define the different areas on most profiles.

The fifth space will always be an "X" for fixed cutter bits, since there are no bearings. This space can be used to distinguish dull grades for fixed cutter bits from dull grades for roller bits.

The measure of the bit gauge is recorded in the sixth space. If the bit is still in gauge, an "I" is used. Otherwise, the amount the bit is undergauge is noted to the nearest $\frac{1}{16}$ of an inch.

The seventh space is for the secondary dull characteristic of the bit, using the same list of codes as was used for the primary dull characteristic. The reason the bit was pulled is shown in the eighth space using the list of codes shown in Figure. 4.7.80.

4.7.3.3 Downhole Tools

Downhole drilling tools are the components of the lower part of a drill string used in normal drilling operations such as the drill bits, drill collars, stabilizers, shock absorbers, hole openers, underreamers, drilling jars as well as a variety of drill stem subs.

As drill bits, drill collars and drill stem subs are discussed elsewhere, this section regards shock absorbers, jars, underreamers, and stabilizers.

4.7.3.4 Shock Absorbers

Extreme vertical vibrations throughout the drill string are caused by hard, broken or changing formations, and the drilling bit chafing against the bottom formation as it rotates.

In shallow wells, the drill string transmits the vibration oscillations all the way to the crown block of the drilling rig. The effects can be devastating as welds fail, seams split, and drill string connections break down under the accelerated fatigue caused by the vibrations. In deep holes, these vibrations are rarely noticed due to elasticity and dampening by the long drill string. Unfortunately, the danger of fatigue still remains and has resulted in many fishing operations.

The drill string vibration dampeners are used to absorb and transfer the shock of drilling to the drill collars where it can be borne without damaging or destroying other drill string equipment. Their construction and design vary with

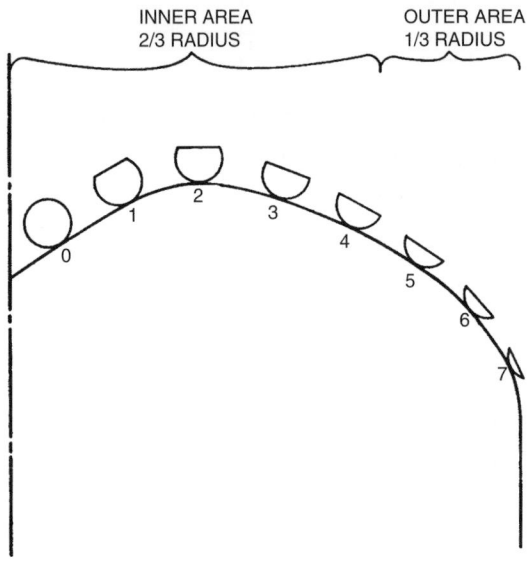

Figure 4.7.81 *Schematics of cutters wear [2] (courtesy of SPE).*

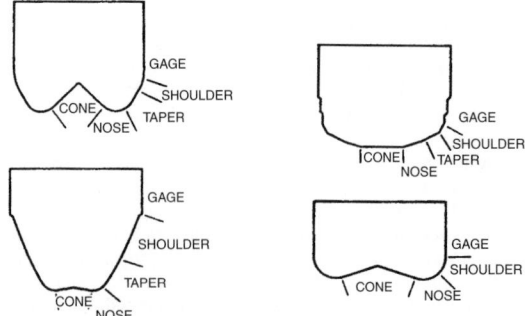

Figure 4.7.82 *Locations of wear on fixed-cutter bit [2]. (Courtesy of SPE).*

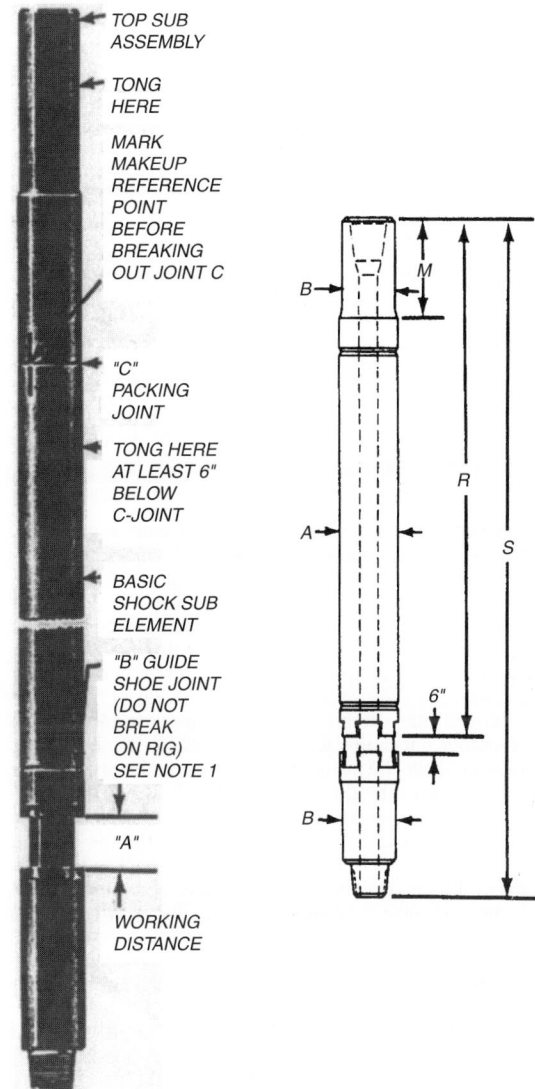

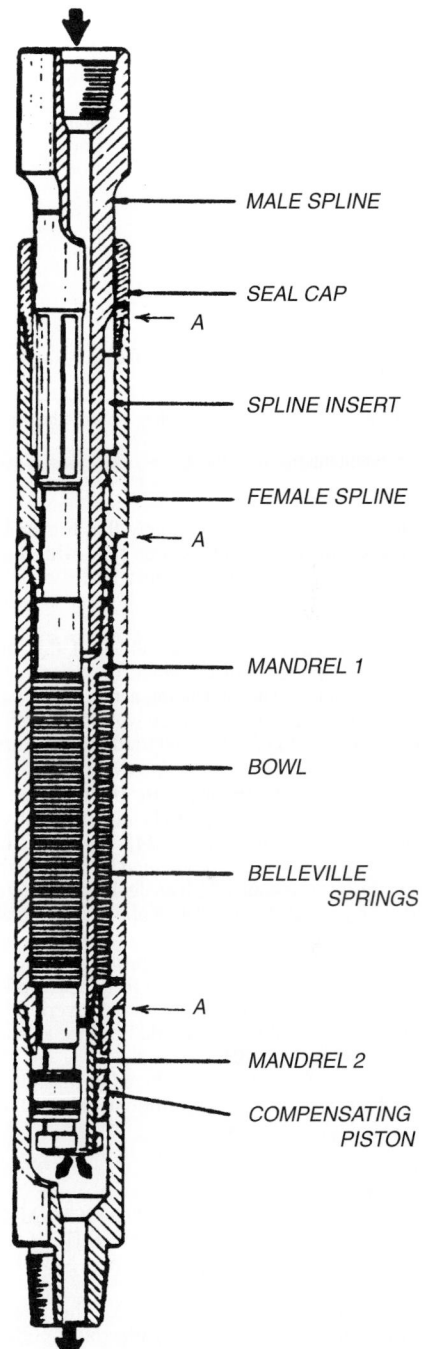

Figure 4.7.84 *Drilco rubber-spring shock dampener. (Courtesy of Smith International, Inc.).*

each manufacturer. To effectively absorb the vibrations induced by the drill bit, an element with a soft spring action and good dampening characteristics is required. There are six basic spring elements used: (1) vulcanized elastometer, (2) elastomeric element, (3) steel wool, (4) spring steel, (5) Belleville steel springs, and (6) gas compression.

Types of Shock Absorbers. Two of the eight commonly used commercial shock absorbers are described below.

Drilco Rubber Type. See Figure 4.7.84 and Table 4.7.13. Shock is absorbed by an elastometer situated between the inner and outer barrels. This shock absorbing element is vulcanized to the barrels. The torque has to be transmitted from the outer into the inner barrel. This tool is able to absorb shocks in axial or in radial directions. There is no need to absorb shocks in the torque because the drill string itself acts like a very good shock absorber so the critical shocks

Figure 4.7.85 *Christensen's Shock Eze[1] (courtesy of Hughes Christensen).*

are in axial directions. These tools cannot be used at temperatures above 200°F. Though they produce a small stroke the dampening effect is good [3].

Christenson Shock Eze. See Figure 4.7.85 [4]. A double action vibration and shock absorber employing Belleville spring elements immersed in oil.

Table 4.7.13 *Drilco Rubber Spring Shock Dampener Model "E"*

Nominal Size Tool (A)	*Suggested Hole Size Recommended for Best Performance	Top Sub Bore	Main Body Bore	End. Conn	OD (B)	S	M	R	Approx. Complete Weight (lb)	C-Joint Center Packing Joint Torque Makeup (ft-lb)
		Body Specifications				Misc. Lengths				
12	$17\frac{1}{2}$ thru 30	$2\frac{13}{16}$	$2\frac{13}{18}$	Specify	To Be	160	38	126	3070	70,000
10	$12\frac{1}{4}$ thru 15	$2\frac{13}{16}$	$2\frac{1}{4}$	Size	Specified	148	23	116	2050	55,000
9	$9\frac{5}{8}$ thru $12\frac{1}{4}$	$2\frac{13}{16}$	$2\frac{1}{4}$	and	Same OD	149	23	116	1635	41,000
8	$8\frac{1}{2}$ thru 11	$2\frac{13}{16}$	$2\frac{1}{4}$	Type	As Drill	145	23	112	1400	35,000
7	$7\frac{1}{2}$ thru 9	$2\frac{1}{4}$	2		Collar	143	23	111	1100	27,000
$6\frac{1}{4}$	$6\frac{3}{4}$ thru 9	$2\frac{1}{4}$	$1\frac{1}{2}$			144	23	111	890	20,000

Note : 1. All dimensions are given in inches, unless otherwise stated
       2. Recommended for optimum tool life.
*Courtesy Smith International, Inc.

The tool features a spline assembly that transmits high torque loads to the bit through its outer tube, while the inner assembly absorbs vibration through a series of steel-disc springs. The spring system works in both suspension and compression.

The high shock-absorbing capabilities of this tool are attained by compression of the stack of springs within a stroke of five inches. The alternating action of the patented spring arrangement provides a wide working range, under all possible conditions of thrust and mud pressure drop.

Placement of Shock Absorber in Drill String. Many operators have their own way of placing shock absorbers in the drill string (see Figure 4.7.86) [4]. In general, the optimum shock absorbing effect is obtained by running the tool as near to the bit as possible. With no deviation expected, the tool should be installed immediately above the bit stabilizer as shown in Figure 4.7.86C.

In holes with slight deviation problems, the shock absorber could be run on top of the first or second string stabilizer. For situations where there are severe deviation problems, the shock absorber should be place as shown in Figures 4.7.86B and 4.7.86D.

For turbine drilling, it is recommended that the shock absorber be placed on top of the first stabilizer above the turbine as in Figure 4.7.86A.

References
 1. Winters, W. J., and H. H. Doiron, "The 1987 IADC fixed cutter bit classification system," SPE/IADC 1642, Proc. 1987 SPE/IADC Drilling Conference New Orleans, March 15–18, 1987, pp. 807–817.
 2. Clark, D. A., et al., "Application of the new IADC dull grading system for fixed cutter bits," SPE/IADC 1645, Proc. 1987 SPE/IADC Drilling Conference, New Orleans, March 15–18, 1987, pp. 851–858.
 3. "Drilco 1986-1987 Products Catalog," Smith International, Inc. 1986-87, Composite Catalog, pp. 1751.
 4. Christensen, *Downhole Tool Technology*, student manual.

4.7.4 Air Hammer Bits
Percussion air hammers have been used for decades in shallow air drilling operations. These shallow operations have been directed at the drilling of water wells, monitoring wells, geotechnical boreholes, and mining boreholes. In the past decade, however, the percussion air hammers have seen increasing use in drilling deep oil and natural gas wells. Percussion air hammers have a distinct advantage over roller cutter bits in drilling abrasive, hard rock formations.

The use of percussion air hammers (or down-the-hole air hammers) is an acceptable option to using rotating tri-cone or single cone drill bits for air and gas drilling operations. The air hammer utilizes an internal piston (or hammer) that is actuated by the compressed air (or other gas) flow inside the drill string. The internal piston moves up and down in a chamber under the action of air pressure applied either below or above the piston through ports in the inside of the air hammer. In the downward stroke, the hammer strikes the bottom of the upper end of the drill bit shaft (via a coupling shaft) and imparts an impact load to the drill bit. The drill bit in turn transfers this impact load to the rock face of the bit. This impact load creates a crushing action on the rock face very similar to that discussed earlier in this section. But in this situation, the crushing action is dynamic and is more effective than the quasi-static crushing action of tri-cone and single cone drill bits. Therefore, air hammer drilling operations require far less WOB as comparable drilling operations using tri-cone or single cone drill bits.

The air hammer is made up to the bottom of the drill string and at the bottom of the air hammer is the air hammer bit. The air hammer drill string must be rotated just like a drill string that utilizes tri-cone or single cone drill bits. The rotation of the drill string allows the inserts (i.e., tungsten carbide studs) on the bit face to move to a different location on the rock face surface. This rotation allows a different position on the rock face to receive the impact load as the upper end of the hammer bit is struck by the hammer. In direct circulation operations, air flow passes through the hammer section, through the drill bit channel and orifices to the annulus, As the air passes into the annulus, the flow entrains the rock cuttings and carries the cuttings to the surface in the annulus. Direct circulation air hammers are available in a wide variety of outside housing diameters (3 inches to 16 inches). These air hammers drill boreholes with diameters from $3\frac{5}{8}$ inches to $17\frac{1}{2}$ inches.

There are also reverse circulation air hammers. These unique air hammers allow air pressure in the annulus to actuate the hammer via ports in the outside housing of the hammer. The reverse circulation air hammer bits are designed with two large orifices in the bit face that allow the return air flow with entrained rock cuttings to flow to the inside of the drill string and then to the surface.

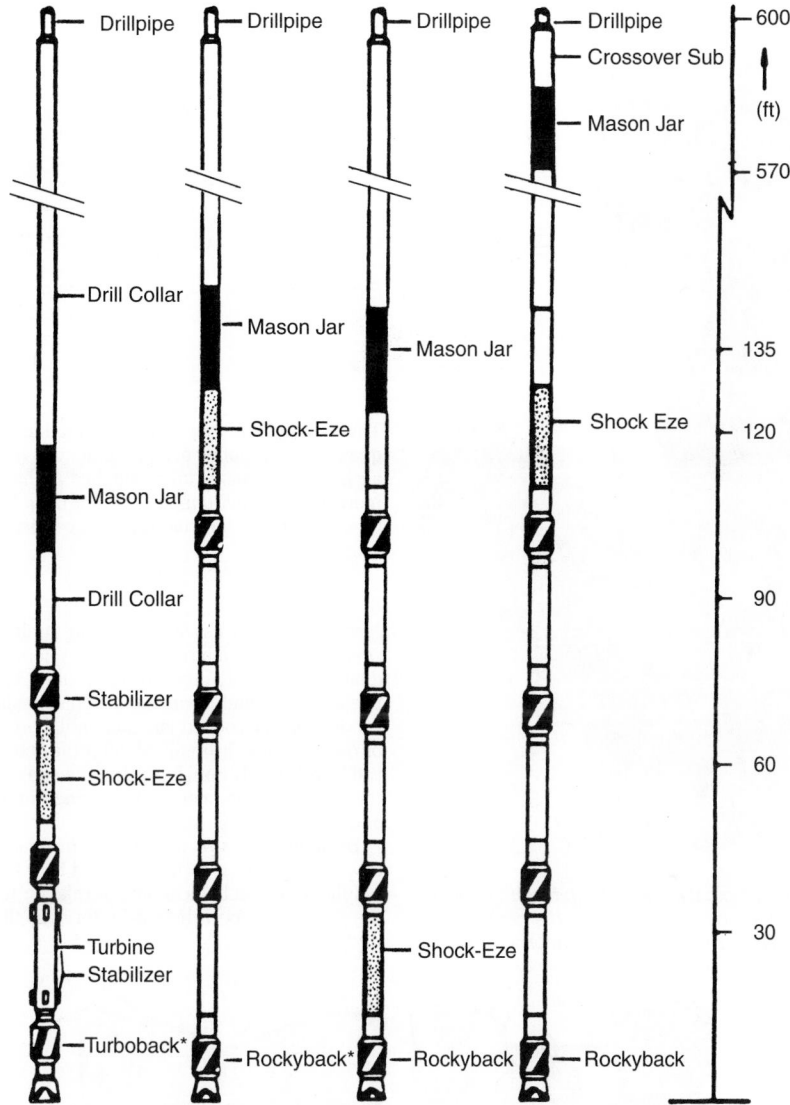

Figure 4.7.86 *Recommended placement of the Shock Eze [1] (courtesy of Hughes Christensen).*

Reverse circulation air hammers are available in larger outside housing diameters (6 inches to 24 inches). These air hammers drill boreholes with diameters from $7\frac{7}{8}$ inches to 33 inches. Figure 4.7.87 shows two typical air hammer bits that would be used with direct circulation air hammers. The larger bit (standing on its shank end) is an $8\frac{5}{8}$ inch diameter concave bit. The smaller bit (laying on its side) is a 6 inch diameter concave bit.

There are five air hammer bit cutting face designs. Figure 4.7.88a shows the profile of the drop center bit and Figure 4.7.88b shows the profile of the concave bit. Figure 4.7.89a shows the profile of the step gauge bit and Figure 4.7.89b. shows the profile of the double gauge bit. Figure 4.7.90 shows the profile of the flat face bit. These five bit cutting face designs are applicable for a variety of drilling applications from non-abrasive, soft rock formations to highly abrasive,

very hard rock formations. The application of these five face designs are shown in Figure 4.7.91.

In the past the air hammer manufacturers have provided the air hammer bits for their specific air hammers. This practice insured compatibility of bit with hammer housing. The increased air hammer use in drilling deep oil and natural gas recovery wells has attracted traditional oil field drill bit manufacturers to fabricate air hammer bits. Although the air hammer bit faces are somewhat uniform in design, the shafts are different for each air hammer manufacturer. The air hammer face and shafts are integral to the bit, thus, manufacturing air hammer bits is complicated. Fortunately, the air hammer has proven in the past decade to be very effective in drilling deep boreholes. This has given rise to competition among traditional drill bit manufacturers to provide improved air hammer bits for deep drilling operations.

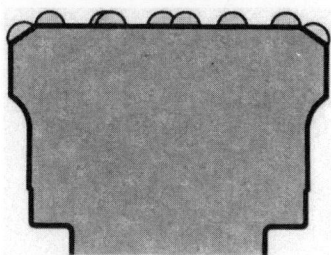

Figure 4.7.90 *Air hammer bit face profile design, flat face bit (courtesy of AB Sandvik Rock Tools).*

This competition has in turn resulted in an increase in the quality and durability of air hammer bits (over the traditional air hammer manufacturer-supplied air hammer bits) in the more hostile environments of the deep boreholes. Operational use of the air hammer will be discussed in detail in section 4.10.

4.7.5 Jars

Jarring is the process of transferring strain energy *(drill pipe stretch)* in the drill string to kinetic energy by releasing the detent in the jar at a given overpull value. A jar is a hollow rod-telescoping cylinder with a detent mechanism that holds the rod in place or cocked position until a predetermined force or overpull valve is obtained. After the detent is released, the rod travels freely until it reaches the end of its stroke, where the jarring action occurs by a sudden impact of a hammer and anvil within the jar.

By definition, the detent is a mechanical device (e.g., catch, latch, collet, spring-operated ball) for positioning and holding one mechanical part in relation to another so that the device can be released by force applied to one of the parts.

Figure 4.7.87 *Two typical air hammer bits with concave face (8$\frac{5}{8}$ inch diameter bit on end, 6 inch diameter on side) (courtesy of Rock Bit International Incorporated).*

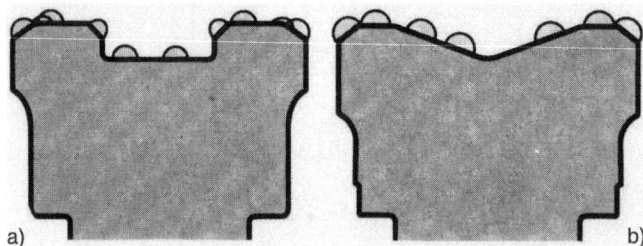

Figure 4.7.88 *Air hammer bit face profile designs, a) drop center bit, and b) concave bit (courtesy of AB Sandvik Rock Tools).*

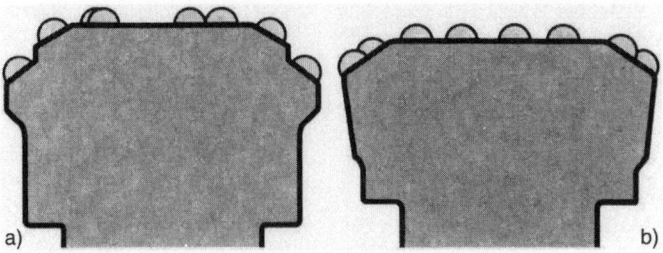

Figure 4.7.89 *Air hammer bit face profile designs, a) step gauge bit, b) double gauge bit (courtesy of AB Sandvik Rock Tools).*

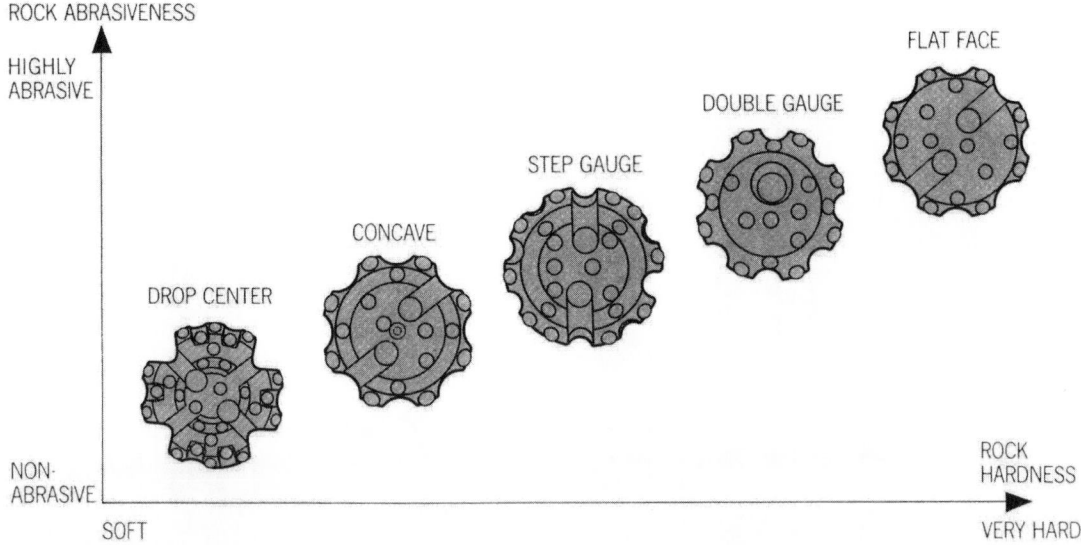

Figure 4.7.91 *Air hammer bit face profile designs and application to rock formation abrasiveness and hardness (courtesy of AB Sandvik Rock Tools).*

4.7.5.1 Type of Jars

There are two general classes of jars: fishing jars and drilling jars.

A fishing jar is used to free stuck drill string and is added to the drill string only when the string becomes stuck.

The drilling jar is used as a part of the drill string, mainly in the bottomhole assembly, (BHA) positioned either in height-weight drill pipe (Figure 4.7.92) or in drill collars (Figure 4.7.93) in most cases. Running drilling jars in the BHA is insurance against loss time and money since the jarring process can begin immediately after the drill string becomes stuck. Note: Jars are not effective if the stuck point is located above the jar.

4.7.5.2 Drilling Jar Design

Drilling jars can be classified as mechanical or hydraulic, depending on the trigging method. A third design, which incorporates a hydraulic section for up jarring and a separate mechanical section for down jarring, is called hydro-mechanical.

Mechanical Jars (Figure 4.7.94). Mechanical jars trigger by some variation of the spring-loaded mechanical detent principle. The detent releases as soon as a predetermined overpull is reached so that the only waiting time is that required to cycle the drawworks and stretch the drill pipe. With most mechanical drilling jars, triggering forces cannot be varied downhole, but those that do allow varied overpull downhole use drill pipe torque to control the triggering overpull. This method normally becomes less predictable as hole drag becomes high.

Hydraulic Drilling Jars (Figure 4.7.95). The operating principle of the conventional hydraulic drilling jar is explained by McGill (SPE 1968). A piston is pulled through a close-fitting bore, causing hydraulic fluid to flow from a high-pressure side to the low-pressure side around the piston. When the piston moves from the close-fitting area of the bore into the area of the larger diameter, the fluid can flow freely around the piston and the jar releases. This provides a time delay, which can be used to stretch the drill pipe to any desired tensile overpull or to apply weight to the jar to jar down within the range of the particular jar. The ability to vary the overpull can be useful in many jarring operations.

This type of hydraulic delay frequently generates heat at a rate that cannot dissipate because this design requires a long metering stroke. This long metering stroke is necessary to allow for changes in the piston/bore radial clearance. The heat build-up in the jar may necessitate periodically stopping the jarring action to allow the jar fluid to cool. To minimize the heat build-up problem, a hydraulic drilling jar with a time delay/mechanical release detent was designed.

The operating principle of the hydraulic delay/mechanical release jar is explained by R. W. Evans (SPE 1990). The metering function of this type of jar is separated from the triggering function by using a separate valve to control the triggering. When jarring, the mandrels move in response of the drawworks while the housing remain stationary. Two pistons that oppose each other to define a high-pressure chamber resist the movement of the mandrels in either direction. The metering orifice allows hydraulic fluid to slowly bleed through the piston until the triggering valves are mechanically opened. This delay or metering time is designed to allow the operator sufficient time to pull to the desired overpull or apply adequate weight to the jar for a downward action before the triggering valves to open. Like the conventional hydraulic jar, both types do not require presetting or adjustment before running in the well. The overpull and downward jarring forces and directions are controlled from the rig floor.

4.7.6 Underreamers

The term *underreaming* has been used interchangeably with *hole opening*. Underreaming is the process of enlarging a hole bore beginning at some point below the surface using a tool with expanding cutters. This permits lowering the tool through the original hole to the point where enlargement of the hole is to begin.

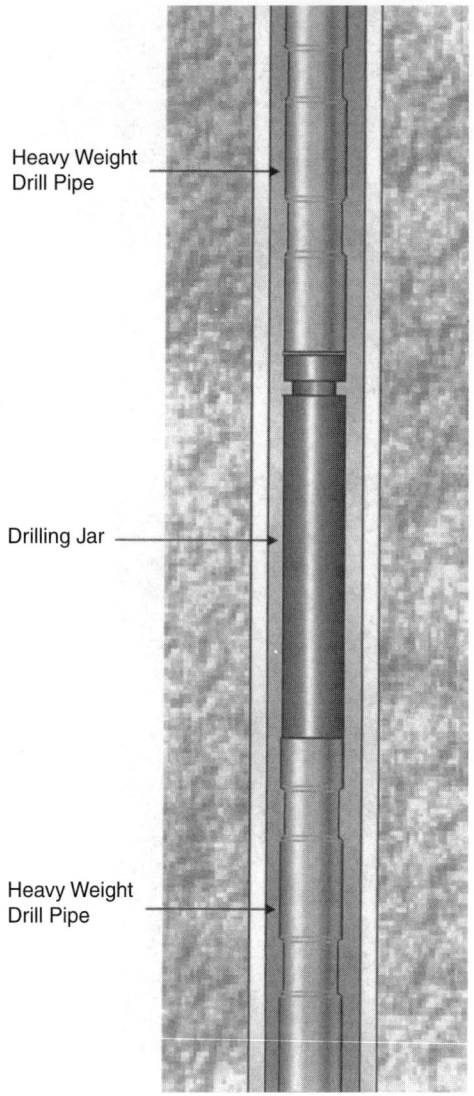

Figure 4.7.92 *Jars positioned in heavy-weight drill pipe.*

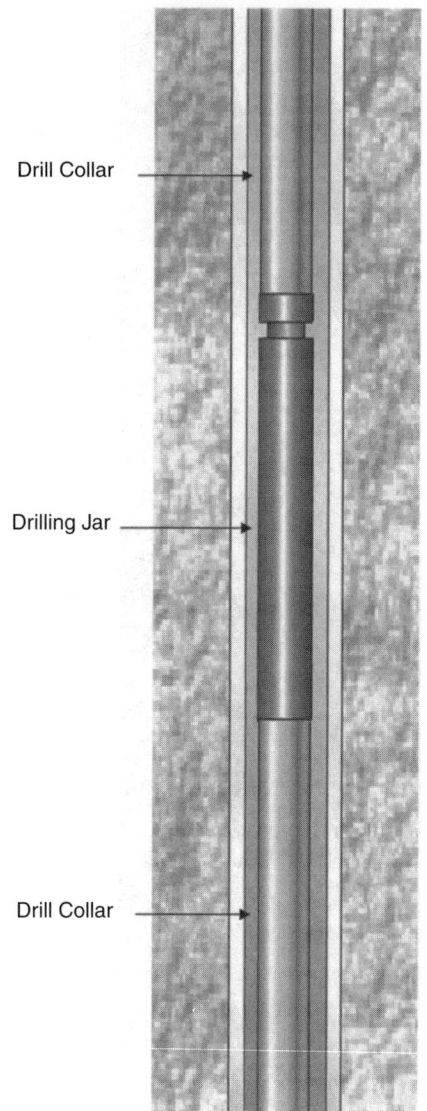

Figure 4.7.93 *Jars positioned in drill collars.*

Hole opening is considered as opening or enlarging the hole from the surface (or casing shoe) downward using a tool with cutter arms at a fixed diameter.

Thus, the proper name for the tools with expandable cutting arms is underreamers. The cutting arms are collapsed in the tool body while running the tool in the hole. Once the required depth is reached, mud circulation pressure moves the cutters opening for drilling operation. Additional pressure drop across the underreamer orifice gives the operator positive indication that the cutter arms are extended fully and the tool is undereaming at full gauge.

Underreamer Design. There are two basic types of underreamers: (1) roller cone rock-type underreamers and (2) drag-type underreamers. The roller cone rock-type underreamers are designed for all types of formations depending upon the type of roller cones installed. The drag-type

underreamers are used in soft to medium formations. Both types can be equipped with a bit to drill and underream simultaneously. This allows for four different combinations of underreamers as shown in Figure 4.7.96. Nomenclature of various under-reamer designs are shown in Figures 4.7.97, 4.7.98, 4.7.99, and 4.7.100.

Underreamer Specifications. Tables 4.7.14 though 4.7.17 [2] show specifications for the four different types of Smith International, Inc. underreamers. Table 4.7.14 shows the specifications for the six standard models of reamaster underreamers. Table 4.7.15 shows the specifications for the nine standard models of drilling-type underreamers, Table 4.7.16 shows the specifications for the ten standard models of Rock-Type underreamers and Table 4.7.17 shows the specifications for the eight standard models of drag-type underreamers.

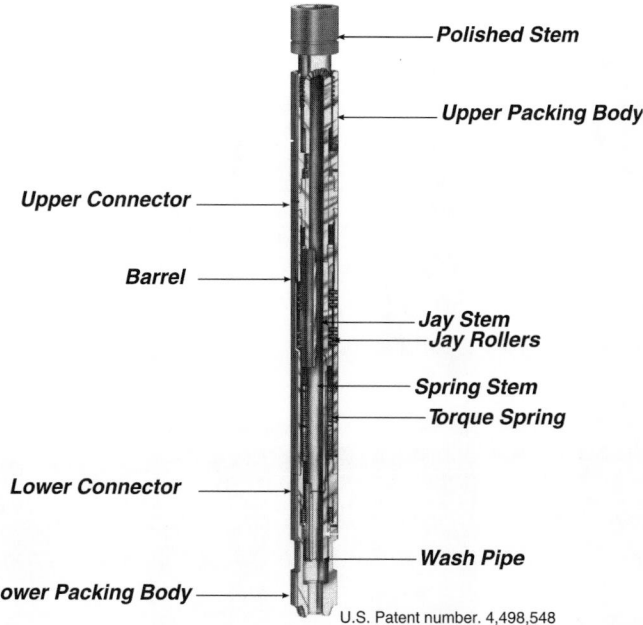

Figure 4.7.94 *Dailey L.I. Mechanical drilling jar (Courtesy of Weatherford International).*

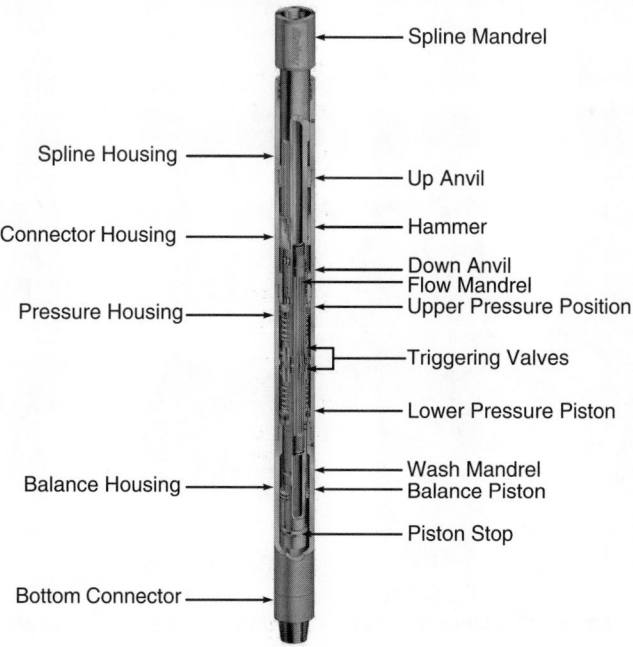

Figure 4.7.95 *Dailey Hydraulic drilling jar (Courtesy of Weatherford International).*

| Reamaster (XTU)
Underreamer | Rock Type
Underreamer | Rock-Drilling
Type
Underreamer | Drag-Type
Underreamer |

Figure 4.7.96 *Types of Underreamers (Courtesy of Smith International, Inc).*

Reamaster Components

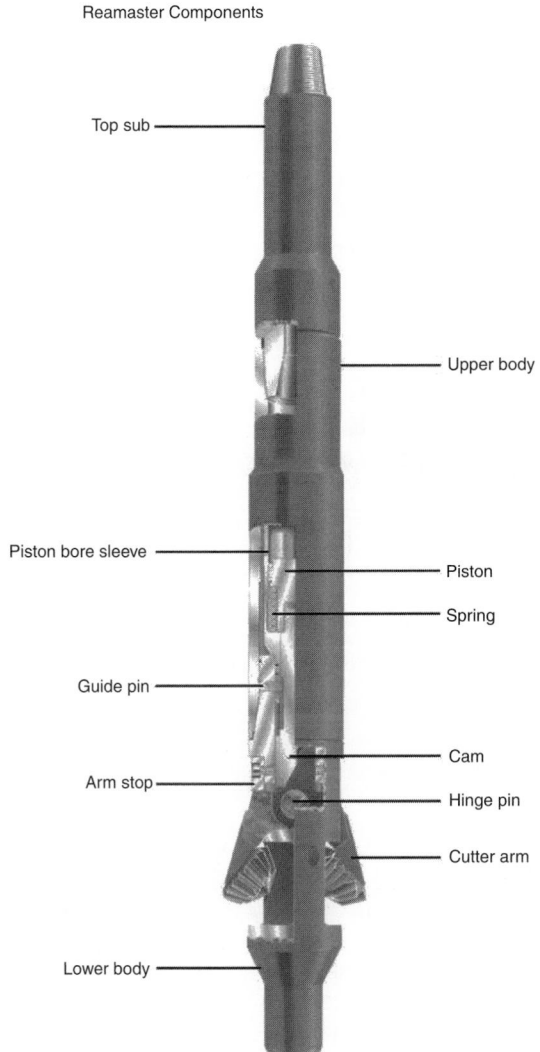

Top sub

Upper body

Piston bore sleeve

Piston

Spring

Guide pin

Cam

Arm stop

Hinge pin

Cutter arm

Lower body

Figure 4.7.97 *Reamaster (XTU) underreamer nomenclature (courtesy of Smith International Inc.).*

Underreamer Hydraulics. Pressure losses across the underreamer nozzles (orifice) are shown in Figures 4.7.101 and 4.7.102 [1]. These pressure drop graphs can be used for pressure losses calculations (given pump output and nozzles) or for orifice (nozzle) selection (given pump output and pressure loss required).

4.7.6.1 Stabilizers

Drill collar stabilizers are installed within the column of drill collars. Stabilizers guide the bit straight in vertical-hole drilling or help building, dropping, or, maintaining hole angle in directional drilling. The stabilizers are used to

1. provide equalized loading on the bit
2. prevent wobbling of the lower drill collar assembly
3. minimize bit walk
4. minimize bending and vibrations that cause tool joint wear
5. prevent collar contact with the sidewall of the hole
6. minimize keyseating and differential pressure.

The condition called "wobble" exists if the bit centerline does not rotate exactly parallel to and on the hole centerline so the bit is tilted.

Stabilizer Design. There are three commonly used stabilizer designs.

Solid-Type Stabilizers (Figure 4.7.103). These stabilizers have no moving or replaceable parts, and consist of mandrel and blades that can be one piece alloy steel (integral blade stabilizer) or blades welded on the mandrel (weld-on blade stabilizer). The blades can be straight or spiral, and their working surface is either hardfaced with tungsten carbide inserts or diamonds [2,3].

Sleeve-Type Stabilizers (Figure 4.7.104). These stabilizers have replaceable sleeve that can be changed in the field. There are two types of sleeve-type stabilizers: the rotating sleeve-type stabilizer (Figure 4.7.104A [2]) and the nonrotating sleeve-type stabilizer (Figure 4.7.104B [4]). Rotating sleeve-type stabilizers have no moving parts and work in the same way as solid-type stabilizers. Nonrotating sleeve-type stabilizers have a nonrotating rubber sleeve supported by the wall of the borehole. The rubber sleeve stiffens the drill collar string in packed hole operations just like a bushing.

Reamers (Figures 4.7.105 and 4.7.106). Reamers are stabilizers with cutting elements embedded in their fins, and are

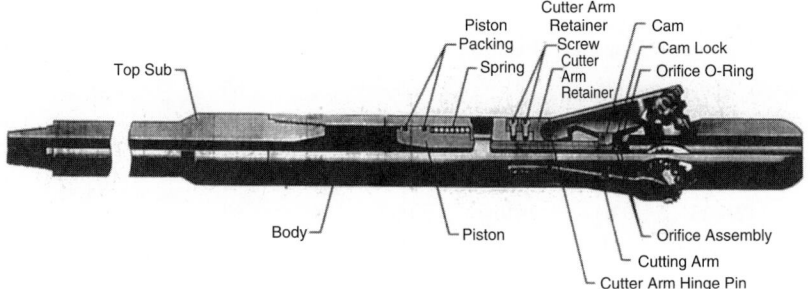

Top Sub

Piston
Packing
Spring

Cutter Arm
Retainer
Screw
Cutter
Arm
Retainer

Cam
Cam Lock
Orifice O-Ring

Body

Piston

Orifice Assembly

Cutting Arm

Cutter Arm Hinge Pin

Figure 4.7.98 *Rock-type underreamer nomenclature (courtesy of Smith International, Inc.).*

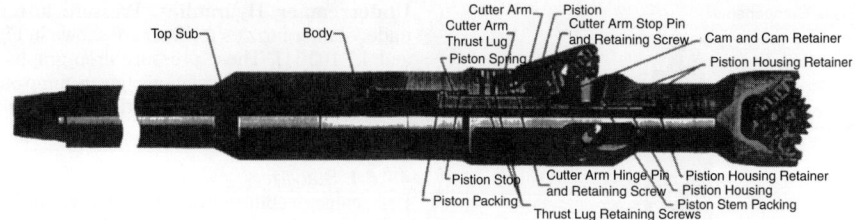

Figure 4.7.99 *Rock-drilling underreamer nomenclature (courtesy of Smith International, Inc.).*

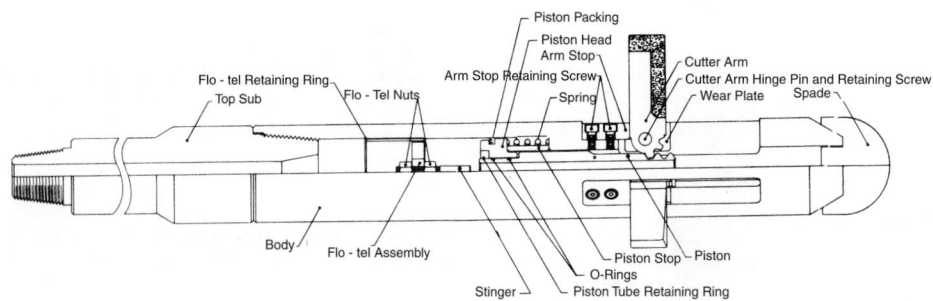

Figure 4.7.100 *Drag-type underreamer nomenclature (open arms) (courtesy of Smith International, Inc.).*

Table 4.7.14 *Reamaster™ (XTU) Underreamer Specification Table*

Tool Series	Opening Diameter	Pilot Hole Size	Body Dia. (Collapsed Diameter)	Fishing Neck Length	Fishing Neck Diameter	Overall Length	Top Pin/Bottom Box Conn. API Reg.	Weight (lb)
5700	$8\frac{1}{2}$, 9	$5\frac{7}{8}$–$6\frac{1}{2}$	$5\frac{3}{4}$	18	$4\frac{3}{4}$	90	$3\frac{1}{2}$	500
7200	$9\frac{7}{8}$, 11, $11\frac{3}{4}$, $12\frac{1}{4}$	$7\frac{1}{2}$–11	$7\frac{1}{4}$	18	$5\frac{3}{4}$	99	$4\frac{1}{2}$	700
8250	$9\frac{7}{8}$, $10\frac{5}{8}$, 11, $12\frac{1}{4}$, $13\frac{1}{2}$	$8\frac{1}{2}$–$9\frac{7}{8}$	$8\frac{1}{4}$	18	$5\frac{3}{4}$	123	$4\frac{1}{2}$	900
9500	$12\frac{1}{4}$, $13\frac{1}{2}$, 15, 16	$9\frac{7}{8}$–$12\frac{1}{4}$	$9\frac{1}{2}$	24	8	136	$6\frac{5}{8}$	1100
11700	14, 15, 16, $17\frac{1}{2}$	$12\frac{1}{4}$–$14\frac{3}{4}$	$11\frac{3}{4}$	20	8	130	$6\frac{5}{8}$	1700
16000	20, 22, 24, 26	$17\frac{1}{2}$–22	16	20	10	140	$8\frac{5}{8}$	3200

Courtesy of Smith International, Inc.

Table 4.7.15 *Drilling-Type Underreamer (DTU) Specification Table*

Tool Series	Standard Opening Diameter	Optional Opening Diameter	Body Diameter	Collapsed Diameter	Fishing Neck Length	Fishing Neck Diameter	Overall Length	Top Pin/Bottom Box Conn. API Reg.	Weight (lb)	Min. Rec. Pilot Hole Diameter
3600	6	$5\frac{1}{2}$ – 6	$3\frac{5}{8}$	$3\frac{5}{8}$	8	$3\frac{3}{8}$	35	$2\frac{3}{8}$	170	$3\frac{3}{4}$
5700	$8\frac{3}{4}$	7 – $8\frac{3}{4}$	$5\frac{3}{4}$	6	18	$4\frac{3}{4}$	70	$3\frac{1}{2}$	360	6
7200	11	9 – 11	$7\frac{1}{4}$	$7\frac{1}{2}$	18	$5\frac{3}{4}$	74	$4\frac{1}{2}$	770	$7\frac{5}{8}$
8200	14	10 – 14	$8\frac{1}{4}$	$8\frac{1}{4}$	18	$5\frac{3}{4}$, 8	79	$4\frac{1}{2}$, $6\frac{5}{8}$	900	$8\frac{1}{2}$
9500	15	12 – 15	$9\frac{1}{2}$	$10\frac{1}{4}$	18	8	82	$6\frac{5}{8}$	1, 150	$9\frac{7}{8}$
11700	$17\frac{1}{2}$	$14\frac{3}{4}$ – 20	$11\frac{3}{4}$	$11\frac{3}{4}$	20	8	96	$6\frac{5}{8}$	1, 670	$12\frac{1}{4}$
15000	$17\frac{1}{2}$, 26	$17\frac{1}{2}$ – 26	$14\frac{3}{4}$	$14\frac{3}{4}$	20	8, 9	97	$6\frac{5}{8}$, $7\frac{5}{8}$	2, 800	$14\frac{3}{4}$
17000	32	24 – 32	17	17	20	9, 10	127	$7\frac{5}{8}$, $8\frac{5}{8}$	3, 800	$17\frac{1}{2}$
22000	36	28 – 36	22	22	20	9, 10	100	$7\frac{5}{8}$, $8\frac{5}{8}$	4, 400	24

Courtesy of Smith International, Inc.

Table 4.7.16 *Rock-Type Underreamer (RTU) Specification Table*

Tool Series	Standard Opening Diameter	Optional Opening Diameter	Body Diameter	Collapsed Diameter	Fishing Neck Length	Fishing Neck Diameter	Overall Length	Top Pin/Bottom Box Conn. API Reg.	Weight (lb)	Min. Rec. Pilot Hole Diameter
3600	6	$4\frac{3}{4}-6\frac{1}{2}$	$3\frac{5}{8}$	$3\frac{3}{4}$	8	$3\frac{3}{8}$	$26\frac{1}{2}$	$2\frac{3}{8}$	175	$3\frac{3}{4}$
4500	$6\frac{1}{2},8\frac{1}{2}$	$6-9$	$4\frac{1}{2}$	$4\frac{5}{8}$	18	$4\frac{1}{4}$	67	$2\frac{7}{8}$	235	$4\frac{3}{4}$
5700	11	$8-11$	$5\frac{3}{4}$	6	18	$4\frac{3}{4}$	$76\frac{1}{2}$	$3\frac{1}{2}$	380	6
6100	12	$11-12$	$6\frac{1}{8}$	$6\frac{1}{8}$	18	$4\frac{3}{4}$	$78\frac{1}{2}$	$3\frac{1}{2}$	380	$6\frac{1}{2}$
7200	14	$9-14$	$7\frac{1}{4}$	$7\frac{3}{8}$	18	$5\frac{3}{4}$	86	$4\frac{1}{2}$	775	$7\frac{5}{8}$
8200	16	$10-16$	$8\frac{1}{4}$	$8\frac{3}{8}$	18	$5\frac{3}{4},8$	89	$4\frac{1}{2},6\frac{5}{8}$	920	$8\frac{1}{2}$
9500	$17\frac{1}{2}$	$12\frac{1}{4}-18$	$9\frac{1}{2}$	$9\frac{3}{4}$	18	8	91	$6\frac{5}{8}$	1, 160	$9\frac{7}{8}$
11700	$17\frac{1}{2}$	$14\frac{3}{4}-22$	$11\frac{3}{4}$	$12\frac{1}{4}$	20	8	91	$6\frac{5}{8}$	1, 670	$12\frac{1}{4}$
15000 LP	26	$17\frac{1}{2}-30$	$14\frac{3}{4}$	$14\frac{3}{4}$	20	8, 9	97	$6\frac{5}{8},7\frac{5}{8}$	2, 800	$17\frac{1}{2}$
22000	32-40	$32-40$	22	22	20	9,10	$124\frac{1}{2}$	$7\frac{5}{8},8\frac{5}{8}$	5, 900	24

Courtesy of Smith International, Inc.

Table 4.7.17 *Drag-Type Underreamer Specification Table*

Tool Series	Body Diameter	Minimum Recommended Pilot Hole Diameter	Standard Expanded Diamter SPX	Servcoloy "S"	Fishing Neck Length	Fishing Neck Diameter(s)	Overall Length	Top Pin Conn. API	Weight (Ib)
3600	$3\frac{5}{8}$	$3\frac{3}{4}$	—	9	8	$3\frac{3}{8}$	$26\frac{1}{2}$	$2\frac{3}{8}$ Reg.	185
4500	$4\frac{1}{2}$	$4\frac{3}{4}$	$6\frac{1}{2},6\frac{3}{4},8\frac{1}{2}$	—	18	$4\frac{1}{4}$	69	$3\frac{1}{2}$ If	230
4700	$4\frac{3}{4}$	5	—	12	18	$4\frac{1}{8}$	67	$2\frac{7}{8}$ Reg.	250
5700	$5\frac{3}{4}$	6	$7\frac{1}{2},8,8\frac{1}{2},12,13$	16	18	$4\frac{3}{4}$	70	$3\frac{1}{2}$ Reg.	350
7200	$17\frac{1}{4}$	$7\frac{5}{8}$	$10,12\frac{1}{4},13,14,15,16$	22	18	$5\frac{3}{4}$	78	$4\frac{1}{2}$ Reg.	750
8200	$8\frac{1}{4}$	$8\frac{1}{2}$	$10,12\frac{1}{4},14,15,16,17$	23	18	$5\frac{3}{4},8$	78	$4\frac{1}{2},6\frac{5}{8}$ Reg.	900
9500	$9\frac{1}{2}$	$9\frac{7}{8}$	$12\frac{1}{4},17\frac{1}{2}$	28	18	8	78	$6\frac{5}{8}$ Reg.	1,100
11700	$11\frac{3}{4}$	$12\frac{1}{4}$	—	36	18	8, 9	86	$6\frac{5}{8},7\frac{5}{8}$ Reg.	1,400

Courtesy Smith International, Inc.

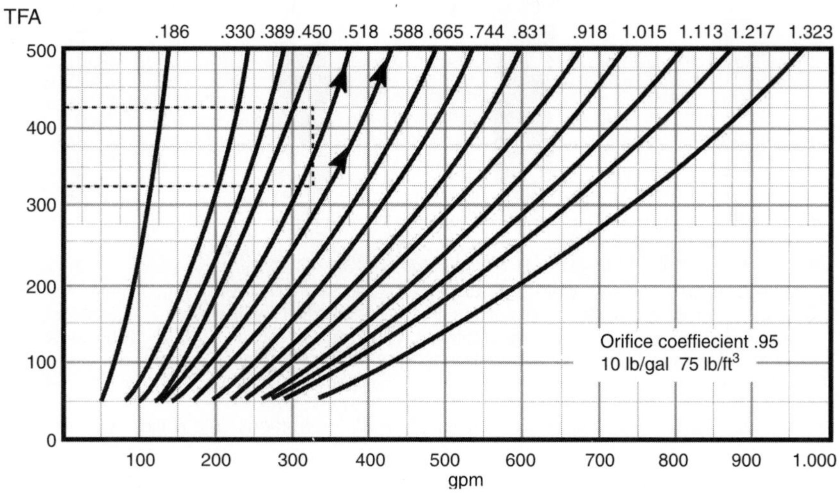

Figure 4.7.101 *Pressure drop across underreamer (reamaster (XTU) and rock-type (DTU) underreamers with three nozzles) (courtesy of Smith International, Inc.).*

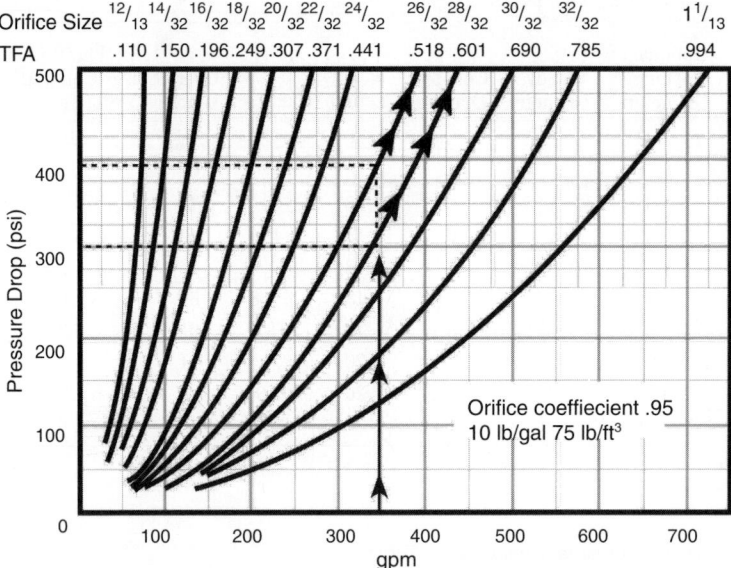

Figure 4.7.102 *Pressure drop across underreamer (rock-type (RTU) and drag-type underreamers with one nozzle) (courtesy of Smith International, Inc.).*

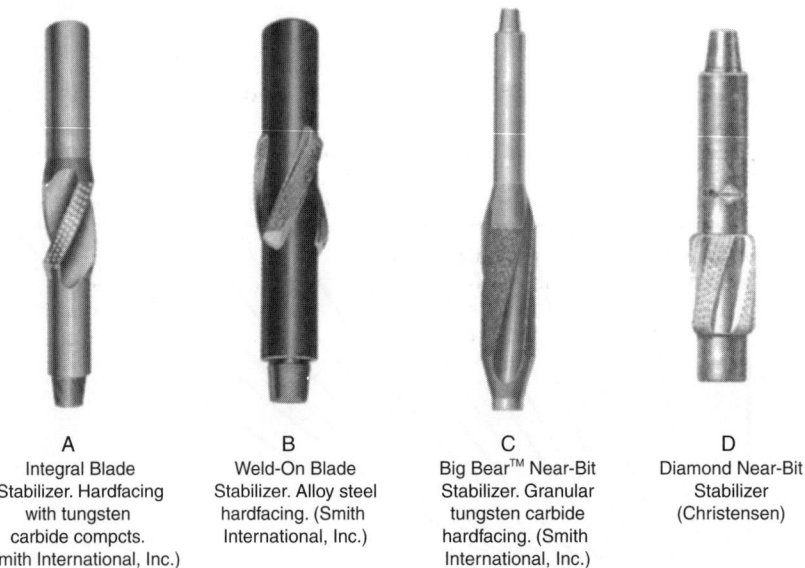

A	B	C	D
Integral Blade Stabilizer. Hardfacing with tungsten carbide compcts. (Smith International, Inc.)	Weld-On Blade Stabilizer. Alloy steel hardfacing. (Smith International, Inc.)	Big Bear™ Near-Bit Stabilizer. Granular tungsten carbide hardfacing. (Smith International, Inc.)	Diamond Near-Bit Stabilizer (Christensen)

Figure 4.7.103 *Solid-type stabilizers (courtesy of Smith International and Baken Hughes INTEQ).*

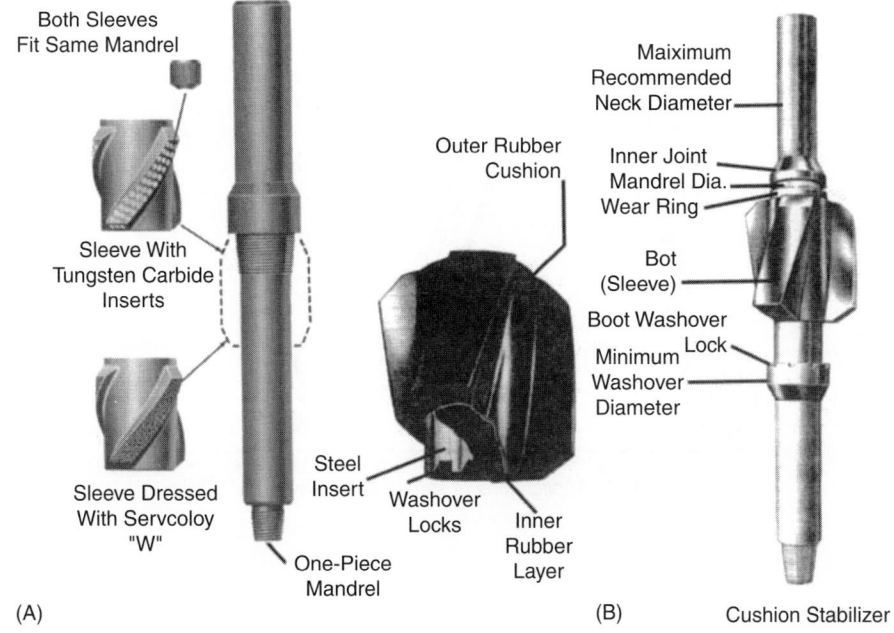

Figure 4.7.104 *Sleeve-type stabilizers. (A) Rotating sleave-type stabilizer (Servco). (B) Grant cushion stabilizers (nonrotating sleeve-type stabilizer) (courtesy of Smith International).*

Figure 4.7.105 *Open bearing design (m-60/62) reamer, 3-point and 6-point types (courtesy of Smith International, Inc.).*

Figure 4.7.106 *Sealed bearing design (Borrox 85) reamer, 3-point and 6-point types (courtesy of Smith International, Inc.).*

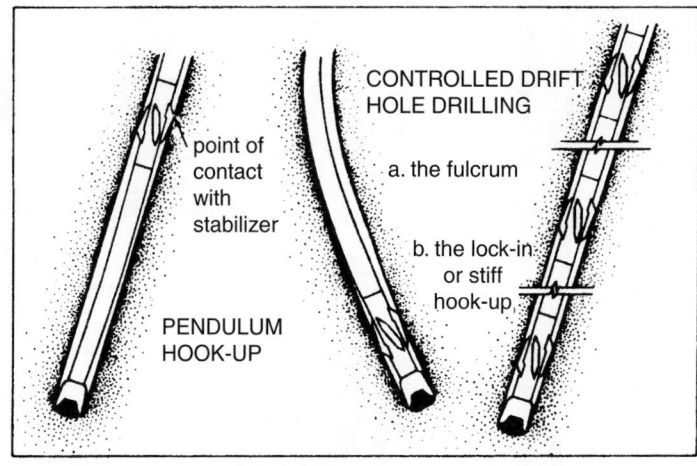

Figure 4.7.107 *Applications of stabilizers in directional drilling.*

used to maintain hole gage and drill out doglegs and keyseats in hard formations. Because of the cutting ability of the reamer, the bit performs less work on maintaining hole gauge and more work on drilling ahead. Reamers can be used as near-bit stabilizers in the bottomhole assembly or higher up in the string. There are basically three types of reamer body:

- Three-point bottom hole reamer. This type of reamer is usually run between the drill collars and the bit to ensure less reaming back to bottom with a new bit.
- Three-point string reamer. The reamer is run in the drill collar string. This reamer provides stabilization of the drill collars to drill a straighter hole in crooked hole country. When run in the string, the reamer is effective in reaming out dog-legs, keyseats and ledges in the hole.
- Six-point bottom hole reamer. This type of reamer is run between the drill collars and the bit when more stabilization or greater reaming capacity is required. Drilling in crooked hole areas with a six-point reamer has proven to be very successful in preventing sharp changes in hole angles.

Reamers are either open bearing design or sealed bearing design with pressure compensation. Open bearing design reamers are used for standard applications. Sealed bearing design reamers are used in more demanding conditions where extended reamer life is required.

Application of Stabilizers Figure 4.7.107 illustrates three applications of stabilizers, pendulum, fulcrum, and lock-in (stiff) hook-up.

The stiff hookup consists of three or more stabilizers placed in the bottom 50 to 60 ft of drill collar string. In mind crooked hole conditions, the stiff hookup will hold the deviation to a minimum. In most cases, deviation will be held below the maximum acceptable angle. In severe conditions, this hookup will slow the rate of angle buildup, allowing more weight to be run for a longer time. This method prevents sudden increases or decreases of deviation, making dog-legs less severe and decreasing the probability of subsequent

keyseats and other undesirable hole conditions. The stiff hookup is beneficial only until the maximum acceptable angle is reached. The pendulum principle should then be used.

To employ pendulum effect in directional drilling, usually one stabilizer is placed in the optimum position in the drill collar string. The position is determined by the hole size, drill collar size, angle of deviation and the weight on the bit. A properly placed stabilizer extends the suspended portion of the drilling string (that portion between the bit and the point of contact with the low side of the hole). The force of gravity working on this extended portion results in a stronger force directing the bit toward vertical so the well trajectory returns to vertical.

To employ fulcrum effect one stabilizer is placed just above the bit and additional weight is applied to the bit. The configuration acts as a fulcrum forcing the bit to the high side of the hole. The angle of hole deviation increases (buildup) as more weight is applied.

To employ a restricted fulcrum effect one stabilizer is placed just above the bit while second stabilizer is placed above the nonmagnetic drill collar. The hookup allows a gradual buildup of inclination with no abrupt changes.

To prevent key-seating one stabilizer is placed directly above the top of drill collars. The configuration prevents drill collars wedging into a key seat during tripping out of the hole.

To prevent differential sticking across depleted sands stabilizers are placed throughout the drill collar string. The area of contact between drill collars and hole is reduced, thus reducing the sticking force.

References
1. "Proven drilling performance," Eastman Christensen 1988-89-General Catalog, pp. 28–29.
2. Smith Services, A Business unit of Smith International, Inc., Remedial Tools Data Handbook, 8th Edition.
3. Christensen, *Downhole Tool Technology*, student manual.
4. Smith Services, A Business unit of Smith International, Inc., Drilling Assembly Handbook, 1998.

4.8 DRILLING MUD HYDRAULICS

4.8.1 Rheological Classification of Drilling Fluids

Experiments performed on various drilling muds have shown that the shear stress-shear rate characteristic can be represented by one of the functions schematically depicted in Figure 4.8.1. If the shear stress-shear rate diagram is a straight line passing through the origin of the coordinates, the drilling fluid is classified as a Newtonian fluid, otherwise it is considered to be non-Newtonian.

The following equations can be used to describe the shear stress-shear rate relationship:

Newtonian fluid

$$\tau = \mu \left(-\frac{dv}{dr} \right) \qquad [4.8.1]$$

Bingham plastic fluid

$$\tau = \tau_y + \mu_p \left(-\frac{dv}{dr} \right)^n \qquad [4.8.2]$$

Power law fluid

$$\tau = K \left(-\frac{dv}{dr} \right)^n \qquad [4.8.3]$$

Herschel and Buckley fluid

$$\tau = \tau_y + K \left(-\frac{dv}{dr} \right)^n \qquad [4.8.4]$$

where τ = shear stress
  v = the velocity of flow
  dv/dr = shear rate (velocity gradient in the direction perpendicular to the flow direction)
  μ = dynamic viscosity
  τ_y = yield point stress
  μ_p = plastic viscosity
  K = consistency index
  n = flow behavior index

The dynamic viscosity (μ), yield point stress(τ_p), plastic viscocity (μ_p), consistency index (K), and flow behavior index (n) are usually determined with the Fann rotational viscometer. The Herschel and Buckley fluid model is not considered further in this text.

4.8.2 Flow Regimes

The flow regime (i.e., whether laminar or turbulent) can be determined using the Reynolds number. In general, the Reynolds number, Re, is calculated in consistent units from

$$Re = \frac{dvp}{\mu} \qquad [4.8.5]$$

where d = diameter of the fluid conduit
  v = velocity of the fluid
  ρ = density of fluid
  μ = viscosity

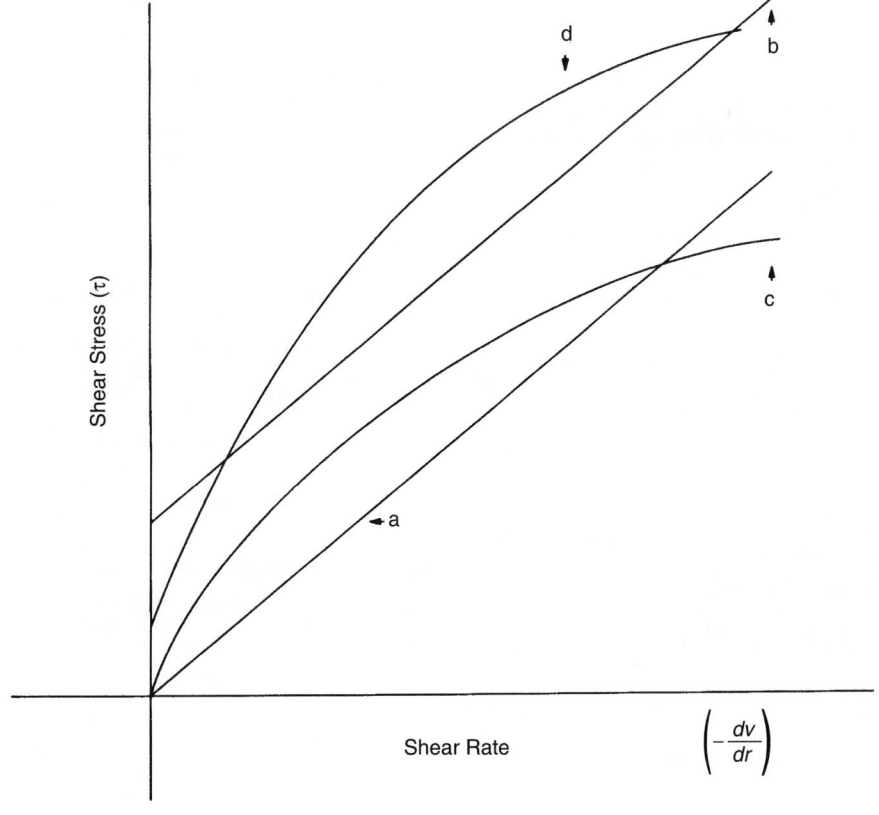

Figure 4.8.1 Shear stress-shear rate diagram: (a) Newtonian fluid, (b) Bingham plastic fluid, (c) Power law fluid, (d) Herschel-Buckley fluid.

In oilfield engineering units

$$Re = \frac{928 d_e \bar{v} \gamma}{\mu} \qquad [4.8.6]$$

where d_e = equivalent diameter of a flow channel (in.)
  $\bar{v}$ = average flow velocity (ft/sec)
  γ = drilling fluid specific weight (lb/gal)
  μ = Dynamic viscosity (cp)

The equivalent diameter of the flow channel is defined as

$$d_e = \frac{4 \text{ (flow cross-sectional area)}}{\text{wetted perimeter}} \qquad [4.8.7]$$

For flow in the i.d. of the workstring, $d_e = d$, and for annular flow, $d_e = d_1 - d_2$,

where d = inside diameter of the pipe (in.)
  d_1 = larger diameter of the annulus (in.)
  d_2 = larger diameter of the annulus (in.)

The flow changes from laminar to turbulent in the range of Reynolds numbers from 2,100 to 4,000 [1]. In laminar flow, the friction pressure losses are proportional to the average flow velocity. In turbulent flow, the losses are proportional to the velocity to a power ranging from 1.7 to 2.0.

The average flow velocity is given by the following equations:

For flow inside circular pipe

$$\bar{v} = \frac{q}{2.45 d^2} \quad \text{(ft/sec)} \qquad [4.8.8]$$

For flow in an annular space between two circular pipes

$$\bar{v} = \frac{q}{2.45 \left(d_1^2 - d_2^2 \right)} \quad \text{(ft/sec)} \qquad [4.8.9]$$

where q = mud flow rate (gpm)

For non-Newtonian drilling fluids, the concept of an effective viscosity* can be used to replace the dynamic viscosity in Equation 4.8.5.

For a Bingham plastic fluid flow in a circular pipe and annular space, the effective viscosities are given as [2].

For pipe flow

$$\mu_e = \mu_p + 6.66 \left(\frac{\tau_y d}{\bar{v}} \right) (Cp) \qquad [4.8.10]$$

For annular flow

$$\mu_e = \mu_p + 4.99 \left(\frac{\tau_y (d_1 - d_2)}{\bar{v}} \right) (Cp) \qquad [4.8.11]$$

For a power law fluid, the following formulas can be used.

For pipe flow

$$\mu_e = \left(\left(\frac{1.6\bar{v}}{d} \right) \left(\frac{3n+1}{4n} \right) \right)^n \left(\frac{300 K d}{\bar{v}} \right) \quad (Cp) \qquad [4.8.12]$$

For annular flow

$$\mu_e = \left(\left(\frac{2.4\bar{v}}{(d_1 - d_2)} \right) \left(\frac{2n+1}{3n} \right) \right)^n \left(\frac{200 K (d_1 - d_2)}{\bar{v}} \right) (Cp) \qquad [4.8.13]$$

The plastic viscosity (μ_p), yield point stress (τ_y), flow behavior index (n), and consistency index (K) are mud rheological properties that are typically calculated from the experiments performed on the drilling mud in a rotational

*Also called equivalent or apparent viscosity in some published works.

viscometer. When these data are available, the following equations can be used:

$$\mu_p = \theta_{600} - \theta_{300} \left(c_p \right) \qquad [4.8.14]$$

$$\tau_y = \theta_{300} - \mu_p \left(\text{lb}/100 \text{ ft}^2 \right) \qquad [4.8.15]$$

$$n = 3.32 \log \frac{\theta_{600}}{\theta_{300}} \qquad [4.8.16]$$

$$K = \frac{\theta_{300}}{(511)^n} \left(\text{lb}/100 \text{ ft}^2 \text{s}^{-n} \right) \qquad [4.8.17]$$

where θ_{600} = viscometer reading at 600 rpm
  θ_{300} = viscometer reading at 300 rpm

Example
Consider a well with the following geometric and operational data:

Casing $9\frac{5}{8}$ in., unit weight = 40 lb/ft, ID = 8.835 in. Drill pipe: $4\frac{1}{2}$ in., unit weight = 16.6 lb/ft, ID = 3.826 in. Drill collars: $6\frac{3}{4}$ in., unit weight = 108 lb/ft, ID = $2\frac{1}{2}$ in. Hole size: $8\frac{1}{2}$ in. Drilling fluid properties: $\theta_{600} = 68$, $\theta_{300} = 68$, density = 10 lb/gal, circulating rate = 280 gpm.

Calculate Reynolds number for the fluid (1) inside drill pipe, (2) inside drill collars, (3) in drill collar annulus, and (4) in drill pipe annulus.

To perform calculation, a Power law fluid is assumed.
Flow behavior index (use Equation 4.8.16) is

$$n = 3.32 \log \frac{68}{41} = 0.729$$

Consistency index (use Equation 4.8.17) is

$$K = \frac{41}{(511)^{0.729}} = 0.433 \text{ lb}/100 \text{ ft}^2 \text{s}^{-0.729}$$

The average flow velocities are
For inside drill pipe (use Equation 4.8.8)

$$\bar{v} = \frac{280}{2.45 (3.826)^2} = 7.807 \text{ (ft/sec)}$$

For inside drill collars

$$\bar{v} = \frac{280}{2.45 (2.25)^2} = 22.58 \text{ (ft/sec)}$$

In drill collar annulus (in open hole) use Equation 4.8.3.

$$\bar{v} = \frac{280}{2.45 (8.5^2 - 6.75^2)} = 4.282 \text{ (ft/sec)}$$

In drill pipe annulus (in the cased hole)

$$\bar{v} = \frac{280}{2.45 (8.835^2 - 4.5^2)} = 1.977 \text{ (ft/sec)}$$

The effective viscosities are
For inside the drill pipe (use Equation 4.8.12)

$$\mu_e = \left(\frac{(1.6)(7.807)}{3.826} \frac{(3)(0.729)+1}{(4)(0.729)} \right)^{0.729}$$
$$\times \left(\frac{(300)(0.433)(3.826)}{7.807} \right) = 160.9 \text{ cp}$$

For inside drill collars

$$\mu_e = \left(\frac{(1.6)(22.575)}{2.25} \frac{(3)(0.729)+1}{(4)(0.729)} \right)^{0.729}$$
$$\times \left(\frac{(300)(0.433)(2.25)}{22.575} \right) = 104.5 \text{ cp}$$

For in drill collar annulus (use Equation 4.8.13)

$$\mu_e = \left(\frac{(2.4)\,(4.282)}{8.5 - 6.75}\, \frac{(2)\,(0.729) + 1}{(3)\,(0.729)} \right)^{0.729}$$

$$\times \left(\frac{(200)\,(0.433)\,(8.5 - 6.756)}{4.282} \right) = 140.1 \text{ cp}$$

In the drill pipe annulus

$$\mu_e = \left(\frac{(2.4)\,(1.977)}{8.835 - 4.5}\, \frac{(2)\,(0.729) + 1}{(3)\,(0.729)} \right)^{0.729}$$

$$\times \left(\frac{(200)\,(0.433)\,(8.835 - 4.5)}{1.977} \right) = 233.6 \text{ cp}$$

Reynolds number (use Equation 4.8.6)
For inside the drill pipe

$$Re = 928\frac{(3.826)\,(7.807)\,(10)}{(160.9)} = 1723$$

For inside drill collars

$$Re = 928\frac{(2.25)\,(22.575)\,(10)}{(104.5)} = 4511$$

In the drill collar annulus

$$Re = 928\frac{(8.5 - 6.75)\,(4.282)\,(10)}{(140.1)} = 496$$

In the drill pipe annulus

$$Re = 928\frac{(8.835 - 4.5)\,(1.977)\,(10)}{(233.6)} = 340$$

4.8.3 Principle of Additive Pressures

Applying the conservation of momentum to the control volume for a one-dimensional flow conduit, it is found that [3].

$$\rho A = \frac{dv}{dt} = -A\frac{dp}{dl} - P_w\tau_w - A\rho g \cos\alpha \qquad [4.8.18]$$

where ρ = fluid density
A = flow area
dv/dt = acceleration (total derivative)
v = flow velocity
τ_w = average wall shear stress
P_w = wetted perimeter
g = gravity acceleration
α = inclination of a flow conduit to the vertical
dP/dl = pressure gradient
l = length of flow conduit

For a steady-state flow, Equation 4.8.18 is often written as an explicit equation for the pressure gradient. This is

$$\frac{dP}{dl} = -\frac{P_w}{A}\tau_w - \rho v\frac{dv}{dl} - \rho g \cos\alpha \qquad [4.8.19]$$

The three terms on the right side are known as frictional, accelerational (local acceleration) and gravitational components of the pressure gradient. Or, in other words, the total pressure drop between two points of a flow conduit is the sum of the components mentioned above. Thus,

$$\Delta P = \Delta P_F + \Delta P_A + \Delta P_G \qquad [4.8.20]$$

where ΔP_F = frictional pressure drop
ΔP_A = accelerational pressure drop
ΔP_G = gravitational pressure drop (hydrostatic head)

Equation 4.8.21 expresses the principle of additive pressures. In addition to Equation 4.8.21, there is the equation of state for the drilling fluid.

Typically, water based muds are considered to be incompressible or slightly compressible. For the flow in drill pipe or drill collars, the acceleration component (ΔP_A) of the total pressure drop is negligible, Equation 4.8.20 can be reduced to

$$\Delta P = \Delta P_F + \Delta P_G \qquad [4.8.21]$$

Equations 4.8.19 through 4.8.22 are valid in any consistent system of units.

Example
The following data are given:

- Pressure drop inside the drill string = 600 psi
- Pressure drop in annular space = 200 psi
- Pressure drop through the bit nozzle = 1,600 psi
- Hole depth = 10,000 ft
- Mud density = 10 lb/gal

Calculate

- bottomhole pressure
- pressure inside the string at the bit level (above the nozzles)
- drill pipe pressure

Because the flow in the annulus is upward, the total bottomhole pressure is equal to the hydrostatic head plus the pressure loss in the annulus. Hydrostatic head (psi) is calculated as follows [1]:

$$P_{hyd} = 0.052\rho L \text{ psi} \qquad [4.8.22]$$

where ρ = density (lb/gal)
L = length of pipe (ft)

Therefore,

$$p_{bottom} = 0.052(10)(10,000) + 200 = 5,400 \text{ psi}$$

Pressure inside the drill string above the nozzle, p_{is} (psi),

$$p_{is} = 5,400 + 1,600 = 7,000 \text{ psi}$$

Drill pipe pressure (also known as standpipe pressure), P_{dp} (psi)

$$p_{dp} = 7,000 + 600 - (0.052)(10)(10,000) = 2,400 \text{ psi}$$

Note that the drill pipe pressure is a sum of all the pressure losses in the circulating system.

4.8.4 Friction Pressure Loss Calculations
4.8.4.1 Laminar Flow
For pipe flow of Bingham plastic type drilling fluid, the following can be used:

$$\Delta p = \frac{\mu_p L v}{1500 d^2} + \frac{\tau_y L}{225 d} \qquad [4.8.23]$$

Corresponding equation for a Power law type drilling fluid is

$$\Delta p = \left[\left(\frac{1.6v}{d} \right)\left(\frac{3n + 1}{4n} \right) \right]^n \frac{KL}{300d} \qquad [4.8.24]$$

For annular flow of Bingham plastic and Power law fluids, respectively,

$$\Delta p = \frac{\mu_p L v}{1000\,(d_1 - d_2)^2} + \frac{\tau_y L}{200\,(d_1 - d_2)} \qquad [4.8.25]$$

and

$$\Delta p = \left[\left(\frac{2.4v}{(d_1 - d_2)} \right)\left(\frac{2n + 1}{3n} \right) \right]^n \frac{KL}{300\,(d_1 - d_2)} \qquad [4.8.26]$$

4.8.4.2 Turbulent Flow

Turbulent flow occurs if the Reynolds number as calculated above exceeds a certain critical value. Instead of calculating the Reynolds number, a critical flow velocity may be calculated and compared to the actual average flow velocity [1].

The critical velocities for the Bingham plastic and Power law fluids can be calculated as follows:

For a Bingham plastic fluid

$$v_c = \frac{1.08\mu_p + 1.08\sqrt{\mu_p^2 + 9.256(d_1 - d_2)^2 \tau_y \rho}}{(d_1 - d_2)} \text{(ft/min)}$$

[4.8.27]

For a power law fluid

$$v_c = \left[\frac{3.878 \times 10^4 K}{\rho}\right]^{1/(2-n)}$$
$$\times \left[\left(\frac{2.4}{d_1 - d_2}\right)\left(\frac{2n+1}{3n}\right)\right]^{n/(2-n)} \text{(ft/min)}$$

[4.8.28]

In the case of pipe flow, for practical purposes, the corresponding critical velocities may be calculated using Equation 4.8.27 and 4.8.28, but letting $d_2 = 0$.

Note: the critical flow velocity is in ft/min instead of the customary ft/sec.

In turbulent flow the pressure losses, Δp (psi), can be calculated from the Fanning equation [1].

$$\Delta p = \frac{f\gamma L v^2}{25.8d}$$

[4.8.29]

where f = Fanning factor
L = length of pipe, ft

The friction factor depends on the Reynolds number and the surface conditions of the pipe. There are numerous charts and equations for determining the relationship between the friction factor and Reynolds number. The friction factor can be calculated by [4].

$$f = 0.046 \, Re^{-0.2}$$

[4.8.30]

Substituting Equation 4.8.7, 4.8.8, and 4.8.30 into Equation 4.8.29 yields [4]

For pipe flow

$$\Delta p = \frac{7.7 \times 10^{-5} \bar{v}^{0.8} q^{1.8} \mu_p^{0.2} L}{d^{4.8}}$$

[4.8.31]

For annular flow

$$\Delta p = \frac{7.7 \times 10^{-5} \bar{v}^{0.8} \mu_p^{0.2} q^{1.8} L}{(d_1 - d_2)^3 (d_1 + d_2)^{1.8}}$$

[4.8.32]

Example

The wellbore, drill string and drilling fluid data from the previous example are used. Casing depth is 4,000 ft. Assuming a drill pipe length of 5,000 ft and a drill collar length of 500 ft, find the friction pressure losses,

For flow inside the drill pipe, the critical flow velocity is

$$v_c = \left[\frac{3.878 \times 10^4 (0.433)}{10}\right]^{1/(2-0.729)}$$
$$\times \left[\left(\frac{2.4}{3.826 - 0}\right)\left(\frac{2(0.729)+1}{3(0.729)}\right)\right]^{0.729/(2-0.729)} \text{(ft/min)}$$

$v_c = 283 \text{ ft/min} = 4.72 \text{ ft/sec}$

Recall $v = 7.807$ ft/sec, because $v > v_c$ the flow is turbulent therefore Equation 4.8.31 is chosen to calculate the pressure

loss p_1.

$$\Delta p_1 = \frac{7.7 \times 10^{-5} (10^{0.8})(280^{1.8})(27^{0.2}) 5000}{3.826^{4.8}} \text{psi}$$

$\Delta p_1 = 190 \text{ psi}$

Note use of equation 4.8.14 to determine the plastic viscosity.

For flow inside the drill collars, it is easy to check that the flow is also turbulent and thus the pressure loss is

$$\Delta p_2 = \frac{7.7 \times 10^{-5} (10^{0.8})(280^{1.8})(27^{0.2}) 500}{2.25^{4.8}} \text{psi}$$

$\Delta p_2 = 243 \text{ psi}$

For annulus flow around the drill collars

To calculate the critical velocity, Equation 4.8.28 is used

$$v_c = \left[\frac{3.878 \times 10^4 (0.433)}{10}\right]^{1/(2-0.729)}$$
$$\times \left[\left(\frac{2.4}{8.5 - 6.75}\right)\left(\frac{2(0.729)+1}{3(0.729)}\right)\right]^{0.729/(2-0.729)} \text{(ft/min)}$$

$v_c = 442 \text{ (ft/min)} = 7.36 \text{ ft/sec}$

The average velocity around the collars is 4.282 ft/sec, as calculated in the previous example. Therefore, the flow is laminar and Equation 4.8.25 is selected to calculate the pressure drop Δp_3

$$\Delta p_3 = \left[\left(\frac{2.4(4.282)}{(8.5-6.75)}\right)\left(\frac{2(0.729)+1}{3(0.729)}\right)\right]^{0.729}$$
$$\times \frac{0.433(500)}{300(8.5-6.75)} \text{psi}$$

$\Delta p_3 = 6.06 \text{ psi}$

For annulus flow around the drill pipe in the open hole section.

It is found from Equation 4.8.28 that the flow is laminar, thus Equation 4.8.25 can be used with the average annular velocity found from Equation 4.8.9.

$$\bar{v} = \frac{280}{(2.45(8.5^2 - 4.5^2))} = 2.19 \text{ ft/sec}$$

$$\Delta p_4 = \left[\left(\frac{2.4(2.19)}{(8.5-4.5)}\right)\left(\frac{2(0.729)+1}{3(0.729)}\right)\right]^{0.729}$$
$$\times \frac{0.433(1000)}{300(8.5-4.5)} \text{psi}$$

$\Delta p_4 = 0.479 \text{ psi}$

For annulus flow around the drill pipe in cased section.

Using Equation 4.8.28 it is determined that the flow is also laminar in the cased hole, therefore Equation 4.8.26 is used again.

$$\Delta p_5 = \left[\left(\frac{2.4(1.977)}{(8.835-4.5)}\right)\left(\frac{2(0.729)+1}{3(0.729)}\right)\right]^{0.729}$$
$$\times \frac{0.433(4000)}{300(8.835-4.5)} \text{psi}$$

$\Delta p_5 = 1.5 \text{ psi}$

Therefore, the total frictional pressure loss is

$$\Delta p_f = \Delta p_1 + \Delta p_2 + \Delta p_3 + \Delta p_4 + \Delta p_5$$

$$\Delta p_f = 94.7 + 243 + 6.06 + .48 + 1.5 = 346 \text{ psi}$$

4.8.4.3 Pressure Loss Through Bit Nozzles

Assuming steady-state, frictionless (due to the short length of the nozzles) drilling fluid flow, Equation 4.8.19 is written

$$\rho v = \frac{\partial v}{\partial 1} = \frac{\partial p}{\partial 1} \qquad [4.8.33]$$

Integrating Equation 4.8.33 assuming incompressible drilling fluid flow (ρ is constant) and after simple rearrangements yields the pressure loss across the bit Δp_b (psi) which is

$$\Delta p_b = \frac{\rho v^2}{2} \qquad [4.8.34]$$

Introducing the nozzle flow coefficient of 0.95 and using field system of units, Equation 4.8.34 becomes

$$\Delta p_b = \frac{\bar{\gamma} v^2}{1,120} \qquad [4.8.35]$$

or

$$\Delta p_b = \frac{\bar{\gamma} v^2}{10,858 A^2} \qquad [4.8.36]$$

where v = nozzle velocity in ft/s
q = flow rate in gpm
$\bar{\gamma}$ = drilling fluid density in lb/gal
A = nozzle flow area in in.^{2}

If the bit is furnished with more than one nozzle, then

$$A = A_1 + A_2 + A_3 + \ldots A_n \qquad [4.8.37]$$

where n is the number of nozzles, and

$$d_{en} = \sqrt{d_1^2 + d_2^2 + \ldots d_n^2} \qquad [4.8.38]$$

where d_{en} is the equivalent nozzle diameter.

Example
A tricone roller rock bit is furnished with three nozzles with the diameters of $\frac{9}{32}$, $\frac{10}{32}$ and $\frac{12}{32}$ in. Calculate the bit pressure drop if the mud weight is 10 lb/gal and flowrate is 300 gpm.
   Nozzle equivalent diameter is

$$d_{en} = \sqrt{\left(\frac{9}{32}\right)^2 + \left(\frac{10}{32}\right)^2 + \left(\frac{12}{32}\right)^2} = 0.5643 \text{ in.}$$

and the corresponding flow area is

$$A = \frac{\pi}{4}(0.5643)^2 = 0.2493 \text{ in.}^2 \qquad [4.8.39]$$

The pressure loss through bit nozzles is

$$\Delta p_b = \frac{(10)(300)^2}{10858(0.2493)^2} = 1334 \text{ psi}$$

References

1. Gatlin, C., *Petroleum Engineering: Drilling and Well Completions*, Prentice-Hall, Inc., Englewood Cliffs, New Jersey, 1960.
2. Bourgoyne, A. T., et al., *Applied Drilling Engineering*, Society of Petroleum Engineers , Richardson, Texas, 1986.
3. White, F. M., *Fluid Mechanics*, McGraw-Hill Book Co., New York, 1979.
4. Moore, P. L., et al., *Drilling Practices Manual*, PennWell Books, Tulsa, Oklahoma, 1974.

4.9 UNDERBALANCED DRILLING AND COMPLETIONS

4.9.1 Definitions

Rotary drilling operations require that a drilling fluid be circulated down the inside of a drill string, through the drill bit nozzles (or open orifices), and up the annulus between the outside of the drill string and the borehole wall (see Sections 4.5 Drilling Muds and Completion Systems, 4.8 Drilling Mud Hydraulics, and 4.14 Well Pressure Control). As the drilling fluid passes through the drill bit nozzles, the fluid entrains the rock cuttings generated by the drill bit advance. In addition to carrying the rock cuttings from the borehole, the drilling fluid cools the drill bit and stabilizes the borehole wall [1].

When drilling for oil and natural gas deposits, the most important function of the drilling fluid is to control the inflow of formation fluids that could enter the wellbore annulus. When the specific gravity of the drilling fluid has been engineered to provide a bottomhole pressure in the borehole annulus that precisely counters the pore pressure of the formation fluids that could enter the wellbore annulus, the drilling operation is called *balanced*. When the specific gravity of the drilling fluid has been engineered to provide a bottomhole pressure in the borehole annulus that is slightly higher than the pore pressure of the formation fluids, the drilling operation is called *overbalanced*, When the specific gravity of the drilling fluid has been engineered to provide a bottomhole pressure in the borehole annulus that is lower than the pore pressure of the formation fluids, the drilling operation is called *underbalanced* [2, 3, 4]. This is sometimes referred to as *flowdrilling*. Having drilled a well using underbalanced techniques, it is important to complete the well underbalanced. If the underbalanced drilling operation is not followed by completions efforts using underbalanced techniques, formation damage may still occur due to poor completions practices.

For a general guideline, when drilling and completing wells into the actual producing rock formations, underbalanced drilling (and completion) techniques are restricted to geologic provinces that have producing formations with pore pressures of the order of approximately 2500 psi or less. From a practical standpoint, this restricts underbalanced drilling and completions to infill production wells in oil and natural gas fields that have significant production history. However, there are numerous drilling operations where underbalanced drilling techniques are utilized in both exploratory and early development wells when upper zones with troublesome rock formations are encountered. Thus, underbalanced drilling and completion techniques have broad applications throughout the oil and natural gas (and geothermal steam and water) resource recovery industry.

4.9.2 Drilling Operations

In oil and natural gas producing wells that have relatively low pore pressures, overbalanced drilling operations created *formation damage* by forcing the drilling mud particles and rock cutting fines into the producing rock formation pores thereby clogging the pores of formation rock in the vicinity of the open borehole. This formation damage reduces the flow of oil and/or natural gas to the wellbore when the well is completed and produced. Great costs are incurred by operating companies to reduce the effects of formation damage in wells that have been drilled with conventional overbalanced methods (e.g., hydraulic fracturing and acidizing).

Producing oil and natural gas geologic provinces become depleted as they are produced. This depletion is in both volume of oil or gas available for production and in reservoir pressure. To reduce the effects of formation damage, infill drilling and completion operations are carried out in these provinces, the new wells are increasingly being drilled and completed using underbalanced drilling techniques. The use of underbalanced drilling and completion techniques is generally confined to the drilling and completion of the rock formations that will produce oil and/or natural gas. Underbalanced drilling operations allow the oil and/or natural gas to flow into the wellbore as drilling process progresses through these producing rock formations. Special surface production equipment must be in place in order to safely collect these produced fluids and control (via metering) the volumetric flow rate of these fluids (together with the drilling fluid) into the wellbore at the bottom of the well. The metering of these volumetric flow rates is critical to the control of any underbalanced drilling operation. These metering measurements are used to continuously adjust the magnitude of the difference between the pore pressure of the openhole exposed producing formation and the pressure at that formation due to the circulating drilling fluid.

4.9.2.1 Air and Gas Drilling

Air and gas drilling refers to the use of compressed air (or other gases) as the circulating drilling fluid for rotary drilling operations. The majority of these drilling operations use compressed air as the circulating drilling fluid. In some oil and natural gas recovery drilling operations it is necessary to drill with a gas that will not support downhole combustion. This is particularly the case when drilling in or near the hydrocarbon producing formations. This objective has been realized by using natural gas or inert atmospheric air (air stripped of most oxygen) as drilling gases. Natural gas from pipelines has been used as a drilling gas since the 1930's. The use of inert air as a drilling gas is a recent technological development [3]. Drilling location deployable equipment units are available to the oil and natural gas resource recovery industry. These units are also called "nitrogen generators." The units strip most of the oxygen content from the compressed air output from standard positive displacement primary compressors (rotary or reciprocation piston compressors). These inert air generating field units have only been in use for a few years.

The basic planning steps for the drilling of a deep well with compressed gases are as follows [4]:

1. Determine the geometry of the borehole section or sections to be drilled with air or other gases (i.e., openhole diameters, the casing or liner inside diameters, and depths).
2. Determine the geometry of the associated drill strings for the sections to be drilled with air or other gases (i.e., drill bit size and type, the drill collar size, drill pipe size, and maximum depth).
3. Determine the type of rock formations to be drilled and estimate the anticipated drilling rate of penetration. Also, estimate the quantity and depth location of any formation water that might be encountered.
4. Determine the elevation of the drilling site above sea level, the temperature of the air during the drilling operation, and the approximate geothermal temperature gradient.
5. Establish the objective of the air (or other gas) drilling operation:

   • To eliminate loss of circulation problems,
   • To reduce formation damage,
   • To allow formation fluids to be produced as the formation is drilled.

6. Determine whether direct or reverse circulation techniques will be used to drill the various sections of the well.

7. Determine the required approximate minimum volumetric flow rate of air (or other gas) to carry the rock cuttings from the well when drilling at the maximum depth.
8. Select the contractor compressor(s) that will provide the drilling operation with a volumetric flow rate of air that is greater than the required minimum volumetric flow rate (use a factor of safety of at least 1.2).
9. Using the compressor(s) air volumetric flow rate to be injected into the well, determine the bottomhole and surface injection pressures as a function of drilling depth (over the interval to be drilled). Also, determine the maximum power required by the compressor(s) and the available maximum derated power from the prime mover(s) (see Section 3.1, Chapter 3).
10. Determine the approximate volume of fuel required by the compressor(s) to drill the well.
11. In the event formation water is encountered, determine the approximate volumetric flow rate of "mist" injection water needed to allow formation water or formation oil to be carried from the well during the drilling operation.
12. Determine the approximate volumetric flow rate of formation water or formation oil that can be carried from the well during the drilling operation (assuming the injection air will be saturated with water vapor at bottomhole conditions).

The circulation system for an air (or other gas) drilling operation is a typical compressible fluid flow calculation problem. In these problems, the pressure and temperature is usually known at the exit to the system. In this case, the exit is at the end of the blooey line. At this exit, the pressure and temperature are the local atmospheric pressure and temperature at the drilling location. The calculation procedures for air drilling circulation problems are to start at these known exit conditions and work upstream through the system. In these calculations, the volumetric flow rate must be assumed or known. If compressors are used to provide compressed air, then the volumetric flow rate is the sum of the outputs of all the primary compressors (see Section 3.4). For direct circulation drilling operations, the equations given in the *"Bottomhole Pressure"* subsection below are applied from the top of the well annulus to each constant cross-section section in sequence starting from the top well. The pressure found at the bottom of each constant cross-section section is used as the initial pressure for the next deeper constant cross-section section until the bottomhole pressure at the bottom of the annulus is determined. In this calculation, all major and minor flow losses should be considered. The minimum volumetric flow rate air (or gas) can be determined using the equations in the *"Bottomhole Pressure"* subsection and the kinetic energy equation given in the *"Minimum Volumetric Flow Rate"* subsection [3,4]. The actual volumetric flow rate to the well must be greater than the minimum by a factor of at least 1.2. Working upstream, the pressure above the drill bit orifices (or nozzles) inside the drill bit is found using the equations in *"Drill Bit Orifices and Nozzles"* subsection. Care must be taken to determine whether the flow through these bit openings is sonic or subsonic. The equations given in the *"Injection Pressure"* subsection are applied from the bottom of the inside of the drill string to each constant cross-section section in sequence starting from the pressure above the drill bit. The pressure found at the top of each constant cross-section section is used as the initial pressure for the pressure at the top of next constant cross-section section until the injection pressure at the top of the inside of the drill string is found.

The methodology outline above has been successful in predicting bottomhole and injection pressures with an accuracy of about 5%. In order to attain this accuracy, it is necessary to consider all major and minor flow losses in the circulating system, and to account for any water or other incompressible fluids being carried from the well by the air or gas drilling fluid.

Bottomhole Pressure. The perfect gas law is used as the basis of the air and gas drilling equations presented. The perfect gas law can be written as (see Section 3.4, "Compressors")

$$\frac{P}{\gamma} = \frac{RT}{S_g} \quad [4.9.1]$$

where

P is pressure (lb/ft^2 absolute, or N/m^2 absolute),
γ is specific weight (lb/ft^3, or N/m^3),
T is absolute temperature (°R, or K),
R is the universal gas constant (53.36 lb-ft/lb-°R, or 29.28 N-m/N-K),
S_g is the specific gravity of the gas.

The equation for the pressure in the air (or gas) flow at the entrance end to the blooey line (just after the Tee from the annulus) can be approximated by

$$P_b = \left[\left(f_b \frac{L_b}{D_b} + K_t + \sum K_v \right) \left(\frac{\dot{w}_g^2 R T_r}{g A_b^2 S_g} \right) + P_{at}^2 \right]^{0.5} \quad [4.9.2]$$

where

P_b is the pressure at the entrance to the blooey line (lb/ft^2 absolute, or N/m^2 absolute).
P_{at} is the atmospheric pressure at the exit to the blooey line (lb/ft^2 absolute, or N/m^2 absolute).
f_b is the Darcy-Wiesbach friction factor for the blooey line.
L_b is the length of the blooey line (ft, or m).
D_b is the inside diameter of the blooey line (ft, or m).
K_t is the minor loss factor for the Tee turn at the top of the annulus.
K_v is the minor loss factor for the valves in the blooey line.
$\dot{w}_g$ is the weight rate of flow of gas (lb/sec, or N/sec).
A_b is the cross-sectional area of the inside to the blooey line (ft^2, or m^2).
T_r is the average temperature of the gas flow in the blooey line (°R, or K)

The approximate value of the K_t can be determined from Figures 4.9.1 and 4.9.2.

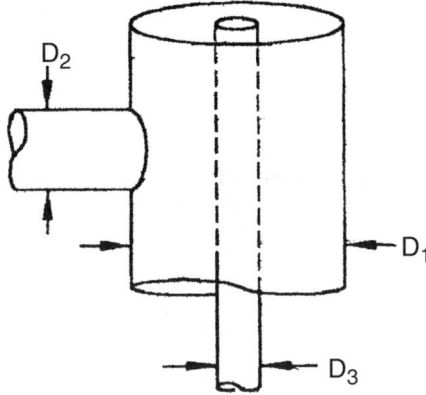

Figure 4.9.1 *Dimensions of the blind Tee at top of annulus.*

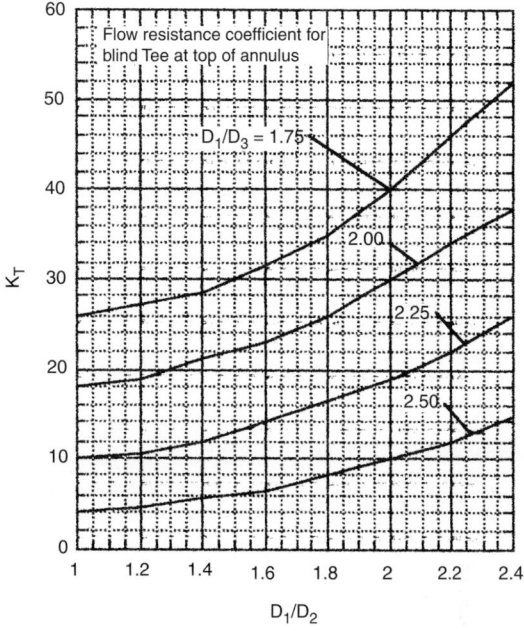

Figure 4.9.2 *Flow resistance coefficient for the blind Tee at top of annulus.*

The friction factor is determined from the empirical von Karman relationship. This relationship is

$$f_b = \left[\frac{1}{2 \log_{10} \left(\frac{D_b}{e} \right) + 1.14} \right]^2 \quad [4.9.3]$$

where e is the approximate absolute roughness of the blooey line inside surface (ft, or m).

For direct (conventional) circulation for air and gas drilling, the bottomhole pressure in the annulus is given by

$$P_{bh} = \left[\left(P_{at}^2 + b_a T_{av}^2 \right) e^{\frac{2a_a H}{T_{av}}} - b_a T_{av}^2 \right]^{0.5} \quad [4.9.4]$$

where

P_{bh} is the bottomhole pressure in the annulus (lb/ft^2 absolute, or N/m^2 absolute). (This pressure can also be the pressure at the bottom of any calculation interval with a uniform cross-sectional flow area.)
P_{at} is the pressure at the top of the annulus (lb/ft^2 absolute, or N/m^2 absolute). (This is the atmospheric pressure when the calculation interval starts at the top of the well However, it is the pressure at the bottom of the calculation interval of the interval above if numerous intervals with various cross-sectional flow areas are being considered.)
T_{av} is the average temperature of the calculation interval (°R, or K).
H is the length (or height) of the calculation interval (ft, or m).

The constants a_a and b_a are

$$a_a = \left(\frac{S_g}{R} \right) \left[1 + \left(\frac{\dot{W}_s}{\dot{W}_g} \right) \right] \quad [4.9.5]$$

and

$$b_a = \frac{f}{2g(D_h - D_p)}\left(\frac{R}{S_g}\right)\frac{\dot{W}_g^2}{\left(\frac{\pi}{4}\right)^2(D_h^2 - D_p^2)^2} \qquad [4.9.6]$$

where

f is the Darcy-Wiesbach (or Moody) friction factor for the annulus (this can be a weighted average if the annulus has dissimilar surfaces).

g is the acceleration of gravity (i.e., 32.2 ft/sec^2, or 9.81 m/sec^2)

$\dot{w}_s$ is the weight rate of flow of the solids (lb/sec, or N/sec).

$\dot{w}_g$ is the weight rate of flow of gas (lb/sec, or N/sec).

D_h is the inside diameter of the annulus borehole (ft, or m).

D_p is the outside diameter of the pipe in the annulus (ft, or m).

The values of $\dot{w}_s$ and $\dot{w}_g$ are determined from

$$\dot{W}_s = \frac{\pi}{4}D_h^2\gamma_w S_s\kappa \qquad [4.9.7]$$

and

$$\dot{W}_g = \gamma_g Q_g \qquad [4.9.8]$$

where

γ_w is the specific weight of the water (62.4 lb/ft^3, or 9810 N/ft^3).

γ_g is the specific weight of the gas (lb/ft^3, or N/ft^3).

S_s is the specific gravity of the solids cuttings.

Q_g is the volumetric flow rate of gas at reference pressure and temperature conditions (ft^3/sec, or m^3/sec).

κ is the drilling rate of penetration (ft/sec, or m/sec).

The compressed air or other gas flowing through the annulus is assumed to be wholly turbulent, therefore, the friction factor is determined from the empirical von Karman relationship for annulus flow. This relationship for the annulus is [5]

$$f = \left[\frac{1}{2\log_{10}\left(\dfrac{D_h - D_p}{e_{av}}\right) + 1.14}\right]^2 \qquad [4.9.9]$$

where e_{av} is the average absolute roughness of the annulus surfaces (ft, or m).

The average annulus open hole absolute roughness can be determined from a surface area weight average relationship between the inside openhole surface area and its roughness and the outside surface area of the drill string and its roughness. The weight average relationship is [4]

$$e_{av} = \frac{e_{oh}D_h^2 + e_p D_p^2}{D_h^2 + D_p^2} \qquad [4.9.10]$$

The openhole borehole inside surface absolute roughness can be approximated from Table 4.9.1. The outside absolute roughness of the outside of the steel pipe is usually taken as 0.00015 ft [5].

Minimum Volumetric Flow Rate. The minimum volumetric flow rate can be approximated by determining the volumetric flow rate that will give a minimum kinetic energy per unit volume in the annulus equal to 3.0 lb-ft/ft^3 (or 141.9 N-m/m^3). This is usually at the section of the annulus that has the largest cross-sectional area. The kinetic energy per unit volume is

$$KE = \frac{1}{2}\frac{\gamma_g}{g}V^2 \qquad [4.9.11]$$

Table 4.9.1 *Openhole Wall Approximate Absolute Roughness for Rock Formation Types [3]*

Rock Formation Types	Surface Roughness (ft)
Competent, low fracture	
Igneous (e.g., granite, basalt)	0.001 to 0.02
Sedimentary (e.g., limestone, sandstone)	
Metamorphic (e.g., gneiss)	
Competent, medium fracture	
Igneous (e.g., granite, basalt)	
Sedimentary (e.g., limestone, sandstone)	0.02 to 0.03
Metamorphic (e.g., gneiss)	
Competent, high fracture	
Igneous (e.g., breccia)	
Sedimentary (e.g., sandstone, shale)	0.03 to 0.04
Metamorphic (e.g., schist)	

where

KE is the kinetic energy per unit volume (lb-ft/ft^3, or N-m/m^3).

V is the velocity of the gas in the annulus (ft/sec, or m/sec).

Drill Bit Orifices or Nozzles. Gas flow through an orifice or nozzle constriction can be either sonic or subsonic. These flow conditions depend upon the critical pressure ratio. The critical pressure ratio is the relationship between the downstream pressure and the upstream pressure. This ratio is a function of the properties of the gas and is given by [5]

$$\left(\frac{P_2}{P_1}\right)_c = \left(\frac{2}{k+1}\right)^{\frac{k}{k-1}} \qquad [4.9.12]$$

where

P_1 is the upstream pressure (lb/ft^2 absolute, or N/m^2 absolute).

P_2 is the downstream pressure (lb/ft^2 absolute, or N/m^2 absolute).

k is the ratio of specific heats of the gas.

The flow conditions through the constriction are sonic if

$$\left(\frac{P_2}{P_1}\right)_c \leq \left(\frac{2}{k+1}\right)^{\frac{k}{k-1}} \qquad [4.9.13]$$

The flow conditions through the constriction are subsonic if

$$\left(\frac{P_2}{P_1}\right)_c \geq \left(\frac{2}{k+1}\right)^{\frac{k}{k-1}} \qquad [4.9.14]$$

If the gas flow through the orifice or nozzle throats is sonic, the pressure above the drill bit inside the drill string can be determined from

$$P_{ai} = \frac{\dot{w}_g T_{bh}^{0.5}}{A_n\left[\left(\dfrac{gkS_g}{R}\right)\left(\dfrac{2}{k+1}\right)^{\left(\frac{k+1}{k-1}\right)}\right]^{0.5}} \qquad [4.9.15]$$

where

P_{ai} is the pressure above the drill bit inside the drill string (lb/ft^2 absolute, or N/m^2 absolute).

A_n is the total flow area of the orifices or nozzles (ft^2, or m^2).

T_{bh} is the temperature of the gas at the bottom of the borehole (°R, or K).

If the gas flow through the orifice or nozzle throats is subsonic, the pressure above the drill bit inside the drill string can be determined from

$$P_{ai} = P_{bh} \left[\frac{\left(\frac{\dot{w}_g}{A_n} \right)^2}{2g \left(\frac{k}{k-1} \right) P_{bh} \gamma_{bh}} + 1 \right]^{\frac{k}{k-1}}$$ [4.9.16]

where γ_{bh} is the specific weight of the gas at bottomhole conditions (lb/ft³, or N/m³).

Injection Pressure. The pressure above the bit at the bottom of the inside of the drill string is used as the initial pressure for a sequence of calculations that determine the pressure at the top of each section of drill string having constant cross-section. The pressure at the top of each section is used as the initial pressure for the section above. This procedure is used until the injection pressure is determined. The injection pressure is obtained from [3]

$$P_{in} = \left[\frac{P_{ai}^2 + b_i T_{av}^2 \left(e^{\frac{2a_iH}{T_{av}}} - 1 \right)}{e^{\frac{2a_iH}{T_{av}}}} \right]^{0.5}$$ [4.9.17]

where

P_{in} is the injection pressure into the inside of the drill string (lb/ft² absolute, or N/m² absolute). (This pressure can also be the pressure at the top of any calculation interval with a uniform cross-sectional flow area.)

The constants a_i and b_i are

$$a_i = \frac{S_g}{R}$$ [4.9.18]

and

$$b_i = \frac{f}{2gD_i} \left(\frac{R}{S_g} \right)^2 \frac{w_g^2}{\left(\frac{\pi}{4} \right)^2 D_i^4}$$ [4.9.19]

where

D_i is the inside diameter of the constant cross-section section (ft, or m).

Water Injection. Water is injected into the volumetric flow rate of air (or other gases) flowing from the compressors to the top of the inside of the drill string for three important reasons:

- Saturate the air or other gas with water vapor sat bottomhole annulus pressure and temperature conditions
- Eliminate the stickiness of the small rock cuttings flour generated by the advance of the drill bit
- Assist in suppressing the combustion of the mixture of produced hydrocarbons and oxygen rich air

Water injection (with additives) is accomplished with a liquid pump that injects water into the compressed gas flow line to the well. The liquid pump draws its water from a liquid suction tank. The typical additives are corrosion inhibitors, polymer, and a foaming agent. The foaming agent causes the gas/liquid mixture to develop into an unstable foam as the mixture passes through the bit orifices. Table 4.9.2 gives a typical additives mixture used to create unstable foam drilling. "Mist drilling" was the old term used to describe the practice of injecting water (with additives) into the compressed gas to the well. Later it was found that a foaming agent improved the hole cleaning effectiveness of water injection. The water injection practice is now known as "unstable foam drilling" [6].

Table 4.9.2 *Typical Approximate Additives Volumes per 20 bbl of Water for Unstable Foam Drilling (actual commercial product volumes may vary)*

Additives	Volume per 20 bbl of Water
Foamer	4.2 to 8.4 gal
Polymer	1 to 2 quarts
Corrosion Inhibitor	0.5 gal

Saturation of Gas. Water is injected into the air or other circulation gases at the surface in order to saturate the gas with water vapor at bottomhole conditions. The reason this is done is to assure that the circulation gas as it flows out of the drill bit orifices into the annulus will be able to carry formation water coming into the annulus as whole droplets. If the gas entering the annulus is not saturated, the gas will use its internal energy to convert the formation water to vapor. This will reduce the capability of the gas to expand as it enters the annulus. If the gas cannot expand properly, it will not create adequate gas flow velocity in the annulus to carry cuttings (and formation water) to the surface. Water saturated gas will carry formation water to the surface as droplets.

The empirical formula for determining the saturation of various gases including air can be found in a variety of chemistry handbooks and other literature. The empirical formula for the saturation pressure of air, p_{sat}, can be written as [7]

$$P_{sat} = 10^{\left[6.39416 - \left(\frac{1750.286}{217.23 + 0.555t_{bh}} \right) \right]}$$ [4.9.20]

where P_{sat} is saturation pressure of the air at annulus bottomhole conditions (psia) and t_{bh} is the temperature of the air at annulus bottomhole conditions (°F). The approximate volumetric flow rate of injected water to an air drilling operation is determined by the relationship between the above saturation pressure and the bottomhole pressure in the annulus, and the weight rate of flow of gas being injected into the top of the inside of drill string. Thus, the flow rate of injected water, q_{iw}, is determined from [8]

$$q_{iw} = \left(\frac{P_{sat}}{P_{bh} - P_{sat}} \right) \left(\frac{3600}{8.33} \right) \dot{w}_g$$ [4.9.21]

where q_{iw} is the volumetric flow rate of injected water (gal/hr) and P_{bh} is the pressure at annulus bottomhole conditions (psia)

Eliminate Stickiness. The next higher volumetric flow rate of injected water is that amount needed to eliminate borehole stickiness. As the drill bit advances in some types of rock formations, the cutting action of the bit creates rock cuttings and a small amount of "rock flour". Rock flour are very small rock particles that act mechanically very much like the flour one cooks with in the kitchen. If a borehole is originally dry, the circulation gas will efficiently carry the rock cutting particles and the rock flour up the annulus to the surface as the drill bit is advanced (this is called "dust drilling"). If a water bearing formation is penetrated, formation water will begin to flow into the annulus. When the water combines with rock flour, the flour particles begin to stick to each other and the borehole wall. This is very much like placing cooking flour in a bowl and putting a small amount of water in with it and mixing it. The cooking flour will become sticky and nearly impossible to work with a spoon. In the open borehole, the slightly wetted rock flour sticks to the nonmoving inside surface of the borehole. Because gas flow eddy currents form just above the top of the drill collars, "mud rings" develop from this sticky rock flour at this location on the borehole wall. These mud rings can build up and create a constriction to the annulus gas flow. This constriction will in turn cause the injection pressure to increase slightly (by 5 to 10 psi) in a matter of a minute or so. This rather sharp increase in pressure should alert the driller that mud rings are forming due to an influx of formation water (or perhaps even crude oil). If mud rings are allowed to continue to form they will begin to resist the rotation of the drill string (since cuttings are not being efficiently remove from the well). This in turn will increase the applied torque at the top of the drill string and increase the danger of a drill string torque failure. Also, the existence of mud rings creates a confined chamber of high pressure gas. If hydrocarbon rock formations are being penetrated, the potential for combustion in the chamber increases. The solution to this operational problem is to begin to inject water into the circulation gas. Additional injected water (above that given by Equations 4.9.20 and 4.9.21) is needed to reduce the stickiness (in much the same way the cooking flour stickiness is reduced by adding more water). The amount of water added to eliminate the mud ring must be determined empirically and will be somewhat unique for each drilling operation.

The procedure for eliminating mud rings is as follows:

1. Begin injecting sufficient water to saturate the gas flow with water vapor.
2. Curtail drilling ahead but continue gas circulation.
3. Bring the rotations of the drill string up to about 100 rpm and lift the drill string up to the top of the drilling mast and lower it several times. This will allow the drill collars to smash into the mud rung structures and break them off the borehole wall.
4. Return to drilling ahead.
5. If the mud rings begin to form again (the injection pressure increases again), increase the water injection flow rate and repeat the above sequence.
6. Continue the above five steps until the volumetric flow rate of injected water reduces the stickiness of the rock flour so that the mud rings no longer form on the open borehole wall (this will be indicted by a return to a nearly constant gas injection pressure).

The typical injected water volumetric flow rate to eliminate rock flour stickiness in a $7\frac{7}{8}$ in. borehole would be of the order of approximately 2 to 10 bbl/hr.

Suppression of Hydrocarbon Combustion. The next higher volumetric flow rate of injected water is that needed to suppress the downhole combustion of hydrocarbons (e.g., downhole fire or explosion). This combustion is created when the advance of the drill bit gives a mixture of drilling air with produced oil, natural gas, and/or coal dust. The ignition spark is easily caused by the action of the steel drill bit cutters on the rock face at the bottom or the sides of the borehole. Figure 4.9.3 gives the ignition (ignition zone) parameters of pressure versus the percent mixture for natural gas mixed with standard atmosphere air (at ASME Standard Conditions) [8,9]. Increasing the water injection volumetric flow rate (with additives, particularly the foam agent) to the borehole wets down rock and steel surfaces reducing the risk of ignition sparks, and creates a bottomhole foam that deprives the combustion process of unlimited source of oxygen. It is thought that the water injection can be successful in suppressing hydrocarbon combustion in vertical wells where the hydrocarbon producing rock formation section has a thickness of approximately 200 ft or less.

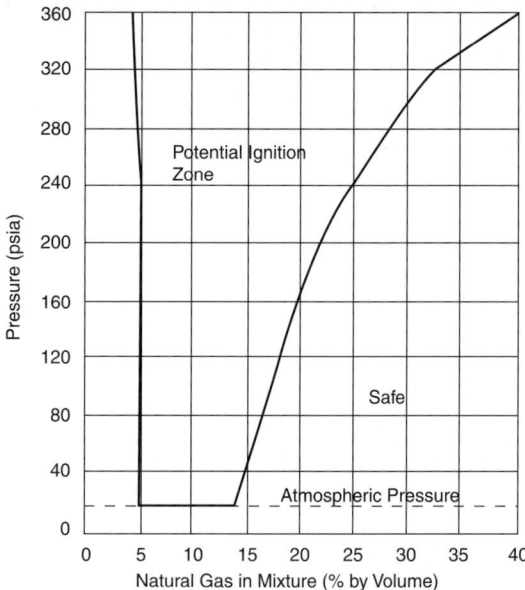

Figure 4.9.3 *Ignition mixture by volume of natural gas and atmospheric air.*

Horizontal drilling in hydrocarbon bearing rock formations is considered quite risky because of the long time exposure of the spark producing drill bit.

In a drilling operation where air is being used as the drilling fluid, as the drill bit is advanced and hydrocarbon bearing rock formations are about to be penetrated, the following steps should be taken to reduce the risk of downhole hydrocarbon combustion:

1. Drilling should be immediately stopped.
2. Air injection should be shut off and the gas flare monitored. If the flare is sustained by hydrocarbons from the well, the operator should note the wetness of the cuttings at the sample catcher (indicting water or distillate in the gas).
3. If the gas flare will not sustain or burn when the air is turned off, air can be turned back on, and water injection operations initiated, followed by careful bit advancement.
4. If the gas flare is sustained and/or the cuttings are wet, precautions must be taken to eliminate any mud rings before drilling forward. If water injection has not commenced, it should be initiated. With the air on and water injection operations underway, the drill string should be raised and lowered while rotating the drill string at approximately 100 rpm to smash any existing mud rings in the hole above the hydrocarbon bearing formations. If the injection pressure has not come down after this attempt to clear the mud rings, repeat the above sequence until the operator is satisfied that any mud rings have been successfully cleared. Once satisfied no existing mud rings are present in the borehole, carefully advance the drill bit through the hydrocarbon bearing formations.

Note that an alternative to the above is to switch the drilling gas to natural gas or liquid nitrogen.

There is a new technology that allows underbalanced drilling operations to be carried out using inert atmosphere.

The technology is the development of large industrial membrane filters that strip most of the oxygen from the compressed air output of the primary compressors [10]. The compressed inert air is either injected directly into the top of the drill string or injected into the top of the drill string after passing through the booster compressor. Details of this technology and its engineering applications are presented in Subsection 4.9.4, "*Compressor and Inert Air Generator Units*," given below.

4.9.2.2 Aerated Drilling (Gasified Fluid Drilling)

Aerated drilling (or gasified fluid drilling) refers to the use of compressed air (or other gases) injected into an incompressible drilling fluid that flows in the annulus. This is accomplished by two basic methods. (1) Drill pipe injection method requires the injection of compressed gas into the incompressible drilling fluid flow as this fluid is injected into the inside of the drill string at the top of the well. This allows the mixture of the gas and incompressible drilling fluid to flow throughout the circulating system. A variation on the drill pipe injection method is the use of a drill string jet sub above the drill collar/drill pipe interface to allow the injected gas to gasify the annulus above the jet sub position in the drill string. This later technique minimizes the aerated flow friction losses inside of drill collars and through drill bit nozzles. (2) Annulus injection method requires the use of parasite tubing installed on the backside of the casing. This allows the upper section of the annulus to be aerated directly from the bottom of the casing.

The basic advantage of aerated drilling is the control of bottomhole pressure in the annulus (via surface adjustments). Through calculations and trial and error adjustments in the field, the inflow of formation fluids into the well can be controlled with the pressure exerted on the bottom of the annulus by the mixture of gas and incompressible drilling fluid. The gas injected into the incompressible drilling fluid is usually air.

Another important application for aerated drilling has been in lost circulation situations. When incompressible drilling fluids are being lost in the annulus to thief formations, aerated drilling techniques have been used successfully to minimize or eliminate the problem. The bubbles created by gasifying the incompressible drilling fluid tend to fill in the fracture or pore openings in the borehole wall. This flow obstruction effect often controls the loss of incompressible drilling fluid to the thief zones.

The basic planning steps for the drilling of a deep well with aerated drilling fluids are as follows [3]:

1. Determine the geometry of the borehole section or sections to be drilled with aerated drilling fluids (i.e., openhole diameters, the casing or liner inside diameters, and depths).
2. Determine the geometry of the associated drill strings for the sections to be drilled with aerated drilling fluids (i.e., drill bit size and type, the drill collar size, drill pipe size, and maximum depth).
3. Determine the type of rock formations to be drilled and estimate the anticipated drilling rate of penetration. Also, estimate the quantity and depth location of any formation water that might be encountered.
4. Determine the elevation of the drilling site above sea level, the temperature of the air during the drilling operation, and the approximate geothermal temperature gradient.
5. Establish the objective of the aerated drilling operation:

   - To allow formation fluids to be produced as the formation is drilled.
   - To control loss of circulation problems,
   - To reduce formation damage.

6. Determine whether direct or reverse circulation techniques will be used to drill the various sections of the well.
7. If underbalanced drilling is the objective, determine the bottomhole pressure limit that must be maintained in order to allow minimal production of formation fluids into the well bore.
8. For either of the above objectives, determine the required approximate volumetric flow rate of incompressible fluid to be used in the aerated fluid drilling operation. This is usually the minimum volumetric flow rate required to clean the rock cuttings from the bottom of the well and transport the cuttings to the surface. In most aerated drilling operations, the incompressible fluid volumetric flow rate is held constant as drilling progresses through the openhole interval (as the gas injection flow rate is increased).
9. Determine the approximate volumetric flow rate of air (or other gas) to be injected with the flow of incompressible fluid into the top of the drill string (or into the annulus) as a function of drilling depth.
10. Using the incompressible fluid and air volumetric flow rates to be injected into the well, determine the bottomhole pressure and the associated surface injection pressure as a function of drilling depth (over the openhole interval to be drilled).
11. Select the contractor compressors that will provide the drilling operation with the appropriate air or gas volumetric flow rate needed to properly aerate the drilling fluid. Also, determine the maximum power required by the compressor(s) and the available maximum derated power from the prime movers (see Section 3.1).
12. Determine the approximate volume of fuel required by the compressor(s) to drill the well.

The circulation system for an aerated drilling operation can be modeled by as a multiphase flow calculation problem. In these problems, the pressure and temperature of the gas is usually known at the exit to the system. In this case, the exit is at the end of the return flow line to the mud tank. At this exit, the pressure and temperature are the local atmospheric pressure and temperature at the drilling location. The calculation procedures for aerated drilling circulation problems are to start at these known exit conditions and work backwards (or upstream) through the system. In these calculations, the volumetric flow rates of both the incompressible fluid and the injected gas must be assumed or known. If compressors are used to provide compressed air, then the volumetric flow rate is the sum of the outputs of all the primary compressors (see Section 3.4). Unlike the air and gas drilling calculations described in the subsection above, most aerated drilling fluid operations are designed on a basis of an incompressible drilling mud (or other liquid phase) circulation rate that can carry the anticipated rock bit generated cuttings to the surface [3]. For direct circulation, the geometry of the well and the physical properties of the drilling mud are used to determine the minimum volumetric flow rate for the incompressible drilling mud in the "*Minimum Volumetric Flow Rate*" subsection below. Once this drilling mud minimum flow rate is known, the gas phase volumetric flow rate to be injected into the well is selected (trial and error) to give the appropriate bottomhole pressure or other drilling conditions (see No. 5 earlier). The equations given in the "*Bottomhole Pressure*" subsection below are used to determine the bottomhole pressure. These equations are applied from the top of the well annulus to each constant cross-section section in sequence starting from the top well.

This is a trial and error calculation of determining the pressure at the bottom of the constant cross-section section of the well annulus. The pressure found at the bottom of each constant cross-section section is used as the initial pressure for the next deeper constant cross-section section until the bottomhole pressure at the bottom of the annulus is determined. In this calculation, all major and minor flow losses should be considered. Working upstream, the pressure above the drill bit orifices (or nozzles) inside the drill bit is found using the equations in the *"Drill Bit Orifices and Nozzle"* subsection. The equations given in the *Injection Pressure* subsection are applied from the bottom of the inside of the drill string to each constant cross-section section in sequence starting from the pressure above the drill bit. The pressure found at the top of each constant cross-section section is used as the initial pressure for the pressure at the top of next constant cross-section section until the injection pressure at the top of the inside of the drill string is found. These are trial and error calculations.

Aerated drilling (or multiphase flow) calculation models are complex and cumbersome to apply and obtain predictions for field operations. The methodology outline above has been used to predict bottomhole and injection pressures with an accuracy of about 15% to 20%.

Minimum Volumetric Flow Rate. Aerated drilling fluid operations are designed using the minimum volumetric flow rate of incompressible drilling mud (or other liquid) that can carry the anticipated bit cuttings to the surface. This design ensures that the borehole will be cleaned even if the injection of air is interrupted. Therefore, the minimum velocity required to carry cuttings form the borehole must be determined for the largest annulus space in the borehole profile. The average incompressible fluid velocity in the annulus is

$$V_f = V_c + V_t \qquad [4.9.22]$$

where

V_f is the minimum fluid velocity (ft/sec, or m/sec).
V_c is the solids critical concentration velocity (ft/sec, or m/sec).
V_t is the terminal velocity of solids (particles) in the drilling fluid (ft/sec, or m/sec).

The solids critical concentration velocity is

$$V_c = \frac{\kappa}{3600\,C} \qquad [4.9.23]$$

where

κ is the drilling rate of penetration (ft/sec, or m/sec)
C is the concentration factor (usually assumed to be 0.04)

The drilling cuttings particle average diameter can be estimated using the following expression

$$D_c = \frac{\kappa}{(60)\,N} \qquad [4.9.24]$$

where

D_c is the average diameter of drill bit cuttings (ft, or m),
N is the average drill bit rotary speed (rpm).

The fluid flow regions are classified as Laminar, transitional, or turbulent. These flow regions can be approximately defined using the non-dimensional Reynolds Number. The Reynolds Number is

$$N_R = \frac{DV}{\nu} \qquad [4.9.25]$$

where

D is the diameter of the flow channel (ft, or m).

V is the velocity of the flow (ft/sec, or m/sec).
ν is the kinematic viscosity of the flowing fluid (ft^2/sec, or m^2/sec).

The generally used empirically derived terminal velocity expressions in English Units are given below. For the laminar region, the expression is

$$V_{t1} = 0.0333 D_c^2 \left(\frac{\gamma_s - \gamma_f}{\mu_e}\right) \qquad 0 < N_R < 2100 \qquad [4.9.26]$$

For the transition region, the expression is

$$V_{t2} = 0.492 D_c \left(\frac{(\gamma_s - \gamma_f)^{\frac{2}{3}}}{(\gamma_f \mu_e)^{\frac{1}{3}}}\right) \qquad 2100 < N_R < 4000 \qquad [4.9.27]$$

For the turbulent region, the expression is

$$V_{t3} = 5.35 \left[D_c \left(\frac{\gamma_s - \gamma_f}{\gamma_f}\right)\right]^{\frac{1}{2}} \qquad N_R > 4000 \qquad [4.9.28]$$

where

V_{t1}, V_{t2}, V_{t3} are terminal velocities (ft/sec).
γ_s is the specific weight of the solids (lb/ft^3).
γ_f is the specific weight of the fluid (lb/ft^3).
μ_e is the effective absolute viscosity (lb-sec/ft^2).

The terminal velocity expressions in SI units are given below. For the laminar region the expression is

$$V_{t1} = 0.0333 D_c^2 \left(\frac{\gamma_s - \gamma_f}{\mu_e}\right) \qquad 0 < N_R < 2100 \qquad [4.9.29]$$

For the transition region, the expression is

$$V_{t2} = 0.331 D_c \left[\frac{(\gamma_s - \gamma_f)^{\frac{2}{3}}}{(\gamma_f \mu_e)^{\frac{1}{3}}}\right] \qquad 2100 < N_R < 4000 \qquad [4.9.30]$$

For the turbulent region, the expression is

$$V_{t3} = 2.95 \left[D_c \left(\frac{\gamma_s - \gamma_f}{\gamma_f}\right)\right]^{\frac{1}{2}} \qquad N_R > 4000 \qquad [4.9.31]$$

where

V_{t1}, V_{t2}, V_{t3} are terminal velocities (m/sec).
γ_s is the specific weight of the solids (N/m^3).
γ_f is the specific weight of the fluid (N/m^3).
μ_e is the effective absolute viscosity (N-sec/m^2).

Equations 4.9.22 to 4.9.31 allow for the determination of the approximate minimum annulus velocity of drilling mud (or other incompressible fluid) that can carry the drill bit cuttings from the borehole. This minimum annulus velocity determination is independent of any gas injection for aerated drilling. Normal aerated drilling operations practice is to have a drilling mud or incompressible fluid volumetric flow rate that will clean the bottom of the well without aeration with injected gas. This is usually the beginning design point for an aerated drilling operation.

Bottomhole Pressure. As in air and gas drilling calculations discussed above, the losses in the entire aerated drilling circulation system must be considered. Thus, aerated drilling calculations are initiated with the known atmospheric pressure at the exit to the surface return flow line. It is assumed that the drilling mud and air mixture will exit the well annulus and enter the surface return flow line to the mud tank (or separator). Knowing the exit pressure, Equation 4.9.30 can be used to determine the upstream return line entrance pressure. This is a trial-and-error calculation procedure:

$$\int_{P_{ex}}^{P_{en}} \frac{dP}{B_s(P)} = \int_0^{L_{sr}} dl \qquad [4.9.32]$$

where

$$B_s\,(P) = \left[\frac{\dot{w}_t}{\left(\frac{P_g}{P}\right)\left(\frac{T_r}{T_g}\right)Q_g + Q_m} \right]$$

$$\times \left\{ \frac{f}{2gD_{sr}} \left[\frac{\left(\frac{P_g}{P}\right)\left(\frac{T_r}{T_g}\right)Q_g + Q_m}{\frac{\pi}{4}D_{sr}^2} \right]^2 \right\}$$

and

$$\dot{w}_t = \dot{w}_g + \dot{w}_m + \dot{w}_s$$

$$\dot{w}_g = \gamma_g Q_g$$

$$\dot{w}_m = \gamma_m Q_m$$

$$\dot{w}_s = \frac{\pi}{4} D_h^2 \gamma_w S_s \kappa$$

where

P_{ex} is the pressure in the aerated fluid as it exits the end of the surface return line (lb/ft^2 or N/m^2 absolute). This pressure is usually assumed to be the atmospheric pressure at the exit.

P_{en} is the pressure in the aerated fluid as it enters the surface return line at the top of the annulus (lb/ft^2, or N/m^2 absolute). Trial and error methods must be used to find the value of this pressure that allows both sides of Equation 4.9.30 to be satisfied.

P_g is the pressure in the gas in the aerated fluid at reference surface conditions (lb/ft^2, or N/m^2 absolute). This is usually assumed to be the atmospheric pressure at the location.

T_g is the temperature of the gas in the aerated fluid at reference surface conditions (lb/ft^2, or N/m^2 absolute). This is usually assumed to be the atmospheric pressure at the location.

T_{av} is the average temperature of the aerated fluid flow in the surface return line (°R, or K),

Q_g is the volumetric flow rate of gas (ft^3/sec, or m^3/sec),

Q_m is the volumetric flow rate of drilling mud (ft^3/sec, or m^3/sec),

D_{sr} is the inside diameter of the surface return flow line (ft, or m),

L_{sr} is the length of the surface return flow line (ft, or m).

The empirical expression for the Darcy-Wiesbach friction factor for flow in the laminar region is [4.9.5]

$$f = \frac{64}{N_R} \qquad [4.9.33]$$

Equation 4.9.31 is valid for $0 \leq N_R \leq 2100$.

The Colebrook empirical expression is used to determine the friction factor for flow in the transition region. This expression is [5]

$$\frac{1}{\sqrt{f}} = -2\log_{10}\left[\frac{\left(\frac{e}{D_{sr}}\right)}{3.7} + \frac{2.51}{N_R\sqrt{f}} \right] \qquad [4.9.34]$$

where e is the absolute roughness of the inside surface of the surface return flow line (ft, or m). Note that Equation 4.9.34 must be solved by trial and error methods. Equation 4.9.34 is valid for $N_R \geq 4000$

The von Karman empirical expression is used to determine the friction factor for flow in the wholly turbulent region. This expression is [5]

$$f = \left[\frac{1}{2\log_{10}\left(\frac{D_{sr}}{e}\right) + 1.14} \right]^2 \qquad [4.9.35]$$

Equation 4.9.35 is valid for $N_R \geq 4000$. In practice, the greater of the f values obtained from Equation 4.9.34 and Equation 4.9.35 must be used in Equations 4.9.30.

For direct (conventional) circulation, the bottomhole pressure in the annulus is given by

$$\int_{P_e}^{P_{bh}} \frac{dP}{B_a\,(P)} = \int_0^H dh \qquad [4.9.36]$$

where

$$B_a\,(P) = \left[\frac{w_t}{\left(\frac{P_g}{P}\right)\left(\frac{T_{av}}{T_g}\right)Q_g + Q_m} \right]$$

$$\times \left\{ 1 + \frac{f}{2g\,(D_h - D_p)} \left[\frac{\left(\frac{P_g}{P}\right)\left(\frac{T_{av}}{T_g}\right)Q_g + Q_m}{\frac{\pi}{4}\,(D_h^2 - D_p^2)} \right]^2 \right\}$$

and

P_{bh} is the bottomhole pressure in the annulus (lb/ft^2 absolute, or N/m^2 absolute). (This pressure can also be the pressure at the bottom of any calculation interval with a uniform cross-sectional flow area.)

P_e is the exit pressure at the top of the annulus (lb/ft^2 absolute, or N/m^2 absolute). (This pressure can also be the pressure at the bottom of any previous calculation interval with a uniform cross-sectional flow area. When several intervals of different cross-sections are being considered, this pressure is the bottomhole pressure calculated in the previous calculation above the interval under consideration. It can also be the entrance pressure to the surface return line.)

T_{av} is the average temperature of the calculation interval (°R, or K).

H is the length (or height) of the calculation interval (ft, or m).

The empirical expression for the Darcy-Wiesbach friction factor for flow in the laminar region is

$$f = \frac{64}{N_R} \qquad [4.9.37]$$

Equation 4.9.35 is valid for $0 \leq N_R \leq 2100$.

The empirical expression for the friction factor for the flow in the transition region is

$$\frac{1}{\sqrt{f}} = -2\log_{10}\left[\frac{\left(\frac{e_{av}}{D_h - D_p}\right)}{3.7} + \frac{2.51}{N_R\sqrt{f}} \right] \qquad [4.9.38]$$

where e_{av} is the average absolute roughness of the annulus surfaces (ft, or m). See Equation 4.9.10 for details. Equation 4.9.38 is valid for $N_R \geq 4000$.

The friction factor is determined from the empirical von Karman relationship. This relationship is

$$f = \left[\frac{1}{2\log_{10}\left(\frac{D_h - D_p}{e_{av}}\right) + 1.14} \right]^2 \qquad [4.9.39]$$

See Equation 4.9.10 for details. Equation 4.9.39 is valid for $N_R \geq 4000$. In practice, the greater of the f values obtained from Equation 4.9.38 and Equation 4.9.39 must be used in Equation 4.9.36.

The solution of Equation 4.9.36 is by trial and error. Over each section of the annulus having constant geometry, the pressure at the bottom of the section (the upper limit in Equation 4.9.36) is selected that allows the left side of Equation 4.9.36 to equal the right side of that equation. This trial and error process is repeated for each geometric section down the annulus until the pressure at the bottom of the annulus is obtained. If the bottom of the annulus pressure (bottomhole pressure) is to be a given known value (for a given depth), then the entire trial and error process must be repeated with adjusted Q_g or Q_m to allow the correct bottomhole pressure to be obtained.

To obtain an initial approximate mixture of Q_g and Q_m for a given bottomhole pressure value, a non-friction solution can be used. Setting $f = 0$ in Equation 4.9.36 gives an expression that can be integrated to yield a closed form solution. Solving this closed form solution for Q_g, gives

$$Q_g = \frac{(\dot{w}_m + \dot{w}_s)\,H - (P_{bh} - P_e)\,Q_m}{\left[P_g\left(\dfrac{T_{av}}{T_g}\right)\ln\left(\dfrac{P_{bh}}{P_e}\right) - \gamma_g H\right]} \qquad [4.9.40]$$

Drill Bit Orifices and Nozzles. The mixture of incompressible fluid and the compressed gas passing through the drill bit orifices or nozzles can be assumed to act as a single phase incompressible fluid. However, this assumption is valid only when the friction losses in the flow through the bit orifices are also assumed to be higher. Thus, borrowing from mud drilling technology, the pressure change through the drill bit is

$$\Delta P_b = \frac{\left(\dot{w}_g + \dot{w}_m\right)^2}{2g\gamma_{mixbh}\,C^2\left(\dfrac{\pi}{4}\right)^2 D_e^4} \qquad [4.9.41]$$

where

ΔP_b is the pressure drop through the drill bit (lb/ft² absolute, or N/m² absolute).
γ_{mixbh} is the specific weight of the fluid mixture under bottomhole conditions (lb/ft³, or N/m³).
C is the loss coefficient for the aerated fluid flow through the drill bit orifices or nozzles (the values for C for aerated fluid flow should be taken as 0.70 to 0.85).
D_e is the effective orifice diameter (ft).

For drill bits with n equal diameter orifices (or nozzles), D_e is

$$D_e = \sqrt{nD_n^2} \qquad [4.9.42]$$

The pressure change obtained from Equation 4.9.38 is added to the bottomhole pressure P_{bh} obtained from Equation 4.9.34. Thus, the pressure above the drill bit inside the drill string, P_{ai}, is

$$P_{ai} = P_{bh} + \Delta P_b \qquad [4.9.43]$$

Injection Pressure. Knowing the pressure at the bottom of the inside of the drill string, Equation 4.9.42 can be used to determine the upstream injection pressure at the top of the inside of the drill string (or at the top of the section of constant cross-section). This is a trial and error calculation procedure. This equation is

$$\int_{P_{in}}^{P_{ai}} \frac{dP}{B_i(P)} = \int_0^H dh \qquad [4.9.44]$$

where

$$B_i(P) = \left[\frac{\dot{w}_g + \dot{w}_m}{\left(\dfrac{P_g}{P}\right)\left(\dfrac{T_r}{T_g}\right)Q_g + Q_m}\right]$$
$$\times \left\{1 - \frac{f}{2gD_i}\left[\frac{\left(\dfrac{P_g}{P}\right)\left(\dfrac{T_r}{T_g}\right)Q_g + Q_m}{\dfrac{\pi}{4}D_i^2}\right]^2\right\}$$

where

P_{in} is the pressure in the aerated fluid as it is injected into the top of the inside of the drill string (lb/ft², or N/m² absolute).

The empirical expression for the Darcy-Wiesbach friction factor for flow in the laminar region is [5]

$$f = \frac{64}{N_R} \qquad [4.9.45]$$

Equation 4.9.43 is valid for $0 \leq N_R \leq 2100$.

The Colebrook empirical expression is used to determine the friction factor for flow in the transition region. This expression is [5]

$$\frac{1}{\sqrt{f}} = -2\log_{10}\left[\frac{\left(\dfrac{e}{D_i}\right)}{3.7} + \frac{2.51}{N_R\sqrt{f}}\right] \qquad [4.9.46]$$

where e is the absolute roughness of the inside surface of the surface return flow line (ft, or m). Note that Equation 4.9.46 must be solved by trial and error methods. Equation 4.9.44 is valid for $N_R \geq 4000$.

The von Karman empirical expression is used to determine the friction factor for flow in the wholly turbulent region. This expression is [5]

$$f = \left[\frac{1}{2\log_{10}\left(\dfrac{D_i}{e}\right) + 1.14}\right]^2 \qquad [4.9.47]$$

Equation 4.9.47 is valid for $N_R \geq 4000$. In practice, the greater of the f values obtained from Equation 4.9.46 and Equation 4.9.47 must be used in Equation 4.9.44.

4.9.2.3 Stable Foam Drilling

From a calculation point of view, stable foam drilling is a special case of the aerated fluid drilling predictive theory given above. In stable foam drilling, a mixture of gas (usually air or nitrogen) and an incompressible fluid (water and a foam agent) is further specified by the foam quality at the top annulus and foam quality at the bottom of the annulus (see Table 4.9.2). This foam quality must be maintained through the annulus in order for a stable foam to exist in this return flow space. Foam quality, Γ, is defined as

$$\Gamma = \frac{Q_g}{Q_s + Q_L} \qquad [4.9.48]$$

where

Q_s is the volumetric flow rate of gas (ft³/sec).
Q_L is the volumetric flow rate of the incompressible fluid (ft³/sec).

The control of the foam quality at the top allows the foam quality at the bottom of the annulus to be calculated. Operationally this control is accomplished by placing a valve on the return flow line from the annulus (back pressure valve).

Upstream of the valve is a pressure gauge and by maintaining a specified back pressure at this position the foam quality at the top of the annulus can be determined. Knowing the foam quality at this position (and the other flow characteristics of the circulating system), the foam quality at any position in the annulus (particularly at the bottom of the annulus) can be determined. The foam quality at the bottom of the annulus (once developed) must be maintained at approximately 0.60 or greater [2,11,12,13]. If the foam quality at the bottom of the annulus drops much below this level, the foam will collapse and the flow will become aerated fluid drilling. To maintain the bottomhole foam quality in the annulus at a magnitude of approximately 0.60 or greater, the foam quality immediately upstream of the back pressure valve must usually be maintained at a magnitude in the range of 0.90 to 0.98.

Stable foam drilling operations can use a variety of incompressible fluids and compressed gases to develop a stable foam. The majority of the operations use fresh water and a commercial foam agent specifically for drilling use (a surfactant) with injected compressed air. Commercial surfactants for drilling can be obtained from a variety of drilling service companies. Recently developed inert atmosphere generators have been used to provide the injected gas for stable foam drilling operations. Using inert atmosphere gas in the stable foam will reduce corrosion of the drill string and the borehole casing and reduce the risk of downhole combustion when drilling through hydrocarbon bearing rock formations.

The basic planning steps for a deep well are as follows:

1. Determine the geometry of the borehole section or sections to be drilled with the stable foam drilling fluids (i.e., open hole diameters, the casing inside diameters, and maximum depths).

2. Determine the geometry of the associated drill string for the sections to be drilled with stable foam drilling fluids (i.e., drill bit size and type, the drill collar size, drill pipe size and description, and maximum depth).

3. Determine the type of rock formations to be drilled in each section and estimate the anticipated drilling rate of penetration.

4. Determine the elevation of the drilling site above sea level, the temperature of the air during the drilling operation, and the approximate geothermal temperature gradient.

5. Establish the objective of the stable foam drilling fluids operation:

   - To drill through loss of circulation formations,
   - To counter formation water entering the annulus (by injecting additional surfactant to foam the formation water in the annulus),
   - To maintain low bottom hole pressure to either preclude fracturing of the rock formations, or to allow underbalanced drilling operations.

6. If underbalanced drilling is the objective, it should be understood that stable foam drilling operations cannot maintain near constant bottomhole annulus pressures.

7. For either of the above objectives, determine the required approximate volumetric flow rate of the mixture of incompressible fluid (with surfactant) and the compressed air (or other gas) to be used to create the stable foam drilling fluid. This required mixture volumetric flow rate is governed by the foam quality at the top of annulus (i.e., return flow line back pressure) and the rock cuttings carrying capacity of the flowing mixture in the critical annulus cross-sectional area (usually the largest cross-sectional area of the annulus). The rock cuttings carrying capability of the stable foam can

be estimated using a minimum kinetic energy per unit volume value in the critical annulus cross-sectional.

8. Using the incompressible fluid and air volumetric flow rates to be injected into the well, determine the bottomhole pressure and the surface injection pressure as a function of drilling depth (over the openhole interval to be drilled).

9. Select the contractor compressor(s) that will provide the drilling operation with the appropriate air or gas volumetric flow rate needed to create the stable foam drilling fluid. Also, determine the maximum power required by the compressor(s) and the available maximum derated power from the prime mover(s).

10. Determine the approximate volume of fuel required by the compressor(s) to drill the well.

Stable foam drilling predictive calculations are carried out in the annulus space only. It is assumed that in deep drilling operations the flow of the mixed gas and incompressible fluid (and foam agent additive) inside the drill string and through the nozzle openings acts as an aerated fluid (governed by Equations 4.9.41 to 4.9.47). However, for flow in the return line and the annulus the flow is stable foam. For modeling stable foam drilling, the basic aerated fluid return line and annulus flow equations (i.e., Equations 4.9.32 to 4.9.39) are modified using the additional limitations imposed by the requirements that the foam qualities be specified at the top and bottom of the annulus. Equations 4.9.32 to 4.9.39 (or Equation 4 9.40) are restricted by Equation 4.9.48 and the specified foam qualities at the top and bottom of the annulus. These restrictions result in the inability to be able to specify the bottomhole annulus pressure for a foam drilling operations. In essence, the bottomhole annulus pressure cannot be controlled. Stable foam flow calculations for the return line and annulus spaces are carried out with the same trial and error methodology that was described above in the aerated drilling subsection.

Given below are the advantages and disadvantages of the stable foam drilling technique. The advantages are as follows:

- The technique does not generally require any additional downhole equipment.
- Nearly the entire annulus is filled with the stable foam drilling fluid, thus, low bottomhole pressures can be achieved.
- Since the bubble structures of stable foam drilling fluids have a high fluid yield point, these structures can support rock cuttings in suspension when drilling operations are discontinued to make connections. Stable foams can have seven to eight times the rock cutting carrying capacity of water.
- Rock cuttings retrieved from the foam at the surface are easy to analyze for rock properties information.

The disadvantages are as follows:

- Bottomhole annulus pressure cannot be specified and maintained. The specified foam qualities at the top and bottom of the annulus and the geometry of the will result in a unique bottomhole annulus pressure for each drilling depth.
- Stable foam fluids injection cannot be continued when circulation is discontinued during connections and tripping. Therefore, it can be difficult to maintain underbalanced conditions during connections and trips.
- Since the injected gas is trapped under pressure inside the drill string by the various string floats, time must be allowed for the pressure bleeddown when making connections and trips. The bleed-down makes it difficult to maintain underbalanced conditions.

- The flow down the inside of the drill string is two phase flow and, therefore, high pipe friction losses are present. The high friction losses result in high pump and compressor pressures during injection.
- The gas phase in the stable foam attenuates the pulses of conventional (measure-while-drilling) MWD systems. Therefore, conventional mud pulse telemetry MWD cannot be used.

Foam Models. In order to adequately model the flow of stable foam flow in the annulus it is necessary to have absolute viscosity that will describe the foam two-phase flow absolute viscosity. Most models are based on the assumption that the two-phase flow absolute viscosity is a strong function of foam quality. Foam is effectively Newtonian at qualities from 0.55 to as high as 0.74 (this foam change over value is dependent upon the type of foam used). In this lower range of foam quality the effective absolute viscosity of the foam can be approximated by

$$\mu_f = \mu_L (1 + 3.6\Gamma) \qquad [4.9.49]$$

where

μ_f is the absolute viscosity of the foam (lb-sec/ft^2).
μ_L is the absolute viscosity of the liquid (incompressible fluid) (lb-sec/ft^2).

For foam qualities greater than the above discussed value (up to 0.98), the effective absolute viscosity of the foam can be approximated by

$$\mu_f = \frac{\mu_L}{1 - \Gamma^n} \qquad [4.9.50]$$

where n is an exponent that depends on the type of foam being used and can vary from 0.33 to 0.49. The higher values of n are more associated with stiff foam (i.e., foam using drilling mud as the incompressible fluid base).

Drilling foam can be modeled to act as a Bingham plastic fluid. However, power law rheology has been also been used [13]. It can not be overstated that the modeling of foam drilling using Equations 4.9.31 to 4.9.50 above (or any other similarly constituted set of equations) will very dependent upon the type incompressible fluid (liquid) used as the bases for the foam and the type of foaming agent added to create the foam.

Bottomhole Pressure. Equations 4.9.32 to 4.9.39 together with the limitations of Equations 4.9.48 to 4.9.50 do not allow the control of the bottomhole pressure during stable foam drilling operations. Thus, stable foam drilling operations are not amenable to situations where precisely controlled bottomhole pressures are necessary for the success of the drilling operations.

Minimum Volumetric Flow Rate. Most stable foam vertical drilling operations are drilled over a depth interval with variable incompressible fluid volumetric flow rates and variable compressible gas volumetric flow rates. These variable volumetric flow rates are necessary in order to keep the annulus surface exit foam quality and the annulus bottomhole quality at predetermined values. The requirement of variable volumetric flow rates versus depth make the control of stable foam drilling operations complicated. In addition to keeping the foam qualities at the top and bottom of the annulus at predetermined values, the control of the flow rates must also assure that the foam flow in the annulus have sufficient rock cuttings carrying capacity to clean borehole as the drill bit is advanced. Stable foam has rock cuttings carrying capabilities during circulation and when circulation is stopped. Because of the variety of foaming agents and base

incompressible fluids that can be used for stable foam operations, the question of a minimum volumetric flow rate is usually determined empirically at the rig floor by the driller during actual operations. Since stable foam circulating fluids create a structure that can support rock cuttings, the flowing velocity in the annulus need not be high relative to air and gas drilling, and aerated fluid drilling. Assuming that the bottomhole annulus foam quality to be greater than approximately 0.60 (to prevent foam collapse), it is likely that successful stable foam drilling operations have the minimum bottomhole kinetic energy per unit volume of the order of 1.0 to 2.0 ft-lb/ft^3.

Drill Bit Orifices and Nozzles. In deep wells the base incompressible fluid (with foam agent additive) and injected gas flow flows down the inside of the drill string as a aerated fluid. Such an aerated mixture can be assumed to pass through the nozzles in much the same manner as an incompressible fluid. Thus, Equations 4.9.41 to 4.9.43 can be used to determine the pressure drop across the drill bit. For stable foam drilling operation, nozzles are usually used in the drill bit to increase the fluid shear as the fluid passes through the bit. This fluid shear aids in the creation of the foam at the bottom of the annulus.

Injection Pressure. The flow condition in the inside of the drill string is two phase (gas and fluid) flow. This aerated fluid flow is modeled by Equations 4.9.44 to 4.9.47. Because of the need to maintain a continuous stable foam in the annulus space during the drilling operation, it will be necessary to continuously increase the volumetric flow rate of both incompressible fluid and gas to the well as the drill bit is advanced. The control of the drilling operation will rest on the reliability of the data from the injection pressure gauge used together with the data from the back pressure gauge on the return line.

4.9.3 Completions Operations

Underbalance drilling operations require that the well completion operations also be underbalanced. In general, underbalanced completion techniques do not have to be used until the drill bit advance approaches the production rock formations (production zones). Most underbalanced wells are completed with either, openhole, well screen, noncemented slotted casing, or slotted tubing strings across the production zones. The placement of well screen, slotted casing, or slotted tubing in wells usually must be accomplished with snubbing and stripping techniques.

- Snubbing is the inserting of tubulars or other downhole tools into a well that is under pressure. In order to maintain a required minimum bottomhole static pressure in the well as tubulars are inserted, the well must be vented (usually through the choke line to the burn pit).
- Stripping is the inserting of tubulars or other downhole tools into a flowing well. Keeping the well flowing as the tubulars are inserted assures that the well is underbalanced.

Section 4.14 "Well Pressure Control" has additional information concerning snubbing and stripping.

Inserting tubulars into open holes through exposed freshly drilled production zones present some unique problems.

4.9.3.1 Sloughing Shales

Since the drilling circulation fluid is not heavy, there is a constant threat of caving and sloughing of the openhole borehole wall. Air and gas drilling operations will have

Table 4.9.3 *Typical approximate additive weights or volumes per 20 bbl of water for controlling sloughing shales (actual commercial product may vary).*

Additives	Weights of Volumes per 20 bbl of Water
Foamer	8.5 gallons
Bentonite	40 lb
CMC	2 lb
Corn starch	5 lb
Soda ash	1 quart

drilling penetration rates that can be twice that of mud drilling operations. This faster drilling penetration rate is an important feature since openhole integrity is very dependent upon the length of time the hole remains open and unsupported by cement and casing.

When drilling with air and gas the shale sequences of rock formations are usually the most susceptible to caving or sloughing. This is due mainly to bedding layered texture of shale and the generally weak bonding between these layers. When these shales are penetrated with a drill bit, the openhole wall surfaces of the exposed shale formations tend to break off and the large fragments and fall into the annulus space between the openhole wall and the drill collar and drill pipe outside surfaces. This sloughing of shale formations can be temporarily controlled by injecting additional additives into water being injected into the circulation air or gas (in addition to those given in Table 4.9.2). Table 4.9.3 gives the formula for these additional additives. This formula has been successfully used in the San Juan Basin, New Mexico, USA.

4.9.3.2 Casing and Cementing

When drilling with air or gas as the circulation fluid, the borehole will be basically dry prior to casing and cementing operations. Therefore, there will be no water or drilling mud in the well to float the casing into the well (making use of buoyancy). This presents some special completions problems for air and gas drilling operations.

When an openhole section of a gas drilled well is to be ceased, the casing with a casing shoe on the bottom of the string is lowered into the dry well. A pre-lush of about 20 bbl of CMC (carboxymethyl-cellulose) treated water must be pumped ahead of the cement. A diaphragm (bottom) plug is usually run between the CMC treated water pre-flush and the cement and a "bumper" plug run at the top of the cement. Fresh water is pumped directly behind the cement and is used to help balance the cement in the annulus between the openhole wall and the outside of the inserted casing. The CMC treated water pre-flush seals the surface of the dry borehole walls prior to the cement entering the open annulus space. This pre-flush limits the rapid hydrating of the cement as it flows from the inside of the casing to the annulus space. Once the casing and cementing operations are properly carried out and the cement successfully sets up in the annulus, it is necessary to remove the water from inside the casing in order to return to air and gas drilling operations (drill out the cement at the casing shoe and continue drilling ahead). There are several safe operational procedures that can be used to remove water from the inside of the casing after a successful cementing operation.

Aerated Fluid Procedure. The aerated fluid procedure is as follows:

1. Run the drill string made up with the appropriate bottomhole assembly and drill bit to a depth a few tens of feet above the last cement plug.

2. Start the mud pump running as slowly as possible, to pump water at a rate of 1.5 to 2.0 bbl/min. This reduces fluid friction resistance to the moving fluids in the circulation system.

3. Bring one compressor and booster on line to aerate the water being pumped to the top of the drill string. The air rate to the well should be about 100 to 150 acfm per barrel of water. If the air volumetric flow rate is too high, the standpipe pressure will exceed the pressure rating of the compressor and the compressor will shut down. Therefore, the compressor must be slowed down until air is mixed with the water going into the drill string.

4. As the fluid column in the annulus (between the inside of the casing and the outside of the drill string) is aerated, the standpipe pressure will drop. Additional compressors can be added (i.e., increasing air volumetric flow rate) to further lighten the fluid column and unload the water form the casing.

5. After the hole has been unloaded, the water injection pumps should be kept in operation to clean the borehole.

6. At this point, begin air or mist drilling. Drill out the cement plug at the bottom of the casing and drill an additional 20 to 100 ft to allow any sloughing walls of the borehole to clean up.

7. Once the hole has been stabilized, stop drilling and blow the hole with air and injected water to eliminate rock cuttings. Continue this drilling and cleaning procedure for 15 to 30 minutes or until the air flow (with injected water) returning to the surface is clean (i.e., shows a fine spray and white color).

8. With the drill bit directly on bottom, continue flowing air with no injected water into the drill string. Air should flow to the well at normal drilling volumetric flow rates until the water and surfactant remaining in the well are swept to the surface.

9. Continuously blow the hole with air for about 30 minutes to an hour.

10. Begin normal air drilling. After 5 to 10 ft have been drilled, the hole should go to dry drilling (although it is sometimes necessary to drill as much as 60 to 90 ft before dry dust appears at the surface). If the hole does not dust after these steps have been carried out, inject another surfactant slug into the air flow to the well. If dry dusting cannot be achieved, unstable foam drilling may be required to complete the air drilling operation.

Gas Lift Procedure. The air lift procedure is as follows:

1. Calculate the lifting capability of the primary and booster compressor on the drilling location. Run the drill string made up with the appropriate bottomhole assembly and drill bit to a depth a few hundred feet above this calculated compressor pressure limit.

2. Start the compressors and force compressed air to the bottom of the drill string and begin aerating the water column in the annulus and flow this aerated water column to the surface (removing this portion of the column from the well).

3. Once this column of water has been removed, shut down the compressors and lower the drill string a similar distance as defined by the lifting capability limit determined in No.1 above. Start up the compressors and remove his next column of water from the well.

4. Continue lowering the drill string in increments and air lifting the entire water column from the well.

5. With the drill bit directly on bottom, continue flowing air into the drill string. Air should flow to the well at normal drilling volumetric flow rates until the water and surfactant remaining in the well are swept to the surface.

6. Continuously blow the hole with air for about 30 minutes to 1 hour.

7. Begin normal air drilling. After 5 to 10 ft have been drilled, the hole should go to dry dust drilling (although it is sometimes necessary to drill as much as 60 to 90 ft before dry dust appears at the surface). If the hole does not dust after these steps have been carried out, inject another surfactant slug into the air flow to the well. If dry dusting cannot be achieved, unstable foam drilling may be required to complete the air drilling operation.

4.9.3.3 Drilling with Casing

There are some new technologies entering the underbalanced drilling and completions operations. These involve using casing (that will be left in the well) as the "drill pipe." Casing while drilling has been used for many decades in shallow water and geotechnical drilling and completions operations. Oil and gas service companies have adopted the drilling with casing concept and developed new technologies to allow this type of drilling to be safely used for deep pressured wells. Figure 4.9.4 shows a downhole deployment valve that can be used at the bottom of the drill pipe/casing string. This flapper type valve allows the back pressure of sensitive production rock formations to be isolated as tools are run through the inside of the drill pipe/casing string. Figure 4.9.5 shows a Wireline Retrievable Float Valve that can be used at the top of the drill pipe/casing string to isolate the gas pressure in the drill string (between the top and bottom valves).

Drilling with casing can utilize a variety of methods to rotate the drill bit (and underreamers). Usually the drill pipe/casing string can be rotated with a rotary table, a top drive swivel, or a hydraulic rotary head. Also a downhole motor can be used to rotate the drill bit. The drilling assembly can be on a separate drill string inside the casing, or attached directly to the bottom of the drill pipe/casing string.

4.9.4 Compressor and Inert Air Generator Units

There are a variety of compressors available commercially. However, the most useful for air and gas drilling operations are the reciprocating piston compressor and the rotary compressor. These compressors are used as primary compressors or as booster compressors. The primary compressor intakes atmospheric air and compresses usually to a pressure of about 200 psig to 350 psig via two or three internal compression stages. The primary compressor can be either a reciprocating piston or a rotary type compressor. If higher pressures are required for the drilling operation, the compressed air from the primary compressor can be passed through a booster compressor for compression up to as high as 2000 psig. Only the reciprocating piston compressor can be used as a booster.

4.9.4.1 Compressor Units

Compressors are rated by their intake volumetric flow rate (of atmospheric air at a specified standard condition (e.g., API, ASME, EU; see Section 3.4). This volumetric flow rate is usually specified by the manufacturer in units of standard ft^3/m (usually written as scfm) or in standard $m^3/minute$ (usually written as scm/m). Be sure to use the manufactures specified standard conditions to make accurate engineering calculations.

Reciprocating Piston. The advantages of the reciprocating piston compressor are

1. Dependable volumetric flow rates at output line back pressures near compressor maximum pressure capability.

Figure 4.9.4 *Downhole deployment valve (Courtesy of Weatherford International Limited).*

2. Compressor will match its output flow rate pressure with the flow line back pressure.
3. More prime mover fuel efficient.

The disadvantages are

1. High capital costs
2. Requires high maintenance
3. Bulky to transport

Rotary. The advantages of the rotary compressor are

1. Low capital costs

drilling mud filled. When drilling with compressed air, the oxygen in the air mixes with methane (or other hydrocarbon gases) to create combustible mixtures. Three basic methods of combustion suppression are used in modern drilling operations. These are using natural gas from a nearby pipeline as the drilling fluid, using inert atmospheric air provided by inert air generator units, and using liquid nitrogen injection (used only for drilling of short distances).

Using natural gas as the drilling fluid essentially eliminates all oxygen from the borehole and, therefore, eliminates the possibility of all fires and explosions. However, using natural gas as a drilling fluid creates new danger problems on the rig floor and in the area around the drilling location. The drilling location fire and explosion problems and the expense of using a marketable gas usually make natural gas uneconomical.

Liquid nitrogen can be mixed with compressor air to reduce the oxygen content of the mixture below the level to support combustion in the event the drilling operation using the mixture encounters hydrocarbons. The expense of this market gas makes liquid nitrogen uneconomical for prolong drilling.

The allowable oxygen content for a mixture of oxygen, nitrogen, and methane is a function of the maximum pressure of the mixture in the borehole during the drilling operation. The minimum oxygen percent of a mixture with nitrogen and methane is given by [2,10]

$$O_{2min} = 13.98 - 1.68 \log_{10} P \qquad [4.9.51]$$

where

O_{2min} is the minimum percent of oxygen content in the mixture (%).
P is the absolute pressure of the mixture (psia).

The minimum percent of oxygen content to support combustion for pressures of less than 3000 psia is approximately 8% or greater. Therefore, to prevent combustion in any drilling or completion operation the oxygen content must be kept below 8%.

4.9.4.3 Inert Air Generator Units
Inert air generator units (or nitrogen generator units) use membrane filtration technology to remove oxygen from atmospheric air. This results in an inert atmospheric gas low in oxygen content [2]. The oxygen percentage in this mixture must be lower than 8% for the mixture to be inert. Because other gas components in atmospheric air are removed in the membrane filtration process, the efficiency of these inert air generator units is a function of the percentage of oxygen content required in the final output of inert atmospheric air exiting the units. This efficiency must be considered in making engineering predictive calculations. The efficiency of the inert air generator unit refers to the ratio of the volumetric flow rate of inert atmospheric air exiting the unit to the volumetric flow rate of atmospheric air entering the unit. The atmospheric air entering the primary compressor is passed to the inert air generator after exiting the compressor. Figure 4.9.6 the typical inert atmospheric air drilling location schematic [10]. Field test data demonstrates a linear relationship between the percentage of oxygen remaining in the output of the inert air generator unit and the percent efficiency of the unit (Figure 4.9.7) [10]. Equation 4.9.52 gives the relationship of percent efficiency of the unit to the oxygen percentage remaining in the inert air output from the unit. Equation 4.9.52 is

$$\% \text{ Efficiency} = (\% \text{ Oxygen}) (3.33) + (33.33) \qquad [4.9.52]$$

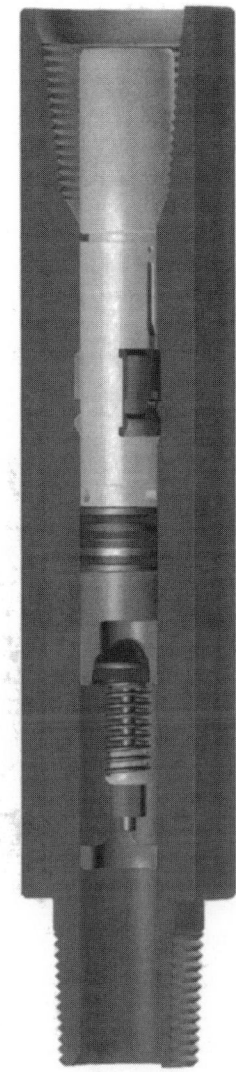

Figure 4.9.5 Wireline retrievable float valve (Courtesy of Weatherford International Limited).

2. Low maintenance
3. Easy to transport and small site footprint on location

The disadvantages are

1. Volumetric flow rates will decrease as flow line back pressure near compressor fixed pressure output
2. Because of fixed output pressure, fuel consumption is the same for all flow line back pressures

4.9.4.2 Allowable Oxygen Content
Compressed air is combustible when mixed with hydrocarbons in the downhole environment. Downhole combustion suppression is extremely important in air and gas drilling applications. Downhole combustion can cause drilling rig and production personnel injury and death, and damage to equipment associated with drilling production operations. Combustion hazards are prevalent in wells that are not

Figure 4.9.6 *Typical compressor and inert air generator unit layout at drilling location. (Courtesy of Weatherford International Limited.)*

where

    % Efficiency is the percent of volumetric flow rate of inert atmospheric air exiting the unit relative to the volumetric flow rate of atmospheric entering the unit.

    % Oxygen is the percent of oxygen remaining in the inert atmospheric air exiting the unit.

Figure 4.9.8 shows the unique Sea Wolf combined primary compressor (on the bottom) and inert air generator (on the top) unit. Such combined units allow for small footprints at land drilling locations and for offshore platforms.

4.9.4.4 Liquid Nitrogen

Liquid nitrogen is used to suppress downhole combustion. The two main reasons for using liquid nitrogen in a drilling operation are that only a short interval needs to be drilled using a non-combustible drilling fluid and that atmospheric air generation units are not available.

    Liquid nitrogen is injected into the compressed atmospheric air from the compressors in order to reduce the overall oxygen content in the resulting flow stream of gas to the well. The objective is to reduce the oxygen content in the resulting flow of gases to the well to a level where combustion cannot be supported. Thus, specifying the desired oxygen content (percent) in the resulting flow of gases to the well, the approximate volumetric flow rate of pure nitrogen to be injected into the compressed air to a well can be obtained from

$$q_n = q_t - \frac{q_t\,(\%O_2)}{0.21} \qquad [4.9.53]$$

where

    q_n is the volumetric flow rate of injected nitrogen gas to the well (scfm).

    q_t is the total volumetric flow rate of gas needed to adequately clean the bottom of the hole for the drilling operation (scfm).

    O_2 is the desired oxygen content in the resulting mixture (%)

The compressor produced volumetric flow rate of air required to drill the well is determined by subtracting the nitrogen flow rate from the total flow rate needed to safely clean the borehole. Note that this calculation is for standard atmospheric conditions (scfm, usually API mechanical equipment standard conditions, see Section 3.4). Therefore, the actual atmospheric conditions at the drilling location must be considered and the volumetric flow rates determined in acfm. Since primary compressor come in specific flow rate segments, care must be taken not to select compressor units that will increase the oxygen content beyond the level desired.

    Pure nitrogen comes to the drilling location in a liquid form (at very low temperatures). Therefore, the above volumetric flow rate of pure nitrogen in scfm must be converted to gallons of liquid nitrogen. This conversion is

One gallon liquid nitrogen = 93.11 scf nitrogen gas   [4.9.54]

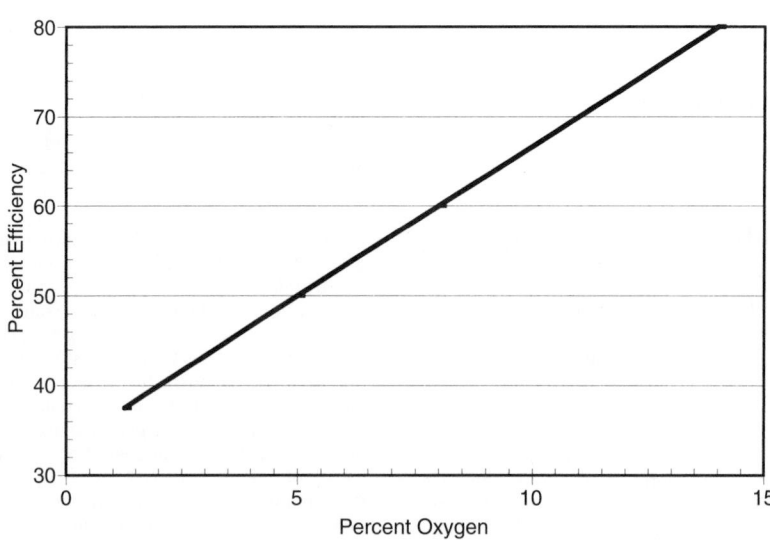

Figure 4.9.7 *Inert air generator unit efficiency versus percentage of oxygen remaining in unit output. (Courtesy of SPE.)*

Figure 4.9.8 *Sea Wolf combined primary compressor and inert air generation unit (Courtesy of Weatherford International Limited).*

4.9.5 Highly Deviated Well Drilling and Completions
Many technologies have been developed through the past decades directed at improving directional drilling and the completions of directional wells using conventional incompressible drilling fluids (e.g., water-based drilling muds and oil-based drilling muds). Air and gas drilling technology has been a small niche area of the drilling industry and, therefore, up until the late 1980's little attention was given to the development of directional drilling technologies for air and gas drilling operations. As the North American oil and gas fields deplete, underbalanced drilling and completion operations have become more important in extending the useful life of these fields. Although there have been some recent development activities to develop air and gas directional drilling technologies, these have not been entirely successful or accepted commercially.

4.9.5.1 Drilling Operations
The mud pulse communication systems between the surface and bottomhole assembly are the major technologies used for MWD and LWD (see Section 4.9.11). No similar technologies have been developed for drilling highly deviated wells with air and gas drilling fluids, aerated drilling fluids, or stable foam drilling fluid. There are two technologies that were developed as possible alternatives to the mud pulse technology for mud drilling operations that have had limited success in air and gas, aerated fluid, and stable foam directional drilling operations. These are electromagnetic earth transmission MWD and wireline steering and logging tools (see Section 4.9.11). One of the most effective methods of making hole using air and gas as the drilling fluid is the downhole air hammer. At present there are no methods for utilizing this drilling tool in deviated wells. Progressive cavity motors have been adapted for use with air and gas, aerated, or stable foam drilling fluids. However, the lack of a reliable

MWD technology is still the primary limitation to directional drilling using air and gas drilling technology.

The risk of downhole fires and explosion exists for both vertical and horizontal drilling operations using air and gas drilling technology. However, the risk is higher when drilling long horizontal boreholes using. This is due to the fact that during a typical horizontal drilling operation, the horizontal interval drilled in the hydrocarbon bearing rock formations is several times longer than in typical vertical interval drilled in a vertical drilling operation (assuming similar hydrocarbon bearing rock formations). Further, the drilling rate of penetration for a horizontal drilling operation will be about half that of vertical drilling (assuming the similar rock type). Thus, this increased exposure time in a horizontal borehole in hydrocarbon bearing rock formation is far greater and, therefore, the risk is far greater. The development of membrane filter technology that basically eliminates the risk of downhole fires and explosions has increased interest in developing more reliable MWD technologies for air and gas drilling operations.

4.9.5.2 Completions Operations
Wells that have been directionally drilled with air and gas drilling technology are completed either open hole completions, or with slotted tubing liners (not cemented). The development of expandable casing and liners holds the prospect of greatly improving completions for highly deviated boreholes drilled with air and gas technology (see Sections 4.16 and 4.18).

References
1. Bourgoyne, A. T., Millheim, K. K., Chenevert, M. E., and Young F. S., *Applied Drilling Engineering*, SPE, First Printing, 1986.

2. *Underbalanced Drilling Manual*, Gas Research Institute, GRI Reference No. GRI 97/0236, 1997.

3. Lyons, W. C., Guo, B., and Seidel, F. A., *Air and Gas Drilling Manual*, McGraw-Hill, 2000.

4. Guo, B., and Ghalambor, A., *Gas Volume Requirements for Underbalanced Drilling Deviated Holes*, PennWell, 2002.

5. Daugherty, R. L., Franzini, J. B., and Finnemore, E. J., *Fluid Mechanics with Engineering Applications*, 8th Edition, McGraw-Hill, 1985.

6. *API Recommended Practice for Testing Foam Agents for Mist Drilling*, API RP-46, lst Edition, November 1966.

7. *Handbook of Chemistry*, McGraw-Hill, 1956.

8. U. S. Bureau of Mines Report Investigations N. 3798.

9. Coward, H. P., and Jones, G. W., "Limits of Flammability of Gases and Vapors," Bureau of Mines Bulletin 503, Washington D.C., 1952.

10. Allan, P. D., "Nitrogen Drilling System for Drilling Applications," SPE Paper 28320, presented at the SPE 69th Annual Technical Conference and Exhibition, New Orleans, Louisiana, September 25–28, 1994.

11. Beyer, A. H., Millhone, R. S., and Foote, R. W., "Flow Behavior of Foam as a Well Circulating Fluid," SPE Paper 3986, Presented at the SPE 47th Annual Fall Meeting, San Antonio, Texas, October 8–11, 1972.

12. Mitchell, B. J., "Test Data Fill Theory Gap on Using Foam as a Drilling Fluid," *Oil and Gas Journal*, September 1971.

13. Kuru, E., Miska, S., Pickell, M., Takach, N., and Volk, M., "New Directions in Foam and Aerated Mud Research and Development," SPE Paper 53963, Presented at the 1999 SPE Latin American and Caribbean Petroleum Engineering Conference, Caracas, Venezuela, April 21–23, 1999.

4.10 DOWNHOLE MOTORS

4.10.1 Background

In 1873, an American, C. G. Cross, was issued the first patent related to a downhole turbine motor for rotating the drill bit at the bottom of a drillstring with hydraulic power [1]. This drilling concept was conceived nearly 30 years before rotary drilling was introduced in oil well drilling. Thus the concept of using a downhole motor to rotate or otherwise drive a drill bit at the bottom of a fluid conveying conduit in a deep borehole is not new.

The first practical applications of the downhole motor concept came in 1924 when engineers in the United States and the Soviet Union began to design, fabricate and field test both single-stage and multistage downhole turbine motors [2]. Efforts continued in the United States, the Soviet Union and elsewhere in Europe to develop an industrially reliable downhole turbine motor that would operate on drilling mud. But during the decade to follow, all efforts proved unsuccessful.

In 1934 in the Soviet Union a renewed effort was initiated to develop a multistage downhole turbine motor [2–5]. This new effort was successful. This development effort marked the beginning of industrial use of the downhole turbine motor. The Soviet Union continued the development of the downhole turbine motor and utilized the technology to drill the majority of its oil and gas wells. By the 1950s the Soviet Union was drilling nearly 80% of their wells with the downhole turbine motors using surface pumped drilling mud or freshwater as the activating hydraulic power.

In the late 1950s, with the growing need in the United States and elsewhere in the world for directional drilling capabilities, the drilling industry in the United States and elsewhere began to reconsider the downhole turbine motor technology. There are presently three service companies that offer downhole turbine motors for drilling of oil and gas wells. These motors are now used extensively throughout the world for directional drilling operations and for some straight-hole drilling operations.

The downhole turbine motors that are hydraulically operated have some fundamental limitations. One of these is high rotary speed of the motor and drill bit. The high rotary speeds limit the use of downhole turbine motors when drilling with roller rock bits. The high speed of these direct drive motors shortens the life of the roller rock bit.

In the 1980s in the United States an effort was initiated to develop a downhole turbine motor that was activated by compressed air. This motor was provided with a gear reducer transmission. This downhole pneumatic turbine has been successfully field tested [6,7].

The development of positive displacement downhole motors began in the late 1950s. The initial development was the result of a United States patent filed by W. Clark in 1957. This downhole motor was based on the original work of a French engineer, René Monineau, and is classified as a helimotor. The motor is actuated by drilling mud pumped from the surface. There are two other types of positive displacement motors that have been used, or are at present in use today: the vane motor and the reciprocating motor. However, by far the most widely used positive displacement motor is the helimotor [2].

The initial work in the United States led to the highly successful single-lobe helimotor. From the late 1950s until the late 1980s there have been a number of other versions of the helimotor developed and fielded. In general, most of the recent development work in helimotors has centered around multilobe motors. The higher the lobe system, the lower the speed of these direct drive motors and the higher the operating torque.

There have been some efforts over the past three decades to develop positive development vane motors and reciprocating motors for operation with drilling mud as the actuating fluid. These efforts have not been successful.

In the early 1960s efforts were made in the United States to operate vane motors and reciprocating motors with compressed air. The vane motors experienced some limited test success but were not competitive in the market of that day [7]. Out of these development efforts evolved the reciprocating (compressed) air hammers that have been quite successful and are operated extensively in the mining industry and have some limited application in the oil and gas industry [8]. The air hammer is not a motor in the true sense of rotating equipment. The reciprocating action of the air hammer provides a percussion effect on the drill bit, the rotation of the bit to new rock face location is carried out by the conventional rotation of the drill string.

In this section the design and the operational characteristics and procedures of the most frequently used downhole motors will be discussed. These are the downhole turbine motor and the downhole positive displacement motor.

4.10.2 Turbine Motors

Figure 4.10.1 shows the typical rotor and stator configuration for a single stage of a multistage downhole turbine motor section. The activating drilling mud or freshwater is pumped at high velocity through the motor section, which, because of the vane angle of each rotor and stator (which is a stage), causes the rotor to rotate the shaft of the motor. The kinetic

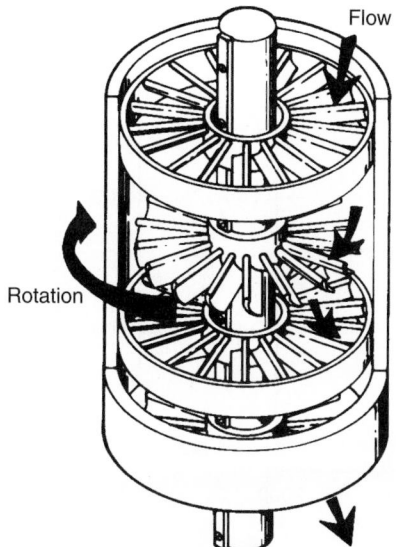

Figure 4.10.1 *Basic turbine motor design principle* (courtesy of Smith International, Inc).

energy of the flowing drilling mud is converted through these rotor and stator stages into mechanical rotational energy.

4.10.2.1 Design

The rotational energy provided by the flowing fluid is used to rotate and provide torque to the drill bit. Figure 4.10.2 shows the typical complete downhole turbine motor actuated with an incompressible drilling fluid.

In general, the downhole turbine motor is composed of two sections: (1) the turbine motor section and (2) the thrust-bearing and radial support bearing. These sections are shown in Figure 4.10.2. Sometimes a special section is used at the top of the motor to provide a filter to clean up the drilling mud flow before it enters the motor, or to provide a by-pass valve.

The turbine motor section has multistages of rotors and stators, from as few as 25 to as many as 300. For a basic motor geometry with a given flowrate, an increase in the number of stages in the motor will result in an increase in torque capability and an increase in the peak horsepower. This performance improvement, however, is accompanied by an increase in the differential pressure through the motor section (Table 4.10.1). The turbine motor section usually has bearing groups at the upper and lower ends of the rotating shaft (on which are attached the rotors). The bearing groups have only radial load capabilities.

The lower end of the rotating shaft of the turbine motor section is attached to the upper end of the main shaft. The drilling fluid after passing through the turbine motor section is channeled into the center of the shaft through large openings in the main shaft. The drill bit is attached to the lower end of the main shaft. The weight on the bit is transferred to the downhole turbine motor housing via the thrust-bearing section. This bearing section provides for rotation while transferring the weight on the bit to the downhole turbine motor housing.

In the thrust-bearing section is a radial support bearing section that provides a radial load-carrying group of bearings that ensures that the main shaft rotates about center even

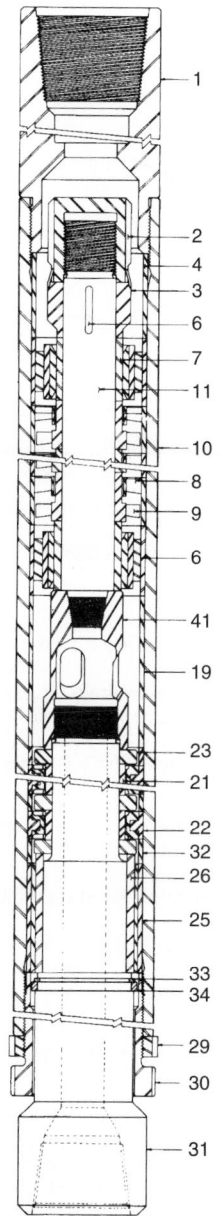

Item No.	Description
1	Top Sub
2	Shaft Cap
3	Lockwasher-Turbine Section
4	Stator Spacer
5	Shaft Key-Turbine Section
6	Intermediates Bearing Body
7	Intermediates Bearing Sleeve
8	Stator - Assembly
9	Rotor -
10	Turbine Housing
11	Turbine Shaft
19	Spacer-Bearing Section
21	Thrust Bearing Sleeve
22	Thrust Bearing Body
23	Thrust Disc
25	Lower Bearing Body
26	Lower Bearing Sleeve
29	Lower Sub Lock Ring
30	Lower Sub
31	Bearing Shaft
32	Lower Bearing Spacer
33	Retaining Ring
34	Catch Ring
40	Float Retainer Ring
41	Shaft Coupling
45	Shaft Cap Lock Screw
46	Eastco Float
	Turbodrill Complete
	[1]Optimal, order by topf sub too
	Repair Accessories
	Rubber Lubricant
	Assembly Compound
	Joint Compound
	Retaining Ring Pliers-External
	Retaining Ring Pliers-Internal
	1/4" Lock Screw Wrench
	Eastco Float Repair Kit

Figure 4.10.2 *Downhole turbine motor design* (courtesy of Baker Hughes Co.).

Table 4.10.1 *Turbine Motor, $6\frac{3}{4}$-in. Outside Diameter, Circulation Rate 400 gpm, Mud Weight 10 lb/gal*

Number of Stages	Torque* (ft-lb)	Optimum Bit speed (rpm)	Differential Pressure (psi)	Horse-power*	Thrust Load (1000 lb)
212	1412	807	1324	217	21
318	2118	807	1924	326	30

*At optimum speed.
Courtesy of Eastman-Christensen.

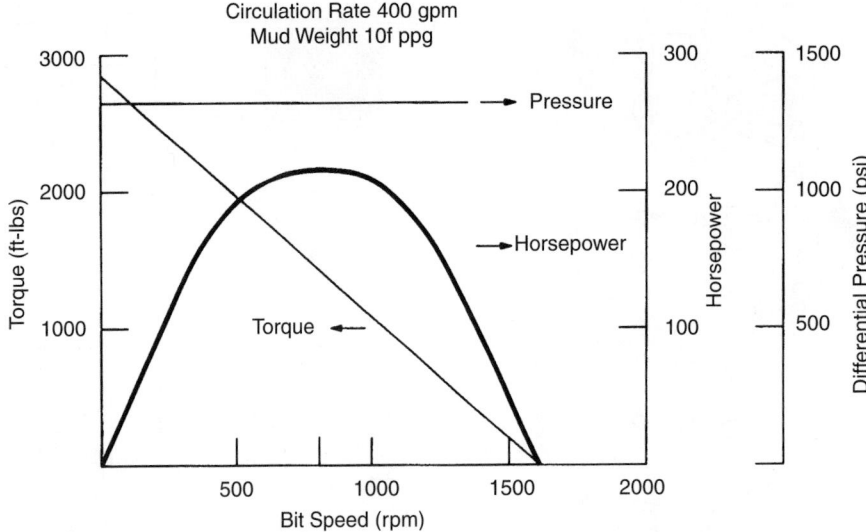

Figure 4.10.3 *Turbine motor, $6\frac{3}{4}$-in. outside diameter, two motor sections, 212 stages, 400 gal/min, 10-lb/gal mud weight (courtesy of Baker Hughes Co.).*

when a side force on the bit is present during directional drilling operations.

There are of course variations on the downhole turbine motor design, but the basic sections discussed above will be common to all designs.

The main advantages of the downhole turbine motor are:

1. Hard to extremely hard competent rock formations can be drilled with turbine motors using diamond or the new polycrystalline diamond bits.
2. Rather high rates of penetration can be achieved since bit rotation speeds are high.
3. Will allow circulation of the borehole regardless of motor horsepower or torque being produced by the motor. Circulation can even take place when the motor is installed.

The main disadvantages of the downhole turbine motor are:

1. Motor speeds and, therefore, bit speeds are high, which limits the use of roller rock bits.
2. The required flowrate through the downhole turbine motor and the resulting pressure drop through the motor require large surface pump systems, significantly larger pump systems than are normally available for most land and for some offshore drilling operations.
3. Unless a measure while drilling instrument is used, there is no way to ascertain whether the turbine motor is operating efficiently since rotation speed and/or torque cannot be measured using normal surface data (i.e., standpipe pressure, weight on bit).
4. Because of the necessity to use many stages in the turbine motor to obtain the needed power to drill, the downhole turbine motor is often quite long. Thus the ability to use these motors for high-angle course corrections can be limited.
5. Downhole turbine motors are sensitive to fouling agents in the mud; therefore, when running a turbine motor steps must be taken to provide particle-free drilling mud.
6. Downhole turbine motors can only be operated with drilling mud.

4.10.2.2 Operations

Figure 4.10.3 gives the typical performance characteristics of a turbine motor. The example in this figure is a $6\frac{3}{4}$-in. outside diameter turbine motor having 212 stages and activated by a 10-lb/gal mud flowrate of 400 gal/min.

For this example, the stall torque of the motor is 2,824 ft-lb. The runaway speed is 1,614 rpm and coincides with zero torque. The motor produces its maximum horsepower of 217 at a speed of 807 rpm. The torque at the peak horsepower is 1,412 ft-lb, or one-half of the stall torque.

A turbine device has the unique characteristic that it will allow circulation independent of what torque or horsepower the motor is producing. In the example where the turbine motor has a 10-lb/gal mud circulating at 400 gal/min, the pressure drop through the motor is about 1,324 psi. This pressure drop is approximately constant through the entire speed range of the motor.

If the turbine motor is lifted off the bottom of the borehole and circulation continues, the motor will speed up to the runaway speed of 1,614 rpm. In this situation the motor produces no drilling torque or horsepower.

As the turbine motor is lowered and weight is placed on the motor and thus the bit, the motor begins to slow its speed and produce torque and horsepower. When sufficient weight has been placed on the turbine motor, the example motor will produce its maximum possible horsepower of 217. This will be at a speed of 807 rpm. The torque produced by the motor at this speed will be 1,412 ft-lb.

If more weight is added to the turbine motor and the bit, the motor speed and horsepower output will continue to decrease. The torque, however, will continue to increase.

When sufficient weight has been placed on the turbine motor and bit, the motor will cease to rotate and the motor is described as being stalled. At this condition, the turbine motor produces its maximum possible torque. Even when the motor is stalled, the drilling mud is still circulating and the pressure drop is approximately 1,324 psi.

The stall torque M_s (ft-lb) for any turbine motor can be determined from [9].

$$M_s = 1.38386 \times 10^{-5} \frac{\eta_h \eta_m n_s \bar{\gamma}_m q^2 \tan\beta}{h} \qquad [4.10.1]$$

where η_h = hydraulic efficiency
  η_m = mechanical efficiency
  n_s = number of stages
  $\bar{\gamma}_m$ = specific weight of mud in lb/gal
  q = circulation flowrate in gal/min
  β = exit blade angle in degrees
  h = radial width of the blades in in.

Figure 4.10.2 is the side view of a single-turbine stage and describes the geometry of the motor and stator.

The runaway speed N_r (rpm) for any turbine motor can be determined from

$$N_r = 5.85 \frac{\eta_v q \tan\beta}{r_m^2 h} \qquad [4.10.2]$$

where η_v = volumetric efficiency
  r_m = mean blade radius in in.

The turbine motor instantaneous torque M (ft-lb) for any speed N (rpm) is

$$M = M_s \left(1 - \frac{N}{N_r}\right) \qquad [4.10.3]$$

The turbine motor horsepower HP (hp) for any speed is

$$HP = \frac{2\pi M_s N}{33,000} \left(1 - \frac{N}{N_r}\right) \qquad [4.10.4]$$

The maximum turbine motor horsepower is at the optimum speed, N_o, which is one-half of the runaway speed. This is

$$N_0 = \frac{N_r}{2} \qquad [4.10.5]$$

Thus, the maximum horsepower HP_{max} is

$$HP_{max} = \frac{\pi M_s N_r}{2(33,000)} \qquad [4.10.6]$$

The torque at the optimum speed M_0 is one-half the stall torque. Thus

$$M_0 = \frac{M_s}{2} \qquad [4.10.7]$$

The pressure drop Δp (psi)/through a given turbine motor design is usually obtained empirically. Once this value is known for a circulation flowrate and mud weight, the pressure drop for other circulation flowrates and mud weights can be estimated.

If the above performance parameters for a turbine motor design are known for a given circulation flowrate and mud weight (denoted as 1), the performance parameters for the new circulation flowrate and mud weight (denoted as 2) can be found by the following relationships:

Torque

$$M_2 = \left(\frac{q_2}{q_1}\right)^2 M_1 \qquad [4.10.8]$$

$$M_2 = \left(\frac{\bar{\gamma}_2}{\bar{\gamma}_1}\right) M_1 \qquad [4.10.9]$$

Speed

$$N_2 = \left(\frac{q_2}{q_1}\right) N_1 \qquad [4.10.10]$$

Power

$$HP_2 = \left(\frac{q_2}{q_1}\right)^3 HP_1 \qquad [4.10.11]$$

$$HP_2 = \left(\frac{\bar{\gamma}_2}{\bar{\gamma}_1}\right) HP_1 \qquad [4.10.12]$$

Pressure drop

$$\Delta p_2 = \left(\frac{q_2}{q_1}\right)^2 \Delta p_1 \qquad [4.10.13]$$

$$\Delta p_2 = \left(\frac{\bar{\gamma}_2}{\bar{\gamma}_1}\right) \Delta p_1 \qquad [4.10.14]$$

Table 4.10.2 gives the performance characteristics for various circulation flowrates for the 212 state, $6\frac{3}{4}$-in. outside diameter turbine motor described briefly in Table 4.10.3 and shown graphically in Figure 4.10.3.

Table 4.10.3 gives the performance characteristics for various circulation flowrates for the 318-stage, $6\frac{3}{4}$-in. outside diameter turbine motor described briefly in Table 4.10.1. Figure 4.10.4 shows the performance of the 318-stage turbine motor at a circulation flowrate of 400 gal/min and mud weight of 10 lb/gal.

The turbine motor whose performance characteristics are given in Table 4.10.2 is made up of two motor sections with 106 stages in each section. The turbine motor whose performance characteristics are given in Table 4.10.3 is made up of three motor sections.

The major reason most turbine motors are designed with various add-on motor sections is to allow flexibility when applying turbine motors to operational situations.

For straight hole drilling the turbine motor with the highest possible torque and the lowest possible speed is of most use. Thus the turbine motor is selected such that the motor produces the maximum amount of power for the lowest possible circulation flowrate (i.e., lowest speed). The high power increase rate of penetration and the lower speed increase bit life particularly if roller rock bits are used.

For deviation control drilling the turbine motor with a lower torque and the shortest overall length is needed.

Example 4.10.1

Using the basic performance data given in Table 4.10.2 for the $6\frac{3}{4}$-in. outside diameter turbine motor with 212 stages, determine the stall torque, maximum horsepower and pressure drop for this motor if only one motor section with 106 stages were to be used for a deviation control operation. Assume the same circulation flow rate of 400 gal/min, but a mud weight of 14 lb/gal is to be used.

Stall Torque. From Table 4.10.2 the stall torque for the turbine motor with 212 stages will be twice the torque value at optimum speed. Thus the stall torque for 10 lb/gal mud weight flow is

$$M_s = 2(1421)$$
$$= 2,842 \text{ ft-lb} \qquad [4.10.15]$$

From Equation 4.10.1, it is seen that stall torque is proportional to the number of stages used. Thus the stall torque for a turbine motor with 106 stages will be (for the circulation flowrate of 400 gal/min and mud weight of 10 lb/gal)

$$M_s = 2,842 \left(\frac{106}{212}\right)$$
$$= 1,421 \text{ ft-lb} \qquad [4.10.16]$$

and from Equation 4.10.9 for the 14-lb/gal mud weight

$$M_s = 1,421 \left(\frac{14}{10}\right)$$
$$= 1,989 \text{ ft-lb} \qquad [4.10.17]$$

Table 4.10.2   Turbine Motor, $6\frac{3}{4}$-in. Outside Diameter, Two Motor Sections, 212 Stages, Mud Weight 10 lb/gal

Circulation Rate (gpm)	Torque* (ft-lb)	Optimum Bit Speed (rpm)	Differential Pressure (psi)	Maximum Horsepower*	Thrust Load (1000 lb)
200	353	403	331	27	5
250	552	504	517	53	8
300	794	605	745	92	12
350	1081	706	1014	145	16
400	1421	807	1324	217	21
450	1787	908	1676	309	26
500	2206	1009	2069	424	32

*At optimum speed.

Table 4.10.3   Turbine Motor, $6\frac{3}{4}$-in. Outside Diameter, Three Motor Sections, 318 Stages, Mud Weight 10 lb/gal

Circulation Rate (gpm)	Torque* (ft-lb)	Optimum Bit Speed (rpm)	Differential Pressure (psi)	Maximum Horsepower*	Thrust Load (1000 lb)
200	529	403	485	40	8
250	827	504	758	79	12
300	1191	605	1092	137	17
350	1622	706	1486	218	23
400	2118	807	1941	326	30
450	2681	908	2457	464	38

*At optimum power.

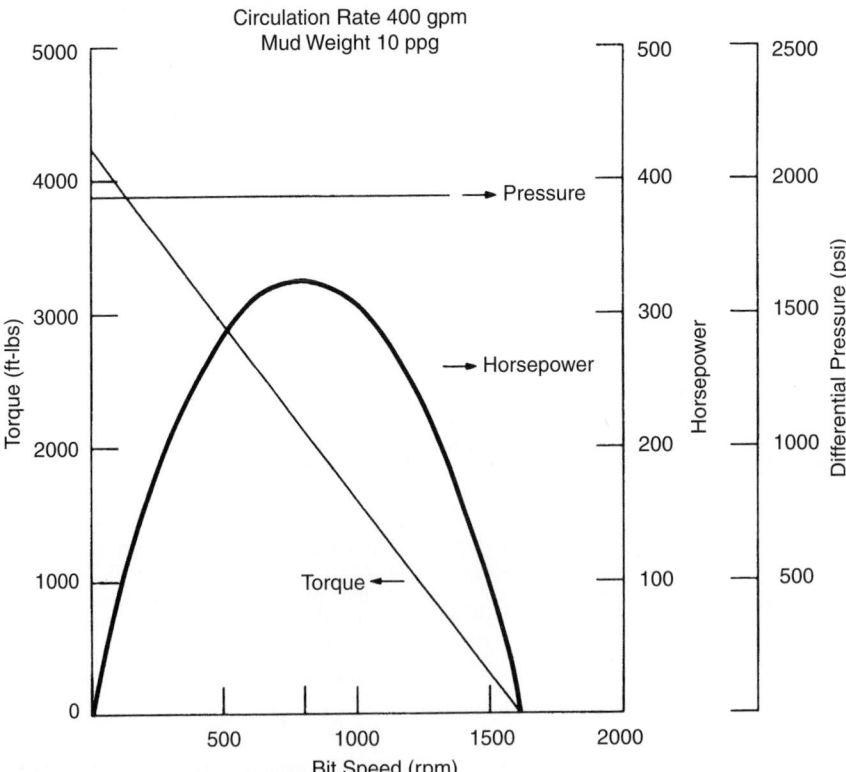

Figure 4.10.4   Turbine motor, $6\frac{3}{4}$-in. outside diameter, three motor sections, 318 stages, 480 gal/min, 10-lb/gal mud weight (courtesy of Baker Hughes Co.).

Maximum Horsepower. From Table 4.10.2 the maximum horsepower for the turbine motor with 212 stages is 217. From Equation 4.10.6, it can be seen that the maximum power is proportional to the stall torque and the runaway speed. Since the circulation flow rate is the same, the runaway speed is the same for this case. Thus, the maximum horsepower will be proportional to the stall torque. The maximum power will be (for the circulation flowrate of 400 gal/min and mud weight of 10 lb/gal)

$$\text{HP}_{\text{max}} = 217\left(\frac{1421}{2842}\right)$$

$$= 108.5 \qquad\qquad [4.10.18]$$

and from Equation 4.10.12 for the 14-lb/gal mud weight

$$\text{HP}_{\text{max}} = 108.5\left(\frac{14}{10}\right)$$

$$= 152 \qquad\qquad [4.10.19]$$

Pressure Drop. Table 4.10.2 shows that the 212-stage turbine motor has a pressure drop of 1,324 psi for the circulation flowrate of 400 gal/min and a mud weight of 10 lb/gal. The pressure drop for the 106 stage turbine motor should be roughly proportional to the length of the motor section (assuming the motor sections are nearly the same in design). Thus, the pressure drop in the 106-stage turbine motor should be proportional to the number of stage. Therefore, the pressure drop should be

$$\Delta p = 1,324\left(\frac{106}{212}\right)$$

$$= 662 \text{ psi} \qquad\qquad [4.10.20]$$

and from Equation 4.10.14 for the 14-lb/gal mud weight

$$\Delta p = 662\left(\frac{14}{10}\right)$$

$$= 927 \text{ psi} \qquad\qquad [4.10.21]$$

The last column the Tables 4.10.2 and 4.10.3 show the thrust load associated with each circulation flowrate (i.e., pressure drop). This thrust load is the result of the pressure drop across the turbine motor rotor and stator blades. The magnitude of this pressure drop depends on the individual internal design details of the turbine motor (i.e., blade angle, number of stages, axial height of blades and the radial width of the blades) and the operating conditions. The additional pressure drop results in thrust, T (lb), which is

$$T = \pi r_m^2 \Delta p \qquad\qquad [4.10.22]$$

Example 4.10.2
A $6\frac{3}{4}$-in outside diameter turbine motor (whose performance data are given in Tables 4.10.2 and 4.10.3) is to be used for a deviation control direction drilling operation. The motor will use a new $8\frac{1}{2}$-in. diameter diamond bit for the drilling operation. The directional run is to take place at a depth of 17,552 ft (measured depth). The rock formation to be drilled is classified as extremely hard, and it is anticipated that 10 ft/hr will be the maximum possible drilling rate. The mud weight is to be 16.2 lb/gal. The drilling rig has a National Supply Company, triplex mud pump Model 10-P-130 available. The details of this pump are given in Table 4.10.4 (also see the section titled "Mud Pumps" for more details). Because this is a deviation control run, the shorter two motor section turbine motor will be used.

Determine the appropriate circulation flowrate to be used for the diamond bit, turbine motor combination and the appropriate liner size to be used in the triplex pump. Also,

Table 4.10.4 *Triplex Mud Pump, Model 10-P-13, National Supply Company, Example 4.10.2*

Input Horsepower, 1300; Maximum Strokes per Minute, 140; Length of Stroke, 10 Inches

	Liner Size (inches)				
	$5\frac{1}{4}$	$5\frac{1}{2}$	$5\frac{3}{4}$	6	$6\frac{1}{4}$
Output per Stroke (gals)	2.81	3.08	3.37	3.67	3.98
Maximum Pressure (psi)	5095	4645	4250	3900	3595

prepare the turbine motor performance graph for the chosen circulation flowrate. Determine the total flow area for the diamond bit.

Bit Pressure Loss. To obtain the optimum circulation flowrate for the diamond bit, turbine motor combination, it will be necessary to consider the bit and the turbine motor performance at various circulation flowrates: 200, 300, 400 and 500 gal/min.

Since the rock formation to be drilled is classified as extremely hard, 1.5 hydraulic horsepower per square inch of bit area will be used as bit cleaning and cooling requirement [10]. The projected bottomhole area of the bit A_b (in.^{2}) is

$$A_b = \frac{\pi}{4}(8.5)^2$$

$$= 56.7 \text{ in.}^2 \qquad\qquad [4.10.23]$$

For a circulation flowrate of 200 gal/min, the hydraulic horsepower for the bit HP_b (hp) is

$$HP_b = 1.5(56.7)$$

$$= 85.05 \qquad\qquad [4.10.24]$$

The pressure drop across the bit Δp_b (psi) to produce this hydraulic horsepower at a circulation flowrate of 200 gal/min is

$$\Delta p_b = \frac{85.05(1,714)}{200}$$

$$= 729 \text{ psi} \qquad\qquad [4.10.25]$$

Similarly, the pressure drop across the bit to produce the above hydraulic horsepower at a circulation flowrate of 300 gal/min is

$$\Delta p_b = \frac{85.05(1,714)}{300}$$

$$= 486 \text{ psi} \qquad\qquad [4.10.26]$$

The pressure drop across the bit at a circulation flowrate of 400 lb/gal is

$$\Delta p_b = 364 \text{ psi} \qquad\qquad [4.10.27]$$

The pressure drop across the bit at a circulation flowrate of 500 gal/min

$$\Delta p_b = 292 \text{ psi} \qquad\qquad [4.10.28]$$

Total Pressure Loss. Using Table 4.10.2 and Equations 4.10.3 and 4.10.14, the pressure loss across the turbine motor can be determined for the various circulation flowrates and the mud weight of 16.2 lb/gal. These data together with the above bit pressure loss data are presented in Table 4.10.5. Also presented in Table 4.10.5 are the component pressure losses of the system for the various circulation flowrates considered. The total pressure loss tabulated in the lower row represents the surface standpipe pressure when operating at the various circulation flowrates.

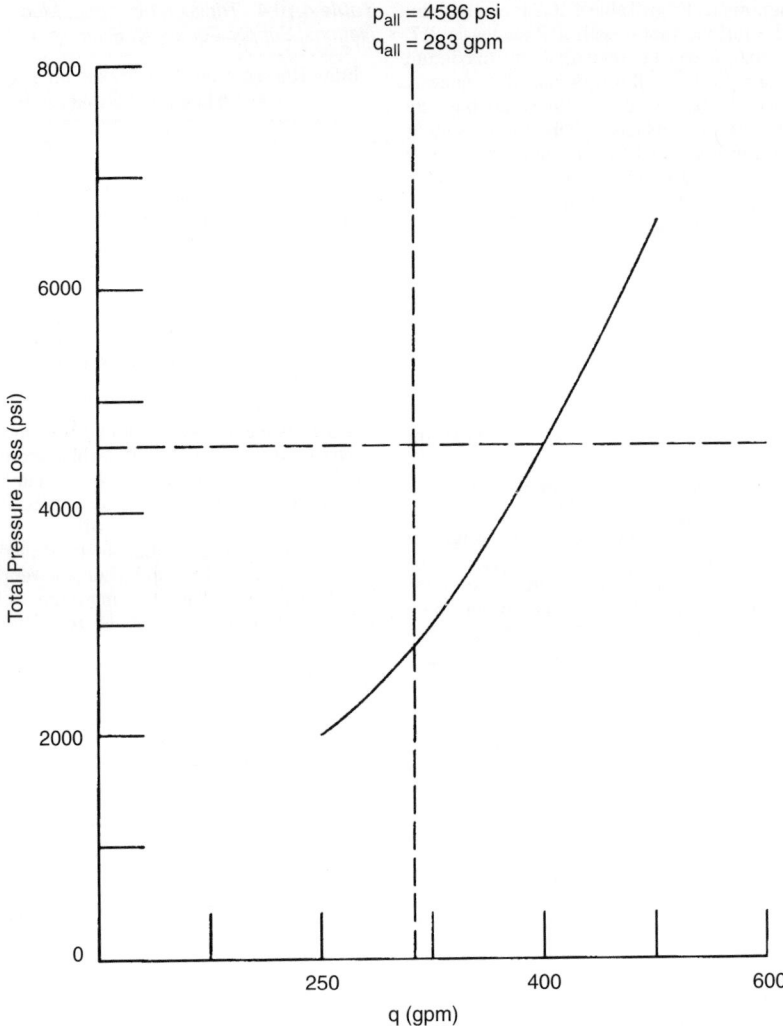

p_{all} = 4586 psi
q_{all} = 283 gpm

Figure 4.10.5 *A $5\frac{3}{4}$-in. liner, total pressure loss versus flowrate, Example 4.10.2 (courtesy of Baker Hughes Co.).*

Table 4.10.5 *Drill String Component Pressure Losses at Various Circulation Flowrates for Example 4.10.2*

Components	Pressure (psi)			
	200 gpm	300 gpm	400 gpm	500 gpm
Surface Equipment	4	11	19	31
Drill Pipe Bore	460	878	1401	2021
Drill Collar Bore	60	117	118	272
Turbine Motor	536	1207	2145	3352
Drill Bit	729	486	364	292
Drill Collar Annulus	48	91	144	207
Drill Pipe Annulus	133	248	391	561
Total Pressure Loss	1970	3038	4652	6736

The maximum pressure available for each liner size will be reduced by a safety factor of 0.90.* The maximum volumetric flowrate available for each liner size will also be reduced by a volumetric efficiency factor of 0.80 and an additional safety factor of 0.90.** Thus, from Table 4.10.4, the allowable maximum pressure and allowable maximum volume, metric flowrates will be those shown in Figures 4.10.5 through 4.10.9, which are the liner sizes $5\frac{1}{4}$, $5\frac{1}{2}$, $5\frac{3}{4}$, 6 and $6\frac{1}{4}$ in., respectively. Plotted on each of these figures are the total pressure losses for the various circulation flowrates considered. The horizontal straight line on each figure is the allowable maximum pressure for the particular liner size. The vertical straight line is the allowable maximum volumetric flowrate for the particular liner size. Only circulation flowrates that are in the lower left quadrant of the figures are practical. The highest circulation flowrate (which produces

Pump Limitations. Table 4.10.4 shows there are five possible liner sizes that can be used on the Model 10-P-130 mud pump. Each liner size must be considered to obtain the optimum circulation flowrate and appropriate liner size.

*This safety factor is not necessary for new, well-maintained equipment.
**The volumetric efficiency factor is about 0.95 for precharged pumps.

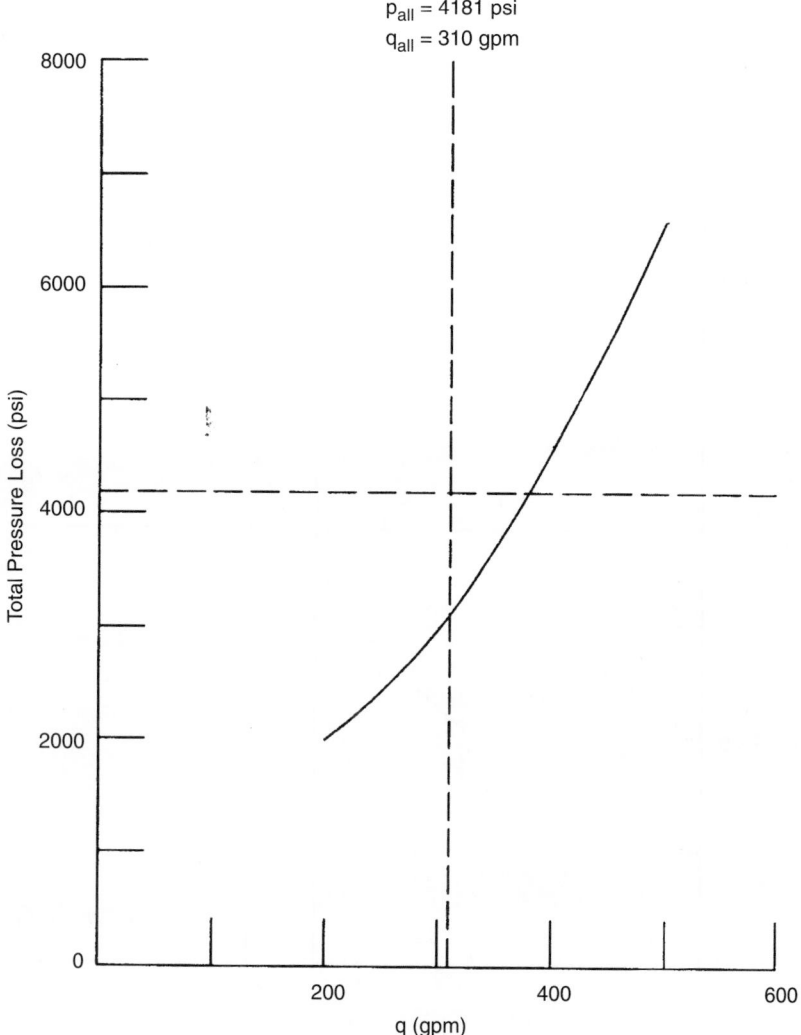

p_{all} = 4181 psi
q_{all} = 310 gpm

Figure 4.10.6 A $5\frac{1}{2}$-in. liner, total pressure loss versus flowrate, Example 4.10.2 (courtesy of Baker Hughes Co.).

the highest turbine motor horsepower) is found in Figure 4.10.7, the $5\frac{3}{4}$ in. liner. This optimal circulation flowrate is 340 gal/min.

Turbine Motor Performance. Using the turbine motor performance data in Table 4.10.2 and the scaling relationships in Equations 4.10.8 through 4.10.14, the performance graph for the turbine motor operating with a circulation flowrate of 340 gal/min and mud weight of 16.2 lb/gal can be prepared. This is given in Figure 4.10.10.

Total Flow Area for Bit. Knowing the optimal circulation flowrate, the actual pressure loss across the bit can be found as before in the above. This is

$$\Delta p_b = \frac{85.05(1,714)}{340}$$

$$= 429 \text{ psi} \qquad [4.10.29]$$

With the flowrate and pressure loss across the bit, the total flow area of the diamond bit A_{tf} (in.^{2}) can be found using [88]

$$\Delta p_b = \frac{q^2 \bar{\gamma}_m}{8795 \left(A_{tf} e^{-0.832 \frac{ROP}{N_b}} \right)^2} \qquad [4.10.30]$$

where ROP = rate of penetration in ft/hr
N_b = bit speed in rpm

The bit speed will be the optimum speed of the turbine motor, 685 rpm. The total flow area A_{tf} for the diamond bit is

$$A_{tf} = \left[\frac{(340)^2 16.2}{8795 (429)} \right]^{1/2} \frac{1}{0.9879}$$

$$= 0.713 \text{ in.}^2 \qquad [4.10.31]$$

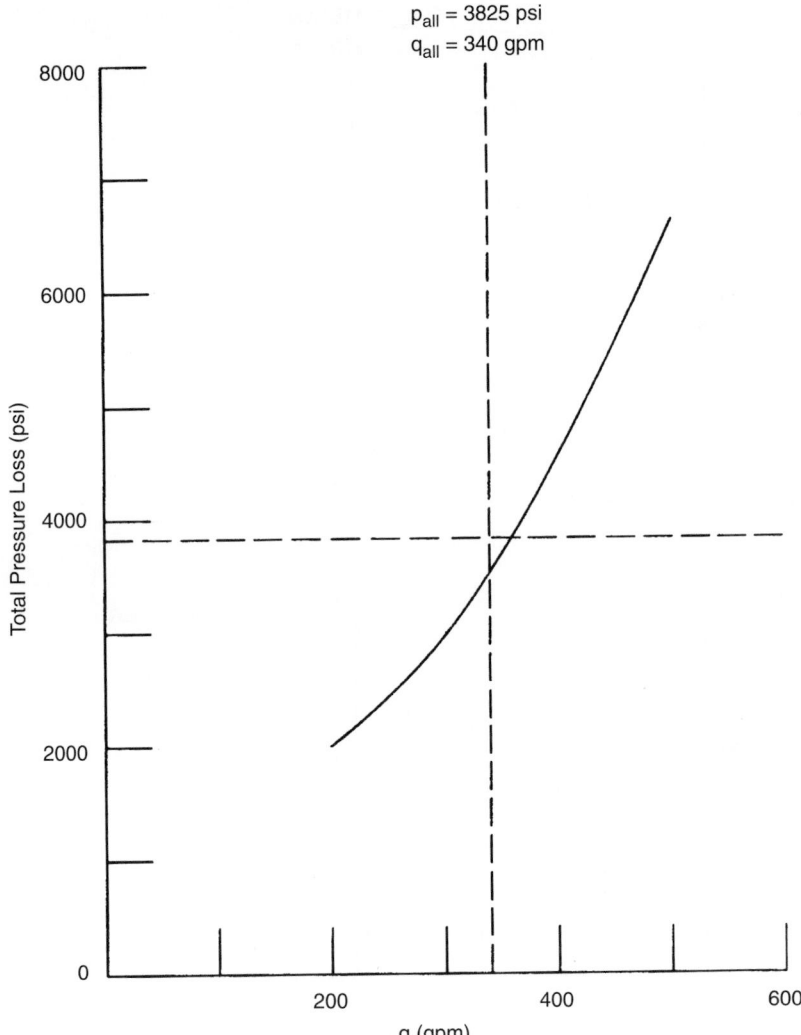

p_{all} = 3825 psi
q_{all} = 340 gpm

Figure 4.10.7 *A 5$\frac{3}{4}$-in. liner, total pressure loss versus flowrate, Example 4.10.2 (courtesy of Baker Hughes Co.).*

4.10.3 Positive Displacement Motor

Figure 4.10.11 shows the typical rigid rotor and flexible elastomer stator configuration for a single chamber of a multichambered downhole positive displacement motor section. All the positive displacement motors presently in commercial use are of Moineau type, which uses a stator made of an elastomer. The rotor is made of rigid material such as steel and is facricated in a helical shape. The activating drilling mud, freshwater, aerated mud, foam or misted air is pumped at rather high velocity through the motor section, which, because of the eccentricity of the rotor and stator configuration, and the flexibility of the stator, allows the hydraulic pressure of the flowing fluid to impart a torque to the rotor. As the rotor rotates the fluid is passes from chamber to chamber (a chamber is a lengthwise repeat of the motor). These chambers are separate entities and as one opens up to accept fluid from the preceding, the preceding closes up. This is the concept of the positive displacement motor.

4.10.3.1 Design

The rotational energy of the positive displacement motor is provided by the flowing fluid, which rotates and imparts torque to the drill bit. Figure 4.10.12 shows the typical complete downhole positive displacement motor.

In general, the downhole positive displacement motor constructed on the Moineau principle is composed of four sections: (1) the dump valve section, (2) the multistage motor section, (3) the connecting rod section and (4) the thrust and radial-bearing section. These sections are shown in Figure 4.10.12. Usually the positive displacement motor has multichambers, however, the number of chambers in a positive displacement motor is much less than the number of stages in a turbine motor. A typical positive displacement motor has from two to seven chambers.

The dump valve is a very important feature of the positive displacement motor. The positive displacement motor does not permit fluid to flow through the motor unless the motor is rotating. Therefore, a dump valve at the top of the motor

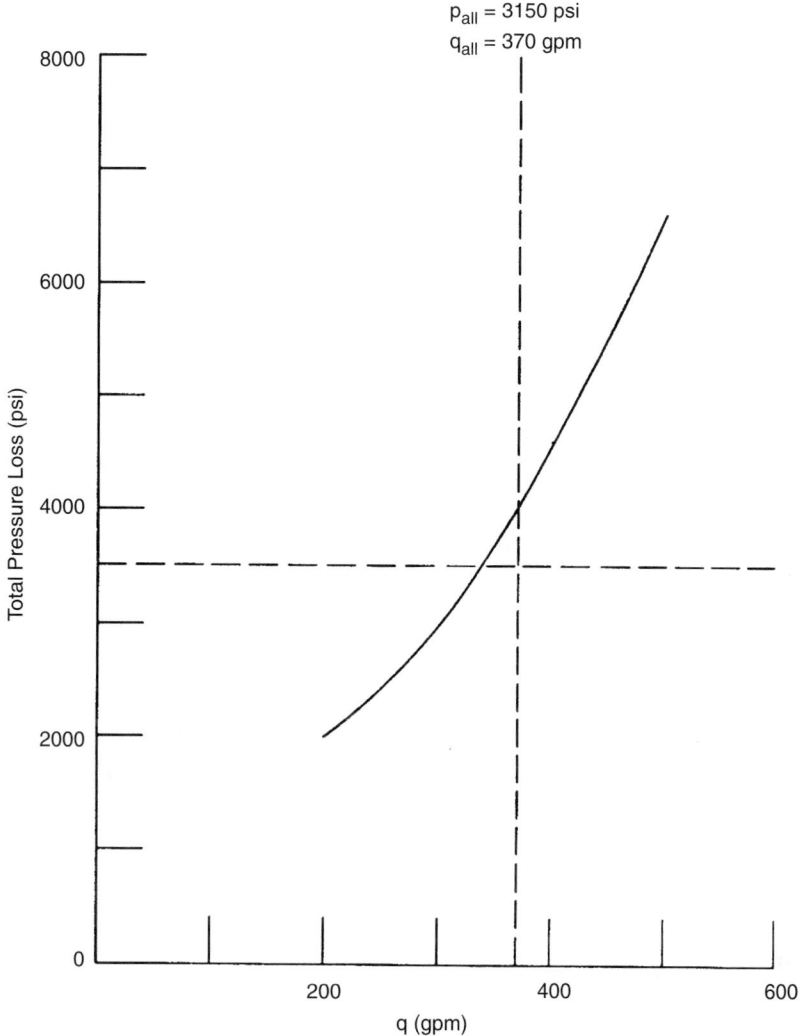

$p_{all} = 3150$ psi

$q_{all} = 370$ gpm

Figure 4.10.8 *A 6-in. liner, total pressure loss versus flowrate, Example 4.10.2 (courtesy of Baker Hughes Co.).*

allows drilling fluid to be circulated to the annulus even if the motor is not rotating. Most dump valve designs allow the fluid to circulate to the annulus when the pressure is below a certain threshold, say below 50 psi or so. Only when the surface pump is operated does the valve close to force all fluid through the motor.

The multichambered motor section is composed of only two continuous parts, the rotor and the stator. Although they are continuous parts, they usually constitute several chambers. In general, the longer the motor section, the more chambers. The stator is an elastomer tube formed to be the inside surface of a rigid cylinder. This elastomer tube stator is of a special material and shape. The material resists abrasion and damage from drilling muds containing cuttings and hydrocarbons. The inside surface of the stator is of an oblong, helical shape. The rotor is a rigid steel rod shaped as a helix. The rotor, when assembled into the stator and its outside rigid housing, provides continuous seal at contact points between the outside surface of the rotor and the inside surface of stator (see Figure

4.10.11). The rotor or driving shaft is made up of n_r lodes. The stator is made up of n_s lodes, which is equal to one lobe more than the rotor. Typical cross-sections of positive displacement motor lobe profiles are shown in Figure 4.10.13. As drilling fluid is pumped through the cavities in each chamber that lies open between the stator and rotor, the pressure of the flowing fluid causes the rotor to rotate within the stator. There are several chambers in a positive displacement motor because the chambers leak fluid. If the first chamber did not leak when operating, there would be no need for additional chambers.

In general, the larger lobe profile number ratios of a positive displacement motor, the higher the torque output and the lower the speed (assuming all other design limitations remain the same).

The rotors are eccentric in their rotation at the bottom of the motor section. Thus, the connecting rod section provides a flexible coupling between the rotor and the main drive shaft located in the thrust and radial bearing section. The main drive shaft has the drill bit connected to its bottom end.

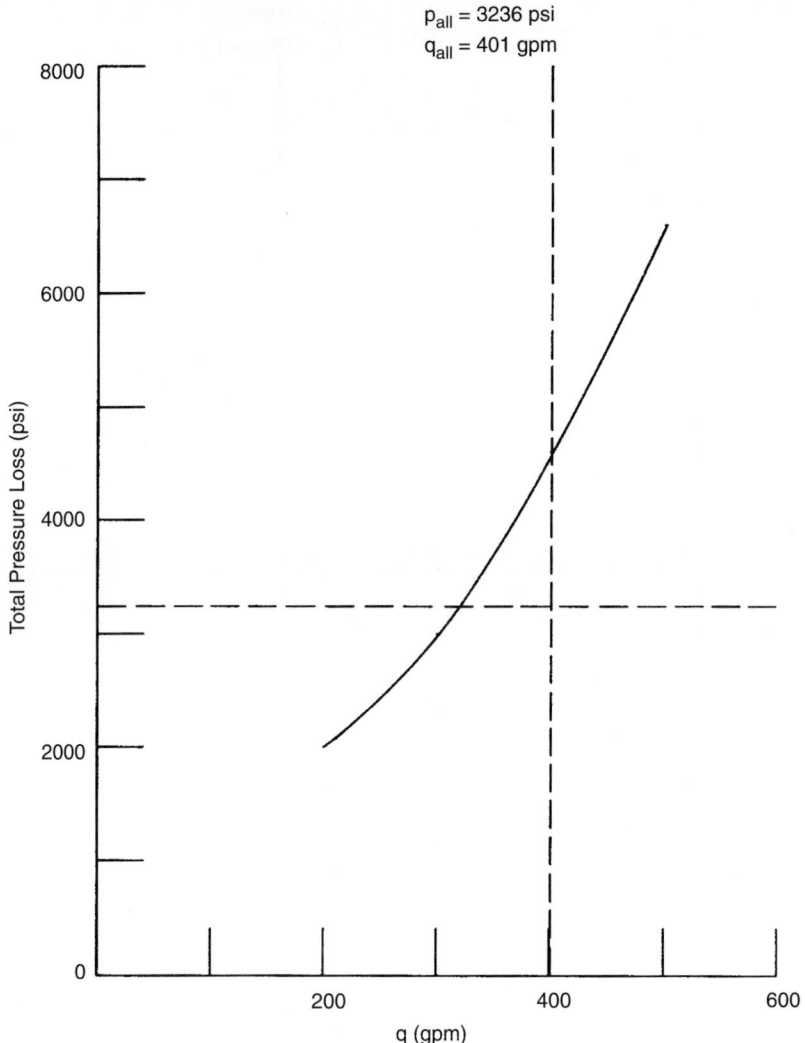

$p_{all} = 3236$ psi
$q_{all} = 401$ gpm

Figure 4.10.9 *A 6¼-in. liner, total pressure loss versus flowrate, Example 4.10.2* (courtesy of Baker Hughes Co.).

The thrust and radial-bearing section contains the thrust bearings that transfer the weight-on-bit to the outside wall of the positive displacement motor. The radial support bearings, usually located above the thrust bearings, ensure that the main drive shaft rotates about a fixed center. As in most turbine motor designs, the bearings are cooled by the drilling fluid. There are some recent positive displacement motor designs that are now using grease-packed, sealed bearing assemblies. There is usually a smaller upper thrust bearing that allows rotation of the motor while pulling out of the hole. This upper thrust bearing is usually at the upper end of thrust and radial bearing section.

There are, of course, variations on the downhole positive displacement motor design, but the basic sections discussed above will be common to all designs.

The main advantages of the downhole positive displacements motor are:

1. Soft, medium and hard rock formations can be drilled with a positive displacement motor using nearly and type of rock bit. The positive displacement motor is especially adaptable to drilling with roller rock bits.

2. Rather moderate flow rates and pressures are required to operate the positive displacement motor. Thus, most surface pump systems can be used to operate these downhole motors.

3. Rotary speed of the positive displacement motor is directly proportional to flowrate. Torque is directly proportional to pressure. Thus, normal surface instruments can be used to monitor the operation of the motor downhole.

4. High torques and low speeds are obtainable with certain positive displacement motor designs, particularly, the higher lobe profiles (see Figure 4.10.13).

5. Positive displacement motors can be operated with aerated muds, foam and air mist.

The main disadvantages of the downhole positive displacement motors are

1. When the rotor shaft of the positive displacement motor is not rotating, the surface pump pressure will rise sharply and little fluid will pass through the motor.

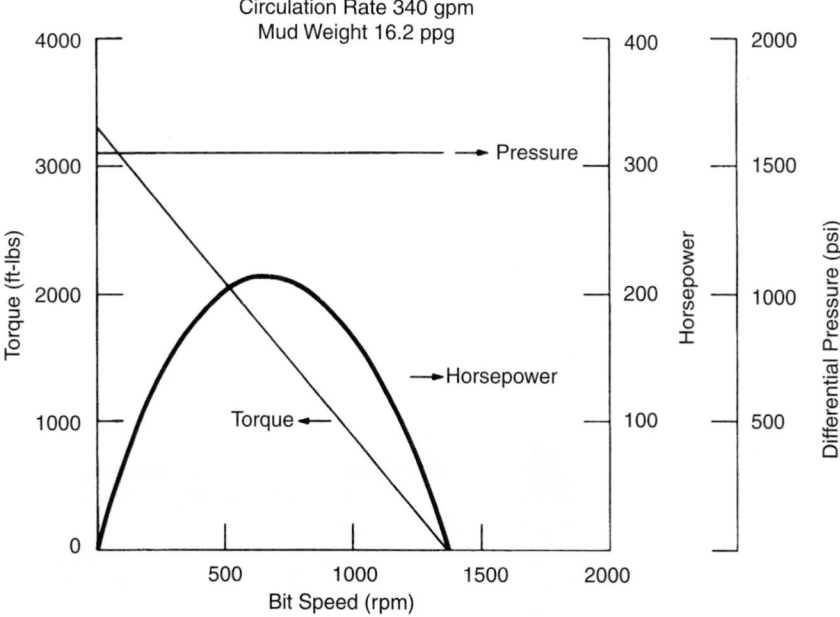

Figure 4.10.10 *Turbine motor, $6\frac{3}{4}$-in. outside diameter, two motor sections, 212 stages, 340 gal/min, 16.2-lb/gal mud weight, Example 4.10.2 (courtesy of Baker Hughes Co.).*

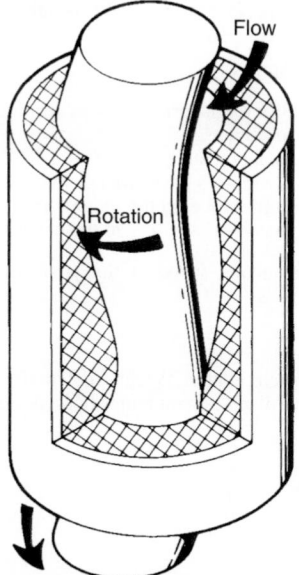

Figure 4.10.11 *Basic positive displacement motor design principle (courtesy of Smith International, Inc.).*

2. The elastomer of the stator can be damaged by high temperatures and some hydrocarbons.

4.10.3.2 Operations

Figure 4.10.14 gives the typical performance characteristics of a positive displacement motor. The example in this figure is a $6\frac{3}{4}$ in. outside diameter positive displacement motor

having five chambers activated by a 400-lb/gal flowrate of drilling mud.

For this example, a pressure of about 100 psi is required to start the rotor shaft against the internal friction of the rotor moving in the elastomer stator (and the bearings). With constant flowrate, the positive displacement motor will run at or near constant speed. Thus, this 1:2 lobe profile example motor has an motor speed of 408 rpm. The torque and the horsepower of the positive displacement motor are both linear with the pressure drop across the motor. Therefore, as more weight is placed on the drill bit (via the motor), the greater is the resisting torque of the rock. The mud pumps can compensate for this increased torque by increasing the pressure on the constant flowrate through the motor. In this example the limit in pressure drop across the motor is about 580 psi. Beyond this limit there will be either extensive leakage or damage to the motor, or both.

If the positive displacement motor is lifted off the bottom of the borehole and circulation continues, the motor will simply continue to rotate at 408 rpm. The differential pressure, however, will drop to the value necessary to overcome internal friction and rotate, about 100 psi. In this situation the motor produces no drilling torque or horsepower.

As the positive displacement motor is lowered and weight is placed on the motor and thus the bit, the motor speed continues but the differential pressure increases, resulting in an increase in torque and horsepower. As more weight is added to the positive displacement motor and bit, the torque and horsepower will continue to increase with increasing differentiated pressure (i.e., standpipe pressure). The amount of torque and power can be determined by the pressure change at the standpipe at the surface between the unloaded condition and the loaded condition. If too much weight is placed on the motor, the differential pressure limit for the motor will be reached and there will be leakage or a mechanical failure in the motor.

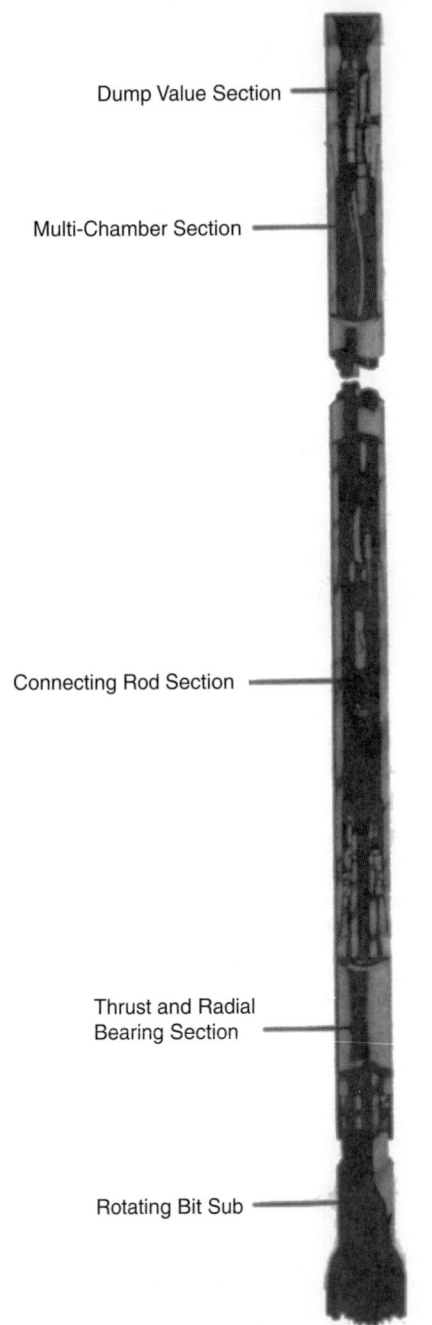

Dump Value Section

Multi-Chamber Section

Connecting Rod Section

Thrust and Radial
Bearing Section

Rotating Bit Sub

Figure 4.10.12 *Downhole positive displacement motor design (courtesy of Smith International, Inc.).*

The rotor of the Moineau-type positive displacement motor has a helical design. The axial wave number of the rotor is one less than the axial wave number for the stator for a given chamber. This allows the formation of a series of fluid cavities as the rotor rotates. The number of stator wave lengths n_s and the number of rotor wave length n_r per chamber are related by [2,9].

$$n_s = n_r + 1 \qquad [4.10.32]$$

The rotor is designed much like a screw thread. The rotor pitch is equivalent to the wavelength of the rotor. The rotor lead is the axial distance that a wave advances during one full revolution of the rotor. The rotor pitch and the stator pitch are equal. The rotor lead and stator lead are proportional to their respective number of waves. Thus, the relationship between rotor pitch t_r (in.) and stator pitch, t_s (in.) is [9]

$$t_r = t_s \qquad [4.10.33]$$

The rotor lead L_r (in.) is

$$L_r = n_r t_r \qquad [4.10.34]$$

The stator lead L_s (in.) is

$$L_s = n_s t_s \qquad [4.10.35]$$

The specific displacement per revolution of the rotor is equal to the cross-sectional area of the fluid multiplied by the distance the fluid advances. The specific displacement s (in.^{3}) is

$$s = n_r n_s t_r A \qquad [4.10.36]$$

Where A is the fluid cross-sectional area (in.^{2}). The fluid cross-sectional area is approximately

$$A \approx 2 n e_r^2 (n_r^2 - 1) \qquad [4.10.37]$$

where e_r is the rotor rotation eccentricity (in.). The special case of a 1:2 lobe profile motor has a fluid cross-sectional area of

$$A \approx 2 e_r d_r \qquad [4.10.38]$$

where d_r is the reference diameter of the motor (in.). The reference diameter is

$$d_r = 2 e_r n_s \qquad [4.10.39]$$

For the 1:2 lobe profile motor, the reference diameter is approximately equal to the diameter of the rotor shaft.

The instantaneous torque of the positive displacement motor M (ft-lb) is

$$M = 0.0133 \, s \, \Delta p \eta \qquad [4.10.40]$$

where Δp = differential pressure loss through the motor in psi

η = total efficiency of the motor. The 1:2 lobe profile motors have efficiencies around 0.80. The higher lobe profile motors have efficiencies that are lower (i.e., of the order of 0.70 or less)

The instantaneous speed of the positive displacement motor N (rpm) is

$$N = \frac{231.016 q}{s} \qquad [4.10.41]$$

where q is the circulation flowrate (gal/min).

The positive displacement motor horsepower HP (hp) for any speed is

$$HP = \frac{q \Delta p}{1,714} \eta \qquad [4.10.42]$$

The number of positive displacement motor chambers n_c is

$$n_c = \frac{L}{t_s} - (n_s - 1) \qquad [4.10.43]$$

where L is the length of the actual motor section (in.).

The maximum torque M_{max} will be at the maximum differential pressure Δp_{max}, which is

$$M_{max} = 0.133 s \Delta p_{max} \eta \qquad [4.10.44]$$

The maximum horse power HP_{max} will also be at the maximum differential pressure Δp_{max}, which is

$$HP_{max} = \frac{q \Delta p_{max}}{1,714} \eta \qquad [4.10.45]$$

It should be noted that the positive displacement motor performance parameters are independent of the drilling

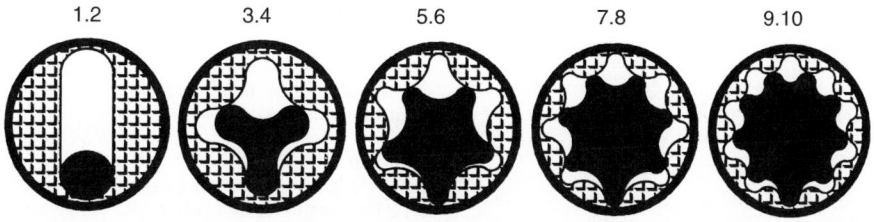

Figure 4.10.13 *Typical positive displacement motor lobe profiles* (courtesy of Smith International, Inc.).

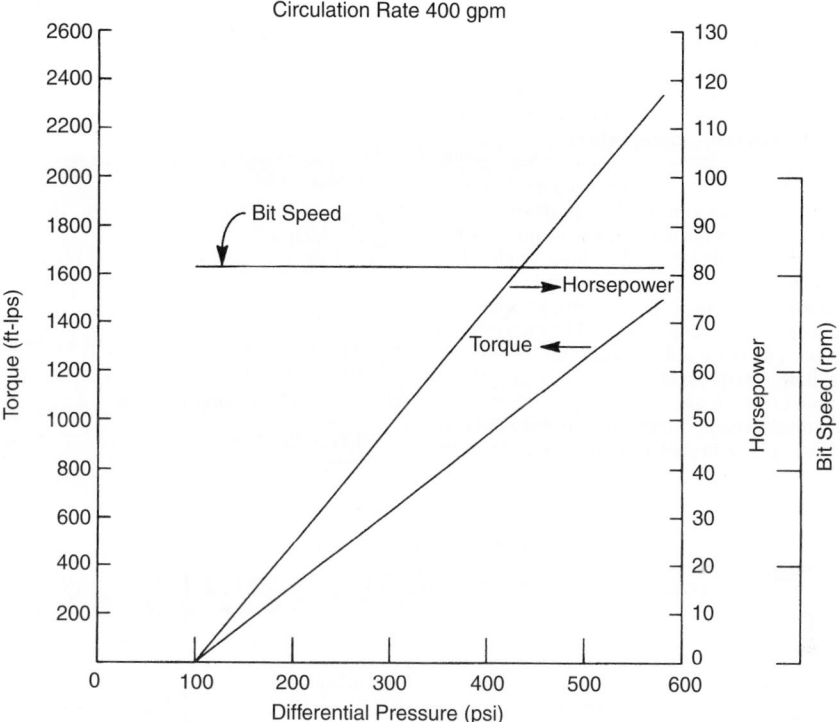

Figure 4.10.14 *Positive displacement motor, $6\frac{3}{4}$-in. outside diameter, 1:2 lobe profile, 400 gal/min, differential pressure limit 580 psi* (courtesy of Baker Hughes Co.).

mud weight. Thus, these performance parameters will vary with motor design values and the circulation flowrate.

If the above performance parameters for a positive displacement motor design are known for a given circulation flowrate (denoted as 1), the performance parameters for the new circulation flowrate (denoted as 2) can be found by the following relationships:

Torque

$$M_2 = M_1 \qquad [4.10.46]$$

Speed

$$N_2 = \left(\frac{q_2}{q_1}\right)N_1 \qquad [4.10.47]$$

Power

$$HP_2 = \left(\frac{q_2}{q_1}\right)HP_1 \qquad [4.10.48]$$

Table 4.10.6 gives the performance characteristics for various circulation flowrates for the 1:2 lobe profile $6\frac{3}{4}$-in. outside diameter positive displacement motor. Figure 4.10.14 shows

Table 4.10.6 *Positive Displacement Motor, $6\frac{3}{4}$-in. Outside Diameter, 1:2 Lobe Profile, Five Motor Chambers*

Circulation Rate (gpm)	Speed (rpm)	Maximum Differential Pressure (psi)	Maximum Torque (ft-lb)	Maximum Horse-power
200	205	580	1500	59
250	255	580	1500	73
300	306	580	1500	87
350	357	580	1500	102
400	408	580	1500	116
450	460	580	1500	131
500	510	580	1500	145

the performance of the 1:2 lobe profile positive displacement motor at a circulation flowrate of 400 gal/min.

Table 4.10.7 gives the performance characteristics for various circulation flowrates for the 5:6 lobe profile,

Table 4.10.7 *Positive Displacement Motor, $6\frac{3}{4}$-in. Outside Diameter, 5:6 Lobe Profile, Five Motor Chambers*

Circulation Rate (gpm)	Speed (rpm)	Maximum Differential Pressure (psi)	Maximum Torque (ft-lb)	Maximum Horse-power
200	97	580	2540	47
250	122	580	2540	59
300	146	580	2540	71
350	170	580	2540	82
400	195	580	2540	94

Courtesy of Eastman-Christensen.

$6\frac{3}{4}$-in. outside diameter positive displacement motor. Figure 4.10.15 shows the performance of the 5:6 lobe profile positive displacement motor at a circulation flow rate of 400 gal/min.

The positive displacement motor whose performance characteristics are given in Table 4.10.6 is a 1:2 lobe profile motor. This lobe profile design is usually used for deviation control operations. The 1:2 lobe profile design yields a downhole motor with high rotary speeds and low torque. Such a combination is very desirable for the directional driller. The low torque minimizes the compensation that must be made in course planning which must be made for the reaction torque in the lower part of the drill string. This reactive torque when severe can create difficulties in deviation control planning. The tradeoff is, however, that higher speed reduces the bit life, especially roller rock bit life.

The positive displacement motor whose performance characteristics are given in Table 4.10.7 is a 5:6 lobe profile motor. This lobe profile design is usually used for straight hole drilling with roller rock bits, or for deviation control operations where high torque polycrystalline diamond compact bit or diamond bits are used for deviation control operations.

Example 4.10.3

A $6\frac{3}{4}$-in. outside diameter positive displacement motor of a 1:2 lobe profile design (where performance data are given in Table 4.10.4) has rotor eccentricity of 0.60 in., a reference diameter (rotor shaft diameter) of 2.48 in. and a rotor pitch of 38.0 in. If the pressure drop across the motor is determined to be 500 psi at a circulation flowrate of 350 gal/min with 12.0 lb/gal, find the torque, rotational speed and the horsepower of the motor.

Torque. Equation 4.10.37 gives the fluid cross-sectional area of the motor, which is

$$A = 2(0.6)(2.48)$$

$$= 2.98 \text{ in.}^2 \qquad [4.10.49]$$

Equation 4.10.37 gives the specific displacement of the motor, which is

$$s = (1)(2)(38.0)(2.98)$$

$$= 226.5 \text{ in.}^3 \qquad [4.10.50]$$

The torque is obtained from Equation 4.10.40, assuming an efficiency of 0.80 for the 1:2 lobe profile motor. This is

$$M = 0.0133(226.5)(500)(0.80)$$

$$= 1205 \text{ ft-lb} \qquad [4.10.51]$$

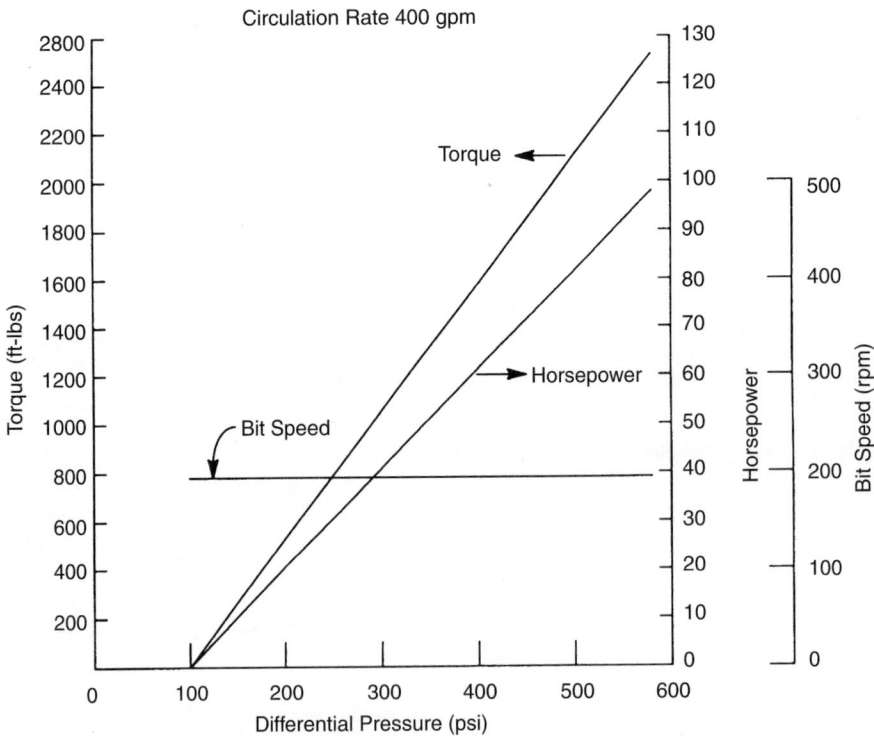

Figure 4.10.15 *Positive displacement motor, $6\frac{3}{4}$-in. outside diameter, 5:6 lobe profile, 400 gal/min, differential pressure limit 580 psi (courtesy of Baker Hughes Co.).*

Speed. The rotation speed is obtained from Equation 4.10.41. This is

$$N = \frac{231.016(350)}{226.5}$$

$$= 357 \text{ rpm} \qquad [4.10.52]$$

Horsepower. The horsepower the motor produces is obtained from Equation 4.10.42. This is

$$HP = \frac{350(500)}{1714}(0.80)$$

$$= 82 \qquad [4.10.53]$$

Planning for a positive displacement motor run and actually drilling with such a motor is easier than with a turbine motor. This is mainly due to the fact that when a positive displacement motor is being operated, the operator can know the operating torque and rotation speed via surface data. The standpipe pressure will yield the pressure drop through the motor, thus the torque. The circulation flowrate will yield the rotational speed.

Example 4.10.4

A $6\frac{3}{4}$-in. outside diameter positive displacement motor (whose performance data are given in Tables 4.10.6 and 4.10.7) is to be used for a deviation control direction drill operation. The motor will use an $8\frac{1}{2}$-in. diameter roller rock bit for the drilling operation. The directional run is to take place at a depth of 10,600 ft (measured depth). The rock formation to be drilled is classified as medium, and it is anticipated that 30 ft/hr will be the maximum possible drilling rate. The mud weight is to be 11.6 lb/gal. The drilling rig has a National Supply Company duplex mud pump Model E-700 available. The details of this pump are given in Table 4.10.8 (also see the section titled "Mud Pumps"). Because this is a deviation control run, the 1:2 lobe profile positive displacement motor will be used since it has the lowest torque for a given circulation flowrate (see Table 4.10.6). Determine the appropriate circulation flowrate to be used for the roller rock bit, positive displacement motor combination and the appropriate liner size to be used in the duplex pump. Also, prepare the positive displacement motor performance graph for the chosen circulation flowrate. Determine the bit nozzle sizes.

Table 4.10.8 *Duplex Mud Pump, Model E-700, National Supply Company, Example 4.10.4*

Input Horsepower, 825; Maximum Strokes per Minute, 65; length of Stroke, 16 Inches

	$5\frac{3}{4}$	6	$6\frac{1}{4}$	$6\frac{1}{2}$	$6\frac{3}{4}$	7
Output per Stroke (gals)	6.14	6.77	7.44	8.13	8.85	9.60
Maximum Pressure (psi)	3000	2450	2085	1780	1535	1260

Bit Pressure Loss. It is necessary to choose the bit pressure loss such that the thrust load created in combination with the weight on bit will yield an on-bottom load on the motor thrust bearings, which is less than the maximum allowable load for the bearings. Since this is a deviation control run and, therefore, the motor will be drilling only a relatively short time and distance, the motor thrust bearings will be operated at their maximum rated load for on-bottom operation. Figure 4.10.16 shows that maximum allowable motor thrust bearing load is about 6,000 lb. To have the maximum weight on bit, the maximum recommended bit pressure loss of 500 psi will be used. This will give maximum weight on bit of about 12,000 lb. The higher bit pressure loss will, of course, give the higher cutting face cleaning via jetting force (relative to the lower recommended bit pressure losses).

Total Pressure Loss. Since bit life is not an issue in a short deviation control motor run operation, it is desirable to operate the positive displacement motor at as high a power level as possible during the run. The motor has a maximum pressure loss with which it can operate. This is 580 psi (see Table 4.10.6). It will be assumed that the motor will be operated at the 580 psi pressure loss in order to maximum the torque output of the motor. To obtain the highest horsepower for the motor, the highest circulation flowrate possible while operating within the constraints of the surface mud pump should be obtained. To obtain this highest possible, or optimal, circulation flowrate, the total pressure losses for the circulation system must be obtained for various circulation flowrates. These total pressure losses tabulated in

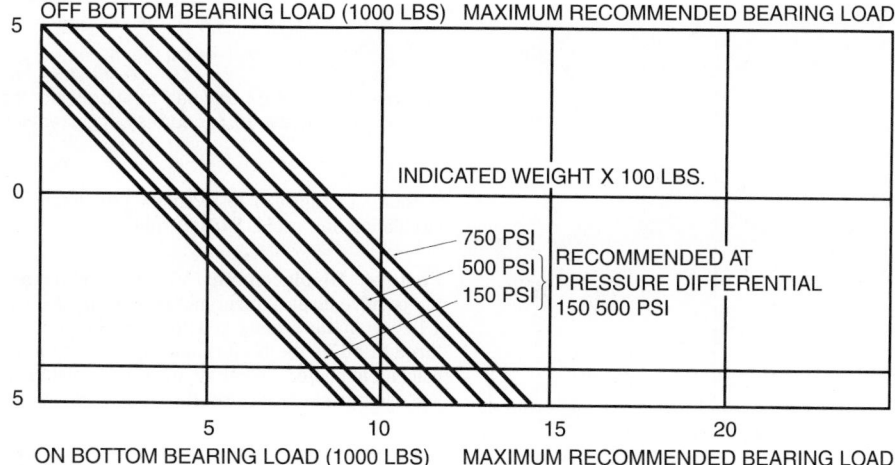

Figure 4.10.16 *Hydraulic thrust and indicated weight balance for positive displacement motor* (courtesy of Smith International, Inc.).

Table 4.10.9 *Drillstring Component Pressure Losses at
Various Circulation Flowrates for Example 4.10.4*

Components	Pressure (psi)			
	200 gpm	300 gpm	400 gpm	500 gpm
Surface Equipment	5	11	19	30
Drill Pipe Bore	142	318	566	884
Drill Collar Bore	18	40	71	111
PDM	580	580	580	580
Drill Bit	500	500	500	500
Drill Collar Annulus	11	25	45	70
Drill Pipe Annulus	32	72	128	200
Total Pressure Loss	1288	1546	1909	2375

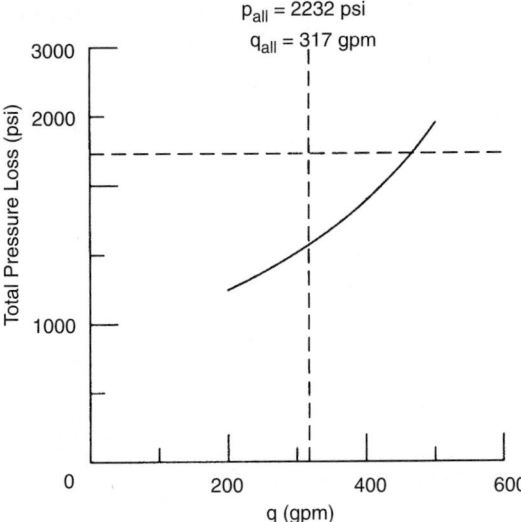

Figure 4.10.18 *A 6-in. liner, total pressure loss vs.
flowrate, Example 4.10.4 (courtesy of Baker Hughes Co.).*

the lower row of Table 4.10.9 represent the surface standpipe
pressure when operating at the various circulation flowrates.

Pump Limitations. Table 4.10.8 shows there are six possible liner sizes that can be used on the Model E-700 mud
pump. Each liner size must be considered to obtain the
optimum circulation flowrate and a appropriate liner size.
The maximum pressure available for each liner size will
be reduced by a safety factor of 0.90. The maximum volumetric flowrate available for each liner size will also be
reduced by a volumetric efficiency factor of 0.80 and an
additional safety factor of 0.90. Thus, from Table 4.10.8,
the allowable maximum pressures and allowable maximum
volumetric flowrates will be those shown in Figures 4.10.17
through 4.10.22 which are the liner sizes $5\frac{3}{4}$, 6, $6\frac{1}{2}$ and 7 in.,
respectively. Plotted on each of these figures are the total
pressure losses for the various circulation flowrates considered. The horizontal straight line on each figure is the
allowable maximum pressure for the particular liner size.
The vertical straight line is the allowable maximum volumetric flowrate for the particular liner size. Only circulation
flowrates that are in the lower left quadrant of the figures
are practical. The highest circulation flowrate (which produces the highest positive displacement motor horsepower)

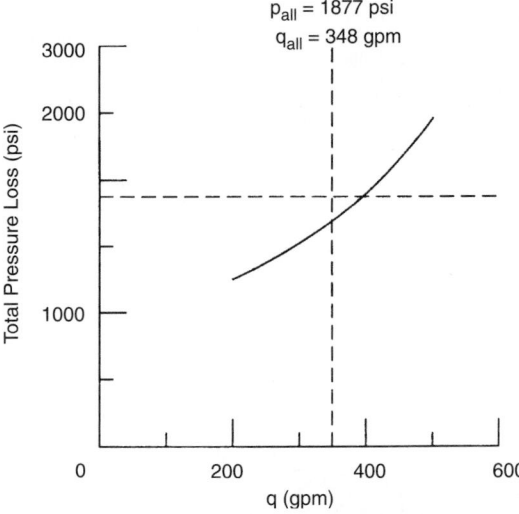

Figure 4.10.19 *A $6\frac{1}{4}$-in. liner, total pressure loss vs.
flowrate, Example 4.10.4 (courtesy of Baker Hughes Co.).*

is found in figure 4.10.19, the $6\frac{1}{4}$-in. liner. The optimal
circulation flow rate is 348 gal/min.

Positive Displacement Motor Performance. Using
the positive displacement motor performance data in Table
4.10.6 and the scaling relationships in Equations 4.10.46
through 4-170, the performance graph for the positive displacement motor operating with a circulation flowrate of
348 gal/min can be prepared. This is given in Figure 4.10.3.

Bit Nozzle Sizes. The pressure loss through the bit must
be 500 psi with a circulation flowrate of 348 gal/min with
11.6-lb/gal mud weight. The pressure loss through a roller
rock bit with three nozzles is (see the section titled "Drilling

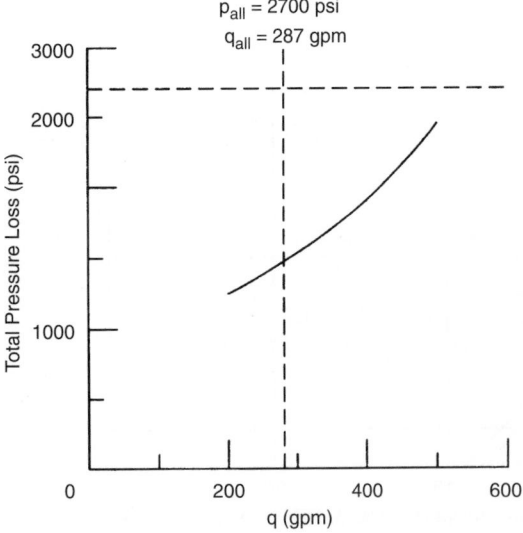

Figure 4.10.17 *A $5\frac{3}{4}$-in. liner, total pressure loss vs.
flowrate, Example 4.10.4 (courtesy of Baker Hughes Co.).*

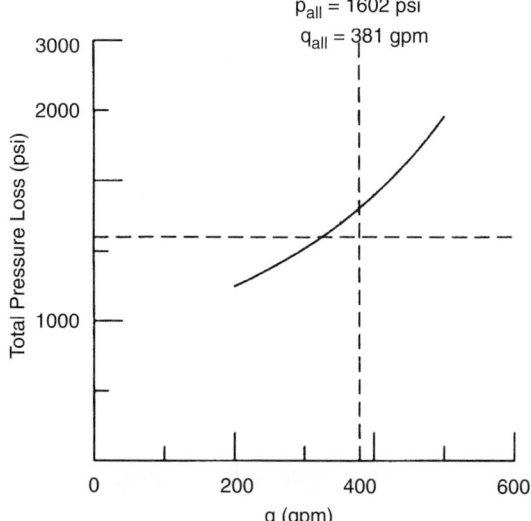

Figure 4.10.20 *A $6\frac{1}{2}$-in. liner, total pressure loss vs. flowrate, Example 4.10.4 (courtesy of Baker Hughes Co.).*

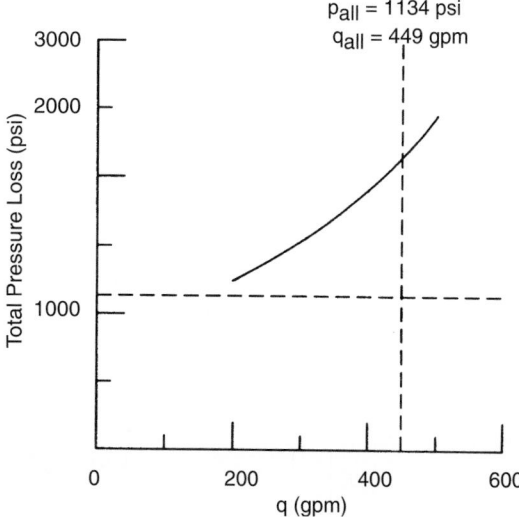

Figure 4.10.22 *A 7-in. liner, total pressure loss vs. flowrate, Example 4.10.4 (courtesy of Baker Hughes Co.).*

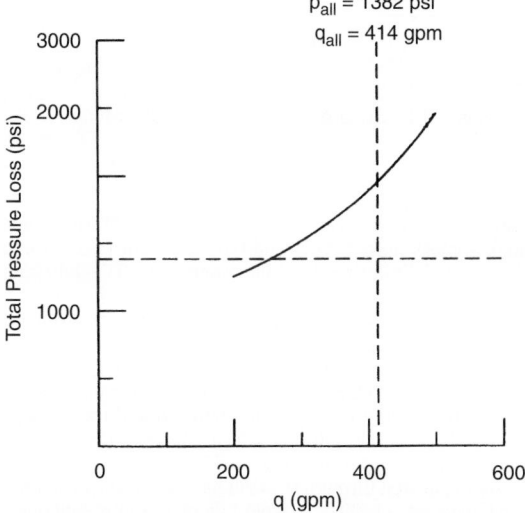

Figure 4.10.21 *A $6\frac{3}{4}$-in. liner, total pressure loss vs. flowrate, Example 4.10.4 (courtesy of Baker Hughes Co.).*

Bits and Downholes Tools")

$$\Delta p_b = \frac{q^2 \bar{\gamma}_m}{7430 C^2 d^4} \qquad [4.10.54]$$

where C = nozzle coefficient (usually taken to be 0.95)
  d_e = hydraulic equivalent diameter in in.

Therefore, Equation 4.10.54 is

$$500 = \frac{(348)^2 (11.6)}{7,430 (0.98)^2 de^4} \qquad [4.10.55]$$

which yields

$$d_e = 0.8045 \text{ in.} \qquad [4.10.56]$$

The hydraulic equivalent diameter is related to the actual nozzle diameters by

$$d_e = \left[ad_1^2 + bd_2^2 + cd_3^2\right]^{1/2} \qquad [4.10.57]$$

where   a = number of nozzles with diameter d_1
  b = number of nozzles with diameter d_2
  c = number of nozzles with diameter d_3
  d_1, d_2 and d_3 = three separate nozzle diameters in in.

Nozzle diameters are usually in 32nds of an inch. Thus, if the bit has three nozzles with $\frac{15}{32}$ of an inch diameter, then

$$d_e = [(3)(0.4688)^2]^{1/2}$$
$$= 0.8120 \text{ in.} \qquad [4.10.58]$$

The above hydraulic equivalent diameter is close enough to the one obtained with Equation 4.10.54. Therefore, the bit should have three $\frac{5}{32}$-in. diameter nozzles.

4.10.4 Down the Hole Air Hammers
4.10.4.1 Design
There are two basic designs for the downhole air hammer. One design uses a flow path of the compressed air through a control rod (or feed tube) down the center hammer piston (or through passages in the piston) and then through the hammer bit. The other design uses a flow path through a housing annulus passage (around the piston) and then through the hammer bit. Figure 4.10.24 shows a schematic of a typical control rod flow design downhole air hammer. The hammer action of the piston on the top of the drill bit shank provides an impact force that is transmitted down the shank to the bit studs, which, in turn, crush the rock at the rock face. In shallow boreholes where there is little annulus back pressure, the piston impacts the top of the bit shank at a rate of from about 600 to 1,700 strikes per minute (depending on volumetric flow rate of gas). However, in deep boreholes where the annulus back pressure is usually high, impacts can be as low as 100 to 300 strikes per minute. This unique air drilling device requires that the drill string with the air hammer be rotated so that the drill bit studs impact over the entire rock face. In this manner an entire layer of the rock face can be destroyed and the drill bit advanced.

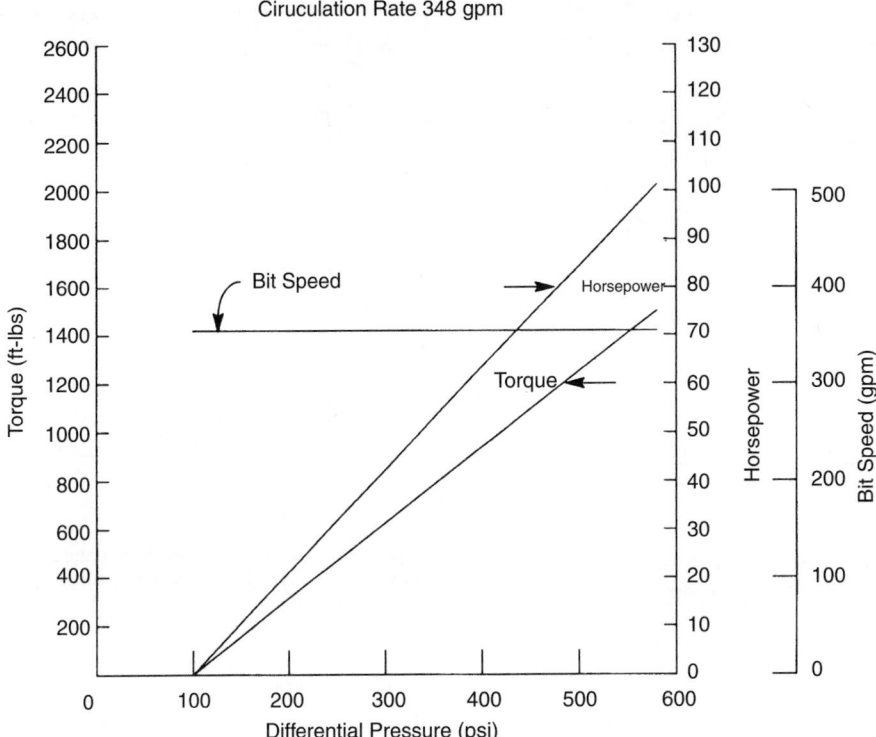

Figure 4.10.23 *Positive displacement motor, $6\frac{3}{4}$-in. outside diameter, 1:2 lobe profile, 348 gal/min, differential pressure limit 580 psi, Example 4.10.4 (courtesy of Baker Hughes Co.).*

Figure 4.10.24 shows the air hammer suspended from a drill string lifted off the bottom (the shoulder of the bit is not in contact with the shoulder of the driver sub). In this position, compressed air flows through the pin connection at the top of the hammer to the bit without actuating the piston action (i.e., to blow the borehole clean). When the hammer is placed on the bottom of the borehole and weight placed on the hammer, the bit shank will be pushed up inside the hammer housing until the bit shoulder is in contact with the shoulder of the driver sub. This action aligns one of the piston ports (of one of the flow passage through the piston) with one of the control rod windows. This allows the compressed air to flow to the space below the piston which in turn forces the piston upward in the hammer housing. During this upward stroke of the piston, no air passes through the bit shank to the rock face. In essence, rock cuttings transport is suspended during this upward stroke of the piston.

When the piston reaches the top of its stroke, another one of the piston ports aligns with one of the control rod windows and supplies compressed air to the open space above the piston. This air flow forces the piston downward until it impacts the top of the bit shank. At the same instant the air flows to the space above the piston, the foot valve at the bottom of the control rod opens and air inside the drill string is exhaust through the control rod, bit shank, and the bit orifices to the rock face. This compressed air exhaust entrains the rock cuttings created by the drill bit for transport up the annulus to the surface. This impact force on the bit allows the rotary action of the drill bit to be very effective in destroying rock at the rock face. This in turn allows the air hammer to drill with low WOB. Typically for a $6\frac{3}{4}$ inch outside diameter air hammer drilling with a $7\frac{7}{8}$ inch air

hammer bit, the WOB can be as low as 1,500 lb. Downhole air hammers must have an oil type lubricant injected into the injected air during the drilling operation. This lubricant is needed to lubricate the piston surfaces as it moves in the hammer housing. Air hammers are used exclusively for vertical drilling operations.

This piston cycle is repeated as the drill string (and thus the drill bit) is rotated and in this manner continuous layers of rock on the rock face are crushed and the cuttings removed. The air flow through the air hammer is not continuous. When the piston is being lifted in the hammer, air is not being exhausted through the drill bit. For example, at a piston impact rate of 600 strikes per minute the air is shut down for about 0.050 seconds per cycle. This is so short a time that the air flow rate through the annulus can be assumed as a continuous flow.

Downhole air hammers are available in housing outside diameters from 3 inches to 16 inches. The 3 inch housing outside diameter hammer can drill a borehole as small as $3\frac{5}{8}$ inches. The 16 inch housing outside diameter hammer can drill a boreholes from $17\frac{1}{2}$ inches to 33 inches. For shallow drilling operations, conventional air hammer bits are adequate. For deep drilling operations (usually oil and gas recovery wells), higher quality oil field air hammer drill bits are required.

There are a variety of manufacturers of downhole air hammers. These manufacturers use several different designs for their respective products. The air hammer utilizes very little power in moving the piston inside the hammer housing. For example, a typical $6\frac{3}{4}$ inch outside diameter air hammer with a 77 lb piston operating at about 600 strikes per minute, will use less than 2 horsepower driving the piston. This is a very

HAMMER

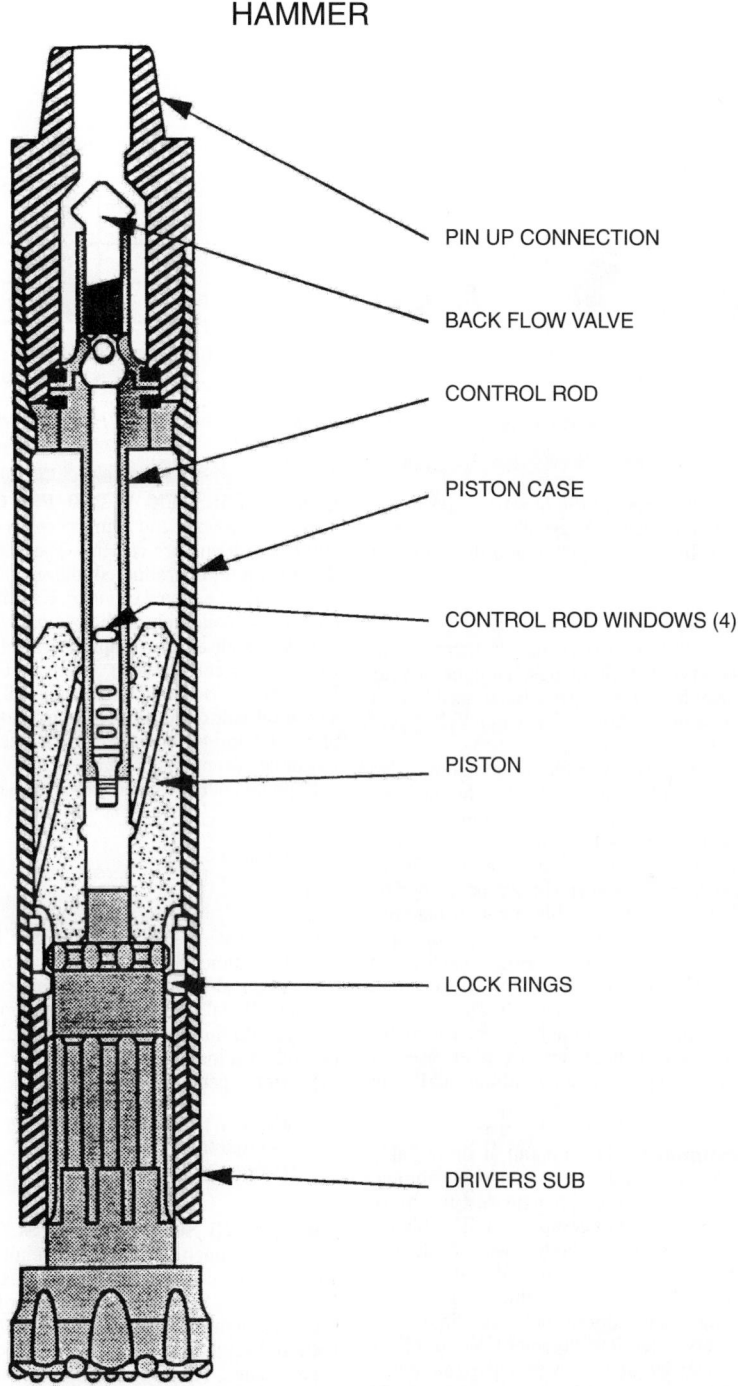

PIN UP CONNECTION

BACK FLOW VALVE

CONTROL ROD

PISTON CASE

CONTROL ROD WINDOWS (4)

PISTON

LOCK RINGS

DRIVERS SUB

Figure 4.10.24 *Schematic cutaway of a typical air hammer* (courtesy of Diamond Air Drilling Services, Inc.).

small amount of power relative to the total needed for the actual rotary drilling operation. Thus, it is clear that the vast majority of the power to the drill string is provided by the rotary table. Therefore, any pressure loss (i.e., energy loss) due to the piston lifting effort can usually be ignored. The major pressure loss in the flow through an air hammer is due

to the flow energy losses from the constrictions in the flow path when the air is allowed to exit the hammer (on the down stroke of the piston). All air hammer designs have internal flow constrictions. These flow constrictions can be used to model the flow losses through the hammer. In most designs these constrictions can be approximately represented by a

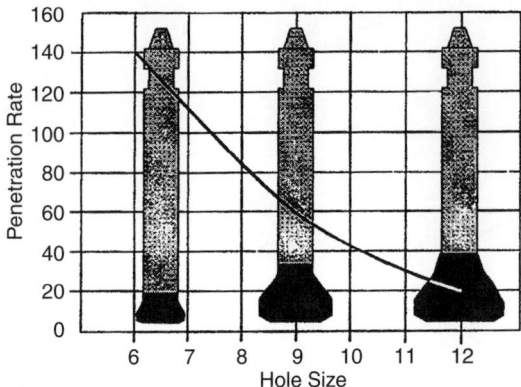

Figure 4.10.25   *Poor Hammer/Bit Ratio. (Courtesy of DrilMaster.)*

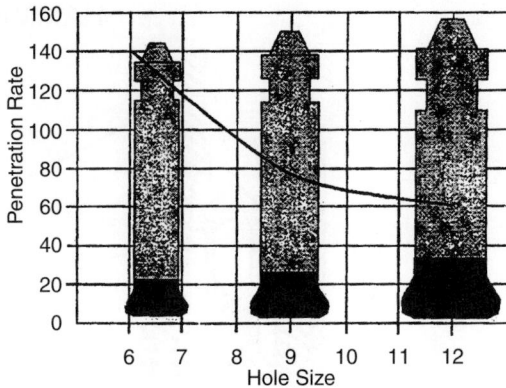

Figure 4.10.26   *Correct Hammer/Bit Ratio. (Courtesy of DrilMaster.)*

set of internal orifice diameters in the flow passages to the drill bit. These internal orifices are usually the ports (and associated channels) through the piston and the orifice at the open foot valve.

4.10.4.2 Operations

Hammer Selection. With such a vast array of hammers on the market it is quite a task to pick the best hammer for your application. Some manufacturers may have up to six different hammers to drill the same size hole. Why so many? Read on!

Size of Hammer. Generally, try to choose a hammer size nearest to the intended hole size. The bigger the hammer diameter, the bigger the piston diameter, the bigger the performance. Figure 4.10.25 shows how rapidly performance drops when using an oversize bit on a small hammer, Figure 4.10.26 shows how performance can be better maintained by going to a bigger hammer, Not only this, a bigger hammer allows a bigger bit shank, better energy transfer resulting in greater strength and lower stresses—more reliability and lower cost per foot. The only reasons hammer selection should not be based entirely on hole size is because of flexibility and capital cost of other hammers. Special hammer sizes such as 5.5 or 11 inch hammers are often used in such situations to maintain performance without sacrificing flexibility.

Hammer Air Consumption. The amount of air available from the air compressor will affect the choice of hammer. Generally, it is best to select a hammer nearest the maximum operating conditions of the compressor. The higher the hammer operating pressure, the better the performance, However, just like your car engine, the manufacturer's ingenuity and technology will play a part in the efficiency and performance of the hammer under different conditions.

Some manufacturers offer low volume (LV) or Mizer hammers. These are usually variations of standard hammers and may contain different pistons, piston cases and/or rigid valves. Another approach by manufacturers is to change the internal port timing and volumes by substituting one part to adjust air consumption characteristics. Apart from the advantage of fewer stocked parts, this gives much greater flexibility and greater opportunity to optimize drilling for different rigs and drilling conditions. The following graphs show performance curves for three typical hammers plotted from manufacturer's data. For example, take three typical cases and show how to estimate operating pressure for each and make a hammer selection.

Case 1: 750SCFM X 250 PSI Compressor. Figure 4.10.27 shows that the compressor produces more air than any of the hammers can use. Hammer 1 is the nearest to the maximum operating conditions of the compressor and will give the best performance. Hammer 2 or 3 will work of course, and could rely on the compressor unloader to control the excess air of the compressor (not a desirable situation with a piston compressor). There are two other reasons why Hammer 2 or 3 should be considered: (1) the lower air consumption will result in lower air velocity and less erosion on the bit and the hammer, and (2) as a means of reducing power requirements and fuel.

Expected performance index:

    Hammer 1   190%
    Hammer 2   104%
    Hammer 3   100%

Case 2: 450 SCFM X 250 PSI. Form a performance aspect, Hammer 3 can be expected to give a far better performance and hold around 220 psi when drilling, The user, for operational reasons, prefers to operate at around 180–200 psi with this specification we would select hammer 2 or possible 3 (Figure 4.10.28).

Expected performance index:

    Hammer 1   190%
    Hammer 2   142%
    Hammer 3   100%

Case 3: 825 SCFM X 350 PSI. According to Figure 4.10.29, Hammer 3 would not use all the air the compressor can produce causing the compressor to partially unload. Hammer 2 on the other hand needs more air than the compressor can make so it will only operate at 320 psi. On paper, both of these hammers could be expected to have the same performance. However, experience dictates that compressors rarely deliver the amount of air to the hammer that is claimed by the stickers on the side, Leaks at drill joints, altitude, blocked intake filters, oil scavenge system, even the weather will effect the air delivery. Knowing this, and the fact that the amount of air might be as little as 660 SCFM, Hammer 3 would be the right choice.

Expected performance index:

    Hammer 1   100%
    Hammer 2   118%
    Hammer 3   136%

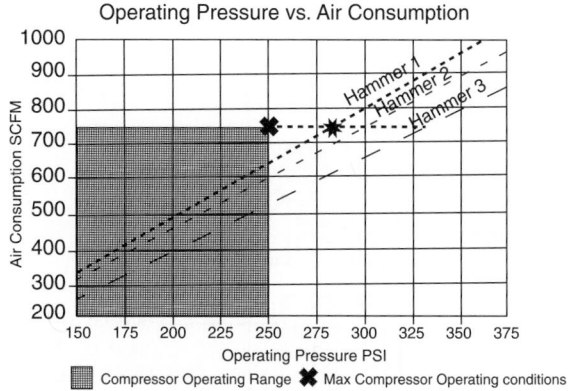

Figure 4.10.27   Case 1: 750 SCFM X 250 PSI Compressor. (Courtesy of DrilMaster.)

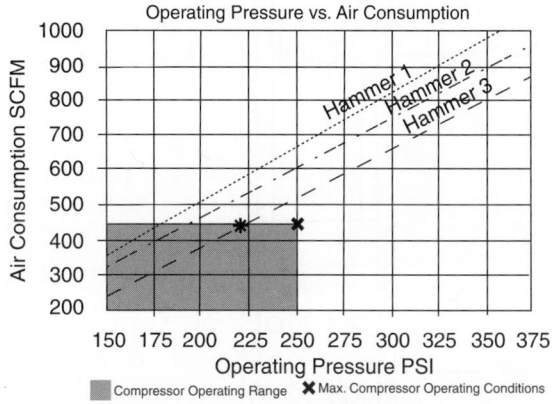

Figure 4.10.28   Case 2: 450 SCFM X 250 PSI Compressor. (Courtesy of DrilMaster.)

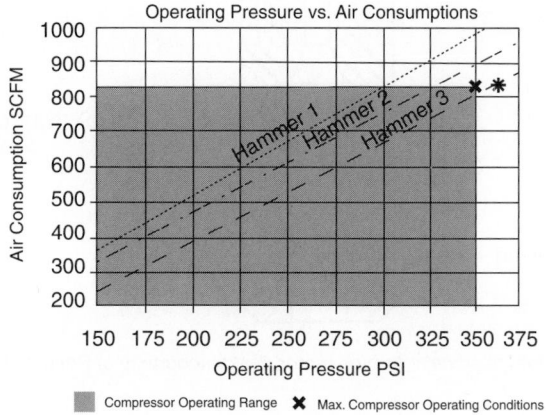

Figure 4.10.29   Case 3: 825 SCFM X 350 PSI Compressor. (Courtesy of DrilMaster.)

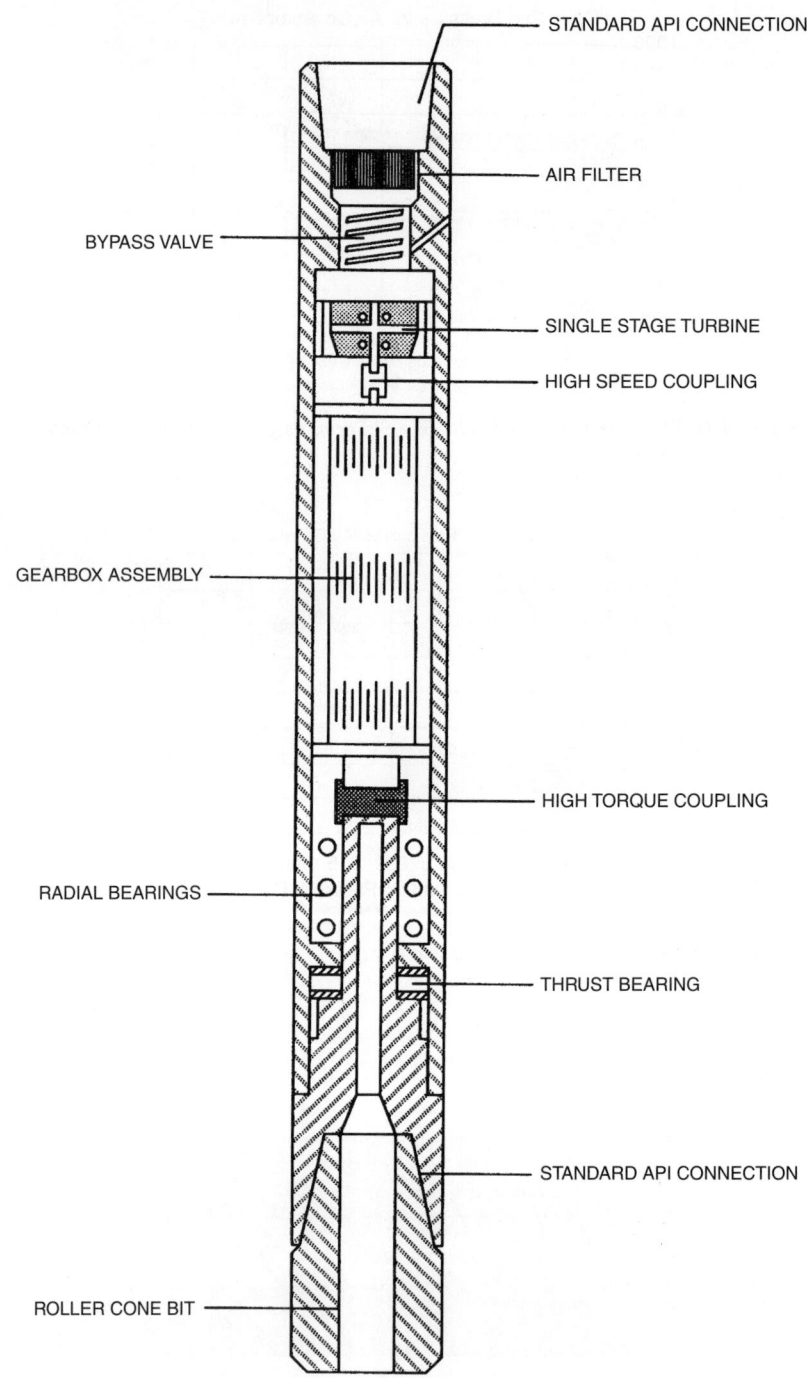

Figure 4.10.30 *Downhole pneumatic turbine motor design* (courtesy of Pneumatic Turbine Partnership).

Heavy Duty Hammers. These hammers are slightly bigger on the outside diameter to permit more wear allowance. Some manufacturers refer to them as "half inch bigger hammers," for example 6.5 inch hammers. In fact, these may not be true 6.5 inch hammers, but the heavy duty hammers with the same internal parts.

Deep Hole Hammers. These hammers are specifically designed to maintain performance where high back pressures are encountered such at high volumes of ground water or very deep holes. Back pressure greatly reduces the impact velocity of the piston eventually causing the piston to cycle but not drill or "wash out". These hammers have a special piston design incorporating a very large bottom volume to reduce the compression ration making them less sensitive to back pressure. DTH hammers are not as efficient as normal drilling because of the extra piston mass and air consumption. A standard hammer can be made to work in high back pressure applications by boosting the air pressure thereby increasing air consumption and improving flushing.

Hammer Characteristics. Every hammer model has its own characteristics, dependent mostly on design and operating pressure. Two values used to describe these characteristics are blow energy and blow frequency (Blows/Minute). How does this effect performance? In basic terms, percussive hammers with a light-blow energy and a high blow frequency drill well in soft, broken formations where a hard-hitting, high-blow energy hammer will bury itself. In hard rock conditions, without a hard-hitting, high-blow energy hammer, power is insufficient to break rock. Great care must be taken when comparing two hammers on the basis of blow energy and blow frequency. Values measure in test beds or calculated may bear no resemblance to balies gained from drilling in actual rock conditions. To make matters more confusing, the bit face configuration and condition of the carbides play a big part in how efficiently energy is transferred to the rock for maximum penetration rates. Use the figures to select a hammer based on air consumption. This will play the biggest part in performance. Hammer and bit characteristics will effect performance, but trial and error is really the only sure way to find out if the performance can be improved.

4.10.5 Special Applications

As it becomes necessary to infill drill the maturing oil and gas reservoirs in the continental United States and elsewhere in the world, the need to minimize or eliminate formation damage will become an important engineering goal. To accomplish this goal, air and gas drilling techniques will have to be utilized (see Section 4.9). It is very likely that the future drilling in the maturing oil and gas reservoirs will be characterized by extensive use of high-angle directional drilling coupled with air and gas drilling techniques.

The downhole turbine motor designed to be activated by the flow of incompressible drilling mud cannot operate on air, gas, unstable foam or stable foam drilling fluids. These downhole turbine motors can only be operated on drilling mud or aerated mud.

Recently, a special turbine motor has been developed to operate on air, gas and unstable foam [5]. This is the downhole pneumatic turbine motor. This motor has been tested in the San Juan Basin in New Mexico and the Geysers area in Northern California. Figure 4.10.30 shows the basic design of this drilling device. The downhole pneumatic turbine motor is equipped with a gear reduction transmission. The compressed air or gas that actuates the single stage turbine motor causes the rotor of the turbine to rotate at very high speeds (i.e., ~20,000 rpm). A drill bit cannot be operated at such speeds; thus it is necessary to reduce the speed with a series of planetary gears. The prototype downhole pneumatic turbine motor has a gear reduction transmission with an overall gear ratio of 168 to 1. The particular version of this motor concept that is undergoing field testing is a 9-in. outside diameter motor capable of drilling with a $10\frac{5}{8}$-in. diameter bit or larger. The downhole pneumatic turbine motor will deliver about 40 hp for drilling with a compressed air flowrate of 3,600 scfm. The motor requires very little additional pressure at the surface to operate (relative to normal air drilling with the same volumetric rate).

The positive displacement motor of the Moineau-type design can be operated with unstable foam (or mist) as the drilling fluid. Some liquid must be placed in the air or gas flow to lubricate the elastomer stator as the metal rotor rotates against the elastomer. Positive displacement motors have been operated quite successfully in many air and gas drilling situations. The various manufacturers of these motors can give specific information concerning the performance characteristics of their respective motors operated with air and gas drilling techniques. The critical operating characteristic of these motors, when operated with unstable foam, is that these motors must be loaded with weight on bit when circulation is initiated. If the positive displacement motor is allowed to be started without weight on bit, the rotor will speed up quickly to a very high speed, thus burning out the bearings and severely damaging the elastomer stator.

References

1. Cross, C. G., "Turbodrill," U.S. Patent 142,992 (September 1873).
2. Tiraspolsky, W., *Hydraulic Downhole Drilling Motors*, Gulf Publishing Company, Editions Technip, Houston, 1985.
3. Ioannesion, R. A., *Osnory-Teorii i Tekhnik: Turbinnogo Bureniya (Foundations of Theory and Technology of Turbodrilling)*, Gostopetekhizdat, Moscow, 1954 (in Russian).
4. Ioannesion, R. A., and Y. V. Vadetsky, "Turbine drilling equipment development in the USSR," *Oil and Gas Journal*, September 28, 1981.
5. Jurgens, R., and C. Marx, "Neve Bohrmotoren für die erdölindustrie (New drilling motors for the petroleum industry)," *Erdol-Erdges Zeitschrift*, April 1979 (in German).
6. Lyons, W. C., et al., "Field testing of a downhole pneumatic turbine motor," Geothermal Energy Symposium, ASME/GRC, January 10–13, 1988.
7. Magner, N. J., "Air motor drill," *The Petroleum Engineer*, October 1960.
8. Downs, H. F., "Application and evaluation of air-hammer drilling in the Permian Basin," *API Drilling and Production Practices*, 1960.
9. Bourgoyne, A. T., et al., *Applied Drilling Engineering*, SPE Textbook Series, Vol. 2, First Printing, SPE Richardson, Texas, 1986.
10. *Drilling Manual*, 11th Edition, IADC, 1992.
11. Winters, W., and T. M. Warren, "Determining the true weight-on-bit for diamond bits," SPE Paper 11950, presented at the 1983 SPE Annual Technical Conference and Exhibition, San Francisco, October 5–8, 1983.

4.11 MWD AND LWD

Most of the cost in a well is expanded during the drilling phase. Any amount of information gathered during drilling can be used to make decisions regarding the efficiency of the process. But the scope and ultimate cost to gather and analyze such information must be offset by a decrease in drilling expenditures, an increase in drilling efficiency and an increase in safety.

As drilling technology moved the pursuit of hydrocarbon resources into higher-cost offshore and hostile environments, intentionally deviated boreholes required information such as azimuth and inclination that could not be derived by surface instruments. Survery instruments, either lowered on a sand line or dropped into the drill pipe for later retrieval, to some degree satisfied the requirements but consumed expensive rig time and sometimes produced questionable results.

For many years researchers have been looking for a simple, reliable *measurement while drilling* technique, referred to by its abbreviation MWD. As early as 1939, a logging while drilling (LWD) system, using an electric wire, was tested successfully but was not commercialized [1, 2]. Mud pulse systems were first proposed in 1963 [3, 4]. The first mechanical mud pulse system was marketed in 1964 by Teledrift for transmitting directional information [5]. In the early 1970s, the steering tool, an electric wire operated directional tool, gave the first real-time measurements while the directional buildup was in progress. Finally, the first modern mud pulse data transmission system was commercialized in 1977 by Teleco [6]. State-of-the-art surveys of the technology were made in 1978 [7], in 1988 [7–10], and in 1990 [11].

A problem with the early MWD mud pulse systems was the very slow rate of data transmission. Several minutes were needed to transmit one set of directional data. Anadrill working with a Mobil patent [12] developed in the early 1980s a continuous wave system with a much faster data rate. It became possible to transmit many more drilling data, and also to transmit logging data making LWD possible. Today, as many as 16 parameters can be transmitted in 16 s. The dream of the early pioneers has been more than fulfilled since azimuth, inclination, tool face, downhole weight-on-bit, downhole torque, shocks, caliper, resistivity, gamma ray, neutron, density, Pe, sonic and more can be transmitted in real time to the rig floor and the main office.

4.11.1 MWD Technology
4.11.1.1 Steering Tool
Up until 1970, all directional drilling was conducted using singleshot and multishot data. The normal procedure was

A. Drill vertically in rotary to the kick-off depth.
B. Kick-off toward the target using a downhole motor and a bent sub to an inclination of approximately 10°.
C. Resume rotary drilling with the appropriate bottomhole assembly to build angle, hold, or drop.

The kick-off procedure required numerous single-shot runs to start the deviation in the correct direction. Since, during this phase, the drillpipe was not rotating a steering tool was developed to be lowered on an electric wireline instead of the single shot. The measurements were then made while drilling.

Measurements by electric cable are possible only when the drillstem is not rotating, hence with a turbine or downhole motor. The steering tool is run in the drillstring and is positioned by a mule shoe and key. The process is identical to the one used in the single-shot measurements. The magnetic orientation sensor is of the flux-gate type and measures the three components of the earth's magnetic field vector in the reference space of the logging tool. Three accelerometers measure the three components of the gravity vector still in the same reference space. These digitized values are multiplexed and transmitted by an insulated electric conductor and the cable armor toward the surface. On the surface a minicomputer calculates the azimuth and the drift of the borehole as well as the angle of the tool face permanently during drilling. In the steering tool system, the computer can also determine the azimuth and slant of the downhole motor underneath the bent sub and thus anticipate the direction that the well is going to take. It can also determine the trajectory followed. Figure 4.11.1 shows the steering tool system.

Naturally for operating the tool, the seal must be maintained at the point where the cable enters the drillstring:

A. At the top of the drillpipes, in which case the logging tool is pulled out every time a new drillpipe length is added on.
B. Through the drillpipe wall in a special sub placed in the drillstring as near the surface as possible, in which case new lengths are added on without pulling out the logging tool.

Figure 4.11.2 shows the typical operation of a steering tool for orienting the drill bit. The electric wireline goes through a circulating head located on top of the swivel. As mentioned previously, the tool has to be pulled out when adding a single.

Figure 4.11.3 shows the same operation using a side-entry sub. With this sub, the electric wireline crosses over from inside the drill pipe to the outside. Consequently, singles may be added without pulling the steering tool out. On the other hand, there is a risk of damaging the cable if it is crushed between the drillpipe tool joints and the surface casing. The wireline also goes through the rotary table and special care must be taken not to crush it between the rotary table and the slips. Furthermore, in case of BHA sticking, the steering tool has to be fished out by breaking and grabbing the electric wireline inside the drillpipes.

Figure 4.11.4 shows the arrangement of the sensors used in a steering tool. There flux-gate-type magnetometers and three accelerometers are positioned with their sensitive axis along the principal axis of the tool. O_x is oriented toward the bent sub in the bent sub/tool axis plane. O_y is perpendicular to O_x and the tool axis. O_z is oriented along the tool axis downward.

The arrangement of Figure 4.11.4 is common to all directional tools based on the earth's magnetic field for orientation: MWD tools or wireline logging tools.

The steering tools have practically been abandoned and replaced by MWD systems, mostly because of the electric wireline. However, the high data rate of the electric wireline (20–30 kbits/s) compared to the low data rate of the MWD systems (1–10 bits/s) make the wireline tools still useful for scientific work.

4.11.1.2 Accelerometers
Accelerometers measure the force generated by acceleration according to Newton's law:

$$F = ma \qquad [4.11.1]$$

where F = force in lb
m = mass in $(lb - s^2)/ft$ (or slugs)
a = acceleration in ft/sec^2

If the acceleration is variable, as in sinusoidal movement, piezoelectric systems are ideal. In case of a constant acceleration, and hence a force that is also constant, strain gages may be employed. For petroleum applications in boreholes,

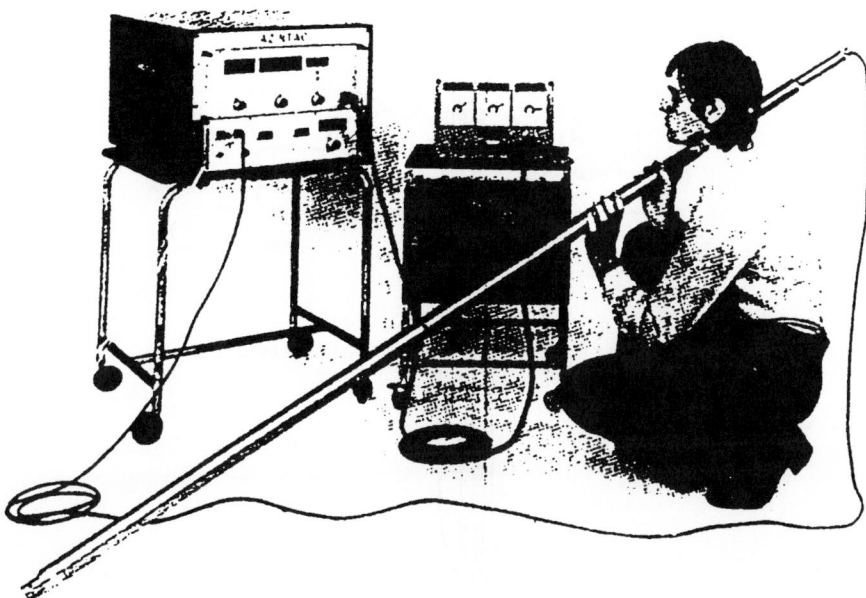

Figure 4.11.1 *Typical steering tool unit with surface panel and driller read-out (courtesy of Institut Francois du Petrole).*

however, it is better to use servo-controlled accelerometers. Reverse pendular accelerometers and "single-axis" accelerometers are available.

Every accelerometer has a response curve of the type shown schematically in Figure 4.11.5. Instead of having an ideal linear response, a nonlinear response is generally obtained with a "skewed" acceleration for zero current, a scale factor error and a nonlinearity error. In addition, the skew and the errors vary with temperature. If the skew and all the errors are small or compensated in the accelerometer's electronic circuits, the signal read is an ideal response and can be used directly to calculate the borehole inclination. If not, "modeling" must be resorted to, i.e., making a correction with a computer, generally placed at the surface, to find the ideal response. This correction takes account of the skew, all the errors, and their variation with temperature. In this case, the accelerometer temperature must be known. The maximum current feedback defines a measurement range beyond which the accelerometer is saturated. Vibrations must be limited in order not to disturb the accelerometer response.

Assume that the accelerometer has the ideal response shown in Figure 4.11.6, with a measurement range of 2 g (32.2 ft/s²). We want to measure 1 g, but the ambient vibration level is ±3 g. In this case, the accelerometer's indications are shaved and the mean value obtained is not 1 g but 0.5 g. The maximum acceleration due to vibrations which are not filtered mechanically, plus the continuous component to be measured, must be less than the instrument's measurement range.

These instruments serve to measure the earth's gravitational field with a maximum value of 1 g. The typical values of the characteristics are:

- Scale factor, 3 mA/g
- Resolution, 10^{-6} g
- Skew, 10^{-3} g
- Service temperature, -55 to $+150°C$

For measurement ranges from 0 to 180°, three accelerometers mounted orthogonally must be used as shown in Figure 4.11.7. The x and y accelerometers are mounted with their sensitive axis perpendicular to the tool axis. The z accelerometer is mounted with its sensitive axis lined up with the tool axis.

Figure 4.11.8 shows the inclination measurement using a triaxial sensor featuring three accelerometers. The three coordinates of earth's gravitational acceleration vector serve to define this vector in the reference frame of the probe. The earth's acceleration is computed as

$$G = \sqrt{G_x^2 + G_y^2 + G_z^2} \qquad [4.11.2]$$

It must be equal to the 32.2 ft/s²; otherwise the accelerometers are not working correctly. When the readings are in g units, G must be equal to one.

For the best accuracy, inclination less than 60° is computed with

$$i = \arcsin \frac{\sqrt{G_x^2 + G_y^2}}{G} \qquad [4.11.3]$$

and for i greater than 60°

$$i = \arccos \frac{G_z}{G} . \qquad [4.11.4]$$

The gravity tool face angle, or angle between the plane defined by the borehole axis and the vertical and the plane defined by the borehole axis and the BHA axis below the bent sub, can also be calculated. Figure 4.11.9 shows the gravity tool face angle. It is readily calculated using the equation

$$TF = \arctan \frac{-G_y}{G_x} \qquad [4.11.5]$$

The gravity tool face angle is used to steer the well to the right, TF>0, or to the left, TF<0.

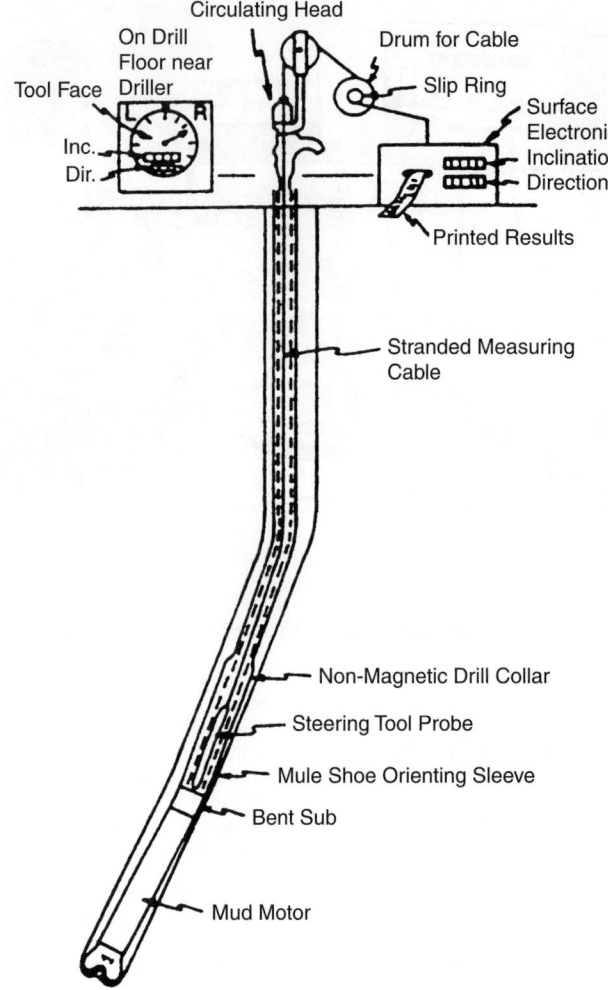

Figure 4.11.2 *Typical operation of a steering tool for orienting the drill bit using a circulating head on the swivel (courtesy of SPE [13]).*

Typical specifications for a gravity sensor are as follows:

Temperature

Operating = 0 to 175°C
Storage = −40 to 175°C
Scale factor = 0.01 V/°C

Power requirement

24 V nominal
Less than 100 mA

Output impedance

10Ω

Mechanical characteristics

Length = 60 cm (24 in.)
Diameter = 3.75 cm (1.5 in.)
Mass = 2 kg (4 lb)
Alignment = ±0.4°

Electrical characteristics

Scale factor = 5 V/g ± 1% (g = 32.2 ft/s^2)
Bias = ±0.005 g @ 25°C
Linearity = ±0.1% full scale

Environmental characteristics

Vibrations = 1.5 cm p-p (peak-to-peak), 10 to 50 Hz
50 g, 50 to 2000 Hz
Shock = 2000 g, 0.5 ms, 0.5 sine

4.11.1.3 Magnetometers

Magnetometers used in the steering tools or MWD tools are of the flux-gate type.

The basic definition of a magnetometer is a device that detects magnetic fields and measures their magnitude and/or direction. One of the simplest types of magnetometers is the magnetic compass. However, due to its damping problems more intricate designs of magnetometers have been developed. The "Hall effect" magnetometer is the least sensitive. The "flux-gate" magnetometer concept is based on the magnetic saturation of an iron alloy core.

If a strip of an iron alloy that is highly "permeable" and has sharp "saturation characteristics" is placed parallel to the earth's magnetic field, as in Figure 4.11.10, some of the lines of flux of the earth's field will take a short cut through the alloy strip, since it offers less resistance to their

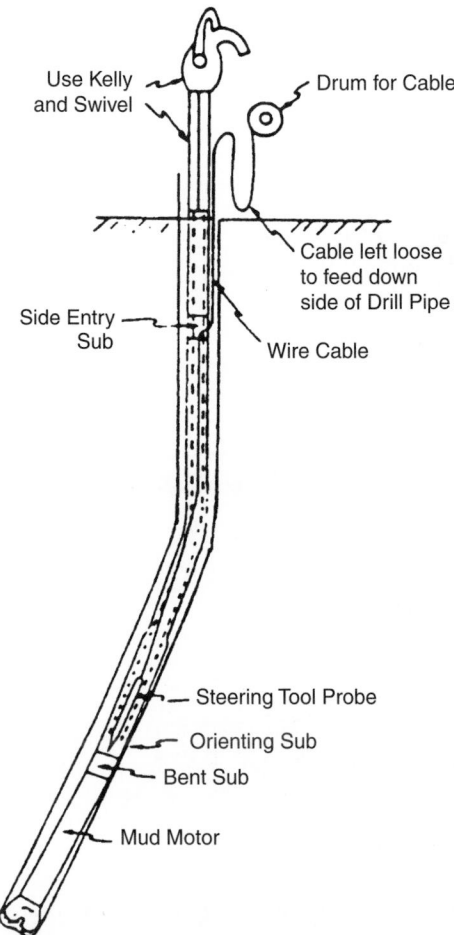

Figure 4.11.3 *Typical operation of a steering tool for orienting the drill bit using a side-entry sub (courtesy of SPE [13]).*

Labels in figure:
Use Kelly and Swivel
Drum for Cable
Cable left loose to feed down side of Drill Pipe
Side Entry Sub
Wire Cable
Steering Tool Probe
Orienting Sub
Bent Sub
Mud Motor

flow than does the air. If we place a coil of wire around the strip, as in Figure 4.11.11 and pass enough electrical current through the coil to "saturate" the strip, the lines of flux due to earth's field will no longer flow through the strip, since its permeability has been greater reduced.

Therefore, the strip of iron alloy acts as a "flux gate" to the lines of flux of the earth's magnetic field. When the strip is not saturated, the gate is open and the lines of flux bunch together and flow through the strip. However, when the strip is saturated by passing and electric current through a coil wound on it, the gate closes and the lines of flux pop out and resume their original paths.

One of the basic laws of electricity, Faraday's law, tells us that when a line of magnetic flux cuts or passes through an electric conductor a voltage is produced in that conductor. If an AC current is applied to the drive winding A-A of Figure 4.11.11 the flux gate will be opening and closing at twice the frequency of the AC current and we will have lines of flux from the earth's field moving in and out of the alloy at a great rate. If these lines of flux can be made to pass through an electrical conductor ("sense winding"), a voltage will be induced each time they pop in or out of the alloy strip. This induced voltage in the sense windings is proportional

to the number of lines of flux cutting through it, and thus proportional to the intensity of that component of the earth's magnetic field that lies parallel to the alloy strip.

When the alloy strip is saturated, a lot of other lines of flux are created that are not shown in Figure 4.11.11. The lines of flux must be sorted out from the lines of flux due to the earth's field to enable a meaningful signal to be produced. A toroidal core as shown in Figure 4.11.12 will enable this separation of lines of flux to be accomplished. The material used for the toroidal core is usually mu metal.

Each time the external lines of flux are drawn into the core, they pass through the sense windings B-B to generate a voltage pulse whose amplitude is proportional to the intensity of that component of the external field that is parallel to the centerline of the sense winding. The polarity, or direction of this pulse, will be determined by the polarity of the external field with respect to the sense windings. When the flux lines are expelled from the core they cut the sense windings in the opposite direction and generate another voltage pulse of the same amplitude but of opposite polarity or sign. Because of this voltage pulse occurring at twice the driving voltage frequency, the flux gate is sometime known as "second harmonic magnetometer".

The main advantages of the flux gate magnetometers are that they are solid state devices much less sensitive to vibration than compasses, they have uniaxial sensitivity, and sensitivity, and they are very accurate.

Typical specifications are:
Temperature

   Operating $= 0$ to $200°C$
   Storage $= -20$ to $200°C$

Power requirement, output in impedance, and *mechanical characteristics* are similar to the accelerometer sensors.
Electrical characteristics

   Alignment $= \pm 0.5°$
   Scale factor $= 5$ V/G $\pm 5\%$
   Bias $= \pm 0.005$ G @ $25°C$
   Linearity $= \pm 2\%$ full scale
   Note: 1 gauss $= 1G = 10^{-4}$ tesla

Environmental characteristics

   Vibrations $= 1.5$ cm p - p, 2 to 10 Hz
   20 g, 10 to 200 Hz
   Shock $= 1000$ g, 0.5 ms, 0.5 sine

Figure 4.11.13 shows the photograph of a Develco high-temperature direction sensor. For all the sensor packages, calibration data taken at 25, 75, 125, 150, 175 and 200°C are provided. Computer modeling coefficients provide sensor accuracy of ± 0.001 G and $\pm 0.1°$ alignment from 0 to 175°C. From 175 to 200 the sensor accuracy is ± 0.003 G and $\pm 0.1°$ alignment.

Example 4.11.1 Steering Tool Measurement–Trajectory Forecast
An interesting problem that can be solved with the steering tool or the MWD measurements is the trajectory forecast when drilling ahead with a given bent sub (constant angle), and a given tool-face angle.

1. After drilling the mud motor length with a given tool face angle and a given bent sub angle, what is the borehole deviation and orientation likely to be at the mud motor depth?

   Using the drawing Figure 4.11.14 find the algorithm to compute these angles.

2. If we change the tool face angle to −30° (turning left), what will be the probable borehole deviation and azimuth after drilling another motor length? Use the same computer program.

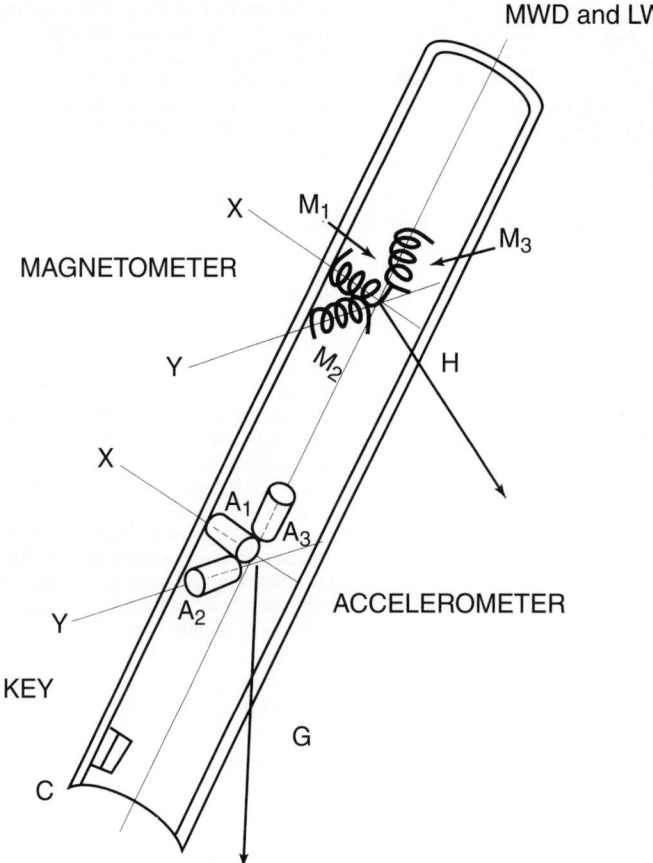

Figure 4.11.4 *Sketch of the principle of the sensor arrangement in a steering tool and any magnetic directional tool (courtesy of SPE [13]).*

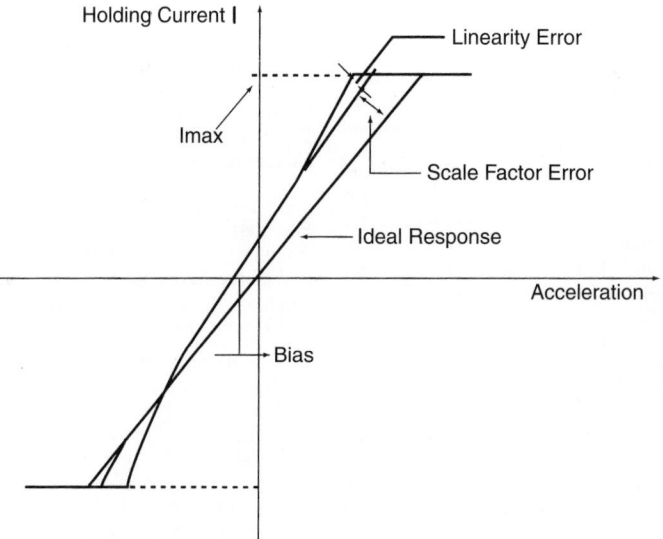

Figure 4.11.5 *Accelerometer response.*

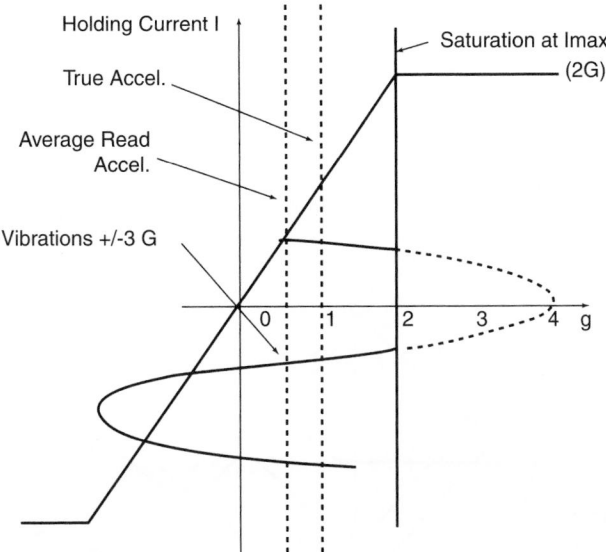

Figure 4.11.6 *Effect of vibrations on an accelerometer response.*

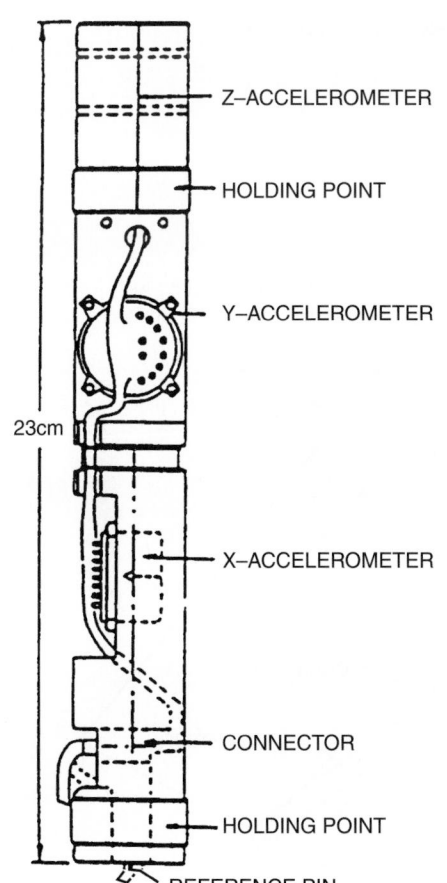

Figure 4.11.7 *Mechanical drawing of the accelerometer section of a directional tool (courtesy of Sunstrand [14]).*

Note: We will assume that the borehole axis is the same as the drill collar axis at the steering tool depth and also that the borehole axis is the same as the mud motor axis at the mud motor depth.

Solution
The same algorithms are used as in Example 3. The vector **Z** $(0, 0, 1)$ is replaced by vector **Z** $(\sin 2° \bullet \cos$ TF, $\sin 2° \bullet \sin$ TF, $\cos 2°)$, This new vector **Z** should be used to compute the new inclination, using the scalar product between vector **G** and new vector **Z**. The new vector **Z** should also be used to compute the new azimuth.

4.11.1.4 Vibrations and Shocks
Measurements while drilling are made with sensors and downhole electronics that must operate in an environment where vibrations and shocks are sometimes extremely severe. A brief study of vibrations and shocks will be made to understand better the meaning of the specifications mentioned earlier.

The vibration frequencies encountered during drilling are well known. They correspond to the rotation of the drill bit, to the passing of the bit rollers over the same hard spot on the cutting face, and to the impact of the teeth. Figure 4.11.15 gives the order of magnitude for frequencies in hertz (60 rpm = 1 Hz). In each type the lowest frequencies correspond to rotary drilling and the highest ones correspond to turbodrilling. Three vibrational modes are encountered:

1. axis vibrations due to the bouncing of the drill bit on the bottom
2. transverse vibrations generally stemming from axial vibrations by buckling or mechanical resonance
3. angular vibrations due to the momentary catching of the rollers of stabilizers

In vertical rotary drilling, the drillpipes are almost axially and angularly free. Therefore, the highest level of axial and angular vibration is encountered for this type of drilling. In deviated rotary drilling, the rubbing of the drill string

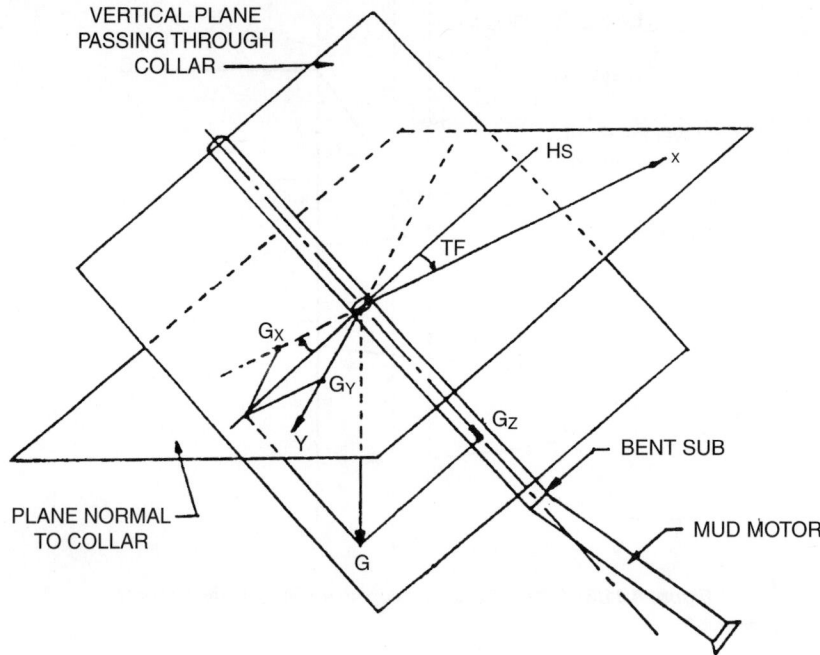

Figure 4.11.8 *Vector diagram of the inclination measurement with three accelerometers.*

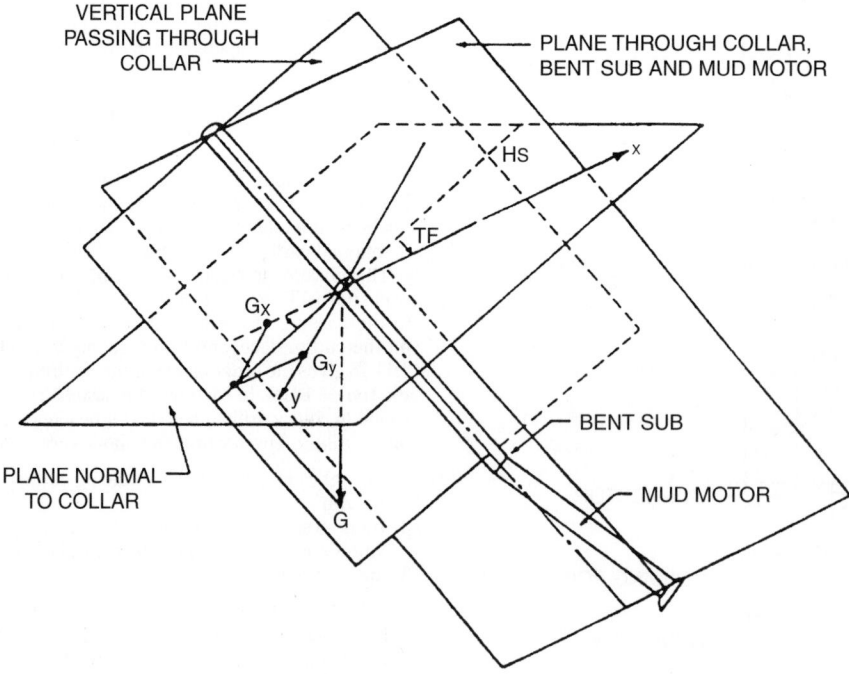

Figure 4.11.9 *Solid geometry representing the gravity tool face angle concept.*

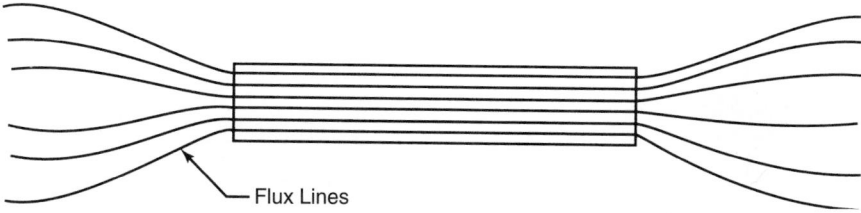

Figure 4.11.10   *Magnetic flux-lines representation in a highly permeable iron alloy core.*

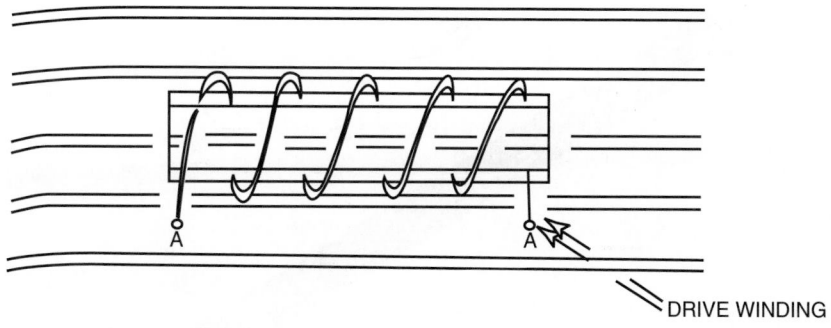

DRIVE WINDING

Figure 4.11.11   *Magnetic flux-lines representation in a highly permeable iron alloy core saturated with an auxiliary magnetic field.*

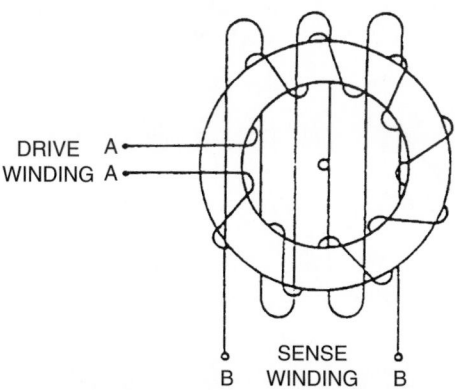

DRIVE   A
WINDING A

SENSE
B   WINDING   B

Figure 4.11.12   *Sketch of principle of a single-axis flux-gate magnetometer.*

on the well wall reduces axial vibrations, but the stabilizers increase angular vibrations. In drilling with a downhole motor, the rubbing of the bent sub on the well wall reduces the amplitude of all vibrations.

Vibrations are characterized by their peak-to peak amplitude at low frequencies or by their acceleration at high frequencies. Assuming that vibration is sinusoidal, the equation for motion is

$$x = \frac{A}{2}\sin 2\pi f \times t \qquad [4.11.6]$$

where x = elongation in m
    A = peak-to-peak amplitude in m
    f = frequency in Hz
    t = time in s

By deriving twice, acceleration becomes

$$a = -\frac{A}{2}(2\pi f)^2 \sin 2\pi f \times t \qquad [4.11.7]$$

Maximum acceleration is thus $a_m = 2A\pi^2 f^2$. For example, peak-to-peak 12 mm at 10 Hz corresponds to $a_m = 11.8$ m/s$^2 = 1.2$ g. Acceleration of gravity is expressed as g.

Very few vibration measurements are described in the literature, but the figures in Figure 4.11.16 can be proposed for vertical rotary drilling. The lower limits correspond to soft sandy formations and the upper limits with heterogeneous formations with hard zones. Table 4.11.1 gives the specifications that the manufacturers propose for several tools.

The shocks that measuring devices are subjected to are generally characterized by an acceleration (or deceleration) and a time span. For example, a device is said to withstand 500 g (5,000 m/s^2) for 10 ms. This refers to a "half-sine." Shock testing machines produce a deceleration impulse having the form shown in Figure 4.11.17.

Measuring devices run inside the drillstring are mainly subject to axial impacts. These shocks come from sudden halts in the mule shoe or from an obstruction in the string. The measuring devices used while drilling are generally subjected to axial and angular impacts caused by the bouncing of the bit on the bottom and by the catching of the rollers and stabilizers on the borehole walls. There is very little information in the literature about measuring impacts while drilling.

Table 4.11.2 gives the specifications compiled by manufacturers for several measuring devices. It can thus be seen that such devices must be equipped with an axial braking system capable of having a stroke of 10 and even 20 cm.

Figure 4.11.13 *Photograph of a high-temperature directional sensor with three accelerometers and three magnetometers (courtesy of Develco [15]).*

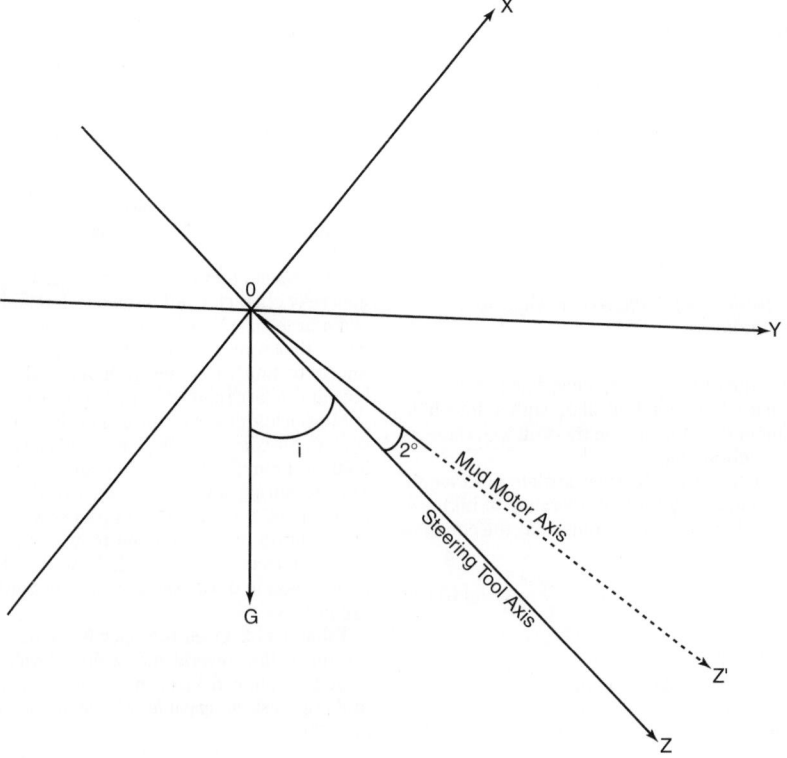

Figure 4.11.14 *Vector diagram showing the mud motor axis as well as the steering tool axis.*

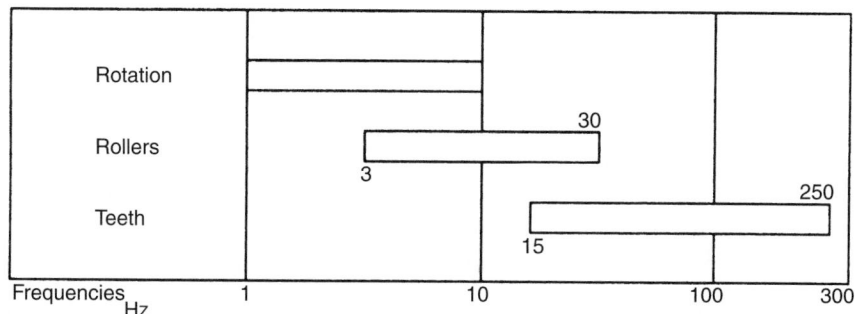

Figure 4.11.15 *Main vibration frequencies encountered while drilling.*

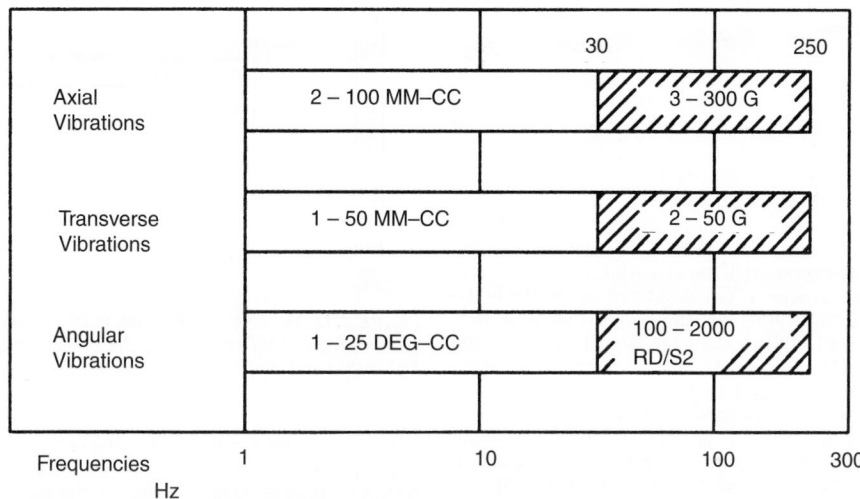

Figure 4.11.16 *Order of magnitude of the vibration amplitudes encountered during drilling.*

Table 4.11.1 *Resistance of Some Directional Tools or Components to Vibrations*

Tool	Low Frequencies	High Frequencies
Azintac (1)	1 mm cm³ (5–50 Hz)	10 g (50–500 Hz)
Drill-director (2)	2 g (5–45 Hz)	5 g (45–400 Hz)
Q-Flex accelerometer		25 g
Develco accelerometer	12.7 mm cm³ (20–40 Hz)	40 g (40–2000 Hz)
4-Gimbal gyroscope	2 g (10–200 Hz)	
2-Axis gyroscope	5 g (10–300 Hz)	
On-shore military specifications		14 g (50–2000 Hz)

(1) Inst. Fr. Du Pet. Trademark
(2) Humphery Inc. Trademark
(The acceleration values correspond to maximum amplitudes)

4.11.1.4.1 Future Developments

Orientation measurements while drilling are practically impossible with gimbal gyroscopes. Two-axis flexible-joint gyroscopes should be able to withstand vibrations and impacts while maintaining a sufficiently accurate heading provided that periodic recalibration is performed by halting drilling and switching on the north seeking mode. In the more distant future, laser or optical-fiber gyroscopes that have been suitably miniaturized should provide a solution.

4.11.1.5 Teledrift and Teleorienter

The first transmission of data during drilling using mud pulses was commercialized by B.J. Hughes Inc. in 1965 under the name of *teledrift* and *teleorienter*. Both tools are purely mechanical. A general sketch of principle is given in Figure 4.11.18. The tool is now operated by Teledrift Inc.

The tool generates at bottom positive pulses by restricting momentarily the flow of mud each time that the mud flow (pumps) is started. The pulses are detected at the

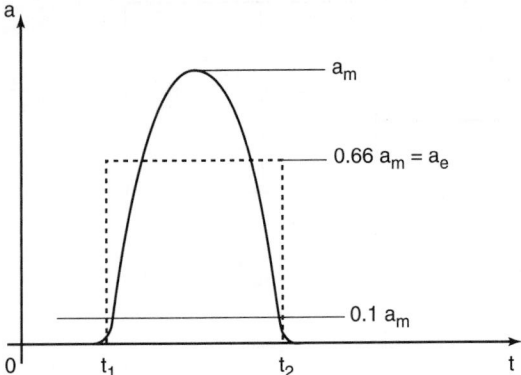

Figure 4.11.17 *Theoretical deceleration variation during a shock or impact.*

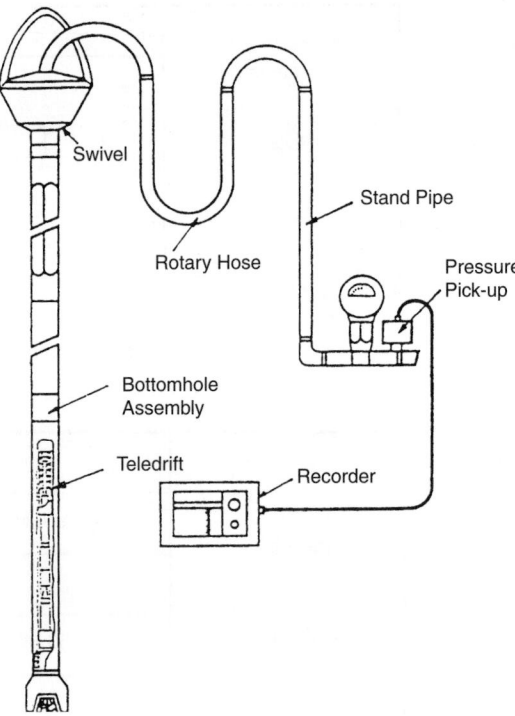

Figure 4.11.18 *Sketch of principle of the teledrift tool or teleorienter tool attached to the drillstring (courtesy of Teledrift, Inc. [16]).*

surface on the stand pipe and recorded as a function of time.

Figure 4.11.19 shows the sketch of principle of the teledrift unit which is measuring inclination.

A pendulum hangs in a conical grooved bore. A spring tends to move the pendulum and the poppet valve upwards when the circulation stops. If the tool is inclined, the pendulum catches the grooves at different levels according to the inclination and stops there. For example, for minimum inclination (Figure 4.11.19B) it stops the poppet valve past the first restriction. In Figure 4.11.19D the poppet valve stops past the seventh restriction due to the high inclination.

When the circulation is started, the poppet valve travels slowly down, generating one pressure pulse when passing each restriction. The measurement range in the standard tool is of 2.5° (also 7° ranges, 1° increments, max. 17°).

Table 4.11.3 gives the inclination angles corresponding to one to seven pulses with three cones. Fifteen cones are available. The maximum measurable angle is 10°. The range must be selected before lowering the drillstring.

The cone can be replaced by a mechanism that senses the angular position rather than the inclination of the drillstring. The tool is then sensitive to the tool face and is called the *teleorienter.*

Figure 4.11.20 shows the read-out display of the driller in a zero tool-face position. Four mud pressure pulses will be recorded each time the pumps are started. In Figure 4.11.21a, a tool-face value of 20° is indicated by three pulses, turning to the right. In Figure 4.11.21b, a tool-face value of −20° is indicated by five pulses, and the borehole is turning left.

These tools have been widely used in the past by the cost conscious operators. The tool could be rented and operated by the rig floor personnel. However, the inclination ranges are limited, only one tool, teledrift or teleorienter, can be used during a trip, and only the tool-face angle is read by the teleorienter, not the azimuth.

These tools are still available but tend to be replaced by the MWD systems.

4.11.1.6 Mud Pressure and EM Telemetry
Two methods are currently used to transmit data from downhole to surface: mud pressure telemetry and electromagnetic earth transmission.

There are three principles for transmitting data by drilling mud pressure:

1. positive pulses obtained by a momentaneous partial restriction of the downhole mud current;

Table 4.11.2 *Resistance of Some Directional Tools or Components to Axial Shocks or Impacts*

Tool	Deceleration (g, G)	Time (ms)	Braking Distance (m)	Initial Velocity (m/s)
Azintac	60	11	0.024	4.35
Drill-director	700	10	0.23	46.2
Q-Flex accelerometer	250	11	0.10	18.15
Develco accelerometer	400	1	0.0013	2.64
4-Gimbal accelerometer	50	10	0.016	3.3
2-Axis gyroscope	100	10	0.033	6.6
Military specifications	30 to 100	10		

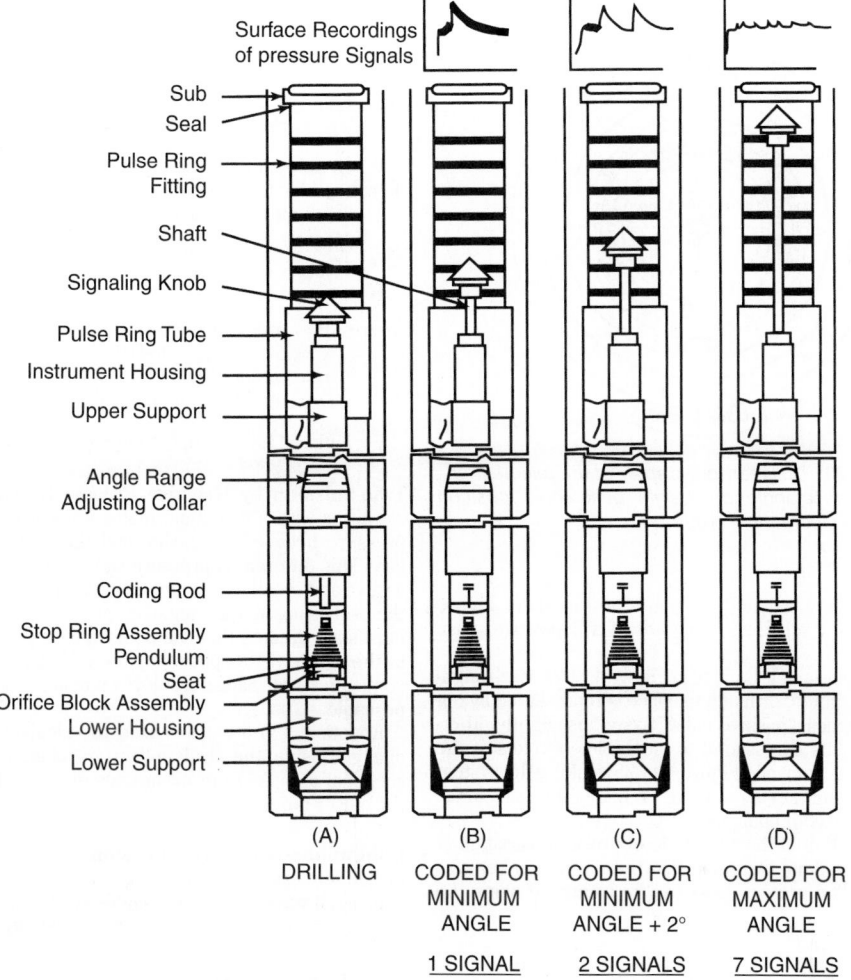

Surface Recordings
of pressure Signals

Sub
Seal
Pulse Ring
Fitting
Shaft
Signaling Knob
Pulse Ring Tube
Instrument Housing
Upper Support
Angle Range
Adjusting Collar
Coding Rod
Stop Ring Assembly
Pendulum
Seat
Orifice Block Assembly
Lower Housing
Lower Support

(A)	(B)	(C)	(D)
DRILLING	CODED FOR MINIMUM ANGLE	CODED FOR MINIMUM ANGLE + 2°	CODED FOR MAXIMUM ANGLE
	1 SIGNAL	2 SIGNALS	7 SIGNALS

Figure 4.11.19 *Teledrift mechanism in various coding positions (courtesy of Teledrift, Inc. [16]).*

Table 4.11.3 *Teledrift Angle Range Settings*

| Angle Range | Deviation Angle in Degrees | | | | | | |
	1 signal	2 signals	3 signals	4 signals	5 signals	6 signals	7 signals
0.5–3.0	0.5	1.0	1.5	2.0	2.5	3.0	3+
1.0–3.5	1.0	1.5	2.0	2.5	3.0	3.5	3.5+
7.5–10.0	7.5	8.0	8.5	9.0	9.5	10.0	10.0+

2. negative pulses obtained by creating a partial and momentaneous communication between the drill string internal mud stream and the annular space at the level of the drill collars;

3. phase changes of a low-frequency oscillation of the drilling mud pressure induced downhole in the drillstring.

Figure 4.11.22 shows sketches of the three systems.

Transmission by Positive Pulses. This system is by far the most common. It is placed in a nonmagnetic drill collar containing sensors of the flux-gate type for measuring the direction of the earth's magnetic field and accelerometers for measuring the gravity vector. An electromagnetic and electronic unit, every time rotation is halted, calculates and memorizes the azimuth, drift and tool face angles. Bottomhole electric power is supplied by an AC generator coupled with a turbine situated on the mud stream in the drill collar, or by batteries.

In rotary drilling a rotation detector triggers angle measurements when the string stops rotating with circulation maintained. With a downhole motor the measurements are

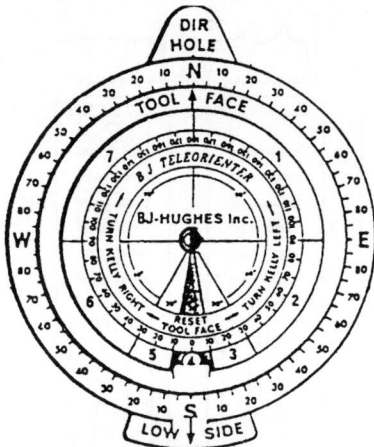

Figure 4.11.20 *Drill read-out display of the teleorienter in a zero tool face angle "go straight, build angle" position (courtesy of Teledrift, Inc. [16]).*

repeated as long as mud continues to circulate. Figure 4.11.23, gives a schematic diagram of a 2 positive pulse generator.

The coding principle is given in Figure 4.11.24. Each angular value of azimuth, drift and tool face is represented by ten bits. The practical "positive pulse" system is slightly different. The "1" bits correspond to incomplete strokes of the poppet valve as shown in Figure 4.11.25, making the system a slow phase-shift keying system. Transmission rates of 0.2 or 0.4 bit/s are commonly used.

The calculation of the amplitude of pressure variation at bottom can be done assuming that the restriction behaves as a choke. The pressure loss can be estimated using the relations

$$\Delta P = \frac{Q^2 \bullet \gamma \bullet 144}{2 \bullet g_c \bullet c^2 \bullet A_0^2} \qquad [4.11.8]$$

where ΔP = pressure loss in psi
$\quad Q$ = flowrate in ft^3/s
$\quad \gamma$ = fluid specific weight in lb/ft^3
$\quad c$ = coefficient assumed to be one
$\quad A_0$ = cross-sectional area of the restriction in in.^{2}
$\quad g_c$ = acceleration of gravity (32.2 ft/s^2)

When using a mud motor, the ΔP due to the restriction must be added to the ΔP due to the motor and the bit nozzles.

The mud motor pressure loss is given by

$$\Delta P = \frac{1714 \bullet W}{\eta \bullet Q} \qquad [4.11.9]$$

where ΔP = pressure loss in psi
$\quad W$ = motor power in HP
$\quad \eta$ = motor efficiency
$\quad Q$ = mud flowrate in gal/min

Formula 4.11.8 will apply to the bit nozzle pressure loss.

Transmission by Negative Pulses. Drilling with a nozzle bit or with a downhole motor introduces a differential pressure between the inside and the outside of drill collars. This differential pressure can be changed by opening a valve and creating a communication between the inside of the drill string and the annular space. In this way, negative pulses are created that can be used to transmit digital data in the same way as positive pulses. Halliburton and other companies are marketing devices using this transmission principle.

Equation 4.11.8 can be used to calculate the pressure change inside the drill collars by changing the cross-sectional area A_0 from bit nozzles only to bit nozzles plus the pulser nozzle.

Continuous-Wave Transmission. Anadrill, a subsidiary of Schlumberger, markets a tool which produces a 12-Hz sinusoidal wave downhole. Ten-bit words representing data are transmitted by changing or maintaining the phase of the wave at regular intervals (0.66 s). A 180° phase change represents a 1, and phase maintenance represents a 0.

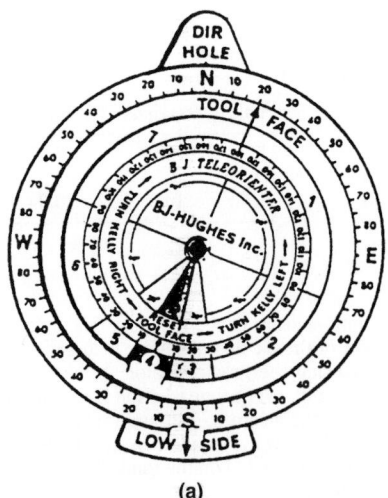

(a)

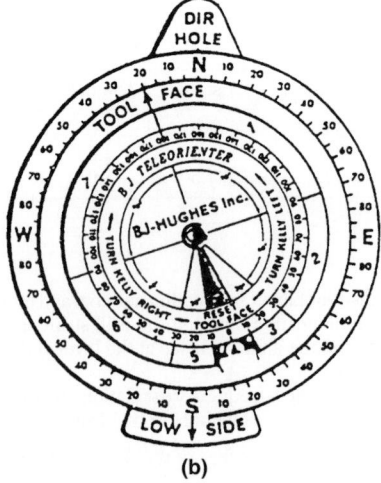

(b)

Figure 4.11.21 *Driller read-out display of the teleorienter: (a) +20° tool-face angle, turning right position; (b) −20° tool-face angle, turning left position (courtesy Teledrift, Inc. [16]).*

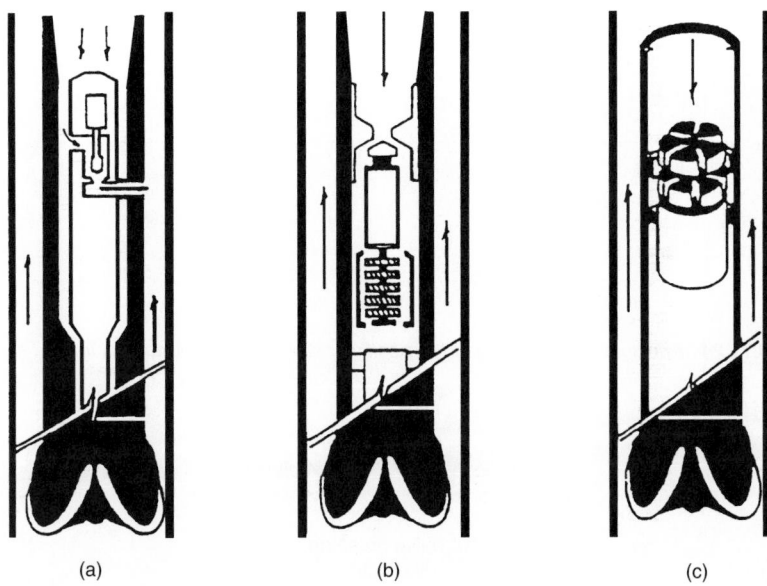

Figure 4.11.22 *Telemetry systems using mud pressure waves: (a) negative pulse system; (b) positive pulse system; (c) continuous wave system.*

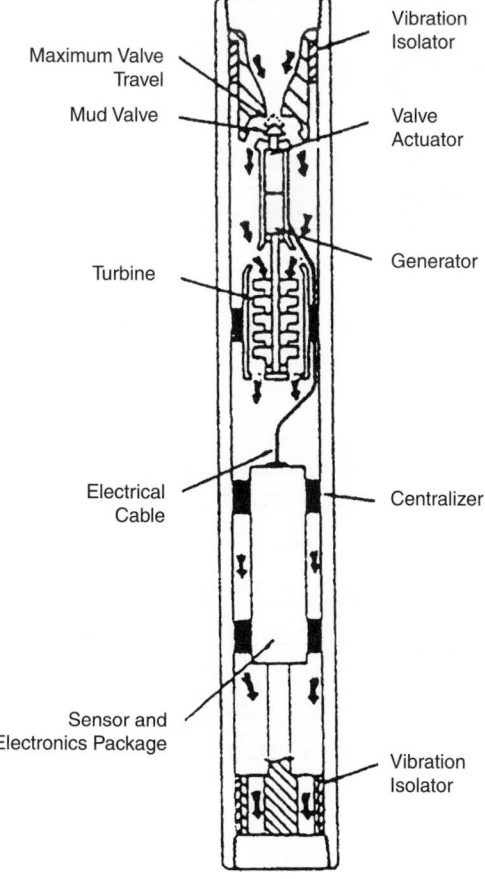

Vibration
Isolator

Maximum Valve
Travel

Mud Valve

Valve
Actuator

Turbine

Generator

Electrical
Cable

Centralizer

Sensor and
Electronics Package

Vibration
Isolator

Figure 4.11.23 *Schematic diagram of the positive pulse system (courtesy of Inteq-Teleco [17]).*

Figure 4.11.26 shows a sketch of principle of the system and of the phase-shift keying technique. Frames of data are transmitted in a sequence. Each frame contains 16 words, and each word has 10 bits. Some important parameters may be repeated in the same frame, for example, in Figure 4.11.27 the torque Tp, the resistivity R and the gamma ray GR, are repeated four times. The weight on bit WOB is repeated twice, and the alternator voltage V_{alt} one time. Note that a synchronization pulse train starts the frame.

The early system was transmitting 1.5 bits/s (4 sine wave to identify one bit). Later systems went to 3 bits/s. Now with a 24-Hz carrier frequency, 6 bits/s can be transmitted.

Then, with data compression techniques (sending only changes for most of the words in the frame and rotating the data), an effective transmission rate of 10 bits/s can be achieved.

The continuous wave technique has a definite advantage over the other techniques: a very narrow band of frequencies is needed to transmit the information. The pulse techniques, on the contrary, use a large band of frequencies, and the various noises, pump noises in particular, are more difficult to eliminate.

In principle, several channels of information could be transmitted simultaneously with the continuous wave technique. In particular, a downward channel to control the tool modes and an upward channel to bring up the information. However, continuous wave transmission requires well maintained pumps and rig circulation system.

Fluidic Pulser System. A new type of pulser is being developed at Louisiana State University. It is based on a patent by A. B. Holmes [107]. The throttling of the mud is obtained by creating a turbulent flow in a chamber as shown in Figure 4.11.28.

A vortex is generated by momentarily introducing a dissymmetry in the chamber. The resulting change in pressure loss can be switched on and off very rapidly. The switching time is approximately 1 ms and the amplitude of the pressure

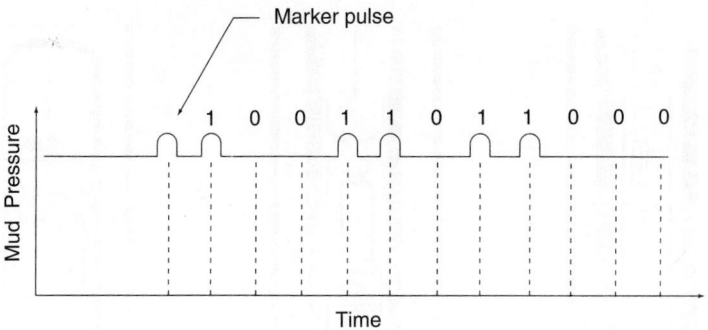

Figure 4.11.24 *Principle of the coding of the positive pulse system (courtesy of Inteq-Teleco [17]).*

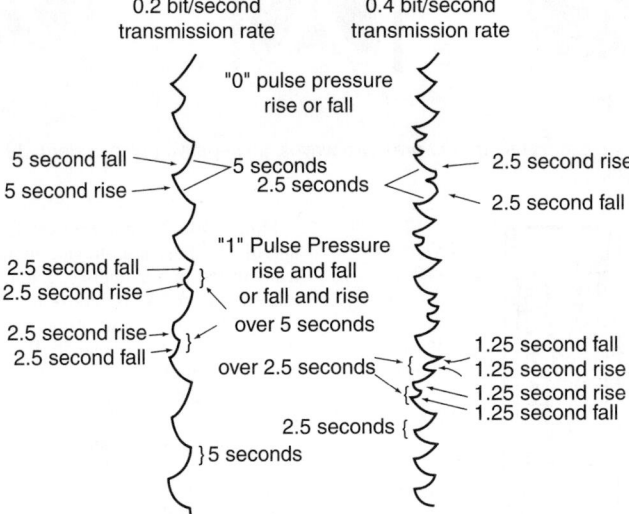

Figure 4.11.25 *Pressure waves used in the practical application of the positive pulse system (courtesy of Inteq-Teleco [17]).*

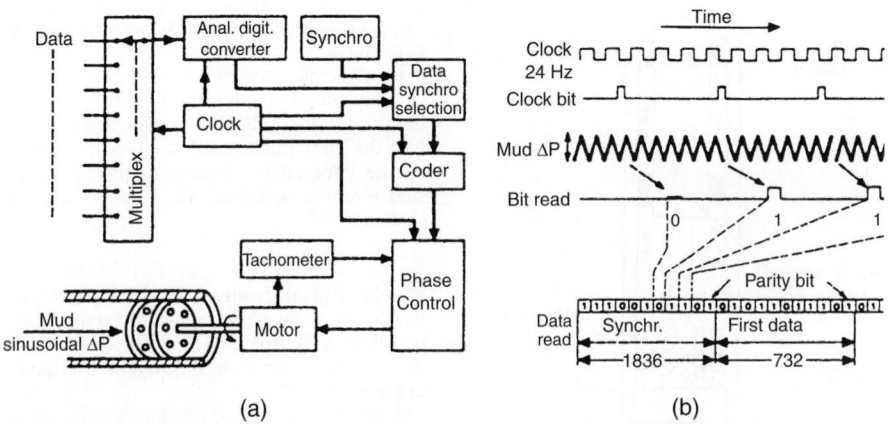

Figure 4.11.26 *Principle of the continuous wave system: (a) sketch of the siren and electronic block diagram; (b) principle of the coding by phase shift keying (courtesy of Anadrill [18]).*

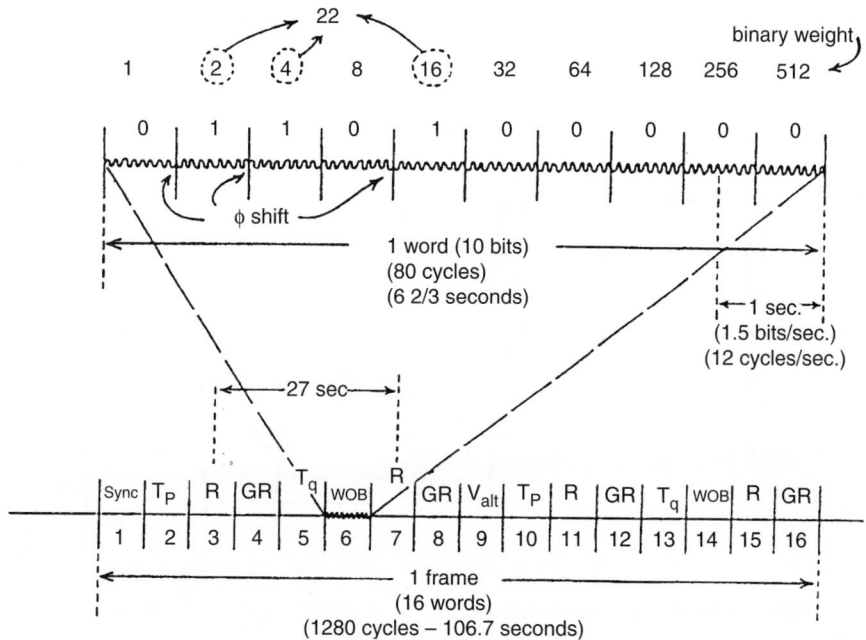

Figure 4.11.27 *Example of a frame of data transmitted by the continuous wave system (courtesy of Anadrill [18]).*

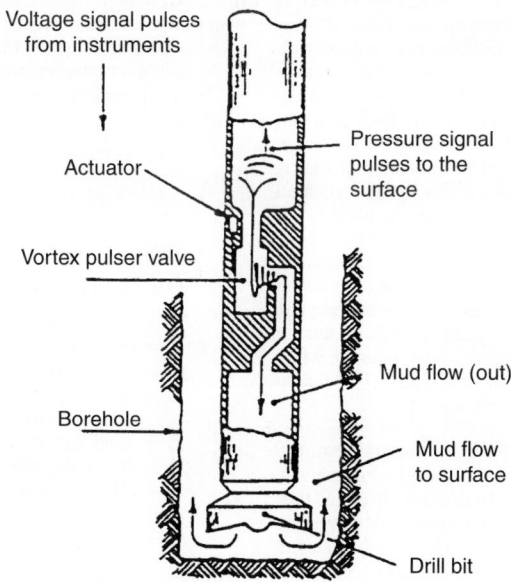

Figure 4.11.28 *Fluidic mud pulser principlet (courtesy of Louisiana State University [19]).*

loss change can be as high as 145 psi (10 bars). The prototype tool can operate up to 20 Hz. Using a continuous wave with two cycles per bit could lead to a rate of 10 bits/s. With a data compression technique, 15 effective bits per second could be transmitted, corresponding to 1.5 data per second.

Surface Detection of the Mud Pressure Signals. The pulse or wave amplitude varies largely according to depth, frequency, mud type and pulse generator device. A typical mud surface pulse amplitude is 1 bar (14.5 psi). In a sine wave transmission the surface amplitude may go as low as

0.1 bar (1.5 psi) rms. The pump noise must be lowered to minimum by the use of properly adjusted dampeners and triplex instead of duplex pumps, the pulse amplitude being about twice as large for the duplex pumps. The pump noise amplitude varies from 0.1 to 10 or more bars (1.5 to 145 psi) with dominant frequencies ranging from 2 to 10 Hz. The rotation speed of the pumps may have to be changed so the noise frequency does not interfere with the measurements. The pressure sensors are generally of the AC-type, which sense only the pressure variations. A common sensor is of the piezoelectric type with a crystal transducer. Generally

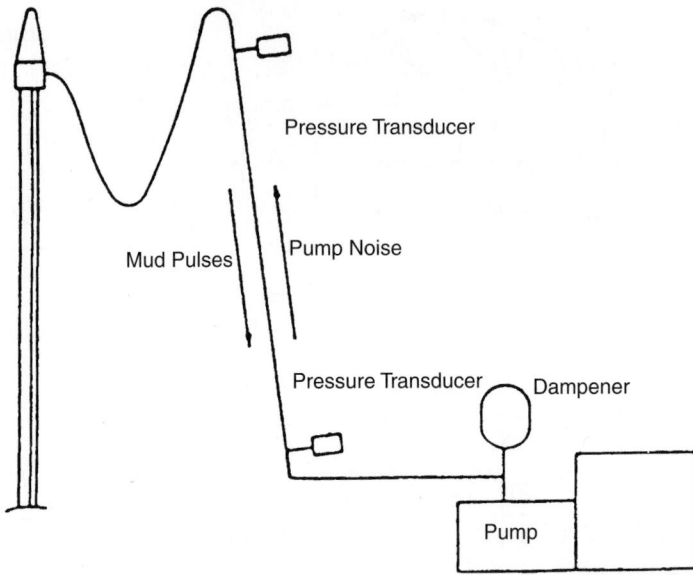

Figure 4.11.29 *Surface pressure transducers location for pump noise elimination.*

a built-in constant current follower amplifier converts the signal to a low impedance voltage. A typical sensitivity is 5 V per 1,000 psi (70 bars) with a maximum constant pressure of 10,000 psi (700 bars). The filtering can be done with digital filters or Fourier transform analyzers. For Fourier transform processing, the signal must be properly analog-filtered and then digitized. Two pressure transducers can be used at different locations on the standpipe, as shown in Figure 4.11.29, to take advantage of the phase shift that is opposite for pump noise and downhole signal. Sophisticated digital cross correlation techniques can then be used.

Downhole Recording. Most MWD service companies offer the possibility of recording the data versus time downhole. The memories available may reach several megabytes, allowing the recording of many parameter values during many hours. This information is particularly valuable when the mud pulse link breaks down. The data can be dumped in a computer, during the following drillpipe.

Retrievable Tools. Retrievable MWD tools similar to the steering tools are available from several service companies. They are generally battery powered and generate coded positive pressure mud pulses or continuous pressure waves the lower part of the tool has a mule shoe that engages in a sub for orientation. Currently tools are available for measuring directional parameters and gamma rays. A typical retrievable tool is shown in Figure 4.11.30.

The benefits of such a tool are apparent in the following instances:

- kickoffs and sidetracks
- correction runs
- high stuck-pipe risk
- high temperature
- slim hole
- low-budget drilling

Velocity and Attenuation of the Pressure Waves. The velocity and attenuation of the mud pulses or waves have been studied theoretically and experimentally. The velocity

depends on the mud weight, mud compressibility, and on the drillpipe characteristics, and varies from 4920 ft/s for a light water-base mud to 3,940 ft/s for a heavy water-base mud. An oil-base mud velocity will vary from 3,940 ft/s for a light mud to 3,280 ft/s for a heavy mud.

The propagation velocity can be calculated using the equation

$$V = \sqrt{\frac{g_c \bullet 144 \bullet B \bullet M}{\gamma(B+M)}} \qquad [4.11.10]$$

and

$$M = E\frac{(a^2-b^2)}{4 \bullet b^2\left(\frac{5}{4}-\lambda\right)+2(1+\lambda)(a^2+b^2)} \qquad [4.11.11]$$

where V = pressure wave velocity in ft/s
g_c = acceleration due to gravity: 32.17 ft/s^2
B = mud bulk modulus in psi (inverse of compressibility)
E = steel Young modulus of elasticity in psi
a = OD of the pipe in in.
b = ID of the pipe in in.
λ = steel Poisson ratio
γ = mud specific weight in lb/ft^3

Figure 4.11.31a and b gives the pressure wave amplitude versus the distance for various typical muds.

Electromagnetic Transmission Systems. One system uses a low-frequency antenna built in the drill collars. This system is a two-way electromagnetic arrangement allowing communication from bottom to surface for data transmission and from surface to bottom to activate or modify the tool mode. At any time the sequence of the transmitted parameters, as well as the transmission rate, can be modified. The tool is battery powered and can work without mud circulation. The receiver is connected between the pipe string and an electrode away from the rig for the bottom to surface mode. This system can be used on-or off-shore in theory. Two tools are available: the directional tool, which transmits

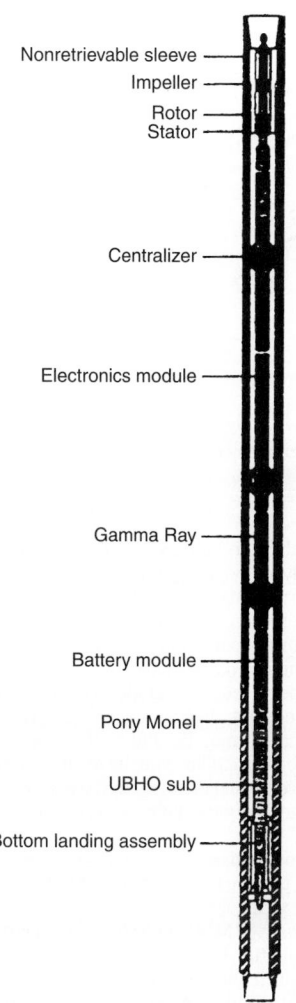

inclination, azimuth, gravity tool face or magnetic tool face, magnetic field inclination and intensity, and the formation evaluation tool, which measures gamma ray and resistivity. The formation evaluation data are stored downhole in a memory that can be interrogated from the surface or transferred to a computer when pulling out.

Figure 4.11.32a gives the attenuation per kilometer as a function of frequency for an average formation resistivity of 10 and 1 $\Omega \bullet m$.

With the downhole power available and the signal detection threshold at surface, Figure 4.11.32b gives the maximum depth that can be reached by the technique as a function of frequency. Assuming that phase-shift keying is used with two cycles per bit, in a 10 $\Omega \bullet m$ area (such as the Rocky mountains) depth of 2 km (6,000 ft) could be reached while transmitting 7 bits/s.

Coding and Decoding. Ten-bit binary codes are used to transmit the information in most techniques. In one technique, the maximum reading to be transmitted is divided ten times. In a word, each bit has the value corresponding to its rank.

Demonstration. Transmit a range of values between 0 and 90°.

Bit	1	1	1	1	1	1	1	1	1	1
Value	45	22.5	11.25	5.62	2.81	1.40	0.70	0.35	0.17	0.08789

Word 1111111111 = 89.91°
Word 1011011001 = 64.06°
Word 0001100111 = 9.04°

In another technique, each bit represents a power of two in a given word. The highest number that can be transmitted is

$$2^9 + 2^8 + 2^7 + 2^6 + 2^5 + 2^4 + 2^3 + 2^2 + 2^1 + 2^0 = 1023$$

as well as zero.

The smallest value that can be transmitted for a full scale of 90° is

$$90/1024 = 0.08789°$$

Each bit has the following numerical value:

Bit	2^9	2^8	2^7	2^6	2^5	2^4	2^3	2^2	2^1	2^0
Value	512	256	128	64	32	16	8	4	2	1

Figure 4.11.30 *Retrievable MWD tool (courtesy of Anadrill [18]).*

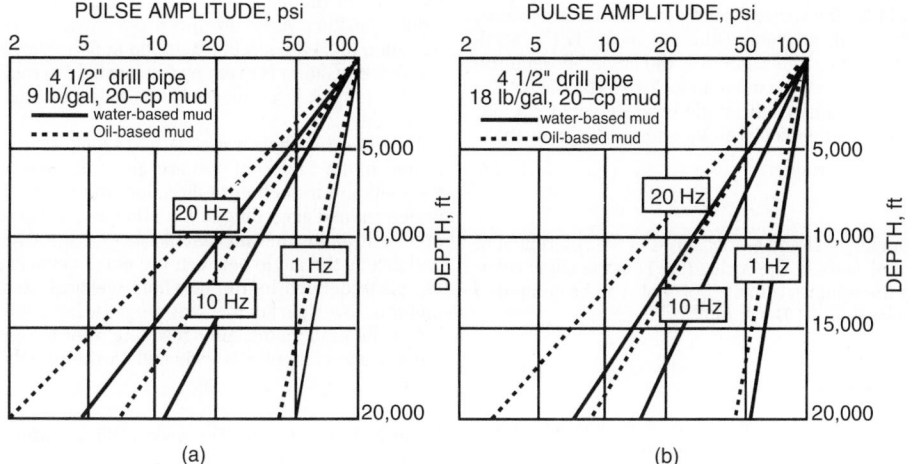

Figure 4.11.31 *Wave amplitude variation as a function of distance in water base mud and in oil-base mud: (a) mud weight, 9 lb/gal; (b) mud weight, 17.9 ft/gal (courtesy of Petroleum Engineer International [20]).*

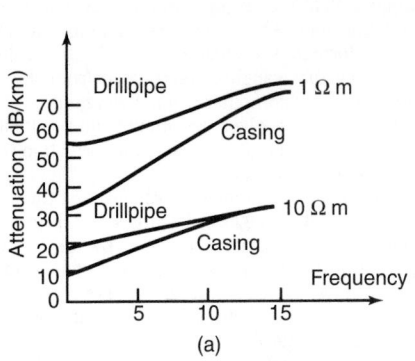 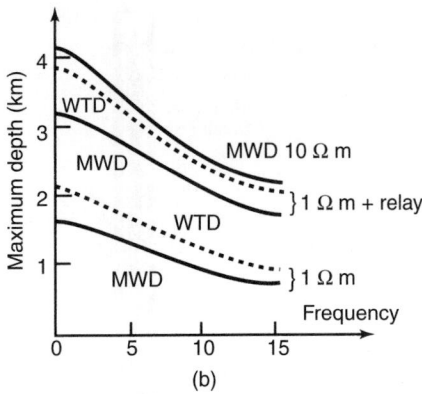

Figure 4.11.32 *Attenuation of electromagnetic signals for 1 and 10 $\Omega \bullet$ m average earth resistivity: (a) attenuation as a function of frequency; (b) maximum depth reached versus frequency (courtesy of Geoservices [21]).*

For example, to transmit 64.06°, the numerical value is

$$64.06/0.08789 = 729$$

We will have one "2^9" bit: $729 - 512 = 217$
one "2^7" bit: $217 - 128 = 89$
one "2^6" bit: $89 - 64 = 25$
one "2^4" bit: $25 - 16 = 9$
one "2^3" bit: $9 - 8 = 1$
one "2^0" bit: $1 - 1 = 0$

The word would be 1011011001 = 729. This is the same binary word as found previously. In each technique the accuracy is 0.08789° for a range of 0 to 90°. The rounding must be done the same way when coding and decoding.

Pressure variation with depth is shown in Figure 4.11.33.

4.11.1.7 Directional Drilling Parameters

With the modern accelerometers and solid-state magnetometers, a complete set of data is available for inclination, tool face and azimuth calculation. Magnetic corrections can be done. Inclination can be calculated with Equations 4.11.3 and 4.11.4. The gravity tool face angle can be calculated with Equation 4.11.5.

Azimuth calculation can be done by using vector analysis. In Figure 4.11.34, the vector $\mathbf{Z}$ represents the borehole axis, vector $\mathbf{H}$ the earth magnetic field and vector $\mathbf{G}$ the vertical or gravity vector. The azimuth is the angle between the vertical planes V_H and V_z counted clockwise starting at V_H. This angle is the same as the angle between vectors $\mathbf{A}$ and $\mathbf{B}$, respectively, perpendicular to V_H and V_z. We know that

$$\mathbf{A} = \mathbf{G} \times \mathbf{H} \text{ (vector product)} \qquad [4.11.12]$$

$$\mathbf{B} = \mathbf{G} \times \mathbf{Z}$$

The components of $\mathbf{H}$ and $\mathbf{G}$ measured in the referential of the MWD tool, and z is the vector (0,0,1) in the same referential. Now the azimuth α of the borehole can be computed with the scalar product $\mathbf{B} \bullet \mathbf{A}$; thus

$$\alpha = \arccos\left(\frac{\mathbf{A} \bullet \mathbf{B}}{|\mathbf{A}| \bullet |\mathbf{B}|}\right) \qquad [4.11.13]$$

Some precautions must be taken to be sure that the correct angle is computed since $\cos(\alpha) = \cos(-\alpha)$.

The MWD sensors are located in a nonmagnetic part of the drill collars. The magnetic collars located several meters away still have an effect by creating a perturbation in the direction of the borehole axis. This introduces an error that is empirically corrected with the single-shot instruments. Since the three components of H are measured, the magnitude of the error vector can be calculated if the module of the nonperturbed earth magnetic vector is known. The corrected dip angle vector can also be computed and compared to the non-perturbed dip angle. The computation should match; if it does not, then a nonaxial perturbation is present. This perturbation may be due to "hot spots," points in the nonmagnetic drill collar that have developed some magnetism, or to external factors such as a casing in the vicinity. Correction techniques have been introduced for the hot spots.

External magnetism due to casing or steel in the well vicinity is used in passive ranging tools for blowout well detection from a relief well.

The accuracy of MWD directional measurements is generally much better than the single- or multishot-type measurements since the sensors are more advanced and the measurements more numerous. The azimuth measurement is made with the three components of the earth magnetic field vector and only with the horizontal component in the case of the single shot or multishot. The accelerometer measurements of the inclination are also more accurate whatever the value of the inclination. The average error in the horizontal position varies from 6 ft per 3,000 ft drilled at no deviation to 24 ft per 3,000 ft drilled at 55° of deviation. The reference position is given by the inertial Ferranti platform FINDS [23]. A large dispersion is noted on the 102 wells surveyed.

When the borehole is vertical and a kickoff must be done, a mud motor and bent sub are generally used. To orient the bent sub in the target direction, the gravity toolface is undetermined according to Equation 4.11.5. Up to about 5° or 6° of deviation the *magnetic tool face* is used. The magnetic tool face is the angle between the north vertical plane and the plane defined by the borehole (vertical) and the mud motor or lower part of the bent sub if the bent sub is located below the mud motor. After reaching 5° or 6° of inclination the surface computer is switched to the gravity tool face mode.

Drilling Parameters. The main drilling parameters measured downhole are

- weight-on-bit
- torque

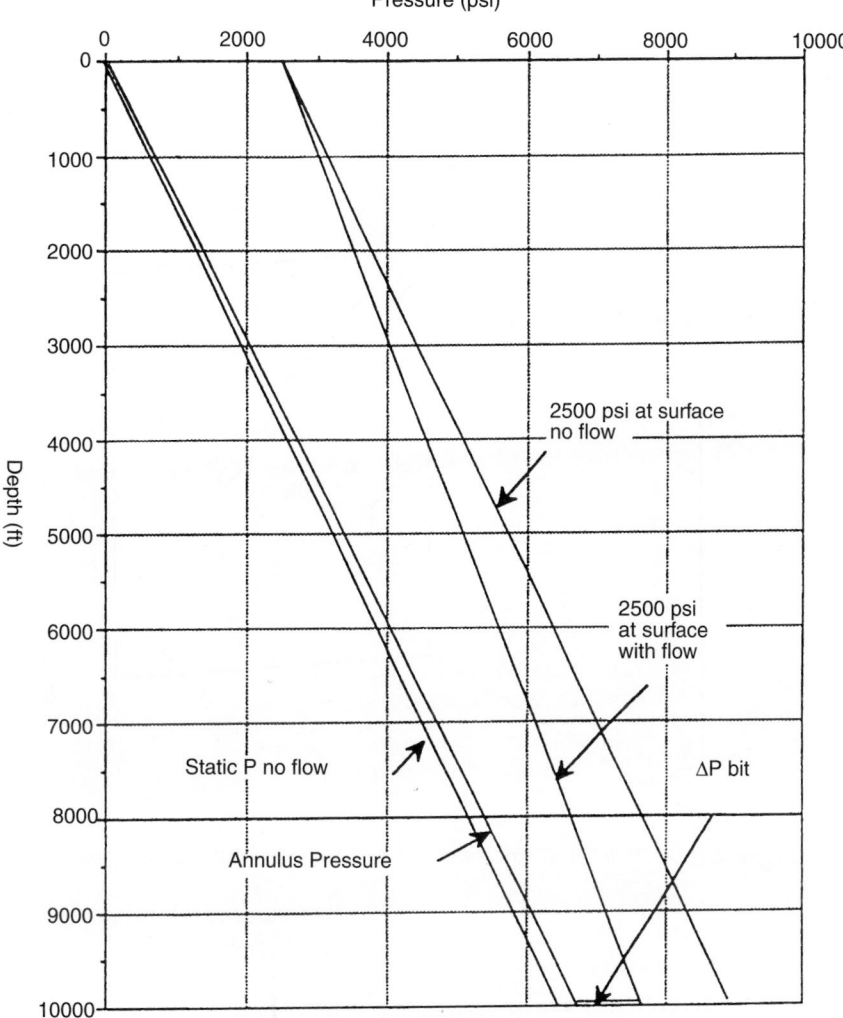

Figure 4.11.33 *Pressure variation with depth: grid with the solution.*

- bending moment
- mud pressure
- mud temperature

Strain gages are usually used for the first four measurements.

Strain Gages. Strain gages are used to measure the strain or elongation caused by the stress on a material. They are usually made of a thin foil grid laid on a plastic support as shown in Figure 4.11.35. They are the size of a small postal stamp and are glued to the structure to be stressed.

The sensitive axis is along the straight part of the conducting foil. When elongated, this conducting foil increases in resistance. The change in resistance is very low. Two gages are usually used and mounted in a Wheatstone bridge. Two more gages not submitted to the strain are also used to compensate for temperature variation.

Weight-on-Bit. Weight-on-bit is usually measured with strain gages attached to a sub subjected to axial load. The axial load is composed of three parts:

1. weight on bit proper
2. end effect due to the differential internal pressure in the drill collar
3. hydrostatic pressure effects

Hydraulic lift must also be taken into account when using diamond bits and PDC bits. The weight-on-bit varies between 0 and 100,000 lb or 0 and 50 ton-force. The end effect is due to the differential pressure between the drill collar internal pressure and the external hydrostatic pressure. This differential pressure acts on the sub internal cross-sectional area.

Torque. The torque is measured by placing two strain gages at 45° on the sub as shown in Figure 4.11.36.

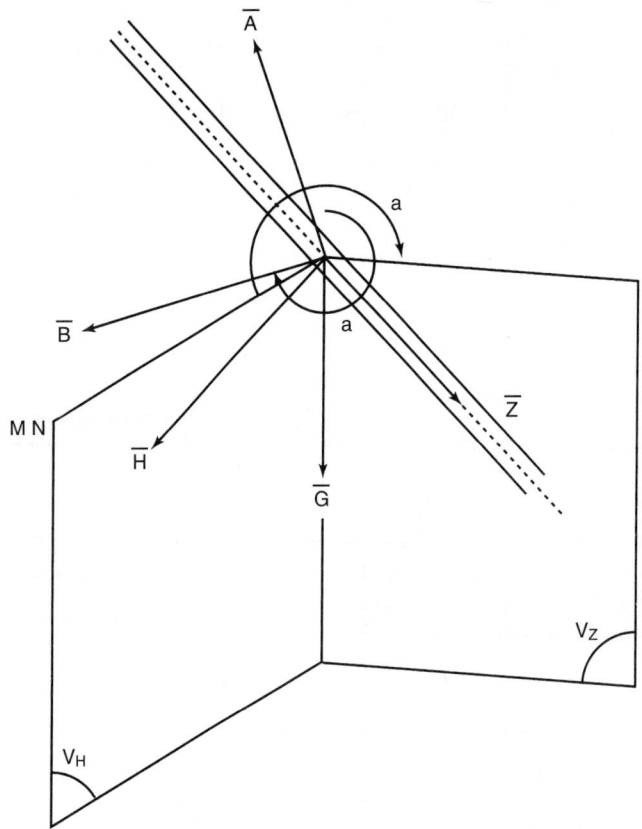

Figure 4.11.34 *Solid geometry sketch of the planes defining the azimuth angle.*

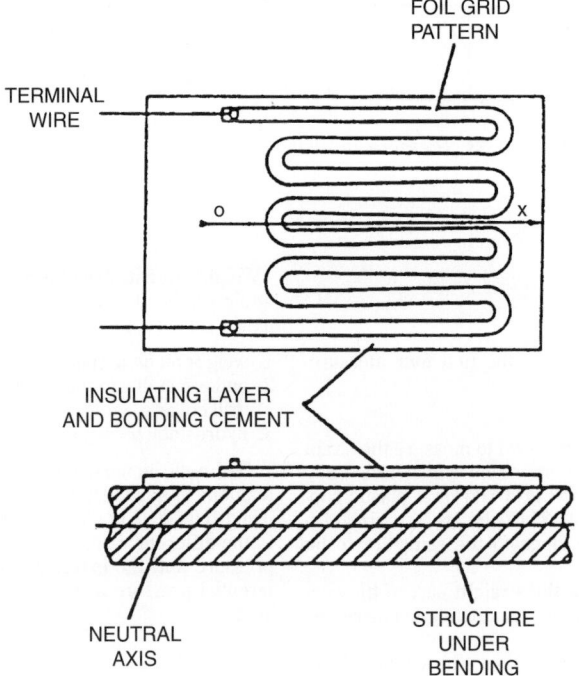

Figure 4.11.35 *Sketch of principle of glued foil strain gage transducers.*

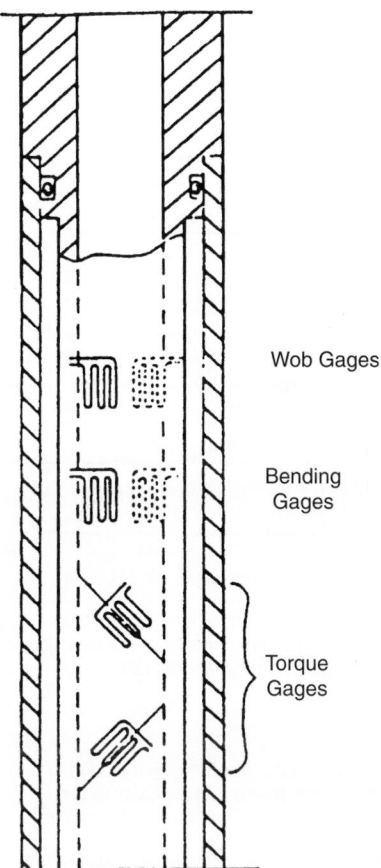

Figure 4.11.36 *Sketch of theoretical strain gage position in a sub to read WOB, torque, and bending moment.*

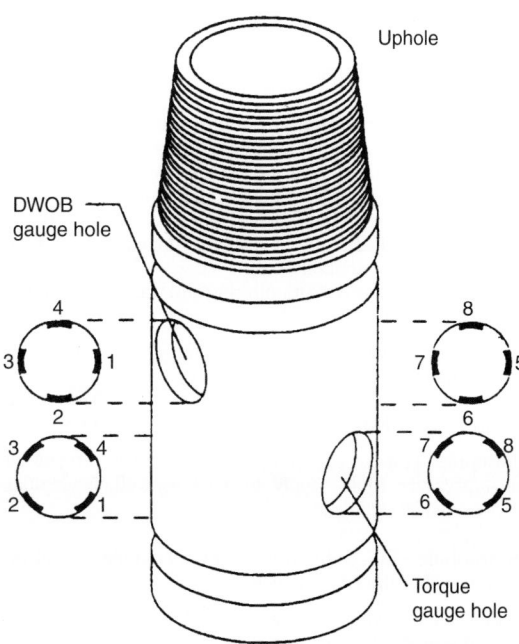

Figure 4.11.37 *Practical design of a drilling parameter sub (courtesy of Anadrill [18]).*

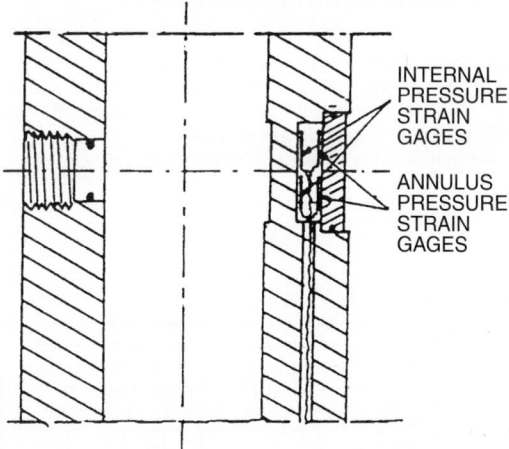

Figure 4.11.38 *Sketch of principle of downhole pressure measurements.*

Using two gages in opposite legs of the bridge will double the sensitivity.

The axial load (compression) gives a uniform stress and strain in the absence of a bending moment. If a bending moment exists, then one side is extended while the other is compressed.

During the rotation an alternative signal for the axial load is superimposed on the DC signal. By filtering, both the axial load and the bending moment can be measured. In practice, the strain gages are placed in holes drilled in the measuring sub as shown in Figure 4.11.37.

Mud Pressure. Internal and external mud pressures are usually measured with strain gages mounted on a steel diaphragm. Figure 4.11.38 shows a sketch of principle.

One steel disphragm is exposed to the internal pressure, the other is exposed to the external pressure. Four gages are normally used. Two of them are sensitive to pressure and temperature, and two are sensitive to the temperature. A Wheatstone bridge is used for detection of the pressure.

Downhole Shocks Measurements. An accelerometer in the MWD telemetry tool measures transverse accelerations, or shocks, that may be damaging for the bottomhole assemblies. When acceleration exceeds a certain threshold, the event is signaled to the surface as being a shock. These events versus time or depth are displayed as shock count. This information is used as a warning against excessive downhole vibrations and to alert the driller to change the rpm or weight on the bit [18].

A simple circuit has been designed to count the number of shocks that the tool experiences above a preset "g" level. The transverse shocks are measured in the range of 2 to 1,000 Hz in excess of the preset level. The level is adjustable and defaults at 25 g's (when no preset level is specified).

Downhole shock measurements are used to

- send alarms of excessive downhole vibration in real-time so that action can be taken to reduce damage to the MWD tools, drill bits, and bottomhole assemblies;

- reduce costly trips to replace damaged equipments;
- improve drilling rate by eliminating counter-productive BHA vibration motion.

Downhole Flowrate Measurement. Anadrill's basic MWD tool can be set up to monitor the alternator voltage being produced by the mud flowing across the MWD turbine downhole. By comparing this voltage to the standpipe pressure and the pump stroke rate, the surface system shows that a washout in the drill string is occurring much quicker than with conventional methods [18].

The downhole flowrate monitoring and washout detection system is used to

- avoid potential twist-offs from extensive drill string washouts;
- determine if the washout is above or below the MWD tool, thus saving rig time when searching for the failure.

Downhole pressure subs may achieve the same result as most pressure tools supply internal as well as annular pressure while circulating.

Bottomhole Gas Detection. Many techniques could be used for bottomhole gas detection:

- mud acoustic velocity (Figure 4.11.39)
- mud acoustic attenuation
- mud specific weight
- mud resistivity
- mud temperature
- annulus noise-level
- annular pressure changes (Figure 4.11.40)

Mud Resistivity. The mud resistivity can be measured only with the water-base muds. It is measured easily with a small microlog-type sensor embedded in the outer wall of the drill collar. Assuming the free gas is dispersed in small bubbles in the mud, the resistivity of the gas cut mud is

$$R_{gcm} = \frac{R_m}{\left(1 - f_g\right)^2} \qquad [4.11.14]$$

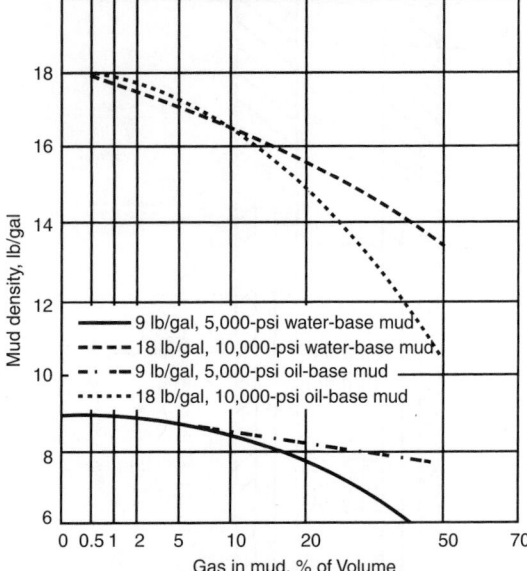

Figure 4.11.40 *Bottomhole mud density in the annulus as a function of the gas content of the mud (courtesy of Petroleum Engineer International [8]).*

where R_{gcm} = gas cut mud resistivity in $\Omega \bullet m$
R_m = gas free mud resistivity in $\Omega \bullet m$
f_g = volumetric gas content (fraction)

The variation is independent of the mud weight, pressure, or temperature, but is sensitive to fluids other than gas, such as oil or saltwater. Figure 4.11.41 shows the resistivity variations for a 1-$\Omega \bullet m$ mud. If we assume that a change of 10% can be detected, then the alarm could be given again for a free gas or oil volumetric concentration of 2 to 5%.

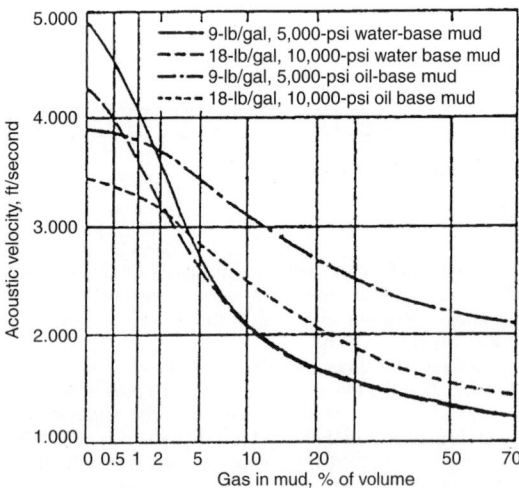

Figure 4.11.39 *Acoustic velocity in the annulus as a function of the gas content in the mud (courtesy of Petroleum Engineer International [8]).*

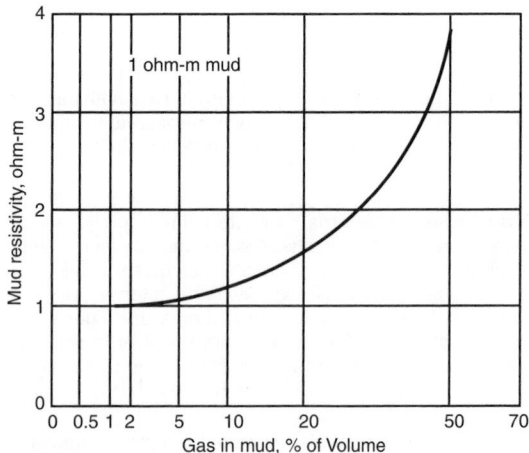

Figure 4.11.41 *Bottomhole mud resistivity in the annulus as a function of the gas content of the mud (courtesy of Petroleum Engineer International [20]).*

Mud Temperature. One can attempt to calculate the variation of the temperature of the mud when it mixes with a gas stream cooled by expansion.

Calculations were made with a 500-gal/min mud flowrate, an expansion from 10,500 to 10,000 psi with an 18-lb/gal mud and also an expansion from 5,500 to 5,000 psi with a 9-lb/gal mud. The temperature decrease of the mud was a few °F up to 50% gas by volume in the mud.

Temperature measurements do not seem to be good gas indicators.

Mud acoustic attenuation and annulus noise level are being investigated. It is expected that attenuation would be very sensitive to free gas concentration.

Example 4.11.2 Drilling Parameters — Drill Collar Pressure Drop
The following data characterize a well during drilling:

depth: 10,000 ft
$4\frac{1}{2}$-in. drillpipes (ID=3.64 in.)
mud specific weight: 12 lb/gal
flowrate: 500 gal/min
three-bit nozzles: $\frac{16}{32}$-in. diameter
mud viscosity: 12 cp
nozzle factor: C=1.0
hole diameter: 8.5 in.

1. Assuming no cutting in the annulus, compute the pressure recorded inside the drill collars downhole and the pressure in the standpipe at surface using the formula given hereafter.
2. A leak develops in the pipe string. The standpipe pressure reading drops to 1,896 psi with the same mud flowrate and the downhole drill collar inside pressure drops to 6,700 psi. What is the flowrate of the leak? What is the area of the leaking hole assuming it is located at 3,000 or 5,000 or 7,000 ft? (Assume that ΔP annulus does not change.)

Equations

1 Hydrostatic pressure is

$$P_H = 0.052 \bullet \gamma \bullet z \qquad [4.11.15]$$

where P_H = hydrostatic pressure in psi
γ = mud specific weight in lb/gal
z = depth in ft

2. Turbulent flow pressure loss in pipe is

$$\Delta P = \frac{\Delta L \bullet \gamma^{0.75} \bullet v^{1.75} \bullet \mu^{0.25}}{1800 \bullet d^{1.25}} \qquad [4.11.16]$$

where ΔP = pressure loss in psi
ΔL = pipe length in ft
γ = fluid specific weight in lb/gal
v = average fluid velocity in ft/s
m = fluid viscosity in cp
d = pipe ID in in.

with

$$v = Q/(2.448 \times d^2)$$

where Q = flowrate in gal/min
d = pipe ID in in.

3. Turbulent flow pressure loss in annulus is

$$\Delta P = \frac{\Delta L \bullet \gamma^{0.75} \bullet v^{1.75} \bullet \mu^{0.25}}{1396 (d_2 - d_1)^{1.25}} \qquad [4.11.17]$$

where (notations same as above)

d_2 = borehole or casing diameter in in.
d_1 = pipe OD in in.

4. Flowrate through a choke, or nozzle, or leak is

$$Q = C \bullet A \bullet \sqrt{\frac{2 \bullet g_c \bullet \Delta P}{144 \bullet \gamma}} \qquad [4.11.18]$$

where Q = flowrate in ft³/s
C = coefficient (0.95 to 1.0)
g_c = acceleration of gravity (32.17ft/s²)
ΔP = pressure loss in psi
γ = fluid specific weight in lb/ft³
A = area in ft²

Solution

1. ΔP drillpipes: 1,590 psi
   ΔP bit nozzles: 718 psi
   ΔP annulus: 74 psi
   Hydrostatic pressure: 6,240 psi
   P inside DC: 7,032 psi
   P standpipe: 2,382 psi
2. Total DP in pipe (friction plus lead): 1,436 psi
   Pressure drop in DC due to leak: 332 psi
   New ΔP across nozzles: 386 psi
   Q through nozzles: 366.5 gal/min
   Q through drillpipes below leak: 366.5 gal/min
   Q through leak: 133.5 gal/min

If leak at 3000 ft:

Pipe ΔP above leak: 477 psi
Pipe ΔP below leak: 646.5 psi
ΔP across leak: 312.5 psi
Leak cross-section: 0.24 in.²

If leak at 5000 ft:

Pipe ΔP above leak: 745 psi
Pipe ΔP below leak: 462 psi
ΔP across leak: 229 psi
Leak cross-section: 0.28 in.²

If leak at 7,000 ft:

Pipe ΔP above leak: 1,272 psi
Pipe ΔP below leak: 277 psi
Sum is more than total ΔP
The leak must be above 7,000 ft

4.11.2 LWD Technology

Logging while drilling has been attempted as early as 1939. The first commercial logs were run in the early 1980s. First gamma ray logs were recorded downhole and transmitted to the surface by mud pulses. Then came the resistivity logs of various types that were also recorded downhole and/or transmitted to the surface. Now, neutron-density and Pe logs are also available. Sonic logs are offered commercially by Schlumberger, Halliburton, and Pathfinder. Newest technologies now include formation pressure (D.S.T. Equivalent) monitoring.

4.11.2.1 Gamma Ray Logs

Gamma rays of various energy are emitted by potassium-40, thorium, uranium, and the daughter products of these two last elements contained in the earth formations surrounding the borehole. These elements occur primarily in shales. The gamma rays reaching the borehole form a spectrum typical of each formation extending from a few keV to several MeV.

The gamma rays are detected today with sodium iodide crystals scintillation counters. The counters, 6 to 112 in.

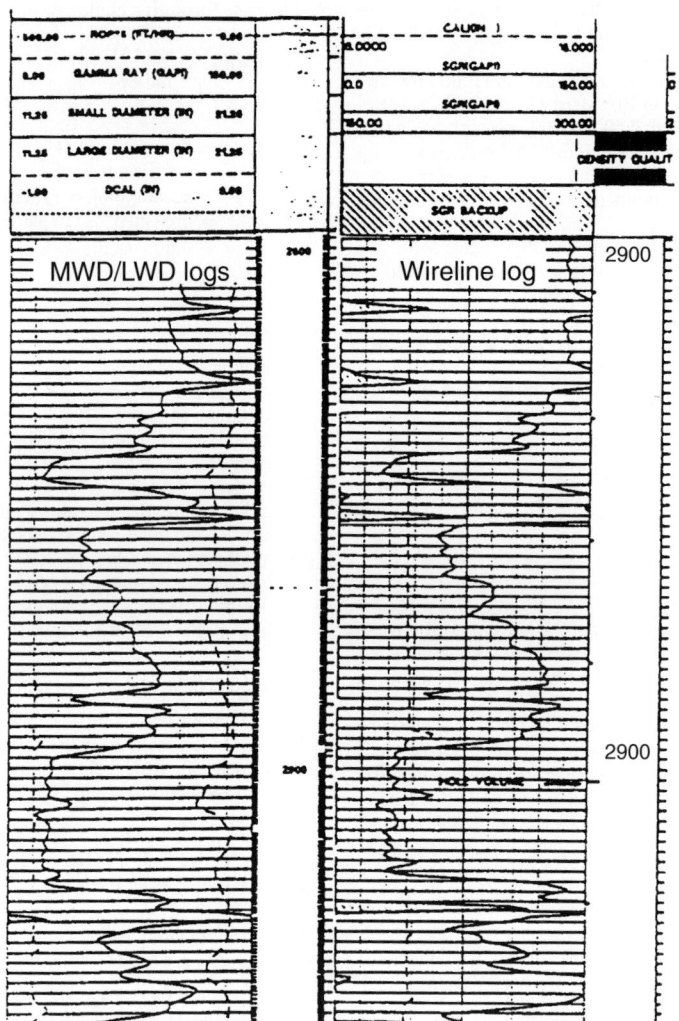

Figure 4.11.42 *Example of good similarity displayed between the MWD gamma ray log and the wireline log.*

long (15 to 30 cm) are shock mounted and housed in the drill collars. Several types of measurements can be made: total gamma rays, direction-focused gamma rays, spectral gamma rays.

Total Gamma Rays. Total gamma ray logs have been run on electric wireline since 1940. The sondes are rather small in diameter (1.5 to 4 in. or 37 to 100 mm). The steel housing rarely exceeds 0.5 in. (12 mm) and a calibration is done in terms of API units, arbitrary units defined in a standard calibration pit located at the University of Houston.

While early MWD total gamma ray tools could not be calibrated in the standard pit since they are too large, newer smaller tools can. Their calibration in API units was difficult because it varies with the spectral content of the radiation. By spectral matching the MWD logs we made to closely resemble the wireline logs. The logs which were recorded by the MWD companies in counts per second (cps) were then recorded in API units.

Another difference between the wireline logs and the MWD logs is the logging speed. With a wireline, the sonde is pulled out at a speed of 500 to 2,000 ft/min (150 to 600 m/min). The time constant used to optimize the effect of the statistical variations of the radioactivity emission, varied from 2 to 6 s. Consequently, the log values are somewhat distorted and inaccurate.

In MWD, the recording speed is the rate of penetration which rarely exceeds 120 to 150 ft/hr or 2 to 2.5 ft/min, two orders of magnitude less than the logging speed. Counters can be made shorter and time constant longer (up to 30 s or more). This results in a better accuracy and a better bed definition. Figure 4.11.42 shows an example of comparison between and MWD gamma ray log and the wireline log ran later.

To summarize, the total gamma ray measurements are used for real-time correlation, lithology identification, depth marker, and kick-off point selection.

Direction-Focused Gamma Rays. It is important to keep the trajectory of horizontal or nearly horizontal wells in the pay zone. By focusing the provenance of the gamma rays it

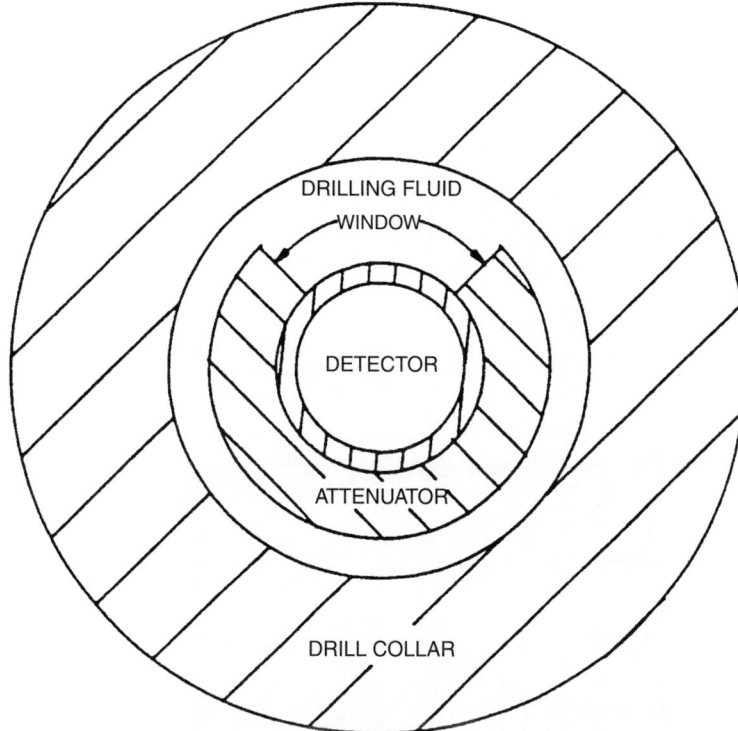

Figure 4.11.43 *Cross-section of an MWD focused gamma ray tool (courtesy of SPWLA [24]).*

is possible to determine if a shale boundary is approached from above or from below.

The tool shown in Figure 4.11.43 has its scintillation detector inserted in a beryllium-copper housing, fairly transparent to gamma rays. A tungsten sleeve surrounds the beryllium-copper housing, with a 90° slot or window running from top to bottom. Figure 4.11.43 is a sketch of the tool cross-section. The center of the window is keyed to the reference axis of the directional sensor. Consequently the directional sensor indicates if the window is pointing up or down.

By rotating the tool, one can differentiate between the level of gamma rays entering from the top and the lower part of the borehole. A sinusoidal response is recorded which depends on the following:

- distance from the bed boundary.
- gamma ray intensity of the bed in which the tool is in.
- the contrast of radioactivity at the boundary.
- the shielding efficiency of the tungsten sleeve.

An example of the log ran is a horizontal borehole as shown in Figure 4.11.44. The depths on the log are along the hole depths. Vertical depths are shown in the higher part of the log with a representation of the true radioactivity of each bed. The following observations can be made:

- Approaching formation bed boundaries are detected by concurrent separation and displacement of the high and low gamma counts. These are shown in Figure 4.11.44 at measured depth intervals (7970–7980 ft) and (8010–8020 ft).
- Radioactive events occur in the measured depth interval (8,100–8,200 ft) with no displacement of the low/high side gamma ray logs. The radioactive events must be perpendicular to the gamma detector and could be indications of vertical natural fractures in the formation.

The disadvantage of relying on such tools is due to the shallow depth of investigation. Gamma detectors rely on statistical measurements by definition.

Spectral Gamma Ray Log. This log makes use of a very efficient tool that records the individual response to the different radioactive minerals. These minerals include potassium-40 and the elements in the uranium family as well as those in the thorium family. The GR spectrum emitted by each element is made up of easily identifiable lines. As the result of Compton effect, the counter records a continuous spectrum. The presence of potassium, uranium and thorium can be quantitatively evaluated only with the help of a computer that calculates in real time the amounts present. The counter consists of a crystal optically coupled to a photomultiplier. The radiation level is measured in several energy windows.

Figure 4.11.45 shows an example of a MWD spectral GR log. On the left track, SGR is the total GR count, and CGR is this total minus the uranium count. On the right side of Figure 4.11.45 the wireline spectral gamma ray in the same interval is displayed. The curves are similar but some differences occur in the amplitude of the three curves.

The main field applications of this log are

1. Clay content evaluation: Some formations may contain nonclayey radioactive materials. Then the curve GR-U or GR-K may give a better clay content estimate.
2. Clay type identification: A plot of thorium versus potassium will indicate what type of clay is present. The thorium/potassium ratio can also be used.
3. Source rock potential of shale: A relation exists between the uranium-to-potassium ratio and the organic carbon content. The source rock potential of shale can thus be evaluated.

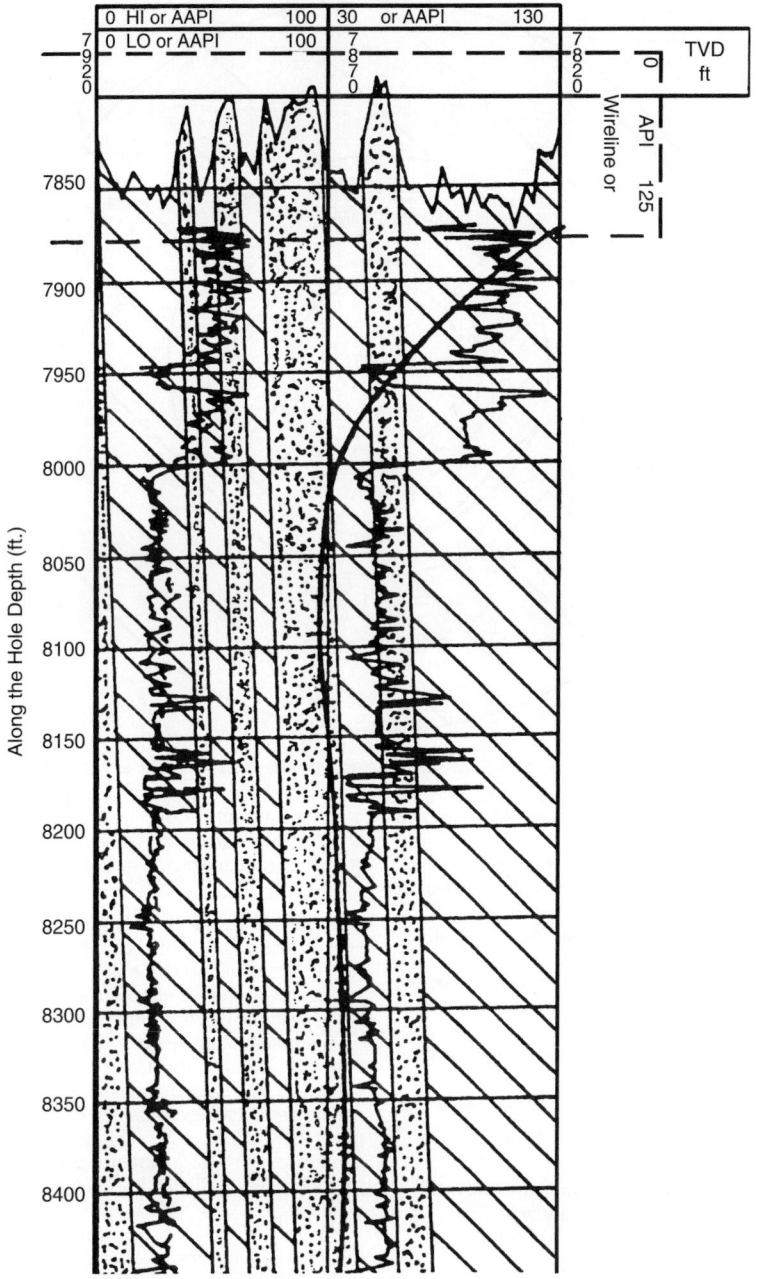

Figure 4.11.44 *MWD focused gamma ray composite log in a horizontal borehole (courtesy of SPWLA [24]).*

4.11.2.2 Resistivity Logs

Four types of resistivity logs are currently run while drilling:

1. Short normal resistivity
2. Focused current resistivity
3. Electromagnetic resistivity
4. Toroidal system resistivity

Short Normal Resistivity (after Anadrill). The short normal (SN) resistivity sub provides a real-time measurement of formation resistivity using a 16-in. electrode device

suitable for formations drilled with water-base muds having a moderate salinity. A total gamma ray measurement is included with the resistivity measurement; an annular bottomhole mud temperature sensor is optional. The short normal resistivity sub schematically shown in Figure 4.11.46 must be attached to the MWD telemetry tools and operate in the same conditions as the other sensors.

Due to the small invasion and the large diameter of the sonde body, a resistivity near the true resistivity of the formation is generally measured. This is particularly true

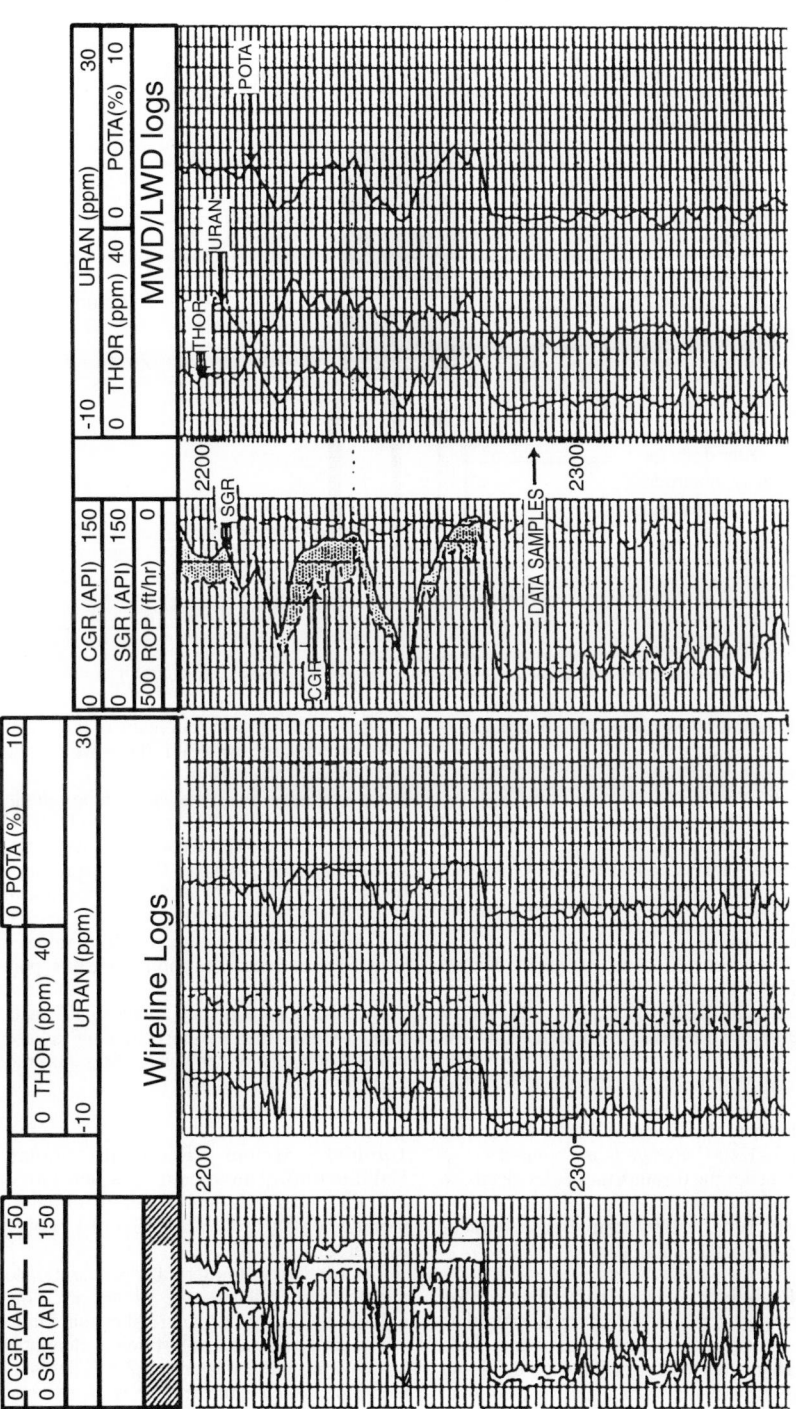

Figure 4.11.45   *Example of natural gamma ray spectral logs recorded while drilling and with a wireline.*

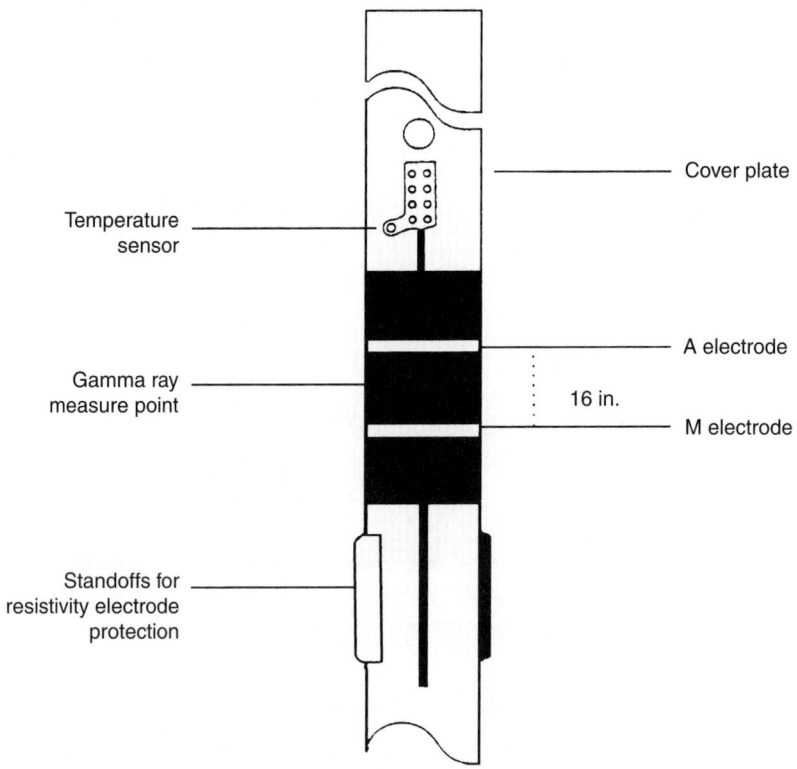

Figure 4.11.46 *Short normal resistivity sub (courtesy Anadrill [25]).*

in shale where no invasion takes place. The main applications are:

- real-time correlation and hydrocarbon identification
- lithology identification for casing point and kick-off point selection
- real-time pore pressure analysis based on resistivity trend in shales
- resistivity range: 0.2 to 100 Ω •m

Focused Current Resistivity. Focused resistivity devices are particularly suited for wells where highly conductive drilling muds are used, where relatively high formation resistivities are encountered and where large resistivity contrasts are expected.

The focused current system employs the guarded electrode design shown in Figure 4.11.47.

The system is similar to the laterolog 3 used in wireline logging. A constant 1-kHz AC voltage is maintained for all electrodes. The current flowing through the center electrode is measured.

The resistivity range is 0.1 to 1000 Ω•m. Beds as thin as 6 in. (15 cm) can be adequately delineated.

Electromagnetic Resistivity. The measurement in electromagnetic resistivity systems is similar to the wireline induction sonde resistivity. The frequency used is 2 MHz instead of 20 kHz. This is due to the drill collars steel that would completely destroy a 20-kHz signal. Early systems had one transmitter coil and two receiver coils. Systems presently in use have two to four transmitters allowing the recording of many curves with different depths of investigation. Figure 4.11.48a shows the CDR, compensated dual resistivity tool of Anadrill.

Figure 4.11.48b is a schematic of the operating principle. Two signals are measured: the wave amplitude reduction and the wave phase shift.

Two values of the resistivity can be calculated. The wave amplitude resistivity (R_{ad}) appears to have a deep investigation radius: 35 to 65 in. according to the formation resistivity. The phase shift resistiviy (R_{ps}) appears to have a shallow investigation radius: 20 to 45 in. An example of tool response is given in Figure 4.11.49.

The deep penetration curve reads a value close to the non-invaded zone resistivity and the shallow penetration curve reads a value much lower than the invaded zone resistivity. The resistivity ranges for an acceptable accuracy are 0.15 to 50 Ω•m for the deep investigation radius (R_{ad}) and 0.15 to 200 Ω•m for the shallow investigation radius (R_{ps}). The vertical resolution is 6 in. (15 cm).

Toroidal System Resistivity (after Gearhart-Halliburton). The system uses one toroidal transmitter operating at 1 kHz and a pair of toroidal receiver coils mounted on the drill collars. Figure 4.11.50 shows a sketch of a toroid.

The winding of the toroid acts as a transformer primary and the drill collar as the secondary. The current lines induced by the drill collar are shown in Figure 4.11.51.

The drill collar acts as a series of elongated electrodes in a way similar to the laterolog 3 wireline sonde. The lower electrode, which is the drill bit, is used to get the "forward" resistivity curve. A lateral resistivity measurement is made between the two toroid receivers. An example of toroid logs is shown in Figure 4.11.52.

The readings of both toroid curves seem to follow closely the ILd and ILm curves.

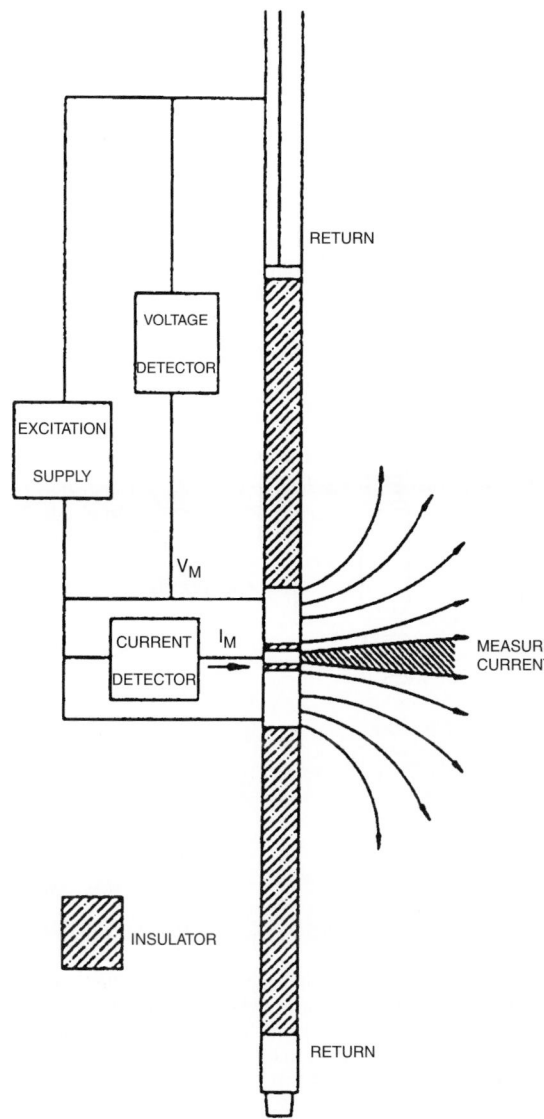

Figure 4.11.47 Block diagram of an LWD focused current system (courtesy of SPE [26]).

Example 4.11.3 Gamma Ray and Resistivity Interpretation
A typical set of logs recorded while drilling is shown in Figure 4.11.53. The wireline caliper is shown in the gamma ray track. Displayed on this attachement are gamma ray, R_{wa} curve, Pe curve, neutron and density curve. The delta-rho curve is the quality curve check for the density log.

1. Draw a lithology description in the depth column.
2. Is the clean formation permeable? Why?
3. Does the porous zone contain hydrocarbons? What type? Give the boundaries.
4. Determine R_w.
5. Compute the hydrocarbon saturation at 8400 ft assuming $a = 1$ and $m = 2$.

Solution

1. 8,450 to 8,434 ft dolomite
   8,434 to 8,430 ft shale

8,430 to 8,426 ft dolomite
8,426 to 8,423 ft shale
8,423 to 8,374 ft dolomite
8,374 to 8,350 ft shale
Rock nature is read on the P_e log.
2. Yes, a mud cake is seen on the caliper log.
3. Yes, the R_{wa} curve increases sharply in the main zone at 8425 ft. Oil from 8425 to 8400 ft. Gas above 8,400 ft. Gas is indicated by a density porous larger than the neutron porosity.
4. $R_w = 0.05\ \Omega \bullet m$; read on R_{wa} curve in the lower porous zone.
5. At 8400 ft, porosity $= 20\%$, $R_{wa} = 0.45$, $F = 25$, $R_t = 11.25\ \Omega \bullet m$

$$S_w = \sqrt{\frac{25 \bullet 0.05}{11.25}} = 0.33 = 33\%$$ [4.11.19]

$$S_{hc} = 67\%$$

4.11.2.3 Neutron-Density Logs
The physics of the measurements made by the MWD neutron-density tools are similar to those of corresponding wireline sondes. A sketch of principle of the Anadrill tool is shown in Figure 4.11.53.

For the neutron porosity measurement, fast neutrons are emitted from a 7.5-curie (Ci) americium-beryllium (Am–Be) source. The quantities of hydrogen in the formation, in the form of water or oil-filled porosity as well as crystallization water in the rock if any, primarily control the rate at which the neutrons slow down to epithermal and thermal energies. Neutrons are detected in near- and far-spacing detectors, located laterally above the source. Ratio processing is used for borehole compensation.

The energy of the detected neutrons has an epithermal component because a high percentage of the incoming thermal neutron flux is absorbed as it passes through 1 in. of drill collar steel. Furthermore, a wrap of cadmium under the detector banks shields them from the thermal neutron arriving from the inner mud channel. This mainly epithermal detection practically eliminates adverse effects caused by thermal neutron absorbers in the borehole or in the formation, such as boron.

Figure 4.11.54 shows a typical comparison of wireline and MWD gamma ray and neutron logs in a borehole in excellent hole conditions. The MWD/LWD log matches the wireline log almost perfectly.

The density section of the tool, also seen in Figure 4.11.53, uses a 1.7 curie (Ci) of 137-cesium (Ce) gamma ray source in conjunction with two gain-stabilized scintillation detectors to provide a high-quality, borehole compensated density measurement.

The tool also measures the photoelectric effect Pe for lithology identification. The density source and detectors are positioned close to the borehole wall in the fin of a full-gage clamp-on stabilizer as seen in Figure 4.11.53. This geometry excludes mud from the path of the gamma rays, greatly reducing borehole effect. In deviated and horizontal wells, the stabilizer may be run under gage for directional drilling purposes. Rotational processing provides a correction in oval holes and yields a differential caliper. Figure 4.11.55 shows a schematic of the tool positions in a borehole.

In the top part of Figure 4.11.55 the stand-off in constant during the rotation. In the oval borehole represented in the lower part of the figure, the stand-off is excessive when the density system is oriented up and the normal $\Delta \rho$ correction is not enough.

Statistical methods are used to measure the density variation as the tool rotates, the stand-off can be estimated and the density corrected. A density caliper can be computed

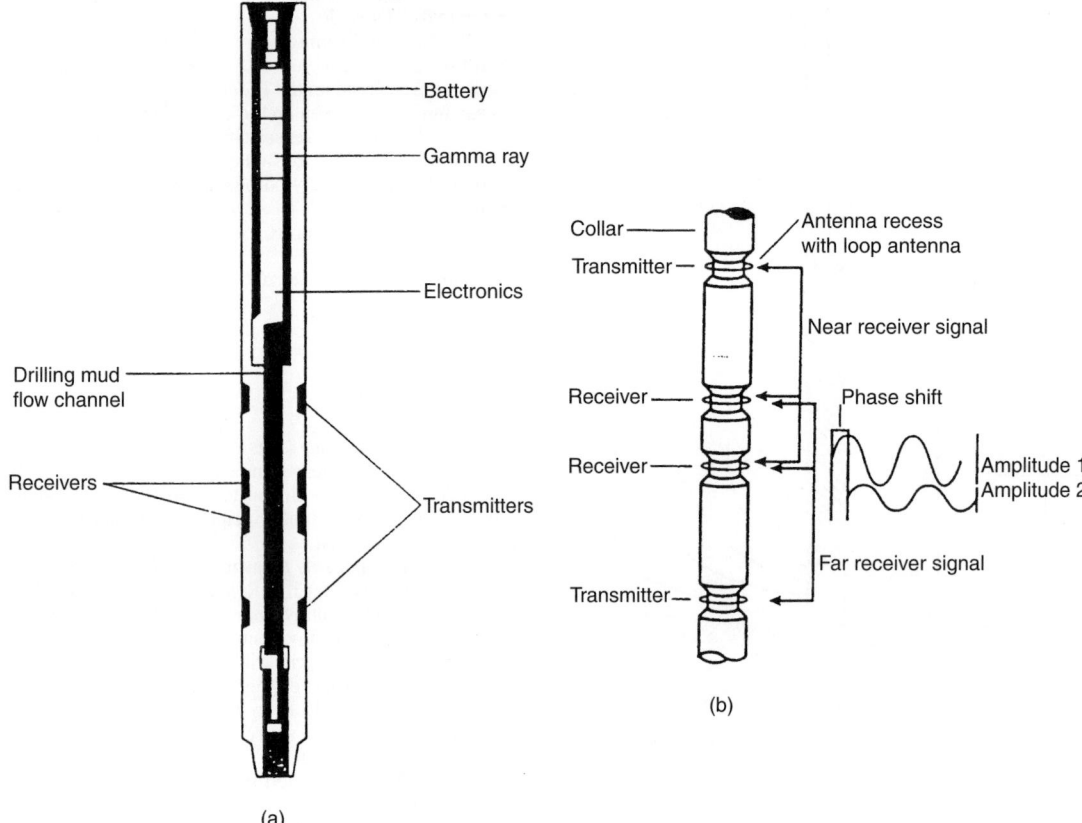

(a)

(b)

Figure 4.11.48 *Compensated dual-resistivity tool: (a) sub design; (b) operating principle (courtesy of Anadrill [25]).*

that works for cavings of 2 in. (5 cm) or less when the tool rotates at a speed ranging from 6 to 150 rpm.

Figure 4.11.56 shows a typical MWD density log compared to a wireline density log. The calipers are also shown. At 1620 ft, the wireline caliper detects a much larger caving because it was run several days later.

Photoelectric (Pe) Curve. The Compton effect (change in gamma ray energy by interaction with the formation electrons) is used for measuring the density of the formation. The energy range is 200 to 450 keV. The photoelectric effect (absorption of a low-energy gamma due to ejection of a low orbital electron from its orbit) is seen for gamma in an energy range of 35 to 100 keV.

By counting the gamma of low energy reaching the first counter a Pe curve sensitive to the nature of the formation can be recorded. A special counter protection fairly transparent to low-energy gamma ray (beryllium) is used. Table 4.11.4 shows the value of Pe for various lithologies.

Two systems are presently used by the MWD service companies concerning the radioactive source installation. One way is to lock the sources in holes in the drill collars. Thus, if the BHA is lost, the sources are left in the formation.

Another way (Anadrill) is to use removable sources. Figure 4.11.57 shows the sources being installed in the tool at surface. The two sources, neutrons and gammas, are mounted on the same flexible shaft. They are moved from

the shield to the sub without being exposed. Furthermore, if the BHA becomes stuck, they can be fished out with an overshot that connects to the fishing head on top of the neutron source possibly lost higher in the BHA, should the wireline operation have difficulties.

Example 4.11.4 Neutron-Density Logs Interpretation
The set of logs in Figure 4.11.58 have all been recorded while drilling in a Gulf Coast sand-shale sequence.

1. Give the boundaries of the clean sand. Are some zones shaly?
2. Give the probable hydrocarbon/water contact. Give the probable nature of the hydrocarbons and the gas/oil contact.
3. Compute the gas saturation at 9692 ft.
4. Can the oil saturation be computed at 9720 ft using the basic formula given in this chapter?

Solution

1. Shaly sand. Cleanest parts: 9679 to 9696 and 9803 to 9808 ft.
2. Hydrocarbon/water contact at 9750 ft with R_{wa} curve. Oil to 9696 ft. Gas above with neutron density.
3. At 9692 ft, $\Phi = 30\%$, $R_t = 8\,\Omega \bullet m$, $R_w = 0.1\,\Omega \bullet m$, $S_w = 3.3\%$, $S_g = 6.7\%$.
4. No, shaly sand formula must be used.

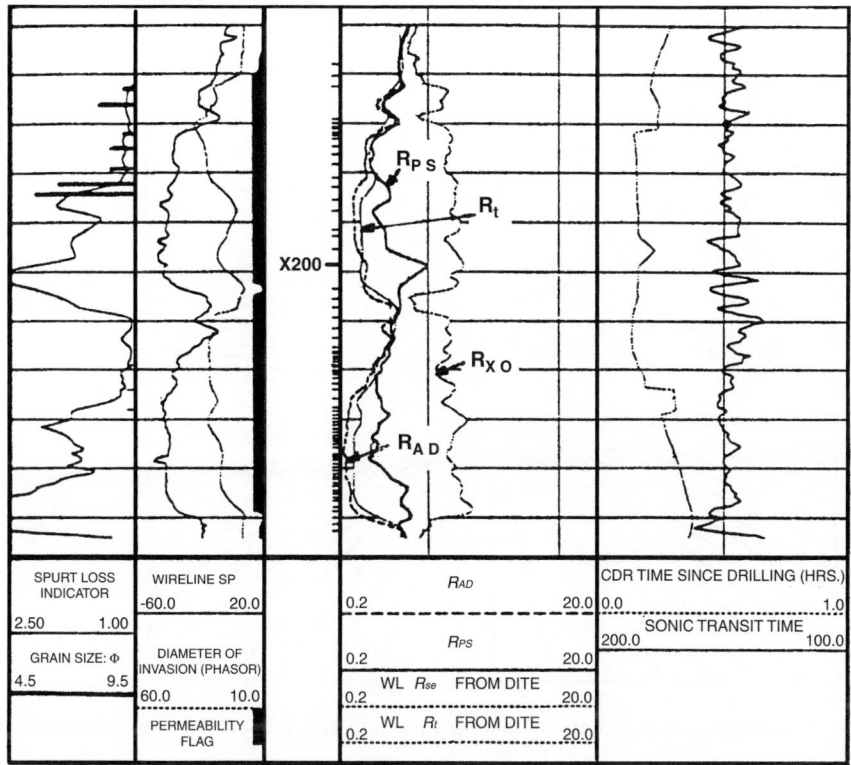

Figure 4.11.49 *Comparison of the compensated dual-resistivity log resistivities run while drilling to the invaded and noninvaded resistivities calculated with wireline phasor induction data. The spurt loss is the ratio R_{ps}/R_{ad} (courtesy of Andrill [25]).*

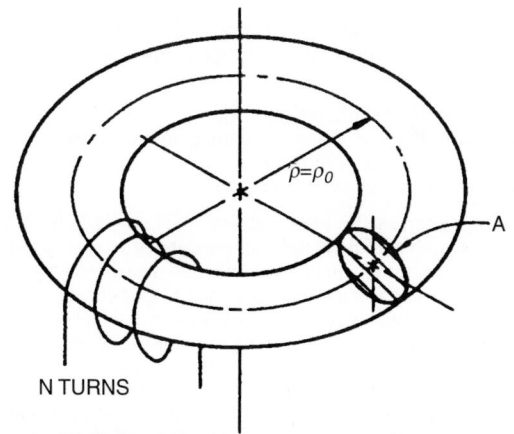

Figure 4.11.50 *Toroid mounted on a drill collar (courtesy of SPWLA [27]).*

Example 4.11.5 Neutron-Density Logs Interpretation
The Gulf Coast logs shown in Figure 4.11.58 have been run in the same interval with MWD/LWD sondes and wireline sondes.

1. Draw the lithology description in the depth column.
2. Looking at the R_{wa} curve, where is the hydrocarbon/water contact?

3. Do we have enough information to know if we have oil or gas?
4. What is the invasion diameter at 8397 ft using the wireline logs? What is the invasion diameter using the MWD/LWD logs?
5. What is needed to compute the hydrocarbon saturation?

Solution

1. Shale-sand sequence:
   Top to 8,374 ft shale
   8,374 to 8,425 ft sand
   8,425 to 8,429 ft shale break
   8,429 to bottom sand
2. Hydrocarbon/water contact: 8,413 ft.
3. No, we need the neutron and density curves.
4. According to chart in Figure 4.11.75, $d_i = 40$ in. with the wireline logs. The d_i cannot be calculated with the MWD/LWD logs since we have only two resistivity curves.
5. We need the porosity. R_w is given by R_{wa} in the lower sand. We also need R_t.

Ultrasonic Caliper and Sonic Log while Drilling. In the ultrasonic caliper sub, 3 ultrasonic sensors are mounted 120° apart (pathfinder), or two ultrasonic sensors are mounted 180° apart on stabilizer blades as shown in Figure 4.11.59.

The sensors function in a pulse-echo mode that allows the direct measurement of stand-off, from short and long axes of the borehole diameter are computed. The vertical resolution

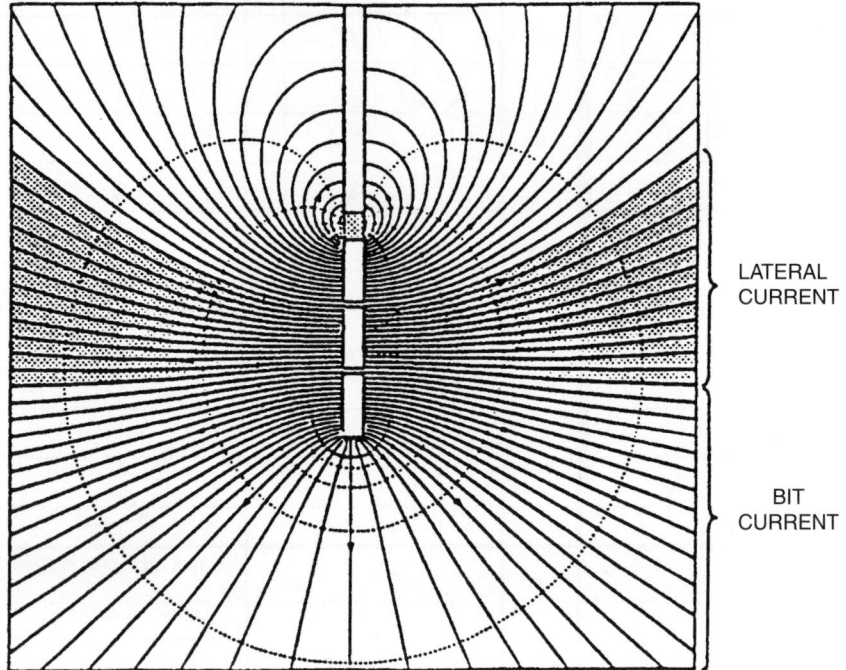

Figure 4.11.51 *Computed current pattern in a homogeneous formation for the MWD toroid system (courtesy of SPWLA [27]).*

is 1 in. (25 mm) and accuracy of the diameter measurement is ± 0.1 in (2.5 mm).

The caliper is used to correct the density and neutron porosity measurements for borehole effects and also can be used as a borehole stability indicator. Figure 4.11.60 shows an example of comparison between the MWD ultrasonic caliper and the four-arm wireline caliper run five days later.

The MWD caliper sub can also be used for downhole detection of free gas in the annulus (gas bubbles, not dissolved gas) through a combination of formation and "faceplate" echo signals. In Figure 4.11.59b a schematic of the system is represented. The faceplate is the interface between the window and the mud. The faceplate echo signal is the echo due to the impedance mismatch between the window and the mud. This echo is affected by the gas content of the mud, with echo amplitude increasing with the gas content. It can be seen in Figure 4.11.59b that concurrently the formation echo decreases.

The smallest amount of gas detectable is about 3% of free gas in volume. Real-time transmission of this information can shorten the time needed to detect gas influxes while drilling. It can help and simplify the kill operations.

Figure 4.11.61 shows an example of drilling in underbalance conditions. Gas influxes are very well outlined.

MWD Sonic. A new LWD tool developed by Anadrill provides sonic compressional Δt measurements in real time and recorded modes. Additional companies offereing sonic measurements are Halliburton and Pathfinder. The tool operates on the same general principles as modern wireline sonic tools. As the drilling operation progresses, the transmitter is actuated and acoustic waves propagating through the mud and formation are detected by the receiver array. Using a downhole processing algorithm, the compressional Δt of the formation is extracted from the waveforms and transmitted uphole in real time via mud telemetry. The compressional Δt and porosity logs are generated, providing an input for lithology identification and overpressure determination.

A successful sonic-while-drilling tool must overcome four major problems:

- suppressing collar arrivals
- transmitter and receiver mounting on drill collars
- interference of drilling noise
- processing sonic waveforms downhole

A diagram of the tool is shown in Figure 4.11.62. The array length and the use of four receivers give a good compromise for compatibility with wireline measurements and spatial aliasing properties. The choice of four receivers also minimizes memory and power requirements, which are both proportional to the number of receivers. The separation between transmitter and receivers is a compromise between a long distance for good signal amplitude and minimum tool cost. This distance is also similar to that used in wireline array tools. The receivers are small, wideband piezoceramic stacks, which have responses similar to wireline receivers.

The battery-powered electronics acquire and store sonic waveforms. Under the control of a downhole microprocessor, the transmitter is fired, and four receiver waveforms are simultaneously digitized at 12 bits and added to a signal stack. The transmitter firing is done in bursts at the rate of 10 Hz, which allows minimum movement while stacking.

The data acquistion rate is generally set so that the sample spacing of the sonic log (the distance between two acquired data points) ranges from 6 in. to 1 ft based on the anticipated drilling rate of penentration (ROP).

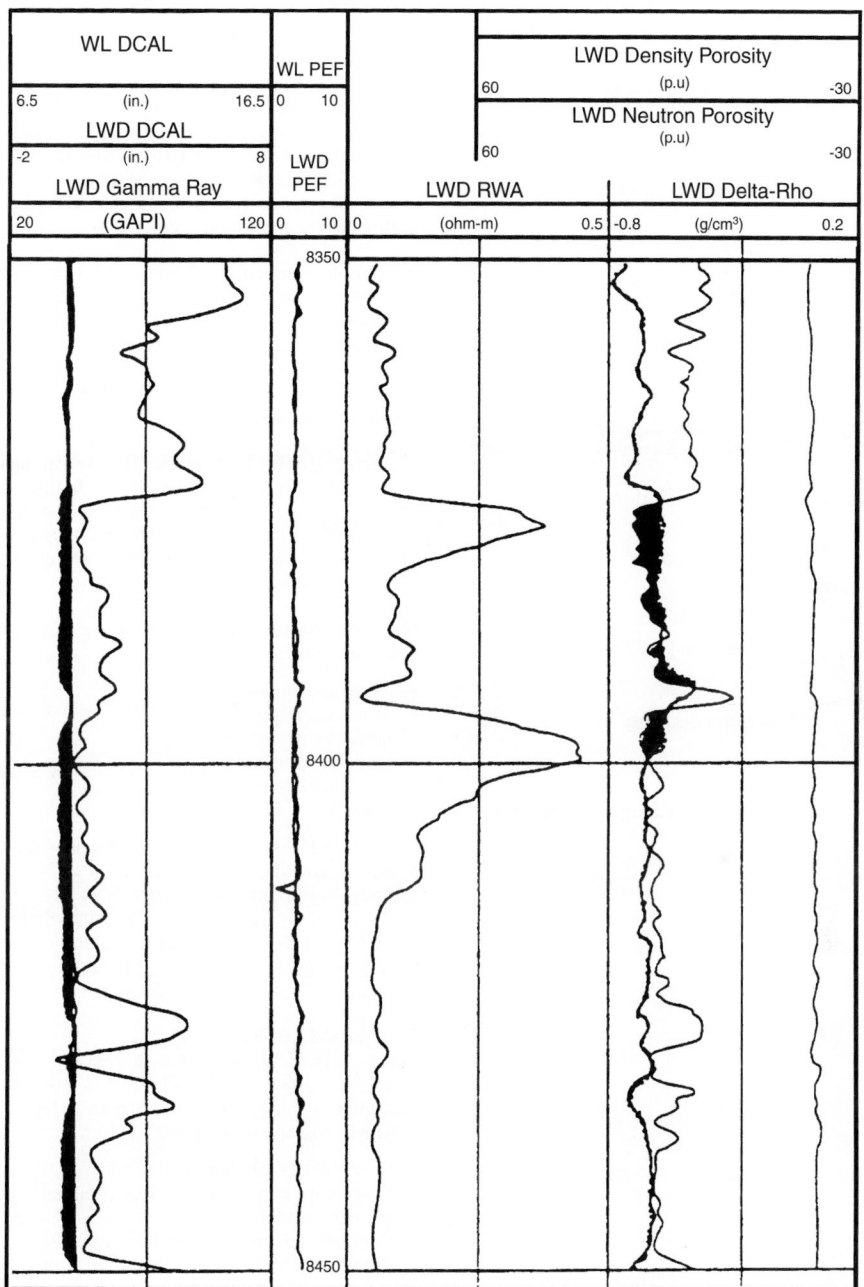

Figure 4.11.52 *LWD logs recorded while drilling [25].*

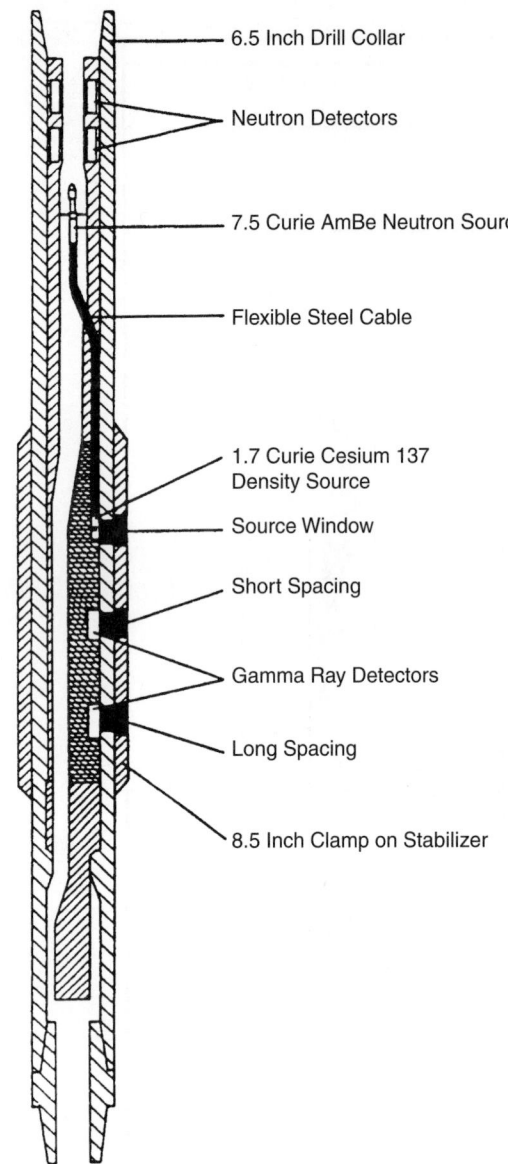

Figure 4.11.53 *Neutron-density sub (courtesy of Anadrill [25]).*

- 6.5 Inch Drill Collar
- Neutron Detectors
- 7.5 Curie AmBe Neutron Source
- Flexible Steel Cable
- 1.7 Curie Cesium 137 Density Source
- Source Window
- Short Spacing
- Gamma Ray Detectors
- Long Spacing
- 8.5 Inch Clamp on Stabilizer

For LWD/wireline sonic log comparison, we highlight the zone from 600 to 1000 ft in Figure 4.11.63. However, the two sonic measurements in the 840–860 ft interval show a major disagreement. This disagreement probably results from the deteriorated hole condition (two large washouts shown on the caliper logs) when the wireline logs were acquired (10 days after drilling).

Of particular interest in Figure 4.11.63 is the shaly sandstone in the 690–720-ft interval. In this zone, the LWD sonic measurements are consistently faster than the wireline measurements. Since the wireline logs were acquired 10 days after drilling, it is likely that shale swelling in the shaly sandstone has taken place. The phenomenon, known as formation alteration, causes the wireline sonic measurements to be slower. In this type of zone, LWD sonic yields a more correct Δt, which will better match surface seismic sections.

4.11.2.4 Measuring While Tripping: Wiper Logs
When the MWD systems are battery powered and have a downhole recording capability or use an electromagnetic telemetry, logging measurements can be repeated each time the bit is pulled out to run into the borehole. This new capability provides a way to map the progression of the filtrate front in the permeable formations. On replace wireline in highly deviated holes after drilling.

Downhole Recording. When the logging measurements are battery powered the logging parameters can be recorded versus time while tripping the drill string. If the depth is simultaneously recorded versus time, the data can be plotted versus depth. Common memory capabilities are of the order of 2 to 10 megabytes. The recording rate in adjusted to obtain about two data sets per foot.

Electromagnetic Telemetry. The electromagnetic telemetry is usually powered downhole with batteries. Parameters such as gamma ray, resistivity and temperature, can be transmitted while tripping up or down. Since a two-way communication is possible, the system can be switched to a "logging only" mode to transmit only the logging information.

Invasion Diameter Versus Time. Many parameters determine the invasion diameter:

- formation porosity
- formation permeability
- mudcake permeability
- mudcake thickness
- differential pressure
- mud filtrate and formation fluid viscosity

Figure 4.11.64 shows two typical cases for a 1-μd, 0.25-in. (6 mm) mudcake, 500 psi (3450 kPa) differential pressure, 20% porosity (a), and 30% porosity (b).

The factor permeability is important only for the low permeabilities, below 1 md. The invasion diameter increases rapidly in the first few days, making the measurements during tripping particularly significant.

Example 4.11.6 Wiper Logs
Figure 4.11.65 shows a set of resistivity logs run in a sand-shale sequence of Gulf Coast. We have one wireline dual induction log, one MWD resistivity log, a wiper-MWD resistivity log and one gamma ray log:

1. Describe the lithology of this zone.
2. How do we know that the cleaner zones marked A and B are permeable?
3. Compare the invasion for the various logs.
4. What is the true resistivity of Zone A and Zone B?

Solution

1. Laminated shaly zone; two fairly clean intervals marked A and B.
2. Invasion causes curve separation for wiper and wireline logs.
3. No invasion in the MWD/LWD log. In the wireline log, using the chart in Figure 4.11.75.
   Sand A: $d_1 = 32$ in.
   Sand B: $d_1 = 30$ in.
   Using the wiper log, we have invasion but it can not be determined quantitatively.
4. In the wireline log (R_{id}),
   Sand A: $R_t = 3.0\ \Omega \cdot m$

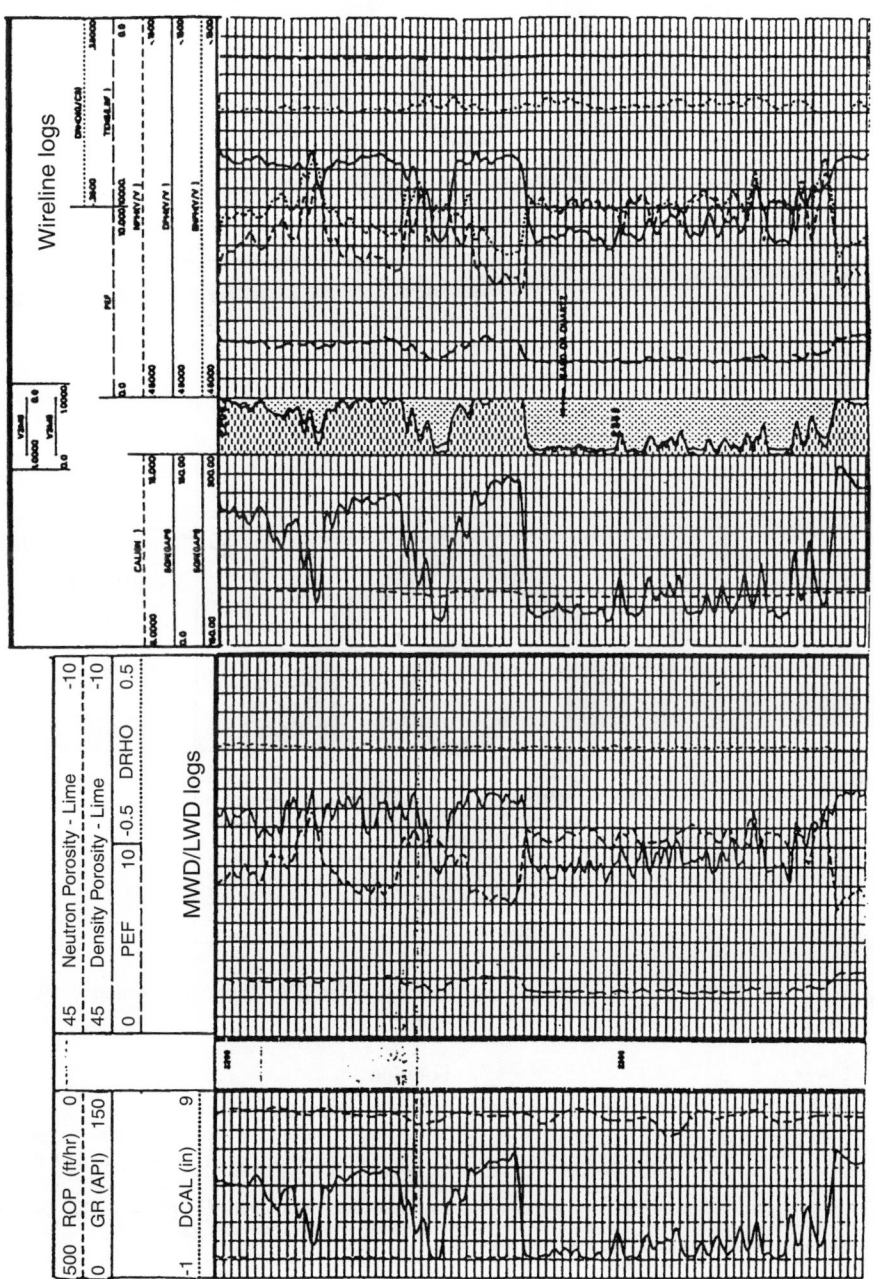

Figure 4.11.54 *Comparison of wireline and LWD gamma ray and neutron logs in a borehole in good conditions.*

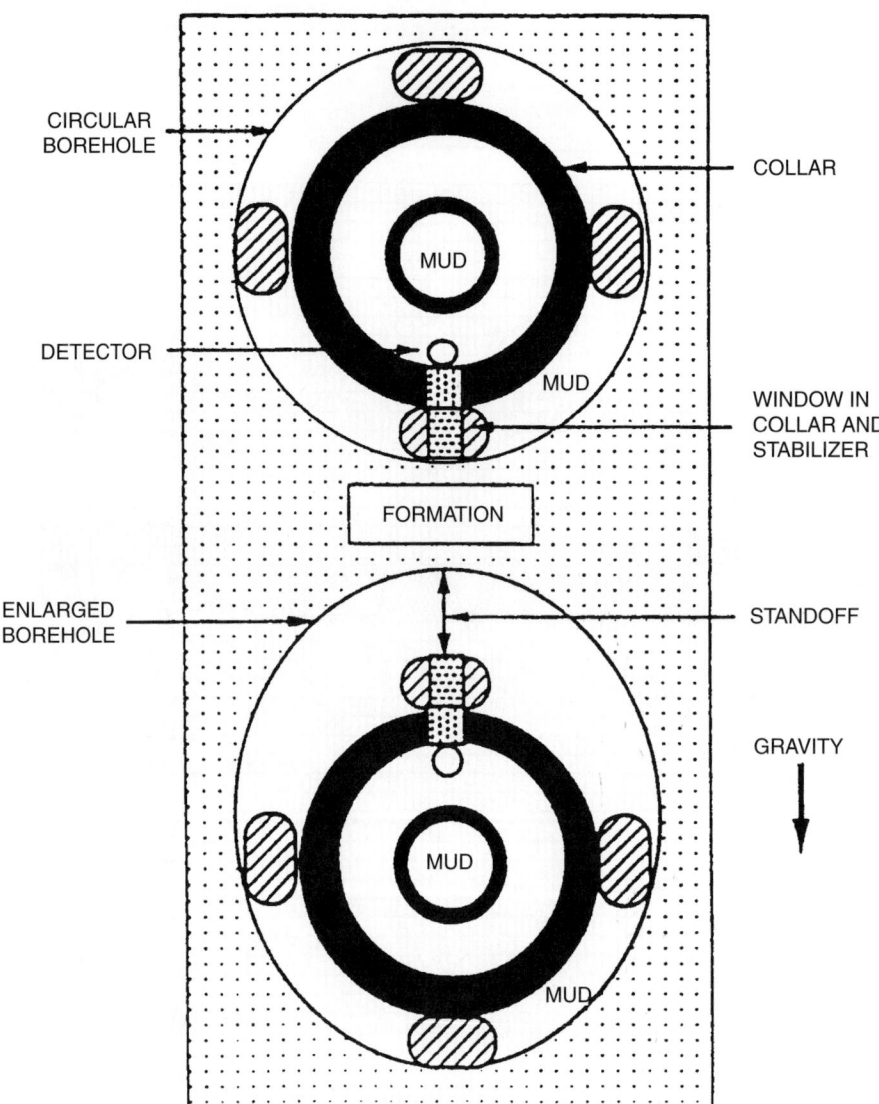

Figure 4.11.55 *Schematic of the density tool position in a borehole (courtesy of Anadrill [30]).*

Sand B: $R_t = 1.8\ \Omega \bullet m$
Using the wiper log (R_{att})
Sand A: $R_t = 3.0\ \Omega \bullet m$
Sand B: $R_t = 1.5\ \Omega \bullet m$

Example 4.11.7 Wiper Logs
Figure 4.11.66 shows a set of MWD/LWD logs recorded in a Gulf Coast well.

1. What are the boundaries of the main sand? What curve is used for lithology?
2. Is there a hydrocarbon/water contact? Where? Which curve(s) will tell us?
3. Is the upper part gas or oil saturated?
4. Determine the porosity at 400 ft.
5. Determine the hydrocarbon saturation at 400 ft.

Solution

1. Top of sand at 383 ft. Bottom of sand at 486 ft.
2. Gamma ray indicates shales and clean formations.
3. Yes, hydrocarbon/water contact at 410 ft. Resistivity and R_{wa}.
4. With chart in Figure 4.11.74, $\Phi = 28\%$. Slightly above sandstone line, probably gas.
5. Saturations: $R_{wa} = 0.01$; $R_t = 6\ \Omega \bullet m$; $\Phi = 28\%$; $S_w = 12\%$; $S_g = 88\%$.

4.11.2.5 Measurements at the Bit
A typical MWD bottomhole assembly used in rotary drilling is as follows from bottom to top: drill bit, stabilizer, resistivity, WOB torque, directional and telemetry system, neutron-density Pe. The typical distances are seen in Figure 4.11.67a. When a mud motor is inserted between the lower stabilizer

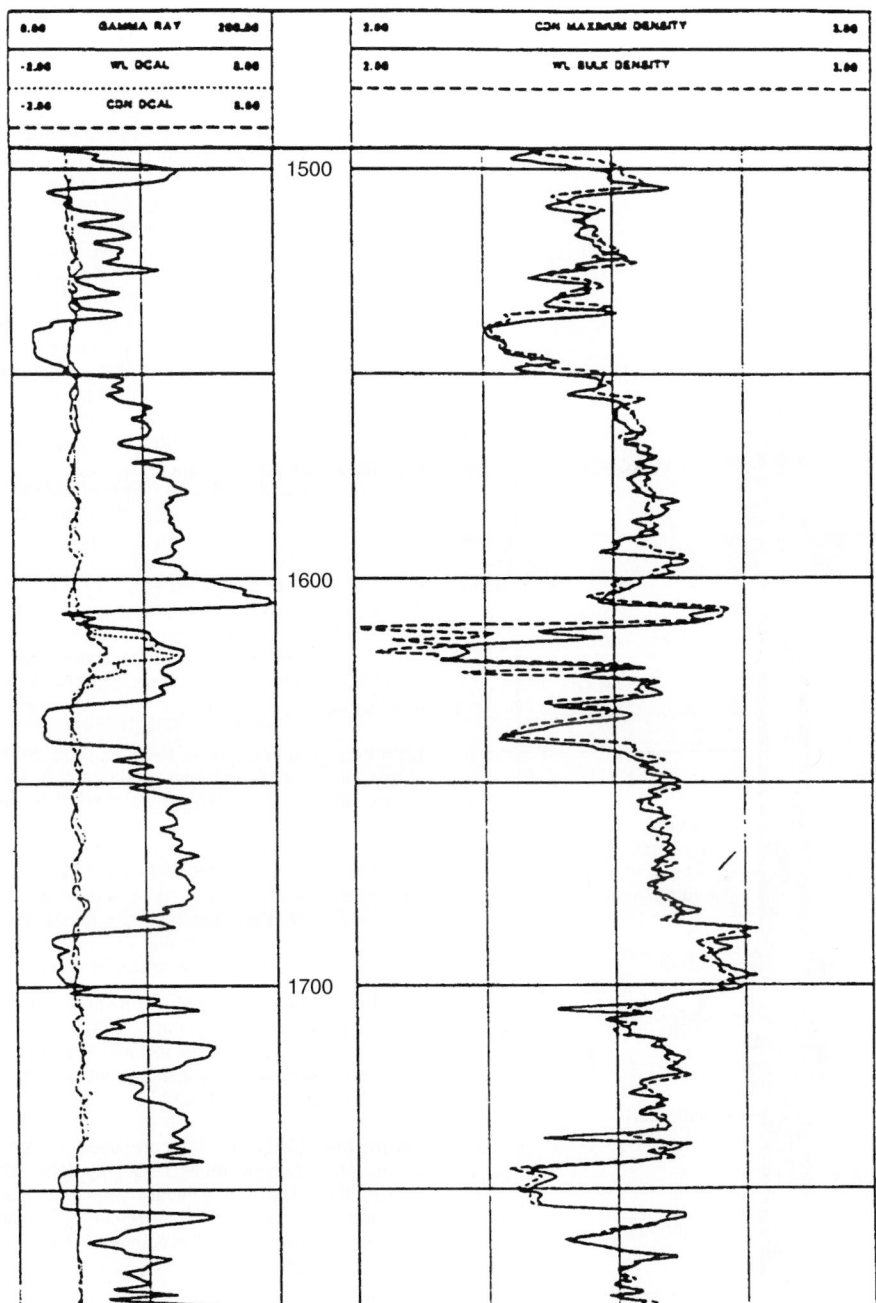

Figure 4.11.56 *Density comparison in a deviated well (courtesy of Anadrill [25]).*

and the drill bit, the distances are increased as shown in Figure 4.11.67b.

The time elapsed between drilling and recording the data at a given depth can be read in Table 4.11.5 for various rates of penetration.

To meet the challenges posed by horizontal drilling in particular, systems have been developed to make the measurements at or near the bit and transmit them to the mud telemetry section of the MWD bottomhole assembly.

The resistivity at the bit tool is similar to the toroidal resistivity tool described in the section titled "Resistivity Logs." As shown in Figure 4.11.68, the Anadrill geosteering package includes below the Power Pack mud motor:

- a surface adjustable bent housing
- a sub for measurement of the bit resistivity, azimuthal gamma ray, inclination, and bit rpm (the same sub contains an electromagnetic transmission system that

Table 4.11.4 *Pe and Rock Matrix Densities for Various Lithologies*

Lithologies	Pe	Rock Density (g/cm^3)
Sandstone	1.81	2.65
Limestone	5.08	2.71
Dolomite	3.14	2.85
Illite	3.45	2.45–2.65
Kaolinite	1.83	2.35–2.50
Smectite	2.04	2.05–2.30

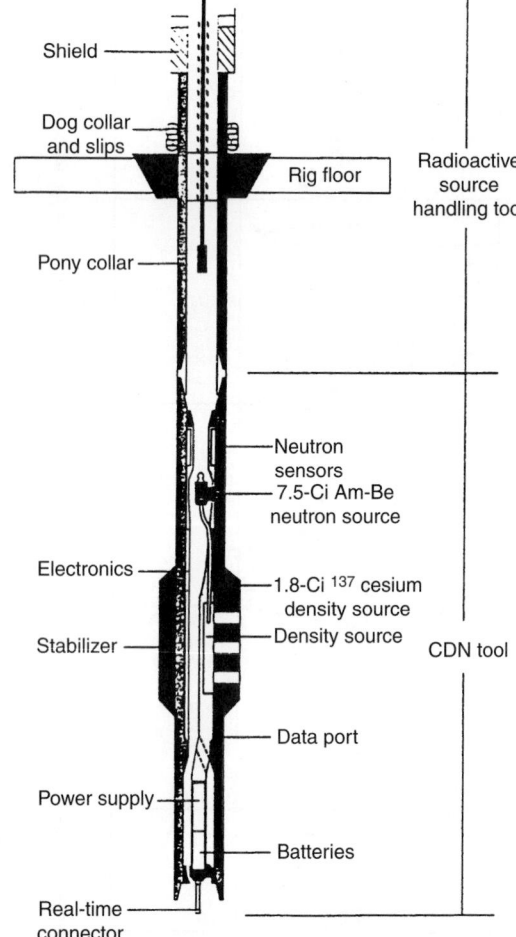

Figure 4.11.57 *Radioactive sources being installed in the neutron-density sub (courtesy of Andrill [25]).*

sends the data to the mud telemetry above the mud motor)

- a $\frac{3}{4}°$ fixed bent-housing section equipped with the bit resistivity toroid and the azimuthal resistivity button
- a stabilizer and bearing section, just above the drill bit

Figure 4.11.69a shows the sketch of principle of the resistivity measurement in water-based mud.

The drill bit resistivity is measured below toroid T_2. An average resistivity is measured between toroid T_1 and T_2. The azimuthal resistivity is measured with button R. In normal buildup drilling, the azimuthal resistivity measures the high side of the borehole and the gamma ray window sees the low side of the borehole. Measurements in other orientations are made by rotating the bottom-hole assembly. In the geosteering configuration, the resolution of resistivity at the bit is 6 ft (1.8 m). When drilling in rotary, the resistivity at the bit (RAB) is also measured with a similar toroidal system as shown in Figure 4.11.69b. The RAB tool has three button electrodes spaced to give azimuthal resistivities with three depths of investigation: 3, 6 and 9 in. (7.6, 15 and 23 cm). A ring electrode is also provided to yield an axial focused resistivity 5 ft above the drill bit. Focusing is improved by a near-bit transmitter. The radius of investigation of the ring electrode is about 12 in. (30 cm) and the vertical resolution about 2 in (5 cm). Figure 4.11.70 is a sketch of the ring electrode measurement. The button electrode measurement is similar with a button instead of the ring.

Both rotary and mud motor systems use an electromagnetic wireline telemetry to relay the data from the near-bit sub to the mud telemetry sub.

4.11.2.6 Basic Log Interpretation

The log interpretation using logging while drilling logs is very similar to the interpretation made with wireline logs. One major difference is that the invasion is usually less important due to the short time elapsed between drilling and logging.

Lithology. Gamma ray is used to differentiate between shales and clean formations. Pe is used to determine the nature of the clean formations: sandstone, limestone or dolomite.

Porosity. Two porosity logs are run today (1994); neutron and density. Soon the sonic will also be available.

With the lithology matching the log scale, and assuming the formation fully invaded by mud filtrate, a neutron porosity and a density porosity can be determined.

- If the two porosities match, the formation is saturated with a liquid, water or oil, and the porosities are true porosities.
- If the neutron porosity is low and the density porosity is high, the formation contains gas and the true porosity can be determined with charts.

Saturation. Resistivity logs are used for saturation estimates. One, two or more resistivity logs with different investigation depths are usually available. The true resistivity of the formation can be estimated. If a water sand has been drilled near the interval of interest, the formation water R_w can be determined by

$$R_w = \frac{R_0}{F} \qquad [4.11.20]$$

where R_0 = true resistivity of the water sand
$\qquad$ F = formation factor

and

$$F = \frac{1}{\Phi^2} \qquad [4.11.21]$$

or

$$F = \frac{0.62}{\Phi^{2.15}} \qquad [4.11.22]$$

where Φ = porosity in fraction

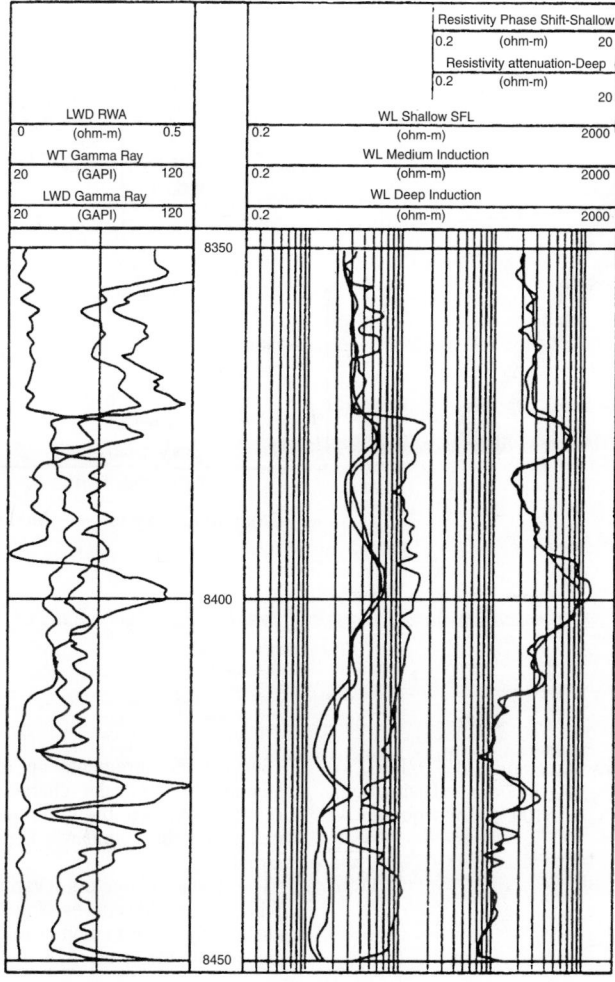

Figure 4.11.58 *Wireline and LWD logs showing the effect of invasion (courtesy of Anadrill [25]).*

Equation 4.11.21 is used in carbonates (limestone and dolomite) and Equation 4.11.22 is used in unconsolidated to medium consolidated sandstones. A third Equation 4.11.23 can be used in highly consolidated sandstones. This is

$$F = \frac{0.81}{\Phi^2} \qquad [4.11.23]$$

If no water sand exists, then an estimate of R_w must be made based on regional knowledge.

Since the true porosity has been determined, formation water saturation can be determined by

$$S_w = \sqrt{\frac{F \cdot R_w}{R_t}} \qquad [4.11.24]$$

where

S_w = formation water saturation
F = formation factor
R_w = formation water resistivity
R_t = true resistivity of the formation

The oil or gas saturation is

$$S_{hc} = 1 - S_w \qquad [4.11.25]$$

where

S_{hc} = hydrocarbon saturation

The volume of hydrocarbons in place at reservoir conditions is

$$Q_{IP} = 775 \cdot \Phi \cdot S_{hc} \cdot h \qquad [4.11.26]$$

where

Q_{IP} = volume of hydrocarbons in bbl/acre
Φ = porosity in fraction
S_{hc} = hydrocarbon saturation in fraction
h = formation thickness in ft.

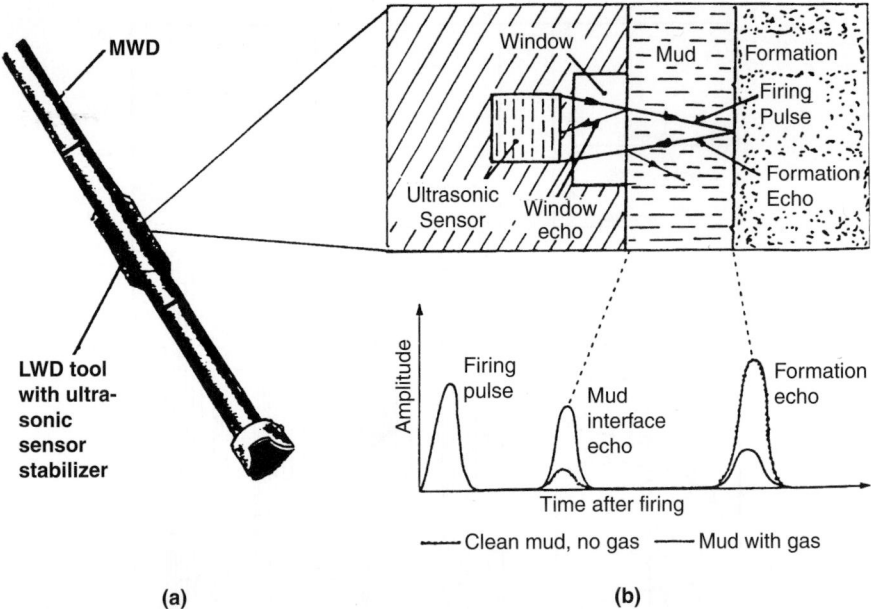

Figure 4.11.59 *MWD ultrasonic caliper: (a) sensor in the stabilizer blades; (b) schematic of the echoes (courtesy of Anadrill [25]).*

Apparent Water Resistivity R_{wa}. Some MWD/LWD log sets display a curve labeled R_{wa}. R_{wa} is computed using Equation 4.11.24 assuming that $S_w = 1$ (100%). Consequently we have

$$R_w = R_{wa} = \frac{R_t}{F} \qquad [4.11.27]$$

Since

$$F = \frac{a}{\Phi^m} \qquad [4.11.28]$$

where

F = formation factor
a = constant depending on the formation, generally 1 or 0.81 or 0.62
m = cementation factor, generally 2 or 2.15

Finally,

$$R_{wa} = \frac{R_t \Phi^m}{a} \qquad [4.11.29]$$

The true porosity Φ is determined with the neutron-density Pe logs. R_t is generally given by the deep investigation resistivity curve. R_{wa} equals R_w in the water formations. It increases rapidly in hydrocarbon saturated formations.

Permeability. Permeable zones can be identified with the resistivity measurements made with different radius of investigation. A departure between the curves of deep and shallow investigation is a qualitative indication of permeability.

The charts mentioned in the section titled "Measuring While Tripping: Wiper Logs" can be used to estimate quantitatively the permeability if several measurements during tripping are made with resistivity devices that can give the invasion diameter.

Surface measurements on the mud can be used to estimate the mudcake characteristics. If the formation pressure is known, the differential pressure can be calculated, and a chart similar to Figures 4.11.64a and b can be plotted.

The invasion diameters at various times should follow one of the permeability curves. Note that the permeability effect is seen only for formations with 1 md or less permeability. Above 1 md, the invasion diameter is dependent mostly on porosity.

Log Samples. Figure 4.11.71 shows samples of gamma ray and spectral gamma ray logs. The boundaries of the clean (not shaly) zone can be seen very clearly:

2,256–2,274 ft shale
2,274–2,356 ft clean zone
2,356–2,388 ft shale

Shale streaks are visible in most of the lower part of the clean section.

A high radioactivity peak can be seen at 2388 ft. Looking at the spectral log it can be seen that this peak is due to a high uranium content. Other radioactive elements' concentration is normal.

Figure 4.11.72 shows sample neutron-density Pe log over the same interval. Figure 4.11.73 shows sample MWD resistivity (left) and wireline dual induction (right) for the same interval.

A high neutron porosity and low density porosity occur in the shale zones: 2256–2274 and 2356–2388 ft. In the clean zone, around 2280–2290 ft, the cleanest part reads $\Phi_n = 18\%$ and $\Phi_d = 13\%$. Plotting the point in the CNL chart of Figure 4.11.74, the rock matrix appears to be a dolomitized limestone and the true porosity is 17%.

(text continued on page 4-355)

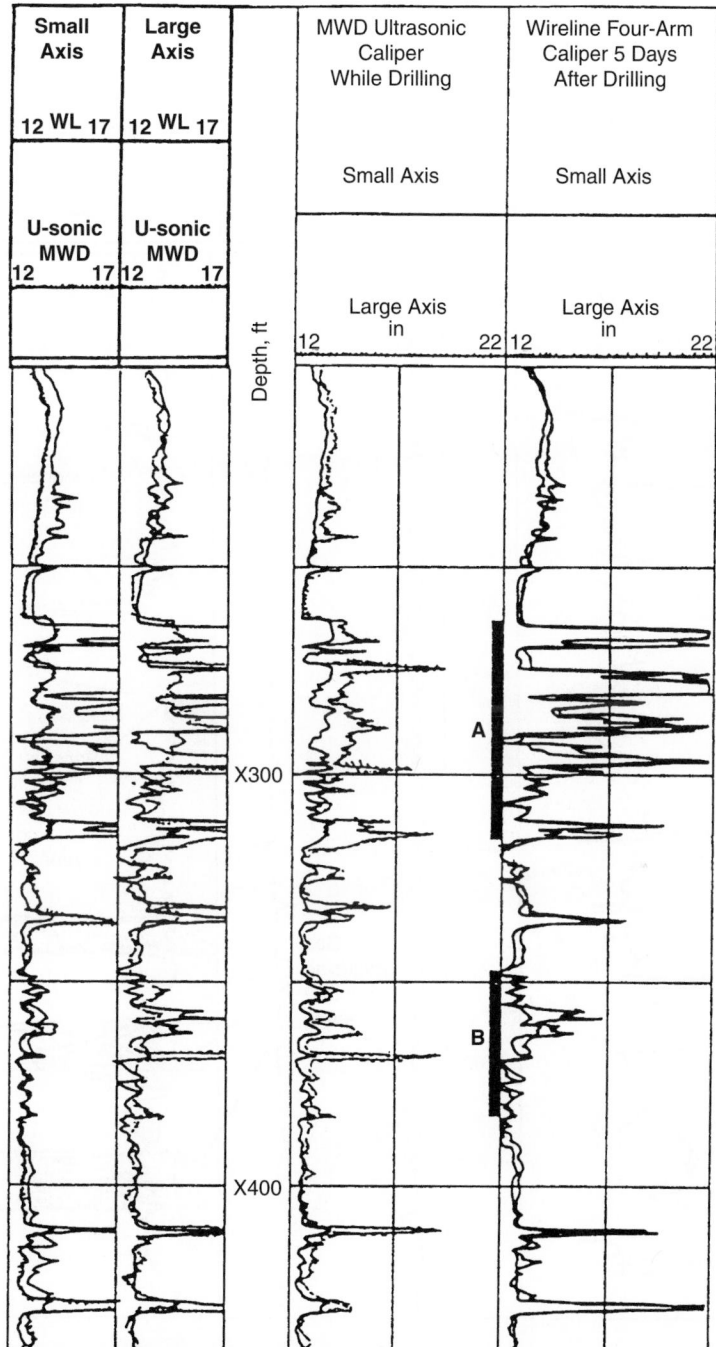

Figure 4.11.60 *Comparison of MWD ultrasonic caliper logs with the 4-arm wireline caliper logs run five days later (courtesy of Anadrill [25]).*

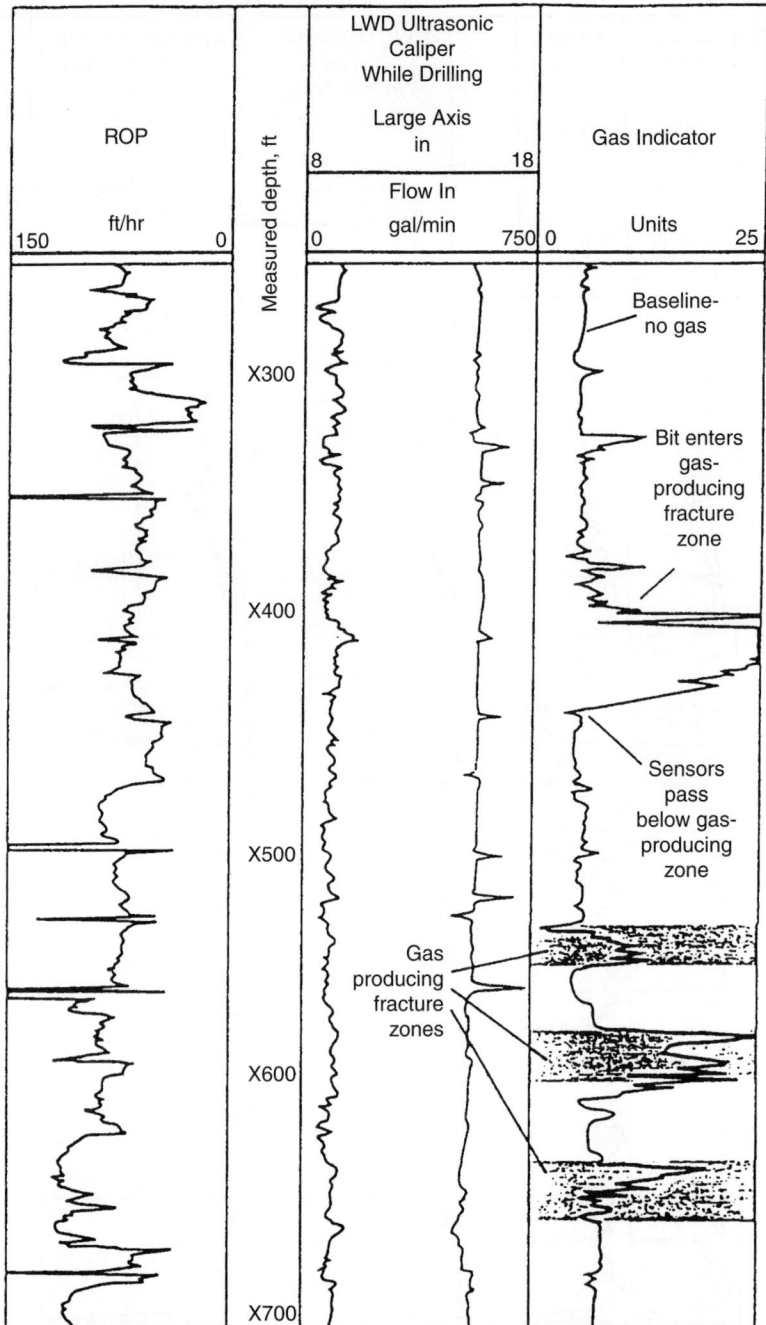

Figure 4.11.61   *MWD ultrasonic caliper and gas influx log (courtesy of Anadrill [25]).*

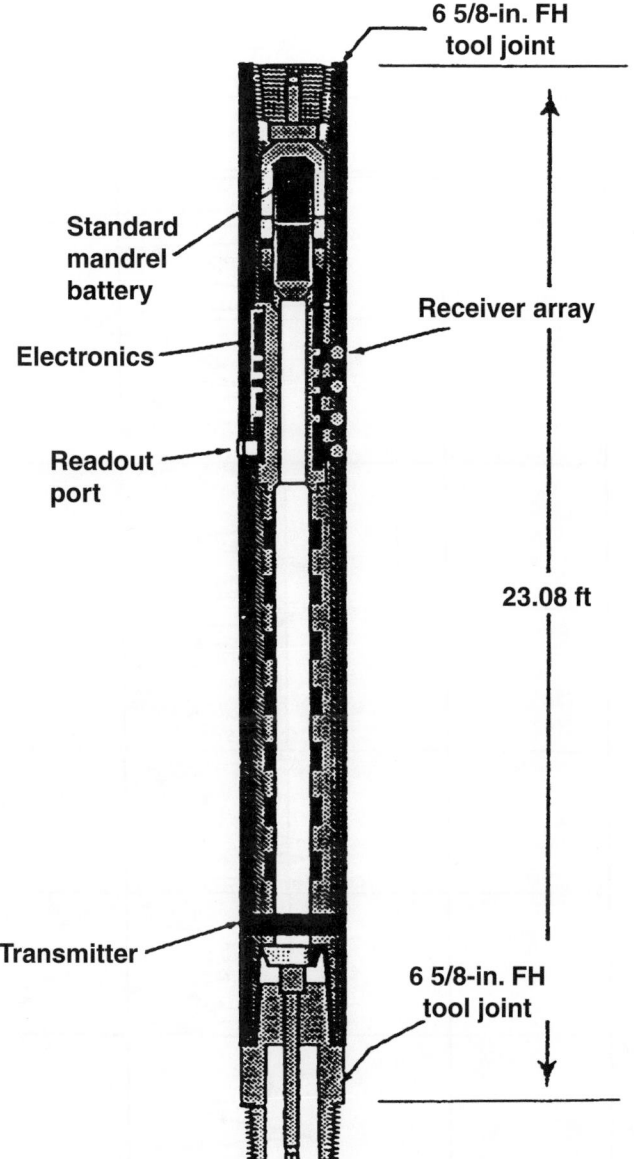

Figure 4.11.62 *Schematic of the LWD sonic tool (courtesy of SPWLA [28]).*

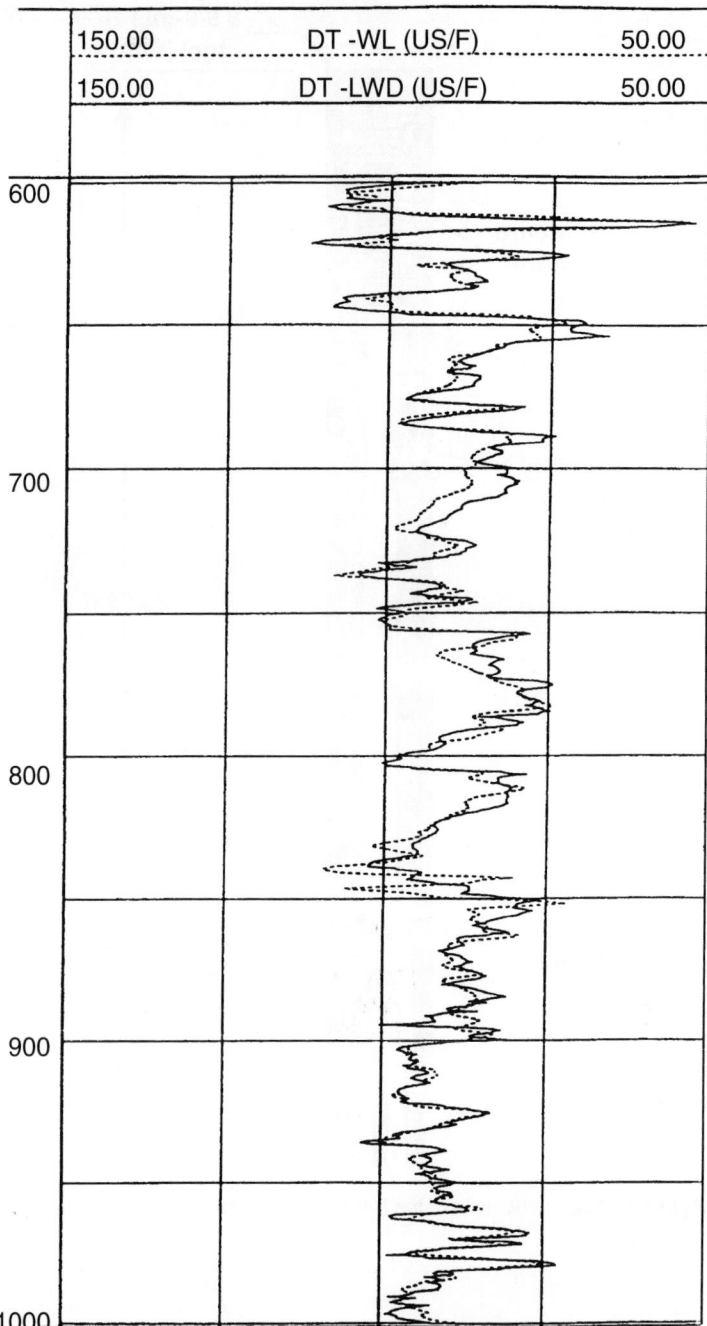

| 150.00 | DT -WL (US/F) | 50.00 |
| 150.00 | DT -LWD (US/F) | 50.00 |

Figure 4.11.63 *Comparison of a wireline sonic log with an LWD sonic log (courtesy of SPWLA [28]).*

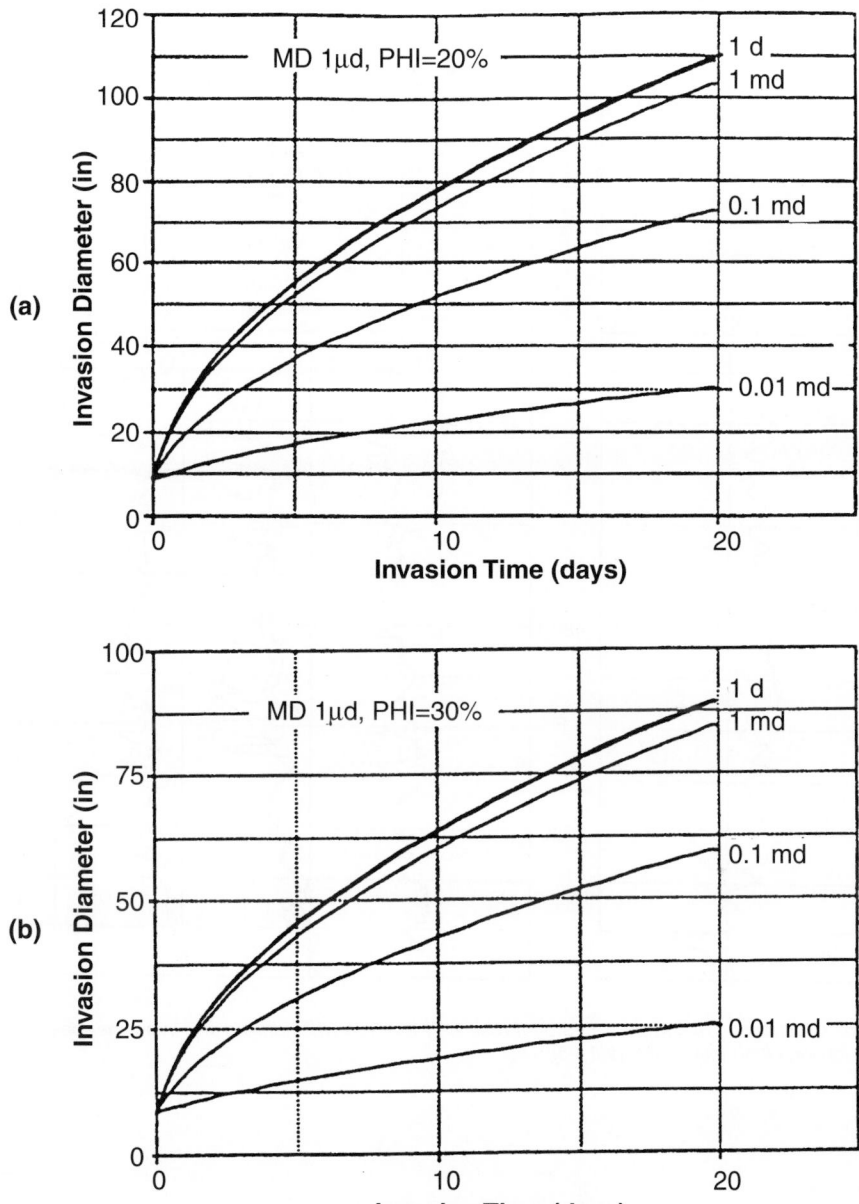

Figure 4.11.64 *Invasion diameter versus invasion time for various formation permeabilities. (a) Filtrate invaded porosity: 20%; mudcake permeability: 1 µd; mudcake thickness: 0.25"; differential pressure: 500 psi. (b) Filtrate invaded porosity: 30%; mudcake permeability: 1 µd; mudcake thickness: 0.25"; differential pressure: 500 psi (courtesy of Louisiana State University [11]).*

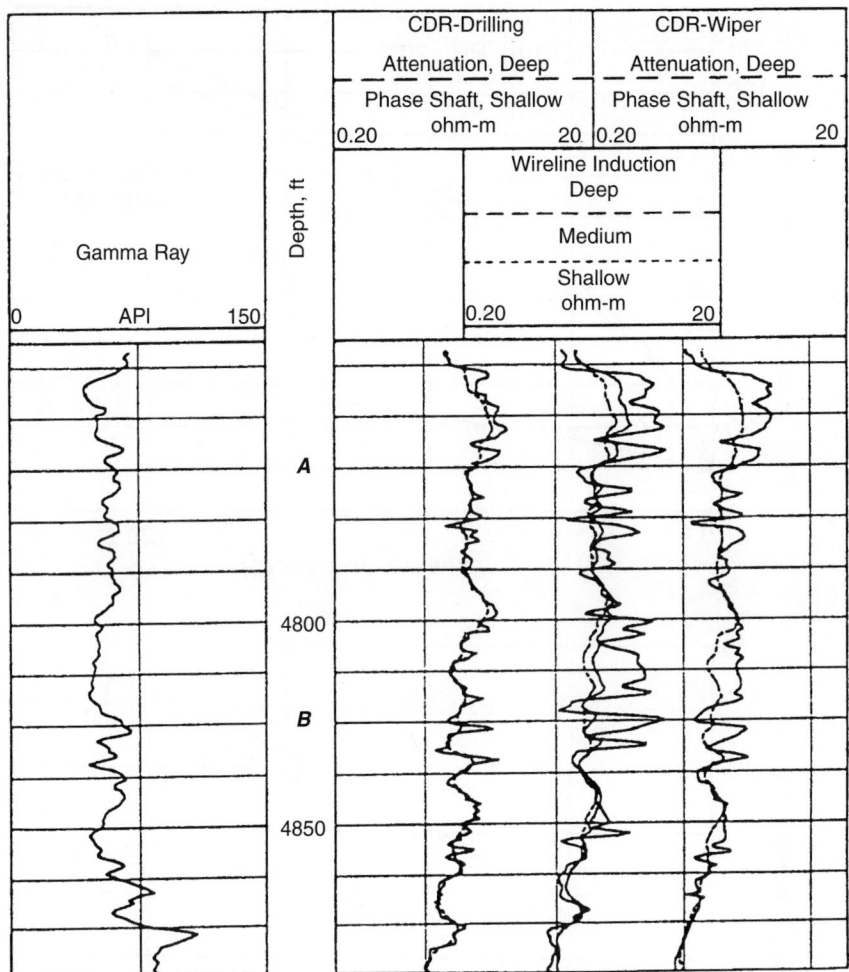

Figure 4.11.65 *Comparison of compensated dual-resistivity logs run while drilling and in a wiper pass with the wireline induction logs (courtesy of Anadrill [25]).*

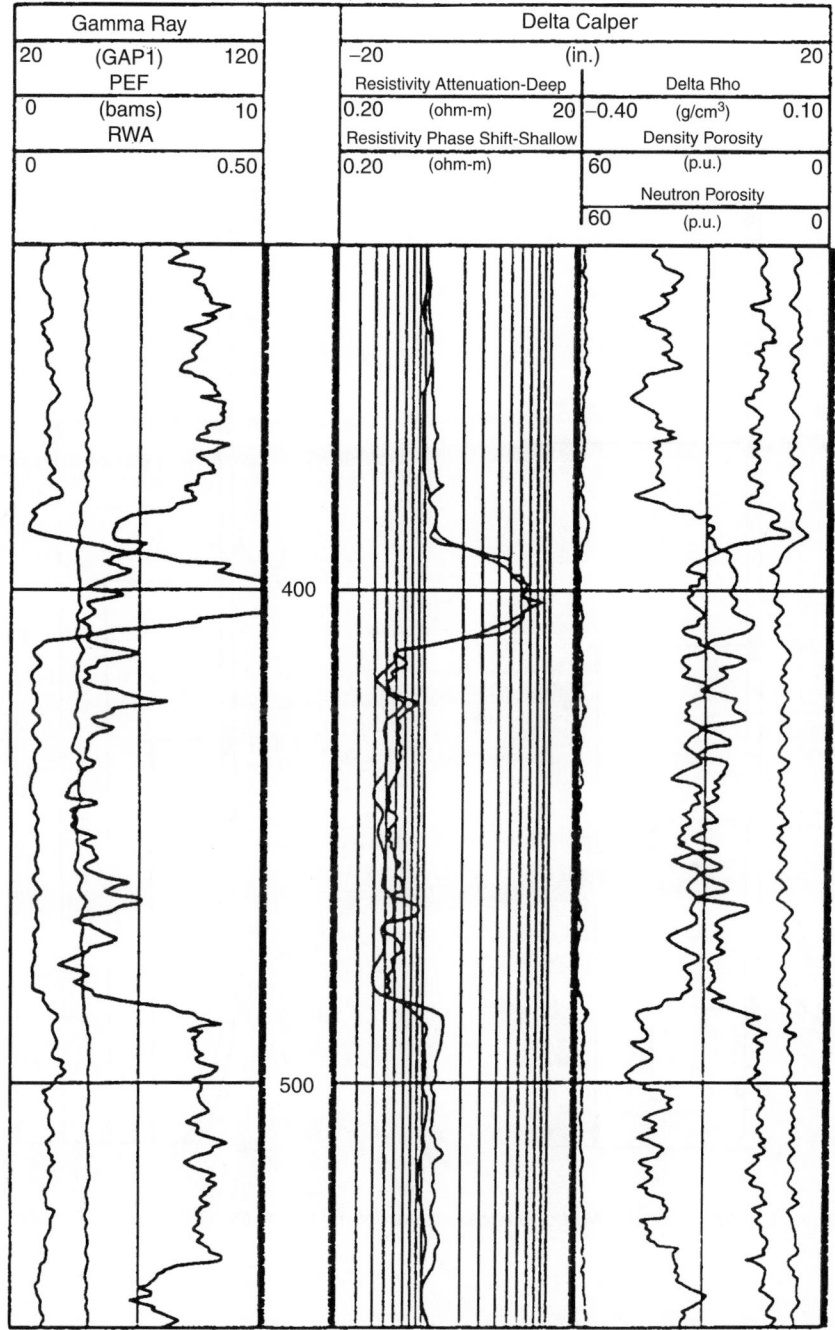

Figure 4.11.66 *LWD resistivity logs recorded during a wiper pass (courtesy of Anadrill [25]).*

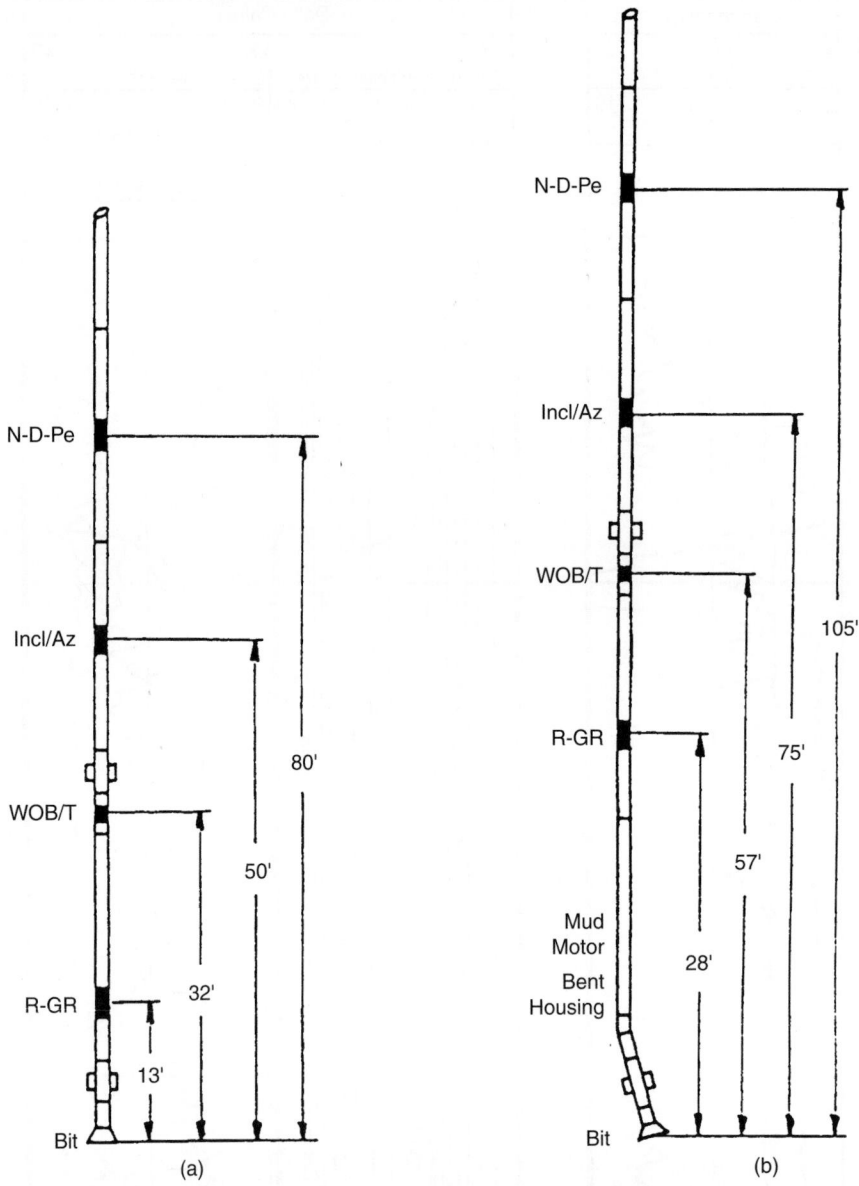

Figure 4.11.67 *Typical MWD bottomhole assemblies: (a) rotary drilling; (b) mud motor drilling.*

Table 4.11.5 *"Time Since Drilling" for Various MWD/LWD Logs with Typical Spacings*

(a) Rotary Drilling				
ROP (ft/hr)	R/GR (hr/min)	WOB/T (hr/min)	INCL/AZ (hr/min)	ND Pe (hr/min)
1	13–0	32–0	50–0	80–0
5	2–36	6–24	10–0	16–0
10	1–18	3–12	5–0	8–0
20	0–39	1–36	2–30	4–0
50	0–15	0–38	1–0	1–36
100	0–8	0–19	0–30	0–48
	13	32	50	80

(b) Mud Motor Drilling				
ROP (ft/hr)	R/GR (hr/min)	WOB/T (hr/min)	INCL/AZ (hr/min)	ND Pe (hr/min)
1	28–0	57–0	75–0	105–0
5	5–36	11–24	15–0	21–0
10	2–48	5–42	7–30	10–30
20	1–24	2–51	3–45	5–15
50	0–34	1–8	1–30	2–6
100	0–17	0–34	0–45	1–3
	28	57	75	105

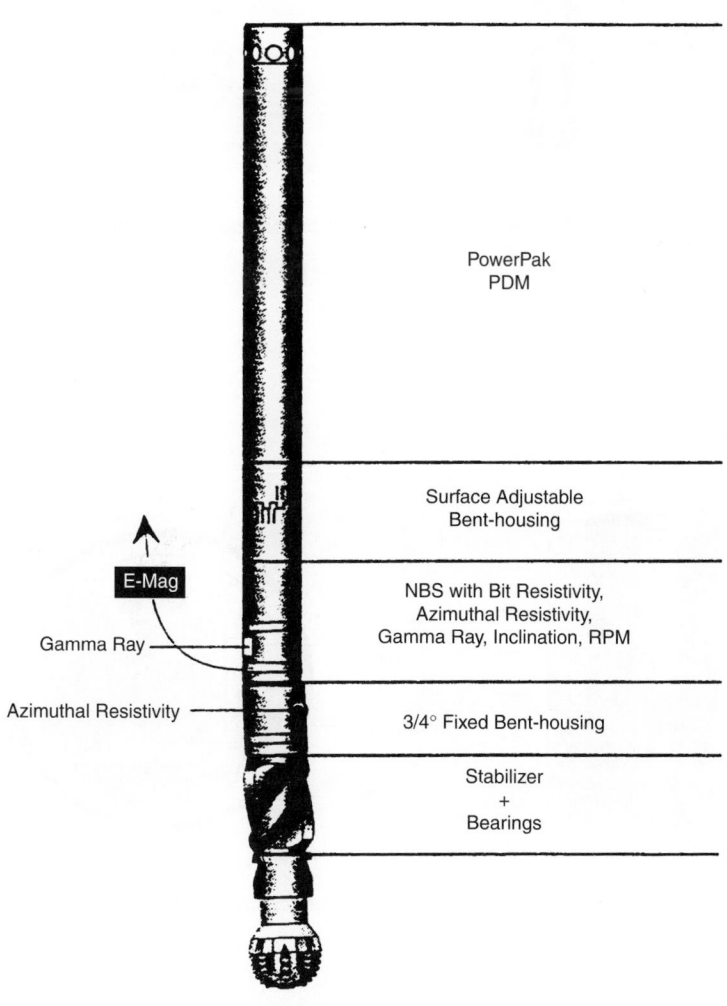

Figure 4.11.68 *Anadrill geosteering tool (courtesy of Anadrill [25]).*

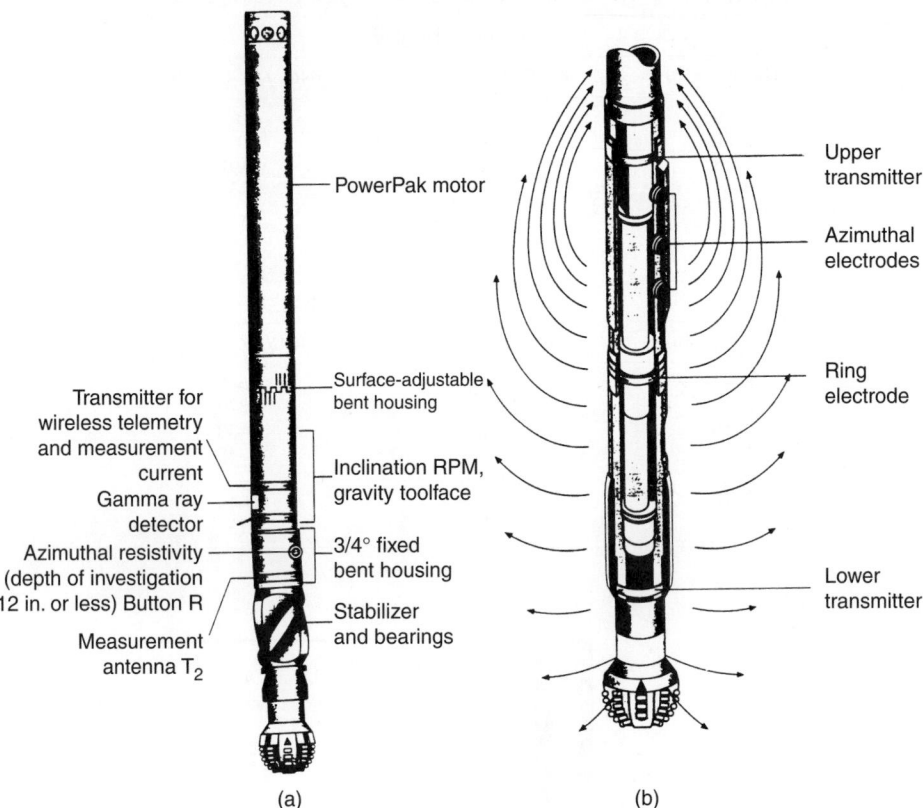

Figure 4.11.69   *Measurement at the drill bit (courtesy of Anadrill [25]).*

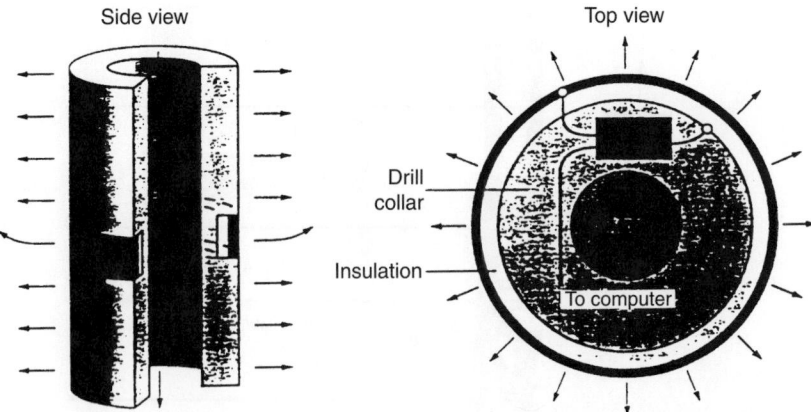

Figure 4.11.70   *Principle of the ring resistivity measurement in the resistivity-at-the-bit tool (courtesy of Anadrill [25]).*

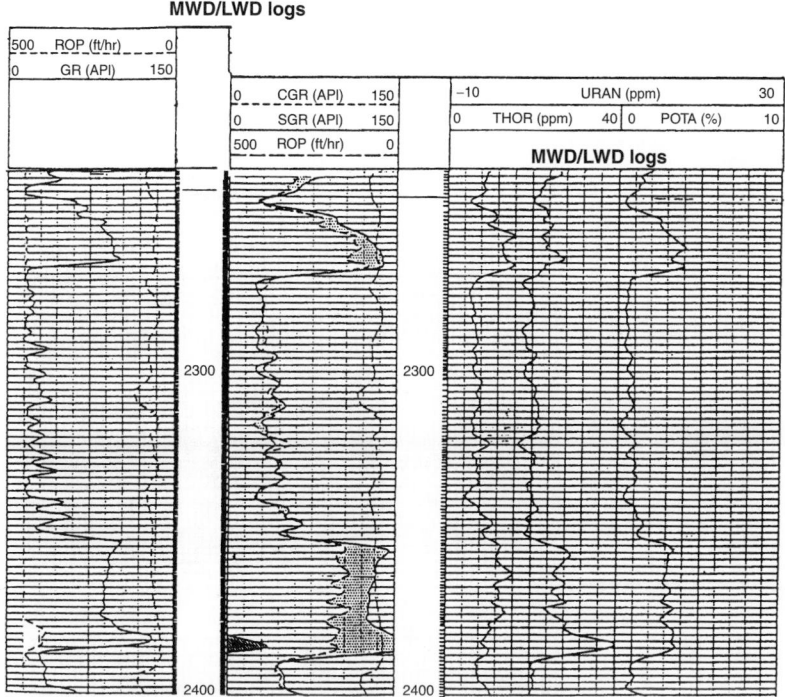

Figure 4.11.71 *Sample of MWD/LWD logs.*

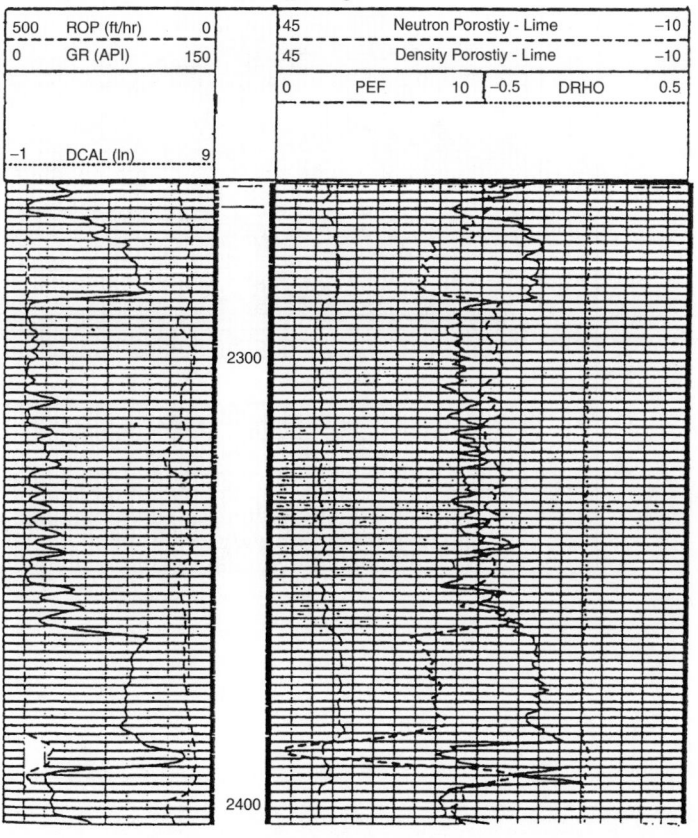

Figure 4.11.72 *Sample of MWD/LWD logs.*

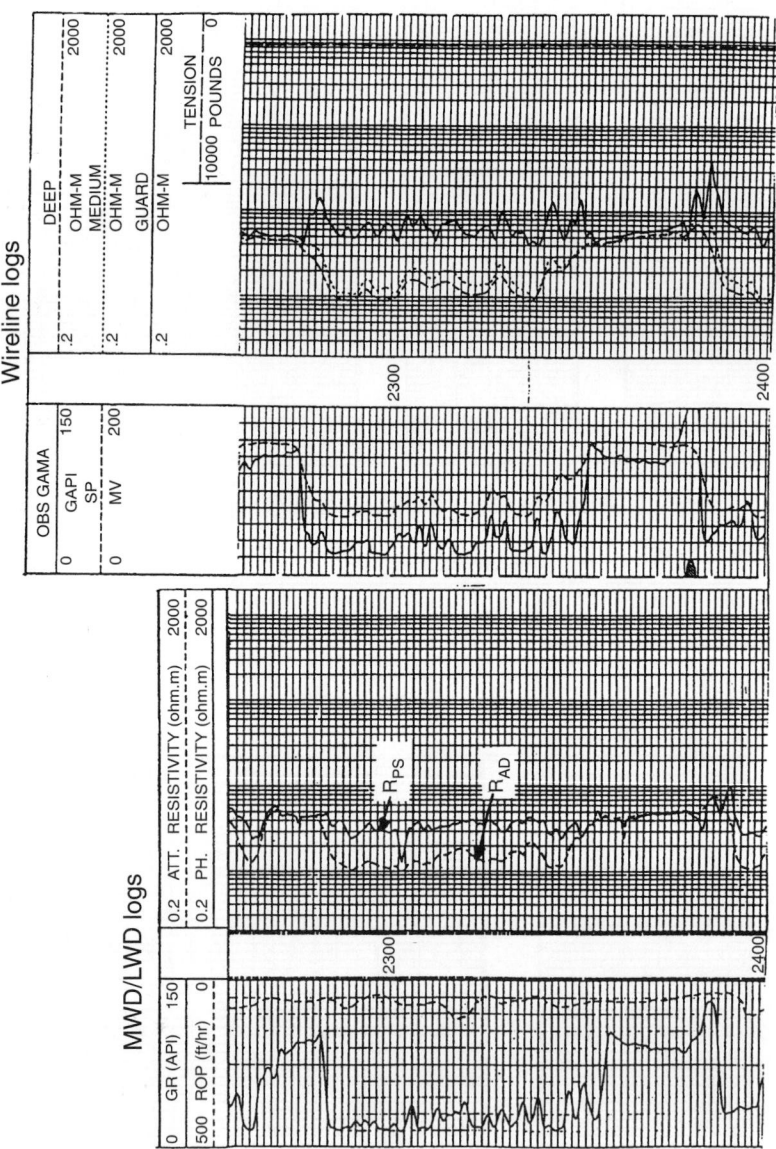

Figure 4.11.73   Sample of MWD/LWD logs.

Porosity and Lithology Determination from
Litho-Density* Log and CNL* Compensated Neutron Log
Liquid-Filled Holes ρ_f = 1.000 g/cc, C_f = 0 ppm

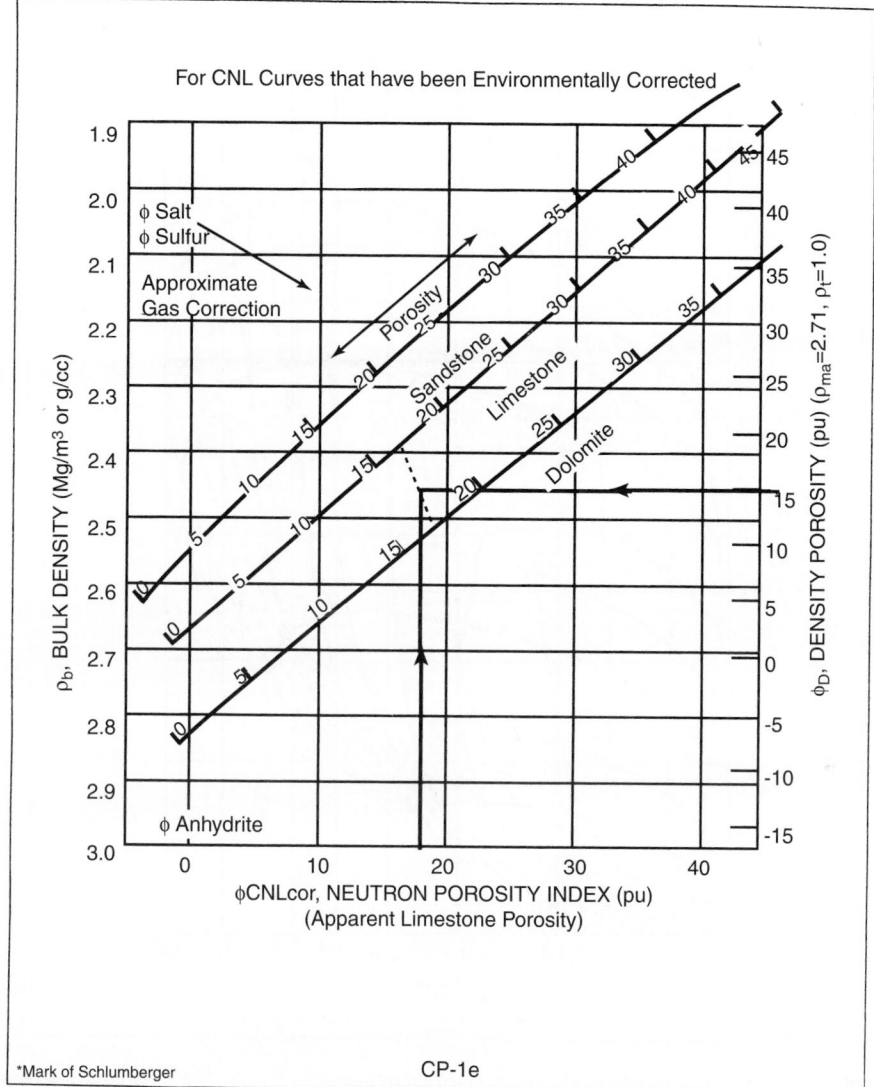

Figure 4.11.74 *Porosity and lithology determination from the Litho-Density* logs and the CNL* Compensated neutron log (courtesy of Schlumberger). (*Trade mark of Schlumberger).*

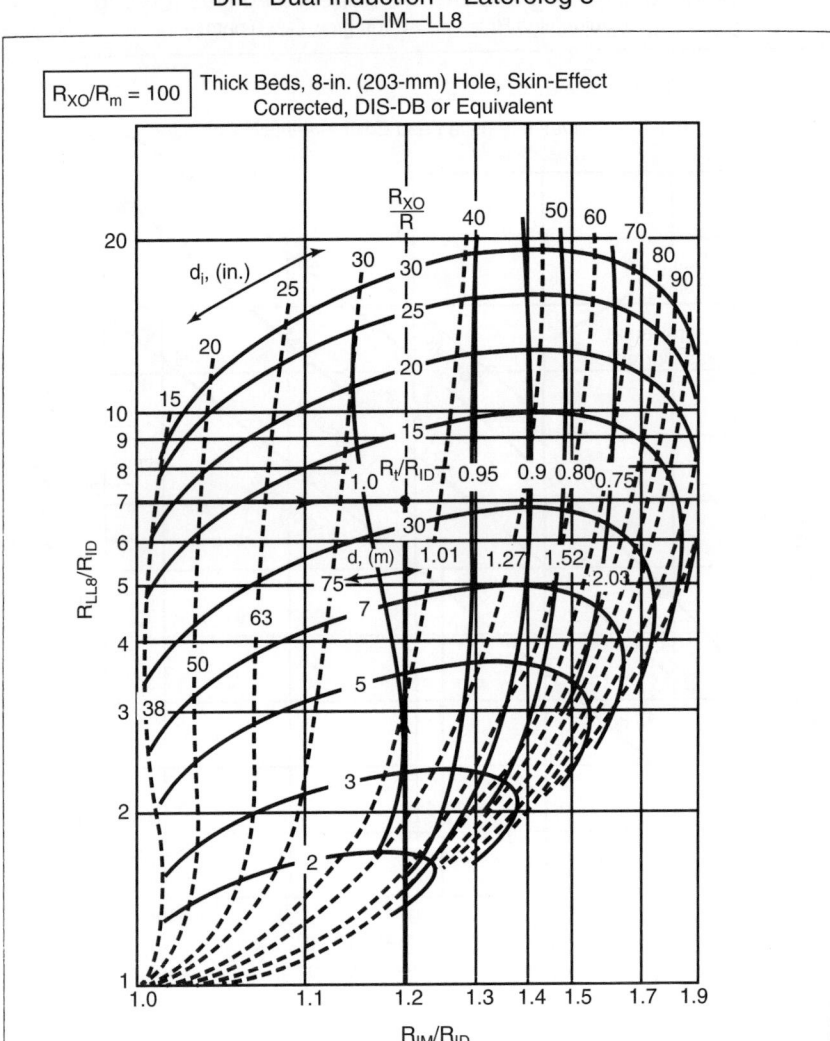

Figure 4.11.75 *Resistivities and invasion determination with the Dual Induction log (DIL*) and the laterolog-8 (courtesy of Schlumberger). (*Trade mark of Schlumberger.)*

(text continued from page 4-340)

Since the deep and shallow curve of the MWD log and the deep, shallow and guard (laterolog) of the wireline log shows a departure, the zone 2274–2356 ft is invaded, consequently permeable. Using the dual induction chart of Figure 4.11.75, we can plot the point at 2290 ft:

$$R_{IM}/R_{ID} = 1.2/1 = 1.2$$

$$R_{LL}/R_{ID} = 7/1 = 7$$

The invasion diameter is 36 in.
Where

$$R_{x0}/R_{ID} = 11$$
$$R_t/R_{ID} = 0.98$$

Consequently:

$$R_t = 0.98\ \Omega \bullet m \cong 1\ \Omega \bullet m$$

$$R_{x0} = 11\ \Omega \bullet m$$

The formation factor is

$$F = 1/\Phi^2 = 1/0.172 = 34.6$$

If we assume a 100% water saturation, the water resistivity is

$$R_w = R_t/F = 0.029\ \Omega \bullet m$$

If an oil or gas zone is present above this water sand, R_t should be computed the same way and S_w calculated.

References

1. Reed, P., "Amerada develops special rotary drill stem for simultaneous electrical logging and drilling," Oil and Gas J., pp. 68–70, Nov. 17, 1939.

2. Mills, B., "Simultaneous and continuous electrical logging and drilling achieve," *Oil Weekly* (now *World Oil*), January 1, 1940.

3. Arps, J. J., "Continuous logging while drilling; A practical reality," SPE 710, SPE Annual Fall Meeting, New Orleans, October 6–9, 1963.

4. Arps, J. J., and J. L., Arps, "The subsurface telemetry problem: A practical solution," *Journal of Petroleum Engineering*, May 1964.

5. Roberts, W. F., and Johnson, H. A., "Systems available for measuring hole direction," Oil and Gas J., V. 76, No. 22, pp. 68–70, 1978.

6. Spinner, T. G., and F. A. Stone, "Mud pulse logging while drilling system design, development and demonstration," IADC/CAODC Drilling Technical Conference, Houston, March 6–9, 1978.

7. "MWD: State of the Art," March 27, Vol. 76, No. 13; April 3, Vol. 76, No. 14; April 17, Vol. 76, No. 16; May 1, Vol. 76, No. 18; May 15, Vol. 76, No. 20; May 29, Vol. 76, No. 22; June 12, Vol. 76, No. 24; July 3, Vol. 76, No. 27; July 17, Vol. 76, No. 29; July 31, Vol. 76, No. 31; 1978.

8. Desbrandes, R., "Status report: MWD technology, Part 1: Data acquisition and downhole recording and processing," *Petroleum Engineering International*, September 1988.

9. Desbrandes, R., "Status report: MWD technology Part 2: Data transmission," *Petroleum Engineering International*, October 1988.

10. Desbrandes, R., "Status report: MWD technology, Part 3: Processing, display and applications," *Petroleum Engineering International*, November 1988.

11. Desbrandes, R., "Invasion diameter and supercharging in time-lapse MWD/LWD logging," *Proceedings MWD Symposium*, Louisiana State University, Baton Rouge, February 26–27, 1990.

12. Patton, B. J., et al., "Development and successful testing of a continuous wave, logging-while-drilling telemetry system," *Journal of Petroleum Engineering*, October 1977.

13. Bourgoyne, A. T., M. E. Chenevert, K. K., Millheim, and F. S. Young, *Applied Drilling Engineering*, SPE Textbook Series, Vol. 2, SPE, 1986.

14. Sunstrand, "Q-flex accelerometers,: brochure, Sundstrand Inc., Redmond, WA, 1985. (Now, a division of Allied Signal Inc., Redmond, WA.)

15. Develco, "Borehole Products," brochure, Develco Inc., Sunnvale, CA, 1985. (Now, a division of Baker-Hughes-Inteq, Houston, TX.)

16. Teledrift, Inc., "Teledrift–Wireless Drift Indicator," Brochure, Teledrift, Inc., Oklahoma City, OK, pp. 4, 1993.

17. "MWD: Measurement-While-Drilling," Inteq-Baker-Hughes, Brochure 602–003, August 1993.

18. "Anadrill Drilling Service Catalog," Anadrill-Schlumberger, AMP-7006, 1993.

19. Holmes, A. B., "Fluidic mud pulser concepts in MWD," *Proceedings MWD Symposium*, Louisiana State University, Baton Rouge, February 26–27, 1990.

20. Desbrandes, R., A. T. Bourgoyne and J. A. Carter, "MWD transmission data rate can be optimized," *Petroleum Engineering International*, June 1987.

21. "Electromagnetic MWD and Directional Drilling Services," Geoserivices Brochure, 1994.

22. McCain, W. D. Jr., *The Properties of Petroleum Fluids*, PennWell Books, Tulsa, OK, 1990.

23. Knox, D. W. J., and J. M. Milne, "Measurement-while-drilling tool performance," SPE 16523, Offshore Europe 1987 Conference, Aberdeen, Proceedings Vol. 1, September 1987.

24. Jan, Y. H., and Harrel, J. W., "MWD directional-focused Gamma Ray-A new tool for formation evaluation and drilling control in horizontal wells," SPWLA 28th Annual Logging Symp., p. A-1-17, 1987.

25. Anadrill-Schlumberger, 1992, *Logging While Drilling*, SMP-9160, Sugar Land, TX.

26. Evans, H. B., Brooks, A. G., Meisner, J. E., and Squire, R. E., "A focused current resistivity logging system for MWD," SPE 16757, 62nd SPE Ann. Tech. Conf. & Exhibition, Dallas, TX, set 27–30, 1987.

27. Gianzero, S. R., Chemali, R., Lin, Y., and Su, S., "A new resistivity tool for Measurement-While-drilling," SPWLA 26th Ann. Logging Symp., pp. A-1-22, 1985.

28. Aron, J. S. K., Chang, R., Dworak, R., Hsu, K., Lau, T., Masson, J-P., Mayes, J., McDaniel, G., Randall, C., and Plona, T. J., "Sonic compressional measure-ment while drilling," SPWLA 31st Ann. Logging Symp., pp. G-1-21, 1994.

29. Pidcock, G., and J. Daudy, "Gulf Canada improves drilling with MWD techniques," *Petroleum Engineering International*, September, Vol. 60, No. 9, 1988.

30. Falconer, I. G., Burgess, T. M., and Sheppared, M. C., "Seaprating Bit and Lithology Effects from Drilling Mechanics data," paper IADC/SPE17191, IACD/SPE Drilling Conf., Dallas, TX, February 1988.

31. Terzaghi, K., 1943, *Theoretical Soil Mechanics*, John Wiley and Sons, New York, NY.

32. Jorden, J. B., and Shirley, O. J., 1966, "Application of drilling performance data to overpressure detection." J. P. T., v. 18, pp. 1387–1394.

33. Rhem, W., and McClendon, R., Measurement of formation pressures from drilling data," SPE 3601, 46th Ann, Fall Meet., New Orleans, LA, October 3–5, 1971.

34. Eaton, B. A., "The equation for pressure prediction from well logs," SPE 5544, SPE 50th Ann. Tech. Conf. & Exhibit, Dallas, Tx, September 28–October 1, 1975.

35. Hottman, C. E., and Johson, R. K., "Estimation of formation pressure from log-derived shales properties," J. P. T., v. 17, pp. 717–722, 1965.

36. Alixant, J-L., "Real-time Effective Stress Evaluation in Shale: Pore Pressure and Permeability Estimation," Ph.D. dissertation, Louisiana State University, p. 210, December 1989.

37. Wyllie, M. R. J., Gregory, A. R., and Gartner, L. W., "Elastic wave velocities in heterogeneous and porous media," Geophysics, v. 21, 1, p. 41, 1956.

38. Raymer, L. L., and Hunt, E. R., "An improved sonic transit time-to-porosity transform," SPWLA Ann. Logging Symp., paper P, 1960.

39. Joshi, S. D., *Horizontal Well Technology*, PennWell Publishing Co., Tulsa, Oklahoma, 1991.

40. Hutchinson, M. W., "Measurement While Drilling and After Drilling by Multiple Service Companies Through Upper Carboniferous Formations at a Borehole Test Facility, Kay Country, Oklahoma," SPE 22735, 66th Ann. Techn. Conf., Dallas, TX, October 6–9, pp. 741–753, 1991.

41. Shi, Ying, "A comparison of MWD and open hole logging in sandstone and limestone," Master of Science Thesis, LSU Petroleum Engineering Department, December 1993.

42. Zhao, Donglin, "The Comparison of MWD and Wireline Logs in Sandstone and Limestone Under Various Borehole Conditions," Master of Science Thesis, LSU Petroleum Engineering Department, May 1994.

4.12 DIRECTIONAL DRILLING

4.12.1 Glossary of Terms used in Directional Drilling

The glossary of terms used in directional drilling [1] has been developed by the API Subcommittee on Controlled Deviation Drilling under the jurisdiction of the American Petroleum Institute Production Department's Executive Committee on Drilling and Production Practice. The most frequently used terms listed below.

Angle of inclination (angle of drift). The angle, in degrees, taken at one at several points of variation from the vertical as revealed by a deviation survey, sometimes called the inclination or angle of deviation.

Angle of twist. The azimuth change through which the drillstring must be turned to offset the twist caused by the reactive torque of the downhole motor.

Anisotrospic formation theory. Stratified or antisotropic formations are assumed to posses different drillabilities parallel and normal to the bedding planes with the result that the bit does not drill in the direction of the resultant force.

Azimuth. Direction of a course measured in a clockwise direction from 0° to 360°; also called bearing.

Back-torque. Torque on a drill string causing a twisting of the string.

Bent sub. Sub used on top of a downhole motor to give a nonstraight bottom assembly. One of the connecting threads in machined at an angle to the axis of the body of the sub.

Big-eyed bit. Drill bit with one large-sized jet nozzle, used for jet deflection.

Bit stabilization. Refers to stabilization of the downhole assembly near the bit; a stabilized bit is forced to rotate around its own axis.

Borehole direction. Refers to the azimuth in which the borehole is heading.

Borehole directional survey. Refers to the measurements of the inclinations, azimuths and specified depths of the stations through a section of borehole.

Bottom-hole assembly (BHA). Assembly composed of the drill bit, stabilizers, reamers, drill collars, subs, etc., used at the bottom of the drillstring.

Bottomhole location. Position of the bottom of the hole with respect to some known surface location.

Bottomhole orientation sub (BHO). A sub in which a free-floating ball rolls to the low side and opens a port indicating an orientation position.

Build-and-hold wellbore. A wellbore configuration where the inclination is increased to some terminal angle of inclination and maintained at that angle to the specified target.

Buildup. That portion of the hole in which the angle of inclination is increased.

Buildup rate. Rate of change (°/100 ft) of the inclination angle in the section of the hole where the inclination from the vertical is increasing.

Clearance. Space between the outer diameter of the tool in question and the side of the drilled hole; the difference in the diameter of the hole and the tool.

Clinograph. An instrument to measure and record inclination.

Closed traverse. Term used to indicate the closeness of two surveys; one survey going in the hole and the second survey coming out of the hole.

Corrective jetting runs. Action taken with a directional jet bit to change the direction or inclination of the borehole.

Course. The axis of the borehole over an interval length.

Course bearing. The azimuth of the course.

Crooked-hole. Wellbore that has been inadvertently deviated from a straight hole.

Crooked-hole area. An area where subsurface formations are so composed or arranged that it is difficult to drill a straight hole.

Cumulative fatigue damage. The total fatigue damage caused by repeated cyclic stresses.

Deflection tools. Drilling tools and equipment used to change the inclination and direction of the drilled wellbore.

Departure. Horizontal displacement of one station from another.

Deviation angle. See "Angle of inclination."

Deviation control techniques
 Fulcrum technique. Utilizes a bending moment principle to create a force on that the bit to counteract reaction forces that are tending to push the bit in a given direction.
 Mechanical technique. Utilizes bottomhole equipment which is not normally a part of the conventional drillstring to aid deviation control. This equipment acts to force the bit to turn the hole in direction and inclination.
 Packed-hole technique. Utilizes the hole wall to minimize bending of the bottomhole assembly.
 Pendulum techniques. The basic principle involved is gravity or the "plumb-bob effect."

Directional drilling contractor. A service company that supplies the special deflecting tools, BHA, survey instruments and a technical representative to perform the directional drilling aspects of the operation.

Direction of inclination. Direction of the course.

Dogleg. Total curvature in the wellbore consisting of a change of inclination and/or direction between two points.

Dogleg severity. A measure of the amount of change in the inclination and/or direction of a borehole; usually expressed in degrees per 100 ft of course length.

Drag. The extra force needed to move the drillstring resulting from the drillstring being in contact with the wall of the hole.

Drainholes. Several high-angle holes drilled laterally form a single wellbore into the producing zone.

Drift angle. The angle between the axis of the wellbore and the vertical (see "Inclination").

Drop off. The portion of the hole in which the inclination is reduced.

Drop-off rate. Rate of change ($°/100$ ft) of the inclination angle in the section of the wellbore that is decreasing toward vertical.

Goniometer. An instrument for measuring angles, as in surveying.

Gyroscopic survey. A directional survey conducted using a gyroscope for directional control, usually used where magnetic directional control cannot be obtained.

Hole curvature. Refers to changes in inclination and direction of the borehole.

Hydraulic orienting sub. Used in directional holes with inclination greater than 6° to find the low side of the hole.

A ball falls to the low side of the sub and restrict an orifice, causing an increase in the circulating pressure. The position of the tool is know with relation to the low side of the hole.

Hydraulically operated bent sub. A deflection sub that is activated by hydraulic pressure of the drilling fluid.

Inclination angle. The angle of the wellbore from the vertical.

Inclinometer. An instrument that measures an angle of deviation from the vertical.

Jet bit deflection. A method of changing the inclination angle and direction of the wellbore by using the washing action of a jet nozzle at one side of the bit.

Keyseat. A condition wherein the borehole is abraded and extended sideways, and with a diameter smaller than the drill collars and bit; usually caused by the tool joints on the drill pipe.

Kickoff point (kickoff depth). The position in the well bore where the inclination of the hole is first purposely increased (KOP).

Lead angle. A method of setting the direction of the wellbore in anticipation of the bit walking.

Magnetic declination. Angular difference, east or west, at any geographical location, between true north or grid north and magnetic north.

Magnetic survey. A directional survey in which the direction is determined by a magnetic compass aligning with the earth's magnetic field.

Measured depth. Actual length of the wellbore from its surface location to any specified station.

Mechanical orienting tool. A device to orient deflecting tools without the use of subsurface surveying instruments.

Methods of orientation
 Direct method. Magnets embedded in the nonmagnetic drill collar are used to indicate the position of the tool face with respect to magnetic north. A picture of a needle compass pointing to the magnets is superimposed on the picture of a compass pointing to magnetic north. By knowing the position of the magnets in the tool, the tool can be positioned with respect to north.
 Indirect method. A method of orienting deflecting tools in which two survey runs are needed, one showing the direction of the hole and the other showing the position of the tool.
 Surface readout. A device on the rig floor to indicate the subsurface position of the tool.
 Stoking. Method of orienting a tool using two pipe clamps, a telescope with a hair line, and an aligning bar to determine the orientation at each section of pipe run in the hole.

Monel (K monel). A nonmagnetic alloy used in making portions of downhole tools in the bottomhole assembly (BHA), where the magnetic survey tools are placed for obtaining magnetic direction information. Monel refers to a family of nickel-copper alloys.

Mud motor. Usually a positive displacement or turbine-type motor, positioned above the bit to provide (power) torque and rotation to the bit without rotating the drillstirng.

Mule shoe. A shaped form used on the bottom of orienting tools to position the tool. The shape resembles a mule shoe or the end of a pipe that has been cut both diagonally and concave. The shaped end forms a wedge to rotate the tool when lowered into a mating seat for the mule shoe.

Multishot survey. A directional survey in which multiple data points are recorded with one trip into the wellbore. Data are usually recorded on rolls of film.

Near-bit stabilizer. A stabilizer placed in the bottomhole assembly just above the bit.

Ouija board (registered trademark of Eastern Whipstock). An instrument composed of two protractors and a straight scale that is used to determine the positioning for a deflecting tool in a inclined wellbore.

Permissible dogleg. A dogleg through which equipment and/or tubulars can be operated without failure.

Pendulum effect. Refers to the pull of gravity on a body; tendency of a pendulum to return to vertical position.

Pendulum hookup. A bit and drill collar with a stabilizer to attain the maximum effect of the pendulum.

Rat hole. A hole that is drilled ahead of the main wellbore and which is of a smaller diameter than the bit in the main borehole.

Reamer. A tool employed to smooth the wall of a wellbore, enlarge the hole, stabilize the bit and straighten the wellbore where kinks and abrupt doglegs are encountered.

Rebel tool (registered trademark of Eastman Whipstock). A tool designed to prevent and correct lateral drift (walk) of the bit tool. It consists of two paddles on a common shaft that are designed to push the bit in the desired direction.

Roll off. A correction in the facing of the deflection tool, usually determined by experience, and which must be taken into consideration to give the proper facing to the tool.

Setting off course. A method of setting the direction of the wellbore in anticipation of the bit walking.

Side track. An operation performed to redirect the wellbore by starting a new hole; at a position above the bottom of the original hole.

Slant hole. A nonvertical hole; usually refers to a wellbore purposely inclined in a specific direction; also used to define a wellbore that is nonvertical at the surface.

Slant rig. Drilling rig specifically designed to drill a wellbore that is nonvertical at the surface. The mast is slanted and special pipe-handling equipment is needed.

Spiraled wellbore. A wellbore that has attained a changing configuration such as a helical form.

Spud bit. In directional drilling, a special bit used to change the direction and inclination of the wellbore.

Stabilizer. A tool placed in the drilling assembly to
  Change or maintain the inclination angle in a wellbore by controlling the location of the contact point between the hole and drill collars.
  Center the drill collars near the bit to improve drilling performance.
  Prevent wear and differential sticking of the drill collars.

Surveying frequency. Refers to the number of feet between survey records.

Target area. A defined area, at a prescribed vertical depth, that is planned to be intersected by the wellbore.

Tool azimuth angle. The angle between north and the projection of the tool reference axis onto a horizontal plane.

Tool high-side angle. The angle between the tool reference axis and a line perpendicular to the hole axis and lying the vertical plane.

Total curvature. Implies three-dimensional curvature.

True north. The direction from any geographical location on the earth's surface to the north geometric pole.

True vertical depth (TVD). The actual vertical depth of an inclined wellbore.

Turbodrill. A downhole motor that utilizes a turbine for power to rotate the bit.

Turn. A change in bearing of the hole; usually spoken of as the right or left turn with the orientation that of an observer who views the well course from the surface site.

Walk (of hole). The tendency of a wellbore to deviate in the horizontal plane.

Wellbore survey calculation method. Refers to the mathematical method and assumptions used in reconstructing the path of the wellbore and in generating the space curve path of the wellbore from inclination and direction angle measurements taken along the wellbore. These measurements are obtained from gyroscopic or magnetic instruments of either the single-shot or multishot type.

Whipstock. A long wedge and channel-shaped piece of steel with a collar at its top through which the subs and drillstring may pass. The face of the whipstock sets an angle to deflect the bit.

Woodpecker drill collar (indented drill collar). Round drill collar with a series of indentations on one side to form an eccentrically weighted collar.

4.12.2 Dogleg Severity (Hole Curvature) Calculations
There are several analytical methods available for calculating dogleg severity:

- Tangential
- Radius of curvature

- Average angle
- Trapezoidal (average tangential)
- Minimum curvature

The tangential and radius of curvature methods are outlined here.

4.12.2.1 Tangential Method [171]
The overall angle change is calculated from

$$\beta = 2\arcsin\sqrt{\sin^2\left(\frac{\Delta v}{2}\right) + \sin^2\left(\frac{\Delta h}{2}\right)\sin^2\left(\frac{V_0 + V_1}{2}\right)}$$

$$[4.12.1]$$

where β = overall angle change (total curvature)
Δh = change of horizontal angle (in horizontal plane)
Δv = change in vertical angle (in vertical plane)
V_0, V_1 = hole inclination angle in two successive surveying stations.

The hole curvature is

$$C = 100\frac{\beta}{L}$$

$$[4.12.2]$$

where L = course length between the surveying stations.

Example 4.12.1
Two surveying measurements were taken 30 ft apart. The readings are as below.

Station 1:

Hole inclination 3°30′
Hole direction, N11°E

Station 2:

Hole inclination 4°30′
Hole direction, N23°E

Find the dogleg severity.

Solution

Change in horizontal angle is
$$\Delta h = 23° - 11° = 12°$$
Change in vertical angle is
$$\Delta v = 4.5° - 3.5° = 1°$$
The overall angle change is

$$\beta = 2\arcsin\sqrt{\sin^2\left(\frac{1.0}{2}\right) + \sin^2\left(\frac{12}{2}\right)\sin^2\left(\frac{3.5 + 4.5}{2}\right)}$$

$$= 1.3°$$

The hole curvature is

$$c = 100\frac{1.3}{30} = 4.33°/100 \text{ ft} = 4°20'/100 \text{ ft}$$

4.12.2.2 Radius of Curvature Method
The dogleg severity is calculated from

$$c = 100\sqrt{a^2\sin^4\phi + b^2}$$

$$[4.12.3]$$

where a = rate of change in direction angle in degrees/ft
b = rate of change in inclination angle in degrees/ft
ϕ = inclination angle in degrees.
The sequence of computations involved is explained in the following example.

Example 4.12.2
From two successive directional survey stations is obtained:

	Station 1	Station 2
Hole inclination angle	30° (ϕ_1)	40° (ϕ_2)
Hole direction angle	N11°E (θ_1)	N18°E (θ_2)

The distance between the stations is 60 ft. Determine the dogleg severity.

Solution

Rate of change in inclination angle is

$$b = \frac{40 - 30}{60} = 0.1667°/\text{ft}$$

Radius of curvature in vertical plane is

$$R_v = \frac{180}{\pi b} = \frac{180}{\pi \times 0.1667} = 363 \text{ ft}$$

Horizontal departure (arc length of projection of wellbore in horizontal plane) is

$$H_d = R_v(\cos\phi_1 - \cos\phi_2) = 363(\cos 30 - \cos 40)$$

$$= 36.29 \text{ ft}$$

Rate of change in hole direction is

$$a = \frac{\theta_2 - \theta_1}{H_d} = \frac{18 - 11}{36.29} = 0.1929°/\text{ft}$$

Hole curvature at the first station is

$$c_1 = 100\sqrt{a^2\sin^4\phi_1 + b^2} = 100\sqrt{0.1929^2 \times \sin^4 30 + 0.1667^2}$$

$$= 17.35°/100 \text{ ft}$$

Hole curvature at the first station is

$$c = 100\sqrt{a^2\sin^4\phi_2 + b^2} = 100\sqrt{0.1929^2 \times \sin^4 40 + 0.1667^2}$$

$$= 18.48°/100 \text{ ft}$$

The average value is

$$c = \frac{17.35 + 18.48}{2} = 17.92°/100 \text{ ft}$$

4.12.2.3 Deflection Tool Orientation
Application of a deflecting tool (e.g., downhole motor with a bent sub) requires determining the orientation of the tool so that the hole takes the desired course. There are three effects to consider when setting a deflection tool.

1. The existing borehole inclination angle
2. The existing borehole direction angle
3. The bent sub angle of the deflection tool itself

These three effects in combination will result in a new dogleg of a wellbore.

The deflection tool orientation parameters can be obtained using the vectorial method of D. Ragland, the Ouija Board, or the three-dimensional mathematical deflecting model.

4.12.2.4 Vectorial Method of D. Ragland
This method is explained by solving two example problems.

Example 4.12.3
Determine the deflection tool-face orientation, tool deflection angle and tool facing change from original course line angle

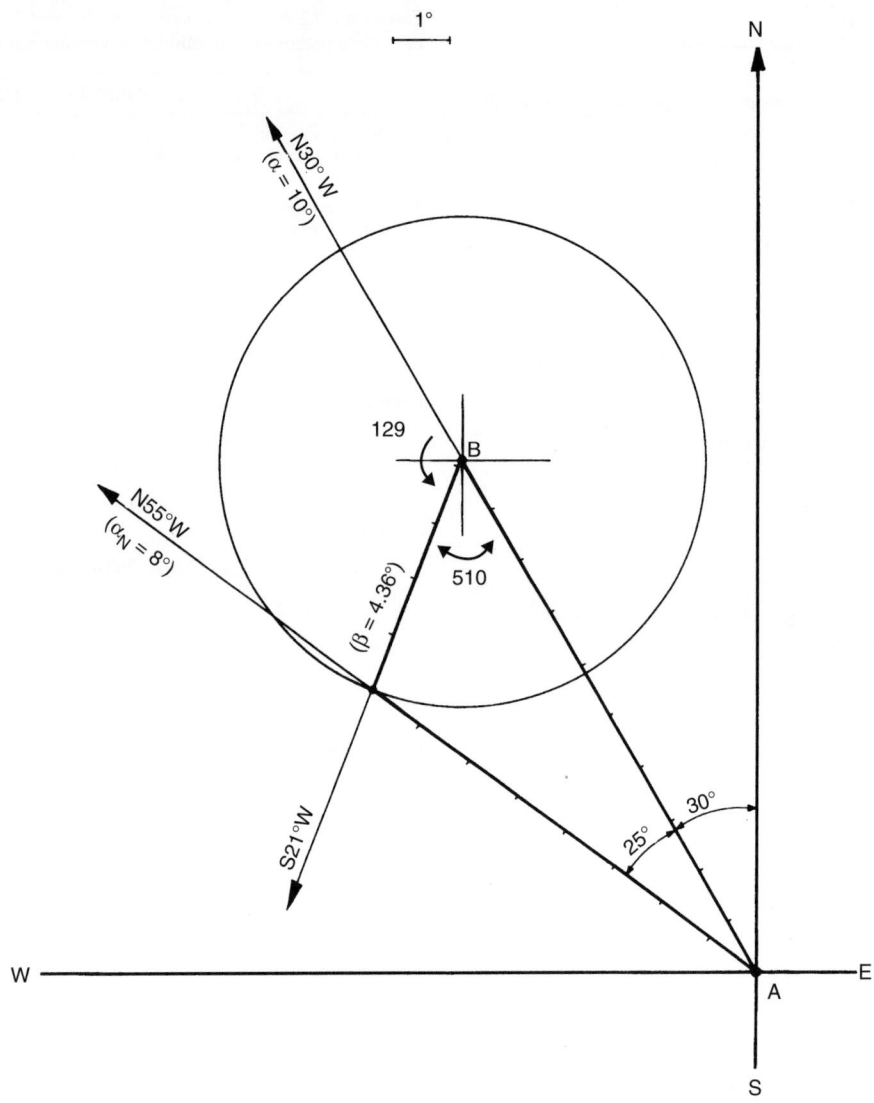

Figure 4.12.1   *Graphic solution of Example 4.12.3.*

if the data are as follows:

> Existing hole inclination angle: 10°
> Existing hole direction: N300W
> Desired change of azimuth: 25° (to the left)
> Desired hole inclination: 8°

Solution

The solution to this problem is shown in Figure 4.12.1. The following steps are involved in preparing a Ragland diagram.

1. Lay a quadrant N-S-W-E.
2. Select a scale for the angles.
3. With a protractor layoff an angle 30° from N to W.
4. Using the selected angle scale find p. B (10° from p. A).

5. With a protractor layoff an angle 25° to the left of line AB.
6. Find p. C using the angle scale (8° from p. A).
7. Describe a circle about p. B with a radius of CB = 4.36° (read from the diagram). This is the desired overall angle change.
8. Read of required tool-face orientation: S21°W.
9. Read off required tool facing change from original course line: 129° to the left.

Example 4.12.4
The original hole direction and inclination were measured to be S50°W and 6°, respectively. It is desired to obtain a new hole direction of S65°W with an inclination angle of 7° after

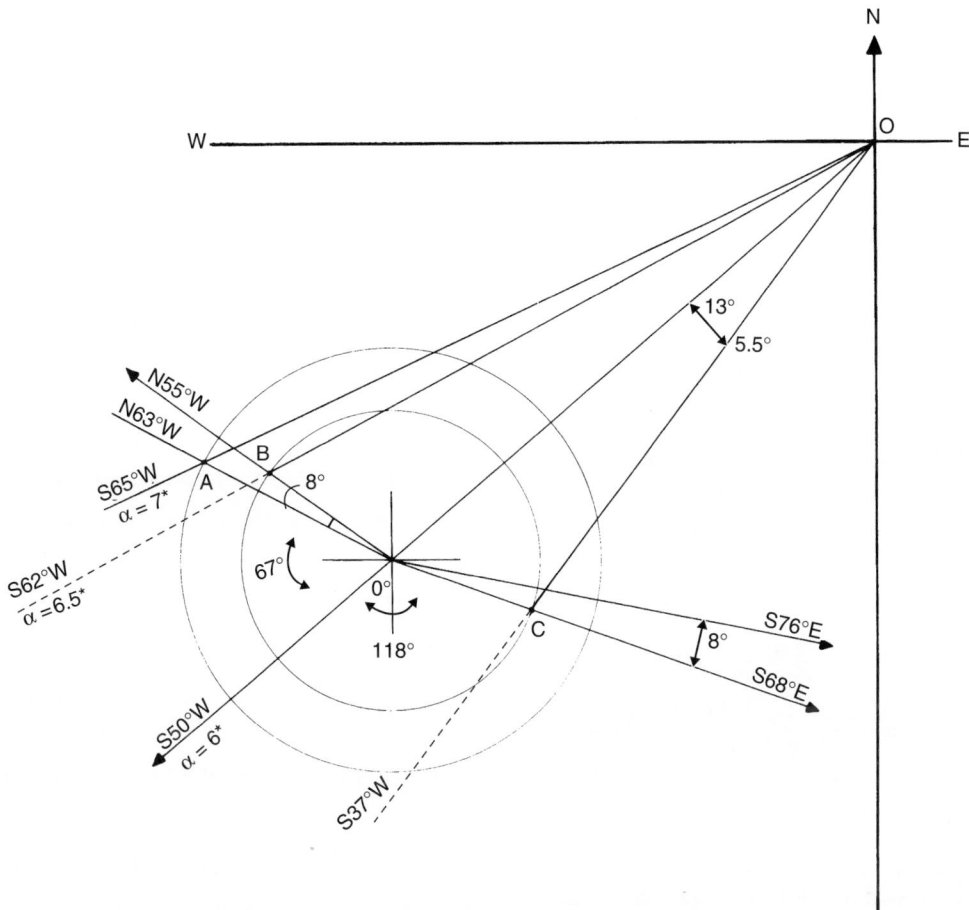

Figure 4.12.2 *Graphical solution of Example 4.12.4. With a downhole motor, there will be a left-hand reaction torque; therefore, the tool should be turned clockwise from the ideal position.*

drilling 90 ft. For this purpose, a whipstock was oriented correctly.

After drilling 90 ft, a checkup measurement was performed that revealed that the hole new direction is S62°W and the new inclination is 6°30′. Determine the expected hole curvature delivered by the whipstock, the magnitude of roll-off for this system and the real hole curvature delivered by the whipstock. How should the whipstock be oriented to drill a hole with a direction of S37°W in 90 ft? What should be the inclination of the hole?

Solution

The solution to this problem is shown in Figure 4.12.2. Steps 1 and 2 are the same as in Example 4.12.3.

1. Layoff an angle 50° from S to W.
2. Draw a line 00′.
3. Draw a line OA that will represent the desired new hole (S65°W = 15° to the right of the current direction). Point A is found at a distance of 7° from point O. Radius OA represents the expected overall angle changed of 2.0°.
4. Read off the original whipstock orientation N63°W (67° to the right of the current hole direction).

5. Draw line OB (direction, S62°W, inclination, 6.5°).
6. Describe a circle about p. O′ with a radius of OB. Radius OB = 1.40° represents the real overall angle change.
7. The angle AO′B is the roll-off angle. Read off the roll-off angle = 8° to the right.
8. Draw line OC (S37°W). The whipstock orientation should be S68°E, and the final angle is 5.5°. If the roll-off effect is not considered, the whipstock direction is S76°E.

4.12.2.5 Three-Dimensional Deflecting Model
A mathematical model has been presented [5] that enables one to analyze and plan deflection tool runs. The model is a set of equations that relate the original hole inclination angle (α), new hole inclination angle (α_N), overall angle change (β), change of direction ($\Delta\varepsilon$), and tool-face rotation from original course direction (γ). These equations are given below:

$$\beta = \arccos\left(\cos\alpha\cos\alpha_N + \sin\alpha\sin\alpha_N\cos\Delta\varepsilon\right) \qquad [4.12.4]$$

$$\alpha_N = \arccos\left(\cos\alpha\cos\beta - \sin\alpha\sin\beta\cos\gamma\right) \qquad [4.12.5]$$

$$\Delta\varepsilon = \arctan\left(\frac{\tan\beta\sin\gamma}{\sin\alpha + \cos\alpha\tan\beta\cos\gamma}\right) \qquad [4.12.6]$$

The above equations can be rearranged to the form suitable for a solution of a particular problem. A hand-held calculator may be used to perform required calculations [6].

Practical usefulness of this model is presented below.

Example 4.12.5

Original hole direction and inclination is N20°E and 10°, respectively. It is desired to deviate the borehole so that the new hole inclination is 13° and direction N30°E in 90 ft. What is the tool orientation and deflecting angle (dogleg) necessary to achieve this turn and build?

Solution

From Equation 4.12.4, the overall angle change is

$$\beta = \arccos[\cos 10 \times \cos 13 + \sin 10 \times \sin 13 \cos(30-20)]$$

$$= 3.59°$$

The dogleg severity (hole curvature) is

$$C = 100\frac{\beta}{L} = 100\frac{3.59}{90} = 3.99°/100 \text{ ft}$$

To obtain the tool orientation we solve Equation 4-267 for γ:

$$\cos\gamma = \arccos\left(\frac{\cos\alpha\cos\beta - \cos\alpha_N}{\sin\alpha\sin\beta}\right)$$

$$= \arccos\left(\frac{\cos 10 \times \cos 3.59 - \cos 13}{\sin 10 \times \sin 3.59}\right) = 38.54°$$

Consequently, the tool-face orientation is 38.54° to the right of the high side of the hole.

Example 4.12.6

Surveying shows that the hole drift angle is 22° and direction S36°W. It is desired to turn the hole of 6° to the right and build angle during next 90ft drilling. For this purpose a deflection tool that can deliver a hole curvature of 3.333°/100 ft will be used. What is the expected new hole inclination angle? What is the required tool direction?

Solution

The expected overall angle change in the interval is

$$\beta = 3.333°/100 \text{ ft} \times 90 \text{ ft} = 3°$$

From Equation 4.12.4, we have

$$\cos\beta = \cos\alpha\cos\alpha_N + \sin\alpha\sin\alpha_N\cos\Delta\varepsilon$$

Considering the coefficients $a = \cos\alpha$ and $b = \sin\alpha\,\cos\Delta\varepsilon$, we obtain

$$\cos\beta = a\cos\alpha_N + b\sin\alpha_N$$

Now consider dividing both sides by $\sqrt{a^2+b^2}$. We obtain

$$\frac{\cos\beta}{\sqrt{a^2+b^2}} = \frac{a}{\sqrt{a^2+b^2}}\cos\alpha_N + \frac{b}{\sqrt{a^2+b^2}}\sin\alpha_N$$

Making $\cos\phi = \frac{a}{\sqrt{a^2+b^2}}$ and $\sin\phi = \frac{b}{\sqrt{a^2+b^2}}$, we have

$$\frac{\cos\beta}{\sqrt{a^2+b^2}} = \cos\phi\cos\alpha_N + \sin\phi\sin\alpha_N = \cos(\phi - \alpha_N)$$

where $\tan\phi = \dfrac{b}{a} = \dfrac{\sin\alpha\cos\Delta\varepsilon}{\cos\alpha}$.

Solving for α_N, we have

$$\alpha_N = \phi - \arccos\left(\frac{\cos\beta}{\sqrt{a^2+b^2}}\right)$$

$$= \arctan\left(\frac{\sin\alpha\cos\Delta\varepsilon}{\cos\alpha}\right)$$

$$- \arccos\left(\frac{\cos\beta}{\sqrt{a^2+b^2}}\right)$$

Using the expressions for a and b, we obtain

$$\alpha_N = \arctan\left(\frac{\sin\alpha\cos\Delta\varepsilon}{\cos\alpha}\right)$$

$$- \arccos\left(\frac{\cos\beta}{\sqrt{\cos^2\alpha + \sin^2\alpha\cos^2\Delta\varepsilon}}\right)$$

Substituting the data of the problem, we obtain for the new angle

$$\alpha_N = \arctan\left(\frac{\sin 22 \times \cos 6}{\cos 22}\right)$$

$$- \arccos\left(\frac{\cos 3}{\sqrt{\cos^2 22 + \sin^2 22 \times \cos^2 6}}\right)$$

$$= 23.88°$$

Using Equation 4.12.5 and solving for γ, we obtain for the tool-face orientation

$$\gamma = \arccos\left(\frac{\cos\alpha\cos\beta - \cos\alpha_N}{\sin\alpha\sin\beta}\right)$$

$$= \arccos\left(\frac{\cos 22 \times \cos 3 - \cos 23.88}{\sin 22 \times \sin 3}\right) = 54.02°$$

Therefore, the tool-face should be oriented at 54.02° to the right of the high side because it is desired to turn the hole to S42°W.

References

1. *Sii Datadrill Directional and Drilling Manual*, Second Revision, December 1983.
2. Lubinski, A., "How severe is the Dogleg?," World Oil, February 1957.
3. Wilson, G. J., "An improved method for computing directional surveys," *Journal of Petroleum Technology*, August 1968.
4. Leblond, A., "Controlled directional drilling," Ecole Nationale Superieure du Petrole at des Moteurs, Reference 19163, April 1971.
5. Millheim, K. K., Gubler, F. H., and Zaremba, H. B., "Evaluating and planning directional wells utilizing post analysis techniques and a three dimensional bottom hole assembly program," SPE Paper 8339, presented at the 54th Annual Fall Technical Conference and Exhibition of the Society of Petroleum Engineers of AIME held in Las Vegas, September 23–26, 1979.
6. Nicholson, J. T., "Calculator program developed for directional drilling," *Oil and Gas Journal*, September 28, 1981.

4.13 SELECTION OF DRILLING PRACTICES

The objective of optimizing drilling practices is to safely deliver a product capable of the highest production capacity in a cost-effective manner. Throughout all phases of well planning and construction, the following ideas must be considered: health, safety, and environment (HSE), production capability, and drilling optimization. These criteria are requisite to delivering a successful drilling outcome. Several choices are normally available to achieve the required outcomes for each criterion, but fewer options, perhaps only one, may be viable to meet the needs of all three.

In today's environment, HSE requirements play an equal role in the planning and implementation of drilling plans. Significant planning is required to protect operating personnel, retain licenses to operate, control costs, and coexist with the environment. Accidents are costly, more so than the up-front costs to mitigate the risk to people, property, and environment.

Production capacity, both rate and volume, are affected by decisions made during planning and construction phases. Proper selection of various cost control options can minimize the negative influences that some of these measures can have on production. Analyzing potential gains in capacity associated with incremental risk and investment is part of the optimization process.

Optimization requires understanding, prioritization, and implementation of the practices most appropriate for the specific well being drilled to accomplish the stated objectives. You must solve potential problems that remove limits or lessen their impact on the outcome. The right answer for one circumstance (e.g., well design, geographical area, hole size) is possibly the wrong answer in another circumstance. Problem solving must be approached with the overall objective in mind while understanding the causes and effects and the relative magnitudes of these effects. No matter how well the planning process is conducted, adjustment in real time is expected and will be required. The parameters of drilling operations parameters must be optimized. Noticing, understanding, and adapting to observed trends will yield significant value to the outcome. Real-time ability to respond to problems during drilling will minimize their impact and not allow these problems to cause other problems, which are often substantially more costly to solve. The days of characterizing drilling practices as simply operational parameters such as weight on bit (WOB), revolutions per minute (rpm), and flow rate are long gone. Drilling optimization is a multivariant analysis that requires diligent effort, but the challenge is rewarding personally and financially.

4.13.1 Health, Safety and Environment

An effective HSE program is characterized by no injury to people, no loss of property, and no harm to the environment. Great HSE performance is an indication of great leadership. It is much more than statistics, although measurements are necessary to facilitate performance improvement. HSE must be considered a core responsibility for all business participants, operators, contractors, and service companies and be accepted by all individuals on a personal basis. It is imperative that all parties are committed from the top management down throughout their organizations.

The main reasons companies in the current era support strong HSE programs include humanitarian reasons (i.e., not hurting people), legal or regulatory requirements, the company's public image, employee morale, and economic reasons (i.e., loss of business or the cost of poor HSE performance).

The modern era of HSE management for oil field operations began with the Piper Alpha disaster in the North Sea, where 167 men lost their lives in one incident. The subsequent investigation, conducted and published by Lord Cullen, contained the framework of HSE management embodied in much subsequent legislation and regulation [1]. This was the birth of safety cases for oil field facilities, including offshore MODU drilling rigs.

One of the key recommendations was that companies should have *safety management systems* (SMSs) that control a company's operations from top to bottom. The systems were recommended to include the elements of ISO 9001 (a standard for management of quality in organizations) and include elements such as management responsibility and commitment, design control, documentation and procedures, process control, control of nonconformance, corrective actions, internal auditing, and training. All the elements of a safety management system are not directly applicable to drilling optimization and are therefore not addressed here. Examples include emergency response plans and oil spill response procedures that companies should have in place as part of their normal procedures.

Generally, each drilling department should have a set of operating guidelines that control the drilling work processes. The operating guidelines are a subset of the SMSs, and the documents may vary from multivolume sets at the major oil companies to much smaller documents for smaller companies. The drilling engineer is responsible for ensuring that all his work complies with these guidance instructions. Variations from these procedures usually will require a higher level of management approval.

The three components of HSE-related drilling issues that are normally considered during drilling operations are discussed separately in the following sections.

4.13.1.1 Health

Health issues related to drilling operations can include industrial hygiene issues related to onsite conditions and may include exposures to drilling fluid components such as oil-base mud fumes and skin contact, highly toxic completion brines, or oils, gas, and toxic materials such as hydrogen sulfide originating from the well. Naturally occurring radioactive materials (NORM) can also be encountered during workover operations, and metals such as mercury are encountered periodically in gas streams and may be found in production separators. An optimized drilling program includes careful consideration of the health impact on employees and workers at any well site.

4.13.1.2 Safety

The general safety culture adapted in the current era was derived from the Dont manufacturing culture and adapted to oil field operations when the company purchased Conoco. Dont was originally in the dynamite manufacturing business, and serious accidents in the past strongly motivated management to adopt a "best practice" approach to safety management. The primary concept is that *all accidents are preventable*. Accidents do not just happen, and with work, resources, and management commitment, accidents can be minimized or eliminated. In most all oil companies and service companies, management personnel leave no doubt that they are committed to providing a safe working environment. A poor safety record leads to the suffering of injured employees, the financial impact of lawsuits, and the loss of business and shareholder support. It is not unusual for service companies to be removed from approved bidding lists if the safety performance is not up to the company's requirements. The second concept generally accepted is that *safety (or HSE) must be considered equally with production and profits* in the decision-making process. "Safety first" is not plausible. If we wanted to only be safe, we would never leave the comfort of our homes in the first place. Safety must be considered as an equal to other factors in the decision-making process to ensure that all jobs meet the company's safety goals.

Safety management is an engineering profession unto itself, with many disciplines and areas of coverage. For the drilling engineer or drilling foreman, safety at the well site is accomplished through sound engineering, formal safety reviews (as appropriate), contracting of reputable firms with strong safety cultures, and appropriate training of all personnel.

Operational risks for drilling operations can vary from simple, shallow onshore jobs to extremely complex offshore operations in remote hostile areas that can be extremely expensive and carry significant risk. Consequently, a "fit for purpose" approach must be considered for the safety management of each operation. A simple onshore job may rely completely on the drilling contract and include relatively simple tools such as a toolbox safety discussion and daily safety meeting. At the other end of the scale, in large offshore production platforms with simultaneous operation of wells and drilling in a remote or hostile area, a large amount of safety engineering may be required. Tools may include a full safety case preparation and hazard and operability (HAZOP) studies [2]. A HAZOP study is an examination procedure. Its purpose is to identify all possible deviations from the way in which a design is expected to work and to identify all the hazards associated with these deviations. When deviations arise that result in hazards, actions are generated that require design engineers to review and suggest solutions to remove the hazard or reduce its risk to an acceptable level. These solutions are reviewed and accepted by the HAZOP team before implementation. HAZOP techniques have been adopted by many countries and are required to provide assurance of safe operations.

It should be understood by all operation personnel that 96% of all accidents are related to unsafe behaviors and that only 4% of accidents are caused by unsafe conditions. Most drilling contractors have adopted policies that focus on encouraging safe behaviors.

4.13.1.3 Environment
Dramatic changes in environmental impact management have occurred continually from the earliest days of the oil field. Common practice in the early days was to produce oil into open pits for storage. Saltwater was routinely dumped into the nearest creek, killing everything in it. Today's best practices include serious management and minimization of all waste streams. Development of projects today may require full environmental impact assessments (EIA) before approval. The first legal requirement in the United States for an EIA was imposed on the Trans-Alaska oil pipeline, which was delayed for years and experienced cost overruns from an initial estimate of $900 million to a final installed cost of $9 billion. Today, most of the East and West coastlines of the United States and Florida are off limits for drilling because of environmental concerns. The Exxon Valdez incident will not leave the collective public memory any time soon and is an example of the negative impact to the environment from oil and gas operations.

Most companies have management systems in place for environmental management of their operations. Drilling personnel are responsible for planning and conducting operations to ensure optimization and compliance.

4.13.2 Production Capacity
Production capacity requires investment. Performance measures of optimization may include one or several of the following: days to pay out, cash flow, profit, finding cost (cost per barrel oil equivalent), and return on investment (ROI). The idea of *diminishing returns* plays a key role in the decision-making process. Delivering rate and reservoir volume carry associated cost and risks. Fundamental to drilling practices optimization is understanding how specific cause-and-effect relationships influence this objective.

Key issues affecting production capability, which therefore must receive a high priority when planning the drilling phase, include permeability (K) and porosity (Φ) of the zone of interest (ZOI). Formation damage resulting from poor planning may be permanent or very costly to correct. The right equipment, resources, knowledge, and skills are prerequisites for optimization.

Several key issues may establish constraints on the drilling program, narrowing choices for the drilling plan. For example, mud types and properties and exposure time to the ZOI may be critical. A special "drill-in" fluid (i.e., drilling mud that can minimize formation damage) may be required. The time and cost to change the drilling mud is deemed high value for achieving the final production capacity versus cost objective. Another example is the determination of the optimum hole and tubular sizes. Hole and tubular sizes must be optimized based on expected initial production (e.g., rate, fluid type, pressure), availability, potential future remediation, and consideration of how production may change throughout the life of the well. The geology and pore pressures determine the number of casing strings required to arrive at the ZOI, and the expected production rate will determine the size of production string that the drilling department must be deliver to the company. In some cases, exploration wells will be designed as expendable and will never be produced. All of these factors influence the spectrum of possible casing plans and potential costs, and they will require evaluation.

Using a "cradle to grave" approach to well design considers the drilling phases and looks at all aspects of the well's life, from initial production through permanent abandonment [3]. Close association is required between the multiple departments (e.g., geology, asset management, drilling, completion, production) of the oil company and the service provider representatives. Communicating and sharing their respective areas of expertise and requirements with the others is paramount to optimization of the outcome.

New technologies can significantly enhance production capacity when properly applied and should not be forgotten. Underbalanced drilling (UBD), expandable sand screens (ESS), expandable drilling liners (EDL), coiled-tubing drilling (CTD), hole enlarging devices (e.g., bi-center drill bits, hole openers), casing drilling, and fiberoptic "intelligent" completions are a few examples.

4.13.3 Well Planning and Implementation
4.13.3.1 Optimum Well Planning
Drilling optimization requires detailed engineering in all aspects of well planning, drilling implementation, and post-run evaluation (Figure 4.13.1) [4,5]. Effective well planning optimizes the boundaries, constraints, learning, nonproductive time, and limits and uses new technologies as well as tried and true methods. Use of *decision support packages*, which document the reasoning behind the decision-making, is key to shared learning and continuous improvement processes. It is critical to anticipate potential difficulties, to understand their consequences, and to be prepared with contingency plans. Post-run evaluation is required to capture learning. Many of the processes used are the same as used during the well planning phase, but are conducted using new data from the recent drilling events.

Depending on the phase of planning and whether you are the operator or a service provider, some constraints will be

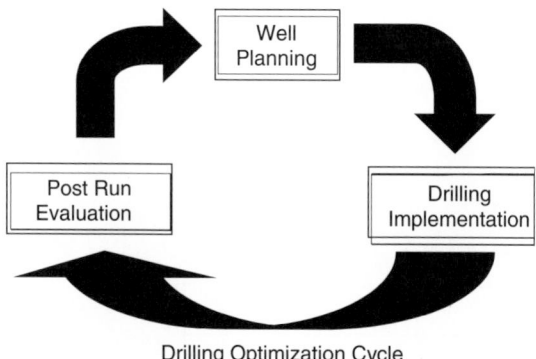

Drilling Optimization Cycle

Figure 4.13.1 *Drilling Optimization Cycle.*

out of your control to alter or influence (e.g., casing point selection, casing sizes, mud weights, mud types, directional plan, drilling approach such as BHA types or new technology use). There is significant value in being able to identify alternate possibilities for improvement over current methods, but well planning must consider future availability of products and services for possible well interventions. When presented properly to the groups affected by the change, it is possible to learn why it is not feasible or to alter the plan to cause improvement. Engineers must understand and identify the correct applications for technologies to reduce costs and increase effectiveness. A correct application understands the tradeoffs of risk versus reward and costs versus benefits.

Boundaries
Boundaries are related to the "rules of the game" established by the company or companies involved. Boundaries are criteria established by management as "required outcomes or processes" and may relate to behaviors, costs, time, safety, and production targets.

Constraints
Constraints during drilling may be preplanned trip points for logs, cores, casing, and BHA or bit changes. Equipment, information, human resource knowledge, skills and availability, mud changeover, and dropping balls for downhole tools are examples of constraints on the plan and its implementation.

The Learning Curve
Optimization's progress can be tracked using learning curves that chart the performance measures deemed most effective for the situation and then applying this knowledge to subsequent wells. Learning curves provide a graphic approach to displaying the outcomes. Incremental learning produces an exponential curve slope. Step changes may be caused by radically new approaches or unexpected trouble. With understanding and planning, the step change will more likely be in a positive direction, imparting huge savings for this and future wells. The curve slope defines the optimization rate.

The learning curve can be used to demonstrate the overall big picture or a small component that affects the overall outcome. In either case, the curve measures the rate of change of the parameter you choose, typically the "performance measures" established by you and your team. Each performance measure is typically plotted against time,

perhaps the chronological order of wells drilled as shown in Figure 4.13.2.

Cost Estimating
One of the most common and critical requests of drilling engineers is to provide accurate cost estimates, or *authority for expenditures* (AFEs). The key is to use a systematic and repeatable approach that takes into account all aspects of the client's objectives. These objectives must be clearly defined throughout the organization before beginning the optimization and estimating process. Accurate estimating is essential to maximizing a company's resources. Overestimating a project's cost can tie up capital that could be used elsewhere, and underestimating can create budget shortfalls affecting overall economics.

Integrated Software Packages
With the complexity of today's wells, it is advantageous to use integrated software packages to help design all aspects of the well. Examples of these programs include

- Casing design
- Torque and drag
- Directional planning
- Hydraulics
- Cementing
- Well control

Decision Support Packages
Decision support packages document the reasoning behind the decisions that are made, allowing other people to understand the basis for the decisions. When future well requirements change, a decision trail is available that easily identifies when new choices may be needed and beneficial.

Performance Measures
Common drilling optimization performance measures are cost per foot of hole drilled, cost per foot per casing interval, trouble time, trouble cost, and AFEs versus actual costs [6].

Systems Approach
Drilling requires the use of many separate pieces of equipment, but they must function as one system. The borehole should be included in the system thinking. The benefit is time reduction, safety improvement, and production increases as the result of less nonproductive time and faster drilling. For example, when an expected average rate of penetration (ROP) and a maximum instantaneous ROP have been identified, it is possible to ensure that the tools and borehole will be able to support that as a plan. Bit capabilities must be matched to the rpm, life, and formation. Downhole motors must provide the desired rpm and power at the flow rate being programmed. Pumps must be able to provide the flow rate and pressure as planned [7].

Nonproductive Time
Preventing trouble events is paramount to achieving cost control and is arguably the most important key to drilling a cost-effective, safe well. Troubles are "flat line" time, a terminology emanating from the days versus depth curve when zero depth is being accomplished for a period of days, creating a horizontal line on the graph. Primary problems invariably cause more serious associated problems. For example, surge pressures can cause lost circulation, which is the most common cause of blowouts. Excessive mud weight can cause differential sticking, stuck pipe, loss of hole, and sidetracking. Wellbore instability can cause catastrophic loss of entire hole sections. Key seating and pipe washouts can cause stuck pipe and a fishing job.

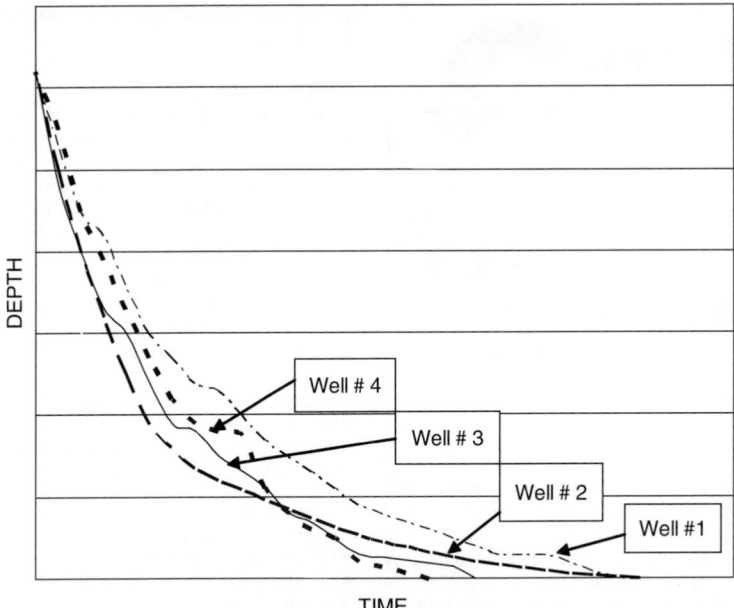

Figure 4.13.2 *Learning Curves.*

When a trouble event leads to a fishing job, "fishing economics" should be performed. This can help eliminate emotional decisions that lead to overspending. Several factors should be taken in to account when determining whether to continue fishing or whether to start in the first place. The most important of these are replacement or lost-in-hole cost of tools and equipment, historical success rates (if known), and spread rate cost of daily operations. These can be used to determine a risk-weighted value of fishing versus the option to sidetrack.

Operational inefficiencies are situations for which better planning and implementation could have saved time and money. Sayings such as "makin' hole" and "turnin' to the right" are heard regularly in the drilling business. These phases relate the concept of *maximizing progress.* Inefficiencies which hinder progress include

- Poor communications
- No contingency plans and "waiting on orders" (WOO)
- Trips
- Tool failure
- Improper WOB and rpm (magnitude and consistency)
- Mud properties that may unnecessarily reduce ROP (spurt loss, water loss and drilled solids)
- Surface pump capacities, pressure and rate (suboptimum liner selection and too small pumps, pipe, drill collars)
- Poor matching of BHA components (hydraulics, life, rpm, and data acquisition rates)
- Survey time

Limits
Each well to be drilled must have a plan. The plan is a baseline expectation for performance (e.g., rotating hours, number of trips, tangibles cost). The baseline can be taken from the learning curves of the best experience that characterizes the well to be drilled. The baseline may be a widely varying estimate for an exploration well or a highly refined measure in a developed field. Optimization requires identifying and improving on the limits that play the largest role in reducing progress for the well being planned. Common limits include

1. **Hole Size.** Hole size in the range of $7\frac{7}{8} - 8\frac{1}{2}$ in. is commonly agreed to be the most efficient and cost-effective hole size to drill, considering numerous criteria, including hole cleaning, rock volume drilled, downhole tool life, bit life, cuttings handling, and drill string handling. Actual hole sizes drilled are typically determined by the size of production tubing required, the required number of casing points, contingency strings, and standard casing decision trees. Company standardization programs for casing, tubing, and bits may limit available choices.

2. **Bit Life.** Measures of bit life vary depending on bit type and application. Roller cones in soft to medium-soft rock often use KREVs (i.e., total revolutions, stated in thousands of revolutions). This measure fails to consider the effect of WOB on bearing wear, but soft formations typically use medium to high rpm and low WOB; therefore, this measure has become most common. Roller cones in medium to hard rock often use a multiplication of WOB and total revolutions, referred to as the WR or WN number, depending on bit vender. Roller cone bits smaller than $7\frac{7}{8}$ in. suffer significant reduction in bearing life, tooth life, tooth size, and ROP. PDC bits, impregnated bits, natural diamond bits, and TSP bits typically measure in terms if bit hours and KREVs. Life of all bits is severely reduced by vibration. Erosion can wear bit teeth or the bit face that holds the cutters, effectively reducing bit life.

3. **Hole Cleaning.** Annular velocity (AV) rules of thumb have been used to suggest hole-cleaning capacity, but each of several factors, including mud properties, rock properties, hole angle, and drill string rotation, must be considered. Directional drilling with steerable systems require "sliding" (not rotating) the drill string during the orienting stage; hole cleaning can suffer drastically at hole angles greater than 50. Hole cleaning in large-diameter holes, even if vertical, is difficult merely because of the fast drilling formations and commonly low AV.

4. **Rock Properties.** It is fundamental to understand formation type, hardness, and characteristics as they relate to drilling and production. From a drilling perspective, breaking down and transporting rock (i.e., hole cleaning) is required. Drilling mechanics must be matched to the rock mechanics. Bit companies can be supplied with electric logs and associated data so that drill bit types and operating parameters can be recommended that will match the rock mechanics. Facilitating maximum production capacity is given a higher priority through the production zones. This means drilling gage holes, minimizing formation damage (e.g., clean mud, less exposure time), and facilitating effective cement jobs.

5. **Weight on Bit.** WOB must be sufficient to overcome the rock strength, but excessive WOB reduces life through increased bit cutting structure and bearing wear rate (for roller cone bits). WOB can be expressed in terms of weight per inch of bit diameter. The actual range used depends on the "family" of bit selected and, to some extent, the rpm used. *Families* are defined as natural diamond, PDC, TSP (thermally stable polycrystalline), impregnated, mill tooth, and insert.

6. **Revolutions per Minute (rpm).** Certain ranges of rpm have proved to be prudent for bits, tools, drill strings, and the borehole. Faster rpm normally increases ROP, but life of the product or downhole assembly may be severely reduced if rpm is arbitrarily increased too high. A too-low rpm can yield slower than effective ROP and may provide insufficient hole cleaning and hole pack off, especially in high-angle wells.

7. **Equivalent Circulating Density (ECD).** ECDs become critical when drilling in a soft formation environment where the fracture gradient is not much larger than the pore pressure. Controlling ROP, reducing pumping flow rate, drill pipe OD, and connection OD may all be considered or needed to safely drill the interval.

8. **Hydraulic System.** The rig equipment (e.g., pumps, liners, engines or motors, drill string, BHA) may be a given. In this case, optimizing the drilling plan based on its available capabilities will be required. However, if you can demonstrate or predict an improved outcome that would justify any incremental costs, then you will have accomplished additional optimization. The pumps cannot provide their rated horsepower if the engines providing power to the pumps possess inadequate mechanical horsepower. Engines must be down rated for efficiency. Changing pump liners is a simple cost-effective way to optimize the hydraulic system. Optimization involves several products and services and the personnel representatives. This increases the difficulty to achieve an optimized parameter selection that is best as a system.

New Technologies

Positive step changes reflected in the learning curve are often the result of effective implementation of new technologies:

1. **Underbalanced Drilling.** UBD is implemented predominantly to maximize the production capacity variable of the well's optimization by minimizing formation damage during the drilling process. Operationally, the pressure of the borehole fluid column is reduced to less than the pressure in the ZOI. ROP is also substantially increased. Often, coiled tubing is used to reduce the tripping and connection time and mitigate safety issues of "snubbing" joints of pipe.

2. **Surface Stack Blowout Preventer (BOP).** The use of a surface stack BOP configurations in floating drilling is performed by suspending the BOP stack above the waterline and using high-pressure risers (typically $13\frac{3}{8}$ in. casing) as a conduit to the sea floor. This method, generally used in benign and moderate environments, has saved considerable time and money in water depths to 6,000 ft [8].

3. **Expandable Drilling Liners.** EDLs can be used for several situations. The casing plan may start with a smaller diameter than usual, while finishing in the production zone as a large, or larger, final casing diameter. Future advances may allow setting numerous casing strings in succession, all of the exact same internal diameter. The potential as a step change technology for optimizing drilling costs and mitigating risks is phenomenal.

4. **Rig Instrumentation.** The efficient and effective application of weight to the bit and the control of downhole vibration play a key role in drilling efficiency. Excessive WOB applied can cause axial vibration, causing destructive torsional vibrations. Casing handling systems and top drives are effective tools.

5. **Real-Time Drilling Parameter Optimization.** Downhole and surface vibration detection equipment allows for immediate mitigation. Knowing actual downhole WOB can provide the necessary information to perform improved drill-off tests [9].

6. **Bit Selection Processes.** Most bit venders are able to use the electric log data (Sonic, Gamma Ray, Resistivity as a minimum) and associated offset information to improve the selection of bit cutting structures. Formation type, hardness, and characteristics are evaluated and matched to the application needs as an optimization process [10,11].

4.13.4 Drilling Implementation

Most of the well drilling cost is time dependent rather than product cost dependent. Time is often the biggest influence. Rigs, boats, many tools, and personnel costs are charged as a function of time. Drilling mud is also discussed in terms of cost per day due to daily maintenance costs. The sum of the daily time-based costs is referred to as the operation's *spread rate*. In floater operations, the rig rate is the big influence. As spread rate increases, it becomes easier to economically justify higher-priced products that will save time. These may include more expensive bits, downhole turbines, rotary steerable systems, or a standard steerable system versus steering tools. The potential rewards of new technologies can be great.

4.13.4.1 Rate of Penetration

It is all about ROP, but how should we define ROP? Drilling optimization relies on minimizing total time. It is understood that we must be safe and must not damage production capacity. Our discussion focuses on drilling processes from spud to TD [12]. First consider each casing interval and then identify any subinterval where a substantially different drilling process would be beneficial or required (e.g., due to formation drillability or pressure changes, hole angle, mud type) and any planned events that would cause a trip. Where trips are required, an automatic opportunity is presented to change the drilling assembly and drilling approach or to replace the current tools with new ones if tool life may not be sufficient to reach the next planned trip point. Alternatively, perhaps a planned trip may be challenged and found to be unneeded.

Discussing drilling "time" implies that the time is used to accomplish an outcome (i.e., creating hole)—hence the concept and measure of ROP. There is more than one measure and definition of ROP. This can create misunderstandings, but there are needs for these various measures. ROP, as historically defined, includes the hours after a bit reaches bottomhole divided by the distance drilled until a decision is made to trip out of the hole. By this definition, ROP includes the time spent actually drilling rock plus any back reaming, taking surveys with the steerable system, and making connections. It does not include time tripping in and out of the hole or circulating before tripping out. The potential for any optimization to occur using this criterion is severely limited.

Overall ROP (sometimes referred to as *effective ROP*) reflects the time to drill an entire interval, start to finish, divided by the distance drilled. This measure includes all time from start to finish, the rotating time, and the nonproductive time, and it is of great importance to the drilling engineer and operator management. The interval may start at drill out and conclude with starting the trip out of the hole to run casing, or it may be a subinterval as defined in the planning stage.

Instantaneous ROP is the ROP being achieved at any point in time. This can be correlated to the bit features, operating parameters, mud type and properties, and formation to optimize choices for each of these parameters in the interval in real time and in future wells. Instantaneous ROP can be studied over the length of each joint (or stand if using a top drive) to assess changes in ROP, usually as a function of hole cleaning. This concept can be taken one step further to compare the ROP of respective joints as drilling progresses to assess bit wear or look for formation changes.

4.13.4.2 Special Well Types
The boundaries, constraints, limits, risks, drilling practices, equipment selection, and costs change significantly, depending on the type of well to be drilled. Because the relative importance and effect of these variables also change, the optimization approach and prioritization become more specific to the particular well type. Common groupings are

- Extended reach drilling (ERD)
- Horizontal
- Deepwater

Extended Reach Drilling
An ERD well is typically defined as one in which the horizontal departure is at least twice its TVD. Be aware that this is two-dimensional thinking and does not account for complex well paths or a degree of difficulty based on the equipment. However, Figure 4.13.3 provides examples of TVD versus horizontal displacement for several groups of drilled assets. ERD wells are expensive and typically are pushing the envelope [13–16]. The "envelope" has been pushed over time, as shown in Figure 4.13.4. Requirements and drilling objectives become very critical. Drill strings are designed differently [17,18], rig capacity may be pushed to the limit, hole cleaning processes and capabilities will be tested, dogleg severity and differential sticking take on heightened sensitivity, and casing wear is a distinct possibility. Rotation of the drill string is still a key component to accomplish hole cleaning. Rotary steerable systems have provided tremendous value and are allowing ERD wells to be drilled longer than ever before.

The key issues are torque, drag, and buckling; hole cleaning; ECDs; rig capability; survey accuracy and target definition; wellbore stability; differential sticking and stuck pipe; well control; casing wear; bit type; BHA type; logistics; and costs.

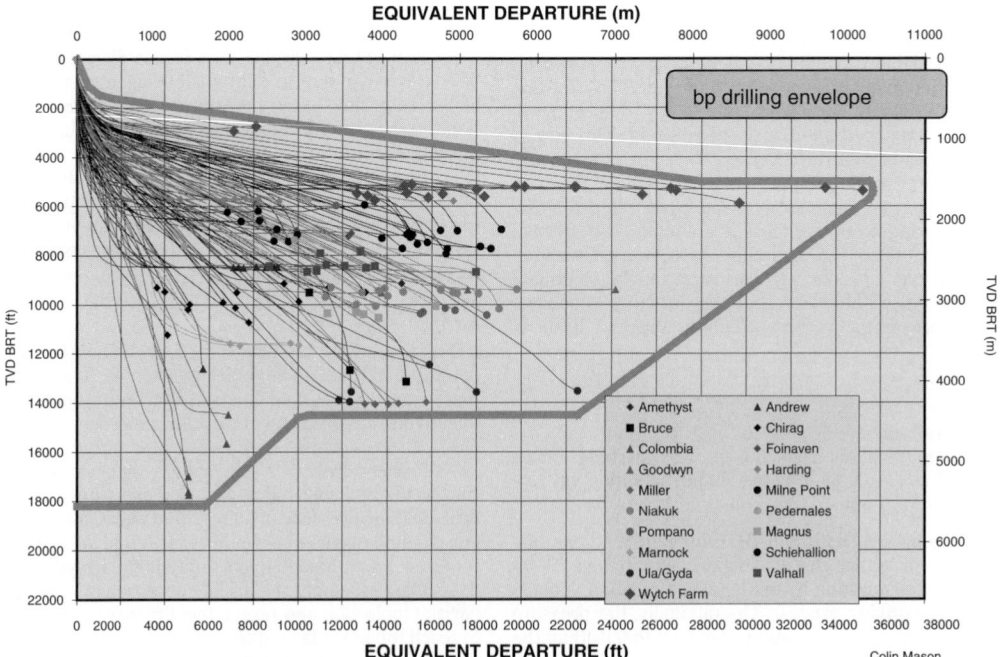

Figure 4.13.3

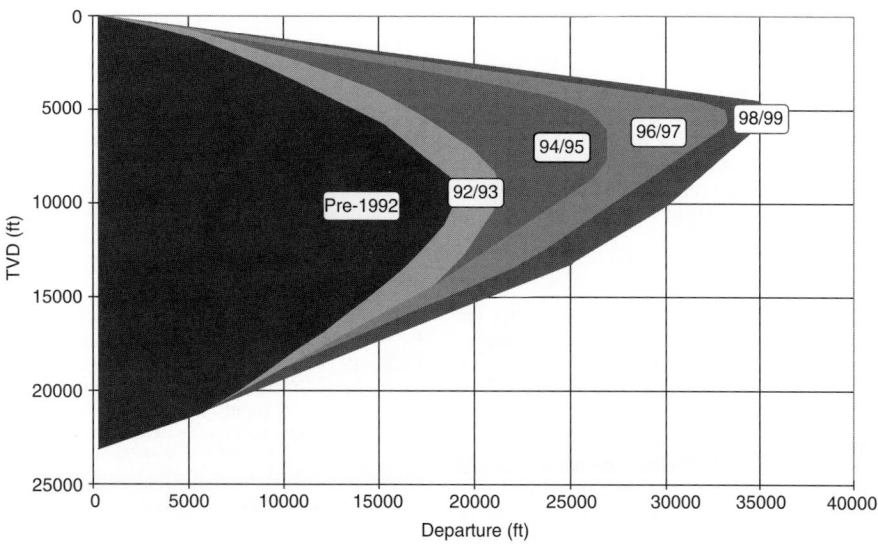

Figure 4.13.4 *Extended-Reach Drilling–Evolution in the 1990s.*

Table 4.13.1 *Well Profiles*

Well Type	Angle Build Rate (°/100 ft)	Feet Drilled to Turn 90°	Curve Radius (ft)	Maximum Lateral Length (ft)	Special Tools Required (?)
Long radius	2–8	1125–4500	1000–3000	8000	No
Medium radius	8–20	450–1125	700–1250	8000	No
Short radius	150–300	30–60	20–40	1000	Yes

Horizontal Wells

Horizontal wells are defined according the rate of build from vertical to horizontal. Choosing one well profile instead of another is based on numerous criteria, including the location of the production zone in relation to the well's surface location, the length of horizontal lateral needed, minimizing cost for footage drilled, and maximizing value. Table 4.13.1 provides a summary of typical ranges of application of each of these technologies, although it seems that as soon as numbers are written, they become obsolete. Product and tool selection must be suited to and optimized for the respective footages, hours, hole angles, and rate of change of hole angles. Large-radius horizontal wells essentially use the same build rates to build the curve as ERD wells use. However, drill collar weight is placed high enough in the drill string so that the collars do not enter the curved build section. In this way, the weight is most effective axially along the drill string, causing less hole drag. Buckling is still a significant planning need. Casing wear is a lesser concern. Mud systems for horizontal laterals normally possess very low low-end viscosity, which maximizes turbulence when pumping, but solids will fall out of solution when pumping stops. Rotation of the drill string is still a key component to accomplish hole cleaning. Laterals are aligned to drill across any expected fractures to add most effectively to the effective permeability.

Deepwater

Until as late as 10 years ago, anything off the continental shelf (500–600 ft) was considered deepwater. However, with recent advances in drilling, exploration, and production technologies, wells are being drilled and produced in depths of 10,000 ft of water. A generally accepted minimum depth of 1,000 ft is common today but will surely change along with the technology. Drilling costs are a major share of the total development cost. A deepwater well may also be an ERD or horizontal profile and therefore possess those sensitivities.

The key issues are rig-buoyed weight and variable deck load capacity; surface stack BOPs; shallow water flows; riser design; hole cleaning; mud temperature cooling; hydrates; low pore pressure; ability to predict pore pressure during planning and in real time; narrow range of pore pressure to fracture gradient; ECD control; wellbore stability; trouble cost; BHA type; logistics; and time cost sensitivity.

New technologies for reducing finding costs include dual density and gradient drilling, composite risers, expandable casing, casing drilling, hole-enlarging devices, and surface stack BOPs.

4.13.4.3 Real Time Optimization Practices

Prudent use of real time information and outcomes during drilling facilitate improved optimization. Real-time decisions must be allowed versus the well plan. This does not, however, mean that all decision-making should be placed on one person.

Drilling Mud

Optimum mud properties control the borehole effectively but minimize the negative effect on instantaneous ROP.

Water loss, spurt loss, and low-gravity solids (i.e., drilled solids) directly reduce instantaneous ROP.

Directional Drilling Techniques

Unnecessary "drilling on the line" (especially on ERD wells) that is dictated by the customer or desired by the directional driller will add unnecessary doglegs, hole drag, time and costs, and hole-cleaning problems to a well. Meaningful criteria specific to the particular well plan should be developed.

Evaluation of the Last Bit

Information can be applied to the next bit run if the formation and the application method are substantially similar for the upcoming need. Bit BHA for the next run may change.

4.13.4.4 Drill-off Tests

Drill-off tests establish the preferred operational parameters in an effort to optimize ROP and bit life. The objective is not necessarily to maximize ROP, because bit life may be severely shortened. The objective is to produce the highest ROP with an acceptable bit and bearing wear rate, if applicable. This is accomplished by using only the magnitudes of rpm and WOB that cause a reasonably sufficient increase in ROP to justify the bit wear that will be caused. New parameters are needed when the formation changes throughout the drilling interval. Sometimes, the ROP is very sensitive to the WOB, and it is important to keep the WOB replaced as it drills off.

Once the WOB and rpm constraints are understood for the downhole tools and possible drill string vibration, the process for performing a drill-off test is as follows:

- Select a WOB believed to be reasonable and then vary the rpm up to the maximum, identifying the lowest rpm that produced the highest ROP.
- Select a reasonable rpm and then vary the WOB through its range, identifying the lowest WOB that produced the highest ROP.

4.13.4.5 Downhole Vibration

The effects of downhole vibration can be disastrous, and recognition, mitigation, and planning for vibration can be critical for success. Drill string failure, downhole tool (DHT) failures, and bit destruction are possible. Less than catastrophic outcomes include reduction in ROP, bit, DHT, and drill string life. A formula for drill string harmonics rule of thumb was removed from the API RP7G recommendations during the 1998 revision because of the inability to accurately predict vibration occurrence. Vibration observed at the surface is not always a model of what is occurring downhole.

Mitigation is accomplished by modifying operational parameters or the drill string and BHA and bit type selection, individually or in combination (e.g., using a downhole motor or turbine). Rig instrumentation or real-time downhole vibration monitoring can be added to recognize drilling vibration. The first step of mitigation operationally involves increasing the rpm. If this does not stop the event, WOB can be lessened. The last choice is to stop and then restart drilling. Damage can be caused very quickly. Quick recognition and mitigation is imperative.

4.13.4.6 Trendology

Drilling must use trend analysis because changes during drilling (over time) tell stories that can and should be used to prevent nonproductive time and to optimize drilling practices. Tabular data or graphical representations can be used to assess the data. The following relationships are predictable, and any unexpected changes are indicators of possible problems.

Standpipe Pressure (SPP) versus Strokes Per Minute (SPM)

An unexpected change may be caused by pump liner failure, drill string and tool washouts, lost nozzles in drill bits, plugged tools, or downhole motor failure.

Torque and Drag

Torque and drag change during the operation. Tracking of these criteria can indicate the borehole condition and hole cleaning. The measurement criteria are pick-up and slack-off weights and free rotating torque. When plotted against time, unexpected and rapid changes are indicators of potential problems. High-angle wells, which typically involve higher daily spread rates and higher risks of occurrence are prime candidates for this level of effort, as are wells with expected hole stability problems.

ROP

Changes in ROP during a bit run should correlate to changes in formation properties (e.g., type, hardness, characteristics), mud properties, or operating parameters (e.g., WOB, rpm, flow rate). An increase in pore pressure can cause a safety incident or borehole instability.

Mud

The mud weight, spurt loss, water loss, and drilled solids all directly affect ROP, hole stability, hole washouts, and future cementing effectiveness.

4.13.5 Post-Run Evaluation

Continuous improvement demands understanding to what extent each variable influences the outcome and how to apply the knowledge. The post-run evaluation process completes the cycle shown in Figure 4.13.1. This should be a continuous improvement cycle.

The data acquired from the well just drilled become the input data for subsequent well planning and drilling stages. These data have become useful information for achieving new learning and greater optimization.

The key criteria that affected the outcome should be the primary focus of evaluation. The data will be used in the specialized "tools" used by operator and service company engineers. End of well (EOW) summary reports are created as a method to communicate the lessons learned.

References

1. The Hon Lord Cullen, "The Public Inquiry into the Piper Alfa Disaster," London, HMSO, 1990.
2. Comer, P. J., Fitt, J. S., and Ostebo, R., "A Driller's HAZOP Method," SPE Paper #15867, 1986.
3. Murchison W. J., Murchison Drilling Schools, Inc., "Drilling Technology for the Man on the Rig," 1998.
4. Devereux, S., "Practical Well Planning and Drilling Manual," PennWell Publishing Company, Tulsa, OK, 1998.
5. Perrin, V. P., Mensa-Wilmot, G., and Alexander, W. L., "Drilling Index: A New Approach to Bit Performance Evaluation," SPE/IADC Paper #37595, 1997.
6. Wolfson, L., Mensa-Wilmot, G., and Coolidge, R., "Systematic Approach Enhances Drilling Optimization and

PDC Bit Performance in North Slope ERD Program" SPE Paper #50557, 1998.

7. Fear, M. J., Meany, N. C., and Evans, J. M., "An Expert System for Drill Bit Selection," SPE/IADC Paper #27470, 1994.

8. "Standard DS-1, Drill Stem Design and Inspection," 2nd Edition, T. H. Hill Associates, Inc., Texas, Houston, March 1998.

9. API RP7G, Recommended Practice for Drill Stem Design and Operating Limits, 16th Edition, 1998.

10. Shanks, E., Schroeder, J., Ambrose, W., and Steddum, R., "Surface BOP for Deepwater Moderate Environment Drilling Operations from a Floating Drilling Unit," Offshore Technology Conference #14265, 2002.

11. Krepp, T. A., and Richardson, B., "Step Improvements Made in Timor Sea Drilling Performance," World Oil, May 1997.

12. Payne, M. L., Cocking, D. A., Hatch, A. J., "Critical Technologies for Success in Extended Reach Drilling, SPE #28293, 1994.

13. Cocking, D. A., Bezant, P. N., and Tooms, P. J., "Pushing the ERD Envelope at Wytch Farm," SPE/IADC Paper #37618, 1997.

14. Mims, M. G., Krepp, A. N., and Williams, H. A., "Drilling Design and Implementation for Extended Reach and Complex Wells," K and M Technology Group, LLC, 1999.

15. Chitwood, J. E., and Hunter, W. A., "Well Drilling Completion and Maintenance Technology Gaps," Offshore Technology Conference #13090, 2001.

16. Fear, M. J., "How to Improve Rate of Penetration in Field Operations," IADC/SPE #35107, 1996.

17. Aldred, W. D., and Sheppard, M. C., "Drillstring Vibrations: A New Generation Mechanism and Control Strategies," SPE #24582, 1992.

18. Belaskie, J. P., Dunn, M. D., and Choo, D. K., "Distinct Applications of MWD, Weight on Bit, and Torque," SPE #19968, 1990.

19. Maidla, E. E., and Ohara, S., "Field Verification of Drilling Models and Computerized Selection of Drill Bit, WOB, and Drillstring Rotation," SPE #19130, 1989.

20. Hill, T. H., Guild, G. J., and Summers, M. A., "Designing and Qualifying Drill Strings For Extended Reach Drilling," SPE #29349, 1995.

21. Saleh, S. T., and Mitchell, B. J., "Wellbore Drillstring Mechanical and Hydraulic Interaction," SPE #18792, 1989.

22. Brett, J. F., Beckett, A. D., Holt, C. A., and Smith, D. L., "Uses and Limitations of Drillstring Tension and Torque Models for Monitoring Hole Conditions," SPE #16664, 1988.

23. Besaisow, A. A., Jan, Y. M., and Schuh, F. J., "Development of a Surface Drillstring Vibration Measurement System," SPE #14327, 1985.

24. Kriesels, P. C., Huneidi, I., Owoeye, O. O., and Hartmann, R. A., "Cost Savings through an Integrated Approach to Drillstring Vibration Control," SPE #57555, 1999.

25. Burgess, T. M., McDaniel, G. L., and Das, P. K. "Improving BHA Tool Reliability With Drillstring Vibration Models: Field Experience and Limitations," SPE #16109, 1987.

4.14 WELL PRESSURE CONTROL

4.14.1 Introduction

Basically all formations penetrated during drilling are porous and permeable to some degree. Fluids contained in pore spaces are under pressure that is overbalanced by the drilling fluid pressure in the well bore. The bore-hole pressure is equal to the hydrostatic pressure plus the friction pressure loss in the annulus. If for some reason the bore-hole pressure falls below the formation fluid pressure, formation fluids can enter the well. Such an event is known as a *kick*. This name is associated with a rather sudden flowrate increase observed at the surface.

A formation fluid influx (a kick) may result from one of the following reasons:

- abnormally high formation pressure is encountered
- lost circulation
- mud weight too low
- swabbing in during tripping operations
- not filling up the hole while pulling out the drillstring
- recirculating gas or oil cut mud.

If a kick is not controlled properly, a blowout will occur. A blowout may develop for one or more of the following causes:

- lack of analysis of data obtained from offset wells
- lack or misunderstanding of data during drilling
- malfunction or even lack of adequate well control equipment

4.14.2 Surface Equipment

A formation gas or fluid kick can be efficiently and safely controlled if the proper equipment is installed at the surface. One of several possible arrangement of pressure control equipment is shown in Figure 4.14.1. The blowout preventer (BOP) stack consists of a spherical preventer (i.e., Hydril) and ram type BOPs with blind rams in one and pipe rams in another with a drilling spool placed in the stack.

A spherical preventer contains a packing element that seals the space around the outside of the drill pipe. This preventer is not designed to shut off the well when the drill pipe is out of the hole. The spherical preventer allows stripping operations and some limited pipe rotation. Hydril Corporation, Shaffer, and other manufactures provide several models with differing packing element designs for specific types of service. The ram type preventer uses two concentric halves to close and seal around the pipe, called pipe rams or blind rams, which seal against the opposing half when there is no pipe in the hole. Some pipe rams will only seal on a single size pipe; 5 in. pipe rams only seal around 5 in. drill pipe. There are also variable bore rams, which cover a specific size range such as $3\frac{1}{2}$ in. to 5 in. that seal on any size pipe in their range.

Care must be taken before closing the blind rams. If pipe is in the hole and the blind rams are closed, the pipe may be damaged or cut. A special type of blind rams that will sever the pipe are called shear blind rams. These rams will seal against themselves when there is no pipe in the hole, or, in the case of pipe in the hole, the rams will first shear the pipe and then continue to close until they seal the well.

A drilling spool is the element of the BOP stack to which choke and kill lines are attached. The pressure rating of the drilling spool and its side outlets should be consistent with BOP stack. The kill line allows pumping mud into the annulus of the well in the case that is required. The choke line side is connected to a manifold to enable circulation of drilling and formation fluids out of the hole in a controlled manner.

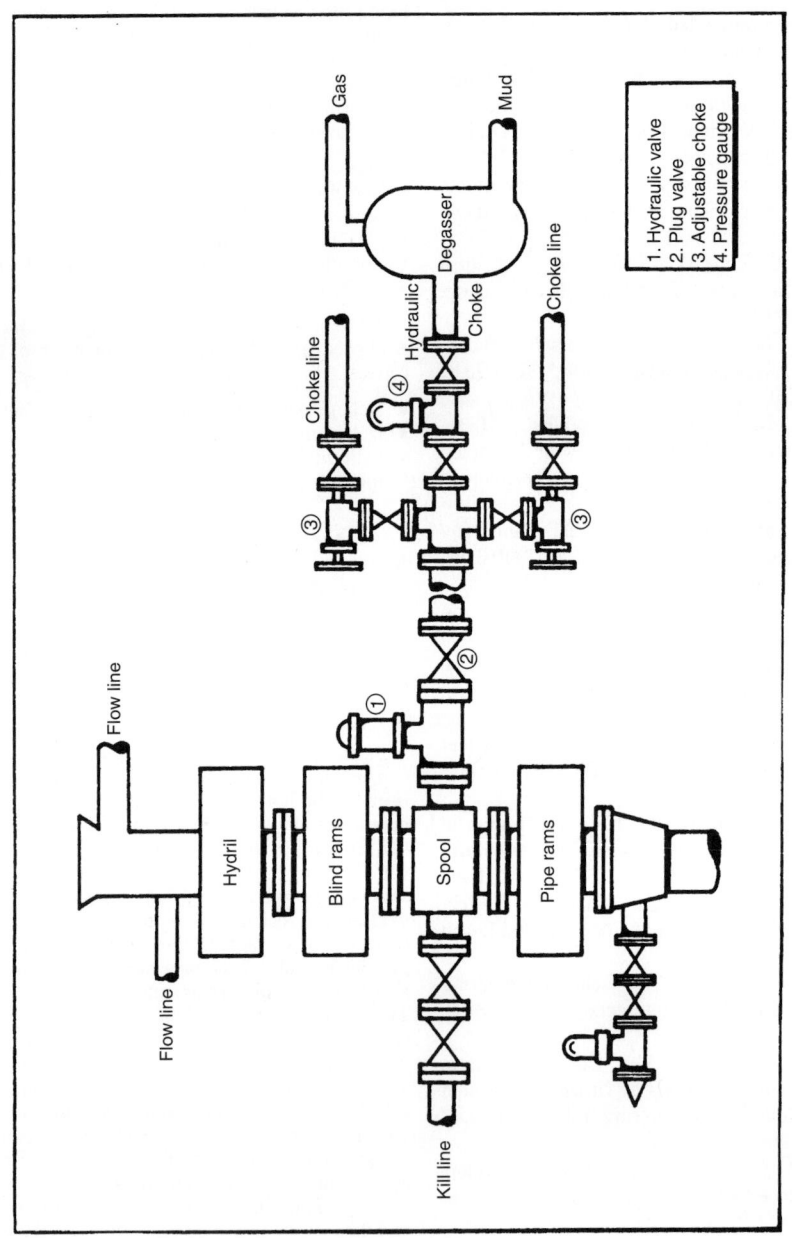

Figure 4.14.1 Pressure control equipment [2].

A degasser is installed on the mud return line to remove any small amounts of entrained gas in the returning drilling fluids. Samples of gas are analyzed using the gas chromatograph.

If for some reason the well cannot be shut in, and thus prevents implementation of regular kick killing procedure, a diverter type stack is used rather, the BOP stack described above. The diverter stack is furnished with a blow-down line to allow the well to vent wellbore gas or fluids a safe distance away from the rig. Figure 4.14.2 shows a diverter stack arrangement.

4.14.3 When and How to Close the Well

While drilling, there are certain warning signals that, if properly analyzed, can lead to early detection of gas or formation fluid entry into the wellbore.

1. *Drilling break.* A relatively sudden increase in the drilling rate is called a drilling break. The drilling beak may occur due to a decrease in the difference between borehole pressure and formation pressure. When a drilling break is observed, the pumps should be stopped and the well watched for flow at the mud line. If the well does not flow, it probably means that the overbalance is not lost or simply that a softer formation has be encountered.
2. *Decrease in pump pressure.* When less dense formation fluid enters the borehole, the hydrostatic head in the annulus is decreased. Although reduction in pump pressure may be caused by several other factors, drilling personnel should consider a formation fluid influx into the wellbore as one possible cause. The pumps should be stopped and the return flow mud line watched carefully.
3. *Increase in pit level.* This is a definite signal of formation fluid invasion into the wellbore. The well must be shut in as soon as possible.
4. *Gas-cut mud.* When drilling through gas-bearing formations, small quantities of gas occur in the cuttings. As these cuttings are circulated up, the annulus, the gas expands. The resulting reduction in mud weight is observed at surface. Stopping the pumps and observing the mud return line help determine whether the overbalance is lost.

If the kick is gained while tripping, the only warning signal we have is an increase in fluid volume at the surface (pit gain). Once it is determined that the pressure overbalance

is lost, the well must be closed as quickly as possible. The sequence of operations in closing a well is as follows:

1. Shut off the mud pumps.
2. Raise the Kelly above the BOP stack.
3. Open the choke line
4. Close the spherical preventer.
5. Close the choke slowly.
6. Record the pit level increase.
7. Record the stabilized pressure on the drill pipe (Stand Pipe) and annulus pressure gauges.
8. Notify the company personnel.
9. Prepare the kill procedure.

If the well kicks while tripping, the sequence of necessary steps can be given below:

1. Close the safety valve (Kelly cock) on the drill pipe.
2. Pick up and install the Kelly or top drive.
3. Open the safety valve (Kelly cock).
4. Open the choke line.
5. Close the annular (spherical) preventer.
6. Record the pit gain along with the shut in drill pipe pressure (SIDPP) and shut in casing pressure (SICP).
7. Notify the company personnel.
8. Prepare the kill procedure.

Depending on the type of drilling rig and company policy, this sequence of operations may be changed.

4.14.4 Gas-Cut Mud

A gas-cut mud is a warning sign of possible formation fluid influx, although it is not necessarily a serious problem. Due to gas expandability, it usually gives the appearance of being more serious than it actually is.

The bottomhole hydrostatic head of gas-cut mud and the expected pit gain can be calculated from the following equations:

$$P_h = P_s + \frac{\gamma_s H}{(1-a_s)} - \frac{P_s a_s}{(1-a_s)} \ln\left(\frac{P_h}{P_s}\right) \qquad [4.14.1]$$

$$V_p = \frac{A_n P_s a_s}{\bar{\gamma}_s} \ln\left(\frac{P_h}{P_s}\right) \qquad [4.14.2]$$

where

$$\gamma_s = \frac{P_s M a_s}{z R T_{av}} + \gamma_m (1-a_s) \qquad [4.14.3]$$

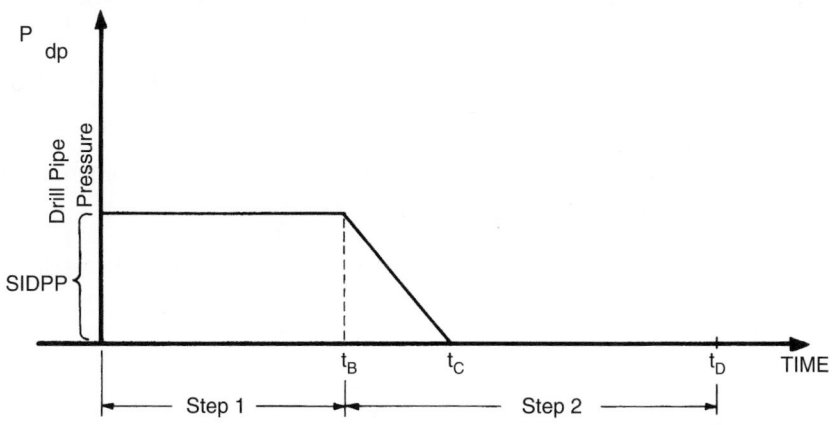

Figure 4.14.2 *Driller's method—Schematic diagram of drill pipe pressure vs. time.*

where P_h = hydrostatic pressure of gas-cut mud in lb/ft^2, abs

P_s = casing pressure at the surface in lb/ft^2, abs

H = vertical hole depth with the gas-cut mud in ft

a_s = gas concentration in mud (ratio of gas to total fluid volume)

$\bar{\gamma}_s$ = gas-cut mud specific weight at surface in lb/ft^3

A_n = cross-sectional area of annulus in ft^2

γ_m = original mud specific weight in lb/ft^3

z = the average gas compressibility factor

M = gas molecular mass

T_{av} = the average fluid temperature in °R

R = universal gas constant in lb·ft

Example 4.14.1

Calculate the expected reduction in bottomhole pressure and pit gain for the data as given below:

Hole size $= 9\frac{5}{8}$ in.
Hole depth $= 10,000$ ft
Drilling rate $= 60$ ft/hr
Original mud density $= 10$ lb/gal
Mud flowrate $= 350$ gal/min
Formation pore pressure gradient $= 0.47$ psi/ft
Porosity $= 30\%$
Water saturation $= 25\%$
Gas saturation $= 75\%$
Gas specific gravity $= 0.6$ (air specific gravity is 1.0)
Surface temperature $= 540°$R
Temperature gradient $= 0.01°$F/ft

Solution

Gas volumetric rate entering the annulus

$$\dot{V} = \frac{\pi(9.625)^2}{4 \times 144} \times 60 \times \frac{1}{60} \times 0.3 \times 0.75$$

$$= 0.1136 ft^3 min = 0.85 gal/min$$

Gas volumetric flowrate at surface (at constant temperature) is

$$\dot{V} = \frac{0.85 \times 4700}{14.7} = 272.6 gal/min$$

The surface pressure is assumed to be 14.7 psia. Gas concentration as surface is

$$a_s = \frac{272.6}{272.6 + 350} = 0.4378$$

Mud weight at surface (assume $z = 1.0$) is

$$\gamma_s = \frac{14.7 \times 144 \times 0.6 \times 29 \times 0.4378}{1,544 \times 540}$$

$$+ 10 \times 7.48(1 - 0.4378)$$

$$= 42.07 lb/ft^3, \text{ or } 5.62 lb/gal$$

Hydrostatic pressure of gas-cut mud is

$$p_h = 14.7 + \frac{0.052 \times 5.62 \times 9,038}{(1 - 0.4378)}$$

$$- \frac{14.7 \times 0.4378}{1 - 0.4378} \ln\left(\frac{p_h}{14.7}\right)$$

Solving the above yields

$$P_h = 4646 psia$$

Hydrostatic head at the bottom of the hole is

$$P_{bh} = 0.052 \times 10(10,000 - 9,038) + 4,646 = 5,146 psia$$

Since the hydrostatic pressure of the original mud is 5,214.7 psia, the reduction in the hydrostatic pressure is

about 69 psi. Because the pore pressure at the vertical depth of 10,000 ft is 4,700 psi, the hydrostatic pressure of the gas-cut mud is sufficient to prevent any formation fluid kick into the hole.

4.14.5 The Closed Well

Upon shutting in the well, the pressure builds up both on the drillpipe and casing sides. The rate of pressure buildup and time required for stabilization depend upon formation fluid type, formation properties, initial differential pressure and drilling fluid properties. In Reference 1, technique is provided for determining the shut-in pressures if the drillpipe pressure is recorded as a function of time. Here we assume that after a relatively short time the conditions are stabilized. At this time we record the shut-in drillpipe pressure (SIDPP) and the shut-in casing pressure (SICP). A small difference between their pressures indicates liquid kick (oil, saltwater) while a large difference is evidence of gas influx. This is true for the same kick size (pit gain).

Assuming the formation fluid does not enter the drillpipe, we know that the SIDPP plus the hydrostatic head of the drilling fluid inside the pipe equals the pressure of the kick fluid (formation pressure). The formation pressure is also equal to the SICP plus the hydrostatic head of the original mud, plus the hydrostatic head of the kick fluid in the annulus.

Thus

$$P_p = 0.052 \bar{\gamma}_m H + SIDPP \qquad [4.14.4]$$

$$= 0.052 \bar{\gamma}_m (H - L) + 0.052 \bar{\gamma}_f L + SICP \qquad [4.14.5]$$

where H = vertical hole depth in ft

$\bar{\gamma}_m$ = original mud specific weight lin in lb/gal

$\bar{\gamma}_f$ = formation fluid specific weight in lb/gal

L = vertical length of the kick in ft

p_p = formation pore pressure in psi

If the hole is vertical the kick length can be calculated as

$$L = \frac{V}{VC_{dc}} \qquad [4.14.6]$$

or

$$L = L_{dc} = \frac{V - L_{dc}VC_{dc}}{VC_{dp}} \qquad [4.14.7]$$

where V = pit gain in bbl

VC_{dc} = annulus volume capacity opposite the drill collars in bbl/ft

L_{dc} = length of drill collar in ft

VC_{dp} = annulus volume capacity opposite the drillpipe in bbl/ft

Equation 4.14.6 is applicable if the kick length is shorter than the drill collars while Equation 4.14.7 is used if the kick fluid column is longer than the drill collars.

From Equations 4.14.4 and 4.14.5 it is found that the required mud weight increase can be calculated from

$$\Delta\gamma_m = \frac{19.23 \times SIDPP}{H} \qquad [4.14.8]$$

The formation fluid density can be obtained from

$$\bar{\gamma}_f = \bar{\gamma}_m - \frac{SICP - SIDPP}{0.052H} \qquad [4.14.9]$$

4.14.6 Kick Control Procedures

There are several techniques available for kick control (kick-killing procedures). In this section only three methods will be addressed.

1. *Driller's method.* First the kick fluid is circulated out of the hole and then the drilling fluid density is raised up to the proper density (kill mud density) to replace the original mud. An alternate name for this procedure is the two circulation method.
2. *Engineer's method.* The drilling fluid is weighted up to kill density while the formation fluid is being circulated out of the hole. Sometimes this technique is known as the *one circulation method.*
3. *Volumetric method.* This method is applied if the drillstring is off the bottom.

The guiding principle of all these techniques is that bottomhole pressure is held constant and slightly above the formation pressure at any stage of the process. To choose the most suitable technique one ought to consider (a) complexity of the method, (b) drilling crew experience and training, (c) maximum expected surface and borehole pressure and (d) time needed to reestablish pressure overbalance and resume normal drilling operations.

4.14.6.1 Driller's Method
The driller's method of controlling a kick is accomplished in two main steps:

Step 1. The well is circulated at half the normal pump speed while keeping the drillpipe pressure constant (see Figure 4.14.2). This is accomplished by adjusting the choke on the mud line so that the bottomhole pressure is constant and above the formation fluid pressure. To maintain a constant bottomhole pressure the formation fluid is allowed to expand, which usually results in a noticeable increase in casing pressure. This step is completed when the formation fluid is out of the hole. At this time casing pressure should be equal to the initial SIDPP if the well could be shut in.

Step 2. When the formation fluid is out of the hole, a kill mud is circulated down the drillpipe. To obtain constant bottomhole pressure, the casing pressure is kept constant (Figure 4.14.3) while the drillpipe pressure drops. Once the kill mud reaches the bottom of the hole the control moves back to the drillpipe side. The drillpipe pressure is maintained constant (almost constant) while the new mud fills the annulus.

Example 4.14.2
Consider the following data:

Vertical hole depth $= 10,000\,\text{ft}$
Hole diameter $= 8\frac{3}{4}\,\text{in.}$
Drillpipe diameter $= 4\frac{1}{2}\,\text{in.}$
Mud density $= 12\,\text{lb/gal}$
Yield point $= 12\,\text{lb/100\,ft}^2$
Circulating pressure at reduced speed $= 800\,\text{psi}$
Shut-in drillpipe pressure $= 300\,\text{psi}$

Calculate the following:

1. Required mud weight to restore the safe overbalance
2. Initial circulating pressure (ICP)
3. Final circulating pressure (FCP)
4. Specific weight of formation fluid

Solution
Mud weight increase to balance formation pressure is

$$\Delta\bar{\gamma}_m = \frac{20 \times 300}{10,000} = 0.6\,\text{lb/gal}$$

Trip margin $\Delta\bar{\gamma}_t$ is

$$\Delta\bar{\gamma}_t = \frac{yp}{6(D_h - D_p)} = \frac{12}{6(8.75 - 4.5)} = 0.47\,\text{lb/gal}$$

Consequently the required mud weight is

$$\bar{\gamma}_m = 12 + 0.6 + 0.47 = 13.1\,\text{lb/gal}$$

Initial circulating pressure is

$$\text{ICP} = 800 + 300 = 1,100\,\text{psi}$$

Final circulating pressure is the system pressure loss corrected for new mud weight. This is

$$\text{FCP} \cong 800\left(\frac{13.1}{12}\right) = 873\,\text{psi}$$

4.14.6.2 Engineer's Method
This method consists of four phases.

Phase 1. During this phase the drilling fluid, weighted to the desired density, is placed in the drillpipe. When the drillpipe is filled with heavier mud, the standpipe pressure is gradually reduced. The expected drillpipe pressure versus the numer

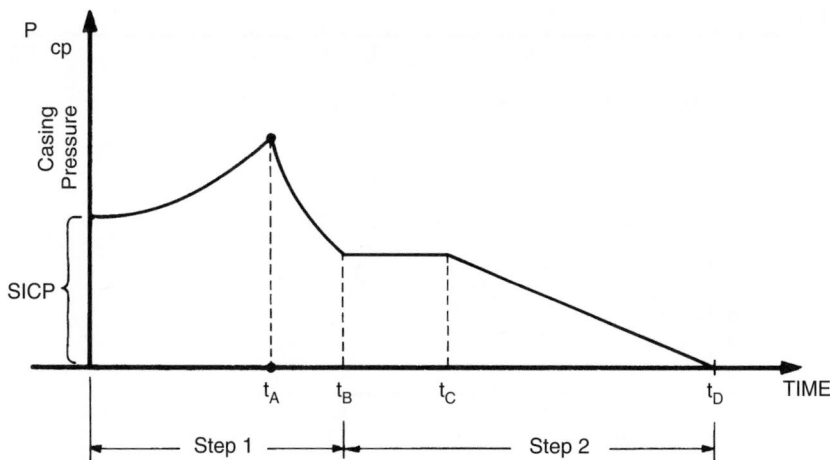

Figure 4.14.3 *Driller's method—Schematic diagram of casing pressure vs. time. t_A = kick fluid out the top of the hole; t_B = kick fluid out of the hole; t_C = kill mud at the bottom of the hole; t_D = killing procedure completed.*

of pump strokes (or time) must be prepared in advance. Only by pumping with a constant number of strokes and simultaneously maintaining the standpipe pressure in accordance with the schedule can one keep the bottomhole pressure constant and above the formation pressure. The annulus pressure at the surface generally rises due to formation fluid expansion, although for some formation fluid the casing pressure may decrease. This depends on phase behavior of the formation fluid and irregularities in the hole geometry.

Phase 2. This phase is initiated when the kill mud begins filling the annulus and is finished when the formation fluid reaches the choke. The standpipe pressure remains essentially constant by proper adjustment of the choke.

Phase 3. The formation fluid is circulated out of the hole whole heavier mud fills the annulus. Again the choke operator maintains the drill pipe pressure constant and constant pumping speed.

Phase 4. During this phase the original mud that follows the kick fluid is circulated out of the hole and a kill mud fills up the annulus. The choke is opened more and more to keep the drill pipe pressure constant. At the end of this phase and safe pressure overbalance is restored.

A qualitative relationship between the drillpipe pressure, casing pressure and circulating time is shown is Figure 4.14.4 and 4.14.5, respectively.

4.14.6.3 Volumetric Method
This method can be used if the kick is taken during tripping up the hole with the bit far from the bottom of the hole. Again the constant bottomhole pressure principle is used to control the situation.

The fundamental principle of this method is equating the pit volume change with the corresponding change in annulus pressure. We write

$$\Delta p_s = \frac{\Delta V_p}{AV} 0.052 \bar{\gamma}_m \qquad [4.14.10]$$

where Δ_{p_s} = change in surface pressure at casing side in psi

ΔV_p = change in pit volume (volume gained) in bbl

$\bar{\gamma}_m$ = mud specific weight in lb/gal

AV = volume capacity of annulus in bbl/ft

Note that the term $(0.052 \gamma_m/AV)$ expresses the expected increase of casing pressure when 1 bbl of pit gain is recorded.

The magnitude of the casing pressure during kick control is

$$P = SICP + \frac{\Delta V_p}{AV} 0.052 \bar{\gamma}_m \qquad [4.14.11]$$

When the pit volume stabilizes, there is equilibrium in the annulus and the kick fluid is out of the hole.

4.14.7 Maximum Casing Pressure
Determination of the maximum expected casing pressure is required for selection of the kick control technique. If the driller's method is used for kick control, the maximum casing pressure $P_{c\,max}$ (psi) is calculated assuming gas influx into the hole. This is

$$P_{c\,max} = \frac{SIDPP}{2} + \left[\left(\frac{SIDPP}{2} \right)^2 + 0.052 \bar{\gamma}_m + \frac{p_p \Delta V_p T_2 z_2}{z_1 T_1 AV} \right]^{0.5}$$

$$[4.14.12]$$

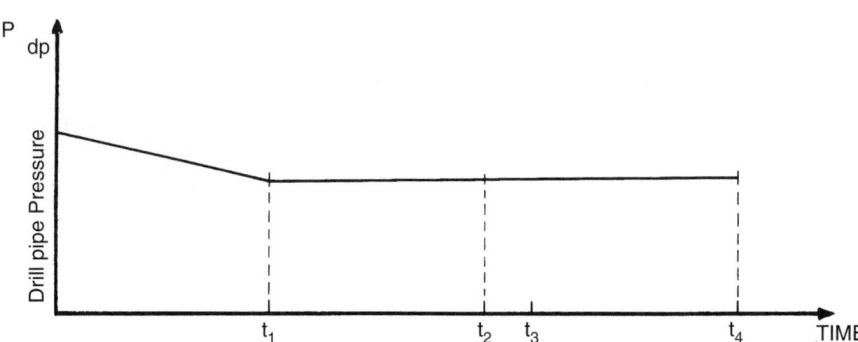

Figure 4.14.4 *Engineer's method—drill pipe pressure vs. time.*

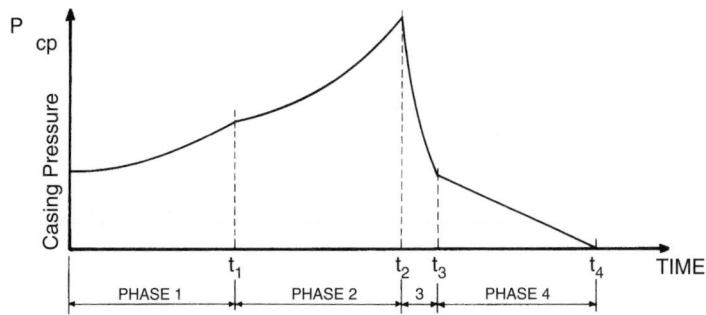

Figure 4.14.5 *Engineer's method—casing pressure vs. time; t_1 = kill mud at the bottom of the hole; t_2 = formation fluid reaches the choke; t_3 = kick fluid out of the hole; t_4 = pressure overbalance is restored.*

where SIDPP = shut-in drillpipe pressure in psi
$\bar{\gamma}_m$ = original mud density specific weight in lb/gal
P_p = formation pressure
ΔV_p = pit gain in bbl
T_2 = gas temperature at surface in °R
T_1 = gas temperature at the bottom of the hole in °R
z_1 = gas compressibility factor at surface conditions
z_2 = gas compressibility factor at the bottom of the hole
AV = volume capacity of annulus in bbl/ft

For the engineer's method the maximum expected casing pressure can be calculated from

$$p_{cmax} = 0.026(\bar{\gamma}_k - \bar{\gamma}_m)L_m + \left([0.026(\bar{\gamma}_k - \bar{\gamma}_m)L_m]^2 \right.$$

$$\left. + 0.052\bar{\gamma}_k \frac{P_p\Delta V_p z_2 T_2}{z_1 T_1 AV} \right)^{0.5} \qquad [4.14.13]$$

where $\bar{\gamma}_k$ = kill mud density
L_m = length of original mud in the annulus at the moment the bubble reaches the top of the hole

Example 4.14.3
Calculate the maximum expected casing pressure for the driller's and engineer's techniques of kick control for the data as below:

Vertical hole depth = 12,500 ft
Original mud weight = 12.0 lb/gal
Shut-in drillpipe pressure = 260 psi
Pit gain = 30 bbl
Volume of mud inside the drillstring = 175 bbl
Annular volume capacity = 0.13 bbl/ft
Gas compressibility ratio $z_1/z_2 = 1.35$
Gas temperature at the surface = 120°F
Well-bore temperature gradient = 1.2°F/100 ft
Surface temperature = 80°F

Solution
Required mud weight for killing the well is

$$\bar{\gamma}_k = 12 + \frac{260}{0.052 \times 12,500} + 0.5 = 12.9 \text{ lb/gal}$$

For trip margin, a 0.5 lb/gal was added arbitrarily. Bottom-hole temperature is

$$T_1 = 80 + 1.2 \times 125 = 230°F$$

Formation pore pressure is

$$p_p = 260 + 0.052 \times 12 \times 12,500 = 8,060 \text{ psi}$$

For the driller's method

$$p_{cmax} = \frac{260}{2} + \left[\left(\frac{260}{2}\right)^2 + 0.052 \right.$$

$$\left. \times 12 \frac{8,060 \times 30 \times 1.35 \times 580}{690 \times 0.13} \right]^{0.5}$$

$$= 1,285 \text{ psi}$$

For the engineer's method

$$p_{cmax} = 0.026(12.9 - 12.0)\frac{175}{0.13}$$

$$+ \left[(0.026 \times 0.9 \times 1,346.1)^2 + 0.052 \right.$$

$$\left. \times 12.9 \frac{8,060 \times 30 \times 1.35 \times 580}{690 \times 0.13} \right]^{0.5}$$

$$= 1,222 \text{ psi}$$

The engineer's method normally results in lower maximum surface pressure than the driller's method. For last field-type calculations, the following equation for the engineer's method can be used:

$$p_{cmax} = 200 \left(\frac{p_p \Delta V_p \bar{\gamma}_k}{AV} \right)^{0.5} \qquad [4.14.14]$$

All symbols in Equation 4.14.14 are the same as those used above.

4.14.8 Maximum Borehole Pressure
When the top of the gas bubble just reaches the casing setting depth, the open part of the hole is exposed to the highest pressure. If this pressure is less than the formation fraction pressure, the kick can be circulated out of the hole safely without danger of an underground blowout.

If the driller's method is used, this maximum pressure p_{bm} can be obtained from

$$p_{bm}^2 - (p_p - 0.052 \times \bar{\gamma}_m \times H + 0.052\bar{\gamma}_m D_1)p_{bm}$$

$$- 0.052\bar{\gamma}_m \frac{\Delta V_p p_p}{AV} = 0 \qquad [4.14.15]$$

Example 4.14.4
The data for this example are the same as the data used in Example 4.14.3.

Casing setting depth = 9,200 ft
Formation fracturing pressure at casing shoe = 7,550 psi

It is required to determine whether or not the formation at the casing shoe will break down while circulating the kick out of the hole. The annulus volume capacity in an open hole is 0.046 bbl/ft.

Solution
Using Equation 4.14.15 we write

$$p_{bm}^2 - (8,060 - 0.052 \times 12 \times 12,500 + 0.052 \times 12 \times 7,550)p_{bm}$$

$$- 0.052 \times 12 \times \frac{30 \times 8,060}{0.046} = 0$$

Solving the above equation yields $p_{bm} = 5,561$ psi. Since the maximum expected pressure in an open hole is less than the formation fracture pressure, the kick can be safely circulated out of the hole.

References
1. Miska, S., F. Luis, and A. Schofer-Serini, "Analysis of the Inflow and Pressure Buildup Under Impending Blowout Conditions," *Journal of Energy Resources Technology*, March 1992.
2. Moore, P. L., *Drilling Practices Manual*, The Petroleum Publishing Co., Tulsa, Oklahoma, 1974.
3. Rehm, B., *Pressure Control in Drilling*. The Petroleum Publishing Co., Tulsa, Oklahoma, 1976.

4.15 FINISHING AND ABANDONMENT

A fish is a part of the drill string that separates from the upper remaining portion of the drill string while the drill string is in the well. This can result from the drill string failing mechanically, or from the lower portion of the drill string becoming stuck or otherwise becoming disconnected from drill string upper portion. Such an event will instigate an operation to free and retrieve the lower portion (or fish) from the well with a strengthened specialized string. Junk is usually described as small items of non-drillable metals that fall or are left behind in the borehole during the drilling, completion, or workover operations. These non-drillable items must be retrieved before operations can be continued [1–3]. The process of removing a fish or junk from the borehole is called *fishing*.

It is important to remove the fish or junk from the well as quickly as possible. The longer these items remain in a borehole, the more difficult these parts will be to retrieve. Further, if the fish or junk is in an open hole section of a well the more problems there will be with borehole stability. There is an important tradeoff that must be considered during any fishing operation. Although the actual cost of a fishing operation is normally small compared to the cost of the drilling rig and other investments in support of the overall drilling operation, if a fish or junk cannot be removed from the borehole in a timely fashion, it may be necessary to sidetrack (directionally drill around the obstruction) or drill another borehole. Thus, the economics of the fishing operation and the other incurred costs at the well site must be carefully and continuously assessed while the fishing operation is underway. It is very important to know when to terminate the fishing operation and get on with the primary objective of drilling a well. Equation 4.15.1 can be used to determine the number of days that should be allowed for a fishing operation.

The number of days, D (days), is

$$D = \frac{V + C_s}{R + C_d} \qquad [4.15.1]$$

where

V is the replacement value of the fish (dollars or other money),

C_s is the estimated cost of the sidetrack or the cost of restarting the well (dollars),

R_f is the cost per day of the fishing tool and services (dollars/day),

C_d is the cost per day of the drilling rig (and appropriate support) (dollars/day).

4.15.1 Causes and Prevention

There are a number of causes for fishing operations. Many of the problems that lead to fishing operations can be prevented by careful operational planning and being very watchful as drilling operations progress for indications of possible future borehole troubles [4]. The major causes of a fishing or junk retrieval operation are

1. *Differential Pressure Sticking*. It is estimated that the cost of stuck pipe in deep oil and gas wells can be approximately 25% of the overall budget. Therefore, a large portion of the fishing tools available were developed for the recovery of stuck pipe. This condition occurs when a portion of the drill string becomes stuck against wall of an open hole section of the borehole. This is due to smooth surface of the drill string (usually the drill collars) has become embedded in the filter cake on the wall. Differential sticking is possible in most borehole operations where drilling mud is being used as the circulation fluid.

Underbalanced operations generally avoid this differential pressure sticking (see Section 4.9). Sticking occurs when the pressure exerted by the mud column is greater than the pressure of the formation fluids. Normally the drill string is differentially stuck when

A. The drill string cannot be rotated, raised or lowered, but circulating pressure is normal.

B. The drill collars are opposite a permeable formation.

C. Sticking was instantaneous when the pipe was stationary after drilling at a higher than normal penetration rate. In some cases a differentially stuck string or bottomhole assembly may be freed by reducing the mud weight.

This will reduce the differential pressure between the column of mud and the permeable zone. However, this procedure should not be used if well control is a problem.

2. *Under Gauge Borehole*. Mud filter cake can build excessively across a low pressure permeable formation when the circulation rate is low, water loss is very high, and there is an extended period between trips. Under these conditions, a drill string or logging tools can become stuck in the under gauge borehole and/or filter cake. Filter cake build up is usually slow and appears as drag on the multichannel recorder, or as an under gauge hole on a caliper survey.

3. *Key seats*. Key seats develop where there is a sudden change in hole deviation or above a washout in a deviated hole. Doglegs above the drill collars are subject to erosion or wear by the drill pipe on the high side of the dogleg. Continuous rotation can slowly cut a groove into the dogleg forming what is known as a key seat. The drill pipe body and tool joints wear a groove in the formation approximately the same diameter as the tool joints. The wear is confined to a narrow groove, because high tension in the drill pipe prevents side ways movement. During a trip out of the hole, the BHA may be pulled into these grooves and the grooves may be too small to allow the BHA to pass through. In this situation no attempt should be made to jar the collars through a key seat. A possible solution to this problem would be to circulate and rotate the drill string and move the string in small increments up through the key seat. All tight spots (over pull and depth of the over pull) should be noted and recorded on the IADC daily report and the drilling recorder. A tight spot that occurs on two successive trips out of the hole with over pull on the second trip greater than the first, is an indication of a key seat forming. A key seat wiper or string reamer should be run on the third trip.

4. *Tapered hole*. Abrasive hole sections will tend to dull bits and thereby reduce bit and stabilizer gauge. Attempting to maximize the length of a bit run in an abrasive formation may prove to be a false economy since an under gauge hole will likely lead to a reaming operation. If the driller fails to ream a new bit to bottom when this situation exists, the bit may jam in the under gauge hole. Proper grading and gauging of the old bit prior to running a new one will prevent this problem. Since this type of sticking usually occurs while tripping in the hole. In such situations, the string should be jarred up immediately. This will usually free the drill string. If jarring does not free the string, it will be necessary to make a back-off and wash over. Wash over procedures will be discussed below.

5. *Object along side the drill string*. Occasionally, an object such as a wrench, bolt, slip or tong part, or a hammer will fall into the hole along side the drill string. Except when the string can be pushed or pulled around the object or the object can be pushed into the wall of the hole, serious

fishing problems can develop. This is especially true when the string gets jammed to one side of the hole in a cased hole. A visual check of all hand and other surface tools is required to see if anything is missing. Never leave the hole unprotected or leave loose objects lying around the rotary area. Jarring might free the string, if not, a short wash over is required using an internal spear to catch the string when it falls.

6. *Inadequate hole cleaning occurs as a result of*
   A. A drill string washout above the bit.
   B. Low circulation rate in a large hole with an unweighted mud system.
   C. Sloughing shale.
   D. Gravel bed in the shallow portion of the hole.
   E. Partial returns.

Indications of sticking due to inadequate hole cleaning are:

A. A significant change in the amount of returns across the shaker before sticking.
B. A decrease in pump pressure or increase in pump strokes followed by an increase in drag while picking up on the pipe (a washout in the drill string).
C. An increase in pump pressure and drag.
D. The inability to circulate if the pipe sticks.
E. Frequent bridges on trips.

Even with the challenging wells being drilled today the incident of parted strings occur less often than decades ago. Improved maintenance, inspection procedures, monitoring systems, materials and coatings are all contributing to a reduction in this fishing operations. The biggest challenge when fishing a parted string is in the interpretation of the condition of the top of the fish. A string may part due to any of the following reasons:

a. A twist off after the drill string has become stuck.
b. A washout.
c. A back lash and subsequent unscrewing of the string.
d. Junk wearing through a tubular.
e. Metal fatigue in the string.

When working shallow without the benefit of a torque limit switch and if the string becomes stuck, the torque can build up very rapidly in the string and cause a twist off. If a work string is in poor condition a twist off can occur at any depth (with or without a torque limit switch). A twist off can be the most difficult type of fishing job due to the possible condition of the top of the fish. Turbulent flow of the circulating drilling fluid can damage a connection and cause a washout in the metal of the connection itself. If such a washout is not detected, the drill string can be weakened in the washed out area resulting in the failure of the string component. A washout in a tubular can in turn washout the formation. This would reduce the annular velocity in the washout area which in turn world diminish hole cleaning. Any time there is a drop in the standpipe pressure that cannot be explained, the string should be pulled immediately.

A string that alternately sticks then releases while drilling forward can result in a build up of torque which, when released, rotates the lower portion of the string at an accelerated rate. The inertia of the lower portion of the string can make the string back off. In a gauged area of the hole, the string could be screwed back together, but in a washed out area it will be necessary to run special tools to engage the fish. This can also occur off bottom while torque is in the string. Junk pushed into a soft formation can later damage tubulars rotating against the junk. It is always better to remove the junk than to push it to the side. Metal fatigue can cause a string to fail under normal operating parameters. Fatigue can be reduced by establishing the working life of

string components and replacing them at the appropriate respective time intervals.

In general, parted strings are easier to fish than stuck pipe. However, if a fish is in an open hole, the likelihood of recovery a fish will diminishes with time. If a fish has a connection is facing up, a screw-in assembly with jars should be run. If the fish top cannot be screwed into, an overshot with a jarring assembly should be run. Different types of fishing and jarring assemblies are shown in Figure 4.15.1. The condition of the bottom of the string pulled after a string parts should give an indication of the condition of the top of the fish and, thus, determine what tools should be used for the fishing job. The piece of fish pulled out of the hole should be a reverse mirror image of what the top of the fish looks like.

4.15.2 Pipe Recovery and Free Point

When a drill string or other tubular becomes stuck in a borehole it is very important that the depth where the pipe is stuck be determined. In most cases this can be accomplished via a simple calculation. The depth to where the drill string is free and where sticking commences is called the free point [2]. This free point can be calculated using measurements taken on the rig floor.

The length of the drill string from the surface to the free point is L_f. This will be less than the overall length of the drill string D. To obtain the L_f, the following procedure is used:

1. An upward force F_1 is applied to the top of the drill string via the drawworks. The force should be slightly greater than the total buoyant weight of the drill string. This ensures that the entire drill string is in tension.
2. With this tension on the drill string, a reference mark is make on the drill string exposed at the top of the string.
3. A greater upward force F_2 is then applied to the drill string. This causes the free portion of the drill string to elastically stretch by an amount ΔL. The stretch (or elastic displacement) is measured by the movement of the original reference mark. The magnitude of F_2 must be limited by the yield stress, or elastic limit stress of the specific pipe steel grade.

Knowing the stretch ΔL and the forces applied F_1 and F_2, Hooke's law (see Section 2.3), the length to the free point is

$$L_f = EA \frac{\Delta L}{(F_2 - F_1)} \qquad [4.15.2]$$

where

E is the Elastic Modulus (Young's Modulus) of steel (i.e., 29×10^6 psi, 200 GPa),
A is the cross-sectional area of the pipe body (ft^2, m^2),
ΔL is the stretch distance (ft, m),
F_1 is the force to place the entire drill string in tension (lb, N),
F_2 is a force greater than F_1 but less that the force limited by the yield stress of the pipe grade (lb, N).

Example 4.15.1
A drill string API $3\frac{1}{2}$ in., 13.3 lb/ft nominal weight, Grade E75 drill pipe is stuck in a 12,000 ft borehole. The driller places 150,000 lb of tension on the top of the drill string above the normal weight of the string and makes a mark on the string at top of the rotary table. The driller then increases the tension to 210,000 lb beyond the weight of the drill string. This later tension will give a maximum stress in the drill pipe body that is less than the 75,000 psi yield stress of Grade E75. The original mark on the drill string shows that the free portion of the drill string has stretched 4 ft due to the additional tension placed on the drill string by the drawworks.

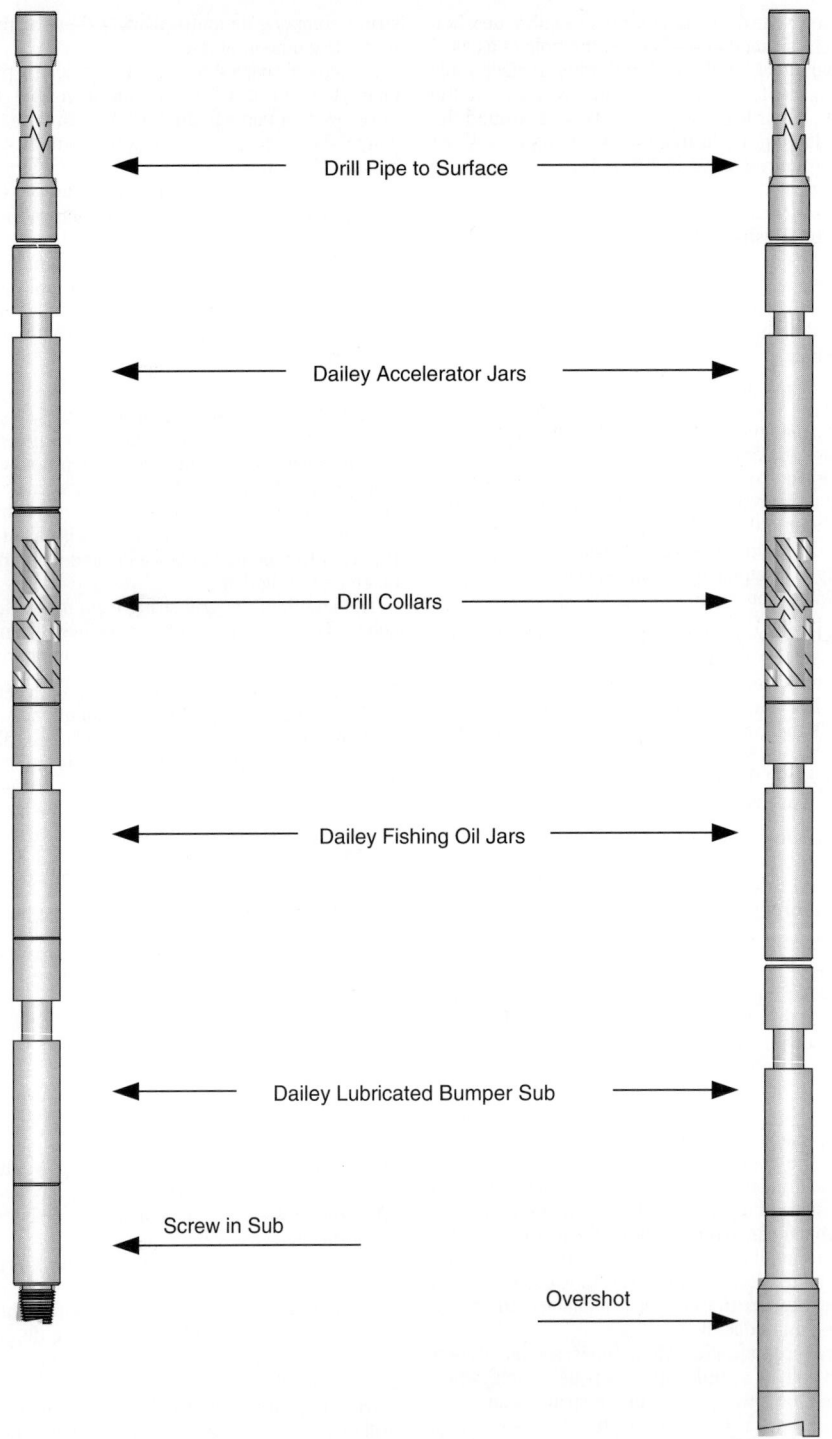

Figure 4.15.1   *Screw in Sub and Overshot Jarring Assemblies. (Courtesy Dailey Fishing Tools)*

The cross-sectional area of the drill pipe body is approximately

$$A = \frac{13.3}{490}$$

$$A = 0.271 \text{ft}^2$$

where the specific weight of steel is 490lb/ft^3. Equation 4.15.1 is

$$L_f = \frac{(29 \times 10^6)(144)(0.0271)(4.0)}{(210000 - 150000)}$$

$$L_f = 7805 \text{ft}$$

There is a special downhole tool that can be used to determine the location where the drill string (or other tubulars) is stuck. This is called a free point tool. Prior to using the free point tool must be calibrated in the free drill string (free pipe). This calibration is accomplished using an anchor system in the free string prior to normal operations being initiated. Once the tool is calibrated, the calibration should not be changed during the course of the operation. The free point tool is calibrated so that 100 units of movement is defined as completely free pipe. When the tool enters an increased cross-sectional area (increased wall thickness) such as Hevi-Wate drill pipe (HWDP), the scale will change so that 70 units of movement is accepted to be completely free pipe (the calibration of the free point tool has not changed). As the tool goes from HWDP into drill collars, there is another scale change. In the drill collars, 40 to 50 units of movement indicates completely free drill collars. A unit of movement in a free point application is a nondefined reference term.

There are four types of anchoring systems used in free point tools today. The oldest is the bow spring anchor. This design has been in use since the late 1950s. The others are the permanent magnet anchoring system, the electromagnet anchoring system, and the motorized anchoring system. All free point tools measure movement between the two anchors (average about 17/1000 of 1 in. in movement). All free point tools read movement between the two anchors in the form of tension on the pipe, compression on the pipe, right hand torque on the pipe, and the newest tools, including the read left hand torque. It is critical that both torque and stretch readings be measured and compared. If the two readings differ, the torque reading is generally more reliable.

The interval between free pipe and stuck pipe is called the transition area. The interval is a length of the borehole. The length of the transition area is an important clue in helping to identify why the pipe is stuck. For example, in free pointing a stuck packer, the transition area would be extremely short if the pipe is free from the surface to the top of the packer. If the pipe were stuck due to fill on top of the packer, the transition are would be spread out over a longer interval. In an open hole a keyseat would result in a very short transition area while differential sticking would result in the transition area being spread out over a much longer interval. When parting the pipe it is important to leave some free pipe looking up (one or two joints of free pipe is recommended). Back-off at least 100 ft. below a casing seat or up inside the casing. Do not back-off in a washed out section or immediately below a drop in hole angle. Always make the back-off at the depth that will facilitate the best chance of future engagement to the fish.

4.15.3 Parting the Pipe

There are seven options to be considered prior to parting the pipe. These are chemical cut, jet cutter, internal mechanical cutter, outside mechanical cutter, multi-string cutter, severing tool, and washover back-off safety joint/washover procedures.

There are five requirements for a back-off to be successful:

1. Free pipe: the connection to be backed off must be free.
2. Torque: the correct amount of left hand torque is required.
3. Weight: the connection being shot must be at neutral weight.
4. Shot placement: the short must be fired across the connection.
5. Shot: use the proper size string shot/prima cord.

4.15.3.1 Chemical Cut

The first chemical cutter was developed by McCullough Tool Company and used in the field in 1957. Today there are several manufactures of chemical cutters. The chemical cutter is lowered inside the pipe (that is to be cut) to a depth of one or two joints above the stuck point. A collar locator is used to correlate depths. Chemical cuts do not require that the pipe be torqued up. This affords a safer operation and is recommended in bad strings of pipe. Sometimes pipe will rotate freely, even though it cannot be pulled from the hole. This makes it impossible to back off the pipe. The chemical cutter utilizes a blast of powerful acid (at high speed and temperature) to make a smooth cut without flare or distortion to the OD or ID of the pipe. It will not damage the outer string of casing or tubing making for easy engagement of the pipe being cut.

4.15.3.2 Jet Cutter

The shaped explosive charges using parabolic geometry were developed after World War I to penetrate thick steel armor. This shaped change was adapted to fit in casing or tubing and became production jet perforating changes that replaced the earlier bullet perforators. Further improvements in the technology allowed the shaped charge concept to be used in a 360° circle design that can be used to completely sever a steel tube. Advantages of the jet cutter are that the jet cutter does not have mechanical slips to set so the condition of the tubular being cut, or what the ID is coated with, has little bearing on the operation of the cutting. Jet cutters are shorter in length than a chemical cutter and greater size ranges are available. The disadvantage of a jet cutter is that the pipe being cut will be deformed and must be dressed off before fishing. Also, adjacent strings could be damaged in multiple tubing completed wells.

4.15.3.3 Internal Mechanical Cutter

The mechanical cutter usually cannot compete with cutters that can be run on wire line due to the cost of trip time. An internal mechanical cutter is shown in Figure 4.15.2. However, if large OD tubulars are being cut a mechanical cutter can be cost effective due to the high cost of large OD chemical and jet cutters. Also a mechanical cutter can be run in conjunction with a spear which allows cutting and retrieving in a single trip. The mechanical cutter is an option that will have merit in several situations. These are shallow depth cuts, large OD tubular cuts, the need to cut and retrieve in a single trip, and in well conditions too adverse for wireline conveyed cutters. The mechanical cutter is lowered into the hole on a tubular string to the point where the cut is to be made. At this point, right hand rotation will allow the friction assembly to unscrew from the mandrel and a gradual lowering of the tool permits the cone to be driven through the slips thereby anchoring the tool in the pipe. As the slips firmly engage, the wedge block forces the knives outward. This action is continued until the pipe is cut. With the cut complete the pipe is raised, the slips disengage, the knives

Figure 4.15.2 *Type "C" Internal Mechanical Cutter.
(Courtesy Weatherford International)*

retracted and the friction assembly returns automatically to
the running in position. A unique feature of this tool is the
automatic nut which allows the resetting and disengaging of
the tool frequently without coming out of the hole.

4.15.3.4 Outside Mechanical Cutter
Figure 4.15.3 shows an external mechanical cutter. Washing
over the stuck pipe is done with washover pipe and a rotary
shoe slightly larger than the cutter to make a gauge run.
When washing over the desired section is completed, the
washover pipe is pulled out of the hole and the rotary shoe
is replaced with the cutter. The cutter and washpipe are then
lowered over the stuck pipe. To operate, the cutter is slowly
raised until the dog assembly engages the joint. The string
is then lowered slightly to reduce excess pressure on the
knives as the cut is started. Rotating the cutter to the right
starts the cut. A slight upward pull and slow uniform rotation

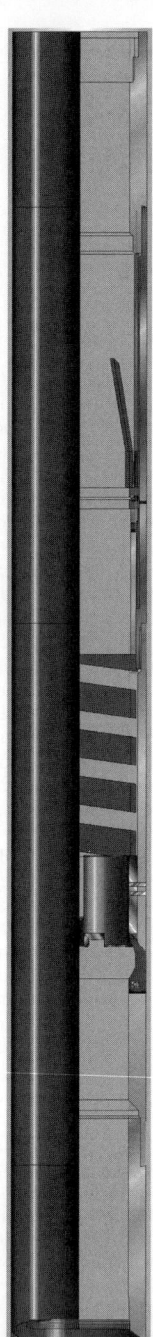

Figure 4.15.3 *Type "A" Outside Mechanical Cutter.
(Courtesy Weatherford International)*

is maintained while the cut is being made. When the cut is
completed, the string is raised, bringing with it the cut off
section of pipe which is held in the cutter and washpipe. At
the surface the cut off section is stripped out of the washpipe
and the process is repeated.

4.15.3.5 Multi-String Cutter
Figure 4.15.4 shows a multi-string cutting assembly. The
multi-string cutter is lowered into the hole on a working

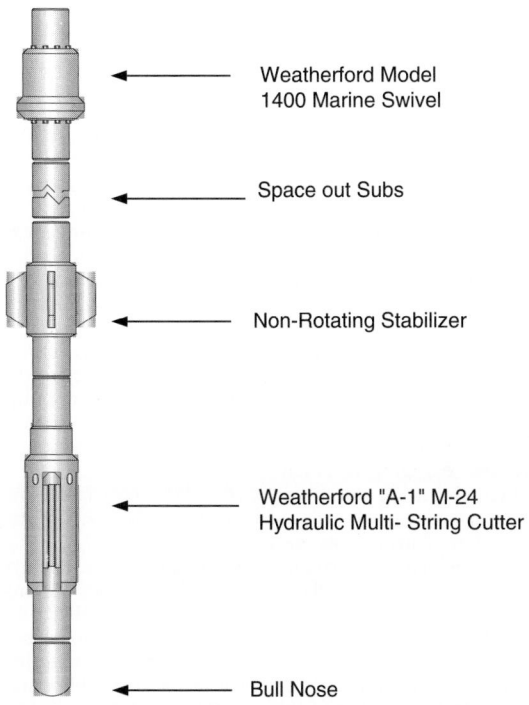

Figure 4.15.4 *Multi-String Cutting Assembly. (Courtesy Weatherford International)*

string. This cutting assembly is used for cutting multiple string of casing cemented together. There are three or four bladed cutter designs and both are hydraulically actuated. The tool is run to the desired depth and rotation is begun. As hydraulic pressure increases the knives are forced into the cutting position and the cut is made. In making the multiple cut, a series of different length knives are used in succession, each enlarging the window cut by the previous set. The tool is easily retracted by easing the hydraulic pressure and picking up on the string. Knives can be quickly exchanged from the rig floor. A well stabilized cutter will do a better downhole cutting job. A space out assembly is required on floating rigs so that the cutter can be accurately repositioned after each knife change. In making single or multipe cuts in large pipe (20 in.) or larger, a replaceable blade stabilizer is used since cutter stabilization is most important for efficient cutting action. On floating rigs, a marine swivel also must be used to stabilize the cutting depth and land properly in the wellhead.

4.15.3.6 Severing Tool

The severing tool is an option for parting pipe, drill collars, casing or other tubing. This tool is so powerful it will often eliminate many fishing options. The 3 in. OD severing tool is rated to sever up to an 11 in. OD drill collar. The explosive material in a severing tool is C-4. If attempting to sever anything other than drill collars, it is recommended that the charge be placed in a tool joint or coupling. The severing tool needs metal mass in order to shatter the drill collar (or other thick walled tubular). If the tool is fired in the pipe body or casing the severing tool may balloon and rail split the pipe wall rather than part it. When the severing tool is fired across a connection on drill pipe or casing, even if the pipe does not

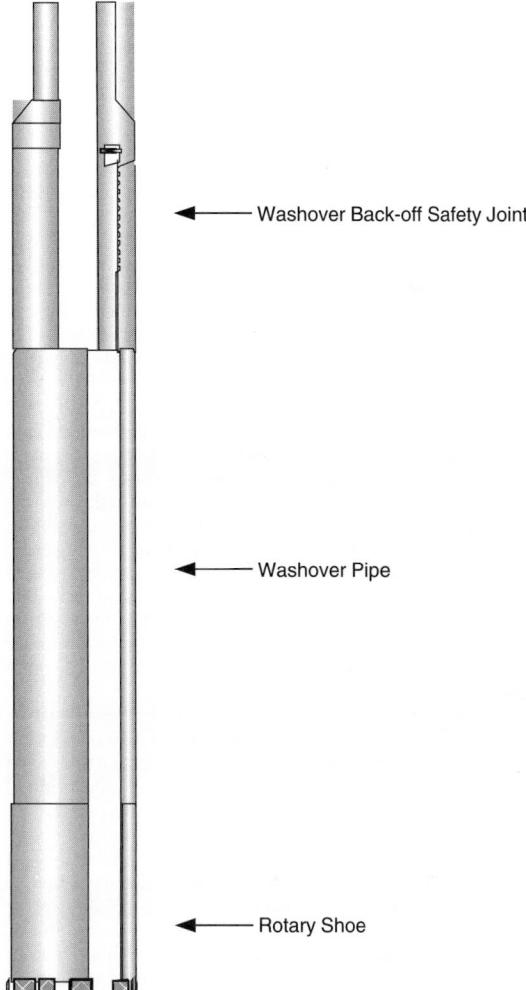

Figure 4.15.5 *Washover Assembly. (Courtesy Weatherford International)*

part, the connection is usually damaged enough so that the pipe can be pulled into.

4.15.3.7 Washover Back-off Safety Joint/Washover Procedures

The washover back-off safety joint is used to cross the wash pipe back to the jarring assembly and allow for washing over, screwing into a fish, and making a string shot back off in one trip. Should the washpipe become stuck, the washover safety joint provides a means to back off at the top of the washpipe with considerably less torque than is required to back off the work string. Due to the high risk of sticking the washpipe, a washover safety joint should be used any time a washover assembly is run in open hole. A fish can be screwed into with the washover safety joint, but several factors should be considered before this is done.

A washover assembly is shown in Figure 4.15.5. A typical washover assembly should consist of the following tools from the bottom up:

- A rotary shoe
- Washpipe

- Washover safety joint
- Bumper jars
- Oil jars
- Drill collars
- Accelerator jar
- Drill pipe or work string

A jarring assembly should always be run with washpipe in case the washpipe becomes stuck. The washover assembly is run into the hole to within 10 ft to the top of the fish. Circulation is established and the washpipe is lowered slowly over the top of the fish. The free portion of the fish standing above the stuck point acts as a guide for the washpipe. Once over the fish, free torque, circulating pressure and all other parameters should be recorded. Use rotary speeds of 40 to 50 rpm in soft formations and higher rotary speeds for harder formation. As with all milling applications, the surface speed should be calculated to insure the correct carbide dressed shoe feed rate. For crushed carbide the recommended feed rate is 150 to 200 surface feet per minute. During this operation, torque and circulating pressure should be closely monitored. The OD of the washpipe in relation to the hole circumference makes sticking the washpipe a real possibility. The string should not be allowed to be static for any period of time. If the torque becomes too great, it is necessary to come out of the hole and lay down about half of the washpipe. If the hole is crooked and the location of the back off was poorly chosen, the top of the fish may be laying under a bend. Shorting of the washpipe may allow it to conform to the curvature of the hole more easily.

When the entire length of the washpipe had washed over the fish, the washpipe can be pulled from the hole and a jarring assembly run back to recover the fish. Special tools exist that allow the washing over, screwing in, and recovering of the fish in one trip. Use of these tools is very dependent on hole conditions and require a seasoned fishing tool operator. If the fish has been engaged but will not pull, a back-off should be made and the washover operation repeated to recover another section of the fish.

When a fish has been washed over and recovered in one trip, the fish must be stripped out of the washpipe. This can be a long and difficult operation. The drill pipe and part of the fishing string is pulled, broken out and set back until the washpipe is in the rotary table and the slips are set around it. The washpipe safety joint with the fish attached is then broken out and raised a few feet. A split slip holder or bowl is then screwed into the threads on the top of the washpipe and split slips are set around the fish to hold it. The fish is stripped out of the washpipe with the elevators. The washpipe can be pulled to change the rotary shoe if necessary. The whole washover process is repeated until all of the fish is recovered. When the last section of the fish is retrieved, the fish is stripped until the bit or stabilizer is near the bottom of the rotary shoe. The fish and washpipe must be pulled, suspended and broken out together. This double stripping job takes about one hour per 100 ft of washpipe. There are tools such as the drill collar spear that can be used to lower the fish and suspend it in the bottom joint of washpipe so that the washpipe can be pulled first, saving a double stripping job.

4.15.4 Jars, Bumper Subs and Intensifiers

Jars, Bumper Subs and Intensifiers are basic components in a fishing assembly. Improved design, new materials, and sophisticated computer programs for jar placement are continually improving jar overall jar performance. The fishing bumper sub is engineered to withstand sustained bumping loads and displacements during fishing, drilling, and workover operations. These tools are designed to permit a 10 to 60 in. vertical strokes downward. The stroke is always available in the tool, whether rotating or not, but the ease of the stroke may be affected by friction within the bumper sub itself. The lubricated bumper sub is better suited for high or sustained circulating pressures. In all bumper subs, an adequate striking surface (to produce impact) must be provided at the limits of the free movement of the sub mechanism. Specially designed splines, torque keys, or torque slippers provide a source of continuous torque transmission. In addition the bores permit full circulation at all times.

By providing an immediate bumping action these tools can help to free drill pipe, drill collars, bits and other tools that have become stuck, lodged or keyseated. To achieve a good impact with a bumper sub, the drill string stretch must be utilized to provide the speed for the impact. This requires dropping the drill or working string until the jar is approximately 50% closed; the string is stopped abruptly at the surface and the downward momentum continues down, supplying the impact. This is a delicate operation and requires a little practice. Another function of the bumper sub is to release engaging tools and provide a "soft touch" when attempting to get over the top of a fish. Whenever a releasing type of fishing tool is to be used, a bumper sub should be run just above it to aid in its release.

The fishing jar is a straight pull operated jar that employs elementary principles of hydraulics and mechanics. Fishing jars can be mechanically or hydraulically actuated. Hydraulic fishing jars are most common. Hydraulic jars require no setting or adjustments before going in the hole, or after the fish has been engaged. The intensity of the jarring blow may be controlled within a wide range of impact load applied by the amount of over pull applied to the string before the jar is tripped. As pull is applied to the jar, oil is forced from one side of the inner body to the other through some type of orifice slots that will meter volume of fluid into storage chambers. This allows for ample time to pull the required stretch in the string.

To minimize the dulling effect on a grapple, hydraulic oil jars should be closed or reset using only sufficient weight to overcome friction in the inner body. During resetting, oil is forced in the opposite direction, unimpeded form one cavity to the other. Some jars require that the resetting be done slowly to optimize the life of the jar. It is possible to jar with torque in the string though it is not recommended. In most cases, the friction created by the torque will slow down the upward movement, reducing the impact loads. When used in fishing operations, an oil jar should be installed below a string of drill collars. For maximum effectiveness, a jar intensifier should be used in the string.

4.15.4.1 Drill Collars in a Jarring Assembly

In a jarring assembly, drill collars provide the mass to deliver the impact. The drill collars can be compared to the weight of head of a conventional hammer. A light hammerhead can deliver a fast impact blow, but the lack of mass reduces inertia, which means the impact is quickly diminished. A heavy hammerhead delivers a slow impact blow, and the high mass increases inertia, increasing the effectiveness of the impact. Too many drill collars can have the same effect as too-heavy a hammerhead. Just as a hammer could be too heavy to swing, the friction produced between the borehole wall and a long string of drill collars could diminish the impact force at the fish.

4.15.4.2 Fluid Accelerator or Intensifier

During conventional jarring operations with either mechanical or hydraulic jars, the intensity of the blow struck is a function of, and proportional to the accelerated movement of the drill string above the jar. This accelerated movement will often be diminished by friction between the string and the wall of the hole. In such cases much of the energy will be lost. Also, at very shallow depths, lack of stretch in the running string causes significant loss in the effectiveness of the acceleration.

The intensifier provides the means to store the required energy immediately above the jar and drill collars. This is done to effectively offset the friction loss of stretch or drag on the string. An important secondary contribution of the intensifier is the use of its contained hydraulic fluid or gas to cushion the shock of the string as it rebounds after each jarring stroke. This reduces the tendency to cause shock damage to the tool and string. Use of the intensifier allows fewer drill collars to be used in specific cases than would otherwise be possible. This is particularly true at shallow operating depths where excessive numbers of drill collars are sometimes used to produce great mass in place of available stretch. However, the use of too many drill collars can be damaging to the tools and the string. The stretch created when over-pulling the drill string provides the energy required to actuate the fishing jar. Because of this energy, oil jars will work without intensifiers but less effectively. Illustrations of Jars, Bumper Subs and Intensifiers are shown in Figure 4.15.6.

4.15.5 Attachment Devices

An attachment or engaging device provides a means to reconnect to the equipment (fish) left in a wall. Engaging devices should have as many as possible of the following features:

1. There should be a means to release the engaging tool from the fish.
2. Full and positive circulation through the fish should be possible.
3. There should be an unobstructed ID that will allow wire line access to the fish.
4. The device should have the highest tensile and torsion yield possible.
5. The device should be able to withstand up and down jarring impact.
6. The tool should allow the application of torque through the work string and into the fish when applying overpull or set down weight or when the engaging tool is at the neutral weight.

In all cases, the preferred method of engaging a fish is screwing in the with a screw in sub, joint of drill pipe, or a drill collar. This will meet all of the above requirements as well as providing the simplest and most fail safe means of solid engagement. Once screwed back together, the screw in point will have as high or higher mechanical properties as the fish below it. Screwing in will allow pump pressures equal to that of the string to be applied to the fish. Other than the jars, the ID of the string will not have changed after screwing in.

Screwing in is the only option when down jars will be used due to the releasing mechanism of most engaging tools. Most engaging tools require over pull to maintain a good bite on the fish, which makes it very difficult to work torque down to the fish. The only drawback to screwing in is that a wire line string shot back-off becomes the surest method of backing off at the screw in point.

4.15.5.1 Cutlip Screw-in Sub

When a fish is partially buried in the wall or partially behind a ledge or even against the high side of the hole, a cut lip sub can be used to pull the connection over to align with the screw-in the assembly. Due to the cutting of the threaded area, this sub is not API approved. If a fish is in a gauged section of a hole in medium to hard rock formations, with low borehole deviation ($\sim 5°$ or less), a joint of drill pipe, or a drill collar may be used to screw into the fish. A cutlip screw-in sub is shown in Figure 4.15.7.

4.15.5.2 Skirted Screw-in Assembly

A skirted screw-in assembly gives an operator all the hole sweeping advantages of an overshot with the strength, torque and circulating ability of a screw-in sub. The skirt can be of an OD that will not allow the skirted assembly to pass the fish. The cut lip makes it possible to pull the fish inside the skirt.

Skirted assemblies are most often made up from a washpipe triple connection bushing and a blank rotary shoe. The shoe can then be cut into the desired shape, cut lip or wallhook. A Weatherford skirted cutlip screw-in assembly is shown in Figure 4.15.8.

4.15.5.3 External Engaging Devices

The second choice for engaging a fish is with an external catch device (when it meets the strength requirements). An external engaging device usually will not restrict the ID of the fishing assembly, thus, leaving wireline options open and allowing full circulation. An external catch device can also be of sufficient OD to eliminate the possibility of passing the fish.

4.15.5.4 Series 150 Releasing and Circulating Overshot

The most-popular external catch device is the Series 150 releasing and circulating overshot. The tool was invented over 60 years ago and has not changed from the original design. This overshot provides the strongest means available to externally engage, pack off, and pull a fish when screwing in is not possible. Its basic simplicity, rugged construction, and availability make this tool the standard for all external catch fishing tools.

The Series 150 is designed to engage, pack off, and grapple (and retrieve) a specific size of tubing, pipe, coupling, tool joint, drill collar or smooth OD tool. Through the installation of the proper parts, these tools may be adapted to engage and pack off a wide range of diameters for each overshot size. Series 150 overshots are available in Full Strength or Slim Hole types. The overshot consists of three outside components: the top sub, bowl and guide. The basic overshot may be addressed with either of two sets of internal parts, depending on whether the fish to be retrieved is near the maximum catch size for the particular overshot. If the fish diameter is near the maximum retrieval for the overshot, a spiral grapple spiral grapple control, and type A packer are used. If the fish is a tubing collar, a type D collar pack-off assembly will be used. If the fish diameter is below the maximum retrieval size (by approximately $\frac{1}{2}$ in.), then basket grapple and mill control packer are used. Both the spiral and basket grapple are the engaging mechanism. The spiral control and basket grapple control locks the torque in the grapple but still allows the grapple to move up and down inside the bowl.

The bowl of the overshot is designed with a helical tapered spiral section within its inside diameter. The gripping mechanism (the grapple) is fitted into this section. When an upward pull is exerted against a fish, a contraction strain is spread evenly over a long section of the fish. The catch range is from $\frac{1}{32}$ to $\frac{1}{16}$ in. above and below the specified catch range. The overshot is engaged by slacking off over the fish, then a straight pull will engage the overshot. As long as there is an over pull, either right or left hand torque may be applied. To release the Bowen overshot, the string is bumped or jarred down on to break the freeze between the overshot bowl and the grapple. Once the freeze has been broken, rotation to the right while slowly picking up on the string will walk the

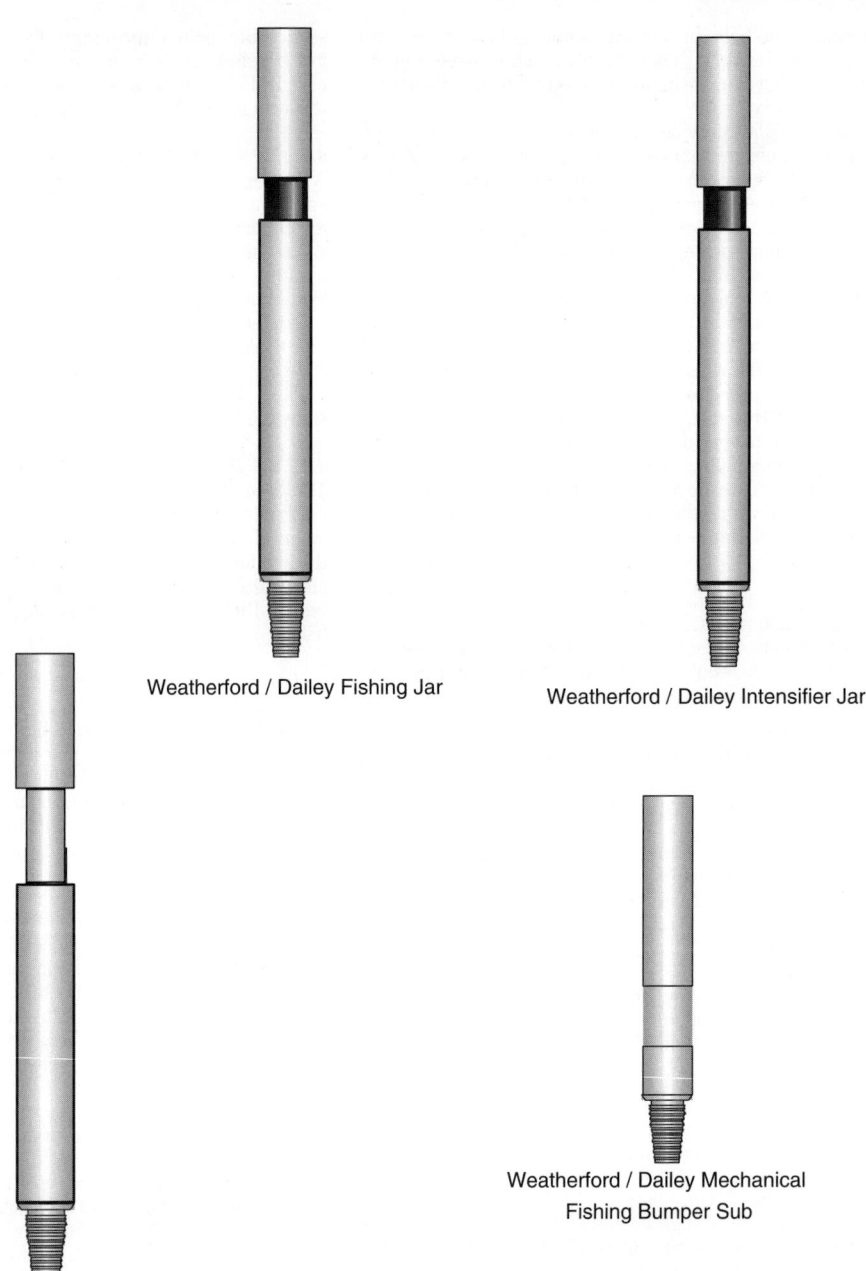

Weatherford / Dailey Fishing Jar

Weatherford / Dailey Intensifier Jar

Weatherford / Dailey Mechanical
Fishing Bumper Sub

Weatherford / Dailey Lubericated Bumper Sub

Figure 4.15.6 *Jars, Bumper Subs and Intensifiers. (Courtesy Weatherford International)*

Figure 4.15.7 *Cutlip Screw in Sub. (Courtesy Weatherford International)*

grapple off the fish. A Series 150 releasing and circulating overshot is shown in Figure 4.15.9.

Tables 4.15.1 and 4.15.2 give specific restrictions for overshot sizes in openhole and cased hole environments, respectively.

4.15.5.5 High-Pressure Pack-Off
The mill control and Type A packers used with the overshot have a limited pressure rating. If high pressures will be required or there are well control concerns, it is advisable to run a high pressure pack-off. The pack-off is run directly above the overshot bowl. Chevron packing is used for sealing. This type of assembly can be used as a permanent tubing patch. The pack-off is good for approximately 5,000 psi.

4.15.5.6 Oversize Cutlip Guide
The oversize guide is used when the fish OD in relation to the hole ID makes it possible for the overshot to bypass the fish. The oversize guide may be used to better utilize inventory by keeping a smaller selection of overshots. But care should be taken not to sacrifice strength to do this. An oversize cutlip guide is shown in Figure 4.15.10.

4.15.5.7 Wallhook Guide
The wallhook guide is used with a bent joint of pipe or hydraulic knuckle joint to sweep a washed out section of hole when a fish cannot be pulled into the overshot with a standard cutlip guide. The assembly is made up so the lip of the wallhook has the maximum possible reach. Once the overshot has passed the top of the fish, the string is slowly rotated until torque indicates the wallhook is madeup to the fish. The string is then elevated until the torque is lost. The fish should be in a position that it can be engaged by

slacking off on the overshot. A wallhook guide is shown in Figure 4.15.10.

4.15.5.8 Hollow Mill Container and Hollow Mill
The hollow mill container with a hollow mill insert is used to dress off the top of a fish that has been damaged. This devise is used to engage the fish in one trip. This accessory is most often used when a fish top is flared after a jet cut or in the event a twist off has occurred. This assembly is run directly below the overshot bowl. A hollow mill container and hollow mill device is shown in Figure 4.15.10.

4.15.5.9 Bowen Series 70 Short Catch Overshot
Series 70 short catch overshots are specifically designed to engage the exposed portion of a fish too short to be caught with a conventional overshots. The short catch overshot consists of a top sub, a bowl, a basket grapple and a basket grapple control. The grapple is inserted in the top end of the overshot and is positioned at the extreme lowered end of the bowl. Spiral grapples, mill control packers and Type A packers cannot be used with a short catch overshot. A Series 70 short catch overshot is shown in Figure 4.15.11.

4.15.5.10 Internal Engaging Devices
If the fish must be engaged internally, releasable spears are used. Spears often restrict the ID of the fishing assembly. There is no pack off system built into the tool but certain accessories may be used to accomplish circulation. The Itco Type Bowen releasing spear is designed to assure positive engagement with a fish. Built to withstand severe jarring and pulling strains, it engages the fish over a large area without damage to the fish. If the fish cannot be pulled, the spear may be released. The Itco type Bowen releasing spear consists of a mandrel, grapple, and a locking ring and nut. The mandrel may be a shoulder type or flush type. The flexible one-piece grapple has an internal helix which mates with the helix on the mandrel. The tang of the grapple rests against a stop on the mandrel when the spear is in the engaged position. A Bowen Itco releasing spear is shown in Figure 4.15.12.

4.15.5.11 Box Taps and Taper Taps
Box taps are used to engage the outside diameter of a fish, usually in a situation where conventional releasing overshots would not be a feasible option. They are designed to tap threads into the steel tubing (or other steel downhole tools) so that the fish can be retrieved. Standard box taps are made with an integral guide on the lower end to insure proper engagement to the fish even when the fish is not centralized in the hole.

Taper taps are used to engage the inside of a fish where conventional releasing spears would not be feasible. Like the box tap, the taper tap is designed to cut threads where no threads existed. Standard taps are machined from high quality heat treated steel, with buttress type threads that are carborized for hardness. They can be ordered with right or left hand wickers. Box and taper taps can be used to fish tubulars, packers, or other types of downhole equipment that cannot be fished by conventional means. They have the widest catch range of any fisting tool. Since all taps are non-releasable, a safety joint should always be run just above them. A box tap and taper tap are shown in Figure 4.15.13.

4.15.6 Fishing For Junk
Junk is all that other devices that may fall into the hole or become detached in the hole that are not tubulars. The type of junk retrieval device to be used in the fishing operation will depend on the properties of the junk (to be milled up or recovered). There are many tools designed to recover junk,

(text continued on page 4-394)

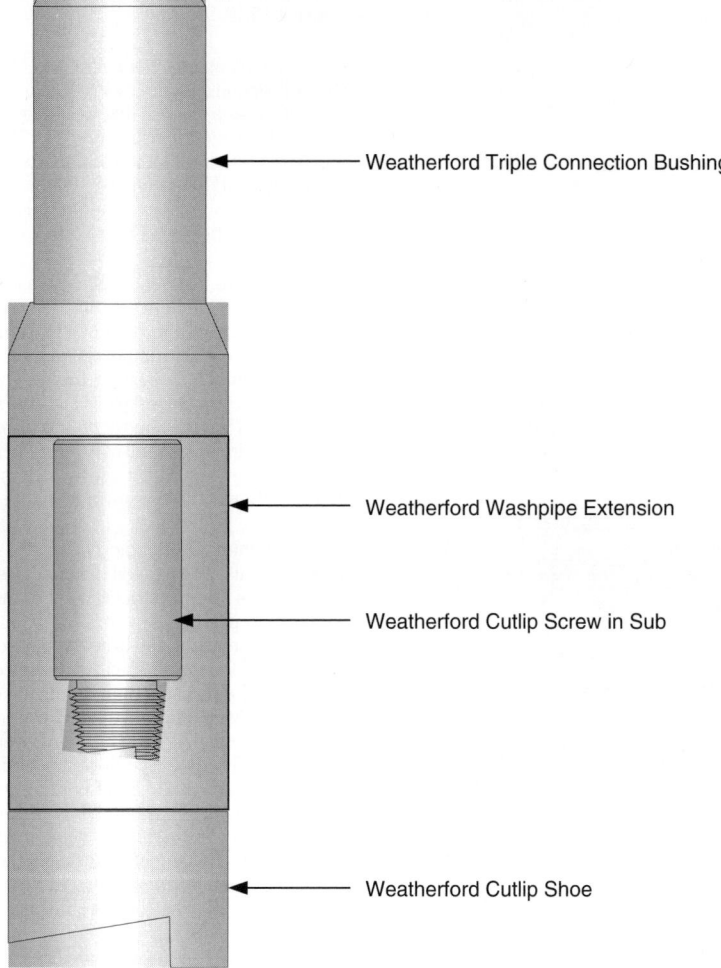

Weatherford Triple Connection Bushing

Weatherford Washpipe Extension

Weatherford Cutlip Screw in Sub

Weatherford Cutlip Shoe

Figure 4.15.8   *Skirted Cutlip Screw in Assembly. (Courtesy Weatherford International)*

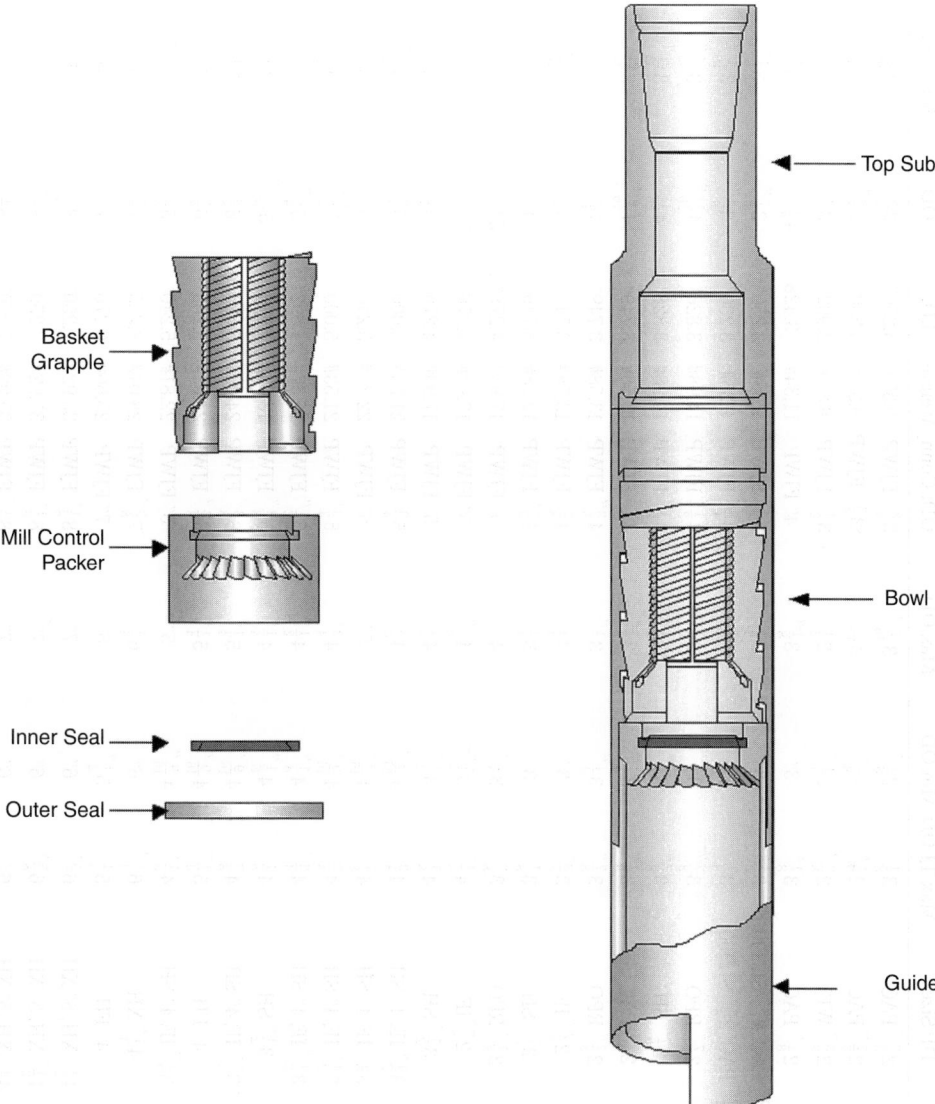

Figure 4.15.9 Series "150" Releasing and Circulating Overshot. (Courtesy Weatherford International)

Table 4.15.1 Recommended Drill Pipe and Drill Collars For Open Hole and Cased Hole

API Casing Size & Wt.	Bit Size	Drill Pipe Recommended In Casing		Drill Pipe Recommended In Open Hole		Drill Collars Recommended In Cased Open Hole		Washpipe Recommended For Fishing			Overshot Recommended For Fishing	
		DP Size	TJ OD	DP Size	Max TJ OD	Min OD	Max OD	OD; Conn	Weight	I.D.	OD	Max Catch
$4\frac{1}{2}''$ 9.50#–11.6	$3\frac{7}{8}''$	$2\frac{7}{8}''$ PAC	$3\frac{1}{8}''$	$2\frac{7}{8}''$ PAC	$3\frac{1}{8}''$	$2\frac{7}{8}''$	$3\frac{1}{16}''$	$3\frac{3}{4}''$ FJWP	9.35#	3.250"	$3\frac{3}{4}''$	$3\frac{1}{16}''$
$4\frac{1}{2}''$ 13.50#	$3\frac{3}{4}''$	$2\frac{7}{8}''$ PAC	$3\frac{1}{8}''$	$2\frac{7}{8}''$ PAC	$2\frac{1}{2}''$	$2\frac{7}{8}''$	$3\frac{1}{16}''$	*$3\frac{3}{4}''$ FJWP	9.35#	3.250"	$3\frac{3}{4}''$	$3\frac{1}{16}''$
$4\frac{1}{2}''$ 15.10#	$3\frac{5}{8}''$	$2\frac{3}{8}''$ WFJ	$2\frac{1}{2}''$	$2\frac{3}{8}''$ WFJ	$2\frac{3}{8}''$	$2\frac{1}{8}''$	$2\frac{1}{2}''$	$3\frac{1}{2}''$ FJWP	8.81#	2.992"	$3\frac{3}{8}''$	$2\frac{1}{2}''$
5" 11.50#–15#	$4\frac{1}{4}''$	$2\frac{7}{8}''$ PAC	$3\frac{1}{8}''$	$2\frac{7}{8}''$ PAC	$3\frac{1}{8}''$	$3''$	$3\frac{1}{8}''$	$4''$ FJWP	11.34#	3.428"	$4\frac{1}{8}''$	$3\frac{1}{8}''$
5" 18#	$4\frac{1}{8}''$	$2\frac{7}{8}''$ PAC, $2\frac{3}{8}''$ SLH90	$3\frac{1}{8}''$	$2\frac{7}{8}''$ PAC, $2\frac{3}{8}''$ SLH90	$3\frac{1}{8}''$	$3''$	$3\frac{1}{8}''$	$3\frac{3}{4}''$ FJWP	9.35#	3.250"	$3\frac{7}{8}''$	$3\frac{1}{8}''$
5" 29.3#–24.2#	$4''$–$3\frac{7}{8}''$	$2\frac{7}{8}''$ PAC, $2\frac{3}{8}''$ SLH90	$3\frac{1}{8}''$	$2\frac{7}{8}''$ PAC, $2\frac{3}{8}''$ SLH90	$3\frac{1}{8}''$	$2\frac{7}{8}''$	$3\frac{1}{16}''$	$3\frac{3}{4}''$ FJWP	9.35#	3.250"	$3\frac{3}{4}''$	$3\frac{1}{16}''$
$5\frac{1}{2}''$ 13#–17#	$4\frac{3}{4}''$	$2\frac{7}{8}''$ RPO	$3\frac{7}{8}''$	$2\frac{7}{8}''$ RPO	$3\frac{7}{8}''$	$3\frac{1}{4}''$	$3\frac{1}{2}''$	$4\frac{1}{8}''$ FJWP	14.98#	3.826"	$4\frac{11}{16}''$	$3\frac{21}{32}''$
$5\frac{1}{2}''$ 13#–17#	$4\frac{3}{4}''$	$2\frac{7}{8}''$ SH	$3\frac{5}{8}''$	$2\frac{7}{8}''$ SH	$3\frac{5}{8}''$	$3\frac{1}{4}''$	$3\frac{1}{2}''$	$4\frac{1}{2}''$ FJWP	14.98#	3.826"	$4\frac{11}{16}''$	$3\frac{21}{32}''$
$5\frac{1}{2}''$ 20#	$4\frac{5}{8}''$	$2\frac{7}{8}''$ PAC	$3\frac{1}{8}''$	$2\frac{7}{8}''$ PAC	$3\frac{1}{8}''$	$3\frac{1}{4}''$	$3\frac{1}{2}''$	$4\frac{3}{8}''$ FJWP	13.58#	3.749"	$4\frac{9}{16}''$	$3\frac{21}{32}''$
$5\frac{1}{2}''$ 20#	$4\frac{5}{8}''$	$2\frac{7}{8}''$ RPO	$3\frac{7}{8}''$	$2\frac{7}{8}''$ RPO	$3\frac{7}{8}''$	$3\frac{1}{4}''$	$3\frac{1}{2}''$	$4\frac{1}{8}''$ FJWP	13.58#	3.749"	$4\frac{9}{16}''$	$3\frac{21}{32}''$
$5\frac{1}{2}''$ 23#–26#	$4\frac{1}{2}''$	$2\frac{7}{8}''$ IF	$3\frac{3}{8}''$	$2\frac{7}{8}''$ IF	$3\frac{3}{8}''$	$3''$	$3\frac{3}{4}''$	$4\frac{3}{8}''$ FJWP	13.58#	3.749"	$4\frac{3}{8}''$	$3\frac{3}{8}''$
$5\frac{1}{2}''$ 23#–26#	$4\frac{1}{2}''$	$2\frac{7}{8}''$ SH	$3\frac{1}{8}''$	$2\frac{7}{8}''$ SH	$3\frac{1}{8}''$	$3''$	$3\frac{3}{4}''$	$4\frac{3}{8}''$ FJWP	13.58#	3.749"	$4\frac{3}{8}''$	$3\frac{3}{8}''$
$6\frac{5}{8}''$ 32#	$5\frac{3}{4}''$	$2\frac{7}{8}''$ RPO	$3\frac{7}{8}''$	$2\frac{7}{8}''$ RPO	$3\frac{7}{8}''$	$3\frac{1}{4}''$	$4\frac{1}{8}''$	$5''$ FJWP	17.93#	4.276"	$5\frac{1}{8}''$	$4\frac{1}{4}''$
$6\frac{5}{8}''$ 32#	$5\frac{3}{4}''$	$2\frac{7}{8}''$ IF	$4\frac{1}{8}''$	$2\frac{7}{8}''$ IF	$4\frac{1}{8}''$	$3\frac{1}{4}''$	$4\frac{1}{8}''$	$5''$ FJWP	17.93#	4.276"	$5\frac{1}{8}''$	$4\frac{1}{4}''$
$6\frac{5}{8}''$ 32#	$5\frac{3}{4}''$	$3\frac{1}{2}''$ SH	$4\frac{1}{8}''$	$3\frac{1}{2}''$ SH	$4\frac{1}{8}''$	$3\frac{1}{2}''$	$4\frac{1}{8}''$	$5''$ FJWP	17.93#	4.276"	$5\frac{1}{8}''$	$4\frac{1}{4}''$
7" 17#–23#–26#	$6\frac{1}{4}''$–$6\frac{1}{8}''$	$3\frac{1}{2}''$ IF, 4" SH	$4\frac{3}{4}''$	$3\frac{1}{2}''$ IF, 4" SH	$4\frac{3}{8}''$	$4\frac{9}{16}''$	$4\frac{3}{4}''$	$5\frac{3}{4}''$ FJWP	21.53#	5.000"	$5\frac{3}{8}''$	$4\frac{3}{4}''$
7" 17#–23#–26#	$6\frac{1}{4}''$–$6\frac{1}{8}''$	$3\frac{1}{2}''$ IF, 4" SH	$4\frac{3}{4}''$–$4\frac{7}{8}''$	$3\frac{1}{2}''$ IF, 4" SH	$4\frac{7}{8}''$	$4\frac{9}{16}''$	$5''$	$6''$ FJWP	22.81#	5.240"	$5\frac{7}{8}''$	$5''$
7" 29#–32#	$6''$	$3\frac{1}{2}''$ IF, 4" SH	$4\frac{3}{4}''$	$3\frac{1}{2}''$ IF, 4" SH	$4\frac{7}{8}''$	$4\frac{9}{16}''$	$5''$	$5\frac{3}{4}''$ FJWP	21.53#	5.000"	$5\frac{3}{8}''$	$4\frac{3}{4}''$
7" 35#	$5\frac{7}{8}''$	$3\frac{1}{2}''$ IF, 4" SH	$4\frac{3}{4}''$	$3\frac{1}{2}''$ IF, 4" SH	$4\frac{3}{4}''$	$4\frac{1}{4}''$	$4\frac{3}{4}''$	$5\frac{1}{2}''$ FJWP	16.87#	4.892"	$5\frac{3}{8}''$	$4\frac{3}{4}''$
7" 38#	$5\frac{3}{4}''$	$3\frac{1}{2}''$ SH	$4\frac{1}{8}''$	$3\frac{1}{2}''$ SH	$4\frac{1}{8}''$	$4\frac{1}{4}''$	$4\frac{1}{4}''$	$5\frac{3}{8}''$ FJWP	18.93#	4.670"	$5\frac{9}{16}''$	$4\frac{1}{4}''$
$7\frac{5}{8}''$ 20#–33.7#	$6\frac{1}{4}''$–$6\frac{5}{8}''$	$3\frac{1}{2}''$ IF, 4" SH	$4\frac{3}{4}''$–$4\frac{7}{8}''$	$3\frac{1}{2}''$ IF, 4" SH	$4\frac{7}{8}''$	$4\frac{9}{16}''$	$5\frac{1}{4}''$	$6\frac{3}{8}''$ FJWP	24.03#	5.625"	$6\frac{3}{8}''$	$5\frac{1}{4}''$
$7\frac{5}{8}''$ 20#–33.7#	$6\frac{1}{4}''$–$6\frac{5}{8}''$	$4''$ FH	$5\frac{1}{4}''$	$4''$ FH	$5\frac{1}{4}''$	$4\frac{9}{16}''$	$5\frac{1}{4}''$	$6\frac{3}{8}''$ FJWP	24.03#	5.265"	$6\frac{3}{8}''$	$5\frac{1}{4}''$
$7\frac{5}{8}''$ 39#	$6\frac{1}{2}''$–$6\frac{3}{8}''$	$3\frac{1}{2}''$ IF, 4" SH	$4\frac{3}{4}''$–$4\frac{7}{8}''$	$3\frac{1}{2}''$ IF, 4" SH	$4\frac{7}{8}''$	$4\frac{9}{16}''$	$5''$	$6''$ FJWP	22.81#	5.240"	$5\frac{7}{8}''$	$5''$
$8\frac{5}{8}''$ 24#–40#	$7\frac{7}{8}''$–7–$5\frac{7}{8}''$	$4\frac{1}{2}''$ XH	$6\frac{1}{4}''$	$4\frac{1}{2}''$ XH	$6\frac{1}{4}''$	$6''$	$6\frac{1}{4}''$	$7\frac{3}{8}''$ FJWP	28.04#	6.625"	$7\frac{3}{8}''$	$6\frac{1}{4}''$
$8\frac{5}{8}''$ 44#–49#	$7\frac{1}{2}''$–$7\frac{3}{8}''$	$4''$ FH	$5\frac{1}{4}''$	$4''$ FH	$5\frac{1}{4}''$	$5\frac{1}{2}''$	$6''$	$7''$ FJWP	25.66#	6.276"	$7\frac{1}{8}''$	$6''$
$9\frac{5}{8}''$ 29.3# 36#	$8\frac{3}{4}''$	$4\frac{1}{2}''$ XH, 5" XH	$6\frac{1}{4}''$–$6\frac{5}{8}''$	$4\frac{1}{2}''$ XH, 5" XH	$6\frac{5}{8}''$	$6''$	$7''$	$8\frac{1}{8}''$ FJWP	35.92#	7.250"	$8\frac{1}{8}''$	$7''$
$9\frac{5}{8}''$ 40#–43.5#	$8\frac{1}{2}''$	$4\frac{1}{2}''$ XH, 5" XH	$6\frac{1}{4}''$–$6\frac{5}{8}''$	$4\frac{1}{2}''$ XH, 5" XH	$6\frac{5}{8}''$	$6''$	$7''$	$8\frac{1}{8}''$ FJWP	35.92#	7.250"	$8\frac{1}{8}''$	$7''$
9.5/8" 47#	$8\frac{1}{2}''$	$4\frac{1}{2}''$ XH, 5" XH	$6\frac{1}{4}''$–$6\frac{5}{8}''$	$4\frac{1}{2}''$ XH, 5" XH	$6\frac{5}{8}''$	$6''$	$7''$	$8\frac{1}{8}''$ FJWP	35.92#	7.250"	$8\frac{1}{8}''$	$7''$
$9\frac{5}{8}''$ 53.5#	$8\frac{3}{8}''$	$4\frac{1}{2}''$ XH, 5" XH	$6\frac{1}{4}''$–$6\frac{5}{8}''$	$4\frac{1}{2}''$ XH, 5" XH	$6\frac{5}{8}''$	$6''$	$7''$	$8\frac{1}{8}''$ FJWP	35.92#	7.250"	$8\frac{1}{8}''$	$7''$

Note: The chart above is a guideline only and should be used as a rule of thumb; the final decision is that of customer; 4" pin-up drill pipe is an option inside 7" casing, discuss with fishing tool manager for guidance.

Table 4.15.2 *Cased Hole Fishing: Recommended Overshot/Washpipe/Cutter*

Casing Size-Wt.	Pipe Size-Wt.	Connection Type O.D.	Overshot Size-Type	Maximum Catch	Washpipe Size	Maximum Washover	External Cutter Type/Size	Maximum Cutter Clearance
$4\frac{1}{2}''$ 9.50–11.60#	$2\frac{3}{8}''$ 4.70#	8RD, $3\frac{1}{16}''$	$3\frac{3}{4}''$ S.H.	$3\frac{1}{16}''$	$3\frac{3}{4}''$	$3\frac{1}{8}''$	Bowen Mech. $3\frac{7}{8}''$	$3\frac{1}{16}''$
$4\frac{1}{2}''$ 13.50#	$2\frac{3}{8}''$ 4.70#	8RD, $3\frac{1}{16}''$	$3\frac{3}{4}''$ S.H.	$3\frac{1}{16}''$	$3\frac{3}{4}''$	$3\frac{1}{8}''$	Turn Down $3\frac{7}{8}''$ ($3\frac{3}{4}''$)	$3\frac{1}{16}''$
$4\frac{1}{2}''$ 15.10#	$2\frac{3}{8}''$ 4.70#	8RD, $3\frac{1}{16}''$	$3\frac{5}{8}''$ X.S.H.	$2\frac{7}{8}''$	$3\frac{1}{2}''$	$2\frac{5}{8}''$	None	—
$5''$ 11.50–23.20#	$2\frac{3}{8}''$ 4.70#	8RD, $3\frac{1}{16}''$	$3\frac{3}{4}''$ S.H.	$3\frac{1}{16}''$	$3\frac{3}{4}''$	$3\frac{1}{8}''$	Bowen Mech. $3\frac{7}{8}''$	$3\frac{1}{16}''$
$5\frac{1}{2}''$ 14.17#	$2\frac{3}{8}''$ 4.70#	8RD, $3\frac{1}{16}''$	$4\frac{11}{16}''$ F.S.	$3\frac{21}{32}''$	$4\frac{1}{2}''$	$3\frac{3}{4}''$	H.E. $4\frac{9}{16}''$	$3\frac{3}{8}''$
$5\frac{1}{2}''$ 14.17#	$2\frac{7}{8}''$ 6.50#	8RD, $3\frac{21}{32}''$	$4\frac{11}{16}''$ F.S.	$3\frac{21}{32}''$	$4\frac{1}{2}''$	$3\frac{3}{4}''$	Bowen $4\frac{11}{16}''$	$3\frac{3}{4}''$
$5\frac{1}{2}''$ 20.00#	$2\frac{3}{8}''$ 4.70#	8RD, $3\frac{1}{16}''$	$4\frac{9}{16}''$ S.H.	$3\frac{21}{32}''$	$4\frac{3}{8}''$	$3\frac{5}{8}''$	Bowen Hyd. $4\frac{9}{16}''$	$3\frac{9}{16}''$
$5\frac{1}{2}''$ 20.00#	$2\frac{7}{8}''$ 6.50#	8RD, $3\frac{21}{32}''$	$4\frac{9}{16}''$ S.H.	$3\frac{21}{32}''$	$4\frac{3}{8}''$	$3\frac{5}{8}''$	Bowen Hyd. $4\frac{9}{16}''$	$3\frac{9}{16}''$
$5\frac{1}{2}''$ 23.00#	$2\frac{3}{8}''$ 4.70#	8RD, $3\frac{1}{16}''$	$4\frac{1}{4}''$ F.S.	$3\frac{1}{8}''$	$4\frac{3}{8}''$	$3\frac{5}{8}''$	Bowen Mech. $3\frac{7}{8}''$	$3\frac{1}{16}''$
$5\frac{1}{2}''$ 23.00#	$2\frac{7}{8}''$ 6.50#	8RD, $3\frac{21}{32}''$	$4\frac{3}{8}''$ S.H.	$3\frac{1}{2}''$	$4\frac{3}{8}''$	$3\frac{5}{8}''$	None	—
$7''$ 20-38#	$2\frac{7}{8}''$ 6.50#	8RD, $3\frac{21}{32}''$	$5\frac{1}{4}''$ F.S.	$4\frac{3}{4}''$	$5\frac{1}{2}''$	$4\frac{3}{4}''$	H.E. $5\frac{9}{16}''$	$4''$
$7''$ 20-38''	$3\frac{1}{2}''$ 9.30#	8RD, $4\frac{1}{4}''$	$5\frac{3}{8}''$ F.S.	$4\frac{3}{4}''$	$5\frac{1}{2}''$	$4\frac{3}{4}''$	Bowen Mech. $5\frac{5}{8}''$	$4\frac{1}{2}''$
$7\frac{5}{8}''$ 26.40–45.30#	$3\frac{1}{2}''$ 9.30#	8RD, $4\frac{1}{2}''$	$5\frac{3}{4}''$ F.S.	$4\frac{3}{4}''$	$5\frac{1}{4}''$	$4\frac{3}{4}''$	Bowen Mech. $5\frac{7}{8}''$	$4\frac{1}{2}''$
$7\frac{5}{8}''$ 26.40–45.30#	$3\frac{1}{2}''$ 13.30#	IF, $4\frac{3}{4}''$	$5\frac{3}{4}''$ F.S.	$4\frac{3}{4}''$	$5\frac{1}{4}''$	$4\frac{3}{4}''$	Bowen Mech. $6\frac{1}{4}''$	$4\frac{3}{4}''$
$8\frac{5}{8}''$ 24-40#	$4''$ 14.00#	FH, $5\frac{1}{4}''$	$7\frac{3}{8}''$ F.S.	$5\frac{3}{4}''$	$7\frac{3}{8}''$	$6\frac{1}{2}''$	Bowen Mech $7\frac{7}{8}''$	$6\frac{1}{4}''$
$8\frac{5}{8}''$ 24-40#	$4\frac{1}{2}''$ 16.60#	XH, $6\frac{1}{4}''$	$7\frac{3}{8}''$ S.H.	$6\frac{1}{4}''$	$7\frac{3}{8}''$	$6\frac{1}{2}''$	Bowen Mech $7\frac{7}{8}''$	$6\frac{1}{4}''$
$8\frac{5}{8}''$ 44-49#	$4''$ 14.00#	FH, $5\frac{1}{4}''$	$7\frac{3}{8}''$ S.H.	$6\frac{1}{4}''$	$7''$	$6\frac{1}{8}''$	None	—
$8\frac{5}{8}''$ 44-49#	$4\frac{1}{2}''$ 16.60#	XH, $6\frac{1}{4}''$	$7\frac{3}{8}''$ S.H.	$6\frac{1}{4}''$	$7\frac{3}{8}''$	$6\frac{1}{2}''$	None	—
$9\frac{5}{8}''$ 36–53.50#	$4\frac{1}{2}''$ 16.60#	XH, $6\frac{1}{4}''$	$8\frac{1}{8}''$ F.S.	$6\frac{5}{8}''$	$8\frac{1}{8}''$	$7\frac{1}{16}''$	Bowen Mech. $8\frac{1}{8}''$	$6\frac{1}{2}''$

Tubing Dimensions	Tube O.D.	Upset O.D.	Coupling O.D.
$2\frac{3}{8}''$ 4.70# 8RD EUE	$2\frac{3}{8}''$	$2\frac{19}{32}''$	$3\frac{1}{16}''$
$2\frac{7}{8}''$ 6.50# 8RD EUE	$2\frac{7}{8}''$	$3\frac{3}{32}''$	$3\frac{21}{32}''$
$3\frac{1}{2}''$ 9.30# 8RD EUE	$3\frac{1}{2}''$	$3\frac{3}{4}''$	$4\frac{1}{2}''$
$4''$ 11.00# 8RD EUE	$4''$	$4\frac{1}{4}''$	$5''$

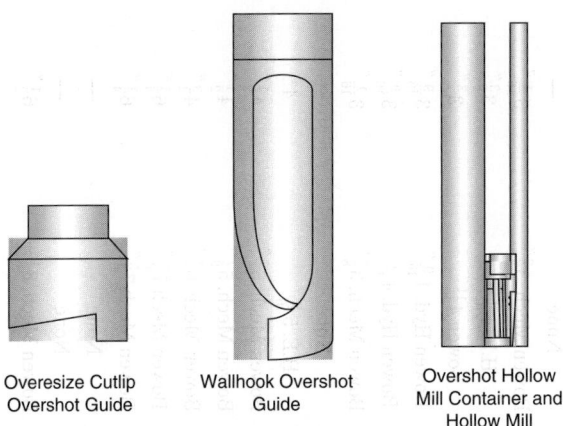

Overesize Cutlip
Overshot Guide

Wallhook Overshot
Guide

Overshot Hollow
Mill Container and
Hollow Mill

Figure 4.15.10 *Oversize Cutlip Guide, Wallhook Guide and Hollow Mill Container with Hollow Mill. (Courtesy Weatherford International)*

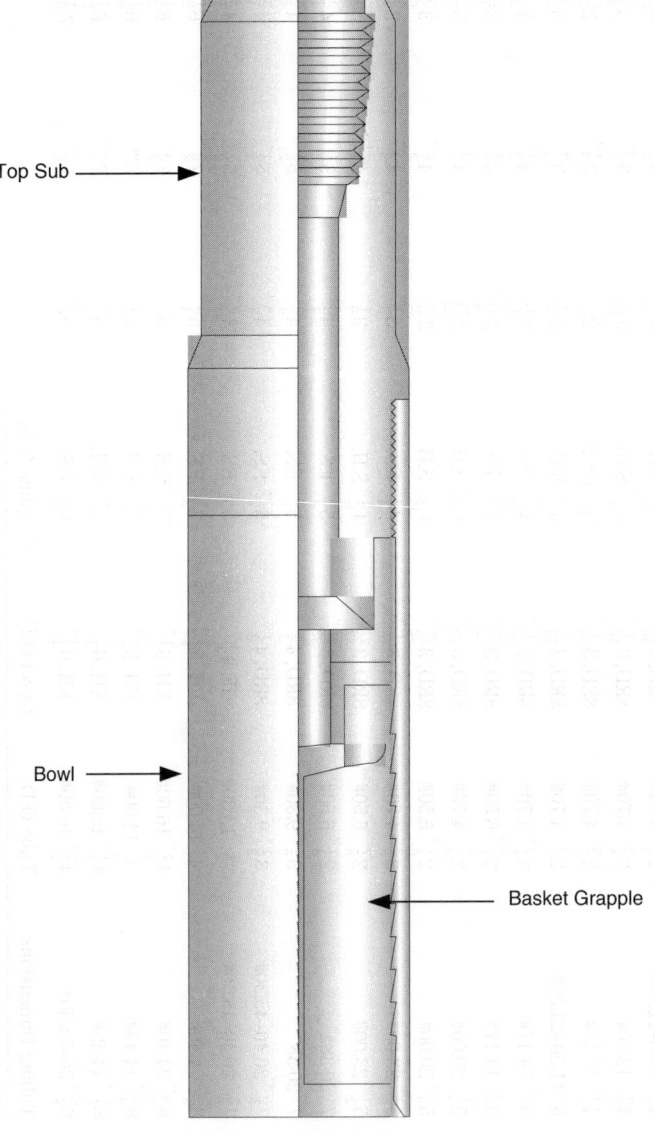

Top Sub

Bowl

Basket Grapple

Figure 4.15.11 *Series "70" Short Catch Overshot. (Courtesy Bowell Tool Co.)*

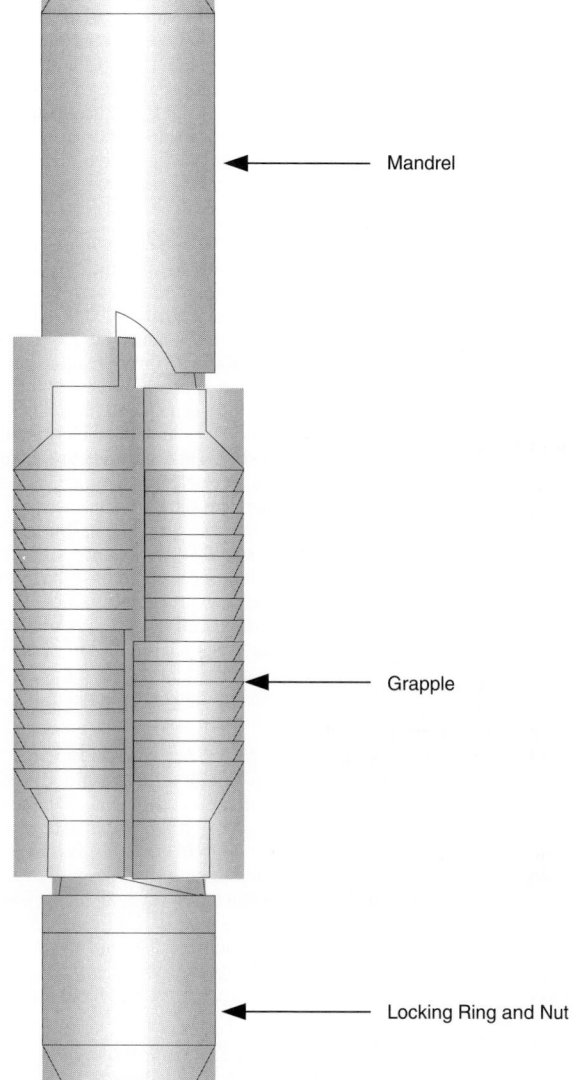

Figure 4.15.12 *ITCO Type Releasing Spear. (Courtesy Bowell Tool Co.)*

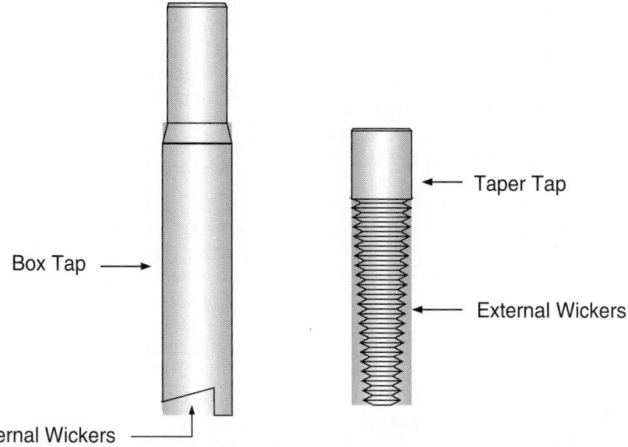

Figure 4.15.13 *Box Tap and Taper Tap. (Courtesy Weatherford International)*

(text continued from page 4-387)

and often an operational practice preference will determine which tool is used.

4.15.6.1 Poor Boy Junk Basket

One of the simplest fishing tools is the poor boy junk basket. It is run in the hole, circulation established, then the tool is slowly rotated down over the fish. If the basket is nearly hole size, its finger like catchers will gather junk toward the center of hole. When weight is applied the basket will bend inward to trap the junk inside. It is most effective for a small solid mass lying loose on bottom, such as a single bit cone. A poor boy junk basket is shown in Figure 4.15.14.

4.15.6.2 Boot Basket

A boot basket is run just above a bit or mill to collect a variety of small pieces of junk. The boot basket design traps junk by producing a sudden decrease in annular velocity when cuttings (with entrained junk) pass the larger OD of the boot and reach the smaller OD of the mandrel and top connection. A boot basket is shown in Figure 4.15.14.

4.15.6.3 Core Type Junk Basket

The core-type junk basket is used to retrieve junk such as bit cones that may or may not be embedded in the wall or bottom of the rock formation. A milling shoe is made up on the bottom of the tool. After it is run nearly to bottom, the mud is circulated at reduced pressure and the tool is slowly rotated and lowered to make contact with the junk. Weight is gradually increased on the shoe as it cuts away the edges of the junk and the formation. The junk is forced into the barrel and a short core is cut. Rotation and circulation are stopped, torque is released from the drill string, and the working string is raised to break the core. Catchers inside the basket hold the core and the junk as it is tripped out of the hole. A core type junk basket is shown in Figure 4.15.14.

4.15.6.4 Jet Powered Junk Baskets and Reverse Circulating Junk Baskets

One of the most common types of junk baskets is the venturi jet basket. The basket is run into the hole and fluid is circulated to clean all cuttings and debris off the junk. A ball is then dropped down the pipe and seated in a valve to produce reverse circulation. The circulating fluid is then jetted outward and downward against the full circumference of the hole where it is deflected in a manner that directs all annular flow below this point into the shoe. Along with drilling fluid (mud), small pieces of junk are sucked up into the barrel where pieces are caught by catcher fingers. A reverse circulating junk basket is shown in Figure 4.15.14.

4.15.6.5 Hydrostatic Junk Baskets

The hydrostatic junk basket derives its actuating forces from circulating fluid pressure differential. The tool and drill string are lowered into the hole nearly empty of fluid. Once on bottom with the junk basket over the junk, application of weight on the string (downward) trips a valve allowing drilling fluids under great hydrostatic pressure to enter the empty drill string carrying debris (junk) with it. The effect is the same as lowering an empty straw, its top covered with a finger, into a glass of water. When the finger is removed, water rushes into the straw. The hydrostatic junk retriever can be opened and closed many times before being removed from the hole.

4.15.6.6 Milling Tools

Junk mills are designed to mill bit cones, reamer blades, or any other junk which may obstruct the wellbore. Many milling tools are fabricated for a specific milling operation. Some mills are manufactured from a solid piece of AISI 4140 heat-treated steel. Others are fabricated by welding the blades and stabilizer pads on to a simple tubular body. Although these tools are quite simple in appearance, the junk mill requires thought and experience to obtain the desired junk removal results. These tools should be designed to be versatile and to withstand spudding, heavy weights, and fast rotation. Factors that affect milling rates and the design of the mill are the type of fish, being milled, the stability of the fish, and its hardness.

4.15.6.7 Mill Designs

The concave design centers the mill over the fish and is less likely to sidetrack. The cone buster junk mill is a very popular type of mill. It is used to mill loose junk, or dress off the top of a fish. The Type P mill is a very aggressive mill. It has a large throat (rather than smaller water courses). The bladed junk mill is used to mill up packers and will cut faster through sand, cement or fill. String mills are used to enlarge windows in a whipstock operation, or to mill out a collapsed casing. Tapered mills are used to dress liner tops, smooth out perforations, and are sometimes run with pilot mills to stabilize and keep the casing clear. Pilot mills are used to mill up casing, drill pipe or liners. They perform better if the fish is stabilized (to reduce vibration and chatter). Pilot milling should be treated as a machining process. Shock subs are usually run directly above the mill to reduce vibration. A section mill is used to cut windows in the casing to facilitate a sidetrack operation. They can also be used to mill out internal casing patches or to cut pipe. Junk, string, taper, pilot, and section mills are shown in Figure 4.15.15.

4.15.6.8 Impression Block

One method that is sometimes used to access the condition of the top of the fish is to run an Impression block. A typical impression block consists of a block of lead that is molded into a circular steel body. The block is made up on drill collars and run into the hole until it just above the fish. Circulation is started to wash all fill off the top of the fish so that a good impression can be obtained. The block is lowered gently to touch the fish and weight is applied. The top of the fish indents the bottom of the soft lead, leaving an impression that can be examined and measured at the surface.

4.15.6.9 Fishing Magnets

Ferrous metallic junk can often be retrieved using a fishing magnet that is made up to the bottom of a working string. This tool is a powerful permanent magnet that has ports for circulation and a nonmagnetic brass sleeve housing to prevent junk from clinging to the side of the magnet. A fishing magnet is lowered into the hole with circulation to wash cuttings off the top of the fish. A cut lip skirt on the bottom helps to rake the junk into the face of the magnet for engagement and also prevents the junk from being knocked off while pulling out of the hole. Do not run excessive weight on a fishing magnet; it is not a drilling tool.

4.15.6.10 Junk Shot

When a large or oddly shaped piece of junk is in the wellbore and it cannot be retrieved using regular junk baskets, a junk shot may be used to break piece into smaller pieces. This junk shot tool contains a shaped charge. This tool must be run on collars and drill pipe to keep the force of its explosion directed downhole (at the junk). The tool is lowered to just above the junk and circulation is established to wash all fill off the junk. When the shaped charge is fired, its downward-directed blow breaks up the junk so that it can be finished with more conventional means.

Poor Boy Junk Basket

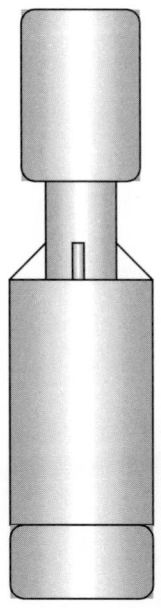

Boot Basket

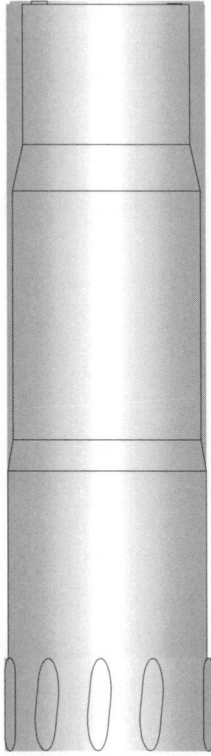

Core Type Junk Basket

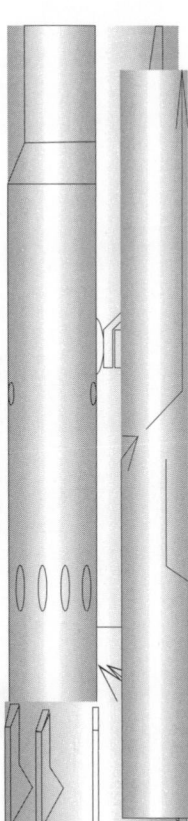

Reverse Circulating Junk Basket

Figure 4.15.14 *Poor Boy, Boot, Core Type and Reverse Circulating Junk Baskets. (Courtesy Weatherford International)*

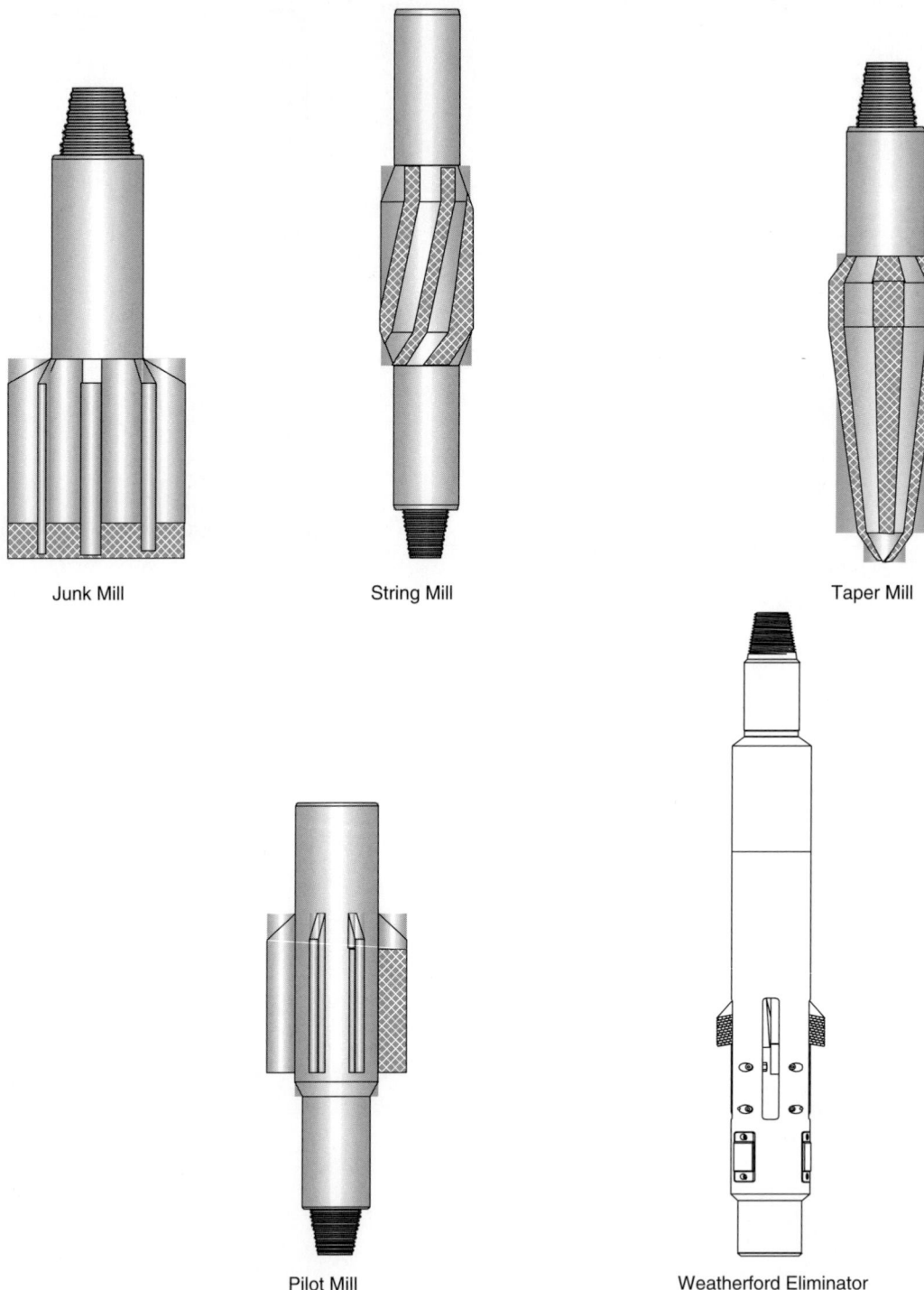

Junk Mill

String Mill

Taper Mill

Pilot Mill

Weatherford Eliminator
Section Mill

Figure 4.15.15   *Junk, String, Taper. Pilot and Section Mills. (Courtesy Weatherford International)*

4.15.7 Abandonment

Well abandonment is a process to permanently shut-in a previously producing well in a safe and environmentally responsible way. This process is applied to both land and offshore wells. Obviously, offshore wells are the most complicated to permanently abandon. To abandon a well the casing is generally cut below the mud line. In the United States and its territorial waters, the depth below the mud line is defined in U.S. Federal Government regulations [6]. Other countries and regions of the world often use similar but different regulation sources. This type of operation was at one time the hardest perform. But with today's new recovery systems and new casing cutting technologies NPT has been greatly reduced, saving the operators thousands of dollars in rig time. Some companies have advanced wellhead recovery systems available on the market. An example of such tool systems is the Weatherford M.O.S.T. Tool System that cuts and retrieves the casing, wellhead, and the temporary guide base in a single trip. These advanced systems when used with a downhole mud motor for cutting can eliminate wellhead damage and confirm cuts without tripping pipe. There are two different types of tools; the deepwater version, and the standard version. The deepwater versions utilize a mud motor and can be used at all water depths and water currents. The standard version has three options for cutting the casing; rotary table, top drive, or a mud motor. When using the rotary table or top drive for cutting a marine swivel must be used. The rotary table or top drive options of cutting the pipe are used primarily for shallow water operations (less than 2,000 ft and water currents less than 3 knots). Using the standard version with a mud motor increases cutting efficiencies (all water depths and water currents). The deepwater and standard versions (using a marine swivel) are shown in Figures 4.15.16 and 4.15.17, respectively.

4.15.8 Wirelines

Wirelines are utilized in a variety of operations in drilling and completions and production operations. During exploration and development drilling operations wireline logging operations, retrieval of bottomhole samples, and placement and retrieval of downhole tools are vital to the assessment of new reservoirs and the continued operation of producing reservoirs. Thus, the subject of wirelines is an important issue in a number sections and subsections throughout this handbook. However, in order not to repeat this subject in the other sections and subsections where wirelines are an important subject, the wireline subject has been concentrated for convenience in this Fishing and Abandonment section.

4.15.8.1 Wireline Construction

There are two basic categories of wirelines. All wirelines are anchored at one end at the surface with a wireline head that can be mounted on the drilling rig floor or on some other auxiliary equipment (e.g., logging truck bed) at the operational site. The free end of the wireline carries a load of some type of downhole tool (a deadload or tool for providing a downhole actuating force).

The first category of wirelines are made up of simple armored cables of various diameters and strengths that have no electrical conductor lines integrated into the overall construction. These wirelines are used to carry loads (e.g., special downhole tools) or otherwise provide actuating forces inside the wellbore. The second category of wirelines are an integrated construction with armored cables (for load carrying) surrounding a variety of electrical conductor lines.

The standard load carrying armored cables are special galvanized improved plow steel wires. The wires of the standard cables are coated with zinc. The tensile strength of each

wire used in the standard cables are in the range of 270,000 to 300,000 psi. These standard load carrying cables are not recommended for operations in H_2S and CO_2 environments.

For operations in H_2S and CO_2 it is necessary to use load carrying armored cables constructed of stainless steel wires. Stainless steel wires have tensile strengths that are slightly below those of the standard wires discussed above (e.g., range of 240,000 to 270,000 psi). Thus, armored cables constructed with stainless steel are rated at slightly lower breaking strengths than the standard armored cables discussed above. Usually H_2S and CO_2 are accompanied by higher downhole temperatures. Higher operating borehole temperatures will result in decreased breaking strength for all armored cables.

All mechanical properties are for room temperature conditions.

4.15.8.2 Electrical Conductors

Most wireline manufactures rate their conductors cables at DC rather than AC. Rating a conductor cables for maximum AC voltage presents difficulties. AC voltage ratings are dependent on frequency and actual wave shape. Such an AC voltage can be characterized by an equivalent DC voltage that will provide the same amount of heating. In many applications, AC power is transmitted as alternating voltages and currents in a since or cosine wave variation in time. Also, 60 hertz is a common power frequency. If this is the method of transmission, then a 707-volt sine wave has 1000 volts peak voltage. Therefore, a 707-volt average voltage at a frequency of 60 hertz would correspond to 1000 volts DC. At higher frequencies there can be more stress on the dielectric due to heating. Generally, the higher the frequency, the lower the breakdown voltage.

All electrical data for resistance and capacitance are nominal values and have been corrected to 20°C.

Proper selection of armored cable materials and conductor insulation materials can allow wirelines to operate up to temperature of 600°F (316°C).

4.15.8.3 Simple Armored Wirelines

This category of wirelines is composed of a variety of armored wirelines that are used in numerous downhole, rig site and/or wellhead operations. These wirelines vary in diamters from $\frac{3}{16}$ in. (4.80 mm) to as high as $\frac{9}{16}$ in. (14.30 mm).

Table 4.15.3 gives the typical construction, dimensions, and mechanical properties for sand lines.

Table 4.15.4 gives the typical construction, dimensions, and mechanical properties for Swablines.

Table 4.15.5 gives the typical construction, dimensions, and mechanical properties of a variety of utility wirelines used in wellhead operations.

Most manufactures of wirelines will construct specially designed armored wirelines for unique downhole operations.

4.15.8.4 Armored Wirelines with Electrical Conductors

This category of wirelines is composed of a variety of armored wirelines that surround electrical conductors.

Table 4.15.6 shows the $\frac{1}{10}$ ths of an inch (2.54 mm) diameter monconductor wireline. The table gives typical construction, dimensions, mechanical properties, and electrical properties.

Table 4.15.7 shows the $\frac{7}{32}$ ths of an inch (5.69 mm) diameter three-conductor wireline. The table gives typical construction, dimensions, mechanical properties, and electrical properties.

(text continued on page 4-404)

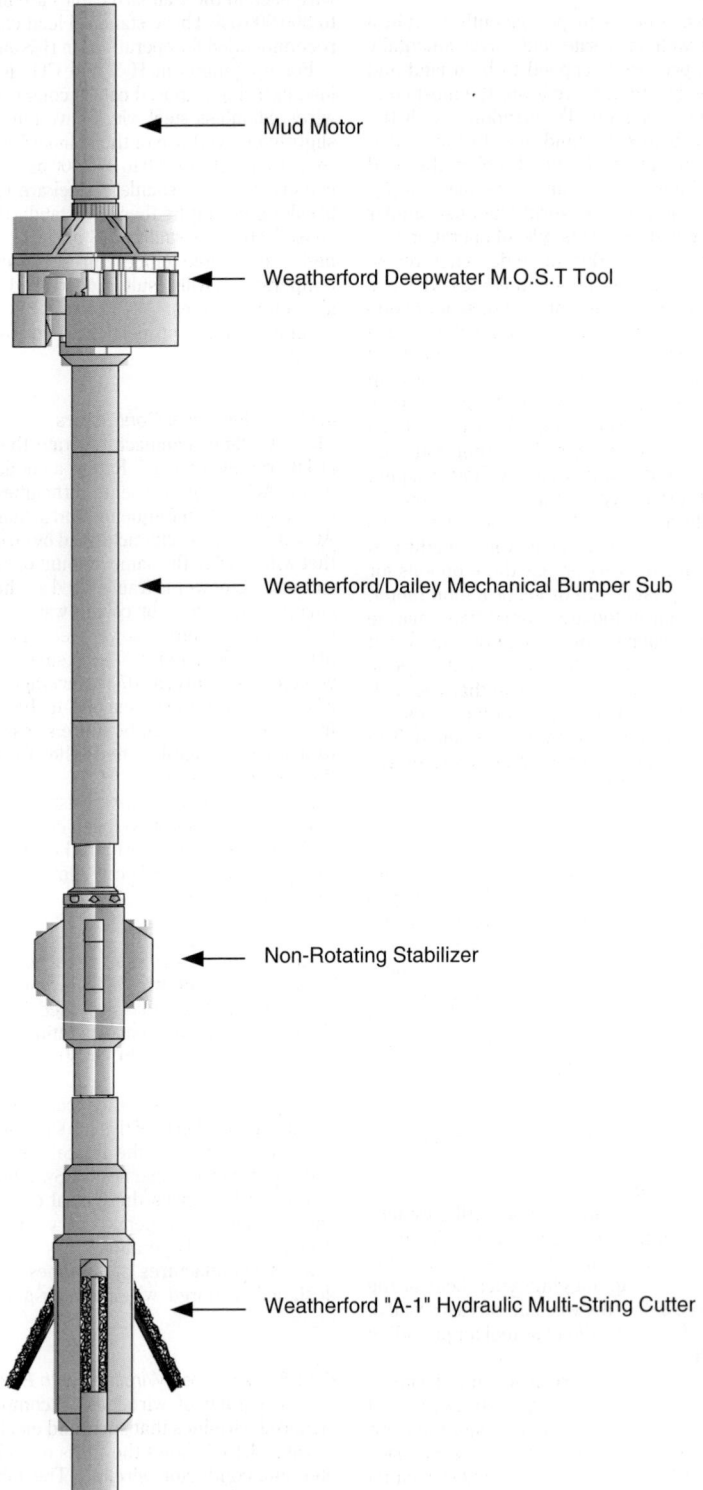

Figure 4.15.16   *Deepwater MOST Tool. (Courtesy of Weatherford International.)*

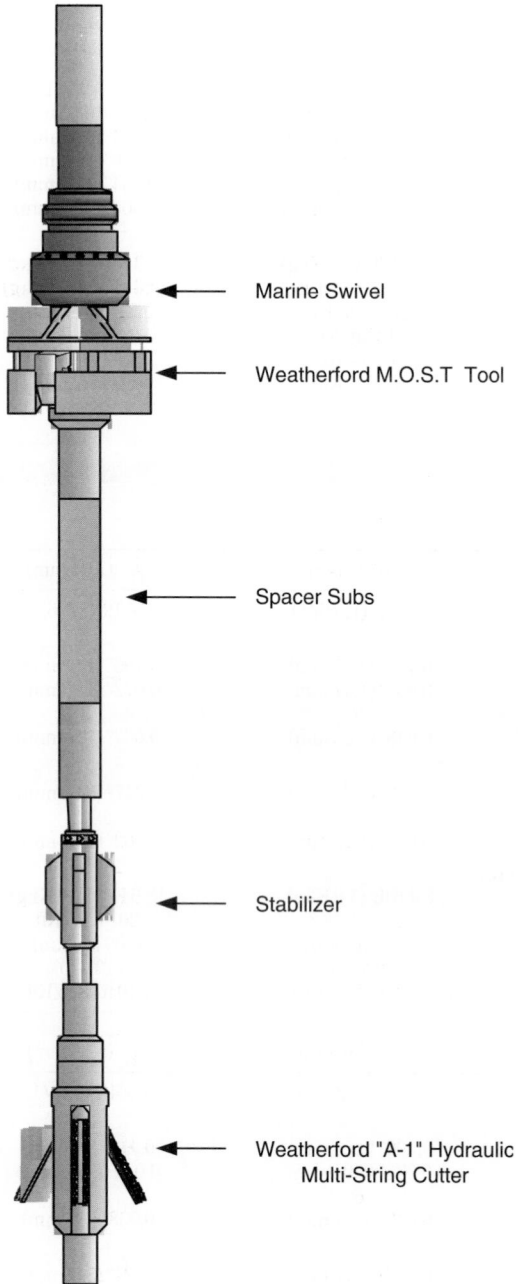

Marine Swivel

Weatherford M.O.S.T Tool

Spacer Subs

Stabilizer

Weatherford "A-1" Hydraulic
Multi-String Cutter

Figure 4.15.17   *Tool with Marine Swivel. (Courtesy Weatherford International.)*

Table 4.15.3 *Sand Lines*

Type (Nominal Diameter)	$\frac{7}{16}$ " (11.11 mm)	$\frac{1}{2}$ " (12.70 mm)	$\frac{9}{16}$ " OS (14.29 mm)
CONSTRUCTION	6×7 (6/1)	6×7 (6/1)	6×7 (6/1)
Polypropylene fiber diameter	0.219″ (5.56 mm)	0.250″ (6.35 mm)	0.290″ (7.40 mm)
Stranded wire diameter	0.142″ (3.61 mm)	0.164″ (4.17 mm)	0.190″ (4.83 mm)
Center wire diameter	0.050″ (1.27 mm)	0.056″ (1.42 mm)	0.066″ (1.68 mm)
Outer wire diameter	0.046″ (1.17 mm)	0.054″ (1.37 mm)	0.062″ (1.57 mm)
Steel wire diameter	0.438″ (11.11 mm)	0.500″ (12.7 mm)	0.582″ (14.8 mm)
MECHANICAL CHARACTERISTICS			
Breaking strength	16.5 Klb (7.5 Kkg)	22.3 Klb (10.1 Kkg)	29.8 Klb (13.5 Kkg)
Breaking strength	17.4 Klb (7.9 Kkg)	23.6 Klb (10.7 Kkg)	31.3 Klb (14.2 Kkg)
Weight	277.5 lb/Kft	375.6 lb/Kft	500.6 lb/Kft
	(413 kg/km)	(559 kg/km)	(745 kg/km)
Diameter tolerance	+ 4% − 0%	+ 4% − 0%	+ 4% − 0%
Stretch coefficient	ft/Kft/Klb	ft/Kft/Klb	ft/Kft/Klb

(Courtesy of Camesa, Inc.)

Table 4.15.4 *Swabline*

Type (Nominal diameter)	$\frac{1}{4}$ " (6.35 mm)	$\frac{5}{16}$ " (7.94 mm)	$\frac{7}{16}$ " (11.11 mm)
CONSTRUCTION (Special)	1×19 (12/6/1)	1×19 (12/6/1)	1×19 (12/6/1)
Inner Layer – Left			
Outside diameter	0.154″ (3.91 mm)	0.196″ (4.98 mm)	0.244″ (6.2 mm)
Center wire diameter	0.058″ (1.47 mm)	0.072″ (1.83 mm)	0.102″ (2.59 mm)
Number of outer diameter	6	6	6
Outer wire diameter	0.048″ (1.22 mm)	0.062″ (1.57 mm)	0.084″ (2.13 mm)
Outer Layer – Right			
Outside diameter	0.250″ (6.35 mm)	0.317″ (8.05 mm)	0.438″ (11.1 mm)
Number wires	12	12	12
Wire diameter	0.048″ (1.22 mm)	0.062″ (1.57 mm)	0.084″ (2.13 mm)
MECHANICAL CHARACTERISTICS			
Breaking strength	8.4 Klb (3.8 Kkg)	13.9 Klb (6.3 Kkg)	25.6 Klb (11.6 Kkg)
Weight	124.3 lb/Kft	207.6 lb/Kft	381.7 lb/Kft
	(185 kg/km)	(309 kg/km)	(568 kg/km)
Diameter tolerance	+ 2% − 0%	+ 2% − 0%	+ 2% − 0%
Stretch coefficient	1.70 ft/Kft/Klb	1.10 ft/Kft/Klb	0.60 ft/Kft/Klb

Type (Nominal diameter)	$\frac{3}{16}$ " (4.76 mm)	$\frac{7}{32}$ " (5.56 mm)	$\frac{1}{4}$ " (6.35 mm)
CONSTRUCTION (Special)	1×16 (9/6/1)	1×16 (9/6/1)	1×16 (9/6/1)
Inner Layer – Left			
Outside diameter	0.100″ (2.54 mm)	0.118″ (3.0 mm)	—
Center wire diameter	0.036″ (0.91 mm)	0.042″ (1.07 mm)	—
Number of outer diameter	6	6	—
Outer wire diameter	0.032″ (0.81 mm)	0.038″ (.97 mm)	—
Outer Layer – Right			
Outside diameter	0.188″ (4.78 mm)	0.224″ (5.69 mm)	—
Number wires	9	9	—
Wire diameter	0.044″ (1.12 mm)	0.054″ (1.37 mm)	—
MECHANICAL CHARACTERISTICS			
Breaking strength	4.6 Klb (2.1 Kkg)	6.8 Klb (3.1 Kkg)	—
Weight	70.6 lb/Kft	104.2 lb/Kft	—
	(105 kg/km)	(155 kg/km)	—
Diameter tolerance	+ 2% − 0%	+ 2% − 0%	—
Stretch coefficient	2.85 ft/Kft/Klb	1.80 ft/Kft/Klb	—

(Courtesy of Camesa, Inc.)

Table 4.15.5 *Utility Lines*

Type (Nominal diameter)	$\frac{3}{16}''$ (4.76 mm)	$\frac{7}{32}''$ (5.56 mm)	$\frac{1}{14}''$ (6.35 mm)
CONSTRUCTION (Normal)	1 × 19 (9/9/1)	1 × 19 (9/9/1)	1 × 19 (9/9/1)
Inner Layer – Left			
Outside diameter	0.112″ (2.84 mm)	0.130″ (3.30 mm)	0.152″ (3.86 mm)
Centre wire diameter	0.056″ (1.42 mm)	0.066″ (1.68 mm)	0.076″ (1.93 mm)
Number of outer wires	9	9	9
Outer wire diameter	0.028″ (.711 mm)	0.032″ (.81 mm)	0.038″ (0.97 mm)
Outer Layer – Right			
Outside diameter	0.188″ (4.78 mm)	0.219″ (5.56 mm)	0.250″ (6.35 mm)
Number wires	9	9	9
Wire diameter	0.050″ (1.27 mm)	0.058″ (1.47 mm)	0.066″ (1.68 mm)
MECHANICAL CHARACTERISTICS			
Breaking strength	6.4 Klb (2.9 Kkg)	8.6 Klb (3.9 Kkg)	11.4 Klb (5.2 Kkg)
Weight	88.6 lb/Kft	127.7 lb/Kft	159 lb/Kft
	(131.8 kg/km)	(190 kg/km)	(237 kg/km)
Diameter tolerance	+2% – 0%	+2% – 0%	+2% – 0%
Stretch coefficient	ft/Kft/Klb	ft/Kft/Klb	ft/Kft/Klb
Type (Nominal diameter)	$\frac{3}{16}''$	$\frac{7}{32}''$	$\frac{1}{4}''$
CONSTRUCTION (Special)	1 × 19 (9/9/1)	1 × 19 (9/9/1)	1 × 19 (9/9/1)
Inner Layer - Left			
Outside diameter	0.108″ (2.74 mm)	0.118″ (3.0 mm)	—
Centre wire diameter	0.052″ (1.32 mm)	0.060″ (1.52 mm)	—
Number of outer wires	9	9	—
Outer wire diameter	0.026″ (0.66 mm)	0.029″ (0.74 mm)	—
Outer Layer - Right			
Outside diameter	0.188″ (4.78 mm)	0.219″ (5.56 mm)	—
Number wires	9	9	—
Wire diameter	0.049″ (1.24 mm)	0.057″ (1.45 mm)	—
MECHANICAL CHARACTERISTICS			
Breaking strength	5.9 Klb (2.7 Kkg)	7.9 Klb (3.6 Kkg)	—
Weight	84 lb/Kft	111.5 lb/Kft	—
	(125 kg/km)	(166 kg/km)	—
Diameter tolerance	+2% – 0%	+2% – 0%	—
Stretch coefficient	1.51 ft/Kft/Klb	ft/Kft/Klb	—

(Courtesy of Camesa, Inc.)

Table 4.15.6   *1/10″ (2.54 mm) Monoconductor 1N10*

PROPERTIES
Cable diameter	$0.101″ + 0.004″ - 0.002″$	$(2.56\,mm + 0.10\,mm - 0.05\,mm)$
Minimum sheave diameter	6″	(16 cm)
Cable stretch coefficient	13.1 ft/Kft/Klb	(14.72 m/Km/5KN)

ELECTRICAL
Maximum conductor voltage	300 VDC	
Conductor AWG rating	24	
Minimum insulation resistance	1,500 MegaΩ/Kft at 500 VDC	(457 MegaΩ/Km at 500 VDC)
Armor electrical resistance	22 Ω/Kft	(72.2 Ω/Km)

MECHANICAL
Cable breaking strength		
Ends fixed	1,000 lb	(4.4 KN) Nominal
Ends free	670 lb	(3.0 KN) Nominal
Maximum suggested working tension	500 lb	(2.2 KN)
Number and size of wires		
Inner armor	$12 \times 0.0140″$	(0.356 mm)
Outer armor	$18 \times 0.0140″$	(0.356 mm)
Average wire braking strength		
Inner armor	42 lb	(0.19 KN)
Outer armor	42 lb	(0.19 KN)

Cable Type	Core Description							Cable Weight		
	Temp Rating	Plastic Insulation	Type Thickness	Copper Construction	Res Typical	Cap. Typical	O.D. Each	In Air	In H$_2$O	Spec. Gravity
	°F (°C)		in. (mm)	in. (mm)	Ω/Kft (Ω/Km)	pf/ft (pf/m)	in. (mm)	lb/Kft (Kg/Km)		
1N10RP	300 (149)	Poly	0.012 (0.305)	7×0.0085 (7×0.216)	21.0 (69.0)	51 (167)	0.049 (1.244)	19 (28)	16 (24)	6.42

The armor wires are high tensile, galvanized extra improved plow steel (GEIPS), and coated with anti-corrosion compound for protection during shipping and storing. Wires are preformed and cables are post tensioned.

Core assembly – Copper strand consists of six wires around one center wire. Conductor resistance is measured at 68 °F. Voids in the copper strand are filled with a water-blocking agent to reduce water and gas migration.

SUPERSEAL – a special pressure seal agent, is applied between armor layers.

The temperature rating assumes a normal gradient for both temperature and weight.

(Courtesy of Camesa, Inc.)

Table 4.15.7 *7/32″ (5.69 mm) 3-CONDUCTOR 3H22*

PROPERTIES

Cable diameter	0.224″ + 0.005″ − 0.002″	(5.69 mm + 0.13 mm − 0.05 mm)
Minimum sheave diameter	13″	(33 cm)
Cable stretch coefficient (nominal)	2.50 ft/Kft/Klb	(2.81 m/Km/5KN)

ELECTRICAL

Maximum conductor voltage	300 VDC	
Conductor AWG rating	23	
Minimum insulation resistance	1,500 MegΩ/Kft at 500 VDC	(457 MegΩ/Km at 500 VDC)
Armor electrical resistance	4.80 Ω/Kft	(15.7 Ω/Km)

MECHANICAL

Cable breaking strength

Ends fixed	4,700 lb	(20.9 KN) Nominal
Ends free	2,880 lb	(12.8 KN) Nominal
Maximum suggested working tension	2,350 lb	(10.5 KN)

Number and size of wires

Inner armor	18 × 0.0220″	(0.559 mm)
Outer armor	18 × 0.0310″	(0.787 mm)

Average wire braking strength

Inner armor	103 lb	(0.46 KN)
Outer armor	204 lb	(0.91 KN)

Cable Type	Temp Rating	Plastic Type	Insulation Thickness	Copper Construction	Res Typical	Cap. Typical	OD Each	Jacket Type	In Air	In H₂O	Spec. Gravity
	°F (°C)		in. (mm)	in. (mm)	Ω/Kft (Ω/Km)	pf/ft (pf/m)	in. (mm)		lb/Kft (Kg/Km)		
3H22RPP	300 (149)	Poly	0.011 (0.279)	7 × 0.0085 (7 × 0.216)	22.5 (73.8)	36 (118)	0.047 (1.194)	Poly	81 (120)	69 (103)	6.91
3H22RXZ	420 (216)	Camtane	0.011 (0.279)	7 × 0.0085 (7 × 0.216)	22.5 (73.8)	35 (115)	0.047 (1.194)	Tefzel	82 (122)	72 (104)	6.82
3H22RTZ	500 (260)	Teflon	0.011 (0.279)	7 × 0.0085 (7 × 0.216)	22.5 (73.8)	40 (131)	0.047 (1.194)	Tefzel	83 (124)	71 (106)	6.81
3H22RTA	550 (288)	Teflon	0.011 (0.279)	7 × 0.0085 (7 × 0.216)	22.5 (73.8)	36 (118)	0.047 (1.194)	PFA	83 (124)	71 (106)	6.81

The armor wires are high tensile. Galvanized extra improved plow steel (GEEIPS), and coated with anti-corrosion compound for protection during shipping and storing. Wires are preformed and cables are post tensioned.

Core assembly – Copper strand consists of six wires around one center wire. Voids in the copper strand are with a water-blocking agent to reduce water and gas migration. Conductor resistance is measured at 68 °F.

The temperature rating assumes a normal gradient for both temperature and weight.

(Courtesy of Camesa, Inc.)

Table 4.15.8 *3/16″ (4.80 mm) 4-Conductor 4H18*

PROPERTIES
Cable diameter	0.189″ + 0.004″ − 0.002″	(4.80 mm + 0.10 mm − 0.05 mm)
Minimum sheave diameter	10″	(25 cm)
Cable stretch coefficient (nominal)	4.25 ft/Kft/Klb	(4.78 m/Km/5KN)

ELECTRICAL
Maximum conductor voltage	300 VDC	
Conductor AWG rating	23	
Minimum insulation resistance	1,500 MegΩ/Kft at 500 VDC	(457 MegΩ/Km at 500 VDC)
Armor electrical resistance	6.70 Ω/Kft	(22.0 Ω/Km)

MECHANICAL
Cable breaking strength		
Ends fixed	3,100 lb	(13.8 KN) Nominal
Ends free	1,900 lb	(8.5 KN) Nominal
Maximum suggested working tension	1,550 lb	(6.9 KN)

Number and size of wires
Inner armor	18 × 0.0185″	(0.470 mm)
Outer armor	18 × 0.0248″	(0.630 mm)

Average wire breaking strength
Inner armor	72 lb	(0.32 KN)
Outer armor	130 lb	(0.58 KN)

Cable Type			Core Description						Cable Weight		
	Temp Rating	Plastic Type	Insulation Thickness	Copper Construction	Res Typical	Cap. Typical	OD Each	Jacket Type	In Air	In H₂O	Spec. Gravity
	°F (°C)		in. (mm)	in. (mm)	W/Kft (W/Km)	pf/ft (pf/m)	in. (mm)		lb/Kft (Kg/Km)		
4H18RPP	300 (149)	Poly	0.0075 (0.191)	7 × 0.0085 (7 × 0.216)	22.5 (73.8)	36 (118)	0.047 (1.194)	Poly	84 (124)	69 (103)	5.73
4H18RXZ	420 (216)	Camtane	0.0075 (0.191)	7 × 0.0085 (7 × 0.216)	22.5 (73.8)	35 (115)	0.047 (1.194)	Tefzel	84 (124)	69 (103)	5.73
4H18RTZ	500 (260)	Teflon	0.0075 (0.191)	7 × 0.0085 (7 × 0.216)	22.5 (73.8)	40 (131)	0.047 (1.194)	Tefzel	86 (127)	71 (105)	5.85
4H18RTA	550 (288)	Teflon	0.0075 (0.191)	7 × 0.0085 (7 × 0.216)	22.5 (73.8)	36 (103)	0.047 (1.194)	PFA	88 (130)	73 (107)	5.94

The armor wires are high tensile, galvanized extra improved plow steel (GEEIPS), and coated with anti-corrosion compound for protection during shipping and storing. Wires are preformed and cables are post tensioned.
Core assembly – Copper strand consists of six wires around one center wire. Voids in the copper strand are with a water-blocking agent to reduce water and gas migration. Conductor resistance is measured at 68 °F.
The temperature rating assumes a normal gradient for both temperature and weight.
(Courtesy of Camesa, Inc.)

(text continued from page 4-397)

Table 4.15.8 shows the $\frac{5}{16}$ th of an inch (8.62 mm) diameter seven-conductor wireline. The table gives typical construction, dimensions, mechanical properties, and electrical properties.

Table 4.15.8 shows the $\frac{3}{16}$ ths of an inch (4.80 mm) diameter four-conductor wireline. The table gives typical construction, dimensions, mechanical properties, and electrical properties.

Most manufactures of wirelines will construct specially designed armored electrical conductor wirelines for unique downhole operations.

4.15.8.5 Wireline Operating and Breaking Strengths

During field operations the line tensions should be no greater than one half of the wireline breaking strength. This will insure that no permanent damage is done to the wireline.

In the tables above the breaking strength for each wireline is given. For stainless steel armored cables or other special construction the manufacturer should be consulted for breaking strength.

4.15.8.6 Wireline Stretching

Many field operations using wirelines requires that the wireline stretch be determined. One common need is the necessity of determine the true depth of the particular logging tool in the well. The true depth can be determined by adding the amount of cable stretch ΔL to the measured wireline length L in the well. The amount of wireline stretch is determined from

$$DL = \left(\frac{KL}{2}\right)[(T + W_t) - 2T_m] \qquad [4.15.3]$$

where

L is the length of wireline in the well (in units of 1,000 ft)

Table 4.15.9 *5/16″ (8.26 mm) 7-Conductor 7H32*

PROPERTIES		
Cable diameter	0.325″ + 0.005″ − 0.002″	(8.26 mm + 0.13 mm − 0.05 mm)
Minimum sheave diameter	18″	(45 cm)
Cable stretch coefficient (nominal)	1.8 ft/Kft/Klb	(2.02 m/Km/5KN)
ELECTRICAL		
Maximum conductor voltage	1000 VDC	
Conductor AWG rating	22	
Minimum insulation resistance	1,500 MegΩ/Kft at 500 VDC	(457 MegΩ/Km at 500 VDC)
Armor electrical resistance	2.3 Ω/Kft	(7.5 Ω/Km)
MECHANICAL		
Cable breaking strength		
Ends fixed	9,500 lb	(42.3 KN) Nominal
Ends free	5,700 lb	(25.4 KN) Nominal
Maximum suggested working tension	4,750 lb	(21.1 KN)
Number and size of wires		
Inner armor	18 × 0.0320″	(0.813 mm)
Outer armor	18 × 0.0445″	(1.130 mm)
Average wire braking strength		
Inner armor	217 lb	(0.97 KN)
Outer armor	420 lb	(1.87 KN)

Cable Type				Core Description						Cable Weight		
	Temp Rating	Plastic Type	Insulation Thickness	Copper Construction	Res Typical	Cap. Typical	OD Each	Tape Type		In Air	In H$_2$O	Spec. Gravity
	°F (°C)		in. (mm)	in. (mm)	Ω/Kft (Ω/Km)	pf/ft (pf/m)	in. (mm)			lb/Kft (Kg/Km)		
7H32RP	300 (149)	Poly	0.013 (0.330)	7 × 0.0100 (7 × 0.254)	15.8 (51.8)	55 (180)	0.056 (1.422)	Dacron		183 (272)	152 (226)	5.86
7H32RZ	420 (216)	Tefzel	0.013 (0.330)	7 × 0.0100 (7 × 0.254)	15.8 (51.8)	67 (220)	0.056 (1.422)	Nomex		188 (280)	157 (234)	6.03
7H32RA	500 (260)	PFA	0.013 (0.330)	7 × 0.0100 (7 × 0.254)	15.8 (51.8)	58 (190)	0.056 (1.422)	Nomex		190 (283)	159 (237)	6.08

The armor wires are high tensile, galvanized extra improved plow steel (GEEIPS), and coated with anti-corrosion compound for protection during shipping and storing. Wires are preformed and cables are post tensioned.
Core assembly – Conductors are bound with conductive tape and voids are filled with conductive paste and string.
Conductors are "Water Blocked" to reduce water and gas migration. Conductor resistance is measured at 68°F.
The temperature rating assumes a normal gradient for both temperature and weight.
Center conductor construction is 7 × 0.0100″. The typical capacitance is decreased by approximately 5 to 10% in comparison to the outer conductors.

T is the surface tension (in units of 1,000 lb)
W$_t$ is the effective weight (in the well fluid) of any downhole tool on the free end of the wireline (in units of 1,000 lb)
T$_m$ is the marked cable tension (usually 1,000 lb)
K is the elastic stretch coefficient (ft/1,000 ft/1,000 lb)

Example 4.15.2
Determine the stretch of 10,000 ft of a monoconductor 7H32RZ wireline with a 300 = lb tool (weight in drilling mud used). The stretch coefficient is 1.8 ft/kft/klb. The wireline has been marked 1000 lb tension. The wirelines tension measured at the surface is 5000 lbs. Using Equation 4.15.3,

$$\Delta L = \left[\frac{1.8(10.0)}{2}\right][(5.0+0.3)-2(1.0)]$$

$$\Delta L = 29.7 \text{ ft}$$

Thus, the true depth as measured by wireline would be 10,000 ft + 29.7 ft = 10,029.7 ft.

References
1. Main, W. C., "Detection of Incipient Drill-pipe Failures," *API Drilling and Production Practices*, 1949.
2. Moore, E. E., "Fishing and Freeing Stuck Drill Pipe," *The Petroleum Engineer*, April 1956.
3. Gatlin, C., *Petroleum Engineering: Drilling and Well Completions*, Prentice-Hall, 1960.
4. Short, J. A., *Fishing and Casing Repair*, PennWell, 1981.
5. McCray, A. W., and Cole, F. W., *Oil Well Drilling Technology*, University of Oklahoma Press, Norman, Oklahoma, 1959.
6. "Well Abandonment and Inactive Well Practices for U. S. Exploration and Production Operations, Environment Guidance Document," *API Bulletin E3*, 1st Edition, January 1993.

4.16 CASING AND CASING STRING DESIGN

4.16.1 Types of Casing

Based on the primary function of the casing string, there are five types of casing to be distinguished.

Stove or Surface Casing

The stovepipe is usually driven to sufficient depth (15–60 ft) to protect loose surface formation and to enable circulation of the drilling fluid. This pipe is sometimes cemented in predrilled holes.

Conductor String

This string acts as a guide for the remaining casing strings into the hole. The purpose of the conductor string is also to cover unconsolidated formations and to seal off overpressured formations. The conductor string is the first string that is always cemented to the top and equipped with casing head and blowout prevention (BOP) equipment.

Surface Casing

This is set deeply enough to protect the borehole from caving-in in loose formations frequently encountered at shallow depths, and protects the freshwater sands from contamination while subsequently drilling a deeper hole. In case the conductor string has not been set, the surface casing is fitted with casing head and BOP.

Intermediate Casing

Also called protection string, this is usually set in the transition zone before abnormally high formation pressure is encountered, to protect weak formations or to case off loss-of-circulation zones. Depending upon geological conditions, the well may contain two or even three intermediate strings. Production string (oil string) is the string through which the well is produced.

Intermediate or production string can be set a liner string. The liner string extends from the bottom of the hole upward to a point about 150–250 ft above the lower end of the upper string.

Casing Program Design

Casing program design is accomplished by two steps. In the first step, the casing sizes and corresponding bit sizes should be determined. In the second step, the setting depth of the individual casing strings ought to be evaluated. Before starting the casing program design, the designer ought to know the following basic information:

- The purpose of the well (exploratory or development drilling)
- Geological cross-sections that should consist of type of formations, expected hole problems, pore and formation's fracture pressure, number and depth of water, oil, gas horizons
- Available rock bits, reamer shoes and casing sizes
- Load capacity of a derrick and mast if the type of rig has already been selected

Before starting the design, it must be assumed that the production casing size and depth of the well has been established by the petroleum engineer in cooperation with a geologist, so that the hole size (rock bit diameter) for the casing may be selected. Considering the diameter of the hole, a sufficient clearance beyond the coupling outside diameter must be provided to allow for mud cake and also for a good cementing job. Field experience shows that the casing clearance should range from about 1.0 in. to 3.5 in. Larger casing sizes require greater value of casing clearance. Once the hole size for production string has been selected, the smallest casing through which a given bit will pass is next determined. The bit diameter should be a little less (0.05 in.) than casing drift diameter. After choosing the casing with appropriate drift diameter, the outside coupling diameter of this casing may be found. Next, the appropriate size of the bit should be determined and the procedure repeated.

Expandable casing technology, expandable drill bits, under-reamers and other tools for optimizations of borehole and/or string designs, are not covered in this section.

Example 4.16.1

The production casing string for a certain well is to consist of 5-in. casing. Determine casing and corresponding bit sizes for the intermediate, surface and conductor string. Take casing data and bit sizes from Table 4.16.1.

Solution

For production hole, select a $6\frac{3}{4}$-in. rock bit. Therefore, the casing clearance $= 6.75 - 5.563 = 1.187$ in.

For intermediate string, select a 7 5/8-in. casing, assuming that wall thickness that corresponds to drift diameters of 6.640 or smaller will not be used. For the 7 5/8-in. intermediate string, use a 9 7/8-in. bit. The casing clearance $= 9.875 - 8.5 = 1.375$ in.

For surface string, select a $10\frac{3}{4}$-in. casing. Note that only unit weights corresponding to drift diameters of 10.036 and 9.894 in. can be used. For the $10\frac{3}{4}$-in. casing, use a $13\frac{3}{4}$-in. bit, so the casing clearance $= 13.75 - 11.75 = 2.0$ in.

For conductor string, select 16 in. casing; the bit size will then be 20 in. and the casing clearance $= 20 - 17 = 3$ in.

Having defined bit and casing string sizes, the setting depths of the individual strings should be determined.

The operation of setting is governed by the principle according to which casing should be placed as deep as possible. However, the designer must remember to ensure the safety of the drilling crew from possible blowout, and to maintain the hole stability, well completion aspects (formation damage) and state regulations.

In general, casing should be set

- Where drilling fluid could contaminate freshwater that might be used for drinking or other household purposes
- Where unstable formations are likely to cave or slough into the borehole
- Where loss of circulation may result in blowout
- Where drilling fluid may severely damage production horizon

Currently, a graphical method of casing setting depth determination is used. The method is based on the principle according to which the borehole pressure should always be greater than pore pressure and less than fracture pressure. (Drilling with borehole pressures lower than pore pressure requires the use of under-balanced drilling technologies not covered in this chapter.)

For practical purposes, a safety margin for reasonable kick conditions should be imposed (Figure 4.16.1). Even when borehole pressure is adjusted correctly, problems may arise from the contact between the drilling fluid and the formation. It depends upon the type of drilling fluid and formation, but in general, the more time spent drilling in an open hole, the greater the possibility of formation caving or sloughing into the borehole. Formation instability may lead to expensive work in the borehole, which influences the time and cost of the drilling operation. To arrest or reduce this problem, special treatment drilling fluids might be used, but these special drilling fluids are expensive. Therefore, the casing and drilling fluid programs depend on each other, and solving the issue of correct casing setting depth evaluation is a rather complicated, optimizing problem.

Table 4.16.1 *Casing Dimensions for Rock Bit Selection*

API Casing Data (Bit Sizes and Clearances)						
Casing Specifications					Recommended Max. Bit Size	
Casing Size D.D. (inches)	Casing Coupling D.D. (inches)	Nominal Weight lb./ft.	Inside Diameter I.D. (inches)	API Drift I.D. (inches)	Roller Cone Bit Size D.D. (inches)	Fixed Cutter Bit Size D.D. (inches)
4.500	5.000	9.50	4.090	3.965	$3\frac{7}{8}$	$3\frac{7}{8}$
4.500	5.000	10.50	4.052	3.927	$3\frac{7}{8}$	$3\frac{7}{8}$
4.500	5.000	11.60	4.000	3.875	$3\frac{7}{8}$	$3\frac{7}{8}$
4.500	5.000	13.50	3.920	3.795	$3\frac{3}{4}$	$3\frac{3}{4}$
5.000	5.563	11.50	4.560	4.435	$3\frac{7}{8}$	$3\frac{7}{8}$
5.000	5.563	13.00	4.494	4.369	$3\frac{7}{8}$	$3\frac{7}{8}$
5.000	5.563	15.00	4.408	4.283	$3\frac{7}{8}$	$3\frac{7}{8}$
5.000	5.563	18.00	4.276	4.151	$3\frac{7}{8}$	$3\frac{7}{8}$
5.500	6.050	14.00	5.012	4.887	$4\frac{3}{4}$	$4\frac{3}{4}$
5.500	6.050	15.50	4.950	4.825	$4\frac{3}{4}$	$4\frac{3}{4}$
5.500	6.050	17.00	4.992	4.767	$4\frac{3}{4}$	$4\frac{3}{4}$
5.500	6.050	20.00	4.778		$4\frac{1}{2}$	$4\frac{1}{2}$
5.500	6.050	23.00	4.670	4.545	$3\frac{7}{8}$	$4\frac{1}{2}$
6.625	7.390	20.00	6.049	5.924	$5\frac{7}{8}$	$5\frac{7}{8}$
6.625	7.390	24.00	5.921	5.798	$4\frac{3}{4}$	$4\frac{3}{4}$
6.625	7.390	28.00	5.791	5.666	$4\frac{3}{4}$	$4\frac{3}{4}$
6.625	7.390	32.00	5.675	5.550	$4\frac{3}{4}$	$4\frac{3}{4}$
7.000	7.656	17.00	6.538	6.413	$6\frac{1}{4}$	$6\frac{1}{4}$
7.000	7.656	20.00	6.456	6.331	$6\frac{1}{4}$	$6\frac{1}{4}$
7.000	7.656	23.00	6.366	6.241	$6\frac{1}{8}$	$6\frac{1}{8}$
7.000	7.656	26.00	6.276	6.151	$6\frac{1}{8}$	$6\frac{1}{8}$
7.000	7.656	29.00	6.184	6.059	6	6
7.000	7.656	32.00	6.094	5.969	$5\frac{7}{8}$	$5\frac{7}{8}$
7.000	7.656	35.00	6.004	5.879	$5\frac{7}{8}$	$5\frac{7}{8}$
7.000	7.656	38.00	5.920	5.795	$4\frac{3}{4}$	$4\frac{3}{4}$
7.625	8.500	20.00	7.125	7.000	$6\frac{3}{4}$	$6\frac{3}{4}$
7.625	8.500	24.00	7.025	6.900	$6\frac{3}{4}$	$6\frac{3}{4}$
7.625	8.500	26.40	6.969	6.844	$6\frac{3}{4}$	$6\frac{3}{4}$
7.625	8.500	29.70	6.875	6.750	$6\frac{3}{4}$	$6\frac{3}{4}$
7.625	8.500	33.70	6.765	6.640	$6\frac{1}{2}$	$6\frac{1}{2}$
7.625	8.500	39.00	6.625	6.500	$6\frac{1}{2}$	$6\frac{1}{2}$
8.625	9.625	24.00	8.097	7.972	$7\frac{7}{8}$	$7\frac{7}{8}$
8.625	9.625	28.00	8.017	7.892	$7\frac{7}{8}$	$7\frac{7}{8}$
8.625	9.625	32.00	7.921	7.795	$6\frac{3}{4}$	$6\frac{3}{4}$
8.625	9.625	36.00	7.825	7.700	$6\frac{3}{4}$	$6\frac{3}{4}$
8.625	9.625	40.00	7.725	7.600	$6\frac{3}{4}$	$6\frac{3}{4}$
8.625	9.625	44.00	7.625	7.500	$6\frac{3}{4}$	$6\frac{3}{4}$
8.625	9.625	49.00	7.511	7.386	$6\frac{3}{4}$	$6\frac{3}{4}$
9.625	10.625	29.30	9.063	8.907	$8\frac{3}{4}$	$8\frac{3}{4}$
9.625	10.625	32.30	9.001	8.845	$8\frac{3}{4}$	$8\frac{3}{4}$
9.625	10.625	36.00	8.921	8.765	$8\frac{3}{4}$	$8\frac{3}{4}$
9.625	10.625	40.00	8.835	8.679	$8\frac{1}{2}$	$8\frac{1}{2}$
9.625	10.625	43.50	8.755	8.599	$8\frac{1}{2}$	$8\frac{1}{2}$

(continued)

Table 4.16.1   (continued)

API Casing Data (Bit Sizes and Clearances)						
Casing Specifications					Recommended Max. Bit Size	
Casing Size D.D. (inches)	Casing Coupling D.D. (inches)	Nominal Weight lb./ft.	Inside Diameter I.D. (inches)	API Drift I.D. (inches)	Roller Cone Bit Size D.D. (inches)	Fixed Cutter Bit Size D.D. (inches)
9.625	10.625	47.00	8.681	8.525	$8\frac{1}{2}$	$8\frac{1}{2}$
9.625	10.625	53.50	8.535	8.379	$8\frac{3}{8}$	$8\frac{3}{8}$
10.750	11.750	32.75	10.192	10.038	$9\frac{7}{8}$	$9\frac{7}{8}$
10.750	11.750	40.50	10.050	9.894	$9\frac{7}{8}$	$9\frac{7}{8}$
10.750	11.750	45.50	9.950	9.794	$9\frac{1}{2}$	$9\frac{1}{2}$
10.750	11.750	51.00	9.850	9.694	$9\frac{1}{2}$	$9\frac{1}{2}$
10.750	11.750	55.50	9.760	9.604	$9\frac{1}{2}$	$9\frac{1}{2}$
10.750	11.750	80.70	9.660	9.504	$9\frac{1}{2}$	$9\frac{1}{2}$
10.750	11.750	65.70	9.560	9.404	$8\frac{3}{4}$	$8\frac{3}{4}$
11.750	12.750	42.00	11.084	10.928	$10\frac{5}{8}$	$10\frac{5}{8}$
11.750	12.750	47.00	11.000	10.844	$10\frac{5}{8}$	$10\frac{5}{8}$
11.750	12.750	54.00	10.880	10.724	$10\frac{5}{8}$	$10\frac{5}{8}$
11.750	12.750	60.00	10.772	10.616	$9\frac{7}{8}$	$9\frac{7}{8}$
13.375	14.375	48.00	12.715	12.559	$12\frac{1}{4}$	$12\frac{1}{4}$
13.375	14.375	54.50	12.615	12.459	$12\frac{1}{4}$	$12\frac{1}{4}$
13.375	14.375	61.00	12.515	12.359	$12\frac{1}{4}$	$12\frac{1}{4}$
13.375	14.375	68.00	12.415	12.259	$12\frac{1}{4}$	$12\frac{1}{4}$
13.375	14.375	72.00	12.347	12.191	11	$10\frac{5}{8}$
16.000	17.000	65.00	15.250	15.062	$14\frac{3}{4}$	$14\frac{3}{4}$
16.000	17.000	75.00	15.124	14.936	$14\frac{3}{4}$	$14\frac{3}{4}$
16.000	17.000	84.00	15.010	14.822	$14\frac{3}{4}$	$14\frac{3}{4}$
18.625	20.000	87.50	17.755	17.567	$17\frac{1}{2}$	$17\frac{1}{2}$
20.000	21.000	94.00	19.124	18.936	$17\frac{1}{2}$	$17\frac{1}{2}$
20.000	21.000	106.50	19.000	18.812	$17\frac{1}{2}$	$17\frac{1}{2}$
20.000	21.000	133.00	18.730	18.542	$17\frac{1}{2}$	$17\frac{1}{2}$
20.000	21.000	169.00	18.376	18.188	$17\frac{1}{2}$	$17\frac{1}{2}$

Example 4.16.2

Suppose that in some area the expected formation pressure gradient is 0.65 psi/ft and formation fracture pressure gradient is 0.85 psi/ft. A gas-bearing formation is expected at depth of 15,000 ft. Assume that a gas kick occurs that, to be removed from the hole, induces a surface pressure of 2,000 psi. The first intermediate casing is set at a depth of 7,200 ft. Determine the setting depth for the second intermediate casing string if required in given conditions. Assume drilling fluid pressure gradient = 0.65 psi/ft.

Solution

The formation fracture pressure line is

$$P_f = (0.85)(D)$$

The borehole pressure line is

$$P_{bh} = 2,000 + (0.65)(D)$$

So

$$(0.85)(D) = 2,000 + (0.65)(D)$$

$$D = \frac{2,000}{0.2} = 10,000 \text{ ft.}$$

The second intermediate casing string is required and must be set at a depth of 10,000 ft.

4.16.2 Casing Data

Special note: Nothing in this specification should be interpreted as indicating a preference by the committee for any material or process, nor as indicating equality between the various materials or processes. In the selection of materials and processes, the purchaser must be guided by his or her experience and by the service for which the pipe is intended.

Casing is classified according to its manner of manufacture, steel grade, dimensions, and weights, and the type of coupling.

The following include excerpts from, and references to API Specification 5CT, Seventh Edition, October 1, 2001.

PRESSURE GRADIENT (psi/ft)

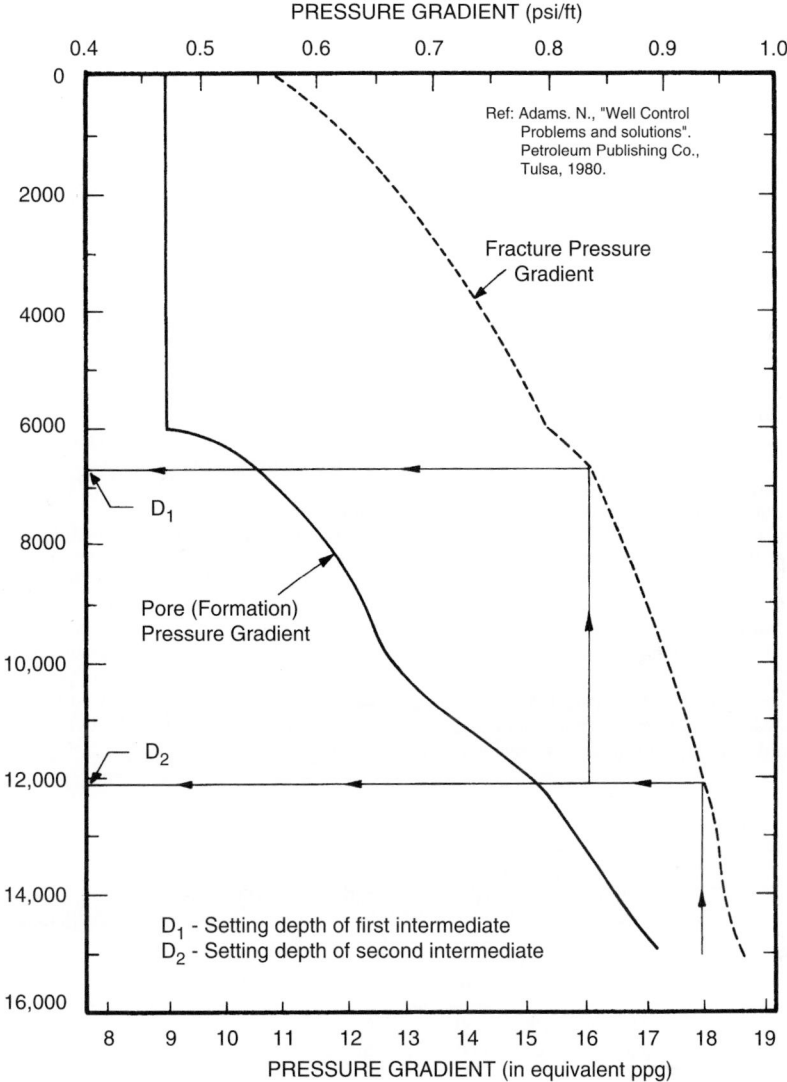

Figure 4.16.1 *Well planning.*

4.16.2.1 Process of Manufacture

Casing and liners shall be seamless or electric welded as defined below and as specified on the purchase.

a. Seamless pipe is defined a wrought steel tubular product made without a welded seam. It is manufactured by hot working steel or, if necessary, by subsequently cold finishing the hot-worked tubular product to produce the desired shape, dimension and properties.

b. Electric-welded pipe is defined as pipe having one longitudinal seam formed by electric-flash welding or electric-resistance welding, without the addition of extraneous metal. The weld seam of electric-welded pipe shall be heat treated after welding to a minimum temperature of 1,000°F (580°C), or processed in such a manner that no untempered martensite remains.

c. Pup-joints shall be made from standard casing or tubing or by machining thick-wall casing, tubing or bar stock.

Cold-drawn tubular products without appropriate heat treatment are not acceptable.

d. Casing and tubing accessories shall be seamless and made from standard casing or tubing, or by machining thick-wall casing, tubing or mechanical tubes, or bar stock or hot forgings.

e. Electric-welded Grade P110 pipe and Grade Q125 casing shall be provided only when the supplementary requirement in API-Specification 5CT Section A.5 (SR11) is specified on the purchase agreement.

f. Grade Q125 upset casing shall be provided only when the supplementary requirement in API-Specification 5CT Section A.4 (SR10) is specified on the purchase agreement.

Further information about heat treatment requirements for the manufacturing process of various grades of pipe steel, as well as straightening methods and traceability requirements are further described in API-Specification 5CT Sections 6.2 through 6.4.2.

4.16.2.2 Material Requirements (Section 7, API Specification 5CT)

Tensile Properties

Product shall conform to the tensile requirements specified in Table 4.16.2. The tensile properties of upset casing and tubing, except elongation of the upset ends, shall comply with the requirements given for the pipe body. In case of dispute, the properties (except elongation) of the upset shall be determined from a tensile test specimen cut from the upset. A record of such tests shall be available to the purchaser.

Yield Strength

The yield strength shall be the tensile stress required to produce the elongation under load specified in Table 4.16.2 as determined by an extensometer.

Tensile Tests

Tensile properties shall be determined by tests on longitudinal specimens conforming to the requirements of Paragraph 10.4 in API Specification 5CT, and ASTM A370: Mechanical Testing of Steel Products, Supplement II, Steel Tubular Products.

Further Inspection and Testing

Requirements for the manufacturing process are described in API Specification 5CT, Sections 7 and 10, and includes such as chemical composition, flattening test of welded pipe, hardness, impact testing and energy absorption, grain size determination, hardenability, Sulfide stress cracking test,

metallographic evaluation, hydrostatic testing, dimensional tests and more.

4.16.2.3 Dimensions, masses, tolerances (section 8, API Specification 5CT)

Labels and Sizes

In the dimensional tables from API-Spec 5CT, pipe is designated by labels, and by size (outside diameter). The outside diameter size of external-upset pipe is the outside diameter of the body of the pipe, not the upset portion.

Dimensions and Masses

Pipe shall be furnished in the sizes, wall thickness and masses (as shown in Tables 4.16.3, 4.16.4) as specified on the purchase agreement except Grades C90, T95 and Q125 which may be furnished in other sizes, masses and wall thickness as agreed between purchaser and manufacturer. All dimensions shown without tolerances are related to the basis for design and are not subject to measurement to determine acceptance or rejection of product.

Diameter

The outside diameter shall be within the tolerances specified below. For threaded pipe, the outside diameter at the threaded ends shall be such that the thread length, L_4, and the full-crest thread length, L_C, are within the tolerances and dimensions specified in API Spec 5B. For pipe furnished non-upset and plain-end and which is specified on the purchase agreement for the manufacture of pup-joints, the non-upset plain-end tolerances shall apply to the full length. (Inside diameter, d, is governed by the outside diameter and mass tolerances)

(text continued on page 4-417)

Table 4.16.2 Tensile and Hardness Requirements. Source: From API Specification 5CT, page 164.

Group	Grade	Type	Total elongation under load %	Yield strength ksi min.	max.	Tensile strength min. ksi	Hardenss[a] max. HRC	HBW/HBS	Specified wall thickness in	Allowable hardness variation[b] HRC
1	2	3	4	5	6	7	8	9	10	11
1	H40		0.5	40	80	60				
	J55		0.5	55	80	75				
	K55		0.5	55	80	95				
	N80	1	0.5	80	110	100				
	N80	Q	0.5	80	110	100				
2	M65		0.5	65	85	85	22	235		
	L80	1	0.5	80	95	95	23	241		
	L80	9Cr	0.5	80	95	95	23	241		
	L80	13Cr	0.5	80	95	95	23	241		
	C90	1.2	0.5	90	105	100	25.4	255	≤ 0.500	3.0
	C90	11	0.5	90	105	100	25.4	255	0.501 to 0.749	4.0
	C90	1.2	0.5	90	105	100	25.4	255	0.750 to 0.999	5.0
	C90	1.2	0.5	90	105	100	25.4	255	≥ 1.000	6.0
	C95		0.5	95	110	105				
	T95	1.2	0.5	95	110	105	25.4	255	≤ 0.500	3.0
	T95	1.2	0.5	95	110	105	25.4	255	0.500 to 0.749	4.0
	T95	1.2	0.5	95	110	105	25.4	255	0.750 to 0.999	5.0
	T95	1.2	0.5	95	110	105	25.4	255	≥ 1.000	6.0
3	P110		0.6	110	140	125				
4	Q125		0.65	125	150	135	b		≤ 0.500	3.0
	Q125		0.65	125	150	135	b		0.500 to 0.749	4.0
	Q125		0.65	125	150	135	b		≥ 0.750	5.0

[a] In case of dispute, laboratory Rockwell C hardness testing shall be used as the referee method.

[b] No hardness limits are specified, but the maximum variation is restricted as a manufacturing control in accordance with section 7.8 and 7.9 of API Spec. 5CT.

Table 4.16.3 Dimensions and Masses for Round Thread, Buttress Thread and Extreme-Line Casing. Source: From API Specification 5CT, page 177, 178

| | | | | | | | | Calculated mass[c] | | | | | |
| | | | | | | | | e_w mass gain or loss due to end finishing[d] lb | | | | | |
	Labels[a]	Outside diameter D in.	Nominal linear mass T&C[b,c] lb/ft	Wall thickness t in.	Inside diameter d in.	Drift diameter in.	Plain end W_{pe} lb/ft	Round thread Short	Round thread Long	Buttress thread Reg. OD	Buttress thread SCC	Extreme-line Stand'd	Extreme-line Optional
1	2	3	4	5	6	7	8	9	10	11	12	13	14
4½	9.50	4.500	9.50	0.205	4.090	3.965	9.41	4.20	—	—	—	—	—
4½	10.50	4.500	10.50	0.224	4.052	3.927	10.24	3.80	—	5.00	2.56	—	—
4½	11.60	4.500	11.60	0.250	4.000	3.875	11.36	3.40	3.80	4.60	2.16	—	—
4½	13.50	4.500	13.50	0.290	3.920	3.795	13.05	—	3.20	4.00	1.56	—	—
4½	15.10	4.500	15.10	0.337	3.826	3.701	15.00	—	2.80	3.20	0.76	—	—
5	11.50	5.000	11.50	0.220	4.560	4.435	11.24	5.40	—	—	—	—	—
5	13.00	5.000	13.00	0.253	4.494	4.369	12.84	4.80	5.80	6.60	2.42	—	—
5	15.00	5.000	15.00	0.296	4.408	4.283	14.88	4.20	5.20	5.80	1.62	4.60	—
5	18.00	5.000	18.00	0.362	4.276	4.151	17.95	—	4.20	4.40	0.22	1.40	—
5	21.40	5.000	21.40	0.437	4.126	4.001	21.32	—	2.95	2.46	-1.72	—	—
5	23.20	5.000	23.20	0.478	4.044	3.919	23.11	—	2.30	2.05	-2.09	—	—
5	24.10	5.000	24.10	0.500	4.000	3.875	24.05	—	1.95	1.24	-2.94	—	—
5½	14.00	5.500	14.00	0.244	5.012	4.887	13.71	5.40	—	—	—	—	—
5½	15.50	5.500	15.50	0.275	4.950	4.825	15.36	4.80	5.80	6.40	2.10	5.80	4.20
5½	17.00	5.500	17.00	0.304	4.892	4.767	16.89	4.40	5.40	5.80	1.50	4.80	3.20
5½	20.00	5.500	20.00	0.361	4.778	4.653	19.83	—	4.40	4.60	0.30	1.40	-0.20
5½	23.00	5.500	23.00	0.415	4.670	4.545	22.56	—	3.20	3.40	-0.90	0.00	-1.60
5½	26.80	5.500	26.80	0.500	4.500	4.375	26.72	—	—	—	—	—	—
5½	29.70	5.500	29.70	0.562	4.376	4.251	29.67	—	—	—	—	—	—
5½	32.60	5.500	32.60	0.625	4.250	4.125	32.57	—	—	—	—	—	—
5½	35.30	5.500	35.30	0.687	4.126	4.001	35.35	—	—	—	—	—	—
5½	38.00	5.500	38.00	0.750	4.000	3.875	38.08	—	—	—	—	—	—
5½	40.50	5.500	40.50	0.812	3.876	3.751	40.69	—	—	—	—	—	—
5½	43.10	5.500	43.10	0.875	3.750	3.625	43.26	—	—	—	—	—	—
6⅝	20.00	6.625	20.00	0.288	6.049	5.924	19.51	11.00	13.60	14.40	2.38	—	—
6⅝	24.00	6.625	24.00	0.352	5.921	5.796	23.60	9.60	12.00	12.60	0.58	3.40	1.80
6⅝	28.00	6.625	28.00	0.417	5.791	5.666	27.67	—	10.20	10.60	-1.42	0.20	-1.40
6⅝	32.00	6.625	32.00	0.475	5.675	5.550	31.23	—	8.80	9.00	-3.02	-1.40	-3.00

(continued)

Table 4.16.3 *(continued)*

1	2 Labels[a]	3 Outside diameter D in.	4 Nominal linear mass T&C[b,c] lb/ft	5 Wall thickness t in.	6 Inside diameter d in.	7 Drift diameter in.	8 Plain end W_{pe} lb/ft	Calculated mass[c] — e_w mass gain or loss due to end finishing[d] lb — 9 Round thread Short	10 Round thread Long	11 Buttress thread Reg. OD	12 Buttress thread SCC	13 Extreme-line Stand'd	14 Extreme-line Optional
7	17.00	7.000	17.00	0.231	6.538	6.413	16.72	10.00	—	—	—	—	—
7	20.00	7.000	20.00	0.272	6.456	6.331	19.56	9.40	—	—	—	—	—
7	23.00	7.000	23.00	0.317	6.366	6.250e	22.65	8.00	10.40	11.00	1.60	6.00	4.20
7	23.00	7.000	23.00	0.317	6.366	6.241	22.65	8.00	10.40	11.00	1.60	6.00	4.20
7	26.00	7.000	26.00	0.362	6.276	6.151	25.69	7.20	9.40	9.60	0.20	2.80	1.00
7	29.00	7.000	29.00	0.408	6.184	6.059	28.75	—	8.00	8.20	-1.20	0.60	-1.20
7	32.00	7.000	32.00	0.453	6.094	6.000e	31.70	—	6.60	6.80	-2.60	-0.60	-2.40
7	32.00	7.000	32.00	0.453	6.094	5.969	31.70	—	6.60	6.80	-2.60	-0.60	-2.40
7	35.00	7.000	35.00	0.498	6.004	5.879	34.61	—	5.60	5.60	-3.80	1.00	-1.80
7	38.00	7.000	38.00	0.540	5.920	5.795	37.29	—	4.40	4.20	-5.20	-0.20	-3.00
7	42.70	7.000	42.70	0.626	5.750	5.625	42.65	—	—	—	—	—	—
7	46.40	7.000	46.40	0.687	5.625	5.500	46.36	—	—	—	—	—	—
7	50.10	7.000	50.10	0.750	5.500	5.375	50.11	—	—	—	—	—	—
7	53.60	7.000	53.60	0.812	5.376	5.251	53.71	—	—	—	—	—	—
7	57.10	7.000	57.10	0.875	5.250	5.125	57.29	—	—	—	—	—	—
7 5/8	24.00	7.625	24.00	0.300	7.025	6.900	23.49	15.80	—	—	—	—	—
7 5/8	26.20	7.625	26.40	0.328	6.969	6.844	25.59	15.20	19.00	20.60	6.21	6.40	4.00
7 5/8	29.70	7.625	29.70	0.375	6.875	6.750	29.06	—	17.40	18.80	4.41	2.60	0.20
7 5/8	33.70	7.625	33.70	0.430	6.765	6.640	33.07	—	15.80	17.00	2.61	0.00	2.40
7 5/8	39.00	7.625	39.00	0.500	6.625	6.500	38.08	—	13.60	14.60	0.21	2.20	4.60
7 5/8	42.80	7.625	42.80	0.562	6.501	6.376	42.43	—	12.01	11.39	3.06	—	—
7 5/8	45.30	7.625	45.30	0.595	6.435	6.310	44.71	—	11.04	11.04	3.36	—	—
7 5/8	47.10	7.625	47.10	0.625	6.375	6.250	46.77	—	10.16	9.23	5.17	—	—
7 5/8	51.20	7.625	51.20	0.687	6.251	6.126	50.95	—	—	—	—	—	—
7 5/8	55.30	7.625	55.30	0.750	6.125	6.000	55.12	—	—	—	—	—	—
7 3/4	46.10	7.750	46.10	0.595	6.560	6.500e	45.54	—	—	—	—	—	—
7 3/4	46.10	7.750	46.10	0.595	6.560	6.435	45.54	—	—	—	—	—	—
8 5/8	24.00	8.625	24.00	0.264	8.097	7.972	23.60	23.60	—	—	—	—	—
8 5/8	28.00	8.625	28.00	0.304	8.017	7.892	27.04	22.20	—	—	—	—	—
8 5/8	32.00	8.625	32.00	0.352	7.921	7.875e	31.13	20.80	27.60	28.30	6.03	13.20	8.80

5/8	32.00	8.625	32.00	0.352	7.921	7.796	31.13	20.80	27.60	28.20	6.03	13.20	8.80
5/8	36.00	8.625	36.00	0.400	7.825	7.700	35.17	19.40	25.60	26.20	4.03	7.60	4.20
5/8	40.00	8.625	40.00	0.450	7.725	7.625e	39.33	—	23.80	24.20	2.03	4.00	0.60
5/8	40.00	8.625	40.00	0.450	7.725	7.600	39.33	—	23.80	24.20	2.03	4.00	0.60
5/8	44.00	8.625	44.00	0.500	7.625	7.500	43.43	—	21.80	22.20	0.03	1.60	-1.80
5/8	49.00	8.625	49.00	0.557	7.511	7.286	48.04	—	19.60	19.80	-2.37	-0.80	-4.20
5/8	32.30	9.625	32.30	0.312	9.001	8.845	31.06	24.40	—	—	—	—	—
5/8	36.00	9.625	36.00	0.352	8.921	8.765	34.89	23.00	32.00	31.00	6.48	—	—
5/8	40.00	9.625	40.00	0.395	8.835	8.750e	38.97	21.40	30.00	29.00	4.48	10.60	7.20
5/8	40.00	9.625	40.00	0.395	8.835	8.679	38.97	21.40	30.00	29.00	4.48	10.60	7.20
5/8	43.50	9.625	43.50	0.435	8.755	8.599	42.73	—	28.20	27.20	2.68	5.40	2.00
5/8	47.00	9.625	47.00	0.472	8.681	8.525	46.18	—	26.60	25.60	1.08	2.20	-1.20
5/8	53.50	9.625	53.50	0.545	8.535	8.500e	52.90	—	23.40	22.40	-2.12	-1.20	-4.60
5/8	53.50	9.625	53.50	0.545	8.535	8.379	52.90	—	23.40	22.40	-2.12	-1.20	-4.60
5/8	58.40	9.625	58.40	0.595	8.435	8.375e	57.44	—	21.50	20.13	-4.40	—	—
5/8	58.40	9.625	58.40	0.595	8.435	8.279	57.44	—	21.50	20.13	-4.40	—	—
5/8	59.40	9.625	59.40	0.609	8.407	8.251	58.70	—	—	—	—	—	—
5/8	64.90	9.625	64.90	0.672	8.281	8.125	64.32	—	—	—	—	—	—
5/8	70.30	9.625	70.30	0.734	8.157	8.001	69.76	—	—	—	—	—	—
5/8	75.60	9.625	75.60	0.797	8.031	7.875	75.21	—	—	—	—	—	—
3/4	32.75	10.750	32.75	0.279	10.192	10.036	31.23	29.00	—	—	—	—	—
3/4	40.50	10.750	40.50	0.350	10.050	9.894	38.91	26.40	—	34.40	7.21	—	—
3/4	45.50	10.750	45.50	0.400	9.950	9.875e	44.26	24.40	—	31.80	4.61	21.20	—
3/4	45.50	10.750	45.50	0.400	9.950	9.794	44.26	24.40	—	31.80	4.61	21.20	—
3/4	51.00	10.750	51.00	0.450	9.850	9.694	49.55	22.60	—	29.40	2.21	18.40	—
3/4	55.50	10.750	55.50	0.495	9.760	9.625e	54.26	20.80	—	27.00	-0.19	15.80	—
3/4	55.50	10.750	55.50	0.495	9.760	9.604	54.26	20.80	—	27.00	-0.19	15.80	—
3/4	60.70	10.750	60.70	0.545	9.660	9.504	59.45	18.80	—	24.40	—	13.00	—
3/4	65.70	10.750	65.70	0.595	9.560	9.404	64.59	16.80	—	22.00	—	—	—
3/4	73.20	10.750	73.20	0.672	9.406	9.250	72.40	—	—	—	—	—	—
3/4	79.20	10.750	79.20	0.734	9.282	9.126	78.59	—	—	—	—	—	—
3/4	85.30	10.750	85.30	0.797	9.156	9.000	84.80	—	—	—	—	—	—
3/4	42.00	11.750	42.00	0.333	11.084	11.000e	40.64	29.60	—	—	—	—	—
3/4	42.00	11.750	42.00	0.333	11.084	10.928	40.64	29.60	—	—	—	—	—

(continued)

Table 4.16.3 (continued)

1	2 Labels[a]	3 Outside diameter D in.	4 Nominal linear mass T&C[b,c] lb/ft	5 Wall thickness t in.	6 Inside diameter d in.	7 Drift diameter in.	8 Plain end W_{pe} lb/ft	9 Round thread Short	10 Round thread Long	11 Buttress thread Reg. OD	12 Buttress thread SCC	13 Extreme-line Stand'd	14 Extreme-line Optional
									Calculated mass[c] e_w mass gain or loss due to end finishing[d] lb				
11 3/4	47.00	11.750	47.00	0.375	11.000	10.844	45.60	27.60	—	35.80	—	—	—
11 3/4	54.00	11.750	54.00	0.435	10.880	10.724	52.62	25.00	—	32.40	—	—	—
11 3/4	60.00	11.750	60.00	0.489	10.772	10.625[e]	58.87	22.60	—	29.60	—	—	—
11 3/4	60.00	11.750	60.00	0.489	10.772	10.616	58.87	22.60	—	29.60	—	—	—
11 3/4	65.00	11.750	65.00	0.534	10.682	10.625[e]	64.03	—	—	—	—	—	—
11 3/4	65.00	11.750	65.00	0.534	10.682	10.526	64.03	—	—	—	—	—	—
11 3/4	71.00	11.750	71.00	0.582	10.586	10.430	69.48	—	—	—	—	—	—
13 3/8	48.00	13.375	48.00	0.330	12.715	12.559	46.02	33.20	—	40.20	—	—	—
13 3/8	54.50	13.375	54.50	0.380	12.615	12.459	52.79	30.80	—	36.80	—	—	—
13 3/8	61.00	13.375	61.00	0.430	12.515	12.359	59.50	28.40	—	33.60	—	—	—
13 3/8	68.00	13.375	68.00	0.480	12.415	12.259	66.17	25.80	—	31.60	—	—	—
13 3/8	72.00	13.375	72.00	0.514	12.347	12.250[e]	70.67	24.20	—	31.60	—	—	—
13 3/8	72.00	13.375	72.00	0.514	12.347	12.191	70.67	24.20	—	—	—	—	—
16	65.00	16.000	65.00	0.375	15.250	15.062	62.64	42.60	—	—	—	—	—
16	75.00	16.000	75.00	0.438	15.124	14.936	72.86	38.20	—	45.60	—	—	—
16	84.00	16.000	84.00	0.495	15.010	14.822	82.05	34.20	—	39.60	—	—	—
16	109.00	16.000	109.00	0.656	14.688	14.500	107.60	—	—	—	—	—	—
18 5/8	87.50	18.625	87.50	0.435	17.755	17.567	84.59	73.60	—	86.40	—	—	—
20	94.00	20.000	94.00	0.438	19.124	18.936	91.59	47.00	61.20	54.80	—	—	—
20	106.50	20.000	106.50	0.500	19.000	18.812	104.23	41.60	54.80	48.40	—	—	—
20	133.00	20.000	133.00	0.635	18.730	18.542	131.45	30.00	40.60	35.20	—	—	—

See Figures 4.16.2, 4.16.3, 4.16.4 and 4.16.5.
[a] Labels are for information and assistance in ordering.
[b] Nominal linear masses, threaded and coupled (col. 4) are shown for information only.
[c] The densities of martensitic chromium steels (L80 Types 9Cr and 13Cr) are less than those of carbon steels. The masses shown are therefore not accurate for martensitic chromium steels. A mass correction factor of 0.989 may be used.
[d] Mass gain or loss due to end finishing. See formula on page 4-417.
[e] Drift diameter for most common bit size. This drift diameter shall be specified on the purchase agreement and marked on the pipe. See 8.10 in API Spec. 5CT for drift requirements.

Table 4.16.4 *Extreme-Line Casing Upset End Dimensions and Masses. Source: From API Specification 5CT, page 180.*

				Pin-and-box outside diameter (turned) −0.010 +0.020 in.		Pin Inside diameter (bored) −0.015 +0.015 in.	Box inside diameter (bored) −0 +0.030 in.	Pin-and-box made-up (power-tight)[b]			Drift dia'r for finished upset member in.	Drift diameter for full length drifting min. in.
Labels		Outside diameter in.	Nominal linear mass, upset and threaded lb/ft					Outside diameter −0.010 +0.020 in.		Inside diameter −0.015 +0.015 in		
				Standard	Optional	Standard and optional	Standard and optional	Standard[c]	Optional[c]	Standard and optional	Standard and optional	Standard and optional
		D		M	M	B	D					
1	2											
1	2	3	4	5	6	7	8	9	10	11	12	13
5	15.00	5.000	15.00	5.360	—	4.208	4.235	5.360	—	4.198	4.183	4.151
5	18.00	5.000	18.00	5.360	—	4.208	4.235	5.360	—	4.198	4.183	4.151
5$\frac{1}{2}$	15.50	5.500	15.50	5.860	5.780	4.746	4.773	5.860	5.780	4.736	4.721	4.653
5$\frac{1}{2}$	17.00	5.500	17.00	5.860	5.780	4.711	4.738	5.860	5.780	4.701	4.686	4.65
5$\frac{1}{2}$	20.00	5.500	20.00	5.860	5.780	4.711	4.738	5.860	5.780	4.701	4.686	4.653
5$\frac{1}{2}$	23.00	5.500	23.00	5.860	5.780	4.619	4.647	5.860	5.780	4.610	4.595	4.545
6$\frac{5}{8}$	24.00	6.625	24.00	7.000	6.930	5.792	5.818	7.000	6.930	5.781	5.766	5.730
6$\frac{5}{8}$	28.00	6.625	28.00	7.000	6.930	5.741	5.768	7.000	6.930	5.731	5.716	5.666
6$\frac{5}{8}$	32.00	6.625	32.00	7.000	6.930	5.624	5.652	7.000	6.930	5.615	5.600	5.550
7	23.00	7.000	23.00	7.390	7.310	6.182	6.208	7.390	7.310	6.171	6.156	6.151
7	26.00	7.000	26.00	7.390	7.310	6.182	6.208	7.390	7.310	6.171	5.156	6.151
7	29.00	7.000	29.00	7.390	7.310	6.134	6.160	7.390	7.310	6.123	6.108	6.059
7	32.00	7.000	32.00	7.390	7.310	6.042	6.069	7.390	7.310	6.032	6.017	5.969
7	35.00	7.000	35.00	7.530	7.390	5.949	5.977	7.530	7.390	5.940	5.925	5.879
7	38.00	7.000	38.00	7.530	7.390	5.869	5.897	7.530	7.390	5.860	5.845	5.795
7$\frac{5}{8}$	26.40	7.625	26.40	8.010	7.920	6.782	6.807	8.010	7.920	6.770	6.755	6.750
7$\frac{5}{8}$	29.70	7.625	29.70	8.010	7.920	6.782	6.807	8.010	7.920	6.770	6.755	6.750
7$\frac{5}{8}$	33.70	7.625	33.70	8.010	7.920	6.716	6.742	8.010	7.920	6.705	6.690	6.640

(continued)

Table 4.16.4 *(continued)*

Labels		Outside diameter in.	Nominal linear mass, upset and threaded lb/ft	Finished pin-and-box dimensions[a]								
				Pin-and-box outside diameter (turned) −0.010 +0.020 in.		Pin Inside diameter (bored) −0.015 +0.015 in.	Box inside diameter (bored) −0 +0.030 in.	Pin-and-box made-up (power-tight)[b]			Drift dia'r for finished upset member in.	Drift diameter for full length drifting min. in.
				M	M	B	D	Outside diameter −0.010 +0.020 in.		Inside diameter −0.015 +0.015 in.		
		D		Standard	Optional	Standard and optional	Standard and optional	Standard[c]	Optional[c]	Standard and optional	Standard and optional	Standard and optional
1	2	3	4	5	6	7	8	9	10	11	12	13
7 5/8	39.00	7.625	39.00	8.010	7.920	6.575	6.602	8.010	7.920	6.565	6.550	6.500
8 5/8	32.00	8.625	32.00	9.120	9.030	7.737	7.762	9.120	9.030	7.725	7.710	7.700
8 5/8	36.00	8.625	36.00	9.120	9.030	7.737	7.762	9.120	9.030	7.725	7.710	7.700
8 5/8	40.00	8.625	40.00	9.120	9.030	7.674	7.700	9.120	9.030	7.663	7.648	7.600
8 5/8	44.00	8.625	44.00	9.120	9.030	7.575	7.602	9.120	9.030	7.565	7.550	7.500
8 5/8	49.00	8.625	49.00	9.120	9.030	7.460	7.488	9.120	9.030	7.451	7.436	7.386
9 5/8	40.00	9.625	40.00	10.100	10.020	8.677	8.702	10.020	10.020	8.665	8.650	8.599
9 5/8	43.50	9.625	43.50	10.100	10.020	8.677	8.702	10.020	10.020	8.665	8.650	8.599
9 5/8	47.00	9.625	47.00	10.100	10.020	8.633	8.658	10.020	10.020	8.621	8.606	8.525
9 5/8	53.50	9.625	53.50	10.100	10.020	8.485	8.512	10.020	10.020	8.475	8.460	8.379
10 3/4	45.50	10.750	45.50	11.460	—	9.829	9.854	11.460	—	9.819	9.804	9.794
10 3/4	51.00	10.750	51.00	11.460	—	9.729	9.754	11.460	—	9.719	9.704	9.694
10 3/4	55.50	10.750	55.50	11.460	—	9.639	9.664	11.460	—	9.629	9.614	9.604
10 3/4	60.70	10.750	60.70	11.460	—	9.539	9.564	11.460	—	9.529	9.514	9.504

See Table 4.16.3 and Figure 4.16.5.

[a] Labels are for information and assistance in ordering.

[b] Nominal linear masses, threaded and coupled (col. 4) are shown for information only.

[c] The densities of martensitic chromium steels (L80 Types 9Cr and 13Cr) are less than those of carbon steels. The masses shown are therefore not accurate for martensitic chromium steels. A mass correction factor of 0.989 may be used.

(text continued from page 4-410)

Tolerances

The following tolerances apply to the outside diameter, D, of pipe:

Label 1	Tolerance on outside diameter, D
< 4−1/2	±0.031 in
≥ 4−1/2	+1% D to − 0,5% D

For upset pipe, the following tolerances apply to the outside diameter of the pipe body immediately behind the upset for a distance of approximately 5.0 in. for sizes 5-1/2 in. and smaller, and a distance approximately equal to the outside diameter for sizes larger than 5-1/2 in. Measurements shall be made with calipers or snap gauges.

Label 1	Tolerance on outside diameter, m_{eu} or L_0
≤3−1/2	+3/32 in. to −1/32 in.
>3−1/2 to ≤5	+7/64 in. to −0.75% D
>5 to ≤8−5/8	+1/8 in. to −0.75% D
>8−5/8	+5/32 in. to −0.75% D

Wall Thickness

Each length of pipe shall be measured for conformance to wall thickness requirements. The wall thickness at any place shall not be less than the tabulated thickness, t, minus the permissible under-tolerance of 12.5%. Wall thickness measurements shall be made with a mechanical caliper, a go/no-go gauge or with a properly calibrated nondestructive testing device of appropriate accuracy. In case of dispute, the measurements determined by the mechanical caliper shall govern.

Mass

Each length of casing shall be weighed separately. The pipe may be weighed plain-end, upset, non-upset, threaded, or threaded and coupled. Threaded-and-coupled pipe may be weighed with the couplings screwed on or without couplings, provided proper allowance is made for the mass of the coupling. Threaded-and-coupled pipe, integral-joint pipe, and pipe shipped without couplings shall be weighed with or without thread protectors if proper allowances are made for the mass of the thread protectors. The masses determined as described above shall conform to the calculated masses as specified herein (or adjusted calculated masses) for the end finish specified on the purchase agreement, within the following stipulated mass tolerances:

Amount	Tolerance
Single lengths	+6,5% to −3,5%
Carload 18 144 kg (40,000 lb) or more	−1,75%
Carload less than 18 144 kg (40,000 lb)	−3,5%
Order items 18 144 kg (40,000 lb) or more	−1,75%
Order item less than 18 144 kg (40,000 lb)	−3,5%

The calculated masses shall be determined in accordance with the following formula:

$$m_L = (m_{pe}L) + e_m$$

where

m_L is the calculated mass of a piece of pipe of length L, in pounds (kilograms)

m_{pe} in the plain-end mass, in pounds per foot (kilograms per meter)

L length of pipe, including end finish (see note on length determination) in ft (meters)

e_m is the mass gain or loss due to end-finishing, in pounds (kilograms). For plain-end non-upset pipe, $e_m = 0$

Note: The densities of martensitic chromium steels (L80 Types 9Cr and 13Cr) are less than those of carbon steels. The masses shown are therefore not accurate for martensitic chromium steels. A mass correction factor of 0,989 may be used.

Length

Pipe shall be furnished in range lengths conforming to Table 4.16.5, as specified on the purchase order. When pipe is furnished with threads and couplings, the length shall be measured to the outer face of the coupling, or if measured without couplings proper allowance shall be made to include the length of coupling. The extreme-line casing and integral joint tubing lengths shall be measured to the outer face of the box end. For pup joints and connectors, the length shall be measured from end to end.

Casing Jointers

If so specified on the purchase order for round-thread casing only, jointers (two pieces coupled to make a standard length) may be furnished to a maximum of 5% of the order, but no length used in making a jointer shall be less than 5 ft.

Coupling

API standards established three types of threaded joints:

1. Coupling joints with rounded thread (Figures 4.16.2 and 4.16.3) (long or short)
2. Coupling joints with asymmetrical trapezoidal thread buttress (Figure 4.16.4)
3. Extreme-line casing with trapezoidal thread without coupling (Figure 4.16.5)

There are also many non-API joints, like Hydril "CTS," Hydril "Super FJ-P," Armco SEAL-LOC, Mannesmann metal-to-metal seal casing and others.

The following are excerpts both from API Specification 5CT, Seventh Edition, October 2001 and API RP 5B1, Fifth Edition, October 1999.

4.16.2.4 Elements of Threads

Threaded connections are complicated mechanisms consisting of many elements that must interact in prescribed fashion to perform a useful function. Each of these elements of a thread may be gauged individually as described in API RP 5B1, 5th Edition, October 1999. The thread elements are defined as

1. *Thread height or depth.* The thread height or depth is the distance between the threaded crest and the thread root normal to the axis of the thread (Figure 4.16.6).
2. *Lead.* For pipe thread inspection purposes, lead is defined as the distance from a point on a thread to a corresponding point on the adjacent thread measured parallel to the thread axis (Figure 4.16.6).
3. *Taper.* Taper is the change of diameter of a thread expressed in in./ft of thread length.

Round Threads (Figure 4.16.6)

The purpose of round top (crest) and round bottom (root) is

A. To improve the resistance of the threads from galling in make-up.
B. To provide a controlled clearance between make-up thread crest and root for foreign particles or contaminants.

Table 4.16.5 *Round-Thread Casing Coupling Dimensions, Masses and Tolerances Source: From API Specification 5CT, page 184.*

| Label 1 | Size[a] Outside diameter | Outside diameter | Minimum length in | | Diameter of recess | Width of bearing face | Mass lb | |
	D in.	W in.	Short N_L	Long N_L	Q^b in.	b in.	Short	Long
1	2	3	4	5	6	7	8	9
$4\frac{1}{2}$	4.500	5.000	$6\frac{1}{4}$	7	$4\frac{19}{32}$	$\frac{5}{32}$	7.98	9.16
5	5.000	5.563	$6\frac{1}{2}$	$7\frac{3}{4}$	$5\frac{3}{32}$	$\frac{3}{16}$	10.27	12.68
$5\frac{1}{2}$	5.500	6.050	$6\frac{3}{4}$	8	$5\frac{19}{32}$	$\frac{1}{8}$	11.54	14.15
$6\frac{5}{8}$	6.625	7.390	$7\frac{1}{4}$	$8\frac{3}{4}$	$6\frac{23}{32}$	$\frac{1}{4}$	20.11	25.01
7	7.000	7.656	$7\frac{1}{4}$	9	$7\frac{3}{32}$	$\frac{3}{16}$	18.49	23.87
$7\frac{5}{8}$	7.625	8.500	$7\frac{1}{2}$	$9\frac{1}{4}$	$7\frac{25}{32}$	$\frac{7}{32}$	27.11	34.46
$8\frac{5}{8}$	8.625	9.625	$7\frac{3}{4}$	10	$8\frac{25}{32}$	$\frac{1}{4}$	35.79	47.77
$9\frac{5}{8}$	9.625	10.625	$7\frac{3}{4}$	$10\frac{1}{2}$	$9\frac{25}{32}$	$\frac{1}{4}$	39.75	56.11
$10\frac{3}{4}$	10.750	11.750	8	—	$10\frac{29}{32}$	$\frac{1}{4}$	45.81	—
$11\frac{3}{4}$	11.750	12.750	8	—	$11\frac{29}{32}$	$\frac{1}{4}$	49.91	—
$13\frac{3}{8}$	13.375	14.375	8	—	$13\frac{17}{32}$	$\frac{7}{32}$	56.57	—
16	16.000	17.000	9	—	$16\frac{7}{32}$	$\frac{7}{32}$	76.96	—
$18\frac{5}{8}$	18.625	20.000	9	—	$18\frac{27}{32}$	$\frac{7}{32}$	119.07	—
20	20.000	21.000	9	$11\frac{1}{2}$	$20\frac{7}{32}$	$\frac{7}{32}$	95.73	126.87

See also Figures 4.16.2 and 4.16.5.
Tolerance on outside diameter W, ± 1% but not greater than ± 1/8 in. Groups 1, 2 and 3.
Tolerance on outside diameter W, ± 1% but not greater than ± 1/8 in., Group 4.
[a] The size designation for the coupling is the same as the size designation for the pipe on which the coupling is used.
[b] Tolerance on diameter of recess, Q, for all groups is 0 to + 0.031 in.

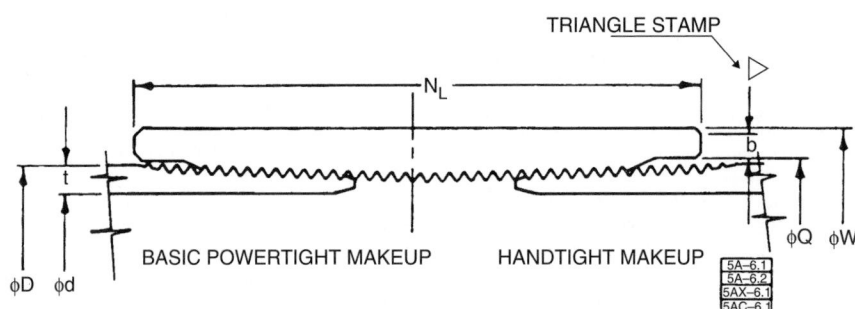

Figure 4.16.2 *D.1—Short round thread casing and coupling.*
From API Spec 5CT 7th Edition, October 2001, p. 136.
See Table 4.16.3 for pipe dimensions, Table 4.16.5 for coupling dimensions and API Spec 5B for L4.

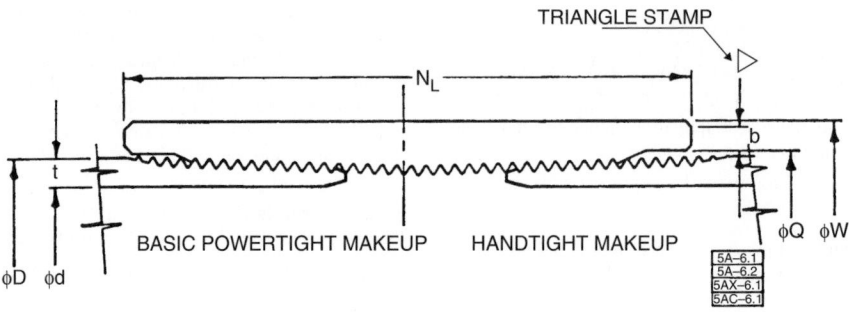

Figure 4.16.3 *Long round thread casing and coupling.*
From API Spec 5CT 7th Edition, October 2001, p. 136.
See Table 4.16.3 for pipe dimensions, Table 4.16.5 for coupling dimensions and API Spec 5B for L4.

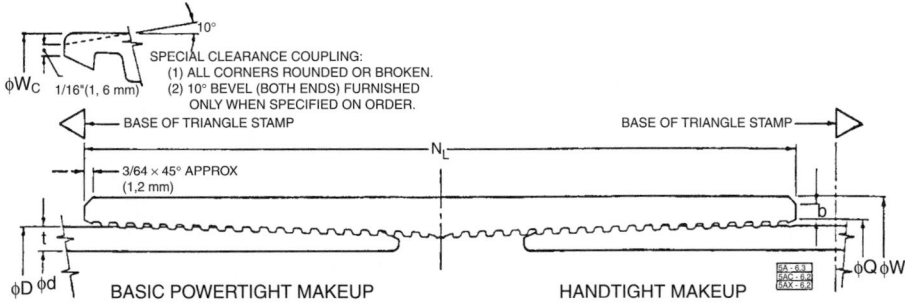

Figure 4.16.4 *Buttress thread casing and coupling.*
From API Spec 5CT 7th Edition, October 2001, p. 137.
See Table 4.16.3 for pipe dimensions, Table 4.16.6 for coupling dimensions and
API Spec 5B for L4.

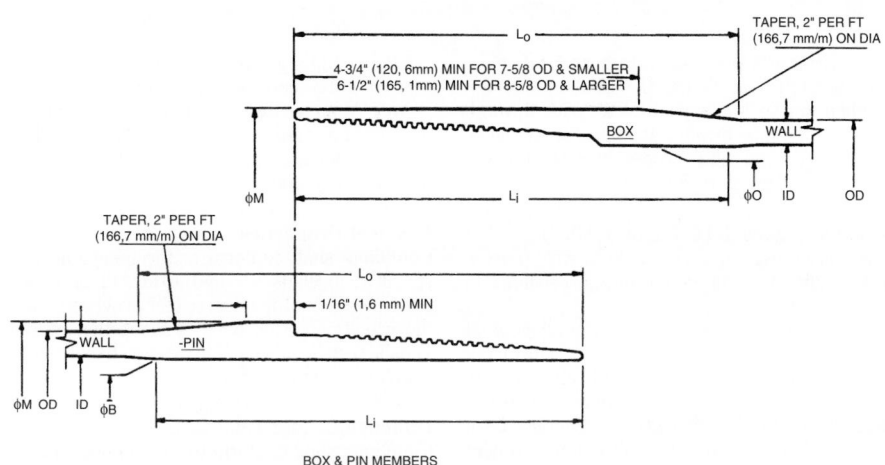

BOX & PIN MEMBERS

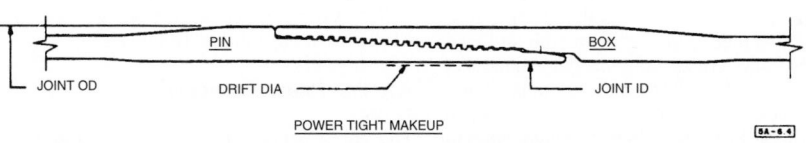

POWER TIGHT MAKEUP

Size	Length of Upset								
	in.	Pin† Min.		in.	Box† Min.		in.	Pin or Box Max	
in. OD	in.	L_i	mm	in.	L_i	mm	in.	L_o	mm
5	$6\frac{5}{8}$		168,3	7		177,8	8		203,2
$5\frac{1}{2}$	$6\frac{5}{8}$		168,3	7		177,8	8		203,2
$6\frac{5}{8}$	$6\frac{5}{8}$		168,3	7		177,8	8		203,2
7	$6\frac{5}{8}$		168,3	7		177,8	8*		203,2*
$7\frac{5}{8}$	$6\frac{5}{8}$		168,3	7		177,8	8		203,2
$8\frac{5}{8}$	8		203,3	$8\frac{3}{4}$		222,2	11		279,4
$9\frac{5}{8}$	8		203,2	$8\frac{3}{4}$		222,2	11		279,4
$10\frac{3}{4}$	8		203,2	$8\frac{3}{4}$		222,2	$12\frac{3}{4}$		323,8

†L_i is the minimum length from end of pipe of the machined diameter B on pin, or machined diameter D plus length of
thread on box, to the beginning of the internal upset runout.

*L_o shall be 9 in. (228,6 mm) max. for 7 in. — 35 lb/ft and 7 in. — 38 lb/ft casing.

Figure 4.16.5 *Extreme-line casing.*
Taken from API Spec 5CT Seventh Edition, October 201, p. 141, 142.
See Table 4.16.4 for pipe dimensions and API Spec 5B for thread details.

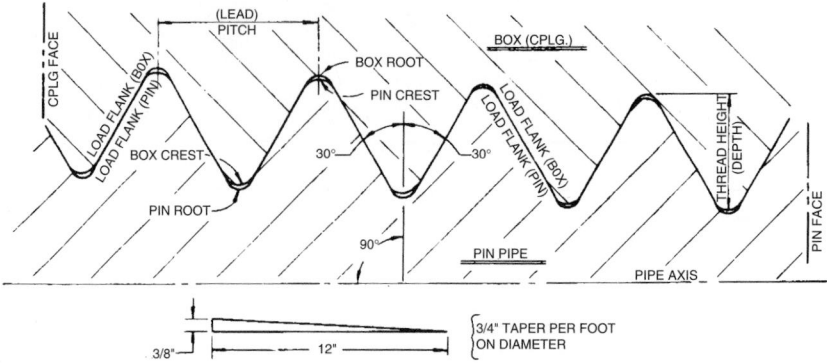

Figure 4.16.6 *Round thread casing and tubing thread configuration. From API RP 5B1, 5th Edition, October 199, p. 6.*

C. To make the crest less susceptible to harmful damage from minor scratches or dents. If sufficient interference is applied during makeup, the leak path through the connection should be through the annular clearance between mated crest and roots. Proper thread compound is necessary to ensure leak resistance.

Buttress Threads (Figure 4.16.7 and 4.16.8)
Buttress threads are designed to resists high axial tension or compression loading in addition to offering resistance to leakage.

The 3° load flank offers resistance to disengagement under high axial tension loading, while the 10° stub flank offers resistance to high axial compression loading. In any event, leak resistance is again accomplished with use of proper thread compound and/or thread coating agents. Leak resistance is controlled by proper assembly (interference) within the perfect thread only.

4.16.2.5 Extreme-Line Casing (Integral Connection)
Extreme-line casing in all sizes uses a modified acme type thread having a 12° included angle between stub and load flanks, and all threads have crests and roots flat and parallel to the axis (Figure 4.16.9). For all sizes, the threads are not intended to be leak resistant when made up. Threads are used purely as a mechanical means to hold the joint members together during axial tension loading. The connection uses upset pipe ends for pin and box members that are an integral part of the pipe body. Axial compression load resistance is primarily offered by external shouldering to the connection or makeup.

Leak resistance is obtained on makeup by interference of metal-to-metal seal between a long radius curved seal surface on the pin member engaging a conical metal seal surface of the box member (Figures 4.16.5 and 4.16.9).

Thread compound is not necessarily a critical agent to ensure leak resistance, but instead is used primarily as an antigalling or antiseizure agent.

Material
Couplings for pipe (both casing and tubing) shall be seamless and, unless otherwise specified on the purchase order, shall be of the same grade as the pipe. Exceptions are stipulated in Sections 9.2 and 9.3 in API Spec 5CT, for grades H-40, J-55, K-55, M65, N-80, and P-110, where couplings of stipulated higher grades are acceptable when specified on the purchase agreement.

When couplings are electroplated, the electroplating process should be controlled to minimize hydrogen absorption.

Note: Most buttress thread couplings will not develop the highest minimum joint strength unless couplings of the next higher order are specified (see API Specification 5CT for more detailed information).

Physical Properties
Couplings shall conform to the mechanical requirements specified in clauses 7 and 10 in API Spec 5CT, including the frequency of testing, re-test provision, etc. A record of these tests shall be open to inspection by the purchaser. Tensile tests shall be made on each heat of steel from which couplings are produced.

Dimensions and Tolerances
Couplings shall conform to the dimensions and tolerances shown in Tables 4.16.5 and 4.16.6. Unless otherwise specified, threaded and coupled casing and tubing shall be furnished with regular couplings.

Note: Couplings inspection procedures are described by API RP 5B1, 5th Edition, October 1999.

4.16.2.6 Thread Protectors
Design
The entity performing the threading shall apply external and internal thread protectors of such design, material and mechanical strength to protect the thread and end of the pipe from damage under normal handling and transportation. External thread protectors shall cover the full length of the thread on the pipe, and internal thread protectors shall cover the equivalent total pipe thread length of the internal thread. Thread protectors shall be of such design and material to inhibit infiltration of dust and water to the threads during transportation and normal storage period. Normal storage period shall be considered as approximately one year. The thread forms in protectors shall be such that the product threads are not damaged by the protectors. Thread protectors are not required for pup-joints and accessories provided they are packaged in a manner that protects the threads.

Material
Protector material shall contain no compounds capable of causing corrosion or promoting adherence of the protectors to the threads and shall be suitable for service temperatures from $-50°F$ to $+150°F$ $(-46°C$ to $+66°C)$.

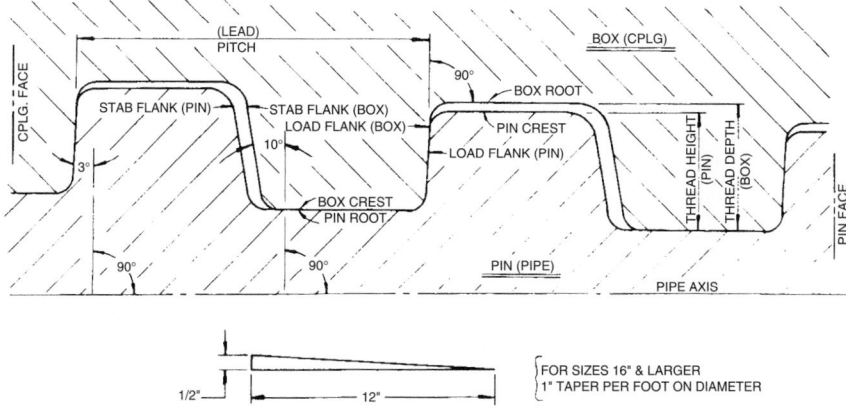

Figure 4.16.7 *Buttress thread configuration for 16-in. OD and larger casing. From API RP 5B1, 5th Edition, October 199, p. 7.*

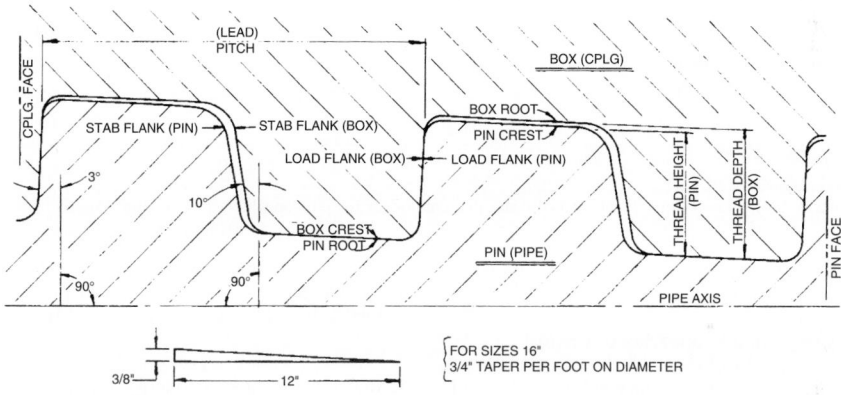

Figure 4.16.8 *Buttress thread configuration for 13 3/8-in. OD and smaller casing. From API RP 5B1, 5th Edition, October 199, p. 7.*

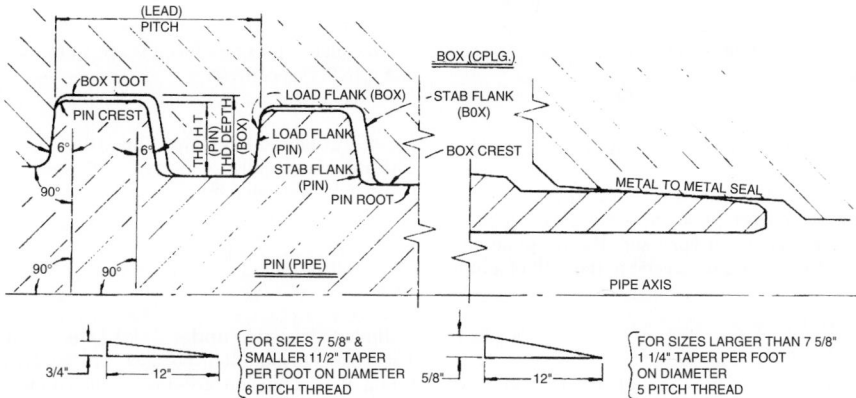

Figure 4.16.9 *Extreme-line casing thread configuration. From API RP 5B1, 5th Edition, October 199, p. 7.*

Table 4.16.6 *Buttress-Thread Casing Coupling Dimensions, Masses and Tolerances. Source: From API Specification 5CT, page 185.*

Label 1	Size[a] Outside diameter D in.	Outside diameter in Regular W in.	Special clearance W_C in.	Minimum length N_L in.	Diameter of counterbore Q^b in.	Width of bearing face B in.	Mass lb Regular	Special clearance
1	2	3	4	5	6	7	8	9
$4\frac{1}{2}$	4.500	5.000	4.875	$8\frac{7}{8}$	4.640	$\frac{1}{8}$	10.12	7.68
5	5.000	5.563	5.375	$9\frac{1}{8}$	5.140	$\frac{5}{32}$	13.00	8.82
$5\frac{1}{2}$	5.500	6.050	5.875	$9\frac{1}{4}$	5.640	$\frac{5}{32}$	14.15	9.85
$6\frac{5}{8}$	6.625	7.390	7.000	$9\frac{5}{8}$	6.765	$\frac{1}{4}$	24.49	12.46
7	7.000	7.656	7.375	10	7.140	$\frac{7}{32}$	23.24	13.84
$7\frac{5}{8}$	7.625	8.500	8.125	$10\frac{3}{8}$	7.765	$\frac{5}{16}$	34.88	20.47
$8\frac{5}{8}$	8.625	9.625	9.125	$10\frac{5}{8}$	8.765	$\frac{3}{8}$	45.99	23.80
$9\frac{5}{8}$	9.625	10.625	10.125	$10\frac{5}{8}$	9.765	$\frac{3}{8}$	51.05	26.49
$10\frac{3}{4}$	10.750	11.750	11.250	$10\frac{5}{8}$	10.890	$\frac{3}{8}$	56.74	29.52
$11\frac{3}{4}$	11.750	12.750	—	$10\frac{5}{8}$	11.890	$\frac{3}{8}$	61.80	—
$13\frac{3}{8}$	13.375	14.375	—	$10\frac{5}{8}$	13.515	$\frac{3}{8}$	70.03	—
16	16.000	17.000	—	$10\frac{5}{8}$	16.154	$\frac{3}{8}$	88.81	—
$18\frac{5}{8}$	18.625	20.000	—	$10\frac{5}{8}$	18.779	$\frac{3}{8}$	138.18	—
20	20.000	21.000	—	$10\frac{5}{8}$	20.154	$\frac{3}{8}$	110.45	—

See also Figures 4.16.6.
Tolerance on outside diameter W, ± 1%
[a] The size designation for the coupling is the same as the size designation for the pipe on which the coupling is used.

Note.   Bare steel protectors shall not be used on Grade L80 Types 9Cr and 13Cr tubulars.

Minimum Performance Properties of Casing
Results of years of field experience have revealed that to reduce the risk of failure, the minimum yield strength should be used instead of average yield strength to determine the performance properties of casing.

Values for collapse resistance, internal yield pressure, pipe body, and joint strength for steel grades as in Table 4.16.2 are given in Table 4.16.7. Table 4.16.7 is directly taken from API Bulletin 5C2, 21st Edition, October 1999. Formulas and procedures for calculating the values in Table 4.16.8 are given in API Bulletin 5C3 6th Edition, October 1, 1994 [152] and are as follows:

Collapse Pressure
(Section 2 of API Bulletin 5C3, 6th edition, October 1, 1994).

Yield Strength Collapse Pressure Formula
The yield strength collapse pressure is not a true collapse pressure but rather the external pressure P_{yp} that generates minimum yield stress Y_p on the inside of the wall of a tube as calculated by

$$P_{yp} = 2Y_p \left[\frac{(D/t) - 1}{(D/t)^2} \right] \qquad [4.16.1]$$

The applicable D/t ratios for field strength collapse pressure are shown in Table 4.16.8.

Plastic Collapse Pressure Formula
The minimum collapse pressure for the plastic range of collapse is calculated by

$$P_p = Y_p \left[\frac{A}{D/t} - B \right] - C \qquad [4.16.2]$$

The formula for minimum plastic collapse pressure is applicable for D/t values as shown in Table 4.16.8. The factors A, B and C are given in Table 4.16.9.

Transition Collapse Pressure Formula
The minimum collapse pressure for the plastic to elastic transition zone P_T is calculated by

$$P_T = Y_p \left| \frac{F}{D/t} - G \right| \qquad [4.16.3]$$

The factors F and G and applicable D/t range for the transition collapse pressure formula are shown in Tables 4.16.8 and 4.16.9, respectively.

Elastic Collapse Pressure Formula
The minimum collapse pressure of the elastic range of collapse is calculated by

$$P_E = \frac{46.95 \times 10^6}{(D/t)((D/t) - 1)^2} \qquad [4.16.4]$$

Collpase Pressure under Axial Tension Stress
The reduced minimum collapse pressure caused by the action of axial tension stress is calculated by

$$P_{CA} = P_{CO} \left[\sqrt{1 - 0.75[(S_A + P_i)/Y_p]^2} \right] - 0.5(S_A + P_i/Y_p) \qquad [4.16.5]$$

Equation 4.16.5 is not applicable if P_{CO} is calculated from the elastic collapse formula.

(text continued on page 4-429)

Table 4.16.7 Minimum Performance of Casing

Column key (1–27):
1 Size: Outside Diameter in. D · 2 Nominal Weight, Threads and Coupling lb per ft · 3 Grade · 4 Wall Thickness in. t · 5 Inside Diameter in. d · 6 Drift Diameter in. · 7 Outside Diameter of Coupling in. W · 8 Outside Diameter Special Clearance Coupling in. W_e · 9 Drift Diameter in. (Extreme Line) · 10 Outside Diameter of Box Power-tight in. M (Extreme Line) · 11 Collapse Resistance psi · 12 Pipe Body Yield Strength 1000 lbs · Internal Yield Pressure at Minimum Yield, psi: 13 Plain End or Extreme Line, Round Thread [14 Short, 15 Long], Buttress Thread Regular Coupling [16 Same Grade, 17 Higher Grade], Special Clearance Coupling [18 Same Grade, 19 Higher Grade] · ★ Joint Strength — 1,000 lbs (Threaded and Coupled): Round Thread [20 Short, 21 Long], Buttress Thread [22 Regular Coupling, 23 Regular Coupling Higher Grade, 24 Special Clearance Coupling, 25 Special Clearance Coupling Higher Grade], Extreme Line [26 Standard Joint, 27 Optional Joint]

1	2	3	4	5	6	7	8	9	10	11	12	13	14	15	16	17	18	19	20	21	22	23	24	25	26	27
4½	9.50	H-40	0.205	4.090	3.965	5.000	—	—	—	2,700	111	3,190	3,190	—	—	—	—	—	77	—	—	—	—	—	—	—
	9.50	J-55	0.205	4.090	3.965	5.000	—	—	—	3,310	152	4,380	4,380	—	—	—	—	—	101	—	—	—	—	—	—	—
	10.50	J-55	0.224	4.052	3.927	5.000	4.875	—	—	4,010	163	4,790	4,790	4,790	4,790	4,790	4,790	4,790	132	162	203	203	203	203	—	—
	11.60	J-55	0.250	4.000	3.875	5.000	4.875	—	—	4,960	184	5,350	5,350	5,350	5,350	5,350	5,350	5,350	154	180	225	225	225	225	—	—
	9.50	K-55	0.205	4.090	3.965	5.000	—	—	—	3,310	152	4,380	4,380	—	—	—	—	—	112	—	—	—	—	—	—	—
	10.50	K-55	0.224	4.052	3.927	5.000	4.875	—	—	4,010	165	4,790	4,790	4,790	4,790	4,790	4,790	4,790	146	180	249	249	249	249	—	—
	11.60	K-55	0.250	4.000	3.875	5.000	4.875	—	—	4,960	184	5,350	5,350	5,350	5,350	5,350	5,350	5,350	170	201	277	277	277	277	—	—
	11.60	C-75	0.250	4.000	3.875	5.000	4.875	—	—	6,130	250	7,290	—	7,290	7,290	7,290	7,290	7,290	—	212	288	288	288	288	—	—
	13.50	C-75	0.290	3.920	3.795	5.000	4.875	—	—	8,170	288	8,460	—	8,460	8,460	8,460	7,490	7,490	—	257	331	320	320	320	—	—
	11.60	L-80	0.250	4.000	3.875	5.000	4.875	—	—	6,350	267	7,780	—	7,780	7,780	7,780	7,780	7,780	—	223	291	—	291	—	—	—
	13.50	L-80	0.290	3.920	3.795	5.000	4.875	—	—	8,540	307	9,020	—	9,020	9,020	9,020	7,990	9,020	—	270	334	—	320	—	—	—
	11.60	N-80	0.250	4.000	3.875	5.000	4.875	—	—	6,350	267	7,780	—	7,780	7,780	7,780	7,780	7,780	—	223	304	304	304	304	—	—
	13.50	N-80	0.290	3.920	3.795	5.000	4.875	—	—	8,540	307	9,020	—	9,020	9,020	9,020	7,990	9,020	—	270	349	349	337	349	—	—
	11.60	C-95	0.250	4.000	3.875	5.000	4.875	—	—	7,010	317	9,240	—	9,240	9,240	9,240	9,240	9,240	—	234	325	325	325	325	—	—
	13.50	C-95	0.290	3.920	3.795	5.000	4.875	—	—	9,630	364	10,710	—	10,710	10,710	10,710	9,490	9,490	—	284	374	374	353	349	—	—
	11.60	P-110	0.250	4.000	3.875	5.000	4.875	—	—	7,560	367	10,690	—	10,690	10,690	10,690	10,690	10,690	—	279	385	385	385	385	—	—
	13.50	P-110	0.290	3.920	3.795	5.000	4.875	—	—	10,670	422	12,410	—	12,410	12,410	12,410	10,990	12,410	—	338	443	443	421	443	—	—
	15.10	P-110	0.337	3.826	3.701	5.000	4.875	—	—	14,320	485	14,420	—	14,420	13,460	14,420	10,990	14,310	—	406	509	509	421	509	—	—
5	11.50	J-55	0.220	4.560	4.435	5.563	5.375	—	—	3,060	182	4,240	4,240	—	—	—	—	—	133	—	—	—	—	—	—	—
	13.00	J-55	0.253	4.494	4.369	5.563	5.375	4.151	5.360	4,110	208	4,870	4,870	4,870	4,870	4,870	4,870	4,870	169	201	252	252	252	252	—	—
	15.00	J-55	0.296	4.408	4.283	5.563	5.375	—	—	5,550	241	5,700	5,700	5,700	5,700	5,700	5,700	5,700	207	246	293	293	287	293	328	—
	11.50	K-55	0.220	4.560	4.435	5.563	5.375	—	—	3,060	182	4,240	4,240	—	—	—	—	—	147	—	—	—	—	—	—	—
	13.00	K-55	0.253	4.494	4.369	5.563	5.375	4.151	5.360	4,140	208	4,870	4,870	4,870	4,870	4,870	4,870	4,870	186	201	309	309	309	309	—	—
	15.00	K-55	0.296	4.408	4.283	5.563	5.375	4.151	5.360	5,550	241	5,700	5,700	5,700	5,700	5,700	5,700	5,700	228	246	359	359	359	359	—	—
	15.00	C-75	0.296	4.408	4.283	5.563	5.375	4.151	5.360	6,970	328	7,770	—	7,770	7,770	7,770	7,770	—	—	295	375	—	364	—	416	—
	18.00	C-75	0.362	4.276	4.151	5.563	5.375	4.151	5.360	10,000	396	9,500	—	9,500	9,290	9,500	6,990	—	—	376	452	—	364	—	446	—
	21.40	C-75	0.437	4.126	4.001	5.563	5.375	4.151	5.360	11,060	470	11,470	—	10,140	9,290	10,140	6,990	—	—	466	510	—	364	—	—	—
	24.10	C-75	0.500	4.000	3.875	5.563	5.375	—	—	13,500	530	13,130	—	10,140	9,920	13,620	6,990	—	—	538	510	—	364	—	—	—
	15.00	L-80	0.296	4.408	4.283	5.563	5.375	4.151	5.360	7,250	350	8,290	—	8,290	8,290	8,290	7,460	8,290	—	295	379	—	364	—	416	—
	18.00	L-80	0.362	4.276	4.151	5.563	5.375	4.151	5.360	10,490	422	10,140	—	10,140	9,910	10,140	7,460	10,140	—	376	457	—	364	—	446	—
	21.40	L-80	0.437	4.126	4.001	5.563	5.375	4.151	5.360	12,760	501	12,240	—	10,810	9,910	10,140	7,460	—	—	466	510	—	364	—	—	—
	24.10	L-80	0.500	4.000	3.875	5.563	5.375	—	—	14,400	566	14,000	—	10,810	9,910	10,140	7,460	—	—	538	510	—	364	—	—	—
	15.00	N-80	0.296	4.408	4.283	5.563	5.375	4.151	5.360	7,230	350	8,290	—	8,290	8,290	8,290	7,460	8,290	—	311	396	396	383	—	437	—
	18.00	N-80	0.362	4.276	4.151	5.563	5.375	4.151	5.360	10,400	422	10,140	—	10,140	9,910	9,910	7,460	10,140	—	396	477	477	383	—	469	—
	21.40	N-80	0.437	4.126	4.001	5.563	5.375	4.151	5.360	12,760	501	12,240	—	10,810	9,910	12,240	7,460	10,250	—	490	537	566	383	—	—	—
	24.10	N-80	0.500	4.000	3.875	5.563	5.375	—	—	14,400	566	14,000	—	10,810	9,910	13,620	7,460	10,250	—	567	537	639	383	—	—	—
	15.00	C-95	0.296	4.408	4.283	5.563	5.375	4.151	5.360	8,090	415	9,840	—	9,840	9,840	9,840	8,850	—	—	326	424	—	402	—	459	—
	18.00	C-95	0.362	4.276	4.151	5.563	5.375	4.151	5.360	12,010	501	12,040	—	12,040	11,770	11,770	8,850	—	—	416	512	—	402	—	493	—
	21.40	C-95	0.437	4.126	4.001	5.563	5.375	4.151	5.360	15,150	595	14,540	—	12,840	11,770	12,240	8,850	—	—	515	563	—	402	—	—	—
	24.10	C-95	0.500	4.000	3.875	5.563	5.375	—	—	17,100	672	16,630	—	12,850	11,770	13,620	8,850	—	—	505	563	—	402	—	—	—
	15.00	P-110	0.296	4.408	5.283	5.563	5.375	4.151	5.360	8,830	481	11,400	—	11,400	11,400	11,400	10,250	11,400	—	388	503	503	479	503	547	—
	18.00	P-110	0.362	4.276	4.151	5.563	5.375	4.151	5.360	13,450	580	13,940	—	13,940	13,620	13,910	10,250	13,940	—	495	606	606	479	606	587	—
	21.40	P-110	0.437	4.126	4.001	5.563	5.375	—	—	17,550	689	16,820	—	14,870	16,820	16,820	10,250	13,980	—	613	671	720	479	613	—	—
	24.10	P-110	0.500	4.000	3.875	5.563	5.375	—	—	10,800	778	19,250	—	14,870	13,620	18,580	10,250	13,080	—	708	671	812	479	613	—	—

4-423

Table 4.16.7 (continued)

Column groups: Columns 1–12 = pipe/coupling dimensions and strengths (Threaded and Coupled / Extreme Line). Columns 13–19 = Internal Yield Pressure at Minimum Yield, psi (Round Thread, Buttress Thread, Special Clearance Coupling). Columns 20–27 = ★ Joint Strength — 1,000 lbs (Threaded and Coupled: Round Thread, Buttress Thread; Extreme Line).

Size OD in. (D)	Nominal Wt, Threads & Coupling lb/ft	Grade	Wall Thick. in. (t)	Inside Dia. in. (d)	Drift Dia. in.	OD Coupling in. (W)	OD Spec. Clearance Coupling in. (Wc)	Drift Dia. in. (Ext. Line)	OD Box Powertight in. (M)	Collapse Resist. psi	Pipe Body Yield 1000 lbs	Plain End or Extreme Line	Round Short	Round Long	Buttress Reg. Same Grade	Buttress Reg. Higher Grade‡	Spec. Clear. Same Grade	Spec. Clear. Higher Grade	Round Short	Round Long	Buttress Reg. Coupling	Buttress Reg. Higher Grade‡	Spec. Clear. Coupling	Spec. Clear. Higher Grade‡	Ext. Line Std Joint	Ext. Line Optional Joint
5½	14.00	H-40	0.244	5.012	4.887	6.050	—	—	—	2,650	161	3,110	3,110	—	—	—	—	—	130	—	—	—	—	—	—	—
	14.00	J-55	0.244	5.012	4.887	6.050	—	—	—	3,120	222	4,270	4,270	—	—	—	—	—	172	—	—	—	—	—	—	—
	15.50	J-55	0.273	4.950	4.825	6.050	5.875	4.653	5.860	4,040	248	4,810	4,810	4,810	4,810	4,810	4,730	4,810	202	217	300	300	300	300	339	339
	17.00	J-55	0.304	4.892	4.767	6.050	5.875	4.653	5.860	4,910	273	5,320	5,320	5,320	5,320	5,320	4,730	5,320	229	247	329	329	318	329	372	372
	14.00	K-55	0.244	5.012	4.887	6.050	—	—	—	3,120	222	4,270	4,270	—	—	—	—	—	189	—	—	—	—	—	—	—
	15.50	K-55	0.275	4.950	4.825	6.050	5.875	4.653	5.860	4,040	248	4,810	4,810	4,810	4,810	4,810	4,730	4,810	222	239	366	366	366	366	429	429
	17.00	K-55	0.304	4.892	4.767	6.050	5.875	4.653	5.860	4,910	273	5,320	5,320	5,320	5,320	5,320	4,730	5,320	252	272	402	402	402	402	471	471
	17.00	C-75	0.304	4.892	4.767	6.050	5.875	4.653	5.860	6,070	372	7,250	—	7,250	7,250	—	6,450	—	—	327	423	—	403	—	471	471
	20.00	C-75	0.361	4.778	4.653	6.050	5.875	4.653	5.860	8,410	437	8,610	—	8,260	8,430	—	6,450	—	—	403	497	—	403	—	497	479
	23.00	C-75	0.415	4.670	4.545	6.050	5.875	4.545	5.860	10,460	497	9,900	—	9,260	8,430	—	6,450	—	—	473	550	—	403	—	549	479
	17.00	L-80	0.304	4.892	4.767	6.050	5.875	4.653	5.860	6,280	397	7,740	—	7,740	7,740	7,740	6,880	7,190	—	338	428	446	403	446	471	471
	20.00	L-80	0.361	4.778	4.653	6.050	5.875	4.653	5.860	8,830	466	9,190	—	9,190	8,990	9,190	6,880	9,190	—	416	503	524	403	524	497	479
	23.00	L-80	0.415	4.670	4.545	6.050	5.875	4.545	5.860	11,160	530	10,560	—	9,880	10,560	10,560	6,880	9,460	—	489	550	596	403	530	549	479
	17.00	N-80	0.304	4.892	4.767	6.050	5.875	4.653	5.860	6,280	397	7,740	—	7,740	7,740	7,740	6,880	7,740	—	348	446	446	424	446	496	496
	20.00	N-80	0.361	4.778	4.653	6.050	5.875	4.653	5.860	8,830	466	9,190	—	9,190	8,990	9,190	6,880	9,190	—	428	524	524	424	524	523	504
	23.00	N-80	0.415	4.670	4.545	6.050	5.875	4.545	5.860	11,160	530	10,560	—	9,880	10,560	10,560	6,880	9,460	—	502	579	596	424	530	577	504
	17.00	C-95	0.304	4.892	4.767	6.050	5.875	4.653	5.860	8,930	471	9,190	—	9,190	9,190	—	8,170	—	—	374	480	—	445	—	521	521
	20.00	C-95	0.361	4.778	4.653	6.050	5.875	4.653	5.860	10,000	554	10,910	—	10,910	10,680	—	8,170	—	—	460	563	—	445	—	549	530
	23.00	C-95	0.415	4.670	4.545	6.050	5.875	4.545	5.860	12,920	630	12,540	—	11,730	10,680	—	8,170	—	—	540	608	—	445	—	606	530
	17.00	P-110	0.304	4.892	4.767	6.050	5.875	4.653	5.860	7,460	546	10,640	—	10,640	10,640	10,640	9,460	10,640	—	445	568	568	530	568	620	620
	20.00	P-110	0.361	4.778	4.653	6.050	5.875	4.653	5.860	11,080	641	12,640	—	12,640	12,360	12,640	9,460	12,640	—	548	667	667	530	667	654	630
	23.00	P-110	0.415	4.670	4.515	6.050	5.875	4.545	5.860	14,520	729	14,520	—	13,580	12,360	14,520	9,460	12,890	—	643	724	759	530	668	732	630
6	20.00	H-40	0.288	6.019	5.924	7.390	—	—	—	2,520	229	3,040	3,040	—	—	—	—	—	184	—	—	—	—	—	—	—
	20.00	J-55	0.288	6.019	5.924	7.234	7.000	5.730	7.000	2,970	315	4,180	4,180	4,180	4,180	4,180	4,060	4,180	245	266	374	374	374	374	—	—
	28.00	J-55	0.352	5.921	5.756	7.390	7.000	5.666	7.000	4,560	382	5,110	5,110	5,110	5,110	5,110	4,060	5,110	314	340	453	453	390	453	477	477
	20.00	K-55	0.288	6.019	5.924	7.234	7.000	5.730	7.000	2,970	315	4,180	4,180	4,180	4,180	4,180	4,060	4,180	267	290	453	453	453	453	—	—
	28.00	K-55	0.352	5.921	5.756	7.390	7.000	5.666	7.000	4,560	382	5,110	5,110	5,110	5,110	5,110	4,060	5,110	342	372	548	548	494	520	605	605
	24.00	C-75	0.352	5.921	5.796	7.390	7.000	5.730	7.000	5,570	520	6,970	—	6,970	6,970	—	5,540	—	—	453	583	—	494	—	605	605
	28.00	C-75	0.417	5.741	5.666	7.390	7.000	5.666	7.000	7,830	610	8,260	—	8,260	8,260	—	5,540	—	—	552	683	—	494	—	648	644
	32.00	C-75	0.475	5.675	5.550	7.390	7.000	5.550	7.000	9,830	688	9,410	—	9,410	9,200	—	5,540	—	—	638	771	—	494	—	717	644
	24.00	L-80	0.352	5.921	5.796	7.390	7.000	5.730	7.000	5,760	555	7,440	—	7,440	7,440	7,440	5,910	7,440	—	473	592	615	494	615	605	605
	28.00	L-80	0.417	5.741	5.666	7.390	7.000	5.666	7.000	8,170	651	8,810	—	8,810	8,810	8,810	5,910	8,120	—	576	693	721	494	650	648	644
	32.00	L-80	0.475	5.675	5.550	7.390	7.000	5.550	7.000	10,320	734	10,040	—	10,040	9,820	10,040	5,910	8,120	—	666	783	814	494	650	717	644
	24.00	N-80	0.352	5.921	5.796	7.390	7.000	5.730	7.000	5,760	555	7,440	—	7,440	7,440	7,440	5,910	7,440	—	481	615	615	520	615	637	637
	28.00	N-80	0.417	5.741	5.666	7.390	7.000	5.666	7.000	8,170	651	8,810	—	8,810	8,810	8,810	5,910	8,120	—	586	721	721	520	650	682	678
	32.00	N-80	0.475	5.675	5.550	7.390	7.000	5.550	7.000	10,320	734	10,040	—	10,040	9,820	10,040	5,910	8,120	—	677	814	814	520	650	756	678
	24.00	C-95	0.352	5.921	5.796	7.390	7.000	5.730	7.000	6,290	659	8,830	—	8,830	8,830	—	7,020	—	—	545	605	—	546	—	688	608
	28.00	C-95	0.417	5.741	5.666	7.390	7.000	5.666	7.000	9,200	773	10,460	—	10,460	10,460	—	7,020	—	—	605	780	—	546	—	716	712
	32.00	C-95	0.475	5.675	5.550	7.390	7.000	5.550	7.000	11,800	872	11,920	—	11,920	11,660	—	7,020	—	—	769	880	—	546	—	793	712
	24.00	P-110	0.352	5.921	5.796	7.390	7.000	5.730	7.000	6,710	763	10,230	—	10,230	10,230	10,230	8,120	10,230	—	641	786	786	650	786	796	796
	28.00	P-110	0.417	5.741	5.666	7.390	7.000	5.666	7.000	10,140	895	12,120	—	12,120	12,120	12,120	8,120	11,080	—	781	922	922	650	832	852	848
	32.00	P-110	0.475	5.675	5.550	7.390	7.000	5.550	7.000	13,200	1,009	13,800	—	13,800	13,500	13,800	8,120	11,080	—	904	1,040	1,040	650	832	944	848

1	2	3	4	5	6	7	8	9	10	11	12	13	14	15	16	17	18	19	20	21	22	23	24	25	26	27
																			★ Joint Strength — 1,000 lbs							
					Threaded and Coupled			Extreme Line				Internal Yield Pressure at Minimum Yield, psi							Threaded and Coupled						Extreme Line	
													Round Thread		Buttress Thread				Round Thread		Buttress Thread					
Size: Outside Diameter in. *D*	Nominal Weight, Threads and Coupling lb per ft	Grade	Wall Thickness in. *t*	Inside Diameter in. *d*	Drift Diameter in.	Outside Diameter of Coupling in. *W*	Outside Diameter Special Clearance Coupling in. W_c	Drift Diameter in.	Outside Diameter of Box Powertight in. *M*	Collapse Resistance psi	Pipe Body Yield Strength 1000 lbs	Plain End or Extreme Line	Short	Long	Regular Coupling		Special Clearance Coupling		Short	Long	Regular Coupling	Regular Coupling Higher Grade:	Special Clearance Coupling	Special Clearance Coupling Higher Grade‡	Standard Joint	Optional Joint
															Same Grade	Higher Grade	Same Grade	Higher Grade								
7	17.00	H-40	0.231	6.578	6.413	7.656	—	—	—	1,450	196	2,310	2,310	—	—	—	—	—	122	—	—	—	—	—	—	—
	20.00	H-40	0.272	6.456	6.331	7.656	—	—	—	1,980	200	2,720	2,720	—	—	—	—	—	176	—	—	—	—	—	—	—
	20.00	J-55	0.272	6.456	6.331	7.656	—	—	—	2,270	316	3,740	3,740	—	—	—	—	—	234	—	—	—	—	—	—	—
	23.00	J-55	0.317	6.346	6.241	7.656	7.375	6.151	7.390	3,270	366	4,360	4,360	4,360	4,360	4,360	3,950	4,360	284	313	432	432	421	432	499	499
	26.00	J-55	0.362	6.276	6.131	7.656	7.375	6.151	7.390	4,320	415	4,980	4,980	4,980	4,980	4,980	3,950	4,980	334	367	490	490	421	490	506	506
	20.00	K-55	0.272	6.456	6.331	7.656	—	—	—	2,270	316	3,740	3,740	—	—	—	—	—	254	—	—	—	—	—	—	—
	23.00	K-55	0.317	6.368	6.241	7.656	7.375	6.151	7.390	3,270	364	4,360	4,360	4,360	4,360	4,350	3,950	4,360	309	341	522	522	522	522	632	632
	26.00	K-55	0.362	6.276	6.135	7.656	7.375	6.151	7.390	4,320	415	4,980	4,980	4,980	4,980	4,980	3,950	4,980	364	401	592	592	533	561	641	641
	23.00	C-75	0.317	6.366	6.211	7.656	7.375	6.151	7.390	3,770	499	5,940	—	5,940	5,940	—	5,380	—	—	416	557	—	533	—	685	632
	26.00	C-75	0.362	6.276	6.151	7.656	7.375	6.151	7.390	5,230	566	6,790	—	6,790	6,790	—	8,380	—	—	489	631	—	533	—	611	641
	29.00	C-75	0.408	6.184	6.059	7.656	7.375	6.059	7.390	6,760	634	7,650	—	7,650	7,650	—	8,380	—	—	562	707	—	533	—	695	674
	32.00	C-75	0.458	6.004	5.069	7.656	7.375	5.969	7.390	8,230	699	8,490	—	8,490	7,930	—	5,380	—	—	633	779	—	533	—	681	674
	35.00	C-75	0.498	6.004	5.879	7.656	7.375	5.879	7.530	9,710	763	9,340	—	8,660	7,930	—	8,380	—	—	701	833	—	533	—	856	781
	38.00	C-75	0.540	5.920	5.795	7.656	7.375	5.795	7.530	10,680	822	10,120	—	8,660	7,930	—	8,380	—	—	767	833	—	533	—	917	781
	23.00	L-80	0.317	6.366	6.241	7.656	7.375	6.151	7.390	3,830	532	6,340	—	6,340	6,340	8,346	5,740	6,340	—	435	565	—	533	—	632	632
	26.00	L-80	0.362	6.276	6.151	7.656	7.375	6.151	7.390	5,410	604	7,240	—	7,240	7,210	7,210	5,740	7,210	—	511	641	—	533	—	641	641
	29.00	L-80	0.408	6.184	6.059	7.656	7.375	6.059	7.390	7,020	676	8,160	—	8,160	8,160	8,160	5,740	7,890	—	587	718	—	533	—	685	674
	32.00	L-80	0.483	6.094	5.368	7.656	7.375	5.969	7.390	8,600	745	9,060	—	9,060	8,160	9,060	5,740	7,890	—	661	791	—	533	—	761	674
	35.00	L-80	0.498	6.004	5.879	7.656	7.375	5.879	7.530	10,180	814	9,900	—	9,240	8,460	9,960	5,740	7,890	—	734	833	—	533	—	856	761
	38.00	L-80	0.590	5.920	5.793	7.656	7.375	6.795	7.530	11,390	877	10,800	—	9,240	8,460	10,800	5,740	7,890	—	801	833	—	533	—	917	761
	23.00	N-80	0.317	6.366	6.241	7.656	7.375	6.151	7.390	3,830	532	6,340	—	6,340	6,340	6,340	5,740	6,340	—	442	588	588	561	588	666	666
	26.00	N-80	0.362	6.276	6.151	7.656	7.375	6.151	7.390	5,410	604	7,240	—	7,240	7,240	7,240	5,740	7,240	—	519	667	667	561	667	675	675
	29.00	N-80	0.408	6.184	6.059	7.656	7.375	6.059	7.390	7,020	676	8,160	—	8,160	8,160	8,160	5,740	7,890	—	507	745	746	561	702	721	709
	32.00	N-80	0.453	6,036	5.069	7.656	7.375	5.969	7.390	8,600	745	9,060	—	9,060	8,460	9,060	5,740	7,890	—	672	823	823	561	702	801	709
	35.00	N-80	0.498	6,006	5.879	7.656	7.375	5.879	7.530	10,180	814	9,960	—	9,240	8,460	9,960	5,740	7,890	—	746	876	898	561	702	805	801
	38.00	N-80	0.540	5.920	5.795	7.656	7.375	5.795	7.530	11,330	877	10,800	—	9,240	8,460	10,800	5,740	7,890	—	814	876	968	561	702	895	801
	23.00	C-95	0.317	6.366	6.711	7.656	7.375	6.151	7.390	4,150	632	7,530	—	7,530	7,530	—	6,810	—	—	605	636	—	589	—	699	699
	26.00	C-95	0.362	6.274	6.151	7.656	7.375	6.151	7.390	5,870	717	8,600	—	8,600	8,600	—	6,810	—	—	593	722	—	589	—	709	709
	29.00	C-95	0.408	6.184	6.059	7.656	7.375	6.059	7.390	7,820	803	9,690	—	9,690	9,690	—	6,810	—	—	683	808	—	889	—	757	744
	32.00	C-95	0.453	6.094	5.969	7.656	7.375	5.959	7.390	9,730	885	10,760	—	10,760	10,050	—	6,810	—	—	768	891	—	589	—	841	744
	35.00	C-95	0.498	6.001	5.879	7.656	7.375	5.879	7.530	11,640	966	11,830	—	10,970	10,050	—	6,810	—	—	853	920	—	589	—	940	841
	38.00	C-95	0.540	5.820	5.795	7.656	7.375	5.795	7.530	13,429	1041	12,820	—	10,970	10,050	—	6,810	—	—	931	920	—	589	—	1,013	841
	26.00	P-110	0.362	6.276	6.151	7.656	7.375	6.151	7.390	6,210	830	9,960	—	9,960	9,960	9,960	7,890	9,960	—	693	853	853	702	853	844	844
	29.00	P-110	0.408	4.184	6.059	7.656	7.375	6.059	7.390	8,310	929	11,220	—	11,220	11,220	11,220	7,890	10,760	—	797	955	955	702	898	902	884
	32.00	P-110	0.453	6.091	5.969	7.656	7.375	5.959	7.390	10,760	1025	12,460	—	12,460	11,640	12,460	7,890	10,760	—	897	1,053	1,053	702	898	1,002	886
	35.00	P-110	0.498	6.004	5.879	7.656	7.375	5.879	7.530	13,010	1119	13,700	—	12,700	11,640	13,700	7,890	10,760	—	996	1,096	1,150	702	898	1,118	1,002
	38.00	P-110	0.540	8.920	5.795	7.656	7.375	5.795	7.530	15,110	1205	14,850	—	12,700	11,640	14,850	7,890	10,760	—	1,087	1,096	1,239	702	898	1,207	1,002
7⅞	24.00	H-40	0.300	7.025	6.900	8.500	—	—	—	2,040	276	2,750	2,750	—	—	—	—	—	212	—	—	—	—	—	—	—
	26.40	J-55	0.328	6.969	6.844	8.500	8.125	6.750	8.010	2,890	414	4,140	4,140	4,140	4,140	4,140	4,140	4,140	316	346	483	483	483	483	558	553
	26.40	K-55	0.328	6.969	6.844	8.500	8.125	6.750	8.010	2,890	414	4,140	4,140	4,140	4,140	4,140	4,140	4,140	342	377	581	581	581	581	700	700
	26.40	C-75	0.328	6.969	6.814	8.500	8.125	6.750	8.010	3,280	564	5.650	—	5,650	5,650	—	5,650	—	—	461	624	—	624	—	700	700
	29.70	C-75	0.375	6.875	6.750	8.500	8.125	6.750	8.010	4,670	611	6,450	—	6,450	6,450	—	6,140	—	—	542	709	—	709	—	700	700
	33.70	C-75	0.430	6.755	6.610	8.500	8.125	6.640	8.010	6,320	725	7,400	—	7,400	7,400	—	6,140	—	—	635	806	—	735	—	766	744
	39.00	C-75	0.500	6.625	6.500	8.500	8.125	6.500	8.010	8,430	839	8,610	—	8,610	8,610	—	6,140	—	—	751	929	—	735	—	851	744
	42.80	C-75	0.562	6.301	6.374	8.500	8.125	--	—	10,240	231	9,670	—	9,670	9,190	—	6,140	—	—	852	1,035	—	735	—	—	—
	47.10	C-75	0.628	6.373	6.250	8.500	8.125	—	—	11,200	1031	10,760	—	9,840	9,190	—	6,140	—	—	953	1,140	—	735	—	—	—
	26.40	L-80	0.328	6.969	6.844	8.500	8.125	6.750	8.010	3,400	602	6,020	—	6,020	6,020	6.020	6,020	6,020	—	482	635	—	635	—	700	700
	29.70	L-80	0.375	6.875	6.730	8.500	8.125	6.750	8.010	4,700	683	6,890	—	6,890	6,890	6,890	6,550	6,890	—	566	721	—	721	—	700	700
	33.70	L-80	0.400	6.765	6.620	8.500	8.125	6.610	8.010	6,560	778	7,900	—	7,900	7,900	7,900	6,550	7,900	—	664	820	—	735	—	766	744
	39.00	L-80	0.500	6.825	6.500	8.500	8.125	6.500	8.010	3.810	895	9,180	—	9,180	9,180	9,180	6,550	9,000	—	786	945	—	735	—	851	744
	42.00	L-80	0.562	6.301	6.374	8.500	8.125	—	—	10,810	928	10,320	—	10,320	9,790	—	6,550	—	—	892	1,053	—	735	—	—	—
	47.10	L-80	0.625	6.375	6.250	8.500	8.125	—	—	12,040	1100	11,480	—	10,490	9,970	—	6,550	—	—	997	1,160	—	735	—	—	—
	26.40	N-80	0.328	6.969	6.844	8.500	8.125	6.750	8.010	3,400	602	6,020	—	6,020	6,020	6,020	6,020	6,020	—	490	659	659	659	659	737	737
	29.70	N-80	0.375	6.875	6.750	8.500	8.125	6.750	8.010	4,700	683	6,890	—	6,890	6,890	6,890	6,550	6,890	—	575	749	749	749	749	737	737
	33.70	N-80	0.430	6.765	4.810	8.500	8.125	6.640	8.010	6,560	778	7,900	—	7,900	7,900	7,900	6,550	7,000	—	674	852	852	773	852	806	784
	39.00	N-80	0.500	6.625	6.500	8.500	8.125	6.500	8.010	5,810	695	9,180	—	9,180	9,180	9,180	6,550	9,000	—	798	981	981	773	967	896	784
	42.80	N-80	0.562	6.301	6.776	8.300	8.125	—	—	10,810	998	10,320	—	10,320	9,700	10,320	6,550	9,000	—	905	1,003	1,093	773	967	—	—
	47.10	N-80	0.623	6.375	6.230	8.500	8.125	—	—	12,010	1100	11,480	—	10,490	9,790	11,480	6,550	9,000	—	1,013	1,205	1,204	773	967	—	—
	26.40	C-95	0.328	6.969	6.814	8.500	8.125	6.750	8.010	3,710	714	7,150	—	7,150	7,150	—	7,150	—	—	560	716	—	716	—	774	774
	29.70	C-95	0.375	6.875	6.750	8.500	8.125	6.720	8.010	5,120	811	8,180	—	8,180	8,180	—	7,780	—	—	659	813	—	812	—	774	774
	33.70	C-95	0.400	6.765	6.610	8.500	8.125	6.640	8.010	7,260	923	9,380	—	9,380	9,380	—	7,780	—	—	772	925	—	812	—	846	823
	38.90	C-95	0.500	6.625	6.500	8.500	8.125	6.500	8.010	9,389	1061	10,900	—	10,900	10,900	—	7,780	—	—	914	1,065	—	812	—	941	823
	42.80	C-95	0.562	6.501	6.725	8.500	8.125	—	—	12,400	1185	12,250	—	12,250	11,620	—	7,780	—	—	1,037	1,187	—	812	—	—	—
	47.10	C-95	0.623	6.375	6.250	8.500	8.125	—	—	14,300	1396	13,630	—	12,460	11,620	—	7,780	—	—	1,159	1,300	—	812	—	—	—
	29.70	P-110	0.375	6.875	6.750	8.500	8.125	6,750	8.010	5,340	940	9,470	—	9,470	9,470	9,470	9,000	9,470	—	769	960	960	960	960	922	922
	33.70	P-110	0.430	6.755	6.410	8.500	8.125	6.600	8.010	7,950	1069	10,860	—	10,860	10,860	10,860	9,000	10,860	—	901	1,093	1,093	967	1,093	1,008	979
	39.00	P-110	0.500	6.425	6.500	8.500	8.125	6.500	8.010	11,140	1231	12,620	—	12,620	12,620	12,620	9,000	12,280	—	1,066	1,258	1,258	967	1,237	1,120	979
	42.80	P-110	0.542	6.501	6.376	8.500	8.125	—	—	13,910	1372	14,190	—	14,190	13,460	14,100	9,000	12,280	—	1,210	1,402	1,402	967	1,237	—	—
	47.10	P-110	0.625	6.375	6.250	8.500	8.125	—	—	16,330	1512	15,780	—	14,430	13,460	15,780	9,000	12,280	—	1,353	1,545	1,545	967	1,237	—	—

Table 4.16.7 (continued)

Spanning column groups: Columns "Outside Diameter of Coupling W", "Outside Diameter Special Clearance Coupling Wc" fall under **Threaded and Coupled**; "Drift Diameter" and "Outside Diameter of Box Power-tight M" fall under **Extreme Line**. Columns 13–19 fall under **Internal Yield Pressure at Minimum Yield, psi**; columns 20–27 fall under **★ Joint Strength — 1,000 lbs** (Threaded and Coupled = Round Thread / Buttress Thread; Extreme Line = Standard Joint / Optional Joint).

Size OD D in.	Nominal Wt, Threads & Coupling lb/ft	Grade	Wall Thick. t in.	Inside Dia. d in.	Drift Dia. in.	OD Coupling W in.	OD Spec. Clear. Coupling Wc in.	Drift Dia. in.	OD Box Power-tight M in.	Collapse Resist. psi	Pipe Body Yield 1000 lbs	Plain End or Extreme Line	RT Short	RT Long	Butt Reg Coupling Same Grade	Butt Reg Coupling Higher Grade	Butt Spec Clear Coupling Same Grade	Butt Spec Clear Coupling Higher Grade	RT Short	RT Long	Butt Reg Coupling	Butt Reg Coupling Higher Grade	Butt Spec Clear Coupling	Butt Spec Clear Coupling Higher Grade	Std Joint	Opt Joint
8⅝	28.00	H-40	0.304	8.017	7.892	9.625	—	—	—	1,640	318	2,470	2,470	—	—	—	—	—	233	—	—	—	—	—	—	—
	32.00	H-40	0.352	7.921	7.796	9.625	—	—	—	2,210	366	2,860	2,860	—	—	—	—	—	279	—	—	—	—	—	—	—
	24.00	J-55	0.264	8.097	7.972	9.625	—	—	—	1,370	381	2,950	2,950	—	3,930	3,930	—	—	244	—	—	—	—	—	—	—
	32.00	J-55	0.352	7.921	7.796	9.625	9.125	7.700	9.120	2,530	503	3,930	3,930	—	3,930	3,930	3,930	3,930	372	417	579	579	579	579	686	686
	36.00	J-55	0.400	7.825	7.700	9.625	9.125	7.700	9.120	3,450	568	4,460	4,460	—	4,460	4,460	4,060	4,460	434	486	654	654	654	654	688	688
	24.00	K-55	0.264	8.097	7.972	9.625	—	—	—	1,370	381	2,950	2,950	—	3,930	3,930	—	—	263	—	—	—	—	—	—	—
	32.00	K-55	0.352	7.921	7.796	9.625	9.125	7.700	9.120	2,530	503	3,930	3,930	—	3,930	3,930	3,930	3,930	402	452	690	690	690	690	869	869
	36.00	K-55	0.400	7.825	7.700	9.625	9.125	7.700	9.120	3,450	568	4,460	4,460	—	4,460	4,460	4,060	4,460	468	526	780	780	780	780	871	871
	36.00	C-75	0.400	7.825	7.700	9.625	9.125	7.700	9.120	4,020	775	6,090	6,090	6,090	6,090	6,090	5,530	5,530	—	648	847	—	839	—	871	871
	40.00	C-75	0.450	7.725	7.600	9.625	9.125	7.600	9.120	5,350	867	6,850	6,850	6,850	7,300	7,300	5,530	5,530	—	742	947	—	839	—	942	886
	44.00	C-75	0.500	7.625	7.500	9.625	9.125	7.500	9.120	6,680	957	7,610	7,610	7,610	7,610	8,120	5,530	5,530	—	834	1,046	—	839	—	1,007	886
	49.00	C-75	0.557	7.511	7.386	9.625	9.125	7.386	9.120	8,200	1059	8,480	8,480	8,480	8,480	9,040	5,530	5,530	—	939	1,157	—	839	—	1,007	886
	36.00	L-80	0.400	7.825	7.700	9.625	9.125	7.700	9.120	4,100	827	6,490	6,490	6,490	6,490	6,490	5,900	5,900	—	678	864	—	839	—	871	871
	40.00	L-80	0.450	7.725	7.600	9.625	9.125	7.600	9.120	5,520	925	7,300	7,300	7,300	7,300	7,300	5,900	5,900	—	776	966	—	839	—	942	886
	44.00	L-80	0.500	7.625	7.500	9.625	9.125	7.500	9.120	6,950	1021	8,120	8,120	8,120	8,120	8,120	5,900	5,900	—	874	1,066	—	839	—	1,007	886
	49.00	L-80	0.557	7.511	7.386	9.625	9.125	7.386	9.120	8,570	1129	9,040	9,040	9,040	9,040	9,040	5,900	5,900	—	983	1,180	—	839	—	1,007	886
	36.00	N-80	0.400	7.825	7.700	9.625	9.125	7.700	9.120	4,100	827	6,490	6,490	6,490	6,490	6,490	5,900	6,490	688	688	895	895	883	895	917	917
	40.00	N-80	0.450	7.725	7.600	9.625	9.125	7.600	9.120	5,520	925	7,300	7,300	7,300	7,300	7,300	5,900	7,300	—	788	1,001	1,001	883	1,001	992	932
	44.00	N-80	0.500	7.625	7.500	9.625	9.125	7.500	9.120	6,950	1021	8,120	8,120	8,120	8,120	8,120	5,900	8,110	—	887	1,105	1,105	883	1,103	1,060	932
	49.00	N-80	0.557	7.511	7.386	9.625	9.125	7.386	9.120	8,570	1129	9,040	9,040	9,040	9,040	9,040	5,900	8,110	—	997	1,222	1,222	883	1,103	1,060	932
	36.00	C-95	0.400	7.825	7.700	9.625	9.125	7.700	9.120	4,360	982	7,710	7,710	7,710	7,710	7,710	7,010	7,010	—	789	976	—	927	—	963	963
	40.00	C-95	0.450	7.725	7.600	9.625	9.125	7.600	9.120	6,010	1098	8,670	8,670	8,670	8,670	8,670	7,010	7,010	—	904	1,092	—	927	—	1,042	979
	44.00	C-95	0.500	7.625	7.500	9.625	9.125	7.500	9.120	7,730	1212	9,640	9,640	9,640	9,640	9,640	7,010	7,010	—	1,017	1,206	—	927	—	1,113	979
	49.00	C-95	0.557	7.511	7.386	9.625	9.125	7.386	9.120	9,690	1341	10,740	10,740	10,740	10,740	10,740	7,010	7,010	—	1,144	1,334	—	927	—	1,113	979
	40.00	P-110	0.450	7.725	7.600	9.625	9.125	7.600	9.120	6,380	1271	10,040	10,040	10,040	10,040	10,040	8,110	10,040	—	1,055	1,288	1,288	1,103	1,288	1,240	1,165
	44.00	P-110	0.500	7.625	7.500	9.625	9.125	7.500	9.120	8,400	1404	11,160	11,160	11,160	11,160	11,160	8,110	11,060	—	1,186	1,423	1,423	1,103	1,412	1,326	1,165
	49.00	P-110	0.557	7.511	7.386	9.625	9.125	7.386	9.120	10,720	1553	12,430	12,430	12,430	12,430	12,430	8,110	11,060	—	1,335	1,574	1,574	1,103	1,412	1,326	1,165
9⅝	32.30	H-40	0.312	9.001	8.845	10.625	—	—	—	1,400	365	2,270	2,270	—	—	—	—	—	254	—	—	—	—	—	—	—
	36.00	H-40	0.262	8.765	8.599	10.625	—	—	10.100	1,740	410	2,560	2,560	—	—	—	—	—	294	—	—	—	—	—	—	—
	36.00	J-55	0.352	8.921	8.765	10.625	10.125	8.599	10.100	2,020	569	3,520	3,520	—	3,520	3,520	3,520	3,520	394	453	639	639	639	639	—	—
	40.00	J-55	0.395	8.835	8.679	10.625	10.125	8.599	10.100	2,570	630	3,950	3,950	—	3,950	3,950	3,950	3,950	455	520	714	714	714	714	770	770
	36.00	K-55	0.352	8.921	8.765	10.625	10.125	8.599	10.100	2,020	564	3,520	3,520	—	3,520	3,520	3,520	3,520	425	489	755	755	755	755	—	—
	40.00	K-55	0.395	8.835	8.679	10.625	10.125	8.379	10.100	2,570	630	3,950	3,950	—	3,950	3,950	3,950	3,950	486	561	843	843	843	843	975	975
	40.00	C-75	0.395	8.835	8.679	10.625	10.125	8.599	10.100	2,980	859	5,390	5,390	5,390	5,390	5,390	4,990	4,990	—	694	926	—	926	—	975	975
	43.50	C-75	0.435	8.755	8.599	10.625	10.125	8.599	10.100	3,750	942	5,930	5,930	5,930	5,930	5,930	4,990	4,990	—	776	1,016	—	934	—	975	975
	47.00	C-75	0.472	8.681	8.525	10.625	10.125	8.525	10.100	4,630	1018	6,440	6,440	6,440	6,440	6,440	4,990	4,990	—	852	1,098	—	934	—	1,032	1,032
	53.50	C-75	0.545	8.535	8.379	10.625	10.125	8.379	10.100	6,380	1166	7,430	7,430	7,430	7,430	7,430	4,990	4,990	—	999	1,257	—	934	—	1,175	1,053
	40.00	L-80	0.395	8.835	8.679	10.625	10.125	8.599	10.100	3,090	916	5,750	5,750	5,750	5,750	5,750	5,320	5,320	—	727	947	—	934	—	975	975
	43.50	L-80	0.435	8.755	8.599	10.625	10.125	8.599	10.100	3,810	1005	6,330	6,330	6,330	6,330	6,330	5,320	5,320	—	813	1,038	—	934	—	975	975
	47.00	L-80	0.472	8.681	8.525	10.625	10.125	8.525	10.100	4,750	1086	6,870	6,870	6,870	6,870	6,870	5,320	5,320	—	893	1,122	—	934	—	1,092	1,032
	53.50	L-80	0.545	8.535	8.379	10.625	10.125	8.379	10.100	6,620	1244	7,930	7,930	7,930	7,930	7,930	5,320	5,320	—	1,047	1,286	—	934	—	1,173	1,053
	40.00	N-80	0.395	8.835	8.679	10.625	10.125	8.539	10.100	3,090	916	5,750	5,750	5,750	5,750	5,750	5,320	5,750	—	737	979	979	979	979	1,027	1,027
	43.50	N-80	0.435	8.755	8.599	10.625	10.125	8.590	10.100	3,610	1005	6,330	6,330	6,330	6,330	6,330	5,320	6,330	—	825	1,074	1,074	983	1,074	1,027	1,027
	47.00	N-80	0.472	8.681	8.525	10.625	10.125	8.525	10.100	4,750	1086	6,870	6,870	6,870	6,870	6,870	5,320	6,870	—	905	1,161	1,161	983	1,161	1,086	1,086
	53.50	N-80	0.545	8.535	8.379	10.625	10.125	8.379	10.100	6,620	1244	7,930	7,930	7,930	7,930	7,930	5,320	7,310	—	1,062	1,329	1,329	983	1,229	1,235	1,109
	40.00	C-95	0.395	8.835	8.679	10.625	10.125	8.599	10.100	3,330	1088	6,820	6,820	6,820	6,820	6,820	6,310	6,310	—	847	1,074	—	1,032	—	1,078	1,078
	43.50	C-95	0.435	8.755	8.599	10.625	10.125	8.599	10.100	4,130	1193	7,510	7,510	7,510	7,510	7,510	6,310	6,310	—	948	1,178	—	1,032	—	1,078	1,078
	47.00	C-95	0.472	8.681	8.525	10.625	10.125	8.525	10.100	5,080	1239	8,150	8,150	8,150	8,150	8,150	6,310	6,310	—	1,010	1,273	—	1,032	—	1,141	1,141
	53.50	C-95	0.545	8.535	8.379	10.625	10.125	8.379	10.100	7,230	1477	9,410	9,410	9,410	9,410	9,410	6,310	6,310	—	1,220	1,458	—	1,032	—	1,297	1,164
	43.50	P-110	0.435	8.755	8.599	10.625	10.125	8.599	10.100	4,430	1381	8,700	8,700	8,700	8,700	8,700	7,310	8,700	—	1,106	1,388	1,388	1,229	1,388	1,283	1,283
	47.00	P-110	0.472	8.681	8.525	10.625	10.125	8.525	10.100	5,310	1403	9,440	9,440	9,440	9,440	9,440	7,310	9,440	—	1,213	1,500	1,500	1,229	1,500	1,358	1,358
	53.50	P-110	0.545	8.535	8.379	10.625	10.125	8.379	10.100	7,930	1710	10,900	10,900	10,900	10,900	10,900	7,310	9,970	—	1,422	1,718	1,718	1,229	1,573	1,544	1,386

4-426

Table columns (numbered 1–27):

1	Size: Outside Diameter, in. D
2	Nominal Weight, Threads and Coupling, lb per ft
3	Grade
4	Wall Thickness, in. t
5	Inside Diameter, in. d
6	Drift Diameter, in.
7	Threaded and Coupled — Outside Diameter Coupling, in. W
8	Threaded and Coupled — Outside Diameter Special Clearance Coupling, in. Wc
9	Extreme Line — Drift Diameter, in.
10	Extreme Line — Outside Diameter of Box Power-tight, in. M
11	Collapse Resistance, psi
12	Pipe Body Yield Strength, 1000 lbs
13	Internal Yield Pressure at Minimum Yield, psi — Plain End or Extreme Line
14	Round Thread Short
15	Round Thread Long
16	Buttress Thread Regular Coupling — Same Grade
17	Buttress Thread Regular Coupling — Higher Grade
18	Special Clearance Coupling — Same Grade
19	Special Clearance Coupling — Higher Grade
20	★ Joint Strength — 1,000 lbs — Round Thread Short
21	Round Thread Long
22	Buttress Thread Regular Coupling
23	Buttress Thread Regular Coupling — Higher Grade
24	Special Clearance Coupling
25	Special Clearance Coupling — Higher Grade
26	Extreme Line — Standard Joint
27	Extreme Line — Optional Joint

1	2	3	4	5	6	7	8	9	10	11	12	13	14	15	16	17	18	19	20	21	22	23	24	25	26	27
10¾	32.75	H-40	0.279	10.192	10.036	11.750	—	—	—	880	367	1,820	1,820	—	—	—	—	—	205	—	—	—	—	—	—	—
	40.50	H-40	0.350	10.050	9.894	11.750	—	—	—	1,420	457	2,280	2,280	—	—	—	—	—	314	—	—	—	—	—	—	—
	40.50	J-55	0.350	10.050	9.894	11.750	11.250	—	—	1,580	629	3,130	3,130	—	3,130	3,130	3,130	3,130	420	—	700	700	700	700	—	—
	45.50	J-55	0.400	9.950	9.794	11.750	11.250	9.794	11.460	2,090	715	3,580	3,580	—	3,580	3,580	3,290	3,580	493	—	796	796	796	796	975	—
	51.00	J-55	0.450	9.850	9.694	11.750	11.250	9.694	11.460	2,700	801	4,030	4,030	—	4,030	4,030	3,290	4,030	565	—	891	891	822	891	1,092	—
	40.50	K-55	0.350	10.050	9.894	11.750	11.250	—	—	1,580	629	3,130	3,130	—	3,130	3,130	3,130	3,130	450	—	819	819	819	819	—	—
	45.50	K-55	0.400	9.950	9.794	11.750	11.250	9.794	11.460	2,090	715	3,580	3,580	—	3,580	3,580	3,200	3,580	528	—	931	931	931	931	1,236	—
	51.00	K-55	0.450	9.850	9.694	11.750	11.250	9.694	11.460	2,700	801	4,030	4,030	—	4,030	4,030	3,290	4,030	606	—	1,043	1,043	1,041	1,043	1,383	—
	51.00	C-75	0.450	9.850	9.694	11.750	11.250	9.694	11.460	3,100	1092	5,490	5,490	—	5,490	—	4,490	—	756	—	1,160	—	1,041	—	1,383	—
	55.50	C-75	0.495	9.760	9.604	11.750	11.250	9.604	11.460	3,950	1196	6,040	6,040	—	6,040	—	4,490	—	843	—	1,271	—	1,041	—	1,515	—
	51.00	L-80	0.450	9.850	9.694	11.750	11.250	9.694	11.460	3,220	1165	5,860	5,860	—	5,860	—	4,790	—	794	—	1,190	—	1,041	—	1,383	—
	55.50	L-80	0.495	9.760	9.604	11.750	11.250	9.604	11.460	4,020	1276	6,450	6,450	—	6,450	—	4,790	—	884	—	1,303	—	1,041	—	1,515	—
	51.00	N-80	0.450	9.850	9.694	11.750	11.250	9.694	11.460	3,220	1165	5,860	5,860	—	5,860	5,860	4,790	5,860	804	—	1,228	1,228	1,096	1,228	1,456	—
	55.50	N-80	0.495	9.760	9.604	11.750	11.250	9.604	11.460	4,020	1276	6,450	6,450	—	6,450	6,450	4,790	6,450	895	—	1,345	1,345	1,096	1,345	1,595	—
	51.00	C-95	0.450	9.850	9.694	11.750	11.250	9.694	11.460	3,490	1383	6,960	6,960	—	6,960	—	5,680	—	927	—	1,354	—	1,151	—	1,529	—
	55.50	C-95	0.495	9.760	9.604	11.750	11.250	9.604	11.460	4,300	1515	7,660	7,660	—	7,660	—	5,680	—	1,032	—	1,483	—	1,151	—	1,675	—
	51.00	P-110	0.450	9.850	9.694	11.750	11.250	9.694	11.460	3,670	1602	8,060	8,060	—	8,060	8,060	6,580	8,060	1,080	—	1,594	1,594	1,370	1,594	1,820	—
	55.50	P-110	0.495	9.760	9.604	11.750	11.250	9.604	11.460	4,630	1754	8,860	8,860	—	8,860	8,860	6,580	8,860	1,203	—	1,745	1,745	1,370	1,745	1,993	—
	60.70	P-110	0.545	9.660	9.504	11.750	11.250	9.604	11.460	5,860	1922	9,760	9,760	—	9,760	9,760	6,580	8,970	1,338	—	1,912	1,912	1,370	1,754	2,000	—
	65.70	P-110	0.595	9.560	9.404	11.750	11.250	9.504	—	7,490	2088	10,650	10,650	—	10,650	10,650	6,580	8,970	1,472	—	2,077	2,077	1,370	1,754	—	—
11¾	42.00	H-40	0.333	11.084	10.928	12.750	—	—	—	1,070	478	1,980	1,980	—	—	—	—	—	307	—	—	—	—	—	—	—
	47.00	J-55	0.375	11.000	10.844	12.750	—	—	—	1,510	737	3,070	3,070	—	3,070	3,070	—	—	477	—	807	807	—	—	—	—
	54.00	J-55	0.435	10.880	10.724	12.750	—	—	—	2,070	850	3,560	3,560	—	3,560	3,560	—	—	568	—	931	931	—	—	—	—
	60.00	J-55	0.489	10.772	10.616	12.750	—	—	—	2,660	952	4,010	4,010	—	4,010	4,010	—	—	649	—	1,042	1,042	—	—	—	—
	47.00	K-55	0.375	11.000	10.844	12.750	—	—	—	1,510	737	3,070	3,070	—	3,070	3,070	—	—	509	—	935	935	—	—	—	—
	54.00	K-55	0.435	10.880	10.724	12.750	—	—	—	2,070	850	3,560	3,560	—	3,560	3,560	—	—	606	—	1,079	1,079	—	—	—	—
	60.00	K-55	0.489	10.772	10.616	12.750	—	—	—	2,660	952	4,010	4,010	—	4,010	4,010	—	—	693	—	1,208	1,208	—	—	—	—
	60.00	C-75	0.489	10.772	10.616	12.750	—	—	—	3,070	1298	5,460	5,460	—	5,460	—	—	—	869	—	1,361	—	—	—	—	—
	60.00	L-80	0.489	10.772	10.616	12.750	—	—	—	3,180	1384	5,830	5,830	—	5,830	—	—	—	913	—	1,399	—	—	—	—	—
	60.00	N-80	0.489	10.772	10.616	12.750	—	—	—	3,180	1384	5,830	5,830	—	5,830	—	—	—	924	—	1,440	1,440	1,440	—	—	—
	60.00	C-95	0.489	10.772	10.616	12.750	—	—	—	3,440	1644	6,920	6,920	—	6,920	—	—	—	1,066	—	1,596	—	—	—	—	—

Table 4.16.7 (continued)

1	2	3	4	5	6	7	8	9	10	11	12	13	14	15	16	17	18	19	20	21	22	23	24	25	26	27
Size: Outside Diameter in. D	Nominal Weight, Threads and Coupling lb per ft	Grade	Wall Thickness in. t	Inside Diameter in. d	Drift Diameter in.	Outside Diameter of Coupling in. W	Outside Diameter Special Clearance Coupling in. Wc	Drift Diameter in. (Extreme Line)	Outside Diameter of Box Power-tight in. M (Extreme Line)	Collapse Resistance psi	Pipe Body Yield Strength 1000 lbs	Plain End or Extreme Line	Round Thread Short	Round Thread Long	Buttress Thread Regular Coupling Same Grade	Buttress Thread Regular Coupling Higher Grade	Buttress Thread Special Clearance Coupling Same Grade	Buttress Thread Special Clearance Coupling Higher Grade	Joint Round Thread Short	Joint Round Thread Long	Joint Buttress Regular Coupling	Joint Buttress Regular Coupling Higher Grade:	Joint Buttress Special Clearance Coupling	Joint Buttress Special Clearance Coupling Higher Grade:	Extreme Line Standard Joint	Extreme Line Optional Joint
13⅜	48.00	H-40	0.330	12.715	12.559	14.375	—	—	—	770	541	1,730	1,730	—	—	—	—	—	322	—	—	—	—	—	—	—
	54.50	J-55	0.380	12.615	12.459	14.375	—	—	—	1,130	853	2,730	2,730	—	2,730	2,730	—	—	514	—	909	909	—	—	—	—
	61.00	J-55	0.430	12.515	12.359	14.375	—	—	—	1,540	962	3,090	3,090	—	3,090	3,090	—	—	595	—	1,025	1,025	—	—	—	—
	68.00	J-55	0.480	12.415	12.259	14.375	—	—	—	1,950	1069	3,450	3,450	—	3,450	3,450	—	—	675	—	1,140	1,140	—	—	—	—
	54.50	K-55	0.380	12.615	12.459	14.375	—	—	—	1,130	853	2,730	2,730	—	2,730	2,730	—	—	547	—	1,038	1,038	—	—	—	—
	61.00	K-55	0.430	12.515	12.359	14.375	—	—	—	1,540	962	3,090	3,090	—	3,090	3,090	—	—	633	—	1,169	1,169	—	—	—	—
	68.00	K-55	0.480	12.415	12.259	14.375	—	—	—	1,950	1069	3,450	3,450	—	3,450	3,450	—	—	718	—	1,300	1,300	—	—	—	—
	68.00	C-75	0.480	12.415	12.259	14.375	—	—	—	2,220	1458	4,710	4,710	—	4,710	—	—	—	905	—	1,496	—	—	—	—	—
	72.00	C-75	0.514	12.347	12.191	14.375	—	—	—	2,590	1558	5,040	5,040	—	5,040	—	—	—	978	—	1,598	—	—	—	—	—
	68.00	L-80	0.480	12.415	12.259	14.375	—	—	—	2,260	1556	5,020	5,020	—	5,020	—	—	—	952	—	1,545	—	—	—	—	—
	72.00	L-80	0.514	12.347	12.191	14.375	—	—	—	2,670	1661	5,380	5,380	—	5,380	—	—	—	1,029	—	1,650	—	—	—	—	—
	68.00	N-80	0.480	12.415	12.259	14.375	—	—	—	2,260	1556	5,020	5,020	—	5,020	—	—	—	963	—	1,585	1,585	—	—	—	—
	72.00	N-80	0.514	12.347	12.191	14.375	—	—	—	2,670	1661	5,380	5,380	—	5,380	—	—	—	1,040	—	1,693	1,693	—	—	—	—
	68.00	C-95	0.480	12.415	12.259	14.375	—	—	—	2,320	1847	5,970	5,970	—	5,970	—	—	—	1,114	—	1,772	—	—	—	—	—
	72.00	C-95	0.514	12.347	12.191	14.375	—	—	—	2,820	1973	6,390	6,390	—	6,390	—	—	—	1,204	—	1,893	—	—	—	—	—
16	65.00	H-40	0.375	15.250	15.062	17.000	—	—	—	670	736	1,640	1,640	—	—	—	—	—	439	—	—	—	—	—	—	—
	75.00	J-55	0.438	15.124	14.936	17.000	—	—	—	1,020	1178	2,630	2,630	—	2,630	2,630	—	—	710	—	1,200	1,200	—	—	—	—
	84.00	J-55	0.495	15.016	14.822	17.000	—	—	—	1,410	1326	2,980	2,980	—	2,980	2,980	—	—	817	—	1,351	1,351	—	—	—	—
	75.00	K-55	0.438	15.124	14.936	17.000	—	—	—	1,020	1178	2,630	2,630	—	2,630	2,630	—	—	752	—	1,331	1,331	—	—	—	—
	84.00	K-55	0.495	15.010	14.822	17.000	—	—	—	1,410	1326	2,980	2,980	—	2,980	2,980	—	—	865	—	1,499	1,499	—	—	—	—
18⅝	87.50	H-40	0.435	17.755	17.567	20.000	—	—	—	*630	994	1,630	1,630	—	—	—	—	—	559	—	—	—	—	—	—	—
	87.50	J-55	0.435	17.755	17.567	20.000	—	—	—	*630	1367	2,250	2,250	—	2,250	—	—	—	754	—	1,329	1,329	—	—	—	—
	87.50	K-55	0.435	17.755	17.567	20.000	—	—	—	*630	1367	2,250	2,250	—	2,250	—	—	—	794	—	1,427	1,427	—	—	—	—
20	94.00	H-40	0.438	19.124	18.936	21.000	—	—	—	*520	1077	1,530	1,530	—	—	—	—	—	581	—	—	—	—	—	—	—
	94.00	J-55	0.438	19.124	18.936	21.000	—	—	—	*520	1480	2,110	2,110	2,110	—	—	—	—	784	907	1,402	1,402	—	—	—	—
	106.50	J-55	0.500	19.000	18.812	21.000	—	—	—	*770	1685	2,410	2,410	2,410	—	—	—	—	913	1,057	1,596	1,596	—	—	—	—
	133.00	J-55	0.635	18.730	18.542	21.000	—	—	—	1,500	2125	3,060	3,060	3,060	—	—	—	—	1,192	1,380	2,012	2,012	—	—	—	—
	94.00	K-55	0.438	19.124	18.936	21.000	—	—	—	*520	1480	2,100	2,110	2,110	—	—	—	—	824	955	1,479	1,479	—	—	—	—
	106.50	K-55	0.500	19.000	18.812	21.000	—	—	—	*770	1685	2,410	2,410	2,410	—	—	—	—	960	1,113	1,683	1,683	—	—	—	—
	133.00	K-55	0.625	18.730	18.542	21.000	—	—	—	1,500	2125	3,060	3,060	3,060	—	—	—	—	1,253	1,453	2,123	2,123	—	—	—	—

‡ For P-110 casing the next higher grade is 150YS, a non-API steel grade having a minimum yield strength of 150,000 psi.

★ Some joint strengths listed in Col. 20 through 27 are greater than the corresponding pipe body yield strength listed in Col. 12.

* Collapse resistance values calculated by elastic formula.

Taken from API Bul 5C2, 17th Edition, March 1980.

Table 4.16.8 *D/t Ranges for Collapse Pressures*

Steel Grade	D/t range for formula (1)	D/t range for formula (2)	D/t range for formula (3)	D/t range for formula (4)
H-40	16.44 & less	16.44 to 26.62	26.62 to 42.70	42.70 & greater
J-K-55	14.80 & less	14.80 to 24.39	24.39 to 37.20	37.20 & greater
C-75	13.67 & less	13.67 to 23.09	23.03 to 32.05	32.05 & greater
L-N-80	13.38 & less	13.38 to 22.46	22.46 to 31.05	31.05 & greater
C-95	12.83 & less	12.83 to 21.21	21.21 to 28.25	28.25 & greater
P-105	12.56 & less	12.56 to 20.66	20.66 to 26.88	26.88 & greater
P-110	12.42 & less	12.42 to 20.29	20.29 to 26.20	26.20 & greater

Table 4.16.9 *Factors for Collapse Pressure Formulas*

Steel Grade	Formula Factor				
	A	B	C	F	G
H-40	2.950	0.0463	755	2.047	0.03125
J-K-55	2.990	0.0541	1205	1.990	0.03360
C-75	3.060	0.0642	1805	1.985	0.0417
L-N-80	3.070	0.0667	1955	1.998	0.0434
C-95	3.125	0.0745	2405	2.047	0.0490
P-105	3.162	0.0795	2700	2.052	0.0515
P-110	3.180	0.0820	2855	2.075	0.0535

(text continued from page 4-422)

Symbols in Equations 4.16.1 to 4.16.5 are as follows:

D = nominal outside diameter in in.
t = nominal wall thickness in in.
Y_p = minimum yield strength of pipe in psi
P_{Yp} = minimum yield strength collapse pressure in psi
P_p = minimum plastic collapse pressure in psi
P_T = minimum plastic/elastic transition collapse pressure in psi
P_E = minimum elastic collapse pressure in psi
P_{CA} = minimum collapse pressure under axial tension stress in psi
P_{CO} = minimum collapse pressure without axial tension stress in psi
S_A = axial tension stress in psi
P_i = internal pressure in psi

Internal Yield Pressure for Pipe
Internal yield pressure for pipe is calculated from Equation 4.16.6. The factor 0.875 appearing in the formula 4.16.6 allows for minimum wall thickness.

$$P_i = 0.875 \left[\frac{2Y_p t}{D} \right] \qquad [4.16.6]$$

where

P_i = minimum internal yield pressure in psi
Y_p = minimum yield strength in psi
t = nominal wall thickness in in.
D = nominal outside diameter in inches.

Internal Yield Pressure for Couplings
Internal yield pressure for threaded and coupled pipe is the same as for plain end pipe, except where a lower pressure is required to avoid leakage due to insufficient coupling strength. The lower pressure is based on

$$P = Y_c \left[\frac{W - d_1}{W} \right] \qquad [4.16.7]$$

where

P = minimum internal yield pressure in psi
Y_c = minimum yield strength at coupling in psi
W = nominal outside diameter of coupling
d_1 = diameter of the root of the coupling thread at the end of the pipe in the powertight position (see API Bulletin 5C3, 6th Edition, October 1994).

Pipe Body Yield Strength (Section 2 of API 5C3)
Pipe body yield strength is the axial load required to yield the pipe. It is taken as the product of the cross sectional area and the specified minimum yield strength for the particular grade of pipe. Values for pipe body yield strength were calculated by means of the following formula:

$$P_Y = 0.7854(D^2 - d^2)Y_p \qquad [4.16.8]$$

where

P_Y = pipe body yield strength in psi
Y_p = minimum yield strength
D = specified outside diameter in in.
d = specified inside diameter in in.

4.16.2.7 Joint Strength (Section 9 of API 5C3)
Round Thread Casing Joint Strength
Round thread casing joint strength is calculated from formulas 4.16.9 and 4.16.10. The lesser of the values obtained from the two formulas governs. Formulas 4.16.9 and 4.16.10 apply both to short and long threads and couplings. Formula 4.16.9 is for minimum strength of a joint failing by fracture, and formula 4.16.10 for minimum strength of a joint failing by thread jumpout or pullout.
The fracture strength is

$$P_j = 0.95 A_{jp} U_p \qquad [4.16.9]$$

The pullout strength is

$$P_j = 0.95 A_{jp} L \left[\frac{0.74 D^{-0.59} U_p}{0.5L + 0.14D} + \frac{Y_p}{L + 0.14D} \right] \qquad [4.16.10]$$

where

P_j = minimum joint strength in lb

A_{jp} = cross - sectional area of the pipe wall under the last perfect thread in in.^{2}

= $0.7854(D-0.1425)^2 - d^2)$ for eight round threads

D = nominal outside diameter of pipe in in.

d = nominal inside diameter of pipe in in.

L = engaged thread length in in.

= $L_4 - M$ for nominal makeup, API Spec 5B

Y_p = minimum yield strength of pipe in psi

U_p = minimum ultimate strength of pipe in psi

Buttress Thread Casing Joint Strength

Buttress thread casing joint strength is calculated from formulas 4.16.11 and 4.16.12. The lesser of the values obtained from the two formulas governs.

Pipe thread strength is

$$P_j = 0.95A_pU_p|1.008 - 0.0396(1.083 - Y_p/U_p)D| \qquad [4.16.11]$$

Casing thread strength is

$$P_j = 0.95A_cU_c \qquad [4.16.12]$$

where

P_j = minimum joint strength in lb

Y_p = minimum yield strength of pipe in lb

U_p = minimum ultimate strength of pipe in psi

U_c = minimum ultimate strength of coupling in psi

A_p = cross-sectional area of plain end pipe in in.^{2}

= $0.7854(D^2 - d^2)$

A_c = cross-sectional area of coupling in in.^{2}

= $0.7854(W^2 - d_i^2)$

D = outside diameter of pipe in in.

W = outside diameter of coupling in in.

d = inside diameter of pipe in in.

d_i = diameter of the root of the coupling thread at the end of the pipe in the powertight position

Extreme-line Casing Joint Strength

Extreme-line casing joint strength is calculated from

$$P_j = A_{cr}U_p \qquad [4.16.13]$$

where

P_j = minimum joint strength in lb

A_{cr} = critical section area of box, pin or pipe, whichever is least, in in.^{2} (see API Bulletin 5C3)

U_p = specified minimum ultimate strength in psi (Table 4.16.9)

4.16.3 Combination Casing Strings

The term *combination casing string* is generally applied to a casing string that is composed of more than one weight per foot, or more than one grade of steel, or both.

4.16.3.1 Design Consideration

Solving the problem of casing string design for known type and size of casing string relies on selection of the most economical grades and weights of casing that will withstand, without failure, the loads to which the casing will be subjected throughout the life of the well.

There are various established methods of designing a technically satisfactory combination casing string. The difference between these methods rely on different design models, different values of the safety factors and different sequences of calculations. There are no commonly accepted methods of combination string design nor accepted values for the safety factors. Some suggestions are offered below; however, the decision is left to the person responsible for the design.

In general, the following loads must be considered: tension, collapse, burst and compression. The reasonably worst working conditions ought to be assumed.

Collapse

The casing must be designed against collapse to withstand the hydrostatic pressure of the fluid behind the casing at any depth, decreased by anticipated pressure inside the casing at the corresponding level. Usually, the maximum collapse pressure to be imposed on the casing is considered to be the hydrostatic pressure of the heaviest mud used to drill to the lending depth of the casing string, acting on empty string. Depending upon design model, it is recommended to use a design factor of 1.0 to 1.2. For example, if it is known that casing will never be empty inside, this fact should be considered for collapse pressure evaluation and selection of the magnitude of safety factor.

Burst

Casing must be designed to resist expected burst pressure at any depth. In burst pressure consideration, it is suggested to consider different design models depending upon the type of casing string.

Conductor String

It is assumed that the external pressure is zero. In any case, the maximum expected internal pressure cannot be greater than fracture pressure at the open hole below the conductor casing shoe; usually, it is the first formation right below the casing shoe. If this pressure is not known (in exploratory drilling), the burst pressure of gas equivalent to 0.9 or 1.0 psi/ft can be assumed. Hydrostatic head due to gas is neglected. For example, if the setting depth of conductor string is 1,000 ft, then the maximum expected burst pressure is even along the string and equal to $(1,100)(1.0) = 1,100$ psi; safety factor = 1.1 to 1.15.

4.16.3.2 Surface and Intermediate Strings

It is suggested to evaluate the burst load based on the internal pressure expected, reduced by the external pressure of the drilling fluid outside the string. Internal pressure is based on the expected bottomhole pressure of the next string with the hole being evacuated from drilling fluid up to a minimum of 50%. In exploratory wells, a reasonable assumption of expected formation pore pressure gradient is required.

Example 4.16.3

Evaluate an expected burst pressure acting on surface casing string in exploratory drilling if setting depth of the next string is 11,000 ft.

Solution

Step 1. Internal pressure in the borehole. Because the next string is set at 11,000 ft, the formation pore pressure gradient is assumed to be 0.65 psi/ft. Thus, the bottomhole pressure (at a depth of 11,000 ft) is $(11,000)(0.65) = 7,150$ psi. Assume 50% of evacuation; thus, $(11,000)(0.5) = 5,500$ ft.

Note: It is assumed that below 5,500 ft, the hole is filled with mud, which exerts a pressure gradient of 0.65 psi/ft. The hole above 5,500 ft is filled with gas, the weight of which is ignored.

The internal pressure at a depth of 5,500 ft is $(5,500)(0.65) = 3,575$ psi. Since the weight of gas is ignored for this type of string, the internal pressure at the top of the hole is also 3,575 psi.

Step 2. External pressure. It is assumed that there is drilling fluid with specific gravity of 1.2 outside the casing.

Thus, the external pressure at surface $= 0.0$ psi and at 5,000 ft is $(5,000)(1.2)(8.34)(0.052) \cong 2,600$ psi.

Step 3. Burst load (Pb). The burst pressure is equal to internal pressure reduced by the external pressure of the drilling fluid outside the casing. Therefore,

at surface: $P_b = 3,575 - 0.0 = 3,573$ psi

at 5,000 ft: $P_b = 3,575 - 2,600 = 987$ psi

The burst pressure line equation is as below:

$$P_b = 3,575 - 0.52(D) \text{ (psi)}$$

$D = $ depth at the hole (ft) (from 0 to 5,000 ft)

Note: For practical purposes, a graphic solution is very advisable.

4.16.3.3 Production String
In exploratory drilling, it assumed that internal pressure acting on the casing is reduced by external saltwater pressure gradient of about 0.5 psi/ft. Internal pressure is based on expected gas pressure gradient. For long strings, the weight of gas is not ignored.

4.16.3.4 Tension Load
The maximum tensile load acting on the casing string is often considered as the static weight of the casing as measured in air.

Casing must be designed to satisfy these equations:

$$P_j = (W)(N_j) \qquad\qquad [4.16.14]$$

$$P_y = (W)(N_p) \qquad\qquad [4.16.15]$$

where

P_j = casing joint strength in lb
P_y = pipe body strength in lb
W = weight (mass) of casing suspended below the cross-section under consideration in lb
N_j = safety factor for joint
N_p = safety factor for pipe body

Safety factors (N_j, N_p) of 1.6 to 2.0, are used and should be applied to the minimum joint tensile strength or the minimum pipe body tensile strength, whichever is the smallest.

4.16.3.5 Compression Load
Under certain conditions, casing can be subjected to the compression load, such as if the weight of the inner strings (conductor or surface casing string) is transferred to the outer string or if the portion of the casing weight is slacked off on the bottom of the hole. This load may result in casing failure and, therefore, must also be considered.

It should be pointed out that hydrostatic pressure does not produce an effective compression and, therefore, is not considered. If the casing is suspended at the top of the hole that is filled with fluid, then the only effect of hydrostatic pressure is reduction of casing weight per foot and the string is effectively under tension.

An example is offered on the following pages of the design procedure to be followed in designing a combination string.

Example 4.16.4
Design a combination casing string if data are as below:

Type of well: exploration well
Type of casing: production for testing purposes
Casing setting depth: 12,000 ft
Casing size: 7 in.

Design model:

Collapse: Assumed external fluid pressure gradient of 0.52 psi/ft and casing empty inside. Safety factor for collapse $= 1.0$. Reduction of collapse pressure resistance due to the axial load is considered.
Burst: Assumed external pressure gradient of saltwater $= 0.465$ psi/ft and formation pore (gas) pressure gradient $= 0.65$ psi/ft. Gas weight is neglected. Safety factor $= 1.1$.
Tension: Casing suspended at the surface. Weight reduction due to buoyancy effect is ignored. Safety factor $= 1.6$.
Compression: casing not subjected to compression load.

The selected coupling should be long with round thread. The available casing grade is N-80 and unit weights as given below.

Steel grade	Unit weight (lb/ft)	Cross-sectional area (in.)	Collapse pressure resistance	Burst pressure resistance	Joint strength (10^3 lb)	Pipe body strength (10^3 lb)
N-80	26.0	7.548	5,410	7,240	519	604
N-80	26.0	8.451	7,020	8,160	597	676
N-80	26.0	9.315	8,600	9,060	672	745

Solution
Part 1. Consider collapse pressure and tension load.

Step 1. Determine the lightest weight of casing to resist collapse pressure for a setting depth of 12,000 ft. Because the maximum collapse pressure is $(12,000)(0.52) = 6,240$ psi, select N-80, 29-lb/ft casing with collapse pressure resistance of 7,020 psi. (Note: assumed safety factor for collapse $= 1.0$.) This is Section 1.

Step 2. The next section (above section 1) is to consist of the next lighter casing, i.e., N-80, 26-lb/ft. This is Section 2. Neglecting the effect of the axial load due to the weight of Section 1 suspended below it, the setting depth of Section 2 is

$$D_2' = \frac{5,410}{(1.0)(0.52)} = 10,403 \text{ ft}$$

Under this assumption, the mass of Section 1 is

$$W_1' = (12,000 - 10,403)(29) = 46,313 \text{ lb}$$

For this axial load, the reduced minimum collapse pressure resistance of Section 2 can be calculated from formula 4-312.
(*Note*: internal pressure P_i for considered case $= 0$.)

$$P_{CA}(2)$$

$$= 5,410 \left[\sqrt{1 - 0.75 \left(\frac{6,136}{80,000} \right)^2} - (0.5)\frac{6,136}{80,000} \right]$$

$$= 5,190 \text{ psi}$$

(*Note*: $S_A = W'/A = 46,313/7.548 = 6,136$ psi)

Step 3. Using the obtained reduced minimum collapse pressure resistance of Section 2, calculate the setting depth of this section:

$$D_2'' = \frac{5,190}{(1.0)(0.52)} = 9,980 \text{ ft}$$

For obtained setting depth of Section 2, the weight of Section 1 is

$$W_1'' = (12,000 - 9,980)(29) = 58,580 \text{ lb}$$

and corresponding reduced minimum collapse pressure resistance of Section 2 is

$$P_{CA}''$$

$$=5,410\left[\sqrt{1-0.75\left(\frac{7,761}{80,000}\right)^2}-(0.5)\frac{7,761}{80,000}\right]$$

$$=5,128\,psi$$

Step 4. The third assumed setting depth of Section 2 is usually taken as a correct setting depth, i.e.,

$$D_2''=\frac{5,128}{(1.0)(0.52)}=9,861\,ft$$

Then, the length and weight of Section 2 is

$$L_1=12,000-9,861=2,128\text{ ft}\text{ and } W_1=62,015\text{ lb}$$

(*Note*: If the next lighter casing were available, then Steps 2 through 4 must be repeated for this casing and that would be Section 3, etc.).

The maximum length of Section 2 is limited by coupling load capacity and is calculated below:

$$L_{2max}=\frac{519,000-(62,015)(1.6)}{(1.6)(26)}=10,090\text{ ft}$$

which is greater than its setting depth (9,861 ft). So, Section 2 extends to the top of the hole. (*Note*: If Section 2 would not cover the entire length of the hole, then the next stronger casing should be applied. That would be Section 3. The setting depth of Section 3 is governed by joint strength, not by collapse pressure.)

Part 2. Check casing string on burst pressure obtained in Part 1 and make necessary corrections.

Step 1. Determine external pressure.
At top: 0.0 psi
At bottom: $(12,000)(0.465)=5,580$ psi.
Step 2. Determine internal pressure
At bottom: $(12,000)(0.65)=7,800$ psi.
At top: 7,800 psi (weight of gas is ignored).
Step 3. Determine burst pressure.
At top: 7,800 psi.
At bottom: $7,800-5,580=2,220$ psi.
Burst pressure line equation is

$$P_b=7,800-\frac{7,800-2,200}{12,000}(D)$$

$$=7,800-(0.456)(D)$$

$$D=\text{hole depth in ft}$$

Graphical solution is presented in Figure 4.16.10.
Step 4. It is apparent that Section 1 is capable of withstanding the expected burst pressure.
Step 5. Section 2 can withstand the expected burst pressure up to the depth calculated below:

$$\frac{7,240}{1.1}=7,800-(0.456)(D)$$

$$D=\frac{7,800-6,581}{0.456}=2,673\,ft$$

Therefore, the length and weight of Section 2 is

$$9,861-2,673=7,188\,ft$$

$$W_2=(7,188)(26)=186,888\,lb$$

To cover the upper part of the hole (2,673 ft), stronger casing must be used.

Step 6. Take the next stronger casing, i.e., N-80, 29 lb/ft with burst pressure resistance of $(8,160)/(1.1)=7,418$ psi. This is Section 3. This casing can be used up to the hole depth of

$$7,418=7,800-(0.456)(D)$$

$$D=\frac{7,800-7,418}{0.456}=837\text{ ft}$$

Then, the length and weight of Section 3 is $2,673-837=1,836$ ft

$$W_3=(1,836)(26)=53,244\,lb$$

To cover the remaining 837 ft of the hole, stronger casing must be used.
Step 7. The next stronger casing is N-80, 32 lb/ft with burst pressure resistance of $(9,060)/(1.1)=8,236$ psi. This is Section 4.

N-80, 32 lb/ft casing is strong enough to cover the remaining part of the hole (*Note*: 8,236 psi > 7,800 psi). Therefore, the length and weight of Section 4 is $837-0.0=837$ ft and $W_4=26,784$ lb.

Part 3. Check casing string on tension obtained in Part 2 and, if necessary, make corrections to satisfy the required magnitude of safety factor.

Step 1. Maximum length of Section 3 due to its joint strength is

$$L_3=\frac{597,000-(1.6)(62,015+186,888)}{(1.6)(29)}$$

$$=4,283\,ft$$

Since the length of Section 3 is 1,836, the safety factor is even greater than required.
Step 2. Maximum length of Section 4 due to its joint strength is

$$L_4=\frac{672,000-(1.6)(62,015+186,888+26,784)}{(1.6)(32)}$$

$$=4,509\,ft$$

Because L_4 is greater than the length obtained in Part 2 (837), it may be concluded that the requirements for tension are satisfied.

A summary of the results obtained is presented in the table below and in Figure 4.16.11.

Section no.	Setting depth (ft)	Length	Weight (lb)	Grade	Unit weight	Coupling
4	837	837	26,784	N-80	32	Long with round thread
3	2,637	1,836	53,244	N-80	29	Long with round thread
2	9,861	7,188	186,888	N-80	26	Long with round thread
1	12,000	2,138	62,015	N-80	29	Long with round thread

4.16.4 Running and Pulling Casing

The following excerpts are taken from API Recommended Practice 5C1, "Care and Use of Casing and Tubing," 18th Edition, May 1999 (Section 4, Running and Pulling Casing).

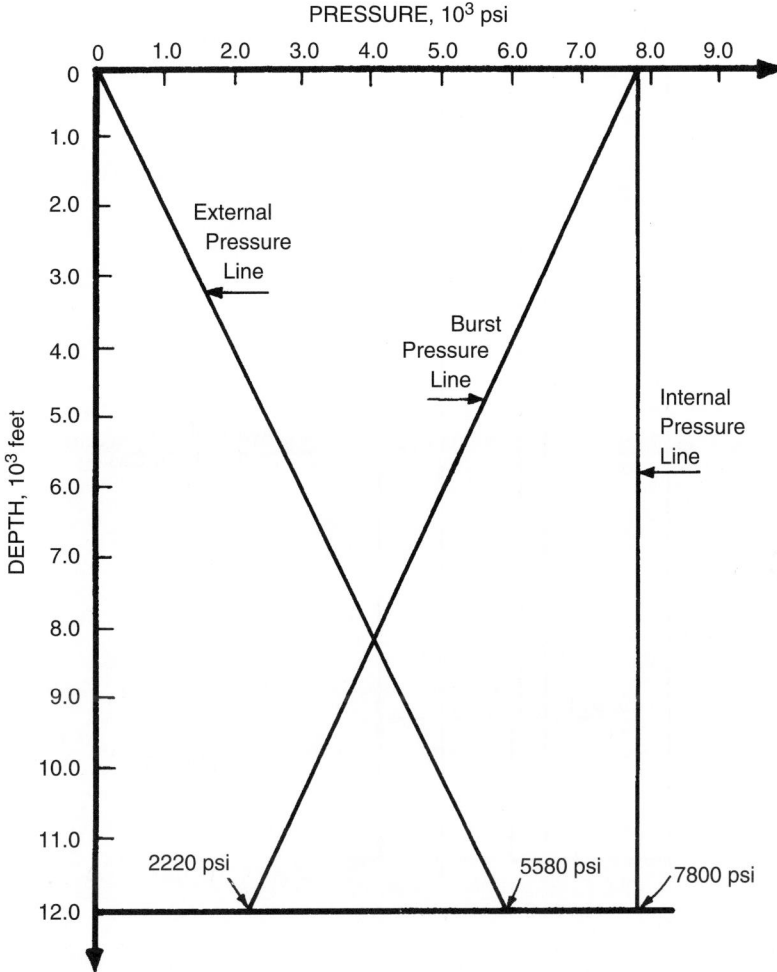

PRESSURE, 10³ psi

Figure 4.16.10 *Graphical representation of burst load determination.*

4.16.4.1 Preparation and Inspection Before Running

4.1.1. New casing is delivered free of injurious defects as defined in API Specification 5CT and within the practical limits of the inspection procedures therein prescribed. Some users have found that, for a limited number of critical well applications, these procedures do not result in casing sufficiently free of defects to meet their needs for such critical applications. Various nondestructive inspection services have been employed by users to ensure that the desired quality of casing is being run. In view of this practice, it is suggested that the individual user:

a. Familiarize himself with inspection practices specified in the standards and employed by the respective mills, and with the definition of "injurious defect" contained in the standards.

b. Thoroughly evaluate any nondestructive inspection to be used by him on API tubular goods to assure himself that the inspection does in fact correctly locate and differentiate injurious defects from other variables that can be and frequently are sources of misleading "defect" signals with such inspection methods.

4.1.2. All casing, whether new, used, or reconditioned, should always be handled with thread protectors in place. Casing should be handled at all times on racks or on wooden or metal surfaces free of rocks, sand, or dirt other than normal drilling mud. When lengths of casing are inadvertently dragged in the dirt, the threads should be recleaned and serviced again as outlined in 4.1.7.

4.1.3. Slip elevators are recommended for long strings. Both spider and elevator slips should be clean and sharp and should fit properly. Slips should be extra long for heavy casing strings. The spider must be level.

Note: Slip and tong marks are injurious. Every possible effort should be made to keep such damage at a minimum by using proper up-to-date equipment.

4.1.4. If collar-pull elevators are used, the bearing surface should be carefully inspected for (a) uneven wear that may produce a side lift on the coupling with danger of jumping it off, and (b) uniform distribution of the load when applied over the bearing face of the coupling.

4.1.5. Spider and elevator slips should be examined and watched to see that all lower together. If they lower

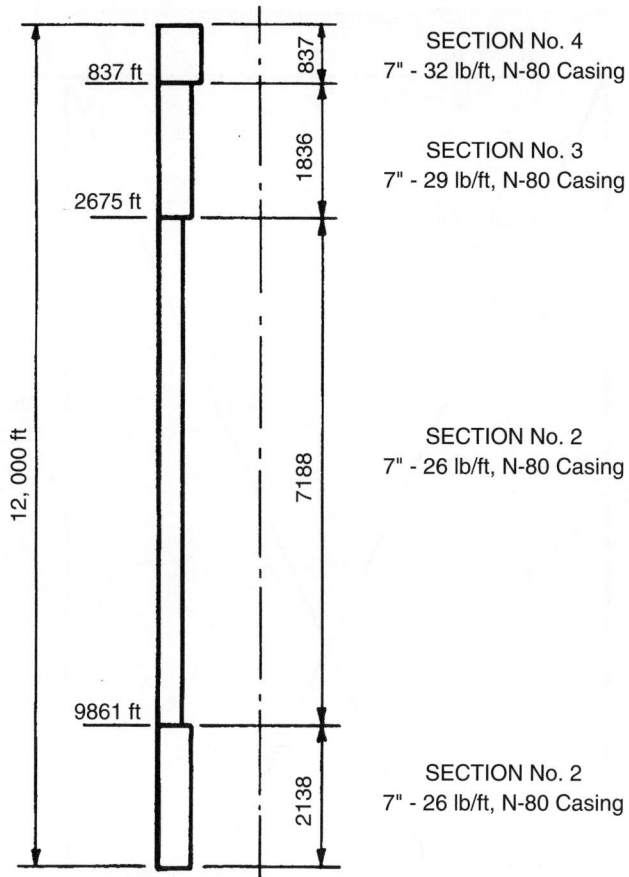

Figure 4.16.11   *Schematic diagram of combination casing string.*

unevenly, there is danger of denting the pipe or badly slip-cutting it.

4.1.6. Care shall be exercised, particularly when running long casing strings, to ensure that the slip bushing or bowl is in good condition. Tongs may be sized to produce 1.5 percent of the calculated pullout strength (API Bulletin 5C3) with units changed to ft-lb (N·m) (150 percent of the guideline torque found in Table 1). Tongs should be examined for wear on hinge pins and hinge surfaces. The backup line attachment to the backup post should be corrected, if necessary, to be level with the tong in the backup position so as to avoid uneven load distribution on the gripping surfaces of the casing. The length of the backup line should be such as to cause minimum bending stresses on the casing and to allow full stroke movement of the makeup tong.

4.1.7. The following precautions should be taken in the preparation of casing threads for makeup in the casing strings:

a. Immediately before running, remove thread protectors from both field and coupling ends and clean the threads thoroughly, repeating as additional rows become uncovered.

b. Carefully inspect the threads. Those found damaged, even slightly, should be laid aside unless satisfactory, means are available for correcting thread damage.

c. The length of each piece of casing shall be measured prior to running. A steel tape calibrated in decimal feet (millimeters) to the nearest 0.01 feet (millimeters) should be used. The measurement should be made from the outermost face of the coupling or box to the position on the externally threaded end where the coupling or the box stops when the joint is made up power tight. On round-thread joints, this position is to the plane of the vanish point on the pipe; on buttress-thread casing, this position is to the base of the triangle stamp on the pipe; and on extreme line casing, this position is to the shoulder on the externally threaded end. The total of the individual lengths so measured will represent the unloaded length of the casing string. The actual length under tension in the hole can be obtained by consulting graphs that are prepared for this purpose and are available in most pipe handbooks.

d. Check each coupling for makeup. If the standoff is abnormally great, check the coupling for tightness. Tighten any loose couplings after thoroughly cleaning the threads and applying fresh compound over entire thread surfaces, and before pulling the pipe into the derrick.

e. Before stabbing, liberally apply thread compound to the entire internally and externally threaded areas. It is recommended that a thread compound that meets the performance objectives of API Bulletin 5A2 be used; however, in special cases where severe conditions are encountered, it is recommended that high-pressure silicone thread compounds as specified in API Bulletin 5A2 be used.

f. Place a clean thread protector on the field end of the pipe so that the thread will not be damaged while rolling pipe on the rack and pulling into the derrick. Several thread protectors may be cleaned and used repeatedly for this operation.

g. If a mixed string is to be run, check to determine that appropriate casing will be accessible on the pipe rack when required according to program.

h. Connectors used as tensile and lifting members should have their thread capacity carefully checked to ensure that the connector can safely support the load.

i. Care should be taken when making up pup joints and connectors to ensure that the mating threads are of the same size and type.

4.16.4.2 Drifting of Casing

4.2.1. It is recommended that each length of casing be drifted for its entire length just before running, with mandrels conforming to API Specification 5CT. Casing that will not pass the drill test should be laid aside.

4.2.2. Lower or roll each piece of casing carefully to the walk without dropping. Use rope snubber if necessary. Avoid hitting casing against any part of derrick or other equipment. Provide a hold-back rope at window. For mixed or unmarked strings, a drift or "jack rabbit" should be run through each length of casing when it is picked up from the catwalk and pulled onto the derrick floor to avoid running a heavier length or one with a lesser inside diameter than called for in the casing string.

4.16.4.3 Stabbing, Making Up, and Lowering

4.3.1. Do not remove thread protector from field end of casing until ready to stab.

4.3.2. If necessary, apply thread compound over the entire surface of threads just before stabbing. The brush or utensil used in applying thread compound should be kept free of foreign matter, and the compound should never be thinned.

4.3.3. In stabbing, lower casing carefully to avoid injuring threads. Stab vertically, preferably with the assistance of a man on the stabbing board. If the casing stand tilts to one side after stabbing, lift up, clean, and correct any damaged thread with a three-cornered file, then carefully remove any filings and reapply compound over the thread surface. After stabbing, the casing should be rotated very slowly at first to ensure that threads are engaging properly and not cross-threading. If spinning line is used, it should pull close to the coupling.

Note: Recommendations in 4.3.4 and 4.4.1 for casing makeup apply to the use of power tongs. For recommendations on makeup of casing with spinning lines and conventional tongs, see 4.4.2.

4.4.4. The use of power tongs for making up casing made desirable the establishment of recommended torque values for each size, weight, and grade of casing. Early studies and tests indicated that torque values are affected by a large number of variables, such as variations in taper, lead, thread height and thread form, surface finish, type of thread compound, length of thread, weight and grade of pipe, etc. In view of the number of variables and the extent that these variables, alone or in combination, could affect the relationship of torque values versus made-up position, it was evident that both applied torque and made-up position must be considered. Since the API joint pullout strength formula in API Bulletin 5C2 contains several of the variables believed to affect torque, using a modified formula to establish torque values was investigated. Torque values obtained by taking 1 percent of the calculated pullout value were found to be generally comparable to values obtained by field makeup tests using API modified thread compound in accordance with API Bulletin 5A2. Compounds other than API modified thread compound may have other torque values. This procedure was therefore used to establish the makeup torque values listed in Table 1. All values are rounded to the nearest 10 ft-lb (10 N·m). These values shall be considered as a guide only, due to the very wide variations in torque requirements that can exist for a specific connection. Because of this, it is essential that torque be related to madeup position as outlined in 4.4.1. The torque values listed in Table 1 apply to casing with zinc-plated or phosphate-coated couplings. When making up connections with tin-plated couplings, 80 percent of the listed value can be used as a guide. The listed torque values are not applicable for making up couplings with PTFE (polytetrafluoroethylene) rings. When making up round thread connections with PTFE rings, 70 percent of the listed values are recommended. Buttress connections with PTFE seal rings may make up at torque values different from those normally observed on standard buttress threads.

Note: Thread galling of gall-prone materials (martensitic chromium steels, 9 Cr and 13 Cr) occurs during movement—stabbing or pulling and makeup or breakout. Galling resistance of threads is primarily controlled in two areas—surface preparation and finishing during manufacture and careful handling practices during running and pulling. Threads and lubricant must be clean. Assembly in the horizontal position should be avoided. Connections should be turned by hand to the hand-tight position before slowly power tightening. The procedure should be reversed for disassembly.

4.16.4.4 Field Makeup

4.4.1. The following practice is recommended for field makeup of casing:

a. For Round Thread, Sizes $4\frac{1}{2}$ through $1\frac{3}{8}$. OD

1. It is advisable when starting to run casing from each particular mill shipment to make up sufficient joints to determine the torque necessary to provide proper makeup. See 4.4.2 for the proper number of turns beyond hand-tight position. These values may indicate that a departure from the values listed in Table 1 is advisable. If other values are chosen, the minimum torque should be not less than 75 percent of the value selected. The maximum torque should be not more than 125 percent of the selected torque.

2. The power tong should be provided with a reliable torque gauge of known accuracy. In the initial stages of makeup, any irregularities of makeup or in speed of makeup should be observed, since these may be indicative of crossed threads, dirty or damaged threads, or other unfavorable conditions. To prevent galling

when making up connections in the field, the connections should be made up at a speed not to exceed 25 rpm.

3. Continue the makeup, observing both the torque gauge and the approximately position of the coupling face with respect to the thread vanish point position.

4. The torque values shown in Tables 1, 2, and 3 have been selected to give recommended makeup under normal conditions and should be considered as satisfactory providing the face of the coupling is flush with the thread vanish point or within two thread turns, plus or minus, of the thread vanish point.

5. If the makeup is such that the thread vanish point is buried two thread turns and 75 percent of the torque shown in Table 1 is not reached, the joint should be treated as a questionable joint as provided in 4.4.3.

6. If several threads remain exposed when the listed torque is reached, apply additional torque up to 125 percent of the value shown in Table 1. If the standoff (distance from face of coupling to the thread vanish point) is greater than three thread turns when this additional torque is reached, the joint should be treated as a questionable joint as provided in 4.4.3.

b. For buttress thread casing connections in sizes $4\frac{1}{2}$ through $13\frac{3}{8}$ OD, makeup torque values should be determined by carefully noting the torque required to make up each of several connections to the base of the triangle; then using the torque value thus established, make up the balance of the pipe of that particular weight and grade in the string.

c. For round thread and buttress thread, sizes 16, $18\frac{5}{8}$, and 20 outside diameter:

1. Makeup of sizes 16, $18\frac{5}{8}$, and 20 shall be to a position on each connection represented by the thread vanish point on 8-round thread and the base of the triangle on buttress thread using the minimum torque shown in Table 1 as a guide. On 8-round thread casing a $\frac{3}{8}$-inch (9.5-millimeter) equilateral triangle is die stamped at a distance of $L_1 + 1/16$ inch (1.6 millimeters) from each end. The base of the triangle will aid in locating the thread vanish point for basic power-tight makeup; however, the position of the coupling with respect to the base of the triangle shall not be a basis for acceptance or rejection of the product. Care shall be taken to avoid cross threading in starting these larger connections. The tongs selected should be capable of attaining high torques [50,000 ft-lb (67,800 N·m)] for the entire run. Anticipate that maximum torque values could be five times the minimum experienced in makeup to the recommended position.

2. Joints that are questionable as to their proper makeup in 4.4.1, item a.5 or a.6 should be unscrewed and laid down to determine the cause of improper makeup. Both the pipe thread and mating coupling thread should be inspected. Damaged threads or threads that do not comply with the specification should be repaired. If damaged or out-of-tolerance threads are not found to be the cause of improper makeup, then the makeup torque should be adjusted to obtain proper makeup (see 4.4.1, item a.1). It should be noted that a thread compound with a coefficient of friction substantially different from common values may be the cause of improper makeup.

4.4.2. When conventional tongs are used for casing makeup, tighten with tongs to proper degree of tightness. The joint should be made up beyond the hand-tight position at least three turns for sizes $4\frac{1}{2}$ through 7, and at least three and one-half turns for sizes $7\frac{5}{8}$ and larger, except $9\frac{5}{8}$, and $10\frac{3}{4}$ grade P110 and size 20 grade J55 and K55, which should be made up four turns beyond hand-tight position. When using a spinning line, it is necessary to compare hand tightness with spin-up tightness. In order to do this, make up the first few joints to the hand-tight position, then back off and spin up joints to the spin-up tight position. Compare relative position of these two makeups and use this information to determine when the joint is made up the recommended number of turns beyond hand tight.

4.4.3. Joints that are questionable as to their proper tightness should be unscrewed and the casing laid down for inspection and repair. When this is done, the mating coupling should be carefully inspected for damaged threads. Parted joints should never be reused without shopping or regauging, even though the joints may have little appearance of damage.

4.4.4. If casing has a tendency to wobble unduly at its upper end when making up, indicating that the thread may not be in line with the axis of the casing, the speed of rotation should be decreased to prevent galling of threads. If wobbling should persist despite reduced rotational speed, the casing should be laid down for inspection. Serious consideration should be given before using such casing in a position in the string where a heavy tensile load is imposed.

4.4.5. In making up the field joint, it is possible for the coupling to make up slightly on the mill end. This does not indicate that the coupling on the mill end is too loose but simply that the field end has reached the tightness with which the coupling was screwed on at the manufacturer's facility.

4.4.6. Casing strings should be picked up and lowered carefully and care exercised in setting slips to avoid shock loads. Dropping a string even a short distance may loosen couplings at the bottom of the string. Care should be exercised to prevent setting casing down on bottom or otherwise placing it in compression because of the danger of buckling, particularly in that part of the well where hole enlargement has occurred.

4.4.7. Definite instructions should be available as to the design of the casing string, including the proper location of the various grades of steel, weights of casing, and types of joint. Care should be exercised to run the string in exactly the order in which it was designed. If any length cannot be clearly identified, it should be laid aside until its grade, weight, or type of joint can be positively established.

4.4.8. To facilitate running and to ensure adequate hydrostatic head to contain reservoir pressures, the casing should be periodically filled with mud while being run. A number of things govern the frequency with which filling should be accomplished: weight of pipe in the hole, mud weight, reservoir pressure, etc. In most cases, filling every six to ten lengths should suffice. In no case should the hydrostatic balance of reservoir pressure be jeopardized by too infrequent filling. Filling should be done with mud of the proper weight, using a conveniently located hose of adequate size to expedite the filling operation. A quick opening and closing plug valve on the mud hose will facilitate the operation and prevent overflow. If rubber hose is used, it is recommended that the quickclosing valve be mounted where the hose is connected to the mud line, rather than at the outlet end of the hose. It is also recommended that at least one other discharge connection be left open on the mud system to prevent buildup of excessive pressure when the quick-closing valve is closed while the pump is still running. A cooper nipple at the end of the mud hose may be used to prevent damaging of the coupling threads during the filling operation.

Note: The foregoing mud fill-up practice will be unnecessary if automatic fill-up casing shoes and collars are used.

4.16.4.5 Casing Landing Procedure
Definite instructions should be provided for the proper string tension, also on the proper landing procedure after the cement has set. The purpose is to avoid critical stresses or excessive and unsafe tensile stresses at any time during the life of the well. In arriving at the proper tension and landing procedure, consideration should be given to all factors, such as well temperature and pressure, temperature developed due to cement hydration, mud temperature, and changes of temperature during producing operations. The adequacy of the original tension safety factor of the string as designed will influence the landing procedure and should be considered. If, however, after due consideration it is not considered necessary to develop special landing procedure instructions (and this probably applies to a very large majority of the wells drilled), then the procedure should be followed of landing the casing in the casing head at exactly the position in which it was hanging when the cement plug reached its lowest point or "as cemented."

4.16.4.6 Care of Casing in Hole
Drill pipe run inside casing should be equipped with suitable drill-pipe protectors.

4.16.4.7 Recovery of Casing
4.7.1. Breakout tongs should be positioned close to the coupling but not too close since a slight squashing effect where the tong dies contact the pipe surface cannot be avoided, especially if the joint is tight and/or the casing is light. Keeping a space of one-third to one-quarter of the diameter of the pipe between the tong and the coupling should normally prevent unnecessary friction in the threads. Hammering the coupling to break the joint is an injurious practice. If tapping is required, use the flat face, never the peen face of the hammer, and under no circumstances should a sledgehammer be used. Tap lightly near the middle and completely around the coupling, never near the end nor on opposite sides only.

4.7.2. Great care should be exercised to disengage all of the thread before lifting the casing out of the coupling. Do not jump casing out of the coupling.

4.7.3. All threads should be cleaned and lubricated or should be coated with a material that will minimize corrosion. Clean protectors should be placed on the tubing before it is laid down.

4.7.4. Before casing is stored or reused, pipe and thread should be inspected and defective joints marked for shopping and regauging.

4.7.5. When casing is being retrieved because of a casing failure, it is imperative to future prevention of such failures that a thorough metallurgical study be made. Every attempt should be made to retrieve the failed portion of the "as-failed" condition. When thorough metallurgical analysis reveals some facet of pipe quality to be involved in the failure, the results of the study should be reported to the API office.

4.7.6. Casing stacked in the derrick should be set on a firm wooden platform and without the bottom thread protector since the design of most protectors is not such as to support the joint or stand without damage to the field thread.

4.16.4.8 Causes of Casing Troubles
The more common causes of casing troubles are listed in 4.8.1 through 4.8.16.

4.8.1. Improper selection for depth and pressures encountered.

4.8.2. Insufficient inspection of each length of casing or of field-shop threads.

4.8.3. Abuse in mill, transportation, and field handling.

4.8.4. Nonobservance of good rules in running and pulling casing.

4.8.5. Improper cutting of field-shop threads.

4.8.6. The use of poorly manufactured couplings for replacements and additions.

4.8.7. Improper care in storage.

4.8.8. Excessive torquing of casing to force it through tight places in the hole.

4.8.9. Pulling too hard on a string (to free it). This may loosen the couplings at the top of the string. They should be retightened with tongs before finally setting the string.

4.8.10. Rotary drilling inside casing. Setting the casing with improper tension after cementing is one of the greatest contributing causes of such failures.

4.8.11. Drill-pipe wear while drilling inside casing is particularly significant in drifted holes. Excess doglegs in deviated holes, or occasionally in straight holes where corrective measures are taken, result in concentrated bending of the casing that in turn results in excess internal wear, particularly when the doglegs are high in the hole.

4.8.12. Wire-line cutting, by swabbing or cable-tool drilling.

4.8.13. Buckling of casing in an enlarged, washed-out uncemented cavity if too much tension is released in landing.

4.8.14. Dropping a string, even a very short distance.

4.8.15. Leaky joints, under external or internal pressure, are a common trouble, and may be due to the following:
   a. Improper thread compound.
   b. Undertonging.
   c. Dirty threads.
   d. Galled threads, due to dirt, careless stabbing, damaged threads, too rapid spinning, overtonging, or wobbling during spinning or tonging operations.
   e. Improper cutting of field-shop threads.
   f. Pulling too hard on string.
   g. Dropping string.
   h. Excessive making and breaking.
   i. Tonging too high on casing, especially on breaking out. This gives a bending effect that tends to gall the threads.
   j. Improper joint makeup at mill.
   k. Casing ovality or out-of-roundness.
   l. Improper landing practice, which produces stresses in the threaded joint in excess of the yield point.

4.8.16. **Corrosion.** Both the inside and outside of casing can be damaged by corrosion, which can be recognized by the presence of pits or holes in the pipe. Corrosion on the outside of casing can be caused by corrosive fluids or formations in contact with the casing or by stray electric current flowing out of the casing into the surrounding fluids or formations. Severe corrosion may also be caused by sulphate-reducing bacteria. Corrosion damage on the inside is usually caused by corrosive fluids produced from the well, but the damage can be increased by the abrasive effects of casing and tubing pumping equipment any by high fluid velocities such as those encountered in some gas-lifted wells. Internal corrosion might also be due to stray electrical currents (electrolysis) or to dissimilar metals in close contact (bimetallic galvanic corrosion).

Because corrosion may result from so many different conditions, no simple or universal remedy can be given for its control. Each corrosion problem must be treated as an individual case and a solution attempted in the light of the known corrosion factors and operating conditions. The condition of the casing can be determined by visual or optical-instrument inspections. Where these are not practical, a casing-caliper survey can be made to determine the condition of the inside surfaces. No tools have yet been designed for determining the condition of the outside of casing in a well. Internal casing caliper surveys indicate the extent, location, and severity of corrosion. On the basis of the industry's experience to date, the following practices and measures can be used to control corrosion of casing:

a. Where external casing corrosion is known to occur or stray electrical current surveys indicate that relatively high currents are entering the well, the following practices can be employed:
1. Good cementing practices, including the use of centralizers, scratchers, and adequate amounts of cement to keep corrosive fluids from contact with the outside of the casing.
2. Electrical insulation of flow lines from wells by the use of nonconducting flange assemblies to reduce or prevent electrical currents from entering the well.
3. The use of highly alkaline mud or mud treated with a bactericide as a completion fluid will help alleviate corrosion caused by sulfate-reducing bacteria.
4. A properly designed cathodic protection system similar to that used for line pipe, which can alleviate external casing corrosion. Protection criteria for casing differ somewhat from the criteria used for line pipe. Literature on external casing corrosion or persons competent in this field should be consulted for proper protection criteria.

b. Where internal corrosion is known to exist, the following practices can be employed:
1. In flowing wells, packing the annulus with fresh water or low-salinity alkaline muds. (It may be preferable in some flowing wells to depend upon inhibitors to protect the inside of the casing and the tubing.)
2. In pumping wells, avoiding the use of casing pumps. Ordinarily, pumping wells, should be tubed as close to bottom as practical, regardless of the position of the pump, to minimize the damage to the casing from corrosive fluids.
3. Using inhibitors to protect the inside of the casing against corrosion.

c. To determine the value and effectiveness of the above practices and measures, cost and equipment-failure records can be compared before and after application of control measures. Inhibitor effectiveness may be checked also by means of caliper surveys, visual examinations of readily accessible pieces of equipment, and water analyses for iron content. Coupons may also be helpful in determining whether sufficient inhibitor is being used. When lacking previous experience with any of the above measures, they should be used cautiously and on a limited scale until appraised for the particular operating conditions.

d. In general, all new areas should be considered as being potentially corrosive and investigations should be initiated early in the life of a field, and repeated periodically, to detect and localize corrosion before it has done destructive damage. These investigations should cover: (1) a complete chemical analysis of the effluent water, including pH, iron, hydrogen sulfide, organic acids, and any other substances that influence or indicate the degree of corrosion. An analysis of the produced gas for carbon dioxide and hydrogen sulfide is also desirable; (2) corrosion rate tests by using coupons of the same materials as in the well; and (3) the use of caliper or optical instrument inspections. Where conditions favorable to corrosion exist, a qualified corrosion engineer should be consulted. Particular attention should be given to mitigation of corrosion where the probable life of subsurface equipment is less than the time expected to deplete a well.

e. When H_2S is present in the well fluids, casing of high yield strength may be subject to sulfide corrosion cracking. The concentration of H_2S necessary to cause cracking in different strength materials is not yet well defined. Literature on sulfide corrosion or persons competent in this field should be consulted.

4.17 WELL CEMENTING

4.17.1 Introduction

Cementing the casing and *liner cementing* (denoted as *primary cementing*) are probably the most important operations in the development of an oil and gas well. The drilling group is usually responsible for cementing the casing and the liner. The quality of these cementing operations will affect the success of follow-on drilling, completion, production and workover efforts in the well [1–3].

In addition to primary cementing of the casing and liner, there are other important well cementing operations. These are *squeeze cementing* and *plug cementing*. Such operations are often called *secondary* or *remedial cementing* [2].

Well cementing materials vary from basic Portland cement used in civil engineering construction of all types, to highly sophisticated special-purpose resin-based or latex cements. The purpose of all of these cementing materials is to provide the well driller with a fluid state slurry of cement, water and additives that can be pumped to specific locations within the well. Once the slurry has reached its intended location in the well and a setup time has elapsed, the slurry material can become a nearly impermeable, durable solid material capable of bonding to rock and steel casing.

The most widely used cements for well cementing are the Portland-type cements. The civil engineering construction industry uses Portland cement and water slurries in conjunction with clean rock aggregate to form concrete. The composite material formed by the addition of rock aggregate forms a solid material that has a compressive strength that is significantly higher than the solid formed by the solidified cement and water slurry alone. The rock of the aggregate usually has a very high compressive strength (of the order of 5,000 to 20,000 psi). The cement itself will have a compressive strength of about 1,000 to 3,000 psi. Therefore, the rock of the aggregate forms a solid material that has a very high compressive strength of the order of 5,000 to 2,000 psi (345 to 1830 bar). The cement itself will have a compressive strength of about 1,000 to 3,000 psi (69 to 207 bar). Therefore, the rock aggregate together with the matrix of solid cement can form a high-strength composite concrete with compressive strengths of the order of 4,000 to 15,000 (276 to 1035 bar).

The well drilling industry does not generally use aggregate with the cement except for silica flour and Ottawa sand. This is mainly due to the tight spacing within a well that precludes the passage of the larger particles of aggregate through the system. Thus, the well drilling industry refers to this material as simply cement. The slurry pumped to wells is usually a slurry of cement and water with appropriate additives. Because of the lack of aggregate, the compressive strength of well cements are restricted to the order of 200 to about 3,000 psi (14 to 207 bar) [1].

Oil well compressive strengths in the range of 500 psi (35 bar) have long been considered acceptable. Cement mixture selection is based on a much broader criteria then compressive strength alone.

4.17.2 Chemistry of Cements

Cement is made of calcareous and argillaceous rock materials that are usually obtained from quarries. The process of making cement requires that these raw rock materials be ground, mixed and subjected to high temperatures.

The calcareous materials contain calcium carbonate or calcium oxide. Typical raw calcareous materials are as follows:

Limestone. This is sedimentary rock that is formed by the accumulation of organic marine life remains (shells or coral). Its main component is calcium carbonate.
Cement rock. This is a sedimentary rock that has a similar composition as the industrially produced cement.
Chalk. This is a soft limestone composed mainly of marine shells.
Marl. This is loose or crumbly deposit that contains a substantial amount of calcium carbonate.
Alkali waste. This is a secondary source and is often obtained from the waste of chemical plants. Such material will contain calcium oxide and/or calcium carbonate.

The argillaceous materials contain clay or clay minerals. Typical raw argillaceous materials are as follows:

Clay. This material is found at the surface of the earth and often is the major component of soils. The material is plastic when wetted, but becomes hard and brittle when dried and heated. It is composed mainly of hydrous aluminum silicates as well as other minerals.
Shale. This sedimentary rock is formed by the consolidation of clay, mud and silt. It contains substantial amounts of hydrous aluminum silicates.
Slate. A dense fine-grained metamorphic rock containing mainly clay minerals. Slate is obtained from metamorphic shale.
Ash. This is a secondary source and is the by-product of coal combustion. It contains silicates.

There are two processes used to manufacture cements: the dry process and the wet process. The dry process is the least expensive of the two, but is the more difficult to control.

In the dry process the limestone and clay materials are crushed and stored in separate bins and their composition analyzed. After the composition is known, the contents of the bins are blended to achieve the desired ultimate cement characteristics. The blend is ground to a mesh size of 100–200. This small mesh size maximizes the contact between individual particles.

In the wet process the clay minerals are crushed and slurried with water to allow pebbles and other rock particles to settle out. The limestone is also crushed and slurried. Both materials are stored in separate bins and analyzed. Once the desired ultimate composition is determined, the slurry blend is ground and then partially dried out.

After the blends have been prepared (either in the dry or wet process), these materials are fed at a uniform rate into a long rotary kiln. The materials are gradually heated to a liquid state. At temperatures up to about 1,600°F the free water evaporates, the clay minerals dehydroxylate and crystallize, and $CaCO_3$ decomposes. At temperatures up to about 1600°F (870°C) the $CaCO_3$ and CaO react with aluminosilicates and the materials become liquids. Heating is continued to as high as 2800°F (1540°C).

When the kiln material is cooled it forms into crystallized clinkers. These are rather large irregular pieces of the solidified cement material. These clinkers are ground and a small amount of gypsum is added (usually about 1.5 to 3%). The gypsum prevents flash setting of the cement and also controls free CaO. This final cement product is sampled, analyzed and stored. The actual commercial cement is usually a blend of several different cements. This blending ensures a consistent product.

There are four chemical compounds that are identified as being the active components of cements.

Tricalcium aluminate ($3CaO \bullet Al_2O_3$) hydrates rapidly and contributes most to heat of hydration. This compound does not contribute greatly to the final strength of the set cement, but it sets rapidly and plays an important role in the early strength development. This setting time can be controlled by the addition of gypsum. The final hydrated product of tricalcium aluminate is readily attacked by sulfate waters. High-sulfate-resistant (HSR) cements have only a 3% or less content of this compound. High early strength cements have up to 15% of this compound.
Tricalcium silicate ($3CaO \bullet SiO_2$) is the major contributor to strength at all stages, but particularly during early stages of curing (up to 28 days). The average tricalcium silicate content is from 40% to maximum of 67%. The retared cements will contain from 40% to 45%. The high early strength cements will contain 60% to 67%.
Dicalcium silicate ($2CaO \bullet SiO_2$) is very important in the final strength of the cement. This compound hydrates very slowly. The average dicalcium silicate content is 25% to 35%.
Tetracalcium aluminoferrite ($4CaO \bullet Al_2O_3 \bullet FeO_3$) has little effect on the physical properties of the cement. For high-sulfate-resistant (HSR) cements, API specifications require that the sum of the tetracalcium aluminoferrite content plus twice the tricaclium aluminate may not exceed a maximum of 24%.

In addition to the four compounds discussed above, the final Portland cement may contain gypsum, alkali sulfates, magnesia, free lime and other components. These do not significantly affect the properties of the set cement, but they can influence rates of hydration, resistance to chemical attack and slurry properties.

When the water is added to the final dry cement material, the hydration of the cement begins immediately. The water is combined chemically with the cement material to eventually form a new immobile solid. As the cement hydrates, it will bond to the surrounding surfaces. This cement bonding is complex and depends on the type of surface to be cemented. Cement bonds to rock by a process of crystal growth. Cement bonds to the outside of a casing by filling in the pit spaces in the casing body [4].

4.17.3 Cementing Principles

There are two basic oil well cementing activities: primary cementing and secondary cementing.

Primary cementing refers to the necessity to fix the steel casing or liner (which is placed in the drilled borehole) to the surrounding formations adjacent to the casing or liner. The purposes of primary cementing are the following:

1. Support vertical and radial loads applied to casing
2. Isolate porous formations from producing zone formations
3. Exclude unwanted subsurface fluids from the producing interval
4. Protect casing from corrosion
5. Resist chemical deterioration of cement
6. Confine abnormal formation pressures

When applied to the various casing or liner strings used in oil well completions, the specific purposes of each string are as follows:

- Conductor casing string is cemented to prevent the drilling fluid from escaping and circulating outside the casing.
- Surface casing string must be cemented to protect fresh-water formations near the surface and provide a structural connection between the casing and the subsurface competent rock formations. This subsurface structural connection will allow the blowout preventor to be affixed to the top of this casing to prevent high-pressure fluids from being vented to the surface. Further, this structural connection will give support for deeper run casing or liner strings.
- Intermediate casing strings are cemented to seal off abnormal pressure formations and cover both incompetent formations, which could cave or slough, and lost circulation formations.
- Production casing string is cemented to prevent the produced fluids from migrating to nonproducing formations and to exclude other fluids from the producing interval.

Cementing operations are carried out with surface equipment specially designed to carry out the primary and secondary cementing operations in oil wells. The key element in any well cementing operation is the recirculating blender (Figure 4.17.1). The recirculating blender has replaced the older jet mixing hopper. The blender provides a constant cement slurry specific weight that could never be achieved by the older equipment. A very careful control of the slurry weight is critical to a successful cementing operation. The recirculating blender is connected to a cement pump that in turn pumps the cement at low circulation rates (and high pressures if necessary) to the cementing head at the top of the casing string (see Figures 4.17.2 and 4.17.3 and Ref. [3]).

The proper amount of water must be mixed with the dry cement product to ensure only sufficient water for hydration of the cement. Excess water above that needed for hydration will reduce the final strength of the set cement and leave voids in the cement column that are filled with unset liquid. Insufficient water for proper hydration will leave voids filled with dry unset cement, or result in a slurry too viscous to pump.

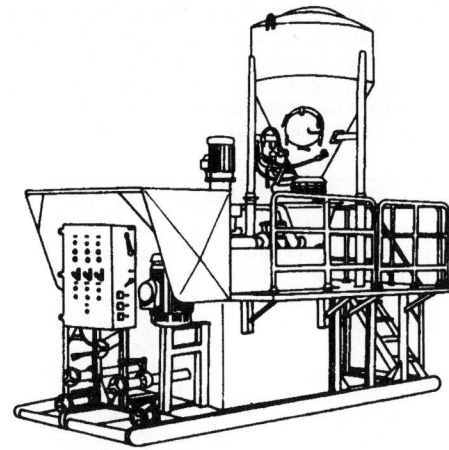

Figure 4.17.1   *Recirculating blender [2].*

The cement is usually dry mixed with the additives that are usually added for a particular cementing application. Often this mixing of additives is carried out at the service company central location. Depending on the application of the well cement, there are a variety of additives that can be used to design the cement slurry characteristics. These are accelerators, retarders, dispersants, extenders, weighting agents, gels, foamers and fluid-loss additives. With these additives the cement slurry and ultimately the set cement can be designed for the particular cementing operation. It is necessary that the engineer in charge of the well carry out the necessary engineering design of the slurries to be used in the well. In addition, the engineer should ensure that the work is carried out in accordance to the design specifications. This critical activity should not be left to the service company technicians.

4.17.4 Standardization and Properties of Cements

The American Petroleum Institute (API) has nine classes of well cements. These are as follows [5]:

Class A: Intended for use from surface to 6,000 ft (1,830 m) depth, when special properties are not required. Available only in ordinary type (similar to ASTM C 150, Type I[1]).

Class B: Intended for use from surface to 6,000 ft (1,830 m) depth, when conditions require moderate to high sulfate resistance. Available in both moderate (similar to ASTM C 150, Type II) and high-sulfate-resistant types.

Class C: Intended for use from surface to 6,000 ft (1,830 m) depth, when conditions require high early strength. Available in ordinary and moderate (similar to ASTM C 150, Type III) and high-sulfate-resistant types.

Class D: Intended for use from 6,000 to 10,000 ft (1,830 to 3,050 m) depth, under conditions of moderately high temperatures and pressures. Available in both moderate and high-sulfate-resistant types.

Class E: Intended for use from 10,000 to 14,000 ft (3,050 to 4,270 m) depth, under conditions of high temperatures and pressures. Available in both moderate and high-sulfate-resistant types.

Class F: Intended for use from 10,000 to 16,000 ft (3,050 to 4,880 m) depth, under conditions of extremely high

---

[1]American Society for Testing and Materials (ASTM).

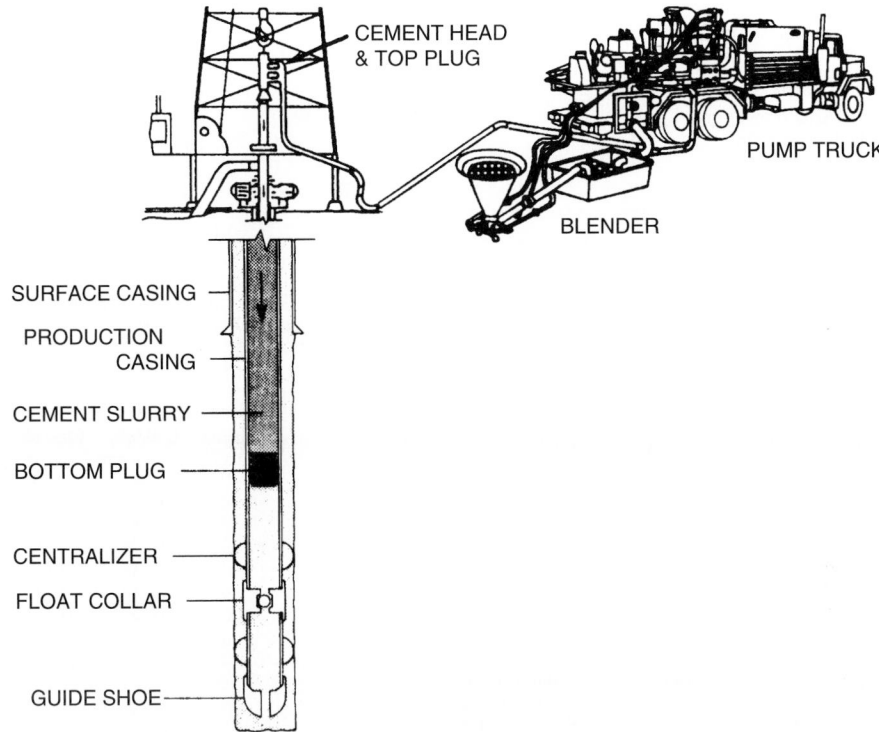

Figure 4.17.2   *Blender, pump truck, cementing head and subsurface equipment [2].*

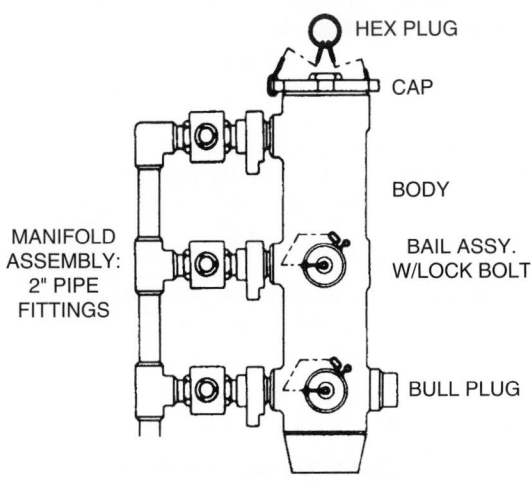

Figure 4.17.3   *Cementing head [2].*

Class H: Intended for use as a basic well cement from surface to 8,000 ft (2,440 m) depth as manufactured, and can be used with accelerators and retarders to cover a wide range of well depths and temperatures. No additions other than calcium sulfate or water, or both, shall be interground or blended with the clinker during manufacture of Class H well cement. Available in moderate and high-(tentative) sulfate-resistant types.

Class J: Intended for use as manufactured from 12,000 to 16,000 ft (3,660 to 4,880 m) depth under conditions of extremely high temperatures and pressures or can be used with accelerators and retarders to cover a range of well depths and temperatures. No additions of retarder other than calcium sulfate or blended with the clinker during manufacture of class J well cement.

The ASTM specifications provide for five types of Portland cements: Types I, II, III, IV and V; they are manufactured for use at atmospheric conditions [6]. The API Classes A, B and C correspond to ASTM Types I, II and III, respectively. The API Classes D, E, F, G, H and J are cements manufactured for use in deep wells and to be subject to a wide range of pressures and temperatures. These classes have no corresponding ASTM types.

Sulfate resistance is an extremely important property of well cements. Sulfate minerals are abundant in some underground formation waters that can come into contact with set cement. The sulfate chemicals, which include magnesium and sodium sulfates, react with the lime in the set cement to form magnesium hydroxide, sodium hydroxide and calcium sulfate. The calcium sulfate reacts with the tricalcium aluminate components of cement to form sulfoaluminate, which causes expansion and ultimately disintegrates to the

temperatures and pressures. Available in both moderate and high-sulfate-resistant types.

Class G: Intended for use from surface to 8,000 ft (2,440 m) depth as manufactured, or can be used with accelerators and retarders to cover a wide range of well depths and temperatures. No additions other than calcium sulfate or water, or both, shall be interground or blended with the clinker during manufacture of Class G well cement. Available in moderate and high-sulfate-resistant types.

Table 4.17.1 *Properties of the Various Classes of API Cements*

Cement Class	A	B	C	D	E	F	G	H
Specific Gravity	3.14	3.14	3.14	3.16	3.16	3.16	3.16	3.16
Surface Area (cm²/gm)	1500	1600	2200	1200	1200	1200	1400	1200
Weight Per Sack, lb	94	94	94	94	94	94	94	94
Bulk Volume, ft³/sk	1	1	1	1	1	1	1	1
Absolute Volume, gal/sk	3.59	3.59	3.55	3.57	3.57	3.57	3.62	3.57

set cement. To increase the resistance of a cement to sulfate attack, the amount of tricalcium aluminate and free lime in the cement should be decreased. Alternatively, the amount of pozzolanic material can be increased in the cement to obtain a similar resistance. The designations of ordinary sulfate resistance, moderate sulfate resistance and high sulfate resistance in the cement classes above indicate decreasing amounts of tricalcium aluminate.

Table 4.17.1 gives the basic properties of the various classes of the dry API cements [5].

4.17.5 Properties of Cement Slurry and Set Cement

In well engineering and applications, cement must be dealt with in both its slurry form and in its set form. At the surface the cement must be mixed and then pumped with surface pumping equipment through tubulars to a designated location in the well. After the cement has set, its structure must support the various static and dynamic loads placed on the well tubulars.

4.17.5.1 Specific Weight

Specific weight is one of the most important properties of a cement slurry. A neat cement slurry is a combination of only cement and water. The specific weight of a neat cement slurry is defined by the amount of water used with the dry cement. The specific weight range for a particular class of cement is, therefore, limited by the minimum and maximum water-to-cement ratios permissible by API standards.

The minimum amount of water for any class of cement is defined as that amount of water that can be used in the slurry with the dry cement that will still produce a neat slurry with a consistency that is below 30 Bearden units of consistency.[2] Note that the minimum water content defined in this manner is much greater than the stoichiometric minimum for cement hydration and setting.

The maximum amount of water for any class of cement is usually defined as the ratio that results in the cement particles remaining in suspension until the initial set of the slurry has taken place. If more than the maximum amount of water is used, then the cement particles will settle in such a manner that there will be pockets of free water within the set cement column.

Table 4.17.2 gives the API recommended optimum water-to-cement ratios and the resulting neat slurry specific weight (lb/gal) and specific volume, or yield (ft³/sack) for the various classes of API cements [2].

Table 4.17.3 gives the maximum and minimum water-to-cement ratios and the resulting neat slurry specific weight and specific volume, or yield for three classes of API cement.

The specific weight of a cement slurry $\bar{\gamma}$ (lb/gal) is

$$\bar{\gamma} = \frac{\text{Cement (lb)} + \text{Water (lb)} + \text{Additive (lb)}}{\text{Cement (gal)} + \text{Water (gal)} + \text{Additive (gal)}} \qquad [4.17.1]$$

[2]This was formally referred to as poise.

Table 4.17.2 *Properties of Neat Cement Slurries for Various Classes of API Cements*

API Class	API Rec'd Water (gal/sk)	API Rec'd Water (gal/sk)	API Rec'd Water (gal/sk)	Percent of Mixing Water
A (Portland)	5.20	15.6	1.18	46
B (Portland)	5.20	15.6	1.18	46
C (High Early)	6.32	14.8	1.32	56
D (Retarded)	4.29	16.46	1.05	38
E (Retarded)	4.29	16.46	1.05	38
F (Retarded)	4.29	16.46	1.05	38
G (Basic)	4.97	15.8	1.15	44
H (Basic)	4.29	16.46	1.05	38

The volumes above are absolute volumes. For example, a 94-lb sack of cement contains 1 ft³ of bulk cement powder, yet the actual or absolute space occupied by the cement particles is only 0.48 ft³.

For dealing with other powdered materials that are additives to cement slurries, the absolute volume must be used. The absolute volume (gal) is

$$\text{Absolute value (gal)} = \frac{\text{Additive material (lb)}}{(8.34\,\text{lb/gal})(\text{S.G. of material})} \qquad [4.17.2]$$

The volume of slurry that results from 1 sack of cement additives is defined as the yield. The yield (ft³/sack) is

$$\text{Yield} = \frac{\text{Cement (gal)} + \text{Water (gal)} + \text{Additives (gal)}}{7.48\,\text{gal/ft}^3} \qquad [4.17.3]$$

where the cement, water and additives are those associated with 1 sack of cement, or 94 lb of cement.

Example 4.17.1

Calculate the specific weight and yield for a neat slurry of Class A cement using the maximum permissible water-to-cement ratio.

Table 4.17.3 gives the maximum water-to-cement ratio for Class A cement as 5.5 gal/sack. Thus using the absolute volume for Class A cement from Table 4.17.1 as 3.59 gal/sack, Equation 4.17.1 is

$$\bar{\gamma} = \frac{94 + 5.5(8.34)}{3.59 + 5.5}$$

$$= 15.39\,\text{lb/gal}$$

The yield determined from Equation 4.17.3 is

$$\text{Yield} = \frac{3.59 + 5.5}{7.48}$$

$$= 1.22\,\text{ft}^3/\text{sack}$$

Table 4.17.3 *Maximum and Minimum Water-to-Cement Ratios for API Classes A, C and E Neat Cement Slurries*

API Class	Maximum Water			Minimum Water		
	Water (gal/sk)	Weight (lb/gal)	Volume (ft^3/sk)	Water (gal/sk)	Weight (pb/gal)	Volume (ft^3/sk)
A	5.5	15.39	1.22	3.90	16.89	1.00
C	7.9	13.92	1.53	6.32	14.80	1.32
E	4.4	16.36	1.07	3.15	17.84	0.90

It is often necessary to decrease the specific weight of a cement slurry to avoid fracturing weak formations during cementing operations. There are basically two methods for accomplishing lower specific weights. These are

1. Adding clay or chemical silicate type extenders together with their required extra water.
2. Adding large quantities of pozzolan, ceramic microspheres or nitrogen. These materials lighten the slurry because they have lower specific gravities than the cement.

When using the first method above great care must be taken not to use too much water. There is a maximum permissible water-to-cement ratio for each cement class. This amount of water can be used with the appropriate extra water required for the added clay or chemical silicate material. Using too much water will result in a very poor cement operation.

It also may be necessary to increase the specific weight of a cement slurry, particularly when cementing through high-pressure formations. There are basically two methods for accomplishing higher specific weights. These are

1. Using the minimum permissible water-to-cement ratio for the particular cement class and adding dispersants to increase the fluidity of the slurry.
2. Adding high-specific-gravity materials to the slurry together with optimal or slightly reduced (but not necessarily the minimum) water-to-cement ratio for the particular cement class.

The first method above is usually restricted to setting plugs in wells since it results in high strength cement that is rather difficult to pump. The second method is used for primary cementing, but these slurries are difficult to design since the settling velocity of the high-specific-gravity additive must be taken into consideration.

4.17.5.2 Thickening Time

It is important that the thickening time for a given cement slurry be known prior to using the slurry in a cementing operation. When water is added to dry cement and its additives, a chemical reaction begins that results in an increase in slurry viscosity. This viscosity increases over time, which will vary in accordance with the class of cement used, the additives placed in the dry cement prior to mixing with water and the temperature and confining pressure in the location where the cement slurry is placed. When the viscosity becomes too large, the slurry is no longer pumpable. Thus, if the slurry has not been placed in its proper location within the well prior to the cement slurry becoming unpumpable, the well and the surface equipment would be seriously damaged.

Thickening time T_t (hr) is defined as the time required for the cement slurry to reach the limit of 100 Bearden

units of consistency.[3] This thickening time must be considerably longer than the time necessary to carry out the actual cementing operation. This can be accomplished by choosing the class of cement that has a sufficiently long thickening time, or placing the appropriate additives in the slurry that will retard the slurry chemical reaction and lengthen the thickening time.

It is necessary that an accurate estimate of the total time be for the actual operation. A safety factor should be added to this estimate. Usually this safety factor is from 30 min (for shallow operations) to as long as 2 hr (for deep complex operations).

The cementing operation time T_o (hr) is the time required for the cement slurry to be placed in the well:

$$T_o = T_m + T_d + T_p + T_s \qquad [4.17.4]$$

where T_m = time required to mix the dry cement (and additives) with water in hr
T_d = the time required to displace the cement slurry (that was pumped to the well as mixing took place) by mud or water from inside the casing in hr
T_p = plug release time in hr
T_s = the safety factor of 0.5 to 2 hr

The mixing time T_m is

$$T_m = \frac{\text{Volume of dry cement}}{\text{Mixing rate}}$$

$$= \frac{V_c(\text{sacks})}{\text{Mixing rate (sack/min)(60)}} \qquad [4.17.5]$$

where V_c is the dry volume of cement (sacks).
The displacement time T_d is

$$T_d = \frac{\text{Volume of fluid required to displace top plug}}{\text{Displacement rate}}$$

$$= \frac{\text{Volume (ft}^3)}{\text{Displacement rate (ft}^3/\text{min)(60)}} \qquad [4.17.6]$$

The cement slurry chosen must have a thickening time that is greater than the estimated time obtained for the actual cementing operation using Equation 4.17.4. Thus, $T_t > T_o$.

Figure 4.17.4 gives a relationship between well depth and cementing operation time and the specifications for the various cement classes. This figure can be used to approximate cementing operation time.

The cement slurry thickening time can be increased or decreased by adding special chemicals to the dry cement prior to mixing with water. Retarders are added to increase the thickening time and thus increase the time when the cement slurry sets. Some common retarders are organic compounds such as lignosulphonate, cellulose derivatives and sugar derivatives. Accelerators are added to decrease

---

[3]70 Bearden units of consistency is considered to be maximum pumpable viscosity.

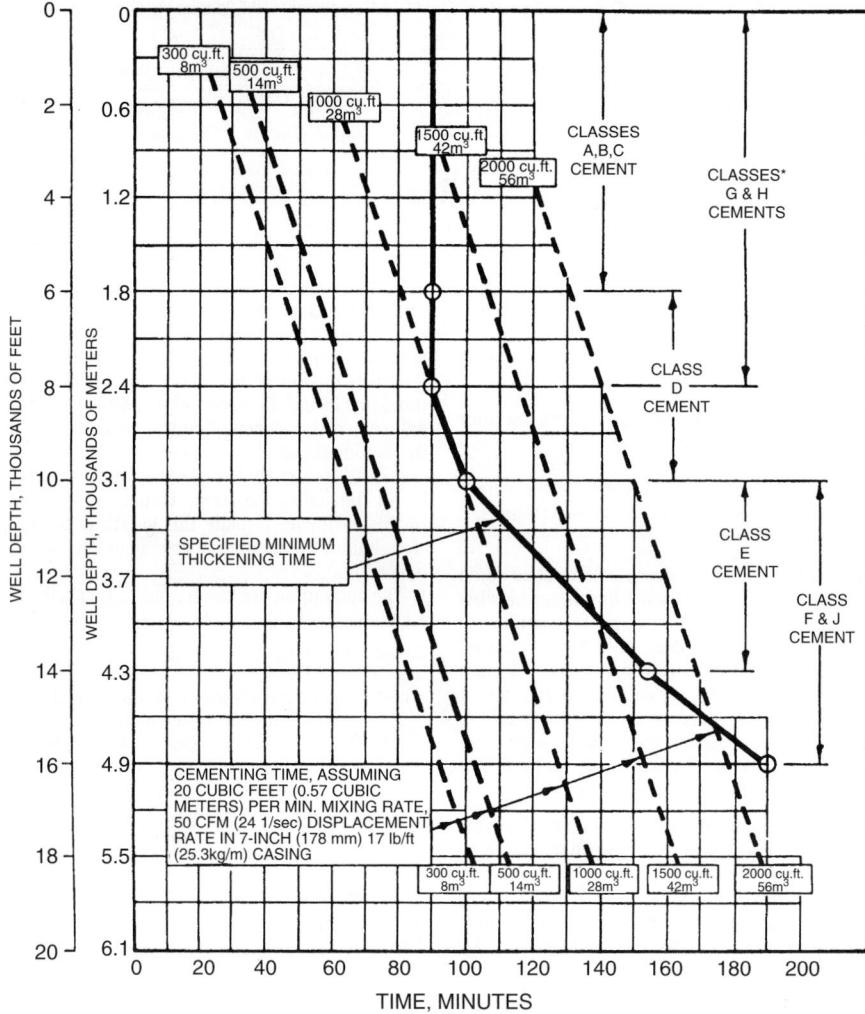

Figure 4.17.4 *Well depth and cementing time relationships [5]. *Specified maximum thickening time—120 minutes.*

the thickening time and thus decrease the time when the cement slurry sets. Accelerators are often used when it is required to have the cement obtain and early compressive strength, usually of the order of 500 psi (35 bar). Early setting cement slurries are used to cement surface casing strings or directional drilling plugs where waiting-on-cementing (WOC) must be kept to a minimum. The most common accelerators are calcium chloride to a lesser extent sodium chloride.

In general the thickening time for all neat cement slurries of all classes of cement will decrease significantly with increasing well environment temperature and/or confining pressure. Thus it is very important that the well extreme temperatures and confining pressures be defined for a particular cementing operation before the cement slurry is designed. Once the well temperature and pressure conditions are known, either from offset wells or from actual well logs during the drilling operation, and the estimated cementing operation time is known, the cement slurry with the required thickening time can be designed. After the initial cement slurry design has been made it is usually necessary to carry out laboratory tests to verify that the actual cement batch mix used will give the required thickening time (and

other characteristics needed). Such tests are usually carried out by the cementing service company laboratory using the operator's specifications.

Example 4.17.2

Calculate the minimum thickening time required for a primary cementing operation to cement a long intermediate casing string. The intermediate casing string is a $9\frac{5}{8}$-in., 53.5-lb/ft casing set in a $12\frac{1}{2}$-in. hole. The string is 12,000 ft in length from the top of the float collar to the surface. The cementing operation will require 1,200 sacks of Class H cement. The single cementing truck has a mixing capacity of 25 sacks per minute. The rig duplex mud pump has an 18-in. stroke (2.5 in. rod) and $6\frac{1}{2}$-in. liners and will be operated at 50 strokes per minute with a 90% volumetric efficiency. The plug release time is estimated to be about 15 min.

The mixing time is obtained from Equation 4.17.5. This is

$$T_m = \frac{1,200}{25(60)}$$

$$= 0.80\,hr$$

The inside diameter of $9\frac{5}{8}$-in., 53.50-lb/ft casing is 8.535 in. (see Section titled Fishing Operations). The internal capacity per unit length of casing is

$$\text{Casing internal capacity} = \frac{\pi}{4}\frac{(8.535)^2}{12}$$

$$= 0.3973 \text{ft}^3/\text{ft}$$

Thus the total internal capacity of the casing is

$$\text{Total volume} = 12,000(0.3973)$$

$$= 4,768 \text{ft}^3$$

The volume capacity of the mud pump per stroke q (ft^3/stroke) (see section titled "Mud Pumps") is

$$q = \left\{ 2\left[\frac{\pi}{4}(6.5)^2\right](18.0) \right.$$

$$\left. + 2\left[\frac{\pi}{4}(6.5)^2 - \frac{\pi}{4}(2.5^2)\right](18.0)\right\}\frac{(0.90)}{(12)^3}$$

$$= 1.1523 \text{ft}^3/\text{stroke}$$

The displacement time is obtained from Equation 4.17.6. This is

$$T_d = \frac{4,768}{50(1.1523)(60)}$$

$$= 1.38 \text{hr}$$

The plug release time is

$$T_p = \frac{15}{60}$$

$$= 0.25 \text{hr}$$

Using a safety factor of 1 hr, the cementing operation time is obtained from Equation 4.17.4. This is

$$T_o = 0.80 + 1.38 + 0.25 + 1.00$$

$$= 3.43 \text{hr}$$

Thus the minimum thickening time for the cement slurry to be used this cementing operation is

$$T_t = 3.43 \text{hr}$$

or 3 hr and 26 min.

4.17.5.3 Strength of Set Cement

A properly designed cement slurry will set after it has been placed in its appropriate location within the well. Cement strength is the strength the set cement has obtained. This usually refers to compressive strength, but can also refer to tensile strength. Cement having a compressive strength of 500 psi (35 bar) is considered adequate for most well applications.

The compressive strength of set cement is dependent upon the water-to-cement ratio used in the slurry, curing time, the temperature during curing, the confining pressure during curing and the additives in the cement. As part of the cement design procedures, samples of the cement slurry to be used in the cementing operation are cured and compression strength tested and often shear bond strength tested. These tests are usually carried out by the cementing company laboratory at the request of the operator.

In the compression test, four or five sample cubes of the slurry are allowed to cure for a specified period of time. The cement cubes are placed in a compression testing machine and the compressive strength of each sample cube obtained experimentally. The average value of the samples is obtained and reported as the compressive strength of the set cement.

In the shear bond strength test, the cement slurry is allowed to cure in the annulus of two concentric steel cylinders. After curing the force to break the bond between the set cement and one of the cylinders (usually the inner one) is obtained experimentally. The shear force, F_s (lb), which can be supported on the inner cylinder, or by the casing, is [4]

$$F_s = 0.969\sigma_c dl \tag{4.17.7}$$

where σ_c = compressive strength of cement in psi
$\quad$ d = outside of the casing in in.
$\quad$ l = length of the cement column in in.

Table 4.17.4 shows the influence of curing time, temperature and confining pressure on the compressive strength of Class H cement [3]. Also shown is the effect of the calcium chloride accelerator. In general, for the ranges of temperature and other values considered, the compressive strength increases with increasing curing time, temperature, confining pressure and the amount of calcium chloride accelerator. In general, the other classes of cement follow the same trend in compressive strength versus curing time, temperature and confining pressure.

Table 4.17.4 *Influence of Time and Temperature on the Compressive Strength of API Class H Cement [3]*

Curing Time (hours)	Calcium Chloride (percent)	Compressive Strength (psi) at Curing Temperature and Pressure of			
		95°F 800 psi	110°F 1,600 psi	140°F 3,000 psi	170°F 3,000 psi
6	0	100	350	1,270	1,950
8		500	1,200	2,500	4,000
12		1,090	1,980	3,125	4,700
24		3,000	4,050	5,500	6,700
6	1	900	1,460	2,320	2,500
8	1	1,600	1,950	2,900	4,100
12	1	2,200	2,970	3,440	4,450
24	1	4,100	5,100	6,500	7,000
6	2	1,100	1,700	2,650	2,990
8	2	1,850	2,600	3,600	4,370
12	2	2,420	3,380	3,900	5,530
24	2	4,700	5,600	6,850	7,400

Table 4.17.5 *Effect of Silica Flour on the Compressive Strength of Class G Cement Cured at Pressure*

Curing Time (days)	Silica Flour (percent)	Compressive Strength (psi), 1 Day at 130°F, All Remaining Days at 700°F
1	0	3525
2		988
4		1012
8		1000
1	40	2670
2		3612
4		3188
8		4588

At curing temperatures of about 200°F (93°C) or greater, the compressive strength of nearly all classes of cement cured under pressure will decrease. This decrease in compressive strength is often denoted as strength retrogression. An increase in cement permeability accompanies this decrease in strength. This strength retrogression usually continues for up to 15 days and, thereafter, the strength level will remain constant. This strength retrogression problem can be solved by adding from 30 to 50% (by weight) of silica flour or silica sand to the cement slurry. This silica additive prevents the strength retrogression and the corresponding increase in permeability of the cement.

Table 4.17.5 shows the effect of silica flour on the compressive strength of Class G cement cured at 700°F (371°C).

Example 4.17.3
For Example 2, determine the total weight that can be supported by the set cement bonded to the $9\frac{5}{8}$-in. casing. Assume the cement slurry yield is 1.05 ft³/sack (see Table 4.17.2) and there are 120 ft of casing below the float collar. A compressive strength of 500 psi is to be used.

The total volume of the cement slurry to be pumped to the well is

$$\text{Volume of slurry} = 1,200(1.05)$$

$$= 1,260 \text{ft}^3$$

The volume of slurry that is pumped to the annulus between the open borehole and the outside of the casing is

$$\text{Volume of annulus} = 1,260 - (0.3973)(120)$$

$$= 1,212 \text{ft}^3$$

The height in the annulus to where the casing is cemented to the borehole well and casing is

$$h = \frac{1,212}{\pi/4[(12.5/12)^2 - (9.625/12)^2]}$$

$$= 3,493 \text{ft}$$

The total weight that can be supported is calculated from Equation 4.17.7. This is

$$F_s = 0.969(500)(9.625)(3,493)(12)$$

$$= 195.5 \times 10^6 \text{lb}$$

4.17.6 Cement Additives
There are many chemical additives that can be used to alter the basic properties of the neat cement slurry and its resulting set cement. These additives are to alter the cement so that it is more appropriate to the surface cementing equipment and the subsurface environment.

Additives can be subdivided into six functional groups that are [8,9]:

1. Specific weight control
2. Thickening and setting time control
3. Loss of circulation control
4. Filtration control
5. Viscosity control
6. Special problems control

4.17.6.1 Specific Weight Control
As in drilling mud design, the cement slurry must have a specific weight that is high enough to prevent high pore pressure formation fluids from flowing into the well. But also the specific weight must not be so high as to cause fracturing of the weaker exposed formations. In general, the specific weight of the neat cement slurry of any of the various API classes of cement are so high that most exposed formations will fracture. Thus, it is necessary to lower the specific weight of nearly all cement slurries. The slurry specific weight can be reduced by using a higher water-to-cement ratio. But this can only be accomplished within the limits for maximum and minimum water-to-cement ratios set by API standards (see Table 4.17.3). The reduction in specific weight is normally accomplished by adding low-specific-gravity solids to the slurry. Table 4.17.6 gives the specific gravity properties of a number of the low-specific-gravity solids that are used to reduce the specific weight of cement slurries.

The most common low-specific-gravity solids used to reduce cement slurry specific weight are bentonite, diatomaceous earth, solid hydrocarbons, expanded perlite and pozzolan. It may not be possible to reduce the cement slurry specific weight enough with the above low-specific-weight materials when very weak formations are exposed. In such cases nitrogen is used to aerate the mud column above the cement slurry to assist in further decreasing the hydrostatic pressure.

Nearly all materials that are added to a cement slurry require the addition of additional water to the slurry. Table 4.17.7 gives the additional requirements for the various cement additives [8].

Bentonite. Bentonite without an organic polymer can be used as an additive to cement slurries. The addition of bentonite requires the use of additional water, thereby, further reducing the specific weight of the slurry (Table 4.17.7). Bentonite is usually dry blended with the dry cement prior to mixing with water. High percentages of bentonite in cement will significantly reduce compressive strength and thickening time. Also, high percentages of bentonite increase permeability and lower the resistance of cement to sulfate attack. At temperatures above 200°F (93°C), the bentonite additive promotes, retrogression of strength in cements with time. Bentonite has been used in 25% by weight of cement. Such high concentrations are not recommended. In general, the bentonite additive makes a poor well cement.

Diatomaceous Earth. Diatomaceous earth has a lower specific gravity than bentonite. Like bentonite this additive also requires additional water to be added to the slurry. This additive will affect the slurry properties similar to the addition of bentonite. However, it will not increase the viscosity as bentonite will do. Diatomaceous earth concentrations as high as 40% by weight of cement have been used. This additive is more expensive than bentonite.

Table 4.17.6 *Physical Properties of Cementing Materials*

Material	Bulk Weight (lbm/cu ft)	Specific Gravity	Weight 3.6[a] Absolute Gal	Absolute Volume gal/lbm	Absolute Volume cu ft/lbm
API cements	94	3.14	94	0.0382	0.0051
Ciment Fondu	90	3.23	97	0.0371	0.0050
Lumnite cement	90	3.20	96	0.0375	0.0050
Trinity Lite-Wate	75	2.80	75.0[e]	0.0429	0.0057
Activated charcoal	14	1.57	47.1	0.0765	0.0102
Barite	135	4.23	126.9	0.0284	0.0038
Bentonite (gel)	60	2.65	79.5	0.0453	0.0060
Calcium chloride, flake[b]	56.4	1.96	58.8	0.0612	0.0082
Calcium chloride, powder[b]	50.5	1.96	58.8	0.0612	0.0082
Cal-Seal, gypsum cement	75	2.70	81.0	0.0444	0.0059
CFR-1[b]	40.3	1.63	48.9	0.0736	0.0098
CFR-2[b]	43.0	1.30	39.0	0.0688	0.0092
DETA (liquid)	59.5	0.95	28.5	0.1258	0.0168
Diacel A[b]	60.3	2.62	78.6	0.0458	0.0061
Diacel D	16.7	2.10	63.0	0.0572	0.0076
Diacel LWL[b]	29.0	1.36	40.8	0.0882	0.0118
Diesel Oil No. 1 (liquid)	51.1	0.82	24.7	0.1457	0.0195
Diesel Oil No. 2 (liquid)	53.0	0.85	25.5	0.1411	0.0188
Gilsonite	50	1.07	32	0.1122	0.0150
HALDAD®-9[b]	37.2	1.22	36.6	0.0984	0.0131
HALDAD®-14[b]	39.5	1.31	39.3	0.0916	0.0122
Hematite	193	5.02	150.5	0.0239	0.0032
HR-4[b]	35	1.56	46.8	0.0760	0.0103
HR-7[b]	30	1.30	39	0.0923	0.0123
HR-12[b]	23.2	1.22	36.6	0.0984	0.0131
HR-L (liquid)[b]	76.6	1.23	36.9	0.0976	0.0130
Hydrated lime	31	2.20	66	0.0545	0.0073
Hydromite	68	2.15	64.5	0.0538	0.0072
LA-2 Latex (liquid)	68.5	1.10	33	0.1087	0.0145
LAP-1 Latex[b]	50	1.25	37.5	0.0960	0.0128
LR-11 Resin (liquid)	79.1	1.27	38.1	0.0945	0.0126
NF-1 (liquid)[b]	61.1	0.98	29.4	0.1225	0.0164
NF-P[b]	40	1.30	39.0	0.0923	0.0123
Perlite regular	8[c]	2.20	66.0	0.0546	0.0073
Perlite Six	38[d]	–	–	0.0499	0.0067
Pozmix® A	74	2.46	74	0.0487	0.0065
Pozmix® D	47	2.50	73.6	0.0489	0.0065
Salt (dry NaCl)	71	2.17	65.1	0.0553	0.0074
Salt (in solution at 77°F with fresh water)					
6%, 0.5 lbm/gal	–	–	–	0.0384	0.0051
12%, 1.0 lbm/gal	–	–	–	0.0399	0.0053
18%, 1.5 lbm/gal	–	–	–	0.0412	0.0055
24%, 2.0 lbm/gal	–	–	–	0.0424	0.0057
Saturated, 3.1 lbm/gal	–	–	–	0.0445	0.0059
Salt (in solution at 140°F with fresh water)					
saturated, 3.1 lbm/gal	–	–	–	0.0458	0.0061
Sand (Ottawa)	100	2.63	78.9	0.0456	0.0061
Silica flour (SSA-1)	70	2.63	78.9	0.0456	0.0061
Coarse silica (SSA-2)	100	2.63	78.9	0.0456	0.0061
Tuf Additive No. 1	–	1.23	36.9	0.0976	0.0130
Tuf-Plug	48	1.28	38.4	0.0938	0.0125
Water	62.4	1.00	30.0	0.1200	0.0160

[a] Equivalent to one 94-lbm sack of cement in volume.

[b] When less than 5% is used, these chemicals may be omitted from calculations without significant error.

[c] For 8 lbm of Perlite regular use a volume of 1.43 gal at zero pressure.

[d] For 38 lbm of Perlite Six use a volume of 2.89 gal at zero pressure.

[e] 75 lbm = 3.22 absolute gal.

Source: Courtesy Halliburton Services.

Table 4.17.7 *Water Requirement of Cementing Materials*

Material	Water Requirements
API Class A and B cements	5.2 gal (0.70 cu ft)/94-lbm sack
API Class C cement (H_1 Early)	6.3 gal (0.84 cu ft)/94-lbm sack
API Class D and E cements (retarded)	4.3 gal (0.58 cu ft)/94-lbm sack
API Class G cement	5.0 gal (0.67 cu ft)/94-lbm sack
API Class H cement	4.3 to 5.2 gal /94-lbm sack
Chem Comp cement	6.3 gal (0.84 cu ft)/94-lbm sack
Ciment Fondu	4.5 gal (0.60 cu ft)/94-lbm sack
Lumnite cement	4.5 gal (0.60 cu ft)/94-lbm sack
HLC	7.7 to 10.9 gal/87-lbm sack
Trinity Lite-Wate cement	7.7 gal (1.03 cu ft)/75-lbm sack (maximum)
Activated charcoal	none at 1 lbm/sack of cement
Barite	2.4 gal (0.32 cu ft)/100-lbm sack
Bentonite (gel)	1.3 gal (0.174 cu ft)/2% in cement
Calcium chloride	none
Gypsum hemihydrate	4.8 gal (0.64 cu ft)/100-lbm sack
CFR-1	none
CFR-2	none
Diacel A	none
Diacel D	3.3 to 7.2 gal /10% in cement (see Lt. Wt. Cement)
Diacel LWL	none (up to 0.7%)
	0.8 to 1.0 gal /1% in cement (except gel or Diacel D slurries)
Gilsonite	2.0 gal (0.267 cu ft)/50 lbm/cu ft
HALAD-9	none (up to 0.5%)
	0.4 to 0.5 gal/sack of cement at over 0.5%
HALAD-14	none
Hematite	0.36 gal (0.048 cu ft)/100-lbm sack
HR-4	none
HR-7	none
HR-12	none
HR-20	none
Hydrated lime	0.153 gal (0.020 cu ft)/lbm
Hydromite	3.0 gal (0.40 cu ft)/100-lbm sack
LA-2 Latex	0 to 0.8 gal/sack of cement
LAP-1 powdered latex	1.7 gal (0.227 cu ft)/1%-in cement
NF-P	none
Perlite regular	4.0 gal (0.535 cu ft)/8 lbm/cu ft
Perlite Six	6.0 gal (0.80 cu ft)/38 lbm/cu ft
Pozmix A	3.6 gal (0.48 cu ft)/74 lbm/cu ft
Salt (NaCl)	none
Sand. Ottawa	none
Silica flour (SSA-1)	1.5 gal (0.20 cu ft)/35% in cement (32.9 lbm)
Coarse silica (SSA-2)	none
Tuf Additive No. 1	none
Tuf Plug	none

Source: Courtesy of Halliburton Services.

Solid Hydrocarbons. Gilsonite (an asphaltite) and coal are used as very-low-specific-gravity solids additives. These additives do not require a great deal of water to be added to the slurry when they are used.

Expanded Perlite. Expanded perlite requires a great deal of water to be added to the slurry when it is used to reduce the specific weight of a slurry. Often perlite as an additive is used in a blend of additives such as perlite with volcanic glass fines, or with pozzolanic materials, or with bentonite. Without bentonite the perlite tends to separate and float in the upper part of the slurry.

Pozzolan. Diatomaceous earth is a type of pozzolan. Pozzolan refers to a finely ground pumice or fly ash that is marketed as a cement additive under that name. The specific gravity of pozzolans is slightly less than the specific gravity of cement. The water requirements for this additive are about the same as for cements. Only a slight reduction in specific weight of a slurry can be realized by using these additives. The cost of pozzolans is very low.

Where very high formations pore pressure are present the specific weight of the cement slurry can be increased by using the minimum water-to-cement ratio and/or by adding high-specific-gravity materials to the slurry. The most common high-specific-gravity solids used to increase cement slurry specific weight are hematite, ilmenite, barite and sand. Table 4.17.6 gives the specific gravity properties of a number of these high-specific-gravity additives.

Hematite. This additive can be used to increase the specific weight of a cement slurry to as high as 19 lb/gal. This is an iron oxide ore with a specific gravity of about 5.02. Hematite requires the addition of some water when it is used as an additive. Hematite has minimal effect on thickening time and compressive strength of the cement.

Ilmenite. This additive has a specific gravity of about 4.67. It is a mineral composed of iron, titanium and oxygen. It requires no additional water to be added to the slurry; thus, it can yield slurry specific weights as high as the hematite additive. Ilmenite also has mineral effect on thickening time and compressive strength of the cement.

Barite. This mineral additive requires much more water to be added to the cement slurry than does hematite. This large amount of added water required will decrease the compressive strength of the cement. This additive can be used to increase the specific weight of a cement slurry to as high as 19 lb/gal.

Sand. Ottawa sand has a low specific gravity of about 2.63. But since no additional water is required when using this additive, it is possible to use sand to increase the cement slurry specific weight. The sand has little effect on the pumpability of the cement slurry. When set the cement will form a very hard surface. Sand used as an additive can be used to increase the specific weight of a cement slurry to as high as 18 lb/gal.

Example 4.17.4
A specific weight of 12.8 lb/gal is required for a Class A cement slurry. It is decided that the cement be mixed with bentonite to reduce the specific weight of the slurry. Determine the weight of bentonite that should be dry blended with each sack of cement. Determine the yield of the cement slurry. Determine the volume (gal) of water needed for each sack of cement.

The weight of bentonite to be blended is found by using Equation 4.17.1. Taking the appropriate data from Tables 4.17.1, 4.17.2, 4.17.6 and 4.17.7 and letting x be the unknown weight of bentonite per sack of cement, Equation 4.17.1 is

$$12.8 = \frac{94 + x + 8.34(5.20 + 0.692x)}{\left(\frac{94}{3.14(8.34)} + \frac{x}{2.65(8.34)} + 5.20 + 0.692x\right)}$$

Solving the above for x gives

$$x = 9.35 \, lb$$

Thus, 9.35 lb of bentonite will need to be added for each sack of Class A cement used.

The yield can be determined from Equation 4.17.3. This is

$$\text{Yield} = \frac{\frac{94}{3.14(8.34)} + \frac{9.35}{2.65(8.34)} + 5.20 + 0.692(9.35)}{7.48}$$

$$= 2.10 \, ft^3/\text{sack}$$

The volume of water needed is

$$\text{Volume of water} = 5.20 + 0.692(9.35)$$

$$= 11.69 \, gal/\text{sack}$$

Example 4.17.5
A specific weight of 18.2 lb/gal is required for a Class H cement slurry. It is decided that the cement be mixed with hematite to increase the specific weight of the slurry. Determine the weight of hematite that should be dry blended with each sack of cement. Determine the yield of the cement slurry. Determine the volume (gal) of water needed for each sack of cement.

The weight of hematite to be blended is found by using Equation 4.17.1. Taking the appropriate data from Tables 4.17.1, 4.17.2, 4.17.6 and 4.17.7 and letting x be the unknown weight of hematite per sack of cement, Equation 4.17.1 is

$$18.2 = \frac{94 + x + 8.34(4.29 + 0.0036x)}{\frac{94}{3.14(8.34)} + \frac{x}{5.02(8.34)} + (4.29 + 0.0036x)}$$

Solving the above for x gives

$$x = 25.8 \, lb$$

Thus, 25.8 lb of hematite will need to be added for each sack of Class H cement used.

The yield can be determined from Equation 4.17.3. This is

$$\text{Yield} = \frac{\frac{94}{3.14(8.34)} + \frac{25.8}{5.07(8.34)} + 4.29 + 0.0036(25.8)}{7.48}$$

$$= 1.15 \, ft^3/\text{sack}$$

The volume of water needed is

$$\text{Volume of water} = 4.29 + 0.0036(25.8)$$

$$= 4.38 \, gal/\text{sack}$$

Example 4.17.6
A specific weight of 17.1 lb/gal is required for a Class H cement slurry. It is decided that the cement be mixed with sand in order to increase the specific weight of the slurry. Determine the weight of sand that should be added with each sack of cement. Determine the yield of the cement slurry. Determine the volume (gal) of water needed for each sack of cement.

The weight of sand to be added is found by using Equation 4.17.1. Taking the appropriate data from Tables 4.17.1, 4.17.2, 4.17.6, and 4.17.7 and letting x be the unknown weight of sand per sack of cement, Equation 4.17.1 is

$$17.1 = \frac{94 + x + 8.34(4.29)}{\frac{94}{3.14(8.34)} + \frac{x}{2.63(8.34)} + 4.29}$$

Solving the above for x gives

$$x = 31.7 \, lb$$

Thus, 31.7 lb of sand will need to be added for each sack of Class H cement used.

The yield can be determined from Equation 4.17.3. This is

$$\text{Yield} = \frac{\frac{94}{3.14(8.34)} + \frac{31.7}{2.63(8.34)} + 4.29}{7.48} = 1.25 \, ft^3/\text{sack}$$

Since no additional water is needed when using sand as the additive, the volume of water needed is

$$\text{Volume of water} = 4.29 \, gal/\text{sack}$$

4.17.6.2 Thickening Setting Time Control
It is often necessary to either accelerate, or retard the thickening and setting time of a cement slurry.

For example, when cementing a casing string run to shallow depth or when setting a directional drilling kick-off plug, it is necessary to accelerate the cement hydration so that the waiting period will be minimized. The most commonly used cement hydration accelerators are calcium chloride, sodium chloride, a hemihydrate form of gypsum and sodium silicate.

Calcium Chloride. This accelerator may be used in concentrations up to 4% (by weight of mixing water) in well having bottomhole tempeatures less than 125°F. This additive is usually available in an anhydrous grade (96% calcium chloride). Under pressure curing conditions calcium chloride tends to improve compressive strength and significantly reduce thickening and setting time.

Sodium Chloride. This additive will act as an accelerator when used in cements containing no bentonite and when in concentrations of about 5% (by weight of mixing water). In concentrations above 5% the effectiveness of sodium chloride as an accelerator is reduced. This additive should be used in wells with bottom-hole temperatures less than 160°F. Saturated sodium chloride solutions act as a retarder. Such saturated sodium chloride solutions are used with cement slurries that are to be used to cement through formations that are sensitive to freshwater. However, potassium chloride is far more effective in inhibiting shale hydration. In general, up to a 5% concentration, sodium chloride will improve compressive strength while reducing thickening and setting time.

Gypsum. Special grades of gypsum hemihydrate cement are blended with Portland cement to produce a cement with reduced thickening and setting time for low-temperature applications. Such cement blends should be used in wells with bottomhole temperatures less than 140°F (regular-temperature grade) or 180°F (high-temperature grade). There is a significant additional water requirement for the addition of gypsum (see Table 4.17.7). When very rapid thickening and setting times are required for low-temperature conditions (i.e., primary cementing of a shallow casing string or a shallow kick-off plug), a special blend is sometimes used. This is 90 lb of gypsum hemihydrate, 10 lb of Class A Portland cement and 2 lb of sodium chloride mixed with 4.8 gal of water. Such a cement slurry and 2 lb of sodium chloride mixed with 4.8 gal of water. Such a cement slurry can develop a compression strength of about 1,000 psi in just 30 min (at 50°).

Sodium Silicate. When diatomaceous earth is used with the cement slurry, sodium silicate is used as the accelerator. It can be used in concentrations up to 7% by weight.

When it is necessary to cement casing or line strings set at great depths, additives are often used in the design of the cement slurry to retard the thickening and setting time. Usually such retarding additives are organic compounds. These materials are also referred to as thinners or dispersants. Calcium lignosulfonate is one of the most commonly used cement retarders. It is very effective at increasing thickening and setting time in cement slurries at very low concentrations (of the order of 1% by weight or less). It is necessary to add an organic acid to the calcium lignosulfonate when high-temperature conditions are encountered.

Calcium-sodium lignosulfonate is a better retarding additive when high concentrations of bentonite are to be used in the design of the cement slurry.

Also sodium tetraborate (borax) and carborymethyl hydroxyethyl cellulose are used as retarding additives.

4.17.6.3 Filtration Control
Filtration control additives are added to cement slurries for the same reason that they are added to drilling muds. However, untreated cement slurries have much greater filtration rates than untreated drilling muds. It is this important to limit the loss of water filtrate from a slurry to a permeable formation. This is necessary for several reasons:

- Minimize hydration of formations containing water sensitive formations
- Limit the increase in slurry viscosity as cement is placed in the well

- Prevent annular bridges
- Allow for sufficient water to be available for cement hydration

Examples of filtration control additives are latex, bentonite with a dispersant and other various organic compounds and polymers.

4.17.6.4 Viscosity Control
The viscosity of the cement slurry will affect the pumping requirements for the slurry and frictional pressure gradient within the well. The viscosity must be kept low enough to ensure that the cement slurry can be pumped to the well during the entire cementing operation period. High viscosity will result in high-pressure gradients that could allow formation fractures.

Examples of commonly used viscosity control additives are calcium lignosulfonate, sodium chloride and some long-chain polymers. These additives also act as accelerators or retarders so care must be taken in designing the cement slurry with these materials.

4.17.6.5 Special Problems Control
There are some special problems in the design of cement slurries for which additives have been developed. These are

- Gel strength additives for the preparation of spacers (usually fragile gel additives).
- Silica flour is used to form a stronger and less permeable cement, especially for high-temperature applications.
- Hydrazine is used to control corrosion of the casing.
- Radioactive tracers are used to assist in assessing where the cement has been placed.
- Gas-bubble-producing compounds to slowly create gas bubbles in the cement as the slurry sets and hardens. It is felt that such a cement will have less tendency to leak formation gas.
- Paraformaldehyde and sodium chromate that counteract organic contaminants left in the well from drilling operations.
- Fibrous materials such as nylon are added to increase strength, in particular, resistance to impact loads.

4.17.7 Primary Cementing
Primary cementing refers to the cementing of casing and liner strings in a well. The cementing of casing or liner string is carried out so that producible oil and gas, or saltwater will not escape from the producing formation to another formation, or pollute freshwater sands at shallower depth.

The running of long casing strings and liner strings to great depths and successfully cementing these strings required careful engineering design and planning.

4.17.7.1 Normal Single-Stage Casing Cementing
Under good rig operating conditions, casing can be run into the hole at the rate from 1,000 to 2,000 ft/hr (300 to 600 m/hr). It is often necessary to circulate the drilling mud in the hole prior to running the casing string. This is to assure that all the cutting and any borehole wall prices have been removed. However, the longer a borehole remains open, the more problems will occur with the well.

Prior to running the casing string into the well the mill varnish should be removed from the outer surface of the casing. The removal of the mill varnish is necessary to ensure that the cement will bond to the steel surface.

The casing string is run into the well with a guide shoe (Figure 4.17.5) or float shoe (Figure 4.17.6). Figure 4.17.4 illustrates a regular pattern guide shoe with a wireline re-entry bevel and no internal components to drill.

Figure 4.17.5 *Regular pattern guide shoe [2]. (Courtesy of Weatherford International Limited.)*

Figure 4.17.6 *Swirl guide shoe [2]. (Courtesy of Weatherford International Limited.)*

Figure 4.17.7 *Ball float collar [2]. (Courtesy of Weatherford International Limited.)*

Figure 4.17.8 *Flapper float collar [2]. (Courtesy of Weatherford International Limited.)*

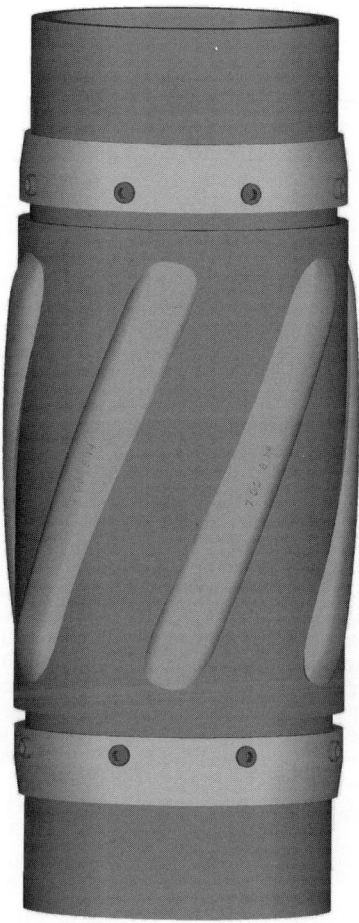

Figure 4.17.9 *Scratcher [2]. (Courtesy of Weatherford International Limited.)*

One to four joints above the guide or float shoe is the float collar. The float collar acts as a back flow valve that keeps the heavier cement slurry from flowing back into the casing string after it has been placed into the annulus between the outside of the casing and the borehole wall. Figures 4.17.7 and 4.17.8 show two typical float collars. Figure 4.17.7 is a plunger style float collar and Figure 4.17.8 is a flapper float collar. Most modern float equipment is PDC drillable and constructed of composites and cement. Typically a float shoe and a float collar with both be run on one string to provide a redundant check valve capability.

Centralizers are placed along the length of the casing string in a density of between 2 per casing joint to 1 every 4 joints. The density of centralizer placement varies dependent on casing to hole size combination, fluid rheology, and well profile and is usually determined using centralizer placement software. Centralizers ensure that the casing is nearly centered in the borehole, thus allowing a more uniform distribution of cement slurry flow around the casing. This nearly uniform flow around the casing is necessary

to remove the drilling mud in the annulus and provide an effective seal.

Centralizers are available in two basic varieties, rigid and bow spring. Figure 4.17.9 shows a typical rigid centralizer and 4.17.10 shows a typical bow spring centralizer. Bow spring centralizers are generally run when casing is being run through a restricted bore hole, into under reamed sections, or into washed out zones. Rigid centralizers perform best when the well internal diameter is constant.

The casing string is lowered into the drilling mud in the well using the rig drawworks and elevators. The displaced drilling mud flows to the mud tanks and is stored there for later use. Once the entire casing string is in place in the borehole, the casing string is left hanging in the elevators through the cementing operation. This allows the casing string to be reciprocated (moved up and down) and possibly rotated as the cement is placed in the annulus. This movement assists the removal of the drilling mud.

While the casing string is hanging in the elevators a cementing head is made up to the upper end of the string (see Figure 4.17.3). The cementing head is then connected with flow liners that come from pump truck (see Figure 4.17.2). The blender mixes the dry cement and additives with water. The high-pressure, low-volume triplex cement pump on the

Figure 4.17.10 *Centralizer [2]. (Courtesy of Weatherford International Limited.)*

pump truck pumps the cement slurry to the cementing head. Usually a preflush or spacer is initially pumped ahead of the cement slurry. This spacer (usually about 20 bbl or 3.2 m³) is used to assist in removing the drilling mud from the annular space between the outside of the casing and the borehole wall.

Figure 4.17.11 gives a series of schematics that show how the spacer and cement slurry displace drilling mud in the well. Two wiper plugs are usually used to separate the spacer and the cement slurry from the drilling mud in the well. The cementing head has two retainer valves that hold the two flexible rubber wiper plugs with two separate plug-release pins (Figure 4.17.11a). When the spacer and the cement slurry are to be pumped to the inside of the casing through the cementing head, the bottom plug-release pin is removed. This releases the bottom wiper plug into the initial portion of the spacer flow to the well (see Figure 4.17.11b). This bottom wiper plug keeps the drilling mud from contaminating the spacer and the cement slurry while they pass through the inside of the casing. When all the cement slurry has passed through the cementing head, the top plug-release pin is removed releasing the top wiper plug into the flow to the well. At this point in the cementing operation the cement

pump begins to pump drilling mud through the cementing head to the well (see Figure 4.17.11a).

When the bottom wiper plug reached the float collar a diaphram is burst at 200 to 400 psi (14 to 28 bar). Typically this pressure event is not witnessed at surface due to the density difference between the cement and the drilling mud. This density differential is almost always large enough to rupture the diaphram. Once the diaphram is broken the spacer string and then into the annulus between the casing and the borehole wall (see Figure 4.17.11d). The spacer and the cement slurry displace the drilling mud below the float collar and the drilling mud in the annulus. The spacer is designed to efficiently displace nearly all the drilling mud in the annulus prior to the cement slurry entering the annulus.

When the top wiper plug reached the float collar, the pump pressure rises sharply. This wiper plug does not have a diaphram and, therefore, no further flow into the well can take place. At this point in the cementing operation the cement pump is usually shut down and the pressure released on the cementing head. The back-flow valve in the float collar stops the heavier cement slurry from flowing back into the inside of the casing string (see Figure 4.17.11e). The volume of spacer and cement slurry is calculated to allow the cement slurry to either completely fill the annulus, or to fill the annulus to a height sufficient to accomplish the objectives of the casing and cementing operation.

Usually a lighter drilling mud is used to follow the heavy cement slurry. In this way the casing is under compression from a higher differential pressure on the outside of the casing. Thus when the cement sets and drilling or production operations continues, the casing will always have an elastic load on the cement-casing interface. This elastic load is considered essential for maintenance of the cement-casing bond and to keep leakage between the cement and casing (i.e., the microannulus) from occurring.

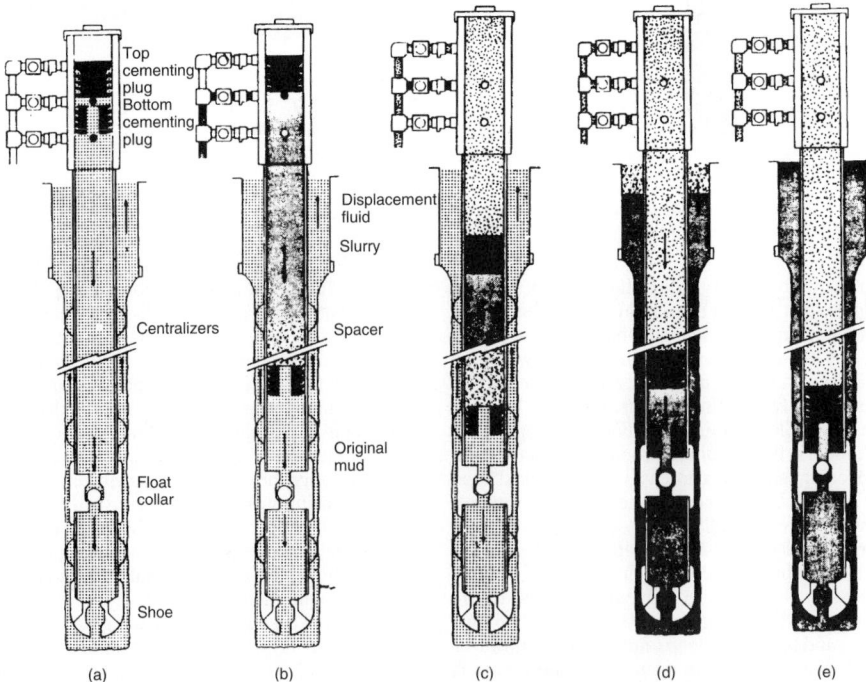

Figure 4.17.11 *Single-stage cementing: (a) circulating mud; (b) pumping spacer and slurry; (c) and (d) displacing; (e) end of job.* •, *plug-releasing pin in; o, plug releasing pin out (Courtesy of Schlumberger).*

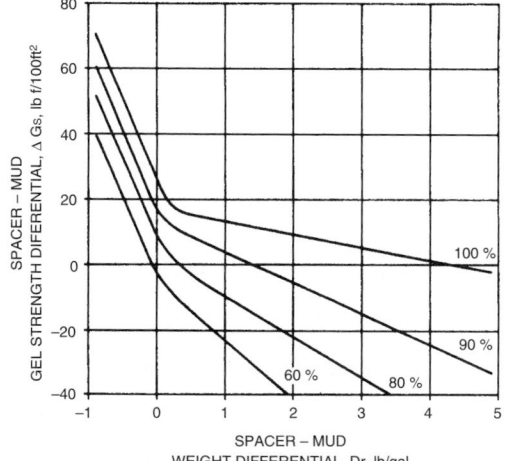

Figure 4.17.12 *Spacer gel and specific weight relationships [9].*

Since the early 1960s there has been a great deal of discussion regarding the desirability of using a low viscosity cement slurry to improve the removal of drilling mud from the annulus [2,3]. More recent studies have shown that low viscosity slurries do not necessarily provide effective removal of drilling mud in the annulus. In fact, there is strong evidence that spacers and slurries should have a higher gel strength and a higher specific weight than the mud that is being displaced [10]. Also, the displacement pumping rate should be low with an annular velocity of 90 ft/min or less [10]. Figure 4.17.12 gives the annular residual drilling mud removal efficiency as a function of differential gel strength and differential specific weight between the drilling mud and the spacer. The most successful spacer is an initial portion of the cement slurry that will give about 200 ft of annulus and that has a higher gel strength and a higher specific weight than the drilling mud it is to displace. The cement slurry spacer should have a gel strength that is about 10 to 15 lb/100 ft^2 greater and a specific weight that is about 2 to 4 lb/gal greater than the drilling mud to be displaced. The spacer should be a specially treated cement slurry.

It makes no sense to use a water or a drilling mud spacer. If such spacers are used, the follow-on normal cement slurry will be just as unsuccessful at removing the water or the drilling mud of the spacer as it would at removing drilling mud if no spacer were used.

The basic steps for the planning and execution of a successful cementing operation are:

1. Condition the well and the drilling mud prior to pulling the drill pipe. Typically at least one hole volume is circulated to remove cutting and debris.
2. Centralize the casing using a placement program.
3. Determine the specific weight and gel strength of drilling mud to be displaced.
4. Estimate the approximate cementing operation time.
5. Run a prefush spacer compatible with the mud and cement.
6. Select the most appropriate API class of cement that meets the depth, temperature, sulfate resistance and other well limitations. Select the cement class that has a natural thickening time that most nearly meets the cementing operation time requirement, or that will require only small amounts of retarding additives.

7. Do not add water loss control additives unless the drilling mud requires it.
8. If the major portion of the cement slurry must have a rather low specific weight, do not use excess water to lower specific weight. Try to use only the optimum recommended water-to-cement ratio. Utilize only additives that will require little or no added water to lower the specific weight of the cement slurry. Avoid using bentonite in large concentrations since it requires a great deal of added water and significantly reduces cement strength. Silicate flour should be used whenever possible to increase compressive strength and decrease permeability of set cement.
9. Design the cement slurry and its initial spacer to have the appropriate gel strength and specific weight. Have the cementing service company run laboratory tests on the cement slurry blend selected. These tests should be run at the anticipated bottomhole temperature and pressure. These tests should be carried out to verify thickening time, specific weight, gel strength (of spacer) and compressive strength.
10. Reciprocate or rotate the casing or liner while conditioning the hole and during the cementing operations. The combination of rotation and reciprocation is widely considered to provide the best results. Rotation is more effective then reciprocation if only one of the two options can be employed. Special cementing hardware may be required to rotate casing.
11. The fluid mechanics should be modeled to provide the optimum pump rates during displacement.
12. The mixed cement slurry should be closely monitored with high-quality continuous specific weight monitoring and recording equipment as the slurry is pumped to the well.
13. When the cementing operation is completed, bleed off internal pressure inside the casing and leave valve open on cementing head.

Example 4.17.7

A 7-in., N-80, 29-lb/ft casing string is to be run in a 13,900-ft borehole. The casing string is to be 13,890 ft in length. Thus the guide shoe will be held about 10 ft from the bottom of the hole during cementing. The float collar is to be 90 ft from the bottom of the casing string. The borehole is cased with $9\frac{5}{8}$–in., 32.30-lb/ft casing to 11,451 ft of depth. The open hole below the casing shoe of the $9\frac{5}{8}$–in. casing is $8\frac{1}{2}$ in diameter. The 7-in. casing string is to be cemented to the top of the borehole. Class E cement with 20% silica flour is to be used from the bottom of the hole to a height of 1,250 ft from the bottom of the borehole. Class G cement with at least 10% silica flour is to be used from 1,250 ft from the bottom to the top of the borehole. The drilling mud in the borehole has a specific weight of 12.2 lb/gal and an initial gel strength of 15 lb/100 ft^2. A space sufficient to give 200 ft of length in the open-hole section will be used. An excess factor of 1.2 will be applied to the Class G cement slurry volume.

1. Determine the specific weight and gel strength for the spacer.
2. Determine the number of cement sacks for each class of cement to be used.
3. Determine the volume of water to be used.
4. Determine the cementing operation time and thus the minimum thickening time. Assume a cement mixing rate of 25 sacks/min. Also assume an annular displacement rate no greter than 90 ft/min while the spacer is moving through the open-hole section and a flowrate of 300 gal/min thereafter. A safety factor of 1.0 hr is to be used.

5. Determine the pressure differential prior to bumping the plug.
6. Determine the total mud returns during the cementing operation.

The basic properties of Class E and Class G cements are given in Table 4.17.1 and 4.17.2.

1. The drilling mud has a specific weight of 12.2 lb/gal and an initial gel strength of 15 lb/100 ft^2. Thus from Figure 4.17.12 it will be desirable to have a cement slurry spacer with a specific weight of at least 15.2 lb/gal and an initial gel strength of about 20 lb/100 ft^2.

The spacer will be designed with Class G cement. Taking the appropriate data from Tables 4.17.6 and 4.17.7, the weight of silica flour to be used is

$$0.10(94) = 9.4 \text{lb/sack}$$

Equation 4.17.1 becomes

$$\bar{\gamma} = \frac{94 + 9.4 + 8.34[4.97 + 0.0456(9.4)]}{\dfrac{94}{3.14(8.34)} + \dfrac{9.4}{2.63(5.34)} + 4.97 + 0.0456(9.4)}$$

$$= 16.4 \text{lb/gal}$$

Thus the spacer will be designed to have a specific weight of 16.4 lb/gal and sufficient fragile gel additive to have an initial gel strength of about 20 lb/100 ft^2.

2. The spacer must have a volume sufficient to give 200 ft of length in the open-hole section. The annular capacity of the open hole is

$$\frac{\pi}{4}\left[\left(\frac{8.5}{12}\right)^2 - \left(\frac{7.0}{12}\right)^2\right] = 0.1268 \text{ft}^3/\text{ft}$$

Equation 4.17.3 gives the yield of the spacer as

$$\text{Yield} = \frac{3.59 + 0.4286 + 4.97 + 0.4286}{7.48} = 1.21 \text{ft}^3/\text{sack}$$

Thus the number of sacks to give that will be 200 ft in length in the open-hole section is

$$\text{Sacks} = \frac{0.1268(200)}{1.21} = 20.96$$

The above is rounded off to the next highest sack, or 21 sacks.

The volume of Class G cement is the sum of the spacer volume, annular volume in the cased portion of the borehole and the applicable annular volume in the open-hole section of the borehole. The annular capacity of the cased portion of the borehole is

$$\frac{\pi}{4}\left[\left(\frac{9.001}{12}\right)^2 - \left(\frac{7.0}{12}\right)^2\right] = 0.1746 \text{ft}^3/\text{ft}$$

Thus the volume of Class G cement slurry to be used to cement the well (excluding spacer, but considering the excess factor) is

$$\text{Volume} = [(2,449 - 1,250)(0.1268) + 11,451(0.1746)]1.2$$

$$= 2,581.66 \text{ft}^3$$

The total number of Class G sacks of cement needed, including the spacer volume, is

$$\text{Total sacks} = \frac{2,581.66}{1.21} + 21 = 2,134 + 21 \approx 2,155$$

The volume of Class E cement is

$$\text{Volume} = \frac{\pi}{4}\left(\frac{8.5}{12}\right)^2(10) + \frac{\pi}{4}\left(\frac{6.184}{12}\right)^2 90$$

$$+ (1,250 - 10)(0.1268)$$

$$= 3.94 + 18.77 + 157.23 = 179.94 \text{ft}^3$$

The specific weight of the Class E cement is determined from Equation 4.17.1. The weight of silica flour to be added is

$$0.20(94) = 18.8 \text{lb}$$

Equation 4.17.1 is

$$\bar{\gamma} = \frac{94 + 18.8 + 8.34[4.29 + 0.0456(18.8)]}{\dfrac{94}{3.14(8.34)} + \dfrac{18.8}{2.63(8.34)} + 4.29 + 0.0456(18.8)}$$

$$= 16.2 \text{lb/gal}$$

Equation 4.17.3 is

$$\text{Yield} = \frac{3.59 + 0.8571 + 4.29 + 0.8573}{7.48} = 1.28 \text{ft}^3/\text{sack}$$

The total number of Class E sacks of cement needed are

$$\text{Total sacks} = \frac{179.94}{1.28} \approx 141$$

3. The volume of water to be used in the cementing operation is

$$\text{Volume} = 2,155(4.97 + 0.4286) + 141(4.29 + 0.8573)$$

$$\approx 11,634 + 726$$

$$\approx 12,360 \text{gal}$$

or about 295 bbl.

4. The total cementing operation time is somewhat complicated since it is desired to reduce the rate of flow when the spacer passes through the open hole section of the well. The mixing time is

$$T_m = \frac{2,155 + 141}{25(60)} = 1.53 \text{hr}$$

During this time the volume of cement slurry pumped to the casing string is

$$\text{Volume pumped} = 2,155(1.21) + 141(1.28)$$

$$= 2,788.1 \text{ft}^3$$

The internal volume of the casing string above the float collar is

$$\text{Internal volume} = 13,800(0.2086)$$

$$= 2,878.7 \text{ft}^3$$

Therefore, the mud pumps are used to fill the remainder of the casing string at a rate of 300 gal/min. Thus, the additional time T_1 to get the bottom plug to the float collar is

$$T_1 = \frac{2,878.7 - 2,788.1}{\left(\dfrac{300}{7.48}\right)(60)} = 0.04 \text{hr}$$

The time to release the plug is

$$T_p = 15/60$$

$$= 0.25 \text{hr}$$

From the time the plug is released the pumping rate is reduced to 85.4 gal/min, which gives a velocity of 90 ft/min in the openhole section of the annulus. The time for the spacer to pass through the open-hole section T_2 (including the 90 ft of internal volume of the casing below the float collar and the 10 ft of open hole below the casing string guide shoe) is

$$T_2 = \frac{3.94 + 18.77 + (2,449 - 10)(0.1268)}{\left(\dfrac{85.4}{7.48}\right)(60)} = 0.49 \text{hr}$$

After this time period, the rate of pumping to the well can be returned to 300 gal/min rate.

Equation 4.17.4 becomes

$$T_0 = 1.53 + 0.04 + 0.25 + 0.49$$

$$+ \frac{2,878.7 - 91 - 332}{\left(\dfrac{300}{7.48}\right)(60)} + 1.0 = 4.33 \text{hr}$$

Therefore, the thickening time for the cement slurries to be used must be greater than 4.33 hr.

5. The pressure differential prior to the top plug reaching the float collar is

$$p = \frac{122.7}{144}(13,900 - 1,250) + \frac{121.2}{144}(1,250 - 90 - 10)$$

$$- \frac{91.3}{144}(13,900 - 90 - 10)$$

$$= 2997 \text{psi}$$

6. The total volume of the mud returns will be

Mud volume = Total volume well without steel

− Steel volume

$$= \frac{\pi}{4}\frac{(8.5)^2}{12}(13,900 - 11,451)$$

$$+ \frac{\pi}{4}\frac{(9.001)^2}{12}(11,451) - \frac{29.0}{490}(13,890)$$

$$= 5,203 \text{ft}^3 \text{ or about } 927 \text{bbl}$$

4.17.7.2 Large-Diameter Casing Cementing

Often when large-diameter casing strings are to be cemented, an alternate method to cementing through inside diameter of the casing is used. This alternative method requires that the inner string of drillpipe be placed inside the large-diameter casing string. The drillpipe string is centralized within the casing and a special stab-in unit is made up to the bottom of the drillpipe string. After the drillpipe string is run into the well through the casing, the special unit on the bottom of the drillpipe string is stabbed into the stab-in cementing collar located a joint or two above the casing string guide shoe, or the unit is stabbed into a stab-in cementing shoe (a combination of the stab-in cementing collar and the guide shoe). Figure 4.17.13 shows the schematic of a large-diameter casing with the inner drillpipe string used for cementing [4]. Figure 4.17.14 shows the stab-in cementing shoe and a stab-in cementing collar. Also shown is a flexible latch-in plug used to follow the cement slurry. Once the stab-in unit of the drillpipe string is seated in the stab-in cementing shoe or cementing collar, a special circulating head is made up to the top of the drillstring. Circulation is established through the circulating head to the drillpipe and up the annular space between the casing and the borehole wall. The stab-in unit may be locked into the cementing collar. The collar will act as a back-flow valve [3]. After the cement slurry has been run, the drillpipe string can be unlocked and withdrawn from the casing.

The advantages of cementing large diameter casing using an inner drillpipe string are:

• Avoids excessive mud contaminations of the cement slurry with drilling mud prior to reaching the annulus
• Allows cement slurry to be added if wash out zones are excessive (avoids top-up job)

When cementing large-diameter casing loss of circulation can occur. The solution to such problems is to recement down the annulus. This technique of cementing is denoted as a top-up job. This type of cementing can be accomplished by running small-diameter tubing called "spaghetti" into the

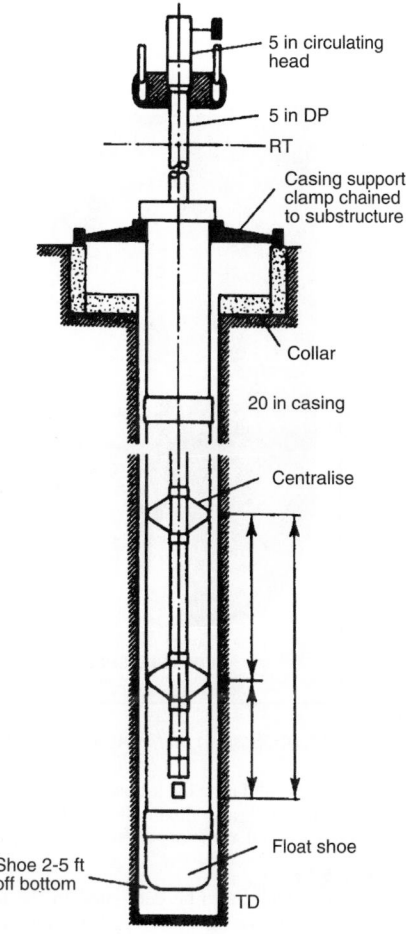

Figure 4.17.13 *Cementing of large-diameter casing with and inner drillpipe string (courtesy of Dowell Schlumberger).*

annulus space from the surface. Under these conditions usually low-specific-weight cement is used so that the formations will not fracture.

The hook load after a cementing operation (but prior to the cement setting) is important to know, particularly when large-diameter casings are to be cemented. Figure 4.17.15 shows the schematic of a casing string with the cement slurry height shown. Prior to the cement slurry setting the hook load V_h (lb) will be

$$V_h = w_c \left\{ 1 - \left[\left(\frac{\gamma_c}{\gamma_s}\right)\frac{h}{1} + \left(\frac{\gamma_m}{\gamma_s}\right)\left(\frac{1-h}{1}\right) \right] \right\} 1$$

$$- A_i[(\gamma_c - \gamma_m)h] \qquad [4.17.8]$$

where w_c = unit weight of the casing in lb/ft
γ_c = specific weight of the cement slurry in lb/ft^3
γ_m = specific weight of the drilling mud in lb/ft^3
γ_s = specific weight of the steel, which is 490 lb/ft^3
1 = length of the casing in ft
h = height the cement slurry is placed in the annular in ft
A_i = internal cross-sectional diameter of the casing in ft^2

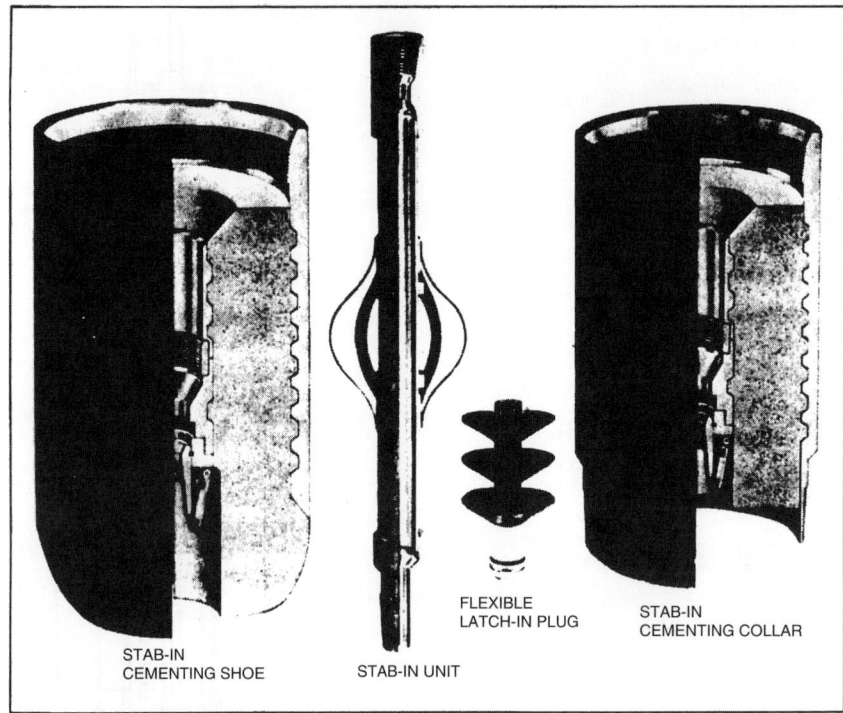

Figure 4.17.14 *Stab-in unit, stab-in cementing shoe, stab-in cementing collar, and flexible latch-in plug [2].*

Equation 4.17.8 is valid for a back-flow valve at the bottom of the casing string such as a float collar. It is also valid for a back-flow valve located at the top of the casing string.

Example 4.17.8

A large-diameter casing is to be cemented to the top in a 2,021-ft borehole. The casing string is 2,000 ft long and has ball float shoe at the bottom of the cement slurry. The borehole is 26 in. in diameter. The casing is to be 20 in., J-55, 94 lb/ft. The drilling mud in the borehole prior to running the casing has a specific weight of 10.0 lb/gal. The cement slurry is to have a specific weight of 17.0 lb/gal. The conventional cementing operation will be utilized, i.e., cementing through the inside of the casing. Determine the hook load after the cementing operation has been completed.

The hook load may be determined from Equation 4.17.8. This is

$$V_u = 94.00 \left\{ 1 + \left[\left(\frac{127.3}{490.00} \right) \frac{2,000}{2,000} + \left(\frac{75.0}{490.0} \right) \frac{0}{2,000} \right] \right\} 2,000$$

$$- \frac{\pi}{4} \left(\frac{19.124}{12} \right)^2 [(127.3 - 75.0)2,000]$$

$$= -69,557 \text{ lb}$$

This answer means that the casing and its contained drilling mud will float when the cementing operation is completed and the back-flow valve is actuated. Thus the casing should be secured down at the well head prior to initiating the cementing operation.

It should be noted that if this cementing operation were to be carried out using an inner drillpipe string to place the cement in the annulus, the above force of buoyancy would be reduced by the buoyed weight of the drillpipe. However, unless very heavy drillpipe were used, the casing and drillpipe would still float on the cement slurry.

As the cement slurry sets the hook load will decrease from its value just after the cementing operation. If the cement slurry sets up to the height (freeze point) indicated

in Figure 4.17.14, and if the casing has been landed as cemented, then the hook load V_h will be approximately

$$V_h = W_c \left(1 - \frac{\gamma_m}{\gamma_s} \right) (1 - h) \qquad [4.17.9]$$

Example 4.17.9

A $5\frac{1}{2}$-in., J-55, 17.00-lb/ft casing string 16,100 ft in length is to be run into a 16,115-ft deep well. The float collar is 30 ft from the bottom of the casing string. The drilling mud in the well has a specific weight of 12.0 lb/gal. The casing string is to be cemented to a height of 4,355 ft above the bottom of the borehole with a cement slurry having a specific weight of 16.4 lb/gal.

1. Determine the approximate hook load prior to the cementing operation.
2. Determine the approximate hook load just after the cement slurry has been run.
3. Determine the approximate hook load after the cement has set.

1. The hook load prior to the cementing operation should read approximately

$$V_h = 17.00 \left(1 - \frac{89.9}{490.0} \right) 16,100 = 223,540 \text{lb}$$

2. The hook load just after the cementing operation can be approximated by Equation 4.17.8. This is

$$V_h = 17.00 \left\{ 1 - \left[\left(\frac{122.7}{490.0} \right) \frac{4,340}{16,100} + \left(\frac{89.9}{490.0} \right) \right. \right.$$

$$\left. \left. \times \left(\frac{16,100 - 4,340}{16,100} \right) \right] \right\} 16,100$$

$$- \frac{\pi}{4} \left(\frac{4.892}{12} \right)^2 [(122.7 - 89.8)4,340]$$

$$= 199,943 \text{ lb}$$

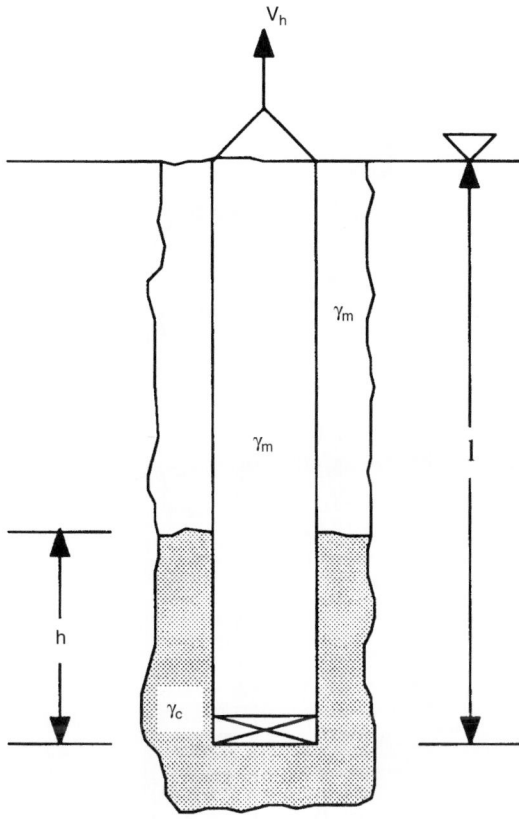

Figure 4.17.15 *Schematic of casing after a cementing operation.*

3. The hook load after the cement has set can be approximated by Equation 4.17.9. This is

$$V_h - 17.00 \left(1 - \frac{89.8}{490.0} \right) (16,100 - 4,340) = 163,282 \text{ lb}$$

4.17.7.3 Multistage Casing Cementing
Multistage casing cementing is used to cement long casing strings. The reasons that multistage cementing techniques are necessary are as follows:

- Reduce the pumping pressure of the cement pumping equipment
- Reduce the hydrostaic pressure on weak formations to prevent fracture
- Selected formations can be cemented
- Entire length of a long casing string may be cemented
- Casing shoe of the previous casing string may be effectively cemented to the new casing string
- Reduces cement contamination

There are methods for carrying out multistage cementing. These are

1. Regular two-stage cementing
2. Continuous two-stage cementing
3. Regular three-stage cementing

Regular two-stage cementing requires the use of stage cementing collar and plugs in addition to the conventional casing cementing equipment. The stage cementing collar is placed in the casing string at near the mid point, or at a position in the casing string where the upper cementing of the casing is to take place. Figure 4.17.16 shows a schematic of a regular two-stage casing cementing operation. The stage cementing collar is a special collar with ports to the annulus that can be opened and closed (sealed off) by pressure operated sleeves (see Figure 4.17.16).

The first stage of cementing (the lower section of the annulus) is carried out similar to a conventional single-stage casing cementing operation. The exception is that a wiper plug is generally not run in the casing prior to the spacer and the cement slurry.[4] During the pumping of the spacer and cement slurry for the lower section of the annulus, the ports on the stage cementing collar are closed. After the appropriate volume of spacer and cement slurry has been pumped to the well (lower section) the first stage plug is released. This plug is pumped to its position on the float collar at the bottom of the casing string with drilling mud or completion fluid as the displacement fluid. This first plug is designed to pass through the stage cementing collar without actuating it. When the first plug is landed on the float collar, there is a pressure rise at the pump. This plug seals the float collar such that further flow throughout the collar cannot take place.

After the first-stage cementing operation has been carried out the opening bomb can be dropped and allowed to fall by gravity to the lower seal of the stage cementing collar. This can be done immediately after the first-stage cementing has taken place, or at some later time when the cement slurry in the lower section of the well has had time to set. In the later case great care must be taken that the first-stage cement slurry has not risen in the annulus to a height above the stage cementing collar. Once the opening bomb is seated, pump pressure is applied that allows hydraulic force to be applied to the lower sleeve of the stage cementing collar. This force shears the lower sleeve retaining pins and exposes the ports to the annulus.

Once the ports are opened the well is circulated until the appropriate drilling mud or other fluid is in the well. The second-stage cementing slurry is mixed and pumped to the well (again without a wiper plug separating the spacer from the drilling mud). This cement slurry passes through the stage cementing float ports to the upper section of the annulus. The closing plug is released at the end of the second-stage cement slurry and is displaced to the stage cementing collar with drilling mud or other fluid. The closing plug seats on the upper sleeve of the stage cementing collar. A pressure of about 1,500 psi causes the retaining pins in the upper sleeve to shear thus forcing the sleeve downward to close the ports in the stage cementing collar.

Continuous two-stage cementing is an operation that requires that the cement slurry be mixed and displaced to the lower and upper sections of the annulus in sequence without stopping to wait for an opening bomb to actuate the stage cementing collar. In this operation the first-stage cement slurry is pumped to the well and a wiper plug released behind it (Figure 4.17.17). Displacing the wiper plug is a volume of drilling mud or completion fluid that will displace the cement slurry out of the casing and fill the inside of the casing string from the float collar at the bottom of the casing string to a height of the stage cementing collar. A bypass insert allows fluid to pass through the wiper plug and float collar after the plug is landed. The opening plug is pumped immediately behind the volume of drilling mud. Immediately behind the opening plug is the second-stage spacer and cement slurry. The opening plug sits on the lower sleeve of the stage cementing collar, opening the ports to the annulus. At the end of the second-stage cement slurry the closing plug is run. This plug sits on

---

[4]A wiper plug may be accommodated with special equipment [2].

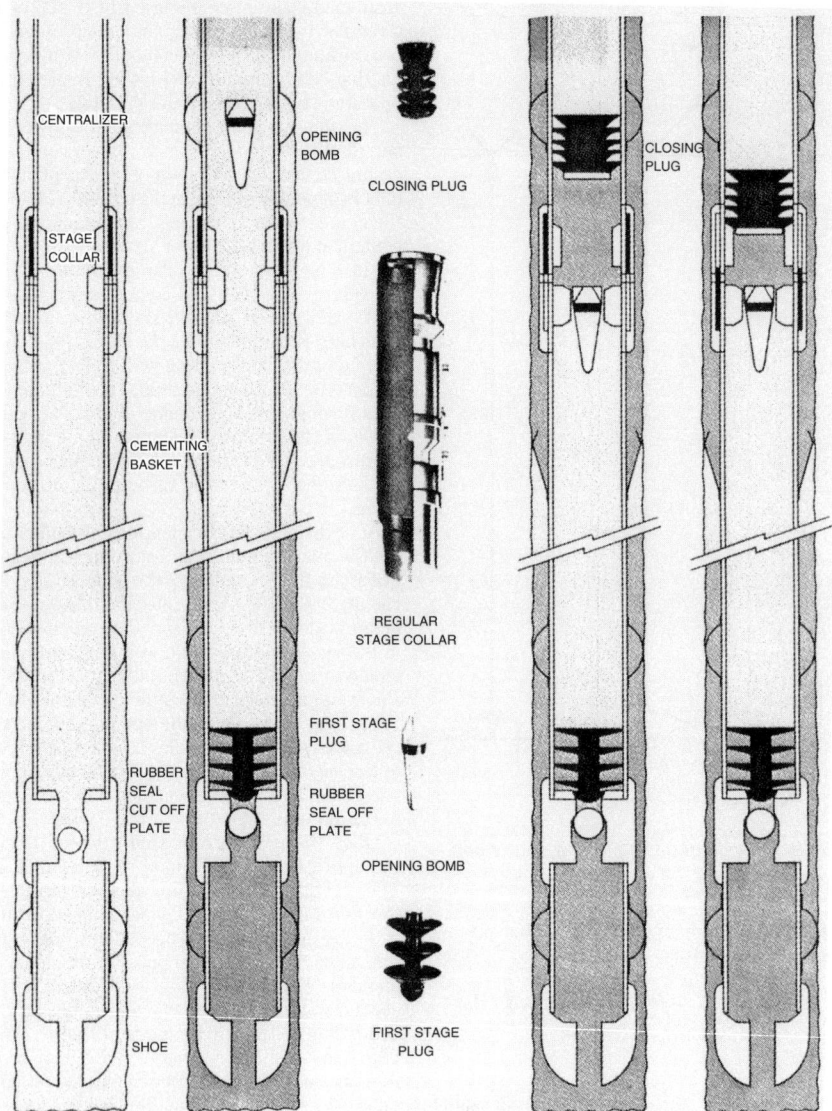

Figure 4.17.16 *Regular two-stage cementing [2].*

the upper sleeve of the stage cementing collar and with hydraulic pressure, closing the ports in the stage cementing collar.

Three-stage cementing is carried out using the same procedure as the regular two-stage cementing operation discussed above. In this case, however, two-stage cementing collars are placed at appropriate locations in the casing string above the float collar. Each stage of cementing is carried out in sequence, the lower annulus section cemented first, the middle annulus section next and the top annulus section last. Each stage of cement can be allowed to set, but great care must be taken in not allowing the lower stage of cement to rise above the stage cementing collar of the next stage above.

4.17.7.4 Liner Cementing

A liner is a short string of casing that does not reach the surface. The liner is hung from the bottom of the previous casing string using a liner hanger that grips the bottom of the previous casing with a set of slips. Figure 4.17.18 shows typical liner types. The liner is run into the borehole on the drillpipe and the cementing operation for the liner is carried out through the same drillpipe. The placing of liners and their cementing operations are some of the most difficult operation in well drilling and completions. Great care must be taken in designing and planning these operations to ensure a seal between the liner and the previous casing.

Figure 4.17.19 shows a typical linear assembly. The liner assembly is made up with the following components:

Float shoe. The float shoe may be placed at the bottom of the liner. This component is a combination of a guide shoe and a float collar.

Landing collar. This is a short sub placed in the string to provide a seat for the casing string.

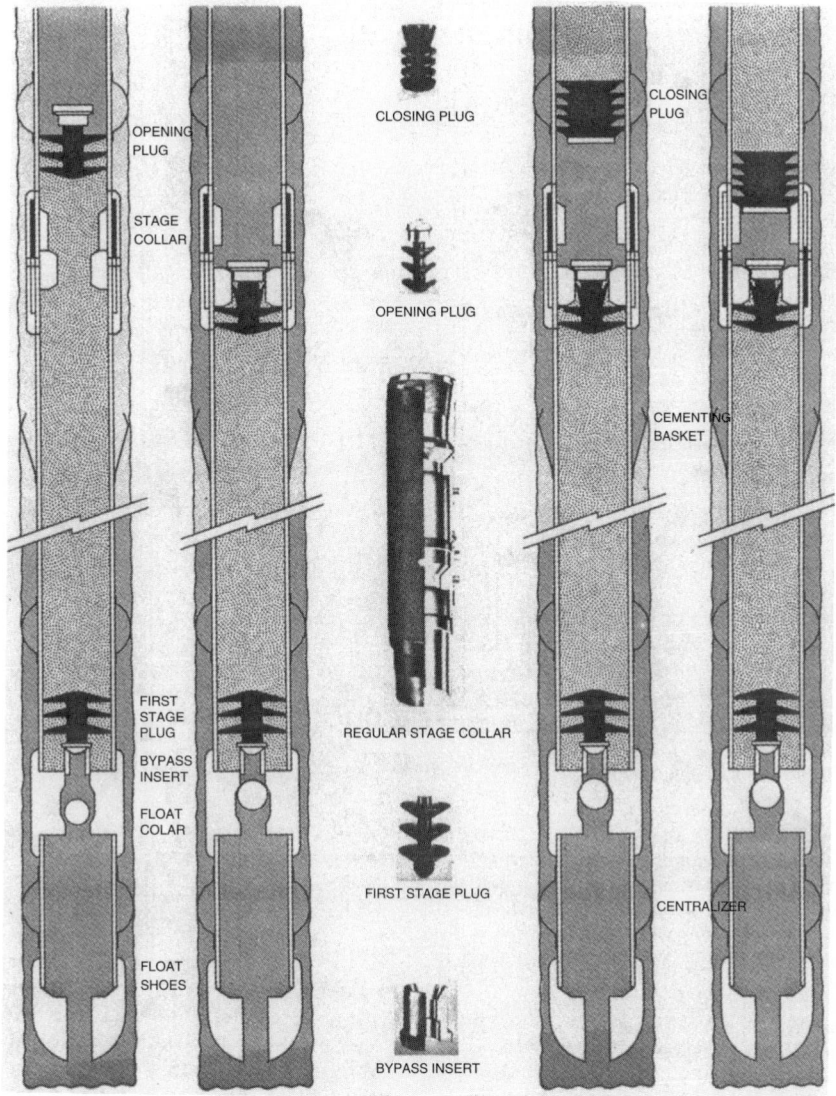

Figure 4.17.17 *Continuous two-stage cementing [2].*

Liner. This is a string of casing used to case off the open hole without bringing the end of the string to the surface. Usually the liner overlaps the previous casing string (shoe) by about 200 to 500 ft. (60 to 150 m.).

Liner hanger. This special tool is installed on the top of the liner string. The top of the liner hanger makes up to the drillpipe on which the entire liner assembly is lowered into the well. Liner hangers can be either mechanically or hydraulically actuated.

The liner hanger is the key element in the running of liners and the follow-on cementing operations. The liner hanger allows the liner to be hung from the bottom section of the previous casing. After the cementing operation the upper part of the liner hanger is retrievable, thus allowing the residual cement above the liner hanger to be cleaned out of the annulus between the drillpipe and the previous casing (by reverse circulation) and the liner left in the well.

The casing joints of the liner are placed in the well in the normal manner. The liner hanger is made up to the top of the liner. The top of the liner hanger is made up to drillpipe and the whole assembly (i.e. liner, liner hanger and drillpipe) is lowered into the well. After the liner is at the desirable location in the well, the drillpipe is connected to the rig pumps and mud circulation carried out. This allows conditioning of the drilling mud in the well prior to the cementing operation and ensures that circulation is possible before the liner is hung and cemented.

The liner hanger is set (either mechanically or hydraulically) and the drillpipe with the upper part of the liner hanger (the setting tool) released. The drillpipe and the setting tool are raised to make sure that the setting tool and the drillpipe can be released from the lower part of the liner hanger and the liner. The drillpipe and the setting tool are lowered to make a tight seal with the lower portion of the liner hanger.

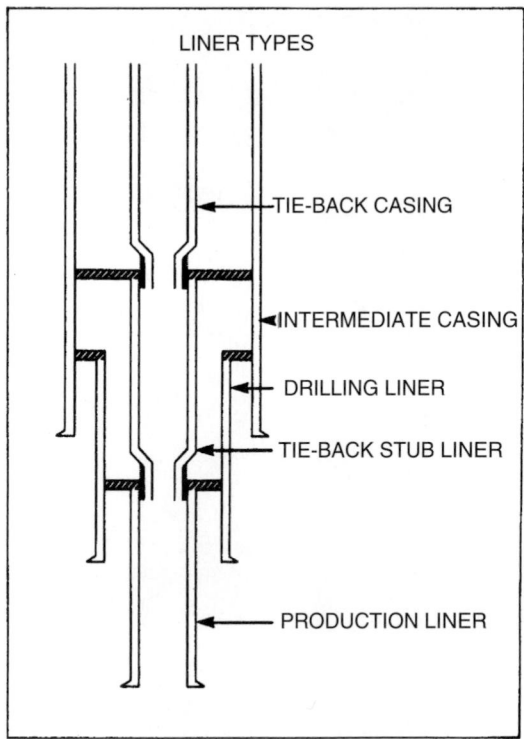

Figure 4.17.18   *Liner types [2].*

After these tests of the equipment have been made, a liner cementing head is made up to the drillpipe at the surface. Figure 4.17.20 shows the liner cementing head and the pump-down plug. The pump-down plug is placed in the liner cementing head, but not released. The cement pump is connected to the liner cementing head and the spacer and cement slurry pumped to the head (no wiper plug is run ahead of the spacer). The pump-down plug is released at the end of the cement slurry and separates the slurry from the follow-on drilling mud. Figure 4.17.21 shows the sequence of the liner cementing operation. The drilling mud displaces the pump-down plug to the liner hanger. As the pump-down plug passes through the liner hanger it latches into the liner wiper plug (see Figure 4.17.19). With increased surface pressure (1200 psi or 83 bar) by the pump the liner wiper by the pump the liner wiper plug with the pump-down plug coupled to it are released from the liner hanger and begin to move again downward. These two coupled plugs eventually seat on the landing collar or on the float collar. Another pressure rise indicates the cement is in place behind the liner.

Once the spacer and the cement slurry have been successfully pumped to their location in the well, the drillpipe and setting tool are released from the lower part of the liner hanger. Also, the liner cementing head is removed from the drillpipe. The drillpipe and setting tool are raised slightly and the excess cement slurry reverse circulated from around the liner hanger area. These steps should be taken immediately after the completion of the cementing operation, otherwise the excess cement slurry between the drillpipe and the previous casing could set and cause later drilling and completions problems if the excess cement is too great. If reverse circulation is not planned, then the excess cement slurry that comes through the liner hanger to the annulus between the

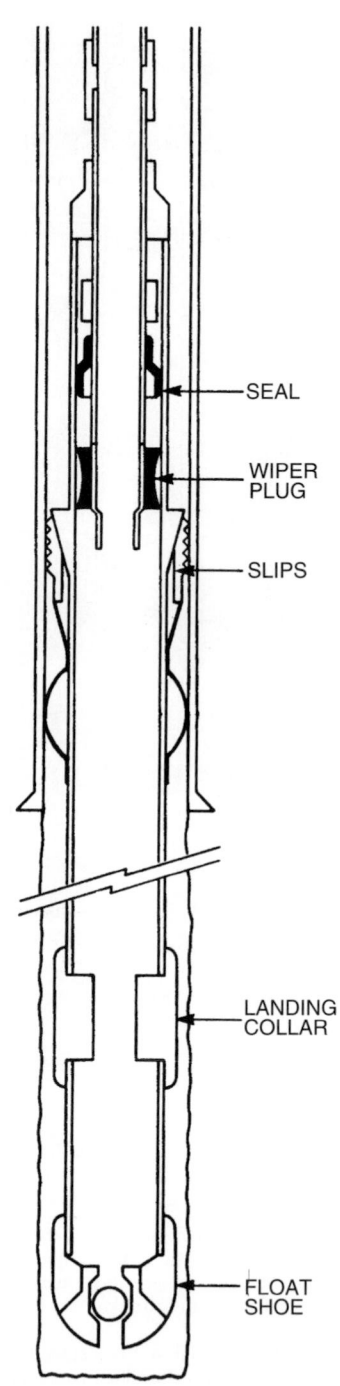

Figure 4.17.19   *Liner assembly [2].*

drillpipe and the previous casing must be kept to a minimum so that it can be easily drilled out after it sets.

The determination of the excess cement slurry should be carefully calculated: too little and the cement seal at the liner hanger is contaminated with drilling mud; too much and there are problems removing it.

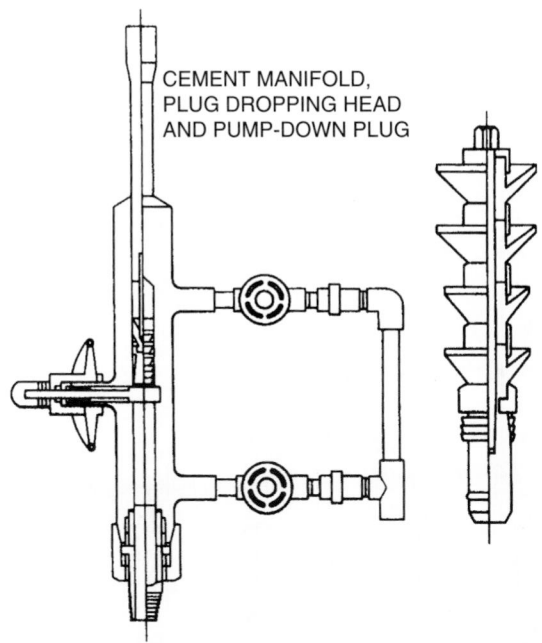

CEMENT MANIFOLD,
PLUG DROPPING HEAD
AND PUMP-DOWN PLUG

Figure 4.17.20 *Liner cementing head [2].*

4.17.8 Secondary Cementing

Secondary or remedial cementing refers to cementing operations that are intended to use cement as a means of maintaining or improving the well's operation. There are two general secondary cementing operations, squeeze cementing and plug cementing.

4.17.8.1 Squeeze Cementing

Squeeze cementing operations utilize the mechanical power of the cement pumps to force a cement slurry into the annular space behind a casing and/or into a formation for the following purposes:

• Repair a faulty primary cementing operation
• Stop loss of circulation in the open hole during drilling operations
• Reduce water-oil, water-gas and gas-oil ratios by selectively sealing certain fluid producing formations
• Seal abandoned or depleted formations
• Repair casing leaks such as joint leaks, split casing, parted casing or corroded casing
• Isolate a production zone by sealing off adjacent non-productive zones

There are numerous squeeze cementing placement methods that utilize many different special downhole tools. In general, the squeeze cementing operation forces the cement slurry into fractures under high pressure, or into casing perforations using low pressure. As a cement slurry is forced into rock fractures, or through casing perforations into the rock formations, the slurry loses part of its mix water. This leaves a filter cake of cement particles at the interface of the fluid and the permeable rock. As the filter cake builds up during the squeeze cementing operation, more channels into the formation are sealed and the pressure increases. If the cement slurry is poorly designed for an intended squeeze cementing operation, the filter cake builds up too fast and the cement pump capability is reached before the cement

slurry has penetrated sufficiently to accomplish its purpose. Thus the squeeze cementing operation slurry should be designed to match the characteristics of the rock formation to be squeeze cemented and the equipment to be used.

There are five important considerations regarding the cement slurry design for a squeeze cementing operation:

1. *Fluid loss control.* The slurry should be designed to match the formation to be squeezed. Low permeability formations should utilize slurries with 100–200 ml/30 min water losses. High Permeability formations should utilize slurries with 50–100 ml/30 min water losses.
2. *Slurry volume.* The cement slurry volume should be estimated prior to the squeeze operation. In general, high-pressure squeeze operations of high-permeability formations that have relatively low fracture strengths will require large volumes of slurry. Low-pressure squeeze operation through perforations will require low volumes.
3. *Thickening time.* High-pressure squeeze operations that pump large volumes in a rather short time period usually require accelerator additives. Low-pressure slow-pumping-rate squeeze operations usually require retarder additives.
4. *Dispersion.* Thick slurries will not flow well in narrow channels. Squeeze cement slurries should be designed to be thin and have low yield points. Dispersive agents should be added to these slurries.
5. *Compressive strength.* High compressive strength is not a necessary characteristic of squeeze cement slurries.

There are basically two squeeze cementing techniques used: the high-pressure squeeze operation and the low-pressure squeeze operation.

High-pressure squeeze cementing operations are utilized where the hydraulic pressure is used to make new channels in the rock formations (by fracturing the rock) and force the cement slurry into these channels.

Low-pressure squeeze cementing operations are utilized where the existing permeability structure is sufficient to allow the cement slurry to efficiently move in formation without making new fracture surfaces with the hydraulic pressure.

Hesitation method of applying pressure is applicable to both high and low-pressure squeeze cementing operations. This method of applying pressure (and thus volume) appears to be more effective than continuous pressure application. The hesitation method is the intermittent application of pressure, separated by a period of pressure leakoff caused by the loss of filtrate into the formation. The leakoff periods are short at the beginning of an operation but get longer as the operation progresses.

Figure 4.17.22 shows a typical squeeze cementing downhole schematic where a retrievable packer is used. The cement slurry is spotted adjacent to the perforations in the casing. The packer raise and set against the casing wall. Pressure is then applied to the drilling mud and cement slurry below the packer forcing the cement slurry into the perforation and into the rock formation or voids in the annulus behind the casing.

Figure 4.17.23 shows how a retrievable packer with a tail pipe can be used to precisely spot the cement slurry at the location of the perforations.

Figure 4.17.24 shows how a drillable cement retainer can be used to ensure that the cement slurry will be applied directly to the perforations.

Most squeeze cementing operations take place in cased sections of a well. However, open-hole packers can be used to carry out squeeze cement operations of thief zones during drilling operations.

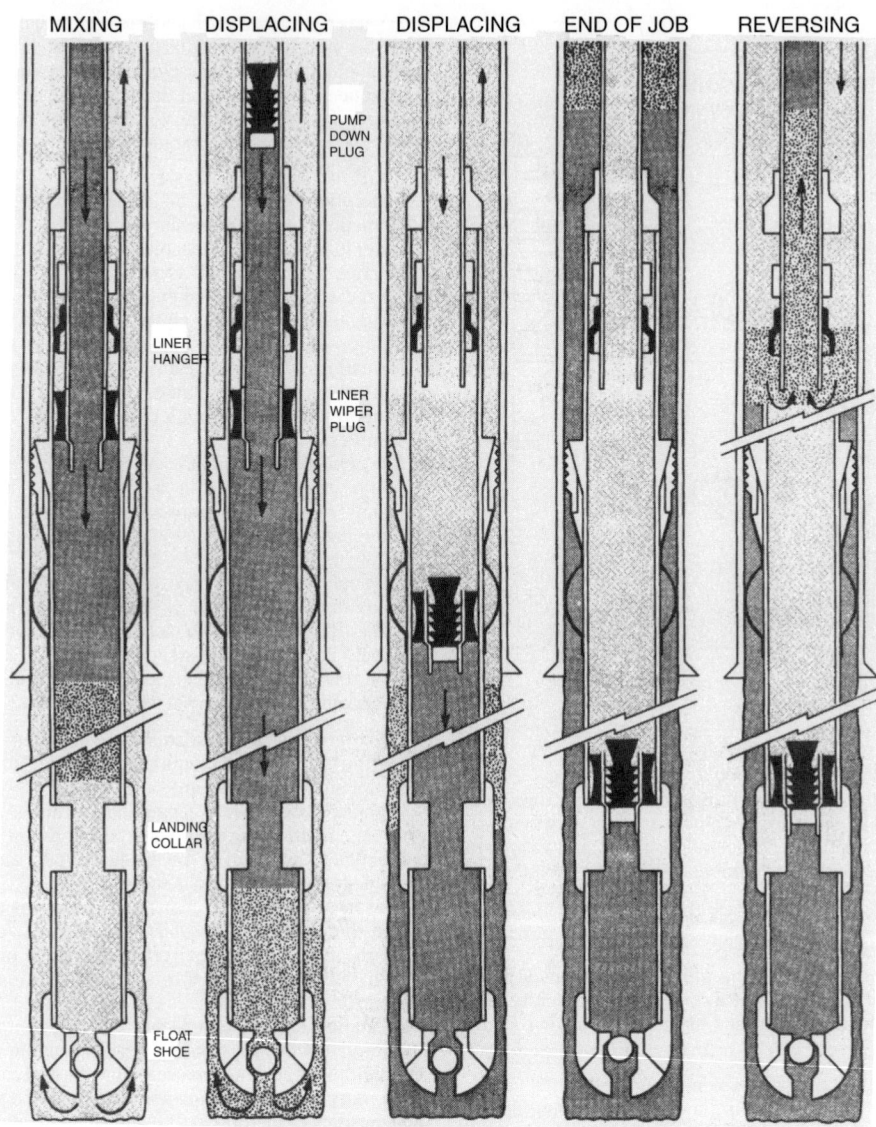

MIXING   DISPLACING   DISPLACING   END OF JOB   REVERSING

PUMP
DOWN
PLUG

LINER
HANGER

LINER
WIPER
PLUG

LANDING
COLLAR

FLOAT
SHOE

Figure 4.17.21   *Liner cementing [2].*

Another technique for carrying out a squeeze cementing operation is the Bradenhead technique. This technique can be used to squeeze a cement slurry in a cased hole or in an open hole. Figure 4.17.25 shows a schematic of the technique. Instead of using a downhole packer, the cement slurry is spotted by drillpipe (or tubing) adjacent to the perforations or a thief zone. After the drillpipe has been raised, the pipe rams are closed over the drillpipe and pump pressure applied to the drilling mud which in turn forces the cement slurry into the perforations or fractures.

The basic squeeze cementing operation procedures are as follows [2]:

1. Lower zones are isolated by a retrievable or drillable bridge plug.
2. Perforations are washed using a perforation wash tool or they are reopened by surging. If there is no danger of

damaging the lower perforations, this operation can be carried out before running the bridge plug, which may be run in one trip with the packer.
3. The perforation wash tool is retrieved and the packer run in the hole with a work string, set at the desired depth and tested. An annular pressure test of 1,000 psi (69 bar) is usually sufficient. The packer is run with or without a tail pipe, depending on the operation to be performed. If cement is to be spotted in front of the perforations, a tail pipe that covers the length of the zone plus 10 to 15 ft (3 to 5 meters) must be run with the packer.
4. An injectivity test is performed using clean, solids-free water or brine. If a low fluid loss completion fluid is in the hole, it must be displaced from the perforations before starting the injecting. This test will give an idea of the permeability of the formation to the cement filtrate.

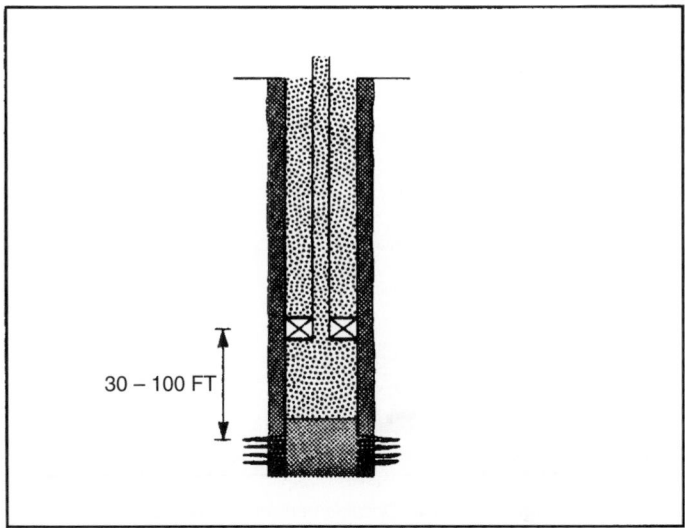

Figure 4.17.22 *Retrievable packer near perforations [2].*

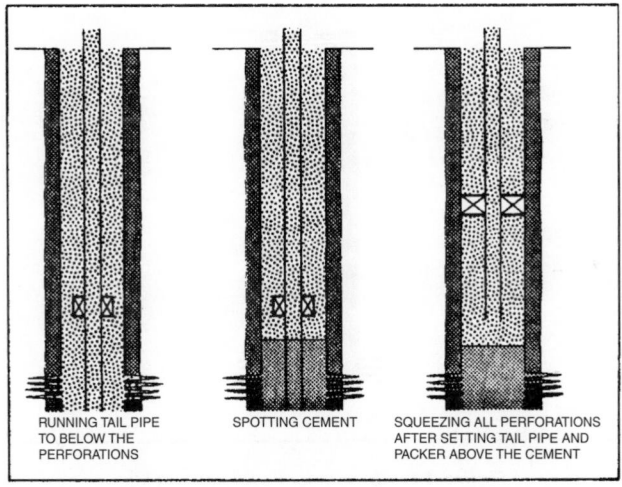

| RUNNING TAIL PIPE TO BELOW THE PERFORATIONS | SPOTTING CEMENT | SQUEEZING ALL PERFORATIONS AFTER SETTING TAIL PIPE AND PACKER ABOVE THE CEMENT |

Figure 4.17.23 *Retrievable packer with tail pipe [2].*

5. A spearhead or breakdown fluid followed by the cement slurry is circulated downhole with the packer by-pass open. This is done to avoid the squeezing of damaging fluids ahead of the slurry. A small amount of back pressure must be applied on the annulus to prevent the slurry fall caused by U tubing. If no tail pipe has been run, the packer by-pass must be closed 2 or 3 bbl (.25 to .33 m^3) before the slurry reaches the packer. If the cement is to be spotted in front of the perforations, with the packer unset, circulation is stopped as soon as the cement covers the desired zone, the tail pipe pulled out of the cement slurry and the packer set at the desired depth at which the packer is set must be carefully decided.

   If tail pipe is run, the minimum distance between perforations and packer is limited to the length of the tail pipe. The packer must not be set too close to the perforations as pressure communication through the annulus above the packer might cause the casing to collapse. A safe setting depth must be decided on after seeing the logged quality of the cement bond. Casing conditions and possible cement contamination limit the maximum spacing between packer and treated zone.

6. Squeeze pressure is applied at the surface. If high-pressure squeezing is practiced, the formation is broken down and the cement slurry pumped into the fractures before the hesitation technique is applied. If low-pressure squeezing is desired, hesitation is started as soon as the packer is set.

7. Hesitation continues until no pressure leak-off is observed. A further test of about 500 psi (35 bar) over the final injection pressure will indicate the end of the injection process. Usually, well-cementing perforations will tolerate pressures above the formation fracture pressure, but the risk of fracturing is increased.

8. Pressure is released and returns are checked. If no returns are noticed, the packer by-pass is opened and

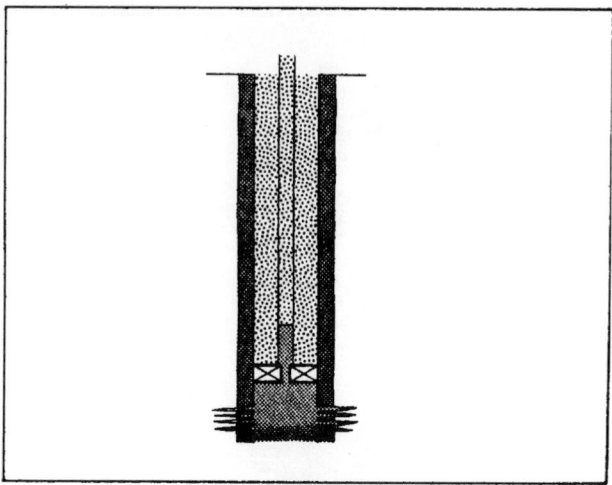

Figure 4.17.24 *Drillable cement retainer [2].*

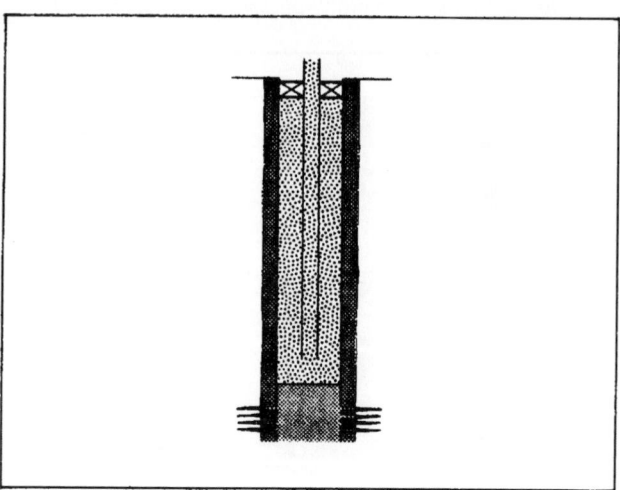

Figure 4.17.25 *Bradenhead squeeze [2].*

excess cement reversed out. Washing off cement in front of perforations can be performed by releasing the packer and slowly lowering the work string during the reversing.

9. Tools are pulled out and the cement is left to cure for the recommended time, usually 4 to 6 hr.

Plug Cementing

The major reasons for plug cementing are:

Abandonment. State regulations have rules on plugging and abandoning wells. Cement plugs are normally used for that purpose (see Figure 4.17.26).

Kick-off plug. Usually an Ottawa sand-cement plug is used to plug off a section of the borehole. This plug uses a hard surface to assist the kick-off procedure (see Figure 4.17.27).

Lost circulation. A cement plug can be placed adjacent to a zone of lost circulation in the hope that the cement slurry will penetrate and seal fractures (see Figure 4.17.28).

Openhole completions. Often in openhole completions it is necessary to shut off water flows, or to provide an anchor for testing tools, or other maintenance operations (Figure 4.17.29).

There are three methods for placing cement plugs.

1. *Balance plug method* is the most commonly used. The cement slurry is placed at the desired depth through the drillpipe or tubing run to that depth. A spacer is placed below and above the slurry plug to avoid contamination of the cement slurry with surrounding drilling mud and to assist in balancing the mud. There are a number of tools available in the market to assist in plug balancing. Most involve drill pipe darts that provide a means of separating fluids, wiping the pipe and provide an indication of the end of displacement.

2. *Dump bailing method* utilizes a bailing device that contains a measure volume of cement slurry. The bailer is run to the appropriate depth on a wireline and releases its load

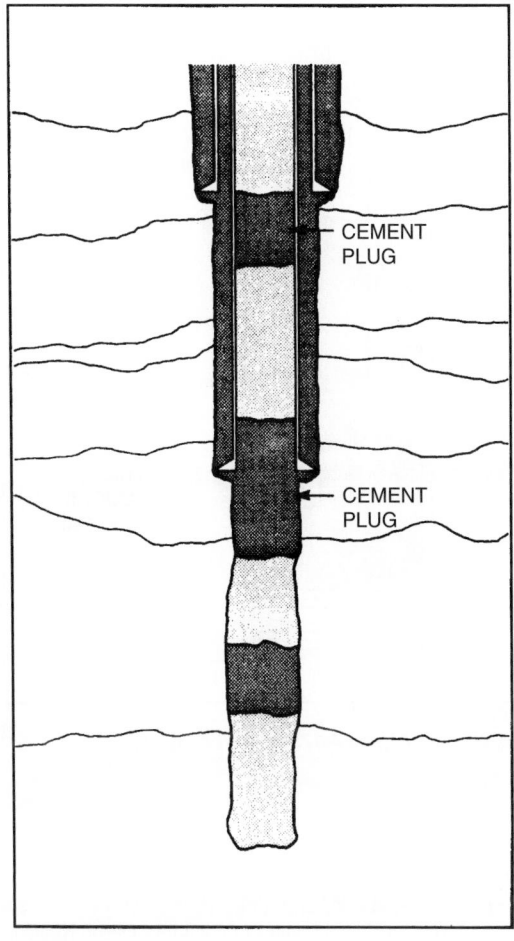

Figure 4.17.26   *Abandonment plugs [2].*

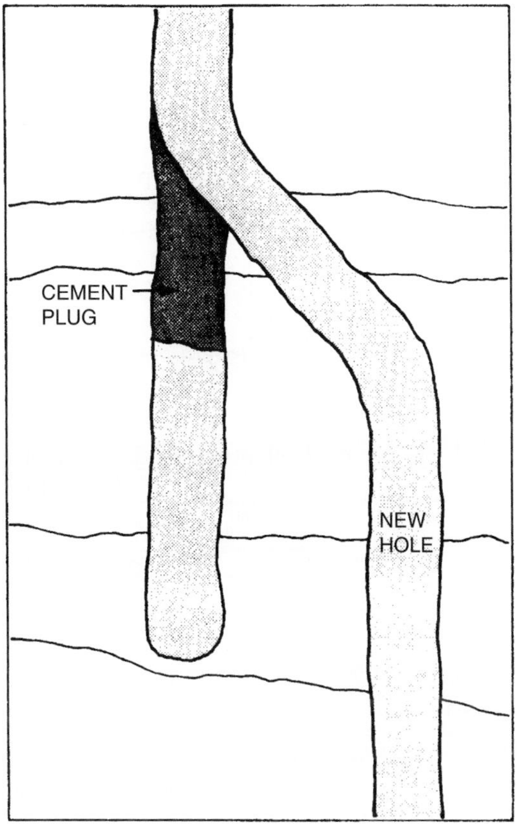

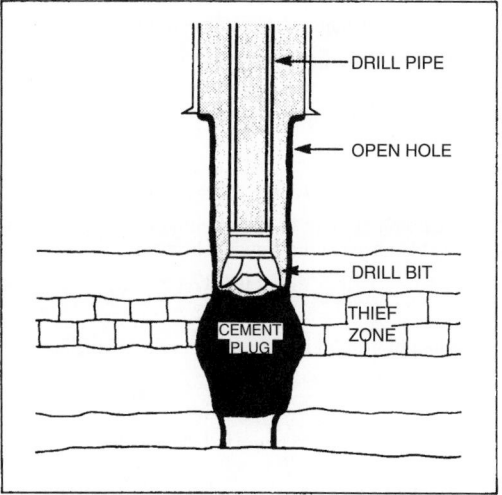

Figure 4.17.27   *Kickoff plug [2].*

upon bumping the bottom or a permanent bridge plug set at the desired depth (see Figure 4.17.30).

Cement plugs fail for the following reasons:

- Lack of hardness (kick-off plug)
- Contaminated cement slurry
- Placed at wrong depth
- Plug migrates from intended depth location due to lack of balancing control

When placing a plug in well the fluid column in the tubing or drillpipe must balance the fluid column in the annulus. If the fluid in the drillpipe is heaviest, then the fluid from the drillpipe (the slurry) will continue down and out the end of the drillpipe and then up into the annulus after pumping has stopped. If the fluid in the annulus is heaviest, then the fluid from the drillpipe will continue downward in the annulus after pumping. The two columns are balanced by using a spacer (or wash) above and below the cement slurry. Since the cement slurry usually has a higher specific weight than the annular fluid, the spacers are normally water.

Example 4.17.10
Balance a 100-sack cement plug of Class C neat cement slurry when 15 bbl of water are to be run ahead of the slurry.

Figure 4.17.28   *Lost circulation plug [2].*

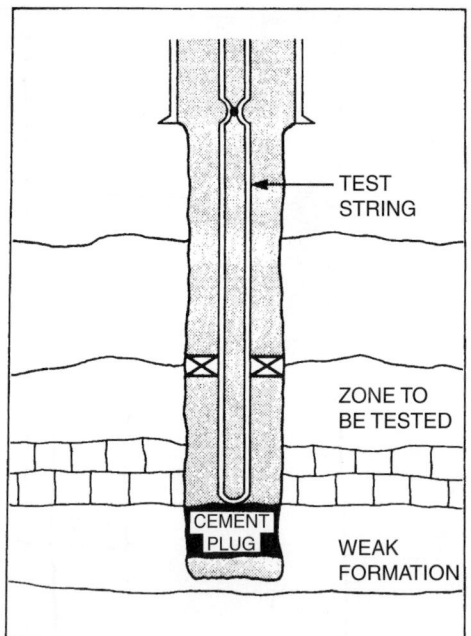

Figure 4.17.29 *Openhole completions plug [2].*

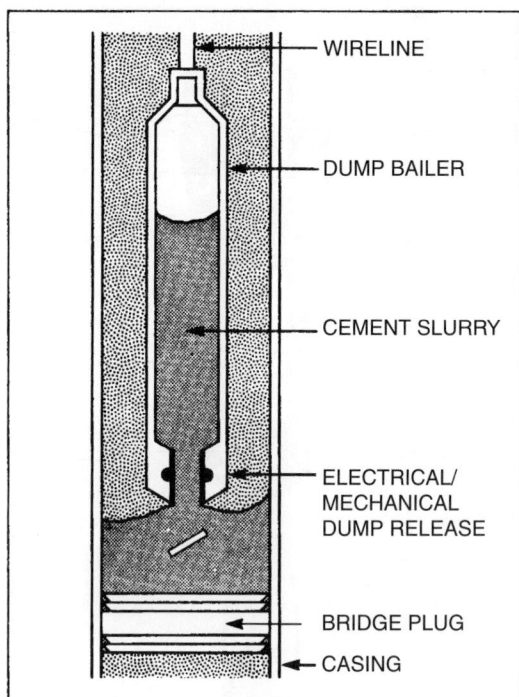

Figure 4.17.30 *Dump bailer [2].*

The plug is to be run through 2-in., 4.716-lb/ft, EUE tubing. The depth of placement is 4,000 ft.

1. Determine the top of the plug.
2. Determine the volume of water to run behind cement slurry to balance the 15 bbl of water run ahead of the slurry.
3. Determine the volume of mud to pump to balance plug.

The annular volume per ft is $0.1997\,\text{ft}^3/\text{ft}$. The tubing volume per ft is $0.0217\,\text{ft}^3/\text{ft}$. The yield of a Class C neat cement slurry is $1.32\,\text{ft}^3/\text{sack}$ (see Table 4.17.2). Thus the volume of the cement slurry is

$$\text{Volume} = 1.32(100) = 132\,\text{ft}^3$$

1. The height H (ft) of the plug is

$$H = \frac{132}{0.1997 - 0.0217} = 742\,\text{ft}$$

Depth to top of plug D (ft) is

$$D = 4,000 - 742 = 3,258\,\text{ft}.$$

2. The amount of water to place behind the plug to balance the 15 bbl ahead of the cement slurry is

$$\text{Volume} = 15\left(\frac{0.0217}{0.1997}\right) = 1.6\,\text{bbl}$$

3. The height h (ft) of trailing water in the tubing is

$$h = \frac{1.6(42)}{(0.0217)(7.48)} = 414\,\text{ft}$$

The volume of mud to pump to balance plug is

$$\text{Volume} = [4,000 - (742 + 414)]\frac{0.0217}{5.6146} = 11\,\text{bbl}$$

References

1. McCray, A. W., and Cole, F. W., *Oil Well Drilling Technology*, University of Oklahoma Press, Norman, 1959.
2. *Cementing Technology*, Dowel Schlumberger Publication. 1984.
3. Smith, D. K., *Cementing*, SPE Publications, Second Printing, 1976.
4. Rabia, H., *Oilwell Drilling Engineering, Principles and Practices*, Graham and Trotman Ltd, 1985.
5. API Specification 10: "API Specification for Materials and Testing for well Cementing," January 1982.
6. ASTM C 150: Standard Specifications for Porland Cement, *Book of ASTM standards*, Part 13, ASTM Publication, 1987.
7. API Standard 10A: "API Specifications for Oil-Well Cements and Cement Additives," April 1969.
8. Bourgoyne, A. T., *Applied Drilling* Engineering, SPE Textbook Series, Vol. 2, 1986.
9. Parker, P. N., Ladd, B. J., Ross, W. M., and Wahl, W. W., "An evaluation of primary cementing technique using low displacement rates," SPE Fall Meeting, 1965, SPE No. 1234.
10. McLean, R. H., Maury, C. W., and Whitaker, W. W., "Displacement mechanics in primary cementing," *Journal of Petroleum Technology*, February 1967.

4.18 TUBING AND TUBING STRING DESIGN

Tubing string is installed in the hole to protect the casing from erosion and corrosion and to provide control for production. The proper selection of tubing size, steel grade, unit weight and type of connections is of critical importance in a well completion program. Tubing must be designed against failure from tensile/compressive forces, internal and external pressures and buckling.

Tubing is classified according to the outside diameter, the steel grade, unit weight (well thickness), length and type of joints. The API tubing list is given in Tables 4.18.1 and 4.18.2.

4.18.1 API Physical Property Specifications

The tubing steel grade data as given by API Specification 5A, 35th Edition (March 1981), Specification 5AX, 12th Edition (March 1982) and Specification 5AC, 13th Edition (March 1982), are listed in Table 4.18.3.

4.18.1.1 Dimensions, Weights and Lengths

The following are excerpts from API Specification 5A, 7th Edition, 2001 "API Specification for Casing, Tubing, and Drill Pipe."

1. *Dimensions and weights.* Pipe shall be furnished in the sizes, wall thicknesses and weights (as shown in Tables 4.18.4, 4.18.5 and 4.18.6) as specified by API Specification 5A: "Casing, Tubing and Drill Pipe."
2. *Diameter.* The outside diameter of 4 in.-tubing and smaller shall be within ± 0.79 mm. Inside diameters are governed by the outside diameter and weight tolerances.
3. *Wall thickness.* Each length of pipe shall be measured for conformance to wall-thickness requirements. The wall thickness at any place shall not be less than the tabulated thickness minus the permissible undertolerance of 12.5%. Wall-thickness measurements shall be made with a mechanical caliper or with a properly calibrated nondestructive testing device of appropriate accuracy. In case of dispute, the measurement determined by use of the mechanical caliper shall govern.
4. *Weight.* Each length of tubing in sizes 1.660 in. and larger shall be weighed separately. Lengths of tubing in sizes 1.050 and 1.315 in. shall be weighed either individually or in convenient lots.

   Threaded-and-coupled pipe shall be weighed with the couplings screwed on or without couplings, provided proper allowance is made for the weight of the coupling. Threaded-and-coupled pipe, integral joint pipe and pipe shipped without couplings shall be weighed without thread protectors except for carload weighings, for which proper allowances shall be made for the weight of the thread protectors.
5. Calculated weights shall be determined in accordance with the following formula:

$$W_L = (Wpe \times L) + e_w$$

   where W_L = calculated weight of a piece of pipe of length L in lb (kg)

   Wpe = plain-end weight in lb/ft (kg/m)

   L = length of pipe, including end finish, as defined below in ft (m)

   e_w = weight gain or loss due to end finishing in lb (kg) (for plain-end pipe, $e_w = 0$)

6. *Length.* Pipe shall be furnished in range lengths conforming to the following as specified on the purchase order: When pipe is furnished with threads and couplings, the length shall be measured to the outer face of the coupling, or if measured without couplings, proper allowance shall

be made to include the length of the coupling. For integral joint tubing, the length shall be measured to the outer face of the box end.

Pipe Range Lengths

	Range 1	Range 2
Total range length*	20–24	28–32
Range length for 100% of carload:[†]		
Permissible variation, max.	2	2
Permissible length, min.	20	28

*By agreement between purchaser and manufacturer, the total range length for Range 1 tubing may be 6.10–8.53 m.

[†] Carload tolerances shall not apply to orders less than a carload shipped from the mill. For any carload of pipe shipped from the mill to the final destination without transfer or removal from the car, the tolerance shall apply to each car. For any order consisting of more than a carload and shipped from the mill by rail, but not to the final destination in the rail cars loaded at the mill, the carload tolerances shall apply to the total order, but not to the individual carloads.

4.18.1.2 Performance Properties

Tubing performance properties, according to API Bulletin 5C2, 18th Edition (March 1982), are given in Table 4.18.7. Formulas and procedures for calculating the values in Table 4.18.7 are given in API Bulletin 5C3, 3rd Edition (March 1980).

4.18.2 Running and Pulling Tubing

The following are excerpts from "API Recommended Practice for Care and Use of Casing and Tubing," API RP 5C1, 12th Edition (March 1981).

4.18.3 Preparation and Inspection before Running

1. New tubing is delivered free of injurious defects as defined in API Standard 5A, 5AC and 5AX, and within the practical limits of the inspection procedures therein prescribed. Some users have found that, for a limited number of critical well applications, these procedures do not result in casing sufficiently free of defects to meet their needs for such critical applications. Various nondestructive inspection services have been employed by users to assure that the desired quality of tubing is being run. In view of this practice, it is suggested that the individual user:
   a. Familiarize her or himself with inspection practices specified in the standards and employed by the respective mills and with the definition of "injurious defect" contained in the standards.
   b. Thoroughly evaluate any nondestructive inspection to be used by him or her on API tubular goods to ensure that the inspection does in fact correctly locate and differentiate injurious defects from other variables that can be and frequently are sources of misleading "defect" signals with such inspection methods.

   Caution: Due to the permissible tolerance on the outside diameter immediately behind the tubing upset, the user is cautioned that difficulties may occur when wraparound seal-type hangers are installed on tubing manufactured on the high side of the tolerance; therefore, it is recommended that the user select the joint of tubing to be installed at the top of the string.
2. All tubing, whether new, used or reconditioned, should always be handled with thread protectors in place. Tubing should be handled at all times on racks or on wooden or metal surfaces free of rocks, sand or dirt other than

Table 4.18.1   *ISO/API Tubing List Sizes, Masses, Wall Thickness, Grade and Applicable End Finish*

Labels			Outside diameter D in.	Nominal linear masses[a,b]			Wall thickness t in.	Type of end-finish[c]						
	2			Non-upset T&C lb/ft	Ext. upset T&C lb/ft	Integ. joint lb/ft					N80			
1	NU T&C	EU T&C	IJ						H40	J55	L80	Type 1,Q	C90[d]	T95[d]	P110
1	2	3	4	5	6	7	8	9	10	11	12	13	14	15	16
1.050	1.14	1.20		1.050	1.14	1.20		0.113	PNU	PNU	PNU	PNU	PNU	PNU	
1.050	1.48	1.54		1.050	1.48	1.54		0.154	PU	PU	PU	PU	PU	PU	PU
1.315	1.70	1.80	1.72	1.315	1.70	1.80	1.72	0.133	PNUI	PNUI	PNUI	PNUI	PNUI	PNUI	
1.315	2.19	2.24		1.315	2.19	2.24		0.179	PU	PU	PU	PU	PU	PU	PU
1.660	2.09		2.10	1.660			2.10	0.125	PI	PI					
1.660	2.30	2.40	2.33	1.660	2.30	2.40	2.33	0.140	PNUI	PNUI	PNUI	PNUI	PNUI	PNUI	
1.660	3.03	3.07		1.660	3.03	3.07		0.191	PU	PU	PU	PU	PU	PU	PU
1.900	2.40		2.40	1.900			2.40	0.125	PI	PI					
1.900	2.75	2.90	2.76	1.900	2.75	2.90	2.76	0.145	PNUI	PNUI	PNUI	PNUI	PNUI	PNUI	
1.900	3.65	3.73		1.900	3.65	3.73		0.200	PU	PU	PU	PU	PU	PU	PU
1.900	4.42			1.900	4.42			0.250			P		P	P	
1.900	5.15			1.900	5.15			0.300			P		P	P	
2.063	3.24		3.25	2.063			3.25	0.156	PI	PI	PI	PI	PI	PI	
2.063	4.50			2.063	4.50			0.225	P	P	P	P	P	P	P
$2\frac{3}{8}$	4.00			2.375	4.00			0.167	PN	PN	PN	PN	PN	PN	
$2\frac{3}{8}$	4.60	4.70		2.375	4.60	4.70		0.190	PNU	PNU	PNU	PNU	PNU	PNU	PNU
$2\frac{3}{8}$	5.80	5.95		2.375	5.80	5.95		0.254			PNU	PNU	PNU	PNU	PNU
$2\frac{3}{8}$	6.60			2.375	6.60			0.295			P		P	P	
$2\frac{3}{8}$	7.35	7.45		2.375	7.35	7.45		0.336			PU		PU	PU	
$2\frac{7}{8}$	6.40	6.50		2.875	6.40	6.50		0.217	PNU	PNU	PNU	PNU	PNU	PNU	PNU
$2\frac{7}{8}$	7.80	7.90		2.875	7.80	7.90		0.276			PNU	PNU	PNU	PNU	PNU
$2\frac{7}{8}$	8.60	8.70		2.875	8.60	8.70		0.308			PNU	PNU	PNU	PNU	PNU
$2\frac{7}{8}$	9.35	9.45		2.875	9.35	9.45		0.340			PU		PU	PU	
$2\frac{7}{8}$	10.50			2.875	10.50			0.392			P		P	P	
$2\frac{7}{8}$	11.50			2.875	11.50			0.440			P		P	P	
$3\frac{1}{2}$	7.70			3.500	7.70			0.216	PN	PN	PN	PN	PN	PN	
$3\frac{1}{2}$	9.20	9.30		3.500	9.20	9.30		0.254	PNU	PNU	PNU	PNU	PNU	PNU	PNU
$3\frac{1}{2}$	10.20			3.500	10.20			0.289	PN	PN	PN	PN	PN	PN	
$3\frac{1}{2}$	12.70	12.95		3.500	12.70	12.95		0.375			PNU	PNU	PNU	PNU	PNU
$3\frac{1}{2}$	14.30			3.500	14.30			0.430			P		P	P	
$3\frac{1}{2}$	15.50			3.500	15.50			0.476			P		P	P	
$3\frac{1}{2}$	17.00			3.500	17.00			0.530			P		P	P	
4	9.50			4.000	9.50			0.226	PN	PN	PN	PN	PN	PN	
4	10.70	11.00		4.000		11.00		0.262	PU	PU	PU	PU	PU	PU	
4	13.20			4.000	13.20			0.330			P		P	P	
4	16.10			4.000	16.10			0.415			P		P	P	
4	18.90			4.000	18.90			0.500			P		P	P	
4	22.20			4.000	22.20			0.610			P		P	P	
$4\frac{1}{2}$	12.60	12.75		4.500	12.60	12.75		0.271	PNU	PNU	PNU	PNU	PNU	PNU	
$4\frac{1}{2}$	15.20			4.500	15.20			0.337			P		P	P	
$4\frac{1}{2}$	17.00			4.500	17.00			0.380			P		P	P	
$4\frac{1}{2}$	18.90			4.500	18.90			0.430			P		P	P	
$4\frac{1}{2}$	21.50			4.500	21.50			0.500			P		P	P	
$4\frac{1}{2}$	23.70			4.500	23.70			0.560			P		P	P	
$4\frac{1}{2}$	26.10			4.500	26.10			0.630			P		P	P	

P = Plain end, N = Non-upset threaded and coupled, U = External upset threaded and coupled, I = Integral joint.
[a] Nominal linear masses, threads and coupling (cols. 2, 3, 4) are shown for information only.
[b] The densities of martensitic chromium steels (L80 types 9Cr and 13Cr) are different from carbon steels. The masses shown are therefore not accurate for martensitic chromium steels. A mass correction factor of 0,989 may be used.
[c] Non-upset tubing is available with regular couplings or special bevel couplings. External-upset tubing is available with regular, special-bevel, or special-clearance couplings.
[d] Grade C90 and T95 tubing shall be furnished in sizes, masses, and wall thicknesses as listed above, or as shown on the purchase agreement.

Table 4.18.2 *ISO/API Tubing List Sizes, Masses, Wall Thickness, Grade and Applicable End Finish*

	Labels				Nominal linear masses[a,b]				Type of end-finish[c]						
		2		Outside diameter	Non-upset	Ext. upset		Wall thick-							
	NU	EU		D	T&C	T&C	Integ. joint	ness t	H40	J55	L80	N80 Type 1,Q	C90[d]	T95[d]	P110
1	T&C	T&C	IJ	mm	kg/m	kg/m	kg/m	mm							
1	2	3	4	5	6	7	8	9	10	11	12	13	14	15	16
1.050	1.14	1.20	–	26.67	1.70	1.79	–	2.87	PNU	PNU	PNU	PNU	PNU	PNU	–
1.050	1.48	1.54	–	26.67	2.20	2.29	–	3.91	PU	PU	PU	PU	PU	PU	PU
1.315	1.70	1.80	1.72	33.40	2.53	2.68	2.56	3.38	PNUI	PNUI	PNUI	PNUI	PNUI	PNUI	–
1.315	2.19	2.24	–	33.40	3.26	3.33	–	4.55	PU	PU	PU	PU	PU	PU	PU
1.660	2.09	–	2.10	42.16	–	–	3.13	3.18	PI	PI	–	–	–	–	–
1.660	2.30	2.40	2.33	42.16	3.42	3.57	3.47	3.56	PNUI	PNUI	PNUI	PNUI	PNUI	PNUI	–
1.660	3.03	3.07	–	42.16	4.51	4.57	–	4.85	PU	PU	PU	PU	PU	PU	PU
1.900	2.40	–	2.40	48.26	–	–	3.57	3.18	PI	PI	–	–	–	–	–
1.900	2.75	2.90	2.76	48.26	4.09	4.32	4.11	3.68	PNUI	PNUI	PNUI	PNUI	PNUI	PNUI	–
1.900	3.65	3.73	–	48.26	5.43	5.55	–	5.08	PU	PU	PU	PU	PU	PU	PU
1.900	4.42	–	–	48.26	6.58	–	–	6.35	–	–	P	–	P	P	–
1.900	5.15	–	–	48.26	7.66	–	–	7.62	–	–	P	–	P	P	–
2.063	3.24	–	3.25	52.40	–	–	4.84	3.96	PI	PI	PI	PI	PI	PI	–
2.063	4.50	–	–	52.40	–	–	–	5.72	P	P	P	P	P	P	P
2³⁄₈	4.00	–	–	60.32	5.95	–	–	4.24	PN	PN	PN	PN	PN	PN	–
2³⁄₈	4.60	4.70	–	60.32	6.85	6.99	–	4.83	PNU	PNU	PNU	PNU	PNU	PNU	PNU
2³⁄₈	5.80	5.95	–	60.32	8.63	8.85	–	6.45	–	–	PNU	PNU	PNU	PNU	PNU
2³⁄₈	6.60	–	–	60.32	9.82	–	–	7.49	–	–	P	–	P	P	–
2³⁄₈	7.35	7.45	–	60.32	10.94	11.09	–	8.53	–	–	PU	–	PU	PU	–
2⁷⁄₈	6.40	6.50	–	73.02	9.52	9.67	–	5.51	PNU	PNU	PNU	PNU	PNU	PNU	PNU
2⁷⁄₈	7.80	7.90	–	73.02	11.61	11.76	–	7.01	–	–	PNU	PNU	PNU	PNU	PNU
2⁷⁄₈	8.60	8.70	–	73.02	12.80	12.95	–	7.82	–	–	PNU	PNU	PNU	PNU	PNU
2⁷⁄₈	9.35	9.45	–	73.02	13.91	14.06	–	8.64	–	–	PU	–	PU	PU	–
2⁷⁄₈	10.50	–	–	73.02	15.63	–	–	9.96	–	–	P	–	P	P	–
2⁷⁄₈	11.50	–	–	73.02	17.11	–	–	11.18	–	–	P	–	P	P	–
3½	7.70	–	–	88.90	11.46	–	–	5.49	PN	PN	PN	PN	PN	PN	–
3½	9.20	9.30	–	88.90	13.69	13.84	–	6.45	PNU	PNU	PNU	PNU	PNU	PNU	PNU
3½	10.20	–	–	88.90	15.18	–	–	7.34	PN	PN	PN	PN	PN	PN	–
3½	12.70	12.95	–	88.90	18.90	19.27	–	9.52	–	–	PNU	PNU	PNU	PNU	PNU
3½	14.30	–	–	88.90	21.28	–	–	10.92	–	–	P	–	P	P	–
3½	15.50	–	–	88.90	23.07	–	–	12.09	–	–	P	–	P	P	–
3½	17.00	–	–	88.90	25.30	–	–	13.46	–	–	P	–	P	P	–
4	9.50	–	–	101.60	14.14	–	–	5.74	PN	PN	PN	PN	PN	PN	–
4	10.70	11.00	–	101.60	–	16.37	–	6.85	PU	PU	PU	PU	PU	PU	–
4	13.20	–	–	101.60	19.64	–	–	8.38	–	–	P	–	P	P	–
4	16.10	–	–	101.60	23.96	–	–	10.54	–	–	P	–	P	P	–
4	18.90	–	–	101.60	28.13	–	–	12.70	–	–	P	–	P	P	–
4	22.20	–	–	101.60	33.04	–	–	15.49	–	–	P	–	P	P	–
4½	12.60	12.75	–	114.30	18.75	18.97	–	6.88	PNU	PNU	PNU	PNU	PNU	PNU	–
4½	15.20	–	–	114.30	22.62	–	–	8.56	–	–	P	–	P	P	–
4½	17.00	–	–	114.30	25.30	–	–	9.65	–	–	P	–	P	P	–
4½	18.90	–	–	114.30	28.13	–	–	10.92	–	–	P	–	P	P	–
4½	21.50	–	–	114.30	32.00	–	–	12.70	–	–	P	–	P	P	–
4½	23.70	–	–	114.30	35.27	–	–	14.22	–	–	P	–	P	P	–
4½	26.10	–	–	114.30	38.84	–	–	16.00	–	–	P	–	P	P	–

P = Plain end, N = Non-upset threaded and coupled, U = External upset threaded and coupled, I = Integral joint.
[a] Nominal linear masses, threads and coupling (cols. 2, 3, 4) are shown for information only
[b] The densities of martensitic chromium steels (L80 types 9Cr and 13Cr) are different from carbon steels. The masses shown are therefore not accurate for martensitic chromium steels. A mass correction factor of 0,989 may be used.
[c] Non-upset tubing is available with regular couplings or special bevel couplings. External-upset tubing is available with regular, special-bevel, or special-clearance couplings.
[d] Grade C90 and T95 tubing shall be furnished in sizes, masses, and wall thicknesses as listed above, or as shown on the purchase agreement.

Table 4.18.3   *Tubing Steel Grade Data*

Steel Grade	Yield Strength, psi		Min. Tensile Strength, psi	Elongation*
	Min.	Max.		
H-40	40,000	80,000	60,000	27
J-55	55,000	80,000	75,000	20
C-75	75,000	90,000	95,000	16
C-95	95,000	110,000	105,000	16
N-80/L-80	80,000	110,000	100,000	16
P-105	105,000	135,000	120,000	15

*The minimum elongation in 2 in. (50.80 mm) should be determined by the following formula:

$$e = 625,000 \frac{A^{0.2}}{U^{0.9}}$$

where e = minimum elongation in 2 in. (50:80 mm) in percent rounded to the nearest $\frac{1}{2}$%

A = cross-sectional area of the tensile specimen in in.^{2}, based on a specified outside diameter or nominal specimen width and specified wall thickness, rounded to the nearest 0.01 or 0.75 in.,^{2} whichever is smaller

U = specified tensile strength in psi.

Table 4.18.4   *Nonupset Tubing — Dimensions and Weights*

1	2	3	4	5	6
				Calculated Weight	
Size: Outside Diameter, in. D	Nominal Weight:[1] Threads and and Coupling, lb. per ft	Wall Thickness in. t	Inside Diameter in. d	Plain End lb/ft w_{pe}	Threads[2] and Coupling lb e_w
*1.050	1.14	0.113	0.824	1.13	0.20
1.315	1.70	0.133	1.049	1.68	0.40
1.660	2.80	0.140	1.380	2.27	0.80
1.900	2.75	0.145	1.610	2.72	0.60
$2\frac{3}{8}$	4.00	0.167	2.041	3.94	1.60
$2\frac{3}{8}$	4.60	0.190	1.995	4.43	1.60
$2\frac{3}{8}$	5.80	0.254	1.867	5.75	1.40
$2\frac{7}{8}$	6.40	0.217	2.441	6.16	3.20
$2\frac{7}{8}$	8.60	0.308	2.259	8.44	2.60
$3\frac{1}{2}$	7.70	0.216	3.068	7.58	5.40
$3\frac{1}{2}$	9.20	0.254	2.992	8.81	5.00
$3\frac{1}{2}$	10.20	0.289	2.922	9.91	4.80
$3\frac{1}{2}$	12.70	0.375	2.750	12.52	4.00
4	9.50	0.226	3.548	9.11	6.20
$4\frac{1}{2}$	12.60	0.271	3.958	12.24	6.00

[1]Nominal weights, threads and coupling (cols. 2), are shown for the purpose of identification in ordering.
[2]Weight gain due to end finishing.
*For information only.
See Figure 4.18.1.
From API Specification 5A, 35th ed. (March 1981).

Table 4.18.5 *External-Upset tubing—Upset end dimensions and Masses for Group 1, 2 and 3 (See Figure 4.18.2)*

				Upset			
Labels[a]		Size Outside diameter in	Normal linear mass, threaded and coupled[b]	Outside diameter[c] +0.0625 0 in.	Length from end of pipe to start of taper[d,c] 0 −1 in.	Length from end of pipe to end of taper[e] in.	Length from end of pipe to start of pipe body[e] max. in.
1	2	D	lb/ft	D_4	L_{eu}	L_e	L_b
1	2	3	4	5	6	7	8
1.050	1.20	1.050	1.20	1.315	$2\frac{3}{8}$	–	–
1.050	1.54	1.050	1.20	1.315	$2\frac{3}{8}$	–	–
1.315	1.80	1.315	1.80	1.469	$2\_$	–	–
1.315	2.24	1.315	1.80	1.469	$2\_$	–	–
1.660	2.40	1.660	2.40	1.812	$2\frac{5}{8}$	–	–
1.660	3.07	1.660	2.40	1.812	$2\frac{5}{8}$	–	–
1.900	2.90	1.900	2.90	2.094	$2\frac{11}{16}$	–	–
1.900	3.73	1.900	2.90	2.094	$2\frac{11}{16}$	–	–
$2\frac{3}{8}$	4.70	2.375	4.70	2.594	4.00	6.00	10.00
$2\frac{3}{8}$	5.95	2.375	5.95	2.594	4.00	6.00	10.00
$2\frac{3}{8}$	7.45	2.375	7.45	2.594	4.00	6.00	10.00
$2\frac{7}{8}$	6.50	2.875	6.50	3.094	$4\frac{1}{4}$	$6\frac{1}{4}$	$10\frac{1}{4}$
$2\frac{7}{8}$	7.90	2.875	7.90	3.094	$4\frac{1}{4}$	$6\frac{1}{4}$	$10\frac{1}{4}$
$2\frac{7}{8}$	8.70	2.875	8.70	3.094	$4\frac{1}{4}$	$6\frac{1}{4}$	$10\frac{1}{4}$
$2\frac{7}{8}$	9.45	2.875	9.45	3.094	$4\frac{1}{4}$	$6\frac{1}{4}$	$10\frac{1}{4}$
$3\frac{1}{2}$	9.30	3.500	9.30	3.750	$4\frac{1}{2}$	$6\frac{1}{2}$	$10\frac{1}{2}$
$3\frac{1}{2}$	12.95	3.500	12.95	3.750	$4\frac{1}{2}$	$6\frac{1}{2}$	$10\frac{1}{2}$
4	11.00	4.000	11.00	4.250	$4\frac{1}{2}$	$6\frac{1}{2}$	$10\frac{1}{2}$
$4\frac{1}{2}$	12.75	4.500	12.75	4.750	$4\frac{3}{4}$	$6\frac{3}{4}$	$10\frac{3}{4}$

[a]Labels are for information and assistance in ordering.
[b]The densities of martensitic chromium steels (L80 9Cr and 13Cr) are different from carbon steels. The masses shown are therefore not accurate for chromium steels. A mass correction factor of 0.989 may be used.
[c]The minimum outside diameter of upset D_4 is limited by the minimum length of full-crest threads. See ISO 10422 or API Spec 5B.
[d]For pup-joints only, the length tolerance on L_{eu} is: $^{+4}_{-1}$ in. The length on L_b may be 4 in longer than specified.
[e]For extended length upsets on external upset tubing, add 1 in to the dimensions in columns 5, 6, and 7.

Table 4.18.6 *Integral Joint Tubing—Upset end dimensions and Masses for Groups 1, 2 (See Figure 4.18.3)*

				Upset dimensions, in.								
				Pin				Box				
Labels		Outside dia.	Nominal linear mass[a] T&C	Outside dia.[b] +0.0625 0	Inside dia.[c] +0.015 0	Length min.	Length of taper min.	Outside diameter +0.015 −0.025	Length min.	Length of taper	Diameter of recess	Width of face min.
1	2	in D	lb/ft	D_4	d_{iu}	L_{iu}	m_{iu}	W_b	L_{eu}	m_{eu}	Q	b
1	2	3	4	5	6	7	8	9	10	11	12	13
1.315	1.72	1.315	1.72	–	0.970	$1\frac{3}{8}$	$\frac{1}{4}$	1.550	1.750	1	1.378	$\frac{1}{32}$
1.660	2.10	1.660	2.10	–	1.301	$1\frac{1}{2}$	$\frac{1}{4}$	1.880	1.875	1	1.723	$\frac{1}{32}$
1.660	2.33	1.660	2.33	–	1.301	$1\frac{1}{2}$	$\frac{1}{4}$	1.880	1.875	1	1.723	$\frac{1}{32}$
1.900	2.40	1.900	2.40	–	1.531	$1\frac{5}{8}$	$\frac{1}{4}$	2.110	2.000	1	1.963	$\frac{1}{32}$
1.900	2.76	1.900	2.76	–	1.531	$1\frac{5}{8}$	$\frac{1}{4}$	2.110	2.000	1	1.963	$\frac{1}{32}$
2.063	3.25	2.063	3.25	2.094	1.672	$1\frac{11}{16}$	$\frac{1}{4}$	2.325	2.125	1	2.156	$\frac{1}{32}$

[a]Nominal linear masses, upset and threaded, are shown for information only.
[b]The minimum outside diameter D4 is limited by the minimum length of full-crest threads. See ISO 10422 or API Spec SB.
[c]The minimum diameter d_{iu} is limited by the drift test.

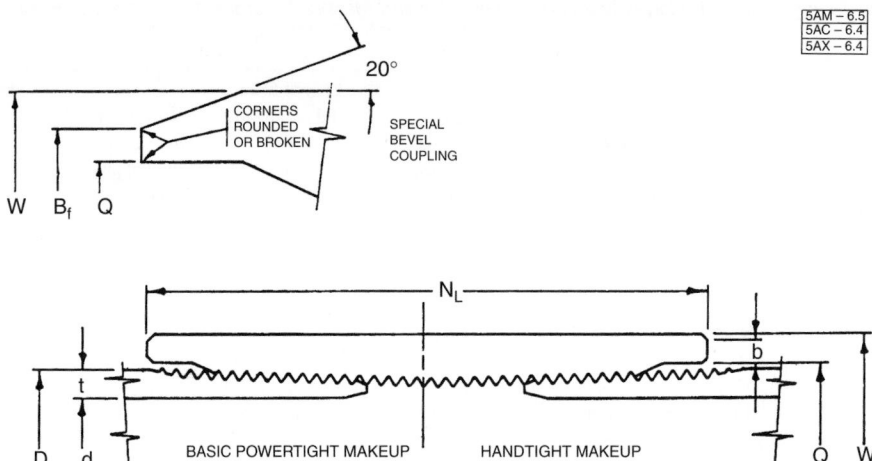

Figure 4.18.1   *Non-upset tubing and coupling [1]. (See Table 4.18.4 for pipe dimensions)*

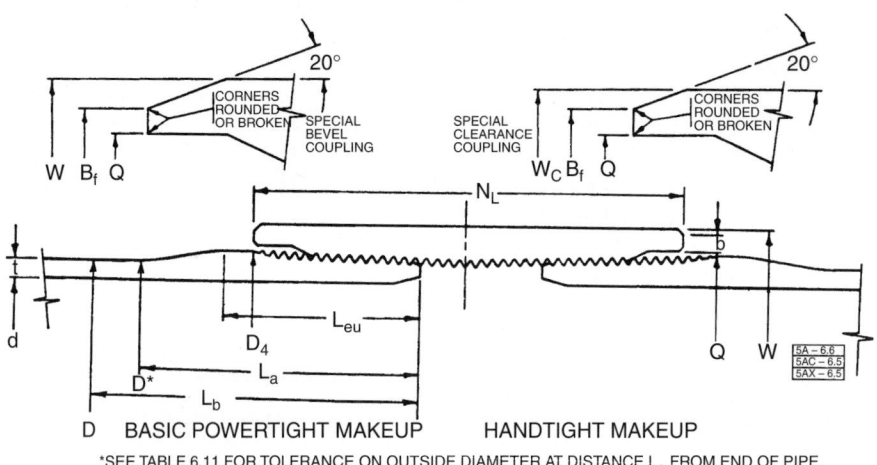

SEE TABLE 6.11 FOR TOLERANCE ON OUTSIDE DIAMETER AT DISTANCE L_a FROM END OF PIPE.

Figure 4.18.2   *External-upset tubing and coupling [1]. (See Table 4.18.5 for pipe dimensions)*

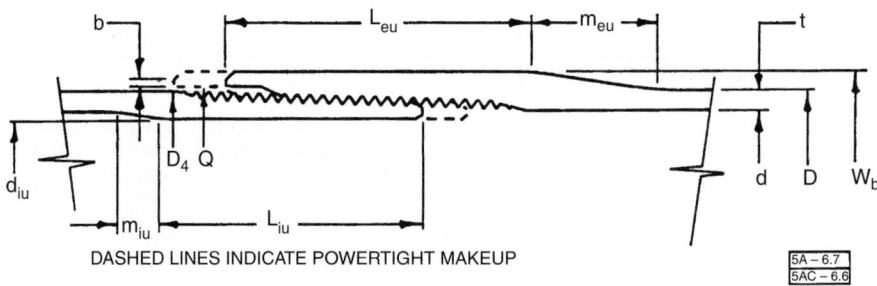

Figure 4.18.3   *Integral-joint tubing [1]. (See Table 4.18.6 for pipe dimensions)*

Table 4.18.7 *Dimensions and Masses for Non-upset, External Upset and Integral Joint Tubing*

Labels[a]				Nominal linear masses[b,c]				Wall thick-ness t in.	Inside dia. d in.	Plain end W_{pe} lb/ft	Calculated mass[c] e_w mass gain or loss due to end finishing[d], lb			
1	2 NU T&C	EU T&C	IJ	Outside dia. D in.	Non-upset T&C lb/ft	Ext. upset T&C lb/ft	Integral joint lb/ft				Non-upset	External upset[e] Regular	Special clearance	Integral joint
1	2	3	4	5	6	7	8	9	10	11	12	13	14	15
1.050	1.14	1.20	–	1.050	1.14	1.20	–	0.113	0.824	1.13	0.20	1.40	–	–
1.050	1.48	1.54	–	1.050	1.48	1.54	–	0.154	0.742	1.48	–	1.32	–	–
1.315	1.70	1.80	1.72	1.315	1.70	1.80	1.72	0.133	1.049	1.68	0.40	1.40	–	0.20
1.315	2.19	2.24	–	1.315	2.19	2.24	–	0.179	0.957	2.17	–	1.35	–	–
1.660	2.09	–	2.10	1.660	–	–	2.10	0.125	1.410	2.05	–	–	–	0.20
1.660	2.30	2.40	2.33	1.660	2.30	2.40	2.33	0.140	1.380	2.27	0.80	1.60	–	0.20
1.660	3.03	3.07	–	1.660	3.03	3.07	–	0.191	1.278	3.00	–	1.50	0.20	–
1.900	2.40	–	2.40	1.990	–	–	2.40	0.125	1.650	2.37	–	–	–	0.20
1.900	2.75	2.90	2.76	1.990	2.75	2.90	2.76	0.145	1.610	2.72	0.60	2.00	–	0.20
1.900	3.65	3.73	–	1.990	3.65	3.73	–	0.200	1.500	3.63	–	2.03	–	–
1.900	4.42	–	–	1.990	4.42	–	–	0.250	1.400	4.41	–	–	–	–
1.900	5.15	–	–	1.990	5.15	–	–	0.300	1.300	5.13	–	–	–	–
2.063	3.24	–	3.25	2.063	–	–	3.25	0.156	1.751	3.18	–	–	–	0.20
2.063	4.50	–	–	2.063	–	–	–	0.225	1.613	4.42	–	–	–	–
$2\frac{3}{8}$	4.00	–	–	2.375	4.00	–	–	0.167	2.041	3.94	1.60	–	–	–
$2\frac{3}{8}$	4.60	4.70	–	2.375	4.60	4.70	–	0.190	1.995	4.44	1.60	4.00	2.96	–
$2\frac{3}{8}$	5.80	5.95	–	2.375	5.80	5.95	–	0.254	1.867	5.76	1.40	3.60	2.56	–
$2\frac{3}{8}$	6.60	–	–	2.375	6.60	–	–	0.295	1.785	6.56	–	–	–	–
$2\frac{3}{8}$	7.35	7.45	–	2.375	7.35	7.45	–	0.336	1.703	7.32	–	–	–	–
$2\frac{7}{8}$	6.40	6.50	–	2.875	6.40	6.50	–	0.217	2.441	6.17	3.20	5.60	3.76	–
$2\frac{7}{8}$	7.80	7.90	–	2.875	7.80	7.90	–	0.276	2.323	7.67	2.80	5.80	3.92	–
$2\frac{7}{8}$	8.60	8.70	–	2.875	8.60	8.70	–	0.308	2.259	8.45	2.60	5.00	3.16	–
$2\frac{7}{8}$	9.35	9.45	–	2.875	9.35	9.45	–	0.340	2.195	9.21	–	–	–	–
$2\frac{7}{8}$	1.050	–	–	2.875	10.50	–	–	0.392	2.091	10.40	–	–	–	–
$2\frac{7}{8}$	11.50	–	–	2.875	11.50	–	–	0.440	1.995	11.45	–	–	–	–
$3\frac{1}{2}$	7.70	–	–	3.500	7.70	–	–	0.216	3.068	7.58	5.40	–	–	–
$3\frac{1}{2}$	9.20	9.30	–	3.500	9.20	9.30	–	0.254	2.992	8.81	5.00	9.20	5.40	–
$3\frac{1}{2}$	10.20	–	–	3.500	10.20	–	–	0.289	2.922	9.92	4.80	–	–	–
$3\frac{1}{2}$	12.70	12.95	–	3.500	12.70	12.95	–	0.375	2.750	12.53	4.00	8.20	4.40	–
$3\frac{1}{2}$	14.30	–	–	3.500	14.30	–	–	0.430	2.640	14.11	–	–	–	–
$3\frac{1}{2}$	15.50	–	–	3.500	15.50	–	–	0.476	2.548	15.39	–	–	–	–
$3\frac{1}{2}$	17.00	–	–	3.500	17.00	–	–	0.530	2.440	16.83	–	–	–	–
4	9.50	–	–	4.000	9.50	–	–	0.226	3.548	9.12	6.20	–	–	–
4	10.70	11.00	–	4.000	–	11.00	–	0.262	3.476	10.47	–	10.60	–	–
4	13.20	–	–	4.000	13.20	–	–	0.330	3.340	12.95	–	–	–	–
4	16.10	–	–	4.000	16.10	–	–	0.415	3.170	15.90	–	–	–	–
4	18.90	–	–	4.000	18.90	–	–	0.500	3.000	18.71	–	–	–	–
4	22.20	–	–	4.000	22.20	–	–	0.610	2.780	22.11	–	–	–	–
$4\frac{1}{2}$	12.60	12.75	–	4.500	12.60	12.75	–	0.271	3.958	12.25	6.00	13.20	–	–
$4\frac{1}{2}$	15.20	–	–	4.500	15.20	–	–	0.337	3.826	15.00	–	–	–	–
$4\frac{1}{2}$	17.00	–	–	4.500	17.00	–	–	0.380	3.740	16.77	–	–	–	–
$4\frac{1}{2}$	18.90	–	–	4.500	18.90	–	–	0.430	3.640	18.71	–	–	–	–
$4\frac{1}{2}$	21.50	–	–	4.500	21.50	–	–	0.500	3.500	21.38	–	–	–	–
$4\frac{1}{2}$	23.70	–	–	4.500	23.70	–	–	0.560	3.380	23.59	–	–	–	–
$4\frac{1}{2}$	26.10	–	–	4.500	26.10	–	–	0.630	3.240	26.06	–	–	–	–

[a]Labels are for information and assistance in ordering.
[b]Nominal linear masses, threaded and coupled (cols. 6, 7 and 8) are shown for information only.
[c]The densities of martensitic chromium (L80 Types 9Cr and 13Cr) are different from carbon steels. The masses shown are therefore not accurate for martensitic chromium steels. A mass correction factor of 0.989 may be used.
[d]Mass gain or loss due to end finishing.
[e]The length of the upset may alter the mass gain or loss due to end finishing.

normal drilling mud. When lengths of tubing are inadvertently dragged in the dirt, the threads should be recleaned.

a. Before running in the hole for the first time, tubing should be drifted with an API drift mandrel to ensure passage of pumps, swabs and packers.

b. Elevators should be in good repair and should have links of equal length.

c. Slip-type elevators are recommended when running special clearance couplings, especially those beveled on the lower end.

d. Elevators should be examined to note if latch fitting is complete.

e. Spider slips that will not crush the tubing should be used. Slips should be examined before using to see that they are working together.
Note: Slip and tong marks are injurious. Every possible effort should be made to keep such damage at a minimum by using proper up-to-date equipment.

f. Tubing tongs that will not crush the tubing should be used on the body of the tubing and should fit properly to avoid unnecessary cutting of the pipe wall. Tong dies should fit properly and conform to the curvature of the tubing. The use of pipe wrenches is not recommended.

g. The following precautions should be taken in the preparation of tubing threads:
  1. Immediately before running, remove protectors from both field end and coupling end and clean threads thoroughly, repeating as additional rows become uncovered.
  2. Carefully inspect the threads. Those found damaged, even slightly, should be laid aside unless satisfactory means are available for correcting thread damage.
  3. The length of each piece of tubing should be measured prior to running. A steel tape calibrated in decimal feet to the nearest 0.01 ft should be used. The measurement should be made from the outermost face of the coupling or box to the position on the externally threaded end where the coupling or the box stops when the joint is made up powertight. The total of the individual length so measured will represent the unloaded length of the tubing string.
  4. Place clean protectors on the field end of the pipe so that the thread will not be damaged while rolling the pipe onto the rack and pulling it into the derrick.
  5. Check each coupling for makeup. If the stand off is abnormally great, check the coupling for tightness. Loose couplings should be removed, the threads thoroughly cleaned and fresh compound should be applied over the entire thread surfaces. Then the coupling should be replaced and tightened before pulling the tubing into the derrick.
  6. Before stabbing, liberally apply thread compound to the entire internally and externally threaded areas. It is recommended that high-pressure, modified thread compound as specified in API Bulletin 5A2: "Bulletin on Thread Compounds" be used except in special cases where severe conditions are encountered. In these special cases it is recommended that high-pressure silicone thread compound as specified in Bulletin 5A2 be used.

h. For high-pressure or condensate wells, additional precautions to insure tight joints should be taken as follows:
  1. Couplings should be removed and both the mill-end pipe thread and coupling thread thoroughly cleaned and inspected. To facilitate this operation,

tubing may be ordered with couplings handling tight, which is approximately one turn beyond hand tight, or may be ordered with the couplings shipped separately.

  2. Thread compound should be applied to both the external and internal threads, and the coupling should be reapplied handling tight. Field-end threads and the mating coupling threads should have thread compounds applied just before stabbing.

4.18.3.1 Stabbing, Making Up and Lowering

1. Do not remove thread protector from field end of tubing until ready to stab.
2. If necessary, apply thread compound over entire surface of threads just before stabbing.
3. In stabbing, lower tubing carefully to avoid injuring threads. Stab vertically, preferably with the assistance of someone on a stabbing board. If the tubing tilts to one side after stabbing, lift up, clean and correct any damaged thread with a three-cornered file, then carefully remove any filings and reapply compound over thread surface. Intermediate supports may be placed in the derrick to limit bowing of the tubing.
4. After stabbing, start screwing the pipe together by hand or apply regular or power tubing tongs slowly. To prevent galling when making up connections in the field, the connections should be made up to a speed not to exceed 25 rpm. Power tubing tongs are recommended for high-pressure or condensate wells to ensure uniform makeup and tight joints. Joints should be made up tight, approximately two turns beyond the hand-tight position, with care being taken not to gall the threads.

4.18.3.2 Field Makeup

1. Joint life of tubing under repeated field makeup is inversely proportional to the field makeup torque applied. Therefore, in wells where leak resistance is not a great factor, minimum field makeup torque values should be used to prolong joint life. Table 4.18.8 contains recommended optimum makeup torque values for nonupset, external upset, and integral joint tubing, based on 1% of the calculated joint pullout strength determined from the joint pull-out strength formula for eight-round thread casing in Bulletin 5C3. Minimum torque values listed are 75% of optimum values, and maximum torque values listed are 125% of optimum values. All values are rounded to the nearest 10 ft-lb. The torque values listed in Table 4.18.8 apply only to tubing with zinc-plated couplings. When making up connections with tin-plated couplings, 80% of the listed value can be used as a guide.
2. Spider slips and elevators should be cleaned frequently, and slips should be kept sharp.
3. Finding bottom should be accomplished with extreme caution. Do not set tubing down heavily.

4.18.3.3 Pulling Tubing

1. A caliper survey prior to pulling a worn string of tubing will provide a quick means of segregating badly worn lengths for removal.
2. Break-out tongs should be positioned close to the coupling. Hammering the coupling to break the joint is an injurious practice.
3. Great care should be exercised to disengage all of the thread before lifting the tubing out of the coupling. Do not jump the tubing out of the coupling.
4. Tubing stacked in the derrick should be set on a firm wooden platform without the bottom thread protector since the design of most protectors is not such as to

Table 4.18.8 *Recommended Tubing Makeup Torque*

1	2	3	4	5	6	7	8	9	10	11	12	13	14
	Nominal Weight lb. per ft.				Torque, ft-lb								
Size: Outside Diameter in.	Threads and Coupling		Integral Joint	Grade	Non-Upset			Upset			Integral Joint		
	Non-Upset	Upset			Opt.	Min.	Max.	Opt.	Min.	Max	Opt.	Min.	Max.
1.050	1.14	1.20	–	H-40	140	110	180	460	350	580	–	–	–
	1.14	1.20	–	J-55	180	140	230	600	450	750	–	–	–
	1.14	1.20	–	C-75	230	170	290	780	590	980	–	–	–
	1.14	1.20	–	L-80	240	180	300	810	610	1010	–	–	–
	1.14	1.20	–	N-80	250	190	310	830	620	1040	–	–	–
1.315	1.70	1.80	1.72	H-40	210	160	260	440	330	550	310	230	390
	1.70	1.80	1.72	J-55	270	200	340	570	430	710	400	300	500
	1.70	1.80	1.72	C-75	360	270	450	740	560	930	520	390	650
	1.70	1.80	1.72	L-80	370	280	460	760	570	950	540	400	680
	1.70	1.80	1.72	N-80	380	290	480	790	590	990	550	410	690
1.660	–	–	2.10	H-40	–	–	–	–	–	–	380	280	480
	2.30	2.40	2.33	H-40	270	200	340	530	400	660	380	280	480
	–	–	2.10	J-55	–	–	–	–	–	–	500	380	630
	2.30	2.40	2.33	J-55	350	260	440	690	520	860	500	380	630
	2.30	2.40	2.33	C-75	460	350	580	910	680	1140	650	490	810
	2.30	2.40	2.33	L-80	470	350	590	940	710	1180	670	500	850
	2.30	2.40	2.33	N-80	490	370	610	960	720	1200	690	520	860
1.900	–	–	2.40	H-40	–	–	–	–	–	–	450	340	560
	2.75	2.90	2.76	H-40	320	240	400	670	500	840	450	340	560
	–	–	2.40	J-55	–	–	–	–	–	–	580	440	730
	2.75	2.90	2.76	J-55	410	310	510	880	660	1100	580	440	730
	2.75	2.90	2.76	C-75	540	410	680	1150	860	1440	760	570	950
	2.75	2.90	2.76	L-80	560	420	700	1190	890	1490	790	590	990
	2.75	2.90	2.76	N-80	570	430	710	1220	920	1530	810	610	1010
2.063	–	–	3.25	H-40	–	–	–	–	–	–	570	430	710
	–	–	3.25	J-55	–	–	–	–	–	–	740	560	920
	–	–	3.25	C-75	–	–	–	–	–	–	970	730	1210
	–	–	3.25	L-80	–	–	–	–	–	–	1010	760	1260
	–	–	3.25	N-80	–	–	–	–	–	–	1030	770	1290
$2\frac{3}{8}$	4.00	–	–	H-40	470	350	590	–	–	–	–	–	–
	4.60	4.70	–	H-40	560	420	700	990	740	1240	–	–	–
	4.00	–	–	J-55	610	460	760	–	–	–	–	–	–
	4.60	4.70	–	J-55	730	550	910	1290	970	1610	–	–	–
	4.00	–	–	C-75	800	600	1000	–	–	–	–	–	–
	4.60	4.70	–	C-75	960	720	1200	1700	1280	2130	–	–	–
	5.80	5.95	–	C-75	1380	1040	1730	2120	1590	2650	–	–	–
	4.00	–	–	L-80	830	620	1040	–	–	–	–	–	–
	4.60	4.70	–	L-80	990	740	1240	1760	1320	2200	–	–	–
	5.80	5.95	–	L-80	1420	1070	1780	2190	1640	2740	–	–	–
	4.00	–	–	N-80	850	640	1060	–	–	–	–	–	–
	4.60	4.70	–	N-80	1020	770	1280	1800	1350	2250	–	–	–
	5.80	5.95	–	N-80	1460	1100	1830	2240	1680	2800	–	–	–
	4.60	4.70	–	P-105	1280	960	1600	2270	1700	2840	–	–	–
	5.80	5.95	–	P-105	1840	1380	2300	2830	2120	3540	–	–	–
$2\frac{7}{8}$	6.40	6.50	–	H-40	800	600	1000	1250	940	1560	–	–	–
	6.40	6.50	–	J-55	1050	790	1310	1650	1240	2060	–	–	–
	6.40	6.50	–	C-75	1380	1040	1730	2170	1630	2710	–	–	–
	8.60	8.70	–	C-75	2090	1570	2610	2850	2140	3560	–	–	–
	6.40	6.50	–	L-80	1430	1070	1790	2250	1690	2810	–	–	–
	8.60	8.70	–	L-80	2160	1620	2700	2950	2210	3690	–	–	–
	6.40	6.50	–	N-80	1470	1100	1840	2300	1730	2880	–	–	–
	8.60	8.70	–	N-80	2210	1660	2760	3020	2270	3780	–	–	–
	6.40	6.50	–	P-105	1850	1390	2310	2910	2180	3640	–	–	–
	8.60	8.70	–	P-105	2790	2090	3490	3810	2860	4760	–	–	–

(continued)

Table 4.18.8 (continued)

1	2	3	4	5	6	7	8	9	10	11	12	13	14
	Nominal Weight lb. per ft.				Torque, ft-lb								
Size: Outside Diameter in.	Threads and Coupling		Integral Joint	Grade	Non-Upset			Upset			Integral Joint		
	Non-Upset	Upset			Opt.	Min.	Max.	Opt.	Min.	Max	Opt.	Min.	Max.
$3\frac{1}{2}$	7.70	–	–	H-40	920	690	1150	–	–	–	–	–	–
	9.20	9.30	–	H-40	1120	840	1400	1730	1300	2160	–	–	–
	10.20	–	–	H-40	1310	980	1640	–	–	–	–	–	–
	7.70	–	–	J-55	1210	910	1510	–	–	–	–	–	–
	9.20	9.30	–	J-55	1480	1110	1850	2280	1710	2850	–	–	–
	10.20	–	–	J-55	1720	1290	2150	–	–	–	–	–	–
	7.70	–	–	C-75	1600	1200	2000	–	–	–	–	–	–
	9.20	9.30	–	C-75	1950	1460	2440	3010	2260	3760	–	–	–
	10.20	–	–	C-75	2270	1700	2840	–	–	–	–	–	–
	12.70	12.95	–	C-75	3030	2270	3790	4040	3030	5050	–	–	–
	7.70	–	–	L-80	1660	1250	2080	–	–	–	–	–	–
	9.20	9.30	–	L-80	2030	1520	2540	3130	2350	3910	–	–	–
	10.20	–	–	L-80	2360	1770	2950	–	–	–	–	–	–
	12.70	12.95	–	L-80	3140	2360	3930	4200	3150	5250	–	–	–
	7.70	–	–	N-80	1700	1280	2130	–	–	–	–	–	–
	9.20	9.30	–	N-80	2070	1550	2590	3200	2400	4000	–	–	–
	10.20	–	–	N-80	2410	1810	3010	–	–	–	–	–	–
	12.70	12.95	–	N-80	3210	2410	4010	4290	3220	5360	–	–	–
	9.20	9.30	–	P-105	2620	1970	3280	4050	3040	5060	–	–	–
	12.70	12.95	–	P-105	4060	3050	5080	5430	4070	6790	–	–	–
4	9.50	–	–	H-40	940	710	1180	–	–	–	–	–	–
	–	11.00	–	H-40	–	–	–	1940	1460	2430	–	–	–
	9.50	–	–	J-55	1240	930	1550	–	–	–	–	–	–
	–	11.00	–	J-55	–	–	–	2560	1920	3200	–	–	–
	9.50	–	–	C-75	1640	1230	2050	–	–	–	–	–	–
	–	11.00	–	C-75	–	–	–	3390	2540	4240	–	–	–
	9.50	–	–	L-80	1710	1280	2140	–	–	–	–	–	–
	–	11.00	–	L-80	–	–	–	3530	2650	4410	–	–	–
	9.50	–	–	N-80	1740	1310	2180	–	–	–	–	–	–
	–	11.00	–	N-80	–	–	–	3600	2700	4500	–	–	–
$4\frac{1}{2}$	12.60	12.75	–	H-40	1320	990	1650	2160	1620	2700	–	–	–
	12.60	12.75	–	J-55	1740	1310	2180	2860	2150	3580	–	–	–
	12.60	12.75	–	C-75	2300	1730	2880	3780	2840	4730	–	–	–
	12.60	12.75	–	L-80	2400	1800	3000	3940	2960	4930	–	–	–
	12.60	12.75	–	N-80	2440	1830	3050	4020	3020	5030	–	–	–

Source: From API Recommended Practice, 5C1, 16th ed. (May 1998).

support the joint or stand without damage to the field thread.

5. Protect threads from dirt or injury when the tubing is out of the hole.
6. Tubing set back in the derrick should be properly supported to prevent undue bending. Tubing that is $2\frac{3}{8}$ in. OD and larger preferably should be pulled in stands approximately 60 ft long or in doubles of range 2. Stands of tubing 1.900-in. OD or smaller and stands longer than 60 ft should have intermediate support.

4.18.3.4 Causes of Tubing Trouble

The more common causes of tubing troubles are as follows:

1. Improper selection for strength and life required.
2. Insufficient inspection of finished product at the mill and in the yard.
3. Careless loading, unloading and cartage.
4. Damaged threads resulting from protectors loosening and falling off.
5. Lack of care in storage to give proper protection.
6. Excessive hammering on couplings.
7. Use of worn-out and wrong types of handling equipment.
8. Nonobservance of proper rules in running and pulling tubing.
9. Coupling wear and rod cutting.
10. Excessive sucker rod breakage.
11. Fatigue that often causes failure at the last engaged thread. There is no positive remedy, but using external upset tubing in place of nonupset tubing greatly delays the start of this trouble.
12. Replacement of worn couplings with non-API couplings.
13. Dropping a string, even a short distance.
14. Leaky joints, under external or internal pressure.
15. Corrosion. Both the inside and outside of tubing can be damaged by corrosion. The damage is generally in the form of pitting, box wear, stress-corrosion cracking, and sulfide stress cracking, but localized attack like corrosion–erosion, ringworm and caliper tracks– can also occur. Since corrosion can result from many causes and influences and can take different forms, no simple and universal remedy can be given for control. Each problem must be treated individually, and the solution must be attempted in light of known factors and operating conditions.
    a. Where internal or external tubing corrosion is known to exist and corrosive fluids are being produced, the following measures can be employed:
    1. In flowing wells the annulus can be packed off and the corrosive fluid confined to the inside of the tubing. The inside of the tubing can be protected with special liners, coatings or inhibitors. Under severe conditions, special alloy steel or glass-reinforced plastics may be used. Alloys do not eliminate corrosion. When H_2S is present in the well fluids, tubing of high-yield strength may be subject to sulfide corrosion cracking. The concentration of H_2S necessary to cause cracking in different strength materials is not yet well defined. Literature on sulfide corrosion or persons competent in this field should be consulted.
    2. In pumping and gas-lifting wells, inhibitors introduced via the casing-tubing annulus afford appreciable protection. In this type of completion, especially in pumping wells, better operating practices can also aid in extending the life of tubing; viz., through the use of rod protectors, rotation of tubing and longer and slower pumping strokes.
    b. To determine the value and effectiveness of the above practices and measures, cost and equipment

failure records can be compared before and after application of control measures.

    c. In general, all new areas should be considered as being potentially corrosive, and investigations should be initiated early in the life of a field, and repeated periodically, to detect and localize corrosion before it has done destructive damage. Where conditions favorable to corrosion exist, a qualified corrosion engineer should be consulted.

4.18.3.5 Selection of Wall Thickness and Steel Grade of Tubing

Tubing design relies on the selection of the most economical steel grades and wall thicknesses (unit weight of tubing) that will withstand, without failure, the forces to which tubing will be exposed throughout the expected tubing life.

Tubing must be designed on:

- Collapse
- Burst
- Tension
- Possibility of permanent corkscrewing

Tubing string design is very much the same as for casing. For shallow and moderately deep holes, uniform strings are preferable; however, in deep wells, a tapered tubing can be desirable. A design factor (safety factor) for tension should be about 1.6. The collapse design factor must not be less than 1.0, assuming an annulus filled up with fluid and tubing empty inside. The design factor for burst should not be less than 1.1.

4.18.3.6 Tubing Elongation/Contraction Due to the Effect of Changes in Pressure and Temperature

During the service life of a well, tubing can experience various combinations of pressures and temperatures that result in tubing length changes. The four basic effects to consider are as follows:

1. Piston effect
2. Helical buckling effect
3. Ballooning and reverse ballooning effect
4. Temperature change effect

The tubing movement due to piston effect is

$$\Delta L_1 = \frac{[(A_p - A_i)\Delta P_i - (A_p - A_o)\Delta P_o]}{EA_s}L \qquad [4.18.1]$$

Tubing movement due to helical buckling is

$$\Delta L_2 = -\frac{r^2 F_f^2}{8EI(W_s + W_i - W_o)} \qquad [4.18.2]$$

where

$$I = \frac{\pi}{64}(OD^4 - ID^4) \qquad F_f = (\Delta P_i - \Delta P_o)A_p$$

Note: If $F_f < 0$, $\Delta L_2 = 0$.

The tubing movement due to ballooning effect is

$$\Delta L_3 = -\frac{v}{E}\frac{\Delta \rho_i - R^2 \Delta \rho_o - \frac{1+2v}{2v}\delta}{R^2 - 1}L^2 - \frac{2v}{E}\frac{\Delta p_i - R^2 \Delta p_o}{R^2 - 1}L \qquad [4.18.3]$$

The tubing movement due to change in temperature is

$$\Delta L_4 = \beta L \Delta T \qquad [4.18.4]$$

Two approaches can be used to handle tubing movement:

1. Provide seals of enough length.
2. Slack off enough weight of tubing to prevent movement.

For practical purposes a combination of the approaches mentioned above can be applicable.

$$\Delta L_5 = \frac{LF_s}{EA_s} + \frac{r^2 F_s^2}{8EI(W_s + W_i - W_o)} \qquad [4.18.5]$$

Notations used in the above equations are

A_i = area corresponding to tubing ID in in.^{2}
A_o = area corresponding to tubing OD in in.^{2}
A_p = area corresponding to packer bore in in.^{2}
A_s = cross-sectional area of the tubing wall in in.^{2}
F_f = fictitious force in presence of no restraint in the packer in lb$_f$
I = moment of inertia of tubing cross-section with respect to its diameter in in.^{4}
L = length of tubing in in.
P_i, P_o = pressure inside and outside the tubing at the packer level respectively in psi
$\Delta P_i, \Delta P_o$ = change in pressure inside and outside of tubing at the packer level in psi
$\Delta p_i, \Delta p_o$ = change in pressure inside and outside of tubing at the surface in psi
R = ratio OD/ID of the tubing
r = tubing and casing radial clearance in in.
$\Delta \rho_i$ = change in density of liquid in the tubing in lbm/in.^{3}
$\Delta \rho_o$ = change in density of liquid in annulus in lbm/in.^{3}
β = coefficient of thermal expansion of the tubing material (for steel, $\beta = 6.9 \times 10^{-6}/°F$)
W_s = average (i.e., including coupling) weight of tubing per unit length in lb/in.
W_i = weight of liquid in the tubing per unit length in lb/in.
W_o = weight of outside liquid displaced per unit length
δ = drop of pressure in the tubing due to flow per unit length in psi/in.
ΔT = change in average tubing temperature in.°F
ν = Poisson's ratio of the tubing material (for steel, $\nu = 0.3$)
E = Young's modulus (for steel, $E = 30 \times 10^6$ psi)

Nomenclature used is in the paper by A. Lubinski et al. API Bulletin 5C2, 20th ed. (May 1987).

Example 4.18.1

Calculate the expected movement of tubing under conditions as specified below. Initially both tubing and annulus are filled with a crude of 30°API. Thereafter, the crude in the tubing is replaced by a 15-lb/gal cement slurry to perform a squeeze cementing operation. While the squeeze cementing job is performed, pressures p_i = 5,000 psi and p_o = 1,000 psi are applied at the surface on the tubing and annulus respectively.

Tubing: $2\frac{7}{8}$ in.; 6.5 lb/ft
Casing: 7 in.; 32 lb/ft (r = 1.61 in.)
A_o = 6.49 in.^{2}, A_i = 4.68 in.^{2}, A_s = 1.81 in.^{2}
Ratio of OD/ID of tubing, R = 1.178
A_p = 8.30 in.^{2}
Length of tubing = 10,000 ft (120,000 in.)
Average change in temperature: −20degF
Pressure drop due to flow is disregarded ($\delta = 0$)

Solution

Pressure changes are the following. At the surface the pressures are Δp_i = 5,000 psi and Δp_o = 1,000 psi. At the packer level, ΔP_i = 9,000 psi and ΔP_o = 1,000 psi. Liquid density changes are $\Delta \rho_i$ = 0.0332 psi/in. and $\Delta \rho_o$ = 0.0 psi/in. The unit weight of tubing is

$$W = W_s + W_i - W_o$$

$$= \frac{6.5}{12} + \frac{1.5 \times 4.68}{231} - \frac{0.876 \times 8.34 \times 6.49}{231} = 0.64 \text{lb/in.}$$

The moment of inertia is I = 1.61 in.^{4} The tubing movement due to piston effect is

$$\Delta L_1 = \frac{120,000}{30 \times 10^6}[(8.3 - 4.68)9,000 - (8.3 - 6.49)1,000]$$

$$= -68.0 \text{in.}$$

The tubing movement due to buckling effect is

$$\Delta L_2 = \frac{(1.61)^2 \times (8.3)^2 (9,000 - 1,000)^2}{8 \times 30 \times 10^6 \times 1.61 \times 0.64}$$

$$= -46.23 \text{in.}$$

The tubing movement due to ballooning effect is

$$L_3 = \frac{0.3}{30 \times 10^6} \frac{0.0332 - 0}{1.178^2 - 1}(120,000)^2 - \frac{2 \times 0.3}{30 \times 10^6}$$

$$\times \frac{5,000 - (1.178^2)1,000}{1.178^2 - 1} 120,000 = -34.69 \text{in.}$$

The temperature effect is

$$\Delta L_4 = -6.9 \times 10^{-6} \times 120,000 \times 20 = -16.56 \text{in.}$$

The total expected tubing movement is

$$\Delta L_6 = \Delta L_1 + \Delta L_2 + \Delta L_3 + \Delta L_4 = -165.48 \text{in.}$$

4.18.3.7 Packer-To-Tubing Force

Certain types of packers permit no tubing motion in either direction. Depending upon operational conditions, a tubing can be landed either in compression (slack off) or tension (pull up). Landing tubing in compression is desirable if the expected tubing movement would produce tubing shortening while landing in tension to compensate the expected tubing elongation.

Restraint of the tubing in the packer results in a packer-to-tubing force. To find the expected packer-to-tubing force, the following sequence of calculations is applicable:

$$F_f = A_p(P_i - P_o) \qquad [4.18.6]$$

$$F_a = (A_p - A_i)P_i - (A_p - A_o)P_o \qquad [4.18.7]$$

$$\Delta L_p = -\Delta L_6 \qquad [4.18.8]$$

$$\Delta L_f = -\frac{LF_f}{EA_s} - \frac{r^2 F_f^2}{8EI(W_s + W_i - W_o)} \qquad [4.18.9]$$

$$\widehat{\Delta L_f} = \Delta L_f + \Delta L_p \qquad [4.18.10]$$

If $\widehat{\Delta L_f}$ is positive, then

$$\widehat{F_f} = -\frac{\widehat{\Delta L_f} EA_s}{L} \qquad [4.18.11]$$

If $\widehat{\Delta L_f}$ is negative, then

$$\widehat{F_f} = \frac{4I(W_s + W_i - W_o)}{A_s r^2}$$

$$\times \left[-L \pm \left(L^2 - \frac{A_s^2 r^2 E \widehat{\Delta L_f}}{2I(W_s + W_i - W_o)} \right)^{0.5} \right] \qquad [4.18.12]$$

and finally

$$F_p = \widehat{F_f} - F_f \qquad [4.18.13]$$

Upon determining the packer to tubing force F_p, the actual force $\widehat{F}_a$ immediately above the packer is given by

$$\widehat{F}_a = F_a + F_p \qquad [4.18.14]$$

The symbols used are

F_a = actually existing pressure force at the lower end of tubing subjected to no restraint in the packer in lb_f

F_f = fictitious force in presence of no restraint in the packer in lb_f

$\widehat{F}_a$ = actually existing force at the lower end of tubing in lb_f

$\widehat{F}_f$ = fictitious force in presence of packer restraint in lb_f

ΔL_6 = overall tubing length change in in.

ΔL_p = length change necessary to bring the end of the tubing to the packer

Other symbols are as previously used.

Example 4.18.2

The operating conditions are the same as those described in Example 1. Assume that the packer does not permit any tubing movement at the packer setting depth (10,000 ft). Since tubing shortening is expected, a 20,000-lb force is slacked off before the squeeze job. Find the tubing-to-packer force.

Solution
Length change due to slack-off force is

$$\Delta L_s = \frac{120,000 \times 20,000}{30 \times 10^6 \times 1.81} + \frac{(1.61)^2 \times (20,000)^2}{8 \times 30 \times 10^6 \times 1.61 \times 0.64}$$

$$= 48.39 \, in.$$

The overall tubing length change is

$$\Delta L_6 = -165.48 + 48.39 = -117.09 \, in.$$

The fictitious and actual forces are

$$F_f = 8.3(12,800 - 4,800) = 66,400 \, lb_f$$

$$F_a = 12,800(8.3 - 4.68) - 4,800(8.3 - 6.49) = 37,648 \, lb_f$$

Note: A positive sign indicates a compressive-type force, a negative sign indicates a tensional force.

$$\Delta L_p = -\Delta L_6 = 115.5 \, in.$$

$$\Delta \widehat{L}_f = -\frac{120,000 \times 66,400}{30 \times 10^6 \times 1.81} - \frac{(1.61^2) \times (66,400^2)}{8 \times 30 \times 10^6 \times 1.61 \times 0.64}$$

$$= -192.9 \, in.$$

$$\Delta \widehat{L}_f = -192.9 + 115.5 = -77.4 \, in.$$

Since $\Delta \widehat{L}_f$ is negative, the force $\widehat{F}_f$ is calculated from Equation 4.18.12:

$$\widehat{F}_f = \frac{4 \times 1.61 \times 0.66}{1.81 \times (1.61^2)} \left[-120,000 + \left((120,000)^2 \right. \right.$$

$$\left. \left. -\frac{(1.81)^2(1.61^2) \times 30 \times 10^6(-77.4)}{2 \times 1.61 \times 0.64} \right)^{0.5} \right] = 30,050.45 \, lb_f$$

Since the force $\widehat{F}_f$ is positive (compressive), a helical buckling of tubing is expected above the packer.

The tubing-to-packer force is

$$F_p = \widehat{F}_f - F_f = 30,050.45 - 66,400 = 36,349.55 \, lb_f$$

4.18.3.8 Permanent Corkscrewing

To ensure that permanent corkscrewing will not occur, the following inequalities must be satisfied:

$$\left[3\left(\frac{P_i - P_o}{R^2 - 1} \right)^2 + \left(\frac{P_i - R^2 P_o}{R^2 - 1} + \sigma_a \pm \sigma_b \right)^2 \right]^{0.5} \le Ym \quad [4.18.15]$$

$$\left[3\left(\frac{R^2(P_i - P_o)}{R^2 - 1} \right)^2 + \left(\frac{P_i - R^2 P_o}{R^2 - 1} + \sigma_a \pm \frac{\sigma_b}{R} \right)^2 \right]^{0.5} \le Ym$$

$$[4.18.16]$$

where

$$\sigma_a - \frac{\widehat{F}_a}{A_s} \qquad and \qquad \sigma_b = \frac{D \times r}{4I} \widehat{F}_f$$

All symbols in Equations 4.18.15 and 4.18.16 are as previously stated.

4.18.4 Packers

A packer is simply a mechanical means of forming a seal between two strings of pipe or between the pipe and the open hole. This generated seal isolates the tubing from the casing or open hole, preventing vertical flow of fluids or gases.

To qualify as a packer, a tool needs only a sealing means and a method of maintaining an opening through the middle. Typically, most packers available today incorporate the following components in their design.

Slips to anchor the packer in the casing or open hole

Cone to support the slips and create a wedge to enhance anchoring capability

Elastomer elements which expand to seal the area between the packer and the casing or open hole

Mandrel which maintains a consistent bore and typically has an ID compatible with the tubular used. Outer body components are supported by the mandrel.

Setting mechanism which initiates the setting sequence when desired during installation.

Typically, packers are identified by their classification (permanent or retrievable), application (production or service), and setting method (mechanical, hydraulic or wireline).

Permanent packers do not have releasing mechanisms and are not retrievable from surface. When a permanent packer is set, the tubing can be removed from the well without releasing the packer. Permanent packers can only be removed by destroying their integrity through milling or drilling.

Retrievable packers are in integral part of the tubing string. The packer is set, released and pulled with the tubing string. The tubing string cannot be removed from the well bore without pulling the packer unless a on/off tool or similar mechanism is used with the packer.

Production packers are designed to meet general or specific well conditions and provide a long operational life in those conditions. They are manufactured from a variety of materials to meet operational life expectancy in different environments. Production packers are used for a number of reasons, which can be summarized into four categories, which are: protecting the casing string, maximum safety and control, energy conservation, and improve productivity.

4.18.4.1 Protecting the Casing

Since the casing string is cemented in the hole, it is considered permanent and non-replaceable. A string of replaceable

tubing is normally run inside the casing to control and provide the conduit necessary to bring the producible fluids to
the surface.

Damaged tubing is relatively easy to remedy as it can be
pulled and replaced. Damage to the casing, on the other
hand, creates a serious problem, which jeopardize the productivity of the well. Casing is difficult and expensive to
repair.

Where a packer is not used, the casing is exposed to produced fluids, which are often corrosive. In order to maintain
a controlled inhibited atmosphere, a packer is necessary to
seal the bottom of the casing annulus, which prevents the
release of produced fluids into the casing annulus.

The casing may also be exposed to pressure differentials
high enough to cause casing damage. The tubing string normally has higher-pressure ratings than the casing. Using a
packer to seal the casing annulus above the productive zone
will confine the high differentials to the tubing string.

4.18.4.2 Safety

A packer acts as a downhole blowout preventer or valve to
provide maximum control of the formation at the formation.

During completion operations, the packer can be used as
a plug to completely shut off the production zone while work
is being done up the hole, or during nippling up operations.
High-pressure zones can be tested or produced without
exposing the casing or wellhead equipment to unnecessary
high pressures.

During producing operations, the packer confines high
pressures to the tubing string where they are more easily controlled. Safety shut-in devices and other accessory
equipment are often installed in the tubing for maximum
control.

4.18.4.3 Energy Conservation

One of the primary reasons for using a smaller tubing string
to produce the formation fluids is to maintain sufficient velocity to take full advantage of available energy. With high
velocities, oil, water and gas are intermingled so that the
gas energy is used, preventing the heavier water from falling
back and killing the well.

When a packer is not used, gas tends to break out into the
annulus, accumulating at the top of the casing. This results
in a loss of energy. A packer funnels all of the production into
the tubing string, so all of the available gas energy is utilized
to lift the fluid.

If two different zones are perforated, interflow between
the zones results in a loss of energy for the more prolific zone
and could artificial lift for both zones. Formation damage
and/or loss of hydrocarbon reserves can also result from
this uncontrolled formation flow.

4.18.4.4 Improve Productivity

A packer allows the casing to be used as an energy storage
area and/or conduit in artificial lift operations; such as, gas
lift and hydraulic pumping.

Sealing between two zones allows both to be produced
simultaneously with maximum control by using two strings
of tubing.

In some cases, it is desirable to perforate a second zone
for production at a later date. To keep this alternate zone
isolated from the producing zone, a packer is set between the
two. With the proper equipment, the zones can be switched
without pulling the tubing string.

Service packers are used to perform specialized functions. When the function is completed, the packer is removed
from the well. Service packers are commonly designed with
higher differential pressure ratings than permanent packers, and are more likely to have their operational capabilities

tested to the fullest extent. Service packers are commonly
used to pressure test casing and repair if a leak exists; temporarily suspend producing intervals; and protect the casing
string from high pressures during hydraulic fracturing.

In a mechanical set application, tubing movement is used
to create the force necessary to set the packer and energize
the element.

In hydraulic set application, a force is generated internally
within the packer by applying fluid pressure to set the slips
and energize the element.

For wireline set applications, an electric wireline pressure
setting tool is used. An electric current ignites a powder
charge to create a gas pressure within the setting tool. The
gas pressure then acts on a hydraulic piston to create a force
sufficient to set the packer slips, energize the element and
shear the setting tool from the packer.

Choosing and successfully installing the best packer for
a job depends on the proper evaluation and analysis of the
forces that may act on the tubing and the packer during the
lifetime of the installation. The effect of the forces can only
be determined by an accurate force analysis. The installation
can proceed with a high degree of confidence if no detrimental effects are shown during analysis. However, if analysis
indicates that damage to the tubing or packer could occur,
reassessment and revision of the installation procedures will
reduce the risk of costly recovery.

The anticipated change in well conditions that may occur
after the packer has been set must be analyzed to evaluate
the forces affecting the packer and tubing string.

Temperature and pressure variations occur within the well
throughout its lifecycle. Changes in these conditions can
lead to a change in length of the tubing string or a change in
force through out the tubing string depending on the type of
packer chosen and the packer to tubing configuration. The
tubing will either lengthen or shorten when temperature,
tubing pressure or annulus pressure is changed in the well.
If the tubing is anchored, the change in force will occur at the
packer as well as over the entire length of the tubing string.
If the tubing is free to move, the tubing string will experience
a net change in length.

The following effects will lead to a change in length or
force:

> Piston effect
> Buckling effect
> Ballooning effect
> Temperature effect

The piston, buckling and ballooning effects result from
pressure changes within the well. The temperature effect
is the result of a change in temperature and is not related
to pressure. Each effect must be calculated separately even
though some of the effects may be related. The magnitude and direction of each effect must then be combined
to determine the total change in force or change in tubing
length.

4.18.4.5 Piston Effect

The piston effect is the result of pressure changes, which
occur inside the tubing string and annulus at the packer.

In retrievable packer installations, pressure changes that
occur inside the tubing at the packer act on the difference in
areas between the packer valve area (A_p) and the tubing ID
area (A_i). Pressure changes that occur in the annulus act on
the difference in areas between the packer valve area (A_p)
and the tubing OD area (A_o).

In permanent packer installations, pressure changes that
occur inside the tubing act on the difference in areas between
the seal bore (packer valve, A_p) and the tubing ID area (A_i).

Pressure changes that occur in the annulus act on the difference in areas between the seal bore (packer valve, A_p) and the tubing OD area (A_o).

The result of the piston effect is either an up or down force at the packer. Upward forces are designated as negative (−) forces, and downward forces are designated as positive (+) forces.

In "anchored" installations (the tubing is not allowed to move), the piston effect leads to a change in force either upward (tension) or downward (compression) at the packer. Upward forces are designated as negative (−), and downward forces are designated as positive (+).

In "free" installations (the tubing-to-packer configuration allows movement of the tubing), the piston effect leads to a change in length either upward (shortening) or downward (lengthening) of the tubing string.

In retrievable packer installations where the tubing is attached to the packer via a coupling on the packer mandrel, the piston effect will always result in a force change on the packer. The resulting force change will either be a change in tubing compression, or a change in tubing tension (either positive or negative) unless an expansion joint is installed in the tubing string above the packer.

If the tubing pressure is greater than the annular pressure, the affected area is calculated on the knife seal OD. If the annular pressure is greater than the tubing pressure, the affected area is calculated on the knife seal ID.

When the packer and tubing string are installed in the well, the hydrostatic pressure in the well acts on the end area of the tubing string. This creates an upward pressure-related effect, resulting in a reduction in the weight of the tubing string as well as a shortening of the overall length of the tubing string. This effect is known as force due to buoyancy or the buoyancy effect. Although the tubing is shortened as a result of the buoyancy effect, the packer-to-tubing relationship is not affected until the packer is set. Once the packer is set and the packer valve is closed, a change in pressure in either the tubing or the annulus will result in a change in the tubing force at the packer, and will affect the tubing-to-packer relationship.

When calculating the piston effect, always consider the tubing-to-packer relationship as "balanced" when the packer is set. Any changes in well condition that occur after the packer is set will affect the packer-to-tubing relationship.

When calculating the force due to the piston effect, it is the change in pressure rather than the absolute pressure that is important. To determine this, both initial and final well conditions need to be known. The initial conditions are those that existed when the packer was set, or when the seal assembly was stung into the packer sealbore. Final conditions are those conditions that are to be expected during production, stimulation, or well work-overs [4].

Use the following formula to calculate piston force:

$$\Delta F_1 = (A_p - A_o)\Delta P_o - (A_p - A_i)\Delta P_i \qquad [4.18.17]$$

where

ΔF_1 = Force change due to Piston Effect, lbs.
A_p = Packer valve area, in^2
A_o = Area of the tubing O.D., in^2
A_i = Area of the tubing I.D., in^2
ΔP_o = Change in the total annular pressure at the packer, psi
ΔP_i = Change in the total tubing pressure at the packer, psi

To calculate the length change due to the Piston Effect (ΔL_1), use this formula:

$$\Delta L_1 = \frac{LF_1}{EA_s} \qquad [4.18.18]$$

where

ΔL_1 = Length change due to Piston Effect, in.
L = Length of the tubing string, in.
F_1 = Force change due to Piston Effect, lbs.
E = Modulus of elasticity, 30×10^6 psi
A_s = Cross sectional area of the tubing, in^2

4.18.4.6 Buckling Effect

The buckling effect is the most difficult of the four basic effects to understand. Tubing buckling will occur due to two different force distributions. One is the compressive force mechanically applied to the end of the tubing, and the other is the internal pressure that results in a force distribution on the ID area of the tubing.

A mechanically applied compressive force acting on the end area of the tubing will cause tubing to buckle. When compressive force is applied to a tubing string that is, essentially, relatively flexible, the tubing will bow outward or will buckle. For example, tubing that is standing in a derrick will buckle or bow outward due to its own weight.

The second force that will cause a tubing string to buckle is pressure. Pressure buckling will occur when the pressure inside a tubing string is greater than the pressure outside the tubing.

The pressure within the tubing string creates a force distribution that acts over the ID area of the tubing, while pressure outside the tubing is creating a force distribution that acts over the OD area of the tubing. Because the pressure inside is higher than the pressure outside, these force distributions produce burst stresses within the tubing. The pressure within the tubing acts perpendicular to the inside tubing wall just as the annulus pressure acts perpendicular to the outside tubing wall. Because of the variance in tubing wall thickness, which can occur during the manufacturing process, the resulting burst stresses cannot be distributed evenly within the tubing. The uneven stress distribution causes the tubing to buckle.

If tubing was manufactured with precisely controlled inside and outside dimensions, the net resulting force due to a pressure differential would be zero, and the tubing would not buckle due to pressure. Dimensional tolerances for oilfield tubulars are outlined in the current edition of API Specification 5CT.

Buckled tubing is defined as tubing that is bowed from its original straight condition. The tubing will buckle or bow outward until it contacts the casing wall at which point it will begin to coil. This coiling of the tubing is referred to as "corkscrewing" the tubing. Corkscrewed tubing is a form of buckled tubing. As long as the stresses that occur because of the buckling do not exceed the yield strength of the tubing, the tubing will return to its original shape when the force is removed. Should the stresses created due to the buckled condition of the tubing exceed the yield strength of the tubing, the tubing will not return to its original shape: It will become permanently corkscrewed.

Buckling will only shorten a tubing string. Pressure buckling will only exert a negligible change in force on a packer and will only occur if the final pressure within the tubing string is greater than the final pressure outside the tubing string.

It should be noted that pressure buckling can occur in a tubing string even if the tubing is in tension throughout its entire length. Although in tension and considered as being in a "straight" condition within the wellbore, the tubing string will buckle due to the uneven force distribution.

Tubing buckling is most severe at the bottom of the tubing string where the pressure is the greatest. It lessens further up the hole until, normally, a point is reached where tubing buckling does not occur. This point is known as the neutral point. Notice that the tubing below the neutral point is buckled, while the tubing above remains straight.

The factors that have the most influence on the amount of tubing buckling that will occur are the radial clearance (r) between the tubing OD and the casing ID., the magnitude of the pressure differential from the tubing ID to the tubing OD, and the size of the packer valve. Because these factors have a direct effect on buckling, as any one of these factors increases, the length change due to buckling will increase.

Use this formula to calculate the length change due to buckling:

$$\Delta L_2 = \frac{r^2 A_p^2 (\Delta P_i - \Delta P_o)^2}{-8EI(W_s + W_i - W_o)} \qquad [4.18.19]$$

where

ΔL_2 = Length change in inches due to buckling
  r = radial clearance between the tubing and the casing, inches
  A_p = Packer valve area, in.^{2}
  ΔP_i = Change in total tubing pressure at the packer, psi
  ΔP_o = Change in total annulus pressure at the packer, psi
  E = Modulus of elasticity, 30×10^6 psi for carbon steels
  I = Moment of inertia of the tubing, in.^{4}
     $I = [(\text{Tubing OD})^4 - (\text{Tubing ID})^4] \times \pi \div 64$
  W_s = Weight of the tubing, lbs./in.
  W_i = Weight of the final fluid displaced in the tubing per unit length, psi
  W_i = (Tubing ID)$2 \times 0.0034 \times$ Final tubing fluid wt., lbs./gal.
  W_o = Weight of the final fluid displaced in the annulus per unit length, psi
  W_o = (Tubing OD)$2 \times 0.0034 \times$ Final annulus fluid wt., lbs./gal.

Use this formula to determine length from packer to neutral point:

$$n = \frac{A_p (P_{ifinal} - P_{ofinal})}{(W_s + W_i - W_o)} \qquad [4.18.20]$$

where

  n = The distance from the Packer to the Neutral Point, in.
  A_p = Packer valve area, in.^{2}
  P_{ifinal} = Total tubing pressure at the packer that will exist for the given conditions, psi.
  P_{ofinal} = Total annulus pressure at the packer that will exist for the given conditions, psi.
  W_s = Weight of the tubing, lbs./in.
  W_i = Weight of the final fluid displaced in the tubing per unit length, psi
  W_i = (Tubing ID)$2 \times 0.0034 \times$ Final tubing fluid wt., lbs./gal.
  W_o = Weight of the final fluid displaced in the annulus per unit length, psi
  W_o = (Tubing OD)$2 \times 0.0034 \times$ Final annulus fluid wt., lbs./gal.

Use this formula for Corrected Length Change due to Buckling:

$$\Delta L_2' = \Delta L_2 \times \frac{L}{n} \times \frac{[2 - (L)]}{n} \qquad [4.18.21]$$

where

$\Delta L_2'$ = Length change due to buckling when neutral point is above wellhead, in.
ΔL_2 = Length change due to buckling, in.
  L = Length of the tubing string, in.
  n = The distance from the packer to the neutral point

4.18.4.7 Ballooning Effect

The third effect, which must be considered in packer installations, is the ballooning effect. Tubing ballooning occurs when pressure is applied to the inside of a tubing string: The pressure differential creates forces within the tubing string, which are trying to burst the tubing. The burst forces created within the tubing string due to the differential pressure causes the tubing to swell.

As the tubing swells, the tubing string will shorten if it is free to move. If the tubing is not free to move, the swelling will create a tension force (−force) on the packer. This shortening of the tubing due to internal differential pressure is referred to as ballooning.

If the pressure differential exists on the outside of the tubing string, the force created is attempting to collapse the tubing. As the tubing tries to collapse, it lengthens if it is free to move. If the tubing is not free to move, it will create a compressive force (+ force) on the packer. This lengthening of the tubing string due to collapse forces is referred to as reverse ballooning.

The ballooning effect is directly proportional to the area over which the pressure is acting, and as a result, the effect of reverse ballooning is slightly greater than that of ballooning.

Unlike the piston and buckling effects, the ballooning effect occurs over the entire length of the tubing string. Ballooning effect calculations are based on changes in the average pressures inside and outside the tubing.

The average pressure is based on the half the sum of the surface pressure plus the pressure at the packer. Increasing the bottom hole pressure, by changing the fluid gradient, would have half the effect of making the same change by applying added surface pressure.

Well conditions can affect the average pressure inside and outside the tubing: The effects due to ballooning and reverse ballooning are calculated together, then the net effect is expressed as either a negative force (ballooning), or a positive force (reverse ballooning), or change in length of the tubing string, depending on the tubing string's ability to move at the packer.

Pressure increases in the tubing tend to swell, or balloon the tubing, and will result in a shortening of the tubing string. Similarly, pressure increases in the casing cause the tubing diameter to contract, resulting in reverse ballooning, and an elongation of the string. As treating pressures increase, and depths become greater, the resultant changes in length can be considerable.

In considering the results of ballooning and reverse ballooning, two point to consider are

1. The pressure that is responsible for ballooning or reverse ballooning is the pressure change in the tubing and/or annulus, and from the conditions that prevailed when the well was completed. The fact that a differential pressure exists between tubing and casing does not necessarily mean that ballooning exists.
2. Pressure calculations are based on changes in the surface tubing or casing pressure (applied) and changes in fluid density (hydrostatic).

Use this formula to calculate the effects due to ballooning:

$$\Delta F_3 = 0.6[(\Delta P_{oa} A_o) - (\Delta P_{ia} A_i)] \qquad [4.18.22]$$

where

ΔF_3 = Ballooning effect force change, lbs.
ΔP_{oa} = Change in average annulus pressure, psi.
A_o = Area of the tubing OD, in.^{2}
ΔP_{ia} = Change in average tubing pressure, psi.
A_i = Area of the tubing ID, in.^{2}

To calculate the length change (DL_3) due to the ballooning effect use this formula:

$$\Delta L_3 = \frac{2L\gamma}{E} \times \frac{(R^2 \Delta P_{oa}) - \Delta P_{ia}}{R^2 - 1} \qquad [4.18.23]$$

where

ΔL_3 = Change in length (in.)
L = Length of the tubing string (in.)
γ = Poisson's ratio: 0.3 for steel
E = Modulus of elasticity: 30×10^6 psi for steel
ΔP_{ia} = Change in average tubing pressure (psi)
ΔP_{oa} = Change in average annulus pressure (psi)
R^2 = Ratio of the tubing OD to the tubing ID (where R = OD ÷ ID)

4.18.4.8 Temperature Effect

The last basic effect is the temperature effect. It is the only one of the four basic effects that is not pressure related: The length and force changes that occur are functions of a change in the average tubing temperature.

The basic principles of expansion and contraction apply when the average temperature of the tubing string is increased or decreased. When the average temperature of the tubing is decreased by injecting cool fluids, the tubing string will shorten in length if it is free to move. If the tubing string is not free to move, then a tension force will be created on the packer.

When the average temperature of the tubing string is increased either by injecting warm fluids, or by producing warm oil and/or gas, the tubing will lengthen, or elongate, if it is free to move. If the tubing is not free to move, a compressive force on the packer will result, caused by the increase in the average tubing temperature. In many packer installations, the temperature effect will be the largest of the four basic effects.

Because the temperature change in a tubing string occurs over the entire length of the tubing string, the change in average tubing temperature must be calculated to determine the resulting change in force or length.

To calculate the average temperature of the tubing the bottom hole temperature and the surface temperature must be known. The following formula is used to calculate the average tubing temperature:

Avg. tbg. temp. (°F)

$$= [(\text{Surface temp. (°F)} + \text{Bottom hole temp. (°F)}) \div 2 \qquad [4.18.24]$$

If reliable temperature data is not available, it is strongly recommended that any and all assumptions made be conservative in nature to prevent possible equipment failure.

Although some assumptions are made, the magnitude and direction of the temperature effect can still be determined. It should be noted, however, that the temperature effect is not felt immediately at the packer. In installations where pressure changes occur within the well, the effects of the pressure change are felt immediately at the packer. When temperature effects occur, it may take from several minutes to several hours for the effects to be felt at the packer. However, it is typically assumed, when performing force analysis calculations, that the resulting effect due to a change in the tubing temperature is an immediate effect.

Treating temperature as an immediate effect allows it to be added to the pressure effects, so that all the effects can be considered, to determine the net effect on the tubing string. In some instances, this assumption can result in indications of potential equipment, or tubular, failure. Each installation should be considered separately to minimize the risks of tubing or equipment failure.

Use this formula to calculate the change in Force due to temperature effect:

$$\Delta F_4 = (207)(As)(\Delta T) \qquad [4.18.25]$$

where

ΔF_4 = Force change due to temperature effect, lbs.
As = Tubing cross sectional area, in.^{2}
ΔT = Change in average temperature, °F.
ΔT = Final average tubing temperature — Initial average tubing temperature

Use this formula to calculate change in tubing length due to temperature effect:

$$\Delta L_4 = (L)(\beta)(\Delta T) \qquad [4.18.26]$$

where

ΔL_4 = Length change due to temperature effect, in.
L = Length of tubing string, in.
β = Coefficient of thermal expansion for steel, 0.0000069 in/in/°F
ΔT = Change in average temperature, °F.

4.18.4.9 Total Effect

The four basic effects plus any mechanically applied forces are used to determine the total effect of the tubing on the packer. The method used to determine the total effect at the packer depends on the type of packer being considered for the installation and the tubing-to-packer relationship.

In packer installations that permit movement of the tubing, the degree and direction of the tubing length change expected (total effect) must be determined after considering each of the four basic effects independently, then collectively.

In packer installations that do not permit movement of the tubing, the degree and direction of the tubing force change expected (total effect) must be determined after considering the four basic effects independently, then collectively.

The total effect is then determined by adding together each basic effect and any mechanically applied effect and can be expressed mathematically.

Use this formula to calculate total length changes with slack off weight applied to packer:

$$\Delta L_{total1} = \Delta L_1 + \Delta L_2 + \Delta L_3 + \Delta L_4 + \Delta L_s \qquad [4.18.27]$$

where

ΔL_{total1} = Total length change, in.
ΔL_1 = Length change of the tubing due to piston effect, in.
ΔL_2 = Length change of the tubing due to buckling effect, in.
ΔL_3 = Length change of the tubing due to ballooning effect, in.
ΔL_s = Length change of the tubing due to slack off force, in.

Use this formula to calculate total length changes with tension applied to packer:

$$\Delta L_{total1} = \Delta L_1 + \Delta L_2 + \Delta L_3 + \Delta L_4 + \Delta L_t \qquad [4.18.28]$$

Table 4.18.9 *Grade Yield and Ultimate Strengths, Ultimate Elongations, and Hardness (courtesy of Precision Tube Technology, Inc.)*

Grade	Min. Yield	Min. Ultimate	Min. Elongation	Hardness Range
HS 70	70 ksi (483 MPa)	80 ksi (552 MPa)	25%	85-94 RH B
HS 80	80 ksi (552 MPa)	88 ksi (607 MPa)	28%	90-98 RH B
HS 90	90 ksi (621 MPa)	97 ksi (669 MPa)	25%	94 RH B-22 RH C
HS 110	110 ksi (758 MP)	115 ksi (793 MPa)	22%	22-28 RH C

where

ΔL_{total1} = Total length change, in.

ΔL_1 = Length change of the tubing due to piston effect, in.

ΔL_2 = Length change of the tubing due to buckling effect, in.

ΔL_3 = Length change of the tubing due to ballooning effect, in.

ΔL_4 = Length change of the tubing due to tension force, in.

When determining length changes of the tubing in an installation where the tubing is stung through the packer, there will always be a change (ΔL_{total}) in the total length of the tubing string if a change in pressure or temperature occurs, and this change must be calculated.

In a packer installation where the tubing is landed on the packer and the change in total length (ΔL_{total}) calculated is a downward movement (a positive value), the total length change will not occur due to the tubing being restrained from downward movement because of its landed configuration. The change in tubing length will result in a change in force (compressive) on the packer.

If the tubing is in a landed condition and the total change in length (ΔL_{total}) calculated is an upward movement (a negative value), the tubing is not restrained and is free to move upward.

If the tubing is latched to the packer, the total length change (ΔL_{total}) will be zero because the tubing is restrained from upward or downward movement and compressive or tension forces will result.

Use this formula to calculate the total force effect with slack off weight applied to the packer:

$$\Delta F_p = \Delta F_1 + \Delta F_3 + \Delta F_4 + \Delta F_s \qquad [4.18.29]$$

where

ΔF_p = Total force change effect, tubing-to-packer, lb

ΔF_1 = Force change due to piston effect, lb

ΔF_3 = Force change due to ballooning effect, lb

ΔF_4 = Force change due to temperature effect, lb

ΔF_s = Force change due to slack off force, lb

Use this formula to calculate the total force effect with tension applied to the packer:

$$\Delta F_p = \Delta F_1 + \Delta F_3 + \Delta F_4 + \Delta F_t \qquad [4.18.30]$$

where

ΔF_p = Total force change effect, tubing-to-packer, lb

ΔF_1 = Force change due to piston effect, lb

ΔF_3 = Force change due to ballooning effect, lb

ΔF_4 = Force change due to temperature effect, lb

ΔF_t = Force change due to tension force, lb

When determining force changes in the tubing string in an installation where the tubing is stung through the packer, the tubing-to-packer force (F_p) will always equal zero because the tubing can never exert a force on the packer.

In a packer installation where the tubing is landed onto the packer and the calculated change in force is a downward force (a positive value), the change in force will result in a compressive force (F_p) on the packer.

If the tubing is landed and the change in force calculated is an upward force (a negative value), the tubing-to-packer force (F_p) will be zero, given that tension cannot be pulled through a landed tubing-to-packer configuration.

If the tubing is latched to the packer, the tubing-to-packer force (F_p) will always result in either compressive or tension forces on the packer.

A sufficient amount of accurate data is important in the complete analysis. The validity of the results depends largely on the accuracy of the data used, and this should weigh heavily on the final selection of the procedure, equipment to be used and the safety factor required.

The formulas used do not intentionally include any safety factor and the safety factor must be added. Magnitude of the safety factor should be based on the severity of the application and the validity of the data used.

When it is necessary to assume some of the data, the company representative or serviceman most familiar with the area should be able to supply the best "guesstimate" and some idea to its accuracy. Considering the guesstimates used, the same personnel should be able to establish the safety margin required and often the amount of detail required in analyzing. Obviously it is not practical to analyze data, which is highly questionable.

4.18.4.10 Coiled Tubing

Coiled tubing is a recent development in tubing technology. Coiled tubing is a continuous length of steel tubing with no joints. In general, the coiled tubing is fabricated to the length needed for a particular well application and rolled onto a reel for transport to a well site for placement in the well. The tubing can be butt-welded or continuously-milled as the tubing is placed on the roll. Coiled tubing is used in a variety of oil and gas well production, completions, and drilling operations. Like conventional tubing, coiled tubing in fabricated in a four material grades. These are HS 70, HS 80, HS 90, and HS 110.

Table 4.18.9 gives the minimum steel strength, elongation and hardness properties for each of the above grades of coiled tubing materials. The same calculation techniques that were discussed above in this section are applicable to the design of coiled tubing strings for placement in oil and gas production wells.

Tables 4.18.10 (USC units) and 4.18.11 (Metric units) give the geometry and performance limitations for the HS 70 grade coiled tubing.

Tables 4.18.12 (USC units) and 4.18.13 (Metric units) give the geometry and performance limitations for the HS 80 grade coiled tubing.

Tables 4.18.14 (USC units) and 4.18.15 (Metric units) give the geometry and performance limitations for the HS 90 grade coiled tubing.

Tables 4.18.16 (USC units) and 4.18.17 (Metric units) give the geometry and performance limitations for the HS 110 grade coiled tubing.

Table 4.18.10 *HS70 Grade USC Units (courtesy of Precision Tube Technology, Inc.)*

DIMENSIONS (Inches)						Nominal Weight	TUBE LOAD BODY (Lbs.)		INTERNAL PRESSURE (psi)		TUBING AREA (sq. in.)		TORSIONAL YIELD (ft.-lbs.)		INTERNAL CAPACITY per 1000 ft.		EXTERNAL DISPLACEMENT per 1000 ft.	
O.D. Specified	(mm)	Wall Specified	(mm)	Wall Minimum	I.D. Calculated	Lbs./ft.	Yield Minimum	Tensile Minimum	Hydro Test 90%	Internal Yield Min	w/min. Wall	Internal Min	Yield	Ultimate	Gallons	Barrels	Gallons	Barrels
1.000	25.4	0.080	2.03	0.076	0.840	0.788	16,200	18,500	9,400	10,500	0.221	0.565	319	344	28.79	0.69	40.80	0.97
1.000	25.4	0.087	2.21	0.083	0.826	0.850	17,500	20,000	10,300	11,400	0.239	0.546	341	370	27.84	0.66	40.80	0.97
1.000	25.4	0.095	2.41	0.090	0.810	0.920	18,900	21,600	11,100	12,300	0.257	0.528	362	395	26.77	0.64	40.80	0.97
1.000	25.4	0.102	2.59	0.097	0.796	0.981	20,100	23,000	11,900	13,300	0.275	0.510	382	420	25.85	0.62	40.80	0.97
1.000	25.4	0.109	2.77	0.104	0.782	1.040	21,400	24,400	12,700	14,200	0.293	0.493	401	443	24.95	0.59	40.80	0.97
1.250	31.8	0.080	2.03	0.076	1.090	1.002	20,600	23,500	7,600	8,400	0.280	0.947	522	555	48.47	1.15	63.75	1.52
1.250	31.8	0.087	2.21	0.083	1.076	1.083	22,300	25,400	8,300	9,200	0.304	0.923	561	599	47.24	1.12	63.75	1.52
1.250	31.8	0.095	2.41	0.090	1.060	1.175	24,100	27,600	9,000	9,900	0.328	0.899	598	642	45.84	1.09	63.75	1.52
1.250	31.8	0.102	2.59	0.097	1.046	1.254	25,800	29,400	9,600	10,700	0.351	0.876	633	683	44.64	1.06	63.75	1.52
1.250	31.8	0.109	2.77	0.104	1.032	1.332	27,400	31,300	10,300	11,400	0.374	0.853	668	724	43.45	1.03	63.75	1.52
1.250	31.8	0.116	2.95	0.111	1.018	1.408	28,900	33,100	11,000	12,200	0.397	0.830	700	764	42.28	1.01	63.75	1.52
1.250	31.8	0.125	3.18	0.118	1.000	1.506	30,900	35,300	11,600	12,900	0.420	0.808	732	802	40.80	0.97	63.75	1.52
1.250	31.8	0.134	3.40	0.128	0.982	1.601	32,900	37,600	12,600	14,000	0.451	0.776	775	856	39.34	0.94	63.75	1.52
1.250	31.8	0.145	3.68	0.138	0.960	1.715	35,200	40,300	13,500	15,000	0.482	0.745	815	907	37.60	0.90	63.75	1.52
1.250	31.8	0.156	3.96	0.148	0.938	1.827	37,500	42,900	14,400	16,000	0.512	0.715	853	956	35.90	0.85	63.75	1.52
1.250	31.8	0.175	4.45	0.167	0.900	2.014	41,400	47,300	16,100	17,900	0.568	0.659	919	1,044	33.05	0.79	63.75	1.52
1.500	38.1	0.095	2.41	0.090	1.310	1.429	29,400	33,500	7,500	8,300	0.399	1.368	893	947	70.02	1.67	91.80	2.19
1.500	38.1	0.102	2.59	0.097	1.296	1.527	31,400	35,800	8,100	9,000	0.428	1.340	949	1,011	68.53	1.63	91.80	2.19
1.500	38.1	0.109	2.77	0.104	1.282	1.623	33,300	38,100	8,600	9,600	0.456	1.311	1,003	1,074	67.06	1.60	91.80	2.19
1.500	38.1	0.116	2.95	0.111	1.268	1.719	35,300	40,300	9,200	10,200	0.484	1.283	1,055	1,135	65.60	1.56	91.80	2.19
1.500	38.1	0.125	3.18	0.118	1.250	1.840	37,800	43,200	9,800	10,800	0.512	1.255	1,106	1,194	63.75	1.52	91.80	2.19
1.500	38.1	0.134	3.40	0.128	1.232	1.960	40,300	46,000	10,600	11,700	0.552	1.215	1,175	1,278	61.93	1.47	91.80	2.19
1.500	38.1	0.145	3.68	0.138	1.210	2.104	43,200	49,400	11,300	12,600	0.590	1.177	1,242	1,358	59.74	1.42	91.80	2.19
1.500	38.1	0.156	3.96	0.148	1.188	2.245	46,100	52,700	12,100	13,500	0.629	1.139	1,305	1,436	57.58	1.37	91.80	2.19
1.500	38.1	0.175	4.45	0.167	1.150	2.483	51,000	58,300	13,600	15,100	0.699	1.068	1,416	1,577	53.96	1.28	91.80	2.19
1.500	38.1	0.190	4.83	0.180	1.120	2.665	54,700	62,600	14,600	16,200	0.746	1.021	1,486	1,668	51.18	1.22	91.80	2.19

1.750	44.5	0.109	2.77	0.104	1.532	39,300	45,000	7,400	8,200	0.538	1.867	1,407	1,492	95.76	2.28	124.95	2.97
1.750	44.5	0.116	2.95	0.111	1.518	41,700	47,600	7,900	8,800	0.572	1.834	1,483	1,579	94.02	2.24	124.95	2.97
1.750	44.5	0.125	3.18	0.118	1.500	44,700	51,100	8,400	9,300	0.605	1.800	1,558	1,665	91.80	2.19	124.95	2.97
1.750	44.5	0.134	3.40	0.128	1.482	47,600	54,400	9,100	10,100	0.652	1.753	1,660	1,784	89.61	2.13	124.95	2.97
1.750	44.5	0.145	3.68	0.138	1.460	51,200	58,500	9,800	10,900	0.699	1.706	1,759	1,901	86.97	2.07	124.95	2.97
1.750	44.5	0.156	3.96	0.148	1.438	54,700	62,500	10,500	11,600	0.745	1.660	1,854	2,014	84.37	2.01	124.95	2.97
1.750	44.5	0.175	4.45	0.167	1.400	60,600	69,300	11,700	13,100	0.831	1.575	2,024	2,221	79.97	1.90	124.95	2.97
1.750	44.5	0.190	4.83	0.180	1.370	65,200	74,500	12,600	14,000	0.888	1.517	2,132	2,356	76.58	1.82	124.95	2.97
1.750	44.5	0.204	5.18	0.195	1.342	69,400	79,300	13,600	15,100	0.953	1.453	2,250	2,506	73.48	1.75	124.95	2.97
2.000	50.8	0.109	2.77	0.104	1.782	45,300	51,800	6,500	7,200	0.619	2.522	1,879	1,979	129.56	3.08	163.20	3.89
2.000	50.8	0.116	2.95	0.111	1.768	48,100	54,900	6,900	7,700	0.659	2.483	1,985	2,097	127.53	3.04	163.20	3.89
2.000	50.8	0.125	3.18	0.118	1.750	51,500	58,900	7,400	8,200	0.698	2.444	2,088	2,213	124.95	2.97	163.20	3.89
2.000	50.8	0.134	3.40	0.128	1.732	55,000	62,800	8,000	8,900	0.753	2.389	2,230	2,375	122.39	2.91	163.20	3.89
2.000	50.8	0.145	3.68	0.138	1.710	59,200	67,600	8,600	9,500	0.807	2.334	2,368	2,534	119.30	2.84	163.20	3.89
2.000	50.8	0.156	3.96	0.148	1.688	63,300	72,300	9,200	10,200	0.861	2.280	2,501	2,690	116.25	2.77	163.20	3.89
2.000	50.8	0.175	4.45	0.167	1.650	70,200	80,300	10,300	11,500	0.962	2.180	2,741	2,975	111.08	2.64	163.20	3.89
2.000	50.8	0.190	4.83	0.180	1.620	75,600	86,400	11,100	12,300	1.029	2.112	2,897	3,163	107.08	2.55	163.20	3.89
2.000	50.8	0.204	5.18	0.195	1.592	80,600	92,100	12,000	13,300	1.106	2.036	3,067	3,372	103.41	2.46	163.20	3.89
2.375	60.3	0.125	3.18	0.118	2.125	61,900	70,700	6,200	6,900	0.837	3.593	3,028	3,181	184.24	4.39	230.14	5.48
2.375	60.3	0.134	3.40	0.128	2.107	66,000	75,500	6,700	7,500	0.904	3.527	3,243	3,421	181.13	4.31	230.14	5.48
2.375	60.3	0.145	3.68	0.138	2.085	71,100	81,300	7,300	8,100	0.970	3.460	3,452	3,656	177.37	4.22	230.14	5.48
2.375	60.3	0.156	3.96	0.148	2.063	76,100	87,000	7,800	8,600	1.035	3.395	3,655	3,886	173.64	4.13	230.14	5.48
2.375	60.3	0.175	4.45	0.167	2.025	84,700	96,800	8,700	9,700	1.158	3.272	4,025	4,313	167.31	3.98	230.14	5.48
2.375	60.3	0.190	4.83	0.180	1.995	91,300	104,300	9,400	10,500	1.241	3.189	4,266	4,595	162.38	3.87	230.14	5.48
2.375	60.3	0.204	5.18	0.195	1.967	97,400	111,300	10,200	11,300	1.335	3.095	4,533	4,913	157.86	3.76	230.14	5.48
2.375	60.3	0.224	5.69	0.214	1.927	106,000	121,100	11,100	12,400	1.453	2.977	4,855	5,301	151.50	3.61	230.14	5.48

Test pressure value equals 90% of internal yield pressure rating. Maximum working pressure is a function of tube condition and is determined by user.

* Available as continuously milled tubing (CM™) or conventional butt-welded tubing sections (W™). All data is for new tubing at minimum strength. Other sizes and wall thicknesses available on request. See individual size sheets for additional wall thicknesses.

Table 4.18.11 *HS70 Grade USC Units (courtesy of Precision Tube Technology, Inc.)*

DIMENSIONS (mm)						Nominal Weight	TUBE LOAD BODY (Newtons)		INTERNAL PRESSURE (kPa)		TUBING AREA (sq. cm)		TORSIONAL YIELD (N-m)		INTERNAL CAPACITY per meter	EXTERNAL DISPLACEMENT per meter
OD Specified (inches)	OD Specified	Wall Specified	Wall Specified (inches)	Wall Minimum	ID Calculated	Kg/m	Yield Minimum	Tensile Minimum	Hydro Test 90%	Internal Yield Min	w/min. Wall	Internal Min	Yield	Ultimate	Liters	Liters
1.000	25.4	2.03	0.080	1.93	21.3	1.17	72,100	82,300	64,800	72,400	1.42	3.64	430	470	0.36	0.51
1.000	25.4	2.21	0.087	2.11	21.0	1.26	77,800	89,000	71,000	78,600	1.54	3.52	460	500	0.35	0.51
1.000	25.4	2.41	0.095	2.29	20.6	1.37	84,100	96,100	76,500	84,800	1.66	3.41	490	540	0.33	0.51
1.000	25.4	2.59	0.102	2.46	20.2	1.46	89,400	102,300	82,100	91,700	1.78	3.29	520	570	0.32	0.51
1.000	25.4	2.77	0.109	2.64	19.9	1.55	95,200	108,500	87,600	97,900	1.89	3.18	540	600	0.31	0.51
1.250	31.8	2.03	0.080	1.93	27.7	1.49	91,600	104,500	52,400	57,900	1.81	6.11	710	750	0.60	0.79
1.250	31.8	2.21	0.087	2.11	27.3	1.61	99,200	113,000	57,200	63,400	1.96	5.95	760	810	0.59	0.79
1.250	31.8	2.41	0.095	2.29	26.9	1.75	107,200	122,800	62,100	68,300	2.12	5.80	810	870	0.57	0.79
1.250	31.8	2.59	0.102	2.46	26.6	1.86	114,800	130,800	66,200	73,800	2.27	5.65	860	930	0.55	0.79
1.250	31.8	2.77	0.109	2.64	26.2	1.98	121,900	139,200	71,000	78,600	2.42	5.50	910	980	0.54	0.79
1.250	31.8	2.95	0.116	2.82	25.9	2.09	128,500	147,200	75,800	84,100	2.56	5.35	950	1,040	0.53	0.79
1.250	31.8	3.18	0.125	3.00	25.4	2.24	137,400	157,000	80,000	88,900	2.71	5.21	990	1,090	0.51	0.79
1.250	31.8	3.40	0.134	3.25	24.9	2.38	146,300	167,200	86,900	96,500	2.91	5.01	1,050	1,160	0.49	0.79
1.250	31.8	3.68	0.145	3.51	24.4	2.55	156,600	179,300	93,100	103,400	3.11	4.81	1,110	1,230	0.47	0.79
1.250	31.8	3.96	0.156	3.76	23.8	2.72	166,800	190,800	99,300	110,300	3.31	4.61	1,160	1,300	0.45	0.79
1.250	31.8	4.45	0.175	4.24	22.9	2.99	184,100	210,400	111,000	123,400	3.67	4.25	1,250	1,420	0.41	0.79
1.500	38.1	2.41	0.095	2.29	33.3	2.12	130,800	149,000	51,700	57,200	2.57	8.83	1,210	1,280	0.87	1.14
1.500	38.1	2.59	0.102	2.46	32.9	2.27	139,700	159,200	55,800	62,100	2.76	8.64	1,290	1,370	0.85	1.14
1.500	38.1	2.77	0.109	2.64	32.6	2.41	148,100	169,500	59,300	66,200	2.94	8.46	1,360	1,460	0.83	1.14
1.500	38.1	2.95	0.116	2.82	32.2	2.55	157,000	179,300	63,400	70,300	3.12	8.28	1,430	1,540	0.81	1.14
1.500	38.1	3.18	0.125	3.00	31.8	2.73	168,100	192,200	67,600	74,500	3.31	8.10	1,500	1,620	0.79	1.14
1.500	38.1	3.40	0.134	3.25	31.3	2.91	179,300	204,600	73,100	80,700	3.56	7.84	1,590	1,730	0.77	1.14
1.500	38.1	3.68	0.145	3.51	30.7	3.13	192,200	219,700	77,900	86,900	3.81	7.59	1,680	1,840	0.74	1.14
1.500	38.1	3.96	0.156	3.76	30.2	3.34	205,100	234,400	83,400	93,100	4.06	7.35	1,770	1,950	0.72	1.14
1.500	38.1	4.45	0.175	4.24	29.2	3.69	226,800	259,300	93,800	104,100	4.51	6.89	1,920	2,140	0.67	1.14
1.500	38.1	4.83	0.190	4.57	28.4	3.96	243,300	278,400	100,700	111,700	4.82	6.59	2,020	2,260	0.64	1.14

44.5	1.750	2.77	0.109	2.64	38.9	2.85	174,800	200,200	51,000	56,500	3.47	12.05	1,910	2,020	1.19	1.55
44.5	1.750	2.95	0.116	2.82	38.6	3.02	185,500	211,700	54,500	60,700	3.69	11.83	2,010	2,140	1.17	1.55
44.5	1.750	3.18	0.125	3.00	38.1	3.23	198,800	227,300	57,900	64,100	3.90	11.61	2,110	2,260	1.14	1.55
44.5	1.750	3.40	0.134	3.25	37.6	3.45	211,700	242,000	62,700	69,600	4.21	11.31	2,250	2,420	1.11	1.55
44.5	1.750	3.68	0.145	3.51	37.1	3.70	227,700	260,200	67,600	75,200	4.51	11.01	2,390	2,580	1.08	1.55
44.5	1.750	3.96	0.156	3.76	36.5	3.96	243,300	278,000	72,400	80,000	4.81	10.71	2,510	2,730	1.05	1.55
44.5	1.750	4.45	0.175	4.24	35.6	4.39	269,500	308,200	80,700	90,300	5.36	10.16	2,740	3,010	0.99	1.55
44.5	1.750	4.83	0.190	4.57	34.8	4.72	290,000	331,400	86,900	96,500	5.73	9.79	2,890	3,190	0.95	1.55
44.5	1.750	5.18	0.204	4.95	34.1	5.02	308,700	352,700	93,800	104,100	6.15	9.37	3,050	3,400	0.91	1.55
50.8	2.000	2.77	0.109	2.64	45.3	3.28	201,500	230,400	44,800	49,600	4.00	16.27	2,550	2,680	1.61	2.03
50.8	2.000	2.95	0.116	2.82	44.9	3.48	213,900	244,200	47,600	53,100	4.25	16.02	2,690	2,840	1.58	2.03
50.8	2.000	3.18	0.125	3.00	44.5	3.73	229,100	262,000	51,000	56,500	4.50	15.77	2,830	3,000	1.55	2.03
50.8	2.000	3.40	0.134	3.25	44.0	3.98	244,600	279,300	55,200	61,400	4.86	15.41	3,020	3,220	1.52	2.03
50.8	2.000	3.68	0.145	3.51	43.4	4.28	263,300	300,700	59,300	65,500	5.21	15.06	3,210	3,440	1.48	2.03
50.8	2.000	3.96	0.156	3.76	42.9	4.58	281,600	321,600	63,400	70,300	5.56	14.71	3,390	3,650	1.44	2.03
50.8	2.000	4.45	0.175	4.24	41.9	5.08	312,200	357,200	71,000	79,300	6.20	14.06	3,720	4,030	1.38	2.03
50.8	2.000	4.83	0.190	4.57	41.1	5.47	336,300	384,300	76,500	84,800	6.64	13.63	3,930	4,290	1.33	2.03
50.8	2.000	5.18	0.204	4.95	40.4	5.83	358,500	409,700	82,700	91,700	7.13	13.13	4,160	4,570	1.28	2.03
60.3	2.375	3.18	0.125	3.00	54.0	4.47	275,300	314,500	42,700	47,600	5.40	23.18	4,110	4,310	2.29	2.86
60.3	2.375	3.40	0.134	3.25	53.5	4.78	293,600	335,800	46,200	51,700	5.83	22.75	4,400	4,640	2.25	2.86
60.3	2.375	3.68	0.145	3.51	53.0	5.14	316,300	361,600	50,300	55,800	6.26	22.32	4,680	4,960	2.20	2.86
60.3	2.375	3.96	0.156	3.76	52.4	5.51	338,500	387,000	53,800	59,300	6.68	21.90	4,960	5,270	2.16	2.86
60.3	2.375	4.45	0.175	4.24	51.4	6.13	376,700	430,600	60,000	66,900	7.47	21.11	5,460	5,850	2.08	2.86
60.3	2.375	4.83	0.190	4.57	50.7	6.61	406,100	463,900	64,800	72,400	8.01	20.57	5,780	6,230	2.02	2.86
60.3	2.375	5.18	0.204	4.95	50.0	7.05	433,200	495,100	70,300	77,900	8.62	19.97	6,150	6,660	1.96	2.86
60.3	2.375	5.68	0.224	5.44	48.9	7.67	471,500	538,700	76,500	85,500	9.37	19.21	6,580	7,190	1.88	2.86

Test pressure value equals 90% of internal yield pressure rating. Maximum working pressure is a function of tube condition and is determined by user.

*Available as continuously milled tubing (CM^{TM}) or conventional butt-welded tubing sections (W^{TM}). All data is for new tubing at minimum strength. Other sizes and wall thicknesses available on request. See individual size sheets for additional wall thicknesses.

Table 4.18.12 HS80 Grade USC Units (courtesy of Precision Tube Technology, Inc.)

OD Specified	OD (mm)	Wall Specified	Wall (mm)	Wall Minimum	ID Calculated	Nominal Weight lb./ft.	TUBE LOAD BODY (Lbs.) Yield Minimum	Tensile Minimum	INTERNAL PRESSURE (psi) Hydro Test 90%	Internal Yield Min	TUBING AREA (sq. in.) w/min. Wall	Internal Min	TORSIONAL YIELD (ft.-lb.) Yield	Ultimate	INTERNAL CAPACITY per 1,000 ft. Gallons	Barrels	EXTERNAL DISPLACEMENT per 1,000 ft. Gallons	Barrels
1.000	25.4	0.080	2.03	0.076	0.840	0.788	18,500	20,300	10,800	12,000	0.221	0.565	365	393	28.79	0.69	40.80	0.97
1.000	25.4	0.087	2.21	0.083	0.826	0.850	20,000	22,000	11,700	13,000	0.239	0.546	390	423	27.84	0.66	40.80	0.97
1.000	25.4	0.095	2.41	0.090	0.810	0.920	21,600	23,800	12,700	14,100	0.257	0.528	414	452	26.77	0.64	40.80	0.97
1.000	25.4	0.102	2.59	0.097	0.796	0.981	23,000	25,300	13,600	15,200	0.275	0.510	437	480	25.85	0.62	40.80	0.97
1.000	25.4	0.109	2.77	0.104	0.782	1.040	24,400	26,800	14,600	16,200	0.293	0.493	458	507	24.95	0.59	40.80	0.97
1.250	31.8	0.080	2.03	0.076	1.090	1.002	23,500	25,900	8,700	9,600	0.280	0.947	597	634	48.47	1.15	63.75	1.52
1.250	31.8	0.087	2.21	0.083	1.076	1.083	25,400	28,000	9,500	10,500	0.304	0.923	641	684	47.24	1.12	63.75	1.52
1.250	31.8	0.095	2.41	0.090	1.060	1.175	27,600	30,300	10,200	11,400	0.328	0.899	683	733	45.84	1.09	63.75	1.52
1.250	31.8	0.102	2.59	0.097	1.046	1.254	29,400	32,400	11,000	12,200	0.351	0.876	724	781	44.64	1.06	63.75	1.52
1.250	31.8	0.109	2.77	0.104	1.032	1.332	31,300	34,400	11,800	13,100	0.374	0.853	763	828	43.45	1.03	63.75	1.52
1.250	31.8	0.116	2.95	0.111	1.018	1.408	33,100	36,400	12,500	13,900	0.397	0.830	800	873	42.28	1.01	63.75	1.52
1.250	31.8	0.125	3.18	0.118	1.000	1.506	35,300	38,900	13,300	14,800	0.420	0.808	836	917	40.80	0.97	63.75	1.52
1.250	31.8	0.134	3.40	0.128	0.982	1.601	37,600	41,300	14,400	16,000	0.451	0.776	885	978	39.34	0.94	63.75	1.52
1.250	31.8	0.145	3.68	0.138	0.960	1.715	40,300	44,300	15,400	17,100	0.482	0.745	931	1,036	37.60	0.90	63.75	1.52
1.250	31.8	0.156	3.96	0.148	0.938	1.827	42,900	47,200	16,500	18,300	0.512	0.715	975	1,093	35.90	0.85	63.75	1.52
1.250	31.8	0.175	4.45	0.167	0.900	2.014	47,300	52,000	18,400	20,400	0.568	0.659	1,050	1,193	33.05	0.79	63.75	1.52
1.500	38.1	0.095	2.41	0.090	1.310	1.429	33,500	36,900	8,600	9,500	0.399	1.368	1,020	1,083	70.02	1.67	91.80	2.19
1.500	38.1	0.102	2.59	0.097	1.296	1.527	35,800	39,400	9,200	10,200	0.428	1.340	1,084	1,156	68.53	1.63	91.80	2.19
1.500	38.1	0.109	2.77	0.104	1.282	1.623	38,100	41,900	9,900	11,000	0.456	1.311	1,146	1,227	67.06	1.60	91.80	2.19
1.500	38.1	0.116	2.95	0.111	1.268	1.719	40,300	44,400	10,500	11,700	0.484	1.283	1,206	1,297	65.60	1.56	91.80	2.19
1.500	38.1	0.125	3.18	0.118	1.250	1.840	43,200	47,500	11,200	12,400	0.512	1.255	1,264	1,365	63.75	1.52	91.80	2.19
1.500	38.1	0.134	3.40	0.128	1.232	1.960	46,000	50,600	12,100	13,400	0.552	1.215	1,343	1,460	61.93	1.47	91.80	2.19
1.500	38.1	0.145	3.68	0.138	1.210	2.104	49,400	54,300	13,000	14,400	0.590	1.177	1,419	1,552	59.74	1.42	91.80	2.19
1.500	38.1	0.156	3.96	0.148	1.188	2.245	52,700	58,000	13,900	15,400	0.629	1.139	1,491	1,641	57.58	1.37	91.80	2.19
1.500	38.1	0.175	4.45	0.167	1.150	2.483	58,300	64,100	15,500	17,300	0.699	1.068	1,618	1,802	53.96	1.28	91.80	2.19
1.500	38.1	0.190	4.83	0.180	1.120	2.665	62,600	68,800	16,700	18,500	0.746	1.021	1,699	1,907	51.18	1.22	91.80	2.19

1.750	*44.5*	**0.109**	*2.77*	0.104	1.532	1.915	45,000	49,500	8,500	9,400	0.538	1.867	1,608	1,705	95.76	2.28	124.95	2.97
1.750	*44.5*	**0.116**	*2.95*	0.111	1.518	2.029	47,600	52,400	9,000	10,000	0.572	1.834	1,695	1,804	94.02	2.24	124.95	2.97
1.750	*44.5*	**0.125**	*3.18*	0.118	1.500	2.175	51,100	56,200	9,600	10,700	0.605	1.800	1,780	1,902	91.80	2.19	124.95	2.97
1.750	*44.5*	**0.134**	*3.40*	0.128	1.482	2.318	54,400	59,900	10,400	11,500	0.652	1.753	1,898	2,039	89.61	2.13	124.95	2.97
1.750	*44.5*	**0.145**	*3.68*	0.138	1.460	2.492	58,500	64,300	11,200	12,400	0.699	1.706	2,011	2,172	86.97	2.07	124.95	2.97
1.750	*44.5*	**0.156**	*3.96*	0.148	1.438	2.662	62,500	68,700	12,000	13,300	0.745	1.660	2,119	2,302	84.37	2.01	124.95	2.97
1.750	*44.5*	**0.175**	*4.45*	0.167	1.400	2.951	69,300	76,200	13,400	14,900	0.831	1.575	2,313	2,538	79.97	1.90	124.95	2.97
1.750	*44.5*	**0.190**	*4.83*	0.180	1.370	3.173	74,500	81,900	14,400	16,000	0.888	1.517	2,437	2,693	76.58	1.82	124.95	2.97
1.750	*44.5*	**0.204**	*5.18*	0.195	1.342	3.377	79,300	87,200	15,500	17,300	0.953	1.453	2,571	2,864	73.48	1.75	124.95	2.97
2.000	*50.8*	**0.109**	*2.77*	0.104	1.782	2.207	51,800	57,000	7,400	8,300	0.619	2.522	2,148	2,261	129.56	3.08	163.20	3.89
2.000	*50.8*	**0.116**	*2.95*	0.111	1.768	2.340	54,900	60,400	7,900	8,800	0.659	2.483	2,268	2,396	127.53	3.04	163.20	3.89
2.000	*50.8*	**0.125**	*3.18*	0.118	1.750	2.509	58,900	64,800	8,400	9,400	0.698	2.444	2,386	2,529	124.95	2.97	163.20	3.89
2.000	*50.8*	**0.134**	*3.40*	0.128	1.732	2.677	62,800	69,100	9,100	10,100	0.753	2.389	2,549	2,715	122.39	2.91	163.20	3.89
2.000	*50.8*	**0.145**	*3.68*	0.138	1.710	2.880	67,600	74,400	9,800	10,900	0.807	2.334	2,706	2,896	119.30	2.84	163.20	3.89
2.000	*50.8*	**0.156**	*3.96*	0.148	1.688	3.080	72,300	79,500	10,500	11,700	0.861	2.280	2,858	3,074	116.25	2.77	163.20	3.89
2.000	*50.8*	**0.175**	*4.45*	0.167	1.650	3.419	80,300	88,300	11,800	13,100	0.962	2.180	3,133	3,400	111.08	2.64	163.20	3.89
2.000	*50.8*	**0.190**	*4.83*	0.180	1.620	3.682	86,400	95,100	12,700	14,100	1.029	2.112	3,310	3,614	107.08	2.55	163.20	3.89
2.000	*50.8*	**0.204**	*5.18*	0.195	1.592	3.923	92,100	101,300	13,700	15,200	1.106	2.036	3,505	3,854	103.41	2.46	163.20	3.89
2.375	*60.3*	**0.125**	*3.18*	0.118	2.125	3.011	70,700	77,800	7,100	7,900	0.837	3.593	3,461	3,635	184.24	4.39	230.14	5.48
2.375	*60.3*	**0.134**	*3.40*	0.128	2.107	3.215	75,500	83,000	7,700	8,600	0.904	3.527	3,707	3,909	181.13	4.31	230.14	5.48
2.375	*60.3*	**0.145**	*3.68*	0.138	2.085	3.462	81,300	89,400	8,300	9,200	0.970	3.460	3,945	4,178	177.37	4.22	230.14	5.48
2.375	*60.3*	**0.156**	*3.96*	0.148	2.063	3.706	87,000	95,700	8,900	9,900	1.035	3.395	4,177	4,442	173.64	4.13	230.14	5.48
2.375	*60.3*	**0.175**	*4.45*	0.167	2.025	4.122	96,800	106,400	10,000	11,100	1.158	3.272	4,600	4,929	167.31	3.98	230.14	5.48
2.375	*60.3*	**0.190**	*4.83*	0.180	1.995	4.445	104,300	114,800	10,800	12,000	1.241	3.189	4,876	5,252	162.38	3.87	230.14	5.48
2.375	*60.3*	**0.204**	*5.18*	0.195	1.967	4.742	111,300	122,400	11,600	12,900	1.335	3.095	5,181	5,614	157.86	3.76	230.14	5.48
2.375	*60.3*	**0.224**	*5.69*	0.214	1.927	5.159	121,100	133,200	12,700	14,100	1.453	2.977	5,548	6,058	151.50	3.61	230.14	5.48

Test pressure value equals 90% of internal yield pressure rating. Maximum working pressure is a function of tube condition and is determined by user.

*Available as continuously milled tubing (CM^{TM}) or conventional butt-welded tubing sections (W^{TM}). All data is for new tubing at minimum strength. Other sizes and wall thicknesses available on request. See individual size sheets for additional wall thicknesses.

Table 4.18.13 HS80 Grade USC Units (courtesy of Precision Tube Technology, Inc.)

DIMENSIONS (mm)						Nominal Weight	TUBE LOAD BODY (Newtons)		INTERNAL PRESSURE (kPa)		TUBING AREA (sq. cm)		TORSIONAL YIELD (N-m)		INTERNAL CAPACITY per meter	EXTERNAL DISPLACEMENT per meter
OD		Wall			ID		Yield	Tensile	Hydro	Internal	w/min.	Internal				
Specified (inches)	OD	Specified (inches)	Wall	Wall Minimum	Calculated	Kg/m	Minimum	Minimum	Test 90%	Yield Min	Wall	Min	Yield	Ultimate	Liters	Liters
25.4	1.000	2.03	0.080	1.93	21.3	1.17	82,300	90,300	74,500	82,700	1.42	3.64	490	530	0.36	0.51
25.4	1.000	2.21	0.087	2.11	21.0	1.26	89,000	97,900	80,700	89,600	1.54	3.52	530	570	0.35	0.51
25.4	1.000	2.41	0.095	2.29	20.6	1.37	96,100	105,800	87,600	97,200	1.66	3.41	560	610	0.33	0.51
25.4	1.000	2.59	0.102	2.46	20.2	1.46	102,300	112,500	93,800	104,800	1.78	3.29	590	650	0.32	0.51
25.4	1.000	2.77	0.109	2.64	19.9	1.55	108,500	119,200	100,700	111,700	1.89	3.18	620	690	0.31	0.51
31.8	1.250	2.03	0.080	1.93	27.7	1.49	104,500	115,200	60,000	66,200	1.81	6.11	810	860	0.60	0.79
31.8	1.250	2.21	0.087	2.11	27.3	1.61	113,000	124,500	65,500	72,400	1.96	5.95	870	930	0.59	0.79
31.8	1.250	2.41	0.095	2.29	26.9	1.75	122,800	134,800	70,300	78,600	2.12	5.80	930	990	0.57	0.79
31.8	1.250	2.59	0.102	2.46	26.6	1.86	130,800	144,100	75,800	84,100	2.27	5.65	980	1,060	0.55	0.79
31.8	1.250	2.77	0.109	2.64	26.2	1.98	139,200	153,000	81,400	90,300	2.42	5.50	1,030	1,120	0.54	0.79
31.8	1.250	2.95	0.116	2.82	25.9	2.09	147,200	161,900	86,200	95,800	2.56	5.35	1,090	1,180	0.53	0.79
31.8	1.250	3.18	0.125	3.00	25.4	2.24	157,000	173,000	91,700	102,000	2.71	5.21	1,130	1,240	0.51	0.79
31.8	1.250	3.40	0.134	3.25	24.9	2.38	167,200	183,700	99,300	110,300	2.91	5.01	1,200	1,330	0.49	0.79
31.8	1.250	3.68	0.145	3.51	24.4	2.55	179,300	197,000	106,200	117,900	3.11	4.81	1,260	1,410	0.47	0.79
31.8	1.250	3.96	0.156	3.76	23.8	2.72	190,800	209,900	113,800	126,200	3.31	4.61	1,320	1,480	0.45	0.79
31.8	1.250	4.45	0.175	4.24	22.9	2.99	210,400	231,300	126,900	140,700	3.67	4.25	1,420	1,620	0.41	0.79
38.1	1.500	2.41	0.095	2.29	33.3	2.12	149,000	164,100	59,300	65,500	2.57	8.83	1,380	1,470	0.87	1.14
38.1	1.500	2.59	0.102	2.46	32.9	2.27	159,200	175,300	63,400	70,300	2.76	8.64	1,470	1,570	0.85	1.14
38.1	1.500	2.77	0.109	2.64	32.6	2.41	169,500	186,400	68,300	75,800	2.94	8.46	1,550	1,660	0.83	1.14
38.1	1.500	2.95	0.116	2.82	32.2	2.55	179,300	197,500	72,400	80,700	3.12	8.28	1,640	1,760	0.81	1.14
38.1	1.500	3.18	0.125	3.00	31.8	2.73	192,200	211,300	77,200	85,500	3.31	8.10	1,710	1,850	0.79	1.14
38.1	1.500	3.40	0.134	3.25	31.3	2.91	204,600	225,100	83,400	92,400	3.56	7.84	1,820	1,980	0.77	1.14
38.1	1.500	3.68	0.145	3.51	30.7	3.13	219,700	241,500	89,600	99,300	3.81	7.59	1,920	2,100	0.74	1.14
38.1	1.500	3.96	0.156	3.76	30.2	3.34	234,400	258,000	95,800	106,200	4.06	7.35	2,020	2,230	0.72	1.14
38.1	1.500	4.45	0.175	4.24	29.2	3.69	259,300	285,000	106,900	119,300	4.51	6.89	2,190	2,440	0.67	1.14
38.1	1.500	4.83	0.190	4.57	28.4	3.96	278,400	306,000	115,100	127,600	4.82	6.59	2,300	2,590	0.64	1.14

44.5	1.750	2.77	0.109	2.64	38.9	2.85	200,200	220,200	58,600	64,800	3.47	12.05	2,180	2,310	1.19	1.55
44.5	1.750	2.95	0.116	2.82	38.6	3.02	211,700	233,100	62,100	69,000	3.69	11.83	2,300	2,450	1.17	1.55
44.5	1.750	3.18	0.125	3.00	38.1	3.23	227,300	250,000	66,200	73,800	3.90	11.61	2,410	2,580	1.14	1.55
44.5	1.750	3.40	0.134	3.25	37.6	3.45	242,000	266,400	71,700	79,300	4.21	11.31	2,570	2,760	1.11	1.55
44.5	1.750	3.68	0.145	3.51	37.1	3.70	260,200	286,000	77,200	85,500	4.51	11.01	2,730	2,950	1.08	1.55
44.5	1.750	3.96	0.156	3.76	36.5	3.96	278,000	305,600	82,700	91,700	4.81	10.71	2,870	3,120	1.05	1.55
44.5	1.750	4.45	0.175	4.24	35.6	4.39	308,200	338,900	92,400	102,700	5.36	10.16	3,140	3,440	0.99	1.55
44.5	1.750	4.83	0.190	4.57	34.8	4.72	331,400	364,300	99,300	110,300	5.73	9.79	3,300	3,650	0.95	1.55
44.5	1.750	5.18	0.204	4.95	34.1	5.02	352,700	387,900	106,900	119,300	6.15	9.37	3,490	3,880	0.91	1.55
50.8	2.000	2.77	0.109	2.64	45.3	3.28	230,400	253,500	51,000	57,200	4.00	16.27	2,910	3,070	1.61	2.03
50.8	2.000	2.95	0.116	2.82	44.9	3.48	244,200	268,700	54,500	60,700	4.25	16.02	3,080	3,250	1.58	2.03
50.8	2.000	3.18	0.125	3.00	44.5	3.73	262,000	288,200	57,900	64,800	4.50	15.77	3,240	3,430	1.55	2.03
50.8	2.000	3.40	0.134	3.25	44.0	3.98	279,300	307,400	62,700	69,600	4.86	15.41	3,460	3,680	1.52	2.03
50.8	2.000	3.68	0.145	3.51	43.4	4.28	300,900	330,900	67,600	75,200	5.21	15.06	3,670	3,930	1.48	2.03
50.8	2.000	3.96	0.156	3.76	42.9	4.58	321,600	353,600	72,400	80,700	5.56	14.71	3,880	4,170	1.44	2.03
50.8	2.000	4.45	0.175	4.24	41.9	5.08	357,200	392,800	81,400	90,300	6.20	14.06	4,250	4,610	1.38	2.03
50.8	2.000	4.83	0.190	4.57	41.1	5.47	384,300	423,000	87,600	97,200	6.64	13.63	4,490	4,900	1.33	2.03
50.8	2.000	5.18	0.204	4.95	40.4	5.83	409,700	450,600	94,500	104,800	7.13	13.13	4,750	5,230	1.28	2.03
60.3	2.375	3.18	0.125	3.00	54.0	4.47	314,500	346,100	49,000	54,500	5.40	23.18	4,690	4,930	2.29	2.86
60.3	2.375	3.40	0.134	3.25	53.5	4.78	335,800	369,200	53,100	59,300	5.83	22.75	5,030	5,300	2.25	2.86
60.3	2.375	3.68	0.145	3.51	53.0	5.14	361,600	397,700	57,200	63,400	6.26	22.32	5,350	5,670	2.20	2.86
60.3	2.375	3.96	0.156	3.76	52.4	5.51	387,000	425,700	61,400	68,300	6.68	21.90	5,660	6,020	2.16	2.86
60.3	2.375	4.45	0.175	4.24	51.4	6.13	430,600	473,300	69,000	76,500	7.47	21.11	6,240	6,680	2.08	2.86
60.3	2.375	4.83	0.190	4.57	50.7	6.61	463,900	510,600	74,500	82,700	8.01	20.57	6,610	7,120	2.02	2.86
60.3	2.375	5.18	0.204	4.95	50.0	7.05	495,100	544,400	80,000	88,900	8.62	19.97	7,030	7,610	1.96	2.86
60.3	2.375	5.69	0.224	5.44	48.9	7.67	538,500	592,500	87,600	97,600	9.37	19.21	7,520	8,210	1.88	2.86

Test pressure value equals 90% of internal yield pressure rating. Maximum working pressure is a function of tube condition and is determined by user.

*Available as continuously milled tubing (CM^{TM}) or conventional butt-welded tubing sections (WTM). All data is for new tubing at minimum strength. Other sizes and wall thicknesses available on request. See individual size sheets for additional wall thicknesses.

Table 4.18.14 HS90 Grade USC Units (courtesy of Precision Tube Technology, Inc.)

OD Specified (mm)	OD (mm)	Wall Specified	Wall (mm)	Wall Minimum	ID Calculated	Nominal Weight lb./ft.	TUBE LOAD BODY (Lbs.) Yield Minimum	Tensile Minimum	INTERNAL PRESSURE (psi) Hydro Test 90%	Internal Yield Min.	TUBING AREA (sq. in.) w/min. Wall	Internal Min.	TORSIONAL YIELD (ft.-lb.) Yield	Ultimate	INTERNAL CAPACITY per 1,000 ft. Gallons	Barrels	EXTERNAL DISPLACEMENT per 1,000 ft. Gallons	Barrels
1.000	25.4	0.087	2.21	0.083	0.826	0.850	22,500	24,200	13,200	14,700	0.239	0.546	439	476	27.84	0.66	40.80	0.97
1.000	25.4	0.095	2.41	0.090	0.810	0.920	24,300	26,200	14,300	15,900	0.257	0.528	466	508	26.77	0.64	40.80	0.97
1.000	25.4	0.102	2.59	0.097	0.796	0.981	25,900	27,900	15,300	17,000	0.275	0.510	491	540	25.85	0.62	40.80	0.97
1.000	25.4	0.109	2.77	0.104	0.782	1.040	27,500	29,600	16,400	18,200	0.293	0.493	515	570	24.95	0.59	40.80	0.97
1.250	31.8	0.087	2.21	0.083	1.076	1.083	28,600	30,800	10,600	11,800	0.304	0.923	721	770	47.24	1.12	63.75	1.52
1.250	31.8	0.095	2.41	0.090	1.060	1.175	31,000	33,400	11,500	12,800	0.328	0.899	769	825	45.84	1.09	63.75	1.52
1.250	31.8	0.102	2.59	0.097	1.046	1.254	33,100	35,700	12,400	13,800	0.351	0.876	814	879	44.64	1.06	63.75	1.52
1.250	31.8	0.109	2.77	0.104	1.032	1.332	35,200	37,900	13,200	14,700	0.374	0.853	858	931	43.45	1.03	63.75	1.52
1.250	31.8	0.116	2.95	0.111	1.018	1.408	37,200	40,100	14,100	15,700	0.397	0.830	900	982	42.28	1.01	63.75	1.52
1.250	31.8	0.125	3.18	0.118	1.000	1.506	39,800	42,900	14,900	16,600	0.420	0.808	941	1,032	40.80	0.97	63.75	1.52
1.250	31.8	0.134	3.40	0.128	0.982	1.601	42,300	45,600	16,200	17,900	0.451	0.776	996	1,100	39.34	0.94	63.75	1.52
1.250	31.8	0.145	3.68	0.138	0.960	1.715	45,300	48,800	17,300	19,300	0.482	0.745	1,048	1,166	37.60	0.90	63.75	1.52
1.250	31.8	0.156	3.96	0.148	0.938	1.827	48,300	52,000	18,500	20,600	0.512	0.715	1,097	1,229	35.90	0.85	63.75	1.52
1.250	31.8	0.175	4.45	0.167	0.900	2.014	53,200	57,300	20,700	23,000	0.568	0.659	1,181	1,342	33.05	0.79	63.75	1.52
1.500	38.1	0.095	2.41	0.090	1.310	1.429	37,700	40,700	9,600	10,700	0.399	1.368	1,148	1,218	70.02	1.67	91.80	2.19
1.500	38.1	0.102	2.59	0.097	1.296	1.527	40,300	43,500	10,400	11,500	0.428	1.340	1,220	1,300	68.53	1.63	91.80	2.19
1.500	38.1	0.109	2.77	0.104	1.282	1.623	42,900	46,200	11,100	12,300	0.456	1.311	1,289	1,380	67.06	1.60	91.80	2.19
1.500	38.1	0.116	2.95	0.111	1.268	1.719	45,400	48,900	11,800	13,100	0.484	1.283	1,357	1,459	65.60	1.56	91.80	2.19
1.500	38.1	0.125	3.18	0.118	1.250	1.840	48,600	52,400	12,500	13,900	0.512	1.255	1,422	1,536	63.75	1.52	91.80	2.19
1.500	38.1	0.134	3.40	0.128	1.232	1.960	51,800	55,800	13,600	15,100	0.552	1.215	1,511	1,643	61.93	1.47	91.80	2.19
1.500	38.1	0.145	3.68	0.138	1.210	2.104	55,600	59,900	14,600	16,200	0.590	1.177	1,596	1,746	59.74	1.42	91.80	2.19
1.500	38.1	0.156	3.96	0.148	1.188	2.245	59,300	63,900	15,600	17,300	0.629	1.139	1,677	1,846	57.58	1.37	91.80	2.19
1.500	38.1	0.175	4.45	0.167	1.150	2.483	65,600	70,700	17,500	19,400	0.699	1.068	1,821	2,028	53.96	1.28	91.80	2.19
1.500	38.1	0.190	4.83	0.180	1.120	2.665	70,400	75,800	18,700	20,800	0.746	1.021	1,911	2,145	51.18	1.22	91.80	2.19

1.750	44.5	**0.109**	2.77	0.104	1.532	1.915	50,600	54,500	9,500	10,600	0.538	1.867	1.809	1,918	95.76	2.28	124.95	2.97
1.750	44.5	**0.116**	2.95	0.111	1.518	2.029	53,600	57,800	10,200	11,300	0.572	1.834	1.907	2,030	94.02	2.24	124.95	2.97
1.750	44.5	**0.125**	3.18	0.118	1.500	2.175	57,400	61,900	10,800	12,000	0.605	1.800	2.003	2,140	91.80	2.19	124.95	2.97
1.750	44.5	**0.134**	3.40	0.128	1.482	2.318	61,200	66,000	11,700	13,000	0.652	1.753	2.135	2,294	89.61	2.13	124.95	2.97
1.750	44.5	**0.145**	3.68	0.138	1.460	2.492	65,800	70,900	12,600	14,000	0.699	1.706	2.262	2,444	86.97	2.07	124.95	2.97
1.750	44.5	**0.156**	3.96	0.148	1.438	2.662	70,300	75,800	13,500	14,900	0.745	1.660	2.384	2,589	84.37	2.01	124.95	2.97
1.750	44.5	**0.175**	4.45	0.167	1.400	2.951	77,900	84,000	15,100	16,800	0.831	1.575	2.602	2,855	79.97	1.90	124.95	2.97
1.750	44.5	**0.190**	4.83	0.180	1.370	3.173	83,800	90,300	16,200	18,000	0.888	1.517	2.741	3,029	76.58	1.82	124.95	2.97
1.750	44.5	**0.204**	5.18	0.195	1.342	3.377	89,200	96,100	17,500	19,400	0.953	1.453	2.893	3,222	73.48	1.75	124.95	2.97
2.000	50.8	**0.109**	2.77	0.104	1.782	2.207	58,300	62,800	8,400	9,300	0.619	2.522	2.416	2,544	129.56	3.08	163.20	3.89
2.000	50.8	**0.116**	2.95	0.111	1.768	2.340	61,800	66,600	8,900	9,900	0.659	2.483	2.552	2,696	127.53	3.04	163.20	3.89
2.000	50.8	**0.125**	3.18	0.118	1.750	2.509	66,300	71,400	9,500	10,500	0.698	2.444	2.684	2,845	124.95	2.97	163.20	3.89
2.000	50.8	**0.134**	3.40	0.128	1.732	2.677	70,700	76,200	10,300	11,400	0.753	2.389	2.867	3,054	122.39	2.91	163.20	3.89
2.000	50.8	**0.145**	3.68	0.138	1.710	2.880	76,100	82,000	11,000	12,300	0.807	2.334	3.045	3,258	119.30	2.84	163.20	3.89
2.000	50.8	**0.156**	3.96	0.148	1.688	3.080	81,300	87,700	11,800	13,100	0.861	2.280	3.216	3,458	116.25	2.77	163.20	3.89
2.000	50.8	**0.175**	4.45	0.167	1.650	3.419	90,300	97,300	13,300	14,800	0.962	2.180	3.525	3,825	111.08	2.64	163.20	3.89
2.000	50.8	**0.190**	4.83	0.180	1.620	3.682	97,200	104,800	14,300	15,900	1.029	2.112	3.724	4,066	107.08	2.55	163.20	3.89
2.000	50.8	**0.204**	5.18	0.195	1.592	3.923	103,600	111,600	15,400	17,100	1.106	2.036	3.943	4,335	103.41	2.46	163.20	3.89
2.375	60.3	**0.125**	3.18	0.118	2.125	3.011	79,500	85,700	8,000	8,900	0.837	3.593	3.894	4,090	184.24	4.39	230.14	5.48
2.375	60.3	**0.134**	3.40	0.128	2.107	3.215	84,900	91,500	8,700	9,600	0.904	3.527	4.170	4,398	181.13	4.31	230.14	5.48
2.375	60.3	**0.145**	3.68	0.138	2.085	3.462	91,400	98,500	9,300	10,400	0.970	3.460	4.438	4,700	177.37	4.22	230.14	5.48
2.375	60.3	**0.156**	3.96	0.148	2.063	3.706	97,900	105,500	10,000	11,100	1.035	3.395	4.699	4,997	173.64	4.13	230.14	5.48
2.375	60.3	**0.175**	4.45	0.167	2.025	4.122	108,900	117,300	11,200	12,500	1.158	3.272	5.175	5,545	167.31	3.98	230.14	5.48
2.375	60.3	**0.190**	4.83	0.180	1.995	4.445	117,400	126,500	12,100	13,400	1.241	3.189	5.485	5,908	162.38	3.87	230.14	5.48
2.375	60.3	**0.204**	5.18	0.195	1.967	4.742	125,200	135,000	13,100	14,500	1.335	3.095	5.829	6,316	157.86	3.76	230.14	5.48

Test pressure value equals 90% of internal yield pressure rating. Maximum working pressure is a function of tube condition and is determined by user.

*Available as continuously milled tubing (CM™) or conventional butt-welded tubing sections (W™). All data is for new tubing at minimum strength. Other sizes and wall thicknesses available on request. See individual size sheets for additional wall thicknesses.

Table 4.18.15 HS90 Grade Metric Units (courtesy of Precision Tube Technology, Inc.)

DIMENSIONS (mm)						Nominal	TUBE LOAD BODY (Newtons)		INTERNAL PRESSURE (kPa)		TUBING AREA (sq. cm)		TORSIONAL YIELD (N-m)		INTERNAL CAPACITY per meter	EXTERNAL DISPLACEMENT per meter
OD Specified	OD Specified (inches)	Wall Specified	Wall Specified (inches)	Wall Minimum	ID Calculated	Weight Kg/m	Yield Minimum	Tensile Minimum	Hydro Test 90%	Internal Yield Min.	w/min. Wall	Internal Min.	Yield Min.	Yield Ultimate	Liters	Liters
25.4	1.000	2.21	0.087	2.11	21.0	1.26	100,100	107,600	91,000	101,400	1.54	3.52	590	650	0.35	0.51
25.4	1.000	2.41	0.095	2.29	20.6	1.37	108,100	116,500	98,600	109,600	1.66	3.41	630	690	0.33	0.51
25.4	1.000	2.59	0.102	2.46	20.2	1.46	115,200	124,100	105,500	117,200	1.78	3.29	670	730	0.32	0.51
25.4	1.000	2.77	0.109	2.64	19.9	1.55	122,300	131,700	113,100	125,500	1.89	3.18	700	770	0.31	0.51
31.8	1.250	2.21	0.087	2.11	27.3	1.61	127,200	137,000	73,100	81,400	1.96	5.95	980	1,040	0.59	0.79
31.8	1.250	2.41	0.095	2.29	26.9	1.75	137,900	148,600	79,300	88,300	2.12	5.80	1,040	1,120	0.57	0.79
31.8	1.250	2.59	0.102	2.46	26.6	1.86	147,200	158,800	85,500	95,200	2.27	5.65	1,100	1,190	0.55	0.79
31.8	1.250	2.77	0.109	2.64	26.2	1.98	156,600	168,600	91,000	101,400	2.42	5.50	1,160	1,260	0.54	0.79
31.8	1.250	2.95	0.116	2.82	25.9	2.09	165,500	178,400	97,200	108,300	2.56	5.35	1,220	1,330	0.53	0.79
31.8	1.250	3.18	0.125	3.00	25.4	2.24	177,000	190,800	102,700	114,500	2.71	5.21	1,280	1,400	0.51	0.79
31.8	1.250	3.40	0.134	3.25	24.9	2.38	188,200	202,800	111,700	123,400	2.91	5.01	1,350	1,490	0.49	0.79
31.8	1.250	3.68	0.145	3.51	24.4	2.55	201,500	217,100	119,300	133,100	3.11	4.81	1,420	1,580	0.47	0.79
31.8	1.250	3.96	0.156	3.76	23.8	2.72	214,800	231,300	127,600	142,000	3.31	4.61	1,490	1,670	0.45	0.79
31.8	1.250	4.45	0.175	4.24	22.9	2.99	236,600	254,900	142,700	158,600	3.67	4.25	1,600	1,820	0.41	0.79
38.1	1.500	2.41	0.095	2.29	33.3	2.12	167,700	181,000	66,200	73,800	2.57	8.83	1,560	1,650	0.87	1.14
38.1	1.500	2.59	0.102	2.46	32.9	2.27	179,300	193,500	71,700	79,300	2.76	8.64	1,650	1,760	0.85	1.14
38.1	1.500	2.77	0.109	2.64	32.6	2.41	190,800	205,500	76,500	84,800	2.94	8.46	1,750	1,870	0.83	1.14
38.1	1.500	2.95	0.116	2.82	32.2	2.55	201,900	217,500	81,400	90,300	3.12	8.28	1,840	1,980	0.81	1.14
38.1	1.500	3.18	0.125	3.00	31.8	2.73	216,200	233,100	86,200	95,800	3.31	8.10	1,930	2,080	0.79	1.14
38.1	1.500	3.40	0.134	3.25	31.3	2.91	230,400	248,200	93,800	104,100	3.56	7.84	2,050	2,230	0.77	1.14
38.1	1.500	3.68	0.145	3.51	30.7	3.13	247,300	266,400	100,700	111,700	3.81	7.59	2,160	2,370	0.74	1.14
38.1	1.500	3.96	0.156	3.76	30.2	3.34	263,800	284,200	107,600	119,300	4.06	7.35	2,270	2,500	0.72	1.14
38.1	1.500	4.45	0.175	4.24	29.2	3.69	291,800	314,500	120,700	133,800	4.51	6.89	2,470	2,750	0.67	1.14
38.1	1.500	4.83	0.190	4.57	28.4	3.96	313,100	337,200	128,900	143,400	4.82	6.59	2,590	2,910	0.64	1.14

44.5	1.750	**2.77**	0.109	2.64	38.9	2.85	225,100	242,400	65,500	73,100	3.47	12.05	2,450	2,600	1.19	1.55
44.5	1.750	**2.95**	0.116	2.82	38.6	3.02	238,400	257,100	70,300	77,900	3.69	11.83	2,590	2,750	1.17	1.55
44.5	1.750	**3.18**	0.125	3.00	38.1	3.23	255,300	275,300	74,500	82,700	3.90	11.61	2,720	2,900	1.14	1.55
44.5	1.750	**3.40**	0.134	3.25	37.6	3.45	272,200	293,600	80,700	89,600	4.21	11.31	2,890	3,110	1.11	1.55
44.5	1.750	**3.68**	0.145	3.51	37.1	3.70	292,700	315,400	86,900	96,500	4.51	11.01	3,070	3,310	1.08	1.55
44.5	1.750	**3.96**	0.156	3.76	36.5	3.96	312,700	337,200	93,100	102,700	4.81	10.71	3,230	3,510	1.05	1.55
44.5	1.750	**4.45**	0.175	4.24	35.6	4.39	346,500	373,600	104,100	115,800	5.36	10.16	3,530	3,870	0.99	1.55
44.5	1.750	**4.83**	0.190	4.57	34.8	4.72	372,700	401,700	111,700	124,100	5.73	9.79	3,720	4,110	0.95	1.55
44.5	1.750	**5.18**	0.204	4.95	34.1	5.02	396,800	427,500	120,700	133,800	6.15	9.37	3,920	4,370	0.91	1.55
50.8	2.000	**2.77**	0.109	2.64	45.3	3.28	259,300	279,300	57,900	64,100	4.00	16.27	3,280	3,450	1.61	2.03
50.8	2.000	**2.95**	0.116	2.82	44.9	3.48	274,900	296,200	61,400	68,300	4.25	16.02	3,460	3,660	1.58	2.03
50.8	2.000	**3.18**	0.125	3.00	44.5	3.73	294,900	317,600	65,500	72,400	4.50	15.77	3,640	3,860	1.55	2.03
50.8	2.000	**3.40**	0.134	3.25	44.0	3.98	314,500	338,900	71,000	78,600	4.86	15.41	3,890	4,140	1.52	2.03
50.8	2.000	**3.68**	0.145	3.51	43.4	4.28	338,500	364,700	75,800	84,800	5.21	15.06	4,130	4,420	1.48	2.03
50.8	2.000	**3.96**	0.156	3.76	42.9	4.58	361,600	390,100	81,400	90,300	5.56	14.71	4,360	4,690	1.44	2.03
50.8	2.000	**4.45**	0.175	4.24	41.9	5.08	401,700	432,800	91,700	102,000	6.20	14.06	4,780	5,190	1.38	2.03
50.8	2.000	**4.83**	0.190	4.57	41.1	5.47	432,300	466,200	98,600	109,600	6.64	13.63	5,050	5,510	1.33	2.03
50.8	2.000	**5.18**	0.204	4.95	40.4	5.83	460,800	496,400	106,200	117,900	7.13	13.13	5,350	5,880	1.28	2.03
60.3	2.375	**3.18**	0.125	3.00	54.0	4.47	353,600	381,200	55,200	61,400	5.40	23.18	5,280	5,550	2.29	2.86
60.3	2.375	**3.40**	0.134	3.25	53.5	4.78	377,600	407,000	60,000	66,200	5.83	22.75	5,650	5,960	2.25	2.86
60.3	2.375	**3.68**	0.145	3.51	53.0	5.14	406,500	438,100	64,100	71,700	6.26	22.32	6,020	6,370	2.20	2.86
60.3	2.375	**3.96**	0.156	3.76	52.4	5.51	435,500	469,300	69,000	76,500	6.68	21.90	6,370	6,780	2.16	2.86
60.3	2.375	**4.45**	0.175	4.24	51.4	6.13	484,400	521,800	77,200	86,200	7.47	21.11	7,020	7,520	2.08	2.86
60.3	2.375	**4.83**	0.190	4.57	50.7	6.61	522,200	562,700	83,400	92,400	8.01	20.57	7,440	8,010	2.02	2.86
60.3	2.375	**5.18**	0.204	4.95	50.0	7.05	556,900	600,500	90,300	100,000	8.62	19.97	7,900	8,560	1.96	2.86

Test pressure value equals 90% of internal yield pressure rating. Maximum working pressure is a function of tube condition and is determined by user.

* Available as continuously milled tubing (CM™) or conventional butt-welded tubing sections (W™). All data is for new tubing at minimum strength. Other sizes and wall thicknesses available on request. See individual size sheets for additional wall thicknesses.

Table 4.18.16 *HS110 Grade USC Units (courtesy of Precision Tube Technology, Inc.)*

DIMENSIONS (Inches)						Nominal	TUBE LOAD BODY (Lbs.)		INTERNAL PRESSURE (psi)		TUBING AREA (sq. in.)		TORSIONAL YIELD (ft.-lb.)		INTERNAL CAPACITY per 1,000 ft.		EXTERNAL DISPLACEMENT per 1,000 ft.	
OD Specified	OD (mm)	Wall Specified	Wall (mm)	Wall Minimum	ID Calculated	Weight lb./ft.	Yield Minimum	Tensile Minimum	Hydro Test 90%	Internal Yield Min	w/min. Wall	Internal Min	Yield	Ultimate	Gallons	Barrels	Gallons	Barrels
1.000	25.4	0.109	2.77	0.104	0.782	1.040	33,000	35,100	19,700	21,900	0.293	0.493	618	684	24.95	0.59	40.80	0.97
1.250	31.8	0.109	2.77	0.104	1.032	1.332	42,200	44,900	15,900	17,700	0.374	0.853	1,030	1,117	43.45	1.03	63.75	1.52
1.250	31.8	0.116	2.95	0.111	1.018	1.408	44,600	47,500	16,900	18,800	0.397	0.830	1,081	1,178	42.28	1.01	63.75	1.52
1.250	31.8	0.125	3.18	0.118	1.000	1.506	47,700	50,800	17,900	19,900	0.420	0.808	1,129	1,238	40.80	0.97	63.75	1.52
1.250	31.8	0.134	3.40	0.128	0.982	1.601	50,700	54,000	19,400	21,500	0.451	0.776	1,195	1,320	39.34	0.94	63.75	1.52
1.250	31.8	0.145	3.68	0.138	0.960	1.715	54,400	57,900	20,800	23,100	0.482	0.745	1,257	1,399	37.60	0.90	63.75	1.52
1.250	31.8	0.156	3.96	0.148	0.938	1.827	57,900	61,700	22,200	24,700	0.512	0.715	1,316	1,475	35.90	0.85	63.75	1.52
1.250	31.8	0.175	4.45	0.167	0.900	2.014	63,800	68,000	24,800	27,600	0.568	0.659	1,417	1,610	33.05	0.79	63.75	1.52
1.500	38.1	0.109	2.77	0.104	1.282	1.623	51,400	54,800	13,300	14,800	0.456	1.311	1,547	1,656	67.06	1.60	91.80	2.19
1.500	38.1	0.116	2.95	0.111	1.268	1.719	54,500	58,000	14,200	15,800	0.484	1.283	1,628	1,751	65.60	1.56	91.80	2.19
1.500	38.1	0.125	3.18	0.118	1.250	1.840	58,300	62,100	15,100	16,700	0.512	1.255	1,706	1,843	63.75	1.52	91.80	2.19
1.500	38.1	0.134	3.40	0.128	1.232	1.960	62,100	66,100	16,300	18,100	0.552	1.215	1,813	1,971	61.93	1.47	91.80	2.19
1.500	38.1	0.145	3.68	0.138	1.210	2.104	66,700	71,000	17,500	19,400	0.590	1.177	1,916	2,095	59.74	1.42	91.80	2.19
1.500	38.1	0.156	3.96	0.148	1.188	2.245	71,100	75,700	18,700	20,800	0.629	1.139	2,013	2,216	57.58	1.37	91.80	2.19
1.500	38.1	0.175	4.45	0.167	1.150	2.483	78,700	83,800	21,000	23,300	0.699	1.068	2,185	2,433	53.96	1.28	91.80	2.19
1.500	38.1	0.190	4.83	0.180	1.120	2.665	84,400	89,900	22,500	25,000	0.746	1.021	2,293	2,574	51.18	1.22	91.80	2.19
1.750	44.5	0.109	2.77	0.104	1.532	1.915	60,700	64,600	11,400	12,700	0.538	1.867	2,170	2,301	95.76	2.28	124.95	2.97
1.750	44.5	0.116	2.95	0.111	1.518	2.029	64,300	68,500	12,200	13,600	0.572	1.834	2,288	2,436	94.02	2.24	124.95	2.97
1.750	44.5	0.125	3.18	0.118	1.500	2.175	68,900	73,400	13,000	14,400	0.605	1.800	2,403	2,568	91.80	2.19	124.95	2.97
1.750	44.5	0.134	3.40	0.128	1.482	2.318	73,500	78,200	14,000	15,600	0.652	1.753	2,562	2,753	89.61	2.13	124.95	2.97
1.750	44.5	0.145	3.68	0.138	1.460	2.492	79,000	84,100	15,100	16,800	0.699	1.706	2,714	2,932	86.97	2.07	124.95	2.97
1.750	44.5	0.156	3.96	0.148	1.438	2.662	84,400	89,800	16,100	17,900	0.745	1.660	2,860	3,107	84.37	2.01	124.95	2.97
1.750	44.5	0.175	4.45	0.167	1.400	2.951	93,500	99,600	18,100	20,100	0.831	1.575	3,122	3,426	79.97	1.90	124.95	2.97
1.750	44.5	0.190	4.83	0.180	1.370	3.173	100,600	107,100	19,500	21,600	0.888	1.517	3,290	3,635	76.58	1.82	124.95	2.97

2.000	50.8	0.109	2.77	0.104	1.782	2.207	69,900	74,500	10,000	11,200	0.619	2.522	2,900	3,053	129.56	3.08	163.20	3.89
2.000	50.8	0.116	2.95	0.111	1.768	2.340	74,200	79,000	10,700	11,900	0.659	2.483	3,062	3,235	127.53	3.04	163.20	3.89
2.000	50.8	0.125	3.18	0.118	1.750	2.509	79,500	84,700	11,400	12,600	0.698	2.444	3,221	3,414	124.95	2.97	163.20	3.89
2.000	50.8	0.134	3.40	0.128	1.732	2.677	84,800	90,300	12,300	13,700	0.753	2.389	3,441	3,665	122.39	2.91	163.20	3.89
2.000	50.8	0.145	3.68	0.138	1.710	2.880	91,300	97,200	13,300	14,700	0.807	2.334	3,653	3,910	119.30	2.84	163.20	3.89
2.000	50.8	0.156	3.96	0.148	1.688	3.080	97,600	103,900	14,200	15,800	0.861	2.280	3,859	4,150	116.25	2.77	163.20	3.89
2.000	50.8	0.175	4.45	0.167	1.650	3.419	108,400	115,400	15,900	17,700	0.962	2.180	4,230	4,590	111.08	2.64	163.20	3.89
2.000	50.8	0.190	4.83	0.180	1.620	3.682	116,700	124,200	17,100	19,000	1.029	2.112	4,469	4,879	107.08	2.55	163.20	3.89
2.375	60.3	0.125	3.18	0.118	2.125	3.011	95,400	101,600	9,600	10,700	0.837	3.593	4,672	4,908	184.24	4.39	230.14	5.48
2.375	60.3	0.134	3.40	0.128	2.107	3.215	101,900	108,500	10,400	11,600	0.904	3.527	5,004	5,277	181.13	4.31	230.14	5.48
2.375	60.3	0.145	3.68	0.138	2.085	3.462	109,700	116,800	11,200	12,400	0.970	3.460	5,326	5,640	177.37	4.22	230.14	5.48
2.375	60.3	0.156	3.96	0.148	2.063	3.706	117,500	125,100	12,000	13,300	1.035	3.395	5,639	5,996	173.64	4.13	230.14	5.48
2.375	60.3	0.175	4.45	0.167	2.025	4.122	130,600	139,100	13,500	15,000	1.158	3.272	6,210	6,654	167.31	3.98	230.14	5.48
2.375	60.3	0.190	4.83	0.180	1.995	4.445	140,900	150,000	14,500	16,100	1.241	3.189	6,582	7,090	162.38	3.87	230.14	5.48
2.625	66.7	0.134	3.40	0.128	2.357	3.574	113,300	120,600	9,400	10,500	1.004	4.408	6,209	6,516	226.66	5.40	281.14	6.69
2.625	66.7	0.145	3.68	0.138	2.335	3.850	122,000	129,900	10,100	11,300	1.078	4.334	6,617	6,970	222.45	5.30	281.14	6.69
2.625	66.7	0.156	3.96	0.148	2.313	4.124	130,700	139,200	10,900	12,100	1.152	4.260	7,014	7,416	218.28	5.20	281.14	6.69
2.625	66.7	0.175	4.45	0.167	2.275	4.590	145,500	154,900	12,200	13,600	1.290	4.122	7,742	8,243	211.17	5.03	281.14	6.69
2.625	66.7	0.190	4.83	0.180	2.245	4.953	157,000	167,100	13,200	14,600	1.383	4.029	8,220	8,793	205.63	4.90	281.14	6.69
2.875	73.0	0.156	3.96	0.148	2.563	4.541	143,900	153,200	9,900	11,000	1.268	5.224	8,541	8,987	268.01	6.38	337.24	8.03
2.875	73.0	0.175	4.45	0.167	2.525	5.059	160,300	170,700	11,200	12,400	1.421	5.071	9,445	10,002	260.12	6.19	337.24	8.03
2.875	73.0	0.190	4.83	0.180	2.495	5.462	173,100	184,300	12,100	13,400	1.524	4.968	10,041	10,680	253.98	6.05	337.24	8.03
3.500	88.9	0.175	4.45	0.167	3.150	6.230	197,400	210,200	9,200	10,200	1.749	7.872	14,447	15,146	404.84	9.64	499.80	11.90
3.500	88.9	0.190	4.83	0.180	3.120	6.733	213,400	227,200	9,900	11,000	1.877	7.744	15,397	16,200	397.16	9.46	499.80	11.90

Test pressure value equals 90% of internal yield pressure rating. Maximum working pressure is a function of tube condition and is determined by user.

*Available as continuously milled tubing (CM^{TM}) or conventional butt-welded tubing sections (W^{TM}). All data is for new tubing at minimum strength. Other sizes and wall thicknesses available on request. See individual size sheets for additional wall thicknesses.

Table 4.18.17 HS110 Grade Metric Units (courtesy of Precision Tube Technology, Inc.)

OD Specified (mm)	OD Specified (inches)	Wall Specified (mm)	Wall Specified (inches)	Wall Minimum (mm)	ID Calculated (mm)	Nominal Weight Kg/m	Tube Load Body Yield Minimum (N)	Tube Load Body Tensile Minimum (N)	Internal Pressure Hydro Test 90% (kPa)	Internal Pressure Yield Min (kPa)	Tubing Area w/min. Wall (sq. cm)	Tubing Area Internal Min. (sq. cm)	Torsional Yield (N-m)	Torsional Ultimate (N-m)	Internal Capacity per meter (Liters)	External Displacement per meter (Liters)
25.4	1.000	2.77	0.109	2.64	19.9	1.55	146,800	156,100	135,800	151,000	1.89	3.18	840	930	0.31	0.51
31.8	1.250	2.77	0.109	2.64	26.2	1.98	187,700	199,700	109,600	122,000	2.42	5.50	1,400	1,510	0.54	0.79
31.8	1.250	2.95	0.116	2.82	25.9	2.09	198,400	211,300	116,500	129,600	2.56	5.35	1,470	1,600	0.53	0.79
31.8	1.250	3.18	0.125	3.00	25.4	2.24	212,200	226,000	123,400	137,200	2.71	5.21	1,530	1,680	0.51	0.79
31.8	1.250	3.40	0.134	3.25	24.9	2.38	225,500	240,200	133,800	148,200	2.91	5.01	1,620	1,790	0.49	0.79
31.8	1.250	3.68	0.145	3.51	24.4	2.55	242,000	257,500	143,400	159,300	3.11	4.81	1,700	1,900	0.47	0.79
31.8	1.250	3.96	0.156	3.76	23.8	2.72	257,500	274,400	153,100	170,300	3.31	4.61	1,780	2,000	0.45	0.79
31.8	1.250	4.45	0.175	4.24	22.9	2.99	283,800	302,500	171,000	190,300	3.67	4.25	1,920	2,180	0.41	0.79
38.1	1.500	2.77	0.109	2.64	32.6	2.41	228,600	243,800	91,700	102,000	2.94	8.46	2,100	2,250	0.83	1.14
38.1	1.500	2.95	0.116	2.82	32.2	2.55	242,400	258,000	97,900	108,900	3.12	8.28	2,210	2,370	0.81	1.14
38.1	1.500	3.18	0.125	3.00	31.8	2.73	259,300	276,200	104,100	115,100	3.31	8.10	2,310	2,500	0.79	1.14
38.1	1.500	3.40	0.134	3.25	31.3	2.91	276,200	294,000	112,400	124,800	3.56	7.84	2,460	2,670	0.77	1.14
38.1	1.500	3.68	0.145	3.51	30.7	3.13	296,700	315,800	120,700	133,800	3.81	7.59	2,600	2,840	0.74	1.14
38.1	1.500	3.96	0.156	3.76	30.2	3.34	316,300	336,700	128,900	143,400	4.06	7.35	2,730	3,000	0.72	1.14
38.1	1.500	4.45	0.175	4.24	29.2	3.69	350,100	372,700	144,800	160,700	4.51	6.89	2,960	3,300	0.67	1.14
38.1	1.500	4.83	0.190	4.57	28.4	3.96	375,400	399,900	155,100	172,400	4.82	6.59	3,110	3,490	0.64	1.14
44.5	1.750	2.77	0.109	2.64	38.9	2.85	270,000	287,300	78,600	87,600	3.47	12.05	2,940	3,120	1.19	1.55
44.5	1.750	2.95	0.116	2.82	38.6	3.02	286,000	304,700	84,100	93,800	3.69	11.83	3,100	3,300	1.17	1.55
44.5	1.750	3.18	0.125	3.00	38.1	3.23	306,500	326,500	89,600	99,300	3.90	11.61	3,260	3,480	1.14	1.55
44.5	1.750	3.40	0.134	3.25	37.6	3.45	326,900	347,800	96,500	107,600	4.21	11.31	3,470	3,730	1.11	1.55
44.5	1.750	3.68	0.145	3.51	37.1	3.70	351,400	374,100	104,100	115,800	4.51	11.01	3,680	3,980	1.08	1.55
44.5	1.750	3.96	0.156	3.76	36.5	3.96	375,400	399,400	111,000	123,400	4.81	10.71	3,880	4,210	1.05	1.55
44.5	1.750	4.45	0.175	4.24	35.6	4.39	415,900	443,000	124,800	138,600	5.36	10.16	4,230	4,650	0.99	1.55
44.5	1.750	4.83	0.190	4.57	34.8	4.72	447,500	476,400	134,500	148,900	5.73	9.79	4,460	4,930	0.95	1.55

44.5	1.750	2.77	2.64	38.9	2.85	0.109	270,000	287,300	78,600	87,600	3.47	12.05	2,940	3,120	1.19	1.55
44.5	1.750	2.95	2.82	38.6	3.02	0.116	286,000	304,700	84,100	93,800	3.69	11.83	3,100	3,300	1.17	1.55
44.5	1.750	3.18	3.00	38.1	3.23	0.125	306,500	326,500	89,600	99,300	3.90	11.61	3,260	3,480	1.14	1.55
44.5	1.750	3.40	3.25	37.6	3.45	0.134	326,900	347,800	96,500	107,600	4.21	11.31	3,470	3,730	1.11	1.55
44.5	1.750	3.68	3.51	37.1	3.70	0.145	351,400	374,100	104,100	115,800	4.51	11.01	3,680	3,980	1.08	1.55
44.5	1.750	3.96	3.76	36.5	3.96	0.156	375,400	399,400	111,000	123,400	4.81	10.71	3,880	4,210	1.05	1.55
44.5	1.750	4.45	4.24	35.6	4.39	0.175	415,900	443,000	124,800	138,600	5.36	10.16	4,230	4,650	0.99	1.55
44.5	1.750	4.83	4.57	34.8	4.72	0.190	447,500	476,400	134,500	148,900	5.73	9.79	4,460	4,930	0.95	1.55
50.8	2.000	2.77	2.64	45.3	3.28	0.109	310,900	331,400	69,000	77,200	4.00	16.27	3,930	4,140	1.61	2.03
50.8	2.000	2.95	2.82	44.9	3.48	0.116	330,900	351,400	73,800	82,100	4.25	16.02	4,150	4,390	1.58	2.03
50.8	2.000	3.18	3.00	44.5	3.73	0.125	353,600	376,700	78,600	86,900	4.50	15.77	4,370	4,630	1.55	2.03
50.8	2.000	3.40	3.25	44.0	3.98	0.134	377,200	401,700	84,800	94,500	4.86	15.41	4,670	4,970	1.52	2.03
50.8	2.000	3.68	3.51	43.4	4.28	0.145	406,100	432,300	91,700	101,400	5.21	15.06	4,950	5,300	1.48	2.03
50.8	2.000	3.96	3.76	42.9	4.58	0.156	434,100	462,100	97,900	108,900	5.56	14.71	5,230	5,630	1.44	2.03
50.8	2.000	4.45	4.24	41.9	5.08	0.175	482,200	513,300	109,600	122,000	6.20	14.06	5,740	6,220	1.38	2.03
50.8	2.000	4.83	4.57	41.1	5.47	0.190	519,100	552,400	117,900	131,000	6.64	13.63	6,060	6,620	1.33	2.03
60.3	2.375	3.18	3.00	54.0	4.47	0.125	424,300	451,900	66,200	73,800	5.40	23.18	6,340	6,650	2.29	2.86
60.3	2.375	3.40	3.25	53.5	4.78	0.134	453,300	482,600	71,700	80,000	5.83	22.75	6,790	7,160	2.25	2.86
60.3	2.375	3.68	3.51	53.0	5.14	0.145	487,900	519,500	77,200	85,500	6.26	22.32	7,220	7,650	2.20	2.86
60.3	2.375	3.96	3.76	52.4	5.51	0.156	522,600	556,400	82,700	91,700	6.68	21.90	7,650	8,130	2.16	2.86
60.3	2.375	4.45	4.24	51.4	6.13	0.175	582,900	618,700	93,100	103,400	7.47	21.11	8,420	9,020	2.08	2.86
60.3	2.375	4.83	4.57	50.7	6.61	0.190	626,700	667,200	100,000	111,000	8.01	20.57	8,930	9,610	2.02	2.86
66.7	2.625	3.40	3.25	59.9	5.31	0.134	504,000	536,400	64,800	72,400	6.48	28.44	8,420	8,840	2.81	3.49
66.7	2.625	3.68	3.51	59.3	5.72	0.145	542,700	577,800	69,600	77,900	6.96	27.96	8,970	9,450	2.76	3.49
66.7	2.625	3.96	3.76	58.8	6.13	0.156	581,400	619,200	75,200	83,400	7.43	27.49	9,510	10,060	2.71	3.49
66.7	2.625	4.45	4.24	57.8	6.82	0.175	647,200	689,000	84,100	93,800	8.32	26.60	10,500	11,180	2.62	3.49
66.7	2.625	4.83	4.57	57.0	7.36	0.190	698,300	743,300	91,000	100,700	8.92	26.00	11,150	11,920	2.55	3.49
73.0	2.875	3.96	3.76	65.1	6.75	0.156	640,100	681,400	68,300	75,800	8.18	33.70	11,580	12,190	3.33	4.19
73.0	2.875	4.45	4.24	64.1	7.52	0.175	713,000	759,300	77,200	85,500	9.17	32.72	12,810	13,560	3.23	4.19
73.0	2.875	4.83	4.57	63.4	8.12	0.190	769,900	819,800	83,400	92,400	9.83	32.05	13,620	14,480	3.15	4.19
89.9	3.500	4.45	4.24	80.0	9.26	0.175	878,000	935,000	63,400	70,300	11.28	50.79	19,590	20,540	5.03	6.12
89.9	3.500	4.45	4.57	79.2	10.01	0.175	949,200	1,010,600	68,300	75,800	12.11	49.96	20,880	21,970	4.93	6.21

Test pressure value equals 90% of internal yield pressure rating. Maximum working pressure is a function of tube condition and is determined by user.

*Available as continuously milled tubing (CM™) or conventional butt-welded tubing sections (W™). All data is for new tubing at minimum strength. Other sizes and wall thicknesses available on request. See individual size sheets for additional wall thicknesses.

4.19 CORROSION IN DRILLING AND WELL COMPLETIONS

Corrosion is a naturally occurring phenomenon where materials interact with the environment resulting in chemical change. This chemical change is a degradation in structure with a few exceptions such as batteries. When this corrosion reaction with metal continues unabated, the final condition is the complete conversion of the metal into metal salts.

With the exception of noble metal such as gold and silver, metals do not normally exist in their free form in nature. The metals are tied up with other elements to form oxides, chlorides, sulfides etc. This can be explained thermodynamically since the metals are moving towards a lower energy state. When we refine metals from ores, we move the metal to a higher energy state; therefore, we must add energy (typically in the form of heat or electricity). When our refined metals react with the environment to form metal salts or compounds, thereby moving to a lower energy state, this is the process of corrosion. This chapter includes a primer into the science of corrosion with a discussion of thermodynamics and electrochemistry.

Though the subject of corrosion is important to all facets of our society, the subject here is the corrosion problems encountered in drilling and well completions. The solutions to these corrosion problems have created a knowledge base from which we can draw on to avoid the many potential pitfalls in drilling and well completion operations.

A wide variety of engineered tools and structures are used in the oil and gas industry. Drilling tools include drill pipe, tool joints, blowout preventers (BOPs), drilling jars, drill collars, stabilizers, and milling/rock bits and cutters. Well completion tools and products include casing, cementing tools, and liner hangers. Each of these tools has different engineering requirements that equates to different metals and conditions. These metals and their condition exhibit varying susceptibilities to the various potential corrosion reactions prevalent in the oilfield drilling and completion environments.

In addition to exposure to atmospheric, coastal marine, and sea environments, the oil and gas industry encounters additional corrosive constituents in oil and gas drilling and production. In the course of drilling and completing wells, we introduce acids and other corrosive agents that further complicate the application of common engineering materials.

This chapter is structured to provide an introduction to the science of corrosion with an expansion of the subject into the specifics that are applicable to drilling and well completion environments. The common techniques to mitigate the effects of corrosion in these oilfield applications are presented.

4.19.1 Corrosion Science
4.19.1.1 Corrosion Terms
To provide a basis for a discussion of corrosion science, a list of definitions of the various terms used in this chapter is provided herein. The sources for these terms are (1) NACE MR0175 [1] or, when the term is not listed in this standard, (2) the ASM Corrosion Volume of ASM's Metals Handbook [2].

Acid
A chemical substance that yields hydrogen ions (H^+) when dissolved in water.

Activation
The changing of a passive surface of a metal to a chemically active state.

Active Potential
The potential of a corroding material.

Alkaline
(1) Having properties of an alkali. (2) Having a pH greater than 7.

Anion
A negatively charged ion that migrates through the electrolyte toward the anode under the influence of a potential gradient.

Anode
The electrode of an electrolytic cell at which oxidation occurs. Electrons flow away from the anode in the external circuit. It is usually at the electrode that corrosion occurs and metal ions enter solution.

Anodic Reaction
Electrode reaction equivalent to a transfer of positive charge from the electronic to the ionic conductor. An anodic reaction is an oxidation process. A common example in corrosion is $Me \rightarrow Me^{n+} + n e^-$.

Austenite
The face-centered crystalline phase of iron base alloys.

Base
A chemical substance that yields hydroxyl ions (OH^-) when dissolved in water.

Carbon Steel
An alloy of carbon and iron containing up to 2% carbon and up to 1.65% manganese and residual quantities of other elements, except those intentionally added in specific quantities for deoxidation (usually silicon and/or aluminum). Carbon steels used in the petroleum industry usually contain less than 0.8% carbon.

Cathode
The electrode of an electrolyte cell at which reduction is the principal reaction. Electrons flow toward the cathode in the external circuit. Typical cathodic processes are cations taking up electrons and being discharged, oxygen being reduced, and the reduction of an element or group of elements from a higher to a lower valence state.

Cathodic Reaction
Electrode reaction equivalent to transfer of negative charge from the electronic to the ionic conductor. A cathodic reaction is a reduction process. A common example in corrosion is: $Ox + n e^- \rightarrow Red$.

Cation
A positively charged ion that migrates through the electrolyte toward the cathode under the influence of a potential gradient.

Chloride Stress Corrosion Cracking
Failure by cracking under the combined action of tensile stress and corrosion in the presence of chlorides and water.

Corrosion Potential (E_{corr})
The potential of a corroding surface in an electrolyte, relative to reference electrode. Also called the rest potential, open-circuit potential, or freely corroding potential.

Current Density
The current flowing to or from a unit area of an electrode surface.

Electrochemical Cell
An electrochemical system consisting of an anode and a cathode in metallic contact and immersed in an electrolyte. The anode and cathode may be different metals or dissimilar areas on the same metal surface.

Electrode Potential
The potential of an electrode in an electrolyte as measured against a reference electrode. The electrode does not include any resistance losses in potential in either the solution or the external circuit. It represents the reversible work to move a unit charge from the electrode surface through the solution to the reference electrode.

Faraday's Law
(1) The amount of any substance dissolved or deposited in electrolysis is proportional to the total electric charge passed. (2) The amounts of different substances dissolved or deposited by the passage of the same electric charge are proportional to their equivalent weights.

Galvanic Corrosion
Accelerated corrosion of a metal because of an electrical contact with a more noble metal or nonmetallic conductor in a corrosive electrolyte.

Galvanic Couple
A pair of dissimilar conductors, commonly metals, in electrical contact.

Gibbs Free Energy
The thermodynamic function $\Delta G = \Delta H - T\Delta S$, where H is enthalpy, T is absolute temperature, and S is entropy. Also called free energy, free enthalpy, or Gibbs function.

Ion
An atom, or group of atoms, that has gained or lost one or more outer electrons and thus carries an electric charge. Positive ions, or cations, are deficient in outer electrons. Negative ions, or anions, have an excess of outer electrons.

Low Alloy Steel
Steel with a total alloying element content of less than about 5%, but more than specified for carbon steel.

Partial Pressure
Ideally, in a mixture of gases, each component exerts the pressure it would exert if present alone at the same temperature in the total volume occupied by the mixture. The partial pressure of each component is equal to the total pressure multiplied by its mole fraction in the mixture. For an ideal gas, the mole fraction is equal to the volume fraction of the component.

Passivation
(1) A reduction of the anodic reaction of an electrode involved in corrosion. (2) The process in metal corrosion by which metals become passive. (3) The changing of a chemically active surface of a metal to a much less reactive state.

pH
The negative logarithm of the hydrogen-ion activity.

Polarization
(1) A reduction of the anodic reaction rate of an electrode involved in corrosion. (2) The process in metal corrosion by which metals become passive. (3) The changing of a chemically active surface of a metal to a much less reactive state.

Pressure-Containing Parts
Those parts whose failure to function as intended would result in a release of retained fluid to the atmosphere. Examples are valve bodies, bonnets, and stems.

Residual Stress
Stress present in a component free of external forces or thermal gradients.

Stainless Steel
Steel containing 10.5% or more chromium. Other elements may be added to secure special properties.

Standard Cubic Foot of Gas
The quantity of a gas occupying one cubic foot at a pressure of one atmosphere or 0.10133 Mpa abs (14.696 psi) and a temperature of 15°C (59°F).

Stress Corrosion Cracking (SCC)
Cracking of metal produced by the combined action of corrosion and tensile stress (residual or applied).

Sulfide Stress Cracking (SSC)
Brittle failure by cracking under the combined action of tensile stress and corrosion in the presence of water and H_2S.

Tubular Component
A cylindrical component (pipe) having a longitudinal hole that is used in drilling/production operations for conveying fluids.

4.19.1.2 Thermodynamics of Corrosion
The phenomena of corrosion can be explained by thermodynamics. An initial assessment of corrosion potential typically includes an examination of the relevant thermodynamics. The following questions can be answered by thermodynamics: Why does corrosion occur? How can we determine whether corrosion is possible? Can we quantify the likelihood and obtain an estimate of the kinetics of the reaction (or corrosion rate)?

In thermodynamics, the energies associated with formation or dissolution of a chemical compound or metal can be found in an expression of this energy known as Gibbs free energy. Gibbs free energy is the combination of enthalpy (H) and entropy (S) and is expressed as follows:

$$\Delta G = \Delta H - T\Delta S$$

The 'T' in the equation is the absolute temperature. The general rule for a reaction to occur without the addition of energy is that:

$$\Delta G < 0$$

A good discussion of thermodynamics as it is related with predicting the chemical reactions associated with corrosion reaction can be found in the ASM Metals Handbook volume on corrosion [1] or Gelling's text [3].

4.19.1.3 Electrochemistry of Corrosion
Another way to express and quantify this energy and predict corrosion tendencies is in electrochemical terms. In aqueous (water) or other polar solvents that permit the

flow of electrons (conductive), a current will flow during a chemical reaction such as corrosion. During a corrosion process/chemical reaction, two or more reactions take place concurrently with the resultant current flow from a higher potential to a lower potential (or higher energy state to a lower energy state). These reactions are either anodic or cathodic. Anodic reactions are those where electrons are discharged and these electrons are used in the cathodic reaction. This is shown schematically in Figure 4.19.1.

Using electrochemistry, we can obtain predictions of equilibrium data for electrochemical reactions. When reactions are steady state (at equilibrium), it has been shown that we can relate Gibbs Free Energy to an equilibrium constant as shown in the following equation [1].

$$\Delta G^0 = -RT \ln K_{eq}$$

where ΔG^0 is Gibbs free energy at equilibrium
R is the gas constant
T is the absolute temperature
K_{eq} is the equilibrium constant

The potential that occurs under steady state (equilibrium) conditions is related to Gibbs free energy in the following manner [4].

$$\Delta G^{0=} -nFE^0$$

where n is the number of electrons transferred in the reaction
F is the Faraday constant (charge of 1 mole of electrons)
E^0 is the equilibrium cell potential

The standard electrode potentials discussed above are for standard concentrations (1.000 M) and, in a real application environment, the concentrations will vary from the standard concentration. The same equation (Nernst equation) listed below permits calculation of electrode potentials at non-standard concentrations or electrode potentials of chemical reactions when reactants or products vary from the standard concentrations. Similarly, non-equilibrium conditions are related as follows:

$$\Delta G = -nFE$$

where n is the number of electrons transferred in the reaction
F is the Faraday constant (charge of 1 mole of electrons)
E is the cell potential

Combining these equations and solving for the equilibrium constant permits the direct relationship with potential:

$$\ln K_{eq} = -nFE^0/RT$$

This equation permits prediction of equilibrium data for electrochemical reactions.

Corrosion is a process of material degradation via a variety of mechanisms. In this discussion of electrochemistry we find direct application to general or weight-loss type corrosion where the mechanism is electrochemical in that the move is towards a lower energy state. To predict whether corrosion is likely to occur, we can combine the individual reactions in the corrosion process. The associated energy with each reaction is combined with all the reactions in the system to yield a total or combined energy. The sign (positive or negative) and the magnitude provide the signpost by which we can predict the likelihood that corrosion will occur.

Taking the corrosion of iron in water as an example, the anodic and cathodic reactions are expressed as follows:

Anodic: $Fe \leftrightarrows Fe^{+2} + 2e^-$ This is the oxidation reaction.

Cathodic: $2H^+ + 2e^- \leftrightarrows H^2$ This is the reduction reaction.

Using standard electrode potentials, [3], the following illustrates the reaction of iron in water:

$$Fe \leftrightarrows Fe^{+2} + 2e^- \qquad E^0 = -0.44V$$
$$H^2 \leftrightarrows 2H^+ + 2e^- \qquad E^0 = 0.00V$$
$$Fe + 2H^+ \leftrightarrows Fe^{+2} + H^2 \quad E^0 = -0.44V$$

In the previous example, the reaction of iron in oxygen free water results in a combined energy (electrode potential) of -0.44 V. This indicates that iron will corrode because the number is negative and relatively large. In another example, we can similarly look at copper in an oxygen free water environment.

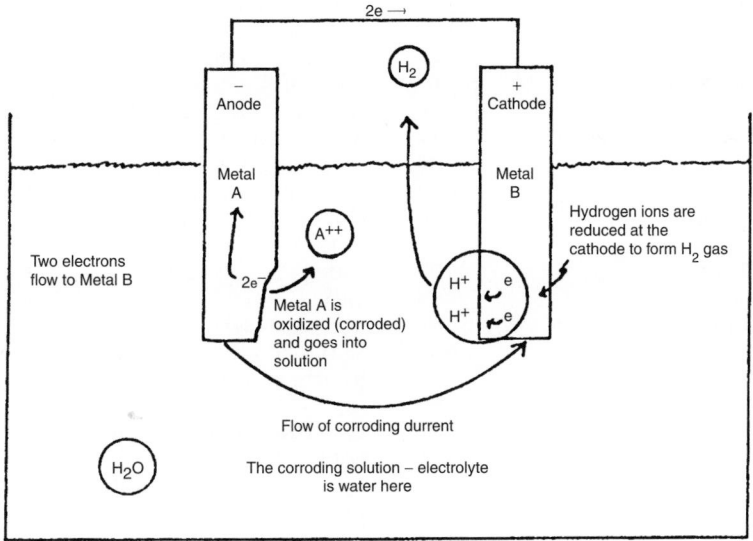

2e $\longrightarrow$

Anode —

Cathode +

H_2

Metal A

Metal B

Two electrons flow to Metal B

A^{++}

$2e^-$

Metal A is oxidized (corroded) and goes into solution

H^+ e

H^+ e

Hydrogen ions are reduced at the cathode to form H_2 gas

Flow of corroding durrent

H_2O

The corroding solution – electrolyte is water here

Figure 4.19.1 *Electrochemical corrosion of metals.*

$$Cu \leftrightarrows Cu^{+2} + 2e^- \qquad E^0 = +0.337V$$
$$H^2 \leftrightarrows 2H^+ + 2e^- \qquad E^0 = 0.00V$$
$$Cu + 2H^+ \leftrightarrows Cu^{+2} + H^2 \quad E^0 = +0.337V$$

This example illustrates the reason why copper is often found free as a metal in nature; the positive number indicates that the reaction will not occur naturally. These examples are in an oxygen free environment, looking at the copper reaction above and introducing oxygen results in the following:

$$Cu \leftrightarrows Cu^{+2} + 2e^- \qquad\qquad E^0 = +0.337V$$
$$\tfrac{1}{2}O_2 + 2H^+ + 2e^- \leftrightarrows H_2O \qquad E^0 = -1.229V$$
$$Cu + \tfrac{1}{2}O_2 + 2H^+ \leftrightarrows Cu^{+2} + H_2O \quad E^0 = -0.892V$$

In an oxygen containing water environment, copper does corrode naturally. From these examples, we know that knowledge of the chemical reaction and the electrode potential yields the probability of a reaction occurring. Table 4.19.1 contains many of these standard electrode potentials [5].

Table 4.19.1 *Standard Reduction Potentials [5]*

Reaction	Standard potential $E°$, V
$H_4 + XeO_6 + 2e^- \rightarrow XeO_3 + 2H_2O$	+3.0
$F_2 + 2e^- \rightarrow 2F^-$	2.87
$O_3 + 2H^+ + 2e^- \rightarrow O_2 + H_2O$	2.07
$H_2O_2 + 2H^+ + 2e^- \rightarrow 2H_2O$	1.776
$Au^+ + e^- \rightarrow Au$	1.691
$MnO_4 + 4H^+ + 3e^- \rightarrow MnO_2 + 2H_2O$	1.695
$2HOCl + 2H^+ + 2e^- \rightarrow Cl_2 + 2H_2O$	1.63
$Mn^{+3} + e^- \rightarrow Mn^{+2}$	1.51
$Au^{+3} + 3e^- \rightarrow Au$	1.498
$MnO_4 + 8H^+ + 5e^- \rightarrow Mn^{+2} + 4H_2O$	1.491
$HOCl + H^+ + 2e^- \rightarrow Cl^- + H_2O$	1.49
$PbO_2 + 4H^+ + 2e^- \rightarrow Pb^{+2} + 2H_2O$	1.455
$ClO_3 + 6H^+ + 6e^- \rightarrow Cl^- + 3H_2O$	1.45
$ClO_4 + 8H^+ + 8e^- \rightarrow Cl^- + 4H_2O$	1.37
$Cl_2 + 2e^- \rightarrow 2Cl^-$	1.359
$ClO_4^- + 8H^+ + 7e^- \rightarrow \tfrac{1}{2}Cl_2 + 4H_2O$	1.34
$Cr_2O_7^{-2} + 14H^+ + 6e^- \rightarrow 2Cr^{+3} + 7H_2O$	1.33
$Au^{+3} + 2e^- \rightarrow Au^+$	1.29
$Tl^{+3} + 2e^- \rightarrow Tl^+$	1.247
$O_3 + H_2O + 2e^- \rightarrow O_2 + 2OH^-$	1.24
$MnO_2 + 4H^+ + 2e^- \rightarrow Mn^{+2} + 2H_2O$	1.23
$O_2 + 4H^+ + 4e^- \rightarrow 2H_2O$	1.229
$MnO_2 + 4H^+ + 2e^- \rightarrow Mn^{+2} + 2H_2O$	1.208
$Pt^{+2} + 2e^- \rightarrow Pt$	1.2
$Cu^{+2} + 2CN^- + e^- \rightarrow Cu(CN)_2^-$	1.12
$Cr^{+8} + 3e^- \rightarrow Cr^{+3}$	1.10
$Br_2 + 2e^- \rightarrow 2Br^-$	1.065
$AuCl^- + 3e^- \rightarrow Au + 4Cl^-$	0.994
$Pd^{+2} + 2e^- \rightarrow Pd$	0.987
$NO_3^- + 4H^+ + 3e^- \rightarrow NO + 2H_2O$	0.96
$2Hg^{+2} + 2e^- \rightarrow Hg_2^{+2}$	0.920
$ClO^- + H_2O + 2e^- \rightarrow Cl^- + 2OH^-$	0.90
$HO_2^- + H_2O + 2e^- \rightarrow 3OH^-$	0.878
$TiO_2 + 4H^+ + 4e^- \rightarrow Ti + 2H_2O$	0.86
$Hg^{+2} + 2e^- \rightarrow Hg$	0.854
$OsO_4 + 8H^+ + 8e^- \rightarrow Os + 4H_2O$	0.85
$Pd^{+2} + 2e^- \rightarrow Pd$	0.83
$\tfrac{1}{2}O_2 + 2H^+ + (10^{-7}M) + 2e^- \rightarrow H_2O$	0.815
$Ag^+ + e^- \rightarrow Ag$	0.799
$Hg_2^{+2} + 2e^- \rightarrow 2Hg$	0.788
$IrCl_8^{-3} + 2e^- \rightarrow Ir + 6Cl^-$	0.77
$Fe^{+3} + e^- \rightarrow Fe^{+2}$	0.771
$NiO_2 + 2H_2O + 2e^- \rightarrow Ni(OH)_2 + 2OH^-$	0.76

Reaction	Standard potential $E°$, V
$Sb^{+5} + 2e^- \rightarrow Sb^{+3}$	0.75
$H_2SeO_3 + 4H^+ + 4e^- \rightarrow Se + 3H_2O$	0.74
$PtCl_4^{-2} + 2e^- \rightarrow Pt + 4Cl^-$	0.73
$O_2 + 2H^+ + 2e^- \rightarrow H_2O_2$	0.682
$Ag_2SO_4 + 2e^- \rightarrow 2Ag + SO_4^{-2}$	0.653
$Te^{+4} + 4e^- \rightarrow Te$	0.63
$PdCl_4^{-2} + 2e^- \rightarrow Pd + 4Cl^-$	0.623
$ClO_3^- + 3H_2O + 6e^- \rightarrow Cl^- + 6OH^-$	0.62
$Hg_2SO_4 + 2e^- \rightarrow 2Hg + SO_4^{-2}$	0.616
$TeO_2 + 4H_2 + 4e^- \rightarrow Te + 2H_2O$	0.593
$Ag_2NO_2 + e^- \rightarrow Ag + NO_2^-$	0.59
$MnO_4^- + 2H_2O + 3e^- \rightarrow MnO_2 + 4OH^-$	0.588
$I_2 + 2e^- \rightarrow 2I^-$	0.536
$Cu^+ + e^- \rightarrow Cu$	0.521
$TeO_4 + 8H^+ + 7e^- \rightarrow Te + 4H_2O$	0.472
$H_2SO_4 + 4H^+ + 4e^- \rightarrow S + 3H_2O$	0.450
$Fe^{+3} + e^- \rightarrow Fe^{+2}$	0.438
$O_2 + 2H_2O + 4e^- \rightarrow 4OH^-$	0.401
$ReO_4^- + 8H^+ + 7e^- \rightarrow Re + 4H_2O$	0.367
$Ag_2O + H_2O + 2e^- \rightarrow 2Ag + 4OH^-$	0.342
$Cu^{+2} + 2e^- \rightarrow Cu$	0.337
$BiO^+ + 2H^+ + 3e^- \rightarrow Bi + H_2O$	0.32
$Re^{+3} + 3e^- \rightarrow Re$	0.3
$Hg_2Cl_2 + 2e^- \rightarrow 2Hg + 2Cl^-$	0.268
$ReO_2 + 4H^+ + 4e^- \rightarrow Re + 2H_2O$	0.26
$As_2O_3 + 6H^+ + 6e^- \rightarrow 2As + 3H_2O$	0.234
$AgCl + e^- \rightarrow Ag + Cl^-$	0.222
$SbO^+ + 2H^+ + 3e^- \rightarrow Sb + 2H_2O$	0.212
$SO_4^{-2} + 4H^+ + 2e^- \rightarrow H_2SO_4 + H_2O$	0.172
$ClO_4^- + H_2O + 2e^- \rightarrow ClO_3^- + 2OH^-$	0.17
$BiCl_4^- + 3e^- \rightarrow Bi + 4Cl^-$	0.168
$Pt(OH)_2 + 2e^- \rightarrow Pt + 2OH^-$	0.16
$Cu^{+2} + e^- \rightarrow Cu^+$	0.158
$Sn^{+4} + 2e^- \rightarrow Sn^{+2}$	0.15
$Sb_2O_3 + 6H^+ + 6e^- \rightarrow 2Sb + 3H_2O$	0.144
$S + 2H^+ + 2e^- \rightarrow H_2S$	0.142
$Hg_2Br_2 + 2e^- \rightarrow 2Hg + 2Br^-$	0.140
$Hg_2O + H_2O + 2e^- \rightarrow 2Hg + 2OH^-$	0.123
$Pd(OH)_2 + 2e^- \rightarrow Pd + 2OH^-$	0.1
$Ir_2O_3 + 3H_2O + 6e^- \rightarrow 2Ir + 6OH^-$	0.1
$HgO + H_2O + 2e^- \rightarrow Hg + 2OH^-$	0.0984
$S_4O_6^{-2} + 2e^- \rightarrow 2S_2O_3^{-2}$	0.09
$AgBr + e^- \rightarrow Ag + Br^-$	0.0713

Table 4.19.1   (continued)

Reaction	Standard potential E°, V	Reaction	Standard potential E°, V
$Sn^{+4}+2e^- \rightarrow Sn^{+2}$	0.070	$2H_2O+2e^- \rightarrow H_2+2OH^-$	−0.828
$Tl_2O_3+3H_2O+4e^- \rightarrow 2Tl^++6OH^-$	0.02	$SiO_2+4H^++4e^- \rightarrow Si+2H_2O$	−0.84
$NO_3^-+H_2O+2e^- \rightarrow NO_2^-+2OH^-$	0.01	$TiO_2+4H^++4e^- \rightarrow Ti+2H_2O$	−0.86
$H_2MoO_4+6H^++6e^- \rightarrow Mo+4H_2O$	0.0	$Fe(OH)_2+2e^- \rightarrow Fe+2OH^-$	−0.877
$2H^++2e^- \rightarrow H_2$	0.000	$SO_4^{-2}+H_2O+2e^- \rightarrow SO_3^{-2}+2OH^-$	−0.92
$AgCN+e^- \rightarrow Ag+CN^-$	−0.02	$Te+2e^- \rightarrow Te^{-2}$	−0.92
$TeO_3^{-2}+3H_2O+4e^- \rightarrow Te+6OH^-$	−0.02	$PO_4^{-3}+2H_2O+2e^- \rightarrow HPO_3^{-2}+3OH^-$	−1.05
$Fe^{+3}+3e^- \rightarrow Fe$	−0.036	$Mn^{+2}+2e^- \rightarrow Mn$	−1.180
$Hg_2I_2+2e^- \rightarrow 2Hg+2I^-$	−0.0405	$CrO_2^-+2H_2O+3e^- \rightarrow Cr+4OH^-$	−1.2
$O_2+H_2O+2e^- \rightarrow HO_2^-+OH^-$	−0.076	$SiF_6^{-2}+4e^- \rightarrow Si+6F^-$	−1.2
$2Cu(OH)_2+2e^- \rightarrow Cu_2O+2OH^-+H_2O$	−0.09	$V^{+2}+2e^- \rightarrow V$	−1.2
$WO_3+6H^++6e^- \rightarrow W+3H_2O$	−0.09	$Zn(OH)_4^{-2}+2e^- \rightarrow Zn+4OH^-$	−1.215
$WO_2+4H^++4e^- \rightarrow W+2H_2O$	−0.12	$H_2GaO_3^-+H_2O+3e^- \rightarrow Ga+4OH^-$	−1.22
$Pb^{+2}+2e^- \rightarrow Pb$	−0.126	$Cr(OH)_3+3e^- \rightarrow Cr+3OH^-$	−1.3
$CrO_4^{-2}+4H_2O+3e^- \rightarrow Cr(OH)_3+5OH^-$	−0.13	$ZrO_2+4H^++4e^- \rightarrow Zr+2H_2O$	−1.43
$Sn^{+2}+2e^- \rightarrow Sn$	−0.136	$Mn(OH)_2+2e^- \rightarrow Mn+2OH^-$	−1.47
$O_2+2H_2O+2e^- \rightarrow H_2O_2+2OH^-$	−0.146	$HfO_2+4H^++4e^- \rightarrow Hf+2H_2O$	−1.57
$AgI+e^- \rightarrow Ag+I^-$	−0.152	$Ti^{-2}+2e^- \rightarrow Ti$	−1.63
$Cu(OH)_2+2e^- \rightarrow Cu+2OH^-$	−0.227	$Al^{+3}+3e^- \rightarrow Al$	−1.622
$Ni^{+2}+2e^- \rightarrow Ni$	−0.250	$HfO^{+2}+2H^++4e^- \rightarrow Hf+2H_2O$	−1.68
$V(OH)_4^++4H^++5e^- \rightarrow V+4H_2O$	−0.25	$HPO_3^{-2}+2H_2O+3e^- \rightarrow P+5OH^-$	−1.71
$Co^{+2}+2e^- \rightarrow CO$	−0.277	$SiO_3+3H_2O+4e^- \rightarrow Si+6OH^-$	−1.73
$Tl^++e^- \rightarrow Tl$	−0.336	$U^{+3}+3e^- \rightarrow U$	−1.79
$In^{+3}+3e^- \rightarrow In$	−0.338	$ThO_2+4H^++4e^- \rightarrow Th+2H_2O$	−1.80
$TlOH+e^- \rightarrow Tl+OH^-$	−0.344	$HPO_2^-+e^- \rightarrow P+2OH^-$	−1.82
$SeO_3^{-2}+3H_2O+4e^- \rightarrow Se+6OH^-$	−0.35	$Be^{+2}+2e^- \rightarrow Be$	−1.847
$PbSO_4+2e^- \rightarrow Pb+SO_4^{-2}$	−0.359	$Th^{+4}+4e^- \rightarrow Th$	−1.90
$Se+2H^++2e^- \rightarrow H_2Se$	−0.36	$Np^{+3}+3e^- \rightarrow Np$	−1.9
$Cu_2O+H_2O+2e^- \rightarrow 2Cu+2OH^-$	−0.361	$Ti^{+3}+e^- \rightarrow Ti^{+2}$	−2.0
$Cd^{+2}+2e^- \rightarrow Cd$	−0.403	$Sc^{+3}+3e^- \rightarrow Sc$	−2.08
$Cr^{+3}+e^- \rightarrow Cr^{+2}$	−0.408	$\frac{1}{2}H_2+e^- \rightarrow H^-$	−2.23
$Fe^{+2}+2e^- \rightarrow Fe$	−0.440	$Nd^{+3}+3e^- \rightarrow Nd$	−2.246
$S+2e^- \rightarrow S^{-2}$ (basic soln.)	−0.447	$Lu^{+3}+3e^- \rightarrow Lu$	−2.26
$Bi_2O_3+3H_2O+6e^- \rightarrow 2Bi+6OH^-$	−0.46	$Be_2O_3^{-2}+3H_2O+4e^- \rightarrow 2Be+6OH^-$	−2.28
$S+H_2O+2e^- \rightarrow HS^-+OH^-$	−0.478	$ZrO(OH)_2+H_2O+4e^- \rightarrow Zr+4OH^-$	−2.32
$H_3PO_3+3H^++3e^- \rightarrow P+3H_2O$	−0.49	$Al(OH)_4+3e^- \rightarrow Al+4OH^-$	−2.33
$Sb+3H^++3e^- \rightarrow H_3Sb$	−0.51	$Mg^{+2}+2e^- \rightarrow Mg$	−2.363
$Ga^{+3}+3e^- \rightarrow Ga$	−0.529	$Y^{+3}+3e^- \rightarrow Y$	−2.37
$As+3H^++3e^- \rightarrow AsH_3$	−0.54	$La^{+3}+3e^- \rightarrow La$	−2.37
$HPbO_2+H_2O+2e^- \rightarrow Pb+3OH^-$	−0.54	$Nd^{+3}+3e^- \rightarrow Nd$	−2.43
$Cr^{+2}+2e^- \rightarrow Cr$	−0.557	$Ce^{+3}+3e^- \rightarrow Ce$	−2.483
$Fe(OH)_3+e^- \rightarrow Fe(OH)_2+OH^-$	−0.56	$H_2BO_3+H_2O+3e^- \rightarrow B+4OH^-$	−2.5
$PbO+H_2O+2e^- \rightarrow Pb+2OH^-$	−0.576	$Ac^{+3}+3e^- \rightarrow Ac$	−2.6
$Nb_2O_5+10H^++10e^- \rightarrow 2Nb+5H_2O$	−0.62	$HfO(OH)_2+H_2O+4e^- \rightarrow Hf+4OH^-$	−2.60
$Ni(OH)_2+2e^- \rightarrow Ni+2OH^-$	−0.66	$ThO_2+2H_2O+4e^- \rightarrow Th+4OH^-$	−2.64
$SbO_2^-+2H_2O+3e^- \rightarrow Sb+4OH^-$	−0.66	$Mg(OH)_2+2e^- \rightarrow Mg+2OH^-$	−2.67
$AsO_2^-+2H_2O+3e^- \rightarrow As+4OH^-$	−0.68	$Na^++e^- \rightarrow Na$	−2.714
$Ag_2S+2e^- \rightarrow 2Ag+S^{-2}$	−0.705	$La(OH)_3+3e^- \rightarrow La+3OH^-$	−2.76
$Ta_2O_5+10H^++10e^- \rightarrow 2Ta+5H_2O$	−0.71	$Ca^{+2}+2e^- \rightarrow Ca$	−2.866
$Co(OH)_2+2e^- \rightarrow Co+2OH^-$	−0.73	$Sr^{+2}+2e^- \rightarrow Sr$	−2.888
$H_3BO_3+3H^-+3e^- \rightarrow B+3H_2O$	−0.73	$Ba^{+2}+2e^- \rightarrow Ba$	−2.906
$Cr^{+3}+3e^- \rightarrow Cr$	−0.744	$Ra^{+2}+2e^- \rightarrow Ra$	−2.916
$Zn^{+2}+2e^- \rightarrow Zn$	−0.763	$Cs^++e^- \rightarrow Cs$	−2.923
$Fe(OH)_3+3e^- \rightarrow Fe+3OH^-$	−0.711	$K^++e^- \rightarrow K$	−2.925
$Se+2e^- \rightarrow Se^{-2}$	−0.78	$Rb^-+e^- \rightarrow Rb$	−2.925
$HSnO_2^-+H_2O+2e^- \rightarrow Sn+3OH^-$	−0.79	$Ba(OH)_2+2e^- \rightarrow Ba+2OH^-$	−2.97
$RuO_2+4H^++4e^- \rightarrow Ru+2H_2O$	−0.8	$Sr(OH)_2+2e^- \rightarrow Sr+2OH^-$	−2.99
$Cd(OH)_2+2e^- \rightarrow Cd+2OH^-$	−0.809	$Ca(OH)_2+2e^- \rightarrow Ca+2OH^-$	−3.02
$ReO_4^-+4H_2O+7e^- \rightarrow Re+8OH^-$	−0.81	$Li^-+e^- \rightarrow Li$	−3.045
$UO_2^{+2}+4H^++7e^- \rightarrow U+2H_2O$	−0.82		

Corrosion rates can be determined experimentally in two ways, first by determining weight loss over a known period of time under specific conditions and second by electrochemical methods. The advantage of the weight loss technique is that the corrosion rate is an average rate over time, the disadvantage is that the effect of natural perturbations to the system are unknown. The advantage of the electrochemical method is that the rate determined is almost instantaneous and the changes in the rate can be measured over time [6].

Fundamentally, Tafel [7] found that the current (I) was equivalent to the rate of a single electrode reaction by the following equation:

$$E = a + b(\log I)$$

where E = potential (volts) measured with respect to a reference electrode

a and b = constants

I = current density of the reaction

The 'b' constant in the equation is known as the Tafel slope. Both anodic and cathodic processes will exhibit a characteristic Tafel constant or slope, often designated b_a and b_c where the 'a' and 'c' subscripts designate anodic and cathodic. To determine the Tafel constants and the corroding (open circuit or unperturbed) current and potential, polarization experiments may be conducted.

As shown earlier, where a natural corrosion process will exhibit at least one anodic and one cathodic electrochemical process, the natural system can be perturbed or polarized in either the anodic or cathodic directions and the effect measured. An example of the resultant relationships from anodic and cathodic polarizations is shown in Figure 4.19.2.

The value of the voltage at the intersection of these Tafel slopes is known as the corrosion potential. The value of the current density at the intersection of these Tafel slopes is known as the corrosion current. From Figure 4.19.2, we can experimentally measure the corrosion potential and the corrosion current density.

The corrosion current density is related to the instantaneous corrosion rate through variations of Faraday's Law, one form of which is presented below [6].

$$\text{Corrosion rate (mils per year)} = (129IM)/(z\rho)$$

where I = corrosion current in ma/cm^2

M = average gram atomic weight

z = number of electrons transferred

ρ = density

In general, corrosion studies, the values obtained from electrochemical measurements can be measured against weight-loss measurements under the same environmental conditions. When a sample of known surface area, density and weight is corroded over a known period of time, the weight loss can be converted into a corrosion rate. One such equation is listed below [8,9].

$$\text{Corrosion rate in mpy (mils per year)} = (534W)/(DAT)$$

where W = weight loss in mg

D = density of metal corroded in g/cm^3

A = area of metal exposed to corrosive environment in square inches

T = exposure time in hours

4.19.1.4 Passivity

In this discussion of corrosion science and electrochemistry, it is necessary to explore the related subject of passivity to explain why materials do not corrode or stop corroding under particular environmental conditions. The preceding electrochemical discussion needs to be carried further to understand this behavior of metals.

The part missing so far in our discussion is the subject of the characteristics of the products of the chemical reactions or corrosion. There are two interrelated attributes that need to be considered that permit a greater understanding of why metals corrode and do not corrode: (1) the solubility

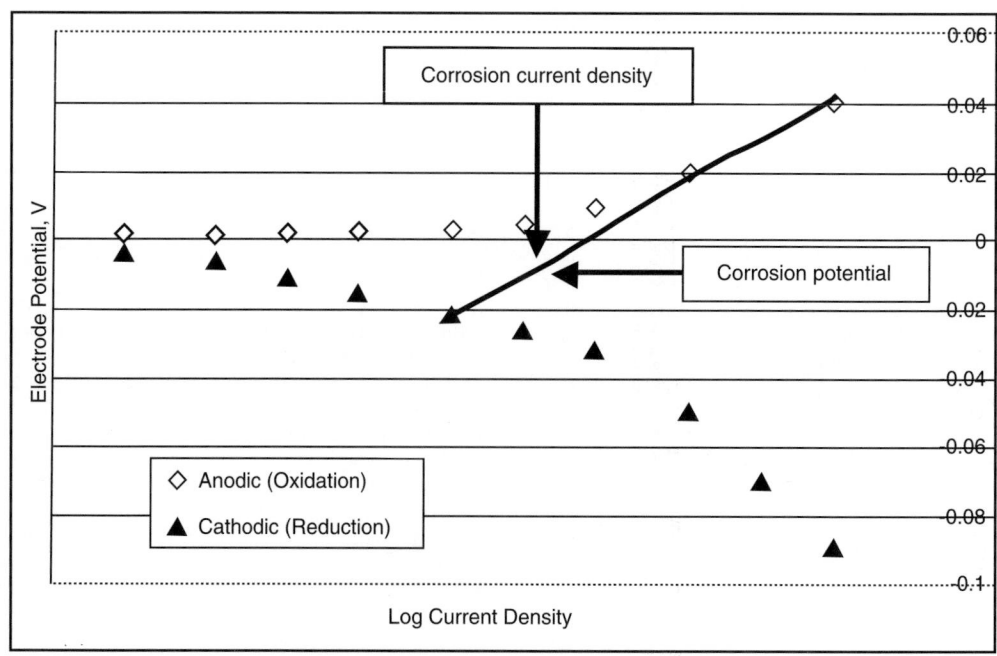

Figure 4.19.2 *Schematic polarization curves.*

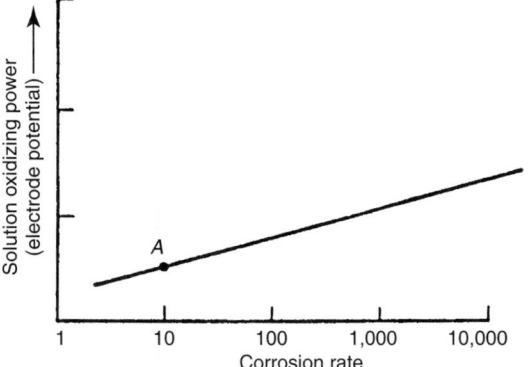

Figure 4.19.3 *Corrosion rate of a metal as a function of solution oxidizing power (electrode potential) [10].*

of the reaction products and (2) the stability of the reaction products.

If, for example, a metal under specific conditions corrodes and produces a stable corrosion product that is insoluble in the environment, then the corrosion product could form a protective film or layer on the metal surface preventing further corrosion by becoming a barrier between the environment and the metal surface. This means that under certain conditions, a normally active metal such as iron, can become inert or perform like a noble metal.

Generally, as the oxidizing power or electrode potential increases, the corrosion rate increases. This relationship is presented in Figure 4.19.3.

This corrosion prediction as illustrated above may not accurately describe the corrosion rates because of the corrosion product solubility and stability. The reality of the corrosion prediction may instead include a potential region that is called passive and reflects the inactive or passive nature of the metal surface. This is demonstrated in Figure 4.19.4.

Note that the x axis is the corrosion rate and these rates can be determined experimentally or via electrochemical

techniques. In electrochemical experiments, the x axis would be experimentally measured as current density as described earlier. The passive region in Figure 4.19.4 is where the corrosion rate decreases as the oxidizing power increases as measured by potential (voltage).

The subject of passivation is not completely understood but there are two commonly accepted theories, *oxide film theory* and the *adsorption theory of passivity*. The oxide film theory relates the passivity to the presence of surface films (primarily oxides) that are diffusion barrier films preventing contact of the metal surface with the environment. The adsorption theory relates the passivity to adsorbed gas that causes the barrier preventing diffusion of the metal ions into the surrounding environment.

One useful tool that assists in defining regions of relative passivity are potential–pH diagrams. These are often commonly referred to as Pourbaix diagrams named after the discoverer of the relationship, Marcel Pourbaix, in 1945 [3]. An example of this type of diagram for iron is presented in Figure 4.19.5.

These diagrams are used to predict whether reactions will occur naturally, the composition of the corrosion products and where these reactions occur and do not occur. We can use the regions where structure-damaging reactions do not occur to prevent corrosion attack; this is accomplished by modifying the pH of the environment or by impressing a potential (cathodic protection).

There have been diagrams produced that describe the corrosion behavior without the fields surface film compositions listed [11]. The general corrosion behavior defined as regions of immunity, passivity and corrosion can be overlaid onto the Potential–pH diagram. Looking at the regions represented in Figure 4.19.5, we obtain a diagram such as that presented in Figure 4.19.6.

From the example in Figures 4.19.5 and 4.19.6, we can readily see the conditions that result in corrosion and those that do not. There is good agreement between the composition of the surface films and the corrosion behavior.

4.19.2 Forms of Corrosion Attack

The degradation of materials may occur from a number of forms of corrosion. The many forms that are encountered

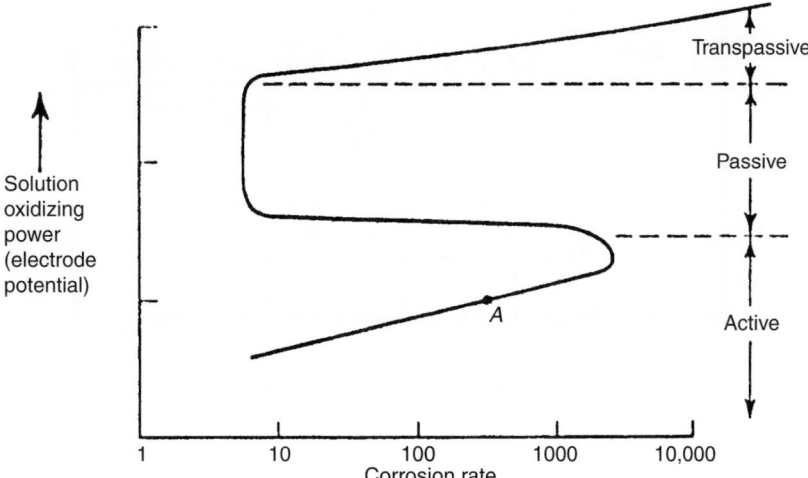

Figure 4.19.4 *Corrosion characteristics of an active passive metal as a function of solution oxidizing power (electrode potential) [10].*

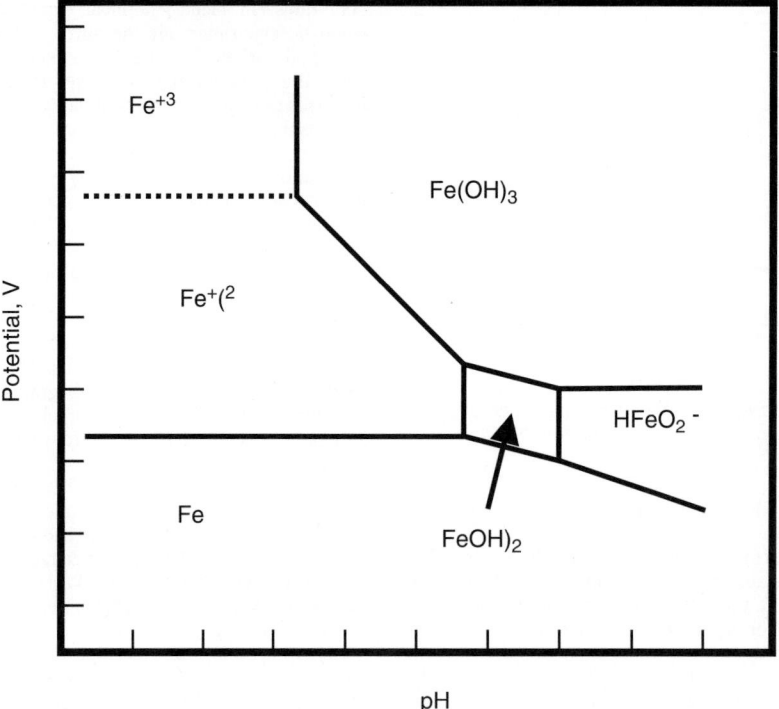

Figure 4.19.5   *Potential – pH diagram for iron.*

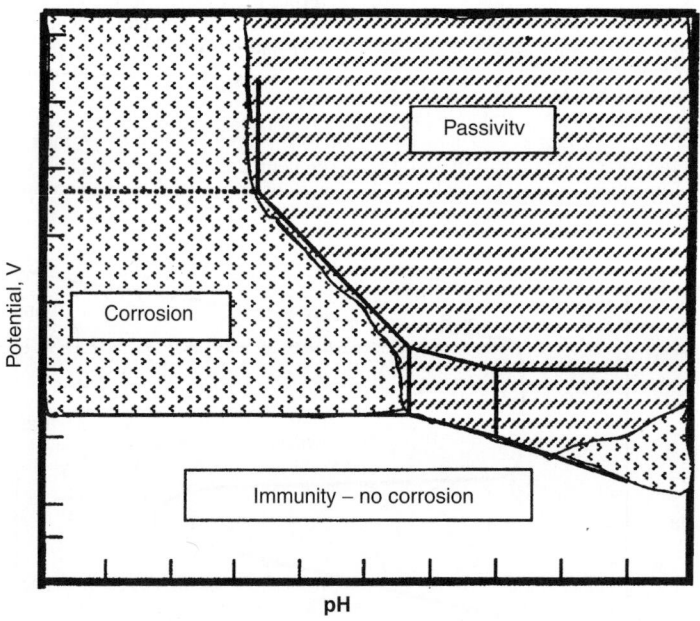

Figure 4.19.6   *Immunity, passivity and corrosion regions overlaid onto potential – pH diagram.*

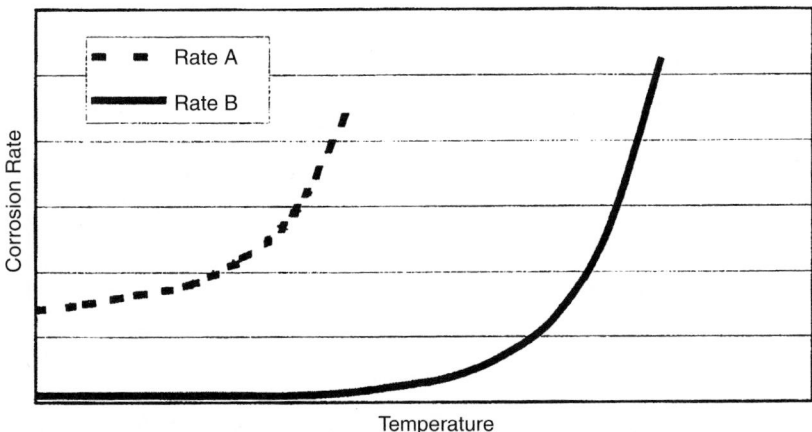

Figure 4.19.7 *Effect of temperature on corrosion rate [12].*

in the oil and gas industry are discussed herein. The forms of attack can be categorized broadly into either a corrosion phenomena, a cracking phenomena or a form of attack that is a combination of a corrosion mechanism and another type or form of metal degradation. Corrosion phenomena include general and localized forms of corrosion. Cracking phenomena is a combination of stress and a corrosion reaction of a susceptible material that results in corrosion cracking. Material degradation may also be caused by the combination of a corrosion mechanism and another type or form of material damage such a fatigue or wear. The combination of these forms of degradation may often be more than additive which indicates that the material damage may be more severe that predicated by only considering one mechanism at a time. These forms of corrosion attack are discussed here.

4.19.2.1 General Corrosion

One of the most common and simplest type of corrosion is general or uniform corrosion. If a material surface degrades in a uniform manner via a chemical reaction without the addition of a non-corrosion mechanism such as wear, then this is general corrosion. This general corrosion is easy to predict using the electrochemical techniques previously discussed and can be easily determined experimentally.

The general corrosion rate as defined by weight loss can be calculated by equations such as the one listed below [8,9].

Corrosion Rate in mpy (mils per year)$=534W/DAT$

where W = weight loss in mg
     D = density of metal corroded in g/cm^3
     A = area of metal exposed to corrosive environment in square inches
     T = exposure time in hours

The most common environmental factors that effect general corrosion are (1) acidity of environment, (2) oxygen and (3) temperature. The temperature is known to have a strong effect on corrosion rates as can be predicted by Faraday's law. The reaction rate with temperature relationship is predicted by the Arhenius relationship where the corrosion rate approximately doubles every 10°C. This doubling (exponential relationship) of the corrosion rate

is explained by the Arhenius Equation [3]. This equation is presented below where k is the reaction rate and T is temperature.

$$k = k^0 e^{(nF/RT)}$$

The relationship between corrosion rate and temperature is illustrated in Figure 4.19.7.

The examples in Figure 4.19.7 are typical for 304 type austenitic stainless steel in sulfuric acid (Rate A) and nitric acid (Rate B). The difference between curves A and B in Figure 4.19.7 was explained by Fontana [12] as being caused by the metal being in the passive state being close to the transpassive region (see Figure 4.19.3) in nitric acid. As the temperature rises, the oxidizing power increases until the stainless becomes transpassive and, once transpassive, the corrosion rate increases exponentially.

The Arhenius relationship is the basis for predictive corrosion rate tests. An example would be weight loss corrosion measurements conducted at lower temperatures being used to predict corrosion rates at higher temperatures. The converse is also true, the life of a structure as predicted by metal loss can be predicted at lower temperature by conducting accelerated corrosion experiments at higher temperatures. This permits the measuring of corrosion rates over days and weeks to predict product or structure life in years. An example of such an experiment to yield predictive corrosion rates is presented in Figure 4.19.8.

Using Figure 4.19.8 as the model, it is possible to have physical measurements of corrosion rates (and, therefore, predicted corrosion loss over a period of time) at high temperatures. For example, the estimated useful life due to general corrosion could be measured to be 120 days, 60 days and 30 days respectively at 70°C, 80°C, and 90°C. The resultant relationship could be used to predict an approximate useful life of 2000 days at 30°C without waiting for 2000 days (almost $5\frac{1}{2}$ years) for test results.

Acidity has a tremendous effect on corrosion in materials. This effect is illustrated below in Figure 4.19.9 for mild steel at two temperatures.

The last factor considered here that is noted for its effect on general corrosion is oxygen, where even minute quantities can have devastating effect on material structures. An example of the effect of oxygen is presented in Figure 4.19.10.

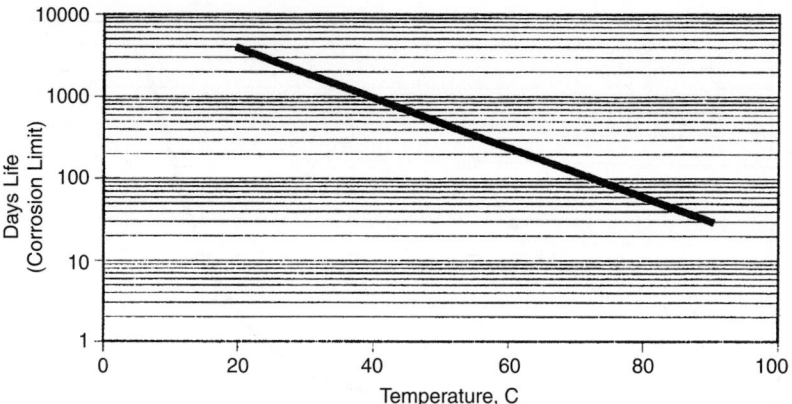

Figure 4.19.8   *Prediction of corrosion limit or life based on idealized corrosion rates.*

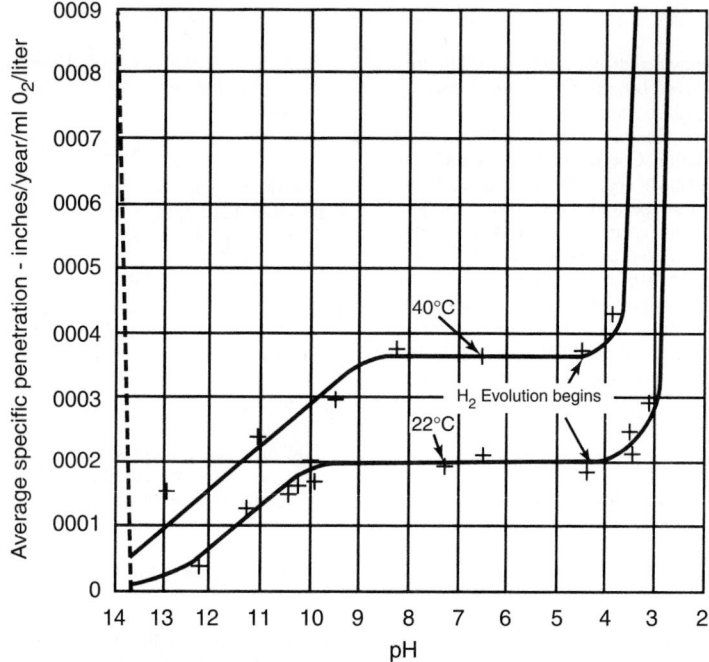

Figure 4.19.9   *Effect of pH on corrosion of mild steel [13].*

Oxygen plays a pivotal role in many forms of corrosion including general corrosion. The primary effect is that a new cathodic reaction is introduced: the oxygen reduction reaction:

$$\frac{1}{2}O_2 + 2H^+ + 2e^- \rightleftarrows H_2O$$

4.19.2.2 Pitting Corrosion

There are several forms of localized corrosion; the first type considered here is pitting corrosion. Pitting is one type of corrosion that is normally manifested in materials that are passive in a particular environment and are, therefore,

considered to be corrosion resistant. Pitting corrosion is a type of localized corrosion that is one of the first that is considered since it limits the applications of many corrosion resistant alloys and is the cause of many product failures. General corrosion, as was previously discussed, is readily understood and predicted. Conversely, pitting corrosion can occur when not expected and can exhibit relatively unpredictable propagation rates.

Pitting corrosion occurs in two stages, pit initiation and pit growth. There have been numerous studies on pitting corrosion [15–17]. The initiation stage has been reported by Gellings [3] to usually be on the order of several months.

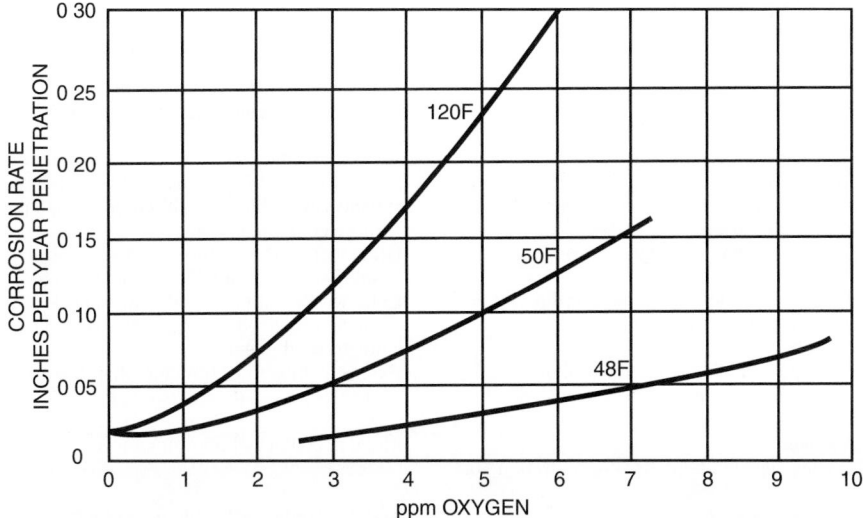

Figure 4.19.10 *Effect of oxygen content on the corrosion rate of steel [14].*

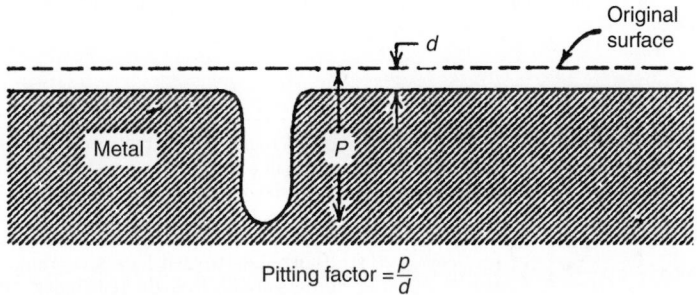

Pitting factor $= \dfrac{p}{d}$

Figure 4.19.11 *Sketch of depest pit with relation to average metal penetration and the pitting factor [9].*

Corrosion pits vary in size, depth and frequency. The shape (size and depth) of pits have been quantified by measurements of the ratio of the depth to the width. A pitting factor is sometimes defined as the ratio of the deepest pit depth to the average pit depth [9], this is presented in the next equation.

$$\text{Pitting factor} = P/d$$

where P = deepest metal penetration
       d = average metal penetration

A pitting factor of 1 would indicate general corrosion where the larger the pitting factor the more severe the pitting corrosion. The application of this pitting factor is demonstrated in Figure 4.19.11.

The insidious problem with pits is that the pit surface could be covered with corrosion products or be as small as a pinhole but with great depth. These features, when pitting occurs in this manner, renders the pitting attack extremely difficult to detect.

From the environmental details, it has been noted that chloride ions and oxidizing agents such as Cu^{+2} and Fe^{+3} salts frequently are associated with pitting. It is also known that hydroxides, chromates and silicates often inhibit the formation of pitting [3].

The material composition and processing history also influence the occurrence and severity of pitting. We know, for example, that austenitic stainless steels are generally known to be susceptible to pitting corrosion. We also know that inhomogeneities such as secondary phases, inclusions, banding, segregation etc., can be the driving factor in pitting initiation. Table 4.19.2 summarizes geometric and material attributes that can lead to localized corrosion.

There have been several relationships developed that relate the composition of iron based corrosion resistant alloys to the relative pitting resistance in the presence of dissolved chlorides and oxygen (salt solutions such as salt water). The PREN relationship that exists in the ISO Standard 15156 [19] is presented below:

$$\text{PREN} = W_{Cr} + 3.3(W_{Mo} + 0.5W_w) + 16W_N$$

where W_{Cr} is the chromium content (% of total composition)
      W_{Mo} is the molybdenum content (% of total composition)
      W_W is the tungsten content (% of total composition)
      W_N is the nitrogen content (% of total composition)

Table 4.19.2 *Geometric Factors and Heterogeneities in Relation to Localized Attack [18]*

System	Metal Area which is Predominantly Anodic*
Metal	
Dissimilar metals in contact.	Metal which is move reactive in a given solution (i.e., metal which has a greater tendency to ionize).
Crevices, deposits on metal surface or any geometrical configuration which result in differences in the concentration of oxygen or other cathodic depolarizers (e.g., CU^{2+}).	Metal in contact with the lower concentration—this follows from considerations of an equivalent reversible cell, although the situation is more complex in practice.
Differences in metallurgical structure	Grain boundaries, more reactive phases (solid solutions, intermetallic compounds, etc.).
Differences in metallurgical condition due to thermal or mechanical treatment.	Cold worked areas anodic to annealed areas, metal subjected to external stress anodic to unstressed metal.
Discontinuities in conducting oxide film or scale or discontinuities in applied metallic or nonmetallic coatings.	Exposed substrate (provide that this is more electrochemically active than the coating).
Environment	
Differences in aeration or concentration of cathodic depolarizers.	Metal in contact with lower concentration.
Differences in velocity.	Metal in contact with solution of higher velocity.
Differences in pH or salt concentration.	Metal in contact with the solution of lower pH or higher salt concentration.

*The table gives a general indication of the area which is likely to be anodic. There are many exceptions, e.g., grain boundaries can be cathodic, the area of metal in contact with a higher salt concentration will be cathodic if the oxygen concentration is higher, etc.

These percentages are based on the mass fraction of chromium in the alloy expressed as a percentage of the total composition. An example of a PREN calculation for a duplex stainless steel such as UNS S399377 with 24.8% nitrogen would yield the following result:

$$PREN = 24.8 + 3.3 \times (2.5 + 0.5 \times 1.1) + 16 \times 0.28$$

$$PREN = 24.8 + 10.1 + 4.5 = 39.4$$

This 39.4 value of the PREN can be used to match up with design criteria or compare with other alloys to rank relative pitting resistance. A note of caution for the PREN relationship presented here is that the relationship was developed for stainless steels and the relationship may not be a good predictor of relative ranking when the alloys being compared includes other types of alloys such as nickel base alloys.

From the above relationship, the higher the PREN number, the higher the resistance to pitting corrosion. Also, we can see the beneficial effect of elements such as chromium, molybdenum, tungsten and nitrogen on pitting corrosion resistance.

To mitigate the effects of pitting corrosion, we can select alloys that exhibit a higher PREN number. Other ways to prevent or minimize pitting attack include modifying the environment (e.g., lowering the pH), increasing the flow velocity (preventing stagnant conditions) and lowering the temperature.

For most metals there is a critical temperature below which pitting does not occur. This temperature is dependent on the material composition and the specific environment. For stainless steels, the critical pitting temperature is related to the PREN number as expressed earlier.

4.19.2.3 Ringworm Corrosion

A type of corrosion that is found in oil and gas production environments that is a form of localized corrosion is ringworm corrosion. This corrosion is related to pitting corrosion and galvanic corrosion. There are localized pits or corrosion attack that occurs circumferentially near the upset of a tubular section such as drill pipe. An example of this type of corrosion is presented in Figure 4.19.12.

Many tubular products such as drill pipe with upset end regions (hot forged) have a region adjacent to the upset that, if not re-heat treated, leaves it in a metallurgical condition that differs greatly from the remainder of the drill pipe structure. This localized condition is more anodic and more susceptible to corrosion attack.

The susceptible metallurgical condition results from thermal gradients of these hot-forming operations. This condition occurs in a narrow band that corrodes preferentially.

4.19.2.4 Crevice Corrosion

Crevice corrosion is a form of localized corrosion attack where the corrosion takes place between adjoining surfaces such as made-up threaded and flanged connections, beneath metal banding and thread protectors, and even beneath scale and corrosion deposits. This type of localized corrosion occurs very similarly to pitting corrosion except that the geometric effects of the crevice influence the initiation and propagation of the attack.

Crevice corrosion is differentiated from pitting corrosion in that the initiation phase is different [20]. Localized variations in the environment cause potential differences and result in local breakdown of passive films and accelerated corrosion. Crevice corrosion has been related to differences in oxygen content commonly known as differential aeration by Gelling [3] and Trethewey and Chamberlain [20]. Oxygen is consumed within the crevice during even low levels of corrosion and the oxygen becomes depleted because the crevice geometry prevents bulk fluid transfer into the crevice (this is a stagnant condition).

In itself, the depleted oxygen does not cause corrosion. Inside the crevice, when even low levels of corrosion occur, there is a corresponding increase in dissolved metal ions and chloride ions (in chloride containing environments).

Figure 4.19.12 *"Ringworm" corrosion—at the inside of the tubing where grain structure near edge of upset portion makes metal susceptible to rapid corrosion [13].*

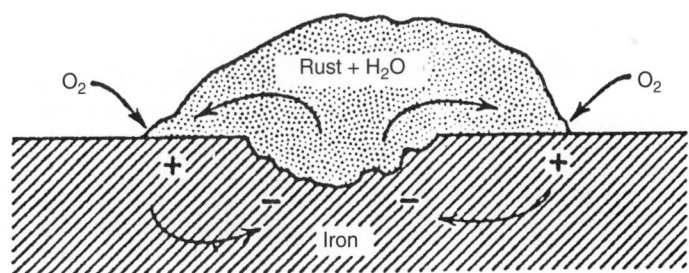

Figure 4.19.13 *Differential aeration cell formed by rust on iron [21].*

These local changes in the environment result in a decrease in the pH locally thus increasing the aggressiveness of the environment with respect of corrosion. These local changes in the environment cause increased corrosion rates and, since the oxygen is depleted, the damaged surface oxide films providing corrosion protection do not reform or are very sluggish to form. This further accelerates the localized corrosion.

Two examples of crevice type corrosion beneath surface scale and in a crevice are presented in Figures 4.19.13 and 4.19.14, respectively.

In figure 4.19.13, there is a crevice and differential aeration cell between the iron rust and the iron basemetal.

Crevice corrosion can be avoided or minimized primarily by changing the geometry (first choice) or by selecting a material that is more resistant to corrosion in the intended environment. The material factors that increase a metal's resistance to pitting corrosion are essentially identical to those factors that increase resistance to crevice corrosion. This means that the equations defined in the discussion on pitting corrosion also apply to crevice corrosion.

The temperature factor in pitting corrosion also has a corollary in crevice corrosion. An increase in temperature results in an increased probability of the occurrence of crevice corrosion. There is also a temperature below which the likelihood of crevice corrosion is very low.

The above discussion infers that the risk of crevice corrosion can be minimized through materials selection for the intended environment. The following, list, adopted from Fontana [12] describes methods and procedures related to design that are used to prevent or minimize crevice

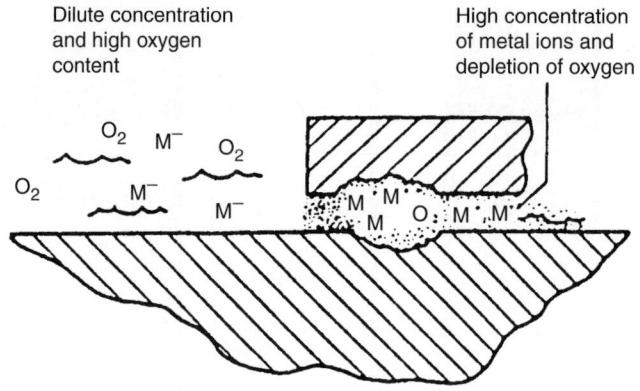

Figure 4.19.14   *Crevice corrosion [9].*

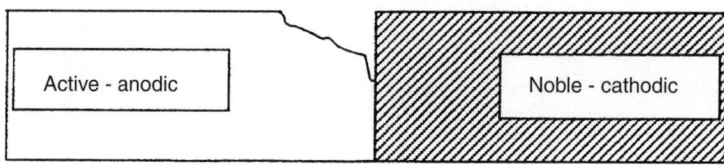

Figure 4.19.15   *Galvanic corrosion.*

corrosion:

1. Use welded butt joints instead of riveted or bolted joints in new equipment. Sound welds and complete penetration are necessary to avoid introducing an additional crevice source.
2. Close crevices by using continuous welds, caulk, braze, solder or paint.
3. Avoid stagnant areas and design for complete drainage.
4. Remove scale and other deposits. Remove solids to prevent deposit formation, a source of crevices.
5. Remove wet packing materials during long shutdowns.
6. Provide as uniform an environment as possible such as using backfill materials in pipeline trenching.
7. Weld instead rolling in tubes in tube sheets.

4.19.2.5 Galvanic Corrosion

Galvanic corrosion occurs as a result of electrical contact between two dissimilar metals in an electrically conductive environment (electrolyte). This is depicted in the schematic presented in Figure 4.19.15.

In Figure 4.19.15, the active metal (anodic) is being corroded starting at the connection or interface with the more noble (cathodic) metal.

The factors affecting galvanic corrosion are related to material and geometry. To predict galvanic corrosion, these two areas need to be examined.

From a materials consideration, the electrochemical (or galvanic) differences between the material being coupled is the characteristic evaluated with the basic benchmark being the relative position on the galvanic scale. The value of the material in question is dependent on the specific environment; for example, this is illustrated in Figure 4.19.16 for the galvanic series in seawater and Table 4.19.3 for the galvanic series in neutral water and soil.

The galvanic series, as can be seen from Figure 4.19.16 in seawater and Table 4.19.3 in neutral water, varies with the environment. The position of a specific material may exhibit a shift in the relative position with respect to other materials from one environment to another. Generally, the greater the galvanic difference between two adjoining materials, the greater the likelihood that galvanic corrosion will occur.

In geometric effects, the three factors are relative surface areas, distance between the different materials and shape effects. The greater the ratio of cathodic (more noble) to anodic (less noble) surface area, the greater the likelihood that galvanic corrosion will occur. The driving force here is the increasing anodic current density with greater ratios resulting in greater polarization. An anodic surface area that is greater than the cathodic area results in small or no accelerated galvanic effects because of the predominant polarization of the more noble material [11].

The closer two dissimilar materials are, the greater the potential for galvanic effects between them. The most severe anticipated galvanic effects are in the less noble material immediately adjacent to the more noble material, this characteristic is depicted in Figure 4.19.15. Another factor that can further reduce potential galvanic effects as it relates to distance is solution conductivity because the current flow through the electrolyte is the relevant feature and decreasing current flow decreases potential galvanic effects.

The geometry enters into the evaluation of the galvanic effects because current flows through the path of least resistance, this means that the current will not readily flow around corners. This and the previously discussed distance relationships illustrate that galvanic effects are adjacent effects and are essentially line of sight.

4.19.2.6 Erosion Corrosion

Erosion corrosion is the combined effect of corrosion under the influence of mechanical action in the form of fluid flow. Passive metals generally rely on corrosion resistance through protective surface films. When these films are

VOLTS: SATURATED CALOMEL HALF-CELL REFERENCE ELECTRODE

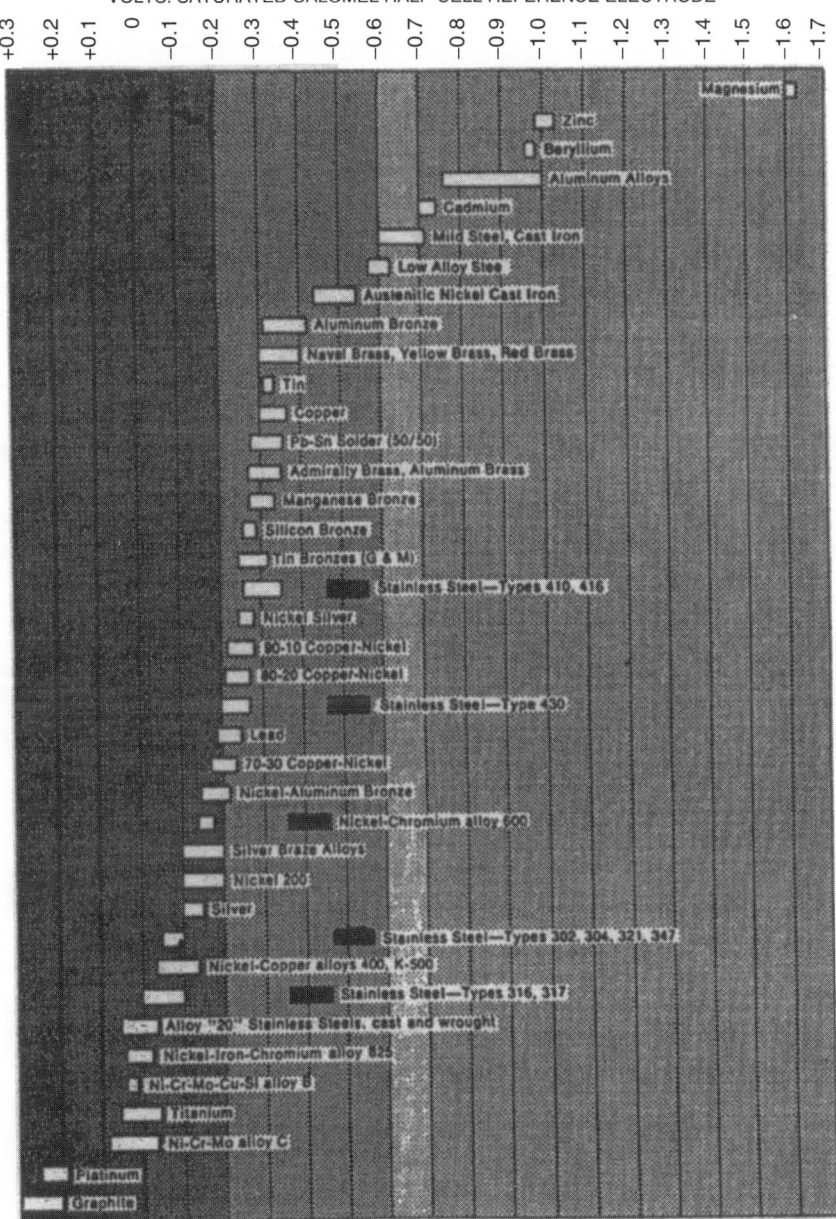

Figure 4.19.16 *Galvanic series in seawater [22].*

disturbed or removed, the surface is active for at least the time required to form a new surface film; the surface could also remain active with corrosion as a result. The continued action of surface film removal through fluid flow results in attack that appears as grooves or waves that exhibit a directional pattern in accordance with the direction of fluid flow. This is illustrated in Figure 4.19.17.

Because erosion corrosion is strongly influenced by fluid flow characteristics, there are critical velocities that are dependent on specific materials and environments below which erosion corrosion does not occur. Corrosion rates of various metals in flowing seawater at three

velocities are presented in Table 4.19.4 to illustrate this phenomenon.

From Table 4.19.4, the effect of flow rate on corrosion rates can be seen. The critical velocity usually occurs at a specific level that is also presented in Figure 4.19.18.

This critical velocity can be estimated by knowing when a specific fluid flow becomes turbulent. The critical velocity can be calculated by the equation presented in API RP 14E [23] which is

$$V_e = \frac{C}{\sqrt{P}}$$

Table 4.19.3 *Galvanic Series in Neutral Water and Soil (after Treseder[22])*

Metals	Volts (note 1)
Commercially pure magnesium	−1.75
Magnesium alloy (6% Al, 3% Zn, 0.15% Mn)	−1.6
Zinc	−1.1
Aluminum alloy (5% Zn)	−1.05
Commercially pure aluminum	−0.8
Mild steel (clean and rust free)	−0.5 to −0.8
Mild steel (rusted)	−0.2 to −0.5
Cast iron (not graphitized)	−0.5
Lead	−0.5
Mild steel in concrete	−0.2
Copper, brass, bronze	−0.2
High silicon cast iron	−0.2
Mill scale on steel	−0.2
Carbon, graphite, coke	+0.3

Note 1: potential values (volts) are in reference to copper/copper sulfate electrode.

The critical (or erosion) velocity is related in the above equation to a constant divided by the square root of the fluid density. Erosion corrosion has been observed on a variety of equipment in flowing fluids including components of pipe and tube (especially in changes of direction), nozzles, production and drilling chokes, valves, pumps, blowers, propellers, turbine blades, compressors, mud motors, and drilling products such as rock bits.

In addition to limiting the application for a material in an environment (velocity limitation), we can also reduce or avoid erosion corrosion by (1) using materials with better corrosion resistance, (2) designing to avoid flow restrictions or abrupt changes in flow direction, (3) minimize protrusions into the flow stream that could change flow rates by introducing turbulent flow, (4) change the environment to reduce the corrosivity, (5) add corrosion inhibitors, (6) modify environment by filtering to remove solids, and (7) use corrosion resistant coatings or hardsurfacings.

4.19.2.7 Cavitation

Cavitation is a type of corrosion damage that is caused by the collapse of vapor bubbles in a liquid on or adjacent to a metal surface. When the vapor bubbles burst, there is energy released in the form of a shock wave that can destroy protective corrosion resistant films on metal surfaces. A schematic of this type of damage is presented in Figure 4.19.19.

Cavitation has been observed in production and drilling chokes, mud motors and drilling mud pumps. Cavitation damage can be minimized or avoided by designing to reduce hydrodynamic pressure differences, coating metals with rubber or plastic and using more corrosion resistant alloys.

4.19.2.8 Fretting Corrosion

Fretting corrosion occurs in contact areas between two surfaces that are subjected to small magnitudes of relative motion in an environment that is corrosive to at least one of the mating surfaces. In a manner very similar to other mechanical interactions with a corrosion mechanism, the surface protective layer is disturbed by the contact causing accelerated localized corrosion. In order for fretting corrosion to occur, the interface must be under stress, relative motion between two surfaces must occur such as in vibrating motion and the stress and vibration must be great enough to deform the surfaces [5]. A schematic of fretting corrosion is presented in Figure 4.19.20.

The mechanism of fretting corrosion involves either the previously noted disruption of the protective surface films

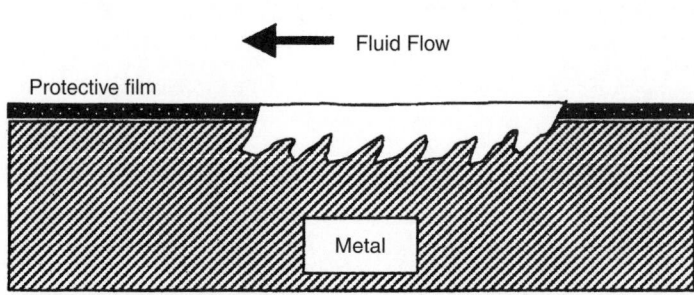

Figure 4.19.17 *Schematic of Erosion Corrosion.*

Table 4.19.4 *Erosion Corrosion Rates in Flowing Seawater (after J.R. Davis[11])*

Material	Typical Corrosion Rates, mg/dm²/day		
	0.3 m/s (1 ft/s)	1.2 m/s (4 ft/s)	8.2 m/s (27 ft/s)
Carbon steel	34	72	254
Cast iron	45	–	270
Silicon bronze	1	2	343
Admiralty bronze	2	20	170
Aluminum bronze	5	–	236
90-10 copper nickel	5	–	99
70-30 copper nickel	<1	<1	39
Monel	<1	<1	4
316 stainless steel	1	<1	<1
Titanium	0	–	0

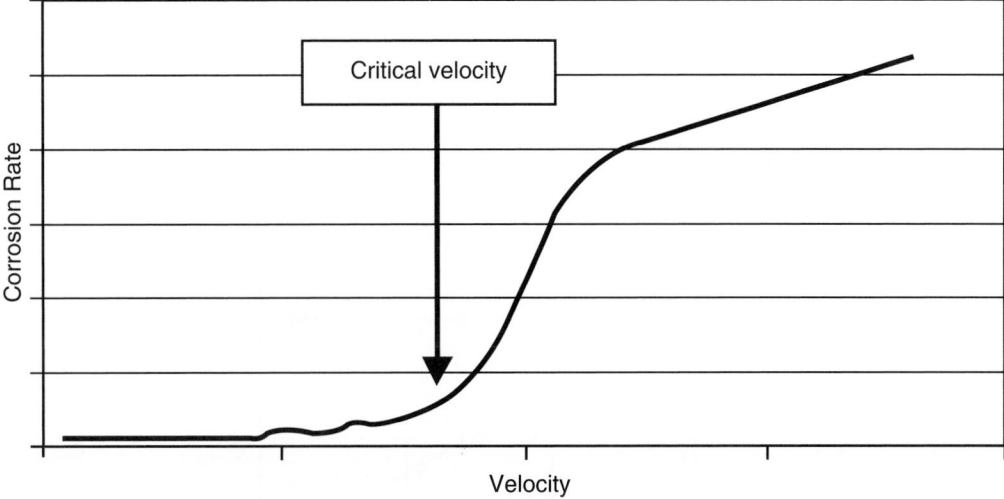

Figure 4.19.18 *Schematic of velocity effect on erosion corrosion.*

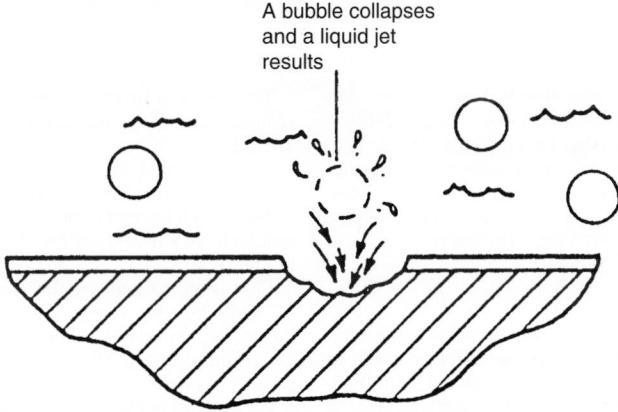

Figure 4.19.19 *Cavitation corrosion [8].*

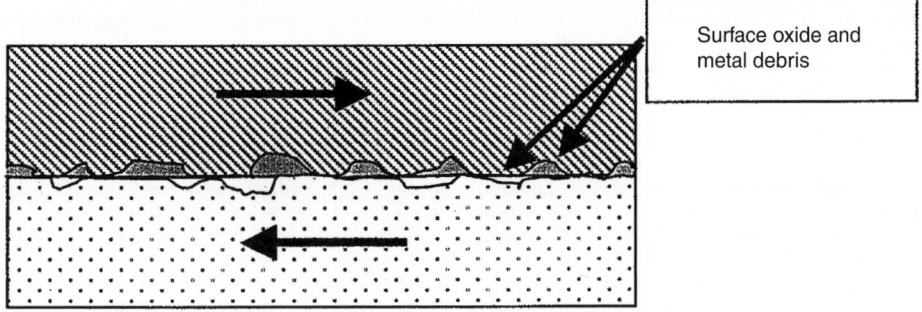

Figure 4.19.20 *Schematic of fretting corrosion (horizontal arrows depict relative motion).*

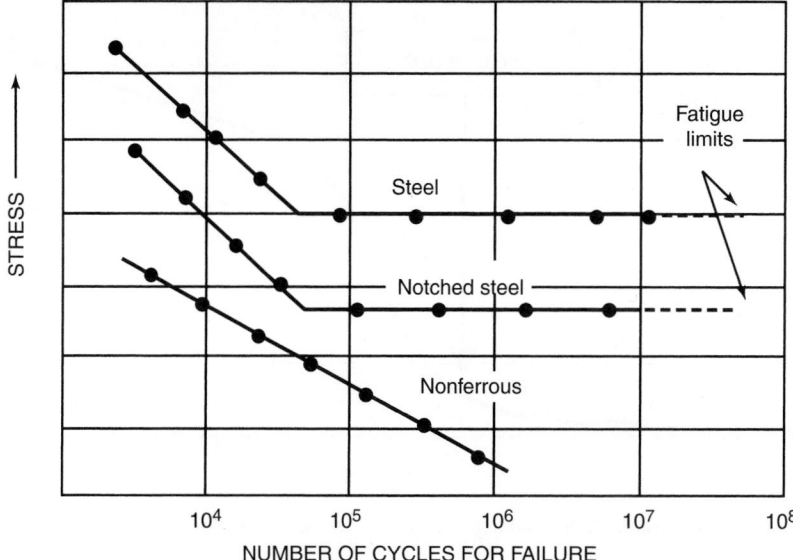

Figure 4.19.21 *Fatigue behavior of ferrous and nonferrous alloys [10].*

or oxidization/corrosion of wear debris. Fretting corrosion may be avoided or reduced by (1) surface coatings, (2) lubrication, (3) increasing surface hardness, (4) using gaskets to absorb vibrational movement, (5) increasing relative motion between surfaces, and (6) decreasing contact stress. These mitigation techniques are the same as those used to reduce the occurrence of adhesive type wear (galling).

4.19.2.9 Corrosion Fatigue

The combination of corrosion and the extended application of cyclical stresses can result in the mechanism known as corrosion fatigue. In fatigue situations for steels under non-corrosive conditions, we find that there is a stress level at which fatigue does not occur; this is known as the endurance limit. Conversely, for non-ferrous alloys, we find that there is no endurance limit. In situations where there are stress concentrations resulting from sharp radii (notches etc.) or material defects such as existing cracks, we discover that the fatigue life is depressed or decreased under all stress conditions. These relationships are typically illustrated through plots of number of cycles to failure against stress, the stress is usually expressed as the highest tensile stress level in alternating or changing stress levels. An example of a life–stress plot for these fatigue relationships in the absence of a corrosion mechanism is presented in Figure 4.19.21.

In moderate to low strength steels, we typically find that this endurance limit is in the range of $\frac{1}{2}$ of the tensile strength [24]. Fatigue can occur as the result of cyclical tensile stresses (lower and higher tensile stresses) as well as stresses reversals where the stress cycles between a zero or compressive stress state and a tensile stress state.

Corrosion has large effects on fatigue resistance. Corrosion mechanisms affect fatigue by simultaneously providing notches/defects that result in earlier fatigue crack initiation and decreasing resistance to fatigue propagation though the cyclical formation and break-down of protective surface films during cyclical stress cycles and the harmful effects of absorbed hydrogen. An example of a corrosion related site that provides a notch is a corrosion pit.

The result of increasing environmental severity and the resultant corrosion on fatigue life is presented in Figure 4.19.22.

One important indicator or measure of environmental severity is the pH or acidity of the environment. In an aerated saltwater environment, increasing acidity or decreasing pH results in concurrent decreases in fatigue life as presented in Figure 4.19.23.

Corrosion fatigue can be minimized or prevented by (1) reducing the stress through design or operation, (2) increasing material tensile strength, (3) selection of a more corrosion resistant alloy, (4) coatings, (5) limiting applications to environments where the potential for corrosion is low, and (6) corrosion inhibition.

4.19.2.10 Stress Corrosion Cracking

Stress corrosion cracking occurs when a susceptible material is subjected to an environment that causes stress corrosion cracking and the application of a tensile stress. There are a set or sets of environmental conditions in which most metals can exhibit stress corrosion cracking. Some examples of specific types of metals and the environmental constituents that can lead to stress corrosion cracking are presented in Table 4.19.5.

From the earlier examples, we can note that material strength level, applied stress, acidity, temperature, and specific agent concentration are frequently important or controlling factors in stress corrosion cracking.

Many mechanisms have been proposed to explain the phenomena of stress corrosion cracking. The identification of the mechanism is complicated because there are a variety of processes and conditions that can cause stress corrosion cracking. For example, we know that stress corrosion cracking can occur as result of either anodic or or cathodic corrosion reactions; it is also possible to have both anodic and cathodic reactions promoting stress corrosion cracking simultaneously. One or more of the following environmental reactions between the material and the environment has been proposed as an essential step in stress corrosion cracking [27]: adsorption of environmental species, surface

Stress, psi x 10^3

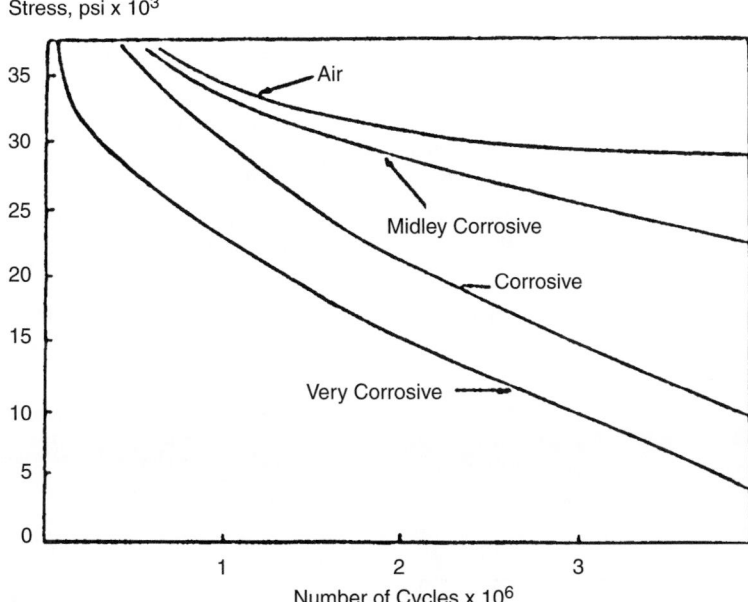

Figure 4.19.22 *Effect of corrosive environment on fatigue life [25].*

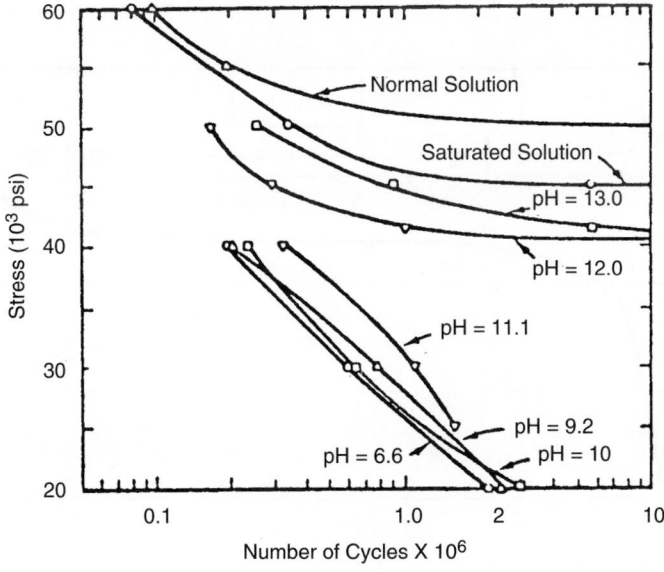

Figure 4.19.23 *Effect of pH on corrosion fatigue in aerated saltwater [26].*

reactions, reactions in the metal ahead of the crack tip and surface films. From these essential reactions, the following potential list of rate-determining steps have been proposed [27]:

1. Mass transport along the crack to or away from the crack tip
2. Reactions in the solution near the crack
3. Surface adsorption at or near the crack tip
4. Surface diffusion
5. Surface reactions

6. Absorption into the bulk
7. Bulk diffusion to the plastic zone ahead of the advancing crack
8. Chemical reactions in the bulk
9. The rate of interatomic bond rupture

In addition to a specific alloy type and the strength/ hardness level listed as a factor in Table 4.19.5, there are other metallurgical variables that include grain size, thermal processing history, additional solid phases that may be present in the matrix or in the grain boundary, concentration

Table 4.19.5 *Metals and Stress Corrosion Cracking (SCC)*

Metal	Environment	Factors that increase risk of SCC
Steels	Hydrogen sulfide	Increasing H_2S, moderate temperatures, more acidic, higher strength/hardness, higher stress levels
Steels	Carbonates	Higher strength
Steels	Chlorides	Higher strength, higher stress levels, more acidic
Brass	Ammonia	Higher strength, higher stress levels
Martensitic stainless steels	Hydrogen sulfide	Increasing H_2S, moderate temperatures, more acidic, higher strength/hardness, higher stress levels
Austenitic and stainless steels	Chlorides	Higher strength, higher chloride, higher stress levels, more acidic, higher temperatures, presence of H_2S
Duplex stainless steels	Chlorides	Higher strength, higher chloride, higher stress levels, more acidic, moderate temperatures, presence of H_2S
Titanium	Alcohol	Higher stress levels, lower water content

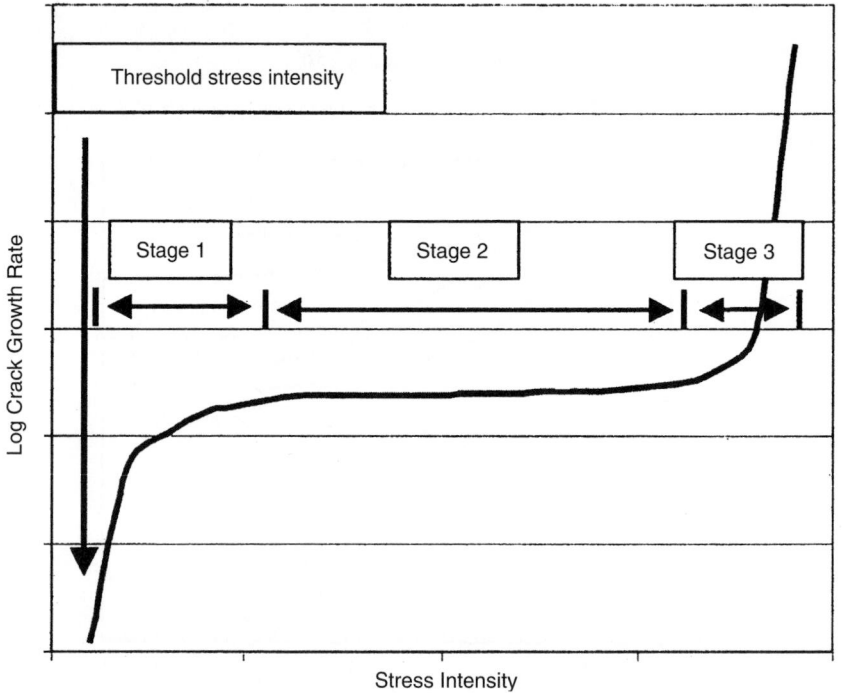

Figure 4.19.24 *Schematic of the stages of stress corrosion cracking.*

of impurities, and segregation of alloying constituents including impurities.

Most stress corrosion cracking typically occurs in three stages: Stage 1 — Initiation, Stage 2 — steady-state crack propagation, and stage 3 — rapid crack propagation or final fracture. The stress below which stress corrosion cracking does not occur is known as the threshold stress intensity. These stages in stress corrosion cracking and the threshold stress intensity are presented in Figure 4.19.24.

For each alloy and condition under a specific set of environmental conditions there is a threshold stress intensity below which stress corrosion cracking will not occur. The stress intensity is effected by bulk applied stress, stress

concentration factors, and residual stresses. Values of this stress intensity have been reported to be as low as 10% of the yield strength [8].

There are a number of ways to prevent stress corrosion cracking. The following list is presented as examples of mitigation techniques.

1. Change the material to one that is more resistant to SCC under the specific environmental conditions.
2. Avoid the metallurgical characteristics of the chosen material that could lead to SCC. These include avoiding susceptible microstructures and phases and reducing material stress levels.

3. Alter the environment to reduce the potential for SCC. Examples of altering the environment include adjusting the pH, limiting or changing the operating temperature, reduction of the environmental agents that promote SCC and the addition of corrosion inhibitors.
4. Apply anodic or cathodic protection to change the potential out of the critical range where SCC occurs. Caution, for SCC, the application of an anodic potential may accelerate the cracking.
5. Use metallic or nonmetallic coatings as barriers to the corrosive environment or that are resistant to the corrosion/stress corrosion cracking mechanisms.
6. Reduce stresses by reduction of operating stresses, avoiding stress concentrations and introducing beneficial compressive stresses.

4.19.2.11 Hydrogen-Induced Cracking
There are many forms of damage that occur as a result of hydrogen. These are typically categorized by the role of stress in the mechanism. The three types of damage that are observed are (1) hydrogen induced cracking (HIC), (2) stress-oriented hydrogen induced cracking (SOHIC), and (3) sulfide stress cracking (SSC). In HIC, the externally applied stress is low and low and does not interact with the cracking mechanism. In SOHIC, this is essentially the same as HIC except that the external stress influences the orientation or direction of the crack propagation. In SSC, the stress is an integral component of the cracking mechanism and does not occur without the application of applied stress.

In HIC, failures can occur at applied stress levels that are a fraction of the yield strength. For example, there have been many well-documented [28] failures due to HIC in low strength pipeline steels in gas transmission lines containing H_2S.

Hydrogen is evolved during corrosion reactions and hydrogen sulfide plays a pivotal role in hydrogen-related damage mechanisms. The hydrogen evolved is mono-atomic or nascent hydrogen and its recombination into molecular hydrogen is inhibited or poisoned by the hydrogen sulfide. The result of this inhibited reaction is greater quantities of the mono-atomic hydrogen that readily diffuses into the metal lattice; the size of the mono-atomic hydrogen is minute and moves easily through metal structures. The reactions of hydrogen sulfide and iron are presented below:

$$Fe + H_2S \rightarrow FeS + 2H$$

$$2H \rightarrow H^2$$

The second reaction above is inhibited or poisoned.

There are several theories that explain the various types of hydrogen damage [1]. Although numerous models and theories exist with respect to HIC, the phenomenon of trapping best quantifies the metal susceptibility to hydrogen damage [29,30]. A crack occurs when the atomic hydrogen migrates to one of these trap sites and then becomes "trapped."

We know that inclusions, especially elongated manganese sulfide inclusions, increase susceptibility to HIC damage. The other known major factor is alloy segregation in the form of banding. Banding is particularly a problem in plate and pipe formed from plate products but can be seen in seamless pipe as well.

To avoid HIC, clean steels with uniform microstructures have the best resistance to this form of cracking. There are standard tests that are used to assess a materials susceptibility to HIC, the most widely accepted standard test method is NACE's TM-0284 [31].

4.19.2.12 Sulfide Stress Corrosion Cracking
Sulfide stress corrosion cracking, also known as sulfide stress cracking (SSC), is a stress corrosion cracking phenomena that occurs in the presence of hydrogen sulfide. Hydrogen sulfide is known to cause or assist in the mechanism of cracking with many alloy types including low carbon steels, low alloy steels, alloy steels, tool steels, martensitic stainless steels, ferritic stainless steels, austenitic stainless steels, duplex stainless steels, nickel-copper alloys, and nickel-base alloys. For many of these materials, there co-exists the potential for chloride stress corrosion cracking. The presence of hydrogen sulfide often increases the severity of the cracking or enlarges the range of environmental conditions where cracking occurs.

There are several competing mechanisms that result in SSC including internal hydrogen stressing, hydrogen embitterment, stress corrosion cracking and combinations of these mechanisms. These competing mechanisms have complicated the identification of a universally accepted molecular mechanism that explains all of the observed phenomena.

The oil and gas industry has been actively developing reservoirs containing H_2S since the early 1950s [32]. The presence of numerous early failures resulted in large and widespread efforts to understand and mitigate the effects of H_2S on materials. The noteworthy characteristics of these failure were (1) they were brittle failures, (2) they occurred in relatively short periods of time, and (3) they were not expected by virtue of previous experience in environments that contained either no H_2S or very low levels of H_2S. Three examples of the brittle nature of these failures are presented in Figure 4.19.25.

The amount of H_2S that causes SSC is conditional on other environmental variables but the historical limit has been listed in NACE MR0175 as 0.05 psi partial pressure of H_2S [2,32]. The environmental factors that influence the existence and extent of SSC include the following:

1. H_2S concentration
2. Acidity (pH)
3. Time
4. Temperature
5. Water content (also gas/oil ratio and solids content)
6. Chloride content
7. Oxygen content
8. Presence of elemental sulfur

The effects of these variables are interrelated and also very much conditional on the specific alloy, its condition and stress state. Though these effects are therefore complicated, we can explore the effects of some of these variables in specific cases illustrating trends. Looking at the combination of time and H_2S concentration effects with material hardness we discover that increasing H_2S decreases the time to failure. Conversely, short time exposure can be successful at greater concentrations of H_2S. These relationships are presented in Figure 4.19.26.

When examining the combined effects of time and acidity (pH), we discover that the time to failure decreases with decreasing pH in H_2S containing environments. The effect of pH with time is presented in Figure 4.19.27.

Another way to illustrate the effects of pH is to examine the effect of pH on the critical stress (Sc); the critical stress is one measure of resistance to cracking and, the higher the number, the greater the resistance. This is presented in Figure 4.19.28.

The relationship between the cracking potential and the pH and H_2S content have lead to a relationship between these and regions or domains of usage that apply to carbon and low alloy steels. These pH-H_2S domains have appeared in EFC

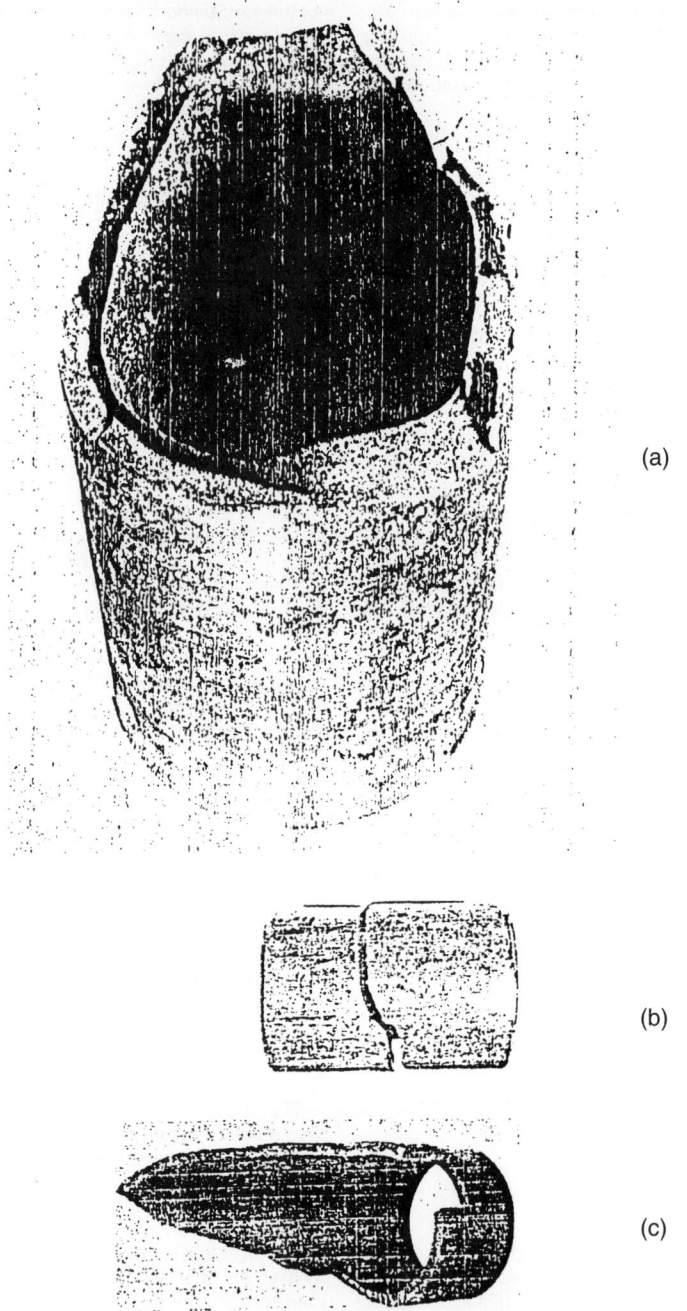

(a)

(b)

(c)

Figure 4.19.25 *SSC failure, drillpipe (a); tubing failure (b); coupling (c) [22,14].*

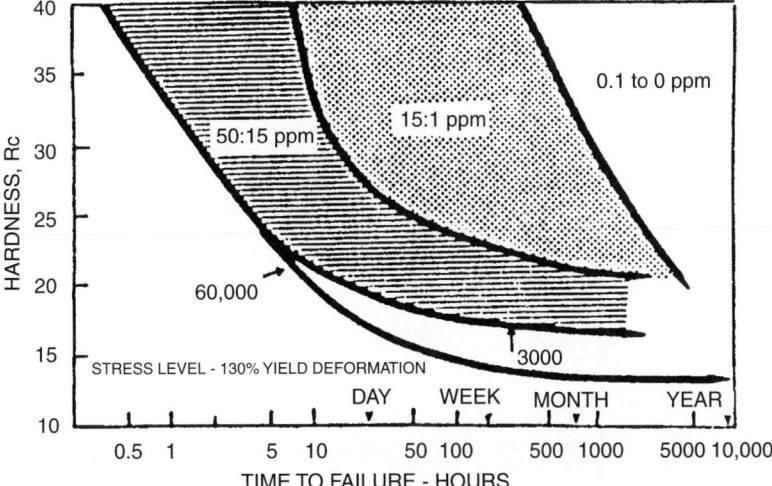

Figure 4.19.26 *Approximate failure time of carbon steel is 5% NaCl and various parts per million of hydrogen sulfide [33].*

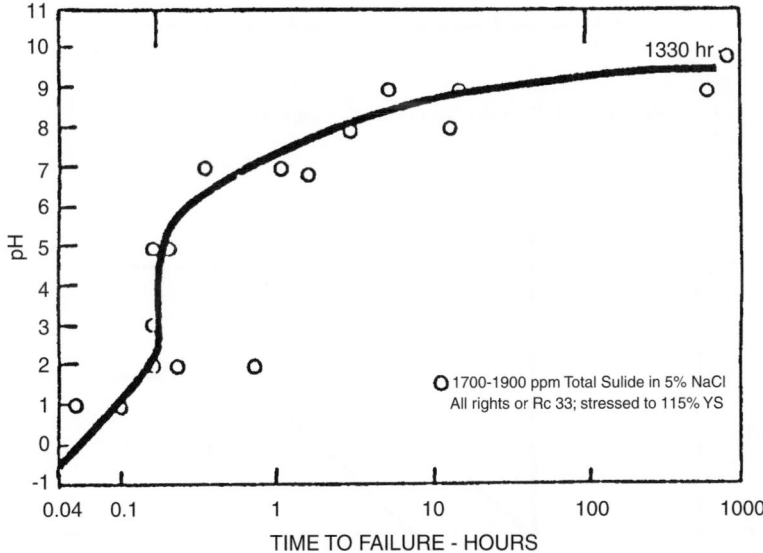

Figure 4.19.27 *Steel fails much less rapidly when pH is above 7 or 8 [33].*

Publication 17 [35] and ISO 15156 [19]. A sketch illustrating these domains is presented in Figure 4.19.29.

Domain 0 in Figure 4.19.29 is considered the region where SSC is not a factor in most carbon and low alloy steels. The environmental severity and, therefore, the care and level of material restriction increases with increasing Domain number. Domain 3 is the region of restrictions that match those historically impose by NACE MR0175 [2].

One of the other major factors that effects materials selection for cracking resistance for both corrosion resistant alloys (CRAs) and carbon and low alloy steels is temperature. For carbon and low alloy steels, it was observed that the incidences of failure seemed to be more prevalent at lower temperatures. An early investigation conducted by Townsend [36] on high-strength wire line steel revealed a maximum susceptibility to cracking near room temperature. The high strength steels are very susceptible to cracking and the tests conducted demonstrated, in relatively short periods of time, the effects of temperature. This relationship between temperature with environmental severity as measure by time to failure is presented in Figure 4.19.30.

From Figure 4.19.30, we can observe that very severe conditions exist for steels in temperatures below about 80°C (175°F). We can also see the effects that this relationship has had on material standards for oil country tubular goods (OCTG) for sour service (H_2S-containing environments). For example, NACE MR0175 [2] and ISO 15156 [19] permit the use of higher strength materials at high temperatures where the environment is not as severe. Table 4.19.6 lists some of these OCTG grades and the minimum temperature they can be safely used.

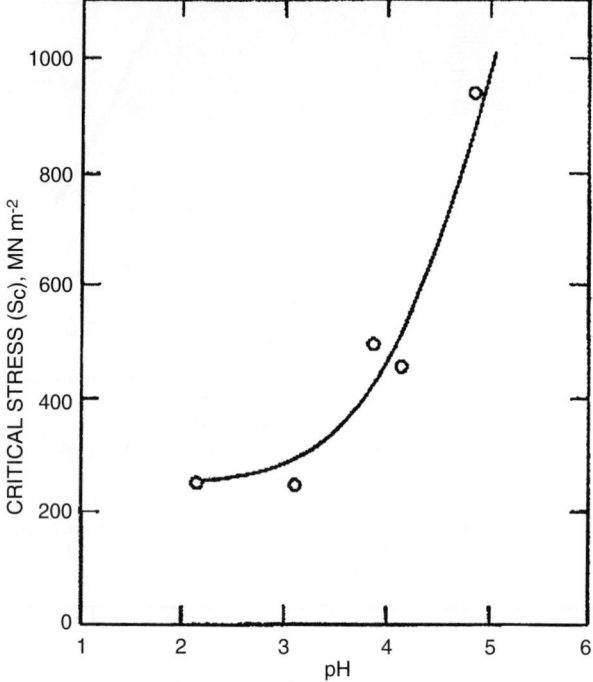

Figure 4.19.28 *Effect of pH on H_2S cracking resistance [34].*

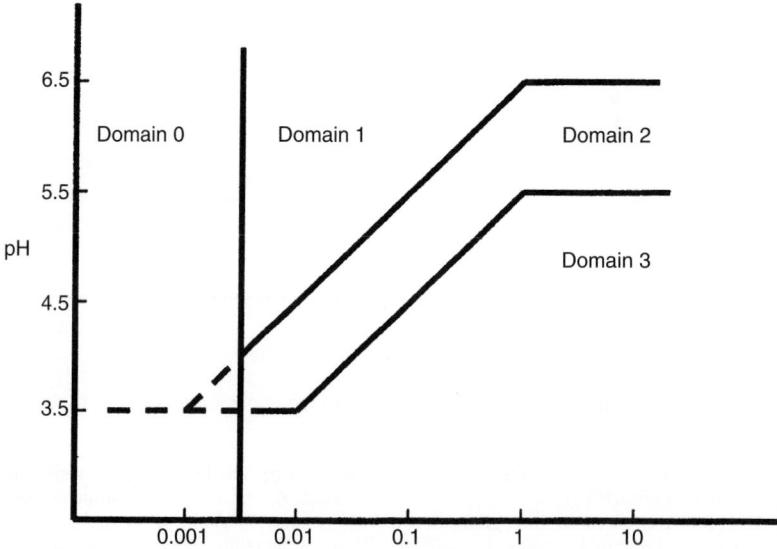

Figure 4.19.29 *pH versus H_2S domain diagram.*

This rationale is supported by laboratory data and experience and permits the use of increasingly less resistant materials in less severe environments as defined by increasing temperature.

One other common variable that effects SSC is the type of oilfield environment that is encountered. Generally, corrosion rates increase with the amount of water present or being produced; the amount of water produced is known as the water cut. The base fluid being gas, oil or brine effects the severity of the environment with respect to cracking. Another early investigation [37] revealed that cracking was much more prevalent in a gas environment as opposed to an oil or brine environment. This relationship is presented in Figure 4.19.31 that correlated the percentage of failures with the hydrogen sulfide content in three distinct phases, brine, oil and gas.

There are numerous other factors that effect SSC such as the presence of oxygen and elemental sulfur that increase

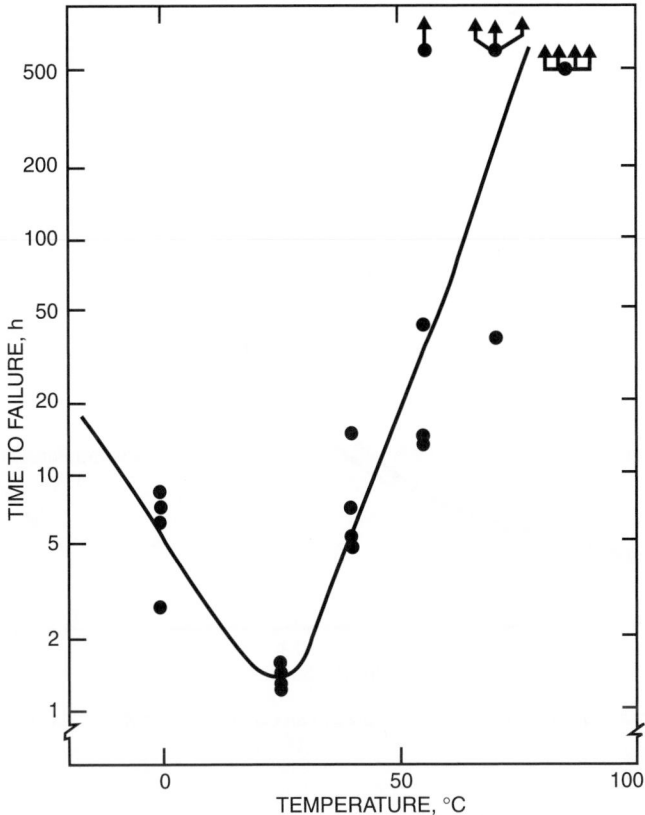

Figure 4.19.30   *Time to failure versus temperature for high strength steel in H_2S [34].*

Table 4.19.6   *OCTG Grade versus Safe Use Temperature in H_2S Containing Environments*

OCTG Grade	Yield Strength Range, MPa (ksi)	Minimum Temperature
L80 type 1	552–655 (80.0–95.0)	No minimum
C90 type 1	621–724 (90.0–105.0)	No minimum
T95 type 1	655–758 (95.0–110.0)	No minimum
C95 type 1	655–758 (95.0–110.0)	65 °C (150 °F)
P110	758–965 (110.0–140.0)	80 °C (175 °F)
Q125 type 1 & 2	862–1034 (125.0–150.0)	107 °C (225 °F)

the severity of the environment. Care must be taken with materials selection when these are encountered.

4.19.2.13 Microbiologically Induced Corrosion (MIC)

Microorganisms are present throughout our environment. These organisms can result in corrosion by producing a deposit on metal surfaces and/or by increasing the acidity/reducing the pH of the environment. This altering of the environment can occur in the bulk or locally.

There are two basic classes of microbes, anaeorbic and aerobic microorganisms. Anaerobic microorganisms flourish in the absence of oxygen such as deep sea or downhole environments. Aerobic microorganisms require oxygen to survive and flourish and are found in all surface environments including salt and fresh waters.

One example of an anaerobic microorganism that is troublesome in the oilfield is the genus Desulovibrio. This is a sulfate-reducing bacteria (SRB) and the resultant corrosion

mechanism involves naturally occurring sulfates to hydrogen sulfide. The resultant acid from the hydrogen sulfide can cause or accelerate corrosion or cause HIC or SSC as discussed previously. A simplistic chemical mechanism illustrating corrosion attack on steel by SRB is presented in Figure 4.19.32.

There are many iron oxidizing types of bacteria such as Gallionalla and Crenothrix that can accelerate corrosion by releasing carbon dioxide (CO_2) as one of the corrosion products as illustrated by the following equation.

$$4FeCO_3 + O_2 + 6H_2O \rightarrow 4Fe(OH)_3 + 4CO_2$$

The iron hydroxide ($Fe(OH)_3$) is not very soluble and will precipitate on the metal surface. These deposits can cause oxygen concentration cells resulting in corrosion beneath the deposits.

In addition to the bacteria examples mentioned previously, there are also molds, yeasts, algae, and protozoa. Molds, yeasts, algae, and protozoa can have a waste product

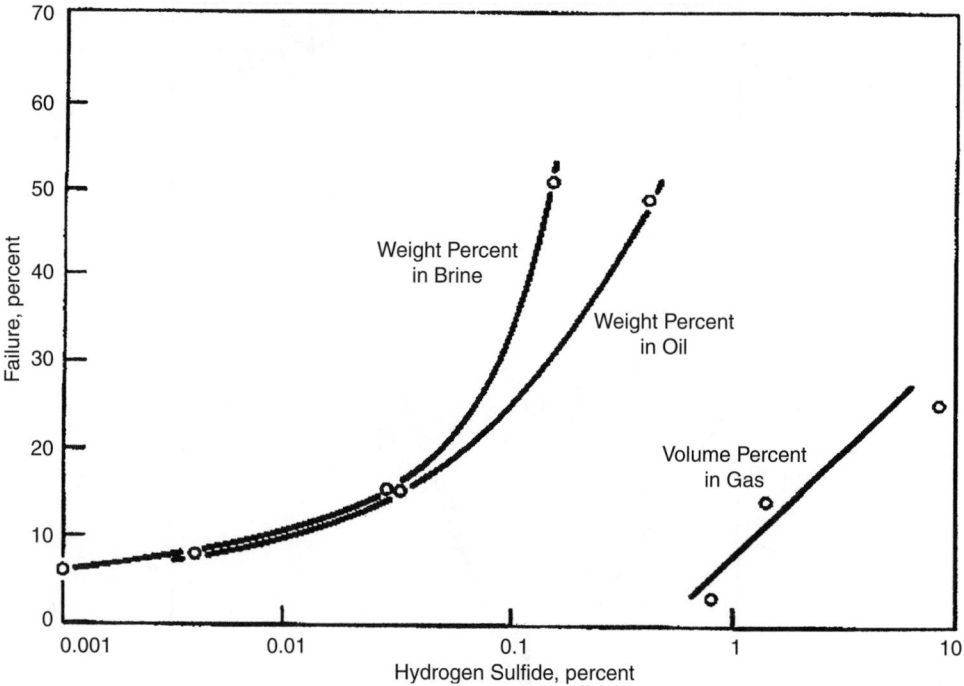

Figure 4.19.31   *Relationship between percentage failures in steels and H_2S content in brine, oil, and gas phases [37].*

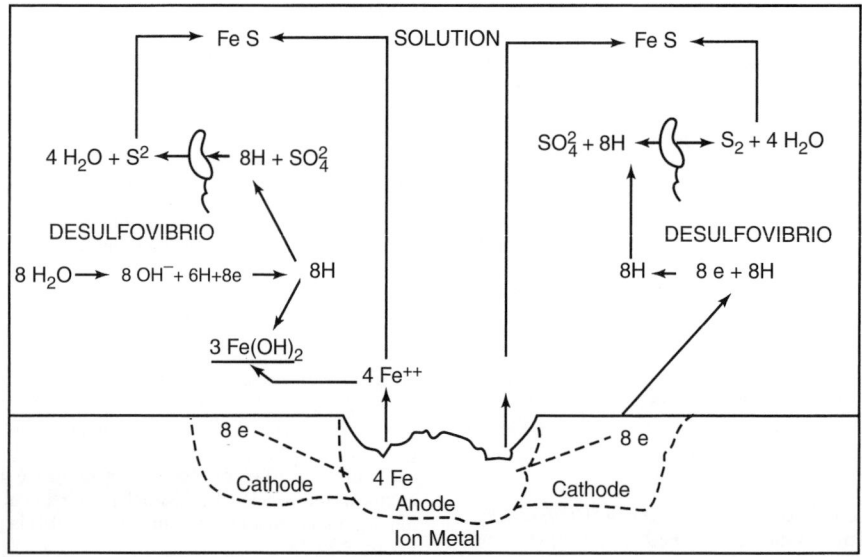

Figure 4.19.32   *Diagram of polarization of local cathode by a film of hydrogen gas bubbles (cathodic area to right of anode is polarized) [38].*

that enhances corrosion or can result in slimes or deposits. Oxygen concentration cells and localized corrosion attack can result beneath these deposits.

There are a number of techniques used to prevent MIC; these involve modifying the environment or coatings. The environment can be adjusted to prevent growth of these microorganisms such as by adjusting pH (Figures 4.19.33 and 4.19.34), removal of nutrients, and the use of biocides.

Coatings and films have been successfully used such as paints, platings, and filming inhibitors.

4.19.3 Oilfield Environments
The concentration of acid gases is related to pH and an illustration of the relationship between them is presented in Figure 4.19.34.

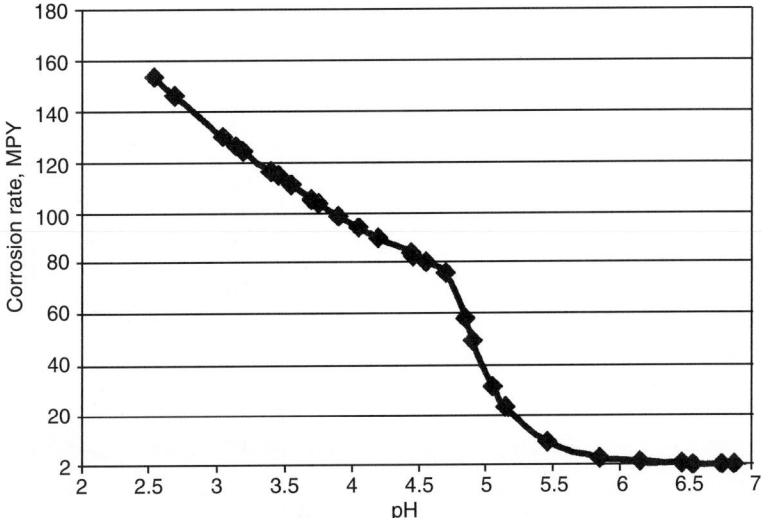

Figure 4.19.33   *Effect of pH on corrosion rate.*

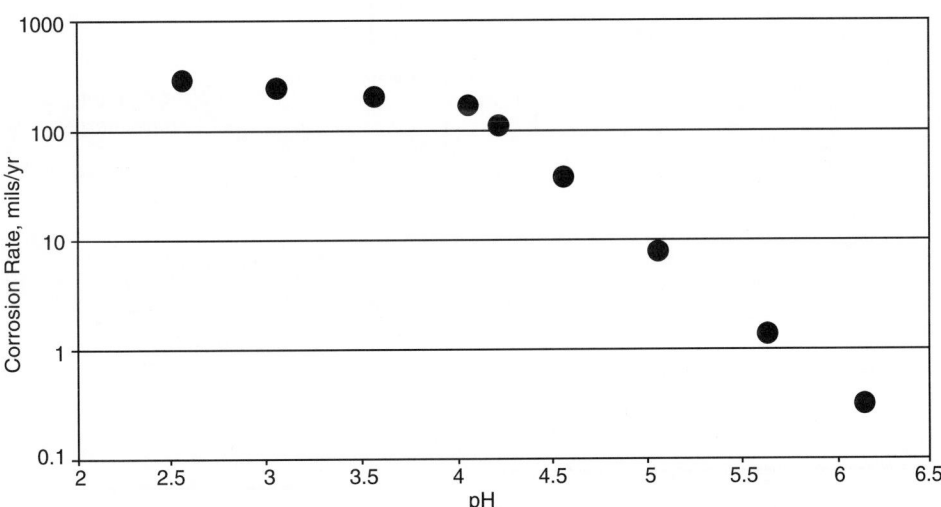

Figure 4.19.34   *Simplified relationship between pH and corrosion rate (steel with CO_2 at 93 °C [200 °F]).*

Carbon dioxide is usually associated more with direct effects on corrosion. Carbon dioxide in water results in carbonic acid (a weak acid) and a CO_2 partial pressure of 1 bar (15 psi) lowers the pH of pure water at 60°C (125°F) to 3.8 [39]. The characteristics of CO_2 corrosion are complex because of the semi-protective nature of the corrosion products. This is contrasted with H_2S where the corrosion products are generally more protective in nature. As discussed previously, the composition and stability of the corrosion products are controlled by pH and temperature.

DeWaard and Milliams [40] published early work in 1975 that became the basis for subsequent models relating corrosion rates to a variety of oilfield conditions. An early nomograph is presented in Figure 4.19.35.

Since that early work, much progress has been made refining these relationships taking into account the stability of protective corrosion product films. The corrosion rate of steels in CO_2-containing oilfield environments as a function

of partial pressure of CO_2 and temperature is presented in Figure 4.19.36.

Hydrogen sulfide also has a similar effect on the corrosion rate of steels. The effect of H_2S concentration on the corrosion rate of steel is presented in Figure 4.19.37.

The effect of H_2S when combined with CO_2 is complicated by the interaction of the sulfide corrosion products with the carbonate corrosion products. The sulfide corrosion products from the H_2S often increase the stability of the surface film thereby reducing the corrosion rate.

4.19.3.1 Corrosion Effects of Temperature, Flow Rates, and Chlorides

In CO_2 corrosion on steels, the effect of temperature is illustrated in Figure 4.19.36. The relationship shown here is only valid for the specific environment and this relationship is strongly influenced by other environmental constituents that cause/effect corrosion such as hydrogen

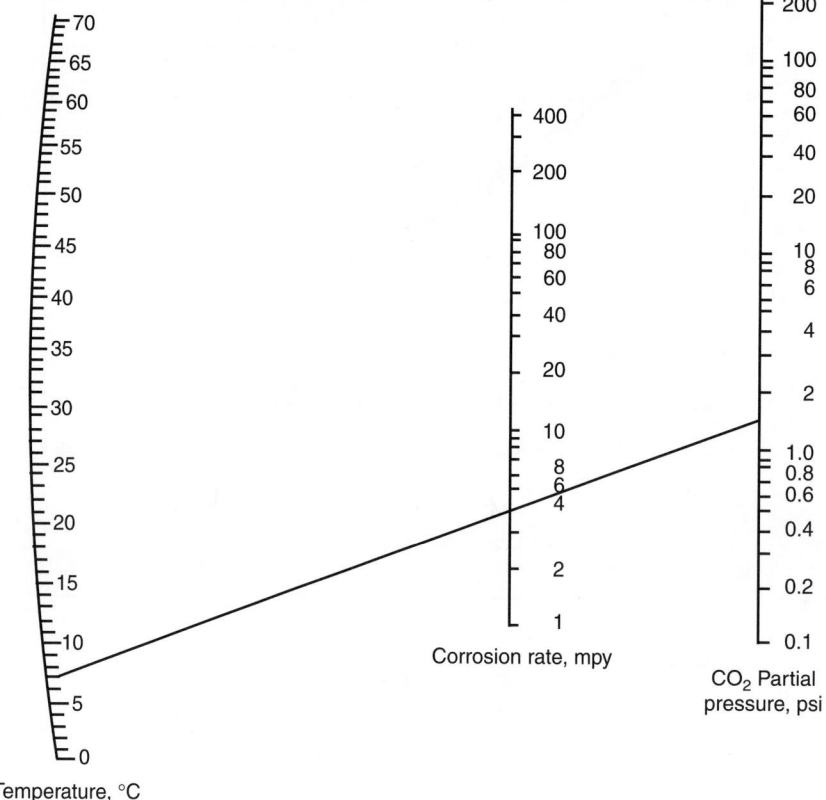

Figure 4.19.35 *Calculation of corrosion rate as a function of temperature and CO_2 [40].*

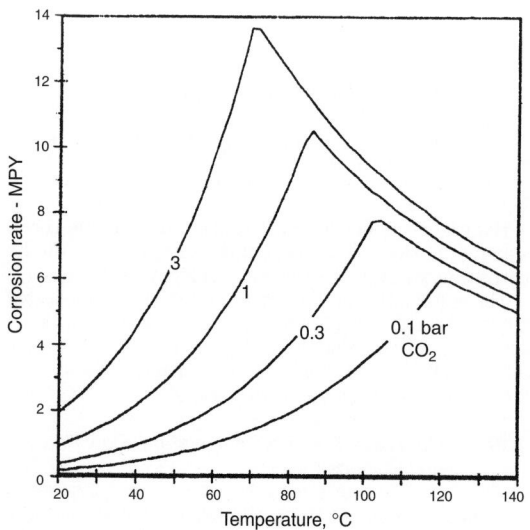

Figure 4.19.36 *Effect of temperature on CO_2 corrosion [41].*

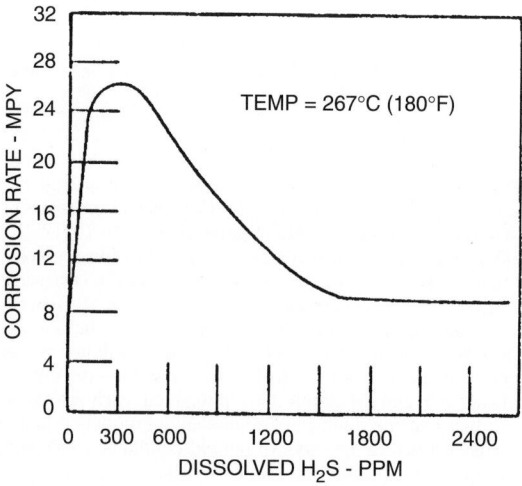

Figure 4.19.37 *Corrosion rate of mild steel in distiled water containing varying concentrations of hydrogen sulfide [26].*

sulfide, sulfates and chlorides that generally increase corrosion rates and carbonates, bicarbonates and alkalis that generally decrease corrosion rates. Chlorides effect general corrosion rates by changing the stability of corrosion products and interference with the formation of protective oxides. The effect of sulfates, chlorides and bicarbonates on steel corrosion is demonstrated by the trends presented in Figure 4.19.38.

At high temperatures, chlorides can also destroy the passivity of many corrosion resistant alloys such as the stainless steels resulting in localized corrosion attack and increased general corrosion rates.

4.19.3.2 Environments Causing Cracking

The oil field has a diverse list of potential environments that can cause cracking as well as corrosion. In production operations, we can have heavy brines, production gases/fluids including H_2S and CO_2, chlorides, organic acids and elemental sulfur. In drilling operations, we can have alkaline (high pH) drilling fluids, brines and the exposure with production fluids. In well stimulation and workover operations, we can have acids (hydrochloric, fluoric and organic) and oxygen. It is this diverse nature that complicates the mitigation of corrosion and environmental cracking.

The risk of environmental cracking is a function of the material and the stresses (service and residual) as well as the severity of the environment. There are a number of oil field environments that can cause cracking in susceptible materials. For example, on one end of the spectrum, we have ultra high strength steels that, if improperly processed could exhibit environmental cracking in mild salt containing environments. These high strength steels can crack in carbonate containing environments and highly alkaline environments. On the other end of the spectrum, we have hot salty gas-water systems that contain elemental sulfur that can cause cracking in many of the normally crack resistant alloys.

High pH drilling fluids can cause SCC in high strength steels. The factors that increase susceptibility (decreasing life) are increasing temperature, increasing stress level and increasing the alkali concentration (increasing pH). The effect of temperature and alkali concentration (NaOH) on SCC cracking of steels is presented in Figure 4.19.39.

The effects of hydrogen sulfide on sulfide stress corrosion and hydrogen induced cracking wee discussed previously. We can add to this by adding a discussion on chlorides, elemental sulfur and oxygen.

The chlorides are known to cause stress corrosion cracking in stainless steels at elevated temperatures. The potential for SCC increases with increasing temperature, stress and chloride content. The potential for SCC can be influenced by the presence of hydrogen sulfide and even small amount of oxygen or elemental sulfur has a dramatic negative effect on susceptibility to SCC.

4.19.3.3 Petroleum Phases (Oil, Gas and Water)

Corrosion occurs as a result of contact with water phases that may contain one or more potential corrodents. The corrosion rates generally increase with increasing water content (also known as water cut). When water is present with a gas phase, the gases dissolve in the water and this acidifies the water. A gas phase that contains a small water fraction could be more corrosive due to its acidified nature.

Other phases, such as brines and oil have varying effects on corrosion. Oil phases typically reduce corrosion rates and when the oil is present in sufficient enough quantities, the metal surfaces in contact with the petroleum fluid(s) will be wetted with oil. This results in an oil film barrier. The corrosion rates (and cracking tendencies) are dramatically reduced when the contacting phase is oil.

Brines have a mixed corrosion relationship. Generally, by themselves, brines may be very corrosive, and corrosion rates may increase with increasing brines content; this is especially true of some of the heavy acid brines such as zinc bromide. The other effect of brines is the reduction of gas solubility; as the brine's salt content increases, the solubility of acid gases decreases. The end result is less acid gas present to lower pH and increase corrosion rates. The effect of brines, gas, and oil on the failure rate of steels with increasing H_2S contents was presented previously in Figure 4.19.31.

4.19.3.4 Injected Fluids (Drilling Fluids, Oxygen and Acids)

Drilling fluids are comprised of many potential constituents. The drilling fluid could be water based, oil based or synthetic based. Brines are a form of water-based fluids. The fluids could have the pH adjusted to high levels with the use of alkalis such as sodium and potassium hydroxide. The role of the drilling fluid on corrosion is related to many factors, including the pH, the chloride content, and the oxygen content.

Two of common failure modes in drill strings are fatigue and corrosion fatigue. An example of a corrosion related affect is increasing chloride contents decrease the fatigue life in drilling fluids. Figures 4.19.40, 4.19.41 and 4.19.42 summarizes the corrosion fatigue tests conducted by Azar [42] that related chloride content to fatigue life in different drilling fluids.

The highest corrosion rates are found in corrosive systems in the presence of even minute quantities of oxygen. Oxygen has anodic and cathodic effects in that it acts as a cathodic depolarizer and an anodic polarizer. Oxygen accelerates corrosion due to other corrodents such as CO_2, H_2S, salt solutions, and injected mineral acids.

In drilling operations, the fluids used in drilling are exposed to air and this leads to oxygen being dissolved in the drilling fluid. This dissolved oxygen with chlorides increases corrosion as previously presented in Figure 4.19.10 and as presented below in Figure 4.19.43.

The effect of oxygen was found to be linear with the range and conditions in Figure 4.19.43 but this is not typical. Oxygen also decreases corrosion fatigue life. Corrosion fatigue tests have been conducted by Azar [42] demonstrating the relationship between oxygen content and fatigue life in drilling muds. This data is presented in Figures 4.19.44 and 4.19.45.

An additional problem exists when the injected oxygen containing fluid mixes with production fluids that may contain acid gases. When H_2S is present, the oxygen could reduce the hydrogen sulfide to elemental sulfur with potential drastically increased environment severity. When combined with CO_2 even at low parts per million (ppm) levels, the corrosion rates increase dramatically. An example of the effect of small oxygen contents on the corrosion rate of steel with CO_2 is presented in Figure 4.19.46.

Acids can result in very rapid corrosion rates and the can render normally passive materials active. When passive materials become active, there can be localized as well as general corrosion. One risk in many downhole anaerobic conditions is that the active pits may not repassivate when the acids are no longer in contact with the metal.

Mineral acids are often injected to stimulate production by cleaning out formations. These acids are typically hydrochloric acid or a mixture of hydrochloric and hydrofluoric acids. These mineral acids are very corrosive at ambient temperatures and the increase in temperature in downhole environments can render the acid attack even more aggressive. The effect of temperature on corrosion rates was previously

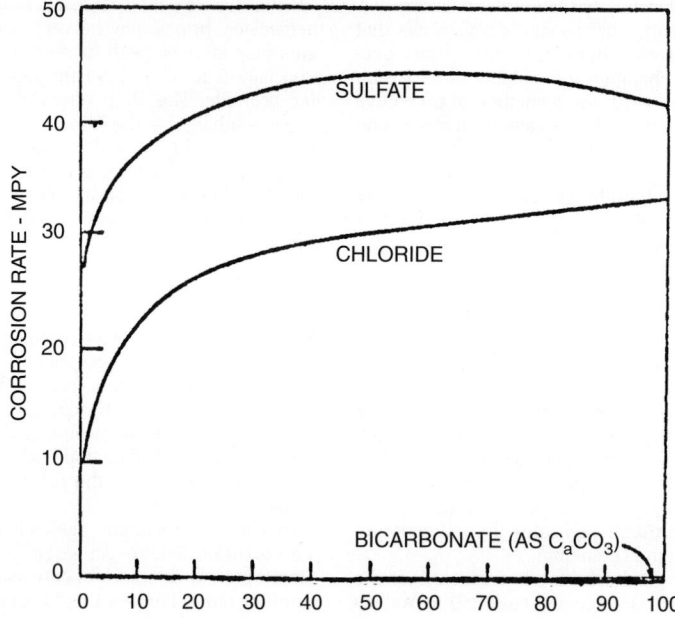

Figure 4.19.38   *Influence of sulfate, chloride and bicarbonate on the corrosion of steel [26].*

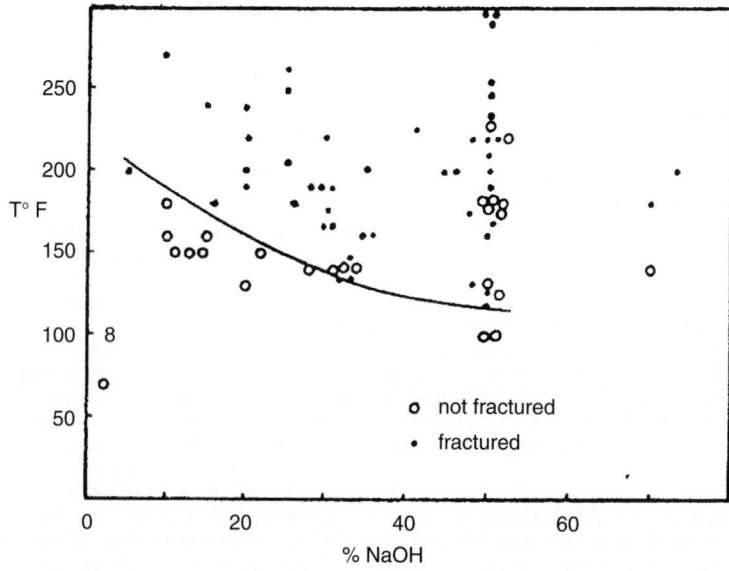

Figure 4.19.39   *Effect of temperature and concentration on alkali SCC cracking of steel [14].*

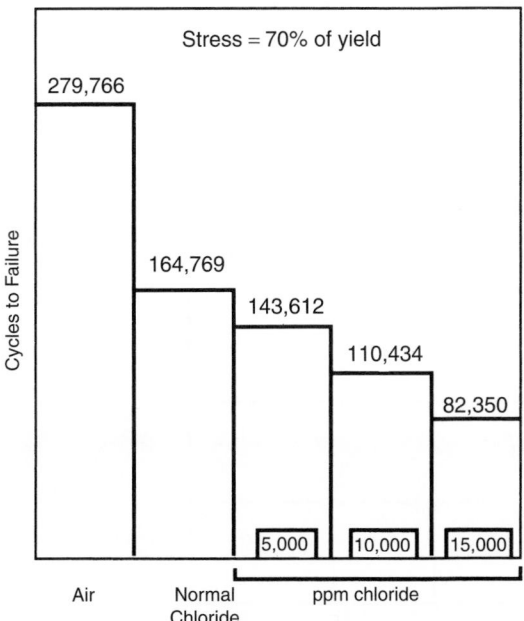

Figure 4.19.40 *Chloride fatigue test—dispersed lignosulfonate mud, 9.5 pH [42].*

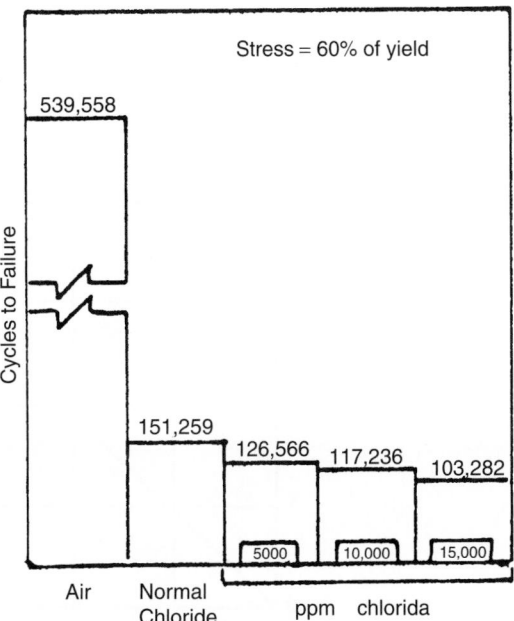

Figure 4.19.42 *Chloride fatigue test—nondispersed bentonite mud, 9.5 pH [42].*

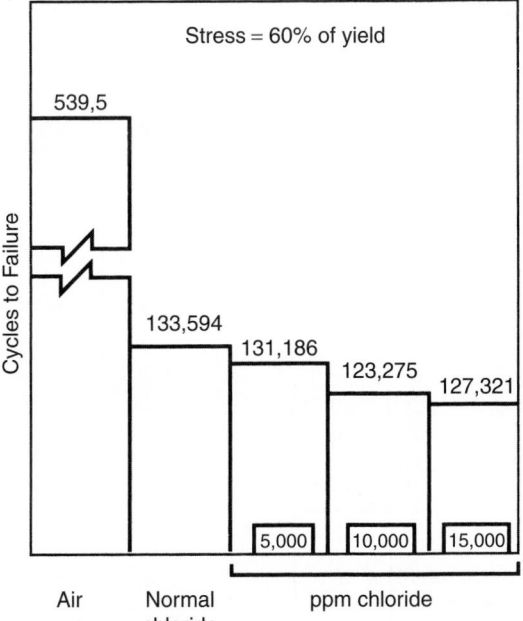

Figure 4.19.41 *Chloride fatigue test—nondispersed bentonite mud, 8.5 pH [42].*

discussed and the relationship can be seen in Figures 4.19.7 and 4.19.8.

These acids are most often inhibited to prevent corrosion on tubing string but the inhibitor used may not be effective in the specific downhole condition, not protective against one of the materials of construction used or may cause corrosion by the flow of spend acid where the inhibitor is no longer present or effective.

4.19.4 Drilling and Well Completion Equipment and Materials

In this segment of the oil and gas industry, the following tools are discussed with the typical materials of construction and types of corrosion problems encountered: drill pipe, collars, tool joints, stabilizers, drilling jars, drill bits, milling tools, mud motors, cementing equipment, casing, tool joints, coiled tubing, blow-out preventers (BOPs), and drilling risers.

4.19.4.1 Drill String Components

The drill string serves the purposes of conducting drilling fluids to the drilling bit, conducting rotational forces to the drill bit, controlling the amount of weight on bit and providing an annulus space between the formation or casing and the drill string to allow for removal of cuttings and drilling fluids. Components that fit into the drill string include drill pipe, drill collars, stabilizers, tool joints, jars, and miscellaneous crossovers and subs. The two basic types of tubulars are the drill pipe itself and drill collars. The drill pipe is composed of the pipe body and either integrally forged (upset) connections or forged and welded-on tool joints.

The typical materials of construction for drill pipe, stabilizers, collars, and tool joints are chromium — molybdenum (Cr–Mo) low alloy steels in the quenched and tempered condition such as specific alloys 4135, 4140, and 4145. There has also been some use of aluminum drill pipe and nonmagnetic collars and subs made from beryllium copper and austenitic type stainless steels.

In drilling environments, the corrosion is typically minimized through the use of pH control, inhibitors and/or the use of natural or synthetic oil based fluids. Corrosion when the controlled drilling environment is functioning as designed is generally not a concern except in some cases of stress corrosion cracking and hydrogen embrittlement

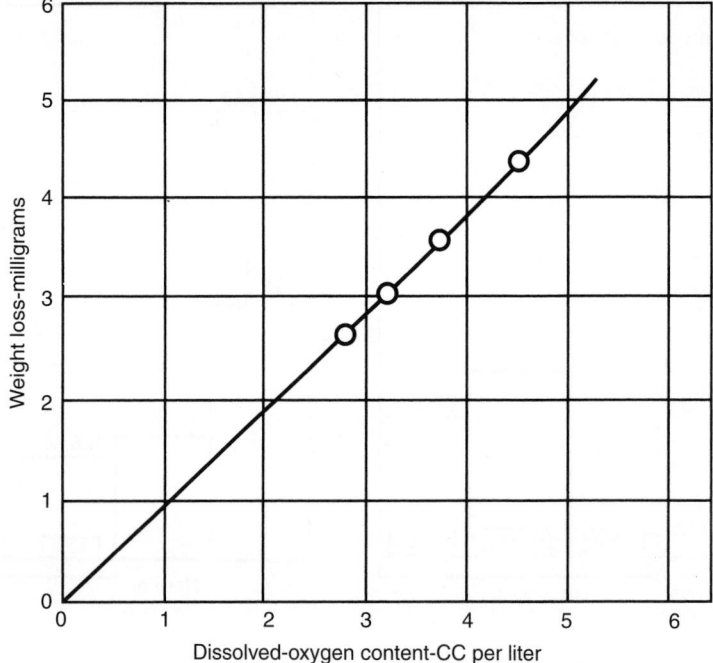

Figure 4.19.43 *Corrosion in sodium chloride solution containing dissoloved oxygen [26].*

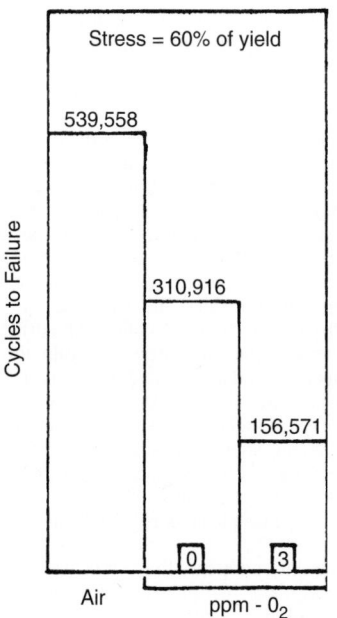

Figure 4.19.44 *Oxygen fatigue test—lignosulfonate mud, 9.5, pH [42].*

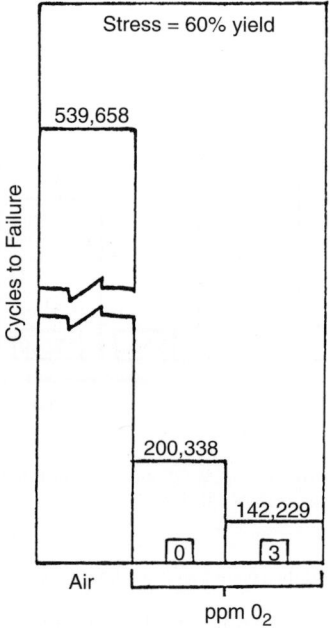

Figure 4.19.45 *Oxygen fatigue test—nondispersed mud, 8.5 pH [42].*

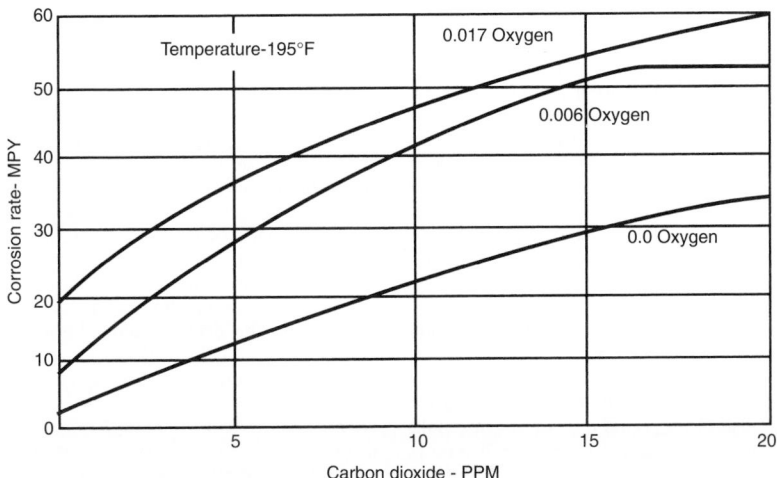

Figure 4.19.46   *Effect of carbon dioxide conentration on corrosion rate [43].*

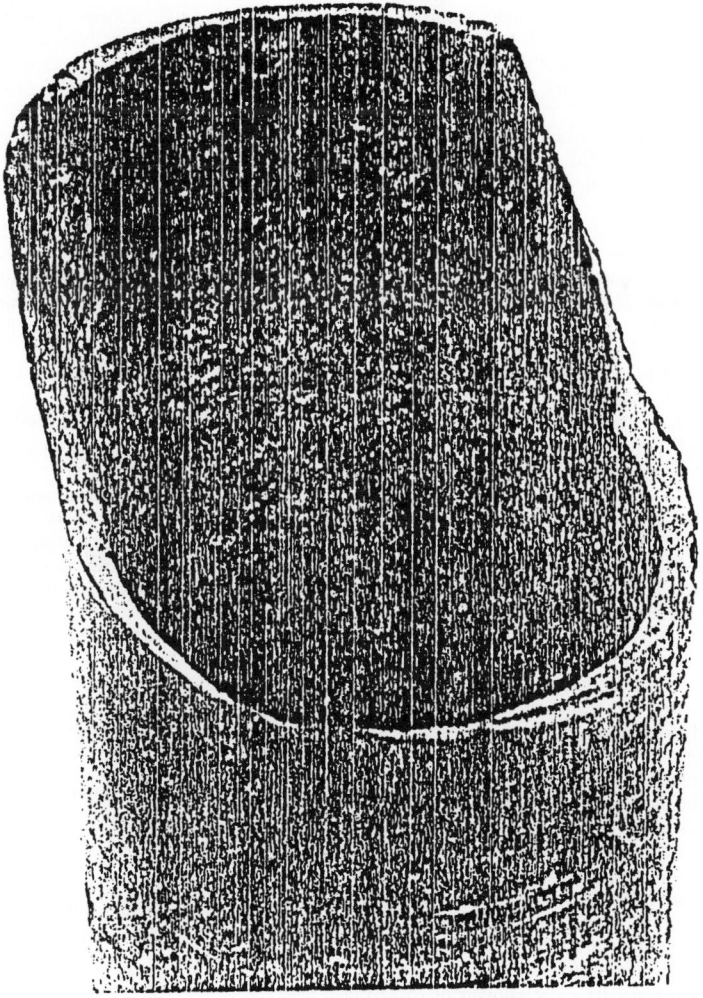

Figure 4.19.47   *Corrosion—fatigue failure of drillpipe caused by internal pitting [14].*

in very high strength steels. An exception is aluminum alloy equipment that may experience accelerated general corrosion in high pH environments as well as low pH environments.

In underbalanced drilling and where drilling fluid control is lost, the drilling equipment may encounter production fluids that may be mixed with lower quantities of drilling fluids. In these cases, accelerated general corrosion rates may be encountered and the risk of environmental cracking increases with increasing concentrations of H_2S and lower pH levels.

As mentioned previously, fatigue and corrosion fatigue are common failure mechanisms in drill string components. When the drilling fluids are not properly controlled, make-up torque not properly applied, when production fluids co-mingle with drilling fluids and when operational circumstances concentrate rotating bending stress (such as rotating a connection for an extended period through a severe bend in the drill string, fatigue failures can result. An example of a corrosion fatigue failure in drill pipe is presented in Figure 4.19.47.

4.19.4.2 Motors, Jars and Drilling Bits

Drilling motors, jars and bits are very complicated pieces of hardware that utilize a variety of materials that serve very specialized functions. These components are grouped together because they are used in conjunction with each other and each utilizes specialized materials with unique types of service. The bits and motors exhibit a relatively short finite life.

Drilling motors, specifically downhole mud motors, work by taking the hydraulic action of the drilling fluid and converting it to a rotational force. Instead of rotating an entire drill string, the mud motor will be at the end of a non-rotating work string or coiled tubing and provide the rotational forces necessary to drill the formation.

The materials in the mud motors are composed of hardened steel rotors and stators. The rotors are typically coated with metallic plating and the stators have a mating (to the rotor) elastomeric member bonded to the ID. There is a non-elastomeric version of the stator/rotor arrangement that utilizes specialized wear components including copper based alloys and cobalt based hardsurfacing alloys.

Jars are composed of high strength materials and their function is to store energy that can be released to create a jar or impact on the bottom hole assembly (BHA) that is usually comprised of the bit, subs and collars. The impact motion of the jar is used to free "stuck" pipe or BHAs. The jars are made from Cr–Mo and Ni–Cr–Mo steels with 4140 and 4340 types being typical. The jars are subjected to fatigue and impact type failures. Corrosion is not an expected failure mode for this product.

Drilling bits may either be the tricone "rotating cone" or the drag "fixed cone" variety. Both types incorporate hard drilling elements that cut into the formation by cutting and fracturing forces aided by hit hydraulics.

The cutting elements of the drilling bits are either the mining grade type of cemented tungsten carbides or natural/synthetic diamonds. The diamonds may be as natural stones or formed into polycrystalline diamond compacts (PDCs). The body of the drag bit may be a matrix head (a high strength braze type material with hard particles embedded in it or a hardened steel head with the cutting elements brazed or press-fit into it. The tricone type of drilling bit is composed of three drilling cones or heads that are attached to three legs or bodies and the assembly is joined together. The steel cones and legs are typically Ni-Cr-Mo carburizing grades of steel such as 4820 and 9310. The bearings that permit bit rotation under extremely high bearings

forces are very highly engineered and utilize such materials as beryllium copper, spinoidal copper alloys, cobalt base wear resistant alloys, hardened tool and stainless steels and cemented tungsten carbides.

Though there are occasionally corrosion failures in these products, the product lives are normally too short to be subjected to the majority of the potential corrosion mechanisms. This is not true if the drilling fluid is not controlled, the action of corrosion in production fluids such as hydrogen sulfide containing gas/water phases can cause cracking in very short periods of time (minutes or hours).

One problem shared by mud motors and drilling bits is the susceptibility to premature failure due to environmental interactions between the mud motor elastomer or tricone bit seal and the drilling fluid. There have been instances of motor and bit failures that have resulted from elastomer property deterioration in drilling fluids.

4.19.4.3 Blow-Out Preventers

Blow-out preventers (BOPs) are used at the surface (or subsea) for pressure control during drilling and work-over operations. Their primary purpose is safety; the key here is to prevent uncontrolled release of production fluids or "blow-outs." The two basic types are ram style and annular style BOPs. They are named for the action of the piston activating the BOP. A ram style utilizes opposing rams that are piston activated whereas the annular type utilizes compression that activates on a spherical or annular type sealing element. Pipe ram or variable ram type BOPs are used to isolate the annulus from the atmosphere. Blind ram BOPs or annular type BOPs too are used to shut off open hole pressure. There is also a shear ram BOP that is designed to sever tubing or drill pipe in the advent of pressure coming up through the tubing or drill pipe. Several BOPs are typically mounted together to form a BOP stack as illustrated in Figure 4.19.48.

A diverter is sometimes used in conjunction with a BOP stack to divert the gas from the annulus away from the stack.

BOPs and diverters are made from low alloy steels, the most common are modifications to 8630 and Cr–Mo steels such as $2\frac{1}{4}$ Cr–1Mo. BOPs may be subjected to corrosive production fluids when used for well control. Since the fluids may contain H_2S, the BOPs are built of materials in the conditions that are resistant to SSC. Corrosion problems with BOPs occur in seal areas that could become corroded by contact with aerated seawater and/or production fluids. The other potential is environmental interactions between the drilling fluid and the BOP sealing elements. There are drilling fluids that cause dramatic property changes in specific elastomers and care must be taken to ensure that the elastomers and the drilling fluids are compatible.

4.19.4.4 Casing and Hangers

Casing is a pipe that is used in the oil and gas industry to ensure a pressure tight connection from the surface to the reservoir being produced. There are many reasons for using casing including the following listed by Petroleum Extension service [44].

1. Prevent cave-in or wash-out of the hole
2. Prevent contamination of freshwater sands by fluids from lower zones
3. Exclude water from the producing formation
4. Confine production to the wellbore
5. Provide a means of controlling well pressure
6. Permit installation of artificial lift equipment for producing the well
7. Provide a flow path for produced fluids

The casing is divided into several types including conductor pipe, surface casing, intermediate casing, liner string

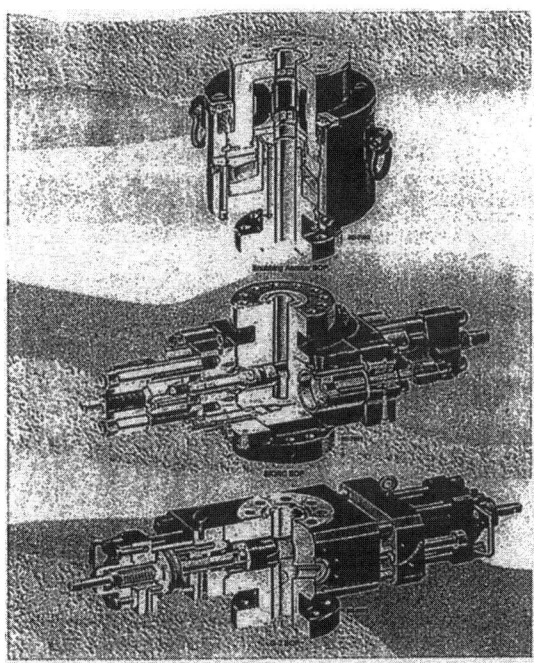

Figure 4.19.48 *Typical arrangement of a BOP stack.*

and production casing. These casing types are shown in Figure 4.19.49.

The casing can be made from a variety of materials ranging from low carbon steels to very highly alloyed nickel base alloys. The alloy and grade selection is based on strength requirements and the anticipated type of corrosion service. Liner hangers and casing are generally made from similar materials though the liner hangers are often made of higher strength corrosion resistant alloys.

The corrosion problems that could occur in casing and liner hangers include SSC, SCC, pitting corrosion, crevice corrosion, and general corrosion.

4.19.4.5 Cementing Equipment

Cementing equipment is used to control the location and flow of cement for cementing operations. This equipment includes a float collar, a guide (cementing) shoe and plugs. The purpose is to lock into place the casing and prevent migration of fluids from one position along the casing to another; this prevents communication between the different downhole formations such as those producing water and hydrocarbons. A schematic of some of this equipment is presented in Figure 4.19.50.

The cementing equipment is composed of materials that are relatively easy to mill or remove by subsequent drilling operations. The materials of construction are often aluminum alloys, cast iron, phenolic plastics and composite materials. There are generally no corrosion problems with these products in drilling fluids and cement. There can be corrosion problems in production environments or acids.

4.19.4.6 Coiled Tubing

Coiled tubing can serve many functions including milling, drilling, injection, production tubing and intervention/workover operations. The coiled tubing is continuous length meaning that there are no joints or connections.

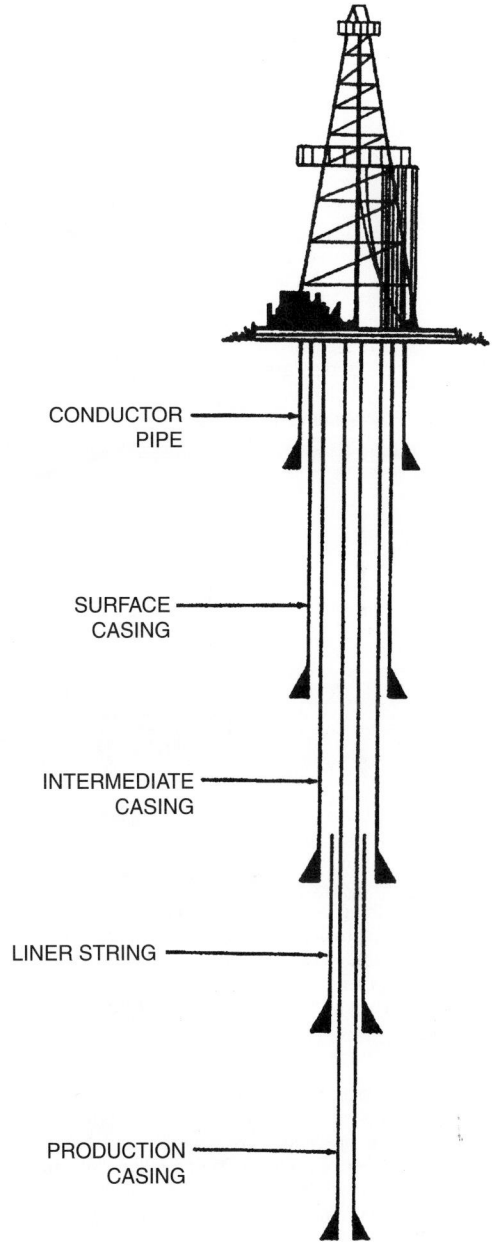

Figure 4.19.49 *The various strings of casing used in a well.*

The coiled tubing is usually made from HSLA type steels with the most popular type being a weathering type of HSLA steel. The strength level of the coiled tubing ranges from about 400 MPa (60 ksi) to about 750 MPa (110 ksi). The weathering type of steel has small intentional additions of chromium, copper and nickel to improve corrosion resistance.

Other materials have been made and used in coiled tubing including 13% chromium stainless steel, 15% chromium stainless steel and titanium.

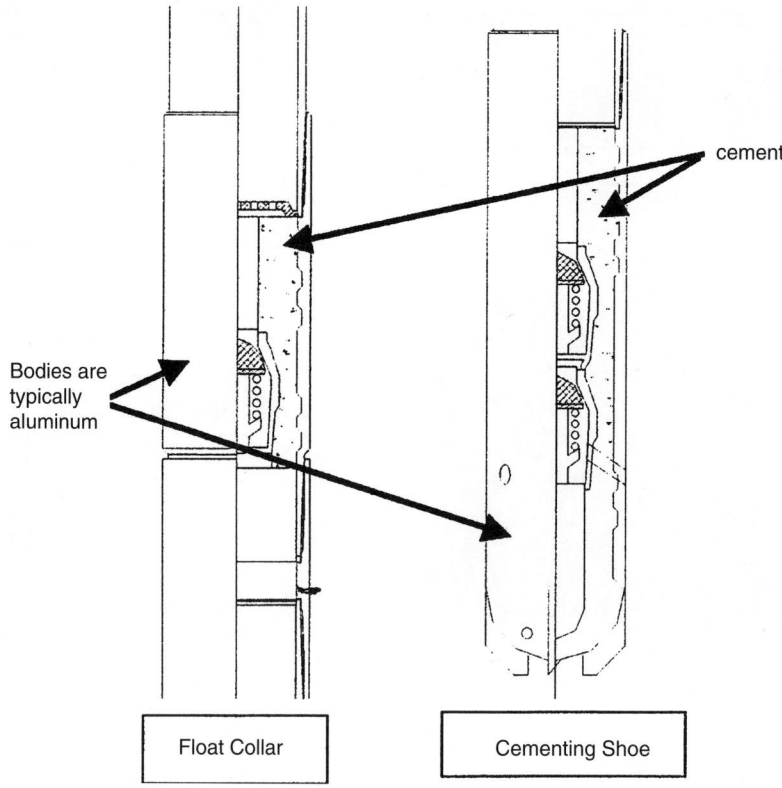

cement

Bodies are
typically
aluminum

Float Collar

Cementing Shoe

Figure 4.19.50 *Cementing equipment.*

The coiled tubing is susceptible to fatigue and corrosion fatigue, pitting corrosion, general corrosion and SSC in the high strength grades of coiled tubing. One of the biggest problems with respect to corrosion and coiled tubing is the pitting corrosion that can occur as a result of acidizing operations.

4.19.4.7 Wireline and Testing Equipment
Wireline is used to convey a large number of different types of tools. Wireline operations are generally quick and completed in a number of hours. Wireline is typically an alloy steel but can be made from corrosion resistant alloys. The wireline is normally very high strength material and some materials can be susceptible to cracking (SSC and SCC). The factors affecting this susceptibility are time, temperature and the contacting environment. Wireline equipment survives in aggressive environments despite it susceptibility because of the typical short duration of the wireline operations.

4.19.4.8 Marine Risers
Marine risers are subjected to offshore environments that range from surface marine, splash zone, and submerged sea exposure. All of the typical marine forms of corrosion are possible including general corrosion, galvanic corrosion, localized corrosion including pitting and crevice corrosion. The spash zone and near surface regions are of particular importance because of the washing motions and oxygen contents.

The marine risers are typically made from line pipe steels (usually high strength low alloy steels known as HSLA steels) such as X65, X70, and X80 and low alloy steels such as 4130.

Another consideration that effects all oilfield equipment, but should be stressed for drilling equipment, is the storage and care of equipment between uses. Other equipment types that more commonly encounter corrosion problems are generally better suited to withstand coastal or offshore storage than drilling equipment. Drilling equipment that functions well in the intended drilling environments may exhibit severe corrosion during storage periods.

4.19.5 Mitigation Techniques
Mitigation techniques include controlled strength level range steels, coatings, inhibitors, corrosion resistant alloys and surfacings. Some of the common mitigation techniques that are applicable to the corrosion mechanism are discussed in the section where the mechanism is discussed. This section summarizes some of these techniques and relates them to drilling and completion equipment.

The selection of the mitigation technique starts with an assessment of the product or tool material and the intended application environment. This assessment can be turned into design criteria whereby the mitigation techniques can be investigated and selected. This process is illustrated in Figure 4.19.51.

4.19.5.1 Materials Selection
In addition to corrosion resistance, the factors that to be considered in materials selection include strength level, product form, impact toughness, meeting code requirements and cost. It is frequently the case that some of the other product

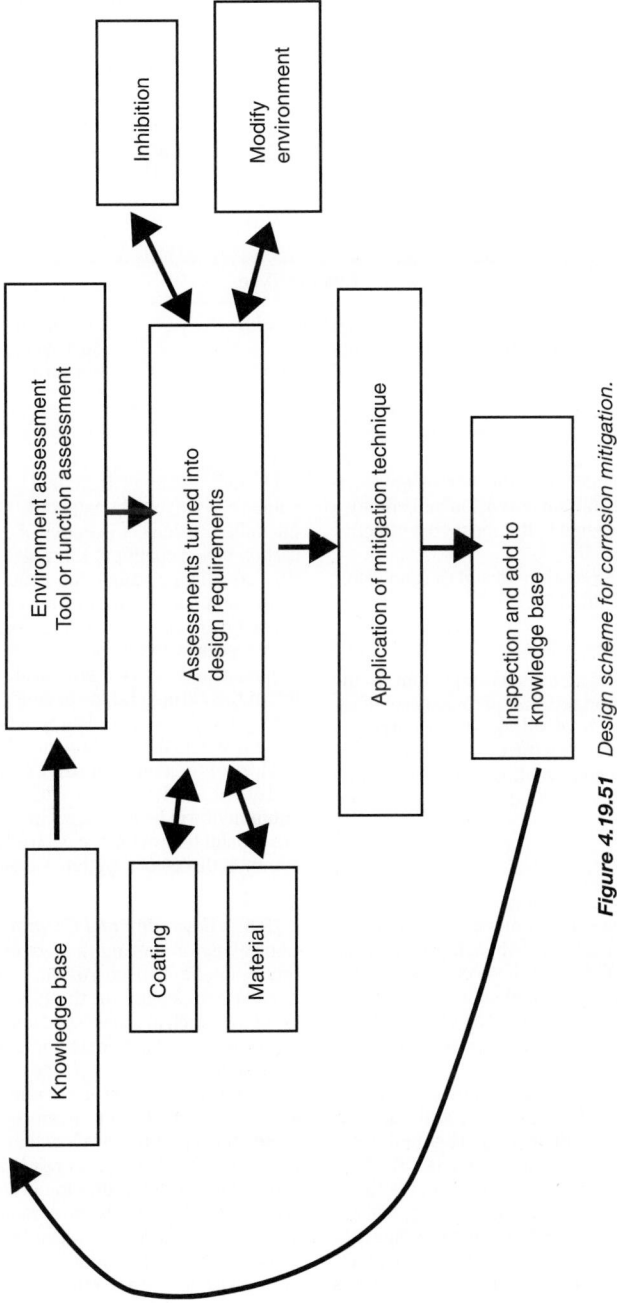

Figure 4.19.51 *Design scheme for corrosion mitigation.*

Table 4.19.7 *Equipment and Materials Choices*

Equipment	Yield Strength, ksi	Material Choices
Drill pipe	75–135 ksi	Cr-Mo steels (4140, 4145)
Tool joints	110 ksi	Cr-Mo steels (4135M)
Collars and Subs	105–120 ksi	Cr-Mo steels, beryllium copper, stainless steel
Jars	110–170 ksi	Ni-Cr-Mo steels (4340, 4330)
Tricone rock bits	110–150 ksi	Ni-Cr-Mo steels (4820, 8620, 9310)
Casing	55–150 ksi	Carbon and low alloys steels, stainless steels
Liner hangers	80–140 ksi	Low alloys steels, stainless steels, nickel base
Blow-out preventers	60–80 ksi	Cr-Mo & Ni-Cr-Mo (8630, 4130)
Cementing equipment	–	Phenolics, aluminum, cast iron
Coiled tubing	70–110 ksi	HSLA steels, stainless steels
Wireline	200 + ksi	Carbon steel (1067), stainless and nickel base
Drilling risers	65 ksi–80 ksi	Carbon, HSLA and low alloy steels

Table 4.19.8 *Typical Three Coat Marine Paint*

Coating Layer	Thickness	Coating Type	Purpose
(1) Prime	0.08 mm	Inorganic zinc	Corrosion protection — galvanic
(2) Intermediate	0.15 mm	Epoxy	Corrosion protection — barrier
(3) Finish	0.10 mm	Polyurethane	Appearance, UV protection — barrier

requirements take precedence over the corrosion resistance. It may also be that permitting corrosion and adding a corrosion allowance to the design is the most cost effective materials design selection.

Some of the specific drilling equipment and the applicable materials are presented in Table 4.19.7.

4.19.5.2 Design

Often the design of the equipment is as important as the material selection in preventing serious corrosion problems. Many types of corrosion such as sulfide stress corrosion (SSC) and stress corrosion cracking (SCC) may be mitigated by reducing the stress level and avoiding stress concentrations. Crevice corrosion may be avoided my preventing tight crevices.

4.19.5.3 Coatings and Surfacings

There are several types of coatings that are used to prevent corrosion in drilling and well completion equipment. These can be broadly classed as metallic coatings, inorganic coatings and organic coatings. With all of these classes, the one feature that has a huge role in the final success rate is the surface preparation. The surface needs to be clean and free of contaminants and exhibit the proper surface profile for the specific coating being applied.

Metallic coatings include (1) hot-dip galvanizing, (2) electroplating of nickel, chromium, copper, tin, and zinc, (3) electroless nickel plating, (4) high velocity (oxy-fuel) spray coatings—also known as HVOF, (5) metallizing (arc or flame spraying), (6) plasma (PVD) and ion implantation, (7) chemical vapor deposition (CVD) coatings, (8) cladding, and (9) weld overlays. The major use of these metallic coatings is to prevent corrosion on carbon and low alloy steels. There are applications of copper plating and hard surfacing for reasons other than corrosion control.

Inorganic coatings includes include (1) surface conversions, (2) inorganic zinc silicate, (3) HVOF ceramic coatings, and (4) flame and arc spray ceramic coatings. The common surface conversions are anodizing used on aluminum, phosphating and oxidizing processes used on steels and chromate surface conversions on steel.

Organic coatings include (1) epoxy paints, (2) acrylic paints, (3) polyurethane paints, (4) organic zinc paints, (5) phenolic coatings, (6) epoxy-phenolic coatings, (7) fluorocarbon based coatings, (8) molybdenum disulfide coatings, (9) and silicone paints and coatings. Generally, where the word paint appears in the above list, the coating consists of at least two components, one a carrier and one the paint. In many cases, with two or more part products that are mixed, the paint undergoes a chemical reaction causing or assisting in the curing and/or drying cycles. The coatings listed above that do not list plant generally require a heat curing cycle that sets the coating onto the substrate.

Most of the common paint systems designed for marine and offshore oil and gas drilling and surface well completion equipment will be coated successively with two or more layers of paint to provide the desired corrosion protection. One common three-coat system is described in Table 4.19.8.

4.19.5.4 Environmental Control

Controlling or altering the factors that effect corrosion can reduce or eliminate corrosion. In drilling fluids, we can monitor the composition of the fluid and make adjustments to avoid too much abrasive sand and adjust the pH to the range where we can avoid corrosion and environmental cracking problems.

We know that corrosion rates generally increase with temperature and the temperature of drilling fluids can be controlled through the use of cooling towers. We also know that injected air (and oxygen) greatly affects the corrosion rates and reduces the life due to corrosion fatigue. The oxygen effects on fatigue life were demonstrated in Figures 4.19.44 and 4.19.45. There could also be entrained acid gases that contaminate the drilling fluid decreasing the pH; a degasser unit can remove these entrained gases.

In Figures 4.19.40, 4.19.41, and 4.19.42, the effects of chlorides on drill string corrosion fatigue were presented. The chloride content can be minimized through freshwater dilution.

The H_2S in the environment can be scavenged through the use of zinc and iron based compounds. There are two types of zinc based scavengers, the slightly water soluble inorganic compounds such as carbonates and the organic zinc-chelate type that is highly water soluble. The zinc carbonate combines with hydrogen sulfide in the following manner:

$$ZnCO_3 + H_2S = ZnS + H_2O + CO_2$$

The zinc sulfide is insoluble and precipitates out of the environment. This type of scavenger works at neutral to basic pH values. These scavengers work well in drilling fluids. The organic zinc-chelates are more expensive but work over a wider pH range including the acidic levels and are typically used in brine type drilling fluids.

The iron-based scavenger used is magnetite. This a very effective H_2S scavenger that works best in acidic conditions; the scavenging rate is slow at high pH levels. The magnetic reacts with H_2S in the following manner:

$$Fe_3O_4 + 6H_2S = 3FeS_2 + 4H_2O + 2H_2$$

Oil based drilling fluids are highly effective in controlling corrosion but their use is restricted by costs and environmental considerations. The oil-based fluids work because the fluids are non-conductive and do not support the flow of electrons necessary for corrosion reactions (there is no electrolyte).

4.19.5.5 Inhibition

An inhibitor is a substance that slows or stops a chemical reaction. When an inhibitor is present in sufficient enough quantity to slow a corrosion reaction, the corrosion is prevented. The basic mechanisms by which inhibitors function are as follows [45]:

1. They absorb onto the corroding material as a thin film.
2. They induce formation of a thick corrosion product, which forms a passive layer.
3. They change the characteristics of the environment either by producing protective particulates or by removing or inactivating an aggressive constituent in the environments.

As a type of material, the inhibitor may be an organic or an inorganic compound. Inorganic inhibitors include chromates (CrO_4^{-2}), silicates (SiO_3^{-2}) and phosphates (PO_4^{-2}). The three basic types of inorganic inhibitors are anodic passivating inhibitors, cathodic inhibitors and cathodic precipitators. There are also mixed inhibitors that suppress both anodic and cathodic reactions.

Anodic inhibitors suppress or prevent anodic reactions by assisting the natural passivation of metal surfaces or by forming films/deposits that isolate the metal from the environment. These anodic inhibitors are sometimes categorized by being either highly oxidizing or by being much less oxidizing. Examples of the highly oxidizing anions include chromates, nitrites and nitrates. Sodium nitrite and sodium chromate are two of the most effective; their effect on the reduction-oxidation (redox) potentials are presented below [46]:

$$2NO_2^- + 6H^+ + 4e^- = N_2O + 3H_2O \text{ where } E^0 = +1.29V$$

$$2CrO_4^{-2} + 10H^+ + 6e^- = Cr_2O_3 + 5H_2O \text{ where } E^0 = 1.33V$$

The low oxidizing anodic inhibitors include phosphates, molybdates, and tungstates. The low oxidizing anions require an external source of oxygen for passivation. These low oxidizing anions are sometimes referred to as dangerous inhibitors because they can actually cause pitting when present in low concentrations. Anodic inhibitors are also considered dangerous because cathodic reactions may not be inhibited on unprotected areas [46].

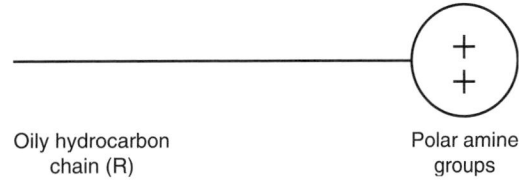

Oily hydrocarbon chain (R)   Polar amine groups

Figure 4.19.52   *A semipolar molecule [47].*

The cathodic inhibitors suppress the cathodic chemical reaction and these are further subgrouped by the way in which the cathodic reaction is affected. The three basic types are cathodic poisons, cathodic precipitators and oxygen scavengers. Of these, oxygen scavengers are the type that are used in drilling and well completion operations.

The oxygen scavenger works by combining with free oxygen, this inhibitor type prevents cathodic depolarization. One common effective oxygen scavenger is sodium sulfite and the reaction to scavenge oxygen is as follows:

$$2Na_2SO_4 + O_2 = 2Na_2SO_4$$

The organic type of inhibitors affect the entire surface of the metal by forming films on the surface. The organic inhibitors are semi-polar in that the make-up of the inhibitor includes a polar (conductive) and a non-polar (non-conductive) component. The semi-polar nature renders one end of the molecule hydrophobic (water repelling) and the other hydrophilic (water loving). This semi-polar molecule is illustrated in Figure 4.19.52.

These semi-polar molecules bond or attach to the metal surface on the polar (conductive) end of the chemical compound. A thin film only a few molecules thick is adsorbed according to the ionic charge of the inhibitor molecule and the charge of the metal surface. These organic inhibitors can preferentially attach to either the cathode or the anode of the system; these inhibitors are termed cationic inhibitors (positively charged) and anionic inhibitors (negatively charged). Amines are examples of organic cationic inhibitors and sulfonates are examples of organic anionic inhibitors. The bonding of these inhibitors is illustrated in Figure 4.19.53.

One advantage of the organic inhibitors is that they are typically self-healing when the inhibitor is present in sufficient concentration; if the film is damaged or removed through flowing conditions, the film reforms. The factors that effect how well the inhibitor works includes:

1. Molecule size
2. Bond strength to the metal surface
3. Type and number of bonding atoms
4. Filming characteristics

The value of the corrosion inhibitor is related to how much the corrosion rate is lowered by use of the inhibitor. This can be calculated by the use of the following formula:

$$E = 100(R_0 - R_i)/R_0$$

where $E =$ inhibitor efficiency
$R_0 =$ rate of corrosion without inhibitor
$R_i =$ rate of corrosion with inhibitor

These inhibitors are used and applied in a number of ways. First, an inhibitor can be used directly on the metal requiring protection such as drill pipe. This is typically sprayed onto the metal surface. Second, the inhibitor can be added to the environment such as additions made to drilling fluids.

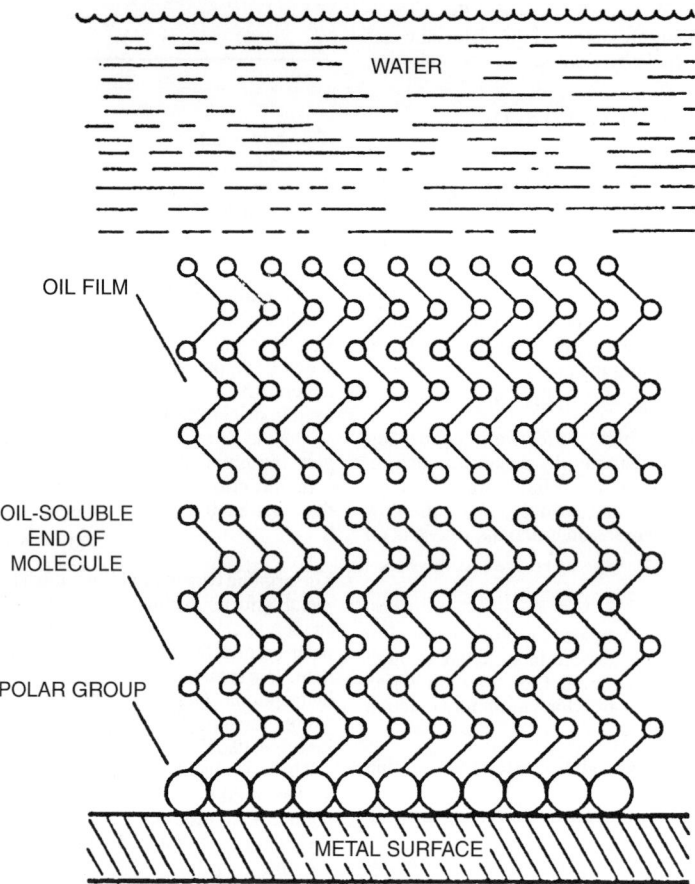

Figure 4.19.53   *Schematic of film-forming inhibitor [48].*

When the inhibitor is added to the environment, it is either added continuously or by batch. When added continuously, a small quantity is added to maintain a minimum concentration in the environment. When added by batch, a much larger concentration is added at intervals that are determined through monitoring.

When inhibition is selected as the means by which corrosion is prevented, the environment should be monitored to ensure a minimum effectiveness of the inhibitor. This is true regardless of the addition method, continuous or batch.

4.19.6 Corrosion Monitoring and Equipment Inspection

One way to combat corrosion is to maintain an effective corrosion-monitoring program to supplement good preventative measures. It is also very important to keep complete records of monitoring programs, control programs and failures that occur.

An effective corrosion control program should be able to detect evidence of corrosion and identify the causes. Therefore, continuous monitoring is essential during drilling operations because the nature of drilling fluid corrosivity changes as the hole is drilled and different formations are penetrated. Several techniques of monitoring corrosion should be used simultaneously whenever possible, and records need to be kept.

4.19.6.1 Linear Polarization Instruments

Linear polarization instruments provide an instantaneous corrosion-rate data, by utilizing polarization phenomena. These instruments are commercially available as two-electrode and three electrode types (Figure 4.19.54). The instruments are portable, with probes that can be utilized at several locations in the drilling fluid circulatory systems. The technique involves monitoring electrical potential of one of the electrodes with respect to one of the other electrodes as a small electrical current is applied. The amount of applied current necessary to change potential (no more than 10 to 20 mV) is proportional to corrosion intensity. The electronic meter converts the amount of current to read out a number that represents the corrosion rate in mpy. Before recording the data, sufficient time should be allowed for the electrodes to reach equilibrium with the environment. The corrosion-rate reading obtained by these instruments is due to corrosion of the probe element at that instant [22].

The limitation of these instruments is that they only indicate overall corrosion rates. Their sensitivity is affected by deposition of corrosion products, mineral scales or accumulation of hydrocarbons. Corrosivity of a system can be measured only if the continuous component of the system is an electrolyte. These devices may not adequately predict localized corrosion.

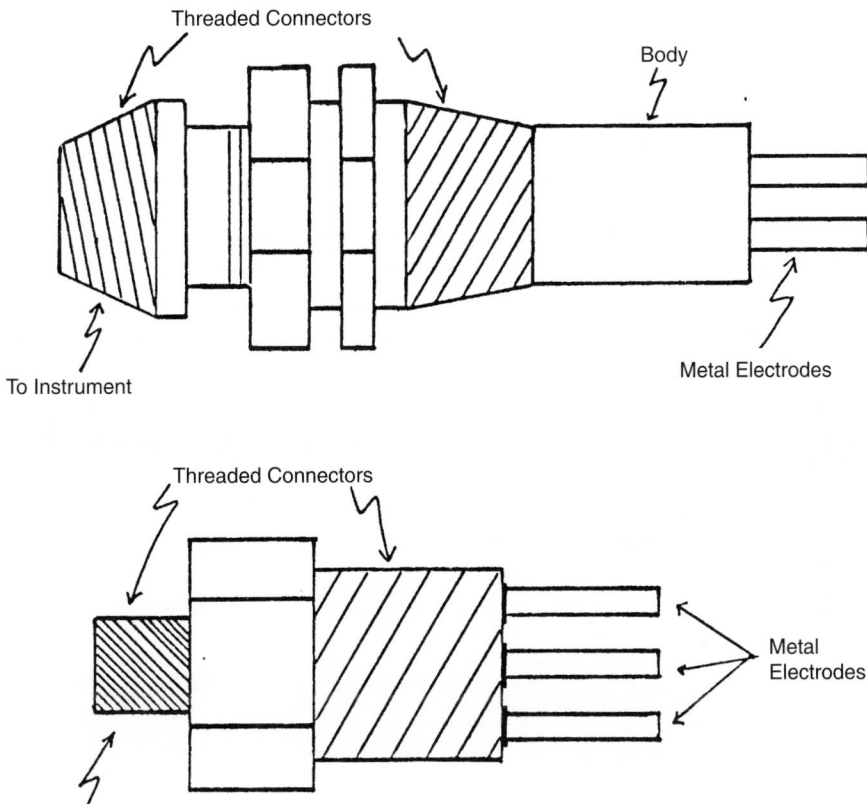

Figure 4.19.54 *Linear polarization instrument probes.*

4.19.6.2 Galvanic Probe

The galvanic probe continuously monitors the corrosion characteristics of the drilling fluid. The probe (Figure 4.19.55) consists of two dissimilar metal electrodes, usually brass and steel. The electrodes are mounted on, but insulated from a threaded high-pressure plug. These electrodes are connected to each other through a DC ammeter capable of detecting microamperes when the probes are immersed in an electrolyte. Enough time is allowed for the electrodes to reach equilibrium and read the current flow through the external loop. The corrosion process occurring on the electrodes generates the current measured [22].

The amount of current flow is a measure of its corrosiveness. The probe generally registers low-current flow (0–10 mA) in slightly corrosive environments. However, high-current flows (40–100 mA) have been recorded in severely corrosive environments. The current intensity generally depends on oxygen concentration of the system, since oxygen depolarizes the brass cathode, thereby continuing the corrosion process of the cell. Among various locations of a surface circulatory system, the instrument can be installed downstream of a deaerator and in the standpipe.

If the instrument indicates current surge in an air-free system, it generally implies hydrogen sulfide contamination, but the galvanic probe is usually best suited to detect corrosion influenced by oxygen contamination. Other limitations

of this instrument are the same as those of linear polarization instruments discussed earlier.

4.19.6.3 Hydrogen Probe

The hydrogen probe (Figure 4.19.56) basically consists of a hollow, thin-walled, steel tube that is sealed on one end and the other end is equipped with a pressure gage. Once mounted in the system, the probe body corrodes. Some of the nascent hydrogen generated by the corrosion process in the presence of hydrogen sulfide diffuses through the tube wall. Once inside the valid space in the tube, the hydrogen atoms combine to form molecular hydrogen gas. As these hydrogen gas molecules are too large to diffuse back through the tube wall, the pressure in the tube rises. The pressure rise in the tube is a function of the amount of hydrogen generated by the corrosion process.

One problem with hydrogen probes do not perform well in aeraed fluids [47].

4.19.6.4 Corrosion Coupons

The most direct method of evaluating the corrosivity of an environment is the use of corrosion coupons. In drilling operations, a drill string corrosion coupon is a ring coupon machined from a section of tubing and sized to fit into the relief groove in the tool joint box.

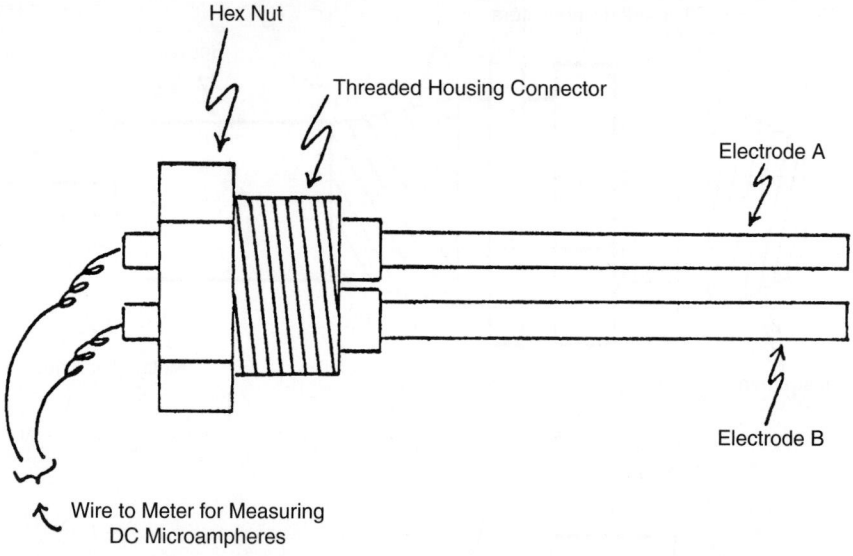

Hex Nut

Threaded Housing Connector

Electrode A

Electrode B

Wire to Meter for Measuring
DC Microampheres

Figure 4.19.55   *Typical galvanic probe.*

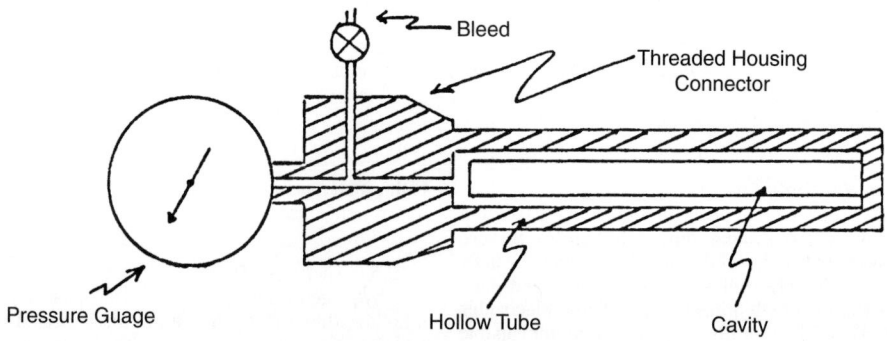

Bleed

Threaded Housing
Connector

Pressure Guage

Hollow Tube

Cavity

Figure 4.19.56   *Hydrogen probe.*

To employ effective control measures, it is very important to determine the type of corrosion attack. Spot analysis of the corrosions film and careful visual examination of the coupon surface can help in determining the type and severity of corrosion attack. Generalized corrosion is represented by continuous attack over the entire surface, but no pitting. The pitting type of corrosion is represented by a high concentration of pits on the coupon surface. This type of corrosion attack is the most serious attack resulting from drilling fluids, as discussed earlier.

Weighing the coupon before and after exposure, and evaluating the change in weight permits a determination of the severity of general corrosion attack. Before installation the coupon must be clean (i.e., free of any corrosion, grease marks, drops of perspiration, etc.) and weighed. After exposure to the system for a known period of time, the coupon is retrieved, visually examined, then cleaned and reweighed within one tenth of a milligram. The difference between the initial and final weights is attributed to corrosion and converted to a corrosion rate knowing the surface area of the corrosion coupon exposed to the environment.

There are several factors such as handling, surface preparation and cleaning, etc., which can affect the results of the tests. The results obtained from this test assume uniform corrosion. Therefore, for proper analysis it becomes very important to include a complete description of the exposed coupons. One of the most important factors is visual inspection of the coupon, describing the form of attack and identifying the corrosion by-product.

In assessing drill string corrosion, there are limitations in using corrosion coupons. The coupon experiences the same downhole conditions as the drill string does and represents the corrosive effects of the downhole environment. However, the coupon is only exposed to the one side (inside) of the drill string and is not subject to the same stresses. Stresses can and do influence corrosion rates and corrosion on the outside of the drill string may not be adequately assessed. The results obtained are only for a certain depth of exposure, while the corrosion may vary appreciably up and down the hole. Finally, the results are not available until the tool containing the coupon is pulled out of the hole.

4.19.7 Recommended Practices

Equipment failure due to corrosion is one problem that inevitably rears its ugly head during drilling operations. Corrosion control is an essential factor in any engineering design and must be considered as early as the initial phase of the operation. An effective and most cost-efficient corrosion control program is imperative for a successful drilling operation. Some recommended practices are as follows [2,14,19,22,49,50].

4.19.7.1 Drilling Operations

1. Use quenched and tempered drill pipe that meets the API requirements for drill pipe.
2. Avoid the use of shrink-type tool joints on drill pipes and drill collars (connectors, in this case). Flash-welded or friction welded tool joints for drill pipes and integral drill-collar joints should be used.
3. Utilize proper make-up torque on tool joints.
4. Use internally, plastic-coated drillpipe at all times. The coating must be holiday-free to be effective.
5. Consider the use of heavy-wall (thick wall) drillpipe in severely corrosive conditions. Heavy walls reduce stress levels and extend the service life of drill pipe.
6. Materials used for tool joints are generally modified 4135 steel. They are heat treated by quenching and tempering at temperatures equal to or above 1150°F to

hardness of Rc 30 to Rc 37 and yield strength ranging from 120,000 psi to 135,000 psi.

7. Most drill collars are also made with materials similar to those used for tool joints. Where possible, use heavy walls are used to reduce stress levels.
8. limit exposure time of the drillstem to the corrosive environment during drillstem tests. A maximum of one hour is recommended.
9. Sustained exposure of drill pipe to operating temperatures above 300°F (149°C) must be limited to a maximum of 10 hr.
10. Run enough drill collars to keep all drill pipe in tension in order to reduce wear and stress on tool joints.
11. Use controlled drilling fluids at all times and minimize contamination by potential corrodents.
12. Use oxygen and hydrogen sulfide scavengers when appropriate to reduce corrosion rates.

4.19.7.2 Transportation

1. Drillpipes and drill collars should be tied down with suitable chains at the bolsters when being transported.
2. Load with all couplings on the same end of the truck.
3. Use thread protectors on tool joints when moving or racking pipes.
4. Avoid chafing a tool-joint shoulders on adjacent joints.
5. Keep load binding chains tight at all times.
6. Avoid damaging coatings as well as the pipe itself.

4.19.7.3 Handling

1. Before unloading, make sure that the thread protectors are tightly in place.
2. Avoid rough handling that can damage the pipe in any way.
3. The pipe must not be dropped at any time.
4. Unload one joint at a time.
5. When rolling down skids, roll pipe parallel to the stack. Do not let the pipe gather any momentum and strike the ends. This can damage threads even when the protectors are in place.
6. Avoid creating knicks or notches on the drillpipe.

4.19.7.4 Chemical Additives

1. When high temperatures above 300°F are expected, do not use sulfur-containing compounds as drilling fluid additives.
2. Use only chemical additives that are compatible with drilling fluids in circulatory systems.
3. Avoid using copper-based compounds such as copper carbonate. Copper can "plate out" on steel and set up galvanic corrosion cells, resulting in accelerated corrosion of the steel.
4. Maintain the pH level of the drilling fluid around 9.5 to reduce corrosion. However, when aluminum drill pipes are in use, the pH must be maintained between 9.5 and 7.
5. Minimize salt contents.
6. Caustics should be dissolved in water before being added. The mixture should be mixed well with mud guns in the pit prior to pumping into the hole.

References

1. ASM International, *Metals Handbook*, Volume 13 — Corrosion, ASM International, Materials Park, Ohio, 1987.
2. NACE Standard MR0175-2002, Sulfide Stress Corrosion Cracking Resistant Materials for Oilfield Equipment," NACE, Houston, Texas, 2002.

3. Gellings, P. J., *Introduction to Corrosion Prevention and Control*, Delft University Press, Delft, The Netherlands, 1985.

4. Denbigh, K., *Principles of Chemical Equilibrium*, 2nd Edition, Cambridge Press, 1981, pp. 133–186.

5. Carter, G. F., *Principles of Physical and Chemical Metallurgy*, ASM International, Materials Park, Ohio, 1979.

6. *Electrochemical Techniques for Corrosion Engineering*, R. Baboian, ed., NACE, Houston, Texas, 1986.

7. Tafe, J., *Z. Physik. Chem.*, Vol. 50, p. 605, 1905.

8. Budinksi, K., *Engineering Materials Properties and Selection*, 2nd Edition, Reston Publishing Co., 1983.

9. Uhlig, H. H., and Reeves, R. W., *Corrosion and Corrosion Control*, 3rd Edition, John Willey and Sons, New York, 1985.

10. Fontana, M. G., and Greene, N. D., *Corrosion Engineering*, McGraw-Hill Book Company, New York, 1967.

11. *Corrosion—Understanding the Basics*, edited by J. R. Davis & Associates, ASM International, Materials Park, Ohio, 2000.

12. Fontana, M. G., *Corrosion Engineering*, McGraw-Hill Book Company, New York, 1986.

13. Chilingarian, G. V., *Drilling and Drilling Fluids*, Elsevier Publishing Co., Amsterdam, The Netherlands, 1983.

14. Craig, B. D., *Practical Oilfield Metallurgy and Corrosion*, 2nd Edition, PennWell Publishing Co., Tulsa, Oklahoma, 1993.

15. Frankenthal, R. P., and Kruger, J., editors, *Passivity of Metals*, Pennington, New Jersey, 1978.

16. Isaacs, H., Bertocci, U., Kruger, J., and Smialowska, S., editors, *Advances in Localized Corrosion*, Proceedings of the 2nd International Conference on Localized Corrosion, NACE, Houston, Texas, 1990.

17. Staehle, R. W., Brown, B. F., Kruger, J., and Agarwal, A., editors, *Localized Corrosion*, Proceedings of the U.R. Evans Conference, Houston, Texas, 1974.

18. Shreir, L. L., *Corrosion*, Volume 1, John Wiley and Sons, New York, 1963.

19. NACE MR0175/ISO 15156-3:2003, NACE, Houston, Texas, 2003.

20. Trethewey, K. R., and Chamberlain, J., *Corrosion for Science and Engineering*, 2nd Edition, Longman Scientific and Technical, London, 1995.

21. "Corrosion Control in Petroleum Production," TPC5, NACE, Houston, Texas, 1984.

22. *NACE Corrosion Engineer's Reference Book*, edited by R. S. Treseder, NACE, Houston, Texas, 1980.

23. *API Recommended Practice 14E*, 5th Edition, Recommended Practice for Design and Installation of Offshore Production Platform Piping Systems (October, 1991).

24. *Metals Handbook — Desk edition*, edited by H. E., Boyer, and T. L., Gail, ASM International, Materials Park, Ohio, 1985.

25. Eeters, L. L., "The Nature of Fatigue," *Journal of Petroleum Technology*, August, 1964.

26. Ostroff, A. G., *Introduction to Oilfield Water Technology*, NACE, Houston, 1979.

27. *Stress Corrosion Cracking*, Edited by R. H. Jones, ASM International, Materials Park, Ohio, 1992.

28. Moore, E. M., and Warga, J. J., *Materials Performance*, pp. 17–23 (June, 1976)

29. *Hydrogen Effects in Metals*, edited by I. M. Bernstein and A. W. Thompson, AIME, 1981.

30. Pressouyre, G. M., and Bernstein, I. M., A Quantitative Analysis of Hydrogen Trapping, *Metallurgical Transactions A*, Vol. 9A, 1978.

31. NACE Standard TM0284-2003, "Evaluation of Pipeline and Pressure Vessel Steels for Resistance to Hydrogen-Induced Cracking," NACE, Houston, Texas, 2003.

32. *H₂S Corrosion in Oil & Gas Production — A Compilation of Classic Papers*, edited by R. N. Tuttle and R. D. Kane, NACE, Houston, Texas, 1981.

33. Hudgins, C. M., et al., "Hydrogen Sulfide Cracking of Carbon and Alloys Steels," *Corrosion*, Vol. 22, pp. 238–251, August, 1966.

34. Treseder, R. S., and Swanson, T. M., "Factors in Sulfide Corrosion Cracking of High Strength Steels," *Corrosion*, Vol. 24, pp. 31–37, February, 1968.

35. "Guidelines on Materials Requirements for Carbon and Low Alloy Steels for H₂S-Containing Environments in Oil and Gas Production," European Federation of Corrosion Publication 16, 2nd Edition, Institute of Metals, London, 2002.

36. Townsend, H. E., *Corrosion*, Vol. 28, pp. 39–45, 1972.

37. Bates, J. F., "Sulfide Cracking of High Yield Strength Steels in Sour Crude Oils," *Materials Protection*, pp. 33–40, January, 1969.

38. "The Role of Bacteria in the Corrosion of Oil Field Equipment," TPC3, NACE, Houston, Texas, 1976.

39. Treseder, R. S., and Tuttle, R. N., *Corrosion Control in Oil and Gas Production*, 1998 edition, CorUPdate, Houston, Texas, 1998.

40. DeWaard, C., and Milliams, D. E., *Corrosion*, Vol. 31, Number 5, 1975.

41. "Predicting CO2 Corrosion in the Oil and Gas Industry," European Federation of Corrosion Publication 13, Institute of Metals, London, 1994.

42. Azar, J. J., "How CO2 and Chlorides Affect Drill Pie Fatigue," Petroleum Engineering International (March, 1979).

43. Deem, H. A., "Plastic Coating of Drill Pipe — Is it Worth the Added Expense?," World Oil (November, 1964).

44. "Casing and Drilling," Unit II Lesson, 4, 2d Edition, Petroleum Extension Service, University of Texas, Austin, Texas, 1982.

45. Nathan, C. C., "Corrosion Inhibitors," NACE, Houston, Texas, 1974.

46. Talbot, D., and Talbot, J., *Corrosion Science and Technology*, CRC Press, New York, 1998.

47. Jones, L. W., *Corrosion and Water Technology for Petroleum Producers*, Oil and Gas Consultants International Inc., Tulsa, Oklahoma, 1988.

48. *Betz Handbook of Industrial Water Conditioning*, Betz Laboratories Inc., Trevose, Pennsylvania, 1980.

49. Moore, W. W., *Fundamentals of Rotary Drilling*, Harcourt Brace Jovanovich Publishing Co., New York, 1981.

50. *Drilling Manual. Manual*, 9th Edition, IADC, 1974.

4.20 ENVIRONMENTAL CONSIDERATIONS FOR DRILLING OPERATIONS

4.20.1 Introduction

Planning for drilling should include environmental considerations. Environmental management at the well site involves thoughtful planning at the onset of exploration or development. In today's world, a project may be postponed or terminated because of these issues. Plans must be developed and permits applied for and received before moving any equipment onto the location. Obtaining the Construction General Permit is the first step in complying with the Storm Water Pollution Prevention Plan (SWPPP) (see Chapter 6, Section 9 details on this federal regulation). After the plans and permits are obtained, the pre-spud meeting, in addition to discussing well depth, casing points, and rig selection, should cover topics pertinent to the environmental management of the drilling and completion operation. Regulating agencies are most concerned with these issues. A site will have particular factors that are imminently apparent and those perhaps not so apparent except to groups exhibiting a certain interest.

It is the role of the regulating agency to protect the public and the public domain from detrimental effects caused by industrial operations. Compliance with the regulator's requirements is usually simple in nature. These requirements are documented and should be followed in a particular order. Concerns disseminated by special interest groups are often unexpected. In an attempt to forego stalling a project by these parties, public disclosure of a project should be made as early as possible, even if it is still pending. A well-informed public disclosure statement released to the immediate community may prevent any future surprises.

4.20.2 Well Site

In planning a drilling operation, the location and access are primarily keyed to environmental decisions. The access and location must be able to maintain the traffic load and mitigate any impact on local resources such as the flora, fauna, and cultural and aesthetic sites. In certain instances, the preparation of an environmental assessment followed by an environmental impact statement may be considered warranted because of proximate

- Wildlife refuges
- Historic or cultural resources
- Recreation areas
- Land containing threatened or endangered species

In the United States, drilling plans are submitted to the Bureau of Land Management (BLM) or state oil and gas commissions in the form of an Application for Permit to Drill (APD), depending on ownership of mineral resources. Other countries have similar requirements. In addition to these agencies, other appropriate surface management agencies may have to be considered, including the BLM, National Park Service, Tribal Authorities, State Environmental Program, and County Environmental Program (see Chapter 6, Table 6.9.1 for a partial listing of federal and state agencies).

In the permitting process, it is important to have the entire operation planned. It is a necessary component of most APDs. In a federal APD, the 13-point surface use plan must address the following items before approval:

1. Location of site and existing roads
2. Planned access roads
3. Location of existing wells
4. Location of existing or proposed production facilities
5. Location and type of water supply
6. Source of construction materials
7. Methods of handling waste disposal
8. Ancillary facilities
9. Well site layout
10. Plans for restoration of the surface
11. Surface ownership
12. Information such as proximity to water, inhabited dwellings, and archeological, historic or cultural sites
13. Certification of liability

Details of the actual drilling program are not considered in detailed in an APD except for the casing and cementing program and how they are designed to protect any underground sources of drinking water (USDW). The bulk of the APD permit is designed to address the impact on surface resources and the mitigating procedures the operator plans to take to lessen these impacts [1].

4.20.3 Environmental Regulations

In the United States, the Environmental Protection Agency (EPA) sets policy concerning environmental regulations. The states are then allowed primacy over jurisdiction of environmental compliance if their regulations are at least as stringent as the federal regulations. Several bodies of government may be involved in environmental decisions concerning the well site. It was at one time common in most oil and gas producing states that the oil and gas division was allowed to administer all matters covering exploration and development. However, because of increased regulations, the responsibility is usually spread among several agencies, with air quality being one that is usually separate from daily oil and gas operations. It is primarily the responsibility of state agencies to protect the integrity of the state's land and water supplies from contamination due to oil and gas activity. The state oil and gas division's environmental regulations usually closely mirror those provided by the EPA.

Congress, in an attempt to promote mineral development in the United States, has exempted most hazardous wastes produced at the well site under the Resource Conservation and Recovery Act (RCRA) Subtitle C regulations. Hazardous waste are listed by inherent characteristics of

- Toxicity
- Ignitability
- Corrosiveness
- Reactivity

Although a number of waste products at the well site are considered characteristic hazardous wastes, some wastes fall under the nonhazardous description. The regulation of these fall under RCRA Subtitle D. Initially, Subtitle D wastes were regulated to control dumping of domestic trash and city runoff. The EPA is considering promulgating regulation of certain oil and gas waste under Subtitle D [2].

Under the RCRA exemption, wastes intrinsically associated with exploration and development of oil and gas do not have to follow Subtitle C regulations for disposal. Under Subtitle C, hazardous wastes must follow strict guidelines for storage, treatment, transportation, and disposal. The cost of handling materials under the Subtitle C scenario is overwhelming. Under the exemption, the operator is allowed to dispose of well site waste in a prudent manner and is not obligated to use licensed hazardous waste transporters and licensed treatment, storage, and disposal facilities (TSDF).

Covered by the exemption are drilling fluids, cuttings, completion fluids, and rig wash. Not covered by the exemption are motor and chain oil wastes, thread cleaning solvents, painting waste, trash, and unused completion fluids.

A waste product, whether exempt or not, should always be recycled if economically possible. Oil-based drilling mud is typically purchased back by the vendor for reuse. Unused chemicals are similarly taken back for resale. Arrangements should be made with the mud company for partial drums or sacks of chemical. Muds also may be used on more than one hole. With the advent of closed-system drilling, the muds be moved off location in the event of a producing well.

If a waste is generated that is a listed or characteristic item, the operator must follow certain guidelines [3]. A listed hazardous waste (e.g., mercury, benzene) is considered hazardous if the concentrations in which they naturally occur above certain limitations (40 CFR 261.31 – 261.33). The listed hazardous waste may not be diluted to achieve a lesser concentration and thus become nonhazardous. A characteristic hazardous waste (40 CFR 261.21 – 261.24) may be diluted to a nonhazardous status.

Most nonexempt, nonacute hazardous waste generated on location is considered a small quantity. In this case, the waste may remain on location for 90 days. At that time, a Department of Transportation (DOT)–licensed motor carrier must transfer the waste to an EPA-certified TSDF for disposal. Appropriate documentation and packaging must conform to regulations. The operator continues to be liable for the waste as denoted by the cradle-to-grave concept [4].

Exempt wastes are usually disposed of on location after gaining permission from the state oil and gas division. Liquid wastes, if not evaporated or fixated on location, are usually injected into Class II injection wells (see Chapter 6, Environmental Considerations). Solid wastes, if not acceptable to local landfills, are remediated onsite or buried in some instances. Table 4.20.1 shows exempt and nonexempt waste, Figure 4.20.1 shows determination of exempt and nonexempt waste, and Figure 4.20.2 shows the possible waste mixtures and their exempt and nonexempt status [5].

Generally, waste that must be produced to complete the work involved in the exploration and production of oil and gas is considered exempt, allowing the operator the option of disposing of the waste in a prudent manner. A nonexempt waste (e.g., radioactive tracer waste) should be avoided if possible. These wastes must be disposed of according to federal and state regulations. Because these regulations are becoming more complicated with time, the operator should consult the primary regulator in questionable circumstances.

4.20.4 Site Assessment and Construction
4.20.4.1 Access and Pad
The road to the wellpad should be constructed to prevent erosion and be of dimensions suitable for traffic. A 16-ft top-running width with a 35-ft bottom width has been shown to sustain typical oil field traffic. The road should be crowned to facilitate drainage, and culverts should be placed at intersections of major runoff (Figure 4.20.3). Before the pad and access is surveyed for final layout, preliminary routes should be studied and walked out. Trained personnel should note any significant attributes of the area such as archeological finds, plants, and animals. Because of this survey, the most favorable access is designated weighing and primarily linked to

- Drilling fluid program
- Periodic operations
- Completions

The rig selection will dictate the basic layout of the pad. Based on the necessary area needed to support its functions, ancillary equipment may be added in space-conservative measures. In addition to the placement of various stationary rig site components, other operations such as logging, trucking, and subsequent completion operations must be provided for. The most environmentally sensitive design will impact the least amount of area and will therefore be the most economic. Potential pad sites and access routes should be laid out on a topographic map before the actual survey. At this time, construction costs can be estimated and compared. Figure 4.20.4 shows such a layout. The cost of building a location includes the cost of reclamation such as any remediation, recontouring, and reseeding of native plants. In the event of a producing well, only that area needed to support the production operations is left in place. The reserve pit and all outlying areas are reclaimed.

4.20.4.2 Rig Considerations
A rig layout diagram provided from the contractor gives the dimensions necessary for planning the drill site. Figure 4.20.5 depicts a layout for a 10,000-ft mobile drilling rig. The main components to the rig are the drive-in unit, substructure, mud pits, pumps, light plant, catwalk, pipe racks, and fuel supply. This type of rig, with a telescoping derrick, may be laid on less than 0.37 acre. The outstanding area is taken by the positioning of the deadmen. The deadmen to which the guy wires are connected are specified by safety considerations for each rig model. A standard drilling rig with a 10,000-ft capacity will have a longer laydown side, as it is measured from the well center. During rig up, the derrick is assembled on the ground and then lifted onto the A legs. The laydown side of the site must allow for this operation. As depicted in Figure 4.20.6, the standard rig exceeds the overall aerial dimensions of the mobile rig at 0.41 acre with the laydown dimension of 130 ft, compared with 90 ft for the telescoping mast unit. A standard drilling rig with a 20,000-ft capacity may run a laydown of 185 ft, with all other dimensions proportionally increased.

With the basic rig layout defined, the ancillary equipment may be determined and laid out adjacent to the structures in place. Access usually is determined where casing, mud, and other equipment may be delivered without disturbance of the infrastructure. When a crane (or forklift) is used, this may mean a simple loop from the rear to the laydown on the working side of the location. A loop is the optimum arrangement whereby multiple truck interference may be avoided. In the event a crane is not available, extra space is needed to accommodate the activity of gin trucks positioning materials. Figure 4.20.7 provides an overall location plan view.

4.20.4.3 Drilling Fluid Considerations
The drilling fluid program defines to some extent a major portion of the pad, including the reserve pit, blow pit, and equipment space. The program may include a closed system or a conventional one. The fluids can be oil-based mud, air, foam, water, or other media. Although the basic rig layout considers most mud drilling activities, it does not figure in mud storage, additional water storage, or air drilling systems. The overall layout may even be reduced in some cases. A closed mud system or air drilling eliminates the requirement for a large reserve pit.

Air Drilling
In the event of air drilling, the blooie line will exit from under the substructure and away from the rig. Depending on the nature of the pay zone, the blooie line will extend different lengths from the rig. For example, a well producing 20 MMCFD in addition to the injection of 2,000 cfm air will require an extension of at least 150 ft because of heat and dust accumulation. The blow pit with berm will often extend another 40 ft beyond that to include both the blow pit and berm. The blow pit is then connected to the reserve pit. This allows the transfer of any injected or produced fluid to a storage area away from the blow pit. The blow pit and the berm should be designed with considerable contingency.

Table 4.20.1 *Exempt and Nonexempt Oil and Gas Production Related Wastes*

Exempt Waste	Nonexempt Waste
Produced water	Unused fracturing fluids or acids
Drilling fluids	Cooling tower cleaning wastes
Drill cuttings	Used well completion/stimulation fluids
Rigwash	Radioactive traces wastes
Geothermal production fluids	Painting wastes
Hydrogen sulfide abatement wastes	Used lubrication oils
Well completion/stimulation fluids	Vacuum truck and drum rinsate containing nonexempt wastes
Basic sediment and water (BS&W) and other tank bottom material from storage facilities that hold product and exempt waste.	Refinery wastes
Accumulated materials such as hydrocarbons, solids, sand and emulsions from production separators, fluid treating vessels, and production impoundments.	Service company wastes such as empty drums, drum rinsate, vacuum truck rinsate, sandblast media, spent solvents, spilled chemicals, and waste acids
Pit sludges and contaminated bottoms from storage or disposal of exempt waste	Waste compressor oil, filters and blowdown
Workover wastes	Used hydraulic oil
Gas plant dehydration and sweetening waste	Waste solvents
Cooling tower blowdown	Caustic or acid cleaners
Spent filters, filter media, and backwash (assuming the filter itself is not hazardous and the residue in it is from an exempt waste stream)	Waste generated in transportation pipeline related pits
Packer fluids	Laboratory wastes
Produced sand	Boiler cleaning wastes
Pipe scale, hydrocarbon solids, hydrates, and other deposits removed from piping and equipment before transportation	Boiler refractory bricks
Hydrocarbon bearing soil contaminated from exempt waste streams	Boiler scrubbing fluids, sludges and ash
Pigging waste from gathering lines	Incinerator wastes
Constituents removed from produced water before it is injected or otherwise disposed of	Industrial wastes from activities other than oil and gas E&P
Waste from subsurface gas storage and retrieval	Pesticide wastes
Liquid hydrocarbons removed from the production stream but not from oil refining	Gas plant cleaning wastes
Gases from the production stream, such as H_2S, CO_2, and volatized hydrocarbons	Drums, insulation and miscellaneous solids
Materials ejected from a producing well during blowdown	Manufacturing wastes
Waste crude from primary operations and production	Contamination from refined products
Light organics volatized from exempt waste in reserve pits, impoundments, or production equipment	
Liquid and solid wastes generated by crude oil and crude tank bottom reclaimers	
Stormwater runoff contaminated by exempt materials	

Although non-E&P wastes generated from crude oil and tank bottom reclamation operations (e.g., waste equipment cleaning solvent) are nonexempt, residuals derived from exempt wastes (e.g., produced water separated from tank bottoms) are exempt.

During operations, the well may be producing oil or natural gas, or both. In the case of oil, the well will have to be mudded up after the well is under control. In the process of getting the well under control, the oil may be sprayed a great distance if the blow pit and berm are not properly constructed. If natural gas is encountered, drilling is usually advanced with a flare in place. The pit and the perm must then be able to protect the surrounding area from fire and to sustain the impact from a steady bombardment of drilling particles. A distance of 30 ft should be allowed from the end of the blooie line to the edge of the berm, with only 5 ft of depth needed at the pit sloping toward the reserve pit built to maintain a flowing velocity of 2 fps. The berm itself should be 20 ft high and composed primarily of 12-ft or larger-diameter stones on the face and

backed with a soil having poor permeability (Figure 4.20.8). A lip should overhang from the top of the berm whereby the ejected materials are diverted back to the pit. The air compressors and boosters are typically located at the rear of the location, with the air piped from that point to the standpipe to alleviate additional piping. Figure 4.20.9 shows an air-drilling layout. Notice that the location may be compacted somewhat because of the reduced reserve pit.

Mud Drilling
The conventional drill site includes mud pits, mud cleaning equipment, water storage, mud storage, pumps, and mixing facilities. Often, the reserve pit associated with the conventional mud system is as large as the leveled location. A large

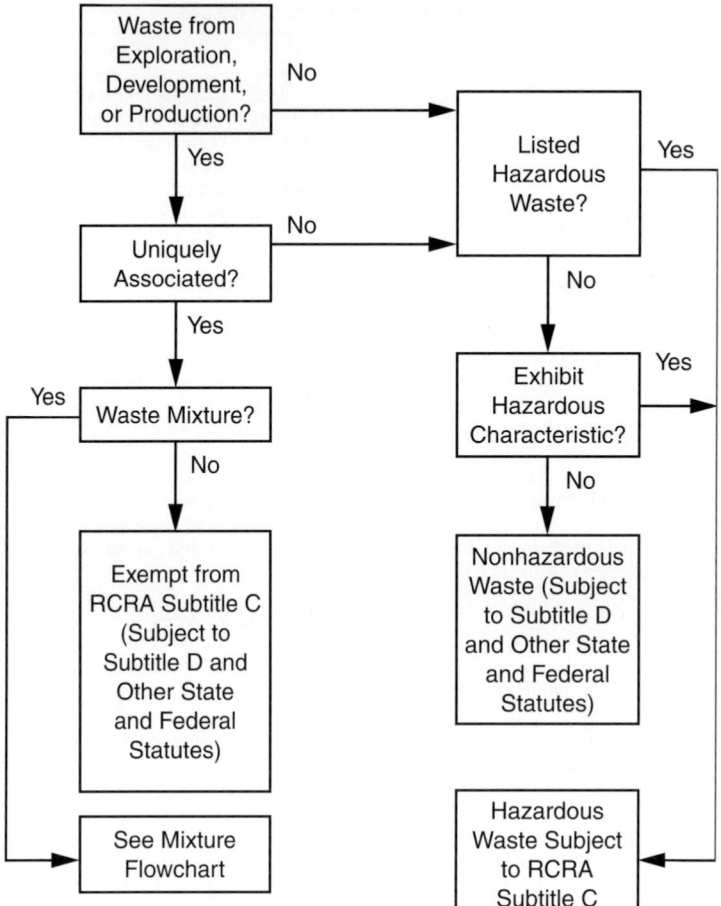

Figure 4.20.1 *Exempt/nonexempt waste flowchart.*

reserve pit allows for contingency, but it also increases cost of reclamation and construction. Generally, a 3-ft freeboard is maintained for safety, and it may be relied on for contingency. A good field estimate for a reserve pit size needed for a conventional operation is

$$V_t = 2(V_s + Vi_I + V_p) + 281D_d + 16.84MV + 3WL \qquad [4.20.1]$$

$$WL = \frac{V_r}{H} \qquad [4.20.2]$$

where

V_t = volume of reserve pit in ft^3
V_s = volume of surface hole (ft^3)
V_i = volume of intermediate hole (ft^3)
V_p = volume of production hole (ft^3)
D_d = forecasted drilling time in days
MV = mud pit volume in bbl
L = pit length: 0.5−0.75 ft of location in ft
W = pit width: 0.25−0.5 ft of pit in ft
H = pit depth ≤ 10 ft

Equation 4.20.1 assumes that for each operation, a volume of cuttings is put into the reserve pit plus a mud volume

equivalent to the circulation system. The additional volumes are attributed to mud dilution and maintenance. An additional contingency is added through 3 ft of freeboard. The reserve pit should be located in an area where capacity may be increased in an emergency or additional fluid trucked away.

Closed Mud System

The closed mud system is the modern solution to an environmentally sensitive drilling operation. The circulation system on the rig is fully self-supporting, requiring only the discharge of drill cuttings. On a simple system, this may only necessitate the addition of a mud cleaner or centrifuge to the drilling rig's conventional mud system. A bid package to the drilling contractor may stipulate a closed system and the requirements, thereby letting the burden of design fall to the contractor, although the liability still rests on the operator. With a closed system, the additional area is sometimes needed within the basic layout for mud cleaning equipment and mud storage. A trench located at the pit side of the mud pits may be allowed then for cutting disposal. In the event of oil-based mud, the cuttings may be collected in a sloped container, where residual fluid is allowed to drain from the cuttings before disposal.

(text continued on page 4-553)

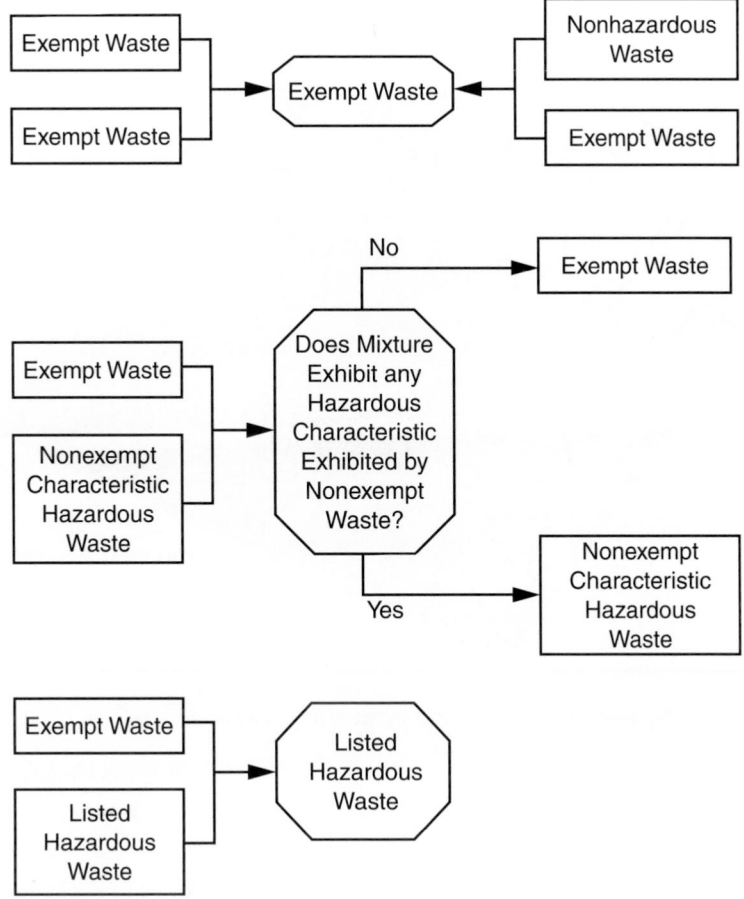

Figure 4.20.2 *Possible waste mixtures and their exempt and nonexempt status.*

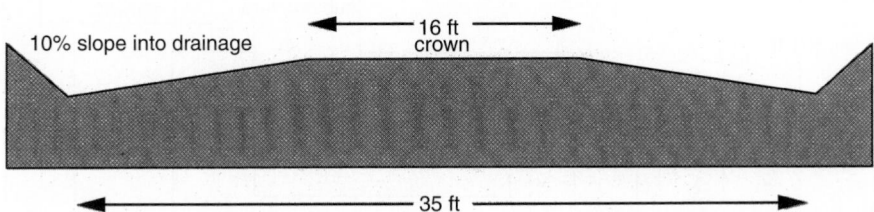

Figure 4.20.3 *Typical oilfield access dimensions.*

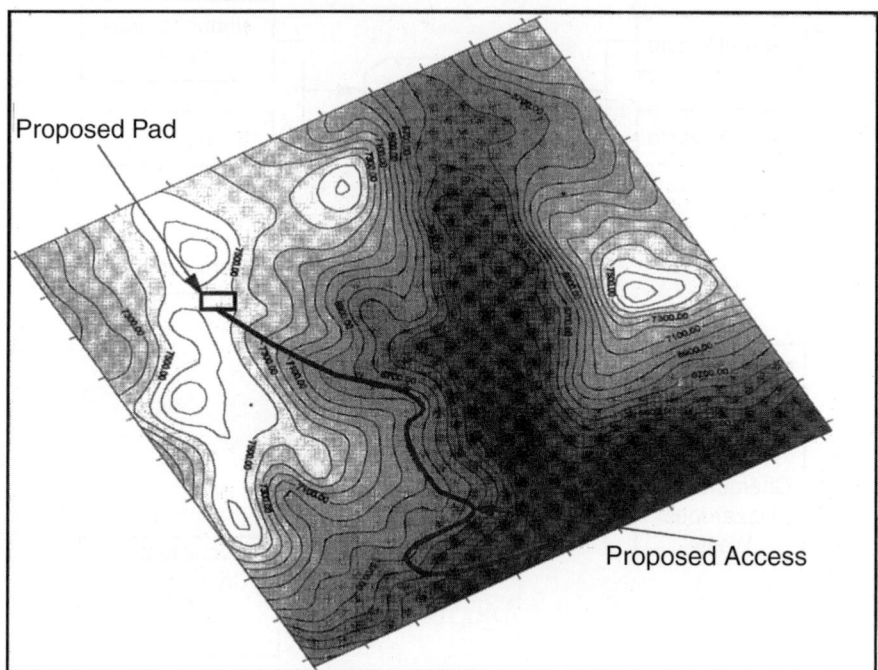

Figure 4.20.4 *Topographic layout of proposed access and pad.*

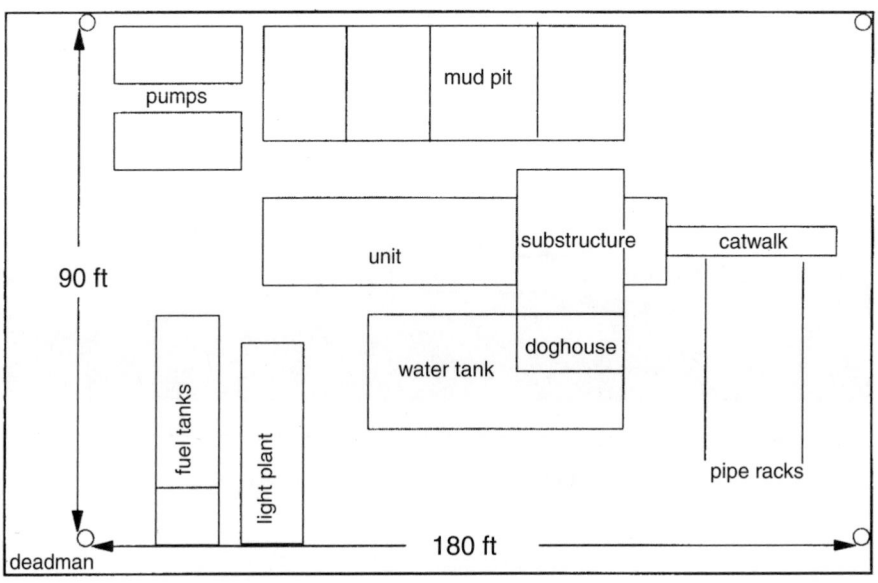

Figure 4.20.5 *Basic mobile 10,000-ft drilling rig layout.*

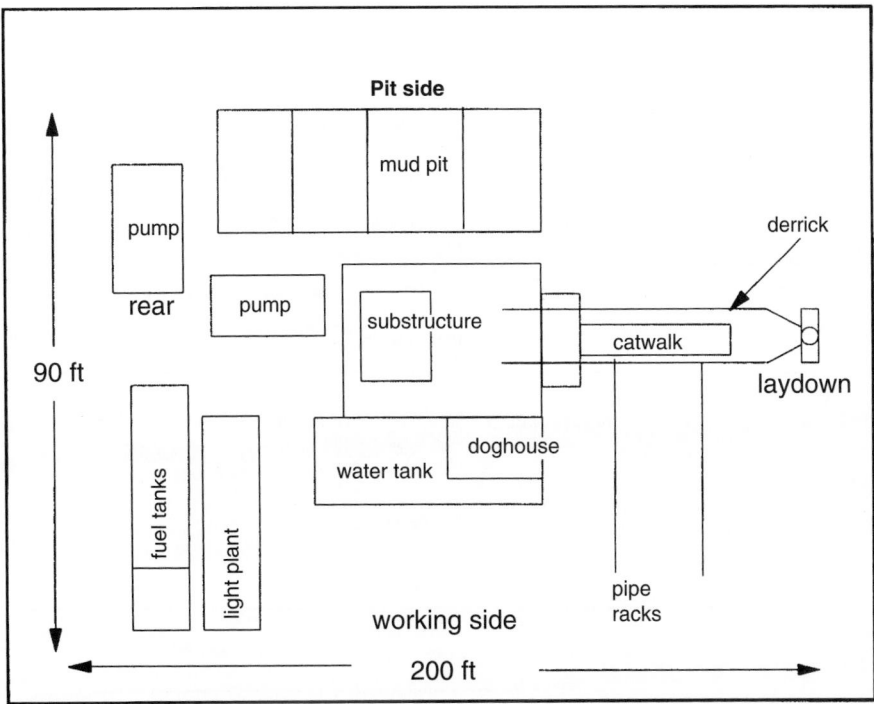

Figure 4.20.6   *Basic layout for drilling rig with standard derrick.*

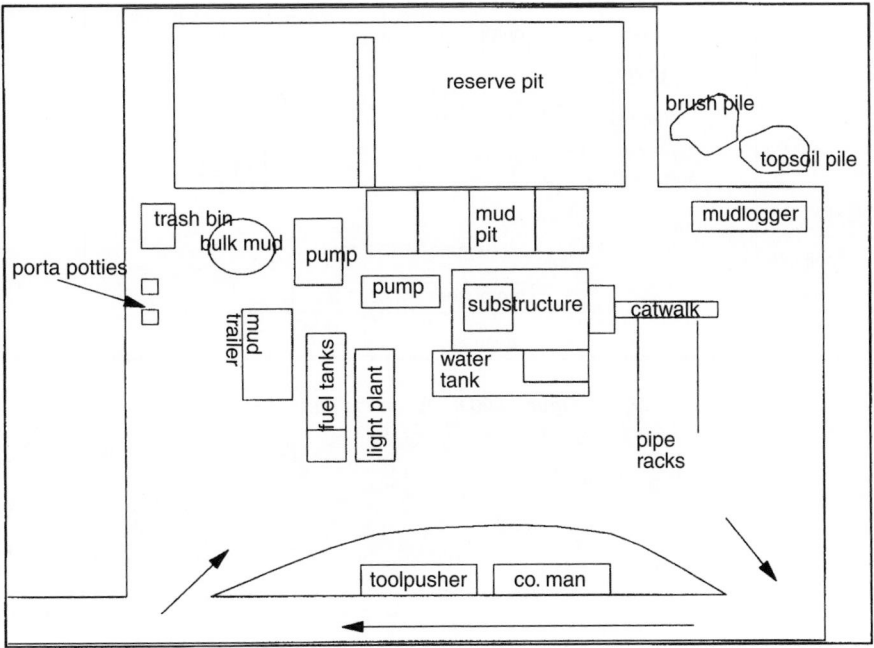

Figure 4.20.7   *Overall location plan.*

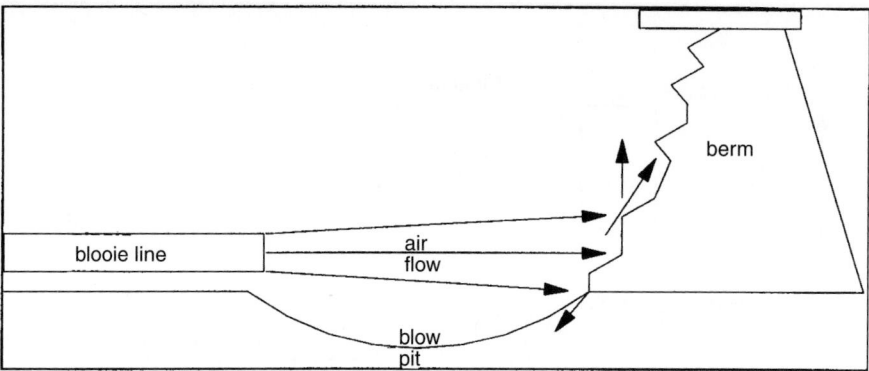

Figure 4.20.8 *Side view of blow pit.*

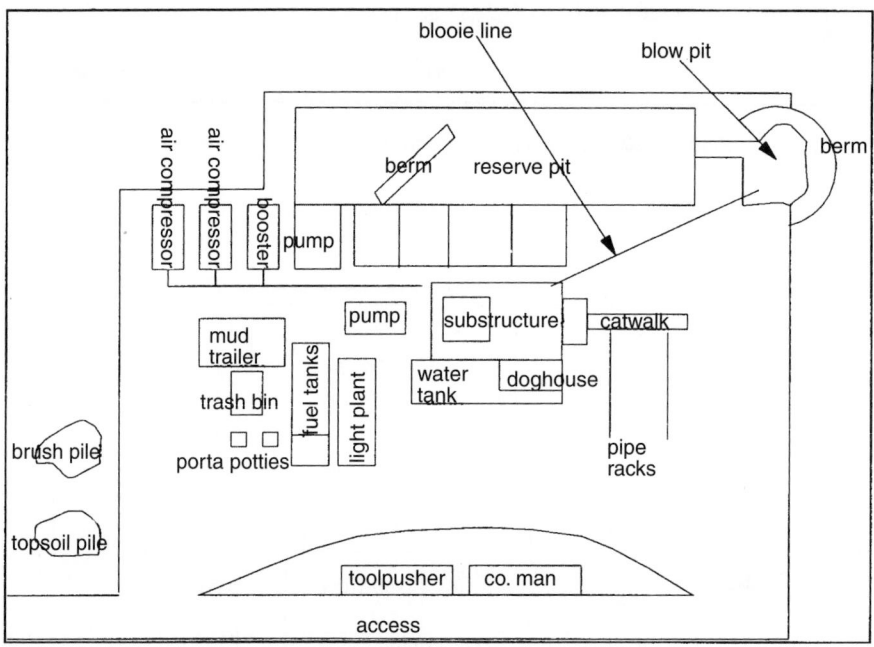

Figure 4.20.9 *Air drilling pad.*

(text continued from page 4-548)

4.20.4.4 Periodic Operations

Periodic operations include cementing, running casing, and logging. Room must be allocated for the storage of casing. Running 10,000 ft of range $3, 5\frac{1}{2}$ in. casing requires a 40×115 ft area. The casing may be stacked, although never in excess of three layers and preferably only two layers. Cementing operations may include placement of bulk tanks in addition to pump trucks and bulk trailers. These are usually located near the water source and rig floor. Laying down drill pipe and logging both necessitate approximately 30 ft of space in front of the catwalk.

4.20.4.5 Completions

If a well proves productive, the ensuing completion operation may require an area in excess of the drilling area. This may mean allocation for frac tank placement, blenders, pump trucks, bulk trucks, and nitrogen trucks. In today's economic climate, the operator should weigh the probability of success, Bayes theorem (Equation 4.20.3), with the cost of constructing and reclaiming an additional area needed for stimulation (Equation 4.20.4). Plans such as these should be considered and proposed in the APD. The operator may then construct the additional space without a permitting delay.

$$P_s = PR(E/R) = \frac{PR(E)PR(R/E)}{PR(E)PR(R/E) + PR(notE)PR(R/notE)}$$
[4.20.3]

$$P_s C_b \le C_a$$
[4.20.4]

where

P_s = probability of success
$PR(E/R)$ = probability of event E given information R
C_b = construction cost of additional space before drilling in $
C_a = construction cost of additional space after drilling in $

4.20.4.6 Pad Construction

Once access and size of the pad have been determined, the areas may be physically flagged to further reduce the amount of dirt moved. Important features of the area should be noted at this time, including

- Depth of water table
- Natural drainage patterns
- Vegetation types and abundance
- Surface water
- Proximate structures

These features are quantified, as well as testing for contamination in soil and water to prevent unnecessary litigation over previous pollution.

In construction of the pad, brush and trees are pushed to one area. Topsoil is then removed from the site and stockpiled for respreading during subsequent seeding operations. The leveled pad is slightly crowned to move fluids collected on the pad to the perimeter, where drainage ditches divert this fluid to the reserve pit. In the event a subsurface pit is not possible, the drainage will run into a small sump. This sump is used as a holding tank for pumping of collected fluids to an elevated reserve pit.

When possible, the reserve pit is placed on the low side of the location to reduce dirt removal. In this event, the pit wall should be keyed into the earth and the summation of forces and moments on the retaining wall calculated to prevent failure. Pits most often fail because of leaking liners undercutting the retaining wall and sliding out. A minimum horizontal-to-vertical slope of 2 is recommended for earthen dams [6]. The pit bottom should be soft filled to prevent liner tears. Other key factors of the location provide for the drainage of all precipitation away from the location. This prevents the operator from unduly managing it as a waste product. Any water landing on the location must be diverted to the reserve pit.

A new scheme for location management had developed whereby wastes are diverted to separate holding facilities according to the hazard imposed by the waste. Separate pits are created to hold rig washing and precipitation wastes, solid wastes, and drilling fluids [7]. The waste is then reused, disposed on site, or hauled away for offsite treatment. The system reduces contamination of less hazardous materials with more hazardous materials, thereby reducing disposal costs.

4.20.5 Environmental Concerns While in Operation

4.20.5.1 Drilling

A blowout primarily consisting of oil presents the greatest environmental hazard while drilling. During normal drilling situations, downhole drilling fluids are usually the greatest potential threat to the environment. In the case of oil-based mud, the cuttings also present a problem though absorption of the diesel base. These cuttings are presently landfarmed, landspread, reinjected, or thermally treated to drive off the hydrocarbons.

Most wells encounter shales when drilling. The oil-based muds are quite effective in reducing the swelling tendencies of the shale in addition to presenting less intrusive invasion characteristics to the reservoir. The oil-based muds may be sold back to the distributor, where they are recycled. An alternative to oil-based muds may be found in using a synthetic-based material [8]. Saline mud is also used to reduce the shale's swelling characteristics. Chlorides may be found to be within toxic levels in the drilling fluid, making its use less desirable. Because the swelling of the clay is attributed to the cation exchange capacity (CEC), the chloride levels may be reduced by replacement with another anion. Calculation of the shale's CEC estimates the amount of cations that can be added to the drilling mud to effectively reduce swelling tendencies. Generally, a multivalent cation is more effective in reducing hydration of the clay. An equilibrium equation is used to define the CEC of a shale (Equation 4.20.5). Figure 4.20.10 shows the CEC relationships among several multivalent cations and sodium in attapulgite clay [9].

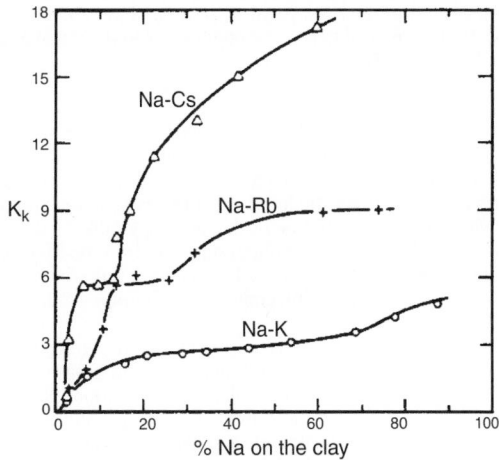

Figure 4.20.10 *Selectivity number K_k vs. exchangeable ion composition on an attapulgite clay.*

Table 4.20.2 *Cation Exchange Capacity of Materials Encountered in Drilling*

Material	mEq/100 g – Moisture Free
Wyoming bentonite	75
Soft shale	45
Kaolinite	10
Drilled cuttings	8

While drilling, the drilled cutting's CEC is usually tested with methylene blue (Equation 4.20.6). Table 4.20.2 shows CEC of several clays encountered in drilling situations [10]. In reducing the amount of any one type cation or anion present in the drilling mud, the environmental risks are reduced. A mixed salt solution containing several salt combinations, each of which is below the toxic limits, may produce the desired effects. Thus,

$$K_k = \frac{(NaX)^c(M)}{(MX)(Na)^c} \qquad [4.20.5]$$

where

NaX = sodium on the shale in mEq
  M = multivalent cation in mEq
  c = charge of multivalent cation

$$MBC = \frac{MB}{M_{shale}} = CEC_{shale} \qquad [4.20.6]$$

where

  MBC = methylene blue capacity in mEq/100 g
  MB = methylene blue capacity in mEq
  M_{shale} = weight of shale in 100 g

There are a variety of toxic chemicals used in the drilling fluid makeup. Chromates and asbestos were once commonly used but are now off the market. A mud inventory should be kept for all drilling additives. Included in the inventory is the Material Safety Data Sheet (MSDS) that describes each material's pertinent characteristics. The chemicals found on the MSDS should be compared with the priority pollutants, and any material should be eliminated if a match is found. The chemicals should also be checked on arrival for breakage and returned to the vendor if defective packaging is found. All mud additives should be housed in a dry area and properly card for to prevent waste. Chemicals should always be mixed in packaged proportions. Wasted chemicals, ejected to the reserve pit by untrained personnel, can present future liabilities to the operator, who is responsible for them from cradle to grave.

4.20.5.2 Rig Practice

All drilling fluids should be contained in the mud pits for reserve pit. The cellar should have a conduit linking it to the reserve pit such that accumulation of mud from connections does not spread over the location. In the case of toxic mud, a bell should be located beneath the rotary table to direct such fluids back into the drilling nipple or into the mud pit in the case of a rotating head. The use of a mud bucket is also recommended in situations where the pipe cannot be shaken dry.

Many operators are now requiring that absorbent or catch pans be placed beneath the drilling rig. Oil and grease leaks or spills are common to the drilling rig operation. Even if the drilling rig is new, leaks and spills are inevitable. Fuel and oil racks should be provide a spill-resistant pour device for direct placement of the lubricant or fuel into equipment. Oil changed on location must be caught in barrels and recycled by the contractor. Pipe dope should be environmentally sound and not a metal type. Thread cleaning on casing is now frequently done with machines that catch the solvent for reuse on subsequent operations.

Drilling contractors should be advised before the operations about what is trying to be accomplished environmentally on location. Contractors know that the standard practices of throwing dope buckets, and everything else into the reserve pit is no longer acceptable, but an occasional drilling crew may not take directives seriously. Because of this, drilling contractors should line item the liabilities associated with imprudent practices.

4.20.5.3 Completions

Inherent to the completion operations are the stimulation fluids used to carry sand or otherwise enhance producing qualities of the well. These stimulations fluids may or may not be toxic in nature. The stimulation fluid, although sometimes batch mixed at the service company's facility, may also be mixed on location. The latter system is preferable unless the service company is willing to let form the operator any liability common to the fluid in case of excess.

Many frac jobs are mini-fraced before the actual operation and the design criteria established before the actual frac job is accomplished. The method helps prevent the screening out of the well beforehand. The screened-out well causes potential problems not only in productivity but also in waste management. The unused mixed chemicals, such as KCl makeup water in frac tanks leftover due to the screenout, must be properly disposed. An alternative to this is the mixing of chemicals on the fly instead of preblending and stocking in frac tanks. Some chemicals, such as KCl in water, must be mixed well in advance on location to attain heightened concentrations. In theses in instances, a properly designed frac job, based on the mini-frac, will allow for some certainty in getting the designed job away. Frac flow back may be introduced to the reserve pit after separation from any hydrocarbons. Current frac fluids are composed of primarily of natural organics such as guar gum, but they may contain other components that may be harmful to the environment. Containment of flowback in a lined reserve pit before disposal is a prudent practice.

Acid jobs may also be designed according to prior investigation, although most service companies accept unused acid back into their facilities. Spent acid flowback may be introduced to the lined reserve pit with little consequence. Often, the residual contains salts such as $CaCl_2$ preciously introduced to the pit. Live acid is occasionally flowed back to surface. This acid may be flowed to the lined pit, given acceptance of the liner material for low pH. The buffer capacity of most drilling fluids is significant, and it is able to assimilate the excess hydrogen ions introduced to the system. The buffer capacity of the fluid may be calculated from Equations 4.20.7 and 4.20.8 [11]. From a water analysis, pH and alkalinity are determined, and the remaining parameters may then be calculated.

$$\beta = \frac{dC}{dpH} \qquad [4.20.7]$$

$$\beta = 2.3\left[\frac{\alpha_1([ALK]-[OH^-]+[H^+])\left([H^+]+\frac{(K_1'K_2')}{[H^+]}+4K_2'\right)}{K'\left(1+\frac{2K_2'}{[H^+]}\right)}\right.$$

$$\left.+[H^+]+[OH^-]\right] \qquad [4.20.8]$$

$$[ALK] = [HCO_3^-] + 2[CO_3^{-2}] + [OH^-] - [H^+] \qquad [4.20.9]$$

$$[Ct] = [H_2CO_3^*] + [HCO_3^-] + 2[CO_3^{-2}] \qquad [4.20.10]$$

$$K_1' = \frac{\gamma[HCO_3^-](H^+)}{[H_2CO_3^*]} \qquad [4.20.11]$$

$$K_2' = \frac{\gamma^4[HCO_3^{-2}](H^+)}{\gamma[HCO_3^*]} \qquad [4.20.12]$$

$$\alpha_1 = \frac{1}{1 + \dfrac{K_2'}{(H*)} + \dfrac{(H^+)}{K_1'}} \qquad [4.20.13]$$

$$Log\gamma = -AZ^2 \left[\frac{[\sqrt{I}]}{[1+\sqrt{I}]} \right] \qquad [4.20.14]$$

where

β = buffer capacity, (equivalents/unit pH change)
C = equivalents of buffer available = assumed equal to [ALK]
[ALK] = equivalents of alkalinity
K = equilibrium coefficients
γ = the monovalent activity coefficient
I = the ionic strength
Z = the charge of the species of interest
A = $1.82 \times 106(DT)^{-1.5}$
D = the dielectric constant for water, 78.3 at 25°C
T = °K
( ) = activity of the ion
[ ] = concentration of the ion

The buffer capacity of the pit fluid is equal to the change in alkalinity of the system per unit change of pH. Figure 4.20.11 shows the buffer intensity (capacity) of a 0.1M carbonate pit fluid [10]. Calculating the initial buffer capacity of the pit fluid allows for prediction of the pH change on introduction of live acid and any addition of buffer, such as sodium bicarbonate, required to neutralize the excess hydrogen ions.

Care should be taken in every stimulation circumstance to allow fluids to drain to the reserve pit. In the completion operation, it is exceedingly difficult to accomplish this because of traffic, and the service company should therefore provide leak-free hoses, lines, and connections. On completion of the job, the hoses should be drained to a common area for holding subsequent to introduction to the reserve pit. Every precaution should be taken to prevent accumulation of fluids on the pad proper, thereby posing a potential risk to groundwater and runoff of location.

As with the drilling operation, the equipment on location providing the completion service may leak oil. The use of absorbents and catch pans is advised.

In the case of produced liquid hydrocarbons and other chemicals spilled during operations, subsequent remediation may be necessary. This section (see also Section 9 of Chapter 6, Environmental Considerations) details some remediation techniques currently employed.

4.20.5.4 Reclamation of the Drill Site

In the event of a dry hole, the reserve pit water usage should be maximized to prepare the mud spacers between plugs. Water in excess of this may be pumped into the hole, including solids. All USDWs must be protected in this event. Once the hole has been properly plugged and the drilling rig removed, the mousehole and rathole should be backed filled immediately to preclude any accidents. Trash is removed from the location and adjacent area and is hauled to permit facilities.

4.20.5.5 Reserve Pit Closure

The reserve pit commonly holds all fluids introduced to the wellbore during drilling and completion operations. This includes the drilling and completion fluids in the event the well is stimulated for production and those cuttings produced during the drilling operation. The reserve pit, on completion of the initial rig site activities, must be reclaimed. On removal of the drilling rig, the reserve pit is fenced to prevent wildlife and livestock from watering. The fence is removed on initiation of reclamation.

The fluids from the reserve pit may be hauled away from location for disposal, reclaimed in situ, or pumped into the wellbore given a dry hole. The operator of the well site

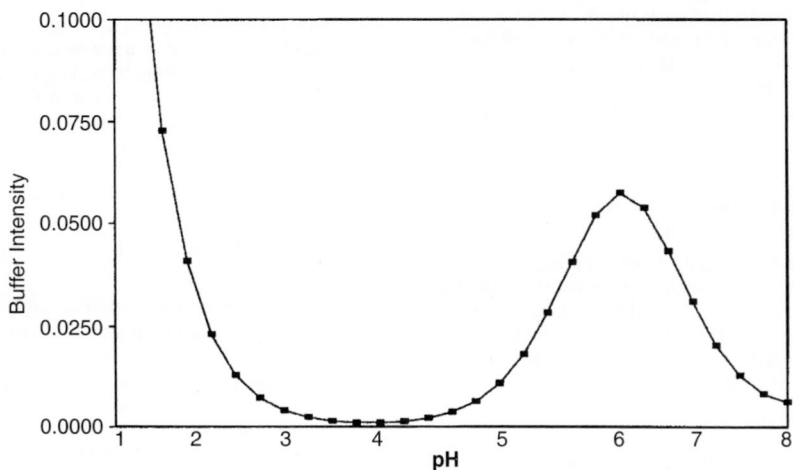

Figure 4.20.11   *Nonideal buffer characteristics of a 0.10 M carbonate reserve pit fluid.*

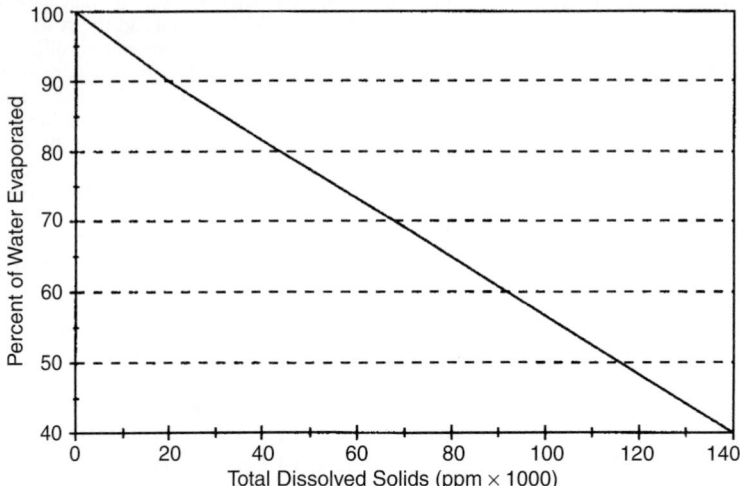

Figure 4.20.12 *Maximum limit of evaporation as defined by TDS (NaCl).*

is responsible for the transportation offsite of the drilling fluids. The fluids may be considered hazardous in nature due to the toxic characteristics of most drilling and completion fluids.

4.20.5.6 Evaporation

Evaporation of the water held in the pit is often the first step in the reserve pit remediation because of economic considerations about trucking and disposal. The evaporation may be mechanically driven or take place naturally. Natural evaporation is very effective in semi-arid regions. The Meyer equation (Equation 4.20.15), as derived from Dalton's law, may be used to estimate the local natural evaporation [6].

$$E = C(e_w - e_a)\psi \qquad [4.20.15]$$

$$\Psi = 1 + 0.1w \qquad [4.20.16]$$

where

E = evaporation rate (in 30 days)
C = empirical coefficient equal to 15 for small shallow pools and 11 for large deep reservoirs
e_w = saturation vapor pressure corresponding to the monthly mean temperature of air for small bodies and monthly mean temperature of water for reservoirs (in.Hg)
e_a = actual vapor pressure corresponding to the monthly mean temperature of air and relative humidity 30 ft above the body of water (in.Hg)
Ψ = wind factor
w = monthly mean wind velocity measured at 30 ft above body of water (miles/hr)

Some mechanically driven systems include heated vessels or spraying of the water to enhance the natural evaporation rate. In heating, the energy needed to evaporate the water is equal to that needed to bring the water to the temperature of vaporization plus the energy required for the evaporation, where for constant volume this is

$$\Delta E = CpdT + \Delta H_{vap} \qquad [4.20.17]$$

The heat capacity of ΔH_{vap} of pure water at 14.7 pisa are commonly taken as 1 btu/lbm (°F) and 970 btu/lbm [12]. Ionic content in the pit fluid raises the energy necessary to

evaporate the fluid. Figure 4.20.12 shows this relationship for brine water containing primarily NaCl [10].

In field evaporative units using natural gas as the fuel source, the primary driving force is the heat supplied to the water. The theoretical evaporation rate for these units may be expressed as

$$\frac{H_c Q_g}{\Delta E \rho_w} = Q_{evap} \qquad [4.20.18]$$

where

H_c = natural gas heating value (btu/mcf)
Q_c = natural gas flow rate to burner (mcfpd)
ρ_w = density of water (ppg)
Q_{evap} = evaporation rate (gpd)

Mechanical efficiency may range from 25% to 75% of the theoretical evaporation rate. Efficiencies may be raised with the application of multi-effect or vapor compression evaporators. The more complicated the systems can seldom be warranted due to the short service offered.

Spray systems rely on forming minuscule droplets of water and allowing the vaporization thereof while in suspension over the reserve pit. Allowance for wind carriage of the droplets beyond the pit must be made to prevent salting damage to the surrounding area. The shear force extended on each droplet in combination with the relative humidity provides the driving force for the operation. Neglecting the shear component, driving force is actual and saturation vapor pressure differential. A derivation of Fick's law may be used to express the molar flux of water in air.

$$Na = \frac{DA}{1000RT} \frac{dV_p}{dx} \qquad [4.20.19]$$

where

Na = moles of water diffused to the air (mol/sec)
D = diffusivity of water in air (0.256 cm^2/secat 25°C, 1 atm)
A = the area perpendicular to the flux (cm^2)
V_p = vapor pressure of water in air (atm)
R = gas law constant (0.0821 atm L/mol °K)
X = the thickness of the film where dV_p exists (cm)

Inspection of Equation 4.20.19 shows that increasing the area of the active water surface will allow for greater evaporation rates. In the case of the spray systems,

$$Na = 0.01256 \frac{nD}{RT} \frac{dV_p}{dx} \qquad [4.20.20]$$

Because of the differences in determining x, the thickness of the film between the two vapor pressures, an overall transfer coefficient is introduced. Based on the two-film theory, the overall transfer coefficient is used. In the case of water evaporation, the gas film is the controlling mechanism and the resulting equation is

$$Na = \frac{Kga}{RT}(V_{Psat} - V_{Pact}) \qquad [4.20.21]$$

where Kga = the overall mass transfer coefficient (t^{-1})

Service companies offering evaporation services can supply the operator with values of Kga maybe used comparatively between all systems for economic analysis.

Using the Meyer equation (Equation 4.20), the evaporative rate from a 5000 ft^2 pit is estimated. The average temperature in an area in the winter is 40°F, and the corresponding saturated vapor pressure is 0.26 in.Hg, with the actual average vapor pressure residing at 0.19 in.Hg. Wind velocity reaches a peak at 40 mph with a time weighted mean velocity of 5 mph, such that the evaporation rate may be estimated as

$$E = 15(0.26 - 0.19)(1 + 0.1(5)) = 1.58 \text{ in./mo or}$$

$$111 \text{ bbl/mo } (0.00083 \text{ mol sec}^{-1})$$

Given this evaporation rate, the overall mass transfer coefficient may then be calculated from Equation 4.20.21:

$$Kga = \frac{NaRT}{(V_{Psat} - V_{Pact})} = 7.94 \text{ sec}^{-1}$$

4.20.5.7 Fixation of Reserve Pit Water and Solids

Another method of reclaiming the reserve pit involves combining water-absorbing materials to the water and mud. Usually, the pit contents are pumped through tanks where sorbent is combined with the pit fluid and solids. The mixture is dried and subsequently buried. Care must be taken with this method such that any harmful containment is immobilized to prevent contamination to the surroundings. Studies have shown that for muds, once most of the water has been evaporated or pulled from the pit, the remainder may be solidified to comply with existing regulations. This may be done with cement, fly ash, pozzalin, or any number of absorbents. Polymers have been developed to handle high pH, salt, and oil contents, for which the previous mixtures fell short. The mixture is then allowed to dry and the bulk mass is then buried. This method requires the forethought on pit construction whereby complete mixing of the slurry is accomplished. If primarily bentonite and water are used, evidence has shown minor or no migration from the pit [13].

$$W = \Sigma M_{at\ldots n} R_{wi\ldots n} \qquad [4.20.22]$$

where

W = water available in pit (bbls)
M_r = mass of absorbent (lbm)
R_w = water required by absorbent (bbl/lbm)

Even though materials such as bentonite can absorb tremendous amounts of water, they cannot solidify to an extent that the pit may be reclaimed. In moist instances, a dozer must be able to walk out to the center of the pit under a load of pushed dirt. In the event the pit materials are wet, the dozer may become mired and unable to complete the work. It is often better to pick a sorbent that will harden sufficiently for this purpose.

4.20.5.8 Final Closure

On elimination of the fluids, the liner to the pit is folded over the residual solids in a way to prevent fluid migration. The liner is then buried in place. The operator may choose to remove the liner contents completely to preclude any future contamination. In the case of a producing well, the location is reclaimed up to the deadmen. The adjacent areas are contoured to provide drainage away from the production facilities. In the case of a dry hole, the entire location is reclaimed to the initial condition. All of the reclaimed are should be ripped to enhance soil conductivity. The topsoil is then spread over the reclaimed area, followed by seeding. Local seed mixtures are broadcast to quicken reintroduction of native plants.

References

1. Bureau of Land Management, "Onshore Oil and Gas Order No. 1: Approval of Operations on Onshore Federal and Indian Oil and Gas Leases," United States Department of Interior, Bureau of Land Management, Washington, D.C., 1983.
2. Fitzpatrick, M., "Common Misconceptions about the RCRA Subtitle C Exemption from Crude Oil and Natural Gas Exploration, Development and Production," Proceedings from the First International Symposium on Oil and Gas Exploration Waste Management Practices, pp. 169–179, 1990.
3. USEPA, "RCRA Information on Hazardous Wastes for Publicly Owned Treatment Works" Office of Water Enforcement Permits, Washington, D.C., 1985.
4. Wentz, C., *Hazardous Waste Management*, McGraw-Hill, New York, 1989.
5. EPA Exemption of Oil and Gas Exploration and Production Wastes from Federal Hazardous Waste Regulations, EPAA530-K-01-004 January 2002 (www.epa.gov/epaoswer/other/oil/oil-gas.pdf).
6. Merritt, F.S., *Standard Handbook for Civil Engineers*, McGraw-Hill, New York, 1983.
7. Pontiff, D., and Sammons, J., "Theory, Design and Operation of an Environmentally managed Pit System," First International Symposium on Oil and Gas Exploration Waste Management Practices, pp. 997-987, 1990.
8. Carlson, T., "Finding Suitable Replacement for Petroleum Hydrocarbons in Oil Muds," SPE Paper 23062, 1992 American Association of Drilling Engineers New Advancements in Drilling Fluids Technology Conference, Houston, TX, 1992.
9. Marshall, C., and Garcia, G., *Journal of Physical Chemistry*, 1959.
10. Bariod, N.L., *Manual of Drilling Fluids Technology*, NL Industries Inc., Houston, 1979.
11. Russell, C., M.S. Thesis, Desalination of Bicarbonate Brine Water: Experimental Finding Leading to an Ion Exchange Process: New Mexico Tech, Socorro, NM, 1994.
12. Engineering Data Book, Gas Processors Suppliers Association, Tulsa, 1981.
13. Grimme, S. J., and Erb, J. E., "Solidification of Residual Waste Pits as an Alternative Disposal Practice in Pennsylvania," Proceedings from the First International Symposium on Oil and Gas Exploration Waste Management Practices, pp. 873–881, 1990.

14. EPA Exemption of Oil and Gas Exploration and Production Wastes from Federal Hazardous Waste Regulations, January, 2002 (http://www.epa.gov/epaoswer/other/oil/oil-gas.pdf).

4.21 OFFSHORE DRILLING OPERATIONS

One of the more remarkable technological accomplishments of the modern era is the ability of mankind to explore for and produce commercial quantities of oil and natural gas in what are often remote and hostile offshore environments. While offshore petroleum operations have existed for more than half a century, major advances in seismic, drilling and production technology have recently pushed the frontiers to ever deeper waters. What a few years ago was considered deep water—about 650 feet—is no longer regarded as extraordinary, and the very definition of deep water has expanded to 3,000 feet and beyond. In the wake of these innovations, offshore provinces have grown to account for about 25% to 30% of all U.S. production and about 30% of estimated undiscovered reserves.

The basic principles of drilling that rule onshore operations also hold true in offshore environments. However, a fundamental difference offshore is that a stable, self-contained platform must be provided for drilling and support equipment. Communication with a well through possibly thousands of feet of water calls for mechanical as well as procedural differences, primarily in well control and platform positioning. Onshore technology can be applied to offshore operations in many instances on bottom-bearing rigs, but the widespread use of floating vessels has necessitated the development of new technology tailored to the offshore environment.

Beyond that, well planning, logistics, and execution take on added risk, expenses, and responsibility when carried out in a distant, often harsh marine environment hundreds of miles from sight of the nearest landmass [1–8].

4.21.1 Drilling Vessels

Drilling vessels predominate offshore exploration and are of two types – either bottom-bearing or floating-type vessels. Water depth is generally the governing factor as to which type of vessel is employed [9].

Bottom-bearing vessels include jackup rigs, submersibles and drilling barges. A jackup rig (Figure 4.21.1) is used for drilling in water depths up to about 400 ft. Jackup rigs are usually towed to their location, where heavy machinery is used to jack the support legs down through the water until they are embedded in the sea floor. The drilling package on a jackup can be raised or lowered on the legs of the vessel to allow for use in various water depths and sea floor soil conditions. These rigs can also be used alongside production platforms for developmental drilling. Jackup rigs thus provide a mobile, bottom-bearing drilling vessel suitable for use in relatively shallow water.

Drilling barges, commonly employed in inland waters and marshes, can be used where water depths do not exceed about 20 ft. These self-contained vessels provide the least expensive drilling option, but have limited applicability because of water depth and environmental limitations.

Submersible rigs likewise are used strictly in shallow water—in depths of 80 ft or less. They have two hulls—the upper containing crew quarters and the drilling rig and the lower filled with air, making the vessel buoyant. Submersibles are towed to their location, where the air in the bottom hull is displaced with sea water for weight and the rig is lowered to the sea floor.

Drillships and semisubmersible rigs are the two dominant types of floating drilling vessels – and can operate in shallow, deep and ultra-deep water.

A drillship (Figure 4.21.2) is a large ocean-going vessel that features a drilling package and large derrick positioned in the middle of its main deck, with a hole or "moonpool" extending down through the hull – through which the drill string is extended through the water and seabed to the target location. Some drillships are designed to work in waters

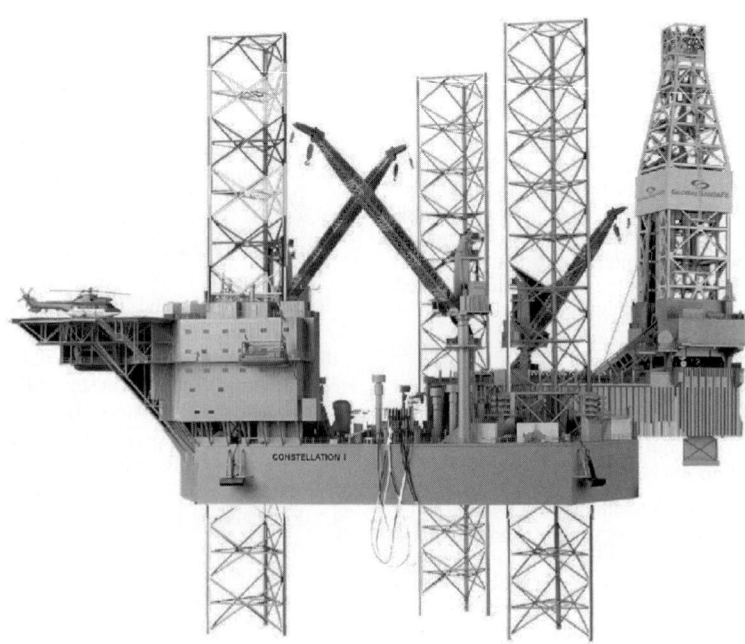

Figure 4.21.1 *Jackup drilling rig.*

Figure 4.21.2 *Drillship.*

Figure 4.21.3 *Semisubmersible drilling rig.*

more than 10,000 ft deep. They are kept on location by complex mooring and/or dynamic positioning systems. Dynamic positioning systems utilize electrically driven propellers, or thrusters, on the underside of the ship's hull, capable of moving the vessel in any direction. They are controlled by the ship's computer system, which uses satellite technology, along with sensors located on the drilling template on the sea floor, to ensure that the ship remains directly above the drill site.

Semisubmersible rigs (Figure 4.21.3)—the most common type of floating drilling vessel—can be used in water up to 10,000 ft deep, and provide a drilling platform generally more stable than does a drillship. Since a major portion of the vessel is submerged, wave effects can be minimized. Mooring systems of semisubmersibles utilize dynamic positioning and/or huge anchors weighing upwards of 15 tons. Ballast can be preferentially shifted within the vessel to provide stability during rough weather.

When choosing an offshore drillsite, the primary considerations are the likelihood of hydrocarbon-laden formations, water depth, the location of shipping lanes, foundation stability (for bottom-bearing vessels), and the possible presence of shallow gas. Seismic surveys and core analysis generally provide suitable information on foundation stability, and seismic studies can indicate the presence of potentially dangerous shallow gas deposits.

4.21.1.1 Marine Riser Systems

The most striking difference between onshore and offshore drilling occurs when the wellhead is located at the sea floor. This configuration makes communication with the well more complex. A marine riser provides communication and circulation capability between the surface and the sea floor, and is used at some point during all offshore drilling. The riser consists of large diameter (17- to 20-in.) steel pipe joints of 50- to 90-foot lengths, with quick-connect couplings. The riser can be connected at the sea floor to a wellhead or to a subsea blowout preventer stack. A diverter system is usually attached at the surface [8–9].

Vertical vessel motion is accommodated by adding a telescoping (slip) joint at the surface. This joint will usually allow up to 50 ft (double amplitude) of vertical vessel motion. Lateral motion is compensated for by the use of ball/flex joint connections at the sea floor and surface.

The riser can be quickly detached from the wellhead or blowout preventer stack to facilitate vessel movement during adverse weather conditions. Figure 4.21.4 shows in more detail a typical riser system that would be used while drilling the intermediate hole from a floating vessel.

A marine riser must be held in tension to prevent the riser from collapsing under its own weight. This can be accomplished by mechanical tensioning devices (Figures 4.21.5, 4.21.6) and by adding buoyant material to the riser pipe at increasing water depths. Many operators drill the conductor hole (approximately, 1,000 ft) without using a riser. When drilling with a floating vessel, the shallow portion of the surface hole is considered expendable, and can easily be abandoned in case of shallow gas flow, deviation difficulties, or other shallow-hole problems. Drilling is accomplished by circulating returns to the sea floor. Once the conductor hole is drilled, casing is run, and a subsea wellhead installed, then the marine riser can be attached and used to drill the remainder of the well.

4.21.1.2 Casing Programs

Casing programs offshore typically consist of 30-in. structural, 20-in. conductor, $13\frac{3}{8}$-in. surface, and $9\frac{5}{8}$-in. intermediate casing strings. Because seawater is in perfect communication with the sediments, pore pressure is increased by increasing water depth, and the margin between fracture pressure and pore pressure is lessened. Also, abnormal pore pressure is found in many offshore geological environments, forcing, fairly stringent casing designs to be used. Several intermediate casing strings or liners are often required to reach the target depth. Most operators prefer to set 7-in. production casing to facilitate larger production tubing and quicker payout [9,10].

4.21.1.3 Well Control

Offshore, well control equipment and associated operations present some differences from that seen and used onshore. In some instances, onshore equipment can be employed, but the offshore environment generally dictates a modification of equipment and procedures. There are several different well configurations used offshore, depending on the type of drilling unit employed and the stage of the drilling operation, and each configuration has specific well control procedures that should be followed. A well may be controlled with a surface blowout preventer stack; a subsea blowout preventer

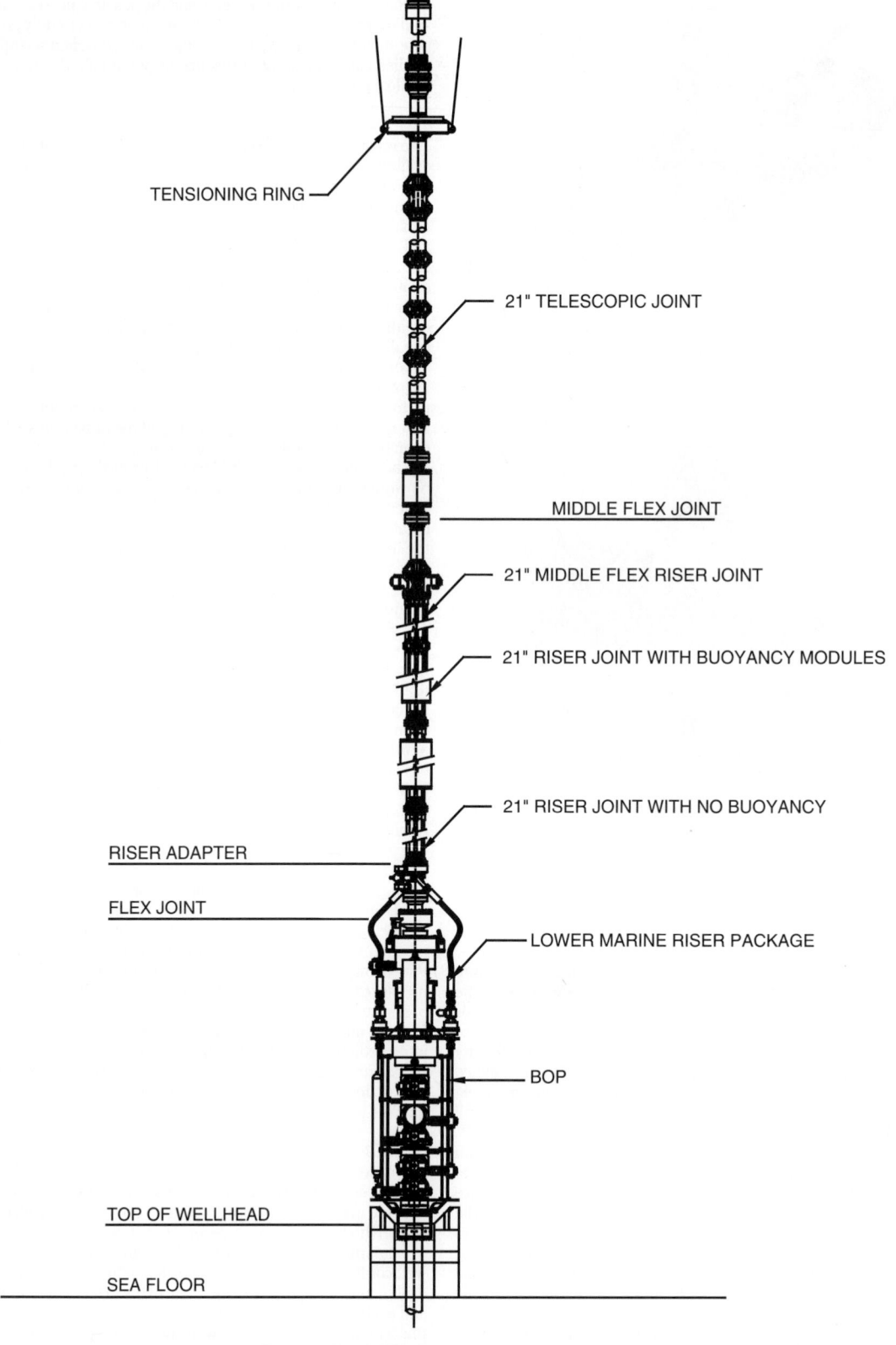

Figure 4.21.4   *Marine riser for floating system.*

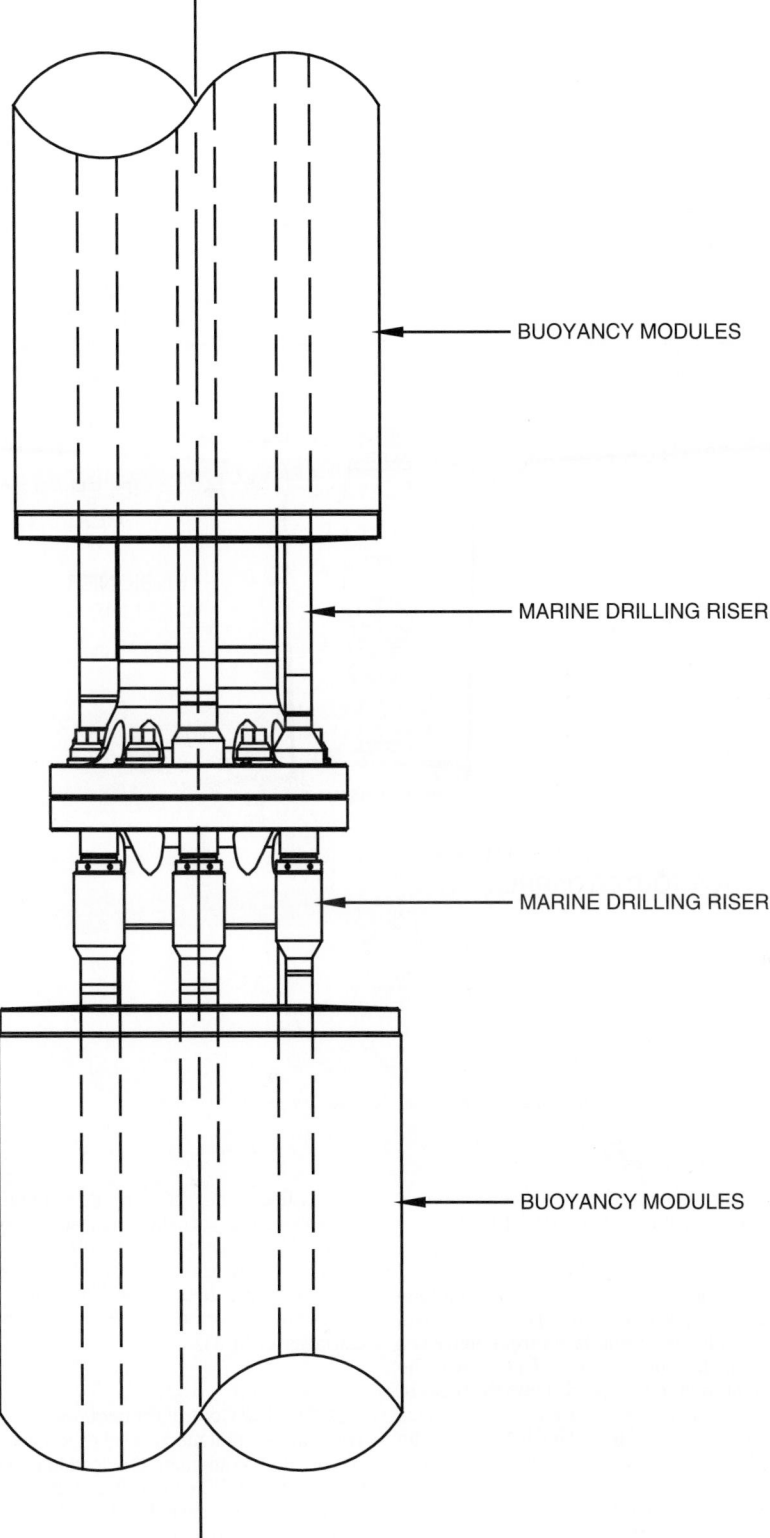

Figure 4.21.5 *Buoyant riser tensioning device.*

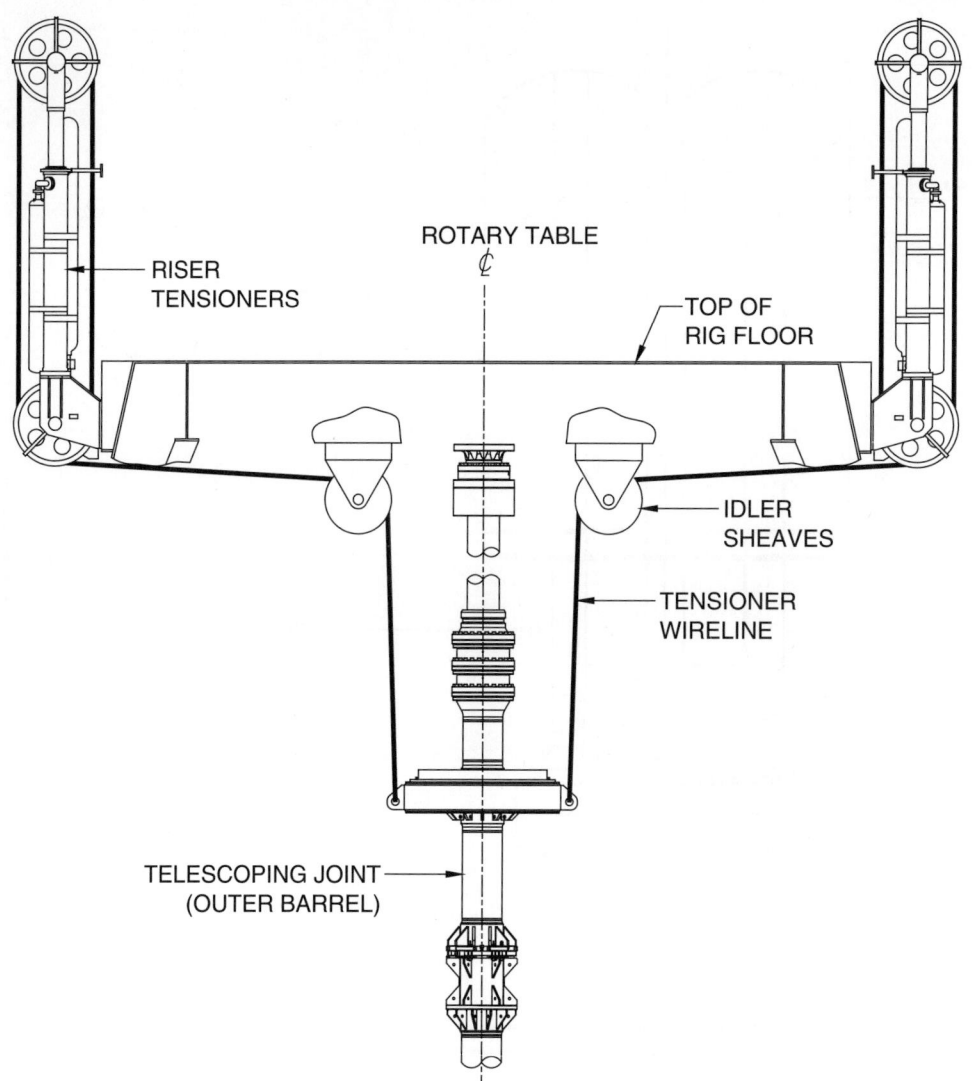

ROTARY TABLE
¢

RISER
TENSIONERS

TOP OF
RIG FLOOR

IDLER
SHEAVES

TENSIONER
WIRELINE

TELESCOPING JOINT
(OUTER BARREL)

Figure 4.21.6   *Mechanical riser tensioning device.*

stack, riser and diverter system; a riser and diverter system with no blowout preventer; or a diverter only [10,11].

4.21.1.4 Surface Blowout Preventer Stacks
When a development platform has been set, or a bottom-bearing drilling rig is used, wells can be drilled with a "land-type" well design. Conductor casing is set from above sea level, through the water, to as much as 1,500 feet subsea. The conductor provides for mud returns, and allows the blowout prevention system to exist in the surface environment. This well design is essentially identical to that onshore, as are the well control procedures that should be employed [10,11].

4.21.1.5 Subsea Blowout Preventer Stacks
Due to the nature of floating drilling operations, the operator does not always have the option of placing the blowout preventers at the surface. By placing the blowout preventer stack at the sea floor and communicating with the well through a detachable marine riser, a well can be shut in at the sea floor. This action removes the drilling vessel from the responsibility of maintaining surface position at all times. Figure 4.21.7 shows a typical subsurface blowout preventer stack. Note the presence of choke and kill lines extending from the surface to the stack. These lines allow fluids to flow out of the well or be pumped into the well, bypassing the marine riser [9–11].

4.21.1.6 Well Control Procedures
There are several mechanical differences between onshore and offshore equipment that lead to procedural differences in well control. When a well is shut in by a subsea blowout preventer, fluids must be circulated down the drillstring, up the casing annulus, and up the choke line to the surface. The marine riser/drillstring annulus is thus bypassed. Because the choke line is of small diameter (3 to $4\frac{1}{2}$ in.) and may be very long, depending on water depth, frictional pressure drop in the choke line may be very large. When

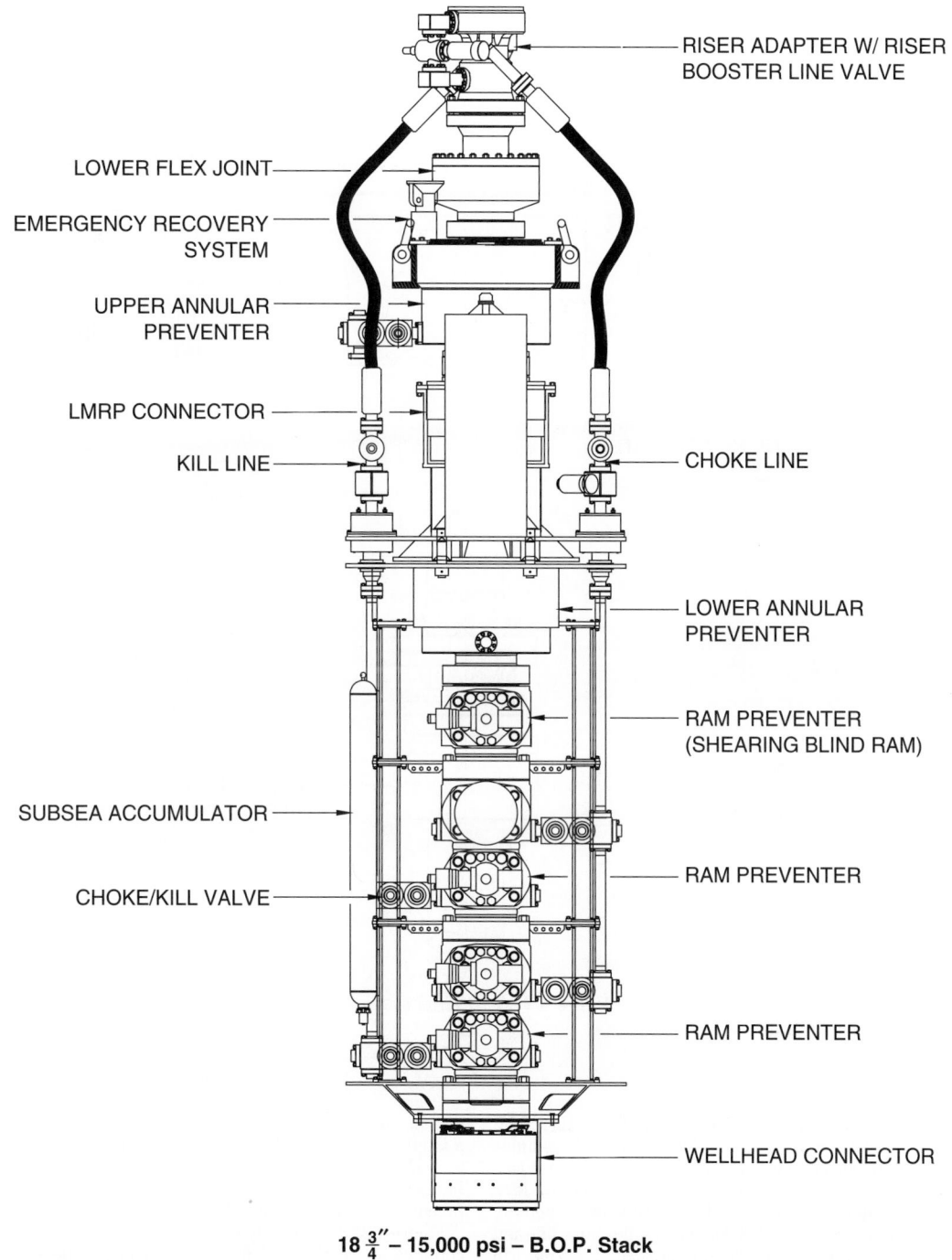

RISER ADAPTER W/ RISER BOOSTER LINE VALVE

LOWER FLEX JOINT

EMERGENCY RECOVERY SYSTEM

UPPER ANNULAR PREVENTER

LMRP CONNECTOR

KILL LINE

CHOKE LINE

LOWER ANNULAR PREVENTER

RAM PREVENTER (SHEARING BLIND RAM)

SUBSEA ACCUMULATOR

RAM PREVENTER

CHOKE/KILL VALVE

RAM PREVENTER

WELLHEAD CONNECTOR

18 $\frac{3}{4}''$ – 15,000 psi – B.O.P. Stack

Figure 4.21.7 *Subsea blowout preventer stack.*

preplanning for well control operations, the kill rate circulating pressure must be gained by circulating the well through the choke line. Neglecting choke line friction in calculations can cause excessive backpressure to be placed on the formation. Choke-line friction is considered when calculating the pump pressure startup schedule. If the circulating pressure has been measured by circulating through the choke line, the remainder of the calculations are performed as for a surface stack. Choke-line friction can be reduced by using larger choke lines, or by circulating the well through both the choke and kill lines [9–11].

When the well is circulated through the choke line, a rapid loss in hydrostatic pressure is encountered when the kick fluid begins to enter the choke line. Hydrostatic pressure is

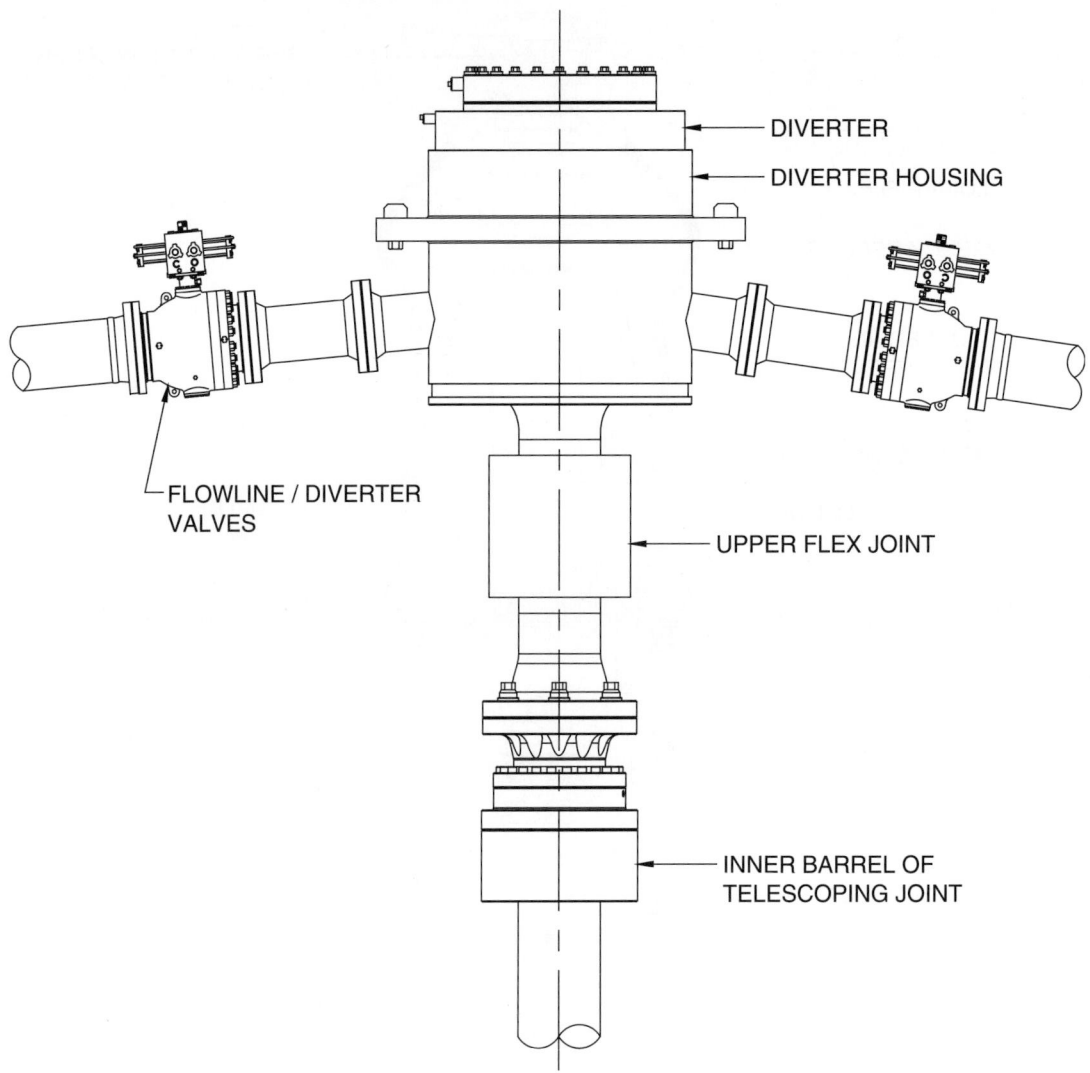

Figure 4.21.8 *Diverter system.*

lost because low-density gas is displacing the drilling mud from the small volume of choke line. Small kick volumes can result in long columns of gas in the choke line. Surface choke response must be rapid enough to prevent new kick fluid from entering the well due to the reduction in bottom-hole pressure. Also, when kill fluid enters the choke line and begins to displace the kick fluid, hydrostatic pressure will increase rapidly, thus increasing bottomhole pressure. Slow response by the choke operator can fracture the formation. A choke operator is faced with an usually demanding job when pumping out a kick through a subsurface blowout preventer.

4.21.1.7 Diverter Systems

Shallow gas hazards are a common potential danger off-shore. Quite often, these hazards can be spotted in seismic analysis, and a surface location chosen to avoid the hazard. However, there is always a risk of encountering a shallow gas flow with insufficient casing in the well to allow a

shut-in. In this instance a diverter system is called on as a safety measure. The ideal function of the diverter system is to allow the well to flow and subside by natural means. In many cases the diverter system simply provides enough time to evacuate the rig [9–11].

Figure 4.21.8 shows a typical diverter system setup. The components of the system are the annular-type preventer, vent lines, the control system, and the conductor or structural casing. The major operational consideration when using a diverter system is to make certain that the valves on the vent lines are fully open before the annular-type preventers are closed. When drilling the conductor or surface hole, shutting in the well may cause a subsurface blowout that is likely to broach to the surface. This cannot be tolerated when drilling from a bottom-supported vessel. Larger diameter vent lines free from turns or bends generally lower the backpressure placed on the well by the diverter system, therefore reducing the risk of a subsurface blowout.

The lack of bends in the lines greatly reduces the erosion potential of the flow. Sufficient conductor casing depth is also very important in preventing a subsurface blowout.

The Minerals Management Service requires a minimum 10-in.-diameter diverter vent line, but many operators are now using as large as 16-in.-diameter vent lines.

While a bottom-supported vessel must divert when shallow gas is encountered, a floating vessel has the additional option of simply abandoning the well. This option has led to the use of riserless systems when drilling the surface hole. However, a dynamic kill provides the only means of controlling the well. A dynamic kill makes use of annular friction as well as a heavier mud to hold backpressure on the formation. If very short wellbores are involved, the dynamic kill rates are usually too large to be practical.

4.21.2 Deepwater Well Planning Considerations

As the offshore industry moves into ever-deeper water to globally explore for and produce oil and natural gas in commercial quantities, particular and complex challenges arise in terms of health, safety and environmental considerations—which are always of the highest priority—as well as well planning and logistics.

In particular, industry experts point out, failures in the design of the structural string and wellhead have proven to be an expensive nemesis. They encourage a very conservative design approach, incorporating stringent safety, environmental and regulatory compliance factors and covering the risks and uncertainties of "worst-case" design assumptions.

Deepwater well planning further calls for special attention to such considerations as well angle, hydrate formation, ocean currents and deepwater hydrostatic pressures and stresses, as well as casing program, drilling fluids and techniques, flow control, kick prevention and response, well testing and intervention, and shut-in and emergency planning.

References

1. Tom Snackelford, Manager, Drilling Services, Global-SantaFe Corporation, Houston, Texas.
2. H.O. Birmingham III, Operations Manager, Deepwater, GlobalSantaFe Corporation, Houston.
3. Aimee M. Dobbs, Multimedia Manager, Challenger Minerals Inc., Houston.
4. Allan R. Beckering, Consultant, GlobalSantaFe Corporation, Houston.
5. Minerals Management Service, U.S. Department of the Interior (www.mms.gov).
6. American Petroleum Institute, Energy Professional (www.api.org).
7. International Association of Drilling Contractors (www.iadc.org).
8. World Petroleum Congress (www.world-petroleum. org).
9. *The Technology of Offshore Drilling, Completion, and Production*, ETA Offshore Seminars, Inc., PennWell, pp. 10, 190, 191, 206, Tulsa, 1976.
10. *Well Control Manual*, Louisiana State University, p. 12-3, Baton Rouge, 19XX *(sic)*.
11. *Blowout Prevention Equipment Systems for Drilling Operations*, 3rd Edition, American Petroleum Institute Recommended Practice 53.

5

Reservoir Engineering

Contributing Authors
Heru Danardatu
F. David Martin
Robert M. Colpitts, P.G.

Contents

Reservoir engineering covers a broad range of subjects including the occurrence of fluids in a gas or oil-bearing reservoir, movement of those or injected fluids, and evaluation of the factors governing the recovery of oil or gas. The objectives of a reservoir engineer are to maximize producing rates and to ultimately recover oil and gas from reservoirs in the most economical manner possible.

This chapter presents the basic fundamentals useful to practical petroleum engineers. Topics are introduced at a level that can be understood by engineers and geologists who are not expert in this field. Various correlations are provided where useful. Newer techniques for improving recovery are discussed.

The advent of programmable calculators and personal computers has dramatically changed the approach of solving problems used by reservoir engineers. Many repetitious and tedious calculations can be performed more consistently and quickly than was possible in the past. The use of charts and graphs is being replaced by mathematical expressions of the data that can be handled with portable calculators or personal computers. Programs relating to many aspects of petroleum engineering are now available. In this chapter, many of the charts and graphs that have been historically used are presented for completeness and for illustrative purposes. In addition, separate sections will be devoted to the use of equations in some of the more common programs suitable for programmable calculators and personal computers.

5.1 BASIC PRINCIPLES, DEFINITIONS, AND DATA

5.1.1 Reservoir Fluids

5.1.1.1 Oil and gas

Reservoir oil may be saturated with gas, the degree of saturation being a function, among others, of reservoir pressure and temperature. If the reservoir oil has dissolved in it all the gas it is capable of holding under given conditions, it is referred to as saturated oil. The excess gas is then present in the form of a free gas cap. If there is less gas present in the reservoir than the amount that may be dissolved in oil under conditions of reservoir pressure and temperature, the oil is then termed undersaturated. The pressure at which the gas begins to come out of solution is called the saturation pressure or the bubble-point pressure. In the case of saturated oil, the saturation pressure equals the reservoir pressure and the gas begins coming out of solution as soon as the reservoir pressure begins to decrease. In the case of undersaturated oil, the gas does not start coming out of solution until the reservoir pressure drops to the level of saturation pressure.

Apart from its function as one of the propulsive forces, causing the flow of oil through the reservoir, the dissolved gas has other important effects on recovery of oil. As the gas comes out of solution the viscosity of oil increases and its gravity decreases. This makes more difficult the flow of oil through the reservoir toward the wellbore. Thus the need is quite apparent for production practices tending to conserve the reservoir pressure and retard the evolution of the dissolved gas. Figure 5.1.1 shows the effect of the dissolved gas on viscosity and gravity of a typical crude oil.

The dissolved gas also has an important effect on the volume of the produced oil. As the gas comes out of solution the oil shrinks so that the liquid oil at surface conditions will occupy less volume than the gas-saturated oil occupied in the reservoir. The number of barrels of reservoir oil at reservoir pressure and temperature which will yield one barrel of stock tank oil at 60°F and atmospheric pressure is referred to as the formation volume factor or reservoir volume factor. Formation volume factors are described in a subsequent section. The solution gas–oil ratio is the number of standard cubic feet of gas per barrel of stock tank oil.

Physical properties of reservoir fluids are determined in the laboratory, either from bottomhole samples or from recombined surface separator samples. Frequently, however, this information is not available. In such cases, charts such as those developed by M.B. Standing and reproduced as Figures 5.1.2, 5.1.3, 5.1.4, and 5.1.5 have been used to determine the data needed [1,2]. The correlations on which the

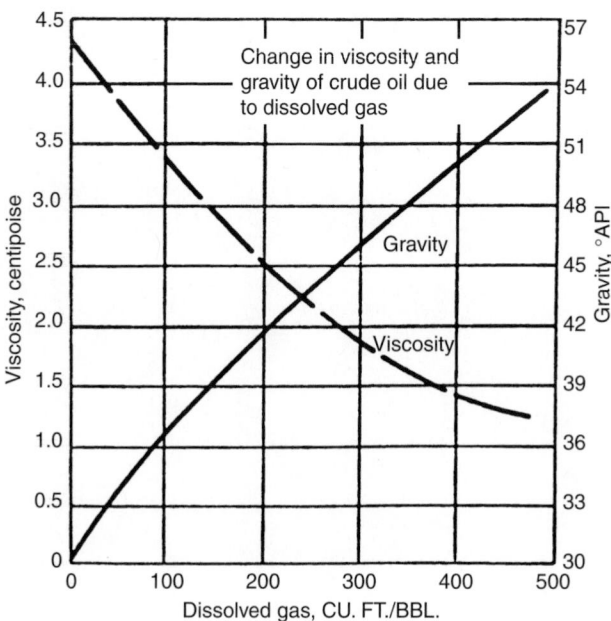

Figure 5.1.1 *Change in viscosity and gravity of crude oil due to dissolved gas.*

EXAMPLE

REQUIRED:
Bubble point pressure at 200°F
of a liquid having a gas-oil ratio of
350 CFB, a gas gravity of 0.75, and
a tank out gravity of 30°API.

PROCEDURE:
Starting at the left side of the
chart, proceed horizontally along line 350
CFB line to a gas gravity of 0.75. From
this point drop vertically to the 30°API
line. Proceed horizontally from the
tank oil gravity scale to the 200°F
line. The required pressure is
found to be 1930 PSIA.

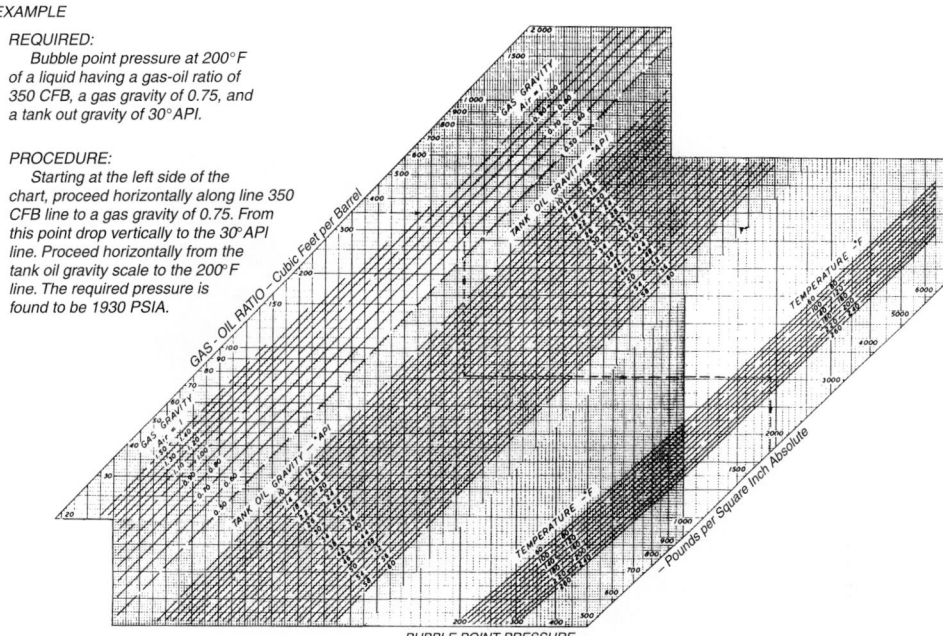

Figure 5.1.2 Bubble point pressure graph [1,2].

EXAMPLE

REQUIRED:
Formation volume at 200°F of a
bubble point liquid having a gas-oil
ratio of 350 CFB, a gas gravity of 0.75,
and a tank oil gravity of 30°API.

PROCEDURE:
Starting at the left side of the chart,
proceed horizontally along the 350 CFB
line to a gas graivty of 0.75. From this
point drop vertically to the 30°API line.
Proceed horizontally from the tank oil
gravity scale to the 200°F line. The
required formation volume is found to be
1.22 barrel per barrel of tank oil.

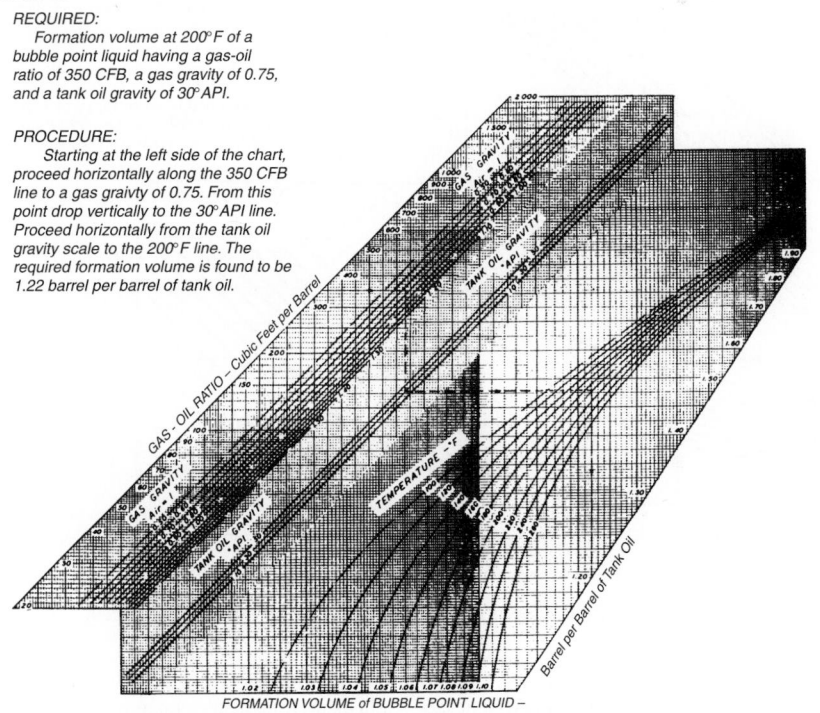

Figure 5.1.3 Formation volume of bubble point liquids [1,2].

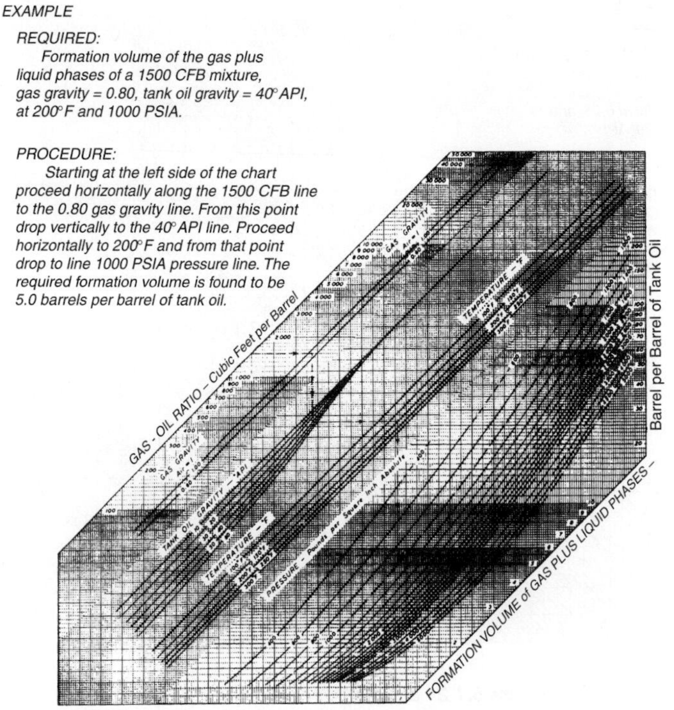

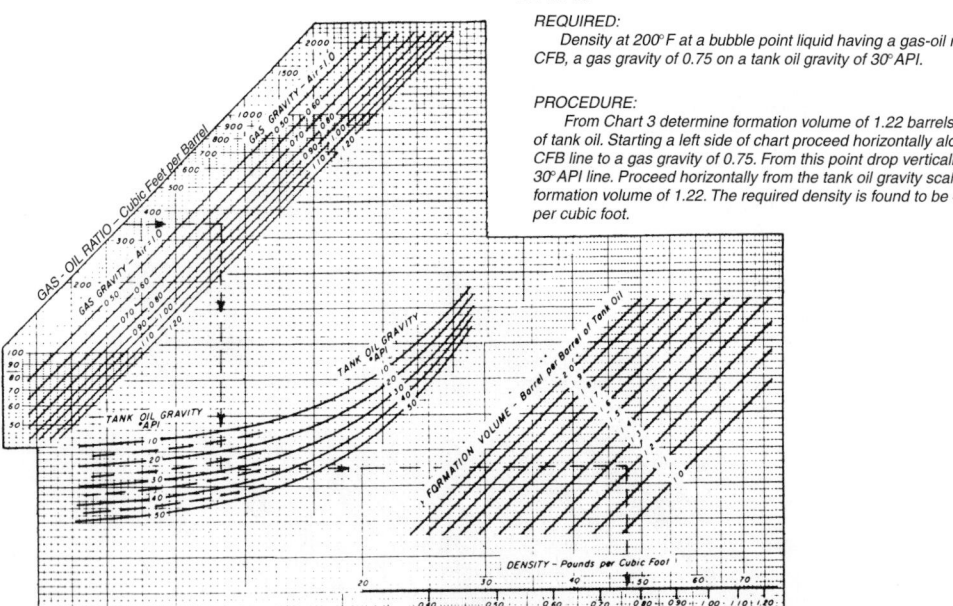

EXAMPLE
REQUIRED:
Formation volume of the gas plus liquid phases of a 1500 CFB mixture, gas gravity = 0.80, tank oil gravity = 40°API, at 200°F and 1000 PSIA.
PROCEDURE:
Starting at the left side of the chart proceed horizontally along the 1500 CFB line to the 0.80 gas gravity line. From this point drop vertically to the 40°API line. Proceed horizontally to 200°F and from that point drop to line 1000 PSIA pressure line. The required formation volume is found to be 5.0 barrels per barrel of tank oil.
GAS - OIL RATIO – Cubic Feet per Barrel
FORMATION VOLUME of GAS PLUS LIQUID PHASES – Barrel per Barrel of Tank Oil

Figure 5.1.4 *Formation volume of gas with liquid phases [1,2].*

EXAMPLE
REQUIRED:
Density at 200°F at a bubble point liquid having a gas-oil ratio at 350 CFB, a gas gravity of 0.75 on a tank oil gravity of 30°API.
PROCEDURE:
From Chart 3 determine formation volume of 1.22 barrels per barrel of tank oil. Starting a left side of chart proceed horizontally along the 350 CFB line to a gas gravity of 0.75. From this point drop vertically to the 30°API line. Proceed horizontally from the tank oil gravity scale to the formation volume of 1.22. The required density is found to be 47.5 pounds per cubic foot.
GAS - OIL RATIO – Cubic Feet per Barrel
GAS GRAVITY – Air = 1.0
TANK OIL GRAVITY °API
FORMATION VOLUME - Barrel per Barrel of Tank Oil
DENSITY – Pounds per Cubic Foot
SPECIFIC GRAVITY – Water at 60°F = 1.0

Figure 5.1.5 *Density and specific gravity of mixtures [1,2].*

charts are based present bubble-point pressures, formation volume factors of bubble-point liquids, formation volume factors of gas plus liquid phases, and, density of a bubble-point liquid as empirical functions of gas-oil ratio, gas gravity, oil gravity, pressure, and temperature. More recent correlations will be presented subsequently.

Until recently, most estimates of PVT properties were obtained by using charts and graphs of empirically derived

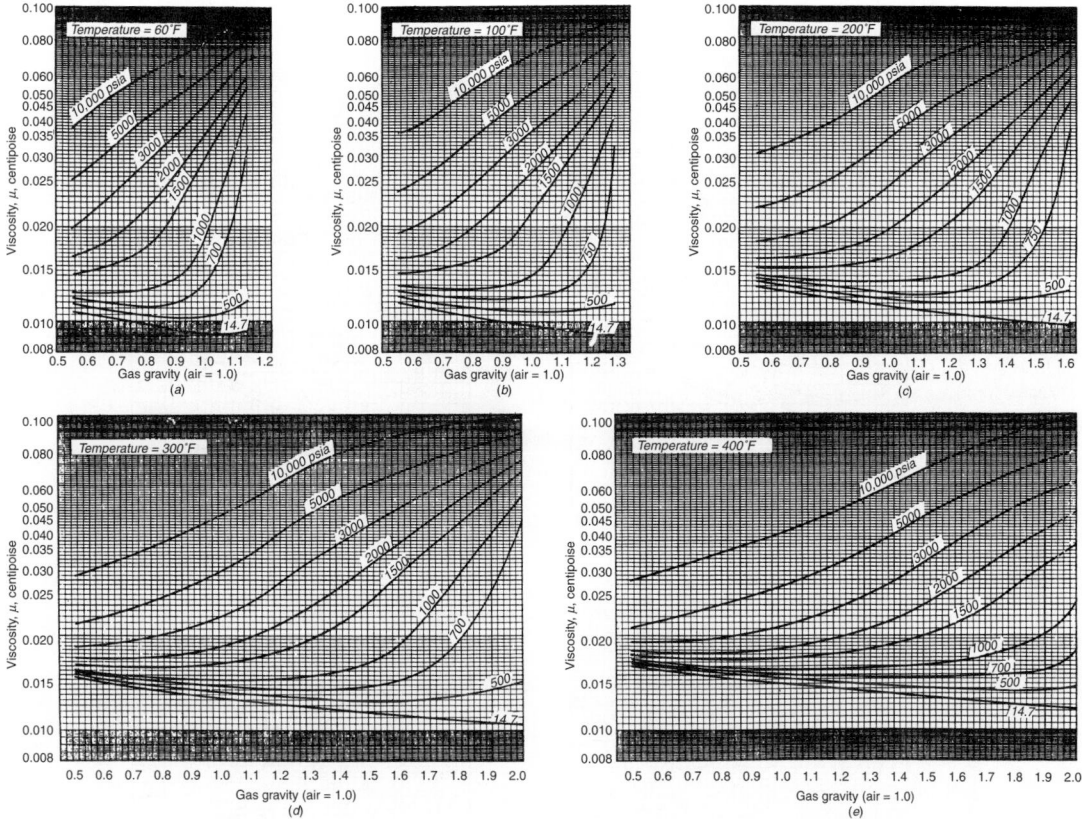

Figure 5.1.6 *Gas viscosity versus gravity at different temperatures [3].*

data. With the development of programmable calculators, graphical data are being replaced by mathematical expressions suitable for computer use. In a later section, the use of such programs for estimating PVT properties will be presented. In the initial sections, the presentation of graphical data will be instructive to gaining a better understanding of the effect of certain variables.

5.1.1.2 Water
Regardless of whether a reservoir yields pipeline oil, water in the form commonly referred to as interstitial or connate is present in the reservoir in pores small enough to hold it by capillary forces.

The theory that this water was not displaced by the migration of oil into a water-bearing horizon is generally accepted as explanation of its presence.

The amount of the interstitial water is usually inversely proportional to the permeability of the reservoir. The interstitial water content of oil-producing reservoirs often ranges from 10% to 40% of saturation.

Consideration of interstitial water content is of particular importance in reservoir studies, in estimates of crude oil reserves and in interpretation of electrical logs.

5.1.1.3 Fluid Viscosities
Gas Viscosity
Viscosities of natural gases are affected by pressure, temperature, and composition. The viscosity of a specific natural

gas can be measured in the laboratory, but common practice is to use available empirical data such as those shown in Figures 5.1.6 and 5.1.7. Additional data are given in the *Handbook of Natural Gas Engineering* [3]. Contrary to the case for liquids, the viscosity of a gas at low pressures increases as the temperature is raised. At high pressures, gas viscosity decreases as the temperature is raised. At intermediate pressure, gas viscosity may decrease as temperature is raised and then increase with further increase in temperature.

Oil Viscosity
The viscosity of crude oil is affected by pressure, temperature, and most importantly, by the amount of gas in solution. Figure 5.1.8 shows the effect of pressure on viscosities of several crude oils at their respective reservoir temperatures [4]. Below the bubble-point, viscosity decreases with increasing pressure because of the thinning effect of gas going into solution. Above the bubble-point, viscosity increases with increasing pressure because of compression of the liquid. If a crude oil is undersaturated at the original reservoir pressure, viscosity will decrease slightly as the reservoir pressure decreases. A minimum viscosity will occur at the saturation pressure. At pressures below the bubble-point, evolution of gas from solution will increase the density and viscosity of the crude oil as the reservoir pressure is decreased further.

Viscosities of hydrocarbon liquids decrease with increasing temperature as indicated in Figure 5.1.9 for gas-free reservoir crudes [5]. In cases where only the API gravity

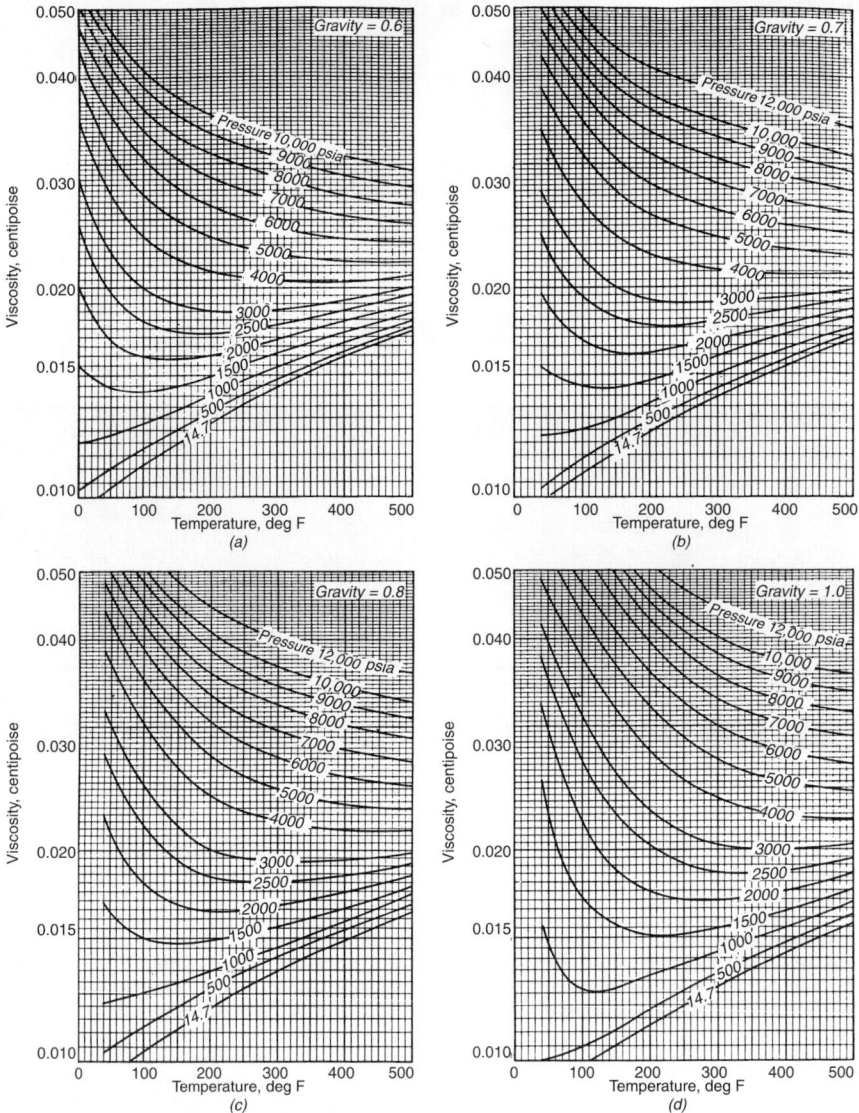

Figure 5.1.7 *Gas viscosity versus temperature at different gravities [3].*

of the stock tank oil and reservoir temperature are known, Figure 5.1.9 can be used to estimate dead oil viscosity at atmospheric pressure. However, a more accurate answer can be obtained easily in the laboratory by simply measuring viscosity of the dead oil with a viscometer at reservoir temperature.

With the dead oil viscosity at atmospheric pressure and reservoir temperature (either measured or obtained from Figure 5.1.9), the effect of solution gas can be estimated with the aid of Figure 5.1.10 [6]. The gas-free viscosity and solution gas-oil ratio are entered to obtain viscosity of the gas-saturated crude at the bubble-point pressure. This figure accounts for the decrease in viscosity caused by gas going into solution as pressure is increased form atmospheric to the saturation pressure.

If the pressure is above the bubble-point pressure, crude oil viscosity in the reservoir can be estimated with

Figure 5.1.11 [5]. This figure shows the increase in liquid viscosity due to compression of the liquid at pressures higher than the saturation pressure. Viscosity of the crude can be estimated from the viscosity at the bubble-point pressure, and the difference between reservoir pressure and bubble-point pressure.

Recent correlations [7] were presented in equation form for the estimation of both dead oil and saturated oil viscosities. These correlations, which are presented in the section on programs for hand-held calculations, neglect the dependence of oil viscosity on composition of the crude. If compositional data are available, other correlations [8–10] for oil viscosity can be used.

Water Viscosity

In 1952, the National Bureau of Standards conducted tests [11] which determined that the absolute viscosity of pure

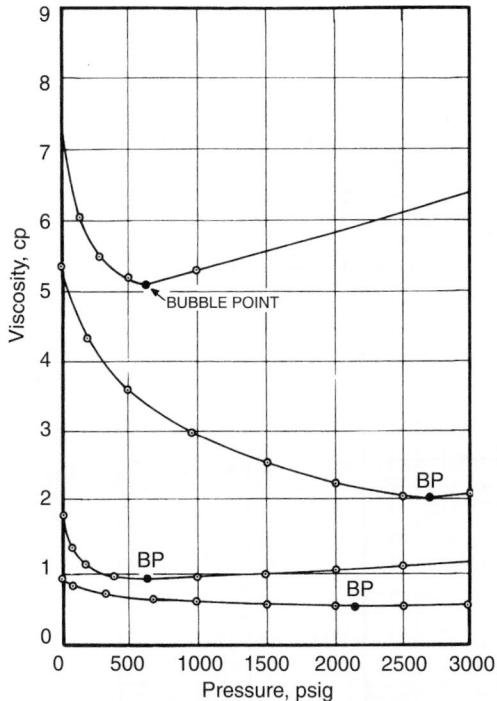

Figure 5.1.8 *Effect of pressure on crude oil viscosities [4].*

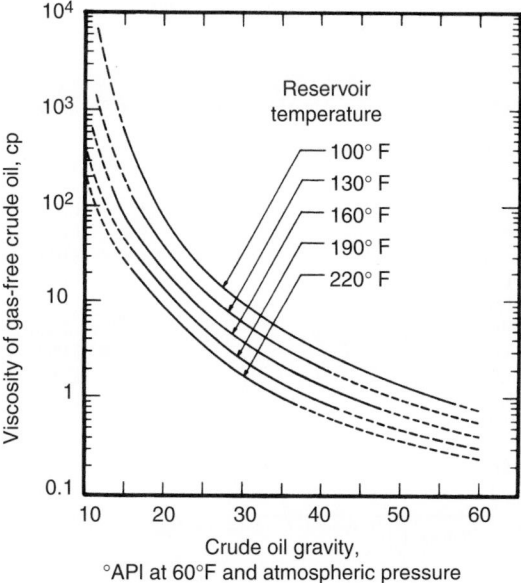

Figure 5.1.9 *Crude oil viscosity as a function of API gravity [5].*

water was 1.0019 cp as compared with the value of 1.005 cp that had been accepted for many years. Effective July 1, 1952, the value of 1.002 cp for the absolute viscosity of water was recommended as the basis for the calibration of viscometers and standard oil samples. Any literature values based on the

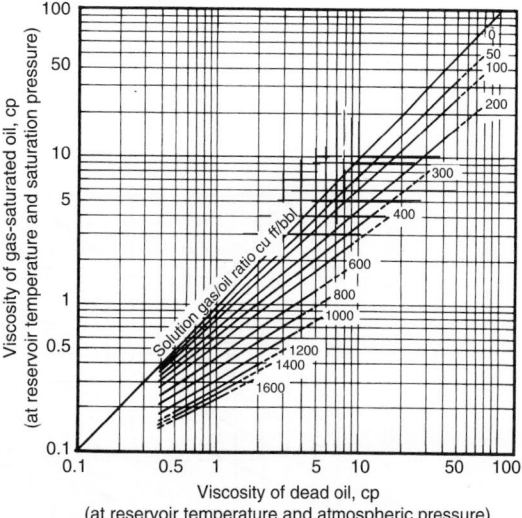

Figure 5.1.10 *Viscosities of gas-saturated crude oils at reservoir temperature and bubble-point pressure [6].*

old standard are in slight error. Water viscosity decreases as temperature is increased as shown in Table 5.1.1.

Although the predominate effect on water viscosity is temperature, viscosity of water normally increases as salinity increases. Potassium chloride is an exception to this generality. Since most oilfield waters have a high sodium chloride content, the effect of this salt on viscosity of water is given in Table 5.1.2.

For temperatures of interest in oil reservoirs ($> 60°F$), the viscosity of water increases with pressure but the effect is slight. Dissolved gas at reservoir conditions should reduce the viscosity of brines; however, the lack of data and the slight solubility of gas in water suggest that this effect is usually ignored. Figure 5.1.12 is the most widely cited data for the effect of sodium chloride and reservoir temperature on water viscosity [13].

5.1.1.4 Formation Volume Factors
These factors are used for converting the volume of fluids at the prevailing reservoir conditions of temperature and pressure to standard surface conditions of 14.7 psia and 60°F.

Gas Formation Volume Factor
The behavior of gas can be predicted from:

$$pV = znRT \qquad [5.1.1]$$

where p = absolute pressure
V = volume of gas
T = absolute temperature
n = number of moles of gas
R = gas constant
z = factor to correct for nonideal gas behavior

For conventional field units, p is in psia, V is in ft³, T is in °R (°F + 460), z is dimensionless, n is in lb moles, and R is 10.73 psia ft³/lb mole °R [14]. The gas formation volume factor, B_g, is the volume of gas in the reservoir occupied by a standard ft³ of gas at the surface:

$$B_g = \frac{V\ ft^3}{5.615\ ft^3/bbl} = \frac{znR(T_R + 460)}{5.615} \qquad [5.1.2]$$

where T_R is the reservoir temperature in °F.

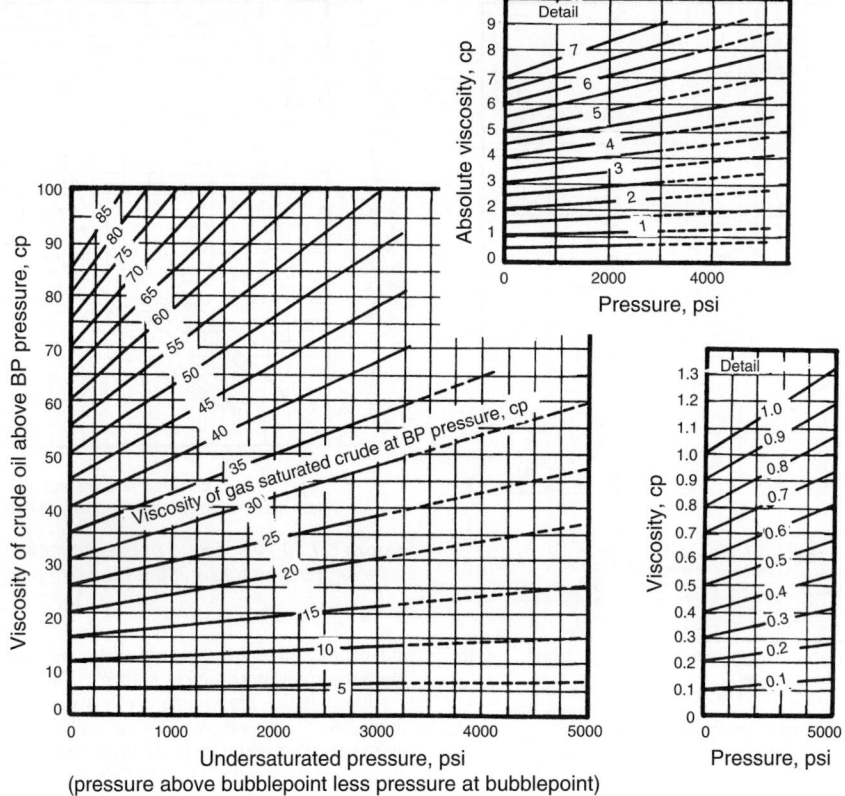

Figure 5.1.11   *Increase in oil viscosity with pressure above bubble-point pressure [5].*

Table 5.1.1   *Viscosity of Pure Water at Various Temperatures*

T, °C	T, °F	Viscosity (cp)
0	32	1.787
10	50	1.307
20	68	1.002
25	77	0.8904
30	86	0.7975
40	104	0.6529
50	122	0.5468
60	140	0.4665
70	158	0.4042
80	176	0.3547
90	194	0.3147
100	212	0.2818

From Reference 12.

Table 5.1.2   *Viscosities of Sodium Chloride Solutions at 68°F*

NaCl (wt %)	Viscosity (cp)
0.1	1.004
0.3	1.008
0.5	1.011
1.0	1.020
1.5	1.028
2.0	1.036
3.0	1.052
4.0	1.068
5.0	1.085
10.0	1.193
15.0	1.351
20.0	1.557
25.0	1.902

From Reference 12.

Since one lb mole is equivalent to $379\,\text{ft}^3$ at 60°F and 14.7 psia [15]:

$$B_g = \frac{\frac{1}{379} \times 10.73}{5.615}\frac{zT_R}{p} = 0.00504\frac{zT_R}{p} \qquad [5.1.3]$$

In this expression, B_g will be in reservoir barrels per standard ft^3 (RB/scf).

Gas formation volume factor can also be expressed in units of reservoir barrels per stock tank barrel or ft^3 of gas at reservoir conditions per ft^3 of gas at standard conditions:

$$B_g = 0.02827(460 + T_R)\frac{z}{p} \qquad [5.1.4]$$

Because the gas formation volume factor can be expressed in so many different units (including the reciprocal of B_g), caution should be exercised when B_g is used. In much of the recent petroleum literature, notably SPE, B_g is expressed in

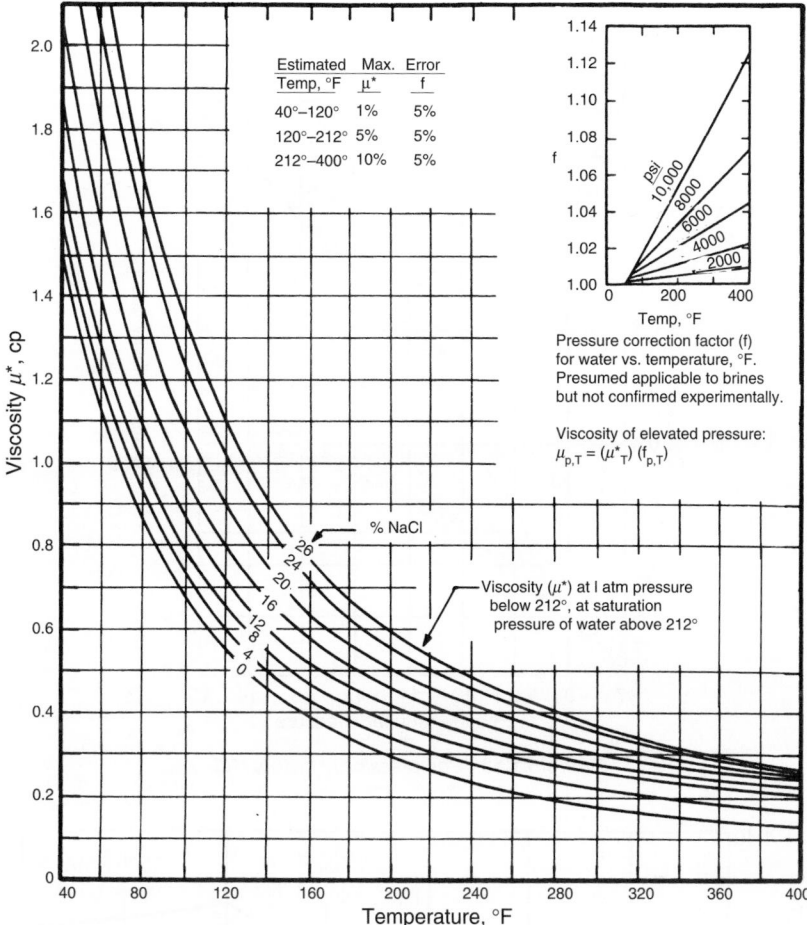

Figure 5.1.12 *Water viscosities for various salinities and temperatures [13].*

RB/scf. If units of ft^3/scf are given, B$_g$ can be divided by 5.615 or multiplied by 0.1781 to get RB/scf.

Gas formation volume factors can be estimated by determining the gas deviation factor or compressibility factor, z, at reservoir pressure, p, and temperature T$_R$ from the correlations of Standing and Katz [16] (Figure 5.1.13). To obtain the z factor, reduced pressure, p$_r$, and reduced temperature, T$_r$, are calculated:

$$p_r = \frac{p}{p_c} \quad\quad [5.1.5]$$

where p$_c$ is the critical pressure and

$$T_r = \frac{T}{T_c} \quad\quad [5.1.6]$$

where T$_c$ is the critical temperature. The critical pressure and temperature represent conditions above which the liquid and vapor phase are indistinguishable.

Compressibility factor and gas formation volume factor can be more conveniently estimated by the use of programs available for hand-held calculators. These programs will be subsequently discussed.

Oil Formation Volume Factor

The volume of hydrocarbon liquids produced and measured at surface conditions will be less than the volume at reservoir conditions. The primary cause is the evolution of gas from the liquids as pressure is decreased from the reservoir to the surface. When there is a substantial amount of dissolved gas, a large decrease in liquid volume occurs. Other factors that influence the volume of liquids include changes in temperature (a decrease in temperature will cause the liquid to shrink) and pressure (a decrease in pressure will cause some liquids to expand). All of these factors are included in the oil formation volume factor, B$_o$, which is the volume of oil in reservoir barrels, at the prevailing reservoir conditions of pressure and temperature, occupied by a stock tank barrel of oil at standard conditions. The withdrawal of reservoir fluids can be related to surface production volumes by obtaining laboratory PVT data with reservoir fluids. Such data include B$_g$ (the gas formation volume factor), B$_o$ (the oil formation volume factor), and R$_s$ (the solution gas-oil ratio which is the volume of gas in standard ft^3 that will dissolve in one stock tank barrel of oil at reservoir conditions).

The formation volume factor is used to express the changes in liquid volume accompanied by changes in pressure. Changes in formation volume factor with pressure for an undersaturated crude are displayed in Figure 5.1.14 [17]. As the initial reservoir pressure decreases, the all-liquid system expands and the formation volume factor increases until the bubble-point pressure is reached. As pressure decreases below the bubblepoint, gas comes out of solutions, the volume of oil is reduced, and B$_o$ decreases. For a saturated

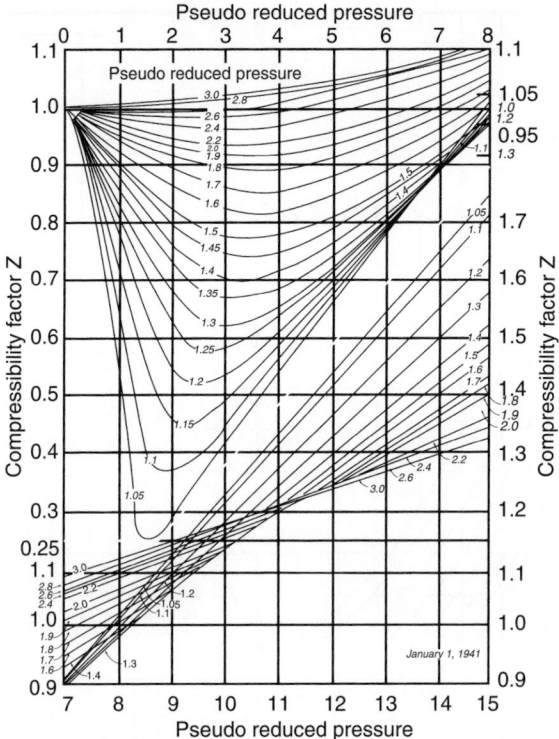

Figure 5.1.13 *Compressibility factors [16].*

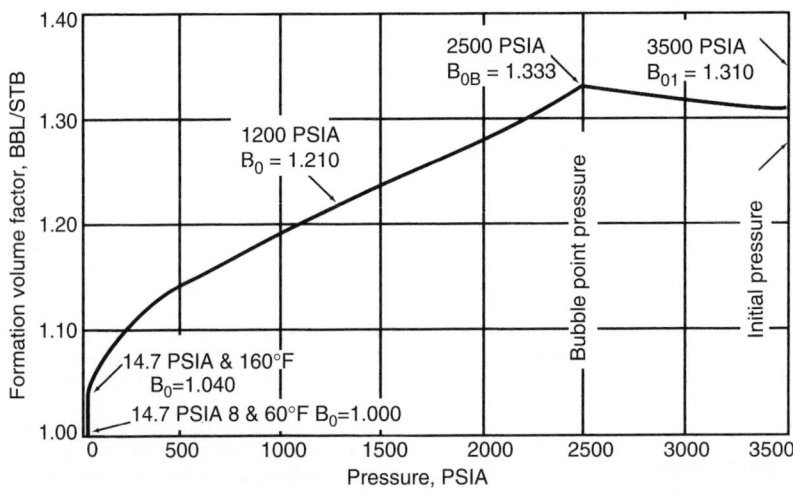

Figure 5.1.14 *Formation volume factor of the Big Sandy Field reservoir oil, by flash liberation at reservoir temperature of 160°F [17].*

crude, the trend would be similar to that observed to the left of bubble-point pressure in Figure 5.1.14.

When two phases exist, the total formation volume factor or 2-phase formation volume factor is [17]:

$$B_t = B_o + B_g(R_{si} - R_s) \qquad [5.1.7]$$

which includes the liquid volume, B_o, plus the gas volume times the difference in initial solution gas-oil ratio, R_{si}, and the solution gas-oil ratio at the specific pressure, R_s. At pressures

above the bubblepoint, R_{si} equals R_s, and the single-phase and 2-phase formation volume factors are identical. At pressures below the bubblepoint, the 2-phase factor increases as pressure is decreased because of the gas coming out of solution and the expansion of the gas evolved.

For a system above the bubblepoint pressure, B_o is lower than the formation volume factor at saturation pressure because of contraction of the oil at higher pressure. The customary procedure is to adjust the oil formation volume

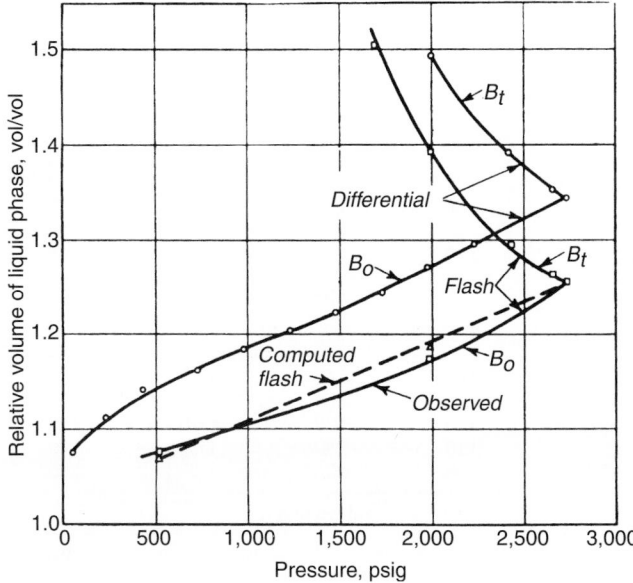

Figure 5.1.15 *Comparison of measured and calculated composite oil volume [19].*

factor at bubble-point pressure and reservoir temperature by a factor that accounts for the isothermal coefficient of compressibility such as [18]:

$$B_o = B_{ob} \exp[-c_o(p - p_b)] \qquad [5.1.8]$$

where B_{ob} is the oil formation volume factor at bubblepoint conditions, p_b is the bubble-point pressure in psi, and c_o is oil compressibility in psi^{-1}.

The basic PVT properties (B_o, R_s, and B_g) of crude oil are determined in the laboratory with a high-pressure PVT cell. When the pressure of a sample of crude oil is reduced, the quantity of gas evolved depends on the conditions of liberation. In the flash liberation process, the gas evolved during any pressure reduction remains in contact with the oil. In the differential liberation process, the gas evolved during any pressure reduction is continuously removed from contact with the oil. As a result, the flash liberation is a constant-composition, variable-volume process and the differential liberation is a variable-composition, constant-volume process. For heavy crudes (low volatility, low API gravity oils) with dissolved gases consisting primarily of methane and ethane, both liberation processes yield similar quantities and compositions of evolved gas as well as similar resulting oil volumes. However, for lighter, highly volatile crude oils containing a relatively high proportion of intermediate hydrocarbons (such as propane, butane, and pentane), the method of gas liberation can have an effect on the PVT properties that are obtained. An example of differences in formation volumes with flash and differential liberation processes can be seen in Figure 5.1.15 [19]. Actual reservoir conditions may be somewhere between these extremes because the mobility of the liberated gas is greater than the oil, the gas is produced at a higher rate, and the oil in the reservoir is in contact with all of the initial solution gas for only a brief period [20]. Since volatile oil situations are uncommon [20], many engineers feel the differential liberation process typifies most reservoir conditions [19]. For reservoir fluids at the bubblepoint when a well is put on production, the gas evolved from the oil as the pressure declines does not flow to the well until the critical gas saturation is exceeded. Since the greatest pressure drop occurs near the wellbore, the critical gas saturation occurs first near the well, especially if the pressure drop is large. In general, differential liberation data are applicable if the reservoir pressure falls considerably below the bubble-point pressure and the critical gas saturation is exceeded in the majority of the drainage area, as indicated by producing gas-oil ratios considerably in excess of the initial solution gas-oil ratio [17]. Flash liberation data may be applicable to reservoirs where there is only a moderate pressure decline below the bubblepoint, as indicated by producing gas-oil ratios not much higher than the initial solution gas-oil ratio, since the liberated gas stays in the reservoir in contact with the remaining oil [17].

Several correlations are available for estimating formation volume factors. Single-phase formation volume factors can be estimated from solution gas, gravity of solution gas, API gravity of the stock tank oil, and reservoir temperature by using the correlations of Standing [1,2]. Figure 5.1.3 provides Standing's empirical correlation of bubble-point oil formation volume factor as a function of the variables mentioned. Total formation volume factors of both solution gas and gas-condensate systems can be obtained from Standing's correlations given in Figure 5.1.4.

Empirical equations have been developed [21] from Standing's graphical data. These equations provide the oil formation volume factor and the solution gas-oil ratio as functions of reservoir pressure [21]:

$$B_o = a\, p^{1.17} + b \qquad [5.1.9]$$

where a is a constant that depends on temperature, oil API gravity and gas gravity and b is a constant that depends on temperature. Values of both constants are given in Table 5.1.3, other values can be interpolated.

Solution gas-oil ratio can be estimated from:

$$R_s = y\, p^{1.17} \qquad [5.1.10]$$

where y is a constant that depends on temperature, gas gravity, and oil gravity. Values of y are provided in Table 5.1.4.

Table 5.1.3 *Values of Constants for Equation 5.1.9*

Oil gravity		Values of a × 10^5					
		T = 120°F			T = 140°F		
°API	Gas gravity:	0.7	0.8	0.9	0.7	0.8	0.9
26		2.09	2.55	3.10	2.03	2.58	3.13
30		2.44	2.98	3.61	2.38	3.01	3.64
34		2.85	3.48	4.21	2.78	3.51	4.24
38		3.33	4.07	4.90	3.26	4.10	4.93
42		3.89	4.75	5.71	3.82	4.78	5.74

Oil gravity		T = 160°F			T = 180°F		
°API	Gas gravity:	0.7	0.8	0.9	0.7	0.8	0.9
26		2.02	2.47	3.02	1.95	2.38	2.91
30		2.33	2.85	3.48	2.27	2.78	3.39
34		2.69	3.29	4.01	2.65	3.24	3.96
38		3.10	3.80	4.62	3.09	3.79	4.61
42		3.58	4.38	5.33	3.60	4.42	5.38

Values of b	
T, °F	b
120	1.024
140	1.032
160	1.040
180	1.048

From Reference 21.

Table 5.1.4 *Values of Constants for Equation 5.1.10*

Oil gravity		T = 120°F			T = 140°F		
°API	Gas gravity:	0.7	0.8	0.9	0.7	0.8	0.9
26		0.0494	0.0577	0.0645	0.0481	0.0563	0.0632
30		0.0568	0.0660	0.0737	0.0550	0.0635	0.0721
34		0.0654	0.0755	0.0842	0.0630	0.0736	0.0823
38		0.0752	0.0864	0.0962	0.0720	0.0841	0.0939
42		0.0865	0.0989	0.1099	0.0824	0.0961	0.1071

Oil gravity		T = 160°F			T = 180°F		
°API	Gas gravity:	0.7	0.8	0.9	0.7	0.8	0.9
26		0.0453	0.0519	0.0591	0.0426	0.0481	0.0543
30		0.0522	0.0597	0.0677	0.0492	0.0557	0.0629
34		0.0601	0.0686	0.0775	0.0567	0.0645	0.0728
38		0.0692	0.0788	0.0887	0.0654	0.0747	0.0842
42		0.0797	0.0906	0.1016	0.0755	0.0865	0.0975

From Reference 21.

Additional correlations suitable for use with programmable calculators are discussed in a later section of this chapter.

Water Formation Volume Factor

The factors discussed that affected B_o also affect the water formation volume factor, B_w. However, gas is only slightly soluble in water so evolution of gas from water has a negligible effect on B_w. Expansion and contraction of water due to reduction of pressure and temperature are slight and offsetting. Hence, B_w is seldom greater than 1.06 [18] and is usually near unity (see Table 5.1.5).

Several correlations for B_w are available, including the effect of gas saturation in pure water and the effect of salinity [23], and the effect of natural gas on B_w as a function of pressure and temperature [24]. However, since B_w is not greatly affected by these variables, only a simplified correction is presented [18]:

$$B_w = (1 + \Delta V_{wp})(1 + \Delta V_{wT}) \qquad [5.1.11]$$

where ΔV_{wp} and ΔV_{wT} are the volume changes caused by reduction in pressure and temperature, respectively. Values of these corrections are given in Figures 5.1.16 and 5.1.17.

5.1.1.5 Fluid Compressibilities

Gas Compressibility

The compressibility of a gas, which is the coefficient of expansion at constant temperature, should not be confused with the compressibility factor, z, which refers to the deviation from ideal gas behavior. From the basic gas equation (see Equation 5.1.2), Muskat [25] provided an expression

Table 5.1.5 *Formation Volumes of Water*

Pressure psia	Formation volumes, bbl/bbl			
	100°F	150°F	200°F	250°F
	Pure water			
5,000	0.9910	1.0039	1.0210	1.0418
4,000	0.9938	1.0067	1.0240	1.0452
3,000	0.9966	1.0095	1.0271	1.0487
2,000	0.9995	1.0125	1.0304	1.0523
1,000	1.0025	1.0153	1.0335	1.0560
Vapor pressure of water	1.0056	1.0187	1.0370	1.0598
Saturation pressure psia	Natural gas and water			
5,000	0.9989	1.0126	1.0301	1.0522
4,000	1.0003	1.0140	1.0316	1.0537
3,000	1.0017	1.0154	1.0330	1.0552
2,000	1.0031	1.0168	1.0345	1.0568
1,000	1.0045	1.0183	1.0361	1.0584

From Reference 22.

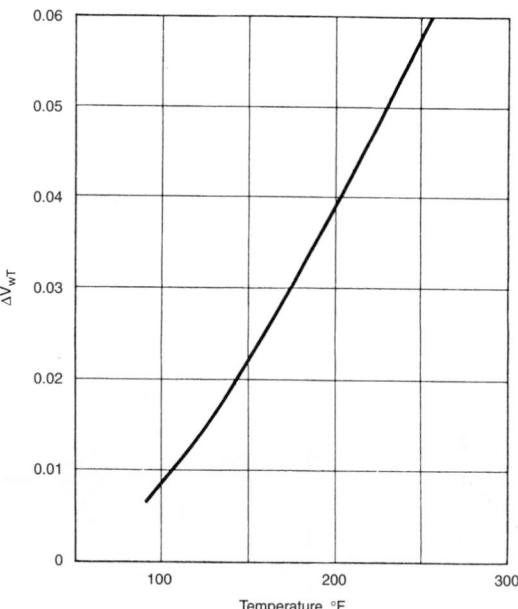

Figure 5.1.17 *Change in water volume due to temperature reduction [18].*

obtained from laboratory PVT data or estimated from the correlations given by Trube [27] (see Figures 5.1.19a and 5.1.19b). Trube defined the pseudo-reduced compressibility of a gas, c_{pr}, as a function of pseudo-reduced temperature and pressure, T_{pr} and p_{pr}, respectively [27]:

$$c_g = \frac{1}{p_{pc}}\left[\frac{1}{p_{pr}} - \frac{1}{z}\left(\frac{dz}{dp_{pr}}\right)T_{pr}\right] \qquad [5.1.13]$$

where p_{pc} is the pseudo-critical pressure (reduced and critical pressures have been defined earlier). Gas compressibility is computed for the pseudo-reduced compressibility from the appropriate figure:

$$c_g = \frac{c_{pr}}{p_{pr}} \qquad [5.1.14]$$

Pseudo-critical pressures and temperatures can be calculated from the mole fraction of each component present in hydrocarbon gas mixture or estimated from Figure 5.1.20 [3].

Oil Compressibility
The compressibility of oil, c_o, can be obtained in the laboratory from PVT data. In the absence of laboratory data, Trube's correlation [28] for compressibility of an undersaturated oil in Figure 5.1.21 can be used in a similar fashion as previously discussed for c_g. Pseudo-critical temperature and pressure can be estimated from Figure 5.1.22 or 5.1.23. With the pseudo-reduced compressibility from Figure 5.1.21, oil compressibility can be estimated:

$$c_o = \frac{c_{pr}}{p_{pc}} \qquad [5.1.15]$$

For conditions below the bubblepoint, dissolved gas must be taken into account. In the absence of laboratory data, the changes in R_s and B_o with changes in pressure can be approximated from Figures 5.1.24 and 5.1.25, which were developed by Ramey [26] from Standing's [1] data:

$$\left(\frac{\partial B_o}{\partial p}\right)_T = \left(\frac{\partial R_s}{\partial p}\right)_T \left(\frac{\partial B_o}{\partial R_s}\right)_T \qquad [5.1.16]$$

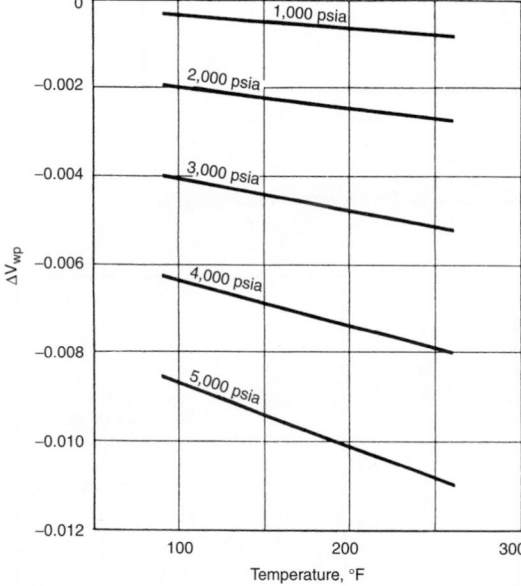

Figure 5.1.16 *Change in water volume due to pressure reduction [18].*

for the coefficient of isothermal compressibility:

$$c_g = \frac{1}{p} - \frac{1}{z}\frac{dz}{dp} \qquad [5.1.12]$$

For perfect gases $(z = 1$ and $dz/dp = 0)$, c_g is inversely proportional to pressure. For example, an ideal gas at 1,000 psia has a compressibility of $1/1,000$ or $1,000 \times 10^{-6}$ psi^{-1}. However, natural hydrocarbon gases are not ideal gases and the compressibility factor, z, is a function of pressure as seen in Figure 5.1.18 [17]. At low pressures, z decreases as pressure increases and dz/dp is negative; thus, c_g is higher than that of an ideal gas. At high pressures, dz/dp is positive since z increases, and c_g is less than that of a perfect gas.

Compared to other fluids or to reservoir rock, the compressibility of natural gas is large; c_g ranges from about $1,000 \times 10^{-6}$ psi^{-1} at 1,000 psi to about 100×10^{-6} psi^{-1} at 5,000 psi [27]. Compressibility of natural gases can be

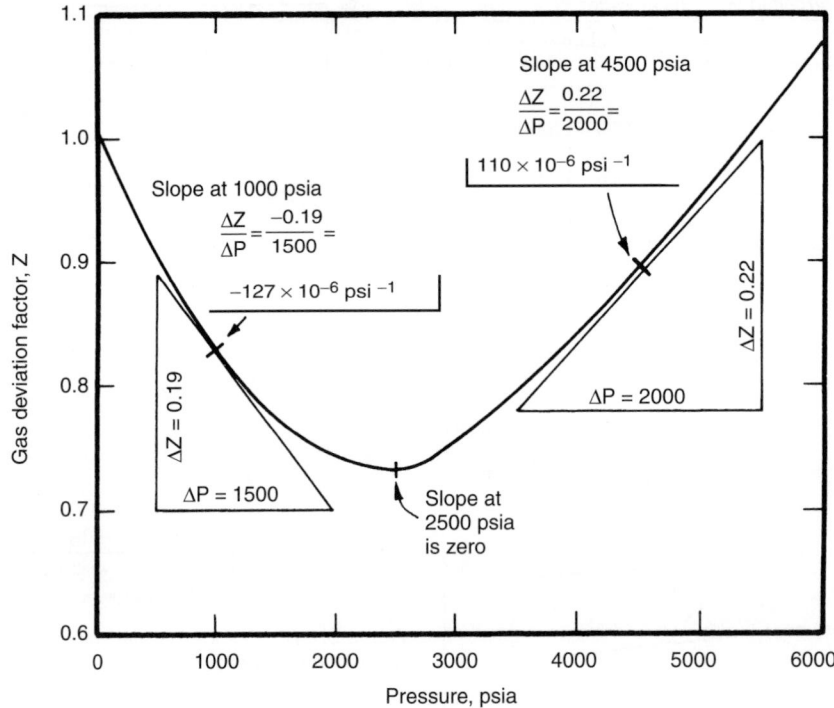

Figure 5.1.18   *Gas compressibility from the gas deviation factor versus pressure [17].*

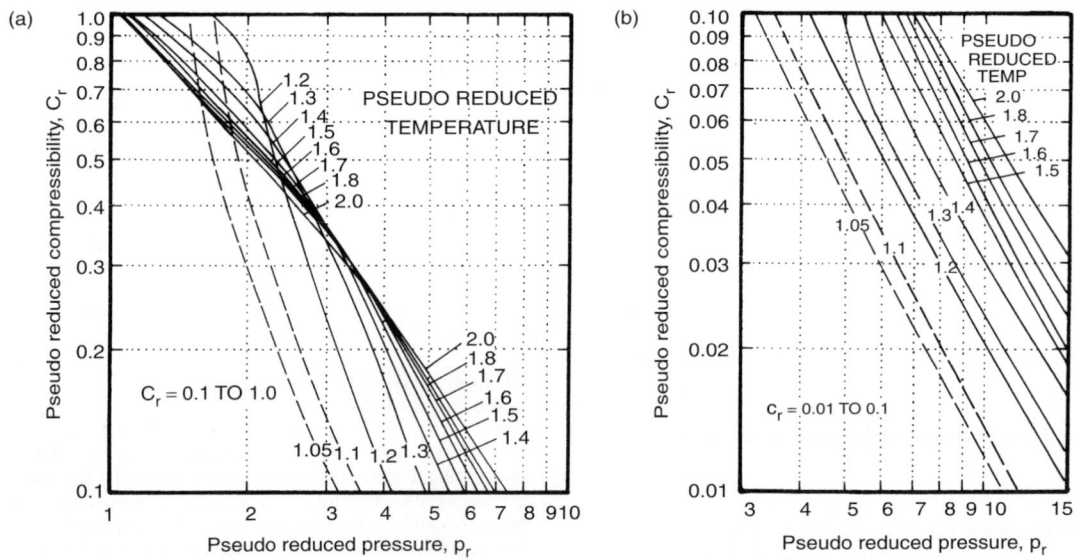

Figure 5.1.19   *(a and b) Variation of reduced compressibility with reduced pressures for various fixed values of reduced temperature [27].*

B_o can be estimated from Figure 5.1.3, and gravities of both oil and gas must be known. Oil compressibility is often on the order of 10×10^{-6} psi^{-1}.

There are several types of petroleum-engineering software equipped with the so-called "PVT data" generator code. Engineers only need fill in the known fluid properties or pressure-temperature data. By choosing the suitable correlation, the software will provide all pertinent PVT data.

Water Compressibility

Although the best approach is to obtain water compressibilities from laboratory PVT tests, this is seldom done and the use of correlations [22] such as are given in Figures 5.1.26

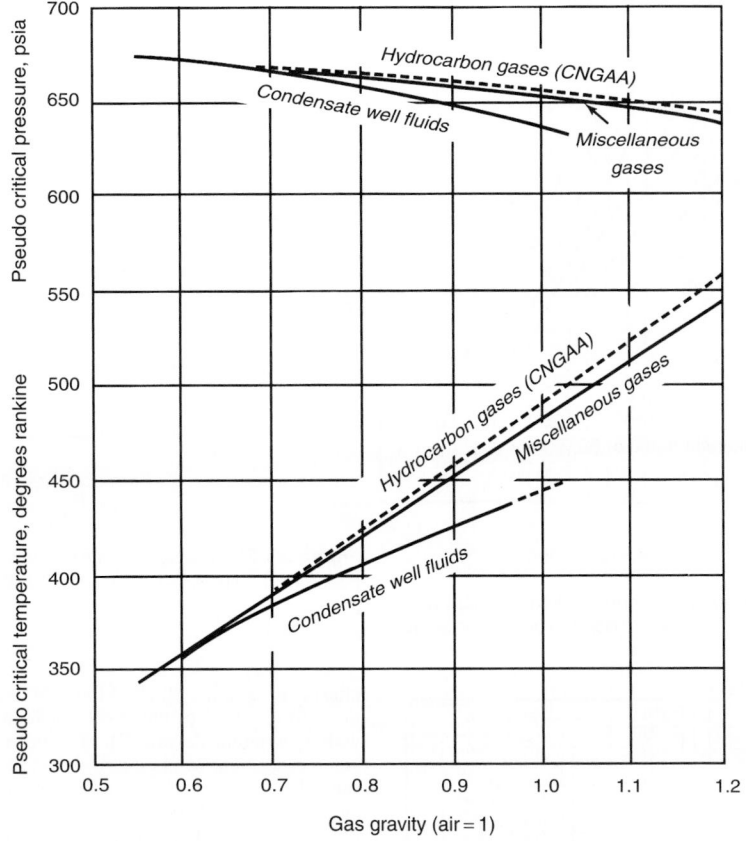

Figure 5.1.20 *Pseudo-critical properties of natural gases [3].*

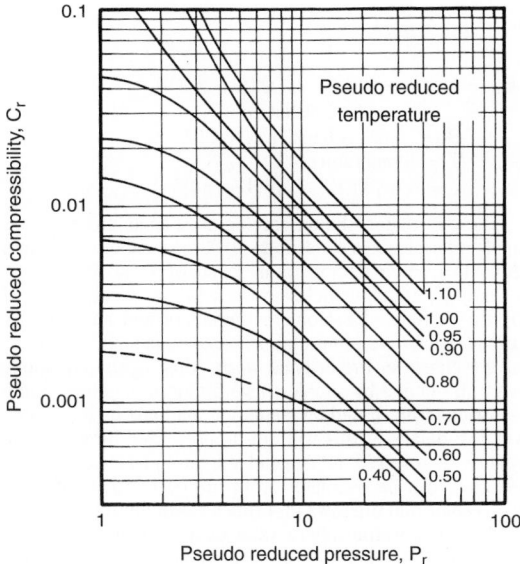

Figure 5.1.21 *Variation of pseudo-reduced compressibility with pseudo-reduced pressures for various fixed values of pseudo-reduced temperature [28].*

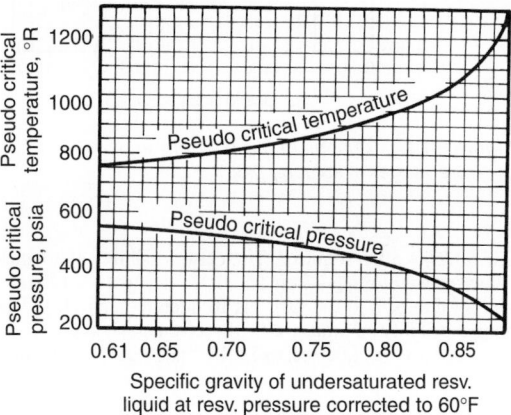

Figure 5.1.22 *Approximate variation of pseudo-critical pressure and pseudo-critical temperature with specific gravity of liquid corrected to 60°F [28].*

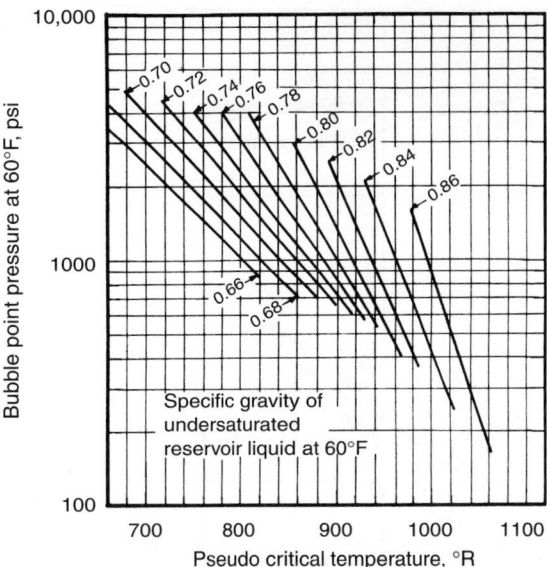

Figure 5.1.23 *Variation of pseudo-critical temperature with specific gravity and bubble point of liquid corrected to 60°F [28].*

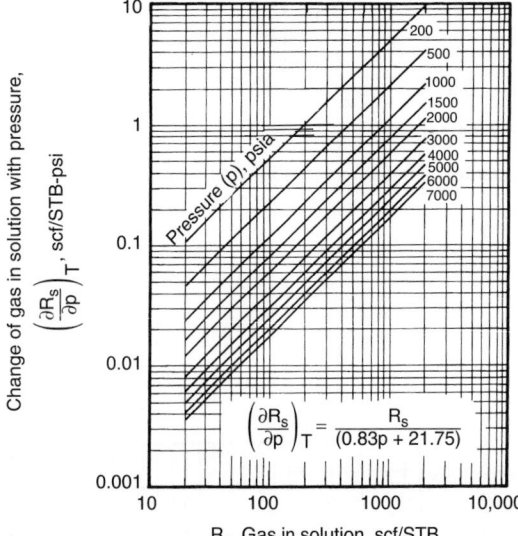

$$\left(\frac{\partial R_s}{\partial p}\right)_T = \frac{R_s}{(0.83p + 21.75)}$$

Figure 5.1.24 *Change of gas solubility in oil with pressure vs. gas in solution [26].*

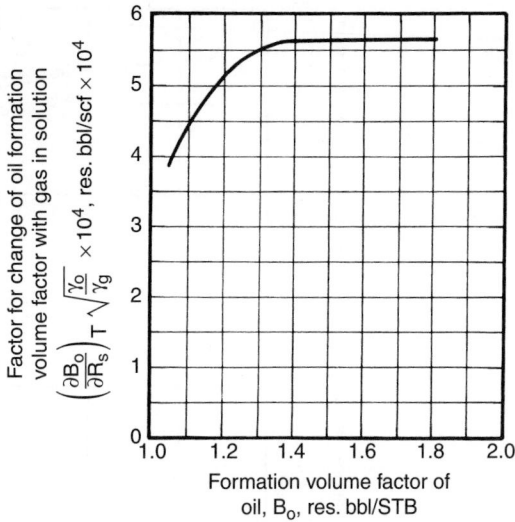

Figure 5.1.25 *Change of oil formation volume factor with gas in solution vs. oil formation volume factor [26].*

and 5.1.27 is often required. The compressibility of nongas-saturated water ranges from 2×10^{-6} psi^{-1} to 4×10^{-6} psi^{-1} and a value of 3×10^{-6} psi^{-1} is frequently used [13]. The compressibility of water with dissolved gas ranges from 15×10^{-6} psi^{-1} at 1,000 psi to 5×10^{-6} psi^{-1} at 5,000 psi [26].

5.1.1.6 Estimation of Fluid Properties with Programmable Calculators and Personal Computers

With the recent widespread use of hand-held programmable calculators and desk-top personal computers, engineers are no longer faced with estimating fluid properties from charts and graphs. Much of the data in the literature have been transmitted to empirical equations suitable for use with programmable calculators or personal computers. In some cases, improved empirical data have been presented recently. This section provides a number of the expressions available for computer use and also provides references for recent books devoted to programs for hand-held calculators. References to some of the software available for personal computers will be given.

Data from Figure 5.1.12 have been used to develop a correlation for water viscosity that can be used on programmable calculators [29]:

$$\mu_w = \left(A + \frac{B}{T}\right) f_{pt} \qquad [5.1.17]$$

where $A = -0.04518 + 0.009313\,(\%NaCl)$
$\qquad\qquad -0.000393\,(\%NaCl)^2$
$\qquad B = 70.634 + 0.09576\,(\%NaCl)^2$
$\qquad T = \text{temperature, }°F$
$\qquad f_{pt} = 1 + [3.5\,E - 12\,p^2(T - 40)]$

The correlation should only be used as an estimate and applies for pressures less than 10,000 psi, salinity less than 26% NaCl, and in a temperature range of 60° to 400°F.

In 1977, Standing's classic work [2] was reprinted [30] by the Society of Petroleum Engineers and an appendix was added by Standing that provides equations for several of the charts in the original work. Most of the equations were developed by simple curve fitting procedures. Some equations were based on computer solutions by other individuals; details of this will not be presented here and the reader is referred to Appendix 2 of Reference 30. Gas viscosity can be estimated from the correlations of Carr, Kobayashi, and Burrows [31] (the basis of Figures 5.1.6 and 5.1.7); first the atmospheric value of gas gravity at reservoir temperature, estimated from gravity and nonhydrocarbon content:

$$\mu_1 = (\mu_1 \text{ uncorrected}) + (N_2 \text{ correction})$$
$$+ (CO_2 \text{ correction}) + (H_2S \text{ correction}) \qquad [5.1.18]$$

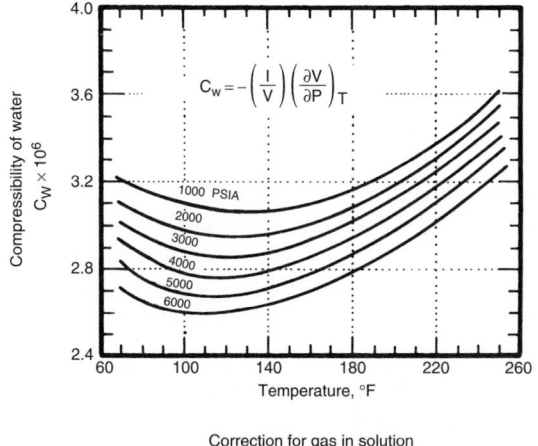

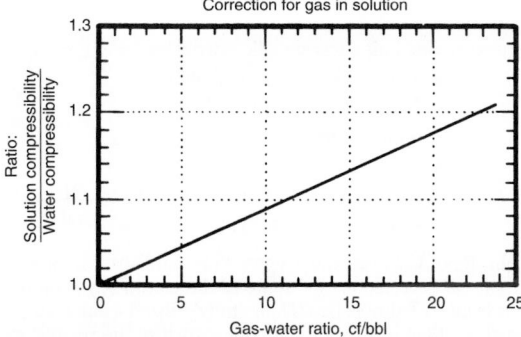

Figure 5.1.26 *Effect of dissolved gas on the compressibility of water [22].*

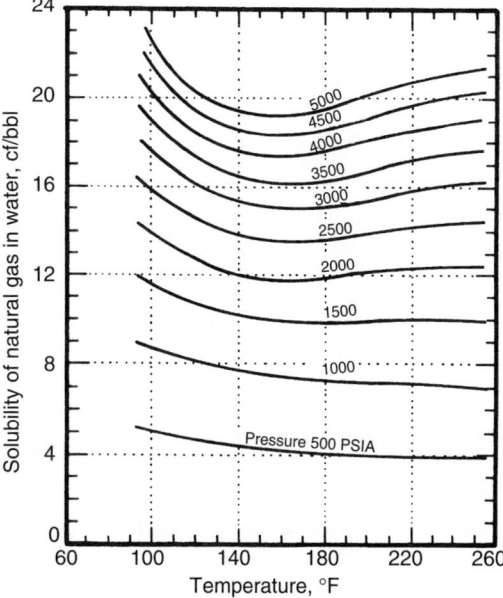

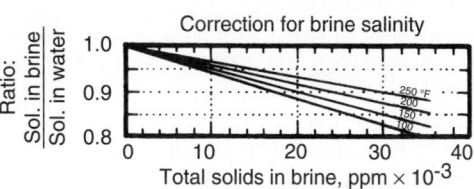

Figure 5.1.27 *Solubility of natural gas in water [22].*

where

$$(\mu_1 \text{ uncorrected}) = [1.709(10^{-5}) - 2.062(10^{-6})\gamma_g]$$
$$\times\, T + 8.188(10^{-3}) - 6.15(10^{-3})\log\gamma_g$$

$$(N_2 \text{ correction}) = y_{N_2}[8.48(10^{-3})\log\gamma_g + 9.59(10^{-3})]$$

$$(CO_2 \text{ correction}) = y_{CO_2}[9.08(10^{-3})\log\gamma_g + 6.24(10^{-3})]$$

$$(H_2S \text{ correction}) = y_{H_2S}[8.49(10^{-3})\log\gamma_g + 3.73(10^{-3})]$$

is adjusted to reservoir conditions by a factor based on reduced temperature and pressure:

$$\ln\left[\left(\frac{\mu_g}{\mu_i}\right)(T_{pr})\right] = a_0 + a_1 p_{pr} + a_2 p_{pr}^2 + a_3 p_{pr}^3$$
$$+ T_{pr}(a_4 + a_5 p_{pr} + a_6 p_{pr}^2 + a_7 p_{pr}^3)$$
$$+ T_{pr}^2(a_8 + a_9 p_{pr} + a_{10}p_{pr}^2 + a_{11}p_{pr}^3)$$
$$+ T_{pr}^3(a_{12} + a_{13}p_{pr} + a_{14}p_{pr}^2 + a_{15}p_{pr}^3)$$
$$[5.1.19]$$

where

$a_0 = -2.462\ 118\ 20E - 00$	$a_8 = -7.933\ 856\ 84E - 01$
$a_1 = 2.970\ 547\ 14E - 00$	$a_9 = 1.396\ 433\ 06E - 00$
$a_2 = -2.862\ 640\ 54E - 01$	$a_{10} = -1.491\ 449\ 25E - 01$
$a_3 = 8.054\ 205\ 22E - 03$	$a_{11} = 4.410\ 155\ 12E - 03$
$a_4 = 2.808\ 609\ 49E - 00$	$a_{12} = 8.393\ 871\ 78E - 02$
$a_5 = -3.498\ 033\ 05E - 00$	$a_{13} = -1.864\ 088\ 48E - 01$
$a_6 = 3.603\ 703\ 20E - 01$	$a_{14} = 2.033\ 678\ 81E - 02$
$a_7 = -1.044\ 324\ 13E - 02$	$a_{15} = -6.095\ 792\ 63E - 04$
	$(= 0.00\ 060\ 957\ 9263)$

A reasonable fit to Beal's correlation (Figure 8 of Reference 5) of gas-free or dead oil viscosity (which is not very precise) is given by Standing:

$$\mu_{oD} = \left[0.32 + \frac{1.8(10^7)}{\text{°API}^{4.53}}\right]\left(\frac{360}{T + 200}\right)^a \qquad [5.1.20]$$

where $a = \text{antilog}\left(0.43 + \frac{8.33}{\text{°API}}\right)$

The dead oil viscosity can then be adjusted for dissolved gas with the correlation by Chew and Connally [6] (Figure 5.1.10) for saturated oil:

$$\mu_{ob} = A(\mu_{oD})^b \qquad [5.1.21]$$

where A and b are functions of solution gas-oil ratio.

Fit of A and b values given in Table 3 of Reference 6 are

$$A = \text{antilog}\{R_s[2.2(10^{-7})R_s - 7.4(10^{-4})]\}$$

$$b = \frac{0.68}{10^{8.62}(10^{-5})R_s} + \frac{0.25}{10^{1.1}(10^{-3})R_s} + \frac{0.062}{10^{3.74}(10^{-3})R_s}$$

The equation for compressibility factors of natural gases (Figure 5.1.13) is:

$$Z = A + (1 - A)/e^B + C\, p_{pr}^D \qquad [5.1.22]$$

where

$$A = 1.39(T_{pr} - 0.92)^{0.5} - 0.36 T_{pr} - 0.101$$

$$B = (0.62 - 0.23T_{pr})p_{pr} + \left[\frac{0.066}{(T_{pr} - 0.86)} - 0.037\right]p_{pr}^2$$

$$+ \frac{0.32}{10^9(T_{pr}^{-1})}p_{pr}^6$$

$$C = (0.132 - 0.32\log T_{pr})$$

$$D = \text{antilog }(0.3106 - 0.49T_{pr} + 0.1824T_{pr}^2)$$

T_{pr} is the dimensionless pseudo-reduced temperature, and p_{pr} is the dimensionless pseudo-reduced pressure. The relationship between formation volume factor of bubble-point liquids and gas-oil ratio, dissolved gas gravity, API oil gravity, and temperature (Figure 5.1.3) is [30]:

$$B_{ob} = 0.9759 + 12(10^{-5})(CN)B_{ob}1.2 \qquad [5.1.23]$$

where

$$(CN)_{B_{ob}} = R_s\left(\frac{\gamma_g}{\gamma_o}\right)^{0.5} + 1.25\,T$$

where

(CN)B_{ob} = bubble-point formation volume factor
        correlating number
R_s = solution gas-oil ratio in ft^3/bbl
γ_g = gas gravity (air = 1)
γ_o = oil specific gravity (water = 1)
T = temperature in °F

The correlation of bubble-point pressure with dissolved gas-oil ratio, dissolved gas gravity, API oil gravity, and temperature (Figure 5.1.2) is [30]:

$$P_b = 18.2[(CN)_{pb} - 1.4] \qquad [5.1.24]$$

where

$$(CN)_{pb} = \left(\frac{R_s}{\gamma_g}\right)^{0.83}[10^{(0.00091T - 0.125°API)}]$$

where (CN)$_{pb}$ is the bubble-point pressure correlating number and the other terms have been previously defined.

Standing also presents equation for; density correction for compressibility of liquids; density correction for thermal expansion of liquids; apparent liquid densities of natural gases; effect of condensate volume on the ratio of surface gas gravity to well fluid gravity; pseudo-critical constants of gases and condensate fluids; pseudo-liquid density of systems containing methane and ethane; and pseudo-critical temperatures and pressures for heptane and heavier.

Beggs and Robinson [7] recently collected PVT data and presented a better estimate of the dead oil viscosity as a function of temperature and oil specific gravity:

$$\mu_{OD} = 10^x - 1 \qquad [5.1.25]$$

where μ_{OD} = viscosity in cp of the gas-free oil at temperature, T, and $X = yT^{-1.163}$, $y = 10^z$, and $Z = 3.0324 - 0.02023\,\gamma_o$, with T in °F and the oil gravity γ in °API. An expression was also given for the saturated oil viscosity, μ, or live oil below the bubblepoint which results from a linear relationship between log μ_{OD} and log μ for a given value of dissolved gas, R_s:

$$\mu = A\,\mu_{OD}^B \qquad [5.1.26]$$

where $A = 10.715\,(R_s + 100)^{-0.515}$

$$B = 5.44(R_s + 150)^{-0.338}$$
$$R_s = \text{scf/STB}$$

In the first book specifically for hand-held calculators, Hollo and Fifadara [32] presented programs for estimating gas deviation factor (based on data of Standing and Katz):

$$Z = 1 + (A_1 + A_2/T_R + A_3/T_R^3)\rho R$$

$$+ (A_4 + A_5/T_R)\rho R^2 + A_6\rho R^2/T_R^3 \qquad [5.1.27]$$

where $\rho R = 0.27\,P_R/ZT_R$ $T_R = T/T_C$ $P_R = P/P_C$
    $A_1 = 0.31506$ $A_2 = -1.0467$ $A_3 = -0.5783$
    $A_4 = 0.5353$ $A_5 = -0.6123$ $A_6 = 0.6815$

A program was also presented to calculate the single-phase formation volume factor using the correlation developed by Standing:

$$B_o = 0.972 + \frac{1.47}{10^4}\left[(R_s)\left(\frac{\gamma_g}{\gamma_o}\right)^{0.5} + 1.25\,T\right]^{1.75} \qquad [5.1.28]$$

where γ_g = solution gas specific gravity
    γ_o = stock tank oil specific gravity
        (141.5/131.5 + °API)
    T = temperature, °F
    R_s = solution gas-oil ratio, scf/STB
    B_o = single-phase formation volume factor, RB/STB
    °API = stock tank oil gravity, °API

In 1980 Vazquez and Beggs [33] published improved empirical correlations for some of the commonly required crude oil PVT properties. Their study utilized a much larger database than was used in previous work so the results are applicable to a wider range of oil properties. The empirical correlations, presented as a function of gas specific gravity, oil API gravity, reservoir temperature, and pressure, are particularly convenient to use with hand-held calculators. Gas gravity was found to be a strong correlating parameter. Since gas gravity depends on gas-oil separation conditions, Vazquez and Beggs chose 100 psig as a reference pressure, which resulted in a minimum oil shrinkage for the separator tests available. Thus, gas gravity must first be corrected to the value that would result from separation at 100 psig:

$$\gamma_{gs} = \gamma_{gp}[1 + 5.912 \times 10^{-5}(\gamma_o)(T)\log(p/114.7)] \qquad [5.1.29]$$

where γ_{gs} = gas gravity (air = 1) that would result from
        separator conditions of 100 psig
    γ_{gp} = gas gravity obtained at separator conditions
        of p and T
    p = actual separator pressure, psia
    T = actual separator temperature, °F
    γ_o = oil gravity, °API

For both dissolved gas and oil formation volume factor, improved correlations were obtained when the measured data were divided into two groups, with the division made at an oil gravity of 30°API. The expression for dissolved gas was presented:

$$R_s = C_1\gamma_{gs}p^{C_2}\exp\{C_3[\gamma_o/(T + 460)]\} \qquad [5.1.30]$$

Values for the coefficients are as follows.

Coefficient	$\gamma_o \leq 30$	$\gamma_o > 30$
C_1	0.0362	0.0178
C_2	1.0937	1.1870
C_3	25.7240	23.9310

For saturated oils (reservoir pressure less than bubble-point), oil formation volume factor was expressed as:

$$B_o = 1 + C_1 R_s + C_2(T-60)(\gamma_o/\gamma_{gs}) + C_3 R_s(T-60)(\gamma_o/\gamma_{gs})$$
$$[5.1.31]$$

The values for the coefficients depend on oil gravity and are given by the following:

Coefficient	$\gamma_o \leq 30$	$\gamma_o > 30$
C_1	4.677×10^{-4}	4.670×10^{-4}
C_2	1.751×10^{-5}	1.100×10^{-5}
C_3	-1.811×10^{-8}	1.337×10^{-9}

Since the oil formation volume of an undersaturated crude depends on the isothermal compressibility of the liquid, the oil formation volume as pressure is increased above bubble-point pressure was calculated from:

$$B_o = B_{ob} \exp[c_o(p - p_b)] \qquad [5.1.32]$$

The correlation for compressibility of oil was given as:

$$c_o = (a_1 + a_2 R_s + a_3 T + a_4 \gamma_{gs} + a_5 \gamma_o)/a_6 p \qquad [5.1.33]$$

where $a_1 = -1,433.0$
$a_2 = 5.0$
$a_3 = 17.2$
$a_4 = -1,180.0$
$a_5 = 12.61$
$a_6 = 105$

Vazquez and Beggs also presented an equation for viscosity of undersaturated crude oils that used the correlations of Beggs and Robinson:

$$\mu_o = \mu_{ob}(p/p_b)^m \qquad [5.1.34]$$

where $m = C_1 p^{C_2} \exp(C_3 + C_4 p)$
$C_1 = 2.6$
$C_2 = 1.187$
$C_3 = -11.513$
$C_4 = -8.98 \times 10^{-5}$

The improved correlations of Vazquez and Beggs were incorporated by Meehan in the development of programs for hand-held calculators. These programs were presented in a series of articles in the *Oil and Gas Journal* [34–35]. Reference 34 contains the programs for calculating gas gravity, dissolved gas-oil ratio, oil formation volume factor, and oil compressibility. Reference 35 contains the program for calculating oil viscosity.

See References 36–40 for a list of books devoted to the use of programs for handheld calculators and personal computers.

5.1.2 Properties of Fluid-Containing Rocks
5.1.2.1 Porosity

The porosity, ϕ, is equal to the void volume of the rock divided by the bulk volume and is expressed as a percent or fraction of the total bulk volume of the rock. Oil-bearing sandstones have porosities which often range from 15% to 30%. Porosities of limestones and dolomites are usually lower.

Differentiation must be made between absolute and effective porosity. Absolute porosity is defined as the ratio of the total pore volume of the rock to the total bulk volume of the rock whereas effective porosity is defined as the ratio of the interconnected pore volume of the rock to the total bulk volume of the rock.

Factors affecting porosity are compactness, character and amount of cementation, shape and arrangement of grains, and uniformity of grain size or distribution.

In problems involving porosity calculations it is convenient to remember that a porosity of one percent is equivalent to the presence of 77.6 barrels of pore space in a total volume of one acre-foot of sand.

5.1.2.2 Pore Volume

The pore volume of a reservoir is the volume of the void space, that is, the porosity fraction times the bulk volume. In conventional units, the pore volume, V_p, in reservoir barrels is:

$$V_p = 7758 \qquad V_b \phi = 7,758 \, A \, h \, \phi \qquad [5.1.35]$$

where V_b is the bulk volume in ac-ft. A is the area in ft^2, h is the reservoir thickness in ft. and ϕ is the porosity expressed as a fraction.

5.1.2.3 Permeability

The permeability of a rock is a measure of the ease with which fluids flow through the rock. It is denoted by the symbol k and commonly expressed in units of darcies. Typical sandstones in the United States have permeabilities ranging from 0.001 to a darcy or more, and for convenience the practical unit of permeability is the millidarcy which equals 0.001 darcy. Some other useful conversion factors are given in Table 5.1.6.

5.1.2.4 Absolute Permeability

If a porous system is completely saturated with a single fluid, the permeability is a rock property and not a property of the flowing fluid (with the exception of gases at low pressure). This permeability at 100% saturation of a single fluid is termed the absolute permeability.

5.1.2.5 Darcy Equation

The Darcy equation relates the apparent velocity, v, of a homogeneous fluid in a porous medium to the fluid viscosity, μ, and the pressure gradient, $\Delta p/L$:

$$v = -\frac{k\Delta p}{\mu L} \qquad [5.1.36]$$

This equation states that the fluid velocity is proportional to the pressure gradient and inversely proportional

Table 5.1.6 *Permeability Conversion Factors*

1 darcy = 1,000 millidarcies; 1 millidarcy = 0.001 darcy

$$= 0.9869233 \, \mu m^2 (1 \, md \cong 10^{-3} \, \mu m^2)$$

$$k = \frac{q\mu}{(A)(\Delta p/L)}$$

1 darcy $= \dfrac{(cc/sec)(cp)}{(cm^2)(atm)/cm}$

$$= 9.869 \times 10^{-9} \, cm^2$$

$$= 1.062 \times 10^{-11} \, ft^2$$

$$= 1.127 \frac{[bbl/(day)](cp)}{(ft^2)(psi)/ft}$$

$$= 9.869 \times 10^{-7} \frac{(cc/sec)(cp)}{cm^2[dyne/(cm^2)(cm)]}$$

$$= 7.324 \times 10^{-5} \frac{[ft^3/(sec)](cp)}{(ft^2)(psi)/ft}$$

$$= 9.679 \times 10^{-4} \frac{[ft^3/(sec)](cp)}{(cm^2)(cm \, water)/cm}$$

$$= 1.424 \times 10^{-2} \frac{[gal/(min)](cp)}{(ft^2)(ft \, water)/ft}$$

From Reference 19.

to fluid viscosity. The negative sign indicates that pressure decreases in the L direction when the flow is taken to be positive. The flow rate, q, is understood to be positive during production and negative during injection. As shown in Table 5.1.6, the Darcy equation can be written as:

$$k = \frac{q\mu L}{A\Delta p} \qquad [5.1.37]$$

Linear Flow

In the Darcy equation for linear displacement:

$$q = \frac{kA(\Delta p)}{\mu L} \qquad [5.1.38]$$

where q = fluid flow rate, cm^3/sec
   A = cross-sectional area of rock perpendicular to flow, cm^2
   p = pressure difference (in atm) across the distance L parallel to flow direction, cm
   μ = viscosity of fluid, cp

A rock has permeability of one darcy if it permits the flow of one cc per second of a one-phase fluid having viscosity of one centipoise under the pressure gradient of one atmosphere per gradient of one atmosphere per centimeter. For liquid flow in a linear porous system, the flow rate q in barrels per day is [20]:

$$q = \frac{1.127\,kA(p_c - p_w)}{\mu L} \qquad [5.1.39]$$

where k is in darcies, A is in ft^2, μ is in cp, L is in ft, and pressures are in psia. For laminar gas flow in a linear system [20]:

$$q = \frac{0.112\,kA(p_c^2 - p_w^2)}{\mu zTL} \qquad [5.1.40]$$

where q is in Mscf/D, T is in $°R(°F + 460)$, z is the dimensionless compressibility factor, and the other terms are in units consistent with Equation 5.1.39.

Radial Flow

Production from or injection into a reservoir can be viewed as flow for a cylindrical region around the wellbore. For the steady-state radial flow of an incompressible fluid [19]:

$$q = \frac{2\pi kh(p_e - p_w)}{\mu \ln(r_e/r_w)} \qquad [5.1.41]$$

where q = volume rate of flow, cc/sec
   k = permeability, darcies
   h = thickness, cm
   μ = viscosity, cp
   p_e = pressure at external boundary, atm
   p_w = pressure at internal boundary, atm
   r_e = radius to external boundary, cm
   r_w = radius to internal boundary, cm
   ln = natural logarithm, base e

For the flow rate, q, in the barrels per day of a liquid [19]:

$$q = \frac{7.08\,kh(p_e - p_w)}{\mu \ln(r_e/r_w)} \qquad [5.1.42]$$

where k is in darcies, h is in ft., pressures are in psia, μ is in cp, and the radii are in consistent units, usually feet. For the laminar flow of a gas in MscfD [20]:

$$q = \frac{0.703\,kh(p_e^2 - p_w^2)}{\mu zT \ln(r_e/r_w)} \qquad [5.1.43]$$

where T is in °R, z is the dimensionless compressibility factor, and the other terms are as defined in Equation 5.1.42.

5.1.2.6 Capacity

Flow capacity is the product of permeability and reservoir thickness expressed in md ft. Since the rate of flow is proportional to capacity, a 10-ft thick formation with a permeability of 100 md should have the same production as a 100-ft thick formation of 10 md, if all other conditions are equivalent.

5.1.2.7 Transmissibility

Transmissibility is flow capacity divided by viscosity or kh/μ with units of md ft/cp. An increase in either reservoir permeability or thickness or a decrease in fluid viscosity will improve transmissibility of the fluid in the porous system.

5.1.2.8 Resistivity and Electrical Conductivity

Electrical conductivity, the electrical analog of permeability, is the ability of a material to conduct an electrical current. With the exception of certain clay minerals, reservoir rocks are nonconductors of electricity. Crude oil and gas are also nonconductors. Water is a conductor if dissolved salts are present so the conduction of an electric current in reservoir rocks is due to the movement of dissolved ions in the brine that occupies the pore space. Conductivity varies directly with the ion concentration of the brine. Thus, the electrical properties of a reservoir rock depend on the fluids occupying the pores and the geometry of the pores.

Resistivity, which is the reciprocal of conductivity, defines the ability of a material to conduct electric current:

$$R = \frac{rA}{L} \qquad [5.1.44]$$

where R is the resistivity in ohm-meters, r is the resistance in ohms, A is the cross-sectional area in m^2, and L is the length of the conductor in meters. As seen in Figure 5.1.28, resistivity of water varies inversely with salinity and temperature [41].

During flow through a porous medium, the tortuous flow paths cause the flowing fluid to travel an effective length, L_e, that is longer than the measured length, L. Some authors have defined this tortuosity, τ, as (L_e/L) while $(L_e/L)^2$ has been used by others.

5.1.2.9 Formation Resistivity Factor

The formation resistivity factor, F_R, is the ratio of the resistivity of a porous medium that is completely saturated with an ionic brine solution divided by the resistivity of the brine:

$$F_R = \frac{R_o}{R_w} \qquad [5.1.45]$$

where R_o is the resistivity (ability to impede the flow electric current) of a brine-saturated rock sample in ohm-m, R_w is the resistivity of the saturating brine in ohm-m, and F_R is dimensionless. The formation resistivity factor, which is always greater than one, is a function of the porosity of the rock (amount of brine), pore structure, and pore size distribution. Other variables that affect formation factor include composition of the rock and confining pressure (overburden).

Archie [42] proposed an empirical formula that indicated a power-law dependence of F_R on porosity:

$$F_R = \phi^{-m} \qquad [5.1.46]$$

where ϕ is porosity and m is a constant (commonly called the cementation factor) related to the pore geometry. The constant, m, was the slope obtained from a plot of F_R vs. porosity on log-log paper. For consolidated, shale free sandstones, the value of m ranged from 1.8 to 2. For clean, unconsolidated sands, m was found to be 1.3, and Archie speculated that m

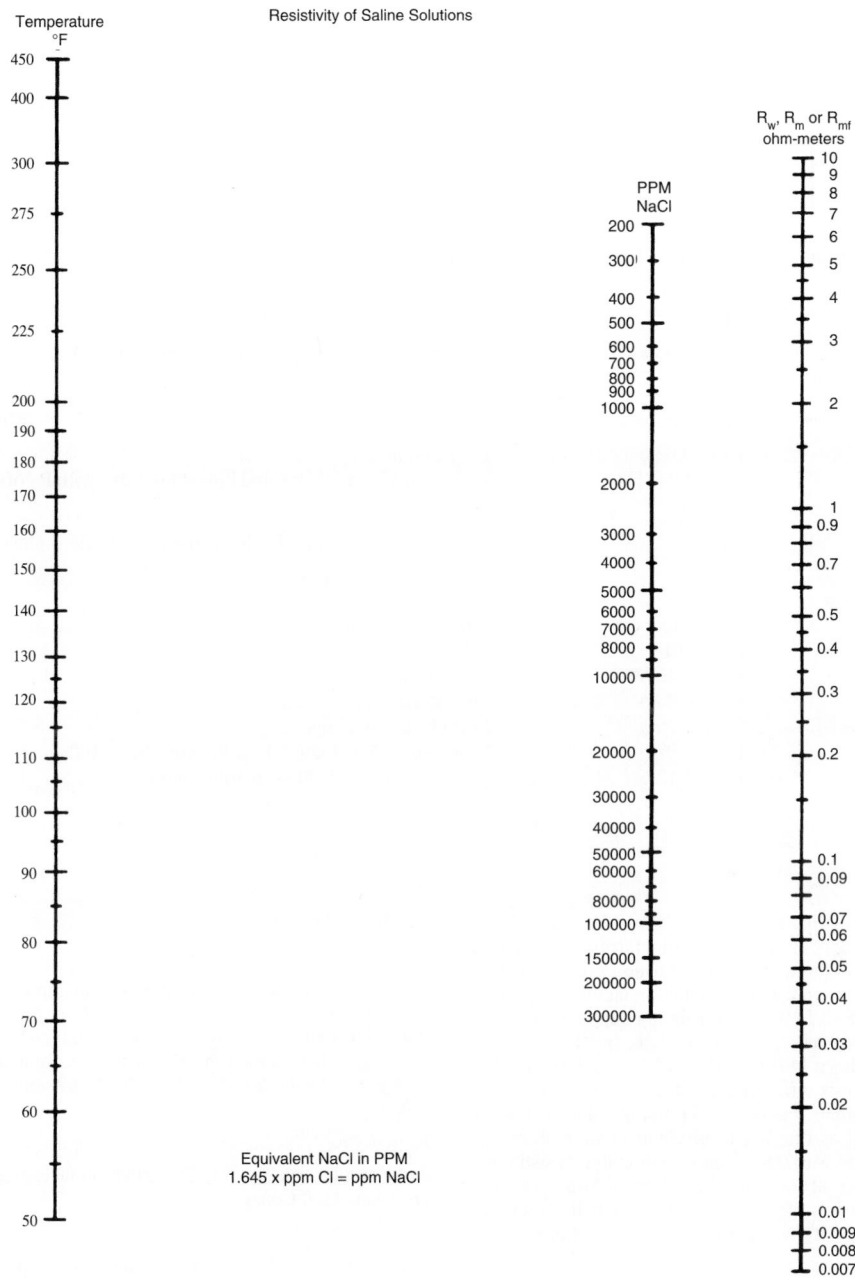

Figure 5.1.28 *Resistivity of saline solutions [41].*

might vary from 1.3 to 2 for loosely or partly consolidated sands. Equations 5.1.45 and 5.1.46 were also combined by Archie to give:

$$R_o = R_w \phi^{-m} \qquad [5.1.47]$$

so that a reasonable estimate of F_R or R_o can be made if the slope, m, is obtained.

Several other correlations [43–55], mostly empirical, between formation factor and porosity have been reported in the literature and these are summarized in Table 5.1.7.

From an analysis of about 30 sandstone cores from a number of different reservoirs throughout the United States, Winsauer et al. [45] presented what is now known as the Humble relation:

$$F_R = 0.62\phi^{-2.15} \qquad [5.1.48]$$

Wyllie and Gregory [46], citing their results and the results of Winsauer et al. [45], proposed that the data for consolidated porous media could be best described empirically by:

$$F_R = a\phi^{-m} \qquad [5.1.49]$$

Table 5.1.7   *Correlations of Formation Resistivity Factor and Porosity*

Source	Relation	Notes
Archie [42]	$F_R = \phi^{-m}$	For consolidated sands, m = 1.8 to 2.5. For unconsolidated sands, m = 1.3
Wyllie and Rose [43]	$F_R = \dfrac{\tau^{1/2}}{\phi}$	Tortuosity $\tau = L_e/L$
Tixier [44]	$F_R = \dfrac{1}{\phi^2}$	For limestone
Winsauer et al. [45]	$F_R = \dfrac{\tau^2}{\phi}$	Theory
	$F_R = \dfrac{\tau^{1.67}}{\phi}$	Experimental (transport time of flowing ions)
	$F_R = 0.62\,\phi^{-2.15}$	Sandstones containing varying amounts of clay
Wyllie/Gregory [46]	$F_R = a\phi^{-m}$	General form of Archie relation
Cornell and Katz [47]	$F_R = \dfrac{L_e^2}{\phi L}$	F_R directly proportional to length and inversely proportional to area
Owen [48]	$F_R = 0.68\,\phi^{-2.23}$	Logs in dolomite, mud filtrate same resistivity as connate water
Hill and Milburn [49]	$F_R = 1.4\phi^{-1.78}$	Results of 450 sandstone and limestone cores with R_w of 0.01 ohm-m
	$F_R = \phi^{-1.93}$	When a = 1
Wyllie/Gardner [50]	$F_R = \dfrac{1}{\phi^2}$	Model of capillary bundle, for conducting wetting phase
Sweeney/Jennings [51]	$F_R = \phi^{-m}$	25 various carbonates
	m = 1.57	Water-wet
	m = 1.92	Intermediate wettability
	m = 2.01	Oil-wet
Carothers [52]	$F_R = 1.45\,\phi^{-1.54}$	Sandstones
	$F_R = 0.85\,\phi^{-2.14}$	Limestones
Porter/Carothers [53]	$F_R = 2.45\,\phi^{-1.08}$	From California logs
	$F_R = 1.97\,\phi^{-1.29}$	From Gulf Coast logs. All sandstones, S_w = 100%
Timur [54]	$F_R = 1.13\,\phi^{-1.73}$	Analysis of over 1,800 sandstone samples
Perez-Rosales [55]	$F_R = 1 + G(\phi^{-m} - 1)$	General theoretical relation
	$F_R = 1 + 1.03(\phi^{-1.73} - 1)$	Theoretical relation for sandstones

Results of a logging study in the Brown Dolomite formation, in which the resistivity of the mud filtrate was the same as the connate water, were used by Owen [48] to establish a relationship between the true formation factor and porosity. Shown in Figure 5.1.29 are the relationships obtained with the equations of Archie, Winsauer, et al., and Tixier.

Hill and Milburn [49] provided data from 450 core samples taken from six different sandstone formations and four different limestone reservoirs. The sandstone formations were described as ranging from clean to very shaly. The formation factor was determined at a water resistivity of 0.01 ohm-m since, at that value, the apparent formation factor approached the true formation factor when the rock contained low-resistivity water. From their data, the following equation was provided:

$$F_R = 1.4\phi^{-1.78} \qquad [5.1.50]$$

An expression was also provided for the case in which the constant, a, in Equation 5.1.49 is taken as unity:

$$F_R = \phi^{-1.93} \qquad [5.1.51]$$

Using 981 core samples (793 sandstone and 188 carbonate), Carothers [52] established a relationship for sands:

$$F_R = 1.45\phi^{-1.54} \qquad [5.1.52]$$

and for limestones:

$$F_R = 0.85\phi^{-2.14} \qquad [5.1.53]$$

As shown in Figure 5.1.30, a relationship was suggested for calcareous sands:

$$F_R = 1.45\phi^{-1.33} \qquad [5.1.54]$$

and for shaly sands:

$$F_R = 1.65\phi^{-1.33} \qquad [5.1.55]$$

Using these data, the nomograph in Figure 5.1.31 was constructed to solve the modified Archie expression (Equation 5.1.49) when it is desired to vary both constants.

Using 1,575 formation factors from California Pliocene well logs, Porter and Carothers [53] presented an in-situ relation:

$$F_R = 2.45\phi^{-1.08} \qquad [5.1.56]$$

and a similar relation for 720 formation factors from Texas-Louisiana Gulf Coast logs:

$$F_R = 1.97\phi^{-1.29} \qquad [5.1.57]$$

This investigation used well log data from sandstone formations known to have water saturations of 100%.

These samples included 569 core plugs (from Alaska, California, Louisiana, Colorado, Trinidad, Australia, and the Middle East) plus 28 samples from Winsauer et al. [45], 362 samples from Hill and Milburn [49], 788 from Carothers [52], and 85 samples from other sources [54].

In a recent paper [55], Perez-Rosales presented the following theoretical expression:

$$F_R = 1 + G(\phi^{-m} - 1) \qquad [5.1.58]$$

and a generalized equation for sandstones:

$$F_R = 1 + 1.03(\phi^{-1.73} - 1) \qquad [5.1.59]$$

Perez-Rosales notes that the previous expressions are fundamentally incorrect since they do not satisfy the requirement that $F_R = 1$ when $\phi = 1$. A graphical comparison

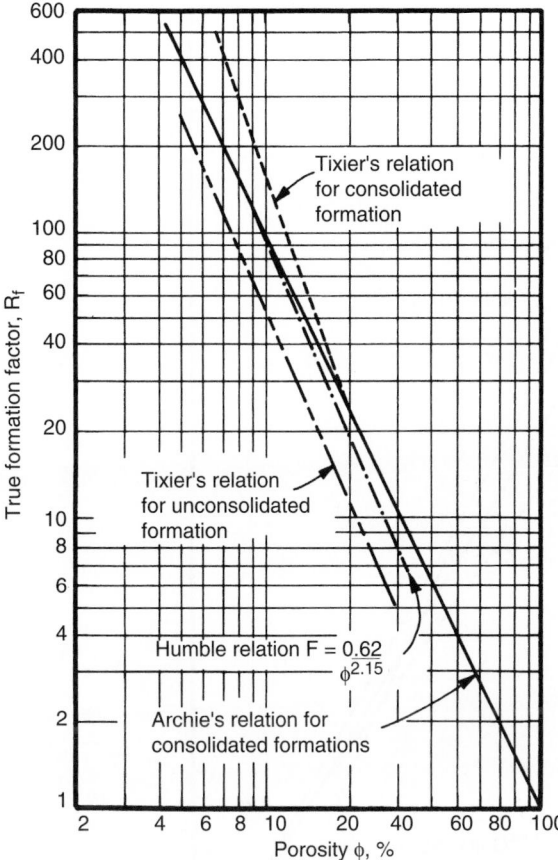

Figure 5.1.29 *Formation factor vs. porosity from logs of Brown Dolomite formation [48].*

of expressions, provided by Perez-Rosales, is shown in Figure 5.1.32 for Equations 5.1.48, 5.1.59, and 5.1.60. In porosity ranges of practical interest, the three expressions yield similar results.

From an analysis of over 1,800 sandstone samples, Timur et al. [54] presented the following expression:

$$F_R = 1.13\phi^{-1.73} \tag{5.1.60}$$

Coates and Dumanoir [56] listed values for the cementation exponent of the Archie equation for 36 different formations in the United States. These data are presented in the following section under "Resistivity Ratio."

In the absence of laboratory data, different opinions have existed regarding the appropriate empirical relationship. Some authors [57] felt that the Archie equation (Equation 5.1.46) with $m = 2$ or the Humble equation (Equation 5.1.48) yields results satisfactory for most engineering purposes, but Equation 5.1.50 may be more valid (these authors point out that the relationship used should be based on independent observations of interest). Another opinion was that, while the Humble relation is satisfactory for sucrosic rocks and the Archie equation with $m = 2$ is acceptable for chalky rocks, in the case of compact or oolicastic rocks the cementation exponent in the Archie equation may vary from 2.2 to 2.5 [58]. Based on the more recent work of Timur et al. [54], it appears that Equation 5.1.60 may be more appropriate as a general expression for sandstones,

if individual formation factor-porosity relationships are not available for specific cases.

Water in clay materials or ions in clay materials or shale act as a conductor of electrical current and are referred to as conductive solids. Results in Figure 5.1.33 show that clays contribute to rock conductivity if low-conductivity, fresh, or brackish water is present [59,60]. The effect of clay on formation resistivity depends on the type, amount, and distribution of clay in the reservoir, as well as the water salinity. Values of m in Equation 5.1.49 for several clays are given in Table 5.1.8 [61].

Other variables that affect resistivity of natural reservoir rocks include overburden pressure and temperature during measurement. The value of the cementation exponent, m, is normally higher at overburden conditions [62], especially if porosity is low or with rocks that are not well-cemented. Thus, F_R increases with increasing pressure. Although the effect of temperature depends on clay content of the sample, F_R tends to increase with increasing temperature, but the effect is not as great as pressure [63,64]. At a fixed pressure, F_R may go through a minimum and then increase as temperature is increased; the combined increase of both temperature and pressure will cause an increase in F_R [64]. Factors that affect the exponent, m, and the coefficient, a, in the modified Archie expression (Equation 5.1.49) are summarized in Tables 5.1.9 and 5.1.10, respectively [65].

To summarize the general relationship between formation resistivity factor and porosity (see Equation 5.1.49), the normal range for the geometric term, a, is 0.6 to 1.4, and the range for the cementation exponent, m, is 1.7 to 2.5 for most consolidated reservoir rocks [62]. Since the exact values depend on pore geometry and composition of the rock, formation factors should be determined with samples of the reservoir rock of interest, under the reservoir conditions of temperature and overburden pressure.

Based on core analyses of 793 sandstone and 188 carbonate samples, Carothers [52] observed different permeabilities and formation factors for samples from the same core even though porosity was identical. Furthermore, permeability generally decreased as formation factor increased. For permeabilities above 10 md, there appeared to be a relation between formation factor, permeability, and lithology. For sandstones, the general relationship was:

$$k = \frac{7 \times 10^8}{F_R^{4.5}} \tag{5.1.61}$$

and the relation for carbonates was:

$$k = \frac{4 \times 10^8}{F_R^{3.65}} \tag{5.1.62}$$

Carothers stated that more data are needed to confirm these observations. Any such relation should be used with caution.

5.1.2.10 Rock Compressibility

The isothermal rock compressibility is defined as the change in volume of the pore volume with respect to a change in pore pressure:

$$c_f = \frac{1}{V_p}\left(\frac{\partial V_p}{\partial p}\right)_T \tag{5.1.63}$$

where c_f is the formation (rock) compressibility with common units of psi^{-1}, V_p is pore volume, p is pressure in psi, and the subscript T denotes that the partial derivative is taken at constant temperature. The effective rock compressibility is considered a positive quantity that is additive to fluid compressibility; therefore, pore volume decreases as fluid pressure decreases [26,66]. Since overburden pressure of a reservoir is essentially constant, the differential

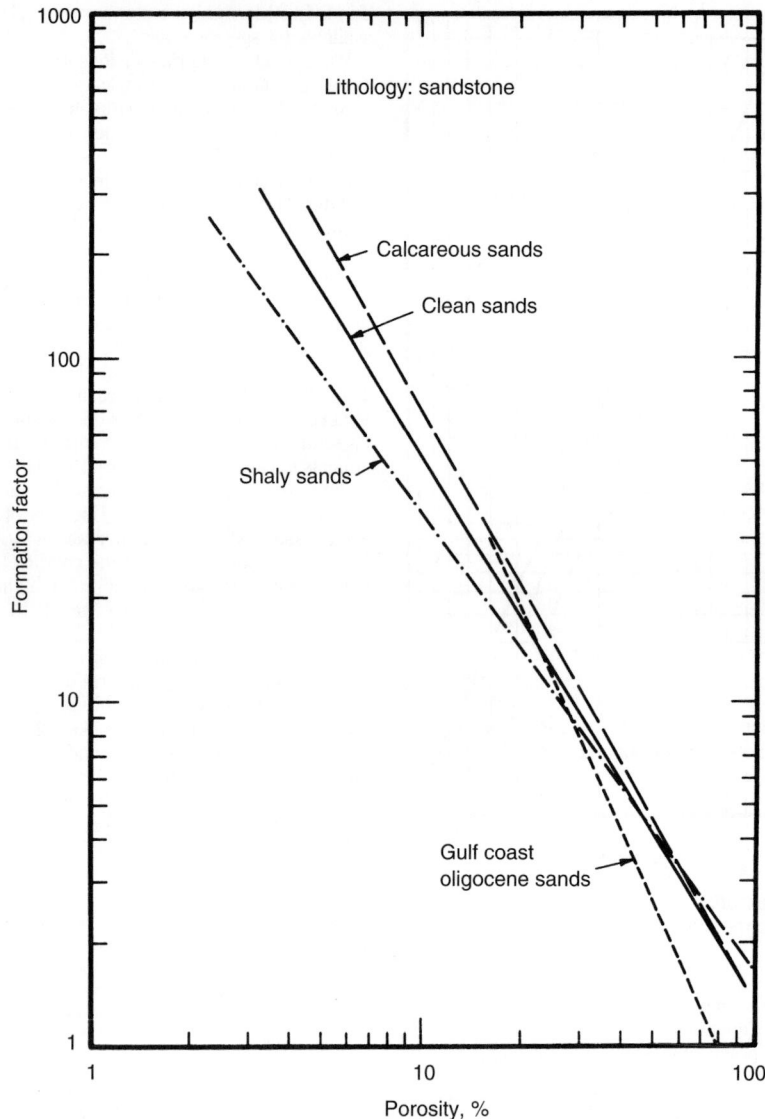

Figure 5.1.30 *Formation factor vs. porosity for clean sands, shaly sands, and calcareous sands [52].*

pressure between the overburden pressure and the pore pressure will increase as the reservoir is depleted. Thus, porosity will decrease slightly, on the order of only one-half percent for a 1,000 psi change in internal fluid pressure [17]. For different reservoirs, porosities tend to decrease as overburden pressure (or depth) increases. Therefore, porosity under reservoir conditions may differ from values determined in the laboratory [67]. For sandstones with 15% to 30% porosity, reservoir porosity was found to be about 1% lower under reservoir conditions; for low porosity limestones, the difference was about 10% [68].

One of the commonly cited correlations between rock compressibility and porosity was developed by Hall [69] (Figure 5.1.34) for several sandstone and limestone reservoirs. All measurements were conducted with an external pressure of 3,000 psi and internal pressures from 0 to 1,500 psi. Fatt [67] found no correlation between

compressibility and porosity, although the porosity range studied (10% to 15%) was very narrow. Van der Knapp [68], citing his measurements and those of Carpenter and Spencer [70], observed a general trend of increasing pore volume compressibility with decreasing porosity. For a particular limestone reservoir, Van der Knapp [68] found that pore compressibility and porosity were related by a simple empirical formula. However, in a more extensive study, Newman [71] suggests that any correlation between pore volume compressibility and porosity does not apply to a wide range of reservoir rocks. As shown in Figure 5.1.35a, Newman's study in limestones showed poor agreement with the correlations of Hall and Van der Knapp. Figures 5.1.35b to 5.1.35d show a comparison of Newman's data with Hall's correlation for consolidated sandstones, friable sandstones, and unconsolidated sandstones. While the general trend of Newman's data on consolidated sandstones (Figure 5.1.35b)

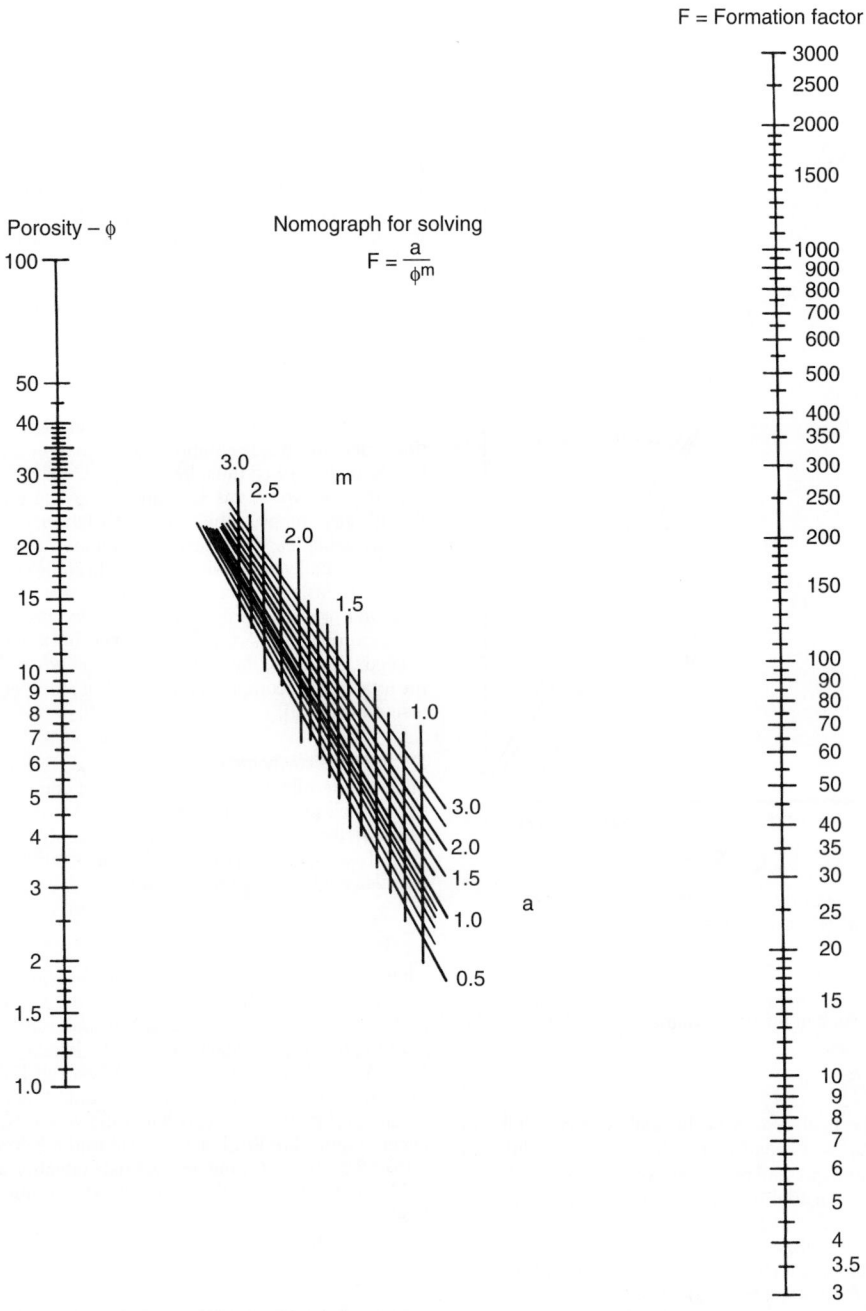

Figure 5.1.31 Nomograph for solving $F_R = a/\phi^m$ [52].

is in the same direction as Hall's correlation, the agreement is again poor. Figure 5.1.35c shows no correlation for Newman's friable sandstones and Figure 5.1.35d for unconsolidated sandstones shows an opposite trend from the correlation presented by Hall. From Newman's data, ranges of compressibilities for various types of reservoir rocks are given in Table 5.1.11. Clearly, formation compressibility

should be measured with samples from the reservoir of interest.

5.1.3 Properties of Rocks Containing Multiple Fluids
5.1.3.1 Total Reservoir Compressibility
The total compressibility of oil- or gas-bearing reservoirs represents the combined compressibilities of oil, gas, water,

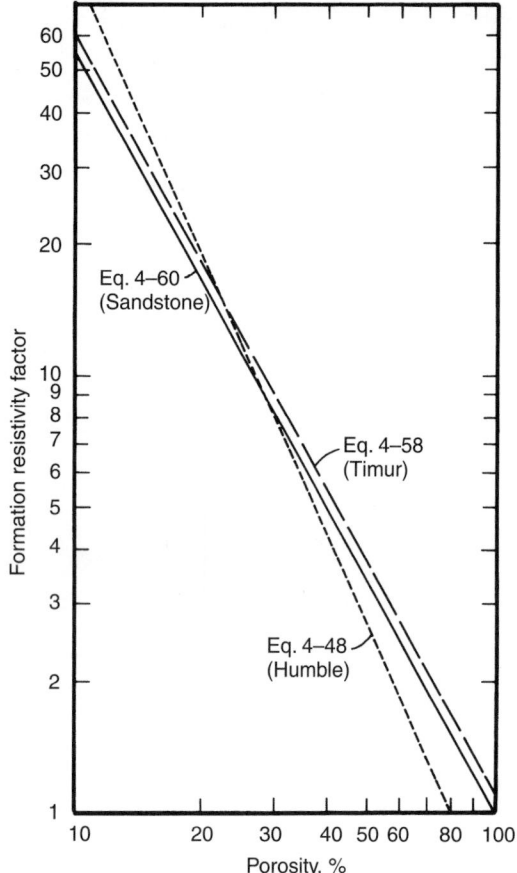

Figure 5.1.32 *Graphical comparison of relationship between formation resistivity factor and porosity [55].*

and reservoir rock in terms of volumetric weighting of the phase saturations:

$$c_t = c_o S_o + c_w S_w + c_g S_g + c_f \qquad [5.1.64]$$

where c_t is the total system isothermal compressibility in vol/vol/psi, c_o, c_w, c_g, and c_f are the compressibilities in psi^{-1} of oil, water, gas, and rock (pore volume), respectively, S is fluid saturation, and the subscripts, o, w, and g refer to oil, water, and gas, respectively.

Based on the treatment by Martin [72], Ramey [26] has expressed volumes in terms of formation volume factors with consideration for gas solubility effects:

$$c_t = S_o \left[\frac{B_g}{B_o} \left(\frac{\partial R_s}{\partial p} \right) - \frac{1}{B_o} \left(\frac{\partial B_o}{\partial p} \right) \right]$$

$$+ S_w \left[\frac{B_g}{B_w} \left(\frac{\partial R_{sw}}{\partial p} \right) - \frac{1}{B_w} \left(\frac{\partial B_w}{\partial p} \right) \right]$$

$$+ S_g \left[-\frac{1}{B_g} \left(\frac{\partial B_g}{\partial p} \right) \right] + c_f \qquad [5.1.65]$$

where p is pressure in psi, R_s is the solubility of gas in oil in scf/STB oil, R_{sw} is the solubility of gas in water in scf/STB water, and B_g, B_o, and B_w are the formation volume factors of gas, oil, and water, respectively.

Fluid and rock compressibilities have been discussed in prior sections of this chapter. Table 5.1.12 provides a summary of these data.

The rock compressibilities in Table 5.1.12 represent a majority of the consolidated sandstone and limestone data from Newman [71] that have porosities in the range of 10% to 30%. Oil compressibility increases as a function of increasing API gravity, quantity of solution gas, or temperature [17]. As pointed out by Ramey [26], when the magnitude of water compressibility is important, the effect of solution gas in the water will be more important. Clearly, the magnitude of gas compressibility will dominate the total system compressibility if gas saturations are high.

In many gas reservoirs, only the gas terms in Equation 5.1.64 may be significant so that the total system compressibility becomes [26]:

$$c_t = c_g S_g \qquad [5.1.66]$$

In certain cases of high pressure and high water saturation, rock and water compressibility may be significant so that Equation 5.1.65 must be used [26].

In oil reservoirs, gas saturations may be low and, even though gas compressibility is much larger than the other compressibilities, each term in Equation 5.1.64 or 5.1.65 should normally be considered [26]. In some cases, not all of the compressibility terms will be important. For example, if reservoir pressure is above the saturation pressure, the gas saturation will be zero [20]. However, if the gas saturation exceeds 2% or 3%, the gas compressibility term dominates the total system compressibility and the other terms become insignificant [20].

5.1.3.2 Resistivity Index
Since crude oil and natural gas are nonconductors of electricity, their presence in reservoir rock increases resistivity. The resistivity index or ratio, I, is commonly used to characterize reservoir rocks that are partially saturated with water and also contain oil or gas, or both:

$$I = \frac{R_t}{R_o} \qquad [5.1.67]$$

where R_t is the resistivity of the rock at some condition of partial water saturation, S_w, and R_o is the resistivity of the rock when completely saturated with water or brine.

Citing the work of Martin et al. [74], Jakosky and Hopper [75], Wyckoff and Botset [76], and Leverett [77], in which the variation in resistivity with water saturation was studied, Archie [42] plotted the resistivity ratio versus S_w on log-log paper (Figure 5.1.36). For water saturations down to about 0.15 or 0.20, the following approximate equation appeared to hold, regardless of whether oil or gas was the nonconducting fluid:

$$S_w = \left(\frac{R_o}{R_t} \right)^{1/n} \qquad [5.1.68]$$

where n has been commonly referred to as the saturation exponent. For clean sands and for consolidated sandstones, the value of n was close to 2.0, so the approximate relation was given by Archie as:

$$S_w = \left(\frac{R_o}{R_t} \right)^{1/2} \qquad [5.1.69]$$

By substituting the equation for R_o (refer to Equation 5.1.47), Archie presented the relationship between water saturation, formation resistivity factor, brine resistivity, and the resistivity of the rock at the given S_w:

$$S_w = \left(\frac{F_R R_w}{R_t} \right)^{1/2} \qquad [5.1.70]$$

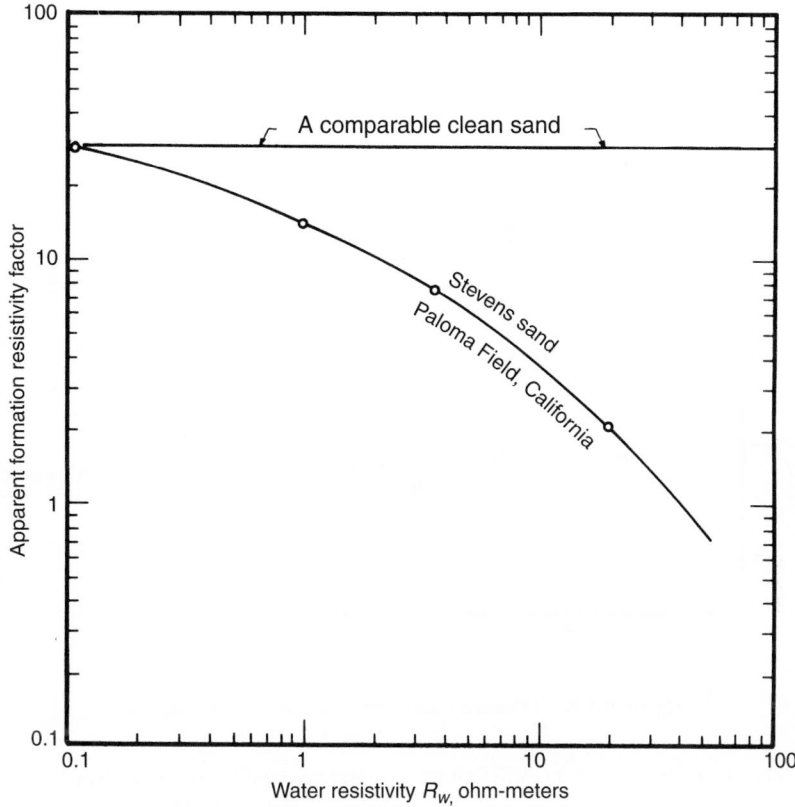

Figure 5.1.33 *Effect of clays on formation resistivity factor [59,60].*

Table 5.1.8 *Values of Exponent in the Archie Equation for Clays*

Mineral particle	Shape factor exponent
Sodium montmorillonite	3.28
Calcium montmorillonite	2.70
Muscovite	2.46
Attapulgite	2.46
Illite	2.11
Kaolinite	1.87

From Reference 61.

Table 5.1.9 *Factors that Influence the m Exponent in Equation 5.1.49 for the Rock–Water Interface*

1. Pore geometry.
   a. Surface-area-to-volume ratio of the rock particle, angularity, sphericity.
   b. Cementation.
   c. Compaction.
   d. Uniformity of mineral mixture.
2. Anisotropy.
3. Degree of electrical isolation by cementation.
4. The occurrence of an open fracture.

From Reference 65.

The more general form of Equation 5.1.70, commonly used, is:

$$S_w = \left(\frac{F_R R_w}{R_t} \right)^{1/n} = \left(\frac{R_w \phi^{-m}}{R_t} \right)^{1/n} \qquad [5.1.71]$$

An even more general form recognizes that the constant, a, in Equation 5.1.49 is not necessarily unity:

$$S_w = \left(\frac{aR_w}{\phi^m R_t} \right)^{1/n} \qquad [5.1.72]$$

The foregoing equations are close approximations in clean formations having a regular distribution of porosity. The accuracy of the equations will not be as good in formations with vugs or fractures.

As shown in Figure 5.1.37, Patnode and Wyllie [59] found that the presence of clays affected the relationship between water saturation and resistivity index. As the water saturation is reduced toward zero, the resistivity approaches that of the clays rather than approaching infinity as with clean sands. Relationships between resistivities and fluid content in the presence of conductive solids have been presented in the literature [78,79].

Early investigations, using data from the Woodbine sand of east Texas, suggested that the saturation exponent, n, may range from 2.3 to 2.7 [80,81]. Wyllie and Spangler [82] presented data (Figure 5.1.38) for several natural and synthetic porous media that showed a variation of the saturation exponent from 1.4 to 2.5. Other investigators found that the

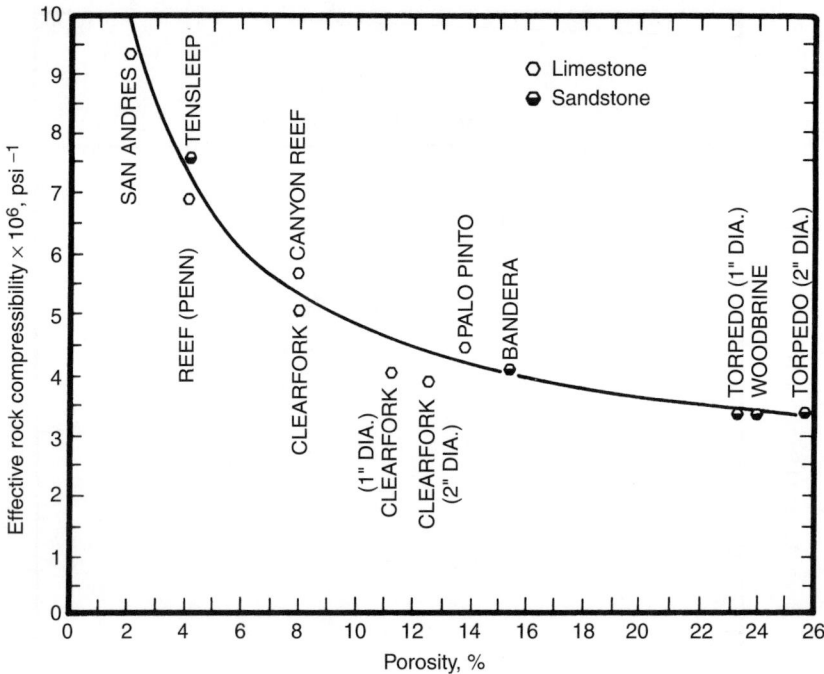

Figure 5.1.34   *Effective rock compressibility vs. porosity [69].*

Table 5.1.10   *Factors that Influence the a Coefficient in Equation 5.1.49*

1. Surface conductance and ionic mobility occurring in water films adsorbed to solid surfaces.
   a. The cation exchange capacity of particular solid materials.
   b. The quantity of water adsorbed to clay particles in the rock framework or within the interstices
2. Salinity of formation water.
3. Wettability relations between particular solid surfaces and hydrocarbons, as they influence cation exchange.
4. The presence and distribution of electrically conductive solid minerals.

From Reference 65.

distribution of fluids within the core sample, at the same water saturation, could affect the resistivity index for both sandstones [83] and for limestones [84]. The exponent, n, was also found to vary depending upon the manner in which the conducting wetting phase saturation was varied [82,85].

Data [56] from a number of different reservoirs for the values of m, the cementation or shape factor, and n, the saturation exponent, in Equation 5.1.71 are presented in Table 5.1.13.

Based on these data and other reasoning, Coats and Dumanoir [56] have proposed that the two exponents, m and n, can be assumed to be equal for water-wet, consolidated reservoirs. Ransome [65] has proposed that the saturation exponent may be a special case of the porosity exponent, and the two exponents may bear certain similarities but not necessarily the same value.

As recently pointed out by Dorfman [86], data in Table 5.1.13 strongly suggest that assuming a saturation exponent of 2 can result in serious errors in the estimation

of water saturation. In low-porosity formations, such as the Cotton Valley sandstone, the saturation was found to vary greatly from the value of 2 [87]. If n is always assumed to be 2, Dorfman contends that many hydrocarbon zones will be overlooked and many water-producing zones could be tested. As related by Hilchie [88], most of the values for the saturation exponent have been obtained at atmospheric conditions and there is the need to obtain laboratory measurements under simulated reservoir pressure and temperature. At atmospheric pressure, the percentage of smaller pores is larger than at reservoir pressure [64], which results in the wrong saturation exponent and a higher value of water saturation [88].

5.1.3.3 Surface and Interfacial Tensions

The term *interface* indicates a boundary or dividing line between two immiscible phases. Types of interfaces include: liquid-gas, liquid-liquid, liquid-solid, solid-gas, and solid-solid. For fluids, molecular interactions at the interface result in a measurable tension which, if constant, is equal to the surface free energy required to form a unit area of interface. For the case of a liquid which is in contact with air or the vapor of that liquid, the force per unit length required to create a unit surface area is usually referred to as the surface tension. Interfacial tension is used to describe this quantity for two liquids or for a liquid and a solid. Interfacial tension between two immiscible liquids is normally less than the surface tension of the liquid with the higher tension, and often is intermediate between the individual surface tensions of the two liquids of interest. Common units of surface or interfacial tension are dynes per centimeter (or the identical ergs/cm^2) with metric units in the equivalent milli-Newton per meter (mN/m).

The surface tension of pure water ranges from 72.5 dynes/cm at 70°F to 60.1 dynes/cm at 200°F in an almost linear fashion with a gradient of 0.095 dynes/cm/°F [25]. Salts in oilfield brines tend to increase surface tension, but

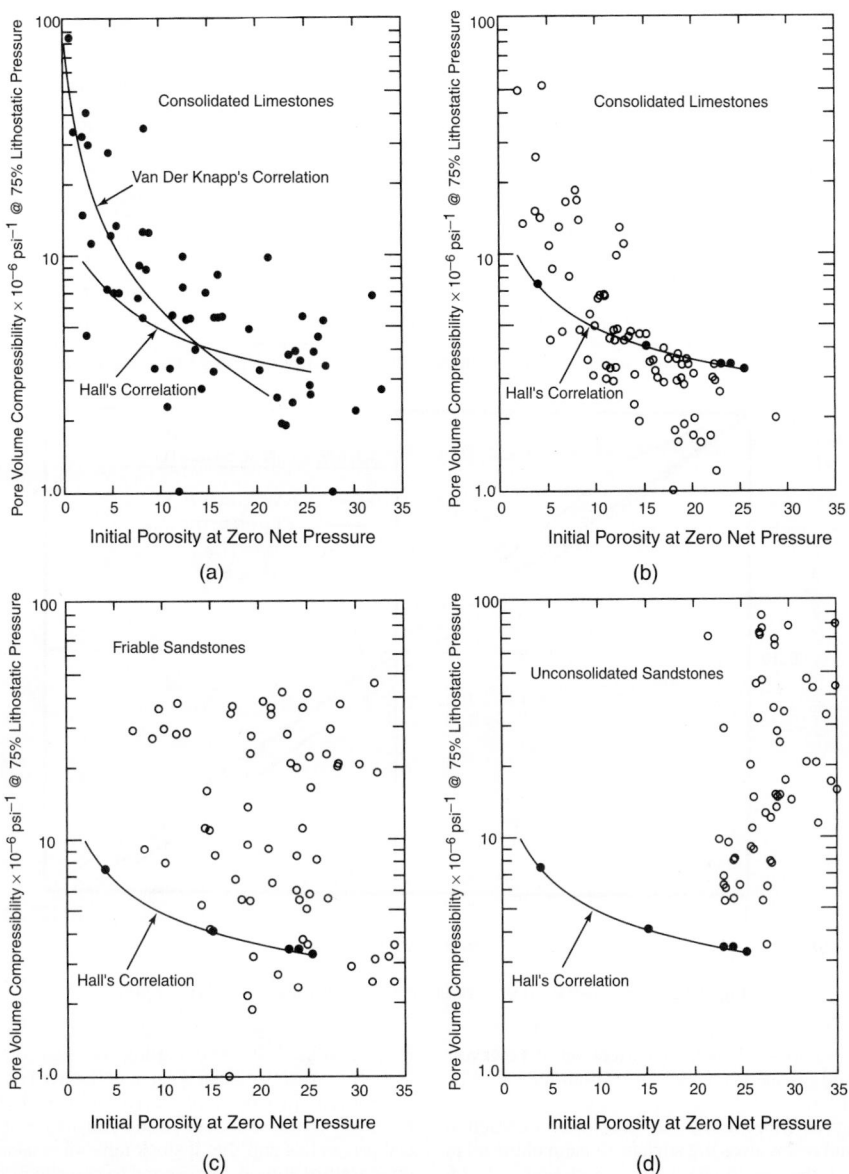

Figure 5.1.35   *Pore-volume compressibility at 75% lithostatic pressure vs. sample porosity [71].*

Table 5.1.11   *Range of Formation Compressibilities*

Formation	Pore volume compressibility, psi^{-1}
Consolidated sandstones	1.5×10^{-6} to 20×10^{-6}
Consolidated limestones	2.0×10^{-6} to 35×10^{-2}
Friable sandstones	2.5×10^{-6} to 45×10^{-6}
Unconsolidated sandstones	5.5×10^{-6} to 85×10^{-6}

From Reference 71.

surface active agents that may dissolve into the water from the oil can lower surface tension. At standard conditions, surface tensions of brines range from 59 to 76 dynes/cm [25]. As shown in Figure 5.1.39, dissolved natural gas reduces surface tension of water as a function of saturation pressure [89].

At a given temperature, surface tension of hydrocarbons in equilibrium with the atmosphere or their own vapor increases with increasing molecular weight (Figure 5.1.40) [90]. For a given hydrocarbon, surface tension decreases with increasing temperature. At 70°F, surface tensions of crude oils often range from 24 to 38 dyne/cm [25].

The presence of dissolved gases greatly reduces surface tension of crude oil as shown in Figure 5.1.41 [91]. Dissolved natural gas reduces surface tension of crude oil more than previously noted for water, but the amount and nature of gas determines the magnitude of the reduction. The direct effect of a temperature increase on reduction of surface tension more than counterbalances the decreased gas solubility at

Table 5.1.12 *Summary of Compressibility Values*

	Compressibility, psi^{-1}	
	Range	Typical value
Consolidated rock*	2×10^{-6} to $\ \ 7 \times 10^{-6}$	3×10^{-6}
Oil [17,73]	5×10^{-6} to 100×10^{-6}	10×10^{-6}
Water (gas-free) [26]	2×10^{-6} to $\ \ 4 \times 10^{-6}$	3×10^{-6}

	Compressibility, psi^{-1}	
	At 1,000 psi	At 5,000 psi
Gas [26]	$1,000 \times 10^{-6}$	100×10^{-6}
Water (with dissolved gas) [26]	15×10^{-6}	5×10^{-6}

*See Figure 5.1.35 (for most of samples having porosities of $20 \pm 10\%$ in Figures 5.1.35a and 5.1.35b).

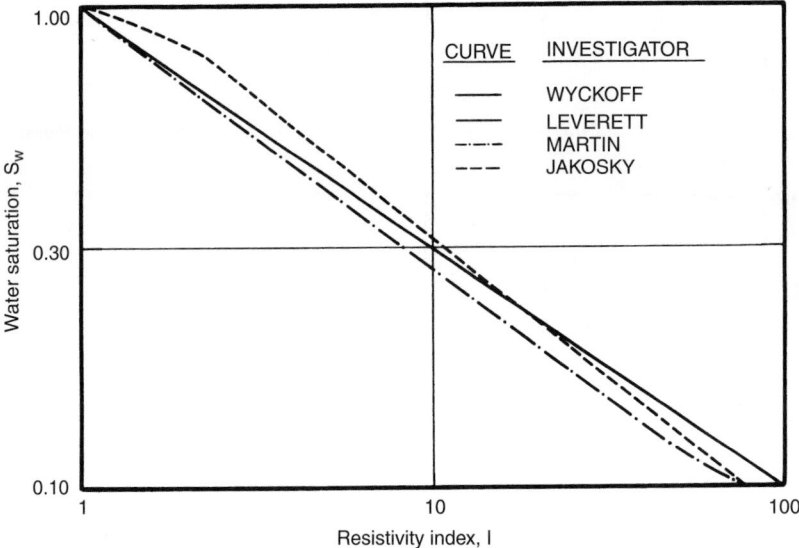

Figure 5.1.36 *Variation of resistivity index with water saturation [42].*

elevated temperatures. Thus, surface tension at reservoir temperature and pressure may be lower than indicated by Figure 5.1.41 [25].

Under reservoir conditions, the interfacial interaction between gas and oil involves the surface tension of the oil in equilibrium with the gas. Similarly, the interaction between oil and water determines the interfacial tension between the crude and brine. Listed in Table 5.1.14 are the surface and interfacial tensions for fluids from several Texas fields [92].

The effect of temperature on interfacial tensions for some oil-water systems is shown in Figure 5.1.42 [92]; the reduction in interfacial tension with increasing temperature is usually somewhat more pronounced than is observed for surface tension. Although no quantitative relation is observed, the general trend suggests lower interfacial tensions for the higher API gravity crudes. However, in studies with a crude oil containing large amounts of resins and asphaltenes, different effects of temperature on interfacial tension were observed when measurements made at aerobic conditions were compared to anaerobic tests [93]. Interfacial tension between the crude and reservoir brine showed a decrease with an increase in temperature under aerobic conditions, whereas at anaerobic conditions, interfacial tension increased with increasing temperatures. This difference in behavior was attributed to oxidation of the stock tank oil

in the aerobic tests. At conditions of reservoir temperature and pressure, interfacial tension of the live reservoir oil was higher than the stock tank oil. The study concluded that live reservoir crude should be used in measurements of interfacial properties and that if stock tank oil is used, at least the temperature should correspond to reservoir conditions.

Figure 5.1.43 shows the effect of dissolved gas and pressure on the interfacial tension of three oil-water systems [89]. For each system, interfacial tension increases as the amount of dissolved gas increases, but drops slightly as pressure is increased above the bubblepoint.

Surface and interfacial tensions are important in governing the flow of fluids in the small capillaries present in oil-bearing reservoirs. The capillary forces in oil or gas reservoirs are the result of the combined effect of surface and interfacial tensions, pore size distribution, pore shape, and the wetting properties of the hydrocarbon/rock system.

5.1.3.4 Wettability and Contact Angle

The contact angle (θ_c), existing between two fluids in contact with a solid and measured through the more dense phase, is a measure of the relative wetting or spreading by a fluid on a solid. A contact angle of zero indicates complete wetting by the more dense phase, an angle of 180° indicates complete

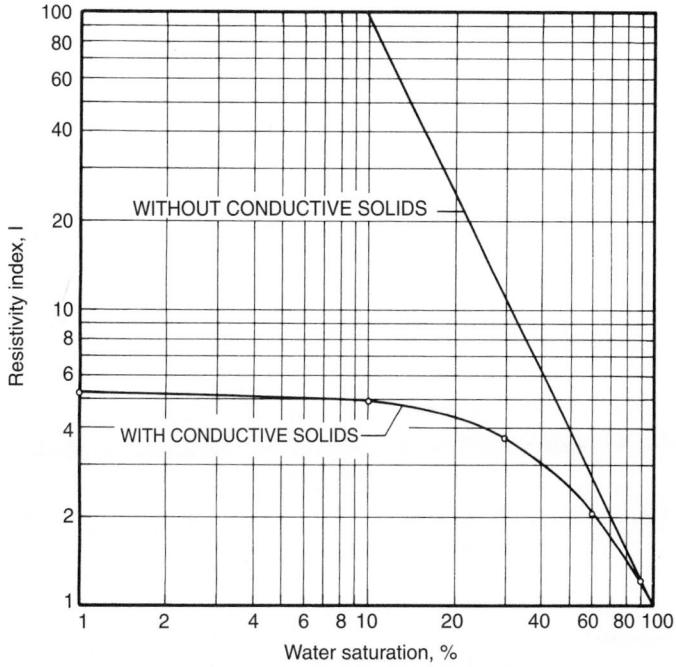

Figure 5.1.37 *Effect of conductive solids on the relationship between resistivity index and water saturation [59].*

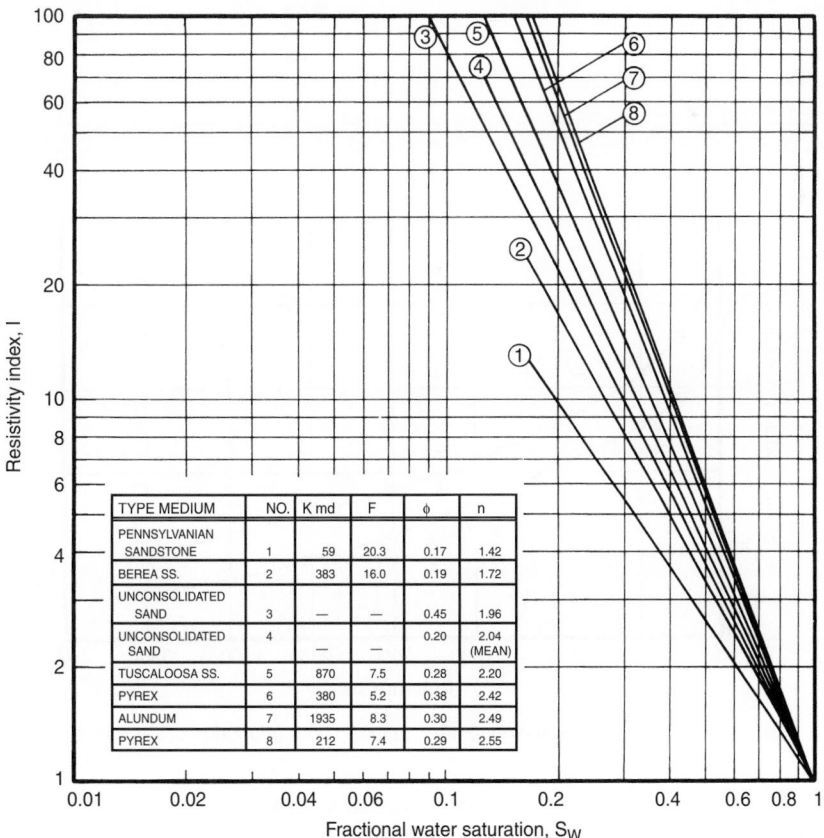

TYPE MEDIUM	NO.	K md	F	φ	n
PENNSYLVANIAN SANDSTONE	1	59	20.3	0.17	1.42
BEREA SS.	2	383	16.0	0.19	1.72
UNCONSOLIDATED SAND	3	—	—	0.45	1.96
UNCONSOLIDATED SAND	4	—	—	0.20	2.04 (MEAN)
TUSCALOOSA SS.	5	870	7.5	0.28	2.20
PYREX	6	380	5.2	0.38	2.42
ALUNDUM	7	1935	8.3	0.30	2.49
PYREX	8	212	7.4	0.29	2.55

Figure 5.1.38 *Relationship between resistivity index and water saturation for several media [82].*

Table 5.1.13   *Values of Constants in Equation 5.1.71*

	Lithology	Avg. m	Avg. n
Ordovician Simpson, W. Texas and E. New Mexico	SS	1.6	1.6
Permian, W. Texas	SS	1.8	1.9
Ellenburger, W. Texas	LS and Dol.	2.0	3.8
Pennsylvanian, W. Texas	LS	1.9	1.8
Viola, Bowie Field, No. Texas	LS	1.77	1.15
Edwards, So. Texas	LS	2.0	2.8
Edwards Lime, Darst Creek, So. Texas	LS	1.94, 2.02	2.04, 2.08
Frio, So. Texas	SS	1.8	1.8
Frio, Agua Dulce, So. Texas	SS	1.71	1.66
Frio, Edinburgh, So. Texas	SS	1.82	1.47, 1.52
Frio, Hollow Tree, So. Texas	SS	1.80, 1.87	1.64, 1.69
Government Wells, So. Texas	SS	1.7	1.9
Jackson, Cole Sd., So. Texas	SS	2.01	1.66
Miocene, So. Texas	Cons. SS	1.95	2.1
	Uncons. SS	1.6	2.1
Navarro, Olmos, Delmonte, So. Texas	SS	1.89	1.49
Rodessa, E. Texas	LS	2.0	1.6
Woodbine, E. Texas	SS	2.0	2.5
Travis Peak and Cotton Valley, E. Texas and W. Louisiana	HD. SS	1.8	1.7
Wilcox, Gulf Coast	SS	1.9	1.8
Annona, No. Louisiana	Chalk	2.0	1.5
Cockfield, So. Louisiana	SS	1.8	2.1
Frio, Chocolate Bayou, Louisiana	SS	1.55–1.94	1.73–2.22
Sparta, So. Louisiana (Opelousas)	SS	1.9	1.6
Nacatoch, Arkansas	SS	1.9	1.3
Pennsylvanian, Oklahoma	SS	1.8	1.8
Bartlesville, Kansas	SS	2.0	1.9
Simpson, Kansas	SS	1.75	1.3
Muddy, Nebraska	SS	1.7	2.0
Lakota Sd., Crook Co., Wyoming	SS	1.52	1.28
Madison, No. Dakota	LS	1.9	1.7
Mississippian, Illinois	LS	1.9	2.0
Mississippian, Illinois	SS	1.8	1.9

After Reference 56.

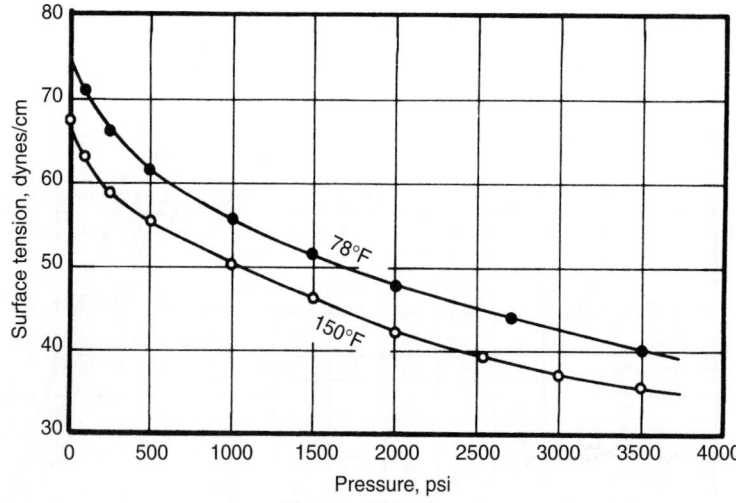

Figure 5.1.39   *Surface tension between water and natural gas as a function of saturation pressure [89].*

wetting of the less dense phase, and an angle of 90° means that neither fluid preferentially wets the solid.

From a combination of Dupre's equation for wetting tension and Young's equation [94], the adhesion tension (τ_A)

can be given as [19,95,97]:

$$\tau_A = \sigma_{os} - \sigma_{ws} = \sigma_{wo} \cos \theta_{wo} \qquad [5.1.73]$$

where σ_{os} is the interfacial tension between the solid and the less dense fluid phase, σ_{ws} is the interfacial tension

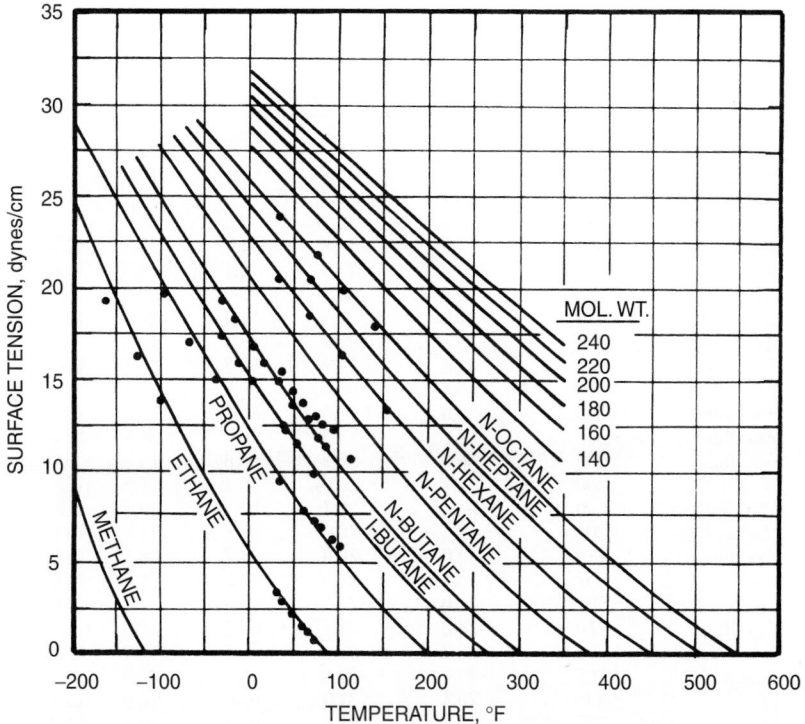

Figure 5.1.40 *Surface tension of several hydrocarbons [90].*

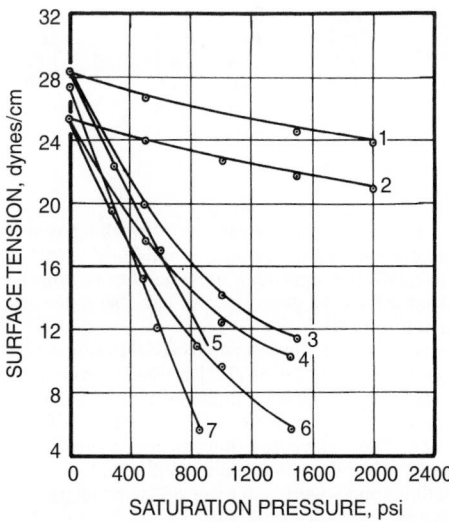

Figure 5.1.41 *Surface tension of several crude oils [91].*

between the solid and the more dense phase, and σ_{wo} is the interfacial tension between the fluids of interest. With gas-oil systems, oil is the more dense phase and is always the wetting phase [96]. With oil-water systems water is almost always the more dense phase, but either can be the wetting phase. For oil and water, a positive adhesion tension ($\theta_c > 90°$) indicates a preferentially water-wet surface,

whereas a negative adhesion tension ($\theta_c < 90°$) indicates a preferentially oil-wet surface. For a contact angle of 90°, an adhesion tension of zero indicates that neither fluid preferentially wets the solid. Examples of various contact angles are depicted in Figure 5.1.44 [96].

The importance of wettability on crude oil recovery has been recognized for many years. This subject is discussed in a subsequent section of this chapter. Although Nutting [98] observed that some producing formations were oil-wet, many early workers considered most oil reservoirs to be water-wet (e.g., References 23, 99, and 100; discussion and comments in Reference 96). From a thermodynamic standpoint, it was felt that pure, clean silica must be wetted by water in preference to any hydrocarbon. In one study [101], no crude oils were tested that had a greater adhesion than pure water. Other results [102] tended to support this contention: capillary pressure tests suggested that all cores tested were water-wet with contact angles ranging from 31° to 80°. However, there are two reasons why these results were obtained [103]: (1) the cores were extracted with chloroform prior to the tests which could have affected the natural wettability, and (2) only receding (decrease in wetting phase saturation) contact angles were measured during the capillary pressure tests. As with capillary pressures, there is a hysteresis in the receding and advancing (increase in wetting phase saturation) contact angles; receding angles are smaller than advancing angles [97]. Bartell and coworkers [95-97] were among some of the first investigators to measure contact angles with crude oil systems that suggested the possibility that oil reservoirs may not be water-wet. Furthermore, they concluded that spontaneous displacement of oil by water should occur only when both advancing and receding angles are less than 90°, and no

Table 5.1.14 *Surface and Interfacial Tensions for Several Texas Fields*

Field	Formation	Depth, ft	Oil Gravity, °API	Surface tension, dynes/cm Oil	Water	Oil-water interfacial tension, dynes/cm 70°F	100°F	130°F
Breckenridge	Marble Falls	3,200	38.2	28.8	67.6	19.0	10.9	
South Bend	Strawn	2,300	36.1	29.9	61.5	29.1	21.4	
Banyon	Austin	2,135	37.0	29.3	72.5	24.4	17.4	9.6
South Bend	Marble Falls	3,900	25.5	31.8	71.4	24.5	16.9	
Banyon	Austin	2,255	37.9	28.9	72.1	16.9	13.6	12.9
Salt Flats	Edwards	2,700	34.9	30.0	73.0	23.0	16.9	16.7
Driscoll	Catahoula	3,929	26.0	32.4	61.4	20.4	16.0	15.5
Wortham	Woodbine	2,800	38.3	29.2	63.2	13.6	7.3	
Wortham	Corsicana	2,200	22.4	33.2	59.6	25.1	16.7	13.2
Mexia	Woodbine	3,000	36.4	30.0	66.2	21.4	19.0	17.6
Powell	Woodbine	3,000	22.9	30.0	66.2	22.6	15.0	
Wortham	Woodbine	2,800	22.2	33.3	66.0	25.8	15.6	
Mexia	Woodbine	3,000	36.6	30.2	66.6	15.0	9.2	
Breckenridge	Marble Falls	3,200	37.7	28.9	70.1	16.2	8.5	10.0
Breckenridge	Marble Falls	3,200	36.6	29.4	74.1	15.5	11.3	
South Bend	Marble Falls	4,200	38.6	28.9	68.1	14.8	10.8	10.1
Van	Woodbine	2,710	33.9	29.0	61.7	18.1	16.2	
Raccoon Bend	Cockfield	3,007	34.1	31.6	69.8	24.7	14.6	14.3
Tomball	Cockfield	5,541	41.6	28.5	62.0	14.1	13.6	
Van	Woodbine	2,710	35.0	28.8	64.1	17.9	15.0	7.8
Saxet	Catahoula	4,308	26.2	32.0	65.2	17.2	11.5	10.8
Saxet	Catahoula	4,308	27.1	32.3	66.5	20.9	16.5	14.1
Pierce Junction	Frio	4,325	29.4	31.0	62.0	16.9	13.9	8.7
Pierce Junction	Frio	4,335	22.2	32.6	64.1	20.7	12.9	2.1
East Texas	Woodbine	3,660	36.5	28.2	68.6	19.7	10.9	9.6
East Texas	Woodbine	3,660	39.5	27.5	70.2	31.4	17.9	13.9
Goose Creek	Pliocene	1,470	14.2	34.1	63.7	24.4	19.5	
Goose Creek	Pliocene	2,040	21.1	33.6	63.5	18.8	15.3	12.5
Goose Creek	Mio-Pliocene	2,560	21.2	33.3	64.2	18.1	12.9	12.5
Talco	Glen Rose	5,000	23.0	31.9	73.9	20.5	18.8	14.8
Big Lake	Ordovician	8,300	42.6	28.5	63.3	18.1	14.8	12.5
Big Lake	Permian	3,000	38.0	27.9	66.2	27.3	18.3	15.7
Crane	Permian	3,500	31.1	29.5	68.2	18.6	14.8	7.8
Echo	Frye (Pa.)	1,950	38.4	27.8	49.5	34.3	24.6	18.6

From Reference 92.

spontaneous imbibition should occur if the two angles are on opposite sides of 90° [97].

A common technique for measuring advancing contact angles using polished mineral crystals is described in the literature [104]. For many crude oil systems, a considerable amount of time may be required before an equilibrium contact angle on a pure mineral is obtained. As shown in Figure 5.1.45 [104], some systems that initially appear to be preferentially water-wet become more oil-wet. Small amounts of polar compounds in some crude oils can adsorb on the rock surfaces and change wettability from preferentially water-wet to more oil-wet [96]. A detailed study on how crude oil components affect rock wettability has been made by Denekas et al. [105]. Imbibition tests have been described to examine wettability of reservoir cores [106,107]. A preferred technique of inferring reservoir wetting from core samples is the Amott method [108] which involves spontaneous imbibition and forced displacement tests; ratios of spontaneous displacement volumes to total displacement volumes are used as an index of wettability. Based on the correlation suggested by Gatenby and Marsden [109], Donaldson et al. [110–112] developed a quantitative indication of wettability, called the U.S. Bureau of Mines (USBM) test, by measuring the areas under capillary pressure curves. The

USBM method has the advantage of working well in the intermediate wettability region. A multitude of techniques for the qualitative indication of wettability that have been proposed will not be described but have been discussed in the literature [133].

In a fairly extensive examination of 55 different reservoirs, Treiber, Archer, and Owens [114] arbitrarily assigned water-wet conditions for contact angles of 0° to 75° and oil-wet conditions for contact angles of 105° to 180° with contact angles of 75° to 105° representing an intermediate (referred to as neutral by others) wettability. With these designations, 27% of the samples were water-wet, 66% were oil-wet, and the remaining 7% were of intermediate wettability. Subsequently, Morrow [115] has defined an intermediate wettability when neither fluid spontaneously imbibes in a "squatters' rights" situation. Morrow found that for contact angles less than 62°, the wetting phase would spontaneously imbibe, and for contact angles above 133°, the nonwetting phase would spontaneously imbibe; therefore, the intermediate wettability condition would be operative for contact angles from 62° to 133°. Using Morrow's guidelines, the data of Treiber, Archer, and Owens indicate that 47% of the samples were of intermediate wettability, 27% were oil-wet, and 26% were water-wet. The distribution in wettability according to lithology is given

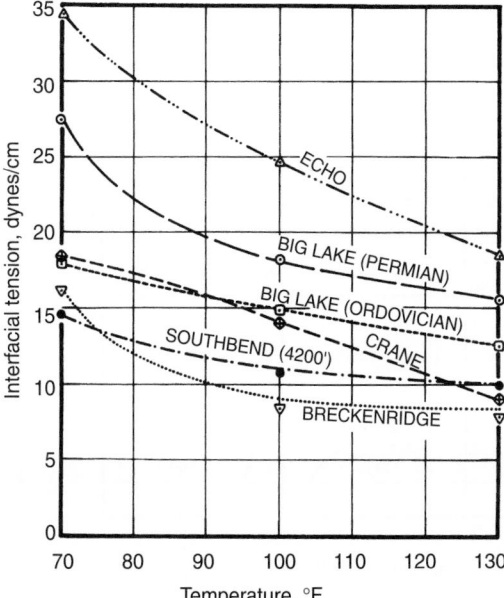

Figure 5.1.42 *Influence of temperature on interfacial tensions for crude oil/water systems [92].*

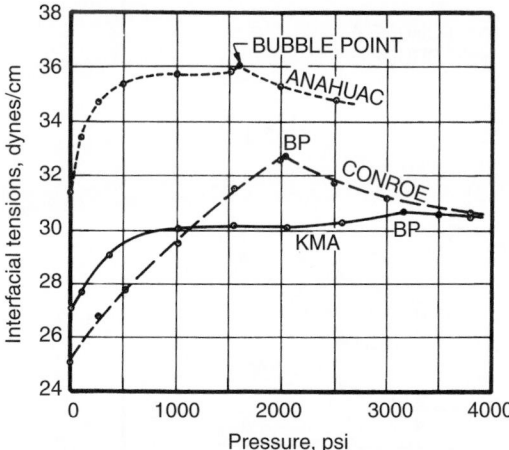

Figure 5.1.43 *Effect of dissolved gas and pressure on interfacial tension between crude oil water [89].*

in Table 5.1.15. In either case, it is apparent that a majority of the samples were not water-wet.

Using 161 core samples representing various carbonate reservoirs, Chilingar and Yen [116] found that 8% were water-wet ($\theta_c > 80°$), 12% were intermediate ($\theta_c = 80° - 100°$), 65% were oil-wet ($\theta_c = 100° - 160°$), and 15% were strongly oil-wet ($\theta_c = 160° - 180°$). The arbitrary definitions of wettability differ from Trieber et al. [114] and Morrow [115], but the distributions appear to be similar to the carbonate data in Table 5.1.15.

In the previous discussion, it was implied that pore surfaces within a reservoir rock are uniformly wetted. The concept whereby a portion of the reservoir surfaces are preferentially oil-wet while other portions are preferentially water-wet was termed fractional wettability by Brown and

Fatt [117] and Fatt and Klikoff [118]. Fractional wetting was believed to be a result of the varying amount of adsorption of crude oil components on the different minerals present in a reservoir. Other evidence [119,120] supported the existence of a heterogenous wetting (also called spotted or Dalmation wettability in the literature). Salathiel [121] introduced the concept of a mixed-wettability condition, a special case of fractional wetting, in which the fine pores and grain contacts are preferentially water-wet and the large pore surfaces are strongly oil-wet. Salathiel concluded that the oil-wet paths can be continuous to provide a means for oil to flow even at very low oil saturations; these results were offered to explain the very good oil recovery noted in some field projects. More recently, Morrow, Lim, and Ward [122] introduced the concept of a speckled wettability in which a rationale is presented whereby oil tends to be trapped in pore throats rather than pore bodies. Speckled wettability mimics behavior of strongly water-wet conditions observed during waterflooding: water breakthrough is abrupt, relative permeability to water at residual oil saturation is low, water is imbibed spontaneously, and oil is not imbibed spontaneously.

When cores are obtained for laboratory tests where wettability is important, precautions must be taken to ensure the wetting preference of the formation is not altered during coring. Mud additives, such as dispersants, weighting agents, lost circulation materials, thinners or colloids, that possess surface-active properties may drastically change core wettability. Surface active agents should be avoided so that the core samples have the same wettability as the reservoir rock. Listed in Table 5.1.16 are the effects of various mud additives on wettability of water-wet and oil-wet cores [107].

In the case of water-wet sandstone or limestone cores, rock-salt, bentonite, carboxymethylcellulose (CMC), and barite had no effect on wettability. However, oil-wet sandstone cores were reversed to a water-wet condition when exposed to CMC, bentonite, or lime solutions. Additional tests with bentonite solutions indicated that wettability of oil-wet cores is not reversed if the solution pH is lowered to a neutral or slightly acidic value. These results suggest that from a wettability standpoint, the best coring fluid is water (preferably formation brine); if bentonite is used, mud pH should be neutral or slightly acidic. If appreciable hydrogen sulfide is suspected in the interval being cored, it may be undesirable to lower pH. In fact, a very alkaline mud (pH 10–12) may be used to keep the sulfide in the ionized state for safety and corrosion considerations. Subsequent work [123] suggests the preferred system to obtain fresh cores is a natural water-base mud with no additives, or a bland mud consisting of bentonite, salt, and CMC. However, recent results [124] conclude that bland muds may not, in fact, be bland. While none of the bland additives altered wettability of water-wet rock samples, all of the components with the exception of bentonite made oil-wet samples significantly less oil-wet. The bland additives that were tested included bentonite, pregelatinized starch, dextrid (an organic polymer), drispac (a polyanionic cellulose polymer), hydroxyethylcellulose, xanthan gum polysaccharide, and CMC. All of the drilling mud components considered to be bland decreased the amount of oil imbibed into a core and increased the amount of water imbibed.

Preventing wettability changes in core material, after it has been recovered at the surface, is equally important so that subsequent laboratory measurements are representative of reservoir conditions. Changes in wettability of core material that occur during handling or storage are usually caused by oxidation of the crude oil, evaporation of volatile components, or decreases in temperature or pressure which cause the deposition of polar compounds, asphaltenes, or heavy hydrocarbon compounds [107,125]. Because of the

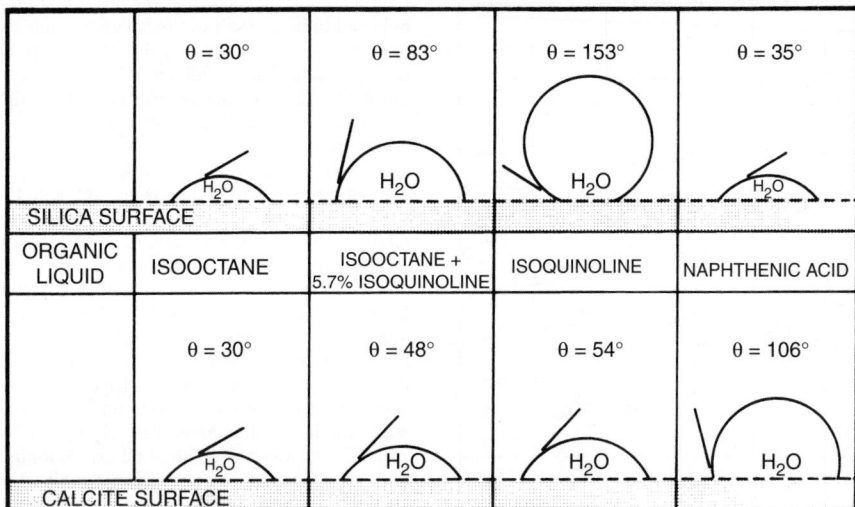

Figure 5.1.44 Examples of contact angles [96].

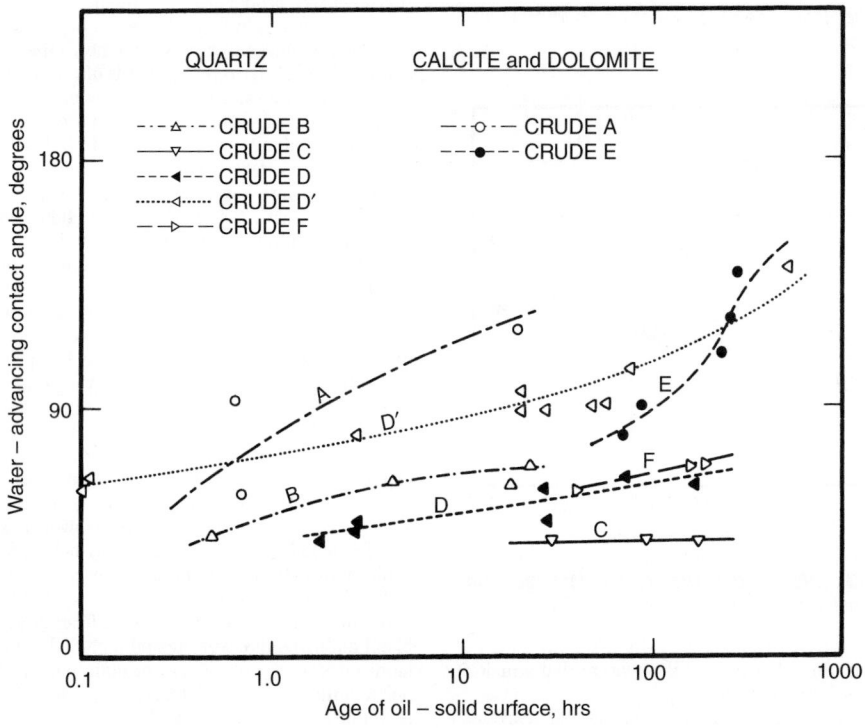

Figure 5.1.45 Change in contact angles with time [104].

complexity of the mechanisms involved, the magnitude and direction of changes in wetting conditions, when reservoir cores are preserved, are not fully understood. Weathering of water-wet cores has been reported to frequently cause the grain surfaces to become oil-wet [107]. In other experiments [126] oil-wet cores changed to water-wet upon contact with air. Therefore it is necessary to preserve core samples at the well-site to ensure that wettability is not altered by contamination, oxidation, or evaporation. Two methods of preserving conventional cores, immediately after they have been removed from the core barrel, will prevent changes in wettability for several months [107]. One method consists of immersing the core in deoxygenated formation brine or suitable synthetic brine (i.e., drilling mud filtrate) and keeping the sample in suitable nonmetallic containers that can be sealed to prevent leakage and the entrance of oxygen. In a

Table 5.1.15 *Wettability of 55 Reservoir Rock Samples*

	Water-wet ($\theta_c < 75°$)*	Intermediate ($\theta_c = 75° - 105°$)*	Oil-wet ($\theta_c > 105°$)*
Sandstones	43%	7%	50%
Carbonate	8%	8%	84%
Total samples	27%	7%	66%
	Water-wet ($\theta_c < 62°$)**	Intermediate ($\theta_c = 62° - 133°$)**	Oil-wet ($\theta_c > 133°$)**
Total samples**	26%	47%	27%

*Based on contact angle/wettability relation suggested by Treiber, Archer, and Owens [114].
**Based on contact angle/wettability relation suggested by Morrow [115].

Table 5.1.16 *Effect of Water-Base Mud Additives on the Wettability of Cores*

Component	Wettability of Test Cores after Exposure to Filtrate		
	Water-Wet Limestone	Water-Wet Sandstone	Oil-Wet Sandstone
Rock-salt	No change	No change	No change
Starch	Slightly less water-wet	Slightly less water-wet	—
CMC	No change	No change	Water-wet
Bentonite	No change	No change	Water-wet
Tetrasodium pyrophosphate	No change	Less water-wet	—
Calcium lignosulfonate	No change	Less water-wet	—
Lime	No change	Slightly more water-wet	Water-wet
Barite	No change	No change	—

From Reference 107.

Table 5.1.17 *Empirical Relative Permeability Equations*

I. Oil-gas relative permeabilities
(for drainage cycle relative to oil)

$$S^* = \frac{S_o}{(1 - S_{iw})}$$

Where S_{iw} is the irreducible water saturation.

	k_{ro}	k_{rg}
A. Unconsolidated sand—well sorted	$(S^*)^{3.0}$	$(1 - S^*)^3$
B. Unconsolidated sand—poorly sorted	$(S^*)^{3.5}$	$(1 - S^*)^2(1 - S^{*1.5})$
C. Cemented sand, oolitic lime, and vugular lime	$(S^*)^{4.0}$	$(1 - S^*)^{2.0}(1 - S^{*2.0})$

II. Water-oil relative permeabilities
(for drainage cycle relative to water)

$$S^* = \frac{S_w - S_{iw}}{(1 - S_{iw})}$$

where S_{iw} is the irreducible water saturation.

	k_{ro}	k_{rw}
A. Unconsolidated sand—well sorted	$(1 - S^*)^{3.0}$	$(S^*)^{3.0}$
B. Unconsolidated sand—poorly sorted	$(1 - S^*)^2(1 - S^{*1.5})$	$(S^*)^{3.5}$
C. Cemented sand, oolitic lime, and vugular lime	$(1 - S^*)^2(1 - S^{*2.0})$	$(S^*)^{4.0}$

From Reference 20.

second method, the cores are wrapped in Saran or polyethylene film and aluminium foil, and then coated with wax or strippable plastic. Cores obtained by either of these methods are referred to as preserved, native-state, or fresh cores, and are preferred for many laboratory tests.

For certain laboratory tests, it may be possible to clean reservoir cores with solvents and resaturate with reservoir fluids to restore the original wetting conditions. Details of preparing such restored-state or extracted cores are discussed subsequently in the section "Coring and Core Analysis." The concept of the method is to clean the core thoroughly until it is water-wet, saturate with reservoir brine, flush with crude oil, and age for over 1,000 hours at reservoir temperature.

Regardless of the method of core handling employed, the rock samples used in the laboratory should have a surface state as close as possible to that present in the reservoir. If preserved cores are used, it is essential they be stored under air-free conditions because exposure to air for as little as 6–8 hours can cause water evaporation and other changes in core properties. If extracted cores are used, drying of the cores can be very critical when hydratable minerals, capable of breaking down at low temperatures are present. Contamination from core holders that contain certain types of rubber sleeves can be prevented by using an inner liner of tubular polyethylene film. Because of the instability of many oilfield waters, it is usually desirable to prepare synthetic brines to prevent core plugging caused by deposition of insolubles.

When possible, tests should be made under reservoir conditions of temperature and pressure using live reservoir oil. This is an improvement over room condition techniques where tests are made at atmospheric conditions with refined laboratory oils. Use of the live crude exposes the rock to compounds present in the oil that might influence wettability, and establishes an environment as close as possible to reservoir conditions. Cores evaluated at atmospheric conditions may be more oil-wet than similar tests at reservoir conditions because of the decreased solubility of wettability-altering compounds at lower temperatures and pressures [107,123]. In a recent contact angle study [93] with calcium carbonate crystals and a crude oil containing 27.3% resins and 2.2% asphaltenes, a complete reversal from a predominantly oil-wet system at lower temperatures to a predominantly water-wet system at higher temperatures was found. While pressure alone had little effect on the wettability of the system, the study speculated that the addition of gas-in-solution with increasing pressure should favor a more water-wet situation than would be indicated from laboratory tests at atmospheric conditions. Even when all precautions have been taken, there is no absolute assurance that reservoir wettability has been duplicated.

5.1.3.5 Capillary Pressure

Curvature at an interface between wetting and nonwetting phases causes a pressure difference that is called capillary pressure. This pressure can be viewed as a force per unit area that results from the interaction of surface forces and the geometry of the system.

Based on early work in the nineteenth century of Laplace, Young, and Plateau (e.g., Reference 94), a general expression for capillary pressure, P_c, as a function of interfacial tension,

σ, and curvature of the interface is [19]:

$$P_c = \sigma \left(\frac{1}{r_1} + \frac{1}{r_2} \right) \qquad [5.1.74]$$

where r_1 and r_2 are the principal radii of curvature at the interface. These radii are not usually measured, and a mean radius of curvature is given by the capillary pressure and interfacial tension.

For a cylindrical vertical capillary, such as a small tube, the capillary pressure for a spherical interface is [19]:

$$P_c = \frac{2\sigma \cos \theta_c}{r} = gh(\rho_1 - \rho_2) \qquad [5.1.75]$$

where r is the radius of the tube, θ_1 is the contact angle measured through the more dense phase that exists between the fluid and the wall of the tube, g is the gravitational constant, ρ is density, h is column height, and the subscripts refer to the fluids of interest. For a fluid that wets the wall of a capillary tube, the attraction between the fluid and the wall causes the fluid to rise in the tube. The extent of rise in the capillary is proportional to the interfacial tension between the fluids and the cosine of the contact angle and is inversely proportional to the tube radius.

An analogous situation can occur during two-phase flow in a porous medium. For example if capillary forces dominate in a water-wet rock, the existing pressure differential causes flow of the wetting fluid to occur through the smaller capillaries. However, if viscous forces dominate, flow will occur through the larger capillaries (from Pouiselle's law, as a function of the 4th power of the radius).

Figure 5.1.46 depicts a typical capillary pressure curve for a core sample in which water is the wetting phase. Variation of capillary pressure is plotted as a function of water saturation. Initially, the core is saturated with the wetting phase (water). The nonwetting phase, oil in this case, is used to displace the water. As shown in the figure, a threshold pressure must be overcome before any oil enters the core. The initial (or primary) drainage curve represents the displacement of the wetting phase from 100% saturation to a condition

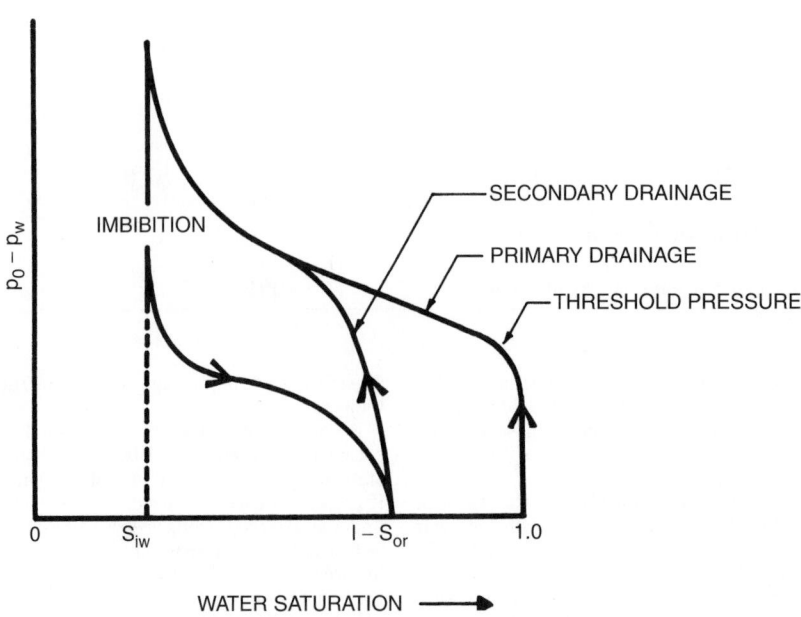

Figure 5.1.46 Example capillary pressure curves.

where further increase in capillary pressure causes little or no change in water saturation. This condition is commonly termed the irreducible saturation, S_{iw}. The imbibition curve reflects the displacement of the nonwetting phase (oil) from the irreducible water saturation to the residual oil saturation. Secondary drainage is the displacement of the wetting phase from the residual oil saturation to the irreducible water saturation. A hysteresis is always noted between the drainage and imbibition curves. Curves can be obtained within the hysteresis loop by reversing the direction of pressure change at some intermediate point along either the imbibition or secondary drainage curve. The nonuniform cross-section of the pores is the basic cause of the hysteresis in capillary pressure observed in porous media. Therefore, capillary pressure depends on pore geometry, interfacial tension between the fluids, wettability of the system (which will be discussed later in this chapter), and the saturation history in the medium.

Leverett [100] introduced a reduced capillary pressure function (subsequently termed the Leverett J function by Rose and Bruce [127]) that was suggested for correlating capillary pressure data:

$$J(S_w) = \frac{P_c}{\sigma \cos \theta_c} \left(\frac{k}{\phi}\right)^{1/2} \qquad [5.1.76]$$

where $J(S_w)$ = the correlating group consisting of the terms of Equation 5.1.75

k = the permeability

ϕ = porosity of the sample

The J function was originally proposed as means of converting all capillary pressure data for clean sand to a universal curve. A series of capillary pressure curves are shown as a function of permeability in Figure 5.1.47 [20]. An example of the J function curve generated from these data is shown in Figure 5.1.48 [20]. While the J function sometimes correlates capillary pressure data from a specific lithology within the same formation, significant variations can be noted for different formations.

Common laboratory methods of measuring capillary pressure include [19]: mercury injection, porous diaphragm or plate (restored state), centrifuge method, and steady-state flow in a dynamic method. While the restored state test is generally considered the most accurate, mercury injection is routinely used. However, it is necessary to correct the mercury injection data for wetting conditions before comparison to results from the restored state test.

A very valuable use of capillary pressure data is to indicate pore size distribution. Since the interfacial tension and contact angle remain constant during a test such as already described, pore sizes can be obtained from capillary pressures. For rocks with more uniform pore sizes, capillary pressure curves will be close to horizontal. The slope of the capillary pressure curve will generally increase with broader pore-size distribution.

If laboratory capillary pressure data are corrected to reservoir conditions, the results can be used for determining fluid saturations. Figure 5.1.49 shows a close agreement in water saturations obtained from capillary pressure and electric logs [48].

Capillary pressure data are helpful in providing a qualitative assessment of the transition zones in the reservoir. A transition zone is defined as the vertical thickness where saturation changes from 100% water to irreducible water for water-oil contact, or from 100% liquid to an irreducible water saturation for gas-oil contact.

5.1.3.6 Effective Permeability

In the previous section, "Absolute Permeability," it was stated that permeability at 100% saturation of a fluid (other than gases at low pressure) is a characteristic of the rock and not a function of the flowing fluid. Of course, this implies that there is no interaction between the fluid and the rock (such as interaction between water and mobile or swelling clays). When permeabilities to gases are measured, corrections must be made for gas slippage which occurs when the

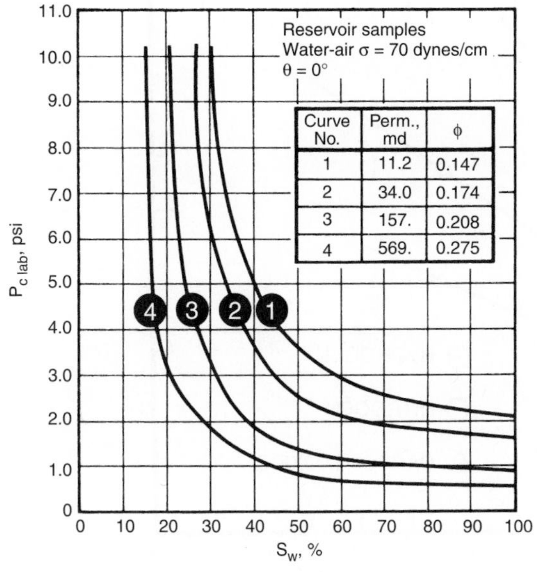

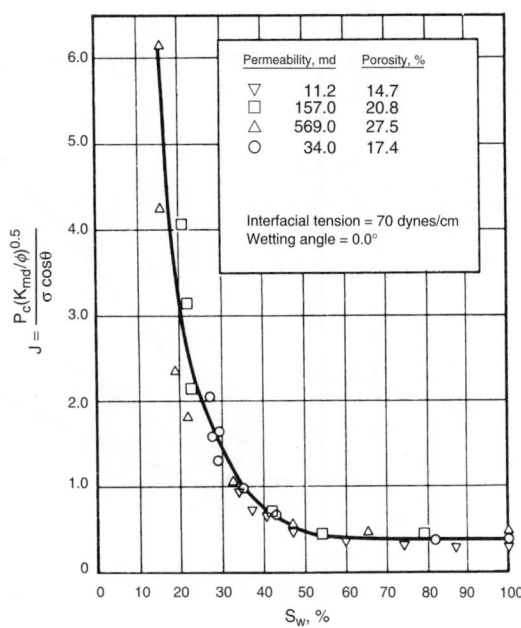

Figure 5.1.47 Capillary pressures of different permeability core samples [20].

Figure 5.1.48 J function plot for data in Figure 5.1.47 [20].

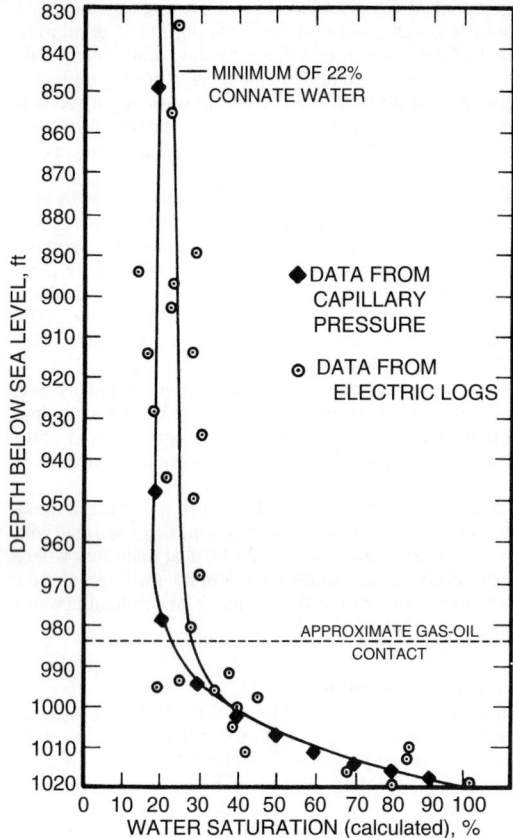

Figure 5.1.49 *Comparison of water saturations from capillary pressure and electric logs [48].*

capillary openings approach the mean free path of the gas. Klinkenberg [128] observed that gas permeability depends on the gas composition and is approximately a linear function of the reciprocal mean pressure. Figure 5.1.50 shows the variation in permeability as a function of mean pressure for hydrogen, nitrogen, and carbon dioxide. Klinkenberg found that by extrapolating all data to infinite mean pressure, the points converged at an equivalent liquid permeability (k_ℓ), which was the same as the permeability of the porous medium to a nonreactive single-phase liquid. From plots of this type, Klinkenberg showed that the equivalent liquid permeability could be obtained from the slope of the data, m, the measured gas permeability, k_g, at a mean flowing pressure $\bar{p}$, at which k_g was observed:

$$k_\ell = \frac{k_g}{1 + (b/\bar{p})} = k_g - \frac{m}{\bar{p}} \qquad [5.1.77]$$

where b is a constant for a given gas in a given medium and is equal to m divided by k_ℓ. The amount of correction, known as the Klinkenberg effect, varies with permeability and is more significant in low permeability formations.

In recent studies [129,130] with very low permeability sandstones, liquid permeabilities were found to be less than gas permeabilities at infinite mean pressure, which is in contrast with the prior results of Klinkenberg. Furthermore, it has been shown [130] that liquid permeabilities decreased with increasing polarity of the liquid. For gas flow or brine flow in low-permeability sandstones, permeabilities were independent of temperature at all levels of confining pressure [130]. The data [130] showed that for a given permeability core sample at a given confining pressure, the Klinkenberg slip factors and slopes of the Klinkenberg plots were proportional to the product of viscosity and the square root of absolute temperature.

As shown in Figure 5.1.51 permeability of reservoir rocks can decrease when subjected to overburden pressure [131]. When cores are retrieved from a reservoir, the confining forces are removed and the rock can expand in all directions

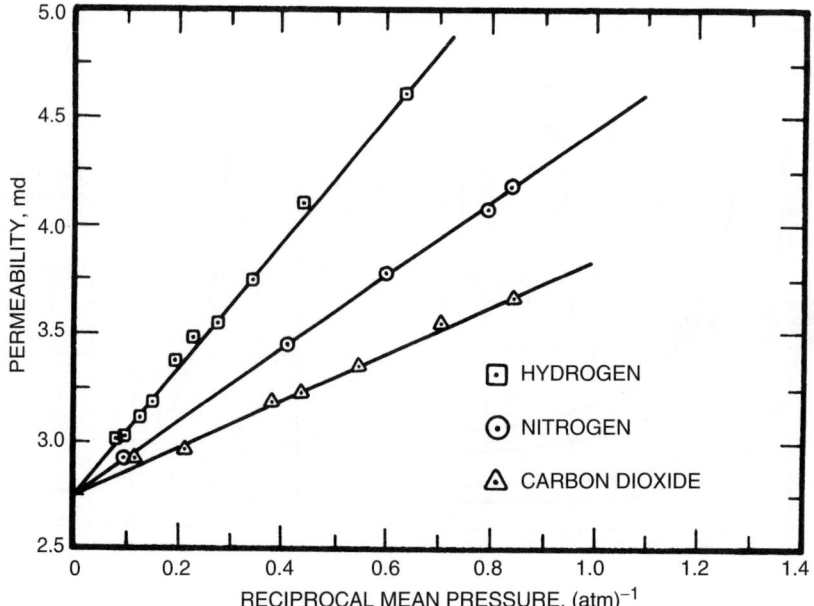

Figure 5.1.50 *Gas slippage in core [128].*

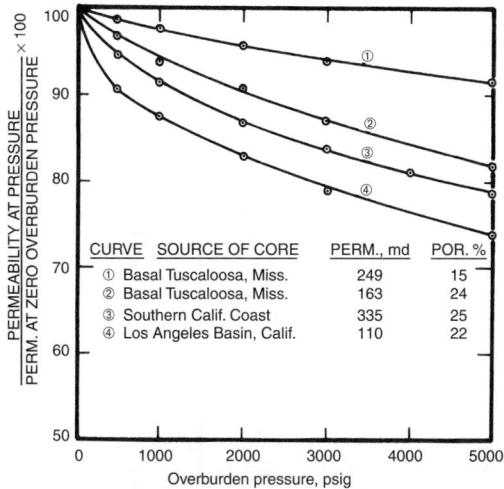

Figure 5.1.51 *Air permeability at different overburden pressures [131].*

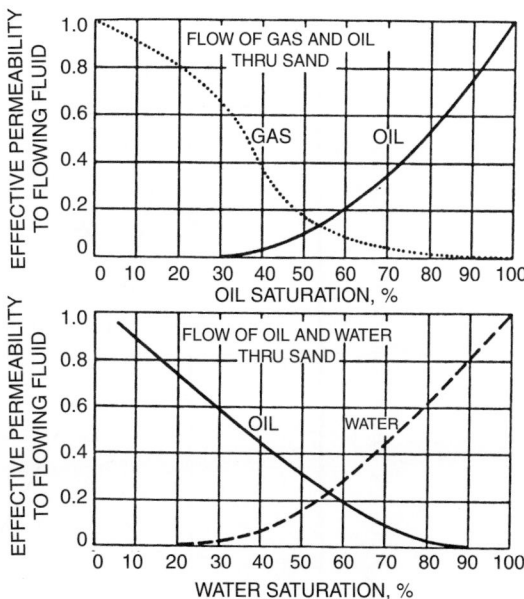

Figure 5.1.53 *Effective permeabilities.*

The fluid system of an oil reservoir consists usually of gas, oil, and water. In the case of such a heterogeneous system, flow of the different phases is a function of fluid saturation in the reservoir by the different phases. The lower the saturation of a certain liquid, as compared to other liquids, the lower the permeability to that liquid. This type of permeability is termed effective permeability and is defined as permeability of the rock to one liquid under conditions of saturation when more than one liquid is present. Typical permeability-saturation relations for oil and gas and for oil and water are shown in Figure 5.1.53.

From the practical point of view, permeability may be considered as a measure of productivity of the producing horizon. Knowledge of permeability is useful in a number of reservoir problems. The concept of effective permeability is of particular importance since it emphasizes a need for production practices, which tend to maintain good permeability of the reservoir to oil.

Thus, the absolute permeability is the permeability measured when the medium is completely saturated with a single fluid. Effective permeability is the permeability to a particular fluid when another fluid is also present in the medium. For example, if both oil and water are flowing, the effective permeability to oil is k_o and that to water is k_w. The sum of the effective permeabilities is always less than the absolute permeability [17]. As noted in the previous section, permeability is commonly expressed in millidarcies (md).

5.1.3.7 Relative Permeability

If the effective permeabilities are divided by a base permeability (i.e., the absolute permeability), the dimensionless ratio is referred to as the relative permeability, namely k_{rg} for gas, k_{ro} for oil, and k_{rw} for water:

$$k_{rg} = \frac{k_g}{k}; \qquad k_{ro} = \frac{k_o}{k}; \qquad k_{rw} = \frac{k_w}{k} \qquad [5.1.78]$$

where k_g, k_o, and k_w are the effective permeabilities to gas, oil, and water, respectively, and k is some base permeability that represents the absolute permeability. For gas-oil two-phase relative permeabilities, the base permeability is often the equivalent liquid permeability. For oil-water two-phase

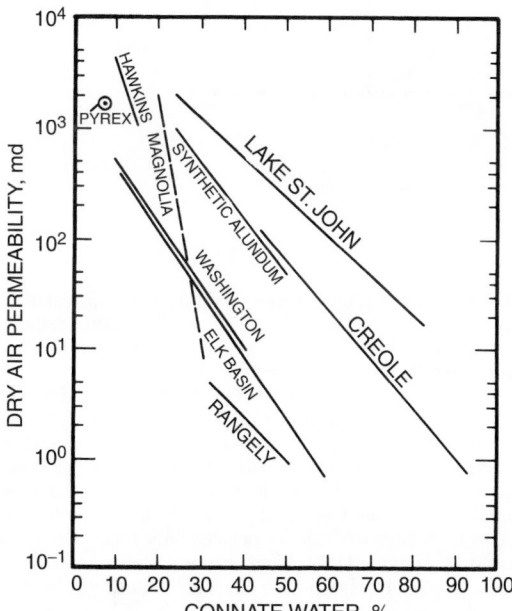

Figure 5.1.52 *Relationship between air permeability and connate water saturation [132].*

which can increase the dimensions of the available flow paths. In reservoirs where this is significant, it is imperative that permeability measured in the laboratory be conducted at the confining pressure that represents the overburden pressure of the formation tested.

As a general trend, air permeability decreases with increasing connate water saturation. Relationships between air permeability and connate water saturation in Figure 5.1.52 show a linear decrease in the logarithm of permeability as a function of water saturation that depends on the individual field [132].

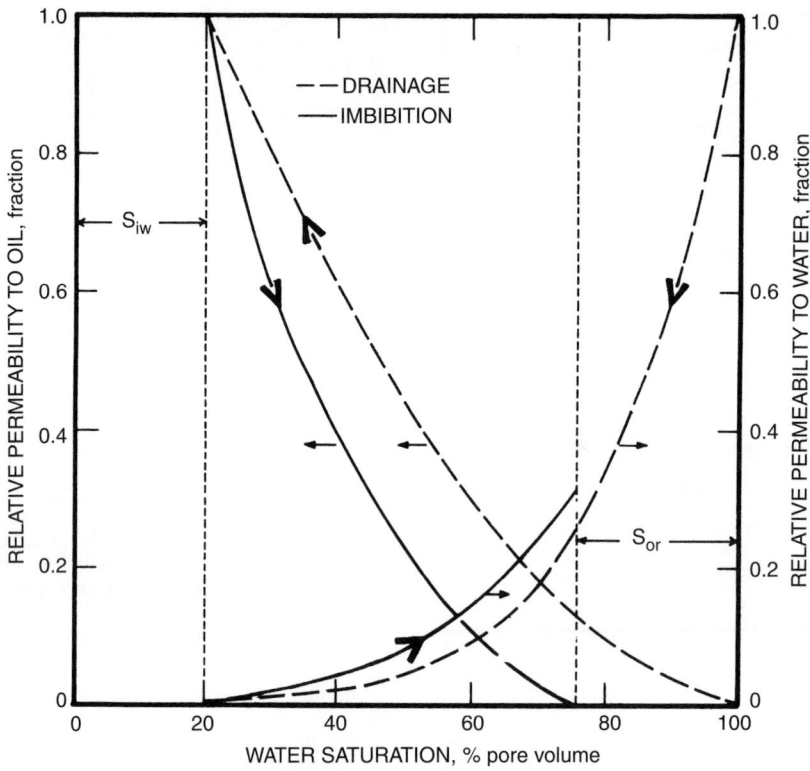

Figure 5.1.54 *Example of water-oil relative permeability data.*

relative permeabilities, three different base permeabilities are often used [133]:

1. The permeability to air with only air present.
2. The permeability to water at 100% S_w.
3. The permeability to oil at irreducible water saturation.

Wyckoff and Botset [76] are generally credited with performing the first gas and liquid relative permeabilities which were conducted in unconsolidated sandpacks in 1936. In these early experiments, a relationship was observed between the liquid saturation of a sand and the permeability to a liquid or gas phase [76,134]. At about the same time, Hassler, Rice, and Leeman [135] measured relative air permeabilities in oil-saturated cores. In 1940, relative permeability measurements were extended to consolidated cores by Botset [136]. Since then, a number of dynamic (fluid displacement or fluid drive) methods [83,137–143] and static (or stationary-phase) methods [144–150] have been proposed to determine relative permeabilities in core samples. In the latter methods, only the nonwetting phase is allowed to flow by the use of a very low pressure drop across the core; hence, this method is applicable only to the relative permeability of the nonwetting phase. The dynamic methods include: (1) steady-state methods in which fluids are flowed simultaneously through a core sample at a fixed gas-oil or water-oil ratio until equilibrium pressure gradients and saturations are achieved, and (2) unsteady-state methods in which an oil-saturated core is flooded with either gas or water at a fixed pressure drop or flow rate so that the average fluid saturation changes result in a saturation gradient. The most popular steady-state procedure is the Penn

State method [83], but the most common dynamic test is the unsteady-state method because of the reduced time requirement. The various methods have been evaluated [139,151] and generally provide similar results.

Based on the initial work of Leverett [100] and Buckley and Leverett [152], Welge [153] was the first to show how to calculate relative permeability ratios in the absence of gravity effects. Subsequently, Johnson, Bossler, and Naumann [154] showed that each of the relative permeabilities could be calculated even when gravity is not neglected. Other calculations of relative permeabilities have been proposed by Higgins [155], Guerrero and Stewart [156,157], and a graphical technique has been presented by Jones and Rozelle [158].

An example of water-oil relative permeability plot vs. water saturation is given in Figure 5.1.54. Several features will be described that pertain to generating relative permeability curves from cores in the laboratory. If a clean, dry core is completely saturated with water, the permeability at 100% S_w should be similar to the equivalent liquid permeability obtained from gas flow measurements at 100% S_g. Exceptions to this generality include some low-permeability systems and other cores that contain clays or minerals that interact with the water used. If a clean core is used, it will probably be strongly water-wet when saturated with brine. As crude oil is injected into the core, the relative permeability to water decreases during the drainage cycle (decreases in wetting phase) while the relative permeability to oil increases. Some water that resides in the nooks and crannies of the pore space cannot be displaced by the oil, regardless of the throughput volume. This water saturation, which does not contribute

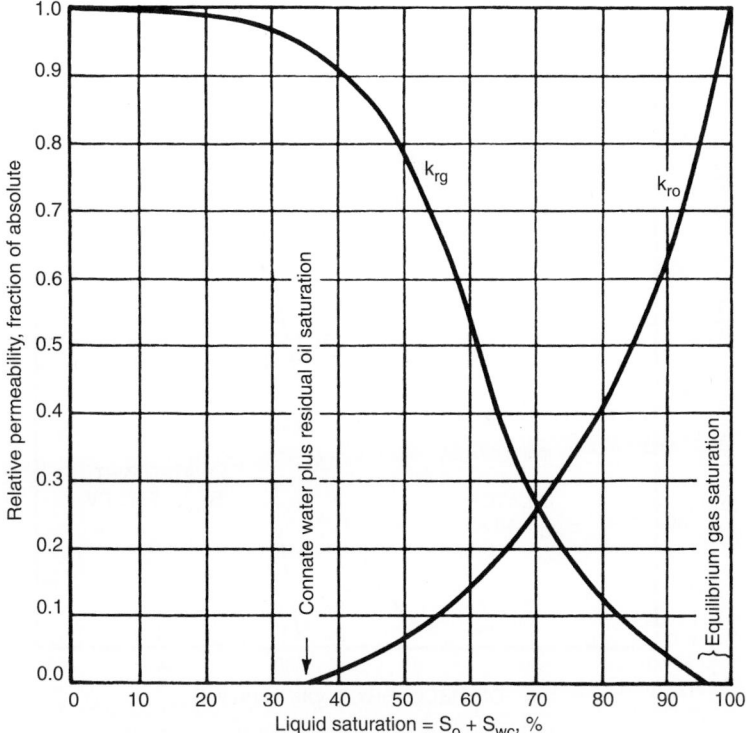

Figure 5.1.55 *Gas-oil relative permeability data [20].*

significantly to occupying the flow paths, is called the irreducible water saturation, S_{iw}. With the core at S_{iw}, there is 100% relative permeability to oil (only oil is flowing) and no permeability to water. At this point, the core can be closed in for about 1,000 hours to allow sufficient time for wettability changes to occur. Then the core is flooded with water in an unsteady-state test, or fixed ratios of water and oil are injected in the steady-state test. If water continues to be the wetting phase, the relative permeability to water (which will be only a function of saturation) will be the same during the drainage and imbibition cycles. (The importance of wettability on relative permeabilities will be discussed in the next section of this chapter.) As the water is injected into the oil-flooded core, k_{rw} increases while k_{ro} decreases. Not all of the oil can be displaced from the core, regardless of the water throughput (at modest flow rates or pressure drops), and this is referred to as the waterflood residual oil saturation, S_{or}.

Similar observations apply to gas-oil relative permeability data as displayed in Figure 5.1.55. Typically, the gas-oil relative permeabilities are plotted against the total liquid saturation, which includes not only the oil but also any connate water that may be present. In the presence of gas, the oil (even if connate water is present) will be the wetting phase in preference to gas. As a result, the k_{ro} curve from gas-oil flow tests resembles the drainage k_{rw} curve from oil-water flow tests. As seen in Figure 5.1.56, the irreducible gas saturation (also called the equilibrium or critical gas saturation) is usually very small. When gas saturation is less than the critical value, gas is not mobile but does impede oil flow and reduces k_{ro}.

Three-phase relative permeabilities pertaining to simultaneous flow of gas, oil, and water have been provided in the

literature [19,50]. Since the occurrence of such three-phase flow in a reservoir is limited [20], relative permeabilities for these conditions will not be discussed in this chapter, and the reader is referred to other sources [19,20,137,140, 159,160].

Based on the work of Corey [150] and Wyllie [23,50], empirical equations have been summarized by Slider [20] to estimate relative permeabilities. These equations permit the estimation of k_{ro} from measurements of k_{rg}. Other empirical equations for estimating two-phase relative permeabilities in consolidated rocks are available in the literature [161].

Early work in unconsolidated sands showed that fluid viscosity or the range of permeability had negligible effects on the relationship between relative permeability and fluid saturation [76,100]. Geffen et al. [141] confirmed that relative permeabilities in cores are not affected by fluid properties provided wettability is not altered. However, others [162,163] have found that viscosity ratio influences relative permeability data when the displacing fluid is non-wetting. For constant wettability conditions, the higher the viscosity of one of the liquids, the lower is the relative permeability of the other liquid [163].

Geffen et al. [141] did cite a number of factors, in addition to fluid saturations, that affect relative permeability results. Because of capillary hysteresis, saturation history was important in that fluid distribution in the pores was altered. Flow rates during laboratory tests need to be sufficiently high to overcome capillary end effects (retention of the wetting phase at the outlet end of the core) [141]. According to Wyllie [23], relative permeability varies because of varying geometry of the fluid phases present, which is controlled by effective pore size distribution, method of obtaining the saturation (saturation history);

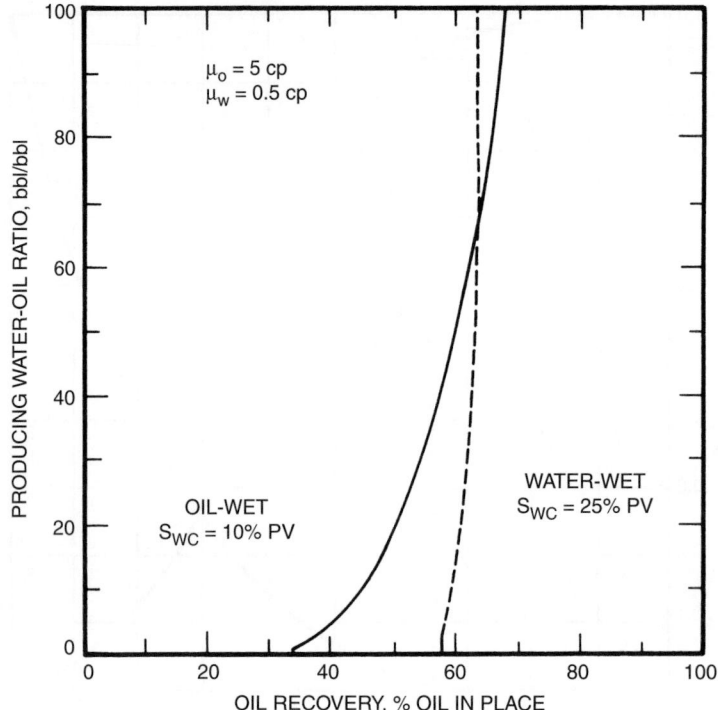

Figure 5.1.56 *Comparison of oil recovery for oil-wet and water-wet cores [133].*

heterogeneity of the core sample, and wettability of the rock-fluid system.

Controversy continues to exist regarding the effect of temperature on relative permeabilities (for example see the discussion and prior citations in References 164 and 165). Miller and Ramey [164] observed no change in relative permeability with temperatures for clean systems, and speculated that for reservoir fluid/rock systems, effects such as clay interactions or pore structure would need to be considered. Honarpour et al. [165] summarized the effects of temperature on two-phase relative permeabilities as measured by various researchers. In field situations, the larger overburden pressure associated with greater depths may be more important in affecting relative permeabilities than the associated temperature increases. As with many other tests, relative permeability measurements should be conducted at reservoir conditions of overburden pressure and temperature with crude oil and brine representative of the formation under study.

5.1.3.8 Effect of Wettability on Fluid-Rock Properties
Oil Recovery and Fluid Saturations
Since a reservoir rock is usually composed of different minerals with many shapes and sizes, the influence of wettability in such systems is difficult to assess fully. Oil recovery at water breakthrough in water-wet cores is much higher than in oil-wet cores [106,110,166–169], although the ultimate recovery after extensive flooding by water may be similar, as shown in Figure 5.1.56. These authors have shown that oil recovery, as a function of water injected, is higher from water-wet cores than from oil-wet cores at economical water-oil ratios. In 1928, Uren and Fahmy [170] observed better recovery of oil from unconsolidated sands that had

an intermediate wettability, and several investigators have suggested better oil recovery from cores of intermediate or mixed wettability [106,108,121,123,171]. More recent evidence by Morrow [172] and Melrose and Brandner [173] suggests that mobilization of trapped oil is more difficult in the intermediate wettability region, but the prevention of oil entrapment should be easier for advancing contact angles slightly less than 90°.

With water-wet cores in laboratory waterfloods, the oil production at water breakthrough ceases abruptly and water production increases sharply. With systems that are not water-wet, water breakthrough may occur earlier, but small fractions of oil are produced for long periods of time at high water cuts. In strongly water-wet systems, the residual oil that is permanently trapped by water resides in the larger pores, whereas in oil-wet systems trapping occurs in the smaller capillary spaces [106,133].

Relative Permeability Characteristics
For a system having a strong wetting preference for either oil or water, relative permeability of the wetting phase is a function of fluid saturation only [76,137,160]. Details of the effect of wettability on relative permeabilities have been discussed by several authors [113,133,174,175]. In a detailed study using fired Torpedo (outcrop sandstone) cores, Owens and Archer [174] changed wettability by adding surface active agents to either the oil or water. Firing of the cores stabilized any clay minerals present and provided more constant internal conditions. Both gas-oil and water-oil relative permeabilities were obtained. Some of the water-oil relative permeability data, all started at the same water saturation and obtained with the Penn State steady-state method, are reproduced in Figure 5.1.57. As the contact angle was

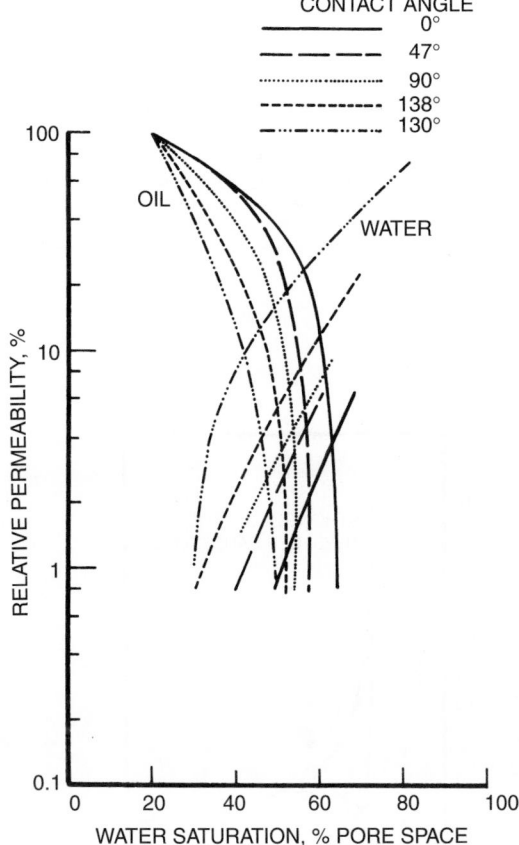

Figure 5.1.57 *Effect of wettability on water-oil relative permeabilities [174].*

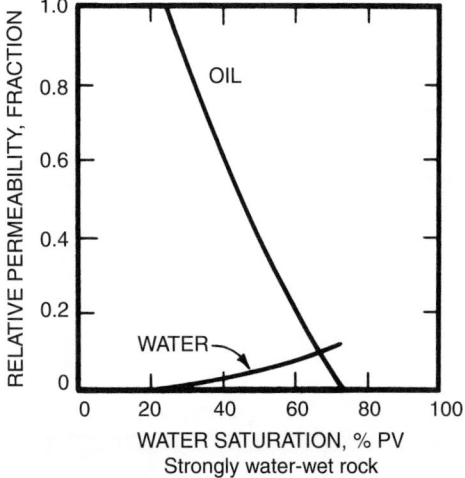

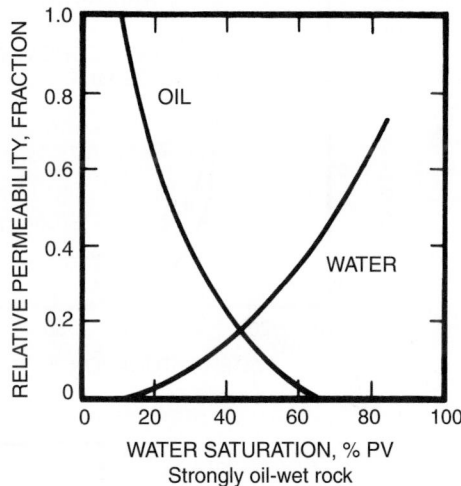

Figure 5.1.58 *Strongly water-wet and strongly oil-wet relative permeability curves [133].*

increased to create more oil-wet conditions the effective permeability to oil decreased. Because of the differences in flow paths for the different wettability conditions, oil-wet systems had lower k_{ro} and higher k_{rw} when compared to water-wet conditions at the same water saturation. As the level of oil-wetting increased, k_{ro} at any saturation decreased whereas k_{rw} increased.

Craig [133] has presented typical relative permeability curves, such as given in Figure 5.1.58, to point out differences in strongly water-wet and strongly oil-wet conditions. Craig suggests that several rules of thumb can help in distinguishing wetting preferences; typical characteristics of water-oil relative permeability curves are given in Table 5.1.18 [133]. Additionally, the strongly oil-wet relative permeability curves tend to have more curvature. Craig suggests the generality that relative permeability curves of intermediate wettability systems will have some of the characteristics of both water-wet and oil-wet conditions. However, as mentioned earlier in the section "Wettability and Contact Angle," a speckled wettability form of intermediate wetting mimics the relative permeability characteristics of strongly water-wet conditions [122].

Capillary Pressure Curves

By convention, oil-water capillary pressure, P_c, is defined as the pressure in the oil phase, p_o, minus the pressure in the water phase, p_w:

$$P_c = p_o - p_w \qquad [5.1.79]$$

Depending on wettability and history of displacement, capillary pressure can be positive or negative. Figure 5.1.59 presents the effect of wettability on capillary pressure as related by Killins, Nielsen, and Calhoun [176]. Drainage and imbibition curves can have similarities, but the capillary pressure values are positive for strongly water-wet and negative for strongly oil-wet conditions. In the intermediate wettability case shown in Figure 5.1.59, the small positive value of threshold pressure during the drainage cycle suggests the sample was moderately water-wet [133]. After the drainage cycle, the sample spontaneously imbibed water until the capillary pressure was zero at a water saturation of 55%. Then, as water pressure was applied, the maximum water saturation of about 88% was achieved. As discussed previously, capillary pressure curves can be used as a criterion of wettability.

Table 5.1.18 *Typical Water-Oil Relative Permeability Characteristics*

	Strongly Water-Wet	Strongly Oil-Wet
Connate water saturation.	Usually greater than 20% to 25% PV.	Generally less than 15% PV, frequently less than 10%.
Saturation at which oil and water relative permeabilities are equal.	Greater than 50% water saturation.	Less than 50% water saturation.
Relative permeability to water at maximum water saturation; i.e., floodout.	Generally less than 30%.	Greater than 50% and approaching 100%.

From Reference 133.

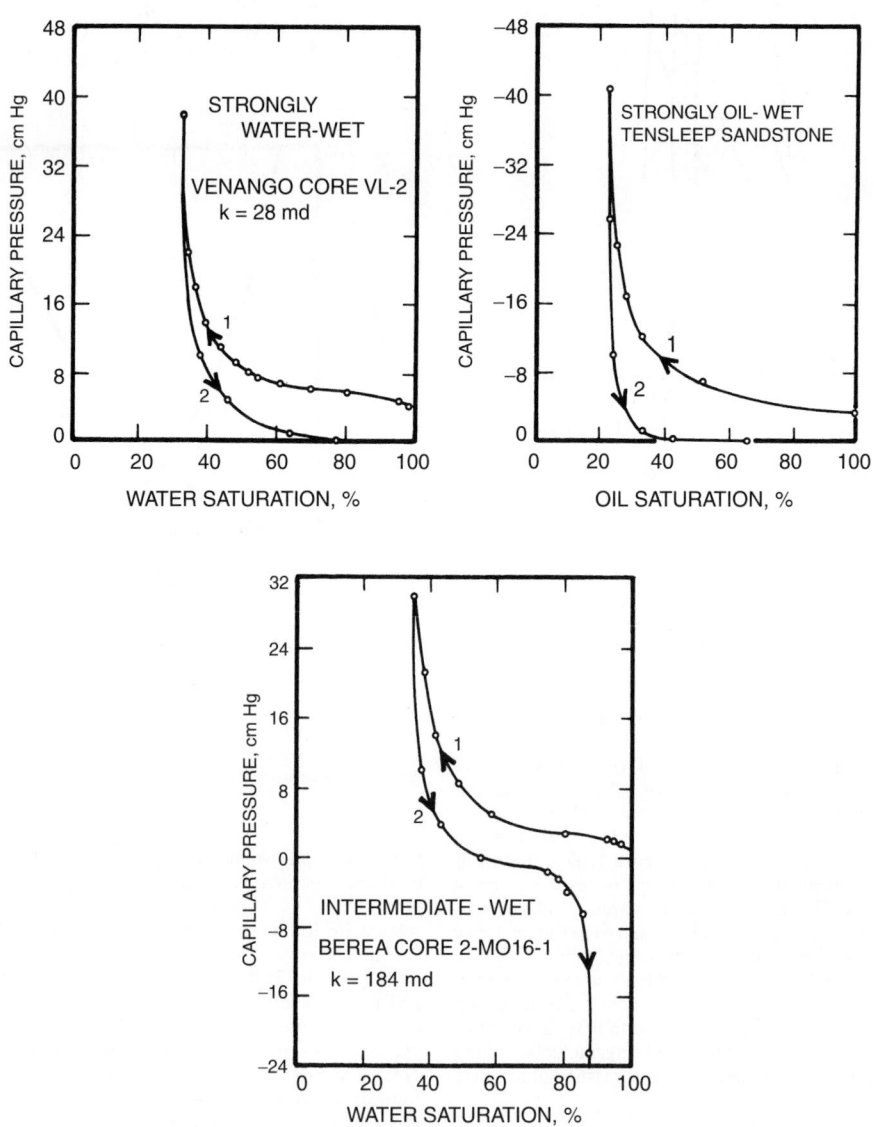

Figure 5.1.59 *Effect of wettability on capillary pressure [176].*

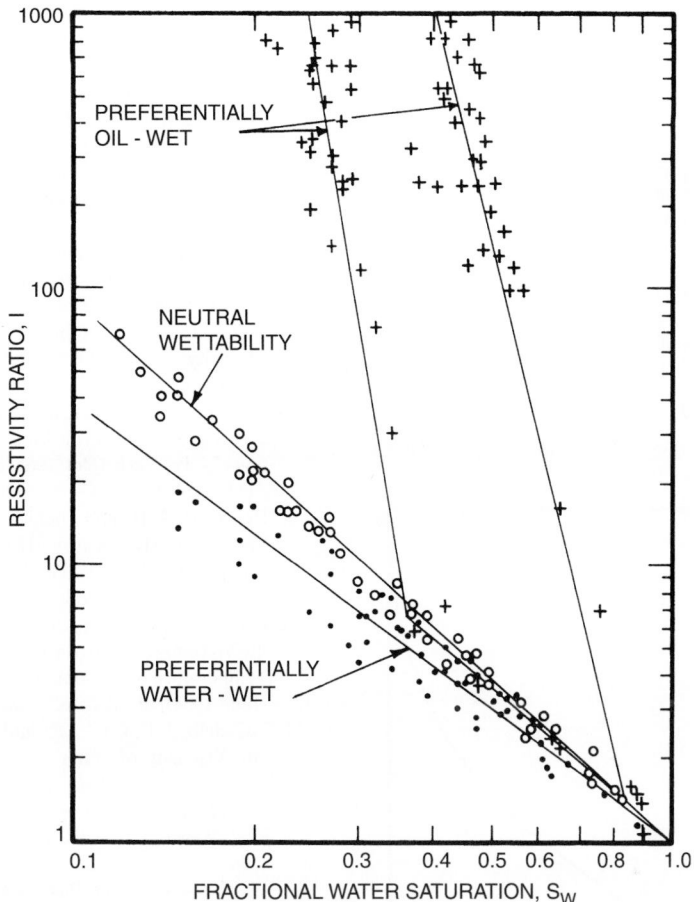

Figure 5.1.60 *Resistivity-saturation relationships for different wettability carbonate cores [51].*

Resistivity Factors and Saturation Exponents

As shown previously in Table 5.1.7, Sweeney and Jennings [51] found that the formation resistivity factor changed when wettability was altered. However, the naphthenic acid they used to alter wettability may have also reduced porosity which could account for the increase in the saturation exponent in Equation 5.1.46. Other investigators [177,178] have found no significant effect of wettability on formation factors. Because of the scarcity of data and the difficulty of altering wettability without affecting other properties, the effect of wettability on formation resistivity remains unclear.

As related by Mungan and Moore [178], three assumptions are made in the Archie saturation expression (Equation 5.1.68): all of the brine contributes to electrical flow; the saturation exponent, n, is constant for a given porous media; and only one resistivity is measured for a given saturation. Since the saturation exponent depends on the distribution of the conducting phase, which is dependent on wettability, the foregoing assumptions are valid only for strongly water-wet conditions. When wettability is altered, the differences in fluid distribution cause variations in the cross-sectional areas of conductive paths and in the tortuous path lengths. These variations affect resistivity, which results in different resistivity-saturation relationships such

as were presented for carbonate cores by Sweeney and Jennings [51] in Figure 5.1.60. The saturation exponent, which is the slope of the lines, was about 1.6 for water-wet cores, about 1.9 for neutral-wet cores, and about 8 for oil-wet cores [51]. Similar data in sandstone cores were provided by Rust [177]; saturation exponents were about 1.7 and 13.5 for water-wet and oil-wet conditions respectively. These differences in oil-wet rocks most likely occur because of the isolation of trapped brine in dendritic fingers or dead-end pores which do not contribute to electrical conductivity.

Table 5.1.19, presenting data given by Mungan and Moore [178], shows that the effect of wettability on saturation exponent becomes more important at low brine saturations. Morgan and Pirson [179] conducted tests on fractionally wetted bead packs in which portions of the beads were water-wet and portions were treated so that they were mildly oil-wet. From their data, plotted in Figure 5.1.61, the saturation exponent increased as the extent of oil-wetting increased.

The foregoing data suggest that unless the reservoir is known to be water-wet, the saturation exponent should be measured with native-state (preferably) or restored cores. If the reservoir is oil-wet and clean cores are used that may be water-wet, the saturation exponents that are obtained can

Table 5.1.19 *Saturation Exponents in Teflon Cores Partially Saturated with Nonwetting Conducting Liquid*

Air-NaCl solution		Oil-NaCl solution	
Brine saturation % PV	n	Brine saturation % PV	n
66.2	1.97	64.1	2.35
65.1	1.98	63.1	2.31
63.2	1.92	60.2	2.46
59.3	2.01	55.3	2.37
51.4	1.93	50.7	2.51
43.6	1.99	44.2	2.46
39.5	2.11	40.5	2.61
33.9	4.06	36.8	2.81
30.1	7.50	34.3	4.00
28.4	8.90	33.9	7.15
		31.0	9.00

From Reference 178.

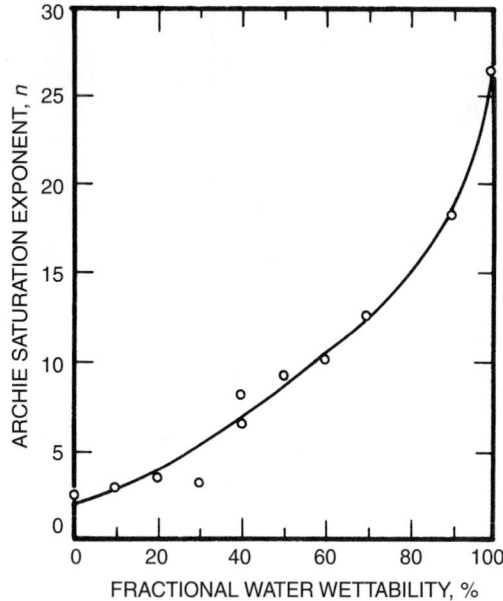

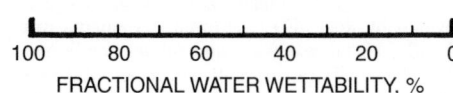

Figure 5.1.61 *Influence of wettability on saturation exponent [179].*

lead to an underestimate of connate water saturation in the formation tested.

References

1. Standing, M. B., "A Pressure-Volume-Temperature Correlation for Mixtures of California Oils and Gases," *Drill. & Prod. Prac.*, API (1947), pp. 275–287.

2. Standing, M. B., *Volumetric and Phase Behavior of Oil Field Hydrocarbon Systems*, Reinhold Publishing Corp., New York (1952).

3. Katz, D. L., et al., *Handbook of Natural Gas Engineering*, McGraw-Hill Book Co., Inc., New York (1959).

4. Hocott, C. R., and Buckley, S. E., "Measurements of the Viscosities of Oils under Reservoir Conditions," *Trans.*, AIME, Vol. 142 (1941), pp. 131–136.

5. Beal, C., "The Viscosity of Air, Water, Natural Gas, Crude Oil and Its Associated Gases at Oil Field Temperature and Pressures," *Trans.*, AIME, Vol. 165 (1946), pp. 94–115.

6. Chew, J. N., and Connally, C. A., "A Viscosity Correlation for Gas-Saturated Crude Oils," *Trans.*, AIME, Vol. 216 (1959), pp. 23–25.

7. Beggs, H. D., and Robinson, J. R., "Estimating the Viscosity of Crude Oil Systems," *J. Pet. Tech.* (Sept. 1975), pp. 1140–1141.

8. Lohrenz, J., Bray, B. G., and Clark, C. R., "Calculating Viscosities of Reservoir Fluids from Their Compositions," *J. Pet. Tech.* (Oct. 1964), pp. 1171–1176; *Trans.*, AIME, Vol. 231.

9. Houpeurt, A. H., and Thelliez, M. B., "Predicting the Viscosity of Hydrocarbon Liquid Phases from Their Composition," paper SPE 5057 presented at the SPE 49th Annual Fall Meeting, Houston, Oct. 6–9, 1974.

10. Little, J. E., and Kennedy, H. T., "A Correlation of the Viscosity of Hydrocarbon Systems with Pressure, Temperature and Composition," *Soc. Pet. Eng. J.* (June 1968), pp. 157–162; *Trans.*, AIME, Vol. 243.

11. Swindells, J. F., Coe, J. R., and Godfrey, T. B., "Absolute Viscosity of Water at 20°C," Res. Paper 2279, *J. Res. Nat'l Bur. Stnds.*, Vol. 48 (1952), pp. 1–31.

12. *CRC Handbook of Chemistry and Physics*, 62nd edition, R. C. Weast (Ed.), CRC Press, Inc., Boca Raton, FL (1982).

13. Matthews, C. S., and Russell, D. G., *Pressure Buildup and Flow Tests in Wells*, Monograph Series, SPE, Dallas (1967), p. 1.

14. Pirson, S. J., *Oil Reservoir Engineering*, McGraw-Hill Book Co., Inc., New York (1958).

15. Kirkbride, C. G., *Chemical Engineering Fundamentals*, McGraw-Hill Book Co., Inc., New York (1947), p. 61.

16. Standing, M. B., and Katz, D. L., "Density of Natural Gases," *Trans.*, AIME, Vol. 146 (1942), pp. 140–149.

17. Craft, B. C., and Hawkins, M. F., *Applied Petroleum Reservoir Engineering*, Prentice-Hall, Inc., Englewood Cliffs (1959).

18. McCain, W. D., *The Properties of Petroleum Fluids*, Petroleum Publishing Co., Tulsa (1973).

19. Amyx, J. W., Bass, D. M., Jr., and Whiting, R. L., *Petroleum Reservoir Engineering*, McGraw-Hill Book Co., Inc., New York (1960).

20. Slider, H. C., *Practical Petroleum Engineering Methods*, Petroleum Publishing Co., Tulsa (1976).

21. Taylor, L. B., personal communication.

22. Dodson, C. R., and Standing, M. B., "Pressure-Volume-Temperature and Solubility Relations for Natural Gas-Water Mixtures," *Drill. & Prod Prac.*, API (1944), pp. 173–179.

23. *Petroleum Production Handbook*, T. C. Frick (Ed.), Vol. II, Reservoir Engineering, SPE, Dallas (1962).

24. Burcik, E. J., *Properties of Petroleum Fluids*, Int'l Human Resources Dev. Corp., Boston (1979).

25. Muskat, M., *Physical Principles of Oil Production*, McGraw-Hill Book Co., Inc., New York (1949).

26. Ramey, H. J., Jr., "Rapid Methods for Estimating Reservoir Compressibilities," *J. Pet. Tech.* (April 1964), pp. 447–454.

27. Trube, A. S., "Compressibility of Natural Gases," *Trans.*, AIME, Vol. 210 (1957), pp. 355–357.

28. Trube, A. S., "Compressibility of Undersaturated Hydrocarbon Reservoir Fluids," *Trans.*, AIME, Vol. 210 (1957), pp. 341–344.

29. "A Correlation for Water Viscosity," *Pet. Eng.* (July 1980), pp. 117–118.

30. Standing, M. B., *Volumetric and Phase Behavior of Oil Field Hydrocarbon Systems*, SPE, Dallas (1977).

31. Carr, N. L., Kobayashi, R., and Burrows, D. B., "Viscosity of Hydrocarbon Gases Under Pressure," *Trans.*, AIME, Vol. 201 (1954), pp. 264–272.

32. Hollo, R., and Fifadara, H., *TI-59 Reservoir Engineering Manual*, PennWell Books, Tulsa (1980).

33. Vazquez, M., and Beggs, H. D., "Correlations for Fluid Physical Property Prediction," *J. Pet. Tech.* (June 1980), pp. 968–970; paper SPE 6719 presented at the 52nd Annual Fall Tech. Conf. and Exhibition, Denver, Oct. 9–12, 1977.

34. Meehan, D. N., "Improved Oil PVT Property Correlations," *Oil & Gas J.* (Oct. 27, 1980), pp. 64–71.

35. Meehan, D. N., "Crude Oil Viscosity Correlation," *Oil & Gas J.* (Nov. 10, 1980), pp. 214–216.

36. Meehan, D. N., and Vogel, E. L., *HP-41 Reservoir Engineering Manual*, PennWell Books, Tulsa (1982).

37. Garb, F. A., *Waterflood Calculations for Hand-Held Computers*, Gulf Publishing Co., Houston (1982).

38. Hollo, R., Homes, M., and Pais, V., *HP-41CV Reservoir Economics and Engineering Manual*, Gulf Publishing Co., Houston (1983).

39. McCoy, R. L., *Microcomputer Programs for Petroleum Engineers, Vol 1. Reservoir Engineering and Formation Evaluation*, Gulf Pub. Co., Houston (1983).

40. Sinha, M. K., and Padgett, L. R., *Reservoir Engineering Techniques Using Fortran*, Int'l Human Resources Dev. Corp., Boston (1985).

41. Smith, H. I., "Estimating Flow Efficiency From Afterflow-Distorted Pressure Buildup Data," *J. Pet. Tech.* (June 1974), pp. 696–697.

42. Archie, G. E., "The Electrical Resistivity Log as an Aid in Determining Some Reservoir Characteristics," *Trans.*, AIME (1942), pp. 54–61.

43. Wyllie, M. R. J., and Rose, W. D., "Some Theoretical Considerations Related to the Quantitative Evaluation of the Physical Characteristics of Reservoir Rock from Electrical Log Data," *Trans.*, AIME, Vol. 189 (1950), pp. 105–118.

44. Tixier, M. P., "Porosity Index in Limestone from Electrical Logs," *Oil & Gas J.* (Nov. 15, 1951), pp. 140–173.

45. Winsauer, W. O., "Resistivity of Brine-Saturated Sands in Relation to Pore Geometry," *Bull.*, AAPG, Vol. 36, No. 2 (1952), pp. 253–277.

46. Wyllie, M. R. J., and Gregory, A. R., "Formation Factors of Unconsolidated Porous Media: Influence of Particle Shape and Effect of Cementation," *Trans.*, AIME, Vol. 198 (1953), pp. 103–109.

47. Cornell, D., and Katz, D. L., "Flow of Gases Through Consolidated Porous Media," *Ind. & Eng. Chem.*, Vol. 45 (Oct. 1953), pp. 2145–2152.

48. Owen, J. D., "Well Logging Study-Quinduno Field, Roberts County, Texas," Symp. on Formation Evaluation, AIME (Oct. 1955).

49. Hill, H. J., and Milburn, J. D., "Effect of Clay and Water Salinity on Electrochemical Behavior of Reservoir Rocks," *Trans.*, AIME, Vol. 207 (1956), pp. 65–72.

50. Wyllie, M. R. J., and Gardner, G. H. F., "The Generalized Kozeny-Carman Equation, Part 2–A Novel Approach to Problems of Fluid Flow," *World Oil* (April 1958), pp. 210–228.

51. Sweeney, S. A., and Jennings, H. Y., Jr., "Effect of Wettability on the Electrical Resistivity of Carbonate Rock from a Petroleum Reservoir," *J. Phys. Chem.*, Vol. 64 (1960), pp. 551–553.

52. Carothers, J. E., "A Statistical Study of the Formation Factor Relation," *The Log Analyst* (Sept.–Oct. 1968), pp. 13–20.

53. Porter, C. R., and Carothers, J. E., "Formation Factor-Porosity Relation Derived from Well Log Data," *The Log Analyst* (Jan.–Feb. 1971), pp. 16–26.

54. Timur, A., Hemkins, W. B., and Worthington, A. W., "Porosity and Pressure Dependence of Formation Resistivity Factor for Sandstones," Proc. Cdn. Well Logging Soc., Fourth Formation Evaluation Symposium, Calgary, Alberta, May 9–12, 1972.

55. Perez-Rosales, C., "On the Relationship Between Formation Resistivity Factor and Porosity," *Soc. Pet. Eng. J.* (Aug. 1982), pp. 531–536.

56. Coates, G. R., and Dumanoir, J. L., "A New Approach to Log-Derived Permeability," *Trans.*, SPWLA 14th Annual Logging Symp. (May 6–9, 1973), pp. 1–28.

57. Amyx, J. W., and Bass, D. M., Jr., "Properties of Reservoir Rocks," Chap. 23 in *Pet. Prod. Handbook*, T. C. Frick and R. W. Taylor (Eds.), SPE, Dallas, 2 (1962) 23/1–23/40.

58. *Log Interpretation–Principles and Applications*, Schlumberger Educational Services, Houston (1972).

59. Patnode, H. W., and Wyllie, M. R. J., "The Presence of Conductive Solids in Reservoir Rocks as a Factor in Electric Log Interpretation," *Trans.*, AIME, Vol. 189 (1950), pp. 47–52.

60. Winn, R. H., "The Fundamentals of Quantitative Analysis of Electric Logs," *Proc.*, Symposium on Formation Evaluation (Oct. 1955), pp. 35–48.

61. Atkins, E. R., and Smith, G. H., "The Significance of Particle Shape in Formation Resistivity Factor-Porosity Relationships," *J. Pet . Tech.* (March 1961); *Trans.*, AIME, Vol. 222, pp. 285–291.

62. Koepf, E. H., "Core Handling-Core Analysis Methods," Chap. 3 in *Determination of Residual Oil Saturation*, Interstate Oil Compact Commission, Oklahoma City, Oklahoma (1978), pp. 36–71.

63. Sanyal, S. K., Marsden, S. S., Jr., and Ramey, H. J., Jr., "The Effect of Temperature on Electrical Resistivity of Porous Media," *The Log Analyst* (March–April 1973), pp. 10–24.

64. Hilchie, D. W., "The Effect of Pressure and Temperature on the Resistivity of Rocks," Ph.D. Dissertation, The University of Oklahoma (1964).

65. Ransom, R. C., "A Contribution Toward a Better Understanding of the Modified Archie Formation Resistivity Factor Relationship," *The Log Analyst* (March–April 1984), pp. 7–12.

66. Earlougher, R. C., Jr., *Advances in Well Test Analysis*, Monograph Series, SPE, Dallas (1977), Vol. 5.

67. Fatt, I., "Pore Volume Compressibilities of Sandstone Reservoir Rock," *J. Pet. Tech.* (March 1958), pp. 64–66.

68. Van der Knapp, W., "Nonlinear Behavior of Elastic Porous Media," *Trans.*, AIME, Vol. 216 (1959), pp. 179–187.

69. Hall, H. N., "Compressibility of Reservoir Rocks," *Trans.*, AIME, Vol. 198 (1953), pp. 309–311.

70. Carpenter, C. B., and Spencer, G. B., "Measurements of Compressibility of Consolidated Oil-Bearing Sandstones," RI 3540, USBM (Oct., 1940).

71. Newman, G. H., "Pore-Volume Compressibility of Consolidated, Friable, and Unconsolidated Reservoir Rocks Under Hydrostatic Loading," *J. Pet. Tech.* (Feb. 1973), pp. 129–134.

72. Martin, J. C., "Simplified Equations of Flow in Gas Drive Reservoirs and the Theoretical Foundation of Multiphase Pressure Buildup Analyses," *Trans.*, AIME, Vol. 216 (1959), pp. 309–311.

73. Calhoun, J. C., Jr., *Fundamentals of Reservoir Engineering*, U. of Oklahoma Press, Norman (1976).

74. Martin, M., Murray, G. H., and Gillingham, W. J., "Determination of the Potential Productivity of Oil-Bearing Formations by Resistivity Measurements," *Geophysics*, Vol. 3 (1938), pp. 258–272.

75. Jakosky, J. J., and Hopper, R. H., "The Effect of Moisture on the Direct Current Resistivities of Oil Sands and Rocks," *Geophysics*, Vol. 2 (1937), pp. 33–55.

76. Wyckoff, R. D., and Botset, H. G., "The Flow of Gas-Liquid Mixtures Through Unconsolidated Sands," *J. Applied Physics*, Vol. 7 (Sept. 1936), pp. 325–345.

77. Leverett, M. C., "Flow of Oil-Water Mixtures Through Unconsolidated Sands," *Trans.*, AIME (1938), pp. 149–171.

78. de Witte, L., "Relations Between Resistivities and Fluid Contents of Porous Rocks," *Oil & Gas J.* (Aug. 24, 1950), pp. 120–132.

79. de Witte, A. J., "Saturation and Porosity From Electric Logs in Shaly Sands," *Oil & Gas J.* (March 4, 1957), pp. 89–93.

80. Williams, M., "Estimation of Interstitial Water from the Electrical Log," *Trans.*, AIME, Vol. 189 (1950), pp. 295–308.

81. Rust, C. F., "Electrical Resistivity Measurements on Reservoir Rock Samples by the Two-Electrode and Four-Electrode Methods," *Trans.*, AIME, Vol. 192 (1952), pp. 217–224.

82. Wyllie, M. R. J., and Spangler, M. B., "Application of Electrical Resistivity Measurements to Problem of Fluid Flow in Porous Media," *Bull.*, AAPG, Vol. 36, No. 2 (1952), pp. 359–403.

83. Morse, R., Terwilliger, P. L., and Yuster, S. T., "Relative Permeability Measurements on Small Core Samples," *Oil & Gas J.* (Aug. 23, 1947), pp. 109–125.

84. Whiting, R. L., Guerrero, E. T., and Young, R. M., "Electrical Properties of Limestone Cores," *Oil & Gas J.* (July 27, 1953), pp. 309–313.

85. Dunlap, H. F., et al., "The Relation Between Electrical Resistivity and Brine Saturation in Reservoir Rocks," *Trans.*, AIME, Vol. 186 (1949), pp. 259–264.

86. Dorfman, M. H., "Discussion of Reservoir Description Using Well Logs," *J. Pet. Tech.* (Dec. 1984), pp. 2195–2196.

87. Wilson, D. A., and Hensel, W. M. Jr., "The Cotton Valley Sandstones of East Texas: a Log-Core Study," *Trans.*, SPWLA, 23rd Annual Logging Symposium, Paper R (July 6–9, 1982), pp. 1–27.

88. Hilchie, D. W., "Author's Reply to Discussion of Reservoir Description Using Well Logs," *J. Pet. Tech.* (Dec. 1984), p. 2196.

89. Hocott, C. R., "Interfacial Tension Between Water and Oil Under Reservoir Conditions," *Pet. Tech.* (Nov. 1938), pp. 184–190.

90. Katz, D. L., Monroe, R. R., and Trainer, R. P., "Surface Tension of Crude Oils Containing Dissolved Gases," *Pet. Tech.* (Sept. 1943), pp. 1–10.

91. Swartz, C. A., "The Variation in the Surface Tension of Gas-Saturated Petroleum with Pressure of Saturation," *Physics*, Vol. 1 (1931), pp. 245–253.

92. Livingston, H. K., "Surface and Interfacial Tensions of Oil-Water Systems in Texas Oil Sands," *Pet. Tech.* (Nov. 1938), pp. 1–13.

93. Hjelmeland, O. S., and Larrondo, L. E., "Investigation of the Effects of Temperature, Pressure, and Crude Oil Composition on Interfacial Properties," *SPE Reservoir Engineering* (July 1986), pp. 321–328.

94. Defay, R., Prigogine, I., Bellemans, A., and Everett, D. H., *Surface Tension and Adsorption*, Longmans, London (1966).

95. Benner, F. C., Riches, W. W., and Bartell, F. E., "Nature and Importance of Surface Forces in Production of Petroleum," *Drill & Prod. Prac.*, API (1938), pp. 442–448.

96. Benner, F. C., and Bartell, F. E., "The Effect of Polar Impurities Upon Capillary and Surface Phenomena in Petroleum Production," *Drill & Prod Prac.*, API (1942), pp. 341–348.

97. Benner, F. C., Dodd, C. G., and Bartell, F. E., "Evaluation of Effective Displacement Pressures for Petroleum Oil-Water Silica Systems," *Drill & Prod. Prac.*, API (1942), pp. 169–177.

98. Nutting, P. G., "Some Physical and Chemical Properties of Reservoir Rocks Bearing on the Accumulation and Discharge of Oil," *Problems in Petroleum Geology*, AAPG (1934).

99. Schilthuis, R. J., "Connate Water in Oil and Gas Sands," *Trans.*, AIME (1938), pp. 199–214.

100. Leverett, M. C., "Capillary Behavior in Porous Solids," *Trans.*, AIME, Vol. 142 (1941), pp. 159–172.

101. Bartell, F. E., and Miller, F. L., "Degree of Wetting of Silica by Crude Petroleum Oils," *Ind. Eng. Chem.*, Vol. 20, No. 2 (1928), pp. 738–742.

102. Slobod, R. L., and Blum, H. A., "Method for Determining Wettability of Reservoir Rocks," *Trans.*, AIME, Vol. 195 (1952), pp. 1–4.

103. Taber, J. J., personal communication.

104. Wagner, O. R., and Leach, R. O., "Improving Oil Displacement Efficiency by Wettability Adjustment," *Trans.*, AIME, Vol. 216 (1959), pp. 65–72.

105. Denekas, M. O., Mattax, C. C., and Davis, G. T., "Effects of Crude Oil Components on Rock Wettability," *Trans.*, AIME, Vol. 216 (1959), pp. 330–333.

106. Moore, T. F., and Slobod, R. L., "The Effect of Viscosity and Capillarity on the Displacement of Oil by Water," *Prod. Monthly* (Aug. 1956), pp. 20–30.

107. Bobek, J. E., Mattax, C. C., and Denekas, M. O., "Reservoir Rock Wettability—Its Significance and Evaluation," *Trans.*, AIME, Vol. 213 (1958), pp. 155–160.

108. Amott, E., "Observations Relating to the Wettability of Porous Rock," *Trans.*, AIME, Vol. 216 (1959), pp. 156–162.

109. Gatenby, W. A., and Marsden, S. S., "Some Wettability Characteristics of Synthetic Porous Media," *Prod. Monthly* (Nov. 1957), pp. 5–12.

110. Donaldson, E. C., Thomas, R. D., and Lorenz, P. B., "Wettability Determination and Its Effect on Recovery Efficiency," *Soc. Pet. Eng. J* (March 1969), pp. 13–20.

111. Donaldson, E. C., et al., "Equipment and Procedures for Fluid Flow and Wettability Tests of Geological Materials," U.S. Dept. of Energy, Bartlesville, Report DOE/BETC/IC-79/5, May 1980.

112. Donaldson, E. C., "Oil-Water-Rock Wettability Measurement," *Proc.*, Symposium of Chemistry of Enhanced Oil Recovery, Div. Pet. Chem., Am. Chem. Soc., March 29–April 3, 1981, pp. 110–122.

113. Raza, S. H., Treiber, L. E., and Archer, D. L., "Wettability of Reservoir Rocks and Its Evaluation," *Prod. Monthly*, Vol. 33, No. 4 (April 1968), pp. 2–7.

114. Treiber, L. E., Archer, D. L., and Owens, W. W., "A Laboratory Evaluation of the Wettability of Fifty Oil-Producing Reservoirs," *Soc. Pet. Eng. J.* (Dec. 1972), pp. 531–540.

115. Morrow, N. R., "Capillary Pressure Correlation for Uniformly Wetted Porous Media," *J. Can. Pet. Tech.*, Vol. 15 (1976), pp. 49–69.

116. Chilingar, G. V., and Yen, T. F., "Some Notes on Wettability and Relative Permeabilities of Carbonate Reservoir Rocks, II," *Energy Sources*, Vol. 7, No. 1 (1983), pp. 67–75.

117. Brown, R. J. S., and Fatt, I., "Measurements of Fractional Wettability of Oil Field Rocks by the Nuclear Magnetic Relaxation Method," *Trans.*, AIME, Vol. 207 (1956), pp. 262–264.

118. Fatt, I., and Klikoff, W. A., "Effect of Fractional Wettability on Multiphase Flow through Porous Media," *Trans.*, AIME, Vol. 216 (1959), pp. 426–432.

119. Holbrook, O. C., and Bernard, G. C., "Determination of Wettability by Dye Adsorption," *Trans.*, AIME, Vol. 213 (1958), pp. 261–264.

120. Iwankow, E. N., "A Correlation of Interstitial Water Saturation and Heterogenous Wettability," *Prod Monthly* (Oct. 1960), pp. 18–26.

121. Salathiel, R. A., "Oil Recovery by Surface Film Drainage in Mixed-Wettability Rocks," *Trans.*, AIME (1973), pp. 1216–1224.

122. Morrow, N. R., Lim, H. T., and Ward, J. S., "Effect of Crude Oil Induced Wettability Changes on Oil Recovery," *SPE Formation Evaluation* (Feb. 1986), pp. 89–103.

123. Rathmell, J. J., Braun, P. H., and Perkins, T. K., "Reservoir Waterflood Residual Oil Saturation from Laboratory Tests," *J. Pet. Tech.* (Feb. 1973), pp. 175–185.

124. Sharma, M. M., and Wunderlich, R. W., "The Alteration of Rock Properties Due to Interactions With Drilling Fluid Components," paper SPE 14302 presented at the SPE 1985 Annual Technical Conference & Exhibition, Las Vegas, Sept. 22–25.

125. Richardson, J. G., Perkins, F. M., Jr., and Osoba, J. S., "Differences in Behavior of Fresh and Aged East Texas Woodbine Cores," *Trans.*, AIME, Vol. 204 (1955), pp. 86–91.

126. Mungan, N., "Certain Wettability Effects in Laboratory Waterfloods," *J. Pet. Tech.* (Feb. 1966), pp. 247–252.

127. Rose, W. R., and Bruce, W. A., "Evaluation of Capillary Character in Petroleum Reservoir Rock," *Trans.*, AIME, Vol. 186 (1949), pp. 127–133.

128. Klinkenberg, L. J., "The Permeability of Porous Media to Liquids and Gases," *Drill. & Prod. Prac.*, API (1941), pp. 200–213.

129. Jones, F. O., and Owens, W. W., "A Laboratory Study of Low-Permeability Gas Sands," *J. Pet. Tech.* (Sept. 1980), pp. 1631–1640.

130. Wei, K. K., Morrow, N. R., and Brower, K. R., "The Effect of Fluid, Confining Pressure, and Temperature on Absolute Permeabilities of Low-Permeability Sandstones," *SPE Formation Evaluation* (Aug. 1986), pp. 413–423.

131. Fatt, I., "The Effect of Overburden Pressure on Relative Permeability," *Trans.*, AIME (1953), pp. 325–326.

132. Bruce, W. A., and Welge, H. J., "The Restored-State Method for Determination of Oil in Place and Connate Water," *Drill. and Prod. Prac.*, API (1947), pp. 166–174.

133. Craig, F. F., Jr., "The Reservoir Engineering Aspects of Waterflooding," Monograph Series, SPE, Dallas, Vol. 3 (1971).

134. Muskat, M., et al., "Flow of Gas-Liquid Mixtures through Sands," *Trans.*, AIME (1937), pp. 69–96.

135. Hassler, G. L., Rice, R. R., and Leeman, E. H., "Investigations on the Recovery of Oil from Sandstones by Gas Drive," *Trans.*, AIME (1936), pp. 116–137.

136. Botset, H. G., "Flow of Gas-Liquid Mixtures through Consolidated Sand," *Trans.*, AIME Vol. 136 (1940), pp. 91–105.

137. Leverett, M. C., and Lewis, W. B., "Steady Flow of Gas-Oil Water Mixtures through Unconsolidated Sands," *Trans.*, AIME, Vol. 142 (1941), pp. 107–116.

138. Krutter, H., and Day, R. J., "Air Drive Experiments on Long Horizontal Consolidated Cores," *Pet. Tech.*, T. P. 1627 (Nov. 1943).

139. Osoba, J. S., et al., "Laboratory Measurements of Relative Permeability," *Trans.*, AIME, Vol. 192 (1951), pp. 47–56.

140. Caudle, B. H., Slobod, R. L., and Brownscombe, E. R., "Further Developments in the Laboratory Determination of Relative Permeability," *Trans.*, AIME, Vol. 192 (1951), pp. 145–150.

141. Geffen, J. M., et al., "Experimental Investigation of Factors Affecting Laboratory Relative Permeability Measurements," *Trans.*, AIME, Vol. 192 (1951), pp. 99–110.

142. Richardson, J. G., et al., "Laboratory Determination of Relative Permeability," *Trans.*, AIME, Vol. 195 (1952), pp. 187–196.

143. Owens, W. W., Parrish, D. R., and Lamoreaux, W. E., "An Evaluation of a Gas-Drive Method for Determining Relative Permeability Relationships," *Trans.*, AIME, Vol. 207 (1956), pp. 275–280.

144. Hassler, G. L., "Method and Apparatus for Permeability Measurements," U. S. Patent No. 2,345,935.

145. Brownscombe, E. R., Slobod, R. L., and Caudle, B. H., "Relative Permeability of Cores Desaturated by Capillary Pressure Method," *Drill & Prod. Prac.*, API (1949), pp. 302–315.

146. Gates, J. I., and Tempelaar-Lietz, W., "Relative Permeabilities of California Cores by the Capillary Pressure Method," *Drill & Prod Prac.*, API (1950), pp. 285–302.

147. Leas, W. J., Jenks, L. H., and Russell, C. D., "Relative Permeability to Gas," *Trans.*, AIME, Vol. 189 (1950), pp. 65–72.

148. Rapoport, L. A., and Leas, W. J., "Relative Permeability to Liquid in Liquid-Gas Systems," *Trans.*, AIME, Vol. 192 (1951), pp. 83–98.

149. Fatt, I., and Dykstra, H., "Relative Permeability Studies," *Trans.*, AIME, Vol. 192 (1951), pp. 249–256.

150. Corey, A. T., "The Interrelation Between Gas and Oil Relative Permeabilities," *Prod. Monthly*, Vol. 19 (Nov. 1954), pp. 38–41.

151. Loomis, A. G., and Crowell, D. C., "Relative Permeability Studies: Gas-Oil and Water-Oil Systems," Bull. 599, U. S. Bureau of Mines, Washington, 1962.

152. Buckley, S. E., and Leverett, M. C., "Mechanism of Fluid Displacement in Sands," *Trans.*, AIME, Vol. 146 (1942), pp. 107–116.

153. Welge, H. J., "A Simplified Method for Computing Oil Recovery by Gas or Water Drive," *Trans.*, AIME, Vol. 195 (1952), pp. 91–98.

154. Johnson, E. F., Bossler, D. P., and Naumann, V. O., "Calculation of Relative Permeability from Displacement Experiments," *Trans.*, AIME, Vol. 216 (1959), pp. 370–372.

155. Higgins, R. V., "Application of Buckley-Leverett Techniques in Oil-Reservoir Analysis," Bureau of Mines Report of Investigations 5568 (1960).

156. Guerrero, E. T., and Stewart, F. M., "How to Obtain a k_w/k_o Curve from Laboratory Unsteady-State Flow Measurements," *Oil & Gas J.* (Feb. 1, 1960), pp. 96–100.

157. Guerrero, E. T., and Stewart, F. M., "How to Obtain and Compare k_w/k_o Curves from Steady-State and Laboratory Unsteady-State Flow Measurements," *Oil & Gas J.* (Feb. 22, 1960), pp. 104–106.

158. Jones, S. C., and Rozelle, W. O., "Graphical Techniques for Determining Relative Permeability from Displacement Experiments," *Trans.*, AIME (1978), pp. 807–817.

159. Corey, A. T., et al., "Three-Phase Relative Permeability," *J. Pet. Tech.* (Nov. 1956), pp. 63–65.

160. Schneider, F. N., and Owens, W. W., "Sandstone and Carbonate Two- and Three-Phase Relative Permeability Characteristics," *Soc. Pet. Eng. J.* (March 1970), pp. 75–84.

161. Honarpour, M., Koederitz, L. F., and Harvey, A. H., "Empirical Equations for Estimating Two-Phase Relative Permeability in Consolidated Rock," *J. Pet. Tech.* (Dec. 1982), pp. 2905–2908.

162. Mungan, N., "Interfacial Effects in Immiscible Liquid-Liquid Displacement in Porous Media," *Soc. Pet. Eng. J.* (Sept., 1966), pp. 247–253.

163. Lefebvre du Prey, E. J., "Factors Affecting Liquid-Liquid Relative Permeabilities of a Consolidated Porous Medium," *Soc. Pet. Eng. J.* (Feb. 1973), pp. 39–47.

164. Miller, M. A., and Ramey, H. J., Jr., "Effect of Temperature on Oil/Water Relative Permeabilities of Unconsolidated and Consolidated Sands," *Soc. Pet. Eng. J.* (Dec. 1985), pp. 945–953.

165. Honarpour, M., DeGroat, C., and Manjnath, A., "How Temperature Affects Relative Permeability Measurement," *World Oil* (May 1986), pp. 116–126.

166. Kinney, P. T., and Nielsen, R. F., "Wettability in Oil Recovery," *World Oil*, Vol. 132, No. 4 (March 1951), pp. 145–154.

167. Newcombe, J., McGhee, J., and Rzasa, M. J., "Wettability versus Displacement in Water Flooding in Unconsolidated Sand Columns," *Trans.*, AIME, Vol. 204 (1955), pp. 227–232.

168. Coley, F. N., Marsden, S. S., and Calhoun, J. C., Jr., "A Study to the Effect of Wettability on the Behavior of Fluids in Synthetic Porous Media," *Prod. Monthly*, Vol. 20, No. 8 (June 1956), pp. 29–45.

169. Jennings, H. Y., Jr., "Surface Properties of Natural and Synthetic Porous Media," *Prod. Monthly*, Vol. 21, No. 5 (March 1957), pp. 20–24.

170. Uren, L. D., and Fahmy, E. H., "Factors Influencing the Recovery of Petroleum from Unconsolidated Sands by Water-Flooding," *Trans.*, AIME, Vol. 77 (1927), pp. 318–335.

171. Kennedy, H. T., Burja, E. O., and Boykin, R. S., "An Investigation of the Effects of Wettability on the Recovery of Oil by Water Flooding," *J. Phys. Chem.*, Vol. 59 (1955), pp. 867–869.

172. Morrow, N. R., "Interplay of Capillary, Viscous and Buoyancy Forces in the Mobilization of Residual Oil," *J. Can. Pet. Tech.* (July–Sept. 1979), pp. 3546.

173. Melrose, J. C., and Brandner, C. F., "Role of Capillary Forces in Determining Microscopic Displacement Efficiency for Oil Recovery by Waterflooding," *J. Can. Pet. Tech.*, Vol. 13 (1974), pp. 54–62.

174. Owens, W. W., and Archer, D. L., "The Effect of Rock Wettability on Oil Water Relative Permeability Relationships," *J. Pet. Tech.* (July 1971), pp. 873–878.

175. McCaffery, F. G., and Bennion, D. W., "The Effect of Wettability on Two-Phase Relative Permeabilities," *J. Can. Pet. Tech.*, Vol. 13 (1974), pp. 42–53.

176. Killins, C. R., Nielsen, R. F., and Calhoun, J. C., Jr.,"Capillary Desaturation and Imbibition in Rocks," *Prod. Monthly*, Vol. 18, No. 2 (Feb. 1953), pp. 30–39.

177. Rust, C. F., "A Laboratory Study of Wettability Effects on Basic Core Parameters," paper SPE 986G, presented at the SPE Venezuelan Second Annual Meeting, Caracas, Venezuela, Nov. 6–9, 1957.

178. Mungan, N., and Moore, E. J., "Certain Wettability Effects on Electrical Resistivity in Porous Media," *J. Can. Pet. Tech.* (Jan.–March 1968), pp. 20–25.

179. Morgan, W. B., and Pirson, S. J., "The Effect of Fractional Wettability on the Archie Saturation Exponent," *Trans.*, SPWLA, Fifth Annual Logging Symposium, Midland, Texas, May 13–15, 1964.

5.2 FORMATION EVALUATION

Formation evaluation, as applied to petroleum reservoirs, consists of the quantitative and qualitative interpretation of formation cores, geophysical well logs, mud logs, flow tests, pressure tests, and samples of reservoir fluids. The goal of the interpretation is to provide information concerning reservoir lithology, fluid content, storage capacity, and producibility of oil or gas reservoirs. The final analysis includes an economic evaluation of whether to complete an oil or gas well and, once completed, an ongoing analysis of how to produce the well most effectively. These interpretations and analyses are affected by geological complexity of the reservoir, rock quality, reservoir heterogeneity, and, from a logistical standpoint, the areal extent and location of the project of interest. In the early stages of development, the purpose of formation evaluation is to define reservoir thickness and areal extent, reservoir quality, reservoir fluid properties, and ranges of rock properties. The key rock properties are porosity, permeability, oil, gas, and water saturations. Because of space limitations and the importance of these properties, methods of measuring porosity, permeability, and fluid saturations will be emphasized.

5.2.1 Coring and Core Analysis

Routine or conventional core analyses refer to common procedures that provide information on porosity, permeability, resident fluids, lithology, and texture of petroleum reservoirs. Table 5.2.1 lists the types of analyses that are obtained and how the results of each analysis are used. Specialized core analyses, such as are listed in Table 5.2.2, are done less often, but are important for specific applications. Routine core analyses can be performed on whole cores or on small plugs that are cut from a larger core. With the exception of petrographic analyses (thin sections, x-ray; scanning electron microscopy, etc.), special core analyses are normally done with core plugs. After a well is drilled and logs are available to identify zones of interest, very small portions of the reservoir can be obtained with percussion sidewall or sidewall drilled cores. Sidewall cores are less expensive and are valuable for petrographic analyses, but are generally not suitable for special core analyses.

The subject of coring and core analysis was summarized in a series of articles [2–10]. An overview article [11] described how core analyses can aid reservoir description. A handbook [12] is available that describes procedures and tools for conventional coring as well as methods for routine core analysis. Procedures for routine core analysis and methods of preserving cores have been recommended by the American Petroleum Institute [13]. Some of the information available in these sources will be highlighted.

5.2.1.1 Coring

Well coring refers to the process of obtaining representative samples of the productive formation in order to conduct a variety of laboratory testing. Various techniques are used to obtain core samples: conventional diamond-bit coring, rubbersleeve coring, pressure coring, sidewall coring, and recovery of cuttings generated from the drilling operation. Conventional coring is normally done in competent formations to obtain full-diameter cores. Rubber sleeve-coring improves core recovery in softer formations. Pressure coring, although relatively expensive, is used to obtain cores that have not lost any fluids during lifting of the core to the surface.

A common problem with all of these techniques is to decide when to core. In many instances, cores from the interval of interest are not obtained because of abrupt stratigraphic changes. A second problem is that, typically, nonproductive intervals of the desired strata are obtained. These intervals did not initially contain a significant amount of hydrocarbon.

5.2.1.2 Core Preservation

The importance of not altering wettability with drilling mud filtrate has been discussed in this chapter in the section entitled "Wettability and Contact Angle." Preventing wettability changes in core material, after it has been recovered at the surface, can be equally important so that subsequent laboratory measurements are representative of formation conditions.

Cores obtained with drilling muds that minimize wettability alteration, and that are protected at the well-site to prevent evaporation or oxidation, are called preserved cores. They are also referred to as fresh cores or native-state cores. Cores that are cleaned with solvents and resaturated with reservoir fluids are called restored-state cores or extracted cores. The restoring process is often performed on nonpreserved or weathered cores, but the same technique could apply to cores that had been preserved.

Table 5.2.1 *Routine Core Analysis Tests*

Type of analysis	Use of results
Porosity	A factor in volume and storage determinations.
Permeability—horizontal and vertical	Defines flow capacity, crossflow, gas and water coning and relative profile capacity of different zones, pay and nonpay zones.
Saturations	Defines presence of hydrocarbons, probable fluid recovery by test, type of recovery, fluid contacts, completion interval.
Lithology	Rock type, fractures, vugs, laminations, shale content used in log interpretation, recovery forecasts, capacity estimates.
Core-gamma ray log	Relates core and log depth.
Grain density	Used in log interpretation and lithology.

From Reference 1.

Table 5.2.2 *Special Core Analysis Tests*

Type of test	Use of results
Capillary pressure	Defines irreducible fluid content, contacts.
Rock compressibility	Volume change caused by pressure change.
Permeability and porosity vs. pressure	Corrects to reservoir conditions.
Petrographic studies	
mineral	Used in log interpretation.
diagenesis	Origin of oil and source bed studies.
clay identification	Origin of oil and log analysis.
sieve analysis	Selection of screens, sand grain size.
Wettability	Used in capillary pressure interpretation and recovery analysis-relative permeability.
Electrical	
formation factor	Used in log interpretation.
resistivity index	
Acoustic velocity	Log and seismic interpretation.
Visual inspection	Rock description and geological study.
Thin sections, slabs	
Air, water, and other liquid permeability	Evaluates completion, workover, fracture and injection fluids; often combined with flood-pot test.
Flood-pot test and waterflood evaluation	Results in values for irreducible saturations, values for final recovery with special recovery fluids such as surfactants, water, and polymers.
Relative permeability	Relative permeability is used to obtain values for effective permeability to
gas-oil	each fluid when two or more fluids flow simultaneously; relative permeability
gas-water	enables the calculation of recovery versus saturation and time while values
water-oil	from flood-pot test give only end-point results.
oil-special fluids	
thermal	

From Reference 1.

Two methods of preserving conventional cores, immediately after they have been removed from the core barrel, will prevent changes in wettability for several months. One method consists of immersing the core in deoxygenated formation brine or suitable synthetic brine (i.e., drilling mud filtrate) and keeping the samples in suitable containers that can be sealed to prevent leakage and the entrance of oxygen. In the second method, the cores are wrapped in Saran or polyethylene film and aluminum foil and then coated with wax or strippable plastic. The second method is preferred for cores that will be used for laboratory determination of residual oil content, but the first method may be preferred for laboratory displacement tests. Plastic bags are often recommended for short-term (2–4 days) storage of core samples. However, this method will not ensure unaltered rock wettability. Air-tight metal cans are not recommended because of the possibility of rust formation and potential leakage.

Cores taken with a pressure core barrel are often frozen at the well-site for transportation to the laboratory. (Cores are left in the inner core barrel.) Normally, the inner barrel containing the cores is cut into lengths convenient for transport. Because of the complexity of the operation, the pressure core barrel is not used as extensively as the conventional core barrel. An alternate procedure involves bleeding off the pressure in the core and core barrel while the produced liquids are collected and measured. Analysis of the depressured core is done by conventional techniques. Fluids collected from the barrel during depressuring are proportionately added to the volumes of liquid determined from core analysis. In this manner a reconstructed reservoir core saturation is provided.

5.2.1.3 Core Preparation
Depending on the type of core testing to be done, core samples may be tested as received in the laboratory or they may be cleaned to remove resident fluids prior to analysis.

Details for cutting, cleaning, and preparing core plugs can be found in API RP-40: Recommended Practice for Core-Analysis Procedure [13], available from API Production Department, 211 North Ervay, Suite 1700, Dallas, TX 75201.

5.2.1.4 Core Analysis
Conventional core analysis procedures are described in detail in API RP-40 and elsewhere [12]. A good discussion on core analysis procedures is in the textbook written by Amyx, Bass, and Whiting [5.1.19].

Porosity
A number of methods [13] are suitable for measuring porosity of core samples. In almost all the methods, the sample is cleaned by solvent extraction and dried to remove liquid. Porosity can be determined by saturating the dry core with brine and measuring the weight increase after saturation. Another common method includes compressing a known volume of gas (usually helium) at a known pressure into a core that was originally at atmospheric pressure. Several other techniques have been used; one of the more common methods is the mercury porosimeter in which pressure on the core plug is reduced and the volume of the expanded air or gas is measured accurately. A summation of fluids technique, which measures and sums the oil, gas and water volumes in a freshly recovered reservoir core sample, is often used for plugs or sidewall samples of non-vuggy consolidated rocks that contain minimum amounts of clay [9].

Equations commonly used for calculation of porosity by gas expansion or compression include:

$$\phi = \left(\frac{V_p}{V_b}\right) 100 \qquad [5.2.1]$$

$$\phi = \left(\frac{V_b - V_{gr}}{V_b}\right) 100 \qquad [5.2.2]$$

where ϕ = porosity expressed as a percent
$\quad V_p$ = pore volume
$\quad V_b$ = bulk volume
$\quad V_{gr}$ = grain volume

All volumes should be in consistent units, commonly cm^3. If pore volume is measured directly in cores that contain vugs (such as some carbonates), Equation 5.2.1 may give erroneously high porosity because the bulk volume may be erroneously low [9]. If bulk volume of vuggy cores is determined by submerging the core in mercury or water, Equation 5.2.2 may yield erroneously low porosity [9]. Thus valid porosity values can only be obtained if bulk volume and grain volume measurements are accurate.

Permeability

The permeability of core plugs is determined by flowing a fluid (air gas, or water) through a core sample of known dimensions. If the absolute permeability is to be determined, the core plug is cleaned so that permeability is measured at 100% of the saturating fluid. Methods of measuring permeability of core plugs are described in API RP-27: Recommended Practice for Determining permeability of Porous Media [14]. Equation 5.1.36 can be used to calculate permeability of core plugs.

Fluid Saturations

Coring procedures usually alter the fluid content of the reservoir rock during the coring process. Drilling fluid is jetted against the formation rock ahead of the coring bit and the core surface as it enters the core barrel; as a result of this flushing action by the drilling mud filtrate, most free gas and a portion of the liquid are displaced from the core. When water base drilling fluid is used, the mud filtrate may displace oil until a condition of residual oil saturation is obtained. Also, this flushing action may result in the fluid content of the core being predominately that of the drilling fluid. When oil base drilling fluid is used, the core sample that is obtained may be driven to an irreducible water saturation.

Factors Affecting Oil Displaced During Coring

During the coring operation, it is important to avoid extreme flushing conditions that could cause mobilization of residual oil [15]. Some of the variables that control the amount of oil flushed from a core by mud filtrate are: borehole-to-formation differential pressure (overbalance), coring penetration rate, core diameter, type of drill bit, drilling mud composition (including particle size distribution), depth of invasion of mud particles into the core, rate of filtrate production (both spurt loss and total fluid loss), interfacial tension of mud filtrate, permeability of the formation (both horizontal and vertical), and nature of the reservoir (uniformity, texture, etc.). In one type of system investigated in the laboratory [16], the amount of oil stripped from cores varied directly with the overbalance pressure, filtration production rate, core diameter and core permeability; it varied inversely with penetration rate. In that system, the overbalance pressure exerted more influence than the other factors. When large pressure gradients exist near the core bit, unintentional displacement of residual oil may occur in coring operations. In this region close to the bit, high velocities caused by this high pressure may mobilize some of the residual oil. Drilling mud composition can affect subsequent laboratory oil displacement tests in core samples by: changing wettability of the reservoir rock, altering interfacial tension of the mud filtrate, being penetrated by mud particles into the zone of interest, and yielding undesirable fluid loss properties. Since fluids with lower interfacial tension contribute to additional oil recovery, whenever possible, the use of mud additives

that lower interfacial tension should be avoided. Greater amounts of residual oil are displaced from cores as the filtrate production rate is increased. High API filter loss or smaller core diameters will generally lead to larger amounts of flushing, but a key factor in the amount of mobilized residual oil is the spurt loss (the rapid loss to the formation that occurs before an effective filter cake is formed). As stated previously, uniformity of the formation being cored will influence the amount of oil that will be displaced. Identical drilling conditions may yield varying results with changes in lithology or texture of the reservoir. In particular, drastic differences may be observed in reservoirs that contain both sandstone and carbonate oil-bearing strata.

Factors Affecting Oil Saturation Changes During Recovery of Cores

Surface oil saturations should be adjusted to compensate for shrinkage and bleedings [123]. Shrinkage is the term applied to the oil volume decrease caused by a temperature change or by a drop in pressure which causes dissolved gases to escape from solution. Shrinkage of reservoir fluids is measured in the laboratory by differentially liberating the samples at reservoir temperature. The formation volume factors are used to adjust surface oil volume back to reservoir temperature and pressure. Gases coming out of solution can cause some oil to flow out of the core even though it may have been flushed to residual oil by mud filtrate. Bleeding is the term applied to this decrease in oil saturation as the core is brought to the surface. Calculations have been proposed [123] to account for shrinkage and bleeding (see the discussion in this chapter entitled "Estimation of Waterflood Residual Oil Saturation").

Measurement of Fluid Saturations

There are two primary methods of determining fluid content of cores; these methods are discussed in API RP-40: Recommended Practice for Core-Analysis Procedure [13]. In the retort or vacuum distillation method, a fluid content still is used to heat and vaporize the liquids under controlled conditions of temperature and pressure. Prior to testing, the gas space in the core is displaced with water. The fluids produced from the still are condensed and measured, and the fluid saturations are calculated. Normally the percent oil and water are subtracted from 100% to obtain the gas saturation; however, considerable error may be inherent in this assumption. The second common method is the distillation-extraction method in which water in the core is distilled, condensed, and accumulated in a calibrated receiving tube. Oil in the core is removed by solvent extraction and the oil saturation is calculated from the weight loss data and the water content data.

Conventional core samples have oil content determined by atmospheric distillation. The oil distilled from a sample is collected in a calibrated receiving tube where the volume is measured. Temperatures up to 1,200°F (about 650°C) are used to distill the oil from the sample which causes some coking and cracking of the oil and the loss of a small portion of the oil. An empirically derived correction is applied to the observed volume to compensate for the loss. Calibration tests are made on each type of oil.

Whole core samples have oil content determined by vacuum distillation. This technique is used to remove oil from the sample without destroying the minerals of the sample. A maximum temperature of 450°F is used. The oil distilled from the sample is collected in a calibrated receiving tube which is immersed in a cold bath of alcohol and dry ice at about −75°C. This prevents the oil from being drawn into the vacuum system. As in the atmospheric distillation method, corrections must be applied to the measured volumes.

The oil content (V_o) divided by the pore volume (V_p) yields the oil saturation (S_o) of a sample in percent of pore space:

$$S_o = \left(\frac{V_o}{V_p}\right) 100 \qquad [5.2.3]$$

Two sources of error are inherent in the retort method. At the high temperatures employed, water of crystallization within the rock is driven off which causes the water saturation to appear to be higher than the actual value. Another error results from the cracking of the oil and subsequent deposition of coke within the pore structure. Thus, a calibration curve should be prepared on various gravity crudes to compensate for the oil lost from the cracking and coking. Both of the above errors will result in a measured oil saturation that is lower than the actual saturation in the rock. Another possible source of error is the liberation of carbon dioxide from carbonate material in the core at elevated temperatures; this would cause a weight loss that can be interpreted as a change in saturation. The solvent extraction method has the disadvantage in that it is an indirect method since only the water removed from the core is measured. However the extraction method has the advantage that the core is usually not damaged and can be used for subsequent tests.

Grain Density and Core Description
Grain density and lithologic descriptions are often provided in data for routine care analysis. Grain density depends on the lithology and composition of the reservoir of interest. Densities of some common minerals found in reservoir rocks are listed in Table 5.2.3 [5.1.41].

Results of Core Analyses from Various Reservoirs
Typical core analyses [5.1.23] of different formations from various states and regions of the U.S. are listed in. Table 5.2.4a to 5.2.4i. In addition to ranges in permeability, porosity, oil saturation, connate water saturation, the depth and thickness of the productive intervals are given.

Special Core Analysis Tests
Special core analysis testing is done when specifically required. Visual inspection and some petrographic studies are frequently done. For sandstones and conglomerates, particle size is often obtained by disaggregating and sieving reservoir rock material. Fractions of the various sizes of grains are determined and described according to the nomenclature in Table 5.2.5. Larger grain size is associated with higher permeability, and very small grain sizes include sit and clay fractions that are associated with lower permeabilities.

Table 5.2.3 *Densities of Common Minerals in Reservoir Rocks*

Material	Matrix density* (gm/cm³)
Sand (consolidated)	2.65
Sand (unconsolidated)	2.65
Limestone	2.71
Dolomite	2.8–2.9
Shale	1.8–2.7
Gypsum	2.32
Anhydrite	2.9–3.0
Halite	2.16

From Reference 5.1.41.
*These figures are averages and may vary from area to area, depending on types and abundance of secondary minerals.

5.2.2 Drill Stem Tests
A drill stem test (DST) is some form of temporary completion of a well that is designed to determine the productivity and fluid properties prior to completion of the well. Although a DST can be performed in uncased hole (open hole) or in cased hole (perforation tests), the open hole test is more common. The tool assembly which consists of a packer, a test valve, and an equalizing valve, is lowered on the drill pipe to a position opposite the formation to be tested. The packer expands against the hole to segregate the mud-filled annular section from the interval of interest, and the test valve allows formation fluids to enter the drill pipe during the test. The equalizing valve allows pressure equalization after the test so the packer can be retrieved. Details of the DST and DST assemblies are described elsewhere [13,19,66] and will only be summarized here. By closing the test valve, a build up in pressure is obtained; by opening the test valve, a decline in pressure is obtained. (Pressure buildup and falloff analyses are discussed subsequently in this section of the chapter.) During the DST, both pressures and flow rates are measured as a function of time.

Interpretation of DST results is often regarded as an art rather than a science. Certainly, a DST can provide a to valuable indication of commercial productivity from a well, provided engineering judgment and experience are properly utilized. Interpretations of various pressure charts are shown in Figure 5.2.1 [13,17] and 5.2.2 [1]; details of interpreting DST data are described in the literature [66,18].

5.2.3 Logging
5.2.3.1 Introduction
This section deals with the part of formation evaluation known as well logging. Well logs are a record versus depth of some physical parameter of the formation. Parameters such as electrical resistance, naturally occurring radioactivity, or hydrogen content may be measured so that important producing characteristics such as porosity, water saturation, pay thickness, and lithology may be determined. Logging instruments (called *sondes*) are lowered down the borehole on armored electrical cable (called a *wireline*). Readings are taken while the tool is being raised up the hole. The information is transmitted uphole via the cable where it is processed by an on-board computer and recorded on magnetic tape and photographic film. In older logging units, downhole signals are processed by analog circuits before being recorded.

Well logging can be divided into two areas: open hole and cased hole. Open hole logging is done after drilling, before casing is set. The purpose of open hole logging is to evaluate all strata penetrated for the presence of oil and gas. Open hole logs give more reliable information on producing characteristics than cased hole logs. Cased hole logs provide information about cement job quality, casing corrosion, fluid flow characteristics, and reservoir performance. In areas where the geologic and producing characteristics of a reservoir are well known, as in development wells, cased hole logs are used for correlation. In recent years, many new open- and cased-hole logs and services have become available, including fluid samplers, sidewall cores, fracture height log, and seismic services. These products, in conjunction with new computer processing techniques, provide the engineer and geologist with an enormous amount of data for any well.

5.2.3.2 Parameters that Can Be Calculated or Estimated from Logs
Porosity
Porosity is defined as the ratio of volume of pores to the total volume of the rock. It occurs as primary (depositional)

(*text continued on page 5-69*)

Table 5.2.4 Typical Core Analyses from Various Reservoirs [5.1.23] (a) Arkansas

Formation	Fluid prod.	Range of prod. depth, ft	Avg. prod. depth, ft	Range of prod. thickness, ft	Avg. prod. thickness, ft	Range of perm. K, md	Avg. perm. K, md	Range of porosity, %	Avg. porosity, %	Range of oil satn., %	Avg. oil satn. %	Range of calc. connate water satn., %	Avg. calc. connate water satn., %
Blossom	C/O*	2,190–2,655	2,422	3–28	15	1.6–8,900	1,685	15.3–40	32.4	1.2–36	20.1	24–55	32
Cotton Vally	C/O	5,530–8,020	6,774	4–79	20	0.6–4,820	333	11.3–34	20.3	0.9–37	13.1	21–43	35
Glen Rose†	O	2,470–3,835	3,052	5–15	10	1.6–5,550	732	17.3–38	23.4	4.0–52	21.0	28–50	38
Graves	C/O	2,400–2,725	2,564	2–26	11	1.2–4,645	1,380	9.8–40	34.9	0.3–29	16.8	19–34	30
Hogg	O	3,145–3,245	3,195	12–33	17	6.5–5,730	1,975	14.4–41	30.9	2.6–56	19.9	26–34	27
Meakin	G/C/O*	2,270–2,605	2,485	2–20	11	3.0–6,525	1,150	17.1–40	31.8	0.6–43	12.9	24–63	43
Nacatoch	C/O	1,610–2,392	2,000	6–45	20	0.7–6,930	142	9.9–41	30.5	0.2–52	4.9	41–70	54
Paluxy	O	2,850–4,890	3,868	6–17	12	5–13,700	1,213	15.1–32	26.9	7.5–49	21.2	28–43	35
Pettit	O	4,010–5,855	4,933	4–19	11	0.1–698	61	6.2–28	15.4	9.1–29	12.7	25–44	30
Rodessa‡	O	5,990–6,120	6,050	8–52	16	0.1–980	135	5.1–28	16.5	0.7–26	14.8	25–38	31
Smackover§	G/C/O	6,340–9,330	8,260	2–74	18	0.1–12,600	850	1.1–34	14.2	0.7–41	12.8	21–50	31
Tokio	C/O	2,324–2,955	2,640	2–19	13	0.5–11,550	2,100	13.6–42	32.1	0.9–57	25.6	17–43	27
Travis Peak	C/O	2,695–5,185	3,275	3–25	10	0.4–6,040	460	9.4–36	24.3	0.5–36	14.3	16–48	36
Tuscaloosa	C/O	3,020–3,140	3,080	4–25	15	0.4–3,760	506	15.6–39	27.3	0.3–53	14.0	31–63	45

From Reference 1.23.
*Indicates fluid produced: G—gas; C—condensate; O—oil.
†Specific zone not indentified locally.
‡Includes data from Mitchell and Gloyd zones.
§Includes data from Smackover Lime and Reynolds zones.

Table 5.2.4 *(continued)* *(b) East Texas Area*

Formation	Fluid prod.	Range of prod. depth, ft	Avg. prod. depth, ft	Range of prod. thickness, ft	Avg. prod. thickness, ft	Range of perm. K, md	Avg. perm. K, md	Range of porosity, %	Avg. porosity, %	Range of oil satn., %	Avg. oil satn., %	Range of calc. connate water satn., %	Avg. calc. connate water satn., %
Bacon	C/O	6,685–7,961	7,138	3–24	11	0.1–2,040	113	1.5–24.3	15.2	2.7–20.6	8.6	9–22	16
Cotton Vally	C	8,448–8,647	8,458	7–59	33	0.1–352	39	6.9–17.7	11.7	1.1–11.6	2.5	13–32	25
Fredericksburg	O	2,330–2,374	2,356	5–8	7	0.1–4.6	1.2	11.9–32.8	23.1	3.3–39.0	20.8	35–43	41
Gloyd	C/O	4,812–6,971	5,897	3–35	19	0.1–560	21	8.0–24.0	14.9	tr–24.3	8.2	16–45	31
Henderson	G/C/O	5,976–6,082	6,020	3–52	12	0.1–490	19	7.0–26.2	15.2	0.8–23.3	10.6	21–44	27
Hill	C/O	4,799–7,668	5,928	3–16	9	0.1–467	70	6.4–32.2	15.6	0.9–26.7	12.2	23–47	33
Mitchell	O	5,941–6,095	6,010	3–43	21	0.1–487	33	7.2–29.0	15.5	1.8–25.9	12.5	15–47	29
Mooringsport	O	3,742–3,859	3,801	4–12	8	0.4–55	5	5.3–19.6	14.6	2.8–26.6	13.8	29–48	40
Nacatoch*	O	479–1,091	743	2–21	12	1.9–4,270	467	13.4–40.9	27.1	0.6–37.4	14.5	24–55	41
Paluxy	O	4,159–7,867	5,413	7–46	27	0.1–9,600	732	6.3–31.1	21.6	2.2–48.7	24.1	22–47	30
Pecan Gap	O	1,233–1,636	1,434	5–20	13	0.5–55	6	16.3–38.1	26.6	3.5–49.8	12.9	30–56	46
Pettit†	G/C/O	5,967–8,379	7,173	2–23	11	0.1–3,670	65	4.5–25.8	14.7	0.9–31.6	9.8	10–35	23
Rodessa	C/O	4,790–8,756	6,765	4–42	17	0.1–1,180	51	2.3–29.0	14.5	tr–25.3	5.3	6–42	23
Sub-Clarksville‡	O	3,940–5,844	4,892	3–25	12	0.1–9,460	599	8.2–38.0	24.8	1.4–34.6	17.9	12–60	33
Travis Peak§	C/O	5,909–8,292	6,551	2–30	11	0.1–180	42	5.6–25.8	15.0	0.1–42.8	12.5	17–38	28
Wolfe City	O	981–2,054	1,517	6–22	13	0.3–470	32	17.1–38.4	27.9	1.5–37.4	15.6	23–68	46
Woodbine	C/O	2,753–5,993	4,373	2–45	14	0.1–13,840	1,185	9.7–38.2	25.5	0.7–35.7	14.5	14–65	35
Young	C	5,446–7,075	6,261	4–33	17	0.1–610	112	4.4–29.8	19.7	tr–4.5	0.8	13–27	21

*Small amount of Navarro data combined with Nacatoch.
†Data for Pittsburg, Potter, and Upper Pettit combined with Pettit.
‡Small amount of Eagleford data combined with Sub-Clarksville.
§Data for Page combined with Travis Peak.

Table 5.2.4 (c) North Louisiana Area

Formation	Fluid prod.	Range of prod. depth, ft	Avg. prod. depth, ft	Range of prod. thickness, ft	Avg. prod. thickness, ft	Range of perm. K, md	Avg. perm. K, md	Range of porosity, %	Avg. porosity, %	Range of oil satn., %	Avg. oil satn., %	Range of calc. connate water satn., %	Avg. calc. connate water satn., %
Annona Chalk	O	1,362–1,594	1,480	15–69	42	0.1–2.5	0.7	14.3–36.4	26.8	6.0–4.0	22	24–40	37
Buckrange	C/O	1,908–2,877	2,393	2–24	13	0.1–2,430	305	13.4–41	31.4	0.7–51	22.6	29–47	35
Cotton Valley[a]	G/C/O	3,650–9,450	7,450	4–37	20	0.1–7,350	135	3.5–34	13.1	0.0–14	3.1	11–40	24
Eagleford[b]	C	8,376–8,417	8,397	9–11	10	3.5–3,040	595	12.8–28	22.9	1.6–28	4.3	...	36
Fredericksburg	G/C	6,610–9,880	8,220	6–8	7	1.6–163	90	12.8–23.1	19.9	1.7–4.3	2.7	35–49	41
Haynesville	C	10,380–10,530	10,420	22–59	40	0.1–235	32	5.5–23.1	13.4	1.1–14.5	5.1	31–41	38
Hosston	C/O	5,420–7,565	6,480	5–15	12	0.4–1,500	140	8.8–29	18.6	0.0–35	8.8	18–37	28
Nacatoch	O	1,223–2,176	1,700	6–12	8	27–5,900	447	25.8–40	31.4	2.5–33	19.5	45–54	47
Paluxy	C/O	2,195–3,240	2,717	2–28	16	0.2–3,060	490	9.6–39	27.2	0.1–48	11.8	23–55	35
Pettit[c]	C/O	3,995–7,070	5,690	3–30	14	0.1–587	26	4.5–27	14.3	0.1–59	15.6	10–43	29
Pine Island[d]	O	4,960–5,060	5,010	5–13	9	0.2–1,100	285	8.5–27	20.6	13.3–37	24.1	16–30	22
Rodessa[e]	G/C/O	3,625–5,650	4,860	6–52	18	0.1–2,190	265	5.1–34	19.1	0.0–31	2.9	21–38	30
Schuler[f]	G/C/O	5,500–9,190	8,450	4–51	19	0.1–3,180	104	3.6–27.4	15.0	0.0–24	4.8	8–51	25
Sligo[g]	C/O	2,685–5,400	4,500	3–21	7	0.1–1,810	158	7.3–35	21.1	0.6–27	9.8	12–47	31
Smackover	C/O	9,980–10,790	10,360	6–55	24	0.1–6,190	220	3.4–23	12.9	1.1–22	7.2	9–47	25
Travis Peak[h]	C/O	5,890–7,900	6,895	7–35	18	0.1–2,920	357	7.0–27	19.4	0.1–35	8.6	26–38	31
Tuscaloosa	G/C/O	2,645–9,680	5,164	4–44	24	0.1–5,750	706	10.7–36	27.6	0.0–37	8.5	31–61	43

[a] Data reported where member formations of Cotton Valley group not readily identifiable.
[b] Data reported as Eutaw in some areas.
[c] Includes data reported as Pettit, Upper Pettit, and Mid Pettit. Sometimes considered same as Sligo.
[d] Sometimes referred to as Woodruff.
[e] Includes data reported locally for Jeter, Hill, Kilpatrick, and Fowler zones.
[f] Includes data reported locally Bodcaw, Vaughn, Doris, McFerrin, and Justiss zones.
[g] Includes data reported as Bridsong-Owens.
[h] Frequently considered same as Hosston.

Table 5.2.4 *(continued) (d) California*

Formation	Area	Fluid prod.	Range of prod. depth, ft	Ave. prod. depth, ft	Range of prod. thickness, ft	Ave. prod. thickness, ft	Range of perm. K, md	Ave. perm. K, md	Range of porosity, %	Ave. porosity, %	Range of oil satn, %	Avg. oil satn, %	Range of total water satn, %	Ave. total water satn, %	Range of calc. water satn, %	Ave. calc. connate water satn, %	Range of gravity, °API	Ave. gravity, °API
Eocene, Lower	San Joaquin Valley[a]	O	6,820–8,263	7,940	–	–	35–2,000	518	14–26	20.7	8–23	14.1	16–51	35	15–49	35	28–34	31
Miocene	Los Angeles Basin and Coastal[b]	O	2,870–9,530	5,300	60–450	165	10–4,000	300	15–40	28.5	6–65	18.8	25–77	50	15–72	36	15–32	26
Miocene, Upper	San Joaquin Valley[c]	O	1,940–7,340	4,210	10–1,200	245	4–7,500	1,000	17–40	28.2	9–72	32[k]	20–68[k]	50[k]	12–62	30	13–34	23
	Los Angeles Basin and Coastal[d]	O	2,520–6,860	4,100	5–1,040	130	86–5,000	1,110	19.5–39	30.8	10–55	25	22–72	44	12–61	30	11–33	21
Miocene, Lower	San Joaquin Valley[e]	O	2,770–7,590	5,300	30–154	76	15–4,000	700	20–38	28.4	4–40	19	25–80	51	14–67	36	15–40	34
	Los Angeles Basin and Coastal[f]	O	3,604–5,610	4,430	20–380	134	256–1,460	842	21–29	24.3	13–20	15.8	32–67	53	27–60	37	34–36	35
Oligocene	San Joaquin Valley[g]	O	4,589–4,717	4,639	–	–	10–2,000	528	19–34	26.3	12–40	22	2–60	43	3–45	30	37–38	38
	Coastal[h]	O	5,836–6,170	6,090	–	–	20–400	107	15–22	19.5	6–17	11.8	19–56	46	15–52	42	–	25
Pliocene	San Joaquin Valley[i]	O	2,456–3,372	2,730	5–80	33	279–9,400	1,250	30–38	34.8	7–43[l]	24.1[l]	33–84	54	10–61	34	18–44	24
	Los Angeles Basin and Coastal[j]	O	2,050–3,450	2,680	–	100	25–4,500	1,410	24–11	35.6	15–80	45	19–54	38	10–40	21	12–23	15

[a] Mainly data from Gatchell zone.
[b] Includes Upper and Lower Terminal, Union Pacific, Ford, "237," and Sesnon zones.
[c] Includes Kernco, Republic, and "26 R" zones.
[d] Includes Jones and Main zones.
[e] Includes "JV," Olcese, and Phacoides zones.
[f] Mainly data from Vaqueros zone.
[g] Mainly data from Oceanic zone.
[h] Mainly data from Sespe zone.
[i] Includes Sub Mulinia and Sub Scalez No. 1 and No. 2 zones.
[j] Includes Ranger and Tar zones.
[k] Oil-base data show high oil saturation (avg. 61 percent) and low water (3–54 percent, avg. 15 percent).
[l] Oil-base data show range 27.6 to 52.4 and avg. of 42.3 percent—not included in above "oil-saturation" values.

Table 5.2.4 (e) *Texas Gulf Coast-Corpus Christi Area**

Formation	Fluid prod.	Range of prod. depth, ft	Avg. prod. depth, ft	Range of prod. thickness, ft	Avg. prod. thickness, ft	Range of perm. K, md	Avg. perm. K, md	Range of porosity, %	Avg. porosity, %	Range of satn., %	Avg. oil satn., %	Range of calc. connate water satn., %	Avg. calc. connate water satn., %	Range of gravity, °API	Avg. gravity, °API
Catahoula	O	3,600–4,800	3,900	1–18	8	45–2,500	670	17–36	30	1–30	14	30–44	36	23–30	29
Frio	C/O	1,400–9,000	6,100	3–57	13	5–9,000	460	11–37	27	2–38	13	20–59	34	23–48	41
Jackson	O	600–5,000	3,100	2–23	9	5–2,900	350	16–38	27	3–32	15	21–70	45	22–48	37
Marginulina	C	6,500–7,300	7,000	5–10	7	7–300	75	14–30	24	1–4	2	20–48	34	55–68	60
Oakville	O	2,400–3,100	2,750	5–35	22	25–1,800	700	21–35	28	9–30	18	32–48	44	23–26	25
Vicksburg	C/O	3,000–9,000	6,200	4–38	12	4–2,900	220	14–32	24	1–17	7	26–54	38	37–65	48
Wilcox	C	6,000–8,000	7,200	30–120	60	1–380	50	15–25	19	0–10	1	22–65	37	53–63	58
Yegua	O	1,800–4,000	3,000	3–21	7	6–1,900	390	22–38	29	4–40	17	14–48	36	20–40	32

*Includes counties in Texas Railroad Commission District IV: Jim Wells, Brooks, Hidalgo, Aransas, San Patricio, Nueces, Willacy, Duval, Webb, Jim Hog, and Starr.

Table 5.2.4 (continued) (f) Texas Gulf Coast-Houston Area

Formation	Fluid prod.	Range of prod. depth, ft	Avg. prod. depth, ft	Range of prod. thickness, ft	Avg. prod. thickness, ft	Range of perm. K, md	Avg. perm. K, md	Range of porosity, %	Avg. porosity, %	Range of oil satn., %	Avg. oil satn., %	Range of total water satn., %	Avg. total water satn., %	Range of calc. water satn., %	Avg. calc. connate water satn., %	Range of gravity, °API	Avg. gravity, °API
Frio	C	4,000–11,500	8,400	2–50	12.3	18–9,200	810	18.3–38.4	28.6	0.1–6.0	1.0	34–72	54	20–63	34		
	O	4,600–11,200	7,800	2–34	10.4	33–9,900	1,100	21.8–37.1	29.8	4.6–41.2	13.5	24–79	52	12–61	33	25–42	36
Marginulina	C	7,100–8,300	7,800	4–28	17.5	308–3,870	2,340	35–37	35.9	0.2–0.8	0.5	33–61	46	14–31	21		
	O	4,700–6,000	5,400	4–10	5.7	355–1,210	490	28.5–37.3	32.6	8.1–21.8	15.3	48–68	59	25–47	36	25–30	28
Miocene	C	2,900–6,000	4,000	3–8	5.5	124–13,100	2,970	28.6–37.6	33.2	0.2–1.5	0.5	55–73	66	23–53	38		
	O	2,400–8,500	3,700	2–18	7.2	71–7,660	2,140	23.5–38.1	35.2	11–29	16.6	45–69	58	21–55	34	21–34	25
Vicksburg	C	7,400–8,500	8,100	1–6	2.0	50–105	86	26.5–31.0	27.1	0–1.5	0.2	66–78	74	53–61	56		
Wilcox	O	6,900–8,200	7,400	3–18	9.3	190–1,510	626	29.5–31.8	30.4	14.4–20.3	15.3	45–55	53	26–36	35	22–37	35
	C	5,800–11,500	9,100	5–94	19.1	3.0–1,880	96	14.5–27.4	19.6	0.2–10.0	1.5	27–62	46	20–54	38	19–42	34
Woodbine	O	2,300–10,200	7,900	3–29	10.0	9.0–2,460	195	16.2–34.0	21.9	4.6–20.5	9.7	32–72	47	20–50	37		
	O	4,100–4,400	4,300	6–13	8.2	14–680	368	23.5–28.7	25.5	10.7–27.4	20.1	34.4–72.7	46	24–59	36	26–28	27
Yegua	G/C	4,400–8,700	6,800	3–63	11.0	24–5,040	750	23.4–37.8	30.7	0.1–15.5	1.2	26–74	57	17–59	33		
	O	3,700–9,700	6,600	2–59	8.5	23–4,890	903	22.9–38.5	31.6	3.5–21.8	11.4	31–73	57	17–53	34	30–46	37
Louisiana Gulf Coast–Lafayette Area																	
Miocene	C	5,200–14,900	11,200	3–98	20.2	36–6,180	1,010	15.7–37.6	27.3	0.1–4.7	1.5	37–79	53	20–74	35		
	O	2,700–12,700	9,000	3–32	11.0	45–9,470	1,630	18.3–39.0	30.0	6.5–26.9	14.3	30–72	51	18–50	32	25–42	36
Oligocene	C	7,300–14,600	9,800	2–80	14.6	18–5,730	920	16.7–37.6	27.7	0.5–8.9	2.3	33–71	51	19–57	32		
	O	6,700–12,000	9,400	2–39	8.3	64–5,410	1,410	22.1–36.2	29.0	5.2–20.0	11.1	34–70	54	23–60	35	29–44	38

Table 5.2.4 (g) Oklahoma-Kansas Area

Formation	Fluid prod.	Range of prod. depth, ft	Avg. prod. depth, ft	Range of prod. thickness, ft	Avg. prod. thickness, ft	Range of perm. K, md	Avg. perm. K, md	Range of perm. K_{90}, md	Avg. perm. K_{90}, md	Range of porosity, %	Avg. porosity, %	Range of oil satn., %	Avg. oil satn., %	Range of total water satn., %	Avg. total water satn., %	Range of calc. connate water satn., %	Avg. calc. connate water satn., %	Range of gravity, °API	Avg. gravity, °API
Arbuckle	G	2,700–5,900	4,500	5.0–37	18.3	3.2–544	131	–	–	9.0–20.9	14.4	0.7–9.4	3.7	34.5–62.7	43.1	28–62	40		
	O	500–6,900	3,500	1.0–65.5	11.8	0.2–1,530	140	0.1–1,270	67.8	2.1–24.3	12.0	5.2–42.3	17.7	20.6–79.3	52.4	20–79	47	29–44	37
	T[b]	800–11,600	3,600	2.0–33	14.3	0.1–354	57	0.1–135	21.8	3.7–23.1	9.2	0–23.8	7.1	37.2–91.9	69.2	37–91	52	42	42
Atoka[c]	G	3,700–3,800	3,700	1.0–9.0	4.0	1.3–609	174	–	–	8.5–17.3	12.9	0–8.1	2.0	36.4–65.2	47.2	32–65	45		
	O	500–4,500	2,600	3.0–16	7.8	0.3–920	144	0.6–2.8	1.7	5.9–28.6	14.5	5.1–35.1	20.7	16.4–61.5	38.7	19–61	37	31–42	38
	T	300–3,700	2,100	2.0–10	6.5	9–166	67.3	–	–	11.9–18.6	14.9	5.8–21.1	12.1	42.7–55.4	47.0	40	40		
Bartlesville	G	700–7,400	2,600	1.5–42	11.4	0.2–36	10.4	5.5	5.5	8.4–21.1	15.6	0–11.1	4.7	23.4–70	54.1	23–68	48		
	O	200–5,700	1,500	1.0–72	14.0	0.2–537	32.7	1.5	1.5	8.5–25.8	17.8	3.3–60.6	18.2	17.4–85.2	44.4	17–72	40	28–42	34
	T	500–2,600	1,200	4–40	14.5	0.1–83	18.2	0.07	0.07	8.5–20.1	14.6	0.9–35.7	12.2	43.9–88	63.5	43–67	54	35	35
Bois D'Arc	G	4,800–5,100	5,000	4–48	19	0.1–43	24.4	0.1–2.2	0.45	3.8–19.8	12.2	0–8.7	4.3	32.9–62.4	42.8	26–62	40		
	O	3,700–7,800	6,500	2.3–50	12.5	0.3–664	36.0	–	–	1.2–19.3	7.2	3.3–25.8	15.0	14.6–58.5	32.4	15–59	32	32–42	40
Booch	G	2,600–3,200	2,900	5–8	6.5	1.4–6.6	4.0	–	–	11.9–14.8	13.4	4.6–8.8	6.7	50–51.3	50.7	50	50		
	O	1,000–3,800	2,600	2–26.5	8.8	0.3–160	19.3	–	–	8.3–21.4	15.6	4.8–49.7	21.5	15.3–60	40.0	15–59	37	29–42	35
	T	2,700–3,300	3,000	4–5	4.5	3.1–13	8.0	–	–	16.9–18.1	17.5	7.4–7.8	7.6	47.3–55.2	51.3	44	44		
Burgess	G	–	1,600	–	20	–	142	–	–	–	14.2	–	6.3	–	37.3	–	35		
	O	300–2,800	1,800	2.5–9	5.8	0.2–104	19	22	22	8.1–22.8	13.2	16.2–33	21.5	19.3–65.4	42.2	19–58	40	31–38	36
First	G	6,800–7,600	7,200	3.0–19.5	11.3	0.6–62	31.3	0.4	0.40	1.5–6.5	4.0	0–7.6	3.8	35.7–71.8	53.8	36–72	54		
Bromide[d]	O	3,700–13,800	8,600	2.0–82	18.7	0.1–2,280	175	0.2–7.4	2.23	1.4–15.7	9.8	3.1–24	11	12.8–67.2	35.4	12–67	34	31–42	40
	T	6,000–13,200	11,500	15–161.3	65.1	0.9–40	18.3	1.40	1.40	1.5–10.9	6.5	0.4–6.8	2.2	29.5–78.8	48.3				
Second	G	6,900–16,200	12,800	20–53.6	37.9	3.4–72	21.4	0.3–0.9	0.60	3.5–14.5	6.8	0–6.9	4.0	28.2–45.7	37.9	28–45	32	42	42
Bromide[e]	O	4,500–11,200	9,000	3.0–6.9	16.2	2.0–585	118	–	–	5.6–11.7	9.3	2.4–24.2	11.5	8.9–44.9	25.1	8–44	25	37–42	41
	T	4,400–13,300	9,700	5–44.5	18.4	0.8–42	12.9	–	–	5.8–11.4	7.4	0–13.6	4.8	21.1–57.6	43.5	40	40		
Burbank	O	1,300–4,500	2,800	3–48	17.3	0.1–226	8.64	–	–	8.4–21.6	15.7	9.3–26.6	15.3	31.5–73.4	47.2	31–73	43	35–41	39
	T	2,800–3,700	3,000	3–19	9.1	0.1–4.8	1.53	–	–	7.1–17.0	13.7	0–15.7	11.2	45.7–80.7	57.8	45–81	51		
Chester	G	4,200–6,700	5,700	2–45	10.9	0.1–269	33.0	0.9–3.5	1.87	2.6–20.7	12.2	0–7.5	1.1	20.9–80.7	46.8	19–81	43		
	O	4,700–6,700	5,700	2–23	8.6	0.1–61	9.11	0–0.5	0.21	2.3–16.0	10.1	7.2–35.9	19.1	17.7–80.8	42.1	17–81	51	38–42	40
	T	4,800–6,100	5,700	4–20.5	10.0	0.1–13	2.38	0.1–5.0	1.18	3.2–17.8	7.7	0–11.1	1.2	40.9–89.2	61.7	40–89	43		
Cleveland[f]	G	2,200–5,700	3,500	2–17	9.0	2.5–338	50.6	–	–	9.8–23.5	16.9	0–7.1	4.1	40–64.4	48.9	30–64	42		
	O	300–6,400	3,200	1–70	13.4	0.1–135	15.4	1.4–2.3	1.85	7.4–24.6	15.2	5.8–35.5	13.1	10.2–74.0	46.7	10–74	44	27–56	42
	T	1,900–3,900	3,100	3–22	7.7	0.1–112	12.9	–	–	11.0–20.4	15.6	0–21.1	7.8	39.9–77.2	55.3	32–77	49		
Deese[g]	G	4,300–11,800	6,500	5–55	19.3	7.8–232	94.1	0.80	0.80	9.8–22.6	16.7	2.2–6.3	3.8	19.1–54.9	42.1	19–49	37		
	O	600–10,000	5,200	2–60.3	11.7	0.4–694	62.8	1.10	1.10	4.7–26.4	17.4	5.9–46.4	20.4	14.0–56.6	37.8	13–57	33	17–42	32
Hoover	G	2,200–6,800	4,000	4–49	16.6	1.9–200	61.8	–	–	11.7–23.4	18.3	0–7.0	0.8	41.1–77.1	53.8	19–76	45		
	O	1,800–2,100	2,000	3–37	11.9	1.3–974	288	–	–	12.7–24.1	19.7	12.6–23.1	16.0	14.6–48.5	40.2	14–47	35	36–42	42
	T	1,900–2,000	2,000	2–17	8.4	55–766	372	–	–	16.7–22.5	20.5	6.6–17.1	14.5	34.8–50.7	42.9	31–42	35		
Hoxbar	G	3,800–8,880	6,300	9–11	10.0	6.4–61	33.7	–	–	13.9–18.2	16.1	0.7–4.4	2.6	40.1–40.6	40.4	34–39	37		
	O	1,000–10,300	4,200	2–63	14.4	0.1–1,620	277	–	–	3.1–29.7	18.5	3.2–48.7	21.4	13.8–68.5	45.1	13–68	39	42	42
	T	2,900–3,000	3,000	3–13	9.3	0.5–31	14.4	–	–	14.3–22.7	18.5	3.3–11.4	6.6	50.5–69.8	57.9			29–42	34

continued

Table 5.2.4 (continued) (g) Oklahoma-Kansas Area

continued

Formation	Fluid prod.	Range of prod. depth, ft	Avg. prod. depth, ft	Range of prod. thickness, ft	Avg. prod. thickness, ft	Range of perm. K, md	Avg. perm. K, md	Range of perm. K_{90}, md	Avg. perm. K_{90}, md	Range of porosity, %	Avg. porosity, %	Range of oil satn., %	Avg. oil satn., %	Range of total water satn., %	Avg. total water satn., %	Range of calc. connate water satn., %	Avg. calc. connate water satn., %	Range of gravity, °API	Avg. gravity, °API
Hunton	O	1,800–9,600	4,600	2–77.3	14.0	0.1–678	34.5	0–77.0	5.24	1–33.8	10.9	1.6–34.5	15.3	16.7–93.4	48.6	17–93	46	24–42	38
	T	2,500–8,700	4,900	2–73	14.7	0.1–48	5.3	0.1–7.9	2.04	1–19.5	7.3	0–61.1	10.6	16.0–88.7	54.5	16–89	48		
Lansing	O	1,900–5,800	3,800	3–16.2	6.5	0.3–390	101	0.3–162	52.3	8.4–16.0	12.2	6.5–28.9	18.1	37.4–68.6	51.9	28–69	49	31–39	37
	T	–	3,300	–	22.0	–	14	–	6.7	–	7.2	–	12.8	–	75.5	–	–		
Layton	G	700–6,100	3,900	4–18	9.3	0.2–210	26.3	–	–	5.1–25.9	14.5	0–7.8	2.4	38.2–83.7	54.1	34–83	47		
	O	500–6,300	2,900	1–57	10.3	0.3–280	54.1	0.5–162	23.3	4.6–27.2	17.8	1.6–37.3	15.3	28–76.3	45.5	23–76	41	30–42	37
	T	1,800–5,700	3,200	3–15.5	7.4	1.1–143	23.8	–	–	14.2–21.3	17.1	0–14.3	6.9	33.2–69.4	45.9	31–69	43		
Marmatom	O	4,300–4,600	4,400	1.5–7.5	4.7	24–105	46.4	0.20	0.20	1.8–21.4	14.0	6.4–16.1	11.7	42.8–66.4	55.5	42–66	53	36–42	40
Misner	G	8,100	8,100	3–14	8.5	37–171	104	–	–	11.0–12.1	11.6	2.1–2.3	2.2	16.9–86.7	21.4	18–22	20		
	O	2,600–6,500	4,300	2–56.5	10.6	0.1–803	89.7	0–2.1	0.62	2.1–20.9	11.9	4.1–41.6	14.8	21.4–51.7	41.5	14–87	38	36–48	42
	T	4,900–6,200	6,000	8–21	15.8	0.1–120	41.8	–	–	1.9–11.3	8.1	0–8.2	4.7	–	33.0	20–51	32		
Mississippi Chat	G	1,800–5,100	4,000	2–34.4	16.1	0.4–516	33.5	0.2–74	13.9	6.5–37.8	21.0	0–6.8	2.4	60.3–93.4	76.7	60–93	77		
	O	800–5,200	3,100	2–48.1	12.2	0.1–361	21.9	0–216	13.7	5.7–39.3	22.3	1.4–30.0	12.9	27.1–94.8	64.0	27–95	58	22–42	35
	T	1,200–5,200	3,900	1–43	10.9	0.2–229	21.3	0–163	14.2	1.5–38.0	18.7	1.1–18.3	7.6	47.4–84.9	71.5	43–85	63		
Mississippi Lime	G	900–8,800	4,600	3–27.1	13.3	0.1–129	22.2	0.1–89	13.2	1.5–23.6	10.3	0–9.3	2.8	22.6–93.5	63.2	22–93	53		
	O	600–6,600	4,100	1.5–95.3	12.0	0.1–1,210	43.5	0.1–185	9.44	1.3–34.1	13.4	2.1–56.5	15.0	16.9–85.3	50.7	16–85	46	22–45	39
	T	400–7,200	4,000	4–70.1	17.4	0.1–135	7.5	0.1–36	4.23	1.1–26.1	9.3	0–41.2	6.9	32.9–94.0	67.6	32–94	61		
McLish	G	3,600–17,000	10,000	14–58	35.3	12–98	48.0	–	–	2.8–9.6	6.7	4.0–14.7	7.8	19.3–76.5	43.9	19–77	44		
	O	1,600–11,200	8,100	3–42	12.2	0.7–157	39.0	–	–	5.5–1.5	11.0	5.1–27.7	13.2	14.8–52.2	32.1	14–52	31	35–48	38
Morrow	G	4,300–9,700	6,100	2–64	11.0	0.1–1,450	115	6.2–8.8	7.5	4.2–24.4	14.8	0–33.0	4.3	29.0–77.0	46.5	16–77	36		
	O	4,100–7,500	5,700	2–37	9.8	0.2–1,840	117	0.3–55	23.1	57–23.2	14.6	0.7–44.5	15.1	23.9–75.5	42.1	16–54	35	33–43	40
	T	5,500–6,900	6,100	3–30	9.5	0.1–410	34.4	0.1–48	28.0	5.5–16.2	11.3	0–15.2	5.0	31.1–90.1	57.2	31–90	38		
Oil Creek	G	7,100–14,000	10,900	14–149	46.3	0.1–132	32.0	–	–	6.1–13.5	9.0	0–6.5	1.6	12.5–40.6	25.2	12–40	24		
	O	5,100–11,700	8,300	3–71	12.6	0.1–615	131	0.2–230	75.6	1.8–23.9	13.1	1.3–29.5	13.0	14.2–76.4	39.1	14–76	34	29–42	36
	T	8,400–13,700	12,300	8–27	15.0	0.1–87	22.1	–	–	5.2–16.1	10.9	0–5.8	2.6	21.7–74.9	46.6	21–74	–		
Oswego	G	4,500–4,600	4,600	8–9	8.5	2.4–151	76.7	–	–	12.0–17.3	14.7	5.1–6.4	5.8	39.8–55.5	47.7	34–55	45		
	O	300–6,300	3,800	3.6–34.1	12.3	0.2–296	27.3	0.1–66	9.24	2.6–21.6	10.1	0–27.1	15.0	16.2–73.4	41.5	15–73	37	35–48	44
	T	1,200–5,800	3,300	2–21	10.6	0.1–117	27.0	0–41	11.5	4.7–20.9	8.7	0–14.5	5.8	41.7–89.7	63.4	42–89	57		
Peru	G	1,200–5,300	3,100	4–17	9.8	3.1–42	15.0	–	–	12.3–17.5	15.6	0.1–7.9	4.1	44.3–59.4	52.5	44–56	51		
	O	200–3,200	1,200	2–42	12.4	0.2–284	20.8	–	–	12.7–33.8	18.7	6.7–36.8	14.7	34.4–73.1	50.6	28–73	44	25–43	36
	T	700–2,500	1,500	4–21	10.3	1.7–804	205	–	–	13.6–24.4	19.2	2.8–25.5	12.0	38.0–60.4	50.7	36–56	51		
Prue	G	3,000–6,600	4,000	5–22	13.8	0.7–42	18.3	–	–	13.8–22.4	17.8	2.3–9.1	5.5	31.4–53.4	42.2	25–49	37		
	O	600–6,700	3,100	2–81	14.6	0.1–254	22.6	–	–	7.6–23.8	17.0	4.7–34.1	16.9	24.4–73.1	41.6	20–72	38	34–46	42
	T	3,000–5,400	3,700	3–18	11.7	0.5–133	42.8	–	–	9.8–23.4	17.5	3.7–34.3	19.0	40.7–60.9	47.1	32–60	36		
Purdy	O	4,200–7,400	4,500	3–30	14.8	7.4–500	182	51–266	179	12.3–18.8	16.7	10.1–27.2	20.0	31.4–58.1	41.5	16–50	38	39–44	41
	T	–	4,200	–	4.8	–	195	–	166	–	17.8	–	13.6	–	56.2	–	36		
Reagan	G	3,500–3,600	3,600	2–13	7.4	1.1–173	39.3	–	–	9.3–12.7	10.8	1.1–7.9	4.2	28.4–68.4	44.4	28–68	40		
	O	2,100–3,700	3,600	1–32	11.0	0.2–2,740	255	–	–	6.9–21.5	13.3	3.0–42.0	14.2	17.5–72.9	32.9	12–72	31	41	41
	T	3,600	3,600	5–7	6.0	19.0–37	38.0	–	–	10.6–12.8	11.7	1.8–10.5	6.2	33.3–46.7	40.0	29–45	29	24–43	38

Table 5.2.4 (g) Oklahoma-Kansas Area[a]

Formation	Fluid prod.	Range of prod. depth, ft	Avg. prod. depth, ft	Range of prod. thickness, ft	Avg. prod. thickness, ft	Range of perm. K, md	Avg. perm. K, md	Range of perm. K_{90}, md	Avg. perm. K_{90}, md	Range of porosity %	Avg. porosity %	Range of oil satn. %	Avg. oil satn. %	Range of total water satn. %	Avg. total water satn. %	Range of calc. connate water satn. %	Avg. calc. connate water satn. %	Range of gravity °API	Avg. gravity °API
Redfork	G	2,300–7,400	4,300	4–19	7.9	0.1–160	23.4	–	–	3.8–21.2	14.5	0–21.7	4.7	16.2–63.6	45.8	16–63	39		
	O	300–7,600	3,100	1–63	10.5	0.1–668	14.2	–	–	6.6–26.1	16.2	5.4–30.8	16.9	29.5–57.7	43.7	27–55	41	32–48	37
	T	1,200–3,800	3,100	2–9	5.3	0–23	6.3	–	–	10.1–18.6	15.3	0.3–36.3	9.9	41.4–69.7	52.6	41–69	49		
Skinner	G	1,000–5,300	3,700	4–29	11.8	0.1–127	27.7	–	–	13.3–19.8	15.7	0–9.9	4.2	30.6–48	40.8	26–47	38		
	O	1,000–5,800	3,200	1–42.5	9.2	0.1–255	20.6	2–6.6	3.30	7.4–21.7	15.3	2.5–29.7	20.1	14.3–78.7	40.3	14–78	38	30–46	36
	T	2,400–4,600	3,400	6–35.9	11.5	0.3–16	6.0	–	2.40	11.7–19.0	15.5	4.9–18.2	8.5	39.9–71.1	52.4	39–71	39		
Strawn	G		1,100		12.0		71.0	–	–		21.3		9.9		61.8		38		
Sycamore	O	1,000–7,400	3,500	2–40.5	12.4	0.1–599	58.1	–	–	8.2–23.5	16.8	5.7–31.1	15.1	28.5–61.5	45.6	22–56	41	31–44	40
	O	2,600–6,700	4,600	2–84	26.4	0.1–3.1	0.67	0.13	0.50	7.2–18.4	13.3	9.2–33.5	21.1	36.0–61.6	45.5	32–62	43	33–36	35
Tonkawa	G	5,000–7,100	5,600	4.42–27.5	9.8	0.3–283	46.7	–	–	11.7–21.4	16.4	0–8.1	2.0	31.6–56.3	44.5	27–56	41		
	O	2,400–5,700	4,800	4.42–28.5	8.7	1.4–278	98.6	8–22	15.0	13.2–22.9	18.4	7.5–16.5	12.5	36.1–78.0	45.0	31–78	38	40–45	43
Tucker	O	2,300–3,100	2,700	4–9	7.0	1.3–406	106	–	–	15.4–18.9	17.1	6.9–17.3	11.4	45.1–52.6	49.0	44–52	45		
Tucker	O	1,300–2,900	2,200	2–14	7.6	2.1–123	36	–	–	12.4–20.3	15.6	7.3–29.8	16.0	35.6–50.1	40.7	33–43	38	29–40	36
	T	2,700–2,900	2,800	8.9–16	12.5	4.3–252	128	53	53	11.8–19.5	15.7	7.1–10.9	9.0	58–64.3	61.2	52–62	52		
Tulip Creek	G	7,200–16,700	13,400	21–268.4	78.1	0.9–24	7.63	0.5–1.0	0.40	2.0–11.9	6.1	0–6.6	4.1	23.7–54.8	33.2	23–55	34		
	O	700–16,800	8,000	2–136	15.3	0.1–1,470	154	0.2–1.8	0.80	2.5–25.0	11.6	3.0–44.5	12.2	10.0–63.0	34.9	9–63	33	49.5	49.5
	T	1,400–12,900	8,600	3–86.5	20	2.0–143	44.6	0.40	0.40	0.7–26.0	11.0	0.7–7.7	2.6	15.9–82.8	45.7	15–82	46	32–50	40
Viola	G	4,300–7,300	5,400	3–73	39.1	3.6–23	10.8	3.40	3.40	8.1–10.1	9.3	1.7–9.4	5.0	19.7–37.2	30.7	19–37	30		
	O	2,100–11,100	4,900	2–111.7	17.2	0.1–1,150	52.3	0.2–186	18.3	1.0–16.1	8.4	3.2–41.0	15.5	24.1–85.5	54.4	24–86	51	28–48	37
	T	2,600–10,300	4,600	2–117	19.6	0.1–997	45.1	0.3–49	4.38	0.6–18.8	7.1	0–33.7	8.6	39.0–90.8	65.7	39–90	58		
Wayside	G	300–2,800	800	3.1–34	10.8	0.2–133	22.2	–	–	13.2–24.9	18.6	8.1–33.8	18.6	29.4–68.0	51.3	28–67	47		
	O	2,800–5,400	4,300	2–35	11.3	0.7–145	72.1	–	–	5.2–15.6	10.8	0.7–8.3	3.6	29.7–60.5	43.9	29–60	44	29–42	35
First Wilcox	O	2,800–7,400	4,900	2–28	10.0	0.2–445	91.3	–	–	5.4–20.5	12.0	3.6–40.5	11.7	15.0–58.2	32.0	14–58	31	33–50	42
	T	3,200–6,100	3,900	1.9–29	7.7	0.3–418	84.1	0.80	0.80	6.8–17.7	10.9	0–16.9	7.9	24.8–63.6	41.7				
Second Wilcox	G	5,000–10,000	6,700	5–28	13.4	0.2–154	76.2	–	–	5.0–15.1	11.2	0–3.8	1.5	17.7–45.8	30.9	17–43	29		
	O	3,700–8,400	6,500	1.3–32	11.3	0.4–2,960	214	–	–	4.2–20.6	12.4	2.9–19.2	10.2	19.0–56.3	36.9	18–56	34	34–42	40
	T	4,700–7,500	6,000	1.5–5	4.4	0.4–756	246	–	–	1.9–20.4	12.9	0–8.4	6.1	41.4–60.5	42.5	40–60	43		
Woodford	O	4,100–5,000	4,600	2.6–30.4	16.2	1.4–250	87.1	2.4–156	79.2	1.9–6.6	4.4	8.3–16.7	11.8	43.0–87.9	60.1	43–87	60	41	41

[a] General geologic sections taken at different points in Oklahoma-Kansas areas indicate some variations in the properties and an appreciable variations in the occurrence and relative depths of many of the more important oil- and/or gas-producing zones, formations, geologic groups, and their members. The general identification of core samples from these producing intervals reflect local conditions or activities significantly. In the development of the average data values, an attempt has been made to combine data originally reported for locally named zones into more generally recognized formations or geologic groups. In some instances (i.e., Deese, Cherokee) data are reported for a major geologic group as well as for some of its individual members. The values designated by the major group name represent areas where the general characteristics permit identification as to the geologic group but not as to group members. In other areas the group members or zones are readily identifiable. The combinations of data and the use of local rather than regional geologic names in some instances are explained in the footnotes.

[b] T Represents transition zone or production of both water and either gas or oil.

[c] Includes data reported as Dornick Hills and Dutcher.

[d] Includes data reported as McClain Country area.

[e] Data reported locally as Bromide Third, Bromide Upper third, and Bromide Lower have been considered as part of the Tulip Creek. Includes Bromide First and Second have been considered as part of the Tulip Creek.

[f] Includes data reported as Cleveland Sand, Cleveland, Lower, and Cleveland Upper.

[g] Includes the numerous zones (Deese First, Second, Third, Fourth, Fifth, Zone A, Zone B, Zone C, and Zone D) reported locally for the Anadarko, Ardmore, and Marietta Basin areas. In northwest Oklahoma, these different zones are normally refereed to as Cherokee. In other areas the zones are frequently identifiable and properties are reported as for Redfork, Bartlesville, etc.

Table 5.2.4 (continued) (h) Rocky Mountain Area

Formation	Fluid prod.	Range of prod. depth, ft	Avg. prod. depth, ft	Range of prod. thickness, ft	Avg. prod. thickness, ft	Range of perm. K, md	Avg. perm. K, md	Range of perm. K90, md	Avg. perm. K90, md	Range of porosity, %	Avg. porosity, %	Range of oil satn., %	Avg. oil satn., %	Range of total water satn., %	Avg. total water satn., %	Range of calc. connate water satn., %	Avg. calc. connate water satn., %	Range of gravity, °API	Avg. gravity, °API
Aneth	O	5,100–5,300	5,200	3.8–23.1	14.0	0.7–34	9.35	0.2–23	6.10	4.4–10.5	8.1	14.5–35.9	25.0	12.5–30.5	23.6	13–31	24	41	41
Boundary	G	5,500–5,600	5,600	8–27	17.5	0.1–2.0	1.05	—	—	4.3–6.5	4.7	4.7	4.7	23.8–35.0	29.4	23–35	29	—	—
butte	O	5,400–5,900	5,600	2–68	16.2	0.1–114	13.3	0.2–33	12.5	5.4–21.6	11.0	4.8–26.7	12.5	9.3–48.8	28.3	7–45	27	40–41	41.1
Cliffhouse	G	3,600–5,800	4,800	2–58	13.7	0.1–3.7	0.94	—	—	7.0–16.2	11.3	0–19.8	4.5	10.2–60.3	36.9	10–59	36	—	—
D Sand	O	4,350–5,050	5,800	7–33	15.0	0–900	192	—	—	8.6–29.5	21.6	8.4–39.5	13.2	—	—	9–48	23	36–42	38
Dakota	G	500–7,100	5,700	2–24	9.5	0.1–710	28.6	33.0	33	7.3–19.6	11.2	0–7.8	3.5	14.8–55.3	40.6	15–52	38	—	40
Desert	O	3,400–7,200	5,600	4–19	7.9	0.1–186	22.3	0.4–2.4	1.13	5.0–23.3	12.7	13.8–35.9	24.4	11.6–44.3	31.0	11–44	29	38–43	40
Frontier sands	O	265–8,295	2,950	8–100	46	0–534	105	—	—	6.3–29.8	20.0	7.6–37.6	15.2	14.8–24.7	19.2	14–25	19	41	41
Gallop	G	1,500–6,900	5,000	5–25	11.6	0.1–324	26.5	0.3–20	10.2	8.5–20.8	13.3	0–25.6	5.7	20.7–59.2	40.0	20–54	33	31–50	39
Hermosa	G	500–6,400	4,600	2–43	12.4	0.1–2,470	48.2	0.1–3.2	0.7	6.9–23.1	12.5	8.5–43.7	25.3	17.2–76.9	35.7	14–77	34	36–42	39
Hospa	O	4,900–7,700	5,600	5–30	14.1	0.1–91	18.6	45.0	45.0	5.5–16.5	10.2	0–6.5	3.0	14.2–45.3	32.7	12–45	32	41–42	40
J Sand	O	5,300–6,000	5,600	3–38.2	15.1	0.1–37	7.32	0–26	4.26	2.7–17.9	8.3	3.9–29.1	10.8	11.6–60.0	35.6	12–60	35	40	40
Madison lime[†]	O	4,800–7,100	5,500	3–17	10.5	0.1–70	18.2	—	—	7.4–11.9	10.5	0.5–23.8	7.5	8.7–49.7	38.1	8–49	37	21.6–30	26
Menefee	G	4,600–5,100	4,800	6–18	13.3	0.7–25	8.63	—	—	6.6–14.8	11.3	0.5–23.8	13.9	32.3–44.8	36.0	31–45	35	—	—
Mesa Verde	G	4,470–5,460	4,900	15–62	25	0–1,795	330	—	—	8.9–32.7	19.6	20.4–29.8	25.0	—	—	6–42	20	36–42	38
Morrison	O	3,400–6,200	4,900	41–450	186	0–1,460	13	—	—	1.8–26.4	11.9	6.0–43.5	17.4	—	—	22–33	27	29–56	42
Muddy	O	5,200–5,700	1,845	7–25	12.7	0.1–20	5.03	—	—	8.7–13.5	11.2	0.3–5.3	1.6	14.5–45.1	27.5	15–43	27	26–42	38
Paradox	O	1,500–6,100	6,900	2–22	10.0	0.1–17	3.57	—	—	10.0–19.8	14.6	0–6.8	3.3	14.5–68.4	42.0	15–64	40	40–43	41
Phoshporia (formerly Embar)	O	700–10,500	4,600	5–100	64	0–126	3.7	—	—	2.0–25.0	8.9	3.0–40.0	22.5	—	—	5–30	21	15–42.3	25.4
Picture Cliffs	G	1,800–5,800	3,400	3–43	16.8	0.1–7.68	1.12	—	—	8.5–25.5	15.7	0–21.1	2.6	23.6–67.5	46.7	23–53	43	55	55
Point Lookout	O	4,300–6,500	5,500	2–101	22.9	0.1–16	1.74	—	2.40	5.6–21.6	10.9	0–9.1	2.9	11.9–55.6	36.7	12–55	36	—	39
Sundance	O	1,100–6,860	3,100	5–100	44	0–1,250	100	—	—	15.0–25.0	19.0	8.0–25.0	17.0	40	40.9	20–49	35	22–63	39
Tensleep	O	600–11,800	4,700	10–200	118	0–2,950	120	—	—	5.0–27.0	13.6	6.0–30.0	23.3	7.0–59	25.7	5–50	19	17–58.5	26.2
Tocito	O	1,400–5,100	4,600	4–58	17.3	0–31	3.36	—	—	12.8–17.8	14.7	11.9–26.6	21.3	40.8–55	46.3	40–55	46	36–40	36

*Not enough wells to justify range of variations.
[†]Data limited to Big Horn Basin

Table 5.2.4 (f) *West Texas-Southeastern New Mexico Areas*

Formation	Group[a]	Range Area[b]	Fluid prod. prod.	Range of prod. depth, ft	Avg. prod. depth, ft	Range of prod. thick-ness, ft	Avg. prod. thick-ness, ft	Range of perm. K, md	Avg. perm. K, md	Range of perm. K_{90}, md	Avg. perm. K_{90}, md	Range of porosity, %	Avg. porosity, %	Range of oil satn., %	Ave. oil satn., %	Range of total water satn., %	Avg. total water satn., %	Range of connate water satn., %	Avg. connate water satn., %	Range of gravity, °API	Avg. gravity, °API
Bend conglomerate	1	1, 2, 4, 5, 6, 8	O	6,000–6,100	6,000	3–22	13.2	4–311	150	—	—	13.8–16.9	15.0	8.1–8.6	8.3	43–64	52	42–62	50	40–42	14
Blinebry	2	3	O	10,300–10,500	10,400	10–28	20.0	1.6–11	5.7	0.9–5.1	2.2	4.0–15.7	10.9	9.5–16.1	11.5	21–41	33	21–39	32	41–45	43
	—	8	G	5,383–5,575	5,480	23–50	36	1.6–3.8	2.4	—	—	10.7–14.8	12.7	2.1–9.1	4.9	34–40	36	31–33	33	39–42	40
Cambrian	—	3	O	5,262–5,950	5,610	4–95	43	0.1–5.3	1.8	0.2–4.2	1.4	3.1–12.5	7.8	9.6–19.3	12.8	29–57	40	27–56	39	39–42	40
Canyon Reef	—	2, 3, 4, 5, 6, 7	O	5,500–6,300	5,900	2.0–95	30.3	0.8–1,130	173	—	—	4.1–16.8	12.0	6.9–21.2	11.8	22–71	39	22–71	38	44–51	48
	—	2, 3, 4, 5, 6, 7	O	4,200–10,400	7,100	4.0–222	36.8	0.6–746	42	0.3–249	17	3.0–21.5	8.9	3.6–39.2	11.6	18.3–73	44	18–73	43	30–47	42
Canyon Sand	—	2, 3, 4, 7	G	—	5,000	—	8.0	—	1.7	—	—	—	15.1	—	5.8	—	46	—	44	37–43	40
Clearfork	1	3, 4, 7	O	3,000–10,000	5,500	3.0–57	16.9	0.1–477	38	—	—	5.5–22.1	14.3	4.8–27.7	13.7	21–72	43	21–72	41		
		2 (part)	G	—	2,400	—	95	—	11	—	7.8	—	13.5	—	5.7	—	50	—	50		
	2	5, 6	O	1,500–6,800	4,400	4.0–180	41	0.1–43	4.6	<0.1–24	2.5	4.1–20.6	9.2	7.5–31.4	15.6	18–84	54	18–84	53	23–42	28
		2 (part)	O	5,400–8,300	6,600	3.0–259	3.3	<0.1–136	5.8	<0.1–109	3.1	1.9–19.4	5.8	5.6–27.1	16.5	22–69	47	21–69	47	28–40	32
Dean	—	2, 4, 7	O	7,700–9,100	8,200	6.0–68	26.2	<0.1–0.3	0.12	—	—	7.5–12.7	10.3	22–44	33.7	20–52	34	19–51	33	37–40	39
Delaware	—	1[c]	G	4,700–5,000	4,800	5.2–39	18.6	1.1–33	12.9	—	—	13.8–21.8	17.9	2.0–10.3	6.0	45–66	53	36–63	49		
	—		O	3,500–5,100	4,200	3.0–52	14.5	0.6–84	24.5	—	—	15.2–25.4	21.0	3.9–15.6	11.2	33–65	49	31–64	42	35–42	40
Devonian	1	2, 5[d]	G	11,200–11,600	11,400	14–117	54	0.5–36	10.5	0.1–1.3	0.5	1.7–5.3	3.3	2.1–6.6	3.7	37–68	53	37–68	53		
			O	11,300–12,300	11,800	8–299	99	0.2–23	4.0	0.1–5.8	0.8	1.3–6.8	4.3	3.3–16.7	9.2	19–53	33	19–53	33	48–52	49
	2	2, 5[e]	G	—	9,200	—	17	—	0.4	—	—	—	6.7	—	5.7	—	62	—	61		
	3	2, 3, 4, 5, 6, 7, 8[f]	O	5,500–9,900	7,700	8–113	34	2.5–50	14.9	0.2–18	7.0	5.5–27.7	15.2	6.8–22.9	11.0	41–76	51	41–76	51	35–46	42
			C	11,000–11,200	11,100	19–34	27	<0.1–2.2	1.1	<0.1–0.9	0.5	2.2–7.7	5.0	3.1–4.8	4.0	45–69	57	45–69	57		
Ellenburger	—	All	O	7,800–12,800	11,200	6.5–954	69	1.0–2,840	177	0.3–1,020	37	1.8–25.2	6.0	5.3–24.6	12.9	22–65	46	22–65	46	36–49	42
Fusselman	—	All	C	4,100–10,600	7,400	11–18	14.3	203–246	225	1.4–54	27.7	3.7–4.6	4.2	0.8–7.6	4.2	47–67	57	47–67	57	37–52	47
	—	All	O	5,500–16,600	10,100	3.0–347	55	0.1–2,250	75	<0.1–396	22.9	1.3–13.8	3.8	1.0–19.2	8.4	40–84	61	40–84	60		
Glorietta (Paddock)[g]	—	All	C	8,700–12,700	10,300	18–51	34	1.2–26	8.4	0.3–1.3	0.9	2.6–3.7	3.3	0.2–3.9	1.7	32–47	40	32–47	40	47–50	48
	—	All	O	9,500–12,500	12,000	8–49	32	0.5–25	10.3	0.2–17	3.9	1.4–10.7	3.3	5.2–16	10.4	25–64	42	26–64	38		
Granite wash	—	3, 4, 6, 7	G	2,200–2,600	2,400	3–44	16.3	4.6–12	5.6	—	9.3	14–18.2	15.0	3.9–4.4	3.7	39–60	51	24–71	47	28–40	33
			O	2,300–6,000	4,300	3–103	22.3	0.4–223	11.5	0.2–126	8.1	5.2–20.9	13.6	3.1–22.1	15.4	24–72	48	24–71	47		
			G	3,000–8,600	4,700	4–8	5.1	11–2,890	477	—	53	12.1–20.4	14.4	2.9–8.7	5.2	39–66	55	39–66	53	40–45	42
			O	2,300–3,400	3,000	2–81	15.6	5–3,290	609	—	30	3.5–26.1	17.7	4.8–22.5	14.7	42–71	54	35–66	49		
Grayburg	1	8	G	3,600–4,200	3,800	3.0–5.0	4.2	1.6–9.3	6.5	—	—	11.1–14.3	12.4	7.1–42	18.6	22–53	39	22–52	38	31–41	36
	2	4, 5, 6, 7	O	2,400–4,500	4,100	3.0–123	27.4	0.5–159	13.7	0.2–48	5.2	7.0–20.0	11.3	6.2–37.9	17.6	26–56	36	25–55	35	23–40	32
			G	4,400	4,400	12–26	20.8	0.6–3.7	2.5	0.3–2.1	1.3	6.3–6.6	6.4	2.4–7.1	4.7	55–68	60	55–68	60		
			O	3,000–4,800	4,400	6–259	45	0.2–118	5.5	0.1–110	2.7	2.7–16.2	7.9	4.8–22.1	13.9	32–84	55	32–84	55		
	3	1, 2, 3	O	1,300–3,900	2,700	4.5–182	50	0.3–1,430	37.7	0.1–228	14.3	5.3–24.3	11.9	8.3–34	18.2	31–78	58	31–78	56	28–35	31

continued

Table 5.2.4 (continued) (l) West Texas–Southeastern New Mexico Areas

Formation	Group[a]	Range Area[b]	Fluid prod.	Range of prod. depth, ft	Avg. prod. depth, ft	Range of prod. thickness, ft	Avg. prod. thickness, ft	Range of perm. K, md	Avg. perm. K, md	Range of perm. K90, md	Avg. perm. K90, md	Range of porosity, %	Avg. porosity, %	Range of oil satn., %	Ave. oil satn., %	Range of total water satn., %	Avg. total water satn., %	Range of calc. connate water satn., %	Avg. calc. connate water satn., %	Range of gravity, °API	Avg. gravity, °API
Pennsylvanian sand (Morrow)[g]	—	2, 3, 4	O	4,100–11,400	9,100	1.7–77	22.3	0.3–462	34.9	0.1–168	14.7	2.7–13.9	7.7	4.7–18.8	9.7	28–58	42	28–58	41	38–47	41
Queen (Penrose)[g]	—	All	G	3,000–3,200	3,100	4.0–29	9.9	10–318	64	—	—	10.7–22.2	16.6	2.6–7.6	7.4	36–62	48	35–58	45		
San Andres	1	8	O	800–4,900	3,500	1.5–38	10.2	0.2–4,190	123	—	1.0	5.7–27.0	17.2	4.2–34.7	15.6	32–68	49	30–66	45	30–42	33
	2	5, 6	O	3,900–4,700	4,500	6–39	18.6	0.3–461	61	0.1–482	53	3.2–14.0	8.5	8.9–33.9	18.7	21–49	36	19–49	36	34–38	37
	3	1, 2, 3, 4, 7	O	4,100–5,300	4,500	4.7–124	40.1	0.3–295	6.9	0.1–208	3.8	3.1–12.8	7.1	4.9–30.6	14.7	26–69	52	25–69	51	30–37	33
Seven Rivers	—	1, 2, 5, 8	O	1,500–5,100	3,300	3.0–197	30.2	0.2–593	9.7	0.2–510	8.4	3.3–25.1	15.5	3.5–24.2	13.2	39–74	58	37–74	56	26–37	32
	—		O	3,600–4,100	3,900	3.0–8.0	5.6	0.6–23	12.2	—	0.3	15.5–16.6	16.0	3.4–9.5	6.4	51–66	56	46–65	54	28–38	32
	—		O	800–4,000	2,600	4.0–136	18.5	0.4–428	51.4	—	8.0	5.9–28.9	16.5	4.2–41.7	16.2	38–70	54	38–61	50	36–42	39
Sprayberry	1	4, 7	O	4,800–8,500	7,100	2.0–59	21.7	0.2–71	6.3	—	4.0	10.1–23.3	15.8	7.0–24.5	15.3	32–68	45	30–67	43	39–47	41
Strawn Lime	—	All	G	—	5,600	11–57	34.4	4.5–310	179	27–189	108	10.9–14.8	12.9	5.5–6.3	5.9	38–39	39	38–39	38		
Strawn Sand	—	2, 3, 4, 7	O	5,200–6,700	5,900	2.0–101	36.7	1.9–196	43	0.6–148	19.1	3.1–12.6	7.2	4.9–28.3	11.2	15–66	44	15–66	43	29–48	42
	—		G	3,800–10,500	7,800	3.0–39	16.8	0.3–42	11.4	—	0.2	2.1–14.2	6.9	1.7–5.2	3.0	48–60	52	48–60	52		
	—		O	1,100–11,300	5,200	2.0–76	15.1	0.2–718	47	0.2–718	11.7	1.0–20.3	12.6	2.7–27.9	12.2	23–77	43	23–77	41		
	—	Others[h]	O	915–7,366	3,938	6–21	14	1.0–400	45	—	—	6.0–27	16.2	5.0–27	14.1	25–60	43	23–59	41	38	38
Tubb	1	1, 2, 5, 8	O	6,100–7,300	6,500	15–43	33.5	0.2–135	27.6	0.1–1.1	0.5	2.5–7.1	4.9	8.5–25.3	12.9	37–64	54	37–64	54	42	42
Wolfcamp (Abo)[g]	1	2	O	—	9,800	—	10.6	—	2.3	—	0.4	—	4.3	—	19.6	—	25	—	25		
	2	4, 5, 6, 7	O	8,400–9,200	8,800	13–129	41.7	2.3–9,410	419	1.5–6,210	274	4.9–18.5	9.9	6.6–16.8	9.7	32–68	44	31–56	44	36–45	40
	3	3 (part)	G	2,500–4,100	3,600	4.0–119	22.5	0.1–1,380	57	—	3.4	7.2–24.5	15.3	0.5–16.1	4.6	30–64	48	26–64	45		
			O	2,400–4,100	3,500	2.0–114	28.0	1.0–1,270	60	—	4.3	5.4–26.3	15.5	1.6–26.8	14.3	32–65	46	29–64	44	40–50	48
	4	8	O	9,000–10,600	9,700	4.5–204	59	0.2–147	20.4	0.2–36	5.4	2.6–12.8	8.1	5.3–23.6	14.4	28–56	54	28–56	39	40–44	42
Yates	—	1, 2, 5, 8	O	1,400–3,500	2,800	3.0–53	10.8	0.2–145	19.3	—	—	12.1–27.4	17.9	1.3–17.0	5.6	43–79	59	36–78	53	27–41	32
	—		G	1,400–4,000	2,300	3.0–66	16.6	1.0–4,000	42.7	—	27.8	2.4–27.0	18.8	3.7–37.3	16.0	31–75	53	31–75	47		

[a] More than one group indicates distinct differences in formation as found in different areas.
[b] Area numbers refer to map of Figure 24-1.
[c] Plus Ward, Pecos, and Southwestern Lea County.
[d] Midland and Ector counties only.
[e] Crane, Ward, Winkler, and Pecos counties only.
[f] Except counties in 4 and 5 above.
[g] Names in parentheses are those commonly used in New Mexico, Area 8.
[h] Archer, Baylor, Jack, Montague, Wichita, Wise, and Young counties.

(text continued from page 5-56)

Table 5.2.5 *Particle Size Definitions*

Material	Particle size, μm	U.S. standard sieve mesh no.
Coarse sand	>500	<35
Medium sand	250–500	35–60
Fine sand	125–250	60–120
Very fine sand	62.5–125	120–130
Coarse silt	31–62.5	—
Medium silt	15.6–31	—
Fine silt	7.8–15.6	—
Very fine silt	3.9–7.8	—
Clay	<2	—

or secondary (diagenetic or solution) porosity. Primary and secondary porosity can be read directly from neutron, density, and sonic logs. These tools do not measure pore volume directly, rather they measure physical parameters of the formation and relate them to porosity mathematically or empirically. Since the sonic tool only records primary (or matrix) porosity, it can be combined with total porosity tools, such as density or a combined neutron and density,

to determine secondary porosity:

$$\phi_{\text{secondary}} = \phi_{\text{total}} - \phi_{\text{sonic}} \qquad [5.2.4]$$

where $\phi_{\text{secondary}}$ = porosity due to vugs and fractures
ϕ_{total} = total porosity as determined from cores, density log, neutron-density crossplot, or local knowledge
ϕ_{sonic} = porosity determined from sonic log.

The combinable magnetic resonance (CMR) tool capable of measuring porosity independent of lithology and measuring movable or bound fluid, which can be used to measure effective porosity in the rock.

Water Saturation

Connate water saturation (S_w) and flushed zone water saturation (S_{xo}) can be calculated from information supplied by well logs.

Connate water saturation is the fraction of pore volume in an undisturbed formation filled with connate water.

$$S_w = \frac{\text{volume of water}}{\text{volume of pores}} \qquad [5.2.5]$$

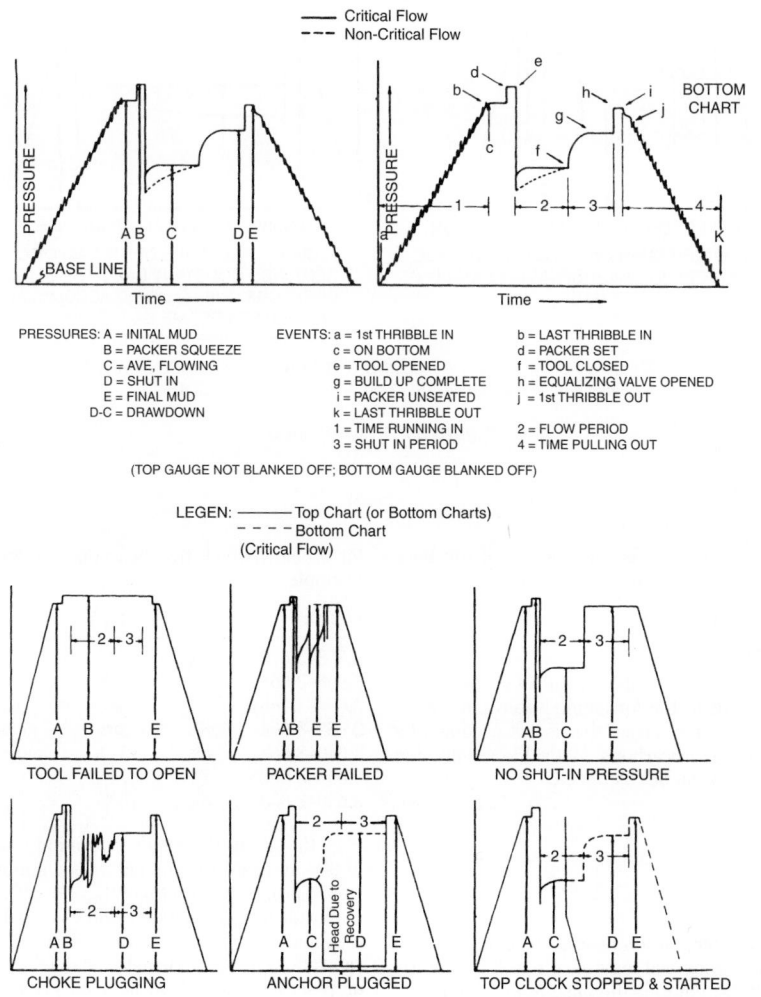

PRESSURES: A = INITAL MUD
B = PACKER SQUEEZE
C = AVE, FLOWING
D = SHUT IN
E = FINAL MUD
D-C = DRAWDOWN

EVENTS: a = 1st THRIBBLE IN
c = ON BOTTOM
e = TOOL OPENED
g = BUILD UP COMPLETE
i = PACKER UNSEATED
k = LAST THRIBBLE OUT
1 = TIME RUNNING IN
3 = SHUT IN PERIOD

b = LAST THRIBBLE IN
d = PACKER SET
f = TOOL CLOSED
h = EQUALIZING VALVE OPENED
j = 1st THRIBBLE OUT

2 = FLOW PERIOD
4 = TIME PULLING OUT

(TOP GAUGE NOT BLANKED OFF; BOTTOM GAUGE BLANKED OFF)

LEGEN: ——— Top Chart (or Bottom Charts)
– – – – Bottom Chart
(Critical Flow)

TOOL FAILED TO OPEN

PACKER FAILED

NO SHUT-IN PRESSURE

CHOKE PLUGGING

ANCHOR PLUGGED

TOP CLOCK STOPPED & STARTED

Figure 5.2.1

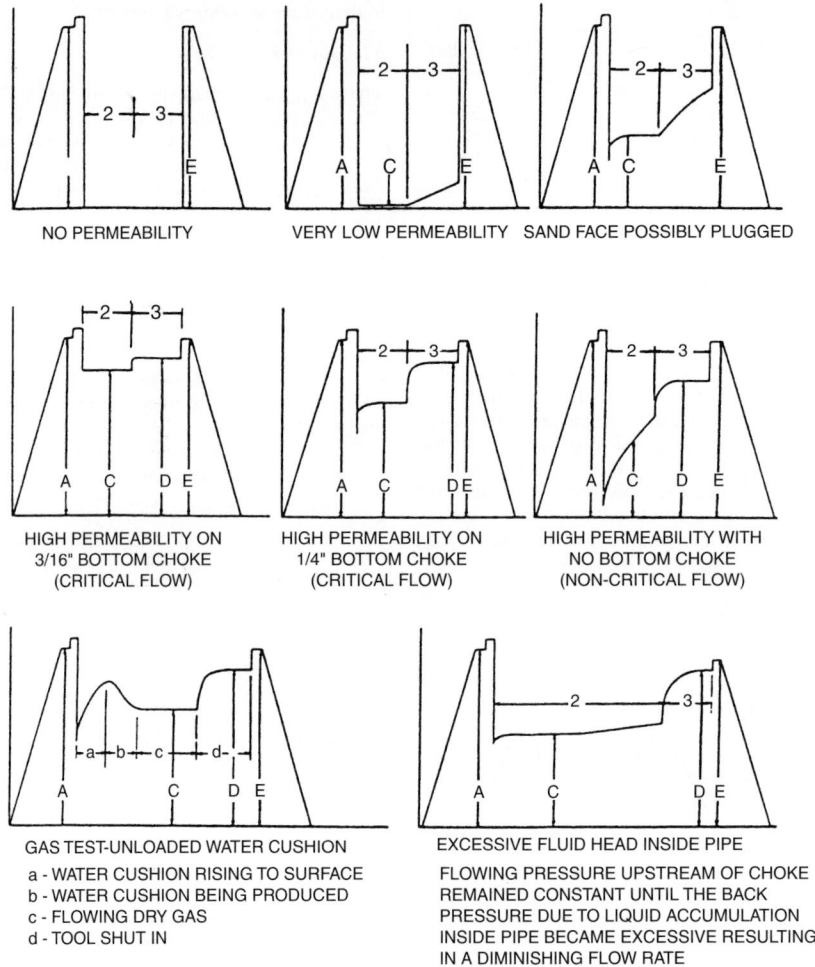

NO PERMEABILITY   VERY LOW PERMEABILITY   SAND FACE POSSIBLY PLUGGED

HIGH PERMEABILITY ON   HIGH PERMEABILITY ON   HIGH PERMEABILITY WITH
3/16" BOTTOM CHOKE   1/4" BOTTOM CHOKE   NO BOTTOM CHOKE
(CRITICAL FLOW)   (CRITICAL FLOW)   (NON-CRITICAL FLOW)

GAS TEST-UNLOADED WATER CUSHION   EXCESSIVE FLUID HEAD INSIDE PIPE

a - WATER CUSHION RISING TO SURFACE   FLOWING PRESSURE UPSTREAM OF CHOKE
b - WATER CUSHION BEING PRODUCED   REMAINED CONSTANT UNTIL THE BACK
c - FLOWING DRY GAS   PRESSURE DUE TO LIQUID ACCUMULATION
d - TOOL SHUT IN   INSIDE PIPE BECAME EXCESSIVE RESULTING
                                         IN A DIMINISHING FLOW RATE

NOTE: THE RUNNING IN AND PULLING OUT PERIODS IN THOSE CHARTS
ARE SHOWN COMPRESSED ON TIME SCALE FOR CLARITY.

Figure 5.2.1 Continued.

Flushed zone saturation (S_{xo}) is the fraction of the pore volume filled with flushing agent (normally drilling fluid).

$$S_{xo} = \frac{\text{volume of flushing agent}}{\text{volume of pores}} \qquad [5.2.6]$$

Prior to penetration by a drill bit, only two fluids are assumed to be present in the formation-water and hydrocarbons. Therefore, all pore space that is not occupied by water is occupied by hydrocarbons. With this assumption hydrocarbon saturation can be calculated:

$$S_h = 1 - S_w \qquad [5.2.7]$$

where S_h = hydrocarbon saturation.

Pay Thickness
The thickness of a hydrocarbon-bearing formation (h_{pay}) is easily determined from well logs once ϕ and S_w cutoffs are established. The S_w cutoff is the maximum value for S_w for a given rock type. The ϕ cutoff is the minimum value for ϕ below which hydrocarbons cannot be produced. For example:

Depth	ϕ	S_w	Comment
3,668–3,670	1%	53%	ϕ too low
3,666–3,668	2%	50%	ϕ too low
3,664–3,666	6%	38%	possible hydrocarbons
3,662–3,664	6%	36%	possible hydrocarbons
3,660–3,662	8%	31%	possible hydrocarbons
3,658–3,660	7%	74%	too wet
3,656–3,658	8%	100%	too wet

In this case, the water saturation cutoff is a maximum of 60% and the porosity cutoff is a minimum of 3%, so this well will have 6 feet of pay ($h_{pay} = 6$ ft). Other factors that may reduce h_{pay} include shaliness, shale streaks, low permeability, and low reservoir pressure. Porosity and water saturation cutoffs are usually established for specific regions or reservoirs based on detailed production and geologic information.

(text continued on page 5-76)

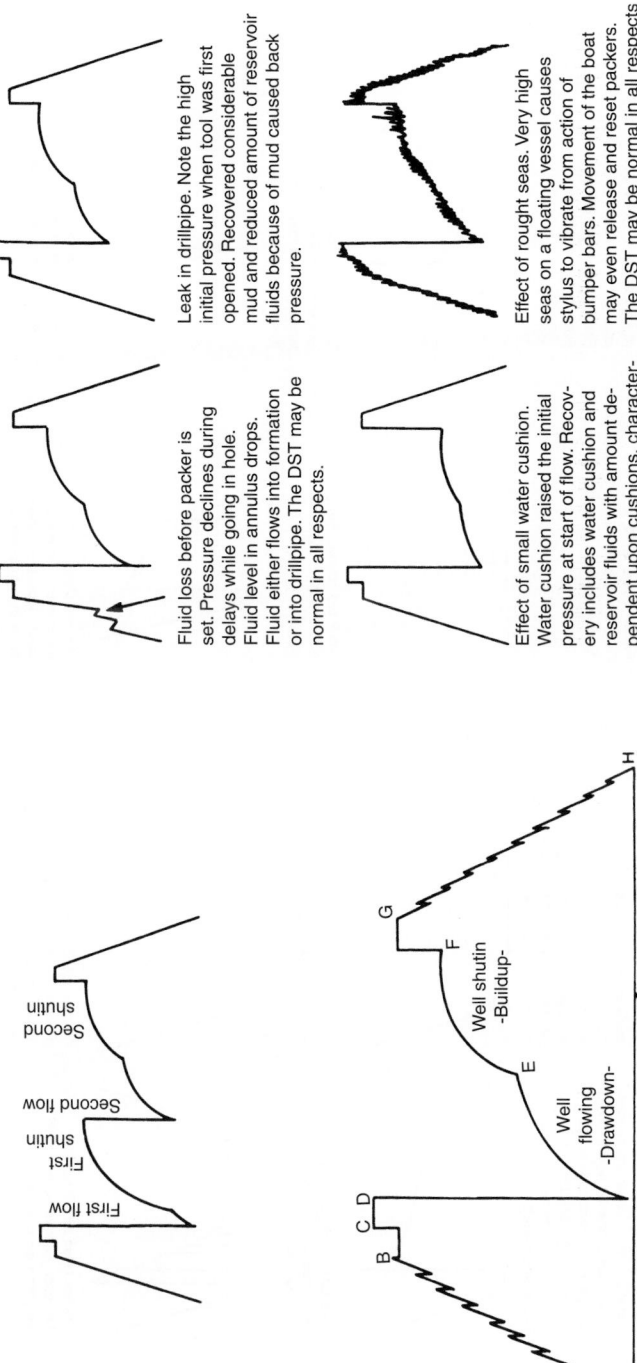

Leak in drillpipe. Note the high initial pressure when tool was first opened. Recovered considerable mud and reduced amount of reservoir fluids because of mud caused back pressure.

Effect of rought seas. Very high seas on a floating vessel causes stylus to vibrate from action of bumper bars. Movement of the boat may even release and reset packers. The DST may be normal in all respects other than the jagged charts.

Fluid loss before packer is set. Pressure declines during delays while going in hole. Fluid level in annulus drops. Fluid either flows into formation or into drillpipe. The DST may be normal in all respects.

Effect of small water cushion. Water cushion raised the initial pressure at start of flow. Recovery includes water cushion and reservoir fluids with amount dependent upon cushions, characteristic of formation and fluids, surface operations, etc.

Figure 5.2.2

Second shutin

Second flow

First shutin

First flow

Well shutin -Buildup-

Well flowing -Drawdown-

Base line

A B C D E F G H

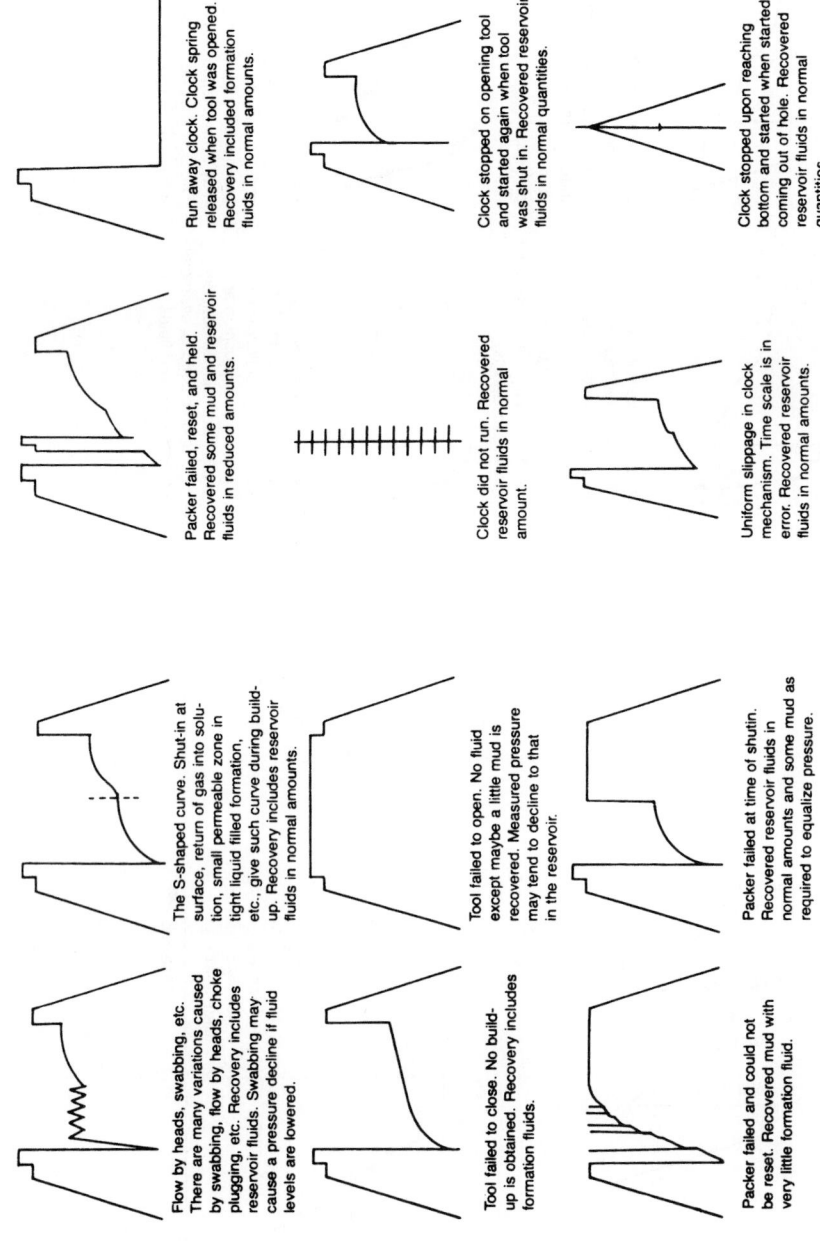

Flow by heads, swabbing, etc. There are many variations caused by swabbing, flow by heads, choke plugging, etc. Recovery includes reservoir fluids. Swabbing may cause a pressure decline if fluid levels are lowered.

The S-shaped curve. Shut-in at surface, return of gas into solution, small permeable zone in tight liquid filled formation, etc., give such curve during build-up. Recovery includes reservoir fluids in normal amounts.

Packer failed, reset, and held. Recovered some mud and reservoir fluids in reduced amounts.

Run away clock. Clock spring released when tool was opened. Recovery included formation fluids in normal amounts.

Tool failed to close. No build-up is obtained. Recovery includes formation fluids.

Tool failed to open. No fluid except maybe a little mud is recovered. Measured pressure may tend to decline to that in the reservoir.

Clock did not run. Recovered reservoir fluids in normal amount.

Clock stopped on opening tool and started again when tool was shut in. Recovered reservoir fluids in normal quantities.

Packer failed and could not be reset. Recovered mud with very little formation fluid.

Packer failed at time of shutin. Recovered reservoir fluids in normal amounts and some mud as required to equalize pressure.

Uniform slippage in clock mechanism. Time scale is in error. Recovered reservoir fluids in normal amounts.

Clock stopped upon reaching bottom and started when started coming out of hole. Recovered reservoir fluids in normal quantities.

Figure 5.2.2 Continued.

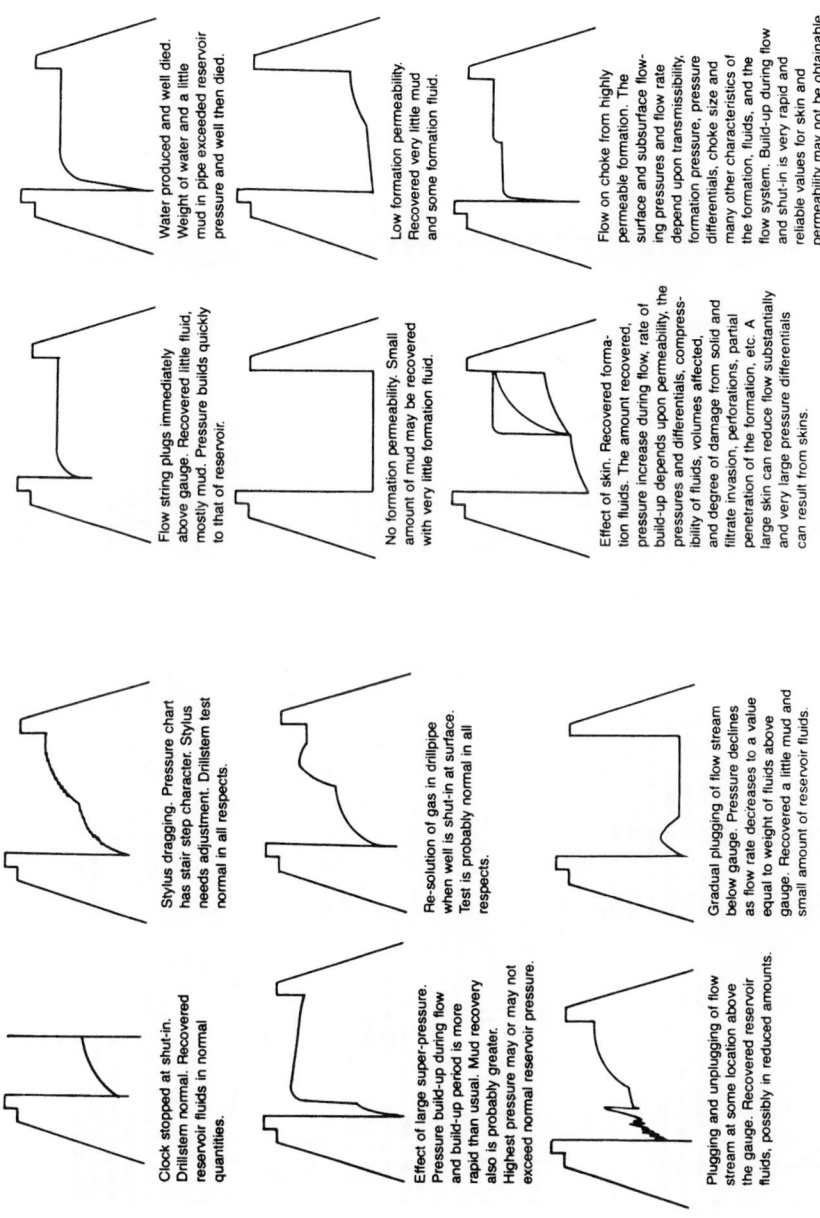

Clock stopped at shut-in. Recovered reservoir fluids in normal quantities.

Stylus dragging. Pressure chart has stair step character. Stylus needs adjustment. Drillstem test normal in all respects.

Water produced and well died. Weight of water and a little mud in pipe exceeded reservoir pressure and well then died.

Flow string plugs immediately above gauge. Recovered little fluid, mostly mud. Pressure builds quickly to that of reservoir.

Effect of large super-pressure. Pressure build-up during flow and build-up period is more rapid than usual. Mud recovery also is probably greater. Highest pressure may or may not exceed normal reservoir pressure.

Re-solution of gas in drillpipe when well is shut-in at surface. Test is probably normal in all respects.

Low formation permeability. Recovered very little mud and some formation fluid.

No formation permeability. Small amount of mud may be recovered with very little formation fluid.

Plugging and unplugging of flow stream at some location above the gauge. Recovered reservoir fluids, possibly in reduced amounts.

Gradual plugging of flow stream below gauge. Pressure declines as flow rate decreases to a value equal to weight of fluids above gauge. Recovered a little mud and small amount of reservoir fluids.

Effect of skin. Recovered formation fluids. The amount recovered, pressure increase during flow, rate of build-up depends upon permeability, the pressures and differentials, compressibility of fluids, volumes affected, and degree of damage from solid and filtrate invasion, perforations, partial penetration of the formation, etc. A large skin can reduce flow substantially and very large pressure differentials can result from skins.

Flow on choke from highly permeable formation. The surface and subsurface flowing pressures and flow rate depend upon transmissibility, formation pressure, pressure differentials, choke size and many other characteristics of the formation, fluids, and the flow system. Build-up during flow and shut-in is very rapid and reliable values for skin and permeability may not be obtainable.

Figure 5.2.2   Continued.

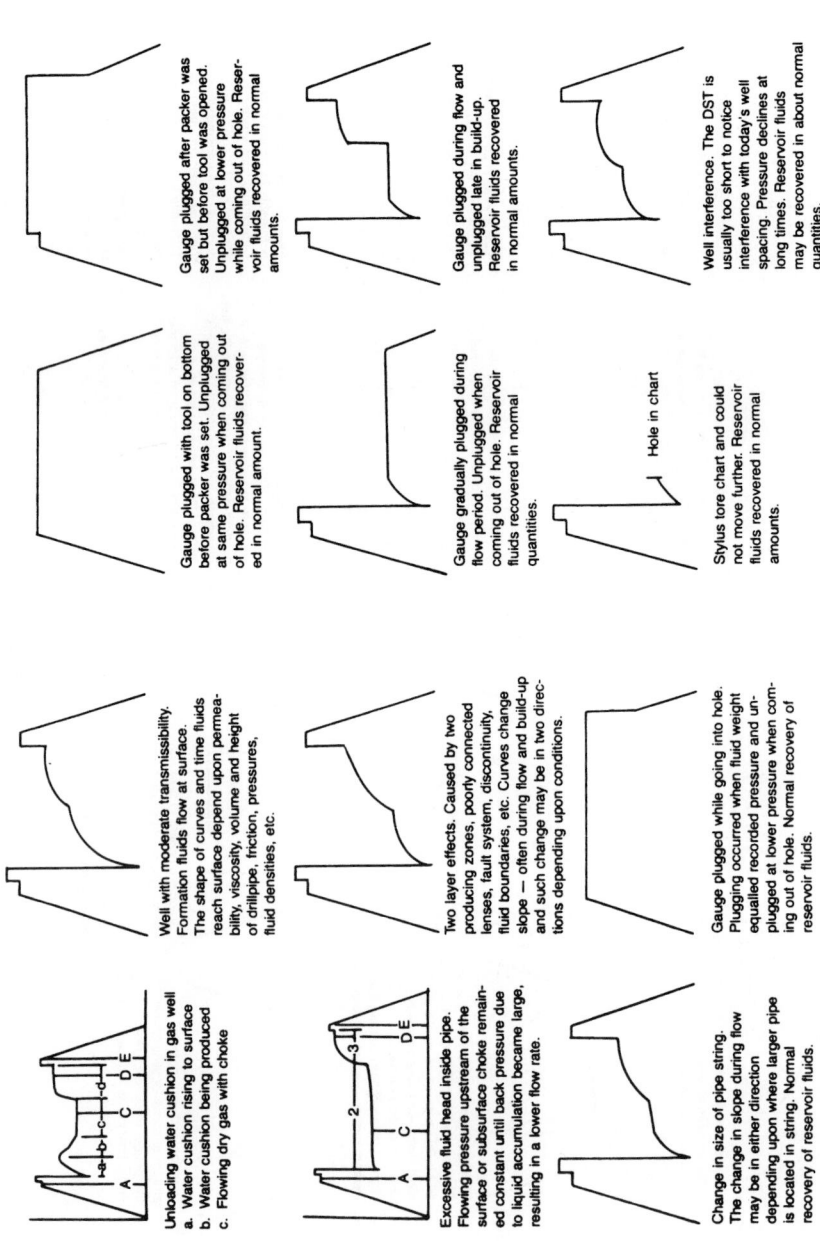

Unloading water cushion in gas well
a. Water cushion rising to surface
b. Water cushion being produced
c. Flowing dry gas with choke

Well with moderate transmissibility. Formation fluids flow at surface. The shape of curves and time fluids reach surface depend upon permeability, viscosity, volume and height of drillpipe, friction, pressures, fluid densities, etc.

Gauge plugged with tool on bottom before packer was set. Unplugged at same pressure when coming out of hole. Reservoir fluids recovered in normal amount.

Gauge plugged after packer was set but before tool was opened. Unplugged at lower pressure while coming out of hole. Reservoir fluids recovered in normal amounts.

Excessive fluid head inside pipe. Flowing pressure upstream of the surface or subsurface choke remained constant until back pressure due to liquid accumulation became large, resulting in a lower flow rate.

Two layer effects. Caused by two producing zones, poorly connected lenses, fault system, discontinuity, fluid boundaries, etc. Curves change slope — often during flow and build-up and such change may be in two directions depending upon conditions.

Gauge gradually plugged during flow period. Unplugged when coming out of hole. Reservoir fluids recovered in normal quantities.

Gauge plugged during flow and unplugged late in build-up. Reservoir fluids recovered in normal amounts.

Change in size of pipe string. The change in slope during flow may be in either direction depending upon where larger pipe is located in string. Normal recovery of reservoir fluids.

Gauge plugged while going into hole. Plugging occurred when fluid weight equalled recorded pressure and unplugged at lower pressure when coming out of hole. Normal recovery of reservoir fluids.

Stylus tore chart and could not move further. Reservoir fluids recovered in normal amounts.

Hole in chart

Well interference. The DST is usually too short to notice interference with today's well spacing. Pressure declines at long times. Reservoir fluids may be recovered in about normal quantities.

Figure 5.2.2 Continued.

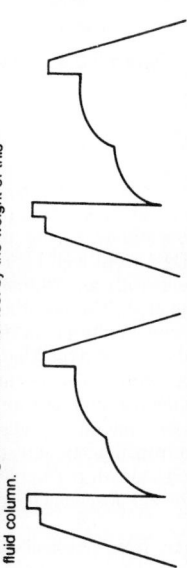

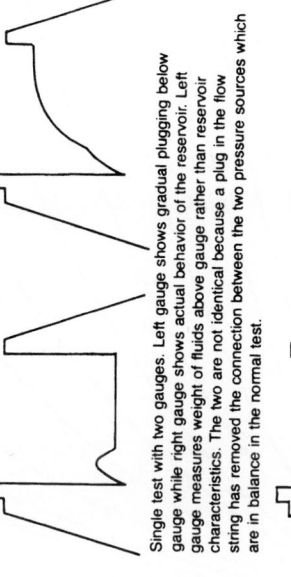

Two tests with the same gauge. Second build-up extrapolated pressure is lower than that of first build-up. Depletion of a small reservoir might be suggested.

Two tests with the same gauge. Character of the curves for the second test differs appreciably from that of the first test. Skin or other parameter in the flow and build-up equations which might be sensitive to flow or shutin has changed between the two tests. Since some fluid entered drillstring during first test, initial pressure of second test will be higher than that of first test by the weight of this fluid column.

Single test with two gauges. The two sets of curves should be identical except for the small difference in pressure related to location of the gauges within the flow string. This is the normal condition to be expected if both gauges are operating properly and no plugging, etc., is occurring.

Single test with two gauges. Left gauge suggests highly permeable formation with little or no skin — maybe negative — while right curve indicates gauge which plugged upon reaching bottom and unplugged when starting out of hole. If right gauge was at bottom, it was probably stuck in debris at bottom of hole.

Single test with two gauges. Left gauge shows gradual plugging below gauge while right gauge shows actual behavior of the reservoir. Left gauge measures weight of fluids above gauge rather than reservoir characteristics. The two are not identical because a plug in the flow string has removed the connection between the two pressure sources which are in balance in the normal test.

Single test with two gauges. Left gauge measures reservoir characteristics while right gauge plugged while going into the well and did not unplug until inspected at surface.

Figure 5.2.2 Continued.

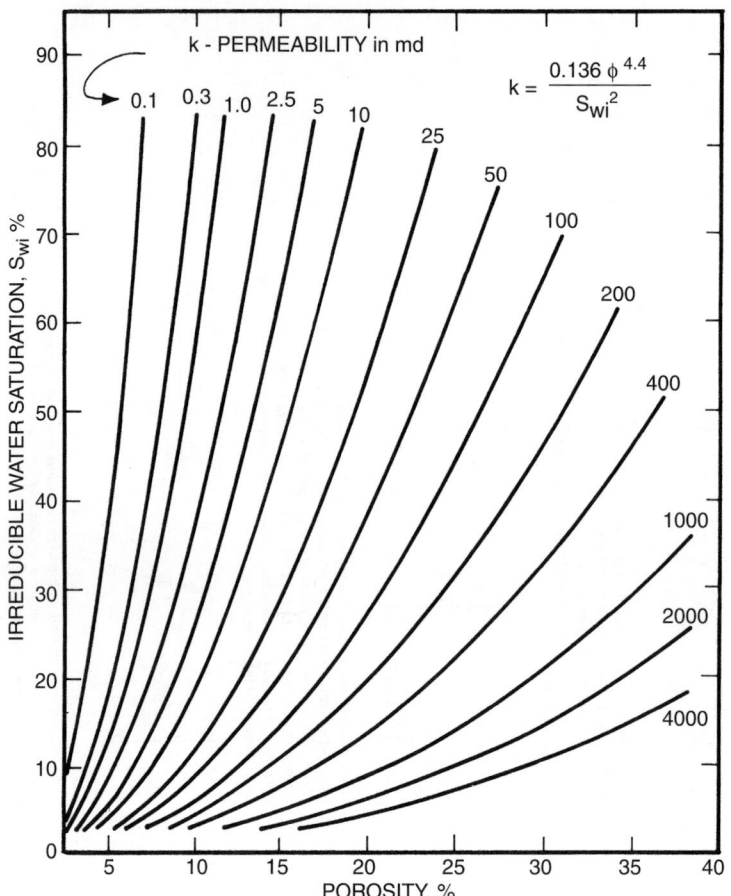

Figure 5.2.3 Timur chart for estimating permeability [19].

(*text continued from page 5-70*)

Lithology

It is often necessary to know the rock type in order to properly design downhole assemblies, casing programs, and completion techniques. Data from well logs can provide the geologist or engineer with an estimate of the lithologic makeup of any formation. The accuracy of this estimate is a function of the complexity of the formation (mineralogic makeup and fluid types) and the kinds of tools used to investigate the rocks. More tools are needed to accurately determine compositions of complex formations. Simple lithologies (three or less minerals, or gas) can be determined with combined neutron, density, and sonic logs. This technique will be discussed later. More complex lithologies can be determined with the aid of special logging tools and computers.

Since well logs infer lithology from physical and chemical parameters, certain rocks will look the same on logs though they differ in their geologic classification. Sandstone, quartz, and chert are all SiO_2 and appear the same on porosity logs. The same is true of limestone and chalk. Dolomite, anhydrite, and salt have very distinct characteristics and are easily distinguished from other rock types. Shales are composed of clay minerals. The type and amount of different clay minerals, which vary widely between shales, can affect their bulk properties.

Permeability

Permeability is one of the essential properties used in evaluation of a potentially producing formation. Unfortunately there are no logging devices that read permeability. This is because permeability is a dynamic property. Most logging tools spend only a few seconds in front of any one point of a formation, therefore it is impossible to measure any time-dependent parameter. There are methods to estimate permeability from well logs, but they are based on general assumptions. From a practical standpoint, log parameters only provide an "order of magnitude" approximation. Several methods of inferring permeability with well logs are discussed where applicable in each section.

Two relationships between porosity and irreducible water saturation (S_{wi}) are used to estimate permeability:

1. The Timur relationship [19] (Figure 5.2.3) for granular rocks (sandstones and oolitic limestones), which generally gives a more conservative estimate of permeability.
2. The Wyllie and Rose relationship [43] modified by Schlumberger [20] (Figure 5.2.4), which generally gives a higher estimate of permeability.

To enter these charts, porosity and irreducible water saturation (S_{wi}) must be known. Porosity can be obtained from

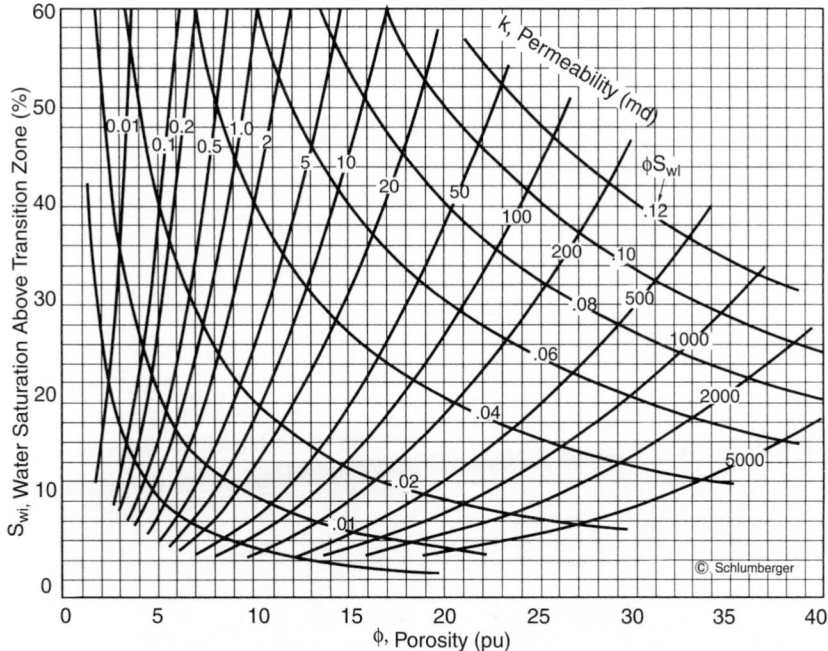

Figure 5.2.4 *Schlumberger chart (after Wyllie and Rose) for estimating permeability [20].*

cores or any porosity device (sonic, neutron, or density). Irreducible water saturation must be found from capillary pressure curves or it can be estimated. The permeabilities from these charts should be considered "order of magnitude" estimates.

5.2.3.3 Influences on Logs

The purpose of well logging is to determine what fluids are in the formation and in what quantity. Unfortunately the drilling process alters the fluid saturations by flushing the pores near the borehole and filling them with the fluid fraction of the drilling mud (mud filtrate). To correct for these influences, the invasion profile must be identified. Figure 5.2.5, an idealized cross-sectional view of the borehole and formations, shows an invasion profile and the appropriate symbols for each part of that profile [20].

Mud Relationships

Since the borehole is filled with mud and the adjacent portion of the formation is invaded with mud filtrate, mud properties must be accurately known so they can be taken into account. Mud has a minor influence on most porosity tools; however, it can have a large effect on the resistivity tools. In general:

$$R_{mc} > R_m > R_{mf} \qquad [5.2.8]$$

This is because the mudcake is mostly clay particles and has very little water associated with it. The clay particles in the mudcake tend to align themselves parallel to the borehole wall, developing a high horizontal resistivity R_h (Figure 5.2.6). Since the mud filtrate is composed only of fluid and has no solids, it will have a lower resistivity. If the resistivity of the mud (R_m) is known, R_{mc} and R_{mf} can be estimated with the following equations:

$$R_{mf} = K_m(R_m)^{1.07} \qquad [5.2.9]$$

$$R_{mc} = 0.69(R_{mf})\left(\frac{R_m}{R_{mf}}\right)^{2.65} \qquad [5.2.10]$$

where $K_m = 0.847$ for 10 lb/gal
0.708   11 lb/gal
0.584   12 lb/gal
0.488   13 lb/gal
0.412   14 lb/gal
0.380   16 lb/gal
0.350   18 lb/gal

These relationships work well for most muds (except lignosulfonate) with resistivities between 0.1 Ω-m and 10 Ω-m.

Another approximation that works well for salt muds is:

$$R_{mf} = 0.75R_m \qquad [5.2.11]$$

$$R_{mf} = 1.5\,R_m \qquad [5.2.12]$$

Temperature Relationships

Mud resistivity is a function of temperature and ion concentration. Since temperature increases with depth due to geothermal gradient, the mud resistivity is lower at the bottom of the hole than at the surface (pits). The temperature of a formation can be found with an equation suggested by Hilchie [21]:

$$T_f = (T_{TD} - T_s)\frac{D_f}{D_{TD}} + T_s \qquad [5.2.13]$$

where T_f = formation temperature (°F)
T_{TD} = temperature at total depth (BHT) (°F)
T_s = average surface temperature (°F)
D_f = formation depth (ft)
D_{TD} = total depth (ft)

Average surface temperatures in various oilfield areas are:
Alberta   40°F
California   65°F
Colorado-Northern New Mexico   55°F

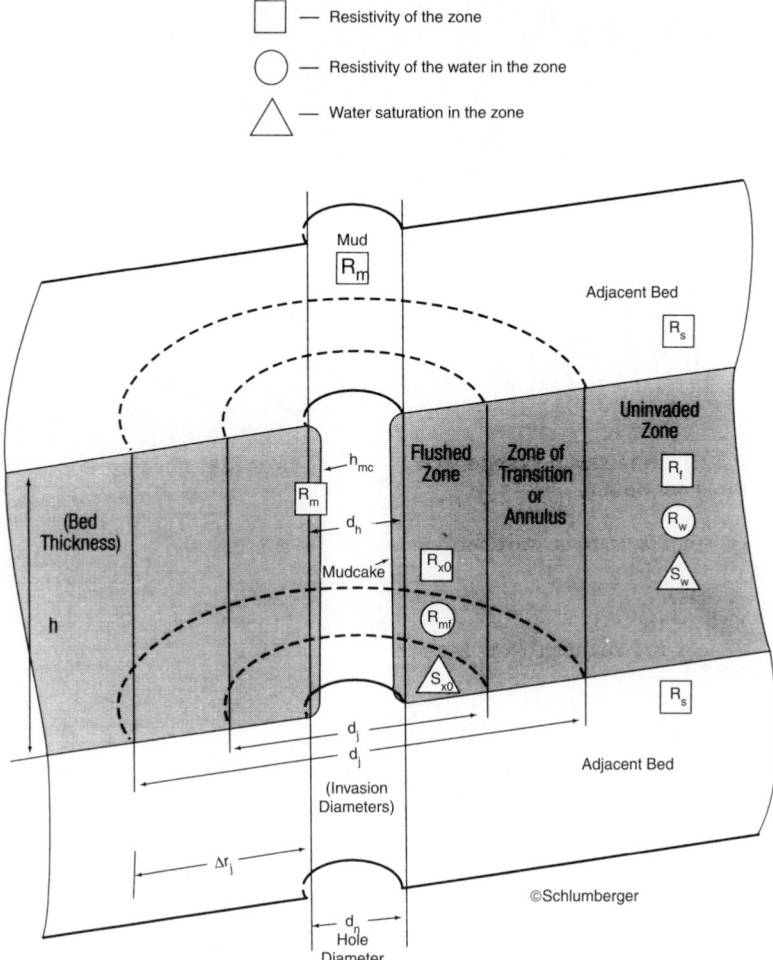

— Resistivity of the zone

— Resistivity of the water in the zone

— Water saturation in the zone

Figure 5.2.5 *Diagram of the borehole environment showing the various zones and their parameters [20].*

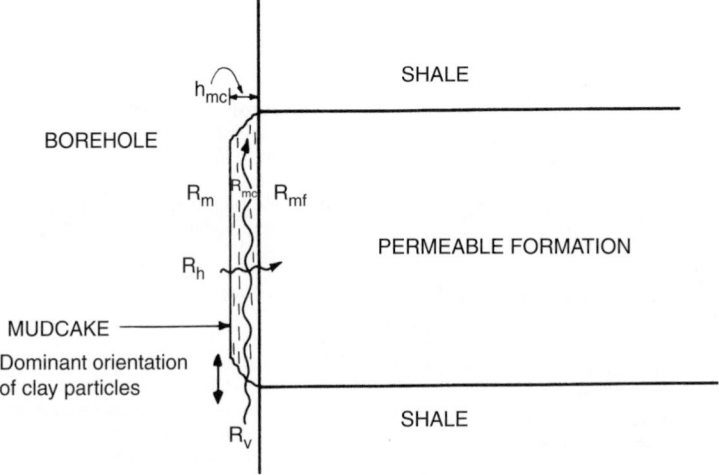

Figure 5.2.6 *Resistivity components on mud cake that develop opposite a permeable formation.*

Temperature Gradient Conversions: 1°F/100 ft = 1.823°C/100 m
1°C/100 m = 0.5486°F/100 ft

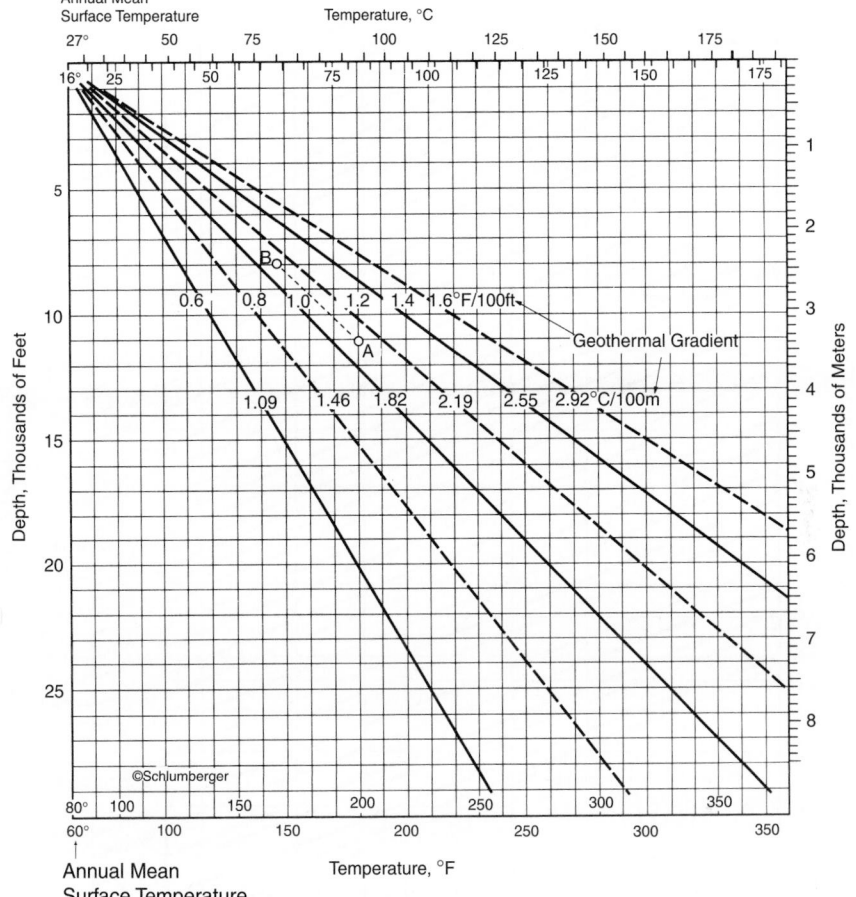

EXAMPLE: Bottom hole temperature, BHT, is 200°F at 11,000 ft (Point A).
Temperature at 8,000 ft is 167°F (Point B).

Figure 5.2.7   *Chart for estimating formation temperature [20].*

Gulf Coast	80°F
Oklahoma	65°F
Permian Basin	65°F
Wyoming	45°F

Figure 5.2.7 solves this equation graphically [20].

Fluid resistivity at any formation depth can be found using the Arps Equation if the resistivity at any temperature and formation temperature are known [20]:

$$R_2 = R_1 \frac{T_1 + 6.77}{T_2 + 6.77} \qquad [5.2.14]$$

where R_2 = fluid resistivity at formation
            temperature, Ω-m
R_1 = fluid resistivity at some temperature, Ω-m
T_2 = formation temperature, °F
T_1 = temperature R_1 was measured at, °F

A monograph that solves this equation graphically is presented in Figure 5.2.8 [20].

5.2.3.4 Openhole Logs and Interpretation
SP (Spontaneous Potential)
The SP log has 4 basic uses: (1) recognition of permeable zones, (2) correlation of beds, (3) determination of R_w, and (4) qualitative indication of shaliness. The SP can only be used in fresh mud and is run with several resistivity tools. The curve is presented in Track 1 and is scaled in millivolts (mV).

The SP log is a record of the naturally occurring electrical currents created in the borehole. These currents or circuits usually occur at bed boundaries and are created by the interaction between fresh drilling mud and salty formation water. The curve represents the potential difference between a stationary electrode on the surface (ground) and a moving electrode in the borehole.

Theory
The total potential (E_t) can be separated into two components: the electrochemical (E_c) and the electrokinetic (E_k). The electrokinetic component is generally very small

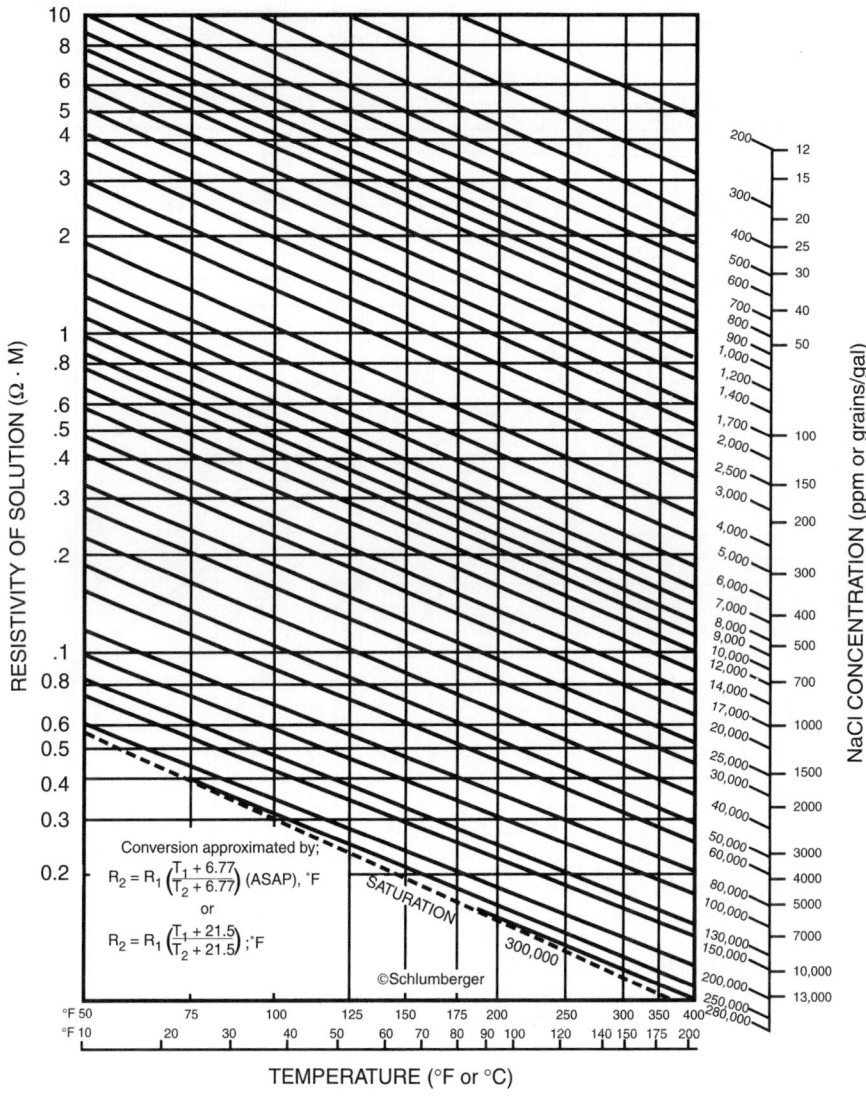

Figure 5.2.8 *Chart for determining salinity, solution resistivity and for converting resistivity to formation temperature [20].*

and is often ignored. It is created when an electrolyte (mud filtrate) flows through a nonmetallic permeable material. The magnitude of E_k is a function of the pressure drop across the material and the resistivity of the electrolyte. The electrokinetic (or streaming) potential is most significant in low pressure (depleted) formations, overbalanced mud conditions and opposite low permeability formations. See Doll's classic paper [201] for more detailed information.

The electrochemical component (E_c) is the sum of the liquid-junction potential (E_{lj}) and the membrane potential (E_m). The liquid junction potential occurs at the interface between fresh mud filtrate and salty formation water. This interface is usually a few inches to a few feet away from the borehole. Only two ions are assumed to be in solution in the mud and formation water: Na^+ and Cl^-. Chloride ions are concentrated in the formation water, and being more mobile than Na^+ ions, move toward lower

concentrations in the borehole (Figure 5.2.9). This creates a net negative charge near the borehole and a current flows toward the undisturbed formation. The liquid-junction potential accounts for about 20% of the electrochemical component.

The membrane (E_m) potential is created at the bed boundary between a permeable bed (sand) and an impermeable bed (shale). The shale acts as an ion-selective membrane, allowing only the smaller Na^+ ion to move through the clay crystal structure from the salty formation water toward the fresh drilling fluid in the bore. This creates a net positive charge along the shale. It also creates a large concentration of negative charges associated with the Cl^- ion in the permeable bed. This phenomena is also shown in Figure 5.2.9. The membrane potential accounts for about 80% of the electrochemical potential. The total effect of these two potentials is a net negative charge within the permeable zone when the connate water is saltier than the mud filtrate.

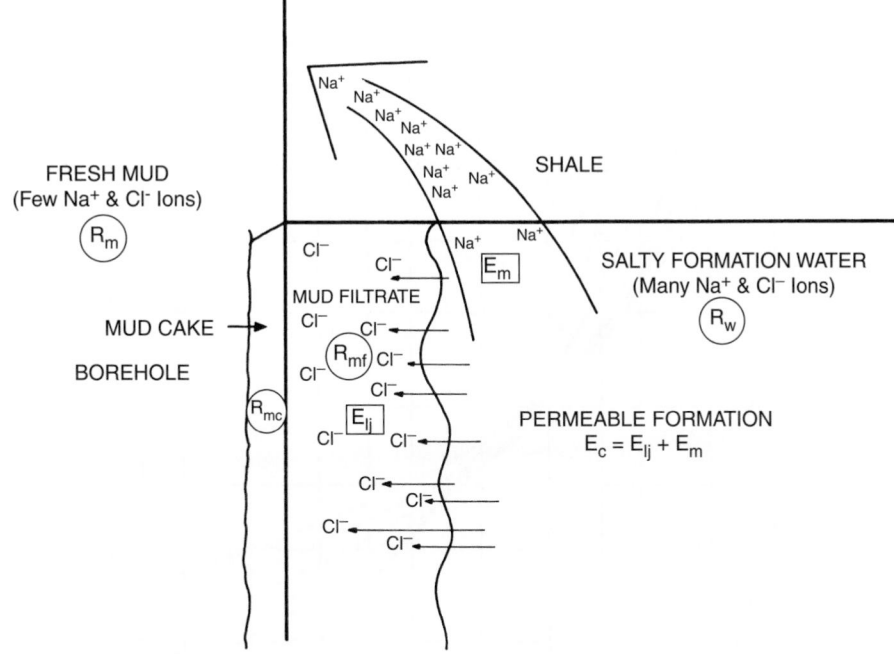

Figure 5.2.9 *Ionic movement that contributes to the development of an SP curve.*

Interpretation

The total electrochemical component of the total potential is what the SP records. It can be calculated with the following equation:

$$E_c = -K \log \frac{a_w}{a_{mf}} \qquad [5.2.15]$$

where $-K = -(0.133\,T + 61)$ (T in °F)
  a_w = chemical activity of formation water
  a_{mf} = chemical activity of drilling mud filtrate

Since the chemical activity of a solution cannot be used, it must be converted to its equivalent electrical resistivity. Chemical activity of a fluid is approximately equal to the inverse of its equivalent electrical resistivity. Conversion to equivalent resistivities makes the equation:

$$E_c = -K \log \frac{R_{mfeq}}{R_{weq}} \qquad [5.2.16]$$

Since E_c is equal to the maximum SP deflection recorded on a log (SSP), Equation 5.2.16 can be rewritten to read:

$$SSP = -K \log \frac{R_{mfeq}}{R_{weq}} \qquad [5.2.17]$$

where SSP is the static (or maximum) spontaneous potential recorded opposite a permeable formation. Since the purpose of an SP log is to find R_{weq} and then R_w, if we know SSP we can solve Equation 5.2.17 for R_{weq}:

$$R_{weq} = \frac{R_{mfeq}}{10^{(-SSP/-K)}} \qquad [5.2.18]$$

Once R_{weq} is known, it is converted to R_w using the chart shown in Figure 5.2.10 [20]. SSP can come directly from the log if the bed is thick and the SP curve reaches a constant value and develops a "flat top." If the curve is pointed or rounded, it must be corrected for bed thickness.

The shape and amplitude of the SP are affected by:

1. Thickness and resistivity of the permeable bed (R_t).
2. Diameter of invasion and resistivity of flushed zone (R_{xo})
3. Resistivity of the adjacent shales (R_s).
4. Resistivity of the mud (R_m).
5. Borehole diameter (d_h)

All of these must be accounted for when examining the SP, and any necessary corrections should be made. To find the magnitude of the SP, take the maximum deflection from the average shale value (shale baseline) to the most negative value. (Figure 5.2.11 shows a curve that needs correction and one that does not.) Bed thickness corrections can be made from Figure 5.2.12 and should *always* increase the magnitude of the SP.

Another use for the SP log is finding permeable zones. Any negative deflection of the cure indicates a potentially permeable zone. The magnitude of the deflection has no relation to the amount of permeability (in millidarcies); it merely indicates that the rock has ionic permeability. No quantitative information on this, parameter can be derived from the SP. Figure 5.2.13 shows an example of permeable and impermeable zones on an SP log.

Interpretation of an SP log follows a few basic rules:

1. If the SP curve is concave to the shale line, the formation is permeable.
2. If the SP curve is convex to the shale line, the formation is impermeable.
3. Constant slope means high resistivity–usually impermeable.
4. High resistivity formations cause the bed boundaries to become rounded.
5. A thin permeable bed does not reach maximum deflection.
6. A thin shale streak does not reach the shale baseline.

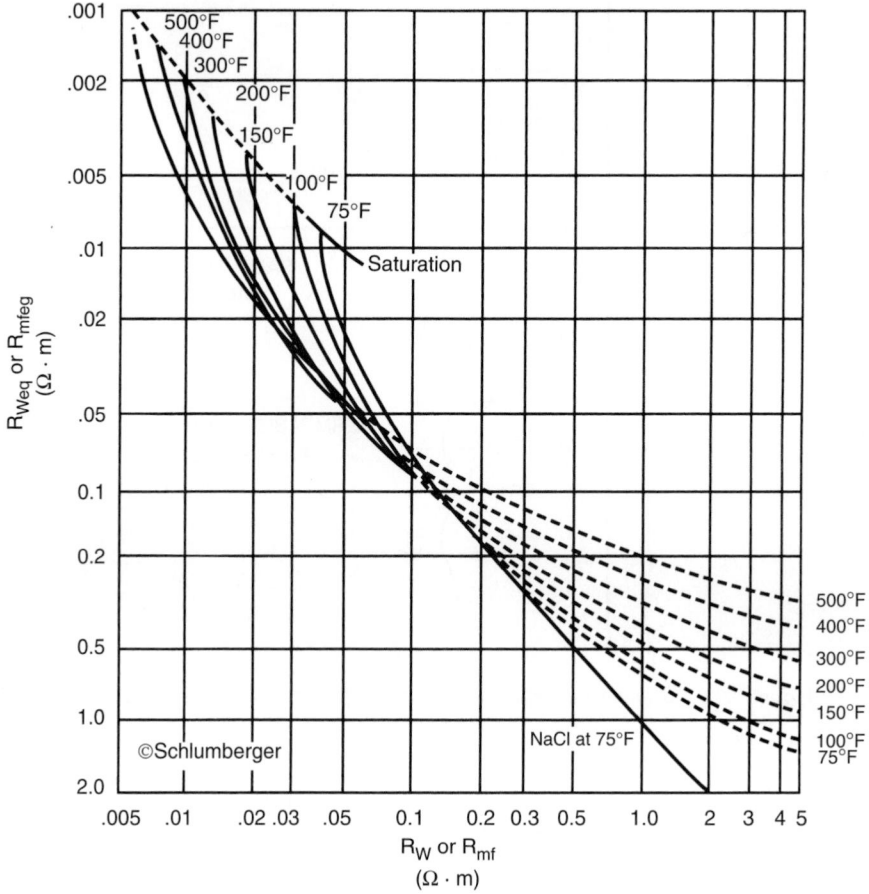

Figure 5.2.10   *Chart for converting R_{mf} to R_{mfeq} and R_{weq} to R_w for SP-R_w calculations [20].*

7. Bed boundaries are picked at the inflection points in clean sands. Bed boundaries should be confirmed with some other log such as the gamma ray.

Resistivity Tools

The purpose of resistivity tools is to determine the electrical resistance of the formation (rock and fluid). Since most formation waters contain dissolved salts, they generally have low resistivities. Hydrocarbons do not conduct electricity, therefore rocks that contain oil and/or gas show high resistivity. This is the way hydrocarbon-bearing zones are differentiated from water zones.

Resistivity tools are divided into three types based on the way measurements are made: (1) non-focused (normal) tools, (2) induction tools, and (3) focused resistivity tools. Microresistivity tools will be treated under a separate heading.

Resistivity tools are further divided by depth of investigation. Tools may read the flushed zone (R_{xo}), the transition zone between the flushed and uninvaded zones (R_i), or the univaded zone (R_t). For purposes of this discussion, all tool names are for Schlumberger equipment. Comparable tools offered by other logging companies are summarized in Table 5.2.6.

Theory Nonfocused (Normal Tools)
The first tools to be used were nonfocused tools [58]. The electrodearrangement is as shown in Figure 5.2.14a [58]. As shown earlier, resistivity, R, is found using Ohm's Law (r = V/I):

$$R = \frac{rA}{L} \qquad [5.1.44]$$

where V = voltage read from meter
  I = current read from meter
  A = surface area, m^2
  L = length, m
  R = resistivity, ohm-m
  r = resistance, ohms

The tools work well in low-resistivity formations with thick beds and in slightly conductive muds.

Inductions Tools
Since a slightly conductive mud is necessary for the normal tools, they cannot be used in very fresh muds or in oil-base muds. The induction tool overcomes these problems by inducing a current into the formation instead of passing it through the mud-filled borehole. Figure 5.2.14b shows a simplified two-coil induction tool [58]. High-frequency alternating current is sent into the transmitter coil. The alternating magnetic field that is created induces secondary in circular paths called ground loops. These currents in

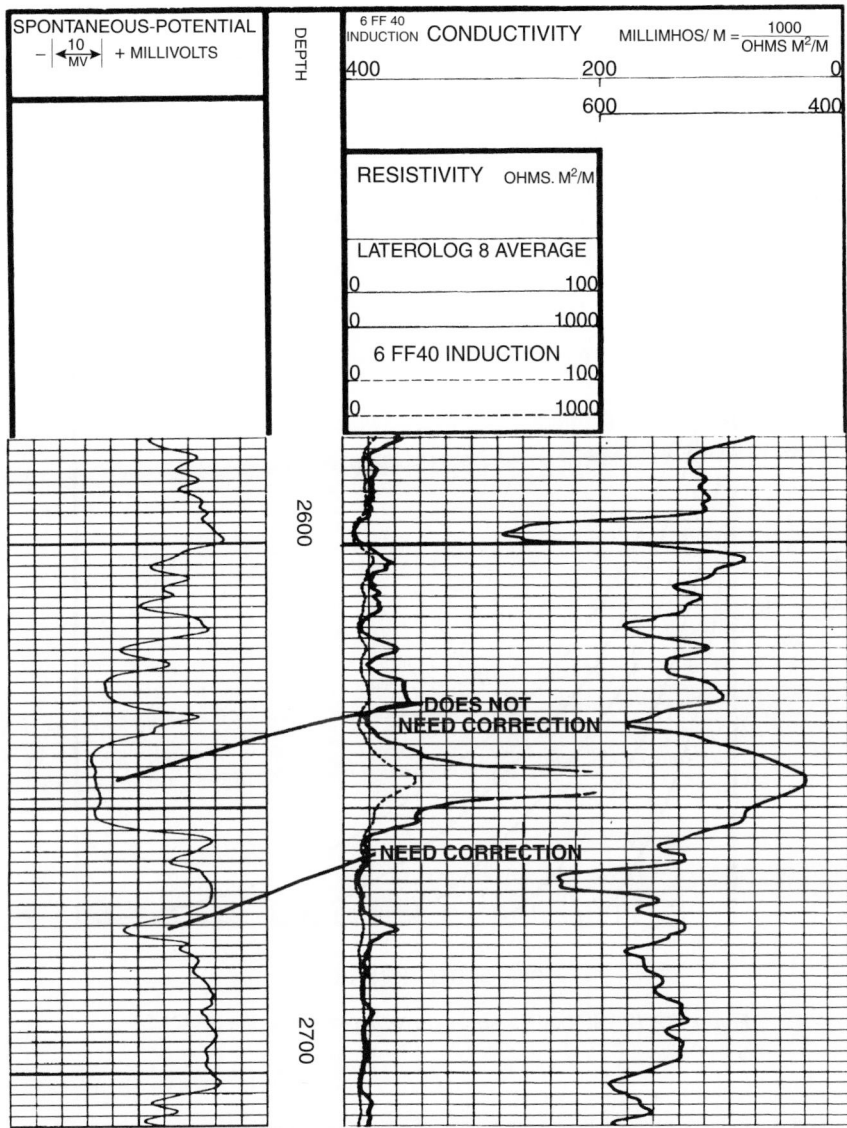

Figure 5.2.11 *Example of an SP curve that requires correction and one that does not.*

turn create a magnetic field which induces currents in the receiver coil. The signal received is proportional to the formation conductivity. Conductivity readings are then converted to resistivity. Additional coils are used to focus the tool so that conductive beds as thin as four feet can be detected. Inductions tools work well in oil-base, foam, air, gas, and fresh mud. The induction tool is unreliable above 500 Ω-m and useless above 1,000 Ω-m or in salt muds. Readings are only considered reasonable below 100 Ω-m and are accurate between 1 and 20 Ω-m when mud is very fresh.

Phasor Induction Tools
Since the early 1960's, induction logging tools have become the principal logging device for fresh, slightly conductive to non-conductive (oil-base) muds. However, these devices are significantly affected by environmental (bore-hole size and mud composition) and geological (bed thickness and

invasion) conditions. Also, high formation resistivities (> 50 to 100 Ω-m) dramatically increase the difference between apparent Rt and true Rt [58A, 199]. In 1986, Schlumberger introduced a new induction log to offset these problems [58A]. This device is known by the trade name Phasor Induction SFL and uses a standard dual induction tool array. The difference between the conventional and phasor devices is in signal processing made possible by miniaturization of computer components. Induction tools all produce two signals: the inphase (R-signal) induction measurement and the quadrature (X-signal) induction signal. The R-signal is what is presented on standard dual induction–SFL log presentations. The R-signal and X-signal measurements are combined during advanced processing in the logging tool itself to produce a log with real time corrections for environmental and geological conditions. Apparent Rt is nearly equal to true Rt in most situations. Vertical resolution of this device is about

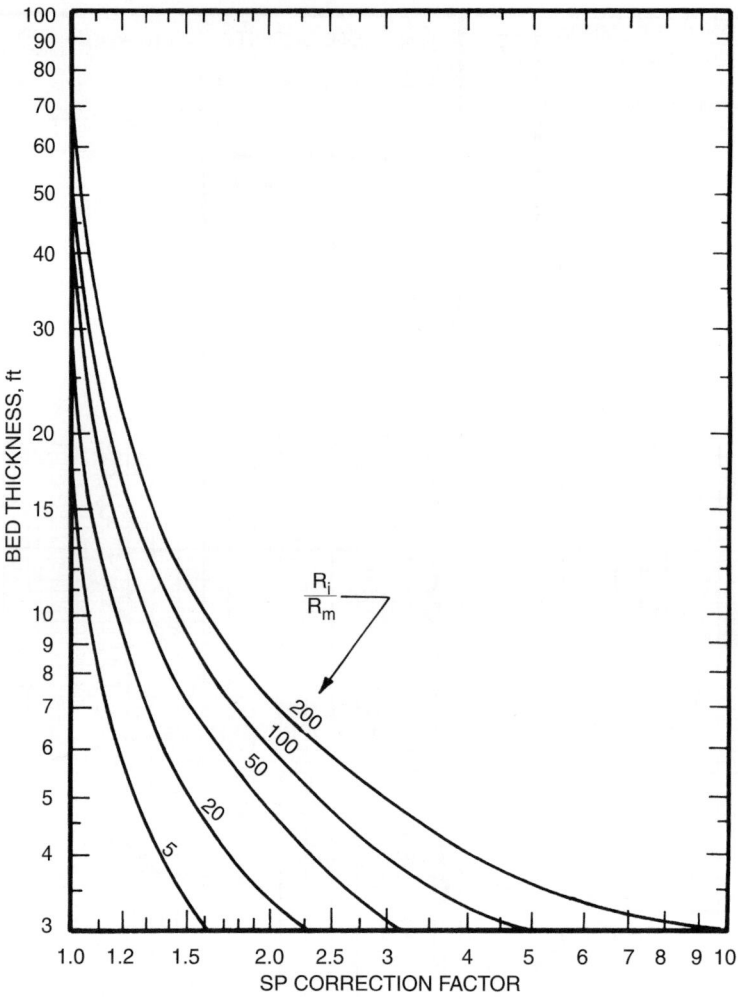

Figure 5.2.12 *SP bed thickness correction chart for the SP [21,22].*

the same as conventional induction tools (about 6 feet), but enhanced and very enhanced resolution phasor tools are available that have vertical resolutions down to 2 feet [58A]. The primary advantages of this tool include much better R_t readings in high resistivity formations (i.e. $> 100\,\Omega$-m) and more accurate readings in salty muds than previously possible.

Focused Resistivity Tools (Laterologs)
The Laterologs are the primary salt-mud resistivity tools. Salt mud presents a problem in that the path of least resistance is within the borehole. Therefore the current must be forced into the formation which has higher resistance. To do this, secondary electrodes (A_1, and A_2) are placed above and below the measuring current-emitting electrode. (A_0). These secondary electrodes emit "focusing" or "guard" currents with the same polarity as the measuring currents. Small monitoring electrodes (M_1, M_2, M'_1 and M'_2) adjust the focusing currents so that they are at the same potential as A_1 and A_2. A sheet of current measuring one to two feet thick is then forced into the formation from A_0. The potential is then measured between M_2 and M_1 and a surface ground.

Since I_0 is a constant, any variation in M_1 and M_2 current is proportional to formation resistivity. Figure 5.2.14c shows the electrode arrangement [58].

Corrections
As previously mentioned, resistivity tools are affected by the borehole, bed thickness, and invaded (flushed) zone. If bed thickness and mud resistivity are known, these effects can be accounted for. Major service companies (Schlumberger, Atlas Wireline, Welex Halliburton) provide correction charts for their tools. Several charts for Schlumberger tools are included in this chapter (Figures 5.2.15 to 5.2.21) [20] and can be used with logs from other service companies. Complete chart books are available from most wireline service companies. Current chart books should be used and can be obtained by calling the appropriate service company or by asking the logging engineer. Correction and interpretation charts for order tools (normals and laterals) are no longer published by service companies, but can be found in a text by Hilchie [23] and in a recent publication [203]. All log readings should be corrected with these charts to obtain accurate R_t and R_{xo} readings.

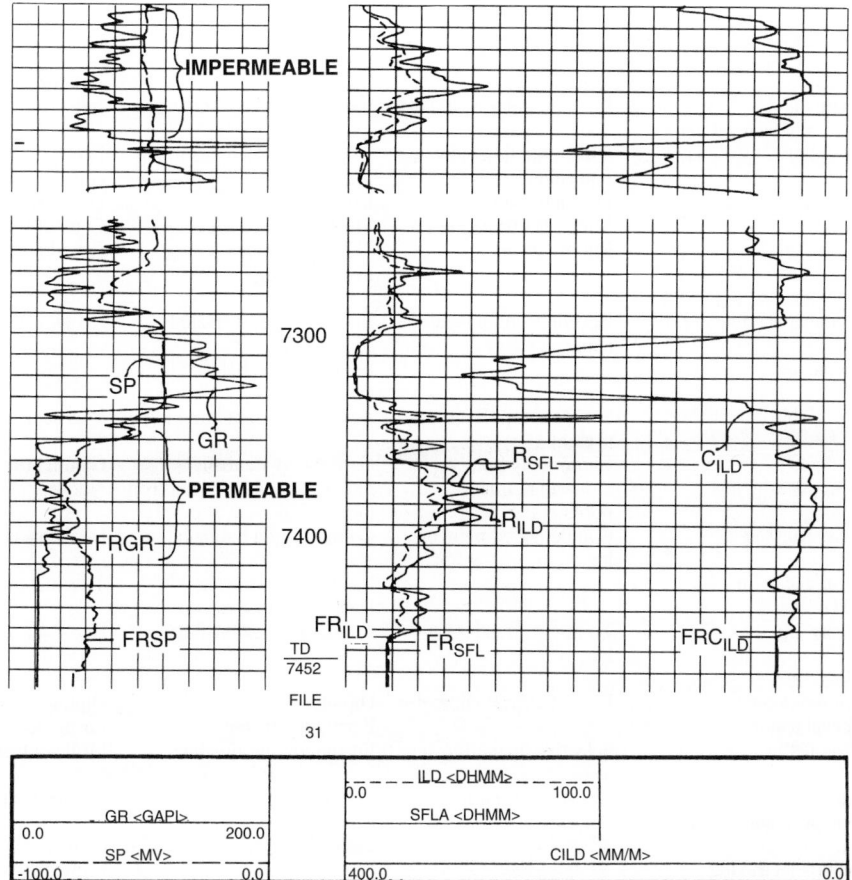

Figure 5.2.13 *Examples of an SP log through an impermeable and a permeable zone.*

The dual induction logs should be corrected for borehole, bed thickness, and invasion effects if three curves are present. The induction-SFL, induction-LL-8, and induction-electric log combinations require an R_{xo} curve to make corrections. If no R_{xo} device is presented, the deep induction curve is assumed to read true R_t. The specific charts included in this text are Figure 5.2.15 for induction log borehole corrections, Figures 5.2.16 and 5.2.17 for induction log bed-thickness corrections, and Figure 5.2.18a to 5.2.18c for invasion corrections (check log to see if an SFL or a LL-8 log was used with the induction curves and select the appropriate Tornado chart). Phasor induction logs only require invasion correction. This is accomplished with the appropriate Tornado Charts (Figures 5.2.18d to 5.2.18k) [20]. An invasion correction Tornado Chart for phasor nduction tools with other frequencies are also available [58a].

The dual laterolog-R_{xo} combination should be corrected for borehole effects; bed thickness corrections are not normally made. Figure 5.2.19 is used to make the borehole corrections of the deep and shallow laterologs, respectively [20]. Figure 5.2.20 is used to correct the micro-SFL logs for mud cake effects [20]. Invasion effects are corrected by using Figure 5.2.21 [20].

Interpretation
With Equation 5.2.20, water saturation can be found using corrected values of R_t derived from the Tornado charts, R_w

can be found from an SP log or chemical analysis, and F_R from a porosity log and Table 5.1.7.

If a porosity log is not available, F_R can be found using Equation 5.1.45. R_o is selected in a water-bearing zone from the deepest reading resistivity curve. Care should be taken to select R_o in a bed that is thick, permeable, and clean (shalefree). The zone should be as close to the zone of interest as possible. R_o can sometimes be selected from below the water/hydrocarbon contact within the zone of interest. Archie's equation (as described earlier) can then be used to calculate water saturation:

$$S_w = \left[\frac{F_R R_w}{R_t} \right]^{1/2}$$ [5.1.70]

If no porosity log or water zone is available, the ratio method can be used to find water saturation. The saturation of the flushed zone (S_{xo}) can be found by:

$$S_{xo} = \left[\frac{FR_{mf}}{R_{xo}} \right]^{1/2}$$ [5.2.19]

When Equation 5.1.70 is divided by Equation 5.2.19, the result is a ratio of water saturations:

$$\frac{S_w^2}{S_{xo}^2} = \frac{R_{xo} R_w}{R_t R_{mf}}$$ [5.2.20]

(*text continued on page 5-102*)

Table 5.2.6 *Service Company Nomenclature*

Schlumberger	Gearhart	Dresser Atlas	Welex
Electrical log (ES)	Electric log	Electrolog	Electric log
Induction electric log	Induction electrical log	Induction electrolog	Induction electric log
Induction spherically focused log			
Dual induction spherically focused log	Dual induction-laterolog	Dual induction focused log	Dual induction log
Laterolog-3	Laterolog-3	Focused log	Guard log
Dual laterolog	Dual laterolog	Dual laterolog	Dual guard log
Microlog	Micro-electrical log	Minilog	Contact log
Microlaterolog	Microlaterolog	Microlaterolog	F_oR_{xo}log
Proximity log		Proximity log	
Microspherically focused log			
Borehole compensated sonic log	Borehole compensated sonic	Borehole compensated acoustilog	Acoustic velocity log
Long spaced sonic log		Long spacing BHC acoustilog	
Cement bond/variable density log	Sonic cement bond system	Acoustic cement bond log	Microseismogram
Gamma ray neutron	Gamma ray neutron	Gamma ray neutron	Gamma ray neutron
Sidewall neutron porosity log	Sidewall neutron porosity log	Sidewall epithermal neutron log	Sidewall neutron log
Compensated neutron log	Compensated neutron log	Compensated neutron log	Dual spaced neutron
Thermal neutron decay time log		Neutron lifetime log	Thermal multigate decay
Dual spacing TDT		Dual detector neutron	
Formation density log	Compensated density log	Compensated densilog	Density log
Litho-density log			
High resolution dipmeter	Four-electrode dipmeter	Diplog	Diplog
Formation interval tester		Formation tester	Formation tester
Repeat formation tester	Selective formation tester	Formation multitester	Multiset tester
Sidewall sampler	Sidewall core gun	Corgun	Sidewall coring
Electromagnetic propagation log			
Borehole geometry tool	X-Y caliper	Caliper log	Caliper
Ultra long spacing electric log			Compensated
Natural gamma ray spectrometry		Spectralog	Spectral natural gamma
Gamma ray spectroscopy tool log		Carbon/oxygen log	
Well seismic tool			
Fracture identification log	Fracture detection log		

From Reference 36.

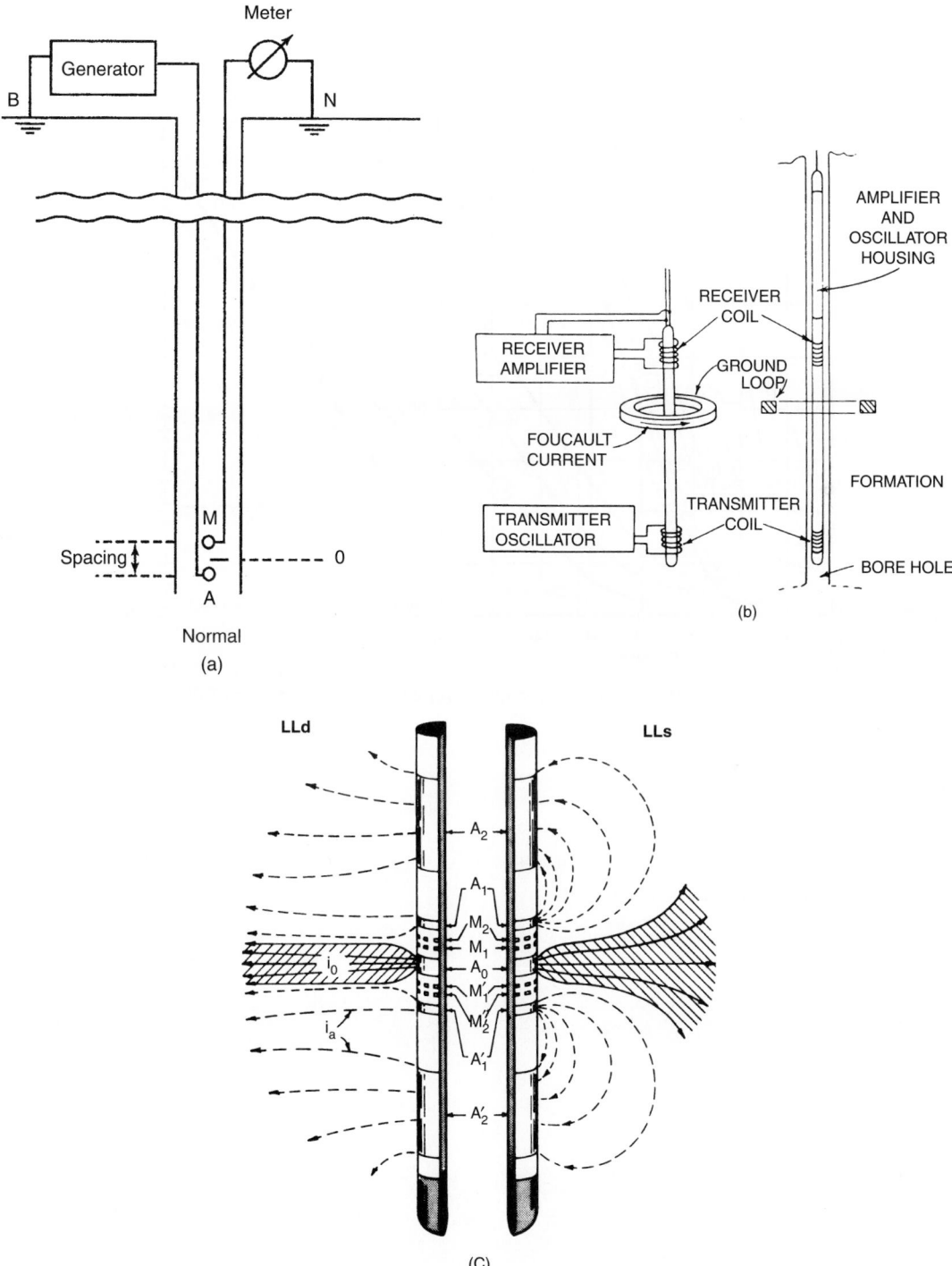

Figure 5.2.14 *(a) Basic electrode arrangement of a normal device used in conventional electric logs. (b) Electrode arrangement for a basic, two coil induction log system. (c) Electrode arrangement and current flow paths for a dual laterolog sonde [5.1.58].*

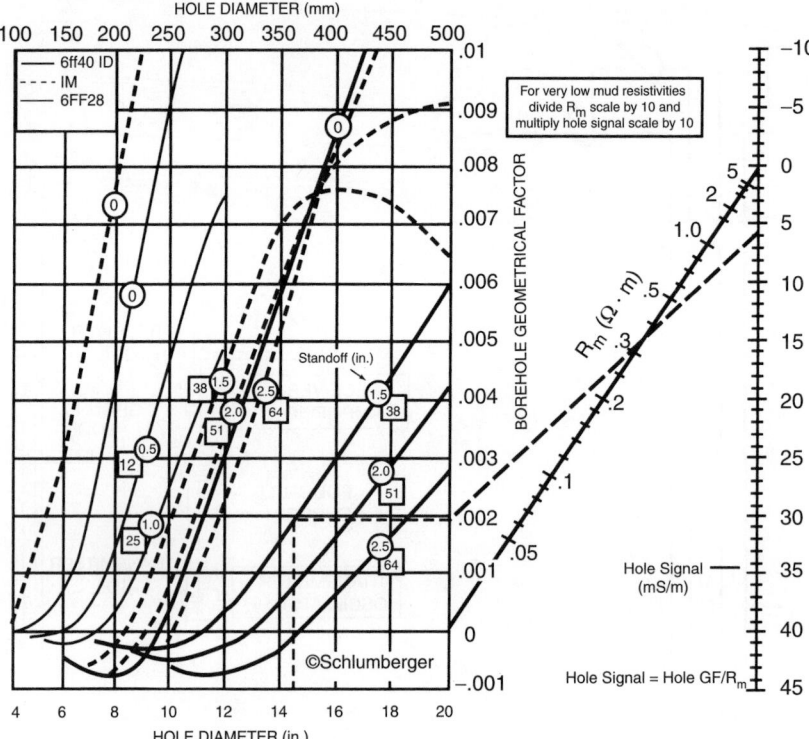

Figure 5.2.15 Borehole correction chart for induction log readings [20].

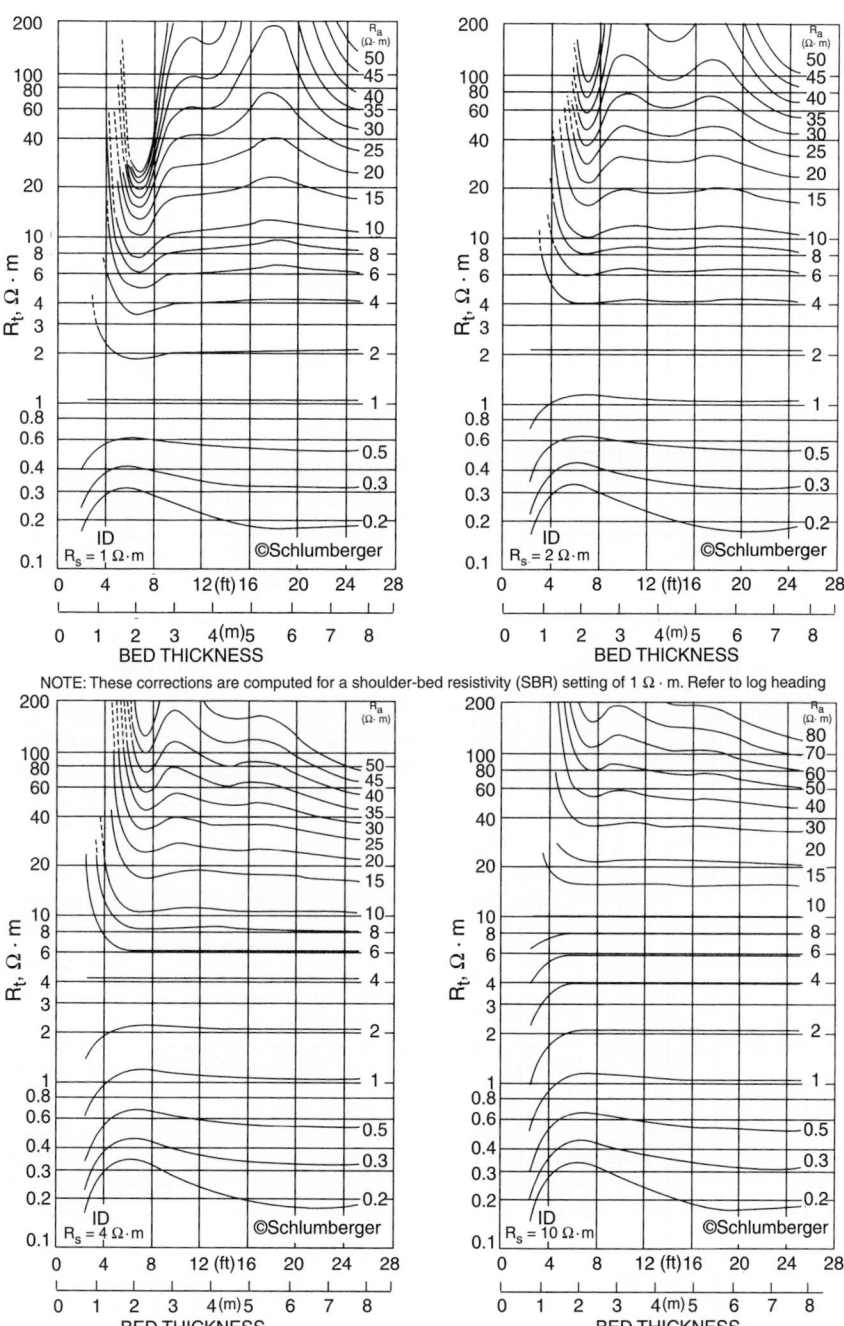

Figure 5.2.16 *Bed thickness correction charts for the deep induction log [20].*

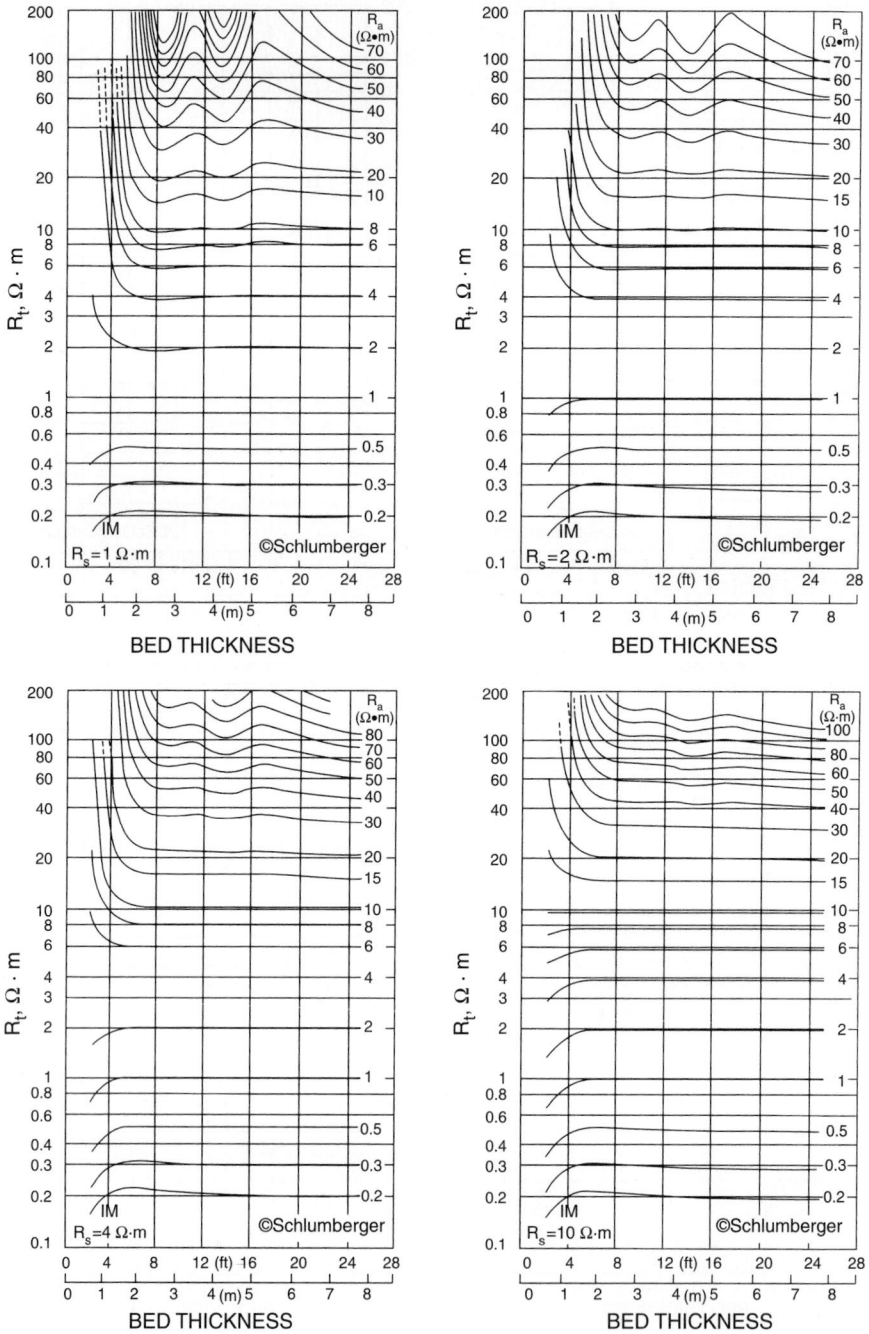

Figure 5.2.17 *Bed thickness correction for the medium induction log [20].*

(a)

Thick Beds, 8-in. (203-mm) Hole, Skin-Effect Corrected,
$R_{xO}/R_m \approx 100$, DIS-DB or Equivalent

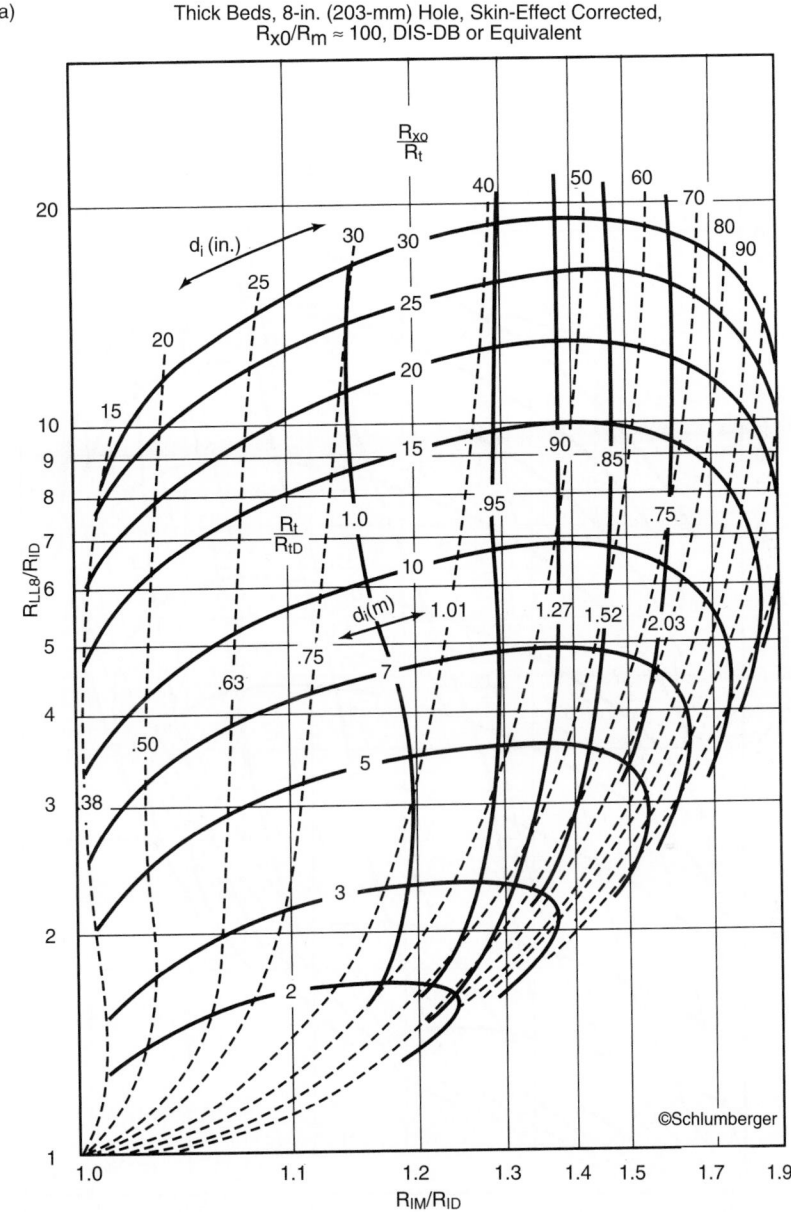

Figure 5.2.18a *Invasion correction charts for the dual induction-laterolog 8 combination [20].*

(b)   Thick Beds, 8-in. (203-mm) Hole, Skin-Effect Corrected,
             $R_{x0}/R_m \approx 20$, DIS-DB or Equivalent

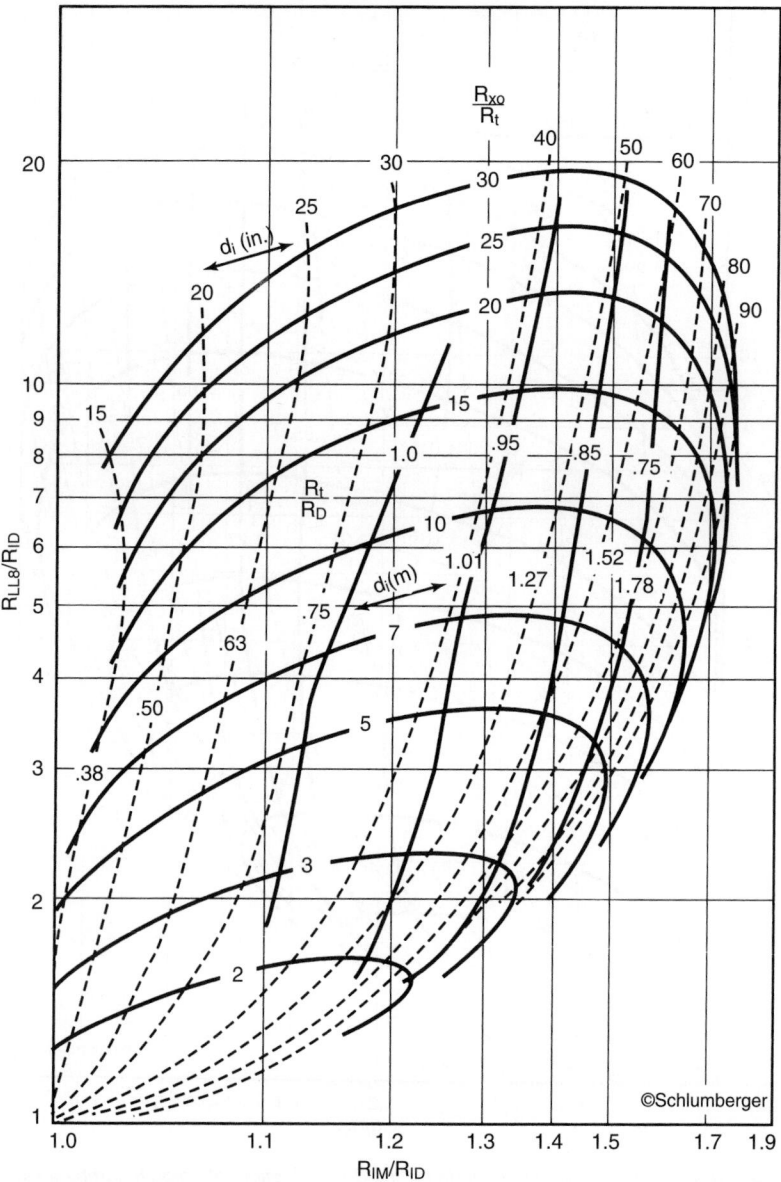

Figure 5.2.18b   *This chart may also be used with dual induction-spherically focused log.*

(c)   Thick Beds, 8-in. (203-mm) Hole, Skin-Effect Corrected,
$R_{x0}/R_m \approx 100$, DIS-EA or Equivalent

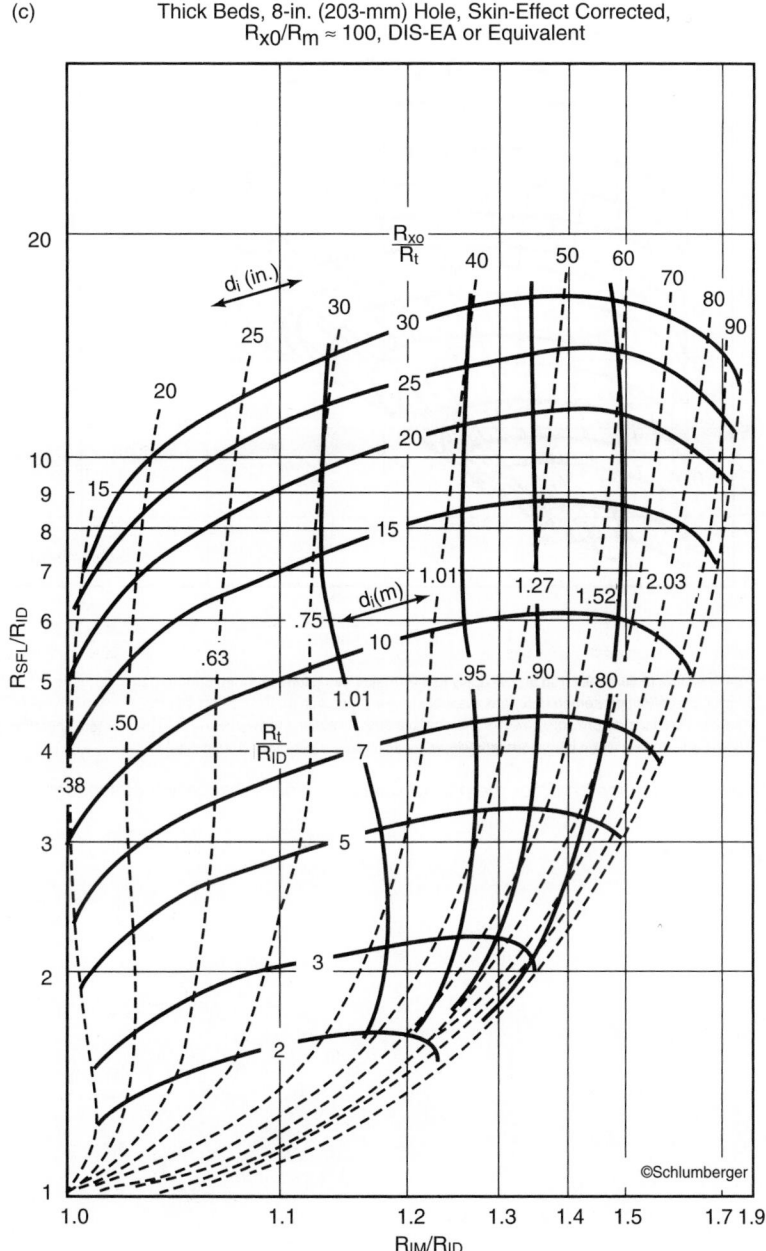

Figure 5.2.18c *Invasion correction chart for the dual induction-spherically focused log combination [20].*

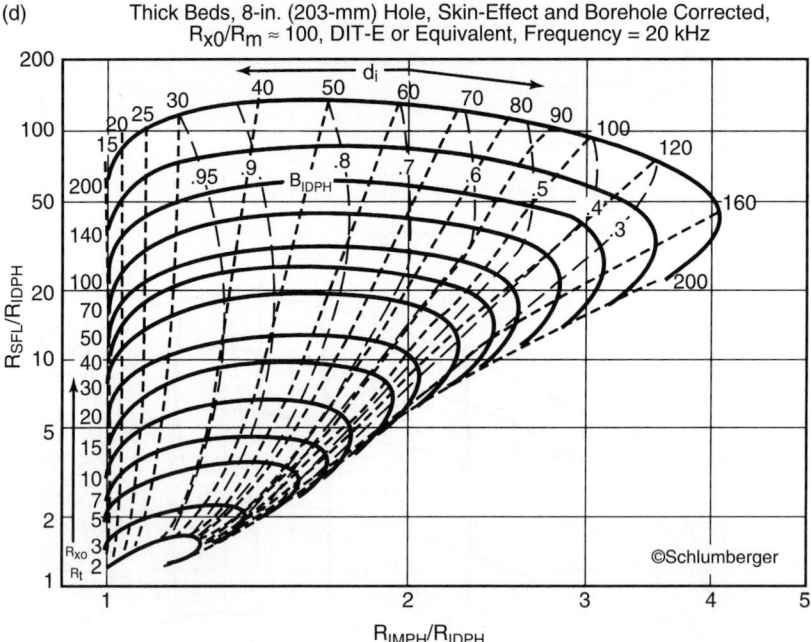

(d)   Thick Beds, 8-in. (203-mm) Hole, Skin-Effect and Borehole Corrected,
$R_{x0}/R_m \approx 100$, DIT-E or Equivalent, Frequency = 20 kHz

These Charts (Figures 5.2.18d and e) apply to the Phasor Induction tool when operated at a frequency of 20 kHz.
Similar charts (not presented here) are available for tool operation at 10 kHz and 40 kHz.
The 20 kHz charts do provide, however, reasonable approximations of R_{x0}/R_t and R_t/R_{IDPH} for tool operation at
10 kHz and 40 kHz when only moderately deep invasion exists (less than 100 inches)

Figure 5.2.18d   Invasion correction chart for the phasor dual induction-spherically focused log combination for
Rxo/Rm = 100 [20].

(e)   Thick Beds, 8-in. (203-mm) Hole, Skin-Effect and Borehole Corrected,
$R_{xo}/R_m \approx 20$, DIT-E or Equivalent, Frequency = 20 kHz

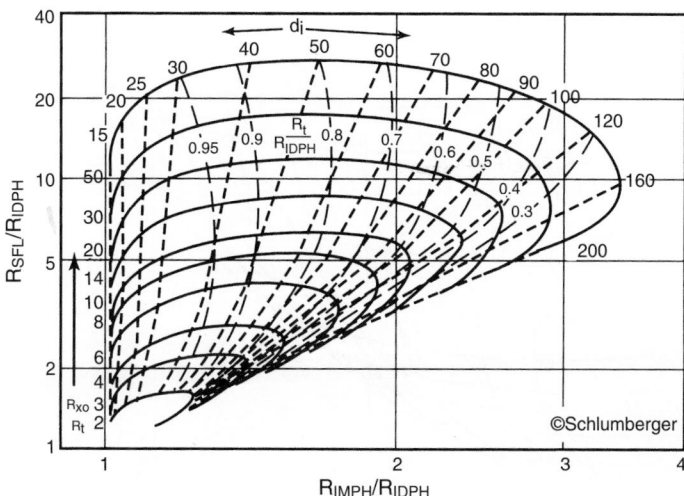

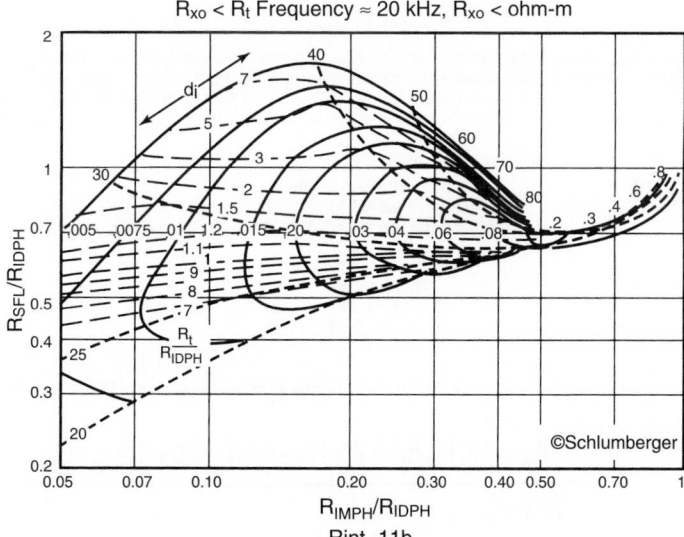

Rint -11b

Figure 5.2.18e *Invasion correction chart for the phasor dual induction-spherically focused log combination for Rxo/Rm = 20 [20].*

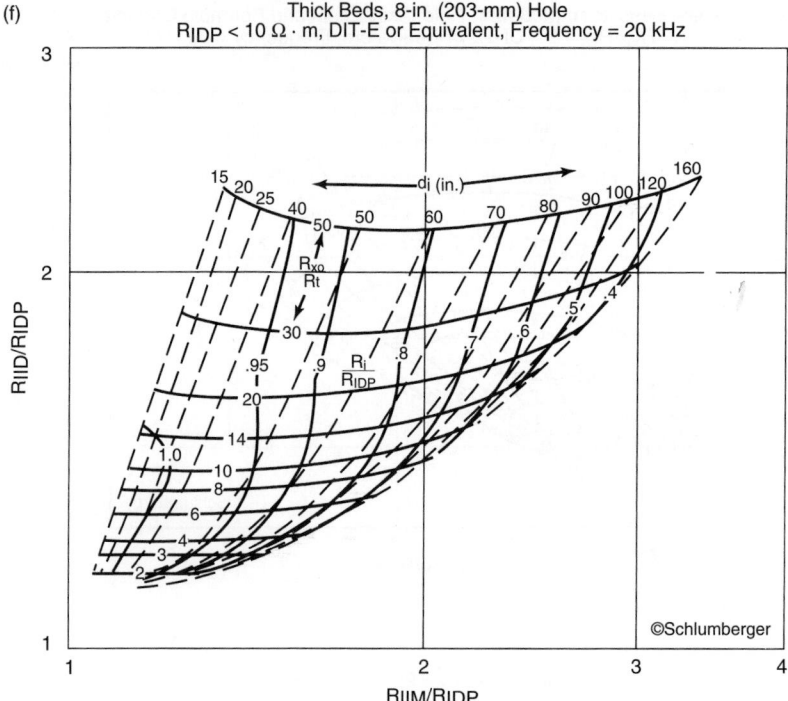

This chart uses the raw, unboosted induction signals and the ID-Phasor value to define the invasion profile in a rock drilled with oil-base mud. To use the chart, the ratio of the raw, unboosted medium induction signal (IIM) and the deep Phasor induction (IDP), is entered in abscissa. The ratio of the raw, unboosted deep induction signal (IID) and the deep Phasor induction (IDP) is entered in ordinate. Their intersection defines d_i, R_{xo}/R_t, and R_t.

EXAMPLE:
$R_{IDP} = 1.6\ \Omega \cdot m$
$R_{IID} = 2.4\ \Omega \cdot m$
$R_{IIM} = 2.4\ \Omega \cdot m$
Giving, $R_{IID}/R_{IDP} = 2.4/1.6 = 1.5$
$R_{IIM}/R_{IDP} = 2.4/1.6 = 1.5$
Therefore, $d_i = 50$ in.
$R_{xo}/R_t = 15$
$R_t/R_{IDP} = 0.94$
$R_t = 0.94\ (1.6)$
$= 1.5\ \Omega \cdot m$

Figure 5.2.18f   *Invasion correction chart for the phasor dual induction log in oil-based mud [20].*

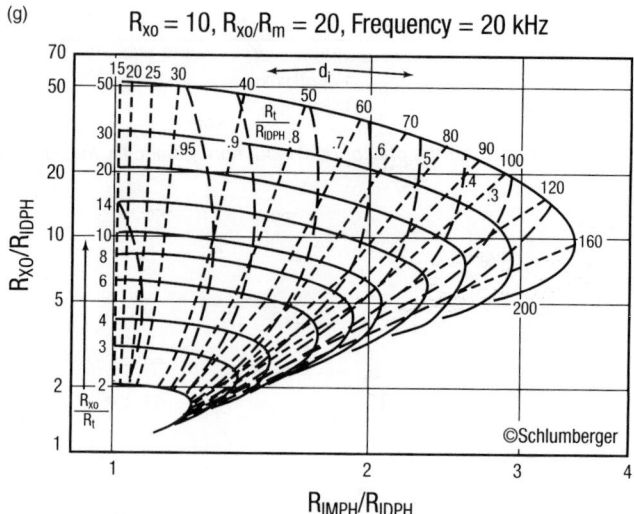

Figure 5.2.18g   *Invasion correction chart for the phasor dual induction-Rxo log combination for Rxo = 10, Rxo/Rm = 20 [20].*

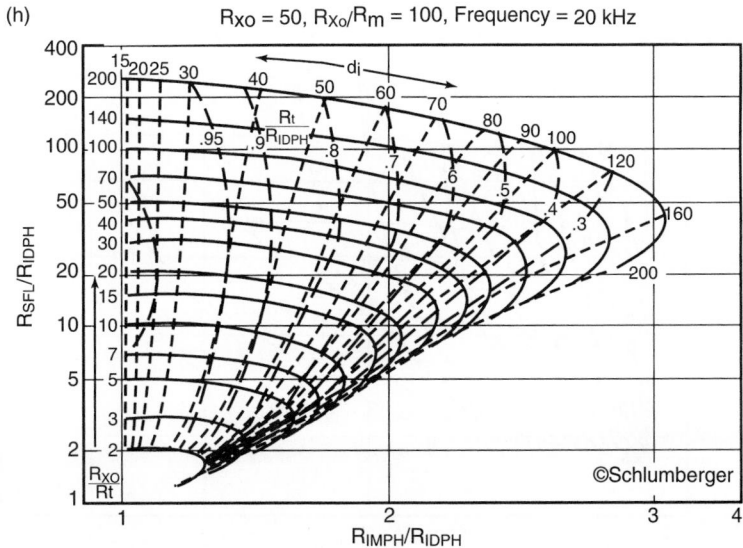

Figure 5.2.18h *Invasion correction chart for the phasor dual induction-Rxo log combination for Rxo = 50, Rxo/Rm = 100 [20].*

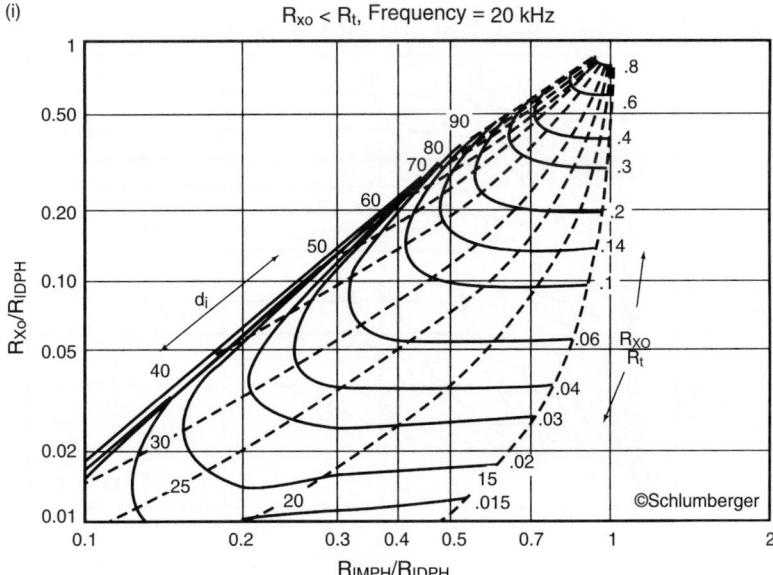

Figure 5.2.18i *Invasion correction chart for the phasor dual induction-Rxo log combination for Rxo < Rt [20].*

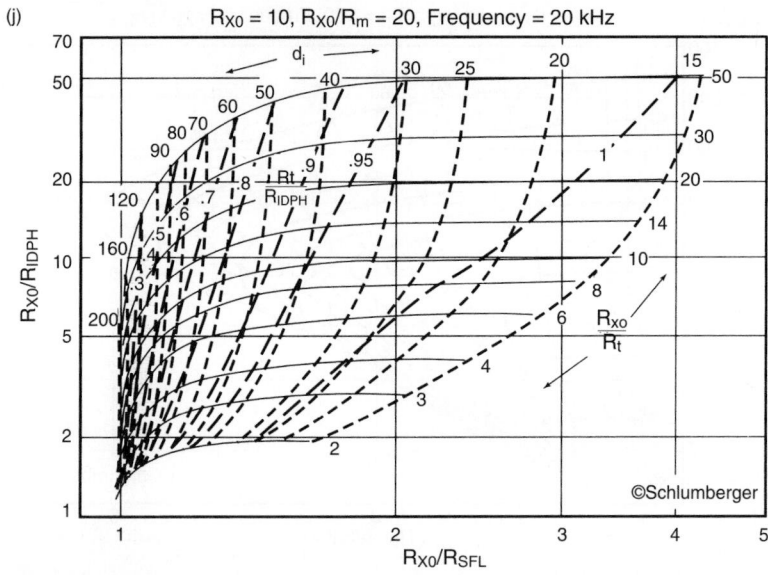

Figure 5.2.18j *Invasion correction chart for the phasor dual induction-spherically focused-Rxo log combination for Rxo = 10, Rxo/Rm = 20 [20].*

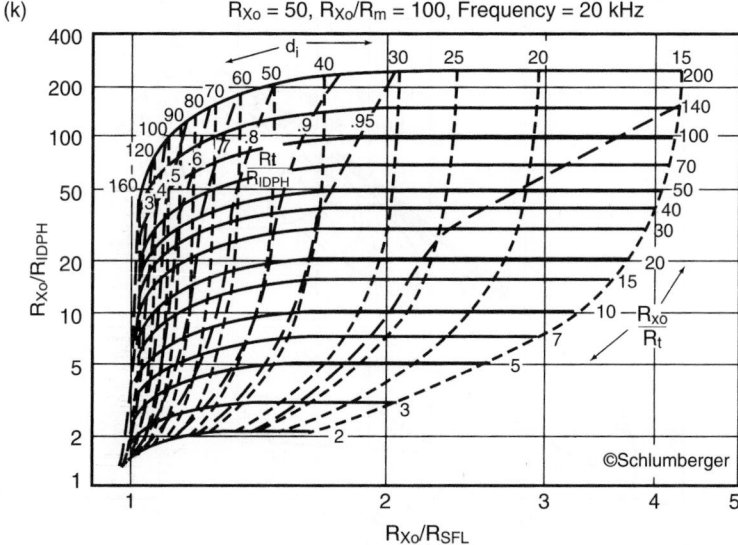

Figure 5.2.18k *Invasion correction chart for the phasor dual induction-spherically focused-Rxo log combination for Rxo = 50, Rxo/Rm = 100 [20].*

(a)

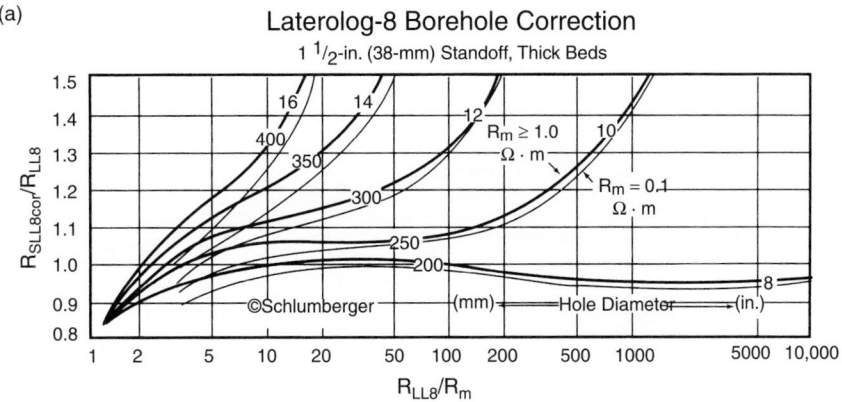

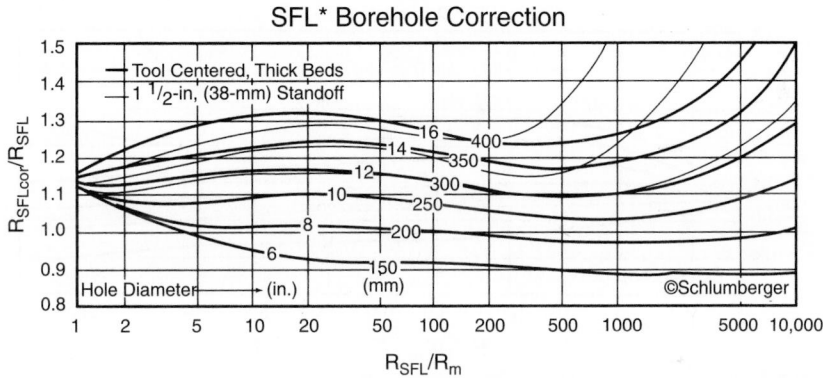

Figure 5.2.19a *Borehole correction charts for the laterolog and spherically focused logs [20].*

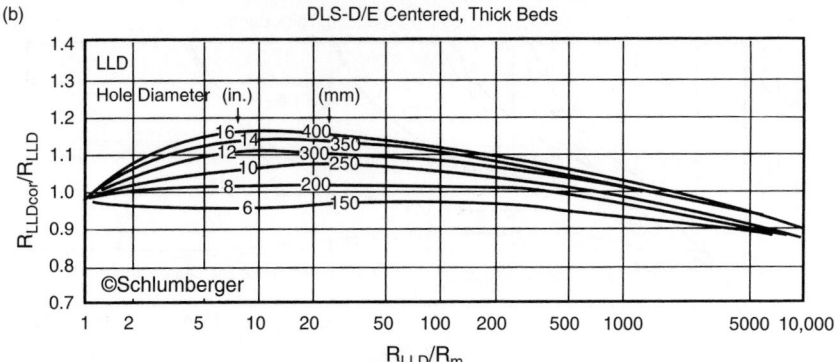

Figure 5.2.19b   *Borehole correction charts for the dual laterolog [20].*

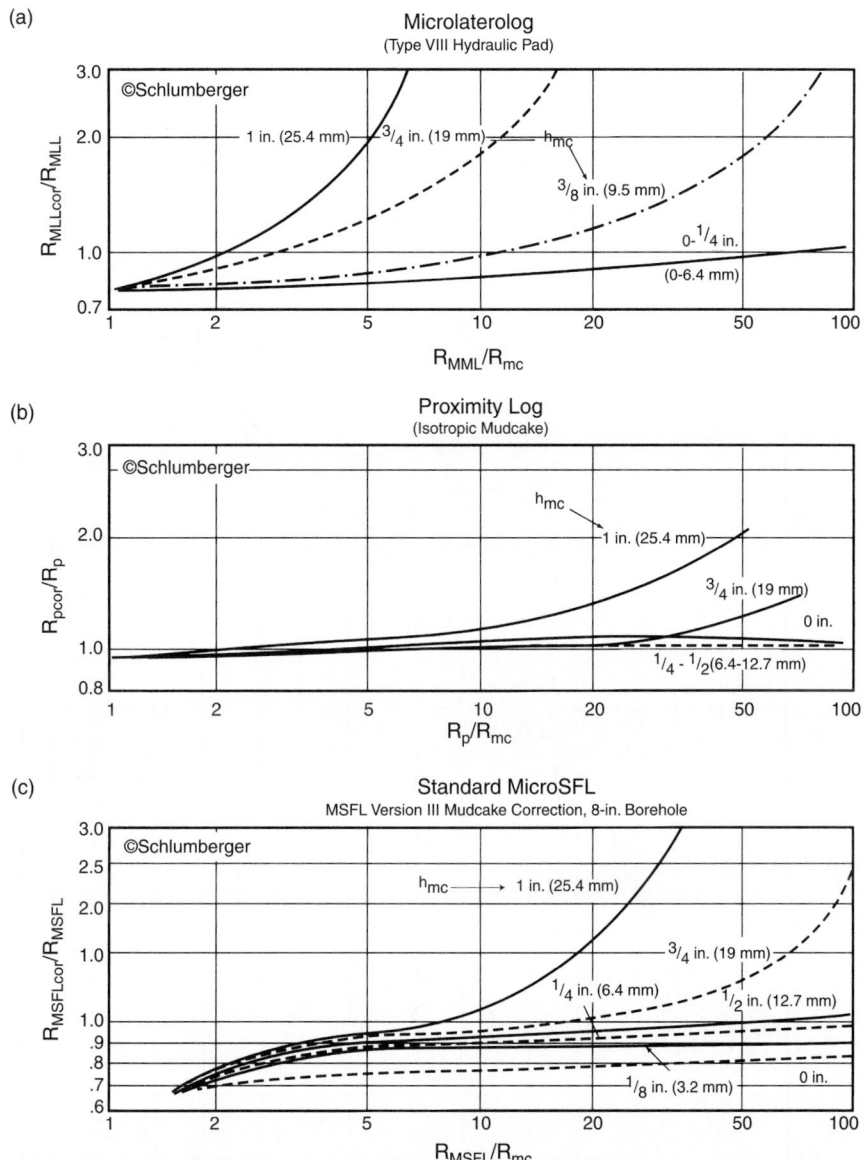

Figure 5.2.20 *Mudcake correction for: (a) microlaterolog, (b) proximity log, and (c) micro SFL log.*

Thick Beds, 8-in. (203-mm) Hole,
No Annulus, No Transition Zone,
Use Data Corrected for Borehole Effect

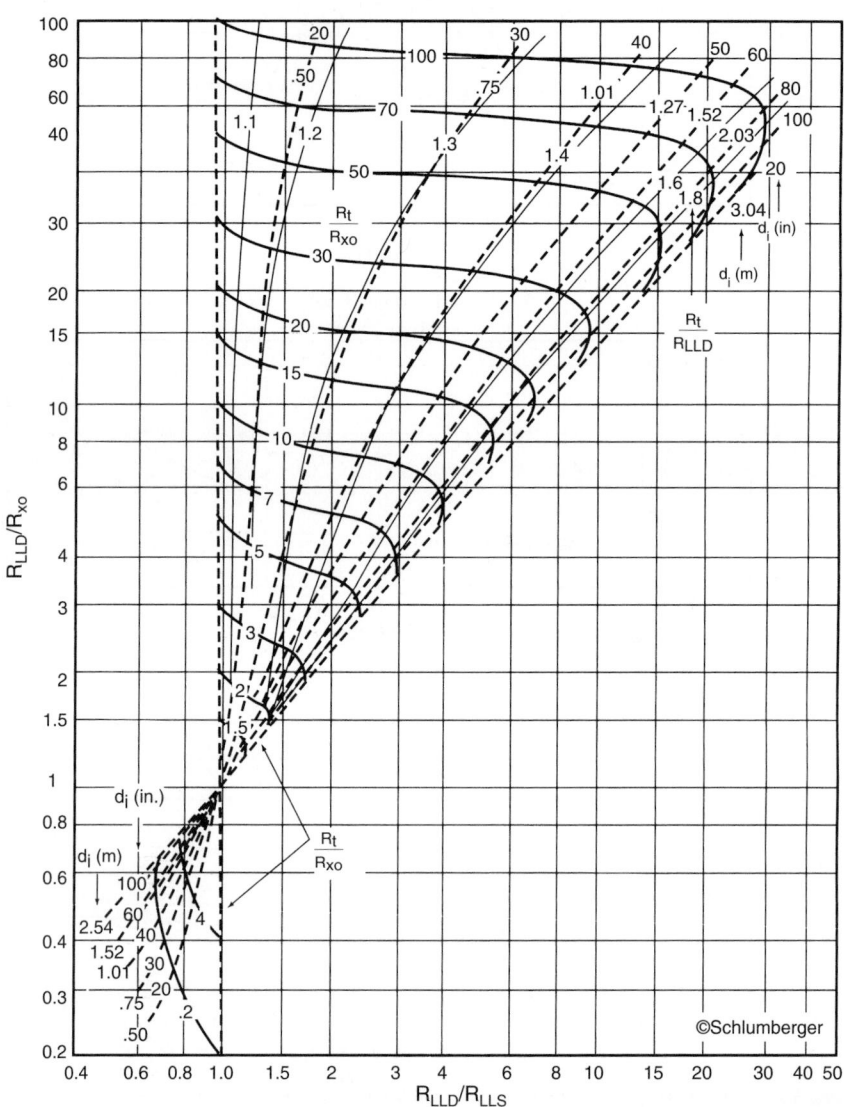

Figure 5.2.21   *Invasion correction chart for the dual laterolog-R_{xo} log combination [20].*

(text continued from page 5-85)

Poupon, Loy, and Tixier [204] found that for "average" residual oil saturations:

$$S_{xo} = S_w^{1/5} \qquad [5.2.21]$$

Substituting Equation 5.2.21 into Equation 5.2.20 and rearranging terms:

$$S_w = \left[\frac{R_{xo}/R_t}{R_{mf}/R_w} \right]^{5/8} \qquad [5.2.22]$$

With this equation, water saturation can be found without knowing ϕ. Note, however, that this interpretation method is based on the assumption that $S_{xo} = S_w^{1/5}$. This relation is for "average" granular rocks and may vary considerably in other rock types. Figure 5.2.22 is a chart [20] that solves Equation 5.2.22.

When two or more resistivity logs with different depths of investigation are combined, permeable zones can be identified. In a permeable zone, the area closest to the borehole will be flushed of its original fluids; mud filtrate fills the pores. If the mud filtrate has a different resistivity than the original formation fluids (connate water), the shallowest-reading resistivity tool will have a different value than the deepest-reading tool (Figure 5.2.23). Many times this difference is significant. The separation of the resistivity curves that result is diagnostic of permeable zones.

Care should be taken not to overlook zones in which curves do not separate. Curve separation may not occur if:

1. The mud filtrate and original formation fluids (i.e., connate water and hydrocarbons) have the same resistivity; both shallow and deep tools will read the same value. This is usually not a problem in oil or gas-saturated rocks.
2. Invasion of mud filtrate is very deep, both shallow and deep tools may read invaded-zone resistivity. This occurs when a long period of time elapses between drilling and logging or in a mud system with uncontrolled water loss.

Microresistivity Tools

Microresistivity tools are used to measure the resistivity of the flushed zone. This measurement is necessary to calculate flushed zone saturation and correct deep-reading resistivity tools for invasion. Microresistivity tools are pad devices on hydraulically operated arms. The microlog and proximity log are the two main fresh-mud microresistivity tools, while the micro SFL and microlaterolog are the two main salt-mud microresistivity tools.

Figure 5.2.24 [58,21,26] shows the electrode pads and current paths for the Microlog (5.2.24a), Micro SFL. (5.2.24b), and proximity log (5.2.24c), and Microlaterolog. Figure 5.2.24d shows the Micro SFL. sonde.

Theory

The microlog makes two shallow nonfocused resistivity measurements, each at different depths. The two measurements are presented simultaneously on the log as the micronormal and microinverse curves. Positive separation (micronormal reading higher than the microinverse) indicates permeability. R_{xo} values can be found by using Figure 5.2.25 [20]. To enter Figure 5.2.25, R_{mc} must be corrected to formation temperature, and mudcake thickness (h_{mc}) must be found. To find h_{mc} subtract the caliper reading (presented in track 1) from the borehole size and divide by two. In washed out or enlarged boreholes, h_{mc} must be estimated. The microlog gives a reasonable value of R_{xo} when mudcake thickness is known. The main disadvantage of this tool is that it cannot be combined with an R_t device, thus a separate logging run is required. The microlog is primarily a fresh mud device and does not work very well in salt-based muds [206]. Generally the mudcake is not thick enough in salt-based muds to give a positive separation opposite permeable zones. The backup arm on the microlog tool, which provides a caliper reading, is also equipped with either a proximity log or a Microlaterolog. The proximity log is designed for fresh muds where thick mudcakes develop opposite permeable formations. There is essentially no correction necessary for mudcakes less than $\frac{3}{4}$ in. If invasion is shallow, the R_{xo} measurement may be affected by R_t because the proximity log reads deeper into the formation than the other microresistivity logs. When the microlog is run with a proximity log, it is presented in track 1 with a microcaliper. The proximity log is presented in tracks 2 and 3 on a logarithmic scale (Figure 5.2.26).

Since its introduction, the microlog (Schlumberger) has become the standard tool for recognizing permeable zones. The theory behind it is similar to using multiple resistivity devices. The tool consists of three electrode buttons on a rubber pad which is pressed against the borehole wall.

In a permeable zone, mud filtrate will enter the formation leaving the clay particles behind on the borehole wall. These clay particles may form a mudcake up to an inch thick. The resistivity of the mudcake is less than the resistivity of the formation saturated with mud filtrate. Two resistivity readings, the microinverse and the micronormal, are taken simultaneously. The microinverse has a depth

of investigation of only an inch; therefore, it reads mostly mudcake (if present). The micronormal has a depth of investigation of 3 to 4 in. and is influenced primarily by fluids in the flushed zone. The difference in resistivity shows up on the log as a separation of the curves with the micronormal reading higher than the microinverse. This is referred to as "positive separation." In impermeable formations, both readings are very high and erratic, and negative separation may occur (micronormal less than microinverse). Shales commonly show negative separation with low resistivities (Figure 5.2.27).

In salt muds, the microlaterolog and micro spherically focused log (MSFL) are used for R_{xo} readings. The microlaterolog is a focused tool with a shallower depth of investigation than the proximity log. For this reason, the microlaterolog is very strongly affected by mudcakes thicker than $\frac{3}{8}$ in. It is presented in tracks 2 and 3 like the proximity log. The MSFL is the most common R_{xo} tool for salt muds. It is a focused resistivity device that can be combined with the dual laterolog, thus providing three simultaneous resistivity readings. Although the depth of investigation is only a few inches, the tool can tolerate reasonably thick mudcakes $\frac{3}{4}$ in. The tool is also available in a slim-hole version. The only disadvantage to this device is that the pad can be easily damaged in rough boreholes.

Interpretation

The saturation of the flushed zone can be found from Equation 5.2.20. R_{mf} must be at formation temperature. Moveable hydrocarbons can be found by comparing S_{xo} and S_w. If $S_w/S_{xo} < 0.7$ then the hydrocarbons in the formation are moveable (this is also related to fluid permeability). If $S_w/S_{xo} > 0.7$, either there are no hydrocarbons or the hydrocarbons present are not moveable.

Gamma Ray Logs

The gamma ray log came into commercial use in the late 1940s. It was designed to replace the SP in salt muds and in air-filled holes where the SP does not work. The gamma ray tool measures the amount of naturally occurring radioactivity in the formation. In general, shales tend to have high radioactivity while sandstone, limestone, dolomite, salt, and anhydrite have low radioactivity. There are exceptions. Recently, tools have been designed to separate gamma rays into their respective elemental sources, potassium (K), thorium (Th), and uranium (U).

Theory

Gamma rays are high-energy electromagnetic waves produced by the decay of radioactive isotopes such as K40, Th, and U. The rays pass from the formation and enter the borehole. A gamma ray detector (either scintillation detector or Geiger-Muller tube) registers incoming gamma rays as an electronic pulse. The pulses are sent to the uphole computer where they are counted and timed. The log, presented in track 1 in Figure 5.2.13, is in API units.

As previously mentioned, there are new gamma ray tools available that determine which elements are responsible for the radioactivity. The incoming gamma rays are separated by energy levels using special energy-sensitive detectors. The data are collected by the computer and analzyed statistically. The log presents total (combined) gamma ray in track 1 and potassium (in %), and uranium and thorium (in ppm) in tracks 2 and 3 (Figure 5.2.27). Combinations of two components are commonly presented in track 1. The depth of investigation of the natural gamma tools is 2–10 in. depending on mud weight, formation density, hole size, and gamma ray energies.

(*text continued on page 5-109*)

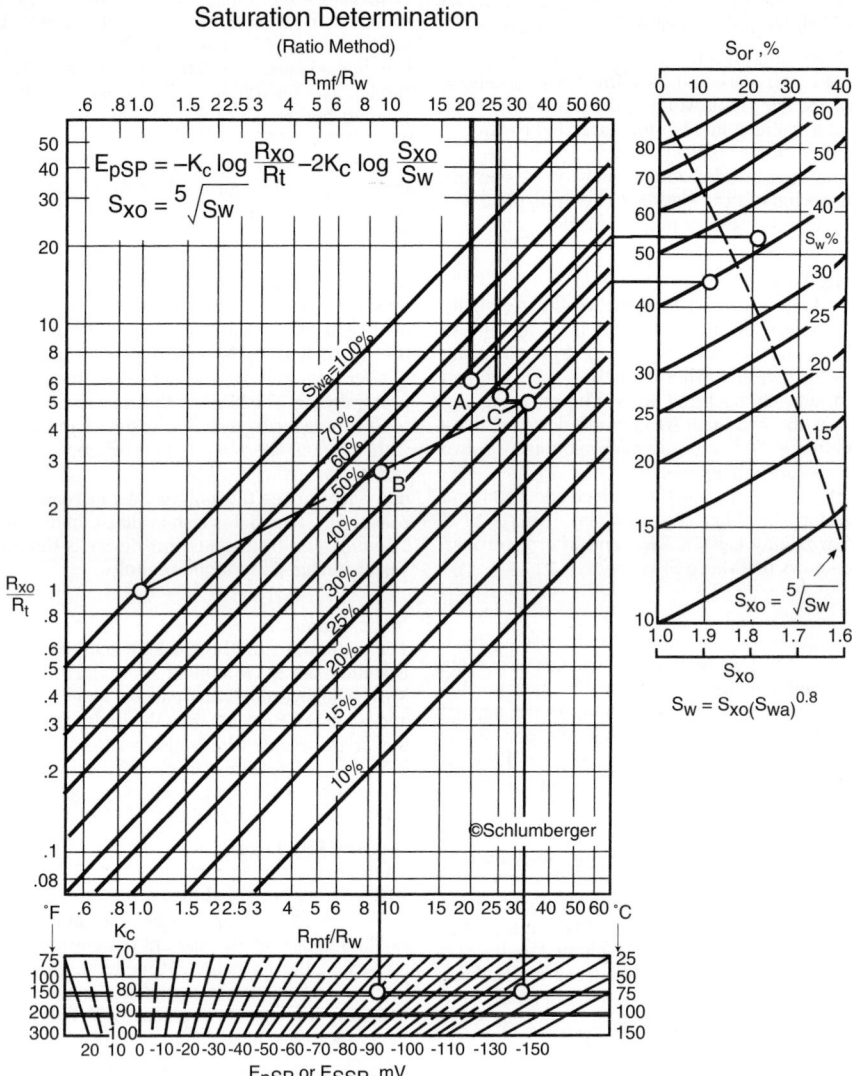

Figure 5.2.22   *Chart for determining water saturation by the ratio method [20].*

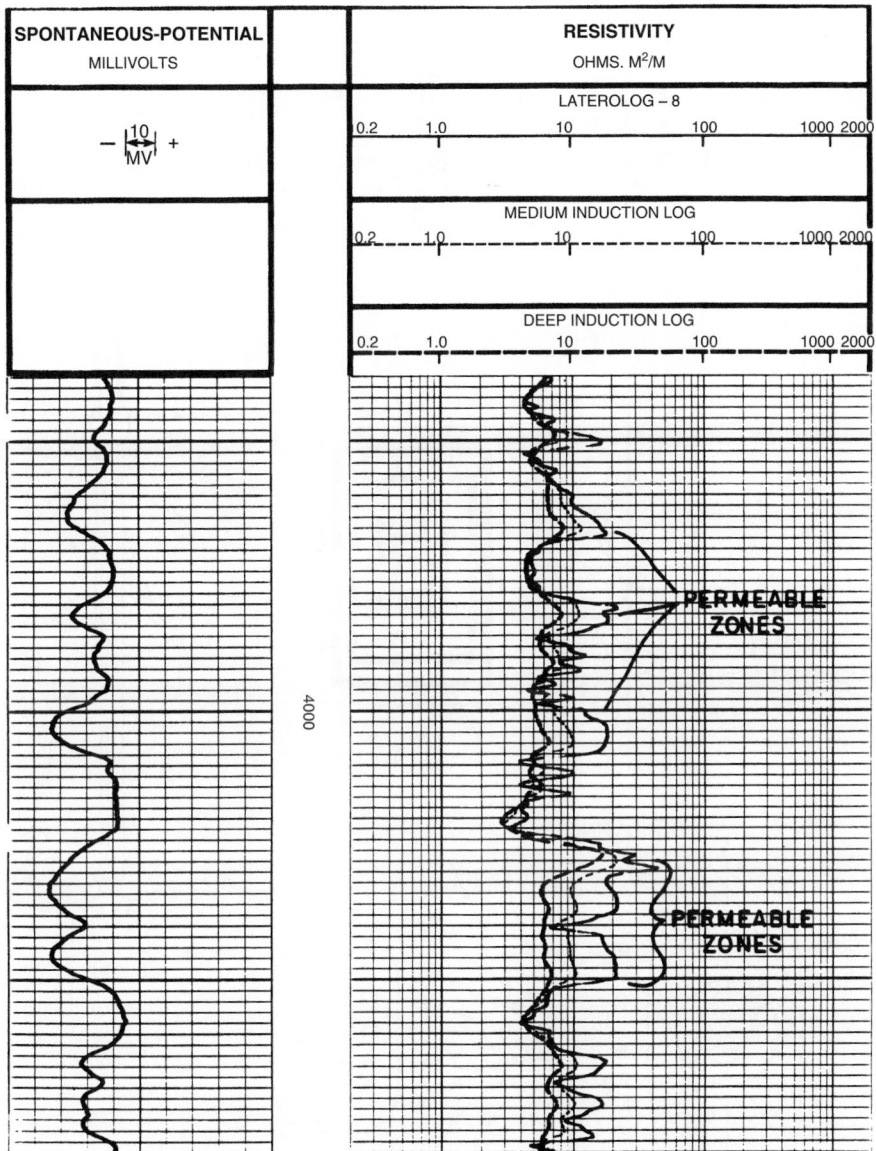

Figure 5.2.23 *Example of resistivity log curve separation as an indication of permeability.*

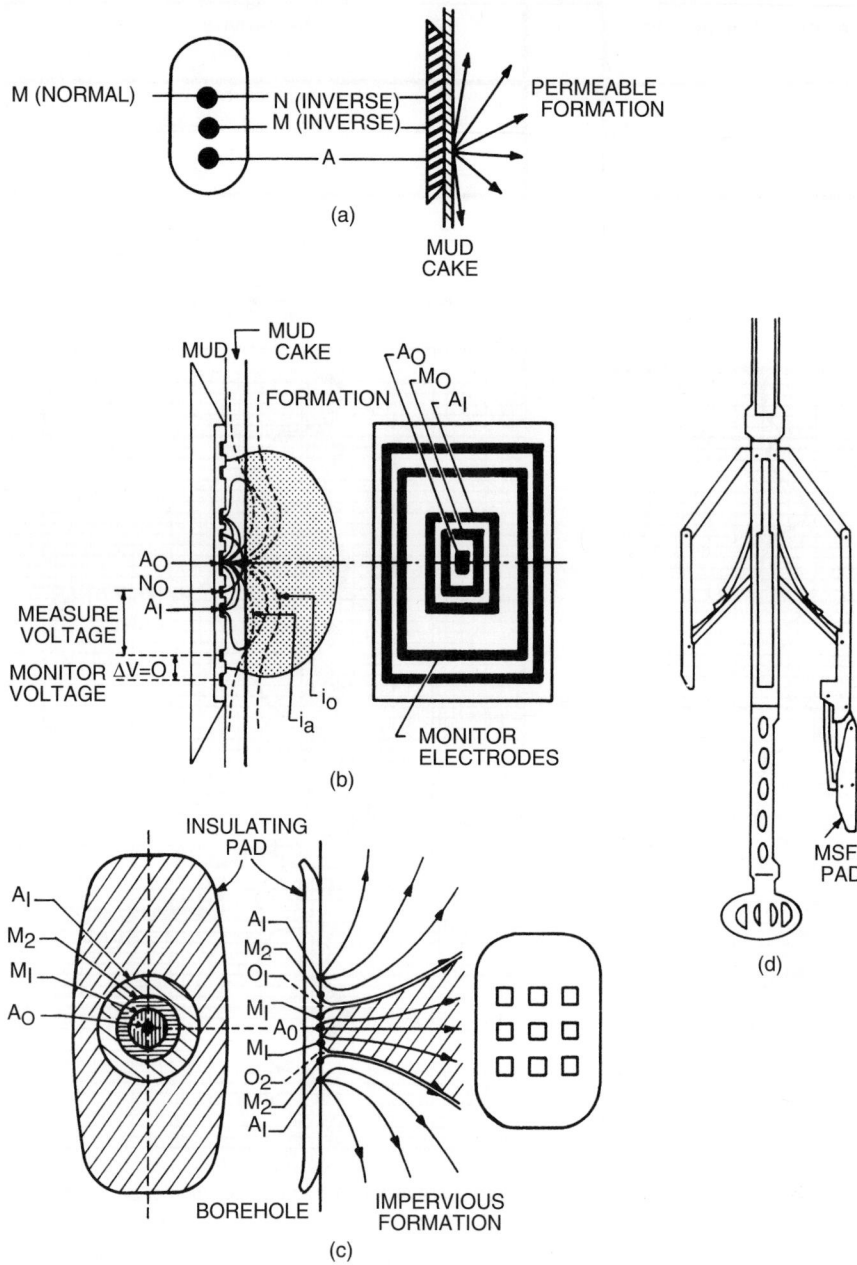

Figure 5.2.24 *Electrode arrangements for various microresistivity devices: (a) microlog; (b) microspherically focused log; (c) microlaterolog; (d) location of the micro pad on the dual laterolog sonde [21,26,58].*

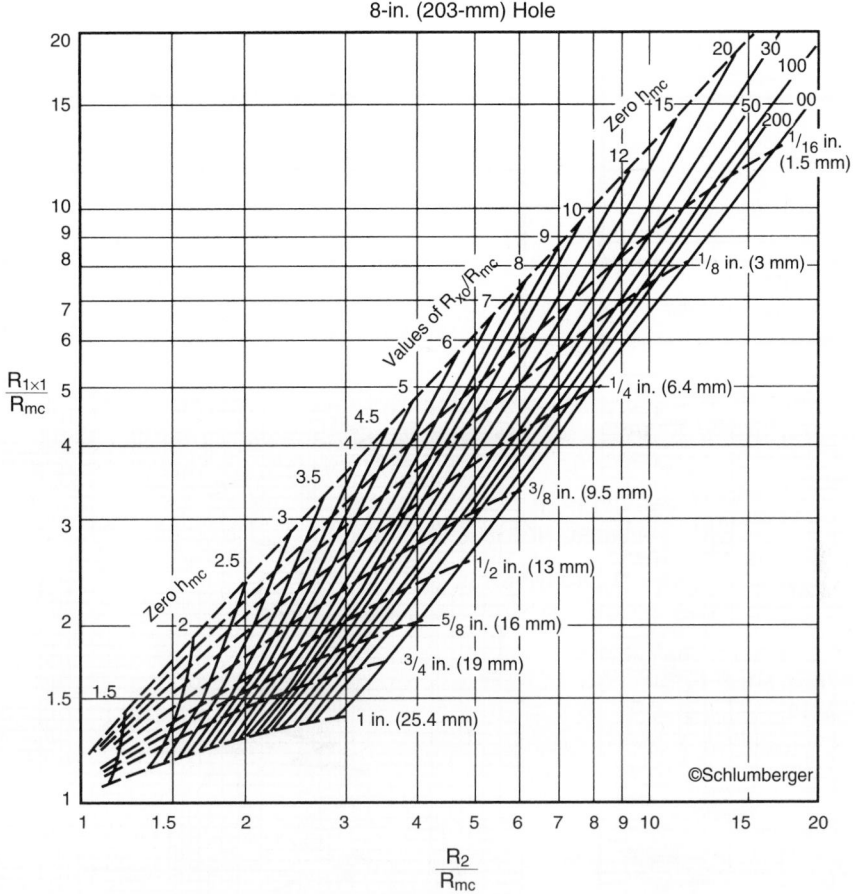

Figure 5.2.25 *Chart for finding Rxo from microlog readings [20].*

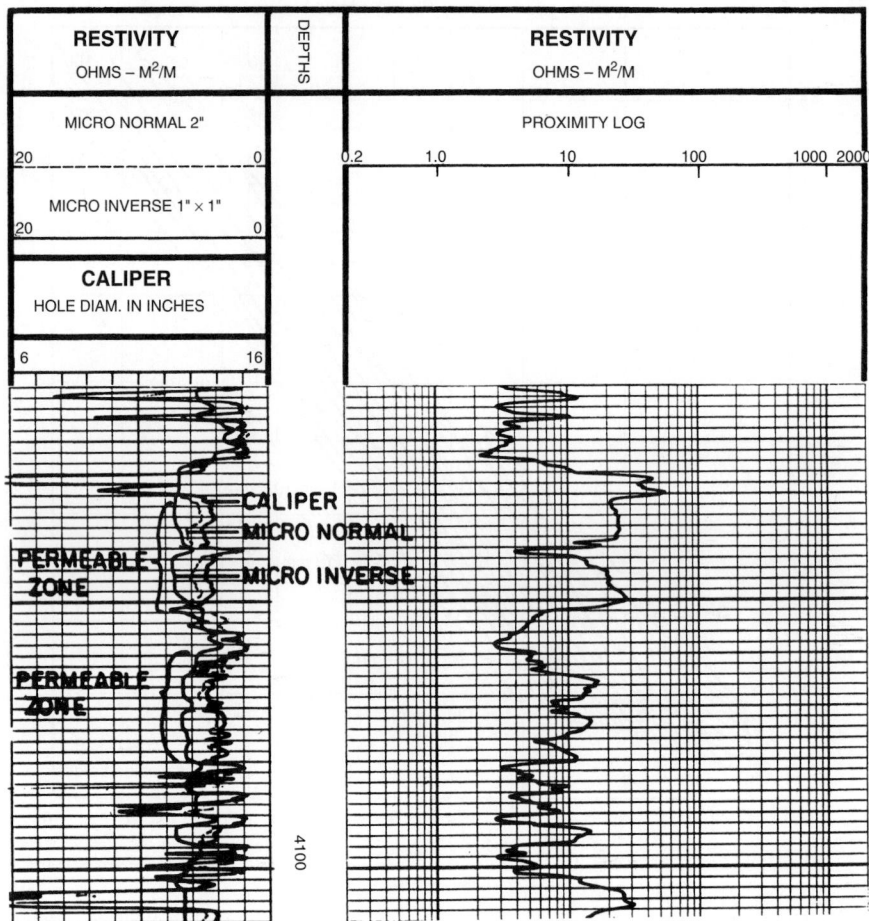

Figure 5.2.26 *Example log showing positive separation of the microlog curves opposite a permeable formation at 4,030 to 4,050 feet.*

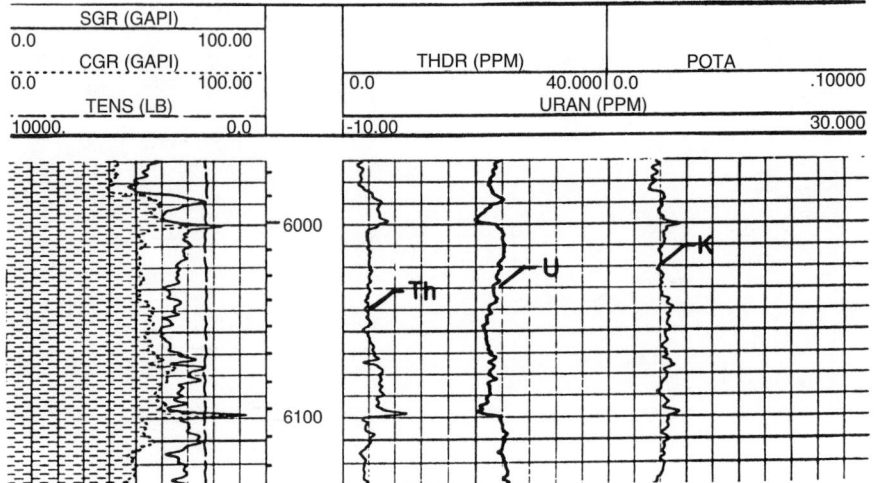

Figure 5.2.27 *Example of a natural gamma spectroscopy log presentation.*

(*text continued from page 5-103*)

Interpretation

The interpretation of a total gamma ray curve is based on the assumption that shales have abundant potassium-40 in their composition. The open lattice structure and weak bonds in clays encourage incorporation of impurities. The most common of those impurities are heavy elements such as uranium and thorium. Thus, shales typically have high radioactivity. Sandstones (quartzose), carbonates and evaporites have strong bonds and generally do not allow impurities. Limestones undergo rearrangement of crystal structure and addition of magnesium to become dolomites. Impurities like uranium (which is very soluble) may enter the crystal lattice during recrystallization. Feldspathic sandstones contain an abundance of potassium-40 and therefore show higher radioactivity than quartzose sandstone. Some evaporite minerals (Such as KCl) contain high amounts of potassium-40 and may appear as shales on the log. Serra et al. [28] provide an excellent discussion of interpretation of the natural or spectral gamma-ray tool.

Sonic (Acoustic) Log

The sonic (acoustic or velocity) total measures the time it takes for a compressional wave to travel through one vertical foot of formation. It can be used to determine porosity (if the lithology is known) and to determine seismic velocities for geophysical surveys when combined with a density log. The sonic log also has numerous cased hole applications.

Theory

A 20 kHz sound wave is produced by the tool and travels through the mud into the formation. The wave travels vertically through the formation. The first arrival of the compressional wave is picked up by a received about one foot away from the transmitter. The wave continues through the formation and is picked up by the far receiver (normally 2-ft below the near receiver). The time difference between the near and far receivers is used to determine formation travel time (Δt). Fractures, vugs, unconsolidated formations, gas-cut mud, lost circulation materials, and rough boreholes can cause sharp increases in Δt, called cycle skips.

Interpretation

Table 5.2.7 shows the velocity and travel time for several commonly encountered oilfield materials. The t_{ma} value in the fourth column is at 0% porosity. Porosity increases travel time. Wyllie and coworkers [208] developed an equation that relates since travel time to porosity:

$$\phi = \frac{\Delta t_{log} - \Delta t_{ma}}{\Delta t_f - \Delta t_{ma}} \qquad [5.2.23]$$

where Δt_{log} = Δt value read from log, μ sec/ft
Δt_{ma} = matrix velocity at 0% porosity, μ sec/ft
Δt_f = 189–190 μ sec/ft (or by experiment)

The Wyllie equation works well in consolidated formations with regular intergranular porosity ranging from 5%–20% [209]. If the sand is not consolidated or compacted, the travel time will be too long, and a compaction correction factor (C_p) must be introduced [208]. The reciprocal of C_p is multiplied by the porosity from the Wyllie equation:

$$\phi = \left(\frac{\Delta t_{log} - \Delta t_{ma}}{t_f - t_{ma}}\right)\left(\frac{1}{C_p}\right) \qquad [5.2.24]$$

The compaction correction factor (C_p) can be found by dividing the sonic porosity by the true (known) porosity. It can also be found by dividing the travel time in an adjacent shale by 100:

$$C_p = \frac{\Delta t_{shale}}{100} C \qquad [5.2.25]$$

where C is a correction factor, usually 1.0 [200]. In uncompared sands, porosities may be too high even after correction if the pores are filled with oil or gas. Hilchie [200] suggests that if pores are oil-filled, multiply the corrected porosity by 0.9; if gas-filled, use 0.7 to find corrected porosity.

Raymer, Hunt, and Gardner [210] presented an improved travel-time to porosity transform that has been adopted by some logging companies. It is based on field observations of porosity versus travel time:

$$\Delta t_{log} = (1 - \phi)^2 \Delta t_{ma} + \phi \Delta t_f \qquad [5.2.26]$$

This relationship is valid up to 37% porosity.

The heavy set of lines in Figure 5.2.28 [20] was derived using the Raymer et al. [210] transform, and the lighter set represents the Wyllie relationship.

The sonic porosity derived from the Wyllie equation (Equation 5.2.23) does not include secondary porosity (vugs and fractures), so it must be cautiously applied in carbonate rocks.

Density Log

The formation density tool measures the bulk electron density of the formation and relates it to porosity. It is a pad device with a caliper arm. The tool is usually run in combination with a neutron log, but it can be run alone.

Theory

The density tool emits medium energy gamma rays from a radioactive chemical source (usually Cs-137). The gamma ray penetrate the formation and collide with electron clouds in the minerals in the rock. With each collision the gamma ray loses some energy until it reaches a lower energy state. This phenomenon is called "Compton scattering." Some gamma rays are absorbed, and a high-energy electron is emitted from the atom. This phenomenon is called the "photoelectric absorption" effect, and is a function of the average atomic weight of each element. Both the Compton-scattered gamma ray and the photo-electrically produced electron return to the borehole where they are detected by

Table 5.2.7 *Matrix Travel Times*

Material	Velocity range ft/sec	Δt range μ sec/ft	Δt commonly used μ sec/ft	Δt at 10% porosity μ sec/ft
Sandstone	18,000–19,500	51.0–55.5	55.0 or 51.0	69.0 or 65.0
Limestone	21,000–23,000	43.5–47.6	47.5	61.8
Dolomite	23,000–24,000	41.0–43.5	43.5	58.0
Salt	15,000	66.7	66.7	—
Anhydrite	20,000	50.0	50.0	—
Shale	7,000–17,000	58.0–142	—	—
Water	5,300	176–200	189	—
Steel casing	17,500	57.0	57.0	—

From References 21 and 36.

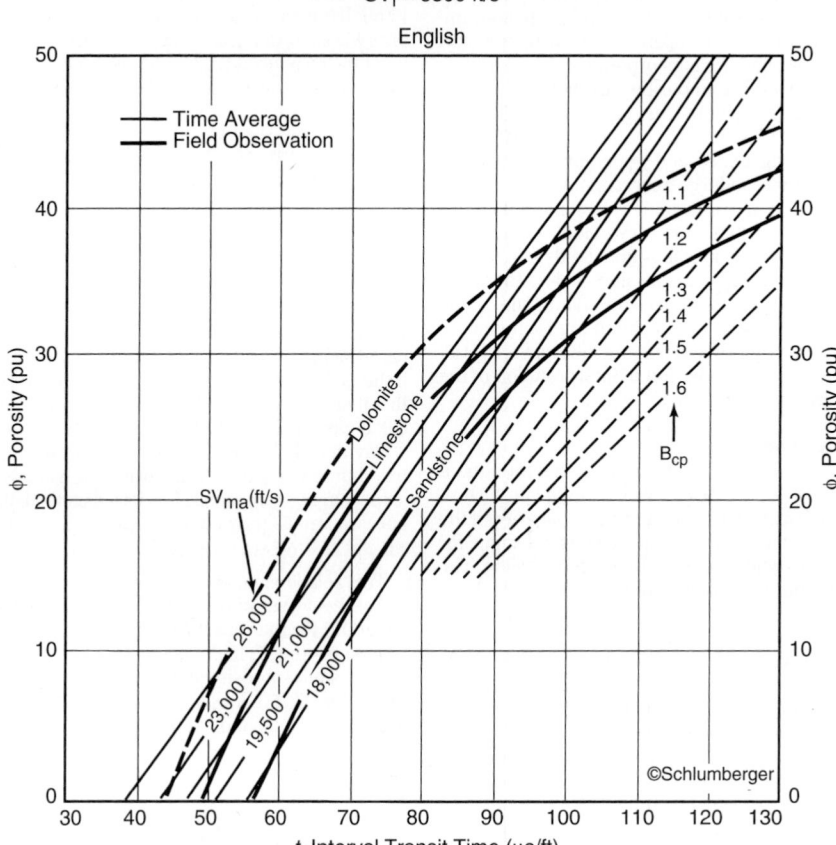

Figure 5.2.28 *Chart for evaluating porosity with a sonic log [20].*

scintillation tubes on the density tool. The main result is that a porous formation will have many returning gamma rays while a nonporous formation will have few returning gamma rays. Each tool has two detectors; one is near the source (short-spacing detector) and another is 1–1.5 ft (35–40 cm) away from the near detector (long-spacing detector).

The long-spacing detector provides the basic value of bulk electron density. The short-spacing detector is used to make a mudcake correction. This correction, made automatically by computer by most service companies, is based on the "spine and ribs" plot (Figure 5.2.29) [5]. The "spine" is the heavy, nearly vertical line from 1.9 to 2.9 g/cc. The ribs are the lighter curved lines trending left to right. The experimental data for constructing the "ribs" are shown in the corners of the plot. Long-spacing count-rates are on the abscissa and short-spacing count-rates are on the ordinate axis. The computer receives data from the sonde and plots it on the chart. If the point falls off the "spine," it is brought back along one of the ribs. Bringing the point back along the "rib" will change the intersection point on the "spine." The correction that is produced is called Δρ and may be either positive or negative depending on the mud properties. Negative Δρ values occur in heavy (barite or iron), weighted muds. Positive values occur in light muds and when the density pad is not flush against the borehole wall as occurs in rough or "rugose" boreholes. The Δρ curve is useful for evaluating the quality of the ρ_b reading. Excursions from 0 that are more than ±0.20 gm/cc on the Δρ curve indicate a poor quality reading.

Interpretation

An equation similar to the Wyllie equation is used to calculate porosity values from bulk density.

$$\phi = \frac{\rho_{ma} - \rho_b}{\rho_{ma} - \rho_f} \qquad [5.2.27]$$

where ρ_{ma} = bulk density of matrix at 0% porosity, g/cc
ρ_b = bulk density from log, g/cc
ρ_f = bulk density of fluid, g/cc

Table 5.2.8 lists commonly used values for ρ_{ma} and ρ_f, and, along with Figure 5.2.30 shows how ρ_f changes with temperature and pressure [58]. As with the sonic tool, an incorrect choice of matrix composition may give negative porosity values.

If a zone is hydrocarbon saturated but not invaded by mud filtrates, the low density of the hydrocarbons will increase the porosity reading to a value that is too high. In this case, Hilchie [21] suggests using the following

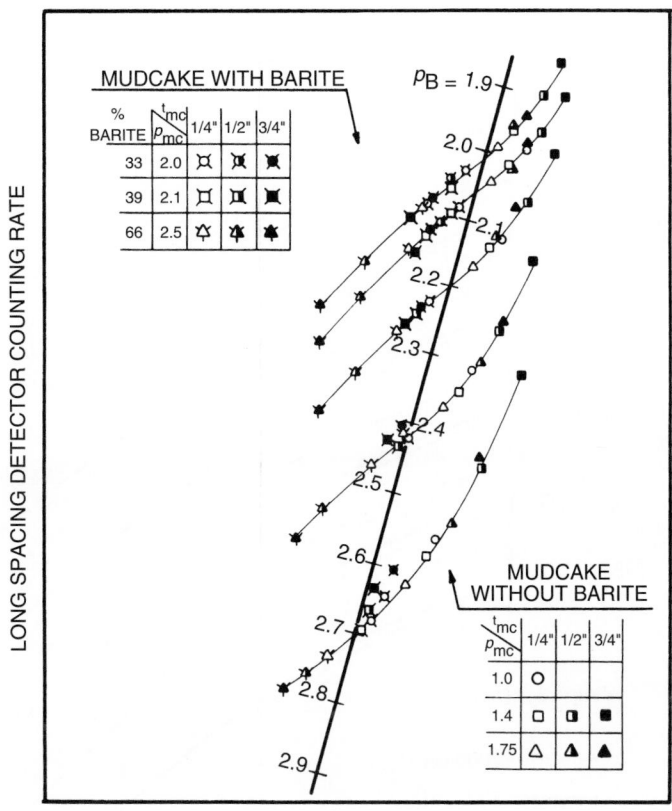

Figure 5.2.29 *Spine and ribs plot used to correct bulk density readings for mudcake effects [58].*

Table 5.2.8 *Bulk Densities Commonly Used for Evaluating Porosity With a Density Log**

Material	ρ_{bulk}	ρ_{log} at 10% porosity (fresh water)
Unconsolidated sand	2.65 g/cc	2.48 g/cc
Silica cemented sand	2.65 g/cc	2.48 g/cc
Calcite cemented sand	2.68 g/cc	2.51 g/cc
Limestone	2.71 g/cc	2.54 g/cc
Dolomite	2.83–2.87 g/cc	2.64–2.68 g/cc
Salt	2.03 g/cc	—
Anhydrite	2.98 g/cc	—
Fresh water	1.0 g/cc	—
Salt water	1.1–1.2 g/cc	—

Fluid Densities for Water (Based on Salinity)**	
Salinity, ppm Nacl	ρ_f, g/cc
0–50,000	1.0
50,000–100,000	1.03
100,000–150,000	1.07
150,000–200,000	1.11
200,000–250,000	1.15
250,000–300,00	1.19

*From Reference 36.
**From Reference 21.

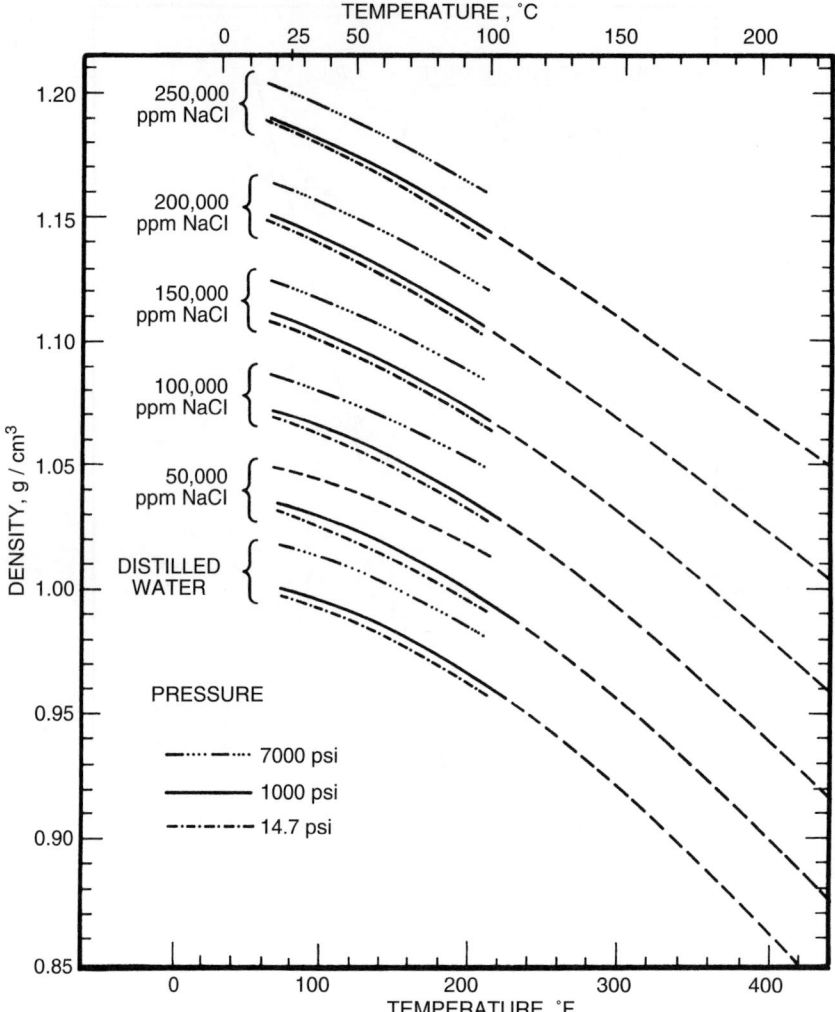

Figure 5.2.30   *Chart showing relationship of water salinity to density and temperature [58].*

equation:

$$\phi = \frac{0.9 \left(\frac{R_w}{R_t}\right)^{1/2} (\rho_w - \rho_h) + \rho_{ma} - \rho_b}{\rho_{ma} - \rho_h} \qquad [5.2.28]$$

where ρ_w = density of formation water, g/cc (estimated from Figure 5.2.30) [58]

ρ_h = density of hydrocarbons, g/cc (from Figure 5.2.31) [199]

Neutron Log
Neutron tools measure the amount of hydrogen in the formation and relate it to porosity. High hydrogen content indicates water (H_2O) or liquid hydrocarbons (C_xH_z) in the pore space. Except for shale, sedimentary rocks do not contain hydrogen in their compositions.

Theory
Neutrons are electrically neutral particles with mass approximately equal to that of a hydrogen atom. High-energy neutrons are emitted from a chemical source (usually AmBe or PuBe). The neutrons collide with nuceli of the formation minerals in elastic-type collisions. Neutrons will lose the most energy when they hit something with equal mass, such as a hydrogen atom. A few microseconds after being released, the neutrons have lost significant energy and enter the thermal state. When in the thermal state, neutrons are captured by the nuclei of other atoms (Cl, H, B). The atom which captures the neutron becomes very excited and emits a gamma ray. The detectors on the tool may detect epithermal neutrons, thermal neutrons or high-energy gamma rays of capture. Compensated neutron tools (CNL) detect thermal neutrons and use a ratio of near-to-far detector counts to determine porosity. Sidewall neutron tools (SNP) detect epithermal neutrons and have less matrix effect (though they are affected by rough boreholes more than the CNL).

Interpretation
Neutron tools are seldom run alone. They are usually combined with a density-porosity tool. Older neutron logs are not presented as porosity but as count rates. Some logs do not specify a scale (Figure 5.2.32), but only which direction

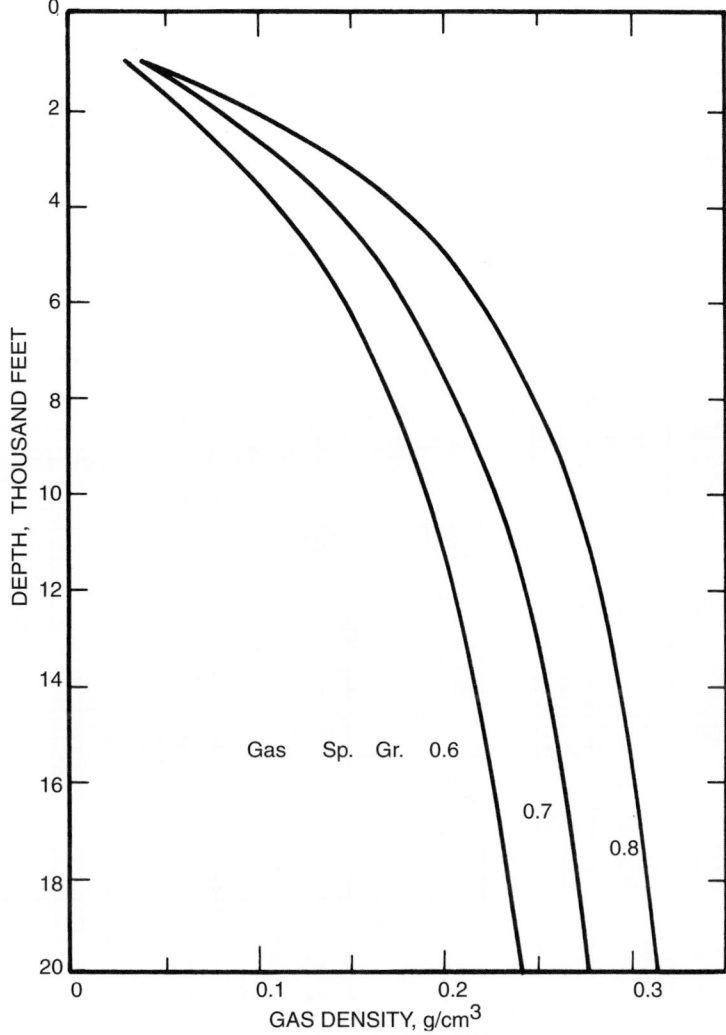

Figure 5.2.31 *Chart for estimating density of gases from reservoir depth and gas specific gravity [21].*

the count rate (or radiation) increases. An increase in radiation indicates lower porosity (less hydrogen). Newer logs present porosity (for a particular matrix, limestone, sandstone or dolomite) directly on the log. Most neutron logs are run on limestone matrix. Figure 5.2.33 corrects the porosity for matrix effect if the log is run on limestone matrix [20]. Neutron logs exhibit "excavation effect" in gas-filled formations. The apparent decrease in porosity is due to the spreading out of hydrogen in gas molecules; gases have less hydrogen *per unit volume* than liquids. Thus the neutron tool sees less hydrogen and assumes less porosity. The magnitude of the effect depends on gas saturation, gas density, and pressure. Care should be taken in using correction charts for neutron tools; each service company has a slightly different design, and the correct chart for the particular tool and service company should be used.

Multiple-Porosity Log Interpretation
As mentioned earlier, the neutron, density, and sonic tools are lithology dependent. If the matrix is incorrectly selected,

porosity may be off as much as 10 porosity units. If two lithology-dependent logs are run simultaneously on the same matrix and presented together, lithology and porosity can be determined. The most common and useful of combinations is the neutron and density. Figures 5.2.34 through 5.2.38 are crossplots of neutron, density, and sonic porosity. The charts are entered with the appropriate values on the ordinate and abscissa. The point defines porosity and gives an indication of matrix. If a point falls between two matrix lines, it is a combination of the two minerals or the neutron-density crossplots. Gas moves the points up and to the left. To correct for the gas effect, move parallel to the gas correction arrow to the assumed lithology. Note that a gassy limestone may look like a sandstone. Shales tend to bring points down and to the right depending on the shale composition. Typically, shaly sandstone will look like a limestone. The sonic-density crossplot is not very helpful in determining porosity or lithology but is extremely useful for determining evaporite mineralogy.

In some areas, neutron and density tools are run on sandstone porosity and therefore cannot be entered in the

(*text continued on page 5-119*)

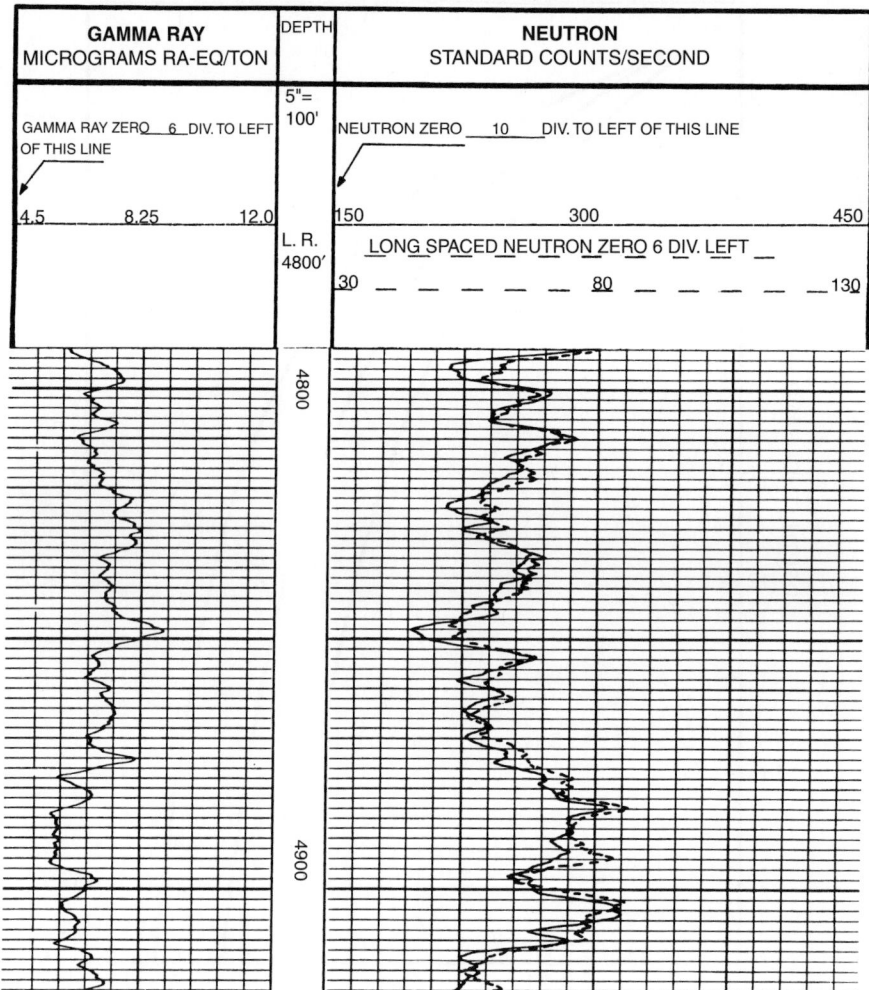

Figure 5.2.32 *Old neutron log presentation.*

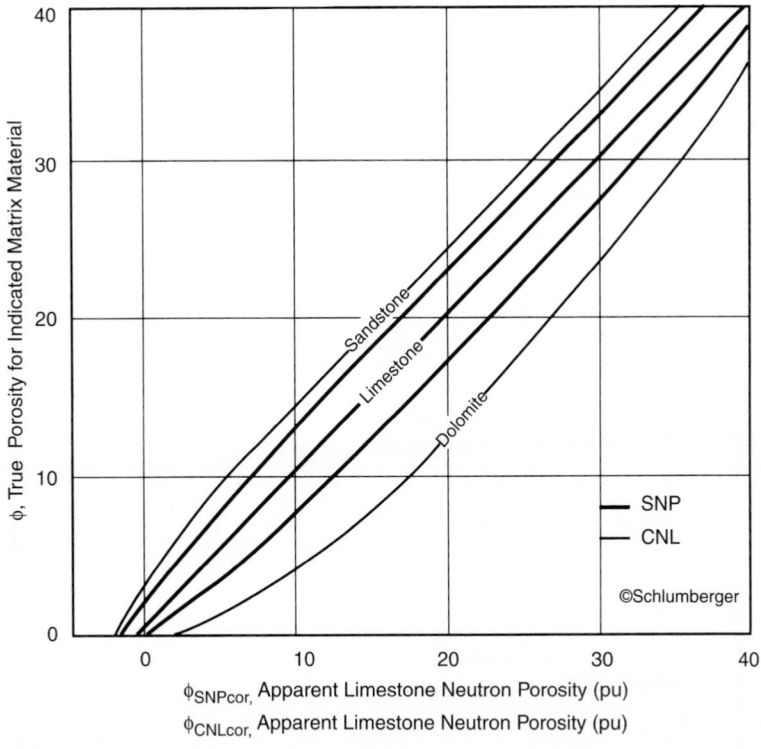

Figure 5.2.33 *Matrix lithology correction chart for neutron porosity logs [20].*

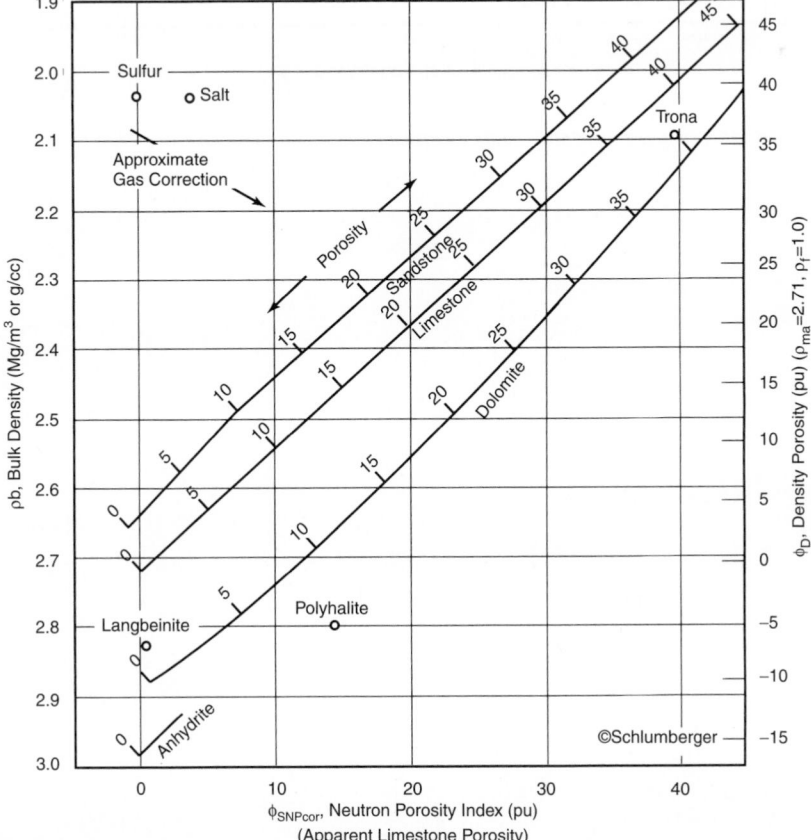

Figure 5.2.34 *Chart for finding porosity and matrix composition from an FDC-SNP log combination in fresh water [20].*

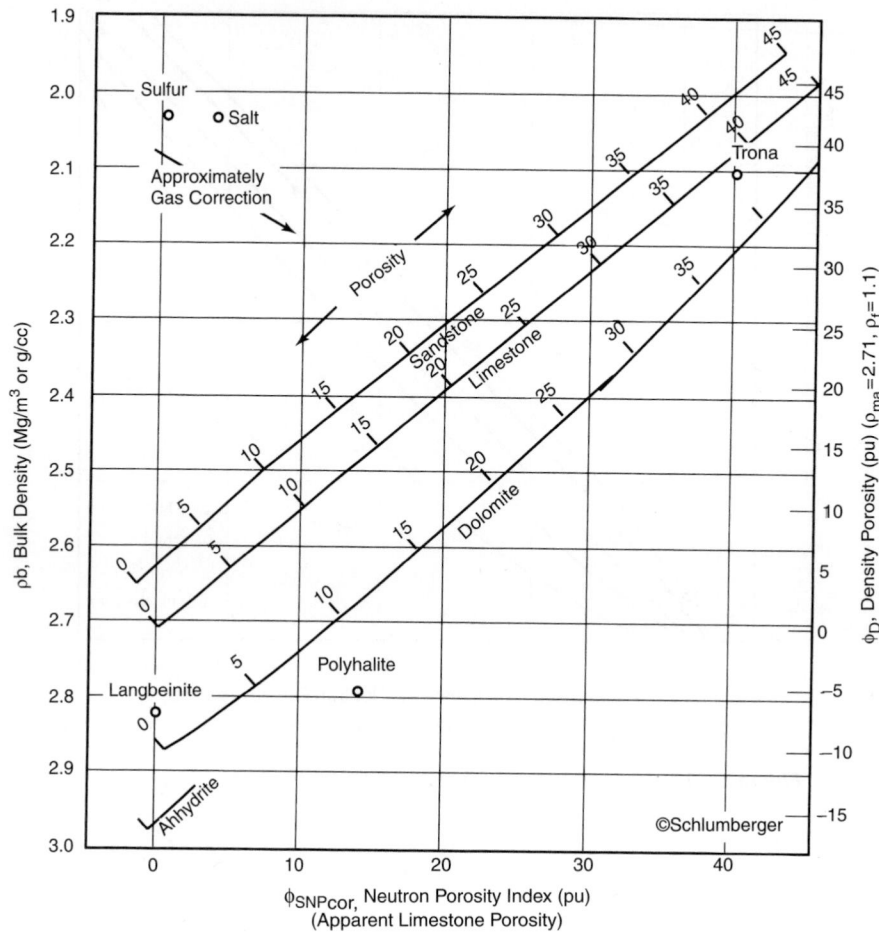

Figure 5.2.35 *Chart for finding porosity and matrix composition from an FDC-SNP log combination in salt water [20].*

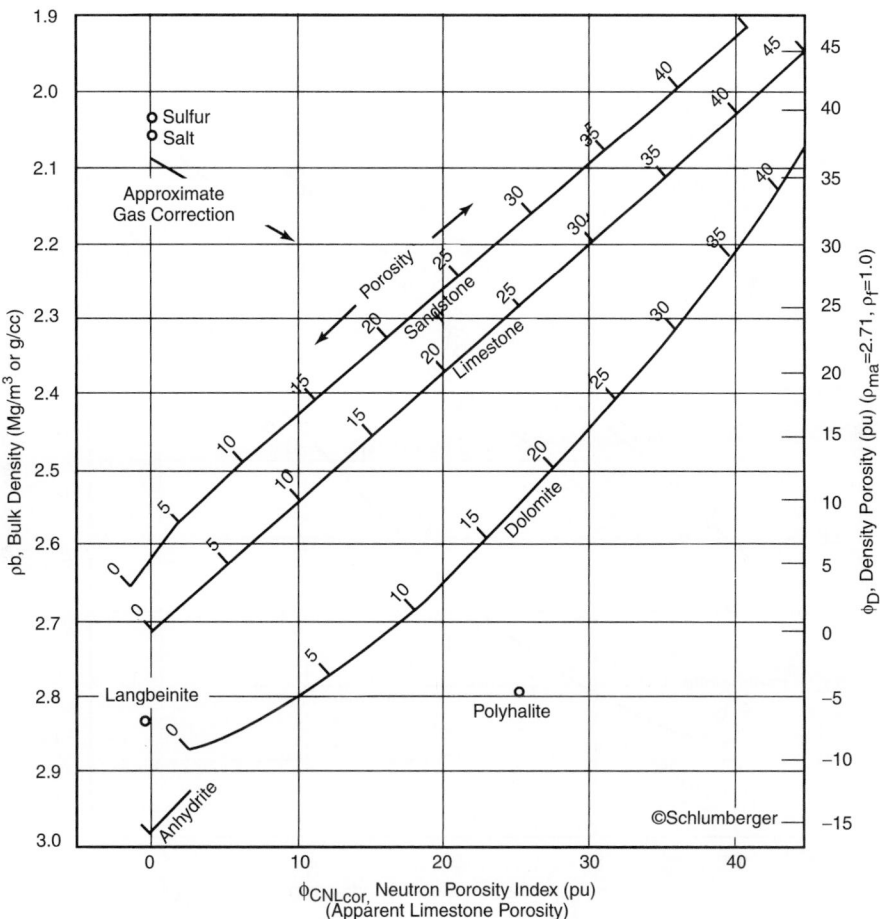

Figure 5.2.36 *Chart for finding porosity and matrix composition from a FDC-CNL log combination in fresh water [20].*

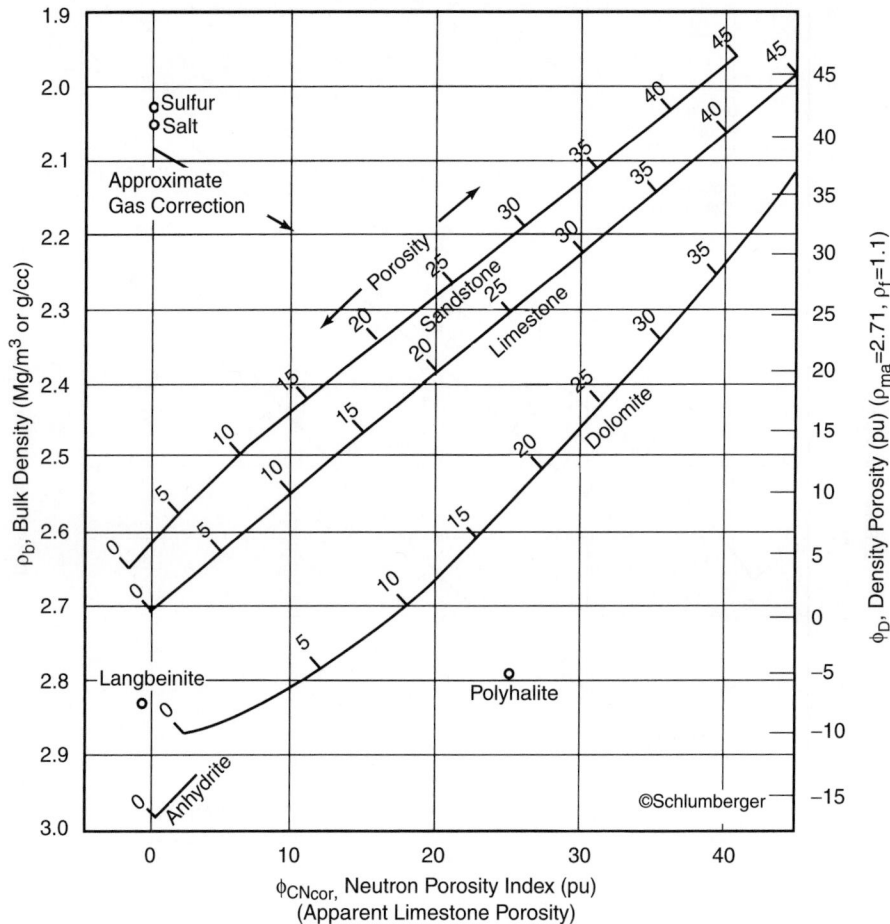

Figure 5.2.37   *Chart for finding porosity and matrix composition from a FDC-CNL log combination in salt water [20].*

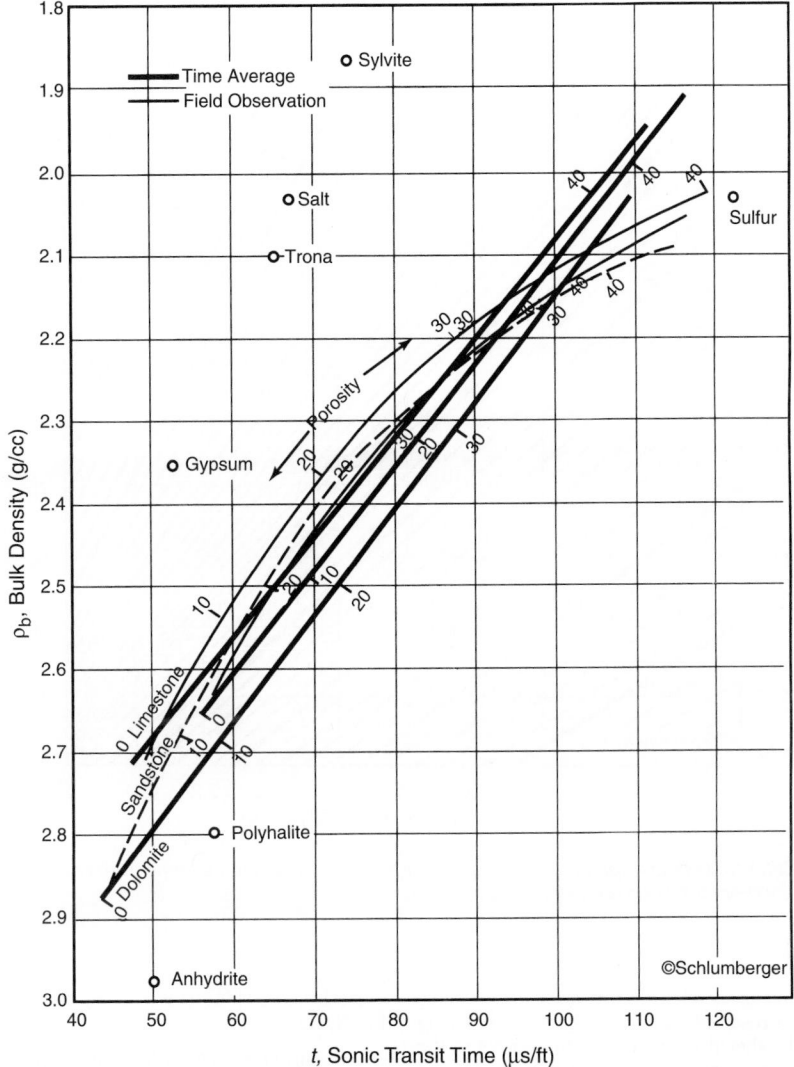

Figure 5.2.38 *Lithology estimation chart for the FDC-sonic log combination in fresh water [20].*

(text continued from page 5-113)

charts directly. To use the neutron-density crossplots when the matrix is not limestone, another method must be applied. Remember that the vertical lines are constant neutron-porosity and the horizontal lines are constant density-porosity. Instead of entering the bottom of sides of the chart, select the appropriate lithology line in the interior of the chart. Draw a horizontal line through the density-porosity and a vertical line through the neutron-porosity. Lithology and porosity are determined at the intersection of these two lines.

Another device that provides good lithology and matrix control is the Lithodensity tool (LDT). It combines a density tool with improved detectors and a P_{ef} curve (photoelectric effect). Combining the P_b, and P_{ef} curve values, an accurate 3- or 4-mineral composition can be determined from the charts in Figures 5.2.39 to 5.2.42. This also provides an excellent way to confirm neutron-density cross-plot interpretations.

Nuclear Magnetic Resonance (NMR)

This log examines the nucleus of certain atoms in the formation. Of particular interest are hydrogen nuclei (protons) since these particles behave like magnets rotating around each other [23]. Hydrogen is examined because it occurs in both water and hydrocarbons.

The log measures fluid by applying a magnetic field, greater in intensity than the earth field, to the formation. Hydrogen protons align themselves with the induced field and when the field is suddenly removed, the protons precess about the earth's magnetic field much like a gyroscope. The nucleus of hydrogen has a characteristic precession rate called the Larmor frequency (∼ 2,100 Hz), and can be identified by sensors on the tool [24]. The nuclei contributing to the total signal occur in the free fluid in the pores; fluid adsorbed on the grains makes no contribution. The signals are then processed in a computer and printed out onto a log.

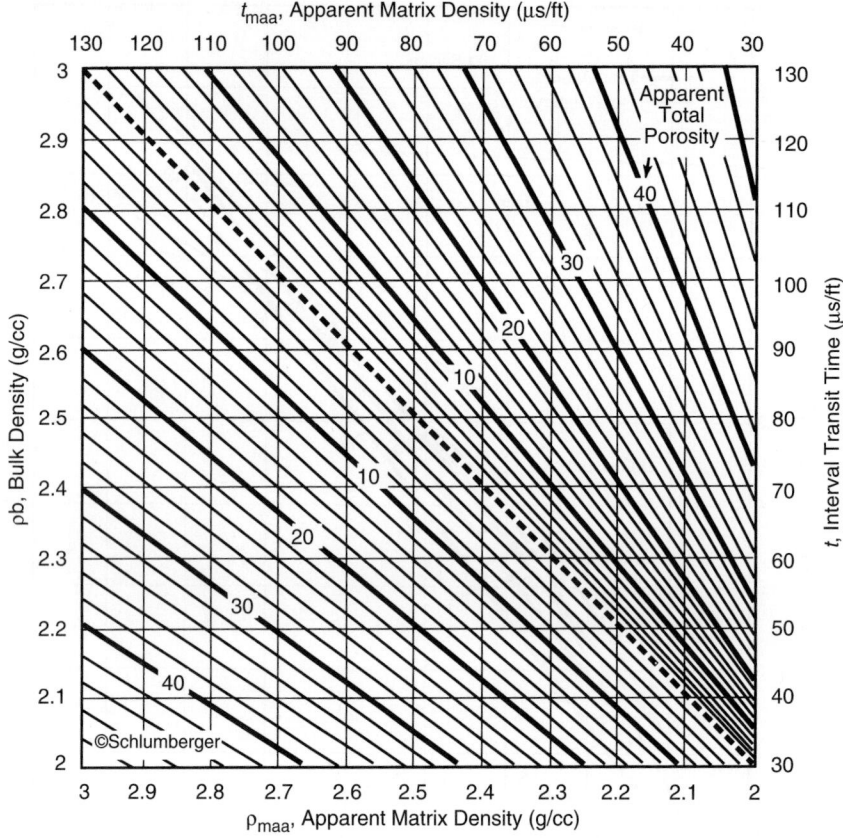

Figure 5.2.39 *Chart for finding apparent matrix density or apparent matrix transit time from bulk density or interval transit time and apparent total porosity [20].*

Normally, proton precession decays along a time constant, T_2. This is a result of each proton precession falling out of phase with other protons due to differences in the local magnetic fields. Moreover, each proton precesses at a slightly different frequency, depending on the kind of fluid it occurs in. This disharmonic relationship makes it possible to differentiate between free water and free oil in a reservoir [34].

Three log modes can be presented:

1. Normal mode—consists of the free fluid index (FFI) in percent obtained from a polarization time of two seconds, the Larmor proton frequency (LFRE), the decay-time constant (or longitudinal relaxation time) of the signal (T_2) and a signal-to-noise ratio (STNR).
2. Continuous mode—gives three free fluid index (FFI) readings taken at polarization times of 100, 200, and 400 ms, respectively, two longitudinal relaxation times (T_1 and T_2), and a signal-to-noise ratio (STNR).
3. Stationary display mode—a signal-stacking mode where eight signals are stacked from each of six polarization times to obtain precise T_1, T_2, and FFI values.

FFI readings yield porosity that is filled with moveable fluid and is related to irreducible water saturation (S_{iw}). Addition of paramagnetic ions to the mud filtrate will disrupt the water portion of the signal and residual oil saturation (S_{or}) can be determined.

Desbrandes [33] summarized the following uses for the NML:

1. Measuring free fluid volume in the pores (ϕ_F).
2. Evaluating permeability by comparing ϕ_F with ϕ_T (total porosity from a neutron-density log combination).
3. Locating intervals at irreducible water saturation by comparing ϕ_F with ϕ_T and R_t, determined with other logs.
4. Determining residual oil saturation by adding paramagnetic ions to the mud filtrate to cancel the water signal and leave the oil signal.

Dielectric Measurement Tools

Dielectric measurement tools examine the formation with high frequency electromagnetic waves (microwaves) rather than high-frequency sound waves (as in the sonic or acoustic logging tools). The way the electromagnetic wave passes through a given formation depends on the dielectric constants (ϵ) of the minerals and fluids contained in the rock.

Two types of tools are available [33]:

1. VHF sondes that have frequencies of 20–47 MHz (found in the deep propagation tool [Schlumberger], and dielectric tools [Dresser-Atlas and Gearhart-Owen], and
2. UHF sondes that have a frequency of 1.1 GHz (found in the electromagnetic propagation tool [Schlumberger]).

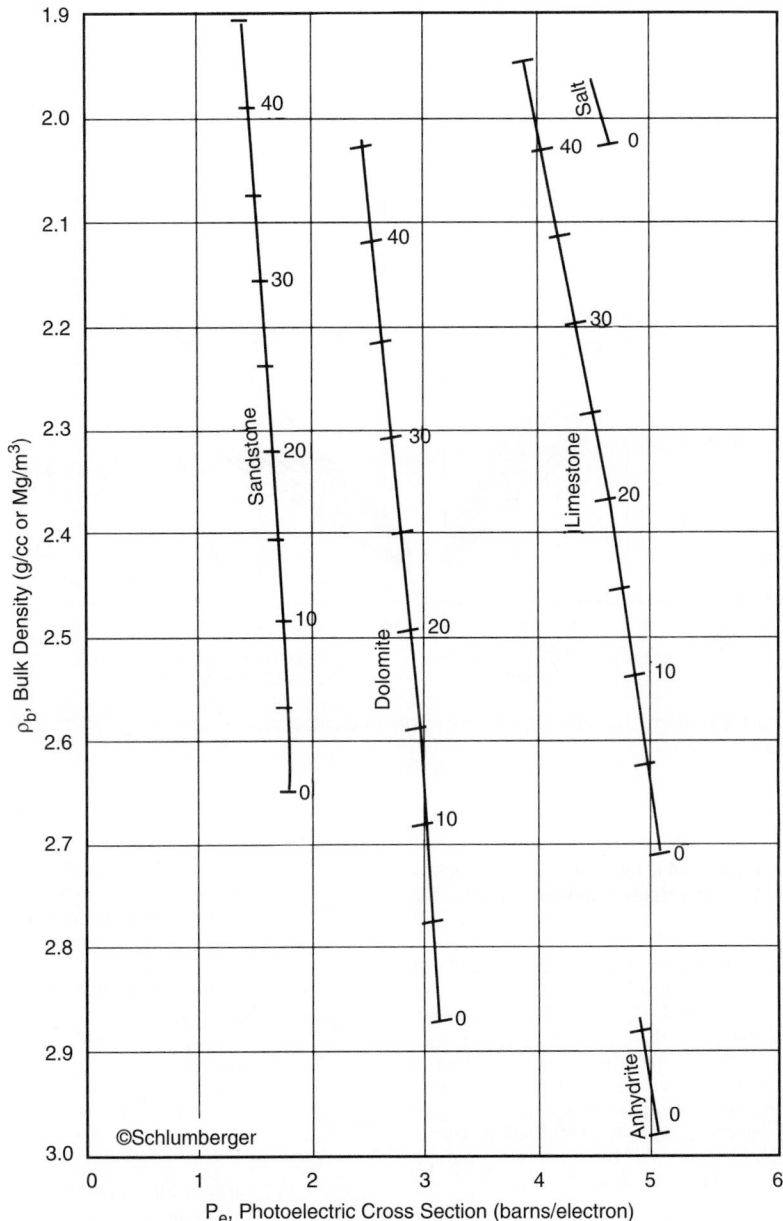

Figure 5.2.40 *Chart for finding porosity and matrix composition from a lithodensity log in fresh water [20].*

The only tool that is currently available is Schlumberger's electromagnetic propagation tool (EPT); the others are still experimental [33].

Theory
The EPT is a sidewall tool that measures the dielectric properties of a formation by passing spherically propagated microwaves into the rock. The tool consists of 2 transmitters (T_1, and T_2) and 2 receivers (R_1, and R_2) mounted in an antenna pad assembly. Its basic configuration is that of a borehole compensation array (much like the borehole

compensated sonic (BHC) log). The transmitter fires a 1.1 GHz electromagnetic wave into the rock around the wellbore. As the wave passes through the rock and fluid there, it is attenuated, and its propagation velocity is reduced. The wave then refracts to the borehole where it is sequentially detected by the two receivers. How much the wave is attenuated is a function of the dielectric permittivity of the formation. Rocks and oil have similar permittivities while water has a very different permittivity. Therefore, the wave responds to the water-filled porosity in the formation, and the response is a function of formation temperature.

The Matrix Identification Plot
ρ_{maa} vs U_{maa}

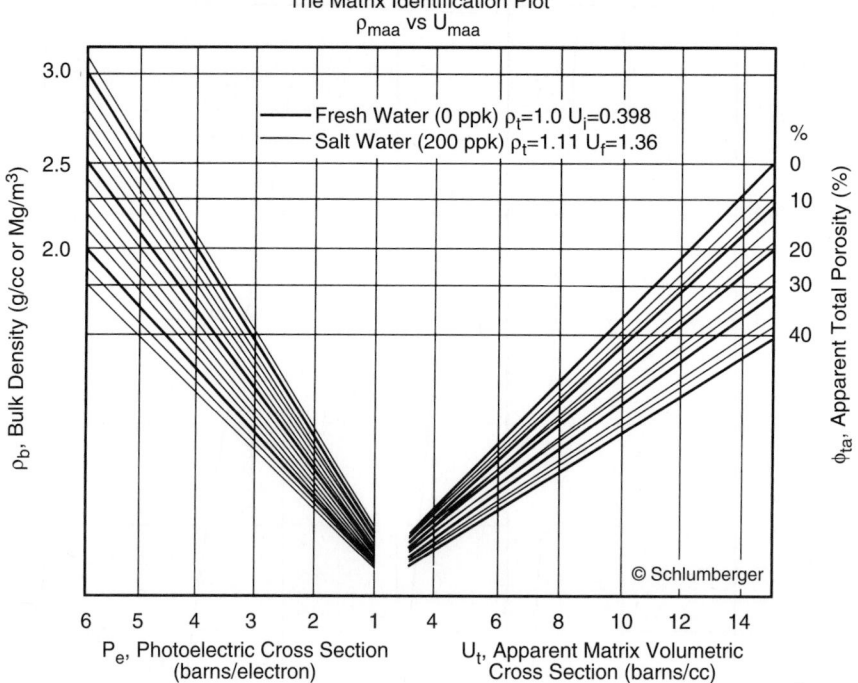

Figure 5.2.41 *Chart for determining apparent matrix volumetric cross section from bulk density and photoelectric cross section [20].*

Since the wave is attenuated by water (and is not too bothered by oil), the log response indicates either R_{xo} (in water-based mud systems) or bulk volume water (in oil-based mud systems).

In order to provide usable values, the velocity of the returning wave is measured and compared to the wave-propagation velocity in free space. The propagation velocity of the formation is then converted into propagation time (T_{pl}). A typical log presentation includes a T_{pl}, curve (in nanoseconds/meter), an attenuation curve (EATT) in decibels/meter, and a small-arm caliper curve (which measures borehole rugosity) recorded in tracks 2 and 3. Figure 5.2.43 shows the basic antenna configuration [35]. Figure 5.2.44 is an example of an EPT log presentation.

The depth of investigation of the tool varies between one and three inches and depends on formation conductivity; high conductivity reduces depth of investigation.

The tool is affected primarily by hole roughness (rugosity) and mud cakes $> \frac{3}{8}$ in. thick. These effects reduce depth of investigation and in extreme situations (i.e., very rough holes and/or very thick mud cakes) keep the tool from reading the formation at all.

Interpretation
The most common way to interpret EPT logs is called the T_{po} *method* [35]. T_{po} in a clean formation is given by:

$$T_{po} = (\phi S_{xo} T_{pfo}) + \phi(1 - S_{xo})T_{phyd} + (1 - \phi)T_{pma} \quad [5.2.29]$$

Rearranging terms and solving for S_{xo}:

$$S_{xo} = \frac{(T_{po} - T_{pma}) + \phi(T_{pma} - T_{phyd})}{\phi(T_{pfo} - T_{phyd})} \quad [5.2.30]$$

where $T_{po} = T_{pl}$ corrected for conductivity losses,
nanosecs/m
$\quad T_{pma} =$ the matrix propagation time, nanosecs/m
$\quad T_{phyd} =$ the hydrocarbon propagation time,
nanosecs/m
$\quad T_{pfo} =$ the fluid propagation time, nanosecs/m

T_{po} can be calculated by

$$T_{po} = T_{pl}^2 - \left(\frac{A_c}{60.03}\right)^2 \quad [5.2.31]$$

where

$$A_c = A_{log} - 50 \quad [5.2.32]$$

($A_{log} =$ EATT curve reading in dB/m)
T_{pfo} is a function of formation temperature (T) and can be found from:

$$T_{pfo} = \frac{20(710 - T/3)}{(440 - T/3)} \quad [5.2.33]$$

T_{pma} is taken from Table 5.2.9 and ϕ is taken from a neutron-density log(ϕ_{ND}). Equation 5.2.30 can be rearranged to find ϕ for a quick-look comparison with other porosity devices (specifically the neutron-density log). By assuming $T_{phyd} \approx T_{pfo}$, T_{phyd} can be eliminated and

$$S_{xo} = \frac{1}{\phi} \frac{(T_{po} - T_{pma})}{(T_{pfo} - T_{pma})} \quad [5.2.34]$$

Since the EPT log measures water-filled porosity, by definition

$$\phi = \phi_{EPT} \frac{(T_{po} - T_{pma})}{(T_{pfo} - T_{pma})} \quad [5.2.35]$$

in a water zone ($R_t = R_o$; $S_w = 100\%$)

(*text continued on page 5-126*)

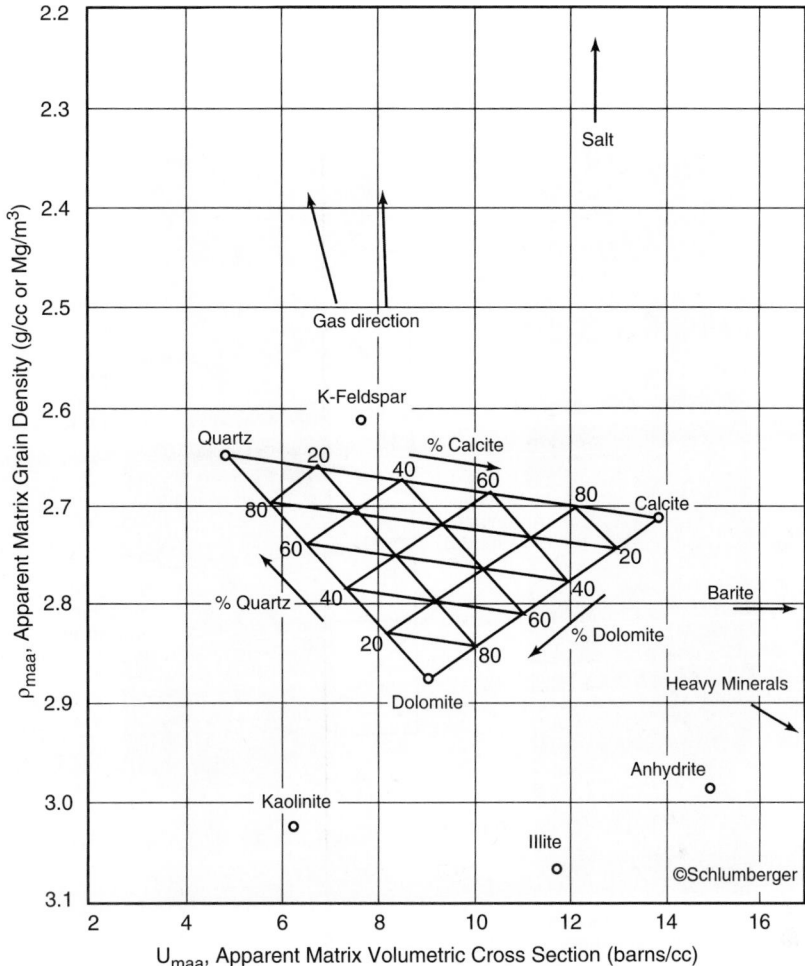

Figure 5.2.42 *Matrix indentification crossplot chart for finding matrix composition from apparent matrix density and apparent matrix volumetric cross section [20].*

Antenna configuration

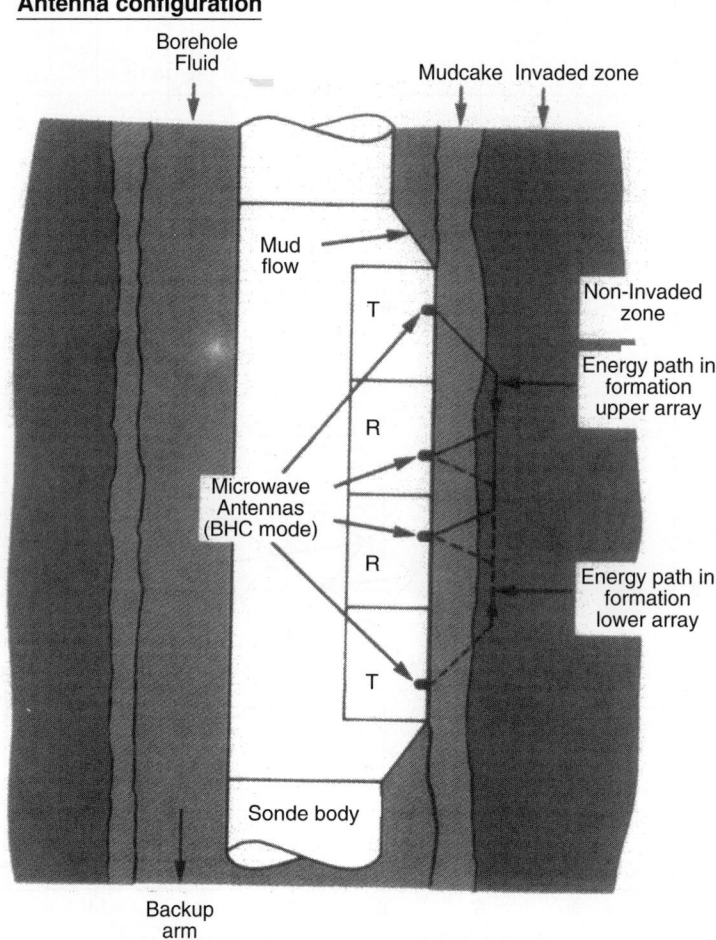

Figure 5.2.43 *Diagram showing antenna-transmitter configuration for an EPT sonde and microwave ray paths through the mudcake and formation adjacent to the borehole [35].*

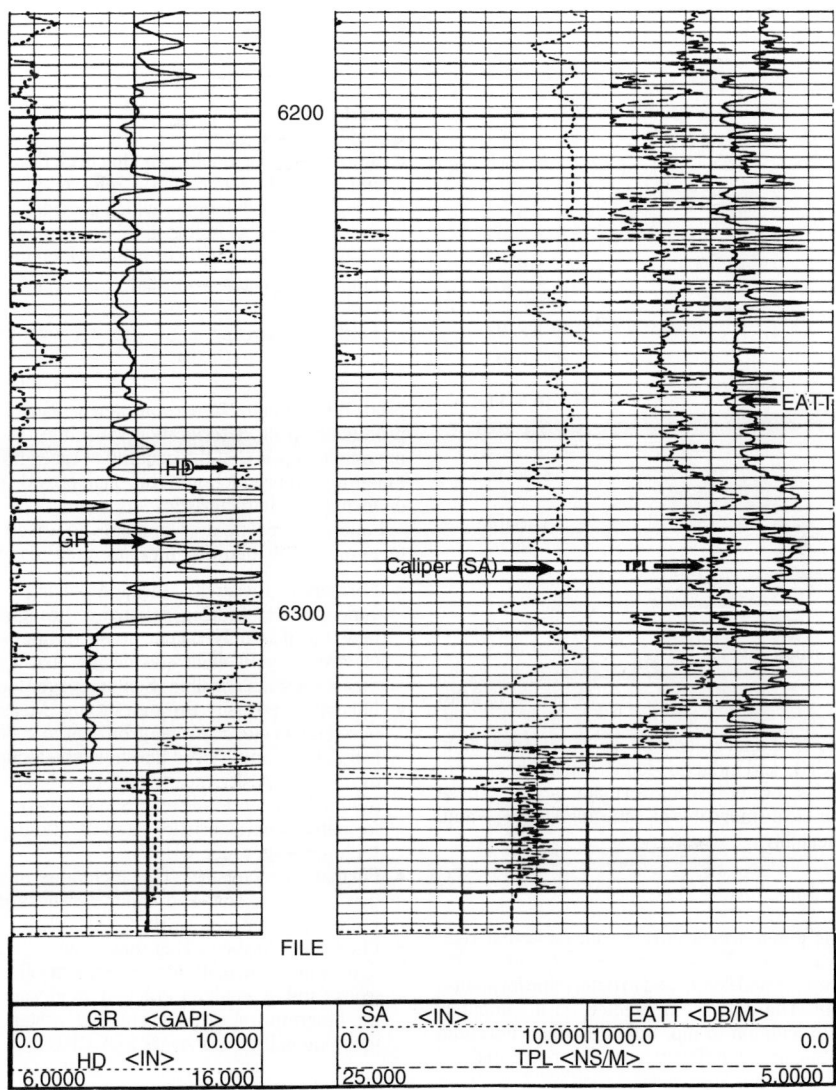

Figure 5.2.44   *EPT log presentation.*

(text continued from page 5-122)

Table 5.2.9 *Matrix Propagation Times with the Electromagnetic Propagation Tool for Various Minerals*

Mineral	T_{pma}, nanosec/m
Sandstone	7.2
Dolomite	8.7
Limestone	9.1
Anhydrite	8.4
Dry colloids	8.0
K-feldspar	7.0–8.2
Muscovite	8.3–9.4
Biotite	7.3–8.2
Talc	7.1–8.2
Halite	7.9–8.4
Siderite	8.8–9.1
Gypsum	6–8
Sylvite (KCl)	7.2–7.3
Limonite	10.5–11.0
Apatite	9.1–10.8
Sphalerite	9.3–9.6
Rutile	31.8–43.5
Petroleum	4.7–5.8
Shale	variable
Fresh water @ 250°C	29.0

From Reference 35.

So, in hydrocarbon zones:

$$S_{xo} = \frac{\phi_{EPT}}{\phi_{ND}} \qquad [5.2.36]$$

Figure 5.2.45 is a quick look at ϕ_{EPT} response compared to FDC, CNL, and induction-log resistivity in gas, oil, fresh water, and saltwater-bearing formations [25]. These responses also indicate moveable oil saturation $(1-S_{xo})$ and, therefore permeability.

5.2.3.5 Special Openhole Logs and Services
Dipmeter
The dipmeter is a four-armed device with pads that read resistivity of thin zones. These four resistivity curves are analyzed by computer and correlated to determine formation dip and azimuth. The dips are presented on a computer-produced log. In addition to dip, hole deviation, borehole geometry, and fracture identification are also presented.

Repeat Formation Tester
The repeat formation tester measures downhole formation pressures. The tool is operated by an electrically driven hydraulic system so that several zones may be pressure tested on the trip into the hole. Once the drawdown pressure and the pressure buildup have been recorded, they can be processed by a computer at the well-site to provide Horner plots from which permeabilities are calculated. Permeabilities from the drawdown test often vary considerably from measured permeabilities and should be considered an order-of-magnitude estimate. This is usually due to a very shallow depth of investigation associated with drawdown tests [33]. The pressure buildup has a better depth of investigation than the drawdown pressure test. Accuracy depends on what type of pressure-wave propagation model is chosen [33,35,36] as well as the compressibility and viscosity of the formation fluids.

Sidewall Cores
After drilling, cores from the side of the borehole can be taken by wireline core guns or drills. Guns are less expensive but do not always recover usable cores. Sidewall drilling

devices have become quite common in the last few years. Up to 20 cores may be cut and retrieved on one trip into the hole.

Cased Hole Logs
Cased hole logs are run to evaluate reservoir performance, casing/cement jog quality, and to check flow rates from producing intervals. The reader is referred to Bateman's book [38] on cased hole logging which provides a more detailed discussion than is possible in this summary.

Cased hole logs can be broadly divided into two classes:

1. Logs that measure formation parameters through the casing.
2. Logs that measure the parameters within and immediately adjacent to the casing.

These logs are all combined to monitor fluids being produced, monitor reservoir performance, and monitor producing-string deterioration with time. They differ from open-hole logs in that the majority of cased hole logs merely monitor fluid production rather than provide extensive data on formation characteristics.

Following recent developments in cased hole logging technology, the following measurements can be made:

- Formation resistivity through casing: Although the concept of measuring resistivity through casing is not new, but it is only recent break throughs in downhole electronics and electrode design that have made the measurements possible [Schlumberger tool: CHFR].
- Formation porosity through casing: Measurements are based on an electronic neutron source instead of a chemical source, and it uses borehole shielding and focusing to obtain porosity measurements that are affected only minimally by borehole environment, casing standoff, and formation characteristics such as lithology and salinity [Schlumberger tool: CHFP].
- Formation accoustics: This technique measures formation compressional and shear slowness in cased wells [Schlumberger tool: DSI].
- Fluid identification: This tool provides a technique for determining formation pressures in old or new cased wells, and it enables efficient fluid sampling without the inherent risks of conventional or standard sampling techniques [Schlumberger tool: CHDT].

Cased Hole Formation Evaluation
Two tools are currently being used to provide formation evaluation in cased holes:

1. Pulsed neutron logs.
2. Gamma spectroscopy tools (GST) logs.

Pulsed Neutron Logs
Pulsed neutron logs are used to monitor changes in fluid content and water saturation with respect to time. Current tools also provide a means of estimating porosity. They are particularly valuable for [38]:

1. Evaluating old wells when old open-hole logs are poor or nonexistent.
2. Monitoring reservoir performance over an extended period of time.
3. Monitoring the progress of secondary and tertiary recovery projects.
4. Formation evaluation through stuck drill pipe (generally a last resort).

EPT Quicklook

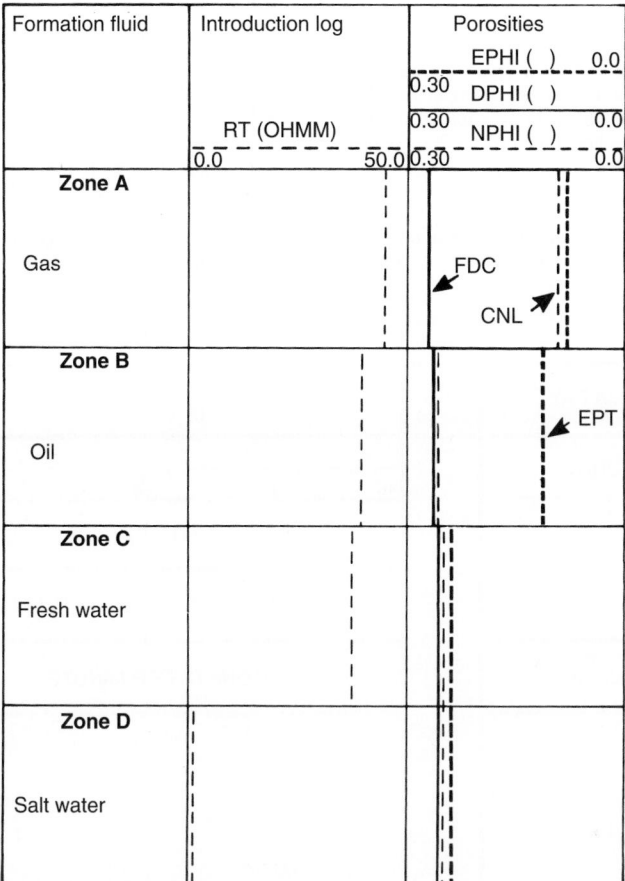

Figure 5.2.45 *EPT quicklook chart comparing curve response of induction resistivity, neutron porosity, density porosity and EPT porosity in water bearing and hydrocarbon-bearing zones [35].*

Theory

A neutron generator that consists of an ion accelerator fires deuterium ions at tritium targets. This produces a burst of 14 keV neutrons which pass through the borehole fluid (must be fresh water), casing, and cement. The burst then forms a cloud of neutrons in the formation which are rapidly reduced to a thermal state by collisions with the atoms in fluids in the rock (made up primarily of hydrogen atoms). Once in a thermal state, they are most liable to be captured by chlorine (or boron). The capture process will produce a gamma ray of capture which is then detected by a scintillometer in the tool. The time it takes for the neutron cloud to die during the capture process is a function of the chlorine concentration in the formation fluid. This is then related to water saturation. Rapid disappearance of the thermal neutron cloud indicates high water saturation. Slower disappearance of the cloud indicates low water saturation (i.e., high hydrocarbons saturations). The rate of cloud decay is exponential and can be expressed by:

$$N = N_0 e^{(-t/\tau)} \qquad [5.2.37]$$

where N = number of gamma rays observed at time t
N_0 = number of gamma rays observed at $t = 0$
t = elapsed time (microseconds)
τ = time constant of the decay process, microseconds

Of most interest is τ since it is strictly a function of the decay rate of the neutron cloud (or rather the slope of the exponential function). From τ the capture cross section, Σ, can be calculated:

$$\Sigma = 4,550/\tau \qquad [5.2.38]$$

The tools that are available to measure τ include:

1. TDT-K (with 3 moveable gates or detectors)
2. DNLL (dual neutron lifetime log) (which uses 2 gates),
3. TDT-M (with 16 fixed gates), and
4. TMD (thermal multigate decay) (which uses 6 gates).

In general, these tools all perform the same function: they measure the decay rate of the neutron cloud in the formation. This is accomplished by using a series of windows to measure near and far-spacing counting rates, as well as background gamma ray rates. The first gates are not triggered until all neutrons in the cloud in the formation are thermalized. At this point neutron capture has started. By using certain gating times and gate combinations, the slope of the straight portion of the decay curve is measured and related to Σ. In addition, the ratio between the short-space-detector and long-space-detector counting rates is also calculated and is related to porosity. (It is similar in principle to the CNL porosity device used in openhole logging).

Log Presentations

Figure 5.2.46 is an example of a DNLL log presentation. Most other TDT logs are presented in a similar way, except that the number of curves varies from company to company. The log shown in this figure consists of 5 curves:

1. Gamma ray curve (Track 1).
2. Gate 1 counting rate (CPM).
3. Gate 2 counting rate (CPM).
4. Ratio curve ($\approx$ CNI. ratio).
5. Sigma curve (Σ).

In addition, the pips located on the left side of track 2 are the corrected casing collar locations.

Gate 1 and Gate 2 show the raw data from the detectors, the ratio curve shows relative hydrogen concentration ($\approx$ water-filled porosity), and the σ curve shows the

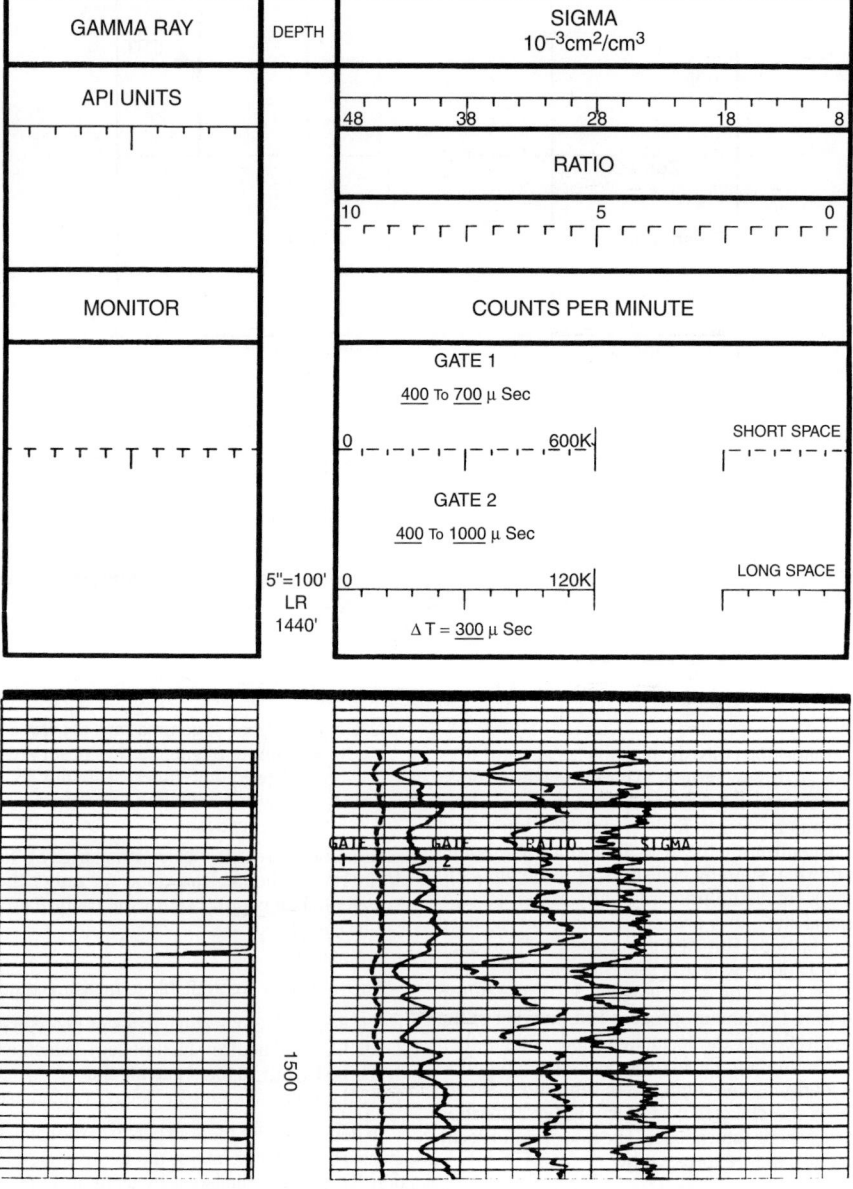

Figure 5.2.46 Dual neutron lifetime log (DNLL) presentation.

capture cross-section. Some logs also show a τ curve, but it is normally omitted [38].

Interpretation

Interpretation of pulsed neutron logs is very straightforward. It relies on knowledge of three parameters (four in hydrocarbon-bearing zones):

1. Σ_{log} (capture units).
2. Σ_{matrix}
3. Σ_{water}
4. $\Sigma_{hydrocarbon}$

According to Schlumberger [41] the log response may be described as:

$$\Sigma_{log} = \Sigma_{ma}(1 - \phi) + \Sigma_w \phi S_w + \Sigma_{hy} \phi (1 - S_w) \qquad [5.2.39]$$

Solving for S_w:

$$S_w = \frac{(\Sigma_{log} - \Sigma_{ma}) - \phi(\Sigma_{hy} - \Sigma_{ma})}{\phi(\Sigma_w - \Sigma_{hy})} \qquad [5.2.40]$$

Porosity (ϕ) can be found either from an openhole porosity log or by combining the ratio curve and Σ_{log}. Figures 5.2.47 to 5.2.50 are charts to find porosity, apparent water salinity, and Σ_w, using this combination. By selecting the chart for the appropriate borehole and casing diameter, these values are easy to determine. Simply enter the proper axes with the log-derived values, and find porosity and Σ_w at the intersection of the two lines. If water salinity and formation temperature are known, use Figure 5.2.50 to find Σ_w.

Estimating Σ_{hy} is another matter. You must first know if the hydrocarbons are oil, methane, or heavier hydrocarbon gases (i.e., propane, butane, pentane). For oil, solution gas-oil ratio and oil gravity (°API) are needed. If the gas is methane, reservoir pressure and temperature are required. For gases other than methane, the specific gravity of the gas can be converted to equivalent methane using Figure 5.2.51, and then Figure 5.2.52 can be entered. Once all of the parameters have been found, Equation 5.2.40 is used to find S_w.

Applications

Pulsed neutron logs are most useful for monitoring changes in water saturation over time while a reservoir is produced. Initially, these logs are run prior to perforating a zone. Subsequent logs are run every few months (or years) depending on production rate and the amount of control desired. Water saturation is calculated for each run using Equation 5.2.40 and subtracted from saturations determined from earlier runs. These values (ΔS_w) show the change in the position of the water table (hydrocarbon-water contact) versus time.

(text continued on page 5-134)

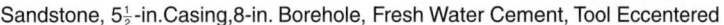
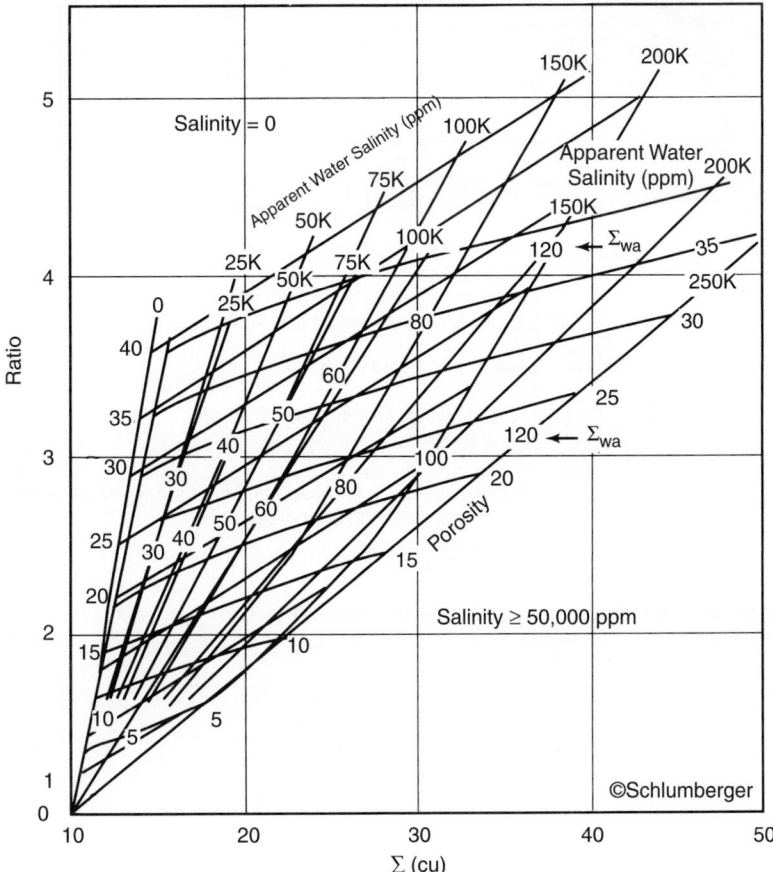

Sandstone, $5\frac{1}{2}$-in.Casing,8-in. Borehole, Fresh Water Cement, Tool Eccentered

Figure 5.2.47 *Chart for determining porosity and apparent water salinity from Σ log and ratio curves in 5 ½-in. casing and an 8-in. borehole [20].*

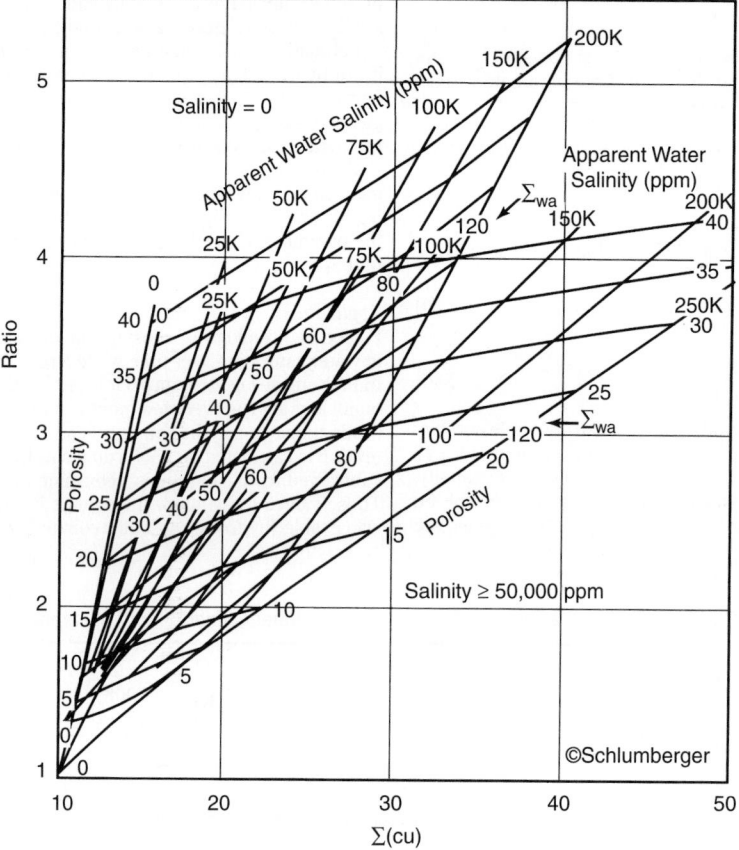

Example: Ratio = 3.1
Σ_{LOG} = 20 cu
Borehole fluid salinity = 80,000 ppm
5 1/2-in. casing cemented in 8 3/4-in. borehole

Thus, from Chart Tcor-3
ϕ = 30%
Apparent water salinity = 50,000 ppm
Σ_{wa} = 40 cu

If this were a clean formation and connate water salinity was
known to be 150,000 ppm, then
$$S_w = \frac{50,000}{150,000} = 33\%$$

Figure 5.2.48 *Chart for determining porosity and apparent water salinity from Σlog and ratio curves in 7-in. casing and a 9-in. borehole [20].*

Sandstone, 7- to 9-in. Casing, 12-in. Borehole, Fresh Water Cement, Tool Eccentered

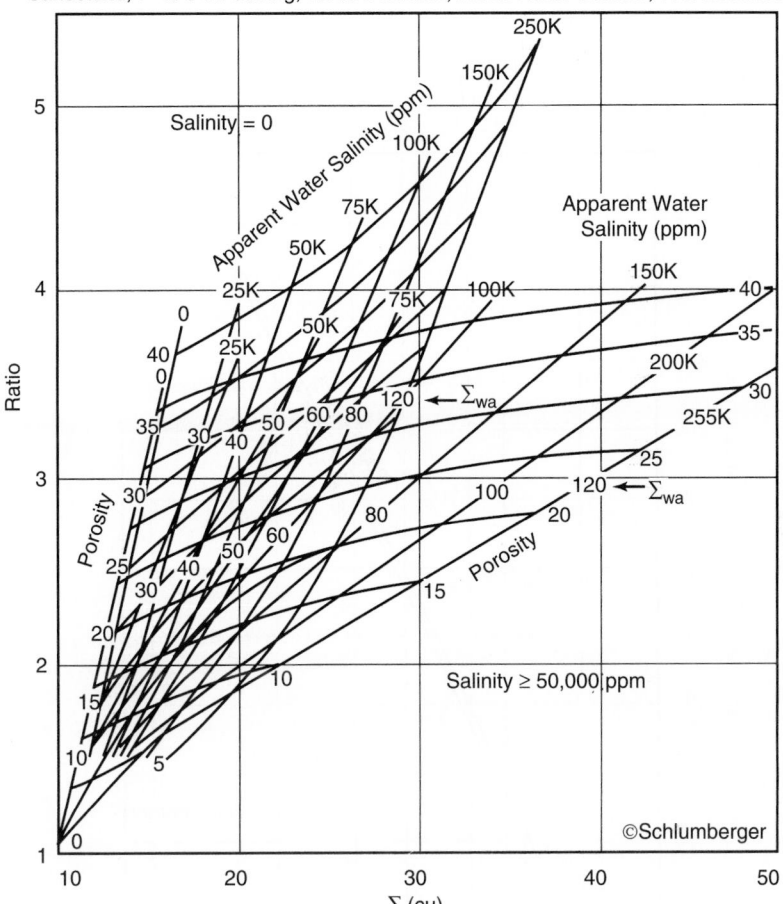

Figure 5.2.49 *Chart for determining porosity and apparent water salinity from $\sum log$ and ratio curves in 7- to 9-in. casing and a 12-in. borehole [20].*

Example: A reservoir section at 90°C temperature and 25 MPa pressure
contains water of 175,000 ppm (NaCl) salinity: 30°API oil with
a GOR of 2000 cu ft/bbl, and methane gas.
Thus, Σ_w = 87 cu (τ = 52 µs)
Σ_o = 19 cu (τ = 240 µs)
Σ_g = 6.9 cu (τ = 660 µs)

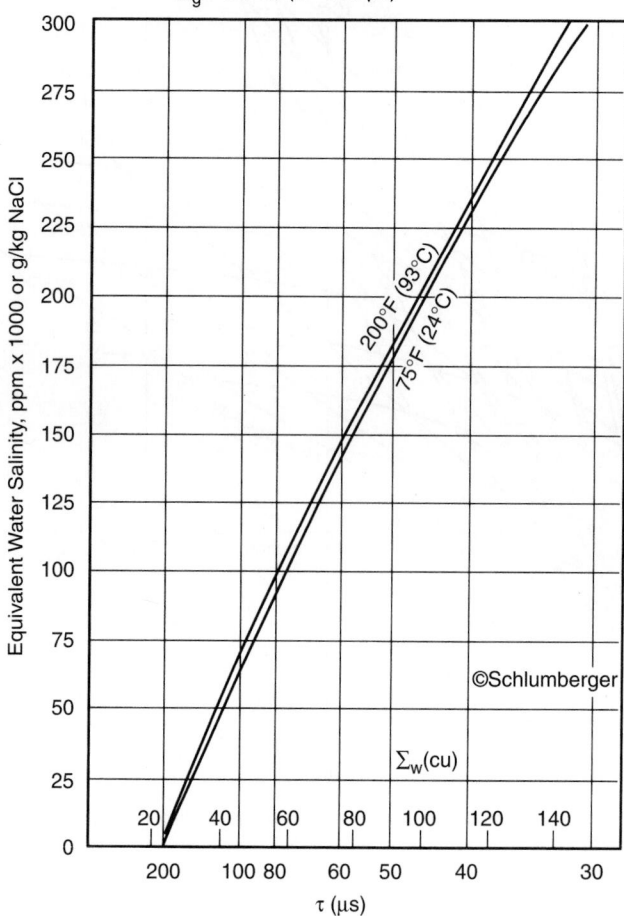

Figure 5.2.50   Chart for finding $\sum_w$ from equivalent water salinity and formation temperature [20].

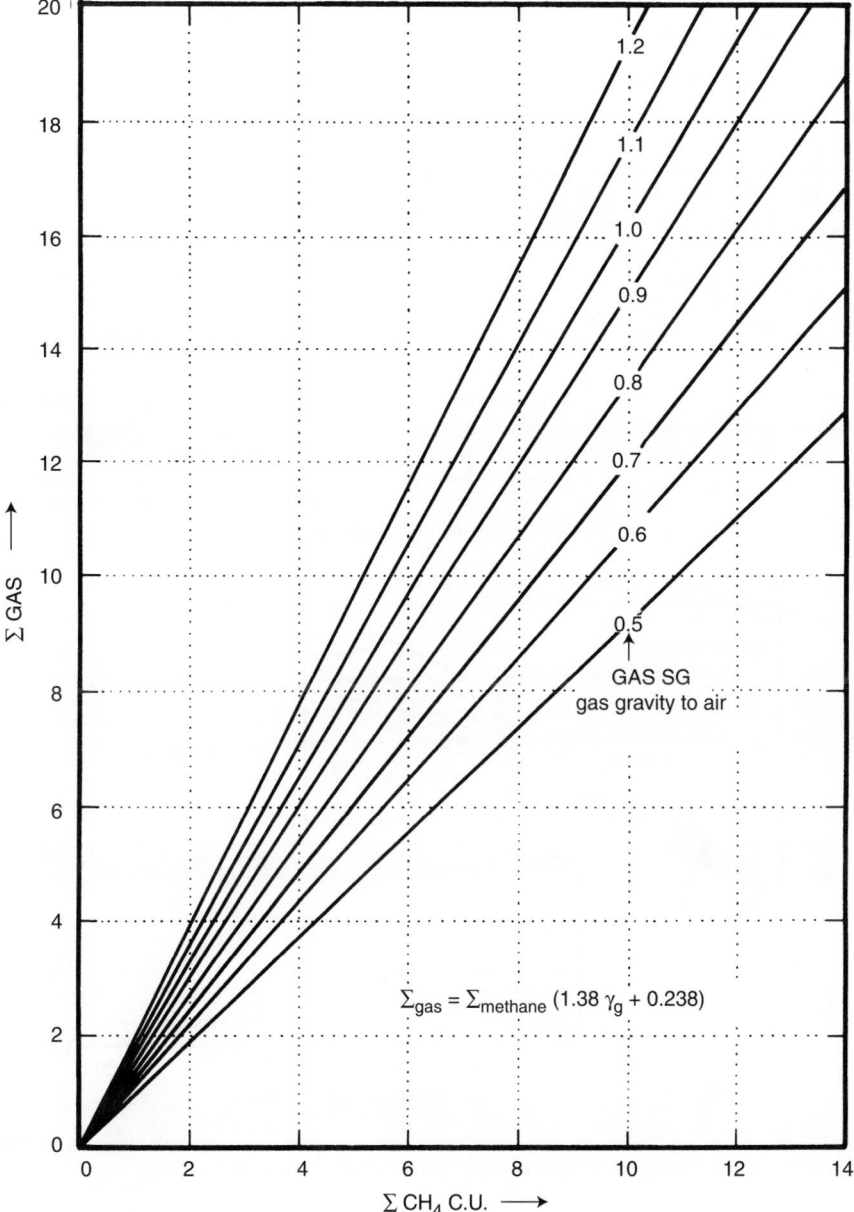

Figure 5.2.51 *Chart for converting $\sum$gas to $\sum$methane [38].*

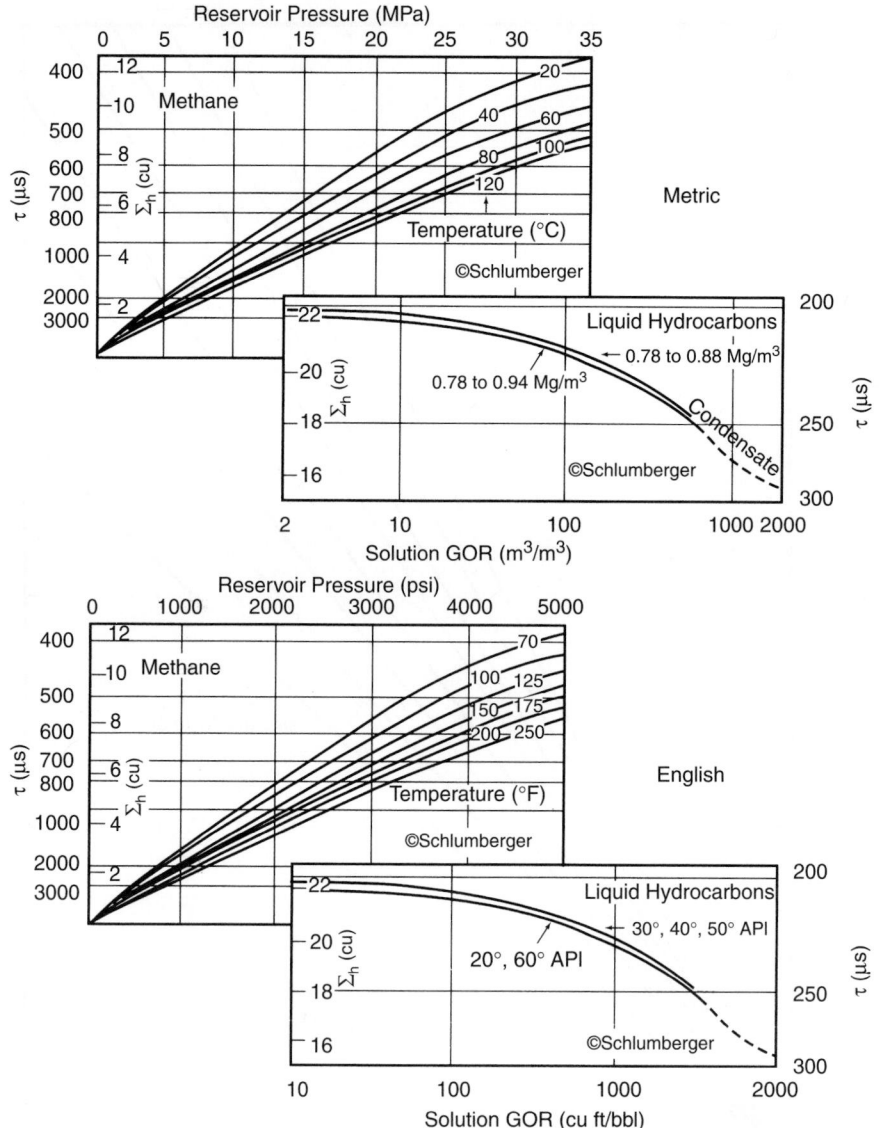

Figure 5.2.52 *Charts for finding Σh (for methane and liquid hydrocarbons) from reservoir pressure and formation temperature (gas) or solution gas-oil ratio (GOR) and API gravity (liquid 7 hydrocarbons) [20].*

(text continued from page 5-129)

Another application is monitoring residual oil saturation in waterflood projects. The procedure outlined by Bateman [38] involves first running a base log (prior to injection). Next, salt water is injected and another log is run. Then fresh water is injected and another log run. If Σ_{brine} and Σ_{fresh} are known, Bateman [38] suggests using:

$$S_o = 1 - \frac{\Sigma_{log(brine)} - \Sigma_{log(fresh)}}{\phi(\Sigma_{brine} - \Sigma_{fresh})} \qquad [5.2.41]$$

to find residual oil saturation (S_{or}) Additional details of estimating S_{or} are given later.

The main problem with using these logs is the presence of shale. Shale normally appears wet, and shale will make a reservoir look like it has higher S_w. Openhole logs and an NGS log are needed to confirm this interpretation although shaly sand corrections can easily be made [38].

Gamma Spectroscopy Tools (GST)
Also known as the carbon-oxygen log, this device has recently been incorporated into pulsed neutron tools to aid in differentiating oil and gas from water. GST tools operate with the same neutron generator as the pulsed neutron devices, but gamma rays returning from the formation are measured.

Two types of gamma rays are produced when neutrons are fired into a formation:

1. Those that result from neutron capture by chlorine and boron.
2. Those that result from inelastic collisions with atoms.

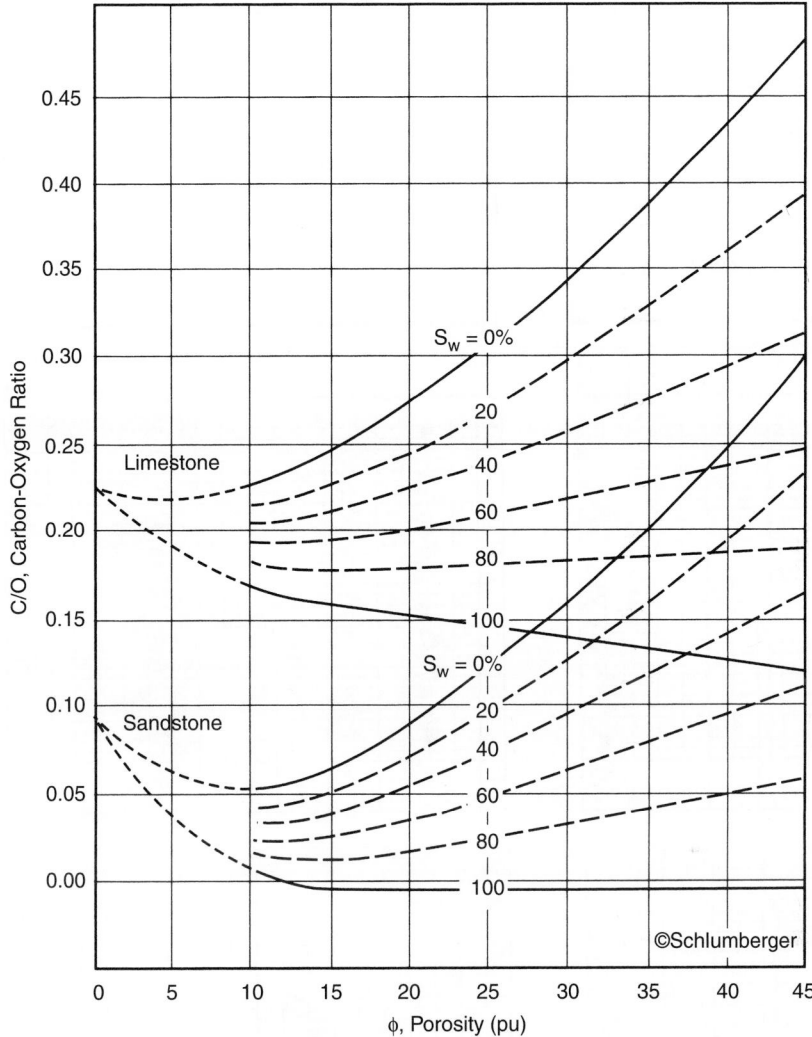

Figure 5.2.53 *Chart for finding water saturation using the carbon-oxygen ratio curve and porosity if matrix lithology is known [20].*

The detector on this tool has energy windows set to receive certain returning gamma rays [39]. The detectors are protected from the fast neutron source by an iron shield, and from returning thermal neutrons by a boron shield.

The energy of the returning gamma rays depends on the atom involved in the collisions. The atoms of interest include carbon, oxygen, silicon, and calcium. Carbon-oxygen ratio is a carbon indicator and when combined with porosity, gives an estimate of water saturation if matrix lithology is known. Figure 5.2.53 is used for this determination. Silicon-calcium ratio is an indicator of matrix and is used to distinguish oil-bearing rock from calcareous rocks (such as limy sands and limestones) [39,40]. Figure 5.2.54 is an example of a carbon-oxygen log.

If capture gamma rays are also detected with separate energy windows, chlorine and hydrogen content can be determined and related to formation water salinity. Figures 5.2.55 and 5.2.56 are used for this purpose. All that is required to estimate salinity of formation waters is knowledge of

borehole fluid salinity, Cl/H ratio, and response mode of the tool. These devices should *not* be confused with the natural gamma spectroscopy log which only measures naturally-occurring gamma rays.

The readings on the GST log are not affected by shale although carbonaceous shales can cause trouble because of the sensitivity to carbon. Usually, however, these effects can be calibrated for or taken into account when this log is interpreted. Much of the interpretation of this type of log is based in regional experience; the analyst should have a good idea of the types of rocks present before trying to make an interpretation. No lithology crossplot charts are presently available to estimate lithology with these logs.

Natural Gamma Spectroscopy
This log operates in the same manner as its openhole counterpart. The main difference is that the log should be calibrated prior to being run in cased holes. No correction charts are currently available for cased hole applications with

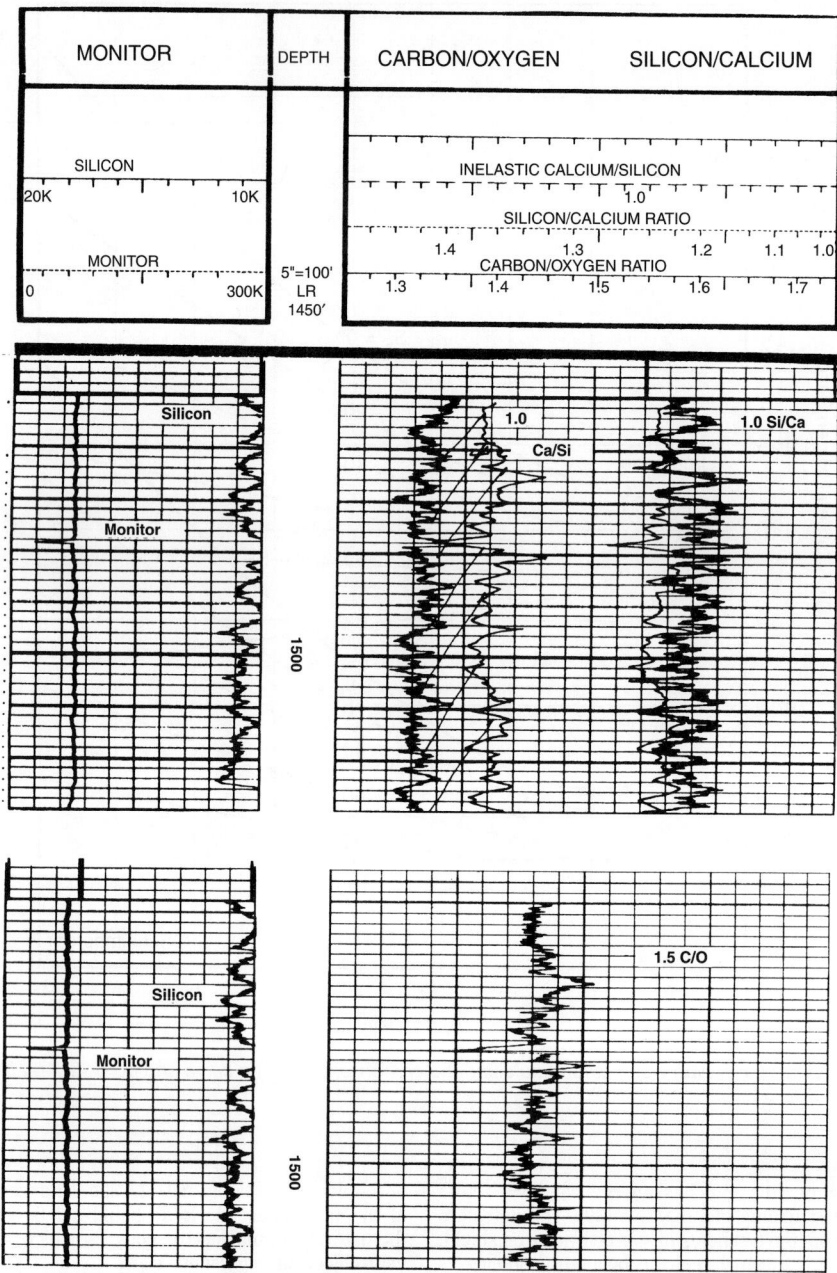

Figure 5.2.54 *Carbon-oxygen and silicon-calcium ratio curves on a carbon-oxygen log.*

this device. Curve presentations are the same as for the open-hole version. Refer to the open-hole section for a discussion of this log.

Cased-Hole Completion Tools

These tools examine cement bond and casing quality. They assure that no leakage or intercommunication will occur between producing horizons, or between water-bearing horizons and producing horizons. The most common completion tools include:

1. Cement bond logs (CBL).
2. Multifingered caliper logs.
3. Electromagnetic inspection logs.
4. Electrical potential logs.
5. Borehole televiewers.

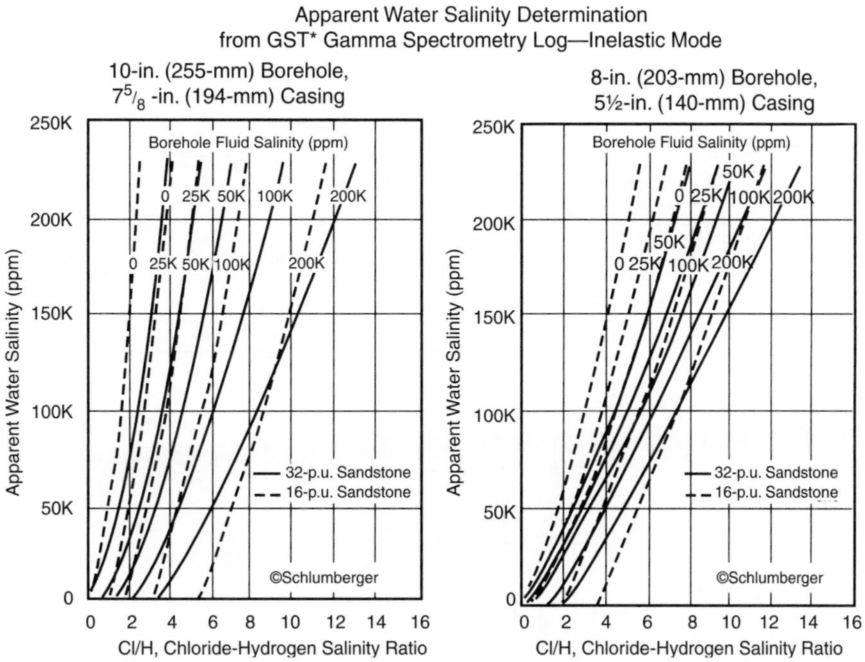

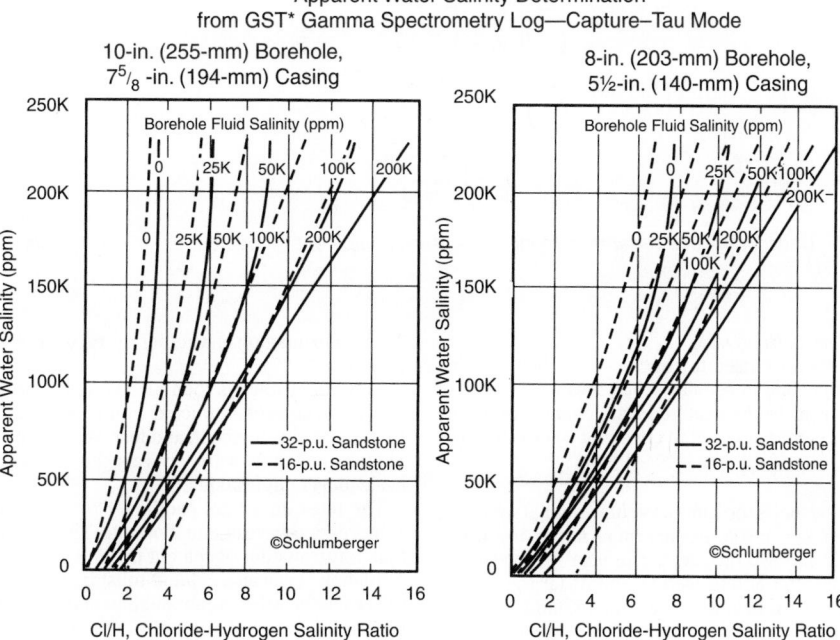

Figure 5.2.55 *Chart for finding apparent water salinity from chlorine-hydrogen ratio and borehole fluid salinity from a GST log (in elastic mode) [20].*

Figure 5.2.56 *Chart for finding apparent water salinity from chlorine-hydrogen ratio and borehole fluid salinity derived from a GST log (Tau-capture mode) [20].*

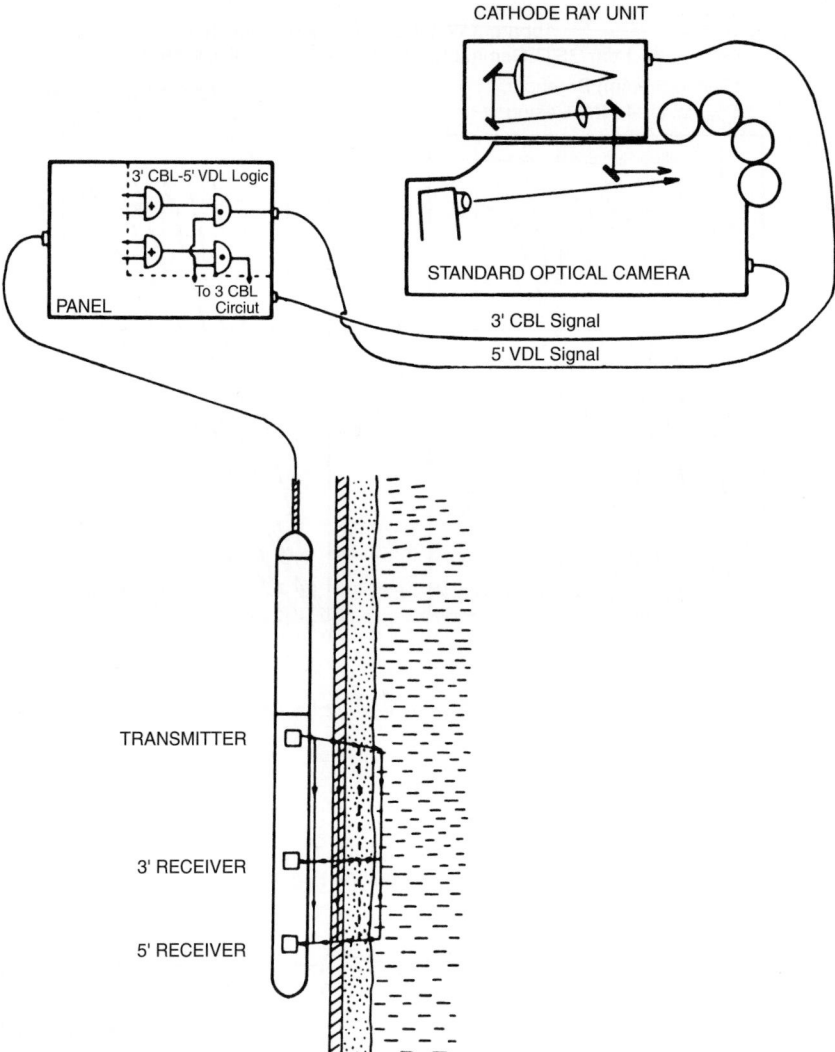

Figure 5.2.57 *Transmitter-receiver arrangement and surface equipment for cement bond log-variable density log combination [41].*

Cement bond logs (CBL-VDL)
Cement bond logs are used to check cement bond quality behind the casing and to estimate compressive strength of the cement. It can also be used to locate channeling in the cement or eccentered pipe and to check for microannulus.

Theory
The cement bond tool is the same as a conventional sonic tool except that the receiver spacings are much larger. It consists of a transmitter and two receivers. The near receiver is 3 ft below the transmitter and is used to find Δt for the casing. The far receiver is 5 ft below the transmitter and is used for the variable density log (VDL) sonic-wave-form output. The operation is the same as a conventional sonic except that the transmit time (one way) is measured. The transmitter is fired and a timer is triggered in both receivers. The wave passes through the fluid in the casing, the casing, and the cement, and into the formation. The near receiver measures the first arrival of the compressional wave and the timer is shut off.

This Δt is a function of whether the casing has cement behind it or not.

The sound wave is then picked up by the lower receiver which recognizes refracted compressional wave arrivals from the casing, cement, and formation, as well as Rayleigh, Stonely, and mud-wave arrivals. Figure 5.2.57 shows the basic tool configurations.

The most important parameter measured by this tool is compressive-wave attenuation-rate. This parameter is a function of the amount of cement present between the pipe and formation. Typically, cement must be at least $\frac{3}{4}$ in. thick on the casing in order for attenuation to be constant [38]. Each part of the log reads different attenuations. The CBL registers attenuation of the compressional wave in the cement and casing which gives an indication of the cement-casing bond-quality. The VDL registers the attenuation of the compressional wave through casing, cement, and formation which gives an indication of acoustic coupling between casing, cement, and surrounding rock. This indicates not only the casing bond quality but also the cement-formation-bond quality.

The basic parameter used to evaluate cement bonds is called the *bond index* and can be calculated by

$$\text{bond index} = \frac{\text{attention in zone of interest (db/ft)}}{\text{attenuation in a well cemented section (db/ft)}}$$
[5.2.42]

Bond index gives a relative way to determine bond quality through any given section of pipe. The minimum value of bond index necessary for a good hydraulic seal varies from region to region, and depends on hole conditions and type of cement used. Ideally, an index of 1.0 indicates excellent pipe-cement bonding; decreasing values show deteriorating conditions which may require squeezing to bring bonding up to acceptable standards. A bond index curve may be presented in track 2.

Log Presentations
Figure 5.2.58 shows a CBL-VDL log. Typically, three curves (and sometimes more, depending on the service company) are presented on the log. Track I contains total travel time. This is total one-way travel time and is a function of the casing size and tool centering. Other curves may be presented in this track, including gamma ray, neutron, and casing collar locator logs. Track 2 contains the cement bond logs amplitude curve. The log is scaled in millivolts and is proportional to the attenuation of the compressional-wave first-arrivals. High attenuation produces low-amplitude values; low attenuation produces high-amplitude values.

The higher the amplitude, the poorer is the casing-to-cement bond. Direction of increasing amplitude is normally indicated by an arrow. Some presentations also include a bond index curve in track 2. Track 3 contains the variable density log (VDL) display. The most common presentation is dark-and light-colored bands that represent the peaks and valleys of the wave train. Figure 5.2.58 shows two possible types of arrivals:

1. Those from the casing which appear as straight, parallel light and dark bonds intermittently broken by small V-shaped spikes which indicate the position of casing collars.
2. Those from the formation which appear as wavy, irregular, and intermittent light and dark bands which represent curve attenuation in the rock surrounding the borehole.

Interpretation
Interpretation of CBL-VDL logs involves recognition of basic curve pattern for determining whether casing is properly bonded or not. These curve patterns are presented in Figures 5.2.59 to 5.2.62. Four basic types of patterns are apparent:

1. Those that show strong casing arrivals only.
2. Those that show strong casing and formation arrivals.
3. Those that show weak casing arrivals and strong formation arrivals.
4. Those that show both weak casing and weak formation arrivals.

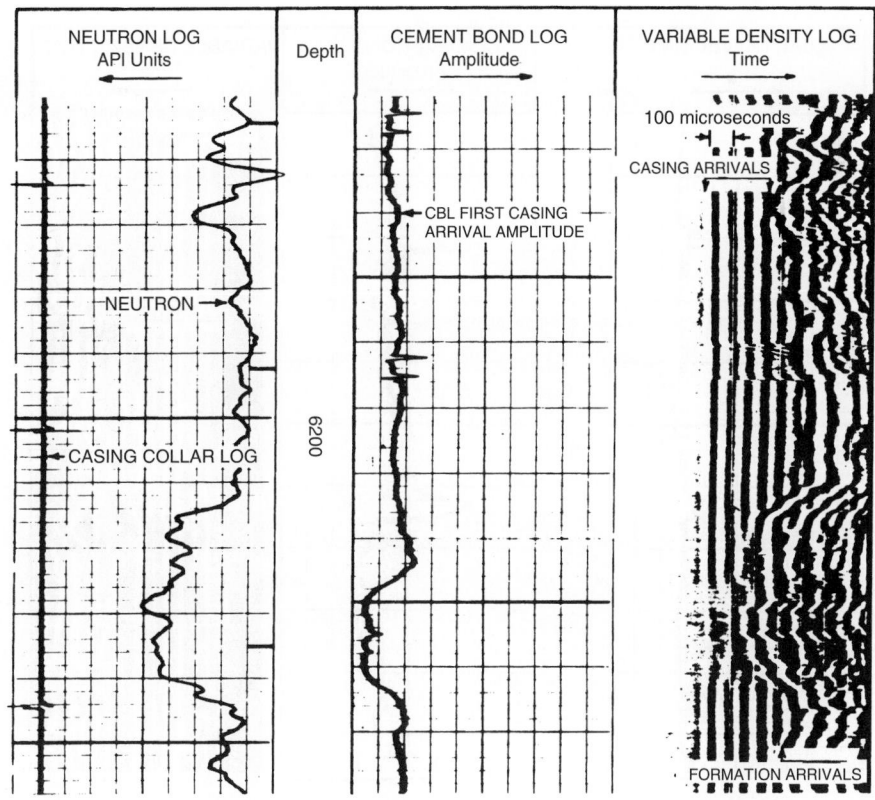

Figure 5.2.58 *Basic CBL-VDL log presentation [41].*

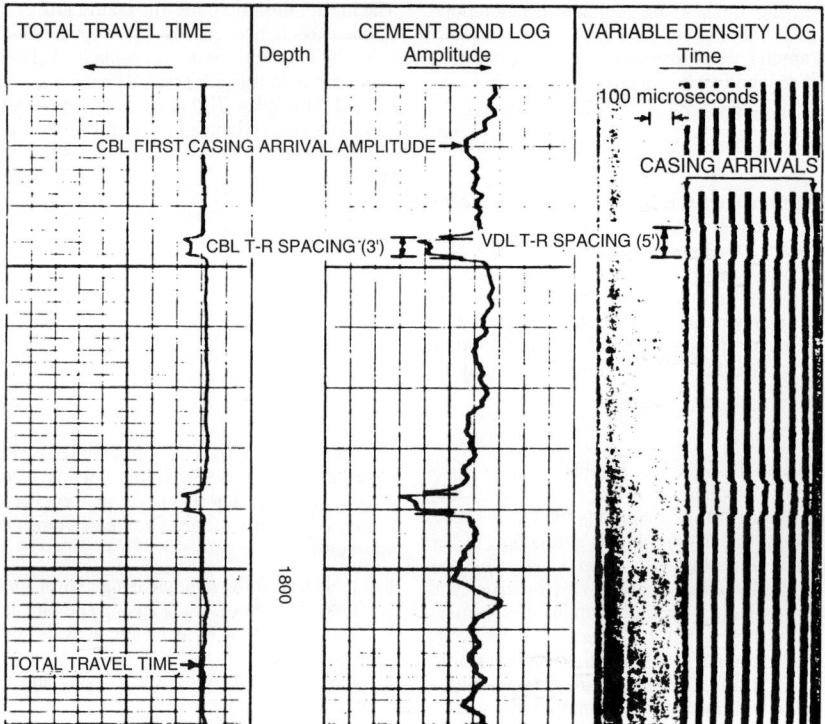

Figure 5.2.59   *CBL-VDL log run in free pipe [41].*

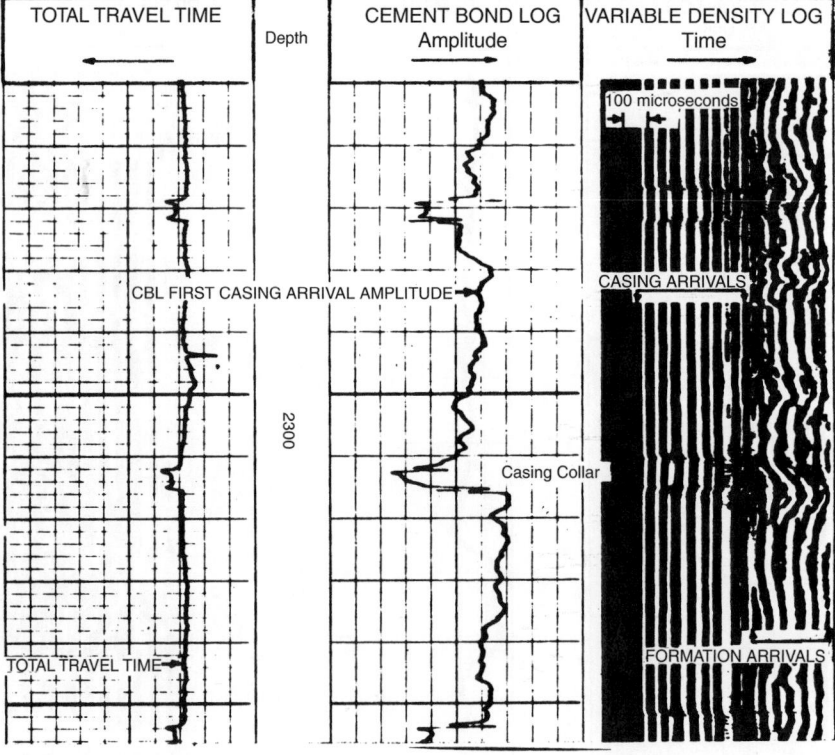

Figure 5.2.60   *CBL-VDL log run in casing eccentered in borehole making contact with the formation [41].*

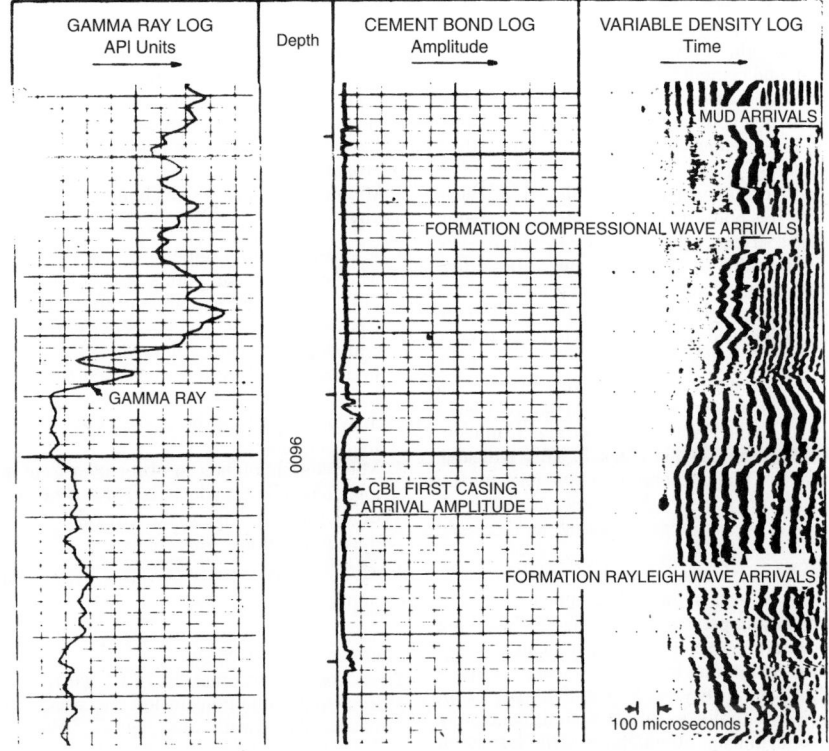

Figure 5.2.61 *CBL-VDL log run in well-bonded casing [41].*

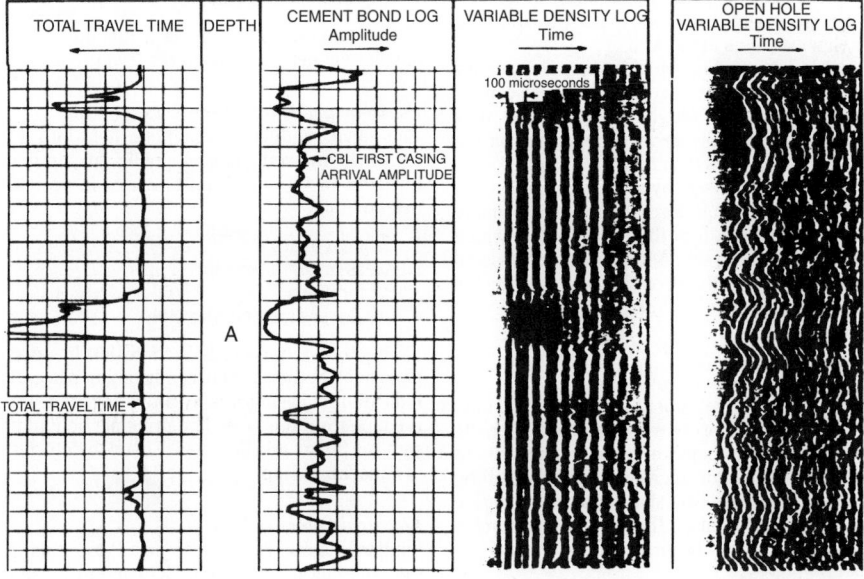

Figure 5.2.62 *CBL-VDL log run in casing with a good cement-casing bond and a poor cement-formation bond [41].*

Strong casing arrivals are shown in Figure 5.2.59 and are characterized by the pronounced casing arrival pattern (straight, alternating light and dark bands). No formation arrivals are present and cement-bond log-amplitude is moderate to high. These indicate free pipe with no cement or cement-casing-formation coupling. A high amplitude curve reading indicates low attenuation, hence no cement in the annulus.

Strong casing and formation arrivals are shown in Figure 5.2.60. This pattern has both the clean, pronounced casing signature as well as a strong, wavy-formation signature. The lack of cement is indicated by the high cement bond log-amplitude (i.e., no cement attenuation). The combination of these signals is interpreted as eccentered casing in contact with the wall of the well-bore. In this situation, proper cementation may be impossible.

Weak-casing and strong-formation arrivals are shown in Figure 5.2.61. This pattern shows no apparent casing or very weak casing patterns and very strong formation-patterns nearly filling the VDL in track 3. This indicates good casing-cement formation bonding, confirmed by the low cement bond log amplitude (high attenuation). Rayleigh and mud wave arrivals are also apparent along the right side of track 3.

Weak-casing and weak-formation arrivals are shown in Figure 5.2.62. This pattern shows what appears to be a slightly attenuated casing pattern but the cement bond log amplitude suggests otherwise. The curve indicates strong to very strong attenuation due to cement. Strongest attenuation occurs at "A" with a very weak formation pattern on the VDL. Comparison with the open-hole VDL (immediately to the right) shows no unusual attenuations of the formation signal. This also confirms poor acoustic coupling between cement and formation with good coupling between casing and cement.

Other possible interpretations for this type of pattern are possible.

1. Gas in the mud can be ruled out by examining long intervals of the log. Generally, this effect will occur over long rather than short sections in the well.
2. Eccentered tool in the casing, which causes destructive interference of compressive-wave first-arrivals, can be confirmed by checking for wiggly casing arrivals or a slight decrease in the casing-arrival time shown on the total-transit-time curve or VDL log [41].
3. Thin cement sheaths, caused by excessive mud cake thickness along a permeable formation, are a problem when cement sheaths are less than $\frac{3}{4}$ in. thick (which allows stronger casing arrivals). At times, the cement and formation have a slight acoustic coupling which gives the VDL a faint or weak formation signal.

Microannulus or Channeling in Cement
Microannulus occurs when the cement is emplaced and the casing is pressurized. When pressure is released after the cement has set, the casing "pops" away from the cement sheath. This generates a gap or microannulus between the casing and the cement. This can also occur if excessive pipe dope or varnish is present on the pipe. Microannulus due to pressurization primarily occurs opposite washed out portions of the borehole.

Channeling occurs when cement is in the annulus but does not completely surround, or is not bonded to, the pipe. This condition will not have proper fluid seal which allows oil, gas or water to pass up the hole outside the casing. Microannulus, on the other hand, may have a proper seal even though a small gap exists. It is very important to be able to distinguish between microannulus and channeling; squeezing may eliminate the channel altogether.

Figure 5.2.63 shows a case of microannulus. Figure 5.2.63a shows strong to weak casing-bond on the CBL amplitude and weak formation-arrivals on the VDL. This indicates poor acoustic coupling between casing, cement, and formation. The weaker the formation signal, the more pronounced the microannulus.

Microannulus can be easily differentiated from channeling by:

1. Pressurizing the pipe and rerunning the CBI-VDL. Microannulus conditions are confirmed by strengthened formation signals on the VDL and decreased CBL amplitude indicating better casing-cement acoustic coupling. Typically, channeling will produce little or no improvement in the signal when the casing is pressurized. Figure 5.2.63b shows a case of stronger formation arrivals indicating the presence of microannulus rather than channeling.
2. Microannulus tends to occur over long intervals of the log; channeling is a localized phenomena. This is a result of microannulus being directly related to pipe expansion during cementing operations.

Once the log has been interpreted, remedial measures can be applied as necessary.

Multifingered Caliper Logs
These logs incorporate up to 64 feelers or scratchers to examine pipe conditions inside the casing. Specifically, they can be combined with other logs to check:

1. Casing collar locations.
2. Corroded sections of pipe.
3. Casing wear.
4. Casing cracks or burstings.
5. Collapsed or crushed casing.
6. Perforations.
7. Miscellaneous breaks.

The number of feelers is a function of pipe diameter; smaller diameter pipe requires fewer feelers on the tool.

Electromagnetic Inspection Logs
This device induces a magnetic field into the casing and measures the returning magnetic flux. In general any disturbance in the flux from readings in normal pipe can be used to find:

1. Casing collars.
2. Areas of corroded pipe.
3. Perforations.
4. Breaks or cracks in the pipe.

This tool only records if corrosion has occurred on the pipe, not whether it is currently taking place. It does give an indication of casing quality and integrity without removing the pipe from the hole. The principle behind this tool is the same as the magna flux device used to detect flaws in metals in a machine shop.

Electrical Potential Logs
Similar in some respects to an SP log, this tool measures the potential gradient of a DC current circulating through a string of casing. This current is applied to provide the casing with cathodic protection thereby preventing casing corrosion; any deviation from a negative field suggests that the pipe is not receiving proper protection and is probably being corroded. Combined with an electromagnetic inspection log, areas currently undergoing corrosion as well as

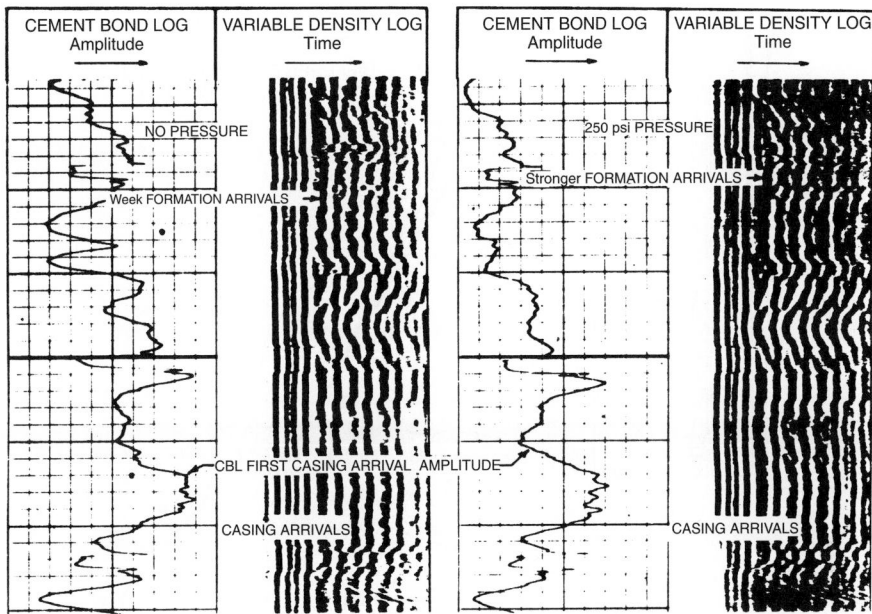

Figure 5.2.63 *CBL-VDL log showing effects of microannulus and the change in signal strength on VDL in pressured and nonpressured pipe [41].*

having a relative amount of damage can be determined with ease.

Borehole Televiewers
This tool incorporates an array of transmitters and receivers to scan the inside of the casing. The signals are sent to the surface where they are analyzed and recorded in a format that gives a picture of the inside of the casing. Any irregularities or cracks in the pipe are clearly visible on the log presentation. This allows engineers to fully scan older pipe and get an idea of the kind and extent of damage that might not otherwise be readable from multifinger caliper, electromagnetic inspection, or electrical potential logs. The main drawback to this device is that it must be run in a liquid-filled hole to be effective.

Production Logs
Production logs are those devices used to measure the nature and behaviour of fluids in a well during production or injection. A Schlumberger manual [42] summarizes the potential benefits of this information:

1. Early evaluation of completion efficiency.
2. Early detection of disturbances which are not revealed by surface measurements (thief zones, channeled cement, plugged perforations, etc.).
3. Detailed information on which zones are producing or accepting fluid.
4. More positive monitoring of reservoir production.
5. Positive identification of encroachment, breakthrough, coring, and mechanical leaks.
6. Positive evaluation of injection efficiency.
7. Essential guidance for remedial workover and secondary or tertiary recovery projects.

The reader is referred to the Schlumberger volume [42] on production log interpretation for examples of various cased-hole-log situations. It is still free upon request.

The types of logs run include

1. Temperature
2. Manometer and gradiomanometer
3. Flow meters
4. Radioactive tracers

Devices that measure water-holdup are also available. These logs can be run singly or in combination on a production combination tool so that a number of parameters may be recorded on the same log sheet.

Temperature Log
A thermometer is used to log temperature anomalies produced by the flow or fluid inside the casing or in the casing annulus. It is used to help determine flow-rates and points of fluid entry or exit, and is, perhaps, most useful for finding fluid movement behind the casing.

Injection Wells
Figure 5.2.64 is the response of the temperature log when fluid is being injected into a reservoir. The sloping portion defines the geothermal gradient the vertical portion defines the zone tasking the water and is a function of the geothermal gradient as well as the injection fluid temperature. Below the sloping position, the temperature/curve rapidly returns to normal formation temperature and the geothermal gradient. The vertical portion of the log clearly indicates where the fluid is leaving the casing.

Production Wells
Figure 5.2.65 is the response of the temperature log when fluid is flowing into a well from perforations in the casing.

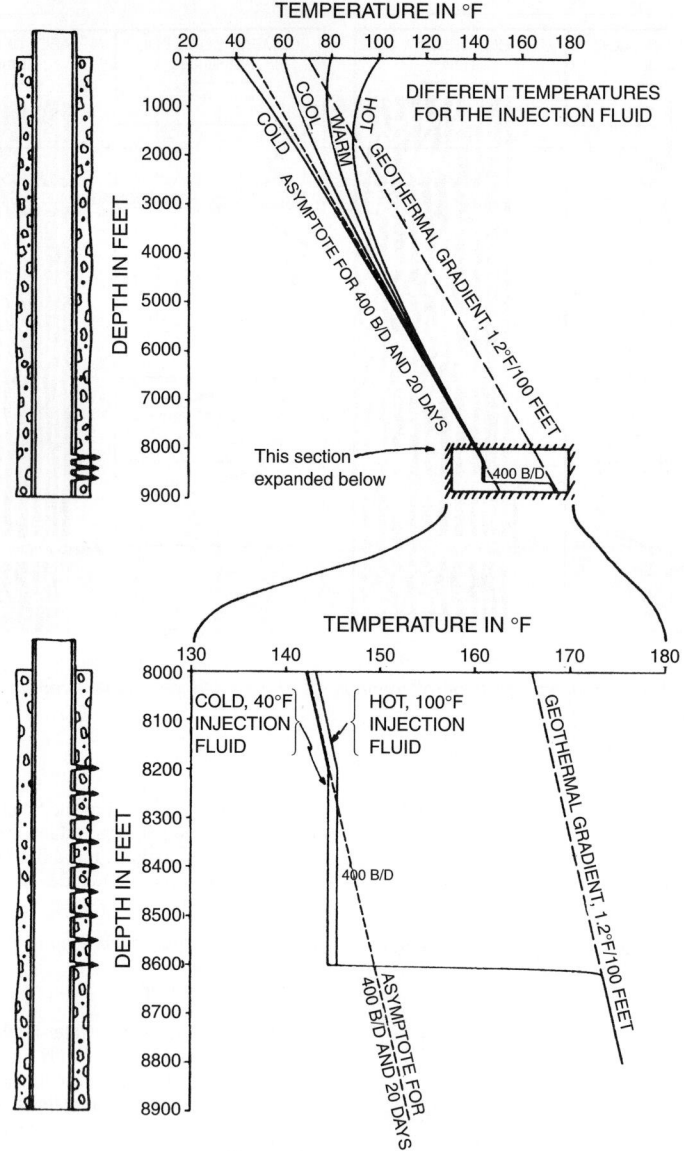

Figure 5.2.64   *The effect of water injection on a temperature log for several injection water temperatures [42].*

Three curves are presented. This figure shows that curve response depends on whether the fluid produced is hotter, the same as, or cooler than the geothermal gradient. If the fluid is hotter or cooler, then the entry point is obvious. If the fluid temperature is the same as the geothermal gradient, the change is so subtle that recognition of the entry point may be very difficult. In this case, a high resolution thermometer may be necessary to pinpoint the fluid entry location.

Flow Behind Casing (Annular Flow)
Figure 5.2.66 is a typical response to annular flow down the outside of the casing in a shut-in well. The figure shows water entering the annulus at about 6,500 ft. Perforations are at ~ 8,500 ft.

In a producing well, the shape of the curve defines the top of the annular space and its relationship to the perforations.

Manometers and Gradiomanometers
Manometers are pressure-sensitive devices used to measure changes in pressure that result from:

1. Leaks in tubing or casing.
2. Fluid inflow through perforations.
3. Gradient measurements in a static mud column.

They are particularly useful for determining pressure opposite a gas-bearing horizon. This value is vital for calculating open-flow potentials in gas wells.

Gradiomanometers are used to check the difference in pressure over a 2-in. interval in a producing well. This is

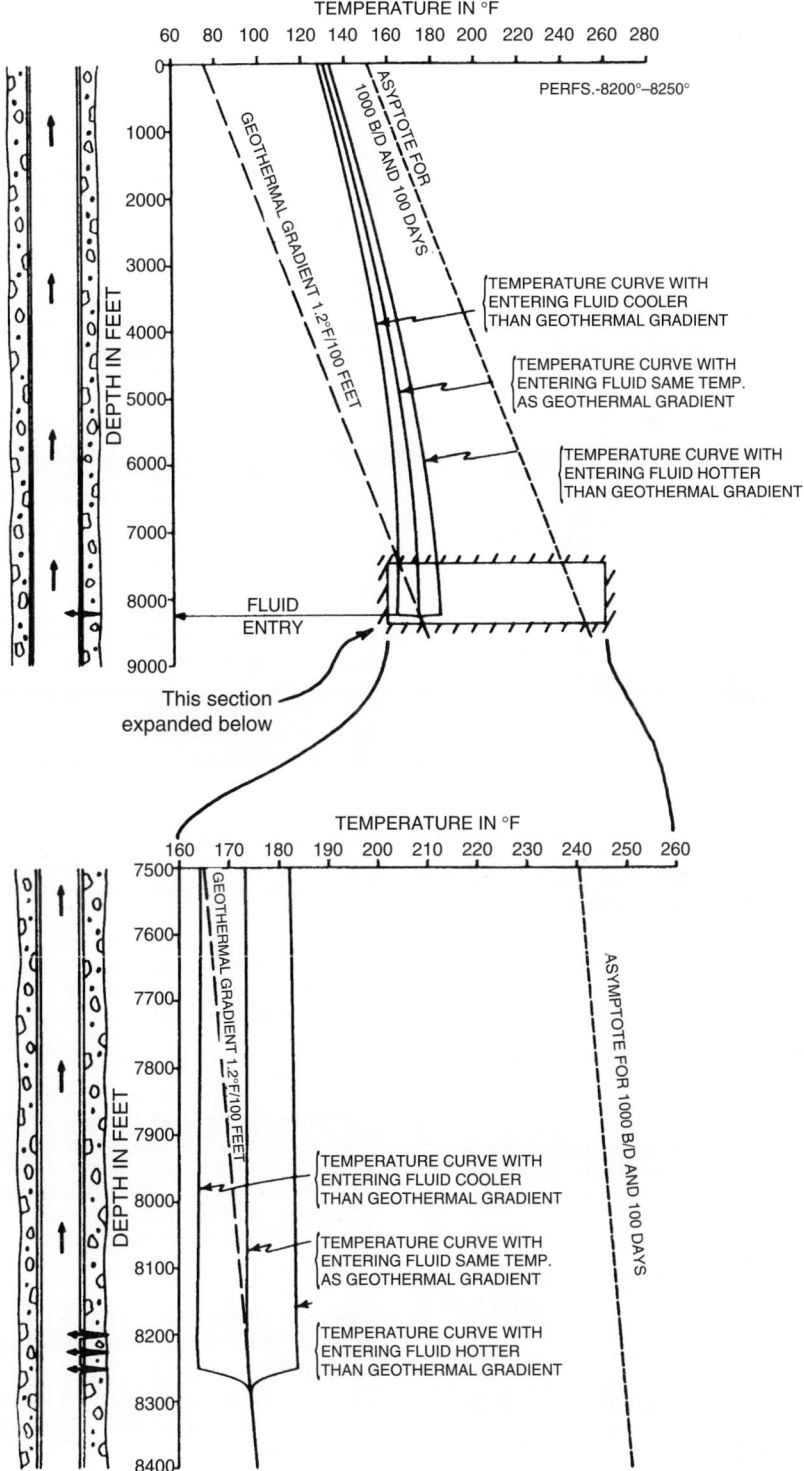

Figure 5.2.65 *Temperature logs in a producing interval for formation fluids with different temperatures [42].*

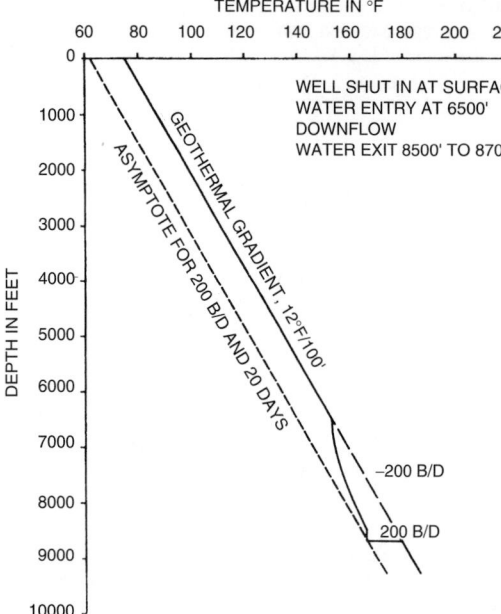

Figure 5.2.66 *Temperature log showing water flow behind the casing. Water is flowing down to the producing interval [42].*

then related to water-holdup in polyphase fluid flow within the casing.

The pressure difference is converted to density and is used to interpret two-phase flow (usually consisting of water as the heavy component and oil as the lighter component). At any given level, the gradiomanometer measures the specific gravity (density) of any fluids entering the borehole. The log reading is related to water holdup and specific gravity by:

$$\rho_{gradiomanometer} = Y_w \rho_w + (1 - Y_w)\rho_o \qquad [5.2.43]$$

where $\rho_{gradiomanometer}$ = specific gravity reading of the gradiomanometer, g/cc

ρ_w = specific gravity of the formation water, g/cc

ρ_o = specific gravity of the oil being produced with the water, g/cc

Y_w = water holdup (or holdup of the heavy phase)

The specific gravity reading is not exclusively dependent on fluid density. Since the fluids are also flowing while measurements are being made, other terms must be added:

$$\rho_{gradiomanometer} = \rho_f(1.0 + K + F) \qquad [5.2.44]$$

where $\rho_{gradiomanometer}$ = specific gravity reading of the instrument, g/cc

ρ_f = specific gravity of the fluid in the casing (oil + water + gas), g/cc

K = a kinetic term

F = a friction term from fluid flowing around the tool

At flow rates less than 2,000 bopd, F is negligible and $\rho_{grad} \approx \rho_f$ [42]. The kinetic term is important when logging from tubing into the casing. Fluid velocity changes become

significant at the change in hole diameter. The change causes a sharp pressure increase on the log. The friction term is important at high flow rates in casing and when logging in small-diameter tubing or casing [42].

Another log similar to the gradiomanometer is the water-holdup meter. The main limitation is that it only reads water holdup and cannot be used if water is not present. It has the advantage over the gradiomanometer where sensitivity is required; small differences in density may not be seen by the gradiomanometer.

Flowmeters
Flowmeters are designed to measure fluid flow in the casing. This measurement is then related to volume of fluid being produced. Three types of flowmeters are available:

1. Fullbore-spinner flowmeter.
2. Continuous flowmeter.
3. Packer flowmeter.

Each is used in certain circumstances and is combinable with other devices so that improved flow rates can be obtained.

Fullbore-Spinner Flowmeter
This device measures velocities of fluid moving up the casing. These velocities are then related to volume of fluid moved with charts available from the various service companies. In general, this tool can be used at flow rates as low as 20 barrels per day in monophase flow situations (usually water). Polyphase flow raises this minimum to 300 barrels per day if gas is present (i.e., oil and water) in $5\frac{1}{2}$ in. casing. This tool is used in wells with hole diameters ranging from $5\frac{1}{2}$ to $9\frac{5}{8}$ in.

Continuous Flowmeter
This tool is similar to the fullbore-spinner flowmeter except that it can be applied to hole diameters between $3\frac{1}{2}$ in. and $6\frac{5}{8}$ in. It has a higher flow threshold (in barrels per day) and should be restricted to use in monophase flow situations (i.e., waterfloods, high-flow-rate gas wells, and high-flow-rate oil wells) [42]. It can be combined with a spinner flowmeter for better flow measurements.

Packer Flowmeter
This is a small spinner-flowmeter with an inflatable packer that can be used in small-diameter tubing ($1^{11}/_{16}$ to $2\frac{1}{8}$ in.) It has an operable flow range from 10 to 1,900 barrels per day an can be applied in low-flow wells as long as measurements are made in the tubing at a sufficient distance above the perforations. Flow measurements are related to volume of fluid flowing the same way found with the other spinner flowmeters.

Radioactive Tracers
Radioactive tracers are combined with cased hole gammaray logs to monitor:

1. Fluid velocities in monophase fluid flow situations where flow velocity is at or near the threshold for spinner flowmeters.
2. Fluid movement behind the casing or to locate channeling in the cement.

Fluid velocity is measured by velocity-shot analysis. A shot of radioactive fluid is injected into the flow stream above two detectors located on a stationary mammary tool. As the radioactive pulse moves down the hole, the amount of time required to move past the two detectors is measured.

This travel time is then related to flow rate in the casing by:

q(B/D)

$$= \frac{\begin{array}{l} \text{spacing(in.)} \times \frac{1(\text{ft})}{12(\text{in.})} \times \frac{\pi}{4}(d_h - d_{\text{tool}})(\text{in.}^2) \times \frac{1(\text{ft.}^2)}{144(\text{in.}^2)} \\ \times 256.5\frac{\text{B/D}}{\text{ft}^3/\text{min}} \end{array}}{\text{time(sec)} \times \frac{1(\text{min})}{60(\text{sec})}}$$

[5.2.45]

where q is flow rate in barrels per day, the spacing between detectors is in inches, the time between detector responses is in seconds, d_h is the hole diameter in in., and d_{tool} is the tool diameter in in.

The main limitation is that slippage and water-holdup factors seriously affect the time reading so this technique cannot be applied in production wells. Moreover, the production of radioactive material is not desirable; therefore, use is mainly restricted to water- or gas-injection wells [42].

Fluid movement behind the casing can be measured with a timed-run radio-active survey. A slug of radioactive fluid is introduced at the bottom of the tubing, and movement is then monitored by successive gamma-ray log runs. Unwanted flow up any channels in the cement can be easily determined and remedial action taken. Again, this technique is mainly applied to water injection wells to monitor flood operations and injection-fluid losses. Figure 5.2.67 is an example of this type of application.

5.2.4 Determination of Initial Oil & Gas in Place
5.2.4.1 Initial Oil in Place
For undersaturated crude, the reservoir contains only connate water and oil with their respective solution gas contents. The initial or original oil in place can be estimated from the volumetric equation:

$$N = \frac{7,758V_b\phi(S_{oi})}{B_{oi}} = \frac{7,758Ah\phi(1 - S_{wi})}{B_{oi}}$$

[5.2.46]

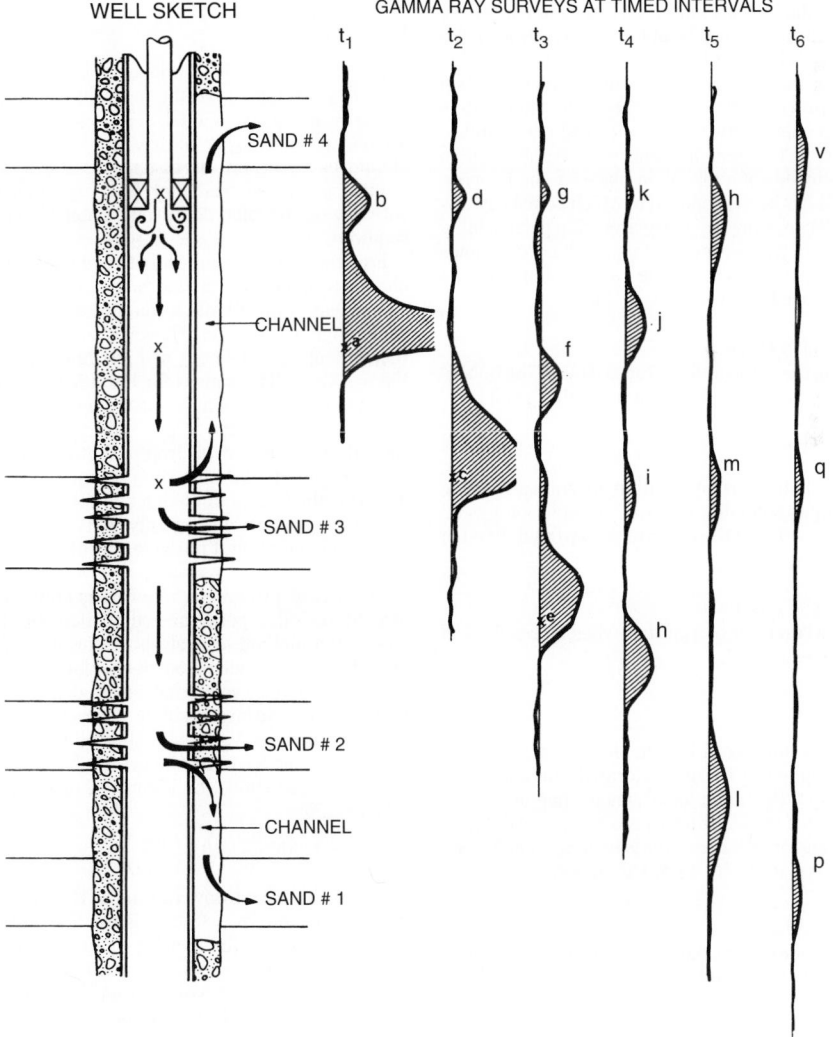

Figure 5.2.67 *Radioactive tracer survey in an injection well where water is flowing behind the casing into another zone [42].*

The constant 7,758 is the number of barrels in each acre-ft, V_b is bulk volume in acre-ft, ϕ is the porosity (ϕV_b is pore volume), S_{oi} is the initial oil saturation, B_{oi} is the initial oil formation volume factor in reservoir barrels per stock tank barrel, A is area in ft^2, h is reservoir thickness in ft, and S_{wi} is the initial water saturation.

In addition to the uncertainty in determining the initial water saturation, the primary difficulty encountered in using the volumetric equation is assigning the appropriate porosity-feet, particularly in thick reservoirs with numerous non-productive intervals. One method is to prepare contour maps of porosity-feet that are then used to obtain areal extent. Another method is to prepare isopach maps of thickness and porosity from which average values of each can be obtained. Since recovery of the initial oil can only occur from permeable zones, a permeability cutoff is used to obtain the net reservoir thickness. Intervals with permeabilities lower than the cutoff value are assumed to be nonproductive. The absolute value of the cutoff will depend on the average or maximum permeability, and can depend on the relationship between permeability and water saturation. A correlation between porosity and permeability is often used to determine a porosity cutoff. In cases in which reservoir cores have been analyzed, the net pay can be obtained directly from the permeability data. When only logs are available, permeability will not be known; therefore a porosity cutoff is used to select net pay. These procedures can be acceptable when a definite relationship exists between porosity and permeability. However, in very heterogeneous reservoirs (such as some carbonates), estimates of initial oil in place can be in error. A technique [43] has been proposed in which actual pay was defined using all core samples above a specific permeability cutoff and apparent pay was defined using all core samples above a specific porosity cutoff; the relationship between these values was used to find a porosity cutoff.

5.2.4.2 Initial Gas In Place

For the foregoing case of an undersaturated oil (at the bubble point with no free gas), the gas in solution with the oil is:

$$G = \frac{7,758 Ah\phi(1-S_{wi})R_s}{B_{oi}} \qquad [5.2.47]$$

where G is the initial gas in solution in standard cubic feet (scf), R_s is gas solubility in the oil or solution gas-oil ratio (dimensionless), and the other terms are as defined in Equation 5.2.46.

5.2.4.3 Free Gas In Place

Free gas within a reservoir or a gas cap when no residual oil is present can be estimated:

$$G = \frac{7,758 V_g \phi(1-S_{wi})}{B_{gi}} \qquad [5.2.48]$$

where 7,758 is the number of barrels per acre-ft, V_g is the pore volume assigned to the gas-saturated portion of the reservoir in acre-ft, B_{gi} is the initial gas formation volume factor in RB/scf, and the other terms are as already defined. (Note: If the formation volume factor is expressed in ft^2/scf, 7,758 should be replaced with 43,560 ft^3/acre-ft).

5.2.5 Productivity Index

The productivity index, J, is a measure of the ability of a well to produce hydrocarbon liquids:

$$J_o = \frac{q_o}{p_e - p_{wf}} \qquad [5.2.49]$$

where q_o is the flow rate of oil in stock-tank barrels of oil per day, p_e is the external pressure in psi, P_{wf} is the flowing bottomhole pressure in psi, and the quantity

$(p_e - p_{wf})$ is referred to as the pressure drawdown. Because the flow rate in this case is in STB/D, the oil productivity index (J_o) has units of STB/D/psi. Since only q and p_{wf} can be measured directly, p_e is commonly replaced with $\bar{p}$ which can be determined from pressure transient testing.

After the well has been shut in for a period of time (usually at least 24 to 72 hours or longer depending on reservoir characteristics), the well is put on production at a low rate with a small choke. The rate of production is recorded as a function of flow time. When the production rate has stabilized, the flow rate is increased by increasing the choke, and flow rate is monitored with time. This process is repeated until a series of measurements has been recorded [19].

In order to attain a stabilized productivity index, a minimum time is required after each individual flow-rate change. This time can be approximated by two equations [18,66]:

$$t_s \cong \frac{380\phi\mu c_t A}{k} \qquad [5.2.50]$$

$$t_s \cong \frac{0.04\phi\mu c_t r_e^2}{k} \qquad [5.2.51]$$

where t_s is the stabilization time in hours, k is the permeability in md, ϕ is the porosity as a fraction, μ is viscosity in cp, c_t is total compressibility in psi^{-1}, A is area in ft^2, and r_e is the external radius in feet which should be based on the distance to the farthest drainage boundary for the well. For large systems or reservoirs with low permeability, very long stabilization times may be required.

Equation 5.2.49 assumes that productivity index does not change with flow rate of time, and in some wells the flow rate will remain proportional to the pressure drawdown over a wide range of flow rates. However, in many wells, the direct relationship is not linear at high flow rates as shown in Figure 5.2.68. The causes for the deviation in the straight-line behavior can include insufficient producing times at each rate, an increase in gas saturation near the wellbore caused by the pressure drop in that region, a decrease in permeability of oil due to the presence of gas, a reduction in permeability due to changes in formation compressibility, an increase in oil viscosity with pressure drop below the bubble point, and possible turbulence at high rates of flow.

A plot of oil production rate versus bottomhole pressure, termed the inflow performance relationship (IPR), was proposed as a method of analysis of flowing and gas-lift wells [44]. Vogel [45] calculated dimensionless IPR curves for solution gas reservoirs that covered a wide range of oil PVT and relative permeability characteristics. From computer simulations, Vogel [45] showed that any solution gas drive reservoir operating below the bubble point could be represented as shown by Figure 5.2.69 or by the following relationship:

$$\frac{q_o}{(q_o)_{max}} = 1 - 0.2\frac{p_{wf}}{\bar{p}} - 0.8\left(\frac{p_{wf}}{\bar{p}}\right)^2 \qquad [5.2.52]$$

where q_o is the oil flow rate in STB/D occurring at bottomhole pressure p_{wf}, $(q_o)_{max}$ is the maximum oil flow rate in STB/D, p_{wf} is the flowing bottomhole pressure, and $\bar{p}$ is the average reservoir pressure. From the well pressure and average reservoir pressure, the ratio of producing rate to maximum oil rate can be obtained; then from the measured production rate, the maximum oil production can be calculated.

Vogel's method handles the problem of a single well test when the permeability near a wellbore is the same as

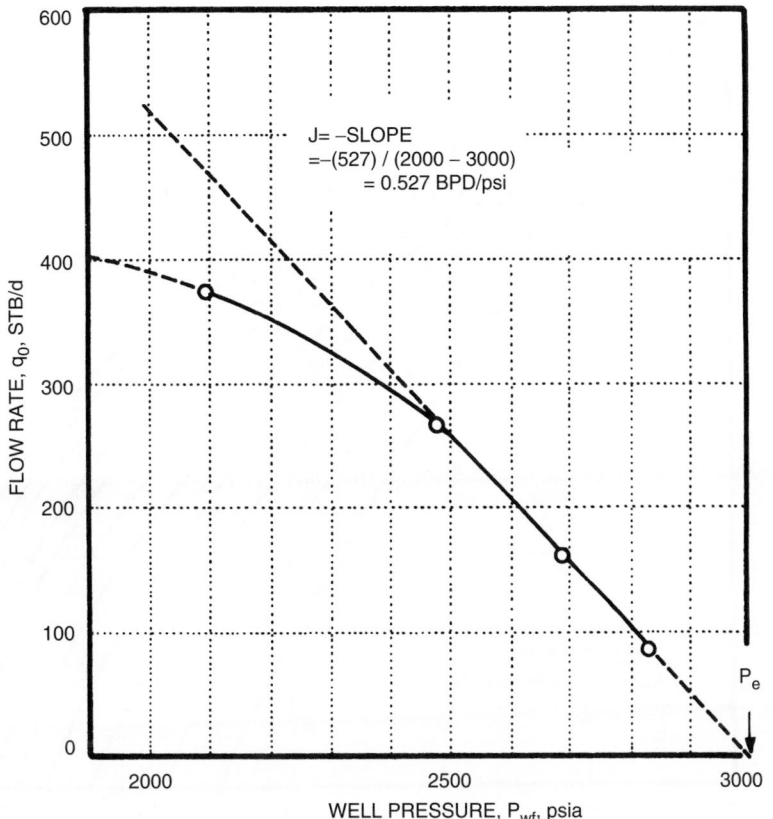

Figure 5.2.68 *Productivity index [18].*

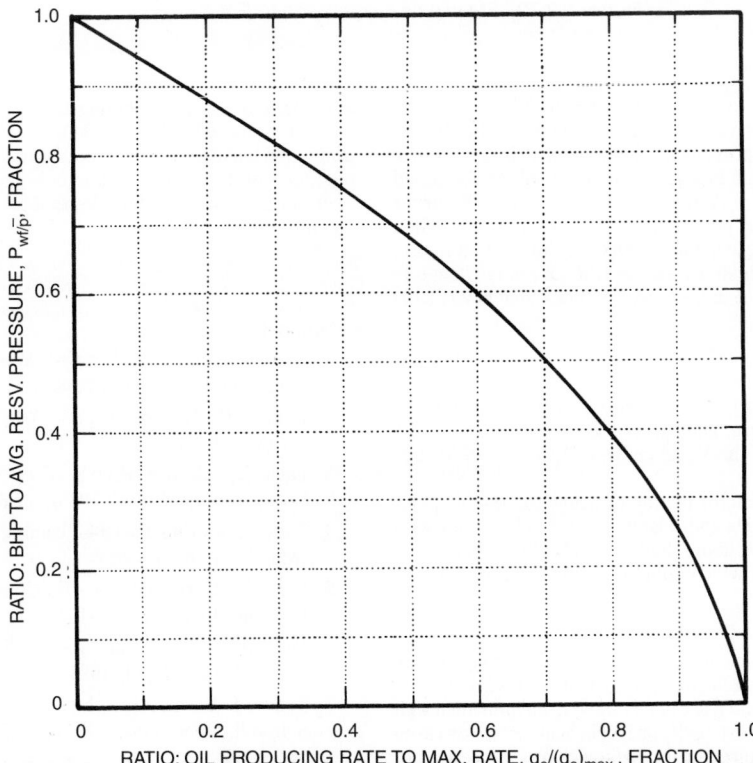

Figure 5.2.69 *Vogel IPR curve [45].*

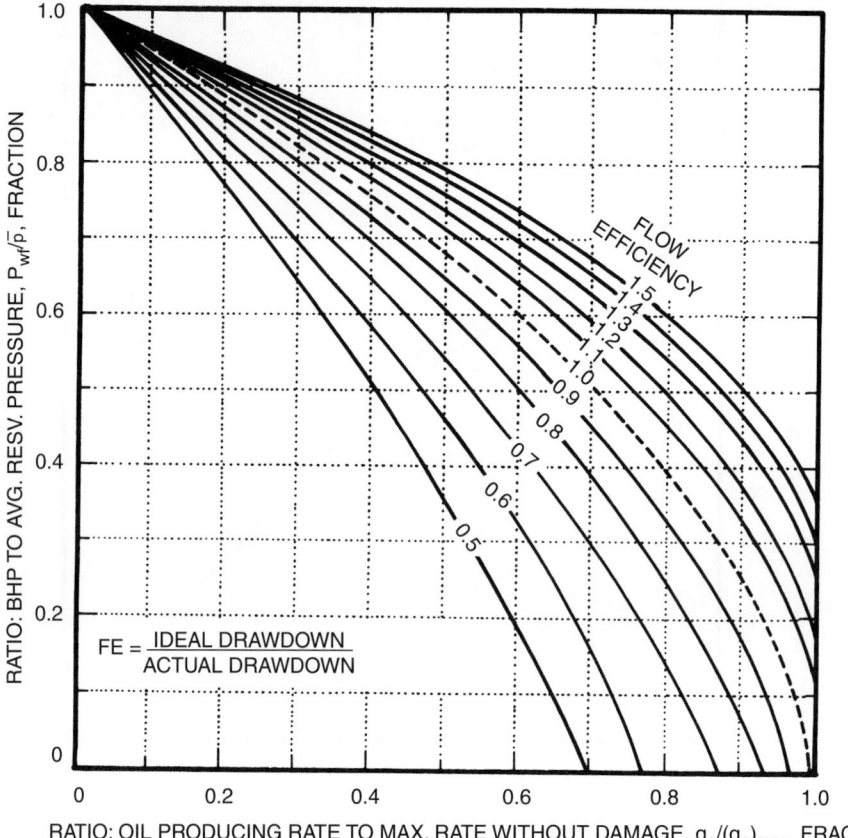

Figure 5.2.70 *Standing's modification to IPR curves [46].*

the permeability throughout the reservoir. When a zone of altered permeability exists near the wellbore, the degree of damage (or improvement) is expressed in terms of a "skin effect" or "skin factor." (Skin effect will be discussed in more detail later.) A modification to Vogel's IPR curves has been proposed by Standing [45] for situations when a skin effect is present (see Figure 5.2.70). In this figure, Standing has provided a series of IPR curves for flow efficiencies between 0.5 and 1.5, where flow efficiency (FE) is defined as:

$$FE = \frac{\bar{p} - p_{wf} - \Delta p_s}{\bar{p} - p_{wf}} \qquad [5.2.53]$$

where $\bar{p}$ is the average reservoir pressure, P_{wf} is the flowing bottomhole pressure, and Δp_s is the pressure drop in the skin region. Thus, the Vogel curve is for a flow efficiency of 1.0.

For reservoir systems operating below the bubble point when fluid properties and relative permeabilities vary with distance from the wellbore, Fetkovich [47] has proposed an empirical equation which combines single-phase and two-phase flow:

$$q_o = J'_o(p_b^2 - p_{wf}^2)^n + J_o(p_e - p_b) \qquad [5.2.54]$$

where p_b is the bubble-point pressure in psia, $\bar{p}$ may be substituted for p_e, J'_o is a form of productivity index and n is an exponent; both J'_o and n are determined from individual well multirate and pressure tests, or isochronal tests. For cases where the data required by the Fetkovich procedure are not

available, a method for shifting the axes of the Vogel plot has been proposed [48]. In this latter method, only one set of production test data (rate and bottomhole flowing pressure) together with the shut-in bottomhole pressure (or average reservoir pressure) and bubble-point pressure are required to construct a reliable IPR.

References

1. Timmerman, E. H., *Practical Reservoir Engineering*, PennWell Books, Tulsa (1982).

2. Keelan, D., "Coring: Part 1—Why It's Done," *World Oil* (March 1985), pp. 83–90.

3. Part, A., "Coring: Part 2—Core Barrel Types and Uses," *World Oil* (April 1985), pp. 83–90.

4. Park, A., "Coring: Part 3—Planning the Job," *World Oil* (May 1985), pp. 79–86.

5. Park, A., "Coring: Part 4—Bit Considerations," *World Oil* (June 1985), pp. 149–154.

6. Park, A., "Coring: Part 5—Avoiding Potential Problems," *World Oil* (July 1985), pp. 93–98.

7. Toney, J. B., and Speights, J. L., "Coring: Part 6—Sidewall Operations," *World Oil* (Aug. 1, 1985), pp. 29–36.

8. Kraft, M., and Keelan, D., "Coring: Part 7—Analytical Aspects of Sidewall Coring," *World Oil* (Sept. 1985), pp. 77–90.

9. Keelan, D., "Coring: Part 8—Plug and Full Diameter Analysis," *World Oil* (Nov. 1985), pp. 103–112.

10. Kersey, D. G., "Coring: Part 9—Geological Aspects," *World Oil* (Jan. 1986), pp. 103–108.

11. Keelan, D. K., "Core Analysis for Aid in Reservoir Description" *J. Pet. Tech.* (Nov. 1982), pp. 2483–2491.

12. Anderson, G., *Coring and Core Analysis Handbook*, Pet. Pub. Co., Tulsa (1975).

13. "API Recommended Practice for Core-Analysis Procedure," API RP 40, API Prod. Dept., Dallas (Aug. 1960).

14. "Recommended Practice for Determining Permeability of Porous Media," API RP 27, third edition, API Prod. Dept., Dallas (Aug. 1956).

15. Stosur, J. J., and Taber, J. J., "Critical Displacement Ratio and Its Effect on Wellbore Measurement of Residual Oil Saturation," paper SPE 5509 presented at the SPE-AIME 50th Annual Fall Meeting, Dallas, Sept. 28–Oct. 1, 1975.

16. Jenks, L. H., et al., "Fluid Flow Within a Porous Medium Near a Diamond Core Bit," *J. Can. Pet. Tech.*, Vol. 7 (1968), pp. 172–180.

17. Black, W. M., "A Review of Drill-Stem Testing Techniques and Analysis," *J. Pet. Tech.* (June 1956), pp. 21–30.

18. Slider, H. C., *Worldwide Practical Petroleum Reservoir Engineering Methods*, PenWell Pub. Co., Tulsa (1983).

19. Timur, A., "An Investigation of Permeability, Porosity, and Residual Water Saturation Relationships for Sandstone Reservoirs," *The Log Analyst* (July–Aug. 1968).

20. *Log Interpretation Charts*, Schlumberger Well Services (1991).

21. Hilchie, D. W., *Applied Openhole Log Interpretation*, second edition, Douglas W. Hilchie, Inc., Golden, CO (1982).

22. Doll, H. G., "The SP Log: Theoretical Analysis and Principles of Interpretation," *Trans., AIME*, Vol. 179 (1948), pp. 146–185.

23. Hilchie, D. W., *Old Electrical Log Interpretation*, Douglas W. Hilchie, Inc., Golden, CO (1979).

24. Frank, R. W., *Prospecting with Old E-Logs*, Schlumberger, Houston (1986).

25. Poupon, A., Loy, M. E., and Tixier, M. P., "A Contribution to Electrical Log Interpretations in Shaly Sands," *J. Pet. Tech.* (June 1954); *Trans., AIME*, pp. 27–34.

26. Suau, J., et al., "The Dual Laterolog-R_{xo} tool," paper SPE 4018 presented at the 47th Annual Meeting, San Antonio (1972).

27. Asquith, G., and Gibson, C., "Basic Well Log Analysis for Geologists," American Association of Petroleum Geologists, Methods in Exploration Series, Tulsa.

28. Serra, O., Baldwin, J., and Quirein, J., "Theory Interpretation and Practical Applications of Natural Gamma Spectroscopy," *Trans.*, SPWLA (1980).

29. Wyllie, M. R. J., Gregory, A. R., and Gardner, L. W., "Elastic Wave Velocities in Heterogeneous and Porous Media," *Geophysics*, Vol. 21, No. 1 (Jan. 1956), pp. 41–70.

30. Merkel, R. H., "Well Log Formation Evaluation," American Association of Petroleum Geologists, Continuing Education Course-Note Series #4.

31. Raymer, L. L., Hunt, E. R., and Gardner, J. S., "An Imporved Sonic Transit Time-To-Porosity Transform," *Trans.*, SPWLA (1980).

32. Wahl, J. S., et al., "The Dual Spacing Formation Density Log," *J. Pet. Tech.* (Dec. 1964), pp. 1411–1416.

33. Desbrandes, R., *Encyclopedia of Well Logging*, Gulf Publishing Co., Houston (1985).

34. Tittman, J., "Geophysical Well Logging," *Methods of Experimental Physics*, Academic Press, Vol. 24 (1986).

35. *Electromagnetic Propagation Tool*, Schlumberger, Ltd., Houston (1984).

36. Bateman, R. M., "Open-Hole Log Analysis and Formation Evaluation," Intl. Human Resources Development Corp. (1985).

37. Bateman, R. M., "Log Quality Control," International Human Resources Development Corp. (1985).

38. Bateman, R. M., "Cased-Hole Log Analysis and Reservoir Performance Monitoring," Intl. Human Resources Development Corp. (1985).

39. Schultz, W. E., and Smith, H. D., Jr., "Laboratory and Field Evaluation of a Carbon/Oxygen (C/O) Well Logging System," *J. Pet. Tech.* (Oct. 1974), pp. 1103–1110.

40. Lock, G. A., and Hoyer, W. A., "Carbon-Oxygen (C/O) Log: Use and Interpretation," *J. Pet. Tech.* (Sept. 1974), pp. 1044–1054.

41. *Cased Hole Applications*, Schlumberger, Ltd., New York (1975).

42. *Production Log Interpretation*, Schlumberger, Ltd., New York (1973).

43. George, C. J., and Stiles, L. H., "Improved Techniques for Evaluating Carbonate Waterfloods in West Texas," *J. Pet. Tech.* (Nov. 1978), pp. 1547–1554.

44. Gilbert, W. E., "Flowing and Gas-lift Well Performance," *Drill. & Prod. Prac.*, API (1955), pp. 126–157.

45. Vogel, J. V., "Inflow Performance Relationships for Solution-Gas Drive Wells," *Trans.*, AIME (1968), pp. 83–92.

46. Standing, M. B., "Inflow Performance Relationships for Damaged Wells Producing by Solution-Gas Drive," *J. Pet. Tech.* (Nov. 1970), pp. 1399–1400.

47. Fetkovich M. J., "The Isochronal Testing of Oil Wells," paper SPE 4529 presented at the SPE 48th Annual Fall Meeting, Las Vegas, Sept. 30–Oct. 3, 1973.

48. Patton, L. D., and Goland, M., "Generalized IPR Curves for Predicting Well Behavior," *Pet. Eng. International* (Sept. 1980), pp. 92–102.

5.3 PRESSURE TRANSIENT TESTING OF OIL AND GAS WELLS

Production rates depend on the effectiveness of the well completion (skin effect), the reservoir permeability, the reservoir pressure, and the drainage area. Pressure transient analysis is a powerful tool for determining the reservoir characteristics required to forecast production rates. Transient pressure data are generated by changing the producing rate and observing the change in pressure with time. The transient period should not exceed 10% of the previous flow or shut-in period. There are a number of methods to generate the transient data available to the reservoir engineer.

Single-well tests such as buildup, falloff, drawdown, injection, and variable-rate describe the isotropic reservoir adjacent to the test well while multiple well tests such as long term interference or short term pulse describe the characteristics between wells. Buildup and falloff tests are most popular because the zero flow rate is readily held constant. Drawdown and injection tests are run less frequently due to problems with maintaining a constant rate. Variable rate tests are useful when wellbore storage is a problem. Multiwell testing for characterizing anisotropic reservoirs has been popularized by the increased use of sophisticated simulation software.

5.3.1 Definitions and Concepts

Several excellent references on well test analyses are available [1.13, 1.66], and a good discussion of difficulties in interpretation of data is available in a recent text. From information in these references several definitions will be given, and the basic concepts of well test analysis will be summarized. More advanced concepts can be found in the foregoing references or in the extensive literature on this subject that has appeared in recent years.

5.3.1.1 Definitions

Transient Region

Flow regimes that occur at different flow times are shown in Figure 5.3.1 for a well flowing at a constant rate. The flowing bottomhole pressure is shown as a function of time on both linear and semilog plots. In the transient region, the reservoir is infinite-acting, and the flowing bottomhole pressure is a linear function of log Δt. This region is amenable to analysis by transient methods, and occurs for radial flow at flow times up to approximately $t = \phi\mu c r_e^2 / 0.00264 k$, where field units are used: t is time in hours, ϕ is porosity as a fraction, μ is viscosity in cp, c is compressibility in psi^{-1}, r_e is the external radius in ft, and k is permeability in md [1.13].

Late-Transient Region

At the end of the transient region and prior to the semisteady-state period, there is a transitional period called the late-transient region (see Figure 5.3.1). There are no simple equations that define this region, but the late-transient period may be very small or practically nonexistent.

Semisteady-State Region

If there is no flow across the drainage boundary and compressibility is small and constant, a semisteady- or pseudosteady-state region is observed in which the pressure declines linearly with time (see Figure 5.3.1). Pressures in the drainage area decrease by the same amount in a given time, and the difference between reservoir pressure and wellbore pressure remains constant during this period. For radial flow, semisteady-state flow conditions start at a flow time of about $t = \phi\mu c r_e^2 / 0.00088 k$, in field units as already specified [1.13]. This region is suitable for reservoir limit tests in which reservoir size and distance to boundaries can be estimated. The most useful test to estimate reservoir limits is the drawdown test [2.18]; interpretation of reservoir limit tests can be difficult as discussed in the literature [2.18].

Steady-State Flow

At a constant flow rate for steady-state flow, the pressure at every point in the reservoir will remain constant with time. This condition is rarely encountered in most well test analyses; steady-state flow may be approached in reservoirs with strong water drives or in cases where reservoir pressure is maintained by gas injection or waterflooding.

Buildup Tests

Pressure buildup tests are conducted by (1) producing an oil or gas well at a constant rate for sufficient time to establish a stabilized pressure distribution, (2) ceasing production by shutting in the well, and (3) recording the resulting increase in pressure. In most cases, the well is shut in at the surface and the pressure is recorded downhole. In pumping wells, buildup tests can be made by: (1) pulling the rods and running a pressure bomb in the tubing, (2) by measuring pressure in the annulus from sonic measurements obtained with an echo-device, or (3) occasionally by using surface-indicating gauges. The pressure buildup curve is analyzed for wellbore conditions such as damage or stimulation and for reservoir properties such as formation permeability, pressure in the drainage area, reservoir limits or boundaries, and reservoir heterogeneities.

Drawdown Tests

Pressure drawdown tests are conducted by: (1) having an oil or gas well shut in for sufficient time to establish a stabilized pressure distribution, (2) putting the well on production at a constant rate, and (3) recording the resulting decrease in bottomhole pressure. An ideal time to run a drawdown test is when the well is initially put on production because in addition to obtaining information on wellbore

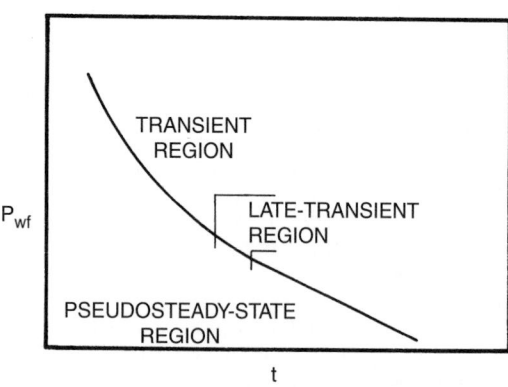

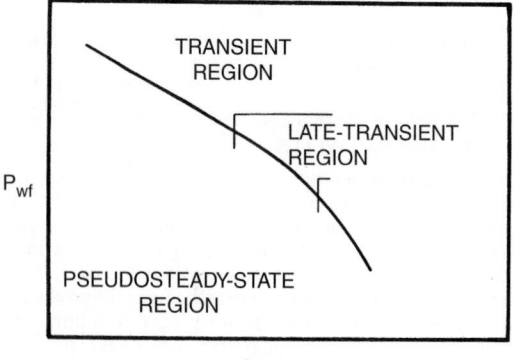

Figure 5.3.1 Flow regimes [1,13].

conditions and formation permeability, estimates of reservoir volume can be made also. A long, constant flow rate is required.

Falloff Tests

Pressure falloff tests are conducted in injection wells and are analogous to the pressure buildup tests in producers. A falloff test consists of (1) injecting fluid at a constant rate, (2) shutting in the well, and (3) recording the decrease in pressure. As long as the mobility ratio between the injected fluid and in-situ fluids is near unity, the analysis of pressure transient tests in injection wells is relatively simple. The equations used in producing well tests are applicable with the exception that the flow is taken to be negative for injection whereas flow is positive for production.

Multiple-Rate Tests

The preceding tests apply to conditions in which the flow rate either has been or is constant. In some cases, maintaining a constant flow rate may not be possible or practical. In other cases, regulatory agencies may require that wells, especially gas wells, be tested at various flow rates. Multiple-rate tests may be conducted at variable flow rates or a series of constant rates, and are applicable to buildup or drawdown tests in producers or falloff tests in injectors. If accurate flow rate and pressure data are obtained, information on permeability, skin, and reservoir pressure can be deduced.

Interference Tests

In the prior tests, the pressure and flow rate applied to only one well at a time. With interference tests, two wells are involved. Interference tests are conducted by producing from or injecting into at least one well and observing the pressure response in at least one shut-in observation well. A change in rate (pressure) at the active producer or injector will cause a pressure interference at the observation well. A special form of multiple-rate testing is the pulse test in which the pressure caused by alternating periods of production (or injection) and shut-in periods is monitored at one or more observation wells. Multiple-rate tests are used to determine if wells are in communication with each other in the same reservoir as well as to provide estimates of formation permeability and the product of porosity and total compressibility.

Miller-Dyes-Hutchinson (MDH) Plot

One of the most useful methods of pressure test analysis is that of Miller, Dyes, and Hutchinson [2]. The MDH method is a plot of bottomhole pressure versus log time on semilog paper. A schematic of an MDH plot for a pressure buildup test is depicted in Figure 5.3.2; the region is identified where MDH and Horner plots are applicable.

Horner Plot

In the Horner plot [3] bottomhole pressure is plotted against $\log(t + \Delta t)/\Delta t$. The Horner method should be applied only to infinite-acting reservoirs; for radial flow, the Horner plot will be a straight line. Several conditions such as boundaries or changes in fluids or fluid properties, can cause the Horner plot to deviate from a straight line (see Figures 5.3.3 and 5.3.4).

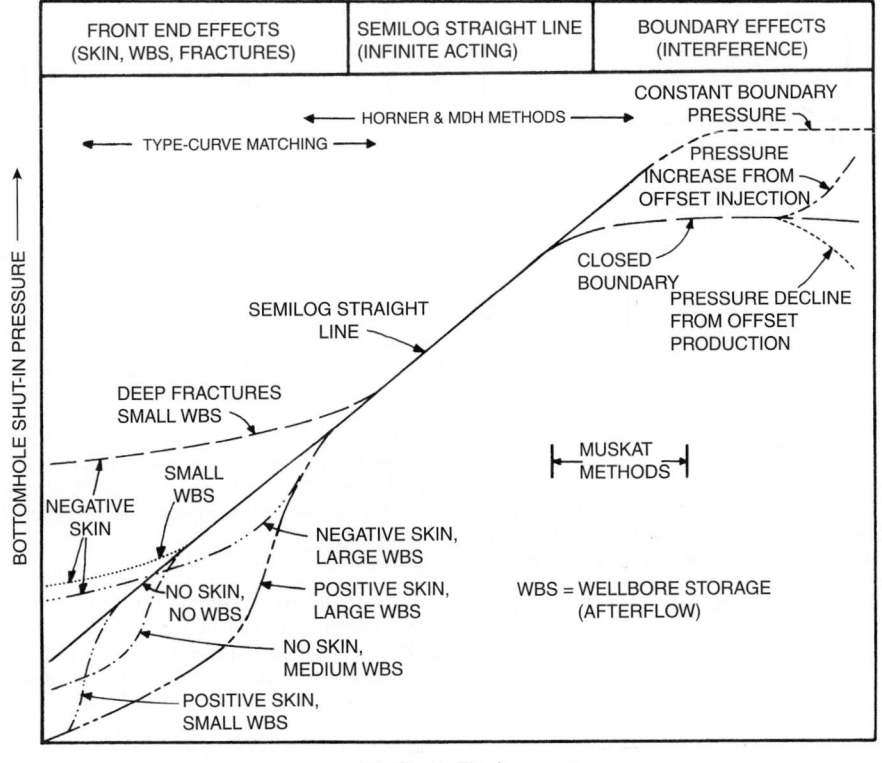

Figure 5.3.2 MDH plot [66].

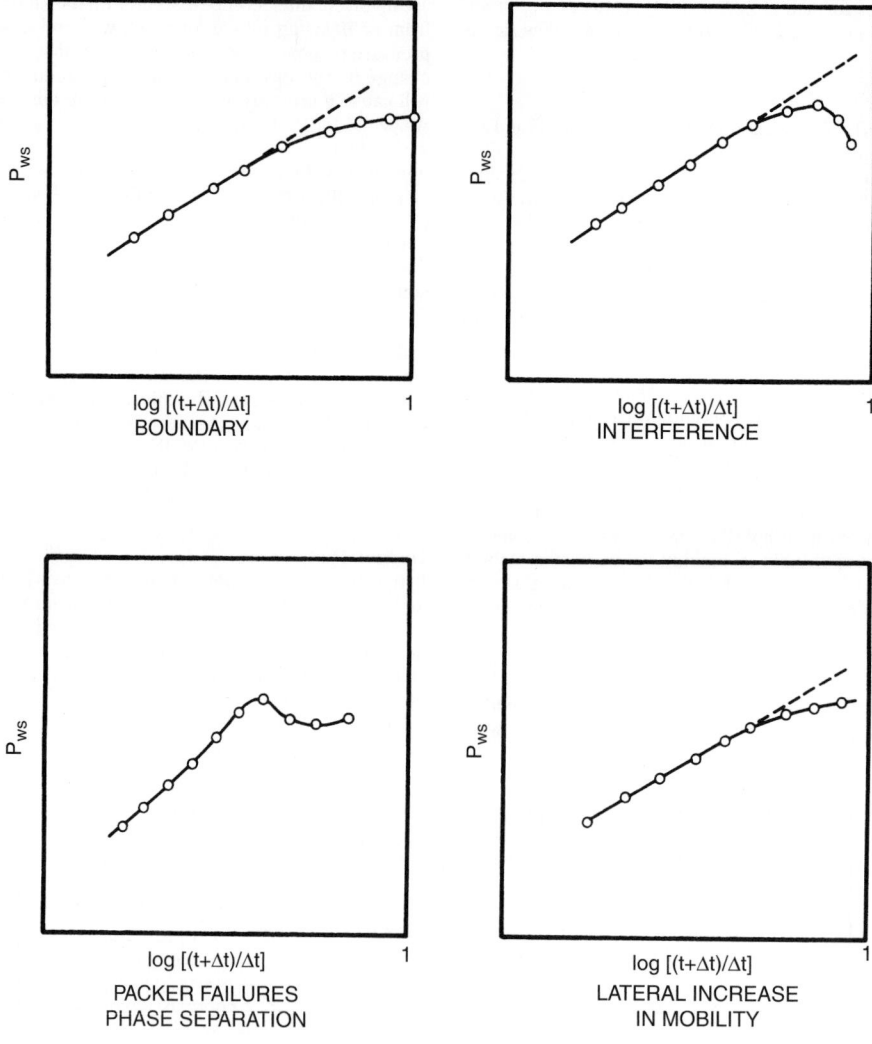

Figure 5.3.3 *Downtrending Horner plots [1,13].*

Skin Factor

The skin effect refers to a zone of altered formation permeability near a wellbore as a result of drilling, completion, or stimulation [4,5]. The extent of altered permeability is expressed in terms of a skin factor, s, which is positive for damage and negative for improvement. Skin factor can range from about -5 for a hydraulically fractured well to a theoretical limit of infinity for a severely damaged or plugged well. A schematic diagram of the pressure distribution near the wellbore of a damaged well is given in Figure 5.3.5. Effects of positive and negative skin are shown for the MDH plot in Figure 5.3.2.

Wellbore Storage

Wellbore storage, also referred to as after flow, wellbore loading or unloading, after production, and after injection, will affect short-time transient pressure behavior. This phenomenon has more of an effect on pressure buildup than drawdown tests, and can be especially important in low-permeability formations or in gas wells. During a buildup test, a well is closed in at the surface, but fluid may continue to flow into the wellbore for some time which causes a lag in the buildup at early times. Various levels of wellbore storage are shown in the MDH buildup plot in Figure 5.3.2. Storage can obscure the transient period thus negating the value of a semi log plot.

5.3.1.2 Concepts

Most techniques used in the analysis of transient tests assume a single well operating at a constant flow rate in an infinite reservoir. At early times, a well transient is like a single well in an infinite reservoir, but at late times, effects of other wells aquifers, or reservoir boundaries can cause the pressure behavior to deviate from the infinite-acting assumption. Other common assumptions include: horizontal flow, negligible gravity effects, a single fluid of small and constant compressibility, a homogeneous and isotropic

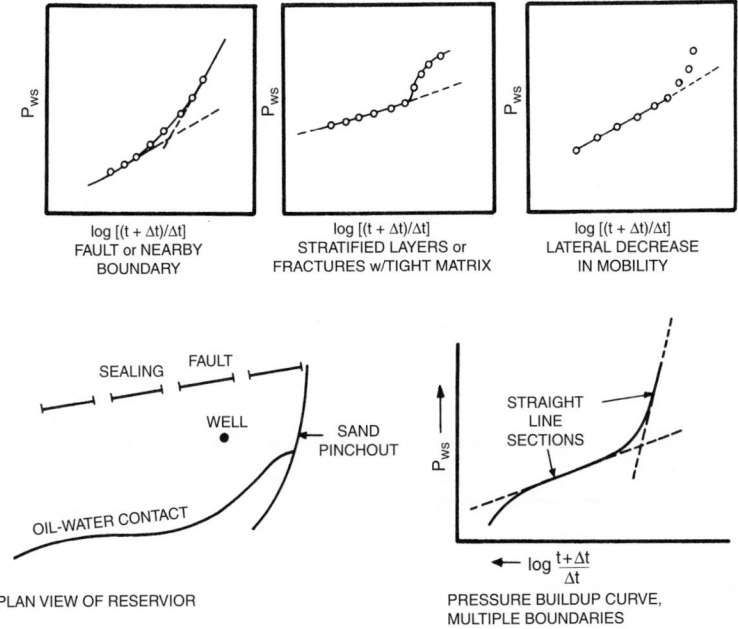

Figure 5.3.4 *Uptrending Horner plots [1,13].*

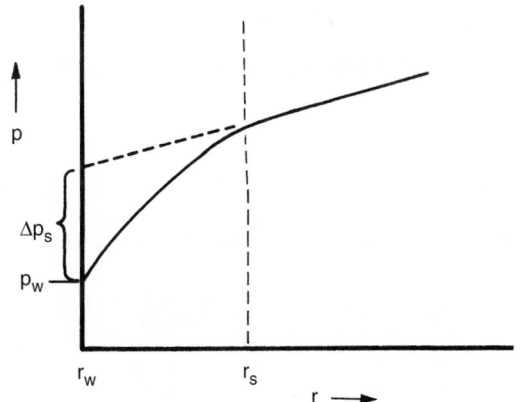

Figure 5.3.5 *Skin region [1].*

porous medium, the Darcy equation is obeyed, and several parameters (including porosity, permeability, viscosity, and compressibility) are independent of pressure.

Pressure transients arriving at the well following a rate change move through three regions on their way to the wellbore. Nearest the wellbore is the early-time region, ETR, where storage and skin effects dominate; next is the middle-time region, MTR, where the formation permeability is determined; and most distant is the late-time region, LTR, where drainage boundaries are sometimes observed (see Figure 5.3.6a for examples of a buildup test and Figure 5.3.6b for a drawdown test) [1]. As discussed earlier, the transient flow region (see Figure 5.3.1) is amenable to analysis by transient flow methods; this region consists of both the ETR and the MTR. The LTR can include the late-transient and the pseudosteady-state or semisteady-state regions. The crux of the analysis involves selecting the proper data to analyze.

Middle-time data will plot as a straight line on semilogarithmic paper. The slope and the intercept of the MTR straight line are used to calculate reservoir permeability, skin factor, and average reservoir pressure. Semilogarithmic straight lines can occur in the ETR and the use of their slopes and intercepts results in unrealistic reservoir characteristics. Typically, improper use of injection well ETR data indicates a tight, fractured reservoir while use of producing well ETR data indicates damaged permeable rock when that is not at all the case.

Flow conditions in the pseudosteady-state LTR occur when transients reach the no-flow drainage boundary during producing-well transient-tests. Flow conditions in the steady-state LTR occur when transients reach the constant pressure boundary in secondary recovery operations. The slope of a Cartesian plot of an LTR drawdown can be useful in determining reservoir limits. Care should be taken to ensure that LTR data are not included in the MTR analysis.

The ETR region is dominated by skin and storage effects close to the wellbore. Storage effects occur when surface and sandface flow rates are unequal such as an injection well that goes on vacuum when shut in or a gas well where gas compresses in the tubing due to afterflow following surface shut in. Conventional pressure transient analysis is dependent on a constant sandface flow rate during the test period and since zero is an easy constant flow rate to maintain, wellbore storage is a frequent problem. The length of the storage period is increased by skin damage since damage acts as an area of reduced permeability around the wellbore.

A logarithmic plot of the change in pressure, $p_i - p_{wf}$, versus the test time, dT, provides a practical means

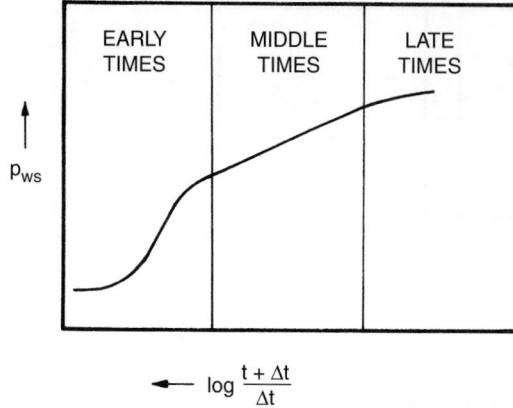

$$\longleftarrow \log \frac{t + \Delta t}{\Delta t}$$

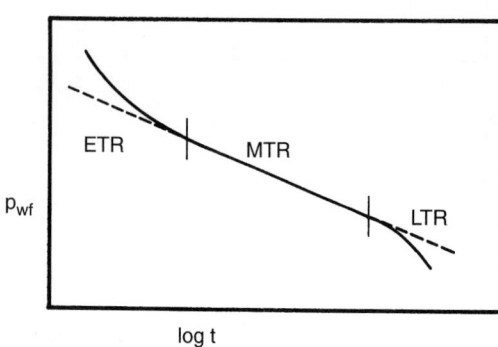

log t

Figure 5.3.6 *Early-, middle-, and late-time regions [1].*

of determining the end of the ETR and the beginning of the MTR semilogarithmic straight line. A logarithmic plot with a unit-slope line (a line with 45° slope) indicates storage effects. The proper MTR semilogarithmic straight line begins at 50 times the end of the unit slope, that is, wellbore storage effects cease at about one and a half log cycles after the disappearance of the unit-slope line.

A logarithmic plot that exhibits a half-slope line (a line with a slope of 26.6°) indicates a fractured wellbore. The proper straight line begins at 10 times the end of the half-slope line if the fractures are unpropped. Pressure drop at the start of the straight line is twice that at the end of the unit-slope line. Injection wells, acid jobs, or naturally fractured reservoirs are typical examples of the uniform flux, unpropped fractures.

A well that has short, propped (infinite capacity) hydraulically induced fractures will exhibit the proper straight line at 100 times the end of the half-slope line. The pressure drop will be about 5 times that at the end of the half-slope line.

Hydraulically stimulated wells in tight formations (< 0.01 md) with long (finite conductivity) fractures never exhibit the proper straight line during a conventional transient test time period. As a practical matter, all production from tight gas wells occurs during the ETR. Type curves or computer simulation are required to successfully analyze this type of ETR data. The ETR can range to hundreds of years in tight gas wells with finite-conductivity fractures.

Use of the logarithmic data plot to determine the start of the semilogarithmic straight line of course means that the

ETR data must be recorded. Pressure changes need to be monitored by the minute and bottomhole pressure at the time of shut-in must be precisely determined. Occasionally a great deal of emphasis is placed on the accuracy of the pressure-measuring equipment when the emphasis should be on the clock.

A problem in determining the initial pressure frequently arises when pressure buildup data from pumping wells are analyzed. A Cartesian plot of the early-time bottomhole pressure versus shut-in time should result in a straight line with the proper initial pressure at the intercept.

5.3.1.3 *Important Pressure Transient Analysis Equations*

permeability, $\quad k = 162.6 \dfrac{q \mu B}{mh}, md$ $\qquad$ [5.3.1]

radius of investigation, $\quad r_i = \left[\dfrac{kt}{948 \phi \mu C_t} \right]^{1/2}, ft$ $\qquad$ [5.3.2]

skin factor (buildup),

$$s = 1.151 \left[\frac{p_{1hr} - p_{wf}}{m} - \log \left(\frac{k}{\phi \mu C_t r_w^2} \right) + 3.23 \right] \qquad [5.3.3]$$

skin factor (drawdown),

$$s = 1.151 \left[\frac{p_i - p_{1hr}}{m} - \log \left(\frac{k}{\phi \mu C_t r_w^2} \right) + 3.23 \right] \qquad [5.3.4]$$

prerssure drop due to skin, $(\Delta p)_s = 0.869 \, ms$ $\qquad$ [5.3.5]

average drainage area pressure,

$$\bar{p} = p_{wf} + \frac{141.2 q \mu B \left(\ln \frac{r_e}{r_w} - \frac{3}{4} - s \right)}{kh} \qquad [5.3.6]$$

average pressure (steady-state),

$$\bar{p} = p_{wf} + \frac{141.2 q \mu B \left(\ln \frac{r_e}{r_w} - \frac{1}{2} - s \right)}{kh} \qquad [5.3.7]$$

flow efficiency,

$$FE = \frac{\bar{p} - p_{wf} - \Delta p_s}{\bar{p} - p_{wf}} \qquad [5.3.8]$$

For these equations, the following nomenclature is applicable:

Rate	q, B/D
Time	t, hr
formation volume factor	B, RB/STB
Viscosity	μ, cp
Thickness	h, ft
Porosity	ϕ, fraction
total compressibility	c_t, psi^{-1}
external drainage radius	r_e, ft
wellbore radius	r_w, ft
semilogarithmic MTR slope	m, psi/cycle
initial reservoir pressure	p_i, psi
flowing bottomhole pressure	p_{wf}, psi
shut-in botomhole pressure	p_{ws}, psi
pressure drop across skin region	Δp_s, psi
MTR semilog intercept	p_{1hr}, psi

In the equation for skin factor during the pressure buildup, P_{ws} is measured just before shutting in the well, and p_{1hr} is obtained from the straight-line portion (extrapolated if necessary) of the buildup curve one hour after shut-in (see Figure 5.3.5a). Similarly, the straight-line portion of the drawdown data must be extrapolated to one hour if the data do not fall on the semilog straight-line (see Figure 5.3.5b).

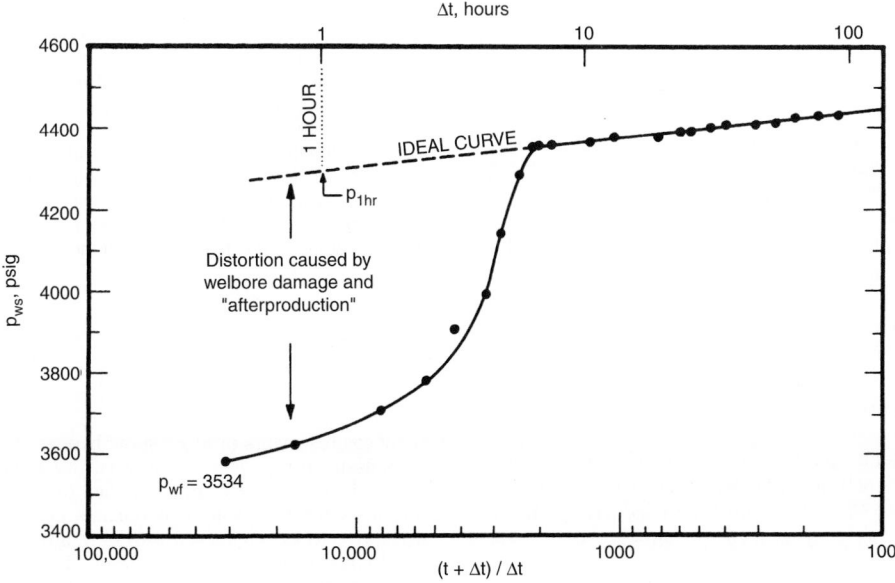

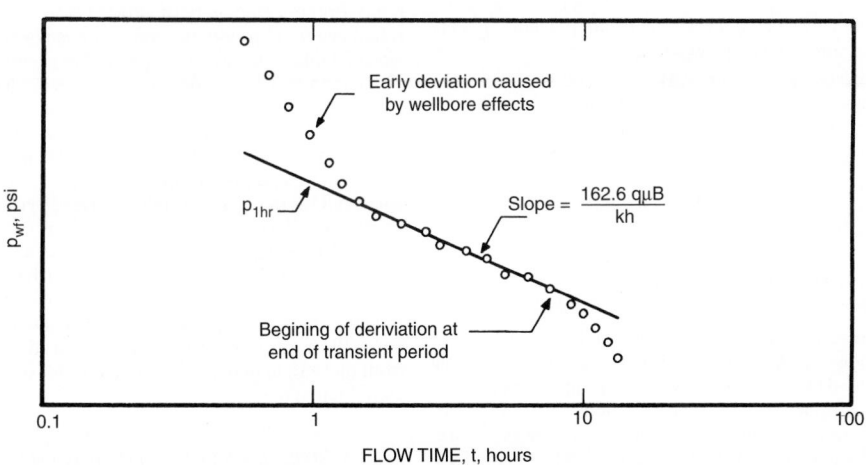

Figure 5.3.7   *Extrapolation of MTR straight line [13].*

5.3.1.4 Type-Curves

While at least one author [2.18] questions the uniqueness of pressure-transient type-curves and others [1.66,2.1] state that type-curves should only be used when conventional techniques cannot be used, curve-matching techniques have recently received more widespread use. In some cases where conventional analyses fail such as when wellbore storage distorts most or all of the data, type-curves may be the only means of interpretation of the pressure data. Type-curves developed by Ramey et al. [6–8] and McKinley [9,10] for pressure buildup and constant-rate drawdown tests and by Gringarten [11] for vertically fractured wells are discussed in a recent monograph [1]. Type-curves are used to estimate formation permeability, damage, and stimulation of the tested well as well as to identify the portion of the data that should be analyzed by conventional

techniques. Families of type-curves for various conditions are available from the SPE book order department in Dallas.

References

1. Lee, J., *Well Testing*, SPE, Dallas (1982).
2. Miller, C. C., Dyes, A. B., and Hutchinson, C. A., Jr., "Estimation of Permeability and Reservoir Pressure From Bottom-Hole Pressure Build-Up Characteristics," *Trans.*, AIME, Vol. 189 (1950), pp. 91–104.
3. Horner, D. R., "Pressure Build-Up in Wells," *Proc.*, Third World Pet. Conf., The Hague (1951) Sec. II, pp. 503–523. Also *Reprint Series, No. 9-Pressure Analysis Methods*, SPE, Dallas (1967), pp. 25–43.

4. van Everdingen, A. F., "The Skin Effect and Its Influence on the Productive Capacity of a Well," *Trans.*, AIME, Vol. 198 (1953), pp. 171–176. Also *Reprint Series, No. 9-Pressure Analysis Methods*, SPE, Dallas (1967), pp. 45–50.

5. Hurst, W., "Establishment of the Skin Effect and Its Impediment to Fluid Flow Into a Well Bore," *Pet. Eng.* (Oct. 1953), B-6 through B-16.

6. Ramey, H. J., Jr., "Short-Time Well Test Data Interpretation in the Presence of Skin Effect and Wellbore Storage," *J. Pet. Tech.* (Jan. 1970), pp. 97–104; *Trans.*, AIME, Vol. 249.

7. Agarwal, R. G., Al-Hussainy, R., and Ramey, H. J., Jr., "An Investigation of Wellbore Storage and Skin Effect in Unsteady Liquid Flow: I. Analytical Treatment," *Soc. Pet. Eng. J.* (Sept. 1970), pp. 279–290; *Trans.*, AIME, Vol. 249.

8. Wattenbarger, R. A., and Ramey, H. J., Jr., "An Investigation of Wellbore Storage and Skin Effect in Unsteady Liquid Flow: II. Finite Difference Treatment," *Soc. Pet. Eng. J.* (Sept. 1970), pp. 291–297; *Trans.*, AIME, Vol. 249.

9. McKinley, R. M., "Wellbore Transmissibility From Afterflow-Dominated Pressure Buildup Data," *J. Pet. Tech.* (July 1971), pp. 863–872; *Trans.*, AIME, Vol. 251.

10. McKinley, R. M., "Estimating Flow Efficiency From Afterflow-Distorted Pressure Buildup Data," *J. Pet. Tech.* (June 1974), pp. 696–697.

11. Gringarten, A. C., Ramey, H. J., Jr., and Raghavan, R., "Unsteady-State Pressure Distributions Created by a Well With a Single Infinite-Conductivity Vertical Fracture," *Soc. Pet. Eng. J.* (Aug. 1974), pp. 347–360; *Trans.*, AIME, Vol. 257.

5.4 MECHANISMS & RECOVERY OF HYDROCARBONS BY NATURAL MEANS

5.4.1 Petroleum Reservoir Definitions [1.17]

Accumulations of oil and gas occur in underground traps that are formed by structural and/or stratigraphic features. A reservoir is the portion of the trap that contains the oil and/or gas in a hydraulically connected system. Many reservoirs are hydraulically connected to water-bearing rocks or aquifers that provide a source of natural energy to aid in hydrocarbon recovery. Oil and gas may be recovered by: fluid expansion, fluid displacement, gravitational drainage, and/or capillary expulsion. In the case of a reservoir with no aquifer (which is referred to as a volumetric reservoir), hydrocarbon recovery occurs primarily by fluid expansion, which, in the case of oil, may be aided by gravity drainage. If there is water influx or encroachment from the aquifer, recovery occurs mainly by the fluid displacement mechanism which may be aided by gravity drainage or capillary expulsion. In many instances, recovery of hydrocarbon occurs by more than one mechanism.

At initial conditions, hydrocarbon fluids in a reservoir may exist as a single phase or as two phases. The single phase may be a gas phase or a liquid phase in which all of the gas present is dissolved in the oil. When there are hydrocarbons vaporized in the gas phase which are recoverable as liquids at the surface, the reservoir is called gas-condensate, and the produced liquids are referred to as condensates or distillates. For two-phase accumulations, the vapor phase is termed the gas cap and the underlying liquid phase is called the oil zone. In the two-phase ease, recovery of hydrocarbons includes the free gas in the gas cap, gas evolving from the oil (dissolved gas), recoverable liquid from the gas cap, and crude oil from the oil zone. If an aquifer or region of high water saturation is present, a transition zone can exist in which the water saturation can vary as a function of vertical depth and formation permeability. Water that exists in the oil- or gas-bearing portion of the reservoir above the transition zone is called connate or interstitial water. All of these factors are important in the evaluation of the hydrocarbon reserves and recovery efficiency.

5.4.2 Natural Gas Reservoirs [1.17]

For reservoirs where the fluid at all pressures in the reservoir or on the surface is a single gaseous phase, estimates of reserves and recoveries are relatively simple. However, many gas reservoirs produce some hydrocarbon liquid or condensate. In the latter case, recovery calculations for the single-phase case can be modified to include the condensate if the reservoir fluid remains in a single phase at all pressures encountered. However, if the hydrocarbon liquid phase develops in the reservoir, additional methods are necessary to handle these retrograde, gas-condensate reservoirs.

5.4.3 Primary Recovery of Crude Oil

Initial crude oil production often takes place by the expansion of fluids which were trapped under pressure in the rock. The expanding fluids may be gas evolving from the oil, an expanding gas cap, a bottom- or edge-water drive, or a combination of these mechanisms. After the initial pressure in the reservoir falls to a low value, the oil no longer flows to the wellbore, and pumps are installed to lift the crude oil to the surface. This mode of oil production is referred to as primary production. Recovery of oil associated with natural reservoir energy varies with producing mechanisms that are broadly classified as: solution-gas or depletion drive, gas cap drive, natural water drive, gravity drainage, and compaction drive. In some reservoirs, production can be attributed mainly to one of the mechanisms; in other cases, production may result from more than one mechanism, and this is referred to as a combination drive.

5.4.3.1 Statistical Analysis of Primary Oil Recovery

Most of the producing mechanisms are sensitive to the rate of oil production; only the solution gas drive mechanism is truly rate-insensitive [2.1]. Primary recoveries are usually reported [2.1] to be less than 25% of the original oil in place by solution gas drive, 30% to 50% of OOIP for water drive, and can exceed 75% of OOIP for gravity drainage in thick reservoirs with high vertical permeabilities. For water drive reservoirs, primary recovery efficiency can be low if the initial water saturation is more than 50%, if permeability is low, or if the reservoir is oil-wet [2.1]. From a recent statistical analysis [1], primary recovery from carbonate reservoirs tends to be lower than for sandstones (see Table 5.4.1 for recoveries by different drive mechanisms). Since primary recoveries tend to be lower for solution gas drive, these reservoirs are usually better candidates for waterflooding and will represent the bulk of the prospective candidates for enhanced oil recovery.

In the United States, much of the primary production involves solution gas reservoirs. Thus, this mechanism will be emphasized in this chapter, but non-U.S. production may involve other mechanisms. The differences in recovery

Table 5.4.1 *Primary Recovery Efficiencies*

Production mechanism	Lithology	State	Average primary recovery efficiency (% OOIP)
Solution gas drive	Sandstones	California	22
		Louisiana	27
		Oklahoma	19
		Texas 7C, 8, 10	15
		Texas 1-7B, 9	31
		West Virginia	21
		Wyoming	25
Solution gas drive	Carbonates	All	18
Natural water drive	Sandstones	California	36
		Louisiana	60
		Texas	54
		Wyoming	36
Natural water drive	Carbonates	All	44

From Reference 1.

mechanisms are important if an engineer is to avoid misapplication of methods; this subject has been addressed in Reference 2.18.

5.4.3.2 Empirical Estimates of Primary Oil Recovery

Several attempts have been made to correlate primary oil recovery with reservoir parameters [1–4]. Based on field data [3] from water-drive reservoirs, a statistical study [4] yielded the following empirical relationship for primary oil recovery:

$$N_p = (0.2719 \log k + 0.25569 S_w + 0.1355 \log \mu_o$$

$$- 15,380\phi - 0.00035h + 0.11403)$$

$$\times \left[7,758 Ah\phi \frac{(1 - S_w)}{B_{oi}} \right] \quad [5.4.1]$$

where N_p is oil production in STB, k is permeability in md, S_w is fractional water saturation, μ_o is oil viscosity in cp, ϕ is fractional porosity, h is pay thickness in ft, A is a real extent in ft^2, and B_{oi} is the initial formation volume factor of oil in reservoir barrels per STB. Based on the first API study [2], correlations were developed for recoverable oil. For solution gas drive reservoirs, the recoverable oil (RO) in stock tank barrels per net acre-ft was:

$$RO = 3,244 \left[\frac{\phi(1 - S_w)}{B_{ob}} \right]^{1.1611} \left[\frac{k}{\mu_{ob}} \right]^{0.0979}$$

$$\times [S_w]^{0.3722} \left[\frac{p_b}{p_a} \right]^{0.174} \quad [5.4.2]$$

For water drive reservoirs, the correlation was:

$$RO = 4,259 \left[\frac{\phi(1 - S_w)}{B_{oi}} \right]^{1.0422} \left[\frac{k\mu_{wi}}{\mu_{oi}} \right]^{0.0770}$$

$$\times [S_w]^{-0.1903} \left[\frac{p_i}{p_a} \right]^{-0.2159} \quad [5.4.3]$$

In the second API study [1], analysis of 116 solution gas drive reservoirs gave the following equation:

$$RO = 6,533 \left[\frac{\phi(1 - S_w)}{B_{ob}} \right]^{1.312} \left[\frac{k}{\mu_{ob}} \right]^{0.0816}$$

$$\times [S_w]^{0.463} \left[\frac{p_b - p_a}{p_b} \right]^{0.249} \quad [5.4.4]$$

However, the second study concluded that none of the *equations developed in either study was statistically appropriate to provide a valid correlation.* Furthermore, no statistically valid correlation was found between oil recovery and definable reservoir parameters. The second study found that when reservoirs were separated by lithology, geographical province, and producing mechanism, the only reasonable correlations that could be developed were between recoverable oil and original oil in place. Even then, the correlations were of poor quality as indicated by Figure 5.4.1 which presents the best correlation for Texas sandstone natural-water-drive reservoirs. The average primary recovery for various groups of reservoirs at the average value of OOIP for each group is listed by production mechanism in Table 5.4.1 [1,5].

In view of the lack of suitable correlations, primary oil recovery for an individual reservoir must be estimated by one of three methods: (1) material balance equations in conjunction with equations for gas-oil ratio and fluid saturations, (2) volumetric equations if residual oil saturation and oil formation volume factor at abandonment are known or estimated, and (3) decline curve analysis, if production history is available. Each of these methods for estimating primary oil recovery and gas recovery, when appropriate, will be discussed in the following sections.

5.4.4 Primary Recovery Factors in Solution-Gas-Drive Reservoirs

Primary recovery from solution-gas-drive reservoirs depends on: type of geologic structure, reservoir pressure, gas solubility, fluid gravity, fluid viscosity, relative permeabilities, presence of connate water, rate of withdrwal, and pressure drawdown. From a statistical study [6,7] the primary recovery factors in Table 5.4.2 were obtained for different oil gravities and solution gas-oil ratios in sands sandstones, limestones, dolomite, and chert. Based on work of the same type in 135 reservoir systems, Wahl [8] presented a series of figures that can be used to estimate primary recovery. One of these figures, for a condition of a 2 cp reservoir oil and a 30% connate water saturation, is reproduced in Figure 5.4.2. To use these figures the following is required: oil viscosity at reservoir conditions, interstitial water saturation, bubble-point pressure, solution gas-oil ratio at the bubble-point pressure, and formation volume factor.

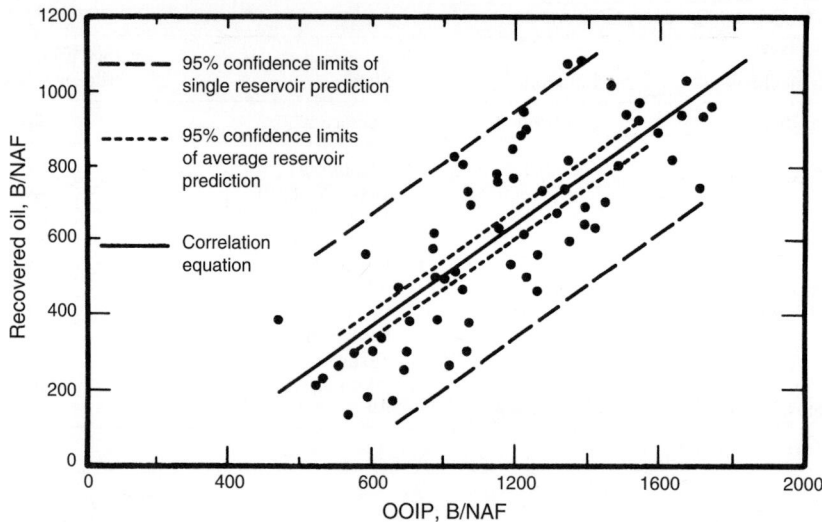

Figure 5.4.1   *Correlation of primary oil recovery for water-drive reservoirs [1].*

Table 5.4.2   *Primary Recovery in Percent of Oil in Place for Depletion-Type Reservoirs*

Oil solution GOR ft³/bbl	Oil gravity °API	Sand or sandstones			Limestone, Dolomite or Chert		
		maximum	average	minimum	maximum	average	minimum
60	15	12.8	8.6	2.6	28.0	4.0	0.6
	30	21.3	15.2	8.7	32.8	9.9	2.9
	50	34.2	24.8	16.9	39.0	18.6	8.0
200	15	13.3	8.8	3.3	27.5	4.5	0.9
	30	22.2	15.2	8.4	32.3	9.8	2.6
	50	37.4	26.4	17.6	39.8	19.3	7.4
600	15	18.0	11.3	6.0	26.6	6.9	1.9
	30	24.3	15.1	8.4	30.0	9.6	(2.5)
	50	35.6	23.0	13.8	36.1	15.1	(4.3)
1,000	15	—	—	—	—	—	—
	30	34.4	21.2	12.6	32.6	13.2	(4.0)
	50	33.7	20.2	11.6	31.8	12.0	(3.1)
2,000	15	—	—	—	—	—	—
	30	—	—	—	—	—	—
	50	40.7	24.8	15.6	32.8	(14.5)	(5.0)

From References 7.

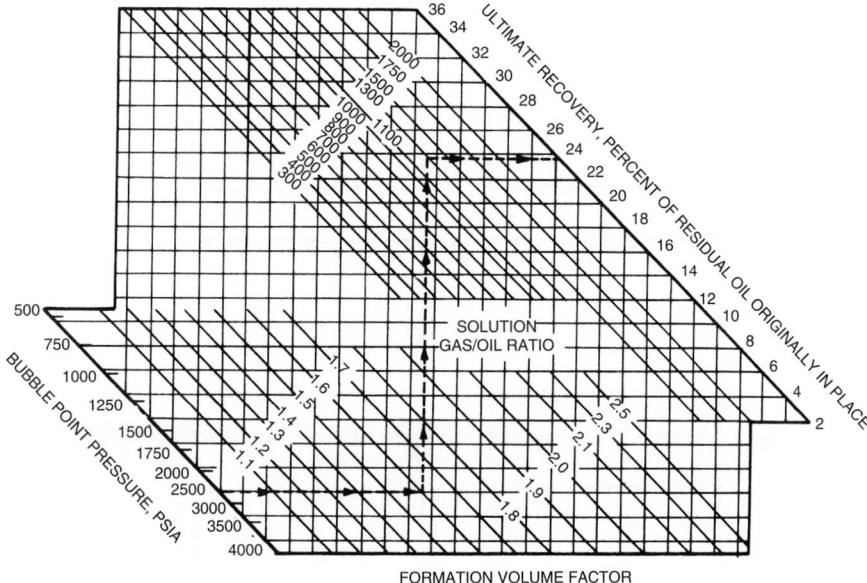

Figure 5.4.2 *Estimates of primary recovery for a solution-gas-drive reservoir [8].*

References

1. "Statistical Analysis of Crude Oil Recovery and Recovery Efficiency," API BUL D14, second edition, API Prod. Dept., Dallas (Apr. 1984).
2. "A Statistical Study of Recovery Efficiency," API BUL D14, API Prod. Dept., Dallas (Oct. 1967).
3. Craze, R. C., and Buckley, S. E., "A Factual Analysis of the Effect of Well Spacing on Oil Recovery," *Drill. & Prod. Prac.*, API (1945), pp. 144–159.
4. Guthrie, R. K., and Greenberger, M. H., "The Use of Multiple-Correlation Analyses for Interpreting Petroleum-engineering Data," *Drill. & Prod. Prac.*, API (1955), pp. 130–137.
5. Doscher, T. M., "Statistical Analysis Shows Crude-Oil Recovery," *Oil & Gas J.* (Oct. 29, 1984), pp. 61–63.
6. Arps, J. J., and Roberts, T. G., "The Effect of the Relative Permeability Ratio, the Oil Gravity, and the Solution Gas-Oil Ratio on the Primary Recovery from a Depletion Type Reservoir," *Trans.*, AIME, Vol. 204 (1955), pp. 120–127.
7. Arps, J. J., "Estimation of Primary Oil Reserves," *Trans.*, AIME, Vol. 207 (1956), pp. 182–191.
8. Wahl, W. L., Mullins, L. D., and Elfrink, E. B., "Estimation of Ultimate Recovery from Solution Gas-Drive Reservoirs," *Trans.*, AIME, Vol. 213 (1958), pp. 132–138.

5.5 MATERIAL BALANCE AND VOLUMETRIC ANALYSIS

Methods of estimating hydrocarbons in place by volumetric methods were discussed. These estimates can be confirmed and future reservoir performance can be predicted with the use of material balance equations. In the most elementary form the material balance equation states that the initial volume in place equals the sum of the volume remaining and the volume produced [1.19]. Material balance equations [1.19]

will be given for different reservoir situations. Nomenclature used in these equations is listed below [1.19,2.18,1].

B_o = oil formation volume factor = volume at reservoir conditions per volume at stock-tank conditions, dimensionless (reservoir barrel per stock-tank barrel, RB/STB)

B_g = gas formation volume factor = volume at reservoir conditions per volume at standard conditions, RB/scf (Note: if the gas formation volume factor is expressed in cu ft/scf, divided by 5.615 or multiply by 0.1781 to get RB/scf)

B_w = water formation volume factor = volume at reservoir conditions per volume at standard conditions, dimensionless (RB/STB)

B_t = $B_o + (R_{si} - R_s)B_g$ = composite oil or total oil formation volume factor = volume at reservoir conditions per volume at standard conditions

c_f = formation (rock) compressibility = pore volume per pore volume per psi

c_w = water compressibility = pore-volume per pore-volume per psi

G = total initial gas in place in reservoir, scf

G_e = cumulative gas influx (encroachment), scf

ΔG_e = gas influx (encroachment) during an interval, scf

G_i = cumulative gas injected, scf

G_p = cumulative gas produced, scf

k_{rg} = relative permeability to gas, dimensionless

k_{ro} = relative permeability to oil, dimensionless

m = GB_{gi}/NB_{oi} = ratio of initial gas-cap/gas-reservoir volume to initial reservoir oil volume

N = initial oil in place, STB

N_p = cumulative oil produced, STB

ΔN_p = oil produced during an interval, STB

p = reservoir pressure, psia

p_i = initial reservoir, pressure, psia

p_r = reduced pressure, dimensionless

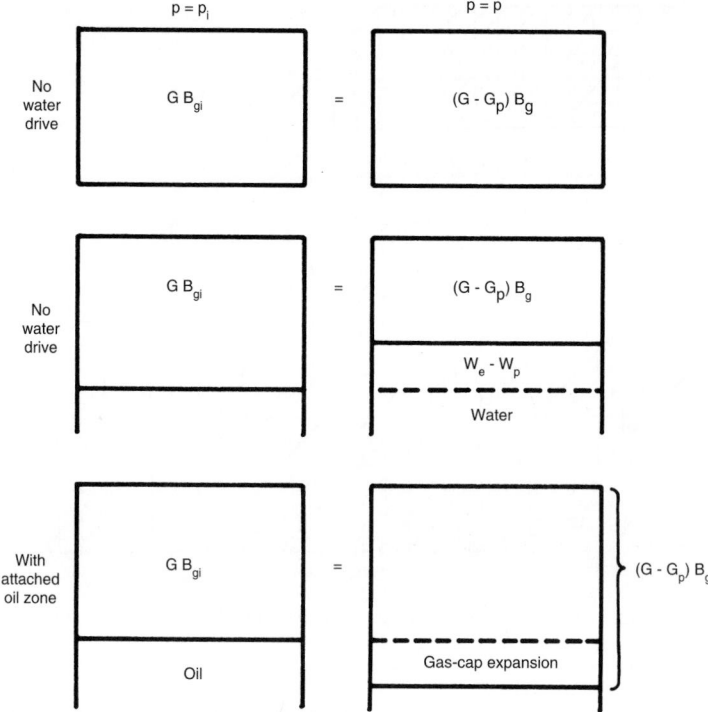

Figure 5.5.1 *Schematic of material balance equations for a dry-gas reservoir [2.18].*

R_s = solution-gas-oil ratio (gas solubility in oil), scf/STB

R_{si} = initial solution-gas-oil ratio, scf/STB

R_{sw} = gas solubility in water, scf/bbl at standard conditions

R_p = cumulative gas-oil ratio, scf/STB

S_o = oil saturation, fraction of pore space

S_g = gas saturation, fraction of pore space

S_{iw} = irreducible or connate water saturation, fraction of pore space

S_w = water saturation, fraction of pore space

S_{wi} = initial water saturation, fraction of pore space

W = initial water in place, reservoir bbl

W_e = cumulative water influx, bbl at standard conditions

W_p = cumulative water produced, bbl at standard conditions

W_i = cumulative water injected, bbl at standard conditions

μ_g = gas viscosity, cp

μ_o = oil viscosity, cp

z = gas deviation factor (compressibility factor), $z = pV/nRT$, dimensionless

i = subscript, initial or original conditions

1 = subscript for conditions at p_1

2 = subscript for conditions at p_2

5.5.1 Material Balance for Gas Reservoirs
5.5.1.1 Material Balance Equations

Reservoirs that contain only free gas are called gas reservoirs. These reservoirs contain a mixture of gaseous hydrocarbons, which may be dry, wet, or condensate gas. Gas reservoirs may be volumetric with no water influx.

For such reservoirs, the material balance is:

$$GB_{gi} = (G - G_p)B_g \qquad [5.5.1]$$

where all terms have been defined earlier in the notation. In the case of a gas reservoir with water encroachment and water production, the material balance equation is:

$$GB_{gi} = (G - G_p)B_g + (W_e - W_p) \qquad [5.5.2]$$

In either equation, the gas formation volume factor can be obtained, as a function of pressure and temperature, as outlined earlier. A schematic representation [2.18] of the material balance equations for dry-gas reservoirs is depicted in Figure 5.5.1.

5.5.1.2 Graphical Form of Material Balance (p/z Plots)

For volumetric gas reservoirs in which there is no water influx and negligible water production, the definition of gas formation volume factor (Equation 5.1.3) can be substituted into Equation 5.5.1, and the resulting equation can be rearranged to give [2.18]:

$$G_p = G - \frac{p}{z}\frac{Gz_i}{p_i} \qquad [5.5.3]$$

where all terms are as defined previously and the subscript i refers to initial conditions. This equation indicates that for a volumetric gas reservoir a plot of cumulative gas production (G_p) in standard cubic feet versus the ratio p/z is a straight line. Within limits of error for average reservoir pressure and cumulative production, this plot is linear as shown in Figure 5.5.2. The straight line can be extrapolated to zero pressure to find the initial gas in place, or can be extrapolated to predict the cumulative production at any future average reservoir pressure. A plot of pressure versus cumulative production is not a straight line because the produced gas is not a perfect

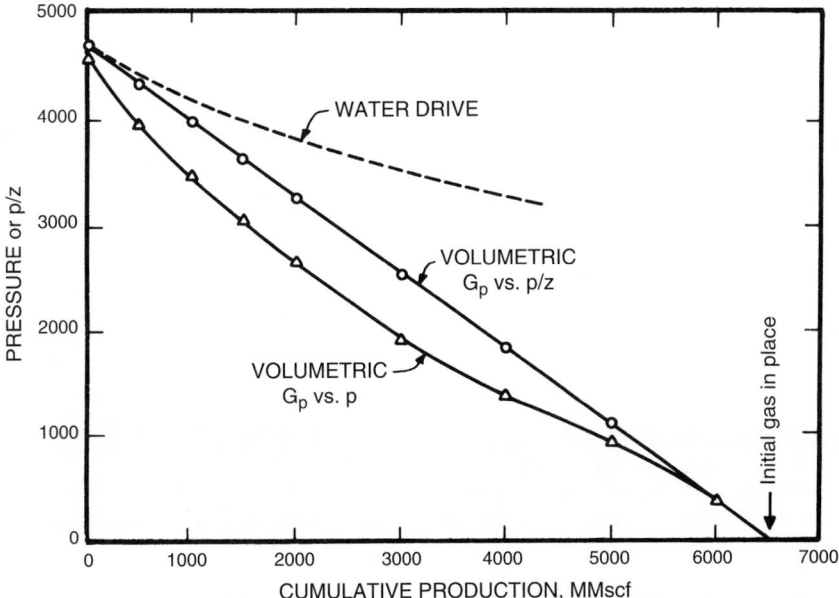

Figure 5.5.2 *A p/z plot [1.17].*

gas. Since the gas deviation factor, z, is a function of pressure, the ratio of p/z can be obtained conveniently from plots of p/z versus p, p_r/z versus z for different reduced temperature, or from computer programs that have gas deviation factors in storage [2.18]. Other graphical interpretations have been suggested [2].

A plot of p/z versus G_p may not be a straight line for several reasons: an unexpected water drive may exist, average reservoir pressure may be inaccurate, or the reservoir pore volume may be changing unpredictably as a result of abnormally high reservoir pressures [2.18]. A water drive reduces relative permeability to gas and increases pressure at abandonment [2.1]. For reservoirs at moderate to high pressure in the absence of a water drive, the recovery efficiency under pressure depletion may range from about 80% to more than 90%; a water drive can reduce recovery to about 60%. To maximize efficiency, water-bearing zones should not be perforated if the water is movable, and production should be at a high rate since water entry is time dependent [2.1].

5.5.2 Material Balance Equations in Oil or Combination Reservoirs

When discovered, a reservoir may contain oil, gas and water that can be intermingled or segregated into zones. As described earlier, recovery may be caused by solution gas drive, water drive, gas cap drive, or a combination of these mechanisms. A general material balance equation should be capable of handling any type of fluid distribution and any drive mechanism.

From the compressibilities given in the first section, water and formation compressibilities are less significant where there is appreciable gas saturation such as in gas reservoirs, gas cap reservoirs, and in undersaturated reservoirs below the bubble point. Because of this and because of the complications they would introduce in already complex equations, water and formation compressibilities are generally neglected, except in undersaturated reservoirs

producing above the bubble point. Gas in solution in the formation is small and also generally neglected. One general material balance equation, the Schilthuis equation, is a volumetric balance stating that the sum of the volume changes in oil, gas, and water must be zero because the reservoir volume is constant.

oil zone	gas cap	water	cumulative oil
[expansion] +	[expansion] +	[influx] =	[zone production]

$$N(B_t - B_{ti}) + \frac{NmB_{ti}(B_g - B_{gi})}{B_{gi}} + W_e = N_p B_t$$

$$+ \left[\begin{array}{c} \text{cumulative gas cap} \\ \text{gas production} \end{array} \right] + \left[\begin{array}{c} \text{cumulative water} \\ \text{production} \end{array} \right]$$

$$+ N_p(R_p - R_{si})B_g + B_w W_p \qquad [5.5.4]$$

All symbols have been defined earlier. Rearranging terms, Equation 5.5.4 can be written:

$$N\left[B_t - B_{ti} + \frac{mB_{ti}(B_g - B_{gi})}{B_{gi}} \right]$$

$$= N_p[B_t + (R_p - R_{si})B_g] - W_e + B_w W_p \qquad [5.5.5]$$

Material balance equations are often expressed in terms of the initial oil in place, N:

$$N = \frac{N_p[B_t + (R_p - R_{si})B_g] - (W_e - B_w W_p)}{B_t - B_{ti} + \frac{mB_{ti}}{B_{gi}}(B_g - B_{gi})} \qquad [5.5.6]$$

This equation applies to a reservoir under any drive mechanism or to fields under the simultaneous influence of dissolved gas drive, water drive, or gas cap drive.

If there is no water drive, $W_e = 0$; therefore:

$$N = \frac{N_p[B_t + (R_p - R_{si})B_g] + B_w W_p}{B_t - B_{ti} + \frac{mB_{ti}}{B_{gi}}(B_g - B_{gi})} \qquad [5.5.7]$$

When there is no original free gas, m = 0; therefore:

$$N = \frac{N_p[B_t + (R_p - R_{si})B_g] - (W_e - B_w W_p)}{B_t - B_{ti}} \qquad [5.5.8]$$

When there is neither an original gas cap nor any water drive, m = 0 and $W_p = 0$; therefore:

$$N = \frac{N_p[B_t + (R_p - R_{si})]B_g}{B_t - B_{ti}} \qquad [5.5.9]$$

Although connate water and formation compressibilities are quite small, relative to the compressibility of reservoir fluids above their bubble points they are significant and account for an appreciable fraction of the production above the bubble point. In cases when compressibilities can be important, Equation 5.5.6 can be written as:

$$N = \frac{N_p[B_t + (R_p - R_{si})B_g] - (W_e - B_w W_p)}{B_t - B_{ti} + (c_f + c_w S_{iw})\Delta p B_{ti}/(1 - S_{iw}) + m B_{ti}(B_g - B_{gi})/B_{gi}} \qquad [5.5.10]$$

Material balance equations for various drive mechanisms and different initial conditions are summarized in Table 5.5.1. Note that in this table B_g is expressed in ft³/scf, and the conversion factor, 0.1781, is the reciprocal of 5.615 ft³ per barrel.

5.5.3 Generalized Material Balance Equation

Many forms of the so-called general material balance equation have appeared in the literature. In the most general form, the material balance equation, expressed in terms of the initial oil in place, is:

$$N = \frac{N_p B_o + B_g(G_p - N_p R_s) - G(B_g - B_{gi}) - (W_e - W_p)}{B_o - B_{oi} + (R_{si} - R_s)B_g + (c_f + c_w S_{iw})\Delta p B_{oi}/(1 - S_{iw})} \qquad [5.5.11]$$

This equation can be obtained by substituting in Equation 5.5.10 the following terms:

$B_i = B_o + (R_{si} - R_s)B_g$; $B_{ti} = B_{oi}$;   and $G_p = N_p R_p$.

A schematic representation [1.19] of the possible changes in fluid distribution is presented in Figure 5.5.3. In order to not omit any of the significant reservoir energies, Slider [2.18] suggests beginning all material balance applications with Equation 5.5.11. With the use of Equation 5.5.11, it is not necessary to list separate equations for the various drive mechanisms and conditions since this one equation can be reduced to any of the individual cases. For reservoirs above the bubble point, some of the terms in Equation 5.5.11 become zero or else cancel out: G = 0 since there is no free gas, G_p, and $N_p R_s$ cancel out since the gas production term is equal to the total gas produced, $R_{si} - R_s$ is zero since the gas in solution at any pressure is equal to the gas originally in solution.

For a reservoir with no initial free gas saturation or no initial gas cap, G = 0. If there is no water encroachment, W_e = 0; however, the water production term should remain, even if there is no water drive, because connate water may be produced when the reservoir pressure declines.

When a reservoir contains free gas, the pore volume expansion or compressibilities of the formation and water are insignificant compared to the free gas terms. Since the gas compressibility is about 100 times the compressibility of the water and formation, a gas saturation of only 1% may provide as much energy as the water and formation compressibility terms. Thus, when the gas saturation is substantial the change in pore volume is insignificant.

5.5.4 Material Balance for Solution-Gas Drive Reservoirs

A schematic representation of material balance equations for solution-gas reservoirs, when the change in pore volume is

Table 5.5.1 *Material Balance Equations*

Gas reservoir with active water drive

$$G = \frac{G_p B_g - 5.615(W_e - W_p)}{B_g - B_{gi}}$$

Gas reservoir; no active water drive ($W_e = 0$)

$$G = \frac{G_p B_g + 5.615 W_p}{B_g - B_{gi}}$$

Oil reservoir with gas cap and active water drive

$$N = \frac{N_p[B_t + 0.1781 B_g(R_p - R_{si})] - (W_e - W_p)}{m B_{oi}\left(\dfrac{B_g}{B_{gi}} - 1\right) + (B_t - B_{oi})}$$

Oil reservoir with gas cap; no active water drive ($W_e = 0$)

$$N = \frac{N_p[B_t + 0.1781 B_g(R_p - R_{si})] + W_p}{m B_{oi}\left(\dfrac{B_g}{B_{gi}} - 1\right) + (B_t - B_{oi})}$$

Initially undersaturated oil reservoir with active water drive (m = 0)

1. Above bubble point

$$N = \frac{\left[N_p(1 + \Delta p c_o) - \dfrac{W_e - W_p}{B_{oi}}\right](1 - S_w)}{\Delta p[c_o + c_t - S_w(c_o - c_w)]}$$

2. Below bubble point

$$N = \frac{N_p[B_t + 0.1781 B_g(R_p - R_{si})] - (W_e - W_p)}{B_t - B_{oi}}$$

Initially undersaturated oil reservoir; no active water drive (m = 0, $W_e = 0$)

1. Above bubble point

$$N = \frac{\left[N_p(1 + \Delta p c_o) - \dfrac{W_p}{B_{oi}}\right](1 - S_w)}{\Delta p[c_o + c_t - S_w(c_o - c_w)]}$$

2. Below bubble point

$$N = \frac{N_p[B_t + 0.1781 B_g(R_p - R_{si})] + W_p}{B_t - B_{oi}}$$

From Reference 2.1.
Note: In this table, B_g is expressed in ft³/SCF.

negligible, is shown in Figure 5.5.4. When these reservoirs are producing above the bubble point or saturation pressure, no gas is liberated and production occurs by expansion of liquids in the reservoir. When reservoir pressure drops below the bubble point, gas is liberated in the reservoir and will be produced with the oil.

5.5.4.1 Liquid Expansion

For some very large reservoirs (often with limited permeability), production may occur for extended periods by expansion of liquids in the reservoir. If production is caused only by liquid expansion, the material balance equation obtained from Figure 5.5.4 for pressures above the bubble

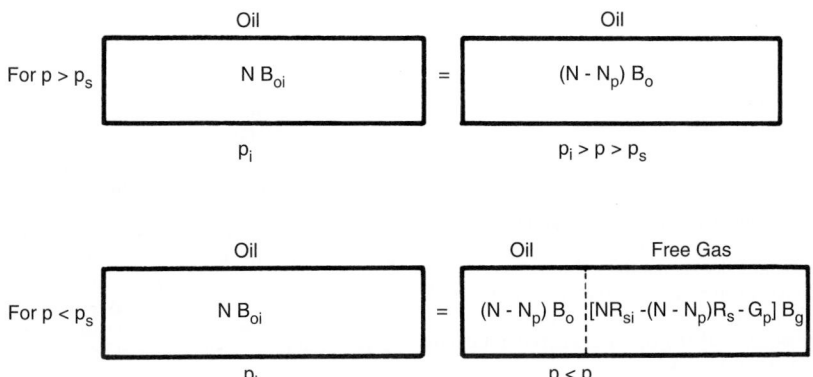

Figure 5.5.4   *Schematic of material balance equations for a solution-gas drive reservoir [2.18].*

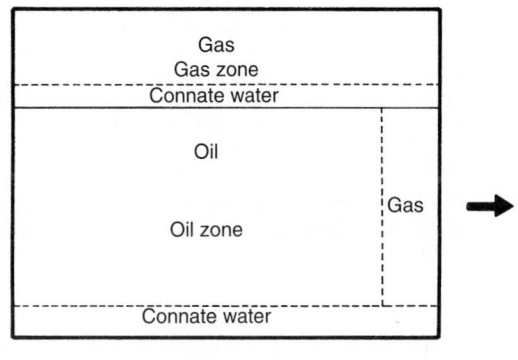

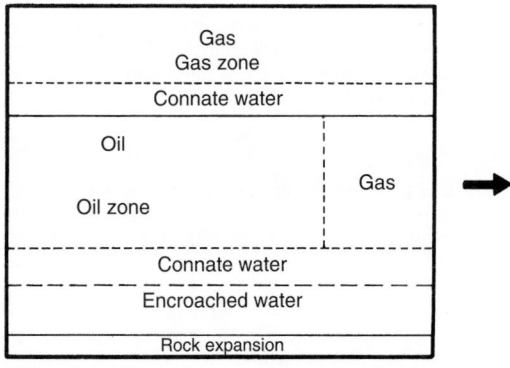

Figure 5.5.3   *Schematic representation of fluid distributions [1.19].*

point is:

$$N = \frac{(N - N_p)B_o}{B_{oi}}$$ [5.5.12]

5.5.4.2 Gas Liberation
When reservoir pressure declines below the bubble-point pressure, the original gas in solution has either been produced as G_p, is still in solution in the oil, $(N - N_p)R_s$, or exists as free gas. For this condition as shown in Figure 5.5.4, the material balance is:

$$N = \frac{N_p B_o + B_g(G_p - N_p R_s)}{B_o - B_{oi} + (R_{si} - R_s)B_g}$$ [5.5.13]

5.5.5 Predicting Primary Recovery in Solution-Gas Drive Reservoirs

Several methods for predicting performance of solution-gas behavior have appeared in the literature. These methods relate pressure decline to gas-oil ratio and oil recovery. Because neither water influx nor gravity segregation is considered, time is not a factor with solution-gas reservoirs, and time must be inferred from the oil in place and production rate [1.17]. The following assumptions are generally made: uniformity of the reservoir at all times regarding porosity, fluid saturations, and relative permeabilities; uniform pressure throughout the reservoir in both the gas and oil zones (which means the gas and oil volume factors, the gas and oil viscosities, and the solution gas will be the same throughout the reservoir); negligible gravity segregation forces; equilibrium at all times between the gas and the oil phases; a gas liberation mechanism which is the same as that used to determine the fluid properties, and no water encroachment and negligible water production.

5.5.5.1 The Schilthuis Method [3]
For solution-gas drive reservoirs where the reservoir pressure is about equal to the saturation pressure and for gas cap drive reservoirs, Equation 5.5.11 can be written [2.18]:

$$N = \frac{N_p B_o + (G_p - N_p R_s)B_g - G(B_g - B_{gi})}{B_o - B_{oi} + (R_{si} - R_s)B_g}$$ [5.5.14]

If this equation is rearranged and solved for cumulative oil produced:

$$N_p = \frac{N[B_o - B_{oi} + (R_{si} - R_s)B_g] + G(B_g - B_{gi}) - B_g G_p}{B_o - B_g R_s}$$

[5.5.15]

In order to predict the cumulative oil production at any stage of depletion, the original oil and gas in place and the initial reservoir pressure must be known. The original pressure establishes values for B_{oi}, B_{gi}, and R_{si}, and the pressure at the given level of depletion establishes values of B_o, B_g, and R_s. Cumulative gas production can be obtained from:

$$G_p = \Sigma R_s \Delta N_p$$ [5.5.16]

A procedure for using this method is [2.18]:

1. At a set value of reservoir pressure, assume the cumulative amount of gas produced, G_p.
2. Calculate N_p by Equation 5.5.15.
3. Calculate S_o using Equation 5.5.31.
4. Determine the relative permeability ratio based on the liquid saturation from data such as that in Figure 5.1.55.
5. Calculate G_p using Equations 5.5.16 and 5.5.17.
6. Compare the calculated G_p with the assumed G_p from Step 1.
7. If the assumed and calculated values do not agree to a satisfactory degree, repeat the calculations from Step 1 for another value of G_p.
8. If the assumed and calculated G_p values agree, return to Step 1 and set a new pressure.

5.5.5.2 The Tarner Method

This [4] is a trial-and-error procedure based on the simultaneous solution of the material balance equation and the instantaneous gas-oil ratio equation. For a pressure drop from p_1 to p_2, the procedure involves a stepwise calculation of cumulative oil produced $(N_p)_2$ and of cumulative gas produced $(G_p)_2$. Several variations in the procedure and equations are possible; the straightforward procedure outlined by Timmerman [2.1] is reproduced here.

1. During the pressure drop from p_1 to p_2, assume that the cumulative oil production increases from $(N_p)_1$ to $(N_p)_2$. At the bubble point pressure, N_p should be set equal to zero.
2. By means of the material-balance equation for $W_p = 0$, compute the cumulative gas produced $(G_p)_2$ at pressure p_2:

$$(G_p)_2 = (N_p)_2(R_p)_2 = N\left[(R_{si} - R_s) - \frac{B_{oi} - B_o}{B_g}\right]$$

$$- (N_p)_2\left(\frac{B_o}{B_g} - R_s\right) \qquad [5.5.17]$$

3. Compute the fractional total liquid saturation $(S_L)_2$ at pressure p_2:

$$(S_L)_2 = S_w + (1 - S_w)\frac{B_o}{B_{oi}}\left[1 - \frac{(N_p)_2}{N}\right] \qquad [5.5.18]$$

4. Determine the k_{rg}/k_{ro} ratio corresponding to the total liquid saturation $(S_L)_2$ and compute the instantaneous gas-oil ratio at p_2:

$$R_2 = R_s + \frac{B_o}{B_g}\frac{\mu_o}{\mu_g}\frac{k_{rg}}{k_{ro}} \qquad [5.5.19]$$

5. Compute the cumulative gas produced at pressure p_2:

$$(G_p)_2 = (G_p)_1 + \frac{R_1 + R_2}{2}[(N_p)_2 - (N_p)_1] \qquad [5.5.20]$$

where R_1 is the instantaneous gas-oil ratio computed at pressure p_1.

Usually three judicious guesses are made for the value $(N_p)_2$ and the corresponding values of $(G_p)_2$ computed by both step 2 and step 5. When the values for $(G_p)_2$ are plotted against the assumed values for $(N_p)_2$, the intersection of the curve indicates the cumulative gas and oil production that will satisfy both equations. In actual application the method is usually simplified further by equating the incremental gas production $(G_p)_2 - (G_p)_1$ rather than $(G_p)_2$ itself. This equality signifies that at each pressure step the cumulative gas, as determined by the volumetric balance is the same as the quantity of gas produced from the reservoir, as controlled by the relative permeability ratio of the rock, which in turn depends on the total liquid saturation.

5.5.5.3 The Muskat Method

The Schilthuis and Tarner forms of material balance have been expressed in integral form. An approach presented by Muskat [5] expresses the material balance in terms of finite pressure differences in small increments. The changes in variables that affect production are evaluated at any stage of depletion or pressure. The assumption is made that values of the variables will hold for a small drop in pressure, and the incremental recovery can be calculated for the small pressure drop. The variables are recalculated at the lower pressure, and the process is continued to any desired abandonment pressure. If the PVT data and the gas-oil relative permeabilities are known at any liquid saturation, the unit recovery by pressure depletion can be computed from a differential form of the material balance equation:

$$\frac{dS_o}{dp} = \frac{\dfrac{S_o B_g}{B_o}\dfrac{dR_s}{dp} + \dfrac{S_o k_{rg}\mu_o}{B_o k_{ro}\mu_g}\dfrac{dB_o}{dp} + (1 - S_o - S_w)B_g\dfrac{d(1/B_g)}{dp}}{1 + \dfrac{k_{rg}\mu_o}{k_{ro}\mu_g}}$$

$$[5.5.21]$$

From the change in saturation at any pressure, the reservoir saturation at that time can be related to the change in oil production and the instantaneous gas-oil ratio. Calculations can be facilitated if the terms in the numerator that are functions of pressure only (B_g, B_o, B_s) are determined for various depletion pressures. Pressure increments of 10 psi or less may be necessary for acceptable accuracy [2.1].

5.5.6 Predicting Primary Recovery in Water-Drive Reservoirs

In the prediction of performance caused by water influx, predictions of water encroachment are made independent of material balance. The extent of water encroachment depends on the characteristics of the aquifer and is a function of the pressure history and time [6]. While several methods are available to predict water drive performance, some [2.18] feel that the theory of unsteady-state compressible flow should be used, and that water encroachment generally follows the constant pressure solution to the radial diffusivity equation. Solutions to the radial diffusivity equation have been provided by van Everdingen and Hurst [7] in terms of dimensionless time, t_D, and a dimensionless fluid flow function, Q_{tD}, which is determined at t_D:

$$t_D = \frac{6.323 \times 10^{-3}kt}{\phi\mu cr^2} \qquad [5.5.22]$$

where k is permeability in md, t is time in days, ϕ is the fractional porosity, c is the compressibility in psi^{-1}, and r is the reservoir radius in ft. Values of Q_{tD} are given in tabular form as a function of t_D in the paper by van Everdingen and Hurst [7] and have been reproduced in several texts [1.17, 2.18]. Because of the length of the tables, they will not be reproduced in this section. The water encroachment, W_e in barrels, can be estimated from:

$$W_e = B\Sigma(dp)(Q_{tD}) \qquad [5.5.23]$$

where dp is the pressure drop is psi, and the constant B is:

$$B = 1.12\phi hcr^2\frac{\theta}{360} \qquad [5.5.24]$$

where h is the reservoir thickness in ft, θ is the angle subtended by the reservoir circumference (θ is 360° for a circular reservoir and θ is 180° for a semicircular reservoir against a fault), and the other terms are as defined above [1.17]. From the slope of cumulative water influx at various times versus the summation term at those times, the aquifer constant, B, can be obtained, and the cumulative water influx for any pressure history can be estimated from

Equation 5.5.23. Plots of Q_{tD} versus t_D for various dimensionless reservoir sizes are also available [1.17,7] and extensions of these data are available as well [2.18].

5.5.7 Volumetric Calculations for Recovery of Gas and Oil

The volumetric equations for original oil and gas in place were given earlier. In this section, volumetric equations will be given for the recovery of gas and oil reservoirs under several common instances.

5.5.7.1 Recovery of Gas

The volume of gas recovered, G_p in scf, from a dry-gas reservoir is:

$$G_p = 7,758Ah\phi(1-S_w)\left(\frac{1}{B_{gi}} - \frac{1}{B_g}\right)$$ [5.5.25]

where 7,758 is the number of barrels per acre-ft, A is the areal extent in acres, h is the reservoir thickness in ft, ϕ is the fractional porosity, S_w is the fractional water saturation, B_{gi} is the initial gas formation volume factor in reservoir barrels per scf, and B_g is the gas formation volume factor in RB/scf at the abandonment pressure. (Note: if the formation volume factors are expressed in ft^3/scf, 7,758 should be replaced with 43,560 ft^3 per acre-ft.

If Equation 5.5.1 is rearranged in terms of the initial gas in place:

$$G = \frac{G_p B_g}{(B_g - B_{gi})}$$ [5.5.26]

The recovery efficiency, E_R, which is the volume of gas recovered divided by the volume of gas initially in place, can be found from:

$$E_R = \frac{B_{gi} - B_g}{B_{gi}}$$ [5.5.27]

For volumetric gas reservoirs (which assumes no change in S_w) recoveries may range from 80% to 90% of the initial gas in place.

For gas reservoirs under water drive, recovery efficiency is:

$$E_R = \frac{(1-S_{wi})B_{gi} - (S_{gr}B_g)}{(1-S_{wi})B_{gi}}$$ [5.5.28]

where S_{gr} is the residual gas saturation and the other terms are as defined above.

5.5.7.2 Recovery of Oil
Solution-Gas or Depletion Drive
Oil recovery by depletion or solution-gas drive is:

$$N_p = 7,758Ah\phi\left[\frac{(1-S_w)}{B_{oi}} - \frac{(1-S_w-S_{gr})}{B_o}\right]$$ [5.5.29]

where B_{oi} is the initial oil formation volume factor and B_o is the oil formation volume factor at abandonment, and the other terms are as previously defined. The ultimate free gas saturation is often estimated from the old data of Arps [4.7] and Craze and Buckley [4.3] in which the average S_g was 30% for a 2.2 cp oil with 400 ft^3/bbl of solution gas. A general rule suggests that for each doubling of solution gas, S_g should be increased 3%; and for each doubling of oil viscosity, S_g should be decreased by 3% [1.14]. As a first approximation, S_{gr} can be taken to be about 0.25 [8].

For a volumetric, undersaturated reservoir,

$$N_p = \frac{N(B_o - B_{oi})}{B_o}$$ [5.5.30]

where N_p is the oil produced in stock tank barrels, N is the initial oil in place in STB and the other terms are as already given. In this case, the oil saturation after any stage of primary production, S_o, can be obtained from:

$$S_o = \frac{(N-N_p)B_o(1-S_w)}{NB_{oi}}$$ [5.5.31]

If calculations are made at the abandonment of primary production, S_o will represent the oil saturation after primary. The gas saturation at any time, S_g, can be found from:

$$S_g = 1 - S_o - S_w$$ [5.5.32]

For a solution-gas drive reservoir with water encroachment [1.73]:

$$N_p = N\left[\frac{B_o - B_{oi}}{B_o} + \frac{W_e}{B_o}\right]$$ [5.5.33]

where W_e is the cumulative water influx in barrels and all other terms are consistent with the prior definitions.

Water-Drive Reservoir
Recovery from a water-drive reservoir (which assumes no appreciable decline in pressure) can be calculated from [1.14]:

$$N_p = 7,758(Ah\phi)\left[\frac{(1-S_w)}{B_{oi}} - \frac{S_{or}}{B_o}\right]$$ [5.5.34]

where S_{or} is the residual oil saturation as a fraction of pore volume. Residual oil saturation can be obtained from cores taken with water base mud or from logs as described earlier. Methods of estimating residual oil saturation after water injection are discussed later. In the absence of data, residual oil saturations are sometimes obtained from the empirical data of Arps [4.7] and Craze and Buckley [4.3] which are given in Table 5.5.2. However, the caution given earlier is repeated here: from a more recent study [4.1] *no statistically valid correlation was found between oil recovery and definable reservoir parameters.*

The recovery efficiency is the case of water influx is given by:

$$E_R = \frac{(1-S_w-S_{or})}{(1-S_w)}$$ [5.5.35]

Table 5.5.2 *Approximation of Residual Oil Saturation*

Reservoir oil viscosity, cp	Residual oil saturation, % PV
0.2	30
0.5	32
1.0	34.5
2.0	37
5.0	40.5
10.0	43.5
20.0	46.5

Average reservoir permeability, md	Deviation of residual oil saturation from viscosity trend, % PV
50	+12
100	+9
200	+6
500	+2
1,000	−1
2,000	−4.5
5,000	−8.5

From Reference 2.7.

Gravity Drainage

For segregation drive reservoirs, Equation 5.5.35 can be used. Residual oil saturations for gravity drainage reservoirs tend to be low (possibly on the order of 0.10) [1.14].

Gas-Cap Drive

For a gas-cap drive reservoir with no water influx, the oil saturation can be estimated from [1.73]:

$$S_o = \left(\frac{N-N_p}{NB_{oi}}\right)(1-S_w)\left[\frac{1}{1-m(B_g-B_{gi})/B_{gi}}\right] \qquad [5.5.36]$$

where m is the ratio of original gas zone to original oil zone in the reservoir and all other quantities have been defined.

References

1. Mayer, E. H., et al., "SPE Symbols Standard," *J. Pet. Tech.* (Dec. 1984), pp. 2278–2332.
2. Ramey, H. J., Jr., "Graphical Interpretations for Gas Material Balance," *J. Pet. Tech.* (*July* 1970), pp. 837–838.
3. Schilthuis, R. J., "Active Oil and Reservoir Energy," *Trans.*, AIME (1936), pp. 35–52.
4. Tarner, J., "How Different Size Gas Caps and Pressure Maintenance Programs Affect Amount of Recoverable Oil," *Oil Weekly* (June 12, 1944), pp. 32–44.
5. Muskat, M., "The Production Histories of Oil Producing Gas-Drive Reservoirs," *Appl. Phys.*, Vol. 16 (March 1945), pp. 147–159.
6. Hurst, W., "Water Influx Into a Reservoir and Its Application to the Equation of Volumetric Balance," *Trans.*, AIME (1943), pp. 57–72.
7. van Everdingen, A. F., and Hurst, W., "The Application of the Laplace Transformation to Flow Problems in Reservoirs," *Trans.*, AIME (Dec. 1949), pp. 305–324.
8. Smith, C. R., Mechanics of Secondary Oil Recovery, Robert E. Krieger Publishing Co., New York (1975).

5.6 DECLINE CURVE ANALYSIS

The conventional analysis of production decline curves for oil or gas production consists of plotting the log of flow rate versus time on semilog paper. In cases for a decline in rate of production, the data are extrapolated into the future to provide an estimate of expected production and reserves.

The empirical relationships for the analysis of production decline curves were first proposed by Arps [1] in which a decline rate, a, was defined as the fractional change in the flow rate, q, with time, t:

$$a = \frac{-dq/dt}{q} \qquad [5.6.1]$$

If time is in days, flow rate in this equation is expressed in terms of stock tank barrels per day in the case of oil and scf per day for gas. Other consistent units of flow rate and time can be used. As shown in Figure 5.6.1, three types of decline can occur: a constant percentage or exponential decline, a hyperbolic decline, and a harmonic decline [4.7]. For the semilog plot, the exponential decline is a straight line whereas the slopes of the hyperbolic and harmonic decline curves decrease with time. For the exponential decline, the drop in production per unit time is a constant fraction of the production rate. For a hyperbolic decline, the decrease in production per unit time as a fraction of the production rate is proportional to a fractional power (between 0 and 1) of the production rate. For a harmonic decline, the decrease in production per unit time as a fraction of the production rate is directly proportional to the rate. Slider [2] presented an

equation for the hyperbolic decline that will reduce to the other types under certain circumstances:

$$\frac{a}{a_i} = \frac{q^n}{q_i^n} \qquad [5.6.2]$$

where a is the decline rate when the production rate is q, and a_i and q_i are the decline rate and production at an initial time. As mentioned above, the exponent, n, is a number between, but not including zero and one for the hyperbolic decline. When n is zero, the decline rate is constant which is the exponential decline. When n is one, the decline rate is proportional to the rate which is the harmonic decline. Several early publications related to decline-curve analysis have appeared in the literature [3–8].

The exponential and hyperbolic types of decline curves are more common than the harmonic decline. The exponential or constant percentage decline is indicative of a homogeneous producing interval where the pressure response has been affected by the outermost reservoir limits [9–10]. The exponential decline may apply to pumping wells that are kept pumped off or gas wells and many oil wells that produce at a constant bottomhole pressure. The hyperbolic decline is indicative of either unsteady-state conditions or pressure response from a variable permeability reservoir [11]. Although frequently encountered, the harmonic decline may be observed with reservoirs that are dominated by gravity drainage [2.18]. Equations for each type of decline-curve are given in Table 5.6.1, and will be discussed for each case.

5.6.1 Exponential Decline

For the exponential or constant percentage decline, the nominal or instantaneous decline rate is:

$$a = \frac{\ell n(q_i/q)}{t} \qquad [5.6.3]$$

and as shown in Table 5.6.1, the rate-time relationship is:

$$q = q_i e^{-at} \qquad [5.6.4]$$

and the relationship between flow rate and cumulative production is:

$$N_p = \frac{q_i - q}{a} \qquad [5.6.5]$$

The annualized (effective) or continuous decline rate, d, is:

$$d = \frac{q_i - q}{q} \qquad [5.6.6]$$

from which the cumulative production, N_p, is:

$$N_p = \frac{q_i - q}{-\ell n(1-d)} \qquad [5.6.7]$$

and the flow rate at time, t, is:

$$q = q_i + N_p \ell n(1-d) \qquad [5.6.8]$$

5.6.2 Hyperbolic Decline

As shown in Table 5.6.1, the time-rate relationship for the hyperbolic decline is:

$$q = \frac{q_i}{(1+na_it)^{1/n}} \qquad [5.6.9]$$

If h is substituted for 1/n, the time-rate relationship is:

$$q = \frac{q_i}{\left(1+\frac{a_it}{h}\right)^h} \qquad [5.6.10]$$

where h is the hyperbolic decline constant.

For the hyperbolic decline, the rate-cumulative production relationship is:

$$N_p = \frac{q_i^n}{a_i(1-n)}(q_i^{1-n} - q^{1-n}) \qquad [5.6.11]$$

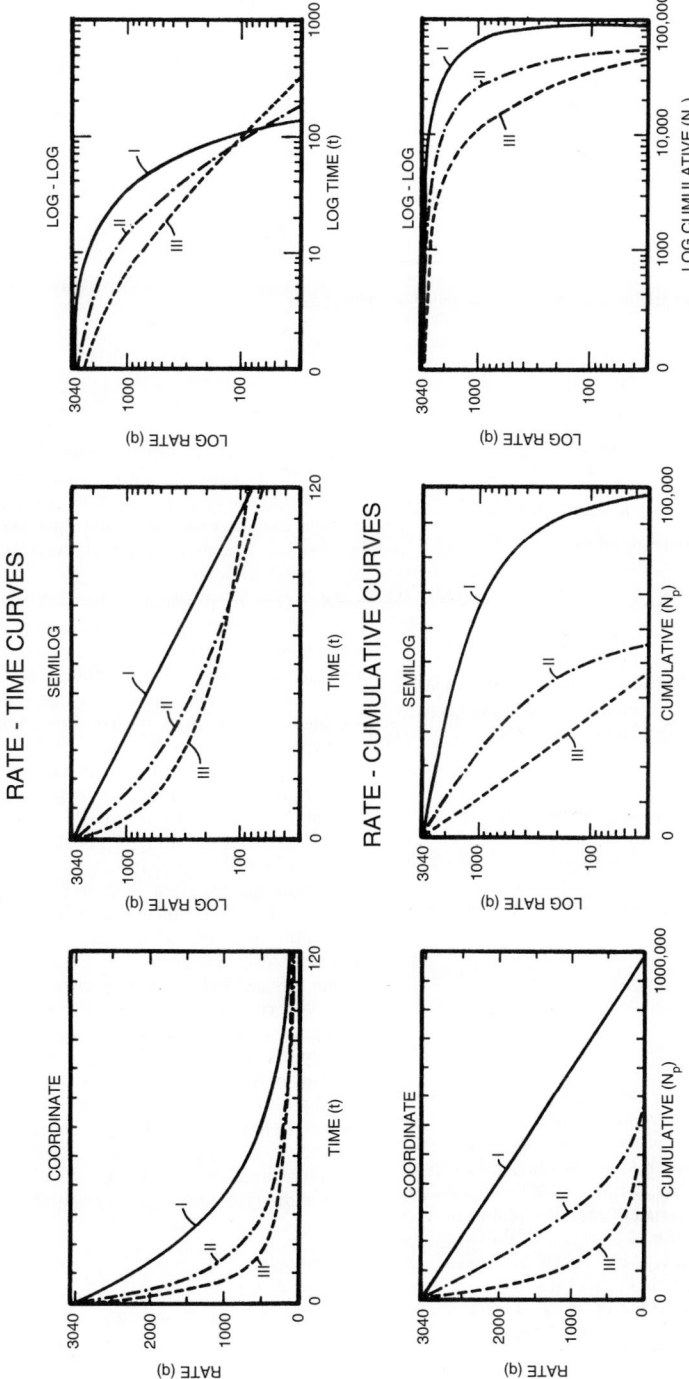

Figure 5.6.1   *Decline curves [4.7].*

Table 5.6.1   *Classification of Production Decline Curves*

Decline type	I. Constant percentage decline	II. Hyperbolic decline	III. Harmonic decline
Basic characteristic	Decline is constant $n = 0$	Decline is proportional to a fractional power (n) of the production rate $0 < n < 1$	Decline is proportional to a production rate $n = 1$
Rate-time relationship	$q = q_i e^{-at}$	$q = \dfrac{q_i}{(1+na_i t)^{1/n}}$	$q = \dfrac{q_i}{1+na_i t}$
Rate-cumulative relationship	$N_p = \dfrac{(q_i - q)}{a}$	$N_p = \dfrac{q_i n}{a_i(1-n)}(q_i^{1-n} - q^{1-n})$ $$q = \dfrac{q_i}{\left(1+\dfrac{a_i t}{h}\right)^h}$$ $N_p = \left[\dfrac{h}{(h-1)}\right]\left(\dfrac{1}{a_i}\right)\left[q_i - q\left(1+\dfrac{a_i t}{h}\right)\right]$	$N_p = \dfrac{q_i}{a_i}\ell n\left(\dfrac{q_i}{q}\right)$

From Reference 1.
a = decline as a fraction of production rate q_i = an initial production rate
t = time n = exponent
a_i = initial decline q = production rate at time t
N_p = cumulative oil production at time t $h = 1/n$

and if h is substituted into this equation:

$$N_p = \left[\frac{h}{(h-1)}\right]\left(\frac{1}{a_i}\right)\left[q_i - q\left(1+\frac{a_i t}{h}\right)\right] \quad [5.6.12]$$

Equation 5.6.10 can be rearranged to:

$$\frac{1}{a} = \frac{1}{a_i} + \frac{t}{h} \quad [5.6.13]$$

which represents a straight line. If the decline rate, $1/a$, is plotted on the y-axis versus the time interval on the x-axis, the intercept at $t = 0$ will yield $1/a_i$, and the slope will yield $1/h$. These values can be substituted into Equation 5.6.10 to give any future estimates of production rates [10].

5.6.3 Harmonic Decline
For a harmonic decline, the time-rate relationship is:

$$q = \frac{q_i}{(1+na_i t)} \quad [5.6.14]$$

and the rate-cumulative production relationship as shown in Table 5.6.1 is:

$$N_p = \frac{q_i}{a_i}\ell n\left(\frac{q_i}{q}\right) \quad [5.6.15]$$

5.6.4 Production Type-Curves
5.6.4.1 Semilog Plots
The complexity of the analysis of hyperbolic decline-curves led to the development of curve-matching techniques. One of the simpler techniques was proposed by Slider [2] with the development of an overlay method to analyze rate-time data. The actual decline-curve data are plotted on transparency paper and compared to a series of semilog plots that represent different combinations of a_i and n. Tabular values needed to plot the hyperbolic type-curves are available [2.18] for values of n from 0.1 to 0.9, in increments of 0.1.

Gentry [12] prepared a series of plots of q_i/q versus $N_p/q_i t$ for different values of n from 0 to 1.0 in increments of 0.1. Using two rates, the cumulative production, and the intervening time, the value of n for a particular hyperbolic decline-curve can be obtained. Gentry provided other curves to estimate a_i, or Equation 5.6.9 can be rearranged:

$$a_i = \frac{(q_i/q)^n - 1}{nt} \quad [5.6.16]$$

With the use of the semilog type-curves, caution must be exercised to ensure that the interval being analyzed is indeed a hyperbolic curve [2.18]. Another problem with the semilog plots is that an exact fit of the data may not be possible; however, the techniques are relatively rapid.

5.6.4.2 Log-log (Fetkovich) Type-Curves
Conventional decline-curve analysis should be used only when mechanical conditions and reservoir drainage remain constant and the well is producing at capacity [2.18]. An advanced approach for decline-curve analysis, which is applicable for changes in pressure or drainage, has been presented by Fetkovich [13,14]. This technique, which is similar to the approach used in pressure testing, involves log-log plots of q/q_i (or q_{Dd}) versus $a_i t$ (or t_{Dd}) for different values of n (see Figure 5.6.2). As shown in this figure, a log-log plot of the dimensionless rate and dimensionless time can identify transient data and/or depletion data, the Arps' equations given in Table 5.6.1 must only be applied to rate-time data that indicate depletion [14]. Use of transient data in the Arps' equations will result in incorrect forecasts that are overly optimistic.

The full-size Fetkovich type-curves can be ordered from the book order department of SPE. The field data are plotted on tracing paper that has the same log-log scale as the full-size type-curves. The log-log plot of flow rate and time can be in terms of barrels/day versus days, barrels/month versus months, or barrels/year versus years, depending on the time interval being studied. Using the best fit on the appropriate type-curve, a match point can be used to obtain q_i and a_i for the actual data. The appropriate equation can then be used to analyze the rate, time, and cumulative production behavior.

References
1. Arps, J. J., "Analysis of Decline Curves," *Trans.*, AIME, Vol. 160 (1945), pp. 228–247.
2. Slider, H. C., "A Simplified Method of Hyperbolic Decline Curve Analysis," *J. Pet. Tech.* (March 1968), pp. 235–236.
3. Chatas, A. T., and Yankie, W. W., Jr., "Application of Statistics to the Analysis of Production Decline Data," *Trans.*, AIME, Vol. 213 (1958), pp. 399–401.

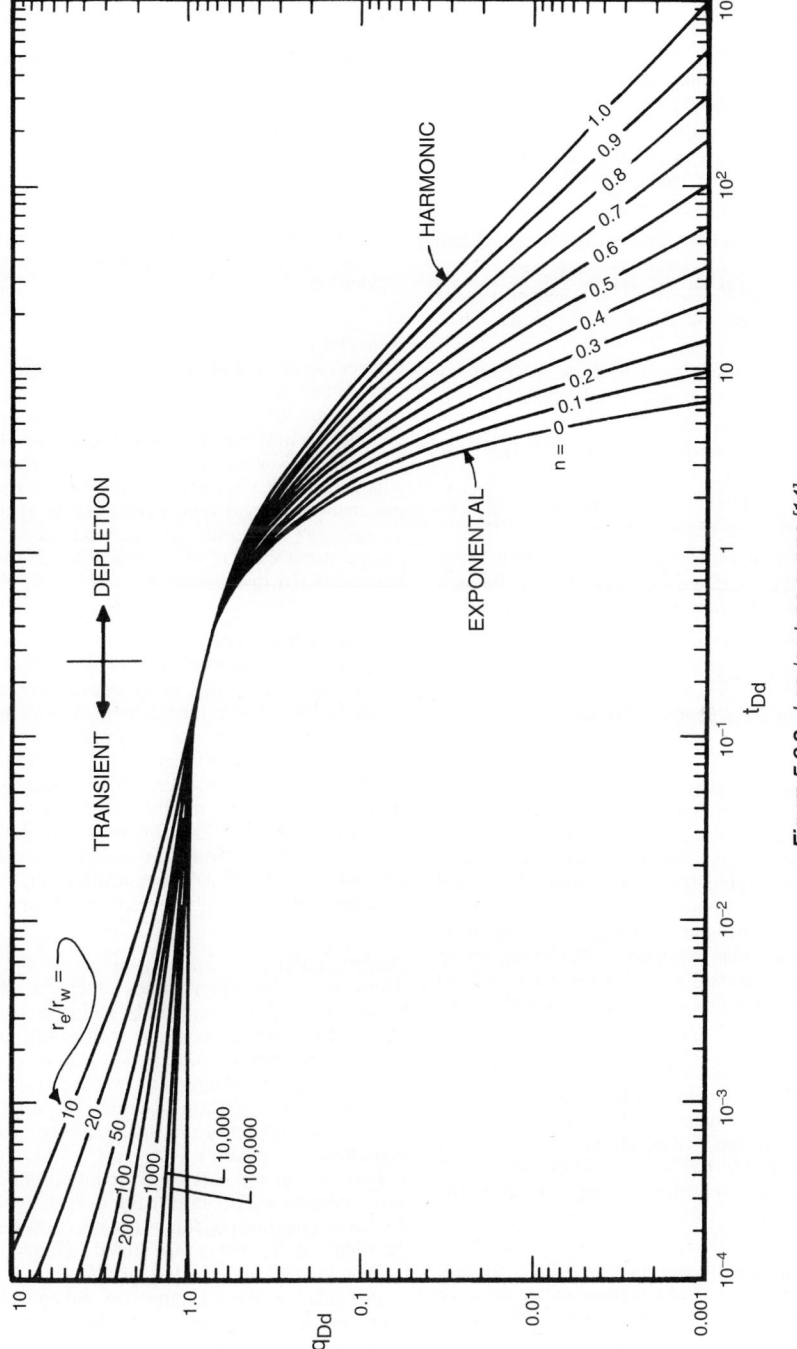

Figure 5.6.2 *Log-log type curves [14].*

4. Gray, K. E., "Constant Per Cent Decline Curve," *Oil & Gas J.*, Vol. 58 (Aug. 29, 1960), pp. 67–74.
5. Gray, K. E., "How to Analyze Yearly Production Data for Constant Per Cent Decline," *Oil & Gas J.* (*Jan.* 1, 1962), pp. 76–79.
6. Brons, F., "On the Use and Misuse of Production Decline Curves," *Prod. Monthly*, Vol. 27, No. 9 (Sept. 1963), pp. 22–25.
7. Brons, F., and Silbergh, M., "The Relation of Earning Power to Other Profitability Criteria," *J. Pet. Tech.* (March 1964), pp. 269–275.
8. Ramsay, H. J., Jr., and Guerrero, E. T., "The Ability of Rate-Time Decline Curves to Predict Production Rates," *J. Pet. Tech.* (Feb. 1969), pp. 139–141.
9. Russell, D. G., and Prats, M., "Performance of Layered Reservoirs with Crossflow-Single-Compressible-Fluid Case," *Soc. Pet Eng. J.* (March 1962), pp. 53–67.
10. Russell, D. G., and Prats, M., "The Practical Aspects of Interlayer Crossflow," *J. Pet Tech.* (June 1962), pp. 589–594.
11. Poston, S. W., and Blasingame, T. A., "Microcomputer Applications to Decline Curve Analysis," *Geobyte* (Summer 1986), pp 64–73.
12. Gentry, R. W., "Decline-Curve Analysis," *J. Pet. Tech.* (Jan. 1972), pp. 3841.
13. Fetkovich, M. J., "Decline Curve Analysis Using Type Curves," *J. Pet. Tech.* (June 1980), pp. 1065–1077.
14. Fetkovich, M. J., et al., "Decline Curve Analysis Using Type Curves: Case Histories," paper SPE 13169 presented at the SPE 1984 Annual Technical Conf. & Exhib., Houston, Sept. 16–19.

5.7 RESERVE ESTIMATES

5.7.1 Definition and Classification of Reserves
5.7.1.1 Definitions [1–3]
Crude Oil

This is defined technically as a mixture of hydrocarbons that existed in the liquid phase in natural underground reservoirs and remains liquid at atmospheric pressure after passing through surface facilities. For statistical purposes, volumes reported as crude oil include: (1) liquids technically defined as crude oil; (2) small amounts of hydrocarbons that existed in the gaseous phase in natural underground reservoirs but are liquid at atmospheric pressure after being recovered from oilwell (casinghead) gas in lease separators[1]; and (3) small amounts of nonhydrocarbons produced with the oil.

Natural Gas

This is a mixture of hydrocarbons and varying quantities of nonhydrocarbons that exist either in the gaseous phase or in solution with crude oil in natural underground reservoirs. Natural gas may be subclassified as follows.

Associated gas is natural gas, commonly known as gas-cap gas, that overlies and is in contact with crude oil in the reservoir.[2]

---

[1]From a technical standpoint, these liquids are termed "condensate"; however, they are commingled with the crude stream and it is impractical to measure and report their volumes separately. All other condensate is reported as either "lease condensate" or "plant condensate" and included in natural gas liquids.

[2]Where reservoir conditions are such that the production of associated gas does not substantially affect the recovery of crude oil in the reservoir, such gas may be classified as nonassociated gas by a regulatory agency. In this event, reserves and production are reported in accordance with the classification used by the regulatory agency.

Dissolved gas is natural gas that is in solution with crude oil in the reservoir.

Nonassociated gas is natural gas that is in reservoirs that do not contain significant quantities of crude oil.

Dissolved gas and associated gas may be produced concurrently from the same wellbore. In such situations, it is not feasible to measure the production of dissolved gas and associated gas separately; therefore, production is reported under the heading of associated-dissolved or casinghead gas. Reserves and productive capacity estimates for associated and dissolved gas also are reported as totals for associated-dissolved gas combined.

Natural Gas Liquids (NGLs)

There are those portions of reservoir gas that are liquefied at the surface in lease separators, field facilities, or gas processing plants. Natural gas liquids include but are not limited to ethane, propane, butanes, pentanes, natural gasoline, and condensate.

Reservoir

A reservoir is a porous and permeable underground formation containing an individual and separate natural accumulation of producible hydrocarbons (oil and/or gas) which is confined by impermeable rock and/or water barriers and is characterized by a single natural pressure system. In most situations, reservoirs are classified as oil reservoirs or as gas reservoirs by a regulatory agency. In the absence of a regulatory authority, the classification is based on the natural occurrence of the hydrocarbon in the reservoir as determined by the operator.

Improved Recovery

This includes all methods for supplementing natural reservoir forces and energy, or otherwise increasing ultimate recovery from a reservoir. Such recovery techniques include (1) pressure maintenance, (2) cycling, and (3) secondary recovery in its original sense (i.e., fluid injection applied relatively late in the productive history of a reservoir for the purpose of stimulating production after recovery by primary methods of flow or artificial lift has approached an economic limit). Improved recovery also includes the enhanced recovery methods of thermal, chemical flooding, and the use of miscible and immiscible displacement fluids.

Reserves [3]

These are estimated volumes of crude oil, condensate, natural gas, natural gas liquids, and associated substances anticipated to be commercially recoverable and marketable from a given data forward, under existing economic conditions, by established operating practices, and under current government regulations. Reserves do not include volumes of crude oil, condensate, or natural gas liquids being held in inventory.

Reserve estimates are based on interpretation of geologic and/or engineering data available at the time of the estimate. Existing economic conditions are prices, costs, and markets prevailing at the time of the estimate. Other assumed future economic conditions may lead to different estimates of recoverable volumes; these volumes are not considered reserves under existing economic conditions constraints, but may be identified as resources.

Marketable means that facilities to process and transport reserves to market are operational at the time of the estimate, or that there is a commitment to install such facilities in the near future, and there is a readily definable market or sales contract. Reserve estimates generally will be revised as reservoirs are produced, as additional geologic and/or engineering data become available, or as economic conditions change.

Natural gas reserves are those volumes which are expected to be produced and that may have been reduced by onsite usage, by removal of nonhydrocarbon gases, condensate or natural gas liquids.

Reserves may be attributed to either natural reservoir or improved recovery methods. Improved recovery includes all methods for supplementing natural reservoir energy to increase ultimate recovery from a reservoir. Such methods include (1) pressure maintenance, (2) cycling, (3) waterflooding, (4) thermal methods, (5) chemical flooding, and (6) the use of miscible and immiscible displacement fluids. Reserves attributed to improved recovery methods usually will be distinguished from those attributed to primary recovery.

All reserve estimates involve some degree of uncertainty, depending chiefly on the amount and reliability of geologic and engineering data available at the time of the estimate and the interpretation of these data. The relative degree of uncertainty may be conveyed by placing reserves in one of two classifications, either proved or unproved. Unproved reserves are less certain to be recovered than proved reserves and may be subclassified as probable or possible to denote progressively increasing uncertainty.

5.7.1.2 Classification of Reserves
Proved Reserves
Attributed to known reservoirs, proved reserves can be estimated with reasonable certainty. In general, reserves are considered proved if commercial producibility of the reservoir is supported by actual production or formation tests. The term proved refers to the estimated volume of reserves and not just to the productivity of the well or reservoir. In certain instances, proved reserves may be assigned on the basis of a combination of core analysis and/or electrical and other type logs that indicate the reservoirs are analogous to reservoirs in the same areas that are producing, or have demonstrated the ability to produce in a formation test. The area of a reservoir considered proved includes (1) the area delineated by drilling and defined by fluid contacts, if any, and (2) the undrilled areas that can be reasonably judged as commercially productive on the basis of available geological and engineering data. In the absence of data on fluid contacts, the lowest known structural occurrence of hydrocarbons controls the proved limit unless otherwise indicated by definitive engineering or performance data.

In general, proved undeveloped reserves are assigned to undrilled locations that satisfy the following conditions: (1) the locations are direct offsets to wells that have indicated commercial production in the objective formation, (2) it is reasonably certain that the locations are within the known proved productive limits of the objective formation, (3) the locations conform to existing well spacing regulation, if any, and (4) it is reasonably certain that the locations will be developed. Reserves for other undrilled locations are classified as proved undeveloped only in those cases where interpretation of data from wells indicates that the objective formation is laterally continuous and contains commercially recoverable hydrocarbons at locations beyond direct offsets.

Reserves that can be produced through the application of established improved recovery methods are included in the proved classification when: (1) successful testing by a pilot project or favorable production or pressure response of an installed program in that reservoir, or one in the immediate area with similar rock and fluid properties, provides support for the engineering analysis on which the project or program is based and (2) it is reasonably certain the project will proceed.

Reserves to be recovered by improved recovery methods that have yet to be established through repeated commercially successful application are included in the proved category only after a favorable production response from the reservoir from either (1) a representative pilot or (2) an installed program that provides support for the engineering analysis on which the project or program is based.

Unproved Reserves
These are based on geologic and/or engineering data similar to that used in estimates of proved reserves, but technical, contractual, or regulatory uncertainties preclude such reserves being classified as proved. Estimates of unproved reserves may be made for internal planning of special evaluations, but are not routinely complied.

Unproved reserves are not to be added to proved reserves because of different levels of uncertainty. Unproved reserves may be divided into two subclassifications: probable and possible.

Probable Reserves
These reserves are attributed to known accumulations and are less certain to be recovered than proved reserves. In general, probable reserves may include (1) reserves that appear to exist a reasonable distance beyond the proved limits of productive reservoirs, where fluid contacts have not been determined and proved limits are established by the lowest known structural occurrence of hydrocarbons, (2) reserves in formations that appear to be productive from core and/or log characteristics only, but that lack definitive tests or analogous producing reservoirs in the area, (3) reserves in a portion of a formation that has been proved productive in other areas in a field, but that is separated from the proved area by faults, (4) reserves obtainable by improved recovery methods and located where an improved recovery method (that has yet to be established through repeated commercially successful operation) is planned but not yet in operation, and where a successful pilot test has not been performed but reservoir and formation characteristics appear favorable for its success, (5) reserves in the same reservoir as proved reserves that would be recoverable if a more efficient primary recovery mechanism were to develop than that assumed in estimating proved reserves, (6) incremental reserves attributable to infill drilling where closer statutory spacing had not been approved at the time of the estimate, and (7) reserves that are dependent for recovery on a successful workover, treatment, retreatment, change of equipment, or other mechanical procedures, when such procedures have not been proved successful in wells exhibiting similar behavior in analogous formations.

Possible Reserves
These are associated with known accumulations and are less certain to be recovered than probable reserves. In general, possible reserves may include (1) reserves indicated by structural and/or stratigraphic extrapolation from developed areas, (2) reserves located where reasonably definitive geophysical interpretations indicate an accumulation larger than could be included within the proved and probable limits, (3) reserves in formations that have favorable log characteristics but questionable productivity, (4) reserves in untested fault segments adjacent to proved reservoirs where a reasonable doubt exist as to whether such fault segment contains recoverable hydrocarbons, (5) incremental reserves attributable to infill drilling that are subject to technical or regulatory uncertainty, and (6) reserves from a planned

improved recovery program that is not in operation and that is in a field in which formation, fluid, or reservoir characteristics are such that a reasonable doubt exists to its success.

Reserve Status Categories

These define the development and producing status of wells and/or reservoirs. They may be applied to proved or unproved (probable or possible) reserves.

Developed Reserves

These are expected to be recovered from existing wells (including reserves behind pipe). Improved recovery reserves are considered developed only after the necessary equipment has been installed, or when the costs to do so are relatively minor. Developed reserves may be subcategorized as producing or nonproducing.

Producing Reserves

These are expected to be recovered from completion intervals open at the time of the estimate and producing to market. Improved recovery reserves are considered to be producing only after an improved recovery project is in operation. Unproved (probable or possible) producing reserves are in addition to proved producing reserves, such as (1) reserves that may be recovered from portions of the reservoir downdip from proved reserves or (2) reserves that may be recovered if a higher recovery factor is realized than was used in the estimate of proved reserves.

Nonproducing Reserves

These include shut-in and behind-pipe reserves. Shut-in reserves are expected to be recovered from completion intervals open at the time of the estimate, but which have not started producing, or were shut in for market conditions of pipeline connection, or were not capable of production for mechanical reasons, and the time when sales will start is uncertain. Behind-pipe reserves are expected to be recovered from zones behind casing in existing wells, which will require additional completion work or a future recompletion prior to the start of production.

Undeveloped Reserves

These are expected to be recovered: (1) from new wells on undrilled acreage, (2) from deepening existing wells to a different reservoir, or (3) where a relatively large expenditure is required to (a) recomplete an existing well, or (b) install production or transportation facilities for primary or improved recovery projects. Undeveloped reserves usually will be distinguished from developed reserves. The ownership status of reserves may change due to the expiration of a production license or contract; when relevant to reserve assignment, such changes should be identified for each reserve classification.

5.7.2 Methods of Estimating Reserves

Method of determining reserves progress from analogy, before a well is drilled, to history after it is plugged and abandoned. The accuracy with which reserves can be estimated progresses along the same path from speculation to history.

5.7.2.1 Analogy

The decision to drill a well is based upon the potential reserves that it will recover. This means that an engineer must be able to predict reserves before a well is drilled. The lack of information about the reservoir restricts the engineering methods available.

Analogy is the only method which can be used without specific well information such as porosity, reservoir thickness, and water saturation. Because analogy employs no specific information about a well, it is the least accurate method of determining reserves. Methods of analogous reserve determination depend on the proximity of similar reserves. The best analogy can be made by taking the median ultimate recovery of a number of wells that are closest and have the same formation and characteristics expected in the proposed well. When ultimate recovery data are not available, volumetric, decline curve or other methods of estimating ultimate recovery may be used.

Unless values of ultimate recovery figures for the group are relatively close, the median ultimate recovery should be calculated by making a normal probability plot. This plot is made by graphing estimated ultimate recovery against the cumulative percent of samples. A best-fit line is drawn through the points and the median is read where the line intersects fifty percent. A straight line indicates a normal distribution; if the line is not straight the distribution is skewed. If there are no similar wells in the area, data from those less similar may be used, but confidence goes down as similarity decreases.

5.7.2.2 Volumetric

If a well is drilled after reserves are determined by analogy, factual information becomes available and reserves can then be determined volumetrically. From log analysis the porosity, water saturation, and productive formation thickness are estimated. A reasonable drainage area is assigned and total hydrocarbons in place are then calculated. When enough wells have been drilled to delineate the field, a subsurface geological contour map showing the subsea sand top and bottom depth, oil-water contact, and gas-oil contact can be prepared. From this map the total areas in acre-feet of each contour are planimetered and graphed as the abscissa against the subsea depth as the ordinate. Lines are then drawn to connect the sand-top points and the sand-bottom-points, the area bounded by the oil-water-contact depth, the sand-top line, the sand-bottom line and the gas-oil-contact line. This area is the gross oil-bearing sand-volume in acre-feet. The area, if present, that is bounded by the gas-oil contact depth, the sand-top-line, the sand-bottom-line and the abscissa is the gross gas-bearing sand-volume in acre-feet. The engineer must determine from core data and/or electric logs the percentage of the gross sand volume that is productive and must then reduce the total acre-feet by that percentage. If there is no subsurface contour map available or if the reservoir is very heterogeneous, an isopach or an isovol map should be constructed. An isopach map is constructed by contouring net sand thickness. This kind of map works well when the reservoir is uniform and when porosity and water saturation are relatively constant. When the water saturation and porosity vary widely from well to well, an isovol map that indicates hydrocarbon thickness is useful. This map is constructed by contouring the value of net pay height multiplied by porosity and by one minus the water saturation. Care should be taken not to rely on the scale provided on the map especially when using xerographed copies, as this and other methods of reproduction can distort one or both axes as much as five percent. A known area such as a section should be measured to calibrate the instrument. Once the contour areas have been planimetered, the net pay volume can be calculated in several ways. If the number of contour intervals is even, the volume can be calculated by Simpson's rule [4]:

$$V_R = 1/3h[(y_o + y_n) + 4(y_1 + y_3 + \cdots + y_{n-1})$$

$$+ 2(y_2 + y_4 + \cdots + y_{n-2})] \qquad [5.7.1]$$

where V_R = reservoir volume, acre-ft
  h = contour interval, ft
  y_o = area on top of sand minus area on base of sand at the highest contour
  y_n = area on top of sand minus area on base of sand at the lowest contour

When the number of contours is uneven, the volume can be found using the slightly less accurate trapezoidal method:

$$V_R = h[1/2(y_o + y_n) + y_1 + y_2 + \cdots + Y_{n-1}] \qquad [5.7.2]$$

Once the reservoir volume is known, the oil in place, N, in stock tank barrels is calculated by an equation similar to that given earlier:

$$N = \frac{7,758V_o(1 - S_w)}{B_o} \qquad [5.7.3]$$

where N = reservoir oil initially in place, STB
  7,758 = number of barrels/acre-ft
  V_o = net producing reservoir volume, acre-ft
  B_o = oil formation volume factor, RB/STB
  S_w = interstitial water saturation, fraction
  ϕ = formation porosity, fraction

Similarly gas-in-place, G, in thousands of standard cubic feet, is estimated by an expression similar to that given earlier:

$$G = \frac{43,560V_g \phi (1 - S_w) p T_s}{z T p_s} \qquad [5.7.4]$$

where G = gas in place, scf
  V_g = gas bearing volume of reservoir, acre-ft
  p_s = standard pressure, psia
  T = reservoir temperature, degrees absolute
  T_s = standard temperature base, degrees absolute
  z = gas deviation factor at reservoir conditions

Volumetric analysis yields the total hydrocarbon content of the reservoir; this figure must be adjusted by a recovery factor to reflect the ultimate recoverable reserves. Recovery factors are based upon empirical correlations, experience, or analogy.

5.7.2.3 Material Balance

If a field development program has been well planned and executed, enough information should be available to calculate reserves by the material balance equation. The material balance equation is derived on the assumption that the reservoir is a homogeneous vessel with uniform porosity, permeability, and fluid properties. The equation accounts for all quantities of materials that enter or leave the vessel. The simplest form of the equation is that initial volume is equal to the volume remaining plus the volume removed. As material is withdrawn from a constant-volume reservoir the pressure declines and remaining material expands to fill the reservoir. Laboratory PVT analysis of the reservoir fluid defines the change in volume per unit pressure drop. Knowing the amount of fluid withdrawn from the reservoir and the drop in pressure one can calculate the corresponding volume of fluid at the original reservoir pressure. The calculated reservoir size should remain constant as fluid is withdrawn and pressure drops. If the calculated reservoir size changes constantly in one direction as the field is produced the assumed production mechanism is probably wrong. Calculations should be repeated assuming different mechanisms until one is found that yields a constant reservoir size. Since Schilthuis [5.3] developed the original material balance equation in 1936 it has been rearranged to solve almost any unknown. The most frequently used forms of the equation are for these types of recovery mechanisms [4]:

1. Oil reservoir with gas cap and active water drive.
2. Oil reservoir with gas cap and no active water drive.
3. Initially undersaturated oil reservoir with active water drive: A. Above bubble point. B. Below bubble point.
4. Initially undersaturated oil reservoir with no active water drive: A. Above bubble point. B. Below bubble point.
5. Gas reservoir with active water drive.
6. Gas reservoir with no active water drive.

The material balance equation, when combined with reliable relative permeability data, can be used to predict future reservoir performance. Many times, reservoirs do not conform to the assumptions made in the material balance equation. Few reservoirs are homogeneous and no reservoirs respond instantaneously to changes in pressure. The precision with which reserves can be calculated or predicted with the material balance equation is affected by the quality of data available and the degree of agreement between the assumptions made in the equation and the actual reservoir conditions.

5.7.2.4 Model Studies

Predicting reservoir performance with the Tarner [5.4] or the Muskat [5.5] method is a long and tedious process and, even with a programmable calculator, the process takes several hours. To resolve the problems caused by the assumptions inherent in the material balance equation, a reservoir would have to be broken up into parts small enough to be considered homogeneous. The material balance equation would then have to be calculated for each part and for each increment of production. This would entail thousands of calculations performed thousands of times and would drastically limit the number of reservoir simulations an engineer could run. Fortunately, computers have cut the required time to a few minutes. Numerical simulators divide the reservoir up into discreet elements, each having the properties and spatial orientation of the associated blocks of a physical reservoir [5]. The simulator treat each block as a small reservoir, and keeps track of fluid entering or leaving the block. When a change in pressure due to injection or withdrawal of fluid occurs, the simulator solves the material balance equation for a number of time steps for each block until equilibrium is reached. Blocks are usually configured so that each well is in an individual block. Time steps are picked so that the required information is resolved without using excessive computer time. Since the simulator keeps track of fluid movement through the reservoir, the output can include a wide variety of parameters. Fluid fronts, saturation changes, pressure distribution, and oil-water contact movement are a few of the things that can be plotted. The three general classifications of simulators are gas, black oil, and compositional. Gas simulators model one or two phases (gas or gas and water). Black oil simulators are designed to model any proportion of gas, oil, and water, and they account for gas going into or out of solution. Compositional simulators are used when PVT data does not adequately describe reservoir behavior such as in condensate reservoirs. These simulators calculate the mass fraction of individual components in each phase and mass transfer between phases as each phase flows at different rates. Most models are run with limited information and must be tuned to properly predict actual reservoir performance. This is done by changing parameters such as relative permeability, porosity, and permeability data until the simulator matches the field history.

5.7.2.5 Production Decline Curves

The most widely used method of estimating reserves is the production rate decline-curve. This method involves extrapolation of the trend in performance. If a continuously changing continuous function is plotted as the dependent variable

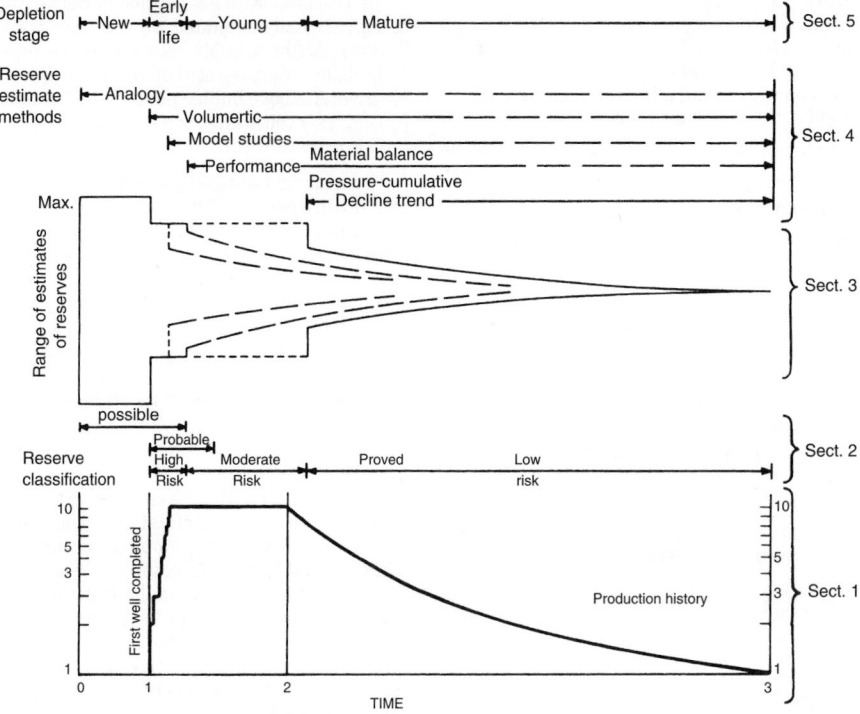

Figure 5.7.1   *Quality of reserves [6].*

against an independent variable, a mathematical or graphical trend can be established. Extrapolation of that trend can then permit a prediction of future performance. For an oil reservoir, the plot of the logarithm of production rate against time is most useful. Although decline-curve analysis is empirical, if care is taken to ensure that production rates are not being affected by such things as the mechanical degradation of equipment or the plugging of the formation by fines or paraffin, the method is reasonably accurate. As discussed, there are three major types of decline curves: constant percentage or exponential, hyperbolic, and harmonic. Although analysis of a large number of actual production decline curves indicates that most wells exhibit a hyperbolic decline with an n value falling between 0 and 0.4, the constant-percentage exponential decline-curve is most widely used. The exponential decline curve is most popular because, when plotted on semilog paper the points make a straight line which is easiest to extrapolate to the economic limit. Now that programmable calculators and personal computers reside at most every engineer's desk, it is easy to punch in the production data and decide which decline curve is best.

5.7.3 Quality of Reserve Estimates

If reserve estimates contained no risk, no dry holes would be drilled. Unfortunately, risk is inversely proportional to knowledge and the least is known before a well is drilled. Hudson and Neuse [6] presented a graphical representation (reproduced in Figure 5.7.1) of reserve estimate quality throughout the life of a property. Section 1 shows the production history of the property. Section 2 indicates the probable risk factor associated with each stage of production from completion to depletion. Section 3 shows the range of

possible reserve estimates through the life of the property. Section 4 shows at what stage of the life of the property each method of reserve estimate method becomes available. The solid line indicates for what period each method is most applicable. Section 5 names the depletion stage of the reservoir.

The quality of reserve estimates throughout the industry should improve now that the SPE definitions for proved reserves are being accepted. When the definitions for probable and possible reserves are approved, uniform factors for risk analysis can be made [7].

References

1. "Society Adopts Proved Reserves Definitions," *J. Pet. Tech.* (Nov. 1981), pp. 2113–2114.
2. "Reserve Definitions Approved," *J. Pet. Tech.* (May 1987), pp. 576–578.
3. Guidelines for Application of the Definitions for Oil and Gas Reserves, Monograph 1, the Society of Petroleum Evaluation Engineers, Houston (1988).
4. Garb, F. A., "Oil and Gas Reserves, Classification, Estimation and Evaluation," *J. Pet. Tech.* (March 1985), pp. 373–390.
5. Crichlow, H. B., *Modern Reservoir Engineering-A Simulation Approach*, Prentice-Hall, Inc., New Jersey (1977).
6. Hudson, E. J., and Neuse, S. H., "Cutting Through the Mystery of Reserve Estimates," *Oil & Gas J.* (March 25, 1985), pp. 103–106.
7. Hudson, E. J., and Neuse, S. H., "Depletion Stage Determines Most Effective Methods for Reserve-Estimate Integrity," *Oil & Gas J.* (April 1, 1985), pp. 80–91.

5.8 SECONDARY RECOVERY

5.8.1 Definitions

5.8.1.1 Secondary and Tertiary Recovery

Primary recovery, as already discussed, refers to the recovery of oil and/or gas that is recovered by either natural flow or artificial lift through a single wellbore. Thus, primary recovery occurs as a result of the energy initially present in the reservoir at the time of discovery. When the initial energy has been depleted and the rate of oil recovery declines, oil production can be increased by the injection of secondary energy into the reservoir. Secondary recovery is the recovery of oil and/or gas that involves the introduction of artificial energy into the reservoir via one wellbore and production of oil and/or gas from another wellbore. Conventional means of secondary recovery include the immiscible processes of waterflooding and gas injection. Currently in the United States, waterflooding is the dominant secondary recovery method in that about half of the oil production is recovered from waterflood projects. After secondary recovery, a substantial amount of oil may remain, and attempts to recover oil beyond primary and secondary recovery are referred to as tertiary recovery. Any method that recovers oil more effectively than plain waterflooding or gas injection is defined as enhanced recovery. The more sophisticated enhanced methods may be initiated as a tertiary process if they follow waterflooding or gas injection, or they may be a secondary process if they follow primary recovery directly. Many of the enhanced recovery projects are implemented after water-flooding. (Enhanced recovery methods are discussed later.)

5.8.1.2 Pressure Maintenance

Pressure maintenance is a secondary recovery process that is implemented early during the primary producing phase before reservoir energy has been depleted. Pressure maintenance projects, which can be accomplished by the injection of either gas or water, will almost always recover more oil reserves than are recoverable by primary producing mechanisms. For example, the return of gas to the formation early in the primary producing history of a field will permit higher rates of oil production.

5.8.2 Gas Injection

Historically, both natural gas and air have been used in gas injection projects, and in some cases nitrogen and flue gases have been injected. Many of the early gas injection projects used air to immiscibly displace crude oil from reservoirs. The injection of hydrocarbon gas may result in either a miscible or immiscible process depending on the composition of the injected gas and crude oil displaced, reservoir pressure, and reservoir temperature. Hydrocarbon miscible injection is considered as an enhanced recovery process and is discussed later.

Although the ultimate oil recovered from immiscible gas injection projects will normally be lower than for waterflooding, gas injection may be the only alternative for secondary recovery under certain circumstances. If permeability is very low, the rate of water injection may be so low that gas injection is preferred. In reservoirs with swelling clays, gas injection may be preferable. In steeply-dipping reservoirs, gas that is injected updip can very efficiently displace crude oil by a gravity drainage mechanism; this technique is very effective in low-permeability formations such as fractured shales. In thick formations with little dip, injected gas (because of its lower density) will tend to override and result in vertical segregation if the vertical permeability is more than about 200 md [5.8]. In thin formations especially if primary oil production has been by solution-gas drive, gas may

be injected into a number of wells in the reservoir on a well pattern basis; this dispersed gas injection operation attempts to bank the oil in a frontal displacement mechanism. In addition to the external gas injection into reservoirs with dip as just described (which may be into a primary or secondary gas cap), a variation called attic oil recovery involves injection of gas into a lower structural position. If there is sufficient vertical permeability, the injected gas will migrate upward to create a secondary gas cap that can displace the oil downward where it is recovered in wells that are already drilled.

Volumetric calculations for gas injection are given in Table 5.8.1. The material balance equation for dispersed gas injection in a solution-gas-drive reservoir where there is no water influx, no water production, and no gas cap can be expressed in terms of the oil produced during an interval [5.8]:

$$\Delta N_p = \frac{(1-N_{p1})\left(\frac{dB_o}{dB_g}-dR_s\right)-d\left[\left(\frac{1}{B_g}\right)B_{ob}\right]}{\left(\frac{B_o}{B_g}-R_s\right)_2+R_{avg}(1-I)} \qquad [5.8.1]$$

where I is the constant fraction of the produced gas which is reinjected into the oil reservoir, B_{ob} is the oil formation volume factor at bubble-point conditions, R is the producing gas-oil ratio, the subscripts 1 and 2 refer to the time increments at p_1 and p_2, the other standard terms are as already defined, and R_{avg} is $(R_1+R_2)/2$. As given in Table 5.8.1, the instantaneous gas-oil ratio, R, is:

$$R=R_s+\frac{B_o k_g \mu_o}{B_g k_o \mu_g} \qquad [5.8.2]$$

where R_s is the solution gas-oil ratio, B is formation volume factor, k is permeability, μ is viscosity, and the subscripts o and g refer to oil and gas, respectively. The relative permeabilities are determined at the total liquid saturation, S_L:

$$S_L=S_w+S_o=S_w+(1-S_w)\frac{(N-N_p)}{N}\frac{B_o}{B_{ob}} \qquad [5.8.3]$$

The simultaneous solution of these three equations will provide estimates of oil produced at any chosen conditions for a dispersed-gas-drive injection project. Additional details can be found in Reference 5.8.

Many of the flow equations and concepts for immiscible gas displacement are similar to those that will be presented later for waterflooding. Because of the importance of water injection processes in U.S. operations, waterflooding concepts will be emphasized.

5.8.3 Water Injection

Water injection processes may be designed to: (1) dispose of brine water, (2) conduct a pressure maintenance project to maintain reservoir pressure when expansion of an aquifer or gas cap is insufficient to maintain pressure, or (3) implement a water drive or waterflood of oil after primary recovery. As mentioned before, waterflooding is the dominant secondary recovery process which accounts for about 50% of the current oil production in U.S. operations. Because of the importance of waterflooding, fluid displacement in waterflooded reservoirs is covered as a separate discussion in a later section.

5.8.4 Spacing of Wells and Well Patterns

5.8.4.1 Spacing of Wells

One section (one sq mile or 5,280 ft by 5,280 ft) is 640 acres. If wells are drilled evenly such that each well is theoretically assigned to drain 40 acres, the 16 wells per section would be as spaced as in Figure 5.8.1. Each 40 acres ($\frac{1}{4}$ mile by $\frac{1}{4}$ mile or 1,320 ft by 1,320 ft) would contain $40\times43,560$ or

Table 5.8.1 *Volumetric Calculations for Gas Injection*

Gas drive field with gas cap

$$\frac{dS_o}{dp} = \frac{\dfrac{B_gS_o}{B_o}\dfrac{dR_s}{dp} + \dfrac{k_g\mu_oS_o}{k_o\mu_gB_o}\dfrac{dB_o}{dp} - \dfrac{(1-S_o-S_w)+m(1-S_w)}{B_g}\dfrac{dB_g}{dp}}{1+\dfrac{k_g\mu_o}{k_g\mu_g}}$$

m= ratio of gas-cap pore volume to oil-zone pore volume, bbl/bbl or ft^3/ft^3. For each 1 bbl of oil zone there will be m bbls gas cap.

Reinjection of all produced gas-pressure maintenance

$$\frac{dS_o}{dp} = \frac{\dfrac{B_gS_o}{B_o}\dfrac{dR_s}{dp} - \dfrac{R_sB_gS_o}{B_o^2}\dfrac{dB_o}{dp} - \dfrac{(1-S_o-S_w)}{B_g}\dfrac{dB_g}{dp}}{1-\dfrac{R_sB_g}{B_o}}$$

Gas reinjection

Producing GOR $R = \dfrac{k_g\mu_oB_o}{k_o\mu_gB_g} + R_s$

Net GOR $R = \left(\dfrac{k_g\mu_oB_o}{k_o\mu_gB_g} + R_s\right)(1-I)$

I = fraction of gas that is reinjected

$$\frac{dS_o}{dp} = \frac{\dfrac{B_gS_o}{B_o}\dfrac{dR_s}{dp} + \left[(1-I)\dfrac{k_g\mu_oS_o}{k_o\mu_gB_o} - \dfrac{IR_sB_gS_o}{B_o^2}\right]\dfrac{dB_o}{dp} - \left[\dfrac{(1-S_o-S_w)+m(1-S_w)}{B_g}\dfrac{dB_g}{dp}\right]}{1+(1-I)\left(\dfrac{k_g\mu_o}{k_o\mu_g}\right) - \dfrac{IR_sB_g}{B_o}}$$

m= ratio of gas-cap volume to oil-zone volume

From References 1.14 and 5.8.

1,742,400 ft^2. The 10-acre region in Figure 5.8.1 would measure 660 ft by 660 ft and would contain 435,000 ft^2. Similarly, a 20-acre region would contain 871,200 ft^2 and would measure $[20 \times 871,200]^{1/2}$ or 933.4 ft by 933.4 ft.

In many parts of the United States, 40-acre spacing or less is common for oil wells, and 160-acre or 320-acre spacing is common for gas wells. Because drilling costs increase considerably with depth, deeper wells may be on larger spacing.

5.8.4.2 Injection Well Placement
Wells may be spaced evenly or unevenly from each other based on surface topology, lease boundaries, regulations, or other factors. Many older fields were developed on irregular spacing. In more recent times, more uniform drilling patterns and well spacing have been used.

In most cases when an injection project is started, primary recovery has been implemented and producing wells will already be in place. For some projects, a number of existing wells will be converted from producers to injectors, and in other cases, new injection wells will be drilled. In either event, the injection well placement must be compatible with the existing wells and should [1.133]: (1) take advantage of known reservoir uniformities or nonuniformities (fractures, directional permeability, regional permeabilities, dip, etc.), (2) provide sufficient fluid injection rate to yield the desired production rate, (3) maximize recovery with a minimum of production of the injected fluid, and (4) in most cases, require a minimum of new wells. Two general types of well locations are common: (1) peripheral or central flooding where the injectors are grouped together, and (2) pattern flooding where certain patterns are repeated throughout the field. The relative location of injectors and producers depends on the geology ad type of reservoir, the volume of reservoir swept, and the time limitations that affect economics. When possible, the injection scheme should take advantage of gravity, i.e., dipping or inclined reservoirs, gas caps, or underlying aquifers.

5.8.4.3 Peripheral or Central Flooding
In peripheral flooding, the injectors are located around the periphery so that the flood progresses toward the center as shown in Figure 5.8.2. When the first row of producers flood out, they are converted to injection status. This type of flood can result in maximum oil recovery with a minimum of produced injectant, and less injectant is required for a given amount of production, but a peripheral flood usually takes longer than a pattern flood. In general, adequate permeability is required to permit movement of fluids at an acceptable rate with the available well spacing. Central flooding is the opposite case in which injectors are located in the center of the field, and the flood progresses outward (see Figure 5.8.2).

A form of peripheral flooding is an end-to-end flood such as that shown in Figure 5.8.2. This type of injection could include the injection of gas into a gas cap or the injection of water into an aquifer. The choice of peripheral or repeating pattern flood is usually made on the basis of: formation permeability, formation dip, area and dimensions of the reservoir, and the initial production response that is acceptable.

5.8.4.4 Pattern Flooding
In pattern flooding, the injectors are distributed among the producers in some repeating fashion. Examples of the common repeating patterns are shown in Figure 5.8.3. Pattern

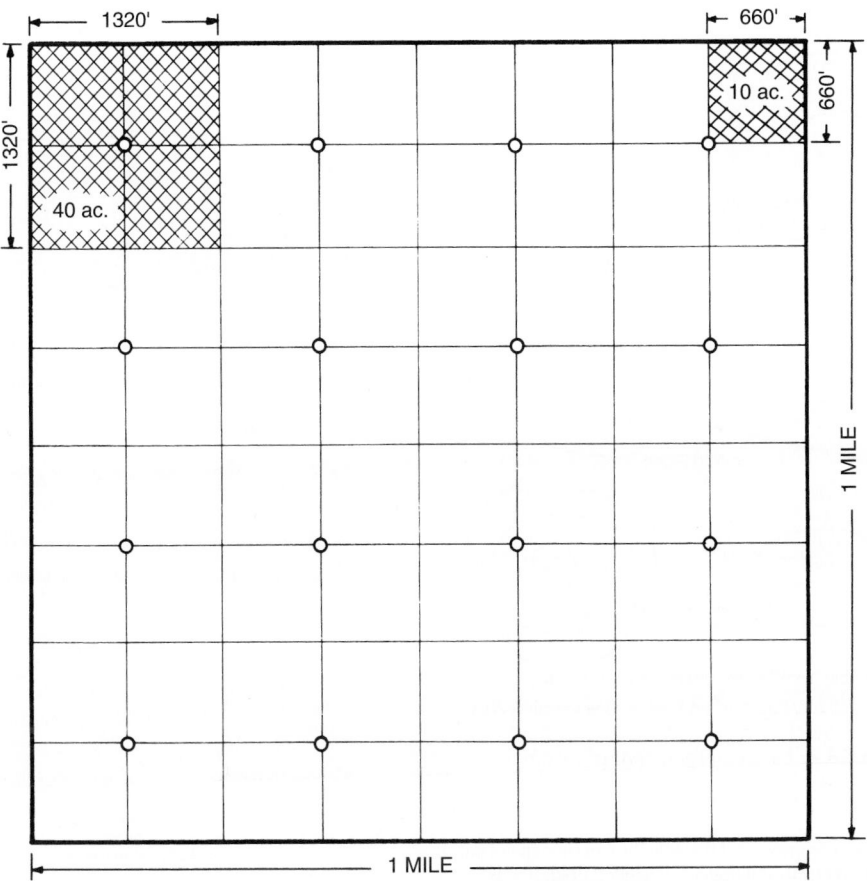

Figure 5.8.1 *Well locations for 40-acre spacing.*

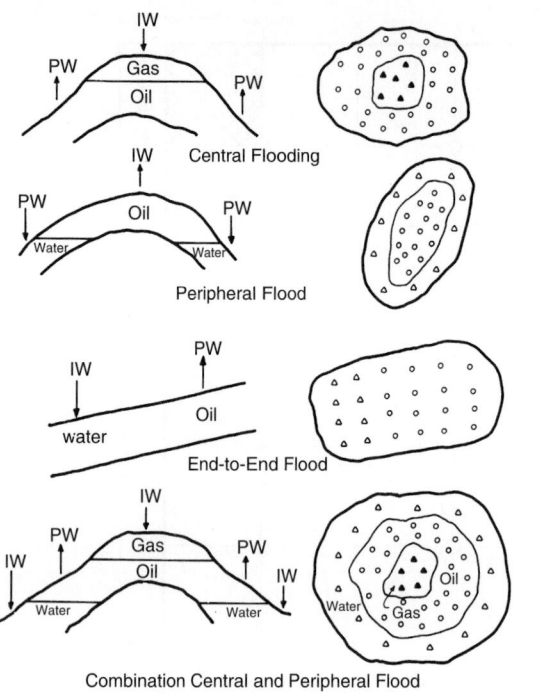

Figure 5.8.2   Peripheral and central flooding.

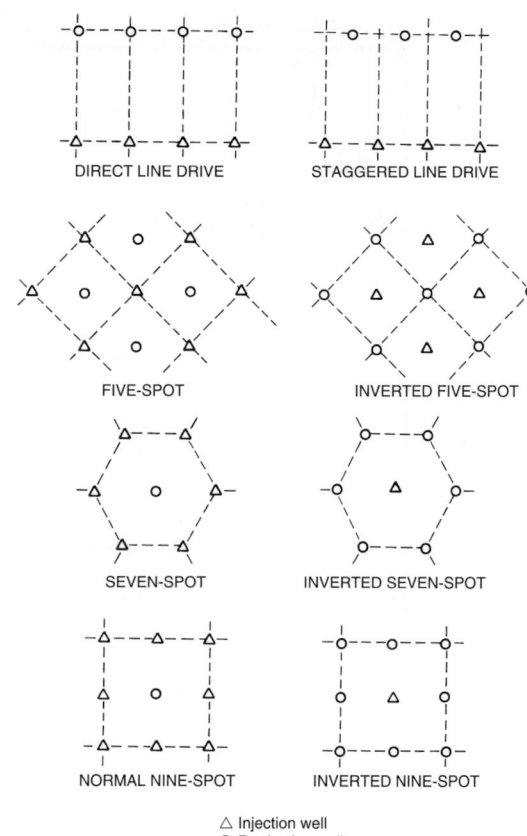

Figure 5.8.3   Well locations in pattern floods.

flooding is very common, and the selection of the type of pattern will depend on circumstances in a given field. If existing wells were drilled on square patterns, 5-spots and 9-spots are common, and both yield similar oil recovery and water-oil ratio performance. If the injected fluid is more mobile than the displacing fluid (which is often the case, especially when oil viscosity is high), a pattern having more producers than injectors may be desired to balance the injection and production rates. In cases where the injected fluid is less mobile or when the formation permeability is low, a pattern having more injectors than producers may be desired. From an inspection of Figure 5.8.3, the ratio of producers to injectors for the various patterns can be determined as given in Table 5.8.2. For either the normal (or regular) 5-spot or the inverted 5-spot (inverted means one injector per pattern), the ratio of producers to injectors is 1:1; in this case, the distinction between normal and inverted is only important if a few patterns are involved, such as for a small pilot flood. For the 7-spot and 9-spot patterns, the distinction between normal and inverted patterns is more important.

There is often confusion between well spacing (or density) and pattern size. As shown in Figure 5.8.4, the pattern area for a 5-spot is twice the well spacing or well density, and the pattern area for a 9-spot is four times the well spacing. When information is given concerning patterns of a given size, the reader is cautioned to find out if well spacing or pattern size is intended.

Dimensions of various distances for 5-spot patterns and 9-spot patterns with different well spacings are given in Table 5.8.3. For the 9-spot pattern, s refers to the shortest distance from a side injection well and the central producer (the opposite for an inverted pattern), and l refers to the longest or lateral distance from injector to producer. For the 5-spot pattern, l is the lateral or straight-line distance from injector to

Table 5.8.2   Well Patterns

Pattern	Ratio of producers to injectors	Drilling pattern
Direct line drive	1	Rectangle
Staggered line drive	1	Offset line of wells
5-spot	1	Square
Normal 7-spot	1/2	Equilateral triangle
Inverted 7-spot	2	Equilateral triangle
Normal 9-spot	1/3	Square
Inverted* 9-spot	3	Square

From Reference 1.133.
*Inverted: One injection well per pattern.

producer, a refers to the distance between wells that are alike, and d refers to the distance between the dissimilar wells. Distances of d and a for the line drive pattern and staggered line drive pattern are shown in Figure 5.8.5, the 5-spot pattern is a special case of the staggered line drive when d/a is 0.5.

Pattern selection is important because it can affect the area swept by the injected fluid. Areal or pattern sweep efficiency is discussed in the section under fluid movement in waterflooded reservoirs, but the principles apply to either water or gas injection.

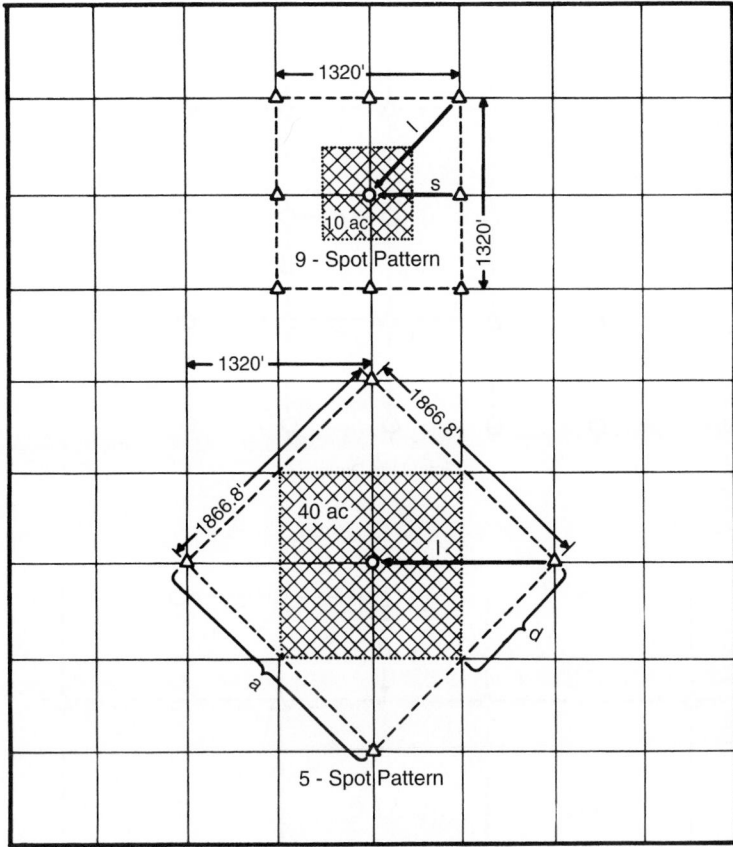

Figure 5.8.4   *Pattern size for 9-spot and 5-spot patterns.*

Table 5.8.3   *Well Distance in 5-Spot and 9-Spot Patterns*

	5-spot pattern			
Well spacing	Pattern size	a (ft)	d (ft)	l (ft)
5 acres	10 acres	660	330	467
10 acres	20 acres	933	467	660
20 acres	40 acres	1,320	660	933
40 acres	80 acres	1,867	933	1,320
	9-spot pattern			
Well spacing	Pattern size		s (ft)	l (ft)
5 acres	20 acres		467	660
10 acres	40 acres		660	933
20 acres	80 acres		933	1,320
40 acres	160 acres		1,320	1,867

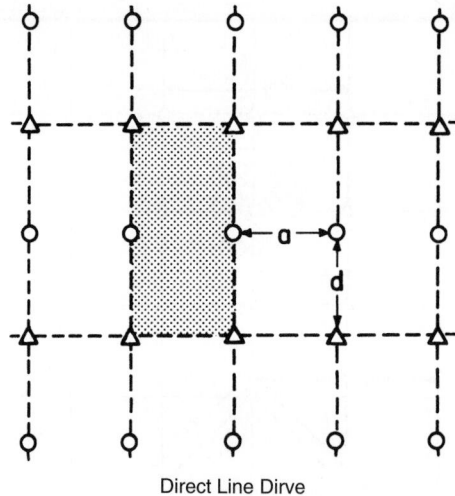

Direct Line Dirve

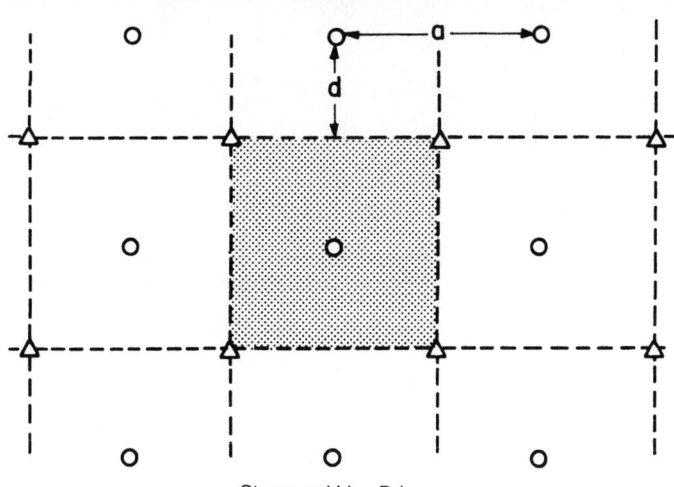

Staggered Line Drive

O Producing Well △ Injection Well ---- Pattern Boundary

Figure 5.8.5 *Distances in line drive patterns [1.25].*

5.9 FLUID MOVEMENT IN WATERFLOODED RESERVOIRS

Many of the principles discussed in this section also apply to immiscible gas injection, primary recovery by gravity drainage, and natural bottom-water drive. However, because of the importance of waterflooding in the United States, the emphasis is placed on fluid movement in waterflooded reservoirs.

The importance of various factors that affect displacement of oil by water were discussed in the first section. In particular, the discussion on the effect of wettability on relative permeability characteristics is important in the understanding of oil displacement during waterflooding.

Several textbooks on waterflooding are available [1.133,5.8, 1–3]. The source most often referred to in this section is the excellent SPE monograph by Craig [1.133]; many of the principles in this monograph are summarized in the Interstate Oil Compact Commission text [2] and in an SPE paper [4]. The text by Smith [5.8] contains many useful aspects of waterflooding, and the recent SPE text [3] contains a more thorough and mathematical treatment of the subject.

5.9.1 Displacement Mechanisms

Under ideal conditions, water would displace oil from pores in a rock in a piston-like manner or at least in a manner representing a leaky piston. However, because of various wetting conditions, relative permeabilities of water and oil are important in determining where flow of each fluid occurs, and the manner in which oil is displaced by water. In addition, the higher viscosity of crude oil in comparison to water will contribute to nonideal displacement behavior. Several concepts will be defined in order that an understanding of displacement efficiencies can be achieved.

5.9.1.1 Buckley–Leverett Frontal Advance

By combining the Darcy equations for the flow of oil and water with the expression for capillary pressure, Leverett [1.100] provided an equation for the fractional flow of water, f_w at any point in the flow stream:

$$f_w = \frac{1 + \dfrac{kk_{ro}}{v_t \mu_o}\left(\dfrac{\partial P_c}{\partial L} - g\Delta\rho\sin\alpha_d\right)}{1 + \dfrac{\mu_w}{\mu_o}\dfrac{k_o}{k_w}} \qquad [5.9.1]$$

where f_w = fraction of water in the flowing stream passing any point in the rock (i.e., the water cut)

k = formation permeability

k_{ro} = relative permeability to oil

k_o = effective permeability to oil

k_w = effective permeability to water

μ_o = oil viscosity

μ_w = water viscosity

v_t = total fluid velocity (i.e., q_t/A)

P_c = capillary pressure = $p_o - p_w$ = pressure in oil phase minus pressure in water phase

L = distance along direction of movement

g = acceleration due to gravity

$\Delta\rho$ = water-oil density differences = $\rho_w - \rho_o$

α_d = angle of the formation dip to the horizontal.

This equation is derived in an appendix in the monograph by Craig [1.133]. Because relative permeabilities and capillary pressure are functions of only fluid saturation, the fractional flow of water is a function of water saturation alone.

In field units, Equation 5.9.1 becomes [1.133]:

$$f_w = \frac{1 + 0.001127\dfrac{kk_{ro}}{\mu_o}\dfrac{A}{q_t}\left(\dfrac{\partial P_c}{\partial L} - 0.433\Delta\rho\sin\alpha_d\right)}{1 + \dfrac{\mu_w}{\mu_o}\dfrac{k_o}{k_w}} \qquad [5.9.2]$$

where permeability is in md, viscosities are in cp, area is in sq ft, flow rate is in B/D, pressure is in psi, distance is in ft, and densities are in g/cc.

In practical usage, the capillary pressure term in Equation 5.9.1 is neglected [1.133]:

$$f_w = \frac{1 - \dfrac{k}{v}\dfrac{k_{ro}}{\mu_o}(g\Delta\rho\sin\alpha_d)}{1 + \dfrac{\mu_w}{\mu_o}\dfrac{k_o}{k_w}} \qquad [5.9.3]$$

and for a horizontal displacement of oil by water, the simplified form of this equation is [1.133]:

$$f_w = \frac{1}{1 + \dfrac{\mu_w}{\mu_o}\dfrac{k_{ro}}{k_{rw}}} \qquad [5.9.4]$$

Examples of idealized fractional flow curves, f_w vs. S_w, are given in Figure 5.9.1 for strongly water-wet and strongly oil-wet conditions [1.133].

Based on the initial work of Leverett [1.100], Buckley and Leverett [1.152] presented equations to describe an immiscible displacement in one-dimensional flow. For incompressible displacement, the velocity of a plane of constant water saturation traveling through a linear system was given by:

$$v = \frac{q}{A\phi}\left(\frac{\partial f_w}{\partial S_w}\right) \qquad [5.9.5]$$

where q is the flow rate in cc/sec (or ft^3/D), A is the cross-sectional area in cm^2 (or ft^2), ϕ is the fractional porosity, v is the velocity or rate of advance in cm/sec (or ft/D), and $(\partial f_w/\partial S_w)$ is the slope of the curve of f_w vs. S_w. This equation states that the rate of advance or velocity of a plane of constant water saturation is directly proportional to the derivative of the water cut at that water saturation. By integrating Equation 5.9.5 for the total time since the start of injection, the distance that the plane of given water saturation moves can be given by:

$$L = \frac{W_i}{A\phi}\left(\frac{\partial f_w}{\partial S_w}\right) \qquad [5.9.6]$$

where W_i is the cumulative water injected and L is the distance that a plane of given saturation has moved.

If L is the distance from injector to producer, the time of water breakthrough, t_{bt}, is given by:

$$t_{bt} = \frac{L}{\dfrac{q}{A\phi}\left(\dfrac{\partial f_w}{\partial S_w}\right)} \qquad [5.9.7]$$

Equation 5.9.6 can be used to calculate the saturation distribution in a linear waterflood as a function of time. According to Equation 5.9.6, the distance moved by a given saturation in a given time interval is proportional to the slope of the fractional flow curve at the saturation of interest. If the slope of the fractional flow curve is graphically obtained at a number of saturations, the saturation distribution in the reservoir can be calculated as a function of time. The saturation distribution can then be used to predict oil recovery and required water injection on a time basis. A typical plot of df_w/dS_w vs. S_w will have a maximum as shown in Figure 5.9.2. However, a problem is that equal values of the slope, df_w/dS_w, can occur at two different saturations which is not possible. To overcome this difficulty, Buckley

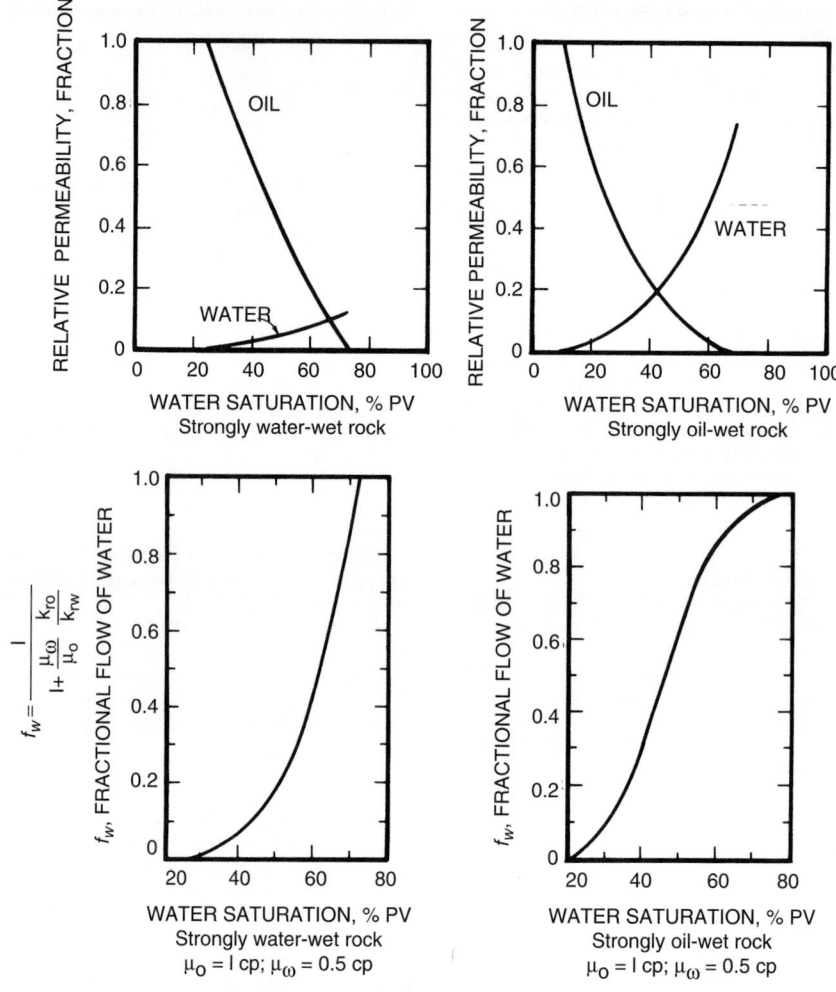

Figure 5.9.1 Effect of water saturation on relative permeabilities and fractional flow of water [1.133].

and Leverett [1.152] suggested that a portion of the saturation distribution curve is imaginary, and that the real curve contains a saturation discontinuity at the front. Since the Buckley-Leverett procedure neglects capillary pressure, the flood front in a practical situation will not exist as a discontinuity, but will exist as a stabilized zone of finite length with a large saturation gradient.

5.9.1.2 Welge Graphical Technique

A more simplified graphical technique was proposed by Welge [153] which involves integrating the saturation distribution from the injection point to the front. The graphical interpretation of this equation is that a line drawn tangent to the fractional flow curve from the initial water saturation (S_{wi}) will have a point of tangency equal to water saturation at the front (S_{wf}). Additionally, if the tangent line is extrapolated to $f_w = 1$, the water saturation will correspond to the average water saturation in the water bank, $\bar{S}_w$. Construction of a Welge plot is shown in Figure 5.9.3. The tangent line should be drawn from the initial water saturation even if that saturation is greater than the irreducible water saturation.

Welge derived an equation that relates the average displacing fluid saturation to the saturation at the producing end of

the system:

$$\bar{S}_w - S_{w2} = Q_i f_{o2} \qquad [5.9.8]$$

where $\bar{S}_w$ = average water saturation, fraction of PV

S_{w2} = water saturation at the producing end of the system, fraction of PV

Q_i = pore volumes of cumulative injected fluid, dimensionless

f_{o2} = fraction of oil flowing at the outflow end of the system

Equation 5.9.8 is important because it relates to three factors of prime importance in waterflooding [1.133]: (1) the average water saturation and thus the total oil recovery, (2) the cumulative injected water volume, and (3) the water cut and hence the oil cut.

Welge also related the cumulative water injected and the water saturation at the producing end:

$$Q_t = \frac{1}{\left(\dfrac{df_w}{dS_w}\right)_{S_{w2}}} \qquad [5.9.9]$$

Thus, the reciprocal of the slope of the tangent line gives the cumulative water influx at the time of water breakthrough. When a value of Q_i and the injection rate are known, the time to reach that stage of the flood can be computed.

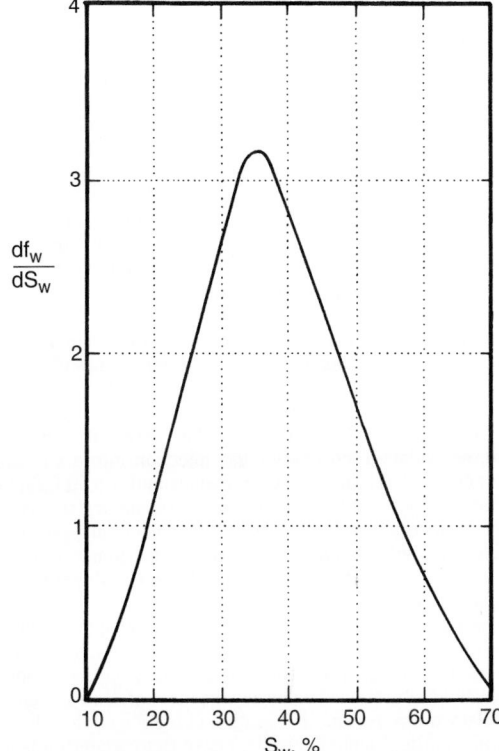

Figure 5.9.2 *Change in fractional flow of water with change in water saturation [1.133].*

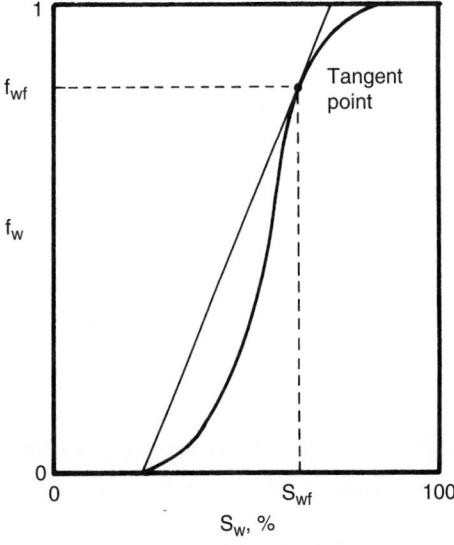

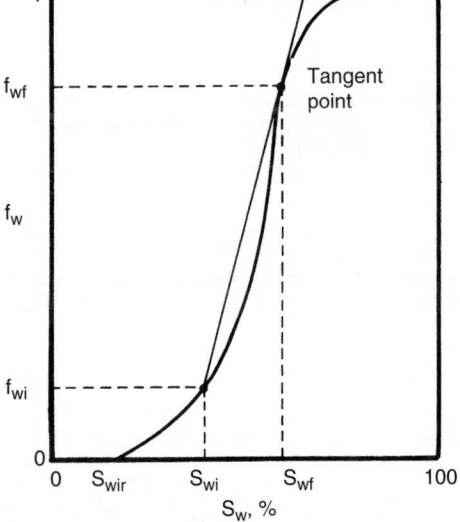

Figure 5.9.3 *Welge graphical plot [1.153].*

For a liquid-filled, linear system, the average water saturation at breakthrough, $\bar{S}_{wbt}$, is:

$$\bar{S}_{wbt} = S_{iw} + \frac{W_i}{A\phi L} \qquad [5.9.10]$$

where S_{iw} is the irreducible or connate water saturation. If Equation 5.9.6 is substituted into Equation 5.9.10:

$$\bar{S}_{wbt} - S_{iw} = \frac{1}{\left(\dfrac{df_w}{dS_w}\right)} = \frac{S_{wf} - S_{iw}}{f_{wf}} \qquad [5.9.11]$$

where S_{wf} is the water saturation at the flood front and f_{wf} is the water cut at the flood front. After breakthrough, water saturation is obtained from Equations 5.9.8 and 5.9.9 where, as mentioned earlier: (1) the tangent point, S_{w2}, represents the water saturation at the producing end of the system, (2) the value of f_w at the point of tangency is the producing water cut, (3) the saturation at which the tangent intersects $f_w = 1.0$ is the average water saturation, and (4) the inverse of the slope of the tangent line is equal to the cumulative injected fluid in pore volumes (Q_i). If connate water is mobile, appropriate corrections need to be made [1.133].

Oil production at breakthrough can be computed from [3]:

$$N_{pbt} = \frac{\phi AL}{B_o}(\bar{S}_{wbt} - S_{iw}) \qquad [5.9.12]$$

After water breakthrough, a number of saturation greater than S_{wf} are selected; the slope of the tangent line and average water saturation are determined for each value of S_w chosen. Oil production after breakthrough is then determined by observing the change in water saturation [3]:

$$\Delta N_p = \frac{\phi AL}{B_o}(\bar{S}_w - \bar{S}_{wbt}) \qquad [5.9.13]$$

The incremental oil production from Equation 5.9.13 can be added to the breakthrough production from Equation 5.9.12, and the resulting total production for the linear system can be listed as a function of S_w, time, or other parameters. If the pore volumes in these equations are in ft^3, divide by 5.615 to get barrels.

5.9.2 Viscous Fingering

A problem often encountered in the displacement of oil by water is the viscosity contrast between the two fluids. The adverse mobility ratios that result promote fingering of water through the more viscous crude oil and can reduce the oil recovery efficiency. An example of viscous fingering is shown in Figure 5.9.4.

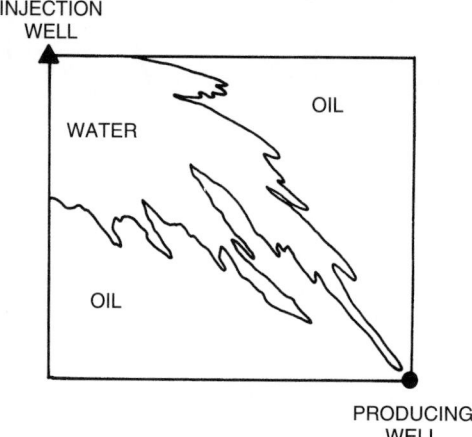

INJECTION
WELL

OIL

WATER

OIL

PRODUCING
WELL

Figure 5.9.4   Viscous fingering.

5.9.3 Mobility and Mobility Ratio

Mobility of a fluid is defined as the ratio of the permeability of the formation to a fluid, divided by the fluid viscosity:

$$\lambda = \frac{k}{\mu} \qquad [5.9.14]$$

where λ = mobility, md/cp
  k = effective permeability of reservoir rock to a given fluid, md
  μ = fluid viscosity, cp

When multiple fluids are flowing through the reservoir, relative permeabilities must be used along with viscosities of the fluids. By convention, the term mobility ratio is defined as the mobility of the displacing fluid divided by the mobility of the displaced fluid. For waterfloods, this is the ratio of water to oil mobilities. Thus the mobility ratio, M, for a waterflood is:

$$M = \frac{k_{rw}/\mu_w}{k_{ro}/\mu_o} = \frac{k_{rw}\mu_o}{k_{ro}\mu_w} \qquad [5.9.15]$$

where k_{rw} and k_{ro}, are relative permeabilities to water and oil, respectively, μ_o is oil viscosity and μ_w is water viscosity. Prior to 1957, there was no accepted definition, and many workers defined mobility ratio as oil to water mobility; in this case, the reciprocal of mobility ratio (as now accepted) must be used. The oil mobility used in Equation 5.9.10 refers to the location in the oil bank ahead of the flood front. For the water mobility, there are several possibilities regarding the location at which the relative permeability should be chosen: at the flood front, at residual oil saturation where only water is flowing (end point), or at some intermediate saturation. Craig [1.133] found a better correlation if the water mobility was determined at the average water saturation behind the flood front at water breakthrough. Thus for the mobility ratio expression, the relative permeability of water is found at the average water saturation at water break-through as determined by the Welge graphical approach. As Craig notes, the mobility ratio of a waterflood will remain constant before breakthrough, but it will increase after water breakthrough corresponding to the increase in water saturation and relative permeability to water in the water-contacted portion of the reservoir. Unless otherwise specified, the term mobility ratio is taken to be the value prior to water breakthrough. As will be discussed later in this section, mobility ratio is important in determining the volume of reservoir contacted by the waterflood.

5.9.4 Recovery Efficiency

Recovery efficiency is the fraction of oil in place that can be economically recovered with a given process. The efficiency of primary recovery mechanisms will vary widely from reservoir to reservoir, but the efficiencies are normally greatest with water drive, intermediate with gas cap drive, and least with solution gas drive. Results obtained with waterflooding have also varied. The waterflood recovery can range from less than the primary recovery to as much as 2.5 times the recovery obtained in some solution-gas drive reservoirs. A recent statistical analysis by the API [4.1] provided the average primary and secondary recovery efficiencies in Table 5.9.1. Generally, primary and ultimate recoveries from carbonate reservoirs tend to be lower than from sandstones. For pattern waterfloods, the average ratio of secondary to primary recovery ranges from 0.33 in California sandstones to greater than one in Texas carbonates. For edge water injection, the secondary-to primary ratio ranged from an average of 0.33 in Louisiana to 0.64 in Texas. By comparison, secondary recovery for gas injection into a gas cap averaged only 0.23 in Texas sandstones and 0.48 in California sandstones. Ultimate primary and secondary recovery performance for different drive mechanisms are given in Table 5.9.2. Solution-gas-drive reservoirs will generally have higher oil saturations after primary recovery, and are usually the better candidates for waterflooding.

Displacement of oil by waterflooding is controlled by fluid viscosities, oil-water relative permeabilities, nature of the reservoir rock, reservoir heterogeneity, distribution of pore sizes, fluid saturations (especially the amount of oil present), capillary pressure, and the location of the injection wells in relation to the production wells. These factors contribute to the overall process efficiency. Oil recovery efficiency (E_R) of a waterflood is the product of displacement efficiency (E_D) and volumetric efficiency (E_V), both of which can be correlated with fluid mobilities:

$$E_R = E_D E_V = E_D E_P E_I \qquad [5.9.16]$$

where E_R = overall reservoir recovery or volume of hydrocarbons recovered divided by volume of hydrocarbons in place at start of project
  E_D = volume of hydrocarbons (oil or gas) displaced from individual pores or small groups of pores divided by the volume of hydrocarbons in the same pores just prior to displacement
  E_P = pattern sweep efficiency (developed from areal efficiency by proper weighting for variations in net pay thickness, porosity, and hydrocarbon saturation): hydrocarbon pore space enclosed behind the injected-fluid front divided by total hydrocarbon pore space of the reservoir or project.
  E_I = hydrocarbon pore space invaded (affected, contacted) by the injection fluid or heat-front divided by the hydrocarbon pore space enclosed in all layers behind the injected fluid

5.9.5 Displacement Sweep Efficiency (E_D)

Factors affecting the displacement efficiency for any oil recovery process are pore geometry, wettability (water-wet, oil-wet, or intermediate), distribution of fluids in the reservoir, and the history of how the saturation occurred. Results are displayed in the relative permeability curves (Figure 5.9.1) from which the flowing water saturation (or conversely the oil saturation) can be obtained at any total fluid saturation. As shown in Figure 5.9.5, displacement efficiencies decrease as oil viscosities increase [1.133].

Table 5.9.1   *Ultimate Recovery (Primary Plus Secondary)*

	Average OOIP B/NAF	Ultimate recovery at Average OOIP B/NAF	Recovery efficiency			
			Primary % OOIP	Secondary % OOIP	Ultimate % OOIP	Secondary to primary ratio
Pattern waterfloods						
California sandstones	1,311	463	26.5	8.8	35.3	0.33
Louisiana sandstones	1,194	611	36.5	14.7	51.2	0.40
Oklahoma sandstones	728	201	17.0	10.6	27.6	0.62
Texas sandstones	942	362	25.6	12.8	38.4	0.50
Wyoming sandstones	774	346	23.6	21.1	44.7	0.89
Texas carbonates	388	123	15.5	16.3	31.8	1.05
Edge water injection						
Louisiana sandstones	1,181	680	41.3	13.8	55.1	0.33
Texas sandstones	897	499	34.0	21.6	55.6	0.64
Gas injection into cap						
California sandstones	909	396	29.4	14.2	43.6	0.48
Texas sandstones	957	412	35.3	8.0	43.3	0.23

From References 4.1 and 4.5.
Recovery Efficiency: Average value of the recoverable oil divided by the average value of the original oil-in-place for the reservoirs in the classification.
OOIP: Original oil-in-place
B/NAF: Barrels per net acre-ft.

Table 5.9.2   *Secondary Recovery Efficiencies*

Secondary recovery method	Lithology	State	Primary plus secondary recovery efficiency (% OOIP)	Ratio of secondary to primary recovery efficiency
Pattern waterflood	Sandstone	California	35	0.33
		Louisiana	51	0.40
		Oklahoma	28	0.62
		Texas	38	0.50
		Wyoming	45	0.89
Pattern waterflood	Carbonates	Texas	32	1.05
Edge water injection	Sandstone	Louisiana	55	0.33
		Texas	56	0.64
Gas cap injection	Sandstone	California	44	0.48
		Texas	43	0.23

From References 4.1 and 4.5.

5.9.6 Volumetric Sweep Efficiency (E_V)

Whereas displacement efficiency considers a linear displacement in a unit segment (group of pores) of the reservoir, macroscopic or volumetric sweep takes into account that fluid (i.e., water) is injected at one point in a reservoir and that other fluids (i.e., oil, water) are produced from another point (Figure 5.9.6). Volumetric sweep efficiency, the percentage of the total reservoir contacted by the injected fluid (often called fluid conformance), is composed of areal (or pattern) efficiency and vertical sweep.

5.9.7 Areal or Pattern Sweep Efficiency (E_P)

Areal sweep efficiency of an oil recovery process depends primarily on two factors: the flooding pattern and the mobilities of the fluids in the reservoir. In the early work on sweep efficiency and injectivity, Muskat and coworkers [1.25,5] presented analytical solutions for direct line drive, staggered line drive, 5-spot, 7-spot, and 9-spot patterns (patterns were discussed earlier; see Figure 5.8.3). Experimental studies on the effect of mobility ratio for different patterns were presented by Dyes, Caudle, and Erickson [6] (5-spot and line drives); Craig, Geffen, and Morse [7], Prats et al. [8], Caudle and Witte [9], and Haberman [10] (5-spot); and Kimbler, Caudle, and Cooper [11] (9-spot). The effect of sweepout beyond the pattern area was studied as well [12,13]. From a mathematical study the breakthrough sweep efficiency of the staggered line drive was presented by Prats [14]. A comparison of the areal sweep efficiency and the ratio d/a is shown in Figure 5.9.7 for direct and staggered line drives [1.25,14], and a review of the early work was provided by Crawford [15].

Areal sweep efficiency at breakthrough for a 5-spot pattern is shown in Figure 5.9.8, and the effect of mobility ratio on areal sweep is shown in Figure 5.9.9. These figures show that areal sweep efficiency is low when mobility ratio is high (note that the data in Figure 5.9.9 from Dyes, Caudle and Erickson are plotted in terms of the reciprocal of mobility ratio as currently defined). Areal sweep efficiencies at

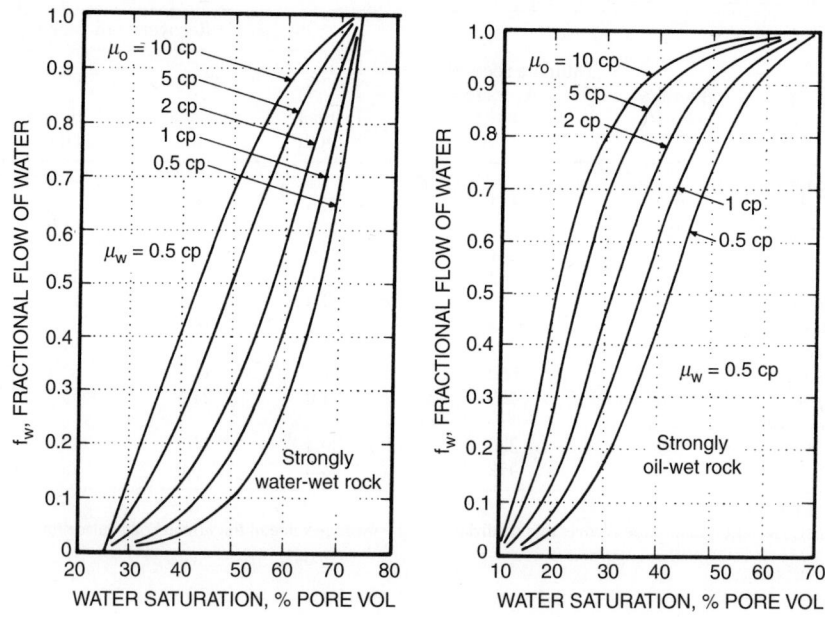

Figure 5.9.5   *Effect of oil viscosity on fractional flow of water [1.133].*

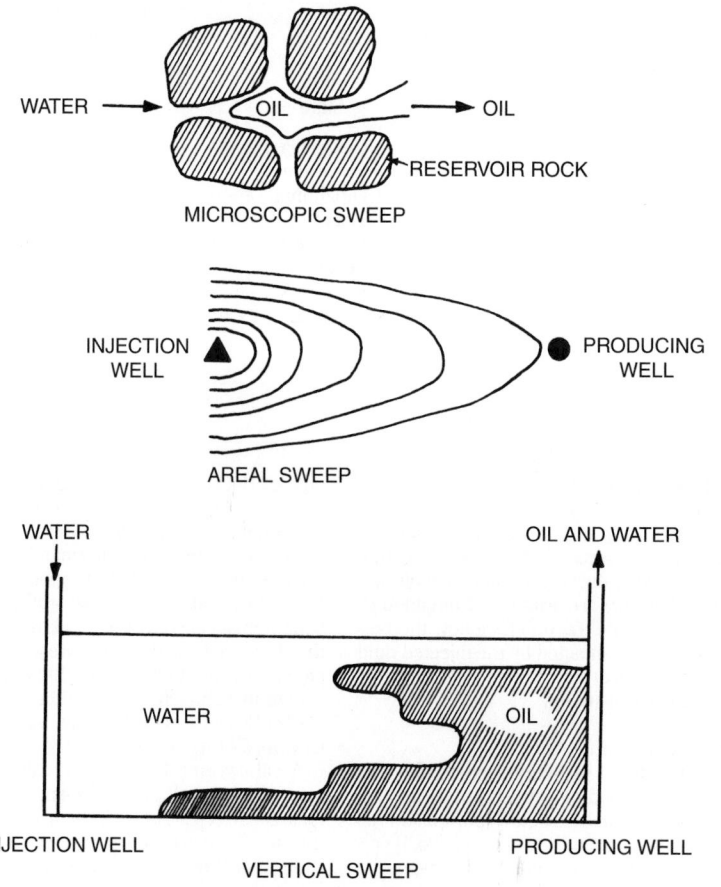

Figure 5.9.6   *Sweep efficiencies.*

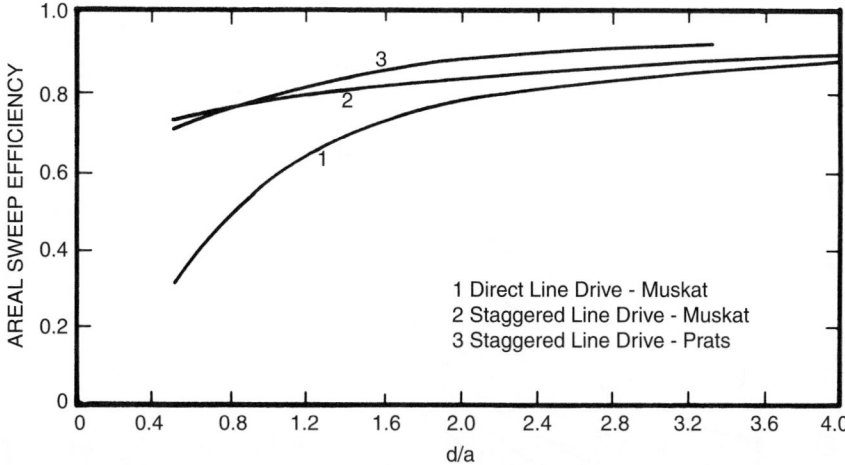

Figure 5.9.7 *Areal sweep efficiencies for direct and staggered line drive patterns [1.25,14].*

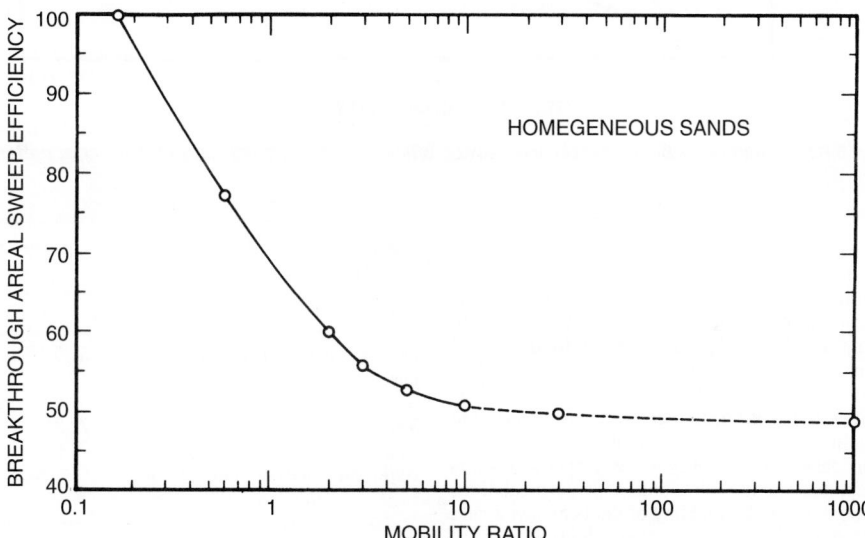

Figure 5.9.8 *Effect of mobility ratio on areal sweep efficiency at breakthrough for a 5-spot pattern [1.133].*

breakthrough, for different patterns and a mobility ratio of one, are summarized in Table 5.9.3 [1.133,2,4].

Areal sweep efficiency is more important for considering rate vs. time behavior of a waterflood rather than ultimate recovery because, at the economic limit, most of the interval flooded has either had enough water throughout to provide 100% areal sweep or the water bank has not yet reached the producing well so that no correction is needed for areal sweep [1.133].

When waterflooding calculations are performed, especially with computers or programmable calculators, the use of equations with adjustable coefficients are very useful. Recently, Fassihi [16] provided correlations for the calculation of areal and vertical sweep efficiencies. For these correlations of areal sweep, the data of Dyes, Caudle, and Erickson [6] were curve-fitted and the resulting equation was:

$$\frac{1-E_P}{E_P} = [a_1 \ln(M+a_2)+a_3]f_w + a_4 \ln(M+a_5)+a_6 \qquad [5.9.17]$$

where E_P is the areal sweep efficiency which is the fraction of the pattern area contacted by water, M is the mobility ratio, and the coefficients are as listed in Table 5.9.4 for the 5-spot, direct line drive, and staggered line drive. These coefficients are valid both before and after breakthrough, and apply to mobility ratios between zero and ten, which is within the range observed in many waterfloods. For the 5-spot pattern, these values of E_p are generally higher than in later experiments, and a correction has been suggested by Claridge [17] that should be multiplied by the E_P from the Dyes et al. data:

$$E_v = \frac{E_P/V_d}{\{M^{0.5}-[(M-1)(1-E_P/V_d)]^{0.5}\}^2} \qquad [5.9.18]$$

where E_V is the volumetric sweep efficiency in a linear displacement, V_d is the displaceable pore volumes injected, and the other terms are as already defined.

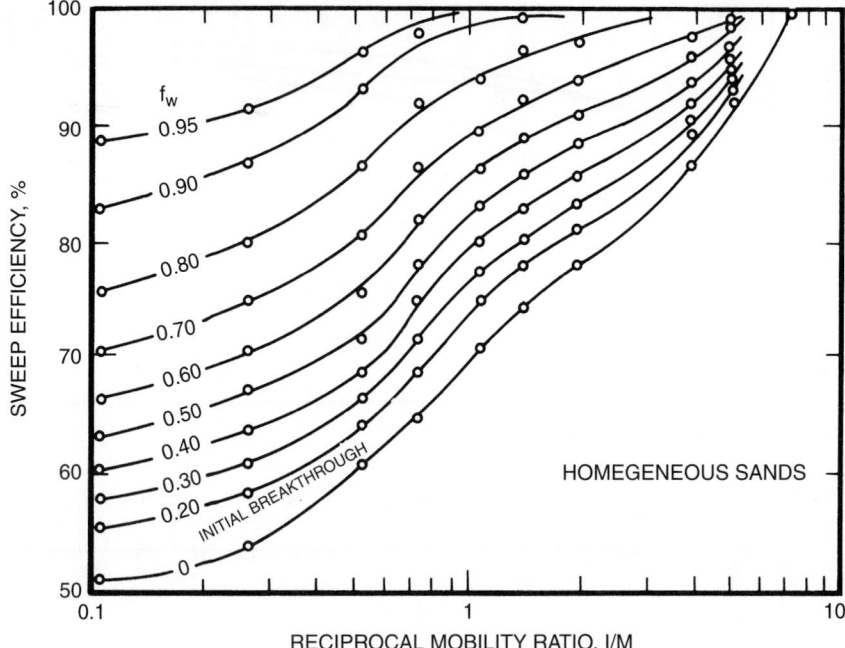

Figure 5.9.9 *Effect of mobility ratio on areal sweep efficiency after breakthrough for a 5-spot pattern [6].*

5.9.8 Vertical or Invasion Sweep Efficiency (E_I)

For well-ordered sandstone reservoirs, the permeability measured parallel to the bedding planes of stratified rocks is generally larger than the vertical permeability. For carbonate reservoirs, permeability (and porosity) may have developed after the deposition and consolidation of the formation; thus the concept of a stratified reservoir may not be valid. However, in stratified rocks, vertical sweep efficiency takes into account the inherent vertical permeability variations in the reservoir. Vertical sweep efficiency of a waterflood depends primarily upon the vertical distribution of permeabilities within the reservoir, on the mobility of fluids involved, and on the density differences between flowing fluids. As a result of nonuniformity of permeabilities in the vertical direction, fluid injected into an oil-bearing formation will seek the paths of least resistance and will move through the reservoir as an irregular front. Consequently, the injected fluid will travel more rapidly in the more permeable zones and will travel less rapidly in the tighter zones. With continued injection, and displacement of some of the resident fluids, the saturation of the injected fluid will become greater in the more permeable areas than in the low-permeability strata. This can cause early breakthrough of injected fluid into the producing wells before the bulk of the reservoir has been contacted. In addition, as the saturation of the injected fluid increases in the highly permeable zones, the relative permeability to that fluid also increases. All of these effects can lead to channeling of the injected fluid, which is aggravated by the unfavorable viscosity ratio common in waterflooding. In many cases, permeability stratification has a dominant effect on behavior of the waterflood.

5.9.9 Permeability Variation

Two methods of quantitatively defining the variation in vertical permeabilities in reservoirs are commonly used. The extent of permeability stratification is sometimes described

Table 5.9.3 *Areal Sweep Efficiency at Breakthrough (M = 1)*

Type of pattern	Ep, % [4]	Ep, % [2]*
Direct line drive (d/a = 1)	57.0	57
Direct line drive (d/a = 1.5)	70.6	—
Staggered line drive	80.0	75 (d/a = 1)
5-spot	72.3	68–72
7-spot	74.0	74–82
9-spot diagonal/directional rate		
0.5	—	49
1	—	54
5	—	69
10	—	78

From Reference 1.133.
*Based on summary of data presented by Craig.

Table 5.9.4 *Coefficients in Areal Sweep Efficiency Correlations*

Coefficient	5-spot	Direct line drive	Staggered line drive
a_1	−0.2062	−0.3014	−0.2077
a_2	−0.0712	−0.1568	−0.1059
a_3	−0.511	−0.9402	−0.3526
a_4	0.3048	0.3714	0.2608
a_5	0.123	−0.0865	0.2444
a_6	0.4394	0.8805	0.3158

From Reference 16.

with the Lorenz coefficient [18] and is often described with the Dykstra–Parsons [19] coefficient of permeability variation.

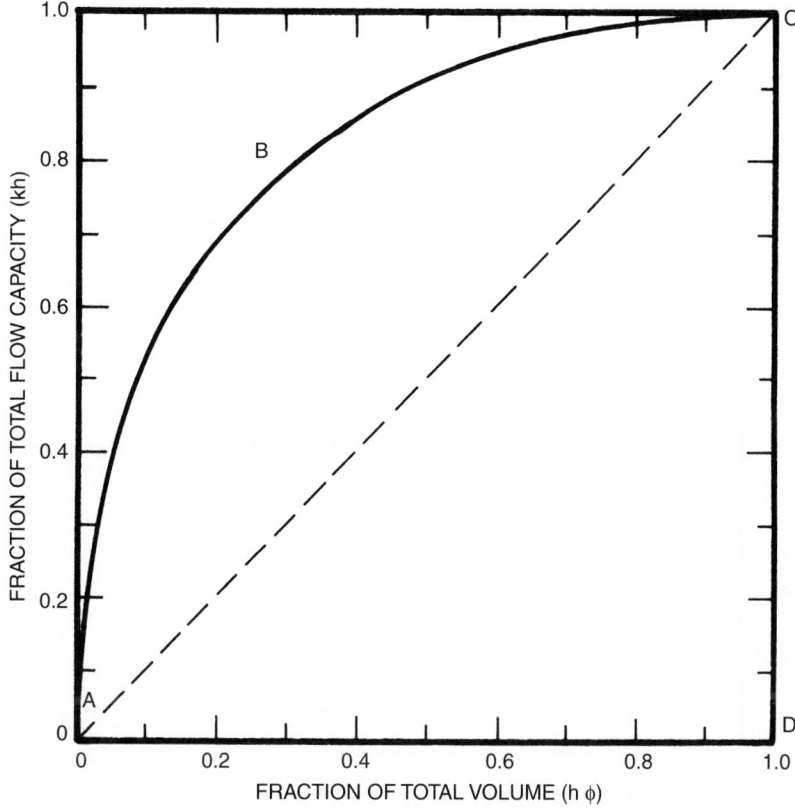

Figure 5.9.10 *Lorenz coefficient plot [1.133].*

5.9.9.1 Lorenz Coefficient
Schmalz and Rahme [18] suggested arranging the vertical distribution of permeabilities from highest to lowest, and plotting the fraction of total flow capacity (kh) versus the fraction of total volume (hϕ). To obtain the Lorenz coefficient (see Figure 5.9.10), the area ABCA is divided by the area ADCA. Values of the Lorenz coefficient can range from zero for a uniform reservoir to a theoretical maximum value of one. However, the Lorenz coefficient is not a unique measure of stratification, and several different permeability distributions can give the same Lorenz coefficient [1.133].

5.9.9.2 Dykstra–Parsons Coefficient of Permeability Variation
The coefficient of permeability variation described by Dykstra and Parsons [19] is also referred to as the permeability variation or permeability variance. This method assumes that vertical permeabilities in a reservoir will have a lognormal distribution. The procedure outlined by Dykstra and Parsons was to: (1) divide permeabilities (usually from core analysis) so that all samples are of equal thickness (often 1 ft), (2) arrange the permeabilities in descending order from highest to lowest, (3) calculate for each sample the percent of samples that have a higher permeability (see example in Table 5.9.5), (4) plot the data from Step 3 on log-probability paper (see Figure 5.9.11) (5) draw the best straight line through data (with less emphasis on points at the extremities, if necessary), (6) determine the permeability at 84.1% probability ($k_{84.1}$) and the mean permeability

at 50% probability (k_{50}), and (7) compute the permeability variation, V:

$$V = \frac{k_{50} - k_{84.1}}{k_{50}} \qquad [5.9.19]$$

Table 5.9.5 *Data for Permeability Variation plot*

Permeability (md)	Percent of samples with greater than stated permeability
950	0
860	5
640	10
380	15
340	20
280	25
210	30
160	35
135	40
130	45
110	50
78	55
65	60
63	65
54	70
40	75
27	80
21	85
20	90
15	95

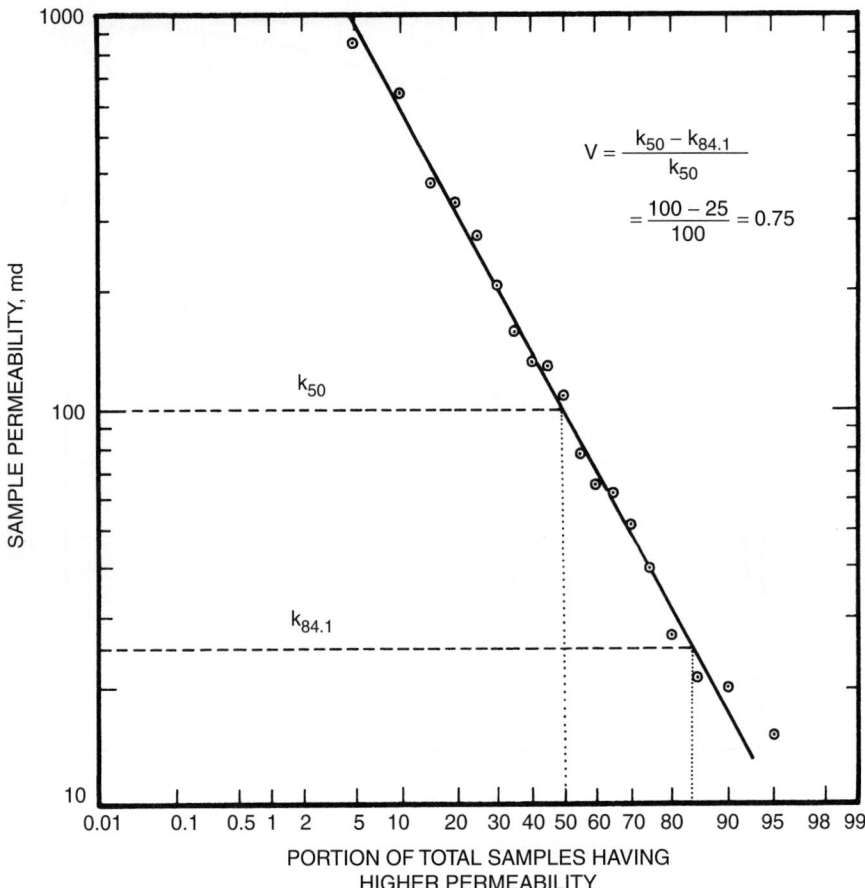

$$V = \frac{k_{50} - k_{84.1}}{k_{50}}$$

$$= \frac{100 - 25}{100} = 0.75$$

Figure 5.9.11 *Dykstra–Parsons plot of permeability variation [1.133,19].*

As with the Lorenz coefficient, the possible values of the Dykstra–Parsons permeability variation range from zero for a uniform reservoir to a maximum value of 1. In some cases, there may be a direct relation between the Lorenz and Dykstra–Parsons coefficients [20], but in many instances a direct relationship with field data will not be observed. Often, insufficient data are available to provide enough samples for adequate analysis, and in some cases, the data may not provide a log-normal distribution. In the remainder of this chapter, the term permeability variation will refer to the Dykstra–Parsons coefficient of permeability variation.

Increasing values of permeability variation indicate increasing degrees of vertical heterogeneity in a reservoir. Permeability variations often range from about 0.5 to 0.8; lower numbers may be observed for relatively uniform reservoirs, and higher numbers may be calculated for very nonuniform reservoirs. Using the data from Dykstra and Parsons, Johnson [21] provided a graphical technique to estimate recovery during an immiscible displacement. One of Johnson's plots is reproduced in Figure 5.9.12 for a producing water-oil ratio (WOR) of 100 which could represent the economic limit for many waterfloods. Lines of constant recovery are given as functions of permeability variation and mobilitiy. Johnson also provided plots for WOR = 1, WOR = 5, and WOR = 25. At any WOR, and increase in vertical permeability variation yielded a lower recovery. As will be discussed later under prediction methods, the

Dykstra–Parsons fractional recovery, R, as a percent of oil in place, must be muliplied by the areal sweep efficiency, Ep, to obtain an estimate of the oil recovered.

As mentioned earlier, correlations for calculating vertical and areal sweep efficiencies were recently provided by Fassihi [16]. The correlating parameter, Y, for vertical coverage, C, is:

$$Y = a_1 C^{a_2} (1 - C)^{a_3} \qquad [5.9.20]$$

where $a_1 = 3.334088568$
$a_2 = 0.7737348199$
$a_3 = -1.225859406$

where the equation for Y was given by deSouza and Brigham [22] in terms of water-oil ratio (WOR), mobility ratio (M), and permeability variation (V):

$$Y = \frac{(WOR + 0.4)(18.948 - 2.499V)}{(M + 1.137 - 0.8094V)^{f(V)}} \qquad [5.9.21]$$

where $f(V) = -0.6891 + 0.9735V + 1.6453V^2$.

These equations are valid for mobility ratios ranging from 0 to 10 and for permeability variations ranging from 0.3 to 0.8.

Based on calculations of WOR vs. oil recovery for a 5-spot pattern, Craig [23] found that there was a minimum number of equal thickness layers required to obtain the same performance as with 100 layers. Table 5.9.6 shows the effect

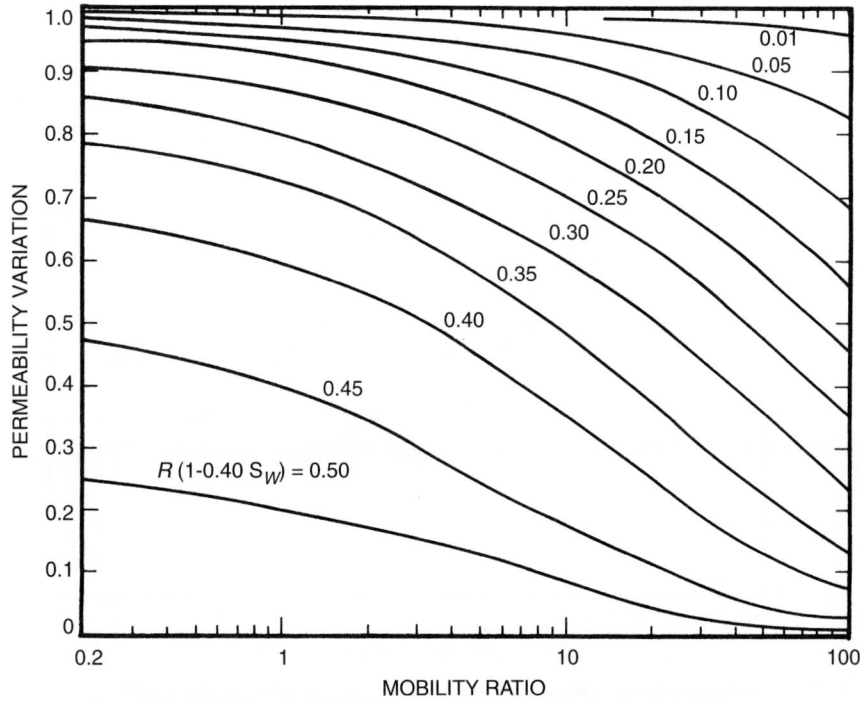

Figure 5.9.12 *Effect of permeability variation and mobility ratio on vertical sweep efficiency at WOR = 100 [21].*

Table 5.9.6 *Minimum Number of Equal-Thickness Layers Required to Obtain Performance of a 100-Layer, 5-Spot Waterflood at Producing WOR's Above 10*

Mobility Ratio	Permeability variation							
	0.1	0.2	0.3	0.4	0.5	0.6	0.7	0.8
0.05	1	1	1	2	4	5	10	20
0.1	1	1	1	2	5	5	10	20
0.2	1	1	2	3	5	5	10	20
0.5	1	1	2	3	5	5	10	20
1.0	1	1	2	3	5	10	10	50
2.0	1	2	3	4	10	10	20	100
5.0	1	3	4	5	10	100	100	100

From Reference 23.

of permeability variation and mobility ratio on the minimum number of layers for WORs above 10. Craig [23] presented similar tables for lower WORs.

A recent analytical extension [24] of the Dykstra–Parsons method allows calculations of total flow rates and flow rates in each layer for both a constant injection rate and for a constant pressure drop. The ability to calculate cumulative injection into a layer allows the incorporation of sweep efficiency of each layer as a function of mobility ratio and displaceable pore volumes injected for the pattern used in the waterflood.

5.9.9.3 Crossflow

In the usual cases where there is vertical communication between the different layers of varying permeabilities, the effect of vertical crossflow must be considered [25,26]. Goddin et al. [26] performed a numerical simulation in a 2-D,

2-layer, water-wet system. For mobility ratios ranging from 0.21 to 0.95, oil recovery with crossflow was between that computed for a uniform reservoir and that for a layered reservoir with no crossflow. Goddin et al. [26] defined a crossflow index, which is a measure of the extent the performance varies from that of a uniform permeability system:

$$\text{crossflow index} = \frac{N_{pcf} - N_{pncf}}{N_{pu} - N_{pncf}} \qquad [5.9.22]$$

where N_{pu} = oil recovery from uniform system with the average permeability
N_{pcf} = oil recovery from layered system with crossflow
N_{pncf} = oil recovery from stratified system with no crossflow

Of the variables investigated, mobility ratio and the permeability ratio of the two layers had the largest effect on crossflow (see Figures 5.9.13 and 5.9.14, respectively). Crossflow was more pronounced at lower mobility ratios or at high ratios of layer permeabilities. The crossflow index of one means that the performance of the layered system with crossflow is identical to the performance of the system with uniform permeability.

Still at issue is the relative importance of mobility ratio and gravity in waterflooding stratified reservoirs [27–31]. For wetting conditions that are not strongly water-wet, additional complications will arise.

5.9.9.4 Estimates of Volumetric Sweep Efficiency

Volumetric sweep efficiency ranges from about 0.1 for very heterogeneous reservoirs to greater than 0.7 for homogeneous reservoirs with good flooding characteristics [3]. For a liquid-filled, 5-spot pattern, Craig [23] found that the volumetric sweep efficiency (E_V) at breakthrough

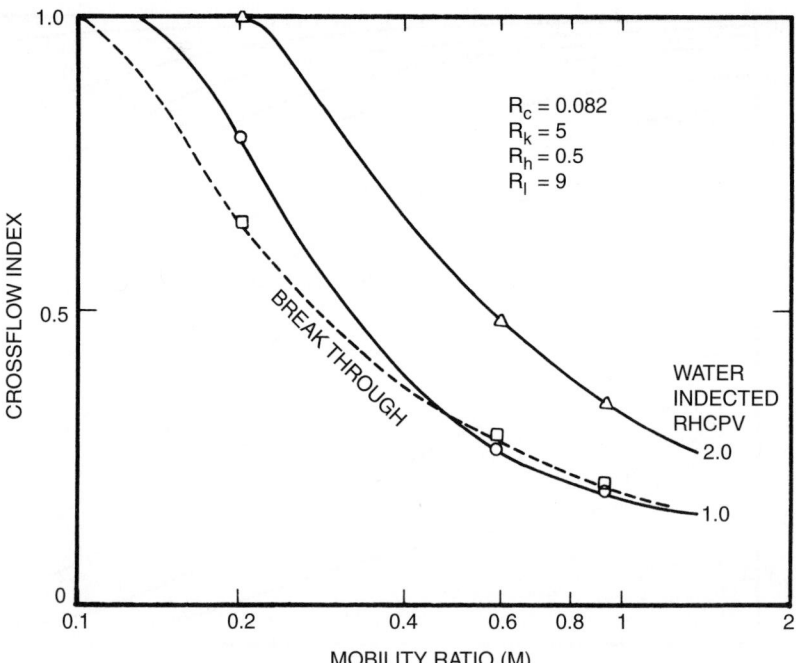

Figure 5.9.13 *Influence of mobility ratio on vertical crossflow [26].*

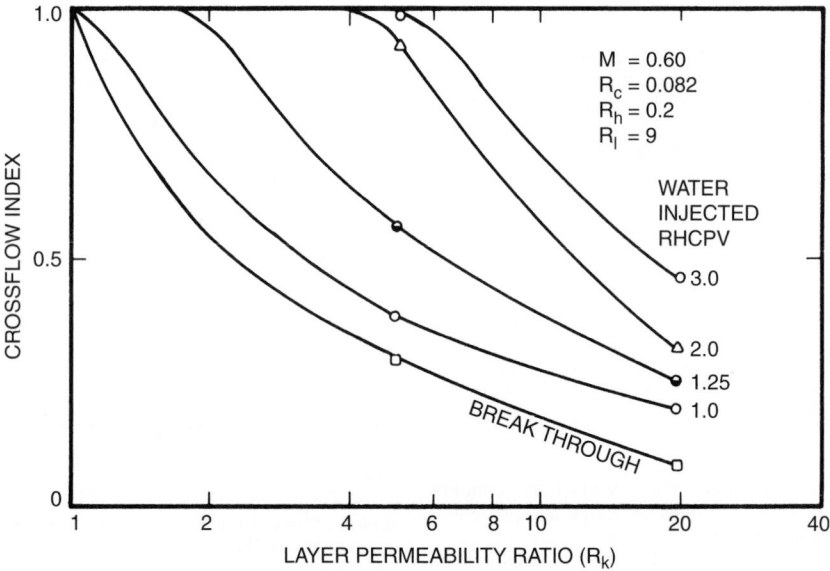

Figure 5.9.14 *Effect of layer permeability ratio on crossflow [26].*

decreases sharply as the permeability variation increases (see Figure 5.9.15). Similar trends were observed for initial gas saturations of 10% and 20%. These data indicated that the major effect of mobility ratio on E_V at breakthrough occurs for mobility ratios ranging from 0.1 to 10.

More recent simulations [32] of 5-spot patterns with a streamtube model yielded the volumetric sweep efficiencies shown in Figures 5.9.16 and 5.1.17 for WORs of 25

and 50, respectively. Mobility ratios of 0.1, 1, 10, 30, and 100 were used. The permeabilities in the 100-layer model were assumed to have a log-normal distribution, and pseudorelative permeability expressions were used. In a companion paper [33] the streamtube model (no crossflow) was compared to the Dykstra–Parsons method (no crossflow) and with a model having the assumption of equal pressure gradient in each layer (with crossflow). The streamtube model

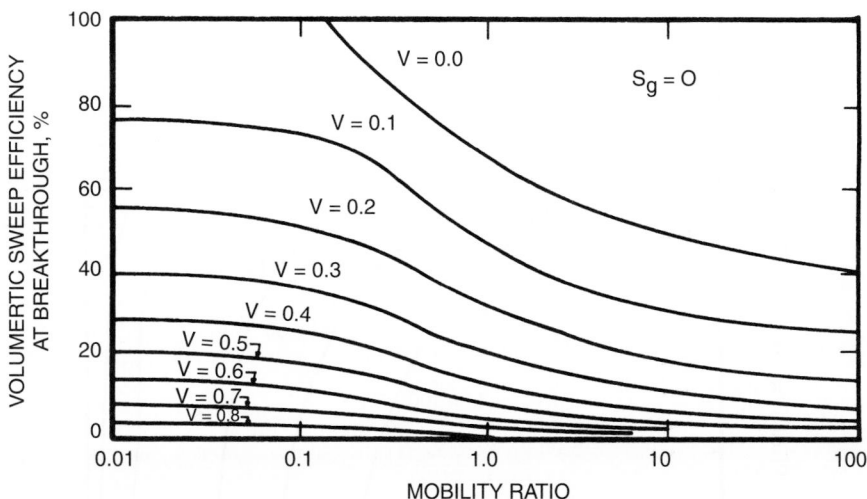

Figure 5.9.15 *Effect of permeability variation and mobility ratio on volumetric sweep at breakthrough [23].*

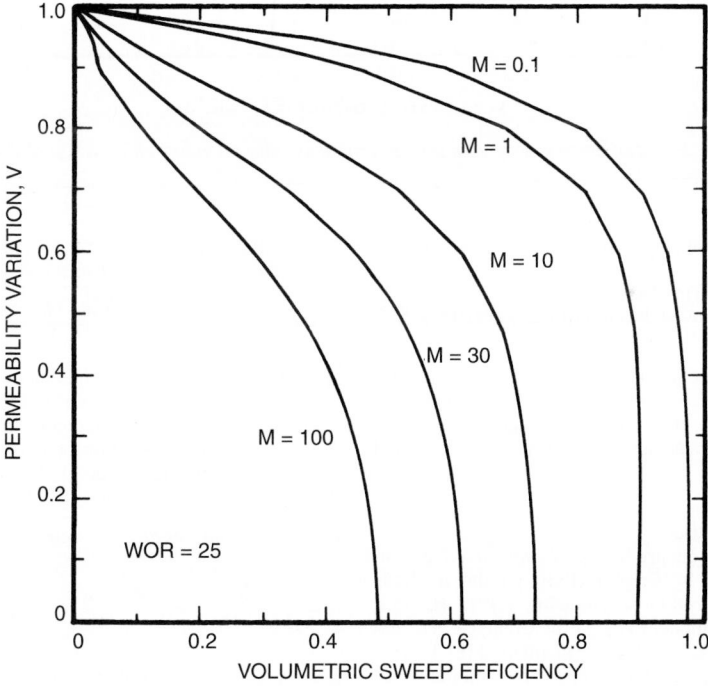

Figure 5.9.16 *Effect of permeability variation and mobility ratio on volumetric sweep at breakthrough [32].*

was more closely described by the model with vertical communication for unfavorable (high) mobility ratios and by the Dykstra–Parsons model for favorable (low) mobility ratios.

5.9.10 Estimation of Waterflood Recovery by Material Balance

Oil recovered by waterflooding, N_{pw}, in STB, can be estimated from [3]:

$$N_{pw} = \frac{7,758 Ah\phi[S_{op} - E_V S_{or} - (1 - E_v)S_{oi}]}{B_o} \qquad [5.9.23]$$

where 7,758 is the number of barrels per acre-ft, A is areal extent of the reservoir in acres, h is reservoir thickness in ft, ϕ is the fractional porosity, S_{op} is the oil saturation at the start of waterflooding, S_{or} is the waterflood residual oil saturation, S_{oi} is the initial oil saturation, B_o is the oil formation volume factor, and E_V is the volumetric sweep or fraction of the reservoir volume swept by the injected water when the economic limit has been reached. In terms of the original oil in place, the waterflood recovery is [3]:

$$N_{pw} = (N - N_p) - N \frac{B_{oi}}{B_o} \left[1 + E_V \left(\frac{S_{or}}{S_{oi}} - 1 \right) \right] \qquad [5.9.24]$$

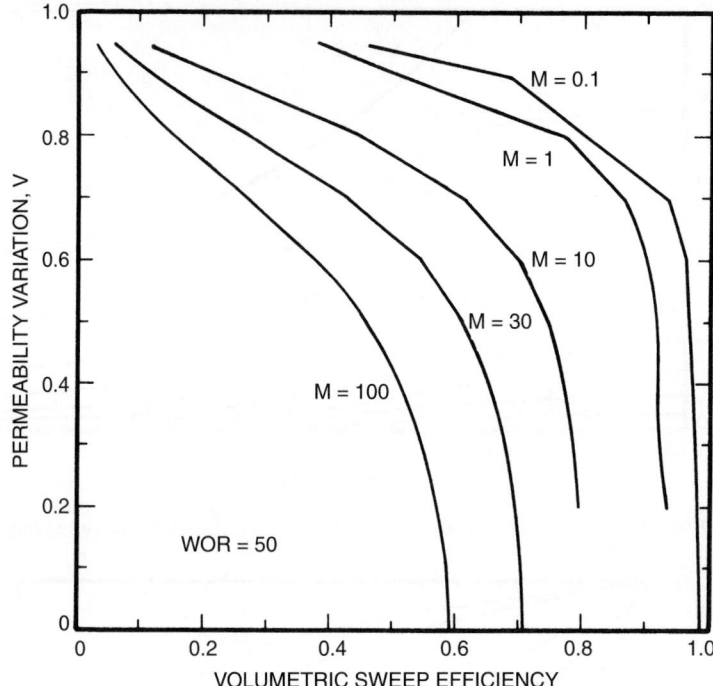

Figure 5.9.17 *Effect of permeability variation and mobility ratio on volumetric sweep at WOR = 50 [32].*

where N_{pw} = oil potentially recoverable by waterflooding, STB

N = initial oil in place, STB

N_p = oil produced during primary operations, STB

B_{oi} = initial FVF

These equations can be altered to include a residual gas saturation, if present. The volumetric sweep efficiency can be estimated from one of the correlations given previously or can be obtained from an analogy from similar water-flood projects.

5.9.11 Prediction Methods

An extensive survey on prediction of waterflood performance was provided by Craig [1.133]. Of the methods reviewed, three appeared most promising: (1) the Higgins-Leighton streamtube model [34], (2) the Craig, Geffen, and Morse model [7], and (3) the Prats et al. method [8]. Discussion of the various prediction methods is beyond the scope of this text, and only two very simple methods will be presented for illustrative purposes. Both the Dykstra–Parsons [19] and Stiles [35] methods are very cursory and, if used, they are normally followed by more extensive evaluations, usually by computer simulation.

For either the Dykstra–Parsons or Stiles methods, the permeabilities are arranged in descending order. For the Dykstra–Parsons method, the permeability variation is determined as described earlier. Two options are then possible: a program [36] for hand-held calculators can be used, or the graphical technique presented by Johnson [21] can be used. The fractional recovery, R, (see Figure 5.9.12 for example) expressed as a fraction of the oil in place when the waterflood is started, must be multiplied by the areal sweep efficiency (for example from Figure 5.9.9) to obtain the waterflood recovery.

Table 5.9.7 *Stiles Method of Calculating Waterflood Performance in Stratified Reservoirs*

$$R = \frac{[k_j h_j + (c_t - c_j)]}{k_j h_t}$$

R = fraction of recoverable oil that has been produced

c_t = total capacity of formation (md-ft)

c_j = mid-ft which have been completely flooded with water

h_t = total net thickness of formation (ft)

h_j = total net thickness flooded (ft)

k_j = permeability of layer just flooded out

Reservoir conditions at P.W.   *Surface conditions*

$$f_w = \frac{Mc_j}{[Mc_j + (c_t - c_j)]} \qquad f'_w = \frac{Ac_j}{[Ac_j + (c_t - c_j)]}$$

f_w = fractional flow of water $A = \dfrac{k_{rw}\mu_o B_o}{k_{ro}\mu_w B_w}$

$$M = \frac{k_{rw}\mu_o}{k_{ro}\mu_w}$$

B_o = oil formation volume factor

B_w = water formation volume factor

From Reference 1.17.

For the Stiles technique, a program [37] for hand-held calculators is available or the procedure summarized in Table 5.9.7 can be used. A straightforward presentation of the Stiles method is in the text by Craft and Hawkins [1.17]. The fractional recovery obtained with the Stiles method is a fraction of the recoverable oil ($S_{op} - S_{or}$) that has been recovered at a given reservoir water cut. Since a water-oil ratio (WOR) is measured at surface conditions, the fractional water cut at reservoir conditions, f_w, is obtained (assuming $B_w = 1.0$) from:

$$f_w = \frac{WOR}{WOR + B_o} \qquad [5.9.25]$$

where B_o is the oil formation volume factor.

5.9.12 Performance Evaluation

Monitoring waterflood performance is crucial to the success of the flood. From a reservoir engineering standpoint, the primary concerns are water injectivity and oil productivity. A few important factors related to these concerns will be summarized.

5.9.13 Injectivity and Injectivity Index

Whereas productivity index was the ability of a well to produce hydrocarbons, injectivity index, I, in B/D/psi, is a measure of the ability of a well to accept fluids [1.17]:

$$I = \frac{q_{sc}}{P_{iwf} - P_e} \qquad [5.9.26]$$

where q_{sc} is the flow rate in B/D at surface conditions, P_{iwf} is the flowing bottomhole pressure in psi, and P_e is the external pressure in psi. Some engineers express injectivity in terms of q_{sc}/p_{iwf} so that when injectivity is given, the reader is cautioned to understand what base pressure was intended. By dividing I by reservoir thickness, a specific injectivity index (specific to one well) can be obtained in B/D/psi/ft. In addition to expressing injectivity in terms of fluid injection rate in B/D, injectivity also is given as B/D/ac-ft and B/D/net ft of producing interval. Values of injectivity depend on properties of the reservoir rock, well spacing, injection water quality, fluid-rock interactions, and pressure drop in the reservoir. Typical values of injectivity are in the range of $8-15$ B/D/net ft or $0.75-1.0$ B/D/net ac-ft. In waterflooding operations, water injection may begin into a reservoir produced by solution-gas-drive in which a mobile gas saturation exists, or injection may begin prior to the development of a mobile gas saturation. In the latter case, the system can be considered filled with liquid.

5.9.13.1 Injectivities for Various Flood Patterns

Analytical expressions for liquid-filled patterns were given by Muskat [1.25] and Deppe [38] for a mobility ratio of one (see Table 5.9.8). While these exact analytical solutions can be developed for steady-state pressure distributions, the equations in Table 5.9.8 cannot be used directly if the mobility is not one. However, the equations are useful in estimating injectivity in limiting conditions. For example, if k and μ are selected for oil at the connate water saturation, an estimate of initial injection rate can be obtained. Then if k and μ are selected for water at residual saturation, an estimate can be made of injectivity at 100% sweep. (These estimates can be useful when equipment is sized for a waterflood). If data on skin factor are available, suitable corrections [2.18,5.8,3] can be inserted in the logarithm term in the denominator in these equations. For unit mobility ratio, the injection rate will remain constant during the flood. If the mobility ratio is more than one, the injection rate increases as more water is injected; if the mobility ratio is less than one, the injection rate decreases. Figure 5.9.18 shows for different mobility ratios, the change in relative injectivity as the water bank extends radially from the injector [23]. Figure 5.9.19 shows, for different mobility ratios, the change in relative injectivity as a 40-acre 5-spot is swept [23].

For water injection into a depletion drive reservoir, several stages can describe the progress of the flood [2.1]. The first stage is the period of radial flow from the start of infection until interference of oil banks from adjacent injectors occurs. The second stage is the period from interference until fill-up of the pre-existing gas space; after fill-up, production response begins. The third stage is the period from fill-up to water breakthrough into the producing wells; water production begins at breakthrough. The fourth and final stage is the period from water breakthrough until floodout. For a 5-spot pattern, the injection rate during fill-up and to the time of interference can be estimated by [3]:

$$i = \frac{0.007082 k_w \left(\dfrac{k_{ro}}{\mu_o}\right) h (p_{iwf} - p_{wf})}{\dfrac{1}{M} \ln \dfrac{r_f}{r_w} + \ln \dfrac{r_{ob}}{r_f}} \qquad [5.9.27]$$

Table 5.9.8 *Injection Rates in Fully Developed Patterns at Unit Mobility Ratio*

Direct line drive $\dfrac{d}{a} \geq 1$	Staggered line drive
$i = \dfrac{0.003541 kh(\Delta p)}{\mu\left[\ln\left(\dfrac{a}{r_w}\right) + 1.571\dfrac{d}{a} - 1.838\right]}$	$i = \dfrac{0.003541 kh(\Delta p)}{\mu\left[\ln\left(\dfrac{a}{r_w}\right) + 1.571\dfrac{d}{a} - 1.838\right]}$
Five-spot	Seven-spot
$i = \dfrac{0.003541 kh(\Delta p)}{\mu\left[\ln\left(\dfrac{d}{r_w}\right) - 0.619\right]}$	$i = \dfrac{0.00472 kh(\Delta p)}{\mu\left[\ln\left(\dfrac{d}{r_w}\right) - 0.569\right]}$
Nine-spot	Nine-spot
$i = \dfrac{0.003541 kh(\Delta p)_{i,c}}{\left(\dfrac{1+R}{2+R}\right)\left[\ln\dfrac{d}{r_w} - 0.272\right]\mu}$	$i = \dfrac{0.00782 kh(\Delta p)_{i,s}}{\left(\dfrac{3+R}{2+R}\right)\left[\ln\dfrac{d}{r_w} - 0.272\right] - \dfrac{0.693}{2+R}\mu}$

R = ratio of producing rate of corner well to side well
$(\Delta p)_{i,c}$ = pressure difference between injection well and corner well, and
$(\Delta p)_{i,s}$ = pressure difference between injection well and side well.

From References 1.25 and 38.
*Units in these equations are B/D, md, ft, psi, and cp.

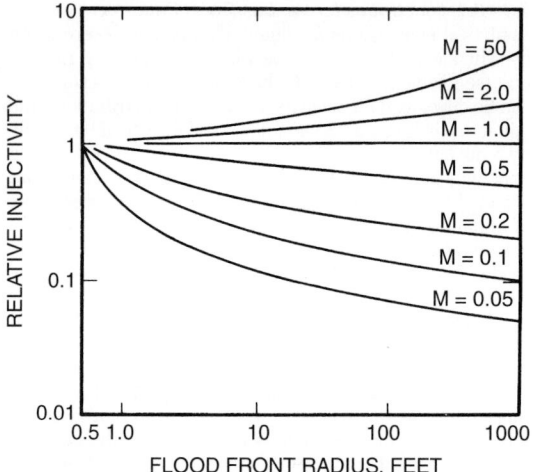

Figure 5.9.18 *Change in injectivity at varying radial distances for different mobility ratios [23].*

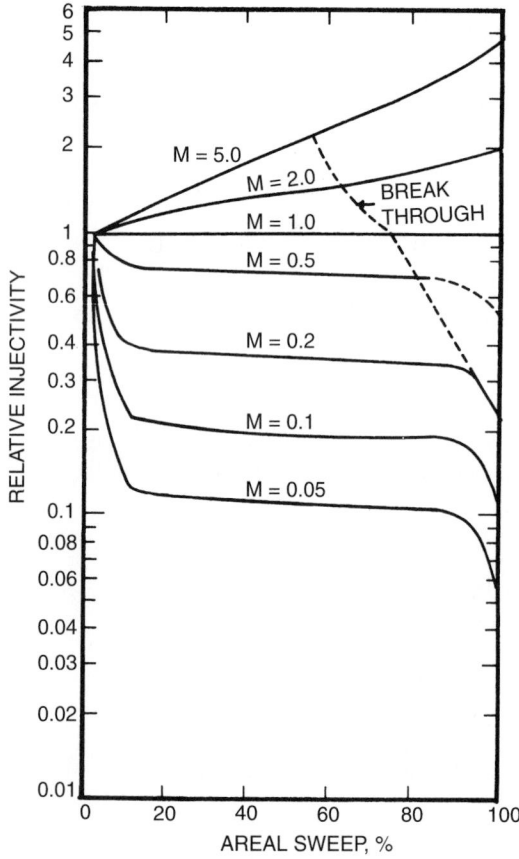

Figure 5.9.19 *Change in injectivity as a function of area swept for different mobility ratios [23].*

where oilfield units of B/D, md, cp, ft, and psi are used, and

r_{ob} = radius of the oil bank, $r_w \leq r_{ob} \leq d/\sqrt{2}$
r_f = radius of the flood-front saturation

Both r_{ob} and r_f can be obtained by a material balance on the injected water:

$$W_i = \pi(r_f^2 - r_w^2)(\bar{S}_w - S_{iw})h\phi \qquad [5.9.28]$$

$$r_f = \sqrt{\frac{W_i}{\pi\phi h(\bar{S}_w - S_{iw})} + r_w^2} \qquad [5.9.29]$$

Since the volume of water injected to fill-up is equal to the volume of gas displaced by the oil bank as the initial gas saturation, S_{gi}, is reduced to the trapped-gas saturation, S_g, a material balance for r_{ob}, is:

$$r_{ob} = \sqrt{\frac{W_i}{\pi\phi h(S_{gi} - S_{gt})} + r_w^2} \qquad [5.9.30]$$

At interference, $r_{ob} = d/\sqrt{2}$, and Equation 5.9.30 can be used to compute the volume of water required to reach interference. Usually, fill-up occurs in a relatively short time after interference, and the volume of water injected at fill-up is:

$$W_{if} = 2d^2\phi h S_{gi} \qquad [5.9.31]$$

At fill-up, r_f is obtained by:

$$r_f = \sqrt{\frac{2d^2 S_{gi}}{\pi(\bar{S}_w - S_{iw})} + r_w^2} \qquad [5.9.32]$$

After interference, the equation in Table 5.9.8 for the 5-spot pattern can be used to estimate injectivity. Additional details of estimating injectivity can be found in several good texts on this subject [2.18,5.8,3].

5.9.13.2 Monitoring Injectivity

Injection well performance can be analyzed and monitored by several means. During and after a period of injection, the pressure transient methods discussed earlier can be used. Additionally, several bookkeeping methods of monitoring injection rates and pressures are quite useful.

Hearn [39] recently proposed a method to analyze injection well pressure and rate data. Permeability is obtained from the slope of a plot of $\Delta p/q_w$ versus the logarithm of cumulative water injected. However, the method can only be used during the initial injection period. After fill-up, $\Delta p/q$, the reciprocal of injectivity index, will cease to be a function of cumulative water injected unless the well experiences damage or is stimulated. In these cases, the plots suggested by Hall [40] are convenient for analysis of the data.

A Hall plot is a graph of cumulative pressure-time versus cumulative water injection. Such plots are useful in observing injection well plugging or any beneficial results of stimulation procedures. An improvement in injectivity is indicated if the slope decreases, whereas plugging is suspected if the slope steepens. Figure 5.9.20 shows an improvement in water injectivity that resulted from a surfactant treatment [41].

The reciprocal of the Hall plot slope is the injectivity index in bbl/D/psi. Effective pressures are obtained by subtracting the static reservoir pressure from the flowing bottomhole pressures [40]:

$$\text{effective pressure} = P_{iwf} - \bar{p} = (p_{wh} + 0.45D - \Delta p_t) - \bar{p} \qquad [5.9.33]$$

where p_{wh} is the wellhead pressure, 0.45 is the hydrostatic pressure gradient in psi/ft, D is the depth to the mid-point of the reservoir, and Δp_t is the pressure drop in the tubing. Although the Hall method assumes that only the wellhead pressure changes with cumulative water injected, the effective pressure drop should be used if permeability capacity,

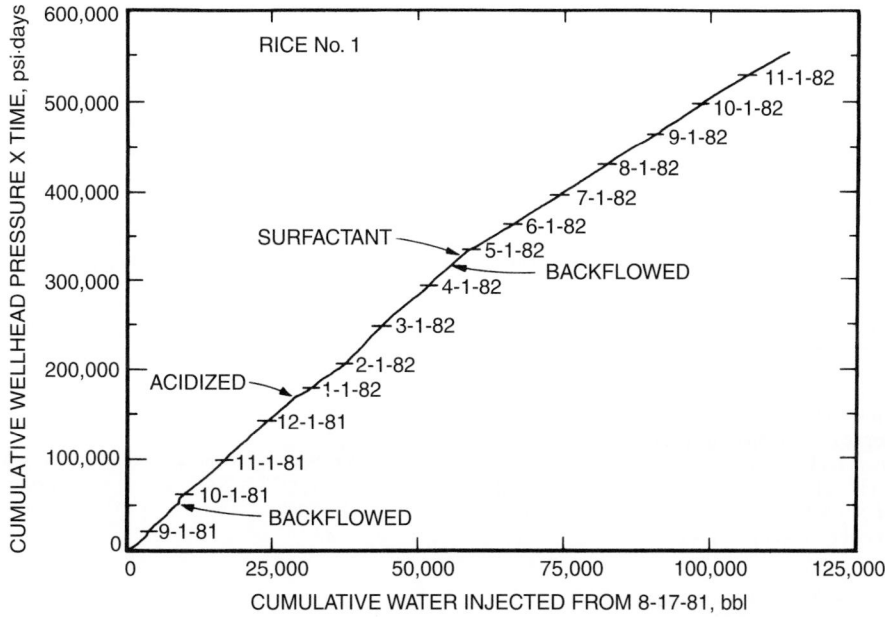

Figure 5.9.20 *Hall plot [40].*

or transmissibility changes, are desired [40]. If the difference between reservoir pressure and the hydrostatic head is more than 15% of the wellhead pressure, serious quantitative errors can result when wellhead pressures are used in the construction of Hall plots; if the difference is less than 15%, only slight errors are caused [1.66].

Earlougher has presented a modified version of the Hall technique in which the slope, m′, is defined as [1.66]:

$$m' = \frac{141.2\mu}{kh}\left[\ln\left(\frac{r_e}{r_w}\right)+s\right] \qquad [5.9.34]$$

The improvement in overall flow efficiency, E_f, can be obtained from [1.66]:

$$\frac{E_f \text{ after}}{E_f \text{ before}} = \frac{m' \text{ before}}{m' \text{ after}} = \frac{I \text{ after}}{I \text{ before}} \qquad [5.9.35]$$

where the injectivity index, I, or Hall slopes are calculated using the pressure difference between the flowing bottomhole pressure and the formation pressure.

5.9.13.3 Production Curves

Plots of waterflood injection and production performance can be presented in a number of ways. For the history of the project, water injection rate, oil production rate, and water-oil ratio or water cut can be plotted vs. time (usually months). The actual water injection and oil production rates can be compared to the predicted rates on a time basis.

Future oil production and ultimate recovery are often extrapolated from graphical methods. One of the more popular methods is a plot of the WOR on a log scale vs. cumulative oil production on a linear scale or a linear plot of the fractional water cut (or percent water produced) vs. cumulative oil produced. Alternatively, the oil-water ratio

can be ploted on a log scale vs. the cumulative production on a linear scale. One of the purposes of these plots is to predict the ultimate oil recovery by extrapolating the curve to some economic limit at which time it becomes no longer profitable to continue the flood. If the operating methods remain relatively unchanged, a method [42] has been proposed for a fully developed waterflood that permits an easy extrapolation of recovery to a given water cut. This latter method consists of a linear plot of E_R, fractional recovery of oil in place, vs. the term $-\{[(1/f_w)-1]-(1/f_w)\}$. This method also provides an estimate of water-oil relative permeabilities.

5.9.13.4 Waterflood Parameters

Important parameters in waterflood operations are the water residual oil saturation, S_{or}, and the relative permeability to water, k_{rw}. A statistical study of these parameters, as well as peak oil rates, was provided by Felsenthal [43]. Data on S_{or} and k_{rw} from core data are listed in Table 5.9.9. Endpoint k_{rw} values were higher in carbonates than in sandstones; for a given lithology, k_{rw} decreased as the absolute permeability decreased.

References

1. Langnes, G. L., Robertson, J. O., Jr., and Chilingar, G.V., *Secondary Recovery and Carbonate Reservoirs*, American Elsevier Publishing Co., New York (1972).

2. *Improved Oil Recovery*, Interstate Oil Compact Commission, Oklahoma City (March 1983).

3. Willhite, G. P., *Waterflooding*, SPE, Richardson, TX (1986).

Table 5.9.9 *Waterflood Parameters*

Waterflood residual oil saturations measured in core test samples

	Sandstone	Carbonate Rock
Mean average S_{or}, % pore space	27.7	26.2
Median average S_{or}, % pore space	26.6	25.2
Number of core samples tested	316	108
Number of source reservoirs	75	20
Standard deviation, % pore space	8.76	8.84

Effect of lithology and K_a on the end points of oil/water relative permeability curves

Permeability Group range of K_a, md	Low (1 to 10)	Medium (11 to 100)	High (101 to 2,000)
Sandstone			
Median k_{rw}^* at S_{or}	0.065	0.133	0.256
Median k_{rw}^{**} at S_{or}	0.033	0.095	0.210
Number of core samples tested	30	213	143
Carbonate rock			
Median k_{rw}^* at S_{or}	0.211	0.357	0.492
Median k_{rw}^{**} at S_{or}	0.179	0.303	0.428
Number of core samples tested	33	45	24

From reference 43.
*Expressed as fraction of k_0 at connate water saturation.
**Expressed as fraction of K_a.

4. Singh, S. P., and Kiel, O. G., "Waterflood Design (Pattern, Rate, and Timing)," paper SPE 10024 presented at the SPE 1982 Intl. Pet. Exhibition & Tech. Symposium, Bejing, China, March 18–26.

5. Muskat, M., and Wyckoff, R. D., "A Theoretical Analysis of Waterflooding Networks," *Trans.*, AIME, Vol. 107 (1934), pp. 62–76.

6. Dyes, A. B., Caudle, B. H., and Erickson, R. A., "Oil Production After Breakthrough-As Influenced by Mobility Ratio," *Trans.*, AIME, Vol. 201 (1954), pp. 81–86.

7. Craig, F. F., Jr., Geffen, T. M., and Morse, R. A., "Oil Recovery Performance of Pattern Gas or Water Injection Operations from Model Tests," *J. Pet. Tech.* (Jan. 1955), pp. 7–15; *Trans.*, AIME, Vol. 204.

8. Prats, M., et al., "Prediction of Injection Rate and Production History for Multifluid Five-Spot Floods," *J. Pet. Tech.* (May 1959), pp. 98–105; *Trans.*, AIME, Vol. 216.

9. Caudle, B. H., and Witte, M. D., "Production Potential Changes During Sweep-Out in a Five-Spot Pattern," *Trans.*, AIME, Vol. 216 (1959), pp. 446–448.

10. Habermann, B., "The Efficiency of Miscible Displacement As a Function of Mobility Ratio," *Trans.*, AIME, Vol. 219 (1960), pp. 264–272.

11. Kimbler, O. K., Caudle, B. H., and Cooper, H. E., Jr., "Areal Sweep-out Behavior in a Nine-Spot Injection Pattern," *J. Pet. Tech.* (Feb. 1964), pp. 199–202.

12. Caudle, B. H., Erickson, R. A., and Slobod, R. L., "The Encroachment of Injected Fluids Beyond the Normal Well Patterns," *Trans.*, AIME, Vol. 204 (1955), pp. 79–85.

13. Prats, M., Hazebroek, P., and Allen, E. E., "Effect of Off-Pattern Wells on the Performance of a Five-Spot Waterflood," *J. Pet. Tech.* (Feb. 1962), pp. 173–178.

14. Prats, M., "The Breakthrough Sweep Efficiency of the Staggered Line Drive," *Trans.*, AIME, Vol. 207 (1956), pp. 361–362.

15. Crawford, P. B., "Factors Affecting Waterflood Pattern Performance and Selection," *J. Pet. Tech.* (Dec. 1960), pp. 11–15.

16. Fassihi, M. R., "New Correlations for Calculation of Vertical Coverage and Areal Sweep Efficiency," *SPE Reservoir Engineering* (Nov. 1986), pp. 604–606.

17. Claridge, E. L., "Prediction of Recovery in Unstable Miscible Flooding," *Soc. Pet. Eng. J.* (April 1972), pp. 143–155.

18. Schmalz, J. P., and Rahme, H. D., "The Variation of waterflood Performance With Variation in Permeability Profile," *Prod. Monthly*, Vol. 15, No. 9 (Sept. 1950), pp. 9–12.

19. Dykstra, H., and Parsons, R. L., "The Prediction of Oil Recovery by Water-flood," *Secondary Recovery of Oil in the United States*, API, Dallas (1950) second edition, pp. 160–174.

20. Warren, J. E., and Cosgrove, J. J., "Prediction of Waterflood Behavior in A Stratified System," *Soc. Pet. Eng. J.* (June 1964), pp. 149–157.

21. Johnson, C. E., Jr., "Prediction of Oil Recovery by Water Flood-A Simplified Graphical Treatment of the Dykstra–Parsons Method," *Trans.*, AIME, Vol. 207 (1956), pp. 345–346.

22. deSouza, A. O., and Bridham, W. E., "A Study on Dykstra–Parsons Curves," Tech. Report 29, Stanford University, Palo Alto (1981).

23. Craig, F. F., Jr., "Effect or Reservoir Description on Performance Predictions," *J. Pet. Tech.* (Oct. 1970), pp. 1239–1245.

24. Reznik, A. A., Emick, R. M., and Panvelker, S. B., "An Analytical Extension of the Dykstra–Parsons Vertical Stratification Discrete Solution to a Continuous, Real-Time Basis," *Soc. Pet. Eng. J.* (Dec. 1984), pp. 643–656.

25. Root, P. J., and Skiba, F. F., "Cross flow Effects Duringan Idealized Process in A Stratified Reservoir," *Soc. Pet. Eng. J.* (Sept. 1965), pp. 229–238.

26. Goddin, C. S., Jr., et al., "A Numerical Study of Waterflood Performance in a Stratified System With Crossflow," *J. Pet. Tech.* (June 1966), pp. 765–771.

27. El-Khatib, N., "The Effect of Crossflow on Waterflooding of Stratified Reservoirs," *Soc. Pet. Eng. J.* (April 1985), pp. 291–302.

28. Collins, H. N., and Wang, S. T., "Discussion of the Effect of Crossflow on Waterflooding of Stratified Reservoirs," *Soc. Pet. Eng. J.* (Aug. 1985), p. 614.

29. El-Khatib, N., "Author's Reply to Discussion of the Effect of Cross flow on Waterflood of Stratified Reservoirs," *Soc. Pet. Eng. J* (Aug. 1985), p. 614.

30. Collins, H. N., and Wang, S. T., "Further Discussion of the Effect of Crossflow on Waterflooding of Stratified Reservoirs," *SPE Reservoir Eng.* (Jan. 1986) p. 73.

31. El-Khatib, N., "Author's Reply to Further Discussion of the Effect of Crossflow on Waterflooding of Stratified Reservoirs," *SPE Reservoir Eng.* (Jan 1986), pp. 74–75.

32. Hirasaki, G. J., Morra, F., and Willhite, G. P., "Estimation of Reservoir Hetrogeneity from Waterflood Performance," SPE 13415, unsolicited paper.

33. Hirasaki, G. J., "Properties of Log-Normal Permeability Distribution for Stratified Reservoirs," SPE 13416, unsolicited paper.

34. Higgins, R. V., ad Leighton, A. J., "A Computer Method to Calculate Two-Phase Flow in Any Irregularly Bounded Porous Medium," *J. Pet Tech.* (June 1962), pp. 679–683; *Trans.* AIME, Vol. 225.

35. Stiles, W. E., "Use of Permeability Distribution in Water Flood Calculations," *Trans.*, AIME, Vol. 186 (1949), pp. 9–13.

36. Garb, F. A., "Waterflood Calculations for Hand-held Computers. Part 8—Using Dykstra-Parsons Methods to Evaluate Stratified Reservoirs," *World Oil* (July 1980), pp. 155–160.

37. Garb, F. A., "Waterflood Calculations for Hand-held Computers. Part 7—Evaluating Flood Performance Using the Stiles Technique," *World Oil* (June 1980), pp. 205–210.

38. Deppe, J. C., "Injection Rates-The Effect of Mobility Ratio, Area Swept, and Pattern," *Soc. pet. Eng. J.* (June 1961), pp. 81–91; *Trans.*, AIME, Vol. 222.

39. Hearn, C. L., "Method Analyzes Injection Well Pressure and Rate Data," *Oil & Gas J.* (April 18, 1983), pp. 117–120.

40. Hall, H. N., "How to Analyze Waterflood Injection Well Performance," *World Oil* (Oct. 1963), pp. 128–130.

41. Martin, F. D., "Injectivity Improvement in the Grayburg Formation at a Waterflood in Lea Country, NM," paper SPE 12599 presented at the 1984 Permian Basin Oil & Gas Recovery Conference, Midland, March 8–9.

42. Ershaghi, I., and Omoregie, O., "A Method for Extrapolation of Cut vs. Recovery Curves," *J. Pet. Tech.* (Feb. 1978), pp. 203–204.

43. Felsenthal, M., "A Statistical Study of Some Waterflood Parameters," *J. Pet. Tech.* (Oct. 1979), pp. 1303–1304.

5.10 ESTIMATION OF WATERFLOOD RESIDUAL OIL SATURATION

It is recognized throughout the industry that there is no single generally accepted method of measuring residual oil [1]. The available methods should be considered on their merits and some combination of methods should always be used as a cross-check. Each method provides somewhat different information about the amount and distribution of residual oil. This section discusses the various methods that are in current use, and recommendations are made as to how residual oil should be measured [2].

5.10.1 Material Balance

Material balance was one of the first and is the most widely used technique employed in estimating oil reserves and depletion. Overall estimates of the amount of in-place and recoverable oil are based mainly on material balance. These calculations are also used as a first screening point to determine if sufficient oil remains after waterflooding for application of tertiary recovery. The quantity of oil remaining in the reservoir, having pore volume V_p, in stock tank barrels, is given by the difference between the initial oil-in-place, N, and the amount of oil produced, N_p. The overall residual saturation $(S_{or})_{MB}$ is based on the volume of oil relative to the reservoir pore space [1].

$$(S_{or})_{MB} = \frac{(N - N_p)B_{or}}{V_p} \qquad [5.10.1]$$

where B_{or} is the oil formation volume factor after waterflooding.

Many of the projects currently being evaluated as tertiary prospects were initially developed twenty to thirty years ago or longer. At that time, methods for estimating hydrocarbon content were not as accurate as methods presently available. In addition, adequate information relating to hydrocarbon volumes may be limited to only a few wells in a field and production data for the field may not always be reliable. In any event, even where material balance might yield a reasonable estimate of the amount of residual oil-in-place, it does not indicate the distribution of oil within the reservoir.

The value of $(S_{or})_{MB}$ will be dependent on both the microscopic displacement efficiency in swept zones and the vertical and horizontal sweep efficiency. It has been suggested that a high ratio of $(S_{or})_{MB}$ to S_{or}, determined by methods applying to the swept zone, indicates that the reservoir still contains an unusually high amount of oil in the unswept region. The reservoir should then receive special consideration as a candidate for obtaining additional oil recovery through infill drilling [1].

In application of tertiary recovery, the residual oil remaining in the swept zone is of most interest because this is the region that will most likely be contacted by a tertiary process. Thus, the material balance, even in more sophisticated forms than the foregoing simple volumetric balance, should only be used for rough screening in evaluating prospects for tertiary recovery. Furthermore, because of the many uncertainties in the measurements used to obtain a material balance, an enhanced recovery prospect should not be discounted on the basis of material balance alone.

When the volume of oil produced is known with reasonable precision, the accuracy of this method depends mainly on the reliability of the original oil-in-place estimate.

5.10.2 Well Test Analyses

In-situ oil saturations can be estimated by combining relative permeability data determined in the laboratory with field well test data such as:

    Production data
    Single well transient tests in producing wells
        Pressure buildup
        Pressure drawdown
        Multiple-rate testing
    Multiple-well transient tests
        Interference tests
        Pulse tests

Details of the methods for developing the well test data can be found in the literature [1.13,1.66,3.1], and were summarized earlier. If sufficient, good quality relative permeability data are available from laboratory tests that simulate downhole conditions, the techniques of correlating well test analyses can give some indication of the reservoir oil saturation. For these techniques to be applicable, numerous assumptions are made: the reservoir interval is homogeneous, horizontal, and isotropic with a small and constant compressibility; fluid properties and saturations are uniform in the formation; relative permeabilities are constant throughout the test area; there are no oil-water or gas-oil contacts; and the tested wells are not affected by other wells outside the test area [1]. The following discussion applies to waterflooded reservoirs with no free gas saturation, but could be extended to more complex systems.

5.10.2.1 Production Data

When the producing oil and water flow rates, formation volume factors, and viscosities are known, relative permeability ratios can be determined:

$$\frac{k_{rw}}{k_{ro}} = \frac{q_w B_w \mu_w}{q_o B_o \mu_o} \qquad [5.10.2]$$

where k_r = relative permeability
       q = flow rate
       B = formation volume factor
       μ = viscosity

and the subscripts w and o refer to water and oil, respectively. Thus, the relative permeability ratio between water and oil can be estimated from the producing water-oil ratio (WOR).

Relative permeabilities determined in the laboratory may be based on any one of the following measures of core permeability: air, water at 100% S_w, or oil at irreducible water saturations. Laboratory-derived permeability ratios can be used to find the water saturation at which the field-derived permeability ratio would occur. In the absence of gas saturation,

$$S_o = 1 - S_w \qquad [5.10.3]$$

Provided core analysis data are available, this method is easy and rapid. However, no information is obtained regarding wellbore damage or specific permeabilities to the flowing fluids. This important information can be obtained from the transient well tests.

5.10.2.2 Transient Tests

Whereas the producing well production data can only provide a relative permeability ratio, transient well testing can provide estimates of reservoir permeabilities to both oil and water (and free gas, if present):

$$k_o = \frac{162.6 q_o B_o \mu_o}{mh} \qquad [5.10.4]$$

$$k_w = \frac{162.6 q_w B_w \mu_w}{mh} \qquad [5.10.5]$$

where k, q, B, and μ are as previously defined and have units of millidarcies, barrels per day, reservoir barrels per stock tank barrel, and centipoise, respectively. The thickness of the interval, h, is in ft and m is the appropriate slope from the Miller-Dyes-Hutchinson (MDH) plot or the Horner plot.

Multiple well or single well tests can be used to estimate the effective permeability to oil by using type-curve matching:

$$k_o = \frac{141.2 q_o B_o \mu_o}{h} \frac{(P_D)_M}{\Delta P_M} \qquad [5.10.6]$$

where $(P_D)_M$ is the dimensionless pressure at the match point for type-curve matching, ΔP_M is the pressure change from transient test data at the match point for type-curve matching, and the previously defined terms are in field units. Laboratory relative permeability curves are then used to find the saturations that correspond to the relative permeabilities obtained from the transient tests.

Multiple well testing can also be used to estimate oil saturation by using the total compressibility from the match point data and:

$$c_t = c_w S_w + c_o S_o + c_g S_g + c_f \qquad [5.10.7]$$

where c_t is system total compressibility; C_f is the formation or pore volume compressibility; c_w, c_o, and c_g are compressibilities of water, oil and gas, all with units of psi^{-1}; and S is the corresponding saturation [3]. Since the gas term is assumed to be zero for the waterflooded case and since $S_w = (1 - S_{or})$,

$$c_t = c_w(1 - S_o) + c_o S_o + c_f \qquad [5.10.8]$$

By rearranging,

$$S_o = \frac{c_t - c_w - c_f}{(c_o - c_w)} \qquad [5.10.9]$$

With a knowledge of oil, water, and formation pore volume compressibilities, and the total compressibility determined from the type-curve matching, reservoir oil saturation can be estimated [1]. Oil saturation based on compressibility is generally less accurate than saturations determined from field relative permeability data and is normally regarded as an approximation.

5.10.2.3 Applicability

Because of the rigid requirements of the assumptions made, and the problems with interpreting the field data, oil saturations obtained from well test analyses are considered rough estimates. The saturation estimate is an overall average for the region of the reservoir influenced by the test. If permeability variations or other conditions cause a variation in the vertical saturation distribution, these techniques will not yield meaningful data. For these techniques to be considered for oil saturation determinations, good laboratory core analysis data are essential. However, because of the low costs and relative ease in conducting the tests, plus the additional important information obtained, well test analyses should be developed along with the other more direct methods of determining residual oil saturations.

5.10.3 Coring and Core Testing

5.10.3.1 Well Coring

Well coring is the process of obtaining representative samples of the productive formation. The choice of depth at which to begin coring can often be a problem. Cores from the regions of interest may not be obtained because of unexpected changes in stratigraphy. There is also the possibility that the region cored will be a nonproductive region which did not contain significant hydrocarbon content initially.

However, analysis and testing of core samples continues to be an important method of determining residual oil [4,5].

Various techniques are used to obtain core samples: conventional diamond-bit coring, rubber-sleeve coring, pressure coring, sidewall coring, and recovery of cuttings generated from the drilling operation. The last two methods are not used for residual oil measurements. Conventional coring is normally done in competent formations to obtain full-diameter cores. In unconsolidated, or poorly consolidated formation, a core barrel containing a rubber sleeve is used. The core sample is held together by the sleeve and its properties during laboratory tests remain reasonably representative of conditions in the formation [4].

Two main problems in coring for determination of residual oil are that further flushing of oil to below-normal waterflood residual can take place around the core bit, and that loss of oil occurs, due mainly to gas expansion, as the core is lifted to the surface.

Flushing During Coring

For a condition where the in-place oil saturation is at its waterflood residual value, no more oil can be produced at normal flow rates. During the coring operation, it is important to avoid extreme flushing conditions that could cause part of the residual oil to be mobilized [2.15]. Some of the variables that control the amount of residual oil flushed from a core by mud filtrate are: borehole-to-formation differential pressure (overbalance), interfacial tension, wettability (the following discussion applies principally to water-wet rocks), core permeability, coring penetration rate, core diameter, type of drill bit, drilling mud composition (including particle size distribution), depth of invasion of mud particles into the core, rate of filtrate production (both spurt loss and total fluid loss), permeability of the formation, and nature of the reservoir (uniformity, texture etc.) [6]. In one type of system investigated in the laboratory, the amount of oil stripped from cores varied directly with the overbalance pressure, filtration production rate, and core permeability; it varied inversely with penetration rate and core diameter. The overbalance pressure is usually the critical variable [2.16].

Overbalance Pressure

Unintentional displacement of residual oil may occur in coring operations when large pressure gradients exist near the core bit. In this region when fluid velocities are high, the resulting viscous forces may become sufficient to overcome the capillary forces that hold the residual oil in place.

Results from an extensive laboratory coring program [2.16] showed a reduction in pre-test residual oil saturation of almost 20 percent to about 60 percent as the pressure gradients varied from about 350 to 1,700 psi/ft. In an evaluation of the same data, other authors [1.123] contend that when analyzing larger diameter cores (4-in. diameter) and considering radial flow, the estimated penetration would result in only a 10% change in residual oil. In addition, they contend that core samples used in retort analysis are usually taken from the center of the core where mud solid penetration into the core would be minimal. However, it is generally recognized that residual oil, which is immobile after normal waterflood operations, can become mobile and be stripped from the core, especially in the region adjacent to the core bit.

The ratio of viscous to capillary forces has been expressed as $\Delta P/L\sigma$, where $\Delta P/L$ is the pressure drop per unit length, and σ is the interfacial tension. In water-wet cores, at least, a critical value of $\Delta p/L\sigma$ must be exceeded before production of residual oil occurs [2.15]. In coring, the overbalance pressure must be kept low in order to minimize flushing. Furthermore, the drilling mud should not contain additives that cause significant reduction in interfacial tension that could mobilize residual oil. The use of dispersants, emulsifiers, lubricants, lost circulation materials, and oil should be avoided. If the overbalance pressure causes the critical displacement ratio to be exceeded, then there will be some displacement of residual oil. Linear displacement tests run in the laboratory show the critical displacement pressure to vary from about 1 (psi/ft)/(dyne/cm) for a 1,000 md sandstone to about 25 for a 100 md sandstone [2.15]. Thus, the permissible overbalance pressure will have significant dependence on the properties of the formation that is being cored (Figure 5.10.1).

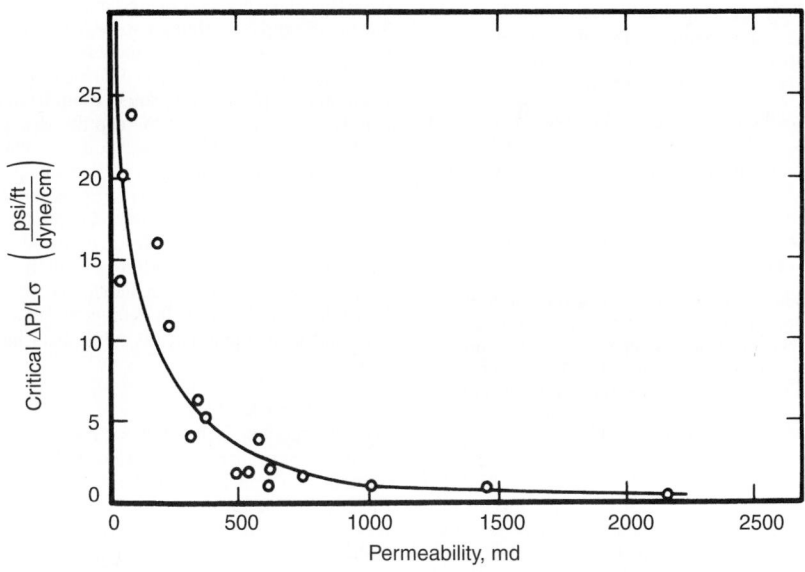

Figure 5.10.1 *Critical capillary number as a function of permeability [2.15].*

Drilling Mud Properties

At bottomhole conditions, API filter loss for water-base muds is often in the range of 5 to 10 cc for 30 minutes, which is sufficient to drive most 3-in. to 4-in. diameter cores to the equivalent of the waterflood residual oil saturation if the region being cored is not already at this condition [1.123]. Higher mud water loss or smaller core diameters can lead to displacement of some of this residual oil. However, only general agreement has been found between API filter loss and the amount of oil stripped from cores. More consistent agreement has been observed between the amount of mobilized residual oil and spurt loss (the rapid fluid loss to the formation that occurs before an effective mud filter cake has built up). Spurt loss has been shown to correlate with solids content and particle size distribution which also influence filtration rates and amount of oil-stripping [2.16]. In general, when taking cores it is always preferable to use a bland water base drilling fluid which contains no oil or surface active materials [1].

Shrinkage and Bleeding

In reservoirs which have been depleted to low pressures and waterflooded to high water-oil ratios, changes in residual saturation in bringing the core to the surface should be fairly minimized. However, in most cases, as the core is raised, gases will come out of solution and can cause residual oil to bleed from the core. The loss of gas causes the oil to shrink.

Corrections for Shrinkage and Bleeding

Shrinkage of residual oil can be estimated from laboratory measurements of shrinkage when the pressure of bottomhole oil samples is lowered. Reduction of temperature will also contribute to shrinkage. The following corrections have been proposed for the effects of shrinkage and bleeding [1.123].

$$(S_{or})_R = (S_{or})_c B_o E \qquad [5.10.10]$$

where $(S_{or})_R$ = average post-waterflood residual oil saturation in the flooded region of the reservoir
$(S_{or})_c$ = average oil saturation from cores
B_o = oil formation volume factor at the time of coring
E = bleeding factor (in the absence of specific bleeding measurements, a value of 1.11 is assumed)

A modification to this calculation has been suggested [7] to compensate for waterflood displacement efficiency by dividing the reservoir residual oil saturation by the conformance factor. In the absence of a reservoir simulator that accounts for reservoir heterogeneity, capillary effects, mobility of oil, and mobility of water, it was suggested the conformance factor be estimated by $(1 - V^2)/M$

where M = mobility ratio or the ratio of the mobility of water at the average water saturation in the reservoir at breakthrough to the mobility of oil in the oil bank ahead of the displacing front
V = the Dykstra–Parsons permeability variation

Thus, the modified calculation would be

$$(S_{or})_R = (S_{or})_c B_o E \frac{M}{1 - V^2} \qquad [5.10.11]$$

However, these corrections, particularly the one for bleeding factor, are only approximate at best and considerable attention has been given to recovering cores at reservoir pressure.

Pressure Core Barrel

Considerable development work has gone into developing a core barrel that will bring cores to the surface without major reduction in reservoir pressure, and thus prevent shrinkage and bleeding. The pressure core barrel is designed to cut the core and seal it within a cylinder before retrieval. Although the cores obtained are small in diameter ($\sim 2\frac{1}{2}$-in.), pressure coring has gained acceptance as one of the best methods of determining residual oil, particularly when information on the vertical profile of the oil saturation is wanting. Success in application of the core barrel depends to a great extent on the skill and experience of the personnel running the operation. The main criterion for success is retrieving cores at or close to formation pressure. Improvements over the years have led to the present success level of about 70% to 80% in consolidated formations [2.3,1,8–13].

Sponge Cores

A recent modification [14] to a conventional core barrel is the incorporation of a porous sponge to collect oil that bleeds from the core. The oil saturation measured by conventional techniques is corrected for the bleeding of oil as measured in the sponge. Oil saturations measured with this technique have approximated the values determined by pressure coring but at a cost that is closer to conventional coring.

5.10.3.2 Core Testing

Laboratory tests to estimate reservoir residual oil can be performed on cores that have been preserved at the wellsite or cores which are extracted with solvent and subsequently restored to reservoir conditions. Cores obtained with drilling muds that minimize wettability alteration, and that are protected at the wellsite to prevent evaporation or oxidation are called preserved cores. Cores that are cleaned with solvents and resaturated with reservoir fluids are called restored-state cores or extracted cores. The restoring process is often performed on nonpreserved or weathered cores, but the same technique could apply to cores that had been preserved.

Core Handling

Conventional Cores. The precautions taken in handling cores once they have been recovered depends mainly on the measurements and tests that are to be performed on them. If the measurements are routine and can be run within a day or two, it is generally considered sufficient to wipe the cores, wrap them in plastic and protect them from exposure to the sun. However, it is better to keep the wrapped cores inside ice boxes to minimize evaporation, especially when longer transit times are anticipated. When maintenance of wettability conditions as they exist in the reservoir is attempted, the most widely preferred method of preservation is to wrap the core in thin sheets of plastic followed by aluminum foil and then isolate the wrapped core from the atmosphere by a coating of wax or plastic. An alternative method of preservation is to store the cores in deoxygenated aqueous solution. This may be formation brine, synthetic brine or mud filtrate [1,15].

Pressure Cores

Special handling is needed for cores obtained using the pressure core barrel. This is normally carried out by the trained crew which assembled the barrel prior to testing. After retrieving the core barrel, drilling fluid is displaced at pressure by gelled kerosene. The complete barrel is then chilled in dry ice for several hours in order to freeze the water in the core sample. The pressure in the core barrel can then be released and an inner metal sleeve containing the core is removed. The core is cut into convenient lengths, of about three to four feet, and kept frozen by means of dry ice during transportation [1].

Measurement of Residual Oil in Recovered Cores

Various techniques are available for determining the oil content of cores. Examples that involve removal of the oil are vacuum distillation, a combination of distillation and solvent extraction (Dean Stark), and high temperature retorting. The Dean Stark method with toluene as solvent is normally used when displacement tests are to be carried out on the extracted cores. In this method, the oil saturation is determined by difference from the amount of water removed from the core [2.12,15,16].

In general, cores obtained with the pressure core barrel under conditions of minimal flushing are needed in order to obtain residual saturations that can be treated with reasonable confidence. Special analytical methods have evolved for treatment of pressure cores. The frozen cores are removed from the metal-containing sleeve and dressed while still frozen. The pressure cores are then allowed to thaw in an inert atmosphere, and volumes of evolved gases are recorded. Next, the free water is distilled from the core. Any remaining oil is removed by a toluene-CO_2 leaching. The amount of oil in the core is determined by adding the volume obtained by distillation to the volume removed during extraction and then making a correction for evolved gas. As a check on extent of penetration of mud filtrate into the core, a tracer can be added to the drilling mud which permits the radial depth of invasion to be estimated. However, filtrate invasion does not necessarily imply flushing of residual oil [1].

Residual Oil from Laboratory Core Floods

Most cores are subjected to cleaning before measurement of permeability and porosity. However, when the preservation of wetting properties is of main interest, displacement tests are run on the cores prior to cleaning.

Another approach to the problem of reservoir wettability is the restored state method. Cleaned cores are saturated with reservoir brine or brine of similar composition. The brine is displaced by reservoir crude to an equivalent connate water saturation. Recontacting the reservoir rock with the reservoir crude is believed to result in adsorption of those components from the crude oil which determined the in-situ wettability and hence restore the system to its original wetting condition.

Relatively little is known about the causes of reservoir wettability and its sensitivity to the numerous variables that may cause the wettability of recovered cores to be changed. It has been shown that wettability can have significant effect on residual oil [1.121,1.25]. This is the main reason why values of residual oil saturation determined by laboratory core flooding tests are treated with caution. Residual saturations determined by laboratory flooding tests are often used in estimating the amount of oil that will be recovered by waterflooding. However, when residual oil saturation is to be determined for evaluation of a tertiary recovery prospect, laboratory flooding tests do not seem to be in favor. In a recent monograph on residual oil determination, results of such tests received a rating of only poor to fair [1]. Nevertheless, there seems to be consensus view in the industry that the problem of wettability as it relates to residual oil has been satisfactorily resolved. However, little attention appears to have been given to restoring cores to their in-situ residual oil saturation at wetting conditions which are representative of the reservoir even though this may be critical to proper laboratory testing of a tertiary process [17,18].

Even though laboratory flooding of reservoir cores may not be a generally acceptable method of determining residual oil, it is considered vital that tertiary processes be tested in the laboratory using these cores.

5.10.4 Tracer Tests for Determining Residual Oil

5.10.4.1 How Tracer Tests Work

The tracer test was conceived by applying principles of chromatographic separation to fluid movement in the reservoir. The outstanding advantage of the tracer test is its ability to investigate a relatively large volume of the formation. It was first suggested that the method could be applied to flow between two wells [19]. The method depends on the effect which the relative solubility of a tracer between oil and water has on the rate at which a pulse of low concentration tracer passes through the formation.

The condition where the reservoir has been flooded out and the oil is immobile is considered later, but the theory can also be applied where both oil and water are flowing, or where the water phase is immobile. If a formation containing residual oil is flooded with a bank of water containing a tracer which is mutually soluble in oil and water, part of the tracer will pass into the oil phase. If there is local equilibrium, the concentration of the tracer in the oil, C_{to}, is related to the concentration in the aqueous phase, C_{tw}, by the distribution coefficient, K_1.

$$K_1 = \frac{C_{to}}{C_{tw}} \qquad [5.10.12]$$

As a result of partitioning, part of the tracer temporarily resides in the immobile residual oil, and the overall velocity of the tracer is less than that of the flowing aqueous phase. The concentration of tracer in the oil together with the oil volume determines the fraction of tracer resident in the oil phase. Since oil volume is directly proportional to oil saturation, the rate at which a pulse of tracer concentration passes through the formation depends on the oil saturation in regions swept by the tracer.

Equilibration of the tracer between the residual oil and water following the tracer bank coupled with dispersion effects determines the shape of a peaked concentration distribution of tracer. Ahead of the peak there is net movement of tracer into the residual oil. Behind the peak there is net movement of tracer molecules from the residual oil back into the water.

From consideration of the way the tracer divides itself between the oil and water phases and the effect of the magnitude of the residual oil saturation on residence time of the tracer, the velocity of the tracer, v_t, is related to the velocity of the associated water, v_w, by

$$v_t = \frac{(1 - S_{or})v_w}{(1 - S_{or}) + K_1 S_{or}} \qquad [5.10.13]$$

from which residual oil is given by

$$S_{or} = \frac{v_w - v_t}{v_w v_t + v_t K_1} \qquad [5.10.14]$$

The long time needed for tracer to move between wells and the broad and uninterpretable residence time distribution that would arise because of streamline and heterogeneity effects for normal well spacings are cited as reasons why the well tracer test may be impractical as a between-well test, although there is still some interest in applying this method to small pilot areas [1].

5.10.4.2 Single-Well Tracer Technique

The single-well or backflow tracer technique to determine residual oil saturation is a recent innovation. Patents [19,20] assigned to Exxon Production Research Company in 1971 described the injection of tracers to measure in-situ oil saturation at distances of typically 10 to 30 ft away from the wellbore of a producing well. The single-well tracer test overcomes many of the difficulties that, in general, make the two-well test impractical.

Test Procedure
The method involves injection of a bank of water containing an alkyl ester as the tracer. The selection of the ester will depend on temperature of the reservoir. For example, ethyl acetate is used in higher temperature reservoirs and n-propyl formate is used at lower temperatures [21]. These tracers are suitable for reservoirs with bottomhole temperatures ranging from about 80°F to 200°F. The ester partially hydrolyzes within the formation to form an acetate and an alcohol, the latter serving as a secondary tracer. Methanol is also added as a material balance tracer to the injected bank and also to the water which is used to push the bank into the formation [21–24].

A desirable volume for the slug is 50–90 barrels per foot of formation. Injection and production rates may limit the size of test that can be conducted in a reasonable period of time. Primary tracer volumes have varied from 40–1,000 barrels while total water volumes have ranged from 175–2,000 barrels [21–24]. Primary tracer concentration has normally been between 0.5–1 volume percent. In an example given by Deans [20], 500 barrels of 1% ethyl acetate and 0.5% methanol in brine is followed by 1,500 barrels of brine containing 0.5% methanol. These quantities are determined by simple volumetric balance based on the desired depth of invasion.

Depending on formation permeability, injection may normally require 1 to 3 days and shut-in time will be 3 to 12 days. After allowing a suitable time period for hydrolysis of about 10% of the ester (usually about 3–12 days), the well is put on production and tracer concentration in the produced water is monitored. Because the alcohol formed by hydrolysis is much more soluble in the aqueous phase than its parent ester, the alcohol is produced ahead of the ester. The greater this separation between the two tracers, the higher the residual oil saturation. The methanol tracer, which is soluble only in water, determines the drift rate in the reservoir and also indicates the fraction of chemical slug that is produced during the test [21–24].

Interpretation
Computer simulation is used to model the injection, reaction, partitioning, production of the tracers and to correct for overall drift of fluids past the wellbore [22]. Fluid drift is mainly caused by injection or production of fluids in the vicinity of the test well. A best fit is obtained for the injected and produced ester and methanol, and parameters given by this fit are used to model the alcohol production for a range of assigned values of residual saturation. The measured alcohol production curve is then compared with the simulated results in order to estimate the residual saturation in the formation. Reservoir heterogeneity and loss of tracer can present problems in interpretation.

Reservoir Heterogeneity
Although the tracer test samples a relatively large pore volume, results will be weighted towards the higher permeability zones. However, this may not be a disadvantage because these zones will normally be swept preferentially by tertiary processes.

The single-well tracer method is not recommended for fractured reservoirs. Fractures cause nonradial flow, which results in tracer profiles that are almost impossible to simulate [23]. Severe permeability variations are also difficult to interpret. Where gross permeability variations exist, it may be necessary to conduct frequent injection profile tests to

determine the intervals that are experiencing high-rate fluid flow [24].

Loss of Tracer
When the chemical tracer is injected into zones that do not subsequently produce fluids, tracer will be lost to the reservoir. In one field test, loss of tracer was estimated at 15% of the injected amount [21]. Conversely, dilution of produced fluids by water from zones that did not receive tracer injection will also present interpretation problems. If high drift-rates cannot be controlled, the slug of water containing chemicals can move so far from the wellbore that tracer profiles may not be well defined. Drift-rate should be less than 1 ft/day [24].

In some cases, it may not be practical to perform tests in wells with large intervals open or in wells with large holdup volumes in the wellbore. In such instances, well workovers may be required [22].

With wells produced by gas lift, corrections are required to account for the loss of part of the more volatile tracers to the gas by stripping action in the wellbore. Tests conducted [23] with ethyl acetate in gas lift wells indicate a loss of about 30%. Appreciable gas accompanying crude oil lifted by other production methods may cause loss of tracer by stripping.

Accuracy
Success in application of the tracer test depends to a considerable extent on the skill and experience of those conducting the test. While there is no absolute measure of success, comparison of measurements with simulated results provides a useful guide as to the reliability and tolerance of the results. In about 10% of the early tests conducted, residual oil saturation could not be determined from results. Oil saturations of 10% to 20% pore volume have been measured with reported accuracies of ±2% to ±4%. Although conventional methods of estimating residual oil cannot be considered an acceptable standard for determining absolute accuracy of the tracer technique, measurements obtained from tracer tests have generally agreed with values obtained from pressure core tests [1,21–24].

Field Application
Logistic considerations require adequate preplanning which means it may be difficult to schedule tests on quick notice [21]. As compared to well logging techniques, considerable time is required to obtain and interpret the data.

5.10.5 Geophysical Well Logging Techniques
Geophysical well logging has the advantage of being an in-situ measurement and is able to give a continuous estimate of residual oil saturation versus depth [25,26]. These features allow the calibration of the measurements in known water-saturated formations. A more detailed discussion of well logging is given earlier. An evaluation of logging techniques for measurement of S_{or} was provided by Fertl [26].

5.10.5.1 Logging Devices
Five measurements that have potential application are:

1. *Electrical resistivity.* Many devices of different depths of investigation are available [1.58,27]. These devices cannot be used in cased holes unless one uses some nonconducting casing.
2. *Pulsed neutron capture.* This name (PNC) covers logs commercially available such as the Dresser Atlas Neutron Lifetime Log (NLL) and the Schlumberger Thermal Decay Time (TDT). The PNC has the virtue of being useful in cased holes.
3. *Carbon-oxygen.* This measurement has the virtues of being directly sensitive to carbon and of working in cased holes.

4. *Nuclear magnetism.* This service is not routinely used but has the unique virtue of being sensitive only to formation fluids.

5. *Dielectric constant.* This service is now routine but is limited to open holes. Its main advantage over resistivity measurements is that water salinity need not be known.

Electrical Resistivity

Resistivity measurements provide a great range of choice as to the volume of formation to be sampled, ranging from a few cubic inches to many cubic feet. Interpretation of the measurements for residual oil saturation requires a determination of the relation between water saturation and resistivity. If the true formation resistivity R_t and the water resistivity R_w are known, then for clean sandstones [1.42],

$$R_t = F_R R_w S_w^n \qquad [5.10.15]$$

where F_R is the formation factor. If after measuring R_t, the fluid around the hole is replaced by water of resistivity R_w' (through chemical flushing followed by displacement with water of resistivity R_w'), then from Equation 5.10.15 it follows that

$$S_w = 1 - S_{or} = \left[\frac{R_t}{R_t'} \frac{R_w'}{R_w} \right]^{1/n} \qquad [5.10.16]$$

Clearly, n, the saturation exponent, must also be known. It is often taken to be $n = 2$. If cores or other logs are available, better estimates of n can be made.

Relations other than Equation 5.10.16 are required for carbonates and shaly sands [28,29].

Resistivity measurements cannot be made in cased holes. This may explain the limited documentation of such methods for residual oil determination in the literature.

Pulsed-Neutron-Capture

The device used for this measurement periodically emits brief bursts of high energy neutrons. Between bursts these neutrons are rapidly reduced in energy and then more slowly absorbed by formation nuclei. It is the rate of this relatively slow absorption that is measured.

Upon neutron absorption most formation nuclei are left in high energy states that decay to ground states through the emission of a characteristic set of gamma rays. That is, the gamma rays emitted have various energies and numbers at these energies that are unique to the capturing nucleus. However, carbon does not participate in this process.

The gamma rays emitted above a fixed energy (the fixed energy used in commercial tools varies from 50 kev to 200 kev) are detected and counted as a function of time. The analysis of this counting rate yields the apparent thermal decay time or equivalently the apparent capture cross-section of the formation. The apparent cross-section can be corrected to true formation cross-section through computation. Certain modes of operation [30] can yield results requiring little or no correction.

The formation capture cross-section Σ_t is related to the cross-sections of the constituents of the fluid-saturated rock (see Equation 5.2.39) by the formula [30,31]:

$$\Sigma_t = \Sigma_{ma}(1 - \phi) + \Sigma_w \phi S_w + \Sigma_h \phi (1 - S_w) \qquad [5.10.17]$$

where Σ_{ma} is the capture cross-section of the rock (including clay or shale), ϕ is the porosity, Σ_w is the cross-section for the water, Σ_h is the cross-section of the hydrocarbons, and S_w is the water saturation. Thus $(1 - S_w)\phi$ is the oil volume per unit volume of formation.

Of course, the desire is to determine the residual oil saturation $(1 - S_w)$ or the residual oil per unit volume, $\phi(1 - S_w)$. Because many of the quantities on the right hand side of Equation 5.10.17 may be unknown, one employs the so-called log-inject-log technique [32–35].

In the porosity is known in addition to Σ_w, then the simplest form of the log-inject-log technique can be used. Here two successive logs are run. In the first, the normal formation water of cross-section Σ_w occupies the pores. The section of interest is then flushed (at low injection rate) with water of cross-section Σ_w' as different from Σ_w as possible, Σ_t and Σ_t' for these two conditions are measured. Then from Equation 5.10.17 it follows that $S_{or} = 1 - S_w$ is

$$S_{or} = 1 - \frac{\Sigma_t' - \Sigma_t}{\phi(\Sigma_w' - \Sigma_w)} \qquad [5.10.18]$$

To obtain accurate values of Σ_t and Σ_t' one should use a very low logging speed, or stationary measurements, or repeat passes that can be averaged.

If porosity is not known, the water flush can be followed by a flush with chlorinated oil [34]. Here the chlorination is adjusted so that the oil has cross-section Σ_w'. Then from the viewpoint of the PNC response, it is just as if we had $S_w = 100\%$. Then Equation 5.10.17 yields

$$\phi = \frac{\Sigma_t'' - \Sigma_t}{\Sigma_w' - \Sigma_w} \qquad [5.10.19]$$

where Σ_t'' is the measured value after the chlorinated oil flush.

The technique just outlined will suffer if Σ_w is small corresponding to low chloride concentration (below 30,000 parts per million). If Σ_w is unknown, it can be controlled by an initial water flush with water of known cross-section.

The most common practice has been to run a PNC log, inject fresh water, and run a second PNC log. Excellent results have been reported for initial water salinities in excess of 30,000 part per million.

Carbon-Oxygen

The oil industry has long sought a logging method that directly measures oil saturation. The carbon-oxygen (C/O) log is the most recent method in this continuing effort. Since the method is sensitive to formation carbon and since oil is largely carbon, the ideal result would yield directly formation oil content [2.39,2.40,36–39].

The C/O log utilizes apparatus similar to that of the PNC log; namely, a pulsed neutron source and a gamma ray detector. The neutron bursts and the detector device are timed to emphasize gamma rays produced by high-energy neutrons scattering off of carbon and oxygen. The gamma rays are not simply counted but are also analyzed for their energy. The gamma rays produced during the neutron burst are primarily inelastic gamma rays, and it is in this time period that carbon contributes gamma rays to the detector. When the neutron source is off, capture gamma rays are detected and analyzed. Carbon does not contribute to the capture-gamma-ray spectrum. The capture gamma-ray spectrum is used to correct for interference in the carbon region of the inelastic spectrum due to calcium, silicon, and oxygen [37].

The device has serious limitations. These include problems with counting statistics, interfering gamma rays, carbonate rocks, and perturbations caused by casing and borehole. Best results are obtained when the tool is stationary for several minutes to overcome counting statistics, and when the formation porosity and oil saturation are high. The measurement may be improved using log-flush-log techniques, but only a few efforts have been documented at this time [26].

Nuclear Magnetism

This method has been thoroughly discussed in the literature [40–42]. The technique involves a polarization of the hydrogen magnetic moments via a large coil (3 ft long and $3\frac{1}{2}$ in. in diameter) carrying a large direct current. The idea is to align the hydrogen magnetic moments along the field created by the coil. This field ideally is at right angles to the earth's magnetic field. After a few seconds (up to about four), this coil-produced field is reduced to zero as quickly as possible. The polarized hydrozen moment then processes about the earth's field at about 2,200 Hz. This includes a voltage into the coil which is detected and processed to yield a measure of the total number of hydrogen nuclei in the formation fluid. This number or concentration is called the free fluid index (FFI). Since oil and water have about the same concentration of hydrogen nuclei, FFI is a measure of porosity as given by the fluid contributing to the signal.

For the detection of residual oil saturation, one relies on the fact that the addition of paramagnetic ions to the formation water will cause the signal from the water to be annulled. The measurement thus utilizes the log-flush technique. A first measurement is made of FFI. The formation is then flushed with water containing paramagnetic ions so that FFI is sensitive only to residual oil. The result is that S_{or} can be estimated from:

$$S_{or} = \frac{FFI'}{FFI}$$ [5.10.20]

This technique requires signal averaging over many repetitions of the basic measurement in order to compensate for the very poor signal to noise ratio.

The method cannot be used in cased holes and requires a fairly large ($>$ 7 in.) open hole. It has a very shallow penetration and highly viscous oils will not contribute to FFI. In this latter case, a separate measure of porosity will be needed. The service is still not widely available but in principle it is the most promising of all logging methods for the determination of S_{or} [43,44].

Dielectric Constant

The dielectric constants of rock and oil are distinctly different from that of water. The dielectric constant of bulk water is about 80 while those of oil and rock are 4 or less. In practice, however, due to polarization effects on heterogeneous media, this difference in dielectric properties is masked unless very high frequencies are employed in the measurement. One study shows that results of the expected order are obtained with use of very high frequencies and, furthermore, the measurements are unaffected by water salinity [45]. Several devices designed in accordance with these ideas have been designed and successfully field tested [46–56].

The use of this measurement for S_{or} determination could also benefit from the log-inject-log procedure. For example, after logging initially, one could displace the water with a fluid having about the same dielectric constant as oil. A further flush with a fluid removing all oil and of known dielectric constant could be made.

To interpret the measurements, one can use an equation similar to Equation 5.10.17:

$$\Delta t_t = \Delta t_{ma}(1-\phi) + \Delta t_w \phi S_w + \Delta t_H(1-\phi)S_H$$ [5.10.21]

where the Δt_t refers to travel time of the electromagnetic energy between the two receivers of the device. Δt_t is the measured parameter while the other Δt symbols refer to formation and fluid travel times. After the first flush, S_{or} is found from:

$$\phi S_{or} = \phi(1-S_w) = \frac{\Delta t_t - \Delta t_t'}{(\Delta t_w - \Delta t_w')}$$ [5.10.22]

After a second flush, ϕ could be determined from an equation similar to Equation 5.10.19. Equation 5.10.22 can be used to obtain the amount of oil per unit volume. If S_{or} is required, a porosity derived from a neutron-density log could be used.

5.10.5.2 Volume of Reservoir Sampled
It is important to know how large a volume of the reservoir is investigated by a given logging method. This is determined

by the vertical resolution and depth of investigation, which have been estimated as follows:

Pulsed neutron capture
   Depth of investigation   ~6 in.
   Vertical resolution   ~1½ ft
Resistivity
   Depth of investigation   ~1 in. to 5 ft
   Vertical resolution   ~1 in. to 5 ft

This large range reflects the large number of resistivity devices available.

Carbon-oxygen
   Depth of investigation   ~2 in.
   Vertical resolution   ~1½ ft
Nuclear magnetism
   Depth of investigation   ~1½ in.
   Vertical resolution   ~3 ft
Dielectric constant
   Depth of investigation   ~1½ in.
   Vertical resolution   ~1½ in.

These measurements are bulk measurements of the formation sampled. Thus, if there are heterogeneities due to fractures, permeability variations, etc., these will result in errors for elemental volumes. However, since for example, a low permeability zone may not contribute to a tertiary recovery process, this feature of the measurements may be an asset rather than a liability.

5.10.5.3 Accuracy of Logging Methods
The estimated uncertainty in residual oil saturations from electric logs is 5% to 10% under optimum conditions and could easily exceed 10% under less favorable conditions [31]. Accuracy of saturations derived from electric logs is generally in the range of ±15 saturation percent [6], which is clearly inadequate for residual oil determinations.

From the laboratory measurements of the C/O ratio, it is known that the accuracy of this technique is poor when porosity is low [2.39]. Oil saturation probably cannot be reliably determined at porosities less than 15%. However, even in a 30% porosity sand, the accuracy of oil saturation measurements would only be ±12.5%. While this uncertainty could be reduced by taking repeated measurements at the same depth, stability of resolution may still be a problem.

The accuracy requirement for the determination of residual oil is so great that the use of the log-inject-log technique is often required. Although this technique is applicable to various logging methods, practical application has been limited to the PNC and NML. The term log-inject-log (L-I-L) refers basically to the fact that the measurements are made with control over the properties of the fluid saturating the rock volume under study. Assumptions with this technique are that the formation is at residual oil saturation; no oil is displaced from the formation during injection the true total cross-section is measured; the formation water is completely displaced within the radius of investigation by the tool; and that shrinkage of oil due to gas stripping during injection is negligible.

With the pulsed-neutron-capture L-I-L technique, accuracy of ±5% oil saturation is obtainable [31,34]. For best results, the contrast in salinities of the two waters should be as great as possible. While this technique has been successfully used, a test in a 100% water saturated formation yielded oil saturations of 40%–60% [35]. These obviously incorrect values may have resulted from incomplete displacement of formation water close to the wellbore by the injected low salinity water.

Accuracy of the nuclear magnetism log depends largely on the signal-to-noise ratio which can be improved by making repeated readings. Newer versions of the NML tool have improved signal-to-noise characteristics. As with other log-inject-log techniques, the possibility of fluid drift in the formation should be considered. Fluid drift should not be sufficient to move injected paramagnetic ions away from the wellbore which would cause erroneously high values of S_{or} [31].

5.10.6 Summary of Methods for Estimating Residual Oil

Economics of primary and secondary recovery processes are usually sufficiently attractive to permit considerable error in the estimation of recoverable reserves. However, for tertiary recovery the amount of oil remaining in a reservoir and its distribution must be known with reasonable confidence. Firstly, a reliable estimate of residual content is extremely important to technical evaluation of field tests. Secondly, the high front-end costs of tertiary processes are such that overestimates of residual oil saturation could have disastrous economic consequence. Thus, a well-planned effort to measure residual oil saturation is a necessity before any tertiary recovery application. Under favorable circumstances, accuracies of ±2% of reservoir pore space can be achieved. In general, accuracies will not be this good, but values within ±5% are considered necessary.

There is no absolute measure of residual oil saturation for a reservoir. When evaluating a tertiary prospect, a combination of methods should be used which provide information on both amount and distribution. Evaluation will normally begin with material balances using information that is already available. Frequently quoted nation-wide estimates of the amount of residual oil that is potentially available for tertiary recovery are based mainly on material balance. Comparisons of material balance with other methods of determining residual show unacceptable scatter, and on average, the material balance gives saturations which are too high by about 9% (by pore volume). This corresponds to about a 30%–50% overestimate in amount of residual oil. It has been suggested that a much higher ratio (say 2 to 1 or more) in residual oil determined by material balance to that given by other methods is an indication that the reservoir contains extensive regions of high oil saturation and may, therefore, be a good prospect for infill drilling.

In addition to material balance, other estimates of residual oil from resistivity logs and laboratory waterflood tests may also be available. However, in general, none of the conventional methods of determining residual oil saturation—analysis of conventional cores, laboratory displacement tests, conventional logging, material balance—are considered sufficiently reliable in themselves. They can provide useful guides as to whether a tertiary prospect should be investigated further.

Over the past 15–20 years there has been increased field testing of a number of more sophisticated techniques: pressure coring, tracer tests, and the various types of log-inject-log procedures. These methods vary in the conditions under which they can be applied and the type of information they provide.

Pressure coring and the sponge core technique provide information on the vertical distribution of residual oil and also have the advantage that the core analysis procedures directly demonstrate the presence of oil. Because of the possibility of flushing, results may tend to be conservatively low. For pressure coring, the time taken to obtain results tends to be longer than other methods because of transportation and the specialized core analysis work that is needed.

After determining residual oil, the extracted cores are used to obtain needed information on reservoir properties, in particular vertical heterogeneity, and they also can be used in laboratory displacement tests. The reduced diameter ($\sim 2\frac{1}{2}$ in.) of pressure cores compared to those obtained by conventional coring is a disadvantage with respect to making laboratory displacement studies. The use of the sponge coring technique has increased because of the lower coring costs, reduced analysis costs, and, since larger diameter cores are obtained, core plugs can be obtained for subsequent conventional or specialized core testing.

The tracer test samples about a half million times more pore volume than the pressure core barrel. Results with the tracer test will be conservatively low because they are usually weighted towards the more permeable zones, where residual oil will tend to be lower. On balance this is probably an advantage because these zones will also tend to be swept more readily during a tertiary process. The tracer test can be used in old wells, but it is important that the well has not been fractured or stimulated severely.

Of the various log-inject-log procedures, the pulsed-neutron-capture method is the most widely tested. It has the advantage that it can be used in cased holes. Problems can arise with borehole rugosity effects and high values of residual oil if displacements are incomplete during the log-inject-log procedure. Results can be affected by flushing because of the small depth of investigation. The method can give accurate results under favorable circumstances. Stabilization of capture-cross-section values for the log-inject-log procedure can take unexpectedly long times and can present problems in interpretation. The nuclear magnetic log has been rated highly as to accuracy, but cannot be used in cased holes and it still has limited commercial availability. Log-inject-log resistivity measurements can give good results under favorable circumstances but are not applicable to cased holes.

5.10.7 Recommended Methods for Assessing Residual Oil

In determining residual oil saturation, at least two reliable methods should be compared. In most cases, a tracer test combined with injectivity profiles should be run in all situations unless there are clear reasons, such as excessive drift, why the tracer test would fail. The second method selected should provide information on vertical distribution of residual oil. The situations of old and new holes will be considered separately.

5.10.7.1 Existing Wells

Considerable cost can be saved if first measurements can be made on existing, preferably watered-out, producing wells. These wells will usually be cased. Old resistivity logs and core analysis data may be of value in estimating oil distribution and related heterogeneity. For more accurate determination of residual oil, the tracer test should be run together with the pulsed neutron capture log. However careful consideration must first be given to the past history to the well. There is no reliable method for determining residual oil if the well has been fractured or subjected to excessive stimulation. Inadvertent fracturing could have occurred during stimulation. Various forms of acidizing can have serious effects on near-borehole rock properties. The operator must also be sure that any injected chemicals which could affect results are absent from the test region, and that the well is clean and can be put on injection.

5.10.8 New Wells

When drilling new wells for residual oil determination, special attention should be given to using a bland drilling mud that contains no additives likely to alter interfacial tension or wetting properties. If full-diameter cores are obtained (which is desirable if economics permit), the sponge coring technique should be used with precautions being taken to ensure that flushing is minimal during coring. It is also recommended that the vertical distribution of residual oil be determined with a suite of open-hole logs that can include resistivity and dielectric constant. In cases where good data on capture cross-section are available, residual oil saturation can be obtained with a pulsed-neutron-capture log. If a log-inject-log scheme is used (for example, with PNC), this should not involve chemical flushing. The tracer test should be run after completion of logging measurements. All of these measurements can be packed up by laboratory displacement tests which should be carried out on preserved cores.

References

1. *Determination of Residual Oil Saturation*, Interstate Oil Compact Commission, Oklahoma City (June 1978).
2. Martin, F. D., Morrow, N. R., and Moran, J., "A Survey of Methods for Estimating Waterflood Residual Oil Saturations," PRRC Report 79–10, Petroleum Recovery Research Center, Socorro (May 1979).
3. Ramey, H. J. Jr., "Interference Analysis for Anisotropic Formations—A Case History," *J. Pet. Tech.* (October 1975), pp. 1290–1298.
4. Jennings, H. J. Jr., and Timur, A., "Significant Contributions in Formation Evaluation and Well Testing," *J. Pet. Tech.* (Dec. 1973), pp. 1432–1442.
5. Donaldson, E. C., and Crocker, M. E., *Review of Petroleum Oil Saturation and Its Determination*, U.S. Dept. of Energy, Bartlesville (1977).
6. Murphy, R., and Owens, W. W., "The Use of Special Coring and Logging Procedures for Defining Reservoir Residual Oil Saturations," *J. Pet. Tech.* (July 1973), pp. 841–850.
7. Kazemi, H., "Determination of Waterflood Residual Oil Saturation from Routine Core Analysis," *J. Pet. Tech.* (Jan. 1977), pp. 31–32.
8. Hagedorn, A. R., and Blackwell, R. J., "Summary of Experience with Pressure Coring," paper SPE 3962 presented at the SPE 47th Annual Fall Meeting, San Antonio, Oct. 8–11, 1972.
9. Bilhartz, H. L. Jr., "Case History: A Pressure Core Hole," paper SPE 6389 presented at the 1977 Permian Basin Oil & Gas Recovery Conference, Midland, March 10–11.
10. Bilhartz, H. L. Jr., and Charlson, G. S., "Coring For In Situ Saturations in the Willard Unit CO_2 Flood Mini-Test," paper SPE 7050 presented at Improved Methods for Oil Recovery of SPE, Tulsa, Apr. 16–19, 1978.
11. Yell, L., "Pressure Coring Yields Valuable Reservoir Data," *Oil & Gas J.* (Oct. 30, 1978), pp. 95–99.
12. Sparks, R. L., "A Technique for Obtaining In Situ Saturations of Underpressured Reservoirs," *J. Pt. Tech.* (Nov. 1982).
13. Hyland, C. R., "Pressure Coring–An Oilfield Tool," paper SPE 12093 presented at the 58th Annual Tech. Conf., San Francisco, Oct. 5–8, 1983.
14. Park, A., "Improved Oil Saturation Data Using Sponge Core Barrel," paper SPE 11550 presented at 1983 Production Operation Symp., Oklahoma City, Feb. 27–March 1.
15. *Fundamentals of Core Analysis, Mod II*, Core Laboratories, Inc.
16. Special Core Analysis and Industrial Water Technology, prepared by Special Core Analysis Studies Section, Core Laboratoires, Inc., April 1972.
17. Grist, D. M., Langley, G. O., and Neustadter, E. L., "The Dependence of Water Permeability on Core Cleaning Methods in the Case of Some Sandstone Samples," *J. Can. Pet. Tech.* (April–June 1975), p. 48.
18. Cuiec, L., "Study of Problems Related to the Restoration of the Natural State of Core Samples," *J. Can. Pet. Tech.* (Oct–Dec. 1977), pp. 68–80.
19. Cooke, C. E. Jr., "Method of Determining Residual Oil Saturation in Reservoirs," U.S. Patent 3,590,923 (July 6, 1971).
20. Deans, H. A., "Method of Determining Fluid Saturations in Reservoirs," U.S. Patent 3,623,842 (Nov. 30, 1971).
21. Sheely, C. Q., "Description of Field Tests to Determine Residual Oil Saturation by Single-Well Tracer Test," paper SPE 5840 presented at the Improved Oil Recovery Symposium, Tulsa, March 22–24, 1976.
22. Tomich, J. F., et al., "Single-Well Tracer Method to Measure Residual Oil Saturation," *J. Pet. Tech.* (Feb. 1973), pp. 211–218.
23. Bragg, J. R., Carlson, L. O., and Atterbury, J. H., "Recent Applications of the Single-Well Tracer Method for Measuring Residual Oil Saturation," paper SPE 5805 presented at the Improved Oil Recovery Symposium, Tulsa, March 22–24, 1976.
24. O'Brien, L. J., Cooke, R. S., and Willis, H. R., "Oil Saturation Measurements at Brown and East Voss Tannehil Fields," *J. Pet. Tech.* (January 1978), pp. 17–25.
25. Fertl, W. H., "Find ROS From Well Logs for Enhanced Recovery Projects," *Oil & Gas J.* (Feb. 12, 1979), pp. 120–127.
26. Fertl, W. H., "Well Logging and Its Applications in Cased Holes," *J. Pet. Tech.* (Feb. 1984), pp. 249–266.
27. *Log Interpretation Fundamentals*, Dressere Atlas Division, Dresser Industries (1975).
28. Waxman, M. H., and Thomas, E. C., "Electrical Conductivities in Oil-Bearing Shaly Sands," *Soc. Pet. Eng. J.* (June 1968), pp. 107–122.
29. Waxman, Mitt, and Smits, L. J. M., "Electrical Conductivities of Oil-Bearing Shaly Sands," *J. Pet. Tech.* (June 1965), pp. 107–122.
30. Richardson, J. E., et al. "Methods for Determining Residual Oil with Pulsed Neutron Capture Logs," *J. Pet. Tech.* (May 1973), pp. 593–606.
31. Wyman, R. E., "How Should We Measure Residual Oil Saturation?" *Bull. Cdn. Pet. Geol.*, Vol. 25, No. 2 (May 1977), pp. 233–270.
32. Murphy, R., and Owens, W. W., "Time-Lapse Logging, A Valuable Reservoir Evaluation Technique," *J. Pet. Tech.* (January 1964), pp. 15–19.
33. Murphy, R., Owens, W. W., and Dauben, D. L., "Well Logging Methods," U.S. Patent 3,757,575 (Sept. 11, 1973).

34. Murphy, R., Foster, G. T., and Owens, W. W., "Evaluation of Waterflood Residual Oil Saturation Using Log-Inject-Log Procedures," *J. Pet. Tech.* (Feb. 1977), 178–186.

35. Bragg, J. R., et al., "A Comparison of Several Techniques for Measuring Residual Oil Saturation," paper SPE 7074 presented at the Fifth Symposium on Improved Oil Recovery, April 16–19, 1978.

36. Culver, R. B., Hopkinson, E. C., and Youmans, A. H., "Carbon-Oxygen (C/O) Logging Instrumentation," *Soc. Pet. Eng. J.* (Oct. 1974), pp. 463–470.

37. *Log Interpretation*, Volume II—Applications, Schlumberger Limited (1974).

38. Fertl, W. H., et al., "Evaluation and Monitoring of Enhanced Recovery Projects in California and Cased-Hole Exploration and Recompletions in West Texas Based on the Continuous Carbon/Oxygen Log," *J. Pet. Tech.* (Jan. 1983), pp. 143–157.

39. Westaway, P., Hertzog, R., and Plasek, R. E., "Neutron-Induced Gamma Ray Spectroscopy for Reservoir Analysis," *Soc. Pet. Eng. J.* (June 1983), pp. 553–564.

40. Robinson, J. D., et al., "Determining Residual Oil with the Nuclear Magnetism Log," *J. Pet. Tech.* (Feb. 1974), pp. 226–236.

41. Loren, J. D., and Robinson, J. D., "Relations Between Pore Size, Fluid and Matrix Properties, and NML Measurements," *Soc. Pet. Eng. J.* (Sept. 1970), pp. 268–278.

42. Herrick, R. C., Couturie, S. H., and Best, D. L., "An Improved Nuclear Magnetism Logging System and its Application to Formation Evaluation," paper SPE 8361 presented at the 1979 SPE Annual Technical Conference and Exhibition, Las Vegas, Sept. 23–26.

43. Neuman, C. H., and Brown, R. J. S., "Application of Nuclear Magnetism Logging to Formation Evaluation," *J. Pet. Tech.* (Dec. 1982), pp. 2853–2862.

44. Neuman, C. H., "Logging Measurement of Residual Oil, Rangely Field, CO," *J. Pet. Tech.* (Sept. 1983), pp. 1735–1744.

45. Poley, J.Ph., Nooteboom, J. J., and deWaal, P. J., "Use of V.H.F. Dielectric Measurements for Borehole Formation Analysis," *The Log Analyst* (May–June 1978), pp. 8–30.

46. Meador, R. A., and Cox, P. T., "Dielectric Constant Logging, A Salinity Independent Estimation of Formation Water Volume," paper SPE 5504 presented at Fall Meeting, Dallas, Sept. 30–Oct. 1, 1975.

47. Calvert, T. J., Rau, R. N., and Wells, L. E., "Electromagnetic Propagation-A New Dimension in Logging." paper SPE 6542 presented at the SPEAIME 52nd Annual Fall Technical Conference, Oct. 9–12, 1977.

48. Freedman, R., and Montaque, D. R., "Electromagnetic Propagation Tool (EPT): Comparison of Log Derived and in situ Oil Saturation in Shaly Fresh Water Sands," paper SPE 9266 presented at the 1980 SPE Meeting, Dallas, Sept. 21–24.

49. Wharton, R. P., et al., "Electromagnetic Propagation Logging: Advances in Technique and Interpretation," paper SPE 9267 presented at 55th Annual Fall Mtg., SPE, Dallas, Sept. 21–24, 1980.

50. Cox, P. T., and Warren, W. F., "Development and Testing of the Texaco Dielectric Log," SPWLA 24th Annual Logging Symp., June 27–30, 1983.

51. Eck, M. E., and Powell, D. E., "Application of Electromagnetic Propagation Logging in the Permian Basin of West Texas," paper SPE 12183 presented at 1983 SPE Annual Mtg., San Francisco, Oct. 5–8.

52. Geng, X., et al., "Dielectric Log-A Logging Method for Determining Oil Saturation," *J. Pet. Tech.* (Oct. 1983), pp. 1797–1805.

53. Dahlberg, K. E., and Ference, M. V., "A Quantitative Test of the Electromagnetic Propagation (EPT) Log for Residual Oil Determination," SPWLA 25th Annual Logging Symposium, New Orleans, June 10–13, 1984.

54. Shen, L. C., Manning, M. J., and Price, J. M., "Application of Electromagnetic Propagation Tool in Formation Evaluation," SPWLA 25th Annual Logging Symposium, New Orleans, June 10–13, 1984.

55. Iskander, M. F., Rattlingourd, S. O., and Oomrigar, J., "A New Electromagnetic Propagation Tool for Well Logging," paper SPE 13189 presented at the 59th Annual Conf., Houston, Sept. 16–19, 1984.

56. Shen, L. C., "Problems in Dielectric-Constant Logging and Possible Routes to Their Solutions," *The Log Analyst* (Nov.–Dec. 1985), pp. 14–25.

5.11 ENHANCED OIL RECOVERY METHODS

5.11.1 Definition

A general schematic of the enhanced oil recovery (EOR) process is depicted in Figure 5.11.1. The more common techniques that are currently being investigated include:

Enhanced Oil Recovery
Chemical Oil Recovery or Chemical Flooding
  Polymer-augmented waterflooding
  Alkaline or caustic flooding
  Surfactant flooding
  —Low tension waterflooding
  —Micellar/polymer (microemulsion) flooding
Hydrocarbon or Gas Injection
  Miscible solvent (LPG or propane)
  Enriched gas drive
  High-pressure gas drive
  Carbon dioxide flooding
  Flue gas
  Inert gas (nitrogen)
Thermal Recovery
  Steamflooding
  In-situ combustion

These procedures are discussed in several texts on the subject [9.2,1–5]. Two studies by the National Petroleum Council [6,7] and several papers summarizing the later study are available [8–11]. The extensive literature on enhanced recovery will not be cited, and the reader is referred to Reference 12 which provides numerous citations and is the basis of the following discussion.

5.11.2 Chemical Flooding

Chemical oil recovery methods include polymer, surfactant/polymer (variations are called micellar-polymer, microemulsion, or low tension waterflooding), and alkaline (or caustic) flooding. All of these methods involve mixing chemicals (and sometimes other substances) in water prior to injection. Therefore, these methods require conditions that are very favorable for water injection: low-to-moderate oil viscosities, and moderate-to-high permeabilities. Hence,

Typical EOR process

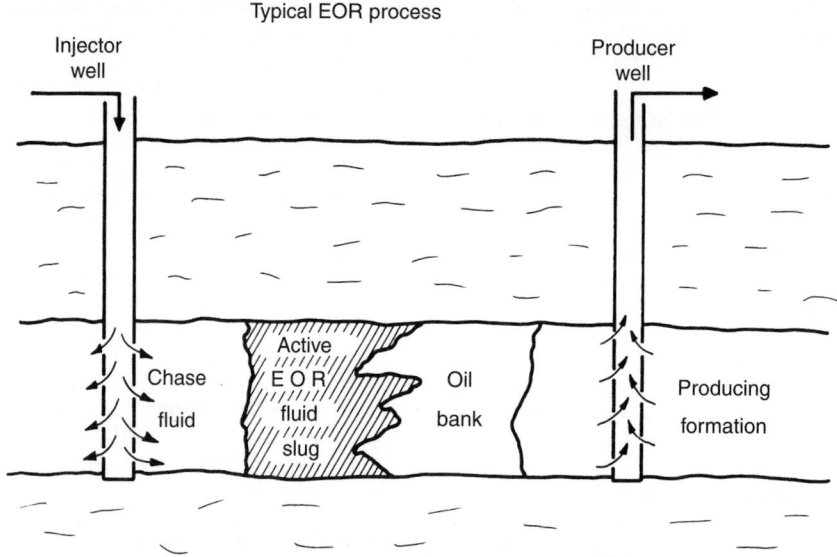

Figure 5.11.1   *General schematic of enhanced oil recovery.*

chemical flooding is used for oils that are more viscous than those oils recovered by gas injection methods but less viscous than oils that can be economically recovered by thermal methods. Reservoir permeabilities for chemical flood conditions need to be higher than for the gas injection methods, but not as high as for thermal methods. Since lower mobility fluids are usually injected in chemical floods, adequate injectivity is required. If previously waterflooded, the chemical flood candidate should have responded favorably by developing an oil bank. Generally, active water-drive reservoirs should be avoided because of the potential for low remaining oil saturations. Reservoirs with gas caps are ordinarily avoided since mobilized oil might resaturate the gas cap. Formations with high clay contents are undesirable since the clays increase adsorption of the injected chemicals. In most cases, reservoir brines of moderate salinity with low amounts of divalent ions are preferred since high concentrations interact unfavorably with the chemicals that are injected.

5.11.2.1 Polymer-Augmented Waterflooding

High mobility ratios cause poor displacement and sweep efficiencies, and result in early breakthrough of injected water. By reducing the mobility of water, water breakthrough can be delayed by improving the displacement, areal, and vertical sweep efficiencies; therefore more oil can be recovered at any given water cut. Thus, the ultimate oil recovery at a given economic limit may be 4%–10% higher with a mobility-controlled flood than with plain water. Additionally, the displacement is more efficient in that less injection water is required to produce a given amount of oil.

The need to control or reduce the mobility of water led to the advent of polymer flooding or polymer-augmented waterflooding. Polymer flooding is viewed as an improved waterflooding technique since it does not ordinarily recover residual oil that has been trapped in pore spaces and isolated by water. However, polymer flooding can produce additional oil over that obtained from waterflooding by improving the displacement efficiency and increasing the volume of reservoir that is contacted. Dilute aqueous solutions of

water-soluble polymers have the ability to reduce the mobility of water in a reservoir thereby improving the efficiency of the flood. Partially hydrolyzed polyacrylamides (HPAM) and xanthan gum (XG) polymers both reduce the mobility of water by increasing viscosity. In addition, HPAM can alter the flow path by reducing the permeability of the formation to water. The reduction in permeability to water that is achieved with HPAM solution can be fairly permanent while the permeability to oil can remain relatively unchanged. The resistance factor is a term that is commonly used to indicate the resistance to flow that is encountered by a polymer solution as compared to the flow of plain water. For example, if a resistance factor of 10 is observed, it is 10 times more difficult for the polymer solution to flow through the system, or the mobility of water is reduced 10-fold. Since water has a viscosity of about 1 cp, the polymer solution, in this case, would flow through the porous system as though it had an apparent or effective viscosity of 10 cp even though a viscosity measured in a viscometer could be considerably lower.

The improvement in areal sweep efficiency resulting from polymer treatment can be estimated from Figure 5.9.9. For example, if the mobility ratio for a waterflood with a 5-spot pattern is 5, the areal sweep efficiency is 52% at breakthrough. If the economic limit is a producing water-oil ratio of 100:1 ($f_w \cong 100/101 = 0.99$), the sweep efficiency at floodout is about 97%. If the polymer solution results in the mobility ratio being lowered to 2, sweep efficiencies are 60% at breakthrough and 100% at the same economic water-oil ratio.

A simplified approach to qualitatively observing the improvement with polymers in a stratified system is illustrated in Figure 5.9.12. For example, if the permeability variation is 0.7, the waterflood mobility ratio is 5, and the initial water saturation is 0.3, the fractional recovery of oil-in-place can be estimated. From the plot, $R(1 - 0.4S_w) = 0.29$, and the fractional recovery, R, is $0.29/[1 - (0.4)(0.3)] = 0.33$. This R needs to be multiplied by the areal sweep efficiency of 0.97 to yield a recovery of 32% of the oil-in-place. If polymers again reduce the mobility ratio to 2 (and if no improvement in permeability variation occurs), a fractional recovery of 0.375

is obtained. Since the areal sweep with the polymer flood is 100%, a recovery of 37.5% of the oil-in-place is estimated. Thus the improvement with polymers is estimated at 0.375–0.32 or 5.5% of the oil-in-place. If the flow distribution with polymer solution lowered the permeability variation (which is not likely), the incremental production could be higher. These calculations are gross oversimplifications of actual conditions and only serve as a tool to show that reducing mobility ratio with polymers can improve the sweep efficiencies.

A properly sized polymer treatment may require the injection of 15%–25% of a reservoir pore volume; polymer concentrations may normally range from 250 to 2,000 mg/L. For very large field projects, millions of pounds of polymer may be injected over a 1–2 year period of time; the project then reverts to a normal waterflood. The polymer flooding literature was reviewed in the late 1970s [13]. Recommendations on the design of polymer floods were recently made available [14].

5.11.2.2 Variations in the Use of Polymers
In-Situ Polymerization
A system is available in which acrylamide monomer is injected and polymerized in the reservoir. Both injection wells and producing wells have been treated.

Crosslinked or Gelled Polymers
Several methods are available for diverting the flow of water in reservoirs with high permeability zones or fracture systems. Some methods are only effective near the injection well while others claim the treatment can be effective at some depth into the reservoir. Both producing wells and injection wells can be treated.

One method is the aluminum citrate process which consists of the injection of a slug of HPAM polymer solution, aluminum ion chelated with citrate ion, and a second slug of polymer. Some of the polymer in the first slug adsorbs or is retained on the surfaces of the reservoir. The aluminum ion attaches to the adsorbed polymer and acts as a bridge to the second polymer layer. This sequence is repeated until the desired layering is achieved. The transport of aluminum ions through the reservoir may be a problem in certain cases, so the effects of the treatment may be limited to near the wellbore.

Another method is based on the reduction of chromium ions to permit the crosslinking of HPAM or XG polymer molecules. A polymer slug containing Cr^{+6} is injected followed by a slug of polymer containing a reducing agent. When the Cr^{+6} is reduced to Cr^{+3}, a gel is formed with the polymer. The amount of permeability reduction is controlled by the number of times each slug is injected, the size of each slug, or the concentrations used. An alternate treatment involves placing a plain water pad between the first and second polymer slugs.

In another variation of the above two methods for HPAM, a cationic polymer is injected first. Since reservoir surfaces are often negatively charged, the cationic polymer is highly adsorbed. When the foregoing sequential treatments are injected, there is a strong attraction between the adsorbed cationic polymer and the anionic polymers that follow. Polymer concentrations used in these variations are normally low, on the order of 250 mg/L. With low molecular weight polymers or if a very stiff gel is desired, polymer concentrations of 1–1.5% are common. The type of polymers are similar to those used in conventional polymer flooding, but the products used for gelation command a higher price.

Methods developed recently, especially for fracture treatments, include Cr^{3+} (acetate)-polyacrylamide, collordal silica, and resorcinol-formaldehyde.

5.11.2.3 Surfactant and Alkaline Flooding
Both alkaline flooding and surfactant flooding improve oil recovery by lowering the interfacial tension between crude oil and the displacing water. With alkaline flooding, the surfactants are generated in situ when the alkaline materials react with crude oil–this technique is normally only viable when the crude oil contains sufficient amount of organic acids to produce natural surfactants. Other possible mechanisms with the caustic materials include emulsification of the oil and alteration in the preferential wettability of the reservoir rock.

With surfactant flooding, surface-active agents are mixed with other compounds (such as alcohol and salt) in water and injected to mobilize the crude oil. Polymer-thickened water is then injected to push the mobilized oil-water bank to the producing wells. Water-soluble polymers can be used in a similar fashion with alkaline flooding. For micellar/polymer flooding, the concentration of polymer used may be similar to the value given for polymer flooding, but the volume of polymer solution may be increased to 50% or more of a reservoir pore volume.

Alkaline Flooding
Alkaline or caustic flooding consists of injecting aqueous solutions of sodium hydroxide, sodium carbonate, sodium silicate or potassium hydroxide. The alkaline chemicals react with organic acids in certain crude oils to produce surfactants in situ that dramatically lower the interfacial tension between water and oil. The alkaline agents also react with the reservoir rock surfaces to alter wettability-either from oil-wet to water-wet, or vice versa. Other mechanisms include emulsification and entrainment of oil or emulsification and entrapment of oil to aid in mobility control. Since an early patent in the 1920s described the use of caustic for improved recovery of oil, much research and some field tests have been conducted. Slug size of the alkaline solution is often 10%–15% PV, concentrations of the alkaline chemical are normally 0.2% to 5%. Recent tests are using large amounts of relatively high concentrations. A preflush of fresh or softened water often precedes the alkaline slug, and a drive fluid (either water or polymer-thickened water) follows the alkaline slug.

Surfactant/Polymer Flooding
Surfactant use for oil recovery is not a recent development. Patents in the late 1920s and early 1930s proposed the use of low concentrations of detergents to reduce the interfacial tension between water and oil. To overcome the slow rate of advance of the detergent, Taber [15] proposed very high concentrations ($\sim 10\%$) of detergent in aqueous solution.

During the late 1950s and early 1960s, several different present-day methods of using surfactants for enhanced recovery were developed. A review of these methods is beyond the scope of this chapter and is available in the literature [16–19]. In some systems, a small slug (> about 5% PV) was proposed that included a high concentration of surfactant (normally 5%–10%). In many cases, the microemulsion includes surfactant, hydrocarbon, water, an electrolyte (salt), and a cosolvent (usually an alcohol). These methods ordinarily used a slug (30%–50% PV) of polymer-thickened water to provide mobility control in displacing the surfactant and oil-water bank to the producing wells. The polymers used are the same as those discussed in the previous section. In most cases, low-cost petroleum sulfonates or blends with other surfactants have been used. Intermediate surfactant concentrations and low concentration systems (low tension waterflooding) have also been proposed. The lower surfactant concentration systems may or may not contain polymer

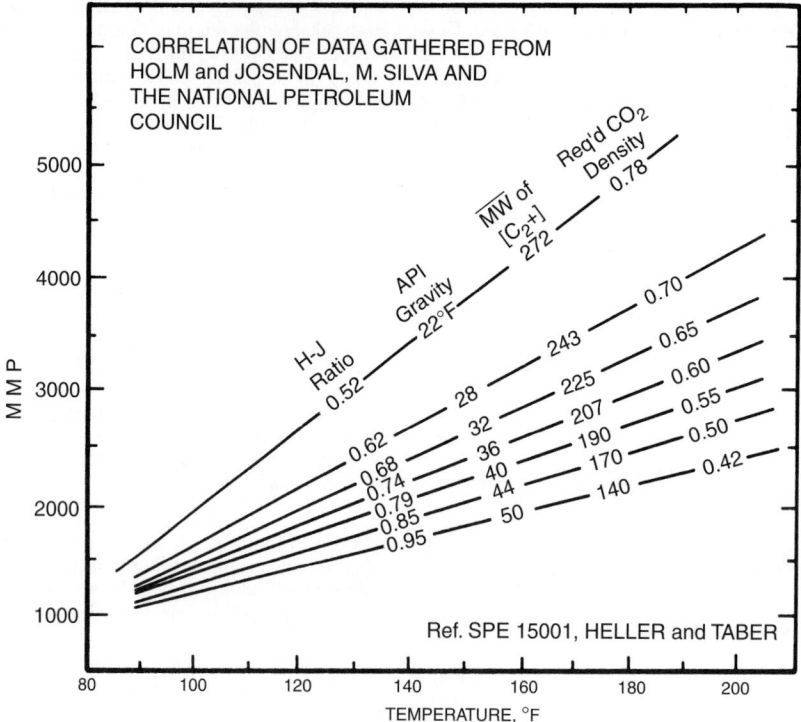

Figure 5.11.2 *Correlations for CO_2 minimum miscibility pressure [23].*

in the surfactant slug, but will utilize a larger slug (30%–100% PV) of polymer solution.

Alkaline/Surfactant/Polymer Flooding
A recent development uses a combination of chemicals to lower process costs by lowering injection cost and reducing surfactant adsorption. These mixtures, termed alkaline/surfactant/polymer (ASP), permit the injection of larger slugs of injectant because of the lower cost.

5.11.3 Gas Injection Methods
5.11.3.1 Hydrocarbon Miscible Flooding
Gas injection is certainly one of the oldest methods utilized by engineers to improve recovery, and its use has increased recently, although most of the new expansion has been coming from the nonhydrocarbon gases [20]. Because of the increasing interest in CO_2 and nitrogen or flue gas methods, they are separated from the hydrocarbon miscible techniques.

Hydrocarbon miscible flooding can be subdivided further into three distinct methods, and field trials or extensive operations have been conducted in all of them. For LPG slug or solvent flooding, enriched (condensing) gas drive and high pressure (vaporizing) gas drive, a range of pressures (and therefore, depths) are needed to achieve miscibility in the systems.

Unless the reservoir characteristics were favorable, early breakthrough and bypassing of large quantities of oil have plagued many of the field projects. In addition, the hydrocarbons needed for the processes are valuable, and there is increasing reluctance to inject them back into the ground when there is some question about the percentage that will be recovered the second time around. Therefore, in the U.S.

in recent years the emphasis has been shifting to less valuable nonhydrocarbon gases such as CO_2, nitrogen, and flue gases. Although nitrogen and flue gases do not recover oil as well as the hydrocarbon gases (or liquids), the overall economics may be somewhat more favorable.

5.11.3.2 Nitrogen and Flue Gas Flooding
As previously mentioned, nitrogen and flue gas (about 87% N2 and 12% CO_2) are sometimes used in place of hydrocarbon gases because of economics. Nitrogen also competes with CO_2 in some situations for the same reason. The economic appeal of nitrogen stems not only from its lower cost on a standard Mcf basis, but also because its compressibility is much lower. Thus, for a given quantity at standard conditions, nitrogen will occupy much more space at reservoir pressures than CO_2 or even methane at the same conditions. However, both nitrogen or flue gas are inferior to hydrocarbon gases (and much inferior to CO_2) from an oil recovery point of view. Nitrogen has a lower viscosity and poor solubility in oil and requires a much higher pressure to generate or develop miscibility. The increase in the required pressure is significant compared to methane and very large (4–5 times) when compared to CO_2. Therefore, nitrogen will not reduce the displacement efficiency too much when used as a chase gas for methane, but it can cause a significant drop in the effectiveness of a CO_2 flood if the reservoir pressures are geared to the miscibility requirements for CO_2 displacements. Indeed, even methane counts as a desirable "light end" or "intermediate" in nitrogen flooding, but methane is quite deleterious to the achievement of miscibility in CO_2 flooding at modest pressures.

5.11.3.3 Carbon Dioxide Flooding

CO_2 is effective for recovery of oil for a number of reasons. In general, carbon dioxide is very soluble in crude oils at reservoir pressures; therefore, it swells the net volume of oil and reduces its viscosity even before miscibility is achieved by the vaporizing gas drive mechanism. As miscibility is approached, both the oil phase and the CO_2 phase (which contains many of the oil's intermediate components) can flow together because of the low interfacial tension and the relative increase in the total volumes of the combined CO_2 and oil phases compared to the water phase. However, the generation of miscibility between the oil and CO_2 is still the most important mechanism, and it will occur in CO_2– crude oil systems as long as the pressure is high enough. This so-called "minimum miscibility pressure" or MMP has been the target of several laboratory investigations and is no longer a mystery. The 1976 NPC report [6] showed that there is a rough correlation between the API gravity and the required MMP, and that the MMP increased with temperature. Some workers have shown that a better correlation is obtained with the molecular weight of the C_5^+ fraction of the oil than with the API gravity. In general the recent work shows that the required pressure must be high enough to achieve a minimum density in the CO_2 phase [21,22]. At this minimum density, which varies with the oil composition, the CO_2 becomes a good solvent for the oil, especially the intermediate hydrocarbons, and the required miscibility can be generated or developed to provide the efficient displacement normally observed with CO_2. Therefore, at higher temperatures, the higher pressures are needed only to increase the CO_2 density to the same value as observed for the MMP at the lower temperature. Figure 5.11.2 shows the variation of minimum miscibility pressure with temperature and oil composition [23].

Although the mechanism for CO_2 flooding appears to be the same as that for hydrocarbon miscible floods, CO_2 floods may give better recoveries even if both systems are above their required miscibility pressures, especially in tertiary floods. Compared to hydrocarbons, CO_2 has a much higher solubility in water, and it has been observed in laboratory experiments to diffuse through the water phase to swell bypassed oil unitl the oil is mobile. Thus, not only are the oil and depth screening criteria easier to meet in CO_2 flooding, but the ultimate recovery may be better than with hydrocarbons when above the MMP. It must be noted, however that this conjecture has not been proved by rigorous and directly comparable experiments.

5.11.4 Thermal Recovery
5.11.4.1 In-Situ Combustion
The theory and practice of in-situ combustion or fireflooding is covered comprehensively in the recent SPE monograph on thermal recovery by Prats [4]. In addition, the continuing evolution of screening criteria for fireflooding [24,25] and steamflooding [26] have been reviewed and evaluated by Chu. A recent appraisal of in-situ combustion was provided by White [27] and the status of oxygen fireflooding was provided by Garon [28].

Part of the appeal of fireflooding comes from the fact that it uses the world's cheapest and most plentiful fluids for injection: air and water. However, significant amounts of fuel must be burned, both above the ground to compress the air, and below ground in the combustion process. Fortunately, the worst part of the crude oil is burned; the lighter ends are carried forward in advance of the burning zone to upgrade the crude oil.

5.11.4.2 Steam Flooding
Of all of the enhanced oil recovery processes currently available, only the steam drive (steamflooding) process is routinely used on a commercial basis. In the United States, a majority of the field testing with this process has occurred in California, where many of the shallow, high-oil-saturation reservoirs are good candidates for thermal recovery. These reservoirs contain high-viscosity crude oils that are difficult to mobilize by methods other than thermal recovery.

In the steam drive process, steam is continuously introduced into injection wells to reduce the viscosity of heavy oil and provide a driving force to move the more mobile oil towards the producing wells. In typical steam drive projects, the injected fluid at the surface may contain about 80% steam and 20% water (80% quality) [6]. When steam is injected into the reservoir, heat is transferred to the oil-bearing formation, the reservoir fluids, and some of the adjacent cap and base rock. As a result, some of the steam condenses to yield a mixture of steam and hot water flowing through the reservoir.

The steam drive may work by driving the water and oil to form an oil bank ahead of the steamed zone. Ideally this oil bank remains in front, increasing in size until it is produced by the wells offsetting the injector. However, in many cases, the steam flows over the oil and transfers heat to the oil by conduction. Oil at the interface is lowered in viscosity and dragged along with the steam to the producing wells. Recoverability is increased because the steam (heat) lowers the oil viscosity and improves oil mobility. As the more mobile oil is displaced the steam zone expands vertically, and the steam-oil interface is maintained. This process is energy-intensive since it requires the use of a significant fraction (25%–40%) of the energy in the produced petroleum for the generation of steam.

In steamflooding, the rate of steam injection is initially high to minimize heat losses to the cap and base rock. Because of reservoir heterogeneities and gravity segregation of the condensed water from the steam vapor, a highly permeable and relatively oil-free channel often develops between injector and producer. Many times this channel occurs near the top of the oil-bearing rock, and much of the injected heat is conducted to the caprock as heat loss rather than being conducted to oil-bearing sand where the heat is needed. In addition, the steam cannot displace oil efficiently since little oil is left in the channel. Consequently, neither the gas drive from the steam vapor nor the convective heat transfer mechanisms work as efficiently as desired. As a result, injected steam will tend to break through prematurely into the offset producing wells without sweeping the entire heated interval.

5.11.5 Technical Screening Guides
In some instances, only one type of enhanced recovery technique is applicable for a specific field condition but, in many instances, more then one technique is possible. The selection of the most appropriate process is facilitated by matching reservoir and fluid properties to the requirements necessary for the individual EOR techniques. A summary of the technical screening guides for the more common EOR processes is given in Table 5.11.1. A distinction is made between the oil properties and reservoir characteristics that are required for each process. Generally, steamflooding is applicable for very viscous oils in relatively shallow formations. On the other extreme, CO_2 and hydrocarbon miscible flooding work best with very light oils at depths that are great enough for miscibility to be achieved. Both steamflooding and in-situ combustion require fairly high permeability reservoirs. Chemical flooding processes (polymer, alkaline, or surfactant) are applicable in low to medium viscosity

Table 5.11.1 Summary of Screening Criteria for Enhanced Recovery Methods

	Oil properties			Oil Saturation	Reservoir characteristics				
	Gravity °API	Viscosity cp	Composition		Formation Type	Net thickness ft	Average permeability md	Depth ft	Temperature °F
Gas injection methods									
Hydrocarbon	>35	<10	High % of C2–C7	>30%PV	Sandstone or carbonate	Thin unless dipping	N.C.	>2,000 (LPG) to >5,000 (H.P. Gas)	N.C.
Nitrogen & flue gas	>24 >35 for N2	<10	High % of C1–C7	>30% PV	Sandstone or carbonate	Thin unless dipping	N.C.	>4,500	N.C.
Carbon dioxide	>26	<15	High % of C2–C12	>30% PV	Sandstone or carbonate	Thin unless dipping	N.C.	>2,000	N.C.
Chemical Flooding									
Surfactant/polymer	>25	<30	Light intermediates desired	>30% PV	Sandstone preferred	>10	>20	<8,000	<175
Polymer	>25	<150	N.C.	Mobile oil Above waterflood residual >10% PV	Sandstone preferred; carbonate possible	N.C.	>10 (normally)	<9,000	<200
Alkaline	13-35	<200	Some organic acids	Above waterflood residual	Sandstone preferred	N.C.	>20	<9,000	<200
Thermal									
Combustion	<40 (10-25 normally)	<1,000	Some asphaltic components	>40–50% PV	Sand or sand stone with high porosity	>10	>100*	>500	>150 preferred
Steamflooding	<25	>20	N.C.	>40–50% PV	Sand or sand stone with high porosity	>20	>200**	300–5,000	N.C.

From Reference 12.
N.C. = Not Critical.
*Transmissibility > 20 md ft/cp
**Transmissibility >100 md ft/cp

oils; depth is not a major consideration except, at great depths, the higher temperature may present problems in the degradation or consumption of some of the chemicals.

Screening guides or criteria are among the first items considered when a petroleum engineer evaluates a candidate reservoir for enhanced oil recovery. A source often quoted for screening criteria is the 1976 National Petroleum Council (NPC) report on Enhanced Recovery [6], which was revised by the NPC in 1984 [7]. Both reports list criteria for six enhanced recovery methods.

Some reservoir considerations apply to all enhanced recovery methods. Because drilling costs increase markedly with depth, shallow reservoirs are preferred, as long as all necessary criteria are met. For the most part, reservoirs that have extensive fractures, gross heterogeneities, thief zones, or are highly faulted should be avoided. Ideally, relatively uniform reservoirs with reasonable oil saturations, minimum shale stringers, and good areal extent are desired.

Implementation of enhanced recovery projects is expensive, time-consuming, and people-intensive. Substantial costs are often involved in the assessment of reservoir quality, the amount of oil that is potentially recoverable, laboratory work associated with the EOR process, computer simulations to predict recovery, and the performance of the project. One of the first steps in deciding to consider EOR is, of course, to select reservoirs with sufficient recoverable oil and areal extent to make the venture profitable.

With any of the processes, the nature of the reservoir will play a dominant role in the success or failure of the process. Many of the failures with EOR have resulted because of unknown or unexpected reservoir problems. Thus, a thorough geological study is usually warranted.

The technique of using cursory screening guides is convenient for gaining a quick overview of all possible methods before selecting the best one for an economic analysis. Common sense and caution must be exercised since the technical guides are based on laboratory data and results of enhanced recovery field trials, and are not rigid guides for applying certain processes to specific reservoirs. Additionally, the technical merits of recent field projects are clouded by various incentive programs that make it difficult to discern true technical applications. Some projects may have been technical misapplications or failures, but economic successes. Certainly, there have been enough technical successes, but economic failures.

Nevertheless, some EOR processes can be rejected quickly because of unfavorable reservoir or oil properties, so the use of preferred criteria can be helpful in selecting methods that may be commercially attractive. If the criteria are too restrictive, some feasible method may be rejected from consideration. Therefore, the guidelines that are adopted should be sufficiently broad to encompass essentially all of the potential methods for a candidate reservoir.

For convenience, brief descriptions of the eight most common enhanced recovery methods are provided in the following sections. These descriptions list the salient features of each method along with the important screening guides. A few general comments are offered here on the relative importance of some individual screening guides to the overall success of the various methods. In addition, we will make some observations on the method itself and its relationship to other enhanced recovery choices that may be available.

5.11.5.1 Hydrocarbon Miscible Flooding [12]
Description
Hydrocarbon miscible flooding consists of injecting light hydrocarbons through the reservoir to form a miscible flood. Three different methods are used. One method uses about 5% PV slug of liquidified petroleum gas (LPG) such as propane, followed by natural gas or gas and water. A second method, called enriched (condensing) gas drive, consists of injecting a 10%–20% PV slug of natural gas that is enriched with ethane through hexane (C_2 to C_6), followed by lean gas (dry, mostly methane) and possibly water. The enriching components are transferred from the gas to the oil. The third method, called high pressure (vaporizing) gas drive, consists of injecting lean gas at high pressure to vaporize $C_2 - C_6$ components from the crude oil being displaced.

Mechanisms
Hydrocarbon miscible flooding recovers crude oil by:

- Generating miscibility (in the condensing and vaporizing gas drive)
- Increasing the oil volume (swelling)
- Decreasing the viscosity of the oil

Technical Screening Guides
Crude oil

Gravity	$> 35°$ API
Viscosity	< 10cp
Composition	High percentage of light hydrocarbons (C_2–C_7)

Reservoir

Oil saturation	$> 30\%$ PV
Type of formation	Sandstone or carbonate with a minimum of fractures and high permeability streaks
Net thickness	Relatively thin unless formation is steeply dipping
Average permeability	Not critical if uniform
Depth	$> 2,000$ ft (LPG) to > 5000 ft (high pressure gas)
Temperature	Not critical

Limitations
- The minimum depth is set by the pressure needed to maintain the generated miscibility. The required pressure ranges from about 1,200 psi for the LPG process to 3,000–5,000 psi for the high pressure gas drive, depending on the oil.
- A steeply dipping formation is very desirable to permit some gravity stabilization of the displacement which normally has an unfavorable mobility ratio.

Problems
- Viscous fingering results in poor vertical and horizontal sweep efficiency.
- Large quantities of expensive products are required.
- Solvent may be trapped and not recovered.

5.11.5.2 Nitrogen and Flue Gas Flooding [12]
Description
Nitrogen and flue gas flooding are oil recovery methods which use these inexpensive nonhydrocarbon gases to displace oil in systems which may be either miscible or immiscible depending on the pressure and oil composition. Because of their low cost, large volumes of these gases may be injected. Nitrogen or flue gas are also considered for use as chase gases in hydrocarbon-miscible and CO_2 floods.

Mechanisms
Nitrogen and flue gas flooding recover oil by:

- Vaporizing the lighter components of the crude oil and generating miscibility if the pressure is high enough.
- Providing a gas drive where a significant portion of the reservoir volume is filled with low-cost gases.

Technical Screening Guides
Crude oil

Gravity	$> 24°$ API ($> 35°$ for nitrogen)
Viscosity	< 10 cp
Composition	High percentage of light hydrocarbons (C_1–C_7)

Reservoir

Oil saturation	$> 30\%$ PV
Type of formation	Sandstone or carbonate with few fractures and high permeability streaks
Net thickness	Relatively thin unless formation is dipping
Average permeability	Not critical
Depth	> 4.500 ft
Temperature	Not critical

Limitations
- Developed miscibility can only be achieved with light oils and at high pressures; therefore, deep reservoirs are needed. ·
- A steeply dipping formation is desired to permit gravity stabilization of the displacement which has a very unfavorable mobility ratio.

Problems
- Viscous fingering results in poor vertical and horizontal sweep efficiency.
- Corrosion can cause problems in the flue gas method.
- The nonhydrocarbon gases must be separated from the saleable produced gas.

5.11.5.3 Carbon Dioxide Flooding [12]
Description
Carbon dioxide flooding is carried out by injecting large quantities of CO_2 (15% or more of the hydrocarbon PV) into the reservoir. Although CO_2 is not truly miscible with the crude oil, the CO_2 extracts the light-to-intermediate components from the oil, and, if the pressure is high enough, develops miscibility to displace the crude oil from the reservoir.

Mechanisms
CO_2 recovers crude oil by:

- Generation of miscibility
- Swelling the crude oil
- Lowering the viscosity of the oil
- Lowering the interfacial tension between the oil and the CO_2-oil phase in the near-miscible regions.

Technical Screening Guides
Crude oil

Gravity	$> 26°$ API (preferably $> 30°$)
Viscosity	< 15 cp (preferably < 10 cp)
Composition	High percentage of intermediate hydrocarbons (C_5–C_{20}), especially C_5–C_{12}

Reservoir

Oil saturation	$> 30\%$ PV
Type of formation	Sandstone or carbonate with a minimum of fractures and high permeability streaks
Net thickness	Relatively thin unless formation is steeply dipping.
Average permeability	Not critical if sufficient injection rates can be maintained.
Depth	Deep enough to allow high enough pressure ($>$ about 2,000 ft), pressure required for optimum production (sometime called minimum miscibility pressure) ranges from about 1,200 psi for a high gravity ($> 30°$ API) crude at low temperatures to over 4,500 psi for heavy crudes at higher temperatures.
Temperature	Not critical but pressure required increases with temperature.

Limitations
- Very low viscosity of CO_2 results in poor mobility control.
- Availability of CO_2.

Problems
- Early breakthrough of CO_2 causes several problems: corrosion in the producing wells; the necessity of separating CO_2 from saleable hydrocarbons; repressuring of CO_2 for recycling; and a high requirement of CO_2 per incremental barrel produced.

5.11.5.4 Surfactant/Polymer Flooding
Description
Surfactant/polymer flooding, also called micellar/polymer or microemulsion flooding, consists of injecting a slug that contains water, surfactant, electrolyte (salt), usually a cosolvent (alcohol), and possibly a hydrocarbon (oil). The size of the slug is often 5%–15% PV for a high surfactant concentration system and 15%–50% PV for low concentrations. The surfactant slug is followed by polymer-thickened water. Concentrations of the polymer often range from 500–2,000 mg/L; the volume of polymer solution injected may be 50% PV, more or less, depending on the process design.

Mechanisms
Surfactant/polymer flooding recovers crude oil by:

- Lowering the interfacial tension between oil and water
- Solubilization of oil
- Emulsification of oil and water
- Mobility enhancement

Technical Screening Guides
Crude oil

Gravity	$> 25°$ API
Viscosity	< 30 cp
Composition	Light intermediates are desirable

Reservoir

Oil saturation	$> 30\%$ PV
Type of formation	Sandstone preferred
Net thickness	> 10 ft
Average permeability	> 20 md
Depth	$<$ about 8,000 ft (see temperature)
Temperature	$< 175°$F

Limitations

- An areal sweep of more than 50% on waterflood is desired.
- Relatively homogeneous formation is preferred.
- High amounts of anhydrite, gypsum, or clays are undesirable.
- Available systems provide optimum behavior over a very narrow set of conditions.
- With commercially available surfactants, formation water chlorides should be $< 20,000$ ppm and divalent ion $(Ca^{++}$ and $Mg^{++}) < 500$ ppm.

Problems

- Complex and expensive system.
- Possibility of chromatographic separation of chemicals.
- High adsorption of surfactant.
- Interactions between surfactant and polymer.
- Degradation of chemicals at high temperature.

5.11.5.5 Polymer Flooding [12]

Description

The objective of polymer flooding is to provide better displacement and volumetric sweep efficiencies during a waterflood. Polymer augmented waterflooding consists of adding water soluble polymers to the water before it is injected into the reservoir. Low concentrations (often 250–2,000 mg/L) of certain synthetic or biopolymers are used; properly sized treatments may require 15%–25% reservoir PV.

Mechanisms

Polymers improve recovery by:

- Increasing the viscosity of water
- Decreasing the mobility of water
- Contacting a large volume of the reservoir

Technical Screening Guides
Crude oil

Gravity	$> 25°$ API
Viscosity	< 150 cp (preferably < 100)
Composition	Not critical

Reservoir

Oil saturation	$> 10\%$ PV mobile oil
Type of formation	Sandstones preferred but can be used in carbonates
Net thickness	Not critical
Average permeability	> 10 md (as low as 3 md in some cases)
Depth	$<$ about 9,000 ft (see temperature)
Temperature	$< 200°$F to minimize degradation

Limitations

- If oil viscosities are high, a higher polymer concentration is needed to achieve the desired mobility control.
- Results are normally better if the polymer flood is started before the water-oil ratio becomes excessively high.
- Clays increase polymer adsorption.
- Some heterogeneities are acceptable, but for conventional polymer flooding, reservoirs with extensive fractures should be avoided. If fractures are present, the crosslinked or gelled polymer techniques may be applicable.

Problems

- Lower injectivity than with water can adversely affect oil production rate in the early stages of the polymer flood.
- Acrylamide-type polymers lose viscosity due to shear degradation or increases in salinity and divalent ions.

- Xanthan gum polymers cost more, are subject to microbial degradation, and have a greater potential for wellbore plugging.

5.11.5.6 Alkaline Flooding [12]

Description

Alkaline or caustic flooding involves the injection of chemicals such as sodium hydroxide, sodium silicate or sodium carbonate. These chemicals react with organic petroleum acids in certain crudes to create surfactants in situ. They also react with reservoir rocks to change wettability. The concentration of the alkaline agent is normally 0.2 to 5%; slug size is often 10% to 50% PV, although one successful flood only used 2% PV, (but this project also included polymers for mobility control). Polymers may be added to the alkaline mixture, and polymer-thickened water can be used following the caustic slug.

Mechanisms

Alkaline flooding recovers crude oil by:

- A reduction of interfacial tension resulting from the produced surfactants
- Changing wettability from oil-wet to water-wet
- Changing wettability from water-wet to oil-wet
- Emulsification and entrainment of oil
- Emulsification and entrapment of oil to aid in mobility control
- Solubilization of rigid oil films at oil-water interfaces (Not all mechanisms are operative in each reservoir.)

Technical Screening Guides
Crude oil

Gravity	$> 13°$ to $35°$ API
Viscosity	< 200 cp
Composition	Some organic acids required

Reservoir

Oil saturation	Above waterflood residual
Type of formation	Sandstones preferred
Net thickness	Not critical
Average permeability	> 20 md
Depth	$<$ about 9,000 ft (see temperature)
Temperature	$< 200°$F preferred

Limitations

- Best results are obtained if the alkaline material reacts with the crude oil: the oil should have an acid number of more than 0.2 mg KOH/g of oil.
- The interfacial tension between the alkaline solution and the crude oil should be less than 0.01 dyne/cm.
- At high temperatures and in some chemical environments, excessive amounts of alkaline chemicals may be consumed by reaction with clays, minerals, or silica in the sandstone reservoir.
- Carbonates are usually avoided because they often contain anhydrite or gypsum, which interact adversely with the caustic chemical.

Problems

- Scaling and plugging in the producing wells.
- High caustic consumption

5.11.5.7 In-Situ Combustion [12]

Description

In-situ combustion or fireflooding involves starting a fire in the reservoir and injecting air to sustain the burning of some of the crude oil. The most common technique is forward combustion in which the reservoir is ignited in an injection well, and air is injected to propagate the combustion front away from the well. One of the variations of this technique is a combination of forward combustion and waterflooding (COFCAW). A second technique is reverse combustion in which a fire is started in a well that will eventually become a producing well, and air injection is then switched to adjacent wells; however, no successful field trials have been completed for reverse combustion.

Mechanisms
In-situ combustion recovers crude oil by:

- The application of heat which is transferred downstream by conduction and convection, thus lowering the viscosity of the crude.
- The products of steam distillation and thermal cracking which are carried forward to mix with and upgrade the crude.
- Burning coke that is produced from the heavy ends of the crude oil
- The pressure supplied to the reservoir by the injected air.

Technical Screening Guides
Crude oil

Gravity	$< 40°$ API (normally 10–25°)
Viscosity	$< 1,000$ cp
Composition	Some asphaltic components to aid coke deposition

Reservoir

Oil saturation	> 500 bbl/acre-ft (or > 40–50% PV)
Type of formation	Sand or sandstone with high porosity
Net thickness	> 10 ft
Average permeability	> 100 md
Transmissibility	> 20 mud ft/cp
Depth	> 500 ft
Temperature	$> 150°$F preferred

Limitations
- If sufficient coke is not deposited from the oil being burned, the combustion process will not be sustained.
- If excessive coke is deposited, the rate of advance of the combustion zone will be slow, and the quantity of air required to sustain combustion will be high.
- Oil saturation and porosity must be high to minimize heat loss to rock.
- Process tends to sweep through upper part of reservoir so that sweep efficiency is poor in thick formations.

Problems
- Adverse mobility ratio.
- Complex process, requiring large capital investment, is difficult to control.
- Produced flue gases can present environmental problems.
- Operational problems such as severe corrosion caused by low pH hot water, serious oil-water emulsions, increased sand production, deposition of carbon or wax, and pipe failures in the producing wells as a result of the very high temperatures.

5.11.5.8 Steamflooding [12]
Description
The steam drive process or steamflooding involves the continuous injection of about 80% quality steam to displace crude oil towards producing wells. Normal practice is to precede and accompany the steam drive by a cyclic steam stimulation of the producing wells (called huff and puff).

Mechanisms
Steam recovers crude oil by:

- Heating the crude oil and reducing its viscosity
- Supplying pressure to drive oil to the producing well

Technical Screening Guides
Crude oil

Gravity	$< 25°$ API (normal range is 10°–25° API)
Viscosity	> 20 cp (normal range is 100–5,000 cp)
Composition	Not critical but some light ends for stream distillation will help

Reservoir

Oil saturation	> 500 bbl/acre-ft (or > 40–50% PV)
Type of formation	Sand or sandstone with high porosity and permeability preferred
Net thickness	> 20 feet
Average permeability	> 200 md (see transmissibility)
Transmissibility	> 100 md ft/cp
Depth	300–5,000 ft
Temperature	Not critical

Limitations
- Oil saturations must be quite high and the pay zone should be more than 20 feet thick to minimize heat losses to adjacent formations.
- Lighter, less viscous crude oils can be steamflooded but normally will not be if the reservoir will respond to an ordinary waterflood.
- Steamflooding is primarily applicable to viscous oils in massive, high permeability sandstones or unconsolidated sands.
- Because of excessive heat losses in the wellbore, steamflooded reservoirs should be as shallow as possible as long as pressure for sufficient injection rates can be maintained.
- Steamflooding is not normally used in carbonate reservoirs.
- Since about one-third of the additional oil recovered is consumed to generate the required steam, the cost per incremental barrel of oil is high.
- A low percentage of water-sensitive clays is desired for good injectivity.

Problems
- Adverse mobility ratio and channeling of steam.

5.11.5.9 Criteria for Gas Injection
For LPG slug or solvent flooding, enriched (condensing) gas drive, and high pressure (vaporizing) gas drive, a range of pressures (and therefore, depths) are needed to achieve miscibility in the systems. Thus, there is a minimum depth requirement for each of the processes as shown earlier (see section on "Hydrocarbon Miscible Flooding"). The permeability is not critical if the structure is relatively uniform; permeabilities of the reservoirs for the current field projects range from less than 1 md to several darcies [29]. On the other hand, the crude oil characteristics are very important. A high-gravity, low-viscosity oil with a high percentage of the C_2–C_7 intermediates is essential if miscibility is to be achieved in the vaporizing gas drives.

As shown earlier under "Nitrogen and Flue Gas Flooding," the screening criteria for flooding with nitrogen or flue gas are similar to those for the high pressure gas drive. Pressure and depth requirements, as well as the need for a very light oil, are even greater if full miscibility is to be realized in the reservoir. The nitrogen and flue gas method is placed between hydrocarbon miscible and CO_2 flooding because the process can also recover oil in the immiscible mode. It can be economic because much of the reservoir space is filled with low cost gas.

Because of the minimum pressure requirement, depth is an important screening criteria, and CO_2 floods are normally carried out in reservoirs that are more than 2,000 ft deep. The oil composition is also important (see section on "Carbon Dioxide Flooding"), and the API gravity exceeds 30° for most of the active CO_2 floods [29]. A notable exception is the Lick Creek, Arkansas, CO_2/waterflood project which was conducted successfully, not as a miscible project, but as an immiscible displacement [30].

5.11.5.10 Criteria for Chemical Methods
For surfactant/polymer methods, oil viscosities of less than 30 cp are desired so that adequate mobility control can be achieved. Good mobility control is essential for this method to make maximum utilization of the expensive chemicals. Oil saturations remaining after a waterflood should be more than 30% PV to ensure that sufficient oil is available for recovery. Sandstones are preferred because carbonate reservoirs are heterogeneous, contain brines with high divalent ion contents, and cause high adsorption of commonly used surfactants. To ensure adequate injectivity, permeability should be greater than 20 md. Reservoir temperature should be less than 175°F to minimize degradation of the presently available surfactants. A number of other limitations and problems were mentioned earlier, including the general requirement for low salinity and hardness for most of the commercially available systems. Obviously, this method is very complex, expen-sive, and subject to a wide range of problems. Most importantly, the available systems provide optimum reduction in interfacial tension over a very narrow salinity range. Preflushes have been used to attempt to provide optimum conditions, but they have often been ineffective.

The screening guidelines and a description of polymer flooding are contained earlier in Section "Polymer Flooding." Since the objective of polymer flooding is to improve the mobility ratio without necessarily making the ratio favorable, the maximum oil viscosity for this method is 100 or possibly 150 cp. If oil viscosities are very high, higher polymer concentrations are needed to achieve the desired mobility control, and thermal methods may be more attractive. As discussed earlier, polymer flooding will not ordinarily mobilize oil that has been completely trapped by water; therefore, a mobile oil saturation of more than 10% is desired. In fact, a polymer flood is normally more effective when started at low producing water-oil ratios [31]. Although sandstone reservoirs are usually preferred, several large polymer floods have been conducted in carbonate reservoirs. Lower-molecular-weight polymers can be used in reservoirs with permeabilities as low as 10 md (and, in some carbonates, as low as 3 md). While it is possible to manufacture even lower-molecular-weight polymers to inject into lower permeability formations, the amount of viscosity generated per pound of polymer would not be enough to make such products of interest. With current polymers, reservoir temperature should be less than 200°F to minimize degradation; this requirement limits depths to about 9,000 ft. A potentially serious problem with polymer flooding is the decrease in the injectivity which must accompany any increase in injection fluid viscosity. If the decreased injectivity is prolonged, oil production rates and project costs can be adversely affected.

Injection rates for polymer solutions may be only 40%–60% of those for water alone, and the reduced injectivity may add several million dollars to the total project costs. Other problems common to the commercial polymers are cited earlier.

Moderately low gravity oils (13°–35° API) are normally the target for alkaline flooding (see section on "Alkaline Flooding"). These oils are heavy enough to contain the organic acids, but light enough to permit some degree of mobility control. The upper viscosity limit (< 200 cp) is slightly higher than for polymer flooding. Some mobile oil saturation is desired, the higher the better. The minimum average permeability is about the same as for surfactant/polymer (> 20 md). Sandstone reservoirs are preferred since carbonate formations often contain anhydrite or gypsum which react and consume the alkaline chemicals. The alkaline materials also are consumed by clays, minerals, or silica; this consumption is high at elevated temperatures so the maximum desired temperature is 200°F. Caustic consumption in field projects has been higher than indicated by laboratory tests. Another potential problem in field applications is scale formation which can result in plugging in the producing wells.

5.11.5.11 Criteria for Thermal Methods
For screening purposes, steamflooding and fireflooding are often considered together. In general, combustion should be the choice when heat losses from steamflooding would be too great. In other words, combustion can be carried out in deeper reservoirs and thinner, tighter sand sections where heat losses for steamflooding are excessive. Screening guides for in-situ combustion are given earlier in Section "In-Situ Combustion." The ability to inject at high pressures is usually important so 500 ft has been retained as the minimum depth, but a few projects have been done at depths of less than 500 ft. Since the fuel and air consumption decrease with higher gravity oils, there is a tendency to try combustion in lighter oils if the fire can be maintained, but no projects have been done in reservoirs with oil gravities greater than 32° API [29].

In summary, if all screening criteria are favorable, fireflooding appears to be an attractive method for reservoirs that cannot be produced by methods used for the lighter oils. However, the process is very complicated and beset with many practical problems such as corrosion, erosion and poorer mobility ratios than steamflooding. Therefore, when the economics are comparable, steam injection is preferred to a combustion drive [4].

Screening criteria for steamflooding are listed earlier in section "Steamflooding". Although steamflooding is commonly used with oils ranging in gravity from 10°–25° API, some gravities have been lower, and there is recent interest in steamflooding light oil reservoirs. Oils with viscosities of less than 20 cp are usually not candidates for steamflooding because waterflooding is less expensive; the normal range is 100–5,000 cp. A high saturation of oil-in-place is required because of the intensive use of energy in the generation of steam. In order to minimize the amount of rock heated and maximize the amount of oil heated, formations with high porosity are desired; this means that sandstones or unconsolidated sands are the primary target, although a steam drive pilot has been conducted in a highly fractured carbonate reservoir in France. The product of oil saturation times porosity should be greater than about 0.08 [26]. The fraction of heat lost to the cap and base rocks varies inversely with reservoir thickness. Therefore, the greater the thickness of the reservoir, the greater the thermal efficiency. Steamflooding is possible in thin formations if the permeability is high. High permeabilities (> 200 md or preferably

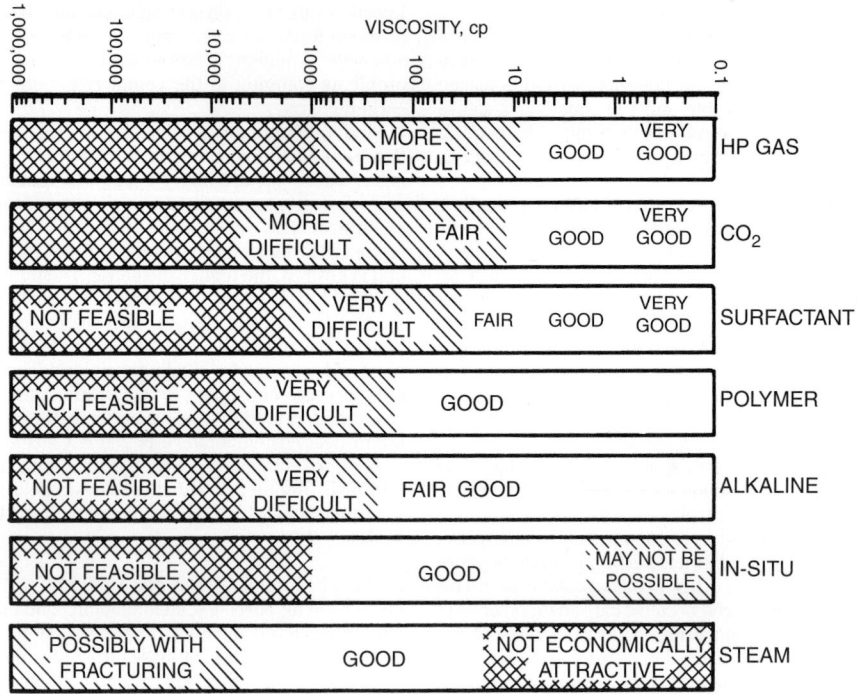

Figure 5.11.3 *Viscosity ranges for EOR processes [12].*

> 500 md) are needed to permit adequate steam injectivity; transmissibility should be greater than 100 md ft/cp at reservoir conditions. Depths shallower than about 300 ft may not permit good injectivity because the pressures required may exceed fracture gradients. Heat losses become important at depths greater than about 2,500 ft. and steamflooding is not often considered at depths greater than 5,000 ft. Downhole steam generators may have potential in deeper formations if operational problems can be overcome.

5.11.5.12 Graphical Representation of Screening Guides
From the summary of screening guides in Table 5.11.1, the viscosity, depth and permeability criteria are presented graphically in Figures 5.11.3 to 5.11.5. The figures have some features which permit the quick application of screening criteria but they cannot replace the table for detailed evaluations. In a sense, the figures present a truer picture than the table because there are few absolutes among the numbers presented as screening guides in the tables. Different authors and organizations may use different parameters for the same process, and most of the guidelines are subject to change as new laboratory and field information evolves. In field applications, there are exceptions to some of the accepted criteria, and the graphs accommodate these nicely. The "greater than" and "less than" designations of the tables can also be displayed better graphically.

The range of values are indicated on the graphs by the open areas, and by cross-hatching along with general words such as "more difficult," "not feasible," etc. The "good" or "fair" ranges are those usually encompassed by the screening parameters in the table. However, the notation of "good" or "very good" does not mean that the indicated process is sure to work; it means simply that it is in the preferred range for that oil or reservoir characteristic.

The influence of viscosity on the technical feasibility of different enhanced recovery methods is illustrated in Figure 5.11.3. Note the steady progression, with increasing viscosity, from those processes that work well with very light oils (hydrocarbon miscible or nitrogen) to oils that are so viscous that no recovery is possible unless mining and extraction are employed.

For completeness, we have included the two "last resort" methods (special steamflooding techniques with shafts, fractures, drainholes, etc., and mining plus extraction) are listed in Figure 5.11.5. These methods are not included in Figures 5.11.4 and 5.11.5 because these unconventional techniques are not considered in most reservoir studies.

Figure 5.11.4 shows that those enhanced recovery processes that work well with light oils have rather specific depth requirements. As discussed, each gas injection method has a minimum miscibility pressure for any given oil, and the reservoir must be deep enough to accommodate the required pressure.

Figure 5.11.5 shows that the three methods that rely on gas injection are the only ones that are even technically feasible at extremely low permeabiliteis. The three methods that use backup waterflooding need a permeability of greater than 10 md in order to inject the chemicals or emulsions and to produce the released oil from the rock. Although most authors show a minimum permeability requirement of 20 md for polymers, we indicate a possible range down as low as 3 md for low molecular weight polymers, especially in some carbonate reservoirs.

The screening guides in the figures can perhaps be summarized by stating a fact well-known to petroleum engineers: oil recovery is easiest with light oil in very permeable reservoirs and at shallow or intermediate depths. Unfortunately, nature has not been kind in the distribution of hydrocarbons,

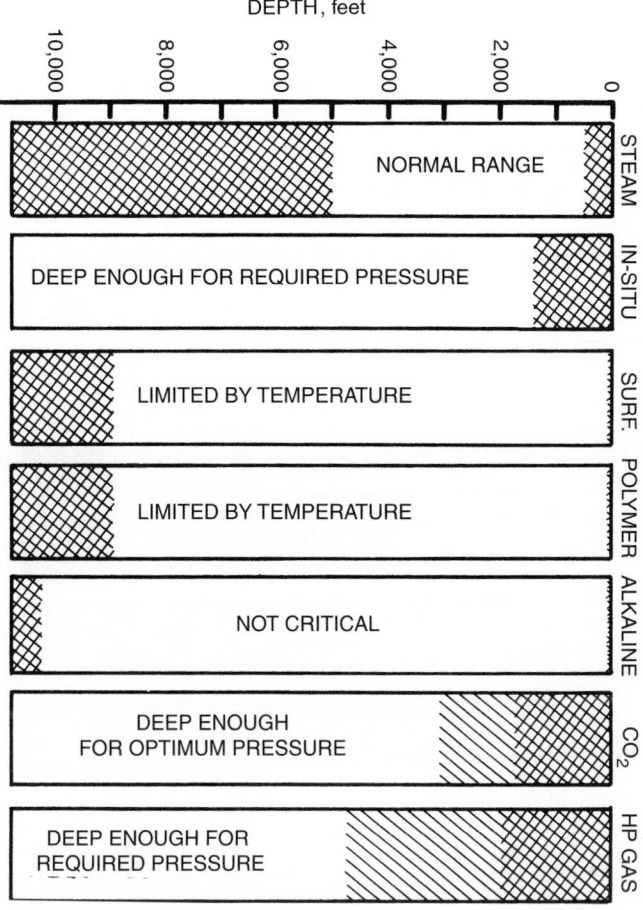

Figure 5.11.4 *Depth requirements for EOR processes [12].*

and it is necessary to select the recovery method that best matches the oil and reservoir characteristics.

5.11.6 Laboratory Design for Enhanced Recovery
5.11.6.1 Preliminary Tests
Water Analysis

A complete water analysis is important to determine the effects of dissolved ions on the EOR processes (especially the chemical methods) or to ascertain any potential water problems such as scale or corrosion that may result when EOR processes are implemented. Water viscosity and density are also measured.

Oil Analysis

Oil viscosity and density are measured as well. A carbon number distribution of the crude may be obtained, especially if CO_2 flooding is being considered.

Core Testing

Routine core analyses, such as porosity, permeability, relative permeabilities, capillary pressure, and waterflood susceptibility tests are normally done by service companies that specialize in these types of tests. Specialized core tests, such as thin sections or scanning electron microscopy, are available to evaluate the relationship between pore structure and

the process being considered. If required, stimulation or injectivity improvement measures can be recommended.

5.11.6.2 Polymer Testing

The desirability of adding polymers is determined by evaluating all available data to assess the performance of normal waterflooding. Any problems such as adverse viscosity ratios or large permeability variations should be identified. If the results of this study indicate that mobility control of the waterflood is warranted, the following laboratory tests are undertaken.

Viscosity Testing

Based on the permeability of the reservoir, relative permeability data, and the desired level of mobility control, polymers of certain molecular weights are selected for testing. Various concentrations of the polymers are dissolved in both the available injection water and in blends of the injection and formation waters. Polymer solutions may be non-Newtonian at certain shear rates, that is, the viscosity decreases at high shear rates (shear-thinning or pseudo-plastic). This shear-thinning behavior is reversible and, if observed in the reservoir, is beneficial in that good injectivity can result from the lower viscosity observed at high shear

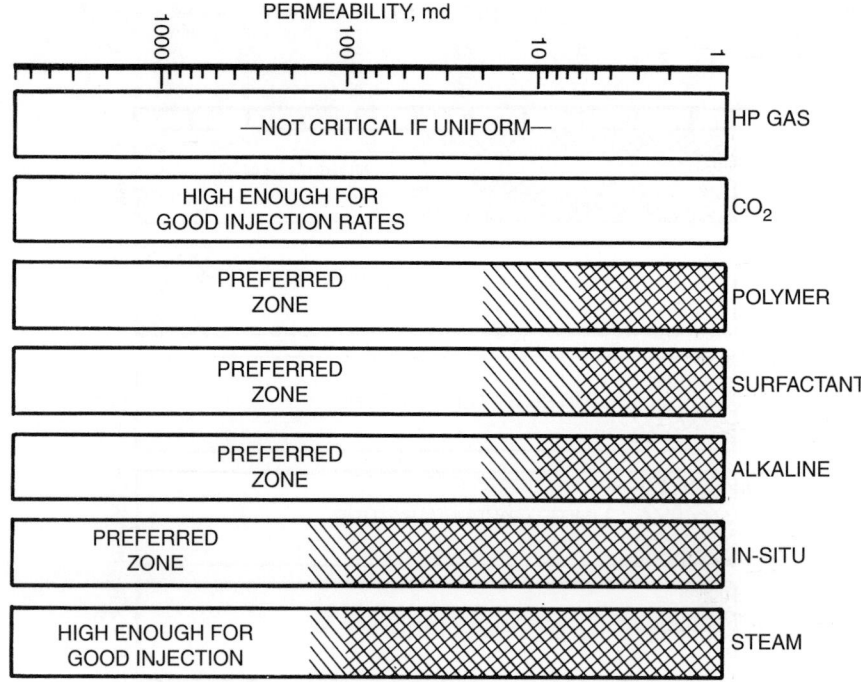

Figure 5.11.5 *Permeability ranges for EOR methods [12].*

rates near the injection well. At the lower shear rates encountered some distance from the injector, the polymer solution develops a higher viscosity. In this testing, it is important to consider not only the viscosity of the injected solution, but, more importantly, the in-situ viscosity that is achieved in the reservoir. Several things can happen that will reduce viscosity when the polymer solution is injected into the reservoir. Reduction in viscosity as a result of irreversible shear degradation is possible at the injection wellbore if the shear rates or shear stresses are large. Once in the reservoir, dilution with formation water or ion exchange with reservoir minerals can cause a reduction in viscosity, and the injected polymer concentration will need to be sufficiently high to compensate for all viscosity-reducing effects.

Polymer Retention

Retention of polymer in a reservoir can result from adsorption, entrapment, or, with improper application, physical plugging. Polymer retention tests are usually performed after a standard waterflood (at residual oil saturation) or during a polymer flood oil recovery test. If polymer retention tests are conducted with only water initially present in the core, a higher level of retention will result from the increased surface area available to the polymer solution in the absence of oil. Effluent samples from the core are collected during both the polymer injection and a subsequent water flush. These samples are analyzed for polymer content. From a material balance, the amount of polymer retained in the core is calculated. Results are usually expressed in lbs per acre-ft. Excessive retention will increase the amount of polymer that must be added to achieve the desired mobility control. The level of polymer retained in a reservoir depends on a number of variables: permeability of the rock, surface area, nature of the reservoir rock (sandstone, carbonate, minerals,

or clays), nature of the solvent for the polymer (salinity and hardness), molecular weight of the polymer, ionic charge on the polymer, and the volume of porosity that is not accessible to the flow of polymer solution. Polymer retention levels often range from less than 100 lb/acre-ft to several hundred lb/acre-ft.

5.11.6.3 Surfactant and Alkali Testing

Laboratory tests consist of measuring the interfacial tension (IFT) between the crude oil and the injected solution (alkaline or surfactant additive). This is usually done with a spinning drop interfacial tensiometer. With surfactants, the requirement for measuring tensions can be minimized by performing vial tests to determine solubilization parameters that can be correlated with IFT. Other tests include determining relative permeabilities, wettability, and total fluid mobilites. Once the optimum conditions are found, results of oil recovery tests with the chemical flood additives are conducted, usually at waterflood residual oil saturation.

5.11.6.4 CO₂ Flooding

For the gas injection projects, the trend in this country is toward the use of carbon dioxide although the full impact of CO_2 flooding will be felt in several years since construction of CO_2 pipelines into the west Texas area was completed in the 1980s. Carbon dioxide flooding is not a truly miscible process; that is, it does not dissolve in all proportions with crude oil. However, CO_2 can extract light to intermediate components out of the crude oil. This CO_2-rich mixture can develop miscibility and effectively displace additional crude oil. The main limitation involved is the very low viscosity of CO_2 that results in fingering of CO_2 through the more viscous crude oil. This causes premature breakthrough of the CO_2 and reduces the amount of oil recovered per unit volume of CO_2.

A prediction of the minimum pressure required to achieve miscibility can be made if the reservoir depth and basic properties of the crude oil are known. Laboratory tests often consist of some means of determining the minimum miscibility pressure, often by observing the oil displacement efficiency by CO_2 in a small-diameter tube (slim tube) packed with sand or glass beads. Carbon number distribution of the crude will be of value in determining if sufficient amounts of the C_5 to C_{12} components are present.

5.11.6.5 Thermal Recovery

Viscosities of very viscous crude oils can be reduced by the use of thermal recovery methods. Fireflooding or in-situ combustion involves starting a fire in the reservoir and injecting air to sustain the burning of some of the crude oil. Heat that is generated lowers the viscosity of the crude oil and results in improved recovery. With the steam drive or steamflooding process, steam is generated on the surface and injected into the injection wells. Some companies are now exploring the use of downhole steam generators in deeper wells where heat loss can be a serious problem. A primary problem with steam flooding is the channeling of steam through thin sections of the reservoir. To combat this problem, several organizations are studying the use of surfactants to create a foam in situ for improving sweep efficiency.

For steamflooding, the most important laboratory tests are, of course, viscosity of the crude oil and permeability of the reservoir core material. To be economically viable, steamfloods must be conducted in thick, very permeable, shallow reservoirs that contain very viscous crude.

References

1. van Poollen, H. K., and Associates, Inc., *Fundamentals of Enhanced Oil Recovery*, PenWell Publishing Co., Tulsa (1980).
2. Stalkup, F. I., Jr., *Miscible Displacement*, SPE Monograph Series, SPE, Dallas, Vol. 8 (1983).
3. Klins, M. A., *Carbon Dioxide Flooding*. International Human Resources Development Corp., Boston (1984).
4. Prats, M., *Thermal Recovery*, Monograph Series, SPE, Dallas, Vol. 7 (1982).
5. Latil, M., et al., *Enhanced Oil Recovery*, Gulf Publishing Co., Houston (1980).
6. Haynes, H. J., et al., *Enhanced Oil Recovery*, National Petroleum Council; Industry Advisory Council to the U.S. Department of the Interior (1976).
7. Bailey, R. E., et al., *Enhanced Oil Recovery*, Natl. Petroleum Council, Washington (June 1984).
8. Broome, J. H., Bohannon, J. M., and Stewart, W. C., "The 1984 Natl. Petroleum Council Study on EOR: An Overview," *J. Pet. Tech.* (Aug. 1986), pp. 869–874.
9. Doe, P. H., Carey, B. S., and Helmuth, E. S., "The Natl. Petroleum Council EOR Study: Chemical Processes," paper SPE 13240 presented at the 1984 Annual Tech. Conf. & Exhib., Sept. 16–19.
10. King, J. E., Blevins, T. R., and Britton, M. W., "The National Petroleum Council EOR Study: Thermal Processes," paper SPE 13242 presented at the 1984 SPE Annual Technical Conf. and Exhib., Houston, Sept. 16–19.
11. Robl, F. W., Emanuel, A. S., and Van Meter, O. E., Jr., "The 1984 Natl. Petroleum Council Estimate of Potential of EOR for Miscible Processes," *J. Pet. Tech.* (Aug. 1986), pp. 875–882.
12. Taber, J. J., and Martin, F. D., "Technical Screening Guides for the Enhanced Recovery of Oil," paper SPE 12069 presented at the SPE 1983 Annual Technical Conf. & Exhib., San Francisco, October 5–8.
13. Chang, H. L., "Polymer Flooding Technology—Yesterday, Today and Tomorrow," *J. Pet. Tech.* (Aug. 1978), pp. 1113–1128.
14. Martin, F. D., "Design and Implementation of a Polymer Flood," Southwestern Petroleum Short Course, *Proc.*, 33rd Annual Southwestern Petroleum Short Course, Lubbock, April 23–24, 1986.
15. Taber, J. J., "The Injection of Detergent Slugs in Water Floods," *Trans.* AIME, Vol. 213 (1958), pp. 186–192.
16. Gogarty, W. B., "Status of Surfactant or Micellar Methods," *J. Pet. Tech.* (Jan. 1976), pp. 93–102.
17. Gogarty, W. B., "Micellar/Polymer Flooding—An Overview," *J. Pet. Tech.* (Aug. 1978), pp. 1089–1101.
18. Gogarty, W. B., "Enhanced Oil Recovery Through the Use of Chemicals—Part 1," *J. Pet. Tech* (Sept. 1983), pp. 1581–1590.
19. Gogarty, W. B., "Enhanced Oil Recovery Through the Use of Chemicals—Part 2," *J. Pet. Tech.* (Oct. 1983), pp. 1767–1775.
20. Taber, J. J., "Enhanced Recovery Methods for Heavy and Light Oils," *Proc.*, International Conference on Heavy Versus Light Oils: Technical Issues and Economic Considerations, Colorado Springs, March 24–26, 1982.
21. Holm, L. W., and Josendal, V. A., "Effect of Oil Composition on Miscible Type Displacement by Carbon Dioxide," *Soc. Pet. Eng. J.* (Feb. 1982), pp. 87–98.
22. Orr, F. M., Jr., and Jensen, C. M., "Interpretation of Pressure-Composition Phase Diagrams for CO_2-Crude Oil Systems," paper SPE 11125 presented at SPE 57th Annual Fall Technical Conf. and Exhibition, New Orleans, Sept. 26–29, 1982.
23. Helier, J. P., and Taber, J. J., "Influence of Reservoir Depth on Enhanced Oil Recovery by CO_2 Flooding," paper SPE 15001 presented at the 1986 SPE Permian Basin Oil & Gas Recovery Conference, Midland, March 13–14.
24. Chu, C., "Current In-Situ Combustion Technology," *J. Pet. Tech.* (Aug. 1983), pp. 1412–1418.
25. Chu, C., "State-of-the-Art Review of Fireflood Field Projects," *J. Pet. Tech.* (Jan. 1982), pp. 19–36.
26. Chu, C., "State-of-the-Art Review of Steamflood Field Projects," *J. Pet. Tech.* (Oct. 1985), pp. 1887–1902.
27. White, P. D., "In-Situ Combustion Appraisal and Status," *J. Pet. Tech.* (Nov. 1985), pp. 1943–1949.
28. Garon, A. M., Kumar, M., and Cala, G. C., "The State of the Art Oxygen Fireflooding," *In Situ*, Vol. 10, No. 1 (1986), pp. 1–26.
29. Leonard, J., "Increased Rate of EOR Brightens Outlook," *Oil & Gas J.* (April 14, 1986), pp. 71–101.
30. Reid, R. B., and Robinson, J. J., "Lick Creek Meakin Sand Unit Immiscible CO_2/Waterflood Project," *J. Pet. Tech.* (Sept. 1981), pp. 1723–1729.
31. Agnew, H. J., "Here's How 56 Polymer Oil-Recovery Projects Shape Up," *Oil & Gas J.* (May 1972), pp. 109–112.

6

Production Engineering

Contributing Authors

Robert P. Badrak
Susan Beck
Ernie Brown
Francisco Ciulla
Tracy Darr van Reet
Ernie Dunn
Michael Economides, Ph.D.
Joel Ferguson
Kazimierz Glowacki, Ph.D.
Bill Grubb
Matthew Hill
Mike Juenke
Reza G. Kashmiri
Joseph V. LaBlanc
Doug LaBombard
Julius P. Langlinais, Ph.D.
Charles Nathan, Ph.D., P.E.
Pudji Permadi, Ph.D.
Jim Pipes
Gary Plisga
Floyd Preston, Ph.D.
Toby Pugh
Bharath N. Rao, Ph.D.
Chris S. Russell, P.E.
Oleg Salzberg
Ron Schmidt
Ardeshir K. Shahraki, Ph.D.

Contents

6.1 PROPERTIES OF HYDROCARBON MIXTURES

This section contains correlation and procedures for the prediction of physical properties of natural gas and oil. Physical constants of single components are given in Table 6.1.1.

6.1.1 Compressibility Factor and Phase Behavior

The compressibility factor Z is a dimensionless factor, independent of the quantity of gas and determined by the character of the gas, the temperature, and the pressure.

$$Z = \frac{PV}{nRT} = \frac{MPV}{mRT} \qquad [6.1.1]$$

	Field Units	SI Units
P = absolute pressure	psia	kPa
V = volume	ft³	m³
n = moles	m/M	m/M
m = mass	lb	kg
M = molecular mass	lb/lb mole	kg/kmole
T = absolute temperature	°R	K
R = universal gas constant	10.73[psia·ft³/ °R·lb mole mole]	8.3145[kPa m³/kmol·K]
ρ = density	slug/ft³	kg/m³

A knowledge of the compressibility factor means that the density ρ is also known from the relationship

$$\rho = \frac{PM}{ZRT} \qquad [6.1.2]$$

6.1.1.1 Compressibility Factor Using the Principle of Corresponding States (CSP)

The following terms are used, P, T, and V, such that

$$P_r = \frac{P}{P_c}, \qquad T_r = \frac{T}{T_c}, \qquad V_r = \frac{V}{V_c} \qquad [6.1.3]$$

where P_r, T_r, and V_r = reduced parameters of pressure, temperature, and volume
P_c, T_c, and V_c = critical parameters of P, T, and V from Table 6.1.1

Compressibility factors of many components are available as a function of pressure in most handbooks [1,2]. In application of CSP to a mixture of gases, pseudocritical temperature (T_{pc}) and pressure (P_{pc}) are defined for use in place of the true T_c and P_c to determine the compressibility factor for a mixture.

$$P_{pc} = \sum y_i P_{ci}, \qquad T_{pc} = \sum y_i T_{ci} \qquad [6.1.4]$$

$$P_{pr} = P/P_{pc}, \qquad T_{pr} = T/T_{pc} \qquad [6.1.5]$$

where subscript i = component in the gas mixture
y_i = mole fraction of component "i" in the gas mixture

For given values of P_{pr} and T_{pr}, compressibility factor Z can be determined from Figure 6.1.1 [1]. If a gas mixture contains significant concentrations of carbon dioxide and hydrogen sulfide, then corrected pseudocritical constants (T'_{pc}, P'_{pc}) are defined as follows:

$$T'_p c = T_{pc} - e \qquad [6.1.6]$$

$$P'_{pc} = \frac{P_{pc} T'_{pc}}{T_{pc} + y_{H_2S}(1 - y_{H_2S})e} \qquad [6.1.7]$$

where

$$e = 120\left[\left(y_{CO_2} + y_{H_2S}\right)^{0.9} - \left(y_{CO_2} + y_{H_2S}\right)^{1.6}\right]$$
$$+ 15\left(y_{H_2S}^{0.5} - y_{H_2S}^{4.0}\right) \qquad [6.1.8]$$

where y_{CO_2}, y_{H_2S} = mole fraction of CO_2 and H_2S in mixture.

6.1.1.2 Direct Calculation of Z Factors

The Hall-Yarborough equation is one of the best:

$$Z = \frac{0.06125P_{pr}t\exp[-1.2(1-t)^2]}{y} \qquad [6.1.9]$$

where P_{pr} = the pseudoreduced pressure
t = the reciprocal, pseudoreduced temperature T_{pc}/T
y = the reduced density which can be obtained as the solution of the equation

$$-0.06125_{pr}t\exp\left[-1.2(1-t)^2\right] + \frac{y+y^2+y^3-y^4}{(1-y)^3}$$
$$-(14.76t - 9.76t^2 + 4.58t^3)y^2$$
$$+(90.7t - 242.2t^2 + 42.4t^3)y^{(2.18+2.82t)} = 0 \qquad [6.1.10]$$

This nonlinear equation can be conveniently solved for y using the simple Newton-Raphson iterative technique.

Good results give cubic equation of state and its modifications. Most significant of the modifications are those of Soave and Peng and Robinson.

The Soave equation (SRK) is [4]:

$$P = \frac{RT}{V-b} - \frac{a}{V(V+b)} \qquad [6.1.11]$$

$$Z^3 = Z^2 + (A - B - B^2)Z - AB = 0 \qquad [6.1.12]$$

where b = $\Sigma x_i b_i$ for a mixture
b_i = $0.08664RT_{ci}/P_{ci}$ for a single component
a = $\Sigma_i\Sigma_j x_i x_j(a_i a_j)^{0.5}(1 - k_{ij})$ for a mixture
a_i = $a_{ci}\alpha_i$ for a single component
a_{ci} = $0.42748(RT_{ci})^2/P_{ci}$
$\alpha_i^{0.5}$ = $1 + m_i(1 - T_{ri}^{0.5})$
m_i = $0.48 + 1.574\omega_i - 0.176\omega_i^2$
A = $aP/(RT)^2$, B = bP/RT

The Peng-Robinson equation (PR) is

$$P = \frac{RT}{V-b} - \frac{a}{V(V+b)+b(V-b)} \qquad [6.1.13]$$

$$Z^3 - (1-B)Z^2 + (A - 2B - 3B^2)Z - (AB - B^2 - B^3) = 0 \qquad [6.1.14]$$

where b = $\Sigma x_i b_i$
b_i = $0.077796RT_{ci}/P_c$
a = $\Sigma_i\Sigma_j x_i x_j(a_i a_j)^{0.5}(1 - k_{ij})$
a_i = $a_{ci}\alpha_i$
a_{ci} = $0.457237(RT_{ci})^2/P_{ci}$
$\alpha_i^{0.5}$ = $1 + m_i(1 - T_{ri}^{0.5})$
m_i = $0.37646 + 1.54226\omega_i - 0.26992\omega_i^2$
A, B as in SRK equation

where P = pressure (absolute units)
T = temperature (°R or K)
R = universal gas constant
Z = compressibility factor
ω = acentric Pitzer factor (see Table 6.1.1)
T_{ci}, P_{ci} = critical parameters (see Table 6.1.1)
k_{ij} = interaction coefficient (= 0 for gas phase mixture)

Both SRK and PR equations are used to predict equilibrium constant K value. See derivation of vapor-liquid equilibrium by equation of state at end of this subsection.

Full description of gas, oil, and water properties are given in Chapter 5, "Reservoir Engineering." Reservoir hydrocarbon fluids are a mixture of hydrocarbons with compositions related to source, history, and present reservoir conditions. Consider the pressure-specific volume relationship for a single-component fluid at constant temperature, below its critical temperature initially hold in the liquid phase at an elevated pressure. This situation is illustrated in Figure 6.1.2[1]. Bubble point and dew point curves in Figure 6.1.2a correspond to the vapor pressure line in Figure 6.1.2b. A locus of bubble points and a locus of dew points that meet at a point C (the critical point) indicate that the properties of liquid and vapor become indistinguishable [6].

Multicomponent systems have different phase behavior than pure component. In the P–T diagram instead of the vapor-pressure line, we have an area limited by saturation line (bubble point+dew point), see Figure 6.1.3[1]. The diagram's shape is more or less the same for two- or three-component systems as for multicomponent systems.

For isothermal production in the reservoir, position A indicates reservoir fluid found as an underposition saturated oil, B indicates reservoir fluid found as a gas condensate, and position C indicates reservoir fluid found as a dry gas. Expansion in the liquid phase to the bubble point at constant temperature is similar to a pure component system. Expansion through the two-phase region does not occur at constant pressure, but is accompanied by a decrease in pressure as the composition of liquid and vapor changes.

6.1.1.3 Classification of Hydrocarbon Fluids
Hydrocarbon fluids usually are classified as to the phase behavior exhibited by the mixture. Figure 6.1.4 shows the pressure–temperature phase diagram of the four general classifications of fluids: dry gas, gas condensate, volatile oil, and black oil. As it can be seen, the source temperature also plays a role in the determination of fluid type. According to MacDonald [7], each type of fluid has composition as given in Table 6.1.2. Sometimes hydrocarbon mixtures are classified as follows: dry gas, wet gas, gas condensate, and black oil (Figure 6.1.4).

In the case of dry gas, a light hydrocarbon mixture existing entirely in gas phase at reservoir conditions and a decline in reservoir pressure will not result in the formation of any reservoir liquid phase; it is a rather theoretical case. Usually gas reservoirs fall into the next group, wet gas.

Gas condensate or retrograde gas system is the case when the critical temperature of system is such that reservoir temperature is between critical and cricondentherm as shown in Figure 6.1.4c. If the pressure is reduced to the cricondenbar pressure, the liquid phase is increasing, but the liquid phase may reevaporate later on. This phenomena — the condensation of liquid upon decrease in pressure — is termed isothermal retrograde condensation. The liquid phase recovered from a condensate system is recovered from a phase that is vapor at reservoir conditions. This is also partly true of volatile oil systems where the vapor phase in equilibrium with the reservoir liquid phase is particularly rich in liquefiable constituents (C_3 to C_{8+}), and a substantial proportion of stock tank liquid may derive from a reservoir vapor phase. We normally do not expect to see retrograde behavior at reservoir pressures below about 2500 psi (17.2 MPa).

Volatile oil systems are those within the two-phase region under reservoir conditions, the vapor stage corresponding to condensate compositions and conditions. Volatile oil is not an apt description because virtually all reservoir fluids are volatile. What is really meant is that the reservoir fluid exhibits the properties of an oil existing in the reservoir at a high temperature near its critical temperature. These properties include a high shrinkage immediately below the bubble point. In extreme cases, this shrinkage can be as much as 45% of the hydrocarbon pore space within 10 psi (0.7 bar) below the bubble point. Near-critical oils have formation factor B_o of 2 or higher; the compositions are usually characterized by 12.5 to 20 mole % heptanes or more.

Ordinary oils are characterized by GOR from 0 to approximately 200 ft^3/bbl (360 m^3/m^3) and with B_0 less than 2. Oils with high viscosity (about 10 cp), high oil density, and negligible gas/oil ratio are called heavy oils. At surface conditions may form tar sands. Oils with smaller than 10 cp viscosity are known as a black oil or a dissolved gas oil system; no anomalies are in phase behavior. There is no sharp dividing line between each group of reservoir hydrocarbon fluid; however, liquid volume percent versus pressure diagram Figure 6.1.5a–e is very useful to understand the subject.

The significant point to be made is that when an oil system exists in intimate contact with an associated gas cap, the bubble point pressure of the oil will be equal to the dew point pressure of the gas cap, and both those values will be equal to the static reservoir pressure at the gas–oil contact, Figure 6.1.6.

6.1.1.4 Reservoir Conditions Phase Behavior
There is one phase flow in reservoir conditions if well flowing pressure P_{wf} is higher than bubble point pressure P_b or dew point pressure P_d and two-phase flow occurs by a wellbore. Reservoir depletion and production consist of two separate processes: flash liberation (vaporization) and differential liberation. A schematic representation of flash vaporization is shown in Figure 6.1.7. At stage 1 reservoir fluid is under reservoir pressure and temperature at known volumes V_t. The pressure in the cell is covered by increasing the space available in the cell for the fluid V_{t2}. This procedure is repeated until a large change in the pressure–volume slope is indicated. The above procedure indicates that flash vaporization is a phase-changing process due to change in pressure and temperature if the mass of reservoir fluid or total composition system remains constant and can be expressed as follows:

$$zF = xL + yV \qquad [6.1.15]$$

where

z = mole fraction of component in a reservoir fluid mixture
F = number of moles of sample at initial reservoir pressure and temperature
x = mole fraction of component in liquid (e.g., P_5)
L = number of moles of equilibrium liquid
y = mole fraction component in gas mixture
V = number of moles of equilibrium gas phase

Equilibrium or flash liberation calculations may be made for reservoir fluid that divides into two phases at any temperature and pressure.

Schematic representation of differential liberation in laboratory conditions is shown in Figure 6.1.8. It begins in the same manner as flash vaporization. The sample is placed in a pressure higher than bubble point pressure. The pressure is lowered until such time that free gas is liberated in the cell. Then for predetermined pressure or volume increments, mercury is withdrawn from the cell, gas is released from solution, and the cell is agitated until the liberated gas is in the equilibrium with the oil. All free gas is ejected from the cell at a constant pressure by injection of mercury. The volume of free gas is displaced, and the oil remaining in

Table 6.1.1 *Physical Constants of Hydrocarbons*

No.	Compound	Formula	Molecular Mass	Boiling point, °C 101.3250 kPa (abs)	Vapor pressure, kPa (abs) 40°C	Freezing point, °C 101.3250 kPa (abs)	Critical constants Pressure, kPa (abs)	Critical constants Temperature, K	Critical constants Volume, m³/kg
	See Note No. →2		1.	2.		3.			
1	Methane	CH₄	16.043	−161.52(28)	(35,000.0)	−182.47ᵈ	4604.0	190.55	0.00617
2	Ethane	C₂H₆	30.070	−88.58	(6000.0)	−182.80ᵈ	4880.0	305.43	0.00492
3	Propane	C₃H₈	44.097	−42.07	1341.0	−187.68ᵈ	4249.0	369.82	0.00460
4	n-Butane	C₄H₁₀	58.124	−0.49	377.0	−138.36	3797.0	425.16	0.00439
5	Isobutane	C₄H₁₀	58.124	−11.81	528.0	−159.60	3648.0	408.13	0.00452
6	n-Pentane	C₅H₁₂	72.151	36.06	115.66	−129.73	3369.0	469.6	0.00421
7	Isopentane	C₅H₁₂	72.151	27.84	151.3	−159.90	3381.0	460.39	0.00424
8	Neopentane	C₅H₁₂	72.151	9.50	269.0	−16.55	3199.0	433.75	0.00420
9	n-Hexane	C₆H₁₄	86.178	68.74	37.28	−95.32	3012.0	507.4	0.00429
10	2-Methylpentane	C₆H₁₄	86.178	60.26	50.68	−153.66	3010.0	497.45	0.00426
11	3-Methylpentane	C₆H₁₄	86.178	63.27	45.73	—	3124.0	504.4	0.00426
12	Neohexane	C₆H₁₄	86.178	49.73	73.41	−99.870	3081.0	488.73	0.00417
13	2,3-Dimethylbutane	C₆H₁₄	86.178	57.98	55.34	−128.54	3127.0	499.93	0.00415
14	n-Heptane	C₇H₁₆	100.205	98.42	12.34	−90.582	2736.0	540.2	0.00431
15	2-Methylhexane	C₇H₁₆	100.205	90.05	17.22	−118.27	2734.0	530.31	0.00420
16	3-Methylhexane	C₇H₁₆	100.205	91.85	16.16	—	2814.0	535.19	0.00403
17	3-Ethylpentane	C₇H₁₆	100.205	93.48	15.27	−118.60	2891.0	540.57	0.00415
18	2,2-Dimethylpentane	C₇H₁₆	100.205	79.19	26.32	−123.81	2773.0	520.44	0.00415
19	2,4-Dimethylpentane	C₇H₁₆	100.205	80.49	24.84	−119.24	2737.0	519.73	0.00417
20	3,3-Dimethylpentane	C₇H₁₆	100.205	86.06	20.93	−134.46	2945.0	536.34	0.00413
21	Triptane	C₇H₁₆	100.205	80.88	25.40	−24.91	2954.0	531.11	0.00397
22	n-Octane	C₈H₁₈	114.232	125.67	4.143	−56.76	2486.0	568.76	0.00431
23	Diisobutyl	C₈H₁₈	114.232	109.11	8.417	−91.200	2486.0	549.99	0.00422
24	Isoctane	C₈H₁₈	114.232	99.24	12.96	−107.38	2568.0	543.89	0.00410
25	n-Nonane	C₉H₂₀	128.259	150.82	1.40	−53.49	2288.0	594.56	0.00427
26	n-Decane	C₁₀H₂₂	142.286	174.16	0.4732	−29.64	2099.0	617.4	0.00424
27	Cyclopentane	C₅H₁₀	70.135	49.25	73.97	−93.866	4502.0	511.6	0.00371
28	Mettylcyclopentane	C₆H₁₂	84.162	71.81	33.85	−142.46	3785.0	532.73	0.00379
29	Cyclohexane	C₆H₁₂	84.162	80.73	24.63	6.554	4074.0	553.5	0.00368
30	Methylcyclohexane	C₇H₁₄	96.189	100.93	12.213	−126.59	3472.0	572.12	0.00375
31	Ethene (Ethylene)	C₂H₄	28.054	−103.77(29)	—	−169.15ᵈ	5041.0	282.35	0.00467
32	Propene (Propylene)	C₃H₆	42.081	−47.72	1596.0	−185.25ᵈ	4600.0	364.85	0.00430
33	1-Butene (Butylene)	C₄H₈	56.108	−6.23	451.9	−185.35ᵈ	4023.0	419.53	0.00428
34	cis-2-Butene	C₄H₈	56.108	3.72	337.6	−138.91	4220.0	435.58	0.00417
35	trans-2-Butene	C₄H₈	56.108	0.88	365.8	−105.55	4047.0	428.63	0.00424
36	Isobutene	C₄H₈	56.108	−6.91	452.3	−140.35	3999.0	417.90	0.00426
37	1-Pentene	C₅H₁₀	70.135	29.96	141.65	−165.22	3529.0	464.78	0.00422
38	1,2-Butadiene	C₄H₆	54.092	10.85	269.0	−136.19	(4502.0)	(444.0)	(0.00405)
39	1,3-Butadiene	C₄H₆	54.092	−4.41	434.0	−108.91	4330.0	425.0	0.00409
40	Isoprene	C₅H₈	68.119	34.07	123.77	−145.95	(3850.0)	(484.0)	(0.00406)
41	Acetylene	C₂H₂	26.038	−84.88ᵉ	—	−80.8ᵈ	6139.0	306.33	0.00434
42	Benzene	C₆H₆	78.114	80.09	24.38	5.533	4898.0	562.16	0.00328
43	Toluene	C₇H₈	92.141	110.63	7.895	−94.991	4106.0	591.80	0.00343
44	Ethylbenzene	C₈H₁₀	106.168	136.20	2.87	−94.975	3609.0	617.20	0.00353
45	o-Xylene	C₈H₁₀	106.168	144.43	2.05	−25.18	3734.0	630.33	0.00348
46	m-Xylene	C₈H₁₀	106.168	139.12	2.53	−47.87	3536.0	617.05	0.00354
47	p-Xylene	C₈H₁₀	106.168	138.36	2.65	13.26	3511.0	616.23	0.00356
48	Styrene	C₈H₈	104.152	145.14	1.85	−30.61	3999.0	647.6	0.00338
49	Isopropylbenzene	C₉H₁₂	120.195	152.41	1.47	−96.035	3209.0	631.1	0.00357
50	Methyl alcohol	CH₄O	32.042	64.54	35.43	−97.68	8096.0	512.64	0.00368
51	Ethyl alcohol	C₂H₆O	46.089	78.29	17.70	−114.1	6383.0	513.92	0.00362
52	Carbon monoxide	CO	28.010	−191.49	—	−205.0ᵈ	3499.0(33)	132.92(33)	0.00332(33)
53	Carbon dioxide	CO₂	44.010	−78.51ᵉ	—	−56.57ᵈ	7382.0(33)	304.19(33)	0.00214(33)
54	Hydrogen sulfide	H₂S	34.076	−60.31	2881.0	−85.53ᵈ	9005.0	373.5	0.00287
55	Sulfur dioxide	SO₂	64.069	−10.02	630.8	−75.48ᵈ	7894.0	430.8	0.00190
56	Ammonia	NH₃	17.031	−33.33(30)	1513.0	−77.74ᵈ	11,280.0	405.6	0.00425
57	Air	N₂+O₂	28.964	−194.2(2)	—	—	3771.0(2)	132.4(2)	0.00323(3)
58	Hydrogen	H₂	2.016	−252.87ᵛ	—	−259.2ᵈ	1297.0	33.2	0.03224
59	Oxygen	O₂	31.999	−182.962ᵛ	—	−218.8ᵈ	5061.0	154.7(33)	0.00229
60	Nitrogen	N₂	28.013	−195.80(31)	—	−210.0ᵈ	3399.0	126.1	0.00322
61	Chlorine	Cl₂	70.906	−34.03	1134.0	−101.0ᵈ	7711.0	417.0	0.00175
62	Water	H₂O	18.015	100.00ᵛ	7.377	0.00	22,118.0	647.3	0.00318
63	Helium	He	4.003	−268.93(32)	—	—	227.5(32)	5.2(32)	0.01436(32)
64	Hydrogen chloride	HCl	36.461	−85.00	6304.0	−114.18ᵈ	8309.0	324.7	0.00222

NOTES

a: Air saturated liquid.

b: Absolute values from weights in vacuum.

c: The apparent values from weight in air are shown for users' convenience and compliance with ASTM-IP Petroleum Measurement Tables. In the United States and Great Britain, all commercial weights are required by law to be weights in air. All other mass data are on an absolute mass (weight in vacuum) basis.

d: At saturation pressure (triple point).

e: Sublimation point.

f: The + sign and number following signify the ASTM octane number corresponding to that of 2,2,4-trimethylpentane with the indicated number of cm³ of TEL added per gal.

g: Determined at 100°C.

h: Saturation pressure and 15°C.

i: Apparent value at 15°C.

j: Average value from octane numbers of more than one sample.

k: Relative density (specific gravity), 48.3°C/15°C (sublimation point; solid C₂H₄/liquid H₂O).

m: Densities of liquid at the boiling point.

n: Heat of sublimation.

p: See Note 10.

s: Extrapolated to room temperature from higher temperature.

t: Gross calorific values shown for ideal gas volumes are not direct conversions of each other using only the gas volume per liquid volume value shown herein. The values differ by the heat of vaporization to ideal gas at 288.15 K.

v: Fixed points on the 1968 International Practical Temperature Scale (IPTS-68).

w: Value for normal hydrogen (25% para, 75% ortho). The value for equilibrium mixture of para and ortho is −0.218; however, in most correlations, 0 is used.

x: Densities at the boiling point in kg/m³ for: Ethane, 546.4; propane, 581.0; propene, 608.8; hydrogen sulfide, 960.0; sulfur dioxide, 1462.0; ammonia, 681.6; hydrogen chloride, 1192.

*: Calculated values.

() Estimated values.

†: Values are estimated using 2nd virial coefficients.

	4.			5.	6.	7.	8.			9.		
	Density of liquid 101.3250 kPa (abs), 15°C						Ideal gas* 101.3250 kPa (abs), 15°C			Specific heat capacity 101.3250 kPa (abs), 15°C c_p kJ/(kg·°C)		
Relative density /15°C^a,b	kg/m³·a (mass in vacuum)	kg/m³·a,c (Apparent mass in air)	m³/kmol	Temperature coefficient of density, at 15°C·, a −1/°C	Pitzer acentric factor, a	Compressibility factor of real gas, Z 101.3250 kPa (abs), 15°C	Relative density Air = 1	Specific volume m³/kg	Volume ratio gas/(liquid in vacuum)	Ideal gas	Liquid	No.
	(300.0)^j	(300.0)^j	(0.05)^i	—	0.0126	0.9981	0.5539	1.474	(442.0)^l	2.204	–	1
81^h.x	357.8^h.x	356.6^h	0.084 04^h	0.002 74^h	0.0978	0.9915	1.0382	0.7863	281.3^h	1.706	3.807	2
83^h	507.8^h.x	506.7^h	0.086 84^h	0.002 11^h	0.1541	0.9810	1.5225	0.5362	272.3^h	1.625	2.476	3
47^h	584.2^h	583.1^h	0.099 49^h	0.002 11^h	0.2015	0.9641	2.0068	0.4068	237.6^h	1.652	2.366(41)	4
37^h	563.2^h	562.1^h	0.1032^h	0.002 14^h	0.1840	0.9665	2.0068	0.4068	229.1^h	1.616	2.366(41)	5
46	631.0	629.9	0.1143	0.001 57	0.2524	0.942^†	2.4911	0.3277	206.8	1.622	2.292(41)	6
50	624.0	623.3	0.1156	0.001 62	0.2286	0.948^†	2.4911	0.3277	204.6	1.600	2.239	7
72^h	596.7^h	595.6^h	0.1209^h	0.001 87^h	0.1967	0.9538	2.4911	0.3277	195.5^h	1.624	2.317	8
44	663.8	662.7	0.1298	0.001 35	0.2998	0.910^†	2.9753	0.2744	182.1	1.613	2.231	9
83	657.7	656.6	0.1310	0.001 40	0.2784	—	2.9753	0.2744	180.5	1.602	2.205	10
94	668.8	667.7	0.1289	0.001 35	0.2741	—	2.9753	0.2744	183.5	1.578	2.170	11
45	653.9	652.8	0.1318	0.001 40	0.2333	—	2.9753	0.2744	179.4	1.593	2.148	12
68	666.2	665.1	0.1294	0.001 35	0.2475	—	2.9753	0.2744	182.8	1.566	2.146	13
86	688.0	686.9	0.1456	0.001 24	0.3494	0.852^†	3.4596	0.2360	162.4	1.606	2.209	14
35	682.8	681.7	0.1468	0.001 22	0.3303	—	3.4596	0.2360	161.1	1.595	2.183	15
21	691.5	690.4	0.1449	0.001 24	0.3239	—	3.4596	0.2360	163.2	1.584	2.137	16
32	702.6	701.5	0.1426	0.001 26	0.3107	—	3.4596	0.2360	165.8	1.613	2.150	17
87	678.0	676.9	0.1478	0.001 30	0.2876	—	3.4596	0.2360	160.0	1.613	2.161	18
77	677.1	676.0	0.1480	0.001 30	0.3031	—	3.4596	0.2360	159.8	1.651	2.193	19
80	697.4	696.3	0.1437	0.001 17	0.2681	—	3.4596	0.2360	164.6	1.603	2.099	20
50	694.4	693.3	0.1443	0.001 24	0.2509	—	3.4596	0.2360	163.9	1.578	2.088	21
73	706.7	705.6	0.1616	0.001 12	0.3961	0.783^†	3.9439	0.2070	146.3	1.601	2.191	22
84	697.7	696.6	0.1637	0.001 17	0.3564	—	3.9439	0.2070	144.4	1.573	2.138	23
66	696.0	694.9	0.1641	0.001 17	0.3041	—	3.9439	0.2070	144.1	1.599	2.049	24
24	721.7	720.6	0.1777	0.001 13	0.4452	—	4.4282	0.1843	133.0	1.598	2.184	25
46	733.9	732.8	0.1939	0.000 99	0.4904	—	4.9125	0.1662	122.0	1.595	2.179	26
08	750.2	749.1	0.09349	0.001 26	0.1945	0.949^†	2.4215	0.3371	252.9	1.133	1.763	27
41	753.4	752.3	0.1117	0.001 28	0.2308	—	2.9057	0.2809	211.7	1.258	1.843	28
38	783.1	782.0	0.1075	0.001 21	0.2098	—	2.9057	0.2809	220.0	1.211	1.811	29
*44	773.7	772.6	0.1269	0.001 13	0.2364	—	3.3900	0.2406	186.3	1.324	1.839	30
	—	—	—	—	0.869	0.9938	0.9686	0.8428	—	1.514	—	31
31^h	522.6^h.x	521.5^h	0.060 69^h	0.003 40^h	0.1443	0.9844	1.4529	0.5619	293.6^h	1.480	2.443	32
19^h	601.4^h	600.3^h	0.093 30^h	0.002 09^h	0.1949	0.9703	1.9372	0.4214	253.4^h	1.483	2.237	33
77^h	627.1^h	626.0^h	0.089 47^h	0.001 76^h	0.2033	0.9660	1.9372	0.4214	264.3^h	1.366	2.241(42)	34
05^h	610.0^h	608.9^h	0.091 98^h	0.001 93^h	0.2126	0.9661	1.9372	0.4214	257.1^h	1.528	2.238	35
10^h	600.5^h	599.4^h	0.093 44^h	0.002 16^h	0.2026	0.9688	1.9372	0.4214	253.1^h	1.547	2.296	36
62	645.6	644.5	0.1086	0.001 60	0.2334	0.948^†	2.4215	0.3371	217.7	1.519	2.241(43)	37
78^h	657^h	656.0^h	0.082 33^h	0.001 76^h	(0.2540)	(0.969)	1.8676	0.4371	287.2^h	1.446	2.262	38
80^h	627.4^h	626.3^h	0.086 22^h	0.002 03^h	0.1971	(0.965)	1.8676	0.4371	274.2^h	1.426	2.124	39
66	686.0	684.9	0.099 30	0.001 55	(0.1567)	0.949^†	2.3519	0.3471	238.1	1.492	2.171	40
5^k	—	—	—	—	0.1893	0.9925	0.8990	0.9081	—	1.659	—	41
50	884.2	883.1	0.088 34	0.001 19	0.2095	0.929^†	2.6969	0.3027	267.6	1.014	1.715	42
23	871.6	870.5	0.1057	0.001 06	0.2633	0.903^†	3.1812	0.2566	223.7	1.085	1.677	43
21	871.3	870.6	0.1219	0.000 97	0.3031	—	3.6655	0.2227	194.0	1.168	1.721	44
50	884.2	883.1	0.1201	0.000 99	0.3113	—	3.6655	0.2227	196.9	1.218	1.741	45
91	868.3	867.2	0.1223	0.000 97	0.3257	—	3.6655	0.2227	193.4	1.163	1.696	46
61	865.3	864.2	0.1227	0.000 97	0.3214	—	3.6655	0.2227	192.7	1.157	1.708	47
15	910.6	909.5	0.1144	0.001 03	0.1997	—	3.5959	0.2270	206.7	1.133	1.724	48
67	866.0	864.9	0.1390	0.000 97	0.3260	—	4.1498	0.1967	170.4	1.219	1.732	49
67	796.0	794.9	0.040 25	0.001 17	0.5648	—	1.1063	0.7379	587.4	1.352	2.484	50
22	791.5	790.4	0.058 20	0.001 07	0.6608	—	1.5906	0.5132	406.2	1.389	2.348	51
93^m	786.6^m (34)	—	0.035 52^m	—	0.0442	0.9995	0.9671	0.8441	—	1.040	—	52
26^h	821.9^h (35)	820.8^h	0.053 55^h	—	0.2667	0.9943	1.5195	0.5373	441.6^h	0.8330	—	53
97^h	789.0^h.x (36)	787.9^h	0.043 19^h	—	0.0920	0.9903	1.1765	0.6939	547.5^h	0.9960	2.08(36)	54
7^h	1396.0^h.x (36)	1395.0^h	0.045 89^h	—	0.2548	0.9801^†	2.2117	0.3691	515.3^h	0.6062	1.359(36)	55
83^h	617.7^h.x (30)	616.6^h	0.027 57^h	—	0.2576	0.9899(30)	0.5880	1.388	857.4	2.079	4.693(30)	56
66^m (36)	855.0^m	—	0.0339^m	—	—	0.9996	1.0000	0.8163	—	1.005	—	57
106^m	71.00^m (37)	—	0.028 39^m	—	0.219^m	1.0006	0.0696	11.73	—	14.24	—	58
20^m (25)	1141.0^m (38)	—	0.028 04^m	—	0.0200	0.9993(39)	1.1048	0.7389	—	0.9166	—	59
93^m (26)	806.0^m (31)	—	0.034 64^m	—	0.0372	0.9997	0.9672	0.8441	—	1.040	—	60
26	1424.5	1423.5	0.049 78	—	0.0737	(0.9875)^† (36)	2.4481	0.3336	475.0	0.4760	—	61
00	999.1	998.0	0.018 03	0.000 14	0.3434	—	0.6220	1.312	1311.0	1.862	4.191	62
251^m	125.0^m (32)	—	0.032 02^m	—	0	1.0005(40)	0.1382	5.907	—	5.192	—	63
538	853.0^x	851.9	0.042 74	0.006 03	0.1232	—	1.2588	0.6485	553.2	0.7991	—	64

[mo]lecular mass (M) is based on the following atomic weights: C = 12.011; [H = 1].008; O = 15.9995; N = 14.0067; S = 32.06; Cl = 35.453.

[Boi]ling point—the temperature at equilibrium between the liquid and vapor [phase]s at 101.3250 kPa (abs).

[Fre]ezing point—the temperature at equilibrium between the crystalline phase [and th]e air saturated liquid at 101.3250 kPa (abs).

[The v]alues for the density and molar volume of liquids refer to the air saturated [liquid] at 101.3250 kPa (abs), except when the boiling

[COM]MENTS

—all dimensional values are reported in SI units, which are derived from the following basic units:
- mass—kilogram, kg
- length—meter, m
- temperature—International Practical Temperature Scale of 1968 (IPTS-68), where 0°C = 273.15 K

derived units are:
- volume—cubic meter, m³
- pressure—Pascal, Pa (1 Pa = N/m²)

Physical constants for molar volume = 22.41383 ± 0.00031
gas constant, $R = 8.31441$ J/(K · mol)
8.31441×10^{-3} m³ · kPa/(K · mol)
1.98719 cal/(K · mol)
1.98596 Btu(IT)/°R· (lb-mol)

Conversion factors
1 m³ = 35.31467 ft³ = 264.1720 gal
1 kg = 2.204623 lb
1 kg/m³ = 0.06242795 lb/ft³ = 0.001 g/cm⁻³
1 kPa = 0.01 bar = 9.869233×10^{-3} atm
= 0.1450377 lb/in²
1 atm = 101.3250 kPa = 14.69595 lb/in²
= 760 Torr
1 kJ = 0.2390057 kcal(thermochemical)
= 0.2388459 kcal(IT)
= 0.9478171 Btu(IT)
see Rossini, F. D. "Fundamental Measures and Constants for Science and Technology"; CRC Press: Cleveland, Ohio, 1974.

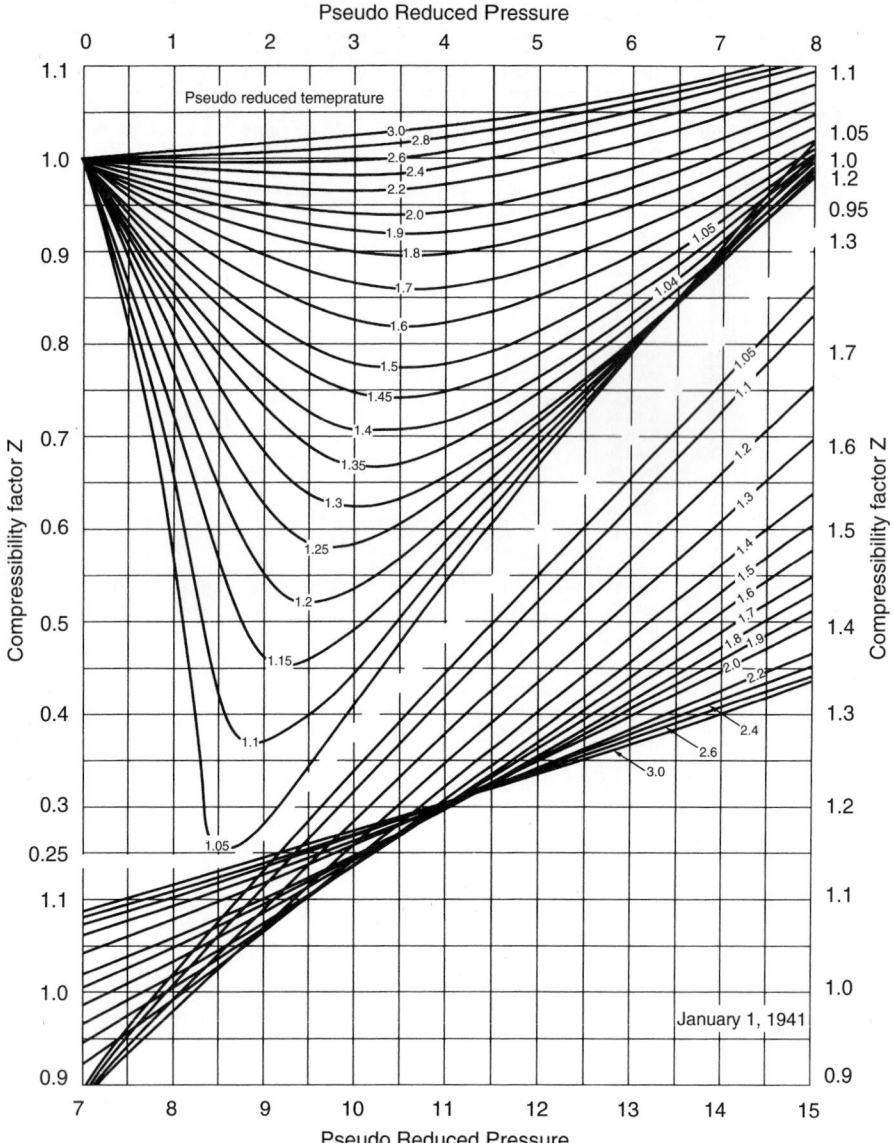

Figure 6.1.1 *Compressibility factor for natural gases [1].*

the cells are thus measured at cell conditions. This procedure is repeated for all the pressure increments until only oil remains in the cell at reservoir temperature and atmospheric pressure.

In contrast to flash vaporization, differential vaporization is undertaken with the decreasing mass participating in the process. Equation 6.1.15 cannot be applied because F is not constant anymore at the given set pressure and temperature.

The question arises: Which process occurs in reservoir conditions? Some specialists, e.g., Moses [8], assume that the reservoir process is a combination of differential and flash. Such statements are incorrect. They produce misunderstanding and confusion. The following is meant to clarify and straighten out this problem:

1. In reservoir conditions, differential process always occurs.

2. In production tubing and surface pipeline flow and in the separators, the flash process takes place (subject to some limiting assumption).

Flash process refers to the conditions in which the mass of the considered system does not vary with changes in pressure and temperature. Flash process in two-phase regions (vaporization or condensation) can be defined in terms of total system composition. The total system composition (z_i) can be measured at any point outside of saturation line, e.g., points A, B, and C (Figure 6.1.3). As a substitution the following treatment can be used; the total system composition in two-phase region flash process remains constant. The flash process may ensue for a composition z_i that separates into two phases for the values of pressures and temperatures inside the saturation curve area. After the temperature and pressure are chosen, all the gas is in equilibrium with all the

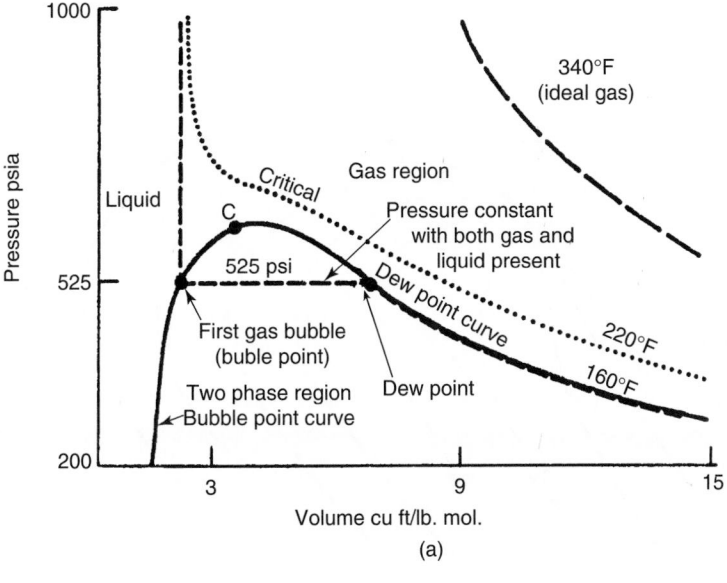

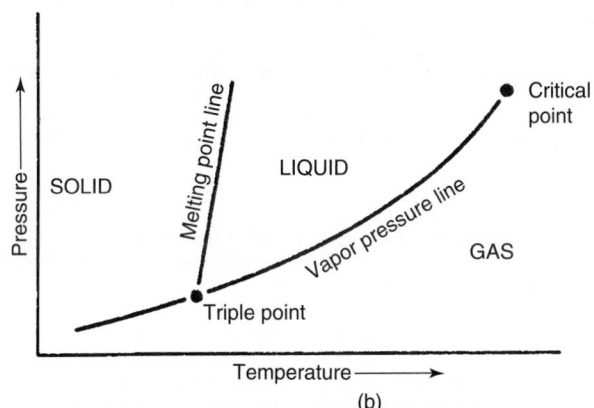

Figure 6.1.2 *Phase diagram for a single component.*

oil. In other words, a change of pressure or temperature, or both, in a flash process can change the equilibrium conditions according to the Gibbs phase rule. This rule provides the number of independent variables that, in turn, define intensive properties. Flash vaporization may be a batch or continuous process. Treating two-phase flow in tubing as a steady state, neglecting the gas storage effect, and gas slippage result in flash process. In a horizontal flow, and in separators, a similar flash process comes about.

The same kind of equilibrium, but with its fluid mass decreasing differentially, is called a *differential process* (liberation or condensation).

In reservoir conditions the hydrocarbon pore volume (HCPV) remains constant if the expansion of interstitial water and rock compressibility are neglected. For such constant HCPV, it must be made clear that differential process occurs always as differential vaporization of differential condensation. Differential vaporization takes place when the reservoir temperature is less than critical temperature of solution ($T_{res} < T_c$), and also it takes place during

retrograde gas reservoir depletion, but only in the region pressure and temperature at which the retrograde liquid is vaporized.

In differential condensation, the oil reservoir pressure is maintained constant or almost constant—for example, by gas injection. Differential condensation can also occur just below the dew point in a gas-condensate reservoir.

Above the bubble point and the dew point curves, the virtual (apparent) value of vaporization and/or condensation is zero, but because the mass of fluid in a depleted reservoir is changing as a result of decreasing pressure, the process could be assumed to be differential. One important statement has to be added: there is no qualitative difference between the reservoir fluid in either differential or flash process, if pressure and temperature fall into the area outside of the saturation curves (Figure 6.1.3). A schematic representation of differential vaporization of oil in reservoir conditions is shown in Figure 6.1.9. As indicated in Figure 6.1.9, six hypothetical cases are distinguished. Study Figure 6.1.9 simultaneously with Figure 6.1.3 and Figure 6.1.10.

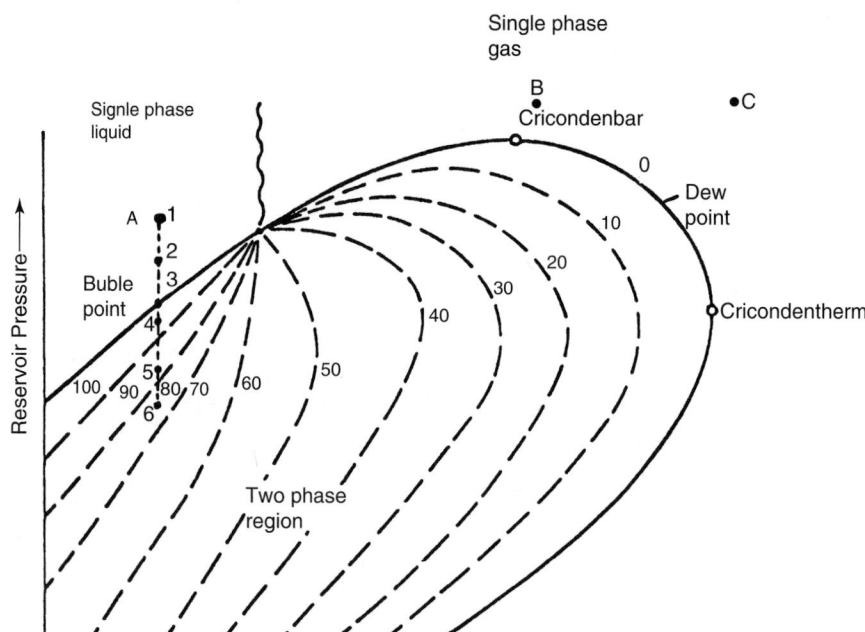

Figure 6.1.3   *Pressure–temperature diagram for reservoir fluids.*

Consider a first sample, in which there is a fixed mass of oil at given temperature, pressure, and HCPV. When the pressure P_1 drops to P_2, the volume of oil increases but the HCPV does not change, so the difference in oil-removed volume equals the total oil production when the pressure changes from P_A to P_2 (sample 2).

The third sample is considered at the bubble point; the oil volume change between P_2 and P_b resembles that between P_1 and P_2. Beginning at P_b, the first gas bubbles are released. Pressure P_4 corresponds to the lowest value of GOR (Figure 6.1.10) and coincides with the highest pressure in a two-phase region, in which only one phase (oil) still flows. Pressure P_4 could be called a *gas flow saturation pressure*. Between P_b and P_4, compositions x_i, y_i, and z_i are changed. HC mass in pore volume is decreasing, so it is the differential process that is contrary to Moses' [8] belief that this is a flash process.

At point 5, HCPV remains constant as in steps 1 to 4, the oil volume has changed and the system is into a two-phase region. An amount of released gas exceeds the gas flow saturation pressure P_4; gas begins to run and is partially removed from the HCPV. This is how two-phase flow is generated. Sample 6 characterizes the same process very close to the bottom-hole area. The reservoir fluid mass difference in steps 1 and 6 equals the total production from an HCPV.

In conclusion, it has been shown that the flash process occurs whenever we are dealing with a closed system or a steady-state flow, e.g., a two-phase flow in vertical tubing, in horizontal pipe flow, and in separators. For any open system, such as a reservoir formation, or for an unsteady-state flow, the differential process is properly describing the quasiequilibrium conditions.

6.1.2 Sampling Process and Laboratory Measurements
The overall quality of the reservoir fluid study and the subsequent engineering calculations based upon that study can be no better than the quality of the fluid samples originally collected during the field sampling process.

Samples representative of the original reservoir can be obtained only when the reservoir pressure is equal to or higher than the original bubble point or dew point.

The pressure drawdown associated with normal production rates will cause two-phase flow near the wellbore if the fluid in the formation was initially saturated or only slightly undersaturated. Relative permeability effects may then cause the material entering the wellbore to be different from the original overall (total) composition fluid existing at the boundary of drainage area. The problem of drawdown in a saturated reservoir cannot be eliminated; therefore, it is necessary to limit the pressure drawdown by reducing the flowrate to the lowest possible stable rate while sampling.

There are two basic methods of sample collection: subsurface (bottomhole) and surface (separator). The fluid sampling method to be used dictates the remainder of the conditioning process. If the bottomhole samples are to be collected, the period of reduced flowrate will generally last from 1 to 4 days, depending on the formation and fluid characteristics and the drainage area affected. After this reduced flowrate period, the well would be shut in and allowed to reach static pressure. The shut-in period would last about 1 day or up to a week or more, depending on formation characteristics. For the case of the saturated reservoir, the shut-in period has the resultant effect of forcing gas into solution in the oil, thus raising the saturation pressure. In some cases, the desired value of P_b is obtained; however, in most cases this value is only approached and the final difference is a function of well productivity, production rate, and fluid properties. At the conclusion of the shut-in period, the well would be properly conditioned and ready for bottomhole sampling. Subsurface sampling is generally not recommended for gas-condensate reservoirs; the same is true for oil reservoirs producing substantial quantities of water. If separator gas and liquid samples are to be collected, the gas and liquid

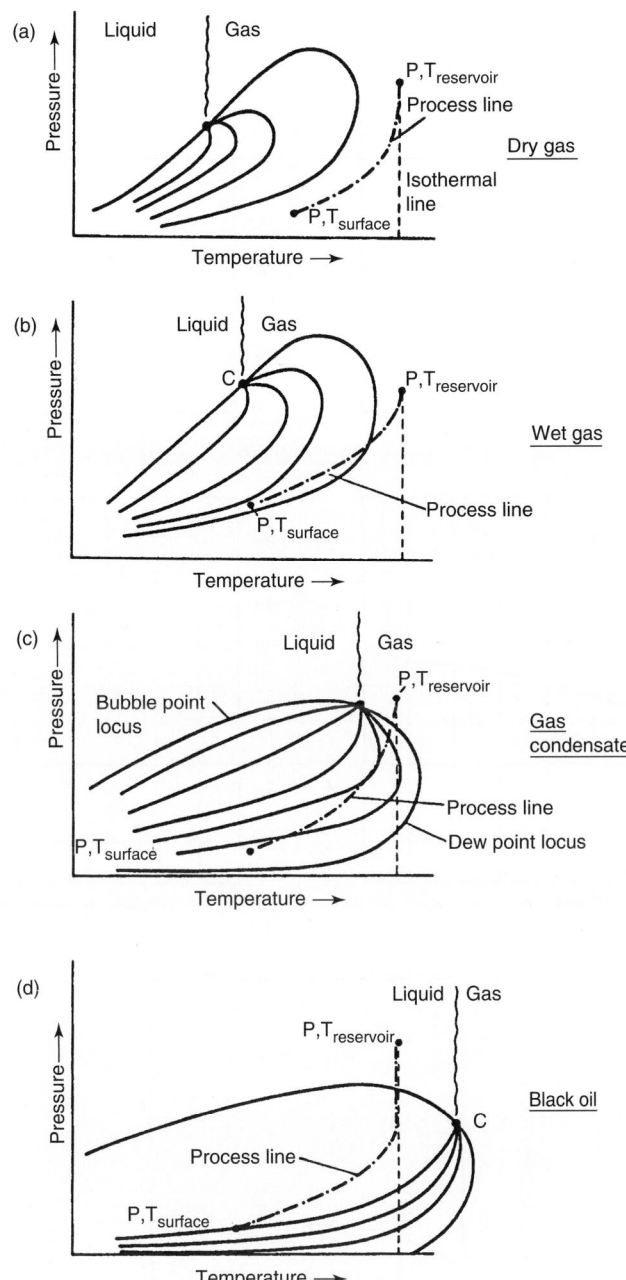

Figure 6.1.4 *Hydrocarbon mixture classification.*

Table 6.1.2 *Typical Compositions of Hydrocarbons Fluids [7]*

	Dry Gas	Wet Gas	Retrograde Condensate	Volatile Oil or Near Critical Oils	Black Oil	Heavy Oil
C_1 (mole %)	90		70	5	30	
C_2–C_6 (mole %)	9		22	30	35	
C_{7+} (mole %)	1		8	15	35	
GOR [ft³/bbl]	∞	150,000	3000–150,000 (10,000+)*	2000–3000 (3000–6000)*	<2000	
[m³/m³]	∞	>27,000	540–27,000	360–540	<360	
API Liquid Gravity	—		40–60	45–70	30–45	<20
Liquid Specific Gravity	—		0.83–0.74	0.8–0.7	0.88–0.8	>0.94

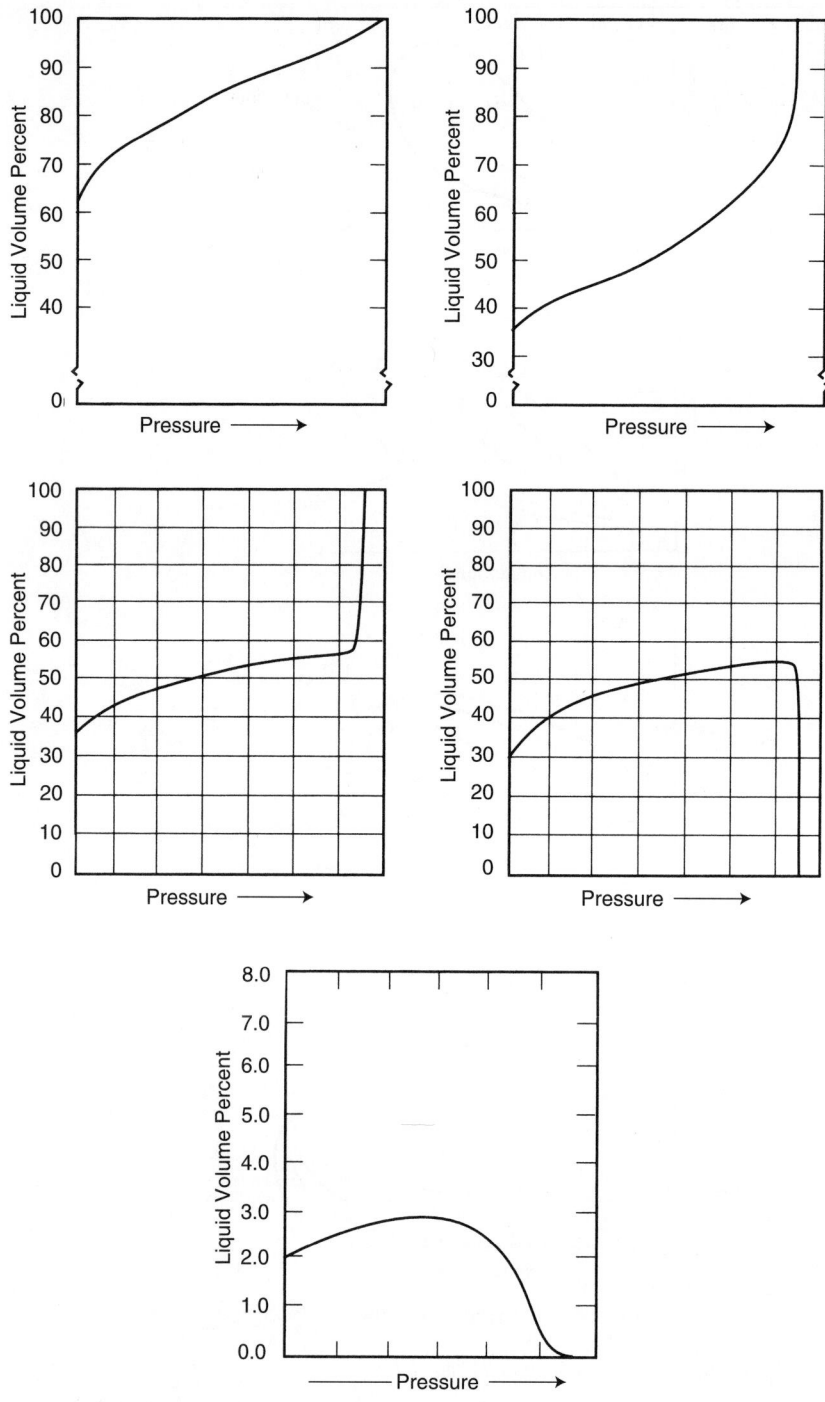

Figure 6.1.5a–e   *Liquid volume percent — pressure diagram for reservoir fluid [6].*

rates must be monitored continually during the period of stable flow at reduced flowrates. A minimum test of 24 hours is recommended, but more time may be needed if the pressure drawdown at the formation has been high. Surface sampling, called *separator sampling*, has wider applications than subsurface sampling, and is the only recommended way of sampling a gas-condensate reservoir, but often can be used with good success for oil reservoirs as well. There are three requirements to successful separator sampling.

1. stable production at a low flowrate
2. accurate measurement of gas and oil flowrates
3. collection of representative samples of first-stage gas and first-stage liquid

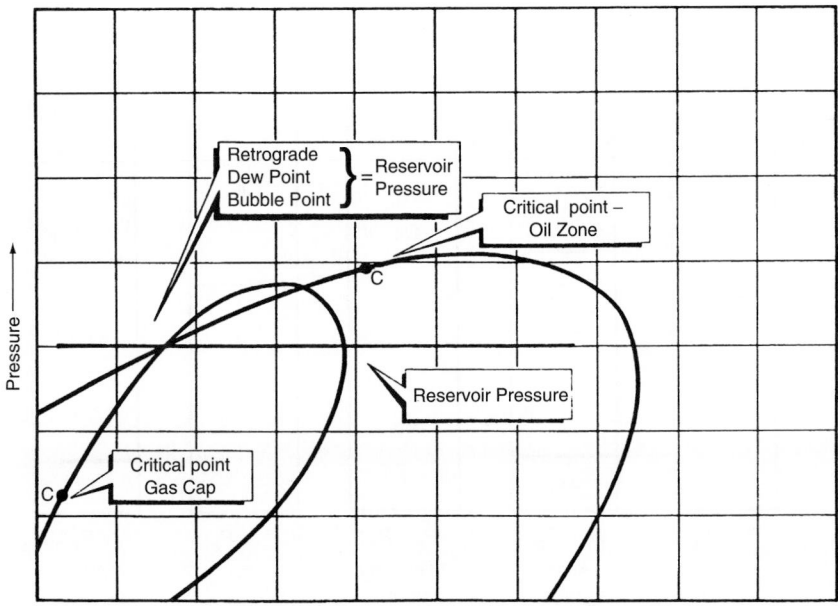

Figure 6.1.6 *Pressure–temperature diagram for gas cap and associated oil [6].*

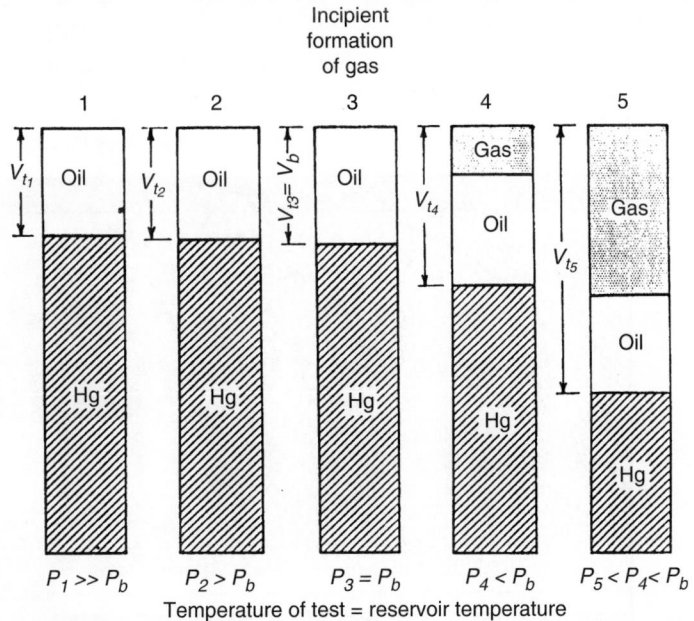

Figure 6.1.7 *Schematic representation of laboratory P–V equilibrium (flash vaporization) [10].*

The above procedure is described in detail in API Standard 811-08800 [9].

The reservoir process is stimulated in the laboratory by flash differential vaporization (Figure 6.1.7 and Figure 6.1.8). Based on both figures, it is possible to prepare the reservoir fluid data for engineering calculations.

In the laboratory, the differential liberation consists of a series—usually 10 to 15—of flash liberations. An infinite series of flash liberations is the equivalent of a true differential liberation. At each pressure level, gas is evolved and measured. The volume of oil remaining is also measured at each depletion pressure. This process is continued to atmospheric pressure. The oil remaining at atmospheric pressure is measured and converted to a volume at 60°F (15.6°C). This final volume is referred to as the residual oil. The volume of oil at each of the higher pressures is divided by the volume of residual oil at 60°F (15.6°C).

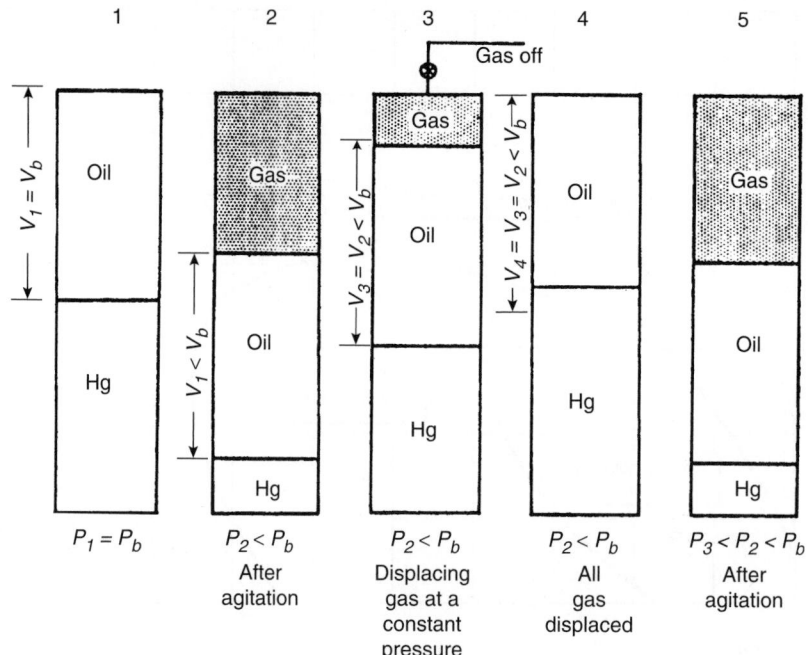

Figure 6.1.8 *Schematic representation of laboratory differential vaporization [10].*

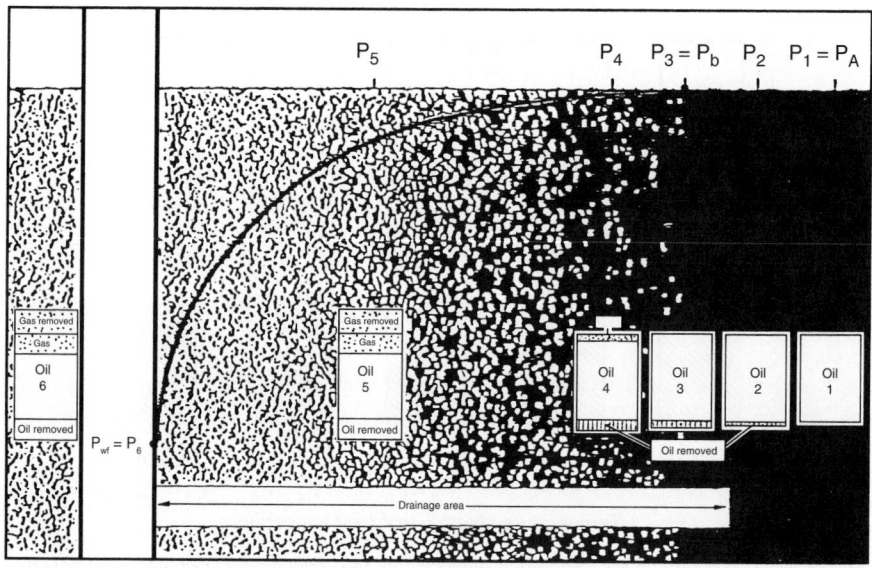

Figure 6.1.9 *Schematic representation of differential vaporization in reservoir conditions.*

Example 6.1.1

Surface separator samples were collected from a well on completion of a 2-hr test on June 8, 1984. The gas/liquid ratio measured on this test was 4565 ft.^{3} of separator gas per barrel of separator liquid and was used as the basis for this recombination. The resultant reservoir fluid exhibited a dew point of 4420 psia at $T_{res} = 285°F$. The reservoir fluid exists as a gas (an undersaturated gas) at $P_{res} = 12,920$ psia.

A constant volume depletion study is also performed on the reservoir fluid. The produced compositions and volumes from the depletion study are used in conjunction with equilibrium constants to calculate cumulative STO and separator

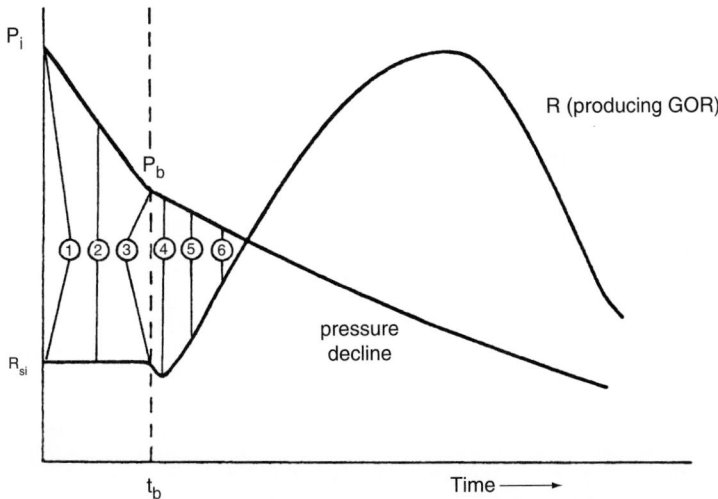

Reservoir parameters	Sample #					
	1	2	3	4	5	6
HCPV	$v_1 = v$	$v_2 = v$	$v_3 = v$	$v_4 = v$	$v_5 = v$	$v_6 = v$
HC Mass in Pore Volume (m)	$m_1 = m$	$m_2 < m_1$	$m_3 < m_2$	$m_4 < m_3$	$m_5 < m_4$	$m_6 < m_5$
Pressure (P)	$P_1 = P_A$	$P_2 < P_1$	$P_b < P_2$	$P_4 < P_b$	$P_5 < P_4$	$P_6 < P_5$
Temperature (T)			assumed constant reservoir temperature			
Total Reservoir Composition z_i	$z_{i1} - z_{iA}$	$z_{i2} - z_{i1}$	$z_{ib} - z_{i1}$	$z_{i4} - z_{i1}$	$z_{i5} \neq z_{i1} \neq z_{i4}$	$z_{i6} \neq z_{i1} \neq z_{i4} \neq z_{i5}$
Gas Composition y_i	$y_{i1} = y_{iA} = 0$	$y_{i2} = y_{iA} = 0$	$y_{i3} = y_{ib} \neq 0$	$y_{i4} = y_{i1} \neq y_{i3}$	$y_{i5} = y_{i1} \neq y_{i3} \neq y_{i4}$	$y_{i6} = y_{i1} \neq y_{i3} \neq y_{i4} \neq y_{i5}$
Oil Composition x_i	$x_{i1} = z_{iA} = z_{i1}$	$x_{i2} = x_{i1} = z_{i1}$	$x_{i3} = x_{ib} \neq x_{i1}$	$x_{i4} \neq x_{i1} \neq x_{i3}$	$x_{i5} \neq x_{i1} \neq x_{i3} \neq x_{i4}$	$x_{i6} \neq x_{i1} \neq x_{i3} \neq x_{i4} \neq x_{i5}$

Figure 6.1.10 *Schematic representation reservoir pressure (P) and GOR vs. time and mass and composition in reservoir differential vaporization process.*

gas recoveries resulting from conventional field separation. Gas plant products in both the primary separator gas and the full well stresses should also be reported.

Sampling Conditions	
Date sampled (on 20/64 choke)	06-08-84 for 1330 hr
Tubing pressure, flowing	9960 psig
Primary separator temperature	95°F
Primary separator pressure	900 psig
Primary separator gas rate (Table 6.1.3)	2229.7 MCF/Day
Liquid rate (2nd stage @ 50 psig)	396 bbl/day
Gas/liquid ratio (GOR)	5631 SCF 1st stg. gas/bbl 50 lb liq.
Shrinkage factor (vol. 50 lb liq./vol. sep. liq.)	0.8108
Gas/liquid ratio (GOR)	4565 SCF 1st stg. gas/bbl 900 lb liq.
Shrinkage factor (vol. S.T. liq./vol. sep. liq.) through 50-lb 2nd stage	0.7445
Pressure base	15.025 psia @ 60°F

Table 6.1.3 *Calculation of Gas Rate [11]*

$\sqrt{HwPf} = 165.6804$	$Hw = 30.0'' \, H_2O,$	$Pf = 915.00 \, psia$
$Fb = 455.0300$	$D = 5.761'',$	$d = 1.50''$
$Fpb = 0.9804$	15.025 psia	
$Fr = 1.0002$	$b = 0.0367$	
$Y2 = 1.0002$	$Hw/Pf = 0.033,$	$d/D = 0.260$
$Fg = 1.1953$	Gravity $= 0.6999,$	$Fg = \sqrt{1/0.6999}$
$Ftf = 0.9680$	Temp. $= 95\,°F,$	$Ftf = \sqrt{520/555}$
$Fpv = 1.0859$	$pTr' = 1.441,$	$pPr' = 1.372$
	$Z = 0.8480,$	$Fpv = \sqrt{1/Z}$

Acid Gas Correction Factor   Epsilon $= 3.57$

$Q = \sqrt{Hw\,Pf} \times Fb \times Fpb \times Fr \times Y2 \times Fg \times Ftf \times Fpv \times 24$

$Q = 2229.7$ MCF/day@15.025   PSIA@60°F

The samples of separator gas and separator liquid were analyzed and the results are reported in Table 6.1.4, showing both composition of each sample and the computed analysis of the well stream based on the GOR in the primary separator. The separator liquid (oil) production was calculated from the measured second-stage production by applying the determined shrinkage factor.

Table 6.1.4 *Hydrocarbon Analysis of Separator Products and Calculated Wellstream [11]*

Component	Separator Liquid Mol %	Separator Liquid Liq. Vol. %	Separator Gas Mol %	Separator Gas GPM @ 15.025 PSIA	Well Stream Mol %	Well Stream GPM @ 15.025 PSIA
Carbon dioxide	0.41	0.19	2.18		1.84	
Nitrogen	0.00	0.00	0.16		0.13	
Methane	22.36	10.16	82.91		71.30	
Ethane	8.54	6.12	8.22	2.243	8.28	2.259
Propane	10.28	7.58	3.82	1.073	5.06	1.421
Iso-Butane	5.69	4.99	1.11	0.372	1.99	0.664
N-Butane	5.11	4.32	0.80	0.256	1.63	0.523
Iso-Pentane	4.84	4.75	0.30	0.113	1.17	0.437
N-Pentane	2.01	1.95	0.17	0.064	0.52	0.193
Hexanes	8.16	8.75	0.22	0.091	1.74	0.731
Heptanes Plus	32.60	51.19	0.11	0.053	6.34	3.907
Total	100.00	100.00	100.00	4.265	100.00	10.135

Calculated specific gravity (Air = 100)= 0.6999 separator gas 1.0853 well stream

Sep. gas heat of combustion (BTU/Cu.ft. @ 15.025 PSIA & = 1206.8 real 60 °F) dry

Sep. gas heat of combustion (BTU/Cu.ft. @ 15.025 PSIA & = 1185.7 water sat. 60 °F) wet

Sep. gas compressibility (@ 1 ATM. & 60 °F) Z = 0.9968

Properties of Heptanes Plus:

Specific gravity = 0.7976
Molecular Weight = 152
Cu.Ft./Gal. = 16.24

Properties of Stock Tank Liquid:

Gravity = 56.8 degrees API @ 60°F

Basis of Recombination:

Separator liquid per MMSCF separator gas = 219.05 bbls

Properties of Separator Liquid:

0.6562 @ 60/60 °F
78.7
25.81 @ 15.025 PSIA & 60 °F

6.1.2.1 Equilibrium Cell Determinations

Following the compositional analyses, portions of the primary separator liquid and gas were physically recombined in their produced ratio in a variable volume, glass-windowed equilibrium cell. Determinations on this mixture were divided into the following main categories.

1. Dew point pressure determination and pressure–volume relations on a constant weight of reservoir fluid at the reservoir temperature: the procedure consisted of establishing equilibrium between gas and liquid phases at a low pressure and measuring volumes of liquid and gas in equilibrium at that pressure. The pressure was raised by the injection of mercury into the cell and phase equilibrium established again at a higher pressure. This procedure was repeated until all of the liquid phase had vaporized, at which point the saturation pressure was observed. The cell pressure was then raised above the dew point pressure in order to determine the supercompressibility characteristics of the single-phase vapor. As a check on all readings and, particularly, to verify the dew point, the cell pressure was incrementally reduced, equilibrium established and volumetric readings made. Reported in Table 6.1.5 are the relative volume relations (Figure 6.1.11) and specific volumes of the reservoir fluid over a wide range of pressures as well as compressibility factors (Figure 6.1.12), the single-phase vapor above the dew point. Reported in Table 6.1.6 are the dew point pressure (Figure 6.1.13) resulting from recombination's

at gas/liquid ratios above and below the ratio measured at the time of sampling.

2. Compositions of the produced wellstream and the amount of retrograde condensation resulting from a stepwise differential depletion. This procedure consisted of a series of constant composition expansions and constant pressure displacements with each displacement being terminated at the original cell volume. The gas removed during the constant pressure was analyzed. The determined compositions (Figure 6.1.14), computed GPM content (Figure 6.1.15), respective compressibility (deviation) factors, and volume of wellstream produced during depletion (Figure 6.1.16) are presented in Table 6.1.7. The volume of retrograde liquid resulting from the gas depletion is shown in Table 6.1.8 (and Figure 6.1.17), both in terms of barrels of reservoir liquid and percentage of hydrocarbon pore space. Show in Table 6.1.7 are the compositions of the gas and liquid remaining in the reservoir after depletion to abandonment pressure.

6.1.2.2 Equilibrium Flash Calculations

The produced compositions and volumes from a depletion study were used in conjunction with equilibrium constants K (derived from a Wilson modified R-K equation of state) to calculate cumulative stock tank liquid and separator gas recoveries resulting from conventional separation. These data are reported in Table 6.1.9 and Figure 6.1.18. Also, gas

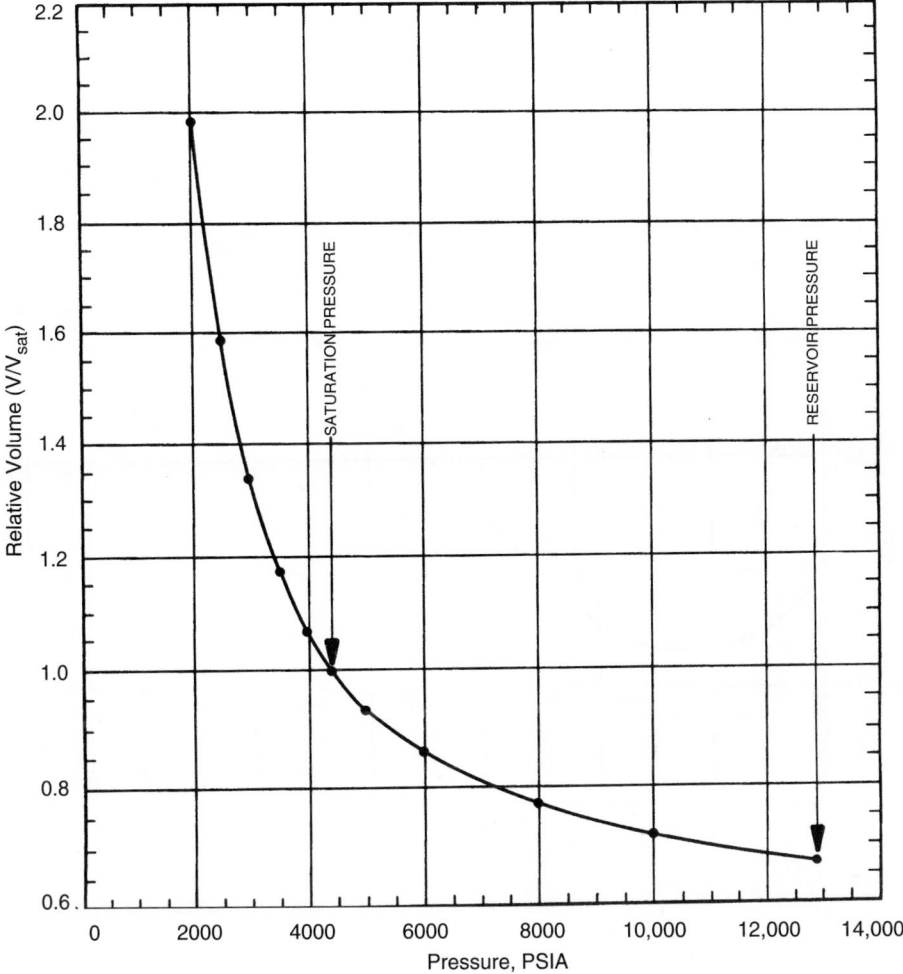

Figure 6.1.11 *Pressure–volume relation of reservoir fluid at 285°F (constant mass expansion) [11].*

plant products in both the primary separator and the full wellstream are attached.

Example 6.1.2

This is a black oil problem. From differential vaporization (Table 6.1.11) and separator test data (Table 6.1.12) discuss the B_o and R_s calculation method.

Figure 6.1.19 and Figure 6.1.20 illustrate laboratory data. These data are reported in a convention other than in Example 6.1.1. The residual oil in the reservoir is never at 60°F (15.6°C), but always at T_{res}. Reporting these data relative to the residual oil at 60°F (15.6°C), gives the relative-oil-volume curve the appearance of an FVF curve, leading to its misuse in reservoir calculations. A better method of reporting these data is in the form of a shrinkage curve. We may convert the relative-oil-volume data in Figure 6.1.19 and Table 6.1.11 to a shrinkage curve by dividing each relative-oil-volume factor B_{od} by the relative oil volume factor at the bubble point, B_{odb}.

The shrinkage curve now has a value of one at the bubble point and a value of less than one at the subsequent pressures below the bubble point, as in Figure 6.1.21. As pressure is reduced and gas is liberated, the oil shrinks. The shrinkage curve describes the volume of this original barrel of oil in the reservoir as pressure declines. It does not relate to a stock tank or surface barrel.

We now know the behavior of the oil in the reservoir as the pressure declines. We must have a way of bringing this oil to the surface through separators and into a stock tank. This process is a flash process. Most reservoir fluid studies include one or more separator tests to simulate this flash process. Table 6.1.12 is a typical example of a set of separator tests. During this test, the FVF is measured. The FVF is the volume of oil and dissolved gas entering the wellbore at reservoir pressure and temperature divided by the resulting stock-tank oil volume after it passes through a separator.

The FVF is B_o; because separators result in a flash separation, we showed a subscript, B_{of}. In most fluid studies, these separator tests are measured only on the original oil at the bubble point. The FVF at the bubble point is B_{ofb}. To make solution-gas-drive or other material-balance calculations, we need values of B_{of} at lower reservoir pressures. From a technical standpoint, the ideal method for obtaining these data is to place a large sample of reservoir oil in a cell, heat it to reservoir temperature, and pressure-deplete it with a differential

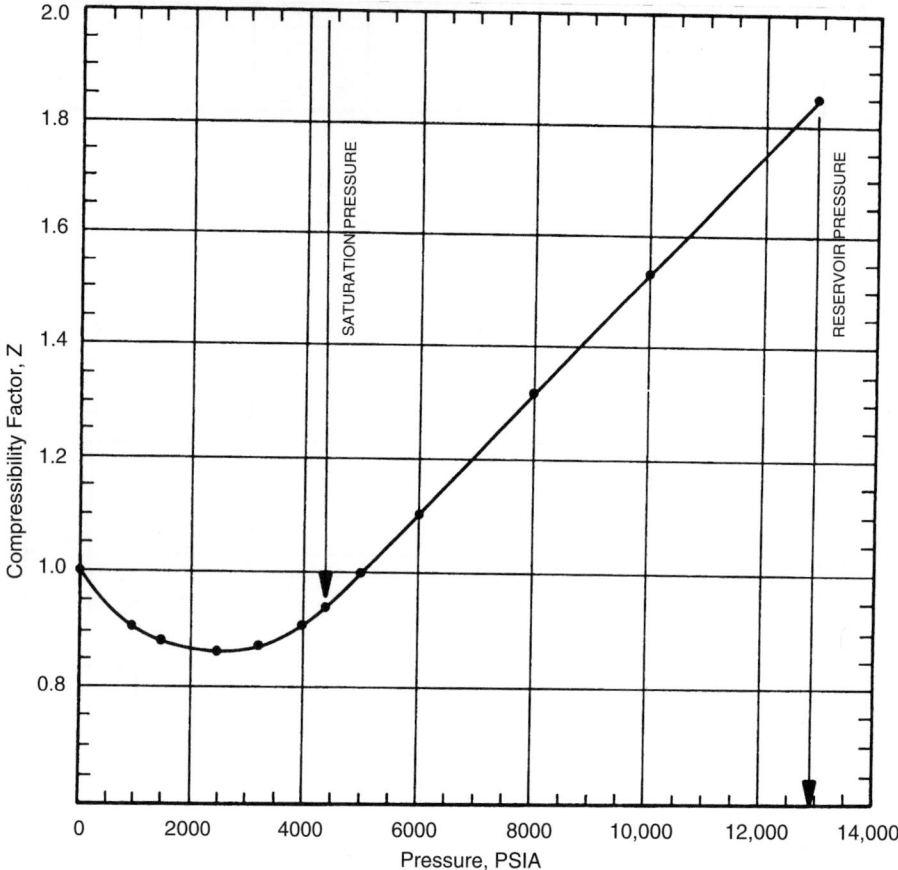

Figure 6.1.12 *Compressibility factor "Z" of wellstream during depletion at 285°F [11].*

process to stimulate reservoir depletion. At some pressure a few hundred psi below the bubble point, a portion of the oil is removed from the cell and pumped through a separator to obtain the flash FVF, B_{of}, at the lower reservoir pressure. This should be repeated at several progressively lower reservoir pressures until a complete curve of B_{of} versus reservoir pressure has been obtained. The process is time consuming and consequently adds to the cost of the study. Most studies include only values of B_{ofb}, the FVF at the bubble point. The values of B_{of} at lower pressures must be obtained by other means.

The method calls for multiplying the flash FVF at the bubble point B_{ofb} by the shrinkage factors at various reservoir pressures obtained earlier. The shrinkage factor was calculated by dividing the relative oil volume factors B_{od} by the relative oil volume factor at the bubble point B_{ofb}. If we combine both calculations, we can start with the differential-relative-volume curve and adjust it to separator or flash conditions by

$$B_o = B_{od} \frac{B_{ofb}}{B_{odb}}$$
[6.1.16]

This calculation is illustrated in Figure 6.1.22.

To perform material-balance calculations, we must also have the separator and stock-tank gas in solution as a function of reservoir pressure. These values are expressed as standard cubic feet per barrel and usually are designated R_{sf}.

The separator test gives us this value at the bubble point, R_{sfb}. As pressure declines in the reservoir, gas is evolved from solution. The amount of gas remaining in solution in the oil is then somewhat less. The differential vaporization tells us how much gas was evolved from the oil in the reservoir: $(R_{sdb} - R_{sd})$, where R_{sdb} is the amount of gas in solution at the bubble point as measured by differential vaporization at the reservoir temperature and R_{sd} is the gas in solution at subsequent pressures.

The units of R_{sdb} and R_{sd} are standard cubic feet per barrel of residual oil. Because we must have the gas in solution in terms of standard cubic feet per barrel of stock-tank oil, this term must be converted to a stock-tank basis. If we divide $(R_{sdb} - R_{sd})$ by B_{odb}, we have the gas evolved in terms of standard cubic feet per barrel of bubble point oil. If we then multiply by B_{ofb}, we will have the gas evolved in terms of standard cubic feet per barrel of stock-tank oil. This expression now is $(R_{sdb} - R_{sd})(B_{ofb}/B_{odb})$. The gas remaining in solution then is $R_s = R_{sfb} - (R_{sdb} - R_{sd})(B_{ofb}/B_{odb})$ standard cubic feet per stock-tank barrel. For every pressure studied during the differential liberation, R_s may be calculated from this equation. This calculation is illustrated in Figure 6.1.23.

It is a fairly common practice to use differential vaporization data for material-balance calculations. Values of B_{od} and R_{sd} are almost always higher than the corresponding values from separator tests; consequently, calculations of OIP and recoverable oil will usually be lower than is correct.

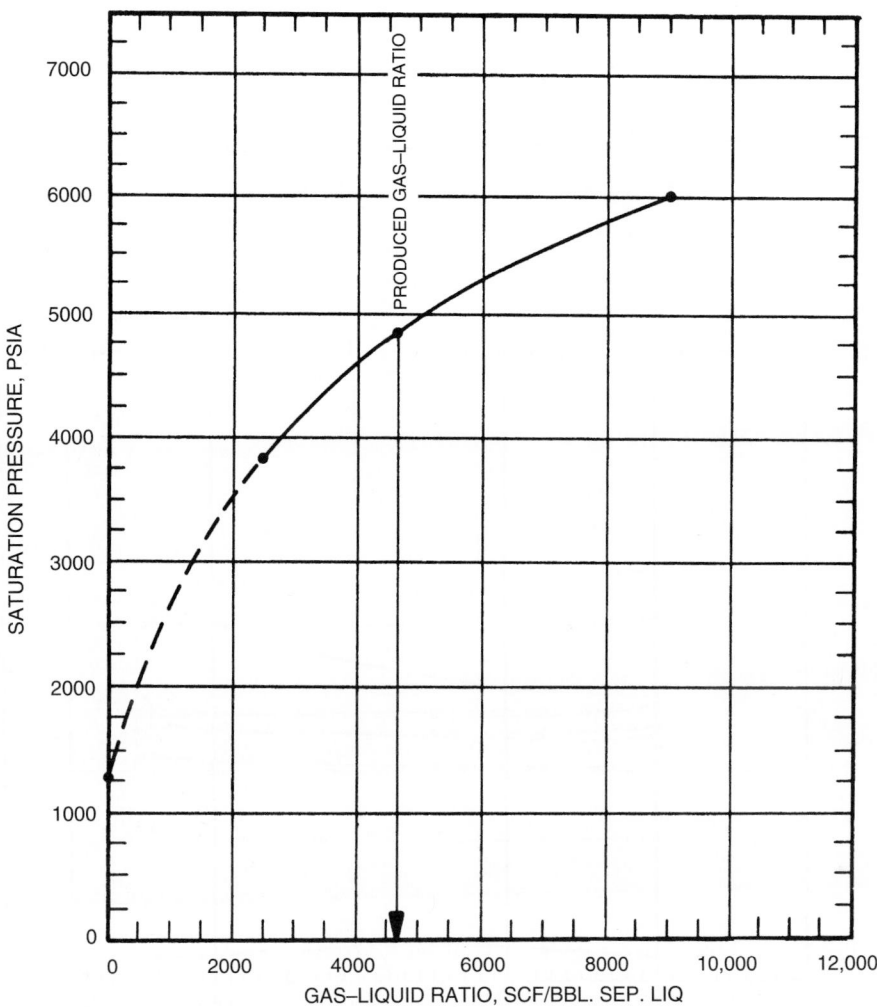

Figure 6.1.13 *Effect of gas/liquid ratio upon saturation pressure at 285°F [11].*

Table 6.1.5 *Pressure–Volume Relation of Reservoir Fluid at 285°F [11]*

Pressure (PSIA)	Relative Volume (V/Vsat)	Specific Volume (Cu.Ft./Lb.)	Retrograde BPMMCF*	Liquid Vol. %**	Deviation Factor (Z)	Calculated Viscosity (Centipoise)
***12,920 Res.	0.6713	0.03635			1.8485	0.0793
10000	0.7201	0.03898			1.5346	0.0674
8000	0.7741	0.04191			1.3197	0.0579
6000	0.8639	0.04677			1.1047	0.0471
5000	0.9373	0.05075			0.9988	0.0412
4420 D.P.	1.0000	0.05414	0.00	0.00	0.9420	0.0374
					(Two Phase)	
4000	1.0677	0.05781	10.58	1.30	0.9102	
3500	1.1764	0.06369	69.83	8.55	0.8775	
3000	1.3412	0.07261	94.17	11.53	0.8575	
2500	1.5879	0.08597	107.92	13.21	0.8460	
2000	1.9831	0.10737	114.27	13.99	0.8453	
1500	2.6605	0.14404	114.27	13.99	0.8505	
1000	4.0454	0.21902	107.92	13.21	0.8621	

*BBLS. per MMSCF of dew point fluid.
**Percentage of hydrocarbon pore space at dew point.
***Extrapolated.

Table 6.1.6 *Observed Saturation Pressures From Stepwise Recombination at 285 °F [11]*

Gas-Liquid Ratio (SCF 1st Stg. Gas) (BBL. 1st Stg. Liq.)	Saturation Pressure (Psia)
9000	6000 Dew Point
4565 (Produced)	4420 Dew Point
2500	3830 Dew Point

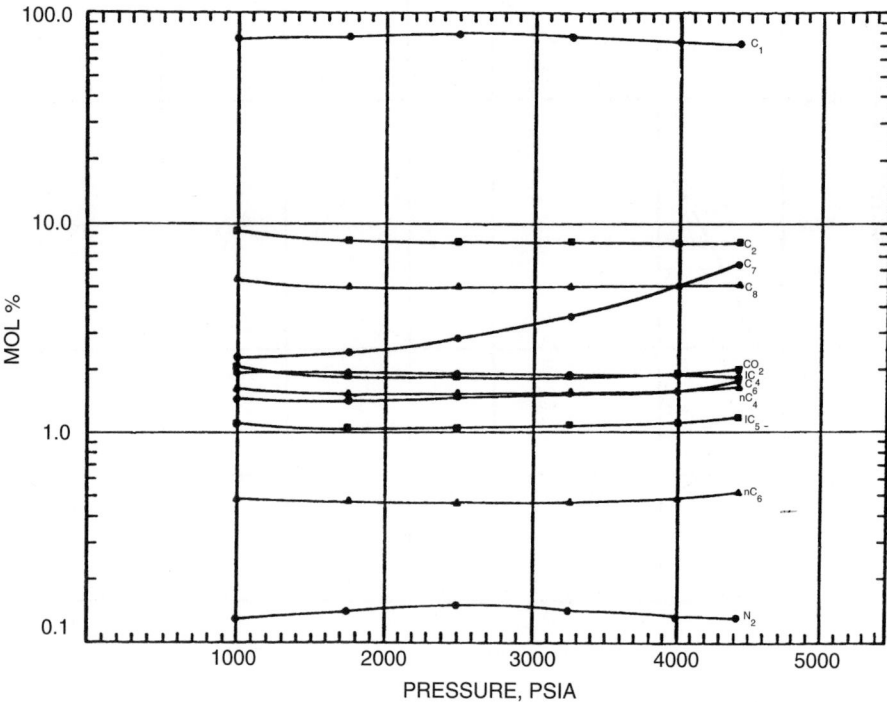

Figure 6.1.14 *Hydrocarbon composite of wellstreams produced during pressure depletion [11].*

The differential vaporization data should be converted to separator flash conditions before use in calculations.

6.1.2.3 Vapor–Liquid Equilibrium Calculations

The basic equilibrium calculations are the bubble point, dew point, and flash (or two-phase equilibrium). In the general flash calculation, the temperature and pressure are usually fixed and L/f is the dependent variable. All equilibrium calculations are based on the definition of the K value, such that

$$K_i = \frac{y_i}{x_i} = \frac{\text{concentration of ``i'' component in vapor phase}}{\text{concentration of ``i'' component in liquid phase}}$$ [6.1.16]

In bubble point calculations, x_i is known and either T or P is fixed. The vapor phase composition (y_i) and the P or T of the system are unknown.

$$y_i = K_i x_i$$ [6.1.17]

Several different values of the dependent variable are assumed. The correct value is the one that yields

$$\sum_{i=1}^{n} K_i x_i = \sum_{i=1}^{n} y_i = 1.0$$ [6.1.18]

An additional requirement when using composition dependent K values is

$$\left| y_i^{m+1} - y_i^m \right| \le \varepsilon, \quad i = 1 \text{ to } n$$ [6.1.19]

where the value of ε is arbitrarily small (10^{-4} to 10^{-6}). This requirement is a consequence of using composition dependent K values. If the composition y_i is not correct, the predicted K, values will not be correct. As a result, the composition of the vapor phase must be stabilized even though the correct value of the dependent variable has been determined. Iterations through the bubble point calculation must be continued until both Equation 6.1.18 and Equation 6.1.19 are satisfied. A logical diagram illustrating the basic bubble calculation is shown in Figure 6.1.24.

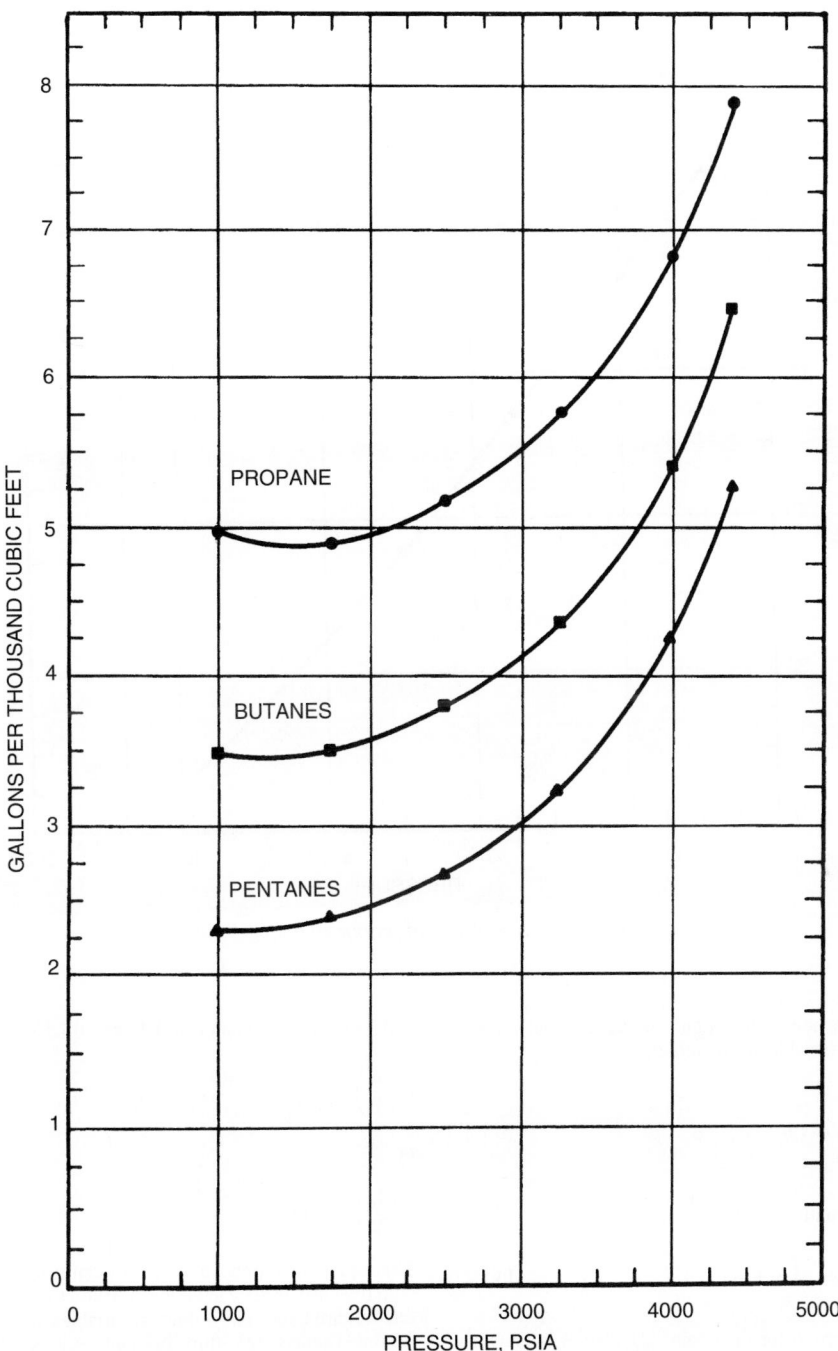

Figure 6.1.15 *GPM content of hydrocarbons produced during depletion at 285°F [11].*

How can we assume initial vapor phase composition? One approach that has been successful is to assume that the mole fraction of the lowest boiling component in the systems is equal to unity with the remaining component mole fractions set to 10^{-6}. Another approach is to get the K value from GPSA [2]. The vapor composition is adjusted after each interaction.

Dew point calculations are the opposite of bubble point calculations: y_i is known and x_i and T or P are to be calculated. The specific equation used in the dew point calculation is

$$x_i = \frac{y_i}{K_i}$$

[6.1.20]

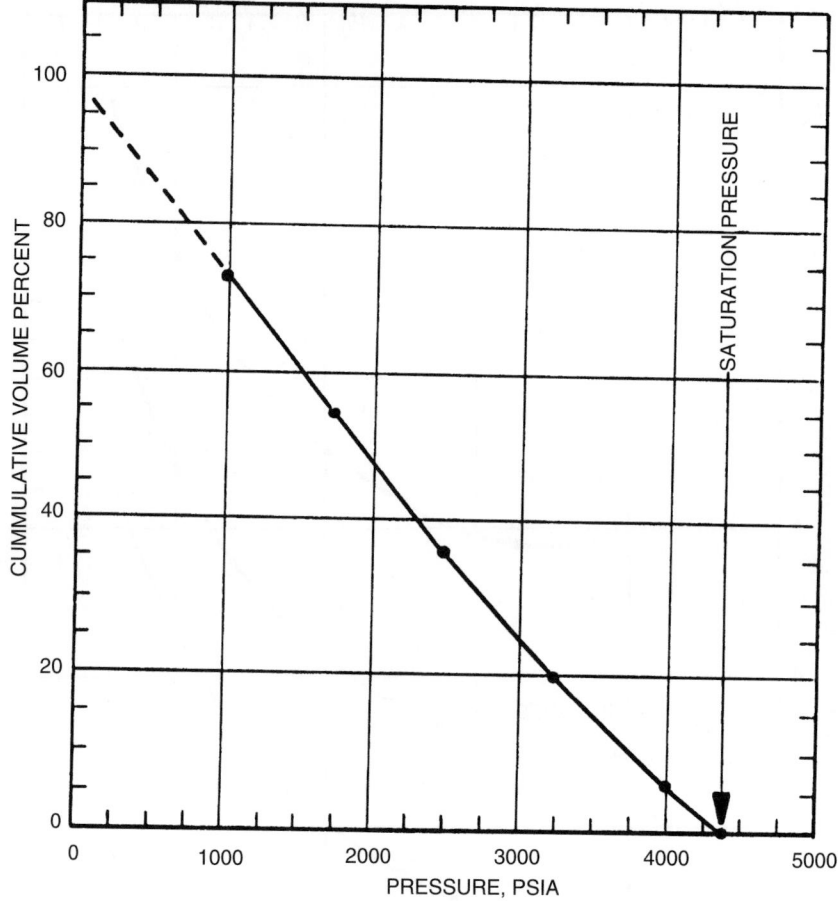

Figure 6.1.16 *Volume of wellstream produced during depletion [11].*

As in the bubble point calculation, several different values of the dependent variable are assumed:

$$\sum_{i=1}^{n} \frac{y_i}{K_i} = \sum_{i=1}^{n} x_i = 1.0 \qquad [6.1.21]$$

subject to the condition that

$$\left| x_i^{m+1} - x_i^m \right| \le \varepsilon, \quad i = 1 \text{ to } n \qquad [6.1.22]$$

A logical diagram for dew point calculation is shown in Figure 6.1.25. In this case x_i is initially unknown and must be assumed. A procedure is to assume that the mole fraction of the highest boiling component in the system is equal to unity. The remaining component mole fractions are set to 10^{-6}. Liquid phase compositions are adjusted by linear combinations of the assumed and calculated value during each iteration. The convergence of algorithms for T and P dependent calculations are given in Figure 6.1.26 and Figure 6.1.27.

The purpose of flash calculations is to predict the composition and amount of the coexisting vapor and liquid phases at a fixed temperature and pressure.

According to Equations 6.1.15 and 6.1.16

$$\phi \left(\frac{L}{P} \right)^m = \sum_{i=1}^{n} \frac{z_i(1 - K_i)}{L/F(1 - K_i) + K_i} = 0.0 \qquad [6.1.23a]$$

or

$$\phi \left(\frac{L}{P} \right)^m = 1.0 \qquad [6.1.23b]$$

This equation is applicable to a wide range of L/F conditions.

In the basic flash calculation, T, P and the overall composition (z_i) are fixed. The unknown variables are x_i, y_i, and L/F. A convergence algorithm that can be used with Equation 6.1.24 is

$$\left(\frac{L}{F} \right)^{m+1} = \left(\frac{L}{F} \right)^m - \frac{\phi \left(\dfrac{L}{F} \right)^m}{\phi \left(\dfrac{L}{F} \right)^{m-1}} \qquad [6.1.24a]$$

where

$$\phi \left(\frac{L}{P} \right)^m = -\sum_{i=1}^{n} \frac{z_i(1 - K_i)^2}{L/F(1 - K_i) + K_i^2} \qquad [6.1.24b]$$

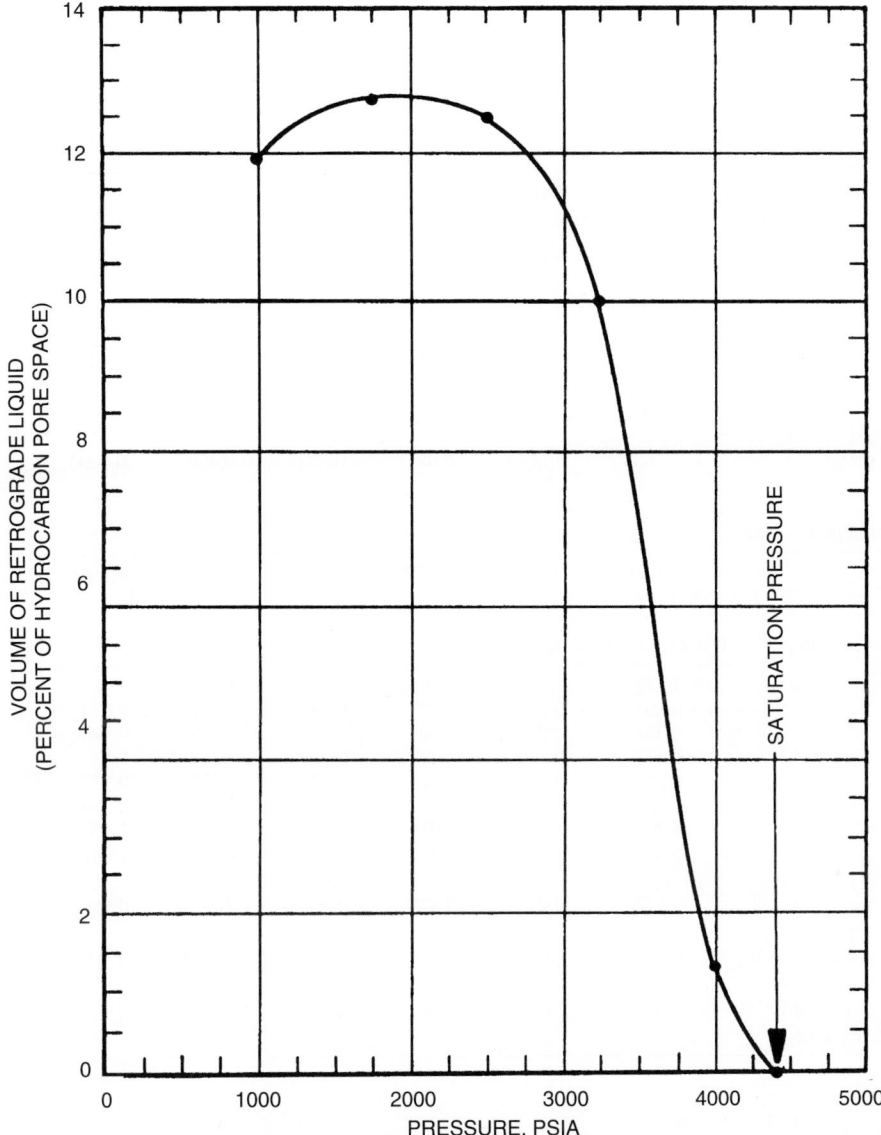

Figure 6.1.17 *Retrograde condensation during depletion at 285°F [11].*

This convergence algorithm is very reliable provided the values of $(L/F)^{m+1}$ are constrained to be valid by material balance considerations:

$$0.0 < (L/F)^{m+1} \leq 1.0 \qquad [6.1.25]$$

The classical bubble point/dew point checks

$$\sum K_i z_i > 1.0 \quad \text{and} \quad \sum z_i/K_i > 1.0 \qquad [6.1.26]$$

(to assure that the mixture is in two-phase region) cannot be conveniently used in most computer equation-of-state–based flash calculations because the K values for a given system are not known until the final solution has been reached. Consequently, the flash calculation (and its convergence algorithm) must be capable of performing "flash calculation" on single-phase systems (subcooled liquids, superheated vapors, and dense gas systems) as well as reliably predicting the amount of vapor and liquid present in a two-phase system. When the above flash equation/convergence algorithm is used on single-phase systems, the final predicted value of L/F will usually be outside the interval described by Equation 6.1.25 unless the material balance constraint is enforced. Should a value of $(L/F)^{m+1}$ outside the limits defined by Equation 6.1.25 be detected in an interaction, we recommend that the value of L/F predicted by Equation 6.1.24a be replaced by the appropriate value described by the following equations:

$$\text{if } (L/F)_{m+1} < 0.0, \quad (L/F)^{m+1} = (L/F)^m/2.0$$

or

$$\text{if } (L/F)^{m+1} > 1.0, \quad (L/F)^{m+1} = [1 + (L/F)^m]/2.0$$

Table 6.1.7   *Depletion Study at 285°F [11]*

Component	Dew Point Fluid 4420	Reservoir Pressure — PSIA					Aband. Liquid 1000
		4000	3250	2500	1750	1000	
Nitrogen	0.13	0.13	0.14	0.15	0.14	0.13	0.04
Carbon dioxide	1.84	1.87	1.90	1.92	1.94	1.93	0.67
Methane	71.30	73.09	74.76	75.69	76.09	74.71	16.61
Ethane	8.28	8.18	8.14	8.12	8.15	9.05	4.76
Propane	5.06	5.01	4.97	4.94	4.93	5.29	4.60
Iso-butane	1.99	1.93	1.89	1.87	1.89	2.02	2.66
N-Butane	1.63	1.59	1.56	1.55	1.54	1.61	2.37
Iso-pentane	1.17	1.12	1.09	1.06	1.06	1.10	2.37
N-Pentane	0.52	0.48	0.47	0.46	0.47	0.48	1.17
Hexanes	1.74	1.59	1.51	1.45	1.40	1.43	5.96
Heptanes Plus	6.34	5.01	3.57	2.79	2.39	2.25	58.77
Totals	100.00	100.00	100.00	100.00	100.00	100.00	100.00
Well stream gravity (Air = 1)	1.0853	1.0043	0.9251	0.8823	0.8595	0.8606	

Properties of heptanes plus

Specific Gravity	0.7976	0.7900	0.7776	0.7652	0.7523	0.7381	0.8181
Molecular Weight	152	145	136	128	120	112	171

GPM content of produced well stream (Gal./MSCF)

Propane	1.421	1.407	1.396	1.387	1.384	1.485	
Iso-Butane	0.664	0.644	0.631	0.624	0.631	0.674	
N-Butane	0.524	0.511	0.502	0.498	0.495	0.518	
Iso-Pentane	0.437	0.418	0.407	0.396	0.396	0.411	
N-Pentane	0.192	0.177	0.174	0.170	0.177	0.177	
Hexanes	0.730	0.667	0.633	0.608	0.587	0.600	
Heptanes plus	3.904	2.992	2.018	1.508	1.232	1.103	
Totals	7.872	6.816	5.761	5.191	4.902	4.968	
Deviation factor "Z" of well stream produced	0.942	0.910	0.876	0.867	0.882	0.911	
Calculated viscosity of well stream produced (CP)	0.0374	0.0317	0.0251	0.0206	0.0174	0.0152	
Well stream produced cumulated percentage	0.00	5.98	20.01	36.71	54.73	72.95	

Table 6.1.8   *Retrograde Condensation During Gas Depletion at 285°F [11]*

	Reservoir Liquid	
Pressure (PSIA)	(BBL./MMSCF of Dew Point Fluid)	(Volume* Percent)
4420 D.P. @ 285°F	0.00	0.00
4000	10.58	1.30
3250	81.47	9.98
2500	101.57	12.46
1750	103.69	12.70
1000	97.34	11.92

*Percent of reservoir hydrocarbon pore space @ dew point.

This procedure eliminates most of the problems associated with flash calculations in single-phase regions and yields excellent results in relatively few iterations inside the two-phase region. Some problems still occur when attempting flash calculations in the dense gas region.

Initial estimates of the phase composition must be made to initiate the flash calculation. Several procedures are available. It was found that a combination of the bubble point/dew point initial phase estimation procedures works quite well [12]. Set the vapor phase mole fraction of the highest component in the system at 1.0 and the liquid phase mole fraction of the heaviest component in the system at 1.0. All other mole fractions are set to 10^{-6}. This procedure is believed to be superior to the technique of basing the initial assumption of the phase composition on some noncomposition-dependent K value estimation procedure, particularly when a wide range of temperatures, pressures, component types, composition ranges, etc., is to be considered.

The estimated vapor and liquid phase compositions must be compared with the calculated phase compositions. Equations 6.1.19 and 6.1.23 describe this checking procedure. If restraints described by these equations for any component (in either phase) are not satisfied, the calculations must be repeated even though an acceptable value for L/F has been determined. Some feel that this detailed checking procedure is unnecessary. It probably is unnecessary for

Table 6.1.9 *Calculated Cumulative Recovery During Depletion [11]*

Cumulative Recovery Per MMSCF of Original Fluid	Reservoir Pressure — PSIA						
	Initial in Place	Dew Point Pressure 4420	4000	3250	2500	1750	1000
Well stream — MSCF	1000.00	0.0	59.80	200.10	367.10	547.30	729.50
Normal temperature separation* stock tank liquid — barrels							
Cumulative produced.	0.0	0.0	7.06	19.79	30.80	41.53	50.84
Remaining in vapor in res.	131.31	131.31	97.16	55.34	34.07	20.89	11.77
Remaining in liquid in res.	0.0	0.0	27.10	56.18	66.44	68.90	68.70
Primary sep. gas — MSCF	794.65	0.0	48.49	166.85	312.88	473.15	635.04
Second stage gas — MSCF	82.58	0.0	4.59	13.72	22.59	31.22	40.51
Stock-tank gas — MSCF	14.75	0.0	0.83	2.56	4.31	6.06	7.95
Total "plant products" in primary separator gas — gallons**							
Propane plus	1822.09	0.0	111.98	392.31	754.17	1161.68	1587.72
Butanes plus	811.70	0.0	50.35	179.85	353.88	554.50	764.40
Pentanes plus	250.79	0.0	15.72	57.75	117.78	189.38	265.12
Total "plant products" in well stream — gallons							
Propane plus	7872.05	0.0	439.19	1321.39	2235.80	3144.93	4043.82
Butanes plus	6451.15	0.0	354.64	1040.24	1722.29	2381.70	3019.14
Pentanes plus	5263.11	0.0	284.57	809.73	1303.52	1760.39	2186.72

*Primary separator at 915 PSIA and 95°F., second stage separator @ 65 PSIA and 70°F, Stock-tank at 15 PSIA and 75°F.
**Recover assumes 100 percent plant efficiency.
All gas volumes calculated at 15.025 PSIA and 60°F and stock-tank liquid measured at 60°F.

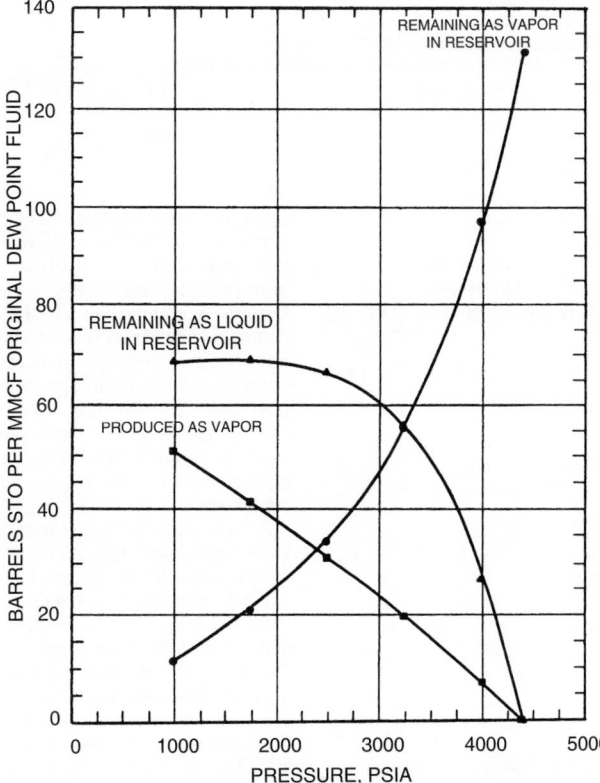

Figure 6.1.18 *Stock-tank liquid production and retrograde condensation during constant volume production at 285°F [11].*

Table 6.1.10 *Calculated Wellstream Yields* [11]*

Pressure, PSIA	4420 D.P.	4000	3250	2500	1750	1000
Gas–Liq. ratio**						
SCF 1st. Stg. gas						
BBL. S.T. Liq.	6052	7899	11,194	14,617	16,842	17,951
BBLS. S.T. Liq.						
MMSCF 1st Stg. gas	165.23	126.60	89.33	68.41	59.38	55.71
BBLS. S.T. Liq.						
MMSCF well stream Fld.	131.31	104.70	76.86	61.50	52.86	49.40

*Primary separator at 915 PSIA and 95°F., Second stage separator at 65 PSIA and 70°F, stock-tank at 15 PSIA and 70°F.
**All gas volumes calculated at 15.205 PSIA and 60°F and stock-tank liquid measured at 60°F.

Table 6.1.11 *Differential Vaporization at 220°F [66]*

Pressure PSIG	Solution Gas/Oil Ratio(1)	Relative Oil Volume(2)	Relative Total Volume(3)	Oil Density gm/cc	Deviation Factor Z	Gas Formation Volume Factor(4)	Incremental Gas Gravity
	R_{sd}	B_{od}					
2620	854	1.600	1.600	0.6562			
2350	763	1.554	1.665	0.6655	0.846	0.00685	0.825
2100	684	1.515	1.748	0.6731	0.851	0.00771	0.818
1850	612	1.479	1.859	0.6808	0.859	0.00882	0.797
1600	544	1.445	2.016	0.6889	0.872	0.01034	0.791
1350	479	1.412	2.244	0.6969	0.887	0.01245	0.794
1100	416	1.382	2.593	0.7044	0.903	0.01552	0.809
850	354	1.351	3.169	0.7121	0.922	0.02042	0.831
600	292	1.320	4.254	0.7198	0.941	0.02931	0.881
350	223	1.283	6.975	0.7291	0.965	0.05065	0.988
159	157	1.244	14.693	0.7382	0.984	0.10834	1.213
0	0	1.075		0.7892			2.039
		@ 60°F. = 1.000					

Gravity of residual oil = 35.1°API @ 60°F.

(1) Cubic feet of gas at 14.65 PSIA and 60°F per barrel of residual oil at 60°F.
(2) Barrels of oil at indicated pressure and temperature per barrel of residual oil at 60°F.
(3) Barrels of oil plus liberated gas at indicated pressure and temperature per barrel of residual oil at 60°F.
(4) Cubic feet of gas at indicated pressure and temperature per cubic foot at 14.65 PSIA and 60°F.

Table 6.1.12 *Separator Test [8]*

Separator Pressure (psig)	Temperature (°F)	GOR, R_{sfb}^*	Stock-Tank Oil Gravity (°API at 60°F)	FVF, B_{ofb}^{**}
50 to 0	75	737		
	75	41	40.5	1.481
		778		
100 to 0	75	676		
	75	92	40.7	1.474
		768		
200 to 0	75	602		
	75	178	40.4	1.483
		780		
300 to 0	75	549		
	75	246	40.1	1.495
		795		

*GOR in cubic feet of gas at 14.65 PSIA and 60°F per barrel of stock-tank oil at 60°F.
**FVF is barrels of saturated oil at 2620 PSIG and 220°F per barrel of stock-tank oil at 60°F.

most problems involving moderate temperature–pressure-composition conditions. However, at extreme condition of temperature, pressure, and composition (low-temperatures, high pressure, high-acid-gas compositions) failure to perform these composition checks will lead to results that are completely incorrect (poor estimates of the phase compositions and incorrect L/F ratios). Unfortunately, the boundary changes in temperature, pressure, or composition can completely alter the difficulty of a given problem. Consequently, careful application of these checks in all calculations is strongly recommended since one can never be sure that a particular problem will not fall into the area of extreme conditions.

A logic diagram illustrating the basic flash calculation is shown in Figure 6.1.28. All the necessary features described earlier are embodied in this diagram.

Flash calculations at fixed L/F and temperature or pressure are frequently necessary. In these calculations, the dependent variable becomes pressure or temperature, and the flash calculation becomes similar in principle to a bubble point or dew point calculation. The flash calculation equation described earlier, Equation 6.1.23, can be coupled with the temperature or pressure adjusting algorithms described forth bubble point/dew point calculations to perform these calculations. Initial estimates of the vapor- or liquid-phase

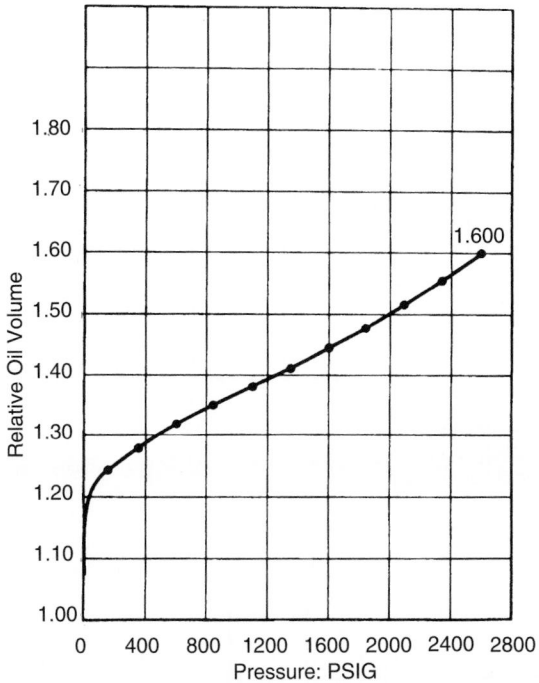

Figure 6.1.19 *Adjustment of oil relative volume curve to separator conditions [6].*

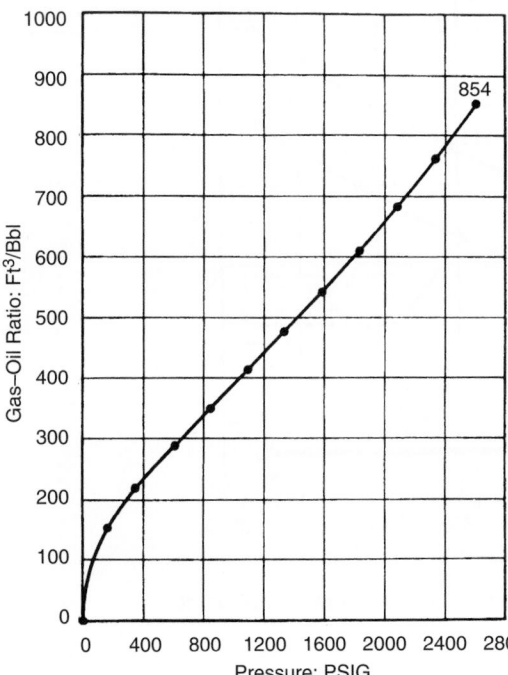

Figure 6.1.20 *Adjustment of gas in solution curve to separator conditions [6].*

compositions must be made, and the approach described in the flash at fixed temperature–pressure conditions can be used quite effectively. The logic diagram for this type of calculation can be deduced from earlier diagrams.

6.1.2.4 Predicting the Properties of Hexane Plus (C_{6+}) Fractions

Physical properties of light hydrocarbons are given in Table 6.1.1. In naturally occurring gas and oil, C_{6+} is unknown and makes a problem. Since the C_{6+} is a combination of paraffins (P), napthenes (N), and aromatics (A) of varying molecular mass (M), these fractions must be defined or characterized in some way. Changing the characterization of C_{6+} fractions present in even small amounts (at 1.0% mole level) can have significant effect on the predicted phase behavior of a hydrocarbon system. The dew point of the gas is heavily dependent upon the heaviest components in the mixture.

The SRK (Equation 6.1.11) and the PR (Equation 6.1.13) require the smallest number of parameters of any of the equations of state. They require the critical temperature, the critical pressure, and the acentric factor. There are many different approaches that can be utilized to predict these parameters for C_{6+} fractions or other mixtures of undefined components.

Some minimum of information must be available on the C_{6+} fraction, usually it is specific gravity (S), average boiling point (T_b), and molecular mass (M) of the fraction.

The following equation is used [14] to estimate the molecular mass (M) of petroleum fractions

$$M = 2.0438 \times 10^2 \exp(0.00218T) \exp(-3.07S)T^{0.118}S^{1.88}$$
$$[6.1.27]$$

where T = mean average boiling point of a petroleum fraction, °R (from ASTM D86 test, Figure 6.1.29)

S = specific gravity, 60°F/60°F

The following equation is to be used to calculate the initial temperature (T_c) of pure hydrocarbons; it is applicable for all families of hydrocarbons:

$$\log T_c = A + B \log S + C \log T_b \qquad [6.1.28]$$

T_c (°R); T_b and S given; A, B, and C as below:

Type Compound	A	B	C
Paraffin	1.47115	0.43684	0.56224
Napthene	0.70612	−0.07165	0.81196
Olefin	1.18325	0.27749	0.65563
Acetylene	0.79782	0.30381	0.79987
Diolefin	0.14890	−0.39618	0.99481
Aromatic	1.14144	0.22732	0.66929

For petroleum factions, physical properties can be predicted more accurately if the fraction of paraffins (P), napthenes (N), and aromatics (A) are known. If θ is a physical property to be predicted and the molecular type fractions are known, a pseudocompound, i.e., a compound having the same boiling point and specific gravity as the fraction, for each molecular type can be defined. These properties can be combined by

$$\theta = P\theta_P + N\theta_N + A\theta_A \qquad [6.1.29]$$

where $P + N + A = 1.0$, are the sum of paraffins, napthenes, and aromatics present in the fraction.

The basic equation for any value of property θ is [15]

$$\theta = aT_b^b S^c \qquad [6.1.30]$$

Correlation constants are given in Table 6.1.13.

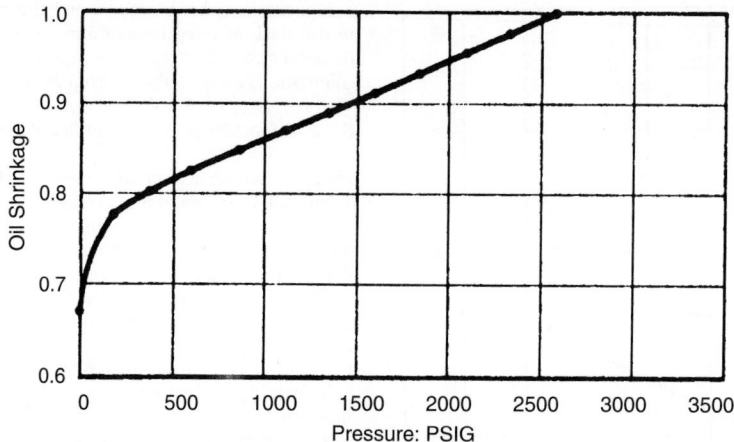

Figure 6.1.21   *Oil-shrinkage curve [8].*

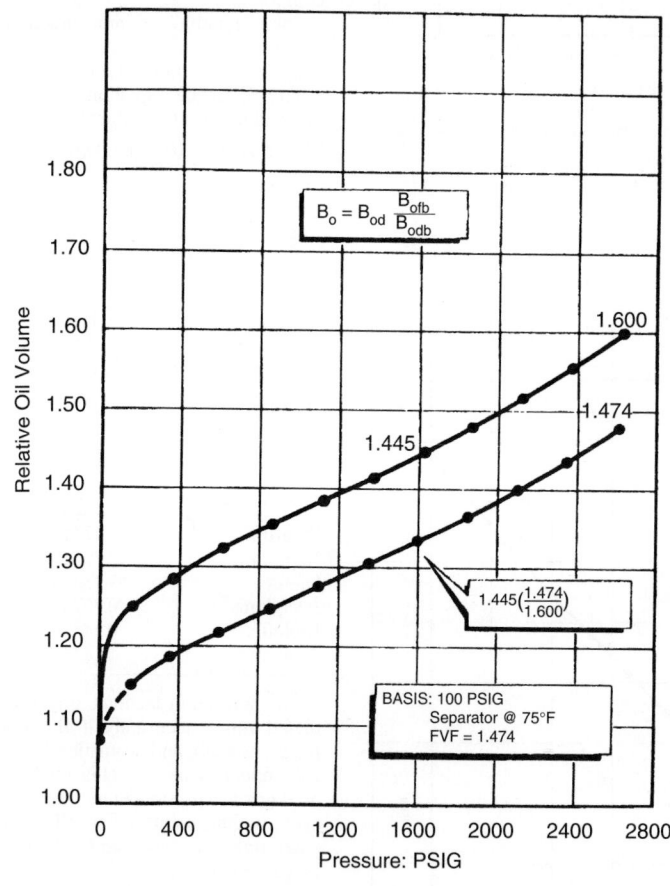

Figure 6.1.22   *Adjustment of oil relative volume curve to separator conditions [6].*

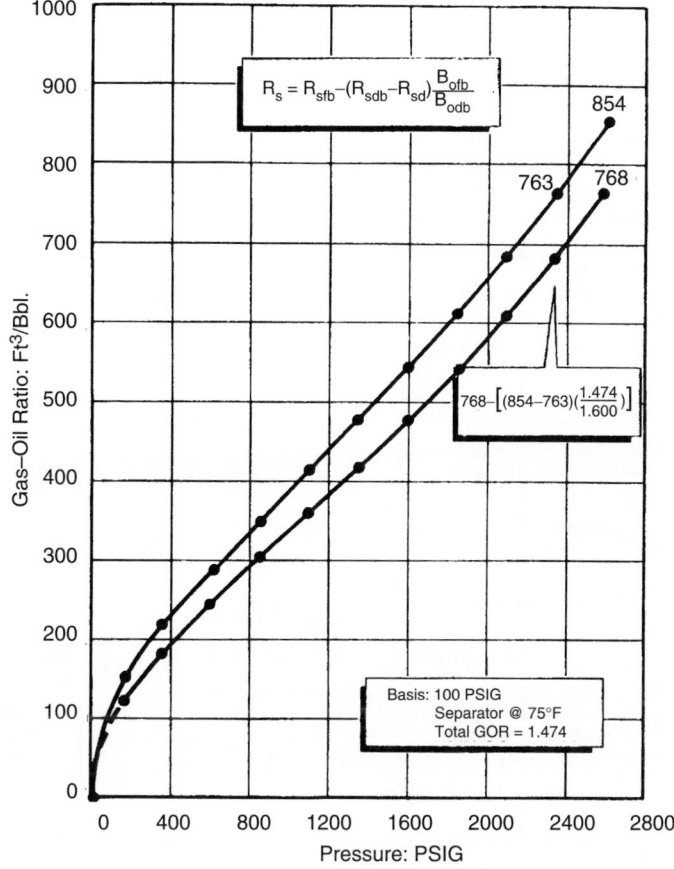

Figure 6.1.23   *Adjustment of gas in solutions curve to separator conditions [6].*

Next, correlation of critical properties and acentric factor (ω) of hydrocarbons and derivatives is developed in terms of M, T_b, and S. Since molecular mass is readily determined by experiments, it is introduced as a correlating variable to obtain more general results. The specific gravity is the ratio of the density of the liquid at 20°C to that of water at 4°C.

The critical properties and acentric factor of C_1 to C_{20} n-alkanes are correlated with M by the following equation:

$$\theta_A = C_1 + C_2 M + C_3 M^2 + C_4 M^3 + C_5/M \qquad [6.1.31]$$

where θ_A represents T_c, $L_n P_c$, V_c, or ωT_c of a n-alkane. The coefficients C_1 to C_5 are reported in Table 6.1.14 for each property. It was additionally correlated S and T_b of the n-alkanes by Equation 6.1.31, and the coefficients are included in Table 6.1.14. The correlated S_A and T_{bA} of the n-alkanes will be required in the perturbation equations to follow as independent variables.

The average absolute deviations (AAD) of the correlations from the American Petroleum Institute project 44 table values are 0.15% for T_c, 1.0% for P_c (excluding methane), 0.8% for V_c, 1.2% for ω, 0.11% for T_b, and 0.07% for S. The specific gravity correlation applies only to C_5–C_{16}, which are the only n-alkanes that are liquids at 20°C.

Properties of the general hydrocarbons and derivatives are correlated as perturbations of those n-alkanes according to the equation

$$\theta = \theta_A + A_1 \Delta S + A_2 \Delta T_b + A_3 (\Delta S)^2 + A_4 (\Delta S)(\Delta T_b) + A_5 (\Delta T_b)^2$$
$$+ A_6 (\Delta S)^3 + A_7 (\Delta S)^2 (\Delta T_b) + A_8 (\Delta S)(\Delta T_b)^2 + A_g (\Delta T_b)^3 \qquad [6.1.32a]$$

with

$$\Delta S = S - S_A \qquad [6.1.32b]$$

$$\Delta T_b = T_b - T_{bA} \qquad [6.1.32c]$$

where S_A and T_{bA} are the gravity and boiling point of the hypothetical n-alkane of the M of the substance of interest and are given by Equation 6.1.31.

The coefficients A_i in Equation 6.1.32a are given by

$$A_i = a_i + b_i M \qquad [6.1.32d]$$

Table 6.1.15 presents the coefficients a_i and b_i that have been determined by fitting Equation 6.1.32d to the properties of a large number of hydrocarbons and derivatives.

State-of-the-art Equation 6.1.32 gives the best results.

The five general categories of experimental data availability for C_{6+} fractions are:

1. *The specific or API gravity of the C_{6+} fraction.* The molecular mass may have been determined experimentally or estimated from some correlation of specific gravity and molecular mass.

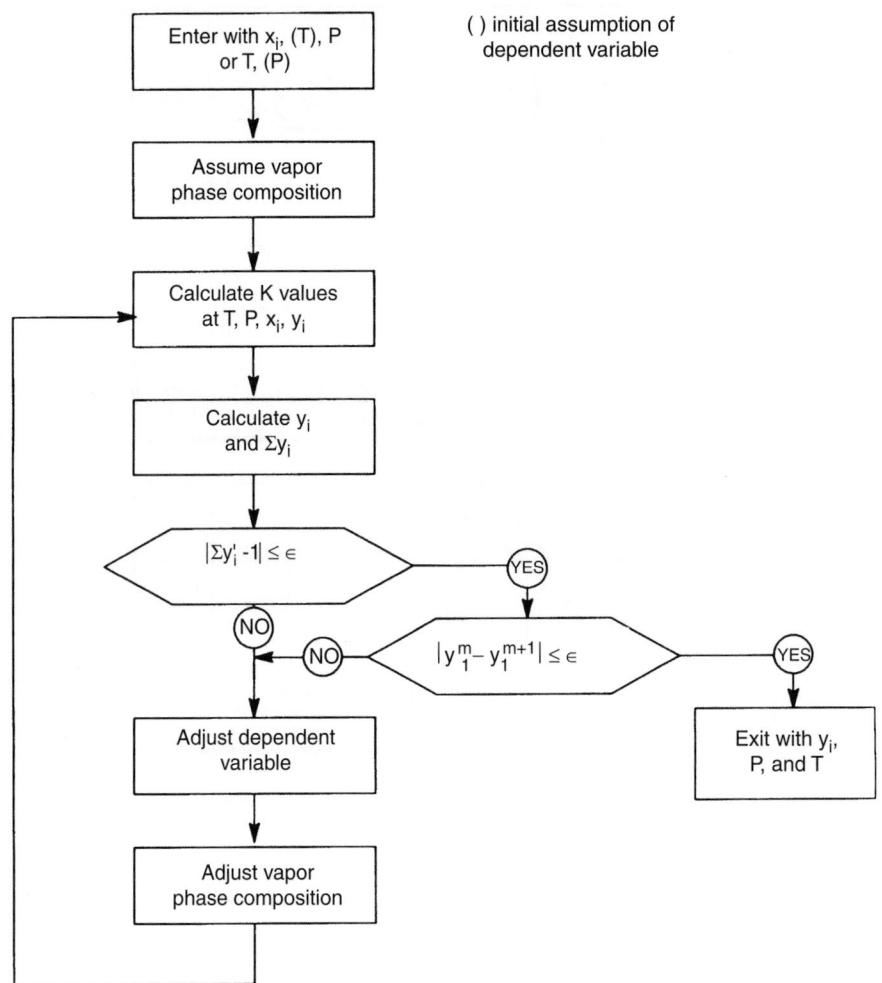

Figure 6.1.24 *Block diagram for bubble point calculations.*

2. *Chromatographic analysis.* The C_{6+} fraction has been analyzed by gas–liquid chromatography. These results may be reported as a series of equivalent n-paraffins up to as high as nC_{30} or as a true boiling point analysis. The specific gravity and/or molecular mass of the fraction may or may not be reported.
3. *ASTM-D158 (or equivalent) analysis.* This analysis is equivalent to a nonrefluxed single-stage batch distillation. Usually the boiling point temperature at seven different points (START, 10%, 30%, 50%, 70%, 90%, END) will be recorded. The points correspond to the volume fraction of the C_{6+} distilled into the receiver vessel. The specific gravity and molecular mass of the total C_{6+} fraction are normally also measured.
4. *A partial TBP analysis.* A true boiling point (TBP) distillation has been performed on the C_{6+} fraction. The TBP distillation is a batch distillation similar to an ASTM distillation but the distillation apparatus contains several trays (usually 10 or more or the equivalent amount of packing) and a high reflux ratio is used. The TBP gives a sharper separation between the subfractions than an

ASTM distillation. Normally, at least five temperatures are reported as a function of liquid volume percent distilled over. Frequently, more than 20 temperatures will be reported. The specific gravity and molecular mass of the total fraction are usually reported.
5. *A complete TBP analysis.* A true boiling point distillation has been performed on the total C_{6+} fraction. The specific gravity and molecular mass have been measured for each of the reported distillate subfractions. Between 5 and 50 temperatures and subfraction properties will be reported.

Table 6.1.15 shows typical information as it may be reported for each of the five categories of C_{6+} characterization. The complete TBP analysis is believed to be the best form of C_{6+} analysis to be used with today's thermodynamic property prediction procedures. Consequently, it is recommended that all noncomplete TBP analyses be converted to this form. This section deals with these conversion techniques. These techniques are based on empirical correlations and, in some cases, experience and judgment. There is also one basic constraint that must be used in these

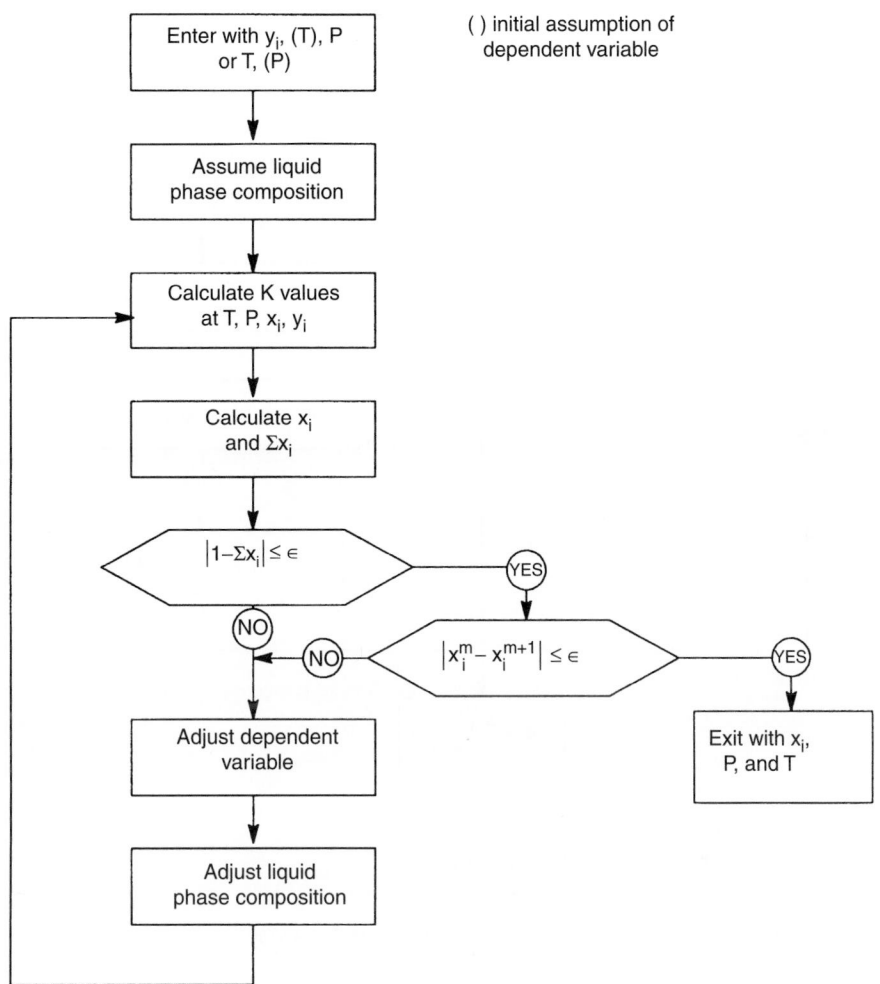

Figure 6.1.25 *Block diagram for dew point calculations.*

conversion techniques — that is, maintenance of volume-mass-molar relationships in the C_{6+} fraction along with consistency in the composition of the total stream. One cannot capriciously change the molecular mass or specific gravity of the total C_{6+} fraction without simultaneously adjusting the reported composition. All of the procedures reported here strive to maintain consistency of the specific gravity, molecular mass and, when possible, the boiling point (s) of the total C_{6+} fraction.

The various procedures for converting noncomplete TBP analyses to complete TBP analyses are illustrated in the following section. A common sample problem is used to illustrate the basic conversion procedure. In addition, the results of several equilibrium calculations are reported for each type of characterization. The gas composition, true boiling point date, gravity, and molecular weight measurements for the C_{7+} fraction are shown in Table 6.1.16. Though the particular system chosen shown C_{7+} as a basis for the heavy and characterization C_{6+} will be used. There are several isomers of hexane, as well as other materials, that can appear in the C_{6+} subfraction. The molecular mass tabulated for the fractions in Table 6.1.16 makes them appear to be normal paraffins.

This, however, is not true and a complete TBP analysis was made on the C_{7+} fraction.

Calculations made based on the different C_{7+} characterizations are compared with experimental values, Table 6.1.17 and Figure 6.1.30. The complete TBP characterization provides the best predictions of the phase behavior and the liquid formation, though there is only a little difference between the full TBP and the partial TBP results. The lumped specific gravity-molecular mass characterization and the lumped n-paraffin characterization give the poorest predictions. All of the characterizations in Table 6.1.18 are in better agreement with experimental values than one would normally expect.

6.1.3 Vapor-Liquid Equilibrium by Equation of State

Prediction of a vapor-liquid mixture is more complicated than prediction of pure component VLE.

The following condition equations for mixture VLE can be derived from classical thermodynamics:

$$T^V = T^L \hspace{3cm} [6.1.33]$$

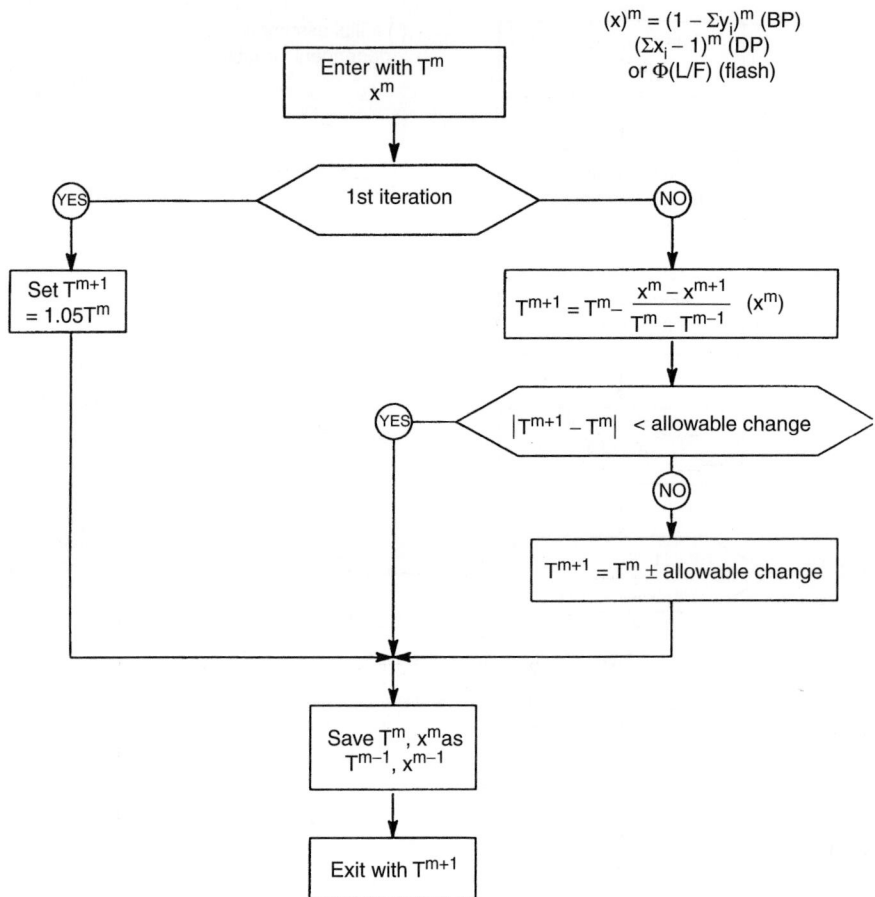

Figure 6.1.26   *Temperature adjustment diagram for equilibrium calculations.*

$$P^V = P^L \qquad [6.1.34]$$

$$f_i^v = f_i^L \qquad [6.1.35]$$

What does "f_i" mean?

The equilibrium constants, or K values, are defined as the ratio of vapor and liquid compositions:

$$K_i = \frac{y_i}{x_i} \qquad [6.1.36]$$

The Gibbs free energy is a property of particular importance because it can be related to the equilibrium state and at the same time can be expressed as a function of T and P:

$$dG_i = -SdT + VdP \qquad [6.1.37]$$

If Equation 6.1.37 is applied to an ideal gas it becomes

$$dG_i = RT\,d(\ln P) \qquad [6.1.38]$$

If we now define a property such that Equation 6.1.38 will apply for all gases under all conditions of temperature and pressure

$$dG_i = RT\,d(\ln f_i) \qquad [6.1.39]$$

f_i is called the fugacity of component "i" and has units of pressure. If Equation 6.1.39 is integrated for an ideal gas:

$$f_i = cP \qquad [6.1.40]$$

"c" is constant and for an ideal gas is equal to 1.0. For real gases the only condition under which the gas will behave ideally is at zero pressure. This can be expressed as

$$\lim_{p \to 0} \frac{f_i}{P} = 1.0 \qquad [6.1.41]$$

The fugacity of a single component in a mixture is defined in a manner similar to Equation 6.1.36

$$dG = RT\,d(\ln f_i) \qquad [6.1.42]$$

and by analogy to the Equation 6.1.39

$$\lim_{p \to 0} \frac{f_i}{x_i P} = 1.0 \qquad [6.1.43]$$

The fugacity is sometimes referred to as a *corrected pressure*. A more valuable parameter for use in correlative procedures would be a variable with characteristics similar to fugacity, but which ranged over a much smaller range of numbers. It is the fugacity coefficient ϕ_i.

For a pure component

$$\phi_i = \frac{f_i}{P} \qquad [6.1.44]$$

For a component in a mixture

$$\phi_i = \frac{f_i}{x_i P} \qquad [6.1.45]$$

The fugacity coefficients are readily calculated form P–V–T data for both, pure component and mixture.

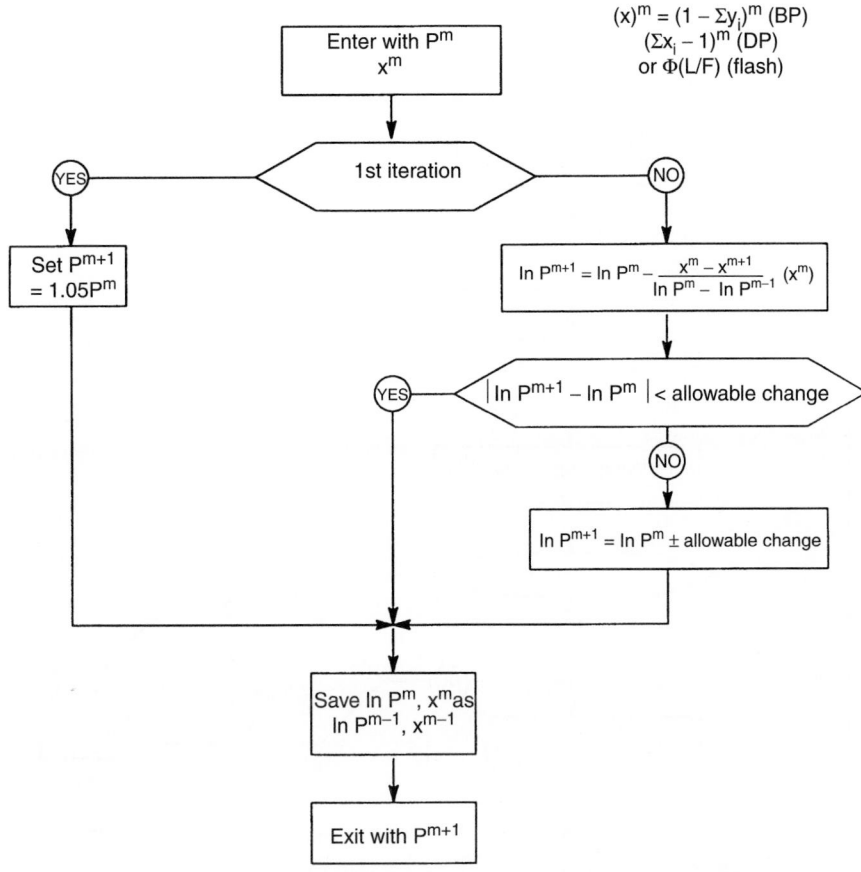

Figure 6.1.27 *Pressure adjustment diagram for equilibrium calculations.*

P–V–T of a fixed composition system can be developed in pressure explicit forms, i.e.,

$$P = \phi(V, T) \qquad [6.1.46]$$

The basic equation of state may be transformed to a compressibility factor Z, but the basic expression given by Equation 6.1.44 still applies. Equilibrium K values are predicted from fugacity, which is related to Gibbs free energy. One of the basic definitions of fugacity gases K values is

$$K = \frac{y_i}{x_i} = \frac{\phi_i^L}{\phi_i^v} = \frac{f_i^L/(Px_i)}{f_i^v/(Py_i)} \qquad [6.1.47]$$

The fugacity coefficient ϕ_i is related to the pressure, volume and temperature by

$$\ln \phi_i = -\frac{1}{RT} \int_V^\infty \left[\left(\frac{\partial P}{\partial n_i} \right)_{T,V,n_j} - \frac{RT}{V} \right] dV - \ln Z \qquad [6.1.48]$$

Applying the SRK equation

$$\ln \phi_i = -\ln\left(Z - \frac{Pb}{RT} \right) + (Z-1)B_i' - \frac{a}{bRT}(A_i' - B_i')\ln\left(1 + \frac{b}{V} \right) \qquad [6.1.49]$$

or

$$\ln \phi_i = -\ln(Z - B) + (Z-1)B_1' - \frac{A}{B}(A_i' - B_i')\ln\left(1 + \frac{B}{Z} \right) \qquad [6.1.50]$$

where

$$B_i = \frac{b_i}{b} \qquad [6.1.51]$$

$$A_i' = \frac{1}{a}\left[2a_i^{0.5} \sum_j^N x_j a_j^{0.5}(1 - k_{ij}) \right] \qquad [6.1.52]$$

Notations are as for Equations 6.1.11 and 6.1.12. For values k_{ij} see Tables 6.1.19 and 6.1.20. The data in Table 6.1.18 was prepared for the Peng-Robinson equation, while the data from Table 6.1.20 was used for the Benedict-Webb-Rubin equation modified by Starling [16].

Applying Peng-Robinson equation

$$\ln \phi_i = -\ln\left(Z - \frac{Pb}{RT} \right) + (Z-1)B_i' - \frac{a}{2^{1.5}bRT}$$

$$\times \left(A_i' - B_i' \right) \ln\left[\frac{V + (2^{0.5} + 1)b}{V - (2^{0.5} - 1)b} \right] \qquad [6.1.53a]$$

or

$$\ln \phi_i = -\ln(Z - B) + (Z-1)B_1' - \frac{A}{2^{1.5}B}$$

$$\times \left(A_i' - B_i' \right) \ln\left[\frac{Z + (2^{0.5} + 1)B}{Z - (2^{0.5} - 1)B} \right] \qquad [6.1.53b]$$

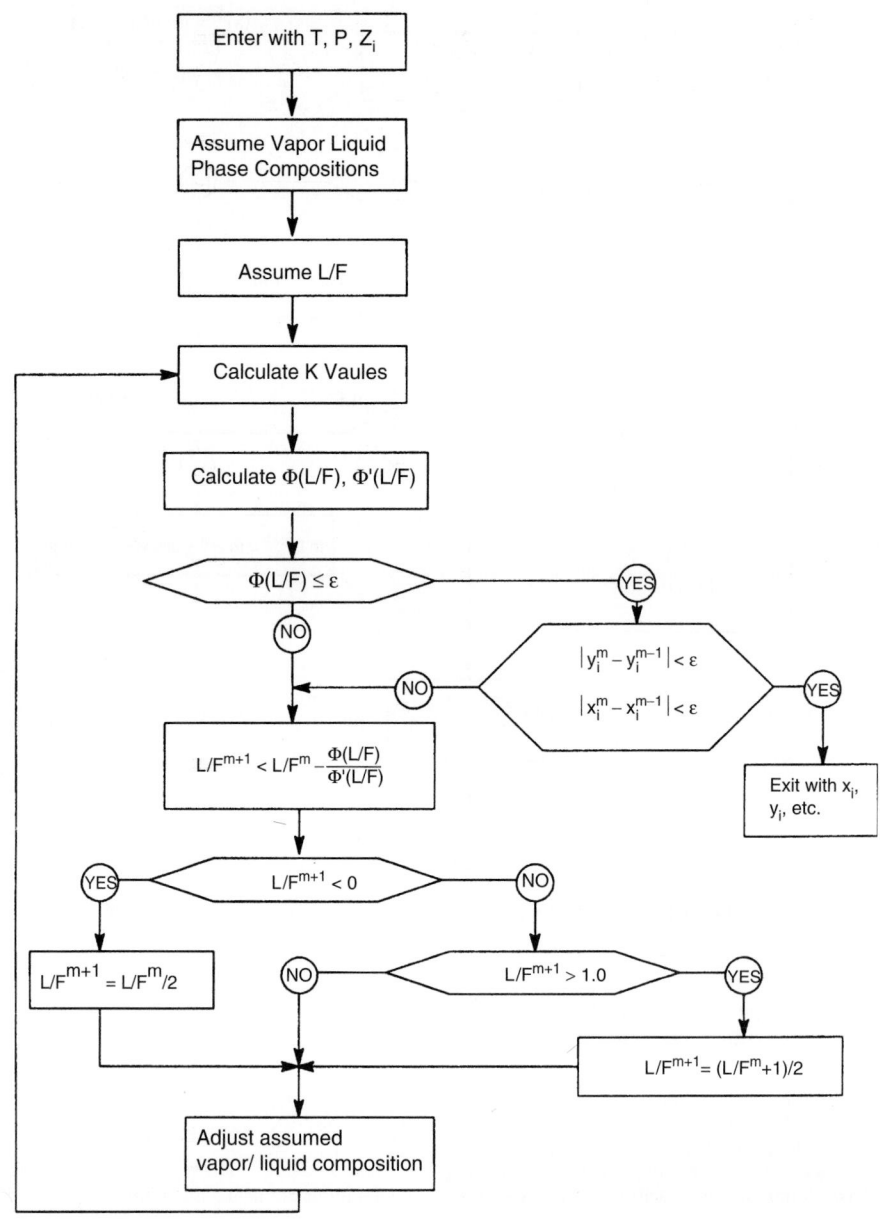

Figure 6.1.28 *Block diagram for two-phase equilibrium calculations.*

where

$$B_i = \frac{b_i}{b}$$ [6.1.54]

$$A_i' = \frac{1}{a}\left[2a_i^{0.5}\sum_{j}^{N}x_j a_j^{0.5}(1-k_{ij})\right]$$ [6.1.55]

Notations are as in Equation 6.1.13 and Equation 6.1.14.

The objectives of any equation-of-state solution method are the reliable and accurate prediction of the volumetric properties of the fluid mixture under consideration. The overall solution procedure is as follows:

- fix the total composition, temperature and pressure
- calculate constant for the equation of state
- solve equation for the volumetric property (specific volume, density or compressibility factor)

When pressure and temperature fall to a two-phase region, the equation must be solved twice, separately for vapor and liquid. The composition of each phase will be different, so the equation of state constants will have to be evaluated for

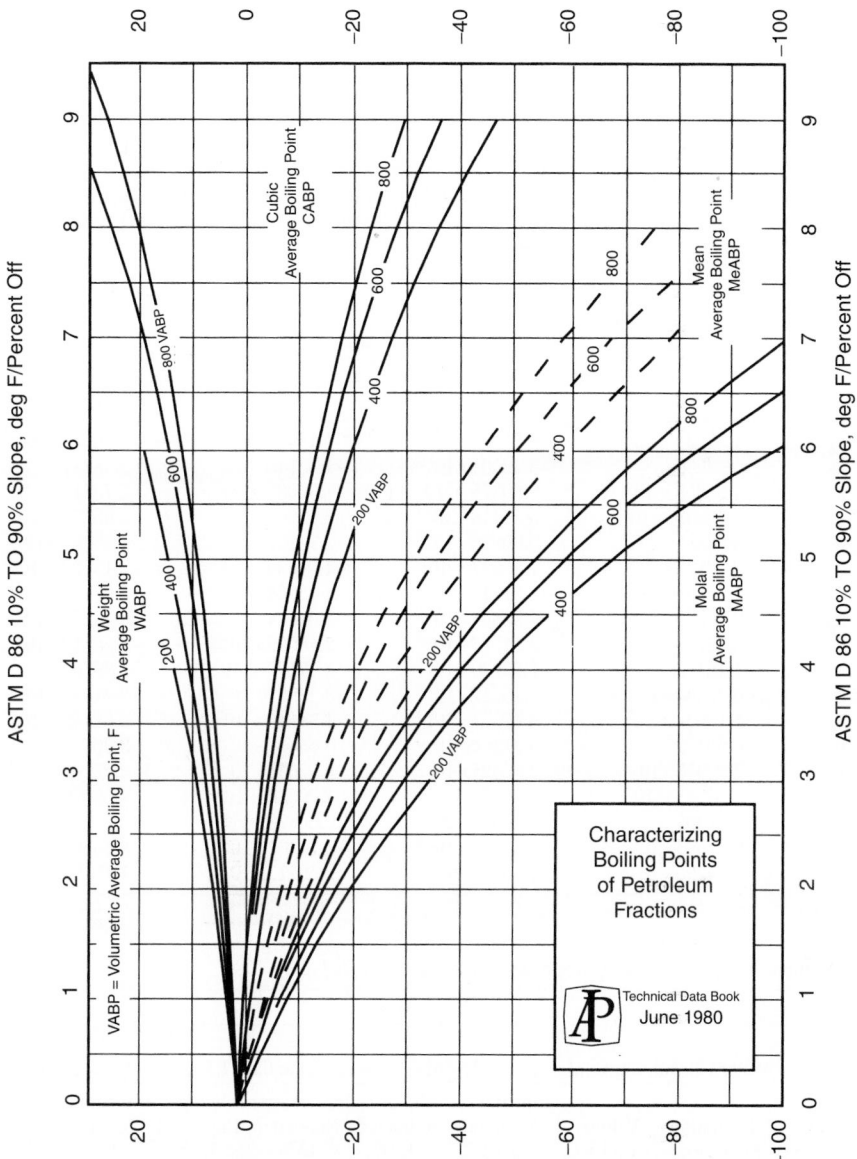

Figure 6.1.29 *Characterizing boiling points of petroleum fractions.*

Table 6.1.13 *Correlation Constants for Equation 6.1.30*

θ	a	b	c
M (Molecular mass)	4.5673×10^{-5}	2.1962	−1.0164
T_c (Critical Temperature °R)	24.2787	0.58848	0.3596
P_c (Critical Pressure, psia)	3.12281×10^9	−2.3125	2.3201
V_c^m (Molar critical volume ft.^{3}/lb mole)	7.0434×10^{-7}	2.3829	−1.683
V_c (Critical Volume ft.^{3}/lb)	7.5214×10^{-3}	0.2896	−0.7666
V (Liquid molar volume at 20°C and 1 atm (cm^3/g mole))	7.6211×10^{-5}	2.1262	−1.8688
ρ (Liquid density g/cc)	0.982554	0.002016	1.0055

Table 6.1.14   Coefficients for θ_A in Equation 6.1.31

θ_A	C_1	C_2	C_3	C_4	C_5
T_c	2.72697×10^2	3.91999	-1.17706×10^{-2}	1.48679×10^{-5}	-2.27789×10^3
$\ln P_c$	1.77645	-1.01820×10^{-2}	2.51106×10^{-5}	-3.73775×10^{-8}	3.50737
V_c	1.54465×10	4.04941	1.73999×10^{-4}	1.05086×10^{-6}	2.99391×10^2
ωT_c	-1.56752×10	1.22751	9.96848×10^{-3}	-2.04742×10^{-5}	-6.90883×10
S	6.64050×10^{-1}	1.48130×10^{-3}	-5.07021×10^{-6}	6.21414×10^{-9}	-8.45218
T_b	1.33832×10^2	3.11349	-7.08978×10^{-3}	7.69085×10^{-6}	-1.12731×10^3

T_c and T_b are in K, P_c in MPa, V_e in cm^3/mol.

Table 6.1.15   Coefficients for Equation 6.1.32d

			θ	
	T_c	$\ln P_c$	V_c	ωT_c
a_1	1.58025×10^3	9.71572	-1.18812×10^3	-1.16044×10^3
a_2	-5.68509	-3.32004×10^{-2}	-1.18745	3.48210
a_3	-1.21659×10^4	-8.60375×10	7.36085×10^3	2.78317×10^4
a_4	7.50653×10	5.50118×10^{-1}	6.83380×10	-2.05257×10^2
a_5	-9.66385×10^{-2}	-9.00036×10^{-4}	-2.12107×10^{-1}	4.55767×10^{-1}
a_6	2.17112×10^4	1.85927×10^2	-4.84696×10^3	-7.13722×10^4
a_7	-1.57999×10^2	-1.51115	-4.12064×10^2	5.08888×10^2
a_8	3.60522×10^{-1}	4.32808×10^{-3}	2.02114	-6.10273×10^{-1}
a_9	-2.75762×10^{-4}	-3.81526×10^{-6}	-2.48529×10^{-3}	-1.68712×10^{-3}
b_1	-1.18432×10	-7.50370×10^{-2}	1.17177×10	1.89761
b_2	5.77384×10^{-2}	3.15717×10^{-4}	-3.48927×10^{-2}	2.41662×10^{-2}
b_3	1.10697×10^2	8.42854×10^{-1}	-1.34146×10^2	-2.67462×10^2
b_4	-6.58450×10^{-1}	-5.21464×10^{-3}	5.63667×10^{-2}	2.06071
b_5	7.82310×10^{-4}	7.87325×10^{-6}	9.52631×10^{-4}	-5.22105×10^{-3}
b_6	-2.04245×10^2	-1.85430	1.80586×10^2	7.66070×10^2
b_7	1.32064	1.36051×10^{-2}	2.56478	-5.75141
b_8	-2.27593×10^{-3}	-3.23929×10^{-5}	-1.74431×10^{-2}	8.66667×10^{-3}
b_9	8.74295×10^{-7}	2.18899×10^{-8}	2.50717×10^{-5}	1.75189×10^{-5}

Table 6.1.16   Sample Analysis for Five Categories of C_{6+} Analysis [12]

				Categories					
1 Specific Gravity		2* Molar Chromatographic		3 ASTM		4 Partial TBP		5 Complete TBP	
Information Reported	Values Reported	Information Reported	Values mol %	Information Reported LV%	Values Reported T, °F	Information Reported LV%	Values Reported T, °F	Information Reported LV%	Values Reported T, °F sp gr mol wt
specific	0.7268	C_6	0.335						
gravity		C_7	0.327	ST	258	ST	155	ST	155
(°API)	(63.2)	C_8	0.341	10	247	S	190	17.52	238 0.745 100
molecular		C_9	0.268	30	283	10	212	33.12	280 0.753 114
mass	104	C_{10}	0.166	50	331	20	246	·	· · ·
				70	403	30	270	·	· · ·
				90	500	40	304	·	· · ·
		C_{33}	0.004	GP	596	50	348	·	· · ·
		C_{34}	0.004			60	380	·	· · ·
		C_{35+}	0.006	specific		70	431	·	· · ·
		specific		gravity °API	0.7867	80	481	·	· · ·
		gravity	?		48.37	90	538	·	· · ·
						95	583	99.73	· · ·
		molecular		molecular		EP	700	EP	698 0.878 310
		mass	?	mass	141.26	Specific			
						gravity	0.7867		
						molecular			
						mass	141.26		

*Chromatographic TBP will be similar to ASTM or partial TBP

Table 6.1.17 *Experimental Data for Illustrative Calculations [12]*

Component	mol %	BPT °F	sp gr	mol wt
C_1	91.35			
C_2	4.03			
C_3	1.53			
iC_4	0.39			
nC_4	0.43			
IC_5	0.15			
nC_5	0.19			
C_6	0.39			
Fraction 7	0.361	209	0.745	100
8	0.285	258	0.753	114
9	0.222	303	0.773	128
10	0.158	345	0.779	142
11	0.121	384	0.793	156
12	0.097	421	0.804	170
13	0.083	456	0.816	184
14	0.069	488	0.836	198
15	0.050	519	0.840	212
16	0.034	548	0.839	226
17	0.023	576	0.835	240
18	0.015	603	0.850	254
19	0.010	628	0.865	268
20	0.006	653	0.873	282
21	0.004	676	0.876	296
22	0.002	698	0.878	310

mol % C_{7+} = 1.540

mol wt C_{7+} = 141.26

sp gr C_{7+} = 0.7867 (43.35 °API)

Phase Behavior Data

dew point at 201°F–3837 psia

Liquid Formation at 201°F

Pressure psia	Bbl/MMSCF	Specific Gravity Liquid
2915	9.07	0.6565
2515	12.44	0.6536
2015	15.56	0.6538
1515	16.98	0.6753
1015	16.94	0.7160
515	15.08	0.7209

both the liquid and the vapor phases. Both SRK and PR are cubic equations, so the solution always gives three roots, as shown in Figure 6.1.31 [17]. However, the P_r–V_r relationship at a given T_r is discontinuous at $V_r = b'_1$, $V_r = b'_2$, and $V_r = b'_3$. We are interested in only $V_r > b'_1$, which in case the SRK equation is equal 0.08664 and 0.077796 for the PR equation. For $V_r > b'_1$ and $T_r > 1.0$, there is only one value of the compressibility factor that will satisfy the equation of state. For $V_r > b'_1$ and $T_r < 1.0$, we will get three values of Z. The largest Z of the vapor Z's is chosen for the vapor and the smallest amount the liquid Z's is chosen for the liquid. However, in an earlier stage of the iterative VLE calculations, it is not uncommon to encounter a single root, mainly because of incorrect compositions [17].

A logic diagram for a trial-and-error solution procedure for cubic equations of state is given in Figure 6.1.32. This diagram shows a traditional Newton-Raphson approach with an interval halving limiting procedure superimposed on it [18]. For purposes of the procedure for locating the boundaries of the liquid- and vapor-phase compressibility factors in

the two-phase region discussion, assume that the equation of state is given in the form

$$Z^3 + \alpha Z^2 + \beta z + \gamma = 0 \qquad [6.1.56]$$

where γ and β are arbitrary constants.

From Figure 6.1.33a–d in the two-phase region the equation of state will have a maximum and a minimum or two points at which the slope of the equation is zero. The value of the compressibility factor at the maximum defines the largest possible value for the liquid-phase compressibility factor. If Equation 6.1.56 is differentiated and set equal to zero the following equation results:

$$Z^2 + \frac{2\alpha Z}{3} + \frac{\beta}{3} = 0 \qquad [6.1.57]$$

Solving the above equation for the values of Z at the maximum and minimum gives

$$Z_{1,2} = \frac{-2\frac{\alpha}{3} \pm \sqrt{4\frac{\alpha^2}{9} - 4\frac{\beta}{3}}}{2} \qquad [6.1.58]$$

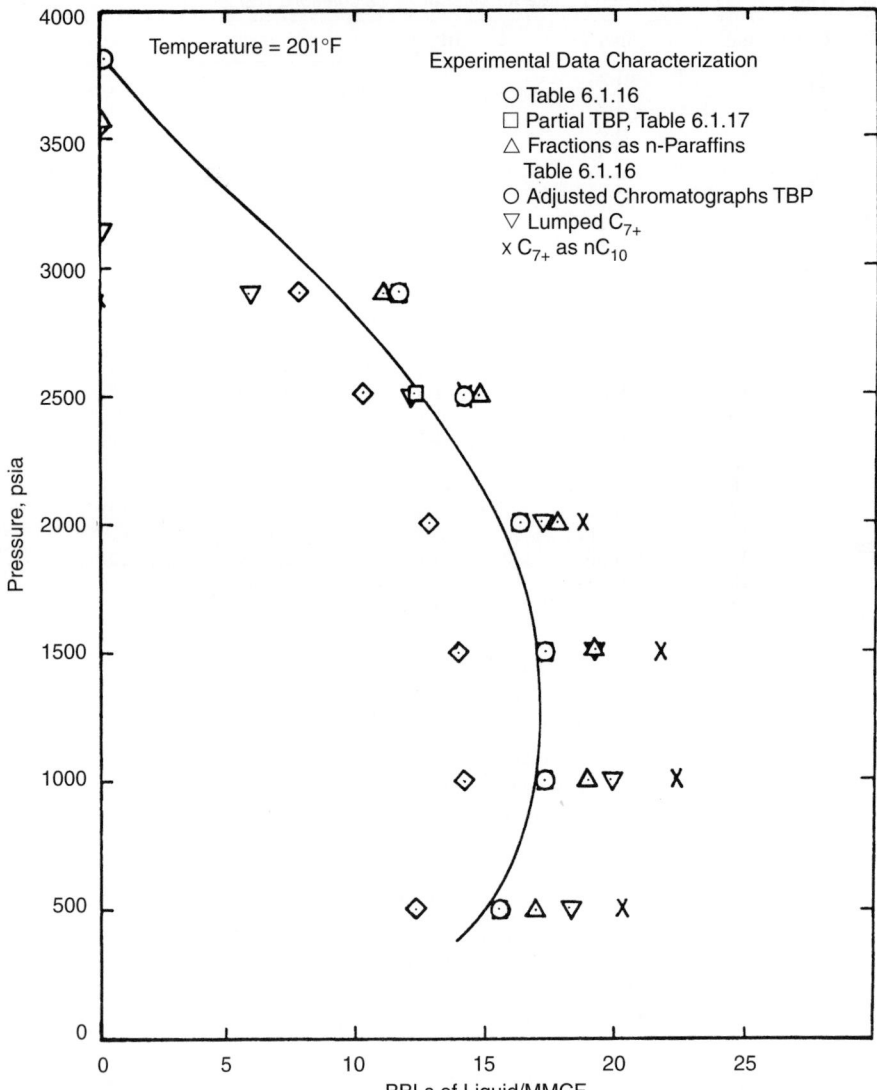

Figure 6.1.30 Effect of C_{6+} Characterization on predicted liquid formation [12].

Table 6.1.18 Effect of C_{7+} Characterization on Predicted Sample Problem Phase Behavior [12]

				C_{7+} Predicted Values for Characterization Chromatographic			
	Experiment Values	C_{7+} from Table 6.1.16	Partial TBP	Fractions as n-paraffins	Fraction from ASTM dest. curve	Lumped C_{7+}	C_{7+} as nC_{10}
Dew Point at 201°F, psig	3822	3824	3800	3583	3553	3150	2877
Amount of liquid at BBL/MMSCF							
2900 psig	9.07	11.57	11.50	11.07	7.37	5.95	0.0
2500 psig	12.44	14.26	14.22	14.89	10.21	12.40	14.09
2000 psig	15.56	16.43	16.44	17.77	12.66	17.09	18.77
1500 psig	16.98	17.48	17.49	19.08	14.00	19.35	21.74
1000 psig	16.94	17.38	17.39	18.99	14.15	19.91	22.25
500 psig	15.08	15.56	15.59	16.98	12.46	18.11	20.32

*Based on 10 equal LV% fractions

Table 6.1.19 *Values of Interaction Parameters* k_{ij} *for Use in the Peng-Robinson Equation*

$$(k_{ij} = 100)$$

Methane	Ethane	Propane	i-Butane	n-Butane	i-Pentane	n-Pentane	n-Hexane	n-Heptane	n-Octane	n-Nonane	n-Decane	Nitrogen	Carbon Dioxide	Hydrogen Sulfide	Toluene	Benzene	Cyclohexane	Water	
0.0	0.0	0.0	0.0	0.0	0.0	0.0	0.0	0.0	0.0	0.0	0.0	3.6	10.0	8.5	4.0	4.0	3.5	50.0	Methane
	0.0	0.0	0.0	0.0	0.0	0.0	0.0	0.0	0.0	0.0	0.0	5.0	13.0	8.4	2.0	2.0	2.0	48.0	Ethane
		0.0	0.0	0.0	0.0	0.0	0.0	0.0	0.0	0.0	0.0	8.0	13.5	7.5	2.0	2.0	2.0	48.0	Propane
			0.0	0.0	0.0	0.0	0.0	0.0	0.0	0.0	0.0	9.5	13.0	5.0	0.0	0.0	0.0	48.0	i-Butane
				0.0	0.0	0.0	0.0	0.0	0.0	0.0	0.0	9.0	13.0	6.0	0.0	0.0	0.0	48.0	n-Butane
					0.0	0.0	0.0	0.0	0.0	0.0	0.0	9.5	5.0	6.0	0.0	0.0	0.0	48.0	i-Pentane
						0.0	0.0	0.0	0.0	0.0	0.0	10.0	5.0	6.5	0.0	0.0	0.0	48.0	n-Pentane
							0.0	0.0	0.0	0.0	0.0	10.0	5.0	6.0	0.0	0.0	0.0	48.0	n-Hexane
								0.0	0.0	0.0	0.0	10.0	5.0	6.0	0.0	0.0	0.0	48.0	n-Heptane
									0.0	0.0	0.0	10.0	11.5	5.5	0.0	0.0	0.0	48.0	n-Octane
										0.0	0.0	10.0	11.0	5.0	0.0	0.0	0.0	48.0	n-Nonane
											0.0	10.0	11.0	4.5	1.0	1.0	1.0	0.0	n-Decane
												0.0	2.0	18.0	18.0	16.0	10.0	0.0	Nitrogen
													0.0	10.0	9.0	7.5	10.0	0.0	Carbon Dioxide
														0.0	0.0	0.0	0.0	0.0	Hydrogen Sulfide
															0.0	0.0	0.0	0.0	Toluene
																0.0	0.0	0.0	Benzene
																	0.0	0.0	Cyclohexane
																		0.0	Water

Table 6.1.20 *Values of Interaction Parameters* k_{ij} *Proposed by Starling*

Methane	Ethylene	Ethane	Propylene	Propane	i-Butane	n-Butane	i-Pentane	n-Pentane	Hexane	Heptane	Octane	Nonane	Decane	Undecane	Nitrogen	Carbon dioxide	Hydrogen Sulfide	
0.0	1.0	1.0	2.1	2.3	2.75	3.1	3.6	4.1	5.0	6.0	7.0	8.1	9.2	10.1	2.5	5.0	5.0	Methane
	0.0	0.0	0.3	0.31	0.4	0.45	0.5	0.6	0.7	0.85	1.0	1.2	1.3	1.5	7.0	4.8	4.5	Ethylene
		0.0	0.3	0.31	0.4	0.45	0.5	0.6	0.7	0.85	1.0	1.2	1.3	1.5	7.0	4.8	4.5	Ethane
			0.0	0.0	0.3	0.35	0.4	0.45	0.5	0.65	0.8	1.0	1.1	1.3	10.0	4.5	4.0	Propylene
				0.0	0.3	0.35	0.4	0.45	0.5	0.65	0.8	1.0	1.1	1.3	10.0	4.5	4.0	Propane
					0.0	0.0	0.08	0.1	0.15	0.18	0.2	0.25	0.3	0.3	11.0	5.0	3.6	i-Butane
						0.0	0.08	0.1	0.15	0.18	0.2	0.25	0.3	0.3	12.0	5.0	3.4	n-Butane
							0.0	0.0	0.0	0.0	0.0	0.0	0.0	0.0	13.4	5.0	2.8	i-Pentane
								0.0	0.0	0.0	0.0	0.0	0.0	0.0	14.8	5.0	2.0	n-Pentane
									0.0	0.0	0.0	0.0	0.0	0.0	17.2	5.0	0.0	Hexane
										0.0	0.0	0.0	0.0	0.0	20.0	5.0	0.0	Heptane
											0.0	0.0	0.0	0.0	22.8	5.0	0.0	Octane
												0.0	0.0	0.0	26.4	5.0	0.0	Nonane
													0.0	0.0	29.4	5.0	0.0	Decane
														0.0	32.2	5.0	0.0	Undecane
															0.0	0.0	0.0	Nitrogen
																0.0	3.5	Carbon dioxide
																	0.0	Hydrogen sulfide

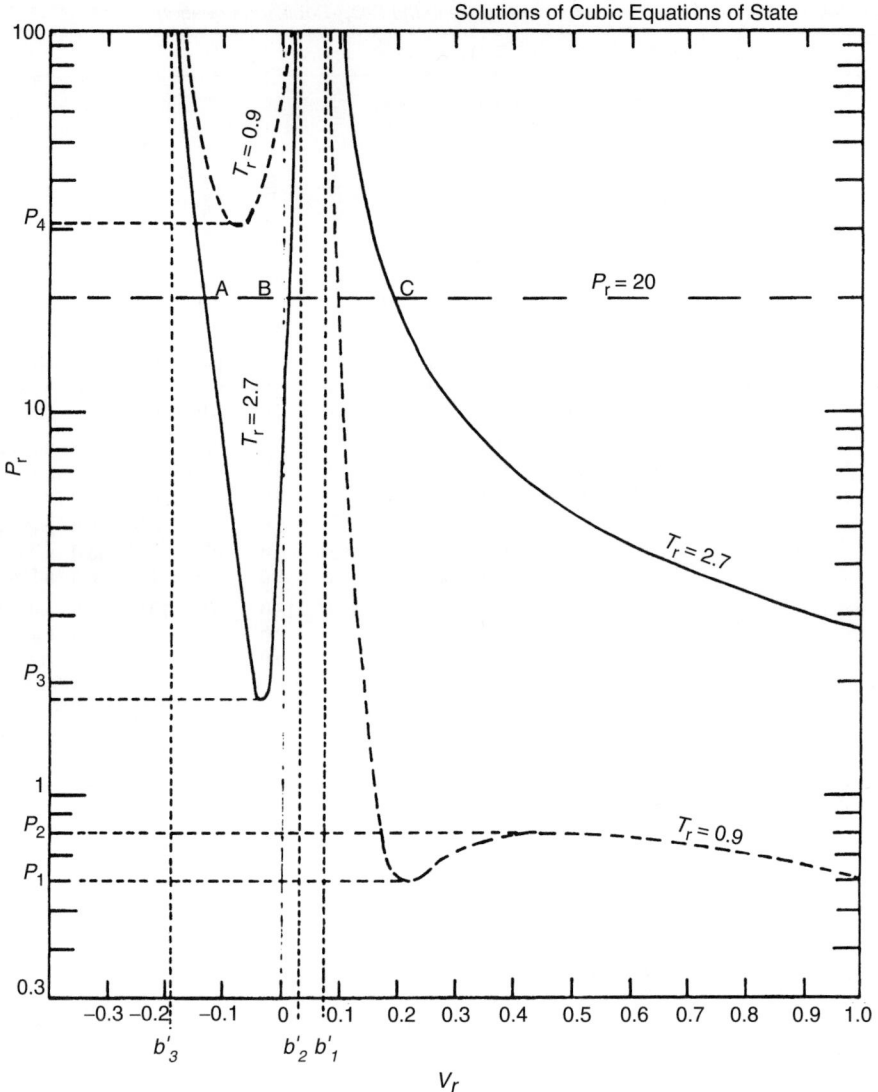

Figure 6.1.31 P_r vs. V_r plot for Peng-Robinson Equation [17].

If the algebraic expression under the square root sign is negative or zero, only one real value of the compressibility factor will satisfy the equation of state (Figure 6.1.33c or d). If, however, the value of the expression under the radical is positive, three real roots exist and limits for the vapor and liquid phase compressibility factors can be determined from Equation 6.1.58. The solutions of Equation 6.1.58 represent the value of Z at the maximum and minimum points of Figure 6.1.33b. The value of the maximum will represent the largest possible value for the liquid compressibility factor and the value at the minimum represents the smallest possible value of the vapor compressibility factor. These limits can then be used with arbitrary values for the other limit to assure that the root obtained is the valid one. The limits thus set up are adjusted at the end of each iteration to narrow the interval of search.

References

1. Katz, D. L., et al., *Handbook of Natural Gas Engineering*, McGraw-Hill Book Co., New York, 1959.

2. *Engineering Data Book*, Vols. 1, 2, 11th Edition, GPSA, Tulsa, Oklahoma, 1998.

3. SPE Reprint No.13, Vol. 1., pp. 233–235.

4. Soave, G., "Equilibrium Constants from a Modified Redlich-Kwong Equation of State," "*Chemical Engineering Science*," 1972.

5. Peng, O. Y., and Robinson, D. B., "A New Two Constant Equation of State," *Industry and Engineering Chemistry Fundamentals*, 1976.

6. *The Phase Behavior of Hydrocarbon Reservoir Fluids*, course given at Core Laboratories Inc., Dallas, Texas.

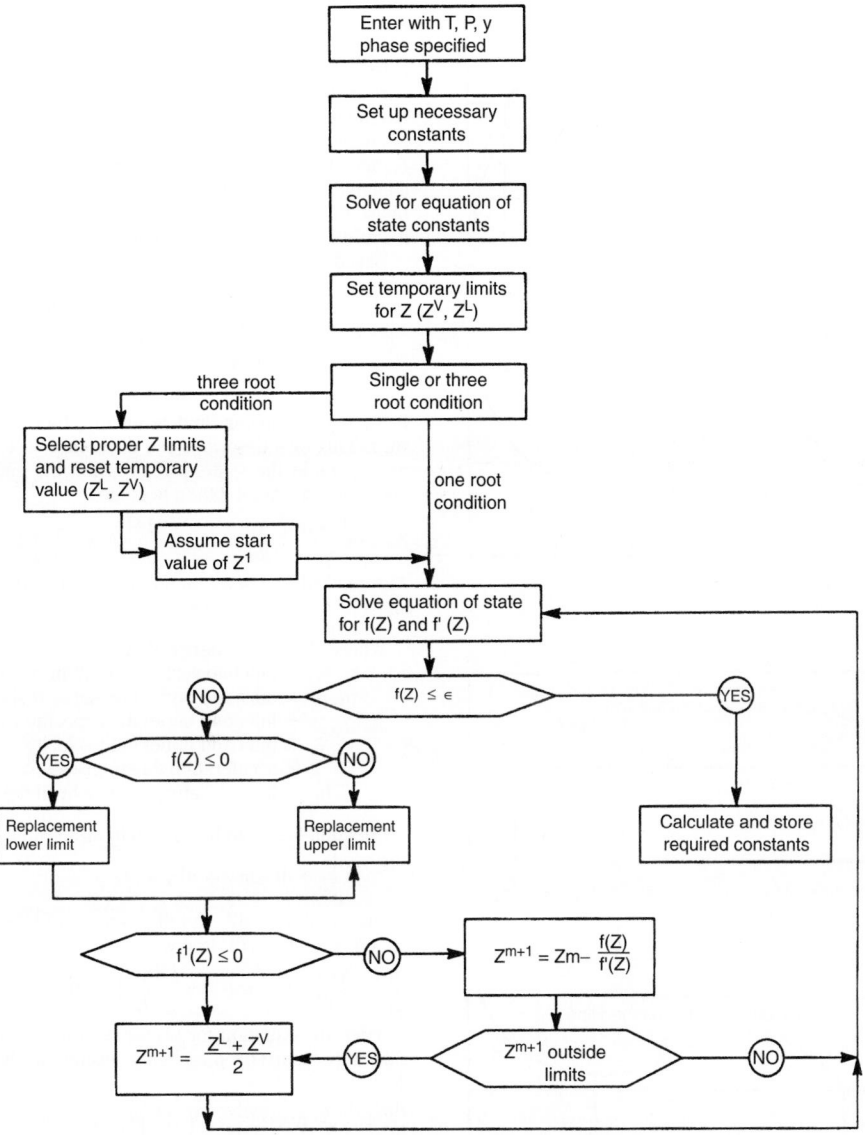

Figure 6.1.32   *Logic diagram for equation-of-state solution.*

7. MacDonald, R. C., "Reservoir Simulation with Interface Mass Transfer," *Report No. UT-71-2, University of Texas, Austin, Texas., 1971.*

8. Moses, P. L., "Engineering Applications of Phase Behavior of Crude Oil and Condensate Systems," *Journal of Petroleum Technology,* July 1986.

9. API 811-08800 Standard RP 44: "Recommended Practice for Sampling Petroleum Reservoir Fluids."

10. Amex, J. W., Bass, D. M., and Whiting, R. L., *Petroleum Reservoir Engineering,* McGraw-Hill Book Co., New York, 1960.

11. Weatherly Laboratories, Inc., *Manual 1984,* Lafayette, Louisiana, 1984.

12. Maddox, R. N., and Erbar, J. H., *Gas Conditioning and Processing,* Campbell CPS, Norman, Oklahoma, 1982.

13. Edmister, W. C., and Lee, B. I., *Applied Hydrocarbon Thermodynamics,* Vol. 1, Gulf Publishing Co., Houston, Texas, 1984.

14. Danbert, T. E., and Danner, R. P., Technical Data Conference and Exhibition of the SPE of AIME, New Orleans, Louisiana, October 3–6, 1976.

15. Danbert, T. E., "Property Predictions," *Hydrocarbon Processing,* March 1980.

16. Starling, E., *Fluid Thermodynamic Properties for Light Petroleum Systems,* Gulf Publishing Co., Houston, Texas, 1973.

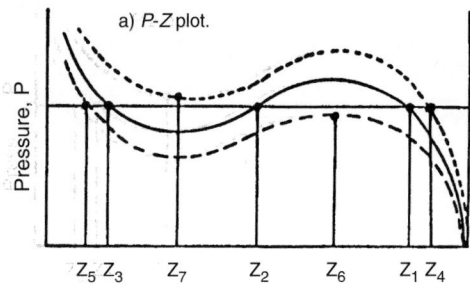

a) *P-Z* plot.

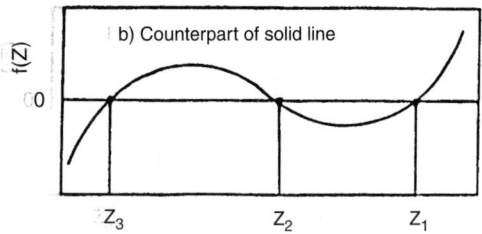

b) Counterpart of solid line

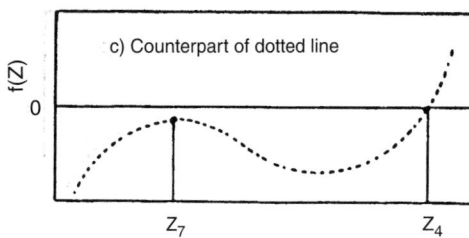

c) Counterpart of dotted line

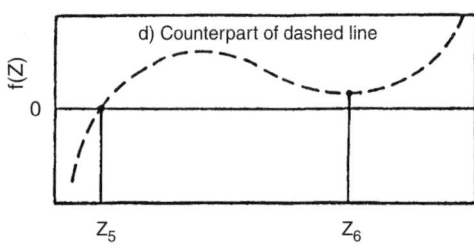

d) Counterpart of dashed line

Figure 6.1.33a–d *Typical behavior of cubic equation-of-state [17].*

17. Asselineau, L., Bogdanic, G., and Vidal, J., *Fluid Phase Equilibrium*, 1979.
18. Gunderson, T., *Computer and Chemical Engineering*, 1982.

6.2 FLOW OF FLUIDS

Fluid is defined as a single phase of gas or liquid or both. Each sort of flow results in a pressure drop. Three categories of fluid flow: vertical, inclined, and horizontal are shown in Figure 6.2.1. The engineer involved in petroleum production

operations has one principal objective to move the fluid from some location in an underground reservoir to a pipeline that may be used to transport it or storage tank. Possible pressure losses in a complete production system and producing pressure profile are shown in Figures 6.2.2 and 6.2.3, respectively. On the way from reservoir to pipeline or storage tank, fluid is changing its temperature, pressure, and, consequently, composition of each phase. In case of a dry gas reservoir, a change in pressure and temperature does not create two-phase flow; also in case of black oil with a very small GOR, it could be assumed that two-phase flow does not occur.

Based on the law of conservation of energy, the total energy of a fluid at any particular point above datum plane is the sum of the deviation head, the pressure head, and velocity head as follows:

$$H = Z_{el} + \frac{144p}{\gamma} + \frac{v^2}{2g} \qquad [6.2.1]$$

In reality, whenever fluid is moving there is friction loss (h_L). This loss describes the difference in total energy at two points in the system. Expressing the energy levels at point 1 versus point 2 then becomes

$$Z_{el1} + \frac{144p_1}{\gamma_1} + \frac{v_1^2}{2g} = Z_{el2} + \frac{144p_2}{\gamma_2} + \frac{v_2^2}{2g} + h_L \qquad [6.2.2]$$

All practical formulas for fluid flow are derived from the above,

where H = total energy of fluid
Z_{el} = pipeline vertical elevation rise (ft)
p_1, p_2 = inlet and outlet pressures (psia)
γ_1, γ_2 = inlet and outlet fluid specific weight
v_1, v_2 = inlet and outlet fluid velocity
g = acceleration due to gravity
h_L = loss of static pressure head due to fluid flow

Equation 6.2.2 can be written in differential form as

$$\frac{dp}{\gamma} + \frac{vdv}{g} + dL\sin\theta + dL_w = 0 \qquad [6.2.3]$$

where $dL\sin\theta = dZ$, and dL_w refers to friction multiplying the equation by γ/dL to give

$$\frac{dp}{dL} = \frac{\gamma vdv}{gdL} + \gamma\sin\theta + \gamma\left(\frac{dL_w}{dL}\right)_f = 0 \qquad [6.2.4]$$

Solving this equation for pressure gradient, and if we consider a pressure drop as being positive in the direction of flow,

$$\frac{dp}{dL} = \gamma\sin\theta + \frac{\gamma vdv}{gdL} + \left(\frac{dp}{dL}\right)_f \qquad [6.2.5]$$

where

$$\left(\frac{dp}{dL}\right)_f = \frac{\gamma dL_w}{dL}$$

Equation 6.2.4 contains three terms that contribute to the total pressure gradient, i.e.:

1. pressure gradient due to elevation

$$\gamma\sin\theta = \left(\frac{dp}{dL}\right)_{el}$$

2. pressure gradient due to acceleration

$$\frac{\gamma vdv}{gdL} = \left(\frac{dp}{dL}\right)_{acc}$$

3. pressure gradient due to viscous forces (friction)

$$\frac{\gamma dL_w}{dL} = \left(\frac{dp}{dL}\right)_f$$

$$\frac{dp}{dL} = \left(\frac{dp}{dL}\right)_{el} + \left(\frac{dP}{dL}\right)_{acc} + \left(\frac{dp}{dL}\right)_f \qquad [6.2.6]$$

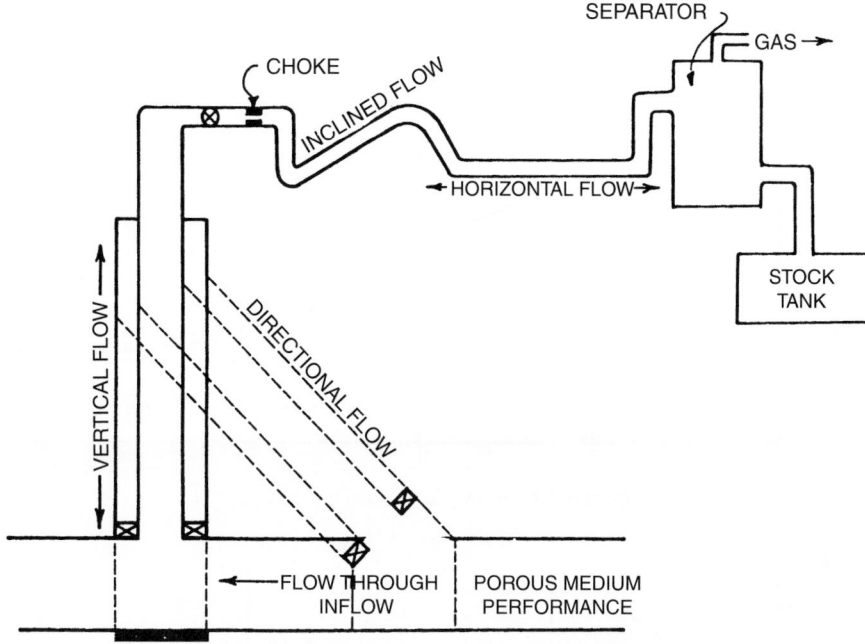

Figure 6.2.1 *Overall production system [1].*

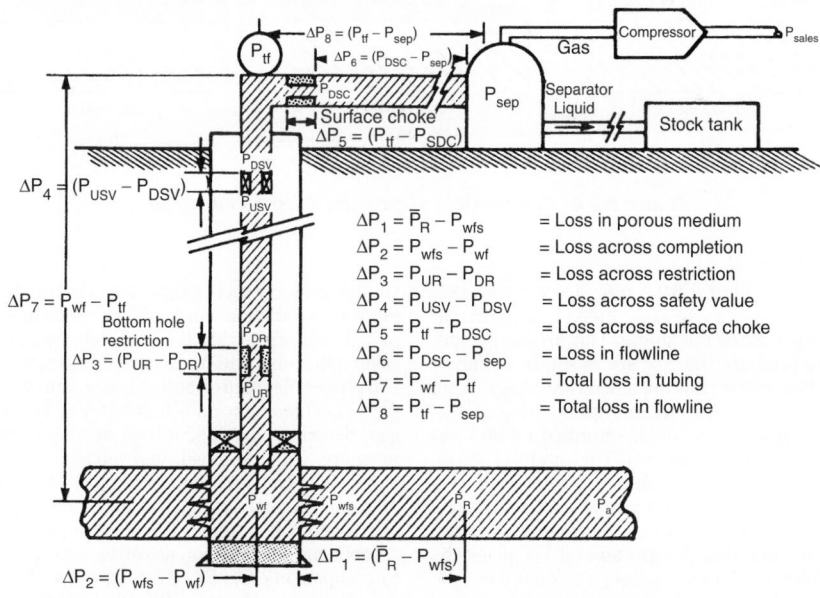

$\Delta P_1 = \bar{P}_R - P_{wfs}$ = Loss in porous medium

$\Delta P_2 = P_{wfs} - P_{wf}$ = Loss across completion

$\Delta P_3 = P_{UR} - P_{DR}$ = Loss across restriction

$\Delta P_4 = P_{USV} - P_{DSV}$ = Loss across safety value

$\Delta P_5 = P_{tf} - P_{DSC}$ = Loss across surface choke

$\Delta P_6 = P_{DSC} - P_{sep}$ = Loss in flowline

$\Delta P_7 = P_{wf} - P_{tf}$ = Total loss in tubing

$\Delta P_8 = P_{tf} - P_{sep}$ = Total loss in flowline

Figure 6.2.2 *Possible pressure losses in complete system [1].*

The acceleration element is the smallest one and sometimes is neglected.

The total pressure at the bottom of the tubing is a function of flowrate and comprises three pressure elements:

1. wellhead pressure—back pressure exerted at the surface form choke and wellhead assembly

2. hydrostatic pressure—due to gravity and the elevation change between wellhead and the intake to the tubing

3. friction losses, which include irreversible pressure losses due to viscous drag and slippage

Figure 6.2.4 illustrates this situation for each single-phase and two-phase flow. Possible pressure losses in a complete

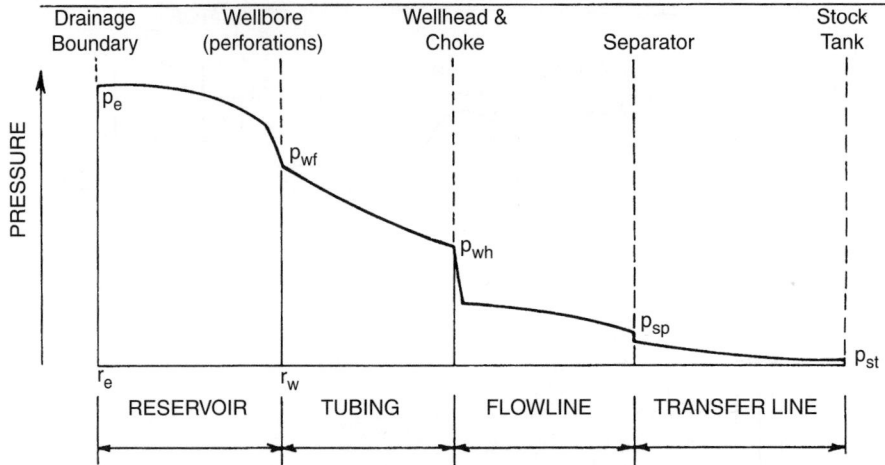

Figure 6.2.3   *Production pressure profile [2].*

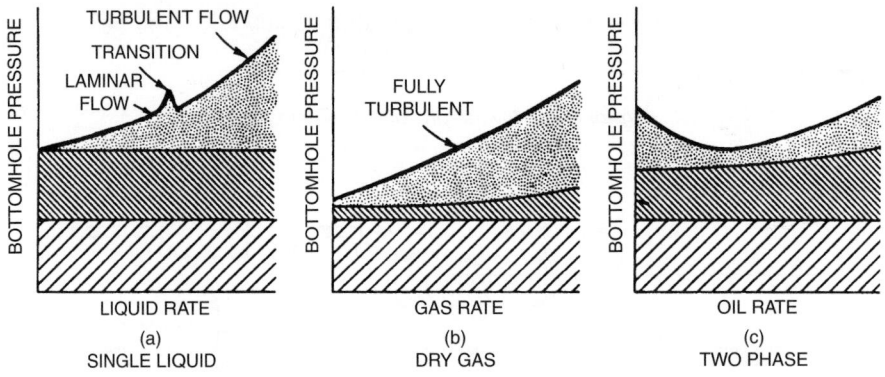

Figure 6.2.4   *Components of pressure losses in tubing [2].*

system are shown in Figure 6.2.2. For a given flowrate, wellhead pressure, and tubing size, there is a particular pressure distribution along the tubing. This pressure-depth profile is called a pressure traverse and is shown in Figure 6.2.5. Gas liberation, gas expansion, and oil shrinkage along the production tubing can be treated as a series of successive incremental states in which saturated oil and gas coexist in equilibrium (flash process). This model is shown in Figure 6.2.6. At (a) the single-phase oil enters the wellbore; (b) marks the first evolution of gas, at the mixture's bubble point; and (c) and (d) show the traverse into the two-phase region. Note that the gas and oil P-T diagrams describing equilibrium phases at points c and d are not the same. This means the composition of equilibrium gas and oil phases changes continuously in the two-phase region. As the two-phase region is entered and gas is liberated, oil and gas phases change in volume and composition, but they are always in a saturated state, the gas at its dew point, and the oil at its bubble point. In Figure 6.2.7 the separation process is shown in forms of the resulting gas and oil.

Engineering analysis of two-phase fluid flow in pipes has focused primarily on the problem of predictive pressure drop, or pressure gradient, from Equation 6.2.6.

In many cases, it has become possible to treat two-phase pipeline problems with empirical numerical techniques that yield reasonably accurate pressure drops. Most of two-phase pipeline simulation currently is performed using "black oil" simulators. A black oil model's validity rests on the assumption that the hydrocarbon mixture is composed of two phases, denoted oil and gas, each with fixed composition. A black-oil model usually treats P–V–T properties (solution gas, densities, and viscosities) as single-value function of pressure. More sophisticated models include the temperature effect on fluid properties as well. The multicomponent or compositional approach is designed for gas condensate and volatile oil systems. These fluids are represented as N-component hydrocarbon mixtures, where N might be equal to components C_1, C_2, C_5, $i\text{-}C_4$, $n\text{-}C_4$, $i\text{-}C_5$, $n\text{-}C_5$, C_6, and C_{7+}. Equations of state (SRK, PR, SBWR) are used to determine physical properties. The term "compositional" implies that the overall or in situ fluid composition varies point by point with the distance as is shown in Figure 6.2.6. When a multicomponent gas–liquid mixture flows through a pipe, the composition, pressure, temperature and liquid holdup distributions are related.

6.2.1 Basic Parameters of Multiphase Flow [1]

Knowledge of the flow regime determines the selection of the appropriate model for pressure gradient and liquid

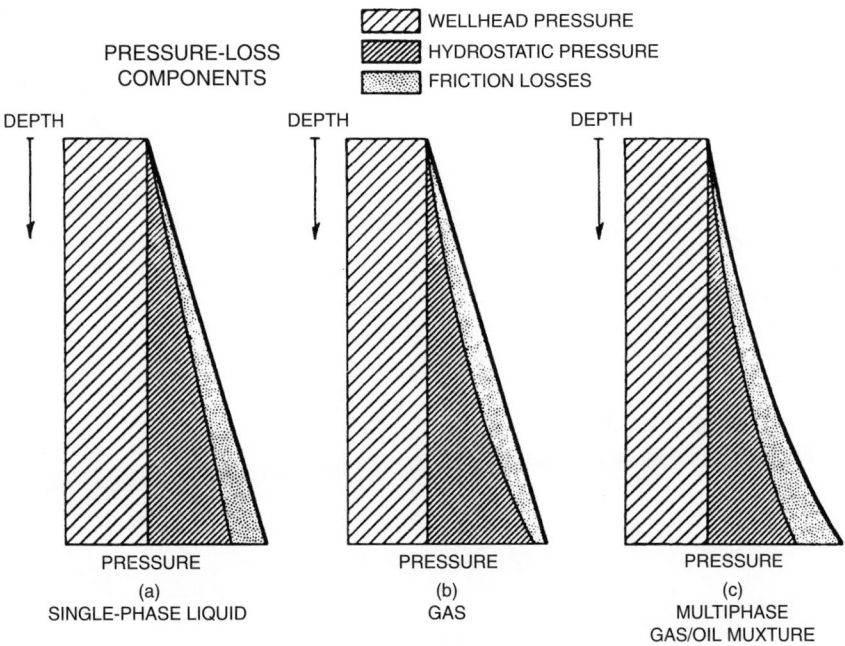

PRESSURE-LOSS
COMPONENTS

▨ WELLHEAD PRESSURE
▧ HYDROSTATIC PRESSURE
▨ FRICTION LOSSES

DEPTH

PRESSURE
(a)
SINGLE-PHASE LIQUID

DEPTH

PRESSURE
(b)
GAS

DEPTH

PRESSURE
(c)
MULTIPHASE
GAS/OIL MUXTURE

Figure 6.2.5 *Pressure traverse for single-phase liquid, gas, and multiphase gas–oil mixture [2].*

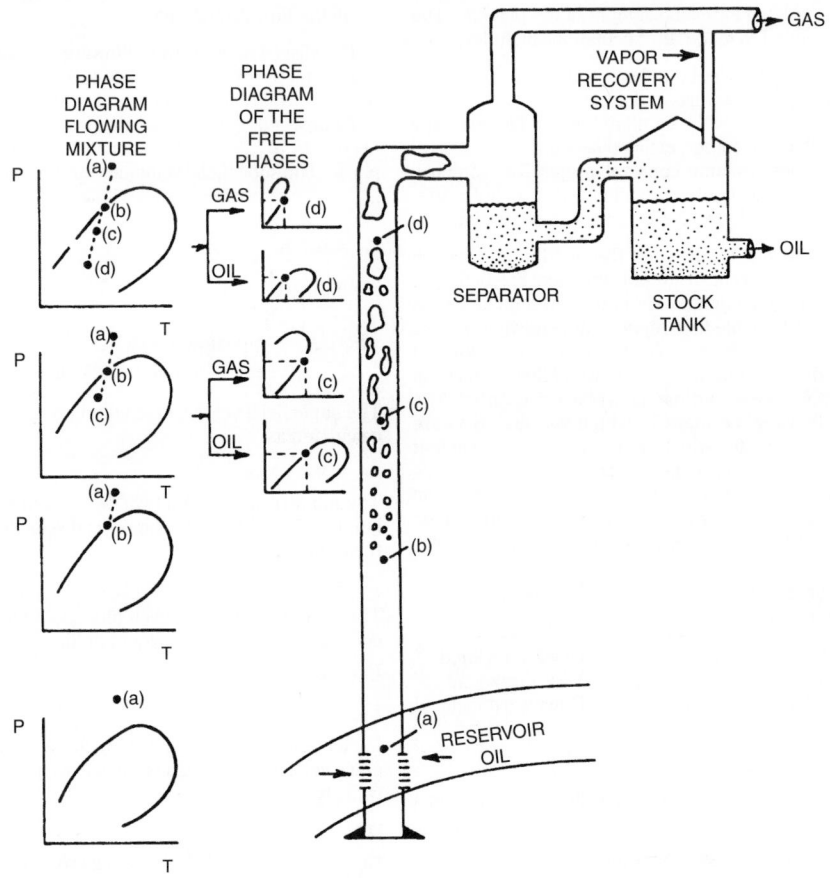

Figure 6.2.6 *Changes in phase behavior in the production tubing [2].*

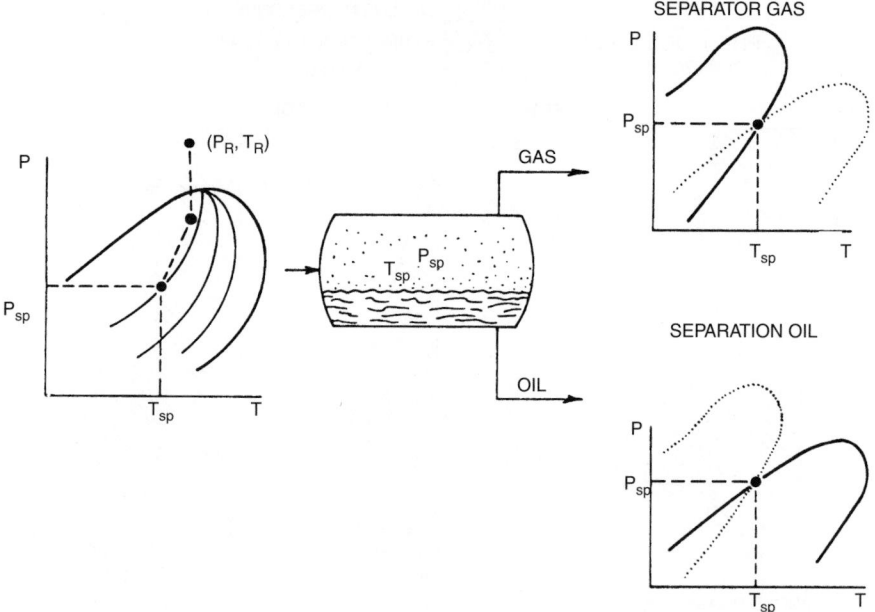

Figure 6.2.7 *Pressure–temperature phase diagram used to describe surface separation [2].*

holdup. The flow regime, pressure gradient, and liquid holdup are calculated for each segment of the pipeline. The information needed to make the calculations includes:

1. pipeline inlet and outlet boundary conditions (liquid and gas flowrates, temperature, and pressure)
2. pipeline geometry, with segments specifications (any riser or well, down-comer, inclined section)
3. fluid properties (assume constant properties, compositional analysis, black oil approaches); this includes gas, oil, and water density; viscosity; and surface tension

It is assumed by flow regime that the distribution of each phase in the pipe is relative to one another. Prediction of flow patterns for horizontal flow is a more difficult task than for vertical flow. Possible flow regimes are shown in Figure 6.2.8. An example of the complexity of two-phase flows in Figure 6.2.8 shows a schematic sequence of flow patterns in vertical pipe. Numerous authors [1–4] have presented flow-pattern and flow-regime maps in which various areas are indicated on a graph for which there are two independent coordinates. Maps are also dependent on ranges of pipe inclination for vertical upward to vertical downward [1]. Selection of the appropriate flow regime map is based solely on the pipe segment inclination θ from the horizontal, as shown below:

Range of Inclination from the Horizontal	Regime Maps
$\theta = 90°$ to $15°$	Upward inclined
$\theta = 15°$ to $-10°$	Near horizontal
$\theta = -10°$ to $-90°$	Downward inclined

6.2.1.1 Flow Regimes

The steps in the determination of the flow regime are as follows:

1. Calculate dimensionless parameters.
2. Refer to flow regime maps laid out in coordinates of these parameters.

3. Determine the flow regime by locating the operating point on the flow regime map.

The discussions in the following sections treat the flow regime maps for vertical upward ($\theta = 90°$); slightly inclined ($\theta = 15°$ to $-10°$) and vertical downward inclinations.

To proceed with the calculations of the flow regime, it is necessary to calculate the superficial velocities for each flow phase. The superficial velocities for the gas, oil and water are

$$v_{sg} = q_g/A_p \qquad [6.2.7a]$$

$$v_{so} = q_o/A_p \qquad [6.2.7b]$$

$$v_{sw} = q_w/A_p \qquad [6.2.7c]$$

where A_p = pipe flow area in ft^2
q = volumetric flowrate at flow conditions in ft^3/s

The superficial velocities of liquid phase (oil and water) are calculated as

$$v_{sL} = v_{so} + v_{sw} \qquad [6.2.7d]$$

The mixture velocity that will be used in some of the calculations is the sum of the superficial velocities of the gas and the liquid phases:

$$v_m = v_{sL} + v_{sg} \qquad [6.2.7e]$$

The average velocity of each phase is related to the superficial velocity through the liquid holdup:

$$U_L = v_L = v_{sL}/H_L \qquad [6.2.7f]$$

$$U_g = v_g = v_{sg}/(1 - H_L) \qquad [6.2.7g]$$

For a homogeneous model, both phases are assumed to have equal velocities and each is equal to a two-phase (or mixture) velocity:

$$v_L = v_g = v_m \qquad [6.2.7h]$$

H_L in Equations 6.2.7f and 6.2.7g refers to liquid holdup.

Liquid holdup is defined as the ratio of the volume of a pipe segment occupied by liquid to the volume of the

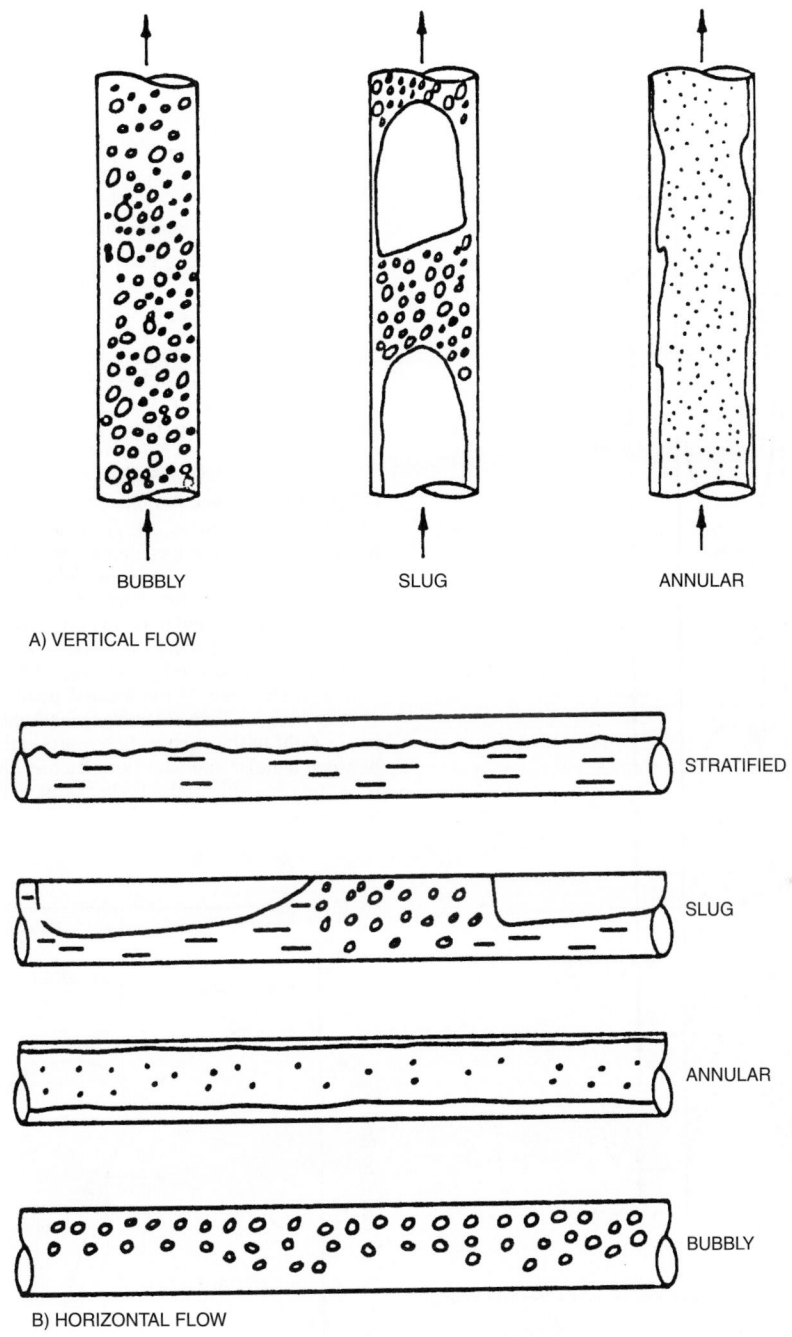

Figure 6.2.8 *Gas–liquid flow regimes.*

pipe segment:

$$H_L = \frac{\text{volume of liquid in a pipe segment}}{\text{volume of pipe segment}} \qquad [6.2.8a]$$

In some cases, e.g., for stratified horizontal flow regime, liquid holdup can be calculated as follows:

$$H_L = \frac{A_L}{A_L + A_g} \qquad [6.2.8b]$$

where A_L = cross-sectional area occupied by liquid (oil and water)

A_g = cross-sectional area occupied by gas

6.2.2 Slightly Inclined Pipes ($-10° < \theta < 15°$)

As can be seen from Figure 6.2.8, there are four flow regimes of interest—stratified, slug, annular, and bubbly—and three flow regime transition zones.

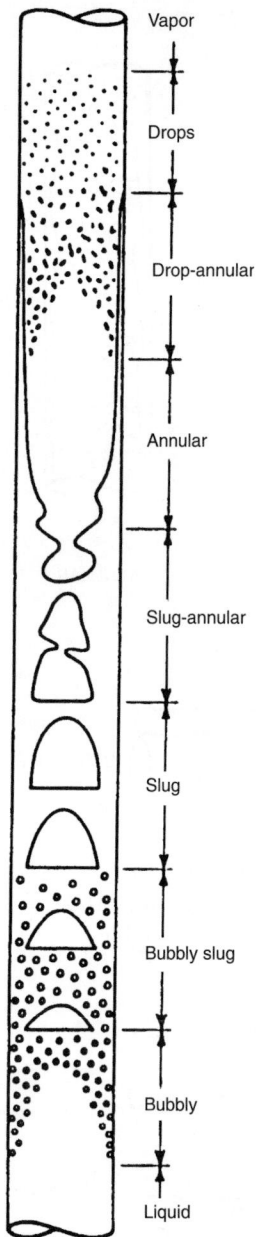

Figure 6.2.9 *Possible sequence of flow patterns in a vertical tube [3].*

6.2.2.1 Step 1. Dimensionless Parameters

1. Martinelli parameter—this is the ratio of the liquid and gas phases as if each flowed alone in the pipe:

$$X = \left[\frac{(dp/dL)_{Ls}}{(dp/dL)_{gs}} \right]^{0.5} = \left(\frac{2f_{hLs}\gamma_L v_{sL}^2/D}{2f_{wgs}\gamma_s v_{sg}^2/D} \right) \quad [6.2.9a]$$

According to standard fluid mechanics book

$$f(^*) = \begin{cases} 0.046/Re^{0.2} & \text{if } Re = \gamma Dv/\mu > 1500 \\ 16/Re & \text{if } Re < 1500 \end{cases} \quad [6.2.9b]$$

*Calculated separately for "ls" and "gs."

2. Gas Froude number (dimensionless gas flowrate):

$$F_g = v_{sg} \left(\frac{\gamma_g}{(\gamma_L - \gamma_g)gD} \right)^{0.5} \quad [6.2.10]$$

3. Turbulence level

$$T = \left[\frac{(dp/dL)_{Ls}}{g(\gamma_L - \gamma_g)\cos\theta} \right] = \left[\frac{2f_{wLs}\gamma_L v_{sL}^2/D}{g(\gamma_L - \gamma_g)\cos\theta} \right] \quad [6.2.11]$$

4. Dimensionless inclination (slope parameter)

$$Y = \left[\frac{g(\gamma_L - \gamma_g)\sin\theta}{(dp/dL)_{gs}} \right] = \left[\frac{g(\gamma_L - \gamma_g)\sin\theta}{2f_{wgs}\gamma_g v_{sg}^2/D} \right] \quad [6.2.12]$$

6.2.2.2 Step 2. Flow Regime Map
The parameter Y is used to select the flow regime map to be used from Figure 6.2.10a–i. This figure presents nine flow regime maps (a to i) for a wide range of dimensionless inclinations, Y, in the range of interest.

6.2.2.3 Step 3. Flow Regime Selection
The flow regime maps are prepared in Froude number–Martinelli parameter–turbulence level (F_g–X–T). Four flow regimes are noted on the maps, doing with the three transition boundaries. Regimes are stratified, slug, annular, and bubbly. The flow conditions for the current pipe segment are located using the X, F_g coordinates. If the located point is in the region labeled "stratified," then the flow regime is indeed stratified. If the located point is outside of the stratified region, then determine whether the point is to the left or right of the vertical line representing the transition between annular and slug flow regimes. If the point X, F_g is on the right side of the vertical line, X,T coordinates are necessary.

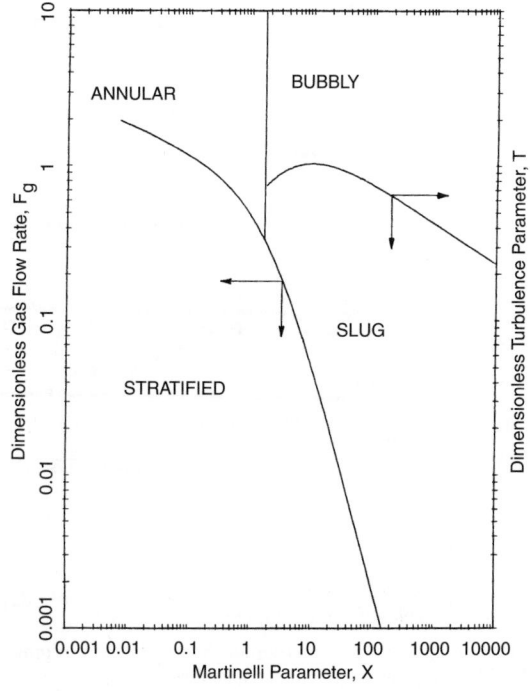

Figure 6.2.10a *Flow regime map for slightly inclined pipes (horizontal) [4].*

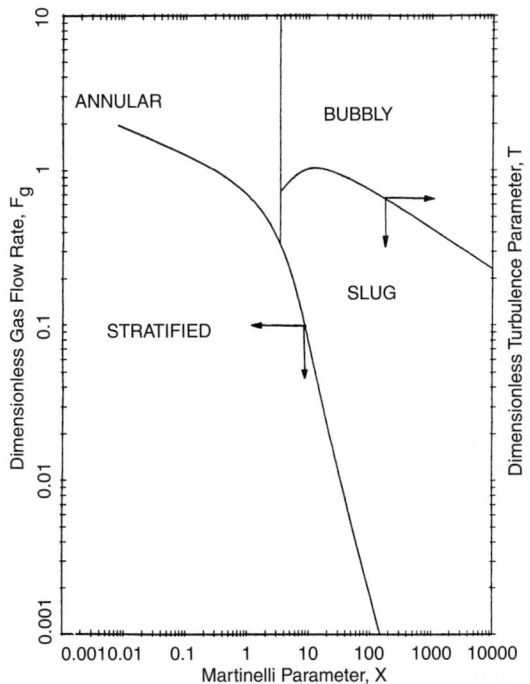

Figure 6.2.10b *Flow regime map for slightly inclined pipes (Y = −100) [4].*

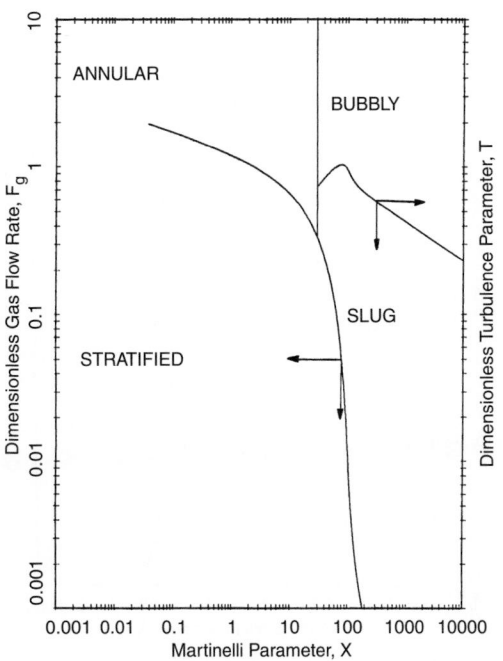

Figure 6.2.10d *Flow regime map for slightly inclined pipes (Y = −10,000) [4].*

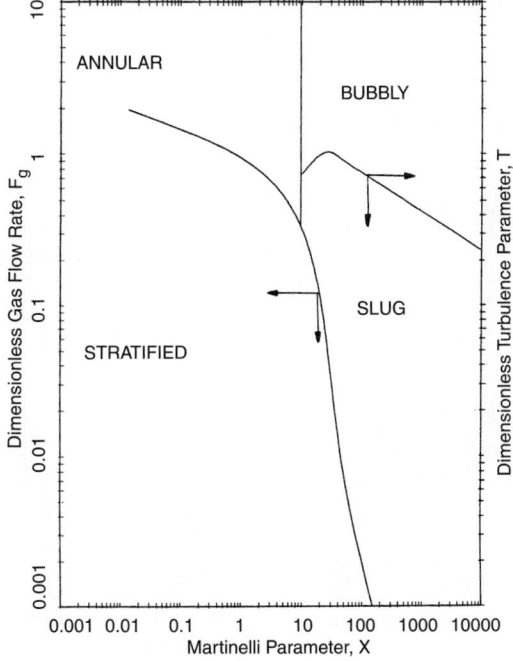

Figure 6.2.10c *Flow regime map for slightly inclined pipes (Y = −1000) [4].*

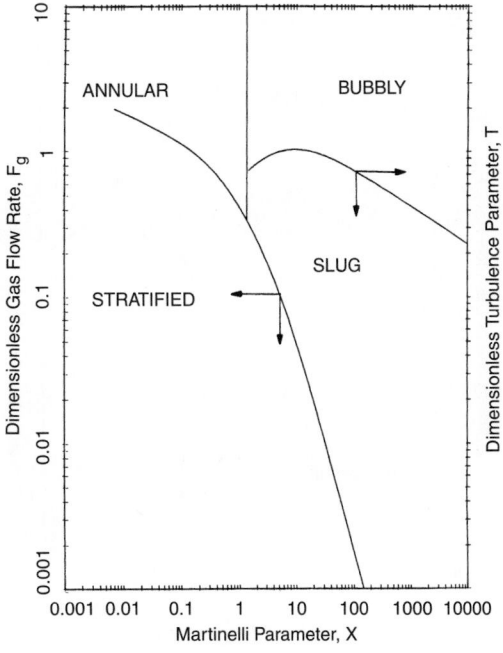

Figure 6.2.10e *Flow regime map for slightly inclined pipes (Y = 10) [4].*

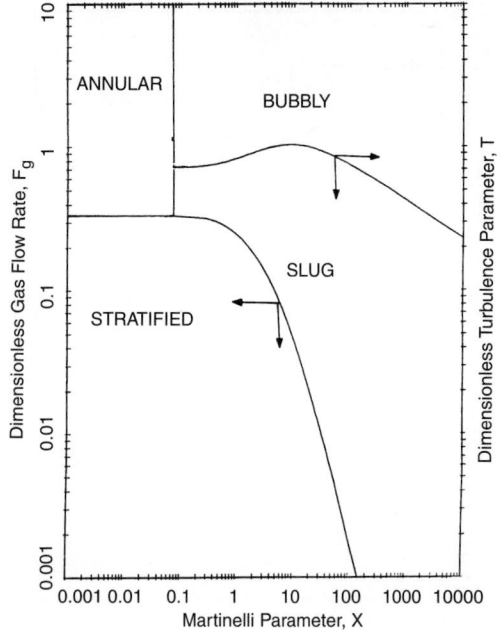

Figure 6.2.10f *Flow regime map for slightly inclined pipes (Y = 30) [4].*

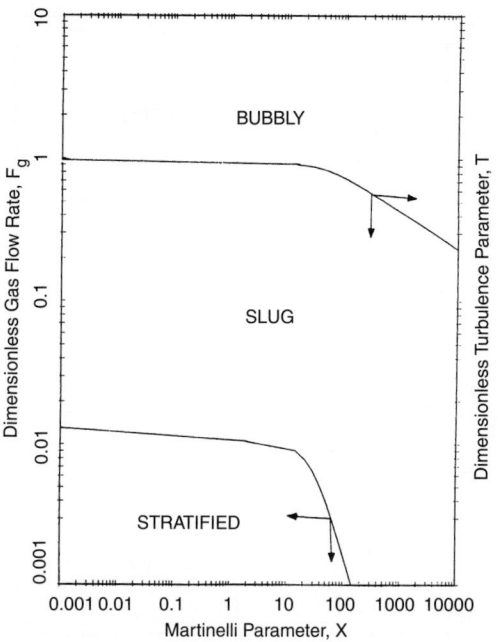

Figure 6.2.10h *Flow regime map for slightly inclined pipes (Y = 1000) [4].*

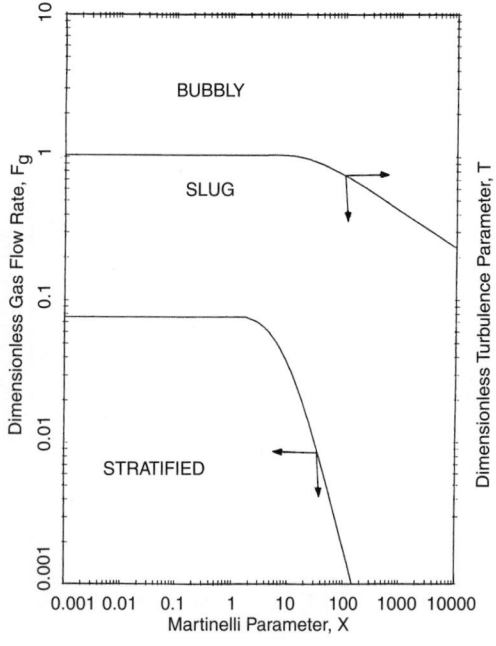

Figure 6.2.10g *Flow regime map for slightly inclined pipes (Y = 100) [4].*

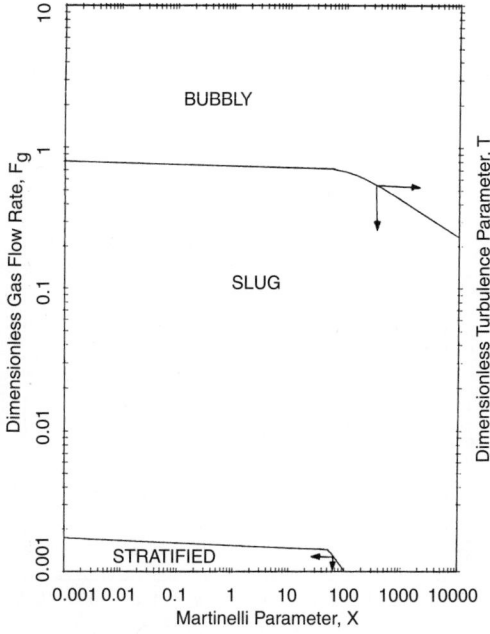

Figure 6.2.10i *Flow regime map for slightly inclined pipes (Y = 10,000) [4].*

6.2.3 Risers and Wells ($\theta = 90°$)

There are three possible flow regimes, including annular, slug, and bubbly, and two regime transitions.

6.2.3.1 Step 1. Dimensionless Parameters

For this regime map, three dimensionless parameters are calculated. The parameters include dimensionless groups, which represent a balance between buoyancy, inertial, and surface tension forces. There are called Kutateladze numbers.

$$K_g = \frac{\gamma_g^{0.5} v_{sg}}{\left[g\sigma_{go}(\gamma_L - \gamma_g)\right]^{0.25}} \qquad [6.2.13]$$

$$K_L = \frac{\gamma_L^{0.5} v_{sL}}{\left[g\sigma_{go}(\gamma_L - \gamma_g)\right]^{0.25}} \qquad [6.2.14]$$

$$\rho^* = \left(\frac{\rho_L}{\rho_g}\right) \qquad [6.2.15]$$

where K_g, K_L = dimensionless gas and liquid flowrates
  ρ_g, ρ_L = density of gas and liquid phases in $lb_m/ft.^3$
  v_{sg}, v_{sL} = superficial gas and liquid velocities in ft./s
  σ_{go} = surface tension between gas–oil phases in $lb_f/ft.$
  ρ^* = dimensionless density ratio
  g = acceleration of gravity in $ft./s^2$

6.2.3.2 Step 2. Flow Regime Map

Once these dimensionless parameters are calculated, then the point with the coordinates (K_g, K_L) is located on the flow regime map in Figure 6.2.11.

6.2.3.3 Step 3. Flow Regime Selection

The boundary for the appropriate density ratio is located for the transition boundary between the slug and bubbly flow regimes. Once the appropriate boundary line is found, then the flow regime is simply bubbly, slug, or annular depending upon the region in which the point falls. This slug to annular transition applies only if the pipe size D is larger than a critical diameter D_{crit} given by

$$D_{crit} = 1.9 \left[\frac{\sigma_{go}(\gamma_L - \gamma_g)}{g\gamma_L}\right]^{0.25} N_L^{(-0.4)} \qquad [6.2.16a]$$

where

$$N_L = \frac{\mu_L}{(\gamma_L \sigma_{go})^{0.5} \left[\sigma_{go}/g(\gamma_L - \gamma_g)\right]^{0.25}} \qquad [6.2.16b]$$

μ_L = liquid phase viscosity in $lb_m/(ft./s)$

Usually, the critical pipe size is about 2 in. for conditions of gas and oil pipelines so that Figure 6.2.11 can be used often. The criterion should be checked each time, however. If $D < D_{crit}$, another method has to be used; see Reference 22, Vol. 3.

6.2.4 Downcomers ($\theta = -90°$)

There are also three possible flow regimes: annular, slug, and bubbly. There are two flow regime transitions to be calculated. Two different maps will be used: one for transition between annular and slug flow regimes, and a second for the transition between the slug and bubbly flow regimes.

6.2.4.1 Annular-Slug Transition

1. Dimensionless parameters X and |Y| from Equations 6.2.9a and 6.2.12, respectively.
2. Flow regime map, see Figure 6.2.12.

3. Flow regime selection. Locate the point with (X, |Y|) coordinates; if the point falls in the region "annular," then the flow regime is annular. If the point falls in region "slug" or "bubbly," then the map from Figure 6.2.13 must be used.

6.2.4.2 Slug-Bubbly Transition

1. Calculate dimensionless parameters K_g, K_L, and ρ^*; use Equations 6.2.14 and 6.2.15, respectively.
2. The appropriate regime boundary for the specific weight ratio γ^* is selected on the regime map with K_g, K_L coordinates.
3. Flow regime selection. Depending upon which region of the map the data point falls into, flow regime is slug or bubbly.

Similarly as for ($\theta = 90°$), Figure 6.2.13 applies for pipes greater than a certain critical diameter.

$$D > D_{crit} = 19 \left[\frac{(\gamma_L - \gamma_g)\sigma_{go}}{g\gamma_L^2}\right]^{0.5} \qquad [6.2.17]$$

The above procedure is valid for $D > 2$ in.

For each flow regime, there is a separate pressure gradient and holdup calculation method.

6.2.5 Stratified Flow Regime

First, the *liquid holdup* H_L and f_i/f_{wg} (friction factor ratio) are calculated. The liquid holdup as a function of X (Equation 6.2.9a), Y (Equation 6.2.12), and (f_i/f_{wg}) is read from Figure 6.2.14a–d. $f_i/f_{wg} = 10$ is recommended to be used as a preliminary estimate. For better accuracy f_i/f_{wg} can be calculated. H_L should be first estimated by the method described above and

$$\frac{f_i}{f_{wg}} = 1.0 \text{ if Equation } 6.2.18b < 1.0 \qquad [6.2.18a]$$

$$\frac{f_i}{f_{wg}} = \left[2 + \frac{0.000025 Re_{Ls}}{D/3.281}\right](1 - H_L)^{5/2}(1 + 75H_L) \qquad [6.2.18b]$$

if Equation 6.2.18b $> (1 + 75H_L)$ $\qquad [6.2.18c]$

(D is in inches).

Figure 6.2.15a–h shows Equation 6.2.18 for pipe diameters ranging from 4 to 36 in. The friction factor ratio (f_i/f_{wg}) can be estimated from the plots in Figure 6.2.15a–h or as a result of calculation.

When the friction factor is determined correctly, Figure 6.2.14 should again be used to estimate a new value of the liquid holdup. If the new value of the liquid holdup obtained is different from determined previously, this new value should be used in Equation 6.2.18 or Figure 6.2.15a–h to refine the estimate for f_i/f_{wg}. This is simply an iteration process.

6.2.5.1 Pressure Gradient

Knowing the liquid holdup, H_L, calculations for the pressure gradients due to friction and gravitational effects are straightforward. First, some geometric parameters are calculated:

1. Dimensionless cross-sectional area occupied by gas (A_G) and by liquid (A_L)

$$A_G = 0.25[\cos^{-1}(2h^* - 1) - (2h^* - 1)\left[1 - (2h^* - 1)^2\right]^{0.5} \qquad [6.2.19a]$$

$$A_L = 0.25[\pi - \cos^{-1}(2h^* - 1) + (2h^* - 1)\left[1 - (2h^* - 1)^2\right]^{0.5} \qquad [6.2.19b]$$

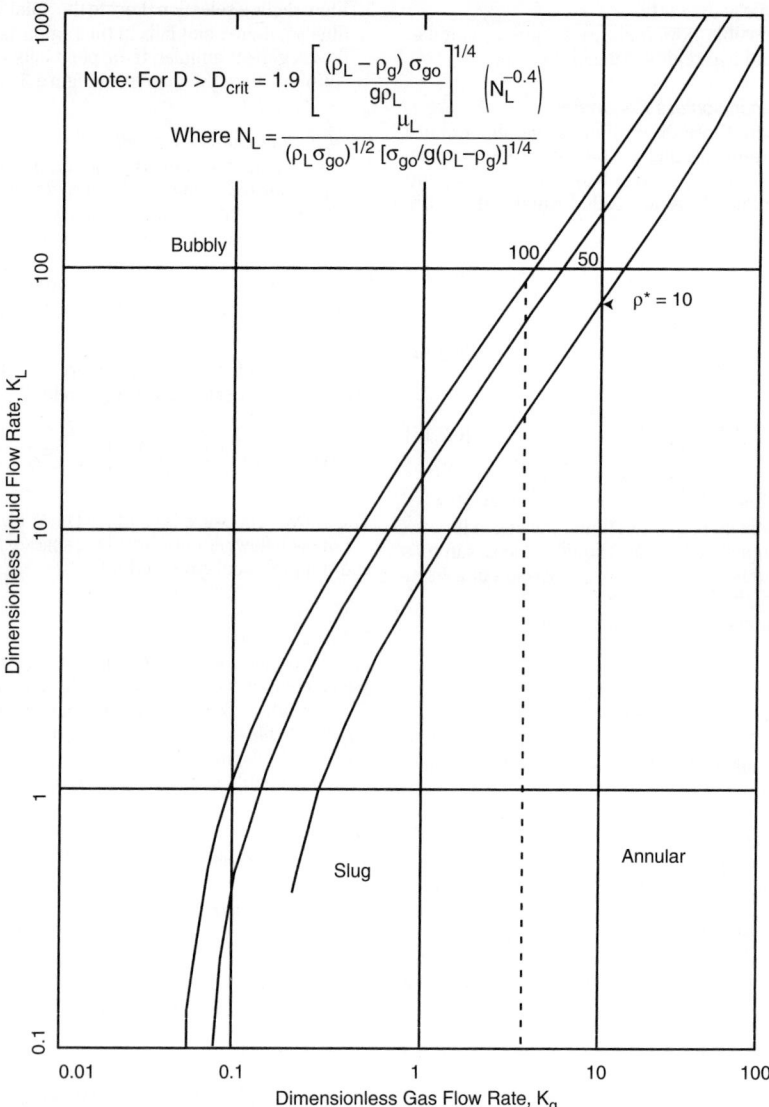

Figure 6.2.11 *Flow regime map for vertical upward inclinations (θ = 90°) [4].*

2. Dimensionless wetted perimeter for gas phase (S_G) and for liquid phase (S_L)

$$S_G = \cos^{-1}(2h^* - 1) \qquad [6.2.20a]$$

$$S_L = \left[\pi - \cos^{-1}(2h^* - 1)\right] \qquad [6.2.20b]$$

3. Dimensionless interfacial length between gas and liquid phases (S_i)

$$S_i = \left[1 - (2h^* - 1)\right]^{0.5} \qquad [6.2.21]$$

4. Dimensionless hydraulic diameter for gas phase (D_G) and liquid phase (D_L)

$$D_G = 4A_G/(S_L + S_i) \qquad [6.2.22a]$$

$$D_L = 4A_L/S_L \qquad [6.2.22b]$$

where h = dimensionless liquid in pipe (= H_L/D)

Dimensionless liquid level h can be expressed in terms of H_L as shown in Figures 6.2.16 to 6.2.19, but in terms of h that approach is much easier for the circular pipe cross-section. Since the liquid holdup is

$$H_L = 4A_L/\pi, \qquad [6.2.23]$$

It is necessary to find h first, using the expression in Equations 6.2.19 to 6.2.22. The value of h must be found numerically, or alternatively, it can be estimated from Figure 6.2.20. After geometric parameters are calculated (Equations 6.2.19 to 6.2.22), the friction factors for the gas and liquid phase has to be calculated:

$$f_{wg} = \begin{cases} 0.046/\mathrm{Re}_g^{0.2} & \text{if } \mathrm{Re}_g = \left(\dfrac{\gamma_g D_G D v_{sg}}{\mu_g(1-h_L)}\right) \geq 1500 \\ 16/\mathrm{Re}_g & \text{if } \mathrm{Re}_g < 1500 \end{cases} \qquad [6.2.24a]$$

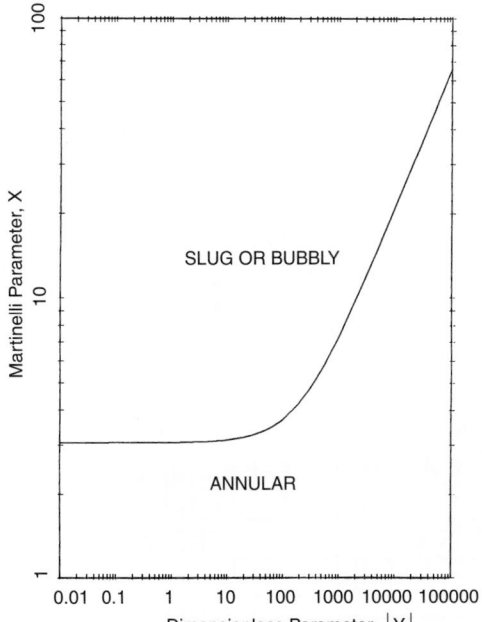

Figure 6.2.12 *Flow regime map for slug-annular transition for vertical downward inclination ($\theta = -90°$) [4].*

$$f_{wL} = \begin{cases} 0.046/Re_L^{0.2} & \text{if } Re_L = \left(\frac{\gamma_g D_L D v_{sL}}{\mu_L H_L}\right) \geq 1500 \\ 16/Re_L & \text{if } Re_L < 1500 \end{cases} \quad [6.2.24b]$$

Then the frictional and gravitational pressure gradients given by Expression 6.2.6 are

$$\left(\frac{dp}{dL}\right)_f = -\left[2f_{wL}\left(\frac{\gamma_L v_{sL}^2}{\pi D}\right)\left(\frac{1}{H_L}\right)^2 (S_L)\right]$$

$$\quad - \left[2f_{wg}\left(\frac{\gamma_g v_{sg}^2}{\pi D}\right)\left(\frac{1}{1-H_L}\right)(S_G)\right] \quad [6.2.25a]$$

$$\left(\frac{dp}{dL}\right)_{el} = \left(\frac{dp}{dZ}\right)\sin\theta = -g\left[H_L\gamma_L + (1-H_L)\gamma_g\right]\sin\theta \quad [6.2.25b]$$

where L = distance in ft.
z = vertical coordinate in ft.
S = pipeline inclination from the horizontal in degree

6.2.5.2 Special Cases for Low and High Liquid Holdup
As $H_L \approx 1.0$ or 0.0, it becomes difficult to determine the liquid holdup accurately. Small errors in the estimation lead to large errors in the frictional portion of the pressure gradient by Equation 6.2.25a; therefore, in such cases as above, it is recommended that the pressure gradient be calculated by the following methods. If the liquid holdup $H_L > 0.99$, then

$$\left(\frac{dp}{dL}\right)_t = -\left[f_{wL}\left(\frac{\gamma_L v_{sL}}{2D}\right)^2\left(\frac{1}{H_L}\right)^2\left(\frac{S_L}{A_L}\right)\right]$$

$$+\left[\left(\frac{f_i}{f_{wg}}\right)f_{wg}\left(\frac{\gamma_g v_{sg}^2}{2D}\right)\left(\frac{1}{1-H_L}\right)^2\left(\frac{S_i}{A_L}\right)\right] - g\gamma_L\sin\theta \quad [6.2.26a]$$

If the liquid holdup $H_L < 0.01$, then

$$\left(\frac{dp}{dL}\right)_t = -\left\{f_{wg}\frac{\gamma_g v_{sg}^2}{2D}\left(\frac{1}{1-H_L}\right)^2\left[\left(\frac{S_G}{A_G}\right)+\left(\frac{f_i}{f_{wg}}\right)\right]\right.$$

$$\left.\times\left(\frac{S_i}{A_G}\right)\right]\right\} - g\gamma_g\sin\theta \quad [6.2.26b]$$

where the gravitational pressure gradient is still the same as in Equation 6.2.25b and the frictional pressure gradient can be determined by subtraction from Equation 6.2.26a or 6.2.26b.

6.2.6 Annular Flow Regime
In the annular flow regime, the extent of liquid entrainment must first be estimated, then the liquid holdup and the pressure gradient can be calculated.

6.2.6.1 Liquid Entrainment E_d
Calculation methods in this area have not been validated and may be poor. First, calculation v_e (critical gas velocity on set of entrainment) is made as follows:

$$v_e = 0.00025\left(\frac{\gamma_L}{\gamma_g}\right)^{0.5}\left(\frac{\sigma_{go}}{\mu_g}\right) \quad [6.2.27]$$

If the value of the $v_{sg} < v_e$, then there is no entrainment $E_d = 0$. If $v_{sg} > v_e$, then the entrainment fraction should be estimated by

$$E_d = 1 - \exp\left[0.23\left(\frac{v_e - v_{sg}}{v_e}\right)\right] \quad [6.2.28]$$

where E_d = mass fraction or volume fraction of the total liquid flow that is in the form of entrained droplets

Equation 6.2.28 is an empirical correlation without experimental basis. Figure 6.2.21 graphically shows the liquid entrainment fraction for various values of the critical entrainment velocity, v_e.

6.2.6.2 Liquid Holdup H_L
After E_d fraction is calculated, the liquid holdup can be estimated. To determine H_L dimensionless specific weight ratio γ_c^* has to be known:

$$\gamma_c^* = \frac{\gamma_c}{\gamma_g} = \left[E_d\left(\frac{\gamma_L}{\gamma_g}\right)\left(\frac{v_{sL}}{v_{sg}}\right)+1\right] \quad [6.2.29]$$

γ_c^* = (density of a gas − droplet mixture in the core of the annular flow)/(density of the gas phase)

with this density ratio, other dimensionless parameters could be defined:

$$X_a = \frac{X}{(\gamma_c^*)^{0.5}} \quad [6.2.30a]$$

$$Y_a = \left(\frac{Y}{\gamma_c^*}\right)\left(\frac{\gamma^* - \gamma_c^*}{\gamma^* - 1}\right) \quad [6.2.30b]$$

where X and Y are the same as the values for the stratified flow regime.

With these parameters, Figure 6.2.22a–d can be used to estimate the liquid holdup in the liquid film H_{Lf} for the annular flow regime.

The total liquid holdup in the annular flow regime is calculated as follows:

$$H_L = H_{Lf} + \left(\frac{E_d v_{sL}}{E_d v_{sL} + v_{sg}}\right) \quad [6.2.31]$$

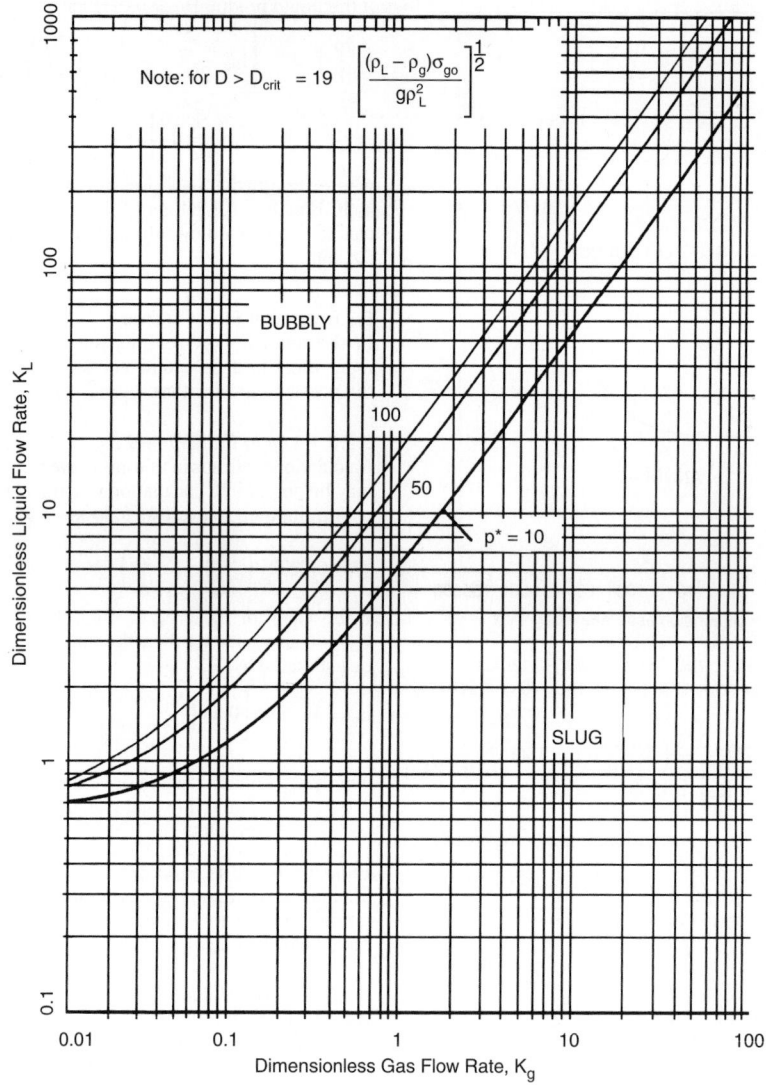

Figure 6.2.13 *Flow regime map for bubble-slug transition for vertical downward inclinations (θ = −90°) [4].*

6.2.6.3 Pressure Gradient

The friction factor for the liquid phase needs to be calculated first.

$$f_{wL} = \begin{cases} 0.046/Re_L^{0.2} & \text{if } Re_L = \left[\dfrac{\gamma_g D v_{sL}(1 - E_d)}{\mu_L}\right] \geq 1500 \\ 16/Re_L & \text{if } Re_L < 1500 \end{cases}$$

[6.2.32]

The fractional and gravitational pressure gradients are

$$\left(\frac{dp}{dL}\right)_f = -\left[2f_{wL}\left(\frac{\gamma_L v_{sL}^2}{D}\right)\left(\frac{1 - E_d}{H_{Lf}}\right)^2\right]$$

[6.2.33a]

and

$$\left(\frac{dp}{dL}\right)_g = g\left[\gamma_L H_{Lf} + (1 - H_{Lf})\gamma_c\right]\sin\theta$$

[6.2.33b]

6.2.6.4 Special Case for Low Liquid Holdup

For low values of the liquid holdup, the approach described below is recommended. If $H_{Lf} < 0.002$, then

$$f_{wg} = \begin{cases} 0.046/Re_g^{0.2} & \text{if } Re_g = \left[\dfrac{\gamma_g D v_{sg}}{\mu_g(1 - H_{Lf})}\right] \geq 1500 \\ 16/Re_g & \text{if } Re_g < 1500 \end{cases}$$

[6.2.34a]

Then the total pressure gradient is

$$\left(\frac{dp}{dL}\right)_{total} = -2(1 + 75H_{Lf})(f_{wg})\left(\frac{\gamma_c v_{sg}^2}{D}\right)\left(\frac{1}{1 - H_{Lf}}\right)^{5/2}$$

$$- g\gamma_c \sin\theta$$

[6.2.34b]

$(dp/dL)_g$ from Equation 6.2.33b could be calculated.

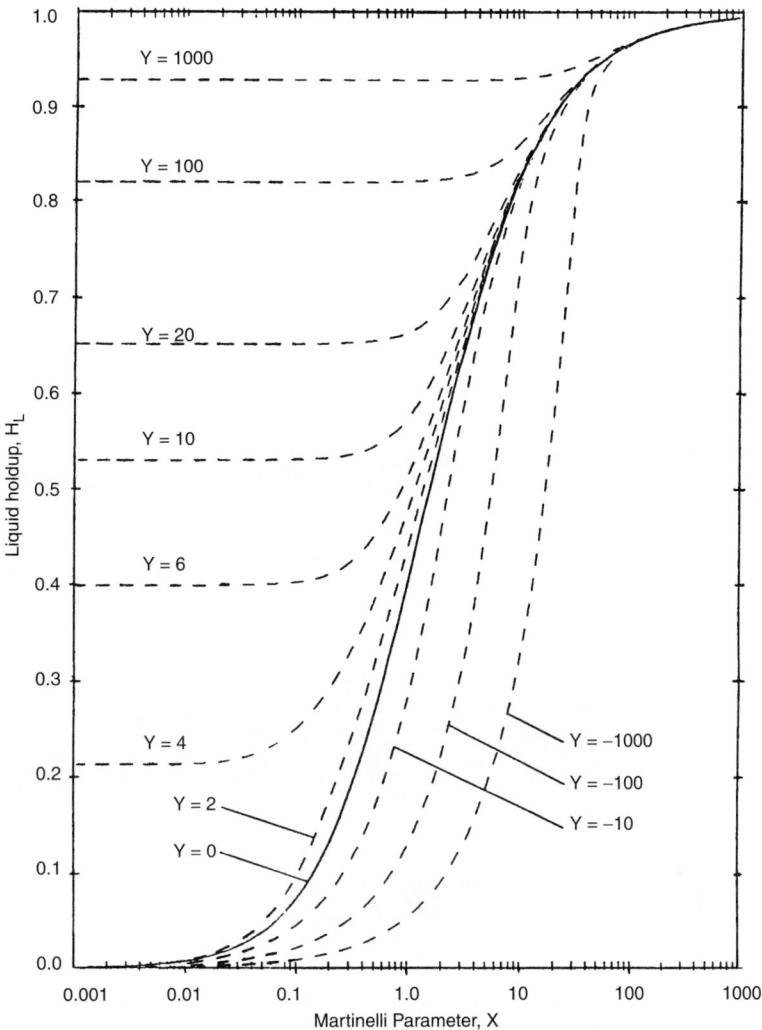

Figure 6.2.14a *Liquid holdup in stratified flow regime ($f_i/f_{wg} = 1$) [4].*

6.2.7 Slug Flow Regime

In slug flow, the liquid slugs tend to contain some gas. The fraction of liquid in the liquid slugs can be estimated by

$$E_{Ls} = \frac{1}{1 + \left(\dfrac{v_m}{28.4}\right)^{1.39}} \qquad [6.2.35]$$

Figure 6.2.23 shows the liquid holdup versus the mixture velocity v_m by this correlation.

6.2.7.1 Slug Velocity

The velocity of the liquid slugs are gas bubbles is determined by

$$v_s = C_o v_m + k \left[\frac{gD(\gamma_L - \gamma_g)}{\gamma_L} \right]^{0.5} \qquad [6.2.36]$$

For C_o and k, see Table 6.2.1.

6.2.7.2 Liquid Holdup

The overall liquid holdup is obtained by

$$H_L = 1 - \frac{v_{sg} + (1 - E_{Ls})(v_s - v_m)}{v_s} \qquad [6.2.37]$$

6.2.7.3 Pressure Gradient

Two cases will be considered, slightly inclined flow and vertical flow. Pressure for slightly inclined flow is a function of an average liquid velocity v_L, a friction factor f_L, and an average slug density ρ_{Ls}.

$$v_L = v_{sL}/H_L \qquad [6.2.38a]$$

$$f_L = \begin{cases} 0.046/Re_L^{0.2} & \text{if} \quad Re_L = \left(\dfrac{\gamma_L D v_L}{\mu_L}\right) \geq 1500 \\[2mm] 16/Re_L & \text{if} \quad Re_L < 1500 \end{cases}$$

$$[6.2.38b]$$

$$\gamma_{sL} = [E_{Ls}\gamma_L + (1 - E_{Ls})\gamma_g] \qquad [6.2.38c]$$

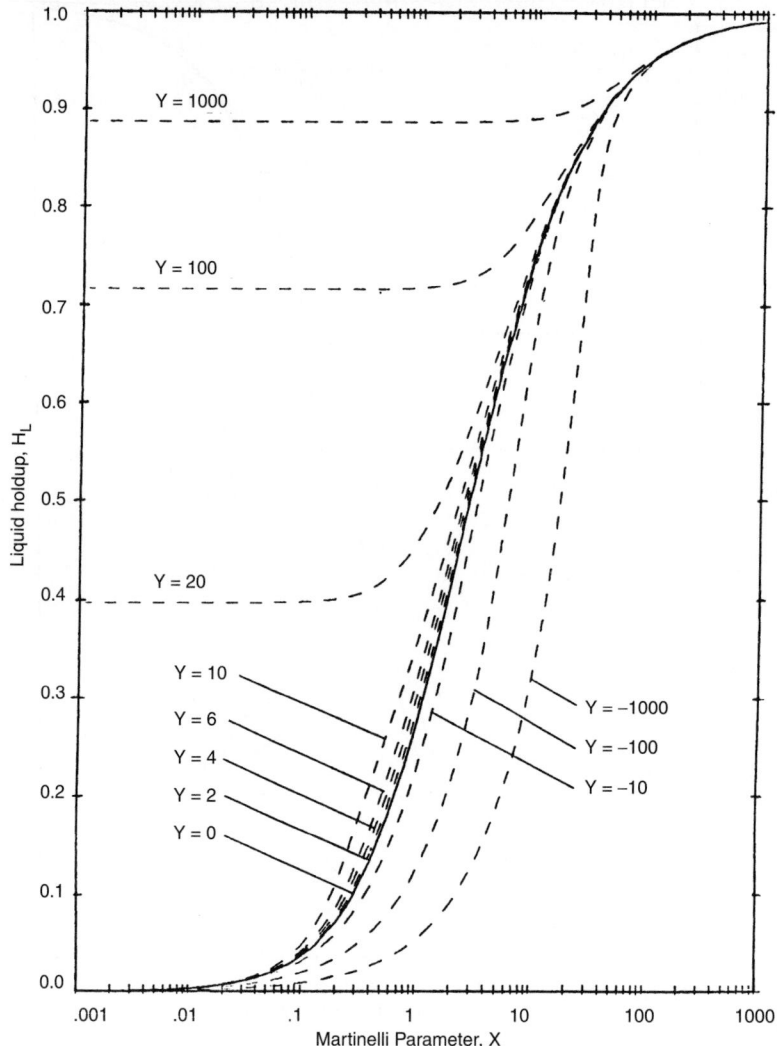

Figure 6.2.14b   *Liquid holdup in stratified flow regime ($f_i/f_{wg} = 5$) [4].*

The frictional pressure gradient is then calculated by

$$\left(\frac{dp}{dL}\right)_f = -\left(\frac{2f_L\gamma_{Ls}v_L^2}{D}\right) \qquad [6.2.39a]$$

and the gravitational pressure gradient is

$$\left(\frac{dp}{dL}\right)_g = -g\left[H_L\gamma_L + (1 - H_L)\gamma_g\right]\sin\theta \qquad [6.2.39b]$$

For vertical flow, the fractional pressure gradient is calculated by

$$\left(\frac{dp}{dL}\right)_f = -\left[2f_m\left(\frac{\gamma_{Ls}v_m^2}{D}\right)H_L\right] \qquad [6.2.40]$$

where γ_{Ls}, from Equation 6.2.38c, with $E_{Ls} = 0.75$

$$H_L = \frac{v_{sL}}{v_L} = \frac{\text{liquid superficial velocity}}{\text{avg. liquid velocity in slug flow}}$$

and f_m, from Equation 6.2.38b, using the mixture velocity v_m (replacing v_L) in the Reynolds number.

6.2.7.4 Optional Correction

The approximation to the pressure gradient above neglects the liquid holdup in the liquid film around the gas bubble. This holdup may be significant for long gas bubbles that occur in wells of gas and oil pipelines. Thus Equations 6.2.40 and 6.2.39b will overpredict the pressure gradient. If greater accuracy is desired, a closer estimate (which may tend to underpredict the pressure gradient) is possible.

The thickness δ^* (dimensionless thickness of liquid film in slug bubble) of the liquid film is estimated and the result used to modify the liquid holdup. To get the liquid film thickness, two dimensionless parameters are first calculated; dimensionless velocity ratio for slug flow in risers N_f and dimensionless velocity ratio for slug flow in risers v^*:

$$N_f = \left[D^3 g(\gamma_l - \gamma_g)\gamma_L\right]^{0.5}/\mu_L \qquad [6.2.41a]$$

$$v^* = v_m\gamma_L^{0.5}/k\left[gD(\gamma_L - \gamma_g)\right]^{0.5} \qquad [6.2.41b]$$

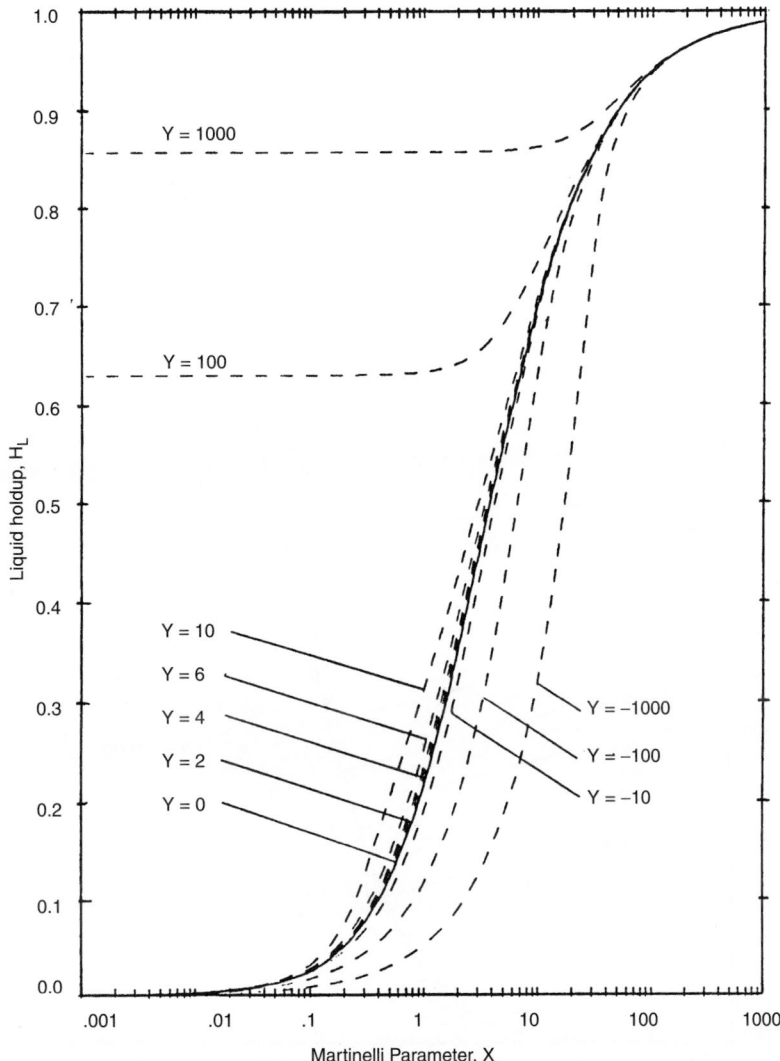

Figure 6.2.14c *Liquid holdup in stratified flow regime ($f_i/f_{wg} = 10$) [4].*

In Figure 6.2.24, the line of constant v^* is found first; then the line of constant N_f is located. The "turbulent film" line is a limiting case for large values of N_f. The intersection of these two lines determines the film thickness δ^*. The liquid holdup calculated by Equation 6.2.37 is then modified by

$$H'_L = 1 - (1 - H_L)/(1 - 2\delta^*)^2 \qquad [6.2.42]$$

This modified value of the liquid holdup is used in Equation 6.2.40 and 6.2.39b to determine the pressure gradient, where the other parameters are calculated as before.

6.2.8 Bubby Flow Regime
The *liquid holdup* is found by a drift-flux model

$$H_L = 1 - \left(\frac{v_{sg}}{C_o v_m + k v_\infty}\right) \qquad [6.2.43a]$$

where the velocity v_∞ is

$$v_\infty = \left[\frac{g\sigma_{go}(\gamma_L - \gamma_g)}{\gamma_L^2}\right]^{0.25} \qquad [6.2.43b]$$

and the parameters C_o and k are determined as below:

Pipe Inclination	C_o	k
$> 0°$	$1.2 - 0.2(\gamma_g/\gamma_L)^{0.5}$	1.4
$0°$	$1.2 - 0.2(\gamma_g/\gamma_L)^{0.5}$	0
$< 0°$	0.9	0

6.2.8.1 Pressure Gradient
The friction factor for the liquid is first calculated by

$$f_{wLm} = \begin{cases} 0.046/Re_{Lm}^{0.2} & \text{if } Re_{Lm} = \left(\frac{\gamma_L D v_m}{\mu_g}\right) \geq 1500 \\ 16/Re_{Lm} & \text{if } Re_{Lm} < 1500 \end{cases} \qquad [6.2.44a]$$

The pressure gradient due to friction in the bubbly flow regime is evaluated from

$$\left(\frac{dp}{dL}\right)_f = \frac{2f_{wLm}v_m(\gamma_g v_{sg} + \gamma_L v_{sL})}{D} \qquad [6.2.44b]$$

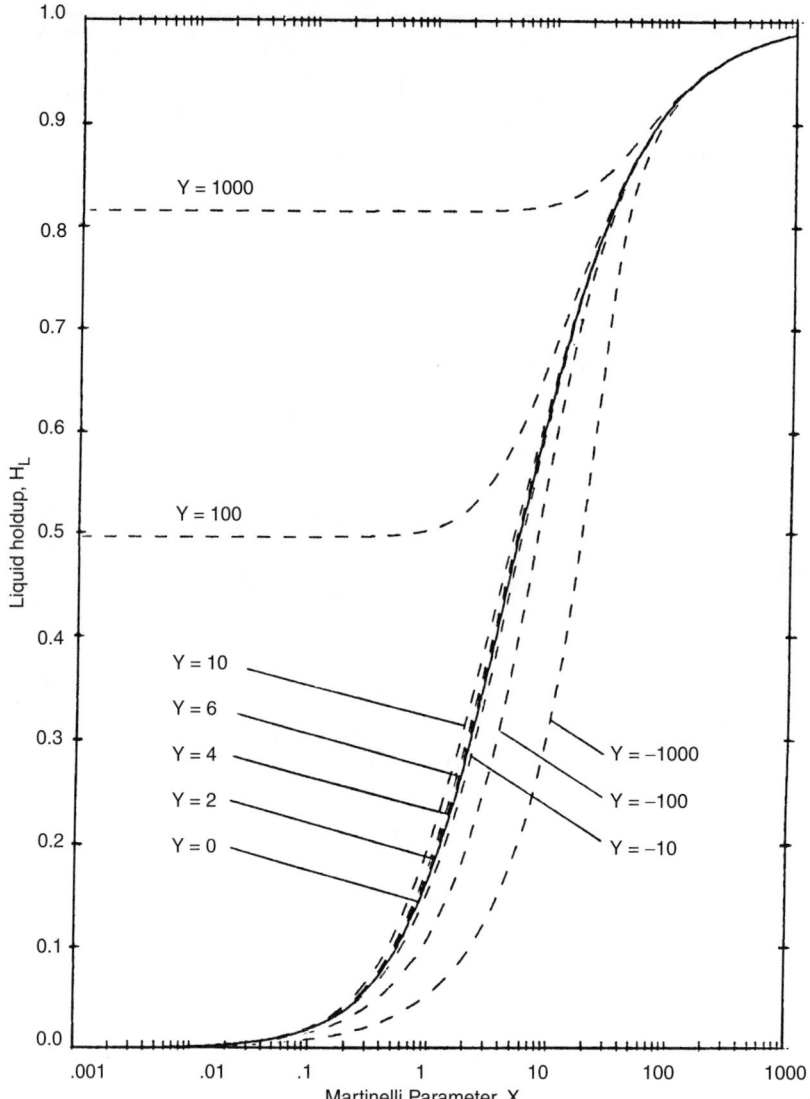

Figure 6.2.14d *Liquid holdup in stratified flow regime (f_i/f_{wg} = 20) [4].*

and the gravitational pressure gradient

$$\left(\frac{dp}{dL}\right)_g = -g\left[H_L\gamma_L + (1-H_L)\gamma_g\right]\sin\theta \qquad [6.2.44c]$$

6.2.9 Correction for Acceleration Effects

The methods used in this section so far neglect the contribution of acceleration effects to the pressure gradient.

The acceleration pressure gradient in pipeline flow is due to the changes in the fluid properties through the pipe segment. These changes include expansion of the gas phase, expansion of the liquid phase, and the changes in quality (due to phase behavior).

To account for acceleration effects, the total pressure gradient should be calculated by:

$$\left(\frac{dp}{dL}\right)_t = \frac{\left(\frac{dp}{dL}\right)_f + \left(\frac{dp}{dL}\right)_g}{1+A} \qquad [6.2.45a]$$

where the parameter A is calculated by the method appropriate for each flow regime as follows:

$$A + \begin{cases} = 0 \quad \text{to neglect acceleration effects} \\[2mm] = (\gamma_g v_{sg} + \gamma_L v_{sL})^2\left[X\left(\frac{dV_g}{dp}\right) + (1-X)\left(\frac{dV_L}{dp}\right)\right] \\ \quad \text{for bubbly or slug flow regime} \\[2mm] = (\gamma_g v_{sg} + \gamma_L v_{sL})^2\left[\frac{X^2}{(1-H_L)}\left(\frac{dV_g}{dp}\right) + \frac{(1-x)^2}{H_L}\left(\frac{dV_L}{dp}\right)\right] \\ \quad \text{for stratified or annular flow regime} \end{cases}$$

$$[6.2.45b]$$

d_g/d_p and d_L/d_p represent the change in specific volume for each phase with a change in pressure as evaluated from the fluid properties. The acceleration portion of the pressure gradient is calculated by

$$(dp/dL)_a = -A(dp/dL)_t \qquad [6.2.45c]$$

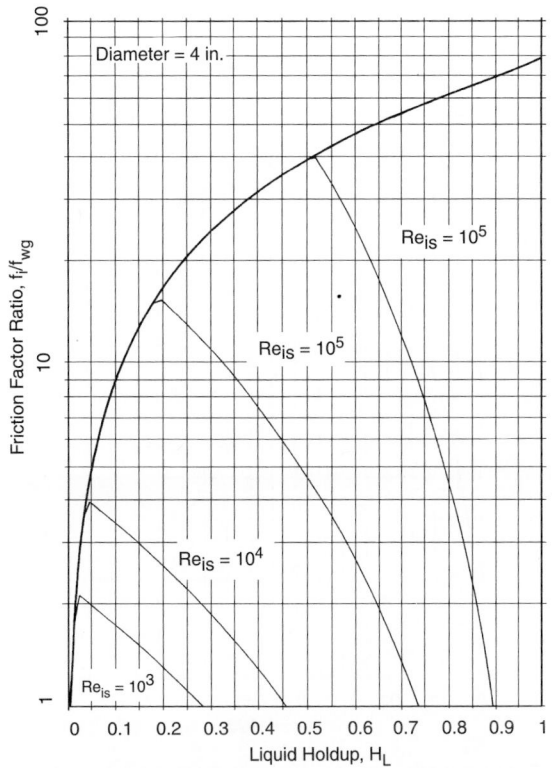

Figure 6.2.15a *Interfacial friction factor ratio for stratified flow regime (D = 4 in.) [4].*

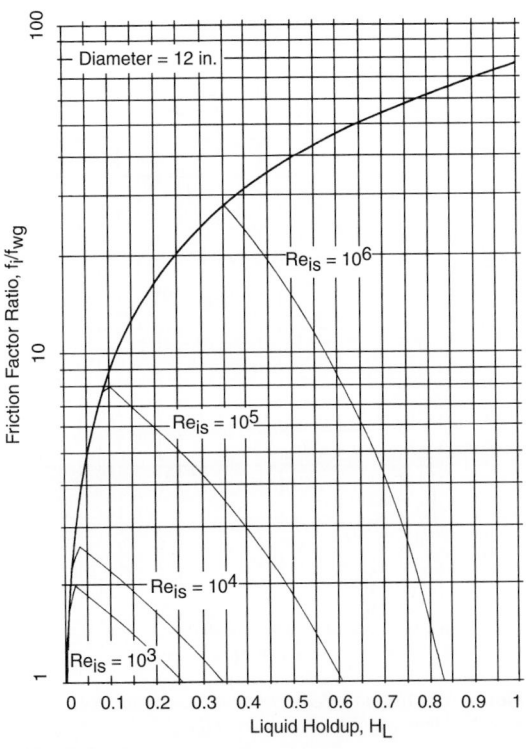

Figure 6.2.15c *Interfacial friction factor ratio for stratified flow regime (D = 12 in.) [4].*

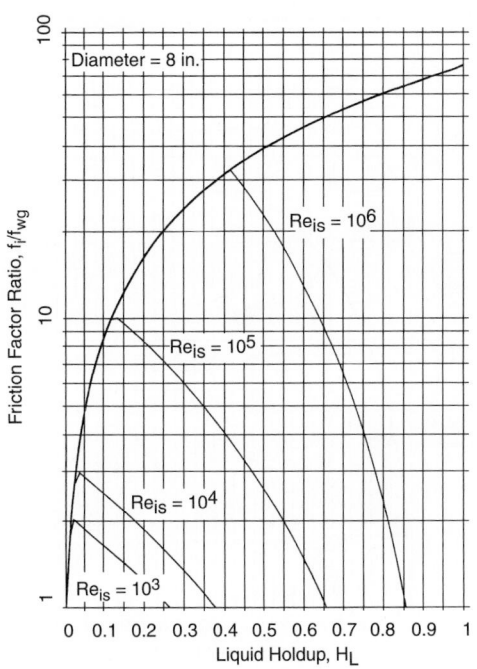

Figure 6.2.15b *Interfacial friction factor ratio for stratified flow regime (D = 8 in.) [4].*

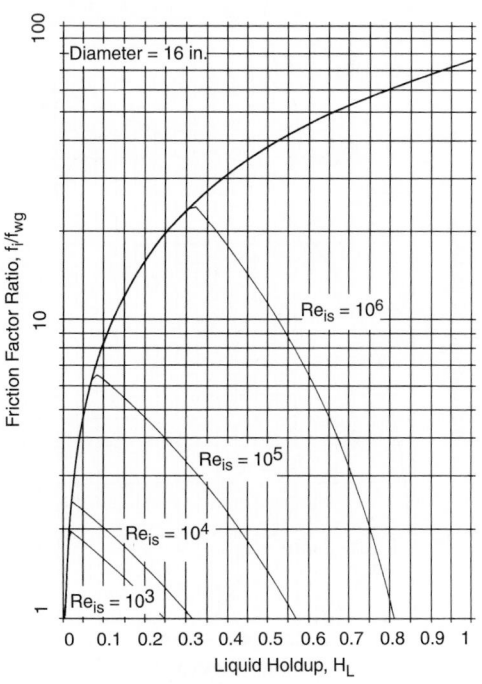

Figure 6.2.15d *Interfacial friction factor ratio for stratified flow regime (D = 16 in.) [4].*

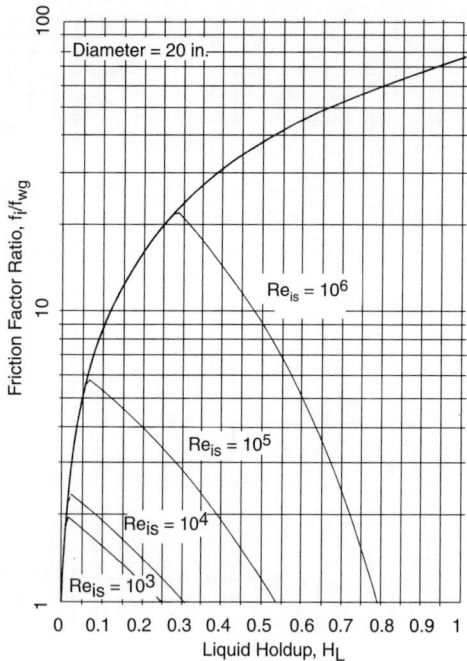

Figure 6.2.15e *Interfacial friction factor ratio for stratified flow regime (D = 20 in.) [4].*

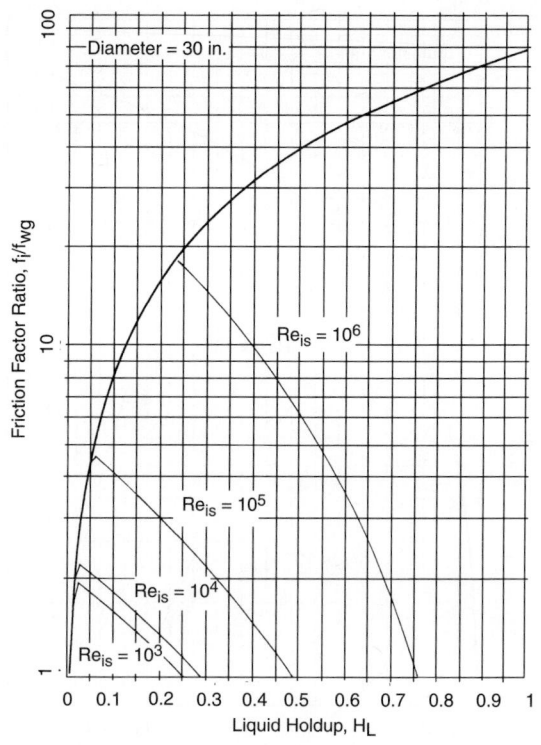

Figure 6.2.15g *Interfacial friction factor ratio for stratified flow regime (D = 30 in.) [4].*

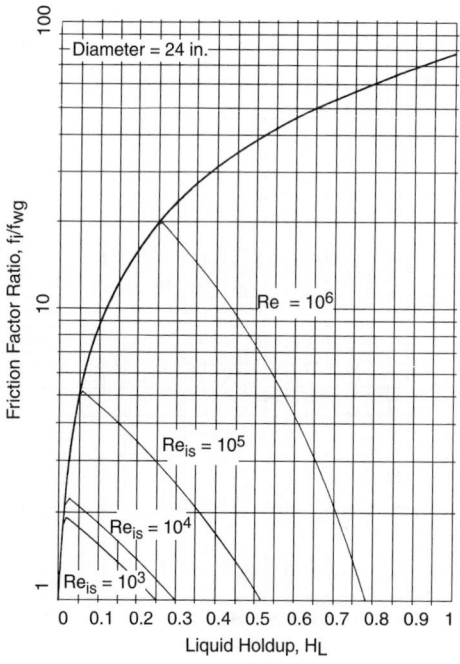

Figure 6.2.15f *Interfacial friction factor ratio for stratified flow regime (D = 24 in.) [4].*

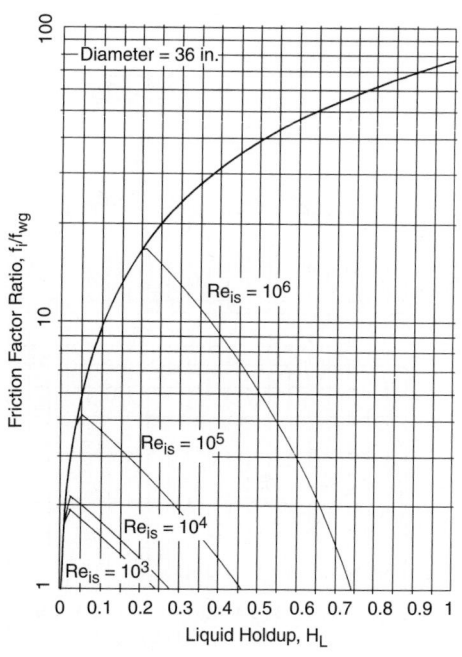

Figure 6.2.15h *Interfacial friction factor ratio for stratified flow regime (D = 36 in.) [4].*

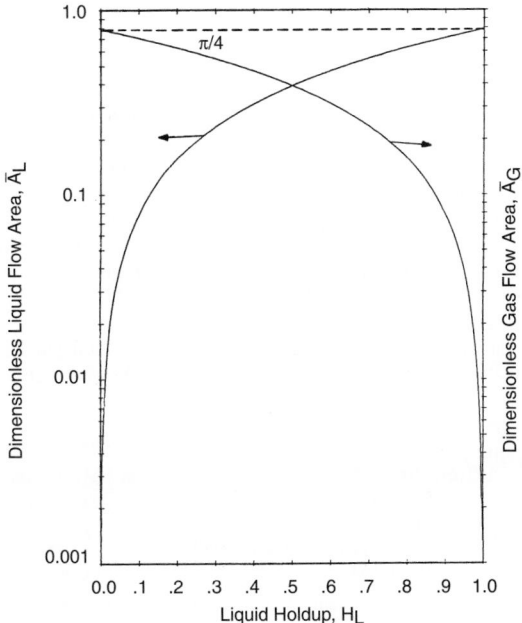

Figure 6.2.16 *Dimensionless cross-sectional areas in stratified flow regimes [4].*

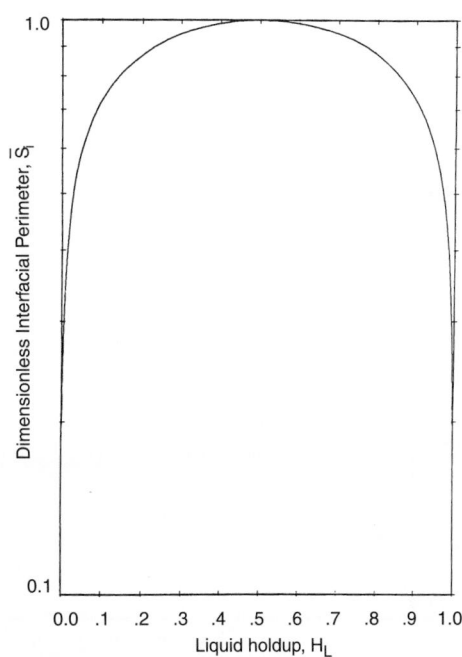

Figure 6.2.18 *Dimensionless interfacial perimeters in stratified flow regimes [4].*

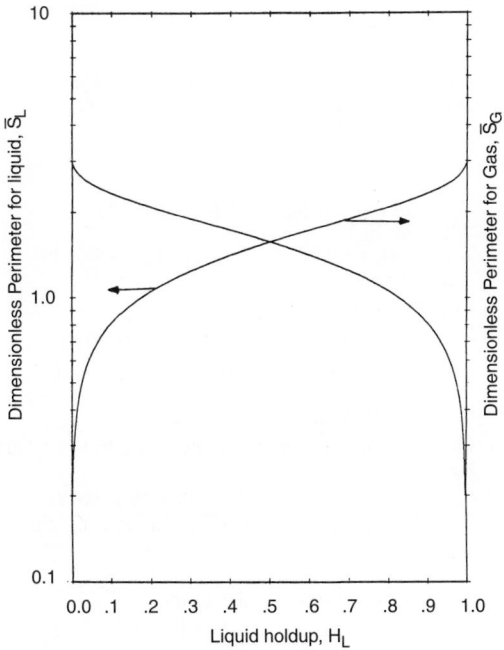

Figure 6.2.17 *Dimensionless wetted perimeters in stratified flow regimes [4].*

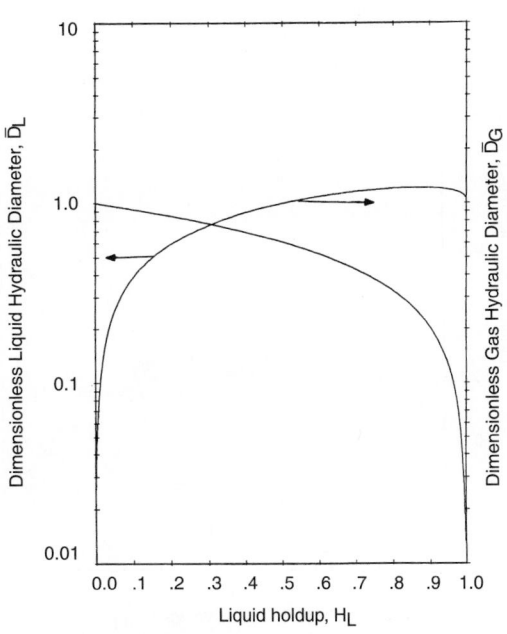

Figure 6.2.19 *Dimensionless hydraulic diameters in stratified flow regimes [4].*

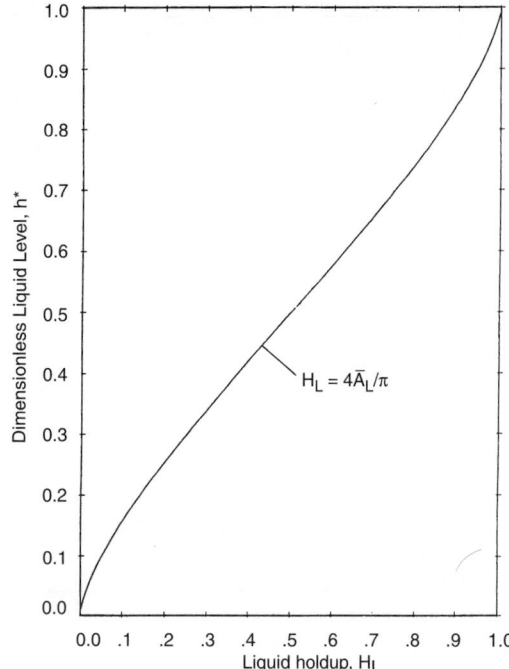

Figure 6.2.20 *Dimensionless liquid level stratified flow regimes [4].*

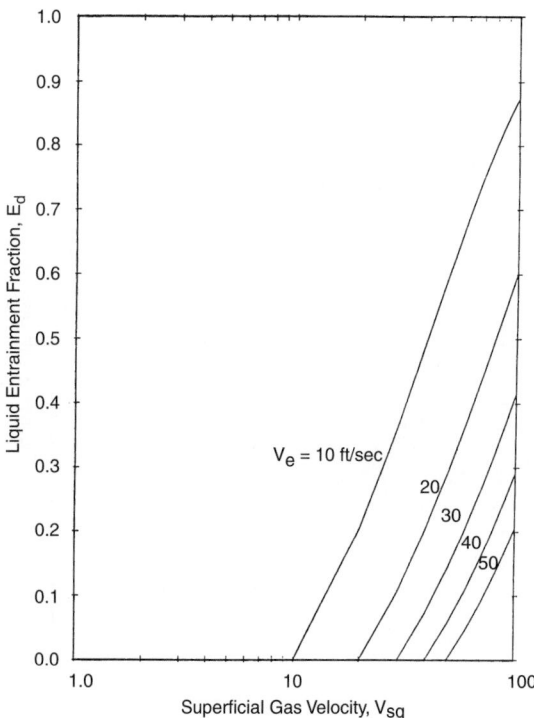

Figure 6.2.21 *Liquid entrainment fraction in annular flow regime [4].*

6.2.10 Limitation
The described-above methods can be applied under the following conditions:

1. flow mass in pipe in constant (steady-state flow)
2. only the gas–liquid phase flow is considered; the liquid phase is treated as an oil
3. the temperature is constant and equal to the average temperature

The above procedure presents the multiphase methods in a simplified form to permit quick prediction to be made. The same problems can be solved based on computer methods that are presented in Volumes 2 and 3 of the AGA project [1].

Example 6.2.1
Calculate the pressure gradient for a horizontal pipeline using the flow regime maps. The following data are given:

$q_g = 40$ MMscf/d
$q_0 = 40,000$ stb/d
ID $= 9$ in.
API(gravity) $= 33°$
$P_{avg} = 2000$ psia
$T_{avg} = 80°F$
S.G. $= 0.75$ at 14.7 psia and T $= 60°F$
$R_p = 990$ scf/bbl

Also calculate missing fluid properties using proper correlations.
dp/dL $=$?

1. Stock-tank oil specific weight γ_{sto}
$$\gamma_{sto} = 62.37[141.5/(131.5 + °API)]$$
$$= 62.37[141.5/(131.5 + 33)]$$
$$= 53.65 \, lb/ft.^3$$

2. Gas specific weight at std conditions γ_{gsc}
$$\gamma_{gsc} = 0.0763\gamma_g = 0.0763(0.75)$$
$$= 0.05723 \, lb/ft.^3$$

3. Pipe flow area
$$A_p = (\pi/4)D^2 = \pi/4(9/12)^2 = 0.4418 \, ft.^2$$

4. Calculate Z, μ_g, μ_0, and GOR at P_{avg} and T_{avg}. From Gas P–V–T Program (see Chapter 5) at p $= 2000$ psia

T $= 80°F$
Z $= 0.685$
$\mu_g = 0.0185$ cp
From Oil P–V–T program at p $= 2000$ psia
T $= 80°F$
$R_p = 990$ scf/bbl
$\mu_0 = 2.96$ cp

5. Calculate γ_g at p_{avg}, T_{avg}, and Z_{avg}
$$\gamma_g = \frac{28.97(\gamma)p_{avg}}{Z_{ave}RT_{avg}} = \frac{28.97(0.75)(2000)}{10.685(10.73)(80 + 460)}$$
$$= 10.95 \, lb/ft.^3$$

6. Calculate vapor superficial velocity, v_{sg} from Equation 6.2.7
$$v_{sg} = \frac{q_g\rho_{gsc}}{A_p\rho_g(24)3600} = \frac{(40 \times 10^6)0.05724}{0.4418 \times 10.95 \times 24 \times 3600}$$
$$= 5.48 \, ft./s$$

7. Calculate liquid superficial velocity v_{sL}
$$v_{sL} = \frac{q_L}{A_p} = \frac{40,000(42)}{(7.481)24(3600)0.4418} = 5.88$$

8. Reynolds number and friction factor based on the superficial velocity, from Equation 6.2.9b

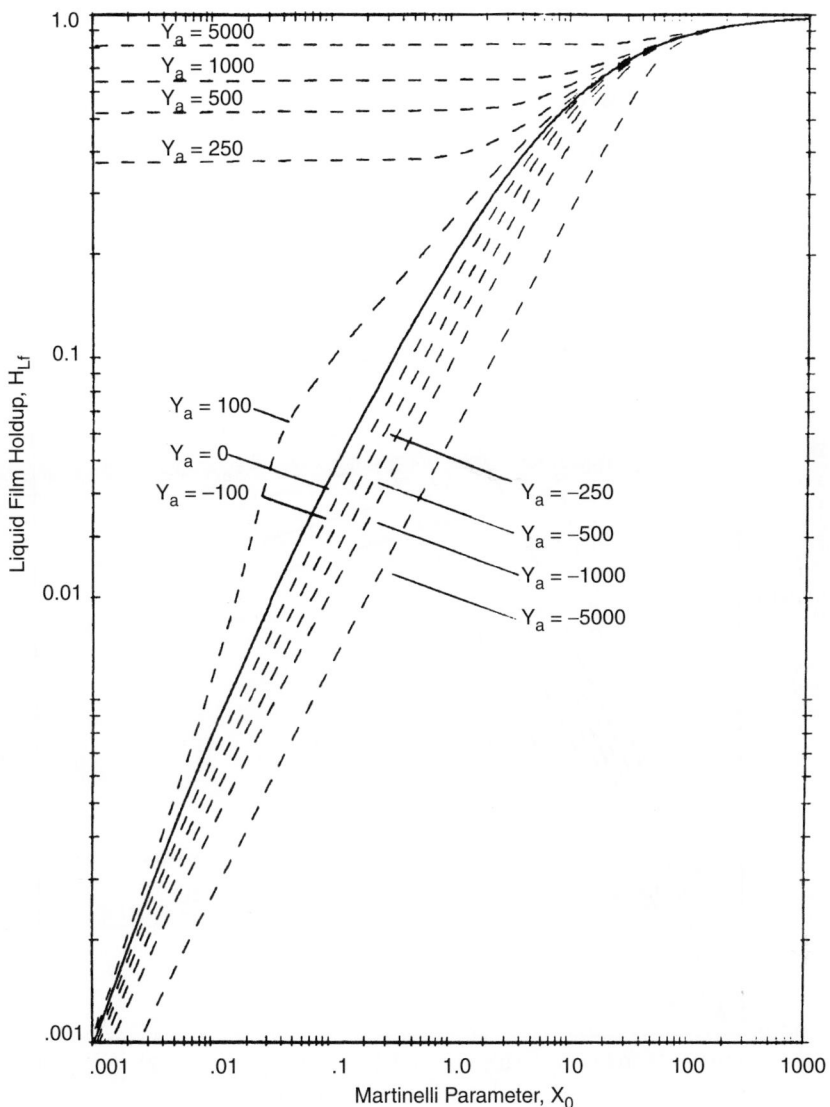

Figure 6.2.22a Liquid holdup in annular flow regime ($E_d = 0$) (expanded Y scale) [4].

a) Liquid

$$R_{eL} = \frac{\gamma_L D v_{sL}}{\mu_L g} = \frac{(53.56)(9/12)(5.88)}{(2.86)2.0886 \times 10^{-5}(32.2)}$$

$$= 118{,}851.7 > 1500$$

$$f_{wLs} = \frac{0.046}{R_{eL}^{0.2}} = \frac{0.046}{(118{,}851.7)^{0.2}} = 4.44 \times 10^{-3}$$

where 2.0886×10^{-5} is a unit conversion factor

b) Gas

$$R_{eg} = \frac{\gamma_g D v_{sg}}{\mu_g g} = \frac{(10.95)(9/12)(5.48)}{(0.0187)(2.0886 \times 10^{-5})(32.2)}$$

$$= 3{,}623{,}080 > 1500$$

$$f_{wgs} = \frac{0.046}{R_{eg}^{0.2}} = \frac{0.046}{(3{,}623{,}080)^{0.2}} = 2.24 \times 10^{-3}$$

9. Martinelli parameter X from Equation 6.2.9a

$$X = \left(\frac{2f_{wLs}\gamma_L v_{sL}^2 / D}{2f_{wgs}\gamma_g v_{sg}^2 / D} \right)^{0.5} = \left[\frac{4.44 \times 10^{-3}(53.65)(5.88)^2}{2.24 \times 10^{-3}(10.95)(5.48)^2} \right]^{0.5}$$

$$= 3.3$$

10. Dimensionless inclination Y from Equation 6.2.12

$$Y = \frac{g(\gamma_L - \gamma_g)\sin\theta}{2f_{wgs}\gamma_g v_g^2 / D}$$

since $\sin\theta = 0$ (horizontal pipe)

$$Y = 0$$

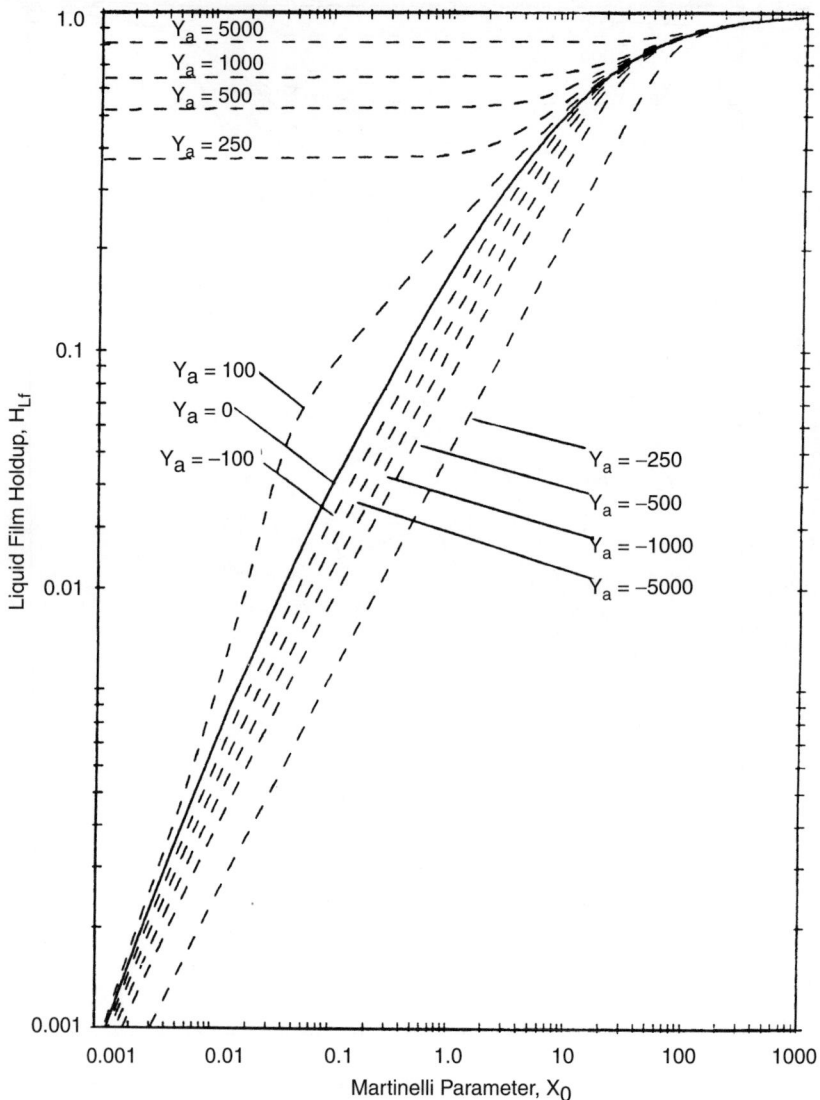

Figure 6.2.22b *Liquid holdup in annular flow regime ($E_d = 0.2$) (expanded Y scale) [4].*

11. Calculate gas Froude number

$$F_g = v_{sg}\left[\frac{\gamma_g}{(\gamma_L - \gamma_g)gD}\right]^{0.5}$$

$$= 5.48\left[\frac{10.95}{(53.65 - 10.95)32.2(9/12)}\right]^{0.5} = 0.56$$

12. Select flow regime for $X = 3.3$ and $Y = 0$. Figure 6.2.10a indicates slug flow.

13. Liquid holdup in the liquid slug, E_{Ls}, from Equation 6.2.7e

$$v_m = v_{sL} + v_{sg} = 5.88 + 5.48 = 11.36 \text{ ft./s}$$

from Equation 6.2.35

$$E_{sL} = \frac{1}{1 + \left(\dfrac{v_m}{28.4}\right)^{1.39}} = \frac{1}{1 + \left(\dfrac{11.36}{28.4}\right)^{1.39}} = 0.781$$

14. Slug velocity v_s from Equation 6.2.36

$$v_s = c_0 v_m + K\left[\frac{gD(\gamma_L - \gamma_g)}{\gamma_L}\right]^{0.5}$$

where $C_0 = 1.3$ and $K = 0$ for $\theta = 0°$

$$v_s = 1.3(11.36) + 0 = 14.768 \text{ ft./s}$$

15. Liquid holdup H_L, Equation 6.2.37,

$$H_L = 1 - \frac{v_{sg} + (1 - E_{Ls})(v_s - v_m)}{v_s}$$

$$= 1 - \frac{5.48 + (1 - 0.781)(14.768 - 11.36)}{14.768} = 0.578$$

16. For $\theta = 0°$, calculate average liquid velocity v_L

$$v_L = \frac{v_{sL}}{H_L} = \frac{5.88}{0.578} = 10.17 \text{ ft./s}$$

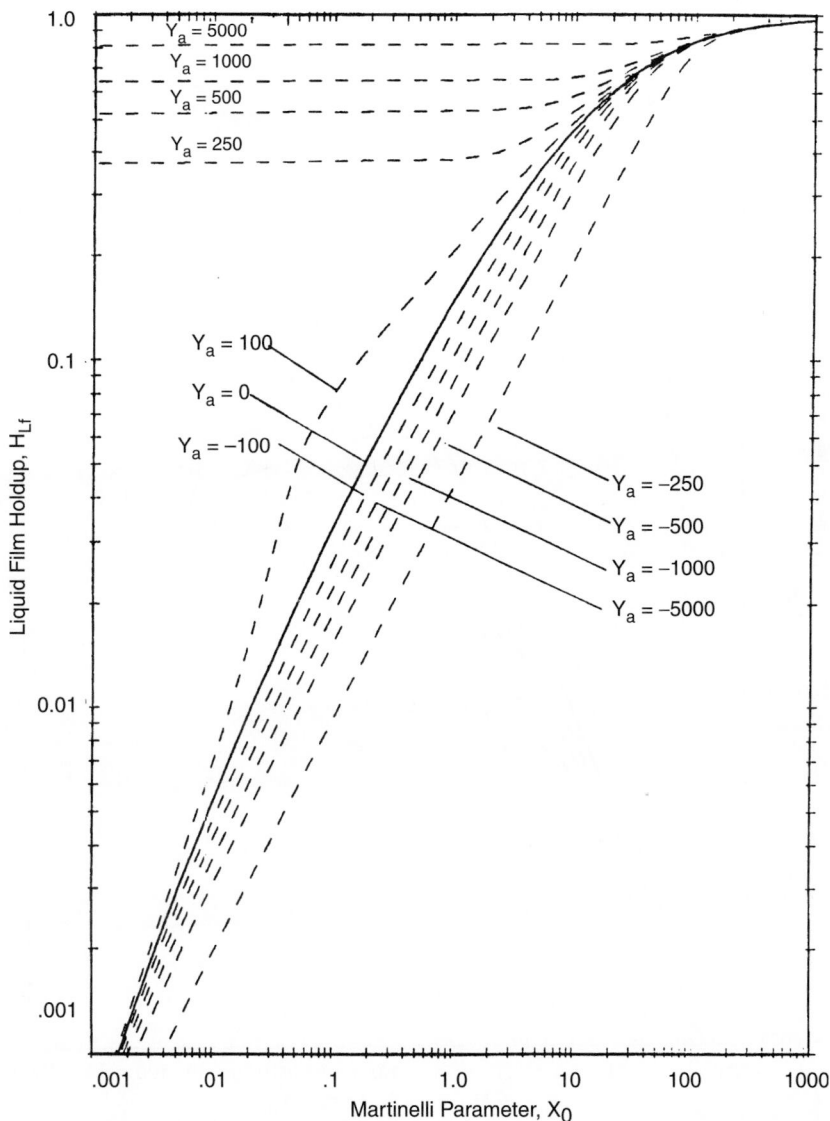

Figure 6.2.22c *Liquid holdup in annular flow regime ($E_d = 0.4$) (expanded Y scale) [4].*

17. Calculate R_{eL} and f_{wL} from Equation 6.2.39b

$$R_{eL} = \frac{\gamma_L D v_L}{\mu_L g} = \frac{53.65(9/12)(10.17)}{12.96(32.2)(2.0886 \times 10^{-5})}$$

$$= 205{,}565 > 1500$$

$$f_{wL} = \frac{0.046}{R_{eL}^{0.2}} = \frac{0.046}{(205{,}564)^{0.2}} = 3.983 \times 10^{-3}$$

18. Average slug density γ_{Ls} from Equation 6.2.39c

$$\gamma_{Ls} = E_{Ls}\gamma_L + (1 - E_{Ls})\gamma_g$$

$$= 0.781(53.65) + (1 - 0.7)(10.95)$$

$$= 44.3 \text{ lb/ft.}^3$$

19. The fractional pressure gradient from Equation 6.2.39a

$$\left(\frac{dp}{dL}\right)_f = -\left(\frac{2f_L \gamma_{Ls} v_L^2}{Dg}\right)$$

$$= -\frac{2(3.983 \times 10^{-3})(44.3)(10.17)^2}{(9/12)(32.2)}$$

$$= -1.51132 \text{ lb/ft.}^3$$

20. The total pressure gradient, since $\theta = 0°$

$$\left(\frac{dp}{dL}\right)_{\text{gravitational}} = 0$$

and

$$\left(\frac{dp}{dL}\right)_{\text{total}} = \left(\frac{dp}{dL}\right)_f = 1.51132 \text{ lb/ft.}^3$$

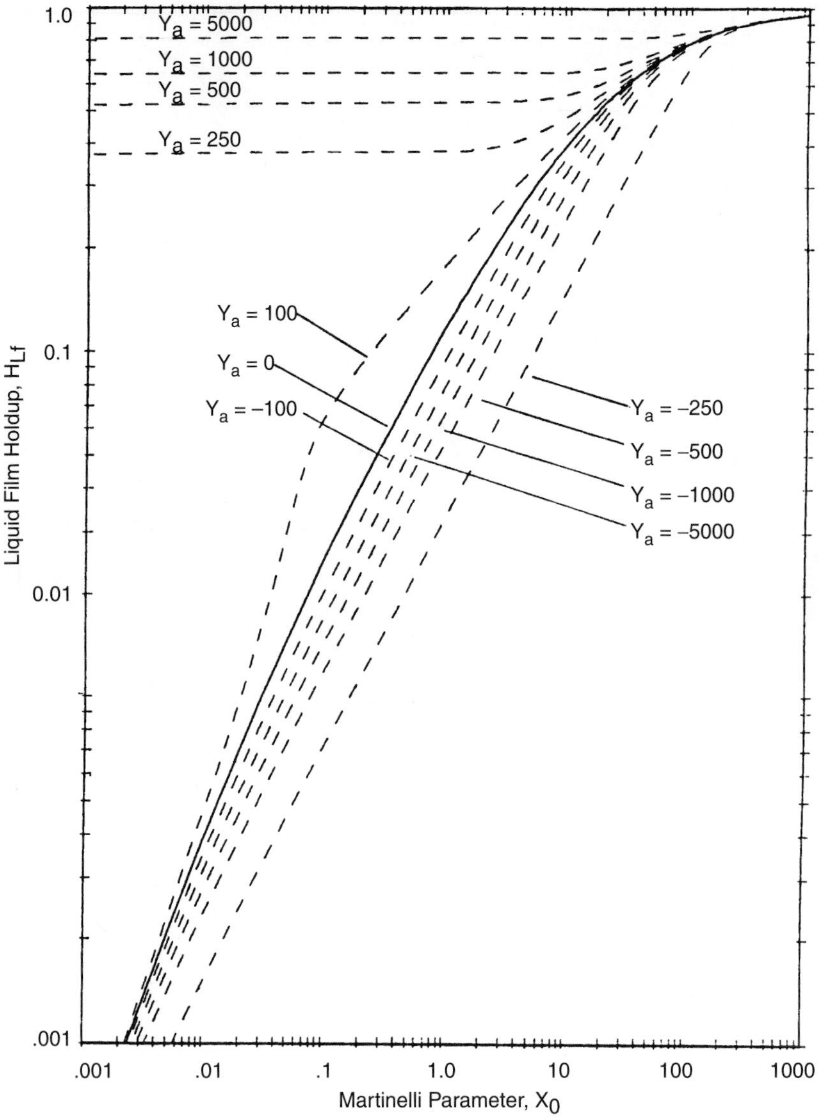

Figure 6.2.22d *Liquid holdup in annular flow regime ($E_d = 0.6$) (expanded Y scale) [4].*

or

$$\left(\frac{dp}{dL}\right)_{total} = -1.0495 \times 10^{-2} \text{ psi/ft.}$$

6.2.11 Empirical Methods

Many empirical correlations have been developed for predicting two-phase flowing pressure gradients, which differ in the manner used to calculate three components of the total pressure gradient (see Equation 6.2.6). Some of these correlations are described below. The range of applicability of these multiphase flow models is dependent on several factors such as tubing size or diameter, oil gravity, gas–liquid ratio and, two-phase flow with or without water-cut. The effect of each of these factors on estimating the pressure profile in a well is discussed for all empirical methods considered [6,7]. A reasonably good performance of the multiphase flow models, within the context of this section, is considered to have a relative error (between the measured and predicted values of the pressure profile) less than or equal to 20%.

6.2.12 The Duns-Ros Method [20,21]

To better understand the initial concept of the Duns-Ros method, Figure 6.2.25 shows a generalized flow diagram. The Dun-Ros correlation is developed for vertical flow of gas and liquid mixtures in wells. This correlation is valid for a wide range of oil and gas mixtures with varying water-cuts and flow regimes. Although the correlation is intended for use with "dry" oil/gas mixtures, it can also be applicable to wet mixtures with a suitable correction. For water contents less than 10%, the Duns-Ros correlation (with a correction factor) has been reported to work well in the bubble, slug (plug), and froth regions. Figure 6.2.26 shows that pressure gradient and holdup also depend significantly on superficial gas velocity.

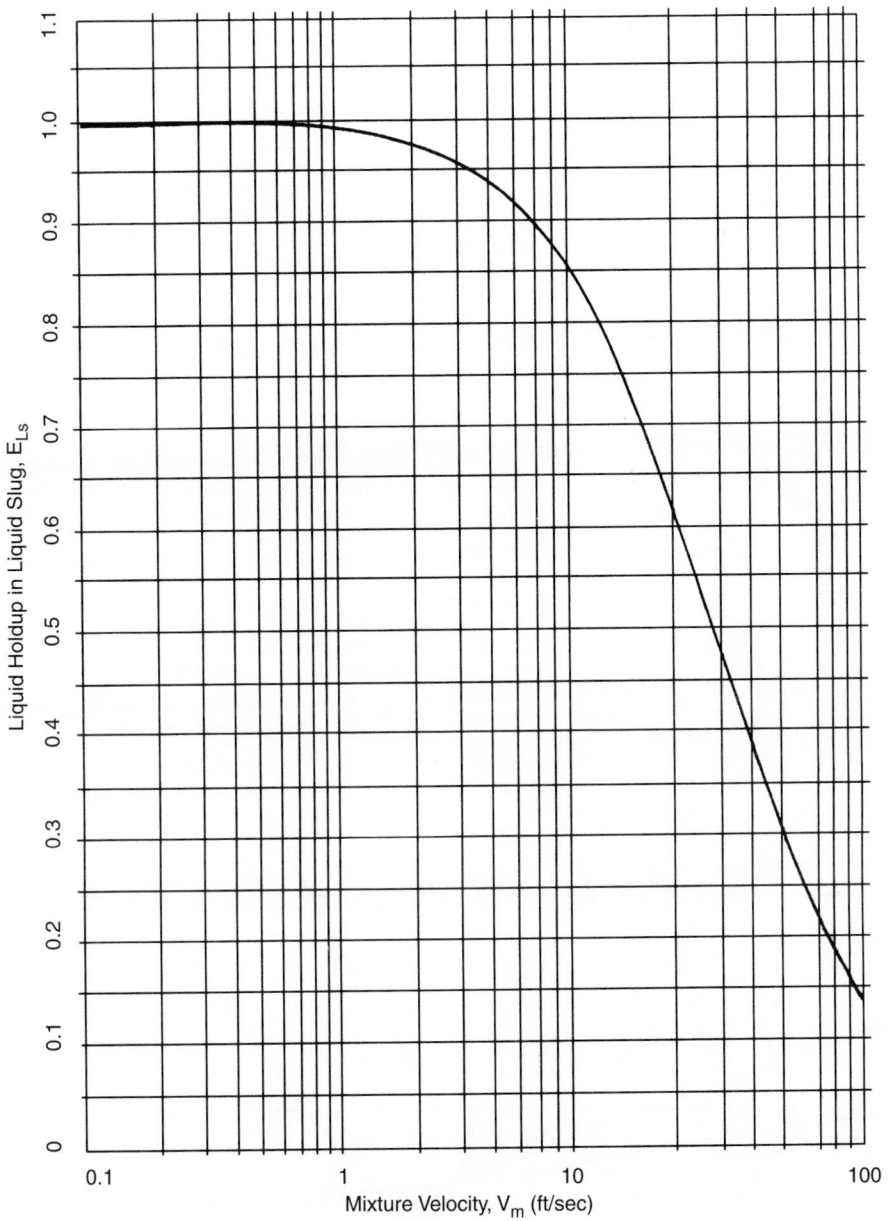

Figure 6.2.23 *Liquid holdup in liquid slug [4].*

At low gas flowrates, the pipe essentially is full of liquid since the gas bubbles are small. Holdup is approximately equal to unity. At liquid rates less than 1.3 ft/s (0.4 m/s), increased gas rate causes the number and size of the bubbles to increase. Ultimately, they combine into plugs that become unstable and collapse at still higher gas concentrations to form slugs. At gas rates greater than 49 ft/s (15 m/s), with the same liquid rate, mist flow is initiated, and gas is in the continuous phase with liquid drops dispersed in it. When the liquid velocity is over 5.3 ft/s (16 m/s), the flow patterns are not as observable. As gas flow increases, no plug flow is observed; flow is turbulent and frothy until some degree of segregation takes place at higher rates. For this degree of liquid loading, mist flow does not occur until gas velocity reaches at least 164 ft/s (50 m/s).

Figure 6.2.27 graphically outlines the flow regime areas. Duns and Ros mathematically defined these areas as functions of the following dimensionless numbers:

$$N_{vg} = v_{sg}A(\gamma_L/\sigma)^{0.25} \qquad \text{gas velocity number} \qquad [6.2.46]$$

$$N_{vL} = v_{sL}A(\gamma_L/\sigma)^{0.25} \qquad \text{liquid velocity number} \qquad [6.2.47]$$

$$N_d = dB(\gamma_L/\sigma)^{0.5} \qquad \text{diameter number} \qquad [6.2.48]$$

$$N_{L\mu} = N_L = \mu_L C(1/\gamma_L\sigma^3)^{0.25} \quad \text{liquid velocity number} \qquad [6.2.49]$$

Table 6.2.1 *Drift-Flux Parameters for Slug Flow Regime*

Pipe Inclination	C_o	k
$80° < \theta \leq 90°$	1.2	0.35
$0° < \theta \leq 80°$	1.3	0.50
$\theta = 0 \leq 1 < 1$	1.3	0
$-80° < \theta < 0°$	1.3	-0.5*
$-90° < \theta < -80°$	0.9	-0.6*

*If the slug velocity becomes small as a result of having the mixture superficial velocity, V_m, very small, then the slug velocity, V_s, should be limited to the mixture velocity.

Any consistent units system may be used, for example,

	English	Metric
ρ_L (liquid density)	slugs/ft.^{3}	kg/m^3
d (diameter of pipe)	ft	M
σ (liquid surface tension)	dyn/cm	dyn/cm
μ_L (liquid velocity)	cp	cp

Equations contain a "g" term that was included into the mist conversion factors A, B, and C.

	English	Metric
A	1.938	3.193
B	120.9	99.03
C	0.1573	0.3146

At high liquid rates, the pressure gradient varied significantly with the gas rate. The various flow regions were divided into three main regions depending on the amount of gas present.

- Region I. The liquid phase is continuous and bubble flow, plug flow, and part of the froth-flow regime exists.
- Region II. In this region the phases of liquid and gas alternate. The region thus covers slug flow and the remainder of the froth flow regime.
- Region III. The gas is in a continuous phase and the mist-flow regime exists.

The different nature of these three main regions necessitates separate correlations for friction and holdup for each region; therefore in principle, six different correlations are to be expected. The identification of flow region is a function of N_{LV}, N_{gv}, L_1, L_2, and N_d. The regions of validity of the correlations are plotted and presented in Figure 6.2.28 as a function of the liquid velocity number N_{Lv} and gas-velocity number N_{gv}. Because N_{Lv} and N_{gv} are directly related to liquid flowrate and gas flowrate, respectively, it can be seen from Figure 6.2.28 that a change in one or both of these rates affects the region of flow.

Duns and Ros suggested the following limits for various flow regions:

- Region I: $0 \leq N_{gv} \leq (L_1 + L_2 N_{Lv})$
- Region II: $(L_1 + L_2 N_{Lv}) < N_{gv}, (50 + 36 N_{Lv})$
- Region III: $N_{gv} > (75 + 84 N_{Lv}^{0.75})$

L_1 and L_2 are functions of N_d, and their relationships are presented in Figure 6.2.29.

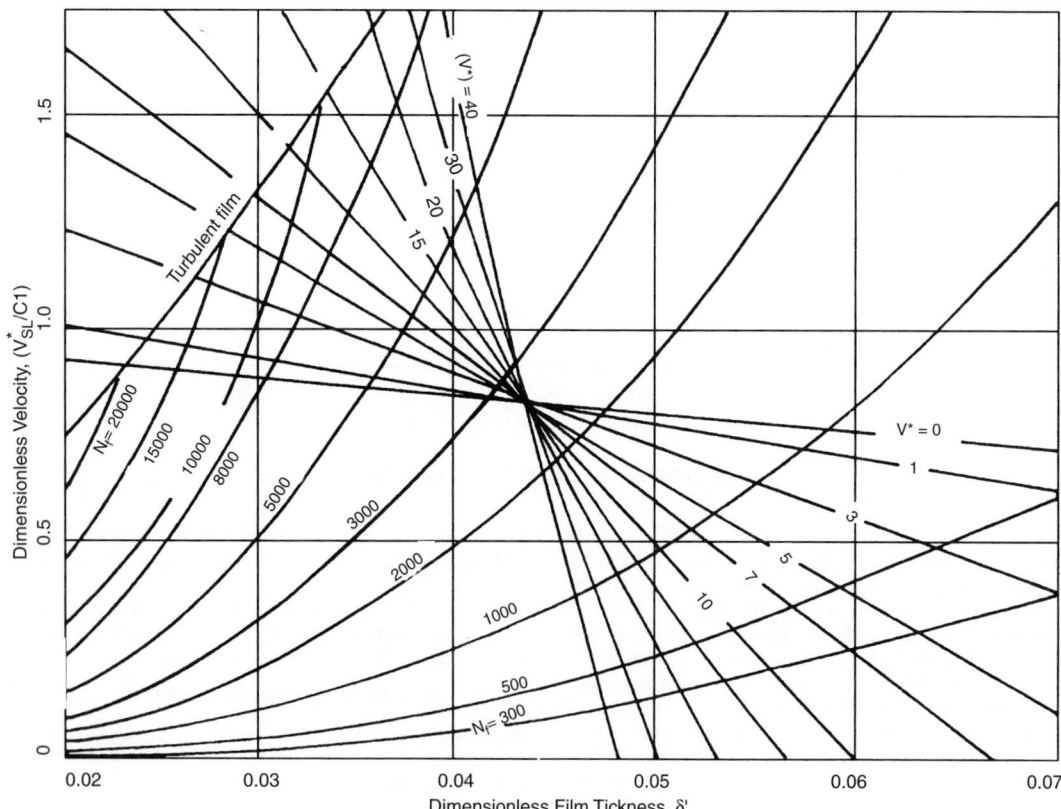

Figure 6.2.24 *Dimensionless film thickness for well in slug flow [4].*

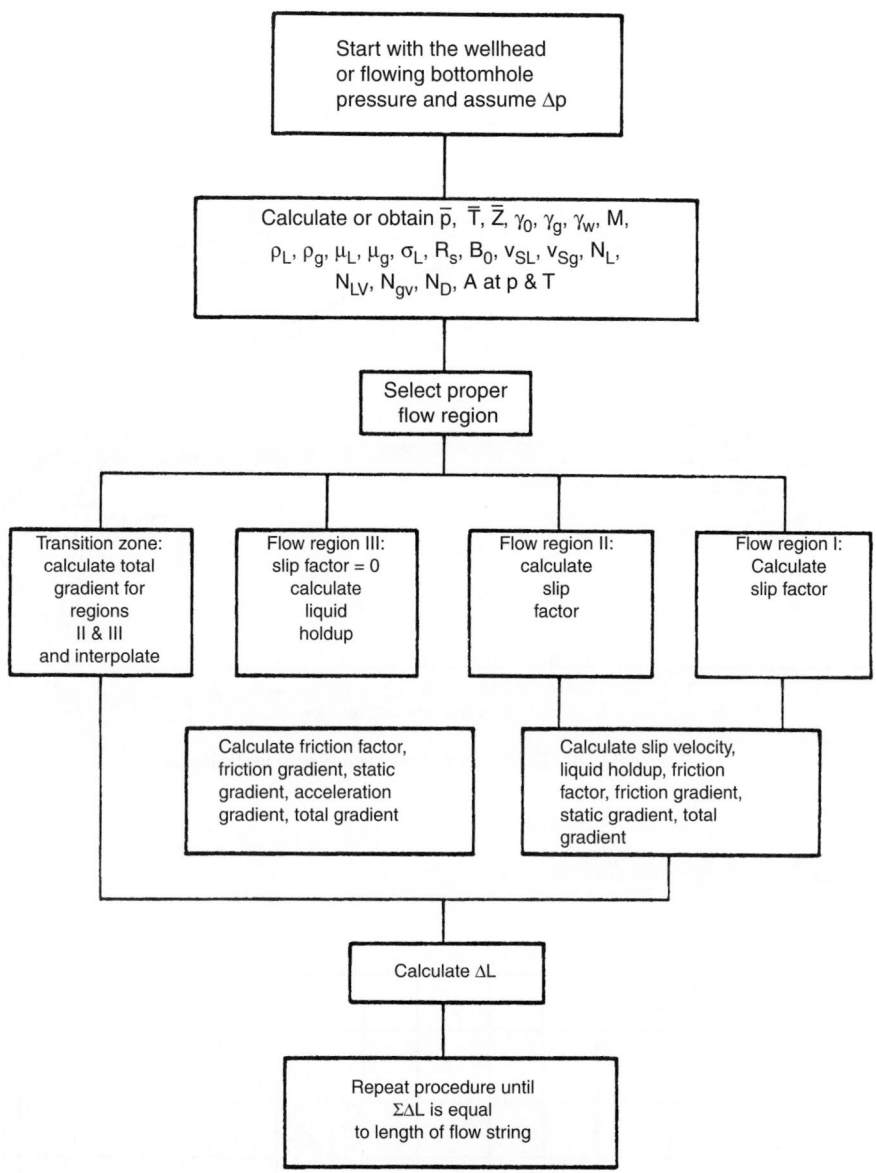

Figure 6.2.25 *Flow diagram for the Duns-Ros method [1].*

It was also found that the liquid holdup is related to the slip velocity, v_s, as follows:

$$v_s = v_{sg}/(1 - H_L) - v_{sL}/H_L \qquad [6.2.50]$$

where v_{sg}, v_{sL} are average gas and average liquid superficial velocities, respectively.

The slip velocity was expressed in dimensionless form as

$$S = v_s(\gamma_L/g\sigma)^{0.25}$$

As soon as S has been determined, v_s, H_L, and, finally, $(dp/dL)_{st}$ can be determined.

Different formulas are used for calculating S in each of the three flow regions. These formulas, which are functions of the four dimensionless numbers, N_{Lv}, N_{gv}, N_d, N_L, are

found in the example below and make use of Figures 6.2.30 and 6.2.31.

Example 6.2.2
Show stepwise procedure for calculation of the pressure traverse by the Duns-Ros method. Apply this procedure to solve the following problem.

Determine the distance ΔL between two pressure points starting surface conditions if $\Delta p = 500$ psig.

Given that tubing size d = 2 in. = 1.995 in. ID

wellhead pressure 1455 psig = p_1

$p_L = p_1 + 500 = 1955$ psig

$p_{bar} = 14.7$ psia

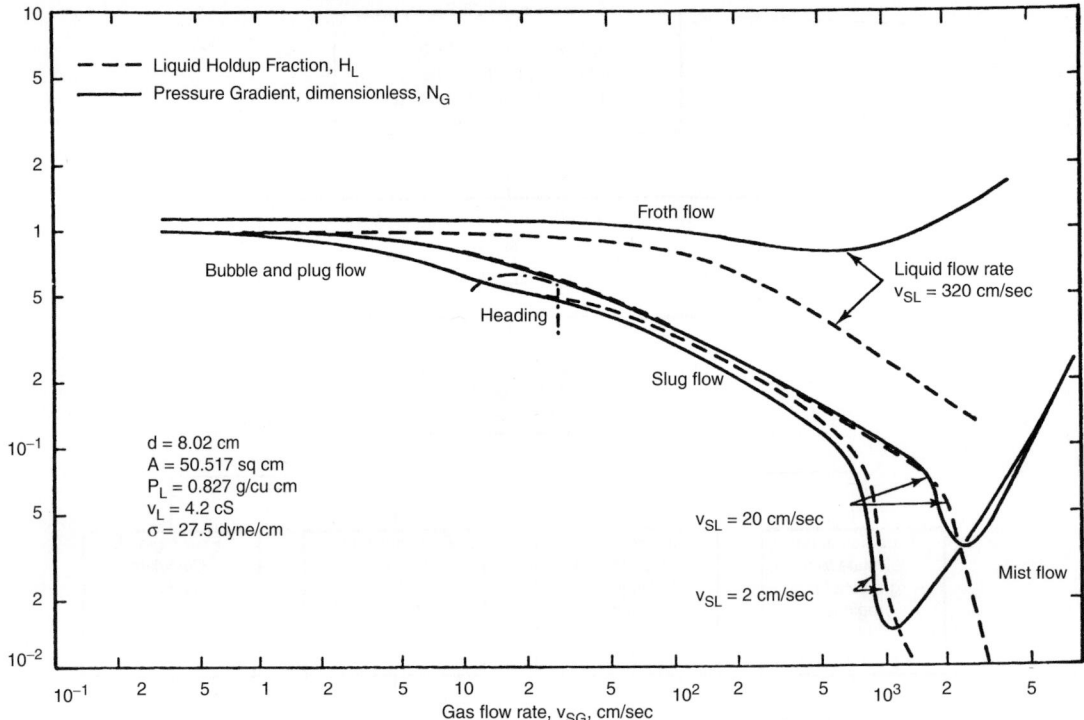

Figure 6.2.26 *Example of two-phase flow in vertical pipe [5].*

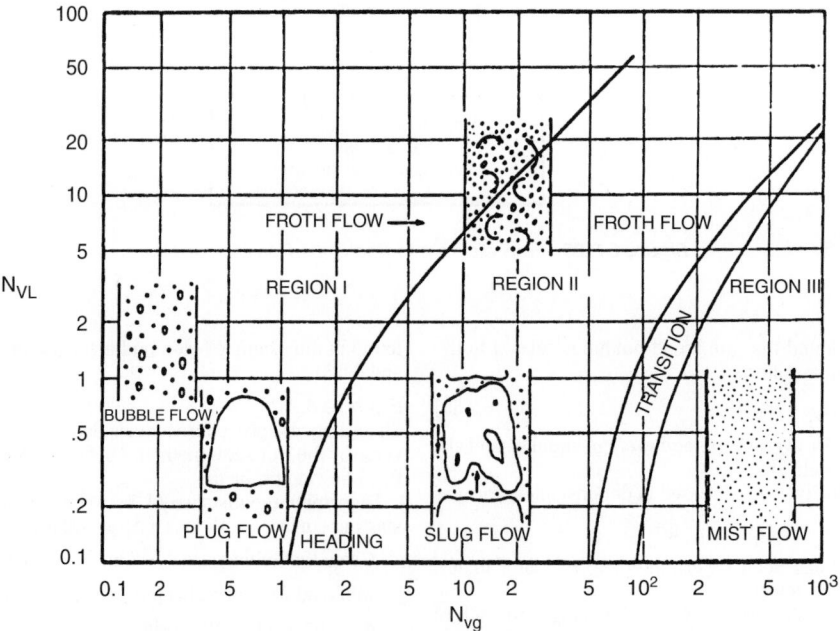

Figure 6.2.27 *Region of occurrence of different flow regimes [5].*

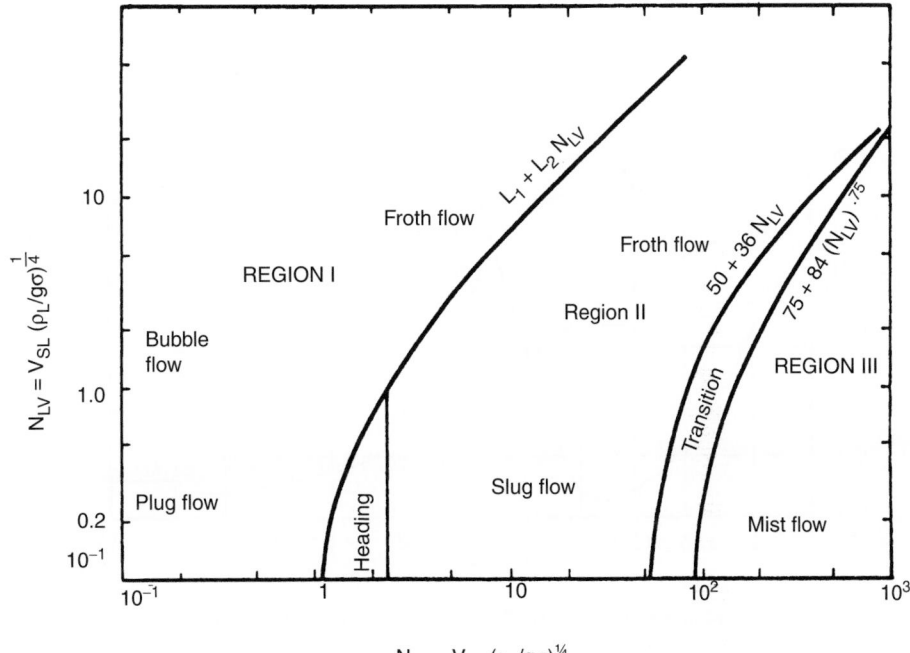

Figure 6.2.28 *Region of validity of Duns-Ros correlation [5].*

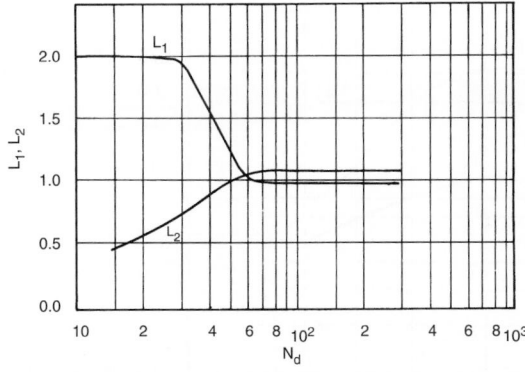

Figure 6.2.29 *L factors versus diameter number, N_d [5].*

$T_1 = 75°F$

$T_2 = 105°F$

$S.G._g = 0.752$

$S.G._o = 54°API$

$q_{oil} = 480$ bpd (std) $q_{water} = 0$ bpd (std)

$q_L = q_w + q_o$

$\mu_g = 0.020$ cp (constant value)

$\sigma_o = 28$ dyn/cm

$GLR = GOR = 3393$ scf/bbl

Choose a commercial steel pipe.

Solutions

1. Determine the specific gravity of the oil:

$$SG_0 = \frac{141.5}{131.5 + °API} = \frac{141.5}{131.5 + 54} = 0.763$$

2. Find the weight associated with 1 bbl of stock-tank liquid:

$$m = SG_0(350)\left(\frac{1}{1 + WOR}\right) + SG_0(350)\left(\frac{WOR}{1 + WOR}\right)$$

$$+ (0.0764)(GLR)(SG_g)$$

$$= 0.763(350)\left(\frac{1}{1 + 0}\right) + 0.763(350)\left(\frac{1}{1 + 0}\right)$$

$$+ 0.0764(3393)(0.752)$$

$$= 462 \text{ lb/stb of oil}$$

3. Determine the specific weight of the liquid phase:

$$\gamma_L = 62.4\left[SG_0\left(\frac{1}{1 + WOR}\right) + SG_w\left(\frac{WOR}{1 + WOR}\right)\right]$$

$$= 62.4\left[0.763\left(\frac{1}{1 + 0}\right) + SG_w\left(\frac{1}{1 + 0}\right)\right]$$

$$= 47.6 \text{ lb/ft.}^3$$

4. Find the average pressure:

$$p_{avg} = \frac{p_1 + p_2}{2} + p_{bar} = \frac{1455 + 1955}{2} + 14.7$$

$$= 1719.7 \text{ psia}$$

5. Find the average temperature:

$$T_{avg} = \frac{T_1 + T_2}{2} = \frac{75 + 105}{2} = 90°F = 550°R$$

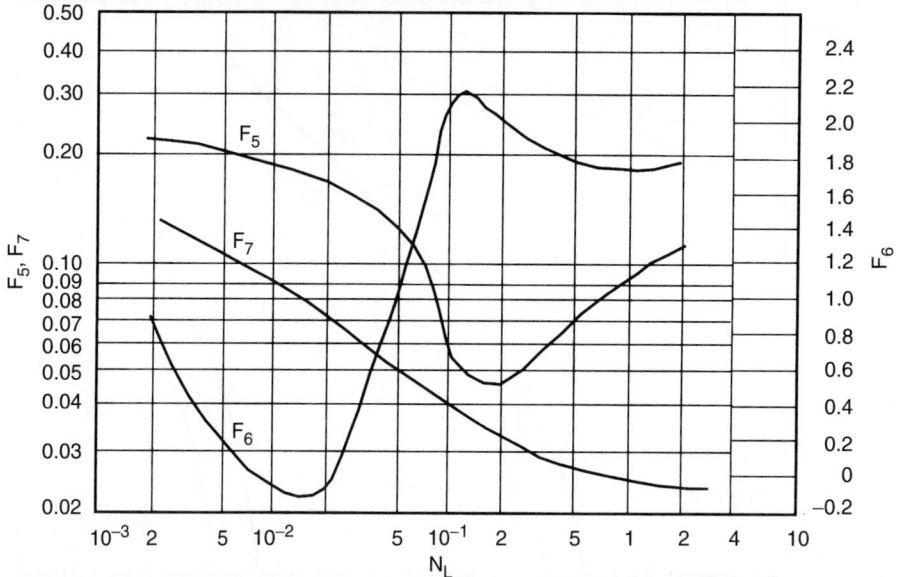

Figure 6.2.30 F_5, F_6, F_7 against viscosity number, N_L [5].

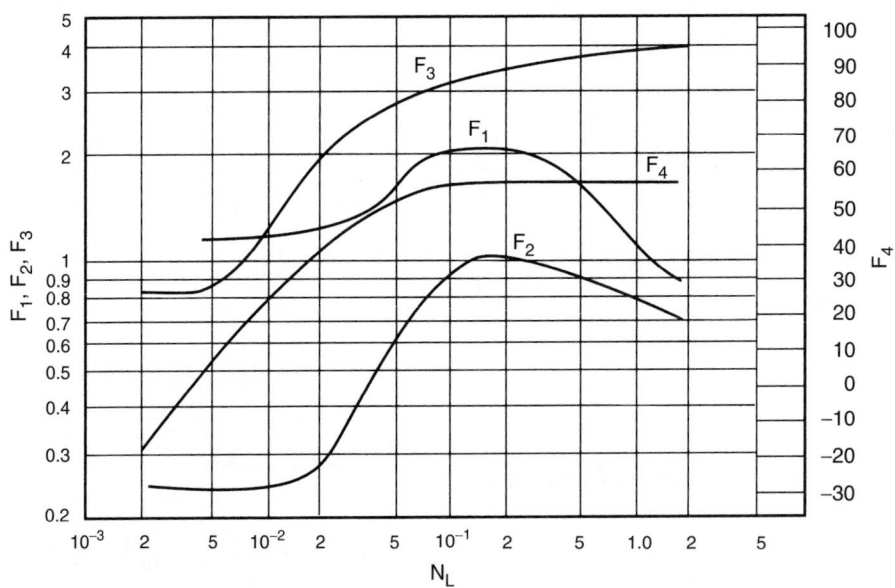

Figure 6.2.31 F_1, F_2, F_3, and F_4 versus viscosity number, N_L [5].

6. Find Z_{avg} for gas phase

$$Z_{avg} = f(T_r, p_r)$$
$$= 0.72$$
$$T_r = T/T_{pc} = 550/394 = 1.4$$
$$p_r = 1719.7/660 = 26$$

7. Find the average specific weight of the gas phase:

$$\gamma_g = SG_g(0.0764)\left(\frac{p}{14.7}\right)\left(\frac{520}{5}\right)\left(\frac{1}{Z}\right)$$
$$= SG_g(0.0764)1/B_g$$
$$= 0.752(0.0764)\frac{1719.7}{14.7}\frac{520}{550}\frac{1}{0.72} = 8823 \text{ lb/ft.}^3$$

8. Find R_s at T_{avg} and p_{avg} (see Chapter 5)

$$R_s = SG_g \left(\frac{p}{18} \times \frac{10^{0.0125(°API)}}{10^{0.00091(T)}} \right)^{1.2048}$$

$T_{avg} = 90°F, \qquad p_{avg} = 1719.7 \text{ psia}$

$$R_s = 0.752 \left(\frac{1719.7}{28} \times \frac{10^{0.0125(54)}}{10^{0.00091(90)}} \right)^{1.2048}$$

$$= 947.3 \text{ scf/stb}$$

9. Calculate the average viscosity of the oil from correlations (see Chapter 5)

$\mu_{od} = 10^X - 1.0$

$X = T^{-1.163} \exp(6.9824 - 0.04658°API)$

$X = 90^{-1.163} \exp(6.9824 - 0.04658 \times 54) = 0.4646$

$\mu_{od} = 1.915 \text{ cP}$

$\mu_{os} = A\mu_{od}^B$

$A = 10.715(R_s + 100)^{-0.515} = 10.715(947.3 + 100)^{-0.515}$

$A = 0.2983$

$B = 5.44(R_s + 150)^{-0.338} = 5.44(947.3 + 150)^{-0.338}$

$B = 0.5105$

$\mu_{os} = 0.2983 \times 1.915^{0.5105} = 0.416 \text{ cp}$

10. Determine the average water viscosity. No water is in the example.

11. Calculate the liquid mixture viscosity:

$$\mu_L = \mu_o \left(\frac{1}{1+WOR} \right) + \mu_w \left(\frac{WOR}{1+WOR} \right) = \mu_o$$

$$= 0.416 \text{ cp}$$

12. Find the liquid mixture surface tension:

$$\sigma_L = \sigma_o \left(\frac{1}{1+WOR} \right) + \mu_w \left(\frac{WOR}{1+WOR} \right) = \sigma_o$$

$$= 28 \text{ dyn/cm}$$

13. Find B_o at p_{avg} and T_{avg}.

$$B_o = 0.972 + 0.00147F^{1.175}$$

$$F = R_s \left(\frac{\gamma_g}{\gamma_o} \right)^{0.5} + 1.25\% \ [T(°F)]$$

$$= 947.3 \left(\frac{0.752}{0.763} \right)^{0.5} + 1.25(90) = 1052.9$$

$B_o = 1.495 \text{ bbl/stb}$

14. Find the turbine flow area A_p:

$$A_p = \frac{\pi d^2}{4} = \frac{\pi}{4} + \left(\frac{1.995}{12} \right)^2 = 0.0217 \text{ ft.}^2$$

15. Find the liquid viscosity number:

$$N_L = 0.1573 \times (0.5) \left(\frac{1}{(28)^3 47.6} \right)^{0.25} = 2.05 \times 10^{-3}$$

16. Find v_{sL} (assume $B_w = 1.0$):

$$v_{sL} = \frac{5.61qL}{86400A_p} \left[B_o \left(\frac{1}{1+WOR} \right) + B_w \left(\frac{WOR}{1+WOR} \right) \right]$$

$$= \frac{5.61(480)}{86400(0.0217)} \left[1.495(1.0) + 1.0 \left(\frac{0}{1+0} \right) \right]$$

$$= 2.147 \text{ ft./s.}$$

17. Find the liquid velocity number:

$$N_{Lv} = 1.938 \, v_{sL} \left(\frac{\gamma_L}{\sigma_L} \right)^{1/4} = 1.938(2.147) \left(\frac{47.6}{28} \right)^{1/4}$$

$$= 4.75$$

18. Find the superficial gas velocity:

$$v_{sg} = \frac{q_L \left[GLR - R_s \left(\frac{1}{1+WOR} \right) \right]}{86,400A_p} \left(\frac{14.7}{p_{avg}} \right)$$

$$\times \left(\frac{T_{avg}}{520} \right) \left(\frac{Z_{avg}}{1} \right)$$

$$= \frac{q_L \left[GLR - R_s \left(\frac{1}{1+WOR} \right) \right] B_g}{86,400A_p}$$

$$= \frac{480 \left[3393 - 947.3 \ \left(\frac{1}{1+0} \right) \right] 14.7 \times 550(0.72)}{84,400(0.0217) \qquad 1719.5 \times 520}$$

$$= 4.08 \text{ ft./s}$$

19. Find the gas velocity number:

$$N_{gv} = 1.938 \, v_{sg} \left(\frac{\gamma_L}{\sigma_L} \right)^{1/4} = 1.938 \times 4.08 \left(\frac{47.6}{28} \right)^{1/4}$$

$$= 9.03$$

20. Find the pipe diameter number:

$$N_d = 120.9d \left(\frac{\gamma_L}{\sigma_L} \right)^{0.5} = 120.9 \left(\frac{1.995}{12} \right) \left(\frac{47.6}{28} \right)^{0.5}$$

$$= 26.2$$

21. Select the proper flow regime from Figure 6.2.3:

$$N_{gv} = 9.03$$

$$N_{Lv} = 4.75$$

These numbers fall in Region II; see Figure 6.2.3.

22. Determine the proper slip factor depending upon the region found in step 21.

a) For Region I: determine the slip factor determination. The slip factor is found by the following formula:

$$S = F_1 + F_2N_{Lv} + F_3' \left(\frac{N_{gv}}{1+N_{Lv}} \right)^2$$

F_1 and F_2 are found in Figure 6.2.31.
$F_3' = F_3 - F_4/N_d$ where F_3 and F_4 are found in Figure 6.2.31. For annular flow N_d is based on the wetted perimeter; thus, $d = (d_c + d_t)$. Region I extends from zero N_{Lv} and N_{gv} up to $N_{gv} = L_1 + L_2N_{Lv}$, where L_1 and L_2 can be found in Figure 6.2.29.

b) For Region II:

$$S = (1 + F_5) \frac{(N_{gv})^{0.982} + F_6'}{(1 + F_7N_{Lv})^2}$$

F_5, F_6 and F_7 can be found in Figure 6.2.30 where $F_6 = 0.029N_d + F_6$. Region II extends from the upper limit of Region I to the transition zone to mist flow given by $N_{gv} = 50 + 36N_{Lv}$.

c) For Region III (mist flow):

$$S = 0$$

$$\text{Therefore, } H_L = \frac{1}{1 + v_{sg}/v_{sL}}$$

This is valid for $N_{gv} > 75 + 84 \, N_{Lv}^{0.75}$. Calculations for Region II from Figure 6.2.5 if $N_L = 2.05 \times 10^{-3}$, then

$$F_5 = 0.218, \qquad F_6 = 0.58, \qquad F_7 = 0.12$$

$$F_6' = 0.029 N_d + 0.58 = 0.029 \times (2.62) + 0.58 = 1.34$$

$$S = (1 + 0.20)\frac{9.03^{0.982} + 1.34}{[1 + (0.12 \times 4.751)]^2} = 4.88$$

23. Determine the slip velocity if in Region I or II:

$$v_s = \frac{S}{1.938(\gamma_L/\sigma_L)^{0.5}}$$

It is Region II; hence

$$v_s = \frac{4.88}{1.938(47.61/28)^{0.5}} = 1.933$$

24. Determine the liquid holdup:

$$H_L = \frac{v_s - v_{sg} - v_{sL} + \left[(v_s - v_{sg} - v_{sL})^2 + 4 v_s v_{sL}\right]^{0.5}}{2 v_s}$$

$$= \Big(1.933 - 4.08 - 2.147 + [(1.933 - 4.08 - 2.147)^2$$

$$+ 4(1.933)(2.147)]^{0.5}\Big) \Big/ \Big(2 \times 1.933\Big)$$

$$= 0.4204 = 0.42$$

This value can be checked

$$v_s = \frac{v_{sg}}{1 - H_L} - \frac{v_{sL}}{H_L} = \frac{4.08}{1 - 0.42} - \frac{2.147}{0.42} = 1.923$$

25. Determine the liquid Reynolds number:

$$N_{RE} = \frac{1488 \gamma_L v_{sL} d}{\mu_L}$$

$$(N_{Re})_L = \frac{1488 \times 47.6 \times 2.147 \times 0.16625}{0.416} = 60,773$$

26. Determine the friction gradient according to the flow region.

a) For Region I and II

$$G_{fr} = 2 f_w \frac{N_{Lv}(N_{Lv} + N_{gv})}{N_d}$$

where $f_w = (f_1)\frac{f_2}{f_3}$

f_1 is found in Figure 6.2.32 and f_2 is found in Figure 6.2.33

The abscissa must be determined in Figure 6.2.33 and is $f_1 R N_d^{2/3}$
where $R = \dfrac{v_{sg}}{v_{sL}}$

$$f_3 = 1 + f_1 (R/50)^{0.5}$$

The friction factor f_w is valid in Regions I and II and covers heading also. It is good from zero N_{Lv} and N_{gv} up to the limit given by $N_{gv} = 50 + 36 \, N_{Lv}$.

b) For Region III:
In mist flow where $N_{gv} > 75 + 84 N_{Lv}^{0.75}$

$$G_{fr} = 2 f_w N_\rho \frac{(N_{gv})^2}{N_d}$$

where $N_\rho = \rho_g/\rho_L$
In Region III f_w is taken as f_1 and may be taken from Figure 6.2.32. For $\varepsilon > 0.05 d$, the value of f_1 is calculated by

$$f_1 = \frac{1}{\left[4 \log_{10}(0.027 \varepsilon/d)\right]^2} + 0.067(\varepsilon/d)^{1.73}$$

For $\varepsilon > 0.05 \, d$, the value of $d - \varepsilon$ should be substituted for d throughout the friction gradient calculation, and also this substitution should be made:

$$v_{sg} = \frac{v_{sg} d^2}{(d - \varepsilon)^2}$$

It is Region II; hence, calculate f_1, f_2, and f_3 such that

$$f_w = (f_1)\frac{f_2}{f_3}$$

From Figure 6.2.32 read f_1, but first determine a value for ε/d. If the value for ε is not known, a good value to use is 0.00015 ft, which is an average value given for commercial sted.

$$\frac{\varepsilon}{d} = \frac{0.00015(12)}{1.995} = 9.02 \times 10^{-4}$$

For given Reynolds number (60,773) and ε/d

$$f_1 = 0.005$$

From Figure 6.2.33

$$\text{where } f_1 \left(\frac{v_{sg}}{v_{sL}}\right) N_d^{2/3} = 0.005 \left(\frac{4.08}{2.147}\right) 26.2^{2/3}$$

$$= 0.0839$$

$$f_2 = 1.01$$

$$f_3 = 1 + f_1 \left(\frac{4.08}{2.147 \times 50}\right)^{0.5} = 1 + 0.005(0.038)^{0.5}$$

$$= 1.001$$

$$f_w = 0.005 \left(\frac{1.01}{1.001}\right) = 0.00505$$

calculate friction gradient G_{fr}:

$$G_{fr} = 2(0.00505)\frac{4.75(4.75 + 9.03)}{26.2} = 0.0252$$

27. Determine the static gradient:

$$G_{st} = H_L + (1 - H_L)\frac{\gamma_g}{\gamma_L}$$

$$= 0.42 + (1 - 0.42)\frac{8.823}{47.6} = 0.5275 \text{ (dimensionless)}$$

28. Determine the total pressure gradient.

a) For Regions I and II:

$$G = G_{st} + G_{fr}$$

b) For Region III (accounting for accelerations):

$$G = \frac{G_{st} + G_{fr}}{1 - (\gamma_L v_{sL} + \gamma_g v_{sg})(v_{sg}/\gamma)}$$

$$= G_{st} + G_{fr} = 0.5275 + 0.0252 = 0.5527$$

29. Convert to gradient in psi/ft:

$$\frac{dp}{dL_{st}} = \frac{G_{st}\gamma_L}{144}, \quad \frac{dp}{dL_{(fr)}} = \frac{G_{fr}\gamma_L}{144}, \quad \text{or} \quad \frac{dp}{dL_{total}} = \frac{G\gamma_L}{144}$$

$$\frac{dp}{dL} = \frac{0.5527(47.6)}{144} = 0.1827 \text{ psi/ft.}$$

30. Determine distance ΔL:

$$\Delta L = \frac{(1955 - 1455)\text{psi}}{0.1827 \text{ psi/ft.}} = 2737 \text{ ft.}$$

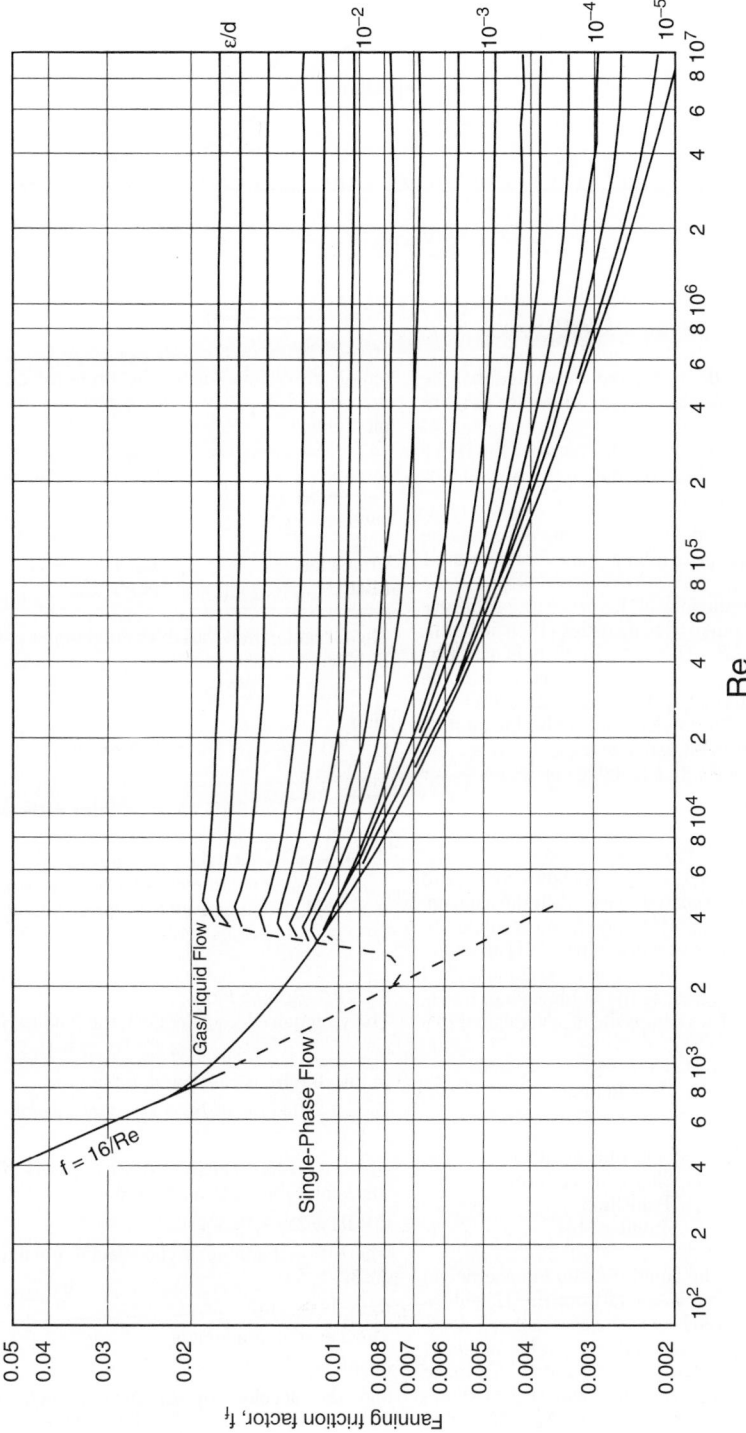

Figure 6.2.32 *Fanning friction factor f_f versus Reynolds number [5].*

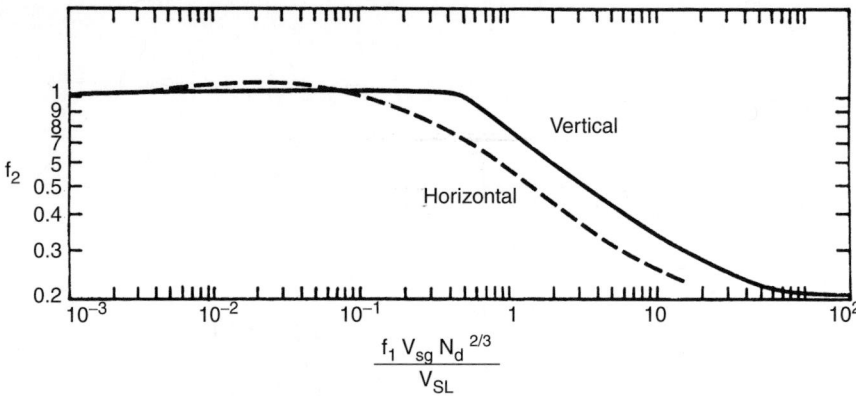

$$\frac{f_1 \, V_{sg} \, N_d{}^{2/3}}{V_{SL}}$$

Figure 6.2.33 *Bubble friction correction [5].*

If water flows together with oil, it is recommended that the calculations be made using the average oil–water mixture properties.

The pressure profile prediction performance of the Duns-Ros method is outlined below in relation to the several flow variables considered [6]:

- *Tubing Size*: In general, the pressure drop is seen to be over predicted for a range of tubing diameters between 1 and 3 in.
- *Oil Gravity*: Good predictions of the pressure profile are obtained for a broad range of oil gravities (13 to 56 °API).
- *Gas-Liquid Ratio (GLR)*: The pressure drop is over predicted for a wide range of GLR. The errors become especially large (>20%) for GLR greater than 5000.
- *Water-Cut*: The Duns-Ros model is not applicable for multiphase flow mixtures of oil, water, and gas. However, the correlation can be used with a suitable correction factor as mentioned earlier.

6.2.13 The Orkiszewski Method [8,9]

This method is recognized for four types of flow pattern, and separate correlations are prepared to establish the slippage velocity and friction for each. This correlation is limited to two-phase pressure drops in a vertical pipe and is an extension of Griffith and Wallis [10] work. The correlation is valid for different flow regimes such as the bubble, slug, transition, and annular mist and is a composite of several methods as shown below:

Method	Flow Regime
Griffith	Bubble
Griffith & Wallis	Slug (density term)
Orkiszewski	Slug (friction term)
Duns & Ros	Transition
Duns & Ros	Annular Mist

It should be noted that the liquid distribution coefficient (hold-up) is evaluated using the data from the Hagedorn and Brown model (discussed later). The correlation is applicable to high-velocity flow range and gas condensate wells in addition to oil wells and has proven accuracy. To make calculations, a computer is preferable, Figure 6.2.34 shows a generalized flow diagram of this method. After assuming a pressure difference and calculating the various required properties, a flow region is selected. Depending on the flow region, the pressure loss calculations—which, in general, include friction and holdup—are made. The vertical

length corresponding to the pressure difference is then determined. The flow regime is found by testing the following limits:

Regime	Limits of Boundary Lines, L
Bubble	$q_g/(q_g + q_l) < L_B$ or $v_{sg}/v_m < L_B$
Slug	$q_g/(q_g + q_l) > L_B$, $N_{vg} < L_s$
Transition	$L_m > N_{vg} > L_s$
Mist	$N_{vg} > L_m$

The foregoing new variables are defined as follows:
Bubble:
$$L_b = 1.071 - (0.2218v_m^2/d) \quad \text{but} \quad \geq 0.13 \qquad [6.2.51]$$
Slug:
$$L_s = 50 + 36N_{vg}q_L/q_g (\text{or } 36N_{vg}v_{sl}/v_{sg}) \qquad [6.2.52]$$
Mist:
$$L_m = 75 + 84(N_{vg}q_L/q_g)^{0.75} \left[\text{or } 84(N_{vg}v_{sl}/v_{sg})^{0.75}\right]$$

where v_t = total fluid velocity ($v_{sl} + v_{sg} = v_m$)
q_t = volumetric total flow ($q_L + q_g$)
N_{vg} = dimensionless velocity influence number
= $v_{sg}(\gamma_L/g\sigma)^{0.25}$

6.2.13.1 Bubble Flow

The γ_m required knowledge of the holdup H_1, such that
$$\gamma_m = \gamma_L H_L + \gamma_g (1 - H_L)$$
In this, the H_L is calculated as follows:
$$H_1 = 1 - 0.5\left[1 + v_m/v_s - (1 + v_m/v_s)^2 - 4v_{sg}/v_s)^{0.5}\right] \qquad [6.2.53]$$

where v_s = slip velocity = 0.8 ft./s (0.244 m/s)
Therefore, the friction gradient is
$$dp/dL = 2f\gamma_L v_L^2/(H_L g_c d) \qquad [6.2.54]$$
where f = Fanning friction factor obtained from Figure 6.2.32
$$R_e = dv_{sL}\gamma_L/\mu_L \qquad [6.2.55]$$
The elevation gradient is
$$dp/dL = \gamma_m F_e \qquad [6.2.56]$$
and the acceleration gradient is negligible. However, Orkiszewski's equation for all these effects is
$$\frac{dp}{dL} = \frac{\gamma_m F_e + 2f\gamma_L v_{sL}^2/(H_L g_c d)}{1 - m_t q_g/(A_p^2 p_{avg} g_c)} \qquad [6.2.57]$$
which is essentially the same as adding the three gradients.

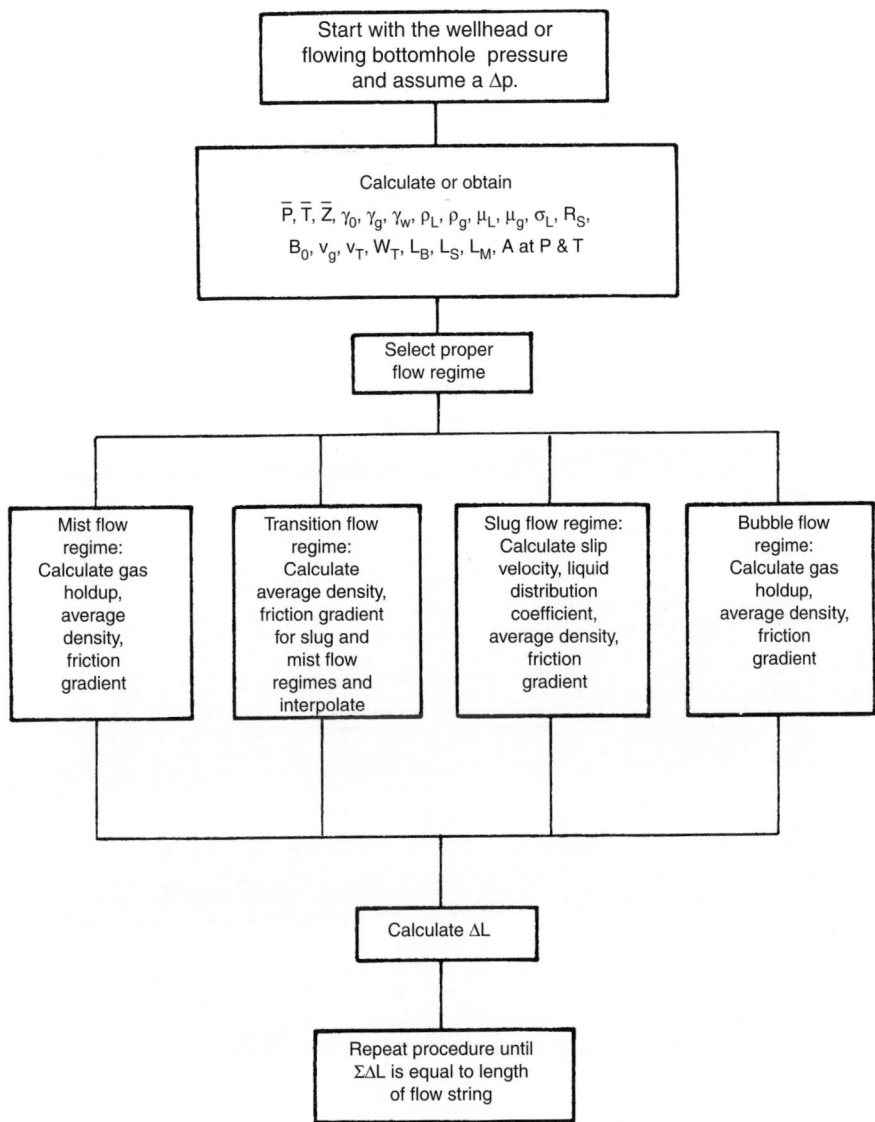

Figure 6.2.34 *Flow diagram for the Orkiszewski method [1,6].*

6.2.13.2 Slug Flow

Slug flow specific weight γ_s is difficult to know and difficult to assume. An attempt was made such that

$$\gamma_s = \frac{m_t + \gamma_L v_s A_p}{q_t + v_s A_p} + \delta \gamma_L \qquad [6.2.58a]$$

in a slightly different term arrangement using velocities, or

$$\gamma_s = \frac{\gamma_L(v_{sL} + v_s) + \gamma_g v_{sg}}{v_m + v_s} + \delta \gamma_L \qquad [6.2.58b]$$

where m_t = total mass/s
$\quad A_p$ = area of pipe
$\quad v_s$ = correlation factor, $C_1 C_2 (gd)^{0.5}$, slip velocity
$\quad \delta$ = liquid distribution coefficient $\qquad [6.2.59]$

C_1 and C_2 are functions of a Reynolds number as follows:

$$C_1 \propto (f) d v_s \gamma_L / \mu_L, \text{ or } Re_b \text{ (Figure 6.2.35)}$$

and

$$C_2 \propto (f) Re_b \quad \text{and} \quad Re_n = d v_t \gamma_L / \mu_L \text{ (Figure 6.2.36)}$$

Since v_s is a dependent variable, it must be found by iteration. A value of v_s is assumed, Re_b is calculated, and C_1 and C_2 are determined. If the calculated value of v_s does not agree with the assumed value, try again. [A good initial try, $v_s = 0.5(gd)^{0.5}$.] For details see Example 6.2.3. Now determine δ as follows:

(a) If $v_m < 10$ (continuous phase is water)

$$\delta = [0.013 \log \mu_L / d^{1.38}] - 0.681 + 0.232 \, \log v_m$$
$$- 0.428 \, \log d \qquad [6.2.60a]$$

(b) If $v_m > 10$ (continuous phase is oil)

$$\delta = [0.0127 \log(\mu_L + 1)/d^{1.415}] - 0.284 + 0.167 \, \log v_m$$
$$+ 0.113 \, \log d \qquad [6.2.60b]$$

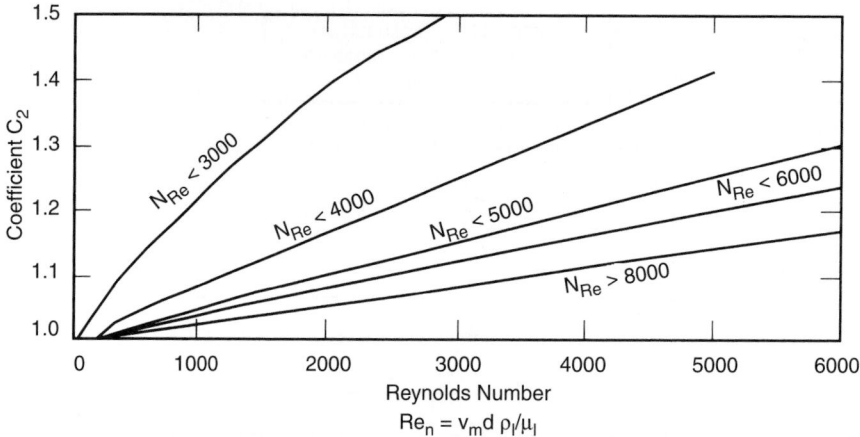

Figure 6.2.35 C_2 constant versus bubble Reynold's number [7].

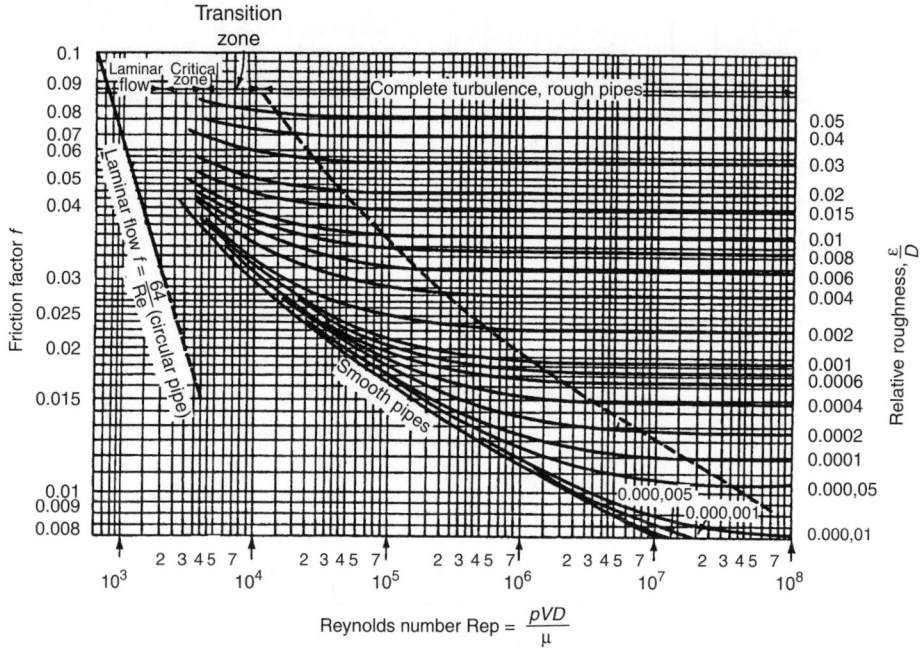

Figure 6.2.36 Friction factor [8].

(c) If $v_m > 10$ (continuous phase is water)

$$\delta = \left[(0.045 \log \mu_L)/d^{0.799}\right] - 0.709 - 0.162 \log v_m$$
$$- 0.888 \log d \qquad [6.2.60c]$$

(d) If $v_m > 10$ (continuous phase is oil) (μ_L in cp)

$$\delta = \left[0.0274 \log(\mu_L + 1)/d^{1.371}\right] + 0.161 + 0.569 \log d + x$$
$$[6.2.60d]$$

where

$$x = -\log v_m\left[(0.01 \log(\mu_L + 1)/d^{1.571}) + 0.397 + 0.63 \log d\right]$$

These constraints apply to δ:

If $v_m < 10$, $\delta \geq -0.065 v_m$

If $v_m > 10$, $\delta \geq (-v_x)(1 - \gamma_s/\gamma_L)/(v_m + v_x)$

Finally, the friction gradient term is

$$\left(\frac{dp}{dL}\right)_f = \frac{2f_f\gamma_L v_m^2}{gd}\left(\frac{v_{sL} + v_s}{v_m + v_s} + \delta\right) \qquad [6.2.61]$$

where f_f is obtained from Figure 6.2.32 using the following:

$$Re = dv_m\gamma_L/\mu_L \qquad [6.2.62]$$

Again, the total pressure gradient includes the elevation (static), friction, and acceleration (negligible) components.

6.2.13.3 Transition Flow
Orkiszewski used linear interpolation between slug and mist.

6.2.13.4 Mist-Flow
The Duns-Ros method is used.

Example 6.2.3

Apply Orkiszewski method to solve Example 6.2.2 with the pipe diameter d = 2 in.

1. Select the starting point as the 1455 psig pressure.
2. The temperature at each point of pressure is given as 75°F at 1455 psig and 105°F at 1955 psig.
3. Δp is equal to 500 psig
4. It is not necessary to assume depth increment since the temperature at 1955 psia is known.
5. The average temperature of the increment is (105 + 75)/2 = 90°F
6. The following calculations are made in order to complete step 7 to determine flow regime.

   a) The average flow conditions are

   $$\Delta p = (1455 + 1955)/2 + 14.7 = 1719.5$$

   $$T = 90°F = 550°R$$

   $$Z = 0.72$$

   $$R_s = 947.3 \text{ scf/stb}$$

   $$B_0 = 1.495 \text{ bbl/stb}$$

   b) The corrected volumetric flowrates are

   $$q_L = 480 \text{ bpd} = 6.4984 \times 10^{-5} \times 480 \text{ scf/s}$$

   $$= 6.4984 \times 10^{-5}(q_0 B_0 + q_w B_w)$$

   $$= 6.4984 \times 10^{-5} \times 480 \times 1.495$$

   $$= 0.0466 \text{ ft.}^3/\text{s}$$

   $$q_g = 3.27 \times 10^{-7} \times (GLR - R_s)q_L(T + 460)/p$$

   $$= 3.27 \times 10^{-7}(0.72)(3397 - 947.3)$$

   $$\times 480(550)/1719.5$$

   $$= 0.08855 \text{ ft.}^3/\text{s}$$

   $$q_t = q_g + q_1 = 0.08855 + 0.0466 = 0.1352 \text{ ft.}^3/\text{s}$$

   c) The corrected weight flowrates are

   $$\dot{w}_L = 4.05(10^{-3})(q_0 SG_0 + q_w SG_w) + 8.85$$

   $$\times (10^{-7})q_L SG_g(R_s)$$

   $$= 4.05(10^{-3})(480 \times 0.763 + 0) + 8.85(10^7)480$$

   $$\times (0.752)947.3$$

   $$= 1.483 + 0.303 = 1.7856 \text{ lb/s}$$

   $$\dot{w}_g = 8.85(10^{-7})q_L g_g(GLR - R_s)$$

   $$= 8.85(10^{-7})480(0.752)(3393 - 947.3)$$

   $$= 0.7813 \text{ lb/s}$$

   $$\dot{w}_t = \dot{w}_L + \dot{w}_g = 1.7856 + 0.7813$$

   $$= 2.567 \text{ lb/s}$$

   d) The corrected specific weights are

   $$\gamma_L = \dot{w}_L/q_L = 1.7856/0.0466 = 38.32 \text{ lb/ft.}^3$$

   $$\gamma_g = \dot{w}_g/q_g = 0.7813/0.08855 = 8.82 \text{ lb/ft.}^3$$

7. Determine the type of flow regime:

   a) Test variables

   $$A_p = 0.0217 \text{ ft.}^2$$

$$v_m = v_t = \frac{q_t}{A_p} = \frac{0.1352}{0.0217} = 6.23 \text{ ft./s}$$

$$\frac{q_g}{q_t} = \frac{0.08855}{0.1352} = 0.655$$

$$N_{vg} = 1.938 \, v_{sg} \left(\frac{\gamma_L}{\sigma_L} \right)^{0.25}$$

$$v_{sg} = \frac{q_g}{A_p} = \frac{0.08855}{0.0217} = 4.081$$

$$N_{vg} = 1.938(4.081) \left(\frac{38.32}{28} \right)^{0.25} = 8.55$$

   b) Boundary limits

   $$d = 2/12 = 0.1662 \text{ ft.}$$

   From Equation 6.2.51

   $$L_B = 1.071 - (0.2218 \, v_m^2/d)$$

   $$= 1.071 - (0.2218 \times 6.23^2/0.1662)$$

   $$= -50.7$$

   Because L_B has such a low value, we must use $L_B = 0.13$ from Equation 6.2.53 $L_s = 50 + 36 N_{vg} q_L/q_g$.

   $$L_s = 50 + 36(8.55)0.0466/0.09955 = 212$$

   These two values, L_B and L_s, indicate that the regime is slug flow.

8. Determine the average density and the friction loss gradient

   a) Slip velocity, v_s

   $$Re_n = dv_m \gamma_L/\mu_L = \big((0.1662)(6.23)(38.32)\big)/$$

   $$\big((0.5 \times 0.000672)\big)$$

   $$= 118,080$$

   Since this value exceeds limits of the graph (Figure 6.2.36) v_s must be calculated using the extrapolation equation.

   $$v_s = 0.5(gd)^{0.5} = 0.5(32.2 \times 0.1662)^{0.5}$$

   $$= 1.155 \text{ (first try)}$$

   $$Re_b = (1.155)(38.32)(0.1662)/(0.5 \times 0.000672)$$

   $$= 21,892$$

   C_1 cannot be read from graph (Figure 6.2.37). To solve this problem, Orkiszewski proposed the following equations:
   if $Re_b \leq 3000$

   $$v_s = (0.546 + 8.74 \times 10^{-6} \times Re_n)(g_c d)^{0.5}$$

   if $Re_b \geq 8000$

   $$v_s = (0.35 + 8.74 \times 10^{-6} Re_n)(g_c d)^{0.5}$$

   if $3000 < Re_b < 8000$

   $$v_s = 0.5F + \big[F^2 + 13.59 \, \mu_L/(\gamma_L d^{0.5})\big]$$

   where $F = (0.251 + 8.74 \times 10^{-6} Re_n)(g_c d)^{0.5}$
   In this example

   $$v_s = 0.35 + 8.74 \times 10^{-6} \times 118,080 \times (32.2 \times 0.1662)^{0.5}$$

   $$= 3.19 \text{ ft./s}$$

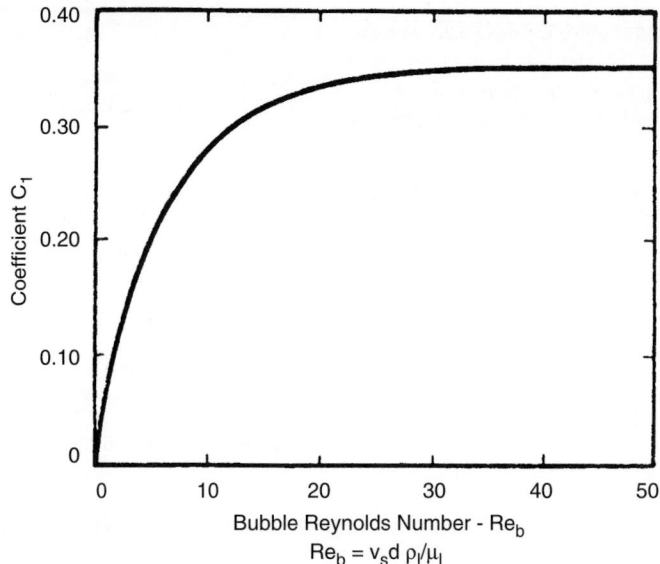

Figure 6.2.37 *C_1 constant versus bubble Reynolds number [7].*

b) Liquid distribution coefficient δ and friction factor f
Equation 6.2.60 is used to evaluate δ since $v_t < 10$
and there is no water.

$$\delta = 0.0127 \log (0.5+1)/0.1662^{1.415} - 0.284$$

$$+ 0.167 \log 6.23 + 0.113 \log (0.1662)$$

$$= -0.211$$

Checking this value

$$-0.211 \geq -0.065 \ (6.23)$$

$$-0.211 \geq -0.405$$

Therefore, δ is okay

$$\varepsilon/d = 0.00015/0.166 = 0.00009$$

and $Re_b = 21892$

$f_f = 0.0065$ (Fanning factor from Figure 6.2.32)

or $f = 0.025$ (friction factor from Figure 6.2.37)

because $f = 4f_f$, so results are consistent.

c) Evaluation of average flowing specific weight from
Equation 6.2.58

$$\gamma_s = \frac{\dot{w}_t + \gamma_L v_s A_p}{q_t + v_s A_p} + \delta\gamma_L$$

$$= \frac{2.57 + 38.32 \times 3.19 \times 0.0217}{0.1352 + (3.19)(0.0217)} + (-0.211)(38.32)$$

$$= 17.45 \ lb_m/ft.^3$$

d) Wall friction loss from Equation 6.2.61

$$\left(\frac{dp}{dL}\right)_f = \frac{2f_f \gamma_L v_m^2}{gd} \left(\frac{v_{sL} + v_s}{v_m + v_s} + \delta\right)$$

$v_{sL} = 2.147$ (see Example 6.2.2)

$$\left(\frac{dp}{dL}\right)_f = \frac{2(0.0063)(38.32)(6.23)^2}{(32.2)(0.1662)}$$

$$\times \left[\frac{2.174 + 3.19}{6.23 + 3.19} + (-0.211)\right]$$

$$\left(\frac{dp}{dl}\right)_f = 1.255\frac{lb}{ft.^3} = 0.0087 \ \text{psi/ft.}$$

$$\left(\frac{dp}{dL}\right)_{total} = \left(\frac{dp}{dL}\right)_{el} + \left(\frac{dp}{dL}\right)_f + \left(\frac{dp}{dL}\right)_{accl}$$

$$\Downarrow$$

$$\text{neglected}$$

$$\left(\frac{dp}{dL}\right)_{el} = \gamma_s \left(\frac{g}{g_c}\right) = 17.45 \frac{32.2}{32.2} \left(\frac{lb_m}{ft.^3} \frac{ft.}{s^2} \frac{lb_f s^2}{lb_m ft.}\right)$$

$$= 17.45 \frac{lb}{ft.^3} = 17.45 \left(\frac{lb_f}{144 \ in.}\right) \frac{1}{ft.}$$

$$\left(\frac{dp}{dL}\right)_{el} = 0.1212 \ \text{psi/ft.}$$

$$\left(\frac{dp}{dL}\right)_{total} = 0.1212 + 0.0087 = \underline{0.1299 \ \text{psi/ft.}}$$

if $\Delta p = 500$ psi since $\Delta L = 500/0.1299 = 3849$ ft.

The performance of Orkiszewski correlation is briefly
outlined below for the flow variables considered [6].

- *Tubing Size*: The correlation performs well between 1-
  and 2-in. tubing sizes. The pressure loss is over predicted
  for tubing sizes greater than 2 in.
- *Oil Gravity*: At low oil gravities (13–30 °API), the correlation over predicts the pressure profile. However,
  predictions are seen to improve as oil °API increases.
- *Gas-Liquid Ratio*: The accuracy of Orkiszewski method
  is very good for gas–liquid ratio up to 5000. The errors
  become large (>20%) for GLR above 5000.
- *Water-cut*: The correlation predicts the pressure drop with
  good accuracy for a wide range of water-cuts.

6.2.14 The Hagedorn-Brown Method [9,12,13]

This correlation was developed using data obtained from a 1500-ft. vertical well. Tubing diameters ranging from 1 to 2 in. were considered in the experimental analysis along with five different fluid types, namely water and four types of oil with viscosities ranging between 10 and 110 cp (@ 80°F). The correlation developed is independent of flow patterns.

The equation for calculating pressure gradient is proposed as

$$\frac{\Delta p}{\Delta L} = \gamma_{hb}(F_e) + \frac{f_f q_L^2 m_o^2}{Ad^5 \gamma_m} + \frac{\left(\gamma_{hb}\Delta[v_m^2/2g]\right)}{\Delta L} \qquad [6.2.63]$$

We have one consistent set of units, where

		English	Metric
p	= pressure	lb/ft.^{2}	kPa
L	= length (height)	ft.	m
f_f	= Fanning friction factor	—	—
q_L	= total liquid flowrate	bbl/day	m^3/s
m_o	= total mass flowing/vol. liquid	slug/bbl	kg/m^3
d	= pipe ID	ft.	m
v_m	= avg. velocity = $v_{sL} + v_{sg}$	ft./sec	m/s
g	= conversion factor, force from mass	32.2 ft./sec^2	9.81 m/sec^2
A	= unit conversion constant	7.41 (10^{10})	8.63 (10^4)
γ_{HB}	= Hagedorn-Brown specific weight		
γ_s	= $v_L H_L + v_g(1 - H_L)$ based on pseudoholdup	lb/ft.^{3}	kg/m^3
γ_m	= avg. two-phase specific weight	lb/ft.^{3}	kg/m^3
V_{sL}	= superficial liquid velocity	ft./s	m/s
V_{sg}	= superficial gas velocity	ft./s	m/s
γ_L	= liquid specific weight	lb/ft.^{3}	kg/m^3
γ_g	= gas specific weight	lb/ft.^{3}	kg/m^3
H_L	= liquid holdup, a fraction		
F_e	= force equivalent	1.0	9.81

The friction factor used in Equation 6.2.63 is found from Figure 6.2.32. Figure 6.2.38 provides a relative roughness number. For this method, the Reynolds number for use with Figure 6.2.32 is

$$R_e = \frac{2.2(10^{-2})q_L m_o}{d\mu_m} \qquad [6.2.64]$$

where

		English	Metric
q	= volumetric flowrate total all fluids	ft^3/s	m^3/s
μ_m	= averaged viscosity, using an equation of the form of Arrhenius:		

$$\mu_m = \mu_L^{H_L} \mu_g^{(1-H_L)}$$

and

$$q_L \times \dot{m}_0 = \dot{w}_t$$

The equation would be solved over finite segments of pipe. Δv_m^2 is the changed velocity at points 1 and 2, the inlet to and outlet from that section. γ_m is the specific weight at the average p and T in the section.

Figure 6.2.39 also contains two empirical correction factors C and ψ. A plot of the data showed that holdup versus viscosity was a series of essentially straight lines. Water was chosen arbitrarily as a base curve (C = 1.0). C then is used

for other viscosity fluids to make the parallel curves coincident. The viscosity correction curve obtained is shown in Figure 6.2.40.

The factor ψ was included to fit some of the data where it was postulated that a transition would occur before mist-flow begins, with gas velocity as the major variable. As gas velocity approached that required for mist-flow, it breaks through the liquid phase and the turbulence produces a liquid "ring," which increases slippage. As velocity increases even further, the shear forces on this ring dissipate it until the primary mechanism is mist flow. Figure 6.2.41 shows the correlation for ψ. In most cases ψ will be equal to 1.0.

The basic correlating equation, 6.2.64 can be converted to a form similar to that for either single flow by allowing $H_L \rightarrow$ 0 for gas or $H_L \rightarrow$ 1.0 for liquids. As gas rate or liquid rate approaches zero, the pressure gradient obtained likewise approaches that for the other single phase. One therefore has a continuous gradient from liquid to two-phase to gas, an important aspect of the model.

Reviewing the foregoing calculation summary, it is necessary to make the calculation for a given diameter pipe and a given flowrate to avoid a trial-and-error solution. One can find R_e and all of the velocity-associated numbers to solve Equation 6.2.64. This would have to be repeated for various pipe size holdup calculations based primarily on data from 1.25-in. (0.031-m) tubing, the correlation in Figure 6.2.39 resulted. Some of the forms in this figure are the output from characterizing numbers and secondary correlations. Four dimensionless characterizing numbers were first proposed by Ros and adapted by others. They are given by Equations 6.2.46 to 6.2.49.

When the liquid stream contains both oil and water, calculate the properties as follows:

the liquid-specific weight γ_L:

$$\gamma_L = \left[\frac{SG_o(62.4) + R_s SG_g(0.0764)/5.614}{B_o}\right]\left(\frac{1}{1+WOR}\right)$$
$$+ \left[SG_w(62.4)\left(\frac{WOR}{1+WOR}\right)\right] \qquad [6.2.65]$$

the total weight associated with 1 bbl of stock tank liquid, w:

$$w = SG_0(350)\left(\frac{1}{1+WOR}\right) + SG_w(350)\left(\frac{WOR}{1+WOR}\right)$$
$$+ (0.0764)(GLR)SG_g \qquad [6.2.66]$$

the weight flowrate $\dot{w}_t$:

$$\dot{w}_t = \dot{w}(q_w + q_o) \qquad [6.2.67]$$

the liquid mixture viscosity μ_L:

$$\mu_L = \mu_o\left(\frac{1}{1+WOR}\right) + \mu_w\left(\frac{WOR}{1+WOR}\right) \qquad [6.2.68]$$

the liquid mixture surface tension σ:

$$\sigma_L = \sigma_o\left(\frac{1}{1+WOR}\right) + \sigma_w\left(\frac{1}{1+WOR}\right) \qquad [6.2.69]$$

the superficial liquid velocity v_{sL} in ft/s (assuming $B_w = 1.0$):

$$v_{sL} = \frac{5.61\,q_L}{86,400A_p}\left[B_o\left(\frac{1}{1+WOR}\right) + B_w\left(\frac{WOR}{1+WOR}\right)\right] \qquad [6.2.70]$$

the superficial gas velocity v_{sg}:

$$v_{sg} = \frac{q_L\left[GLR - R_s\left(\frac{1}{1+WOR}\right)\right]}{86,400\,A_p}\left(\frac{14.7}{p}\right)\left(\frac{T}{520}\right)\left(\frac{Z}{1}\right) \qquad [6.2.71]$$

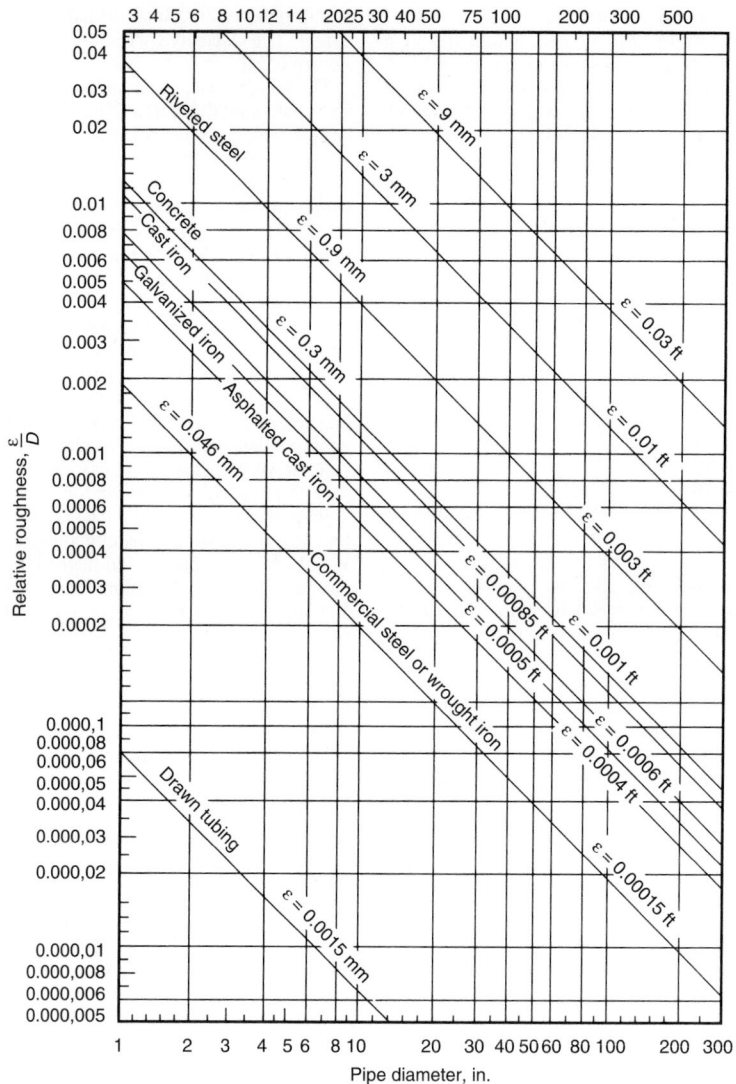

Figure 6.2.38 *Relative roughness for various kinds of pipe [8].*

Mixture specific weight is calculated by both using the Hagedorn-Brown holdup correlation and assuming no slippage. The higher value is then used.

If bubble flow is the dominant regime, the pressure gradient is used in the same way as in the Orkiszewski approach (step 7).

Example 6.2.4
Solve the problem in Example 2 using the Hagedorn-Brown method:

$p_1 = 1455$ psig
$p_i = 1955$ psig
$d = 1.995$ in ID ($A_p = 0.0217$)
$p_{bar} = 14.7$ psia
$T_1 = 75°F$
$T_2 = 105°F$
$SG_g = 0.752$

$SG_0 = 54°API$
$q_0 = 480$ bpd
$q_w = 0$
$\mu_s = 0.020$ cp
$\sigma_0 = 28$ dyn/cm
$GLR = 3393$ scf/bbl

Solution
1. $p = (1455 + 1955)/2 + 14.7 = 1179.7$
2. $T = (75 + 105)/2 = 90°F$
3. $SG_0 = 0.763$ from Example 6.2.2
4. Total mass flowing from volume liquid-$m_o = 462$ lb/stb of oil from Example 6.2.2
5. Solution gas/oil ratio R_s and oil formation volume factor B_o

$$R_s = 947.3 \text{ scf/stb}$$

$$B_o = 1.495 \text{ bbl/stb}$$

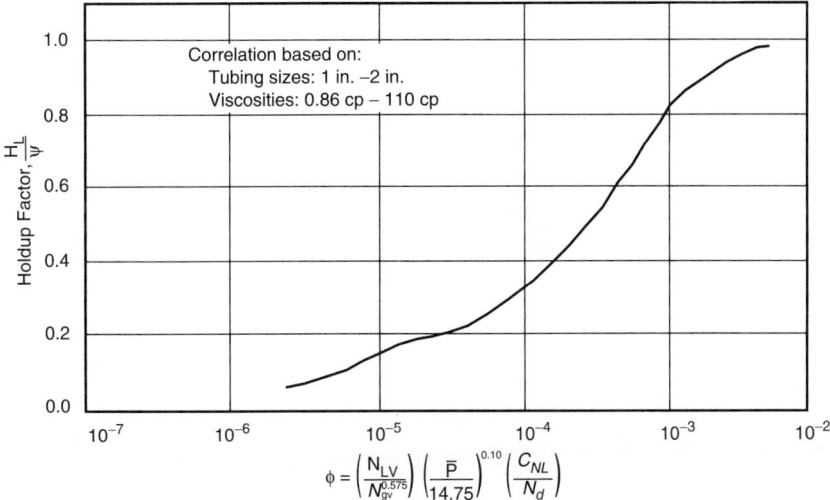

$$\phi = \left(\frac{N_{LV}}{N_{gv}^{0.575}}\right)\left(\frac{\bar{P}}{14.75}\right)^{0.10}\left(\frac{C_{NL}}{N_d}\right)$$

Figure 6.2.39 *Holdup factor correlation [9].*

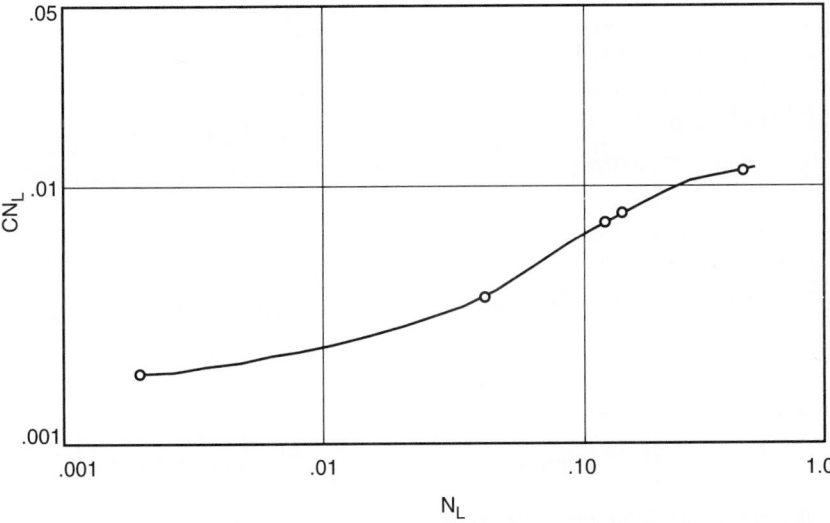

Figure 6.2.40 *Correlation for viscosity number coefficient [10].*

6. Liquid specific weight γ_L and gas specific weight γ_g from Equation 6.2.65

$$\gamma_L = \frac{(0.763)62.4 + 947.3(0.752)(0.0764)}{(1.495)\qquad\qquad(5.614)} = 38.33 \text{ lb/ft.}^3$$

$$\gamma_g = 8.823 \text{ lb/ft.}^3$$

7. Oil viscosity, μ_{os} and oil surface tension, σ_0

$$\mu_{os} = 0.5 \text{ cp}$$

$$\sigma_o = \sigma_L = 28 \text{ dyn/cm}$$

8. $N_L = 0.1573(0.5)(1/(38.33 \times 28^3))^{0.25} = 2.6 \times 10^{-3}$ from Equation 6.2.49
9. $CN_L = 0.0022$ (from Figure 6.2.40)
10. $v_{sL} = (5.61 \times 480)/(86,400 \times 0.0217)1.495 = 2.147$ ft./s from Equation 6.2.70

11. $N_{vL} = 1.938(2.147)(38.33/28)^{0.25} = 4.5$ from Equation 6.2.65
12. $v_{sg} = 4.08$ ft./s from Equation 6.2.71
13. $N_{gv} = 1.938 \times 4.08(38.33/28)^{0.25} = 8.55$ from Equation 6.2.46
14. Check the flow regime; calculate A and B:

$$A = L_B = 1.071 - [0.2218(4.08 + 2.147)^2]/0.1662$$

$$= -50.68 \text{ from Equation 6.2.51}$$

The minimum limit for L_B is 0.13. To assume 0.13

$$B = v_{sg}/(v_{sL} + v_{sg}) = 4.08/(4.08 + 2.147) = 0.655$$

Since $B - A = 0.616 - 0.13 = 0.486$, the difference is positive; so continue the Hagedorn-Brown procedure. In case B-A is negative, use the Orkiszewski method to find flow regime.

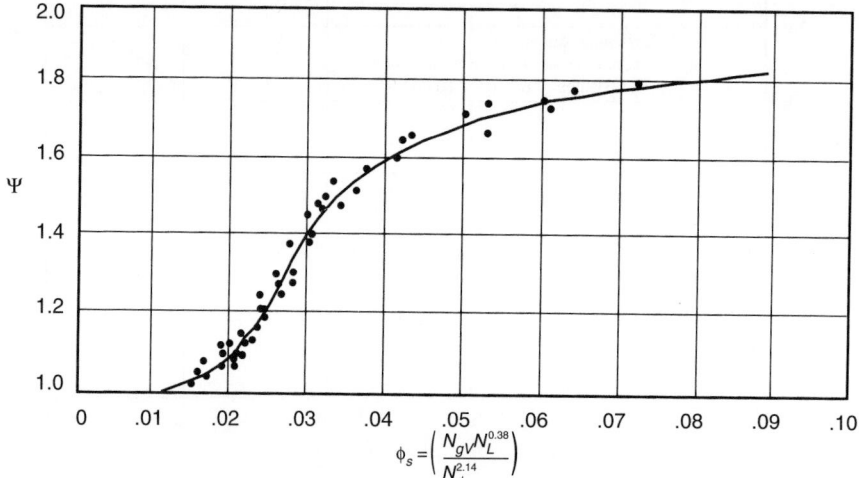

Figure 6.2.41   *Correlation for secondary factor correction [10].*

15. $N_d = 120.9(0.1662)(38.33/28)^{0.5} = 23.5$

16. Calculate holdup correlating function ϕ

$$\phi = \frac{N_{Lv}}{N_{gv}^{0.575}} \left(\frac{p}{14.7}\right)^{0.1} \left(\frac{CN_L}{N_d}\right)$$

$$= \frac{4.5}{8.55^{0.575}} \left(\frac{1719.7}{14.7}\right)^{0.1} \left(\frac{0.0022}{23.5}\right) = 0.000197$$

17. Now from Figure 6.2.39, $H_L/\psi = 0.42$

18. Calculate secondary correlation factor ϕ_s

$$\phi_s = \frac{N_{gv}N_L^{0.38}}{Nd^{2.14}} = \frac{8.55 \times (2.6 \times 10^{-3})0.38}{(23.5)^{2.14}} = 0.001036$$

19. From Figure 6.2.41, $\psi = 1.0$

20. Liquid holdup $H_L = (H_L/\psi)(\psi) = 0.42 \times 1.0 = 0.42$

21. The two-phase Reynolds number $(Re)_{tp}$ is

$$Re_{tp} = \frac{2.2 \times 10^{-2} \times 221{,}760}{(0.1662)0.5^{0.42}0.02^{0.58}} = 379{,}760$$

22. Relative roughness $\varepsilon/d = 0.00015/0.1662 = 0.0009$

23. Fanning friction factor from Figure 6.2.32

$$f_f = 0.00510$$

24. Calculate the average two-phase specific weight γ_m using two methods:

a) $\gamma_m = \gamma_{HB} = \gamma_L H_L + \gamma_g(1 - H_L)$

$$= 38.33(0.42) + 8.823(0.58) = 21.22 \text{ lb/ft.}^3$$

b) $\gamma_m = \dfrac{\dot{w}}{v_m}$

$$= \frac{350SG_0 + 0.0764SG_g(GOR) + 350SG_w(WOR)}{5.61B_0 + 5.61(WOR) + (GOR - R_s)B_g}$$

$$= \frac{350 \times 0.763 + 0.0764(0.752)3393 + 0}{5.61(1.495) + (3393 - 947.3) \times (14.7/1719.7)(550/520)(0.72)}$$

$$= 19.006 \text{ lb/ft.}^3.$$

Use 21.22 lb/ft.³ as a proper value.

25. Calculate

a) $Z_1, B_{01}, R_{s1}, v_{sL2}$ and v_{sg1} at T_2, p_1 (repeat steps 6, 7, 9, 16, 18)

$$T_1 = 75°; p_1 - 1455 \text{ psig}$$

$$\left.\begin{array}{l} T_r = (75 + 460)/394 = 1.358 \\ p_r = 1455/660 = 2.2 \end{array}\right\} \quad Z_1 = 0.71$$

$$R_{s1} = 0.752\left[(1455/18) \times 10^{0.0125(54)}/10^{0.00091(75)}\right]1.2048$$

$$= 804.4 \text{ scf/stb}$$

$$B_{01} = 0.972 + 0.000147F^{1.175} = 1.403 \text{ bbl/stb}$$

where $F = 804.4(0.752/0.763)^{0.5} + 1.25(75) = 892.3$

$$v_{sL1} = \left(\frac{5.61 \times 480}{86{,}400 \times 0.0217}\right)(1.403) = 2.02 \text{ ft./s}$$

$$v_{sg1} = \frac{[480(3393 - 804.4)14.7 \times 535 \times 0.71]}{86{,}400 \times 0.0217 \times 1{,}455 \times 520}$$

$$= 4.89 \text{ ft./s}$$

b) $Z_2, B_{02}, R_{s2}, v_{sL2},$ and v_{sg2} at T_2, p_2

$$T_2 = 105°F \qquad p_2 = 1955$$

$$\left.\begin{array}{l} T_r = (105 + 460)/394 = 1.434 \\ p_r = 1955/660 = 2.962 \end{array}\right\} \quad Z_2 = 0.73$$

$$R_{s2} = 0.752\left[(1955/18) \times 10^{0.0125(54)}/10^{0.00091(105)}\right]^{1.2048}$$

$$= 1064.5 \text{ scf/stb}$$

26. Calculate the two-phase velocity at both P_q and P_2:

$$V_{m1} = V_{sL1} + V_{sg1} = 2.02 + 4.89 = 6.91$$

$$V_{m2} = V_{sL2} + V_{sg2} = 2.262 + 3.368 = 5.63$$

27. Determine value $\Delta(V_m^2)$

$$\Delta(V_m^2) = (6.91^2 - 5.63^2) = 16.05$$

28. From Equation 6.2.64

$$\Delta L = \frac{\Delta p - \gamma_{HB}\Delta\left(\dfrac{V_m^2}{2g_c}\right)}{\gamma_{HB}(F_e) + (f_f q_L^2 m_0^2/Ad^5\gamma_m)}$$

$$\Delta p = 500 \text{ psi} = 500 \times 144 \text{ lb/ft.}^2$$

$$\gamma_{HB} = \gamma_m = 21.22, \qquad A = 7.41 \times 10^{10}$$

$$\Delta L = \frac{500(144) - 21.22\dfrac{16.05^2}{2 \times 32.2}}{21.22(1) + \dfrac{[0.0051(480)^2(462)^2]}{[7.41 \times 10^{10}(0.1662)^2 21.22]}}$$

$$= (72,000 - 84.88)/(21.22 + 0.005774) = 3388 \text{ ft.}$$

or

$$\frac{\Delta p}{\Delta L} = \frac{500}{3388} = 0.1476 \text{ psi/ft.}$$

The performance of the Hagedorn-Brown method is briefly outlined below [6].

- *Tubing Size:* The pressure loses are accurately predicted for tubing size between 1 and 1.5 in., the range in which the experimental investigation was conducted. A further increase in tubing size causes the pressure drop to be over predicted.
- *Oil Gravity:* The Hagedorn-Brown method is seen to over predict the pressure loss for heavier oils (13–25 °API) and under predict the pressure profile for lighter oils (45–56 °API)
- *Gas-Liquid Ratio:* The pressure drop is over predicted for gas–liquid ratio greater than 5000.
- *Water-cut:* The accuracy of the pressure profile predictions is generally good for a wide range of water-cuts.

6.2.15 The Beggs-Brill Method [2,9,15]

The Beggs and Brill correlation is developed for tubing strings in inclined wells and pipelines for hilly terrain. This correlation resulted from experiments using air and water as test fluids over a wide range of parameters given below:

gas flowrate 0 to 300 Mscfd
liquid flowrate 0 to 30 gal/min
average system pressure 35 to 95 psia
pipe diameter 1 and 1.5 in.
liquid holdup 0 to 0.870
pressure gradient 0 to 0.8 psi/ft.
inclination angle −90° to +90° also horizontal flow patterns

A flow diagram for calculating a pressure traverse in a vertical well is shown in Figure 6.2.42. The depth increment equation for Δ_L is

$$\Delta L = \frac{\Delta p\left(1 - \dfrac{\gamma_t v_t v_{sg}}{gp}\right)}{\gamma_t \sin\theta + \dfrac{f_t G_t v_t}{2gd}} \qquad [6.2.72]$$

where γ_t = two-phase specific weight in lb/ft.^{3}
v_t = two-phase superficial velocity
$\quad(v_t = v_{sL} + v_{sg})$ in ft./s
f_t = two-phase friction factor
G_t = two-phase weight flux rate (lb/s·ft.^{2})

A detailed procedure for the calculation of a pressure traverse is following:

1. Calculate the average pressure and average depth between the two points:

$$p = (p_1 + p_2)/2 + 14.7$$

2. Determine the average temperature T at the average depth. This value must be known from a temperature versus depth survey.
3. From P–V–T analysis or appropriate correlations, calculate $R_s, B_0, B_w, \mu_0, \mu_w, \mu_h, \sigma_0, \sigma_{ws},$ and Z at T and p_s.
4. Calculate the specific gravity of the oil SG_0:

$$SG_0 = \frac{141.5}{131.5 + API}$$

5. Calculate the liquid and gas densities at the average conditions of pressure and temperatures:

$$\gamma_L = \gamma_0\left(\frac{1}{1 + WOR}\right) + \gamma_w\left(\frac{1}{1 + WOR}\right) = \gamma_0 f_0 + \gamma_w f_w$$

$$\gamma_0 = \frac{350 SG_0 + 0.0764 R_s SG_g}{5.615}$$

$$\gamma_w = \frac{250\, SG_w}{5.615\, B_w}$$

$$\gamma_g = \frac{0.0764 SG_g p(520)}{(14.7)(T + 460)Z}$$

6. Calculate the *in situ* gas and liquid flowrates.

$$q_g = \frac{3.27 \times 10^{-7} Z q_0 (R - R_s)(T + 460)}{P}$$

$$q_L = 6.49 \times 10^{-5}(q_0 B_0 + q_w B_w)$$

7. Calculate the in situ superficial gas, liquid and mixture velocities:

$$v_{sL} = q_L/A_p$$

$$v_{sg} = q_g/A_p$$

$$v_t = v_{sL} + v_{sg}$$

8. Calculate the liquid, gas, and total weight flux rates:

$$G_L = \gamma_L v_{sL}$$

$$G_t = G_L + G_g$$

$$G_g = \gamma_s v_{sg}$$

9. Calculate the input liquid content (no-slip holdup):

$$\lambda = \frac{q_L}{q_L + q_g}$$

10. Calculate the Froude number N_{FR}, the liquid viscosity, μ_L, the mixture viscosity μ_m and the liquid surface tension σ_L:

$$N_{FR} = \frac{v_t^2}{gd}$$

$$\mu_L = \mu_0 f_0 + \mu_w f_w$$

$$\mu_t = \mu_L \lambda + \mu_g(1 - \lambda)(6.72 \times 10^{-4})$$

$$\sigma_L = \sigma_0 f_0 + \sigma_w f_w$$

11. Calculate the no-slip Reynolds number and the liquid velocity number:

$$(N_{Re})_{ns} = \frac{G_t d}{\mu_t}$$

$$N_{LV} = 1.938 v_{sL}\left(\frac{\gamma_L}{\sigma_L}\right)^{0.25}$$

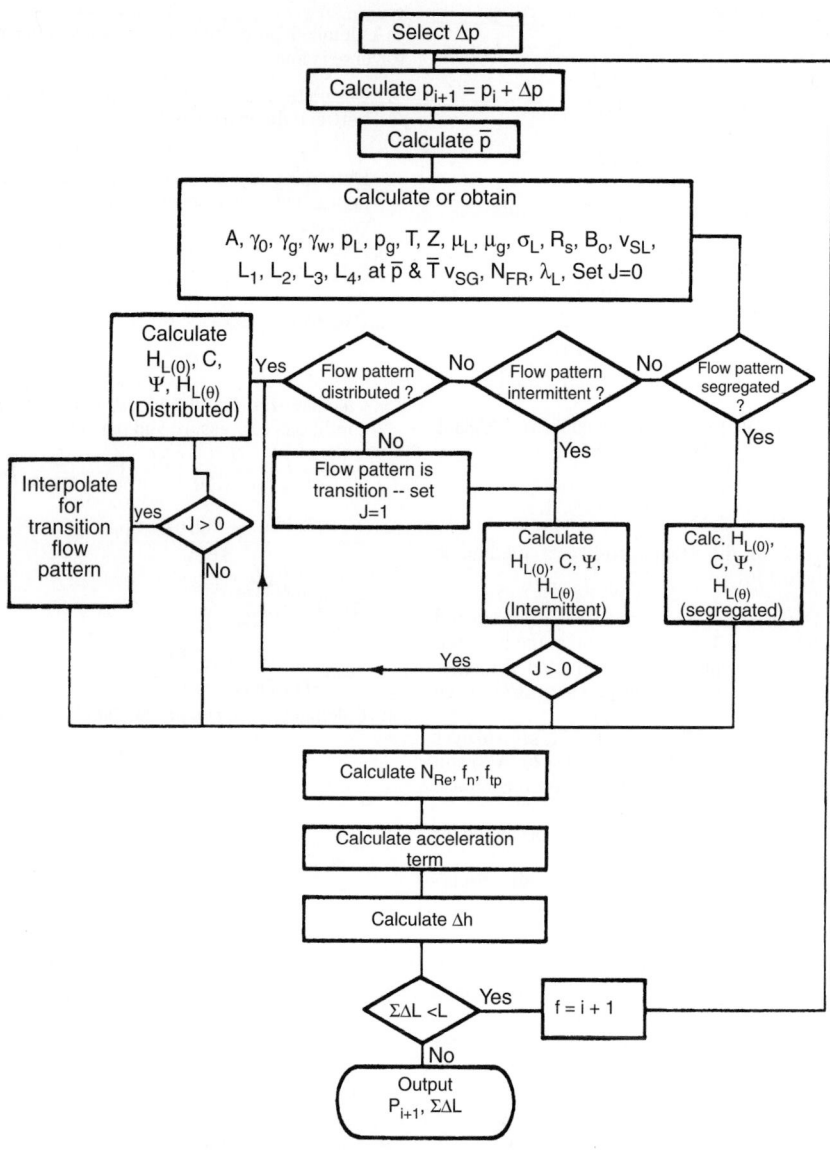

Figure 6.2.42 *Flow diagram for the Beggs-Brill method [1].*

12. To determine the flow pattern that would exist if flow were horizontal, calculate the correlating parameters, L_1, L_2, L_3, and L_4:

$$L_1 = 316\lambda^{0.0302}$$

$$L_3 = 0.10\lambda^{-1.4516}$$

$$L_2 = 0.0009252\lambda^{-2.4684}$$

$$L_4 = 0.5\lambda^{-6.738}$$

13. Determine flow pattern using the following limits:
Segregated:

$$\lambda < 0.01 \quad \text{and} \quad N_{FR} < L_1$$

or

$$\lambda \geq 0.01 \quad \text{and} \quad N_{FR} < L_2$$

Transition:

$$\lambda \geq 0.01 \quad \text{and} \quad L_2 < N_{FR} < L_3$$

Intermittent:

$$0.01 \leq \lambda < 0.4 \quad \text{and} \quad L_3 < N_{FR} < L_1$$

or

$$\lambda \geq 0.4 \quad \text{and} \quad L_3 < N_{FR} \leq L_4$$

Distributed:

$$\lambda < 0.4 \quad \text{and} \quad N_{FR} \geq L_1$$

or

$$\lambda \geq 0.4 \quad \text{and} \quad N_{FR} > L_4$$

14. Calculate the horizontal holdup $H_L(O)$:

$$H_L(O) \frac{a\lambda^b}{N_{FR}^c}$$

where a, b and c are determined for each flow pattern from the following table:

Flow pattern	a	b	c
Segregated	0.98	0.4846	0.0868
Intermittent	0.845	0.5351	0.0173
Distributed	1.065	0.5824	0.0609

15. Calculate the inclination correction factor coefficient:

$$C = (1 - \lambda) \ln \left(d \lambda^e N_{LV}^f N_{FR}^g \right)$$

where d, e, f, and g are determined for each flow condition from the following table:

Flow Pattern	d	e	f	g
Segregated uphill	0.011	−3.768	3.539	−1.614
Intermittent uphill	2.96	0.305	−0.4473	0.0978
Distributed uphill	No correction		$C = 0$	
All flow patterns downhill	4.70	−0.3692	0.1244	−0.5056

16. Calculate the liquid inclination correction factor:

$$\psi = 1 + C\left[\sin(1.8\theta) - 0.333 \sin^3(1.8\theta) \right] = 1 + 0.3C$$

for vertical well

17. Calculate the liquid holdup and the two-phase density:

$$H_L(\theta) = H_L(0)\psi$$

$$\rho_t = \rho_L H_L + \rho_g (1 - H_L)$$

18. Calculate the friction factor ratio:

$$f_t/f_{ns} = e_s$$

where

$$S = \frac{\ln(y)}{-0.0523 + 3.182 \ln(y) - 0.8725 [\ln(y)]^2 + 0.01853 \, [\ln(y)]^4}$$

$$y = \lambda/[H_L(\theta)]^2$$

S becomes unbounded at a point in the interval $1 < y < 1.2$; and for y in this interval, the function S is calculated from

$$S = \ln(2.2y - 1.2)$$

19. Calculate the no-slip friction factor:

$$f_{ns} = 1/\left\{ 2 \log[N_{Rens}/(4.5223 \log N_{Rens} - 3.8215)] \right\}^2$$

or

$$f_{ns} = 0.0056 + 0.5/(N_{Rens})^{0.32}$$

20. Calculate the two-phase friction factor:

$$f_t = f_{ns}/(f_t/f_{ns})$$

21. Calculate ΔL. If the estimated and calculated values for ΔL are not sufficiently close, the calculated value is taken as the new estimated value and the procedure is repeated until the values agree. A new pressure increment is then chosen and the process is continued until the sum of the ΔL's is equal to the well depth.

Example 6.2.5
Solve the problem in Example 6.2.2 using the Beggs-Brill method.

Solution
1. $p = 1719.7$ psia
2. $T = 90°F$
3. $R_s = 947.3$ scf/stb $B_0 = 1.495$ bbl/stb
   $\mu_{os} = 0.5$ cp, $\sigma_0 = 28$ dyn/cm, $Z = 0.72$
4. $SG_0 = 0.736, \gamma_g = 8.823$ lb/ft.^{3}
5. $\gamma_0 = 38.32$ lb/ft.^{3} (from Example 6.2.3)
6. $q_g = 0.08855$ ft.^{3}/s
   $q_L = 0.0466$ ft.^{3}/s
7. $A_p = 0.0217$ ft.^{2}
   $v_{sL} = q_L/A_p = 2.147$ ft./s, $v_{sg} = 4.081$ ft./s
8. Calculate the liquid, gas, and total weight flux rates:

$$G_L = \gamma_L v_{sL}, \quad G_g = \gamma_g v_{sg}$$

$$G_t = G_L + G_g = 38.32 \times 2.147 + (8.823) \times 4.081$$

$$= 118.3 \, \text{lb}/(\text{s·ft.}^2)$$

9. Calculate the input liquid (no-slip holdup):

$$\lambda = \frac{q_L}{q_L + q_g} = \frac{0.0466}{0.0466 + 0.8855} = 0.3448 = 0.345$$

10. The Froude number, viscosity, and surface tension

$$N_{FR} = \frac{v_t^2}{gd} = \frac{6.23^2}{32.174 \times 0.1662} = 7.26$$

$$\mu_L = \mu_o f_o + \mu_w f_w = 0.5(1.0) + \mu_w(0.0) = 0.5$$

$$\mu_t = \left(6.72 \times 10^{-4} \right)\left[0.5 \times 0.345 + 0.02(1 - 0.345) \right]$$

$$= 1.164 \times 10^{-4}$$

$$= 0.0001164 \, \text{lb}_m/(\text{ft./s})$$

$$\sigma_L = \sigma_o f_o + \sigma_w f_w = 28 \times 1.0 = 28 \, \text{dyn/cm}$$

11. Calculate the no-slip Reynolds number and the liquid velocity number:

$$(N_{Re})_{ns} = \frac{G_t d}{\mu_t} = \frac{118.3(0.1662)}{0.000464} = 168,884$$

$$N_{LV} = 1.938 \times 2.147(38.32/28)^{0.25} = 4.5$$

12. Determine the flow pattern that would exist if flow were horizontal:

$$L_1 = 316\lambda^{0.302} = 316 \times (0.345)^{0.302} = 229.14$$

$$L_2 = 0.0009252(0.345)^{-2.4684} = 1.2796 \times 10^{-2}$$

$$L_3 = 0.10\lambda^{-1.4516} = 0.10(0.345)^{-1.4516} = 0.4687$$

$$L_4 = 0.5\lambda^{-6.738} = 0.5(0.345)^{-6.738} = 650.3$$

13. Determine flow pattern:

$$0.4 > \lambda > 0.01 \quad \text{and} \quad L_2 < N_{FR} < L$$

The flow pattern is intermittent.

14. Calculate the horizontal holdup:

$$H_L(O) = 0.845(0.345)^{0.5351}/7.26^{0.0173} = 0.462$$

15. Calculate the inclination correction factor coefficient:

$$C = (1 - 0.345) \ln \left(2.96 \times 0.345^{0.305} 4.5^{-0.4473} 7.26^{0.0978} \right)$$

$$= 0.18452$$

16. Calculate the liquid holdup inclination correction factor:

$$\psi = 1 + C\left[\sin(1.8 \times 90) - 0.333 \sin^3(1.9 \times 90) \right]$$

$$= 1 + C(0.309 - 0.009826) = 1 + 0.3C$$

$$= 1 + 0.3(0.18452) = 1.055$$

17. Calculate the liquid holdup and the two-phase density:

$$H_L(90) = H_L(O)\psi = 0.462 \times 1.055 = 0.4876$$

$$\gamma_t = \gamma_L H_L + \gamma_g(1 - H_L)$$

$$= 38.32(0.4876) + 8.823(1 - 0.4876)$$

$$= 23.2 \text{ lb/ft.}^3$$

18. Calculate the friction factor ratio:

$$y = \left[0.345/(0.4876)^2\right] = 1.451, \ln 1.451 = 0.3723$$

$$f_t/f_{ns} = \exp\left[0.3723/(-0.0523 + 3.182 \times 0.3723)\right.$$

$$\left. - (0.8725 \times 0.3723^2 + 0.1853 \times 0.3723^4)\right]$$

$$= \exp(0.3723/1.0188) = 3^{0.36796} = 1.4447$$

19. Calculate the no-slip friction factor:

$$f_{ns} = 1/\left\{2\log\left[N_{Rens}/(4.5223 \log N_{Rens} - 3.8215)\right]\right\}^2$$

$$= 1/36.84 = 0.0271$$

20. Calculate the two-phase friction factor:

$$f_t = f_{ns}(f_t/f_{ns}) = 0.0271(1.4447) = 0.0391$$

21. Determine the distance ΔL for $\Delta p = 500$ psi from Equation 6.2.72

$$\Delta L = \frac{500\left[1 - \dfrac{23.2(6.23)4.081}{32.174 \times 1719.7}\right]144}{23.2(1.0) + \dfrac{0.0391(118.3)6.23}{2(32.174)0.1662}} = \frac{72,000(0.9893)}{25.894}$$

$$= 2750 \text{ ft}$$

and

$$\frac{\Delta p}{\Delta L} = \frac{500}{2750} = 0.18 \text{ psi/ft.}$$

Example 6.2.6
Solve Example 6.2.1 using the Beggs-Brill method:
$q_g = 40$ MMscf/d, $p_{avg} = 2000$ psia
$q_0 = 40,000$ stb/d, $T_{avg} = 80°F$
ID = 9 in. $SG_g = 0.75$ at $p = 14.7$ psia in $T = 60°F$
$R_p = 990$ scf/bbl

Solution
1. $SG_0 = 141.5/131.5 + API = 0.86$
2. Calculate R_s, B_0, μ_g, Z_g at p_{avg} and T_{avg}:

$$Z_g = 0.685$$

$$\mu_g = 0.0184 \text{ cp}$$

$$R_s = 477 \text{ scf/stb}$$

$$B_0 = 1.233 \text{ rb/stb}$$

$$\mu_0 = 2.96 \text{ cp}$$

3. Calculate γ_0 and γ_g at average parameters:

$$\gamma_0 = \frac{350(0.86) + 0.0764(477)(0.75)}{5.614(1.233)} = 47.42 \text{ lb/ft.}^3$$

$$\gamma_g = \frac{0.0764\gamma_g p(520)}{(14.7)(T + 460)Z_g} = \frac{0.0764(0.73)(2000)(520)}{(14.7)(80 \times 460)(0.685)}$$

$$= 10.96 \text{ lb/ft.}^3$$

4. Calculate the in situ gas and liquid flowrates,

$$q_g = \frac{3.27 \times 10^{-7}Z_g q_0(R - R_s)(T + 460)}{p}$$

$$= \frac{3.27 \times 10^{-7}(0.685)(40,000)(990 - 477)(80 + 540)}{2000}$$

$$= 1.241 \text{ ft.}^3/s$$

$$q_L = 6.49 \times 10^{-5}(q_0 B_0 + q_w B_w)$$

$$= 6.49 \times 10^{-5}[40,000(1.233) + 0]$$

$$= 3.201 \text{ ft.}^3/s$$

5. Calculate A_p:

$$A_p = \frac{\pi}{4}D^2 = \frac{\pi}{4}\left(\frac{9}{12}\right)^2 = 0.4418 \text{ ft.}^2$$

6. Calculate the in situ superficial gas, liquid, and mixture velocities:

$$v_{sL} = q_L/A_p = 3.201/0.4418 = 7.25 \text{ ft./s}$$

$$v_{sg} = q_g/A_p = 1.241/0.4418 = 2.81 \text{ ft./s}$$

$$v_m = v_{sL} + v_{sg} = 7.25 + 2.81 = 10.06 \text{ ft./s}$$

7. Calculate the liquid, gas, and total mass flux rates:

$$G_L = \rho_L v_{sL} = (47.42)(7.25) = 343.6 \text{ lb/(s} \cdot \text{ft.)}$$

$$G_g = \rho_g v_{sg} = (10.96)(2.81) = 30.79 \text{ lb/(s} \cdot \text{ft.)}$$

$$G_m = G_L + G_g = 343.6 + 30.8 = 374.4 \text{ lb/(s} \cdot \text{ft.)}$$

8. Calculate the no-slip holdup:

$$\lambda = \frac{q_L}{q_L + q_s} = \frac{3.201}{32 + 1.241} = 0.72$$

9. Calculate the Froude number N_{FR}, the mixture viscosity μ_m, and surface tension σ_L:

$$N_{FR} = \frac{v_m^2}{gd} = \frac{(10.06)^2}{(32.2 \times 9)} = 4.186$$

$$\mu_m = 6.27 \times 10^{-4}[\mu_L\lambda + \mu_g(1 - \lambda)]$$

$$= 6.27 \times 10^{-4}[2.96(0.72) + 0.0184(0.28)]$$

$$= 1.44 \times 10^{-3} \text{ lb/(ft./s)}$$

$$\sigma_L = 37.5 - 0.257(\text{API}) = 37.5 - 0.257(33)$$

$$= 29.0 \text{ dyn/cm}$$

10. Calculate the non-slip Reynolds number and the liquid velocity number:

$$Re_{NS} = \frac{G_m d}{\mu_m} = \frac{(374.4)(9)}{(1.44 \times 10^{-3})12} = 195,000$$

$$N_{LV} = 1.938 \, v_{sL}\left(\frac{\gamma L}{\sigma L}\right)^{0.25} = 1.938(7.25)(47.42/29)^{0.25}$$

$$= 15.88$$

11. Calculate $L_1, L_2, L_3,$ and L_4:

$$L_1 = 316\lambda^{0.302} = 316(0.721)^{0.302} = 286$$

$$L_2 = 0.000952\lambda^{-2.4684} = 0.0009252(0.721)^{-2.4684}$$

$$= 0.0021$$

$$L_3 = 0.10\lambda^{-1.4516} = 0.10(0.721)^{-1.4516} = 0.161$$

$$L_4 = 0.5\lambda^{-6.738} = 0.5(0.721)^{-6.738} = 4.53$$

12. Determine flow pattern:

Since $0.721 \geq 0.4$ and $L_3 < N_{FR} \leq L_4$

Flow is intermittent.

13. Calculate the horizontal holdup $H_L(O)$:

$$H_L(O) = a\lambda^b/N_{FR}^c = 0.845 \times 0.721^{0.5351}/7.186^{0.0173}$$

$$= 0.692$$

14. Calculate ψ and $H_L(O)$ and two-phase specific weight:

Since $\theta = 0°$, $\lambda = 1 + 0 = 1$

$$H_L(0°) = H_L(0)\lambda = 0.692$$

$$\gamma_t = \gamma_L H_L + \gamma_g(1 - H_L) = 47.42(0.692)$$

$$+ 10.96(1 - 0.692)$$

$$= 36.19 \, \text{lb/ft.}^3$$

15. Calculate the friction factor ratio:

$$y = \frac{\lambda}{H_L^2} = \frac{0.721}{(0.692)^2} = 1.506$$

$$\ln(y) = 0.4092$$

$$S = \ln(y)/\left[-0.0523 + 3182 \ln y - 0.8725(\ln y)^2 \right.$$

$$\left. + 0.01853(\ln y)^4\right]$$

$$= 0.3706$$

$$f_t/f_{ns} = e^s = e^{0.3706} = 1.449$$

16. Calculate the non-slip friction factor f_{ns}:

$$f_{ns} = 1/\{2\log[\text{Re}_{ns}/(4.5223 \log \text{Re}_{ns} - 38,215)]\}^2$$

$$= 1/(2\log[195,000/(4.5223 \log 195,000 - 3.8215)])^2$$

$$= 0.01573$$

17. Calculate the two-phase friction factor:

$$f_t = f_{ns}(f_t/f_{ns}) = 0.01573(1.449) = 0.0227$$

18. Calculate the pressure gradient:

$$\frac{\Delta L}{\Delta p(144)} = \frac{1 - \dfrac{\gamma_t v_m v_{sg}}{gP}}{\gamma_{tp} \sin \theta + \dfrac{f_t G_m v_m}{2gd}}$$

$$= \frac{1 - \dfrac{(36.19)(10.06)(2.81)}{(32.2)2,000(144)}}{36.19(1)(0) + \dfrac{0.0227(374.4)(10.06)}{2(32.2)(9/12)}}$$

$$= 0.5646$$

$$\frac{\Delta L}{p} = 81.3 \quad \text{or} \quad \frac{\Delta p}{\Delta L} = 1.23 \times 10^{-2} \, \text{psi/ft.}$$

because pressure is decreasing in flow direction to proper value of $\Delta p/\Delta L = -1.23 \times 10^{-2}$ psi/ft.

The performance of the correlation is given below [7].

- *Tubing Size*: For the range in which the experimental investigation was conducted (i.e., tubing sizes between 1 and 1.5 in.), the pressure losses are accurately estimated. Any further increase in tubing size tends to result in an over prediction in the pressure loss.
- *Oil Gravity*: A reasonably good performance is obtained over a broad spectrum of oil gravities.
- *Gas–Liquid Ratio*: In general, an over predicted pressure drop is obtained with increasing GLR. The errors become especially large for gas–liquid ratio above 5000.

- *Water-Cut*: The accuracy of the pressure profile predictions is generally good up to about 10% water-cut.

6.2.16 Mechanistic models

In the past two decades, mechanistic models have been the focus of many investigations to improve the flow behavior predictions in multiphase flow. These models rely more on the theory or mechanisms (or phenomena) in multiphase flow rather than solely on experimental work. Empiricism is still used in a mechanistic approach to predict certain flow mechanisms or provide closure relationships. One of the preliminary works on mechanistic modeling of flow pattern transition and steady, upward, and gas-liquid flow in vertical tubes is presented by Taitel et al. [16]. This flow pattern prediction was later modified by Barnea et al. [17] to account for a whole range of pipe inclinations. These two works have provided the basis for predicting flow patterns by defining transition boundaries between bubble, slug, churn, and annular flows. Although many mechanistic models predict flow behavior for a single flow pattern, only a few comprehensive mechanistic models such as Hasan and Kabir [18,19] and Ansari et al. [20] have been published in the literature to model all four flow pattern transitions in multiphase flow. Both these mechanistic the models have enjoyed considerable success in multiphase flow predictions and have generally been accepted in the petroleum industry. The Ansari et al. model [20] is relatively more complex, and the reader is referred to the original papers [20,21] for a complete presentation of the model. Only the details of the Hasan and Kabir model are presented here.

6.2.17 Hasan and Kabir model [18,19]

Hasan and Kabir developed a mechanistic model to predict multiphase flow pressure gradients and wellbores for the same four flow patterns discussed in the empirical approach.

The bubble/slug transition: The bubble/slug flow transitions across at a void fraction of 0.25 and is expressed in terms of superficial gas velocity.

$$v_{sg} = \frac{\sin \theta}{4 - C_o}(C_o v_{sL} + v_s) \qquad [6.2.73]$$

where C_o is the flow parameter written as

$$C_o = \begin{cases} 1.2 \text{ if } d < 4 \text{ in} \\ 2.0 \text{ if } d > 4 \text{ in} \end{cases} \qquad [6.2.74]$$

In equation 6.2.73, V_s is the bubble-rise or slip velocity and is given by Harmathy [22] expression as

$$v_s = 1.53\left[\frac{g\sigma_L(\rho_L - \rho_g)}{\rho_L^2}\right]^{1/4} \qquad [6.2.75]$$

The transition to slug flow occurs at superficial gas velocities greater than that given by equation 6.2.73. The slug flow transition is also checked by finding the rise velocity of a Taylor bubble as

$$v_{TB} = 0.35\sqrt{\frac{g\sigma_L(\rho_L - \rho_g)}{\rho_L^2}}\sqrt{\sin \theta}(1 + \cos \theta)^{1.2} \qquad [6.2.76]$$

It should be noted that v_{TB} is dependent on the pipe diameter. In smaller diameter pipes, when Taylor bubble velocity is less than in the slip velocity, the rising smaller bubbles approached the back of the Taylor bubble, coalesce with it, and increase its size thus causing a transition to slug flow.

Dispersed bubble/slug/churn flow transitions: Taitel et al. [16] proposed an equation for the transition to dispersed-bubble flow and is given by

$$v_m^{1.12} = 4.68d^{0.48}\sqrt{\frac{g\sigma_L(\rho_L - \rho_g)}{\rho_L^2}}\left(\frac{\sigma_L}{\rho_L}\right)^{0.6}\left(\frac{\rho_L}{\mu_L}\right)^{0.08} \qquad [6.2.77]$$

However, as the gas void fraction exceeds 0.52, bubble coalescence cannot be prevented, and transition to either slug, churn, or annular flow must occur.

Slug/churn transitions is predicted from

$$\rho_g V_{sg}^2 = 0.0067(\rho_L V_{sL}^2)^{1.7} \text{ if } \rho_L V_{sL}^2 < 50$$
$$[6.2.78a]$$

and

$$\rho_g V_{sg}^2 = 17.1 \log_{10}(\rho_L V_{sL}^2) - 23.2 \text{ if } \rho_L V_{sL}^2 \geq 50$$
$$[6.2.78b]$$

Annular-flow transition: The transition to annular flow is given by

$$v_{sg} = 3.1 \left[\frac{g\sigma_L(\rho_L - \rho_g)}{\rho_g^2} \right]^{1/4} \qquad [6.2.79]$$

This criterion is partly based on the gas velocity required to prevent fallback of entrained liquid droplets in the gas stream.

Bubble and dispersed bubble flow: The liquid hold up for the bubble and dispersed bubble flow is given by

$$H_L = 1 - \frac{v_{sg}}{C_o v_m + v_s} \qquad [6.2.80]$$

The total pressure gradient is estimated by Equation 6.2.81, and it is used in conjunction with the mixture density, ρ_m, once the hold up is calculated.

$$\left(\frac{dp}{dL}\right)_t = \rho_m g \sin\theta + \frac{2f\rho_m v_m^2}{d} + \rho_m v_m \frac{dv_m}{dL} \qquad [6.2.81]$$

where $\rho_m = H_L\rho_L + (1 - H_L)\rho_g$

As usual, the friction factor is determined from the Moody diagram shown in Figure 6.2.32. Mixture Reynolds number for determining the friction factor is given by the Equation 6.2.82.

$$Re_m = \frac{\rho_L v_m d}{\mu_L} \qquad [6.2.82]$$

Slug and churn flow: The slip velocity for expressions for slug and churn flow is given by Equation 6.2.83

$$v_s = 0.35 \sqrt{\frac{gd(\rho_L - \rho_g)}{\rho_L}} \sqrt{\sin\theta}(1 + \cos\theta)^{1.2} \qquad [6.2.83]$$

A value of 1.2 and 1.15 are used for C_o in Equation 6.2.80 for slug and churn flow, respectively. The friction pressure gradient is calculated from Equation 6.2.81. Because of the chaotic nature of the flow and difficulty in analyzing the churn flow regime, predictions in this flow regime are less accurate.

Annular flow: The liquid hold up for the central core in annular flow is given by

$$H_{LC} = \frac{Ev_{sL}}{Ev_{sL} + v_{sg}} \qquad [6.2.84]$$

Here $H_{LC} = 1 - f_{gc}$ where f_{gc} is the gas void fraction and E is the liquid entrainment in the central core. The liquid entrainment in the core is calculated by the following expressions

$$\left\{ \begin{array}{l} E = 0.0055\left[(v_{sg})_{crit} \times 10^4\right]^{2.86} \text{ if } (v_{sg})_{crit} \times 10^4 < 4 \\ E = 0.857\log_{10}\left[(v_{sg})_{crit} \times 10^4\right] - 0.20 \text{ if } (v_{sg})_{crit} \times 10^4 > 4 \end{array} \right\}$$
$$[6.2.85]$$

$(v_{sg})_{crit}$ is given by

$$(v_{sg})_{crit} = \frac{v_{sg}\mu_g}{\sigma_L}\left(\frac{\rho_g}{\rho_L}\right)^{1/2} \qquad [6.2.86]$$

The friction pressure gradient is calculated from

$$\left(\frac{dp}{dL}\right)_f = \frac{2f_C\rho_C}{d}\left(\frac{v_{sg}}{f_{gc}}\right)^2 \qquad [6.2.87]$$

where

$$f_C = \frac{0.079(1 + 75H_{LC})}{N_{Reg}^{0.25}} \qquad [6.2.88]$$

and

$$\rho_C = \frac{v_{sg}\rho_g + v_{sL}\rho_L E}{v_{sg} + v_{sL}E} \qquad [6.2.89]$$

Wallis [3] presented a simple equation for gas void fraction in the core given by

$$f_{gc} = (1 + X^{0.8})^{-0.378} \qquad [6.2.90]$$

where X is the Lockhart-Martinelli parameter given in terms of the gas mass fraction x_g as

$$X = \left[\frac{(1 - x_g)}{x_g}\right]^{0.9}\left(\frac{\mu_L}{\mu_g}\right)^{0.1}\sqrt{\frac{\rho_L}{\rho_g}} \qquad [6.2.91]$$

Example 6.2.7
Solve Example 6.2.2 using the Hasan-Kabir mechanistic model.

The bubble/slug transition is checked using Equation 6.2.73. The slip velocity is first found Equation 6.2.75

$$v_s = 1.53\left[\frac{g\sigma_L(\rho_L - \rho_g)}{\rho_L^2}\right]^{1/4}$$

$$v_s = 1.53\left[\frac{32.17^2(0.001918)(47.6 - 8.823)}{47.6^2}\right]^{1/4}$$

$$= 0.656 \text{ ft/s}$$

Using $C_o = 1.2$, V_{sg} is calculated from Equation 6.2.73 as

$$v_{sg} = \frac{\sin\theta}{4 - C_o}(C_o v_{sL} + v_s)$$

$$v_{sg} = \frac{1}{4 - 1.2}(1.2 \times 2.147 + 0.656)$$

$$= 1.1544 \text{ ft/s}$$

Since $V_{sg} > 0.656$ ft/s, bubble flow does not exist. Now, check dispersed-bubble transition (Equation 6.2.77)

$$v_m^{1.12} = 4.68\left(\frac{d}{12}\right)^{0.48}\sqrt{\frac{g(\rho_L - \rho_g)}{g_c\sigma_L}}\left(\frac{g_c\sigma_L}{\rho_L}\right)^{0.6}\left(\frac{\rho_L}{g_c\mu_L}\right)^{0.08}$$

$$\left(\frac{dP}{dL}\right)_f = \frac{2f\rho_L v_m^2 H_L}{d}$$

$$= \frac{2 \times 0.004 \times 47.6 \times 6.227^2 \times 0.5025}{0.1662}$$

$$= 1.38645 \text{ lb}_f/\text{ft}^3 = 0.009628 \text{ psi/ft}$$

Thus, the total pressure gradient

$$\frac{dP}{dL} = 0.1965 + 0.009628 = 0.20618 \text{ psi/ft}$$

$$\Delta L = \frac{500}{0.20618} = 2425.07 \text{ ft}$$

In general, the performance of the Hasan-Kabir mechanistic model is good for a wide range of tubing sizes, oil gravity, gas–liquid ratio, and water-cut.

6.2.18 Summary
In this work, attention was paid only to six methods: flow regime maps, the Duns-Ros, Orkiszewski, Hagedorn-Brown, Beggs-Brill, and Hasan-Kabir methods. They are the most often used. A comprehensive discussion of the many other multiphase models published in the literature, including the

ones discussed here, can be found in a recently published monograph [23]. As alluded to in the preceding discussions, the pressure profile in a pipe is influenced by a number of parameters such as tubing size and geometry, well inclination, and fluid composition and properties (which in turn depend on pressure and temperature). Therefore, the empirical methods presented above should be used with caution, keeping in mind the experimental range of validity for which the correlation have been developed. For example, the Hagedorn-Brown, Orkiszewski, and Duns-Ros correlations have all been developed for vertical wells, and their use in deviated wells may or may not yield accurate results. In addition, the Duns-Ros correlation is not applicable for wells with water-cut and should be avoided for such cases. Similarly, the Beggs-Brill method, primarily developed for inclined wells, is applicable to multiphase flows with or without water cut. In general, mechanistic models such as the Hasan-Kabir have a wide range of applicability because of their limited reliance on empiricism.

Over the years, numerous modifications to the above discussed empirical models have been proposed to improve their multiphase flow predictions and to extend their range of applicability. Some of these modifications are discussed in the monograph by Brill and Mukherjee [23] and are used extensively in many commercial software programs. The monograph also presents a comparative study of the performance of both empirical and mechanistic models with measured field data from a well databank for a variety of cases. Although some methods/models perform better than others for certain cases, no single method performs accurately for all cases despite development of new mechanistic models. Therefore, it is strongly recommended that the model limitations and validity be kept in mind before a particular model/correlation is selected in the absence of other relevant information such as measured field data.

Reference

1. Brown, K. E., *The Technology of Artificial Lift Methods*, Vol. 1, Pennwell Books, Tulsa, Oklahoma, 1977.
2. Golan, M., and Whitson, C. H., *Well Performance*, Prentice Hall, Englewood Cliffs, New Jersey, 1991.
3. Wallis, G. B., *One-Dimensional Two-Phase Flow*, McGraw-Hill Book Co., New York, 1969.
4. Crowley, Ch. J., Wallis, G. B., and Rothe, P. H., *State of the Art Report on Multiphase Methods for Gas and Oil Pipelines*, Vols. 1–3, AGA Project, PR-172-609, December 1986.
5. Duns, Jr. H., and Ros, N. C. J., "Vertical Flow of Gas and Liquid Mixtures in Wells," *Proc. Sixth World Pet. Congress*, Frankfurt, Section II, 22-PD6, June 1963.
6. Lawson, J. D., and Brill, J. P., "A Statistical Evaluation of Methods Used to Predict Pressure Losses for Multiphase flow in Vertical Oilwell Tubing," *J. Pet. Tech.*, August 1974.
7. Vohra, I. R., Robinson, J. R., and Brill, J. P., "Evaluation of Three New Methods for Predicting Pressure Losses in Vertical Oilwell Tubing," *J. Pet. Tech.*, August 1974.
8. Orkiszewski, J., "Predicting Two-Phase Pressure Drops in Vertical Pipe," *J. Pet. Tech.*, June 1967.
9. Byod, O. W., *Petroleum Fluid Flow Systems*, Campbell Petroleum Series, Norman, Oklahoma, 1983.
10. Griffith, P. G., and Wallis, G. B., "Two-Phase Slug Flow," *ASME Journal of Heat Transfer*, 1967.
11. Moody, L. F., "Friction Factors in Pipe Flow," *Transactions of ASME*, November 1941.
12. Hagedorn, A. R., and Brown, K. E., "The Effect of Liquid Viscosity in Vertical Two-Phase Flow," *Journal of Petroleum Technology*, February 1964.
13. Hagedorn, A. R., and Brown, K. E., "Experimental Study of Pressure Gradients Occurring During Continuous Two-Phase Flow in Small Diameter Vertical Conduits," *J. Pet. Tech.*, April 1965.
14. Barker, A., Nielson, K., and Galb, A., "Pressure Loss Liquid-Holdup Calculations Developed," *Oil & Gas Journal*, March 1988.
15. Beggs, H. D. and Brill, J. P., "A Study of Two-Phase Flow in Inclined Pipes," *J. Pet. Tech.*, May 1973.
16. Taitel, Y. M., Barnea, D., and Dukler, A. E., "Modeling Flow Pattern Transitions for Steady Upward Gas-Liquid Flow in Vertical Tubes," *AIChE J.* 26, 1980.
17. Barnea, D., Shoham, O., and Taitel, Y., "Flow Pattern Transition for Vertical Downward Two-Phase Flow," *Chem, Eng. Sci.* 37, 1982.
18. Hasan, A. R., and Kabir, C. S., "A Study of Multiphase Flow Behavior in Vertical Wells," *SPEPE 263 Trans.*, AIME, May 1998.
19. Hasan, A. R., and Kabir, C. S., "Predicting Multiphase Flow Behavior in a Deviated Well," *SPEPE 474*, November 1988.
20. Ansari, A. M., Sylvester, N. D., Shoham, O., and Brill, J. P., "A Comprehensive Mechanistic Model for Two-Phase Flow in Wellbores," *SPEPF Trans.*, AIME, May 1994.
21. Ansari, A. M., Sylvester, N. D., Sarica, C., et al. "Supplement to paper SPE 20630, A Comprehensive Mechanistic Model for Upward Two-Phase Flow in Wellbores," paper SPE 28671, May 1994.
22. Harmathy, T. Z., "Velocity of Large Drops and Bubbles in Media of Infinite or Restricted Extent," *AIChE J.* 6, 1960.
23. Brill, J. P., and Mukherjee, H.: "Multiphase Flow in Wells," SPE Monograph Volume 17, Henry L. Doherty Series, Richardson, Texas, 1999.

6.3 NATURAL FLOW PERFORMANCE

The most important parameters that are used to evaluate performance or behavior of petroleum fluids flowing from an upstream point (in reservoir) to a downstream point (at surface) are pressure and flow rate. According to basic fluid flow through reservoir, production rate is a function of flowing pressure at the bottomhole of the well for a specified reservoir pressure and the fluid and reservoir properties. The flowing bottomhole pressure required to lift the fluids up to the surface may be influenced by the size of the tubing string, choke installed at downhole or surface and pressure loss along the pipeline.

In oil and gas fields, the flowing system may be divided into at least four components:

1. reservoir
2. wellbore
3. chokes and valves
4. surface flowline

Each individual component, through which reservoir fluid flow, has its own performance and, of course, affects each other. A good understanding of the flow performances is very important in production engineering. The combined performances are often used as a tool for optimizing well production and sizing equipment. Furthermore, engineering and economic judgments can depend on good information on the well and reasonable prediction of the future performances.

As has been discussed in previous sections, hydrocarbon fluids produced can be either single phase (oil or gas) or two phases. Natural flow performance of oil, gas, and the mixture will therefore be discussed separately. Some illustrative examples are given at the end of each subsection.

6.3.1 Inflow Performance

Inflow performance represents behavior of a reservoir in producing the oil through the well. For a heterogeneous reservoir, the inflow performance might differ from one well to another. The performance is commonly defined in terms of a plot of surface production rate (stb/d) versus flowing bottomhole pressure (P_{wf} in psi) on cartesian coordinate. This plot is defined as inflow performance relationship (IPR) curve and is very useful in estimating well capacity, designing tubing string, and scheduling an artificial lift method.

For single-phase liquid flow, radial flow equation can be written as (for oil)

1. semi–steady-state condition

$$q_0 = 0.00708 k_0 h \frac{(P_r - P_{wf})}{\overline{\mu}_0 \overline{B}_0 \left(\ln \frac{r_e}{r_w} - \frac{3}{4} + s \right)} \qquad [6.3.1a]$$

2. steady-state condition

$$q_0 = 0.00708 k_0 h \frac{(P_r - P_{wf})}{\overline{\mu}_{\overline{B}_0} \left(\ln \frac{r_e}{r_w} - \frac{1}{2} + s \right)} \qquad [6.3.1b]$$

where
q_0 = surface measured oil rate in stb/d
k_0 = permeability to oil in md
h = effective formation thickness in ft.
P_r = average reservoir pressure in psia
P_{wf} = flowing bottomhole pressure in psia
$\overline{\mu}_0$ = oil viscosity evaluated at $\frac{(P_r + P_{wf})}{2}$ in cp
$\overline{B}_0$ = oil formation volume factor evaluated at $\frac{(P_r + P_{wf})}{2}$ in bbl/stb
r_e = drainage radius in ft.
r_w = wellbore radius in ft.
s = skin factor, dimensionless

Assuming all parameters but P_{wf} are constants in the equations above, it is also clear the flowrate q_0 is linearly proportional to flowing pressure P_{wf}. Therefore, for laminar flow the plot q_0 versus P_{wf} on a cartesian coordinate must be linear. This is illustrated in Figure 6.3.1. Strictly speaking, it shows the behavior of single-phase liquid flowing over the range of P_{wf}. In actual cases, however, straightline IPR may be shown by reservoirs producing at P_r and P_{wf} above the bubble point pressure P_b, and by strong water-drive reservoirs.

Productivity index, usually denoted by the symbol J, is commonly expressed in practice for well performance. It is mathematically defined as

$$J = \frac{q_0}{P_r - P_{wf}} \qquad [6.3.2]$$

where J is in stb/d/psi. The term $(P_r - P_{wf})$ is called *pressure drawdown*. Equation 6.3.1a or 6.3.1b can be rearranged to be used in estimating well productivity index.

By knowing reservoir pressure P_r, it is possible to construct an oil IPR curve from a single flow test on a well. Or, because of the linearity of liquid IPR curves, by conducting a two-point flow test (two different flowrates while measuring the flowing bottomhole pressure) on a well, the static reservoir pressure can be determined.

The equations discussed above are derived from the laminar Darcy's law. In a case where turbulent flow occurs, a modified equation should be used. The occurrence of turbulence at the bottomhole may indicate too few open perforations or too narrow fracture in fractured well or other incorrect completion method applied. All these bring about

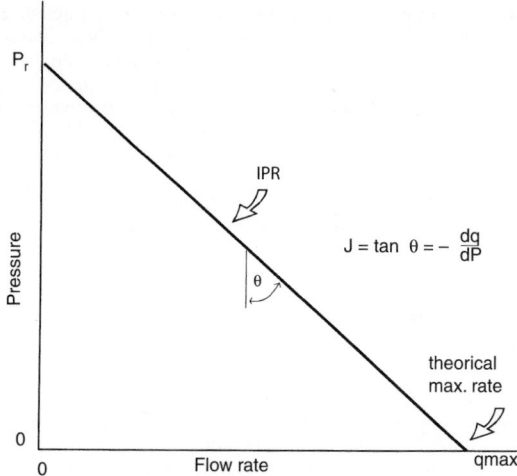

Figure 6.3.1 Inflow performance relationship of single-phase oil reservoirs.

inefficient production operation because the high drawdown encountered results in insufficient flowrate. The symptom may be analyzed using the correlation of Jones et al. [1]:

$$\frac{\Delta P}{q_0} = C + D q_0 \qquad [6.3.3]$$

where

$$C = \frac{\mu_0 B_0}{0.00708 k_0 h} \left(\ln \frac{r_e}{r_w} - \frac{3}{4} + s \right) \qquad [6.3.4]$$

is called the laminar flow coefficient, and

$$D = \frac{9.08 \times 10^{-13} \beta B_0^2 \gamma_0}{4 \pi^2 h^2 r_w} \qquad [6.3.5]$$

is the turbulence coefficient, with β = the turbulence factor in ft.$^{-1}$ and γ_0 = oil specific weight in lb/ft.^{3} and other terms are the same as in the previous equations. The magnitude of the turbulent factor is in the order of 10^{-6} and 10^{-8} and is usually negligible when compared with the laminar flow coefficient in most oil wells. But if this is not the case, plot $(\Delta P/q)$ versus q on a cartesian coordinate paper. If the flow is fully laminar, then the plot has a slope of zero. But when turbulence is measurable, the plot has non-zero positive slope, which also means that the productivity decreases as flowrate increases.

6.3.1.1 Predicting Future Oil Well IPR

Pertaining to our problem here dealing with single-phase oil flow in reservoirs, we always assume that gas does not develop over the whole range of flowing pressure at downhole. The consequence is that the following equations are valid for wells that produce only oil (and water).

Recalling the radial flow equation for oil (Equation 6.3.1a,b for instance), we obtain

$$q_{0max} = \frac{0.00708 k_0 h P_r}{\overline{\mu}_0 \overline{B}_0 \left(\ln \frac{r_e}{r_w} - \frac{3}{4} + s \right)} \qquad [6.3.6]$$

where q_{0max} is a theoretical possible maximum flowrate when $P_{wf} \approx 0$.

Assuming no changes in producing interval, skin factor and drainage radius occur during a period of time from present to the future, and also $\bar{\mu}_o$ and $\bar{B}_o$ are nearly constant over the whole range of pressure, the future possible maximum flowrate is:

$$(q_{0_{max}})_f = (q_{0_{max}})_p \times \frac{(k_{ro} \times P_r)_f}{(k_{ro} \times P_r)_p} \qquad [6.3.7]$$

Because no gas develops in the reservoir, the relative permeability to oil can be a function of water saturation. Figure 6.3.2 suggests the possibility of changes in oil inflow performance curves with time ($t_2 > t_1$).

6.3.1.2 Tubing Performance
A tubing performance may be defined as the behavior of a well in giving up the reservoir fluids to the surface. The performance is commonly showed as a plot of flowrate versus flowing pressure. This plot is called the tubing performance relationship (TPR). For a specified wellhead pressure, the TPR curves vary with diameter of the tubing. Also, for a given tubing size, the curves vary with wellhead pressure. Figure 6.3.3 shows the effect of tubing size and wellhead pressure [2].

For single-phase liquid flow, pressure loss in tubing can be determined using a simple fluid flow equation for vertical pipe, or using some graphical pressure loss correlations where available with GLR $= 0$.

Tubing performance curves are used to determine the producing capacity of a well. By plotting IPR and TPR on the same graph paper, a stabilized maximum production rate of the well can be estimated. Figure 6.3.3 shows the combined plots for determining the flowrate. The larger the diameter of tubing, the higher the flowrate that can be obtained. But there is a critical diameter limiting the rate, even lowering the well capacity. For a specified tubing size, the lower the wellhead pressure, the higher the production rate.

6.3.1.3 Choke Performance
A choke can be installed at the wellhead or downhole to control natural flow or pressure. Chokes are widely used in oil fields. Several reasons in installing chokes are to regulate production rate, to protect surface equipments from slugging, to avoid sand problem due too high drawdown, or to control flowrate in order to avoid water or gas coning.

There are two types of wellhead choke that are commonly used, positive chokes and adjustable chokes. A positive choke has a fixed size in diameter so that it must be replaced to regulate production rate. An adjustable choke permits gradual changes in the size of the opening.

Placing a choke at the wellhead can mean fixing the wellhead pressure and thus flowing bottomhole pressure and production rate. For a given wellhead pressure, by calculating pressure loss in the tubing, the flowing bottomhole pressure can be determined. If reservoir pressure and productivity index of the well are known, the flowrate can then be determined using Equation 6.3.2.

The rate of oil flowing through a choke (orifice or nozzle) depends upon pressure drop in the choke, the inside diameters of pipe and choke, and density of the oil. For incompressible fluids, the Equation 6.3.8 may be used to estimate the flowrate of oil:

$$q_0 = 10,285CA\sqrt{\frac{\Delta P}{\gamma_o}} \qquad [6.3.8]$$

where q_0 = oil rate in bbl/day
   C = flow coefficient as function of diameter ratio and Reynolds number (see Figure 6.3.4)
   A = cross-sectional area of choke in in.^{2}
   ΔP = pressure drop across the choke in psi
   γ_o = oil specific weight in lb/ft.^{3}

In installing a choke, the downstream pressure of the choke is usually 0.55 of the upstream pressure, or even less to ensure no change in flowrate or upstream pressure. This condition is called a sonic flow. A subsonic flow occurs when the upstream pressure or flowrate is affected by a change in downstream pressure.

6.3.1.4 Flowline Performance
After passing through a choke installed at the wellhead, the oil flows through flowline to a separator. If the separator is far from the wellhead and the pressure loss in the flowline cannot be neglected, pressure-flowrate relationship for flowline can be generated similar to tubing performance curves. Usually the separator pressure is specified. Then by using pressure gradient curves available for horizontal pipes or using a simple horizontal fluid flow equation, the wellhead pressure or downstream pressure of the choke or intake pressure of the flowline can be determined as a function of flowrate. This pressure-flowrate plot is useful in sizing the flowline. Figure 6.3.5 illustrates the relationship between the wellhead pressure and flowrate for some different flowline diameters. This plot is called *flowline performance curve.*

By plotting TPR in term of wellhead pressure for various tubing sizes and flowline performance curves on the same graph (Figure 6.3.6), selecting tubing string-flowline combination for a well can be established based on the pipe's availability, production scheme planned, and economic consideration.

Example 6.3.1
Determination of oil inflow performance.

Suppose two flowrates are conducted on an oil well. The results are as follows:

	Test 1	Test 2
q_o, stb/d	200	400
P_{wf}, psi	2400	1800

Gas/oil ratios are very small. Estimate the reservoir pressure and productivity index of the well, and also determine the maximum flowrate.

Solution
1. Plot the two data points on a Cartesian graph paper (q_o versus P_{wf}), see Figure 6.3.7. Draw a straight line through these two points, the intersection with ordinate is the estimated reservoir pressure which is about 3000 psi.
2. The productivity index is

$$J = \frac{200}{3,000 - 2,400} = 0.333 \text{ stb/d/psi}$$

3. The theoretical maximum flowrate is:

$$q_{max} = 0.333(3,000 - 0) = 1,000 \text{ stb/d}$$

Example 6.3.2
The well illustrated in Example 6.3.1 has a vertical depth of 4100 ft. Tubing string of $2\frac{3}{8}$ in. has been installed. The flowing pressure at wellhead is 210 psi. What is the stabilized oil rate achieved?

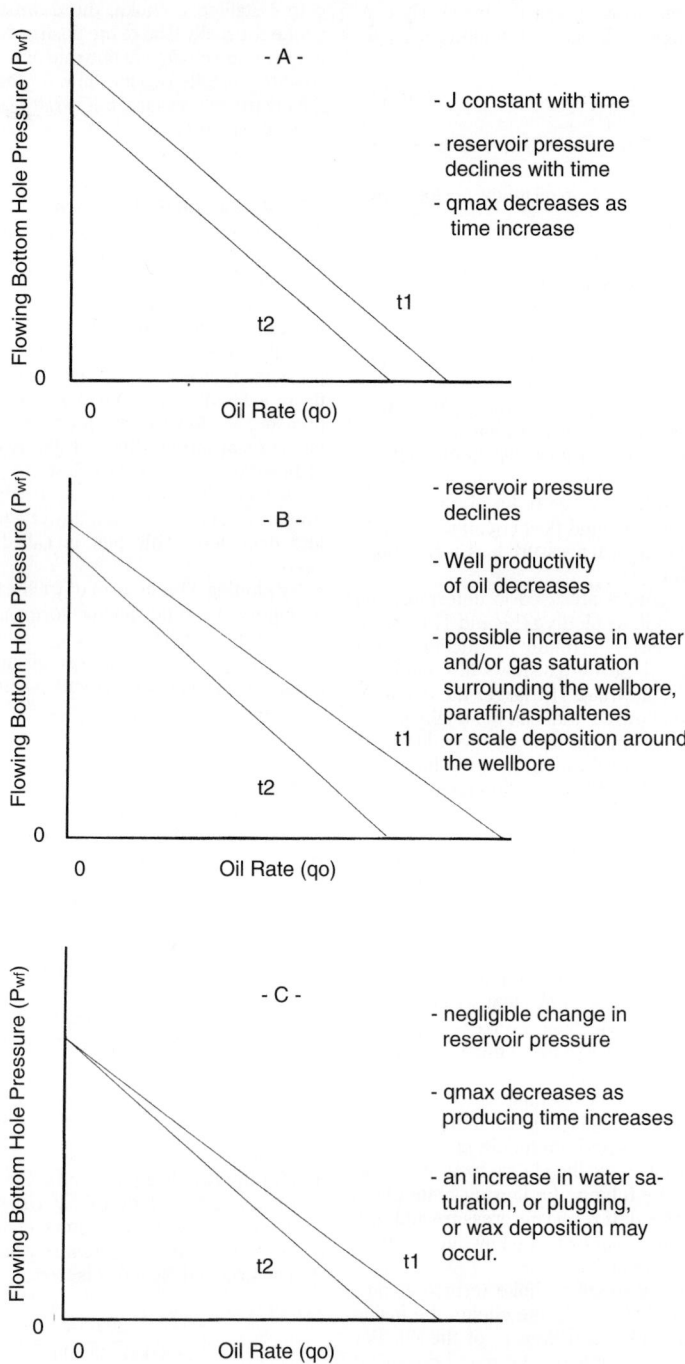

Figure 6.3.2 *Possibility of changes in wells productivity.*

Solution
Using Gilbert's correlation, we obtain:

q, stb/d	P_{wf}, psi
200	1700
400	1695
600	1690

Plot these values of q and P_{wf} on the same graph paper used for solving problem Example 6.3.1 above (Figure 6.3.7). What we get is the TPR curve intersecting the IPR curve. This intersection represents the stabilized rate achieved, which is 435 stb/d.

Example 6.3.3
The oil well of Example 6.3.1 is producing 25° API oil. If a positive choke with diameter of 14/64 in. is installed at the

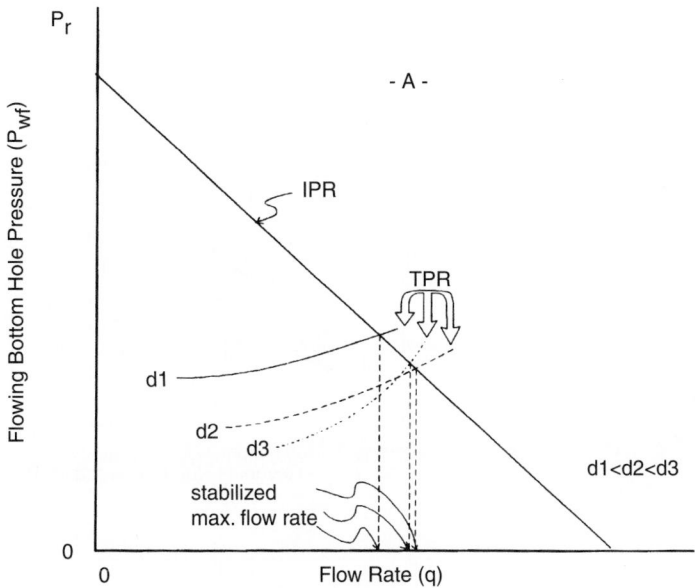

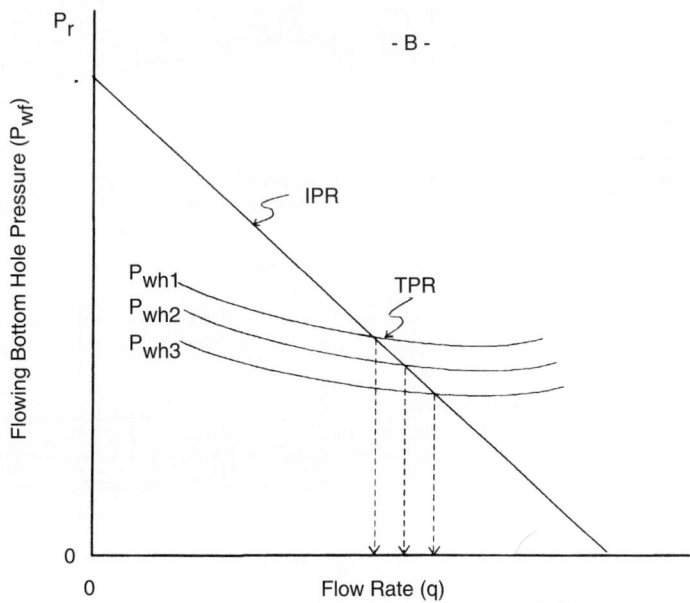

Figure 6.3.3 *Effects of tubing size and tubing head pressure on a well productivity.*

wellhead, determine the pressure at the downstream of the choke.

Solution

$$\gamma_0 = \frac{141.5}{131.5 + 25} \times 62.4 = 56.42 \text{ lb/ft.}^3$$

Rearranging Equation 6.3.8 and assuming that $C \approx 1.0$

$$\Delta P = 56.42 \times \left(\frac{435}{10{,}285(1.0)(0.03758)} \right)^2 = 72 \text{ psi}$$

$$P_{downstream} = (210 - 72) \text{ psi} = 138 \text{ psi}$$

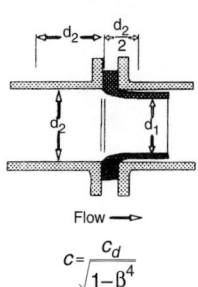

$$c = \frac{c_d}{\sqrt{1-\beta^4}}$$

Example: The flow coefficient C for a diameter ratio β of 0.60 at a Reynolds number of 20,000 (2×10^4) equal 1.03

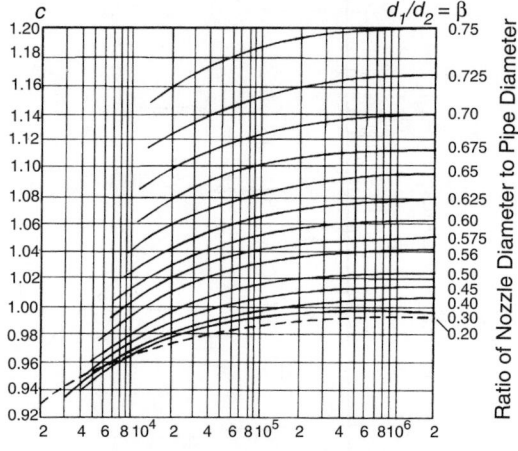

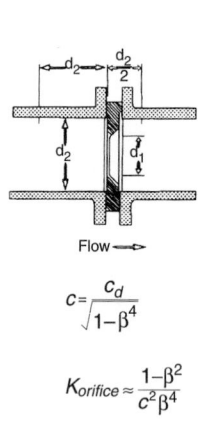

$$c = \frac{c_d}{\sqrt{1-\beta^4}}$$

$$K_{orifice} \approx \frac{1-\beta^2}{c^2\beta^4}$$

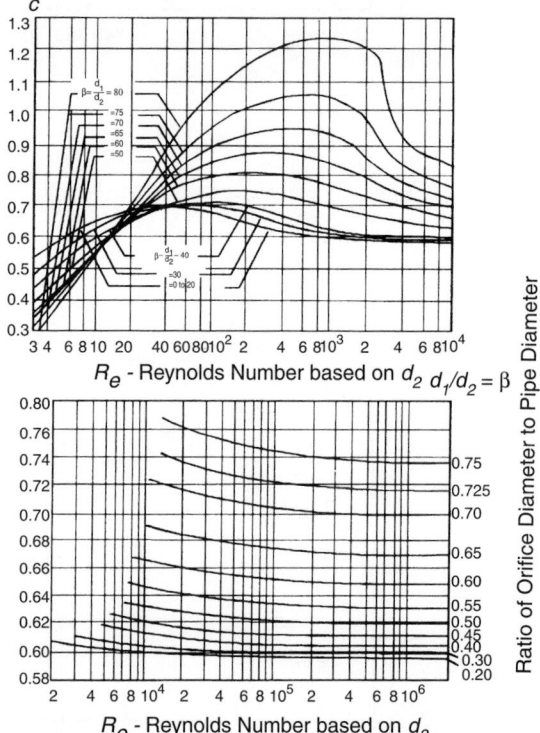

Figure 6.3.4 *Flow coefficient versus diameter ratio and Reynolds number [2].*

6.3.1.5 Gas Flow Performances

As for oil wells, performance curves characterizing a gas production system are very useful tools used to visualize and graphically predict the effects of declining reservoir pressure, changes in tubular size, increasing water production, or installing gas compressors.

6.3.1.6 Gas Inflow Performance

A mathematical expression commonly used to relate gas flowrate and flowing bottomhole pressure is

$$q = C(P_r^2 - P_{wf}^2)^n \qquad [6.3.9]$$

where
- q = gas flowrate in Mscf/d
- P_r = shut-in reservoir pressure in psia
- P_{wf} = flowing bottomhole pressure in psia
- C = stabilized performance coefficient, constant
- n = numerical exponent, constant

Equation 6.3.9 was first introduced by Rawlins and Schellhardt [3] in 1935 and is known as the back-pressure equation. From gas well test data, plotting q versus ($P_r^2 - P_{wf}^2$) on a log-log graph will give a straight line passing through the

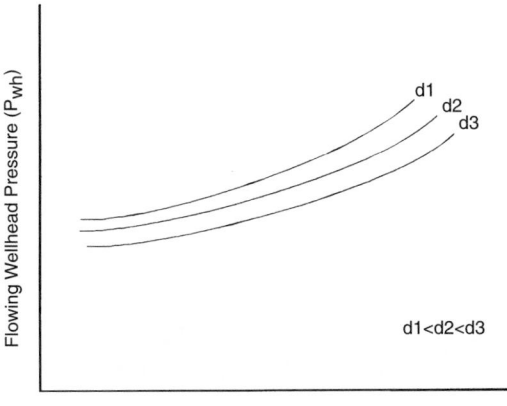

Figure 6.3.5 *Flowline performance curves for different flowline diameters.*

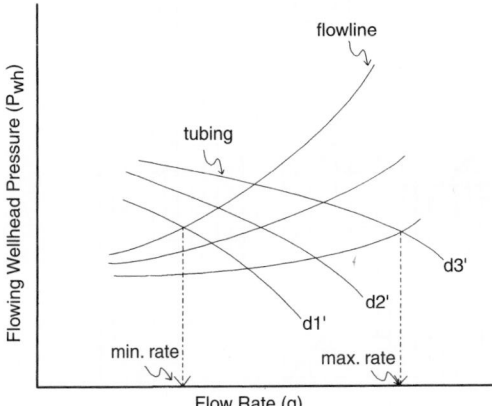

Figure 6.3.6 *Combined tubing-flowline performance curves.*

data points, see Figure 6.3.8. This plot was made based on a stabilized four-point test.

The information that can be obtained from this plot is the absolute open flow potential (AOFP) of the well. This is defined as the theoretical maximum flowrate when flowing pressure at the sand face is zero.

Determination of the exponent n and the coefficient C is given here using Figure 6.3.8 as follows:

1. Choose two values of q arbitrarily but separated one cycle from each other.
2. Read corresponding values of $(P_r^2 - P_{wf}^2)$.
3. Calculate

$$n = 1/\text{slope} = \frac{\log q_2 - \log q_1}{\log(P_r^2 - P_{wf}^2)_2 - \log(P_r^2 - P_{wf}^2)_1}$$

Choosing $q_1 = 1$ gives $(P_r^2 - P_{wf}^2)_1 = 1.5$, and $q_2 = 10$ gives $(P_r^2 - P_{wf}^2)_2 = 20.7$
Then

$$n = \frac{\log 10 - \log 1}{\log 20.7 - \log 1.5} = 0.877$$

4. Rearranging Equation 6.3.9, we obtain:

$$C = \frac{q}{(P_r^2 - P_{wf}^2)^n} = \frac{10,000}{(20.7 \times 10^6)^{0.877}}$$

$$= 3.84 \times 10^{-3} \text{ Mscf/d/psia}^{2n}$$

The AOFP of the well can then be calculated as:

$$\text{AOFP} = q_{max} = 3.84 \times 10^{-3}(3887^2 - 0^2)0.877$$

$$= 7585 \text{ Mscf/d}$$

$$= 7.585 \text{ MMscf/d}$$

From the graph, the AOFP = 7.6 MMscf/d

The IPR curve can be constructed by using the deliverability equation above. By taking some values of P_{wf} arbitrarily, the corresponding q's can be calculated. The IPR curve for the example is shown in Figure 6.3.9.

For situations where multipoint tests cannot be run due to economic or other reasons, single-point data can be used to generate the IPR curve provided that a shut-in bottom-hole pressure is known. Mishra and Caudle [4] proposed a simple method for generating a gas IPR curve from just a single-point test data. Employing the basic gas flow in term of pseudo-pressure function, they developed a dimensionless IPR curve to be used as a reference curve. As an alternative, the dimensionless IPR equation to the best-fit curve is introduced

$$\frac{q}{q_{max}} = \frac{5}{4}\left(1 - 5^{\frac{m(P_{wf})}{m(P_r)} - 1}\right) \qquad [6.3.10]$$

where q_{max} = AOFP in Mscf/d
$m(P_{wf})$ = pseudo-pressure function for real gas and defined as

$$m(P_{wf}) = 2\int_{P_b}^{P_{wf}}\left(\frac{P}{\mu Z}\right)dP$$

$$m(P_r) = 2\int_{P_b}^{P_r}\left(\frac{P}{\mu Z}\right)dP \qquad [6.3.11]$$

where μ = gas viscosity (function of P at isothermal condition) in cp
Z = gas compressibility factor, dimensionless

The use of pseudo-pressure function is quite complex. A numerical integration technique, however, can be applied to this problem. A detail example in applying this numerical technique can be found in *Fundamentals of Reservoir Engineering* [5].

More recently, Chase and Anthony [6] offered a simpler method that is a modification to the Mishra-Caudle method. The method proposed involves substitution of real pressure P or P^2 for the real gas pseudopressure function $m(P)$. The squared pressure P^2 is used for pressures less than approximately 2100 psia, and the relevant equation is

$$\frac{q}{q_{max}} = \frac{5}{4}\left(1 - 5^{\left(\frac{P_{wf}^2}{P_r^2} - 1\right)}\right) \qquad [6.3.12]$$

The real pressures P is suggested for pressures greater than approximately 2900 psia. By having the average reservoir pressure P_r, and a single-point test data P_{wf} and q, it is possible to determine the AOFP and to generate the inflow performance curve:

$$\frac{q}{q_{max}} = \frac{5}{4}\left(1 - 5^{\left(\frac{P_{wf}}{P_r} - 1\right)}\right) \qquad [6.3.13]$$

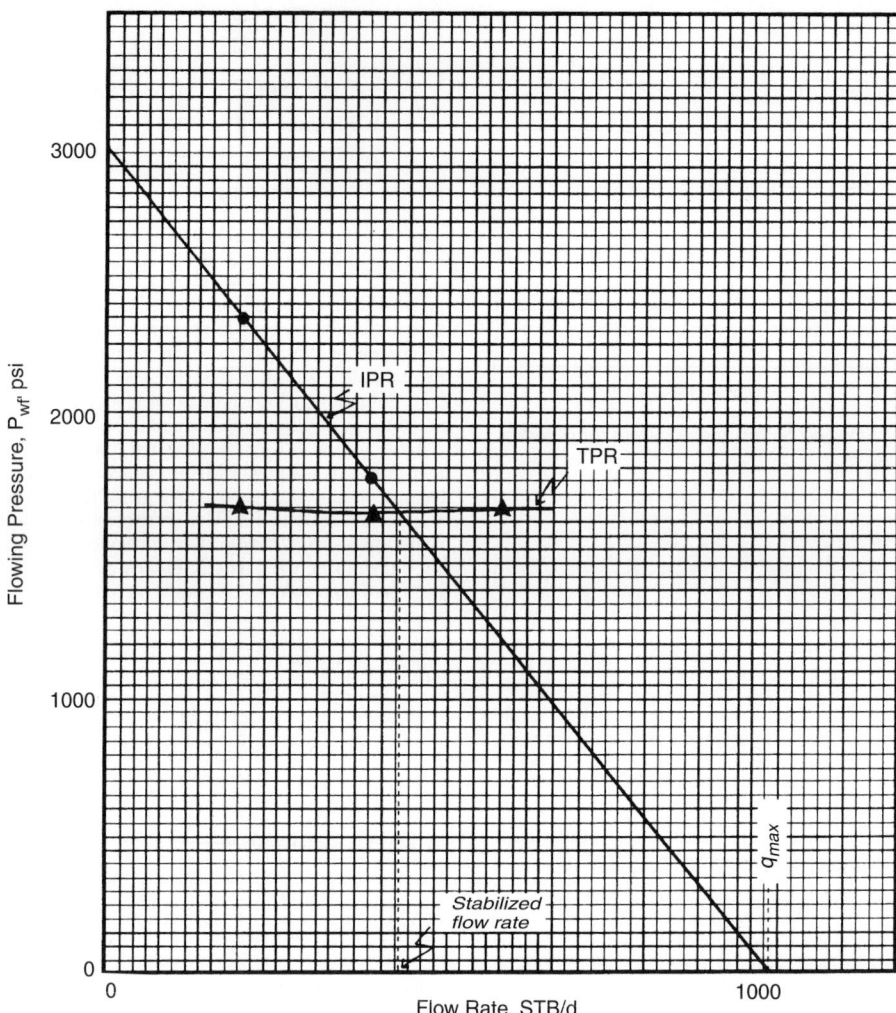

Figure 6.3.7   *Pressure-rate relationship for examples A and B on pages 542 to 544.*

For pressures ranging for 2100 to 2900 psia, the original Mishra-Caudle's technique is recommended.

6.3.1.7 Low-Permeability Well Tests

The requirement of the back-pressure method of testing is that the data be obtained under stabilized conditions. That means that the coefficient C of Equation 6.3.9 is constant with time. This coefficient depends on reservoir characteristics, extent of drainage radius, and produced fluid characteristics.

Wells completed in highly permeable formations stabilize quickly. As demand for gas increased over the years, wells were completed in less permeable formations. In wells of this type, the stabilization period may be very long. Therefore, methods were needed that would permit testing of this type of well without undue waste of time.

In 1955 M. H. Cullender described the isochronal method for determining flow characteristics [7]. The method is based on the assumption that the slope of performance curves of gas wells, exponent n of Equation 6.3.9, is independent of the drainage area. It is established almost immediately after the well is opened. However, the performance coefficient C decreases with time as the radius of drainage recedes from the well. When the radius reaches the boundary of the reservoir or the area of interference of another well, C becomes a constant and the flow is stabilized.

Under the method the well is opened, the flow and pressure data are obtained at specific time intervals without changing the rate of flow. The well is then closed in until the shut-in pressure is reached, approximately the same as at the beginning of the first test. The well is then opened and produced at a different rate, and the pressure and flow data are collected. This procedure is repeated as many times as desired.

Plotting of these data on log-log paper results in a series of parallel lines, the slope of which gives the coefficient n. This is illustrated in Figure 6.3.10. Relationship of coefficient C and time for a gas well is illustrated in Figure 6.3.11.

From these test and theoretical considerations, different procedures have been developed that permit prediction of

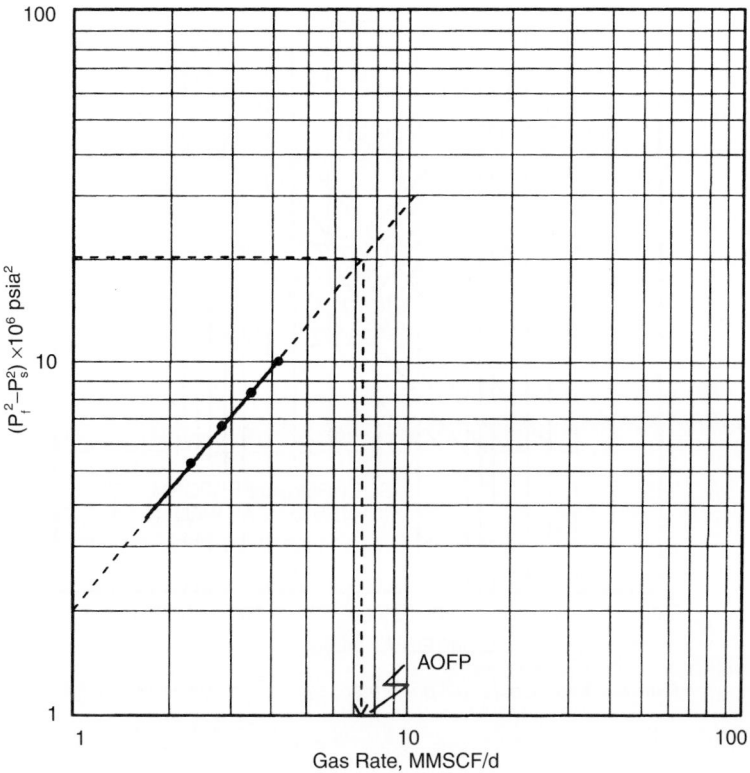

Figure 6.3.8 *Stabilized four-point test and open flow potential of a gas well.*

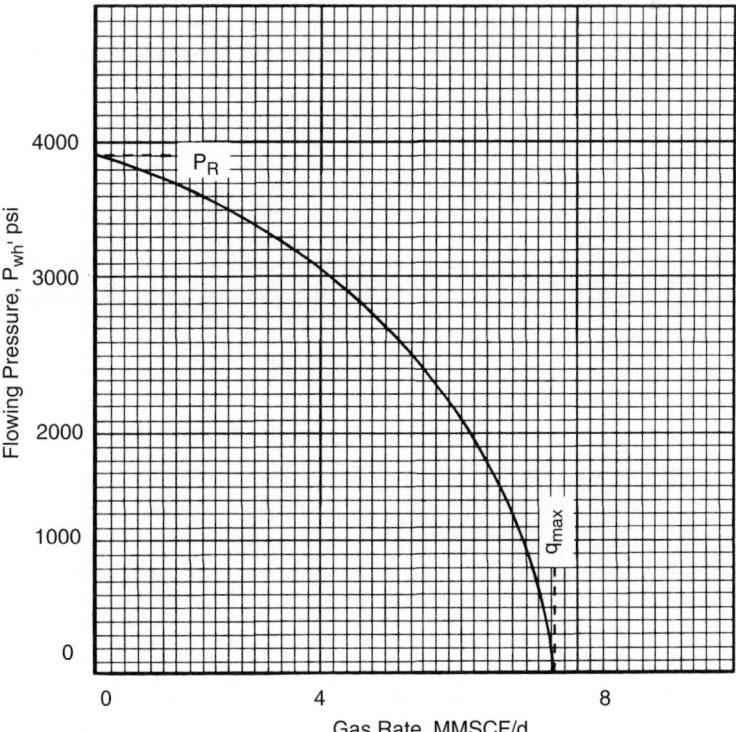

Figure 6.3.9 *Inflow performance relationship for a gas reservoir.*

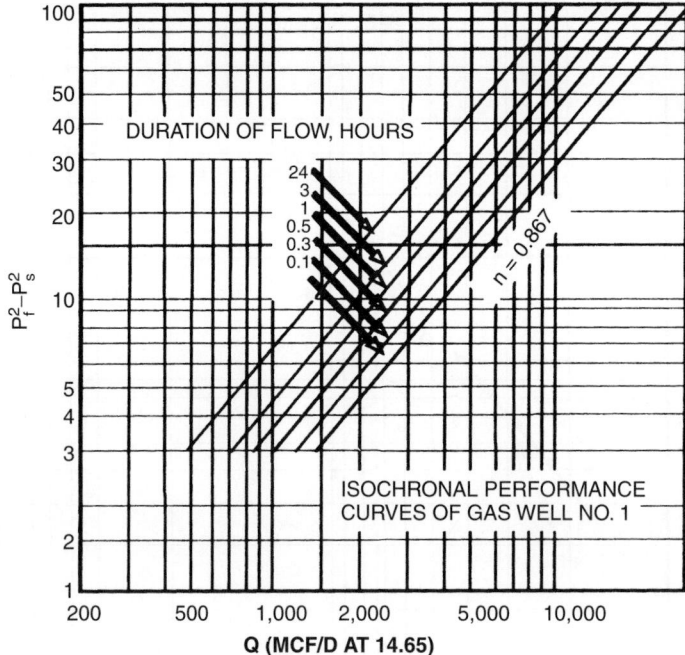

Figure 6.3.10   *Isochronal performance curves of gas well no. 1.*

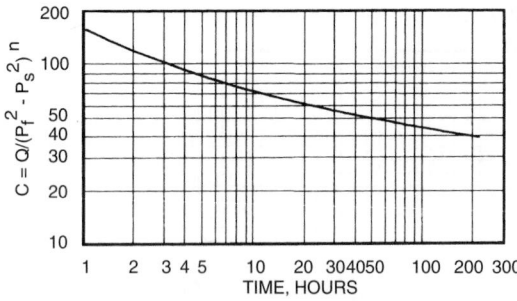

Figure 6.3.11   *Relationship of coefficient of performance and time of gas well no. 3.*

the coefficients of performance of gas wells produced from low-permeability formations.

6.3.1.8 Predicting Future IPR

Predicting the wells' deliverability is important to be able to plan some changes required to maintain the production capacity. Here simple but reliable methods are introduced to forecasting future inflow performance gas wells.

Accompanying Equation 6.3.10, Mishra and Caudle presented an empirical equation for predicting gas wells productivity [4]. The equation is:

$$\frac{(q_{max})_f}{(q_{max})_p} = \frac{5}{3}\left(1 - 0.4^{\frac{m(P_r)_f}{m(P_r)_p}}\right) \qquad [6.3.14]$$

where subscripts f and p refer to future and present time, respectively. Later, Chase and Anthony [6] also proposed

the simplified form of Equation 6.3.14, by substituting real pressure for pseudo-pressure function.

To estimate the AOFP of a gas well, one does not have to run well tests as discussed above. An alternative method is to calculate bottomhole pressure, static and flowing, without running a pressure gage down into the well, by knowing pressures at the wellhead. The calculation of flowing bottomhole pressures will be discussed later. Below, the equations for calculating static bottomhole pressure are given.

One of the most common methods in estimating static bottomhole pressure of a gas well is that of Cullender and Smith, which treats the gas compressibility factor as a function of depth [8]. If we divide the well by equal length, one can calculate the static bottomhole pressure as follows:

1. Having a knowledge of static wellhead pressure, pressure at midpoint of the well is calculated by trial and error:

$$0.0375 G_g \frac{H}{2} = (P_{ms} - P_{ws})(I_{ms} - I_{ws}) \qquad [6.3.15]$$

where G_g = gas gravity (air = 1.0)
   H = well depth in ft.

$$I_{ms} = (T_{ms} \times Z_{ms})/P_{ms} \qquad [6.3.16]$$

where T_{ms} = absolute temperature at midpoint in °R
   P_{ms} = pressure at midpoint (assumed to calculate Z_{ms}) in psia
   Z_{ms} = gas compressibility factor evaluated at T_{ms} and P_{ms}
   P_{ws} = static wellhead pressure in psia

$$I_{ws} = \frac{T_{ws}Z_{ws}}{P_{ws}} \qquad [6.3.17]$$

where T_{ws} = absolute temperature at wellhead in °R
Z_{ws} = gas compressibility factor evaluated at T_{ws} and P_{ws}

The problem here is to calculate P_{ms}. If calculated $P_{ms} \approx P_{ms}$ assumed to determine Z_{ms}, then calculation of bottomhole pressure is the next step. If not, use calculated P_{ms} to determine new Z_{ms} and again use Equations 6.3.16 and 6.3.17 to calculate a new P_{ms}. Repeat this procedure until calculated P_{ms} is close to assumed P_{ms}.

2. The same procedure is used, and the equation for static bottomhole pressure P_{bs} is:

$$0.0375G_g \frac{H}{2} = (P_{bs} - P_{ms})(I_{bs} + I_{ms}) \qquad [6.3.18]$$

3. The very last step is to apply Simpson's rule to calculate P_{bs}:

$$0.0375G_gH = \frac{P_{bs} - P_{ws}}{3}(I_{ws} + 4I_{ms} + I_{bs}) \qquad [6.3.19]$$

For wells producing some liquids, the gas gravity G_g in Equations 6.3.15 and 6.3.18 must be replaced by γ_{mix}:

$$G_{mix} = \frac{G_g + 4584G_o/R}{1 + 132800G_o/RM_o} \qquad [6.3.20]$$

where G_{mix} = specific gravity of mixture (air = 1.00)
G_g = dry gas gravity
G_o = oil gravity
R = surface producing gas–oil ratio in scf/stb
M_o = molecular weight of oil in lbm/lb-mole

The G_{mix} is then used to determine the pseudocritical properties for calculation of the compressibility factor.

If water production is quite significant, the following equation may be used [9]:

$$G_{mix} = \frac{G_g + 4584\left(\dfrac{G_0}{R} + \dfrac{1}{R_w}\right)}{1 + 132,800\left(\dfrac{G_0}{RM_0} + \dfrac{1}{18R_w}\right)} \qquad [6.3.21]$$

where R_w = producing gas/water ratio in scf/stb

6.3.1.9 Tubing Performance
In a gas well, tubing performance can be defined as the behavior of the well in producing the reservoir gas through the tubing installed. At a specified surface pressure, the flowing bottomhole pressure can be calculated by using an equation for vertical flow of gas. Katz presented the equation that is simple but valid only for dry gas [10]:

$$q_g = 200,000\left(\frac{sD^5(P_{wf}^2 - e^sP_{wh}^2)}{G_g\overline{T}\overline{Z}Hf(e^s - 1)}\right)^{0.5} \qquad [6.3.22]$$

where q_g = gas flowrate in scf/d
D = diameter of tubing in in.
P_{wf} = bottomhole flowing pressure in psia
P_{wh} = wellhead flowing pressure in psia
G_g = gas gravity (air = 1.0)
$\overline{T}$ = average temperature in °R
$\overline{Z}$ = average gas compressibility factor
H = vertical depth in ft.
f = friction factor = $\{2\log[3.71/(\varepsilon/D)]\}^{-2}$
ε = absolute pipe roughness, $\cong 0.0006$ in.

$$s = 0.0375G_gH/\overline{TZ} \qquad [6.3.23]$$

The average temperature used in the Equation 6.3.22 is simply the arithmetic average between wellhead temperature

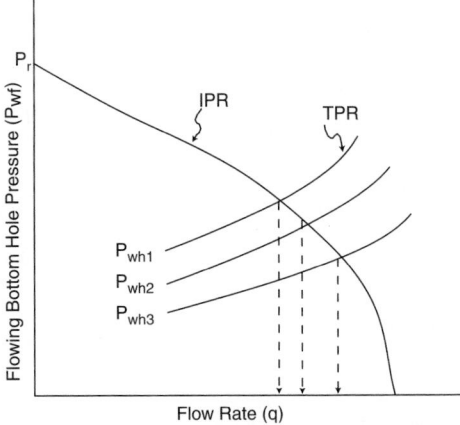

Figure 6.3.12 *Effect of well head pressure on gas well deliverability.*

and bottomhole temperature. The gas compressibility factor $\overline{Z}$ is evaluated at the average temperature and the arithmetic average between the flowing wellhead and bottomhole pressures. This method is a trial-and-error technique, but one or two iterations is usually sufficiently accurate.

By knowing all parameters in Equation 6.3.22 but q_g and P_{wf}, the TPR can then be constructed. The use of TPR here is the same as discussed previously for oil wells. Figure 6.3.12 shows an idea of the effect of wellhead pressure on a well deliverability. A decrease in P_{wh}, thus an increase in flowrate, can be done by changing a choke/bean diameter to a bigger one.

For a specified wellhead pressure, the flowing bottomhole pressure can be estimated as a function of flowrate and tubing diameter. Equation 6.3.22 can be used to do this.

In many cases, gas wells produce some liquids along with the gas itself. The equation for dry gas should be modified to account for the liquid content. One of the modifications presented in the literature was made by Peffer, Miller, and Hill [9]. For steady-state flow and assuming that the effects of kinetic energy are negligible, the energy balance can be arranged and written as:

$$\frac{G_gH}{53.34} = \int_{P_{wh}}^{P_{wf}} \frac{\left(\dfrac{P}{TZ}\right)}{F^2 + \left(\dfrac{P}{TZ}\right)^2}dP \qquad [6.3.24]$$

where P_{wf} = flowing bottomhole pressure in psia
P_{wh} = flowing wellhead pressure in psia

$$F^2 = \frac{2666.5fQ^2}{d^5} \qquad [6.3.25]$$

d = inside diameter of tubing in in.
f = friction factor, dimensionless
Q = flowrate, MMscf/d

Evaluation of friction factor f depends on the stream of fluids in the well. For single-phase (dry gas) and fully developed turbulent flow with an absolute roughness of 0.0006 in., Cullender and Smith [8] suggested the use of:

$$F = \frac{0.10797Q}{d^{2.612}} \quad \text{for} \quad d < 4.277 \text{ in.} \qquad [6.3.26]$$

and

$$F = \frac{0.10337Q}{d^{2.582}} \quad \text{for} \quad d > 4.277 \text{ in.} \qquad [6.3.27]$$

When some liquids are present in the flowing stream Peffer et al. suggested the use of an apparent roughness of 0.0018 instead of using the absolute roughness of 0.0006. Also, adjustment in specific gravity of the fluids should be made by using Equation 6.3.20 or Equation 6.3.21. Applying these adjustments, the method of Cullender and Smith [8] may be used for a wide range of gas-condensate well condition.

However, whenever the Reynold's number for a specific can be calculated and pipe specifications are available, the friction factor can then be easily determined to be used for pressure loss calculations. Equation 6.3.28 and Equation 6.3.29 can be used to calculate the Reynold's number and friction factor, respectively.

$$R_e = \frac{20011 G_g Q}{\mu_g d} \qquad [6.3.28]$$

$$\frac{1}{\sqrt{f}} = 2.28 - 4 \log\left(\frac{\varepsilon}{d} + \frac{21.28}{R_e^{0.9}}\right) \qquad [6.3.29]$$

where R_e = Reynolds number
G_g = gas gravity (air = 1.00)
Q = gas flowrate in MMscf/d
μ_g = gas viscosity in cp
d = inside diameter of pipe in in.
ε = absolute roughness in in.
f = friction factor, dimensionless

For a well divided in equal lengths the upper half of the well has a relation:

$$37.5 G_g \frac{H}{2} = (P_{MF} - P_{WH})(I_{MF} + I_{WH}) \qquad [6.3.30]$$

where P_{MF} = flowing pressure at midpoint in psia
P_{WH} = flowing wellhead pressure in psia
$I = (P/TZ)/[F^2 + (P/TZ)^2]$
F = shown in Equation 6.3.34
H = well depth in ft.

and the lower half of the well has the relation:

$$37.5 G_g H = (P_{WF} - P_{MF})(I_{WF} + I_{MF}) \qquad [6.3.31]$$

where P_{WF} = flowing bottomhole pressure in psia.

After trial and error as previously discussed, Simpson's rule applies:

$$37.5 G_g H = \frac{P_{WF} - P_{WH}}{3}(I_{WH} + 4I_{MF} + I_{WF}) \qquad [6.3.32]$$

6.3.1.10 Choke Performance

Chokes or beans are frequently installed in gas wells. These restrictions can be at the surface or at the subsurface. A surface choke is usually installed for:

1. regulating production rate
2. maintaining sufficient back pressure to avoid sand production
3. protecting surface equipment from pressure surge
4. preventing water coning
5. obeying regulatory bodies

Subsurface restrictions can be a tubing safety valve, a bottomhole choke, or a check valve. A tubing safety valve functions to stop flowstream whenever the surface control equipment is damaged or completely removed. A bottomhole choke is installed if low wellhead pressure is required or freezing of surface control equipment and lines is expected. A check valve is installed to prevent backflow of an injection well. Basically, there are two types of flow conditions: subsonic or sub-critical flow and sonic flow. The criteria to

distinguish subsonic from sonic flow has been discussed previously in the section titled "Choke Performance."

For subsonic flow, the following equation given by Nind can be used to calculate the gas flowrate [11].

$$Q = 1248 C A P_u \left\{ \frac{k}{(k-1)G_g T_u} \left[\left(\frac{P_d}{P_u}\right)^{2/k} - \left(\frac{P_d}{P_u}\right)^{\frac{k+1}{k}} \right] \right\}^{0.5} \qquad [6.3.33]$$

where Q = gas flow rate in Mcf/d
C = discharge coefficient, ≈ 0.86
A = cross-sectional area of choke or restriction in in.^{2}
P_u = upstream pressure in psia
P_d = downstream pressure in psia
G_g = gas gravity (air = 1.00)
T_u = upstream temperature in °R
k = specific heat ratio, C_p/C_f

Equation 6.3.33 shows that subsonic flow is affected by upstream and downstream pressure.

In critical or sonic flow, gas flowrate depends only on upstream pressure as shown as:

$$Q = 879 C A P_u \left[\frac{k}{G_g T_u} \left(\frac{2}{k+1}\right)^{\frac{k+1}{k-1}} \right]^{0.5} \qquad [6.3.34]$$

where Q = flowrate in Mcf/d
C = discharge coefficient
A = choke area in in.^{2}
P_u = upstream pressure in psia
G_g = gas gravity (air = 1.00)
T_u = upstream temperature in °R
k = specific heat ratio

The discharge coefficient C can be determined using Figure 6.3.4 by having a knowledge of diameter ratio β and Reynold's number. Reynold's number may be calculated using the following equation:

$$R_e = \frac{20.011 Q G_g}{\mu_g d} \qquad [6.3.35]$$

where Q = gas flowrate in Mcf/d
G_g = gas gravity (air = 1.00)
μ_g = gas viscosity, evaluated at upstream pressure and temperature in cp
d = internal diameter of pipe (not choke) in in.

Gas flow through restriction (orifice) may also be estimated using Figure 6.3.13. For conditions that differ from chart basis, correction factors are required. A gas throughput read from the chart must be multiplied by the proper correction factor,

$$Q = \text{Gas throughput} \times 0.0544 \sqrt{G_g T}, \text{ Mcf/d} \qquad [6.3.36]$$

where T is the absolute operating temperature in °R.

6.3.1.11 Flowline Performance

For a single-phase gas flow, pressure-rate relation may be obtained from a known Weymouth equation:

$$Q = 433.49 \frac{T_b}{P_b} \left[\frac{(P_u^2 - P_d^2)d^{16/3}}{G_g \overline{T} \overline{Z} L} \right]^{0.5} E \qquad [6.3.37]$$

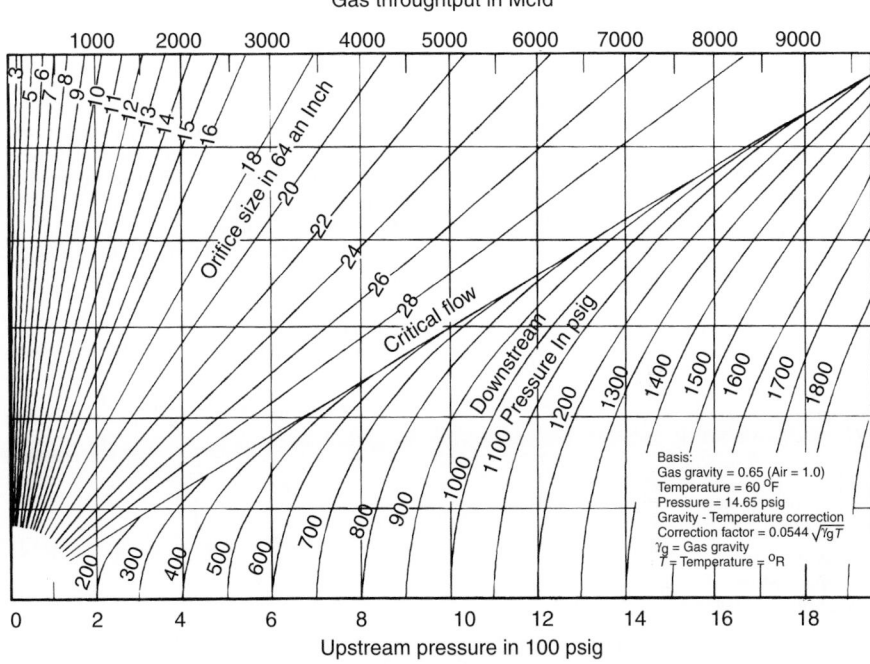

Figure 6.3.13 *A correlation for gas flow through orifice [12].*

where Q = gas flowrate in scf/d
T_b = base temperature in °R
P_b = base pressure in psia
P_u = upstream pressure in psia
P_d = downstream pressure in psia
d = inside diameter of pipe in in.
G_g = gas gravity (air = 1.00)
$\overline{T}$ = average flow line temperature in °R
$\overline{Z}$ = average gas compressibility factor
L = pipe length in mi
E = pipe line efficiency, fraction

or the modified Panhandle (Panhandle B) equation (for long lines):

$$Q = 737 \left(\frac{T_b}{P_b}\right)^{1.02} \left(\frac{P_u^2 - P_d^2}{\overline{T}\,\overline{Z}\,L\,G_g^{0.961}}\right)^{0.510} d^{2.530} E \qquad [6.3.38]$$

with the terms and units the same as the Weymouth Equation 6.3.37.

The pipeline efficiency E depends on flow stream and pipeline conditions. A gas stream may contain some liquids; the higher the liquid content, the lower the line efficiency. The pipeline may be scaled, or condensate-water may accumulate in low spots in the line. Ikoku [12] presents the information about line efficiency as shown below.

Type of line	Liquid Content	E (gal/MMscf)
Dry-gas field	0.1	0.92
Casing-head gas	7.2	0.77
Gas and condensate	800	0.60

This gives an idea in estimating E for a particular condition.

Example 6.3.4
A gas well was flowed at a rate of 7.20 MMscf/d. The stabilized sandface pressure at the end of the test was 1155 psia, and the current average reservoir pressure was estimated to be 1930 psia. Determine the following parameters using the modified Mishra-Caudle [4] method:

1. AOFP at current conditions (P_r = 1,930 psia)
2. deliverability at P_{wf} = 1,000 psia
3. AOFP at a future average pressure P_{rf} = 1,600 psia
4. deliverability at a future P_{wf_f} = 1,155 psia

Solution
The readers may refer to the original paper [4]. Here the modified one is given

(a) For a pressure less than 2100 psia, Equation 6.3.12 is used:

$$\frac{q}{q_{max}} = \frac{5}{4}\left[1 - 5^{\left(\frac{P_{wf}^2}{P_r^2} - 1\right)}\right]$$

$$AOFP = q_{max} = \frac{7.2 \text{ MMscf/d}}{\frac{5}{4}\left[1 - 5^{\left(\frac{1155^2}{1930^2} - 1\right)}\right]} = 8.94 \text{ MMscf/d}$$

(b) $q = 8.94 \times 10^6 \times \dfrac{5}{4}\left[1 - 5^{\left(\frac{1000^2}{1930^2} - 1\right)}\right]$

$= 7.73$ MMscf/d

(c) $\dfrac{(AOFP)_f}{(AOFP)_p} = \dfrac{5}{3}\left[1 - 0.4^{\frac{P_{rf}^2}{P_{rp}^2}}\right]$

$$(AOFP)_f = 8.94 \times 10^6 \times \tfrac{5}{3}\left(1 - 0.4\tfrac{1600^2}{1930^2}\right)$$

$$= 6.96 \text{ MMscf/d}$$

(d) Again use Equation 6.3.12:

$$q_f = (AOFP) \times \frac{5}{4}\left[1 - 5^{\left(\frac{P_{wf}^2}{P_r^2}\right)} - 1\right]_f$$

$$= 6.96 \times 10^6 \times \frac{5}{4}\left[1 - 5^{\left(\frac{1155}{1600}\right)^2 - 1}\right]$$

$$= 4.675 \text{ MMscf/d}$$

Example 6.3.5

The following data are obtained from a producing gas well:

Gas gravity = 0.65
Well depth (vertical) = 6000 ft.
Wellhead temperature = 570°R
Formation temperature = 630°R
Flowing wellhead pressure = 1165 psia
Flowrate = 10.0 MMscf/d
Tubing ID = 2.441 in.
Pseudo-critical temperature = 374°R
Pseudo-critical pressure = 669 psia
Absolute roughness = 0.00065

Calculate the following bottomhole pressure using Equation 6.3.22.

Solution

Rearranging Equation 6.3.22, we obtain:

$$P_{wf} = \left(e^s P_{wh}^2 + \frac{q_g^2}{4 \times 10^{10}} \times \frac{G_g \overline{T}\overline{Z}Hf(e^s - 1)}{SD^5}\right)^{0.5}$$

Assume $P_{wf} = 2000$ psia:

$$\overline{P} = \frac{2000 + 1165}{2} = 1582.5 \text{ psia}$$

$$T = \frac{570 + 630}{2} = 600°R$$

$$\left.\begin{array}{l} pP_r = \dfrac{1582.5}{669} = 2.365 \\[2mm] PT_r = \dfrac{600}{374} = 1.064 \end{array}\right\} \rightarrow \overline{Z} = 0.848$$

$$S = 0.0375(0.65)(6,000)/(600)(0.848) = 0.287$$

$$f = \left\{2\log\left[371/\left(\frac{0.00065}{2.441}\right)\right]\right\}^{-2}$$

$$P_{wf} = \left\{e^{0.287}(1,165)^2 + \frac{(10 \times 10^6)^2}{4 \times 10^{10}}\right.$$

$$\times \frac{(0.65)(600)(0.848)(6,000)}{(0.287)(2.441)^5}$$

$$\left.\times \left[0.0146(e^{0.287} - 1)\right]\right\}^{0.5}$$

$$= 1,666 \text{ psia } (\ll 2,000 \text{ psia assumed})$$

Assume 2nd $P_{wf} = 1,666$ psia:

$$\overline{P} = \frac{1,666 + 1,165}{2} = 141.5 \text{ psia}$$

$$\left.\begin{array}{l} pP_r = \dfrac{141.5}{669} = 2.116 \\[2mm] pT_r = 1.604 \end{array}\right\} \rightarrow \overline{Z} = 0.857$$

$$S = 0.0375(0.65)(6,000)/(600)(0.857) = 0.284$$

$$P_{wf} = 1666.7 \text{ psia } (\cong 1,666 \text{ psia assumed})$$

Example 6.3.6

Suppose the gas well in Example 6.3.5 is produced through a flowline of 2.5 in. diameter and 1250 ft long. The average operating temperature is 100°F. Additional data given are specific heat ratio, k = 1.3, and the estimated gas viscosity at the operating condition, $\mu_g = 0.0131$ cp.

Find the positive choke size required for critical condition, and the pressure at downstream of the flowline.

Solution

(1) For a critical flow, Equation 6.3.34 can be used.

$$Q = 879 \times C \times A \times P_u \left[\frac{k}{G_g T_u}\left(\frac{2}{k+1}\right)^{\frac{k+1}{k-1}}\right]^{0.5}$$

$$= 879 \times C \times \frac{\pi}{4}d_{ch}^2 \times 1165$$

$$\times \left[\frac{1.3}{0.65 \times 570}\left(\frac{2}{1.3+1}\right)^{\frac{1.3+1}{1.3-1}}\right]^{0.5}$$

$$= 27,881Cd_{ch}^2 \text{ (Mscf/d)}$$

$$R_e = \frac{20 \times 10,100 \times 0.65}{0.013 \times 2.5} = 4.01 \times 10^6$$

Assume a diameter choke, such that $d_{ch} = 0.7$ in. Use Figure 6.3.4 to find coefficient C:

$$\beta = \frac{d_{ch}}{d_{pipe}} = \frac{0.7}{2.5} = 0.28$$

For $R_e = 4.01 \times 10^6$ and $\beta = 0.28$, C = 0.998, so

$$Q = 27,881 \times 0.998 \times (0.7)^2 = 13.63 \text{ MMscf/d}$$

If we assume another choke size, let $d_{ch} = 0.6$ in.:

$$\beta = \frac{0.6}{2.5} = 0.24 \rightarrow C = 0.997$$

$$Q = 27,881 \times 0.997 \times (0.6)^2 = 10.0 \text{ MMscf/d}$$

(2) For a critical flow, we may assume $(P_d)_{choke} = 0.5(P_u)_{choke}$. Let's take $T_b = 530°R$ and $P_b = 15$ psia. Then using Equation 6.3.37 with E = 0.92,

$$P_d^2 = P_u^2 - \left(\frac{Q/E}{433.4\,T_b/P_b}\right)^2 \times \frac{G_g \overline{T}\overline{Z}L}{d^{16/3}}$$

$$= \left(\frac{1165}{2}\right)^2 - \left(\frac{10^7/0.92}{15,317}\right)^2 \times \frac{(0.65)(560)(\overline{Z})\left(\frac{1250}{5280}\right)}{(2.5)^{16/3}}$$

$$= 339,306 - 327,422\overline{Z}$$

Assume that $P_d = 200$ psia to find $\overline{Z}$ such that:

$$\overline{P} = \frac{200 + (0.5 \times 1165)}{2} = 391.2 \text{ psia}$$

$$\left.\begin{array}{l} pP_r = \dfrac{391.25}{669} = 0.585 \\[2mm] pT_r = \dfrac{560}{374} = 1.497 \end{array}\right\} \rightarrow \overline{Z} = 0.939$$

$$P_d^2 = 339,306 - 327,422 \times 0.939$$

$$P_d = 178.5 \text{ psia}(\neq P_{d_{assumed}} = 200 \text{ psia})$$

Assume now that $P_d = 178.5$ psia such that:

$$\overline{P} = \frac{178.5 + (0.5 \times 1165)}{2} = 380.5 \text{ psia}$$

$$\left.\begin{array}{l} pP_r = \dfrac{380.5}{669} = 0.569 \\[2mm] pT_r = 1.497 \end{array}\right\} \rightarrow \sim \overline{Z} = 0.940$$

$$P_d^2 = 339{,}306 - 327{,}422 \times 0.940$$

$$P_d = 177.5 \text{ psia} (\cong P_{d_{assumed}} = 178.5 \text{ psia})$$

6.3.2 Two-Phase Flow Performance
6.3.2.1 Two-Phase Inflow Performance
When a reservoir pressure is below the bubble point pressure, the simple equation of inflow performance (e.g., the productivity index is constant) is no longer valid, because at this condition the oil flowrate will decline much faster at increasing drawdown than would be predicted by Equation 6.3.1a or Equation 6.3.2. An illustrative comparison of the two types of IPR is shown in Figure 6.3.14.

6.3.2.2 Vogel's Method
The well-known inflow performance equation for two-phase flow has been proposed by Vogel [13]. The equation is,

$$\frac{q_0}{q_{max}} = 1 - 0.2\frac{P_{wf}}{P_r} - 0.8\left(\frac{P_{wf}}{P_r}\right)^2 \qquad [6.3.39]$$

which fits a general dimensionless IPR shown in Figure 6.3.15. The reference curve and Equation 6.3.39 is valid for solution-gas drive reservoir with reservoir pressures below the bubble point. The formation skin effect is not taken into account. The method is originally developed with the flowing

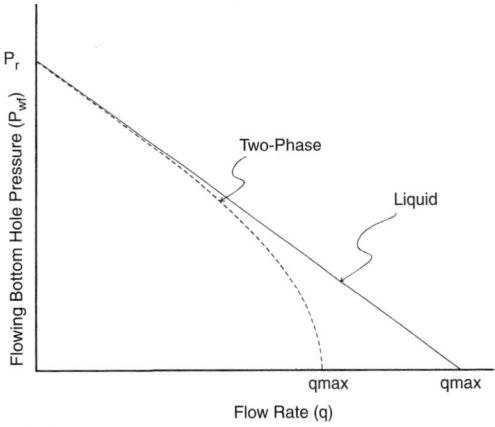

Figure 6.3.14 *Illustrative comparison of liquid and two-phase IPR curves.*

efficiency FE = 1.0. However, for a given well with any FE known, Equation 6.3.39 or the reference curve may be used to generate the IPR curve.

For reservoir pressures above the bubble point but with flowing pressures below the bubble point, the constant J equation and the Vogel equation can be combined to estimate the IPR curves. The equation is

$$q_0 = q_b + (q_{max} - q_b)\left[1 - 0.2\frac{P_{wf}}{P_r} - 0.8\left(\frac{P_{wf}}{P_r}\right)^2\right] \qquad [6.3.40]$$

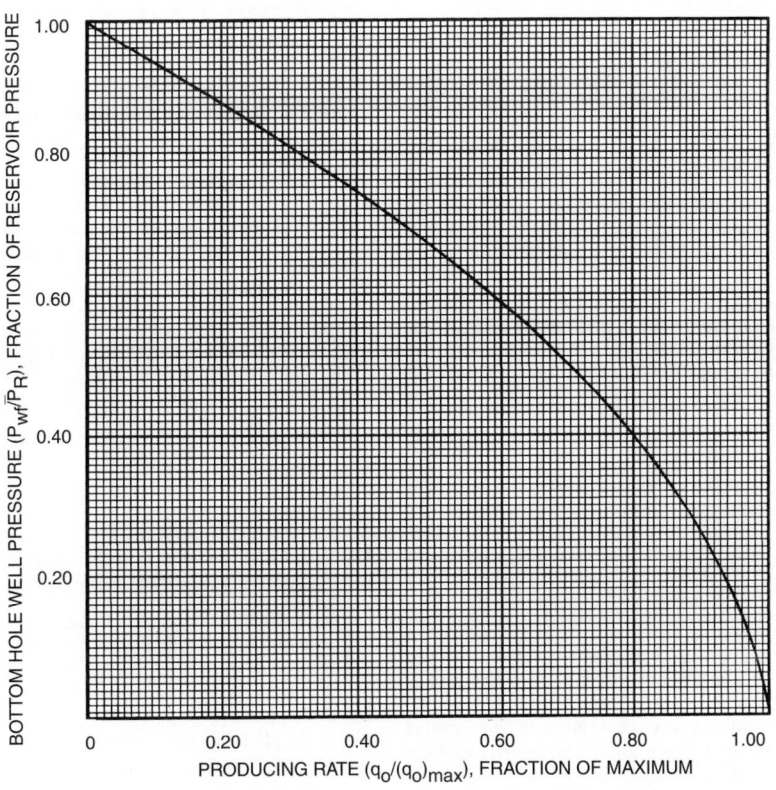

Figure 6.3.15 *A general dimensionless IPR for solution gas drive reservoirs [13].*

The maximum flowrate q_{max} is calculated using the following equation:

where q_o = oil flowrate in stb/d
  q_{max} = the theoretical maximum flowrate when $P_{wf} = 0$ in stb/d
  q_b = oil flowrate at $P_{wf} = P_b$ in stb/d
  P_b = bubble point pressure in psia
  P_{wf} = flowing bottomhole pressure in psia
  P_r = average reservoir pressure in psia

$$q_{max} = q_b + \frac{J \times P_b}{1.8} \qquad [6.3.41]$$

The productivity index J is determined based on the following bottomhole pressure of the test:

1. For $(P_{wf})_{test} > P_b$, then

$$J = \frac{(q_o)_{test}}{P_r - (P_{wf})_{test}} \qquad [6.3.42]$$

2. For $(P_{wf})_{test} < P_b$, then

$$J = \frac{(q_o)_{test}}{P_r - P_b + (P_b \times M/1.8)} \qquad [6.3.43]$$

where $M = (1 - 0.2(P_{wf})/P_b) - 0.8(P_{wf}/P_r)^2$
  $P_{wf} = (P_{wf})_{test}$

The q_b is calculated using Equation 6.3.2 with $P_{wf} = P_b$

6.3.2.3 Fetkovich Method

Analyzing isochronal and flow–afterflow multipoint back-pressure tests conducted on oil wells, Fetkovich [14] found that back-pressure curves for oil wells followed the same form as for gas wells; that is

$$q_o = J_0'(P_r^2 - P_{wf}^2)^n \qquad [6.3.44]$$

where J_0' = back-pressure curve coefficient, stb/d/(psia)^{2n}
  n = back-pressure curve exponent or exponent of inflow performance curve

The plot of q_o versus $(P_r^2 - P_{wf}^2)$ on log-log paper is considered as good as was obtained from gas well back-pressure tests. Conducting a multipoint back-pressure test on a well, Equation 6.3.44 can be used to predict the IPR curve for the well.

Figure 6.3.16 shows the comparison of IPR's for liquid, gas and two-phase (gas and liquid). Fetkovich reported that Vogel's equation yields n = 1.24 (Figure 6.3.17).

For reservoir pressures above bubble point pressures, the inflow performance curves can be constructed using the following equation:

$$q_o = J_0'(P_b^2 - P_{wf}^2)^n + J(P_r - P_b) \qquad [6.3.45]$$

The maximum flowrate of a well can be determined using the following equation:

$$q_{max} = q_0 \left[1 - \left(\frac{P_{wf}}{P_r}\right)^2 \right]^n \qquad [6.3.46]$$

6.3.2.4 Modified Standing's Method

Vogel's reference curve is originally derived for undamaged or unformulated wells. In other words, the curve is only valid for wells with skin factor s = 0. Later, Standing [15] presented a set of companion curves that can be used to predict IPR curves for damaged or stimulated wells. His method is based on the definition of single-phase flow efficiency. In fact, solution-gas drive reservoirs producing oil at $P_{wf} < P_b$ and/or $P_r < P_b$ have inflow performance of two-phase flow.

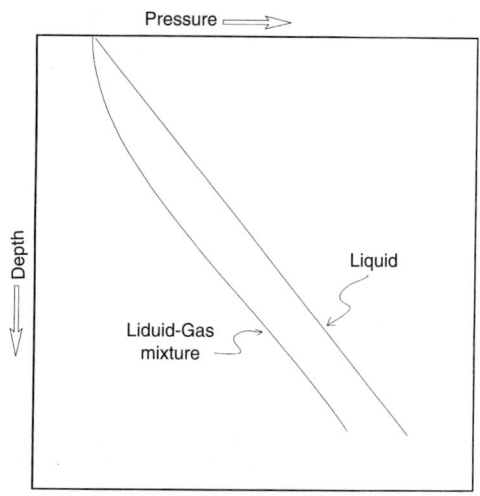

Figure 6.3.16 *Pressure gradients of flowing liquid and liquid–gas mixture.*

The IPR of this type of reservoirs have been shown to have quadratic forms as suggested by Vogel and Fetkovich.

Camacho and Raghhavan [16] found that Standing's definition of flow efficiency is incorrect to be used in two-phase flow behavior. It is suggested that the definition of flow efficiency must also reflect the quadratic form of the inflow performance equation. This is expressed by

$$FE = \frac{(1 + VP_{wf}'/P_r)(1 - P_{wf}'/P_r)}{(1 + VP_{wf}/P_r)(1 - P_{wf}/P_r)} \qquad [6.3.47]$$

where FE = flowing efficiency
  V = quadratic curve factor (V = 0.8 for Vogel and V = 1.0 for Fetkovich)
  P_{wf}' = ideal flowing bottomhole pressure (e.g., when skin factor s = 0) in psia
  P_{wf} = actual flowing bottomhole pressure in psia
  P_r = average reservoir pressure in psia

The flowrate when FE ≠ 1.0 can be calculated using

$$\frac{q_0}{q_{max}^{FE=1.0}} = FE \left[1 + V \left(\frac{P_{wf}}{P_r}\right) \right] \left[1 - \left(\frac{P_{wf}}{P_r}\right) \right] \qquad [6.3.48]$$

where $q_{max}^{FE=1.0}$ = maximum flow rate for undamaged/unstimulated well or when FE = 1.0

6.3.2.5 Predicting Future IPR

Predicting future well deliverability is frequently needed in most oil fields. Some of the many reasons are:

(1) to estimate when the choke should be changed or adjusted to maintain the production rate
(2) to predict well capability and evaluate if the tubing has to be changed
(3) to do planning for selecting future artificial lift methods
(4) to do planning for reservoir pressure maintenance or secondary recovery projects

Some prediction methods available in the literature are discussed.

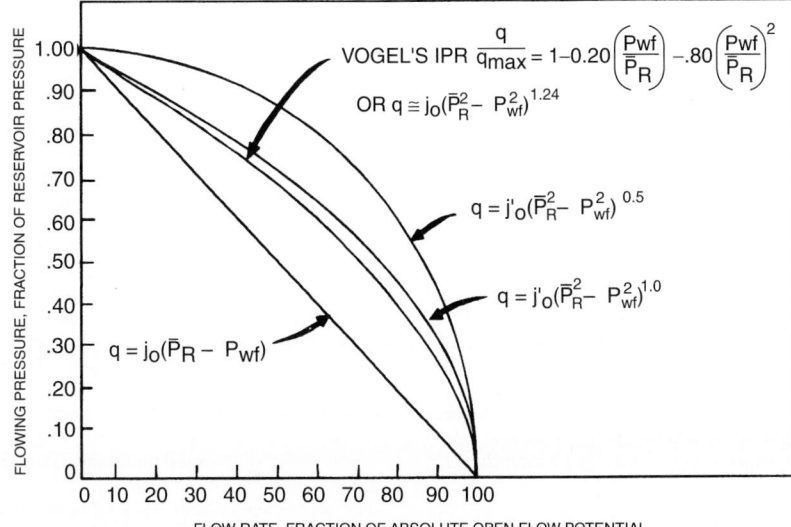

Figure 6.3.17 *Inflow performance relationship for various flow equations [14].*

6.3.2.5.1 Standing's Method

The method has been developed based on Vogel's equation, the definition of productivity index, and the assumption that the fluid saturation is to be the same everywhere in the reservoir. Three basic equations presented are:

$$J_p^* = \frac{1.8(q_{max})_p}{(P_r)_p} \qquad [6.3.49]$$

$$J_f^* = J_p^* \left\{ \frac{[K_{ro}/(\mu_o B_o)]_f}{[k_{ro}/(\mu_o B_o)]_p} \right\} \qquad [6.3.50]$$

$$(q_{max})_f = (J_f^* \cdot P_{rf})/1.8 \qquad [6.3.51]$$

where

J^* = productivity index at zero drawdown in stb/d/psi
k_{ro} = relative permeability to oil, fraction
μ_o = oil viscosity in cp
B_o = oil formation volume factor in bbl/stb

and subscript p and f refer to present and future conditions, respectively. The relative permeability to oil is at corresponding oil saturation in the reservoir. A method for determining oil saturation and k_{ro} may be found in the reservoir engineering chapter. In the example presented by Standing, Tarner's method was used and k_{ro} was evaluated using Corey-type relationship:

$$K_{ro} = \left(\frac{S_o - S_{or}}{1 - S_{wc} - S_{or}} \right)^4 \qquad [6.3.52]$$

where

S_{or} = residual oil saturation, fraction
S_{wc} = connate water saturation, fraction

6.3.2.5.2 Combined Fetkovich-Vogel Method

Fetkovich suggested that a future well deliverability may be estimated by the relation $J_f'/J_p' = (P_r)_f/(P_r)_p$. Recalling Equation 6.3.44 and taking $n = 1$, we can write

$$\frac{(q_{max})_f}{(q_{max})_p} = \left[\frac{(P_r)_f}{(P_r)_p} \right]^3 \qquad [6.3.53a]$$

or

$$(q_{max})_f = (q_{max})_p \times \frac{(P_r)_f^3}{(P_r)_p^3} \qquad [6.3.53b]$$

After calculating the maximum flowrate using Equation 6.3.53b, the inflow performance curve into the future can be constructed using Vogel's equation. This method is valid only for undamaged wells.

6.3.2.5.3 Unified Method

Recently Kelkar and Cox proposed a new method for predicting future IPR [18]. This method is a result of unification of some methods discussed previously. The relationship suggested can be applied to any of the reference methods. Two sets of data points (each at different average reservoir pressure) are required to predict the future inflow performance curve. The procedure is as follows:

1. Calculate the maximum flowrate (q_{max} or Q_{max}) for both tests conducted using the reference method (Vogel, Fetkovich or Modified Standing).
2. Calculate J^*:

$$J_1^* = \frac{(q_{max})_1}{(P_r)_1} \quad \text{and} \quad J_2^* = \frac{(q_{max})_2}{(P_r)_2} \qquad [6.3.54]$$

3. Determine constants A' and B' as

$$A' = \frac{J_1^* - J_2^*}{(P_r^2)_1 - (P_r^2)_2} \qquad [6.3.55]$$

$$B' = \frac{\dfrac{J_1^*}{(P_r^2)_1} - \dfrac{J_2^*}{(P_r^2)_2}}{\dfrac{1}{(P_r^2)_1} - \dfrac{1}{(P_r^2)_2}} \qquad [6.3.56]$$

4. Calculate the maximum flowrate of the corresponding future pressure $(P_r)_f$

$$(q_{max})_f = A'(P_r^3)_f + B'(P_r)_f \qquad [6.3.57]$$

5. Construct the future inflow performance curve using the reference inflow performance equation used in Step 1 above for reservoir pressure $(P_r)_f$ and the maximum flowrate calculated in Step 4.

6.3.2.6 Tubing Performance

The problem of simultaneous flow of oil, gas, and water through the vertical tubing of an oil well is complex. The fluid is a compressible mixture, its density increasing with depth. The gradient line has a distinct curvature (see Figure 6.3.16). Along the gradient line of a given well, different flow regimes occur, which may range from a mist flow in the region of low pressures to a single-phase flow at the pressures when all gas is in the solution.

The knowledge of tubing performance of flowing wells is important for efficient operations. Present and future performance of the wells may be evaluated. This may suggest changes in operating practices and equipment to prolong the flowing life of a well. Figures 6.3.18 and 6.3.19 show the idea of the effects of tubing size and a change in IPR on a well performance, respectively.

For a given wellhead pressure, flowing bottomhole pressure varies with production rate. Plotting these two flowing parameters on a Cartesian coordinate will give a curve called *tubing performance relationship* (TPR). By plotting the TPR and IPR of an oil well on a graph paper, the stabilized production capacity of the well is represented by the intersection of the two curves (see Figure 6.3.18).

To construct a TPR curve for a given well, the fluids and well geometry data should be available. Section 6.2 "Flow of Fluids" provides some good multiphase flow correlations that can be used. In cases where these data and accessibility to computer are limited, a graphical flowing gradients correlation is needed. In fact, many improved graphical correlations covering a broad range of field conditions are available in the literature. The readers may refer to Brown [19] to get a complete set of flowing gradient curves. Sometimes a different company has a different set of curves. Although the best correlation is available, a particular field condition might have specific well characteristics such as salt or asphaltene deposition in the tubing or severe emulsification that may bring about a higher pressure drop than would be estimated from existing graphical correlations. The discrepancy might be used for analyzing the wells. So, it is recommended that a good multiphase flow equation be used instead of a graphical correlation.

For convenience, however, a set of working curves developed by Gilbert [20] is presented for illustrating their use in solving well performance problems. The curves are shown in Figures 6.3.20 through Figure 6.3.29, and available only for small flowrate (50 to 600 stb/d) and tubing sizes of 1.66, 1.90, 2.375, 2.875, and 3.50 in.

The procedure how to use these flowing gradient curves in production engineering problems is given in the example later.

6.3.2.7 Use of Vertical Pressure Gradients

In the preceding section of single-phase flow performances, the functions of vertical flow performance curves have been discussed. The following is a more detailed discussion on the applications of pressure gradients in analyses of flowing well performance. Accurate well test data, obtained under stabilized flow conditions, are needed for such analyses.

6.3.2.7.1 Subsurface Data

With flowing tubing known and well test data available, the flowing bottomhole pressure for a given rate of production can be determined by calculating the flowing pressure gradient to the bottom of the well.

If the static bottomhole pressure is known, the productivity index of the well can be determined from one production rate by determining the flowing bottomhole pressure for the rate of production.

If only surface data are available, the productivity index of the well may be estimated by determining the flowing bottomhole pressures for two or more rates of production.

6.3.2.7.2 Tubing Size

As stated, the size of the tubing is one of the important parameters affecting the pressure gradients. For low velocity, the slippage of gas by the liquid contributes to the pressure losses. For high velocities, friction becomes the controlling factor. Between these two extremes, there is a range of velocities giving the optimum gradient at the inlet of tubing in the bottom of the well.

If the future range of expected rates and gas/oil ratios can be estimated, selection of the tubing size can be made, which would assure operation within the efficient range of the gradients, with the resulting increase in the flowing life of the well. Such selection can be made by calculating gradients for different tubing sizes for a given set of conditions.

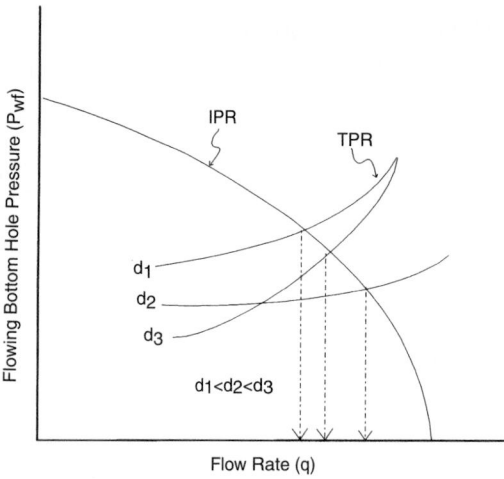

Figure 6.3.18 *Effect of tubing size on oil well deliverability.*

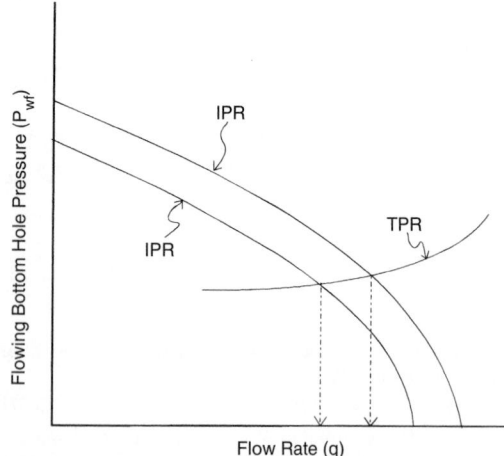

Figure 6.3.19 *Effect of changes in inflow performance on oil well productivity.*

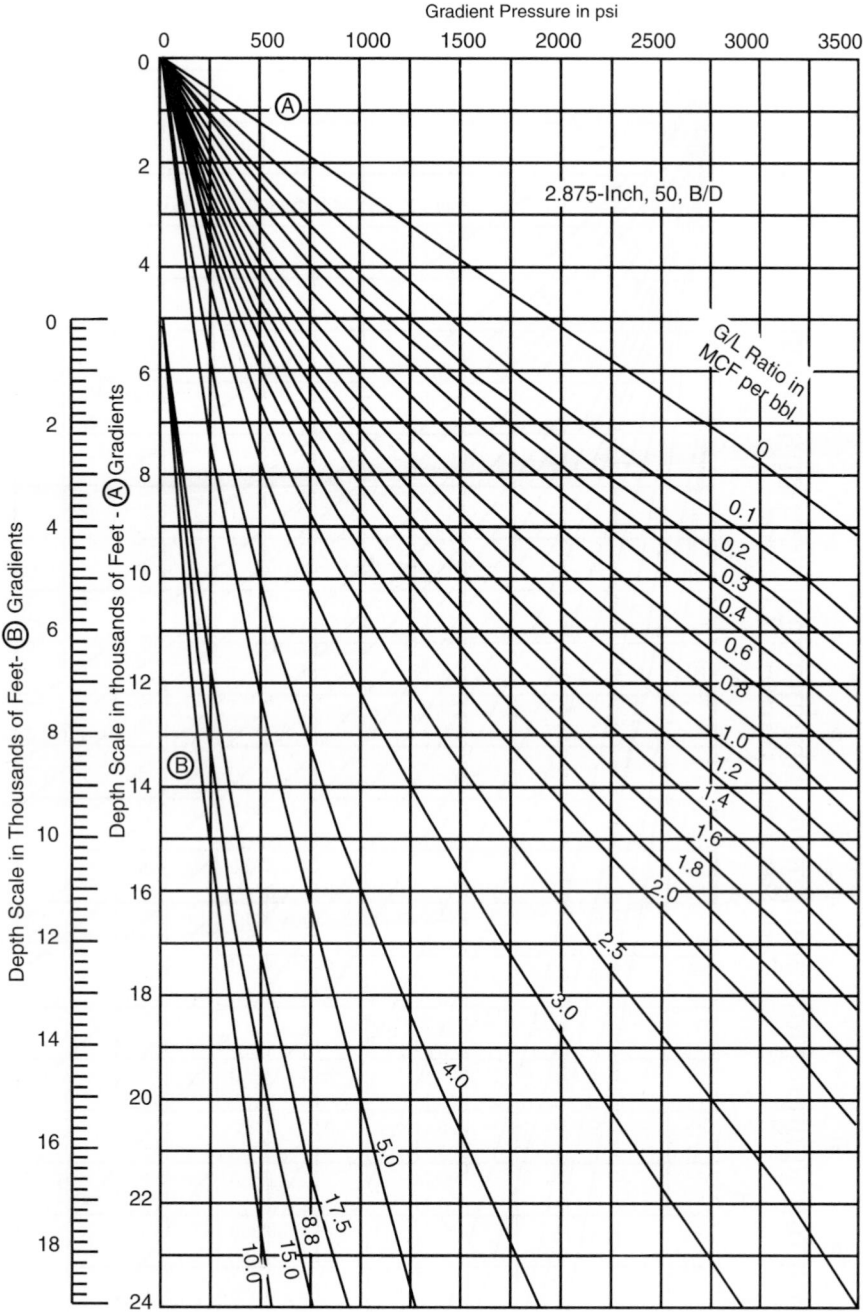

Figure 6.3.20 *Flowing pressure gradients for 2.875-in. tubing with rate of 50 bpd.*

6.3.2.8 Water Content

As the gas/oil ratio decreases, the flowing gradients, of course, increase, other conditions being equal. This is clearly illustrated in Figure 6.3.20 for example. As the water content of the produced fluid increases, the overall gas/liquid ratio decreases. If the future behavior of the water content increase can be estimated, future behavior of the well can be evaluated.

6.3.2.9 Wellhead Pressure

To a degree, the wellhead pressure is controllable because it depends, among other things, on the size, length, and geometry of flow lines and on the separator pressure. At the same time, the wellhead pressure has a marked effect on the slope characteristics of the gradient because of the question of density which is involved. For a given reduction of the wellhead pressure, the reduction of the bottomhole flowing pressure

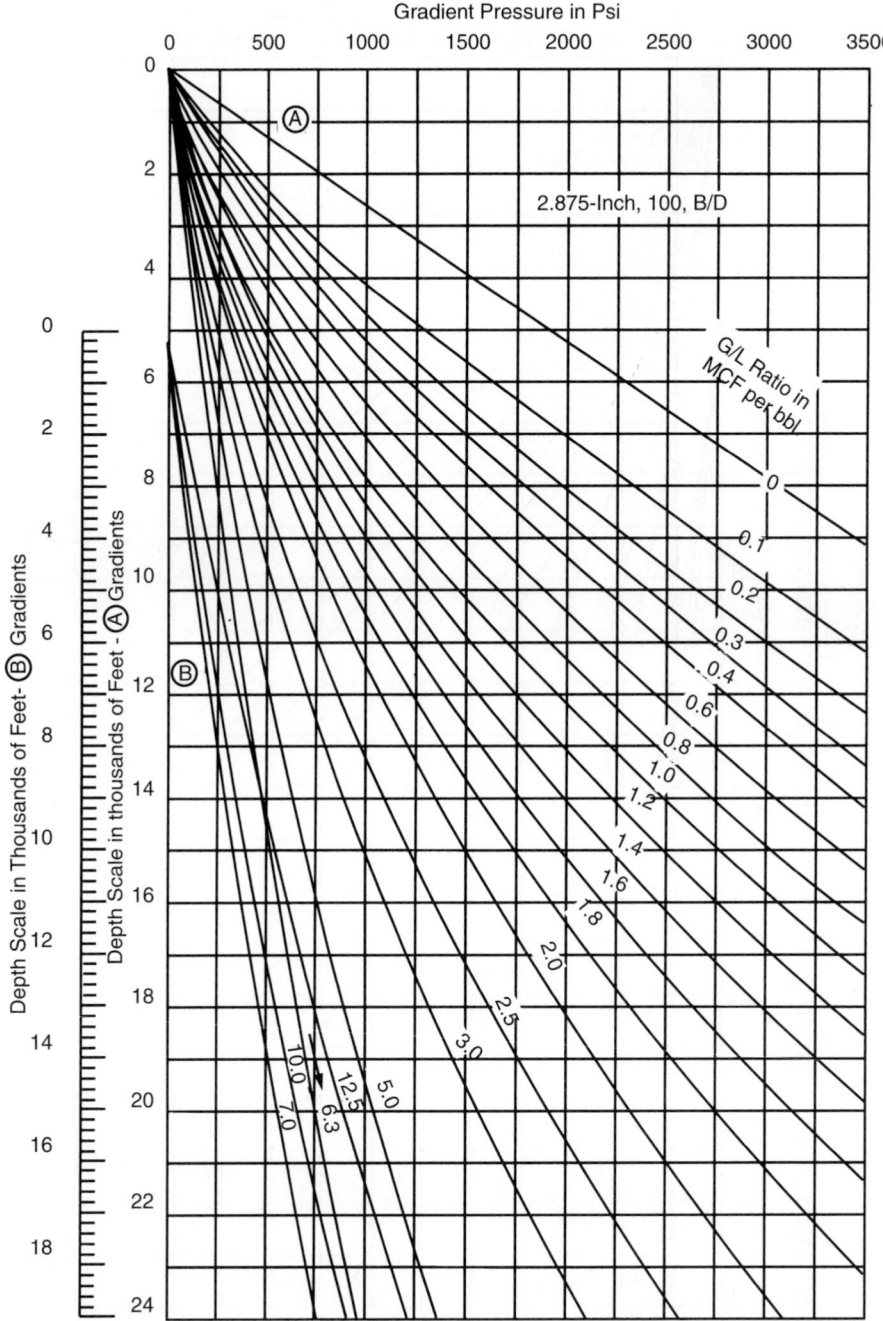

Figure 6.3.21 *Flowing pressure gradients for 2.875-in. tubing with rate of 100 bpd.*

may be substantially higher. This is particularly true in case of high-density flow.

The reduction of the bottomhole flowing pressure should result in increase of production. It can be seen from the above that the degree of this increase cannot be estimated from the surface data alone. It can be estimated if information is available also on the flowing gradient and the productivity index of the well.

6.3.2.10 Predicting the Flowing Life

The natural flow of an oil well continues as long as there exists a proper balance between two conflicting pressure requirements at the bottom of the well. First, this pressure must be sufficiently high to sustain the vertical lift. Second, it must be sufficiently low to create a pressure differential that permits reservoir fluids to enter the well.

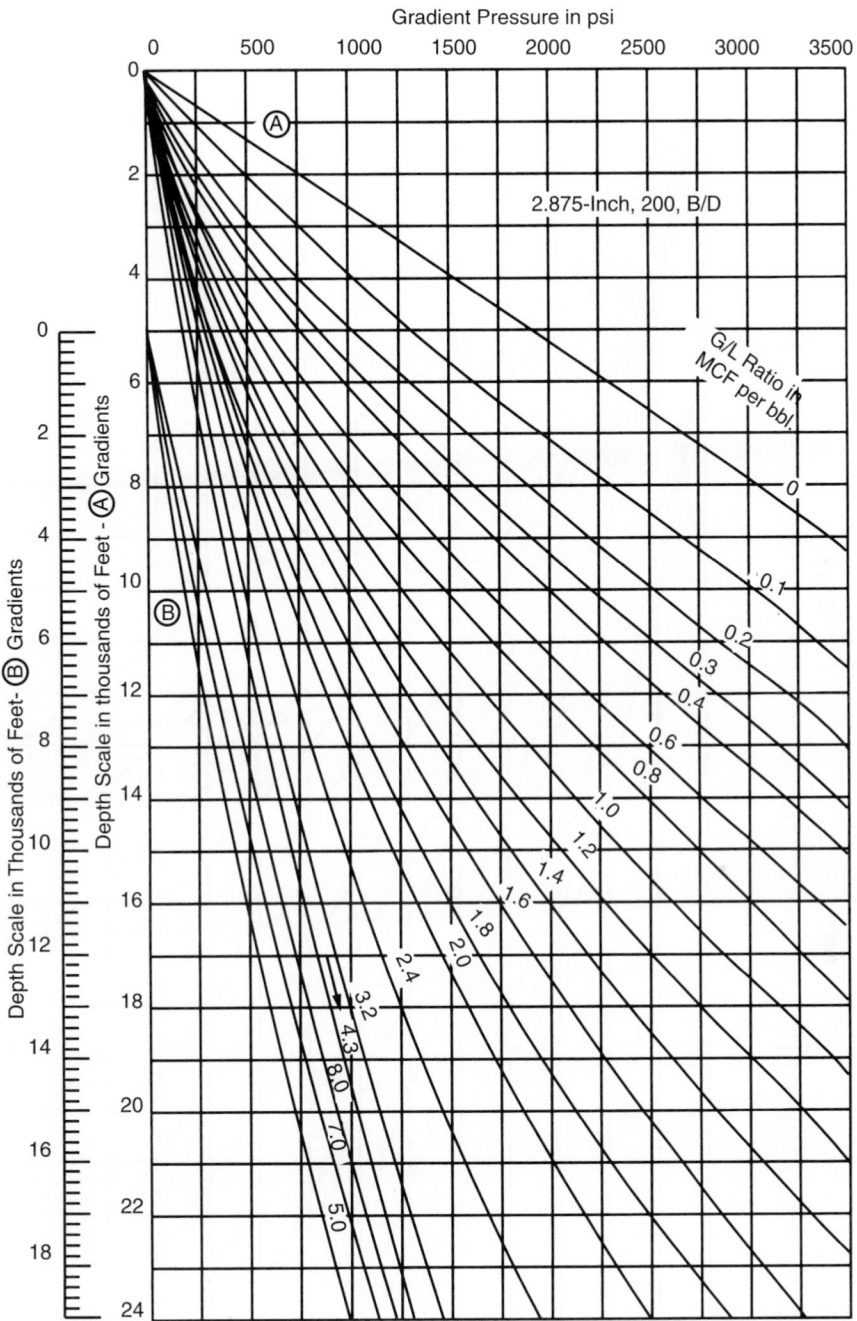

Figure 6.3.22 *Flowing pressure gradients for 2.875-in. tubing with rate of 200 bpd.*

This balance may be destroyed by either one or a combination of two sets of conditions:

1. increase in flowing pressure gradients for any of the reasons mentioned above, increases in the lift pressure requirements above the point needed for maintaining the pressure differential with the available reservoir energy.

2. the declining reservoir energy is not able to maintain this differential for the required vertical lift pressure.

In either case, the natural flow of the well either declines to an uneconomical rate or ceases completely.

The uses of the flowing pressure gradients discussed above may be applied to estimating the length of the flowing life of a well. Additional information needed are the estimate

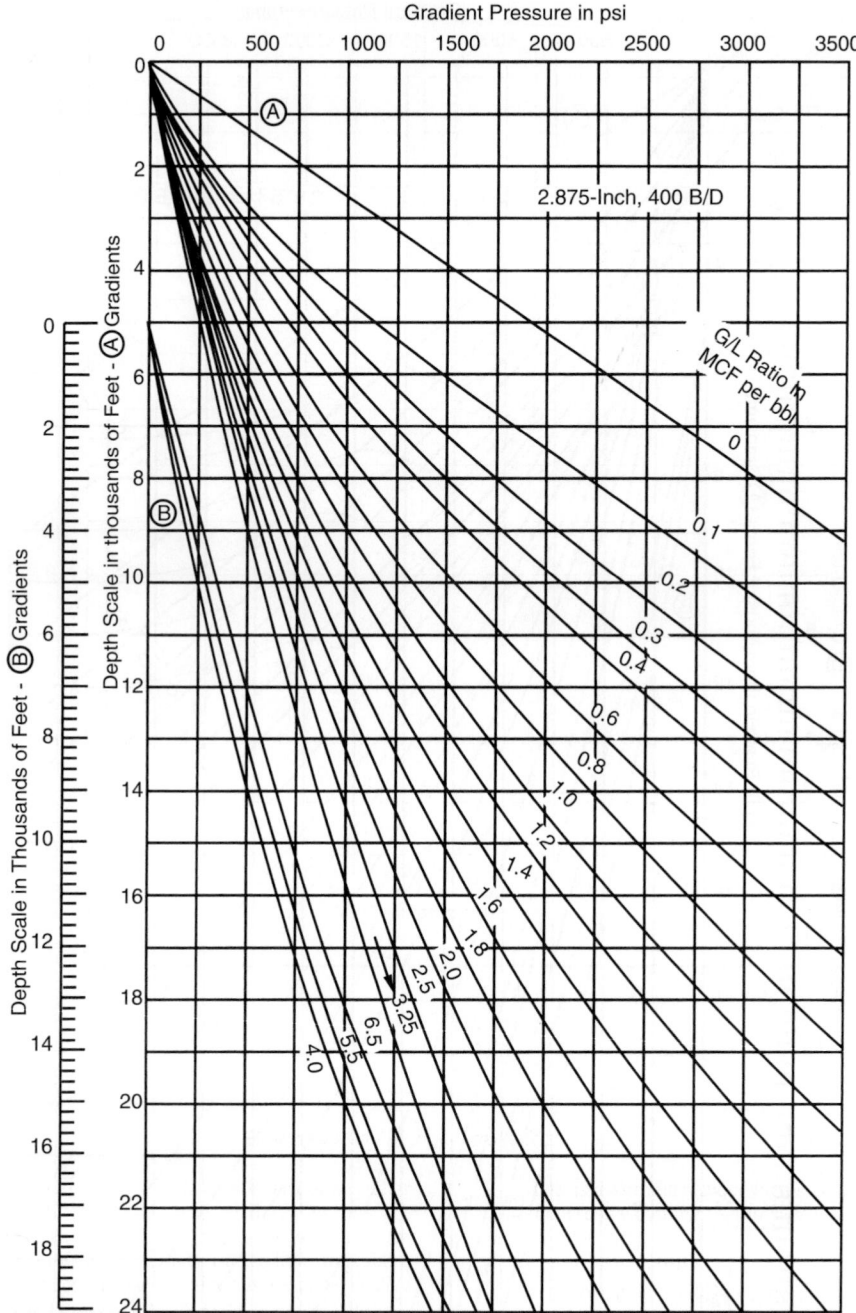

Figure 6.3.23 *Flowing pressure gradients for 2.875-in. tubing with rate of 400 bpd.*

of the static bottomhole pressure of water encroachment, and of gas/oil ratio behavior at different future stages of the cumulative production of the well. Graphic methods have been developed for making such estimates.

6.3.2.11 Choke Performance
The reasons involved in installing chokes in oil and gas fields have been mentioned before. The graphical correlations and

empirical or semiempirical equations used for single-phase oil or gas are not valid for two-phase conditions.

The correlations for multiphase flow through chokes have been published, but not one of them gives satisfactory results for all ranges of operating conditions (flow parameters). Theoretically, the correlations are developed with the assumption that the simultaneous flow of liquid and gas is under critical flow conditions so that when

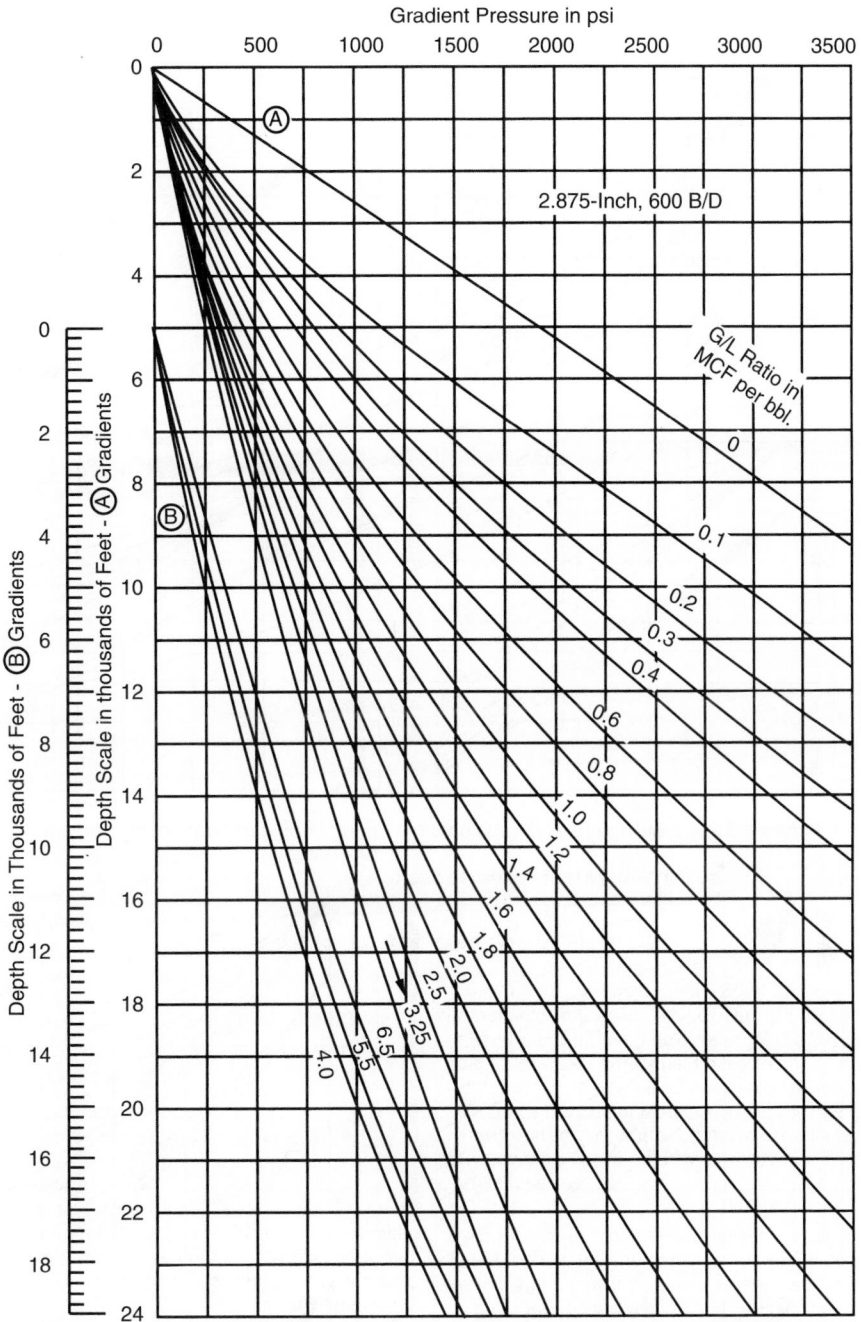

Figure 6.3.24 *Flowing pressure gradients for 2.875-in. tubing with rate of 600 bpd.*

an oil well choke is installed fluctuations in line pressure is gradually increased, there will be no change in either the flowrate or the upstream pressure until the critical–subcritical flow boundary ($P_{downstream} \cong 0.5 - 0.55\,P_{upstream}$) is reached.

In more oil fields, the most popular correlations are of Gilbert [20] and Poetmann and Beck [21]. Ashford [22] also developed a correlation for multiphase flow through chokes.

6.3.2.12 Gilbert's Correlation

The equation developed to estimate a parameter of fluid flow through the orifice is:

$$P_{wh} = \frac{435R^{0.546}q}{d^{1.89}} \qquad [6.3.58]$$

where P_{wh} = wellhead pressure in psig
$\;\; R$ = gas/liquid ratio in Mscf/stb

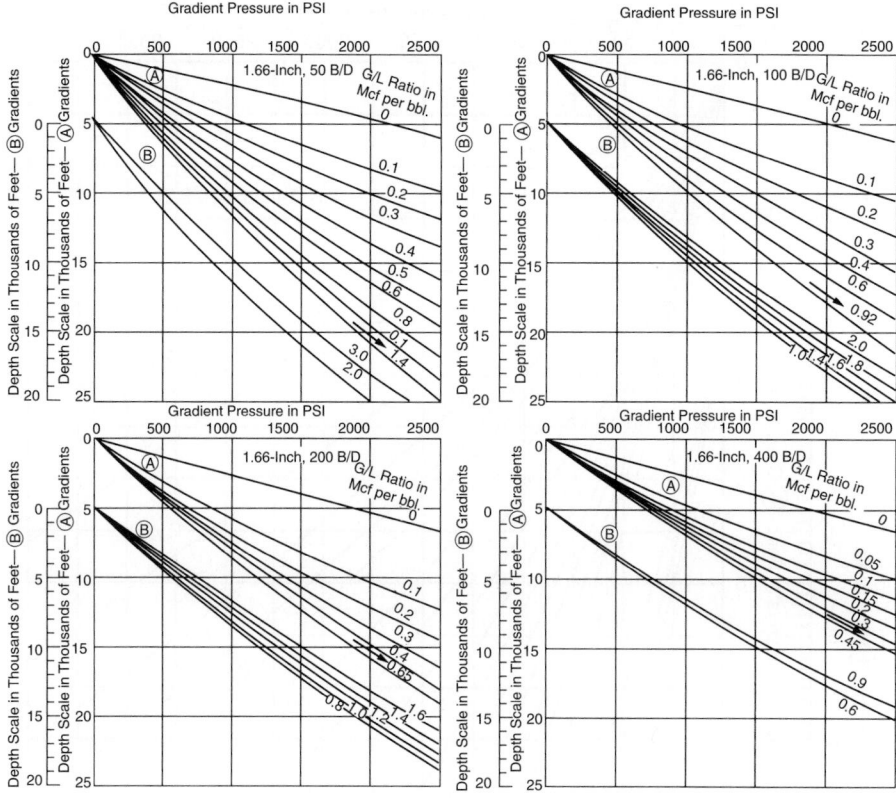

Figure 6.3.25 *Flowing pressure gradients for 1.66-in. tubing with rate of 50–400 bpd.*

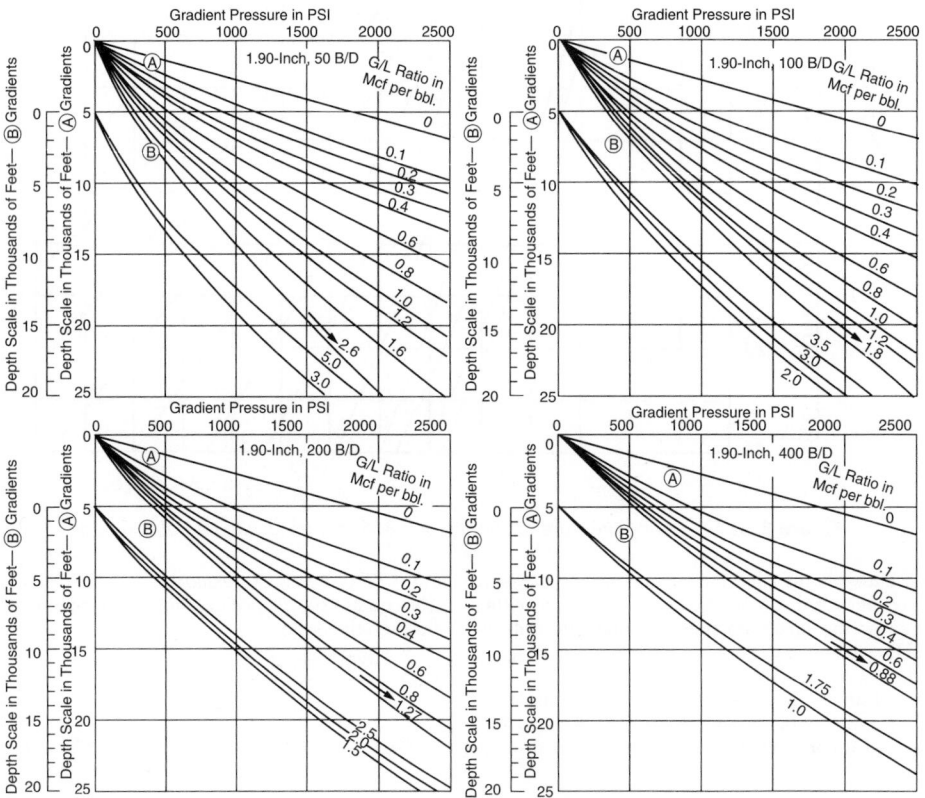

Figure 6.3.26 *Flowing pressure gradients for 1.90-in. tubing with rate of 50–400 bpd.*

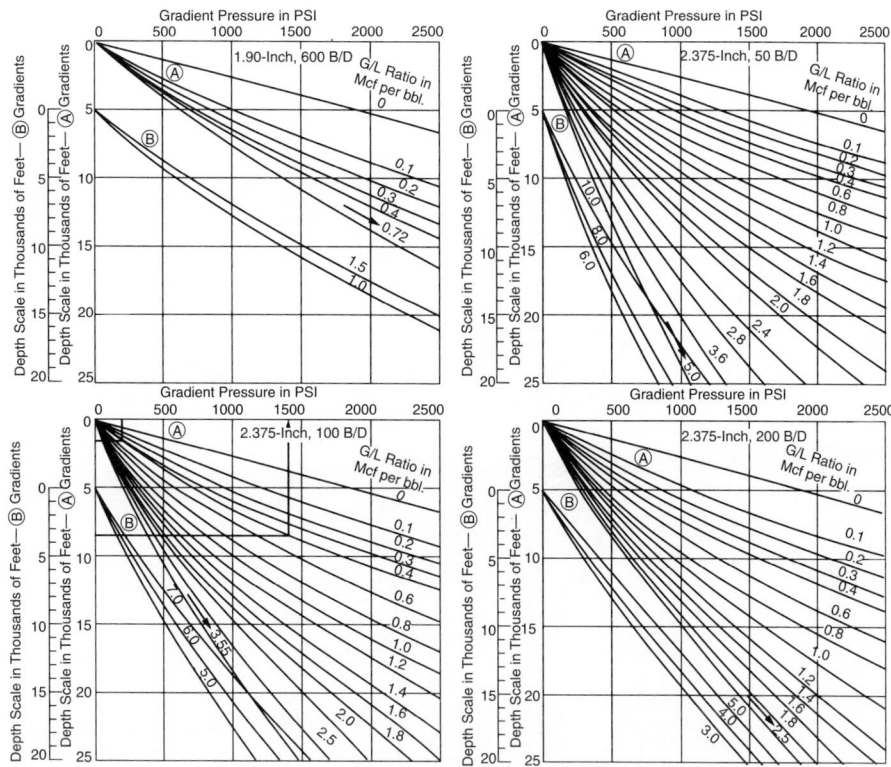

Figure 6.3.27 *Flowing pressure gradients for 1.90-in. tubing with rate of 600 bpd and 2.375-in. tubing with rate of 50–200 bpd.*

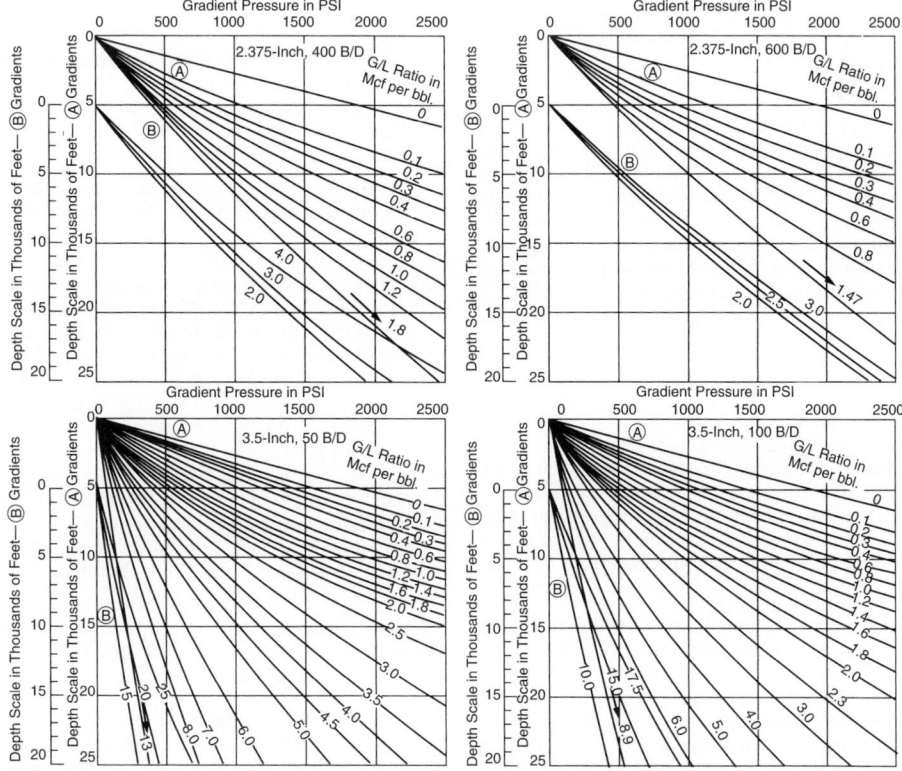

Figure 6.3.28 *Flowing pressure gradients for 2.375-in. tubing with rate of 400–600 bpd and 3.5-in. tubing with rate of 50–100 bpd.*

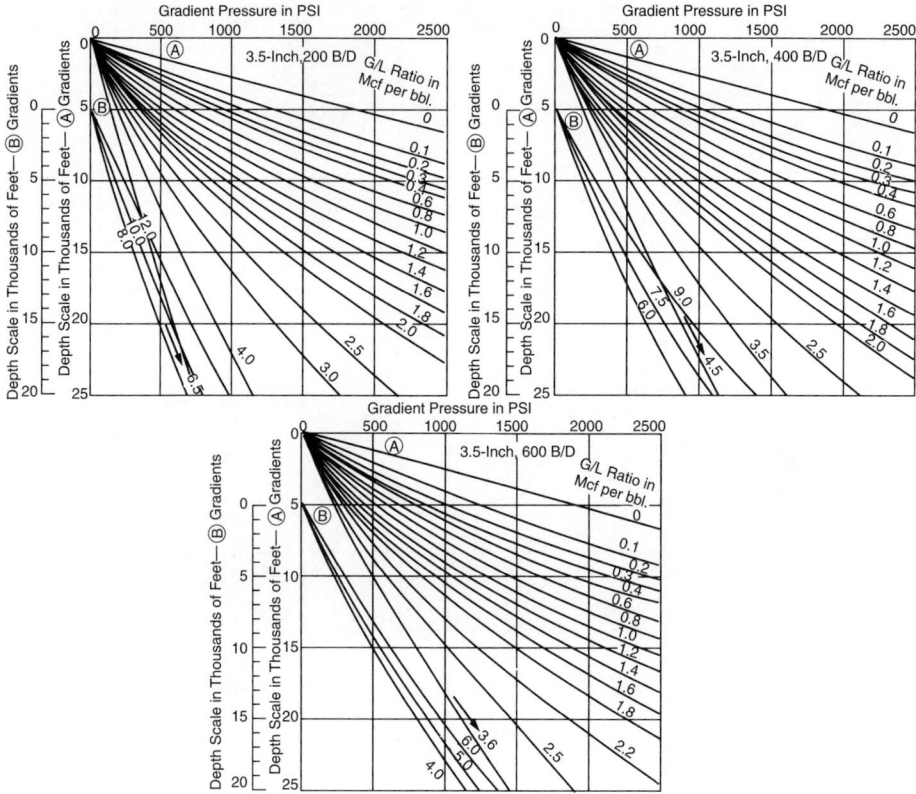

Figure 6.3.29 *Flowing pressure gradients for 3.5-in. tubing with rate of 200–600 bpd.*

q = gross liquid rate in stb/d
d = choke (bean) size in 1/64 in.

This equation is derived using regularly reported daily individual well production data from Ten Section Field in California. Gilbert noted that an error of $\frac{1}{128}$ in. in bean size can give an error of 5 to 20% in pressure estimates. In the type of formula used, it is assumed that actual mixture velocities through the bean exceed the speed of sound, for which condition the downstream, or flow line, pressure has no effect upon the tubing pressure. Thus, the equation applies for tubing head pressure of at least 70% greater than the flow line pressure.

6.3.2.13 Poetmann–Beck's Correlation
F. H. Poetmann and R. L. Beck [21] developed charts for estimating flow of oil and gas through chokes. The charts shown in Figures 6.3.30 through 6.3.32, relate the variables of gas/oil ratio, oil production rate, tubing pressure, and choke size. With three of the variables known, the fourth can be determined. The charts are valid only under the following conditions:

1. The flow is a simultaneous, two-phase flow of oil and gas. The charts are not valid if water is present.
2. The flow through the choke is at the critical flow conditions; that is, at acoustic velocity. This occurs when the downstream (flow line) pressure is 0.55 or less of the upstream (tubing) pressure. Under such conditions the rate of flow is not affected by downstream pressure.

Actually, this last limitation is of small practical significance since chokes are usually selected to operate at critical flow conditions so that the well's rate of flow is not affected by changes in flowline pressure.

The manner of use of these charts is as follows:

1. The 20°, 30°, and 40° API gravity charts can be used for gravity ranges of 15° to 24°, 25° to 34°, and 35° and up, respectively.
2. When the starting point is the bottom scale, a vertical line is drawn to intersection with tubing pressure curve, then a horizontal line to intersection with the choke size, and then vertical line to the upper scale. Reverse procedure is used when the upper scale is the point of beginning.
3. Performance of a given size choke for a given gas/oil ratio can be plotted. Such a plot would show relationship between different rates and corresponding tubing pressures for a given choke.
4. Free gas in tubing can be estimated by use of charts. For instance, in case of Figure 6.3.32 for 2250 ft.^{3}/bbl gas/oil ratio, 1265 psi tubing pressure and 6/64 in. choke, the rate was found to be 60 bbl/d. For 1265 psi tubing pressure the solution gas is 310 ft.^{3}/bbl. Therefore, the free gas is $2250 - 310 = 1940$ ft.^{3}/bbl.

The results obtained by use of these charts compared very favorably with observed data obtained on 108 wells covering a wide range of conditions.

6.3.2.14 Ashford's Correlations
Ashford [22] developed a model for multiphase flow through a choke by applying the polytropic expansion theory.

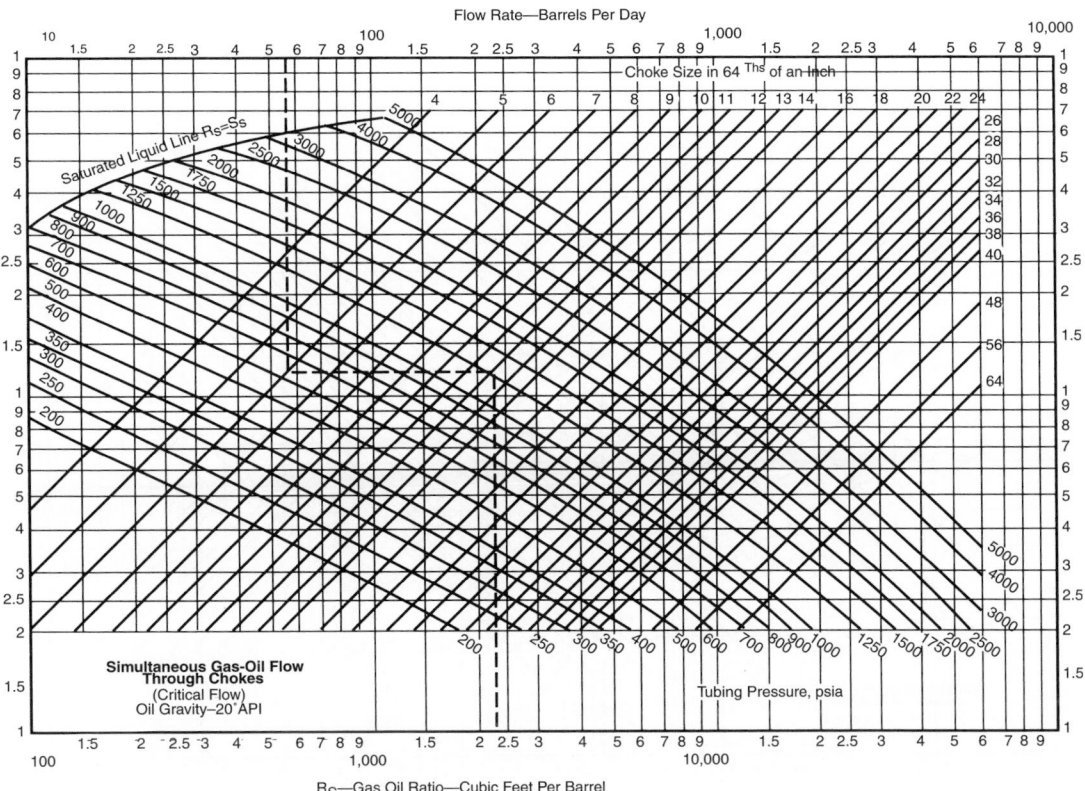

Figure 6.3.30 *A correlation for two-phase flow through chokes with oil gravity of 20°API.*

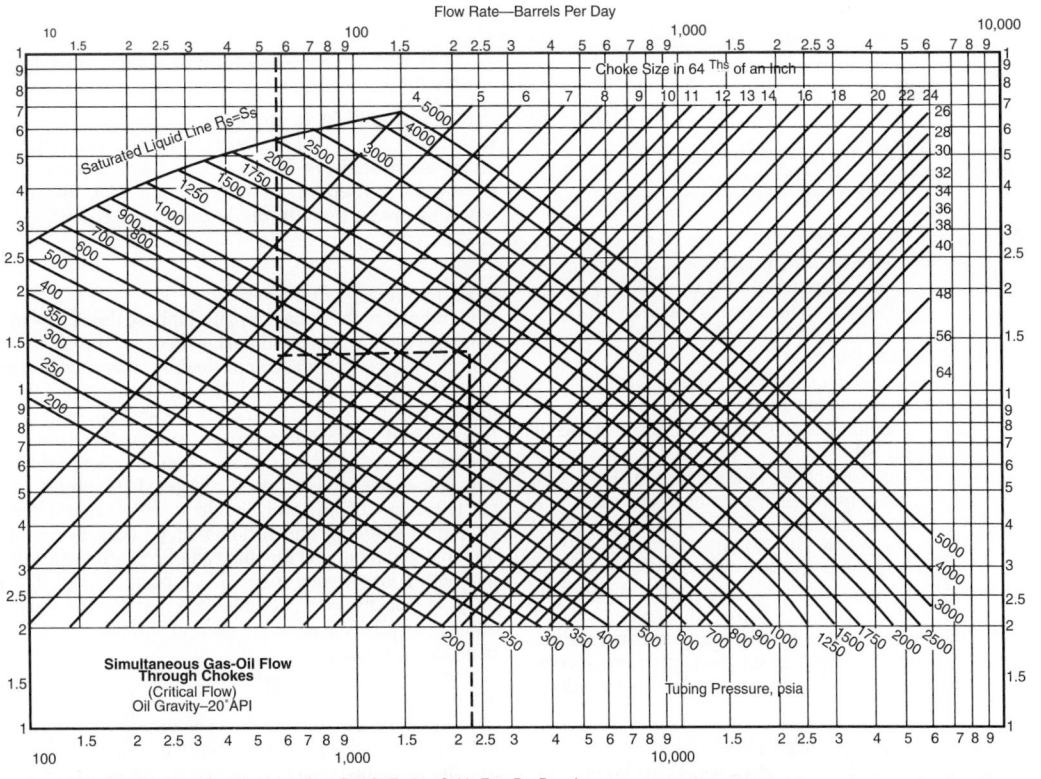

Figure 6.3.31 *A correlation for two-phase flow through chokes with oil gravity of 30°API.*

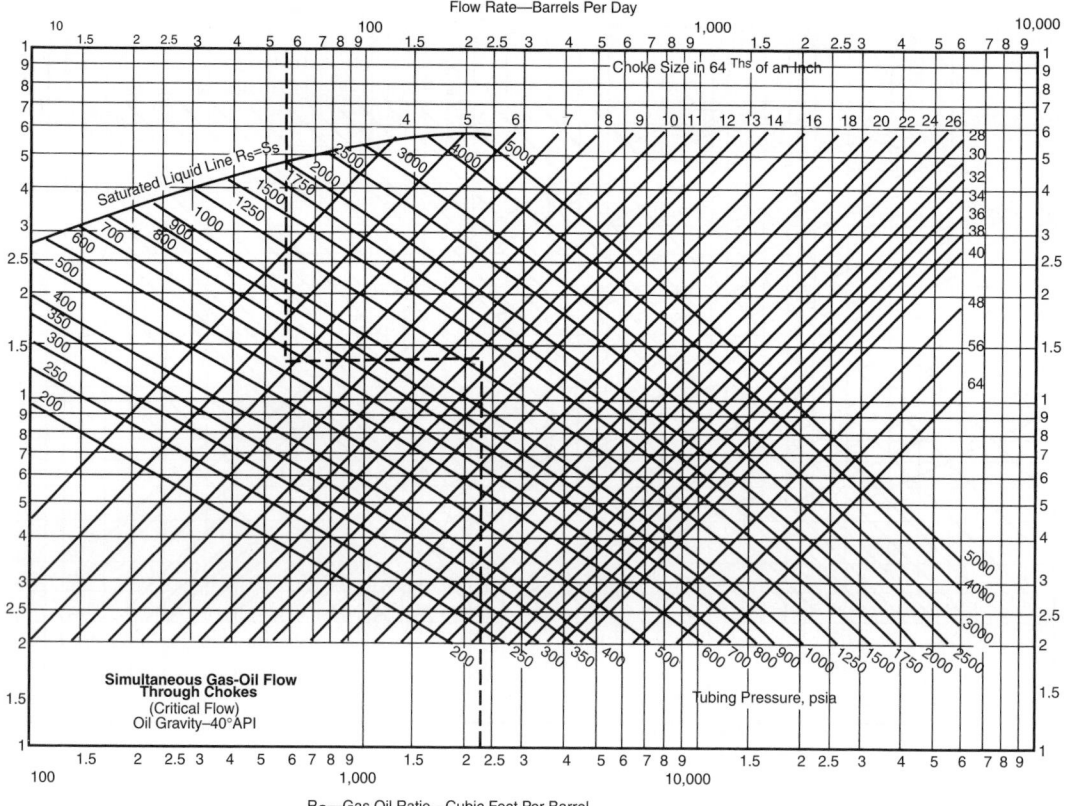

Flow Rate—Barrels Per Day

R_S—Gas Oil Ratio—Cubic Feet Per Barrel

Figure 6.3.32 *A correlation for two-phase flow through chokes with oil gravity of 40° API.*

The final form of his equations is:

$$q_0 = 1.53 \frac{Cd^2P_u}{(B_o + WOR)^{1/2}} \times \frac{[(T_uZ_u(R - R_s) + 151\,P_u)]^{1/2}}{T_uZ_u(R - R_s) + 111P_1}$$

$$\times \frac{[G_o + 0.000217G_gR_s + (WOR)G_w]^{1/2}}{G_o + 0.000217G_gR + (WOR)G_w} \qquad [6.3.59]$$

Where q_o = oil flowrate in stb/d
C = orifice discharge coefficient
d = choke diameter in 1/64 in.
P_u = upstream choke pressure in psia
B_o = oil formation volume factor in bbl/stb
WOR = water/oil ratio
R = producing gas-oil ratio at standard condition in scf/stb
R_s = solution gas-oil ratio at choke conditions in scf/stb
T_u = upstream choke temperature in °R
Z_u = gas compressibility factor evaluated at upstream conditions
G_g = gas gravity (air = 1.00)
G_o = oil gravity (water = 1.00)
G_w = formation water gravity (water = 1.00)

Ashford stated that once C (discharge coefficient) has been fairly well defined for a given production province or operation the Equation 6.3.59 may be used in a conventional manner to evaluate:

1. flowrates arising from changes in choke sizes.
2. wellhead pressures arising from changes in choke sizes.

3. choke sizes necessary to achieve a given wellhead pressure for a known liquid rate.

If C is unknown, a value of 1.0 may be used to obtain a reasonable estimate of choke performance. Later, based on an extensive study, Sachdeva et al., [23] recommended that C = 0.75 be used for a choke configuration involving an elbow upstream from the choke and C = 0.85 be used for a choke free of upstream disturbances.

6.3.2.15 Flowline Performance

Understanding the behavior of multiphase fluids flow in horizontal or inclined pipe is important because the efficiency of a producing system is accomplished by analyzing all components through which the fluids flow. In the analysis a flowline may be considered as a restriction because higher pressure loss resulted. For instance, for a given set of fluids data 2.5-in. line causes higher pressure loss when compared with 3 and 4-in. line, so one tends to take the larger size to produce more oil. The diameter, however, should not be oversized because additional slugging and heading may occur. Some operators just add a parallel line instead of replacing the current line with a larger size. It should be remembered that production capacity, pipes availability, separator pressure, and other constraints may be involved in judging a final design of producing system.

The knowledge of pressure flowrate relationship is very useful in designing an efficient flowing system. The procedure used to generate a flowline performance for a given set of fluids data and a given diameter of the pipe is similar with the one for single-phase. No pressure loss correlations for horizontal or inclined pipes are given. The reader can, however, find some good correlations in Section 6.2 titled "Flow of Fluids."

Example 6.3.7

Consider an oil reservoir producing at an average reservoir pressure below $P_r = 2400$ psia, which is the bubble point pressure. A single-point flow test conducted on a well at stabilized condition resulted in $q = 500$ stb/d and $P_{wf} = 250$ psi. Measured GOR is 400 scf/stb. The well of total depth of 7000 ft. is produced through $2\frac{3}{8}$-in. tubing. Find (1) the maximum rate possible assuming $\dot{P}_{wh} = 200$ psi, and (2) the productivity index for the condition corresponding to the maximum possible flowrate.

Solution (assume FE = 1.0)

(1)

1. Using Vogel's equation, calculate q_{max}

$$\frac{500}{q_{max}} = 1 - 0.2\left(\frac{250}{2,400}\right) - 0.8\left(\frac{250}{2400}\right)^2$$

$$q_{max} = 515 \text{ stb/d}(= \text{oil AOFP})$$

2. Choose some values of P_{wf} and determine corresponding q's using Vogel's equation

$$q = 515\left[1 - 0.2\left(\frac{P_{wf}}{2400}\right) - 0.8\left(\frac{P_{wf}}{2400}\right)^2\right]$$

P_{wf}, psi	q, stb/d
500	476
800	435
1200	361
1500	290
2000	143
2400	0

Plot P_{wf} versus q to get the IPR curve (see Figure 6.3.33).

3. Let's select Gilbert's working curves shown in Figures 6.3.26 and 6.3.27 for $2\frac{3}{8}$-in. tubing and choose the curves with rates of 100, 200 and 400 stb/d.
4. Determine the equivalent length depth of $P_{wf} = 200$ psi. This is done by tracing down a vertical line through pressure point of 200 psi at zero depth until the line of GLR = 400 scf/stb is found and read the depth. This is the equivalent depth. Add this equivalent depth to the actual well depth. Then find the pressure at this total equivalent depth for rate and GLR given. For all three rates chosen, we can obtain

q, stb/d	P_{wf}, psi
100	1410
200	1370
400	1310
(600)	(1400)

5. Plot these values on the same graph made in Step 2 above.
6. Read the coordinate points (q, P_{wf}) of the intersection between the two curves (IPR and TPR), see Figure 6.3.33. It is obtained that the maximum possible rate is 332 stb/d.

(2) The productivity index J of two-phase fluids may be defined as:

$$J = -dq/dP_{wf} = -\left[-0.2\frac{q_{max}}{P_r} - 1.6\frac{q_{max}}{P_r}(P_{wf})\right]$$

$$= 0.2 \times \frac{515}{(2400)} + 1.6 \times \frac{515}{2400^2}(1325)$$

$$= 0.232 \text{ stb/d/psi}$$

Example 6.3.8

Suppose we have more test data for Example 6.3.7. The additional data given are FE = 0.7. (1) What is the AOFP when the skin effect is removed (FE = 1.0)? (2) What is the actual maximum potential (AOFP) of the well? (3) Determine the maximum possible rate. (4) Determine the maximum possible rate if no damage occurred.

Solution

1. Using Equation 6.3.48, for V = 0.8,

$$\frac{500}{q_{max}^{FE=1}} = 0.7\left[1 + 0.8\left(\frac{250}{2400}\right)\right]\left[1 - \left(\frac{250}{2400}\right)\right]$$

$$q_{max}^{FE=1} = 736 \text{ stb/d}$$

2. The maximum potential will occur when $P_{wf} = 0$,

$$q_{max}^{FE=1} = q_{max}^{FE=1} \times 0.7\left[1 + 0.8\left(\frac{0}{2400}\right)\right]\left(1 - \frac{0}{2400}\right)$$

$$= 515.2 \text{ stb/d}$$

3. The maximum rate possible is 332 stb/d (already determined in Example 6.3.7).
4. Construct an IPR with FE = 1.0. Vogel's equation is valid now,

$$q_0 = 736\left[1 - 0.2\left(\frac{P_{wf}}{2400}\right) - 0.8\left(\frac{P_{wf}}{2400}\right)^2\right]$$

P_{wf}, psi	q, stb/d
500	680.0
800	621.5
1200	515.2
1500	414.0
2000	204.4

Plotting P_{wf} versus q, we can determine that q = 477 stb/d (Figure 6.3.33).

Example 6.3.9

An oil well is produced at $P_{wf} > P_b$. Data given are:

H = 5000 ft.
d = 2.375 in.
$P_r = 1500$ psi
J = 0.4 stb/d/psi
GLR = 0.8 Mscf/stb

Find the production rate of liquid and gas if (a) the bean size is $\frac{22}{64}$ in. or (b) the bean size is $\frac{30}{64}$ in. Assume critical flow conditions and use Gilbert's equation.

Solution

Since the well is operated above the reservoir bubble point, we can treat the inflow performance as a linear one:

$$P_{wf} = P_r - \frac{q}{J} = 1500 - \frac{2}{0.4}$$

Chose some values of q (e.g., 100, 200, and 400 stb/d) and use the gradient curves for 2.375-in. tubing. Gilbert's curves

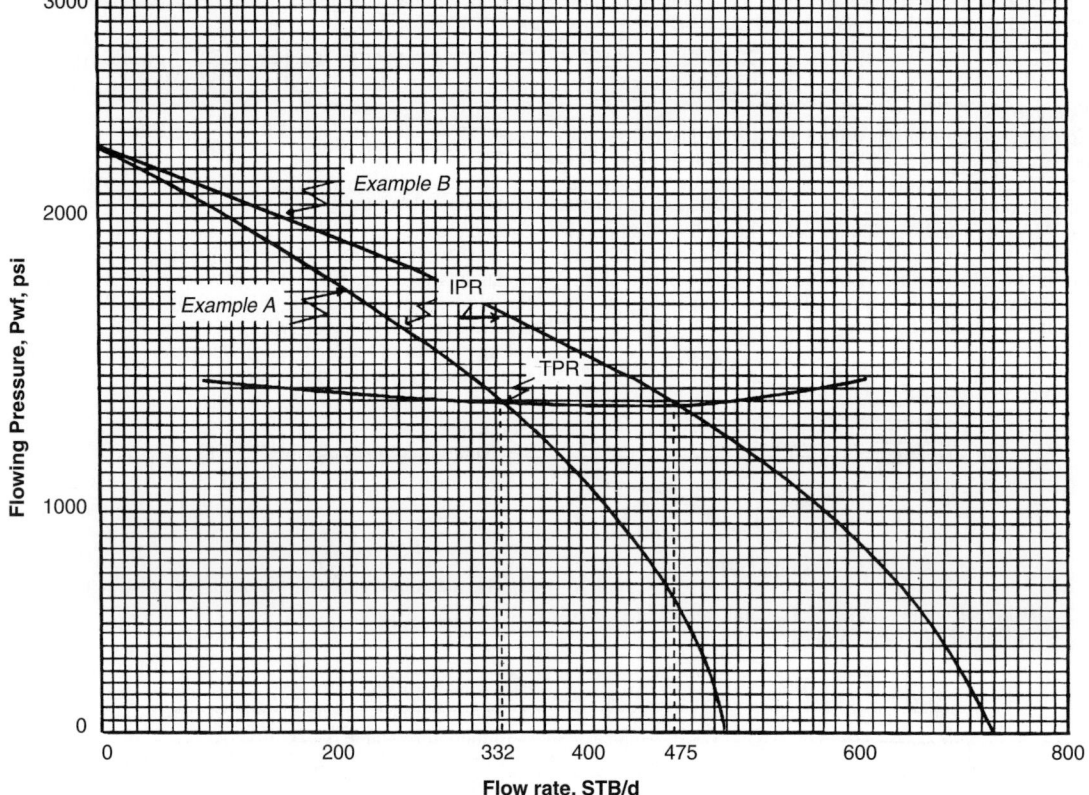

Figure 6.3.33 *Pressure rate relationship for Example 6.3.7.*

are used here for convenience. For a given P_{wf}, P_{wh} is found with a procedure opposite to that for determining P_{wf} for a known P_{wh} (solution to Example 6.3.7, (1), Step 4). Doing so, we can get the following:

q, stb/d	P_{wf}, psi	P_{wh}, psi
100	1250	540
200	1000	400
400	500	40

Plot q versus P_{wh} (shown in Figure 6.3.34). Generate other q versus P_{wh} for the two choke sizes using:

$$P_{wh} = 435R^{0.546}q/S^{1.89}$$

Where (a) $q = (22^{1.89} \times P_{wh})/(435 \times 0.8^{0.546})$, or (b) $q = (30^{1.89} \times P_{wh})/(435 \times 0.8^{0.546})$

These are straight line equations through the origin. Draw the choke performance curves (see Figure 6.3.34). From this figure, we can determine that:

1. installing a $\frac{22}{64}$-in. choke:

$$q_{liquid} = 264 \text{ stb/d}$$

$$q_{gas} = 264 \times 0.8 \text{ M} = 0.211 \text{ MMscf/d}$$

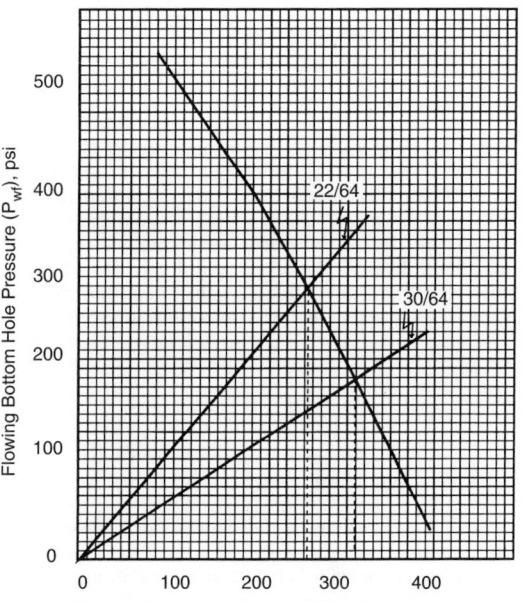

Figure 6.3.34 *Pressure rate relationship for Example 6.3.9.*

2. installing a 30/64-in. choke:

$$q_{liquid} = 319 \text{ stb/d}$$

$$q_{gas} = 0.255 \text{ MMscf/d}$$

Example 6.3.10
Constructing an IPR curve of an oil well is sometimes difficult in terms of data available. The present reservoir pressure is not available or measured. Some oil companies do not want to lose their production caused by closing the well to measure the static reservoir pressure. A practical means of overcoming this problem is to test the well at two different rates while measuring the flowing bottomhole pressure.

This technique does not require expensive, time-consuming tests. The pressure measurements can be very accurate, using a subsurface gage on flowing wells; they can be obtained with surface-recording downhole gage; or they can simply be obtained with casing pressure and fluid level shots, depending on the well condition.

A typical two-point test datum is taken from a well in the Judy Creek Beaverhill Lake 'A' Pool producing under a solution gas drive [24].

	Test 1	Test 2
q, stb/d	690.6	470
P_{wf}, psi	1152.0	1970
P_b, psi	2290.0	2290

The task is to estimate the static reservoir pressure and to construct IPR for this well.

Solution

$$\left(\frac{q}{q_{max}}\right)_1 = 1 - 0.2\left(\frac{1152}{2290}\right) - 0.8\left(\frac{1152}{2290}\right)^2 = 0.697$$

$$\left(\frac{q}{q_{max}}\right)_2 = 1 - 0.2\left(\frac{1970}{2290}\right) - 0.8\left(\frac{1970}{2290}\right)^2 = 0.236$$

$$q_c = (470 - 690.6)/(0.236 - 0.697) = 478.5 \text{ stb/d}$$

$$= (q_{max} - q_b)$$

Flowrate at bubble point,
$$q_b = 470 - (0.236)(478.5) = 357 \text{ stb/d}$$

$$J_b = 1.8 \times (478.5/2290) = 0.376$$

$$P_r = 2290 + (357/0.376) = 3239 \text{ psia}$$

The calculated reservoir pressure from the two-point test was 3239 psia, which is consistent with static pressures measured in this area of the 'A' Pool.

Figure 6.3.35 is the constructed IPR. A third test was run to verify this curve, with a rate of 589 stb/d at 1568 psia. This point fell essentially on the curve generated by the original two-test points.

Example 6.3.11
A reservoir with a back-pressure curve slope $(1/n) = 1.12$ has the following two flow tests:

	$\bar{P}$, psia	q_o, stb/d	P_{wh}, psia
Test 1	2355.9	335.1	1300
Test 2	2254.9	245.8	13

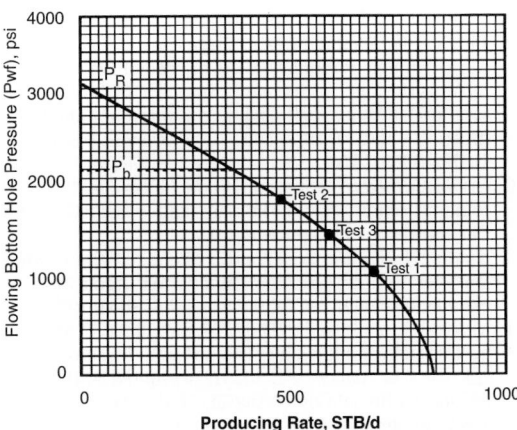

Figure 6.3.35 *Pressure-rate relationship for Example 6.3.10.*

Calculation of future IPR:

- Using Fetkovich's method:

$$q_{max} = \frac{q_0}{\left[1 - \left(\frac{P_{wf}}{P}\right)^2\right]^n}$$

$$q_{max\,1} = \frac{335.1}{\left[1 - \left(\frac{1300}{2355.4}\right)^2\right]^{0.893}} = 463.5 \text{ stb/d}$$

$$q_{max2} = \frac{254.8}{\left[1 - \left(\frac{1300}{2254.9}\right)^2\right]^{0.893}} = 365.5 \text{ stb/d}$$

- Calculate J^*:

$$J_1^* = \frac{463.5}{2355.4} = 0.197 \text{ stb/d/psi}$$

$$J_2^* = \frac{365.5}{2254.9} = 0.162 \text{ stb/d/psi}$$

- Calculate A' and B':

$$A' = \frac{0.197 - 0.162}{(2355.4)^2 - (2254.9)^2} = 7.554 \times 10^{-8}$$

$$B' = \frac{\dfrac{0.197}{(2355.4)^2} - \dfrac{0.162}{(2254.9)^2}}{\dfrac{1}{(2355.4)^2} - \dfrac{1}{(2254.9)^2}} = -0.222$$

- Calculate maximum future rate (for instance, a future reservoir pressure of 1995 psia):

$$q_{max} = 7.554 \times 10^{-8}(1995)^3 + (-0.222)(1,995)$$

$$= 157 \text{ stb/d}$$

- The future IPR curve can then be predicted using the equation

$$q_o = 157\left[1 - \left(\frac{P_{wf}}{1995}\right)^2\right]^{0.893}$$

References

1. Jones, L. G., Blount, M., and Glaze, O. H., "Use of Short Term Multiple Rate Flow Tests to Predict Performance of Wells Having Turbulence," Paper SPE 6133, prepared for the 51st Annual Fall Technical Conference and Exhibition of the SPE of AIME, New Orleans, Louisiana, October 3–6, 1976.
2. Crane Company Industrial Products Group, "Flow of Fluid Through Valves, Fittings and Pipe," Chicago, Illinois, Technical Paper no. 410.
3. Rawlins, E. L., and Schellhardt, M. A., "Back-Pressure Data on Natural Gas Wells and Their Application to Production Practices," Monograph 7, USBM, 1936.
4. Mishra, S., and Caudle, B. H., "A Simplified Procedure for Gas Deliverability Calculations Using Dimensionless IPR Curves," paper SPE 13231, presented at the 59th Annual Technical Conference and Exhibition, Houston, Texas, September 16–19, 1984.
5. Dake, L. P., *Fundamentals of Reservoir Engineering*," Elsevier Scientific Publishing Co., Amsterdam, 1978.
6. Chase, R. W., and Anthony, T. M., "A Simplified Method for Determining Gas Well Deliverability," paper SPE 14507, presented at the SPE Eastern Regional Meeting, Morgantown, West Virginia, November 6–8, 1985.
7. Cullender, M. H., "The Isochronal Performance Method of Determining Flow Characteristics of Gas Wells," *Transactions of AIME*, 1955.
8. Cullender, M. H., and Smith, R. V., "Practical Solution of Gas-Flow Equations for Wells and Pipelines with Large Temperature Gradients," *Transactions of AIME*, 1956.
9. Katz, D. L., et al., *Handbook of Natural Gas Engineering*, McGraw-Hill Book Co., New York, 1959.
10. Peffer, J. W., Miller, M. A., and Hill, A. D., "An Improved Method for Calculating Bottomhole Pressure in Flowing Gas Wells with Liquid Present," paper SPE 15655, presented at the 61st Annual Technical Conference and Exhibition of the SPE, New Orleans, Louisiana, October 5–8, 1986.
11. Nind, T. E. W., *Principles of Oil Well Production*, 2nd Edition, McGraw-Hill Book Company, New York, 1981.
12. Ikoku, C. U., *Natural Gas Production Engineering*, John Wiley & Sons, Inc., New York, 1984.
13. Vogel, J. V., "Inflow Performance Relationships for Solution-Gas Drive Wells," *Journal of Petroleum Technology*, January 1968.
14. Fetkovich, M. J., "The Isochronal Testing of Oil Wells," paper SPE 4529, prepared for the 48th Annual Fall Conference and Exhibition of the SPE of AIME, Las Vegas, Nevada, September 30–October 3, 1973.
15. Standing, M. B., "Inflow Performance Relationships for Damaged Wells Producing by Solution-Gas Drives," *Journal of Petroleum Technology*, November 1970.
16. Camacho, V., Raghavan, R. G., "Inflow Performance Relationships for Solution-Gas Drive Reservoirs," *Journal of Petroleum Technology*, May 1989.
17. Standing, M. B., "Concerning the Calculation of Inflow Performance of Wells Producing from Solution-Gas Drive Reservoirs," *Journal of Petroleum Technology*, September 1971.
18. Kelkar, B. G., and Cox, R., "Unified Relationship to Predict Future IPR Curves for Solution-Gas Drive Reservoirs," paper SPE 14239, prepared for the Annual Technology Conference and Exhibition of the SPE of AIME, Las Vegas, Nevada, September 22–25, 1985.
19. Brown, K. E., *The Technology of Artificial Lift Methods*, Vols. 3a and 3b, Petroleum Publishing Co., Tulsa, Oklahoma, 1980.
20. Gilbert, W. E., "Flowing Gas Well Performance," *Drilling and Production Practices*, 1954.
21. Poettmann, F. E., and Beck, R. L., "New Charts Developed to Predict Gas-Liquid Flow Through Chokes," *World Oil*, March 1963.
22. Ashford, F. E., "An Evaluation of Critical Multiphase Flow Performance Through Wellhead Chokes," *Journal of Petroleum Technology*, August 1974.
23. Sachdeva, R., Schmidt, Z., Brill, J. P., and Blais, R. M., "Two-phase flow through chokes," paper SPE 15657, presented at the 61st Annual Technical Conference and Exhibition of the SPE, New Orleans, Louisiana, October 5–8, 1986.
24. Richardson, J. M., and Shaw, A. H., "Two-Rate IPR testing—A practical production tool," *The Journal of Canadian Petroleum Technology*, March–April, 1982.

6.4 ARTIFICIAL LIFT

6.4.1 Sucker Rod Pumping

6.4.1.1 Pumping System Components

The sucker rod pumping system consists of a pumping unit, a sucker rod string, and a subsurface pump.

The pumping unit is the surface component that supplies the power and reciprocating motion for the operation of the subsurface pump. The sucker rod string connects the pumping unit to the subsurface pump, which is reciprocating, piston type pump. Each of these components of the pumping system will be discussed in detail.

6.4.1.2 Pumping Unit Operation

The prime mover is the power source of the pumping unit, which can be an electric motor or an engine that usually runs on well gas. The prime mover drives the gear reducer, which rotates the crankshaft. The crankshaft drives the crank and pitman, which causes the walking beam to oscillate about the saddle bearing and imparts a reciprocating motion to the sucker rods. A sketch of a common pumping unit is shown in Figure 6.4.1.

6.4.1.3 Pumping Unit Types and Specifications

Four types of beam pumping units are recognized. The classification is based on where the fulcrum is placed (class I or class III) and how they are counterbalanced (air, crank, or beam). Four types are shown in Figure 6.4.2. In a class I lever system (also called a conventional pumping unit), the fulcrum is near the center of the walking beam and the pitman applies lifting force by pulling downward at the rear of the walking beam. The class I lever system with phased crank counterbalance is the latest in pumping unit designs. In this unit, the fulcrum is near the center of the walking beam similar to a conventional unit; however, the phased crank arms allow the unit to have a larger part of the crank rotation dedicated to the upstroke (186° or longer). This longer upstroke often results in lower torque loads or higher lifting capacities. Examples of this type of unit are the Weatherford American Maximizer II or the Lufkin Reverse Mark. In a class III lever system, the lifting force of the pitman is applied upward near the center of the beam. These class III units are also referred to as units with front mounted geometry. The Lufkin Mark II is one such unit. A variation on the class III system that normally has a crank-type counterbalance has instead, a piston and cylinder filled with compressed air as the counterbalance.

Conventional pumping units in smaller sizes can have the counterbalance weights mounted at the rear of the walking beam (beam-balanced units) or mounted on the crank arm (crank-balanced unit). Larger conventional units are all crank balanced. The very largest units (gear reducer sizes 1824 or larger) are almost always air balanced.

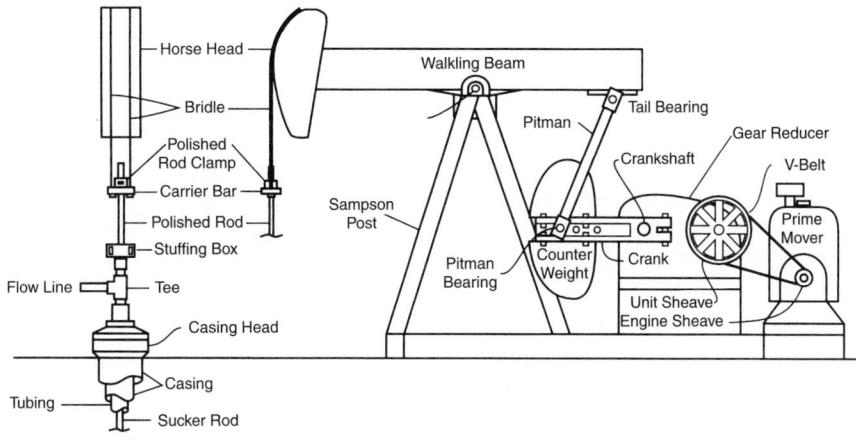

Figure 6.4.1 *Surface equipment of a sucker rod pumping installation [60].*

Table 6.4.1 lists the specifications for American Petroleum Institute (API) standard units. Some manufacturers have additional intermediate sizes not shown in Table 6.4.1. Not all manufacturers make all sizes. The specification is a three-part code; i.e., a unit rating of 160-173-64 identifies a pumping unit having a gear reducer rating of 160,000 in.-lb of torque, a structural load-bearing capacity as measured at the polished rod of 17,300 lb force and a maximum stroke length of 64 in.

Each pumping unit size generally has two to four stroke lengths at which it can operate. The stroke length can be changed by changing the crank hole location. One should refer to the pumping unit manufacture to determine the available stroke lengths for a given unit.

In addition to the unit classifications discussed above, several special purpose pumping units have been designed and used widely in meeting specific challenges. These include:

- Units able to achieve stroke lengths of 288 in. and longer, which operate at lower strokes per minute, reducing rod reversals, down-hole friction, and dynamic loading (Figure 6.4.2a).
- Low profile designs for use under irrigation systems and areas where visual impact needs to be minimized such as parks and residential areas. (Figure 6.4.2b).
- Self-lowering units, which are also used with overhead irrigation systems.
- Trailer mounted units for temporary production requirements and well testing.
- Hydraulic/pneumatic units for use where downstroke rod friction is a concern. (Figure 6.4.2c).

6.4.2 Prime Movers

Both internal combustion engines and electric motors have been used as prime movers for pumping units. Internal combustion engines operating on gas have typically been used in areas where ample lease gas in available and wells can be pumped continuously. New products have been introduced recently that allow for automatic intermittent operation of gas engines on pumping units. Inflatable engine clutches, and clutches operated by screw actuators as well as engines that can start smaller pumping units under load allow lease automation equipment to be utilized on wells powered by these gas engines. Where sufficient electric power is available, electric motors typically offer the operator lower equipment and maintenance costs.

why do we think this is not true ???

6.4.2.1 Electric Motor Types

The National Electrical Manufacturers Association (NEMA) recognizes four electric motor designs that are typically used with oilfield pumping units.

1. NEMA C 4% slip; fair starting torque
2. NEMA D 5% to 8% slip; higher starting torque than a NEMA C
3. NEMA D extended 8% to 13% slip
4. Special purpose ultra-high slip typically up to 30% slip; superior starting, electrical, and mechanical characteristics.

Slip is the differences in motor speed between its synchronous (unloaded) and loaded condition divided by the unloaded speed. An induction motor that has a synchronous speed of 1200 rpm and a loaded speed of 1050 rpm has a slip of

$$\frac{1200 - 1050}{1200} \times 100 = 12.5\% \text{ slip}$$

this is only for induction motors

Most oilfield pumping unit electric prime movers are three-phase induction motors. Single-phase motors are restricted to pumping units used for shallow, low-volume wells requiring minimal horsepower.

6.4.2.2 Internal Combustion Engines

Internal combustion engines designed to run on natural gas or propane and serve as pumping unit prime movers can be classified based on their speed, strokes per cycle and the number of cylinders. Slow-speed engines are those having a crankshaft rpm of 750 rpm or less. High-speed engines have speeds up to 2000 rpm; these are multi-cylinder engines. A relatively new engine package employs a small horsepower (up to 18 hp); high RPM twin cylinder four-stroke engine, intermediate sheave set (jackshaft), and a simple control for automatically starting and stopping the engine.

6.4.2.3 Selecting Prime Mover Size

Several empirical equations are in use for scaling theoretical or hydraulic horsepower rating such as obtained from beam pumping system design procedures (see the section titled "Selection of Beam Pumping Unit Installation") to brake or prime mover horsepower. These have been discussed by Brown [3] and Curtis and Showalter [4]. These empirical

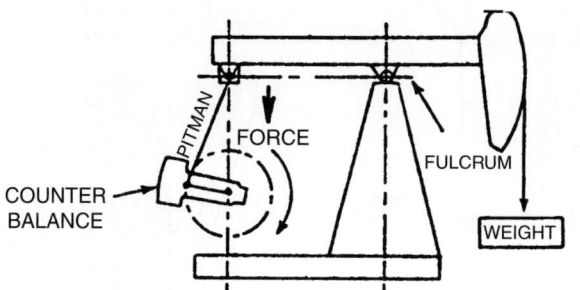

CLASS I LEVER SYSTEM - CONVENTIONAL UNIT.

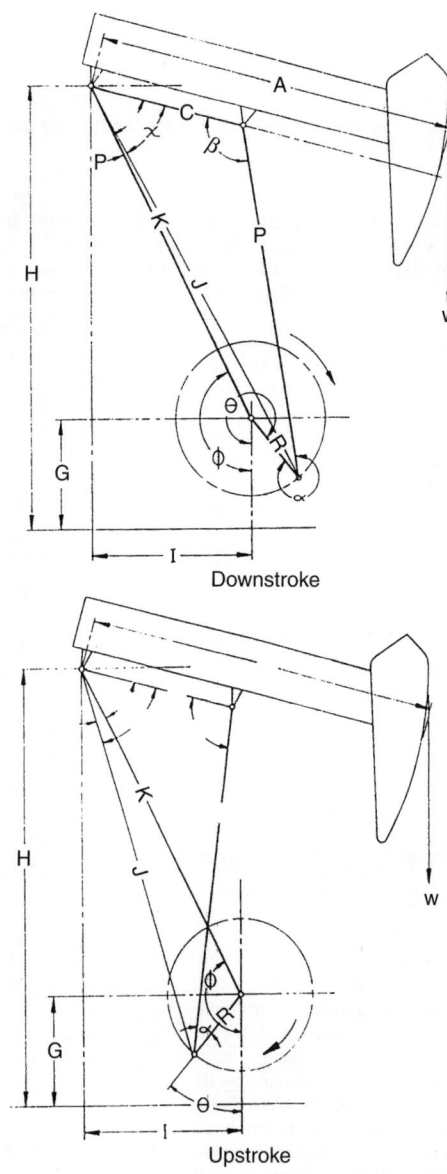

Figure 6.4.2 *Simplified lever diagrams of Class I and Class III pumping units [2].*

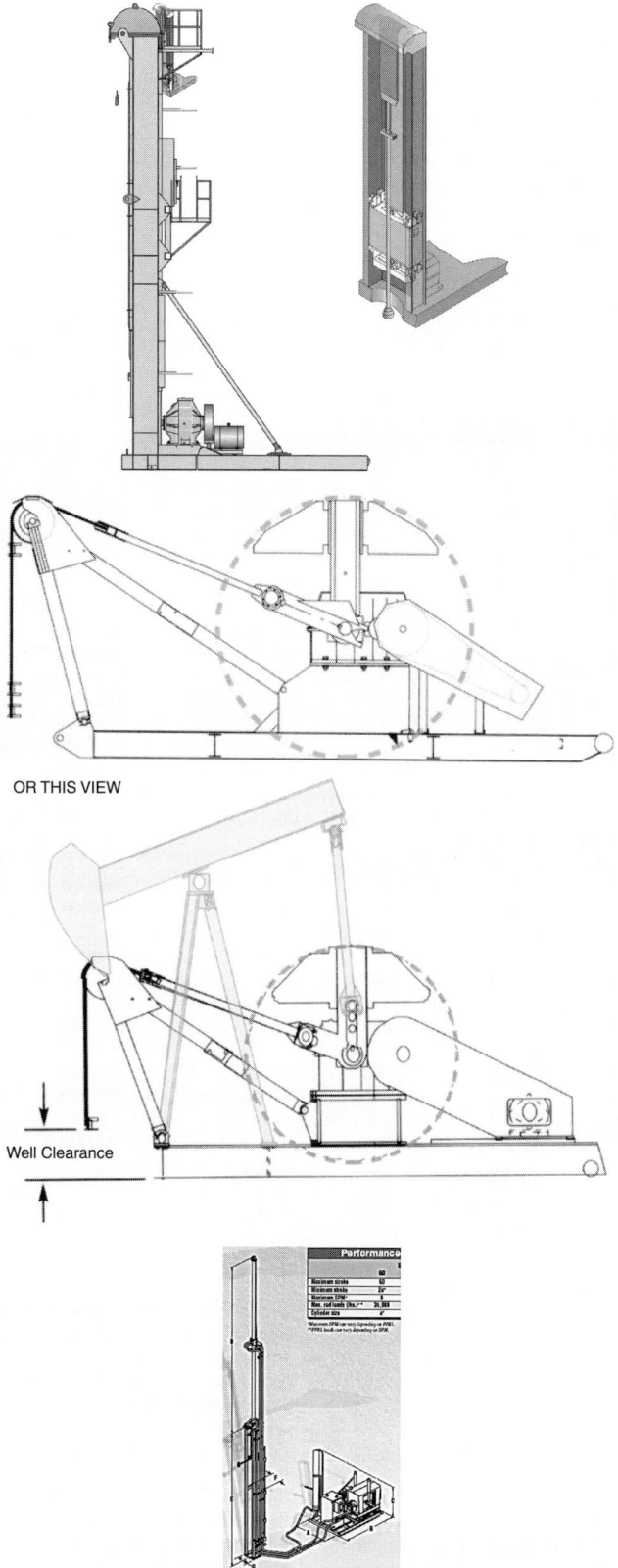

OR THIS VIEW

Well Clearance

Figure 6.4.2 *(continued)*.

Table 6.4.1 *Pumping Unit Size Ratings [3]*

1	2	3	4	1	2	3	4
Pumping Unit Size	Reducer Rating, in.-lb	Structure Capacity, lb	Max. Stroke Length, in.	Pumping Unit Size	Reducer Rating, in.-lb	Structure Capacity, lb	Max. Stroke Length, in.
6.4–32–16	6400	3200	16	320–213–86	320,000	21,300	86
6.4–21–24	6400	2100	24	320–256–100	320,000	25,600	100
				320–305–100	320,000	30,500	100
10–32–24	10,000	3200	24	320–213–120	320,000	21,300	120
10–40–20	10,000	4000	20	320–256–120	320,000	25,600	120
				320–256–144	320,000	25,600	144
16–27–30	16,000	2700	30				
16–53–30	16,000	5300	30	456–256–120	456,000	25,600	120
				456–305–120	456,000	30,500	120
25–53–30	25,000	5300	30	456–365–120	456,000	36,500	120
25–56–36	25,000	5600	36	456–256–144	456,000	25,600	144
25–67–36	25,000	6700	36	456–305–144	456,000	30,500	144
				456–305–168	456,000	30,500	168
40–89–36	40,000	8900	36				
40–76–42	40,000	7600	42	640–305–120	640,000	30,500	120
40–89–42	40,000	8900	42	640–256–144	640,000	25,600	144
40–76–48	40,000	7600	48	640–305–144	640,000	30,500	144
				640–365–144	640,000	36,500	144
57–76–42	57,000	7600	42	640–305–168	640,000	30,500	168
57–89–42	57,000	8900	42	640–305–192	640,000	30,500	192
57–95–48	57,000	9500	48				
57–109–48	57,000	10,900	48	912–427–144	912,000	42,700	144
57–76–54	57,000	7,600	54	912–305–168	912,000	30,500	168
				912–365–168	912,000	36,500	168
80–109–48	80,000	10,900	48	912–305–192	912,000	30,500	192
80–133–48	80,000	13,300	48	912–427–192	912,000	42,700	192
80–119–54	80,000	11,900	54	912–470–240	912,000	47,000	240
80–133–54	80,000	13,300	54	912–427–216	912,000	42,700	216
80–119–64	80,000	11,900	64				
114–133–54	114,000	13,300	54	1280–427–168	1,280,000	42,700	168
114–143–64	114,000	14,300	64	1280–427–192	1,280,000	42,700	192
114–173–64	114,000	17,300	64	1280–427–216	1,280,000	42,700	216
114–143–74	114,000	14,300	74	1280–470–240	1,280,000	47,000	240
114–119–86	114,000	11,900	86	1280–470–300	1,280,000	47,000	300
160–173–64	160,000	17,300	64	1824–427–192	1,824,000	42,700	192
160–143–74	160,000	14,300	74	1824–427–216	1,824,000	42,700	216
160–173–74	160,000	17,300	74	1824–470–240	1,824,000	47,000	240
160–200–74	160,000	20,000	74	1824–470–300	1,824,000	47,000	300
160–173–86	160,000	17,300	86				
				2560–470–240	2,560,000	47,000	240
228–173–74	228,000	17,300	74	2560–470–300	2,560,000	47,000	300
228–200–74	228,000	20,000	74				
228–213–86	228,000	21,300	86	3648–470–240	3,648,000	47,000	240
228–246–86	228,000	24,600	86	3648–470–300	3,648,000	47,000	300
228–173–100	228,000	17,300	100				
228–213–120	228,000	21,300	120				

equations are all essentially of the form

$$HP_{pm} = \frac{q \times D}{PMF} \qquad [6.4.1]$$

where HP_{pm} = primer mover in hp
  q = oil and water daily production in bbls/d
  d = net lift of liquid in ft.
  PMF = prime mover factor

This prime mover factor-PMF consists of two components: a units conversion factor that accepts the other terms of equation 6.4.1 in the given units and converts the numerator product to horsepower; and a component that increases the horsepower to account for energy losses such as friction due to the cyclic nature of the prime mover load when applied to a pumping unit. The influence of these extra effects is not easily estimated. Thus an empirical adjustment is made based on experience. Another factor affecting motor size can be the trend towards longer stroke lengths on conventional beam type pumping units. The prime mover must be large enough to start the unit without the assistance of system inertia. Conventional beam pumping units with very long stroke lengths relative to their gearbox size, can exacerbate this initial loading to the point that although a given size motor is large enough to run the unit, it is not large enough to start it.

Sophisticated computer predictive programs mentioned later in this text, can, in many cases, more accurately predict motor loading for a given set of operating parameters.

If the lifted fluid has a specific gravity of 1.0 and is lifted with 100% efficiency, the PMF would be 135,735. The computed horsepower would be the theoretical or hydraulic horsepower. Because of the need for a prime mover rating considerably above the theoretical, various manufacturers use PMFs from 1.8 to 3.0 times smaller than 135,735. Curtis and showalter indicate that for high slip NEMA D motors or slow-speed internal combustion engines, the recommended prime mover horsepower can be obtained from the following equation:

$$HP_{pm} = \frac{q \times D}{56,000}$$ [6.4.2]

For high-speed internal combustion engines or normal slip NEMA C motors, the prime mover horsepower suggested is

$$HP_{pm} = \frac{q \times D}{45,000}$$ [6.4.3]

*High RPM and lower
NEMA rating increase
required power*

Example Problem
Determine the prime mover horsepower for lifting 100 bpd of oil and water production having a composite specific gravity of 1.0. The net lift is 10,000 ft. Assume a slow-speed internal combustion engine or a high slip NEMA D motor will be selected.

Solution
Use Equation 6.4.2, such that

$$HP_{pm} = \frac{\dfrac{100 \text{ bbl}}{d}(10,000 \text{ ft.})}{54,000} = 17.9 \text{ hp}$$

The actual engine or motor would be selected, with the nearest larger name-plate rating at 20 hp.

6.4.3 Sucker Rods
6.4.3.1 Types, Sizes and Grades
Sucker rods are solid, high-grade rods, which are run inside the tubing string to connect the pumping unit at the surface to the pump at the bottom of the well. There are two types of sucker rods specified by API 11B; steel sucker rods and fiber reinforced plastic (FRP) sucker rods. Steel sucker rods are 25 or 30 ft. in length and are available in diameters of 5/8, 3/4, 7/8, 1, and $1\frac{1}{8}$ in. FRP sucker rods are 25, 30, or 37.5 ft. in length and are available in diameters of 3/4, 7/8, 1, and $1\frac{1}{4}$ in. Sucker rods having lengths less than 25 ft. are called pony rods. Pony rods are usually manufactured in lengths of 2, 4, 6, 8 and 10 ft. Figure 6.4.3 shows the general shape and dimensions specified by API for steel sucker rod and pony rod ends; and Figure 6.4.4 shows the general dimensions and tolerances for FRP sucker rods and pony rods.

Steel sucker rods are available in three API grades, K, C, and D, which are suitable for most applications. API specifies the chemical composition for each grade. Most manufacturers also offer a high-strength grade not specified by API. FRP sucker rod ends are available in five API grades, A, B, K, C, and D.

Steel sucker rods are heat-treated, using various methods, in order to achieve the specified yield and tensile strengths. Today's heat-treating methods for sucker rods include normalize, normalize & temper, quench & temper, and case hardening. Although the, normalize & temper method is adequate for many applications, the quench & temper method offers some advantages over other methods which would be important in many applications. These advantages include a higher yield to tensile ratio, finer grain structure, and higher toughness.

Sucker rods are connected with an internally threaded coupling, which is specified by API 11B. Couplings are available in two classes, class T (or through hard) and class SM (or spray metal) and are available in three styles, sucker rod, polished rod, and sub. Sucker rod couplings have the same thread in each end and cannot be used on polished rods. Polished rod couplings have the same thread in each end. Subcouplings have different threads in each end and are used to connect different size rods. Subcouplings have polished rod threads and therefore can be used on sucker rods or polished rods. Table 6.4.2 and Figure 6.4.5 show the general dimensions of sucker rod coupling, polished rod couplings, and sub couplings.

6.4.3.2 Selection of API Steel Sucker Rods
The principal consideration in selection of API steel sucker rods are the range of stress, the level of stress that the rods will experience and the degree of corrosiveness of the environment in which they will operate. Rods should be selected to be able to withstand not only the maximum stress that they will experience but also the range of stress. This requires the use of the modified Goodman diagram described in "Allowable Rod Stress and Stress Range."

Chemical inhibition programs can be effective in overcoming corrosion. Refer to Chapter 7 for a discussion of corrosion mitigation procedures. No existing grade of rods will withstand all possible corrosion conditions and some treatment may be needed even for mild corrosion environments.

For nonsour (so-called sweet crudes) one should select the lowest grade rod which meets the stress and stress range conditions.

Where sulfide stress cracking exists, one should use a rod with a hardness less than 23 HRC. Grade C rods meet this requirement. The stress and stress range must be checked to see if the C rods also meet this stress criterion. D grade rods are not recommended for sour crude environments unless an effective corrosion inhibition program is applied.

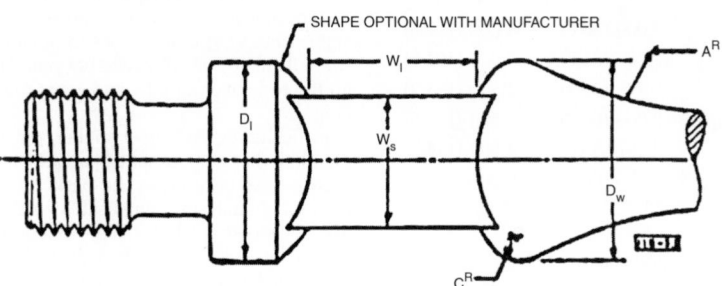

Figure 6.4.3 *General dimensions for steel sucker rod pin ends.*

Nominal Size of Rod Body +0.015 (+0.38)	Rod Pin Size	Nominal Diameter of Pin P	Outside Diameter of Pin Shoulder +0.005 −0.010 (+0.127 −0.0254)	Width of Wrench Square W_s	Length of Wrench Square W_1	Maximum Diameter of End Fitting D_E	Maximum Length of End Fitting L	Maximum Diameter of Extension X	Length[1] of Pin-and-Pin Sucker Rod +2.0(+51)	Length[1] of Pin-and-Pin Pony Rod +2.0(+51)
0.750 (19.05)	$\frac{5}{8}$	$\frac{15}{16}$ (23.81)	1.250 (31.75)	1 (25.40)	$1\frac{1}{4}$ (31.75)	D_1	(4)	(2)	296, 356, 446 (7518, 9042, 11,328)	32, 68, 104, 212 (813, 1727, 2642, 5385)
0.875 (22.23)	$\frac{3}{4}$ (19.05)	$1\frac{1}{16}$ (26.99)	1.500 (38.10)	$1\frac{3}{16}$ (33.34)	$1\frac{1}{4}$ (31.75)	D_1	(4)	(2)	296, 356, 446 (7518, 9042, 11,328)	32, 68, 104, 212 (813, 1727, 2642, 5385)
1.000 (25.40)	$\frac{7}{8}$ (22.23)	$1\frac{3}{16}$ (30.16)	1.625 (41.28)	$1\frac{3}{16}$ (33.34)	$1\frac{1}{4}$ (31.75)	D_1	(4)	(2)	296, 356, 446 (7518, 9042, 11,328)	32, 68, 104, 212 (813, 1727, 2642, 5385)
1.250 (31.75)	1 (25.40)	$1\frac{3}{8}$ (34.93)	2.00 (50.80)	$1\frac{1}{2}$ (38.10)	$1\frac{3}{8}$ (41.28)	D_1	(4)	(2)	296, 356, 446 (7518, 9042, 11,328)	32, 68, 104, 212 (813, 1727, 2642, 5385)

Notes :
All dimensions in inches (mm).
[1]Minimum length exclusive of fillet.
[2]The extension is that portion of the rod body or that portion of that end fitting which is immediately adjacent to the smaller end of the elevator taper. If this section of the end fitting is longer than 0.25 inch (6.3mm) the maximum outside diameter shall not be more than 0.200 inch (5.08) larger than the diameter of the road body. If this section of the end fitting is 0.25 inch (6.3mm)
[3]The length of sucker rods and peny rods shall be measured from contact face of pin shoulder to contact face of pin shoulder.
[4]Not to exceed 10 inches (254mm) exclusive of extension, if used.

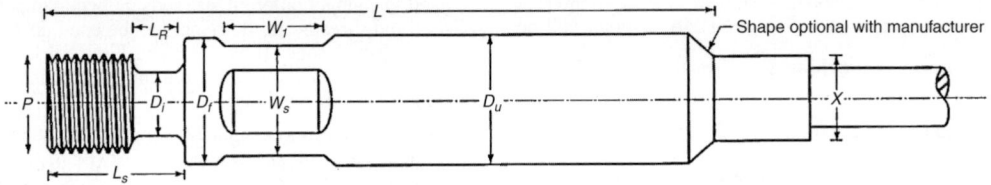

Notes:
See Table 6.
See Section 8 for details of shoulder connection.

Figure 6.4.4   *General dimensions for frp sucker rod pin ends [63].*

Table 6.4.2 *Dimensions for Couplings and Subcouplings*

Nominal Coupling Size[1]	Outside Diameter (W) +0.005 (+0.13) −0.010 (−0.25)	Length (N_L) +0.062 (+1.57) −0.000 (−0.00)
$\frac{5}{8}$ (15.9) S.H[2]	1.250 (31.8)	4.000 (101.6)
$\frac{5}{8}$ (15.9)	1.500 (38.1)	4.000 (101.6)
$\frac{3}{4}$ (19.1) S.H.	1.500 (38.1)	4.000 (101.6)
$\frac{3}{4}$ (19.1)	1.625 (41.3)	4.000 (101.6)
$\frac{7}{8}$ (22.2) S.H.	1.625 (41.3)	4.000 (101.6)
$\frac{7}{8}$ (22.2)	1.812 (46.0)	4.000 (101.6)
1 (25.4) S.H.	2.000 (50.8)	4.000 (101.6)
1 (25.4)	2.187 (55.6)	4.000 (101.6)
$1\frac{1}{8}$ (28.6) S.H.	2.375 (60.3)	4.500 (114.3)
1 (25.4) S.H. Sub Coupling	2.000 (50.8)	4.500 (114.3)
1 (25.4) Sub Coupling	2.187 (55.6)	4.500 (114.3)
$1\frac{1}{8}$ (28.6) Sub Coupling	2.375 (60.3)	5.000 (127.0)

Notes:
See Table 6.
See Section 8 for details of shoulder connection
All dimensions in in. (mm).
[1] Size of coupling is same as corresponding sucker rod size.
[2] S.H. is reduced outside diameter coupling known as slim hole.

Grade K rods are available when other grades have not performed satisfactorily.

6.4.3.3 Allowable Rod Stress and Stress Range

A string of sucker rods when in normal operation is subject to alternating high and low stress because of the nature of the pumping cycle. On the upstroke, the rods bear a load that includes their own weight, the weight of the fluid they are lifting, friction and effects of acceleration. On the downstroke, the rods carry some friction load and the load their own weight diminished by effects of deceleration. The ratio of upstroke to downstroke load and hence ratio of upstroke to downstroke stress can be 2 to 1 or often 3 to 1 or more. This cycle of alternating high and low stress occurs at a frequency at least equal to the pumping speed. A unit pumping at 20 strokes per minute goes from high stress to low stress every 3 s or 10,500,000 cycles per year. This process repeated on the rods over months and years can easily lead to metal fatigue.

In designing API steel sucker rod strings, it is recommended that the modified Goodman diagram shown in Figure 6.4.6 be used as the basis for stress analysis. This method of analysis compensates for the deleterious effects of cyclic stress and helps to prevent premature metal fatigue caused by cyclic stress. The key applicable terms for the diagram and the associated equation are given in the diagram. However, more explanation is given below for the terms of greatest interest. The diagram can be reduced to an equation

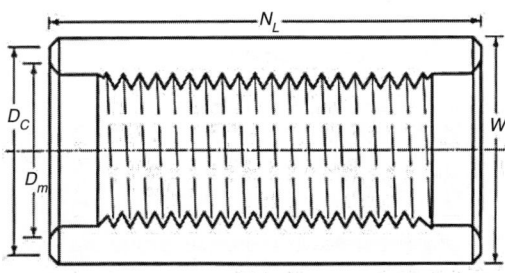

Sucker-rod coupling (do not use on polished rod)

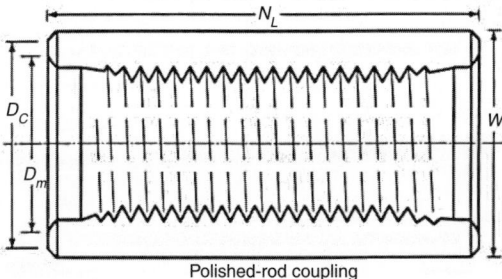

Polished-rod coupling

Subcoupling

Figure 6.4.5 *Sucker rod couplings, polished rod couplings and subcouplings [6].*

as follows:

$$S_A = (T/4 + 0.5625 \times S_{min}) \times SF \qquad [6.4.4]$$

The equation in the form above can be used in either English or metric units. An example is given below. The terms in the equation have the following significance:

S_A = Maximum allowable stress in the rods for a given value of minimum stress, S_{min} and service factor (see definitions below).

T = Minimum tensile stress for rods of a given API rating. Refer to Table 6-27 for these minimum tensile strength values. (Example: for C rods, $T = 90,000$ psi.)

S_{min} = Minimum stress to be experienced by the rods in the pump cycle.

SF = Service factor. This is a factor that adjusts, usually downward, the estimated allowable maximum stress to account for corrosive conditions. See Table 6.4.4 for suggested service factors for API grade C and D rods.

Equation 6.4.4 can be written in forms applicable to specific API rod grades as follows:

Grade C rods:

$$S_A = (22,500 + 0.5625 \times S_{min}) \times SF \qquad [6.4.5]$$

Grade D rods:

$$S_A = (28,750 + 0.5625 \times S_{min}) \times SF \qquad [6.4.6]$$

$$S_A,\ S_{min} = \text{lbf/in.}^2 \text{ or psi}$$

SF is dimensionless.

Example 6.4.1

Given : A 77 Grade D rod string operating in a saltwater-crude oil environment has a minimum polished rod load measured with a dynamometer of 15,000 lbf.

Desired : Maximum allowable stress and load for this rod string.

Solution

Step 1: A 77-rod string has a cross-section area of 0.601 in.^{2} (See Table 6.4.2). Use this area to compute rod stress from rod load.

$$S_{min} = \frac{\text{Min. load}}{\text{Rod area}} = \frac{15,000 \text{ lbf}}{0.601 \text{ in.}^2} = 24,960 \text{ psi}$$

Step 2: Determine the service factor. From Table 6.4.4, for a grade D rod, the service factor for saltwater (brine) service is recommended to be 0.90.

Step 3: Use the Goodman diagram equation to compute S_A. Use Equation 6.4.6 for grade D rods.

$$S_A = (28,750 + 0.5625 \times S_{min}) \times SF$$

$$= (28,750 + 0.5625 \times 24,960) \times 0.90$$

$$= 38,510 \text{ psi}$$

Step 4: Compute maximum allowable loads by multiplying S_A by rod area.

$$\text{Max allowable load} = S_A \times \text{rod area}$$

$$38,510 \text{ psi} \times 0.601 \text{ in}^2 = 23,140 \text{ lbf}$$

Service factors should be considered as guidelines and not as highly precise universal parameters. Experience from a given area should be used to determine proper service factors for given types of corrosive environments.

6.4.4 Reciprocating Rod Pumps
6.4.4.1 The Sucker Rod Pump
Sucker rod pumps, also called "plunger" pumps, consist of the following basic components as shown in Figure 6.4.7. A cylinder or barrel, a piston or plunger, a traveling valve that is attached to the plunger, a standing valve that is attached to the bottom of the barrel, and a device for locking the pump in place called an anchor or hold-down.

The pump is placed at the bottom of the well at a depth that allows the pump to remain covered by fluid during the pumping operation. The pumping cycle starts with an upward stroke of the sucker rods, in which the plunger and traveling valve are pulled up through the barrel. The traveling valve closes, the standing valve opens and fluid enters into the barrel to occupy the space displaced by the upward moving plunger. During this motion, the fluid on top of the plunger is forced upward in the tubing towards the surface. On the downward stroke of the plunger, the traveling valve opens and the standing valve closes trapping the fluid above it. The downward motion of the plunger through this trapped fluid forces it to pass through the open traveling valve to the top of the plunger. With each stroke of the pump, more fluid is added to the tubing until the column of fluid reaches the surface and flows into the flow line.

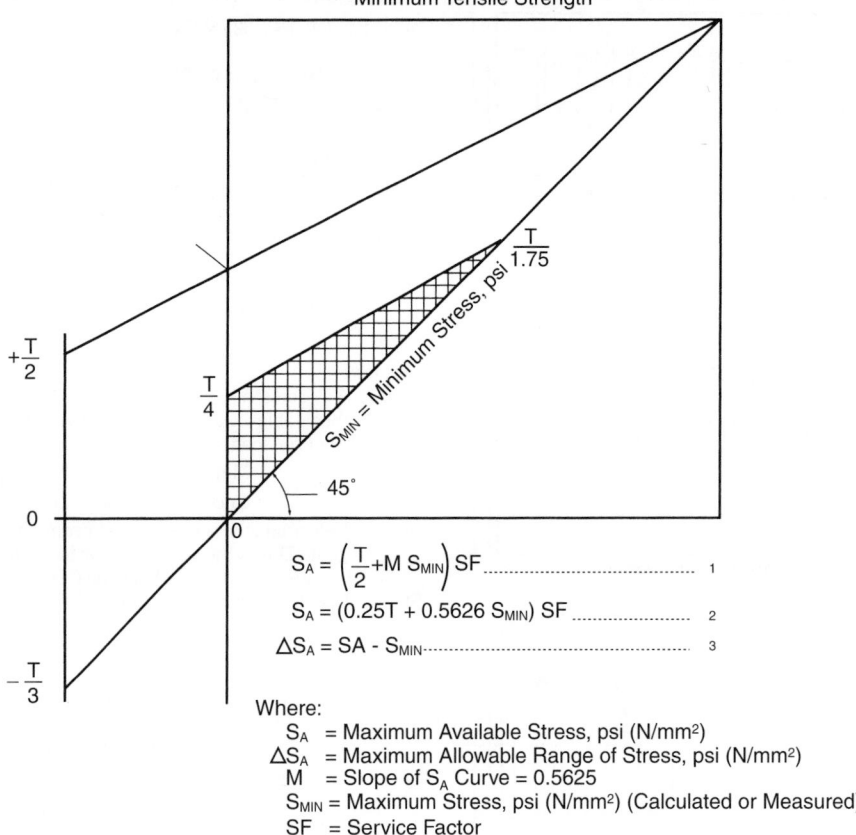

Minimum Tensile Strength

$$S_A = \left(\frac{T}{2} + M\, S_{MIN}\right) SF \dotfill 1$$

$$S_A = (0.25T + 0.5626\, S_{MIN})\, SF \dotfill 2$$

$$\Delta S_A = SA - S_{MIN} \dotfill 3$$

Where:
S_A = Maximum Available Stress, psi (N/mm²)
ΔS_A = Maximum Allowable Range of Stress, psi (N/mm²)
M = Slope of S_A Curve = 0.5625
S_{MIN} = Maximum Stress, psi (N/mm²) (Calculated or Measured)
SF = Service Factor
T = Minimum Tensile Strength, pso(N/mm²)

Figure 6.4.6 *Modified Goodman diagram for allowable stress and range of stress in noncorrosive service [7].*

Sucker rod pump are generally classified as tubing pump and insert or rod pumps. The major difference between the tubing pump and rod pump is that the rod pump can easily be pulled for repairs, while the tubing pump barrel is an

Table 6.4.3 *Chemical and Mechanical Properties of Sucker Rods*

API Grade	Chemical Composition
K	AISI 46XX Series Steel
C	AISI 10XX Series Steel
	AISI 15XX Series Steel
D Carbon	AISI 10XX Series Steel
	AISI 15XX Series Steel
D Alloy	AISI 41XX Series Steel
D Special	Special*

*Special alloy shall be any chemical composition that contains a combination of nickel, chromium, and molybdenum that total a minimum of 1.15 percent alloying content.

API Grade	Minimum Yield 0.2% Offset psi (Mpa)	Minimum Tensile psi (Mpa)	Maximum Tensile psi (Mpa)
K	60,000 (414)	90,000 (620)	115,000 (793)
C	60,000 (414)	90,000 (620)	115,000 (793)
D	85,000 (586)	115,000 (793)	140,000 (965)

Table 6.4.4 *Suggested Service Factors for API Grade C and Grade D Rods in Corrosive Conditions [2]*

Service	API C	API D
Noncorrosive	1.00	1.00
Salt water	0.90	0.65
Hydrogen sulfide	0.70	0.50

integral part of the tubing string which must be pulled for barrel repair or replacement.

All sucker rod pumps require a device to hold them in place in the well. This device is the anchor or hold-down. There are two standard types of hold-downs, cup type and mechanical type. These hold-downs may be found on top or on bottom of the pump depending on the type of pump and the requirements. Some special pumps have a hold-down in the center of the pump, while others may have a hold-down at the top and at the bottom.

The choice between a tubing pump and a rod pump for a given installation is an important one. A comparison of the advantages and disadvantages for rod and tubing pumps as well as top and bottom hold-downs is presented in Table 6.4.5. Additional information is available in reference [8].

6.4.4.2 Pump Designations
The API presents a standardized notation, shown in Figure 6.4.8, for designating sucker rod pumps.

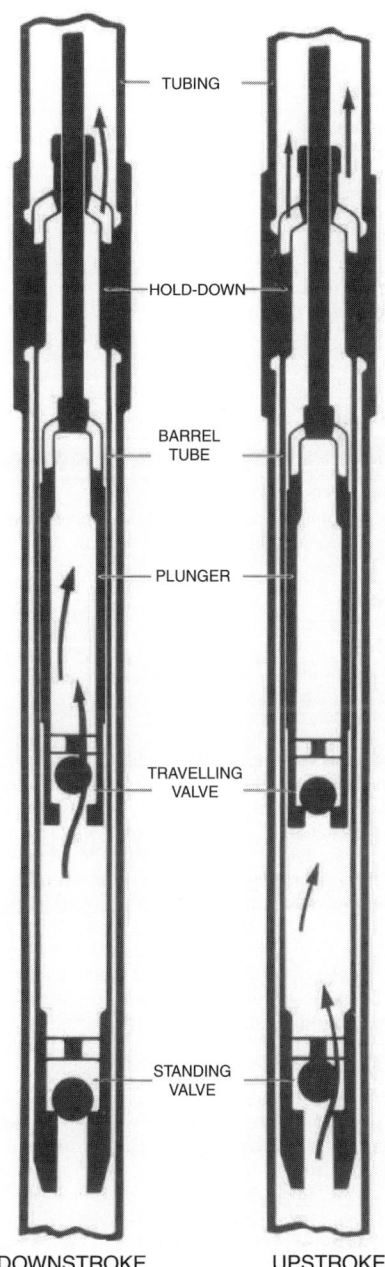

TUBING

HOLD-DOWN

BARREL
TUBE

PLUNGER

TRAVELLING
VALVE

STANDING
VALVE

DOWNSTROKE   UPSTROKE

Figure 6.4.7 *Sucker rod pump.*

6.4.4.3 Tubing Pumps

Because of their rugged design, the tubing pump is considered the heavy-duty pump of the industry. Tubing pumps have a great production capacity than any API insert pump for the same tubing size. Their inherent design, however, makes them more difficult to install and retrieve than insert pumps and their capacity for large fluid volumes makes them impractical for deep wells. The weight of the large volume fluid column is too much for the strength limits of existing sucker rods.

The barrel of the tubing pump is attached directly to the tubing at or near the lower end of the tubing. A seating nipple is placed below the pump barrel to accept the hold-down and seat the pump. After the pump barrel is in place, the standing valve assembly and plunger assembly is run into the well. The lower end of the plunger assembly has a standing valve puller to which the standing valve is attached. During the pump installation process, when the standing valve assembly reaches the seating nipple the standing valve is locked in place with the hold-down. The rods are rotated counterclockwise to disconnect the standing valve puller from the standing valve. The plunger is then raised sufficiently to allow minimum clearance on the down-stroke. This spacing must be adjusted after initial pumping begins to account for the fluid load causing additional rod stretch.

6.4.4.4 Rod Pumps

A rod pump has the same basic parts as the tubing pump, but the barrel is not an integral part of the tubing as in the tubing pump; therefore, all the components of the pump can be pulled from the tubing using the rod string.

The rod pump is more versatile than the tubing pump. It has a much higher compression ratio when spaced properly, making it a better choice for gassy wells. The rod pump can be used at much greater depths than the tubing pump, but is not capable of pumping the large capacities. Rod pumps can be equipped with top or bottom hold-downs or both for a variety of applications.

6.4.5 Selection of Well Equipment
6.4.5.1 Pump Submergence

The energy to fill the pump should be supplied by the formation rather than by a high fluid head in the casing tubing annulus if maximum production is desired. Where feasible, pump intake should be below the perforations or as close above them as possible.

6.4.5.2 Gas Separation

Gas evolution within the pump barrel or below the standing valve can cause significant reduction in pump volumetric efficiency. Where gas evolution is a problem, some type of downhole gas separator (commonly called a *gas anchor*) should be used.

6.4.5.3 Sand Control

Where sand production exists, some method of sand control, such as gravel packs, screens, or formation bonding agents, should be used to maximize pump life.

6.4.5.4 Common Pump Problems and Solutions

Several frequently occurring operating problems and their solutions are discussed in Reference [8] and are summarized here.

6.4.5.5 Corrosion

The principal corroding agents in pumping wells are carbon dioxide, hydrogen sulfide, oxygen and brine, either singly or in combination. The failure mechanisms that these agents can create are sulfide stress cracking, corrosion fatigue, erosion corrosion, stress corrosion, galvanic corrosion, pitting and wear abrasion.

When corrosive conditions exist, the metallurgy of the pump and downhole accessories should be chosen to resist the corrosion. Chemical inhibitors can be helpful but will not prevent corrosion in all cases. Such inhibitors protect casing, tubing, and rods better than the pump parts. Thus, corrosion resistance here is best obtained through proper selection of pump materials for construction. Refer to Chapter 7 for more details on corrosion mitigation.

Table 6.4.5 *Comparison of Pump Types and Anchor Types*

Tubing Pumps	
Advantages	Disadvantages
• Largest displacement • Strongests	• Must pull tubing to remove pump • Poor for gassy wells • Large rod load

Rod Pumps Traveling Barrel, Bottom Anchor	
Advantages	Disadvantages
• Sand kept in motion near barrel • Good for intermittent pumpers • Strong plunger-rod connection • Open cages allow low pressure drop • Low pressure drop across barrel	• Less desirable for low static level wells • Long pull tubes can become bowed

Stationary Barrel, Bottom Anchor	
Advantages	Disadvantages
• Generally suited for deep wells • Useful for low static level wells • Useful for gassy wells	• Hazardous to use in sandy wells • Not recommended for intermittently produced wells

Stationary Barrel, Top Anchor	
Advantages	Disadvantages
• Useful for sandy wells • Useful for some low fluid level gassy wells	• Thin wall pump barrels can burst in deep wells

6.4.5.6 Fluid Pound

A fluid condition exists when insufficient liquid enters the pump barrel on the pump upstroke. Above the liquid, a gas space is created between the traveling valve and the standing valve. During the next downstroke, the traveling valve does not open until the plunger hits the liquid. This impact sends a strong shock wave up the rods to the surface. In some instances the shock can be carried to the gear reducer of the beam pumping unit. When this pounding or shock is repeated with each pump cycle and for significant periods of time, major structural damage can occur in the downhole and surface equipment. This damage can include:

a. gear tooth failure in pumping units
b. damage to the pumping unit base structure
c. rod fatigue and failure
d. accelerated wear on traveling valve cages, balls and seats
e. accelerated wear on tubing threaded connections

Fluid pound can be caused by one of two conditions: (1) a "pumpset-off" condition of the well in which the well cannot produce as much fluid as the pump can lift to the surface, and (2) restricted intake to the pump because of mechanical blockage at the pump.

The pumped-off condition can be prevented by slowing down the pump, using a smaller size pump or shorter stroke length, or putting the pump on a percentage timer if the prime mover is electric. Various pump-off controllers are available to sense when a well is nearing a pumped-off condition. The controllers then shut off the pump for a predetermined period.

If the fluid pound is caused by restricted fluid entry, none of the above alternative remedies are appropriate. Only proper pump servicing will remove the restriction.

6.4.5.7 Gas Pound

Gas pound has a behavior and observed effect somewhat similar to that of fluid pound but has a different cause. It occurs when no "pumped-off" condition exists. There may or may not be an intake flow restriction. The cause of gas pound is occurrence of excessive gas in the pump barrel on the pump upstroke. It can be caused by inadequate downhole gas separation or gas breakout due to oil flow through flow restrictions in the standing valve or dip tube. A gassy crude can evolve significant gas if it is forced at high velocity through a flow restriction. If the gas pound is caused by excessive free gas and there is ample flow area in the standing valve port and cages, then a better gas anchor is needed. If flow restrictions exist, then more open cages and large-diameter dip tubes should be tried.

6.4.5.8 Sand Accumulation

Although proper choice of metallurgy for downhole pump barrels, plungers, valves and cages will help in alleviating a sand accumulation problem, sand control methods should be evaluated also.

6.4.5.9 Scale Formation

Chemical scale deposits can occur in some producing wells, and in extreme cases be so severe as to seriously restrict flow of oil through the pump and tubing. Chemical scale inhibition agents are available that will either eliminate or significantly reduce the effect of scale deposition. Because scale deposition is a cumulative effect, treatment must follow a fixed and uninterrupted treatment regime. Refer to Chapter 7 for a complete discussion of the causes and control procedures for scale deposition.

6.4.5.10 Computation Procedure

Selection of the proper equipment for a new well follows, an orderly sequence based on computation of pumping unit performance. The method suggested here is that recommended by API [9]. Computer programs are available to perform the calculations described here for manual use. The API RP11L calculation procedure is intended for design of conventional

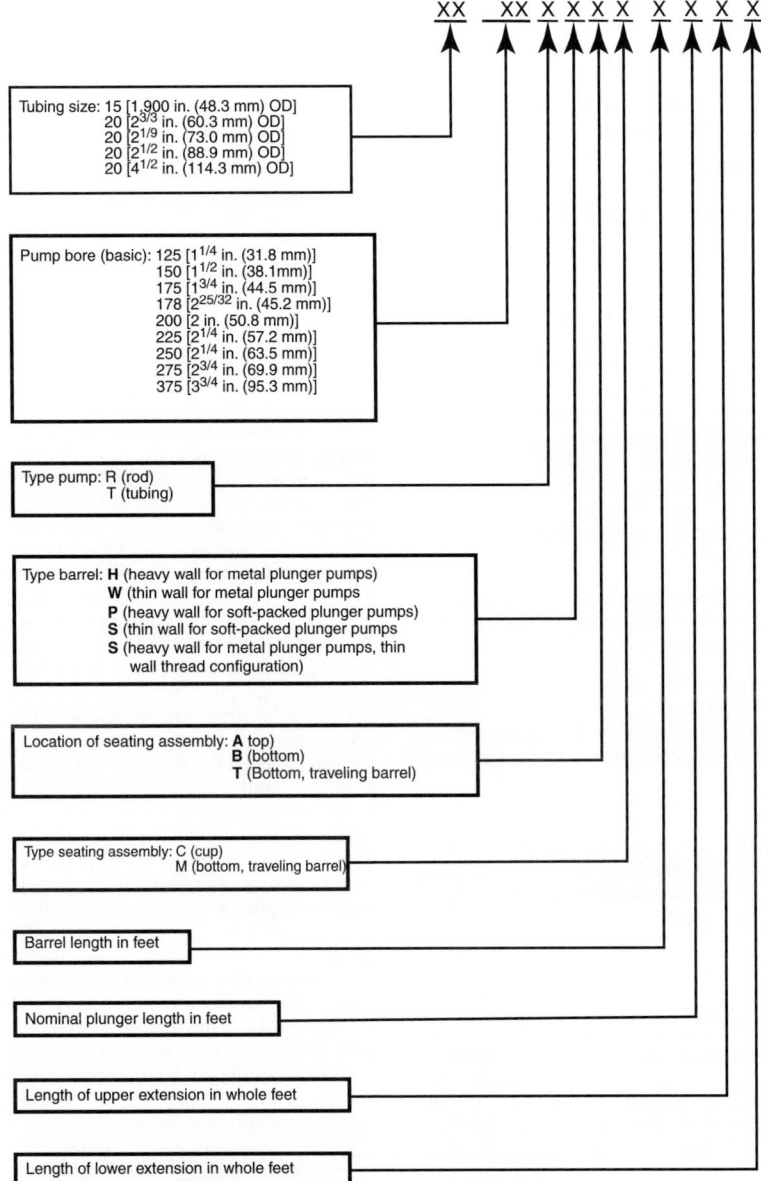

Figure 6.4.8 *API pump designations [8].*

geometry (class I) units pumping in non-deviated wells at depths of less than 12,000 ft. The method is based on use of API grade steel sucker rods.

Various commercially available computer programs, such as Rodstar and S-rod, make complete calculations for a wider variety of pumping unit types (class I, class I with Phased crank, and class III units) rod types (steel rods, continuous rod, and fiberglass rods) depth and inclination of the hole. Some of these programs bases their design on dynamic analysis of the elastic behavior of the rods and tubing and can include inertial effects of the surface equipment. With correct input as to well conditions and equipment characteristics, these wave equation programs can make a very precise estimate of loads and stresses in the pumping system and lead to an optimal design.

6.4.5.11 Required Information
Before the API RP11L calculation procedure can be started, the following information must be available for the well.

a. expected oil and water production, in bbl/d, and their specific gravities
b. casing and tubing sizes
c. anchoring depth (if any) for tubing
d. pump setting depth
e. fluid level during production (when pumping)

The design considerations in proper selection of each of these items are discussed separately in the following sections.

6.4.5.12 Expected Fluid Production

Expected production depends upon how much the pump in the well can lift and on how much oil the formation can produce. The volume of oil that can theoretically be displaced or moved by the pump is called the *pump displacement* PD and is given by the following equation:

$$PD = A_p \times S_p \times N \times Con \qquad [6.4.7]$$

The volume of oil actually produced at the surface is given by

$$Q = E_v \times PD \qquad [6.4.8]$$

where

PD = theoretical displacement of the pump in bpd
A_p = cross section area of plunger in in.^{2}
S_p = effective stroke length of the plunger in in.
N = pumping speed in strokes/min
Con = a units conversion constant, 0.1484
Q = rate of oil and/or water produced at the surface in bpd
E_v = volumetric efficiency of pump displacement process (a fraction)

To simplify the calculations, the terms A_p and Con are frequently lumped together to give a pump constant, K. Pump displacement is then computed from

$$PD = K \times S_p \times N \qquad [6.4.9]$$

Table 6.4.6 gives pump constants K for various sizes of plungers and can be used to compute pump displacement.

Example 6.4.2

Given: A $1\frac{1}{2}$-in. plunger is being used to pump oil at 16 strokes per minute and an effective downhole stroke of 51.5 in. The pump volumetric efficiency is estimated to be 70%.

Desired: Daily surface production of oil, bpd.

Table 6.4.6 *Pump Factors or Constants [57]*

Plunger diam in.	Area of plunger sq. in. (A_p)	Pump Constant (K)
$\frac{5}{8}$	0.307	0.046
$\frac{3}{4}$	0.442	0.066
$\frac{15}{16}$	0.690	0.102
*1	0.785	0.117
*$1\frac{1}{16}$	0.886	0.132
$1\frac{1}{8}$	0.994	0.148
*$1\frac{1}{4}$	1.227	0.182
*$1\frac{1}{2}$	1.767	0.262
*$1\frac{3}{4}$	2.405	0.357
*$1\frac{25}{32}$	2.488	0.370
*2	3.142	0.466
*$2\frac{1}{4}$	3.976	0.590
*$2\frac{1}{2}$	4.909	0.728
*$2\frac{3}{4}$	5.940	0.881
*$3\frac{3}{4}$	11.045	1.640
*$4\frac{3}{4}$	17.721	2.630

*API sizes

Solution

Step 1. Using data from Table 6.4.6 and Equation 6.4.7 we obtain pump displacement:

$$PD = A_p \times S_p \times N \times Con$$

$$= (1.767 \text{ in.}^2) \times (51.5 \text{ in.}) \times (16 \text{ spm}) \times 0.1484$$

$$= 216 \text{ bpd}$$

Step 2. Determine surface production from Equation 6.4.8.

$$Q = E_v \times PD$$

$$= 0.70 \times 216 \text{ bpd} = 151 \text{ bpd}$$

An alternate solution using pump constants is

$$PD = K \times S_p \times N$$

$$= (0.262)(51.5)(16) = 216 \text{ bpd}$$

Volumetric efficiency Ev is normally between 70% and 100%. In unusual circumstances where a well is partially flowing while pumping, the Ev can be above 100% (flumping wells). Several factors contribute to decrease Ev below 100%. These include slippage past the plunger on the upstroke and gas evolution from the oil as the oil enters the pump barrel and as gas evolves from the oil as it rises up the tubing. This latter effect is accounted for by the oil formation volume factor. An oil with a formation volume factor of 1.15 requires 1.15 barrels of oil at reservoir condition to yield 1.0 barrels at the surface. This factor, alone, without any pump slippage would give an Ev of $(1/1.15) \times 100 = 87\%$.

The pump can displace the computed pump displacement PD only if the formation can produce this much oil. Optimal performance is obtained when the pump production at the surface matches the ability of the formation to produce, or when the pump produces at the statutory limit on production or the limit determined by prudent reservoir operation.

The estimate of the formation's ability to produce should be based on one or more well tests whenever possible. A productivity index or inflow performance relation should be used to determine the volume of fluid which the formation can produce under various conditions of flowing bottom hole pressure. A method widely used for solution gas drive fields producing oil, gas, and only a small amount of water, or none is based on Vogel's curve [58]. The curve is reproduced here as Figure 6.4.9. This method needs only a single a single point production test, wherein reservoir shut in pressure is measured and total liquid production (oil and water) and flowing bottomhole pressure are measured. The equation for this curve is given below as

$$\frac{q_0}{(q_0)_{max}} = 1 - 0.2\frac{(Pwf)}{(Pr)} - 0.8\frac{(Pwf)^2}{(Pr)^2} \qquad [6.4.10]$$

where q_0 = total liquid production per unit of time at flowing bottomhole pressure Pwf
$(q_0)_{max}$ = maximum possible flow per unit of time $Pwf = 0$
Pwf = flowing bottomhole pressure
Pr = average reservoir pressure

The equation is used to compute the expected flowrate q_0 at any given flowing bottomhole pressure Pwf. Values of Pr and $(q_0)_{max}$ are constants that must be determined from the flow test. Pr is measured during the flow test and $(q_0)_{max}$ is computed using Equation 6.4.10 and the flowrate data.

An example calculation is given below:

Given: A reservoir flow test indicated that the well flowed 100 bpd when the flowing bottomhole pressure was 1500 psia and the reservoir average pressure was 2000 psi.

Desired: Determine the flowrate in bp/d when the flowing bottomhole pressure is 250 psia.

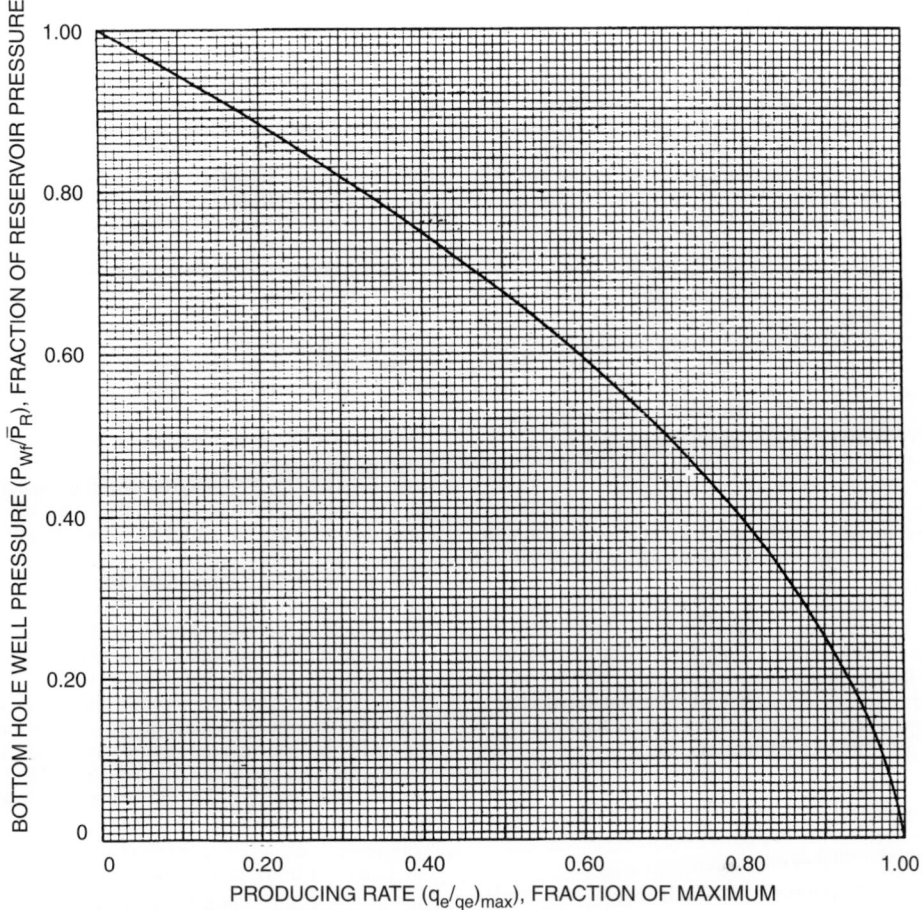

Figure 6.4.9 *Inflow performance relationships for solution gas drive reservoirs [10].*

Solution
Step 1: Compute $(q_0)_{max}$ using Equation 6.4.10, such that

$$\frac{(q_0)}{(q_0)_{max}} = 1 - 0.2\frac{(Pwf)}{(Pr)} - 0.8\frac{(Pwf)^2}{(Pr)^2}$$

$$\frac{100}{(q_0)_{max}} = 1 - 0.2\frac{(1500)}{(2000)} - 0.8\frac{(1500)^2}{(2000)^2}$$

$$(q_0)_{max} = 250 \text{ bbl/d}$$

Step 2. Use Pr and $(q_0)_{max}$ with desired Pwf in Equation 6.4.10 to compute q_0, such that

$$\frac{q_0}{250} = 1 - 0.2\frac{(250)}{(2000)} - 0.8\frac{(250)^2}{(2000)^2}$$

$$q_0 = 241 \text{ bpd}$$

Estimation of water production should be done in company with determination of the oil production.

Where the size of the selected pumping unit is critical or where little data are available, it may be desirable to use a test pumping unit to determine not only flow capacity but also working fluid levels.

6.4.5.13 Casing and Tubing Size and Tubing Anchor
Casing size is usually determined long before the decision is made to install a beam pumping unit. Frequently, tubing is also already in place. However, its diameter and weight and weight per foot should be known. If tubing is already in place, it is not cost effective to change out the tubing for a large size as a means of installing a larger pump. Where a pump is needed that is larger than possible with the existing tubing, a casing pump should be investigated.

Tubing anchors have the advantage of keeping the tubing from buckling and oscillating during the pump cycle. Not only does this reduce wear on the tubing, but, as will be seen in the API RP11L calculations, anchors increase the net effective bottomhole stroke over that for unanchored tubing. Anchors can create problems in wells with severe scale, sand, or corrosion conditions as these conditions may so damage the anchor that it cannot be released to allow tubing removal.

6.4.5.14 Pump Setting Depth
Whenever possible, the pump should be set at a depth in the well where pump intake pressure is above the bubble point of the oil as it is produced. This would suggest locating the pump one to three tubing joints below the perforations. Sometimes this is not possible because of total well depth,

nor desirable because of sand production. Alternatively, the pump should be placed out of and above the turbulence zone near the perforations (three to six joints above the perforations). In any case, the pump should have significant submergence, 50 to 100 ft of fluid, to assist in rapid fluid fillage of the pump barrel. For low productivity wells, this amount of submergence may not be either possible or economically feasible but the consequence will be that wells with lower submergence may experience pump-off and some type of pump-off controller would be needed.

6.4.5.15 Expected Pump Fluid Level
This is one of the needed parameters about which there will be much uncertainty in a new installation. If a well test and an inflow performance relationship is available for the well, one can possibly convert flowing bottomhole pressure to flowing liquid height. Such a calculation would be fairly accurate for nearly dead oils or for wells producing mostly water. An alternative, conservative design approach is to assume that the flowing liquid level is at the pump depth. Once a unit is in operation, one of the commercially available acoustical well sounders can be used to measure the working fluid level. This value would be useful in any comparison of the well's current performance with that predicted by the API RP11L procedure.

6.4.5.16 Preliminary Design Data
In addition to selecting the data listed in the section titled "Required Information" which items are generally constant for a given pumping installations, the user of the API RP11L calculation procedure must select initial or tentative values for each of the following four parameters:

a. plunger diameter
b. pumping speed
c. stroke length
d. sucker rod string design

The following paragraphs give guidelines for proper selection of these parameters.

6.4.5.17 Plunger Diameter
For a given size tubing there is a maximum pump size for each type of pump. Refer to Table 6.4.7. An additional guide to plunger size selection for a given fluid volume and pumping depth is found in Table 6.4.8.

6.4.5.18 Pumping Speed
For any given selection of stroke length there is a maximum recommended pumping speed. If this speed is exceeded,

the rods are likely to "float" or go into compression on the downstroke. Rods are normally under tension on the downstroke as well as the upstroke. If they are put into compression, they will buckle, causing wear on rods and tubing. The range of stress will be large, thus reducing the Goodman diagram maximum allowable stress. The alternating tension and compression will cause severe stress on rod threads and coupling and accelerate rod parting from fatigue and accelerated effects of corrosion. Dynamometer tests (to be discussed later) can detect floating rods. The maximum pumping speed for given stroke lengths is given in Figures 6.4.10, 6.4.11, and 6.4.12. These figures are for three widely used types of pumping units.

6.4.5.19 Stroke Length
Stroke length is a primary variable in determining pumping unit size. Because of the small number of stroke lengths (usually two to four) available for any pumping unit, it is wise to select an available stroke length, then use this length in Equation 6.4.7 to determine an appropriate first estimate of strokes per minute. The selected stroke and speed combination should be checked to see that the rods do not float. Use Figure 6.4.10, 6.4.11, or 6.4.12, whichever is appropriate.

6.4.5.20 Sucker Rod String Selection
For shallow wells, less than 2000 ft most pumping installations will use sucker rods of the same diameter from the top of the well to the pump. Since the load on the rods is as its greatest at the top of the rod string on the upstroke, for single size rod strings one needs only to check the stress at this point using the Goodman diagram to see if the rods are satisfactory. For deeper wells, one usually uses a larger-diameter rod near the surface and smaller-diameter rods further down the well. Such multiple-size rod strings are called *tapered rods*. A coding system for designating the sizes has been adopted. A 76-rod string consists of $\frac{7}{8}$-in. rods near the top and $\frac{6}{8}$- (or $\frac{3}{4}$-) in. rods near the bottom. An 88-rod string has 1-in. (i.e., eight $\frac{1}{8}$-s) rods throughout its length. A 75-rod string has $\frac{7}{8}$-in. rods near the top, $\frac{6}{8}$- (or $\frac{3}{4}$-) in. rods near the middle and $\frac{5}{8}$-in. rods near the plunger. This is called a three-way tapered rod string. Four-way tapers are also possible. This coding scheme does not specify what percent of each rod size is used in the string.

Two approaches have been used in determining the percent of each rod size. In the first, one starts at the plunger depth and adds rods of the smallest size (for example $\frac{5}{8}$-in. rods in a 75-rod string design). These are added until the stress at the top of these rods is the maximum allowable according to some criterion such as the Goodman diagram. At that point, one changes to the next large size ($\frac{3}{4}$-in. rods in the current 75-rod string example). One then continues to add rods until the stress at the top of these rods meets the same maximum stress criterion. At this point, one then changes to the largest size and adds rods until either the top of the hole is reached (a successful design) or until the maximum stress criterion is met but the top of the hole has not been reached. In this latter case, one does not have an acceptable design. One would need either to change to a larger-diameter string such as an 86-rod design or go to a four-way taper (85-rod string in the present instance), adding 1-in. rods to the top of the string.

The above maximum stress method leads to the lightest rod string. It is not as conservative as the next or balanced stress method. In the balanced stress method, the percentages of each size rod in the 75 rod combination, for example, are adjusted until the same stress exists at the top of each size of rods. Usually this stress is less than the maximum allowable stress. The rod string selected by this method is

Table 6.4.7 *Maximum Pump Size and Type [2]*

Pump type	Tubing size, in.			
	1.900	$2\frac{3}{8}$	$2\frac{7}{8}$	$3\frac{1}{2}$
Tubing one-piece, thin-wall barrel (tw)	$1\frac{1}{2}$	$1\frac{3}{4}$	$2\frac{1}{4}$	$2\frac{3}{4}$
Tubing one-piece, heavy-wall barrel (th)	$1\frac{1}{2}$	$1\frac{3}{4}$	$2\frac{1}{4}$	$2\frac{3}{4}$
Tubing linear barrel (tl)	—	$1\frac{3}{4}$	$2\frac{1}{4}$	$2\frac{3}{4}$
Rod one-piece, thin-wall barrel (rw)	$1\frac{1}{2}$	$1\frac{1}{2}$	2	$2\frac{1}{2}$
Rod one-piece, heavy wall barrel (rh)	$1\frac{1}{16}$	$1\frac{1}{4}$	$1\frac{3}{4}$	$2\frac{1}{4}$
Rod liner barrel (rl)	—	$1\frac{1}{4}$	$1\frac{3}{4}$	$2\frac{1}{4}$

Table 6.4.8 *Pump Plunger Sizes Recommended for Optimum Conditions [67]*

Net lift of fluid ft.	Fluid production—Barrels per day—80 pct efficiency									
	100	200	300	400	500	600	700	800	900	1000
2000	$1\frac{1}{2}$	$1\frac{3}{4}$	2	$2\frac{1}{4}$	$2\frac{1}{2}$	$2\frac{3}{4}$	$2\frac{3}{4}$	$2\frac{3}{4}$	$2\frac{3}{4}$	$2\frac{3}{4}$
	$1\frac{1}{4}$	$1\frac{1}{2}$	$1\frac{3}{4}$	2	$2\frac{1}{4}$	$2\frac{1}{2}$				
3000	$1\frac{1}{2}$	$1\frac{3}{4}$	2	$2\frac{1}{4}$	$2\frac{1}{2}$	$2\frac{1}{2}$	$2\frac{3}{4}$	$2\frac{3}{4}$	$2\frac{3}{4}$	$2\frac{3}{4}$
	$1\frac{1}{4}$	$1\frac{1}{2}$	$1\frac{3}{4}$	2	$2\frac{1}{4}$	$2\frac{1}{4}$	$2\frac{1}{2}$			
4000	$1\frac{1}{4}$	$1\frac{3}{4}$	2	$2\frac{1}{4}$	$2\frac{1}{4}$	$2\frac{1}{4}$	$2\frac{1}{4}$	$2\frac{1}{4}$		
		$1\frac{1}{2}$	$1\frac{3}{4}$	2	2					
5000	$1\frac{1}{4}$	$1\frac{3}{4}$	2	2	$2\frac{1}{4}$	$2\frac{1}{4}$				
		$1\frac{1}{2}$	$1\frac{3}{4}$	$1\frac{3}{4}$	2					
6000	$1\frac{1}{4}$	$1\frac{1}{2}$	$1\frac{3}{4}$	$1\frac{3}{4}$						
		$1\frac{1}{4}$	$1\frac{1}{2}$							
7000	$1\frac{1}{4}$	$1\frac{1}{2}$			In this tabulation surface pumping strokes up to 74 in. only are considered.					
	$1\frac{1}{8}$	$1\frac{1}{4}$								
8000	$1\frac{1}{4}$									
	$1\frac{1}{8}$									

somewhat heavier than the selected by the first method and is generally to be preferred.

The API, using this latter approach, has given rod size and percent length of rods for available rod combinations and plunger diameter. These percentages appear in columns 6 through 11 of Table 6.4.9. For example, a 76-rod string using a 1.50-in. plunger would have 33.8% $\frac{7}{8}$-in. rods and 66.2% $\frac{3}{4}$-in. rods. An 85-rod string with 1.75-in. plunger would have 29.6%, 30.4% 29.5% and 10.5% of 1-in., $\frac{7}{8}$-in., $\frac{3}{4}$-in. and $\frac{5}{8}$-in. rods, respectively.

6.4.5.21 API RP11L Calculations

The calculation sequence to design a pumping unit installation is best explained through an example calculation. Assume that the following data are available for a pumping well.

1. desired surface production, 150 bpd
2. estimated pumping volumetric efficiency, 85%
3. pump setting depth, 5000 ft
4. fluid level during pumping, 4500 ft
5. tubing is 2 in. nominal, not anchored
6. pumping speed, 16 spm
7. stroke length at surface, 54 in.
8. plunger diameter, 1.50 in.
9. fluid specific gravity, 0.90
10. sucker rod design (API)

Desired calculated results are:

1. effective downhole stroke, in.
2. pump displacement, bpd
3. peak polished rod load, lbf

4. minimum polished rod load, lbf
5. peak torque, lbf-in
6. polished rod horsepower, hp
7. counterbalance effect, lbf measured at polished rod

The calculation sequence is shown in Figure 6.4.13. Blank calculation sheets similar to Figure 6.4.13 can be obtained from API. The stepwise sequence for performing the calculations is described below.

Step 1. Record on the calculation sheet the factors Wr, E_r, F_0, from Table 6.4.9 and E_t from Table 6.4.9 for the selected rod design (calculation lines 1–4).

Step 2. Using the data recorded in Step 1 and the given data, compute the nondimensional parameters F_0/Sk_r, N/N_0, N/N_0, and the term $1/k_t$ (calculation lines 5–11).

Step 3. Compute the effective stroke length S_p and the pump displacement PD (calculation lines 12–14). For line 12, one needs to get S_p/S from Figure 6.4.14 using the previously computed values for F_0/Sk_r and N/N_0.

Step 4. Compare the computed pump displacement with that desired. The calculated value of PD is 175 bbl/d. Use Equation 6.4.8 to compute Q and compare it to the desired Q:

$$Q = EV \times PD$$

$$= (0.85) \times 175 \text{ bpd} = 149 \text{ bpd}$$

This is close enough to the desired production to proceed to the next step of the calculations.

Calculation line 14 is a crucial intermediate checkpoint. If the production is close to the desired

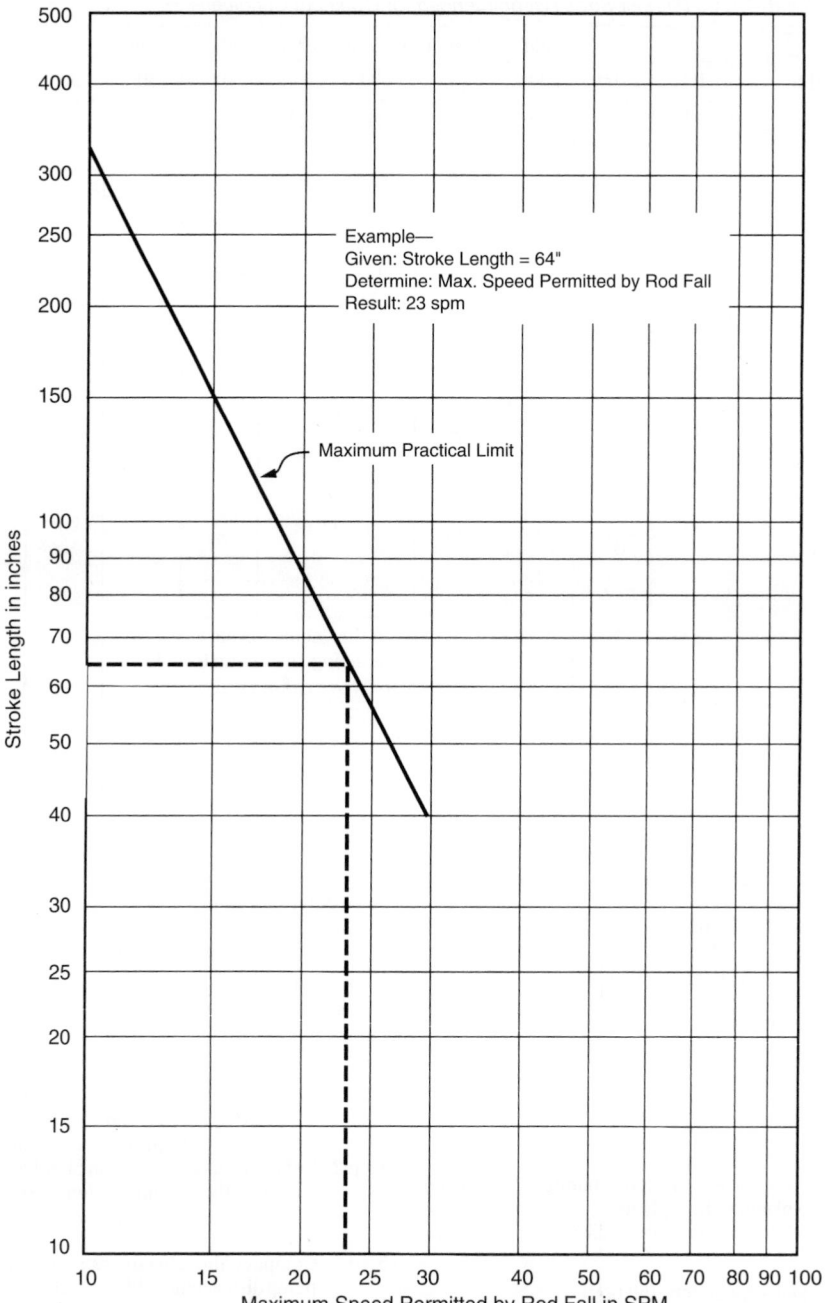

Figure 6.4.10 *Maximum practical pumping speed, conventional unit [2].*

production, then one should proceed with calculation lines 15 through 27. However, if production is considerably different from that desired, one should make different selections for some of the eight parameters that are listed on the calculation sheet as known or assumed data. Parameters that allow greatest change in pump displacement are stroke length and plunger diameter. Stroke lengths should be selected that are available from the manufacturers' catalogues, stroke and speed combination should not allow rods to float, and the plunger diameter should be one possible for the given tubing size. When relatively small adjustments are needed, one usually changes pumping speed first, then changes stroke length if necessary.

In making calculations for lines 5 through 11, one should try to select input data that will keep F_o/Sk_r less than about 0.35 and N/N_o less than about 0.5 for optimal pumping performance.

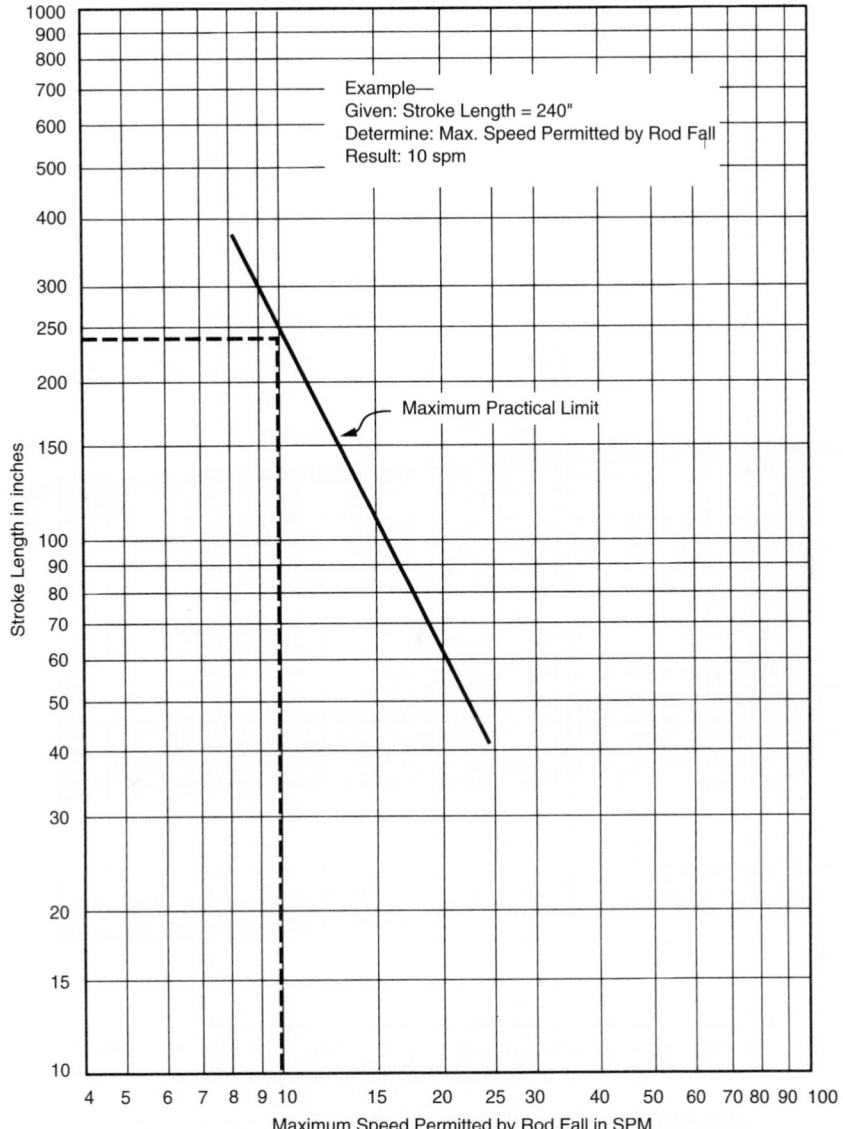

Figure 6.4.11 *Maximum practical pumping speed, air balanced unit [2].*

Step 5. Compute nondimensional terms for calculation lines 15 through 17.

Step 6. Using the previously computed values for F_o/Sk_r, N/N_0', and N/N_0 record from Figures 6.4.15 through 6.4.18 the values needed in lines 18 through 22. Note the special correction procedure needed for Ta from Figure 6.4.19 when Wrf/Sk is different from 0.30.

Step 7. Compute the five operating characteristics of the pumping unit. This step is the final one in the calculation sequence if the operating characteristics are acceptable. However, because the overall task is to design not just a feasible pumping unit but rather to seek an optimal design, it may be necessary to select other alternative values for the assumed data and then repeat the calculation. It is the repetitive nature of the calculation that makes it attractive for a computer solution.

6.4.5.22 Testing the Design

The five operating characteristics of the pumping unit (lines 23 through 28) are used to specify more completely the selected unit. The operating characteristics will allow one to:

1. determine whether the rods are overstressed
2. specify feasible specifications for pumping unit such as:

   a. Peak torque
   b. Structural capacity
   c. Stroke length
   d. Prime mover horsepower

An explanation and definition of each of these five operating characteristics of calculation lines 23 through 27 is given below. This followed by a description of how to use these operating characteristics to choose feasible specifications for the pumping unit operating characteristics. The operating

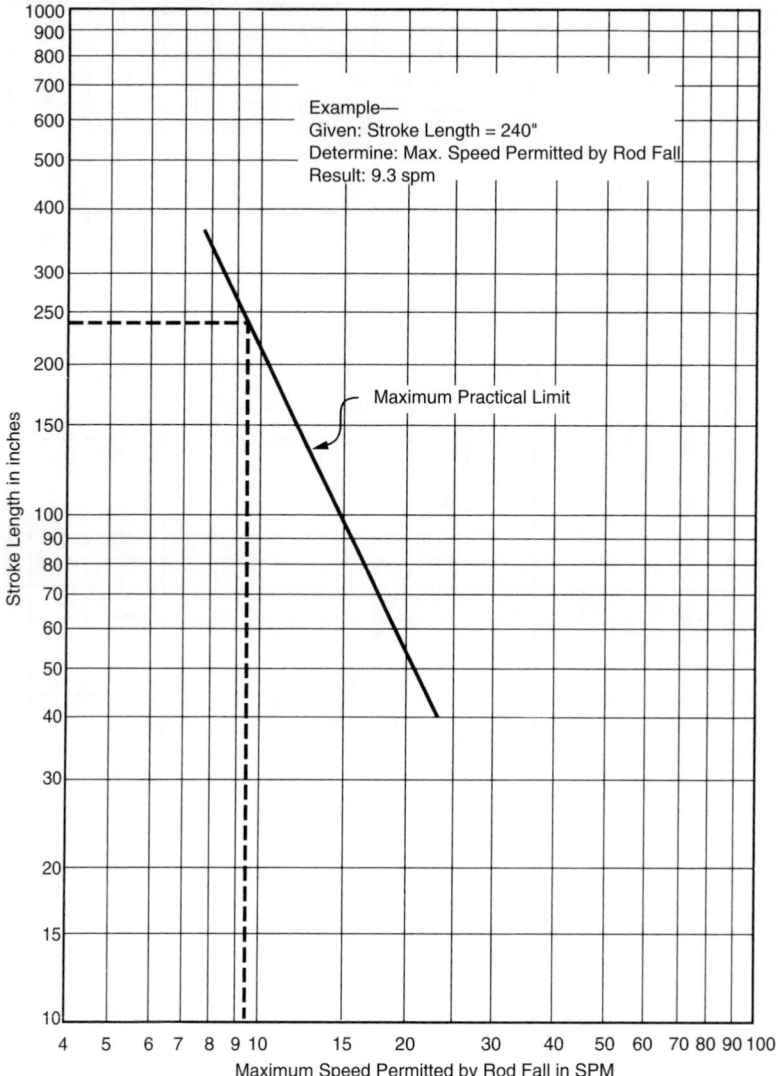

Figure 6.4.12 *Maximum practical pumping speed, Mark II unit [2].*

characteristics of lines 23 through 27 have the following significance.

  PPRL = peak polished rod load in pounds. This is the maximum load on the polished rod during the upstroke.

  MPRL = minimum polished rod load in pounds. This is the maximum load on the polished rod during the downstroke.

  PT = peak torque in inch-pounds. This is the maximum torque generated on the gear reducer during the pumping cycle.

  PRHP = theoretical horsepower needed by the pumping unit.

  CBE = Counterbalance effect. This is the pounds of counterbalance effect measured at the polished rod needed to properly counterbalance the unit. It is *not* equal to the pounds of counterbalance weights needed. See discussion below.

6.4.5.23 Rod Stress

One should use the PPRL and MPRL in a stress analysis of the rod string to determine whether the rods are overstressed. For this, the Goodman diagram is used. The procedure is explained using the above calculation example.

Step 1. Convert the rod loads (PPRL and MPRL) to stress by dividing by the rod cross section area.

  Since PPRL and MPRL are computed at the polished rod, one uses the cross-section area of the largest rod ($\frac{7}{8}$-in. diameter in this case). The cross-section areas of the rods are listed in Table 6.4.3. For $\frac{7}{8}$-in. rods, the cross-section area is 0.601 in.^{2}. The stress's are

$$\frac{\text{PPRL}}{\text{Rod area}} = \frac{14{,}356\ \text{lb}}{0.601\ \text{in.}^2} = 23{,}887\ \text{lb/in.}^2$$

$$\frac{\text{MPRL}}{\text{Rod area}} = \frac{5249\ \text{lb}}{0.601\ \text{in.}^2} = 8734\ \text{lb/in.}^2$$

Table 6.4.9 *Rod and Pump Data [9]*

1	2	3	4	5	6	7	8	9	10	11
Rod* no.	Plunger diam., inches	Rod weight lb per ft	Elastic constant, in. per lb ft E_r	Frequency Factor, F_e	Rod string, % of each size					
					$1\frac{1}{8}$	1	$\frac{7}{8}$	$\frac{3}{4}$	$\frac{5}{8}$	$\frac{1}{2}$
44	All	0.726	1.990×10^{-6}	1.000	—	—	—	—	—	100.0
54	1.06	0.906	1.668×10^{-6}	1.138	—	—	—	—	44.6	55.4
54	1.25	0.929	1.633×10^{-6}	1.140	—	—	—	—	49.5	50.5
54	1.50	0.957	1.584×10^{-6}	1.137	—	—	—	—	56.4	43.6
54	1.75	0.990	1.525×10^{-6}	1.122	—	—	—	—	64.6	35.4
54	2.00	1.027	1.460×10^{-6}	1.095	—	—	—	—	73.7	26.3
54	2.25	1.067	1.391×10^{-6}	1.061	—	—	—	—	83.4	16.6
54	2.50	1.108	1.318×10^{-6}	1.023	—	—	—	—	93.5	6.5
55	All	1.135	1.270×10^{-6}	1.000	—	—	—	—	100.0	—
64	1.06	1.164	1.382×10^{-6}	1.229	—	—	—	33.3	33.1	33.5
64	1.25	1.211	1.319×10^{-6}	1.215	—	—	—	37.2	35.9	26.9
64	1.50	1.275	1.232×10^{-6}	1.184	—	—	—	42.3	40.4	17.3
64	1.75	1.341	1.141×10^{-6}	1.145	—	—	—	47.4	45.2	7.4
65	1.06	1.307	1.138×10^{-6}	1.098	—	—	—	34.4	65.6	—
65	1.25	1.321	1.127×10^{-6}	1.104	—	—	—	37.3	62.7	—
65	1.50	1.343	1.110×10^{-6}	1.110	—	—	—	41.8	58.2	—
65	1.75	1.369	1.090×10^{-6}	1.114	—	—	—	46.9	53.1	—
65	2.00	1.394	1.070×10^{-6}	1.114	—	—	—	52.0	48.0	—
65	2.25	1.426	1.045×10^{-6}	1.110	—	—	—	58.4	41.6	—
65	2.50	1.460	1.018×10^{-6}	1.099	—	—	—	65.2	34.8	—
65	2.75	1.497	0.990×10^{-6}	1.082	—	—	—	72.5	27.5	—
65	3.25	1.574	0.930×10^{-6}	1.037	—	—	—	88.1	11.9	—
66	All	1.634	0.883×10^{-6}	1.000	—	—	—	100.0	—	—
75	1.06	1.566	0.997×10^{-6}	1.191	—	—	27.0	27.4	45.6	—
75	1.25	1.604	0.973×10^{-6}	1.193	—	—	29.4	29.8	40.8	—
75	1.50	1.664	0.935×10^{-6}	1.189	—	—	33.3	33.3	33.3	—
75	1.75	1.732	0.892×10^{-6}	1.174	—	—	37.8	37.0	25.1	—
75	2.00	1.803	0.847×10^{-6}	1.151	—	—	42.4	41.3	16.3	—
75	2.25	1.875	0.801×10^{-6}	1.121	—	—	46.9	45.8	7.2	—
76	1.06	1.802	0.816×10^{-6}	1.072	—	—	28.5	71.5	—	—
76	1.25	1.814	0.812×10^{-6}	1.077	—	—	30.6	69.4	—	—
76	1.50	1.833	0.804×10^{-6}	1.082	—	—	33.8	66.2	—	—
76	1.75	1.855	0.795×10^{-6}	1.088	—	—	37.5	62.5	—	—
76	2.00	1.880	0.785×10^{-6}	1.093	—	—	41.7	58.3	—	—
76	2.25	1.908	0.774×10^{-6}	1.096	—	—	46.5	53.5	—	—
76	2.50	1.934	0.764×10^{-6}	1.097	—	—	50.8	49.2	—	—
76	2.75	1.967	0.751×10^{-6}	1.094	—	—	56.5	43.5	—	—
76	3.75	2.039	0.722×10^{-6}	1.078	—	—	68.7	31.3	—	—
76	3.75	2.119	0.690×10^{-6}	1.047	—	—	82.3	17.7	—	—
77	All	2.224	0.649×10^{-6}	1.000	—	—	100.0	—	—	—
85	1.06	1.883	0.873×10^{-6}	1.261	—	22.2	22.4	22.4	33.0	—
85	1.25	1.943	0.841×10^{-6}	1.253	—	23.9	24.2	24.3	27.6	—
85	1.50	2.039	0.791×10^{-6}	1.232	—	26.7	27.4	26.8	19.2	—
85	1.75	2.138	0.738×10^{-6}	1.201	—	29.6	30.4	29.5	10.5	—
86	1.06	2.058	0.742×10^{-6}	1.151	—	22.6	23.0	54.3	—	—
86	1.25	2.087	0.732×10^{-6}	1.156	—	24.3	24.5	51.2	—	—
86	1.50	2.133	0.717×10^{-6}	1.162	—	26.8	27.0	46.3		
86	1.75	2.185	0.699×10^{-6}	1.164	—	29.4	30.0	40.6		
86	2.00	2.247	0.679×10^{-6}	1.161	—	32.8	33.2	33.9		
86	2.25	2.315	0.656×10^{-6}	1.153	—	36.9	36.0	27.1		
86	2.50	2.385	0.633×10^{-6}	1.138	—	40.6	39.7	19.7		
86	2.75	2.455	0.610×10^{-6}	1.119	—	44.5	43.3	12.2		
87	1.06	2.390	0.612×10^{-6}	1.055	—	24.3	75.7	—	—	—
87	1.25	2.399	0.610×10^{-6}	1.058	—	25.7	74.3	—	—	—

(continued)

Table 6.4.9 *Rod and Pump Data [9] (continued)*

1	2	3	4	5	6	7	8	9	10	11
Rod* no.	Plunger diam., inches	Rod weight lb per ft	Elastic constant, in. per lb ft E_r	Frequency Factor, F_e	Rod string, % of each size					
					$1\frac{1}{8}$	1	$\frac{7}{8}$	$\frac{3}{4}$	$\frac{5}{8}$	$\frac{1}{2}$
87	1.50	2.413	0.607×10^{-6}	1.062	—	27.7	72.3	—	—	—
87	1.75	2.430	0.603×10^{-6}	1.068	—	30.3	69.7	—	—	—
87	2.00	2.450	0.598×10^{-6}	1.071	—	33.2	66.8	—	—	—
87	2.25	2.472	0.594×10^{-6}	1.075	—	36.4	63.6	—	—	—
87	2.50	2.496	0.588×10^{-6}	1.079	—	39.9	60.1	—	—	—
87	2.75	2.523	0.582×10^{-6}	1.082	—	43.9	56.1	—	—	—
87	3.25	2.575	0.570×10^{-6}	1.084	—	51.6	48.4	—	—	—
87	3.75	2.641	0.556×10^{-6}	1.078	—	61.2	38.8	—	—	—
87	4.75	2.793	0.522×10^{-6}	1.038	—	83.6	16.4	—	—	—
88	All	2.904	0.497×10^{-6}	1.000	—	100.0	—	—	—	—
96	1.06	2.382	0.670×10^{-6}	1.222	19.1	19.2	19.5	42.3	—	—
96	1.25	2.435	0.655×10^{-6}	1.224	20.5	20.5	20.7	38.36	—	—
96	1.50	2.511	0.633×10^{-6}	1.223	22.4	22.5	22.8	32.3	—	—
96	1.75	2.607	0.606×10^{-6}	1.213	24.8	25.1	25.1	25.1	—	—
96	2.00	2.703	0.578×10^{-6}	1.196	27.1	27.9	27.4	17.6	—	—
96	2.25	2.806	0.549×10^{-6}	1.172	29.6	30.7	29.8	9.8	—	—
97	1.06	2.645	0.568×10^{-6}	1.120	19.6	20.0	60.3	—	—	—
97	1.25	2.670	0.563×10^{-6}	1.124	20.8	21.2	58.0	—	—	—
97	1.50	2.707	0.556×10^{-6}	1.131	22.5	23.0	54.5	—	—	—
97	1.75	2.751	0.548×10^{-6}	1.137	24.5	25.0	50.4	—	—	—
97	2.00	2.801	0.538×10^{-6}	1.141	26.8	27.4	45.7	—	—	—
97	2.25	2.856	0.528×10^{-6}	1.143	29.4	30.2	40.4	—	—	—
97	2.50	2.921	0.515×10^{-6}	1.141	32.5	33.1	34.4	—	—	—
97	2.75	2.989	0.503×10^{-6}	1.135	36.1	35.3	28.6	—	—	—
97	3.25	3.132	0.475×10^{-6}	1.111	42.9	41.9	15.2	—	—	—
98	1.06	3.068	0.475×10^{-6}	1.043	21.2	78.8	—	—	—	—
98	1.25	3.076	0.474×10^{-6}	1.045	22.2	77.8	—	—	—	—
98	1.50	3.089	0.472×10^{-6}	1.048	23.8	76.2	—	—	—	—
98	1.75	3.103	0.470×10^{-6}	1.051	25.7	74.3	—	—	—	—
98	2.00	3.118	0.468×10^{-6}	1.055	27.7	72.3	—	—	—	—
98	2.25	3.137	0.465×10^{-6}	1.058	30.1	69.9	—	—	—	—
98	2.50	3.157	0.463×10^{-6}	1.062	32.7	67.3	—	—	—	—
98	2.75	3.180	0.460×10^{-6}	1.066	35.6	64.4	—	—	—	—
98	3.25	3.231	0.453×10^{-6}	1.071	42.2	57.8	—	—	—	—
98	3.75	3.289	0.445×10^{-6}	1.074	49.7	50.3	—	—	—	—
98	4.75	3.412	0.428×10^{-6}	1.064	65.7	34.3	—	—	—	—
99	All	3.676	0.393×10^{-6}	1.000	100.0	—	—	—	—	—
					$1\frac{1}{4}$	$1\frac{1}{8}$	1	$\frac{7}{8}$	$\frac{3}{4}$	$\frac{5}{8}$
107	1.06	2.977	0.524×10^{-6}	1.184	16.9	16.8	17.1	49.1	—	—
107	1.25	3.019	0.517×10^{-6}	1.189	17.9	17.8	18.0	46.3	—	—
107	1.50	3.085	0.506×10^{-6}	1.195	19.4	19.2	19.5	41.9	—	—
107	1.75	3.158	0.494×10^{-6}	1.197	21.0	21.0	21.2	36.9	—	—
107	2.00	3.238	0.480×10^{-6}	1.195	22.7	22.8	23.1	31.4	—	—
107	2.25	3.336	0.464×10^{-6}	1.187	25.0	25.0	25.0	25.0	—	—
107	2.50	3.435	0.447×10^{-6}	1.174	26.9	27.7	27.1	18.2	—	—
107	2.75	3.537	0.430×10^{-6}	1.156	29.1	30.2	29.3	11.3	—	—
108	1.06	3.325	0.447×10^{-6}	1.097	17.3	17.8	64.9	—	—	—
108	1.25	3.345	0.445×10^{-6}	1.101	18.1	18.6	63.2	—	—	—
108	1.50	3.376	0.441×10^{-6}	1.106	19.4	19.9	60.7	—	—	—
108	1.75	3.411	0.437×10^{-6}	1.111	20.9	21.4	57.7	—	—	—
108	2.00	3.452	0.432×10^{-6}	1.117	22.6	23.0	54.3	—	—	—
108	2.25	3.498	0.427×10^{-6}	1.121	24.5	25.0	50.5	—	—	—
108	2.50	3.548	0.421×10^{-6}	1.124	26.5	27.2	46.3	—	—	—
108	2.75	3.603	0.415×10^{-6}	1.126	28.7	29.6	41.6	—	—	—

(continued)

Table 6.4.9 *Rod and Pump Data [9] (continued)*

1	2	3	4	5	6	7	8	9	10	11
Rod* no.	Plunger diam., inches	Rod weight lb per ft	Elastic constant, in. per lb ft E_r	Frequency Factor, F_e	Rod string, % of each size					
					$1\frac{1}{4}$	$1\frac{1}{8}$	1	$\frac{7}{8}$	$\frac{3}{4}$	$\frac{5}{8}$
108	3.25	3.731	0.400×10^{-6}	1.123	34.6	33.9	31.6	—	—	—
108	3.75	3.873	0.383×10^{-6}	1.108	40.6	39.5	19.9	—	—	—
109	1.06	3.839	0.378×10^{-6}	1.035	18.9	81.1	—	—	—	—
109	1.25	3.845	0.378×10^{-6}	1.036	19.6	80.4	—	—	—	—
109	1.50	3.855	0.377×10^{-6}	1.038	20.7	79.3	—	—	—	—
109	1.75	3.867	0.376×10^{-6}	1.040	22.1	77.9	—	—	—	—
109	2.00	3.880	0.375×10^{-6}	1.043	23.7	76.3	—	—	—	—
109	2.25	3.896	0.374×10^{-6}	1.046	25.4	74.6	—	—	—	—
109	2.50	3.911	0.372×10^{-6}	1.048	27.2	72.8	—	—	—	—
109	2.75	3.930	0.371×10^{-6}	1.051	29.4	70.6	—	—	—	—
109	3.25	3.971	0.367×10^{-6}	1.057	34.2	65.8	—	—	—	—
109	3.75	4.020	0.363×10^{-6}	1.063	39.9	60.1	—	—	—	—
109	4.75	4.120	0.354×10^{-6}	1.066	51.5	48.5	—	—	—	—

*Rod No. shown in first column refers to the largest and smallest rod size in eighths of an inch. For example, Rod No. 76 is a two-way taper of $\frac{7}{8}$ and $\frac{6}{8}$ rods. Rod No. 85 is a four-way taper of $\frac{3}{8}$, $\frac{7}{8}$, $\frac{6}{8}$, and $\frac{5}{8}$ rods. Rod No. 109 is a two-way taper of $1\frac{1}{4}$ and $1\frac{1}{8}$ rods. Rod No. 77 is a straight string of $\frac{7}{8}$ rods, etc.

Fluid Level, H = 4500 ft. Pumping Speed, N = 16 SPM Plunger Diameter, D = 1.50 in.
Pump Depth, L = 5000 ft. Length of Stroke, S = 54 in. Spec. Grav. of Fluid, G = 0.9

Tubing Size 2 in. Is it anchored? Yes, No Sucker Rode 33.8% − $\frac{7}{8}''$ & 66.2% − $\frac{3}{4}''$

Record Factors.

1. $W_r = 1.833$
2. $E_r = .804 \times 10^{-6}$

3. $F_e = 1.082$
4. $E_t = 307 \times 10^{-6}$

Calculate Non-Dimensional Variables:

5. $F_e = .340 \times G \times D^2 \times H = .340 \times 0.9 \times 2.25 \times 4500 = 3098$ lbs.
6. $1/k_r = E_r \times L = 0.804 \times 10^{-6}\ 5000 = 4.020 \times 10^{-3}$ in/lb.
7. $Sk_r = S \div 1/k_r = 54 \div 4.020 \times 10^{-3} = 13,433$ lbs.
8. $F_e/Sk_r = 3098 \div 13,433 = .231$

9. $N/N_e = NL + 245,000 = 16 \times 5000 \div 245,000 = .326$
10. $N/N'_e = N/N_e \div F_e = .326 \div 1.082 = .301$
11. $1/k_r = E_t \times L = .307 \times 10^{-6} \times 5000 = 1.535 \times 10^{-3}$ in/lb.

Solve for S_p and PD:

12. $S_p/S = .86$
13. $S_p = [(S_p/S) \times S] − [F_e \times 1/k_t] = [.86 \times 54] − [3098 \times 1.535 \times 10^{-3}] = 41.7$ ln.
14. $PD = 0.1166 \times S_p \times N \times D^2 = 0.1166 \times 41.7 \times 16 \times 2.25 = 175$ barrels per day

   If the calculated pump displacement fails to satisfy known or anticipated requirements, appropriate adjustments must be made in the assumed data and steps 1 through 14 repeated. When the calculated pump displacement is acceptable, proceed with the Sign Calculation.

Determine Non-Dimensional Parameters P:

15. $W = W_r \times L = 1.833 \times 5000 = 9165$ lbs.
16. $W_{rf} = W[1 − (.128G)] = 9165[1 − (.128 \times .9)] = 8110$ lbs.

17. $W_{rt}/Sk_r = 8110 \div 13,433 = .604$

Record Non-Dimensional Factors:

18. $F_1/Sk_r = .465$
19. $F_2/Sk_r = .213$

20. $2T/S^2k_r = .37$
21. $F_3/Sk_r = .29$

22. $T_a = .997$

Solve for Operating Characteristics:

23. $PPRL = W_{rf} + [(F_1/Sk_r) \times Sk_r] = 8110 + [.465 \times 13,433] = 14,356$ lbs.
24. $MPRL = W_{rt} − [(F_2/Sk_r)] = 8110 − [.22 \times 13,433] = 5249$ lbs.
25. $PT = (2T/S^2k_r) \times Sk_r \times S/2T_a = .37 \times 13,433 \times 27 \div .997 = 133,793$ lb inches
26. $PRHP = (F_3/Sk_r) \times Sk_r \times S \times N \times 2.53 \times 10^{-5} = .29 \times 13,433 \times 54 \times 16 \times 2.53 \times 10^{-6} = 8.5$
27. $CBE = 1.06(W_{rt} + 1/2F_e) = 1.06 \times (8110 + 1549) = 10,239$ lbs.

Figure 6.4.13 *Example design calculations — conventional sucker rod pumping system [57].*

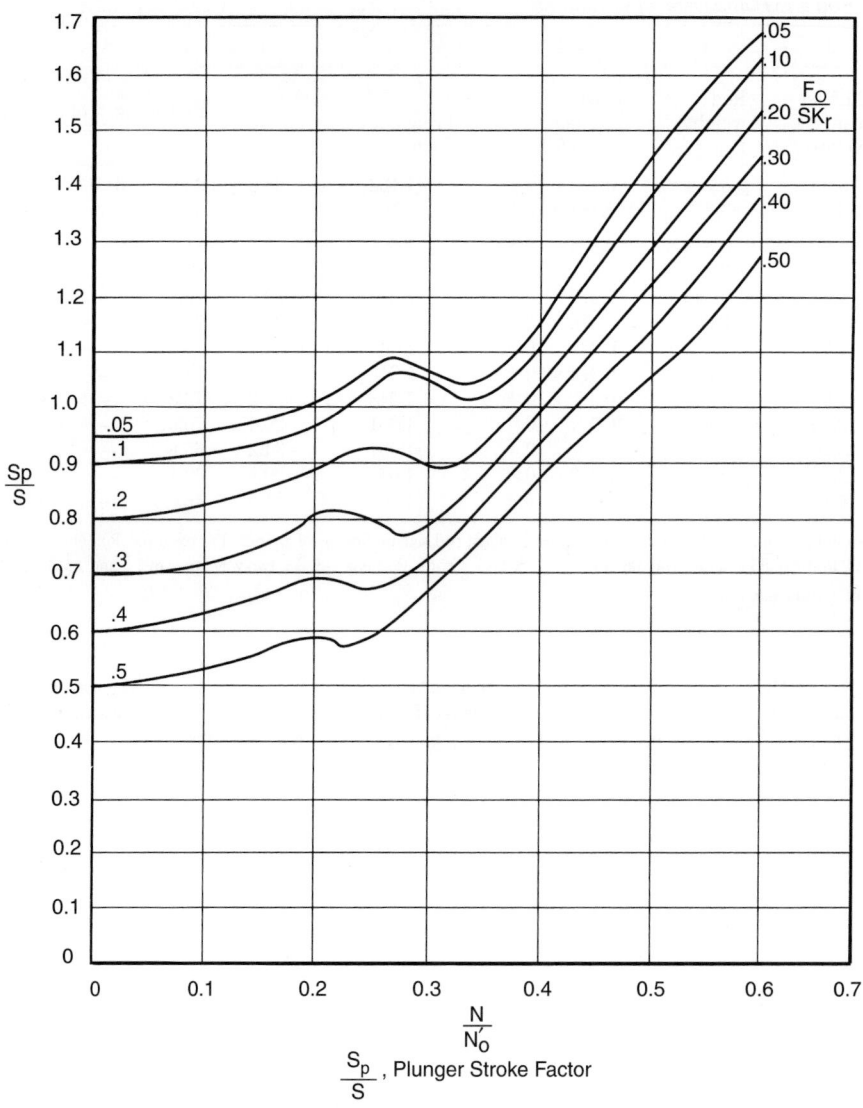

Figure 6.4.14 *Plunger stroke factor [9].*

Step 2. Select the rod grade and tensile strength. Refer to Table 6.4.3. Let us assume that the rods are grade C. These have a minimum tensile strength of 90,000 lbf/in.[2] Minimum rather than maximum tensile strength is used to give a more conservative design and to reduce the possibility of rod failure.

Step 3. Apply the modified Goodman diagram equation. Refer to "Allowable Rod Stress and Stress Range" for a discussion of this equation. Thus

$$S_A = (T/4 + 0.5625 \times S_{min}) \times SF$$

where T = minimum tensile strength in psi
S_{min} = minimum tensile stress experience by the rod
S_F = service factor
S_A = allowable maximum stress in rod in psi

Here T = 90,000 psi for a grade C rod
S_{min} = 8734 psi from Step 2.

Let us assume a service factor of 1.0, i.e., normal noncorrosive conditions (no H_2S, CO_2 or brine)

Step 4. Compute S_A:

$$S_A = (T/4 + 0.5625 \times S_{min}) \times SF$$

$$= (90,000/4 + 0.5625 \times 8734) \times 1.0 = 27,412 \text{ psi}$$

Step 5. Compare allowable maximum stress S_A with the design maximum.

The API R11L design gives a maximum stress of 23,887 psi. The modified Goodman diagram allowable stress is 27,412 psi. Thus this rod string will not be overstressed. The design stress is $(23,887/27,412) \times 100 = 87\%$ of the maximum stress. The actual design stress should not be more than 95% of the Goodman diagram allowable stress. If this percent value is much less than 50%, a lighter rod string could be used. The comparison of actual to allowable stress in this step is critically dependent upon the value chosen for the service factor. In this example, a service of 0.80 would mean that the allowable

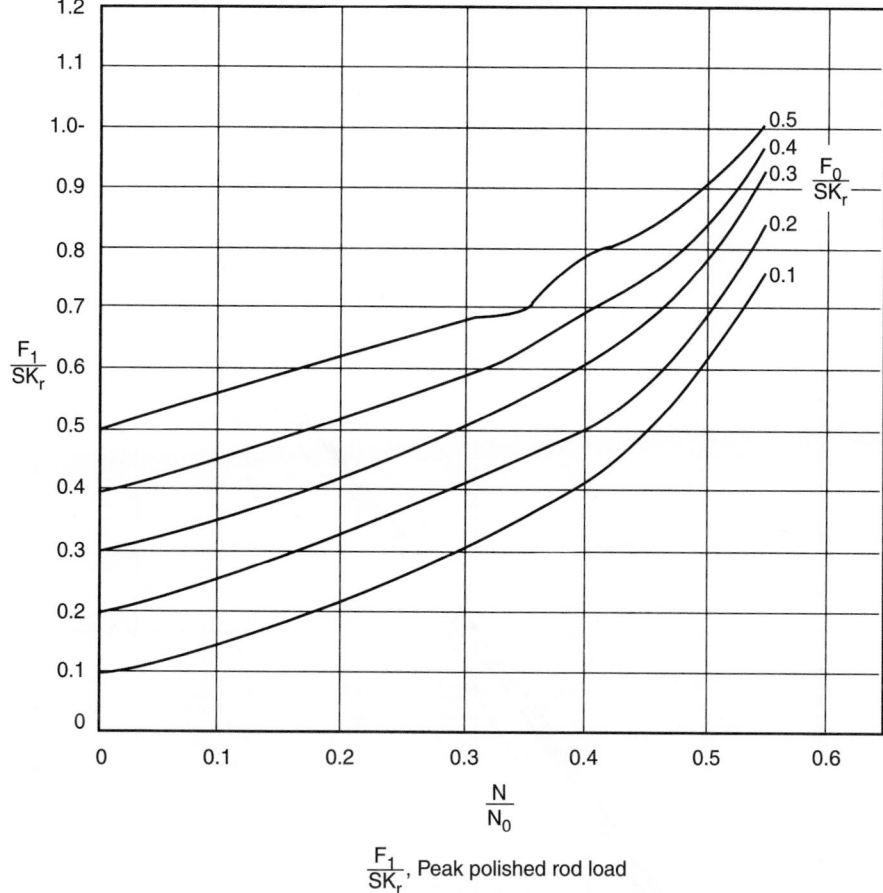

$\dfrac{F_1}{SK_r}$, Peak polished rod load

Figure 6.4.15 *Peak polished rod load factor [9].*

stress was

$$S_A = (90,000/4 + 0.5625) \times 0.80 = 21923 \text{ psi}$$

which gives a percentage of the allowable stress of

$$\frac{23,887 \text{ psi}}{21,923 \text{ psi}} \times 100 = 109\%$$

and so the rods would be overstressed.

6.4.5.24 Peak Torque
The calculated peak torque for this example design is 133,793 lb-in. The unit selected must have a gear reducer torque rating larger than this computed value. Refer to Table 6.4.1 for available ratings. Any unit having a rating of 160,000 lb-in. or larger would serve. Normally one would choose a 160,000-lb-in. unit rather than a 228,000 or 320,000-lb-in. unit. Normal notation is to refer to a 160,000-lb-in. unit as a 160 unit.

6.4.5.25 Structural Capacity
Structural capacity is based on the peak polished rod load (PPRL). In each torque rating class, several structural capacities are available. One chooses a structural capacity rating larger than the PPRL. A 17,300-lb structural capacity unit would be selected here.

6.4.5.26 Stroke Length
Although a stroke length of 54 in. was initially chosen, one now needs to see of the selected pumping unit with

160,000 lb-in. of torque and 17,300 lbf structural capacity is available in a 54-in. stroke. The API unit closest to that desired is 160-173-64. The 64 refers to the *largest* stroke available in this size. One would need to check manufacturers' catalogues to see what smaller strokes are possible for a 160-173-64 unit. If, for instance, a 53-in. stroke were available as a second stroke length setting, one would need to increase the pumping speed slightly to overcome the reduction in stroke length.

6.4.5.27 Prime Mover Horsepower
The computed polished rod horsepower rating PRHP is a theoretical horsepower value and is less than the rated capacity of any prime mover that should be selected. The value also does not indicate whether one should use an electrical motor or an internal combustion engine. Refer to "Selecting Prime Mover Site" for a discussion of prime mover selection criteria. If a slow-speed internal combustion engine or a high slip NEMA D electric motor were selected, the prime mover horsepower would be computed by Equation 6.4.2

$$HP_{pm} = \frac{q \times D}{PMF}$$

$$= \frac{(150 \text{ bpd}) \times (4500 \text{ ft})}{56,000} = 12.0$$

Note that the actual lifting depth, 4500 ft, and the surface flow rate 150 bpd were used. The efficiency given as 85%

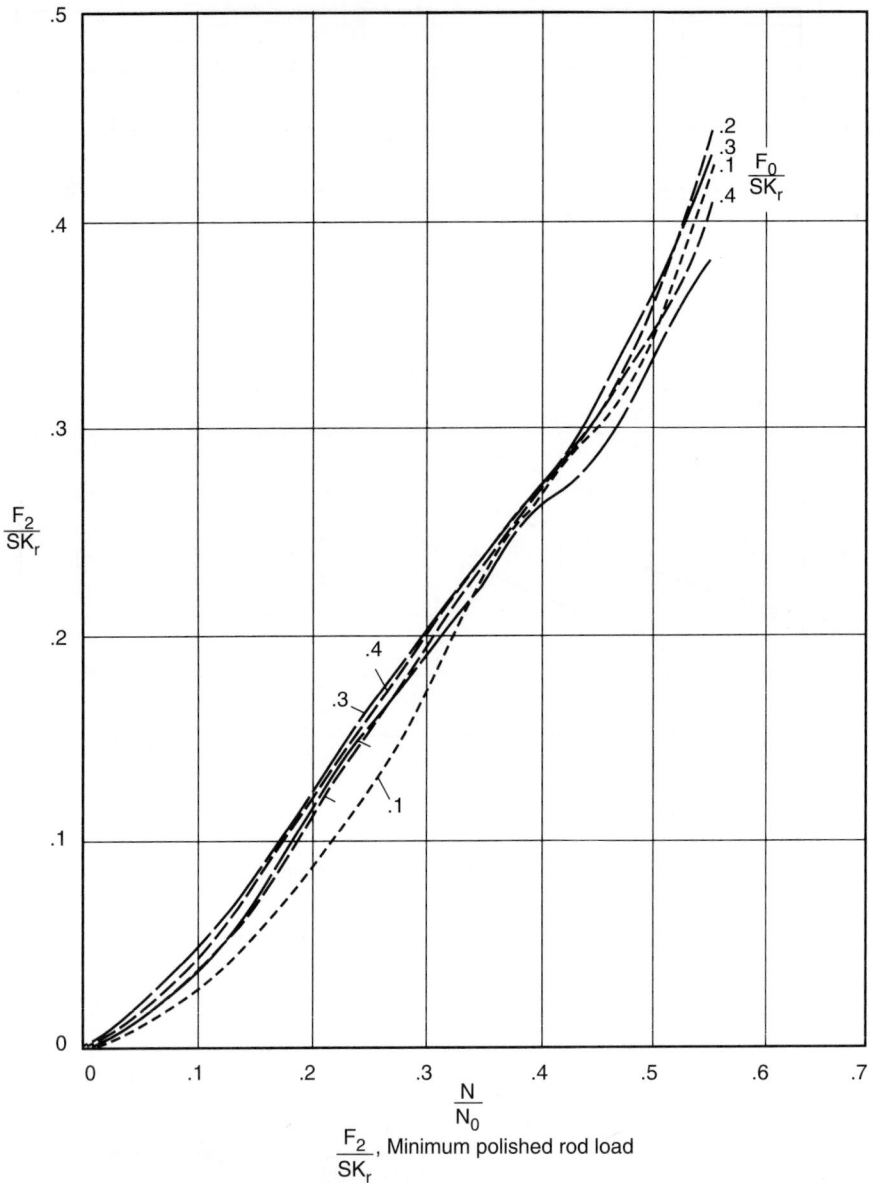

Figure 6.4.16 *Minimum polished rod load factor [9].*

is incorporated into the value selected for PMF. One would choose the next size larger available nameplate rating for the engine or motor to serve as prime mover.

6.4.6 Analyzing Existing Pumping Units
Several different procedures are available to test an existing pumping installation. Because causes of malfunction are many and because the observed consequences of many malfunctions can be the same, the process of identifying one or more problems than can exist can be complex. In most instances, the most rapid and accurate analysis will occur where the largest amount of precise data are available. The benefit of the extra data comes in less work over expense and less lost production due to downtime.

6.4.6.1 Dynamometers
An extremely valuable tool for analysis of well performance is the well dynamometer. Several manufacturers provide these devices that measure polished rod load versus polished rod position. The new electronic dynamometers also allow rod position and rod load to be recorded versus time. If an electric prime mover is used, some new dynamometer systems also provide a graph of motor current versus rod position and versus time. This later graph is a simple and rapid way to determine proper counterbalance. The load measuring part of the dynamometer is attached to the polished rod so that the load can be sensed and sent to a recorder. A companion part of the dynamometer attached to the walking beam senses the polished rod position and sends it to the same recorder. The graph produced is called a *dynagraph* or more

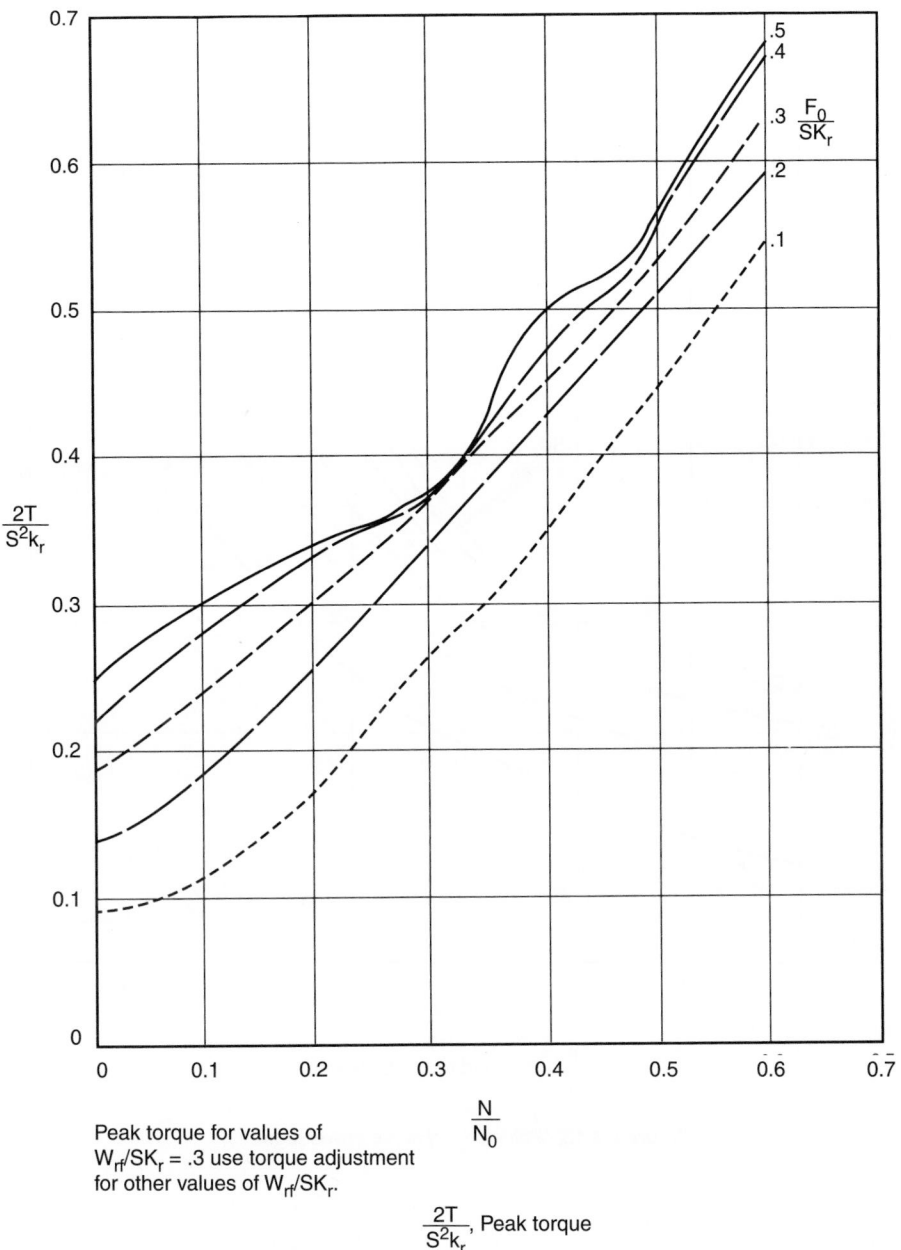

Peak torque for values of
W_{rf}/SK_r = .3 use torque adjustment
for other values of W_{rf}/SK_r.

$\dfrac{2T}{S^2k_r}$, Peak torque

Figure 6.4.17 *Peak torque factor [9].*

commonly a *dynamometer* or *dynagraph card*. A typical card is shown in Figure 6.4.20.

Force × distance equals work. The dynamometer measures force on the polished rod and distance the polished rod moves. The dynamometer thus records work done during one pump cycle. Because power is the rate of doing work, and because the time to create one pumping cycle, creating one dynagraph card is known, the rate of doing work, namely the hosepower at the polished rod can be estimated from the dynamometer card.

The rods, fluid, and tubing (if unanchored) constitute an elastic system whose motion under the action of the reciprocating pump can be very complex. The dynamometer records the changing load versus position at one point in the system, i.e., the polished rod. To understand the behavior of the system at points other than at the point of measurement is very difficult in many installations. Older references to the pumping system as following simple harmonic motion represent a gross oversimplification for moderate to deep wells. For such wells, the only reasonably precise estimation

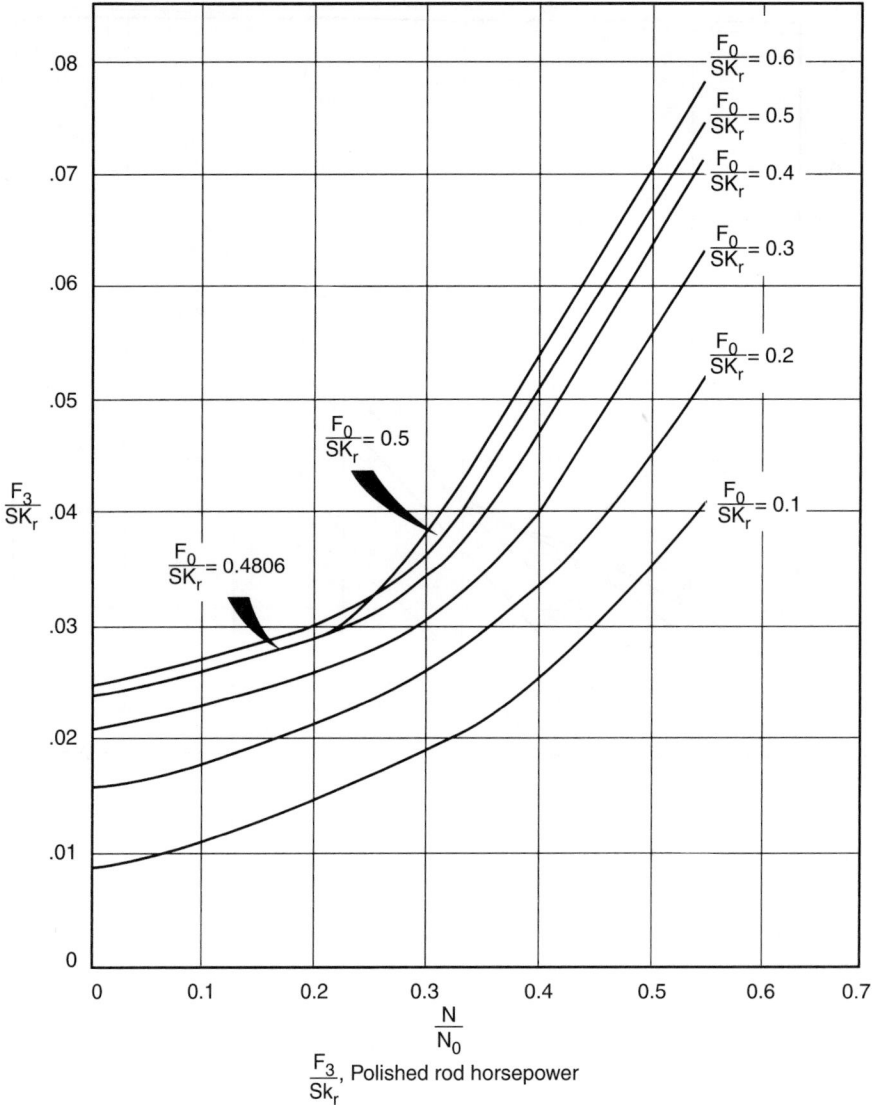

Figure 6.4.18   *Polished rod horse power factor [9].*

of downhole behavior based on dynamometer cards can be obtained through computer programs based on solutions to the wave equation such as described by Gibbs [11]. However, for troubleshooting of wells, several types of problems can easily be identified through qualitative analysis of the appearance of the dynamometer card. Other problems can be identified trough simple computations using dynamometer card data. In any program of troubleshooting of wells with a dynamometer, it is always helpful to have a dynamometer card for the well when it is operating normally. The card when taken during a troubleshooting test should be run under identical conditions of pumping speed, stroke length and dynamometer constant settings as when the well was running normally. In the following paragraphs, some of the more common and easily identified problems will be described where these problems can be identified through card shape.

6.4.6.2 Fluid Pound
Refer to the section titled "Fluid Pound" for a discussion of the causes and cure for fluid pound. Its presence is indicated by a relatively steep drop in the load somewhere along the downstroke portion of the dynagraph. The farther that this drop occurs to the left, the farther the plunger has moved down the pump barrel before it hits liquid. A typical example of fluid pound appears in Figure 6.4.21. A more severe example of fluid pound is given in Figure 6.4.22.

6.4.6.3 Gas Pound
This effect is similar to that of fluid pound and cannot always be positively distinguished from fluid pound based solely on card shape. Gas pound shows a more gradual load decrease on the downstroke relative to fluid pound. An example of gas pound is shown in Figure 6.4.23.

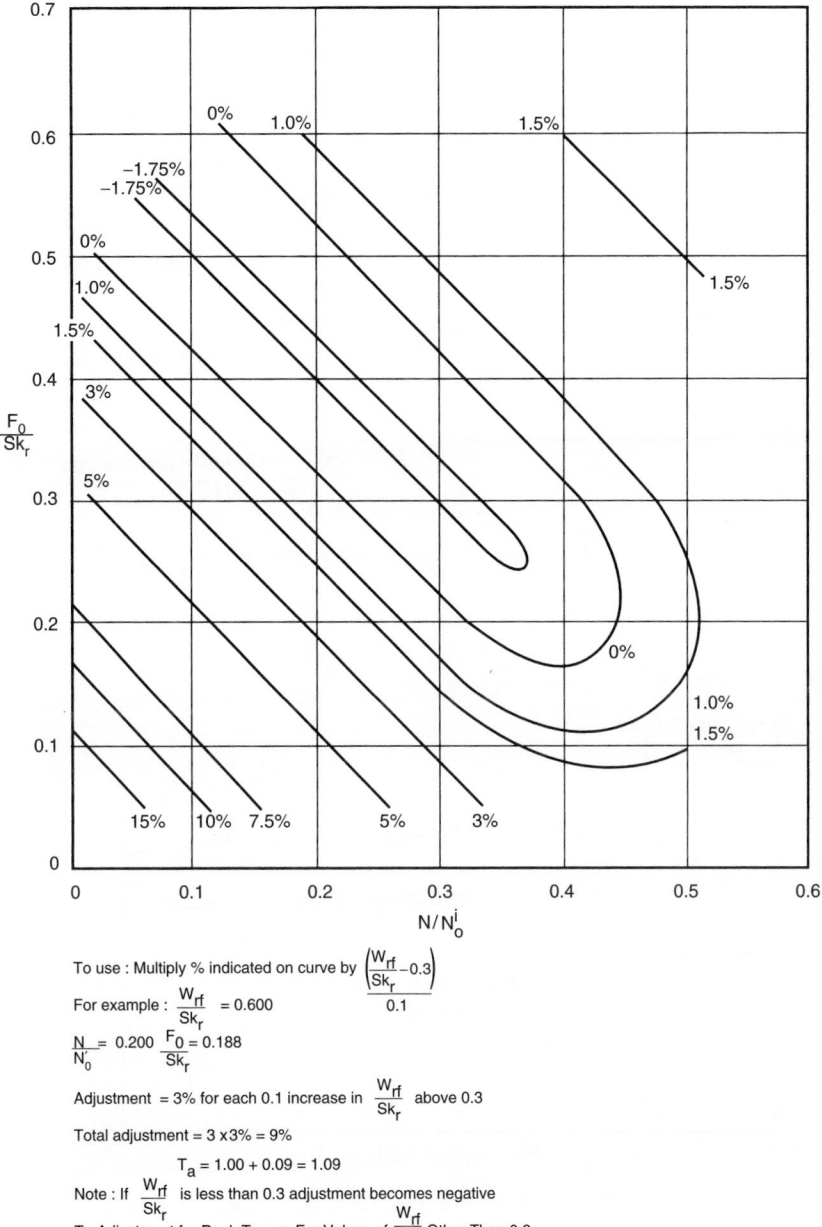

To use : Multiply % indicated on curve by $\left(\dfrac{\frac{W_{rf}}{Sk_r} - 0.3}{0.1}\right)$

For example : $\dfrac{W_{rf}}{Sk_r} = 0.600$

$\dfrac{N}{N_o'} = 0.200 \quad \dfrac{F_0}{Sk_r} = 0.188$

Adjustment $= 3\%$ for each 0.1 increase in $\dfrac{W_{rf}}{Sk_r}$ above 0.3

Total adjustment $= 3 \times 3\% = 9\%$

$\qquad T_a = 1.00 + 0.09 = 1.09$

Note : If $\dfrac{W_{rf}}{Sk_r}$ is less than 0.3 adjustment becomes negative

T_a Adjustment for Peak Torque For Values of $\dfrac{W_{rf}}{Sk_r}$ Other Than 0.3

Figure 6.4.19 *Adjustment for peak torque factor [9].*

6.4.6.4 Well Pumping Off

Under normal conditions, exactly the same dynagraph is traced on each pump cycle. However, when a condition of fluid pound exists and the drop off a load occurs later and later in the downstroke cycle, then a well pumped off condition exists. The well cannot produce as much as the pump is attempting to lift to the surface so the pump barrel fills less with each successive cycle. This condition is shown in Figure 6.4.24.

6.4.6.5 Additional Qualitative Dynamometer Tests

Analysis of problems wells is enhanced if a few extra simple tests are made while the dynamometer is in place. These tests include (1) traveling valve check, (2) standing valve check and (3) counterbalance effect measurement. Each is described below and typical diagnostic examples are shown.

For the traveling valve check, leave the dynamometer in place recording a stable card. Stop the pumping unit about

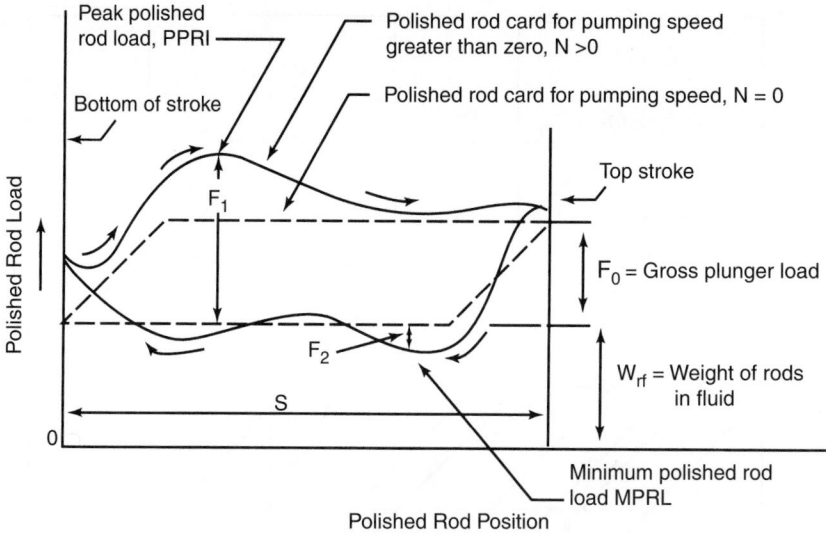

Figure 6.4.20 *Dynamometer card [9].*

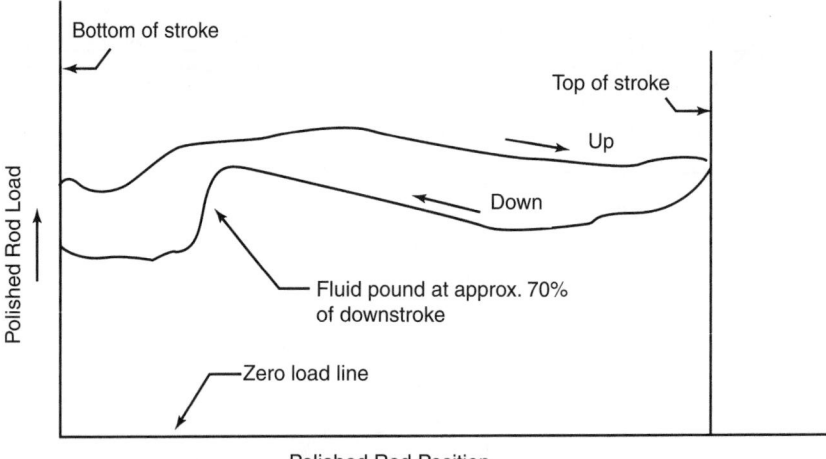

Figure 6.4.21 *Example of fluid pound [12].*

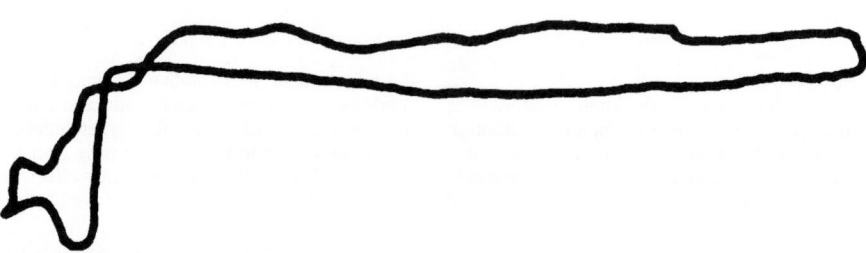

Figure 6.4.22 *Severe case of fluid pound [2].*

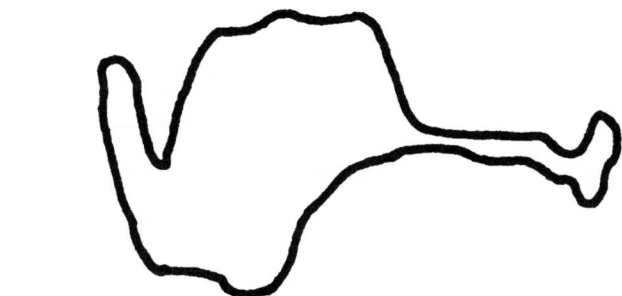

Figure 6.4.23   *Example of gas pound [2].*

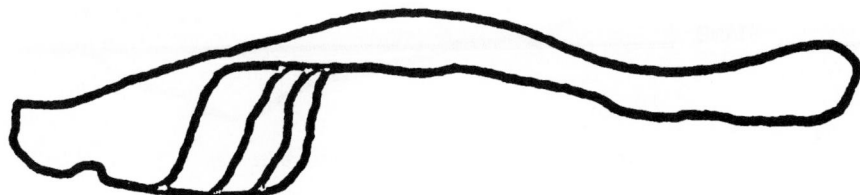

Figure 6.4.24   *Example of well pumping off [2].*

half way through the upstroke and pull the polished rod position indicator cord sharply four or five times at 1-s intervals. A nonleaking traveling valve will show a single horizontal line on the card. A leaking valve will appear as in Figure 6.4.25. The constant line across the middle of the diagram labeled "Traveling Valve" is the line that would appear for a nonleaking valve. However, when the valve leaks, fluid above the valve moves downward through the leak and exerts a pressure on the standing valve and reduce the load on the traveling valve and plunger. In the case of a severe leak, the entire fluid load will bear on the standing valve and the dynamometer will indicate only the buoyant weight of the rods.

The standing valve check is done in a manner similar to that for the traveling valve but on the downstroke rather than the upstroke. At approximately half way through the downstroke, stop the unit and the pull the polished rod position indicator cord sharply four or five times at 1-s intervals. If the traveling valve holds and the standing check valve leaks, the dynamometer card will appear as in Figure 6-136. When the standing leaks, the fluid in the barrel, instead of going up through the traveling valve, will go downward through the leaky standing valve. This fluid motion will cause fluid to move downward through the traveling valve causing it to close. The dynamometer will then sense an increase in load on the rods. This is the cause for the rise shown on the example card in Figure 6.4.26.

The counterbalance effect measurement will indicate whether a pumping unit is properly counterbalanced. Proper counterbalancing is important to reduce power consumption and to ensure long life to the prime mover and pumping unit gear reducer. To measure the effect in pounds that the counterweights have at the polished rod, install the dynamometer as when preparing a dynagraph, then when a stable card has been obtained, stop the unit where the maximum counterbalance is operative, i.e., where the weights are acting vertically downward on the upstroke. Install a polished rod clamp above the stuffing box, then chain the clamp to the casing head so that the polished rod will not move upward when the brake is released. Now release the brake and the pull the polished rod indicator cord. The line created on the dynagraph occurs at a load equal to the counterbalance effect (CBE). An example of a card for a properly balanced well is shown in Figure 6.4.27. On this figure the normal, nonleaking standing valve (SV) and traveling valve (TV) check lines are shown to indicate that effectively counterbalanced units have the CBE line approximately half way between the TV and SV lines.

6.4.7 Gas Lift

The technique of gas lift is a very important one, particularly in those cases in which downhole pumps cannot be used, such as wells on offshore platforms and deviated holes. Gas lift utilizes the energy of compression in the gas to decrease the hydrostatic gradient in a liquid column and thus cause the column to reach the surface. This is accomplished with the use of gas lift valves (active as pressure-regulating valves) located between the tubing and casing annulus. Gas is compressed and fed into the casing annulus and passed into the tubing at a regulated rate by a gas lift valve. Then, the combination of gas and well fluid is carried to the surface. Here the fluids are separated and the low-pressure gas is either recompressed for reuse as gas lift gas or compressed for the sales line if excessive gas is produced. At the heart of this artificial lift technique is the gas lift valve, which acts as a pressure regulator, attempting to maintain either a constant casing pressure or a constant tubing pressure, depending on the construction of the valve (Fig 6.4.28).

Gas lift is utilized in one of two ways: (1) by continuous gas injection into the tubing causing a continuous flow of reservoir fluid or (2) by rapid injection of very large quantities into the tubing, causing a slug of fluid in the tubing to be carried to the surface. The valve then closes, awaiting another column of fluid to build in the tubing. The first method is called *continuous lift* and is used in wells with production capacity such that it will flow continuously. The

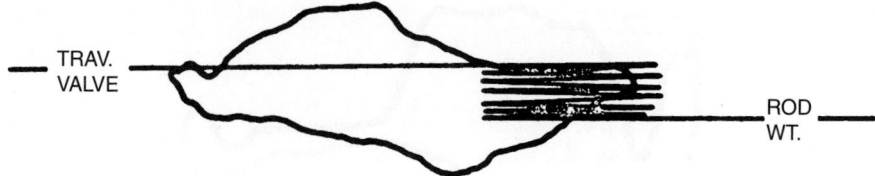

Figure 6.4.25   *Traveling valve with severe leak [2].*

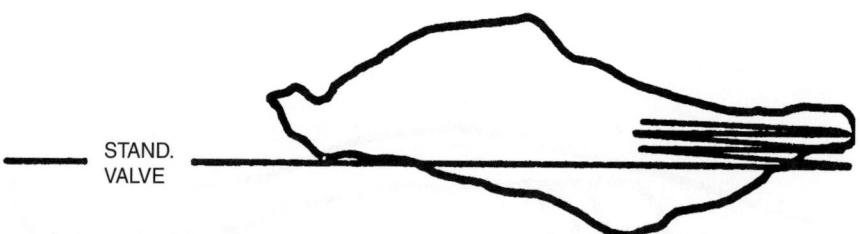

Figure 6.4.26   *Standing valve with severe leak [2].*

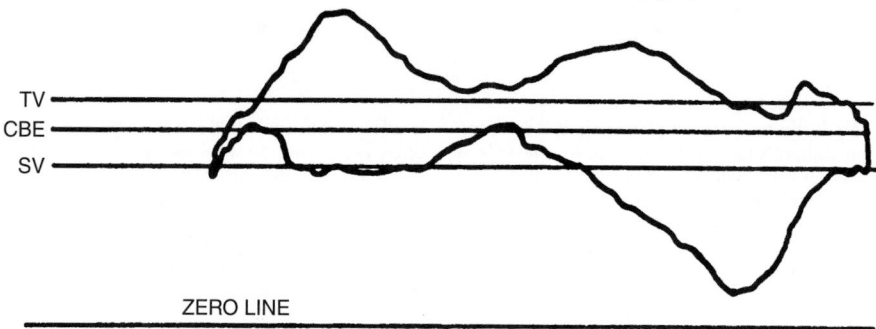

Figure 6.4.27   *Effective counterbalance [2].*

second method refers to intermittent lifting of wells where the reservoir is of very limited productivity. Quite obviously, the intermittent lift method is inefficient and requires specialized surface equipment to handle the large rates involved

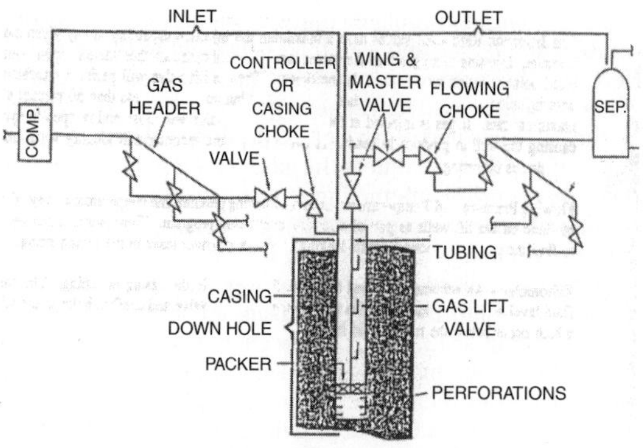

Figure 6.4.28   *Typical gas lift system.*

whenever a slug of production reaches the surface. Because of the limited use of intermittent gas lift, this treatment will be confined to continuous gas lift. Continuous lift can be accomplished with casing-operated valves, tubing-operated valves, or valves that have characteristics of both. These will be described later.

The next section will describe concepts necessary for gas lift design and then will give example designs. First, some knowledge of the productivity of the reservoir is necessary to determine the required drawdown on the perforations. If there is too little drawdown, the well underproduces. If there is too large, a drawdown and the perforations (gravel packs, etc.) could be damaged. Consequently, an engineering knowledge of the production capabilities of similar wells is required.

Another tool needed by the designer is a reliable or industry accepted computer software program for generating a single gradient curve or a comprehensive set of gradient curves for multiphase flow of oil, water and gas in vertical pipes. These will be described in a later section. Also, an engineer designing gas lift installations needs a working concept of valve mechanics. This topic is also covered in a later section. As with computer software for gradient curves, there is available computer software for assisting the designer with determining gas lift mandrel depths, gas lift valve calculations or set pressures, and analyzing gas lift valve performance under dynamic flowing conditions.

After a working knowledge of these topics is covered, an example design of both tubing-operated and casing-operated valves will be presented. These are offered as simple examples under certain assumptions. They can be elaborated upon and changed under certain conditions. In other words, there are many sound engineering variations leading to working designs. A trained representative of a reputable gas lift service company should be consulted for further design techniques.

6.4.7.1 Inflow Performance

Inflow performances of a reservoir can be defined as the functional relationship between the flowing bottomhole pressure and the resulting flowrate. Several models are available, such as straight line (productivity index, or PI), Vogel's method, the Fetkovich method, and others [13].

First, the straight line, or PI, is given as the ratio of the total barrels of fluid produced to the pressure drawdown across the reservoir (drawdown is defined as the average reservoir pressure minus flowing bottomhole pressure), or

$$PI = \frac{q(bpd)}{\delta P(psi)}$$

This value can be obtained by direct measurement on a particular well by physically measuring the flowing and shut-in bottomhole pressure while also noting the surface flowrate. This number is valid only while the oil/water ratio for the well remains fairly constant. The effect of changing oil/water ratios can be seen below. If measured values are not available, one can estimate the PI with the approximate relationship obtained from the Darcy equation. This is done by utilizing a typical relative permeability relationship and correlations for viscosity and formation volume factors:

$$PI = HK_a \left(\frac{K_{ro}}{\mu_o B_o} + \frac{K_{rW}}{\mu_W B_W} \right)$$

where K_a = absolute permeability of the rock in darcies
K_{ro} = relative permeability to oil
K_{rW} = relative permeability to water
B = formation volume factor
μ = viscosity
H = reservoir thickness in ft.

A second method is that developed by Vogel, originally obtained for volumetric reservoirs at bubble point pressure where resulting gas saturations around the vicinity of the wellbore cause a decrease in the relative permeability. The relationship derived is given as

$$\frac{q}{q_0} = 1 - 0.2 \left(\frac{P_w}{P_b} \right) - 0.8 \left(\frac{P_w}{P_b} \right)^2$$

where q = flowrate in bpd
q_0 = flowrate at maximum drawdown in bpd
P_w = flowing bottomhole pressure of interest in psig
P_b = bubble point and reservoir pressure in psig

With this method, a measured value of flowing bottomhole pressure, average reservoir pressure and the associated flowrate is required to define q_0. Once the equation is thus normalized, it can then be used to predict the well's performance at any drawdown.

A third method occasionally used is that of Fetkovich, where flowrate is given by

$$q = C(P_r - P_w)^n$$

where C and n are constants and must be determined by conducting two separate flow tests. The parameter P_r is the average reservoir pressure.

Example 6.4.3

A well flowing 50% water (viscosity = 0.5 cp and FVF = 1) and 50% oil (viscosity = 4 cp and FVF = 1.5) from a 20-ft.-thick reservoir. Core analysis indicates the absolute permeability to be 300 md. Estimate the PI of this well.

Answer

Assuming typical relative permeability relationships, we estimate the relative permeability of the reservoir to oil to be 0.25 and to water of 0.25. This yields a PI = 3.25 bfpd/psi. Note that when this well produced all oil, the PI would be estimated at 1 bopd/psi.

If this well is tested at a flowrate of 200 bopd and 200 bwpd, with a shut-in bottomhole pressure of 4500 psig, and a flowing bottomhole pressure of 4350 psig, what is the calculated PI? What is the expected production rate for a drawdown of 1000 psi by the PI method?

Answer

PI = 2.67 bfpd/psi, implying a rate of 2670 bfpd for a 1000-psi drawdown.

If a well is tested at 400 bopd for a drawdown of 150 psi, what does the Vogel method predict for a 1000-psi drawdown? The reservoir pressure (also the bubble point pressure) is 4500 psi.

Answer

q_0 = 6767 bopd, and yields a flowrate of 2439 bopd for a 1000 psi-drawdown.

A second test on this well shows a production rate of 1000 bopd for a 500-psi drawdown. What flowrate does the Fetkovich method predict for the 1000-psi drawdown?

Answer

C = 0.006 and n = 0.79; flowrate = 1704 bfpd at the 1000-psi drawdown.

Neither of these calculation techniques can be considered preferable; each depends on the engineer's experience and the actual reservoir conditions.

6.4.7.2 Gradient Curves

The first concept needed in gas lift design is that of gradient curves. For the case of single fluid flow in pipes, one can calculate the total pressure difference between two points with well-defined mathematical equations, requiring only a friction factor empirical relationship for the case of turbulent flow. Even in the case of non-Newtonian fluids, appropriate assumptions have been made enabling the calculation of pressure difference with a single equation for pipe or annular flow. In the case of multiphase flow involving gas–liquid mixtures, a single equation is not possible. Pressure variation along the length of the pipe causes gas volume changes. In the case of oil–gas flow, the oil may liberate solution gas into the flow stream. Consequently, a change of the *in situ* gas–liquid fraction implies a constantly changing hydrostatic component of pressure, a constantly changing friction factor and, possibly, an acceleration component of pressure change.

The only method of calculating pressure differences for multiphase flow is to resort to computer numerical evaluation utilizing incrementing type algorithms. Starting conditions in the first increment are used to estimate conditions in the next increment and so on until the total pressure difference is calculated. Calculation techniques are available in the literature such as Poettmann and Carpenter, Hagedorn

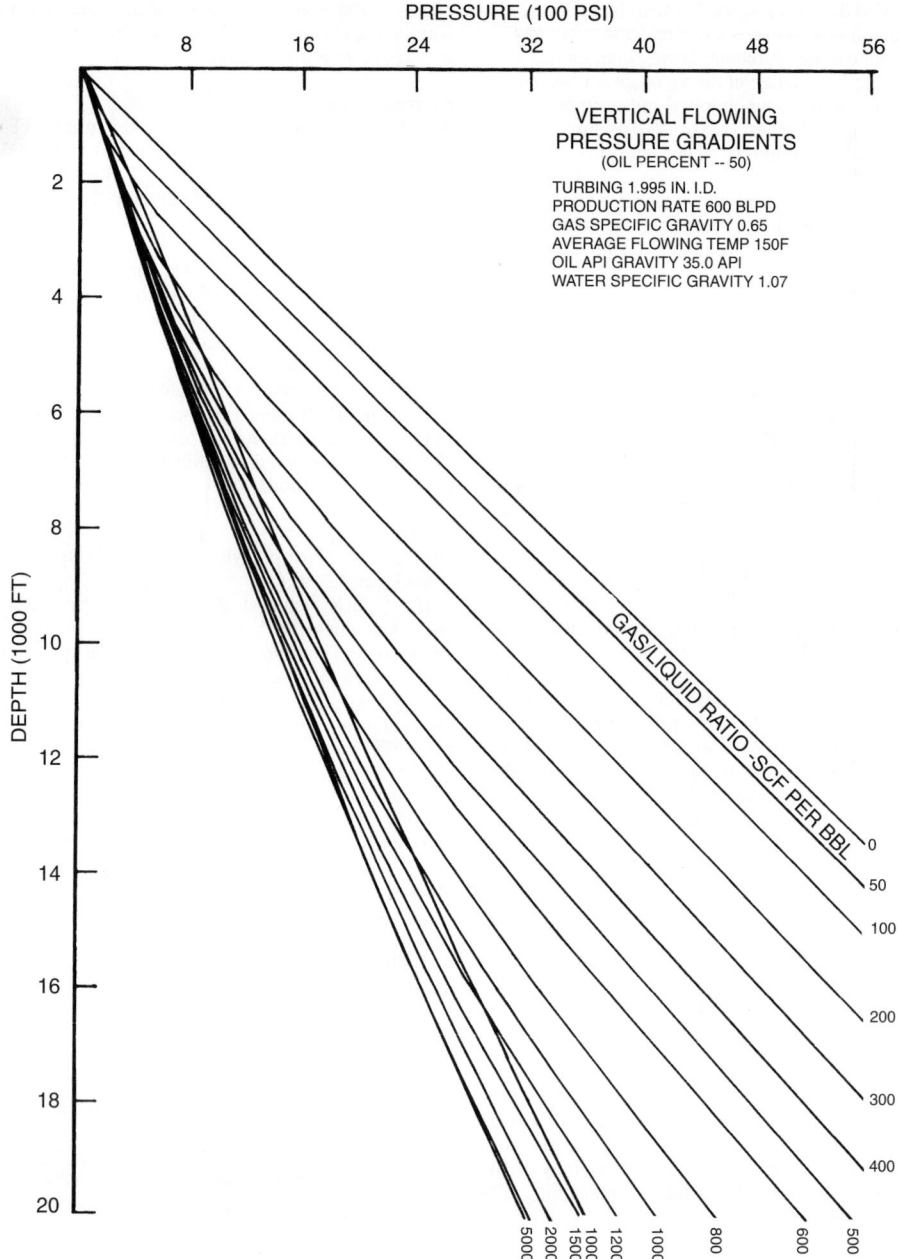

Figure 6.4.29 *Example of vertical flowing pressure gradient for multiphase flow.*

and Brown, Orkiszewski, Beggs and Brill, Duns and Ros, and many others [14]. The reader is encouraged to use appropriate computer programs with which to obtain gradient curves for design work. The appropriate choice of calculation technique is dependent on such factors as the geographical area the wells are located, etc. The author will refer to the gradient curves given by K. Brown in *The Technology of Artificial Lift Methods* [13].

Gradient curves for multiphase flow in vertical pipes are used in one of three ways: (1) to predict the pressure at the bottom of a measured interval if the pressure at the top of the interval, the gas/liquid ratio, and the liquid flowrate are

known; (2) to predict the pressure at the top of a measured interval if the pressure at the bottom of the interval, the gas/liquid ratio, and the liquid flowrate are known; or (3) to predict the gas/liquid ratio required if the pressures at both ends of a measured interval and the liquid flowrate are known.

First, one must realize that the vertical axis on typical gradient curves (see Figure 6.4.29) does not represent absolute depth but rather a relative vertical length. The user must determine the appropriate section of a gradient curve to use for his or her particular problem. For (1) above, one first identifies the appropriate gradient curve and then locates the pressure of interest at the top of the interval, noting the vertical ordinate for that point. The interval length of interest is then added to this vertical ordinate. This identifies the appropriate interval of interest. The pressure at the interval bottom is the predicted pressure. In the case of (2) above, this process is reversed. For (3) above, we imagine two vertical lines drawn at the end pressures. These lines intersect the entire family of gradient curves, but only one curve (perhaps an interpolated one) has the prescribed vertical interval. This process is a trial-and-error one until the correct curve is identified. Of course, a computer program with which to calculate gradient curves would be somewhat easier to use.

Example 6.4.4
Three example problems will be worked below for a well with 2-in. tubing (1.995″ ID) flowing all oil at a rate of 600 bpd and a GLR of 400 scf/bbl. It is assumed that the gas and oil properties given by Brown [13; 3a, page 194] are appropriate (see Figure 6.4.29).

1. If the well is flowing with a surface pressure of 400 psig, what is the pressure at a depth of 5000 ft.? At 10,000 ft.? The answers are 1520 psig; 3280 psig.
2. If the same well is flowing with a bottomhole pressure of 4000 psig at a depth of 8000 ft., what is the pressure at the surface? At 4000 ft.? The answers are 1240 psig; 2520 psig.
3. The same well is flowing with an unknown gas/liquid ratio. The surface pressure is 600 psig and a known pressure of 1600 psig at a depth of 5000 ft. What is the flowing gas/liquid ratio? The answer is a pressure difference of 1000 psi over the 5000-ft interval implies approximately 600 scf/bbl.

6.4.7.3 Valve Mechanics
Gas lift valves, by necessity, are constructed as shown in Figure 6.4.30 and 6.4.31, where all components must be built into a small cylindrical-shaped tube. The diameter and length will vary according to size restrictions imposed by tubing size, mandrel size, etc. Gas Lift Valves are available in two different configurations:

- Wireline retrievable valves–installed in sidepocket mandrels and will not require pulling the tubing string or completion string for repair or recalibration on gas lift valves (example shown in Figure 6.4.32).
- Tubing retrievable valves–installed on conventional mandrels and can only be repaired or recalibrated by removing the tubing or completion string (example shown in Figure 6.4.33).

A typical valve must have a closing force (provided by a spring, gas pressured chamber, or both of these), an opening force (provided by a metal bellows upon which either tubing or casing pressure acts) and a flow-controlling orifice.

If tubing pressure is exerted against the bellows in the open position, this valve is referred to as a fluid-operated or production pressure — operated (PPO) gas lift valve. If casing pressure in applied on the bellows in the open position, then we have a casing-operated or injection pressure — operated (IPO) gas lift valve.

Gas lift valves operate similarly to pressure regulators. Note that a casing-operated valves acts to maintain a set casing pressure. If casing pressure increases, the valve opens further, attempting to relieve the additional pressure. Conversely, a pressure decrease causes the valve to pinch down in an attempt to decrease gas flow and thereby maintain casing pressure.

On the other hand, a tubing-operated valve attempts to maintain a set tubing pressure. An increase in tubing pressure opens the valve to allow additional gas into the tubing in an attempt to lighten the fluid column above. If tubing pressure decreases, then the valve pinches down in an attempt to increase the tubing fluid gradient by reducing the gas flow into the tubing. There are several variations on this arrangement such as balanced valves, etc., but these will not be covered here.

The force balance equations (see Craft, Holden, and Graves [1]) give the following relationships for a casing-operated valve that is open and on the verge of closing:

$$P_{vc} = P_{bT} + S_t(1 - A_v/A_b)$$

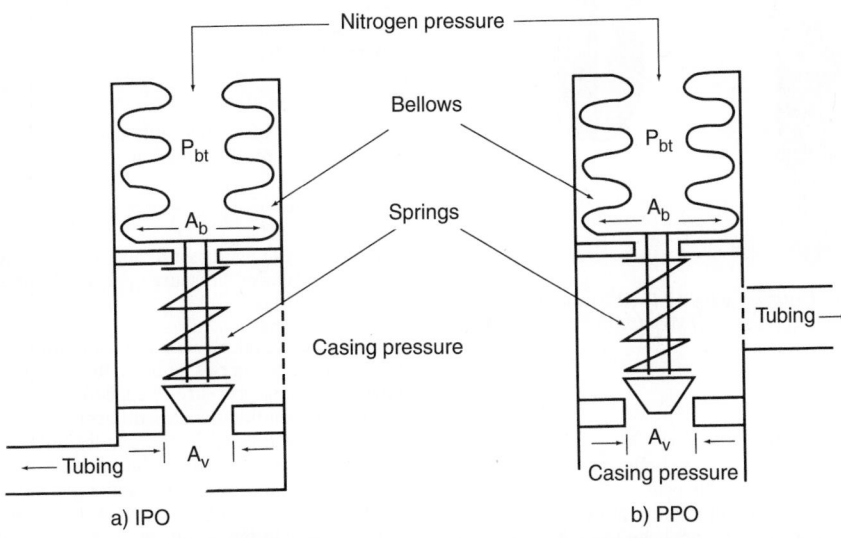

Figure 6.4.30 Simplified drawings of IPO and PPO valves.

Model R -1, IPO
Gas Lift Valve

Model RF -1, PPO
Gas Lift Valve

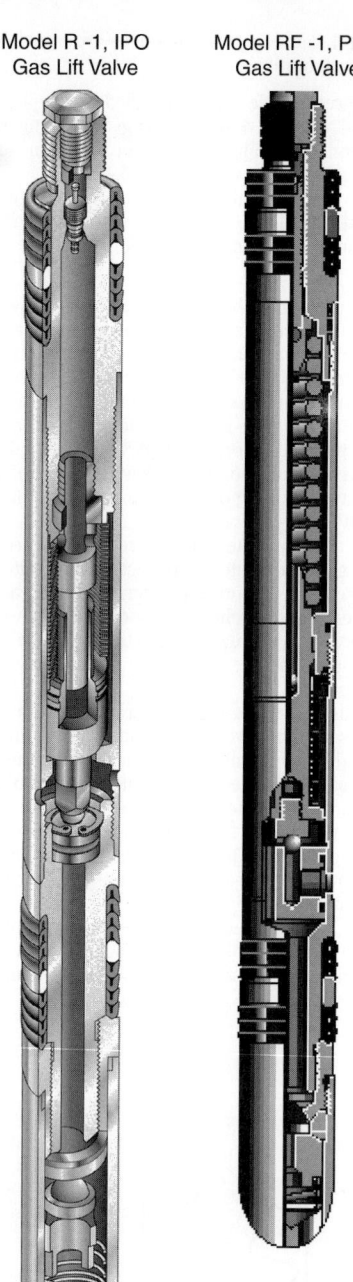

Figure 6.4.32 *Example of sidepocket [wireline retrievable gas lift valve] mandrel (valve installs into mandrel pocket—valve not shown). Courtesy of weatherford-PS.*

In the case of a valve fully closed and on the verge of opening, we have the equation

$$P_{v0} = \frac{P_{bT}}{(1 - A_v/A_b)} + S_t - \left[\frac{A_v/A_b}{(1 - A_v/A_b)} \right] P_t$$

where P_t = tubing pressure opposite the valve in psig
P_{v0} = casing pressure opposite the valve at opening in psig

Note that P_{v0} and P_{vc} are not equal due to the fact that the casing pressure is applied to the valve stem during flow and tubing pressure is applied to the valve stem during closed conditions. This difference is called the *valve spread* and is utilized in the gas lift unloading process to ensure that valves above the lifting valve are closed. Hence, a casing-operated valve will have a gas passage somewhat as shown below in Figures 6.4.34. For exact valve performance characteristics, one would have to consult the valve manufacturer or utilize the computer software program,

Figure 6.4.31 *Cutaway view of 1" IPO model R-1 and 1" PPO model RF-1 (courtesy of Weatherford-Production Systems).*

where A_v = area of the valve port
A_b = effective area of the bellows
S_t = spring constant in psig
P_{bT} = bellows pressure temperature T in psig
P_{vc} = casing pressure opposite the valve at closing in psig.

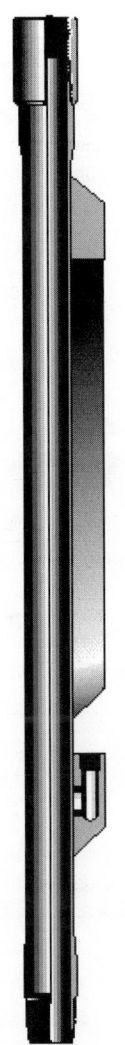

Figure 6.4.33 *Example of conventional gas lift valve mandrel (valve screws on the lower lug adapter—valve not shown). Courtesy of Weatherford-PS.*

Valve Performance Clearinghouse (VPC) 2002 (current version for year 2003). The VPC Program is a proprietary software program available only to VPC members. This gas lift valve performance correlation program is based on tested valve performance and incorporates valve parameters

such as loadrate, maximum effective travel, flow coefficients, dynamic test data. Other gas lift valve performance correlations available are Thornhill-Craver (use for orifice valves only), Winkler-Eads, Bertovic (developed for PPO valves only), API Simplified, TUALP (Tulsa University Artificial Lift Project), and VPC/TUALP. Typical flow performance curves, as predicted by VPC correlation, for 1" OD IPO and 1" OD PPO valves are illustrated in Figure 6.4.35 and in Figure 6.4.36 respectively. In Figure 6.4.37, the predicted flow performance for same IPO valve is compared with different correlations. This shows the reason for not using a square-edged orifice table or correlation (Thornhill-Craver) to select the correct port size in a gas lift valve.

In the case of a tubing-operated valve, the situation is somewhat different in that a valve spread is meaningless. One would calculate a valve tubing closing pressure at a tubing pressure greater than the valve tubing opening pressure. This is impossible; thus the actual closing pressure for a tubing operated valve is determined experimentally in a test rack at the shop; however, the valve closing equation above for a casing operated valve is very close if we substitute tubing pressure for P_{vc}.

Since the bellows volume can be considered constant, the bellows pressure at any temperature is related to that at shop conditions of 60°F by the equation

$$\frac{P_{bT}}{Z_T(460 + T)} = \frac{P_{b60}}{Z_{60}(460 + 60)}$$

The gas deviation factor for nitrogen is given by Sage and Lacy [15] or Craft, Holden and Graves [1]. However, Z factors for nitrogen are very close to 1.0 and only deviate from that value by up to 5%.

Example 6.4.5
A casing-operated valve is to be run at a depth of 2000 ft., where the operating temperature is expected to be 95°F. The valve has an A_v/A_b ratio of 0.1, and a spring effect of 200 psig. What nitrogen pressure at depth is required in the bellows if the closing pressuring desired at depth is 900 psig? What nitrogen pressure must be placed in the valve at 60°F?

Answer
P_{bT} = 720 psig; P_{b60} = 683 psig using a Z factor of 1.006 for nitrogen at 95°F.

One must also calculate the pressure increase for static gas columns due to its own density. A good approximation for methane is

$$\delta P = 0.25(P_{wh}/100)(D/100)$$

For example, if the available casing surface pressure P_{wh} is 1000 psig, then the casing pressure will increase 25 psi/1000 ft. of depth. This can be rewritten as

$$P_a = P_{wh}(1 + D/4.0 \times 10^4)$$

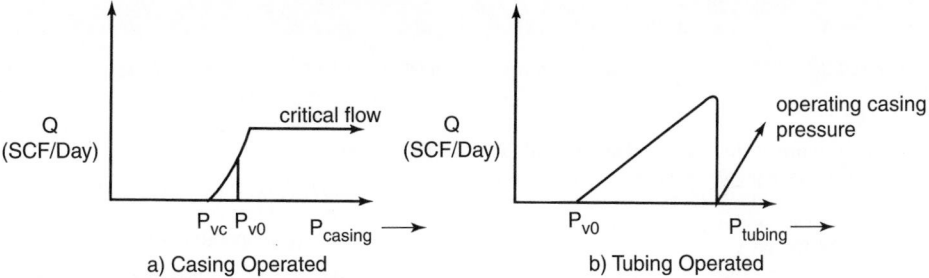

Figure 6.4.34 *Gas passage characteristics for casing and tubing operated valves.*

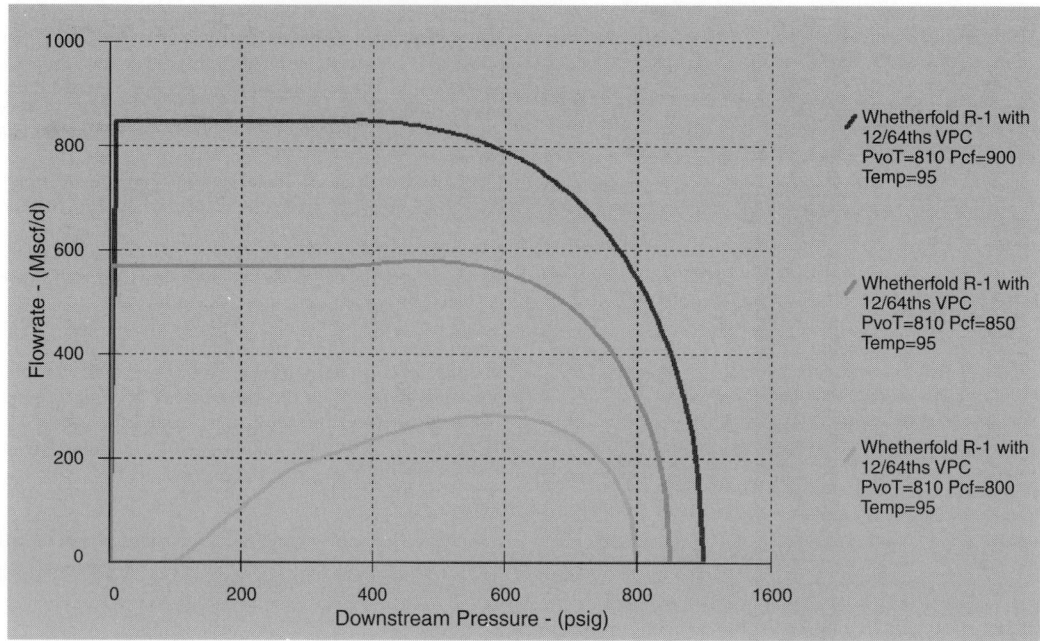

Figure 6.4.35 *VPC correlation — flow performance on IPO model R-1 (courtesy of Weatherford-PS).*

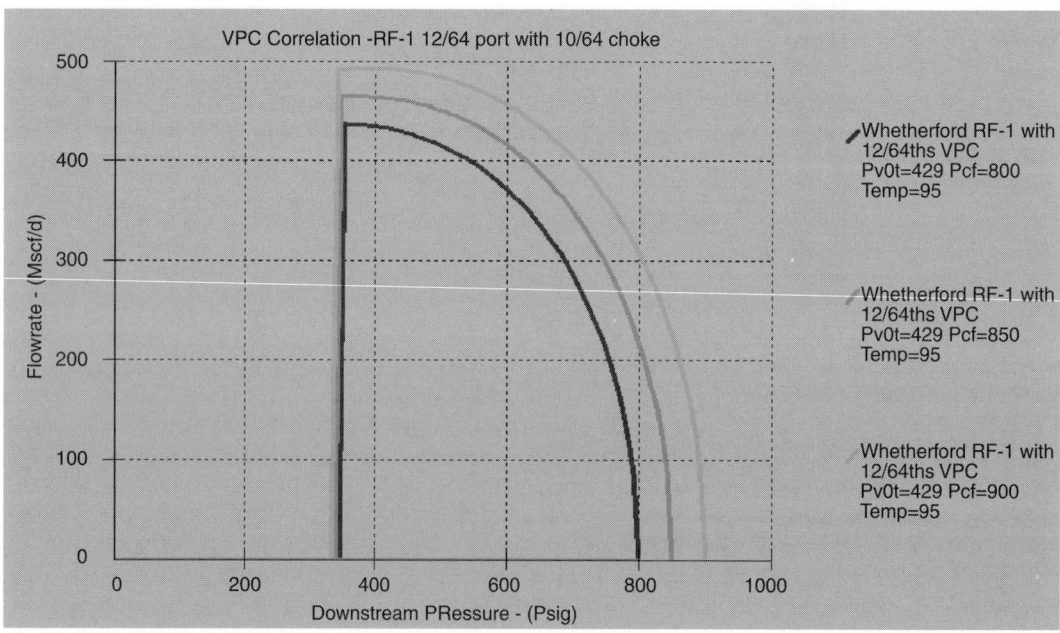

Figure 6.4.36 *VPC correlation — flow performance on PPO model RF-1 (courtesy of Weatherford-PS).*

where P_B is the pressure at the bottom of an interval D in feet, and P_{wh} is the surface pressure, in psia. Typical gas lift situations have a large annular casing cross-section and the gas flow to the valve can be thought of as a static column. For greater accuracy, we have

$$P_B = P_{wh} \exp(0.01875\sigma D / Z_{avg} T_{avg})$$

where

σ = gas specific gravity (air = 1.0)
Z_{avg} = average Z factor over the interval
T_{avg} = average temperature of the gas column in °R

For gas flow in small pipes, frictional pressure losses may be critical. Buthod and Whitely [7] have developed the

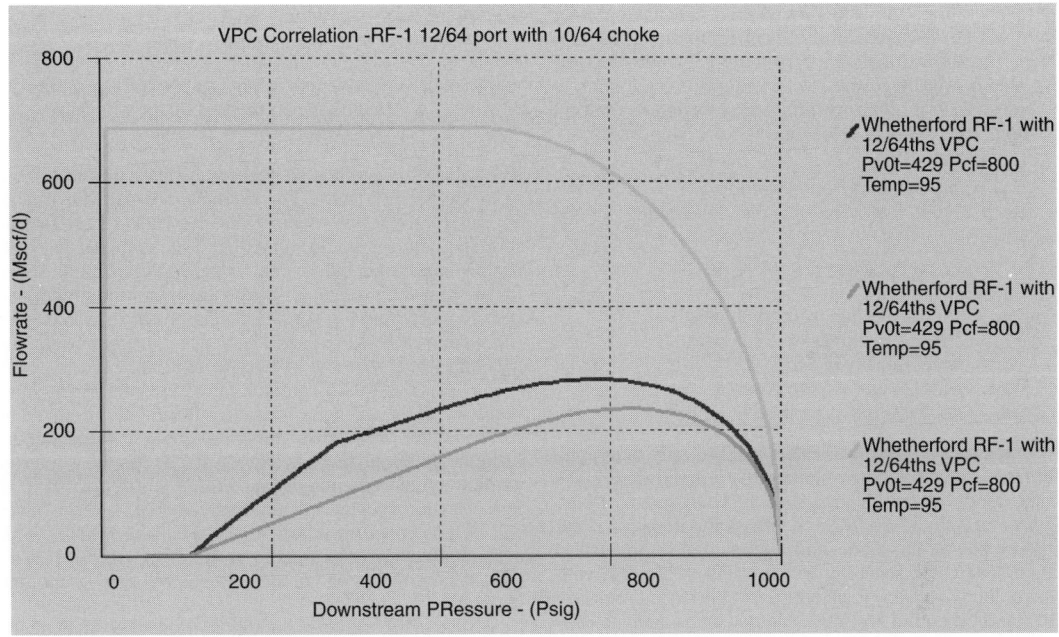

Figure 6.4.37 *Comparison of Correlations — IPO model R-1 under same conditions (courtesy of Weatherford-PS).*

following equation

$$P_B^2 = B(P_{wh}^2 - A) + A$$

where

$B = \exp(0.0376\sigma D / Z_{avg} T_{avg})$
$A = 1.51 \times 10^{-5} (q T_{avg} Z_{avg})^2 / d^{5.23}$
d = equivalent flow channel diameter in in.
q = gas flow rate in Mcf/d

For example, 300 Mcf/d of gas flow down a 1-in. ID tubing ($\sigma = 0.65$, $T_{avg} = 560°R$, and $Z_{avg} = 1$), with a surface pressure of 1000 psig, will have a pressure at a 2000-ft depth of 1025 psig as compared to the expected 1050 psig for a static gas column.

6.4.7.4 Unloading the Well

A phenomenon one must address in gas lift design is that of unloading the casing annulus. Unloading simply means removing all the casing (packer) fluid left in the well between the tubing and casing at the time of completion and replacing it with gas down to the point of injection. First, if drilling mud or other abrasive fluid is in the casing annulus, it must be replaced with clean water by using a circulating pump. In the case of tubing retrievable valves, this task is easily accomplished at the time the pulling rig is on location. For wireline retrievable valves, a mandrel below the expected point of gas injection should be opened, and clean water circulated until the packer fluid is completely replaced. The accepted practice for circulating fluid through the pocket and pocket seal bores in a sidepocket gas lift mandrel is to use a circulating valve or dump kill/shear orifice valve as illustrated in Figure 6.4.38. By using these special service valves, the pressure integrity of the sidepocket mandrel is maintained.

It is very likely that the desired point of gas injection will be deeper than can be unloaded with a single valve. Imagine a U-tube that is filled with liquid on both sides and has a gas pressure applied to one side. The fluid will move from

Figure 6.4.38 *Cutaway view of dump kill valve, model RDDK-1 (courtesy of Weatherford-Production Systems.).*

one side to the other, spilling over at the top, and continuing until the pressure at the bottom on both sides become equal. If the U-tube is too long, this process will stop whenever the pressures equalize. For example, if the desired depth of injection is at 5000 ft. and the casing fluid is formation water of density given by 0.465 psi/ft., a gas pressure at the valve of 2325 psig would be required to unload the well with a single valve. Since this gas pressure is excessive, the conclusion is that one simply cannot unload the well with a single valve.

By this logic, several unloading valves will be needed. There are two methods for locating the unloading valves and the operating valve. The first method, Single Operating Depth, will be discussed in the following paragraphs with details. The second method, "Bracketing Envelope" Operating Depth, will be covered with general comparisons to the first method. In both methods, the location of the first unloading valve depth is determined by the same steps.

6.4.7.5 Single Operating Depth Method

The depth of the first valve can be determined by the equation

$$P_{ts} + \sigma_w D_1 = P_{cs} + \sigma_g D_1$$

where P_{ts} = surface tubing pressure in psig
P_{CS} = surface casing pressure in psig
σ_w = water density in psi/ft
σ_g = average gas density in psi/ft
D_1 = depth to first valve in ft.

This equation can be solved either algebraically or graphically (see Figure 6.4.39) by plotting the left and right sides independently, finding the intersection. This is done by plotting pressure on the *X-axis* and depth on the negative *Y-axis* (standard plot for gas lift design). For example, if the well is being unloaded to a separator whose pressure is 50 psi, with a packer fluid of 0.5 psi/ft, and the available casing pressure is 1000 psig, then the first valve can be run as deep as 2000 ft. Of course, a more shallow depth would ensure successful unloading to the first valve, but would cause the deepest possible point of lift.

Once this first valve have been uncovered, it can be used to reduce the hydrostatic pressure in the tubing above that point by gas injection. Reduction of the hydrostatic pressure simply means that a second valve can be run below and the casing gas pressure will push the packer fluid down, through the second valve so it can be lifted to the surface by the first valve. Again, the depth to the second valve is determined by

a pressure balance condition given by

$$P_{ts} + \sigma_{min}D_1 + \sigma_w(D_2 - D_1) = P_{CS} + \sigma_g D_2$$

where σ_{min} = the average minimum two-phase gradient that can be developed by the first valve in psi/ft.

The value of P_{CS} to be used here may differ from that used to locate the first valve. For example, a design for casing-operated valves will require that the casing pressure be reduced. Therefore, the value to be used is the surface casing pressure available for that valve. In the example above for the first valve, assume the average minimum gradient to be 0.15 psi/ft. and the surface casing pressure available for the second valve is 975 psig. The deepest depth to which the second valve can be run is therefore 3421 ft.

This process would continue to locate subsequent valve depths until the casing pressure (right side of the equation) can no longer produce the minimum gradient. Note that the distance between valves is decreasing. Typically, whenever a minimum incremental distance (on the order of 400 ft. or less) or the expected lift depth is reached, several valves will be positioned, if possible, at this distance apart. This accounts for any errors made in the design and increases the probability that one of these valves will be the optimum lift depth.

At this point, another definition is needed, that of the deepest point of injection. This depth is defined by the equation

$$FBHP - \sigma_{ave}(D_T - D_{iv}) = P_{cs\ lift} + D_{iv}\sigma_g$$

where FBHP = flowing bottomhole pressure in psig
σ_{ave} = average flowing gradient in the tubing from the perforations to the lift valve in psi/ft.
D_T = total depth of the well to the perforations in ft.

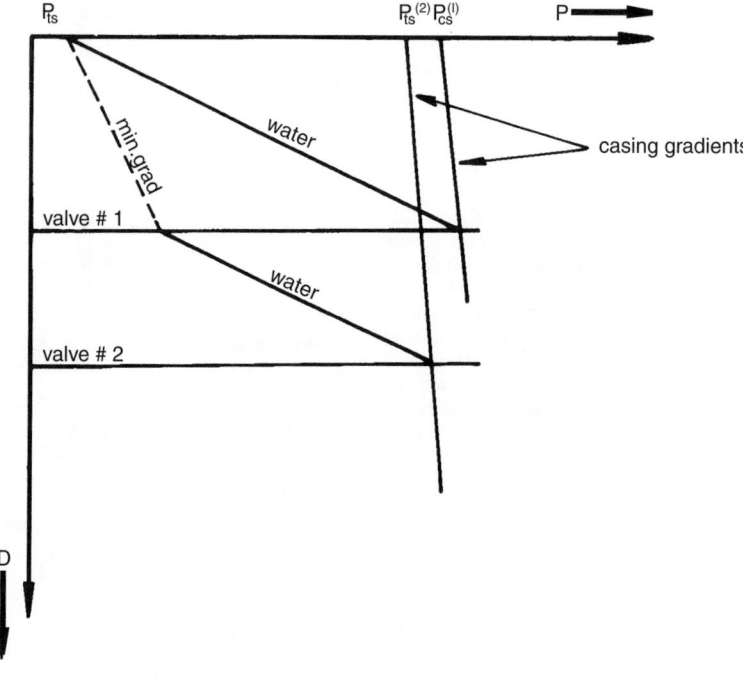

Figure 6.4.39 *Locating the unloading valve depths.*

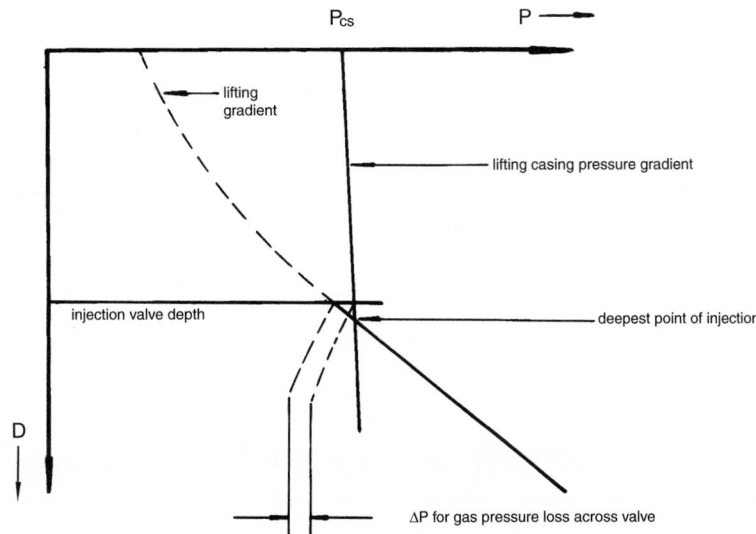

Figure 6.4.40 *Locating the injection valve depth by single operating depth method.*

D_{iv} = depth to the injection valve in ft.
$P_{cs \, lift}$ = surface casing pressure operating the lift valve in psig

The average flowing gradient in the tubing can be estimated to be the average density of the flowing liquids. This assumes that essentially no gas is liberated from solution until the liquid is above the lift valve. If, on the other hand, gas is liberated, then a straight line approximation is no longer valid, and a gradient curve should be used to approximate the second term on the left side of the equation above. Graphically, we can solve the above expression as shown in Figure 6.4.40.

6.4.7.6 "Bracketing Envelope" Operating Depth Method
The purpose of gas lift is injecting high-pressure gas into a well with the result of reducing the bottom hole pressure and increasing the fluid recovery. The deepest gas injection depth as possible will optimize the fluid recovery and minimize the volume of injection gas requirements. Traditionally, the gas lift design engineer solved for a single gas injection depth, as illustrated in Figure 6.4.40. From experience, this method would limit the potential production from the well, because the gas lift valves were not spaced deep enough or were spaced too far apart due to inaccurate information (such as errors in flow correlations, well performance [PI or IPR], changing reservoir conditions, changing surface operating parameters).

Weatherford's (McMurry-Macco) gas lift design engineers developed a better approach for spacing mandrels within an operating area. This method can account for errors in the original design information. The "bracketing envelope" method controls the differential between the injection pressure and the flowing tubing pressure in the operating area. This differential pressure is termed as the "operating differential", and the well's deliverability can be modeled based upon PI by available software programs (specifically Well Evaluation Model–WEM). As shown in Figure 6.4.41, with the operating differential (Pc-Pt) decreasing, the production rate from the well increases. This becomes more significant the higher the PI. Gas lift valve spacing is directly proportional to the operating differential selected.

The smaller operating differential, the closer the valve spacing. Therefore, the operating differential (Pc-Pt) may be controlled through the valve spacing. This fact makes it necessary to attempt to define the unloading area and the operating area. A wider spacing may be utilized in the unloading area reserving more valves for the operating area. Once the operating area has been identified the concept of the bracketing envelope can be utilized to maintain the proper operating differential and to offset any errors introduced through the inherent inaccuracies of the multiphase flow correlations used to profile tubing pressures or other design parameters.

A bracketing area tolerance must be selected. This tolerance should be based on the reliability of the information available. A small error tolerance may be used with good data. Questionable or unknown data justifies using a larger error tolerance. This will result in a more conservative design. Experience has indicated a potential 10% to 20% error associated with the use of the curves. Computer generated curves are more accurate than the published curves but inaccuracies still exist. The bracketing area is defined in the following manner (10% was selected for this example):

1. Multiply the wellhead pressure by the bracketing error percent

$$Pts = 120 \, Psi$$

$$120 \times 10\% = 12 \, Psi$$

Subtract and add this amount to the wellhead pressure and mark at the zero depth on the graphical design sheet, reference Figure 6.4.41.

$$P1 = 120 + 12 = 132 \, psi$$

$$P2 = 120 - 12 = 108 \, psi$$

2. Find the tubing pressure at the indicated point of injection or POI. This depth is determined by same method used for selecting the deepest point of injection, as illustrated in Figure 6.4.40. When using the bracketing envelope method, this depth locates the center of the operating area or envelope. For this current example, the tubing pressure at 3850 ft is 700 psi. Multiply this pressure by

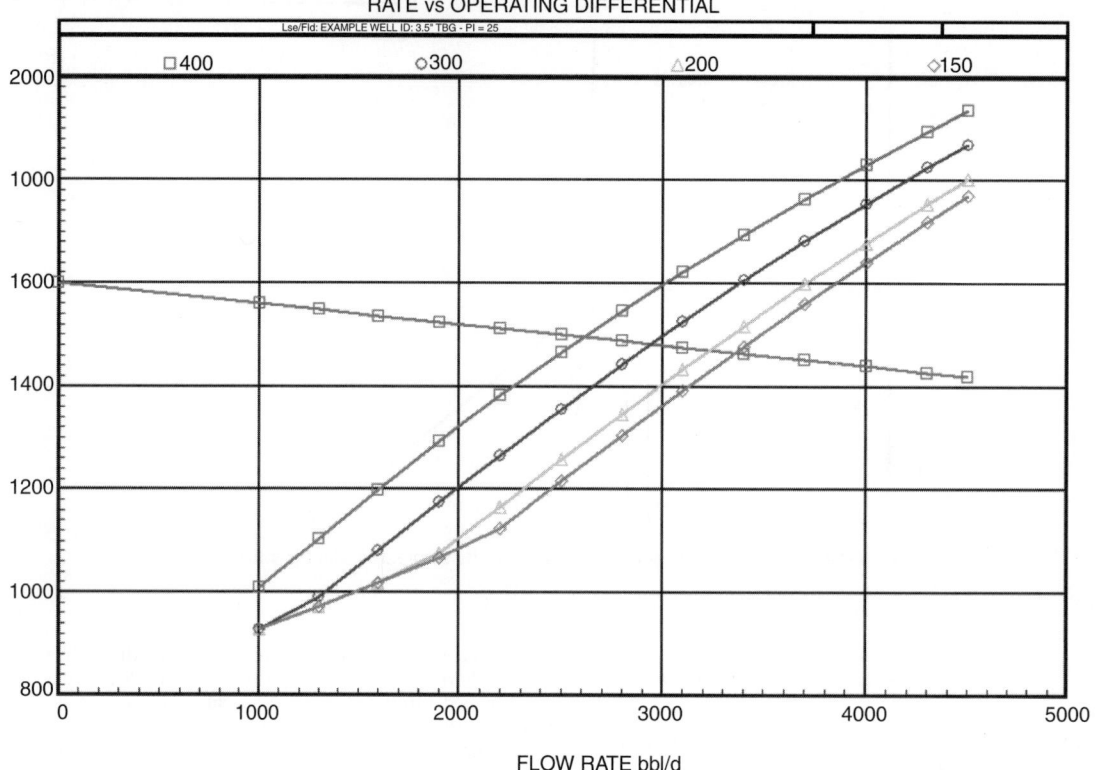

Figure 6.4.41 *Example of operating differential or gas lift valve spacing depths versus potential well production capacity.*

the bracketing area percent that was selected.

$$700 \text{ psi} \times 10\% = 70 \text{ psi}$$

Subtract and add this amount to the tubing pressure at the point of injection.

$$P3 = 700 + 70 = 770 \text{ psi}$$

$$P4 = 700 - 70 = 630 \text{ psi}$$

Mark these two pressures at the point of injection.

3. Connect a line from P1 (132 psi) to P3 (770 psi). At the intersection of this line and the operating differential pressure line, P5, is the approximate depth at which the bracketing spacing should begin or the top of the bracketing envelop. For this example, the top of the bracketing envelope is 3300 ft.

4. Connect a line from P2 to P4 until it intersects the differential pressure line. At the intersection is the depth for the bottom of the bracketing envelope.

Since the operating differential was selected at 150 psi (from our WEM modeling), all mandrels from the top of the bracket at 3300 ft. to the bottom of the bracket at 4600 ft. will be spaced 323 ft. apart.

$$\text{Bracketing spacing} = \text{Operating Differential (psi)} /$$

$$\text{Kill fluid gradient (psi/ft.)}$$

$$= 150 \text{ psi}/0.465 \text{ psi/ft.}$$

$$= 323 \text{ ft.}$$

For convenience, 350 ft. spacing would be used. The graphical solution for defining the bracketing envelope and bracketing spacing is illustrated in Figure 6.4.35.

Now, the logic of the unloading process must be explained. First, the casing operated valve design will be examined. An adjustable gas flow valve at the surface at the entrance to the casing annulus is required to have the valves function properly. Note that once lift is achieved, opening the surface casing valve will cause an increase in the casing pressure and will thereby cause the lift valve to open (allowing more gas into the tubing) since a casing-operated valve attempts to maintain a preset value of pressure. Conversely, pinching down on the surface valve causes the lift valve to reduce the gas passage. This is true unless the valve is fully opened and flow through the orifice is critical.

In the unloading process, gas pressure is applied to the casing annulus, causing water to be pushed through the valves until the first valve is uncovered and begins passing gas. To maintain a constant casing pressure, the surface valve is set to a gas flowrate that is essentially equal to the gas passage rate of the first valve that is now uncovered. Remember, packer fluid is being pushed through lower valves into the tubing and is being lifted to the surface by the first valve. Once the second valve is uncovered, gas is passed into the tubing by both valves. However, since the surface valve is set to a rate equal to one valve, the casing pressure will drop. The second valve closes when the casing pressure drops to the set closing pressure of the first valve. The casing pressure will then remain at this new, lower valve as long as the surface valve has a flowrate equal to that of the second valve. By manipulating the surface valve, a pumper must at this time ensure that the casing pressure remains at this new value, below the opening pressure of the first valve.

This process is repeated when the third valve is uncovered, and the casing pressure is again lowered, consequently closing the second valve. This continues until all the unloading valves have been closed and the lifting valve is the only valve injecting gas into the tubing. Note that if the well supplies formation fluid at such a rate that the next lower valve cannot be uncovered, the well is flowing, which is the desired result. If conditions become favorable in the future, the next valve may be uncovered and the well will begin lifting from the lower valve.

The use of casing-operated valves implies that one has temporarily available gas at a higher than lift pressure for the unloading process, or the well will be lifted with less than available casing pressure, thereby lowering efficiency. However, due to the operating characteristics of casing valves, the pumper, by manipulating the surface valve, has some surface control of the lifting valve. This design could not be used in the case of a dual lift, that is, when there are two tubing strings in a single wellbore.

Tubing-operated valves are sometimes referred to as automatic valves since the unloading process proceeds without pumper assistance. These valves respond primarily to tubing pressure and one causes the valves to respond to tubing pressure only by maintaining a constant casing pressure. The design process is done graphically, and requires first that the injection valve depth be identified, and then that the expected lifting gradient be determined. Since the flowing gradient–injection valve intersection determines a pressure and depth, and the flowing surface tubing pressure is known, the lifting gradient can be obtained from a family of gradient curves as described above in the *gradient curve* section or by a computer program.

The depth to the first valve is obtained as described above. The valve closing pressure is now dictated to be a value as close to, yet greater than, the lifting gradient; that is, to the right of the lifting gradient on a graphical design. This condition is required such that when this lifting gradient is achieved by the lifting valve, all valves above will be closed. The process starts with gas pressure applied to the casing annulus, causing the water to be pushed through the valves and up the tubing to the surface. When the first valve is uncovered, it will lighten the fluid column above because a temporary lifting gradient from the first valve is created when water enters the tubing from lower valves. This temporary gradient will become lighter and approach the final lifting gradient but not reach it due to the valve closing tubing pressure. Consequently, the depth to the second valve is found by the condition where

$$P_{ts} + P_{TG} + P_{safety} + \sigma_w(D_2 - D_1) = P_{CS} + \sigma_g D_2$$

where P_{TG} = total pressure difference of temporary gradient in psi
P_{safety} = small margin to ensure flow through valve in psi

Once the depth to the second valve is determined, its valve closing pressure is also chosen at a tubing pressure that is larger than the lifting gradient at that depth. Again, a temporary gradient is found from the surface unloading pressure to the second valve, and the above process is repeated for the next valve depth. This process continues until a transfer can be made to the injection valve found earlier.

Fluid-operated valves are typically used whenever minimal pumper surveillance is desired. Also, in the case of dual completions, the most popular choice is fluid-operated valves. However, another valve option would be "balanced gas lift valves". These type valves can be controlled with

either casing annulus pressure or tubing pressure. Therefore, the gas lift designer can used either fluid-operated or balanced gas lift valves, which allow maximum utilization of the available casing pressure to the operating depth. The efficiency of the gas lift design increases by lowering the injection depth. However, a string of fluid-operated or balanced valves usually require more unloading valves than a string of casing-operated valves.

Tubing-operated valves are typically used whenever minimal pumper surveillance is desired. Also, in the case of dual completions, tubing-operated valves are the only choice. Tubing-operated valves allow maximum utilization of the available casing pressure, thereby increasing efficiency. However, a string of tubing-operated valves usually require more unloading valves than a string of casing operated valves.

6.4.7.7 Tubing Operated Valves Example

Now that all the tools have been developed, a comprehensive example using tubing operated valves (see Figure 6.4.42 and 6.4.43) will be presented. The well is 10,000 ft. deep ($2\frac{7}{8}$-in. tubing) with a shut-in bottomhole pressure of 4000 psig and a measured PI of 1.33 bfpd/psi. The well flows 20% oil (30° API) and 80% water (specific gravity of 1.07). The surface pressure during lift is expected to be 50 psig, and the available gas pressure is 1000 psi (approximately 0.025 psi/ft. increase due to density). The bottomhole temperature is 200°F and the expected flowing temperature is 100°F. The design flowrate is 800 bfpd.

The flowing bottomhole pressure is 3400 psi. The expected gradient below the injection valve is 0.45 psi/ft. This information indicates the lowest possible point of gas injection to be at 4950 ft., where casing pressure is equal to the flowing tubing gradient. Consequently, the design point of injection is chosen at approximately 4700 ft. where a differential of 100 psi is available between the tubing gradient and casing pressure. This differential allows for the pressure loss for gas flow through the valve and a safety margin for design error.

The packer fluid is assumed to be 0.5 psi/ft. density. An actual value can be used if it is known. However, since lease water is typically used, a high estimate is used as a design factor. This implies that the first valve can be run as deep as 2000 ft. In the case of mandrels already run in the well, any depth above 2000 ft. would work. The second valve depth is determined by first finding the design GLR curve from the lowest valve. By trial and error, the GLR curve connecting the surface tubing pressure of 50 psi and the design point of injection-flowing gradient intersection of 1020 psi at 4700 ft. is found to be approximately 210 scf/bfpd. This GLR curve indicates pressure of 350 psi at 2000 ft. Consequently, the closing pressure of the first valve must be greater than 350 psi. The actual number is dependent on engineering judgment and the manufacturer of the valve used. In this example, we choose 425 psig, or 75 psi greater than the GLR curve.

A column of water will exist between the first and second valve, and a temporary GLR curve will exist between the first valve and the surface. The temporary gradient must be at least as heavy as 425 psig at the first valve since this is the closing pressure of the first valve. For a design factor, we will choose 475 psi, which we will call the transfer point. Adding a water column of 0.5 psi/ft. to this 475 psi, we see that this tubing pressure is equal to the casing pressure at a depth of 3200 ft. Note that once the second valve begins admitting gas into the tubing, the gradient will become lighter and begin closing the first valve. During the time when the gradient from valve two is somewhat heavy, the first valve will be open and assist in reducing the gradient. Also, if no

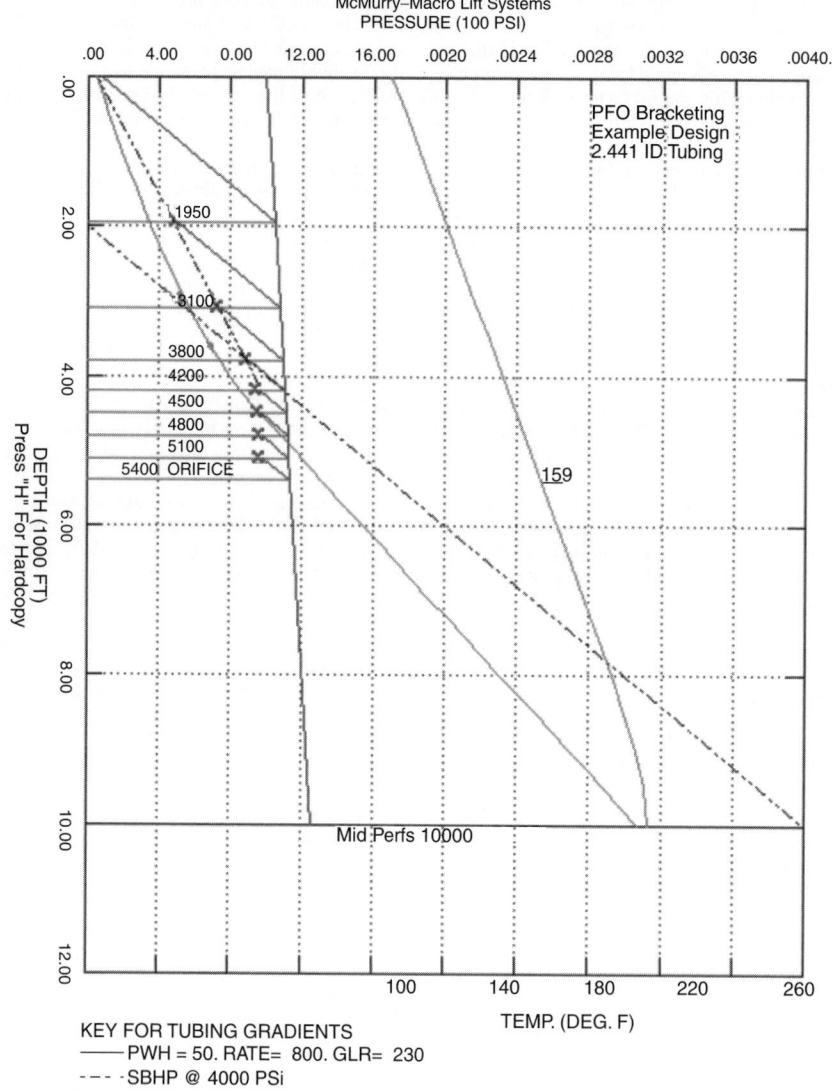

Figure 6.4.42 *PFO Bracketing.*

fluid is in the casing, such as with a well shut-in after it has been unloaded, the deepest valve that can open and allow gas into the tubing will do so. This would then start a process similar to unloading in that the opened valve lightens the gradient, allowing a deeper valve to begin passing gas and thus close the valve above. By definition, this condition must exist somewhere in the well; we simply cannot gas lift the well if the liquid column is below or unable to open the lowest valve.

The lifting gradient has a pressure of 600 psig at the second valve at 3200 ft. A typical closing pressure of 650 psi is chosen for the second valve, and the liquid column transfer point from the third valve is chosen at 700 psi. This gives a depth of 4000 ft for the third valve. The gradient pressure at the third valve is 815 psig, giving a closing pressure of 860 psig and a transfer point of 900 psig. The fourth valve is therefore located at 4440 ft. with a gradient pressure of 860 psig. The valve closing pressure is taken as 950 psig and the transfer point as 985 psig.

This unloading process continues uncovering the next lowest valve and closing the valve above until a reservoir drawdown is achieved and formation fluid is lifted. In the example, formation fluid will be lifted by the third valve during the unloading process, but will be in small volumes since it will be on the order of perhaps 400 bfpd according to our PI information. If the PI used is correct and the lower valves are sized to input the correct gas flowrate, the well will continue to increase the drawdown until the last valve is lifting the well. If we can expect the well to produce 50 Mcfpd from the formation, then an injection rate of approximately 118 Mcfpd is required to bring the GLR up to 210 scf/bf. The valve chosen must be capable of varying gas flowates from the 118 Mcfpd to 168 Mcfpd by throttling to accommodate those situations where the formation gas does not materialize, such as in the case of heading, etc. Its closing pressure will be less than 1000 psi, the value of the final lifting gradient at the valve depth.

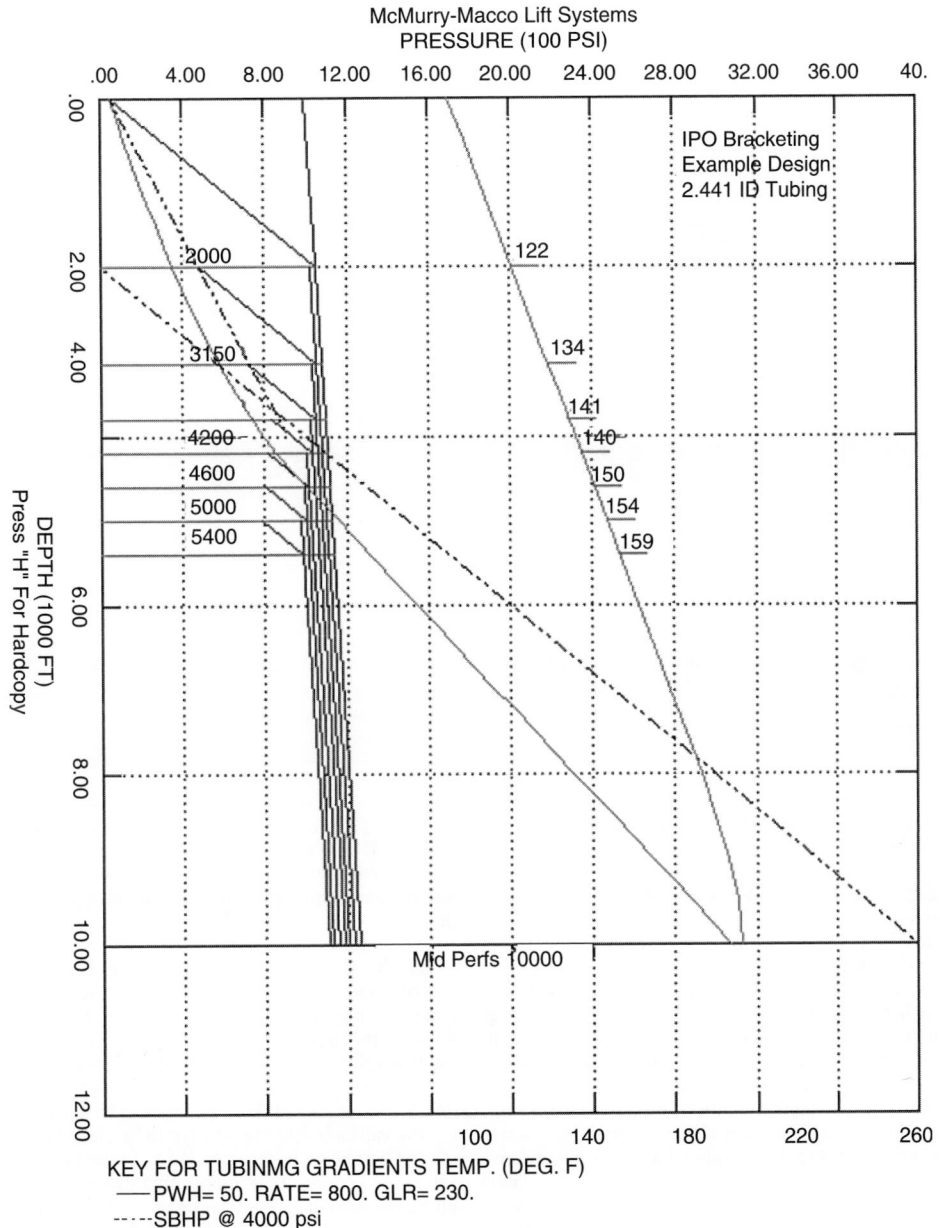

Figure 6.4.43 IPO Bracketing.

Note that if our PI information were incorrect, a GLR gradient would indeed establish itself. For example, if the actual PI is higher, then the well will lift from a more shallow valve, perhaps the third or fourth. This would be a heavier gradient than the 210 GLR designed, but would establish itself at a greater liquid flowrate determined by the available gas and GLR actually established. If, on the other hand, the actual PI is less than expected, the well will unload to the deepest valve and establish a flow rate of less liquid with the available gas passage rate at a steeper lifting gradient. From experience, this variance, the predicated PI and the actual PI in the well completion, is one reason the gas lift design engineer would prefer the "bracketing envelope" design method. The designer must be aware of both of these possibilities since a greater than expected liquid flowrate can be damaging to the formation (gravel pack, etc.) as can a greater than expected drawdown across the formation.

A malfunctioning gas lift design would require testing to determine the difficulty. For example, a casing fluid level determination as used in pumping wells can be useful in cases when the unloading process is not successful. Also, a flowing pressure survey (pressures taken while the well is lifting at various points in the tubing) can determine the actual point of gas injection. Pressure charts on the tubing and casing can sometimes pinpoint lift problems. Furthermore, a temporary increase, if available, of the casing

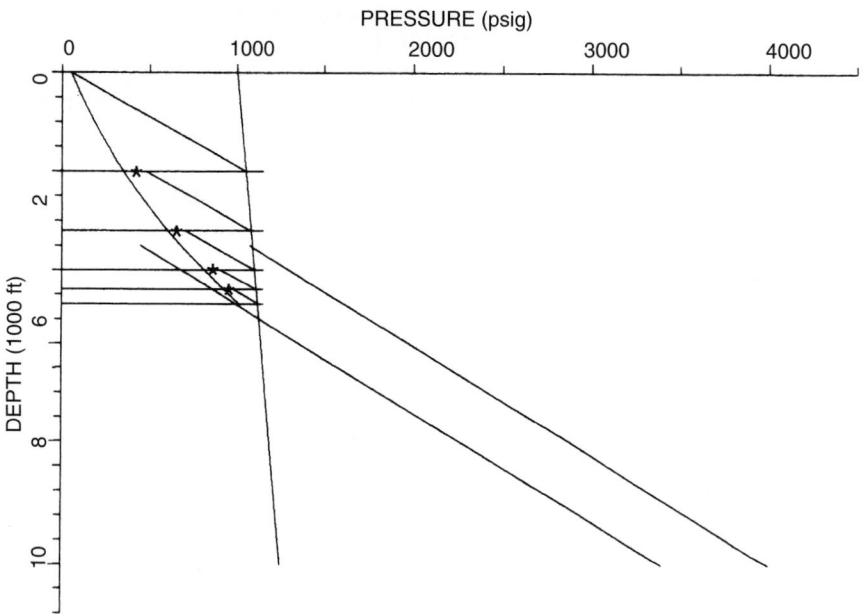

Figure 6.4.44 *Example design fluid-operated values and single operating depth method.*

pressure will help ensure a successful unloading of the well. After unloading, the casing pressure can be reduced to the designed value.

6.4.7.8 Casing Operated Valves Example

The same well described in the tubing-operated valves example will now be designed for gas lift with casing-operated valves. Several design techniques are available for casing-operated valves and one of these will be presented here. Further, an important assumption must be made concerning the casing pressure available. In this example, it is assumed that the 1000-psi lift gas is the maximum available pressure, that is, casing pressure cannot be increased, even temporarily. This assumption implies that the available casing pressure must be able to open all valves, even when there is no packer fluid in the casing annulus. This is the case whenever the well is shut-in (no packer fluid in the casing annulus) and the unloading valves are used to again start the well flow.

Some repetitive calculations are required in order to close the upper valves. The number of repetitive calculations will be reduced as the designer gains more experience in design methods and well characteristics within the field. Another option available to the designer is the application of a "spacing safety factor" during his graphical design. This spacing safety factor ranges from 25 to 50 psi, as illustrated in Figure 6.4.44. First, casing pressure traverses are drawn on the pressure versus depth (see Figure 6.4.45). Lines spaced 50 psi apart are drawn for reference. The static tubing pressure and flowing tubing pressure lines from the perforations are also placed on the graph.

The depth to the first valve is found in the same way as the tubing operated valves at 2000 ft. Because of the condition mentioned above, the opening pressure of the first valve is chosen to be 1050 psig, or the casing pressure at depth. If the valves chosen have an A_v/A_b of 0.067, the closing pressure of the valve at depth is 980 psig. This is an observed surface

casing pressure of approximately 930 psig. Also, if the valve has a 200-psi spring effect, the required bellows pressure at depth is 793 psig, or a test rack pressure of 702 psig nitrogen pressure must be loaded into the bellows at 60°F. This surface casing pressure of 930 psig becomes the operating pressure of the second valve.

The depth to the second valve is chosen as the depth at which the U-tube pressures are equal: that of the casing (corresponding to a surface casing pressure of 930 psig) and the tubing. The tubing pressure is again a water gradient between the first and second valves and a GLR gradient from the first valve to the surface. A convenient gradient to use is the design lifting gradient from the deepest valve. Since the depth is not known at this point, it must be decided. By knowing the valves below the first are to be closed by a decrease of 30 psi surface casing pressure, we determine that four valves are needed. This implies that the final surface casing lifting pressure is 870 psig and the design depth of injection is approximately 4400 ft. Using the Hagedorn-Brown correlation, the design GLR curve is found to be approximately 230 scf/bbl.

The logic of using this GLR curve as the gradient above the valves is that during unloading, the valves can certainly achieve this condition since this is the design lift condition. Consequently, using a transfer point of 400 psig and a 0.5 psig/ft water density, the second valve should be run at 3250 ft. where the casing pressure at depth is 1025 psig. The closing pressure of the second valve, therefore, is taken to be 995 psig (surface casing pressure of 920). This casing pressure becomes the operating pressure for the third valve, which has a value of 1010 psi at the third valve at a depth of 4000 ft. This valve depth is found by using a transfer point of 635 psig at the second valve. The closing pressure of the third valve is therefore 980 psig (a surface casing pressure of 870 psig). Since this is the last valve to close, the operating pressure for the lift valve is a surface casing pressure of 870 psig.

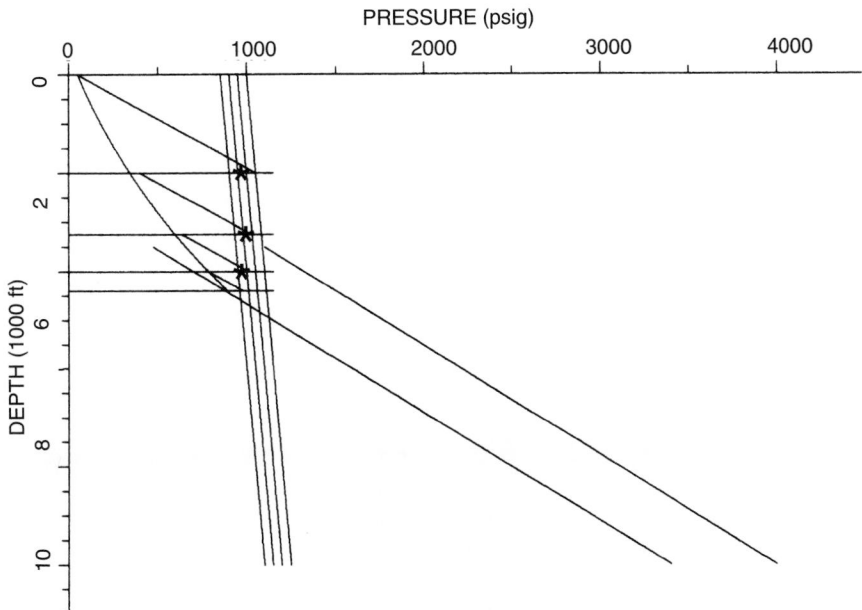

Figure 6.4.45 *Example design using casing-operated (IPO) values and single operating depth method.*

The closing pressure of the valve at 4400 ft. must be chosen such that the gas rate through the valve achieves the design GLR curve of 230 scf/bbl. If the formation can furnish 50 Mcfpd, then the valve must supply an additional 134 Mcfpd. This setting would depend on the type of valve and manufacturer used.

With an expected surface flowing temperature of 100°F and a Bht of 200°F, the design temperatures at the valves are 120, 132.5, 140, and 144°F. The bellows pressures needed in the second and third valves are therefore 808 and 793 psig. The surface rack nitrogen pressure needed for proper operation is 700 and 678 psig. By the choke equation [60], the orifice required in the lifting valve should be taken as $\frac{8}{64}$-in., ensuring that excessive flowrates are prevented since this represents a maximum flowrate of 150 Mcf/d. A better solution for the designer would be using the dynamic flow characteristics of the particular gas lift valve and model the flowrate performance with available software programs, as discussed in "Valve Mechanics."

6.4.8 Fundamentals of Hydraulic Pumping
6.4.8.1 Definition
A hydraulic pumping system such as is being marketed today takes liquid from a power fluid reservoir on the surface, puts it through a reciprocating multiplex piston pump or surface electrical submersible pump to increase its pressure, and then injects the pressurized liquid down hole through a tubing string. At the bottom of the injection tubing string, the pressurized liquid is directed into the nozzle of a jet pump or the hydraulic engine of a piston pump, both of which have been set below the producing fluid level.

6.4.8.1.1 Types of Operating Systems
There are two basic operational types of hydraulic pump systems: the open power fluid system and the closed power fluid system. In an open power fluid (OPF) system the operating power fluid mixes with the produced fluid while down hole

and both fluids are returned to the surface in a commingled state. In a close power fluid (CPF) system the production and operating power fluids are never allowed to intermix throughout the entire system. The closed power fluid system is used with piston pumps only.

6.4.8.1.2 OPF System
In an OPF system only two down-hole fluid conduits are required: one, to contain the pressurized power fluid and direct it to the pump, and another (normally the annulus) to contain both the spent power fluid and produced fluid and to return them to the surface (Figure 6.4.46). Since this is the simpler and more economical of the two systems, it is by far the most common system used.

Besides the simplicity and economical advantage of the OPF system, the intermingling of the power fluid and the produced fluid has some other inherent advantages.

First, the circulated power fluid is ideal for carrying chemical additives. Corrosion, scale, and paraffin inhibitors can be added to extend the lift of the subsurface equipment. Also, if there are emulsions downhole in the produced fluids, emulsion breakers can be added to the power fluid stream.

Second, the commingled power fluid has a diluting effect. Where highly corrosive production fluids are being lifted, the clean power fluid reduces the concentration of the corrosive elements by about 50%. Where extremely viscous oil is produced, the injected power fluid has a diluting effect and can often reduce the viscosity enough to make lifting the heavy crude more practical.

Third, in production fluids with a high paraffin content, the OPF system allows the circulation of heated fluids or dissolving agents through the power fluid lines to remove waxy build ups which may hinder or halt production.

6.4.8.1.3 CPF System (Piston Pumps Only)
In a closed power fluid system (Figure 6.4.47), an extra tubular string is required both down hole and on the surface. One string is for carrying the production to the tank battery and

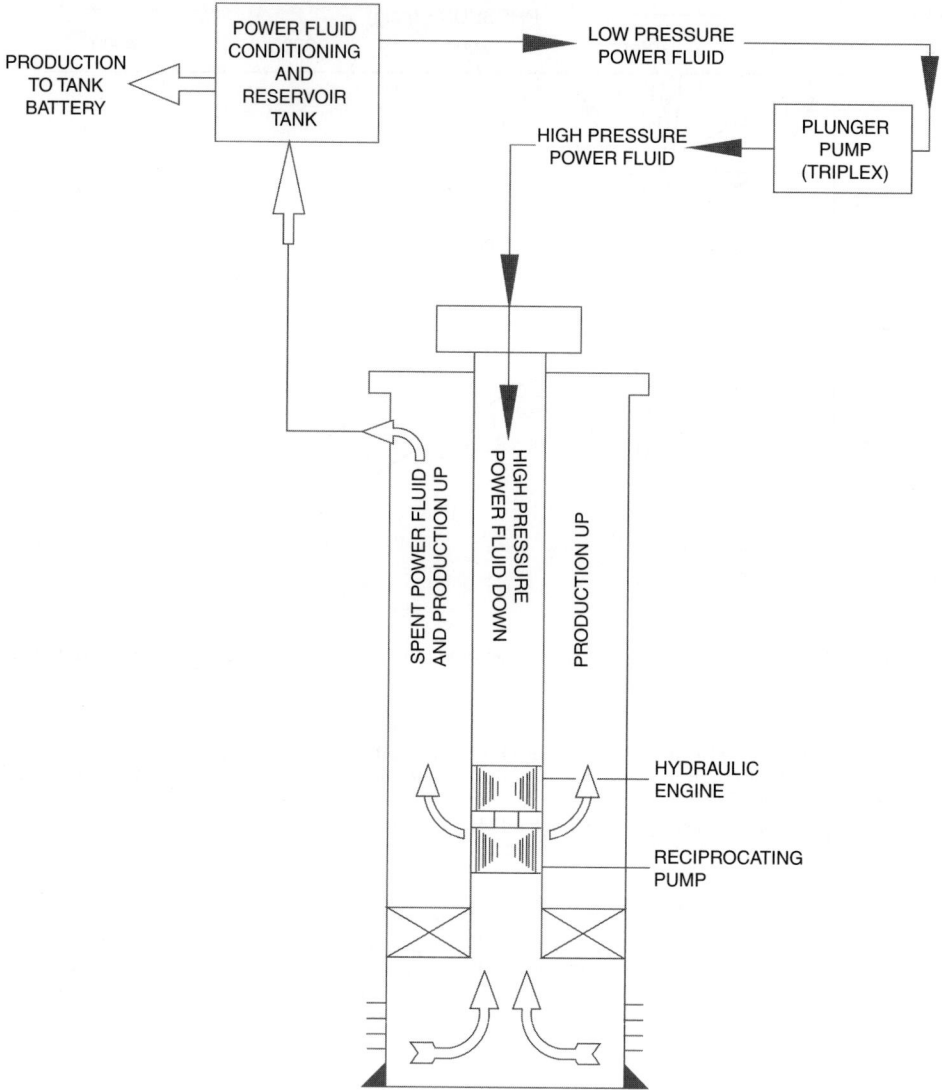

Figure 6.4.46   *Open power fluid system.*

the other to return the spent power fluid to the power fluid tank for repressurization and recirculation.

This requirement for an additional tubing string, plus the associated complication of the bottomhole design, makes the closed power fluid system more expensive than the open system. For this reason, the CPF system is less popular and less used than the open power fluid configuration.

Since the produced fluids and power fluids are always separated, the closed system finds some advantage where the produced fluids are extremely abrasive or corrosive and where inhibitors are not used. A closed system allows the use of less exotic materials in the engine end of the pump and may extend the lift of both the pump and the surface power fluid facility.

A closed system may also find some slight preference on marine platforms and in industrial or residential installations where available space is scarce and at a premium. Because the power fluid conditioning and reservoir tank only needs to be large enough to provide an adequate volume of power fluid to feed the multiplex pump, the size of the power fluid

tank is relatively small and almost all of the produced fluid can be put directly into the flow line.

In most down-hole pumps designed for use in a closed power fluid system, the, pump end is lubricated by the power fluid. However, approximately 10% of the power fluid is lost into the production fluid so this amount of make-up fluid must be fed back into the power fluid system from the production line.

It must also be realized that even in a closed system, the power fluid can not remain clean. First, all of the pipes, fitting, pumps, tanks, etc. are not completely free of contaminated materials. Second, when a liquid containing any solid material is leaked through a close fit clearance (such as slippage past the engine piston) the solid material will tend to be held back.

6.4.8.2 Types of Subsurface Pump Systems
There are three basic subsurface pump systems: the free-type, the fixed-type, and the wireline-type.

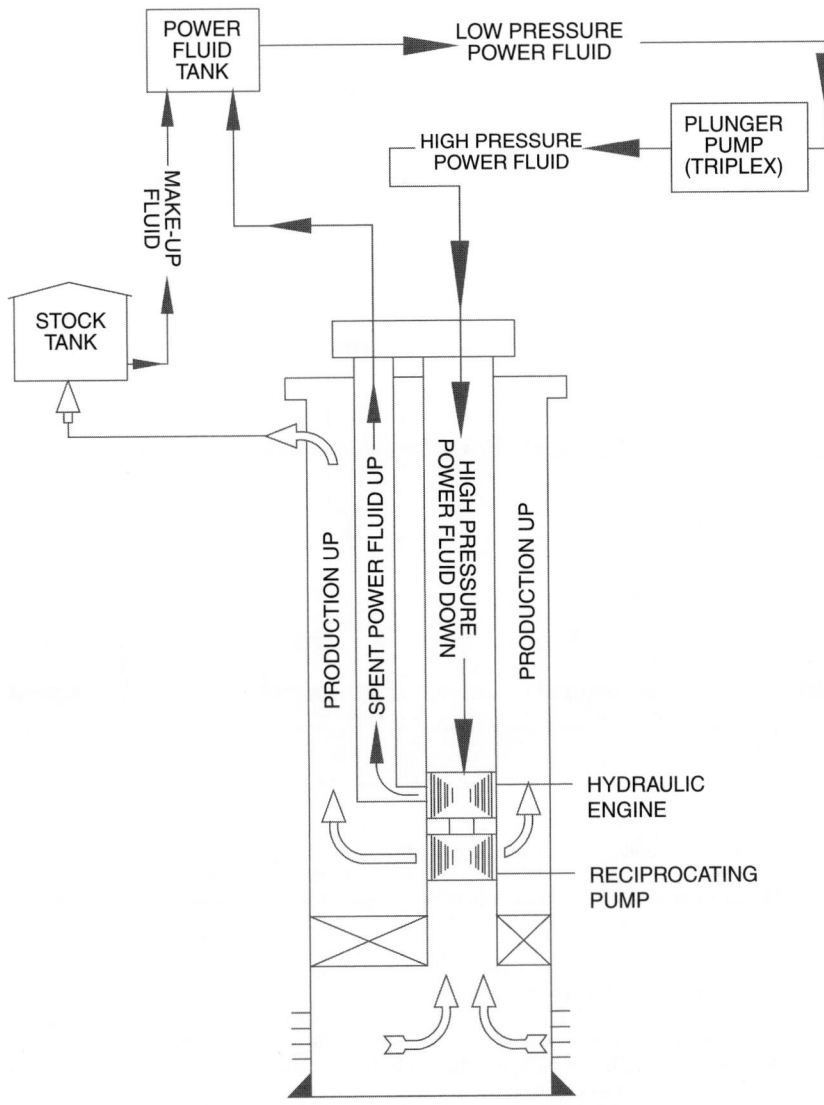

Figure 6.4.47 *Closed power fluid system.*

The free-type does not require a pulling unit to run or retrieve the pump. The pump is placed inside the power fluid tubing string and is "free" to be circulated by the power fluid to bottom and back out again (Figure 6.4.48).

In a fixed-type system the subsurface pump is attached to the power fluid tubing and is run in the hole as an integral part of the tubing string. The pump must be run or pulled with a pulling unit (Figure 6.4.49).

In a wireline-type system (for jet pumps only), the subsurface pump is installed in a sliding sleeve, straddling a gas-lift mandrel, or straddling a chemical-injection mandrel. The pump is run into the well or pulled from the well on a wireline. The pump can be operated in standard flow or reverse flow.

6.4.8.2.1 Free-Pump Systems
There are two main types of free-pumping installation designs: the casing-free design and the parallel-free design.

6.4.8.2.2 Casing-Free Installations
The casing free design is the least complicated and least expensive free pump design. It consists of a single tubing string, a bottom-hole assembly, and a packer. During operation, power fluid is circulated down the tubing string where it operates the subsurface hydraulic pump and mixes with the produced liquids and gas. This mixture of spent power fluid and produced fluids is then returned to the surface via the casing annulus.

Figure 6.4.50 shows a casing-free design for an open power fluid system. In this design all of the produced gas must pass through the pump. This effects the liquid displacement efficiency of a piston pump in direct relation to the amount of gas being produced. Jet pump performance can actually improve due to the presence of gas. Because of the simplicity and cost benefits of the casing-free open power fluid system, there are more hydraulic pumps installed of this design than any other type installation.

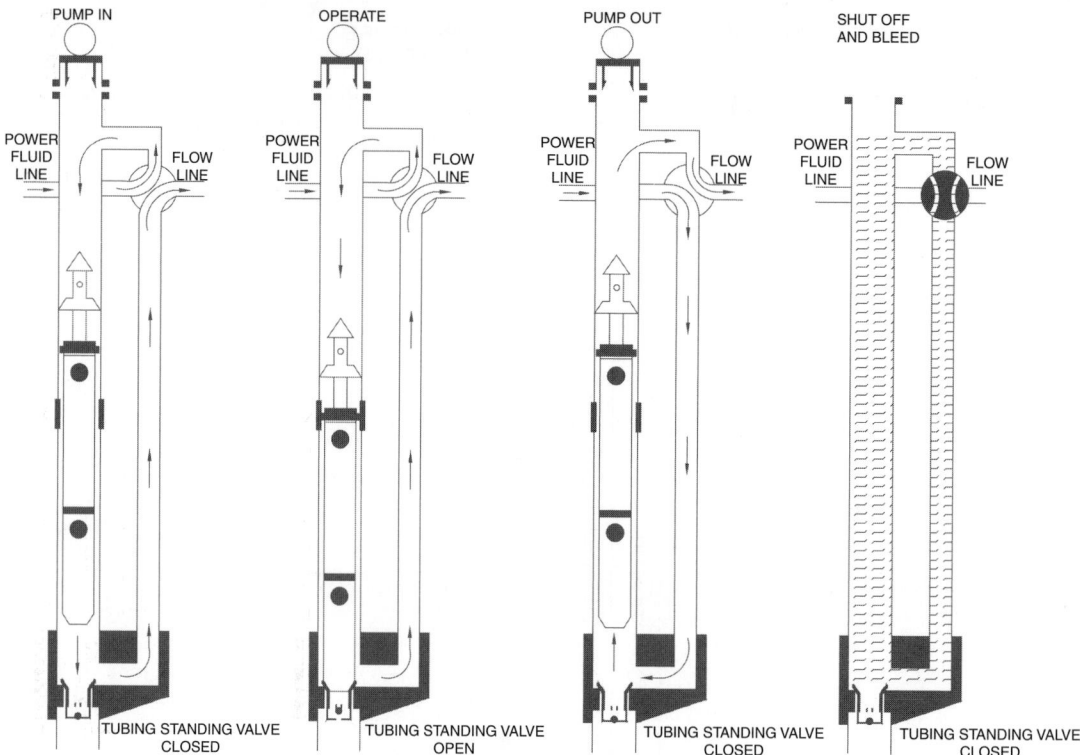

Figure 6.4.48 *Installation/removal of free style pump.*

Figure 6.4.51 shows some variations of the casing-free design.

The casing free OPF with gas vent can be used where displacement efficiency is severely affected by a high GOR. In this design, an auxiliary parallel string is run into a dual packer below the pump in order to vent gas to the surface. The production plus spent power fluid, returns by the casing annulus.

Another variation in installation design in shown in Figure 6.4.52. This shows a safety valve installed below the hydraulic pump. The safety valve is kept open by power fluid pressure. If a disaster strikes at the surface, power fluid pressure will be released and the spring-loaded safety valve will shut in the well at the packer.

6.4.8.2.3 Parallel Free Installations

The parallel free design for an open power fluid system incorporates two strings of tubing and a bottom-hole assembly, but no packer (Figure 6.4.53). The bottomhole assembly is attached to the main string and has a landing bowl which receives the spear that is on the bottom of the parallel string.

During production operations, the pressurized power fluid is pumped down the main tubing string where it operates the pump. It then mixes with the produced fluid and this mixture returns up the parallel string to the surface. This subsurface design permits gas to be vented up the casing annulus.

The drawback to this design is the extra tubing string required. Usually the parallel string is smaller and must carry both the production and spent power fluid. This situation can create higher return friction losses and result in a need for more horsepower. The maximum size of both strings is dictated by the casing diameter. This also restricts the size of the pump, which can be run and the volume of fluid, which can be lifted.

Figure 6.4.54 shows two designs used with closed power fluid installations. Three parallel tubing strings are required in both of these designs: a power fluid string, a power fluid return string and a production string.

In parallel-free type installations, the main string should always be anchored to minimize tubing stretch as an unanchored string could actually unseat the parallel string(s) and disable it (them) from functioning as a return conduit(s).

6.4.8.2.4 Fixed-Pump Systems

Fixed-pump installations are like sucker rod tubing-pump installations, electric submersible pump installations, or conventional gas lift installations in that the bottomhole pump is attached to and run to bottom on the tubing string. These are considered to be "permanent-type" installations because a pulling unit is required to change out the pump.

There may be several reasons for selecting a fixed-pump but the main reasons would be to lift large volumes. Since the physical size of these pumps is not limited by tubing inside diameter, larger pumps can be used which are able to produce larger volumes than those produced by a free style pump.

6.4.8.2.5 Fixed-Insert Installations

In a fixed-insert installation (or fixed-concentric installation) a large tubing string is run to bottom. The pump is then run on a string of macaroni tubing inside the main tubing string and seated in a seating shoe (Figure 6.4.49). In this design, the macaroni string carries the pressurized power fluid to the pump. The exhausted power fluid plus the produced fluids are carried up the tubing-tubing annulus. This allows gas to be vented up the casing annulus to the surface.

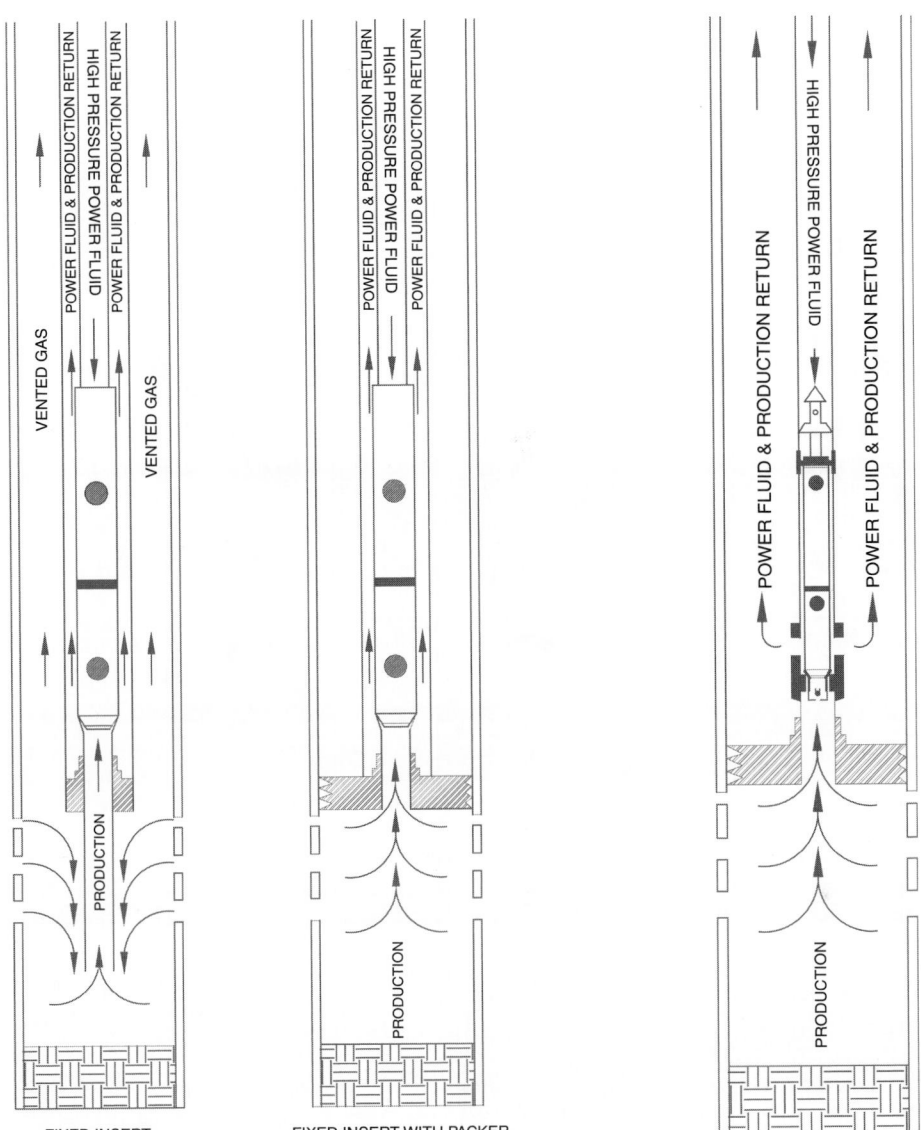

Figure 6.4.49 *Fixed insert open power fluid installations.*

Figure 6.4.50 *Casing free, open power fluid installation.*

A variation of the fixed-insert design would involve a packer set below the pump and bottomhole assembly. This will cause all of the produced gas to go through the hydraulic pump but isolates the perforated zone below the pump from the rest of the well. This design would be advantageous when bad casing up the hole is a problem, or when another zone above the packer is being produced.

6.4.8.2.6 Fixed-Casing Installations

In fixed-casing installations, the pump is run on the tubing string with a packer below the pump (Figure 6.4.55). The tubing string carries the pressurized power fluid down to the pump. The spent power fluid and the production are returned through the casing annulus.

This type of installation is generally used where high production rates require that large pumps be run but any produced gas must pass through the hydraulic pump.

6.4.8.2.7 Bottomhole Assemblies

Bottomhole assemblies (BHAs) are housings where hydraulic pumps are located while they are operating. They have an internal seal sleeve that matches with a seal on the outside diameter of the pump when the pump is landed.

The casing-type BHA and the parallel BHA both provide the same sealing/operating functions, however, only a single tubing string is run with a packer for the casing type. In a stab-in parallel installation, the main tubing string is run by itself. The macaroni string is run separately and stabbed into the shoe assembly. The screw-in type requires more time to install and the use of parallel tubing string clamps.

Figure 6.4.56 shows some BHAs for single displacement casing free pumps.

6.4.8.2.8 Retrievable Tubing Standing Valve

A retrievable tubing standing valve is also required when using a free-type pump (Figure 6.4.57). This standing valve

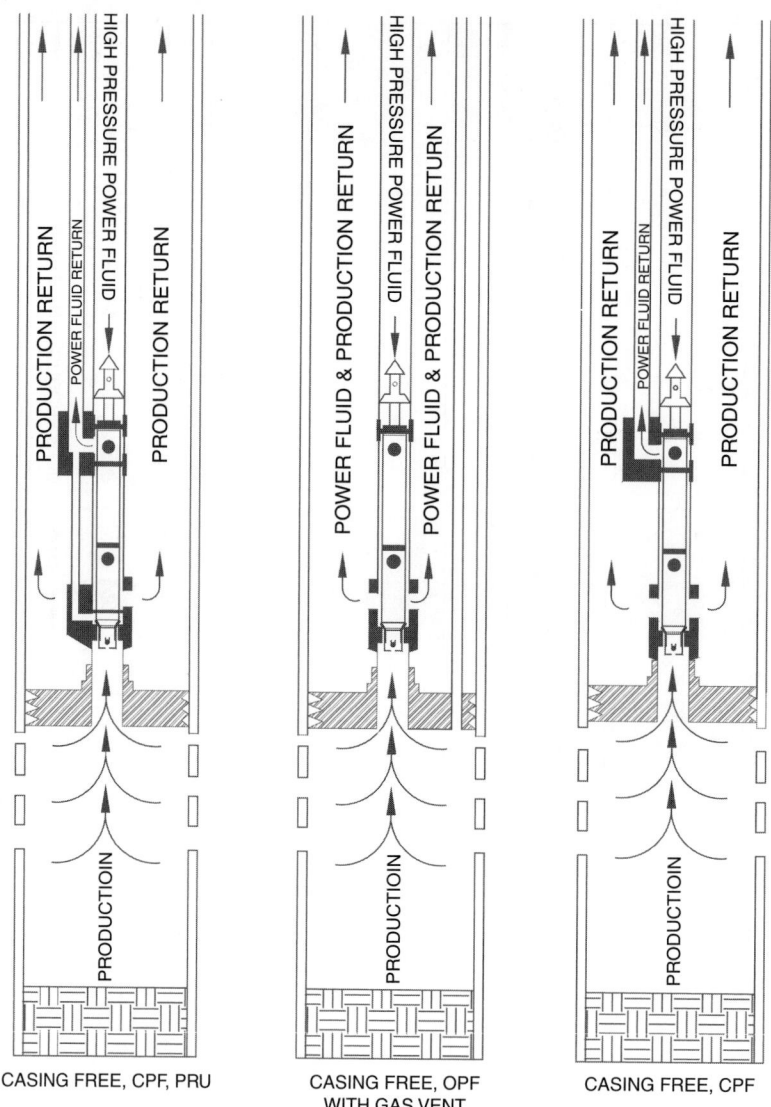

Figure 6.4.51 *Variations of casing free installations.*

lands in the lower end of the bottomhole assembly and the pump sits on top of it during operation. The purpose of the standing valve is to prevent the power fluid from traveling out the bottom of the tubing string when installing or retrieving the subsurface pump.

It is designed to be retired with a wireline tool for servicing if necessary.

6.4.8.2.9 Cleaning the System

In a free-type installation, the system can be cleaned by, circulating fluid after installing the tubing standing valve and before installing the pump.

In a fixed type installation, however, this is not possible so a circulating valve, starting filter assembly and check valve are normally used. The circulating valve, when run, is placed in the circulating position. This permits fluid to be circulated down the tubing string, but it is prevented from going to

the pump. When the circulating time is finished, a ball is dropped down the tubing string and lands in the circulating valve. Pressure causes a sleeve in the valve to shift which shuts off the circulation ports and permits the power fluid to flow through a starting filter and on to the hydraulic pump.

6.4.8.3 Power Water Versus Power Oil

Arguments can be made for and against the use of either water or oil as a power fluid.

6.4.8.3.1 Power Oil

Power oil has a natural lubricity that water does not. This is especially important when operating piston pumps due to their close tolerance clearances. Also, oil has some compressibility as opposed to water. This means that neither the surface piston pump nor the subsurface pump is exposed

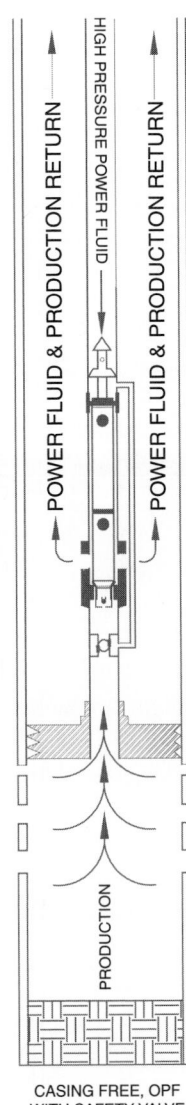

Figure 6.4.52 *Casing free installation with subsurface safety valve.*

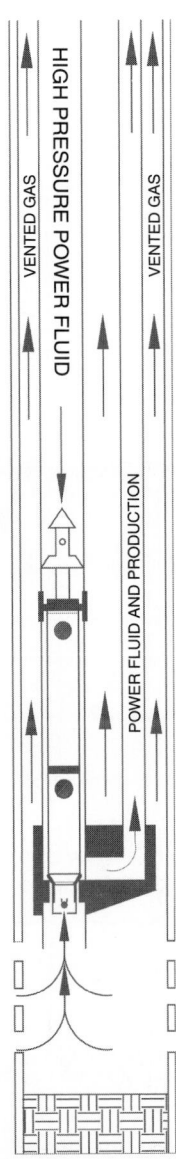

Figure 6.4.53 *Parallel free, open power fluid installation.*

to as much "fluid hammer" with oil as with water, and the service life of the equipment is usually longer on oil.

Significant drawbacks to power oil are the potential fire hazard, which is not present with the use of power water, and pollution damage in the event of a line break.

Another issue is in the potential revenue the operator has tied up in his power oil system, particularly in a central surface system with a power oil tank. For example, if the system has a 750-barrel power oil tank, this represents 750 barrels of marketable oil the operator cannot sell.

6.4.8.3.2 Power Water

During recent years an increased number of hydraulic systems have changed from using power oil to using power water. Many of them were changed to power water because of ecological reasons, code restrictions, town site locations, or increased watercuts or because the produced crude oil had a high viscosity.

A high-viscosity power fluid can mean excessive friction losses in the system. This in turn increases the operating pressure and, consequently the horsepower requirements for lifting the well. Therefore, in some instances it would be prohibitive to use the produced crude as a power fluid.

Water, because of its low viscosity, can be used in these instances. There are hydraulic installations where produced water is heated and used as a power fluid (and as a diluent or thinner) for the 8° gravity, crude being pumped.

Water, having low lubrication qualities, sometimes requires a chemical additive for lubrication when using hydraulic piston pumps. Frequently, the chemicals used will include oxygen inhibitors and agents to combat corrosion. These are easily added at the multiplex suction via a chemical pump. Improvements in lubricants, surface, and subsurface equipment, designed in recent years, has greatly enhanced the use of power water.

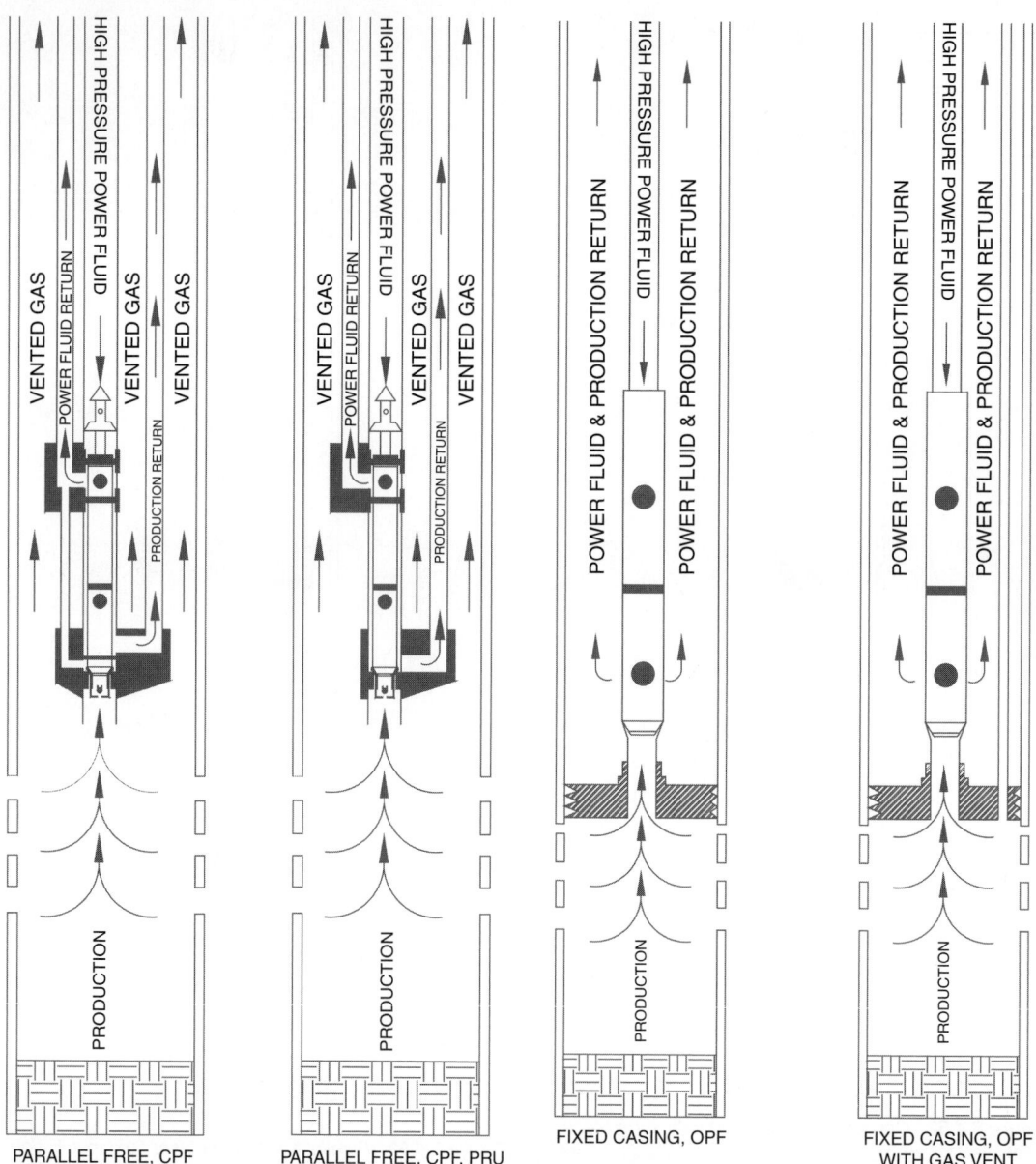

Figure 6.4.54 *Parallel free, closed power fluid installation.*

Figure 6.4.55 *Fixed casing, open power fluid installations.*

Aside from the usual chemical treatment for corrosion and scale depositions, salt crystals will occasionally be a problem in power water. This problem can usually be solved by having a fresh water blanket on the bottom of the power fluid tank, or by injecting fresh water into the power fluid.

Surface multiplex pump modifications for converting from oil service to water service are limited mainly to the fluid end of the pump and its plungers/lines. For example, a power oil fluid end can be made of ductile iron or forged steel but a fluid end for power water will be of aluminum/bronze to resist the corrosion effects of water. It is now common to use the aluminum/bronze fluid end for both power oil and power water. Metal-to-metal pistons and liners can be used for power oil service, but metal pistons against soft packing are needed for water service. As with the fluid end metal-lurgy, it is now common to use the soft packing for both water and oil service.

6.4.8.4 Inflow Performance for Artificial Lift
6.4.8.4.1 Inflow Performance
The relationship between production rate and producing bottomhole pressure is called *inflow performance*, which is the ability of a well to give up fluids.

In any artificial lift method, including hydraulic and jet pumping, the pumping system must be designed to provide the extra energy necessary to lift the production to the surface at the desired (or achievable) rate. Thus, the producing bottomhole pressure for an established production rate is

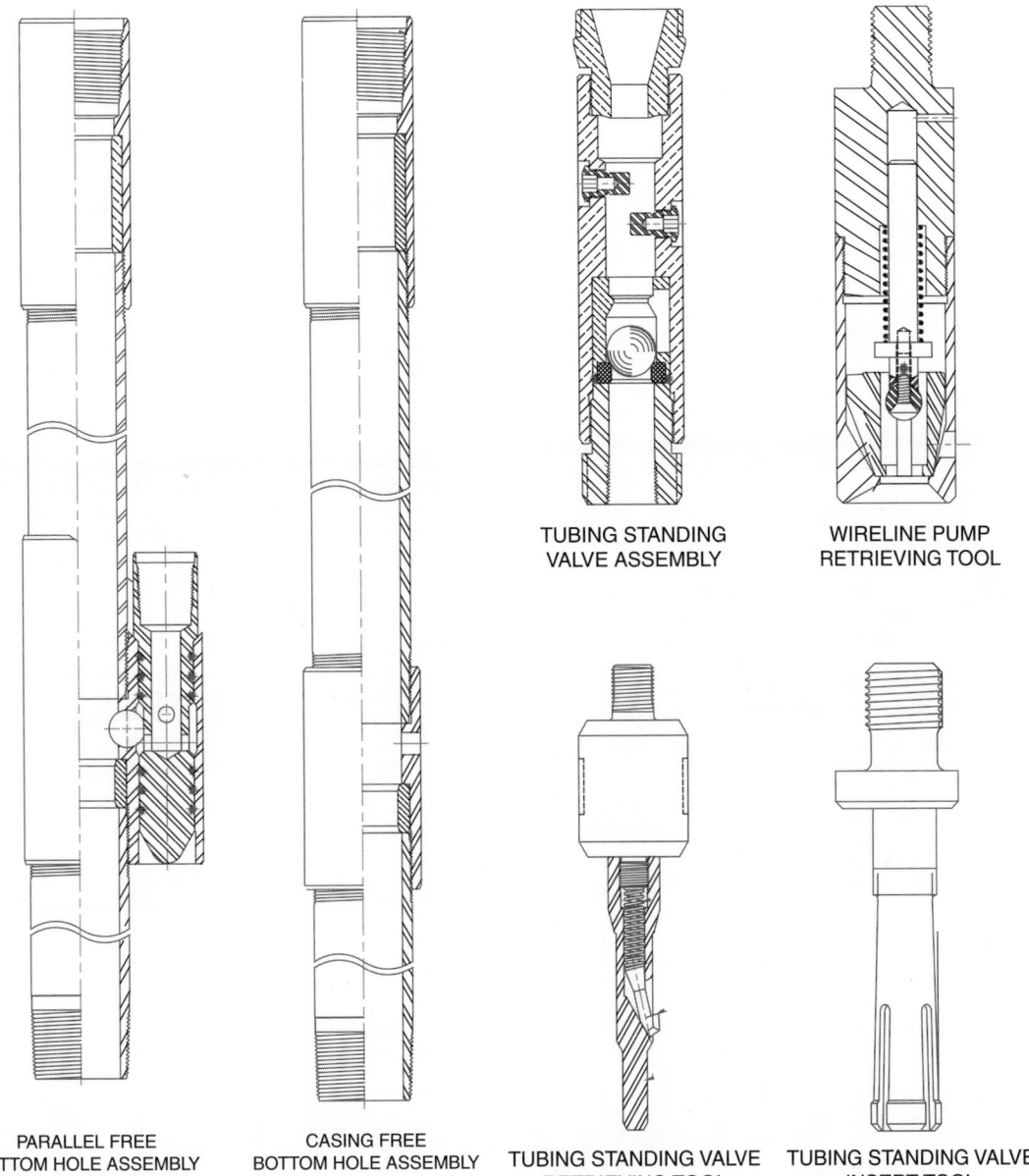

**TUBING STANDING
VALVE ASSEMBLY**

**WIRELINE PUMP
RETRIEVING TOOL**

**TUBING STANDING VALVE
RETRIEVING TOOL**

**TUBING STANDING VALVE
INSERT TOOL**

**PARALLEL FREE
BOTTOM HOLE ASSEMBLY**

**CASING FREE
BOTTOM HOLE ASSEMBLY**

Figure 6.4.56 *Free style bottomhole assemblies.*

Figure 6.4.57 *Retrievable tubing standing valve and accessories.*

one of the primary pieces of input data required in the design of the installation.

Knowing how much fluid can be produced at one particular bottomhole pressure, however, does not tell the whole story. It also needs to be known in order to have a complete IPR is:

- How much was the botomhole pressure lowered to achieve the current production rate (draw down)
- How much more (or less) will it need to be lowered in order to produce at a different rate

For a given well, at a given time, there is only one producing bottomhole pressure associated with a specified production rate. To predict accurately the performance of any pump in a well, it is necessary to know the producing bottomhole pressure at the desired production rate as determined by the well's inflow performance. The two most commonly used methods for this making this prediction are Productivity Index (PI) and Inflow Performance Relationship curve. Several different methods are used to estimate the flowing bottomhole pressure for a given production rate including the straight line method (PI), J.V. Vogel's method [17] and others [18].

6.4.8.4.2 Productivity Index

PI is defined as the production rate in barrels per day (bbl/d), which can be achieved for each psi that the bottomhole pressure is reduced. It has long been used for estimating well production capacity.

H.H. Evinger and M. Muskat first introduced the concept of the PI relationship, shown in the following equation, in 1942 [19]. However, as was suggested by Evinger and Muskat, the producing rate does not always change with producing bottom-hole pressure along a straight line as the PI equation suggests.

The equation that is typically used for PI is:

$$PI = \frac{\left[\begin{array}{c} \text{Change in} \\ \text{Oil Production} \end{array}\right] + \left[\begin{array}{c} \text{Change in} \\ \text{Water Production} \end{array}\right]}{\text{(Change in pressure)}}$$

$$= \frac{\text{Change in Total Production Rate (bbl/d)}}{\text{(SBHP} - \text{PBHP)(psi)}} \qquad [6.4.11]$$

$$= \frac{\text{Production Rate}}{\text{Drawn Down}} = \text{bbl/d/psi}$$

Provided the data is available, an improved equation derived from Darcy's radial flow equation [18] is:

$$PI = HK_A \left(\frac{K_{RO}}{\mu_o B_o} + \frac{K_{RW}}{\mu_W B_W} \right) \qquad [6.4.11a]$$

where K_A = absolute permeability of the rock in darcies
K_{RO} = relative permeability to oil
K_{RW} = relative permeability to water
B = formation volume factor
μ = viscosity
H = reservoir thickness

The above values are normally obtained through measuring shut-in bottomhole pressure and a flowing bottomhole pressure with its associated flow rate (provided the flowing data is not close to the static case). As can be seen, both oil and water phases are represented in the above equation. A more accurate estimation of PI is possibly by using data for the individual phases as opposed to considering the flow as a single phase.

A given PI is valid only for the conditions for which its measurements were made, i.e. percentages of water and oil. As those values change, the accuracy of the PI value is reduced and a new value should be calculated or new measurement made.

6.4.8.4.3 Inflow Performance Relationships
In 1968, J. V. Vogel [17] offered a solution to the problem of determining an inflow performance curve for a solution gas drive well where the producing bottomhole pressure is below the bubble point pressure.

Vogel's work resulted in the construction of a reference curve which could be used as a general solution for predicting production rates and bottomhole pressures. This curve is known as the "General IPR Curve by Vogel," Figure 6.4.58.

$$\frac{Q_{actual}}{Q_{max}} = 1 - 0.2 \left(\frac{P_w}{P_b} \right) - 0.8 \left(\frac{P_w}{P_b} \right)^2 \qquad [6.4.12]$$

where Q_{actual} = flow rate (bpd)
Q_{max} = flow rate at maximum draw down (bpd)
PBHP = flowing bottomhole pressure (psi)
SBHP = static bottomhole pressure (psi)

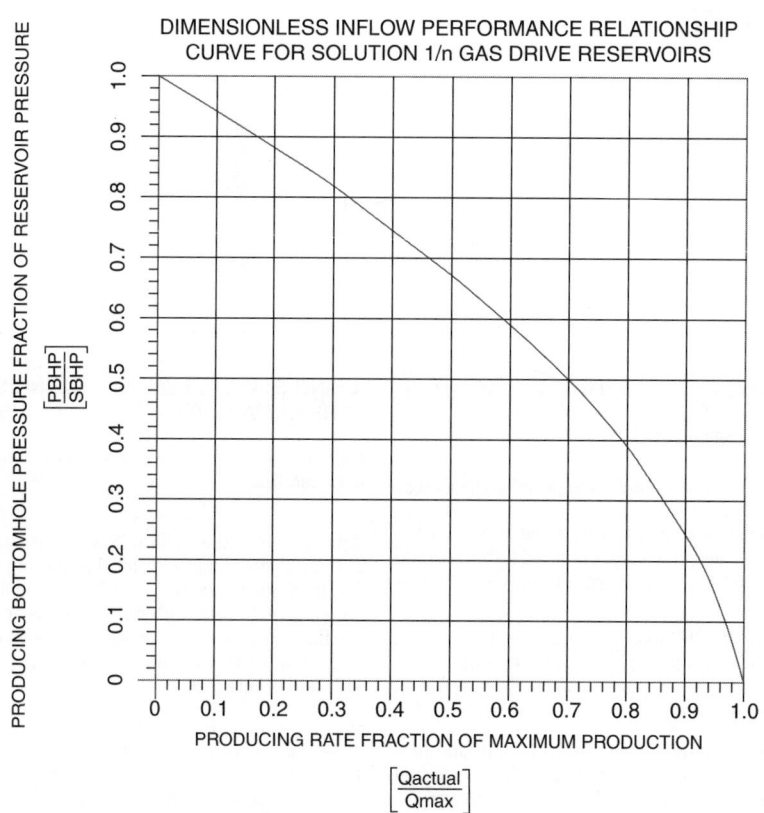

Figure 6.4.58 General IPR curve by Vogel.

This is a general or "unity" reference curve, which is easy to use for predicting producing rates and pressures below bubble point pressure. It can be used when data from only one production test are available provided it is not close to the static case.

Standing [20] proposed the following equation for estimating the bubble point pressure of the oil:

$$P_b = 18 \left(\frac{GOR_s}{SPGR_{gas}} \right) \left[\frac{10^{0.00091 \; BHT}}{10^{0.0125 \; API}} \right]^{0.83} \qquad [6.4.13]$$

where GOR_s = Dissolved gas–oil ratio
$SPGR_{gas}$ = Specific gravity of gas
BHT = Bottom hole temperature in °F
API = API degree of oil
Following is a sample IPR calculation for a well (Figure 6.4.58).

Given: SBHP = 1878
         PBHP = 1543
         Production Rate (Q) = 154 Barrels/d
         Desired Production Rate = 414 Barrels/p
Find: PIP at 414 Barrels/d

1) Determine PBHP/SBHP 1543/1878 = 0.82
2) From Figure 2-1, find Q/Q_{max} = 0.30
3) Calculate Q_{max} = 154/0.30
4) Find Q/Q_{max} = 414/513 = 0.80
5) From Figure 2-1 obtain PBHP/SBHP = 0.38
6) Calculate PBHP at desired production rate = 0.38 × 1878 = 713
7) 713 PSI is pump intake pressure at 414 Barrels/d

In 1970, M. B. Standing [21] proposed a companion chart to go along with Vogel's which accounted for damaged or improved wells. To determine and predict rates and pressures in damaged or improved (reworked) wells, more than one set of test data is required.

6.4.8.5 Jet Pumps

A Jet Pump is a venturi-type device where high-pressure fluid is caused to accelerate to a high velocity and thereby create a low-pressure area into which reservoir fluids will flow. The jet pump has no moving parts. A typical downhole jet pump is shown in Figure 6.4.59. A closer look at the effective mechanism is shown in Figure 6.4.60. These pumps require a high-pressure fluid pump on the surface, and this fluid is circulated down the well, typically the tubing. However there are reverse flows available and are used in most offshore applications. The flows through the jet causing a low-pressure area, and the power fluid produced fluid mixture are brought to the surface by a second conduit, usually the casing annulus. Once installed, the jet pump can be repaired or replaced by simply reversing the circulation and circulating the pump up the tubing to the surface. The repairs can be made a well site then the pumps can pump back into position in the bottom hole assembly.

Jet pumps can produce high volumes and handle free gas very well. However, jet pumps are not as efficient as positive displacement pumps. Which means jet pumps require higher surface horsepower. A thorough coverage of the application of jet pumps is given by, Petrie et al [22]. This type of pump is very useful in certain situations, for example, where high production rates are desired or needed, and, locations where other forms of lift cannot be used, such as populated areas, offshore platforms or remote locations. Deviated wells or wells that produce sand or where gas lift is not an option.

Since fluid density, gas and viscosity effects are variables needed in the calculations to simulate a jet pump performance the calculations are complex and require iterative solutions. Here, the computer enters the picture. Weatherford Artificial Lift Systems has the latest in software design and can furnish characteristic plots of the form show in Figure 6.4.61. This plot is for a particular pump size and well since the PI of that well is needed as well as tubing length and diameter, fluid type, and fluid properties. However, once this plot has been generated, and engineer can predict the power fluid and surface pressure requirements for a desired production rate. Because the pump size is known, the intake pressure generated is predicted as well. Because PI must be known to design any lifting mechanism, the well's performance will follow the PI line overlaid on the graph. Note that the cavitation zone is also shown, which the pump can operate but throat becomes a choke or damage occurs to the throat portion of the pump.

6.4.8.5.1 Jet Pump Performance Equations

The following calculations, performed in order, will provide the information necessary to analyze the performance of a jet pump for any well application. (The user may wish to use more sophisticated calculation techniques in place of some of the equations listed below.)

1. Specify jet pump configuration.
   Determine nozzle area (A_N) and area ratio (R)
2. Specify production rate and associated pump intake pressure.
$$Q_{PROD} \quad \text{and} \quad PIP$$
3. Initialize down-hole power fluid pressure.
$$P_{PWR} = 5000 + GRAD_{PWR} \times DEPTH \qquad [6.4.14]$$
4. Compute oil specific gravity ($SPGR_{OIL}$)
$$SPGR_{OIL} = \frac{141.5}{131.5 + °API} \qquad [6.4.15]$$
5. Compute oil gradient ($GRAD_{OIL}$)
$$GRAD_{OIL} = 0.433 \times SPGR_{OIL} \qquad [6.4.16]$$
6. Compute oil viscosity (μ_{OIL}) for average wellbore temperature
$$T = LOG \left(\frac{BHT + WHT}{2} + 460 \right) \qquad [6.4.17]$$
$$C = 3.55(6.328 - T) + 11.53 \times SPGR_{OIL} - 9.254 \qquad [6.4.18]$$
$$E = e^C \qquad [6.4.19]$$
$$\mu_{OIL} = (e^E - 0.6)SPGR_{OIL} \qquad [6.4.20]$$
7. Compute water viscosity (μ_{WTR}) for average wellbore temperature
$$T_{AVG} = \frac{WHT + BHT}{2} \qquad [6.4.21]$$
$$T = 1.003 - 0.01479 \times T_{AVG} + 0.00001928 \times T_{AVG}^2 \qquad [6.4.22]$$
$$\mu_{WTR} = e^T$$
8. Compute gradient of production ($GRAD_{PROD}$) using percentage of water in production (WC)
$$GRAD_{PROD} = GRAD_{WATER} \times WC + GRAD_{OIL}(1 - WC) \qquad [6.4.23]$$
9. Compute Solution GOR (GOR_s) using bottomhole temperature and degrees API of the oil
$$GOR_s = SPGR_{GAS} \left[\left(\frac{PIP}{18} \right) \left(\frac{10^{0.0125 \; API}}{10^{0.00091 \; BHT}} \right) \right]^{1.2048} \qquad [6.4.24]$$

TYPICAL SINGLE SEAL JET PUMP

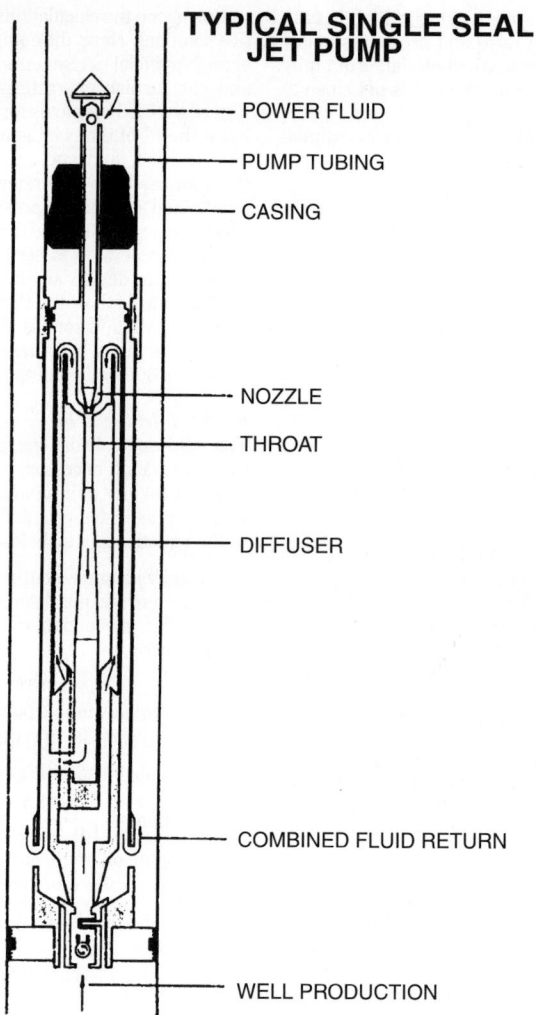

- POWER FLUID
- PUMP TUBING
- CASING
- NOZZLE
- THROAT
- DIFFUSER
- COMBINED FLUID RETURN
- WELL PRODUCTION

Figure 6.4.59

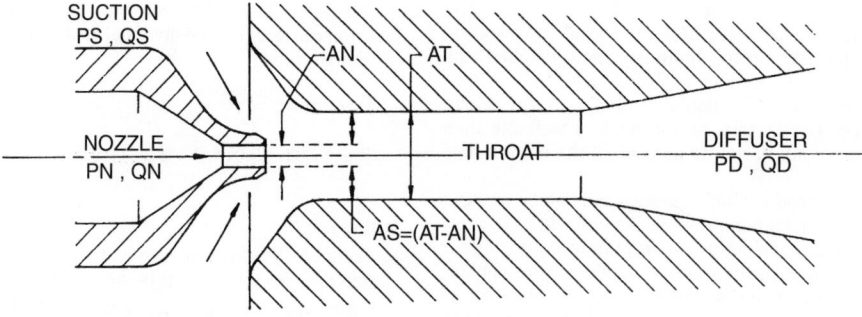

Figure 6.4.60

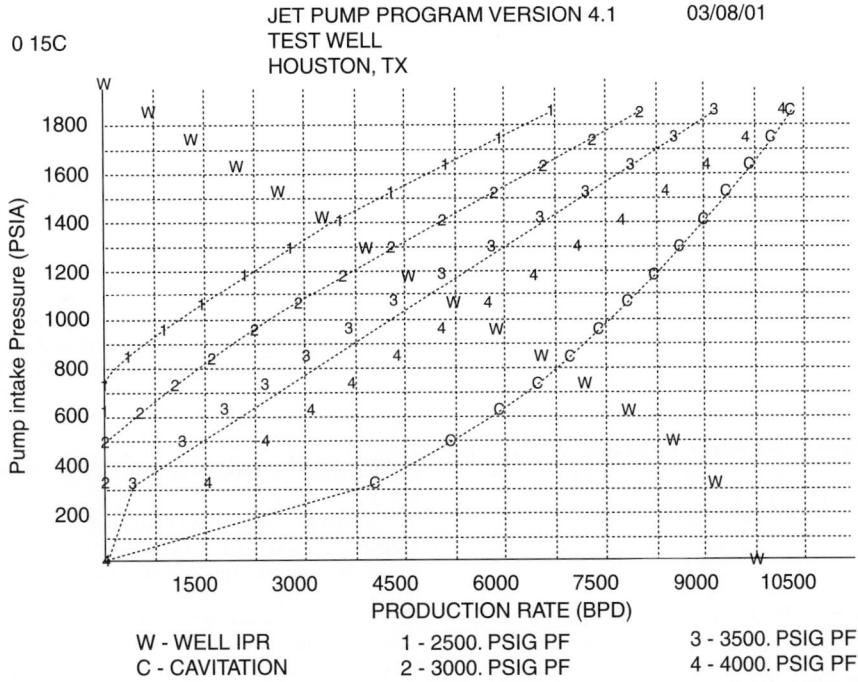

Figure 6.4.61 *Production chart.*

10. Compute F and oil formation volume factor (B_o)

    Where $GOR_s \leq GOR$:

    $$F = GOR_s \left(\frac{SPGR_{GAS}}{SPGR_{OIL}} \right)^{0.5} + 1.25\, BHT \qquad [6.4.25]$$

    If $GOR_s > GOR$

    $$F = GOR \left(\frac{SPGR_{GAS}}{SPGR_{OIL}} \right)^{0.5} + 1.25\, BHT \qquad [6.4.26]$$

11. Compute oil formation volume factor (B_o)

    $$B_o = 0.972 + 0.000147\, F^{1.175} \qquad [6.4.27]$$

12. Compute gas compressibility factor (Z)

    $$BZ = \frac{BHT + 460}{203,320} \qquad [6.4.28]$$

    $$CZ = 0.0694 - 17.6\, BZ \qquad [6.4.29]$$

    $$DZ = \frac{0.022 \times PIP}{14.65} - 3.5 \qquad [6.4.30]$$

    $$EZ = 214 \times BZ \qquad [6.4.31]$$

    $$Z = CZ \times (DZ)^2 + EZ + 0.15 \qquad [6.4.32]$$

13. Compute gas formation volume factor (B_g)

    $$B_g = \frac{0.0283 \times Z \times (BHT + 460)}{PIP} \qquad [6.4.33]$$

14. Compute two-phase formation volume factor (B_T)

    For $GOR_S \geq GOR$:

    $$B_T = B_O \qquad [6.4.34]$$

    For $GOR_S < GOR$:

    $$B_T = B_O + \frac{B_g \times (GOR - GORs)}{5.6146} \qquad [6.4.35]$$

15. Compute Volumetric Efficiency (EFF_V) using percentage of water in production (WC)

    $$EFF_V = \frac{1}{WC + (1 - WC) \times B_T} \qquad [6.4.36]$$

16. Compute nozzle constant (K_Q).

    $$K_Q = 1264 \times A_N \sqrt{\frac{0.433}{GRAD_{PWR}}} \qquad [6.4.37]$$

17. Compute power fluid volume (Q_{PWR}).

    $$Q_{PWR} = K_Q \sqrt{P_{PWR} - PIP} \qquad [6.4.38]$$

18. Compute momentum ratio (M)

    $$M = \frac{GRAD_{PROD} \times Q_{PROD}}{GRAD_{PWR} \times Q_{PWR} \times EFF_V} \qquad [6.4.39]$$

19. Check for solution convergence.
    If change in M is less than 0.001, go to step 30.

20. Compute return fluid volume (Q_{RTN}).

    $$Q_{RTN} = Q_{PROD} + Q_{PWR} \qquad [6.4.40]$$

21. Compute return fluid gradient ($GRAD_{RTN}$).

    $$GRAD_{RTN} = \frac{GRAD_{PWR} \times Q_{PWR} + GRAD_{PROD} \times Q_{PROD}}{Q_{RTN}} \qquad [6.4.41]$$

22. Compute water fraction of return (WC$_{RTN}$).
For power oil:

$$WC_{RTN} = \frac{WC \times Q_{PROD}}{Q_{RTN}} \qquad [6.4.42]$$

For power water:

$$WC_{RTN} = \frac{Q_{PWR} + WC \times Q_{PROD}}{Q_{RTN}} \qquad [6.4.43]$$

23. Compute return fluid viscosity (μ_{RTN}).

$$\mu_{RTN} = \mu_{WTR} \times WC_{RTN} + \mu_{OIL}(1 - WC_{RTN}) \qquad [6.4.44]$$

24. Compute pressure loss due to friction (PL$_{FN}$)
Turbulent Flow Only
Reynold's Number (Re) and Pressure Loss due to friction (PL):

$$Re = \frac{213 \times Q \times G(D_1 - D_2)}{\mu(D_1^2 - D_2^2)^2} \qquad [6.4.45]$$

$$PL = \frac{2.644 \times 10^{-5} \times Q^2 \times L \times G \times f}{(D_1 - D_2)(D_1^2 - D_2^2)^2 \left(\dfrac{D_1}{D_1 - D_2}\right)^{0.1}} \qquad [6.4.46]$$

Where:

f = friction factor = $\dfrac{0.236}{R_e^{0.21}}$

G = Gradient (psi/ft.)
L = Length (feet)
Q = FlowRate (BBL/d)
μ = Viscosity (c.p.)
D_1 = Casing I.D. (Annular Flow)
D_1 = Tubing I.D. (Tubing Flow)
D_2 = Tubing O.D. (Annular Flow)
D_2 = 0   (Tubing Flow)

$$PL = \frac{202 \times 10^{-8} \times L}{(D_1 - D_2)(D_1^2 - D_2^2)\left(\frac{D_1}{D_1-D_2}\right)^{0.1}} \left[\frac{D_1^2 - D_2^2}{D_1 - D_2}\right]^{0.21}$$

$$\times \left[\frac{\mu}{G}\right]^{0.21} G \times Q^{1.79} \qquad [6.4.47]$$

25. Compute pressure loss in return fluid (PL$_{RTN}$)
A flowing gradient correlation is preferred for this calculation if the GLR > 10.
However, if one is not available or GLR = 10, then see pressure loss equation under step 24.

26. Compute pump discharge pressure (P$_{DIS}$).

$$P_{DIS} = P_{WELLHEAD} + GRAD_{RTN} \times DEPTH + PL_{RTN} \qquad [6.4.48]$$

If a flowing gradient is used, then replace right side of equation with results from correlation.

27. Compute pressure ratio (N).

$$N = \frac{2R + (1 - 2R)\dfrac{M^2 R^2}{(1-R)^2} - (1 + k_{34})R^2(1 + M)^2}{1 + k_1 - \text{numerator}} \qquad [6.4.49]$$

Where:

k_1 = 0.03
k_{34} = 0.2

28. Compute down-hole power fluid pressure (P$_{PWR}$).

$$P_{PWR} = \frac{P_{DIS} - PIP}{N} + P_{DIS} \qquad [6.4.50]$$

29. Go to step 17.
30. Compute cavitation limit (M$_{LIM}$).

$$M_{LIM} = \frac{1 - R}{R}\sqrt{\frac{PIP - 1}{1.3(P_{PWR} - PIP)}} \qquad [6.4.51]$$

31. Check for cavitation.
    a. If M > M$_{LIM}$, pump is cavitating, stop.
    b. If M < M$_{LIM}$, go to step 30.

32. Compute pressure loss down tubing (PL$_{FN}$).
See pressure loss equations in step 24.

33. Compute multiplex pressure (P$_{TPLX}$)

$$P_{TPLX} = P_{PWR} - GRAD_{PWR} \times DEPTH + PL_{FN} \qquad [6.4.52]$$

34. Compute hydraulic horsepower (HHP).

$$HHP = 0.000017 \times P_{TPLX} \times Q_{PWR} \qquad [6.4.53]$$

As one can see, the process to size a jet pump properly becomes is complex, but the computer programs that are now available have made performing these calculations much faster and easier. Weatherford Artificial Lift Systems, which now has the Oilmaster, Guiberson, and Kobe product lines can furnish this software and service.

6.4.9 Hydraulic Piston Pumps
6.4.9.1 General

The subsurface production pump is the heart of a hydraulic pumping system. This pump is driven by a reciprocating hydraulic engine piston, which is connected to the production pump piston (Figure 6.4.62). While the stroke length of the pump is fixed, various pump bore sizes are available for different volume and depth requirements. A wide operating speed range adds further flexibility to meet production rate requirements.

The basic components of all hydraulic pumps include an engine piston and cylinder, an engine reversing valve (which controls piston motion by directing the power fluid to the appropriate areas), and the pump piston and barrel. Conventional valves (usually balls and seats) control the production fluid intake and discharge. The arrangement of these components in the pump is based on the specific designs of the individual pumps.

The two most common pump end designs are:

• The "single-acting" production pump end, which displaces produced fluid only during the upstroke or only during the downstroke, and
• The "double-acting" production pump end, which displaces produced fluid on both the upstroke and downstroke.

The engine end may be designed to displace equal volumes of power fluid on each up and down stroke (i.e., a double displacement pump), or to displace a greater power fluid volume during one or the other of the two strokes (i.e., a single displacement pump).

Generally, the power fluid used to drive these pumps is clean crude oil or clean water drawn from the top of a settling tank or from a well site unit such as the Unidraulic®. In most installations, the spent power fluid is mixed with the produced well fluids at the pump and both come to the surface together.

The power required to return the spent power fluid to the surface is essentially what is required to overcome mechanical and fluid friction, and the difference in piston areas as seen by the power fluid and return fluid. The reason is that the static pressure head of the incoming power fluid at the pump tends to balance the static pressure head of the returning power fluid column.

6.4.10 Operation
6.4.10.1 General

As stated above, hydraulic subsurface piston pumps are composed of two basic sections — a hydraulic engine and a piston

Model 220 Pump

Figure 6.4.62 *Model 220 pump.*

are eliminated. These include:

- rod stretch
- the difficulty in adjusting the stroke length of the pump in order to eliminate or at least gas locking problems
- load limitations imposed by depth
- excessive rod/tubing friction and wear caused by crooked holes and deviated wells

6.4.10.2 Single Displacement Pump

The pump in Figure 6.4.57 is a single displacement pump, and its operation is as follows. Additional pump models are available that utilize different designs but the concept holds true for all of them.

The hydraulic engine end of the Oilmaster 220® series is basically composed of an engine barrel, an engine piston, and a reversing valve mechanism. The reversing valve and pilot valve assembly are mounted in stationary positions above the engine piston. The bottom of the engine piston is exposed to the power fluid on both the up and down strokes. During the down stroke, the reversing valve is in the down position and the pilot valve is in the up position. A dashpot on the bottom of the pump piston halts the downward motion of the downstroke of the engine/pump piston traveling assembly. During the downstroke, the reversing valve is held in place by the flow of the power fluid around it. At the end of the downstroke, the fluid flow is halted and the pressure forces on the reversing valve cause it to shift to the up position. At this time, the pilot valve is hydraulically shifted to the down position and opens a pathway, which exposes the top of the reversing valve to discharge pressure. With the reversing valve in the up position, a pathway is opened which exposes the top of the engine piston to discharge pressure. The resulting pressure imbalance creates a net force up on the engine piston. At the end of the upstroke, the pilot valve is pushed up, which reopens the pathway for the top of the reversing valve to be exposed power fluid again. This moves the reversing valve down, which then allows the top of the engine piston to be exposed to power fluid. Because there is no middle rod on top of the engine piston, more area on top is exposed to the power fluid than on the bottom and this results in a net force downward.

As stated above, the arrangement of the pump end is the same as with a sucker rod pump in that there is a barrel, a piston, a piston traveling valve, and a standing valve. As was also noted above, the sucker rod string is replaced with a column of high-pressure power fluid, which supplies the energy needed to move the engine piston. As the engine piston moves upward, the pump piston also moves upward, causing the barrel chamber under the pump piston to fill with production fluid. When the engine piston moves downward, the pump piston also makes a down stroke, displacing the production fluid in the pump barrel from below the pump piston to above it.

6.4.10.3 Double Displacement Pump

The pump in Figures 6.4.63 and 6.4.64 is a double displacement pump and its operation is as follows. As is the case with the single displacement pumps, additional pump models are available that utilize different designs but the concept holds true for all of them.

When the pump starts the upstroke, the reversing valve is positioned at the top of the valve body. Power fluid enters the pump through a port in the center of the valve body. The reversing valve directs the power fluid into the barrel above the reversing valve housing and under the upper piston. Because the pressure below the upper piston is greater than the pressure above it, the result is an upward movement of the piston.

pump. They are directly connected with a middle rod. Therefore, as the engine piston moves upward, the pump piston also moves upward causing the barrel chamber under the pump piston to fill with production fluid. When the hydraulic engine makes a down stroke, the pump piston also makes a down stroke displacing the production fluid in the pump barrel.

The arrangement of the pump end is the same as with a sucker rod pump in that there is a barrel, a piston, a piston traveling valve and a standing valve. However, since there is no mechanical linkage to the surface because the sucker rod string is replaced by a column of high pressure power fluid, many of the limitations imposed by sucker rod pumping

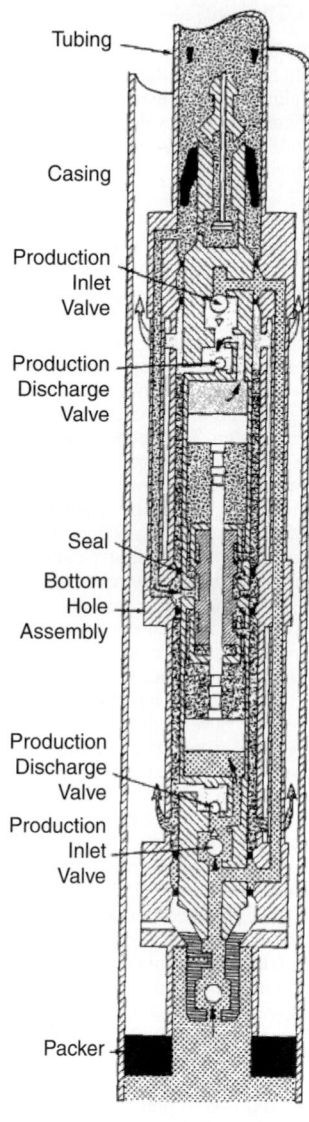

Figure 6.4.63 *PL-II double displacement pump, upstroke.*

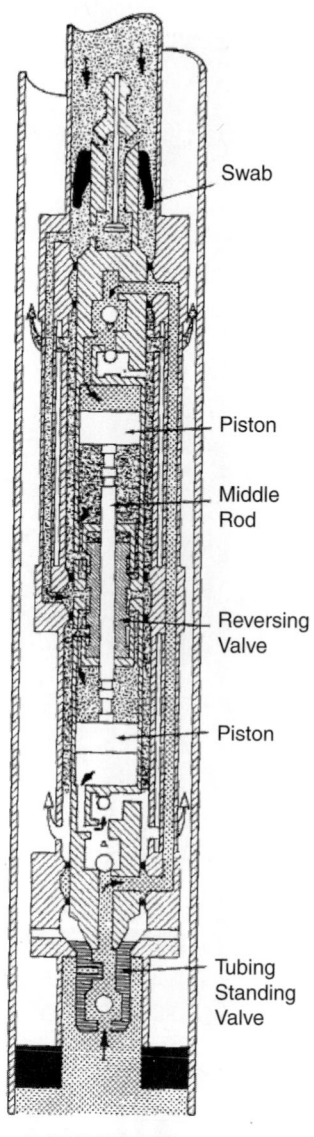

Figure 6.4.64 *PL-II double displacement pump, downstroke.*

The fluid above the upper piston, which is production that entered the upper barrel during the downstroke, enters the valve section of the pump between the inlet valve and discharge valve. The inlet valve is closed and the discharge valve is opened, allowing the production to exit the pump and bottomhole assembly in route to the surface.

The lower piston is also moving up, causing the spent power fluid above that piston to enter the lower end of the valve body. This spent power fluid passes by the reversing valve as it leaves the pump and returns to the surface.

Below the lower piston, the pressure is reduced inside the lower barrel and between the lower valves by the upward movement. This reduced pressure opens the inlet valve, and the higher pressure on the return side closes the discharge valve. This allows production to enter the lower barrel so it can subsequently be discharged on the downstroke.

The middle rod of the pump moves through the center of the reversing valve. As the pump reaches the end of the upstroke, a reduced O.D. section of the middle rod enters the I.D. of the reversing valve. This undercut creates a path that allows the top of the reversing valve to be exposed to the power fluid. With a higher pressure above, the reversing valve is made to shift to the bottom of the reversing valve body. This shift causes the power fluid to be directed to the lower half of the pump and the downstroke begins.

On the downstroke, production in the lower barrel is discharged. Production also enters the pump below the lower production inlet valve and is directed to the upper valve section through side tubes in the bottomhole assembly. This fluid then enters the upper barrel, above the piston, to be subsequently discharged on the next upstroke.

The upper piston is also moving down, causing the spent power fluid below it to enter the upper end of the valve body. This spent power fluid passes by the reversing valve as it leaves the pump and returns to the surface.

Another upstroke begins as the pump reaches the end of the downstroke. Again, a reduced O.D. section of the middle rod enters the I.D. of the reversing valve. This undercut creates a path that allows the bottom of the reversing valve to be exposed to the power fluid. With a higher pressure below, the reversing valve is made to shift to the top of the reversing valve body. This shift causes the power fluid to be directed to the upper half of the pump and the upstroke begins.

Because the power fluid is cleaned and chemically treated at the surface, it provides good chemical protection to virtually all of the pump.

6.4.10.4 Piston Velocity
A common cause of failure among all piston pumps is excessive piston velocity on the displacement stroke. This usually occurs when a pumped-off or fluid pound situation occurs.

The pump in Figure 6.4.62 uses the velocity of the power fluid going around the reversing valve to control the downstroke velocity. When this velocity is excessive, the reversing valve moves to reduce the flow-by area and this action reduces the downward speed (velocity) of the engine piston. The downward speed of pump piston matches the downward speed of the engine piston.

The pump in Figures 6.4.63 and 6.4.64 controls piston velocity through a series of sensing holes in the reversing valve. These holes sense whenever the flow of power fluid through the reversing valve/reversing valve body is excessive and directs the reversing valve to restrict the flow.

6.4.10.5 Fluid Separation
Obviously, there must be some means of keeping high-pressure power fluid and low-pressure return fluid separated both inside and outside of the pump.

Internally, a close-fitting metal-to-metal seal around the middle rod accomplishes separation. This seal prevents the power fluid inside the pump from bypassing the pump engine and mixing with the produced fluids. A leak or loss of this seal results in reduced pump speed and engine efficiency.

Externally, separation is accomplished by elastomeric seals on the outside diameter of the pump and one or more seal collars, which are part of the bottomhole assembly. This seal prevents the power fluid outside of the pump from bypassing the pump engine and mixing with the produced fluids. A leak or loss of this seal also results in reduced pump speed and engine efficiency.

6.4.10.6 Piston Size
The engine piston diameter for a given size of hydraulic pump will always be the same. The pump piston, however, can be reduced by several sizes from the size of the engine piston. For example, a $2\frac{1}{2}$-in. hydraulic pump (the maximum size which can be used in $2\frac{1}{2}$-in. tubing) will always have a 2-in. diameter hydraulic engine piston. The pump piston, however, can have a diameter of 2, $1\frac{3}{4}$, $1\frac{5}{8}$, $1\frac{1}{2}$, $1\frac{1}{4}$, or $1\frac{1}{16}$ in. The displacement rate chart (Table 6.4.10) shows the maximum pump end displacement rates for different pump piston sizes of the single displacement Powerlift® I pumps. The displacement rate chart (Table 6.4.11) shows the maximum pump end displacement rates for different pump piston sizes of the double displacement Powerlift® II pumps.

Varying the piston size primarily permits two things:

- Sizing the pump end to the actual well requirements
- Sizing the pump to have the lowest operating pressure possible.

(The smaller the pump piston size in relation to the size of the engine piston, the lower the pressure required to operate the hydraulic unit.)

6.4.11 Selecting A Pump
6.4.11.1 Pump/Engine Ratio
The pump/engine ratio (P/E) is an important factor to consider in selecting a pump because of its relationship to the surface pump that is providing the high-pressure power fluid. The P/E is determined by dividing the engine piston area (A_E) less the middle rod area (A_{MR}) into the pump piston area (A_p),

$$P/E = \frac{A_P}{A_E - A_{MR}} \qquad [6.4.54]$$

A *higher* P/E requires a *lower* power fluid volume and *higher* multiplex pump pressure. A *lower* P/E requires a *higher* power fluid volume and *lower* multiplex pump pressure.

Ensure that pump selection calculations are done carefully. If the P/E is too low, the increase in power fluid volume will cause increased friction loss in the system, resulting in higher multiplex pump pressure.

6.4.11.2 Hydraulic Piston Pump Calculations
The following calculations, performed in order, will provide the information necessary to determine the proper piston pump for any well application. (The user may wish to use more sophisticated calculation techniques in place of some of the equations listed below.)

1. Compute oil-specific gravity (SPGR$_{OIL}$)

$$SPGR_{OIL} = \frac{141.5}{131.5 + API} \qquad [6.4.55]$$

2. Compute oil gradient (GRAD$_{OIL}$)

$$GRAD_{OIL} = 0.433 \times SPGR_{OIL} \qquad [6.4.56]$$

3. Compute solution GOR (GOR$_s$) using pump intake pressure, bottomhole temperature and °API of the oil

$$GOR_s = SPGR_{GAS} \left[\left(\frac{PIP}{18} \right) \left(\frac{10^{0.0125 \, API}}{10^{0.00091 \, BHT}} \right) \right]^{1.2048}$$

[6.4.57]

4. Compute value of F for use in computing oil formation volume factor (B$_o$)

For GOR$_s \leq$ GOR :

$$F = GOR_s \left(\frac{SPGR_{GAS}}{SPGR_{OIL}} \right)^{0.5} + 1.25 \, BHT \qquad [6.4.58]$$

For GOR$_s >$ GOR :

$$F = GOR \left(\frac{SPGR_{GAS}}{SPGR_{OIL}} \right)^{0.5} + 1.25 \, BHT \qquad [6.4.59]$$

5. Compute oil formation volume factor (B$_o$)
$$B_o = 0.972 + 0.000147 \, F^{1.175}$$

6. Compute gas compressibility factor (Z)

$$BZ = \frac{BHT + 460}{203,320} \qquad [6.4.60]$$

$$CZ = 0.0694 - 17.6 \, BZ \qquad [6.4.61]$$

$$DZ = \frac{0.022 \times PIP}{14.65} - 3.5 \qquad [6.4.62]$$

$$EZ = 214 \times BZ \qquad [6.4.63]$$

$$Z = CZ \times (DZ)^2 + EZ + 0.15 \qquad [6.4.64]$$

Table 6.4.10 Powerlift® Signal Displacement Pump

Type	Pump Size	Nom. Pump Size	Piston Diameter (in) Engine	Piston Diameter (in) Pump	O.D. Pump (in)	Piston Area (sq in) Engine	Piston Area (sq in) Pump	Middle Rod Area (sq in)	Stroke Length (in)	Displacement (bbl/day) SPM Engine	Displacement (bbl/day) SPM Pump	Pump Displ. at Max SPM	Speed Range	P/E
FREE	$2 \times 1\frac{5}{8} \times 1\frac{1}{16}$	2	$1\frac{5}{8}$	$1\frac{1}{16}$	$1\frac{7}{8}$	2.074	0.887	0.358	49	15.08	6.45	225	35	0.52
	$2 \times 1\frac{5}{8} \times 1\frac{1}{4}$	2	$1\frac{5}{8}$	$1\frac{1}{4}$	$1\frac{7}{8}$	2.074	1.227	0.358	49	15.08	8.92	312	35	0.72
	$2 \times 1\frac{5}{8} \times 1\frac{1}{2}$	2	$1\frac{5}{8}$	$1\frac{1}{2}$	$1\frac{7}{8}$	2.074	1.767	0.358	49	15.08	12.85	450	35	1.03
	$2 \times 1\frac{5}{8} \times 1\frac{5}{8}$	2	$1\frac{5}{8}$	$1\frac{5}{8}$	$1\frac{7}{8}$	2.074	2.074	0.358	49	15.08	15.08	528	35	1.21
FREE	$2\frac{1}{2} \times 2 \times 1\frac{1}{16}$	$2\frac{1}{2}$	2	$1\frac{1}{16}$	$2\frac{1}{4}$	3.142	0.887	0.358	66	30.77	8.69	191	22	0.32
	$2\frac{1}{2} \times 2 \times 1\frac{1}{4}$	$2\frac{1}{2}$	2	$1\frac{1}{4}$	$2\frac{1}{4}$	3.142	1.227	0.358	66	30.77	12.02	264	22	0.44
	$2\frac{1}{2} \times 2 \times 1\frac{1}{2}$	$2\frac{1}{2}$	2	$1\frac{1}{2}$	$2\frac{1}{4}$	3.142	1.767	0.554	66	30.77	17.30	467	27	0.68
	$2\frac{1}{2} \times 2 \times 1\frac{5}{8}$	$2\frac{1}{2}$	2	$1\frac{5}{8}$	$2\frac{1}{4}$	3.142	2.074	0.554	66	30.77	20.30	547	27	0.80
	$2\frac{1}{2} \times 2 \times 1\frac{3}{4}$	$2\frac{1}{2}$	2	$1\frac{3}{4}$	$2\frac{1}{4}$	3.142	2.405	0.554	66	30.77	23.56	636	27	0.93
	$2\frac{1}{2} \times 2 \times 2$	$2\frac{1}{2}$	2	2	$2\frac{1}{4}$	3.142	3.142	0.554	66	30.77	30.77	831	27	1.21
FREE	$2\frac{1}{2} \times 1\frac{5}{8} \times 1\frac{1}{16}$	$2\frac{1}{2}$	$1\frac{5}{8}$	$1\frac{1}{16}$	$2\frac{1}{4}$	2.074	0.887	0.358	49	15.08	6.45	225	35	0.52
	$2\frac{1}{2} \times 1\frac{5}{8} \times 1\frac{1}{4}$	$2\frac{1}{2}$	$1\frac{5}{8}$	$1\frac{1}{4}$	$2\frac{1}{4}$	2.074	1.227	0.358	49	15.08	8.92	312	35	0.72
	$2\frac{1}{2} \times 1\frac{5}{8} \times 1\frac{1}{2}$	$2\frac{1}{2}$	$1\frac{5}{8}$	$1\frac{1}{2}$	$2\frac{1}{4}$	2.074	1.767	0.358	49	15.08	12.85	450	35	1.03
	$2\frac{1}{2} \times 1\frac{5}{8} \times 1\frac{5}{8}$	$2\frac{1}{2}$	$1\frac{5}{8}$	$1\frac{5}{8}$	$2\frac{1}{4}$	2.074	2.074	0.358	49	15.08	15.08	528	35	1.21
FREE	$3 \times 2\frac{1}{2} \times 1\frac{3}{4}$	3	$2\frac{1}{2}$	$1\frac{3}{4}$	$2\frac{3}{4}$	4.909	2.405	0.866	60	43.71	21.42	643	30	0.59
	$3 \times 2\frac{1}{2} \times 2$	3	$2\frac{1}{2}$	2	$2\frac{3}{4}$	4.909	3.142	0.866	60	43.71	27.98	839	30	0.78
	$3 \times 2\frac{1}{2} \times 2\frac{1}{4}$	3	$2\frac{1}{2}$	$2\frac{1}{4}$	$2\frac{3}{4}$	4.909	3.976	0.866	60	43.71	35.41	1062	30	0.98
	$3 \times 2\frac{1}{2} \times 2\frac{1}{2}$	3	$2\frac{1}{2}$	$2\frac{1}{2}$	$2\frac{3}{4}$	4.909	4.909	0.866	60	43.71	43.71	1311	30	1.21
FIXED	$2\frac{1}{2} \times 1\frac{5}{8} \times 1\frac{1}{16}$	$2\frac{1}{2}$	$1\frac{5}{8}$	$1\frac{1}{16}$	$2\frac{1}{4}$	2.074	0.887	0.358	66	20.32	8.69	235	27	0.52
	$2\frac{1}{2} \times 1\frac{5}{8} \times 1\frac{1}{4}$	$2\frac{1}{2}$	$1\frac{5}{8}$	$1\frac{1}{4}$	$2\frac{1}{4}$	2.074	1.227	0.358	66	20.32	12.02	325	27	0.72
	$2\frac{1}{2} \times 1\frac{5}{8} \times 1\frac{1}{2}$	$2\frac{1}{2}$	$1\frac{5}{8}$	$1\frac{1}{2}$	$2\frac{1}{4}$	2.074	1.767	0.358	66	20.32	20.32	467	27	1.03
	$2\frac{1}{2} \times 1\frac{5}{8} \times 1\frac{5}{8}$	$2\frac{1}{2}$	$1\frac{5}{8}$	$1\frac{5}{8}$	$2\frac{1}{4}$	2.074	2.074	0.358	66	20.32	17.31	549	27	1.21
											20.32			

Special Powerlift® I Single-Displacement Pump

Type	Pump Size	Nom. Pump Size	Piston Diameter (in) Engine	Piston Diameter (in) Pump	O.D. Pump (in)	Piston Area (sq in) Engine	Piston Area (sq in) Pump	Middle Rod Area (sq in)	Stroke Length (in)	Displacement (bbl/day) SPM Engine	Displacement (bbl/day) SPM Pump	Pump Displ. at Max SPM	Speed Range	P/E
FREE	$2\frac{1}{2} \times 1\frac{5}{8} \times 1\frac{1}{2}$	2	$1\frac{5}{8}$	$1\frac{1}{2}$	$1\frac{7}{8}$	2.074	1.767	0.554	45.6	14.04	11.96	478	40	1.16
	$2\frac{1}{2} \times 1\frac{5}{8} \times 1\frac{5}{8}$	2	$1\frac{5}{8}$	$1\frac{5}{8}$	$1\frac{7}{8}$	2.074	2.074	0.554	45.6	14.04	14.04	561	40	1.36
FREE	$2\frac{1}{2} \times 2 \times 1\frac{3}{4}$	$2\frac{1}{2}$	2	$1\frac{3}{4}$	$2\frac{1}{4}$	3.142	2.405	0.866	66	30.77	23.56	825	35	1.06
	$2\frac{1}{2} \times 2 \times 2$	$2\frac{1}{2}$	2	2	$2\frac{1}{4}$	3.142	3.142	0.866	66	30.77	30.77	1078	35	1.38

Table 6.4.11 *Powerlift® II Double Displacement Pump*

| Pump Size | Nom. Pump Size | Piston Diameter (in) | | Pump O.D. (in) | Piston Area (sq in) | | Middle Rod Area (sq in) |
		Engine	Pump		Engine	Pump	
$2\frac{1}{2} \times 1\frac{1}{4}$	$2\frac{1}{2}$	1.885	1.250	$2\frac{1}{4}$	2.791	1.227	0.355
$2\frac{1}{2} \times 1\frac{1}{2}$	$2\frac{1}{2}$	1.885	1.500	$2\frac{1}{4}$	2.791	1.767	0.355
$2\frac{1}{2} \times 1\frac{7}{8}$	$2\frac{1}{2}$	1.885	1.885	$2\frac{1}{4}$	2.791	2.791	0.355
$2 \times 1\frac{1}{16}$	2	1.572	1.063	$1\frac{7}{8}$	1.941	0.887	0.248
$2 \times 1\frac{1}{4}$	2	1.572	1.250	$1\frac{7}{8}$	1.941	1.227	0.248
$2 \times 1\frac{9}{16}$	2	1.572	1.572	$1\frac{7}{8}$	1.941	1.941	0.248

| Pump Size | Stroke Length (in) | Displacement (BPD per SPM) | | Pump Displacement BPD at Maximum SPM | Speed Range | P/E |
		Engine	Pump			
$2\frac{1}{2} \times 1\frac{1}{4}$	24.00	17.69	8.74	918	105	0.503
$2\frac{1}{2} \times 1\frac{1}{2}$	24.00	17.69	12.59	1322	105	0.725
$2\frac{1}{2} \times 1\frac{7}{8}$	60.25	43.97	50.00	2500	50	1.146
$2 \times 1\frac{1}{16}$	21.00	12.10	5.53	597	108	0.524
$2 \times 1\frac{1}{4}$	21.00	12.10	7.65	826	108	0.725
$2 \times 1\frac{9}{16}$	52.00	26.35	30.00	1560	52	1.147

7. Compute gas formation volume factor (B_g)

$$B_g = \frac{0.0283 \times Z \times (BHT + 460)}{PIP} \qquad [6.4.65]$$

8. Compute two-phase formation volume factor (B_T)

For $GOR_s \geq GOR$:

$$B_T = B_O \qquad [6.4.66]$$

For $GOR_s < GOR$:

$$B_T = B_O + \frac{B_g \times (GOR - GOR_s)}{5.6146} \qquad [6.4.67]$$

9. Compute volumetric efficiency (EFF)

$$EFF_V = \frac{1}{CW + (1 - CW) \times B_T} \frac{STBF}{BBL} \qquad [6.4.68]$$

10. Compute required pump end displacement (DISP)

$$DISP = \frac{Q_{PROD}}{(EFF_V) \times (EFF_{PUMP\,END})} \qquad [6.4.69]$$

11. Select pump size
    Pump end displacement (DISP) should be less than 75% of the pump displacement at maximum strokes per minute.
12. Calculate strokes per minute (SPM)

$$SPM = \frac{DISP}{(PUMP\ END\ DISP/STROKE/MINUTE)} \qquad [6.4.70]$$

13. Calculate power fluid volume (Q_{PWR})

$$Q_{PWR} = \frac{SPM \times (ENGINE\ END\ DISP/STROKE/MINUTE)}{EFF_{ENGINE}} \qquad [6.4.71]$$

14. Compute return fluid volume (Q_{RTN})

$$Q_{RTN} = Q_{POWER\ FLUID} + Q_{PRODUCTION} \qquad [6.4.72]$$

15. Compute oil viscosity (μ_{OIL}) using average wellbore temperature

$$T = LN\left(\frac{BHT + WHT}{2} + 460\right) \qquad [6.4.73]$$

$$C = 3.55(6.328 - T) + 11.53\ SPGR_{OIL} - 9.254 \quad [6.4.74]$$

$$E = e^C \qquad [6.4.75]$$

$$\mu_{OIL} = (e^E - 0.6)SPGR_{OIL} \qquad [6.4.76]$$

16. Compute water viscosity (μ_{WTR}) using average wellbore temperature

$$T_{AVG} = \frac{BHT + WHT}{2} \qquad [6.4.77]$$

$$T = 1.003 - 0.01479\ T_{AVG} + 0.00001928\ T_{AVG}^2 \quad [6.4.78]$$

$$\mu_{WTR} = e^T \qquad [6.4.79]$$

17. Compute pressure loss in power fluid (PL_{PWR})
    Reynold's number

$$RN = \frac{213QG(D_1 - D_2)}{\mu(D_1^2 - D_2^2)} \qquad [6.4.80]$$

Turbulent flow pressure loss

$$PL = \frac{2.644 \times 10^{-5}Q^2LGf}{(D_1 - D_2)(D_1^2 - D_2^2)^2\left(\dfrac{D_1}{D_1 - D_2}\right)^{0.1}} \qquad [6.4.81]$$

where:

$$f = \text{friction factor} = \frac{0.236}{RN^{0.21}}$$

G = Gradient(psi/ft.)

L = Length (ft.)

Q = FlowRate (bbl/d)

μ = Viscosity (c.p.)

D_2 = Tubing O.D. (annular flow)

D_2 = 0 (tubing flow)

D_1 = Casing I.D. (annular flow)

D_1 = Tubing I.D. (tubing flow)

$$PL = \frac{202 \times 10^{-8} \times L}{(D_1 - D_2)(D_1^2 - D_2^2)^2 \left(\frac{D_1}{D_1 - D_2}\right)^{0.1}} \left[\frac{D_1^2 - D_2^2}{D_1 - D_2}\right]^{0.21}$$

$$\times \left[\frac{\mu}{G}\right]^{0.21} G \times Q^{1.79} \qquad [6.4.82]$$

18. Compute return water fraction (WC$_{RTN}$)

For power water :

$$WC_{RTN} = \frac{PRODUCED\ WATER + Q_{PWR}}{Q_{RTN}} \qquad [6.4.83]$$

For power oil : $WC_{RTN} = \dfrac{PRODUCED\ WATER}{Q_{RTN}}$

$$[6.4.84]$$

19. Compute gradient of return fluid (GRAD$_{RTN}$)

$$GRAD_{RTN} = WC_{RTN} \times GRAD_{WATER}$$

$$+(1 - WC_{RTN})GRAD_{OIL} \qquad [6.4.85]$$

20. Determine viscosity of return fluid (μ$_{RTN}$)

For $WC_{RTN} > 0.6 : \mu_{RTN} = \mu_{WATER}$ $\qquad [6.4.86]$

21. Compute pressure loss in return fluid (PL$_{RTN}$)

A flowing gradient correlation is preferred for this calculation if the GLR > 10. However, if one is not available or GLR = 10, then see pressure loss equations under Step 17.

22. Compute pump discharge pressure (P$_{DIS}$)

$$P_{DIS} = PRESSURE_{WELL\ HEAD} + GRAD_{RTN}$$

$$\times DEPTH + PL_{RTN} \qquad [6.4.87]$$

If a flowing gradient correlation is used, then replace right side of equation with results from correlation.

23a. Compute pressure loss in pump in Figure 6.4.20 and its BHA (PL$_{PUMP}$)

$$PL_{PUMP} = 973 \left(\frac{SPM}{SPM_{MAX}}\right)^{1.355} \qquad [6.4.88]$$

23b. Compute Pressure Loss in Pump in Figure 6.4.20 and its BHA (PL$_{PUMP}$)

$$PL_{PUMP} = 204.86 \left(\frac{SPM}{SPM_{MAX}}\right)^5 + 204.56 \left(\frac{SPM}{SPM_{MAX}}\right)^4$$

$$+ 368.18 \left(\frac{SPM}{SPM_{MAX}}\right)^3 + 336.02 \left(\frac{SPM}{SPM_{MAX}}\right)^2$$

$$+ 652.24 \left(\frac{SPM}{SPM_{MAX}}\right) + 0.344 \qquad [6.4.89]$$

24. Compute multiplex pressure (P$_{TPLX}$)

$$P_{TPLX} = P_{DIS} \times \left(\frac{P}{E} + 1\right) - PIP \times \frac{P}{E} + PL_{POWER}$$

$$+ PL_{PUMP} - (DEPTH \times GRAD_{POWER}) \quad [6.4.90]$$

25. Compute hydraulic horsepower (HHP)

$$HHP = 0.000017 \times P_{TPLX} \times Q_{POWER} \qquad [6.4.91]$$

6.4.12 Surface Power Fluid Conditioning Systems

The purpose of a surface power fluid conditioning system is to provide a constant and adequate supply of suitable power fluid to operate the subsurface pumps. The success and economical operation of any hydraulic pumping fluid installation is, to a large extent, dependent on the effectiveness of the surface conditioning system in supplying clean power fluid for surface power and down-hole pump system.

The presence of gas, solids, or abrasive materials in the power fluid will seriously affect the operation and wear life of the subsurface pump as well as the surface power unit. Therefore, the primary objective in conditioning crude oil or water for use as power fluid is to make it as free of gas and solids as is practical. In addition to removing gas and solid material, chemical treatment of the power fluid at the surface will also increase the wear life of the pumping equipment.

There are two types of power fluid conditioning systems for hydraulic pump installations: the central power fluid conditioning system and the well-site, self-contained power fluid conditioning system (Unidraulic®).

6.4.12.1 Central Power Fluid Conditioning System

A central power fluid conditioning facility (Figure 6.4.65) is one in which the power fluid for one or more wells is treated to remove gas and solids at a large centralized facility.

Figure 6.4.66 shows a typical power fluid treating system which has been proven through years of design experience. This power fluid treating system design assumes that the normal lease separators and heater treaters have delivered "stock tank" oil to the treating facility which is essentially free of gas.

The power fluid-settling tank in this system is usually a 24-ft.-high, three-ring, bolted steel tank. A tank of this height generally will provide an adequate head for gravity flow of fluid from the tank to the intake of the charge pump. The diameter of the power fluid tank is dictated by the amount of power fluid required.

The basic purpose of the power fluid settling tank is to allow separation of solids from the power fluid, which the lease separator has not removed in the continuous flow system. The clean fluid is supplied to the surface pump, where it is pressurized and then used to operate the down-hole pump.

In a tank of static fluid, all foreign material in the fluid, which is heavier than the fluid would fall or settle out to the bottom. Some of the particles, such as fine sand would fall more slowly than the heavier solids. These factors, plus viscosity-related resistance factors influence the rate of separation. In time, however, all solids and heavier liquids would settle out leaving a layer of clean fluid.

In an actual power fluid system it is not practical, nor is it necessary, to furnish tank space to allow settling under perfectly still conditions. Sufficient settling can be accomplished when the upward flow through the settling tank is maintained at a velocity just slower than the velocity at which the contaminating material will fall. It has been found through tests and experience that an upward velocity of 1 ft./hr is low enough to provide gravity separation of entrained particles in most crude oils.

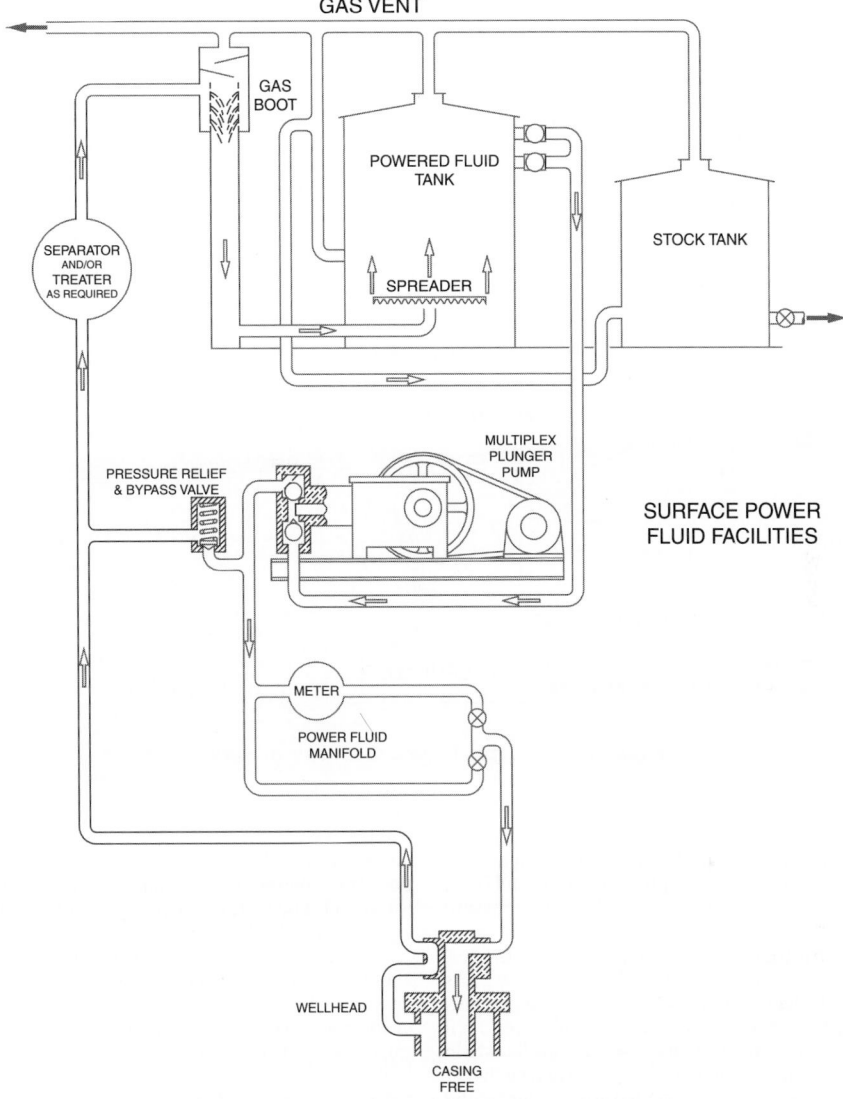

Figure 6.4.65 *Central power fluid conditioning facility.*

6.4.13 Weatherford Unidraulic® Unit

The Weatherford Unidraulic® Fluid Conditioning Unit is a unitized surface equipment package for hydraulic pumping applications. The unit provides complete fluid conditioning as well as a surface pump to provide pressurized power fluid to the downhole pump Figure 6.4.67.

The Unidraulic® is designed for continuous operation and will perform satisfactorily when properly set, adjusted and maintained. Periodic inspections and repairs will be required. The end-user of the equipment must determine the intervals of inspections.

6.4.13.1 Vertical Accumulator

This is an ASME-coded vessel that is available is multiple sizes and working pressure ratings. Some of the sizes/ratings are 30 in. in diameter and 72 in. high with a 200 psi working pressure rating, 36 in. in diameter and 36 in. high with a 200 psi working pressure rating and 36 in. in

diameter and 96 in. high with a 650 psi working pressure rating.

Produced fluid and spent power fluid flow from the wellhead to the accumulator. After entering the accumulator, the fluids pass over a series of baffles in order to separate the free gases from the liquids as well as to remove the majority of trapped gases. The gases plus liquids are then directed to the cyclone desander(s). This vessel also serves as a surge chamber to buffer high pressure kicks from the well so that those surge forces are kept from the cyclone desander(s) and the horizontal fluid conditioning vessel. This vessel also has a safety relief valve that is preset to open at the maximum working pressure.

6.4.13.2 Fluid Conditioning Vessel

This is also an ASME coded vessel that is available is multiple sizes and working pressure ratings. Some of the sizes/ratings are 42 in. in diameter and 141 in. long with

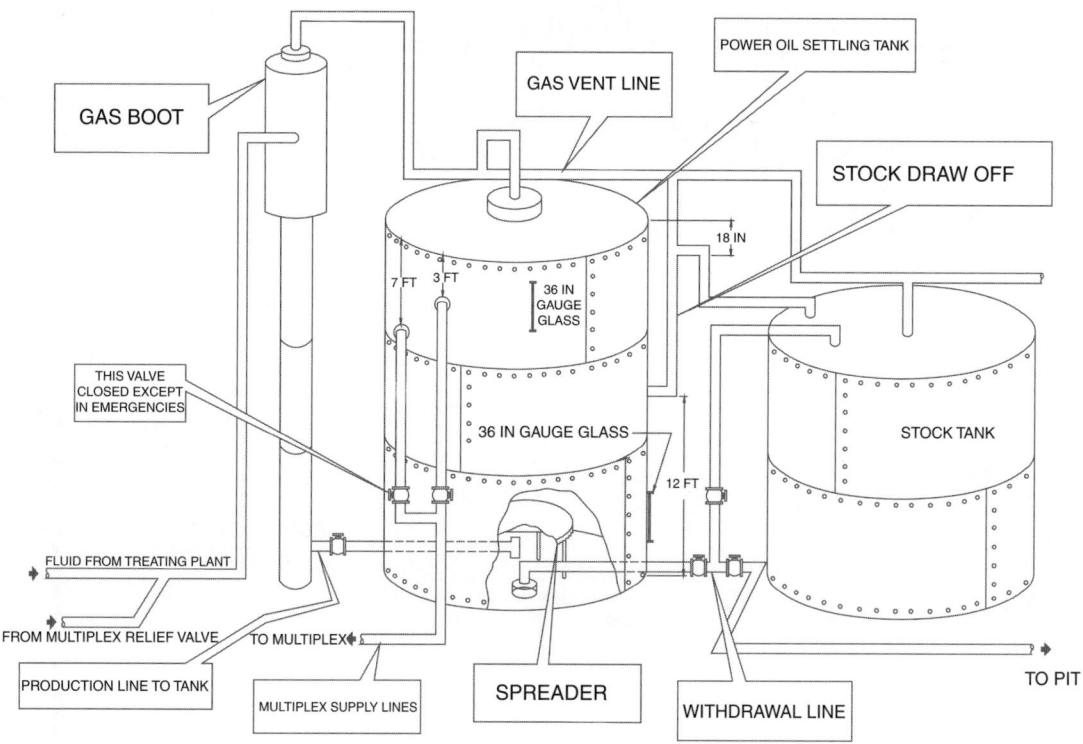

GAS BOOT

GAS VENT LINE

POWER OIL SETTLING TANK

STOCK DRAW OFF

18 IN

7 FT   3 FT

36 IN
GAUGE
GLASS

THIS VALVE
CLOSED EXCEPT
IN EMERGENCIES

36 IN GAUGE GLASS

STOCK TANK

12 FT

FLUID FROM TREATING PLANT

FROM MULTIPLEX RELIEF VALVE   TO MULTIPLEX

PRODUCTION LINE TO TANK

MULTIPLEX SUPPLY LINES

SPREADER

WITHDRAWAL LINE

TO PIT

Figure 6.4.66 *Typical power fluid treating system.*

a 130 psi working pressure rating, 60 in. in diameter and 240 in. long with a 150 psi working pressure rating and 60 in. in diametr and 240 in. long with a 500 psi working pressure rating.

After the fluids leave the desander(s), they enter the vessel, where they pass over a series of baffles. There free gas is separated from the fluid flow and sent to the production line. The remaining fluid, which consists of oil and water, then passes over additional baffles that cause the two liquids to separate from each other. Another baffle is located such that any sediment in the vessel is prevented from entering the power fluid outlets and going to the triplex. By appropriately adjusting the outlets, either oil or water can be sent to the triplex for use as the power fluid, and the vessel made to fill with the desired power fluid (either oil or water). An auxiliary pump is available to charge the intake of the triplex pump if it is needed, but the pressure in the vessel, which is usually flowline pressure, is almost always sufficient to meet that requirement. If the flowline pressure is too low, then a back-pressure valve can be installed between the horizontal vessel and the flowline in order to increase the pressure in the horizontal vessel. This vessel also has a safety relief valve that is preset to open at the maximum working pressure.

By design, the retention time of the power fluid in the conditioning vessel is not sufficient for any appreciable setting of solids particles from the fluids, so most solids separation must be done by the cyclone desander. This is what makes proper installation, sizing, monitoring, and operation of the cyclone desander so important to the entire system operation.

The purpose of the tanks is to supply the multiplex pump with sufficient fluid. If the liquid level in the horizontal vessel is allowed to drop to a point that the pump draws gas, the pump will cavitate and cause damage. Therefore, as protection for the pump, an oversupply of fluid must be directed to the vessel. The fluid not required by the multiplex pump is automatically discharged to the flow line.

This over supply of fluid is also important in maintaining a proper oil/water interface. If the supply is insufficient, the interface level will move up to the pump suction connection on the vessel. The interface often carries iron sulphite, which can cause pump damage.

6.4.13.3 Power Fluid Pump
Fluid from the conditioning vessel is supplied to the surface power fluid pump (normally a multiplex pump), which provides pressurized liquid to operate the subsurface hydraulic pump. An electric motor, diesel, or gas engine powers the power fluid pump.

6.4.13.4 Control Panel
The control panel is normally a standard oil-field type Murphy panel with indicators and limit controls for safety, start up, operation, and shut down.

6.4.13.5 Cyclone Desander
The cyclone desander is considered to be the very heart of the Unidraulic® Unit. Without excellent solids separation, the result will be unnecessarily short pump runs and excessive multiplex maintenance. It must be sized to affect maximum solid particle separation and furnish essentially clean fluid for the surface power fluid pump and the downhole hydraulic pump.

A cyclone desander, microstrainer, filter, centrifuge, or similar device is used to remove solids such as sand, silt, dirt, or drilling mud, from the produced fluid or from any

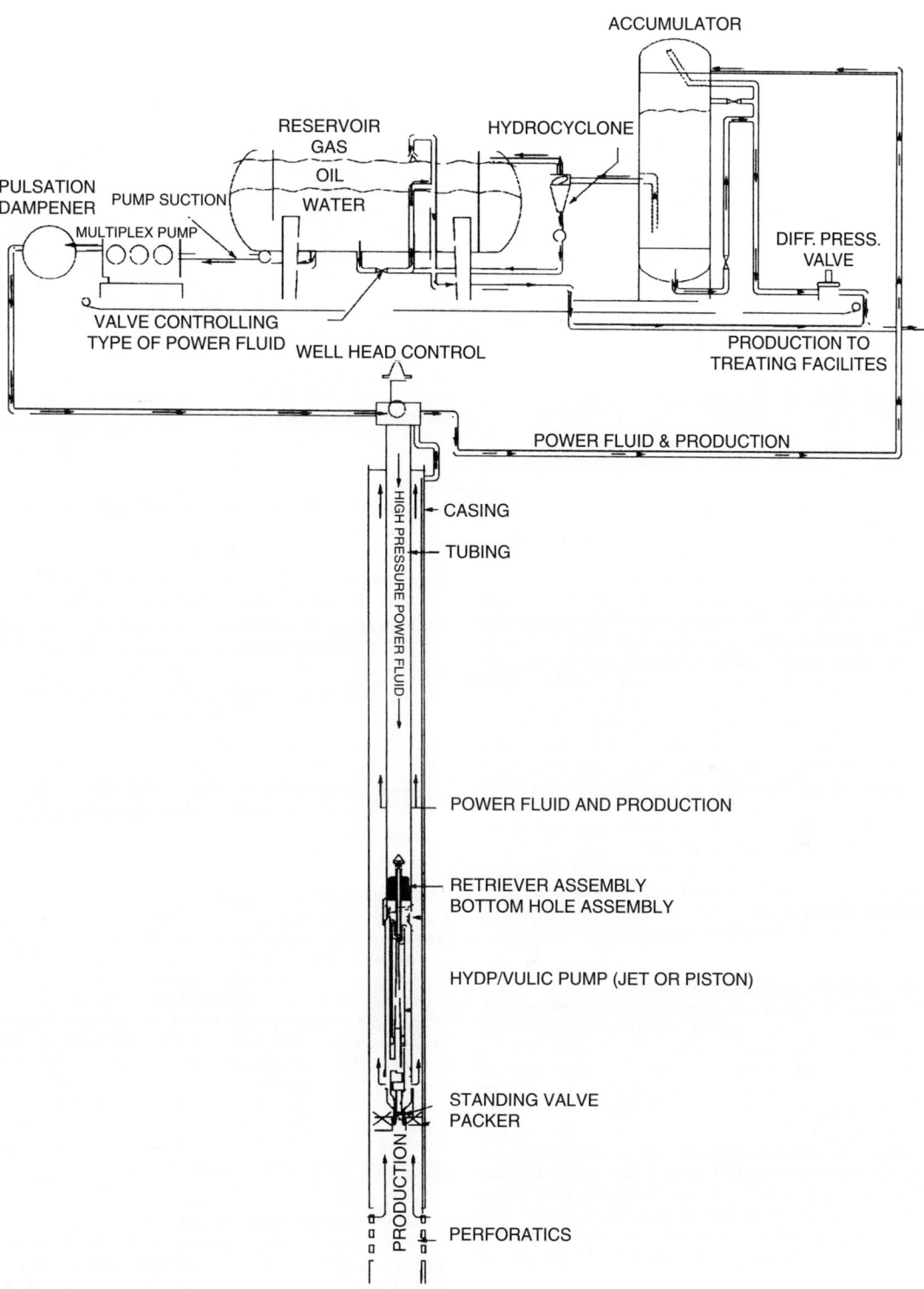

Figure 6.4.67

Table 6.4.12 *Approximate Size Equivalents*

Inches	Microns	Screen Mesh	Common Materials
.00004	1		Clay, Colloids
.0002	5		Bentonite, Talcum Powder, Fine Silt, Barites, Red Blood Corpuscles, Fine Cement Dust, Cocoa
.0004	10	1470	
.0018	44	325	Pollen, Milled Flour, Coarse Silt, Fine Human Hair
.002	53	270	
.003	74	200	
.004	105	140	API Sand, Coarse Human Hair, Table Salt
.006	149	100	
.012	297	50	Coarse Sand
.20	500	18	
.040	1000	18	
.125	3125		Foreign Matter

fluid that is subsequently used to operate a hydraulic pump. The desander should function to provide a fluid clean enough to prevent excessive wear in the surface multiplex pump and the subsurface (downhole) hydraulic pump.

Surface pumps, subsurface piston pumps, and hydraulic engines used to drive the piston pumps are the components of an hydraulic pumping system most susceptible to wear because they contain close fitting metallic parts moving next to or rubbing against each other. Jet pumps are also subject to wear when operated with contaminated fluids.

Most subsurface piston pumps and hydraulic engines have a piston/cylinder radial clearance of 0.001 in. Other parts of the pumps have metal-to-metal seals which have clearances of less than 0.001 in. Solid contaminants in the fluid cause wear leading to excessive clearance of mating parts. This, in turn, leads to inefficient operation of the components and eventual replacement.

Table 6.4.12 may be used as a guide for estimating sizes of particulate matter which might enter the cyclone desander cleansing unit for the majority of oil wells which are candidates for hydraulic pumping.

6.4.13.6 Removal of Suspended Solids

The volume of suspended solids such as sand, drilling mud, and scale in one barrel of produced fluid may be 100 lb or more when a pump is first installed in a well. More typical is a well with particles in the 10- to 500-micron size range and a total content of suspended solids of 100 lb or less per 1000 barrels entering the cyclone desander inlet.

With reasonable operating efficiency, the cyclone desander should allow no more than 5% of the total suspended solids to pass on to the clean fluid reservoir. Of these solids, none should be larger than 25 microns (.001 diameter). No more than 20% of the total suspended solids entering the desander should be larger than 10 microns (.0004 diameter) and the remaining 80% should be 10 microns or less.

Ideally, the cyclone desander should function to prevent suspended solids not larger than one micron (.00004-in. diameter) from entering the clean fluid reservoir. Realistically, 10 microns is closer to the maximum size, which should be adopted as standard.

When fluid volume exceeds 2000 barrels per day, two cyclone desanders operating in parallel may be required.

6.4.13.7 Pressure Drop

Particle separation in a cyclone desander is related to the pressure drop from the inlet to the outlet of the cyclone desander.

Generally the greater the pressure drop across the cyclone desander (up to 50 psi) the cleaner the fluid. However, a pressure drop greater than 50 psi will carry the ligher solids upward giving a solids concentration which is too high even if the particle size is in the acceptable size range.

The recommended pressure drop across the cyclone desander is 35 to 50 psi with the optimum being 40 to 45 psi. If the pressure drop across the cyclone desander gets below 35 psi, excessively large particles will be passed to the conditioning vessel. The volume of fluid passing through it and the cyclone desander's internal parts — the vortex and feed nipple, primarily determines the pressure drop. In operation, it is easier to change the vortex and/or feed nipple of the cyclone desander to maintain the desired pressure drop rather than to change the volume of fluid passing through the cyclone desander.

References

1. Vogel, J. V., "Inflow Performance Relationship for Solution Gas Drive Wells," Journal of Petroleum Technology, January, 1968, pp. 83–93.
2. Brown, K. E., "The Technology of Artificial Lift Methods," Volume 1, Petroleum Publishing Company, 1977.
3. Evinger, H. H. and Muskat, M., "Calculation of Theoretical Productivity Factor," Transactions AIME (1942) 146, 126.
4. Standing, M. B., "a Pressure-Volume-Temperature Correlation for Mixtures of California Oils and Gases," Drilling and Production Practices, API, 1947, p. 275.
5. Standing, M. B., "Inflow Performance Relationships for Damaged Wells Producing by Solution Gas Drive," Journal of Petroleum Technology, November, 1970, pp. 1399–1400.

6.4.14 Introduction Electrical Submersible Pumps

The electrical submersible pump (ESP) is considered the high volume and depth choice among lift systems. ESP systems require very little surface space and can be installed in highly deviated wells for both onshore and offshore applications. Adaptable to a variety of wellbore configurations, ESP systems can be used in wells with $4\frac{1}{2}$-in. casing and larger, and are excellent candidates for vertical wells with doglegs (crooked holes). Typically, these systems do best in high volume wells where gas, solids, and excessive pressures are not present in great volumes. ESP systems are highly efficient beginning at 1000 BPD, although can be operated as low as 200 BPD, require very little maintenance and are resistant to corrosive downhole environments.

6.4.14.1 ESP Application Considerations

The ESP system have the ability to operate in a large range of applications up to 13,000'+ TVD and flow rates up to 60,000 BPD. Typical operating ranges can be found in (Figure 6.4.68). ESPs are designed individually and specifically for each application. Careful attention must be paid each design choosing the proper equipment for any given well. Operating and ESP system outside of its defined range can result in less than desirable performance including possibly shorten the run-life of the system. Additional design information is presented in "ESP Product Design."

Pump recommended operating range for Manufacturer A.

Model	Housing OD (inches)	BEP @ 60Hz (Bls/day)	Maximum HP (HP)	Recommended Operating Range @ 60 Hz (Bls/day)	
				Minimum	Maximum
400-365	4.00	365	94	200	500
400-450	4.00	450	94	200	650
400-675	4.00	675	94	400	1000
400-850	4.00	850	94	600	1300
400-1200	4.00	1200	94	800	1500
400-1600	4.00	1600	94	1100	2100
400-1800	4.00	1800	94	1100	2100
400-2050	4.00	2050	94	1200	2400
400-2600	4.00	2600	256	1500	3750
400-3750	4.00	3750	256	2000	5800
400-4200	4.00	4200	256	3200	6000
400-5500	4.00	5500	256	3500	7500
400-5600	4.00	5600	256	3600	7600
513-1600	5.13	1600	375	1100	2200
513-1750	5.13	1750	375	1100	2250
513-2500	5.13	2500	375	1700	3900
513-3400	5.13	3400	375	2200	5000
513-3600	5.13	3600	375	2200	5000
513-5700	5.13	5700	375	3500	8000
513-8000	5.13	8000	637	5500	10000
513-9500	5.13	9500	637	6000	13000
538-4700	5.38	4700	375	3500	7000
538-7000	5.38	7000	375	5000	10500
538-8500	5.38	8500	375	5000	11000
538-11000	5.38	11000	637	8000	16000

Figure 6.4.68 ESP system design.

6.4.14.2 ESP Product Identification
ESP systems are divided into subsurface and surface components. Both are required for any installation. Beginning downhole, each component is defined below. A typical ESP installation schematic is found in (Figure 6.4.69).

6.4.14.2.1 Submersible Electric Motors
The ESP system's prime mover is the submersible motor. The motor is a two-pole, three-phase, squirrel-cage induction type. Motors run at a nominal speed of 3500 rpm in 60-Hz operation (2917 rpm in 50-Hz). Motors are filled with a highly refined mineral oil that provides dielectric strength, bearing lubrication, and thermal conductivity.

The standard motor thrust bearing is a fixed-pad type. Its purpose is to support the thrust load of the motor rotors. Other types are used in high-temperature applications (above 250°F). Heat generated by motor operation is transferred to the well fluid as it flows past the motor housing. A minimum fluid velocity of 1 ft./sec is recommended to provide adequate cooling. Because the motor relies on the flow of well fluid for cooling, a standard ESP should never be set at or below perforations or producing zone unless the motor is shrouded.

Motors are manufactured in four different diameters (series) as 3.75, 4.56, 5.40, and 7.38 inches. Thus, motors can be used in casing sizes as small as 4.50 inches.

60 Hz horsepower capabilities range from a low of 7.5 HP in 3.75-inch series to a high of 1200 HP in the 7.38-inch series. Motor construction may be a single section or several "tandem" sections bolted together to reach a specific horsepower. Motors are selected on the basis of the maximum OD that can be run easily in a given casing size and specific horsepower required.

6.4.14.2.2 Seal Section
The seal section's primary purpose is to isolate the motor oil from the well fluid while bottomhole pressure (BHP) and the motor's internal pressure are balanced. There are two types of seal section design — the positive seal and the labyrinth path. The positive seal design relies on an elastic, fluid-barrier bag to allow for the thermal expansion of motor fluid in operation, while still isolating the well fluid from the motor oil. The labyrinth path design uses the specific gravity of the well fluid and motor oil to prevent the well fluid from entering the motor. This is accomplished by allowing the well fluid and motor oil to communicate through tube paths connecting segregated chambers.

The seal section performs four basic functions:

1. Connects the pump to the motor by connecting both the housing and drive shafts
2. Houses a thrust bearing to absorb pump shaft axial thrust

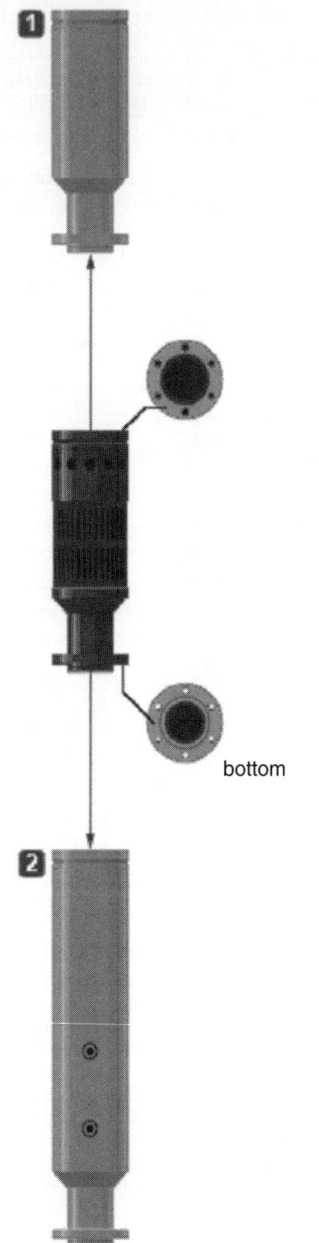

bottom

Figure 6.4.69 *Typical ESP installation.*

3. Isolates motor oil from well fluid while allowing wellbore-motor pressure equalization
4. Allows thermal expansion of motor oil resulting from operating heat rise and thermal contraction of the motor oil after shutdown

6.4.14.2.3 Pump Intake/Gas Separator

There are three types of pump intakes: the standard intake, the static gas separator, and the rotary gas separator.

The amount of free gas by volume at pump intake conditions should be no more than 10% to 15% by volume. The

standard intake has several fairly large ports, allowing fluids to flow into the lower section of the pump and enter the bottom stage in the pump. Some models are equipped with a screen to keep large debris out of the pump.

The static gas separator is a reverse-flow intake. This type separator will, in most cases, separate up to 20% (by volume) free gas. The fluid moves up along the outside diameter of the gas separator and then must reverse its flow as it enters the perforated holes of the separator. The fluid changes direction and must go back down to the pickup impeller. A portion of the gas breaks out and travels up into the annulus. The remaining fluid goes through the pickup impeller, changing directions and moves up into the lower portion of the pump, to first stage of the pump.

The rotary gas separator will separate free gas with an efficiency of up to 90% under some conditions. The rotary gas separator should be used where the free gas available at the intake exceeds 20% (VLR of 0.20), unless there is abrasive material in the fluid. Then the use of rotary separator must be carefully considered. Even though the rotary gas separator is very efficient, there can still be cases where the pump will gas lock. If the pump intake pressure gets so low that slug flow develops, then there will be moments of time when nothing but gas is present at the intake. Only during this interval will gas enter the pump.

6.4.14.2.4 Multistage Centrifugal Pump

The ESP pump is a multistage centrifugal pump type. A stage consists of an impeller and a diffuser. The impeller is keyed to the shaft and rotates at the rpm of the motor. Centrifugal force causes the fluid to move from the center (or eye) of the impeller outward. These forces impart kinetic or velocity energy to the fluid.

The diffuser is stationary, and its function is to direct the fluids to flow efficiently from one impeller to another and to convert a portion of the velocity (kinetic) energy into pressure (potential) energy.

The stages (an impeller–diffuser combination) are placed onto a keyed shaft and then loaded into a steel housing (pipe or tube). When the threaded head and base are screwed into the housing, they compress against the outside edge of the diffuser. It is this compression that holds the diffusers stationary. If for any reason this compression is lost, then the diffusers would be free to rotate. This rotation would cause the pump to lose almost all of its ability to produce any head (or lift).

The impellers incorporate a fully enclosed curved vane design, the maximum efficiency of which is a function of impeller design and type, which generally increases when the volumetric capacity of that design. The operating efficiency is a function of the percent of design capacity at which the impeller is operated.

The fluid enters the impeller at the eye. The vanes in the impeller create channels through which the fluid is directed. The size of the impeller (or the volume between the upper and lower shroud) determines the volume per unit time (or fluid rate) that can be produced.

There are two types of impellers used in oil well submersible pumps. These are the radial flow and the mixed flow. The radial stages generally range from 200 to 2500 BFPD. The radial stage is a flat stage and is the most efficient design for these lower flow rates. The mixed-flow stage ranges from approximately 1700 up to 40,000 BFPD.

Through the use of the corrosion-resistant materials, cast Ni-resist (high nickel-iron) or molded nonmetallic polyphenylene-sulfide (ryton) impellers and diffusers with K-monel shafting, pump wear, and corrosion can be minimized. Most stages are Ni-resist. However, unless otherwise specified, the housings, heads, and bases of the pumps, protectors, and motors will be carbon steel. In corrosive applications the equipment should be corrosion coated, or special ferritic steel housing should be specified. These multistage pumps may be assembled as a floater or fixed-impeller compression pump design, depending on how the axial thrust of the pump is handled.

6.4.14.2.5 Surface and Subsurface Measurement Instrumentation Equipment

6.4.14.2.5.1 Switchboard The switchboard is basically a motor control device. Voltage capability ranges from 600 to 4800 volts, on standard switchboards. All enclosures are NEMA 3R, which is suitable for virtually all, outdoor applications. The switchboards range in complexity from a simple motor starter/disconnect switch to an extremely sophisticated monitoring/control device.

There are two major construction types: electromechanical and solid state. *Electromechanical construction switchboards* provide overcurrent/overload protection through three magnetic inverse time-delay contact relays with push-button, manual reset. Undercurrent protection is provided by silicone-controlled rectifier (SCR) relays. These features provide protection against downhole equipment damage caused by conditions such as pump-off, gas lock, tubing leaks, and shut-off operations. The *solid-state switchboards* incorporate the motor controller (as the Vortex and Eliminator). The purpose of the motor controller is to protect the downhole unit by sensing abnormal power service and shutting down the power supply if current exceeds or drops below preset limits. This is accomplished by monitoring each phase of the input power cable to the downhole motor. A valuable switchboard option is the recording ammeter. Its function is to record, on a circular strip chart, the input amperage to the downhole motor. The ammeter chart record shows whether the downhole unit is performing as designed or whether abnormal operating conditions exist. Abnormal conditions can occur when a well's inflow performance is not matched correctly with pump capability or when electric power is of poor quality. Abnormal conditions that are indicated on the ammeter chart record are primary line voltage fluctuations, low amps, high amps, and erratic amps.

6.4.14.2.5.2 Variable Speed Drives The variable-speed drive (VSD) is a highly sophisticated switchboard-motor controller. A VSD performs three distinct functions. It varies the capacity of the ESP by varying the motor speed, protects downhole components from power transients, and provides "soft-start" capability. Each of these functions is discussed in more detail below. A VSD changes the capacity of the ESP by varying the motor speed. By changing the voltage frequency supplied to the motor and thus motor rpm, the capacity of the pump is also changed in a linear relationship. Thus, well production can be optimized by balancing flow performance with pump performance. This applies to both long-range reservoir changes and short-term transients such as those associated with high-GOR wells. This may eliminate the need to change the capacity of a pump to match changing well conditions, or it may mean longer run life by preventing cycling problems. This capability is also useful in determining the productivity of new wells by documenting pressure and production values over a range of drawdown rates. The change in voltage frequency can be made manually or automatically. The VSD can operate automatically in a "closed loop" mode with a programmable controller and pressure-sensing instrument.

The VSD also protects the downhole motor from poor-quality electricity power. The VSD is relatively insensitive to incoming power balance and regulation while providing closely regulated and balanced output. The VSD will not put

transients out to the downhole motor but it can be shut down or damaged by such transients. Given the choice, most operators prefer to repair surface installation equipment rather than pull and run downhole equipment. Within limits, the VSD upgrades poor quality electric power by "rebuilding." The VSD takes a given frequency and voltage AC input, converts the AC to DC, and then rebuilds the DC to an AC waveform.

The soft-start capability of a VSD provides two major benefits. First, it reduces the startup drain on the power system. Second, the strain on the pump shaft (and its associated components) is significantly reduced when compared with that of a standard start. This capability is valuable in gassy or sandy wells. In some cases, slowly ramping a pump up to operating speed may avoid pump damage.

6.4.14.2.5.3 Pressure-Sensing Instrument The pressure-sensing instrument provides the operator with precise downhole pressure and temperature data. This instrument has two components: the downhole transducer/sending unit, and a surface readout unit. The downhole transducer/sending unit connects electrically and bolts to the base of the motor. Both pressure and temperature are transmitted from the transducers/sending unit to the surface readout through the motor windings and the power cable on a DC carrier signal. The transducer receives operating power from the motor's neutral winding. This allows the operation of the pressure-sensing instrument even when the motor is not running.

The main function of the pressure-sensing instrument unit is to determine the producing potential of a well. This is accomplished by determining both static and dynamic reservoir pressures. By correlating the change in pressure with a given producing rate, a well's inflow performance can be accurately quantified. This in turn will allow equipment selection that optimizes well production.

6.4.14.2.6 Surface and Subsurface Power Cable

Electric power is supplied to the downhole motor by a special submersible three-phase cable. There are two cable configurations: flat (or parallel) and round. Round construction is used except where casing clearance requires the lower profile of flat construction. The standard range of conductor sizes is 1/0 to 6 AWG (American wire gauge). This range meets virtually all motor's amperage requirements. Almost all conductors are copper.

Mechanical protection is provided by armor made from galvanized steel or, in extremely corrosive environments, Monel. Unarmored cable is used in low-temperature (less than 180°F) wells with a static BHP of less than 1500 psi.

Cable is constructed with three individual conductors, one for each power phase. Each conductor is enclosed by insulation and sheathing material. The thickness and composition of the insulation and sheathing determines the conductor's resistance to current leakage, its maximum temperature capability, and its resistance to permeation by well fluid and gas.

6.4.14.2.6.1 Motor Flat Cable The *motor flat cable* is the lowest section of the power cable string. The motor flat cable has a lower profile than standard flat power cable so that it can run the length of the pump, seal, and intake sections in limited clearance situations. The motor flat cable is manufactured with a special terminal, designed to allow entry of electric power into the motor while sealing the well fluid entry.

6.4.14.2.7 Casing Wellhead

The ESP wellhead or tubing support is used as a limited pressure seal. The wellhead provides a pressure tight pack-off around the tubing and power cable. High-pressure wellheads, up to 5000 psi, use an electrical power feed to prevent gas migration through the cable. Wellheads are manufactured to fit standard casing sizes from $4\frac{1}{2}$ to $10\frac{3}{4}$ in.

6.4.14.3 Surface Control Panel
6.4.14.3.1 Surface Junction Box

A *junction box* connects the power cable from the switchboard to the well's power cable. The junction box is necessary to vent to the atmosphere of any gas that may migrate up the power cable from the well. This prevents accumulation of gas in the switchboard that can result in an explosive and unsafe operating condition. *A junction box is required on all ESP installations.*

6.4.14.3.2 Surface Transformer

The ESP system involves three different transformer configurations: three single-phase transformers, three-phase standard transformers, or three-phase autotransformers. Transformers generally are required because primary line voltage does not meet the downhole motor voltage requirement. Oil-immersed self-cooled (OISC) transformers are used in land-based applications. Dry type transformers are sometimes used in offshore applications that exclude oil-filled transformers.

6.4.14.4 Introduction ESP System Design

The sizing of an ESP system, as with any other artificial lift method, is not an exact and universal procedure due to the different application conditions where the method is to be used. Designing an efficient ESP is not a complicate task, but reliable and accurate information must be handled for the calculation process in order to guarantee the appropriate selection of components.

The data requirements for selection of an ESP are categorized as mechanical data, production data, fluid data, and power supply. The following is a brief description of each:

Mechanical Data

1. Casing size and weight
2. Tubing size, weight, and thread
3. Well depth (both measured and true vertical)
4. Perforation depth (both measured and true vertical)
5. Unusual conditions such as tight spots, doglegs, and deviation from true vertical at desired setting depth

Production Data

1. Current and desired production rate
2. Oil production rate
3. Water production rate
4. GOR, free gas, solution gas, and gas bubble point
5. Static BHP and fluid level
6. Producing BHP and stabilized fluid level
7. Bottomhole temperature
8. System backpressure from, flowlines, separator, and wellhead choke

Fluid Data

1. Oil API gravity, viscosity, pour point, paraffin content, sand, and emulsion tendency
2. Water specific gravity, chemical content, corrosion potential, and scale-forming tendency
3. Gas specific gravity, chemical content, and corrosion potential

4. Reservoir FVF, bubble point pressure, and viscosity/temperature curve

Power Supply

1. Voltage available and frequency
2. Capacity of the service
3. Quality of service (spikes, sags, etc.)

6.4.14.4.1 Determining Reservoir Inflow Capacity (Productivity Index)

The reservoir inflow capacity will be governed by the inflow performance relationship (IPR) curve. This curve shows the flow rate associated with each bottomhole flowing pressure for a specific reservoir condition. Depending on how stable the reservoir static pressure (Pws) is, this information could be considered for a long period of time (if PI is high) or only for current well condition (if PI is low).

There are several methodologies to calculate this IPR curve: the straight line method being the most used (if flowing pressure, Pwf, is higher than the bubble point pressure, Pb) and the Vogel method (if Pwf is lower than Pb).

1. If the bottomhole flowing pressure (Pwf) is higher than or equals to the bubble point pressure (Pb), no free gas is present at the reservoir so compressibility of the liquid is insignificant. Under this assumption, a straight line (or constant PI) behavior could be considered for the relation between Pwf and flow rate (Q):

$$PI = \frac{Q}{Pws - Pwf} \qquad [6.4.92]$$

2. If Pwf is lower than Pb, free gas will be liberated from the solution. This means that PI will decrease while the pressure decreases. Under these conditions, the method of Vogel is one of the most appropriate procedures to establish the relationship between Pwf and Q. Following, the equation:

$$\frac{Q}{Q\max} = 1 - 0.2 \cdot \left(\frac{Pwf}{Pws}\right) - 0.8 \cdot \left(\frac{Pwf}{Pws}\right)^2 \qquad [6.4.93]$$

Qmax will be the maximum flow rate that reservoir can produce when Pwf is equal to zero (maximum drawdown).

6.4.14.4.2 Determining Fluid Properties at Pump Intake Condition

Pressure and temperature conditions along the wellbore change depending on specific production conditions and the mechanical configuration of the well. Due to these changes, produced fluid properties also change affecting not only their physical characteristics but also their relative volumes. The relationship between pressure, volume, and temperature is known as PVT properties of fluids. The best way to attain these properties is with laboratory analysis. Another, perhaps the common way, is with PVT correlation such as Standing, Vasquez & Beggs, Lasater, Lee, Baker & Swerdloff, Dodson & Standing, etc.

Determining fluid properties as fluid specific gravity and viscosity at pump intake conditions is very important because they have a strong influence on the pump performance curve. Fluid specific gravity will affect both the head capacity and the horsepower requirement of the pump, typically reduced for fluids lighter than water (specific gravity less than 1.00). Viscosity has a different effect on the pump, increasing the horsepower requirement (up to 2.5 times) and reducing displacement (up to 40%) and head capacities (up to 30%), as it increases. Therefore, knowledge of these two parameters is extremely important to select the appropriate pump.

PVT properties will also help to determine the equivalent volumes of oil, gas, and water produced by the well at pump intake conditions. Following is a detailed calculation procedure to determine such volumes.

6.4.14.4.2.1 Oil Volume at Pump Intake First, gas solubility, Rs, should be calculated using following relation:

$$Rs = \gamma_g \left[\frac{PIP + 14.7}{18 \times 10^{(0.0009 - (T+460) - 0.0125 \text{ API})}} \right]^{1.204} \qquad [6.4.94]$$

Then, the oil volumetric factor, Bo, using:

$$Bo = 0.972 + 0.000147 \times \left(Rs \cdot \left(\frac{\gamma_g}{\gamma_o} \right)^{0.5} + 1.25T \right)^{1.175} \qquad [6.4.95]$$

with $\gamma_o = \dfrac{141.5}{131.5 + \text{API}}$ \qquad [6.4.96]

Finally, calculation of oil volume at intake, Vo, is made using:

$$Vo = Qo \times Bo \qquad [6.4.97]$$

6.4.14.4.2.2 Water Volume at Pump Intake Volume of water at pump intake conditions (Vw) can be assumed as equal to the water flow rate at stock conditions (Qw) because of its relative insignificant compressibility.

6.4.14.4.2.3 Free Gas Volume at Pump Intake First, gas compressibility factor, z, is calculated using the following relation:

$$z = A + B \cdot Pr + (1 - A) \cdot e^{-C} - H \cdot \left(\frac{Pr}{10} \right)^4 \qquad [6.4.98]$$

where : $Tr = (T + 460)/(175 + 307\gamma_g)$

$Pr = Pa/(701 - 47\gamma_g)$

$A = -0.101 - 0.36Tr + 1.3868 \cdot (Tr - 0.919)^{0.5}$

$B = 0.021 + 0.0425/(Tr - 0.65)$

$C = Pr(D + E \cdot Pr + F \cdot Pr^4)$

$D = 0.6222 - 0.224Tr$

$E = 0.0657/(Tr - 0.86) - 0.037$

$F = 0.32e^{(-19.53(Tr-1))}$

$H = 0.122e^{(-11.3(Tr-1))}$

Second, gas volumetric factor, Bg, is calculated as:

$$Bg = \frac{0.0283 \cdot z \cdot (T + 460)}{PIP + 14.7} \qquad [6.4.99]$$

Then, free gas volume at intake, Vg, is obtained using:

$$Vg = 0.17811 \cdot Qo \cdot (GOR - Rs) \cdot Bg \qquad [6.4.100]$$

Once the oil, water, and gas volumes at pump intake are calculated, the total fluid volume, Vt, to be handled by the pump should be calculated as:

$$Vt = Vo + Vw + Vg \qquad [6.4.101]$$

The fraction of free gas content, Fg, will be:

$$Fg = \frac{Vg}{Vt} \qquad [6.4.102]$$

If Fg is higher than 10% with radial-flow impeller pumps or 15% with mixed-flow impeller pumps, the use of a gas separator is recommended in order to minimize gas interference at the pump and an eventual premature failure of the pump.

6.4.14.4.3 Determining Total Dynamic Head
The Total Dynamic Head (TDH) could be defined as the differential pressure (or energy) that the pump must supply to get the desired flow rate at the tank. This differential pressure is defined by the pump discharge pressure (function of surface pressure, flow losses through tubing string, and weight of liquid column inside the tubing) and the pump intake pressure (function of the reservoir inflow performance). The best way to estimate the discharge pressure is by using multiphase flow correlations that consider elevation, acceleration, and friction forces as Duns and Ros, Orkizsewski, Beggs and Brill, Ansari, etc. The intake pressure could be calculated as a static column above the perforations, using as a reference the bottomhole flowing pressure corresponding to a specific flow rate.

The following is a simplified calculation procedure that assumes a single-phase flow pattern into the tubing string. This single-phase fluid will be a liquid; properties are equal to the average properties of current produced fluids (water, oil, and gas).

1. Calculation of fluid specific gravity, γf.

$$\gamma_f = \frac{(\gamma_o \cdot Vo) + (\gamma_w \cdot Vw) + (\gamma_g \cdot Vg)}{Vt} \qquad [6.4.103]$$

2. Calculation of net suction head, Hs (in feet):

$$Hs = \frac{PIP}{(0.433 \cdot \gamma_f)} \qquad [6.4.104]$$

3. Calculation of equivalent vertical head, Hd (in feet):

$$Hd = Hsd - Hs$$

4. Calculation of surface back-head, Pd (in feet):

$$Pd = \frac{P_{surface}}{(0.433 \cdot \gamma_f)} \qquad [6.4.105]$$

5. The friction losses in the tubing string (Ft) could be estimated using the Hazen-Williams correlation, which is shown graphically on Table 6.4.13 (formula is also shown).

6. Calculation of Total Dynamic Head, TDH (in ft.):

$$TDH = Hd + Pd + Ft \qquad [6.4.106]$$

6.4.14.4.3.1 Selection of Pump, Motor, and Seal Section Selection of a specific pump involves identifying a pump of the largest possible diameter that can be run in the well. The pump should have the target capacity (Vt) within its recommended operating range and close to its Best Efficient Point (BEP). The initial pump capacity selection can be made using as reference the Table 6.4.14. Use the recommended operating range as a guide.

The individual pump curve should then be reviewed to determine the optimal producing range and the proximity of the design-producing rate to the pump's BEP (see Table 6.4.15 as a reference). It is very important to choose a producing rate that is in the recommended capacity range of the specific pump. When a pump operates outside this range, premature failure can result.

Once the pump is chosen, the number of stages (N_{stages}) required can be calculated using the head per stage (H_{stage}) reading from the pump performance curve, as follows:

$$N_{stages} = \frac{TDH}{H_{stage}} \qquad [6.4.107]$$

The horsepower required by the pump design can then be calculated. To accomplish this, the horsepower required per stage is read from the specific pump performance curve. The required motor horsepower (BHP_{motor}) is determined by multiplying the horsepower required per stage (BHP_{stage}) by the number of design stages (N_{stages}). The performance

Table 6.4.13 *Friction Losses as Hazen-Williams correlation*

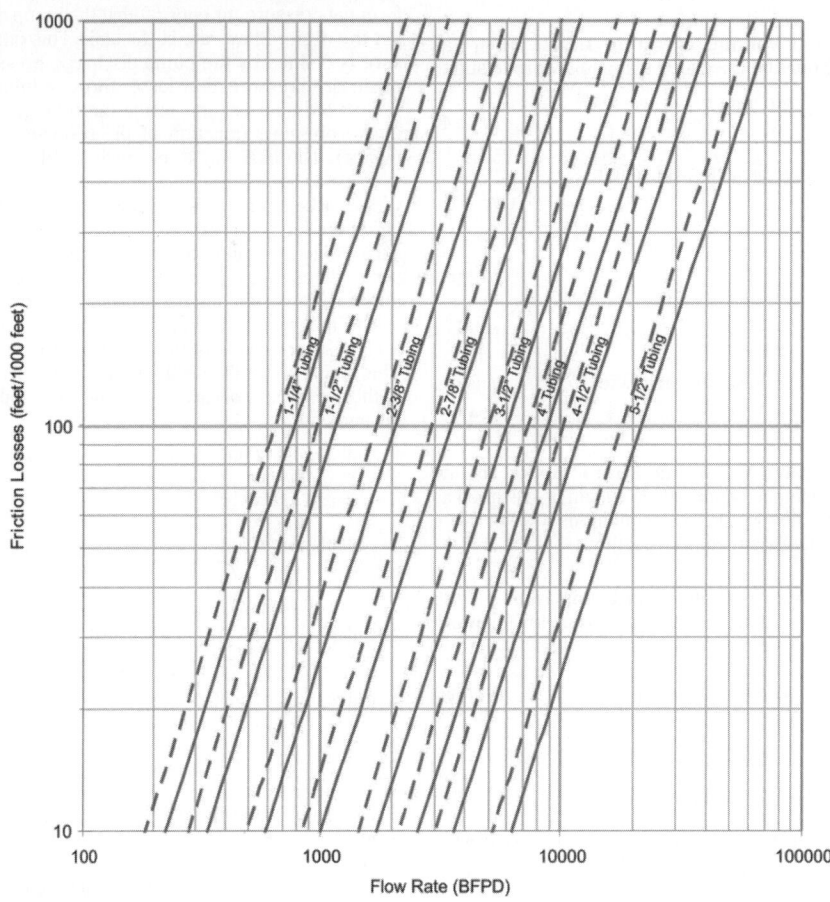

Based on Hazen William Formula:

$$F = \frac{2.083 \left(\dfrac{100}{C}\right)^{1.85} \left(\dfrac{Q}{34.3}\right)^{1.85}}{ID^{4.8655}} \qquad [6.4.108]$$

Where:
F: Friction Losses (feet/1000 feet)
C: Pipe Coefficient (Old = 100, New = 120)
Q: Flow Rate (BFPD)
ID: Tubing Inside Diameter (inches)

---

curve horsepower data apply only to specific gravity 1.0 fluids. For other fluids (other specific gravities), the water horsepower also must be multiplied by the specific gravity of the fluid pumped (γ_f). Thus, the following equation for the motor horsepower calculation:

$$BHP_{motor} = BHP_{stage} \cdot N_{stages} \cdot \gamma_f \qquad [6.4.109]$$

Once the design motor horsepower is determined, specific motor selection is based on setting depth, casing size, and motor voltage. Although the cost of the motor is generally unrelated to voltage, overall ESP system cost may be reduced by, using higher-voltage motors in deep applications. This lower cost can sometimes occur because a higher voltage can lower the cable conductor size required. A smaller conductor size, lower-cost cable can more than offset the increased cost of a higher-voltage switchboard. Setting depth is a major variable in motor selection because of starting and voltage drop losses that are a function of the motor amperage and cable conductor size.

The seal section selection variables are pump and motor series (sizes), motor horsepower, and well temperature. Normally the seal section is the same series as the pump and motor. Large horsepower motors (150 HP and larger) may require a larger oil capacity, so a positive seal double-bag model or a tandem labyrinth model is used. Temperature will limit the type of elastomer used in positive seal types (single- or double-bag) as 250°F for HSN elastomers, 300°F for Viton, and 350°F for Aflas. Higher temperatures could be handle by labyrinth type seal sections.

Finally, in order to ensure the appropriate selection of pump and motor, the following precautions must be taken:

1. *Pump, Motor, and Seal Section Sizes:* Check for outside diameter of these elements and confirm that they can be run into the specific casing size. This information is typically provided by manufacturers

2. *Pump Housing Limit:* Pumps are provided with two different types of housing material. Maximum pressure to

Table 6.4.14 *Pump Recommended Operating Range*

Model	Housing OD (inches)	BEP @ 60 Hz (Bls/day)	Maximum HP (HP)	Recommended Operating Range @ 60 Hz (Bls/day)	
				Minimum	Maximum
400–365	4.00	365	94	200	500
400–450	4.00	450	94	200	650
400–675	4.00	675	94	400	1000
400–850	4.00	850	94	600	1300
400–1200	4.00	1200	94	800	1500
400–1600	4.00	1600	94	1100	2100
400–1800	4.00	1800	94	1100	2100
400–2050	4.00	2050	94	1200	2400
400–2600	4.00	2600	256	1500	3750
400–3750	4.00	3750	256	2000	5800
400–4200	4.00	4200	256	3200	6000
400–5500	4.00	5500	256	3500	7500
400–5600	4.00	5600	256	3600	7600
513–1600	5.13	1600	375	1100	2200
513–1750	5.13	1750	375	1100	2250
513–2500	5.13	2500	375	1700	3900
513–3400	5.13	3400	375	2200	5000
513–3600	5.13	3600	375	2200	5000
513–5700	5.13	5700	375	3500	8000
513–8000	5.13	8000	637	5500	10000
513–9500	5.13	9500	637	6000	13000
538–4700	5.38	4700	375	3500	7000
538–7000	5.38	7000	375	5000	10500
538–8500	5.38	8500	375	5000	11000
538–11000	5.38	11000	637	8000	16000

Table 6.4.15 *Typical Pump Performance Curves, Fixed Speed*

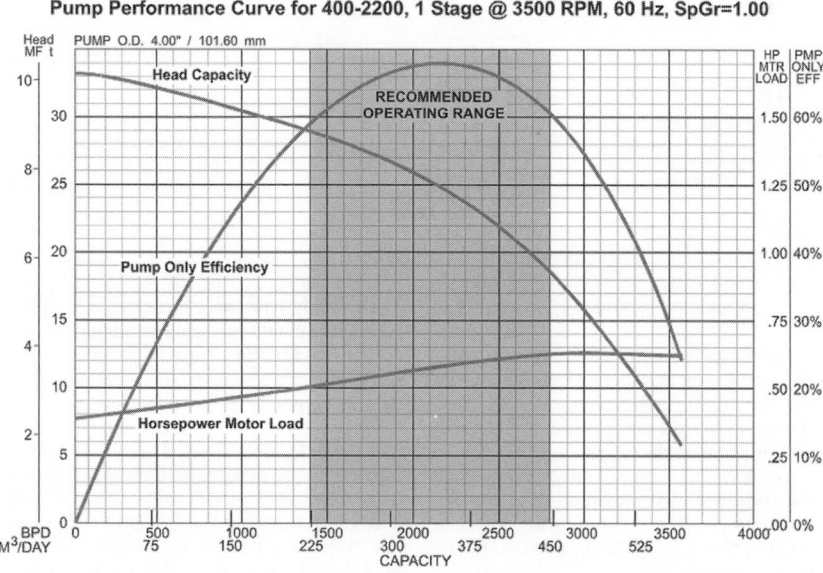

Pump Performance Curve for 400-2200, 1 Stage @ 3500 RPM, 60 Hz, SpGr=1.00

be supported by the housing will be at the start-up of the equipment, which means, at zero flow rate. To calculate this value, read the pump head at 0 BPD from the pump performance curve (when the pump head curve cross the left-vertical axis) and multiply it by the number of stages. Then, find the equivalent pressure to this maximum head

and check the limit of each type of housing provided for that specific pump model. Manufacturers typically provide this information.

3. *Pump, Motor and Seal Shaft Limit:* Check for horsepower limits of pump, motor and seal section shafts in order to determine if a standard or high strength material

Table 6.4.16 *Fluid Velocity Passing the Motor*

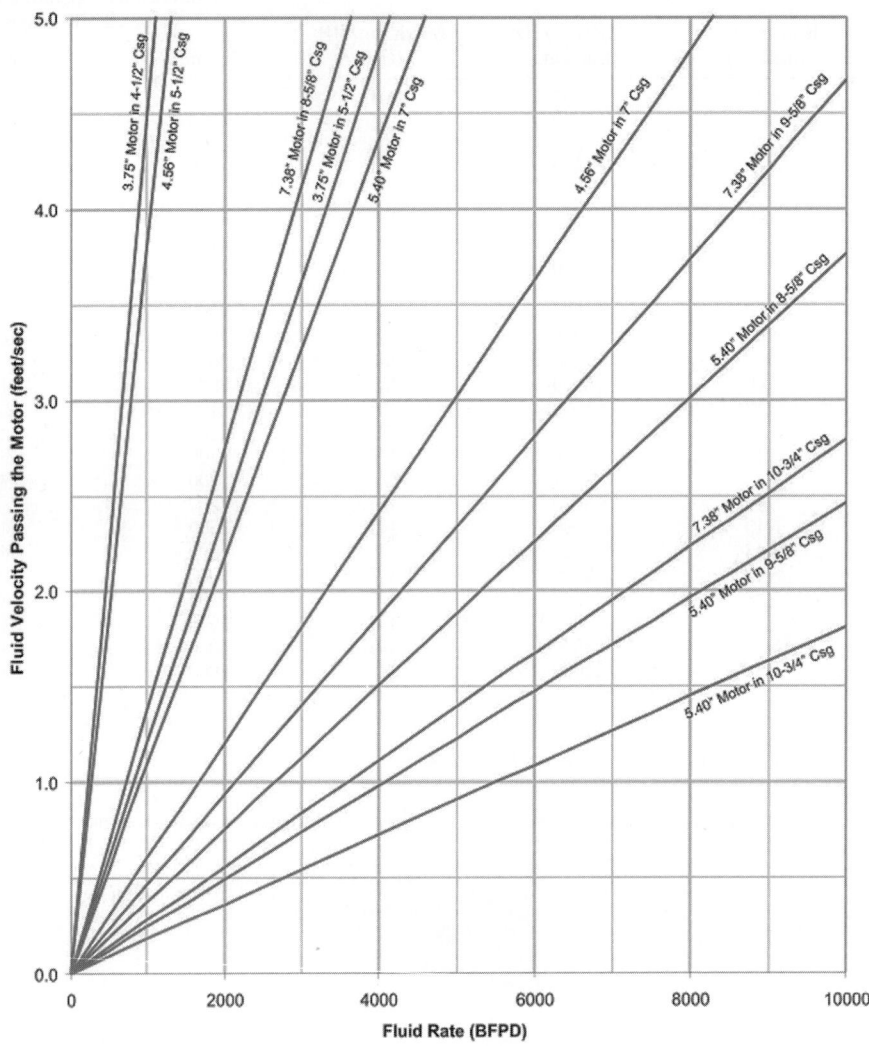

must be used. Manufacturers typically provide this information.

4. *Seal Thrust Bearing Capacity:* Check for maximum axial load that thrust bearing can support by multiplying the differential pressure through the pump by the shaft cross-sectional area, axial load on the shaft is obtained. Information provided by manufacturers.

5. *Fluid Passing the Motor:* To guarantee enough fluid cooling capacity, the recommended minimum fluid velocity passing the motor must be 1 ft/sec. Knowing the flow rate, casing size and motor series (size) such velocity could be estimated using the Table 6.4.16.

6.4.14.4.4 Selection of Downhole Power Cable
Selection of proper type and size of downhole power cable will depend on several well conditions: bottomhole ambient temperature, motor load current, casing and tubing sizes, pump setting depth, well fluid and environment (presence of H_2S, Co_2, free gas, additives, etc.), and power cost considerations.

The cable type must be selected in order to match environmental conditions where ambient temperature and fluid composition are the main parameters to be considered. Table 6.4.17 shows a summary of different cable specifications as well as a guide for their application ranges.

Once the type of cable is selected, Table 6.4.18 must be used to determine the voltage drop. Using motor nominal current as the maximum, the voltage drop per 1000 ft. can be read per each size of cable. In order to get a reasonable leakage, a maximum of 30 V/1000 ft. is accepted. If voltage drop is higher than such limit, a bigger size cable must be selected. Note that Table 6.4.18 works with conductor temperature and not with ambient temperature.

Also, note that Table 6.4.19 has been built at 77°F of conductor temperature. To correct such temperature to the ambient bottomhole condition, the read value in Table 6.4.18 must be multiplied by the correction factor read on Table 6.4.19. Again, results cannot exceed 30 V/1000 ft.

Finally, verify that voltage at the motor during start-up conditions is not too low. Divide the total cable voltage drop

Table 6.4.17 *Downhole Power Cable Information for Manufacturer A.*

Model	Shape	Insulation	Jacket	Armor	Temperature Rating	Application
DW205 R	Round	Polypropilene	Nitrile	Galvanized or Armor	205°F	Maximum operating downhole temperatures of 170°F. Good
DW205 FL	Flat	Polypropilene	Nitrile	Galvanized or Armor	205°F	for mild environments. Low cost cable. Flat configuration allows downhole equipment to fit in smaller casing.
DW300 R	Round	EPDM	Nitrile	Galvanized or Armor	300°F	Maximum operating downhole temperatures of 265°F. Good
DW300 RE	Round	EPDM	EPDM	Galvanized or Armor	300°F	for moderate (Nitrile Jacket) to harsh (EPDM Jacket) environments. Most popular cable in the industry.
DW400 R	Round	EPDM	EPDM	Galvanized or Armor	400°F	Maximum operating downhole temperatures of 365°F. Good
DW450 FL	Flat	EPDM	EPDM w/ Lead Sheath	Galvanized or Armor	450°F	for harsh environments.

Table 6.4.18 *Cable Voltage Drop*

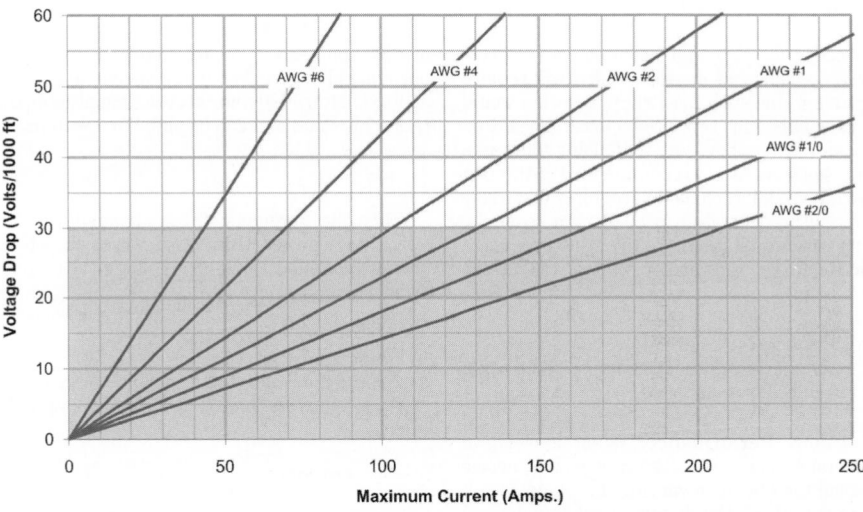

by the surface voltage required (motor nominal voltage plus cable voltage drop). Enter the result in horizontal axis of Table 6.4.20 and meet the red line. Read the corresponding value on vertical axis and, if result is higher than 40%, the motor rating is OK. If not, use a motor with lower nameplate voltage.

6.4.14.4.5 Selection of Switchboard
All applications, except where VSDs are used, will require a surface switchboard or control panel. Selection of such switchboard will depend on voltage and amperage requirements. The surface voltage will be the result of adding motor nameplate voltage plus cable voltage losses. Amperage will be equal to motor nameplate current.

In 60-Hz frequency applications, switchboards are available in 600, 1500, 2500, 3600, and 4800 V with different amperage capacities. Use of motor controller (as vortex

or eliminator) must be determined in order to finalize the switchboard selection process.

6.4.14.4.6 Selection of Transformers
Distribution of service transformers is electrically rated by input/output voltage and KVA (KVA is the abbreviation for Kilo-Volts-Amperes or thousand of Volts-Amperes that is a measure of apparent power). The minimum required total transformer KVA rating can be found using the following formula for three phase operation:

$$KVA = \frac{V_{surface} \cdot I_{mn} \cdot \sqrt{3}}{1000} \qquad [6.4.110]$$

When using a single auto-transformer or three-phase transformer, the calculated KVA value must not exceed the transformer's rating. Three single-phase transformers have a total KVA rating of three times their individual rating.

Table 6.4.19 *Cable Voltage Drop Correction Factor*

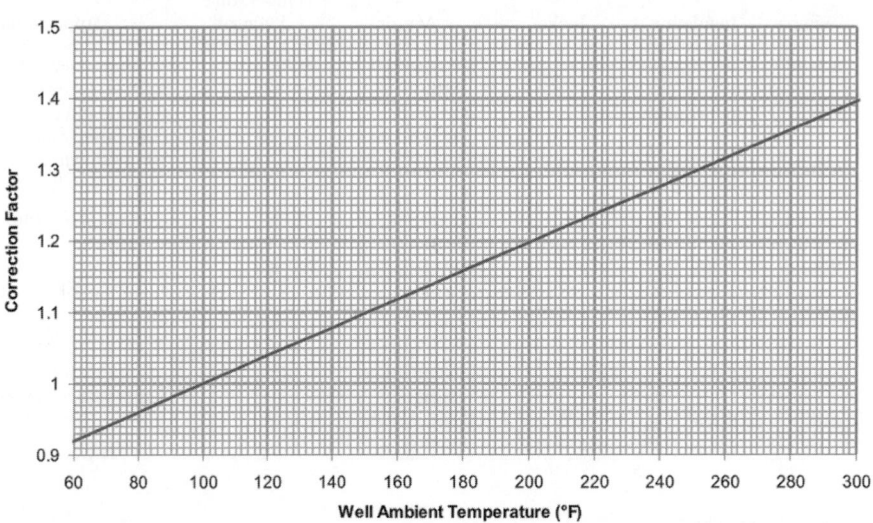

Transformer sizing for normal installations is relatively simple; however, special cases do arise which will require transformer derating. One such case where derating would be necessary is where only two single-phase transformers must be used. (Note however that using only two single-phase transformers is not recommended). An open Wye-Wye or open Wye-Delta connection requires that transformers total rating be decreased to 86.6%. For example, another derating case would be where the primary voltage is 12,470 V, and the three, transformer primary ratings are 14,400/24,940 Y.

6.4.14.4.7 Design with VSD
The following important considerations must be taken when applying a VSD to an ESP system:

6.4.14.4.7.1 Pump Performance: When applying a VSD to a submersible pump installation, it is first necessary to understand the effects of varying the pump speed. Pump performance is affected by changes in rotational speed (known as centrifugal pump's affinity laws).

When the speed (N) is changed, the flow (Q) varies directly as the speed:

$$\frac{Q_1}{Q_2} = \frac{N_1}{N_2} \qquad [6.4.111]$$

When the speed is changed, the head (H) varies directly as the square of the speed:

$$\frac{H_1}{H_2} = \left(\frac{N_1}{N_2}\right)^2 \qquad [6.4.112]$$

When the speed is changed, the brake horsepower required by the pump (BHP) varies directly as the cube of the speed:

$$\frac{BHP_1}{BHP_2} = \left(\frac{N_1}{N_2}\right)^3 \qquad [6.4.113]$$

6.4.14.4.7.2 Motor Performance: The rotational speed of an AC motor varies directly with the change in frequency or cycles per second. A normal 60 cycles input to a two-pole electric motor produces a rotational synchronous speed

of 3500 rpm. Any other frequency (F) will produce a proportional change in rotational speed; i.e., 30 cycles equals approximately 1750 rpm, 90 cycles equal approximately 5250 rpm. This statement can be simplified with the formula:

$$\frac{F_1}{F_2} = \frac{N_1}{N_2} \qquad [6.4.114]$$

When the frequency (F) is changed, the motor output brake horsepower (BHP_{motor}) varies directly as the frequency; provided a constant voltage (V)-to-frequency ratio is maintained:

$$\frac{BHP_{motor1}}{BHP_{motor2}} = \frac{F_1}{F_2} \quad \text{provided} \quad \frac{F_1}{F_2} = \frac{V_1}{V_2} \qquad [6.4.115]$$

Assuming the voltage-to-frequency ratio is constant, the motor full current load (I) will remain approximately constant:

$$I_1 = I_2 \quad \text{provided} \quad \frac{F_1}{F_2} = \frac{V_1}{V_2} \qquad [6.4.116]$$

It is important to note that variable speed drives generally maintain a constant voltage-to-frequency ratio over a limited frequency range. Therefore, the transformer located between the drive and the motor may require a change to the transformer ratio to maintain the required constant voltage of frequency ratio (maintain constant flux density). Be sure to consult the VSD manufacturer to obtain the equipment limitations and select a transformer that is capable of operating within the required range.

6.4.14.4.7.3 Operating Range: Because the pump will operate at a head/capacity that intersects, the system required head/capacity, definite speed limitations (both high and low) need to be established to prevent premature failures. Be sure the speed range allows the pump to perform within the recommended pump operating range.

1. A speed that is too low may result in an insufficient flow past the motor to maintain adequate cooling (a minimum of 1 ft./s is recommended).
2. A speed that is too low may also result in the unit operating at shut-in (zero flow). Operating in this condition

Table 6.4.20 *Cable Voltage at Start-Up*

Single Stage Variable Frequency Pump Performance Curve for 400-2200

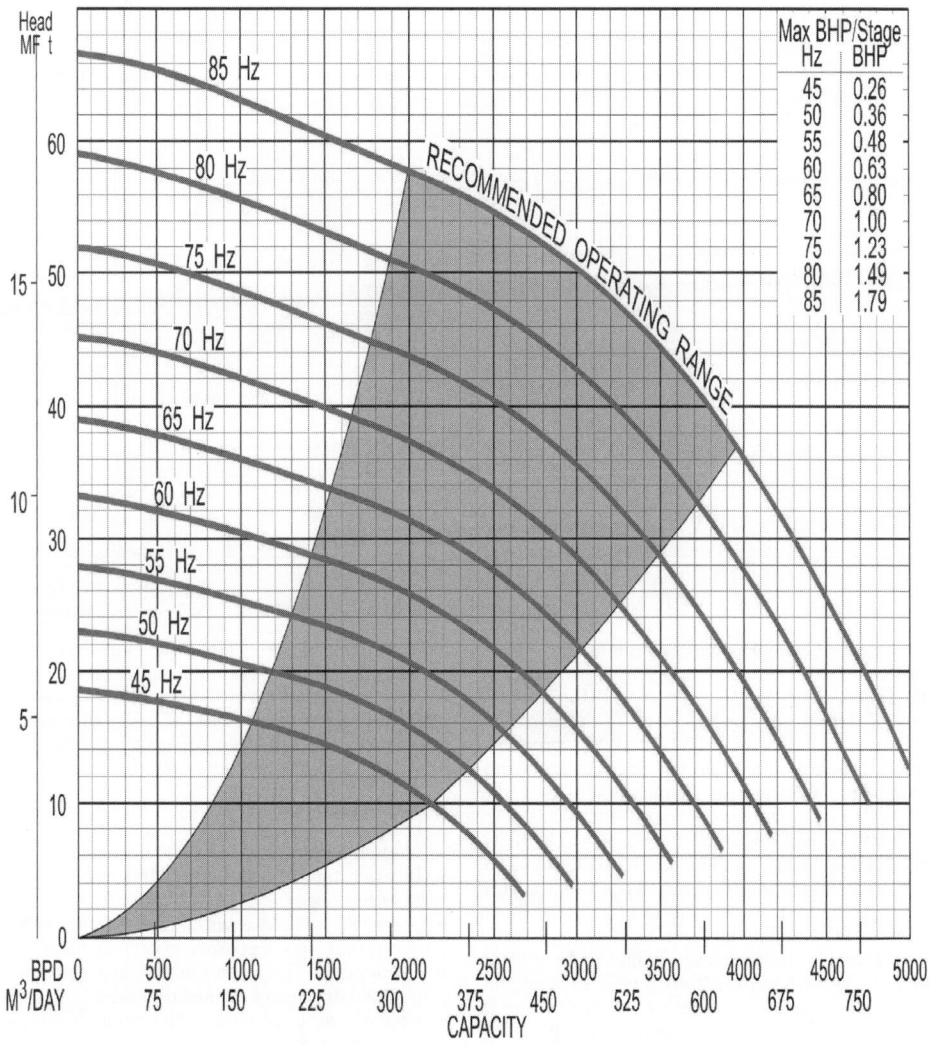

will result in an unit failure within a very short period of time. In addition to the motor having insufficient flow past it for cooling, the pump is adding energy to the fluid in the form of heat that compounds the problem. Operating in this condition also results in extreme hydraulic down thrust, which will reduce the life of the thrust washers, thus reducing the unit run life.

3. A speed that is too high can result in a motor overload condition. Since the pump brake horsepower varies as the cube of the speed, extreme care must be taken when selecting the required motor horsepower rating (the motor must be sized for the largest anticipated load).

Table 6.4.21 shows a typical variable speed curve for an ESP.

References

1. Bradley, Howard B., "Petroleum Engineer Handbook," Chapter 7, Society of Petroleum Engineers, 1987.
2. Brown, Kermit E., "The Technology of Artificial Lift Methods," Volume 1, Chapter 1, Petroleum Publishing Company, 1977.
3. Brown, Kermit E., "The Technology of Artificial Lift Methods," Volume 2b, Chapter 4, Petroleum Publishing Company, 1980.
4. Divine, David, "Electric Submersible Pumping," David Divine, 1982.

Table 6.4.21 *Typical Pump Performance Curves, Variable Speed*

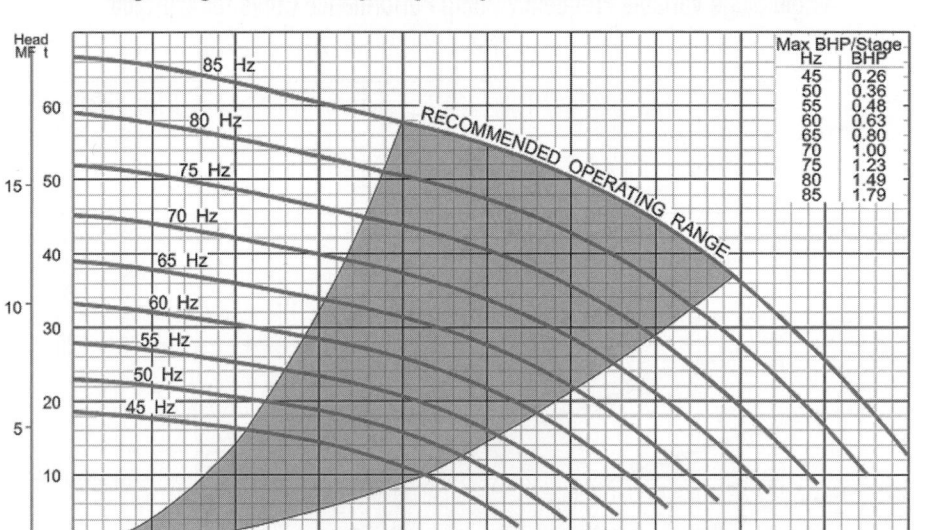

Single Stage Variable Frequency Pump Performance Curve for 400-2200

6.4.15 Progressing Cavity Pumping Systems
6.4.15.1 Introduction
In the late 1920s, Rene Moineau developed the concept for a series of helical gear pumps. One of these pumps took the form of what is now known as the progressing cavity (PC) pump, screw pump, or Moineau pump. The PC pump has been utilized in many industries in a wide variety of applications since its licensing. It has been used as a pump in just about every industry: chemical, coal, food, metal working, mining, paper textile, tobacco, water, waste treatment, and petroleum. Since the first serious efforts to apply them as a method of artificial lift in the early 1980s, PC pumping systems have experienced a gradual emergence in the petroleum industry. As of 2002, from Alaska to South America, from Russia to Australia, from the interior of Africa to secluded mountain spas in Japan, it is estimated that there are approximately 50,000 wells worldwide operating on-shore and off-shore with these systems.

PC pumping systems possess unique characteristics that can make them preferable over other artificial lift systems. Their most important characteristic is high overall system efficiency. PC pumping systems typically exhibit overall efficiencies of 50% to 70%, which is higher than any of the other major artificial lift types. Some additional advantages of PC pumping systems include:

- The ability to produce high viscosity fluids, large solids concentrations, and moderate percentages of free gas
- Low internal shear rates, which limit fluid emulsification by agitation
- No valves or reciprocating parts to clog, gas lock, or wear
- Low capital and power costs
- Simple installation and operation as well as low maintenance
- Low profile and low noise surface equipment.

PC pumping systems also have some disadvantages compared with other forms of artificial lift. The most prominent of these are limitations with respect to pump capacity, lift, and elastomer compatibility. The list below summarizes the current application limitations and major operational difficulties associated with PC pumping systems:

- Limited production rates: maximum = 800 m^3/d (5000 bbls/d)
- Limited lift: maximum = 3000 m (9800 ft.)
- Limited temperature capability: maximum = 170°C (330°F)
- Sensitivity to fluid environment: elastomer may swell or deteriorate upon exposure to certain fluids
- Tendency of pump stator to sustain permanent damage if pumped dry even for short periods
- Lack of user experience with system design, installation and operation

These limitations are rapidly being overcome with the development of new products and improvements in materials and equipment design. In many applications, PC pumping systems can provide the most economic (and in some cases the only) means of artificial lift when configured and operated properly.

6.4.15.1.1 Major Applications
Presently the major applications for PC pumping systems include:

Heavy Oil

- <18 API gravity
- 500 to 15,000 cps viscosity
- 300 to 800 m (1000–2500 ft.)
- Up to 70 m^3/d (440 bbls)

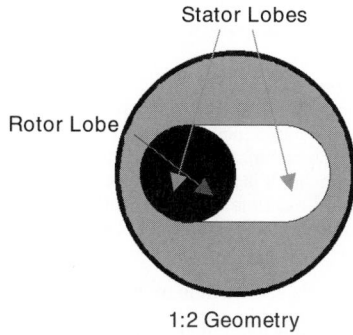
1:2 Geometry

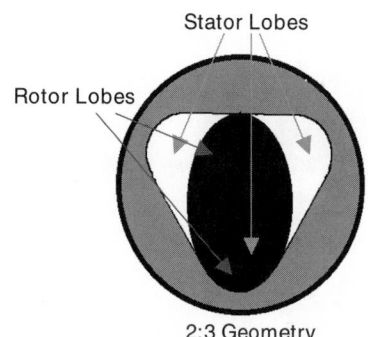
2:3 Geometry

Figure 6.4.70 *Lobe ratio.*

- Sand cuts up to 85%
- Water cuts up to 100%
- Low aromatics, GOR
- H_2S and CO_2 possible

Medium Oil — 18 to 30 API Gravity

- <500 cps viscosity
- 600 to 1400 m (2000–4500 ft.)
- Up to 500 m^3/d (3100 bbls)
- <2% Sand cuts
- Water cuts up to 100%
- Greater possibilities of aromatics, higher GORs, H_2S, and CO_2

Light Oil 30 API Gravity

- <20 cps viscosity
- >1000 m (3200 ft.)
- up to 500 m^3/d (3100 bbls)
- Trace sand
- Water cuts up to 100%
- High possibility of aromatics, high GORs, H_2S, and CO_2

Water

- Water source wells
- Gas water separation
- Coalbed methane extraction
- 800 m^3/d (5000 bbls)
- Depth 900 m (3000 ft.)
- Abrasive cuts up to 70%
- Low aromatics
- High GORs, H_2S, and CO_2 possible
- Mature steam drive operations

6.4.15.1.2 Pumping Principles
PC pumps are positive displacement pumps, which consist of two primary parts, *a rotor* and *a stator*. The geometry of the assembly is such that it constitutes two or more series of lenticular, spiral, separate cavities. When the rotor rotates inside the stator, the cavities move spirally from on end of the stator to the other, from inlet to outlet, creating the pumping action. As each cavity progresses from the inlet to the outlet, the pressure increases from the inlet pressure to the outlet pressure, ideally in a linear fashion.

Most PC pumping systems are rod-driven with the stator run into the well on the bottom of the production tubing, and the rotor connected to the bottom of the rod string. To turn the rotor downhole, the rod string is rotated at surface by means of a hydraulic or direct drive power transmission.

6.4.15.1.3 Pump Geometry
The pump geometry is governed by four parameters:

1. Lobe ratio
2. Offset or eccentricity
3. Stator pitch
4. Rotor diameter

Lobe Ratio: The stator will always have one more lobe than the rotor. Most PC pumps are of the 1:2 geometry. The 1 represents the number of lobes on the rotor; 2 represents the number of lobes on the stator. Most multilobe pumps will be of a 2:3 geometry (Figure 6.4.70).

Cavity Formation: At any cross-section, the number of cavities is equal to the number of lobes on the stator, i.e., 1:2 geometry there are 2 cavities 180° apart. Cavities are one stator pitch in length. One cavity starts where the other ends. The pitch of the rotor is one-half that of the stator.

Rotor Movement: Rotor rotates about its own axis in a clockwise direction at the pump speed. Rotor rotates eccentrically about the stator axis in a counter-clockwise direction at the pump speed times the number of lobes on the rotor.

Offset or Eccentricity: *Eccentricity* is the radius of the circle the rotor head traces about the stator centreline. For a 1:2 geometry pump, the eccentricity is the difference between the rotor head centreline and rotor centre.

Cavity Area: The cavity area for a 1:2 geometry pump is equal to the rotor diameter multiplied by the stator pitch and four times the eccentricity.

$$\text{Cavity Area} = D \times 4e \times \text{stator pitch}$$

Pump Displacement: The *pump displacement* is a product of the rotor minor diameter, pump eccentricity, and stator pitch multiplied through by a constant.

$$V = Cedp$$

where : $V = m^3$/d/rpm
$\quad C = 5.7\ e\text{-}6$
$\quad d = $ rotor minor diameter (mm)
$\quad e = $ eccentricity(mm)
$\quad p = $ stator pitch (mm)

2ndCavity
3rdCavity
1stCavity

Figure 6.4.71 shows a typical progressing cavity pumping system configuration.

Typical Progressing Cavity Pumping System Configuration

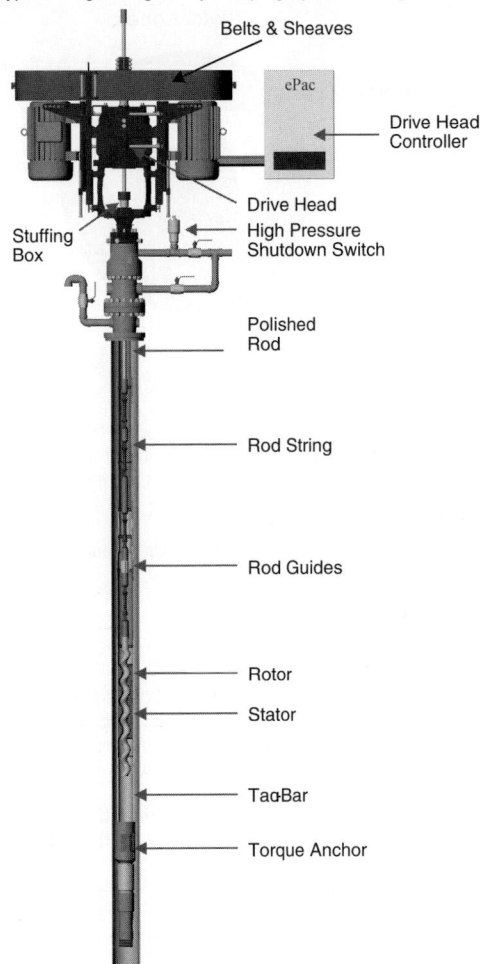

Figure 6.4.71 *Typical progressing cavity pumping system configuration.*

6.4.15.2 How Progressing Cavity Pumps Develop Pressure

The combination of the maximum pressure that can be developed in a single cavity and the number of cavities in a PC pump defines its pressure capability. The pressure that can be supported by each cavity is a function of the interference fit between the rotor and stator and the properties of the produced fluid. In general, as the rotor/stator interference fit and fluid viscosity increases the pressure capability of each cavity increases. Differential pressure between the cavities causes fluid to leak across the rotor/stator seal lines from the high-pressure cavity to the low-pressure cavity. Fluid slip is directly related to the amount of differential across the pump and is independent of speed. With conventional sizing, each cavity can support approximately 35 psi (240 kPa) of pressure differential.

The example in Figure 6.4.72a illustrates that if the rotor/stator interference fit is the same for two pumps, by adding more cavities we increase the pressure capability of that pump.

The example in Figure 6.4.72b illustrates that if the if two pumps have the same number of cavities, by increasing the

rotor/stator interference fit the pressure differential those cavities can support increases.

6.4.15.2.1 Pump Efficiency or "Slip"

The most common method to measure pump performance is by volumetric efficiency. Volumetric efficiency can be defined as the benchmark flow rate @ 0 head minus the "Slip" @ rated head. It is represented as a percentage.

The following example is an illustration of a typical PC pump performance curve. The curve is for a PC pump model 10–1200 where:

10 = pump displacement in $m^3/d/100$ rpm

1200 = pump rates lift in meters of head based on fresh water. To convert this to a pressure, multiply the meters of lift by the fresh water gradient.
(i.e., 1200 m × 9.81 kPa/m = 11,788 kPa).

If the pressure differential across the pump continues to increase, it will eventually exceed the rotor/stator interference fit. When this occurs, a percentage of fluid from the low-pressure cavity will not be produced into the high-pressure cavity resulting in a reduction if flow. With no pressure differential across the pump, the rotor/interference fit is not compromised, and 100% of the fluid is produced from the low- pressure cavity to the high-pressure cavity (i.e., 100% volumetric efficiency at 0 head). In this example it can be seen that at the rated lift of the pump there is 20% slip (80% volumetric efficiency). If the differential pressure continues to increase, a pressure will eventually be reached at which no flow occurs. At this point the slip is 100% and, in the following example, occurs when the pump has a pressure differential of 1800 m of head across it. This is also the maximum pressure the progressing cavity pump can develop under this particular set of test conditions (Figure 6.4.74).

The following example illustrates that if pump 1 and pump 2 have the same rotor/stator interference fit, by adding more cavities, the pump efficiency will be greater because there is less pressure differential between the cavities (Figure 6.4.75a).

To optimize pump life, we try to find a balance between the lift capacity of the pump versus rotor/stator interference fit to achieve a desired pump efficiency for a specific set of well parameters. To extend pump life, we want to add seal lines so the pump is less loaded. By reducing the pressure differential across the pump, we no longer need to have as tight of a rotor/stator interference fit to achieve the desired pump efficiency. This is illustrated below by the two graphical examples (Figure 6.4.75b).

6.4.15.2.2 Slip as a Function of Pump Volumetric Efficiency

Slip is directly affected by the pressure differential across the pump. The more pressure differential across a pump for a given set of condition, the higher the slip will be. Therefore, it is safe to say as the pressure differential across the pump increases, the volumetric efficiency will decrease for a given set of conditions.

Even though slip remains constant for a given differential pressure, the pump volumetric efficiency can be significantly impacted depending on what the pump rotational speed is. The effect of the pump rotational speed on volumetric efficiency is illustrated in the following example.

Example
- PCP Model: **10–1200**
- Test Speed: **300 rpm**
- Pump Efficiency: **70%**

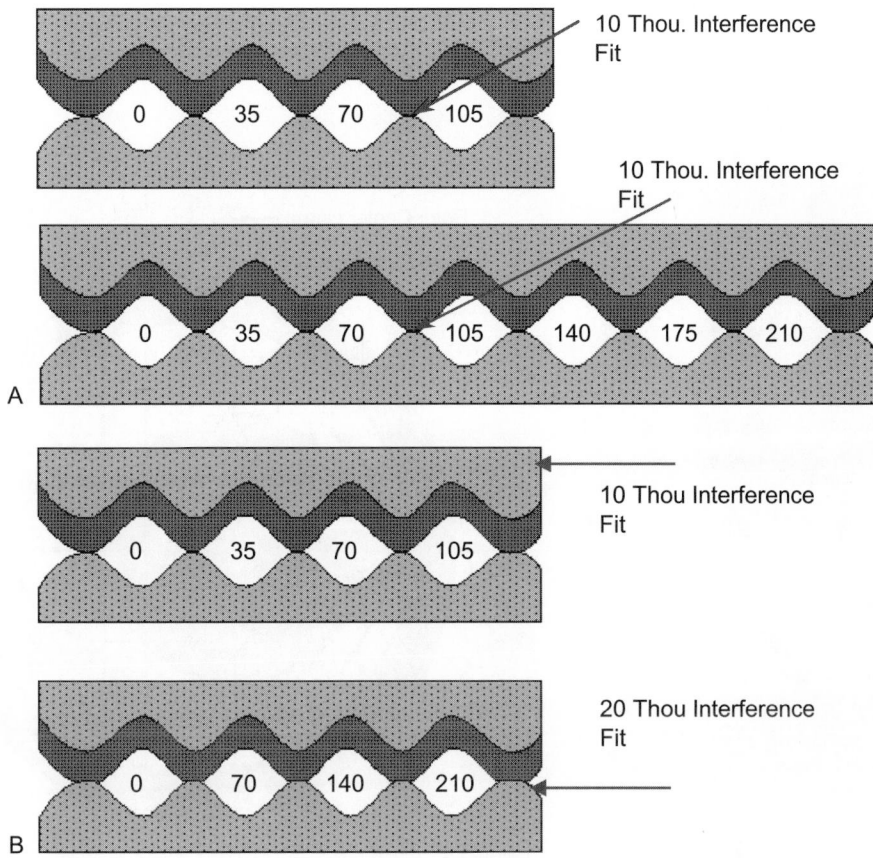

Figure 6.4.72

Theoretical production at 100% efficiency

$$= 30 \text{ m}^3/\text{d} \ (10 \text{ m}^3/\text{d}/100 \text{ rpm} \times 300 \text{ rpm} \times 100\%)$$

- Fluid rate at measured efficiency = 21 m^3/d (30 m^3/d × 70%)

$$\text{Slip} = 9 \text{ m}^3/\text{d at rated lift or 1200 m of}$$

$$\text{head } (30 \text{ m}^3/\text{d} - 21 \text{ m}^3/\text{d})$$

Now if the pump speed is reduced to 150 rpm, but the pressure differential across the pump remains the same the theoretical rate is:

- 10 m^3/d/100 rpm ×150 rpm = 15 m^3/d; however, because since the pressure differential is still the same, the slip remains constant at 9 m^3/d. Therefore, the new pump efficiency is:

$$(15 \text{ m}^3/\text{d} - 9 \text{ m}^3/\text{d})/15 \text{ m}^3/\text{d} = 40\%@150 \text{ rpm}$$

Now if the pump speed is increased to 400 rpm but the pressure differential across the pump remains the same the theoretical rate is:

- 10 m^3/d/100 rpm ×400 rpm = 40 m^3/d; however, because since the pressure differential is still the same, the slip remains constant at 9 m^3/d. Therefore, the new pump efficiency is:

$$(40 \text{ m}^3/\text{d} - 9 \text{ m}^3/\text{d})/40 \text{ m}^3/\text{d} = 78\%@400 \text{ rpm}$$

6.4.15.2.3 Pressure Stage
Although not commonly used anymore, some operators still refer to stages when defining the pressure rating of a PC pump. The problem with this terminology is that not all manufacturers define a pump stage by the same set of parameters.

As previously discussed, a pump cavity can support approximately 35 psi (240 kPa) pressure differential with a standard rotor/stator interference fit. Each stator pitch has two cavities 180° out of phase. Therefore, by definition one stator pitch would have a differential pressure rating of 70 psi (480 kPa). However, most manufacturers have defined a pump stage as three cavities or 1–1/2 stator pitches. Therefore, by this standard a pump stage would be equal to approximately 100 psi (690 kPa). By this definition if you had an 18-stage pump, the pump would be rated for 1800 psi (12,400 kPa) of pressure differential.

6.4.15.3 Elastomers
The weakest link in progressing cavity pumping systems is the elastomer. Elastomer failures often occur as a result of the chemical or physical breakdown that is induced by the downhole environment and produced fluid properties.

Elastomer performance is dependent on its chemical and mechanical properties. The most commonly referenced mechanical properties are:

- *Hardness (Shore A):* very important as it is used to characterize the relationship between the rotor/stator interference fit and the resulting sealing force
- *Tear Strength:* the next most important property as it is used to measure the resistance to tearing and is directly related to fatigue and abrasion resistance

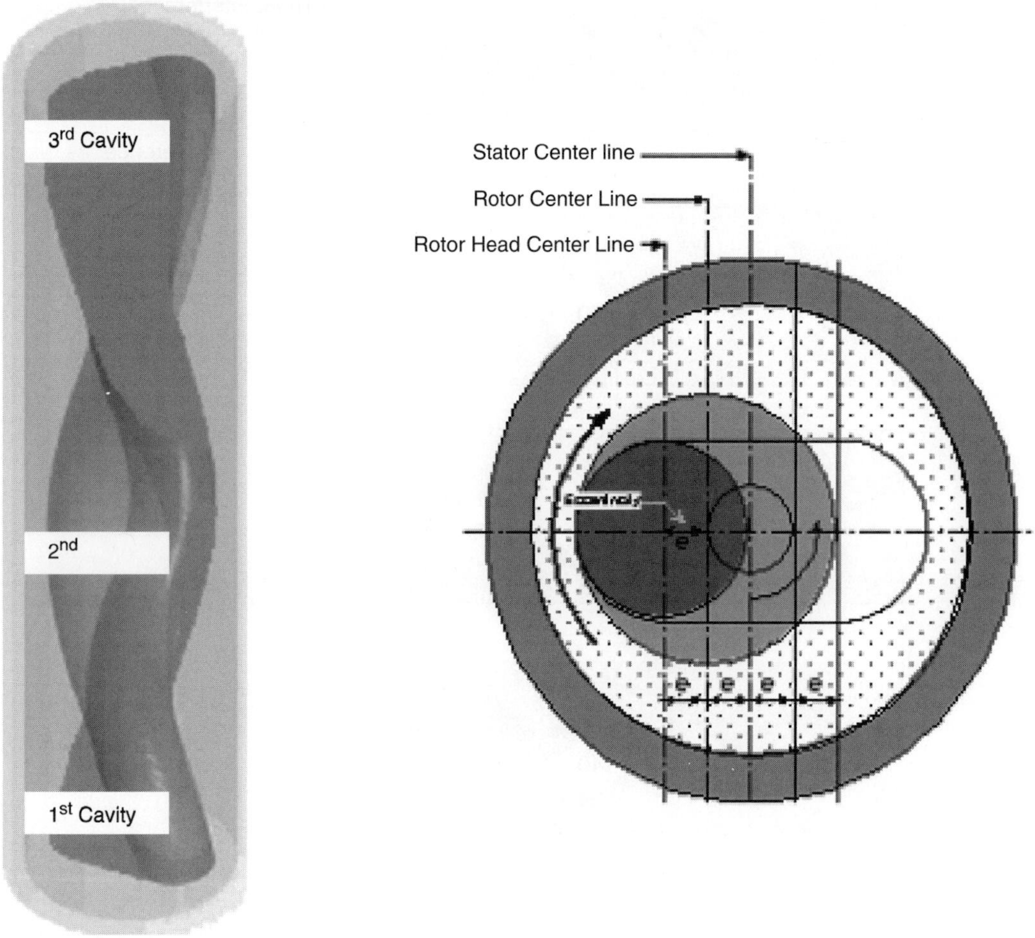

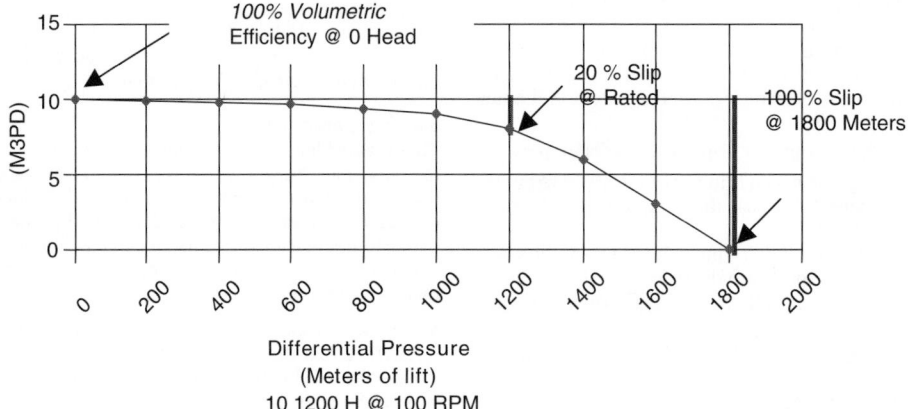

3rd Cavity
2nd
1st Cavity
Stator Center line
Rotor Center Line
Rotor Head Center Line
e
e e e e

Figure 6.4.73 *How progressing cavity pumps develop pressure.*

100% Volumetric
Efficiency @ 0 Head
20 % Slip
@ Rated
100 % Slip
@ 1800 Meters
(M3PD)
15
10
5
0
0
200
400
600
800
1000
1200
1400
1600
1800
2000
Differential Pressure
(Meters of lift)
10 1200 H @ 100 RPM

Figure 6.4.74 *Differential pressure (meters of lift) 10–1200 HNBR @ 100 rpm 10–1200 @ 100 rpm.*

- *Tensile Strength*
- *Abrasion Resistance*
- *Resilience*

Although not classed as a mechanical property, equally important is an elastomer's resistance to liquid, gas swell, and temperature degradation. Elastomers are very sensitive to temperature, and in most cases, the mechanical properties will deteriorate significantly with increasing temperature. The rate of fluid swell will also increase with higher temperatures. Chemical resistance is normally evaluated through compatibility testing with the produced fluid.

In PC pumping systems, there are basically three types of elastomers. The formula for an elastomer is generally proprietary, and as a result, there is no naming standard. However, there are certain generic names common to the manufacturers, but the properties of the elastomers can vary significantly. For example, there are approximately 20 different ingredients such as fillers, plasticizers, curatives, and carbon black. Of each of these ingredients, there are a number of different types, which can lead to an infinite number of formulations. Almost all of the PC pump elastomers use some variation of a synthetic nitrile rubber.

Nitrile (NBR)—Trade Name Buna The majority of elastomers are classified as a conventional nitrile rubber (NBR). The base polymer is an emulsion copolymerisation of butadiene with acrylonitrile. NBR elastomers are typically classed as a medium nitrile (Buna) or high nitrile.

Medium Nitrile (Buna): The medium nitrile has ACN content between 30% and 40%. It is sulphur or peroxide cured at elevated temperatures. For these reasons it is not recommended for applications that contain high levels of H_2S or temperatures that exceed 100°C (212°F), due to the advanced vulcanization and surface hardening that occurs.

Medium nitrile elastomers typically have the best mechanical properties, are the easiest to mold, and generally are the most economical.

High Nitrile: High-nitrile elastomers have ACN content between 40% and 50%. Increasing the ACN content increases the polarity, which improves the elastomers resistance to nonpolar oils and solvents such as *benzene, toluene, and xylene*. However, increasing the ACN content generally results in a decline in certain mechanical properties.

Saturated Nitrile (HNBR) NBRs often contain a large degree of unsaturation in the form of double and triple bonds. Unlike a single bond, which is more stable, the unsaturated bonds are more susceptible to chemical attach and/or additional cross-linking. The unsaturation is why temperature, H_2S, and some chemicals limit NBRs. To decrease the amount of unsaturated bonds, the nitrile polymer is put through a hydrogenation process, which makes it more stable. The resulting elastomer is *commonly referred to as*

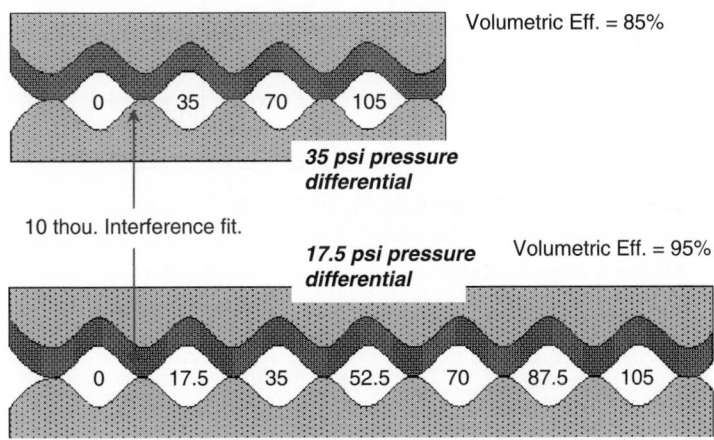

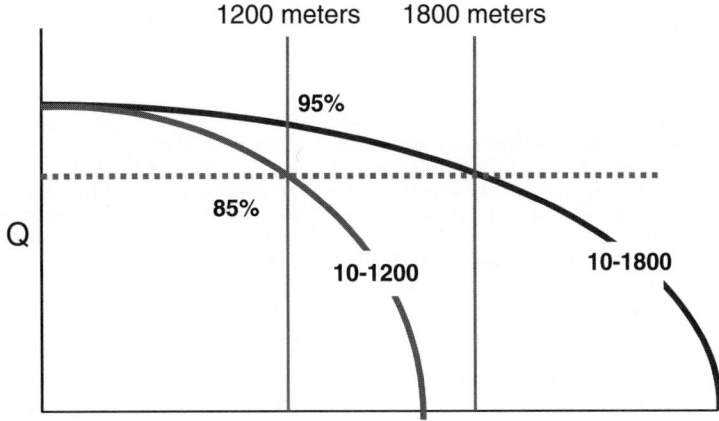

Figure 6.4.75 Pump lift capacity effect on slip (1200 and 1800 m).

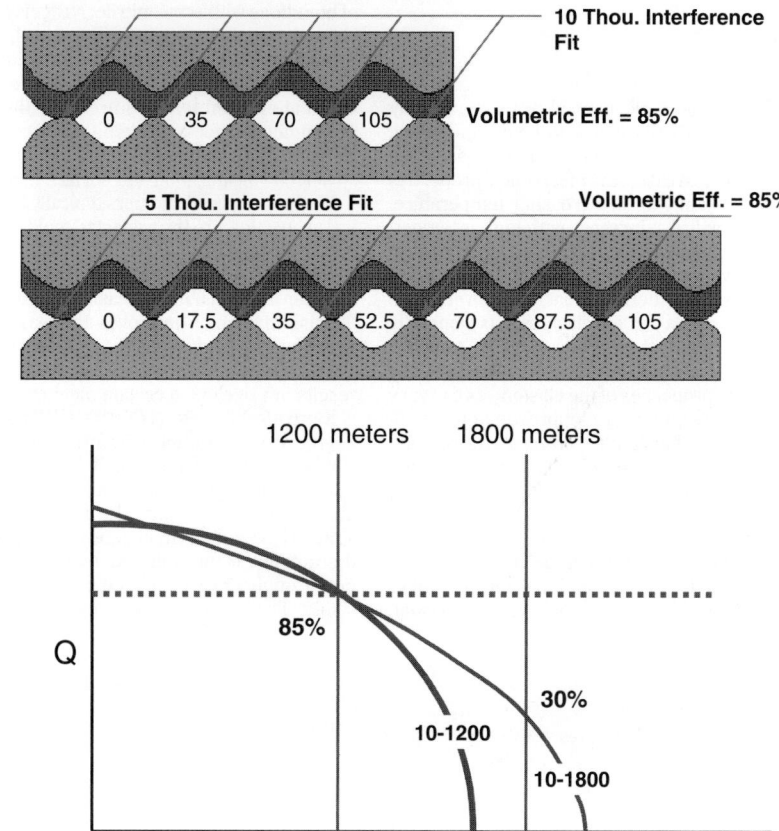

Figure 6.4.75 *Continued.*

Table 6.4.22 *General Elastomer Selection Guide*

	Elastomer Type			
Characteristics	Buna	High Nitrile	Hydrogenated	Viton
Mechanical Properties	*Excellent*	*Good*	*Good*	*Poor*
Abrasive Resistance	*Very Good*	*Good*	*Good*	*Poor*
Aromatic Resistance	*Good*	*Very Good*	*Good*	*Excellent*
H_2S Resistance	*Good*	*Good*	*Excellent*	*Excellent*
Water Resistance	*Very Good*	*Good*	*Excellent*	*Excellent*
Temperature Limit**	*95C (203°F)*	*95C (203°F)*	*135°C (275°F)*	*150°C (302°F)*

**The internal operating temperature of the pump due to the friction heat generated by the rotor/stator inteference fit can be significantly higher than the reservoir fluid temperature.

a saturated nitrile or hydrogenated nitrile (HNBR). Mechanical properties are similar to a NBR, but the big advantage is in the increased heat resistance, 125°C (257°F), and improved tolerance to chemicals and H_2S. HNBR elastomers are difficult to mold and typically cost four times as much for an equivalent volume of NBR.

Fluroelastomers (FKM) — Trade Name VITON Fluorine is used to saturate the carbon bond. The carbon fluorine bond is extremely strong, which gives it superior heat and chemical resistance to NBR elastomers. FKM elastomers contain low level of fillers and additives but a high mount of fluro-polymer. The low level of fillers and additives results in extremely poor mechanical properties.

FKM elastomers (VITON) have minimal fluid swell with most oilfield fluids, including aromatics, and have the potential to be used in temperatures up to 200°C (400°F). They have been somewhat successful in light oil applications as long as the pump is not subjected to excessive mechanical loading. FKM is extremely hard to mold, and the costs can be several hundred times that compared to that of NBR, making it uneconomical in most applications (Table 6.4.22).

6.4.15.3.1 Produced Fluid and Wellbore Effects on Elastomers
All most all fluids produced through the pump to some degree will impact the pump performance and mechanical properties. Elevated temperatures and exposure to some chemicals, such as amine-based corrosion inhibitors, also impact the elastomer.

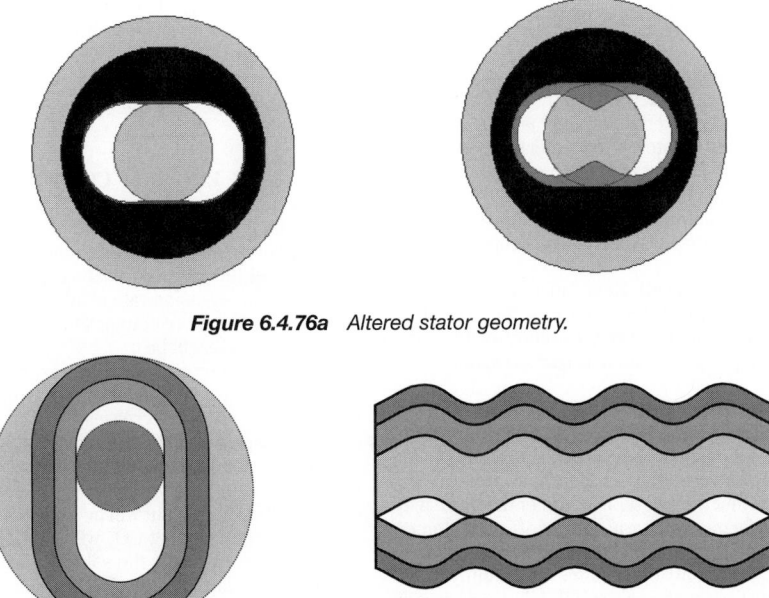

Figure 6.4.76a *Altered stator geometry.*

Figure 6.4.76b *Wellbore effects on elastomers — softening & hardening.*

The most common effects that we see occur to elastomers when exposed to the above environments are:

- Swelling
- Softening
- Shrinkage
- Hardening
- Blistering

6.4.15.3.1.1 Fluid and Gas Swell The elastomer is a permeable membrane and is therefore susceptible to absorption of fluid and gas by a diffusion process that results in overall increase in mass and volume. Fluid absorption will continue until equilibrium between the produced fluid and elastomer is achieved. Depending on the fluid properties and wellbore conditions, this can occur in weeks to years. In some cases the absorption rate is so slow that equilibrium may not be reached within the pump's life. Fluid swell typically results in the deterioration of the elastomers mechanical and chemical properties.

Oilfield fluids that commonly cause swell are:

- Higher API gravity oils that contain single-ring aromatics (i.e., benzene, toluene, xylene).
- Water
- Gas
- Chemicals (i.e., amine-based corrosion inhibitors)

Elastomers that have high compatibility with a fluid will swell more than a fluid that is not compatible. Fluid compatibility varies depending on the type of elastomer. For example nonpolar fluids such as aromatics will swell lower in polar elastomers such as high nitrile NBR and swell higher in lower polar elastomers such as conventional NBR (Buna) elastomers.

Swelling can cause an excessively tight rotor/stator interference fit. A certain amount of torque is required to overcome the rotor/stator interference fit. In some cases, because of swelling, it is not uncommon to have the fit become so tight that the torque required to overcome the fit

exceeds what is available with the surface drive configuration. A tight pump can cause accelerated hysteresis that may eventually result in a premature pump failure. Badly swollen stators can generally not be reused. However, if swelling is minor the stator can sometime be sized with a smaller rotor and reused. The internal operating temperature of the pump due to the friction heat generated by the rotor/stator interference fit can be significantly higher than the reservoir fluid temperature.

The amount of swell or expansion can be expressed as a percentage of the elastomer thickness with the thicker elastomer sections swelling comparatively more to thinner sections. Therefore the amount of change is dependent on the cross-sectional profile of the stator elastomer and since this profile in nonuniform the resultant swell or expansion is also non-uniform. The altered stator geometry will result in lower pump volumetric efficiencies, Figure 6.4.76a. Figure 6.4.76b is an example of a uniform-thickness pump. Having a uniform elastomer thickness means the elastomer also swells or expands in a uniform fashion. Properly sizing the rotor for aggressive applications is therefore much simpler.

6.4.15.3.1.2 Softening Certain fluids that attack the chemical make-up of the elastomer will breakdown some of the mechanical properties, which results in softening of the elastomer. Amine-based corrosion inhibitors are a good example of a chemical that will cause severe softening of an elastomer.

6.4.15.3.1.3 Hardening Heat, pressure, time, and curatives such as sulphur are used in the manufacturing vulcanization process of the elastomer. As an oilfield wellbore can typically have all of the above, it is not uncommon to see the elastomer hardness increase over time in any particular application. However, wells that contain a high mole% of H_2S may aggressively advance vulcanize the elastomer. As the elastomer hardens, it becomes more prone to fatigue and flex cracking–type failures due to the deterioration of these particular mechanical properties. HNBR elastomers have been formulated to reduce the effects of H_2S.

6.4.15.3.1.4 Shrinkage Lower-molecular-weight paraffin (C3–C30) can extract plasticizers from NBR elastomers, which will cause an overall reduction in elastomer volume. Shrinkage will reduce the rotor/stator interference, resulting in a reduction in pump volumetric efficiency. Wells that generally exhibit waxing problems correlate with elastomer shrinkage. Extraction of the plasticizers results in a deterioration of the elastomer's mechanical properties.

6.4.15.3.1.5 Thermal Swell Heat causes overall expansion. Operating at elevated temperatures will result in advanced vulcanization and the deterioration of the elastomer's mechanical properties such as *tear strength*.

6.4.15.3.1.6 Gas Swell The amount of gas that can enter the elastomer is dependent on the permeation rate, diffusion process, and solubility of the gas in the elastomer. The diffusion process is governed by temperature, pressure differential, and elastomer thickness. The higher the diffusivity rate, the easier it is for gas to move in and out of the elastomer. The permeation rate is determined mainly by the gas size, shape, and polarity. Solubility determines the quantity of gas that will absorb into the elastomer. Gas solubility increases as the molecular weight of the gas increases; i.e., CO_2 and H_2S have higher molecular weight than methane and therefore are more soluble.

6.4.15.3.1.7 Explosive Decompression When an elastomeric compound is exposed to high pressure for a period sufficient for gas molecules to diffuse into the compound, a subsequent rapid reduction in pressure can cause internal fracturing of the rubber. Internal fracturing may be in the form of cracks or blisters. This phenomenon is known as *explosive decompression (ED)*.

ED is the result of a three-stage process:

- *Stage One*, gas diffusion into the elastomer.
- *Stage Two*, the migration of gas molecules, immediately after decompression, to *areas of flow energy state (LES)* within the elastomer.
- *Stage Three*, is crack growth and subsequent blister formation in the *areas of LES*.

For blister growth the *critical pressure* inside the gas bubble must be at or above the *fracture or tearing strength* of the elastomer. The critical pressure required for crack growth or blistering increases with the square root of the elastic modulus and critical tear strength of the material.

As the distance below the surface of the elastomer increases, the critical pressure required to initiate crack or blister growth increases. This would mean that you would expect to see less damage toward the center of the material than at the surface; however, the opposite is the normal. Observations of failed pieces often show that ED damage is generally greater in the thicker sections. This is explained by the ability of the gas molecules to easily diffuse out of thin sections of elastomer or the surface layer as opposed to migrating out from thicker sections.

6.4.15.3.2 Other Effects on Elastomers
6.4.15.3.2.1 Gas Temperature At higher temperatures gas solubility and diffusion coefficients change to increase the rate at which gas can escape from the elastomer. However there is a counterbalancing effect offset by the decrease in the elastomer *tear strength* with increasing temperature. Depending on the elastomer the net effect of ED in higher temperature applications could be less.

6.4.15.3.2.2 Gas Pressure Gas pressure correlates directly to the pressure within the *areas of LES*. Therefore the higher the operating pressures of the PC pump the greater the gas pressure within the *areas of LES*. Higher gas pressure will create more energy for crack and blister growth.

6.4.15.3.2.3 Effects of CO_2 When exposed to a gas, most PC pump elastomers will become saturated with that gas resulting in a volumetric increase in the elastomer. The solubility of gas in the elastomer increases with the molecular weight of the gas, temperature, polarity, and differential pressure. The diffusion rate is influenced by its size, shape, and polarity. The more complex the size and shape the lower the diffusion rate will be.

As the temperature and pressure of an application increases the time required for a gas to reach equilibrium within the elastomer decreases.

If we compare CO_2 to methane (CH_4) and CO_2 molecule has a greater molecular weight and polarity as well as a more complex molecular size and shape. Therefore, CO_2 will have a lower diffusion rate but will be more soluble than methane. The result of this is that CO_2 will take longer to diffuse into the elastomer; however, the solubility will be greater, resulting in higher volumetric swell. The impact of this is that when a rapid pressure drop occurs across the pump *(i.e., due to pump stoppage resulting in a fluid dump scenario or when the pump is retrieved to surface)*, it will take longer for the CO_2 to diffuse out of the elastomer than methane. This is supported by observations on dissected pumps, with similar well conditions, in which the observed ED damage to the elastomer was more catastrophic when the gas mixture contained CO_2.

The resulting rapid reduction in pressure causes the gas trapped inside the elastomer to expand at a tremendous rate, which results in severe swelling, cracking and blistering. A small decrease in pressure results in a significant increase in gas volume. This does not mean that the gas within the elastomer will occupy this volume. Factors such as the elastomer thickness and mechanical properties work as a closed vessel to restrict the expansion of gas. As noted previously in order for crack or blister growth to occur the *critical pressure* within the *areas of LES* must exceed the *fracture or tear strength* of the elastomer.

6.4.15.3.2.4 Minimizing the Effects of Explosive Decompression The degree of damage sustained to an elastomer as a result of ED is governed by several conditions:

1. Pressure drop across the elastomer face
2. Free gas production through the pump
3. Time take for the reduction in pressure to occur
4. Concentration of gas within the elastomer
5. Tear energy of the elastomer
6. Mechanical properties of the elastomer

A PC pumping system designed to address the effect of CO_2 on the elastomer has the potential to significantly reduce the impact of ED. Design criteria that should be considered, as they pertain to the aforementioned conditions, for CO_2 applications are:

6.4.15.4 Drive Heads
Progressing cavity pumping systems are extremely flexible in terms of their ability to work in diverse range of applications. However, their success in a particular application requires that the pumping equipment be properly matched to the well conditions and operating environment. Because PC pumping systems function in an integrated manner, it is also important that the individual pieces of equipment

are configured so that they operate together effectively as a system.

When purchasing a PC pumping unit, the design considerations play an important role in the safe operations of the unit. The application designer must establish the maximum surface loading and ensure the equipment being purchased will operate within its maximum limit. To choose the best drive for the application the following should be considered:

- Depth of well
- Fluid level
- Pump size
- Size of sucker and polished rod
- Operating torque
- Speed range
- Prime mover (electric or gas driven motor or engine)

The loading or work done by a PC pumping system is primarily determined by the following:

- Pump take pressure
- Pump discharge pressure
- Rod string/tubing friction
- Internal pump friction

Several other factors influence these parameters and they usually change throughout the life of an oil well.

Other factors may include:

- Fluid contaminants
- Pressure, volume, temperature properties the production medium, reservoir characteristics, tubing roughness
- Side wall stress as a result of tubing/rod contact
- Tubing and casing pressure fluctuations
- Interference pump fit between rotor/stator

Computer aided software tools are available to help determine these factors. Calculations must be made before purchasing a unit. When predicting load limits, a safety factor of at least 10% should be considered to account for error associated with the reliability of assumptions or the accuracy of mathematical correlations.

Operating loads may change through the operating cycle life.

After the initial start up, actual operating loads should be confirmed to verify design expectations. If the system loading exceeds the design, changes to operating parameters or equipment are necessary. Control devices that will protect the unsupervised system and provide the operator with an indication of equipment condition are equally important. The PC drive systems could include instrumentation to monitor and control the following: braking mechanism temperature, backspin speed, operating torque, and hydraulic fluid levels.

The following items are recommended best practices outlined to reduce hazards while working around PC pumping equipment. Production personnel should consider:

- All new installations require a design input data sheet be filled out, ensuring that a qualified person within the producing company has reviewed the design of the unit being installed.
- A file should be maintained for each PC pump system in the field. The file should include the following information:

  1. **Equipment Specifications**
     Equipment specifications should include the maximum allowable: forward speed, backspin speed, drive head fluid temperature, hydraulic fluid levels, pressures, braking mechanism temperature, horsepower/torque, thrust loading, and motor fan speed. It should also include the power factor and motor efficiency, nameplate volt/amps, and motor speed.

  2. **Sheave and Belt Information**
     Sheave dimensions, belt types and sizes, maximum-rated sheave rpm.

  3. **Sizing Data**
     Normal operating horsepower at the motor, start-up horsepower requirements and torque, normal operating torques and operating speed, production volumes, bottomhole flowing pressure, static reservoir pressure, maximum tubing head pressure, and thrust loading.

  4. **Commissioning/Start-up Measurements**
     Start-up values at design speed such as: starting volts/amps, running volt/amps, calculated starting and running horsepower and torque, forward and backspin rpm hydraulic fluid level, braking mechanism temperature, production volumes, annular fluid level data, and polished rod stock up amount.

  5. **Limit Set Points-Control Devices (if applicable)**
     Shut down settings for the following: hydraulic drive head fluid levels, maximum braking temperature, backspin speed, forward speed, running torque, start-up torque, and low torque shutdown.

  6. **Equipment/Operating Changes**
     A list of any changes to operating parameters or equipment. After changes are made, the operator must ensure the system is still operating within the design parameters.

- The operator must ensure that the equipment is operating within its design after every PC work over.
- The production operator should review the design predictives to ensure the equipment will operate within the calculated parameters.
- The operator must be familiar with the limit shutdowns for the equipment. The operator can refer to the operator's manual for those maximum limit values.
- The actual loads should be checked at least every 6 months to ensure the operating loads fall within the design parameters.
- For right angle drive assemblies, the drive assembly should be installed perpendicular to the control panel to ensure that in the event of a sheave failure, debris would be deflected away from personnel standing at the control panel. Caution must also be used in the placement of other equipment on the lease, to ensure that a sheave failure would not adversely effect other equipment or personnel.

Reprinted from Canadian Association of Petroleum Producers (CAPP) Safety Alert Guide

6.4.15.5 Hydraulic Drives and Power Units

Progressing cavity pumps are often driven by hydraulic drive systems. Hydraulic drives and power units are particularly used when variable speed is required. Also, with gas engine as the prime mover, electrical power is not required on site. All hydraulic power units featured automatic torque control and recoil as standard features.

The hydraulic power unit, or "skid" powers the hydraulic drive head. The skid has a prime mover, which drives a hydraulic pump. Fluid flows from the pump then drives the hydraulic motor on the drive head, which in turn drives the rod string.

When a well is shut down, the rods release torsional energy stored in them, and fluid may drain through the pump, turning the rods in reverse. There is potential for the rods to spin back at a high rate of speed, causing a safety hazard if spinning parts (sheaves, for example) should fail. The backspin may also cause damage to the hydraulic motor.

A backspin retarder is generally built into every hydraulic power unit system. A check valve is in the hydraulic lines on the pressure side of the motor. When the drive is turning in the forward direction, the check valve is forced open by the fluid flowing into the motor. When the drive spins in reverse, the check valve is forced closed and drains the fluid at a slow controlled rate of speed. In order to ensure that there is no loss of suction in the motor, the backspin speed is very slow (Figure 6.4.77).

6.4.15.5.1 Right-Angle Drive
Right-angle drives typically will have an internal gear reduction system. Although not limited to them, the most common internal gear ratios are 2:1 and 5:1. This allows the drive to operate at lower speeds. The limitation with this style of drive is typically in the amount of torque that can be transmitted through the pinion gears and as such the drive is generally delegated to lower horsepower applications.

By design, the right-angle drive can be driven by a gas engine and therefore is a popular option for applications in which the well site is not electrified motor.

6.4.15.5.2 Direct Drive
Direct drives generally do not incorporate any type of internal gear reduction. Power transmission and speed control is accomplished directly through a set of belts and sheaves connected to the motor.

6.4.15.5.3 Recoil Control
Rod string backspin occurs whenever a surface driven PC pumping system shuts down due to either routine operator intervention or automatic power cut-off in overload cases in which the pump seizes or sands up. When the power supply to the drive is lost or interrupted, the torsional strain and fluid energy that is stored in the production system during pumping is released. The release of this energy will cause the surface equipment and drive string to accelerate in the direction opposite to its normal operating mode. If not suitably restrained by the surface drive braking equipment, backspin speeds can increase to the point where severe equipment damage results. The most serious type of incident (due to its potential to cause personnel injury) is the fragmentation and subsequent "explosion" radially outward of drive head sheaves.

Backspin events can be divided into two different types: (1) routine shutdown of the pumping system (i.e., normal shutdown), and (2) when the pump rotor seizes within the stator (i.e., seized pump scenario). In a normal shutdown, the fluids in the production tubing drains back to the well through the pump causing both the pump and drive system to rotate backward. This continues until the fluid column in the annulus and the pump friction balance the fluid energy in the tubing. This can take anywhere from several minutes to hours depending on the circumstances. Although the recoil event resulting from a normal shutdown can result in equipment damage due to excessive energy input, there is usually limited potential for a catastrophic failure as long as the brake is initially functional. More likely to produce a catastrophic failure is the seized pump scenario. In the instance, shutdown occurs with the drive string is usually loaded to an extremely high torque. Strain energy (twists) stored in the drive string tends to be released rapidly resulting in a short but high-speed recoil.

There are five major types of recoil devices used in progressing cavity pumping systems.

1. Disc brake
2. Fixed orifice
3. Centrifugal
4. Vane
5. Hydro-dynamic

6.4.15.5.3.1 Disc Brake: Features a hydraulic braking system like on vehicle, which consists of a pump, manifold and a disk brake. The self-powered braking system engages when the rod string begins to run in the reverse direction. A belt connects the main shaft (rod string) with the hydraulic pump, which produces pressure that activates the caliper or brake pads. The pressure of the system proportionately varies with the speed of the main shaft therefore adjusting the braking force (Figure 6.4.78).

6.4.15.5.3.2 Fixed Orifice: In hydraulic systems, the hydraulic motor at the wellhead will force the hydraulic fluid backwards through the system when the rods recoil. A check valve is in the hydraulic lines on the pressure side of the motor. When the drive is turning in the forward direction, the check valve is forced open by fluid flowing into the motor. When he drive spins in reverse, the check valve is forced closed and drains the fluid at a slow rate of speed (Figure 6.4.79).

6.4.15.5.3.3 Centrifugal: The technology used is similar to that of a drum brake except that centrifugal force, created during backspin, assisted by the mechanical design of a clutch is used to throw brake shoes against a stationary housing. Because the braking system relies on centrifugal force the braking mechanism will not activate until the polished rod speed reaches a specified speed. An example of this is the Weatherford MG drive head, and once an approximate speed of 250 rpm is reached, centrifugal force overcomes the spring tension and braking is activated. In turn, once the speed of the brake is reduced to a speed of under the approximated 250 rpm, the spring tension retracts the brake shoes and the drive is allowed to spin with out a brake unless the speed returns to approximately 250 rpm (Figure 6.4.80).

6.4.15.5.3.4 Vane: Drive head braking resistance is developed using a vane style hydraulic pump that is activated in the recoil or shutdown mode through a normally overrunning clutch. Clutch engagement occurs during backspin as a result of rollers in the outer portion of the clutch locking onto an inner clutch hub that is located on the drive head shaft. Engagement of the clutch causes the vane pump rotor within the drive head to rotate with in its housing. Rotation of the rotor produces braking torque by forcing the drive head lubricant oil through a restriction. Baking torque generated in the vane style pump is transferred from the rotor through the clutch and into the drive shaft and polished rod.

The brake torque generated is proportional to the pumping speed (i.e., recoil speed) and changes as a function of the lubricating oil viscosity and the nature of the restriction. The restriction occurs between the upper and lower sealing surfaces on the rotating vane pump rotor and the stationary bottom rotor housing and upper thrust plate (Figure 6.4.81).

6.4.15.5.3.5 Hydrodynamic: Much like the torque converter in a vehicle, braking is accomplished by forcing vanes to rotate within a viscous fluid.

6.4.15.5.3.6 Safe Operation of Wellhead Drives
When undergoing a controlled or uncontrolled shut down of a progressing cavity pumping system, it is critical to understand the recoil speed, if left unrestricted, can exceed the surface drive component (brake/belts/sheaves) ratings and cause that component to fail. The worst type of failure from a personal safety standpoint is to exceed the rim speed of he sheaves casing the sheave to explode.

HYDRAULIC DRIVE HEAD and POWER UNIT (GAS PRIME MOVER)

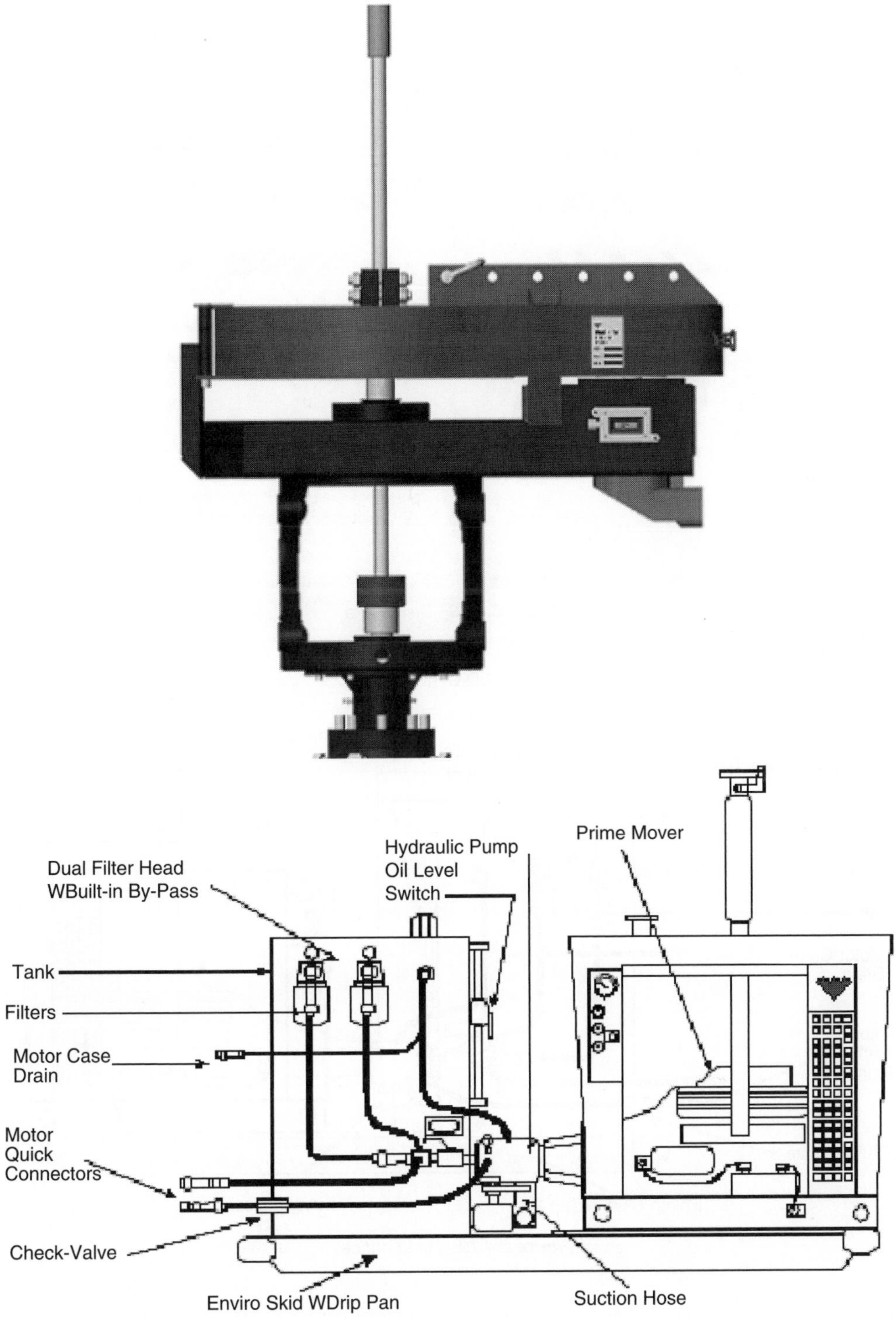

Figure 6.4.77   Drive heads.

DISC BRAKE

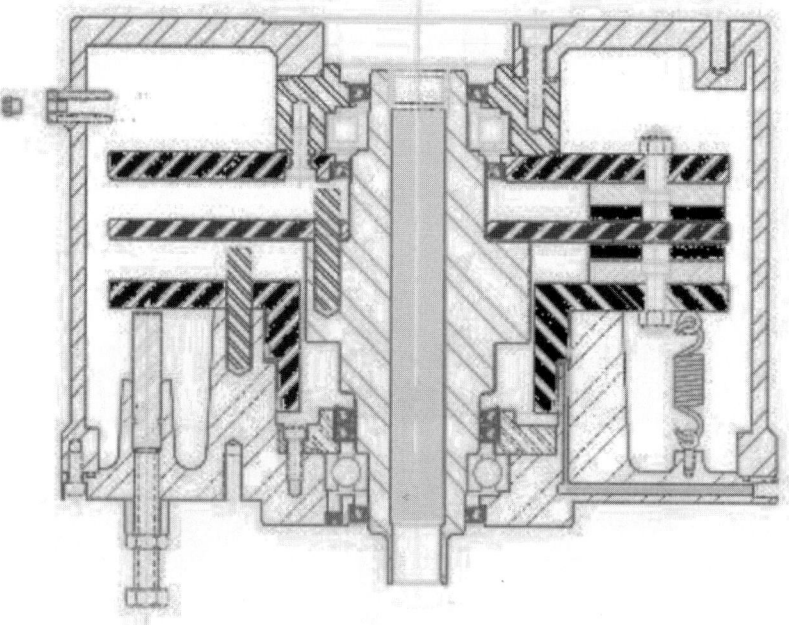

Figure 6.4.78   *Disc brake.*

FIXED ORIFICE

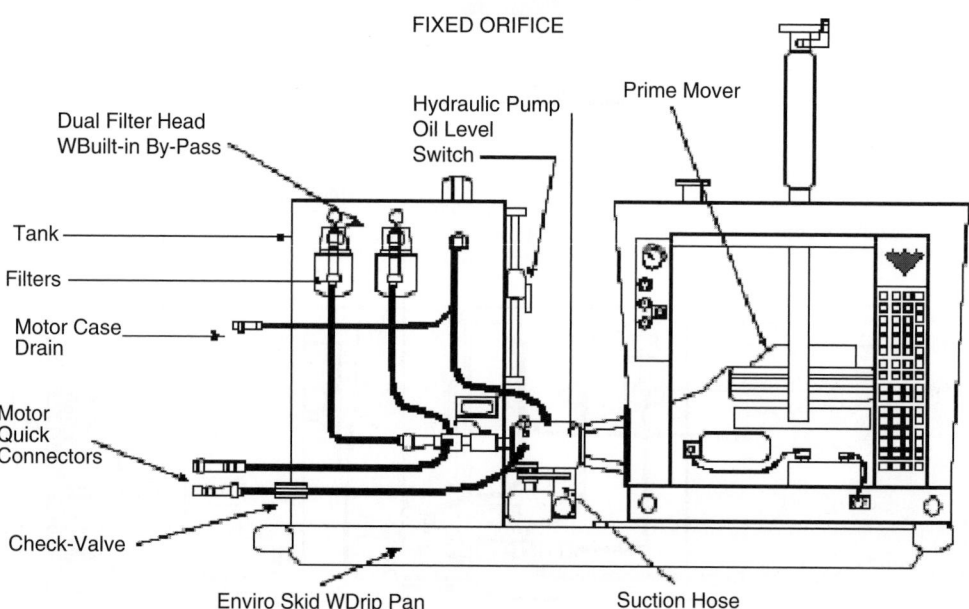

Figure 6.4.79   *Fixed orifice.*

Remember that the recoil control unit of any surface drive should be considered as a safety device to control the release of torque from the rod string during shut down, bit it not the only means of safe operation. For improved overall safety in operation and shut down of progressing cavity surface drive systems, the follow steps should be considered.

Properly designed surface equipment

- Do not exceed manufacturers recommended specifications for torque and speed ratings
- Design the belts and sheaves for the prime mover horsepower using correct safety factors.
- Use sheaves that are in new condition only.

CENTRIFUGAL

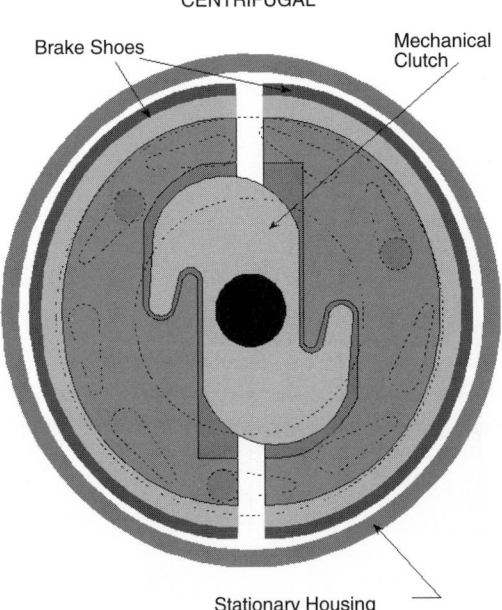

Figure 6.4.80 *Centrifugal.*

- Use a straight polished rod with minimum stick up.
- Use PC pump polished rod clamps only (not conventional clamps).

Improved stability of the surface drive in higher torque/speed applications.

- Flange mount drives.

Limit the input torque to the polished rod.

- Maximum torque should not exceed the surface drive ratings.
- Maximum torque generated by an electric motor can be as high as 250% of the rated torque so a high amp shut down must be properly set.
- Maximum torque should not exceed the torque rating of the sucker rod string.

History shows that throughout the industry most of the serious incidents with any PC pump drive heads have occurred when the pump is seized downhole and PC pumping system has been torqued up in an attempt to free the pump. In a direct electric application, this input torque is typically limited only by the maximum horsepower the motor can deliver and the rod strings ability to transmit torque down to the pump. Unknown to most operators is the fact that all electric motors have the ability to deliver over 250% of the full load nameplate horsepower. This potential applied torque can far exceed the published structural and braking limits of the drive head.

In new installations, the best way to limit the maximum torque input is to restrict the motor size and sheave ratio in order to limit the maximum torque that can potentially be applied. Under certain circumstances, the potential exists for a recoil event to occur where the stored energy in the rod string and the fluid column may be greater than the drive head is designed to control.

This potential is based on the sheave ratio and the connected horsepower in direct electric motor applications and ability of the drive head recoil control device to safely handle the recoil.

6.4.15.6 Belt and Sheave Selection Considerations

A variety of sizes, styles and lengths of belts and sheaves are available. Due to the numerous alternatives and complexity of selecting the correct belts and sheaves it is recommended that the procedure in a power transmission products manual be followed. The following is a simplistic example of the steps necessary to select the right belt and sheave combination for your application:

6.4.15.6.1 Determine the Size and Speed of Electric Motor

- Estimate the desired polished rod speed and maximum torque the system will operate under. Example — Desired speed of 200 rpm and a maximum torque of 900 ft.lbs.
- From Table 6.0.1 — Torque/Horsepower/Speed Relationships, select the electric motor that best meets the desired speed and torque.
- According to the table, a 40-hp, 1200-rpm, 460-V, 60-hz, three-phase motor produces 175 ft.lbs of torque at full load ratings. To develop a polished rod speed of 200 rpm a 6:1 sheave ratio is required (1200/200). This assumes a 1:1 drive ratio drive. With a 6:1 sheave ratio; 1050 ft.lbs (175 × 6) of torque can be applied to the polished rod (assuming the belts and sheave will transfer it). Torque will vary with changes to line voltage, line current, and line frequency. With the addition of a variable frequency drive (VFD), motor speeds can be increased 50%, and then the maximum polished rod speed becomes 300 rpm in this example. Note that motor torque decreases as motor frequency increases above rated speed.

6.4.15.6.2 Sheave Selection

- Sheaves are made with either cast iron or ductile iron. Cast iron has a maximum sheave rim velocity of 65,000 ft./min. Depending on he situation; recoil speeds in excess of this can be reached. A spoked wheel ductile iron sheave is rated for a maximum speed of 10,000 ft./min and a solid wheel ductile sheave for 15,000 ft./min.

6.4.15.6.3 Belt Selection

- There are various styles and sizes of V belts available for PC pumping systems. These include:

  - classical, sizes "A," "B," "C," and "D"
  - narrow, sizes "3V," "5V," and "8V"
  - joint or banded
  - wrapped
  - cogged

In this example, with a small driver sheave of "the amount of" "rubber on the sheave" is limited and a cogged belt with transfer more power than the other styles.

- Using a Power Transmission Products Manual the horsepower each belt transfers can be determined; and knowing the Horsepower requirements, calculating the number of belts is simple. Ensure that the combined horsepower rating of all the V belts exceeds the design horsepower of the system.
- If the amount of power transferred by V belts is a concern than a high torque drive (HTD) belt and corresponding sprockets are recommended. This combination provides a positive slip-proof engagement when the belt teeth mesh with the sprocket grooves.

6.4.15.7 Installation Procedures

The procedures listed below are intended to form a basic framework for a typical installation of a progressing cavity

VANE

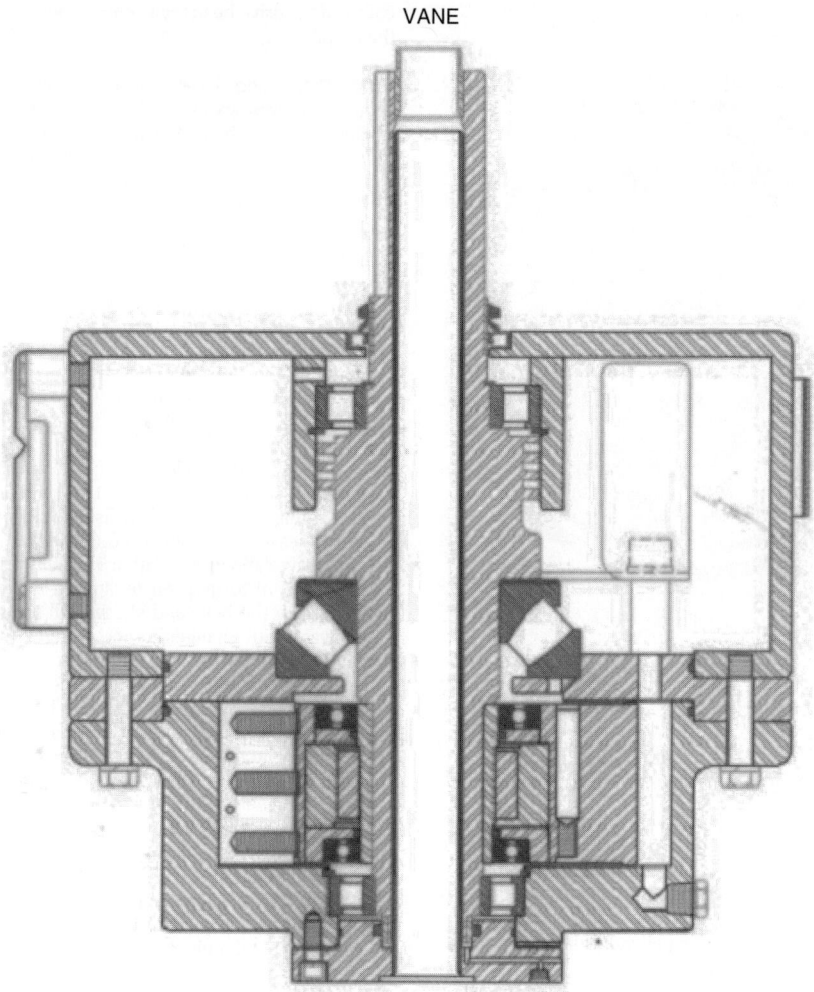

Figure 6.4.81   *Vane.*

pumping systems. Special applications or equipment may necessitate changes in thee procedures. In these cases, a technical representative can provide assistance in formulating modified installation procedures.

6.4.15.7.1 Preinstallation Procedures
1. Clean out the well, removing any solid (e.g., sand, coal fines) build-up in the bottom of the wellbore. It is recommended that a minimum of 2 or 3 m below the pump be clear. Solids build-up should also be at least 2m below the bottom of the perforations.
2. Ensure all proper crossovers and changeovers are on location to attach the stator to the tubing string and the rotor to the rod string. It is recommended that swages be avoided due to their restricted drift and potential for thread failure.
3. Ensure that various lengths of the appropriate size pony rods are available to allow proper space out of the rotor and polish rod.
4. If a pump seating nipple (PSN) is going to be used, verify clearance against the major diameter of the rotor. Information for *Weatherford pumps* provided in Table 7.1.1 (Recommended Pup-Joint and Pony rod Connection) can assist with this.

5. The rotor and stator are matched for optimum pump efficiency in a particular application. Ensure the rotor serial number located on the rotor head (top) corresponds with the last set of numbers on the stator.
6. Measure the distance between the top of the tag bar pin and the bottom of the elastomer. This is required for the space out procedures described in "Space-out Guidelines."

6.4.15.7.2 Stator Installation Procedures
1. Attach any accessories (i.e., gas anchor, no-turn tool, tail joint) to bottom of the stator.
2. Attach stator to first tubing joint with appropriate fittings. Depending on the pump model being used, it may be convenient to attach a pup joint directly on top of the stator to reduce the assembly length and simplify handling. If the production tubing is of smaller diameter then the pump stator, run larger diameter tubing above the pump and then use a crossover to down size. It is recommended that any down size in tubing be done at least 1.2 m (4 ft.) above the stator. Failure to do this may restrict the eccentric rotor movement and produce fatigue failures in the rod string directly above the pump.

3. Run in the production tubing keeping an accurate tally including the stator and fittings. Ensure that all connections are cleaned, inspected, doped and torqued to optimum levels according the guidelines in API RP5C1 ("Recommended Practice for Care and Use of Casing, Tubing and Drill Pipe"). Recommended tubing make-up torque is included for common sizes and grades of tubing in Table 6.4.23 — Suggested Torque Make-up for Tubing. The stator should be considered as equivalent to Grade J55 tubing. Proper makeup is critical because the rotor normally rotates clockwise in the stator, which tends to loosen improperly tightened tubing connections. Make sure the backup wrench is on the tubing below the collar.

4. Positioning of the pump intake is critical to ensure optimal pump performance and life. The most suitable pump location is dependent on the nature of the application. In most cases, the pump should be positioned with its intake near or below the bottom the perforations to facilitate natural separation of the gas from the produced fluid. However, in wells producing significant quantities of sand, it may be preferable to locate the pump above the perforations to limit problems with sanding equipment in. For unique applications, a technical representative can provide detailed pump landing recommendations on an individual application basis. Note that except in special circumstances, the use of extended length tail joints is not recommended.

6.4.15.7.3 Rotor Installation Procedures

1. Count out the number of rods required to equal the total tubing tally plus one extra rod to ensure sufficient length of rod string. Be sure to include the length of the rotor. Check all rods for evidence of damage, wear, and corrosion. It is recommended that rods that are permanently deformed (i.e., yielded) due to excessive torque not be used in applications with high loads or fluctuating loads. Use APIRP11BR Section 6 ("Recommended Practice for Care and Handling of Sucker Rods") as a guide to grade the rod. Replace all suspect or substandard rods.

2. Care should be taken while handling the rotor on surface to avoid damaging the pin threads or the finished surface. The rotor should be supported in a manner as to prevent excessive bending that may cause permanent damage.

3. Liberally apply grease over the full length of the rotor to provide initial lubrication. Attach the rotor to the first sucker rod. A transition coupling may be required if the rotor pin and sucker rod threads are not the same size. It is recommended that one standard rod without guides (i.e., 7.6 m (25 ft.)) be placed above the pump rotor before any pony rods are added to the string. This is because the positioning of large centralizers or guides a short distance above the pump may restrict the orbital motion of the rotor causing damage to the pump. *If pony rods are required directly above the rotor they should be limited to a 0.3 m (1 ft.) machined pony rod.*

4. Run in all but the last few sucker rods making them up as outlined in APIRP11BR Section 5 ("Recommended Practice for Care and Handling of Sucker Rods"). Note that it is essential that each connection is made up to its recommended torque during installation because tightening under loading damages the connection and may result in a premature failure. It is recommended that if significant sand production is anticipated, a shear be installed at least two full sucker rod lengths above the rotor to facilitate rod string removal in the event of a sanded pump. It is also good practice to record details of the rod string configuration, especially when different types of rod centralizers and rod guides are being used.

6.4.15.8 6.4.5.8.4 Space Out Guidelines

1. Record the weight of the rod string before to running the last few sucker rods.

2. Continue to run the remaining sucker rods taking care to lower them very slowly to avoid damaging the stator elastomer. The rod string will normally rotate as the rotor enters the stator.

3. Continue to lower the rod string until the rotor comes to rest on the tag bar at the bottom of the stator. This bottoming out will be indicated by a reduction in rod weight to near zero. Pull up and lower the rotor again to reconfirm that it is positioned on the tag bar.

4. With the rotor landed on the tag bar, pull up slowly on the rod string until full rod weight (measured in step 1 above) is restored. At this point all slack is taken out of the rod string and the rotor is slightly above the tag bar.

5. Lift the rod string the distance from the tag bar to the bottom of the stator rubber. This tag bar distance was measured in step 6 under "Stator Installation Procedures."

   The space-out procedures also need to compensate for rod stretch due to pump loading. This distance can be obtained from the manufacturer.

   To accommodate for rod stretch during operation the rods should be lifted up by the amount determined above. Note that in some cases it may be necessary to space-out an incremental amount to account for thermal expansion of the rod string.

6. The rotor is now correctly landed in the stator and the length of the sucker rods in the hole must be adjusted to allow for the polish rod. The rotor should now be in the operating position. Do not lift the rotor from this position. Landing with least amount of rotor sticking out of the stator minimizes downhole vibration.

7. Measure the height of the drive head and add the length of the drive assembly to the operating position. This becomes the clamping point. Allow and additional 6 in. to 12 in. for polish rod stick up above the clamping point.

8. Remove enough of the sucker rods to make room for the polish rod. The difference between the sucker rod removed versus the length of polish rod added must be made up with pony rods. It is recommended that the polish rod be long enough to allow the rotor to be pulled totally out of the stator without removing the surface equipment. Not only is this important to allow a proper flush-by procedure, but it is also a safety consideration in well servicing operations.

9. Attach the polish rod to the top pony rod using a polish rod coupling. To ensure proper sealing of the stuffing box, it is critical that the polish rod is straight and does not have significant wear or corrosion. It is recommended that the polish rod not be used to move or lift the drive head because of the potential to cause permanent bending of the polish rod. DO NOT leave wrench marks on the polish rod.

10. Confirm the pump space-out (steps 3 to 7 above) with the modified string configuration. The *target* mark should be within 15 cm (6 in.) of the top of the polish rod. If it is not, make the necessary adjustments to the pony rod lengths. When properly positioned, install a clamp on the polish rod adjacent to the top of the flow tee. It is also recommended that any pony rods be buried at least two full sucker rod lengths down the rod string to prevent surface vibration problems.

6.4.15.9 Safety Warning

Polished rod stick-up of more than 46 cm (18 in.) above the top of the drive clamp can cause severe damage during recoil.

Table 6.4.23 *Suggested Make-up Torque For Tubing*

Size		Weight			Makeup Torque	
Sl	Imperial	Sl	Imperial	Grade	Sl	Imperial
60.3 mm	2375 in	7.0 kg/m	4.7 lbs/ft	H40	1340 N·m	990 ft·lbs
60.3 mm	2375 in	7.0 kg/m	4.7 lbs/ft	J55	1750 N·m	1290 ft·lbs
60.3 mm	2375 in	7.0 kg/m	4.7 lbs/ft	C75	2310 N·m	1700 ft·lbs
60.3 mm	2375 in	7.0 kg/m	4.7 lbs/ft	L80	2390 N·m	1760 ft·lbs
60.3 mm	2375 in	7.0 kg/m	4.7 lbs/ft	N80	2450 N·m	1800 ft·lbs
60.3 mm	2375 in	7.0 kg/m	4.7 lbs/ft	C90	2610 N·m	1920 ft·lbs
60.3 mm	2375 in	7.0 kg/m	4.7 lbs/ft	P-105	3080 N·m	2270 ft·lbs
73.0 mm	2875 in	9.7 kg/m	6.5 lbs/ft	H40	1700 N·m	1250 ft·lbs
73.0 mm	2875 in	9.7 kg/m	6.5 lbs/ft	J55	2230 N·m	1650 ft·lbs
73.0 mm	2875 in	9.7 kg/m	6.5 lbs/ft	C75	2940 N·m	2170 ft·lbs
73.0 mm	2875 in	9.7 kg/m	6.5 lbs/ft	L80	3050 N·m	2250 ft·lbs
73.0 mm	2875 in	9.7 kg/m	6.5 lbs/ft	N80	3120 N·m	2300 ft·lbs
73.0 mm	2875 in	9.7 kg/m	6.5 lbs/ft	C90	3340 N·m	2460 ft·lbs
73.0 mm	2875 in	9.7 kg/m	6.5 lbs/ft	P-105	3940 N·m	2910 ft·lbs
88.9 mm	3500 in	13.8 kg/m	9.3 lbs/ft	H40	2340 N·m	1730 ft·lbs
88.9 mm	3500 in	13.8 kg/m	9.3 lbs/ft	J55	3090 N·m	2280 ft·lbs
88.9 mm	3500 in	13.8 kg/m	9.3 lbs/ft	C75	4080 N·m	3010 ft·lbs
88.9 mm	3500 in	13.8 kg/m	9.3 lbs/ft	L80	4240 N·m	3130 ft·lbs
88.9 mm	3500 in	13.8 kg/m	9.3 lbs/ft	N80	4330 N·m	3200 ft·lbs
88.9 mm	3500 in	13.8 kg/m	9.3 lbs/ft	C90	4650 N·m	3430 ft·lbs
88.9 mm	3500 in	13.8 kg/m	9.3 lbs/ft	P-105	5490 N·m	4050 ft·lbs
101.6 mm	4000 in	16.4 kg/m	11.0 lbs/ft	H40	2630 N·m	1940 ft·lbs
101.6 mm	4000 in	16.4 kg/m	11.0 lbs/ft	J55	3470 N·m	2560 ft·lbs
101.6 mm	4000 in	16.4 kg/m	11.0 lbs/ft	C75	4600 N·m	3390 ft·lbs
101.6 mm	4000 in	16.4 kg/m	11.0 lbs/ft	L80	4780 N·m	3530 ft·lbs
101.6 mm	4000 in	16.4 kg/m	11.0 lbs/ft	N80	4880 N·m	3600 ft·lbs
101.6 mm	4000 in	16.4 kg/m	11.0 lbs/ft	C90	5250 N·m	3870 ft·lbs
114.3 mm	4500 in	19.0 kg/m	12.8 lbs/ft	H40	2930 N·m	2160 ft·lbs
114.3 mm	4500 in	19.0 kg/m	12.8 lbs/ft	J55	3870 N·m	2860 ft·lbs
114.3 mm	4500 in	19.0 kg/m	12.8 lbs/ft	C75	5130 N·m	3780 ft·lbs
114.3 mm	4500 in	19.0 kg/m	12.8 lbs/ft	L80	5340 N·m	3940 ft·lbs
114.3 mm	4500 in	19.0 kg/m	12.8 lbs/ft	N80	5450 N·m	4020 ft·lbs
114.3 mm	4500 in	19.0 kg/m	12.8 lbs/ft	C90	5870 N·m	4330 ft·lbs

USEFUL FORMULAS

Hydraulic Torque

$$T_{hydraulic} = CVP_{net}$$

where:

$T_{hydraulic}$ = Hydraulic Pump Torque (N·m or ft·lbs)
   C = Constant (SI:0.111 Imperial: 8.97×10^{-2})
   V = Pump Displacement (m^3/day/rpm or bbls/day/rpm)
   P_{net} = Net Lift (kPa or psi)

Total Torque

$$T_{total} = T_{hydraulic} + T_{friction}$$

where:

   T_{total} = Total Pump Torque (N·m or ft·lbs)
$T_{hydraulic}$ = Hydraulic Pump Torque (N·m or ft·lbs)
   $T_{friction}$ = Pump Friction Torque (N·m or ft*lbs)

Pump Power

$$P_{pump} = CT_{total}N$$

where:

   C = Constant (SI: 1.05×10^4 Imperial: 1.91×10^4)
P_{pump} = Pump Power (kW or hp)
T_{total} = Total Pump Torque (N·m or ft·lbs)
   N = Pump Rotational Speed (rpm)

Polished Rod Torque

$$T_{polishrod} = \frac{CI_{line}VN_{motor}pfN_{pt}}{N}$$

where:

$T_{polishrod}$ = Polish Rod Torque (N·m or ft·lbs)
   C = Constant (SI: 16.495×10^4 Imperial: 12.113×10^4)
   I_{line} = Line Current (A)
   V = Line-to-Line Voltage (V)
   N_{motor} = Nominal Motor Efficiency (%)
   pf = Motor Power Factor
   N_{pt} = Power Transmission System Efficiency (%)
   N = Polished Rod Rotational Speed (rpm)

Power (mechanical)

$$P = \frac{TN}{5252}$$

where:

   P = Power (hp)
   T = Torque (ft·lbs)
   N = Rotational Speed (rpm)

Rod Torque

$$T_{rod} = T_{hydmotor}D_{ratio}$$

where:

   T_{rod} = Rod Torque (ft·lbs)
   $T_{hydmotor}$ = Hydraulic Motor Torque (ft·lbs)
   D_{ratio} = Drive Ratio

Specific Gravity

$$Sg = \frac{141.5}{(API + 131.5)}$$

where:

Sg = Specific Gravity
API = Fluid Gravity in API°

Liquid Column Pressure

$$P_{head} = SgG_{water}H$$

where:

P_{head} = Head Pressure (kPa or psi)
Sg = Specific Gravity
G_{water} = Water Gradient (SI: 9.81 kPa/m
Imperial:0.433 psi/ft)
H = Fluid Height (m or ft)

Specific Gravity

$$Sg = \frac{p}{p_{water}}$$

where:

Sg = Specific Gravity
p = Fluid density
p_{water} = Water Density (SI:1000 kg/m³
Imperial:62.2 lbs/ft³)

Belt Length

$$L = 2C + \frac{\pi(D + d)}{2} + \frac{(D - d)^2}{4C}$$

where:

L = Belt Length (in)
C = Center Distance (in)
D = Driven Sheave Pitch Diameter (in)
D = Driver Sheave Pitch Diameter (in)

Net Lift

$$P_{net} = (P_{thp} + P_{tbgliquid} + P_{tbglosses}) - (P_{chp} + P_{csggas} + P_{csgliquid})$$

where:

P_{net} = Net Lift (kPa or psi)
P_{thp} = Tubing Head Pressure
$P_{tbgliquid}$ = Tubing Liquid Column Pressure
P_{chp} = Casing Head Pressure
P_{csggas} = Casing Gas Column Pressure
$P_{csgliquid}$ = Casing Liquid Column Pressure

Hydraulic Motor Torque

$$T_{hydmotor} = \frac{M_{disp}P}{75}$$

where:

$T_{hydmotor}$ = Hydraulic Motor Torque (ft·lbs)
M_{disp} = Hydraulic Motor Displacement (in³)
P = System Pressure (psi)

- Electrical Output HP = (RPM × Torque (Ft − lbs)/5252
- Electrical Output Hp = (Amps × Volts × Motor Efficiency × Motor P.F. × 1.73)/746

- Hydraulic Torque (Ft − lbs) = $\Big(($System Press. (psi) ×Hyd. Motor Displ (in³) × Sheave Ratio)$\Big) \Big/ \Big((2 × 3.1416 × 12$ in/ft)$\Big)$

- Hydraulic Horse Power = $\Big(($System Press. (psi) ×Hyd. Pump Displ. (in³) ×Hyd. Pump rpm)$\Big) \Big/ \Big(395,934\Big)$

- Fresh Water Gradient = 0.433 psi/foot
= 9.8 Kpa/meter
- Specific Gravity = 141.5/(API 131.5)
- Head Pressure (kPa) = Fluid Density (kg/m³) × depth (meters) × 0.00981

UNIT CONVERSIONS AND EQUIVALENTS

S.I.	S.I. to Imperial multiply by	Imperial to S.I. multiply by	Imperial
Celsius degrees (°C)	1.800 + 32	0.5556 − 17.8	Fahrenheit degrees (°F)
cubic metre (m³)	6.29	0.159	barrel (bbls)
cubic metre (m³)	0.03531	28.32	thousand cubic feet (mcf)
standard cubic metre gas per standard cubic metre oil (m³/m³)	5.618	0.178	standard cubic foot gas per standard barrel oil (scf/STB)
kilogram (kg)	2.205	0.454	pound-mass (lbs)
kilogram-metre squared (kg·m²)	23.70	0.0422	pound-mass-feet squared (lbs·ft²)
kilogram per cubic metre (kg/m³)	0.0624	16.02	pound-mass per cubic foot (lbs/ft³)
kilojoules (kJ)	0.9479	1.055	British thermal unit (Btu)
kilo Newton (kN)	0.2248	4.448	thousand pound-force (kip)
kilopascal (kPa)	0.145	6.895	pound-force per square inch (psi)
kilowatt (kW)	1.341	0.7457	horsepower (hp)
Litre (l)	0.220	4.54	US gallon (g)
mega Pascal (mPa)	0.145	6.895	thousand pound-force per square inch (ksi)
metre (m)	3.281	0.3048	foot (ft)
millimetre (mm)	0.0394	25.4	inch (in)
Newton (N)	0.2248	4.448	pound-force (lbf)
Newton-metre (N·m)	0.7376	1.356	foot-pound-force (ft·lbf)

6.5 STIMULATION AND REMEDIAL OPERATIONS

Reservoir stimulation deals with well productivity, and a successful stimulation treatment requires the accurate identification of the parameters controlling the well production. As a result, causes of impaired production must be identified. Furthermore, it must be determined whether or not a particular treatment can improve production. This is the very first step of the stimulation job design.

Darcy's law in its simplest form is adequate to study the issue. For oil, a familiar expression (for steady-state and in a radial reservoir) is written as

$$q = \frac{kh(p_e - p_{wf})}{141.2\,B\mu(\ln r_e/r_w + s)} \qquad [6.5.1]$$

where q = well flowrate (STB/d)
  k = reservoir permeability in md
  h = reservoir thickness in ft.
  p_e = outer boundary reservoir pressure in psi
  p_{wf} = flowing bottomhole pressure in psi
  B = formation volume factor in resbbl/STB
  μ = fluid viscosity in cp
  r_e = equivalent drainage radius in ft.
  r_w = well radius in ft.
  s = skin effect

Each of the variables on the right-hand side of Equation 6.5.1 affects well productivity, and certain actions may favorably change its impact. Of particular interest to the stimulation engineer are the permeability and the skin effect. Both of these variables may be obtained from a pretreatment pressure transient test. Ignorance of these two variables would result in an inappropriate or less than optimum stimulation treatment; in addition, a posttreatment analysis and job evaluation would be impossible.

As can be easily seen from Equation 6.5.1, a low value of the permeability (tight reservoir) or a high value of the skin effect (damaged or badly completed well) would result in low well productivity. There is virtually nothing practical that can be done to the permeability, although certain investigators have erroneously suggested that hydraulic fracturing increases the effective reservoir permeability. A hydraulic fracture is a superimposed structure on a reservoir that remains largely undisturbed outside of the fracture. The fracture, however, can greatly improve the well productivity by creating a large contact surface between the well and the reservoir. The production improvement results from effectively increasing the wellbore radius.

Matrix stimulation is generally intended to reduce a large skin effect resulting from permeability damage around the wellbore during completion or production. Often, there is confusion in distinguishing matrix acidizing (a form of matrix stimulation) from acid fracturing. The latter requires that the treatment is done at formation fracturing pressure, and it relies on a residual *etched* width of the created fracture. The two methods of stimulation are applicable to entirely different types of formations: matrix acidizing is applied to high-permeability reservoirs, whereas acid fracturing is appropriate for low-permeability, acid-soluble reservoirs such as carbonates.

All stimulation practices adjust the skin effect, either by improving a negative component (fracturing) or by reducing a positive value caused by damage (matrix stimulation) [1].

6.5.1 Fracturing

Hydraulic fracturing of petroleum reservoirs is a reasonably new activity, spanning 40 years. The understanding of fracture propagation, its geometry, and direction is even newer, and addition to the body of knowledge of fracturing as a reservoir stimulation treatment is a very active process.

A classic concept introduced in 1957 [2] concluded that fractures are "approximately perpendicular to the axis of least stress." The stress field can be decomposed into three principal axes: a vertical and two horizontal, which are unequal. For most reservoirs the minimum stress is horizontal, resulting in vertical hydraulic fractures.

6.5.1.1 Stress Distribution

In a sedimentary environment, the vertical stress σ_v is equal to the weight of the overburden and can be calculated from

$$\sigma_v = \frac{1}{144}\int_0^H \rho\,dH \qquad [6.5.2]$$

where ρ = density of each layer in lb/ft.^{3}
  H = thickness of overburden in ft.

Equation 6.5.2 can be evaluated using a density log. In its absence, a value of 1.1 psi/ft. can be used as a reasonable approximation.

A porous medium, containing fluid, is subjected to an *effective* stress, rather than the absolute stress given by Equation 6.5.2. The effective stress σ' is related to the pore pressure p by

$$\sigma' = \sigma - \alpha p \qquad [6.5.3]$$

where α is Biot's "poroelastic" constant and varies from 0 to 1. For most petroleum reservoirs, it is equal to 0.7. It is important that the concept of the effective stress is understood. An implication is that in a propped hydraulic fracture, the effective stress on the proppant is greatest during production ($p = p_{wf}$) and must be considered in the proppant selection.

Although the absolute and effective overburden stress can be computed via Equations 6.5.3 and 6.5.4, the two principal horizontal stresses are more complicated and their determination requires either field or laboratory measurements.

In a tectonically inactive formation, the elastic properties of the rock (Poisson ratio) may be used to relate the effective vertical stress with the effective minimum horizontal stress

$$\sigma'_{H,min} = \frac{v}{1 - v}\sigma'_v \qquad [6.5.4]$$

where v = Poisson ratio.

For sandstone formations, the Poisson ratio is approximately equal to 0.25, leading to a value of $\sigma'_{H,min}$ approximately equal to $\frac{1}{3}\sigma'_v$. For most shales the Poisson ratio is larger, leading to abrupt changes in the horizontal stress profile. This variation, which can envelope a sandstone reservoir because of overlaying and underlaying shales, is the single most important reason for fracture height containment.

6.5.1.2 Vertical versus Horizontal Fractures

Consider Figure 6.5.1. Graphed are the three principal stresses, σ_v, $\sigma_{H,min}$, $\sigma_{H,max}$. The first is given by Equation 6.5.2, whereas $\sigma_{H,min}$ can be extracted from Equation 6.5.4. The maximum horizontal stress $\sigma_{H,max}$ can be considered as equal to $\sigma_{H,min}$ plus some tectonic component σ_{tec}. If the original ground surface remains in place, then $\sigma_{H,min}$ is less than σ_v, leading always to a vertical fracture which would be perpendicular to $\sigma_{H,min}$. However, if the present ground surface has been the result of massive glaciation and erosion, as depicted in Figure 6.5.1, the overburden is reduced. Because the horizontal stresses are "locked" in place, there exists a critical depth, shallower of which the minimum horizontal stress is no longer the minimum stress. In such a case, a

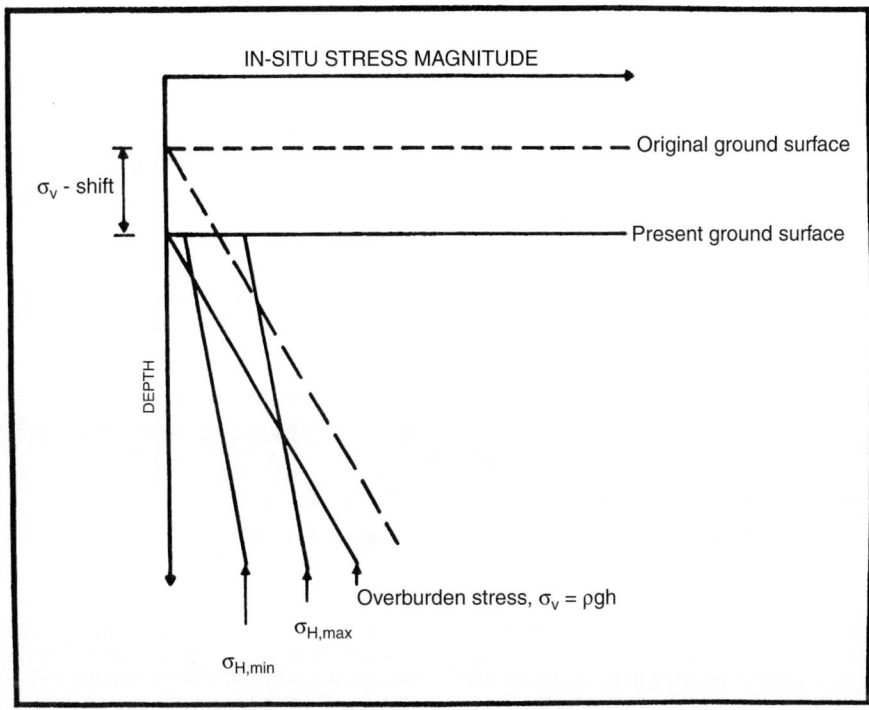

Figure 6.5.1 *Stress profiles (vertical and two horizontal stresses). Glaciation and erosion reduced the overburden, enabling a critical depth above which horizontal fractures may be generated. Below, only vertical fractures, normal to the minimum horizontal stress, are generated.*

horizontal fracture will be created in the reservoir. This has been observed in a number of shallow reservoirs.

The definition of *principal* stress direction implies that all shear stresses vanish. Thus, when a vertical well is drilled, usually it coincides with a principal stress direction. This is not the case when a deviated or horizontal well is drilled (unless, in the latter case, the well is drilled in the direction of one of the principal horizontal stresses). However, for the mass of deviated wells that are drilled from platforms or drilling pads, their direction implies a nonvanishing shear stress [3]. The implications for fracturing are substantial. A deviated well requires a higher fracture initiation pressure. Furthermore, the production performance of a fractured deviated well is impaired. This point will be addressed in a later subsection.

6.5.1.3 Pressure Related to Fracturing

The pressure signature created during the pumping of a fracturing treatment, and the associated pressure decline after injection, contains a significant amount of information relating to the fracture itself as well as the reservoir where the fracture has been placed. Applying specialized analysis of pressure during the various stages of the fracturing process provides a powerful technique for developing a comprehensive understanding about this process. Analysis during pumping provides a qualitative indicator of fracture growth, as well as estimates of several primary fracture parameters. The analysis of pressure decline after pumping can be broken into two distinct periods, with the early period being dominated by pressure falloff that is related to the fracture closing, whereas late time is governed by the transient pressure response of the reservoir. Figure 6.5.2

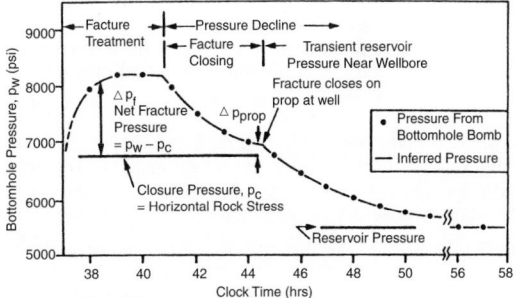

Figure 6.5.2 *Expected pressure response during a hydraulic fracture treatment [6].*

shows the pressure response during the three stages of fracture evolution: growth, closure, and afterclosure reservoir response.

Pressure measured during pumping provides an indication of the fracture growth process. The primary diagnostic tool for this period is the slope of the log-log plot of net pressure (i.e., the fracturing pressure above the reference closure pressure) versus pumping time. Figure 6.5.3 demonstrates how the slope of the log-log plot is used to characterize the fracture geometry.

The pressure response during fracture closure is governed largely by the rate of fluid loss. The analysis of pressure during this period estimates the fluid efficiency and

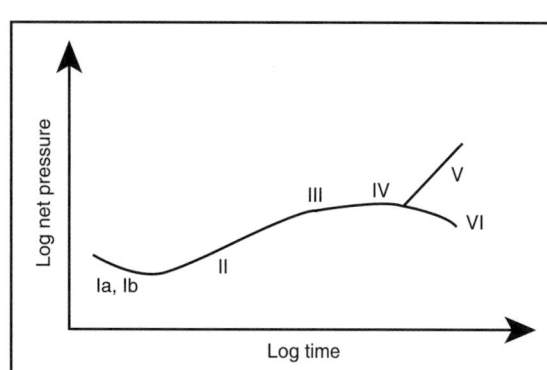

Interpretation of log-log plot fracture pressure slopes		
Propagation Type	Log-Log Slope	Interpretation
Ia	$-\frac{1}{2}$ to $-\frac{1}{3}$	Frac Height growing faster than length; KGD
Ib	$-\frac{1}{2}$ to $-\frac{1}{3}$	Frac Height growing faster than length; Radial
II	$\frac{1}{2}$ to $\frac{1}{4}$	Confined Frac Height. Length growing faster than height; PKN
III	Reduced from II	Controlled height growth Stress-sensitive fissure
IV	0	Height growth through pinch point Fissure dilation T-shaped fracture
V	≥ 1	Restricted extension
VI	Negative following IV	Uncontrolled height growth
Note: $n = 0.5$		

Figure 6.5.3 *Interpretation of log-log plot fracture pressure slopes.*

the leakoff coefficient. These parameters are determined from a plot of the pressure decline versus a specialized function of time, commonly referred to as the *G*-plot. This specialized plot provides the fracturing analog to the Horner plot for well testing.

The final fracturing pressure analysis pertains to the evaluation of pressure after fracture closure. The pressure response during this period loses its dependency on the mechanical response of an open fracture and is governed by the transient pressure response within the reservoir. This transient results from fluid loss during fracturing and can exhibit either linear flow or a long-term radial response. Each of these flow patterns can be addressed in a manner analogous to conventional well test analysis for a fixed-length conductive fracture. The afterclosure period characterizes the reservoir's production potential.

6.5.1.4 Breakdown Pressure (Fracture Initiation Pressure)
The fracture initiation pressure is estimated via Terzaghi's criterion [4], giving an upper bound for the value of the breakdown pressure p_b, such that

$$p_b = 3\sigma_{H,min} - \sigma_{H,max} + T_0 - p \qquad [6.5.5]$$

where T_0 = tensile strength of the formation (psi)

Hence, hydraulic fracturing describes the tensile failure of the rock and Equation 6.5.5 can be used during a fracture calibration treatment to calculate the horizontal stress components.

6.5.1.5 Determination of Closure Pressure
Closure pressure is defined as the pressure when the fracture width becomes zero. In a homogenous reservoir and where $\sigma_{H,min}$ is the smallest stress, the closure pressure is approximately equal to this value. Nolte [5, 6] pioneered the analysis of the pressure response during fracture calibration treatments and the calculation of important fracture variables.

The pressurization/pressure decline stages shown in Figure 6.5.2 can be refined and used in accordance with Nolte's analysis to calculate the closure pressure and, as will be shown in the next subsection, the leakoff coefficient and fracturing fluid efficiency. The closure pressure is not exactly

the minimum horizontal stress. With very little fluid leakoff (i.e., not upsetting the pore pressure in Equation 6.5.3) and with a contained fracture height, the closure pressure is very near the minimum horizontal stress. Otherwise, the closure pressure is a bulk variable taking into account fluid leakoff and especially horizontal stress heterogeneities along the fracture area. If the injected fluid is minimized, then with both leakoff and fracture height migration also minimized, the closure pressure is approximately equal to the effective minimum horizontal stress.

The pump-in/flowback test, which can be done as the first peak in Figure 6.5.2, has been devised to allow for the estimation of the closure pressure. The test involves injecting fluid, normally treated water, at rates (e.g., 5 to 10 bpm) and volumes (e.g., 30 to 50 bbl) sufficient to create a fracture.

Of particular importance is the flowback period. This must be done at rates between $\frac{1}{3}$ and $\frac{1}{4}$ of the injection rate. The flowback rate must be held constant via a regulating valve exactly to prevent any flowrate transients to mask the pressure response. During this flowback period, the interpretation is qualitative and based on a deduction of the ongoing closure process and should have two distinctly different periods:

- while the fracture is closing
- after the fracture is closed

The pressure profile would then have two regions reflecting the two different phenomena. These two regions would be separated by a clear change in slope, and the pressure corresponding to this inflection point is the *fracture closure pressure.*

Another method for determining fracture closure pressure is called the "equilibrium test". This is an injection test similar to the conventional pump-in/shut-in/decline, with one exception: instead of shutting in the well, the fluid continues to be injected at a small rate, as illustrated in Fig. 6.5.4. The treating pressure will initially decline as in the conventional shut-in decline because the new injection rate will approach or even be less than the leak-off rate. The fracture volume and the pressure will decrease with time as more fluid leaks off than is injected. This will result in the fracture volume being sufficiently reduced, and therefore, the fracture length will recede as the fracture approaches closure.

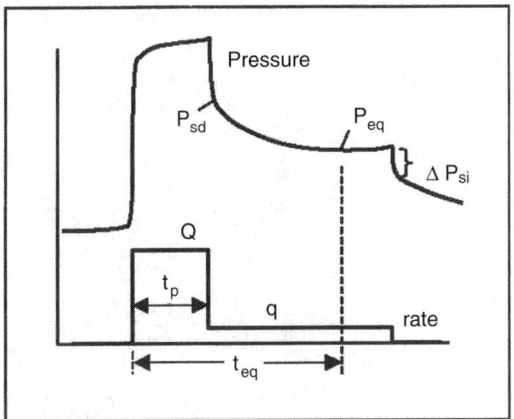

Figure 6.5.4 *Equilibrium test.*

The leak-off rate will decrease with time, and eventually, the leak-off rate and the injection rate q become equal. At that time, the fracture volume will stop decreasing, and the wellbore pressure will flatten out and start increasing. The minimum pressure when rate equilibrium is reached will be called the *equilibrium pressure*.

The equilibrium pressure P_{eq} is an upper bound of the closure pressure P_c. By subtracting the instantaneous pressure change at the final shut-in, ΔP_{si}, any remaining pressure relating to friction or tortuosity is removed. The corrected equilibrium pressure, $P_{eq} - \Delta P_{si}$, differs from closure pressure only by the net pressure in the fracture, which should be relatively small because the rate q is small and therefore provides a direct approximation of closure pressure.

6.5.1.6 Fracturing Pressure Decline Analysis

Castillo [8] extended Nolte's techniques for pressure decline analysis. He introduced a time function $G(\Delta t_D)$, which, graphed against pressure during the closing period of the fracture calibration treatment, forms a straight line.

Figure 6.5.5 is a graphical depiction after Castillo [8]. First, the necessity of an independent determination of the closure pressure (e.g., through a pump-in/flowback test) becomes obvious. The closure pressure would mark the *end* of the straight line, an important point of concern when dealing with real field data. Castillo's time function is given by

$$G(\Delta t_D) = \frac{16}{3\pi}\left[(1 + \Delta t_D)^{3/2} - \Delta t_D^{3/2} - 1\right] \text{ (upper bound)}$$

$$[6.5.6]$$

A second expression, for lower bound, is given in Castillo [8]. The dimensionless time Δt_D is simply the ratio of the closing time Δt and the injection time t_p.

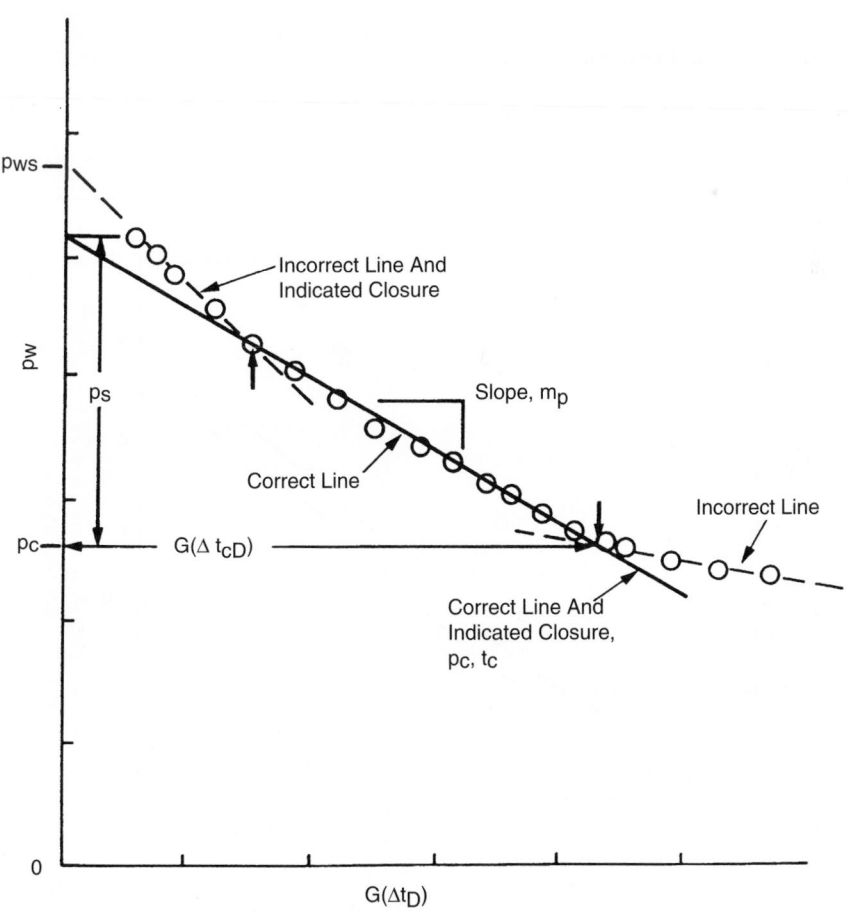

Figure 6.5.5 *Pressure decline analysis [8].*

The slope of the straight line in Figure 6.5.5, m_p, is

$$m_p = \frac{\pi C_L r_p \sqrt{t_p}}{2c_f} \qquad [6.5.7]$$

where C_L = leakoff coefficient in ft./min$^{1/2}$
r_p = ratio of the leakoff height (h_p) to the fracture height (h_t)
t_p = pumping time in min
c_f = fracture compliance in ft./psi

The fracture compliance is model dependent. The Perkins and Kern (PKN) model is given by

$$c_f = \frac{\pi \beta h_f}{2E'} \qquad [6.5.8]$$

where

$$\beta = \frac{2n' + 2}{2n' + 3 + a} \qquad [6.5.9]$$

and

$$E' = \frac{E}{1 - v^2} \qquad [6.5.10]$$

(Fracturing models will be described later.)
In Equations 6.5.9 and 6.5.10

n' = fluid power law exponent
a = viscosity degradation coefficient (0, linear degradation; 1, zero viscosity at the tip)
E = Young's modulus in psi
v = Poisson ratio

Nolte [6] contains all pertinent equations for fracturing pressure decline analysis.

Another important variable that can be extracted from pressure decline analysis is the fluid efficiency. The independently determined closure pressure identifies not only the end of the straight line in Figure 6.5.5 but also the fracture

closure time, corresponding to Δt_{cD}. This dimensionless closure time can then be used with Figure 6.5.6 to estimate the fluid efficiency. This is a particularly important variable and allows the determination of the total fluid requirements and the ratio of "pad" volume to the proppant carrying fluid.

Recently Mayerhofer, Economides, and Nolte [9, 10] investigated the stress sensitivity of crosslinked polymer filtercakes in an effort to decouple the components of fracturing pressure decline. Fracturing fluid leakoff can be regarded as a linear flow from the fracture into the reservoir. Therefore, a new approach to analyze the pressure decline of a fracturing treatment is visualized.

The concept of individual pressure drops in series, constituting the overall pressure drop between the fracture and the reservoir, can be used and is given by

$$\Delta p_{total} = \Delta p_{cake} + \Delta p_{iz} + \Delta p_{res} \qquad [6.5.11]$$

where Δp_{total} = pressure drop across the filtercake
Δp_{iz} = pressure drop within the polymer invaded zone
Δp_{res} = pressure drop in the reservoir

Figure 6.5.7 is a conceptual diagram of the individual zones and the corresponding pressure drops.

The effects of stress-sensitive filtercake leakoff were described by the hydraulic filtercake resistance, which is defined (with Darcy's law) as [9, 11]

$$R = \frac{l_c}{k_c} = \frac{\Delta p_{cake} A}{\mu q} \qquad [6.5.12]$$

The dimensionless resistance was also introduced:

$$R_D = \frac{\Delta p q_{ref}}{q \Delta p_{ref}} \qquad [6.5.13]$$

which is a rate-normalized pressure at any time with respect to a reference value.

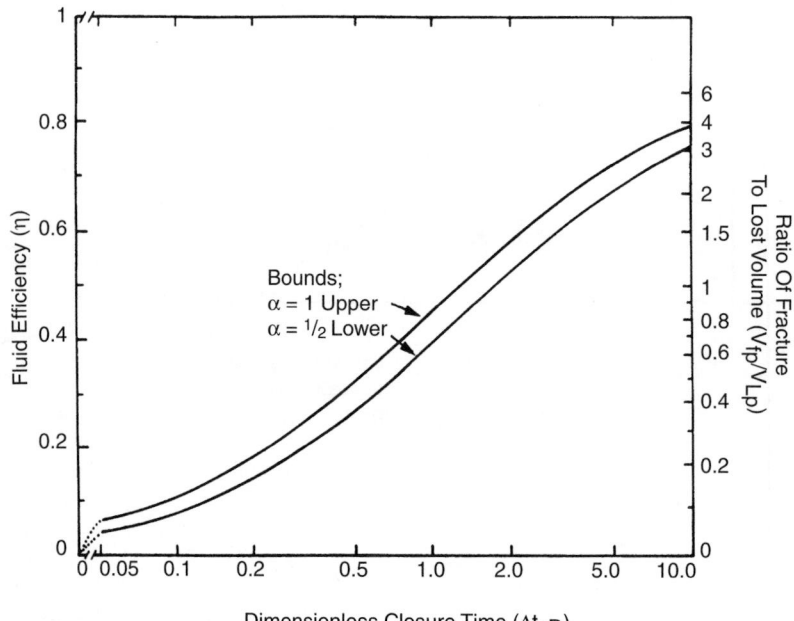

Figure 6.5.6 *Fracturing fluid efficiency correlation. The dimensionless closure time Δt_{cD} is the ratio of the closure time to the pumping time [6].*

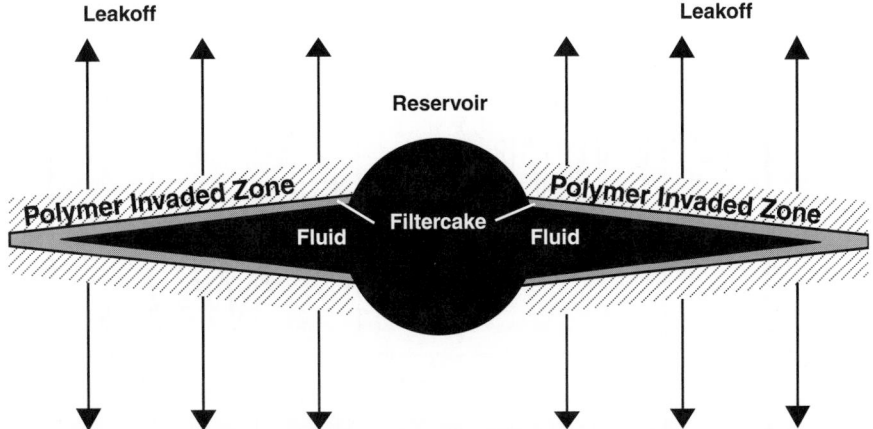

Figure 6.5.7 *Hydraulic fracture with filtercake and invaded zone.*

It was found [9, 10] that polymer filtercakes behave as viscoelastic bodies. The Kelvin or Voight model, which is a mechanical analog commonly used in linear viscoelastic theory, was found to be an appropriate model for analyzing the relation between differential pressure across the filtercake and the dimensionless resistance:

$$R_D(t) = \frac{1}{\mu} \int_0^t \sigma(t - \tau)e^{-\tau/\lambda}\, d\tau \qquad [6.5.14]$$

where $\sigma(t - \tau)$ = change of the differential pressure
$\quad\quad\quad \lambda$ = retardation time
$\quad\quad\quad \mu$ = viscosity

The viscoelastic filtercake relaxation, which was described by Equation 6.5.14, and the additional cake increase are the essential features during closure [9, 10]. Figure 6.5.8 shows, in a plot of R_D versus Δp, the dominance of the stress-sensitive relaxation of the filtercake deposited and compressed during pumping over the additional cake increase.

The stress-sensitive filtercake resistance is equivalent to a skin-effect and can therefore be incorporated as a component of the linear flow from the fracture into the reservoir.

6.5.1.7 Pressure Interpretation After Fracture Closure

Another application of pressure evaluation pertains to the pressure response after fracture closure. The pressure during this period reflects the transient reservoir response to fracturing and is independent of the mechanisms governing fracture propagation. Its character is determined entirely by the response of a reservoir disturbed by the fluid-leakoff process. During this period, the reservoir may initially exhibit formation linear flow, followed by transitional behavior and finally long-term pseudoradial flow.

The afterclosure response is similar to the behavior observed during a conventional well test of a propped fracture. It therefore supports an evaluation methodology analogous to the established principles of pressure transient evaluation. The afterclosure period provides information that is traditionally determined by a standard well test (i.e., transmissibility and reservoir pressure). It completes a chain of fracture pressure analysis that provides a continuum of increasing data for developing a unique characterization of the fracturing process.

6.5.1.8 Properties of Fracturing Fluids

The expected functions of the fracturing fluid are to initiate and propagate the fracture and to transport the proppant with minimum leakoff and minimum treating pressure.

Fluid viscosity is thus critical. An ideal fracturing fluid should have relatively low viscosity in the tubing (sufficient to carry proppant through the surface equipment but low enough to avoid unnecessary friction pressure losses), and high viscosity within the fracture where a large value can provide bigger fracture width and transport the proppant efficiently down the fracture. However, what a high-viscosity fracturing fluid does, inadvertently, is to plug the high permeability of the propped fracture, creating a highly unfavourable mobility. A mechanism to reduce the viscosity after the job to a very low value is then necessary. How are these apparently contradictory demands accomplished?

A typical, water-base fracturing fluid consists of water and a thickening polymer such as guar or one of the guar derivatives such as hydroxypropyl guar (HPG) or carboxymethylhydroxypropyl guar (CMHPG). The polymer concentrations could vary from 15 to 80 lb/1000 gal, depending on the required viscosity.

Such polymer solutions produce viscous fluids at ambient temperatures. At reservoir conditions these solutions thin substantially. To overcome this limitation, crosslinking agents are used to significantly increase the effective molecular weight of the polymer, thereby increasing the viscosity of the solution. Several different metal ions can be used as crosslinkers with borate compounds and transition metal complexes such as titanium (Ti[IV]), zirconium (Zr[IV]), and aluminum (Al[III]) being the most common. The reaction of these crosslinkers is often delayed so that the substantial increase in viscosity takes place near the perforations. This delay reduces tubing friction pressure and improves the long-term stability of the viscous fluid.

Recently polymer-free, water-base fracturing fluids have been developed using viscoelastic surfactants (VESs). Surfactants, when added to water, will associate into structures called micelles. The surfactants used to create VES fluids form rodlike micelles that will associate with one another at a sufficient concentration, and the resulting hindered movement causes the fluid to become both viscous and elastic.

In addition to water-based fluids, there are other types of fracturing fluids. Oil-based fluids were the first fluids to be

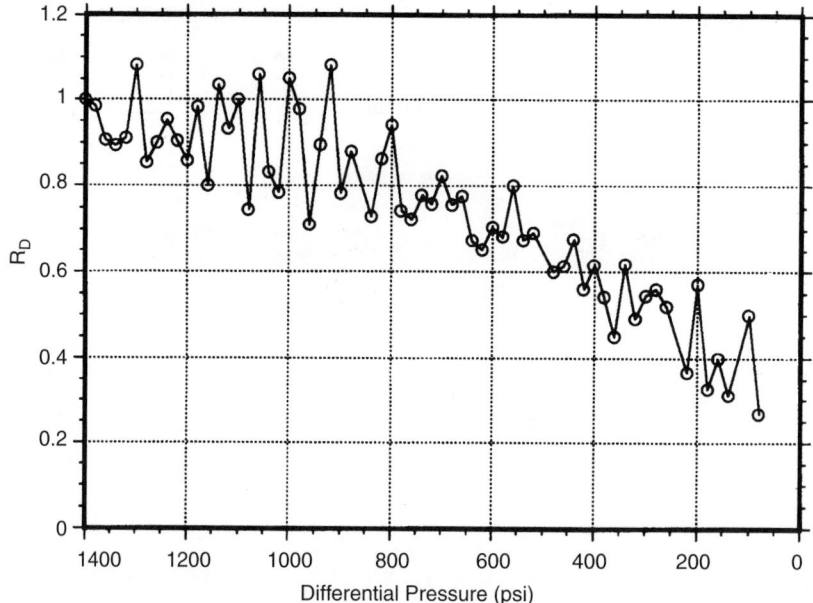

Figure 6.5.8 Stress-sensitive dimensionless resistance [9].

used. They can be thickened via an associative mechanism using an aluminum-phosphate ester polymer chain. However, oil-based fluids are expensive to use and dangerous to handle due to their flammability. Hence, they are applied to formations that are perceived to be particularly water sensitive. Multiphase fluids such as emulsions (oil and water) and foams (gas and water) have been used widely.

The most common emulsion is composed of 67% hydrocarbon (internal) phase and 33% water (external) viscosified phase. Emulsions are very viscous fluids, providing good proppant transport but resulting in high friction pressures and high fluid costs. They also thin significantly, and thus, they cannot be used in hot wells.

Foams (a mixture of liquid and gas stabilized with a surfactant) are particularly popular as fracturing fluids because the contained gas provides a very rapid cleanup following the treatment. The gas phase of foam is generally pumped at a concentration ranging from 52% up to 95%. Stable dispersions of gas in liquid can be prepared with qualities less than 52%, but they can be used effectively as energized fluids even though they do not meet the true definition of a foam. Viscosifying the liquid phase with a polymer is an effective method for increasing the stability and viscosity of foams. Both nitrogen and CO_2 are commonly used as the gas phase. Nitrogen is pumped as a gas and is readily available throughout the oilfield. Nitrogen is considered completely nondamaging, as it is an inert gas and is not very soluble in liquids. CO_2 is pumped as a liquid and therefore creates a denser foam that aids in lowering surface treating pressures because of the increased hydrostatic head in the wellbore. CO_2 is also much more soluble in water or oil, requiring more gas to saturate the liquid and create the foam. The solubility of the CO_2 can aid in cleanup although the CO_2 saturated fluid will naturally have a lower surface tension. Economically, nitrogen will show benefits in shallow wells, whereas CO_2 will become the gas of choice at deeper depths.

For all fracturing fluids, the viscous properties of the fluid must be reduced or "broken" after the treatment.

Unfortunately, the polymer chains concentrate throughout the treatment as the base liquid leaks off to the formation. The concentrated polymer is very difficult to completely break down, even in the presence of breakers. Different combinations of crosslinkers and polymers can be more resistant than others, leading to only partial decomposition, which can result in significant residue and therefore damage to the proppant pack. This phenomenon can lead to substantial proppant pack permeability damage with devastating effects on the fractured well performance.

For polymer fluids, breakers such as oxidative compounds (e.g., peroxydisulfates) or enzymes (e.g., hemicellulase) are used to reduce the length of the polymer chains and their molecular weight. These same breakers will also be added to emulsions and foams to degrade the polymer stabilizers in the water phase. VES fluids are broken by simply destabilizing the rodlike micelle structure. This may be accomplished by simply producing well fluids, either hydrocarbons or formation water, and mixing these fluids with the treating fluid. External breakers may also be added to the VES fluids to accelerate the viscosity degradation and facilitate cleanup immediately after the treatment.

The effects of the type of breaker, its quantity, and the mode of application are shown in Figure 6.5.9. Encapsulated breakers are desirable because they become active only when the fracturing treatment is over. Early breaker polymer reaction is detrimental because it degrades the needed viscous properties of the fluid, whereas minimizing or eliminating the breaker is particularly problematic because it may lead to permanent proppant-pack permeability impairment. Encapsulated breakers allow very high breaker concentrations to be run that can minimize fluid damage, while preventing premature viscosity degradation [13].

Foams, emulsions, and VES fluids are generally considered to be less damaging to the proppant pack than conventional polymer-based fluids. Foams and emulsions use significantly less polymer due to the addition of the

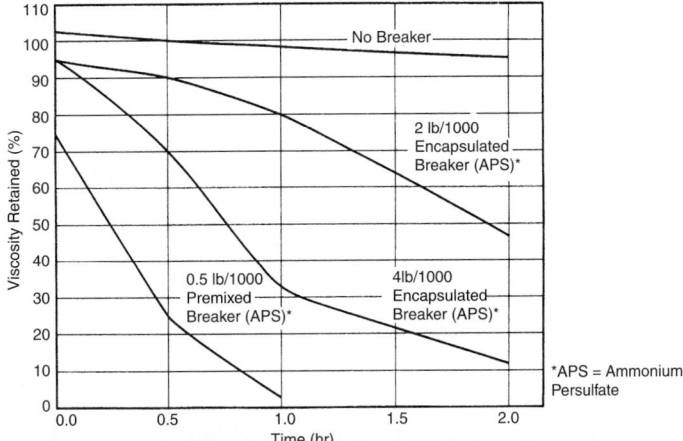

Figure 6.5.9 *Viscosity degradation of fracturing fluids with and without encapsulated breakers [13].*

gas or hydrocarbon phase. With less polymer present, the potential for damage is minimized. VES fluids contain no solids, so once the viscosity is reduced, there will virtually no permeability reduction to the proppant pack.

The selection of what type of fluid is based on the evaluation of numerous fluid performance parameters, including viscosity and stability at temperature, fluid efficiency or the ability to control fluid loss to the formation, minimal friction pressure losses during dumping, fluid damage to the proppant pack, compatibility with the formation rock and the formation fluids, environmental friendliness, and cost. Each of these performance parameters must be weighed against the others in order to select the optimum fracturing for any given reservoir. Each of these parameters can be economically evaluated against the others during the design process using net present value simulations as described in a subsequent section.

6.5.1.9 Proppants

During the execution of the fracture treatment, the imposed hydraulic pressure holds the fracture open. However, when the pumping stops, it is up to the injected particulates to hold open or prop the fracture. The propped fracture width and thus the amount of proppant required will be addressed in the design subsection. However, two other variables are important in the determination of the proppant pack permeability: the proppant strength and the grain size. For a given stress under which the proppant pack will be subjected, the maximum value of fracture permeability can be estimated. Bauxite, a high strength proppant, and ISP (Intermediate Strength Proppant, a synthetic material) maintain a large portion of their permeability at high stresses. Sand, however, experiences more than a magnitude permeability reduction when the stress increases from 4000 to 8000 psi. Resin coatings can be applied to sand to increase the crush resistance and therefore the associated permeability. Understanding proppant permeability at a given stress is important in the selection of a proppant because, although sands are less costly, they *crush* readily, and therefore higher-strength, but more costly, proppants are more suitable at higher stresses. At lower stresses the permeability provided by sand may be sufficient.

Proppant size is also important. Larger grain sizes result in larger fracture permeability. However, larger sizes are more susceptible to crushing as stresses increases, and the relative reduction in the pack permeability is much larger in the larger-size proppants. Reference [14] contains a number of correlations for size and size distribution effects on proppant pack permeability.

These permeabilities are maximum values. As mentioned earlier, fracture permeability damage is caused by unbroken polymer residue, which is by far the biggest culprit. Thus, although proppant strength and size selection can be done using formation strength criteria, damage due to fracturing fluid residue must be controlled. Otherwise, additional damage factors, as high as 80% to 90%, can be experienced after the stress-induced permeability impairment is accounted for.

It is also important to ensure that the proppant remains trapped inside the fracture during clean up and production. This can be a major problem in high flow-rate wells and where fluid drag forces dislodge and carry proppant out of the fracture. This can also be exacerbated in wide fractures (6 or more proppant grains) where a stable bridging arch is difficult to maintain regardless of closure stress. In most cases, proppant flowback does not reduce well production, but the proppant that does flowback can have a detrimental wear effect on the production equipment and may require the use of separators in the production line.

Several techniques have been used to control proppant flowback: forced closure, resin flush, the use of curable-resin–coated proppants, and fiber technology. *Forced closure* is a procedure in which fluid flowback begins immediately at the end of pumping. The theorized benefits of forced closure are that a "reverse" screenout takes place at the perforations (i.e., the fracture width closes to below that required for a stable arch) and that the fracture closes before the proppant has a chance to settle in the fracture.

The *resin flush technique* involves pumping a curable resin into the fracture at the end of the job. The resin coats the proppant in the fracture near the wellbore and forms a bond that glues the individual proppant grains together while still maintaining most of the permeability.

A curable resin coating may also be applied to sand or other types of proppants to prevent the flowback of proppants near the wellbore. The curable-resin–coated proppants are mixed and pumped in the later stages of the treatment, and the well is shut in for a period of time to allow the resin to bind the proppant particles together. Under sufficient closure stress, shut-in time, and temperature, the resin-coated proppant cures into a consolidated, but permeable, proppant pack that resists flowback.

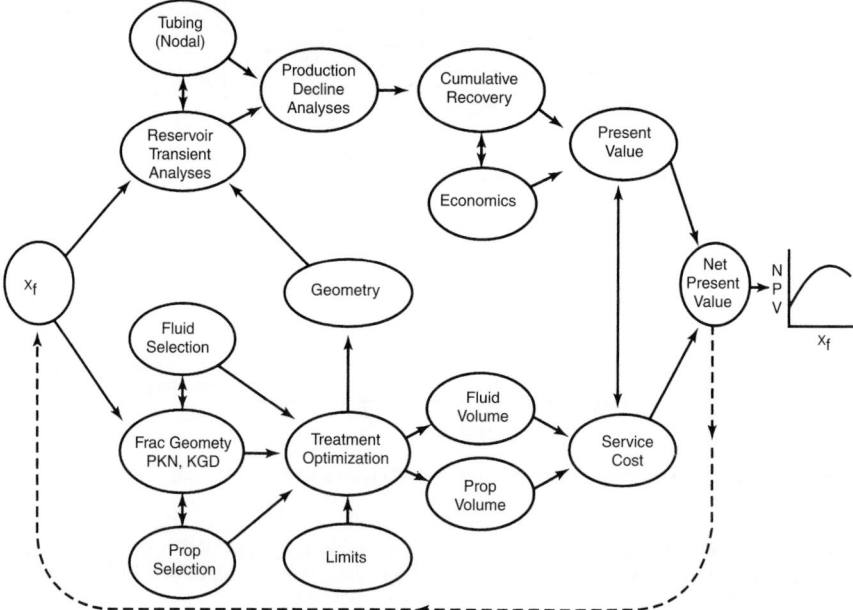

Figure 6.5.10 *The components of the fracture net present value (NPV) calculation [15].*

Fiber technology holds the proppant in the fracture during production without the use of chemical curing reactions. Providing a physical mechanism of random fiber reinforcement prevents proppant flowback while allowing more flexibility in the flowback design.

6.5.1.10 Propped Fracture Design

The previous information was intended to serve as background information for propped fracture design. Meng and Brown [15] and Balen *et al.* [16] presented the concept and applications of the net present value (NPV) as a systematic approach to fracture design. Others have also outlined similar schemes. The complexity of the various design components and their interrelationships invariably require an economic criterion for meaningful comparisons of design options and fracture sizes.

Figure 6.5.10 contains the steps and components for optimizing fracture design. First, a fracture half-length x_f is selected. This is done incrementally, with each new fracture half-length longer than the previous (e.g., by 100 ft).

At first, let's follow the lower branch on Figure 6.5.10. For a given formation, the lithology, temperature, and reservoir fluids would dictate the choice of the fracturing fluid while the state of stress and the desired fractured performance would point toward the proppant selection. A fracture propagation simulator may then describe the fracture geometry. There are several types of simulators including fully three-dimensional (3D), planar 3D (PL3D), pseudo 3D (P3D) (coupled 3D fracture and two dimension [2D] fluid flow), and the classic analytical 2D models. The latter include the PKN model (Perkins and Kern [17]; Nordgren [18]) and the KGD model (Khristianovich and Zheltov [19]; Geertsma and de Klerk [20]).

The higher the complexity of the simulation, the higher the demand for appropriate data and the longer the simulation time. For the purpose of this exercise, the elegant, analytical PKN and KGD models will be used. A depiction of

the PKN geometry is given in Figure 6.5.11. For this simulation two limits are necessary: the fracture height, h_f, and the maximum allowable or available treating pressure.

Thus the fracture geometry can be determined and, with the imposed limits, a treatment optimization may be done. The average width of a fracture of half-length x_f can be obtained from

$$\overline{w} = 0.3 \left[\frac{q_i \mu (1-\nu) x_f}{G} \right]^{1/4} \left(\frac{\pi}{4} \gamma \right) \qquad [6.5.15]$$

where q_i = injection rate in bpm
μ = fracturing fluid viscosity in cp
G = elastic shear modulus in psi
γ = geometric factor = 0.75

The bracketed expression on the right hand side of Equation 6.5.16 gives the maximum width value (at the wellbore), whereas the multiplier in the parentheses provides a geometrically averaged width.

For the KGD model, depicted in Figure 6.5.12, the analogous expression is

$$\overline{w} = 0.29 \left[\frac{q_i \mu (1-\nu) x_f^2}{G h_f} \right]^{1/4} \left(\frac{\pi}{4} \right) \qquad [6.5.16]$$

In both Equations 6.5.16 and 6.5.17, the relationship between viscosity or injection rate and the fracture width is quarter root, indicating that to double the width a 16-fold increase in either of these two variables is needed.

The calculation of fracture width, the assumed fracture height and half-length, and the fluid leakoff lead toward a simple material balance

$$q_i t_i = \frac{\overline{w} A_f}{2} + \frac{(K_L C_L A_f r_p)\sqrt{t_i}}{2} \qquad [6.5.17]$$

where the lefthand side is equal to the fluid volume injected, the first term on the righthand side is the volume of the

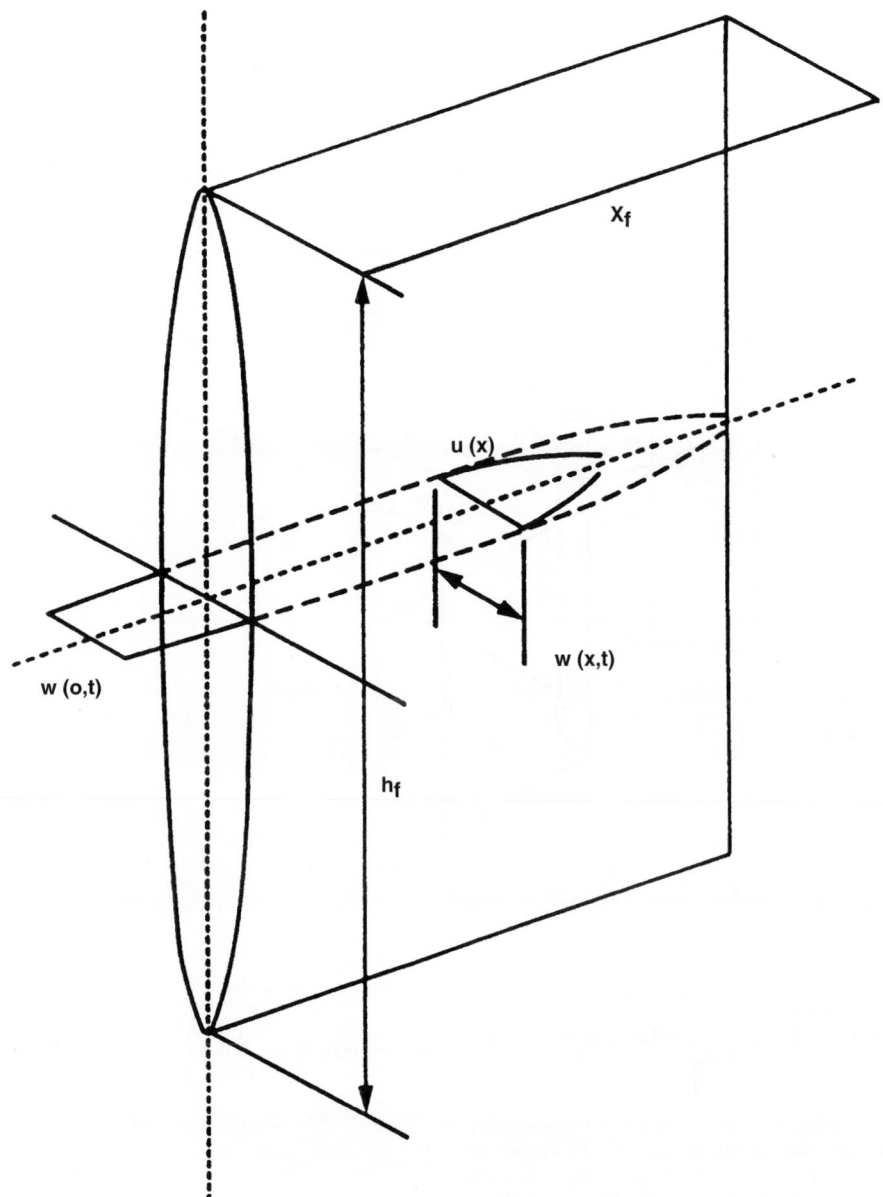

Figure 6.5.11 *The geometry of the Perkins and Kern [17] and Nordgren [18] model (PKN).*

fracture created, and the second term represents the fluid leakoff. The latter is a square root of time relationship.

The area A_f is total fracture area and is equal to $4x_fh_f$. All other variables in Equation 6.5.17 have been described earlier except K_L, which is a multiplier to the leakoff coefficient and is applicable during pumping. Nolte [5] has shown that

$$K_L = \frac{8}{3}\eta + \pi(1 - \eta)$$
[6.5.18]

where η = fluid efficiency

is a good approximation.

Equation 6.5.17 is a quadratic relationship for the square root of the injection time t_i, and can be solved readily. Thus,

the product q_it_i can be calculated, representing the total amount of fluid V_i required to generate a fracture of the calculated geometric and leakoff features.

The pad volume has been related to the total volume injected, V_i, in Reference [6], such that

$$\text{Pad volume} = V_i \frac{(1 - \eta)}{(1 - \eta)}$$
[6.5.19]

The next item is to calculate the proppant volume and its injection schedule. The latter is given by

$$c_p(t) = c_f \left(\frac{t - t_{pad}}{t_i - t_{pad}} \right)$$
[6.5.20]

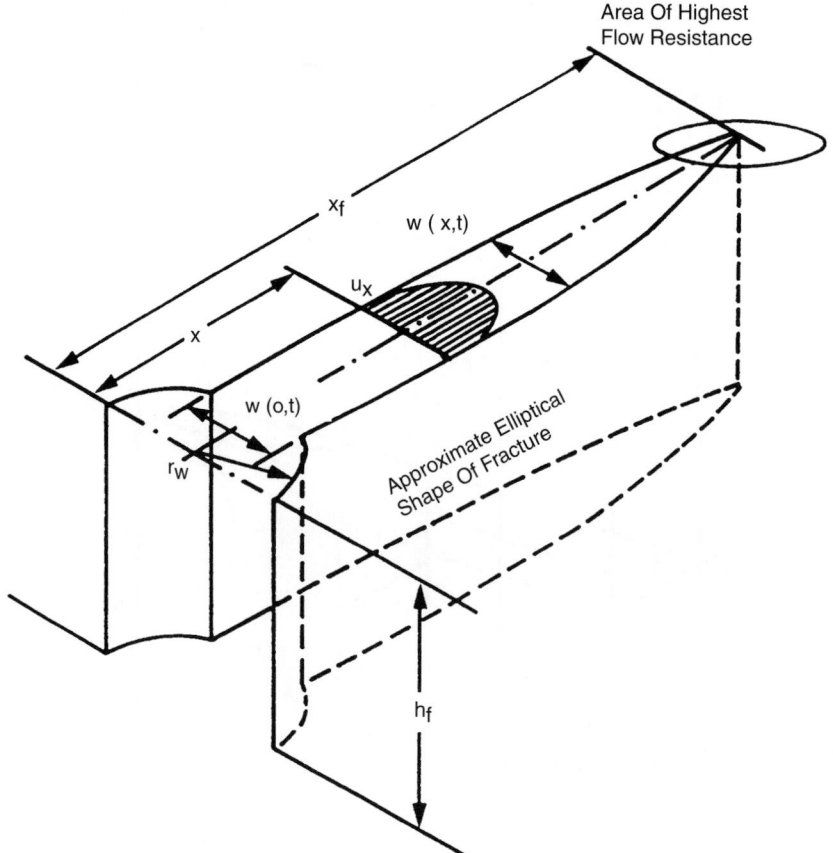

Figure 6.5.12 *The geometry of the Khristianovich and Zheltov [19] and Geertsma and de Klerk [20] model (KGD).*

where $c_p(t)$ = slurry concentration (lb/gal added) as a function of time

c_f = desired slurry concentration at the end of the job (lb/gal absolute)

t_{pad} = time for pad injection

The continuous proppant addition described by Equation 6.5.20 (and shown schematically in Figure 6.5.13, opposed to the classic "stairstep" proppant addition) from t_{pad} to t_i can provide the total amount of proppant injected. This amount, divided by the fracture area, can provide the proppant concentration within the fracture C_p given in lb/ft.^{3} The propped width may then be calculated from

$$w = \frac{12C_p}{(1 - \phi_p)\rho_p}$$ [6.5.21]

where ϕ_p = proppant pack porosity in fraction

ρ_p = proppant density in lb/ft.^{3}

Finally, the net fracturing pressure to generate a half length equal to x_f may be calculated for the PKN model (for a Newtonian fluid)

$$\Delta p_f = 0.0254 \left[\frac{G^3 q_i \mu x_f}{(1 - v)^3 h_f^4} \right]^{1/4}$$ [6.5.22]

and for the KGD model

$$\Delta p_f = 0.050 \left[\frac{G^3 q_i \mu}{(1 - v)^3 h_f x_f^2} \right]^{1/4}$$ [6.5.23]

The treatment pressure is then simply

$$p_f = \sigma_{H,min} + \Delta p_f$$ [6.5.24]

Equations 6.5.22 and 6.5.23 show the classic pressure patterns that may reveal during execution the type of fracture that is generated. For the PKN model and for a constant fracture height (Equation 6.5.22), there is an increasing pressure profile for increasing fracture length, whereas for the KGD model (Equation 6.5.23), there is a decreasing pressure profile for a propagating fracture.

Assuming also that the treating pressure must be kept below a certain level to avoid migration, then a PKN-type fracture can be propagated at constant net pressure. This can be accomplished by decreasing the fluid viscosity or the injection rate as can be concluded from Equation 6.5.22. Given the choice, treatment optimization has shown that reduction in viscosity is more desirable than a reduction in injection rate.

The calculation of the fluid volume (Equation 6.5.16) and proppant mass requirements (Equation 6.5.20), injection rates, and treatment pressures (Equation 6.5.24) lead to the calculation of the cost to create the fracture.

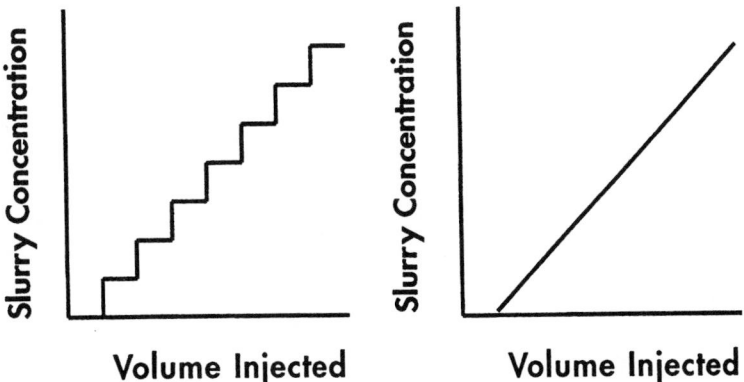

Figure 6.5.13 *Continuous vs. "stairstep" proppant addition. The continuous pressure addition is superior and is described by Equation 6.5.20.*

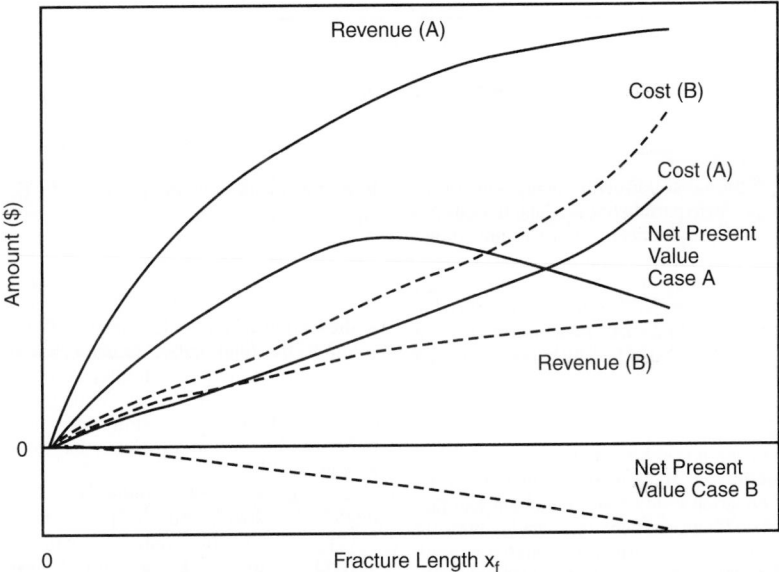

Figure 6.5.14 *Case studies of the NPV design procedure. Case A is positive, Case B is negative.*

The second branch in Figure 6.5.10 may now be addressed. The calculated propped fracture width w (Equation 6.5.21), the fracture half-length x_f, the proppant pack permeability (stress- and damage-adjusted proppant) k_f, and the known reservoir permeability lead to the dimensionless fracture conductivity

$$F_{CD} = \frac{k_f w}{k x_f} \qquad [6.5.25]$$

Cinco-Ley *et al.* [21] presented the solution for the performance of finite-conductivity fractures. Their type curves are in the standard form of a log-log graph of dimensionless pressure p_D against dimensionless time t_{Dxf} for a range F_{CD}s. Hence, any real time corresponds to a dimensionless time, and the dimensionless pressure value is related to the real flowrate and real pressure drop. Thus, the expected performance can be calculated. No-flow boundary conditions for various reservoir geometries have been given in

Reference [15]. The flowrate decline curve can then be coupled with the tubing performance resulting in wellhead rates.

It is very easy step to integrate these rates into the cumulative production. The incremental cumulative production (above the one the unstimulated well could deliver) multiplied by the unit price of the hydrocarbons (oil or gas) and discounted to time zero is the present value of the incremental revenue.

If the cost to perform the job, calculated from the first branch in Figure 6.5.10, is subtracted from the present value of the incremental revenue, it results in the NPV.

A plot of the construction is given in Figure 6.5.14. The fracture half-length is graphed against the NPV. Optimum fracture design corresponds to the maximum NPV. Two case studies are graphed: case A, which provides a positive NPV, and case B, in which the incremental revenue does not recover the stimulation cost. In this case hydraulic fracturing should not be done.

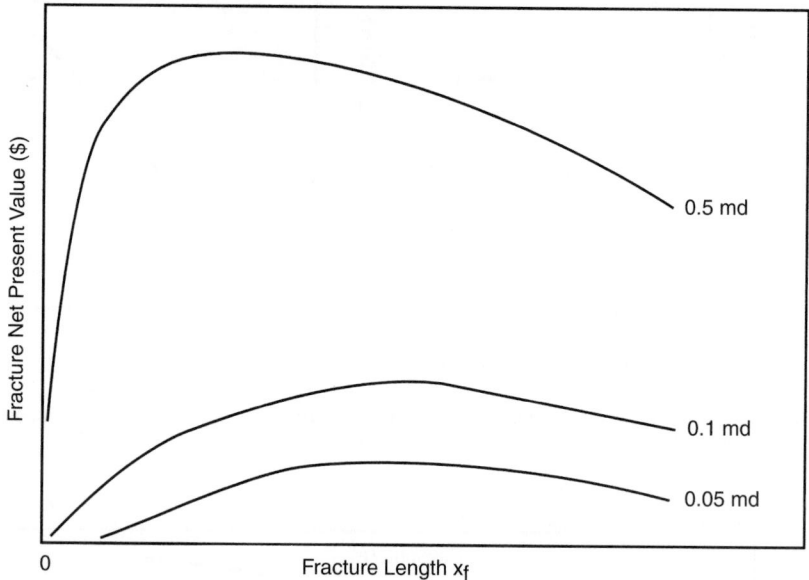

Figure 6.5.15 *NPV fracture design parametric study of reservoir permeability.*

Potentially, one of the most powerful applications of the NPV approach is to perform parametric studies. If a variable (e.g., permeability or slurry concentration) is not known or determined, then a sensitivity analysis can be done to assess the impact of its variation on the optimum design. Large spread in the maximum NPV could be considered as the "cost of ignorance." In that case, measurement of the unknown variable via some type of testing is warranted. An example parametric study with the reservoir permeability is shown in Figure 6.5.15.

6.5.1.11 Fracture Propagation Modeling
Common to all descriptions of fracture propagation are (1) a material balance equation relating volume change to flow across the boundaries, (2) an elasticity relationship between fracture aperture and net pressure, (3) a fluid-flow equation relating the flowrate with the pressure gradient in the fracture, and (4) a tip propagation criterion.

6.5.1.11.1 Width Equations
The theory of linear elasticity provides solutions to idealized problems. One of the them, the pressurized crack problem, deals with a crack (a straight line) of length 2b that is in an infinite plane. The stress σ acting far from the crack and normal to its direction is compressive, trying to close it. On the other hand, a pressure p is acting against the stress, trying to open the crack from the inside. If the net pressure, $\Delta p = p - \sigma$, is positive, the crack will be open and its shape will be elliptic. The maximum width is given by

$$w_{max} = \frac{(1 - v)}{G}(2b)\Delta p \qquad [6.5.26]$$

(The equations, unless otherwise stated, are written in a coherent system of units in this section.) Several models originating from Perkins and Kern [17] figure the hydraulically induced fracture as a constant height channel obeying Equation 6.5.26 in every vertical cross-section, with 2b replaced by h_f. In such a channel of elliptical shape (with a width significantly less than the height), a Newtonian fluid

having constant flowrate q is driven by the pressure drop

$$\frac{dp}{dx} = -\frac{64\mu q}{\pi w^3 h_f} \qquad [6.5.27]$$

The elasticity relation [6.5.26] and the fluid-flow equation [6.5.27] are combined to establish a relation between width at the wellbore and fracture length. To obtain a closed form solution, the fluid leakoff is neglected at this stage of the model development. In addition, zero width (zero net pressure) is assumed at the tip. With these assumptions the created width profiles are similar, and hence, a constant multiplier can be used to transform the wellbore width into the average width.

Historically, the KGD model [19, 20] preceded the PKN model. It is also based on the pressurized crack solution but applied in the horizontal direction. There are additional differences, that is, the pressure drop equation is written for a channel with rectangular cross section and the existence of a nonwetted (nonpressurized) zone is assumed near the tip.

The two widely used width equations were presented earlier as Equations 6.5.15 and 6.5.16. The PKN equation has been considered as a suitable approximation for long fractures (compared to height), whereas the KGD has been suggested as more appropriate for fractures with significant height compared to length.

A width equation such as Equation 6.5.15 or 6.5.16 leaves only one degree of freedom for the geometry of the fracture. If we give (directly or indirectly) the fracture volume, the shape is determined. The additional information on the fracture volume is provided through the material balance.

6.5.1.11.2 Material Balance
Material balance suggests that the injected fluid either generates fracture volume or leaks off. In describing leakoff, Carter [22] applied two important assumptions. In his formulation the following was presented:

$$\frac{V_L}{A_L} = 2C_L\sqrt{t} + S_p \qquad [6.5.28]$$

where V_L = volume of the fluid leaked off
$\quad\quad A_L$ = area available for leakoff
$\quad\quad C_L$ = leakoff coefficient
$\quad\quad S_P$ = spurt loss coefficient

The first term represents decreasing intensity of the fluid leakoff with time elapsed, and the second term is an additional volume that is lost at the very moment of opening (the spurt loss). In addition, Carter assumed that *from the point of view of the material balance,* the fracture geometry can be well approximated by a constant rectangular cross-section, with the only dimension changing with time being the length. He wrote the material balance for a unit time interval in the form

$$q_i = 4 \int_0^t \frac{C_L}{\sqrt{t - \tau}} h_f \frac{dx_f}{d\tau} d\tau + 2h_f w \frac{dx_f}{d\tau} \qquad [6.5.29]$$

where q_i = opening time

The solution of the above integral equation is

$$x_f(t) = \frac{q_i w}{8\pi h_f C_L^2} \left[\exp(\alpha^2) \mathrm{erf}\, c(\alpha) + \frac{2\alpha}{\sqrt{\pi}} - 1 \right] \qquad [6.5.30]$$

where

$$\alpha = \frac{2C_L \sqrt{\pi t}}{w} \qquad [6.5.31]$$

(Note that x_f is the length of one wing, and q_i is the injection rate for two wings.) The complementary error function erf $c(x)$ is available in the form of tables or computer algorithms.

If we want to apply the above equation, we have to decide how to estimate the constant width in this relation. It is the sum of the average width and the spurt width,

$$w = \overline{w} + 2S_p \qquad [6.5.32]$$

where, strictly speaking, the averaging should be done not only in space but also in time.

The combination of Equation 6.5.30 with one of the width equations results in a system that is completely determined if either the length or the injection time is given. For the solution, simple iterative methods can be applied.

The use of Equation 6.5.30 is somewhat complicated. The material balance would take a simple form if the volume leaked off could be determined. Unfortunately, it depends on the opening time *distribution,* which in turn reflects the history of the fracture growth process. A simple version of the material balance is derived assuming some plausible bounds on the distribution of the opening time. Then the fluid lost during pumping (in the two wings) $2V_L$ is bounded by [5]

$$\pi > \frac{2V_L}{2C_L x_f h_f \sqrt{t}} > \frac{8}{3} \qquad [6.5.33]$$

The upper bound corresponds to low fluid efficiency and the lower bound to high fluid efficiency. At any medium efficiency, we can use a linear interpolation between these two bounds. Therefore, Equation 6.5.30 can be replaced by

$$x_f = \frac{(1 - \eta)q_i t}{\left[\frac{8}{3}\eta + \pi(1 - \eta) \right] 2h_f C_L \sqrt{t}} \qquad [6.5.34]$$

where the fluid efficiency is computed from

$$\eta = \frac{2x_f h_f \overline{w}}{q_i t} \qquad [6.5.35]$$

The combination of Equation 6.5.34 with one of the width equations is another short-cut 2D model. Essentially the same concept was used in the design section (see Equations 6.5.17 and 6.5.18).

6.5.1.11.3 Detailed Models

Clearly, the short-cut 2D models are based on several approximations, some of those being contradictory. For example, the geometric picture behind the Carter equation (and behind the upper and lower bounds) would require a fracture propagating with a constant width. The PKN or KGD width equations, on the other hand, give width changing in time as well.

Nordgren [18] presented a constant-height model in the form of a partial differential equation that contains coherent assumptions on the geometry. Kemp [23] showed the correct tip boundary condition for Nordgren's equation. Interestingly, the numerical solution does not differ much from the one of the PKN models. The main reason is that in both the detailed Nordgren model and the PKN versions, the fracture tip propagation rate is controlled by the linear velocity of the fluid at the tip. In other words, in these models there is no mechanism to hamper the opening of the fracture faces once the fluid arrives there. This latter statement is valid also for the different KGD variations.

Appearance of irregular pressure profiles and posttreatment observation of fracture height growth initiated a departure from the ideal geometry assumptions. This generated higher dimensional models and prompted the introduction of improved calculation procedures. The two most important concepts are the vertical distribution of the (minimum horizontal) stress and the fracture toughness.

Most of the researchers agree that stress distribution is the major factor controlling the height growth of hydraulically induced fractures [24]. Building this concept into a PL3D or P3D model, a more realistic fracture shape can be computed. The fracture is contained in the pay layer if the minimum principal stress is significantly higher in the neighboring layers. On the other hand, if the stress in the neighboring layers is only moderately higher than in the pay layer, then a limited height growth is predicted. The PL3D and P3D models differ in how detailed the computation of the height is and to what degree it is coupled with the fluid flow equation [25]). Although the significance of the vertical distribution of the stress is well understood, the usefulness of this concept is somewhat limited by the fact that the necessary data are often lacking. (In fact, even the value of the minimum horizontal stress in the pay layer might be uncertain within a range of several hundred psi.)

There is less consensus in the usefulness of the concept of fracture toughness. This material property is defined as the critical value of the stress intensity factor necessary to initiate the rupture. The stress intensity factor is a quantity having the dimension pressure (i.e., stress) multiplied by the square root of length. Its value increases with both the net pressure and the size of the fracture. Several investigators have arrived at the conclusion that within the physically realistic range of the fracture toughness, its influence on fracture propagation is not significant.

P3D models are used routinely nowadays for the design of fracturing treatments, in real time during the actual treatment, and for postjob evaluation. There are two broad categories of P3D models: cell-based and lumped:

- *Lumped* models assume that the fracture consists of two half-ellipses of variable half-heights, joined along a horizontal line in the fracture length direction. At each time step, the fracture length tip and top and bottom tips are calculated as part of the solution.
- *Cell-based* models assume that the fracture is divided into a number of PKN cells along the fracture length direction. At each time step, the fracture length and height of each cell is computed as part of the solution.

Regardless of the numerical scheme employed, P3D models are more powerful than the simpler PKN-type models because they allow limited height growth as part of the solution, thereby expanding the range of treatments that can be designed and monitored.

A major drawback of the P3D models is that these solutions are all based on the concept of averaging reservoir properties over the fracture height, thereby limiting the range of treatments that can be designed. PL3D models remove this restriction because they employ a 2D mesh to describe the fracture footprint, that is, a mesh that allows variations in fluid pressure and fracture width along the fracture length and height directions. PL3D models are much more powerful and can model complicated geometric configurations, including runaway height growth situations, pinch points, concave sections on the fracture perimeter, and indirect vertical fracturing. However, they are computationally very expensive compared with P3D models and are currently only used in a limited number of treatment designs that involve more complex fracturing behavior.

There are two classes of PL3D models available:

- models based on a *moving* mesh (usually constructed with triangular elements)
- models based on a *fixed* mesh (usually constructed with rectangular elements)

The moving-mesh models are desirable because they provide good resolution at both early and late times during the injection and consume a relatively small number of elements, making them computationally fairly efficient. However, remeshing is required as the fracture footprint changes its shape, resulting in accumulative interpolation errors. These errors can become significant, especially in situations involving layered reservoirs. The fixed-mesh models suffer from poor resolution at early times and can become computationally expensive at later times once many elements become activated. However, they do not suffer from errors in mass balance, unless the mesh is coarsened at later times to reduce the number of active elements and improve Computer Processing Unit (CPU) times.

There are also a limited number of noncommercial "truly" 3D models available that allow nonplanar fracture growth (i.e., limited twisting and turning of the fracture). However, these models are currently prohibitively expensive to exercise, even for research purposes, and do not generally address transverse or longitudinal shear failure that will naturally arise as soon as nonplanar fracture growth is allowed.

Many fracturing treatments are performed in reservoirs that exhibit highly nonlinear or plastic-like material behavior, such as in the Gulf of Mexico. The current linear elastic models (whether P3D, PL3D, PKN, KGD, or radial) need to be adapted to cope with these plastic deformations. Fracturing treatments performed in such soft formations are expected to generate more fracture width and different pressure responses compared with fractures injected in competent rock. Possible remedies include the implementation of a fracture growth criterion in current models that are based on continuum damage mechanics theory [27] or the theory of plasticity [28].

6.5.1.11.4 Evaluation of Fracture Design

Successful stimulation is when the optimum design treatment is performed and the posttreatment flowrate coincides with the one forecasted. Figure 6.5.16 shows a posttreatment well performance showing a good agreement with the predicted flowrate from the designed fracture length. If the two deviate and especially if posttreatment performance is far below expectations, then an evaluation procedure should be implemented.

Primarily, two items should be examined:

- Fracture height migration—this can be done via a posttreatment temperature or radioactive log.
- Fracture permeability reduction—this could be the result of proppant pack damage or a choke (overdisplacement

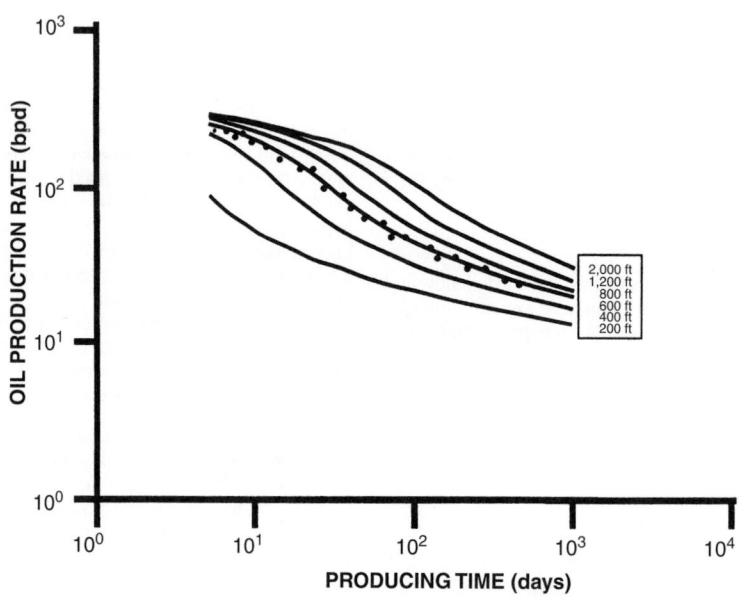

Figure 6.5.16 *Posttreatment fractured well performance and comparison with predicted flowrate.*

or other reasons that reduce the contact between well and fracture). Assessment of the geometric and conductivity characteristics of the fracture can be done via a postttreatment pressure transient test using the model outlined in Reference [21].

6.5.1.12 Acid Fracturing

Acid fracturing is a stimulation process in which acid, usually hydrochloric acid (HCl), is injected into a limestone ($CaCO_3$) or dolomite ($CaMgCO_3$) formation at a pressure sufficient to fracture the formation. As the acid flows along the fracture, portions of the fracture face are dissolved. The flowing acid tends to etch in a nonuniform manner, creating conductive channels that will remain after the fracture closes and ultimately will provide the necessary fracture conductivity. The length of the etched fracture is determined by the acid type, strength, volume, leakoff parameters, reaction rate, and spending rate.

There are certain comparative advantages and disadvantages of acid fractures (versus propped fractures) in carbonate reservoirs.

- **Acid fractures have placement advantages in highly fissured reservoirs** in which proppant placement is difficult and often leads to screenouts due to the high leakoff of fracturing fluid to the associated fissures.
- In very low permeability formations in which long fracture lengths are required, propped fractures will be preferred. In acid fracturing, the acid-based fracturing fluid is continually spending and therefore is consumed as the fracture penetrates deeper into the reservoir. This results in fracture half-lengths that may be limited to a few hundred feet, especially as the temperature of the reservoir increases [30].
- Operationally, acid fracturing has several advantages over propped fracturing, including 1) the addition of proppant is eliminated, thereby greatly simplifying the pumping execution process; 2) problems related to screenouts are eliminated; 3) proppant cleanout from the wellbore after the treatment is not necessary; and 4) no proppant production problems are encountered when production begins.

The maximum fracture conductivity $k_f w$ for an acid fracture is given by

$$k_f w (md/in.) = 9.39 \times 10^{13} (w^3/12) \qquad [6.5.36]$$

However, this value is affected by two variables: the effective stress (σ' and the rock embedment strength S_{rock}. Nierode and Kruk [31] proposed fracture conductivity correlations based on experimental data.

$$k_f w = C_1 e - c_2 \sigma' \qquad [6.5.37]$$

where

$$C_1 = 1.77 \times 10^8 w^{2.47} \qquad [6.5.38]$$

and

$$C_2 = (13.9 - 1.3 \, \mathbf{In} \, S_{rock}) \times 10^{-3} \qquad [6.5.39]$$

for

$$S_{rock} < 20,000 \, psi \qquad [6.5.40]$$

and

$$C_2 = (3.8 - 0.28 \, \mathbf{In} \, S_{rock}) \times 10^{-3} \qquad [6.5.41]$$

for

$$S_{rock} > 20,000 \, psi \qquad [6.5.42]$$

Ben-Naceur and Economides [32] have presented acid fracture performance type curves for a range of effective stresses (3,000 to 9,000) and rock embedment strengths (30,000 to 200,000).

6.5.1.13 Acid Systems and Placement Techniques

The effective length of an acid fracture is limited by the distance live acid travels along the fracture before becoming spent or lost to the matrix to fluid loss. Controlling fluid loss is a difficult challenge because the acid is continuously reacting with the fracture face and increasing the fracture face permeability. Flow channels, called wormholes, will develop, which will greatly increase the effective area where leakoff can take place.

Controlling leakoff becomes a challenge due to the acid continually reacting with the fracture face and removing any filtercake that is laid down, plus the formation of wormholes. To overcome this problem, many acid fracture treatments are design to pump alternating stages of nonreactive viscous pad fluid and acid. The initial viscous pad promotes viscous fingering of the acid that follows. The selective acid flow increases penetration and tends to create etched flow channels that provide good conductivity. Subsequent stages of the viscous fluid will reestablish any filtercake removed by the acid and also flow into and plug off the excessive leakoff associated with wormholes. Crosslinked acid systems have also shown to be very effective at controlling excessive leakoff. These systems are designed to crosslink upon acid spending, which limits the creation of wormholes [33].

Acid and oil emulsion fluids have been designed for use in high temperature wells to reduce the acid reaction rate, which aids in getting live acid to the tip of the fracture. These emulsions are generally oil–external-phase and provide the retarding properties because the oil phase physically separates the acid from the reactive carbonate surface [34].

6.5.1.14 Fracturing of Deviated and Horizontal Wells

Horizontal wells are rapidly emerging as a major type of well completions. There are three meaningful comparisons of performance between a vertical and a horizontal well in the same reservoir:

- Open-hole or fully perforated completions (i.e., no well configuration skin effects) in the vertical and horizontal cases.
- A vertical well with a vertical hydraulic fracture and a fully completed horizontal well.
- A vertical well with a vertical hydraulic fracture and a horizontal well with one or more vertical hydraulic fractures.

Only the second and third comparisons are of interest to the fracturing engineer. For a reservoir with permeability isotropy in the horizontal plane ($k_x = k_y$), but accounting for permeability anisotropy in the vertical plane, the relationship for equal productivity indexes is [35]:

$$r'_{wD} x_f = \frac{2 r_{ev}/L}{\left(a + \sqrt{a^2 - (L/2)^2}\right) \left(\frac{\beta h}{(\beta+1) r_w}\right)^{\beta h/L}} \qquad [6.5.43]$$

where P is the vertical-to-horizontal permeability anisotropy ratio and is defined as

$$\beta = \sqrt{k_H/k_v} = \sqrt{(k_x k_y)^{0.5}/k_v} \qquad [6.5.44]$$

Also, the large axis of the drainage ellipse a is given by

$$a = \frac{L}{2} \left\{ 0.5 + \left[0.25 + \left(\frac{r_{eH}}{L/2} \right)^4 \right]^{0.5} \right\}^{0.5} \qquad [6.5.45]$$

Equation 6.5.43 is for a fully producing along its length, undamaged, horizontal well. It is also valid for reservoirs in which the horizontal permeability is the same in all directions. However, in highly anisotropic formation, while the direction of the *hydraulic* fracture is likely to be normal to

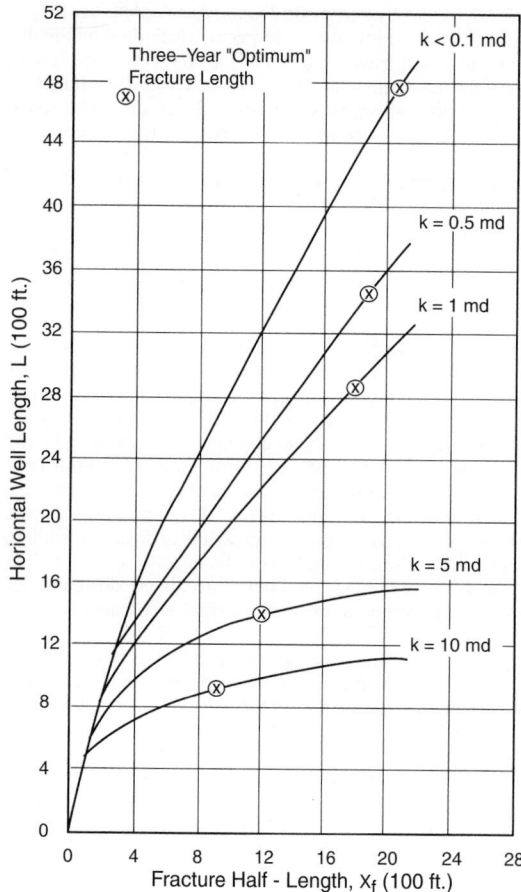

Figure 6.5.17 *Fracture half-length in vertical well and required horizontal well length for equal productivity index (βh = 150 ft.).*

the smallest permeability [36], the horizontal well can be drilled normal to the maximum permeability. In such a case, the situation will be considerably different from the one for a horizontally isotropic reservoir.

Equation 6.5.43 can be used for a comparison of the performance of an unfractured horizontal well of length L with a fractured vertical well with a fracture half-length x_f.

Figure 6.5.17 is such a comparison for five permeabilities and βh = 150 ft. The optimum fracture lengths are also marked. This graph shows the required horizontal well lengths to deliver the same productivity index. Clearly, although detailed economic calculations on the costs versus benefits of all options may be undertaken, it is evident that in reservoirs where vertical wells are usually fracture stimulated (k < 1 md), horizontal wells without fracture stimulation are not viable alternatives. The required horizontal well lengths would be far more expensive than the cost of a vertical well with a hydraulic fracture at the indicated optimum fracture half-length. For higher permeability reservoirs, the required horizontal well lengths are more attractive. However, in all cases, these horizontal wells must be fully stimulated to remove damage and allow flow contributions from the entire well length. Production logging often indicates that less than one-half of the wellbore length contributes to production. A well with this type of limited producing interval would require the horizontal well lengths to be more than doubled from those reported above.

However, reservoirs with βh ≤ 25 ft are normally naturally fractured. Although the theoretically optimum fracture half-lengths may require certain equivalent horizontal well lengths, actual fracture executions may not deliver the desired fracture lengths. Screenouts because of excessive leakoff or the opening of fissures normal to the hydraulic fracture trajectory may prevent the creation of these lengths. Thus, horizontal wells may be good alternatives to vertical wells with propped fractures in cases in which fracture length will be limited.

Finally, *horizontal* permeability anisotropy is exceptionally important. In moderately to highly anisotropic formations, horizontal wells can be good substitutes for fractured vertical wells.

Of particular importance is the third comparison, which allows the possibility to drill the horizontal well either in the direction that would result in transverse fractures or in the direction that would result in largely longitudinal (parallel) fractures. A comprehensive review of the issue has been presented by McLennan *et al.* [37].

The drilling of a highly deviated or horizontal well (that does not coincide with a principal stress axis) results in a nonvanishing shear stress component. This was addressed in an earlier subsection, and it frequently implies additional pressure requirements for fracture initiation. The fracturing pressure of horizontal wells is affected by the stress concentrations near the well given by [38].

$$
\begin{matrix}
\sigma_{xx} \\
\sigma_{yy} \\
\sigma_{zz} \\
\tau_{yz} \\
\tau_{xz} \\
\tau_{xy}
\end{matrix} =
\begin{vmatrix}
1 & 0 & 0 \\
0 & \sin^2 a & \sin^2 a \\
0 & \cos^2 a & \cos^2 a \\
0 & -\sin a \cos a & -\sin a \cos a \\
0 & 0 & 0 \\
0 & 0 & 0
\end{vmatrix}
\begin{matrix}
\sigma_v \\
\sigma_{H,min} \\
\sigma_{H,min}
\end{matrix} \quad [6.5.46]
$$

and at the borehole wall by

$$\sigma_{rr} = p_w \quad [6.5.47]$$

$$\sigma_{\theta\theta} = (\sigma_{xx} + \sigma_{yy} - p_w) - 2(\sigma_{xx} - \sigma_{yy})\cos(2\theta) - 4\tau_{xy}\sin(2\theta) \quad [6.5.48]$$

$$\sigma_{zz} = \sigma_{zz} - 2\nu(\sigma_{zz} - \sigma_{yy})\cos(2\theta) - 4\nu\tau_{xy}\sin(2\theta) \quad [6.5.49]$$

$$\tau_{r\theta} = \tau_{rz} = 0$$

$$\tau_{\theta z} = 2(-\tau_{xz}\sin\theta + \tau_{yz}\cos\theta) \quad [6.5.50]$$

Figure 6.5.18 presents the theoretical fracturing pressure of horizontal wells at different deviation angles from the fracture direction in a reservoir. Also, six actual wells with multiple (in certain cases, more than ten) fracturing treatments are indicated. The agreement between predicted and observed values is excellent.

Because in highly deviated wells the direction of fracture initiation is likely to differ from the ultimate direction of propagation, this turning of the fracture direction can have major implications during fracture execution (screenouts as a result of inadequate width "around a bend" and a choked fracture for the production after the treatment).

A most attractive element is the option to drill a horizontal well in the direction of minimum horizontal stress (i.e., transverse fractures will be initiated) or in the direction of maximum horizontal stress (i.e., a longitudinal fracture will be initiated). In Reference [37], guidelines are given to decide on either of these two options.

The base case is a vertical well with a vertical fracture. If FCD > 10, then a horizontal well with transverse orthogonal fractures is indicated. If FCD < 2, a horizontal well with a colinear, longitudinal, fracture is indicated. For FCD values between 2 and 10, a more detailed calculation is needed. This is a gain outlined in Reference [37].

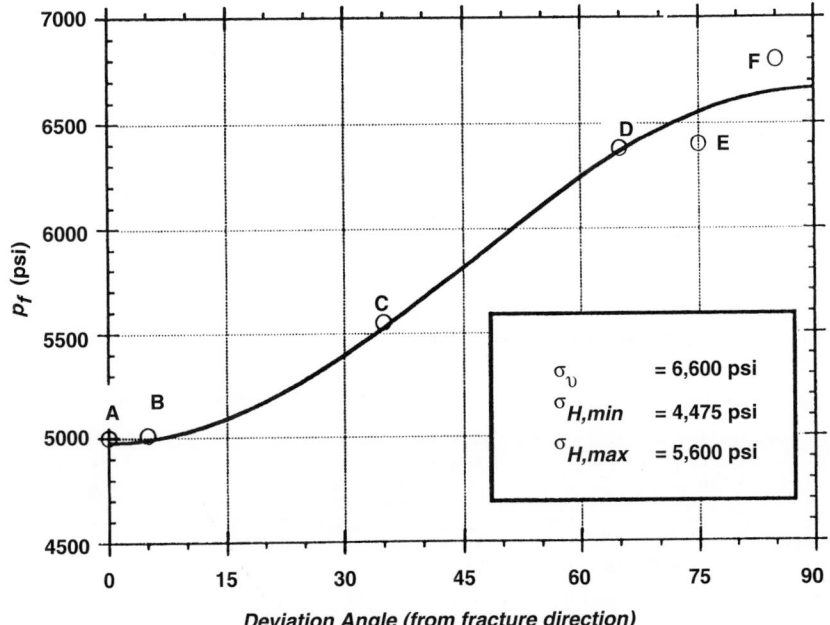

Figure 6.5.18 *Fracturing pressure of six arbitrarily oriented horizontal wells [38].*

In all cases the decision to drill a horizontal well instead of a vertical well must be done on the basis of NPV calculations. The incremental performance, if any, must cover the incremental cost of drilling the horizontal well.

For transverse fractures, each individual fracture is penalized by a skin effect describing the inefficient contact between well and fracture [35]. Then the sum of the individual NPVs is optimized. Each fracture is assessed a share of the drainage area and a portion of the incremental drilling costs.

6.5.2 Matrix Stimulation

Matrix stimulation is the process of injecting a fluid, either an acid or a solvent, into a well at a pressure below the fracturing pressure. For sandstone reservoirs this type of treatment will simply attempt to remove the damage and restore the natural permeability, whereas in carbonates, the damage is not only removed but may also be bypassed by creating new highly conductive channels called wormholes.

Any damage around the near wellbore region may severely reduce the production or injection rate of a well. This damage, which may have a variety of origins, may also be of different type and nature and can be characterized as either natural or induced. Natural damages include fines migration, swelling clays, water-formed scales, and organic deposits, such as paraffins or asphaltenes. Induced damages form after the injection or leakoff of a fluid and include plugging from solids or polymers, wettability changes, water blocks, emulsions, and bacterial growth.

Flow impairment around the wellbore has been described by a positive skin effect as has been introduced in Equation 6.5.1. A common representation of the skin effect due to damage is given by Hawkins' formula:

$$S = \left(\frac{k}{k_S} - 1\right) \ln \frac{r_s}{r_w} \qquad [6.5.51]$$

where k_s = permeability of damage zone in md
r_s = extent of damage zone in ft.

The skin effect, influencing the well performance (and the one obtained from a pressure transient test), is a multicomponent variable including mechanical effects, phase, and rate-dependent effects, along with the near-wellbore damage. Thus, the total skin effect s_t is

$$s_t = s_d + s_{c+\theta} + s_p + \sum ps \qquad [6.5.52]$$

where

S_d = skin due to damage
$s_{c+\theta}$ = skin effect due to partial penetration and slant [39]
s_p = skin effect due to perforations [40]
Σps = rate and phase-dependent pseudoskin factors.

The $s_{c+\theta}$ and s_p can be calculated as outlined in References [39] and [40], respectively, whereas Σps can be usually quantified by variations (and correlation) of the producing pressure and rate. This exercise is necessary for every well in order to isolate the damage-induced skin effect, which is the only one that can be removed via matrix treatments. The other components of the skin effect can be large, and their quantification is necessary in the design and evaluation of matrix stimulation treatments, but they will not be removed or reduced due to the stimulation treatment. The type of fluid used in a treatment often depends on the damage being addressed. Acids can be used when plugging is a problem, but solvents are used for organic deposits. In sandstone reservoirs, knowing the damage mechanism is especially important, because the damage must be removed to regain matrix permeability. In carbonate rocks, damage identification is less critical because of the new flow channels created to bypass the damage.

Because not all damage is acid-removable, an identification of the different types of damage and their treatments are shown in Figure 6.5.19 from Reference [41]. Only damage caused by scales or silts and clays (under headings number 4 and 7) can be removed by acid.

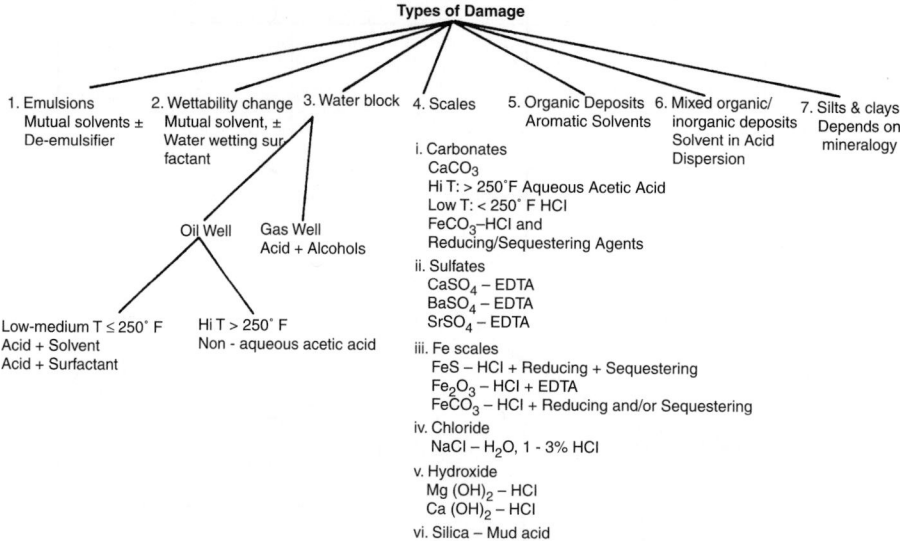

Figure 6.5.19 *Types of damage and suggested matrix treatments.*

Sandstone acidizing and the appropriate fluid selection depends on mineralogy, temperature and nature of damage. HCl or organic acid can effectively remove carbonate, iron, and hydroxide scales while chelating agents such as EDTA will be required to dissolve most sulfates. If the damage can be characterized as relating to silts and clays, a blend of HCl and hydrofluoric acid will be required. This acid blend is commonly referred to as *mud acid* and can be run at varying ratios using different concentrations where the hydrofluoric acid will range from 0.5% to 3% and the HCl will have a concentration of 3% to 13.5%. Figure 6.5.20 from Reference 42 outlines a treatment selection guide for damage removal in sandstone reservoirs.

Fluid selection guidelines for preflush fluids

Mineralogy	Permeability		
	> 100 md	20 to 100 md	< 20 md
< 10% silt and < 10% clay	15% HCl	10% HCl	7.5% HCl
> 10% silt and > 10% clay	10% HCl	7.5% HCl	5% HCl
> 10% silt and < 10% clay	10% HCl	7.5% HCl	5% HCl
< 10% silt and > 10% clay	10% HCl	7.5% HCl	5% HCl

Note: Selection guidelines for all temperatures

Guidelines when clay content contains chlorite/glauconite
    For 4% to 6% chlorite/glauconite, use < 20-md guidelines with 5% acetic acid.
    For > 6% to 8% chlorite/glauconite, do not use HCl; use 10% acetic acid preflush to mud acid plus 5% acetic acid.
    For > 8% chlorite/glauconite, do not use HCl; use 10% acetic acid and organic mud acid.

Guidelines when clay content contains zeolites
    For < 2% zeolite, use 5% acetic acid in all fluids containing HCl and preflush with 10% acetic acid.
    For > 2% to 5% zeolite, *do not* use HCl preflush; use 10% acetic acid preflush and overflush to mud acid containing 10% acetic acid.
    For > 5% zeolite, *do not* use HCl in any system; use 10% acetic acid preflush and overflush to organic acid prepared from 10% citric acid/HF.

Fluid selection guidelines for mud acid fluids

Mineralogy	Permeability		
	> 100 md	20 to 100 md	< 20 md
< 10% silt and < 10% clay	12% HCl–3% HF	8% HCl–2% HF	6% HCl–1.5% HF
> 10% silt and > 10% clay	13.5% HCl–1.5% HF	9% HCl–1% HF	4.5% HCl–0.5% HF
> 10% silt and < 10% clay	12% HCl–2% HF	9% HCl–1.5% HF	6% HCl–1% HF
< 10% silt and > 10% clay	12% HCl–2% HF	9% HCl–1.5% HF	6% HCl–1% HF

Note: Selection guidelines for all temperatures
Guidelines when clay content contains chlorite/glauconite
    For 4% to 6% chlorite/glauconite, use < 20-md guidelines with 5% acetic acid.
    For >6% to 8% chlorite/glauconite, do not use HCl; use 10% acetic acid preflush to mud acid plus 5% acetic acid.
    For >8% chlorite/glauconite, do not use HCl; use 10% acetic acid and organic mud acid.

Guidelines when clay content contains zeolites
    For <2% zeolite, use 5% acetic acid in all fluids containing HCl and preflush with 10% acetic acid.
    For >2% to 5% zeolite, *do not* use HCl preflush; use 10% acetic acid preflush and overflush to mud acid containing 10% acetic acid.
    For >5% zeolite, *do not* use HCl in any system; use 10% acetic acid preflush and overflush to organic acid prepared from 10% citric acid/HF.

6.5.2.1 Matrix Acidizing Design
Reference 43 lists the major steps necessary for a matrix acidizing design:

1. Ensure that the well is damaged and that a high skin effect is not mechanically induced.
2. Establish the nature of damage.
3. Determine the appropriate fluid for the treatment (see Figure 6.5.19 for general fluid selection and Figure 6.5.20 for acidizing treatments).
4. Calculate maximum rate and pressure to avoid unintentional formation fracturing.

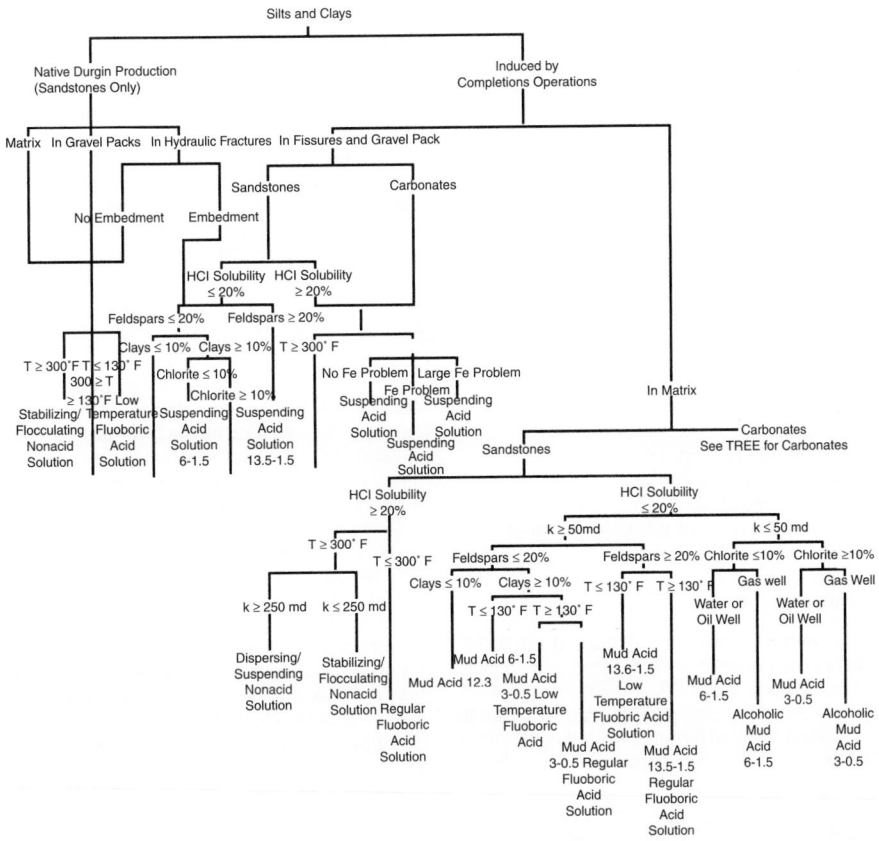

Figure 6.5.20 *Simulated and measured pressure response during a matrix acidizing treatment [42].*

5. Determine volume requirements for treatment.
6. If the formation is multilayered or if it presents a substantial vertical permeability anisotropy, determine a placement technique (diversion).
7. Evaluate treatment for its success (or failure) and consider these findings in future, similar treatments.

Of the above, items 4, 5, 6, and 7 will be outlined.

6.5.2.1.1 Rate and Pressure Limits for Matrix Treatments

A major step in designing a matrix stimulation job is to determine the conditions for the matrix flow regime. To avoid fracturing the formation, the bottomhole injection pressure must not exceed the fracturing pressure. The maximum allowable injection rate that does not fracture the formation is derived from Darcy's radial flow equations. Equation 6.5.53 is a simplified inflow performance relationship that can be used as a guideline for treatment design [44].

$$q_{max} = \frac{4.917 \times 10^{-6} kh(p_{fg} \times d) - \Delta p_s - p_r}{\mu B \left(\ln \frac{r_s}{r_w} + s \right)} \qquad [6.5.53]$$

where q_{max} = maximum injection rate (bbl/min)
k = undamaged permeability (md)
h = net height of the formation (ft)
p_{fg} = fracture gradient (psi/ft)
d = TVD to mid-perforations (ft)
p_s = safety margin (typically 200 to 500 psi) (psi)
p_r = average reservoir pressure (psi)

m = viscosity of injection fluid (cp)
B = formation volume factor (res bbl/STB)
r_s = drainage radius (ft)
r_w = wellbore radius (ft)
s = skin value (dimensionless)

The maximum allowable surface pressure can be determined using Equation 6.5.54.

$$p_s = p_{fg} \times d - p_h + p_f \qquad [6.5.54]$$

where p_s = maximum surface pressure (psi).
p_{fg} = fracture gradient (psi/ft.)
d = TVD to mid-perforations (ft.)
p_h = hydrostatic pressure (psi).
p_f = pipe friction pressure (psi).

Friction pressure can be determined through Reynolds number calculations, by applying values from friction pressure charts, or by measuring at the wellsite with pretreatment injection tests; however, friction is nearly negligible for many matrix treatments due to the low pumping rates of the treatment.

6.5.2.1.2 Fluid Volume Requirements

For sandstones, a matrix acidizing treatment usually consists of a preflush, an acid, and an overflush volume. The preflush provides a separation between conate water and hydrofluoric acid to help prevent the formation of damaging sodium and potassium fluosilicates and reacts with carbonate minerals to prevent their reaction with the mud acid. The reaction of mud acid with carbonate minerals will not only create a calcium fluoride precipitate that is potentially damaging, but it also spends the mud acid, limiting the strength and effectiveness of the acid when contacting the silts and fines that are damaging the well. Depending on the carbonate and clay contents of the formation and its permeability, 5% to 15% HCl or 5% to 10% acetic acid is used. The volume of the preflush needs to be large enough to completely dissolve the carbonate minerals to a depth into the reservoir that ensures the mud acid will not contact these minerals. The volume of required mud acid can be estimated by using a derivation of the Hawkins equation that accounts for both damage and stimulation [45].

$$s = \frac{k}{k_s} \ln \frac{r_s}{r_i} + \frac{k}{k_i} \ln \frac{r_i}{r_w} - \ln \frac{r_s}{r_w} \qquad [6.5.55]$$

where s = skin value (dimensionless)
  k = undamaged permeability (md)
  k_s = damaged permeability (md)
  k_i = stimulated or improved permeability (md)
  r_w = wellbore radius (ft)
  r_s = damage radius (ft)
  r_i = stimulated radius (ft)

Experience has shown that mud acid volumes of 125 to 200 gal/ft. of formation will generally provide acceptable results [36].

The overflush volume, ranging from 50 gal/ft. to one to five times the volume of the acid formulation, is usually either 3% to 10% HCl, NH_4Cl, or a light hydrocarbon, such as diesel. For a gas well, nitrogen may be used.

For carbonate reservoirs, the flow and reaction of acid within the formation is signficantly more complicated. Viscous fingering, but particularly special instabilities called *wormholes*, are generated. The quantification of the wormholing phenomenon as it pertains to the skin effect reduction was presented by Daccord *et al.* [46]:

$$-\frac{1}{d} \ln \left(1 + AcN_{Pe}^{1/3} \frac{bV}{\pi h \phi r_w^d} \right) \qquad [6.5.56]$$

where d = fractal dimension (= 1.6 for mass-transfer–limited kinetics and = 2 for surface-reaction kinetics)
  Ac = acid capacity number [46]
  N_{Pe} = Peclet number
  V = volume of acid injected.

6.5.2.1.3 Diversion In Matrix Acidizing

The placement of treating fluids into all zones of interest, especially those that are damaged, is a significant challenge. Successful matrix treatments depend on the uniform vertical distribution of the treatment fluid over the entire production (or injection) interval. When acid is pumped into a well, it naturally tends to flow into the highest permeability or zone with the least amount of damage (or both) and through the open perforations. Improper placement of acid between regions of different injectivities can be avoided by using diversion techniques.

The earliest diverting techniques relied on mechanical means to restrict flow and included inflatable packers and ball sealers. Chemical diverters became popular not only because they could be pumped concurrently with the matrix stimulation fluid, but because they were also designed to clean up once the well went to production. Chemical diverters such as sized rock salt (water soluble) and benzoic acid

flakes (highly hydrocarbon soluble and slowly water soluble at high temperatures) effectively divert fluid stages by temporarily bridging and plugging the perforation tunnels. Hydrocarbon resins (hydrocarbon soluble) can be made into particulates of various sizes and can effectively provide diversion by creating a filtercake on the formation. Selection of the correct particle size is based on the permeability and therefore the associated pore throat size, of the formation. Resin diverters injected preferably just prior to the acid injection or with the acid itself are deposited on the sandface, creating a diverter cake of a resistance R_{cake} and thus a temporary skin factor s_{cake} [11]:

$$s_{cake} = \frac{2\pi kh}{A} R_{cake} \qquad [6.5.57]$$

where A is the area of flow and R_{cake} is experimentally determined. The amount of diverter deposited on the sandface increases as long as the layer takes a disproportionate amount of fluid. Cake deposition ends when the flowrate is distributed to another zone. Figure 6.5.21 is a depiction of matrix stimulation injection without and with a diverter.

More recently foams and fluids made from VESs have shown to be effective diverting agents. These systems enter into the matrix porosity of the reservoir, where they form a viscous structure that inhibits further flow into the zone. Viscoelastic diverting fluids are increasingly popular because of improved stability, being operationally simple, and not requiring specialized pumping equipment compared with foams. Both of these systems provide the benefit of helping the acid to selectively treat the appropriate zone. They are inherently less stable in hydrocarbon bearing zones (thus allowing acid penetration) and they are very stable in water zones (thus diverting acid away from these zones and minimizing the stimulation effects) [47].

6.5.2.1.4 Matrix Stimulation in Horizontal Wells

To evaluate a proper matrix stimulation treatment in horizontal wells, a description of the damage profile along and normal to the horizontal well trajectory is necessary. Frick and Economides [48] have shown that the distribution of formation damage surrounding a horizontal well is neither radial nor is it evenly distributed along the wellbore.

During drilling and well completion, mud filtrate and completion fluids penetrate the pay zone. Because the exposure time of the formation to drilling and completion fluids is longer at the horizontal section nearer the vertical section, the shape of damage distribution along the wellbore will be a truncated cone, with the larger base near the vertical section of the well. This profile of damage is evident also during production, because the pressure gradient normal to the well nearer the vertical section is (usually) the largest.

Permeability anisotropy, represented by β, generates an elliptical shape of damage distribution normal to the well. The geometrical shape of a truncated elliptical cone results in an expression for the skin effect around a horizontal well [48]:

$$s'_{eq} = \left(\frac{k}{k_s} - 1 \right) \ln \left[\frac{1}{(\beta) + 1} \sqrt{\frac{4}{3} \left(\frac{a_{H,max}^2}{r_w^2} + \frac{a_{H,max}}{r_w} + 1 \right)} \right]$$
$$[6.5.58]$$

where $a_{H,max}$ is the horizontal half-axis of the larger base of the elliptical cone of damage, as shown in Figure 6.5.22 along with cross-sections of damage for various β. This expression is analogous to Hawkins' formulation for vertical wells and can be used in the usual manner, implying a steady-state pressure drop.

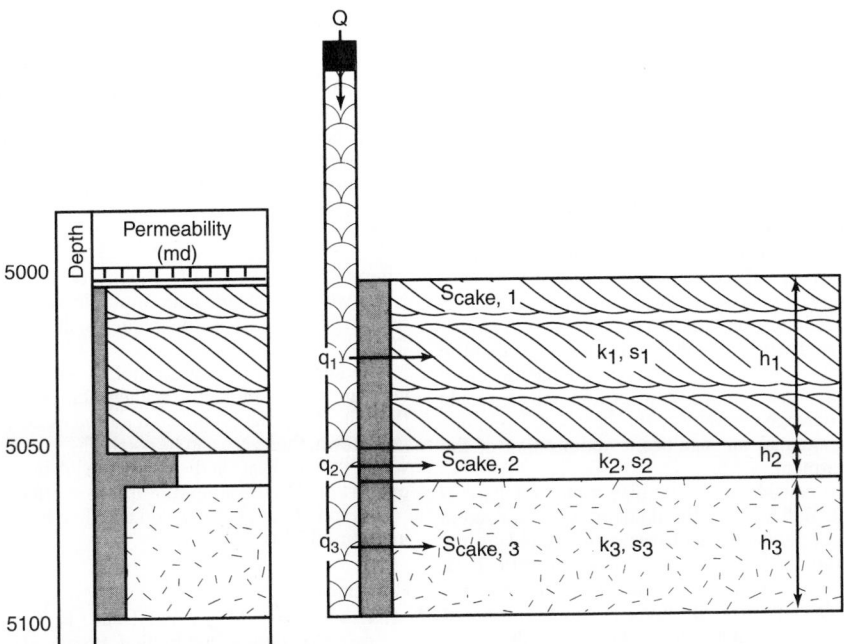

Figure 6.5.21 *Fluid distribution for matrix stimulation injection in a multi-layered reservoir. On the left, without diverters; on the right, with diverters.*

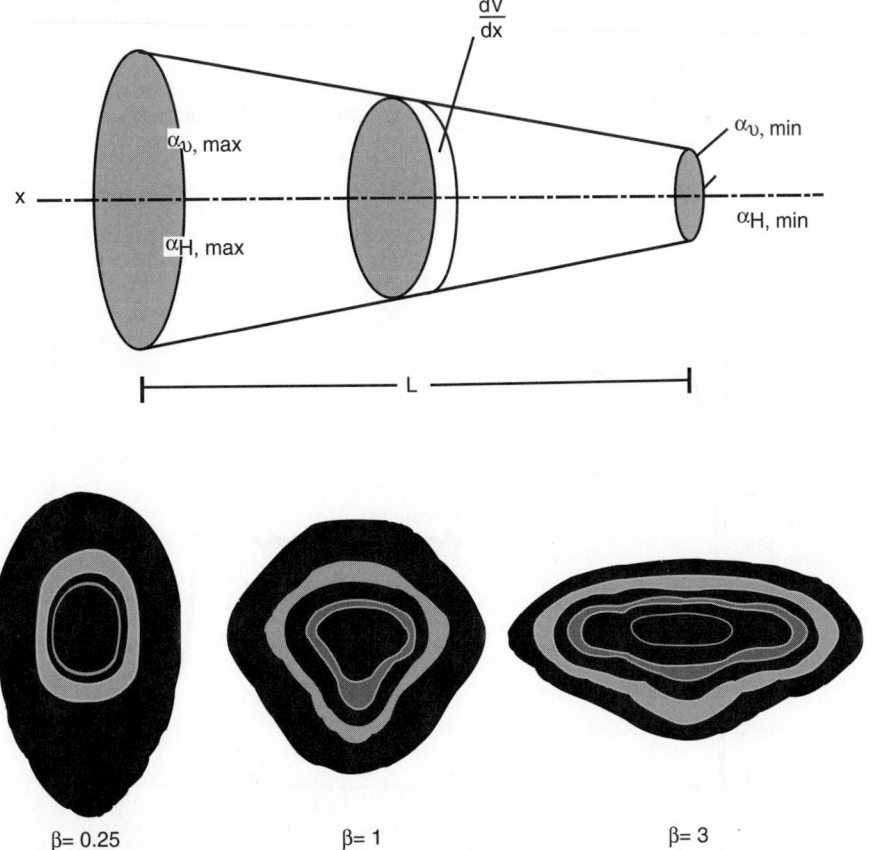

Figure 6.5.22 *Shape of damage along and normal to a horizontal well [48].*

Stimulation Considerations. Horizontal wells have such long exposed intervals that stimulation fluid volumes per unit length used in vertical wells are not practical. While pumping as much as 150 gal/ft. is routine in vertical wells, such coverage implies 300,000 gal of acid in a 2000-ft. horizontal well. Pumping at 1 bpm, these volumes would require 120 hours of pumping.

A further constraint is proper placing of acid (diversion): acid has a tendency to extend existing flow paths. Thus, acid thief zones are either natural or created, and there is therefore a need for substantial optimization of matrix stimulation in terms of technical and economic considerations.

When planning for a partial removal of the damage, the distribution of the stimulation fluid is a crucial issue. A simple calculation of posttreatment skin effect determines that the distribution of the stimulation fluid should mimic the shape of damage. Deliberate blanking of horizontal well segments and complete damage removal in the perforated sections offer another potential for significant optimization of the stimulation treatment.

The previous considerations require a methodology of acid placement. Bull-heading the fluids into the horizontal portion is not recommended. Such treatments stimulate only a minor portion of the horizontal well, usually near the heal. Coiled tubing is a proper tool for the distribution of stimulation fluids. After the coiled tubing is run to the toe of the well, the stimulation fluid can be pumped as the coiled tubing is concurrently withdrawn. The rate of coiled tubing withdrawal can be calculated and depends on the injection rate and volumetric coverage. In addition, new jetting tools placed on the end of the coiled tubing have proven to be very effective at ensuring acid coverage over long intervals [49].

6.5.2.2 Evaluation of Matrix Acidizing Treatments

Paccaloni [50] was the first to use instantaneous pressure and rate values to compute the changing skin effect at any time during the treatment and thus evaluate the progress of the job. He used a steady-state pressure response and defined a "damage ratio" **DR**:

$$\mathbf{DR} = \frac{\ln(r_e/r_w) + s}{\ln(r_e/r_w)} \qquad [6.5.59]$$

The method allows the estimation of both the original (damage) skin effect as well as its evolution.

A new technique, which does not use a steady-state assumption, was introduced by Prouvost and Economides [51]. The method uses reservoir transients during acid injection to simulate pressure response and compares it with measured values. The difference between these values is attributed to the changing skin effect, namely,

$$\Delta s = \frac{kh\Delta p_{departure}}{141.2qB\mu} \qquad [6.5.60]$$

where: $\Delta p_{departure}$ = difference between simulated pressure (with a constant nonremovable skin effect) and the measured value. The technique is illustrated on Figure 6.5.23.

The simulated pressure requires a knowledge of reservoir and well parameters, namely the permeability k, and the skin effect that cannot be removed by acid. If the latter is not known, then zero can be used. The two pressure curves would then be apart at the end of the job by a constant value proportional to the nonremovable skin effect.

A graph of the evolving skin can then be drawn as shown in Figure 6.5.24.

References

1. Economides, M. J., and Nolte, K. G., (Eds.), *Reservoir Stimulation*, 2nd Edition. Prentice-Hall, Englewood Cliffs, 1989.
2. Hubbert, M. K., and Willis, D. G., "Mechanics of Hydraulic Fracturing," *Transactions of AIME, 1957.*
3. Roegiers, J-C., and Detournay, E., "Considerations on Failure Initiation in Inclined Boreholes," Proceedings 29th Symposium on Rock Mechanics, Minnesota University, 1957.
4. Terzaghi, K., "Die Berechnung der Durchlassigkeitsziffer des Tones aus dem Verlauf der Hydrodynamischen Spannungsercheinungen," *Sber. Akad. Wiss., Wein,* 1923.
5. Nolte, K. G., "Determination of Fracture Parameters from Fracturing Pressure Decline," SPE 8341, 1979.
6. Nolte, K. G., " A General Analysis of Fracturing Pressure Decline with Application to Three Models," SPEFE, December 1986.

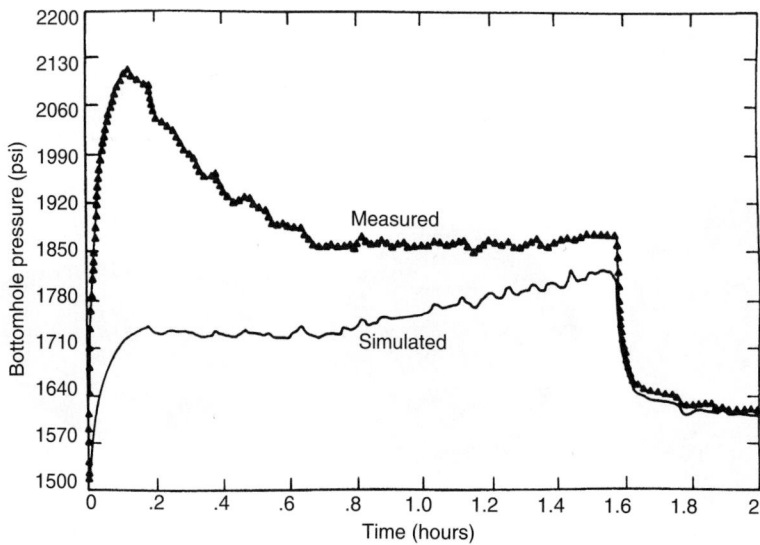

Figure 6.5.23 *Simulated and measured pressure response during a matrix acidizing treatment [51].*

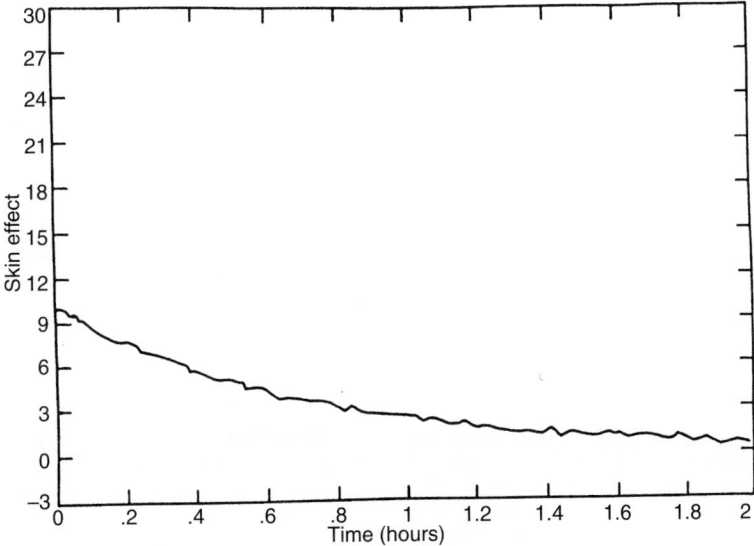

Figure 6.5.24 *Evolution of skin effect [51].*

7. Weng, X., Pandy, V., and Nolte, K. G., "Equilibrium Test: A Method for Closure Pressure Determination," SPE 78713, presented at Mechanic Conference, Irving, Texas, October 20–23, 2002.

8. Castillo, J.L., " Modified Fracture Pressure Decline Analysis Including Pressure-Dependent Leakoff," SPE 16417, 1987.

9. Mayerhofer, M. J., Economides, M. J., and Nolte, K. G., "Experimental Study of Fracturing Fluid Loss," CIM/AOSTRA 91-92 presented at the Annual Technical Conference of the Petroleum Society of CIM and AOSTRA, Banff, 1991.

10. Mayerhofer, M. J., Economides, M. J., and Nolte, K. G., "An Experimental and Fundamental Interpretation of Filtercake Fracturing Fluid Loss," SPE 22873, 1991.

11. Doerler, N., and Prouvost, L. P., "Diverting Agents: Laboratory Study and Modeling of Resultant Zone Injectivities," SPE 16250, 1987.

12. Samuel, M., *et al.*: "Polymer Free Fluid for Fracturing," SPE 38622 presented at the Annual Technical Conference and Exhibition, San Antonio, Texas, October 5, 1997.

13. Gulbis, *et al.* "Encapsulated Breaker for Aqueous Polymeric Fluids," SPE 19433, 1990.

14. Montgomery, C. T., and Steanson, R. E., "Proppant Selection: The Key to Successful Fracture Stimulation," *Journal of Petroleum Technology*, December 1985.

15. Meng, H-Z., and Brown, K. E., "Coupling of Production Forecasting, Fracture Geometry Requirements and Treatment Scheduling in the Optimum Hydraulic Fracture Design," SPE 16435, 1987.

16. Balen, R. M., Meng, H-Z., and Economides, M. J., "Applications of the Net Present Value (NPV) in the Optimization of Hydraulic Fractures," SPE 18541, 1988.

17. Perkins, T. K., and Kern, L. B., "Widths of Hydraulic Fractures," *Journal of Petroleum Technology*, September 1961.

18. Nordgren, R. P., "Propagation of Vertical Hydraulic Fractures," *SPEJ*, August 1972.

19. Khristianovich, S. A., and Zheltov, Y. P., "Formation of Vertical Fractures by means of Highly Viscous Liquid,"

Proceedings 4th World Petroleum Congress, Section II, 1959, pp. 579–586.

20. Geertsma, J., and de Klerk, R., "A Rapid Method of Predicting Width and Extent of Hydraulically Induced Fractures," *Journal of Petroleum Technology*, December 1969.

21. Cinco-Ley, H., Samaniego, F., and Dominquez, N., "Transient Behaviour for a Well with a Finite Conductivity Vertical Fracture," *SPEJ*, August 1978.

22. Carter, R. D., Appendix to "Optimum Fluid Characteristics for Fracture Extension," by G. C. Howard and C. R. Fast, *Drilling and Production Practices*, 1957.

23. Kemp, L. F., "Study of Nordgren's Equation of Hydraulic Fracturing." *SPEJ*, August 1990.

24. Simonson, E. R., Abu-Sayed, A. S., and Clifton, R. J., "Containment of Massive Hydraulic Fractures," *SPEJ*, February 1978.

25. Ben-Naceur, K., "Modeling of Hydraulic Fractures," in *Reservoir Stimulation*, 2nd Edition, M. J. Economides, and K. G. Nolte (Eds.), Prentice-Hall, Englewood Cliffs, 1989.

26. Palmer, I. D., and Veatch, Jr., R. D., "Abnormally High Fracturing Pressures in Step-Rate Tests," SPEPE August 1990; and *Transactions of AIME*.

27. Valko, P., and Economides, M. J., "A Continuum Damage Mechanics Model of Hydraulic Fracturing," *Journal of Petroleum Technology*, March 1993.

28. Van Dam, D. B., Papanastasiou, P., and Pater, C. J., "Impact of Rock Plasticity on Hydraulic Fracture Propagation and Closure," SPE 63712 presented at the 2000 SPE Annual Technical Conference and Exhibition, Dallas, Texas, October 1–4, 2000.

29. Valko, P., and Economides, M. J., "Applications of a Continuum Damage Mechanics Model to Hydraulic Fracturing," SPE 25887, 1993.

30. Nierode, D.E., Williams, B. B., and Bombardieri, "Prediction of Stimulation from Acid Fracturing Treatments," *Journal of Canadian Petroleum Technology*, October–December 1972.

31. Nierode, D. E., and Kruk, K. F., "An Evaluation of Acid Fluid-Loss Additives, Retarded Acids, and Acidizing Fracture Conductivity," SPE 4549, 1973.

32. Ben-Naceur, K., and Economides, M. J., "The Effectiveness of Acid Fractures and Their Production Behaviour," SPE 18536, 1988.

33. Robert, J. A., and Crowe, C. W., "Carbonate Acidizing Design," in *Reservoir Stimulation*, 3rd Edition, M. J. Economides and K. G. Nolte (Eds.), John Wiley & Sons Ltd., West Sussex, England, 2000.

34. Navarrete, R. C., Holms, B. A., McConnell, S. B., and Linton, D. E., "Emulsified Acid Enhances Well Production in High-Temperature Carbonate Formations," SPE 50612, 1998.

35. Mukherjee, H., and Economides, M. J., "A Parametric Comparison of Horizontal and Vertical Well Performance," SPE 18303, 1988.

36. Ben-Naceur, K., and Economides, M. J., "Production from Naturally Fissured Reservoirs Intercepted by a Vertical Hydraulic Fracture," *SPEFE*, December 1989.

37. McLennan, J. D., Roegiers, J-C., and Economides, M. J., "Extended Reach and Horizontal Wells," in *Reservoir Stimulation*, 2nd Edition, M. J. Economides and K. G. Nolte (Eds.), Prentice-Hall, Englewood Cliffs, NJ, 1989.

38. Owens, K. A., Andersen S. A., and Economides, M. J., "Fracturing Pressures for Horizontal Wells," 1992.

39. Cinco-Ley, H., Ramey, Jr., H. J., and Miller, F. G., "Pseudoskin Factors for Partially Penetrating Directionally Drilled Wells," SPE 5589, 1975.

40. Karakas, M., and Tariq, S., "Semi-Analytical Productivity Models for Perforated Completions," SPE 18271, 1988.

41. Piot, B. M., and Lietard, O. M., "Nature of Formation Damage," in *Reservoir Stimulation*, 2nd Edition, M. J. Economides and K. G. Nolte (Eds.,), Prentice-Hall, Englewood Cliffs, NJ, 1989.

42. McLeod, H. O., and Norman, W. D., "Sandstone Acidizing," in *Reservoir Stimulation*, 3rd Edition, M. J. Economides and K. G. Nolte (Eds,), John Wiley & Sons Ltd., West Sussex, England, 2000.

43. Gidley, J. L., Ryan, J. C., and Mayhill, T. D., "Study of the Field Application of Matrix Acidizing," SPE 5693, 1976.

44. Williams, B. B., Gidley, J. L., and Schecter, R. S., (Eds.), *Acidizing Fundamentals*. Society of Petroleum Engineers, Dallas, Texas, 1979.

45. Ayoub, J. A., personal communication, 1990.

46. Daccord, G., Touboul, E., and Lenormand, R., "Carbonate Acidizing: A Quatitative Study of Wormholing Phenomenon," SPE 16887, 1987.

47. Chang, F.F, Acock, A.M., Geoghagan, A., and Huckabee, P. T., "Experience in Acid Diversion in High Permeability Deep Water Formations Using Visco-Elastic-Surfactant," SPE 68919, 2001.

48. Frick, T. P., and Economides, M. J., "Horizontal Well Damage Characterization and Removal," SPE 21795, 1991.

49. Ali, S. A., Irfan, M., Rinaldi, D., Malik, B. Z., Tong, K. K., and Ferdiansyah, E., "Case Study: Using CT-Deployed Scale Removal to Enhance Production in Duri Steam Flood, Indonesia," SPE 74850, 2002.

50. Paccaloni, G., "New Method Proves Value of Stimulation Planning," *The Oil & Gas Journal*, November 1979.

51. Prouvost, L. P., and Economides, M. J., " Applications of Real-Time Matrix Acidizing Evaluation Method," SPE 17156, 1988.

6.6 OIL AND GAS PRODUCTION PROCESSING SYSTEMS

6.6.1 Surface Production/Separation Facility

The purpose of the surface production facility (Figures 6.6.1 and 6.6.2) is:

- To separate the wellstream into its three fundamental components—gases, liquids, and solid impurities
- To remove water from the liquid phase
- To treat crude oil to capture gas vapors
- To condition gas

This section is a discussion of the design, use, function, operation, and maintenance of common facility types. In this section facilities will be referred to as the following:

- Oil production facilities as *batteries*
- Primary gas production facilities as *gas units*
- Water handling/disposal as *brine stations*

Identification of the service of each facility and associated equipment is also required. Is the service natural gas, crude oil, produced water, multiphase? Is it sweet (little to no H_2S) or sour (H_2S present)? H_2S or sour service is ambiguous at best without the following:

- National Association of Corrosion Engineers (NACE) MR-01-75 addresses material selection to avoid sulfide stress cracking. Threshold concentrations for application are in the Scope section of this document. In general, MR-01-75 applies to wet gas steams (where free water is present) when operating pressure exceeds 50 psig and

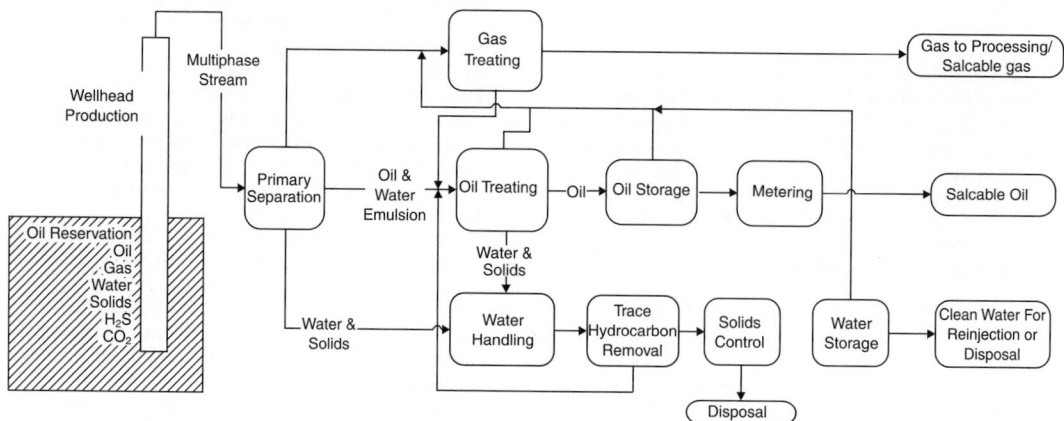

Figure 6.6.1 *Typical oil production process system.*

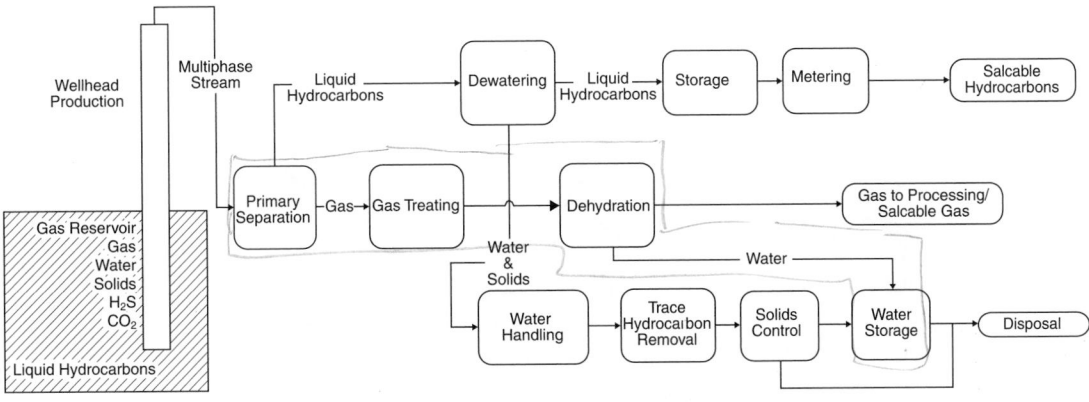

Figure 6.6.2 *Typical gas production process system.*

H_2S is present, or to multiphase systems when pressures exceed 350 psig and H_2S is present. The Minerals Management Service (MMS) cautions that even though materials listed may be resistant to sulfide stress corrosion environments, they may not be suitable for use in chloride stress cracking environments [1].

- Where H_2S concentration exceeds 100 ppm, The Texas Railroad Commission Rule 36 [2] and New Mexico OCD Rule 118 [3] apply. Both these rules address public safety. Other states have similar rules.
- Many oil and gas companies also have recommended standards for various levels of H_2S concentration, so be aware that these may also need to be taken into account.

Before anything is designed, construction codes need to be reviewed as well. Some of the most referenced codes are:

- American National Standards Institute (ANSI) B31.8 "Gas Transmission and Distribution Piping Systems" is used for general oil field construction. ANSI B31.8 dictates that the qualification of welders and the quality of fit-up, taking, welding, and radiographic inspection for noncritical service should conform to API Standard 1104 "Standard for Field Welding of Pipelines".
- ANSI B31.3 "Chemical Plant and Petroleum Piping" applies to all process piping in gas plants and compressor station installation with pressures over 750 psig.
- Title 49 of the Code of Federal Regulations (CFR) covers Department of Transportation (DOT) regulations in parts 191, 192, and 195 [4–6].
- As always, be aware that various other state and federal requirements may apply. Refer to Table 6.9.1 for a partial listing of applicable federal and state regulatory agencies.

Figure 6.6.3 is an example process flow diagram of surface production equipment more commonly referred to as a *battery*.

From a technical point of view, production equipment can be divided into four major groups:

- pressure vessels
- storage tanks
- prime movers (pumps and compressors)
- piping

6.6.1.1 Nomenclature of Separating Pressure Vessels

Pressure vessels in the oilfield are used for separating well fluids produced from oil and gas wells into the gaseous and liquid components. Vessels are separated into two groups by

the American Society of Metallurgical Engineering (ASME), the "governing" body of pressure vessel standards and codes [7]. These are low-pressure, less than or equal to 125 psig, or high-pressure, greater than 125 psig. Some of the more common types of vessels used for separation in oil and gas production are:

- *Separator*—a vessel used to separate a mixed-phase stream into gas and liquid phases that are "relatively" free of each other. Other terms used are scrubbers, knock-outs, line drips, and descanters. The design of these vessels may be either low or high pressure.
- *Scrubber or Gas Knockout*—a vessel designed to handle streams of high gas/liquid ratios. The liquid is generally entrained as mist in the gas or is free, flowing along the pipe wall. These vessels usually have a small liquid collection section. The terms scrubber and knockout are used interchangeably. Typically, these are high-pressure vessels.
- *Three–Phase Separator*—a vessel used to separate gas and two immiscible liquids of different densities (e.g., gas, oil, and water). Typically, these vessels are low pressure but can just as easily be designed for high pressures.
- *Liquid-Liquid Separator or Free Water Knockout*—two immiscible liquid phases can be separated using the same principles as for gas and liquid separators. These vessels are designed to operate at much lower velocities and pressures than the gas-liquid separators; they are designed to take advantage of the physical properties of the liquids, and inasmuch vertical vessels may be more effective than horizontal.
- *Filter Separator*—These separators usually have two compartments. The first contains filter/coalescing elements. As the gas flows through the elements, the liquid particles coalesce into larger droplets, and when the droplets reach sufficient size, the gas flow causes them to flow out of the filter elements into the center core. The particles are then carried into the second compartment of the vessel (containing a vane-type or knitted wire mesh mist extractor) where the larger droplets are removed. A lower barrel or boot may be used for surge or storage of the removed liquid.
- *Line Drip*—Typically used in pipelines with very high gas/liquid ratios to remove only free liquid from the gas stream and not necessarily all the liquid from a gas stream. Line drips provide a place for free liquids to separate and accumulate. Typically, these are high-pressure vessels.

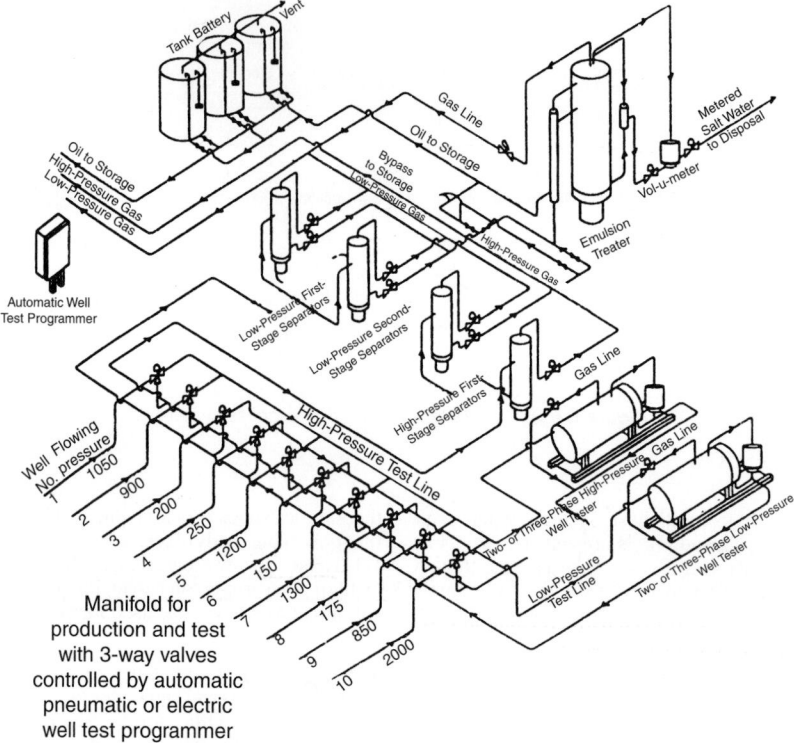

Figure 6.6.3 *Typical surface production equipment.*

- *Slug Catcher*—A separator, which is designed to absorb sustained inflow of large liquid volumes at irregular intervals. Usually found on gas-gathering systems or other two-phase pipeline systems. A slug catcher may be a single large vessel or a manifold system of pipes. Typically, these are high-pressure vessels.
- *Flash Tank (Chamber, Trap, or Vessel)* — A conventional oil and gas separator operated at low pressure, with the liquid from a higher-pressure separator being "flashed" into it. This flash chamber is quite often the second or third stage of separation, with the liquid being discharged from the flash chamber to storage.
- *Expansion Vessel*—A first-stage separator on a low-temperature or cold-separation unit. This vessel may be equipped with a heating core to melt hydrates, or a hydrate-preventative liquid (such as glycol) may be injected into the well fluid just before expansion into this vessel.
- *Desanders*—A high-pressure vessel that is installed on some gas wells to "catch" any sand that is produced along with the gas. This sand may be from a frac job, or it may be from the formation. The pressure drop at the desander allows the majority of this sand to drop out at a controlled location rather than in the lines, at measurement points, or in production equipment.

All new pressure vessels should be constructed per ASME code. It is up to the purchaser if the vessel is to be certified and registered by the National Board of Boiler and Pressure Vessel Inspectors [8]; in critical service, this may be worth the extra time and expense to obtain a National Board Number and the associated paperwork. It is imperative that all the construction drawings, steel mill reports, and the U1A form that is provided by the ASME Code Shop/Fabricator be filed

in a Central Filing System. Those records are required to be maintained and accessible as part of the Code requirements. Any subsequent repair or alterations made to the vessel are to be documented and filed as well. It is suggested that a unique company number be assigned to each vessel and cold stamped into a vessel leg or skirt, as nameplates can be damaged or removed over time, especially if equipment is moved from one location to another. This unique number allows for maintaining the records from cradle to grave for each vessel.

Utilizing used equipment is a matter of timing and economics. The cost of new versus used needs to be taken into account on all projects. The cost of used equipment needs to include inspection, repair, and certification costs as well as the cost of the vessel itself.

All equipment should be upgraded and tested to a level that meets regulations and industry standards (i.e., API) for operability and safety. Used pressure vessels should conform to API 510 standards [9].

The design and operation of various vessels are discussed later in this section.

6.6.2 Pressure Vessel Relief Systems

All pressure vessels are to be equipped with a pressure relief valve set to open at a pressure no greater than the maximum allowable working pressure (MAWP) of that vessel and sized to prevent pressure from rising more than 10% or 3 psi, whichever is greater, above the MAWP at full open flow. For vessels with multiple relief valves, 16% or 4 psi, whichever is greater, above the MAWP is allowed. This provision **DOES NOT** apply to relief valve/rupture disk combinations. Pressure vessels operating at pressures above 125 psig may be equipped with a rupture disk or a second (redundant) relief valve. The rupture disk should be designed to rupture at not more than 15% above the MAWP and sized to handle the maximum flow through the vessel. Relief valves and rupture disks should be installed in the gaseous phase of a vessel, and be placed so that there is no restriction between the valve and the vessel. A locking handle full-opening ball valve can be installed below the relief valve but must remain locked in the open position. The installation of this isolation valve is so that the relief valve can be tested in place or replaced without having to take the vessel out of service to do so. Vessels that operate in a fluid-packed state should be equipped with an appropriately sized liquid service relief valve.

6.6.2.1 Product Storage
Tanks used in oil and gas production operations are all atmospheric. Gas plants and refineries use the pressurized tanks/vessels.

- *Atmospheric*—Atmospheric tanks are designed and equipped for storage of contents at atmospheric pressure. This category usually employs tanks of vertical cylindrical configuration that range in size from small shop welded to large field erected tanks. Bolted tanks, fiberglass tanks, and occasionally rectangular welded tanks are also used for atmospheric storage. API standards relating to tankage are listed in Table 6.6.1 [10]. Note: 1,000 bbl tanks are the largest size that can be shop welded and transported to location. Internally coated, welded steel tanks should normally be used in hydrocarbon or brine service. Tank decks should be connected to ensure proper internal coating. Operations preferences and service conditions should be considered when selecting flat or cone bottom tank. Fiberglass tanks are typically used in brine water service rather than for hydrocarbon due to fire safety. It is important to always ground the fluid in a fiberglass tank, bond and ground metal parts, and gas blanket the vapor space to minimize oxygen entrainment.

Table 6.6.1 *API standards — tanks (partial listing) [10]*

Standard Number	Title
API Spec 12B	Specification for Bolted Tanks for Storage and Production Liquids
API Spec 12D	Specification for Field Welded Tanks for Storage and Production Liquids
API Spec 12F	Specification for Shop Welded Tanks for Storage and Production Liquids
API Spec 12P	Specification for Fiberglass Reinforced Plastic Tanks
API 12R 1	Recommended Practice: Setting, Maintenance, Inspection, Operation and Repair in Production Service
API RP 620	Recommended Practice: Design and Construction of Large, Welded, Low-Pressure Storage Tanks
API RP 650	Recommended Practice: Welded Steel Tanks for Oil Storage
API Spec 653	Specification for Tank Inspection, Repair, Alteration and Reconstruction
API RP 2000	Recommended Practice: Venting Atmospheric and Low-Pressure Storage Tanks: Non-Refrigerated and Refrigerated

- Low Pressure (0 to 2.5 psig or 0 to 17 kPa)—Low-pressure tanks are normally used for storage of intermediates and products that require a low internal gas pressure. These are generally cylindrical in shape with flat or dished bottoms and sloped or domed roofs. Low-pressure storage tanks are usually of welded design; however, bolted tanks are often used for operating pressures near atmospheric. Many refrigerated storage tanks operate around 0.5 psig. These tanks are built to API Standards 12D, 12F, 620, or 650.
- Medium Pressure (2.5 to 15 psig or 17 to 103 kPa)—Medium-pressure tanks are normally used for storage of higher volatility intermediates and products. The shape may be cylindrical with flat or dished bottoms and sloped or domed roofs. Medium-pressure tanks are usually of welded design. Welded spheres may also be used, particularly for pressure at or near 15 psig. These tanks are built to API Standard 620.
- High Pressure (above 15 psig or 103 kPa)—High-pressure tanks/vessels are generally used for the storage of refined products or fractionated components at pressures greater than 15 psig. They are welded and may be of cylindrical or spherical configuration. Because they are above 15 psig, they are designed and constructed per ASME code.

Gas processing industry liquids are commonly stored underground, in conventional or solution-mined caverns. For more details about product storage options, see Table 6.6.2.

6.6.2.1.1 Walkways and stairways
Engineering controls should be installed on new tanks and walkways so that breathing equipment is not required to do routine work. Engineering controls include mechanical or electronic tank gauges that eliminate the need for daily opening of the thief hatch to measure the tank level. All readouts should be at ground level. Walkways and stairways should be constructed to API standards and be wide enough so that a person with a self-contained breathing apparatus (SCBA) can have egress without restriction (minimum of 30 in. [0.8 m] wide). Walkways should be quoted complete with all necessary supports. Engineers should carefully review standard walkways and stairways provided by manufacturers, as some do not meet the structural requirements of API or OSHA.

6.6.2.1.2 Tank Venting
All tank venting designs should be in accordance to API 2000 Standard. Oil and gas production operation tanks should be provided with both normal and emergency venting. Normal venting is typically addressed with deadweight pressure/vacuum valves. The weight is determined by the design of the tank. Tanks used for production operations typically have vent valves set at 4 oz. Emergency venting is addressed with gauge hatch covers and, if necessary, additional weighted covers. The normal operation vents must work in conjunction with the gas blanket and the vapor recovery unit (VRU) (more information on VRUs is found in "Compressors"), if installed. Frangible deck joints are recommended on larger tanks. These should be constructed according to API 650 with full penetration welds to facilitate internal coating of the tank.

6.6.2.1.3 Gas Blankets
Gas blankets should be used in order to minimize the possibility of an explosive atmosphere in water tanks, to prevent oxygenation of produced water, and to prevent air in the intake of a VRU if the controls fail. Gas blankets must work in coordination with the VRU and normal tank venting. Normal relief pressures should be at the widest range permissible

Table 6.6.2 *Product Storage Options [11]*

	Atmospheric	Low Pressure (0–2.5 psig)	Medium Pressure (2.5–15 psig)	High Pressure (Above 15 psig)	Underground
Crude oil	X	X	X	–	X
Condensate	X	X	X	X	X
Refined oil	X	X	–	–	X
Gasoline	X	X	X	–	X
Butane	–	X_R	X_R	X	X
Propane	–	X_R	X_R	X	X
Raw NGLs	–	X_R	X_R	X	X
Ethane	–	X_R	X_R	X	X
Petrochemicals	–	X_R	X_R	X	X
Natural gas	–	–	–	X	X
LNG	–	X_R	X_R	X	–
Treating agents	X	X	–	–	–
Dehydration liquids	X	X	–	–	–
Chemicals	X	X	X	–	–
Solids	X	–	–	–	–
Produced water	X	–	–	–	–

X_R, refrigerated only.

by tank design. The source of gas for a gas blanket is typically a residue gas system, which also provides "clean" gas for fired equipment or may come off of the gas scrubber in the facility.

6.6.2.2 Prime Movers
6.6.2.2.1 Centrifugal Pumps
Centrifugal pumps are the most commonly used pumps in production operations. Single-stage ANSI pumps driven by small-horsepower electric motors are used for transfer of fluids as well as charge pumps upstream of larger high-pressure water injection pumps. ANSI pumps are preferred in production operations because all similar pumps meeting ANSI specifications will have identical baseplate, as suction/discharge dimensions, regardless of the manufacturer. In addition, this allows for greater future utility and ease of maintenance and replacement.

Materials should be specified according to service and design life requirements. Carbon steel housing may be suitable for sweet oil service, but in corrosive service, alloys such as 316 stainless steel (SS) are justified. When sizing a pump, it is necessary to take into account current and future operating conditions in order to provide for flexibility after installation. It is much easier and is more economic to install a larger impeller in a pump than it is to replace the motor and/or pump. The engineer needs to find a balance between conservative pump size and being able to keep the pump operating at or near its best efficiency. As operating conditions fall too far to one side or the other, surge effects and cavitation can occur.

Multistage centrifugal pumps are commonly used for water injection. These pumps can move large volumes of water at relatively high pressure (less than 2,000 psi). Because of the corrosive service, these pumps are typically made of 316 SS. They may be horizontal spilt case or vertical "can" design. If at all possible, it is suggested that there be a spare pump set up to switch to if problems arise with the primary pump.

6.6.2.2.2 Reciprocating Positive Displacement Pumps
In a reciprocating pump, a volume of liquid is drawn into the cylinder through the suction valve on the intake stroke

and is discharged under positive pressure through the outlet valves on the discharge stroke. The discharge from a reciprocating pump is pulsating and changes only when the speed of the pump is changed. This is because the intake is always a constant volume. Often an air chamber is connected on the discharge side of the pump to provide a more even flow by evening out the pressure surges. Reciprocating positive displacement (PD) pumps are most often used for water injection or disposal. They are designed for high pressures and low volumes as compared to the multistage centrifugal pumps. Typically, the fluid end is 316 SS due to the corrosive nature of the fluids handled. As always, know the service conditions before specifying the material. When sizing a PD pump, it is suggested that the initial (design) speed be less than 90% of the manufacturer's rated speed. This will enhance the reliability of the machine and allows some flexibility for increasing flow rate if necessary. By changing the speed and the plunger diameters, a fluid end can address a range of flow rates and pressures.

Reciprocating PD pumps are available in four designs:

- *Simplex* — This pump has a single liquid cylinder that forces liquid out through the top outlet on both the in and the out stroke (here up and down). This basic type of pump might be used for air pumps, feed pumps for the furnace, fire, bilge, and fuel oil service.
- *Duplex* — These are similar to the simplex pump, having two pistons instead of one, providing smoother operation. Duplex pumps have no "dead spots" because one or the other steam piston is always under force of steam (or compressed air). The two pistons are about 1/4 cycle out of synchronization with each other.
- *Triplex* — These are similar to the duplex pumps, having three pistons or plungers. This is the most commonly used pump in drilling and well servicing. They are also widely used for disposal wells or small waterfloods.
- *Quintuplex* — These are similar to the triplex pumps, having five pistons or plungers.

Piping design should carefully address the suction-side pipe hydraulics. It is critical to provide adequate Net Positive Suction Head (NPSH) to avoid cavitation and vibration problems. The use of suction and discharge pulsation dampeners is strongly recommended. Suction lines should be sized to limit the velocity to less than 5 ft./s, and discharge piping should include a relief valve and check valve upstream of the discharge block valve. The relief valve should be piped back to a tank to avoid spills.

Vibration is of major concern with PD pumps; mechanical vibration and structural stress can result in failures and spills, and so pulsation dampeners, valves, etc. should be supported. OSHA requires guards to be placed over plungers and belts.

6.6.2.2.3 Compressors

Compression in the oil field ranges from low-pressure sliding vane compressor to 2500+ high-pressure integral units. Compressors are used to boost the pressure from the wellhead or the facility to sales line pressure, to inject gas back into the reservoir for pressure maintenance or enhanced oil recovery (CO_2, nitrogen, etc.), or boost the pressure from the wellhead or facility to a central facility. Depending upon the size and the field conditions/location, they may be powered by electric motors, gas/internal combustion or by turbine engines. VRUs are the most common use of compression in surface oil production. The larger compressors are addressed in "Compressor Stations".

VRUs are installed for the economic recovery of tank vapors, compliance with Air Emission Regulations (Federal and/or State), and for safety considerations. Emissions can be estimated or measured prior to installation. The Texas Natural Resource Commission has allowed 90 days of temporary operation under Standard Exemption 67 [12] to determine actual emissions. The majority of the VRUs on the market today use sliding vane rotary compressors. In the low-pressure applications, a blower-type compressor should also be considered.

6.6.2.3 Piping Guidelines
6.6.2.3.1 Materials

The most commonly used materials in oil and gas operations are carbon steel, SS (304 or 316 families), fiberglass, and polyethylene. Each one has a place when properly matched to the application. In most cases, the decision on which one to use is based on economics (material costs as well as the coating and ditching costs), corrosion resistance, land use, support requirements, proximity to the public, and impact of failures.

Much of what is addressed is risk-based analysis.

- *Carbon steel* is the most commonly used material. Most onshore operations use electric fusion welded (EFW), electric resistance welded (ERW), or grade B seamless pipe. The EFW and ERW pipe are considered acceptable for most applications and are cheaper than seamless pipe. EFW is typically only available in 16-in. or greater diameter. DOT-regulated lines cannot contain ERW pipe.

  Bare carbon steel can be used for liquid hydrocarbon, steam, and gas service. If internally coated, it can also be used in water and acid service.
- *SS* is used when increased corrosion protection is needed. The most commonly used ones in the oil field are 316 and 304. The low-carbon versions should be used if it is to be welded. The 316L and 304L do not become sensitized during welding; hence, do not require postweld heat-treating. 316 SS is preferred in brine/salt water service, and 304 SS is commonly used in wet CO_2 service.
- *Fiberglass* is commonly used where corrosion, either internal or external, is of concern. It is available in a wide variety of diameters and pressure ratings. It is primarily used in water systems (water injection and gathering systems as well as inplant piping at injection stations). Because of the effect of sunlight (UV) on the resins used

in fiberglass pipe, it should be buried, painted, or otherwise protected from the elements. If buried, thrust blocks should be considered at the elbows to counteract the hydraulic forces.
- *Polyethylene* is commonly used due to its low cost, low weight, ease of installation, corrosion resistance, and flow characteristics. For the most part, it is not recommended for pressure above 100 psig or temperatures above 100°F, although there has been a lot of progress made in the characteristics. Poly pipe is most often used for gas gathering and flowlines, but it should be derated per manufacturer's recommendations when used in wet hydrocarbon service. Because of its composition, it should not be used to transport flammable fluids above ground.

6.6.2.3.2 Buried versus Above Ground

One of the first of these items to be addressed is whether the pipe is to be above ground or buried. OSHA requires that all nonmetallic or low-melting-point pipe carrying flammable materials to be buried. And whenever nonmetallic pipe is buried, detection tape or trace cable should be buried above the pipe to aid in locating the pipe in the future. Buried pipe may be more or less costly than aboveground systems. One should take into account the relative cost of the pipe, access, operational preferences, fire, freeze, mechanical protection, and pipe supports. It is recommended to externally coat buried pipe to avoid external corrosion. Cathodic protection may be used instead of or in conjunction with external coating systems.

It is common to paint aboveground pipe for aesthetics as well as to provide some corrosion protection (this includes painting fiberglass pipe to protect it from UV degradation). Aboveground piping should be labeled for the type of service and carrier fluid. ASME A13.1 "Scheme for the Identification of Piping Systems" can be referred to where color-coding and method of identification is required.

6.6.2.3.3 Internal Coatings and Linings

Solids production and erosion can be detrimental to some internal coatings, so these as well as the specific service and considerations of environmental risk associated with a leak should be evaluated. In most cases, internal coating should be considered for all steel piping systems. There are many different products available; the ones most commonly found in the oil field are cement lining, plastic coating, and extruded polyethylene.

6.6.2.3.4 Piping Connections

The following parameters need to be considered when determining what connections are best suited for a project:

- operations preference and experience (You *have* to get "buy-in" from the field personnel)
- operating pressure of the systems
- hydrocarbon content in the produced fluid
- H_2S concentrations in the product stream
- impact on surrounding area in the event of a failure
- expected life of the project

6.6.2.3.5 Flanged versus Threaded

Welded/flanged construction may provide the following advantages:

- less susceptible to mechanical damage (especially in vibrating service)
- greater structural joint strength
- more resistant to fatigue failure
- better hydrocarbon containment in the event of fire
- lower fugitive emissions
- make-ups are more precise than threaded

Threaded construction may provide the following advantages:

- less expensive to install
- easier and less expensive to repair
- no hot work involved
- quicker to disassemble

6.6.2.3.6 Valves
Types of valves commonly used in oil and gas production

- *Gate Valves* — A gate is a linear-motion valve that uses a typically flat closure element perpendicular to the process flow, which slides into the flow stream to provide shut-off. With a gate valve, the direction of fluid flow does not change, and the diameter through which the process fluid passes is essentially equal to that of the pipe. Hence, gate valves tend to have minimal pressure drop when opened fully. Gate valves are designed to minimize pressure drop across the valve in the fully opened position and stop the flow of fluid completely. In general, gate valves are not used to regulate fluid flow. A gate valve is closed when a tapered disk of a diameter slightly larger than that of the pipe is lowered into position against the valve seats. The valve is fully open when the disk is pulled completely out of the path of the process fluid into the neck [13]. Typically, these are used on pipelines, on gas wells, and as plant block valves.
- *Globe Valves* — A globe valve is a linear-motion valve characterized by a body with a longer face-to-face that accommodates flow passages sufficiently long enough to ensure smooth flow through the valve without any sharp turns. Globe valves are widely used to regulate fluid flow in both on/off and throttling service. The amount of flow restriction observed with valve disk (or globe) location is relative to the valve seat. The valve seat and stem are rotated 90° to the pipe. The direction of fluid flow through the valve changes several times, which increases the pressure drop across the valve. In most cases, globe valves are installed with the stem vertical and the higher-pressure fluid stream connected to the pipe side above the disk, which helps to maintain a tight seal when the valve is fully closed. Traditionally the valve disk and seat were both metal, although some modern designs use an elastomer disk seal. These valves are inexpensive and simple to repair [13].
- *Plug Valves* — Plug valves are similar to gates and work particularly well in abrasive service. These are available either lubricated or nonlubricated.
- *Ball Valves* — Ball valves are quarter-turn, straight-through flow valves that use a round closure element with matching rounded elastometric seats that permit uniform sealing stress. These valves are limited to moderate temperature service (below 250°C) by the plastic seats that create a seal around the ball. The type of seat can vary with the valve pressure rating and materials of construction. Some valve seats are a single molded form, while other valve seats with higher-pressure ratings often incorporate a Trunion design where each face of the ball is sealed separately. They have found applications in flow control, as well as on/off use in isolating a pipe stream. The pressure drop across the valve in a fully open position is minimal for a full-port design. However, with restricted-port designs the pressure drop can be significant [13].
- *Butterfly Valves* — The butterfly valve is a quarter-turn rotary-motion valve that uses a round disk as the closure element. The sealing action of a butterfly valve is achieved by rotating a disk of approximately the same diameter, as the pipe from a globe valve is a function of the position

in line with fluid flow to a position perpendicular to flow. The axial length of these valves is less than any other valve, which in cases in which flange faces are used with large pipe sizes (greater than 10 in.), these valves have the lowest initial cost. If resilient seats or piston rings on the disk are used, these valves can be sealed by relatively low operating torque on the valve stem. This sealing action is assisted by the fluid-pressure distribution that tends to close the valve. This same hydraulic unbalance requires that a latching device or worm gear be installed to prevent unwanted closure of manually operated valves. Although butterfly valves are used for low-pressure drop applications, the pressure drop across the valve is quite high with large flow rates compared with a gate valve [13]. Lug-type body is preferred in hydrocarbon applications because there is less bolt length exposed in a fire. The lug type is also easier to flange up than the wafer type because uneven tightening of the bolts in wafer type can cause uneven pull up and leaking. High-performance butterfly valves are commonly used for throttling in both low and high-pressure applications. A SS stem with a SS or a plastic coated disc is suggested for produced water service.
- *Check Valves* — Check valves commonly used in production service are the swing, piston, or split disk styles. The swing check is used in low-velocity flows where flow reversals are infrequent. Piston checks with or without spring assists are used in high–flow rate streams with frequent reversals (i.e., compressor or high-pressure pump discharges). Split disk checks are typically used in low-pressure applications in which a minimum pressure drop across the valve is required (i.e., transfer/charge pump discharges). Because of the service, SS internals are preferred over plastic-coated in any service where they come in contact with produced water to minimize corrosion failures.
- *Relief Valve* — A pressure relief valve is a self-operating valve that is installed in a process system to protect against overpressurization of the system. Relief valves are designed to continuously modulate fluid flow to keep pressure from exceeding a preset value. There are a wide variety of designs, but most resemble diaphragm valves, globe valves, or swing check valves. With many of these designs, a helical spring or hydraulic pressure is used to maintain a constant force that acts on the backside of the valve disk or diaphragm causing the valve to normally be closed. When the force exerted by the process stream (i.e., fluid pressure) on the valve disk is greater than the constant force exerted by the spring, the valve opens, allowing process fluid to exit the valve until the fluid pressure falls below the preset value. These valves can be preset to a specific relief pressure, or they may be adjustable [13].

Materials used in valve construction are as varied as their applications. The following are the ones commonly used in the oil field. The material used in the valve body is addressed first. Valves carrying flammable fluids should be of steel or ductile iron.

Malleable iron should not be used in sour service because it does not meet NACE standards. Brass and bronze are suitable for fresh water and low-pressure air service. Nickel aluminum bronze and 316 SS are most often used in produced water applications. 304 SS is preferred for wet CO_2, whereas regular steel is acceptable for dry CO_2. Aluminum is acceptable for tank vent valves.

6.6.2.3.7 Valve trim
Elastomer selection is critical. The most commonly used elastomer in brine water and hydrocarbon service is Buna-N

(nitrile). The limitation is that it should not be used when temperatures exceed 200°F for prolonged periods. A 90-derometer peroxide-cured Buna is the first choice for use in CO$_2$ or H$_2$S service. Viton is good to about 350°F and is resistant to aromatics, but should not be used with amines or high pressure CO$_2$. Teflon is good to somewhere between 250°F to 350°F, depending on the application, and has exceptionally good chemical resistance characteristics. Teflon is not a true elastomer because it has no physical memory.

6.6.2.4 Pressure Vessel Design — Phase Separation [11, 14–16]

Practical separation techniques for liquid particles in gases are discussed. The principles used to achieve physical separation of gas and liquids or solids are momentum, gravity settling and coalescing. Any separator may employ one or more of those principles, but the fluid phases must be immiscible and have different densities for separation to occur.

6.6.2.4.1 Momentum

Fluid phases with different densities will have different momentum. If a two-phase stream changes direction sharply, greater momentum will not allow the particles of the heavier phase to turn as rapidly as the lighter fluid, so separation occurs. Momentum is usually employed for bulk separation of the two phases in a stream.

6.6.2.4.2 Gravity Settling

Liquid droplets will settle out of a gas phase if the gravitational forces acting on the droplet are greater than the drag force of the gas flowing around the droplet (Figure 6.6.4). These forces can be described mathematically. Drag force (F) on a liquid droplet in a gas stream is determine from

$$F = C_D A_p \gamma_g \frac{V_t^2}{2g} \qquad [6.6.1]$$

Hence,

$$V_t = \left(\frac{2Fg}{C_D A_p \gamma_g} \right)^{0.5}$$

and from Newton's second law

$$F = \frac{w}{g}g \quad \text{and} \quad \text{if} \quad w = \frac{\gamma_L - \gamma_g}{\gamma_L} W_p$$

$$V_t = \left[\frac{2gW_p (\gamma_t - \gamma_g)}{\gamma_t \gamma_g A_p C_D} \right]^{0.5} = \left[\frac{4gD_P (\gamma_t - \gamma_g)}{3\gamma_g C_D} \right]^{0.5} \qquad [6.6.2]$$

where

F = drag force in lb
V_t = terminal velocity in ft./s
g = acceleration due to gravity in ft./s^2
W_P = weight of particle in lb
$\gamma_{L,g}$ = liquid and gas-phase specific weights in lb/ft.^{3}
A_P = particle cross-sectional area in ft.^{2}
D_p = droplet diameter in ft.
C_D = drag coefficient of particle (dimensionless)

The drag coefficient has been found to be a function of the shape of the particle and the Reynolds number of the flowing gas. For the purpose of this equation, particle shape is considered to be a solid, rigid sphere.

Reynolds number is defined as

$$Re = \frac{1,488D_p V_t \gamma_g}{\mu} \qquad [6.6.3]$$

Where μ = viscosity (cp)

In this form, a trial-and-error solution is required since both particle size D_p and terminal velocity V_t are involved. To avoid trial and error, values of the drag coefficient are presented in Figure 6.6.5 as a function of the product of drag coefficient C_D times the Reynolds number squared; this technique eliminates velocity from expression.

The abscissa of Figure 6.6.5 is represented by

$$C_D(Re)^2 = \frac{(0.95)(10^8)\gamma_g D_p^3(\gamma_t - \gamma_g)}{\mu^2} \qquad [6.6.4]$$

For production facility design (turbulent flow), the following formula for drag coefficient is proper:

$$C_D = \frac{24}{Re} + \frac{3}{(Re)^{0.5}} + 0.34 \qquad [6.6.5]$$

and if D_p is expressed in micrometers $\rightarrow d_m$

$$V_t = 0.0119 \left[\left(\frac{(\gamma_t - \gamma_g)}{\gamma_g} \right) \frac{d_m}{C_D} \right]^{0.5} \qquad [6.6.6]$$

Equations 6.6.5 and 6.6.6 can be solved by an iterative solution as follows

1. Write the equation for laminar flows ($C_D = 0.34$)

$$V_t = 0.0204 \left[\frac{(\gamma_t - \gamma_g)}{\gamma_g} d_m \right]^{0.5}$$

2. Calculate $Re = 0.0049(\gamma_g d_m V_t / \mu)$
3. From Equation 6.6.5 calculate C_D
4. Recalculate V_t using Equation 6.6.6. Go to step 2

The above technique is proper assuming that known diameter drops are removed (e.g., 100 μm).

6.6.3 Separator Design and Construction

There are three types of separators: vertical, horizontal (single and double tube), and sometimes spherical (Tables 6.6.3 to 6.6.5).

There are many vessel design software packages out on the market today, and everyone has a favorite. When the requirements are input and the program is run, one is then armed with the information needed to talk to the vendors about a bid specification or to look at what is available within

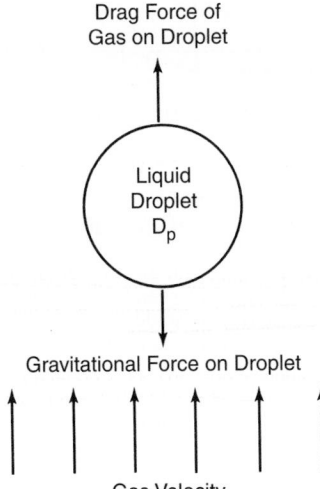

Forces on Liquid Droplet in Gas Stream
Drag Force of
Gas on Droplet

Liquid
Droplet
D_p

Gravitational Force on Droplet

Gas Velocity

Figure 6.6.4 *Forces on liquid input in gas system [11].*

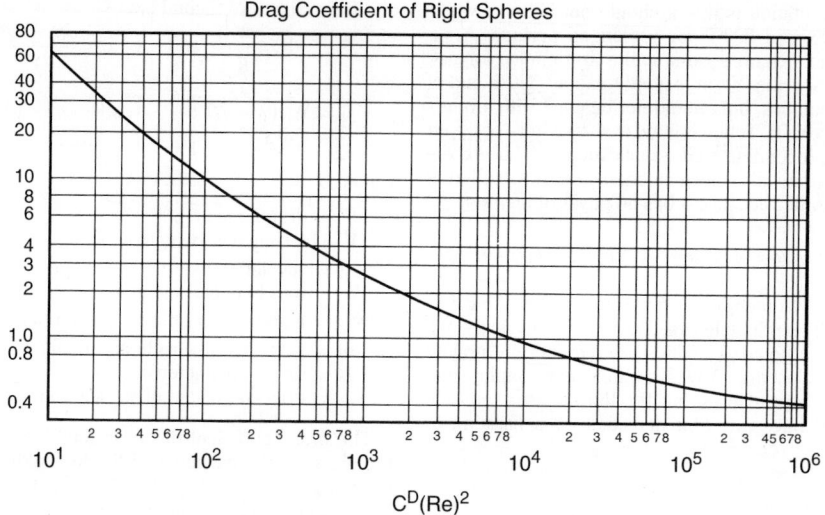

Figure 6.6.5 *Drag coefficient of a rigid sphere [11].*

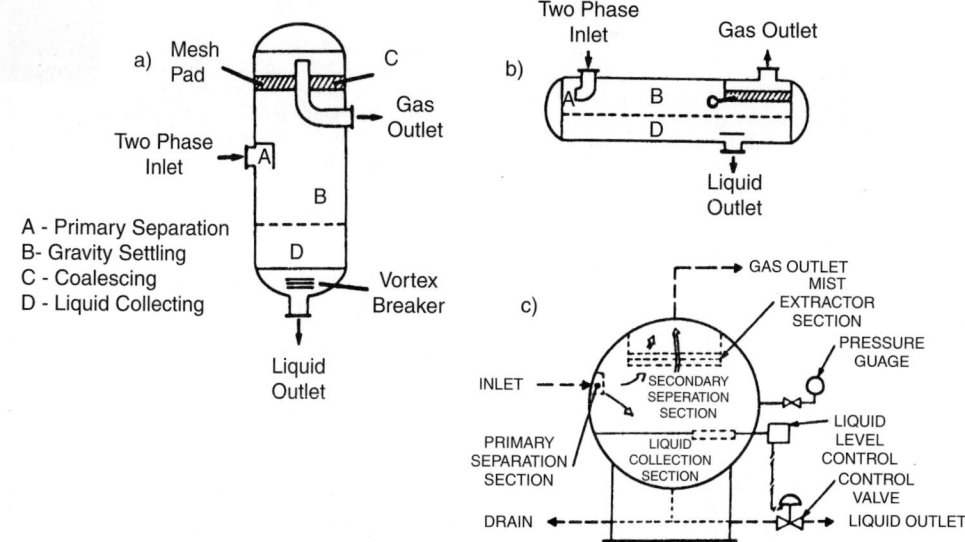

Figure 6.6.6 *Gas liquid separators: (a) vertical, (b) horizontal, (c) spherical.*

the company's used equipment inventory. When purchasing a new vessel the data and requirements should always be provided to the supplier and reviewed by engineering on both sides to obtain the best vessel fit for each application.

Vertical separators are usually selected when the gas/liquid ratio is high or total gas volumes are low. In this sort of vessel, the fluids enter the vessel striking a diverting plate that initiates primary separation. Liquid removed by the inlet diverter falls to the bottom of the vessel. The gas moves upward, usually passing through a mist extractor to remove suspended mist, and then flows out. Liquid is removed by reservoir in the bottom. Mist extractors can significantly reduce the required diameter of vertical separators. However, if the production stream is dirty, the mist extractor will frequently plug and it may be removed.

Subsequent scrubbers may need to be added or bypass lines installed to allow equipment to be cleaned with relative ease.

Horizontal separators are most often efficient for large volumes of total fluids and when large amounts of dissolved gases are present with the liquid. The greater liquid surface area provides for optimum conditions for releasing gas from the liquid.

Spherical separators are rarely used in production operations.

6.6.4 Vertical Separators
The following calculations are presented as a guide to the design and sizing of two-phase separators. Sizing should be based on the maximum expected instantaneous rate.

Table 6.6.3 *Standard Vertical Separators: Size and Working Pressure Ratings*

Size (Dia. × Ht.)	Working Pressure (psi)								
16″ × 5′		230	500	600	1000	1200	1440	1500	2000
16″ × 7 1/2′		230	500	600	1000	1200	1440	1500	2000
16″ × 10′		230	500	600	1000	1200	1440	1500	2000
20″ × 5′		230	500	600	1000	1200	1440	1500	2000
20″ × 7 1/2′		230	500	600	1000	1200	1440	1500	2000
20″ × 10′		230	500	600	1000	1200	1440	1500	2000
24″ × 5′	125	230	500	600	1000	1200	1440	1500	2000
24″ × 7 1/2′	125	230	500	600	1000	1200	1440	1500	2000
24″ × 10′		230	500	600	1000	1200	1440	1500	2000
30″ × 5′		230	500	600	1000	1200	1440	1500	2000
30″ × 7 1/2′		230	500	600	1000	1200	1440	1500	2000
30″ × 10′	125	230	500	600	1000	1200	1440	1500	2000
36″ × 5′	125								
36″ × 7 1/2′	125	230	500	600	1000	1200	1440	1500	2000
36″ × 10′	125	230	500	600	1000	1200	1440	1500	2000
36″ × 15′		230	500	600	1000	1200	1440	1500	2000
42″ × 7 1/2′		230	500	600	1000	1200	1440	1500	2000
42″ × 10′		230	500	600	1000	1200	1440	1500	2000
42″ × 15′		230	500	600	1000	1200	1440	1500	2000
48″ × 7 1/2′		230	500	600	1000	1200	1440	1500	2000
48″ × 10′	125	230	500	600	1000	1200	1440	1500	2000
48″ × 15′	125	230	500	600	1000	1200	1440	1500	2000
54″ × 7 1/2′		230	500	600	1000	1200	1440	1500	2000
54″ × 10′		230	500	600	1000	1200	1440	1500	2000
54″ × 15′		230	500	600	1000	1200	1440	1500	2000
60″ × 7 1/2′		230	500	600	1000	1200	1440	1500	2000
60″ × 10′	125	230	500	600	1000	1200	1440	1500	2000
60″ × 15′	125	230	500	600	1000	1200	1440	1500	2000
60″ × 20′	125	230	500	600	1000	1200	1440	1500	2000

Separators all have "high" working pressure

Table 6.6.4 *Standard Horizontal Separators: Size and Working Pressure Ratings*

Size (Dia.×Ht.)	Working Pressure (psi)								
12 3/4″ × 5′		230	500	600	1000	1200	1440	1500	2000
12 3/4″ × 7 1/2′		230	500	600	1000	1200	1440	1500	2000
12 3/4″ × 10′		230	500	600	1000	1200	1440	1500	2000
16″ × 5′		230	500	600	1000	1200	1440	1500	2000
16″ × 7 1/2′		230	500	600	1000	1200	1440	1500	2000
16″ × 10′		230	500	600	1000	1200	1440	1500	2000
20″ × 5′		230	500	600	1000	1200	1440	1500	2000
20″ × 7 1/2′		230	500	600	1000	1200	1440	1500	2000
20″ × 10′		230	500	600	1000	1200	1440	1500	2000
24″ × 5′	125	230	500	600	1000	1200	1440	1500	2000
24″ × 7 1/2′	125	230	500	600	1000	1200	1440	1500	2000
24″ × 10′	125	230	500	600	1000	1200	1440	1500	2000
24″ × 15′		230	500	600	1000	1200	1440	1500	2000
30″ × 5′	125	230	500	600	1000	1200	1440	1500	2000
30″ × 7 1/2′	125	230	500	600	1000	1200	1440	1500	2000
30″ × 10′	125	230	500	600	1000	1200	1440	1500	2000
30″ × 15′		230	500	600	1000	1200	1440	1500	2000
36″ × 7 1/2′		230	500	600	1000	1200	1440	1500	2000
36″ × 10′	125	230	500	600	1000	1200	1440	1500	2000
36″ × 15′	125	230	500	600	1000	1200	1440	1500	2000
36″ × 20′		230	500	600	1000	1200	1440	1500	2000
42″ × 7 1/2′		230	500	600	1000	1200	1440	1500	2000
42″ × 10′		230	500	600	1000	1200	1440	1500	2000
42″ × 15′		230	500	600	1000	1200	1440	1500	2000
42″ × 20′		230	500	600	1000	1200	1440	1500	2000
48″ × 7 1/2′		230	500	600	1000	1200	1440	1500	2000
48″ × 10′	125	230	500	600	1000	1200	1440	1500	2000
48″ × 15′	125	230	500	600	1000	1200	1440	1500	2000

(continued)

Table 6.6.4 Standard Horizontal Separators: Size and Working Pressure Ratings (continued)

Size (Dia.×Ht.)			Working Pressures (psi)							
48″ × 20′		230	500	600	1000	1200	1440	1500	2000
54″ × 7 1/2′		230	500	600	1000	1200	1440	1500	2000
54″ × 10′		230	500	600	1000	1200	1440	1500	2000
54″ × 15′		230	500	600	1000	1200	1440	1500	2000
54″ × 20′		230	500	600	1000	1200	1440	1500	2000
60″ × 7 1/2′		230	500	600	1000	1200	1440	1500	2000
60″ × 10′	125	230	500	600	1000	1200	1440	1500	2000
60″ × 15′	125	230	500	600	1000	1200	1440	1500	2000
60″ × 20′	125	230	500	600	1000	1200	1440	1500	2000

Table 6.6.5 Standard Spherical Separators: Size and Working Pressure Ratings

Size (O.D.)			Working Pressures (psi)							
24″		230	500	600	1000	1200	1440	1500	2000	3000
30″		230	500	600	1000	1200	1440	1500	2000	3000
36″		230	500	600	1000	1200	1440	1500	2000	3000
41″	125									
42″		230	500	600	1000	1200	1440	1500	2000	3000
46″	125									
48″		230	500	600	1000	1200	1440	1500	2000	3000
54″	125									
60″		230	500	600	1000	1200	1440	1500	2000	3000

For practical purposes, for vertical separators, Equation 6.6.2 is written as

$$V_t = K \left(\frac{\gamma_L - \gamma_g}{\gamma_g} \right)^{0.5} \qquad [6.6.7]$$

where

V_t = terminal velocity of liquid droplets or maximum allowable superficial velocity of gas in ft./s

K = a constant depending on design and separating conditions in ft./s. See Table 6.6.6 [11].

Example 6.6.1

A vertical gravity separator is required to handle 10 MMscfd of 0.6 specific gravity gas at an operating pressure of

Table 6.6.6 Typical K-Factor Values in Equation 6.6.7

Separator Type	K factor (ft./s)
Horizontal (w/vertical pad)	0.40 to 0.50
Vertical or Horizontal (w/horizontal pad)	0.18 to 0.35
@Atm. pressure	0.35
@300 psig	0.33
@600 psig	0.30
@900 psig	0.27
@1500 psig	0.21
Spherical	0.20 to 0.35
Wet steam	0.25
Most vapors under vacuum	0.20
Salt and caustic evaporators	0.15

1. K = 0.35 @ 100 psig (subtract 0.01 for every 100 psi above 100 psig).
2. For glycol and amine solutions, multiply K by 0.6 to 0.8.
3. Typically use $\frac{1}{2}$ of the above K values for approximate sizing of vertical separators without woven wire de-misters.
4. For compressor suction scrubbers and expander inlet separators multiply K by 0.7 to 0.8.

1,000 psia and a temperature of 60°F. Liquid flow = 2,000 bpd of 40° API oil, μ_g = 0.014 cp.

Solution
1. Find K from Table 6.6.6 where K = 0.26 ft./s.
2. Calculate minimum diameter, such that if

$$V_t = K \left(\frac{\gamma_L - \gamma_g}{\gamma_g} \right)^{0.5}$$

then

$$V_g = \frac{Q_A}{A_g} \frac{ft.^3/s}{ft.^2}$$

$$A_g = \frac{\pi}{4} D^2$$

Gas-specific weight is $\gamma_g = (PM/ZRT)$
$M = 0.6 \times 29 = 17.4$
$P = 1,000$ psia
$T = 520°R$
$R = 10.73$ ft.$^3 \cdot$ psia $\cdot$ 1b$_{mol}^{-1}$ R^{-1} Z = 0.84 (for given P, T, M)

$$\gamma_g = \frac{1,000 \times 17.4}{0.84 \times 10.73 \times 520} = 3.713 \text{ lb/ft.}^3$$

Liquid-specific weight is

$$\gamma_L = 62.4 \times SG_0 = 62.4 \frac{141.5}{131.5 + 40} = 51.48 \text{ lb/ft.}^3$$

Weight rate of flow w (lb/s) is

$$\dot{w} = \frac{Q(scf)17.4(lb_m/lb\ mol)}{379.4(scf/lb\ mol) \times (24 \times 3,600)s}$$

$$= \frac{10 \times 10^6 \times 17.4}{379.4 \times 86,400} = 5.3 \text{ lb/s}$$

$$Q_A = \frac{w}{\gamma_g} = \frac{5.3 \, lb_m/s}{3{,}713 \, lb_m/ft.^3} = 1.43 \, ft.^3/s$$

$$V_g = \frac{Q_A}{A_g} = \frac{143 \times 4}{\pi D^2} = \frac{1.82}{D^2}$$

$$V_t = V_g$$

$$\frac{1.82}{D^2} = 0.26 \left(\frac{51.48 - 3.713}{3.713} \right)^{0.5} = 0.99 \, ft./s$$

$$D = 1.39 \, ft. = 16.76 \, in.$$

Minimum diameter is 20 in. (see Table 6.6.3). If

$$V_t = \left[\frac{4gD_p(\gamma_L - \gamma_g)}{3\gamma_g C_D} \right]^{0.5}$$

D_P is usually 100 to 150 μm.
Assume $D_P = 150 \, \mu m = 150 \times 0.00003937/12 = 4.92 \times 10^{-6} \, ft.$
From Equation 6.6.4

$$C_D(Re)^2 = \frac{0.95 \times 10^8 \times 3.713(0.000492)^3(51.48 - 3.713)}{(0.014)^2}$$

$$= 10{,}238$$

From Figure 6.6.5

$$C_D = 0.99$$

$$V_t = \left(\frac{4 \times 32.174 \times 0.000492(51.48 - 3.713)}{(0.014)^2} \right)^{0.5}$$

$$= 0.52 \, ft./s$$

$$V_g = \frac{1.82}{D^2} = 0.52 \, ft./s$$

$D = 1.87 \, ft. = 22.5 \, in.$, and from Table 6.6.3, the minimum diameter is 24 in.
Assume $D_p = 100 \, \mu m = 100 \times 0.00003937/12 = 0.000328 \, ft.$

$$C_D(Re)^2 = \frac{0.95 \times 10^8 \times 3.713(0.000328)^3(51.48 - 3.713)}{(0.014)^2}$$

$$= 2{,}592$$

From Equation 6.6.5

$$C_D = 1.75$$

$$V_t = \frac{(4 \times 32.174 \times 0.000328 \times 48.307)}{3 \times 3.713 \times 1.75} = 0.1046 \, ft./s$$

$$V_g = \frac{1.82}{D^2} = 0.1046 \, ft./s$$

$$D = 4.17 \, ft. = 50 \, in.$$

The maximum allowable superficial velocity calculated from the factors in Table 6.6.6 is for separators normally having a wire mesh mist extractor (Figure 6.6.7). This rate should allow all liquid droplets larger than 10 μm to settle out of the gas. The maximum allowable superficial velocity should be considered for other types of mist extractors.

Further calculations refer to $D = 20 \, in.$ (separator with a mist extractor). Diameters $D = 24 \, in.$ and $D = 50 \, in.$ refer to separator without a mist extractor.

3. Calculate liquid level h. A certain liquid storage is required to ensure that the liquid and gas reach equilibrium at separator pressure. This is defined as "retention time" where liquid is retained in the vessel assuming plug flow. The retention time is thus the volume of the liquid storage in the vessel divided by the liquid flowrate.

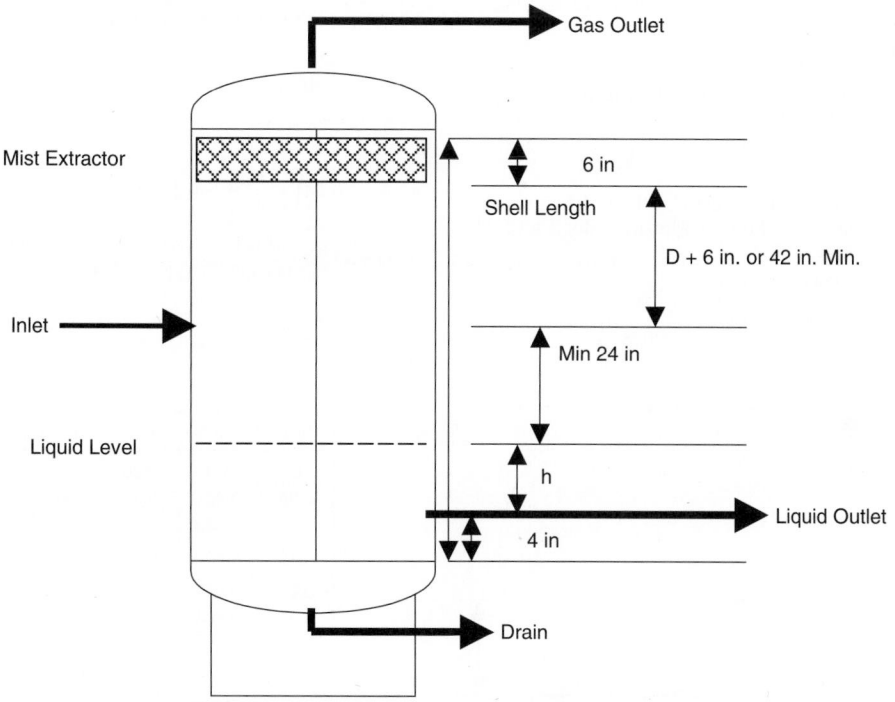

Figure 6.6.7 *The geometry of a vertical separator [15].*

Basic design criteria for liquid retention time in two-phase separators are generally as follows [17]:

Oil Gravities	Minutes (Typical)
Above 35° API	1
20–30° API	1 to 2
10–20° API	2 to 4

$$t = \frac{\text{Vol.}}{Q} \frac{\text{ft.}^3}{\text{ft.}^3/\text{s}}$$

$$\text{Vol.} = \frac{\pi D^2 h}{4\ 12} = \frac{\pi d^2 h}{4(144)12} = 4.545 \times 10^{-4} d^2 h$$

$D = 20$ in.

$\text{Vol.} = 0.1818h$

$Q_L = \text{bpd}$

$$Q = Q_L \times 5.615 \left(\frac{\text{ft.}^3}{\text{barrel}}\right) \times \left(\frac{\text{day}}{24\,\text{hr}}\right)$$

$$\times \left(\frac{\text{hr}}{3,600\,\text{s}}\right) = 0.000065 Q_L$$

$$t = \frac{0.1818h}{0.000065 Q_L} = 2,797 \frac{h}{Q_L}$$

Assume $t_r = 1$ min (API $= 40°$)
So,

$$h = \frac{t_r Q_L}{2,797} = \frac{2,000}{2,797} = 0.715\,\text{ft.} = 8.6\,\text{in.}$$

4. Calculate seam-to-seam length (L_{ss}). The seam-to-seam length of the vessel should be determined from the geometry of the vessel once a diameter and weight of liquid volume are known.

$$L_{ss} = \frac{h + 76}{12}\,(\text{in.}) \quad \text{or} \quad L_{ss} = \frac{h + D + 40}{12}$$

$$= \frac{8.6 + 76}{12} = 7.05\,\text{ft.}$$

5. Compute slenderness ratio (L_{ss}/D), which is usually in the range 3 to 4, such that

$$L_{ss}/D = \frac{7.05 \times 12}{20} = 4.23$$

6. Choosing the diameter because (L_{ss}/D) > 4, let's assume higher diameter $D = 24$ (next higher diameter from Table 6.6.3).

$$\text{Vol} = 0.2618h$$

$$t = \frac{0.2618h}{0.000065 Q_L} = 4,028 \frac{h}{Q_L}$$

If

$$t = 1\,\text{min}$$

then

$$h = \frac{2,000}{4,028} = 0.5\,\text{ft.} = 6\,\text{in.}$$

$$L_{ss} = \frac{6 + 76}{12} = 6.83\,\text{ft.} = 82\,\text{in.}$$

$$\frac{L_{ss}}{D} = \frac{82}{24} = 3.4$$

Proper size of separator is 24 in. $\times\ 7\frac{1}{2}$ ft.

6.6.5 Horizontal Separator

In the case of horizontal separators, the gas drag force does not directly oppose the gravitational settling force. The thru droplet velocity is assumed to be the vector sum of the vertical terminal velocity and the horizontal gas velocity. The minimum length of the vessel is calculated by assuming the time for the gas to flow from the inlet to the outlet is the same as for the droplets to fall from the top of the vessel to the surface of the liquid. In calculating the gas capacity of horizontal separators, the cross-sectional area of that portion of the vessel occupied by liquid (at maximum level) is subtracted from the total vessel cross-sectional area, as shown in Figure 6.6.8.

Separators can be any length, but the ratio L_{ss} to D of a vessel is usually in the range of 2:1 to 4:1, so that

$$V_t = K \left(\frac{\gamma_L - \gamma_g}{\gamma_g}\right) \left(\frac{L}{10}\right)^{0.56} \text{(ft./s)} \qquad [6.6.8]$$

Sometimes separators without mist extractors are sized using Equation 6.6.7 with a constant K of typically 1/2 of that used for vessels with mist extractors.

Example 6.6.2
Solve Example 6.6.1 in the case of horizontal separator with mist extractor if F = 0.564 and 0.436

1. Find K value from Table 6.6.6 where K = 0.3
2. The gas capacity constraint is determined from

$$V_g = \frac{Q}{A_g}$$

$$A_g = 0.564 \times \frac{\pi}{4} D^2 = 0.443 D^2$$

$$Q = Q_g \times \frac{10^6\,\text{scf}}{\text{MMscf}} \times \frac{\text{day}}{24\,\text{hr}} \times \frac{\text{hr}}{3,600\,\text{s}} \times \frac{14.7}{P} \times \frac{TZ}{520}$$

$$= 0.327 \frac{TZ}{P} Q_g$$

h/D	F	h/D	h/D F
0	1.0	0.30	0.748
0.05	0.981	0.35	0.688
0.10	0.948	0.40	0.626
0.15	0.906	0.45	0.564
0.20	0.858	0.50	0.500
0.25	0.804	0.55	0.436

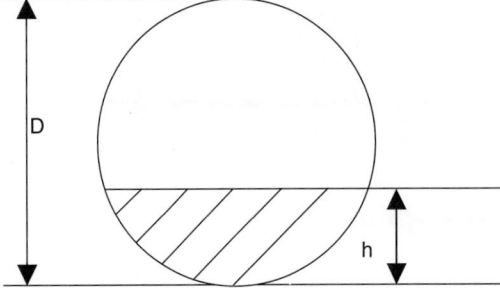

Figure 6.6.8 *The fraction of the total area available for gas flow in a horizontal separator.*

$$V_g = \frac{0.327TZQ_g}{0.443D^2P} = 0.74\frac{TZQ_g}{D^2P}$$

The residence time of the gas (t_g) has to be equal to the time required for the droplet to fall to the gas–liquid interface t_r:

$$t_g = \frac{L_{eff}}{V_g}\frac{\text{Effective length for separator}}{\text{Gas velocity}} = \frac{L_{eff}D^2P}{0.74TZQ_g}$$

$$t_d = \frac{D-h}{V_t}$$

If $F = 0.564$, then $h/D = 0.45$, where $h = 0.45D$ and $t_d = 0.55D/V_t$. If $L_{ss} = L_{eff} + D$ for gas capacity, then

$$V_t = K\left(\frac{\gamma_L - \gamma_g}{\gamma_g}\right)^{0.5}\left(\frac{L_{eff} + D}{10}\right)^{0.56}$$

and

$$t_d = \frac{0.55D}{0.3\left(\frac{\gamma_L - \gamma_g}{\gamma_g}\right)^{0.5}\left(\frac{L_{eff} + D}{10}\right)^{0.56}}$$

where $t_g = t_d$ and

$$\frac{L_{eff}D^2P}{0.74TZQ_g} = \frac{0.55D}{0.3\left(\frac{\gamma_L - \gamma_g}{\gamma_g}\right)^{0.5}\left(\frac{L_{eff} + D}{10}\right)^{0.56}}$$

$$D^2L_{eff} = \frac{6D}{(L_{eff} + D)^{0.56}}$$

From the above constraint for given D, L_{eff} could be calculated, but gas capacity does not govern.

3. Liquid capacity constraint is calculated from

$$t_r = \frac{\text{Vol.}}{Q}$$

where Vol. $= (\pi D^2/4)\,(1 - F)\,L_{eff} = 0.343D^2L_{eff}$
$Q = Q_L \times 5.615\,(\text{ft.}^3/\text{barrel}) \times (\text{day}/24\,\text{hr})$
$\times\,(\text{hr}/3{,}600\,\text{s}) = 0.000065\,Q_L$
$t_r = 0.343D^2L_{eff}/0.000065\,Q_L$
$D^2L_{eff} = (t_r \times 0.000065\,Q_L)\,/0.343$

If t_r is in minutes, then $D^2L_{eff} = 0.01137\,t_r\,Q_L$.

4. Assume retention time $t_r = 1$ min. Compute combination D (Table 6.6.4) and L_{eff}, such that

$D = 2$ ft. $L_{eff} = 5.7$ ft. $L_{ss} = \frac{4}{3}\,L_{eff} = 7.6$ ft.
$D = 2.5$ ft. $L_{eff} = 3.6$ ft. $L_{ss} = \frac{4}{3}\,L_{eff} = 4.9$ ft.

5. Compute slenderness (L_{ss}/D) $= 3.8$ and 1.96, respectively. $D = 2$ ft. (24 in.) is a proper size.

A second case is where $F = 0.436$. The gas capacity constant is determined from

$$V_g = \frac{Q}{A_g}\quad\text{where}\quad Q = 0.327\frac{TZQ_g}{P}$$

$$A_g = 0.436 \times \frac{\pi}{4}D^2 = 0.342D^2$$

$$V_g = \frac{0.327TZQ_g}{0.342D^2P} = 0.96\frac{TZQ_g}{D^2P}$$

$$t_g = \frac{L_{eff}}{V_g} = \frac{L_{eff}D^2P}{0.96TZQ_g}$$

$$t_d = (D-h)\,/V_t\quad\text{if}\quad F = 0.436,\quad \frac{h}{D} = 0.55,\quad h = 0.55D$$

$$t_d = \frac{0.45D}{V_t}\quad\text{if}\quad V_t = k \times 3.587\left(\frac{L_{ss}}{10}\right)^{0.56}$$

$L_{ss} = L_{eff} + D$ for gas capacity, such that
$$V_t = 0.988K\,(L_{eff} + D)^{0.56}\quad\text{where}\quad K = 0.3$$

$$t_d = \frac{1.52D}{(L_{eff} + D)^{0.56}}$$

$$t_g = t_d$$

$$\frac{L_{eff}D^2P}{0.96TZQ_g} = \frac{1.52D}{(L_{eff} + D)^{0.56}}$$

Substituting Q_g, P, T, Z gives

$$L_{eff}D^2 = \frac{6.4D}{(L_{eff} + D)^{0.56}}$$

Liquid capacity constraint is determined from

$$t_r = \frac{\text{Vol.}}{Q}$$

where Vol. $= (\pi D^2/4)\,(1 - F)\,L_{eff} = 0.443D^2L_{eff}$
$Q = 0.000065\,Q_L$ if t_r is in min
$t_r = 0.443D^2L_{eff}/0.000065\,Q_L$
$D^2L_{eff} = 0.0088\,t_r\,Q_L$

Assume retention time $t_r = 1$ min and $D = 2$ ft.

$$L_{eff} = \frac{0.0088 \times 1 \times 2{,}000}{4}\,4.4\text{ ft.}$$

$$L_{ss} = \frac{4}{3} \times 4.4 = 5.9\text{ ft.}$$

$$D = 2.5$$

$$L_{eff} = \frac{0.0088 \times 1 \times 2{,}000}{2.5 \times 2.5} = 2.8\text{ ft.}$$

$$L_{ss} = \frac{4}{3} \times 2.8 = 3.8\text{ ft.}$$

As you can see, the liquid level has significant meaning for separator length. Usually, the fraction of the total area F available for gas flow is equal to 0.5. The liquid level control placement of a horizontal separator is more critical than in a vertical separator and the surge space is somewhat limited.

6.6.6 Vessel Internals [18]

The proper selection of internals can significantly enhance the operation of separators. Proprietary internals often are helpful in reducing liquid carryover at design conditions. Nevertheless, they cannot overcome an improper design or operation at off-design conditions.

Production equipment involving the separation of oil and gas often uses impingement-type mist-extraction elements. This element is usually of the vane type or of knitted wire.

The vane type consists of a labyrinth formed with parallel metal sheets with suitable liquid collection "pockets." The gas, in passing between plates, is agitated and has to change direction a number of times. Obviously, some degree of centrifugation is introduced, for as the gas changes direction the heavier partials tend to be thrown to the outside and are caught in the pockets provided.

Coalescence of small particles into those large enough to settle by gravity is provided by two mechanisms: agitation and surface. The surface of the element is usually wet, and small particles striking it are absorbed. Inasmuch as the pockets are perpendicular to the gas flow, the liquid thus formed does not have to flow against the gas. Consequently, small compact units have a large capacity.

As the plates are placed closer together and more pockets are provided, greater agitation, centrifugal force, and collection surface are provided, but the pressure drop is increased correspondingly. Thus, for a given flow rate, the collection efficiency is normally some function of the pressure drop.

In the average application, this pressure drop varies from 1.2 to 10 in. (3 to 25 cm) of water. Because of this pressure drop and to prevent gas bypassing the extractor, a liquid-collection pas incorporating a liquid seal is necessary for the liquid to drain properly.

Increased use has been made of mist extractors composed of a knitted wire mesh supported on a lightweight support. This material has given generally favorable results and has a low installed cost.

The element consists of wire knitted into a pad having a number of unaligned, asymmetrical openings. Although similar in appearance to filter media, its action is somewhat different. The latter are rather dense and have small openings. This knitted wire, on the other hand, has about 97% to 98% free voids and collects the particles primarily by impingement.

The material is available in single wound units of varying thicknesses in diameter up to 35 in. (90 cm) or in laminated strips for insertion through manholes in large process vessels.

The principle of separation is similar to that of the vane-type unit. The gas flowing through the pad is forced to change direction a number of times, although centrifugal action is not so pronounced. Impingement is the primary mechanism.

A liquid particle striking the metal surface, which it does not "wet," flows downward where adjacent wires provide some capillary space. At these points, liquid collects and continues to flow downward. Surface tension tends to hold these drops on the lower face of the pad until they are large enough for the downward force of gravity to exceed that of the upward gas velocity and surface tension.

Efficiency is a function of the number of targets presented. This may be accomplished by increasing the pad thickness, changing wire diameter, or the closeness of the weave.

The wire mesh normally used falls within the following range:

- Wire diameter—0.003 to 0.011 in. (0.0076 to 0.028 cm)
- Void volume—92% to 99.4%
- Specific weight—3 to 33 lb/ft.^{3} (48 to 529 kg/m^3)
- Surface area—50 to 600 ft.^{2}/ft.^{3} (164 to 1970 m^2/m^3)

The most commonly used wire has a void volume of 97% to 98%, a bulk-specific weight approximately 12 lb/ft.^{3} (192 kg/m^3), a surface area of 100 to 125 ft.^{2}/ft.^{3} (328 to 410 m^2/m^3), with a wire diameter of 0.011 in. (0.028 cm). A pad thickness of 4 to 6 in. (10 to 15 cm) is sufficient for most separator applications, although thicknesses up to 3 ft. (0.9 m) have been reported. In separator service, 4 to 6 in. (10 to 15 cm) will normally suffice.

Any common metal may be used in these units, including carbon steel, SS, aluminum, monel, etc. The pressure drop is a function of the entrainment load, the pad design, and gas velocity but will not exceed 3 cm of water in the average insulation. Because of this small pressure drop, the elements do not have to be "held down" and are normally only wired to the support grid to prevent shifting unless surging flow is anticipated.

Experience has shown that the support grid should contain at least 90% free area in order to eliminate any restrictions to liquid drainage. The pads are light in weight so that a light angle-iron support is adequate. When both liquid and solids are present, a portion of the latter obviously will be scrubbed out. If only dry solids are present, the efficiency of this design is substantially less. Many vessel carryover problems are encountered. Foaming is a major culprit and requires more than simply better mist extraction. Most such problems develop by default. The vendor automatically uses the standard sizing curves and equipment, and the buyer assumes this will be good enough.

With glycols, amines, and similar materials, which tend to foam, a dual mist extractor would normally be specified—the lower one being of the vane type and the upper one being a wire mesh. A space of 6 to 12 in. (15 to 30 cm) would be left between them. The vane type will handle large volumes of liquid but is relatively inefficient on small droplets. It, therefore, serves as a bulk removal device (and helps to coalesce foam). The wire mesh, which has limited liquid capacity, may therefore operate more effectively.

When using the vane-type mist extractor, one must be careful that the pressure drop across it does not exceed the height above the liquid level if a downcomer pipe is used. Otherwise, liquid will be "sucked out" overhead. The downcomer pipe can become partially plugged to accentuate the problem. Two wire mesh pads may be used in like fashion, with the first being used as a coalescer. As a rule of thumb, the coalescer pad should have about half the free space area of the second pad. Any wire mesh pad should be installed so that the flow is perpendicular to the pad face (pad is horizontal in a vertical vessel).

With materials like glycol and amine, which wet metal very well, Teflon-coated mesh may prove desirable. Remember, the liquid must be nonwetting in order to stay as droplets that can run down the wires and coalesce into bigger droplets. A wetting fluid will tend to "run up" the wires. It has been noted that centrifugal force is an integral part of the separation process. The standard oil and gas separator may have an inlet that utilizes centrifugal force to separate the larger droplets. The same principle is used in some mist extractor elements except that higher velocities are needed to separate the smaller droplets. The velocity needed for separation is a function of the particle diameter, particle and gas specific weight and the gas viscosity.

With a given system, the size of particle collected is inversely proportional to the square root of the velocity. Consequently, the success of a cyclonic mist extractor is dependent upon the velocity attained. Furthermore, the velocity needed to separate a given size of particle must increase, as the density of the particle becomes less. In addition to producing the necessary velocity, the mist extractor must provide an efficient means of collecting and removing the particles collected in order to prevent re-entrainment.

One common type of equipment is often called a *steam separator* because it has been widely used to separate condensate and pipe scale in steam systems. Normally, a relatively small vessel imparts a high velocity to the incoming gas and then makes the gas change direction radically to prevent re-entrainment. In general, it will separate particles 40 microns and larger very efficiently.

Another type uses the same principle but, in addition, forces the gas to pass through a labyrinth that introduces impingement effects and forces the gas to change direction a number of times. This is, in reality, a combination type and is relatively efficient. The general performance characteristics are the same as efficient mist extractors of other types. Some, however, are complex and relatively expensive.

6.6.7 Oil–Water–Gas Separation

Two liquids and gas or oil–water–gas separation can be easily accomplished in any type of separator by installing either special internal baffling to construct a water lap or crater siphon arrangement or by use of an interface liquid level control. A three-phase feature is difficult to install in a spherical separator due to the limited internal space available. With three-phase separation, two liquid level controls and two liquid dump valves are required. Figures 6.6.9 to 6.6.11 [15]

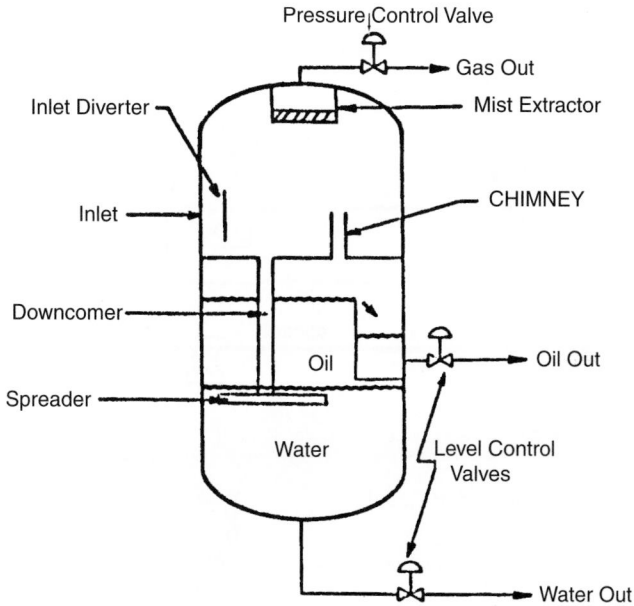

Figure 6.6.9 *Three-phase vertical separator.*

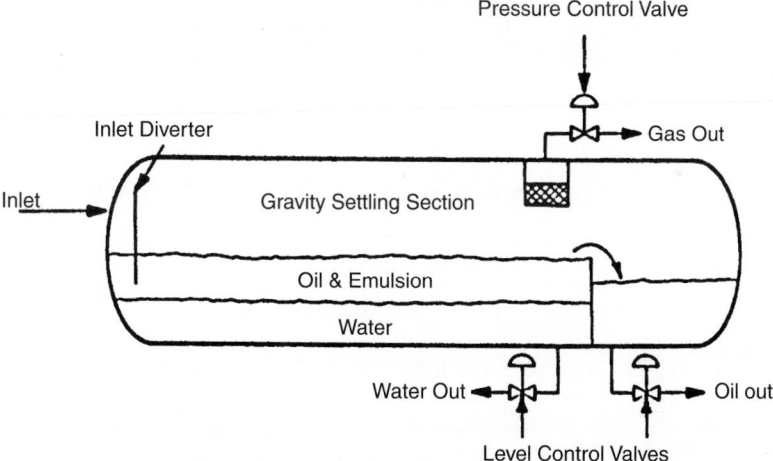

Figure 6.6.10 *Three-phase horizontal separator.*

illustrate schematics of some three-phase separators. The basic design aspects of three-phase separation have been covered under "Sizing of Two Phase Oil–Gas Separators". Regardless of shape, all three-phase vessels must meet the following requirements:

- Liquid must be separated from the gas in a primary separating section.
- Gas velocity must be lowered to allow liquids to drop out.
- Gas is then scrubbed through an efficient mist extractor.
- Water and oil must be diverted to a turbulence-free section of the vessel.
- Liquids must be retained in the vessel long enough to allow separation.
- The water–oil interface must be maintained.
- Water and oil must be removed from the vessel at their respective outlets.

Sizing a three-phase separator for water removal is mainly a function of retention time; required retention time is related to the volume of the vessel, the amount of liquid to be handled and the relative specific gravities of the water and oil. The effective retention volume in a vessel is that portion of the vessel in which the oil and water remain in contact with one another. As far as oil–water separation is concerned, once either substance leaves the primary liquid section, although it may remain in the vessel in a separate compartment it cannot be considered as part of the retention volume. There are two primary considerations in specifying retention time:

- Oil settling time to allow adequate water removal from oil
- Water settling time to allow adequate oil removal from water

Basic design criteria for liquid retention time in three-phase separators are given in Table 6.6.7 [11].

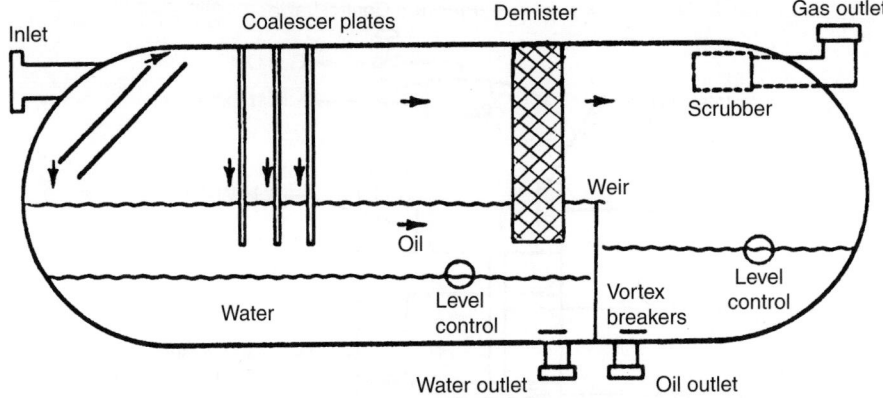

Figure 6.6.11   *Three-phase horizontal separator with coalescer plates.*

Table 6.6.7   *Typical Retention Time for Liquid-Liquid Separation*

Type of Separation	Retention Time
Hydrocarbon/water separators	
Above 35° API hydrocarbons	3 to 5 min
Below 35° API hydrocarbons	
100°F and above	5 to 10 min
80°F	10 to 20 min
60°F	20 to 30 min
Ethylene glycol/hydrocarbon	20 to 60 min
separators (cold separators)	
Amine/hydrocarbon separators	20 to 30 min
Coalescers, hydrocarbon/water	
separators	
100°F and above	5 to 10 min
80°F	10 to 20 min
60°F	20 to 30 min
Caustic/propane	30 to 45 min
Caustic/heavy gasoline	30 to 90 min

Example 6.6.3
Determine the size of a vertical separator to separate 6,300 bpd of oil from its associated gas, water 500 bpd (SG = 0.75) at a pressure of 300 psig, and a temperature with SG_g=1.03, 100°F. The oil has a density of 49.7 lb/ft.^{3} and solution gas/oil ratio of 580 scf/STB at 60°F and 14.7 psia.

1. The gas capacity constraint $V_g = V_t$, such that

$$V_g = \frac{Q}{A_g} \quad Q = \frac{R_s \times Q_0}{86,400} \times \frac{14.7}{P} \times \frac{TZ}{520}$$

$$P = 300 \quad T = 560 \quad Z = 0.94 \text{ (from chart)}$$

$$Q = 2.1 \text{ ft.}^3/s$$

$$A_g = \frac{\pi D^2}{4} = 0.785 D^2$$

$$V_t = K \left(\frac{\gamma_0 - \gamma_g}{\gamma_g} \right) \quad K = 0.33 \text{ (from Table 6.6.6)}$$

$$\gamma_0 = 49.7 \text{ lb/ft.}^3$$

$$\gamma_g = \frac{PM}{ZRT} = \frac{300 \times 29 \times 0.75}{0.94 \times 10.73 \times 560} = 1.155 \text{ lb/ft.}^3$$

$$\left(\frac{\gamma_0 - \gamma_g}{\gamma_g} \right)^{0.5} = \left(\frac{49.7 - 1.155}{1.155} \right)^{0.5} = 6.48$$

$$V_t = 2.14 \text{ ft./s}$$

$$D = 1.117 \text{ ft.} = 13.4 \text{ in.}$$

2. Calculate minimum diameter from requirement for water droplets to fall through oil layer. Use 500-μm droplets if no other information is available.
   For μ_0 =assume 3.0 cp.

$$V_t = V_0$$

According to Stokes law

$$V_t = 2660 \frac{(\gamma_w - \gamma_0)}{\mu} D_P^2$$

$$V_t(\text{ft./s}); \quad \gamma_w, \gamma_0(\text{lb/ft.}^3); \quad D_p(\text{ft.}); \quad \mu(\text{cp})$$

$$V_0 = \frac{Q}{A} \quad Q(\text{ft./s}); \quad A(\text{ft.}^2); \quad Q_0(\text{bpd})$$

$$Q = 6.49 \times 10^{-5} Q_0$$

$$A = \frac{\pi D^2}{4} = 0.785 D^2 \quad \text{Where D is in ft.}$$

$$V_0 = \frac{6.49 \times 10^{-5} Q_0}{0.785 D^2} = 8.27 \times 10^{-5} \frac{Q_0}{D^2}$$

$$8.27 \times 10^{-5} \frac{Q_0}{D^2} = 2660 \left(\frac{\gamma_w - \gamma_0}{\mu_0} \right) D_p^2$$

$$D_p = 500 \text{ μm} = 500 \times 3.2808 \times 10^{-6} \text{ft.}$$

$$D_p = 1.64 \times 10^{-3} \text{ft.}$$

$$D = \left[\frac{8.27 \times 10^{-5} Q_0 \mu_0}{2600 \, (\gamma_w - \gamma_0) \, D_P^2} \right]^{0.5} = 4.15 \text{ ft.} = 49.8 \text{ in.}$$

3. Calculate the total weight of oil and water in separator $(h_o + h_w)$

$$t_r = \text{Vol.}/Q$$

$$\text{Vol.} = (\pi D^2/4) h$$

$$Q = 6.49 \times 10^{-5} Q_L \quad \text{If t is in min}$$

$$t_r \, 60 = \frac{\pi D^2 h}{4 \times 6.49 \times 10^{-5} Q_L} \quad Q_L(\text{bpd})$$

$$D^2 h = 4.958 \times 10^{-3} Q_L t_r$$

For two-phase separator design

$$D_2 h_0 = 4.958 \times 10^{-3} Q_0 (t_r)_0$$

$$D_2 h_w = 4.958 \times 10^{-3} Q_w (t_r)_w$$

$$h_0 + h_w = 4.958 \times 10^{-3} \frac{Q_0 (t_r)_0 + Q_w (t_r)_w}{D^2}$$

where h_0 = height of oil pad in ft.
h_w = height from water outlet to interface in ft.

Choose a nominal diameter from Table 6.6.3, considering minimum diameter in pts 2.54 in (4.5 ft.) is a proper choice. $(t_r)_0 = (t_t)_w = 5$ min from Table 6.6.7.

$$h_0 + h_w = 4.958 \times 10^{-3} \frac{500 \times 5 + 6{,}300 \times 5}{(4.5)^2} = 8.32 \text{ ft.}$$

If $D = 5$ ft. $h_0 + h_w = 6.74$ ft.

4. Calculate seam-to-seam length and slenderness ratios for $D = 4.5$ ft.

$$L_{ss} = h_0 + h_w + \frac{76}{12} = 8.32 + 6.33 = 14.6 \text{ ft.}$$

$$\frac{L_{ss}}{D} = \frac{14.6 \text{ ft}}{4.5} = 3.2 \quad \text{for} \quad D = 5 \text{ ft.}$$

$$L_{ss} = 6.74 + \frac{76}{12} = 13 \text{ ft.}$$

$$\frac{L_{ss}}{D} = \frac{13 \text{ ft.}}{5} = 2.6$$

Vertical three-phase separators have slenderness ratios on the order of 15 to 3. In this case, both diameters are acceptable.

Example 6.6.4
Determine the size of a horizontal separator for data given in Example 6.6.3 and $F = 0.5$.

1. Gas capacity constraint: $t_g = t_d$

$$V_g = \frac{Q}{A_g}$$

$$Q = \frac{R \times Q_0}{86{,}400} \times \frac{14.7}{P} \times \frac{TZ}{520} = 2.1 \text{ ft.}^3/\text{s}$$

$$A_g = \frac{\pi D^2}{4} \times F = 0.393 D^2$$

$$V_g = \frac{2.1 \text{ft.}^3/\text{s}}{0.393 D^2} = \frac{5.34}{D^2}$$

$$t_g = \frac{L_{eff}}{V_g} = \frac{L_{eff} D^2}{5.34}$$

$$t_d = \frac{D/2}{V_t} = \frac{D}{2V_t}$$

$$V_t = K \left(\frac{\gamma_L - \gamma_g}{\gamma_g} \right)^{0.5} \left(\frac{L_{ss}}{10} \right)^{0.56}$$

$$= 0.33 \left(\frac{49.7 - 1.155}{1.155} \right)^{0.5} \left(\frac{L_{eff} + D}{10} \right)^{0.56}$$

$$= 0.589 (L_{eff} + D)^{0.56}$$

$$\frac{L_{eff} D^2}{5.34} = \frac{D}{0.589 (L_{eff} + D)^{0.56}}$$

$$L_{eff} D^2 = \frac{9.07 D}{(L_{eff} + D)^{0.56}}$$

For assumed D, calculate L_{eff}:

D (ft.)	L_{eff} (ft.)
4.0	0.93
4.5	0.79
5.0	0.72

2. Calculate maximum oil pad thickness $(h_0)_{max}$. Use $500 \, \mu m$ droplet if no other information is available $t_w = t_0$:

$$t_w = \frac{h_0}{V_t} \quad V_t = 2{,}660 \frac{(\gamma_w - \gamma_0)}{\mu} D_p^2$$

$$t_w = \frac{h_0 \mu}{2{,}660 \, (\gamma_w - \gamma_0) \, D_p^2}$$

$$t_0 = 60 (t_r)_0 \quad t_r (\text{min})$$

$$\frac{h_0 \mu}{2{,}660 \, (\gamma_w - \gamma_0) \, D_p^2} = 60 (t_r)_0$$

$$h_0 = \frac{159{,}600 \, (\gamma_w - \gamma_0) \, D_p^2 \, (t_r)_0}{\mu}$$

This is the maximum thickness the oil pad can be and still allow the water droplets to settle out in time $(t_r)_0$. If $D_p = 500 \, \mu m = 500 \times 3.2808 \times 10^{-6} \text{ ft.} = 1.64 \times 10^{-3} \text{ ft.}$

$$h_0 = \frac{0.43 \, (\gamma_w - \gamma_0) \, (t_r)_0}{\mu}$$

For a given retention time (from Table 6.6.7, is equal to 5 min) and a given water retention time (also 5 min), the maximum oil pad thickness constraint establishes a maximum diameter.

$$\gamma_w = 64.3 \text{ lb/ft.}^3 \quad \gamma_0 = 49.7 \text{ lb/ft.}^3 \quad \mu = 3 \text{ cp}$$

$$(h_0)_{max} = 10.46 \text{ ft.}$$

3. Calculate the fraction of the vessel cross-sectional area occupied by the water phase (see Figures 6.6.8 and 6.6.12) $A, A_w, A_0 \text{ (ft.}^2)$:

$$Q(\text{ft.}^3/\text{s}) \quad t(\text{s}) \quad L_{eff}(\text{ft.})$$

$$A = \frac{Q_t}{L_{eff}}$$

$$Q = 6.49 \times 10^{-5} Q_0, \quad \text{also } Q = 6.49 \times 10^{-5} Q_w$$

$$t_0 = 60 (t_r)_0 \quad t = 60 (t_r)_w$$

$$A_0 = 3.89 \times 10^{-3} \frac{Q_0 \, (t_r)_0}{L_{eff}} \quad A_w = 3.89 \times 10^{-3} \frac{Q_w \, (t_r)_w}{L_{eff}}$$

$$A = 2(A_0 + A_w)$$

Hence

$$\frac{A_w}{A} = \frac{3.89 \times 10^{-3} Q_w \, (t_r)_w \, L_{eff}}{2 L_{eff} 3.89 \times 10^{-3} \left[Q_w \, (t_r)_w + Q_0 \, (t_r)_0 \right]}$$

$$= \frac{0.5 Q_w \, (t_r)_w}{Q_w \, (t_r)_w + Q_0 \, (t_r)_0}$$

4. From Figure 6.6.12, determine the coefficient "d," and calculate D_{max} for oil pad thickness constraint:

$$\frac{A_w}{A} = \frac{0.5(5)500}{500(5) + 6{,}300(5)} = 0.037$$

$$d = \frac{h_0}{D} = 0.43$$

$$D_{max} = \frac{(h_0)_{max}}{d} = \frac{10.46}{0.43} = 24.3 \text{ ft.}$$

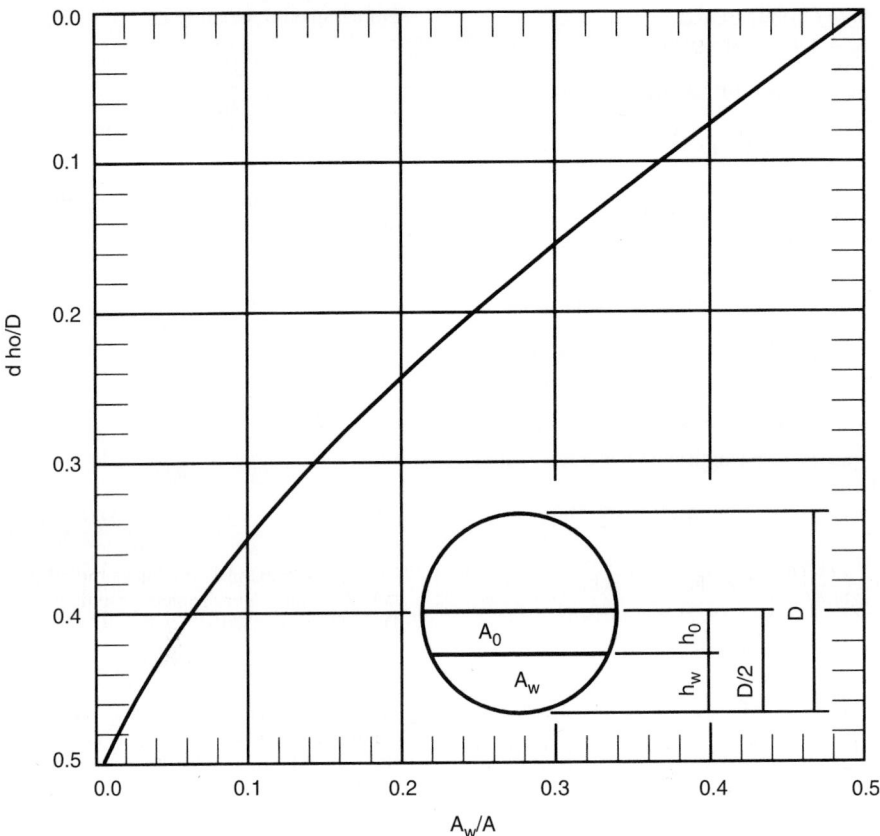

Figure 6.6.12 Coefficient "d" for a cylinder half filled with liquid [15].

D_{max} depends on Q_0, Q_w, $(t_r)_0$, and $(t_r)_w$.

5. Liquid retention constraint:

$$t = \frac{\text{Vol.}}{Q}$$

$$\text{Vol.} = \frac{1}{2}\left(\frac{\pi D^2}{4}L_{eff}\right) = 0.393 D^2 L_{eff}$$

$$(\text{Vol.})_0 = 0.393 D^2 L_{eff}\left(\frac{A_0}{A_{liq}}\right)$$

$$(\text{Vol.})_w = 0.393 D^2 L_{eff}\left(\frac{A_w}{A_{liq}}\right)$$

Q_0 and Q_w are in bpd:

$$Q = 6.49 \times 10^{-5} Q_0 \quad Q = 6.49 \times 10^{-5} Q_w$$

$$t_0 = \frac{0.393 D^2 L_{eff}\left[\dfrac{A_0}{A_{liq}}\right]}{6.49 \times 10^{-5} Q_0}$$

$$\frac{Q_0 t_0}{D^2 L_{eff}} = 6,055\frac{A_0}{A_{liq}}, \quad \text{also} \quad \frac{Q_w t_w}{D^2 L_{eff}} = 6,055\frac{A_w}{A_{liq}}$$

$(t_r)_0$ and $(t_r)_w$ are in min:

$$\frac{Q_0\,(t_0)_0}{D^2 L_{eff}} = 101\frac{A_0}{A_{liq}} \quad \frac{Q_w\,(t_r)_w}{D^2 L_{eff}} = 101\frac{A_w}{A_{liq}}$$

Adding by sides

$$\frac{Q_0\,(t_r)_0 + Q_w\,(t_r)_w}{D^2 L_{eff}} = 101\frac{A_0 + A_w}{A_{liq}} = 101$$

$$D^2 L_{eff} = 0.0099[Q_0(t_r)_0 + Q_w(t_r)_w]$$

for given $(t_r)_0$ and $(t_r)_w = 5$ min, and

$$Q_0 = 6,300 \text{ bpd} \quad Q_w = 500 \text{ bpd}$$

$$D^2 L_{eff} = 0.0099[500(5) + 6,300(5)] = 336.6$$

6. Compute combinations of D and L_{eff}

$$L_{ss}, \quad \text{where} \quad L_{ss} = L_{eff}\frac{4}{3} \quad \text{and} \quad \frac{L_{ss}}{D}$$

D (ft.)	L_{eff} (ft.)	L_{ss}	L_{ss}/D
5.0	13.5	18.0	3.5
5.5	11.1	14.8	2.7
4.5	16.6	22.2	4.9

For three-phase horizontal separators, slenderness ratio is in the range of 3 to 5 usually, so D = 4.5 and 5 ft. are proper choices. As one can see, liquid retention constraint limits three-phase separator size; gas capacity, and oil pad thickness do not govern.

6.6.8 Two-Stage Separation Systems

In high-pressure gas-condensate separation systems, it is generally accepted that a stepwise reduction of the pressure on the liquid condensate will appreciably increase the recovery of stock tank liquids. The calculation of the actual performance of the various separators in a multistage system can be made, using the initial wellstream composition and the operating temperatures and pressures of the various stages. Theoretically, three or four stages of separation would increase the liquid recovery over two stages: the net increase over two-stage separation will rarely pay out the cost of the second and/or third separator. Therefore, it has been generally accepted that two stages of separation plus the stock tank are the most optimum considered. The actual increase in liquid recovery for two-stage separation over single stage may vary from 3% to 15% (sometimes even more) depending upon the wellstream composition, operating pressures and temperatures.

The optimum high stage or first separator operating pressure is generally governed by the gas transmission line pressure and operating characteristics of the well. This will generally range in pressure from 600 to 1200 psia (41.4 to 82.7 bar). For each high or first-stage pressure, there is an optimum low-stage separation pressure that will afford the maximum liquid recovery. This operating pressure can be determined from an equation based on equal pressure ratios between the stages.

$$R = \left(\frac{P_1}{P_s}\right)^{0.5} \quad \text{or} \quad P_2 = \frac{P_1}{R} = P_s R^{n-1} \qquad [6.6.9]$$

where R = pressure ratio
n = number of stages
P_1 = first-stage separator pressure in psia
P_2 = second-stage separator pressure in psia
P_s = stock tank pressure in psia

Figure 6.6.13 illustrates a schematic flow diagram of a typical high-pressure well production equipment installation [19]. The basic equipment is illustrated for two-stage separation of the high-pressure stream. From the wellhead, the high-pressure wellstream flows through a high-pressure separator and indirect heater gas production unit. In this unit the inlet stream is heated prior to choking to reduce the wellstream pressure to sales line pressure. This is done to prevent the formation of hydrates in the choke or downstream of the choke in the separator or sales line. From the indirect heater, the wellstream passes to the high-pressure

separator where the initial separation of the high-pressure gas stream and produced well fluids occur.

From the high-pressure separator, the gas flows through an orifice meter and to the sales gas line. The liquid from the high-pressure separator passes through a diaphragm valve where the pressure is reduced and it is discharged to a low-pressure flash separator. In the low-pressure flash separator that would operate at approximately 100 psi (6.9 bar), a second separation occurs between the liquids and the lighter hydrocarbons in the liquids. The gas released from the low-pressure flash separator is returned back to the high-pressure unit, where it may be used for both instrument and fuel gas for the indirect heater. As illustrated in Figure 6.6.13, a secondary makeup line is shown from the high-pressure separator, which would provide additional makeup gas for the instrument gas and fuel gas, if not enough gas was released from the low-pressure separator. However, typically more gas is released than is required, and the additional low-pressure gas may be sold in a low-pressure gas gathering system and/or used for other utility purposes, such as fuel for company compressor engines or other fired equipment on that lease. This may be for reboilers on dehydrators or acid gas sweetening units, etc. From the low-pressure flash separator, the liquid is discharged though another diaphragm valve into a storage tank that is generally operated at atmospheric pressure.

Example 6.6.5
Perform flash vaporization calculation to determine the increase recoveries that would be seen in both low-pressure flash gas as well as increased liquid recoveries in the storage tank. High-pressure gas $SG_g = 0.67$, the high-pressure separator pressures from 500 to 1,000 psi, the low-pressure separator $P = 100$ psi, and the storage tank pressure $P_s = 14.7$ psia. Temperature for all vessels is the same, 70°F.

Results
Figure 6.6.14 illustrates the gas produced from the low-pressure flash separator for the above-described wellstream at various high-pressure operating pressures (line pressure). The gas produced from the low-pressure flash separator in Mcf per year may be read from ordinate, based on the high-pressure gas stream flow rate in MMscfd and the high-pressure separator operating pressure.

Figure 6.6.15 illustrates the increase in stock tank liquid recovery that would be achieved using the low-pressure flash separator. This chart is also based on high-pressure gas flow rate in MMscfd and the high-pressure separator operating

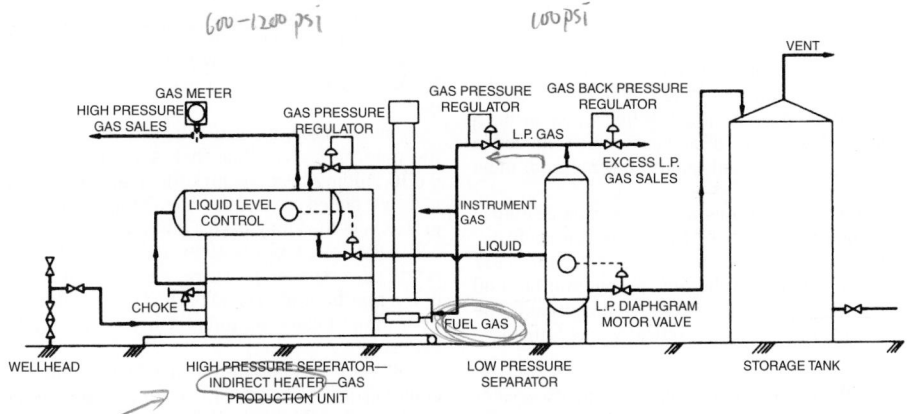

Figure 6.6.13 *Two-stage separation system.*

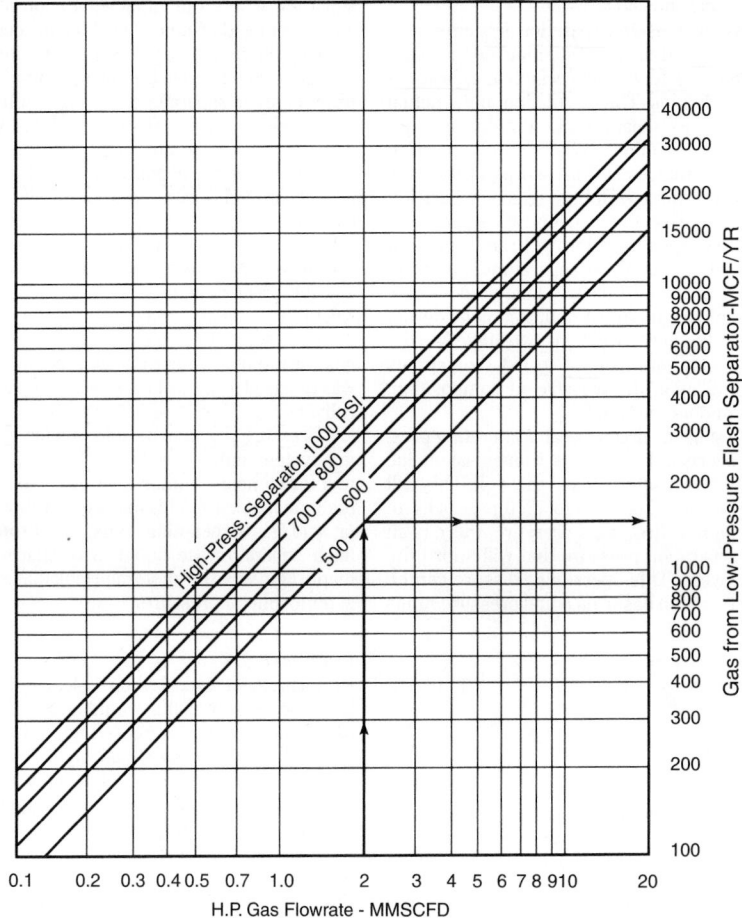

Figure 6.6.14 *Low-pressure gas from flash separator [20].*

pressure. The increase in stock tank liquid recovery may be read from the chart in bbl/year.

This additional recovery not only gives profit but also prevents the unneeded waste of precious hydrocarbon energy that would normally be vented out the stock tank using only single-stage separation.

6.6.9 Crude Oil Treating Systems

Water content of the untreated oil may vary from 1% to over 95%. Purchasers, depending on local conditions, accept a range from 0.2% to 3% of water in oil. When water forms a stable emulsion with the crude oil and cannot be removed in conventional storage tanks, emulsion-treating methods must be used.

An *emulsion* is a heterogeneous liquid system consisting of two immiscible liquids with one of the liquids intimately disposed in the form of droplets in the second liquid (the water remaining is less than 10% of the oil). A common method for separating water–oil emulsion is to heat the stream. The use of heat in treating crude oil emulsions has four basic benefits:

1. Heat reduces the viscosity of the oil, resulting in a greater force during collision of the water droplets.
2. Heat increases the droplets' molecular movement.

3. Heat can enhance the action of treating chemicals, causing the chemicals to work faster to break the film surrounding the droplets of the dispersed phase of the emulsion.
4. Heat may increase the difference in density between the oil and the water, thus accelerating settling.

In general, at temperatures below 180°F (82°C), the addition of heat will increase the difference in density. Most light oils are treated below 180°F. For heavy crudes (below 20°API), heat may have a negative effect on difference in density.

In some cases, increased heat may cause the density of the water to be less than that of oil, as is shown in Figure 6.6.16. Adding heat changes the quality of the oil. The light ends are boiled off, and the remaining liquid has a lower API gravity and thus may have a lower value. Figures 6.6.17 and 6.6.18 illustrate typical gravity and volume losses for 33°API crude oil versus temperature. The molecules leaving the oil phase may be vented or sold. Heat can be added to the liquid by a direct heater, an indirect heater, or any type of heat exchanger.

A direct fired heater is one in which the fluid to be heated comes in direct contact with the immersion-type heating tube or element of the heater. These heaters are generally used when large amounts of heat input are required and to heat

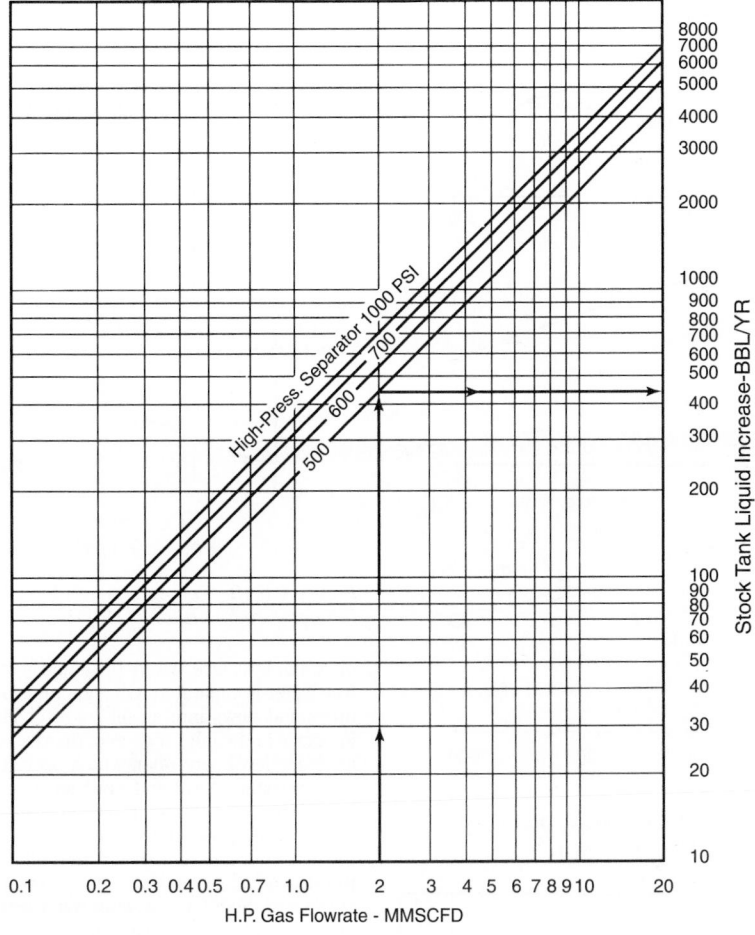

Figure 6.6.15 *Stock tank liquid increase with flash separator [20].*

Are these same
as Fig 6.6.13 ?
and heater treater?

low-pressure noncorrosive liquids. These units are normally constructed so that the heating tube can be removed for cleaning, repair, or replacement.

An indirect fired heater is one in which the fluid passes through the pipe coils or tubes immersed in a bath of water, oil, salt, or other heat transfer medium that, in turn, is heated by an immersion-type heating tube similar to that used in the direct fired heater. Those heaters generally are used for high-corrosive fluids and are more expensive than direct fired heaters.

Heat exchangers are very useful where waste heat is recovered from an engine, turbine, or other process stream or where fired heaters are prohibited.

6.6.9.1 Treating Equipment

All devices that accelerate the separation of two phases when the natural retention time is too long for commercial application, usually over 10 min, are called *treaters.* Emulsion treaters use some combination of heat, electricity, chemicals, retention time, and coalescence to separate oil and water. Treaters are designed as either vertical or horizontal vessels. The vertical heater is shown in Figure 6.6.19. Flow enters the top of the treater into a gas separation section. This section can be small if the treater is located downstream of the separator. The liquids flow through a downcomer to the base of the treater, which serves as a free-water knockout; the

bottom section can be small. If the total wellstream is to be treated, this section should be sized for sufficient retention time to allow the free water to settle out. This minimizes the amount of fuel gas needed to heat the liquid stream rising through the heating section. The oil and emulsion flows upward around the fire tubes to a coalescing section, where sufficient retention time is provided to allow the small water droplets to coalesce and to settle to the water section. Treated oil flows out the oil outlet. The oil level is maintained by pneumatic or lever-operated dump valves.

It is necessary to prevent stream from being formed on the fire tubes. This can be done by employing the "40° rule." That means that the operating pressure is kept equal to the pressure of the saturated steam at a temperature equal to the operating temperature plus 40°F (4.5°C). The normal full-load temperature difference between the fire tube wall and the surrounding oil is approximately 30°F, allowing 10°F for safety; the 40° rule will prevent flashing of steam on the wall of the heating tube.

A low-pressure vertical flow treater, of large diameter, is called a "gunbarrel" (Figure 6.6.20). Most gunbarrels are unheated, although it is possible to provide heat by heating the incoming stream external to the tank, installing heat coils in the tank, or circulating water to an external heater in a closed loop as shown in Figure 6.6.21. A heated gunbarrel emulsion treater is shown in Figure 6.6.22.

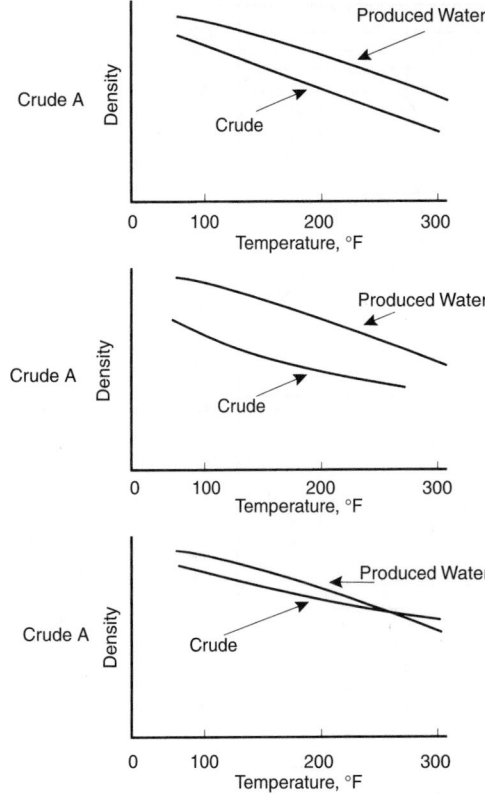

Figure 6.6.16 *Relationship of specific gravity with temperature for these three crude oils [13].*

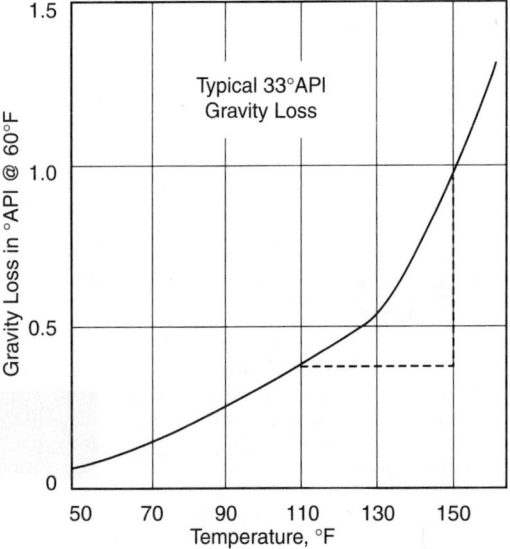

Figure 6.6.17 *API gravity loss versus temperature for crude oil [13].*

For higher flow, inlet horizontal treaters are normally preferred. A typical design of a horizontal treater is shown in Figure 6.6.23. Flow enters the front section of the treater where gas is flashed. The liquid flows downward to near the

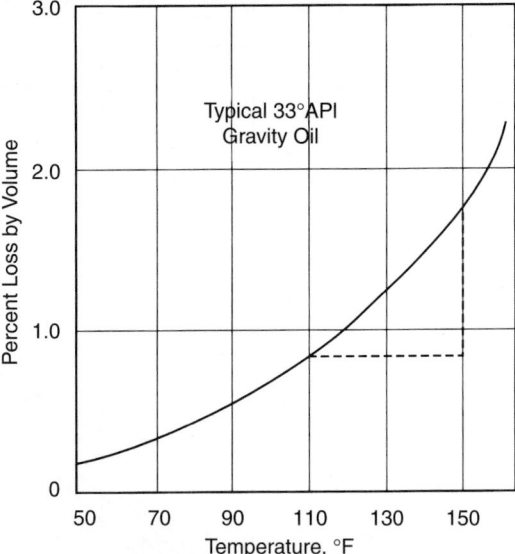

Figure 6.6.18 *Percentage loss by volume versus temperature for crude oil [13].*

oil–water interface, where the liquid is water-washed and the free water is separated. Oil and emulsion rises past the fire tubes and flows into an oil surge chamber. The oil–water interface in the inlet section of the vessel is controlled by an interface level controller, which operates a dump valve for the free water. The oil and emulsion flows through a spreader into the back or coalescing section of the vessel, which is fluid packed. The spreader distributes the flow evenly throughout the length of this section. Treated oil is collected at the top through a collection device used to maintain uniform vertical flow of the oil. Coalescing water droplets fall counter to the rising oil. The oil–water interface level is maintained by a level controller and dump valve for this section of the vessel. A level control in the oil surge chamber operates a dump valve on the oil outlet line regulating the flow of oil out the top of the vessel and maintaining a liquid-packed condition in the coalescing section. Gas pressure on the oil in the surge section allows the coalescing section to be liquid-packed. The inlet section must be sized to handle separation of the free water and heating of the oil. The coalescing section must be sized to provide adequate retention time for coalescence to occur and to allow the coalescing water droplets to settle downward counter to the upward flow of the oil.

6.6.9.2 Heat Input Requirements [21, 22]

The required heat input for an insulated vessel can be calculated as

$$Q_{th} = \dot{w}c\Delta T$$

where

Q_{th} = heat required in Btu/hr
$\dot{w}$ = weight flowrate in lb/hr
c = specific heat constant at average temperature [Btu/(lb°F)] = 0.5 for oil and = 1.0 for water
Δt = temperature increase, assuming a water weight of 350 lb/bbl

$$Q_{th} = \frac{350}{24}q_0(SG_0)(0.5)\Delta T + \frac{350}{24}q_w(SG_w)(1.0)(\Delta T)$$

$$= 14.6\Delta T(0.5q_0 SG_0 + q_w SG_w)$$

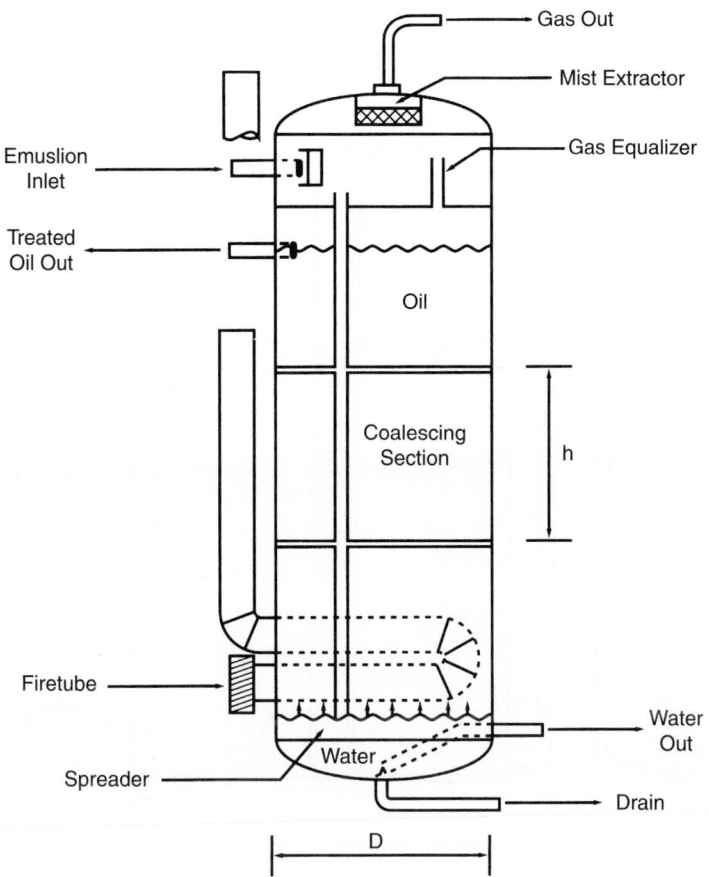

Figure 6.6.19 *Vertical treater [15].*

If heat loss is assumed to be 10% of the heat input, then

$$Q = \frac{Q_{th}}{0.9} = 16.2\Delta T(0.5q_0SG_0 + q_wSG_w) \qquad [6.6.10]$$

An alternative way is to employ the basic heat transfer equations, which are used in indirect heat sizing as follows:

$$Q = V_0AT_m \quad \text{or} \quad A = \frac{Q}{V_0T_m} \qquad [6.6.11]$$

where Q_0, Q_w = heat required in Btu/hr
A = total heat transfer area (coil area) in ft.^{2}
T_m = log mean temperature difference in °F

For low-pressure oil, about 35° API and water liquid, streams, the heat required may be determined from the following equation:

$$Q = q_t[6.25 + 8.33(X)]\Delta T \qquad [6.6.12]$$

where q_t = total liquid flowrate in bpd
X = decimal water content in liquid
ΔT = the difference between inlet and outlet temperature °F

The overall film or heat transfer coefficient for high-pressure gas streams may be found from Figure 6.6.24 using the gas flowrate and tube size selected. The overall film or heat transfer coefficient for water may be found from Figure 6.6.25 and the coefficient for oil from Figure 6.6.26, based on liquid flowrate and tube size. For liquid streams that are a mixture of oil and water, the overall coefficient may be averaged and calculated as

$$V_{0(mix)} = V_{0(oil)} + \left[V_{0(water)} - V_{0(oil)}\right]X \qquad [6.6.13]$$

These film or heat-transfer coefficients are based on clean tubes; in other words, no allowance is made for any fouling factors. If any fouling is to be expected, excess coil area should be allowed in the heater selection.

$$T_m = \frac{GTD - LTD}{\ln\left(\dfrac{GTD}{LTD}\right)} \qquad [6.6.14]$$

where T_m = log mean temperature difference in °F
GTD = greater temperature difference
= water bath temperature − inlet fluid temperature
LTD = least temperature difference
= water bath temperature − outlet fluid temperature in °F

A water bath temperature must be assumed for the calculations as mentioned before. Usually 180°F is the maximum designed temperature recommended for indirect water bath heaters.

The coil area (A) is required for indirect heaters (Figure 6.6.27) [23] can be calculated from the basic heat transfer equation after all the above factors have been determined.

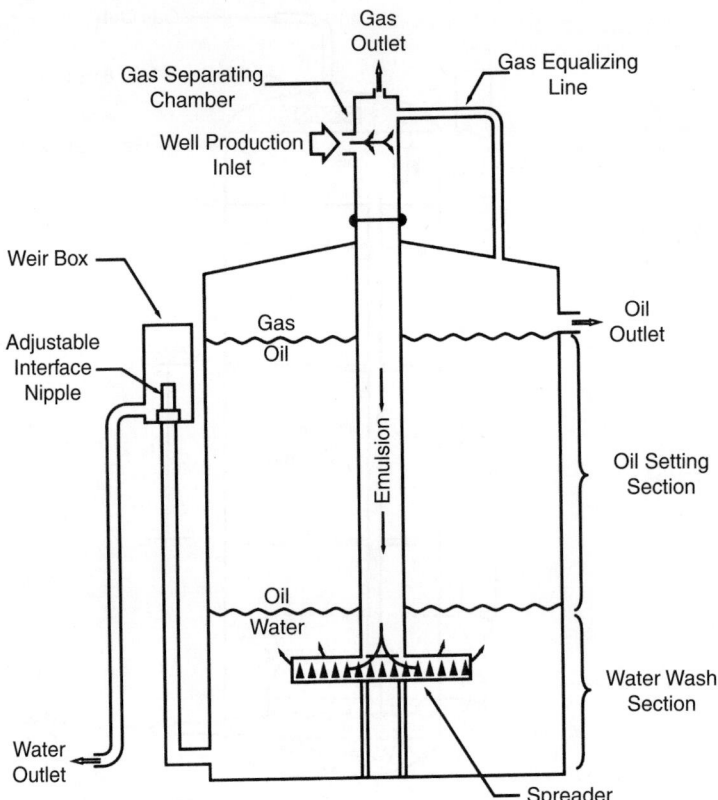

Figure 6.6.20   *Low-pressure settling tank with internal flume [20].*

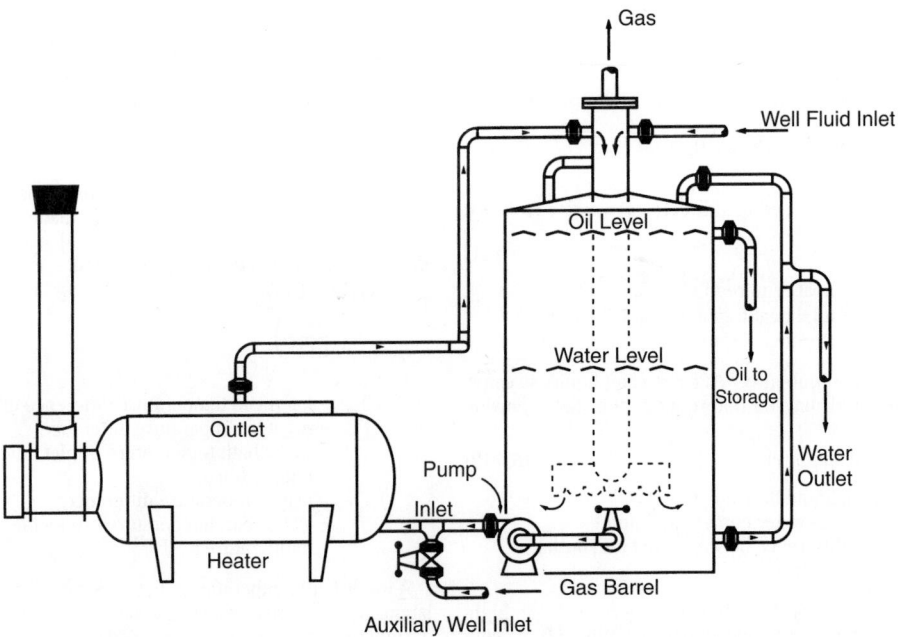

Figure 6.6.21   *Heater and gunbarrel is forced circulation method of heating [20].*

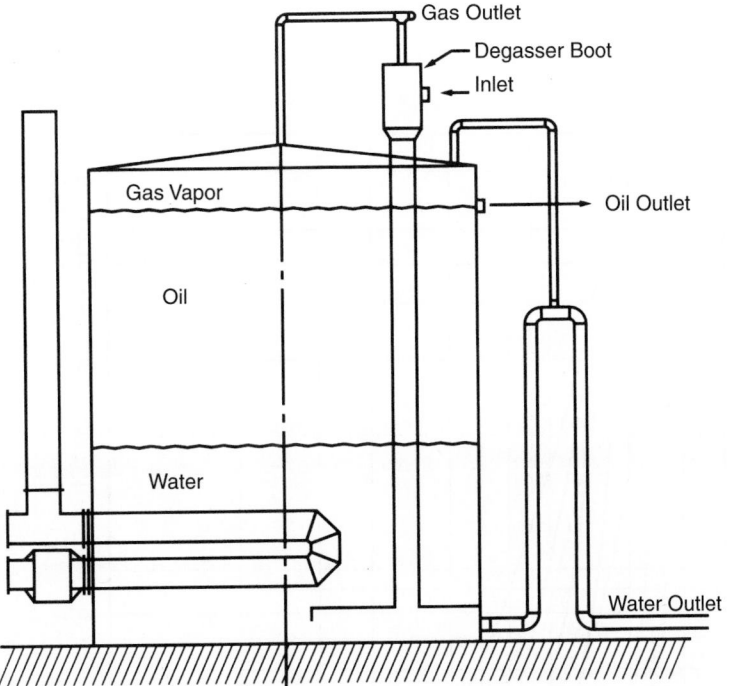

Figure 6.6.22 *Heated gunbarrel emulsion treater [20].*

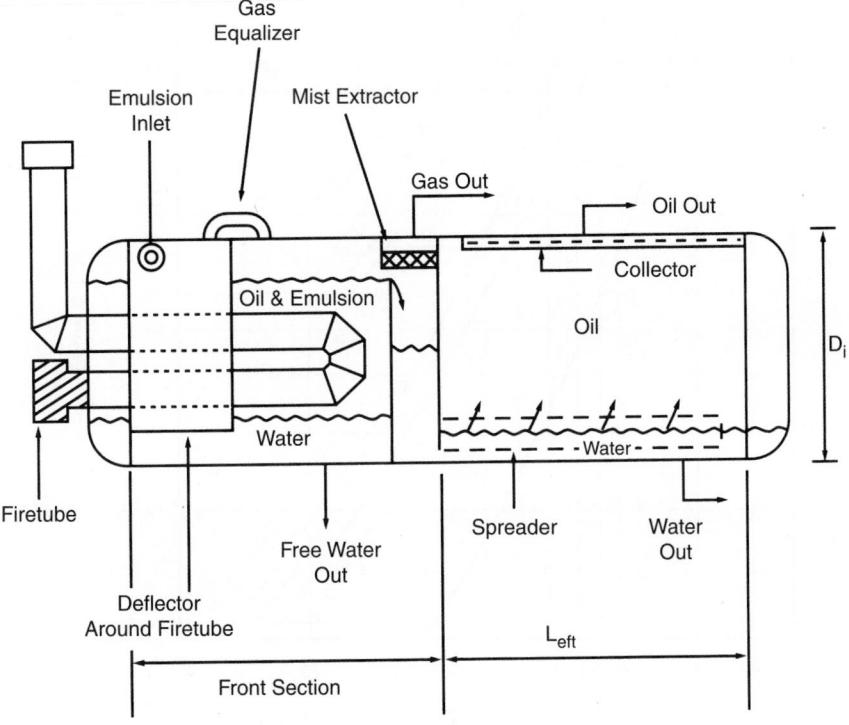

Figure 6.6.23 *Horizontal heater-treater [20].*

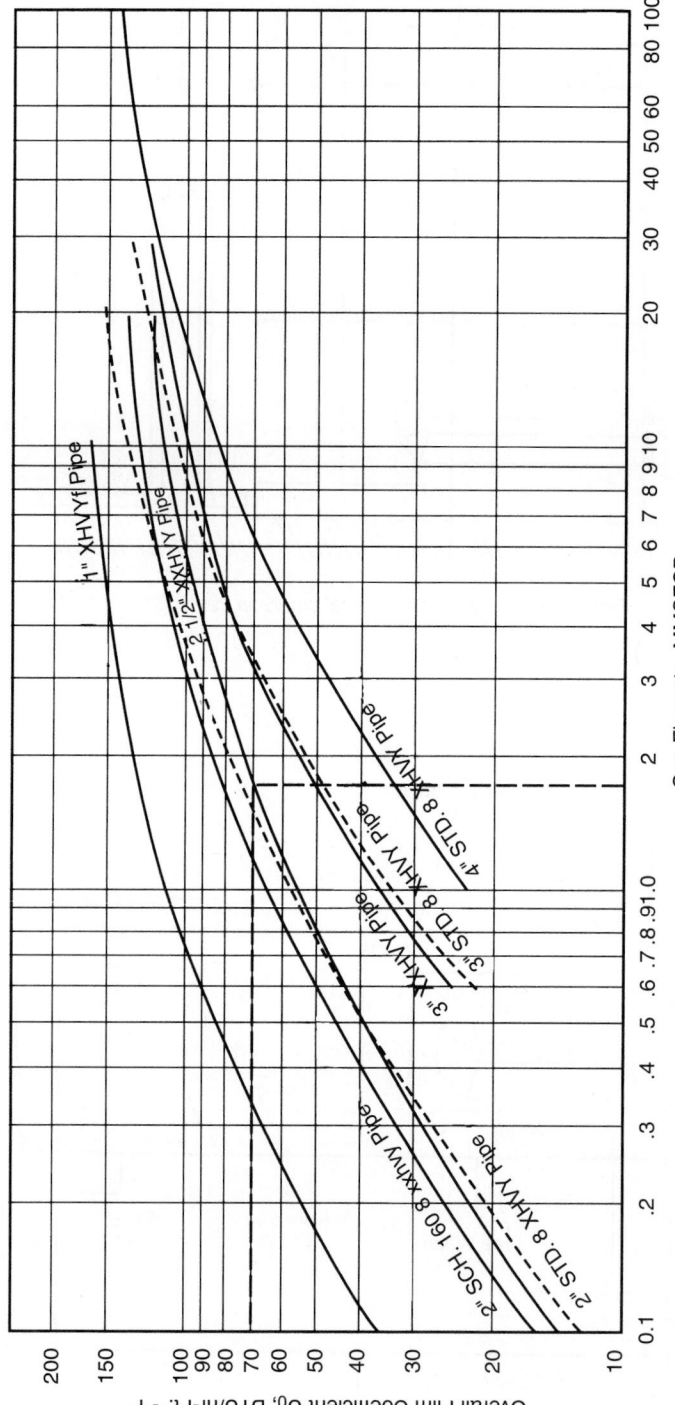

Figure 6.6.24 *Overall film coefficient for natural gas in indirect heaters [22].*

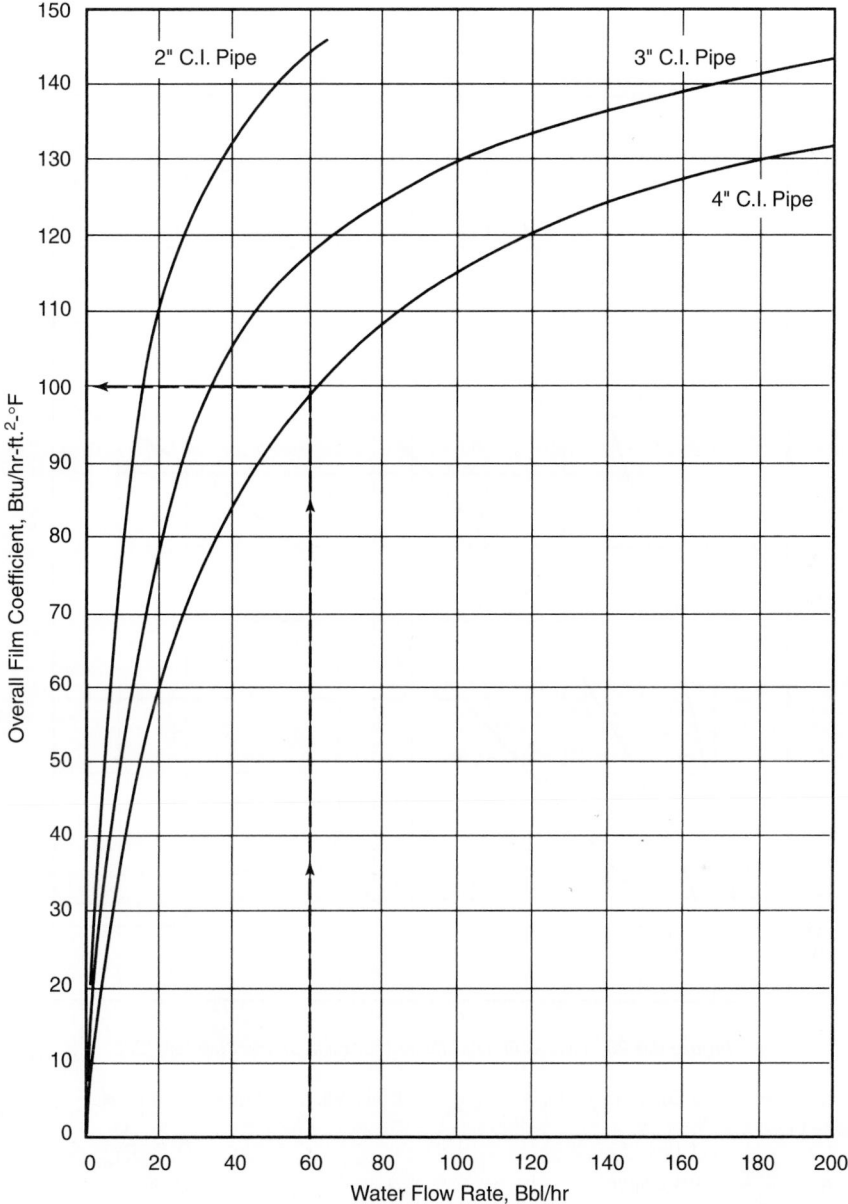

Figure 6.6.25 *Overall film coefficient for water in indirect heaters [22].*

An indirect heater then may be selected from the standard models listed in Tables 6.6.8 and 6.6.9 [22] based on the heat and the coil area required.

By selecting a heater with a larger heat capacity and coil area than that calculated, sufficient excess will be provided to allow for heat loss from the vessel and any fouling that may occur within the tubes, and will allow the heater to be operated at less than the maximum design water bath temperature.

Example 6.6.6
For a given date, calculate the heater size such that

    Oil flowrate = 2,500 bpd
    Water flowrate = 1,400 bpd

Inlet temperature = 60°F
Outlet required temperature = 100°F
Coil size required = 3 in. cast iron

Heat required, from Equation 6.6.12, is

$$Q = q_t(6.25 + 8.33X)T$$

$$q_t = 2,500 + 1,400 = 3,900 \text{ bpd}$$

$$X = \frac{1,400}{3,900} = 0.36 = 36\%$$

$$\Delta T = 110 - 60 = 40°F$$

$$Q = 3,900[6.25 + 8.33(0.36)]40 = 1,442,813 \text{ Btu/hr}$$

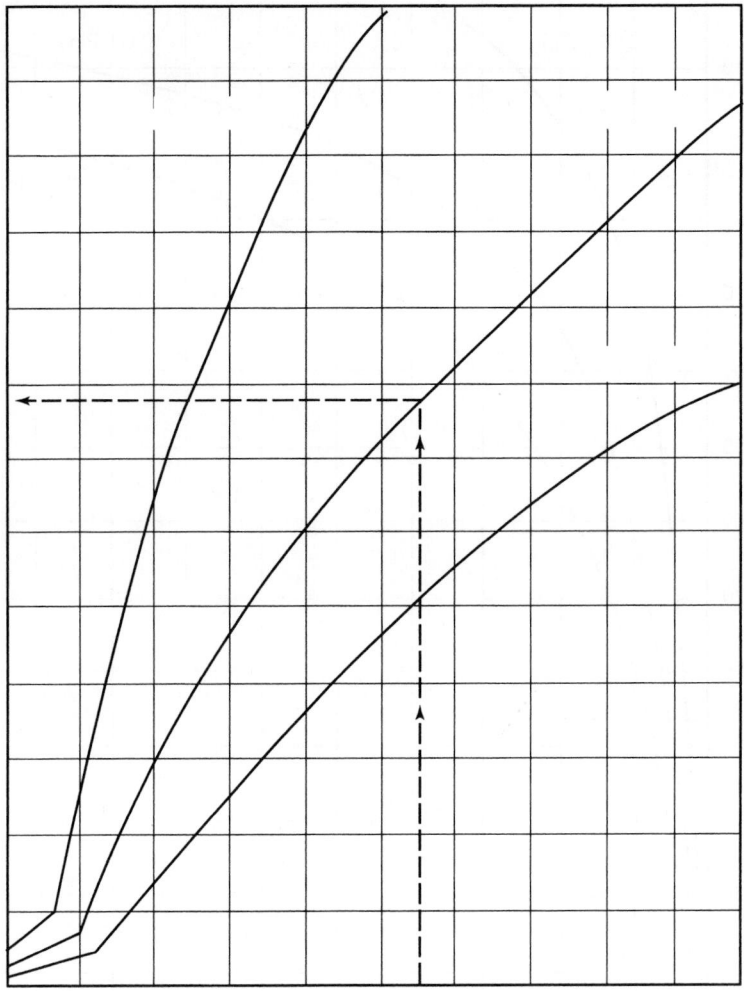

Figure 6.6.26 *Overall film coefficient for oil in indirect heaters [22].*

Heat transfer coefficient, from Equation 6.6.13, is

$$V_{0(mix)} = V_{0(oil)} + \left[V_{0(water)} - V_{0(oil)}\right]X$$

Oil flowrate is $q_0 = \dfrac{2,500}{24} = 104.2 \text{ bphr}$

From Figure 6.6.26, $V_{0(oil)} = 35.7 \text{ Btu/(hr ft.}^2 \text{ °F)}$.

Water flowrate is $q_w = \dfrac{1,400}{24} = 58.4 \text{ bphr}$

from Figure 6.6.25, $V_{0(water)} = 117.5 \text{ Btu/(hr ft.}^2 \text{ °F)}$.

$V_{0(mix)} = 35.7 + (117.5 - 35.7)0.36 = 65.15 \text{ Btu/(hr ft.}^2 \text{ °F)}$
Log mean temperature difference, from Equation 6.6.14, is

$$T_m = \frac{GTD - LTD}{\ln \dfrac{GTD}{LTD}}$$

$GTD = 180 - 60 = 120°F \quad LTD = 180 - 100 = 80°F$

$$T_m = \frac{40}{\ln \dfrac{120}{80}} = 98.7°F$$

Coil area, from Equation 6.6.11, is

$$A = \frac{Q}{V_0 T_m}$$

where

$$A = \frac{1,442,813}{117.5 \times 98.7} = 124.4 \text{ ft.}^2$$

From Table 6.6.9

> Heater size: 60 in. × 12 ft.
> Firebox capacity: 1,500,000 Btu/hr
> Coil data: 243-in. cast iron tubes
> Coil area: 238.5 ft.²

Example 6.6.7
Size a horizontal treater for given data:

> Oil gravity: 33°API, $SG_0 = 0.86$ at 60°F
> Oil flowrate: 6,000 bpd
> Inlet oil temperature: 100°F
> $SG_w = 1.03$.

Assume that 80% of the cross-sectional area is effective, retention time is 15 min, and treating temperature is 120°F

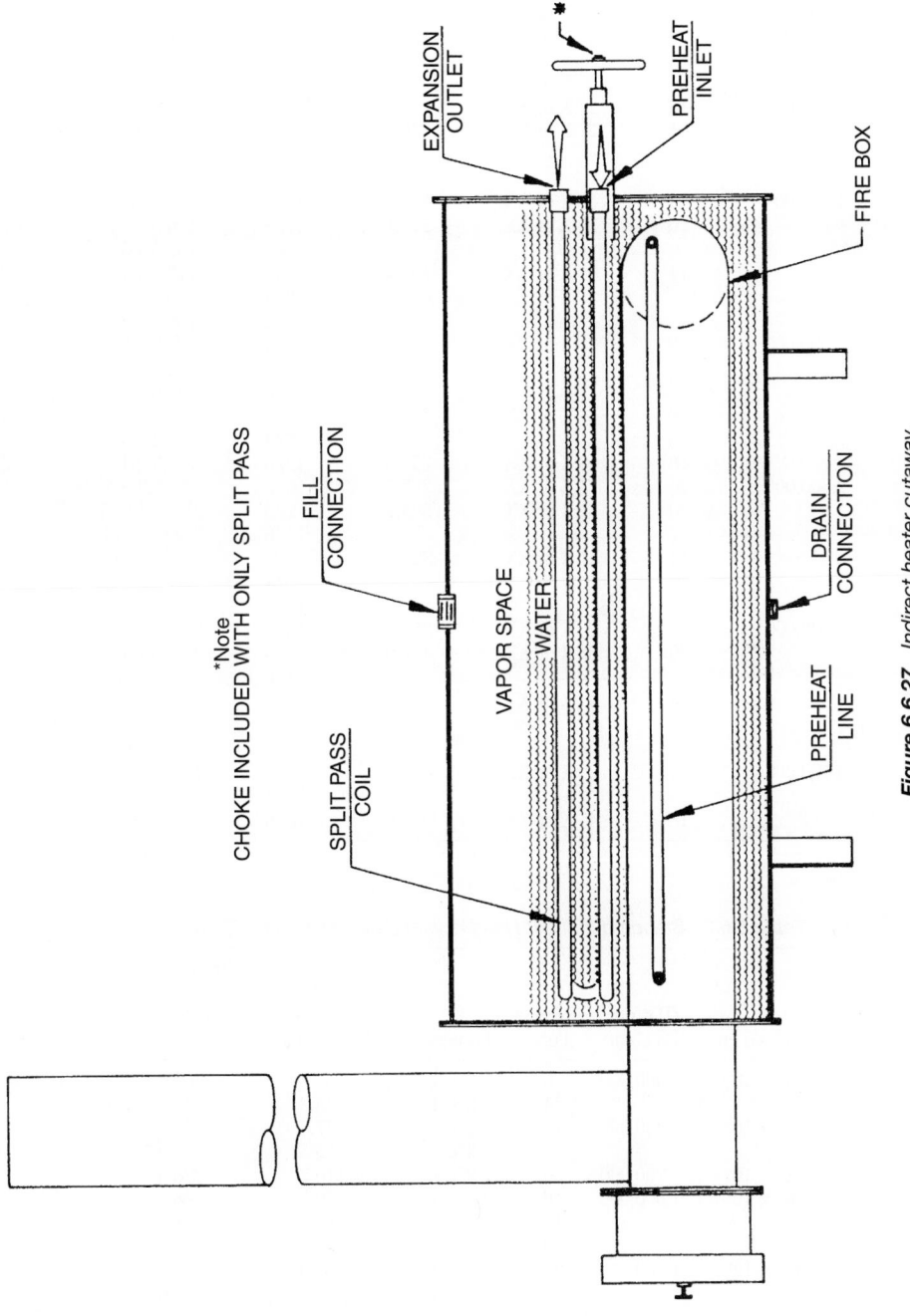

Figure 6.6.27 *Indirect heater cutaway.*

Table 6.6.8 *Standard Indirect Heaters with Steel Coils [22]*

Size Dia × Len.	Firebox Rating BTU/hr	No. of Tubes	Tube Size	Coil Area Sq ft	Split Pass Tubes Split	Split Areas	Equivalent Len. Of Pipe for Press. Drop, ft.
24″ × 3′	100,000	Spiral	1″ X	21.5	–	–	93.2
30″ × 6′	250,000	14	1″ X	26.2	8/6	15.0/11.2	96.4
		8	2″ X or XX	28.9	4/4	14.5/14.4	68.3
30″ × 10′	500,000	8	2″ X or XX	48.8	4/4	24.4/24.4	100.3
36″ × 10′	750,000	14	2″ X or XX	85.3	8/6	48.7/36.6	177.7
					10/4	60.9/24.4	
		8	3″ X or XX	72.3	4/4	36.1/36.2	112.7
48″ × 10′	1,000,000	18	2″ X or XX	109.7	12/6	73.1/36.6	229.3
		14	3″ X or XX	126.4	8/6	72.2/54.2	200.7
					10/4	90.3/36.1	
					12/2	108.3/18.1	
60″ × 12′	1,500,000	24	2″ X or XX	176.1	16/8	117.4/58.7	354.7
		20	3″ X or XX	217.1	12/8	130.3/86.8	328.7
					14/6	152.0/65.1	
		14	4″ X or XX	196.4	8/6	112.2/84.2	243.8
					10/4	140.3/56.1	
					12/2	168.3/28.1	
72″ × 12′	2,000,000	38	2″ X or XX	278.7	20/18	146.7/132.0	563.3
					28/10	205.4/73.3	
					30/8	220.0/58.7	
		30	3″ X or XX	325.6	22/8	238.8/86.8	495.4
					26/4	282.2/43.4	
		18	4″ X or XX	252.5	12/6	168.3/84.2	315.1
60″ × 20′	2,500,000	24	2″ X or XX	295.5	16/8	197.0/98.5	546.7
		20	3″ X or XX	363.7	12/8	218.2/145.5	488.7
					14/6	254.6/109.1	
		14	4″ X or XX	328.4	8/6	187.7/140.7	355.8
					10/4	234.6/93.8	
					12/2	281.5/46.9	
60″ × 24′	3,000,000	24	2″ X or XX	355.2	16/8	236.8/118.4	642.7
		20	3″ X or XX	437.0	12/8	262.2/174.8	568.7
					14/6	305.9/131.1	
		14	4″ X or XX	394.3	8/6	225.3/169.0	411.8
					10/4	281.6/112.7	
					12/2	338.0/56.3	
72″ × 24′	4,000,000	30	3″ X or XX	655.4	22/8	480.6/174.8	855.4
					26/4	568.0/87.4	
		18	4″ X or XX	507.0	12/6	338.0/169.0	531.1

Table 6.6.9 *Standard Indirect Heaters with Cast Iron Coils [22]*

Size Dia × Len.	Rating BTU/hr	No. of Tubes	Tube Size	Coil Area sq ft (outside tubes surface)	Equivalent Len. of Pipe for Press. Drop, ft.
30″ × 10′	500,000	6	2″C.I.	33.6	65.4
		4	3″C.I.	33.3	47.7
36″ × 10′	750,000	14	2″C.I.	77.4	154.0
		8	3″C.I.	65.5	97.9
48″ × 10′	750,000	20	2″C.I.	110.2	220.5
48″ × 10′	1,000,000	18	2″C.I.	99.3	198.4
		12	3″C.I.	97.8	148.2
		18	3″C.I.	146.2	223.5
60″ × 12′	1,500,000	24	2″C.I.	162.0	312.8
		24	3″C.I.	238.5	346.9
		16	4″C.I.	209.6	250.5
72″ × 12′	2,000,000	38	2″C.I.	256.0	496.0
		38	3″C.I.	377.1	550.7
		20	4″C.I.	261.8	314.1

or 150°F. Then oil viscosity = μ_0 = 5.5 cp at 100°F, 4 cp at 120°F, and 2.5 cp at 150°F.

1. Settling equation:

$V_t = V_0$ (terminal velocity of water = velocity of oil)

Flow around settling oil drops in water or water drops in oil is laminar, and thus Stokes law governs:

$$V_t = 2{,}600 \frac{\gamma_w - \gamma_0}{\mu_0} D_p^2 \text{(ft./s)}$$

$$\gamma_w, \gamma_0 (\text{lb/ft.}^3) \quad D_p(\text{ft.}) \quad \mu_0(\text{cp})$$

$$V_0 = \frac{Q}{A}$$

$$Q = 6.49 \times 10^{-5} Q_0$$

$$Q(\text{ft.}^3/s) \quad A(\text{ft.}^2) \quad Q_0(\text{bpd})$$

$A = D \times L_{eff}$ is the highest cross-sectional area.

$$V_0 = \frac{6.49 \times Q_0}{10^5 DL_{eff}} = 6.5 \times 10^{-5} \frac{Q_0}{DL_{eff}}$$

$$2{,}600 \frac{\gamma_w - \gamma_0}{\mu_0} D_p^2 = 6.5 \times 10^{-5} \frac{Q_0}{DL_{eff}}$$

$$DL_{eff} = 2.44 \times 10^{-8} \frac{Q_0 \mu_0}{D_p^2 (\gamma_w - \gamma_0)} \quad \text{where } D_p \text{ is in ft.}$$

or

$$DL_{eff} = 2{,}267 \frac{Q_0 \mu_0}{D_m^2 (\gamma_w - \gamma_0)} \quad \text{where } D_m \text{ is in } \mu m$$

The diameter of water droplets to be settled from the oil (μm) is a function of viscosity of the oil (cp) and, according to Arnold and Stewart [15], can be expressed as

$$d_m = 500(\mu_0)^{-0.675}$$

$$\text{at } 100°F \ d_m = 500(5.5)^{-0.675} = 158 \ \mu m$$
$$120°F \ d_m = 500(4)^{-0.675} = 196 \ \mu m$$
$$150°F \ d_m = 500(2.5)^{-0.675} = 270 \ \mu m$$

$$\gamma_m = 1.03 \times 62.4 = 64.3 \ \text{lb/ft.}^3$$

Assume the same for temperatures 60°F, 100°F, 120°F, and 150°F.

$$SG_0 \text{ at } 60°F = \frac{141.5}{API + 131.5} = \frac{141.5}{32.8 + 131.5} = 0.860$$

$$SG_0 \text{ at } 100°F = \frac{141.5}{(API - loss) + 131.5} = \frac{141.5}{32.8 + 131.5}$$
$$= 0.861$$

where "loss" from Figure 6.6.17 is $0.3 - 0.1 = 0.2$:

$$SG_0 \text{ at } 120°F = \frac{141.5}{32.7 + 131.5} = 0.862$$

$$SG_0 \text{ at } 150°F = \frac{141.5}{32.2 + 131.5} = 0.864$$

Because specific gravity of oil as a function of temperature changes in a small range, assume constant values for oil and water specific weights:

$$\gamma_w = 64.3 \ \text{lb/ft.}^3$$

$$\gamma_0 = 53.7 \ \text{lb/ft.}^3$$

$$\gamma_w - \rho_0 = 106 \ \text{lb/ft.}^3$$

	$T = 100°F$	$T = 120°F$	$T = 150°F$
$\gamma_w = 64.3 \ (\text{lb/ft.}^3)$	10.6	10.6	10.6
$\mu_0 \ (\text{cp})$	5.5	4.0	2.5
$d_m \ (\mu m)$	158.0	196.0	269.0

Calculate D versus L_{eff}

a. $DL_{eff} = 2{,}267 \times \dfrac{6{,}000 \times 4}{(196)^2 10.6} = 133.6$ if $T = 120°F$

b. $DL_{eff} = 2{,}267 \times \dfrac{6{,}000 \times 25}{(-296)^2 10.6} = 44.3$ if $T = 150°F$

D (ft.)	L_{eff} (a)	L_{eff} (b)
2.0	66.8	22.2
4.0	33.4	11.1
4.5	30.0	9.9
5.0	26.7	8.9
6.0	22.3	7.4
9.0	14.8	4.9
15.0	8.9	3.0
30.0	4.5	1.5

2. Calculate the retention time, such that

$$t = \text{Vol.}/Q \quad \text{Vol.} = (\pi D^2/4) L_{eff} \times 0.8 = 0.63 D^2 L_{eff}$$

$$Q = 6.49 \times 10^{-5} Q_0$$

$$t = \frac{0.63 D^2 L_{eff}}{6.49 \times 10^{-5} Q_0} = 9{,}681 \frac{D^2 L_{eff}}{Q_0}$$

$$D^2 L_{eff} = 0.000103 Q_0 t$$

$(t_r)_0$ is in min

$$t = 60(t_r)_0$$

$$D^2 L_{eff} = 0.00618 Q_0 (t_r)_0$$

If the retention time $(t_r)_0 = 15$ min, then

$$D^2 L_{eff} = 0.00618 \times 6{,}000 \times 15 = 556.2$$

D (ft.)	2	4	4.5	5	6	9	15	30
L_{eff} (ft.)	139	24	27.5	22.2	15.5	6.9	2.5	0.6

If the retention time $(t_r)_0 = 20$ min, then

$$D^2 L_{eff} = 0.00618 \times 6{,}000 \times 20 = 741.6$$

D (ft.)	2	4	5	6	9	15	30
L_{eff} (ft.)	185	46	29.7	20.6	9.2	3.3	0.8

Plot Figure 6.6., D versus L_{eff} gives the solution.
For settling temperature 150°F, the minimum effective length has to be 2.7 ft. if $(t_r)_0 = 20$ min and 3.4 ft. if $(t_r)_0 = 15$ min. At lower settling temperature, L_{eff} should be 24 and 32 ft. for $(t_r)_0 = 20$ min and 15 min, respectively.

3. The heat required, from Equation 6.6.10, is

$$Q = 16.2 \Delta T \ (0.5 q_0 SG_0 + q_w \ SG_w)$$

or

$$Q_{th} = 14.6 \Delta T \ (0.5 q_0 SG_0 + q_w \ SG_w)$$

$$q_w = 0.1\% \text{ of oil flow rate}$$

$$\Delta T = 150 - 100 = 50°F \text{ and } 120 - 100 = 20°F$$

If $\Delta T = 50°F$, then

$Q_{thI} = 14.6 \times 50 \times (0.5 \times 6{,}000 \times 0.861 + 600 \times 1.03)$

$\quad = 2{,}336{,}730 \text{ Btu/hr}$

if $\Delta T = 20°F$, then

$Q_{thII} = 14.6 \times 20 \times (0.5 \times 6{,}000 \times 0.861 + 600 \times 1.03)$

$\quad = 934{,}692 \text{ Btu/hr}$

4. Conclusions are as follows. Treating temperature plays a more important role than retention time. For $T = 150°F$ and $L_{eff} = 7.3$ ft., any diameter above retention time curves is correct. The diameter of the front section has to be the same as the coalescing section.

An economical solution would be a 6×20 ft. for the coalescing section and a 2.5 Btu/hr firebox rating.

References

1. Department of Interior—Minerals Management Service, 30 CFR Part 20, *Oil and Gas and Sulfur Operations in the Outer Continental Shelf*, 1995, http://www.mms.gov/federalregister/PDFs/docref.pdf.
2. Texas Administrative Code Title 16, Part 1, Chapter 3, Rule §3.36, *"Oil, Gas or Geothermal Resource Operation in Hydrogen Sulfide Areas,"* April, 1995, http://www.rrc.state.tx.us/rules/16ch3.html.
3. New Mexico Oil Conservation Division, Rulebook, Rule 118, NMAC Citation Number 19 NMZC 15.3.118, http://www.emnrd.state.nm.us/ocd/ocdrules/oil&gas/rulebook/RULEBOOK.DOC.
4. Title 49 CFR Part 191—Transportation of Natural and Other Gas by Pipeline; Annual Reports, Incident Reports, and Safety-Related Condition Reports, http://www.access.gpo.gov/nara/cfr/waisidx/49cfr191.html.
5. Title 49 CFR Part 192—Transportation of Natural and Other Gas by Pipeline: Minimum Federal Safety Standards, http://www.access.gpo.gov/nara/cfr/waisidx_02/49cfr192_02.html.
6. Title 40 CFR Part 195—Transportation of Hazardous Liquids by Pipeline, http://www.access.gpo.gov/nara/cfr/waisidx_02/49cfr195_02.html.
7. ASME Pressure Vessel Code, http://www.asme.org/bpvc/.
8. National Board of Boiler and Pressure Vessel Inspectors, http://www.nationalboard.org/.
9. API 510—Pressure Vessel Inspection Code: Maintenance, Inspection, Rating, Repair and Alteration, 8th Edition, June 1997.
10. API Standards and Technical Publications, http://api-ep.api.org/filelibrary/ACF47.pdf.
11. *Engineering Data Book,* Vols. 1 and 2, 10th Edition, GPSA, Tulsa, OK, 1987.
12. Texas Natural Resource Conservation Commission–Standard Exemption 67, http://www.tnrc.state.tx.us/permitting/airperm/nsr_permits/oldselist/889list/62-72.htm#67.
13. e-Valves.com. E-Commerce Support for Valves, http://www.evalves.org/
14. Perry, R. H., and Green, D., *Perry's Chemical Engineers Handbook,* 6th Edition, McGraw-Hill, Inc., New York, 1984.
15. Arnold, K., and Steward, M., *Surface Production Operation,* Gulf Publishing Co., Houston, Texas, 1986.
16. Robertson, J. A., and Clayton, T. C., *Engineering Fluid Mechanics,* Houghton-Mifflin Co., Boston, 1985.
17. API Specification 12J, 6th Edition, API, Washington, DC, June 1, 1988.
18. Campbell, J. M., *Gas Conditioning and Processing,* Vols. 1 and 2, Norman, OK, 1984.
19. Sivalls, C. R., *Oil and Gas Separation Design Manual,* Sivalls, Inc., Odessa, Texas.
20. Bradley, B. H., *Petroleum Engineers Handbook,* Society of Petroleum Engineers, Richardson, Texas, 1987.
21. API Specification 12K, 6th Edition, API, Washington, DC, June 1, 1988.
22. Sivalls, C. R., Technical Bulletin 113, Sivalls, Inc., Odessa, Texas.
23. Sivalls, C. R., *Indirect Heaters Design Manual*, Sivalls, Inc., Odessa, Texas.

6.7 GAS PRODUCTION ENGINEERING

Quality specifications for natural gas are individually negotiated and prescribed in the contract between the purchaser or the pipeline companies and the producer. Gas contracts usually contain the following basic considerations:

- Minimum, maximum, and nominal delivery pressure
- Water dew point or water content
- Maximum condensable hydrocarbon content or hydrocarbon dew point
- Minimum heating value
- Contaminants such as H_2S, carbon disulfide, mercaptans, etc.
- Composition
- Maximum delivery pressure

Above quality parameters together with the price and quantity are the fundamental factors, and they determine production equipment required [1].

6.7.1 Gas–Water Systems and Dehydration Methods

Water vapor is the most common contaminant in natural gas, usually in the range of 8,000 to 10,000 ppm by volume (400 to 500 lb_M water vapor/MMscf of gas), while the pipeline specifications restrict the water content to a value no greater than 120 to 160 ppm by volume (6 to 8 lb_m water vapor/MMscf). Most gas sales contracts specify a maximum value for the amount of water vapor allowable in the gas. Typical values are 7 lb/MMscf in the southern United States, 4 lb/MMscf in the northern United States, and 2 to 4 lb/MMscf in Canada. These values correspond to dew points of approximately 32°F for 7 lb/MMscf, 20°F for 4 lb/MMscf, and 0°F for 2 lb/MMscf in a 1,000-psi gas line. In order to design and operate dehydration processes, a reliable estimate of the water content of natural gas is essential.

Figure 6.7.1 is based on experimental data and shows the solubility of water in sweet hydrocarbon liquids. In sour hydrocarbon liquids, water solubility can be substantially higher [2].

Liquid water and water vapor are removed from natural gas to:

- Prevent formation of hydrates in processing and transmission systems
- Meet a water dew point requirement of the sales gas contract
- Prevent corrosion

6.7.1.1 Water Content of Natural Gases

The water content of a gas is a function of pressure, temperature, composition, and the salt content of the free water [2]. The effect of composition increases with pressure for acid gases. For lean, sweet natural gases containing over 70% methane and small amounts of "heavy ends," pressure temperature correlation's are suitable. Figure 6.7.1 is an example.

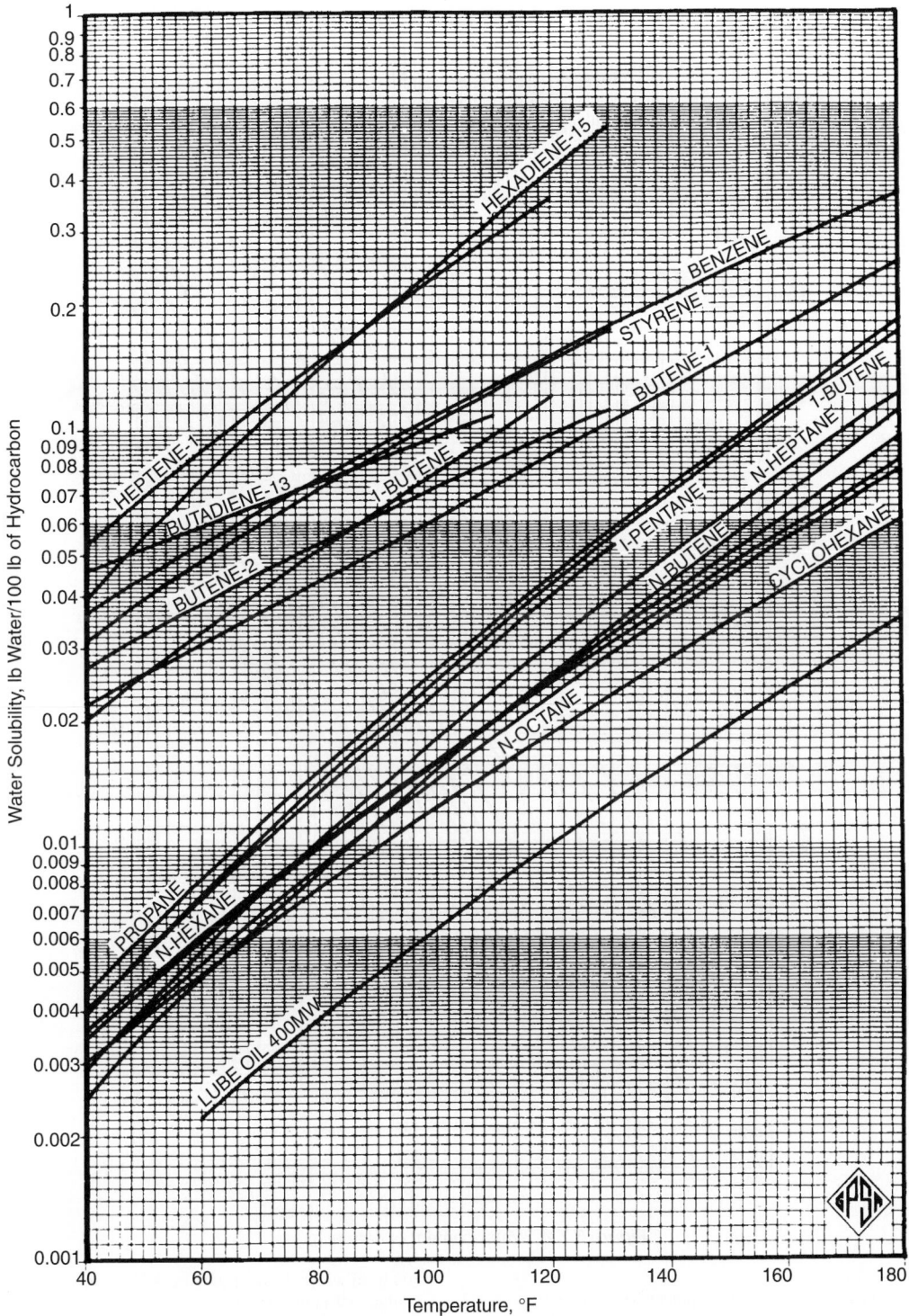

Figure 6.7.1 *Solubility of water in liquid hydrocarbons.*

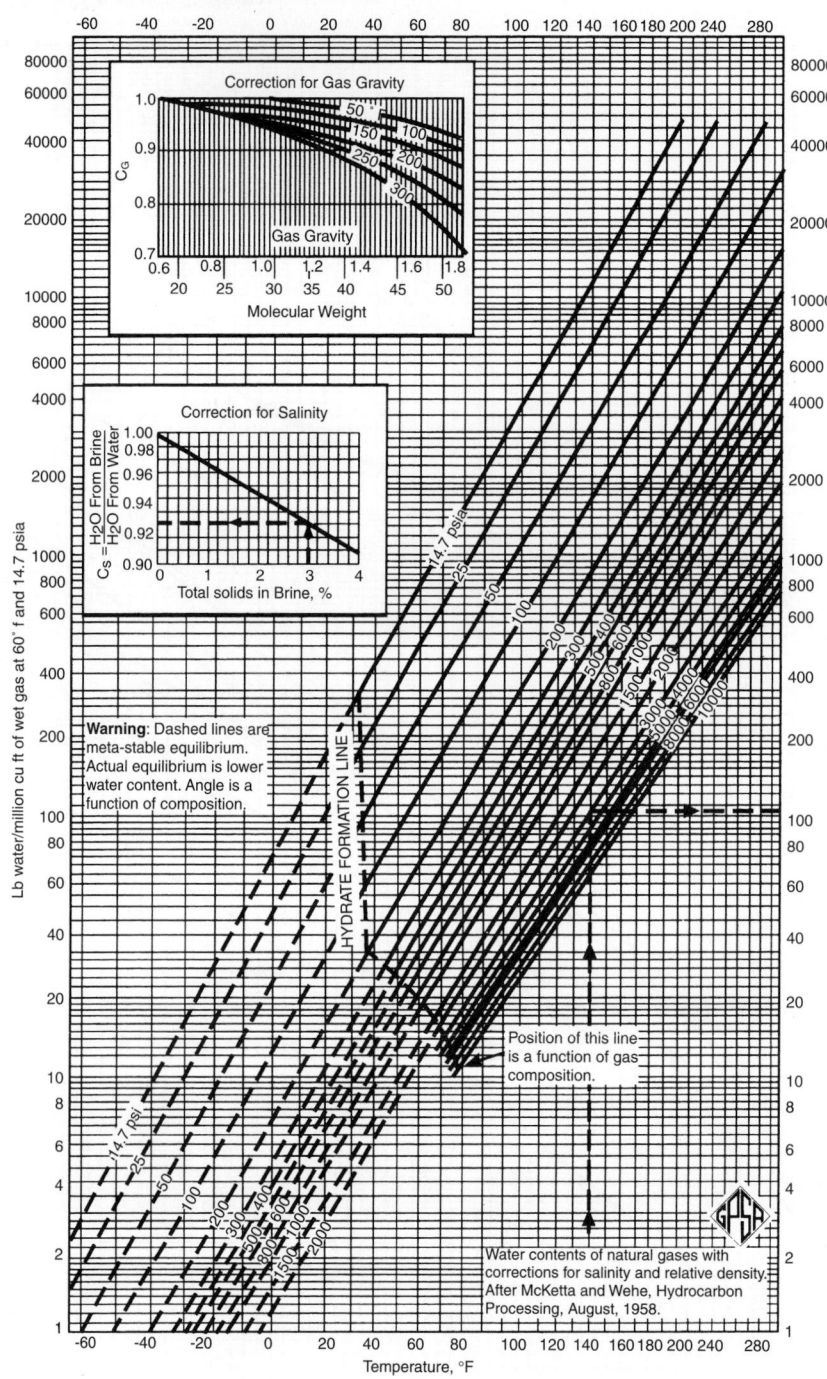

Figure 6.7.2 *Water content of natural gases.*

Example 6.7.1

A natural gas $SG_g = 0.9$ is in contact with brine in the reservoir. The brine contains 30,000-ppm solids (approximately 3% NaCL). The pressure of the gas is 3,000 psia, and the temperature is 150°F. How much water is in the gas in pounds of water per million cubic feet of gas?

1. Follow dashed lines on Figure 6.7.2 from the general chart of Figure 6.7.2 at 3000 psia and 150°F gas would contain

105 lb of water/MMscf of 0.6 gas if the gas had been in contact with pure water.

2. Correction for gas gravity from 0.6 to 0.9. From the "correction chart for gravity" follow the dashed line from the abscissa where the desired gravity is 0.9, vertically to the 150°F isotherm. Read horizontally, the correction factor for gas gravity C_g is 0.98.

3. Correction for salinity of brine; see the second correction chart on the general Figure 6.7.2. At brine salinity of 3% go

vertically to the correction line, and read on the ordinate the ordinate value of $C_s = 0.93$

The final answer for water content W:

$$W = 105 \times C_g \times C_s = 105 \times 0.98 \times 0.93$$

$$= 95.6 \text{ lb water in the gas}$$

Experimental value for this gas was 96.2, which is satisfactory accuracy.

The McKetta and Wehe [2] chart (Figure 6.7.2) is not explicit for temperatures below the hydrate formation line. Because of this, it is better to use Campbell's [1] correlation (Figure 6.7.2) for water content of sweet gases. To determine the moisture within the pressure range of 1 to 690 bar (14.7 to 10,000 psia) and the temperature range of $-40°C$ ($-40°F$) to $110°C$ ($230°F$), the following analytical expression is used:

$$W = \frac{A}{P} + B \quad \text{if} \quad SG_g = .06$$

or

$$W = \frac{A}{P} + B \times C_G \times C_s \quad \text{if} \quad SG_g > 0.6 \qquad [6.7.1]$$

where A, B = constants; see Table 6.7.1
SG_g = specific gravity of gas or relative density
W = water content in g/m^3
P = gas pressure in atm

Table 6.7.1 *Coefficients A and B in Equation [4]*

Temp., °C.	A	B	Temp., °C.	A	B
−40	0.1451	0.00347	+32	36.10	0.1895
−38	0.1780	0.00402	+34	40.50	0.207
−36	0.2189	0.00465	+36	45.20	0.224
−34	0.2670	0.00538	+38	50.80	0.242
−32	0.3235	0.00623	+40	56.25	0.263
−30	0.3930	0.00710	+42	62.70	0.285
−28	0.4715	0.00806	+44	69.25	0.310
−26	0.5660	0.00921	+46	76.70	0.335
−24	0.6775	0.01043	+48	85.29	0.363
−22	0.8090	0.01168	+50	94.00	0.391
−20	0.9600	0.01340	+52	103.00	0.422
−18	1.1440	0.01510	+54	114.00	0.454
−16	1.350	0.01705	+56	126.00	0.487
−14	1.590	0.01927	+58	138.00	0.521
−12	1.868	0.021155	+60	152.00	0.562
−10	2.188	0.02290	+62	166.50	0.599
−8	2.550	0.0271	+64	183.30	0.645
−6	2.990	0.3035	+66	200.50	0.691
−4	3.480	0.03380	+68	219.00	0.741
−2	4.030	0.03770	+70	238.50	0.793
0	4.670	0.04180	+72	260.00	0.841
+2	5.400	0.04640	+74	283.00	0.902
+4	6.225	0.0515	+76	306.00	0.965
+6	7.150	0.0571	+78	335.00	1.023
+8	8.200	0.0630	+80	363.00	1.083
+10	9.390	0.0696	+82	394.00	1.148
+12	10.720	0.767	+84	427.00	1.205
+14	12.390	0.0855	+86	462.00	1.250
+16	13.940	0.0930	+88	501.00	1.290
+18	15.750	0.1020	+90	537.50	1.327
+20	17.870	0.1120	+92	582.50	1.327
+22	20.150	0.1227	+94	624.00	1.405
+24	22.80	0.1343	+96	672.0	1.445
+26	25.50	0.1453	+98	725.0	1.487
+28	28.70	0.1595	+100	776.0	1.530
+30	32.30	0.1740	+110	1093.0	2.620

When natural gases contain substantial quantities of acid gases as H_2S and/or CO_2, the water content of such a sour natural gas mixture can be considerably higher that the chart for sweet gas would indicate, especially at pressure above 70 bar (1,000 psia). Charts on Figure 6.7.6 and Figure 6.7.3 expand the pressure and temperature ranges for determining the water vapor content for sour gases to data given for sweet gases [2].

Example 6.7.2
Determine the amount of water that will drop out in a plant inlet separator from a gas stream that consists of 16% H_2S, 13% CO_2, and 71% hydrocarbons.

$$T_{res} = 212°F, \quad P = 4675 \text{ psia}, \quad T_{sep} = 80°F,$$

$$\text{and} \quad P_{sep} = 1000 \text{ psia}.$$

1. Reduce two-acid components CO_2 and H_2S to H_2S^*:

$$H_2S^* \text{ of pseudocomposition} = H_2S + 0.75 \times CO_2 \times 13$$

$$= 25.75\% = 16 + .075$$

2. Read water content from chart if $H_2S^* = 25.75$ (from Figure 6.7.5)

$$W = 1.36 \text{ lb/MMscf} \quad \text{at} \quad 3000 \text{ psia} \quad \text{and} \quad 212°F$$

$$= 1.13 \text{ bbl/MMscf} \quad \text{at} \quad 6000 \text{ psia} \quad \text{and} \quad 212°F$$

By logarithmic interpolation with a pocket calculator (or by a plot on log-log paper), the water content at 4,675 psia is determined as follows:

$$\frac{\log 4,675 - \log 6,000}{\log 3,000 - \log 6,000} = \frac{\log x - \log 1.13}{\log 1.36 - \log 1.13}$$

$$\log x = 0.0820$$

$$x = 1.21 \text{ bbl/MMscf}$$

3. The water content at the separator:
If $P = 1,000$ psia, $T = 80°F$, $W = 0.11$ bbl/MMscf
4. Water drop out in the separator

$$1.21 - 0.11 = 1.10 \text{ bbl/MMscf}$$

6.7.1.2 Measurement of Water Content of Natural Gas
The first step in designing gas production equipment is to determine the water content of the gas. Many methods of measuring the amount of water in natural gases have been developed to fit various operations [1, 4]. No single method of analysis can be used under all conditions (Table 6.7.2) [5].

Dew point sensors are devices for moisture detection utilizing the physical properties of water and the laws of physics and chemistry to effect a measurement. Some of them are applicable to gas samples only. Other instruments can be used to monitor moisture in both liquid and gas samples (Table 6.7.3) [5]. Specifications for water content measurement are given in GPA Publication 2140 [2]. These include the valve freeze method, the Bureau of Mines dew point tester, and the cobalt bromide method. Cobalt bromide color change occurs at about 25 to 30 ppm.

The dew point tester permits the visual determination of the temperature at which water will condense from a gas onto a silvered mirror (Figure 6.7.8) [6], which is significant because it represents the actual equilibrium saturation temperature of the gas for temperatures above the hydrate-formation value. And, at temperature below the equilibrium hydrate-formation conditions, it measures a reproducible metastable equilibrium condition between gas and liquid water.

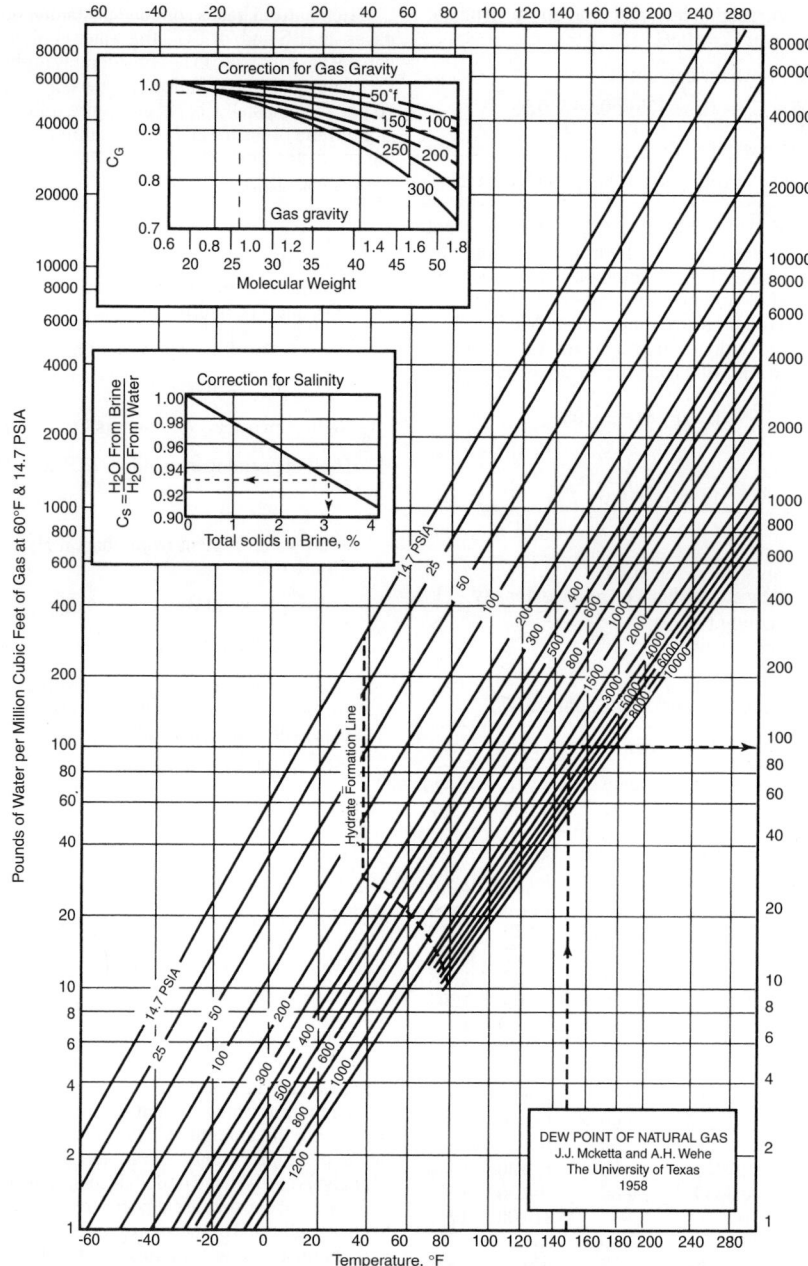

Figure 6.7.3 *Water content of sweet natural gas, LB PER MMscf (at 60°F and 14.7 psia) [1].*

Figure 6.7.8 presents a sectional view of the apparatus without lead lines and refrigerant source. Gas entering the apparatus through valve **A** is deflected by nozzle **B** so as to strike the cooled mirror **C**. The mirror is cooled through the copper cooling rod **F** by the evaporation of a refrigerant, such as propane, CO_2 or some liquid gas in chiller **G**. Pressure gages and a bulb thermometer are used to record the pressure and temperature condition for the inception of condensation or fog formation on the silvered mirror.

In the absence of interfering substances, the accuracy of the determination, given an experienced operator, is reported to be ±0.1°C down to 0°C and ±0.3°C from 0 to −18°C.

A silicon chip hygrometer [7] makes use of a tiny silicon chip to sense the presence of moisture. It can operate in environments ranging from −40°C to 45°C with no effect on accuracy, and gas flow rates from 50 to 1,500 cm^3/min. It can also operate under static or vacuum conditions.

Example 6.7.3

A natural gas dew point is measured at −4°F at a pressure of 14.7 psia. Express this water content in terms of ppm and vpm if the relative density (specific gravity) of the gas is equal to 0.7.

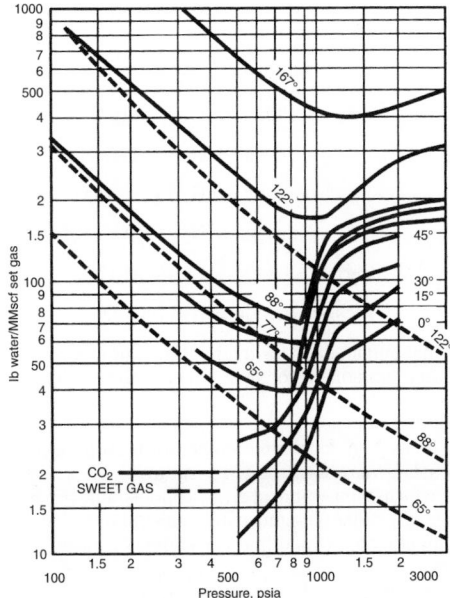

Figure 6.7.4   *Water content of CO_2 [2].*

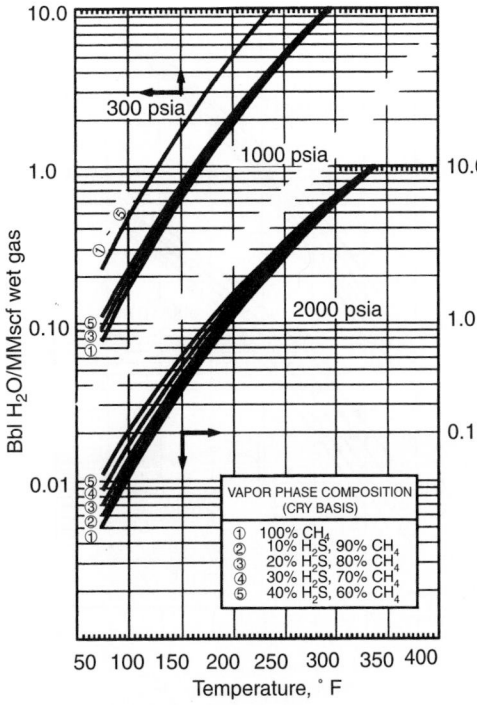

Figure 6.7.6   *Calculated water content of acid gas mixtures to 2000 psia [4].*

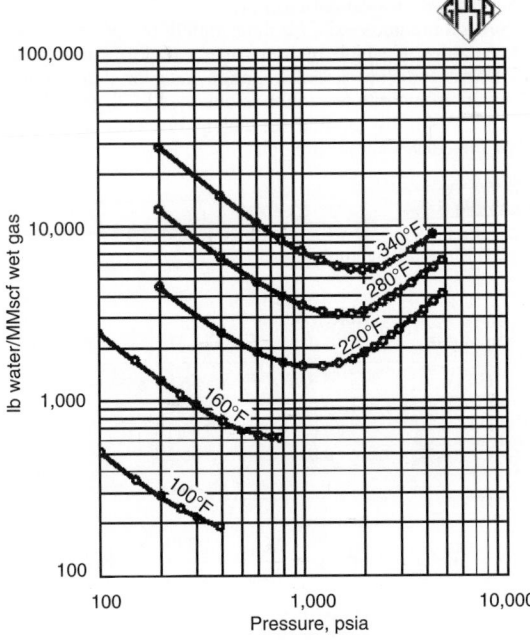

Figure 6.7.5   *Water content of H_2S [2].*

Table 6.7.2   *Methods of Measuring Water Content*

Method	Measurement
Electrolysis for water	Electrolysis current of sample is measured.
Dielectric constant change	Capacitance of a sample is measured.
Electric impedance	Electric impedance of the vapor of a sample is measured.
Piezoelectric crystals	Frequency of crystal with sample moisture is measured.
Heat absorption	Energy absorption and deposition of a sample is measured.
Infrared absorption	Infrared electromagnetic radiation absorption of a sample is measured.
Microwave absorption	Microwave electromagnetic radiation absorption of a sample is measured.

(1) Water content W, in (g/m^3) and $(lb/MMscf)$, from Equation 6.7.1

$$W = \frac{A}{P} + B \times C_G \times C_s \quad -4°F = -20°C$$

14.7 psia = 1.03 atm

A = 0.960   B = 0.0134 from Table 6.7.1

$$W = \frac{0.96}{1.03} + 0.0134 = 0.945 \text{ g/m}^3 \quad \text{if} \quad SG_g = 0.6$$

$$= 58.9 \text{ lb/MMscf} \quad \text{if} \quad SG_g = 0.6$$

for $SG_g = 0.7$ $C_G = 1.0$ $C_s = 1.0$

W = 58.9 lb. water per MMscf

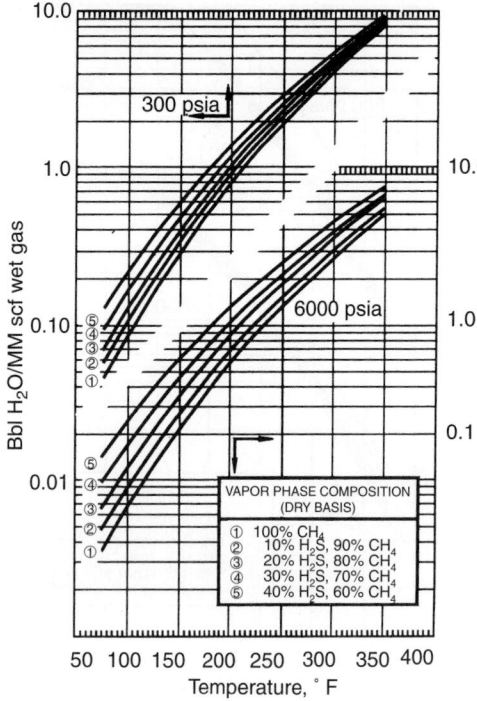

Figure 6.7.7 *Calculated water content of acid gas mixtures to 6000 psia [4].*

(2) Water content in terms of parts per million by weight (ppm)

$$1 \text{ lb mol} = 0.7 \times 29 = 20.3 \text{ lb}$$

$$= 379.3 \text{ ft.}^3 \text{ at } 14.7 \text{ psia and } 60°F$$

$$\frac{58.9 \text{ lb}}{10^6 \text{ scf}} = \frac{58.9 \text{ lb}}{2,636.43 \text{ lb mol}} = \frac{58.9 \text{ lb}}{2,636.43 \times 20.3 \text{ lb}}$$

(3) Water content in terms of parts per million by volume (vpm)

$$\frac{58.9 \text{ lb}}{10^6 \text{ scf}} = \frac{58.9/18.00 \,(\text{lb/lb mol})}{10^6 \text{ scf}} = \frac{3.2722 \text{ lb mol}}{10^6 \text{ scf}}$$

$$= \frac{3.2722 \text{ lb mol} \times 379.4 \text{ scf/lb mol}}{10^6 \text{ scf}} = 1,241 \text{ vpm}$$

(4) Water content W, from Figure 6.7.3 is 57.6 lb/MMscf

6.7.2 Gas Hydrates

The amount of water soluble in a natural gas vapor is limited to temperature and pressure at the dew point. If water will condense in the pipeline and accumulate in sufficient quantities, hydrate can be formed. Gas hydrates are a form of calthrate first discovered by Davy in 1810. A calthrate is any compound wherein guest molecules are entrapped in a cage structure composed of host molecules. With natural gas water molecules as shown in Figure 6.7.9 form hydrates. Contained within each lattice is a series of cavities or cages that must be occupied by enough guest molecules to stabilize this lattice crystal. The guest molecules provide stability to the lattice structure in the same manner that liquid in the pores of a subsurface sand prevents subsidence. Not all cages must be full. Therefore, there is no specific formula

for hydrates. The crystal has an "ice like" appearance, which can be represented as:

$$\text{Methane} = CH_4 \cdot 6H_2O$$

$$\text{Ethane} = C_2H_6 \cdot 8H_2O$$

$$\text{Propane} = C_3H_8 \cdot 17H_2O$$

$$\text{Isobutane i} = C_4H_{10} \cdot 17H_2O$$

$$\text{Nitrogen} = N_2 \cdot 6H_2O$$

$$\text{Carbon Dioxide} = CO_2 \cdot 6H_2O$$

$$\text{Hydrogen Sulfide} = H_2S \cdot 6H_2O$$

Normal butane does form a hydrate, but it is very unstable. Other components of a natural gas mixture do not form hydrate.

Hydrates normally form when a gas stream is cooled below its hydrate formation temperature. At high pressure these solids may form at temperatures well above 32°F. Hydrate formation is almost always undesirable because the crystals may cause plugging of flow lines, chokes, valves, and instrumentation; reduce line capacities; or cause physical damage.

The two major conditions that promote hydrate formation are, the gas being at the appropriate pressure and temperature and the gas being at or below its water dew point with "free water" present. As pressure increases, the hydrate formation temperature also increases. If there is no free water present, hydrates cannot form [4].

Several methods exist for determining the pressure and temperature at which hydrates begin to form: graphical, analytical and experimental. Rough data for determining the start of hydrate formation is obtained by the graphical method. With this method, for gas mixtures not containing H_2S, the curves shown in Figure 6.7.10 [2].

Example 6.7.4
Find the pressure at which hydrate forms at $T = 50°F$ for a gas with the following composition:

Component	Mole fraction in gas	M	lb/mole of mixture
C_1	0.88	16.04	14.12
C_2	0.09	30.07	2.71
C_3	0.02	44.10	0.88
C_4	0.01	58.12	0.58
			18.29

$$\text{Relative Density} = \frac{M_{gas}}{M_{air}} = \frac{20.08}{28.97} = 0.693$$

From Figure 6.7.10 at 50°F

$$P = 320 \text{ psia for } SG_g = 0.7$$

To solve technological design problems, it is simple to use an analytical method to express the relationship between pressure and temperature of hydrate formation. Usually, equations of such dependence are given in the form of $\log P = at + \beta$, that is, when the dependence of P versus T has a linear character in semilog coordinates. However, as some experiments show [8], such dependence frequently does not have a linear character and may be more accurately expressed by:

$$\text{Log } P = a(t + kt^2) + \beta$$

Table 6.7.3 *Summary of Moisture Detector Features*

Type	Range	Sample Phase	Sample System Required	Remarks
Electrolytic hygrometer	0–10 to 0–1,000 ppm	Clean gas. Special sampling for liquids	Yes	Sample flow must be constant
Change of capacitance	0–10 to 0–1,000 ppm	Clean gas or liquid	Yes	Sample temperature must be constant
Impedance type	0–20,000 ppm	Clean gar or liquid	For Liquids	Sample temperature of liquids must be constant
Piezoelectric type	0–5 to 0–25,000 ppm	Clean gas only	Yes	
Heat of adsorption type	0–10 to 0–5,000 ppm	Clean gas or liquid. Special sampling for liquids	Yes	Sample flow must be constant
Infrared absorption	0–0.05 to 0–50%	Liquids and slurries	Yes	
Microwave absorption	0–1 to 0–90%	Liquids, slurries and pastes	No	

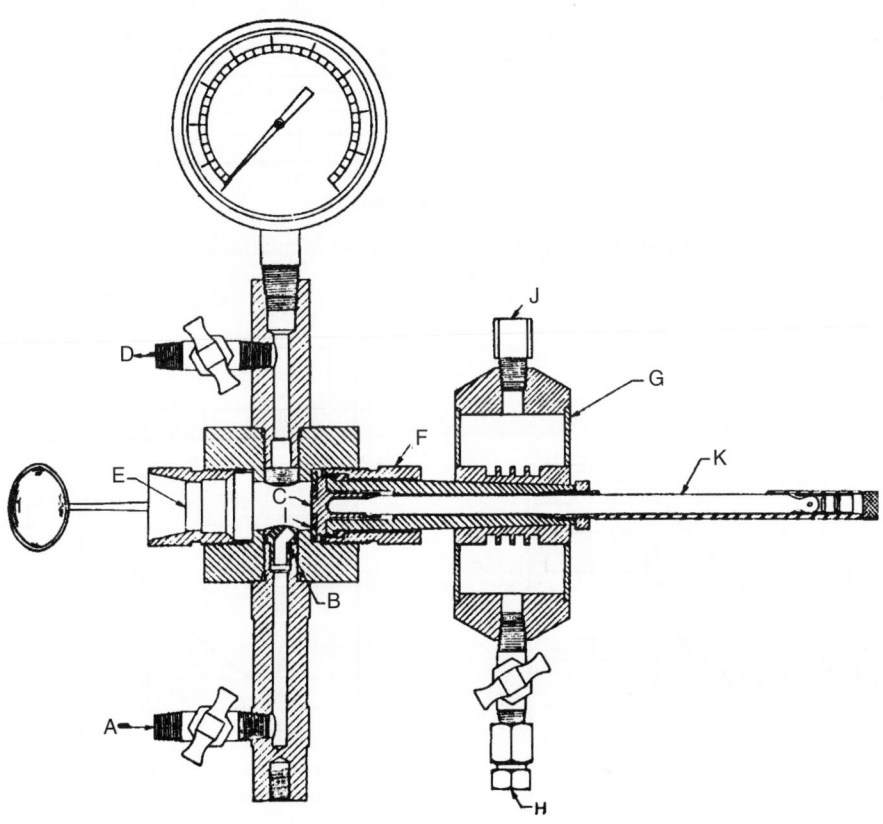

Figure 6.7.8 *Bureau of Mines dew point tester.*

where P = pressure in atm
SG$_g$ = relative density (specific gravity)
a, β, k = coefficients
t = Temperature in °C or T = temperature in K

Example 6.7.5
Find the pressure at which hydrate forms at T = 50°F for the gas from Example 6.7.4 using an analytical equation:

$$SG_g = 0.693$$

$$\log P = \beta + 0.0497\,(t + kt^2) \qquad [6.7.2]$$

β = 0.85 k = 0.002 from Figure 6.7.11 [8]

t = 50°F = 10°C

$$\log P = 0.85 + 0.0497[10 + 0.002 \times (10)^2]$$

$$\log P = 1.0335$$

P = 20.02 atm = 294 psia

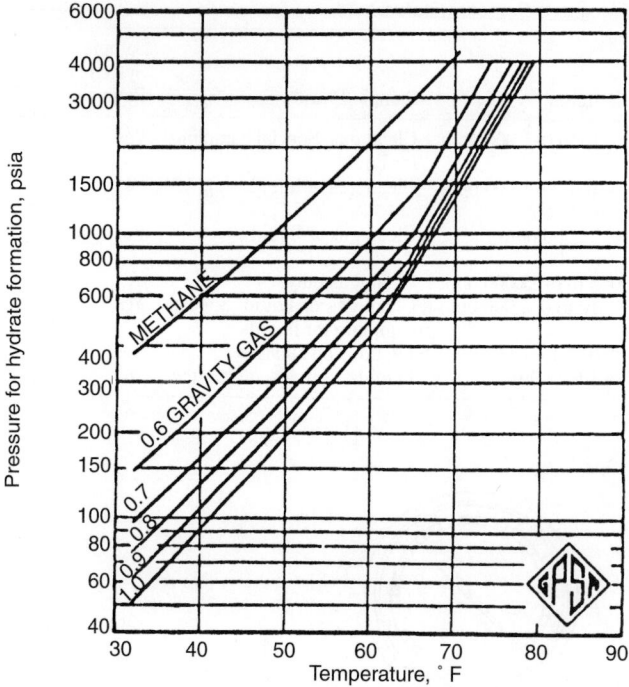

Figure 6.7.9 *Schematic of natural gas hydrate lattice.*

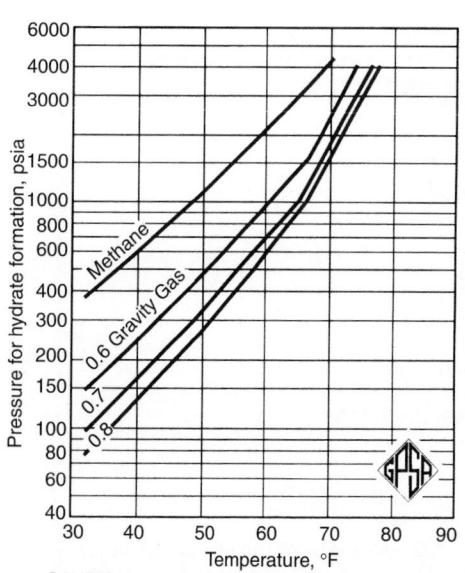

CAUTION: Figures 6.7.10, 6.7.12, and 6.7.13 should only be used for first approximations of hydrate formation conditions. For more accurate determination of hydrate conditions make calculations with K$_{v-s}$.

Figure 6.7.10 *Pressure temperature curves for predicting hydrate formation.*

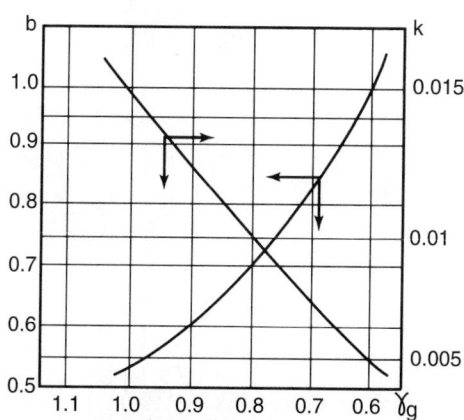

Figure 6.7.11 *Dependence of coefficients β and k on the relative density (SG$_g$) [8].*

Using Figure 6.7.10 we got ≈ 320 psia, it is a worst a value overestimate of 10%.

Example 6.7.6
The gas in Example 6.7.5 is to be expanded from 1,500 psia to 500 psia. What is the minimum initial temperature that will permit the expansion without hydrate formation?
   From Figure 6.7.12 [2] for SG$_g$ = 0.6   T$_{min}$ = 100°F
   From Figure 6.7.13 [2] for SG$_g$ = 0.7   T$_{min}$ = 125°F
   The 1500-psia-pressure line and the 500-psia final pressure line intersect just above the 100°F curve on Figure 6.7.13. Approximately 112°F is the minimum initial temperature.

Table 6.7.4 *Equations for the Relationship between Pressure and Temperature of Hydrate Formation for Several Gases [8]*

Gas and its relative density	Temp. Interval °C	Equations
CH_4	0 to −11	$\ln P = 5.6414 - 1154.61/T$
	0 to +23	$\ln P = 1.415 + 0.417(T + 0.01T^2)$
	+24 to +47	$\ln P = 1.602 + 0.0428T$
C_2H_6	0 to −10	$\ln P = 6.9296 - 1694.86/T$
	0 to +14.5	$\ln P = 0.71 + 0.0547T$
C_3H_8	0 to −12	$\ln P = 5.4242 - 1417.93/T$
	0 to +8.5	$\ln P = 0.231 + 0.0576T$
CO_2	0 to −6	$\ln P = 13.4238 - 3369.1245/T$
	0 to +9.8	$\ln P = 1.08 + 0.056T$
H_2S	−32 to +29.6	$\ln P = 2.844 + 0.0466T$
C_nH_{2n+2} 0.6 to 1.0	0 to +25	$\ln P = \beta + 0.0497(T + kT^2)$

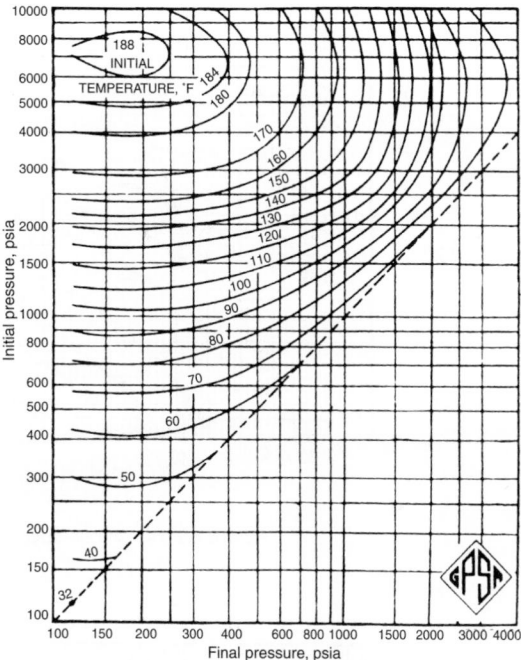

Figure 6.7.13 *Permissible expansion of 0.7 specific gravity natural gas without hydrate formation [2].*

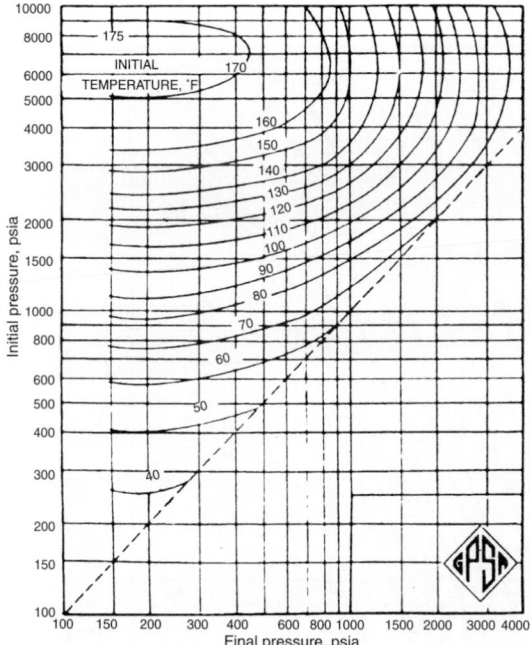

Figure 6.7.12 *Permissible expansion of 0.6 specific gravity natural gas without hydrate formation [2].*

Example 6.7.7

How far may a 0.6 specific gravity (relative density) gas at 2,000 psia and 100°F be expanded without hydrate formation?

On Figure 6.7.11 find the intersection of 2,000 initial pressure line with the 100°F initial temperature curve. Read on the *X-axis*, the permissible final pressure of 1100 psia.

Example 6.7.8

How far may a 0.6 gravity gas at 2,000 psia and 140°F be expanded without hydrate formation?

On Figure 6.7.11, the 140°F initial temperature curve does not intersect the 2,000-psia initial pressure line. Therefore, the gas may be expanded to atmospheric pressure without hydrate formation.

WARNING: Figures 6.7.12 and 6.7.13 should only be used for first approximations of hydrate conditions.

6.7.3 Hydrate Prediction Based on Composition for Sweet Gases

Several correlation's have proven useful for predicting hydrate formation of sweet gases and gases having minimal amounts of CO_2 and H_2S. The most reliable ones require a gas analysis. The Katz method [9, 10] utilizes vapor solid equilibrium constants defined by Equation 6.7.2,

$$K_{v-s} = Y/X_s \quad \text{and} \quad \sum Y_i/K = 1.0 \qquad [6.7.3]$$

The applicable K-value correlation's for the hydrate forming molecules (CH_4, C_2H_6, C_3H_8, i–C_4H_{10}, n–C_4H_{10}, CO_2 and H_2S) are shown in Figure 6.7.14 through 6.7.20. All molecules too large to form hydrate, e.g., nitrogen and helium have a K-value of infinity.

Example 6.7.9

Calculate the temperature for hydrate formation at 435 psia for a gas with the following composition (%):

$$N_2\text{-}5, \; C_1\text{-}78, \; C_2\text{-}6, \; C_3\text{-}3, \; iC_4\text{-}1,$$

$$H_e\text{-}1, \; CO_2\text{-}4, \; C_5\text{-}2.$$

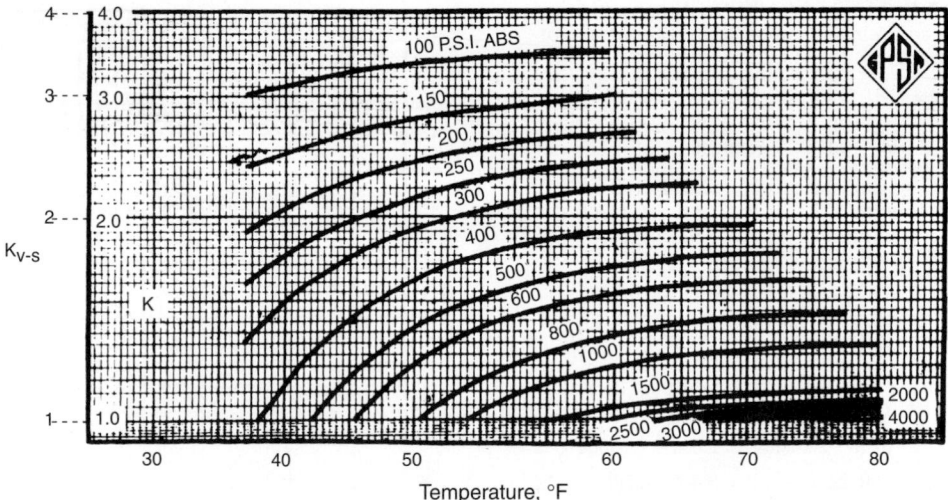

Figure 6.7.14 *Vapor–solid equilibrium constants for methane.*

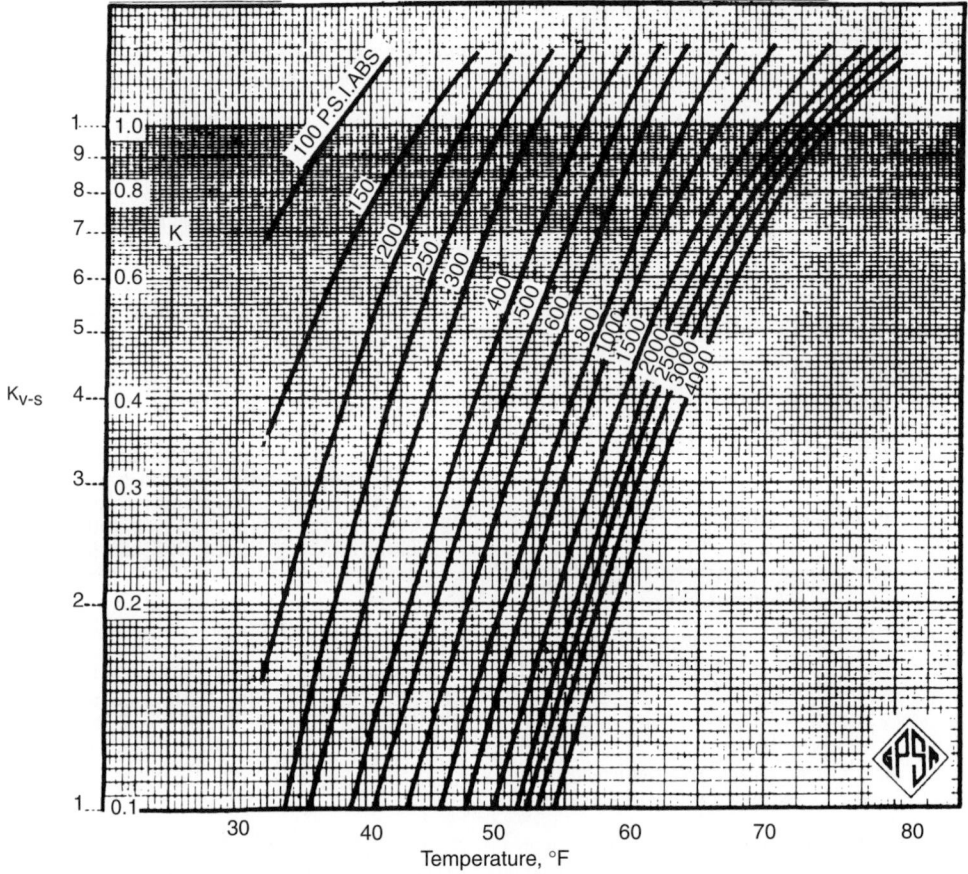

Figure 6.7.15 *Vapor–solid equilibrium constants for ethane.*

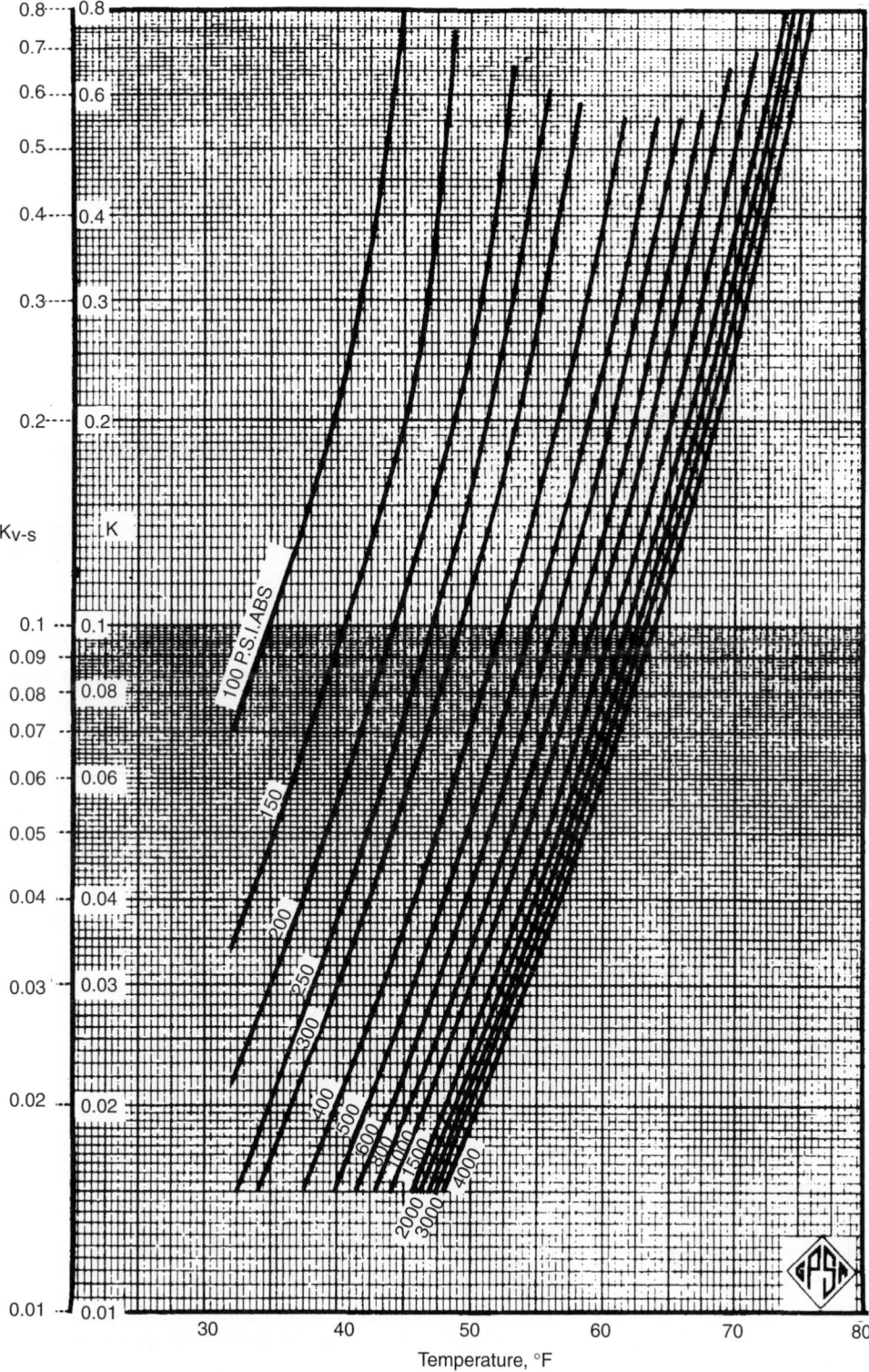

Figure 6.7.16 *Vapor–solid equilibrium constants for propane.*

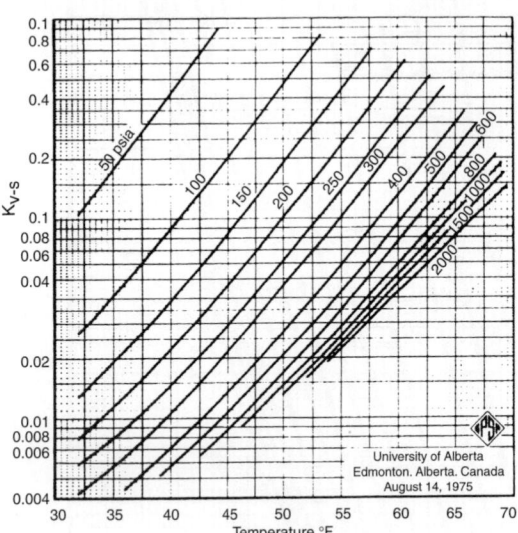

Figure 6.7.17 *Vapor–solid equilibrium constants for iso-butane.*

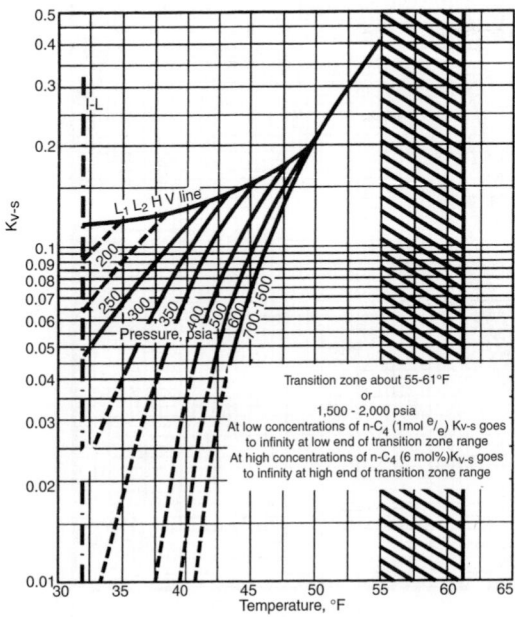

Figure 6.7.18 *Vapor–solid equilibrium constants for n-butane.*

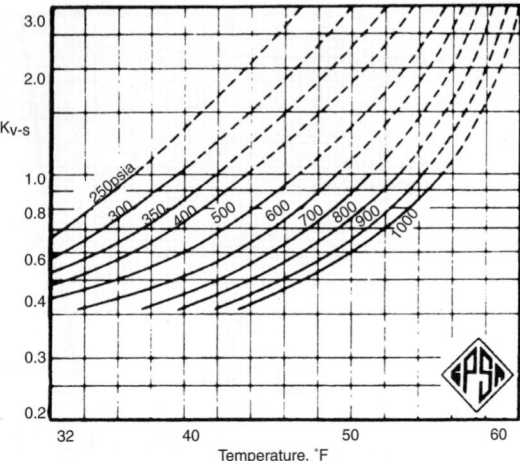

Figure 6.7.19 *Vapor–solid equilibrium constants for CO_2.*

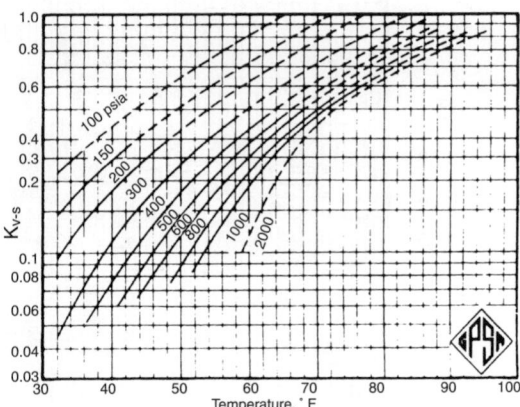

Figure 6.7.20 *Vapor–solid equilibrium constants for H_2S.*

		$t = 59°F$		$t = 50°F$		$t = 54°F$	
	Mole Fraction	y_i		y_i		y_i	
Comp	(y_I)	K	K	K	K	K	K
N_2	0.05	Inf	0	Inf	0	Inf	0
C_1	0.78	1.8	0.433	1.65	0.473	1.74	0.448
C_2	0.06	1.3	0.046	0.475	0.126	0.74	0.081
C_3	0.03	0.27	0.110	0.066	0.454	0.047	0.213
i-C_4	0.01	0.08	0.125	0.026	0.384	0.047	0.213
He	0.01	Inf	0	Inf	0	Inf	0
CO_2	0.04	~ 5	0.008	1.7	0.02	~ 3	0.011
C_{5+}	0.02	Inf	0	Inf	0	Inf	0
		$\sum K = 0.722$		$\sum K = 1.457$		$\sum K = 1.003$	

Temperature at which hydrate will form is about 54°F.
Using Equation 6.7.2 gives 435 psia = 30.6 atm

$$\log 30.6 = 0.8 + 0.0497(t + 0.0077t^2)$$

$$0.000383t^2 + 0.0497t - 0.686 = 0$$

$$t = 12.57°C = 54.6°F$$

Both results are close enough.

Example 6.7.10

The gas with the composition below is at 3,500 psia and at 150°F. What will be the hydrate conditions when the gas is expanded?

Comp	Mole Fraction	Initial Pressure (psia)	Initial Temperature (°F)	Final Pressure (psia)	Final Temperature (°F)
C_1	0.09267	3,500	150		
C_2	0.529	3,500	150	300	38
C_3	0.0138	3,500	150	400	45
iC_4	0.0018	3,500	150	500	52
nC_4	0.0034	3,500	150	600	58
nC_5	0.0014	3,500	150	700	64
	= 1.000				

Solution

Step 1. Make several adiabatic flash calculations at different pressures and plot on a pressure versus temperature graph; see Figure 6.7.21 [2].

Step 2. Assume some temperatures (40°F, 50°F, and 60°F) and predict the hydrate pressure for this gas using the vapor-solid K data. Plot the results on Figure 6.7.21.

$$(y/K) = 1 \quad at \quad 277 \text{ psia} \quad for \quad T = 40°F$$

$$(y/K) = 1 \quad at \quad 452 \text{ psia} \quad for \quad T = 50°F$$

$$(y/K) = 1 \quad at \quad 964 \text{ psia} \quad for \quad T = 60°$$

Step 3. The intersection of the lines in Figure 6.7.21 is the point at which hydrates start to form; in this example, 500 psia and 52°F.

The constants for H_2S shown in Figure 6.7.20 should be satisfactory at concentrations up to 15–20-mol% H_2S in the gas.

At the concentration of 30% or greater, hydrates mat form at about the same conditions as for pure H_2S.

Slightly better results than the K-charts method will generally give computer solutions, which have been developed for hydrate prediction applying P-V-T equations of state.

6.7.4 Hydrate Inhibition [2]

The formation of hydrates can be prevented by dehydrating to prevent a free-water phase or by inhibiting hydrate formation in the free-water phase. Dehydration is usually preferable, but inhibition can often be satisfactory.

Inhibition utilizes injection of one of the glycols or methanol to lower the hydrate formation temperature for a given pressure.

Ethylene, diethylene, and triethylene glycols have been used for glycol injection. The most popular has been ethylene glycol because of its lower cost, lower viscosity, and lower solubility in liquid hydrocarbons.

Physical properties of the most common glycols are given in Table 6.7.5 [2].

Estimation of properties for glycol–water mixtures can be achieved by a weight fraction average of the appropriate glycol curve and the water curve shown on each of the figures.

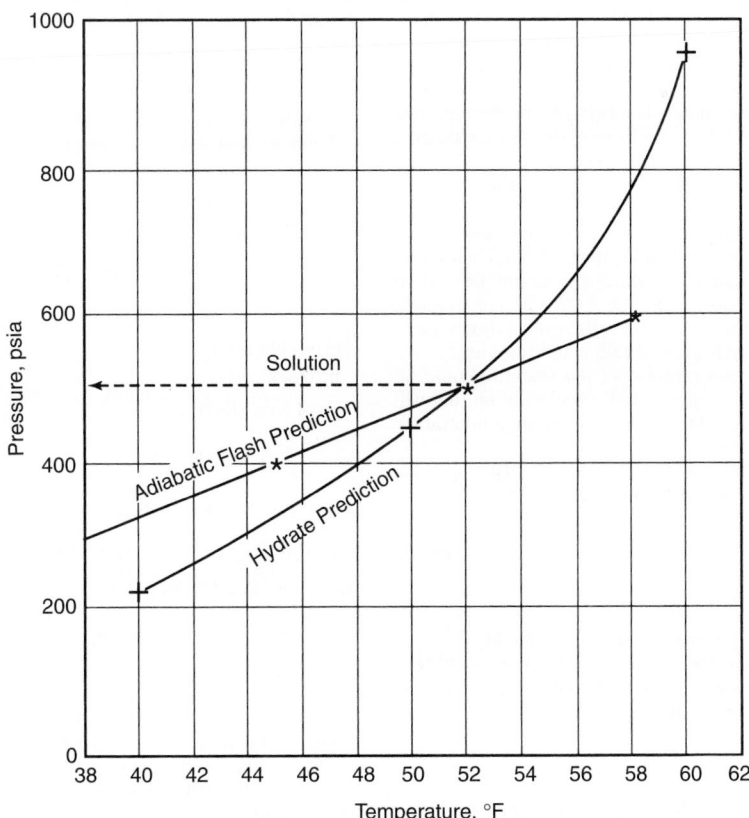

Figure 6.7.21 *Solution sketch for Example 6.7.10.*

Table 6.7.5 *Physical Properties of Selected Glycols and Methanol*

Formula	Ethylene Glycol $C_2H_6O_2$	Diethylene Glycol $C_4H_{10}O_3$	Triethylene Glycol $C_6H_{14}O_4$	Tetraethylene Glycol $C_6H_{18}O_5$	Methanol CH_3OH
Molecular Weight	62.1	106.1	150.2	194.2	32.04
Boiling Point* at 760 mm Hg, °F	387.1	472.6	545.9	597.2	148.1
Boiling Point* at 760 mm Hg, °C	197.3	244.8	285.5	314	64.5
Vapor Pressure at 77°F (25°C) mm Hg	0.12	< 0.01	< 0.01	< 0.01	120
Density					
(g/cc) at 77°F (25°C)	1.110	1.113	1.119	1.120	0.790
(g/cc) at 140°F (60°C)	1.085	1.088	1.092	1.092	
Pounds Per Gallon at 77°F (25°C)	9.26	9.29	9.34	9.34	6.59
Freezing Point, °F	8	17	19	22	−144.0
Pour Point, °F	—	−65	−73	−42	
Viscosity in centipoise					
at 77°F (25°C)	16.5	28.2	37.3	44.6	0.52
at 140°F (60°C)	4.68	6.99	8.77	10.2	
Surface Tension at 77°F (25°C), dynes/cm	47	44	45	45	22.5
Refractive Index at 77°F (25°C)	1.430	1.446	1.454	1.457	0.328
Specific Heat at 77 °F (25°C) Btu/(lb · °F)	0.58	0.55	0.53	0.52	0.60
Flash Point, °F (PMCC)	240	255	350	400	53.6
Fire Point, °F (C.O.C.)	245	290	330	375	

Note: These properties are laboratory results on pure compounds or typical of the products, but should not be confused with, or regarded as, specifications.

*Glycols decompose at temperatures below their atmospheric boiling point. Approximate decomposition temperatures are:

Ethylene Glycol   329°F   Triethylene Glycol   404°F
Diethylene Glycol   328°F   Tetraethylene Glycol   460°F

To allow determination of mixture properties at lower temperatures, the pure glycol and water property curves have been extrapolated below their freezing points.

To be effective, the inhibitor must be present at the very point where the wet gas is cooled to its hydrate temperature. For example, in refrigeration plants glycol inhibitors are typically sprayed on the tube-sheet faces of the gas exchangers so that they can flow with the gas through the tubes. As water condenses, inhibitors are present to mix with the water and prevent hydrates.

Glycol and its absorbed water are separated from the gas stream possibly along with liquid hydrocarbons. The glycol–water solution and liquid hydrocarbons can emulsify when agitated or when let down together from a higher to lower pressure. Careful separator design will allow nearly complete recovery of the glycol for regeneration and recycle.

The regenerator in a glycol injection system should be operated to produce a regenerated glycol solution that will have a freezing point below the minimum temperature encountered in the system.

Lowering the hydrate freezing point in natural gas systems by an antifreeze compound maybe calculated by:

$$d = \frac{K_H(I)}{(100 - I)M_I} \qquad [6.7.4]$$

where

d = depression of gas hydrate freezing point in °F
K_H = constant for methanol $= 2335$, for glycols $= 4000$
I = minimum inhibitor concentration in the free water %
M_I = molecular mass of solute inhibitor

Example 6.7.11

Estimate the methanol (MeOH) injection rate required to prevent hydrate formation in 2 MMscfd of 0.6 specific gravity (relative density) natural gas at 800 psia and 40°F. The gas is water saturated at 1,000 psia and 100°F. No hydrocarbons are condensed.

Solution

From Figure 6.7.2:
Water content at 100°F and 1,000 psia $= 60$ lb/MMscf
Water content at 40°F and 800 psia $= 10.5$ lb/MMscf
Hydrate temperature of gas $= 57.5$°F (Figure 6.7.10)

$$D = 57.5° - 40° = 17.5°F$$

From Equation 6.7.4:

$$17.5 = (2,335)(I)/[(100 - I)32]$$

$$I = 19.4\%$$

From Figure 6.7.22 [2]:

$$C\frac{\text{lb MeOH/MMscf}}{19.4 \text{ wt\% MeOH}} = 0.975$$

So $\dfrac{19.4}{0.975}$ lb MeOH/MMscf → required for gas phase

$19.9 \times 2 + 39.8$ lb MeOH/day

MeOH in liquid $= 99$ lb/day $(0.194)/(1.0 - 0.199) = 23.98$

Minimum MeOH injection rate $= 23.98 + 39.8$

$$= 63.8 \text{ lb/day}$$

6.7.5 Gas Dehydration

In those situations in which inhibition is not feasible or practical, dehydration must be used. Both liquid and solid desiccants may be used, but economics favor liquid desiccant dehydration when it will meet the required dehydration specification.

Liquid desiccant dehydration equipment is simple to operate and maintain. It can easily be automated for unattended

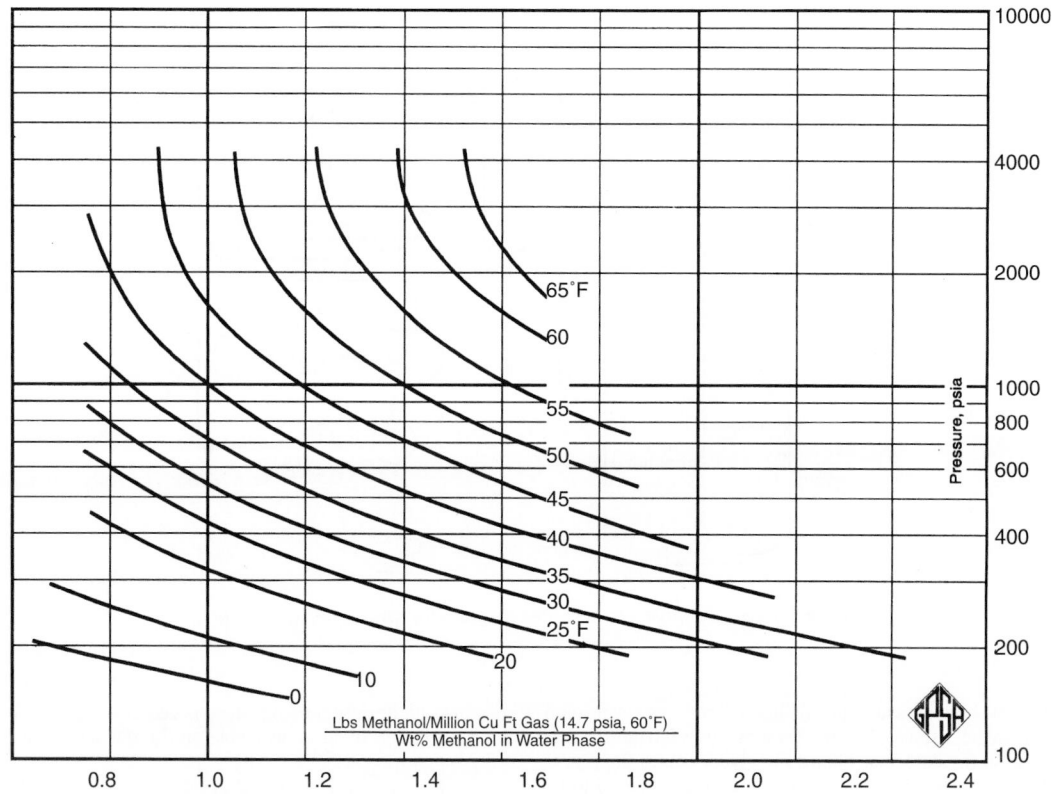

Figure 6.7.22 *Ratio of methanol vapor composition to methanol liquid composition [2].*

operation, for example, glycol dehydration at a remote production well. Liquid desiccants can be used for sour gases, but additional precautions in the design are needed due to the solubility of the acid gases in the desiccant solution.

Solid desiccants are normally used for extremely low dew point specifications as required for expander plants to recover liquid hydrocarbons.

The more common liquids in use for dehydrating natural gas are triethylene glycol (TEG), diethylene glycol (DEG), and tetraethylene glycol (TREG). In general, glycols are used for applications where dew point depressions of the order of 15 to 50°C (59 to 120°F) are required. TEG is the most commonly used glycol.

For the following description of the process and flow through a typical glycol dehydration unit refer to the schematic flow diagram as shown in Figure 6.7.23 [2].

The wet inlet gas stream first enters the unit through a separate vertical inlet gas scrubber. In this scrubber any liquid accumulations in the gas stream are removed. The inlet scrubber is normally provided with a tangenital inlet diverter that affects a circular flow of the well fluids around the wall of the vessel for centrifugal separation. The wet gas then passes out of the top of the inlet scrubber through a high-capacity, high-efficiency, stainless steel wire mesh mist eliminator that allows for virtually no liquid carryover. The separated well fluids drain into a quiet settling chamber in the bottom of the vessel and are discharged through a diaphragm-operated motor valve by liquid level control. The vertical inlet gas scrubbers may be equipped for either a two-phase (oil–gas) operation or a three phase (oil–gas–water) operation.

The wet gas leaves the top of the inlet scrubber and passes to the vertical glycol–gas contactor. The gas enters the bottom of this vessel and flows upward through the contact medium countercurrent to the glycol flow. The contact medium in the glycol–gas contactor may be valve-type trays or bubble cap trays. The operation is the same in that the liquid glycol flows down through the packing and the gas vapor flows up through the packing contacting the glycol. In trayed columns, the gas contacts the glycol on each tray as it passes through the vessel and the glycol absorbs the water vapor from the gas stream. Above the top tray in the contactor is a space for entrainment settling where most of the entrained glycol particles in the gas stream will settle out. A high-efficiency mist eliminator in the top of the contactor vessel will remove any glycol not settling out. The dry gas then leaves the contactor tower at the top.

The incoming dry lean glycol from the surge tank is cooled in a heat exchanger before it enters the contactor for a maximum contacting efficiency. The lean concentrated glycol is picked up from the surge tank by the glycol pump and is pumped at the contactor operating pressure through the heat exchanger and into the top of the contactor column. The dry glycol enters the contactor on the top of the tray. The dry glycol flows downward through the contactor vessel by passing across each tray and spilling over the wire box on the tray, and then passing down through a downcomer to the next tray. By this concurrent flow of gas and glycol, the driest incoming glycol on the top is in contact with the driest outgoing gas for maximum dehydration of the gas stream. The bottom tray downcomer is fitted with a seal to hold the liquid seal on the trays.

The wet-rich glycol that has now absorbed the water vapor from the gas stream leaves the bottom of the glycol-gas contactor column and passes through a high-pressure glycol

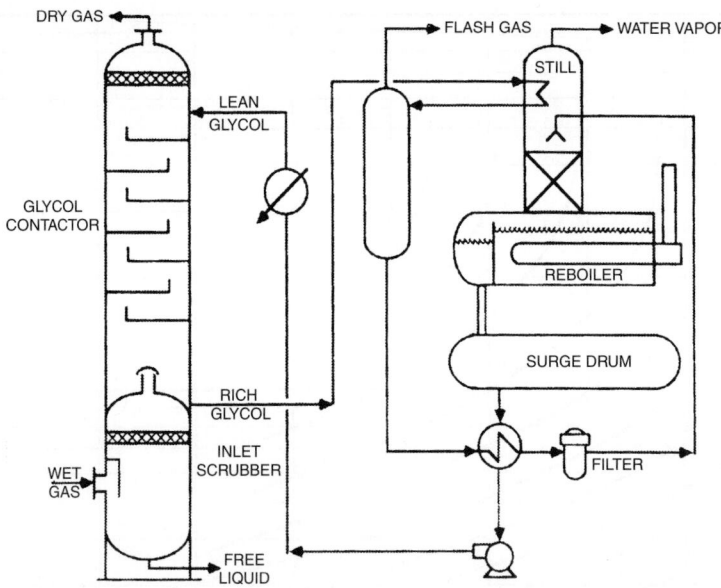

Figure 6.7.23 *Example process flow diagram for glycol dehydration unit.*

filter. The high-pressure glycol filter will remove any foreign solid particles that may have been picked up from the gas stream in the contactor before the glycol enters the power side of the glycol pump. This is generally considered to be the ideal location for primary filtration of the glycol stream.

From the glycol filter, water-rich glycol flows through the condenser coil, flashes off gas in the flash tank, and flows through the glycol–glycol heat exchanger to the regenerator portion of the unit. The warmed water-rich glycol enters the lower part of the stripping still column that is packed with ceramic saddles and is insulated. An atmospheric reflux condenser is integral with the stripping still at the top of the still, plus some water vapor to provide the adequate reflux required for the stripping column. This reflux condenser is also packed with ceramic saddles to assure that all the vapor to be vented will come in contact with the cool wall of the condenser.

The wet glycol, after entering the stripping still column will flow downward toward the reboiler contacting hot rising glycol vapors, water vapors, and stripping gas. The water vapor has a lower boiling point than glycol; therefore, any rising glycol vapors will be condensed in the stripping still and be returned to the reboiler section. In the reboiler, the glycol must travel a substantially horizontal path along the firebox to reach the liquid overflow exit at the opposite end. Here in the reboiler the glycol is heated to between 175°C and 200°C to remove enough water vapor to reconcentrate it to 99.5% or more. For extra dry glycol (99% plus), it may be necessary to add some stripping gas to the reboiler.

The warm wet glycol stream flows from the reboiler to a low-pressure surge tank, Next, the regenerated glycol flows through the glycol heat exchangers for cooling and is recirculated to the contactor by the glycol pump.

6.7.5.1 Dehydrator Design [1]

TEG dehydrators utilizing tray or packed column contactors may be sized from standard models by using the following procedures and associated graphs and tables. Custom-design glycol dehydrators for specific applications

may also be designed using these procedures. The following information must be available on the gas stream to be dehydrates:

- gas flowrate (MMscmd or MMscfd)
- composition or specific gravity of gas
- maximum working and operating pressure (bar or psig)
- gas inlet temperature (°C or °F)
- water dew point or water content required of the outlet gas (kg/MMscm or lb/MMscf)

From these, one can calculate:

1. the minimum concentration of TEG in the lean solution entering the top of the absorber required to meet outlet gas water specification
2. the lean (dry) TEG circulation rate required to absorb from the gas the needed amount of water
3. the total heat load on the reboiler

To obtain the answer, it is necessary to have a vapor–liquid equilibrium correlation for TEG–water system.

The *minimum lean (dry) TEG concentration*, a TEG absorber, is essentially isothermal. The heat of the solution is about 21 kJ/kg (91 BTU/lb) of water absorbed in addition to the latent heat. But, the mass of water absorbed plus the mass of TEG circulated is trivial to the mass of gas, so the inlet gases temperature controls. The temperature rise seldom exceeds 2°C except when dehydrating at a pressure below 10 bar (145 psia).

In Figure 6.7.24 [2] diagonal line represent %TEG in a TEG-water mixture entering the top of the absorber (glycol contactor).

Example 6.7.12

What equilibrium water dew point could be obtained at 80°F with a lean glycol solution containing 99.5 wt% TEG?

In Figure 6.7.23 locate 80°F on the abscissa; go vertically to the 99.5% line and the horizontally to the ordinate. The answer is −26°F.

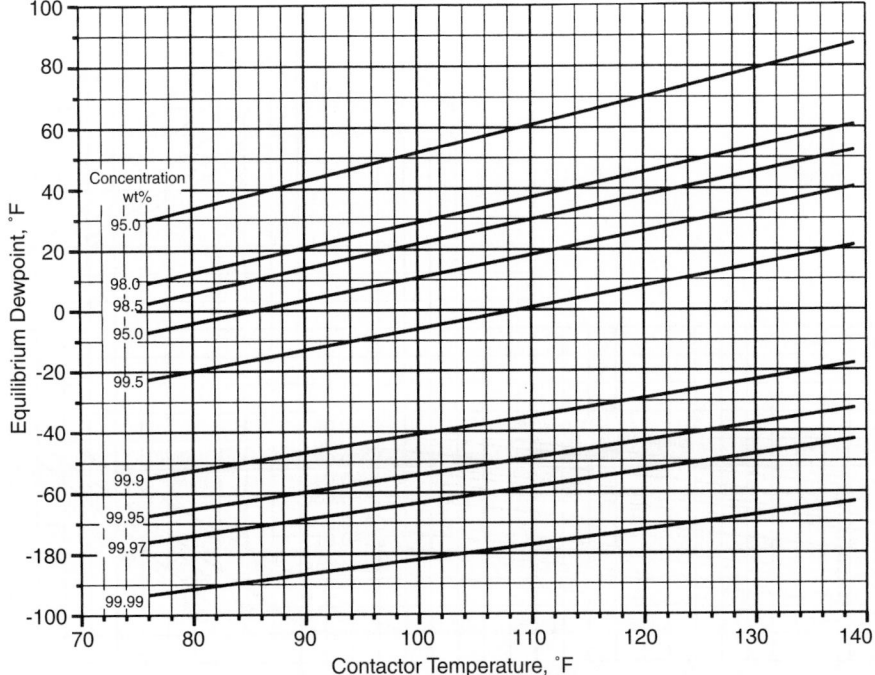

Figure 6.7.24 *Equilibrium H_2O dewpoint versus temperature at various TEG concentrations.*

It is theoretical water dew point, which could be attained in a test cell, but not in a real absorber. The gas and TEG are not in contact for a long enough time to reach equilibrium.

Practical tests show that a well-designed, properly operated unit will have an actual water dew point $7 \pm 1.5°C$ ($10°F$ to $15°F$) higher than the equilibrium dew point.

The procedure for calculation of minimum glycol concentration is as follows:

1. Establish the desired outlet dew point needed from sales contract specification or from minimum system temperature.
2. Subtract the approach ($10°F$ to $15°F$) from the desired outlet dew point to find the corresponding equilibrium water dew point.
3. Enter the value in the second step on the ordinate of Figure 6.7.24 and draw a horizontal line.
4. Draw a vertical line from the inlet gas temperature on the abscissa.
5. The intersection of the lines in Steps 3 and 4 establishes minimum lean TEG concentration required to obtain the water dew point on Step 1.

If water content is specified or calculated in mass per unit gas volume, a water content-pressure dew point temperature correlation is required (see Figure 6.7.3) [1].

Example 6.7.13
The gas sales contract specifies outlet water content of $5\text{-lb}/10^6$ scf at a pressure of 1,000 psia. The inlet gas temperature is $100°F$. What minimum TEG concentration is required?

For $5\text{-lb}/10^6$ scf and 1,000 psia, the equivalent dewpoint from Figure 6.7.3 is $25°F$. Using $13°F$ approach the equilibrium dew point is $12°F$. From Figure 6.7.23 at $12°F$ and $100°F$ contact temperature wt% TEG = 98.4%

A given lean TEG concentration is produced in the reboiler and still column (regenerator) section by control of reboiler temperature, pressure, and the possible use of stripping gas. So long as no stripping gas is used, the concentration of the lean TEG leaving the reboiler is independent of the rich TEG entering. When stripping gas is used, a water material balance around the absorber finds the concentration of rich TEG leaving the absorber.

By definition:

Wt% rich TEG

$$= \frac{(100)\ \text{wt lean TEG}}{\text{Wt lean TEG} + \text{wt water absorbed} + \text{wt water in lean TEG}} \quad [6.7.5]$$

The weight quantities in this equation may be found per unit of time (or per unit of gas flow). In any case, the values used depend on circulation rate. This rate depends on dew point requirements, lean TEG concentration, and amount of absorber contact and economics.

Economics dictates a rather low circulation rate. This rate usually will be a 7.5 to 22.5 L (2 to 6 gal) TEG solution per pound of absorbed water from the gas. The minimum rate governed by the rate required for effective gas-liquid contact in the absorber; the maximum is limited by economics. Because regeneration takes place at low pressure, calculations are simple. Figure 6.7.25 [11] has been used to predict regenerator performance based on Equation 6.7.5.

The minimum wt% of rich TEG on the bottom abscissa is found from Figure 6.7.25. Neglecting the small amount of water in the lean TEG, the rich TEG concentration can be determined from the following:

$$\text{Rich TEG} = \frac{8.34(SG_{gL})(\text{lean TEG})}{8.34(SG_{gL}) + 1/L_w} \quad [6.7.6]$$

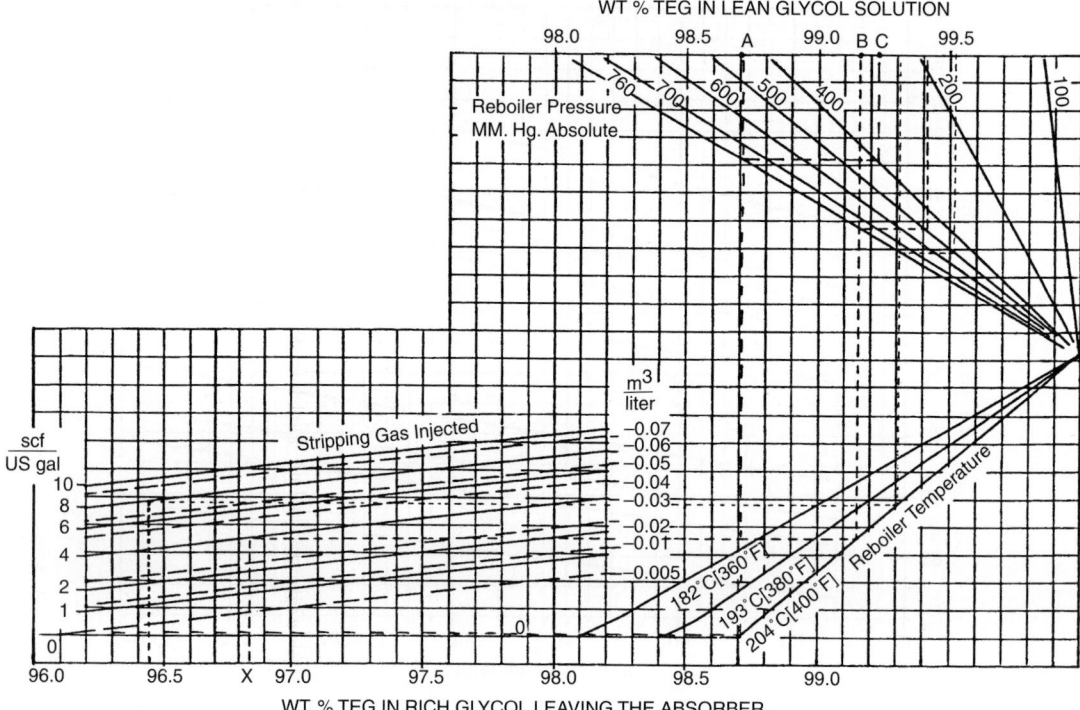

WT % TEG IN LEAN GLYCOL SOLUTION

WT. % TEG IN RICH GLYCOL LEAVING THE ABSORBER.

Figure 6.7.25 *Nomograph for estimating regenerator performance as a function of pressure, reboiler temperature [11].*

where rich TEG = wt% in rich TEG solution
  8.34 = water specific weight in lb/gal
  (SG_{gL}) = relative density of lean TEG solution at operating temperature of contactor contactor
Lean TEG = wt% TEG in lean TEG solution
  L_w = glycol to water circulation rate in gal TEG/lb H_2O

The diagonal lines in the lower left part of Figure 6.7.25 represent various amounts of stripping gas. Degree of regeneration depends on temperature regeneration, pressure and quantity of stripping gas. The general procedure for using nomograph Figure 6.7.25 is as follows:

1. *Atm. Pressure, No Stripping Gas*
   wt% rich glycol is not a variable. Proceed vertically from 0 stripping gas and temperature line intersection. Read 98.8 wt% TEG at 204°C; 98.4 wt% at 193°C.
2. *Atm. Pressure, Stripping Gas*
   a. Proceed vertically from B to temperature line and then horizontally
   b. Proceed vertically from X
   c. Intersection of two lines from (X) to (B) fixes amount of stripping gas.
3. *Vacuum, No Stripping Gas*
   a. Proceed vertically from the intersection of 0 gas line and temperature line to atmospheric line (760 mm H_g).
   b. Proceed horizontally from a point in (a) to pressure line necessary to fix value of point (c).

If both stripping gas and vacuum are used, procedure 2 and 3 are combined.

Lean TEG (99.0% to 99.9%) is available from most glycol reconcentrators. A value of 99.5% lean TEG is adequate for most design considerations.

Example 6.7.14
A 96.45-wt% rich glycol enters a regenerator using 8 scf of stripping gas per gallon of glycol solution. Reboiler temperature is 400°F. Find wt% TEG in lean glycol solution under atmospheric pressure and 400-mmHg absolute pressure.
From a bottom line for the rich glycol at 96.45-wt% flow to stripping gas, we get an injected value of 8 scf/gal. Proceed to 400°F and then vertically to get a 99.31 wt% if atmospheric pressure is used. If a vacuum is employed and the absolute pressure is 400 mmH$_g$, the low glycol concentration is 99.52 wt%.

$$L = \frac{L_w(W_r)G}{24} \tag{6.7.7}$$

where $W_r = \dfrac{(W_I - W_o)G}{24}$ [6.7.8]

where L = glycol circulation rate in gal/hr
  L_w = glycol to water circulation rate in gal TEG/lb H_2O
  W_i = water content of inlet gas in lb H_2O/MMscf
  G = gas flowrate in MMscfd
  W_r = water removed from gas in lb/hr
  W_o = outlet water content in lb/hr

The *required heat load*, for the reboiler, can be estimated from

$$Q_t = 2,000(L) \tag{6.7.9}$$

where Q_t = total heat load on the reboiler in Btu/hr

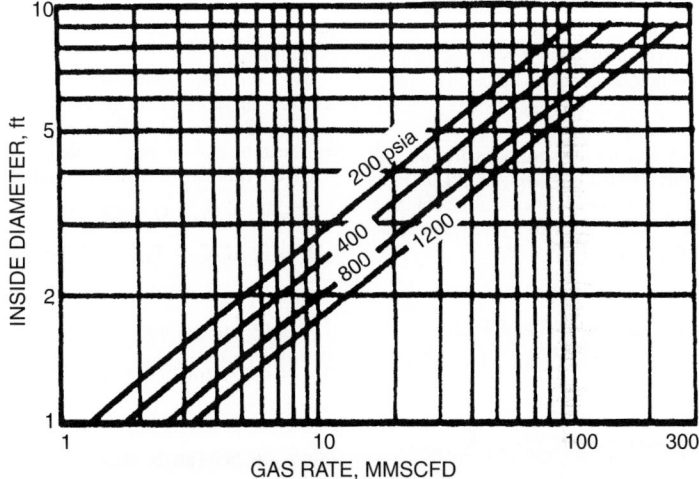

Figure 6.7.26 *Glycol contactor capacity [11].*

The above formula for determining the required reboiler heat load is an approximation that is accurate enough for most high-pressure glycol dehydrator sizing.

A more detailed determination of the required reboiler heat load may be made from the following procedure:

$$Q_t = Q_1 + Q_w + Q_r + Q_h \qquad [6.7.10]$$

$$Q_1 = L\rho_{gL}C(T_2 - T_1) \qquad [6.7.11]$$

$$Q_w = W_r \times 970.3\,(G) \qquad [6.7.12]$$

$$Q_r = 0.25\,Q_w \qquad [6.7.13]$$

$Q_h = 5,000$ to $20,000$ Btu/hr depending on the reboiler size

where Q_1 = sensible heat required for glycol in Btu/hr
Q_w = heat of vaporization required for water in Btu/hr
Q_r = heat of vaporize reflux water in still in Btu/hr
Q_h = heat loss from reboiler and stripping still in Btu/hr
ρ_{gl} = glycol specific weight in lb/gal
970.3 = heat of vaporization of water at 212°F, 14.7 psia in Btu/lb
T_2, T_1 = outlet and inlet temperatures in °F
L, W$_I$, W$_o$, G = as in Equation 6.7.7.

The size of the major components of a glycol dehydration unit may be estimated using the following procedures. The diameter of the inlet gas scrubber, may be estimated using techniques "Surface Oil production Systems". The diameter of the glycol contactor can be estimated from Figure 6.7.25 [11].

The height of the glycol contactors is based on the number of trays used in the column. The number of trays may be estimated using Figures 6.7.26, 6.7.27, and 6.7.28 [2], which are based on the dew point depression required and the glycol circulation rate. Typical field dehydration units use a value of 9 to 10 L (2.5 to 3 gal) of TEG per pound of water. Regardless of the type of tray, a spacing of 55 to 60 cm (22 to 24 in.) is recommended. Therefore, the total height of the contactor tower will be based on the number of trays required plus and additional 1.8 to 3 m (6 to 10 ft.) to allow space for a vapor

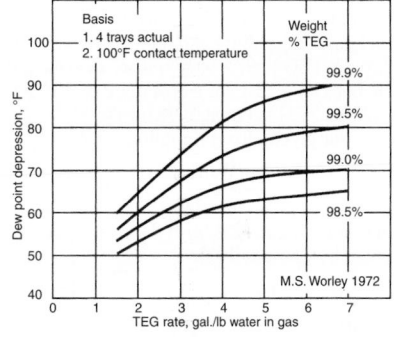

a) four actual trays

Figure 6.7.27 *Approximate glycol flow rates, four actual trays [11].*

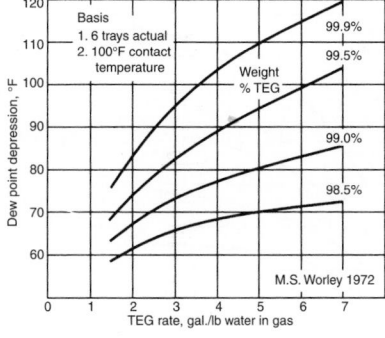

b) six actual trays

Figure 6.7.28 *Approximate glycol flow rates, six actual trays [11].*

disengagement area above the top tray and an inlet gas area at the bottom of the column.

Bubble cap trays are normally used in glycol dehydrators to facilitate low liquid loading and to offer large turndown.

Example 6.7.15

Design the tower and reboiler for a glycol dehydration unit for these given data: gas flow rate 0.7 MMscf/hr at 1,000 psia and temperature of 100°F, and an exit dew point gas of 20°F. The inlet gas is fully water saturated with a molecular mass of 20.3. For gas reboiler $T_2 = 400°F$, $T_1 = 280°F$.

Solution

Inlet water content (see Figures 6.7.1 and 6.7.2)

$$T = 100°F \quad W_I = 60 \text{ lb/MMscf} \quad \text{for} \quad M = 17.4$$

$$P = 1,000 \text{ psia}$$

$$W_{I(\text{at } M = 20.3)} = W_{I(\text{at } M = 17.4)} \times C_G = 60 \times 0.99$$

$$= 59.4 \text{ lb/MMscf}$$

Outlet water content:

$$T = 20°F \quad W_o = 4.2 \text{ lb/MMscf} \quad \text{for} \quad M = 17.4$$

$$P = 1,000 \text{ psia}$$

$$W_{o(\text{at } M = 20.3)} = 4.2 \times 1 = 4.2 \text{ lb/MMscf}$$

Concentration of lean glycol (see Figure 6.7.21)

For 100°F inlet gas temperature and outlet dew point 20°F, theoretical value for TEG is 97.9 wt%, but we decrease this point 10 to 15°F. Let's take 12°F for a dew point of 8°F, actual concentration of lean TEG is equal to 98.6 wt%.

Circulation rate for glycol, from Equation 6.7.7:

$$L = \frac{L_w(W_r)G}{24}$$

L_w is assumed to be 3 gal TEG/lb H_2O; a more accurate procedure is given by Campbell [1].

$$W_r = W_I - W_o = 59.4 - 4.2 = 55.2 \text{ lb/MMscf}$$

$$G = 0.7 \text{ MMscf/hr} = 16.8 \text{ MMscfd}$$

$$L = \frac{3 \times 55.2 \times 16.8}{24} = 115.9 \text{ gal/hr}$$

From Figure 6.7.29, find actual tray numbers if:

- dew point depression is $100 - 20 = 80°F$
- lean glycol concentration is 98.6 wt%
- assume the $L_w = 3$ gal/lb

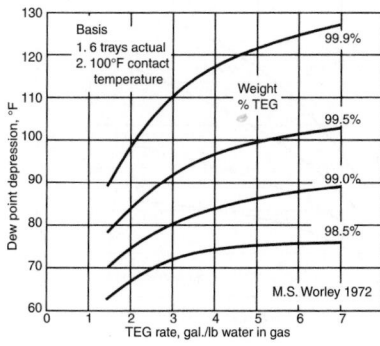

c) eight actual trays

Figure 6.7.29 *Approximate glycol flow rates, eight actual trays [11].*

Eight actual trays contactor with $L_w = 3$ gal/lb, 80°F dew point depression and 99.1 wt% lean TEG satisfy these conditions.

Next the inside diameter of the contactor can be found from Figure 6.7.25 [11] for Q = 16 MMscfd, P = 1,000 psia and D = 3 ft. Based on this diameter scrubber model DHT-3610 can be chosen from Table 6.7.6 [11].

Glycol reboiler duty from Equation 6.7.10:

$$Q_t = Q_1 + Q_w + Q_r + Q_h$$

$$Q_1 = L\rho_{gl}C(T_2 - T_1) = 115.9 \text{ gal/hr} \times 9.26 \text{ gal/hr}$$

$$\times 0.665 \text{ Btu/lb°F} \times (400 - 280)$$

$$= 85,644 \text{ Btu/hr}$$

$$Q_w = 970.3 \text{ Btu/lb} \times 55.2 \text{ lb/MMscf} \times 16.8 \text{ MMscfd} \times \tfrac{1}{2}$$

$$= 37.492 \text{ Btu/hr}$$

$$Q_h = 10,000 \text{ Btu/hr}$$

$$Q_t = 142,509 \text{ Btu/hr}$$

If SG = 0.7 and T = 100°F, calculate the gas capacity of the gas-glycol contactor selected for the specific operating conditions:

$$G_o = G_s(C_t)(C_g) \qquad [6.7.14]$$

where

G_o = gas capacity of contactor at operating conditions in MMscfd

G_s = gas capacity of contactor at SG = 0.7 and T = 100° T, based on operating pressure in MMscfd

C_t = correction factor for operating temperature (Table 6.7.5a)

C_g = correction factor for gas specific gravity (Table 6.7.5b)

For more accurate calculation procedure, see Sivalls' *Glycol Dehydration Design Manual* [11].

Table 6.7.6 *Correction Factors for Gas Capacity for Trayed Glycol-Gas Contactors in Equation 6.7.14 [11]*

(a)	
Operating Temp. °F	Correction Factor C_t
40	1.07
50	1.06
60	1.05
70	1.04
80	1.02
90	1.01
100	1.00
110	0.99
120	0.98

(b)	
Gas Specific Gravity	Correction Factor C_g
0.55	1.14
0.60	1.08
0.65	1.04
0.70	1.00
0.75	0.97
0.80	0.93
0.85	0.90
0.90	0.88

Table 6.7.7 *Vertical Inlet Scubbers Specifications [11]*

Model No.	Size O.D.	Nominal W.P. psig	Nominal Gas Capacity MMSCFD[1]	Intel & Gas Outlet Conn	Std Oil Valve	Shipping Weight lb.
VS-162	18"	250	1.8	2"	1"	900
VS-202	20"		2.9	3"	1"	1000
VS-242	24"		4.1	3"	1"	1200
VS-302	30"		6.5	4"	1"	1400
VS-362	36"		9.4	4"	1"	1900
VS-422	42"		12.7	6"	2"	2600
VS-482	48"		16.7	6"	2"	3000
VS-542	54"		21.2	6"	2"	3500
VS-602	60"		26.1	6"	2"	4500
VS-165	16"	500	2.7	2"	1"	1000
VS-205	20"		4.3	3"	1"	1300
VS-245	24"		6.1	3"	1"	2100
VS-305	30"		9.3	4"	1"	2700
VS-365	36"		13.3	4"	1"	3800
VS-425	42"		18.4	6"	2"	4200
VS-485	48"		24.3	6"	2"	5000
VS-545	54"		30.6	6"	2"	5400
VS-605	60"		38.1	6"	2"	7500
VS-166	16"	600	3.0	2"	1"	1100
VS-206	20"		4.6	3"	1"	1400
VS-246	24"		6.3	3"	1"	2200
VS-306	30"		9.8	4"	1"	2800
VS-366	36"		14.7	4"	1"	3900
VS-426	42"		20.4	6"	2"	4500
VS-486	48"		27.1	6"	2"	5100
VS-546	54"		34.0	6"	2"	6000
VS-606	60"		42.3	6"	2"	5100
VS-1610	16"	1000	3.9	2"	1"	1100
VS-2010	20"		6.1	3"	1"	1600
VS-2410	24"		8.8	3"	1"	2500
VS-3010	30"		13.6	4"	1"	3200
VS-3610	36"		20.7	4"	1"	4400
VS-4210	42"		27.5	6"	2"	6300
VS-4810	48"		36.9	6"	2"	8400
VS-5410	54"		46.1	6"	2"	9700
VS-6010	60"		57.7	6"	2"	14500
VS-1612	16"	1200	4.2	2"	1"	1150
VS-2012	20"		6.5	3"	1"	1800
VS-2412	24"		10.0	3"	1"	2600
VS-3012	30"		15.3	4"	1"	3400
VS-3612	36"		23.1	4"	1"	4700
VS-4212	42"		31.0	6"	2"	6700
VS-4812	48"		40.5	6"	2"	8500
VS-5412	54"		51.4	6"	2"	11300
VS-6012	60"		62.3	6"	2"	14500
VS-1614	16"	1440	4.8	2"	1"	1500
VS-2014	20"		6.7	3"	1"	2100
VS-2414	24"		11.2	3"	1"	2800
VS-3014	30"		17.7	4"	1"	3900
VS-3614	36"		25.5	4"	1"	5400
VS-4214	42"		34.7	6"	2"	7800
VS-4814	48"		45.3	6"	2"	9200
VS-5414	54"		56.1	6"	2"	12900
VS-6014	60"		69.6	6"	2"	16000

1. Gas capacity based on 100°F, 0.7 sp gr, and vessel working pressure.

Example 6.7.16

The sketch to Example 6.7.16 illustrates gas well surface facilities to reduce water content in the gas stream. The well stream arrives at the well head at 2,000 psia and 123°F and separates water in two vessels. The first separator works at 123°F and P = 2,000 psia, the second under pressure, P = 800 psia.

Calculate the following:

1. Temperature in gas separator (approximately).
2. Heat removed or added (if one is necessary) by heat exchanger before J-T expansion.
3. Does the well produce water (in liquid phase) in reservoir condition? If the well does produce water, how much does it produce?

What is the temperature in the gas separator? We calculate

$$527 \text{ vpm change in } \frac{\text{lb water}}{10^6 \text{ scf}}$$

$$X = \frac{\text{lb mole } H_2O \times 379.4 \text{ scf/mole}}{10^6 \text{ scf}} = 527 \times 10^6$$

$$X = \frac{527}{379.4} = 1.389 \text{ mole}$$

$$= (1.389 \text{ mole} \times 18 \text{ lb } H_2)/\text{lb mole} = 25 \text{ lb } H_2O/10^6 \text{ scf}$$

From Figure 6.7.1 for water content of 25 lb/10^6 scf and P = 800 psia, dew point = 67°F

Is heat added or removed?

Use any H-S diagram for natural gas with $SG_g = 0.6$. Starting from T = 67°F and 800 psia at constant enthalpy (J-T expansion) to P = 2,000 psia gives T = 123°F, so assuming constant enthalpy, no heat exchange is necessary.

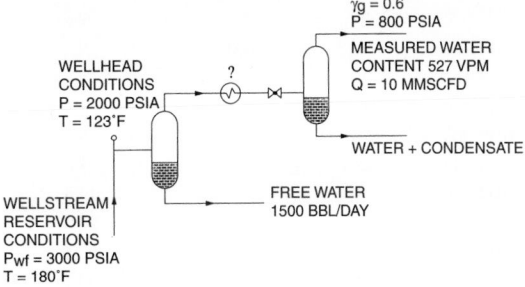

Does the well produce water in reservoir condition? The conditions are as follows:

Dissolved water in gas at reservoir conditions: T = 180°F and P = 3,000 psia, W = 195 lb/10^6 scf.

Dissolved water in gas in separator number 1: T = 123°F and P = 2,000 psia, W = 70 lb/10^6 scf.

Hence, an amount of water from the gas stream is (195 to 70)10 = 1,250 lb/d. Separator number 1 produces 1,500 lb/day water, which is more than can be condensed from the gas, which means the well produces water in liquid phase at a rate of 250 lb/d.

6.7.6 Other Considerations

Under conventional dehydration condition, 40% to 60% of methanol in feed gas to a glycol dehydrator will be absorbed by the TEG. This will add additional heat duty on the reboiler and additional vapor load on the regenerator.

Glycol losses can be defined as mechanical carryover from the contactor, plus vaporization from the contactor and regenerator and spillage. Glycol losses, exclusive of spillage, range from 0.05 gal/MMscf for high-pressure low-temperature gases to as much as 0.30 gal/MMscf for low-pressure high-temperature gases. Excessive losses usually result from foaming in the absorber or regenerator.

Glycol losses in CO_2 dehydration systems can be significantly higher than is natural gas systems particularly at

Table 6.7.8 *Tray-Type Glycol-Gas Contactors [11]*

Model No.	Size O. D.	Nominal W.P. psig	Nominal Gas Capacity MMSCFD[1]	Gas Inlet & Outlet Size	Glycol Inlet & Outlet Size	Glycol Cooler Size	Shipping Weight lb.
DHT-122	$12\frac{3}{4}$"	250	1.5	2"	$\frac{1}{2}$"	2" × 4"	800
DHT-162	16"		2.4	2"	$\frac{3}{4}$"	2" × 4"	900
DHT-182	18"		3.2	3"	$\frac{3}{4}$"	3" × 5"	1100
DHT-202	20"		4.0	3"	1"	3" × 5"	1400
DHT-242	24"		6.1	3"	1"	3" × 5"	2000
DHT-302	30"		9.9	4"	1"	4" × 6"	2400
DHT-362	36"		14.7	4"	$1\frac{1}{2}$"	4" × 6"	3200
DHT-422	42"		19.7	6"	$1\frac{1}{2}$"	6" × 8"	4400
DHT-482	48"		26.3	6"	2"	8" × 8"	6300
DHT-542	54"		32.7	6"	2"	6" × 8"	7700
DHT-602	60"		40.6	6"	2"	6" × 8"	9500
DHT-125	$12\frac{3}{4}$"	500	2.0	2"	$\frac{1}{2}$"	2" × 4"	1000
DHT-165	16"		3.2	2"	$\frac{3}{4}$"	2" × 4"	1200
DHT-185	18"		4.3	3"	$\frac{3}{4}$"	3" × 5"	1500
DHT-205	20"		5.3	3"	1"	3" × 5"	1700
DHT-245	24"		8.3	3"	1"	3" × 5"	2900
DHT-305	30"		13.1	4"	1"	3" × 5"	3900
DHT-365	36"		19.2	4"	$1\frac{1}{2}$"	4" × 6"	6000
DHT-425	42"		27.4	6"	$1\frac{1}{2}$"	6" × 8"	7700
DHT-485	48"		35.1	6"	2"	6" × 8"	10000
DHT-545	54"		44.5	6"	2"	6" × 8"	12000
DHT-605	60"		55.2	6"	2"	6" × 8"	15300
DHT-126	$12\frac{3}{4}$"	600	2.2	2"	$\frac{1}{2}$"	2" × 4"	1100
DHT-166	16"		3.4	2"	$\frac{3}{4}$"	2" × 4"	1300
DHT-186	18"		4.5	3"	$\frac{3}{4}$"	3" × 5"	1600
DHT-206	20"		5.5	3"	1"	3" × 5"	1800
DHT-246	24"		8.5	3"	1"	3" × 5"	3000
DHT-306	30"		14.3	4"	1"	3" × 5"	4000
DHT-366	36"		21.2	4"	$1\frac{1}{2}$"	4" × 6"	6300
DHT-426	42"		29.4	6"	$1\frac{1}{2}$"	6" × 8"	8400
DHT-466	48"		39.2	6"	2"	6" × 8"	11300
DHT-546	54"		49.3	6"	2"	6" × 8"	13400
DHT-606	60"		61.3	6"	2"	6" × 8"	16500
DHT-1210	$12\frac{3}{4}$"	1000	2.7	2"	$\frac{1}{2}$"	2" × 4"	1300
DHT-1610	16"		4.3	2"	$\frac{3}{4}$"	2" × 4"	1600
DHT-1810	18"		5.5	3"	$\frac{3}{4}$"	3" × 5"	2100
DHT-2010	20"		7.3	3"	1"	3" × 5"	2600
DHT-2410	24"		11.3	3"	1"	3" × 5"	4200
DHT-3010	30"		18.4	4"	1"	3" × 5"	5500
DHT-3610	36"		27.5	4"	$1\frac{1}{2}$"	4" × 6"	8500
DHT-4210	42"		37.1	6"	$1\frac{1}{2}$"	6" × 8"	11800
DHT-4810	48"		49.6	6"	2"	6" × 8"	16200
DHT-5410	54"		62.0	6"	2"	6" × 8"	20200
DHT-6010	60"		77.5	6"	2"	6" × 8"	26300
DHT-1212	$12\frac{3}{4}$"	1200	3.0	2"	$\frac{1}{2}$"	2" × 4"	1500
DHT-1612	16"		4.7	2"	$\frac{3}{4}$"	2" × 4"	1900
DHT-1812	18"		6.0	3"	$\frac{3}{4}$"	3" × 5"	2300
DHT-2012	20"		7.8	3"	1"	3" × 5"	3000
DHT-2412	24"		12.0	3"	1"	3" × 5"	4900
DHT-3012	30"		20.1	4"	1"	3" × 5"	6400
DHT-3612	36"		29.8	4"	$1\frac{1}{2}$"	4" × 6"	10000
DHT-4212	42"		41.4	6"	$1\frac{1}{2}$"	6" × 8"	13100
DHT-4812	48"		54.1	6"	2"	6" × 8"	18400

(continued)

Table 6.7.8 *Tray-Type Glycol-Gas Contactors [11] (continued)*

Model No.	Size O. D.	Nominal W.P. psig	Nominal Gas Capacity MMSCFD[1]	Gas Inlet & Outlet Size	Glycol Inlet & Outlet Size	Glycol Cooler Size	Shipping Weight lb.
DHT-5412	54"		68.4	6"	2"	6" × 8"	23500
DHT-6012	60"		85.0	6"	2"	6" × 8"	29000
DHT-1214	$12\frac{3}{4}$"	1440	3.1	2"	$\frac{1}{2}$"	2" × 4"	1800
DHT-1614	16"		4.9	2"	$\frac{3}{4}$"	2" × 4"	2200
DHT-1814	18"		0.5	3"	$\frac{3}{4}$"	3" × 5"	2800
DHT-2014	20"		8.3	3"	1"	3" × 5"	3500
DHT-2414	24"		13.3	3"	1"	3" × 5"	5800
DHT-3014	30"		22.3	4"	1"	3" × 5"	7500
DHT-3614	36"		32.8	4"	$1\frac{1}{2}$"	4" × 6"	11700
DHT-4214	42"		44.3	6"	$1\frac{1}{2}$"	6" × 8"	14400
DHT-4814	48"		58.3	6"	2"	6" × 8"	20000
DHT-5414	54"		74.0	6"	2"	6" × 8"	25800

Gas capacity based on 100°F, 0.7 sp gr and contactor working pressure.

pressure above 900 psia. This is due to the solubility of TEG in dense phase CO_2.

Glycol becomes corrosive with prolonged exposure to oxygen. A dry gas blanket on the glycol surge tank will help eliminate oxygen absorption.

There are several enhanced glycol concentration processes that are used for obtaining TEG purity higher than 98.7 wt%, which is the TEG purity obtained by reboiling at 400°F and atmospheric pressure. The processes are listed on Table 6.7.10 [2].

6.7.6.1 Solid Desiccant Dehydration (Adsorption)

There are several solid desiccants, which possess the physical characteristic to adsorb water from natural gas. These desiccants are generally are used in dehydration systems consisting of two or more towers and associated regeneration equipment. See Figure 6.7.30 [2], for a simple two-tower system.

The continuous process requires two (or more) vessels, with one tower onstream adsorbing water from the gas stream while the other tower is being regenerated and cooled. Generally the bed is designed to be online for 8 to 24 hr. When the bed is taken off line, hot gas is used to heat the sieve to 230°C to 290°C (450°F to 550°F). The regeneration gas is returned to the process after it has been cooled and the free water removed. The towers are switched before the onstream tower becomes saturated.

Adsorption describes any process wherein molecules from the gas are held on the surface of a solid by surface forces. Solid desiccants that possess a total area of 500,000 to 800,000 m^2/kg (2,400,00 to 3,900,000 $ft.^2$/lb) are used as adsorbants, and desiccants in commercial use fall into one of three categories.

Alumina — a manufactured or natural occurring form of aluminum oxide that is activated by heating.

Gels — Aluminum or silica gels manufactured and conditioned to have an affinity for water.

Molecular Sieves — Manufactured or naturally occurring aluminosilicates exhibiting a degree of selectivity based on crystalline structure in their adsorption of natural gas constituents.

Alumina is a hydrated form of alumina oxide (Al_2O_3). It is used for gas and liquid dehydration and will give outlet dew points in the range of −100°F. Less heat is required to regenerate alumina than for molecular sieve, and the regeneration temperature is lower.

Table 6.7.9 *Typical Liquid Desiccant Properties*

	Ethylene Glycol	Diethylene Glycol	Triethylene Glycol
Molecular Weight	62.07	106.12	150.17
Specific Gravity @ 68°F	1.1155	1.1184	1.1255
Specific Weight, lb/gal.	9.292	9.316	9.375
Boiling Point @ 760 MMHg, °F	387.7	474.4	550.4
Freezing Point, °F	9.1	18.0	24.3
Surface Tension @ 77°F, dynes/cm	47.0	44.8	45.2
Heat of Vaporization @ 760 MMHg, BTU/lb	364	232	174

Table 6.7.10 *Enhanced Glycol Regeneration Processes*

	TEG Conc. wt%	Water Dew Point Depression Possible, °F
Vacuum	99.2 to 99.9	100 to 150
COLDFINGER®	99.96	100 to 150
DRIZO®	99.99+	180 to 220
Stripping Gas	99.2 to 99.98	100 to 150

Silica gel is a generic name for a gel manufactured from sulfuric acid and sodium silicate. It is essentially pure silicon dioxide, S_iO_2. When used for dehydration, silica gel will give outlet dew points of approximately −70° to −80°F.

Molecular sieves are a class of aluminosilicates and possess the highest capacity, will produce the lowest water dew points, and can be used to sweeten and dry gases and liquids. Their equilibrium water capacity is much less dependent on adsorption temperature and relative humidity. They are usually more expensive.

Molecular sieve dehydrators are commonly used ahead of NGL recovery plants where extremely dry gas is required. Cryogenic NGL plants designed to recover ethane produce

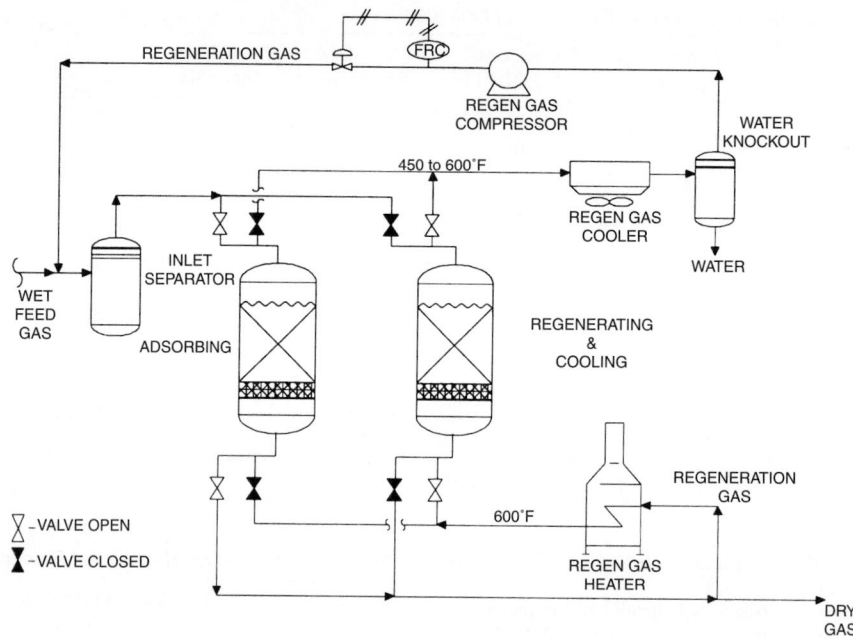

Figure 6.7.30 *Example solid desiccant dehydrator twin-tower system.*

Table 6.7.11 *Typical Desiccant Properties*

Desiccant	Shape	Bulk Density, lb/ft^3	Particle Size	Heat Capacity, Btu/(lb·°F)	Approx. Minimum Moisture Content of Effluent Gas (ppmw)
Alumina Gel Alco H-151	Spherical	52	1/4"	0.20	5–10
Activated Alumina Alcoa F-1	Granular	52	1/4"–8 mesh		0.1
Silica Gel Sorbead®-R	Spherical	49	4–8 mesh	0.25	5–10
Silica Gel Sorbead®-H	Spherical	45	3–8 mesh	0.25	5–10
Mole Sieve Davison-4A	Spherical	42–45	4–8 mesh or 8–12 mesh	0.24	0.1
Mole Sieve Linde-4A	Extruded Cylinder	40–44	1/8" or 1/16"	0.24	0.1

very cold temperatures and require a very dry feed gas to prevent formation of hydrates. Dehydration to approximately 1 ppmw is possible with molecular sieves. Typical solid desiccant properties are given in Table 6.7.11 [2].

Because solid desiccant units cost more to buy and operate than glycol units, their use is usually limited to applications such as very sour gases, very low water dew point requirements, simultaneous control of water, and hydrocarbons dew points, and in certain cases such as oxygen-containing gasses. In cryogenic plants, solid desiccant dehydration usually is preferred over methanol injection to prevent hydrate and ice formation. Solid desiccants are also often used for the drying and sweetening on NGL liquids. Any host source can be used to heat the regeneration gas including waste heat from engines and turbines. Heat is a major operating cost and is a major design consideration.

6.7.6.2 Solid Desiccant Dehydrator Design
The allowable superficial velocity through the bed is the first parameter that must be estimated. The pressure drop through the bed is related to the superficial velocity by a modified Ergun equation:

$$\frac{P}{L} = B\mu V + C\rho V^2 \qquad [6.7.15]$$

where

$\quad$ P = pressure drop in psi
$\quad$ L = length of packed bed in ft.
$\quad$ μ = viscosity in cp
$\quad$ V = Superficial velocity in ft./min
$\quad$ ρ = density in lb/ft.^{3}
$\quad$ B, C = constants as below

Particle Type	B	C
1/8 in bead	0.0560	0.0000889
1/8 in extrude	0.0722	0.000124
1/16 in bead	0.152	0.000136
1/16 in extrude	0.238	0.000210

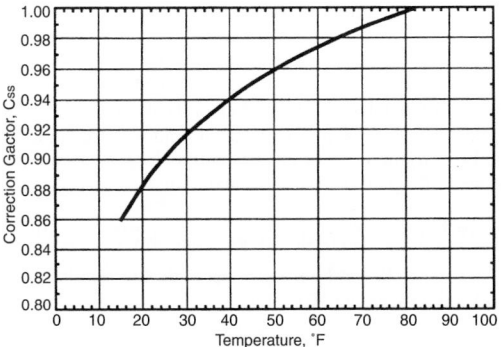

Figure 6.7.31   Mole sieve capacity correlation for undersaturated gas.

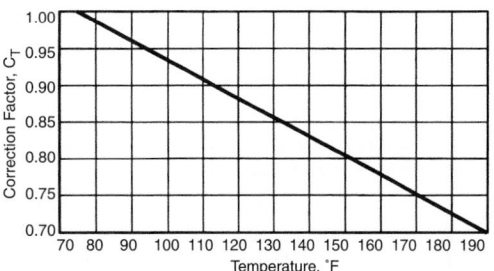

Figure 6.7.32   Mole sieve capacity correlation for temperature.

Figure 6.7.31 [2], was derived from this modified Ergun equation by assuming a gas composition and setting P/L equal to 0.333 psi/ft. The design pressure drop through the bed should be about 5 psi. A design pressure drop higher than 8 psi is not recommended. Remember to check the pressure drop after the bed height has been determined. Once the allowable superficial velocity is estimated, calculate the bed diameter:

$$D = \frac{4(ACFM)}{3.14(V)} \qquad [6.7.16]$$

where

D = diameter in ft.
ACFM = actual cubic feet per minute

The next step is to choose a cycle time and calculate the pounds of sieve required. Eight- to 12-hr cycles are common. Cycles greater than 12 hr may be justified, especially if the feed gas is not saturated. Long cycles mean less regeneration and longer sieve life, but larger beds and additional capital investment.

During the adsorption cycle, the bed can be thought of as operating within three zones. The top zone is called the saturation or equilibrium zone. The desiccant in this zone is in equilibrium with the wet inlet gas. The middle or mass transfer zone (MTZ) is where the water content of the gas is reduced from saturated to <1 ppm. The bottom zone is unused desiccant and is often called the active zone. If the bed operates too long in adsorption, the mass transfer zone begins to move out the bottom of the bed causing "breakthrough." At breakthrough, the water content of the outlet gas begins to increase and will eventually reach feed gas content when the mass transfer zone is completely displaced.

Unfortunately, both the water capacity and the rate at which the molecular sieve adsorbs water change as the molecular sieve's age. The object of the design is to install enough sieves so that 3 to 5 years into the life of the sieve the mass transfer zone will be at the bottom of the bed at the end of the adsorption cycle.

In the saturation zone, the molecular sieve is expected to hold approximately 13 lb of water per 100 lb sieve. This capacity needs to be adjusted when the gas is not water saturated or the temperature is above 75°F. See Figures 6.7.32 and 6.7.33 [2] for the correction factors.

To determine the pounds of molecular sieve required in the saturation zone, calculate the amount of water to be removed during the cycle and divide by the sieve

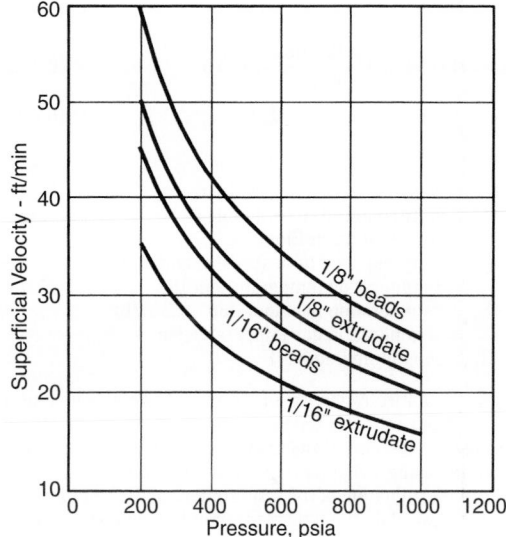

Figure 6.7.33   Allowable superficial velocity for molecular sieve dehydrator.

capacity:

$$Ss = \frac{W_r}{(0.13)(C_{ss})(C_T)} \qquad [6.7.17]$$

$$L_s = \frac{(S_s)(\text{bulk density})(4)}{(3.14)(D^2)} \qquad [6.7.18]$$

where

S_s = amount of molecular sieve required in the saturation zone lb/hr
W_r = water removed in lb/hr
C_{ss} = saturation correction factor for sieve
C_T = temperature correction factor
L_s = length of packed bed saturation zone in ft.

Bulk density is 42 to 45 lb/ft.³ for spherical particles and 40 to 44 lb/ft.³ for extruded cylinders.

Even though the MTZ will contain some water, the saturation zone is calculated assuming it will contain all the water to be removed.

Refer to Table 6.7.11 for bulk densities of common desiccants.

The length of the MTZ can be calculated as follows:

$$L_{MTZ} = (V/35)^{0.35}(Z) \qquad [6.7.19]$$

where

$Z = 1.70$ for 1/8 in sieve
$\quad = 0.85$ for 1/16 in sieve

The total bed height is the summation of the saturation zone and the MTZ heights. Approximately 6-ft. free space above and below the bed is needed.

6.7.7 Regeneration Calculations

The first step is to calculate the total heat required to desorb the water and heat the desiccant and vessel. A 10% heat loss is assumed

$$Q_w = (1800 \text{ Btu/lb})(\text{lbs of water on bed}) \qquad [6.7.20]$$

$$Q_{si} = (\text{lb of sieve})\frac{0.24 \text{ Btu}}{\text{lb°F}}(T_{rg} - T_i) \qquad [6.7.21]$$

$$Q_{st} = (\text{lb of steel})\frac{0.12 \text{ Btu}}{\text{lb °F}}(T_{rg} - T_i) \qquad [6.7.22]$$

$$Q_{hl} = (\text{heat loss}) = (Q_w + Q_{si} + Q_{st})(0.10) \qquad [6.7.23]$$

where

Q_w = desorption of water heat duty in Btu/hr
Q_{si} = duty required to heat mole sieve to regeneration temperature in Btu
Q_{st} = duty required to heat vessel and piping to regeneration temperature in Btu
Q_{hl} = regeneration heat loss duty in Btu/hr
T_{rg} = regeneration gas temperature in °F
T_i = inlet temperature °F

For the entire regeneration cycle, only about one-half of the heat put into the regeneration gas is utilized. This is because by the end of the cycle the gas is leaving the bed about the same temperature at which it enters.

For determination of the requirement of the regeneration gas rate, calculate the total regeneration load from:

$$Q_{tr} = (2.5)(Q_w + Q_{si} + Q_{st} + Q_{hl})$$

$$= \text{total regeneration heat duty in Btu.} \qquad [6.7.24]$$

The heating time is usually 50% to 60% of the total regeneration time, which must include a cooling period. Figure 6.7.32 shows a typical temperature profile for a regeneration cycle (heating and cooling). For 8-hr adsorption cycles, the regeneration normally consists of $4\frac{1}{2}$ hr heating, 3 hr of cooling and $\frac{1}{2}$ hour of standby and switching. For longer cycles the heating time can be lengthened as long as a minimum pressure drop of 0.01 psi/ft. is maintained

$$m_{rg}(\text{lb/hr}) = \frac{Q_{tr}}{(C_p)(600°\text{F} - T_i)(\text{heating time})} \qquad [6.7.25]$$

Figure 6.7.33 can be used to estimate the required minimum velocity to meet 0.01 psi/ft.

6.7.8 General Comments

The regeneration cycle frequently includes depressuring/repressuring to match the regeneration gas pressure and/or to maximize the regeneration gas volume to meet the velocity criterion. Some applications, termed *pressure swing adsorption*, regenerate the bed only with depressurization and sweeping the bed with gas just above atmospheric pressure.

Bottom bed support typically includes three to five layers of inert ceramic balls in graduated sizes (smallest on top). On top of the bed, a holddown screen is provided, again covered with a layer of ceramic balls. In some cases a layer of less expensive desiccant can be installed on the top bed to catch contaminants such as free water, glycol, hydrocarbons, or amines. This may extend bed life.

Since solid desiccants can produce dust, 1-μm filters are frequently installed at the outlet of the dehydration unit to protect downstream equipment.

Operating performance must be monitored periodically to adjust cycle length to ensure adequate dehydration is obtained. The size of the unit and the frequency of regeneration cycles also affect the timing of the performance test.

6.7.9 Gas Flow Measurement

6.7.9.1 Orifice-Meter Measurement

The most common method for measuring volumes is the differential measurement device, the orifice meter, which

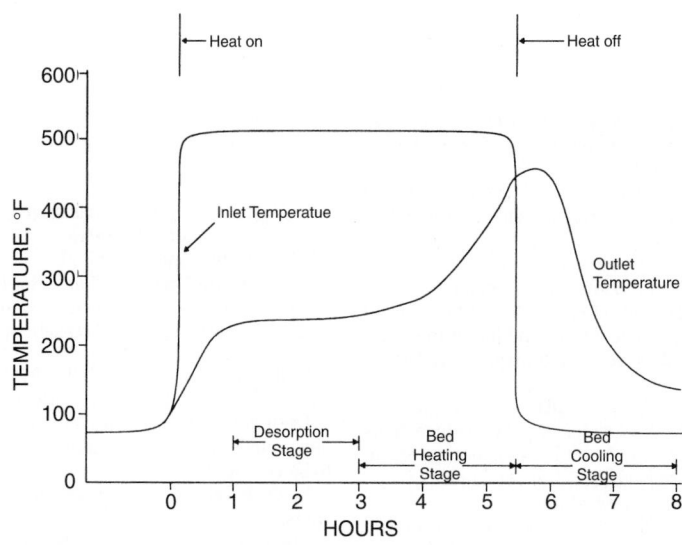

Figure 6.7.34 *Inlet and outlet temperature during typical solid desiccant bed regeneration cycle [2].*

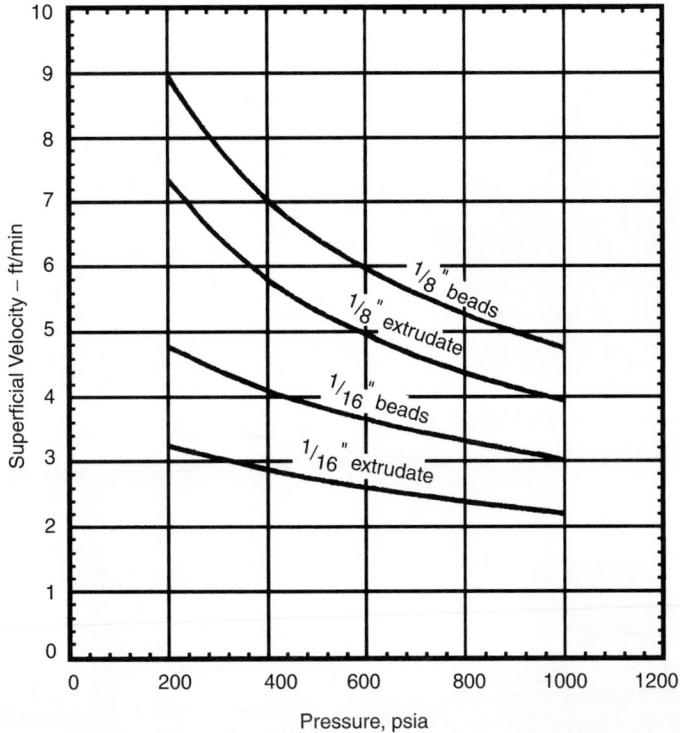

Figure 6.7.35 *Minimum regeneration velocity for mole sieve dehydrator [2].*

is widely accepted for use in measuring liquids and vapors. A correctly installed and maintained orifice may provide an overall accuracy within plus or minus 2%.

The procedures presented here for calculating flow by use of an orifice meter are designed to provide approximate solutions using a hand calculator or equivalent, and do not include the rigorous, iterative solution procedures required when using the Reader-Harris/Gallagher flow equation recommended for accurate, custody transfer calculations with computing equipment. The procedures for applying the Reader-Harris/Gallagher equation can be found in Chapter 14.3 of API *Manual of Petroleum Measurement Standards* published after 1994.

The API *Manual of Petroleum Measurement Standards* is the source of a large portion on the information presented below (Table 6.7.12). Chapter 14 is of particular interest to gas processors because it applies specifically to the measurement of gas and liquefied gas products. Chapter 14 is further divided as follows:

Table 6.7.12 *API Chapter 14, Measurement Subsections*

Chapter	Title
14.3	Concentric, Square-Edged Orifice Meters (ANSI/API 2350) (GPA 8185) (AGA Report No. 3)
14.4	Converting Mass of Natural Gas Liquids and Vapors to Equivalent Liquid Volumes (GPA 8173)
14.5	Calculation of Gross Heating Value, Specific Gravity, and Compressibility of Natural Gas Mixtures from Compositional Analysis (GPA 2172)
14.6	Installing and Proving Density Meters
14.7	Mass Measurement of Natural Gas Liquids (GPA 8182)
14.8	Liquefied Petroleum Gas Measurement

The orifice meter consists of a static pressure and differential pressure recoding gages connected to an orifice flange or orifice fitting. The orifice meter tube (meter run) consists of upstream and downstream sections of pipe for which size and tolerance have been determined through calculation, which conform to specifications set forth in ANSI/API 2530 (GPA 8185) (Table 6.7.15).

The orifice plate is held perpendicular to flow by flanges or a fitting. Bore, circumference, edge sharpness, and other tolerances must meet specification as set forth in ANSI/API 2530 (GPA 8185).

Orifice Flanges (Figure 6.7.36) — When slip-on or screwed orifice flanges are used, the end of the pipe shall extend through the flanges so that there is no recess greater than $\frac{1}{4}$ in. between the end of the pipe and the orifice plate. It is preferred that no recess exists.

When weld neck flanges or weldend orifice fittings are used, the average inside diameter of the section of pipe connected to the inlet side of the fitting or flange should equal the inside diameter of the fitting or flange within the tolerance given in Figure 6.7.36a.

When flanged orifice fittings are used, they should be aligned with the inside diameter of the flange on the meter tube so there are no sharp edges or offsets at the flange connection. Where separable gaskets are used to seal on orifice plate in an orifice holder, care must be taken to assure proper alignment and prevent gasket extension into the flow steam.

Orifice flanges require that the line be shut down and depressured in order to inspect or change the orifice plate.

Single Chamber Orifice Fitting (Figure 6.7.36b) — This fitting also requires that the line be shut down and depressured in order to inspect or change the orifice plate. However, this fitting does not require breaking apart the flanges. Instead, the bolts are loosened on the cover plate and the cover

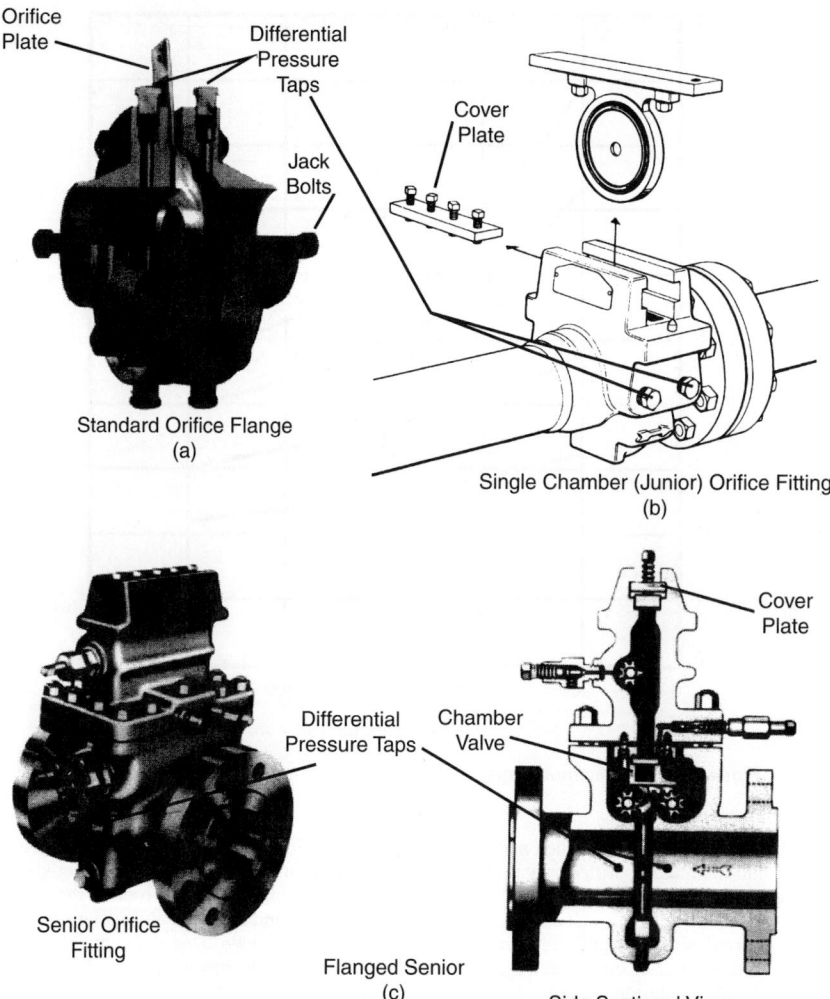

Figure 6.7.36a–c *Orifice plate holders.*

plate removed. The orifice plate holder and orifice plate are then removed from the fitting. These fittings provide precise alignment of the orifice plate.

Senior Orifice Fitting (Figure 6.7.36c) — This fitting allows the removal and inspection of an orifice plate while the line remains under pressure. It allows the orifice plate holder and orifice plate to be raised into the upper cavity for the fitting by the use of a crank handle. A valve is then closed to separate the upper cavity from the lower cavity of the fitting. The upper cavity is then depressured, the top cover plate removed, and the orifice plate cranked out.

In this section, certain terms are used which are peculiar to the measurement industry. Many of the terms are almost self-explanatory, or their general use in the industry makes it somewhat unnecessary to define them. However, in order to be specific with regard to these basic quantities, it is advisable to list the following definitions of terms commonly used in orifice meter measurement.

6.7.9.2 Definitions

Differential Taps — For meters using "flange taps," the center of the upstream pressure tap is placed 1 in. from the upstream face of the orifice plate. The center of the downstream pressure tap is placed one inch from the downstream face of the orifice plate. In locating the pressure tap hole in the flange, an allowance of 1/16 in. is normally made for the gasket thickness. The location of the pressure tap holes in the flanges is subject to the tolerance indicated on Figure 6.7.37. For meters using "pipe taps," the upstream pressure tap is placed two and one half times the *actual* inside pipe diameter from the upstream face of the orifice plate and the downstream pressure tap is placed eight times the *actual* inside pipe diameter from the downstream face of the orifice plate. The location of these pressure tap holes is subject to a tolerance 10 times that necessary for flange taps.

Special care should be used in selecting the values of the basic orifice flow factor since the orifice factors for the two kinds of taps are not the same.

Differential pressure — Differential pressure is the difference between two pressures. The differential pressure across an orifice meter tube is the difference between the pressure at the upstream tap before the gas passes through the orifice and the pressure at the downstream tap after it has passed through the orifice. It is commonly measured by

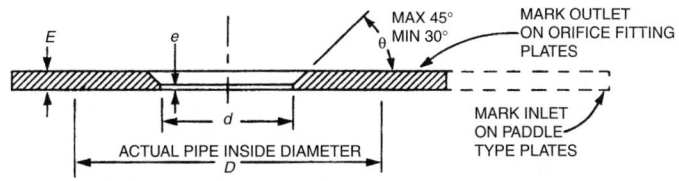

MAX 45°
MIN 30°
MARK OUTLET ON ORIFICE FITTING PLATES
MARK INLET ON PADDLE-TYPE PLATES
ACTUAL PIPE INSIDE DIAMETER

Nominal Inside Diameter, in inches										
	2	3	4	6	8	10	12	16	20	24	30
Published	1.687	2.624	3.152			9.562			18.812		
Inside	1.939	2.900	3.438	4.897 / 5.761	7.981	10.020	11.938	15.000	19.000	23.000	29.000
Diameter	2.067	2.300 / 3.068	3.826 / 4.026	5.187 / 6.065	7.625 / 8.071	10.136 / 11.374	12.090	14.688 / 15.250	19.250	22.624 / 23.250	28.750 / 29.250

Orifice Plate Thickness, E, in inches										
	2	3	4	6	8	10	12	16	20	24	30
Minimum	0.115	0.115 / 0.115	0.115 / 0.115	0.115 / 0.115	0.115 / 0.115	0.115 / 0.175	0.175	0.175 / 0.175	0.240	0.240 / 0.240	0.370 / 0.370
Maximum	0.130	0.130 / 0.130	0.130 / 0.130	0.163 / 0.192	0.254 / 0.269	0.319 / 0.379	0.398	0.490 / 0.500	0.505	0.505 / 0.505	0.562 / 0.578
Recommended	0.125	0.125 / 0.125	0.125 / 0.125	0.125 / 0.125	0.125 / 0.125	0.250 / 0.250	0.250	0.375 / 0.375	0.375	0.375 / 0.375	0.500 / 0.500

Maximum Orifice Edge Thickness, e, in inches

Orifice Diameter d	e ≤ d/8	2.067	2.300	3.068	3.826	4.026	5.187	6.065	7.625	8.071	10.136	11.374	12.090	14.688	15.250	19.250	22.624	23.250	28.750	29.250
0.250	1/32	x1/32	x1/32	1/32	1/32	1/32														
0.375	3/64			x3/64	3/64	3/64														
0.500	1/16				x1/16	1/16	1/16	1/16												
0.625	5/64					x5/64	5/64	5/64												
0.750	3/32						x3/32	3/32												
0.875	7/64							x7/64	7/64	7/64										
1.000	1/8								1/8	1/8	1/8									
1.125	9/64								x9/64	9/64	9/64									
1.250	5/32									x5/32	5/32	5/32	5/32							
1.375	11/64										11/64	11/64	11/64							
1.500	3/16										x3/16	3/16	3/16	3/16	3/16					
1.625	13/64											13/64	13/64	13/64	13/64					
1.750	7/32											x7/32	7/32	7/32	7/32					
1.875	15/64												x15/64	15/64	15/64					
2.000	1/4													1/4	1/4	1/4				
2.250	9/32													x9/32	9/32	9/32				
2.375	19/64														x19/64	19/64	19/64	19/64		
2.500	5/16															5/16	5/16	5/16		
2.750	11/32															11/32	11/32	11/32		
2.875	23/64															23/64	23/64	23/64	23/64	23/64
3.000	3/8															x3/8	3/8	3/8	3/8	3/8
3.250	13/32																13/32	13/32	13/32	13/32
3.500	7/16																x7/16	7/16	7/16	7/16
3.625	29/64																	x29/64	29/64	29/64
3.750	15/32																		15/32	15/32
4.000	1/2																		1/2	1/2
4.250	17/32																		17/32	17/32
4.500	9/16																		x9/16	9/16
4.625	37/64																			
4.750	19/32																			x50/54
5.000	5/8																			

Notes:

1. The maximum edge thickness is defined by e ≤ D/50 or e ≤ d/8, whichever is smaller.
2. Orifice edge thickness marked with x in this table is the maximum for that particular meter tube diameter and is applicable to all larger orifice diameters for that meter tube diameter.
3. Orifice diameters smaller than those marked x are defined by e ≤ d/8.
4. Orifice plates of which the edge thickness meets the value e ≤ D/30 need not be rebeveled unless reconditioning is required for other reasons.
5. All dimensions are in inches. For ease in machining, the next smaller value of e in even multiples of 1/16 or 1/32 inch may be used where e is given in 1/64ths.
6. Orifices used to measure dual directional flows must not be beveled. Where e exceeds the above limits, the flow constant F_b may be subject to higher uncertainty.

Figure 6.7.37 *Orifice plate dimensions.*

means of a liquid in a U-tube. The upstream pressure is transmitted to one leg of the U-tube and the downstream pressure to the other leg of the U-tube. This will cause the liquid in the two legs of the U-tube to assume different levels. The difference in these levels is the differential pressure measured in inches of the liquid that may be used in the U-tube. Mercury is generally used in commercial orifice meters of this type. The meter is calibrated so that this differential in inches of mercury is converted into inches of water, h_w, and recorded in those units on the meter chart.

Another method of obtaining differential pressure is by using a "dry," or bellows-type manometer. This instrument measures the difference in pressure by the movement of metallic bellows, or diaphragm, against the resistance of a calibrated range spring (Table 6.7.14).

Atmospheric Pressure —The pressure acting upon each square in of the earth's surface is equal to the weight of a column of the atmosphere one square inch in cross-section, extending vertically upward to its limit – an unknown height. The atmospheric pressure is usually measured by a barometer in inches of mercury column; such measurements are generally corrected to the height of an equivalent column of mercury at 32°F.

Average values of atmospheric pressure for various elevation above sea level are given in the following table:

Table 6.7.13 *Elevation versus Pressure*

Elevation in feet	Inches Mercury	Pressure psia	Elevation in feet	Inches Mercury	Pressure psia
0	30.00	14.73	2000	27.98	13.74
100	29.90	14.69	2500	27.48	13.5
200	29.80	14.64	3000	26.97	13.25
300	29.70	14.59	3500	26.47	13.00
400	26.60	14.54	4000	25.96	12.75
500	29.49	14.49	4500	25.46	12.5
600	29.39	14.44	5000	24.95	12.25
800	29.19	14.34	5500	24.45	12.01
1000	28.99	14.24	6000	23.94	11.76
1500	28.49	13.99	6500	23.44	11.51

The average atmospheric pressure in the localities of older gas fields is about 14.4 psia, and that the average figure for an assumed average atmospheric pressure is still generally used in many gas calculations.

Static or Line Pressure — Static pressure is usually measured by a coiled metal tube with a flattened cross section (pressure spring). The pressure of the gas in the line is admitted to and acts on the inside of the tube and the atmospheric pressure acts upon the outside of the tube. Therefore the pressure spring measures the difference between the pressure in the line and the atmospheric pressure. The pressure in the line may be greater or may be less than the atmospheric pressure.

The static pressure may be taken from either the upstream or downstream pressure; however, in certain instances the use of pressure between the upstream and downstream pressure simplifies the calculations. When intermediate pressure connections are used, the recorded static pressure is less than the upstream pressure and greater than the downstream pressure.

Absolute Pressure — Absolute pressure is the pressure above the absolute zero, or above a perfect vacuum. The atmospheric pressure is always expressed as an absolute pressure. If the pressure of a gas in a line is greater than the atmospheric pressure, the atmospheric pressure in psia is added to the gage pressure in psig to obtain the absolute pressure of the gas in the line. If the gage pressure is 40 psig (above atmospheric pressure) and the atmospheric pressure is 14.4 psia, the absolute pressure, p_f, is 54.4 psia.

If the line pressure is less than atmospheric pressure, the absolute pressure of the gas in the line in psia is equal to the atmospheric pressure in psia minus the gage pressure in psig. If the gage pressure is 4.0 psia (below atmospheric pressure) and the atmospheric pressure is 14.4 psia, the absolute pressure, p_f, of the gas in the line is 10.4 psia.

It is quite common to measure pressures below atmospheric pressure with gages calibrated in inches of mercury "vacuum." In such cases it is necessary to convert the "vacuum" from inches of mercury into pounds per square inch by multiplying by the conversion factor 0.491; for example, 10 in. of mercury "vacuum" = 4.91 psig (10 × 0.491) below the absolute pressure of the atmosphere.

Pressure Base — The pressure base is the pressure at which the cubic foot is the unit of measurement according to the contract. The basic orifice flow factors referred to in this handbook were calculated for a pressure base of

14.73 psia. If measurement at any other pressure is desired, then correction must be made for the effect of the change to that base. Frequently base pressures are indicated as gage pressures — as a certain number of "ounces." Such an indication would, of course, be above an assumed average atmospheric pressure. In order to definitely specify the basis for the measurement, the base pressure must an absolute pressure. For example, a pressure base of four ounces (above an assumed atmospheric pressure of 14.4 psia) may be written as 14.65 psia. Likewise, a pressure base of eight ounces (above the assumed average atmospheric pressure of 14.7 psia), may be expressed at 15.2 psia. The use of the absolute value is definite and leaves no doubt as to the value of the absolute pressure base. The use of any expression of an actual or an assumed value of the atmospheric pressure is eliminated.

Absolute Temperature — On the Fahrenheit thermometer scale, the absolute temperature is expressed in degrees above and below an arbitrary zero, which is 32°, the freezing point of water. The absolute zero is 460° below the zero of the Fahrenheit thermometer scale. The absolute temperature is the temperature above the absolute zero and is obtained by adding 460° to the reading of the thermometer. If the reading of the Fahrenheit thermometer is 60°, the corresponding absolute temperature is 60° added to 460° or 520°F absolute. If the reading is −20°F, the absolute temperature is −20°F +460° or 440°F absolute.

Temperature Base — The temperature base is the temperature at which the cubic foot is the unit of measurement according to the contract. The orifice flow factors referred to in this handbook were calculated for a temperature base condition of 520°F absolute (60°F). If measurement at any other temperature base is desired, then correction must be made for the effect of the change to that base.

Specific Gravity — The specific gravity of a gas is the weight of a cubic foot of gas compared to the weight of a cubic foot of dry air under the same pressure and temperature conditions. If the specific gravity of a gas is 2.0, it's twice as heavy as air, or if the specific gravity of a gas is 0.6, it is six-tenths as heavy as air.

Discharge Coefficient — The discharge coefficient is the ratio of the average velocity through the orifice divided by the theoretical velocity through the orifice. The theoretical velocity is the velocity each particle would assume if it fell through a height equal to the drop in pressure between the taps expressed in feet of fluid of the same density at the flowing fluid. The term "fluid" applies to a gas or liquid.

Referring to the well known formula,

$$V_f = 2gh$$

where

V_f = velocity in feet per second
g = Acceleration of gravity in feet per second per second
h = differential or drop in pressure in feet head of flowing fluid

Let us assume that air is measured through an orifice, the differential is 1.00 in. of water, the pressure is 14.4 psia and g is 32.17. One inch of water is equal to a head of 69.5 ft. of air at a pressure of 14.4 psia.

The theoretical velocity is then;

$$V_f = 2 \times 32.17 \times 69.5 = 66.9 \text{ feet per second}$$

Viscosity — Viscosity may be defined simply as the stickiness of a liquid or gas. It is that property of a liquid or gas, which tends to prevent one particle from moving faster in a stream than another particle, or it is the resistance to change of velocity between adjacent particles. Molasses is generally

Table 6.7.14 Orientation Table for Flow Sensors

Type of Design	Accuracy	Rangeability	Approx. Straight Pipe-Run Requirement (Upstream Diam./Downstream Diam.)	Pressure Loss Thru Sensor
Elbow Taps	5–10*	3:1 ①	② 25/10	N
Jet Deflection	2*	25:1	② 20/5	M
Laminar Flowmeters	½–5*	10:1	15/5	H
Magnetic Flowmeters	¼**–2*	10:1	N	N
Mass Flowmeters	2**	100:1	N	N
Metering Pumps	½b–1*	20:1	N	—
Orifice (Plate or Integral Cell)	½–2*	3:1 ①	② 20/5	H
Pitot Tubes	2–5*	3:1 ①	② 40/10	L
Positive Displacement Gas Meters	½–1***	200:1	N	M
Positive Displacement Liquid & Steam Meters	¼–1**	10:1	N	H
Solids Flowmeters	¼**–4*	20:1	5/3	—
Target Meters	1–5*	4:1	15/5	H
Thermal Meters	1–2*	20:1	5/3	A
Turbine Flowmeters	¼**	10:1	10/5	H
Ultrasonic Flowmeters Transit Doppler	1** 2–3*	20:1 20:1	② 20/5 15/5	N N
Variable Area Flowmeters	½–10**	5:1	N	A
Venturi Tubes and Flow Nozzles	¼–3*	3:1 ①	② 20/5	M
Vortex Flowmeters Shedding Precession	½–1** 1**	10:1 ② 10:1 ●	15/5 15/5	H H
Weirs, Flumes	2–5*	100:1		M

Legend:

L = Limited
SD = Some Designs
H = High
A = Average
M = Minimal
N = None
SR = Square Root

– – – – = Non Standard Range

* = ± % Full Scale
** = ± % Rate
*** = ± % Registration

① The data in this column is for general guidance only.
② Inherent Rangeability of Primary Device is substantially greater than shown. Value used reflects limitation of differential pressure sensing device, when maximum readout accuracy is desired.
③ Pipe size establishes the upper limit.
④ Practically unlimited with the probe type design.
⑤ Must be conductive.
⑥ Can be re-ranged.
⑦ Varies with upstream disturbance.
⑧ Depends on application conditions.

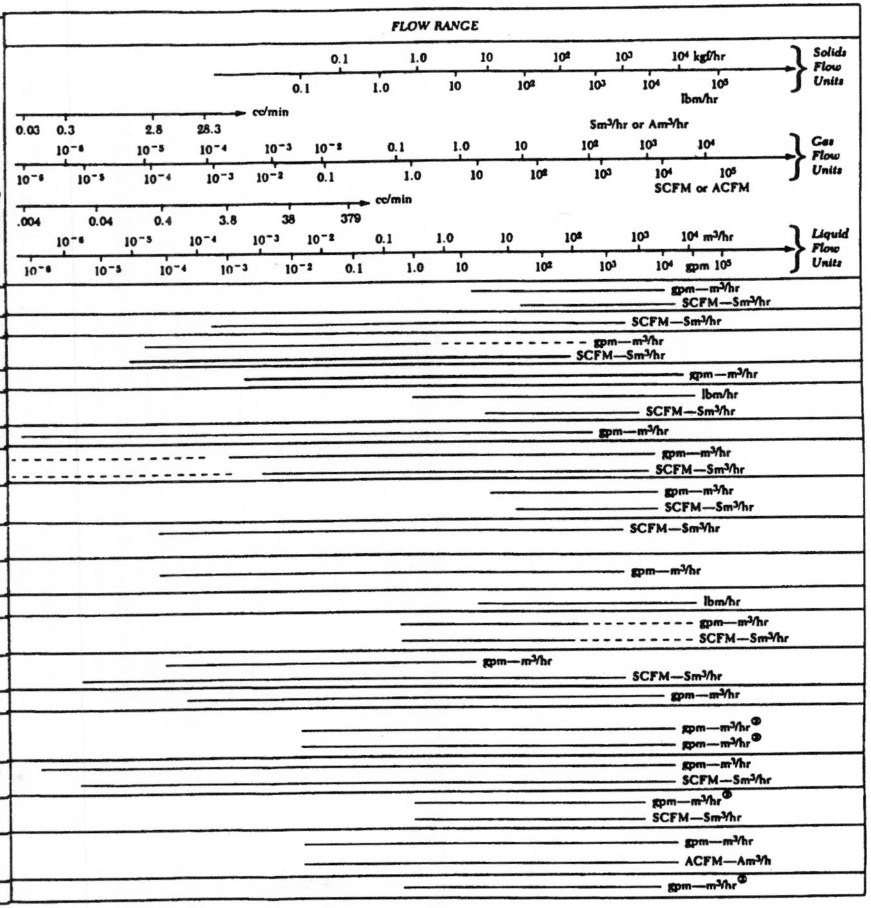

Table 6.7.15 *Minimum Motor Tube Lengths in Terms of Pipe Diameters and Beta Ratio*

Installation Figure	Dimension	Beta Ratio				
		0.5	0.6	0.67	0.70	0.75
6-225	A	25.0	30.0	36.0	38.6	43.5
	A'	10.2	12.2	14.2	15.3	17.5
	B	3.8	4.1	4.2	6.3	4.5
	C	5.0	5.5	6.2	6.4	7.0
	C'	5.2	6.7	8.0	8.9	10.5
6-226	A	20.8	25.0	28.8	31.0	35.2
	A'	10.0	11.4	12.8	13.5	15.0
	B	3.8	4.0	4.2	4.3	4.5
	C	5.0	5.5	6.2	6.5	7.2
6-227	C'	5.0	5.9	6.6	7.0	7.8
	A	10.0	13.8	17.4	19.0	22.0
	A'	9.0	10.3	11.7	12.3	13.8
	B	3.8	4.0	4.2	4.3	6.5
	C	5.0	5.5	6.2	6.5	7.1
	C'	4.0	4.8	5.5	5.8	6.7
6-228	A	6.9	9.3	12.5	13.9	16.7
	B	3.8	4.0	4.2	4.3	4.5
6-229	A	7.5	9.7	11.8	12.1	13.6
	B	3.8	4.1	4.2	4.3	4.5

Use for all pipe sizes. Based on flange taps. For pipe taps and 2 diameters to A, A'', and C and 8 pipe diameters to B.

known as a viscous liquid. Its viscosity is greater than that of water. The viscosity of gasoline is less that that of water. The viscosity of gas is a small fraction of the viscosity of water. The viscosity of a liquid decreases with increases in temperature, but the viscosity of a gas increases with an increase in temperature.

Viscosity references are commonly given in the form of Saybolt seconds, which is the time in seconds required for 60 cc of the fluid to pass through the standard capillary tube of a Saybolt Universal viscosimeter. This viscosity index gives an indication from which the kinematic viscosity can be determined – the kinematic viscosity being the ratio of the absolute viscosity to the density of the fluid. Saybolt second readings are not dependable for fluids less viscous than water, and should not be used for values lower than about 32 s. For fluids having lower viscosities, the viscosity is usually indicated as a decimal fraction of the viscosity of water at 68°F.

Reynold's Number—Reynold's number is an index by which discharge coefficients for various flowing conditions can be correlated. This "number" takes into account the individual way in which discharge coefficients varies with the orifice and meter tube diameters, the fluid viscosity and density, and the rate of flow through the orifice.

For any given orifice installation the discharge coefficient is constant for various flowing conditions, which will produce the same Reynold's number when measuring any gas or liquid through that installation.

Both the discharge coefficient and the value of the Reynold's number are dimension-less numbers. For this reason, corresponding values for these quantities are identical whether obtained with English units of measurement or with metric units—if they are consistently used.

The actual value of the Reynold's number, R_d, can be obtained from the basic equation:

$$R_d = \frac{\text{Velocity} \times \text{Diameter} \times \text{Density}}{\text{Absolute Viscosity}}$$

6.7.9.3 Various Conditions Affecting Measurement Accuracy

Measurement is very important for both the producers and the gas users; gas meter are, in fact, the cash registers for gas companies. At $3 per 1,000 ft.[3], a 1% error on 100 MMcfd of standard heating value gas is $3,000/d. A 1/10% error, better than the capability of current technology, is still equal to $109,500 per year.

The accuracy of measurement techniques, depends on:

Condition of the Orifice Edge—The greatest source of error in the primary measuring element is probably the possible deviation from the specification that the upstream edge of the orifice plate be square and sharp. A slight, almost imperceptible, rounding of the orifice edge can produce a considerable increase in the discharge coefficient, which results in low measurement. This is especially true with the smaller orifices in the smaller line sizes, since the effect of the edge imperfection is relative. A wire-edge burr, or fin, on the orifice edge is also undesirable since it can alter the flow pattern of the stream from that corresponding to proper measurement.

Condition of Meter Tube—Some error can be introduced as a result of a variation of the finish of the inside of the meter tube. The accepted orifice discharge coefficients were obtained from tests with meter tubes constructed of commercial iron pipe with inside surface roughness corresponding to such commercial item. Too smooth and inside surface can introduce an error in the measurement, just as an error can be produced by too rough a surface—because either condition constitutes a deviation from the conditions under which the accepted discharge coefficients were determined. Measurement would be high with too smooth and inside surface, and low with too rough and inside surface. *Report No. 3* [13], permits the use of bored meter tubes but specifies that the finish of the inside surface of the tube correspond to that of commercial pipe. The boring of meter tubes is desirable in those instances where the pipe from which the meter

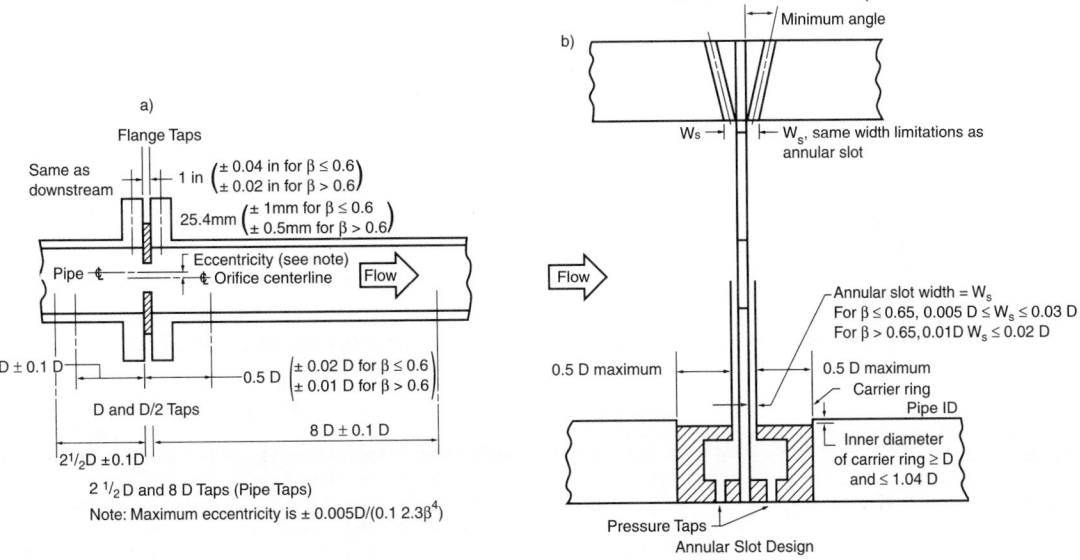

Figure 6.7.38 *Pressure-tap spacing.*

tube is to be made is "out-of-round" beyond permissible limits.

Pulsation — The effect of pressure and velocity pulsation's in the vicinity of the orifice constitutes a very indefinite phase in the measurement of gas with an orifice meter. This pulsation can be of a low frequency form such as might result from reciprocating compressors, undamped pressure regulators, chatting valves, or liquid surging back and forth at low points in the line. It might also be a high frequency pulsation caused by resonance of the pipelines themselves. The effect of the pulsation's of lower frequency probably have the greater effect on the measurement; however, no conclusive information is at present available by which the pulsation error can be completely correlated with pulsation frequency or with wave form and amplitude of those pulsations. This problem has been recognized; considerable research is underway, and more is contemplated, to determine means of either eliminating the effect of pulsations or determining the degree of error, which the pulsation might produce.

In some instances, the effect of pulsation has been largely reduced by installing restrictions in the line, which produce pressure drops up to 10% of the line pressure. In many cases, resistances, which create much smaller pressure drops, have been found satisfactory. In other instances, installing volume tanks, baffles, or dampeners in the line has minimized the effect. When any of these devices are used, they should be installed between the meter tube and the probable source of the pulsation. It has also been found advantageous to decrease the line size, thereby employing orifices with larger diameter ratios and measuring the gas with a higher differential pressure across the orifice plate.

A word of caution should be introduced regarding operations, which might be employed to obtain a smoother chart record — such as pinching gage lines or overdamping a meter. None of these operations eliminate the basic error corresponding to the pulsation in the line. Pinching gage lines valves might increase the error in the measurement. Under these conditions, a considerable error might exists without any visible indication on the instrument. Probably the best advice would be to install the meter at a point as remote as possible from the source of pulsation.

Effect of Water Vapor — In the measurement of gas containing moisture in the vapor state, the effect of moisture depends largely on the specific gravity of the gas. Natural gas is quite dry and its specific gravity is usually quite close to that of water vapor 0.62. For this reason the only appreciable correction would be a direct volume correction based upon the partial pressure of the water vapor, p_v, at flowing conditions. Under these conditions, the volume of dry gas could be determined as follows:

$$Q_{h(dry)} = Q_{h(moist)} \times \frac{p_f - p_v}{p_f}$$

In the measurement of gas with an orifice meter, it is the actual density of the flowing fluid, which must be considered. If the specific gravity has been obtained on a dry basis, and if the gas is measured with some moisture content, it would be necessary to correct the specific gravity to that corresponding to the flowing mixture. The flow rate should be calculated bases upon this corrected specific gravity; subsequently, a volume correction should also be made, based upon the partial pressure of the water vapor at flowing conditions.

Wet Gas Measurement — The effect of liquid in the gas stream on measurement is a problem that has never been completely solved. Various arrangements of meter tubes, gage line piping, and drip pots have been used in an effort to minimize the hazard resulting from liquid accumulating ahead of the orifice plate at low points in the gage line or in the chambers of the meter manometer. An accumulation of liquid ahead of the orifice plate disturbs the normal flow pattern and alters the discharge coefficients for the orifice, liquid trapped in the gage line distorts the differential pressure and causes the manometer to give an incorrect indication. *Bellows-type* manometers with gage lines attached to the bottom of the manometer chambers are self-draining and, therefore, eliminate that portion of the error corresponding to liquid, which may find its way into the manometer.

6.7.10 Flow Measurement

In the measurement of gas by orifice meter, the chart contains records of the differential and the static pressures. From these records the quantity of gas measured is determined by use of the formula:

$$Q_h = C'(h_w p_f)^{0.5} \tag{6.7.26}$$

where

Q_h = rate of flow at base conditions in cubic feet per hour

C' = Orifice flow constant. It is the rate of flow in cu. Ft. per hr. at base conditions when the pressure extension, $h_w p_f = 1.000$

h_w = differential in inches of water

p_f = static pressure in psia

$(h_w p_f)^{0.5}$ = pressure extension

Orifice Flow Constant, C' — In the text of this publication, this term is called the "orifice constant," for the sake of brevity. Its value is obtained by multiplying various factors together, as expressed by the formula:

$$C' = F_b \times F_{pb} \times F_{tb} \times F_g \times F_{tf} \times F_r \times Y \times F_{pv} \times F_m \tag{6.7.27}$$

where

F_b = basic orifice flow factor
  • Flange taps (Table 6.7.16)
  • Pipe taps (Table 6.7.17)
F_{pb} = pressure base factor (Table 6.7.18)
F_{tb} = temperature base factor (Table 6.7.19)
F_g = specific gravity factor (Table 6.7.20)
F_{tf} = flowing temperature factor (Table 6.7.21)
F_r = Reynold's number factor
  • Flange taps (Table 6.7.22)
  • Pipe taps (Table 6.7.23)
Y = expansion factor
  • Flange taps, pressure upstream (Table 6.7.24)
  • Pipe taps, pressure upstream (Table 6.7.25)
  • Flange taps, pressure downstream (Table 6.7.26)
  • Pipe taps, pressure downstream (Table 6.7.27)
  • Flange taps, mean pressure (Table 6.7.28)
F_{pv} = supercompressibility factor (Figure 6.1.1)
F_m = manometer factor (Table 6.7.30)

The basic orifice factor, F_b depends upon the type of pressure taps and the pipe and orifice diameters. F_b can be calculated from an analytical equations as;

$$F_b = 338.17\, d^2 K_o \tag{6.7.28}$$

where

d = actual diameter of the orifice in in.

K_o = coefficient of discharge for an infinite Reynold's number (which will be the minimum value for any particular orifice and pipe size)

and

$$K_o = \frac{K_e}{1 + \dfrac{15E}{1{,}000{,}000\, d}}$$

where $E = d(830 - 500\beta + 9{,}000\beta^2 - 4{,}200\beta^3 + B)$
  $B = 530/D^{0.5}$ (for flange taps)
  $= 875/D + 75$ (for pipe taps)

$$K_e = 0.5993 + \frac{0.007}{D} + 0.364 + \frac{0.076}{D^{0.5}}\beta^4 + 0.4\left(1.6 - \frac{1}{D}\right)^5$$

$$\times \left(0.07 + \frac{0.5}{D}\right)\beta^{5/2} - 0.009 = \frac{0.034}{D}[0.5 - \beta]^{3/2}$$

$$+ \frac{65}{D^2} + 3\,[\beta - 0.7]^{5/2} \quad \text{(for flange taps)}$$

$$= 0.5925 + \frac{0.0182}{D} + 0.440 - \frac{0.06}{D}\beta^2 + 0.935 + \frac{0.225}{D}\beta^5$$

$$+ 1.35\,\beta^{14} + \frac{1.43}{D^{0.5}}(0.25 - \beta)^{5/2} \quad \text{(for pipe taps)}$$

where D = the actual internal diameter in in.
  β = the orifice diameter ratio expressed as d/D

F_b should be rounded to four decimal places
  In practice, Tables 6.7.16 and 6.7.17 are used.

Basic Orifice Flow Factor, F_b — This term is called the "orifice factor" for brevity. The orifice factor is based upon the conditions: pressure base, $p_b = 14.73$ psia; temperature base, $T_b = 60°F$, (520°F absolute); specific gravity, $G = 1.000$; flowing temperature, $T_f = 60°F$ (520°F absolute); ad under the conditions where the Reynold's number is infinite and the expansion factor is unity. The value of this factor depends upon: the location of the differential taps; the diameter of the orifice, d; and upon the internal diameter of the pipe, D (Tables 6.7.16 and 6.7.17).

Pressure Base Factor, F_{pb} — The orifice factors were calculated to give gas volumes at a base pressure, p_b, of 14.73 psia. If measurement at any other base pressure is desired, then a base factor must be applied. These factors, and the equations representing the factors, are shown on Table 6.7.18.
  The pressure base factor, F_{pb} can be calculated by

$$F_{pb} = \frac{14.73}{P_b} \tag{6.7.29}$$

Temperature Base Factor, F_{tb} — The orifice factors were calculated to give gas volumes at a base temperature, T_b, of 60°F (520°F absolute). If measurement at any other temperature base is desired, than a temperature base factor must be applied. These factors, and the equations representing the factors are shown on Table 6.7.19.
  The temperature base factor, F_{tb} can be calculated by

$$F_{tb} = \frac{T_b}{519.67} \tag{6.7.30}$$

Specific Gravity Factor, F_g — The orifice factors were calculated to give gas volumes based upon the flowing gas having a specific gravity G of 1.000. For gases having a specific gravity other than 1.000, a specific gravity factor must be applied. These factors and the equations representing the factors are shown on Table 6.7.20.
  The real gas density (or specific gravity) factor F_g is to be applied to change from a real gas relative density of 1.0 to the real gas density G_r (or SG_g) of the flowing gas;

$$F_{gr} = \frac{1^{0.5}}{G_r} \tag{6.7.31}$$

where $G_r = G_i Z_b(\text{air})/Z_b(\text{gas})$ and $G_i = M_{gas}/M_{air}$

Flowing Temperature Factor, F_{tf} — The orifice factors where calculated, assuming that the gas flowing through the meter tubes at a temperature, T_f, of 60°F (520°F absolute). If measurement is to be made at any other flowing temperature, then a flowing temperature factor must be applied. These factors and the equation representing the factor are shown on Table 6.7.21.

(text continued on page 6-342)

Table 6.7.16 *Basic Orifice Factor, F_b, Flange Taps [14]*

Orifice Diameter Inches	Pipe Sizes, Nominal and Published Inside Diameters, in inches								
	2			3			4		
	1.687	1.939	2.067	2.300	2.624	2.900	3.068	3.152	3.438
0.250	12.696	12.708	12.711	12.714	12.712	12.708	12.705	12.703	12.697
0.375	28.475	28.440	28.428	28.411	28.394	28.382	28.376	28.373	28.364
0.500	50.780	50.588	50.523	50.436	50.358	50.314	50.293	50.285	50.260
0.625	80.099	79.510	79.313	79.054	78.820	78.689	78.627	78.600	78.525
0.750	117.11	115.62	115.14	114.52	113.99	113.70	113.56	118.50	113.33
0.875	162.99	159.56	158.48	157.13	156.01	155.41	155.15	155.03	154.71
1.000	219.86	212.47	210.23	207.44	205.19	204.05	203.55	203.33	202.76
1.125	291.16	276.20	271.71	266.36	262.09	259.95	259.04	258.66	257.64
1.250	386.36	353.59	345.14	335.13	327.43	323.64	322.04	321.37	319.61
1.375		448.60	433.51	415.76	402.25	395.81	393.09	391.98	389.04
1.500			542.29	510.87	488.09	477.37	472.97	471.15	466.40
1.625				623.93	587.00	569.66	562.60	559.74	552.32
1.750					701.53	674.46	663.43	658.98	647.56
1.875					835.32	793.90	777.20	770.46	753.18
2.000						930.67	906.03	896.08	870.61
2.125						1091.2	1052.5	1038.2	1001.5
2.250							1223.2	1199.9	1147.7
2.375									1311.7
2.500									1498.5

Orifice Diameter Inches	Pipe Sizes, Nominal and Published Inside Diameters, in inches								
	4		6				8		
	3.826	4.026	4.897	5.187	5.761	6.065	7.625	7.981	8.071
0.250	12.688	12.683							
0.375	26.354	28.349							
0.500	50.235	50.226	50.198	50.192	50.183	50.180			
0.625	78.452	78.423	78.340	78.323	78.298	78.289			
0.750	113.15	113.08	112.87	112.82	112.75	112.72			
0.875	154.40	154.27	153.88	153.78	153.63	153.57	153.35	153.32	153.31
1.000	202.21	201.99	201.35	201.20	200.96	200.86	200.46	200.40	200.38
1.125	256.70	256.34	255.31	255.08	254.72	254.57	253.99	253.90	253.87
1.250	318.04	317.45	315.83	315.49	314.95	314.73	313.92	313.78	313.75
1.375	386.46	385.52	383.00	382.49	381.71	381.38	380.25	380.07	380.03
1.500	462.28	460.80	456.94	456.17	455.04	454.58	453.03	452.79	452.73
1.625	545.90	543.62	537.78	536.66	535.04	534.39	532.28	531.96	531.88
1.750	637.85	634.41	625.74	624.11	621.80	620.90	618.03	617.61	617.51
1.875	738.77	733.70	721.05	718.72	715.45	714.20	710.33	709.78	709.65
2.000	849.33	842.14	824.01	820.72	816.15	814.43	809.24	808.52	808.35
2.125	970.97	960.50	935.00	930.40	924.09	921.73	914.80	913.87	913.66
2.250	1104.7	1089.9	1054.4	1048.1	1039.5	1036.3	1027.1	1025.9	1025.6
2.375	1252.1	1231.7	1182.9	1174.3	1162.6	1158.4	1146.2	1144.7	1144.3
2.500	1415.1	1387.2	1320.9	1309.4	1293.9	1288.2	1272.3	1270.3	1269.9
2.625	1595.6	1558.2	1469.3	1454.0	1433.5	1426.1	1405.4	1402.9	1402.3
2.750	1797.2	1746.7	1629.0	1608.8	1582.1	1572.4	1545.7	1542.5	1541.8
2.875		1955.5	1801.0	1774.7	1740.0	1727.5	1693.4	1689.3	1688.4
3.000		2195.0	1986.7	1952.7	1907.9	1891.9	1848.6	1843.5	1842.4
3.125			2187.2	2143.7	2086.4	2066.1	2011.6	2005.2	2003.8
3.250			2404.3	2349.1	2276.6	2250.9	2182.6	2174.7	2172.9
3.375			2639.5	2570.3	2479.1	2446.9	2361.9	2352.1	2349.9
3.500			2895.6	2808.6	2695.1	2655.0	2549.8	2537.7	2535.1
3.625			3180.9	3066.0	2925.7	2876.1	2746.6	2731.9	2728.6
3.750				3346.4	3172.2	3111.2	2952.7	2934.8	2930.9
3.875				3658.9	3435.8	3361.6	3168.4	3147.0	3142.2
4.000					3718.3	3628.3	3394.4	3368.6	3362.9
4.250					4354.9	4216.7	3879.5	3842.4	3834.2
4.500						4901.0	4412.9	4360.6	4349.1
4.750							5000.8	4928.2	4912.4
5.000							5650.2	5551.2	5529.6
5.250							6369.4	6236.5	6207.5
5.500							7171.1	6992.2	6953.7
5.575								7830.2	7778.0
6.000									8707.2

(continued)

Table 6.7.16 Basic Orifice Factor, F_b, Flange Taps [14] (continued)

Orifice Diameter Inches	Pipe Sizes, Nominal and Published Inside Diameters, in inches								
	10			12			16		
	9.562	10.020	10.136	11.374	11.938	12.090	14.688	15.000	15.250
1.000	200.20								
1.125	253.56	253.49	253.47						
1.250	313.32	313.21	313.18	312.95	312.86	312.84			
1.375	379.45	379.30	379.27	378.95	378.83	378.79			
1.500	451.96	451.77	451.73	451.31	451.15	451.11	450.54	450.49	
1.625	530.88	530.64	530.58	530.05	529.85	529.80	529.07	529.00	528.95
1.750	616.22	615.92	615.84	615.18	614.92	614.85	613.95	613.87	613.80
1.875	708.00	707.62	707.53	706.70	706.38	706.30	705.19	705.09	705.00
2.000	806.25	805.78	805.66	804.63	804.25	804.15	802.80	802.67	802.57
2.125	911.00	910.40	910.26	909.00	908.53	908.41	906.79	906.63	906.51
2.250	1022.3	1021.5	1021.4	1019.8	1019.2	1019.1	1017.2	1017.0	1016.8
2.375	1140.1	1139.2	1139.0	1137.1	1136.4	1136.2	1133.9	1133.7	1133.5
2.500	1264.5	1263.4	1263.2	1260.9	1260.0	1259.8	1257.1	1256.9	1256.7
2.625	1395.6	1394.3	1393.9	1391.2	1390.2	1389.9	1386.7	1386.4	1386.2
2.750	1533.4	1531.7	1531.4	1528.0	1526.8	1526.5	1522.7	1522.4	1522.1
2.875	1678.0	1675.9	1675.5	1671.4	1670.0	1669.7	1665.2	1664.8	1664.5
3.000	1829.5	1826.9	1826.4	1821.5	1819.8	1819.4	1814.2	1813.7	1813.4
3.125	1987.8	1984.8	1984.1	1978.2	1976.2	1975.7	1969.6	1969.1	1968.7
3.250	2153.3	2149.5	2148.7	2141.6	2139.2	2138.6	2131.5	2130.9	2130.4
3.375	2325.8	2321.3	2320.2	2311.8	2308.9	2308.3	2300.0	2299.3	2298.7
3.500	2505.7	2500.2	2498.9	2488.8	2485.4	2484.6	2475.0	2474.1	2473.5
3.625	2692.9	2686.3	2684.8	2672.7	2668.7	2667.8	2656.5	2655.6	2654.9
3.750	2887.7	2879.7	2878.0	2863.5	2858.9	2857.8	2844.7	2843.6	2842.8
3.875	3090.2	3080.7	3078.6	3061.5	3056.0	3054.7	3039.4	3038.2	3037.3
4.000	3300.7	3289.4	3286.8	3266.5	3260.1	3258.5	3240.9	3239.5	3238.4
4.250	3746.3	3730.3	3726.8	3698.6	3689.7	3687.6	3663.9	3662.0	3660.6
4.500	4226.3	4204.2	4199.3	4160.5	4148.5	4145.6	4114.0	4111.6	4109.8
4.750	4742.9	4712.9	4706.3	4653.6	4637.4	4633.6	4591.6	4588.5	4586.1
5.000	5298.9	5258.6	5249.7	5179.2	5157.5	5152.4	5097.3	5093.2	5090.2
5.250	5897.8	5843.8	5832.0	5738.7	5710.1	5703.3	5631.5	5626.2	5622.4
5.500	6543.6	6472.0	6456.4	6334.1	6296.8	6288.1	6194.9	6188.2	6183.2
5.750	7240.6	7147.1	7126.7	6967.3	6919.2	6908.0	6788.3	6779.8	6773.5
6.000	7994.1	7873.1	7846.9	7640.8	7579.2	7564.9	7412.5	7401.7	7393.8
6.250	8809.8	8655.0	8621.3	8357.8	8279.1	8260.9	8068.5	8054.9	8045.0
6.500	9694.5	9498.3	9455.5	9121.6	9021.9	8998.9	8757.5	8740.5	8728.1
6.750	10656	10409	10355	9935.9	9810.7	9781.8	9480.6	9459.6	9444.2
7.000	11713	11394	11327	10805	10649	10613	10239	10213	10194
7.250		12468	12381	11733	11540	11496	11035	11003	10980
7.500		13656	13541	12726	12489	12435	11870	11831	11803
7.750				13789	13500	13434	12745	12698	12664
8.000				14928	14579	14498	13664	13607	13566
8.250				16160	15730	15633	14629	14560	14511
8.500				17508	16963	16845	15642	15560	15501
8.750					18297	18148	16707	16610	16539
9.000						19566	17827	17711	17627
9.250							19005	18869	18770
9.500							20245	20086	19969
9.750							21553	21366	21229
10.000							22931	22713	22554
10.250							24385	24132	23948
10.500							25924	25628	25415
10.750							27567	27211	26961
11.000							29332	28900	28599
11.250								30711	30347

(continued)

Table 6.7.16 *Basic Orifice Factor, F_b, Flange Taps [14] (continued)*

| Orifice Diameter Inches | Pipe Sizes, Nominal and Published Inside Diameters, in inches | | | | | | | | |
| | 20 | | | 24 | | | 30 | | |
	18.812	19.000	19.250	22.624	23.000	23.250	28.750	29.000	29.250
2.000	801.42	801.37	801.31						
2.125	905.13	905.08	905.00						
2.250	1015.2	1015.1	1015.0						
2.375	1131.6	1131.5	1131.4	1130.2	1130.1	1130.0			
2.500	1254.4	1254.3	1254.2	1252.8	1252.7	1252.6			
2.625	1383.6	1383.5	1383.3	1381.7	1381.6	1381.5			
2.750	1519.1	1519.0	1518.9	1517.0	1516.8	1516.7			
2.875	1661.1	1660.9	1660.8	1658.6	1658.4	1658.3	1656.0		
3.000	1890.4	1809.3	1809.1	1806.6	1806.4	1806.3	1803.7	1803.6	1803.5
3.125	1964.2	1964.0	1963.7	1961.0	1960.8	1960.6	1957.7	1957.6	1957.5
3.250	2125.3	2125.1	2124.8	2121.8	2121.5	2121.3	2118.0	2117.9	2117.8
3.375	2292.9	2292.7	2292.4	2288.9	2288.6	2288.4	2284.7	2284.6	2284.5
3.500	2466.9	2466.7	2466.3	2462.5	2462.1	2461.9	2457.8	2457.7	2457.5
3.625	2647.4	2647.1	2646.7	2642.4	2642.0	2641.8	2637.2	2637.1	2636.9
3.750	2834.3	2834.0	2833.5	2828.8	2828.3	2828.0	2823.0	2822.9	2822.7
3.875	3027.7	3027.3	3026.8	3021.5	3021.0	3020.7	3015.2	3015.0	3014.8
4.000	3227.6	3227.2	3226.6	3220.7	3220.2	3219.8	3213.8	3213.5	3213.3
4.250	3646.8	3646.3	3645.6	3638.4	3637.7	3637.3	3630.0	3629.8	3629.5
4.500	4092.2	4091.6	4090.7	4081.9	4081.1	4080.6	4071.8	4071.5	4071.2
4.750	4563.9	4563.0	4562.0	4551.2	4550.2	4549.6	4539.3	4538.9	4538.5
5.000	5062.0	5061.0	5059.7	5046.5	5045.3	5044.6	5032.4	5031.9	5031.5
5.250	5586.7	5585.5	5583.9	5567.9	5566.5	5565.6	5551.1	5550.6	5550.1
5.500	6138.7	6136.9	6134.9	6115.4	6113.8	6112.7	6095.7	6095.1	6094.5
5.750	6717.2	6715.4	6713.0	6689.3	6687.3	6686.1	6666.0	6665.4	6664.7
6.000	7323.6	7321.3	7318.3	7289.6	7287.3	7285.8	7262.3	7261.5	7260.7
6.250	7957.7	7954.9	7951.3	7916.6	7913.8	7912.1	7884.5	7883.6	7882.7
6.500	8620.2	8616.7	8612.4	8570.5	8567.1	8565.0	8532.7	8531.6	8530.6
6.750	9311.3	9307.1	9301.8	9251.3	9247.4	9244.9	9207.0	9205.8	9204.6
7.000	10032	10027	10020	9959.5	9954.8	9951.9	9907.6	9906.2	9904.8
7.250	10782	10776	10768	10695	10690	10686	10634	10633	10631
7.500	11563	11555	11546	11459	11452	11448	11388	11386	11384
7.750	12375	12366	12354	12251	12243	12238	12168	12165	12163
8.000	13218	13208	13194	13071	13062	13056	12974	12972	12969
8.250	14095	14082	14066	13921	13910	13903	13808	13805	13802
8.500	15006	14991	14972	14800	14787	14779	14668	14665	14662
8.750	15951	15933	15911	15709	15694	15685	15556	15552	15549
9.000	16932	16912	16885	16648	16631	16620	16471	16467	16463
9.250	17951	17927	17896	17619	17598	17586	17414	17409	17404
9.500	19008	18980	18944	18621	18597	18583	18385	18379	18374
9.750	20105	20072	20030	19656	19629	19612	19383	19377	19371
10.000	21244	21205	21157	20724	20692	20673	20411	20404	20397
10.250	22427	23382	22326	21826	21790	21767	21467	21459	21451
10.500	23655	23604	23539	22962	22921	22896	22552	22543	22534
10.750	24932	24872	24798	24135	24088	24059	23667	23656	23646
11.000	26259	26190	26104	25345	25291	25258	24811	24800	24788
11.250	27638	27559	27461	26593	26531	26493	25986	25973	25960
11.500	29072	28983	28870	27879	27809	27766	27192	27177	27162
11.750	30564	30462	30334	29206	29127	29078	28429	28411	28395
12.000	32118	32002	31856	30575	30485	30430	29697	29678	29659
12.500	35420	35271	35085	33445	33331	33260	32331	32307	32284
13.000	39006	38818	38582	36503	36358	36268	35100	35069	35040
13.500	42916	42674	42376	39794	39581	39468	38007	37969	37933
14.000	47248	46922	46524	43243	43016	42875	41060	41013	40969
14.500				46961	46680	46506	44264	44207	44153
15.000				50937	50593	50379	47628	47558	47491

(continued)

Table 6.7.16   *Basic Orifice Factor, F_b, Flange Taps [14] (continued)*

Orifice Diameter Inches	Pipe Sizes, Nominal and Published Inside Diameters, in inches								
	20			24			30		
	18.812	19.000	19.250	22.624	23.000	23.250	28.750	29.000	29.250
15.500				55196	54775	54515	51161	51076	50995
16.000				59763	59252	58936	54873	54770	54672
16.500				64706	64062	63972	58774	58651	58533
17.000					69290	68794	62877	62729	62588
17.500							67195	67019	66850
18.000							71741	71532	71332
18.500							76530	76284	76048
19.000							81581	81291	81013
19.500							86910	86570	86246
20.000							92537	92142	91763
20.500							98493	98027	97586
21.000							104848	104285	103755
21.500								110986	110343

Table 6.7.17   *Basic Orifice Factor, F_b, Pipe Taps [14]*

Orifice Diameter Inches	Pipe Sizes, Nominal and Published Inside Diameters, in inches								
	2			3				4	
	1.687	1.939	2.067	2.300	2.624	2.900	3.068	3.152	3.438
0.250	0.1106	0.1091	0.1087	0.1081	0.1078	0.1078	0.1079	0.1079	0.1081
0.375	0.0890	0.0878	0.0877	0.0879	0.0888	0.0898	0.0905	0.0908	0.0918
0.500	0.0758	0.0734	0.0729	0.0728	0.0737	0.0750	0.0758	0.0763	0.0778
0.625	0.0694	0.0647	0.0635	0.0624	0.0624	0.0634	0.0642	0.0646	0.0662
0.750	0.0676	0.0608	0.0586	0.0559	0.0546	0.0548	0.0552	0.0555	0.0568
0.875	0.0684	0.0602	0.0570	0.0528	0.0497	0.0488	0.0488	0.0489	0.0496
1.000	0.0702	0.0614	0.0576	0.0522	0.0473	0.0452	0.0445	0.0443	0.0443
1.125	0.0709	0.0635	0.0595	0.0532	0.0469	0.0435	0.0422	0.0417	0.0407
1.250		0.0650	0.0617	0.0552	0.0478	0.0434	0.0414	0.0406	0.0387
1.375			0.0629	0.0575	0.0496	0.0443	0.0418	0.0408	0.0379
1.500				0.0590	0.0518	0.0461	0.0431	0.0418	0.0382
1.625					0.0539	0.0482	0.0450	0.0435	0.0392
1.750					0.0554	0.0504	0.0471	0.0456	0.0408
1.875						0.0521	0.0492	0.0477	0.0427
2.000						0.0532	0.0508	0.0495	0.0448
2.125							0.0519	0.0509	0.0467
2.250									0.0483
2.375									0.0494

Orifice Diameter Inches	Pipe Sizes, Nominal and Published Inside Diameters, in inches								
	4		6				8		
	3.826	4.026	4.897	5.187	5.761	6.065	7.625	7.981	8.071
0.250	0.1084	0.1085							
0.375	0.0932	0.0939							
0.500	0.0800	0.0810	0.0850	0.0862	0.0883	0.0893			
0.625	0.0685	0.0697	0.0747	0.0762	0.0789	0.0802			
0.750	0.0590	0.0602	0.0655	0.0672	0.0703	0.0719			
0.875	0.0513	0.0524	0.0575	0.0592	0.0625	0.0642	0.0716	0.0730	0.0734
1.000	0.0453	0.0461	0.0506	0.0523	0.0556	0.0573	0.0652	0.0668	0.0672

(continued)

Table 6.7.17 *Basic Orifice Factor, F_b, Pipe Taps [15] (continued)*

Pipe Sizes, Nominal and Published Inside Diameters, in inches								
Orifice	4		6				8		
Diameter Inches	3.826	4.026	4.897	5.187	5.761	6.065	7.625	7.981	8.071
1.125	0.0408	0.0412	0.0448	0.0464	0.0495	0.0512	0.0593	0.0609	0.0613
1.250	0.0376	0.0377	0.0401	0.0414	0.0442	0.0458	0.0538	0.0555	0.0560
1.375	0.0358	0.0353	0.0363	0.0373	0.0397	0.0412	0.0489	0.0506	0.0511
1.500	0.0350	0.0341	0.3340	0.0341	0.0360	0.0372	0.0445	0.0462	0.0466
1.625	0.0351	0.0336	0.0313	0.0315	0.0329	0.0339	0.0405	0.0421	0.0425
1.750	0.0359	0.0340	0.0300	0.0298	0.0304	0.0311	0.0369	0.0384	0.0388
1.875	0.0372	0.0349	0.0293	0.0287	0.0285	0.0290	0.0338	0.0352	0.0355
2.000	0.0388	0.0363	0.0292	0.0281	0.0273	0.0273	0.0311	0.0323	0.0327
2.125	0.0407	0.0380	0.0297	0.0281	0.0265	0.0262	0.0288	0.0299	0.0301
2.250	0.0427	0.0398	0.0305	0.0285	0.0261	0.0256	0.0268	0.0277	0.0280
2.375	0.0445	0.0417	0.0316	0.0293	0.0262	0.0253	0.0252	0.0259	0.0261
2.500	0.0461	0.0435	0.0330	0.0304	0.0267	0.0254	0.0239	0.0244	0.0246
2.625	0.0472	0.0450	0.0345	0.0317	0.0274	0.0258	0.0230	0.0232	0.0233
2.750		0.0462	0.0362	0.0332	0.0284	0.0265	0.0224	0.0224	0.0224
2.875			0.0379	0.0348	0.0295	0.0274	0.0220	0.0218	0.0218
3.000			0.0395	0.0364	0.0306	0.0285	0.0219	0.0214	0.0213
3.125			0.0410	0.0380	0.0323	0.0297	0.0220	0.0213	0.0211
3.250			0.0422	0.0394	0.0338	0.0311	0.0223	0.0214	0.0212
3.375			0.0433	0.0408	0.0353	0.0325	0.0228	0.0217	0.0214
3.500				0.0419	0.0367	0.0339	0.0235	0.0221	0.0218
3.625				0.0428	0.0381	0.0354	0.0243	0.0227	0.0224
3.750					0.0393	0.0367	0.0252	0.0234	0.0230
3.875					0.0404	0.0380	0.0262	0.0243	0.0238
4.000					0.0413	0.0391	0.0273	0.0252	0.0248
4.250							0.0297	0.0273	0.0268
4.500							0.0321	0.0296	0.0290
4.750							0.0344	0.0320	0.0314
5.000							0.0364	0.0342	0.0336
5.250							0.0381	0.0361	0.0356
5.500								0.0377	0.0373

Pipe Sizes, Nominal and Published Inside Diameters, in inches								
Orifice	10			12			16		
Diameter Inches	9.562	10.020	10.136	11.374	11.938	12.090	14.688	15.000	15.250
1.000	0.0728								
1.125	0.0674	0.0691	0.0695						
1.250	0.0624	0.0641	0.0646	0.0687	0.0704	0.0708			
1.375	0.0576	0.0594	0.0599	0.0643	0.0661	0.0666			
1.500	0.0532	0.0550	0.0555	0.0601	0.0620	0.0625	0.0697	0.0705	
1.625	0.0490	0.0509	0.0514	0.0561	0.0580	0.0585	0.0662	0.0670	0.0676
1.750	0.0452	0.0471	0.0476	0.0523	0.0543	0.0548	0.0628	0.0636	0.0642
1.875	0.0417	0.0436	0.0440	0.0488	0.0508	0.0513	0.0595	0.0603	0.0610
2.000	0.0385	0.0403	0.0407	0.0454	0.0475	0.0480	0.0563	0.0571	0.0578
2.125	0.0355	0.0373	0.0377	0.0423	0.0443	0.0449	0.0532	0.0541	0.0548
2.250	0.0329	0.0345	0.0349	0.0394	0.0414	0.0419	0.0503	0.0512	0.0519
2.375	0.0305	0.0320	0.0324	0.0367	0.0387	0.0392	0.0475	0.0485	0.0492
2.500	0.0283	0.0298	0.0301	0.0342	0.0361	0.0366	0.0449	0.0458	0.0466
2.625	0.0265	0.0277	0.0281	0.0319	0.0337	0.0342	0.0424	0.0433	0.0441
2.750	0.0248	0.0260	0.0263	0.0298	0.0316	0.0320	0.0400	0.0409	0.0417
2.875	0.0234	0.0244	0.0246	0.0279	0.0295	0.0300	0.0378	0.0387	0.0394
3.000	0.0222	0.0230	0.0233	0.0262	0.0277	0.0281	0.0356	0.0365	0.0372
3.125	0.0212	0.0218	0.0220	0.0246	0.0260	0.0264	0.0336	0.0345	0.0352
3.250	0.0205	0.0209	0.0210	0.0232	0.0245	0.0249	0.0317	0.0326	0.0333
3.375	0.0199	0.0201	0.0202	0.0220	0.0232	0.0235	0.0300	0.0308	0.0314
3.500	0.0195	0.0195	0.0196	0.0210	0.0220	0.0223	0.0283	0.0291	0.0297
3.625	0.0193	0.0191	0.0191	0.0200	0.0209	0.0212	0.0268	0.0275	0.0281
3.750	0.0192	0.0188	0.0188	0.0193	0.0200	0.0202	0.0254	0.0261	0.0267
3.875	0.0193	0.0187	0.0186	0.0187	0.0192	0.0194	0.0240	0.0247	0.0253
4.000	0.0195	0.0187	0.0186	0.0182	0.0185	0.0187	0.0228	0.0235	0.0240

(continued)

Table 6.7.17 *Basic Orifice Factor, F_b, Pipe Taps [15] (continued)*

Orifice Diameter Inches	Pipe Sizes, Nominal and Published Inside Diameters, in inches								
	10			12			16		
	9.562	10.020	10.136	11.374	11.938	12.090	14.688	15.000	15.250
4.250	0.0203	0.0192	0.0189	0.0176	0.0176	0.0177	0.0207	0.0213	0.0217
4.500	0.0215	0.0200	0.0197	0.0175	0.0172	0.0171	0.0190	0.0194	0.0198
4.750	0.0230	0.0212	0.0208	0.0178	0.0171	0.0170	0.0176	0.0180	0.0183
5.000	0.0248	0.0228	0.0223	0.0185	0.0175	0.0172	0.0166	0.0168	0.0171
5.250	0.0267	0.0245	0.0239	0.0195	0.0181	0.0178	0.0160	0.0161	0.0162
5.500	0.0287	0.0263	0.0257	0.0207	0.0190	0.0186	0.0156	0.0156	0.0156
5.750	0.0307	0.0282	0.0276	0.0221	0.0202	0.0197	0.0155	0.0154	0.0153
6.000	0.0326	0.0302	0.0295	0.0236	0.0215	0.0210	0.0157	0.0154	0.0153
6.250	0.0343	0.0320	0.0314	0.0253	0.0230	0.0224	0.0161	0.0157	0.0154
6.500	0.0358	0.0336	0.0331	0.0270	0.0246	0.0240	0.0167	0.0162	0.0159
6.750		0.0351	0.0346	0.0288	0.0262	0.0256	0.0175	0.0169	0.0164
7.000		0.0363	0.0359	0.0304	0.0279	0.0272	0.0184	0.0177	0.0172
7.250				0.0320	0.0295	0.0288	0.0195	0.0187	0.0181
7.500				0.0334	0.0310	0.0304	0.0206	0.0198	0.0191
7.750				0.0347	0.0325	0.0318	0.0219	0.0209	0.0202
8.000					0.0338	0.0332	0.0232	0.0222	0.0214
8.250					0.0349	0.03444	0.0246	0.0235	0.0227
8.500							0.0260	0.0249	0.0240
8.750							0.0273	0.0262	0.0253
9.000							0.0286	0.0276	0.0267
9.250							0.0299	0.0288	0.0280
9.500							0.0311	0.0301	0.0292
9.750							0.0322	0.0312	0.0304
10.000							0.0332	0.0323	0.0315
10.250							0.0341	0.0333	0.0326
10.500								0.0341	0.0335

Orifice Diameter Inches	Pipe Sizes, Nominal and Published Inside Diameters, in inches								
	20			24			30		
	18.812	19.000	19.250	22.624	23.000	23.250	28.750	29.000	29.250
2.000	0.0663	0.0667	0.0672						
2.125	0.0635	0.0639	0.0645						
2.250	0.0609	0.0613	0.0618						
2.375	0.0583	0.0588	0.0593	0.0658	0.0665	0.0669			
2.500	0.0558	0.0562	0.0568	0.0635	0.0642	0.0646			
2.625	0.0534	0.0538	0.0544	0.0613	0.0620	0.0624			
2.750	0.0510	0.0515	0.0520	0.0591	0.0598	0.0603			
2.875	0.0488	0.0492	0.0498	0.0570	0.0577	0.0582	0.0669		
3.000	0.0466	0.0470	0.0476	0.0549	0.0556	0.0561	0.0650	0.0654	0.0657
3.125	0.0445	0.0449	0.0455	0.0529	0.0536	0.0541	0.0632	0.0636	0.0639
3.250	0.0425	0.0429	0.0435	0.0509	0.0517	0.0521	0.0615	0.0618	0.0622
3.375	0.0405	0.0410	0.0416	0.0490	0.0497	0.0502	0.0597	0.0601	0.0604
3.500	0.0387	0.0391	0.0397	0.0471	0.0479	0.0484	0.0580	0.0584	0.0587
3.625	0.0369	0.0373	0.0379	0.0453	0.0461	0.0466	0.0563	0.0567	0.0571
3.750	0.0352	0.0356	0.0362	0.0436	0.0444	0.0449	0.0547	0.0551	0.0554
3.875	0.0336	0.0340	0.0346	0.0419	0.0427	0.0432	0.0530	0.0534	0.0538
4.000	0.0320	0.0324	0.0330	0.0403	0.0411	0.0416	0.0515	0.0519	0.0523
4.250	0.0291	0.0295	0.0301	0.0372	0.0380	0.0385	0.0484	0.0488	0.0492
4.500	0.0265	0.0269	0.0274	0.0343	0.0351	0.0356	0.0455	0.0459	0.0463
4.750	0.0242	0.0246	0.0250	0.0316	0.0324	0.0329	0.0427	0.0431	0.0435
5.000	0.0221	0.0225	0.0229	0.0292	0.0299	0.0303	0.0401	0.0405	0.0409

(continued)

Table 6.7.17 *Basic Orifice Factor, F$_b$, Pipe Taps [15] (continued)*

| Orifice Diameter Inches | Pipe Sizes, Nominal and Published Inside Diameters, in inches | | | | | | | | |
| | 20 | | | 24 | | | 30 | | |
	18.812	19.000	19.250	22.624	23.000	23.250	28.750	29.000	29.250
5.250	0.0203	0.0206	0.0210	0.0269	0.0276	0.0280	0.0376	0.0380	0.0384
5.500	0.0188	0.0190	0.0194	0.0248	0.0255	0.0259	0.0352	0.0356	0.0360
5.750	0.0175	0.0177	0.0180	0.0230	0.0236	0.0240	0.0330	0.0334	0.0338
6.000	0.0164	0.0165	0.0168	0.0213	0.0218	0.0222	0.0309	0.0313	0.0317
6.250	0.0155	0.0156	0.0158	0.0197	0.0203	0.0206	0.0289	0.0293	0.0297
6.500	0.0148	0.0149	0.0151	0.0184	0.0189	0.0192	0.0271	0.0274	0.0278
6.750	0.0144	0.0144	0.0145	0.0172	0.0176	0.0179	0.0253	0.0257	0.0261
7.000	0.0141	0.0141	0.0141	0.0162	0.0166	0.0168	0.0237	0.0241	0.0244
7.250	0.0140	0.0140	0.0139	0.0153	0.0156	0.0159	0.0223	0.0226	0.0229
7.500	0.0140	0.0140	0.0139	0.0146	0.0149	0.0150	0.0209	0.0212	0.0215
7.750	0.0143	0.0141	0.0140	0.0140	0.0142	0.0144	0.0196	0.0199	0.0202
8.000	0.0146	0.0144	0.0142	0.0136	0.0138	0.0139	0.0185	0.0187	0.0190
8.250	0.0151	0.0149	0.0146	0.0133	0.0134	0.0135	0.0174	0.0177	0.0179
8.500	0.0157	0.0154	0.0151	0.0132	0.0132	0.0132	0.0165	0.0167	0.0170
8.750	0.0163	0.0160	0.0157	0.0131	0.0130	0.0130	0.0156	0.0158	0.0161
9.000	0.0171	0.0168	0.0163	0.0131	0.0130	0.0130	0.0149	0.0151	0.0153
9.250	0.0180	0.0176	0.0171	0.0133	0.0131	0.0130	0.0142	0.0144	0.0146
9.500	0.0189	0.0185	0.0180	0.0136	0.0133	0.0132	0.0137	0.0138	0.0140
9.750	0.0198	0.0194	0.0189	0.0139	0.0136	0.0134	0.0132	0.0133	0.0135
10.000	0.0209	0.0204	0.0198	0.0143	0.0140	0.0138	0.0129	0.0129	0.0130
10.250	0.0219	0.0214	0.0208	0.0149	0.0144	0.0142	0.0126	0.0126	0.0127
10.500	0.0230	0.0225	0.0219	0.0154	0.0150	0.0147	0.0123	0.0124	0.0124
10.750	0.0241	0.0236	0.0229	0.0161	0.0155	0.0152	0.0122	0.0122	0.0122
11.000	0.0252	0.0247	0.0240	0.0168	0.0162	0.0158	0.0121	0.0121	0.0121
11.250	0.0263	0.0258	0.0251	0.0175	0.0169	0.0165	0.0121	0.0121	0.0121
11.500	0.0273	0.0268	0.0261	0.0183	0.0176	0.0172	0.0122	0.0122	0.0121
11.750	0.0284	0.0278	0.0272	0.0191	0.0184	0.0180	0.0124	0.0123	0.0122
12.000	0.0293	0.0288	0.0282	0.0200	0.0192	0.0188	0.0126	0.0124	0.0123
12.500	0.0312	0.0307	0.0301	0.0218	0.0210	0.0205	0.0131	0.0130	0.0128
13.000	0.0327	0.0323	0.0318	0.0236	0.0228	0.0222	0.0139	0.0137	0.0135
13.500				0.0255	0.0246	0.0240	0.0148	0.0146	0.0143
14.000				0.0272	0.0264	0.0258	0.0159	0.0156	0.0153
14.500				0.0289	0.0280	0.0275	0.0172	0.0168	0.0165
15.000				0.0304	0.0296	0.0291	0.0185	0.0181	0.0177
15.500				0.0318	0.0311	0.0306	0.0199	0.0194	0.0191
16.000					0.0323	0.0319	0.0213	0.0209	0.0205
16.500							0.0228	0.0223	0.0219
17.000							0.0242	0.0238	0.0233
17.500							0.0257	0.0252	0.0248
18.000							0.0270	0.0266	0.0261
18.500							0.0283	0.0279	0.0275
19.000							0.0296	0.0292	0.0288
19.500							0.0307	0.0303	0.0299
20.000							0.0317	0.0313	0.0310

Table 6.7.18   *Pressure-Base Factors, F_{pb}*

$$F_{pb} = 14.73 \div \text{base pressure, psia}$$

Pressure base, psia	Factor F_{pb}	Pressure base, psia	Factor F_{pb}
14.4..	1.0229	15.2 (8 oz. above 14.7).............	0.9691
14.65 (4 oz. above 14.4).........	1.0055	15.325 (10 oz. above 14.7).....	0.9612
14.73..	1.0000	15.4 (1 psi above 14.4)...........	0.9565
14.9 (8 oz. above 14.4)...........	0.9886	15.7 (1 psi above 14.7)...........	0.9382
14.95 (4 oz. above 14.7).........	0.9853	16.4 (2 psi above 14.4)...........	0.8982
15.025 (10 oz. above 14.4).....	0.9804	16.7 (2 psi above 14.7)...........	0.8820

Table 6.7.19   *Temperature-Base Factors, F_{tb}*

$$F_{tb} = \frac{460 + \text{temperature base } °F}{520}$$

Temperature base, °F	Factor F_{tb}	Temperature base, °F	Factor F_{tb}	Temperature base, °F	Factor F_{tb}
45	0.9712	65	1.0096	85	1.0481
50	0.9808	70	1.0192	90	1.0577
55	0.9904	75	1.0288	95	1.0673
60	1.0000	80	1.0385	100	1.0769

Table 6.7.20   *Specific-Gravity Factors, F_g*

$$F_g = \sqrt{\frac{1.0000}{\gamma_g}}$$

Specific gravity γ_g	Factor F_g	Specific gravity γ_g	Factor F_g	Specific gravity γ_g	Factor F_g	Specific gravity γ_g	Factor F_g
0.500	1.4142	0.675	1.2172	0.850	1.0847	1.05	0.9759
0.505	1.4072	0.680	1.2127	0.855	1.0815	1.06	0.9713
0.510	1.4003	0.685	1.2082	0.860	1.0783	1.07	0.9667
0.515	1.3935	0.690	1.2039	0.865	1.0752	1.08	0.9623
0.520	1.3868	0.695	1.1995	0.870	1.0721	1.09	0.9578
0.525	1.3801	0.700	1.1952	0.875	1.0690	1.10	0.9535
0.530	1.3736	0.705	1.1910	0.880	1.0660	1.11	0.9492
0.535	1.3672	0.710	1.1868	0.885	1.0630	1.12	0.9449
0.540	1.3608	0.715	1.1826	0.890	1.0600	1.13	0.9407
0.545	1.3546	0.720	1.1785	0.895	1.0570	1.14	0.9366
0.550	1.3484	0.725	1.1744	0.900	1.0541	1.15	0.9325
0.555	1.3423	0.730	1.1704	0.905	1.0512	1.16	0.9285
0.560	1.3363	0.735	1.1664	0.910	1.0483	1.17	0.9245
0.565	1.3304	0.740	1.1625	0.915	1.0454	1.18	0.9206
0.570	1.3245	0.745	1.1586	0.920	1.0426	1.19	0.9167
0.575	1.3188	0.750	1.1547	0.925	1.0398	1.20	0.9129
0.580	1.3131	0.755	1.1509	0.930	1.0370	1.21	0.9091
0.585	1.3074	0.760	1.1471	0.935	1.0342	1.22	0.9054
0.590	1.3019	0.765	1.1433	0.940	1.0314	1.23	0.9017
0.595	1.2964	0.770	1.1396	0.945	1.0287	1.24	0.8980

(continued)

Table 6.7.20 *Specific-Gravity Factors, F_g (continued)*

$$F_g = \sqrt{\frac{1.0000}{\gamma_g}}$$

Specific gravity γ_g	Factor F_g	Specific gravity γ_g	Factor F_g	Specific gravity γ_g	Factor F_g	Specific gravity γ_g	Factor F_g
0.600	1.2910	0.775	1.1359	0.950	1.0260	1.25	0.8944
0.605	1.2856	0.780	1.1323	0.955	1.0233	1.26	0.8909
0.610	1.2804	0.785	1.1287	0.960	1.0206	1.27	0.8874
0.615	1.2752	0.790	1.1251	0.965	1.0180	1.28	0.8839
0.620	1.2700	0.795	1.1215	0.970	1.0153	1.29	0.8805
0.625	1.2649	0.800	1.1180	0.975	1.0127	1.30	0.8771
0.630	1.2599	0.805	1.1146	0.980	1.0102	1.31	0.8737
0.635	1.2549	0.810	1.1111	0.985	1.0076	1.32	0.8704
0.640	1.2500	0.815	1.1077	0.990	1.0050	1.33	0.8671
0.645	1.2451	0.820	1.1043	0.995	1.0025	1.34	0.8639
0.650	1.2403	0.825	1.1010	1.00	1.0000	1.35	0.8607
0.655	1.2356	0.830	1.0976	1.01	0.9950	1.36	0.8575
0.660	1.2309	0.835	1.0944	1.02	0.9901	1.37	0.8544
0.665	1.2263	0.840	1.0911	1.03	0.9853	1.38	0.8513
0.670	1.2217	0.845	1.0879	1.04	0.9806	1.39	0.8482

Table 6.7.21 *Flowing-Temperature Factors, F_{tf}*

$$F_{tf} = \sqrt{\frac{520}{460 + \text{actual flowing temperature}}}$$

°F	Factor	°F	Factor	°F	Factor	°F	Factor	°F	Factor	°F	Factor
1	1.0621	21	1.0398	41	1.0188	61	0.9990	81	0.9804	110	0.9551
2	1.0609	22	1.0387	42	1.0178	62	0.9981	82	0.9795	120	0.9469
3	1.0598	23	1.0376	43	1.0168	63	0.9971	83	0.9786	130	0.9388
4	1.0586	24	1.0365	44	1.0157	64	0.9962	84	0.9777	140	0.9309
5	1.0575	25	1.0355	45	1.0147	65	0.9952	85	0.9768	150	0.9233
6	1.0564	26	1.0344	46	1.0137	66	0.9943	86	0.9759	160	0.9158
7	1.0552	27	1.0333	47	1.0127	67	0.9933	87	0.9750	170	0.9085
8	1.0541	28	1.0323	48	1.0117	68	0.9924	88	0.9741	180	0.9014
9	1.0530	29	1.0312	49	1.0107	69	0.9915	89	0.9732	190	0.8944
10	1.0518	30	1.0302	50	1.0098	70	0.9905	90	0.9723	200	0.8876
11	1.0507	31	1.0291	51	1.0088	71	0.9896	91	0.9715	210	0.8810
12	1.0496	32	1.0281	52	1.0078	72	0.9887	92	0.9706	220	0.8745
13	1.0485	33	1.0270	53	1.0068	73	0.9877	93	0.9697	230	0.8681
14	1.0474	34	1.0260	54	1.0058	74	0.9868	94	0.9688	240	0.8619
15	1.0463	35	1.0249	55	1.0048	75	0.9859	95	0.9680	250	0.8558
16	1.0452	36	1.0239	56	1.0039	76	0.9850	96	0.9671	260	0.8498
17	1.0441	37	1.0229	57	1.0029	77	0.9840	97	0.9662	270	0.8440
18	1.0430	38	1.0219	58	1.0019	78	0.9831	98	0.9653	280	0.8383
19	1.0419	39	1.0208	59	1.0010	79	0.9822	99	0.9645	290	0.8327
20	1.0408	40	1.0198	60	1.0000	80	0.9813	100	0.9636	300	0.8272

Table 6.7.22 *b Values for Reynold's Number Factor, F_r, Flange Taps [15]*

Orifice Diameter Inches	Pipe Sizes, Nominal and Published Inside Diameters, in inches								
	2			3				4	
	1.687	1.939	2.067	2.300	2.624	2.900	3.068	3.152	3.438
0.250	0.0878	0.0911	0.0926	0.0951	0.0979	0.0999	0.1010	0.1014	0.1030
0.375	0.0677	0.0709	0.0725	0.0755	0.0792	0.0820	0.0836	0.0844	0.0867
0.500	0.0562	0.0576	0.0588	0.0612	0.0647	0.0677	0.0694	0.0703	0.0730
0.625	0.0521	0.0505	0.0506	0.0516	0.0541	0.0566	0.0583	0.0591	0.0618
0.750	0.0537	0.0485	0.0471	0.0462	0.0470	0.0486	0.0498	0.0505	0.0528
0.875	0.0596	0.0507	0.0477	0.0445	0.0429	0.0432	0.0439	0.0443	0.0460
1.000	0.0678	0.0559	0.0515	0.0458	0.0417	0.0403	0.0402	0.0403	0.0411
1.125	0.0763	0.0630	0.0574	0.0495	0.0427	0.0396	0.0386	0.0383	0.0380
1.250	0.0826	0.0707	0.0645	0.0550	0.0457	0.0408	0.0388	0.0381	0.0365
1.375		0.0773	0.0716	0.0614	0.0501	0.0435	0.0406	0.0394	0.0365
1.500			0.0773	0.0679	0.0555	0.0474	0.0437	0.0420	0.0378
1.625				0.0736	0.0613	0.0522	0.0477	0.0457	0.0402
1.750					0.0670	0.0575	0.0524	0.0501	0.0434
1.875					0.0718	0.0628	0.0574	0.0549	0.0474
2.000						0.0676	0.0624	0.0598	0.0517
2.125						0.0715	0.0669	0.0645	0.0563
2.250							0.0706	0.0686	0.0607
2.375									0.0649
2.500									0.0683

Orifice Diameter Inches	Pipe Sizes, Nominal and Published Inside Diameters, in inches								
	4		6				8		
	3.826	4.026	4.897	5.187	5.761	6.065	7.625	7.981	8.071
0.250	0.1047	0.1054							
0.375	0.0894	0.0907							
0.500	0.0763	0.0779	0.0836	0.0852	0.0880	0.0892			
0.625	0.0653	0.0669	0.0734	0.0753	0.0785	0.0801			
0.750	0.0561	0.0577	0.0644	0.0664	0.0701	0.0718			
0.875	0.0487	0.0502	0.0567	0.0587	0.0625	0.0643	0.0723	0.0738	0.0742
1.000	0.0430	0.0442	0.0500	0.0520	0.0557	0.0576	0.0660	0.0676	0.0680
1.125	0.0388	0.0396	0.0444	0.0462	0.0498	0.0516	0.0602	0.0619	0.0623
1.250	0.0361	0.0364	0.0399	0.0414	0.0447	0.0464	0.0549	0.0566	0.0571
1.375	0.0347	0.0345	0.0363	0.0375	0.0403	0.0419	0.0501	0.0518	0.0522
1.500	0.0345	0.0336	0.0336	0.0344	0.0367	0.0381	0.0457	0.0474	0.0479
1.625	0.0354	0.0339	0.0318	0.0322	0.0337	0.0348	0.0418	0.0435	0.0439
1.750	0.0372	0.0350	0.0308	0.0306	0.0314	0.0323	0.0383	0.0399	0.0403
1.875	0.0398	0.0369	0.0305	0.0298	0.0298	0.0302	0.0353	0.0367	0.0371
2.000	0.0430	0.0396	0.0306	0.0296	0.0287	0.0288	0.9327	0.0340	0.0343
2.125	0.0467	0.0427	0.0318	0.0300	0.0281	0.0278	0.0304	0.0315	0.0318
2.250	0.0507	0.0463	0.0334	0.0310	0.0281	0.0274	0.0286	0.0295	0.0297
2.375	0.0548	0.0501	0.0354	0.0324	0.0286	0.0274	0.0271	0.0278	0.0280
2.500	0.0589	0.0540	0.0378	0.0343	0.0295	0.0279	0.0259	0.0264	0.0265
2.625	0.0627	0.0579	0.0406	0.0365	0.0308	0.0287	0.0251	0.0253	0.0254
2.750	0.0660	0.0615	0.0436	0.0391	0.0324	0.0299	0.0246	0.0245	0.0246
2.875		0.0648	0.0468	0.0419	0.0344	0.0314	0.0244	0.0240	0.0240
3.000		0.0674	0.0500	0.0448	0.0366	0.0332	0.0245	0.0238	0.0237
3.125			0.0533	0.0479	0.0390	0.0353	0.0248	0.0239	0.0237
3.250			0.0565	0.0510	0.0416	0.0376	0.0254	0.0242	0.0240
3.375			0.0594	0.0541	0.0443	0.0400	0.0263	0.0248	0.0244
3.500			0.0621	0.0570	0.0471	0.0426	0.0274	0.0255	0.0251
3.625			0.0643	0.0597	0.0499	0.0452	0.0286	0.0265	0.0260
3.750				0.0621	0.0527	0.0479	0.0301	0.0276	0.0271
3.875				0.0641	0.0554	0.0505	0.0317	0.0289	0.0283
4.000					0.0579	0.0531	0.0334	0.0304	0.0297
4.250					0.0621	0.0579	0.0372	0.0338	0.0330
4.500						0.0619	0.0414	0.0375	0.0366
4.750							0.0457	0.0415	0.0405
5.000							0.0500	0.0457	0.0446
5.250							0.0539	0.0497	0.0487
5.500							0.0575	0.0535	0.0525
5.575								0.0569	0.0560
6.000									0.0589

(continued)

Table 6.7.22 *b Values for Reynold's Number Factor, F_r, Flange Taps [15] (Continued)*

	Pipe Sizes, Nominal and Published Inside Diameters, in inches								
Orifice Diameter Inches	10			12			16		
	9.562	10.020	10.136	11.374	11.938	12.090	14.688	15.000	15.250
1.000	0.0738								
1.125	0.0685	0.0701	0.0705						
1.250	0.0635	0.0652	0.0656	0.0698	0.0714	0.0718			
1.375	0.0588	0.0606	0.0610	0.0654	0.0671	0.0676			
1.500	0.0545	0.0563	0.0567	0.0612	0.0631	0.0635	0.0706	0.0713	
1.625	0.0504	0.0523	0.0527	0.0573	0.0592	0.0597	0.0670	0.0678	0.0684
1.750	0.0467	0.0485	0.0490	0.0536	0.0555	0.0560	0.0636	0.0644	0.0650
1.875	0.0432	0.0450	0.0455	0.0501	0.0521	0.0526	0.0604	0.0612	0.0618
2.000	0.0401	0.0418	0.0423	0.0469	0.0488	0.0493	0.0572	0.0581	0.0587
2.125	0.0372	0.0389	0.0393	0.0438	0.0458	0.0463	0.0542	0.0551	0.0558
2.250	0.0346	0.0362	0.0366	0.0410	0.0429	0.0434	0.0514	0.0523	0.0529
2.375	0.0322	0.0337	0.0341	0.0383	0.0402	0.0407	0.0487	0.0495	0.0502
2.500	0.0302	0.0315	0.0319	0.0359	0.0377	0.0382	0.0461	0.0469	0.0476
2.625	0.0283	0.0296	0.0299	0.0336	0.0354	0.0358	0.0436	0.0445	0.0452
2.750	0.0267	0.0278	0.0281	0.0316	0.0332	0.0337	0.0413	0.0421	0.0428
2.875	0.0254	0.0263	0.0266	0.0297	0.0312	0.0317	0.0390	0.0399	0.0406
3.000	0.0242	0.0250	0.0252	0.0280	0.0294	0.0298	0.0370	0.0378	0.0385
3.125	0.0233	0.0239	0.0241	0.0265	0.0278	0.0282	0.0350	0.0358	0.0365
3.250	0.0226	0.0230	0.0231	0.0251	0.0263	0.0266	0.0331	0.0339	0.0346
3.375	0.0222	0.0223	0.0224	0.0239	0.0250	0.0253	0.0314	0.0322	0.0328
3.500	0.0219	0.0218	0.0218	0.0229	0.0238	0.0241	0.0298	0.0305	0.0311
3.625	0.0218	0.0214	0.0214	0.0220	0.0228	0.0230	0.0282	0.0290	0.0295
3.750	0.0218	0.0213	0.0212	0.0213	0.0219	0.0221	0.0268	0.0275	0.0281
3.875	0.0221	0.0213	0.0211	0.0208	0.0212	0.0213	0.0255	0.0262	0.0267
4.000	0.0225	0.0215	0.0213	0.0204	0.0206	0.0207	0.0243	0.0249	0.0254
4.250	0.0238	0.0222	0.0219	0.0200	0.0198	0.0198	0.0223	0.0228	0.0232
4.500	0.0256	0.0236	0.0232	0.0201	0.0195	0.0194	0.0206	0.0210	0.0213
4.750	0.0280	0.0254	0.0249	0.0207	0.0196	0.0194	0.0193	0.0196	0.0198
5.000	0.0307	0.0277	0.0270	0.0217	0.0202	0.0199	0.0184	0.0185	0.0187
5.250	0.0337	0.0303	0.0295	0.0232	0.0213	0.0208	0.0178	0.0178	0.0179
5.500	0.0370	0.0332	0.0323	0.0249	0.0226	0.0221	0.0176	0.0174	0.0174
5.750	0.0404	0.0363	0.0354	0.0271	0.0243	0.0237	0.0176	0.0174	0.0172
6.000	0.0439	0.0396	0.0386	0.0294	0.0263	0.0255	0.0180	0.0176	0.0173
6.250	0.0473	0.0429	0.0418	0.0320	0.0285	0.0277	0.0186	0.0181	0.0177
6.500	0.0506	0.0462	0.0451	0.0348	0.0309	0.0300	0.0195	0.0188	0.0183
6.750	0.0536	0.0494	0.0483	0.0376	0.0335	0.0325	0.0206	0.0198	0.0192
7.000	0.0563	0.0523	0.0513	0.0406	0.0362	0.0352	0.0220	0.0210	0.0203
7.250		0.0550	0.0541	0.0435	0.0390	0.0379	0.0235	0.0224	0.0216
7.500		0.0573	0.0564	0.0464	0.0418	0.0407	0.0252	0.0240	0.0230
7.750				0.0491	0.0446	0.0434	0.0271	0.0257	0.0247
8.000				0.0517	0.0473	0.0461	0.0291	0.0276	0.0264
8.250				0.0541	0.0499	0.0487	0.0312	0.0296	0.0283
8.500				0.0561	0.0523	0.0512	0.0334	0.0317	0.0304
8.750					0.0544	0.0534	0.0357	0.0338	0.0325
9.000						0.0553	0.0380	0.0361	0.0346
9.250							0.0402	0.0383	0.0368
9.500							0.0425	0.0406	0.0390
9.750							0.0448	0.0428	0.0412
10.000							0.0469	0.0450	0.0434
10.250							0.0490	0.0471	0.0455
10.500							0.0509	0.0491	0.0476
10.750							0.0526	0.0509	0.0495
11.000							0.0542	0.0526	0.0513
11.250								0.0541	0.0529

(continued)

Table 6.7.22 b Values for Reynold's Number Factor, F_r, Flange Taps [15] (Continued)

Orifice Diameter Inches	Pipe Sizes, Nominal and Published Inside Diameters, in inches								
	20			24			30		
	18.812	19.000	19.250	22.624	23.000	23.250	28.750	29.000	29.250
2.000	0.0667	0.0671	0.0676						
2.125	0.0640	0.0644	0.0649						
2.250	0.0614	0.0617	0.0623						
2.375	0.0588	0.0592	0.0597	0.0659	0.0665	0.0669			
2.500	0.0563	0.0567	0.0573	0.0636	0.0642	0.0646			
2.625	0.0540	0.0544	0.0549	0.0614	0.0620	0.0624			
2.750	0.0516	0.0521	0.0526	0.0592	0.0599	0.0603			
2.875	0.0494	0.0498	0.0504	0.0571	0.0578	0.0582	0.0664		
3.000	0.0473	0.0477	0.0483	0.0551	0.0557	0.0562	0.0645	0.0649	0.0652
3.125	0.0452	0.0456	0.0462	0.0531	0.0538	0.0542	0.0627	0.0631	0.0634
3.250	0.0432	0.0437	0.0442	0.0511	0.0518	0.0523	0.0610	0.0613	0.0616
3.375	0.0413	0.0418	0.0423	0.0492	0.0500	0.0504	0.0592	0.0596	0.0599
3.500	0.0395	0.0399	0.0405	0.0474	0.0481	0.0486	0.0575	0.0579	0.0582
3.625	0.0377	0.0382	0.0387	0.0457	0.0464	0.0468	0.0559	0.0562	0.0566
3.750	0.0361	0.0365	0.0370	0.0439	0.0447	0.0451	0.0542	0.0546	0.0550
3.875	0.0345	0.0349	0.0354	0.0423	0.0430	0.0435	0.0526	0.0530	0.0534
4.000	0.0329	0.0333	0.0339	0.0407	0.0414	0.0419	0.0511	0.0515	0.0518
4.250	0.0301	0.0305	0.0310	0.0376	0.0383	0.0388	0.0481	0.0485	0.0488
4.500	0.0275	0.0279	0.0284	0.0348	0.0355	0.0359	0.0452	0.0456	0.0460
4.750	0.0252	0.0256	0.0260	0.0322	0.0328	0.0333	0.0425	0.0429	0.0432
5.000	0.0232	0.0235	0.0239	0.0297	0.0304	0.0308	0.0399	0.0403	0.0407
5.250	0.0214	0.0217	0.0220	0.0275	0.0281	0.0285	0.0374	0.0378	0.0382
5.500	0.0199	0.0201	0.0204	0.0254	0.0260	0.0264	0.0351	0.0355	0.0359
5.750	0.0186	0.0188	0.0191	0.0236	0.0241	0.0245	0.0329	0.0333	0.0336
6.000	0.0175	0.0177	0.0179	0.0219	0.0224	0.0228	0.0308	0.0312	0.0315
6.250	0.0167	0.0168	0.0170	0.0204	0.0209	0.0212	0.0289	0.0292	0.0296
6.500	0.0161	0.0162	0.0163	0.0191	0.0195	0.0198	0.0270	0.0274	0.0277
6.750	0.0157	0.0157	0.0158	0.0179	0.0183	0.0185	0.0253	0.0257	0.0260
7.000	0.0155	0.0155	0.0154	0.0169	0.0172	0.0174	0.0237	0.0241	0.0244
7.250	0.0155	0.0154	0.0153	0.0161	0.0163	0.0165	0.0223	0.0226	0.0229
7.500	0.0157	0.0155	0.0154	0.0154	0.0156	0.0157	0.0209	0.0212	0.0215
7.750	0.0160	0.0158	0.0156	0.0148	0.0150	0.0151	0.0197	0.0199	0.0202
8.000	0.0166	0.0163	0.0160	0.0145	0.0145	0.0146	0.0185	0.0187	0.0190
8.250	0.0172	0.0169	0.0165	0.0142	0.0142	0.0142	0.0175	0.0177	0.0179
8.500	0.0181	0.0177	0.0172	0.0141	0.0140	0.0140	0.0165	0.0167	0.0170
8.750	0.0190	0.0186	0.0181	0.0141	0.0140	0.0139	0.0157	0.0159	0.0161
9.000	0.0201	0.0196	0.0190	0.0143	0.0141	0.0140	0.0150	0.0151	0.0153
9.250	0.0213	0.0208	0.0201	0.0146	0.0143	0.0141	0.0144	0.0145	0.0146
9.500	0.0226	0.0221	0.0213	0.0150	0.0146	0.0144	0.0138	0.0139	0.0141
9.750	0.0241	0.0234	0.0226	0.0155	0.0150	0.0147	0.0134	0.0135	0.0136
10.000	0.0256	0.0249	0.0240	0.0161	0.0155	0.0152	0.0130	0.0131	0.0132
10.250	0.0271	0.0264	0.0255	0.0168	0.0162	0.0158	0.0128	0.0128	0.0128
10.500	0.0288	0.0280	0.0270	0.0176	0.0169	0.0165	0.0126	0.0126	0.0126
10.750	0.0305	0.0297	0.0287	0.0185	0.0177	0.0173	0.0125	0.0125	0.0125
11.000	0.0323	0.0314	0.0303	0.0195	0.0186	0.0181	0.0125	0.0125	0.0124
11.250	0.0340	0.0332	0.0320	0.0205	0.0196	0.0191	0.0126	0.0125	0.0124
11.500	0.0358	0.0349	0.0338	0.0216	0.0207	0.0201	0.0127	0.0126	0.0125
11.750	0.0376	0.0367	0.0355	0.0228	0.0218	0.0211	0.0130	0.0128	0.0127
12.000	0.0394	0.0385	0.0373	0.0241	0.0230	0.0223	0.0133	0.0131	0.0129
12.500	0.0430	0.0420	0.0408	0.0268	0.0255	0.0248	0.0141	0.0138	0.0136
13.000	0.0463	0.0454	0.0442	0.0296	0.0283	0.0274	0.0151	0.0148	0.0145
13.500	0.0494	0.0485	0.0474	0.0326	0.0311	0.0302	0.0164	0.0161	0.0157
14.000	0.0520	0.0513	0.0503	0.0356	0.0341	0.0331	0.0179	0.0175	0.0171
14.500				0.0386	0.0370	0.0361	0.0197	0.0192	0.0187
15.000				0.0415	0.0400	0.0390	0.0215	0.0210	0.0205

(continued)

Table 6.7.22 *b Values for Reynold's Number Factor,* F_r, *Flange Taps [15] (Continued)*

Orifice Diameter Inches	Pipe Sizes, Nominal and Published Inside Diameters, in inches								
	20			24			30		
	18.812	19.000	19.250	22.624	23.000	23.250	28.750	29.000	29.250
15.500				0.0444	0.0429	0.0418	0.0236	0.0230	0.0224
16.000				0.0470	0.0456	0.0446	0.0257	0.0251	0.0244
16.500				0.0494	0.0481	0.0472	0.0280	0.0273	0.0266
17.000					0.0503	0.0495	0.0303	0.0296	0.0288
17.500							0.0327	0.0319	0.0312
18.000							0.0351	0.0343	0.0335
18.500							0.0375	0.0366	0.0358
19.000							0.0398	0.0390	0.0382
19.500							0.0421	0.0413	0.0405
20.000							0.0443	0.0435	0.0427
20.500							0.0463	0.0455	0.0448
21.000							0.0482	0.0475	0.0468
21.500								0.0492	0.0486

Table 6.7.23 *b Values for Reynolds Number Factor,* F_r, *Pipe Taps*

$$F_r = 1 + \frac{b}{\sqrt{h_w p_f}}$$

Orifice Diameter, d_0, (in.)	Internal Diameter of Pipe, d_i, in.								
	2			3				4	
	1.689	1.939	2.067	2.300	2.626	2.900	3.068	3.152	3.438
0.250	0.1105	0.1091	0.1087	0.1081	—	—	—	—	—
0.375	0.0890	0.0878	0.0877	0.0879	0.0888	0.0898	0.0905	0.0908	0.0918
0.500	0.0758	0.0734	0.0729	0.0728	0.0737	0.0750	0.0758	0.0763	0.0778
0.625	0.0693	0.0647	0.0635	0.0624	0.0624	0.0634	0.0642	0.0646	0.0662
0.750	0.0675	0.0608	0.0586	0.0559	0.0546	0.0548	0.0552	0.0555	0.0568
0.875	0.0684	0.0602	0.0570	0.0528	0.0497	0.0488	0.0488	0.0489	0.0496
1.000	0.0702	0.0614	0.0576	0.0522	0.0473	0.0452	0.0445	0.0443	0.0443
1.125	0.0708	0.0635	0.0595	0.0532	0.0469	0.0435	0.0422	0.0417	0.0407
1.250	—	0.0650	0.0616	0.0552	0.0478	0.0434	0.0414	0.0406	0.0387
1.375	—	—	0.0629	0.0574	0.0496	0.0443	0.0418	0.0408	0.0379
1.500	—	—	—	0.0590	0.0518	0.0460	0.0431	0.0418	0.0382
1.625	—	—	—	—	0.0539	0.0482	0.0450	0.0435	0.0392
1.750	—	—	—	—	0.0553	0.0504	0.0471	0.0456	0.0408
1.875	—	—	—	—	—	0.0521	0.0492	0.0477	0.0427
2.000	—	—	—	—	—	0.0532	0.0508	0.0495	0.0448
2.125	—	—	—	—	—	—	0.0519	0.0509	0.0467
2.250	—	—	—	—	—	—	—	—	0.0483
2.375	—	—	—	—	—	—	—	—	0.0494

(continued)

Table 6.7.23 *b Values for Reynolds Number Factor,* F_r, *Pipe Taps (Continued)*

$$F_r = 1 + \frac{b}{\sqrt{h_w p_f}}$$

Orifice Diameter, d_0, (in.)	Internal Diameter of Pipe, d_i, in. 4		6				8		
	3.826	4.026	4.897	5.189	5.761	6.095	7.625	7.981	8.071
0.500	0.0799	0.0810	0.0850	—	—	—	—	—	—
0.625	0.0685	0.0697	0.0747	0.0762	0.0789	0.0802	—	—	—
0.750	0.0590	0.0602	0.0655	0.0672	0.0703	0.0718	—	—	—
0.875	0.0513	0.0524	0.0575	0.0592	0.0625	0.0642	0.0716	0.0730	0.0733
1.000	0.0453	0.0461	0.0506	0.0523	0.0556	0.0573	0.0652	0.0668	0.0662
1.125	0.0408	0.0412	0.0448	0.0464	0.0495	0.0512	0.0592	0.0609	0.0613
1.250	0.0376	0.0377	0.0401	0.0413	0.0442	0.0458	0.0538	0.0555	0.0560
1.375	0.0358	0.0353	0.0363	0.0373	0.0397	0.0412	0.0489	0.0506	0.0510
1.500	0.0350	0.0340	0.0334	0.0340	0.0360	0.0372	0.0445	0.0462	0.0466
1.625	0.0351	0.0336	0.0313	0.0315	0.0329	0.0339	0.0404	0.0421	0.0425
1.750	0.0358	0.0340	0.0300	0.0298	0.0304	0.0311	0.0369	0.0384	0.0388
1.875	0.0371	0.0349	0.0293	0.0287	0.0285	0.0290	0.0338	0.0352	0.0355
2.000	0.0388	0.0363	0.0292	0.0281	0.0273	0.0273	0.0311	0.0323	0.0327
2.125	0.0407	0.0380	0.0297	0.0281	0.0265	0.0262	0.0288	0.0298	0.0301
2.250	0.0427	0.0398	0.0305	0.0285	0.0261	0.0258	0.0268	0.0277	0.0280
2.375	0.0445	0.0417	0.0316	0.0293	0.0262	0.0253	0.0252	0.0259	0.0261
2.500	0.0460	0.0435	0.0330	0.0304	0.0267	0.0254	0.0239	0.0244	0.0246
2.625	0.0472	0.0450	0.0345	0.0317	0.0274	0.0258	0.0230	0.0232	0.0233
2.750	—	0.0462	0.0362	0.0331	0.0284	0.0265	0.0224	0.0224	0.0224
2.875	—	—	0.0379	0.0347	0.0295	0.0274	0.0220	0.0218	0.0218
3.000	—	—	0.0395	0.0364	0.0308	0.0285	0.0219	0.0214	0.0213
3.125	—	—	0.0410	0.0380	0.0323	0.0297	0.0220	0.0213	0.0211
3.250	—	—	0.0422	0.0394	0.0338	0.0311	0.0223	0.0214	0.0212
3.375	—	—	0.0432	0.0408	0.0353	0.0325	0.0228	0.0216	0.0214
3.500	—	—	—	0.0419	0.0367	0.0339	0.0235	0.0221	0.0218
3.625	—	—	—	0.0428	0.0381	0.0354	0.0243	0.0227	0.0224
3.750	—	—	—	—	0.0393	0.0367	0.0252	0.0234	0.0230
3.875	—	—	—	—	0.0404	0.0380	0.0262	0.0243	0.0238
4.000	—	—	—	—	0.0413	0.0391	0.0273	0.0252	0.0246
4.250	—	—	—	—	—	—	0.0296	0.0273	0.0268
4.500	—	—	—	—	—	—	0.0321	0.0296	0.0290
4.750	—	—	—	—	—	—	0.0344	0.0320	0.0314
5.000	—	—	—	—	—	—	0.0364	0.0342	0.0336
5.250	—	—	—	—	—	—	0.0381	0.0361	0.0356
5.500	—	—	—	—	—	—	—	0.0377	0.0372

(continued)

Table 6.7.23 b Values for Reynolds Number Factor, F_r, Pipe Taps (Continued)

$$F_r = 1 + \frac{b}{\sqrt{h_w p_f}}$$

Orifice Diameter, d_0, (in.)	Internal Diameter of Pipe, d_i, in.								
	10			12			16		
	9.564	10.020	10.136	11.376	11.938	12.090	14.688	15.000	15.250
1.000	0.0728	—	—	—	—	—	—	—	
1.125	0.0674	0.0690	0.0694	—	—	—	—	—	
1.250	0.0624	0.0641	0.0646	0.0687	0.0704	0.0708	—	—	—
1.375	0.0576	0.0594	0.0599	0.0643	0.0661	0.0666		—	—
1.500	0.0532	0.0550	0.0555	0.0601	0.0620	0.0625	0.0697	0.0705	—
1.625	0.0490	0.0509	0.0514	0.0561	0.0580	0.0585	0.0662	0.0670	0.0676
1.750	0.0452	0.0471	0.0476	0.0523	0.0543	0.0548	0.0628	0.0636	0.0642
1.875	0.0417	0.0436	0.0440	0.0488	0.0508	0.0513	0.0594	0.0603	0.0610
2.000	0.0385	0.0403	0.0407	0.0454	0.0475.	0.0480	0.0563	0.0572	0.0578
2.125	0.0355	0.0372	0.0377	0.0423	0.0443	0.0449	0.0532	0.0541	0.0548
2.250	0.0329	0.0345	0.0349	0.0394	0.0414	0.0419	0.0503	0.0512	0.0519
2.375	0.0305	0.0320	0.0324	0.0367	0.0387	0.0392	0.0475	0.0484	0.0492
2.500	0.0283	0.0298	0.0301	0.0342	0.0361	0.0366	0.0449	0.0458	0.0466
2.625	0.0265	0.0277	0.0281	0.0319	0.0337	0.0342	0.0424	0.0433	0.0440
2.750	0.0248	0.0260	0.0262	0.0298	0.0316	0.0320	0.0400	0.0409	0.0417
2.875	0.0234	0.0244	0.0246	0.0279	0.0295	0.0300	0.0378	0.0387	0.0394
3.000	0.0222	0.0230	0.0232	0.0262	0.0277	0.0281	0.0356	0.0365	0.0372
3.125	0.0212	0.0218	0.0220	0.0244	0.0260	0.0264	0.0336	0.0345	0.0352
3.250	0.0204	0.0209	0.0210	0.0232	0.0245	0.0249	0.0317	0.0326	0.0332
3.375	0.0199	0.0201	0.0202	0.0220	0.0232	0.0235	0.0300	0.0308	0.0314
3.500	0.0195	0.0195	0.0196	0.0210	0.0220	0.0222	0.0283	0.0291	0.0297
3.625	0.0193	0.0191	0.0191	0.0200	0.0209	0.0212	0.0268	0.0275	0.0281
3.750	0.0192	0.0188	0.0188	0.0193	0.0200	0.0202	0.0254	0.0261	0.0267
3.875	Q.0193	0.0187	0.0186	0.0187	0.0192	0.0194	0.0240	0.0247	0.0253
4.000	0.0195	0.0187	0.0186	0.0182	0.0185	0.0187	0.0228	0.0235	0.0240
4.250	0.0203	0.0192	0.0189	0.0176	0.0176	0.0177	0.0207	0.0213	0.0217
4.500	0.0215	0.0200	0.0197	0.0175	0.0172	0.0171	0.0190	0.0194	0.0198
4.750	0,0230	0.0212	0.0208	0.0178	0.0171	0.0170	0.0176	0.0180	0.0182
5.000	0.0248	0.0228	0.0223	0.0185	0.0174	0.0173	0.0166	0.0168	0.0170
5.250	0.0267	0.0244	0.0239	0.0195	0.0181	0.0178	0.0160	0.0161	0.0162
5.500	0.0287	0.0263	0.0257	0.0207	0.0190	0.0186	0.0156	0.0156	0.0156
5.750	0.0307	0.0282	0.0276	0.0221	0.0202	0.0197	0.0155	0.0154	0.0153
6.000	0.0326	0.0302	0.0295	0.0231	0.0215	0.0210	0.0157	0.0154	0.0153
6.250	0.0343	0.0320	0.0316	0.0253	0.0230	0.0224	0.0161	0.0157	0.0154
6.500	0.0358	0.0336	0.0331	0.0270	0.0246	0.0239	0.0167	0.0162	0.0159
6.750	—	0.0351	0.0346	0.0288	0.0262	0.0256	0.0174	0.0169	0.0164
7.000	—	0.0363	0.0359	0.0304	0.0279	0.0272	0.0184	0.0177	0.0172
7.250	—	—	—	0.0320	0.0295	0.0288	0.0195	0.0187	0.0181
7.500	—	—	—	0.0334	0.0310	0.0304	0.0206	0.0198	0.0191
7.750	—	—	—	0.0347	0.0325	0.0318	0.0219	0.0209	0.0202
8.000	—	—	—	—	0.0338	0.0332	0.0232	0.0222	0.0214
8.250	—	—	—	—	0.0349	0.0344	0.0246	0.0235	0.0227
8.500	—	—	—	—	—	—	0.0259	0.0248	0.0240
8.750	—	—	—	—	—	—	0.0273	0.0262	0.0253
9.000	—	—	—	—	—	—	0.0286	0.0276	0.0267
9.250	—	—	—	—	—	—	0.0299	0.0288	0.0280
9.500	—	—	—	—	—	—	0.0311	0.0300	0.0292
9.750	—	—	—	—	—	—	0.0322	0.0312	0.0304
10.000	—	—	—	—	—	—	0.0332	0.0323	0.0315
10.250	—	—	—	—	—	—	0.0341	0.0333	0.0326
10.500	—	—	—	—	—	—	—	0.0341	0.0335

(continued)

Table 6.7.23   b Values for Reynolds Number Factor, F_r, Pipe Taps (Continued)

$$F_r = 1 + \frac{b}{\sqrt{h_w p_f}}$$

Internal Diameter of Pipe, d_i, in.

Orifice Diameter, d_o, (in.)	20			24			30		
	18.814	19.000	19.250	22.626	23.000	23.250	28.628	29.000	29.250
2.000	0.0663	0.0667	0.0672		—	—		—	—
2.125	0.0635	0.0639	0.0644	—				—	—
2.250	0.0609	0.0613	0.0618	—	—	—	—	—	
2.375	0.0583	0.0588	0.0593	0.0658	0.0665	0.0669	—		
2.500	0.0558	0.0562	0.0568	0.0635	0.0642	0.0646			—
2.625	0.0534	0.0539	0.0544	0.0613	0.0620	0.0624	—	—	—
2.750	0.0510	0.0515	0.0520	0.0591	0.0598	0.0603	—	—	—
2.875	0.0488	0.0492	0.0498	0.0570	0.0577	0.0582	0.0667		
3.000	0.0466	0.0470	0.0476	0.0549	0.0556	0.0561	0.0649	0.0654	0.0657
3.125	0.0445	0.0449	0.0455	0.0529	0.0536	0.0541	0.0630	0.0636	0.0639
3.250	0.0425	0.0429	0.0435	0.0509	0.0516	0.0521	0.0613	0.0616	0.0622
3.375	0.0406	0.0410	0.0416	0.0490	0.0497	0.0502	0.0595	0.0601	0.0604
3.500	0.0387	0.0391	0.0397	0.0471	0.0479	0.0484	0.0578	0.0584	0.0587
3.625	0.0369	0.0373	0.0379	0.0454	0.0461	0.0466	0.0561	0.0567	0.0571
3.750	0.0352	0.0356	0.0362	0.0436	0.0444	0.0449	0.0545	0.0550	0.0554
3.875	0.0336	0.0340	0.0346	0.0419	0.0427	0.0432	0.0528	0.0534	0.0538
4.000	0.0320	0.0324	0.0330	0.0403	0.0411	0.0416	0.0513	0.0518	0.0522
4.250	0.0291	0.0295	0.0301	0.0372	0.0380	0.0385	0.0482	0.0488	0.0492
4.500	0.0265	0.0269	0.0274	0.0343	0.0351	0.0356	0.0453	0.0459	0.0463
4.750	0.0242	0.0246	0.0250	0.0316	0.0324	0.0328	0.0425	0.0431	0.0435
5.000	0.0221	0.0225	0.0229	0.0292	0.0299	0.0303	0.0399	0.0405	0.0409
5.250	0.0203	0.0206	0.0210	0.0269	0.0276	0.0280	0.0374	0.0380	0.0384
5.500	0.0188	0.0190	0.0194	0.0248	0.0255	0.0259	0.0350	0.0356	0.0360
5.750	0.0175	0.0177	0.0180	0.0230	0.0236	0.0240	0.0328	0.0334	0.0338
6.000	0.0164	0.0165	0.0168	0.0212	0.0218	0.0222	0.0307	0.0313	0.0317
6.250	0.0155	0.0156	0.0158	0.0197	0.0202	0.0206	0.0287	0.0293	0.0297
6.500	0.0148	0.0149	0.0150	0.0184	0.0189	0.0192	0.0269	0.0274	0.0278
6.750	0.0143	0.0144	0.0145	0.0172	0.0176	0.0179	0.0252	0.0257	0.0260
7.000	0.0141	0.0141	0.0141	0.0162	0.0166	0.0168	0.0236	0.0241	0.0244
7.250	0.0140	0.0140	0.0139	0.0153	0.0156	0.0158	0.0221	0.0226	0.0229
7.500	0.0140	0.0140	0.0139	0.0146	0.0148	0.0150	0.0207	0.0212	0.0215
7.750	0.0142	0.0141	0.0140	0.0140	0.0142	0.0144	0.0195	0.0199	0.0202
8.000	0.0146	0.0144	0.0142	0.0136	0.0138	0.0138	0.0183	0.0187	0.0190
8.250	0.0151	0.0148	0.0146	0.0133	0.0134	0.0132	0.0173	0.0177	0.0179
8.500	0.0156	0.0154	0.0151	0.0132	0.0132	0.0130	0.0164	0.0167	0.0169
8.750	0.0163	0.0160	0.0157	0.0131	0.0130	0.0130	0.0155	0.0158	0.0161
9.000	0.0171	0.0168	0.0163	0.0131	0.0130	0.0130	0.0148	0.0151	0.0153
9.250	0.0180	0.0176	0.0171	0.0133	0.0131	0.0130	0.0142	0.0144	0.0146
9.500	0.0189	0.0185	0.0180	0.0136	0.0133	0.0132	0.0136	0.0138	0.0140
9.750	0.0198	0.0194	0.0189	0.0139	0.0136	0.0134	0.0132	0.0133	0.0134
10.000	0.0209	0.0204	0.0198	0.0143	0.0140	0.0138	0.0128	0.0129	0.0130
10.250	0.0219	0.0214	0.0208	0.0148	0.0144	0.0142	0.0125	0.0126	0.0127
10.500	0.0230	0.0225	0.0219	0.0154	0.0150	0.0147	0.0123	0.0124	0.0124
10.750	0.0241	0.0236	0.0229	0.0160	0.0155	0.0152	0.0122	0.0122	0.0122
11.000	0.0252	0.0247	0.0240	0.0168	0.0162	0.0158	0.0121	0.0121	0.0121
11.250	0.0263	0.0261	0.0251	0.0175	0.0169	0.0165	0.0122	0.0121	0.0121
11.500	0.0273	0.0268	0.0262	0.0183	0.0176	0.0172	0.0122	0.0121	0.0122
11.750	0.0284	0.0278	0.0272	0.0191	0.0184	0.0180	0.0124	0.0123	0.0122
12.000	0.0293	0.0288	0.0282	0.0200	0.0192	0.0190	0.0126	0.0124	0.0123
12.500	0.0312	0.0307	0.0301	0.0218	0.0210	0.0204	0.0132	0.0130	0.0128
13.000	0.0327	0.0323	0.0318	0.0236	0.0228	0.0222	0.0140	0.0137	0.0135
13.500	—	—		0.0254	0.0246	0.0240	0.0150	0.0146	0.0143
14.000	—	—	—	0.0272	0.0264	0.0258	0.0161	0.0156	0.0153

(continued)

Table 6.7.23   b Values for Reynolds Number Factor, F_r, Pipe Taps (Continued)

$$F_r = 1 + \frac{b}{\sqrt{h_w p_f}}$$

Internal Diameter of Pipe, d_i, in.

Orifice Diameter, d_o, (in.)	20			24			30		
	18.814	19.000	19.250	22.626	23.000	23.250	28.628	29.000	29.250
14.500	—	—	—	0.0289	0.0280	0.0275	0.0173	0.0168	0.0165
15.000	—	—	—	0.0304	0.0296	0.0291	0.0186	0.0181	0.0177
15.500	—			0.0318	0.0311	0.0306	0.0200	0.0194	0.0190
16.000		—	—		0.0323	0.0318	0.0215	0.0209	0.0204
16.500					—	—	0.0230	0.0223	0.0219
17.000	—					—	0.0244	0.0238	0.0233
17.500	—	—	—	—	—	—	0.0259	0.0252	0.0248
18.000						—	0.0272	0.0266	0.0261
18.500	—			—		—	0.0286	0.0279	0.0275
19.000		—	—				0.0298	0.0292	0.0288
19.500				—	—	—	0.0309	0.0303	0.0299
20.000	—		—		—		0.0318	0.0313	0.0310

F_d	2			3				4			
	1.689	1.939	2.067	2.300	2.626	2.900	3.068	3.152	3.438	3.826	4.026
0.10	0.1295	0.1209	0.1173	0.1118	0.1058	0.1017	0.0996	0.0986	0.0957	0.0924	0.0909
0.11	0.1253	0.1167	0.1132	0.1077	0.1016	0.0976	0.0954	0.0945	0.0915	0.0882	0.0867
0.12	0.1212	0.1126	0.1090	0.1035	0.0975	0.0934	0.0913	0.0903	0.0874	0.0841	0.0826
0.13	0.1172	0.1086	0.1051	0.0996	0.0935	0.0895	0.0874	0.0864	0.0835	0.0802	0.0787
0.14	0.1134	0.1049	0.1013	0.0958	0.0898	0.0858	0.0837	0.0827	0.0797	0.0765	0.0750
0.15	0.1098	0.1013	0.0978	0.0923	0.0863	0.0823	0.0801	0.0792	0.0762	0.0729	0.0715
0.16	0.1065	0.0980	0.0944	0.0889	0.0829	0.0789	0.0768	0.0758	0.0729	0.0696	0.0682
0.17	0.1033	0.0948	0.0912	0.0856	0.0798	0.0758	0.0737	0.0727	0.0698	0.0665	0.0650
0.18	0.1001	0.0916	0.0881	0.0827	0.0767	0.0727	0.0706	0.0696	0.0667	0.0634	0.0620
0.19	0.0973	0.0888	0.0853	0.0799	0.0739	0.0699	0.0678	0.0669	0.0639	0.0607	0.0592
0.20	0.0945	0.0861	0.0825	0.0771	0.0712	0.0672	0.0651	0.0642	0.0613	0.0580	0.0566
0.21	0.0919	0.0835	0.0800	0.0746	0.0687	0.0647	0.0626	0.0617	0.0588	0.0555	0.0541
0.22	0.0894	0.0811	0.0776	0.0722	0.0663	0.0623	0.0603	0.0593	0.0564	0.0532	0.0518
0.23	0.0872	0.0788	0.0753	0.0700	0.0641	0.0602	0.0581	0.0571	0.0543	0.0510	0.0496
0.24	0.0851	0.0767	0.0733	0.0679	0.0621	0.0581	0.0561	0.0551	0.0523	0.0490	0.0476
0.25	0.0831	0.0748	0.0714	0.0660	0.0602	0.0563	0.0542	0.0533	0.0504	0.0472	0.0458
0.26	0.0812	0.0730	0.0695	0.0642	0.0584	0.0545	0.0524	0.0515	0.0487	0.0455	0.0441
0.27	0.0796	0.0713	0.0679	0.0626	0.0568	0.0530	0.0509	0.0500	0.0471	0.0440	0.0426
0.28	0.0781	0.0699	0.0665	0.0612	0.0554	0.0516	0.0495	0.0486	0.0458	0.0426	0.0412
0.29	0.0766	0.0685	0.0651	0.0598	0.0541	0.0502	0.0482	0.0473	0.0445	0.0413	0.0399
0.30	0.0753	0.0672	0.0638	0.0586	0.0529	0.0490	0.0470	0.0461	0.0433	0.0401	0.0388
0.31	0.0741	0.0661	0.0627	0.0575	0.0518	0.0480	0.0460	0.0451	0.0423	0.0391	0.0378
0.32	0.0730	0.0650	0.0616	0.0564	0.0508	0.0470	0.0450	0.0441	0.0413	0.0382	0.0368
0.33	0.0720	0.0640	0.0606	0.0555	0.0499	0.0461	0.0441	0.0432	0.0405	0.0374	0.0360
0.34	0.0712	0.0632	0.0599	0.0548	0.0492	0.0455	0.0435	0.0426	0.0398	0.0368	0.0354
0.35	0.0705	0.0626	0.0593	0.0542	0.0486	0.0449	0.0429	0.0420	0.0393	0.0362	0.0349
0.36	0.0699	0.0620	0.0587	0.0537	0.0481	0.0444	0.0424	0.0416	0.0388	0.0358	0.0345
0.37	0.0693	0.0615	0.0582	0.0532	0.0477	0.0440	0.0421	0.0412	0.0385	0.0355	0.0341
0.38	0.0689	0.0611	0.0578	0.0529	0.0474	0.0437	0.0418	0.0409	0.0382	0.0352	0.0339
0.39	0.0684	0.0607	0.0575	0.0525	0.0471	0.0435	0.0415	0.0407	0.0380	0.0350	0.0337
0.40	0.0681	0.0604	0.0572	0.0523	0.0469	0.0433	0.0414	0.0405	0.0379	0.0349	0.0336
0.41	0.0678	0.0603	0.0571	0.0522	0.0468	0.0432	0.0414	0.0405	0.0379	0.0349	0.0336
0.42	0.0677	0.0602	0.0570	0.0521	0.0468	0.0433	0.0414	0.0405	0.0379	0.0350	0.0337
0.43	0.0676	0.0601	0.0570	0.0522	0.0469	0.0434	0.0415	0.0402	0.0381	0.0352	0.0339
0.44	0.0675	0.0601	0.0570	0.0522	0.0470	0.0435	0.0417	0.0408	0.0382	0.0354	0.0341

(continued)

Table 6.7.23 b Values for Reynolds Number Factor, F_r, Pipe Taps (Continued)

$$F_r = 1 + \frac{b}{\sqrt{h_w p_f}}$$

Internal Diameter of Pipe, d_i, in.

	2			3				4			
F_d	1.689	1.939	2.067	2.300	2.626	2.900	3.068	3.152	3.438	3.826	4.026
0.45	0.0675	0.0602	0.0571	0.0524	0.0472	0.0437	0.0419	0.0410	0.0385	0.0357	0.0344
0.46	0.0676	0.0603	0.0572	0.0526	0.0474	0.0440	0.0422	0.0413	0.0388	0.0360	0.0347
0.47	0.0676	0.0604	0.0574	0.0528	0.0476	0.0442	0.0424	0.0416	0.0391	0.0363	0.0351
0.48	0.0677	0.0606	0.0576	0.0530	0.0479	0.0446	0.0428	0.0420	0.0395	0.0367	0.0355
0.49	0.0678	0.0608	0.0578	0.0532	0.0482	0.0449	0.0431	0.0423	0.0399	0.0372	0.0359
0.50	0.0680	0.0611	0.0581	0.0536	0.0486	0.0453	0.0436	0.0428	0.0404	0.0377	0.0365
0.51	0.0682	0.0613	0.0584	0.0539	0.0490	0.0458	0.0440	0.0433	0.0409	0.0382	0.0370
0.52	0.0684	0.0615	0.0587	0.0543	0.0494	0.0462	0.0445	0.0437	0.0413	0.0387	0.0375
0.53	0.0687	0.0619	0.0590	0.0547	0.0499	0.0467	0.0450	0.0442	0.0419	0.0393	0.0381
0.54	0.0689	0.0622	0.0594	0.0551	0.0503	0.0472	0.0455	0.0448	0.0424	0.0398	0.0387
0.55	0.0692	0.0625	0.0598	0.0555	0.0508	0.0477	0.0460	0.0453	0.0430	0.0404	0.0393
0.56	0.0694	0.0628	0.0601	0.0559	0.0513	0.0482	0.0466	0.0458	0.0435	0.0410	0.0399
0.57	0.0696	0.0632	0.0605	0.0563	0.0517	0.0487	0.0471	0.0464	0.0441	0.0416	0.0405
0.58	0.0699	0.0635	0.0608	0.0567	0.0522	0.0492	0.0476	0.0469	0.0447	0.0422	0.0412
0.59	0.0701	0.0638	0.0612	0.0571	0.0527	0.0497	0.0482	0.0474	0.0453	0.0428	0.0418
0.60	0.0704	0.0641	0.0615	0.0575	0.0532	0.0502	0.0487	0.0480	0.0458	0.0434	0.0424
0.61	0.0705	0.0643	0.0618	0.0578	0.0535	0.0506	0.0491	0.0484	0.0463	0.0439	0.0429
0.62	0.0706	0.0645	0.0620	0.0581	0.0539	0.0510	0.0495	0.0489	0.0468	0.0444	0.0434
0.63	0.0707	0.0648	0.0623	0.0585	0.0543	0.0515	0.0500	0.0493	0.0473	0.0450	0.0440
0.64	0.0708	0.0649	0.0625	0.0587	0.0546	0.0518	0.0504	0.0497	0.0477	0.0454	0.0444
0.65	0.0709	0.0651	0.0627	0.0590	0.0549	0.0522	0.0508	0.0501	0.0481	0.0459	0.0449
0.66	0.0708	0.0651	0.0628	0.0591	0.0551	0.0525	0.0511	0.0504	0.0485	0.0463	0.0453
0.67	0.0708	0.0653	0.0629	0.0594	0.0554	0.0528	0.0514	0.0508	0.0489	0.0467	0.0458
0.68	0.0708	0.0653	0.0630	0.0595	0.0556	0.0531	0.0517	0.0511	0.0492	0.0471	0.0462
0.69	0.0705	0.0652	0.0629	0.0595	0.0557	0.0532	0.0518	0.0512	0.0494	0.0473	0.0464
0.70	0.0704	0.0651	0.0629	0.0595	0.0558	0.0533	0.0520	0.0514	0.0496	0.0476	0.0467

	6				8			10			12		
F_d	4.897	5.189	5.761	6.065	7.625	7.981	8.071	9.564	10.020	10.136	11.376	11.938	12.090
0.10	0.0845	0.0836	0.0821	0.0814	0.0784	0.0778	0.0777	0.0757	0.0752	0.0751	0.0739	0.0734	0.0733
0.11	0.0803	0.0795	0.0779	0.0772	0.0742	0.0736	0.0735	0.0716	0.0710	0.0709	0.0697	0.0692	0.0691
0.12	0.0763	0.0755	0.0739	0.0732	0.0702	0.0696	0.0695	0.0676	0.0670	0.0669	0.0657	0.0652	0.0651
0.13	0.0725	0.0717	0.0701	0.0694	0.0664	0.0658	0.0657	0.0638	0.0632	0.0631	0.0619	0.0614	0.0613
0.14	0.0689	0.0680	0.0665	0.0658	0.0628	0.0622	0.0621	0.0601	0.0596	0.0595	0.0583	0.0578	0.0577
0.15	0.0655	0.0646	0.0631	0.0624	0.0594	0.0588	0.0587	0.0567	0.0562	0.0561	0.0549	0.0544	0.0543
0.16	0.0623	0.0614	0.0599	0.0591	0.0561	0.0556	0.0554	0.0535	0.0530	0.0528	0.0516	0.0512	0.0510
0.17	0.0592	0.0583	0.0568	0.0561	0.0531	0.0525	0.0524	0.0504	0.0499	0.0498	0.0486	0.0481	0.0480
0.18	0.0563	0.0554	0.0539	0.0532	0.0502	0.0496	0.0495	0.0475	0.0470	0.0469	0.0457	0.0452	0.0451
0.19	0.0536	0.0527	0.0512	0.0505	0.0475	0.0469	0.0468	0.0448	0.0443	0.0442	0.0430	0.0425	0.0424
0.20	0.0511	0.0502	0.0487	0.0479	0.0449	0.0444	0.0442	0.0423	0.0418	0.0416	0.0404	0.0400	0.0398
0.21	0.0486	0.0477	0.0462	0.0455	0.0425	0.0419	0.0418	0.0398	0.0393	0.0392	0.0380	0.0375	0.0374
0.22	0.0464	0.0455	0.0440	0.0433	0.0403	0.0397	0.0396	0.0376	0.0371	0.0370	0.0358	0.0353	0.0352
0.23	0.0444	0.0435	0.0420	0.0412	0.0382	0.0377	0.0375	0.0356	0.0351	0.0350	0.0338	0.0333	0.0332
0.24	0.0425	0.0416	0.0401	0.0393	0.0364	0.0358	0.0357	0.0337	0.0332	0.0331	0.0319	0.0314	0.0313
0.25	0.0407	0.0399	0.0383	0.0376	0.0346	0.0341	0.0339	0.0320	0.0315	0.0313	0.0301	0.0297	0.0295
0.26	0.0391	0.0382	0.0367	0.0360	0.0330	0.0324	0.0323	0.0303	0.0298	0.0297	0.0285	0.0280	0.0279
0.27	0.0377	0.0368	0.0353	0.0345	0.0315	0.0310	0.0309	0.0289	0.0284	0.0283	0.0271	0.0266	0.0265
0.28	0.0364	0.0355	0.0340	0.0332	0.0303	0.0297	0.0296	0.0276	0.0271	0.0270	0.0258	0.0253	0.0252
0.29	0.0352	0.0343	0.0328	0.0321	0.0291	0.0286	0.0284	0.0265	0.0260	0.0258	0.0246	0.0242	0.0240
0.30	0.0342	0.0333	0.0318	0.0311	0.0281	0.0275	0.0274	0.0255	0.0249	0.0248	0.0236	0.0231	0.0230
0.31	0.0333	0.0324	0.0309	0.0302	0.0272	0.0266	0.0265	0.0245	0.0240	0.0239	0.0227	0.0222	0.0221
0.32	0.0325	0.0317	0.0301	0.0294	0.0264	0.0259	0.0257	0.0238	0.0233	0.0232	0.0220	0.0215	0.0214
0.33	0.0319	0.0310	0.0295	0.0288	0.0258	0.0252	0.0251	0.0231	0.0226	0.0225	0.0213	0.0208	0.0207
0.34	0.0314	0.0305	0.0290	0.0283	0.0253	0.0247	0.0246	0.0226	0.0221	0.0220	0.0208	0.0203	0.0202

(continued)

Table 6.7.23 b Values for Reynolds Number Factor, F_r, Pipe Taps (Continued)

$$F_r = 1 + \frac{b}{\sqrt{h_w p_f}}$$

Internal Diameter of Pipe, d_i, in.

	6				8			10			12	
F_d	4.897	5.189	5.761	6.065	7.625	7.981	8.071	9.564	10.020	10.136	11.376	11.938	12.090
0.35	0.0310	0.0301	0.0286	0.0278	0.0294	0.0243	0.0242	0.0222	0.0217	0.0216	0.0204	0.0199	0.0198
0.36	0.0306	0.0298	0.0283	0.0275	0.0246	0.0240	0.0239	0.0219	0.0214	0.0213	0.0201	0.0196	0.0195
0.37	0.0305	0.0296	0.0281	0.0274	0.0244	0.0239'	0.0237	0.0218	0.0213	0.0212	0.0200	0.0195	0.0194
0.38	0.0304	0.0285	0.0280	0.0273	0.0244	0.0238	0.0237	0.0217	0.0212	0.0211	0.0199	0.0195	0.0193
0.39	0.0305	0.0296	0.0281	0.0274	0.0244	0.0239	0.0238	0.0218	0.0213	0.0212	0.0200	0.0195	0.0194
0.40	0.0306	0.0296	0.0283	0.0276	0.0246	0.0240	0.0239	0.0220	0.0215	0.0213	0.0202	0.0197	0.0196
0.41	0.0309	0.0300	0.0285	0.0278	0.0248	0.0243	0.0241	0.0222	0.0217	0.0216	0.0204	0.0199	0.0198
0.42	0.0312	0.0303	0.0288	0.0281	0.0252	0.0246	0.0245	0.0226	0.0221	0.0219	0.0208	0.0203	0.0202
0.43	0.0316	0.0308	0.0293	0.0286	0.0256	0.0251	0.0249	0.0230	0.0225	0.0224	0.0212	0.0207	0.0206
0.44	0.0321	0.0313	0.0298	0.0291	0.0261	0.0256	0.0254	0.0235	0.0230	0.0229	0.0217	0.0213	0.0211
0.45	0.0327	0.0319	0.0304	0.0297	0.0267	0.0262	0.0261	0.0241	0.0236	0.0235	0.0223	0.0219	0.0217
0.46	0.0334	0.0326	0.0311	0.0304	0.0274	0.0269	0.0267	0.0248	0.0243	0.0242	0.0230	0.0226	0.0224
0.47	0.0341	0.0333	0.0318	0.0311	0.0282	0.0276	0.0275	0.0256	0.0251	0.0249	0.0238	0.0233	0.0232
0.48	0.0350	0.0341	0.0326	0.0319	0.0290	0.0284	0.0283	0.0264	0.0259	0.0258	0.0246	0.0242	0.0240
0.49	0.0358	0.0349	0.0335	0.0328	0.0299	0.0293	0.0292	0.0273	0.0268	0.0267	0.0255	0.0250	0.0249
0.50	0.0368	0.0359	0.0344	0.0337	0.0308	0.0303	0.0302	0.0283	0.0278	0.0276	0.0265	0.0260	0.0259
0.51	0.0378	0.0369	0.0354	0.0347	0.0318	0.0313	0.0312	0.0293	0.0288	0.0287	0.0275	0.0270	0.0269
0.52	0.0388	0.0380	0.0365	0.0358	0.0329	0.0324	0.0323	0.0304	0.0299	0.0298	0.0286	0.0281	0.0280
0.53	0.0399	0.0391	0.0376	0.0369	0.0340	0.0335	0.0334	0.0315	0.0310	0.0309	0.0297	0.0293	0.0291
0.54	0.0410	0.0402	0.0387	0.0380	0.0351	0.0346	0.0345	0.0326	0.0321	0.0320	0.0308	0.0304	0.0303
0.55	0.0422	0.0413	0.0399	0.0392	0.0363	0.0358	0.0357	0.0338	0.0333	0.0332	0.0321	0.0316	0.0315
0.56	0.0433	0.0425	0.0411	0.0404	0.0375	0.0370	0.0369	0.0350	0.0345	0.0344	0.0333	0.0328	0.0327
0.57	0.0446	0.0438	0.0423	0.0416	0.0388	0.0383	0.0381	0.0363	0.0358	0.0357	0.0346	0.0341	0.0340
0.58	0.0458	0.0450	0.0436	0.0429	0.0401	0.0395	0.0394	0.0376	0.0371	0.0370	0.0358	0.0354	0.0353
0.59	0.0472	0.0463	0.0449	0.0442	0.0414	0.0409	0.0408	0.0389	0.0384	0.0383	0.0372	0.0367	0.0366
0.60	0.0484	0.0476	0.0462	0.0455	0.0427	0.0422	0.0421	0.0402	0.0398	0.0397	0.0385	0.0381	0.0380
0.61	0.0497	0.0489	0.0474	0.0468	0.0440	0.0435	0.0433	0.0415	0.0411	0.0409	0.0398	0.0394	0.0393
0.62	0.0510	0.0502	0.0488	0.0481	0.0402	0.0448	0.0447	0.0429	0.0424	0.0423	0.0412	0.0407	0.0406
0.63	0.0523	0.0515	0.0501	0.0494	0.0466	0.0461	0.0460	0.0442	0.0437	0.0436	0.0425	0.0421	0.0420
0.64	0.0535	0.0527	0.0513	0.0507	0.0479	0.0474	0.0473	0.0455	0.0451	0.0449	0.0438	0.0434	0.0433
0.65	0.0548	0.0540	0.0526	0.0520	0.0492	0.0487	0.0486	0.0468	0.0464	0.0463	0.0452	0.0447	0.0446
0.66	0.0560	0.0552	0.0538	0.0532	0.0505	0.0500	0.0499	0.0481	0.0476	0.0475	0.0465	0.0460	0.0459
0.67	0.0572	0.0564	0.0551	0.0544	0.0517	0.0512	0.0511	0.0494	0.0489	0.0488	0.0477	0.0473	0.0472
0.68	0.0584	0.0576	0.0563	0.0556	0.0530	0.0525	0.0523	0.0506	0.0502	0.0500	0.0490	0.0486	0.0484
0.69	0.0595	0.0587	0.0574	0.0567	0.0541	0.0536	0.0535	0.0518	0.0513	0.0512	0.0502	0.0497	0.0496
0.70	0.0606	0.0598	0.0585	0.0579	0.0552	0.0548	0.0546	0.0529	0.0525	0.0524	0.0513	0.0509	0.0508
0.71	0.0616	0.0608	0.0595	0.0589	0.0563	0.0558	0.0557	0.0540	0.0535	0.0534	0.0524	0.0520	0.0519
0.72	0.0625	0.0618	0.0605	0.0599	0.0573	0.0568	0.0567	0.0550	0.0546	0.0545	0.0535	0.0530	0.0529
0.73	0.0635	0.0627	0.0614	0.0608	0.0583	0.0578	0.0577	0.0560	0.0556	0.0555	0.0545	0.0541	0.0540
0.74	0.0643	0.0636	0.0623	0.0617	0.0592	0.0587	0.0586	0.0570	0.0565	0.0564	0.0554	0.0550	0.0549
0.75	0.0650	0.0643	0.0630	0.0624	0.0599	0.0594	0.0594	0.0577	0.0573	0.0572	0.0562	0.0558	0.0557

(continued)

Table 6.7.23 b Values for Reynolds Number Factor, F_r, Pipe Taps (Continued)

$$F_r = 1 + \frac{b}{\sqrt{h_w p_f}}$$

Internal Diameter of Pipe, d_i, in.

	16			20			24			30		
F_d	14.688	15.000	15.250	18.814	19.000	19.250	22.626	23.000	23.250	28.628	29.000	29.250
0.10	0.0706	0.0705	0.0704	0.0690	0.0689	0.0688	0.0680	0.0679	0.0678	0.0669	0.0669	0.0668
0.11	0.0665	0.0663	0.0662	0.0648	0.0647	0.0647	0.0638	0.0637	0.0637	0.0628	0.0627	0.0627
0.12	0.0624	0.0622	0.0621	0.0607	0.0607	0.0606	0.0597	0.0596	0.0596	0.0587	0.0586	0.0586
0.13	0.0585	0.0584	0.0582	0.0569	0.0568	0.0567	0.0558	0.0558	0.0557	0.0548	0.0548	0.0547
0.14	0.0549	0.0548	0.0546	0.0532	0.0531	0.0530	0.0522	0.0521	0.0520	0.0511	0.0511	0.0511
0.15	0.0514	0.0512	0.0511	0.0497	0.0497	0.0496	0.0487	0.0486	0.0486	0.0477	0.0476	0.0476
0.16	0.0481	0.0479	0.0478	0.0464	0.0464	0.0463	0.0454	0.0453	0.0453	0.0444	0.0443	0.0443
0.17	0.0450	0.0448	0.0447	0.0433	0.0433	0.0432	0.0423	0.0423	0.0422	0.0413	0.0413	0.0412
0.18	0.0420	0.0419	0.0417	0.0404	0.0403	0.0402	0.0394	0.0393	0.0392	0.0383	0.0383	0.0383
0.19	0.0393	0.0391	0.0390	0.0376	0.0376	0.0375	0.0366	0.0366	0.0365	0.0356	0.0356	0.0355
0.20	0.0367	0.0365	0.0364	0.0350	0.0350	0.0349	0.0340	0.0340	0.0339	0.0330	0.0330	0.0329
0.21	0.0343	0.0341	0.0340	0.0326	0.0326	0.0325	0.0316	0.0316	0.0315	0.0306	0.0306	0.0305
0.22	0.0320	0.0318	0.0317	0.0304	0.0303	0.0302	0.0294	0.0293	0.0293	0.0284	0.0283	0.0283
0.23	0.0299	0.0298	0.0296	0.0283	0.0282'	0.0281	0.0273	0.0272	0.0272	0.0263	0.0262	0.0262
0.24	0.0280	0.0278	0.0277	0.0264	0.0263	0.0262	0.0254	0.0253	0.0253	0.0244	0.0243	0.0243
0.25	0.0262	0.0261	0.0260	0.0246	0.0246	0.0245	0.0236	0.0236	0.0235	0.0226	0.0226	0.0226
0.26	0.0246	0.0244	0.0243	0.0230	0.0229	0.0228	0.0220	0.0219	0.0219	0.0210	0.0210	0.0209
0.27	0.0231	0.0230	0.0229	0.0215	0.0215	0.0214	0.0206	0.0205	0.0204	0.0196	0.0195	0.0195
0.28	0.0218	0.0217	0.0216	0.0202	0.0202	0.0201	0.0193	0.0192	0.0191	0.0183-	0.0182	0.0182
0.29	0.0206	0.0205	0.0204	0.0191	0.0190	0.0189	0.0181	0.0180	0.0180	0.0171	0.0171	0.0170
0.30	0.0196	0.0194	0.0193	0.0180	0.0179	0.0179	0.0170	0.0170	0.0169	0.0161	0.0160	0.0160
0.31	0.0187	0.0185	0.0184	0.0171	0.0170	0.0170	0.0161	0.0161	0.0160	0.0152	0.0151	0.0151
0.32	0.0179	0.0177	0.0176	0.0163	0.0162	0.0162	0.0153	0.0153	0.0152	0.0144	0.0143	0.0143
0.33	0.0172	0.0170	0.0169	0.0156	0.0156	0.0155	0.0147	0.0146	0.0146	0.0137	0.0137	0.0136
0.34	0.0166	0.0165	0.0164	0.0151	0.0150	0.0150	0.0142	0.0141	0.0140	0.0132	0.0131	0.0131
0.35	0.0162	0.0161	0.0160	0.0147	0.0146	0.0145	0.0137	0.0137	0.0136	0.0128	0.0127	0.0127
0.36	0.0159	0.0157	0.0156	0.0144	0.0143	0.0142	0.0134	0.0134	0.0133	0.0125	0.0124	0.0124
0.37	0.0157	0.0155	0.0154	0.0141	0.0141	0.0140	0.0132	0.0132	0.0131	0.0123	0.0122	0.0122
0.38	0.0155	0.0154	0.0153	0.0140	0.0140	0.0139	0.0131	0.0130	0.0130	0.0122	0.0121	0.0121
0.39	0.0155	0.0154	0:0153	0.0140	0.0139	0.0139	0.0131	0.0130	0.0130	0.0122	0.0121	0.0121
0.40	0.0156	0.0154	0.0153	0.0141	0.0140	0.0139	0.0132	0.0131	0.0130	0.0122	0.0122	0.0122
0.41	0.0157	0.0155	0.0154	0.0142	0.0142	0.0141	0.0133	0.0132	0.0132	0.0124	0.0124	0.0123
0.42	0.0159	0.0158	0.0157	0.0144	0.0144	0.0143	0.0136	0.0135	0.0134	0.0126	0.0126	0.0126
0.43	0.0162	0.0161	0.0160	0.0148	0.0147	0.0146	0.0139	0.0138	0.0138	0.0130	0.0129	0.0129
0.44	0.0166	0.0164	0.0163	0.0151	0.0151	0.0150	0.0143	0.0142	0.0141	0.0134	0.0133	0.0133
0.45	0.0170	0.0169	0.0168	0.0156	0.0155	0.0155	0.0147	0.0146	0.0146	0.0138	0.0138	0.0137
0.46	0.0175	0.0174	0.0173	0.0161	0.0160	0.0160	0.0152	0.0151	0.0151	0.0143	0.0143	0.0143
0.47	0.0180	0.0179	0.0178	0.0166	0.0166	0.0165	0.0158	0.0157	0.0157	0.0149	0.0149	0.0148
0.48	0.0186	0.0185	0.0184	0.0172	0.0172	0.0171	0.0164	0.0163	0.0163	0.0155	0.0155	0.0154
0.49	0.0192	0.0191	0.0190	0.0178	0.0178	0.0177	0.0170	0.0169	0.0169	0.0162	0.0161	0.0161
0.50	0.0199	0.0198	0.0197	0.0185	0.0185	0.0184	0.0177	0.0176	0.0176	0.0169	0.0168	0.0168
0.51	0.0206	0.0205	0.0204	0.0193	0.0192	0.0191	0.0184	0.0184	0.0183	0.0176	0.0176	0.0175
0.52	0.0213	0.0212	0.0211	0.0200	0.0199	0.0199	0.0192	0.0191	0.0191	0.0183	0.0183	0.0183
0.53	0.0221	0.0220	0.0219	0.0208	0.0207	0.0207	0.0200	0.0199	0.0199	0.0191	0.0191	0.0191
0.54	0.0229	0.0227	0.0226	0.0215	0.0215	0.0214	0.0208	0.0207	0.0207	0.0199	0.0199	0.0199
0.55	0.0237	0.0235	0.0234	0.0224	0.0223	0.0223	0.0216	0.0215	0.0215	0.0208	0.0207	0.0207
0.56	0.0244	0.0243	0.0242	0.0232	0.0231	0.0231	0.0224	0.0223	0.0223	0.0216	0.0216	0.0215
0.57	0.0253	0.0251	0.0250	0.0240	0.0239	0.0239	0.0232	0.0232	0.0231	0.0224	0.0224	0.0224
0.58	0.0261	0.0260	0.0259	0.0248	0.0248	0.0247	0.0241	0.0240	0.0240	0.0233	0.0233	0.0232
0.59	0.0269	0.0268	0.0267	0.0256	0.0256	0.0255	0.0249	0.0248	0.0248	0.0241	0.0241	0.0241
0.60	0.0277	0.0276	0.0275	0.0265	0.0264	0.0264	0.0257	0.0257	0.0256	0.0250	0.0249	0.0249
0.61	0.0284	0.0283	0.0282	0.0272	0.0272	0.0271	0.0265	0.0265	0.0264	0.0258	0.0257	0.0257
0.62	0.0292	0.0291	0.0290	0.0280	0.0280	0.0279	0.0273	0.0272	0.0272	0.0266	0.0265	0.0265
0.63	0.0299	0.0298	0.0297	0.0288	0.0287	0.0287	0.0281	0.0280	0.0280	0.0273	0.0273	0.0273
0.64	0.0306	0.0305	0.0304	0.0295	0.0295	0.0294	0.0288	0.0288	0.0287	0.0281	0.0281	0.0280
0.65	0.0313	0.0312	0.0312	0.0302	0.0302	0.0301	0.0295	0.0295	0.0294	0.0288	0.0288	0.0288
0.66	0.0320	0.0319	0.0318	0.0309	0.0308	0.0308	0.0302	0.0301	0.0301	0.0295	0.0295	0.0295
0.67	0.0326	0.0325	0.0324	0.0315	0.0315	0.0314	0.0309	0.0308	0.0308	0.0302	0.0302	0.0302
0.68	0.0332	0.0331	0.0330	0.0322	0.0321	0.0321	0.0315	0.0315	0.0314	0.0308	0.0308	0.0308
0.69	0.0337	0.0336	0.0335	0.0327	0.0326	0.0326	0.0320	0.0320	0.0320	0.0314	0.0314	0.0313
0.70	0.0342	0.0341	0.0340	0.0332	0.0332	0.0331	0.0326	0.0325	0.0325	0.0319	0.0319	0.0319

Table 6.7.24 *Expansion Factor, Y_1, Flange Taps; Static Pressure Taken from Upstream Taps [12]*

h_o/P_n Ratio	$\beta = d/D$												
	.1	.2	.3	.4	.45	.50	.52	.54	.56	.58	.60	.61	.62
0.0	1.000	1.000	1.000	1.000	1.000	1.000	1.000	1.000	1.000	1.000	1.000	1.000	1.000
0.1	0.9989	0.9989	0.9989	0.9988	0.9988	0.9988	0.9988	0.9988	0.9988	0.9988	0.9987	0.9987	0.9987
0.2	0.9977	0.9977	0.9977	0.9977	0.9976	0.9976	0.9976	0.9976	0.9975	0.9975	0.9975	0.9975	0.9974
0.3	0.9966	0.9966	0.9966	0.9965	0.9965	0.9964	0.9964	0.9963	0.9963	0.9963	0.9962	0.9962	0.9962
0.4	0.9954	0.9954	0.9954	0.9953	0.9953	0.9952	0.9952	0.9951	0.9951	0.9950	0.9949	0.9949	0.9949
0.5	0.9943	0.9943	0.9943	0.9942	0.9941	0.9940	0.9940	0.9939	0.9938	0.9938	0.9937	0.9936	0.9936
0.6	0.9932	0.9932	0.9931	0.9930	0.9929	0.9928	0.9927	0.9927	0.9926	0.9925	0.9924	0.9924	0.9923
0.7	0.9920	0.9920	0.9920	0.9919	0.9918	0.9916	0.9915	0.9915	0.9914	0.9913	0.9912	0.9911	0.9910
0.8	0.9909	0.9909	0.9908	0.9907	0.9906	0.9904	0.9903	0.9902	0.9901	0.9900	0.9899	0.9898	0.9897
0.9	0.9898	0.9897	0.9897	0.9895	0.9894	0.9892	0.9891	0.9890	0.9889	0.9888	0.9886	0.9885	0.9885
1.0	0.9886	0.9886	0.9885	0.9884	0.9882	0.9880	0.9879	0.9878	0.9877	0.9875	0.9874	0.9873	0.9872
1.1	0.9875	0.9875	0.9874	0.9872	0.9870	0.9868	0.9867	0.9866	0.9864	0.9863	0.9861	0.9860	0.9859
1.2	0.9863	0.9863	0.9862	0.9860	0.9859	0.9856	0.9855	0.9853	0.9852	0.9850	0.9848	0.9847	0.9846
1.3	0.9852	0.9852	0.9851	0.9849	0.9847	0.9844	0.9843	0.9841	0.9840	0.9838	0.9836	0.9835	0.9833
1.4	0.9841	0.9840	0.9840	0.9837	0.9835	0.9832	0.9831	0.9829	0.9827	0.9825	0.9823	0.9822	0.9821
1.5	0.9829	0.9829	0.9828	0.9826	0.9823	0.9820	0.9819	0.9817	0.9815	0.9813	0.9810	0.9809	0.9808
1.6	0.9818	0.9818	0.9817	0.9814	0.9811	0.9808	0.9806	0.9805	0.9803	0.9800	0.9798	0.9796	0.9795
1.7	0.9806	0.9806	0.9805	0.9802	0.9800	0.9796	0.9794	0.9792	0.9790	0.9788	0.9785	0.9784	0.9782
1.8	0.9795	0.9795	0.9794	0.9791	0.9788	0.9784	0.9782	0.9780	0.9778	0.9775	0.9772	0.9771	0.9769
1.9	0.9784	0.9783	0.9782	0.9779	0.9776	0.9772	0.9770	0.9768	0.9766	0.9763	0.9760	0.9758	0.9756
2.0	0.9772	0.9772	0.9771	0.9767	0.9764	0.9760	0.9758	0.9756	0.9753	0.9750	0.9747	0.9745	0.9744
2.1	0.9761	0.9761	0.9759	0.9756	0.9753	0.9748	0.9746	0.9744	0.9741	0.9738	0.9734	0.9733	0.9731
2.2	0.9750	0.9749	0.9748	0.9744	0.9741	0.9736	0.9734	0.9731	0.9729	0.9725	0.9722	0.9720	0.9718
2.3	0.9738	0.9738	0.9736	0.9732	0.9729	0.9724	0.9722	0.9719	0.9716	0.9713	0.9709	0.9707	0.9705
2.4	0.9727	0.9726	0.9725	0.9721	0.9717	0.9712	0.9710	0.9707	0.9704	0.9700	0.9697	0.9694	0.9692
2.5	0.9715	0.9715	0.9713	0.9709	0.9705	0.9700	0.9698	0.9695	0.9692	0.9688	0.9684	0.9682	0.9680
2.6	0.9704	0.9704	0.9702	0.9698	0.9694	0.9688	0.9686	0.9683	0.9679	0.9675	0.9671	0.9669	0.9667
2.7	0.9693	0.9692	0.9691	0.9686	0.9682	0.9676	0.9673	0.9670	0.9667	0.9663	0.9659	0.9656	0.9654
2.8	0.9681	0.9681	0.9679	0.9674	0.9670	0.9664	0.9661	0.9658	0.9654	0.9650	0.9646	0.9644	0.9641
2.9	0.9670	0.9669	0.9668	0.9663	0.9658	0.9652	0.9649	0.9646	0.9642	0.9638	0.9633	0.9631	0.9628
3.0	0.9658	0.9658	0.9656	0.9651	0.9647	0.9640	0.9637	0.9634	0.9630	0.9626	0.9621	0.9618	0.9615
3.1	0.9647	0.9647	0.9645	0.9639	0.9635	0.9628	0.9625	0.9622	0.9617	0.9613	0.9608	0.9605	0.9603
3.2	0.9636	0.9635	0.9633	0.9628	0.9623	0.9616	0.9613	0.9609	0.9605	0.9601	0.9595	0.9593	0.9590
3.3	0.9624	0.9624	0.9622	0.9616	0.9611	0.9604	0.9601	0.9597	0.9593	0.9588	0.9583	0.9580	0.9577
3.4	0.9613	0.9612	0.9610	0.9604	0.9599	0.9592	0.9589	0.9585	0.9580	0.9576	0.9570	0.9567	0.9564
3.5	0.9602	0.9601	0.9599	0.9593	0.9588	0.9580	0.9577	0.9573	0.9568	0.9563	0.9558	0.9554	0.9551
3.6	0.9590	0.9590	0.9587	0.9581	0.9576	0.9568	0.9565	0.9560	0.9556	0.9551	0.9545	0.9542	0.9538
3.7	0.9579	0.9578	0.9576	0.9570	0.9564	0.9556	0.9553	0.9548	0.9543	0.9538	0.9532	0.9529	0.9526
3.8	0.9567	0.9567	0.9564	0.9558	0.9552	0.9544	0.9540	0.9536	0.9531	0.9526	0.9520	0.9516	0.9513
3.9	0.9556	0.9555	0.9553	0.9546	0.9540	0.9532	0.9528	0.9524	0.9519	0.9513	0.9507	0.9504	0.9500
4.0	0.9545	0.9544	0.9542	0.9536	0.9529	0.9520	0.9516	0.9512	0.9506	0.9501	0.9494	0.9491	0.9487

h_o/P_n Ratio	$\beta = d/D$												
	.63	.64	.65	.66	.67	.68	.69	.70	.71	.72	.73	.74	.75
0.0	1.000	1.000	1.000	1.000	1.000	1.000	1.000	1.000	1.000	1.000	1.000	1.000	1.000
0.1	0.9987	0.9987	0.9987	0.9987	0.9987	0.9987	0.9986	0.9986	0.9986	0.9986	0.9986	0.9986	0.9986
0.2	0.9974	0.9974	0.9974	0.9974	0.9973	0.9973	0.9973	0.9973	0.9972	0.9972	0.9972	0.9971	0.9971
0.3	0.9961	0.9961	0.9961	0.9960	0.9960	0.9960	0.9959	0.9959	0.9958	0.9958	0.9958	0.9957	0.9957
0.4	0.9948	0.9948	0.9948	0.9947	0.9947	0.9946	0.9946	0.9945	0.9945	0.9944	0.9943	0.9943	0.9942
0.5	0.9935	0.9935	0.9934	0.9934	0.9933	0.9933	0.9932	0.9931	0.9931	0.9930	0.9929	0.9929	0.9928
0.6	0.9923	0.9922	0.9921	0.9921	0.9920	0.9919	0.9918	0.9918	0.9917	0.9916	0.9915	0.9914	0.9913
0.7	0.9910	0.9909	0.9908	0.9907	0.9907	0.9906	0.9905	0.9904	0.9903	0.9902	0.9901	0.9900	0.9899
0.8	0.9897	0.9896	0.9895	0.9894	0.9893	0.9892	0.9891	0.9890	0.9889	0.9888	0.9887	0.9886	0.9884
0.9	0.9884	0.9883	0.9882	0.9881	0.9880	0.9879	0.9878	0.9877	0.9875	0.9874	0.9873	0.9871	0.9870
1.0	0.9871	0.9870	0.9869	0.9868	0.9867	0.9865	0.9864	0.9863	0.9861	0.9860	0.9859	0.9857	0.9855

(continued)

Table 6.7.24 *Expansion Factor, Y_1, Flange Taps; Static Pressure Taken from Upstream Taps [12] (Continued)*

h_o/P_n Ratio						$\beta = d/D$							
	.63	.64	.65	.66	.67	.68	.69	.70	.71	.72	.73	.74	.75
1.1	0.9858	0.9857	0.9856	0.9854	0.9853	0.9852	0.9851	0.9849	0.9848	0.9846	0.9844	0.9843	0.9841
1.2	0.9845	0.9844	0.9843	0.9841	0.9840	0.9838	0.9837	0.9835	0.9834	0.9832	0.9830	0.9828	0.9826
1.3	0.9832	0.9831	0.9829	0.9828	0.9827	0.9825	0.9823	0.9822	0.9820	0.9818	0.9816	0.9814	0.9812
1.4	0.9819	0.9818	0.9816	0.9815	0.9813	0.9812	0.9810	0.9808	0.9806	0.9804	0.9802	0.9800	0.9728
1.5	0.9806	0.9805	0.9803	0.9802	0.9800	0.9798	0.9796	0.9794	0.9792	0.9790	0.9788	0.9786	0.9783
1.6	0.9793	0.9792	0.9790	0.9788	0.9787	0.9785	0.9783	0.9781	0.9778	0.9776	0.9774	0.9771	0.9769
1.7	0.9780	0.9779	0.9777	0.9775	0.9773	0.9771	0.9769	0.9767	0.9764	0.9762	0.9760	0.9757	0.9754
1.8	0.9768	0.9766	0.9764	0.9762	0.9760	0.9758	0.9755	0.9753	0.9751	0.9748	0.9745	0.9743	0.9740
1.9	0.9755	0.9753	0.9751	0.9749	0.9747	0.9744	0.9742	0.9739	0.9737	0.9734	0.9731	0.9728	0.9725
2.0	0.9742	0.9740	0.9738	0.9735	0.9733	0.9731	0.9728	0.9726	0.9723	0.9720	0.9717	0.9714	0.9711
2.1	0.9729	0.9727	0.9725	0.9722	0.9720	0.9717	0.9715	0.9712	0.9709	0.9706	0.9703	0.9700	0.9696
2.2	0.9716	0.9714	0.9711	0.9709	0.9706	0.9704	0.9701	0.9698	0.9695	0.9692	0.9689	0.9685	0.9682
2.3	0.9703	0.9701	0.9698	0.9696	0.9693	0.9690	0.9688	0.9685	0.9681	0.9678	0.9675	0.9671	0.9667
2.4	0.9690	0.9688	0.9685	0.9683	0.9680	0.9677	0.9674	0.9671	0.9668	0.9664	0.9661	0.9657	0.9653
2.5	0.9677	0.9675	0.9672	0.9669	0.9666	0.9663	0.9660	0.9657	0.9654	0.9650	0.9646	0.9643	0.9639
2.6	0.9664	0.9662	0.9659	0.9656	0.9653	0.9650	0.9647	0.9643	0.9640	0.9636	0.9632	0.9628	0.9624
2.7	0.9651	0.9649	0.9646	0.9643	0.9640	0.9637	0.9633	0.9630	0.9626	0.9622	0.9618	0.9614	0.9610
2.8	0.9638	0.9636	0.9633	0.9630	0.9626	0.9623	0.9620	0.9616	0.9612	0.9608	0.9604	0.9600	0.9595
2.9	0.9625	0.9623	0.9620	0.9616	0.9613	0.9610	0.9606	0.9602	0.9598	0.9594	0.9590	0.9585	0.9581
3.0	0.9613	0.9610	0.9606	0.9603	0.9600	0.9596	0.9592	0.9588	0.9584	0.9580	0.9576	0.9571	0.9566
3.1	0.9600	0.9597	0.9593	0.9590	0.9586	0.9583	0.9579	0.9575	0.9571	0.9566	0.9562	0.9557	0.9552
3.2	0.9587	0.9584	0.9580	0.9577	0.9573	0.9569	0.9567	0.9561	0.9557	0.9552	0.9547	0.9542	0.9537
3.3	0.9574	0.9571	0.9567	0.9564	0.9560	0.9556	0.9552	0.9547	0.9543	0.9538	0.9533	0.9528	0.9525
3.4	0.9561	0.9558	0.9554	0.9550	0.9546	0.9542	0.9538	0.9534	0.9529	0.9524	0.9519	0.9514	0.9508
3.5	0.9548	0.9545	0.9541	0.9537	0.9533	0.9529	0.9524	0.9520	0.9515	0.9510	0.9505	0.9500	0.9494
3.6	0.9535	0.9532	0.9528	0.9524	0.9520	0.9515	0.9511	0.9506	0.9501	0.9496	0.9491	0.9485	0.9480
3.7	0.9552	0.9518	0.9515	0.9511	0.9506	0.9502	0.9497	0.9492	0.9487	0.9482	0.9477	0.9471	0.9465
3.8	0.9509	0.9505	0.9502	0.9497	0.9493	0.9488	0.9484	0.9479	0.9474	0.9468	0.9463	0.9457	0.9451
3.9	0.9496	0.9492	0.9488	0.9484	0.9480	0.9475	0.9470	0.9465	0.9460	0.9454	0.9448	0.9442	0.9436
4.0	0.9483	0.9479	0.9475	0.9471	0.9465	0.9462	0.9457	0.9451	0.9446	0.9440	0.9434	0.9428	0.9422

h_o/P_n Ratio						$\beta = d/D$							
	.1	.2	.3	.4	.45	.50	.52	.54	.56	.58	.60	.61	.62
0.0	1.000	1.000	1.000	1.000	1.000	1.000	1.000	1.000	1.000	1.000	1.000	1.000	1.000
0.1	1.0007	1.0007	1.0006	1.0006	1.0006	1.0006	1.0006	1.0006	1.0006	1.0006	1.0005	1.0005	1.0005
0.2	1.0013	1.0013	1.0013	1.0013	1.0012	1.0012	1.0012	1.0012	1.0011	1.0011	1.0011	1.0011	1.0010
0.3	1.0020	1.0020	1.0020	1.0019	1.0019	1.0018	1.0018	1.0018	1.0017	1.0017	1.0016	1.0016	1.0016
0.4	1.0027	1.0027	1.0026	1.0026	1.0025	1.0024	1.0024	1.0023	1.0023	1.0022	1.0022	1.0021	1.0021
0.5	1.0033	1.0033	1.0033	1.0032	1.0031	1.0030	1.0030	1.0029	1.0029	1.0028	1.0027	1.0027	1.0026
0.6	1.0040	1.0040	1.0040	1.0039	1.0038	1.0036	1.0036	1.0035	1.0034	1.0034	1.0033	1.0032	1.0032
0.7	1.0047	1.0047	1.0046	1.0045	1.0044	1.0043	1.0042	1.0041	1.0040	1.0039	1.0038	1.0038	1.0037
0.8	1.0054	1.0053	1.0053	1.0052	1.0050	1.0049	1.0048	1.0047	1.0046	1.0045	1.0044	1.0043	1.0042
0.9	1.0060	1.0060	1.0060	1.0058	1.0057	1.0055	1.0054	1.0053	1.0052	1.0050	1.0049	1.0048	1.0048
1.0	1.0067	1.0067	1.0066	1.0065	1.0063	1.0061	1.0060	1.0059	1.0058	1.0056	1.0055	1.0054	1.0053
1.1	1.0074	1.0074	1.0073	1.0071	1.0069	1.0067	1.0066	1.0065	1.0063	1.0062	1.0060	1.0059	1.0058
1.2	1.0080	1.0080	1.0080	1.0078	1.0076	1.0073	1.0072	1.0071	1.0069	1.0068	1.0066	1.0065	1.0064
1.3	1.0087	1.0087	1.0086	1.0084	1.0082	1.0080	1.0078	1.0077	1.0075	1.0073	1.0071	1.0070	1.0069
1.4	1.0094	1.0094	1.0093	1.0091	1.0089	1.0086	1.0084	1.0083	1.0081	1.0079	1.0077	1.0076	1.0074
1.5	1.0101	1.0101	1.0100	1.0097	1.0095	1.0092	1.0090	1.0089	1.0087	1.0085	1.0082	1.0081	1.0080
1.6	1.0108	1.0107	1.0106	1.0104	1.0101	1.0098	1.0096	1.0095	1.0093	1.0090	1.0088	1.0087	1.0085
1.7	1.0114	1.0114	1.0113	1.0110	1.0108	1.0104	1.0103	1.0101	1.0099	1.0096	1.0094	1.0092	1.0091
1.8	1.0121	1.0121	1.0120	1.0117	1.0114	1.0111	1.0109	1.0107	1.0104	1.0109	1.0099	1.0098	1.0096
1.9	1.0128	1.0128	1.0126	1.0123	1.0121	1.0117	1.0115	1.0113	1.0110	1.0108	1.0105	1.0103	1.0102
2.0	1.0135	1.0134	1.0133	1.0130	1.0127	1.0123	1.0121	1.0119	1.0116	1.0114	1.0110	1.0109	1.0107

(continued)

Table 6.7.24 *Expansion Factor, Y_1, Flange Taps; Static Pressure Taken from Upstream Taps [12] (Continued)*

h_o/P_n Ratio	$\beta = d/D$												
	.1	.2	.3	.4	.45	.50	.52	.54	.56	.58	.60	.61	.62
2.1	1.0142	1.0141	1.0140	1.0136	1.0134	1.0129	1.0127	1.0125	1.0122	1.0119	1.0116	1.0114	1.0112
2.2	1.0148	1.0148	1.0147	1.0143	1.0140	1.0136	1.0133	1.0131	1.0128	1.0125	1.0122	1.0120	1.0118
2.3	1.0155	1.0155	1.0154	1.0150	1.0146	1.0142	1.0140	1.0137	1.0134	1.0131	1.0127	1.0126	1.0124
2.4	1.0162	1.0162	1.0160	1.0156	1.0153	1.0148	1.0146	1.0143	1.0140	1.0137	1.0133	1.0131	1.0129
2.5	1.0169	1.0168	1.0167	1.0163	1.0159	1.0154	1.0152	1.0149	1.0146	1.0142	1.0139	1.0137	1.0134
2.6	1.0176	1.0175	1.0174	1.0170	1.0166	1.0161	1.0158	1.0155	1.0152	1.0148	1.0144	1.0142	1.0140
2.7	1.0182	1.0182	1.0180	1.0176	1.0172	1.0167	1.0164	1.0161	1.0158	1.0154	1.0150	1.0148	1.0146
2.8	1.0189	1.0189	1.0187	1.0183	1.0179	1.0173	1.0170	1.0167	1.0164	1.0160	1.0156	1.0154	1.0151
2.9	1.0196	1.0196	1.0194	1.0189	1.0185	1.0180	1.0177	1.0173	1.0170	1.0166	1.0162	1.0159	1.0157
3.0	1.0203	1.0203	1.0201	1.0196	1.0192	1.0186	1.0183	1.0180	1.0176	1.0172	1.0167	1.0165	1.0162
3.1	1.0210	1.0210	1.0208	1.0203	1.0198	1.0192	1.0189	1.0186	1.0182	1.0178	1.0173	1.0170	1.0168
3.2	1.0217	1.0216	1.0214	1.0209	1.0205	1.0198	1.0195	1.0192	1.0188	1.0184	1.0179	1.0176	1.0173
3.3	1.0224	1.0223	1.0221	1.0216	1.0211	1.0205	1.0202	1.0198	1.0194	1.0189	1.0184	1.0182	1.0179
3.4	1.0230	1.0230	1.0228	1.0223	1.0218	1.0211	1.0208	1.0204	1.0200	1.0195	1.0190	1.0187	1.0184
3.5	1.0237	1.0237	1.0235	1.0229	1.0224	1.0217	1.0214	1.0210	1.0206	1.0201	1.0196	1.0193	1.0190
3.6	1.0244	1.0244	1.0242	1.0236	1.0231	1.0224	1.0220	1.0216	1.0212	1.0207	1.0202	1.0199	1.0196
3.7	1.0251	1.0251	1.0248	1.0243	1.0237	1.0230	1.0226	1.0222	1.0218	1.0213	1.0207	1.0204	1.0201
3.8	1.0258	1.0258	1.0255	1.0249	1.0244	1.0236	1.0233	1.0229	1.0224	1.0219	1.0213	1.0210	1.0207
3.9	1.0265	1.0264	1.0262	1.0256	1.0250	1.0243	1.0239	1.0235	1.0230	1.0225	1.0219	1.0216	1.0213
4.0	1.0272	1.0271	1.0269	1.0263	1.0257	1.0249	1.0245	1.0241	1.0236	1.0231	1.0225	1.0222	1.0218

h_o/P_n Ratio	$\beta = d/D$												
	.63	.64	.65	.66	.67	.68	.69	.70	.71	.72	.73	.74	.75
0.0	1.000	1.000	1.000	1.000	1.000	1.000	1.000	1.000	1.000	1.000	1.000	1.000	1.000
0.1	1.0005	1.0005	1.0005	1.0005	1.0005	1.0004	1.0004	1.0004	1.0004	1.0004	1.0004	1.0004	1.0004
0.2	1.0010	1.0010	1.0010	1.0010	1.0009	1.0009	1.0009	1.0009	1.0008	1.0008	1.0008	1.0008	1.0007
0.3	1.0015	1.0015	1.0015	1.0014	1.0014	1.0014	1.0013	1.0013	1.0013	1.0012	1.0012	1.0011	1.0011
0.4	1.0021	1.0020	1.0020	1.0019	1.0019	1.0018	1.0018	1.0017	1.0017	1.0016	1.0016	1.0015	1.0014
0.5	1.0026	1.0025	1.0025	1.0024	1.0024	1.0023	1.0022	1.0022	1.0021	1.0020	1.0020	1.0019	1.0018
0.6	1.0031	1.0030	1.0030	1.0029	1.0028	1.0028	1.0027	1.0026	1.0025	1.0025	1.0024	1.0023	1.0022
0.7	1.0036	1.0036	1.0035	1.0034	1.0033	1.0032	1.0032	1.0031	1.0030	1.0029	1.0028	1.0027	1.0026
0.8	1.0042	1.0041	1.0040	1.0039	1.0038	1.0037	1.0036	1.0035	1.0034	1.0033	1.0032	1.0030	1.0029
0.9	1.0047	1.0046	1.0045	1.0044	1.0043	1.0042	1.0041	1.0040	1.0038	1.0037	1.0036	1.0034	1.0033
1.0	1.0052	1.0051	1.0050	1.0049	1.0048	1.0047	1.0045	1.0044	1.0043	1.0041	1.0040	1.0038	1.0037
1.1	1.0057	1.0056	1.0055	1.0054	1.0053	1.0051	1.0050	1.0049	1.0047	1.0046	1.0044	1.0042	1.0041
1.2	1.0062	1.0061	1.0060	1.0059	1.0058	1.0056	1.0055	1.0053	1.0052	1.0050	1.0048	1.0046	1.0044
1.3	1.0068	1.0066	1.0065	1.0064	1.0062	1.0061	1.0059	1.0058	1.0056	1.0054	1.0052	1.0050	1.0048
1.4	1.0073	1.0072	1.0070	1.0069	1.0067	1.0066	1.0064	1.0062	1.0060	1.0058	1.0056	1.0054	1.0052
1.5	1.0078	1.0077	1.0076	1.0074	1.0072	1.0070	1.0069	1.0067	1.0065	1.0063	1.0060	1.0058	1.0056
1.6	1.0084	1.0082	1.0081	1.0079	1.0077	1.0075	1.0073	1.0071	1.0069	1.0067	1.0065	1.0062	1.0060
1.7	1.0089	1.0088	1.0086	1.0084	1.0082	1.0080	1.0078	1.0076	1.0074	1.0071	1.0069	1.0066	1.0064
1.8	1.0094	1.0093	1.0091	1.0089	1.0087	1.0085	1.0083	1.0080	1.0078	1.0076	1.0073	1.0070	1.0068
1.9	1.0100	1.0098	1.0096	1.0094	1.0092	1.0090	1.0088	1.0085	1.0083	1.0080	1.0077	1.0074	1.0071
2.0	1.0195	1.0103	1.0101	1.0099	1.0097	1.0095	1.0092	1.0090	1.0087	1.0084	1.0081	1.0078	1.0075
2.1	1.0111	1.0109	1.0106	1.0104	1.0102	1.0100	1.0097	1.0094	1.0092	1.0089	1.0086	1.0083	1.0079
2.2	1.0116	1.0114	1.0112	1.0109	1.0107	1.0104	1.0102	1.0099	1.0096	1.0093	1.0090	1.0087	1.0083
2.3	1.0121	1.0119	1.0117	1.0114	1.0112	1.0109	1.0106	1.0104	1.0101	1.0098	1.0094	1.0091	1.0087
2.4	1.0127	1.0124	1.0122	1.0120	1.0117	1.0114	1.0111	1.0108	1.0105	1.0102	1.0098	1.0095	1.0091
2.5	1.0132	1.0130	1.0127	1.0125	1.0122	1.0119	1.0116	1.0113	1.0110	1.0106	1.0103	1.0099	1.0095
2.6	1.0138	1.0135	1.0133	1.0130	1.0127	1.0124	1.0121	1.0118	1.0114	1.0111	1.0107	1.0103	1.0099
2.7	1.0143	1.0140	1.0138	1.0135	1.0132	1.0129	1.0126	1.0122	1.0119	1.0115	1.0111	1.0107	1.0103
2.8	1.0148	1.0146	1.0143	1.0140	1.0137	1.0134	1.0131	1.0127	1.0124	1.0120	1.0116	1.0112	1.0107
2.9	1.0154	1.0151	1.0148	1.0145	1.0142	1.0139	1.0136	1.0132	1.0128	1.0124	1.0120	1.0116	1.0111
3.0	1.0160	1.0157	1.0154	1.0150	1.0147	1.0144	1.0140	1.0137	1.0133	1.0129	1.0124	1.0120	1.0116

(continued)

Table 6.7.24 *Expansion Factor, Y_1, Flange Taps; Static Pressure Taken from Upstream Taps [12] (Continued)*

h_o/P_n Ratio	$\beta = d/D$												
	.63	.64	.65	.66	.67	.68	.69	.70	.71	.72	.73	.74	.75
3.1	1.0165	1.0162	1.0159	1.0156	1.0152	1.0149	1.0145	1.0141	1.0137	1.0133	1.0129	1.0124	1.0120
3.2	1.0170	1.0167	1.0164	1.0161	1.0158	1.0154	1.0150	1.0146	1.0142	1.0138	1.0133	1.0128	1.0124
3.3	1.0176	1.0173	1.0170	1.0166	1.0163	1.0159	1.0155	1.0151	0.0147	1.0142	1.0138	1.0133	1.0128
3.4	1.0181	1.0178	1.0175	1.0171	1.0168	1.0164	1.0160	1.0156	1.0151	1.0147	1.0142	1.0137	1.0132
3.5	1.0187	1.0184	1.0180	1.0177	1.0173	1.0169	1.0165	1.0160	1.0156	1.0151	1.0146	1.0141	1.0136
3.6	1.0192	1.0189	1.0186	1.0182	1.0178	1.0174	1.0170	1.0165	1.0161	1.0156	1.0151	1.0146	1.0140
3.7	1.0198	1.0195	1.0191	1.0187	1.0183	1.0179	1.0175	1.0170	1.0165	1.0160	1.0155	1.0150	1.0144
3.8	1.0204	1.0200	1.0196	1.0192	1.0188	1.0184	1.0180	1.0175	1.0170	1.0165	1.0160	1.0154	1.0148
3.9	1.0209	1.0206	1.0202	1.0198	1.0194	1.0189	1.0185	1.0180	1.0175	1.0170	1.0164	1.0159	1.0153
4.0	1.0215	1.0211	1.0207	1.0203	1.0199	1.0194	1.0190	1.0185	1.0180	1.0174	1.0169	0.0163	1.0157

Table 6.7.25 *Expansion Factor, Y_1, Pipe Taps; Static Pressure Taken from Upstream Taps [14]*

h_o/P_n Ratio	$\beta = d/D$										
	.1	.2	.3	.4	.45	.50	.52	.54	.56	.58	60
0.0	1.000	1.000	1.000	1.000	1.000	1.000	1.000	1.000	1.000	1.000	1.000
0.1	0.9990	0.9989	0.9988	0.9985	0.9984	0.9982	0.9981	0.9980	0.9979	0.9978	0.9977
0.2	0.9981	0.9979	0.9976	0.9971	0.9968	0.9964	0.9962	0.9961	0.9959	0.9957	0.9954
0.3	0.9971	0.9968	0.9964	0.9956	0.9952	0.9946	0.9944	0.9941	0.9938	0.9935	0.9931
0.4	0.9962	0.9958	0.9951	0.9942	0.9936	0.9928	0.9925	0.9921	0.9917	0.9913	0.9908
0.5	0.9952	0.9947	0.9939	0.9927	0.9919	0.9910	0.9906	0.9902	0.9897	0.9891	0.9885
0.6	0.9943	0.9937	0.9927	0.9913	0.9903	0.9892	0.9887	0.9882	0.9876	0.9870	0.9862
0.7	0.9933	0.9926	0.9915	0.9898	0.9887	0.9874	0.9869	0.9862	0.9856	0.9848	0.9840
0.8	0.9923	0.9916	0.9903	0.9883	0.9871	0.9857	0.9850	0.9843	0.9835	0.9826	0.9817
0.9	0.9914	0.9905	0.9891	0.9869	0.9855	0.9839	0.9831	0.9823	0.9814	0.9805	0.9794
1.0	0.9904	0.9895	0.9878	0.9854	0.9839	0.9821	0.9812	0.9803	0.9794	0.9783	0.9771
1.1	0.9895	0.9884	0.9866	0.9840	0.9823	0.9803	0.9794	0.9784	0.9773	0.9761	0.9748
1.2	0.9885	0.9874	0.9854	0.9825	0.9807	0.9785	0.9775	0.9764	0.9752	0.9739	0.9725
1.3	0.9876	0.9863	0.9842	0.9811	0.9791	0.9767	0.9756	0.9744	0.9732	0.9718	0.9702
1.4	0.9866	0.9853	0.9830	0.9796	0.9775	0.9749	0.9737	0.9725	0.9711	0.9696	0.9679
1.5	0.9857	0.9842	0.9818	0.9782	0.9758	0.9731	0.9719	0.9705	0.9690	0.9674	0.9656
1.6	0.9847	0.9832	0.9805	0.9767	0.9742	0.9713	0.9700	0.9685	0.9670	0.9652	0.9633
1.7	0.9837	0.9821	0.9793	0.9752	0.9726	0.9695	0.9681	0.9666	0.9649	0.9631	0.9610
1.8	0.9828	0.9811	0.9781	0.9738	0.9710	0.9677	0.9662	0.9646	0.9628	0.9609	0.9587
1.9	0.9818	0.9800	0.9769	0.9723	0.9694	0.9659	0.9643	0.9626	0.9608	0.9587	0.9565
2.0	0.9809	0.9790	0.9757	0.9709	0.9678	0.9641	0.9625	0.9607	0.9587	0.9566	0.9542
2.1	0.9799	0.9779	0.9745	0.9694	0.9662	0.9623	0.9606	0.9587	0.9566	0.9544	0.9519
2.2	0.9790	0.9768	0.9732	0.9680	0.9646	0.9605	0.9587	0.9567	0.9546	0.9522	0.9496
2.3	0.9780	0.9758	0.9720	0.9665	0.9630	0.9587	0.9568	0.9548	0.9525	0.9500	0.9473
2.4	0.9770	0.9747	0.9708	0.9650	0.9613	0.9570	0.9550	0.9528	0.9505	0.9479	0.9450
2.5	0.9761	0.9737	0.9696	0.9636	0.9597	0.9552	0.9531	0.9508	0.9484	0.9457	0.9427
2.6	0.9751	0.9726	0.9681	0.9621	0.9581	0.9534	0.9512	0.9489	0.9463	0.9435	0.9404
2.7	0.9742	0.9716	0.9672	0.9607	0.9565	0.9516	0.9493	0.9469	0.9443	0.9414	0.9381
2.8	0.9732	0.9705	0.9659	0.9592	0.9549	0.9498	0.9475	0.9449	0.9422	0.9392	0.9358
2.9	0.9723	0.9695	0.9647	0.9578	0.9533	0.9480	0.9456	0.9430	0.9401	0.9370	0.9335
3.0	0.9713	0.9684	0.9635	0.9563	0.9517	0.9462	0.9437	0.9410	0.9381	0.9348	0.9312
3.1	0.9704	0.9674	0.9623	0.9549	0.9501	0.9444	0.9418	0.9390	0.9360	0.9327	0.9290
3.2	0.9694	0.9663	0.9611	0.9534	0.9485	0.9426	0.9400	0.9371	0.9339	0.9305	0.9267
3.3	0.9684	0.9653	0.9599	0.9519	0.9469	0.9408	0.9381	0.9351	0.9319	0.9283	0.9244
3.4	0.9675	0.9642	0.9587	0.9505	0.9452	0.9390	0.9362	0.9331	0.9298	0.9261	0.9221
3.5	0.9665	0.9632	0.9574	0.9490	0.9436	0.9372	0.9343	0.9312	0.9277	0.9240	0.9198
3.6	0.9656	0.9621	0.9562	0.9476	0.9420	0.9354	0.9324	0.9292	0.9257	0.9218	0.9175
3.7	0.9646	0.9611	0.9550	0.9461	0.9404	0.9336	0.9306	0.9272	0.9236	0.9196	0.9152
3.8	0.9637	0.9600	0.9538	0.9447	0.9388	0.9318	0.9287	0.9253	0.9216	0.9175	0.9129
3.9	0.9627	0.9590	0.9526	0.9432	0.9372	0.9301	0.9263	0.9233	0.9195	0.9153	0.9106
4.0	0.9617	0.9579	0.9514	0.9417	0.9356	0.9283	0.9249	0.9213	0.9174	0.9131	0.9083

(continued)

Table 6.7.25 Expansion Factor, Y_1, Pipe Taps; Static Pressure Taken from Upstream Taps [14] (Continued)

h_o/P_n Ratio	$\beta = d/D$									
	.61	.62	.63	.64	.65	.66	.67	.68	.69	.70
0.0	1.000	1.000	1.000	1.000	1.000	1.000	1.000	1.000	1.000	1.000
0.1	0.9976	0.9976	0.9975	0.9974	0.9973	0.9972	0.9971	0.9970	0.9969	0.9968
0.2	0.9953	0.9951	0.9950	0.9948	0.9947	0.9945	0.9943	0.9941	0.9938	0.9935
0.3	0.9929	0.9927	0.9925	0.9923	0.9920	0.9917	0.9914	0.9911	0.9907	0.9903
0.4	0.9906	0.9903	0.9900	0.9897	0.9893	0.9890	0.9986	0.9881	0.9876	0.9871
0.5	0.9882	0.9879	0.9875	0.9871	0.9867	0.9862	0.9857	0.9851	0.9845	0.9389
0.6	0.9859	0.9854	0.9850	0.9845	0.9840	0.9834	0.9828	0.9822	0.9814	0.9806
0.7	0.9835	0.9830	0.9825	0.9819	0.9813	0.9807	0.9800	0.9792	0.9784	0.9774
0.8	0.9811	0.9806	0.9800	0.9794	0.9787	0.9779	0.9771	0.9762	0.9753	0.9742
0.9	0.9788	0.9782	0.9775	0.9768	0.9760	0.9752	0.9742	0.9733	0.9722	0.9710
1.0	0.9764	0.9757	0.9750	0.9742	0.9733	0.9724	0.9714	0.9703	0.9691	0.9677
1.1	0.9741	0.9733	0.9725	0.9716	0.9707	0.9696	0.9685	0.9673	0.9660	0.9645
1.2	0.9717	0.9709	0.9700	0.9690	0.9680	0.9669	0.9657	0.9643	0.9629	0.9613
1.3	0.9694	0.9685	0.9675	0.9664	0.9653	0.9641	0.9628	0.9614	0.9598	0.9581
1.4	0.9670	0.9660	0.9650	0.9639	0.9627	0.9614	0.9599	0.9584	0.9567	0.9548
1.5	0.9646	0.9636	0.9625	0.9613	0.9600	0.9586	0.9571	0.9554	0.9536	0.9516
1.6	0.9623	0.9612	0.9600	0.9587	0.9573	0.9558	0.9542	0.9525	0.9505	0.9484
1.7	0.9599	0.9587	0.9575	0.9561	0.9547	0.9531	0.9514	0.9495	0.9474	0.9452
1.8	0.9576	0.9563	0.9550	0.9535	0.9520	0.9503	0.9485	0.9465	0.9443	0.9419
1.9	0.9552	0.9539	0.9525	0.9510	0.9493	0.9476	0.9456	0.9435	0.9412	0.9387
2.0	0.9529	0.9515	0.9500	0.9484	0.9467	0.9448	0.9428	0.9406	0.9381	0.9355
2.1	0.9505	0.9490	0.9475	0.9458	0.9440	0.9420	0.9399	0.9376	0.9351	0.9323
2.2	0.9481	0.9466	0.9450	0.9432	0.9413	0.9393	0.9371	0.9346	0.9320	0.9290
2.3	0.9458	0.9442	0.9425	0.9406	0.9387	0.9365	0.9342	0.9317	0.9289	0.9258
2.4	0.9434	0.9418	0.9400	0.9381	0.9360	0.9338	0.9313	0.9287	0.9258	0.9226
2.5	0.9411	0.9393	0.9375	0.9355	0.9333	0.9310	0.9285	0.9257	0.9227	0.9194
2.6	0.9387	0.9369	0.9350	0.9329	0.9307	0.9282	0.9256	0.9227	0.9196	0.9161
2.7	0.9364	0.9345	0.9325	0.9303	0.9280	0.9255	0.9227	0.9198	0.9165	0.9129
2.8	0.9340	0.9321	0.9300	0.9277	0.9253	0.9227	0.9199	0.9168	0.9134	0.9097
2.9	0.9316	0.9296	0.9275	0.9252	0.9227	0.9200	0.9170	0.9138	0.9103	0.9064
3.0	0.9293	0.9272	0.9250	0.9226	0.9200	0.9172	0.9142	0.9108	0.9072	0.9032
3.1	0.9269	0.9248	0.9225	0.9200	0.9173	0.9144	0.9113	0.9079	0.9041	0.9000
3.2	0.9246	0.9223	0.9200	0.9174	0.9147	0.9117	0.9084	0.9049	0.9010	0.8968
3.3	0.9222	0.9199	0.9175	0.9148	0.9120	0.9089	0.9056	0.9019	0.8979	0.8935
3.4	0.9199	0.9175	0.9150	0.9122	0.9093	0.9062	0.9027	0.8990	0.8948	0.8903
3.5	0.9175	0.9151	0.9125	0.9097	0.9067	0.9034	0.8999	0.8960	0.8918	0.8817
3.6	0.9151	0.9126	0.9100	0.9071	0.9040	0.9006	0.8970	0.8930	0.8887	0.8839
3.7	0.9128	0.9102	0.9075	0.9045	0.9013	0.8979	0.8941	0.8900	0.8856	0.8806
3.8	0.9104	0.9078	0.9050	0.9019	0.8987	0.8951	0.8913	0.8871	0.8825	0.8774
3.9	0.9081	0.9054	0.9025	0.8993	0.8960	0.8924	0.8884	0.8841	0.8794	0.8742
4.0	0.9057	0.9029	0.9000	0.8968	0.8933	0.8896	0.8856	0.8811	0.8763	0.8710

Table 6.7.26 *Expansion Factor, Y_2, Flange Taps; Static Pressure Taken from Downstream Taps [14]*

Pipe Sizes, Nominal and Published Inside Diameters, in inches

Orifice Diameter Inches	2			3			4		
	1.687	1.939	2.067	2.300	2.624	2.900	3.068	3.152	3.438
0.250	12.850	12.813	12.800	12.782	22.765	12.754	12.748	12.745	12.737
0.375	29.362	29.098	29.006	28.883	28.772	28.711	28.682	28.670	28.635
0.500	53.713	52.817	52.482	52.020	51.594	51.354	51.244	51.197	51.065
0.625	87.237	84.920	84.085	82.924	81.802	81.143	80.837	80.704	80.334
0.750	132.29	126.87	124.99	122.45	120.08	118.67	118.00	117.70	116.87
0.875	192.87	181.02	177.09	171.93	167.26	164.58	163.31	162.76	161.17
1.000	275.73	251.11	243.28	233.30	224.61	219.77	217.53	216.55	213.79
1.125	392.50	342.99	327.99	309.44	293.87	285.49	281.67	280.03	275.43
1.250		466.00	438.00	404.53	377.50	363.41	357.13	354.45	347.04
1.375			583.98	524.69	478.89	455.83	445.75	441.49	429.84
1.500				679.11	602.80	565.80	549.95	543.32	525.41
1.625					755.89	697.45	672.96	662.83	635.77
1.750					947.87	856.39	819.07	803.79	763.53
1.875						1050.4	994.01	971.22	912.00
2.000						1290.7	1205.6	1171.9	1085.5
2.125							1465.1	1415.0	1289.7
2.250									1532.0
2.375									1822.9

Pipe Sizes, Nominal and Published Inside Diameters, in inches

Orifice Diameter Inches	4		6				8		
	3.826	4.026	4.897	5.187	5.761	6.065	7.625	7.981	8.071
0.250	12.727	12.723							
0.375	26.599	28.585							
0.500	50.937	50.887	50.740	50.707	50.653	50.629			
0.625	79.976	79.837	79.438	79.351	79.219	79.164			
0.750	116.05	115.73	114.81	114.62	114.32	114.20			
0.875	159.58	158.94	157.11	156.72	156.13	155.89	155.11	154.99	154.97
1.000	211.03	209.92	206.63	205.92	204.85	204.41	203.01	202.80	202.76
1.125	270.91	269.10	263.71	262.52	260.72	259.99	257.62	257.28	257.20
1.250	339.88	337.06	328.73	326.87	324.03	322.87	319.10	318.57	318.44
1.375	418.80	414.51	402.07	399.32	395.09	393.34	387.63	386.81	386.63
1.500	508.77	502.39	484.21	480.26	474.21	471.70	463.40	462.20	461.93
1.625	611.12	601.81	575.75	570.19	561.74	558.25	546.62	544.93	544.54
1.750	727.55	714.17	677.39	669.69	658.09	653.34	637.52	635.20	634.67
1.875	860.19	841.21	790.00	779.49	763.79	757.41	736.36	733.25	732.53
2.000	1011.7	985.07	914.59	900.39	879.40	870.95	843.36	839.31	838.37
2.125	1185.4	1148.4	1052.3	1033.4	1005.6	994.54	958.80	953.61	952.40
2.250	1385.4	1334.5	1204.7	1179.6	1143.2	1128.9	1063.0	1076.4	1074.9
2.375	1617.2	1547.4	1373.4	1340.5	1293.2	1274.6	1216.3	1206.0	1206.1
2.500	1887.7	1792.3	1560.5	1517.5	1456.5	1432.8	1359.2	1348.9	1346.5
2.625	2206.1	2076.0	1768.3	1712.6	1634.4	1604.3	1512.1	1499.3	1496.4
2.750		2407.0	1999.9	1928.1	1828.3	1790.4	1675.4	1659.7	1656.1
2.875			2258.6	2166.5	2040.0	1992.3	1849.9	1830.7	1826.3
3.000			2548.6	2431.0	2271.2	2211.6	2036.1	2012.7	2007.4
3.125			2875.3	2725.3	2524.3	2450.2	2234.7	2206.4	2200.0
3.250			3244.9	3054.0	2801.9	2710.0	2446.6	2412.5	2404.8
3.375			3665.7	3422.4	3106.9	2993.4	2672.6	2631.7	2622.4
3.500				3837.6	3443.1	3303.1	2913.7	2864.8	2853.7
3.625				4308.1	3814.5	3642.4	3171.2	3112.8	3099.6
3.750					4226.4	4014.9	3446.1	3376.7	3361.1
3.875					4685.0	4425.2	3739.9	3657.7	3639.3
4.000					5197.9	4878.5	4054.3	3957.1	3935.3
4.250							4751.6	4616.7	4586.7
4.500							5554.8	5369.1	5328.1
4.750							6485.5	6231.3	6175.4
5.000							7571.6	7224.5	7148.8
5.250							8850.5	8376.6	8274.2
5.500								9724.0	9565.4

(continued)

Table 6.7.26 *Expansion Factor, Y_2, Flange Taps; Static Pressure Taken from Downstream Taps [14] (Continued)*

Orifice Diameter Inches	Pipe Sizes, Nominal and Published Inside Diameters, in inches								
	10			12			16		
	9.562	10.020	10.136	11.374	11.938	12.090	14.688	15.000	15.250
1.000	202.16								
1.125	256.23	256.01	255.96						
1.250	316.90	316.57	316.49	315.82	315.57	315.51			
1.375	384.30	383.80	383.68	382.67	382.31	382.22			
1.500	458.53	457.80	457.64	456.17	455.65	455.53	453.93	453.79	
1.625	539.73	538.70	538.47	536.40	535.67	535.49	533.28	533.09	532.94
1.750	628.05	626.63	626.30	623.46	622.46	622.22	619.20	619.94	618.74
1.875	723.63	721.72	721.28	717.45	716.12	715.79	711.75	711.41	711.14
2.000	826.66	824.14	823.56	818.51	816.75	816.32	811.01	810.55	810.21
2.125	937.31	934.04	933.29	926.74	924.46	923.90	917.03	916.45	916.01
2.250	1055.8	1051.6	1050.7	1042.3	1039.4	1038.7	1029.9	1029.2	1028.6
2.375	1182.3	1177.0	1175.8	1165.3	1161.7	1160.8	1149.7	1148.8	1148.1
2.500	1316.9	1310.5	1309.0	1296.0	1291.4	1290.3	1276.6	1275.4	1274.5
2.625	1460.1	1452.2	1450.4	1434.4	1428.8	1427.4	1410.5	1409.1	1408.0
2.750	1611.8	1602.3	1600.1	1580.7	1573.9	1572.2	1551.7	1550.0	1548.6
2.875	1772.6	1761.1	1758.4	1735.2	1727.0	1725.0	1700.2	1698.1	1696.5
3.000	1942.6	1928.8	1925.6	1897.9	1888.2	1885.8	1856.1	1853.6	1851.7
3.125	2122.2	2105.8	2102.0	2069.1	2057.6	2654.7	2019.6	2016.6	2014.4
3.250	2311.8	2292.3	2287.8	2249.0	2235.4	2232.1	2190.7	2187.2	2184.6
3.375	2511.7	2488.7	2483.4	2437.8	2421.9	2418.0	2369.7	2365.6	2362.5
3.500	2722.5	2695.4	2689.2	2635.8	2617.2	2612.7	2556.5	2551.7	2548.1
3.625	2944.5	2912.8	2905.5	2843.2	2821.6	2816.4	2751.5	2745.9	2741.7
3.750	3178.4	3141.3	3132.8	3060.3	3035.4	3029.3	2954.6	2948.2	2943.4
3.875	3424.6	3381.4	3371.5	3287.5	3258.8	3251.8	3166.0	3158.7	3153.2
4.000	3683.9	3633.6	3622.2	3525.2	3492.1	3484.0	3385.8	3377.5	3371.3
4.250	4244.2	4176.9	4161.7	4033.1	3989.6	3979.1	3851.7	3841.0	3832.9
4.500	4865.6	4776.3	4756.2	4587.4	4530.9	4517.3	4353.5	4339.9	4329.7
4.750	5555.6	5438.0	5411.6	5191.9	5119.1	5101.6	4893.0	4875.9	4863.0
5.000	6323.1	6169.4	6135.0	5851.1	5757.9	5735.6	5472.1	5450.6	5434.5
5.250	7178.8	6979.1	6934.6	6570.1	6451.7	6423.4	6092.7	6066.0	6046.0
5.500	8135.5	7877.4	7820.2	7354.9	7205.3	7169.7	6757.2	6724.3	6699.6
5.750	9208.8	8876.6	8803.3	8212.4	8024.4	7979.8	7468.2	7427.8	7397.6
6.000	10418	9991.5	9898.0	9150.7	8915.6	8860.0	8228.7	8179.4	8142.5
6.250	11786	11240	11121	10179	9886.4	9817.5	9041.9	8981.9	8937.2
6.500	13344	12644	12493	11309	10946	10860	9911.4	9836.9	9784.9
6.750		14231	14038	12552	12103	11998	10842	10754	10689
7.000		16035	15790	13925	13371	13242	11837	11732	11654
7.250				15446	14763	14605	12902	12777	12684
7.500				17135	16295	16102	14044	13894	13784
7.750				19021	17986	17750	15269	15091	14959
8.000					19861	19572	16583	16372	16216
8.250					21948	21594	17996	17746	17562
8.500							19517	19221	19004
8.750							21157	20807	20551
9.000							22927	22515	22214
9.250							24842	24357	24004
9.500							26917	26347	25932
9.750							29173	28502	28015
10.000							31630	30840	30269
10.250							34316	33384	32714
10.500								36161	35373

(continued)

Table 6.7.26   *Expansion Factor, Y₂, Flange Taps; Static Pressure Taken from Downstream Taps [14] (Continued)*

Orifice Diameter Inches	Pipe Sizes, Nominal and Published Inside Diameters, in inches								
	20			24			30		
	18.812	19.000	19.250	22.624	23.000	23.250	28.750	29.000	29.250
2.000	806.73	806.59	806.42						
2.125	911.54	911.37	911.15						
2.250	1022.9	1022.7	1022.5						
2.375	1141.0	1140.7	1140.4	1136.8	1136.5	1136.3			
2.500	1265.7	1265.4	1265.0	1260.6	1260.2	1260.0			
2.625	1397.2	1396.8	1396.3	1390.9	1390.5	1390.2			
2.750	1535.5	1535.0	1534.4	1527.9	1527.4	1527.0			
2.875	1680.7	1680.1	1679.4	1671.6	1670.9	1670.5	1663.7		
3.000	1832.8	1832.1	1831.2	1821.9	1821.1	1820.6	1821.6	1812.3	1812.1
3.125	1991.9	1991.1	1990.0	1979.0	1978.1	1977.5	1968.0	1967.7	1967.4
3.250	2158.1	2157.1	2155.8	2142.9	2141.8	2141.1	2130.0	2129.7	2129.3
3.375	2331.4	2330.3	2328.8	2313.6	2312.3	2311.5	2298.6	2298.2	2297.8
3.500	2512.0	2510.6	2508.9	2491.2	2489.8	2488.8	2473.9	2473.4	2472.9
3.625	2699.8	2698.3	2696.3	2675.8	2674.1	2673.0	2655.8	2655.2	2654.7
3.750	2895.1	2893.3	2891.0	2867.4	2865.5	2864.2	2844.4	2843.7	2843.1
3.875	3097.8	3095.7	3093.1	3066.1	3063.8	3062.4	3039.7	3039.0	3038.3
4.000	3308.1	3305.7	3302.7	3271.9	3269.3	3267.7	3241.8	3241.0	3240.2
4.250	3751.8	3748.8	3744.9	3705.1	3701.8	3699.7	3666.5	3665.4	3664.4
4.500	4226.9	4223.1	4218.2	4167.7	4163.5	4160.8	4118.7	4117.4	4116.1
4.750	4734.3	4729.5	4723.4	4660.2	4654.9	4651.5	4598.9	4597.2	4595.5
5.000	5274.7	5268.8	5261.3	5183.1	5176.6	5172.4	5107.3	5105.2	5103.1
5.250	5849.2	5842.0	5832.8	5737.2	5729.2	5724.1	5644.3	5641.8	5639.3
5.500	6458.8	6150.1	6438.8	6323.1	6313.4	6307.2	6210.4	6207.3	6204.3
5.750	7104.7	7094.2	7080.6	6941.5	6929.8	6922.4	6806.0	6802.3	6798.7
6.000	7788.2	7775.5	7759.3	7593.1	7579.2	7570.3	7431.5	7427.1	7422.8
6.250	8510.8	8495.6	8476.2	8278.6	8262.2	8251.7	8087.5	8082.2	8077.1
6.500	9273.8	9255.8	9232.7	8999.1	8979.7	8967.3	8774.3	8768.1	8762.1
6.750	10079	10058	10031	9755.4	9732.6	9718.1	9492.6	9485.4	9478.4
7.000	10929	10903	10871	10548	10522	10505	10243	10234	10226
7.250	11824	11795	11757	11380	11349	11329	11026	11016	11007
7.500	12768	12733	12689	12250	12214	12191	11841	11830	11819
7.750	13763	13722	13671	13161	13119	13093	12691	12678	12666
8.000	14811	14764	14704	14113	14065	14036	13575	13560	13546
8.250	15915	15861	15791	15109	15055	15020	14494	14477	14461
8.500	17079	17016	16935	16150	16088	16048	15448	15429	15411
8.750	18306	18233	18140	17238	17166	17121	16439	16418	16398
9.000	19600	19515	19408	18374	18293	18241	17468	17444	17421
9.250	20964	20867	20744	19561	19468	19410	18535	18508	18482
9.500	22404	22293	22151	20801	20695	20629	19642	19612	19582
9.750	23925	23797	23635	22096	21976	21901	20789	20755	20722
10.000	25531	25384	25199	23448	23313	23228	21977	21939	21902
10.250	27229	27062	26850	24861	24708	24612	23208	23165	23124
10.500	29026	28834	28593	26337	26165	26057	24482	24435	24389
10.750	30928	30710	30435	27879	27686	27564	25802	25749	25698
11.000	32944	32695	32382	29492	29274	29137	27168	27109	27053
11.250	35082	34799	34444	31177	30933	30780	28582	28517	28454
11.500	37353	37031	36627	32941	32667	32495	30045	29973	29904
11.750	39766	39401	38942	34786	34479	34286	31559	31480	31403
12.000	42336	41921	41400	36717	36374	36158	33126	33038	32953

(continued)

Table 6.7.26 *Expansion Factor, Y_2, Flange Taps; Static Pressure Taken from Downstream Taps [14] (Continued)*

| Orifice Diameter Inches | Pipe Sizes, Nominal and Published Inside Diameters, in inches | | | | | | | | |
| | 20 | | | 24 | | | 30 | | |
	18.812	19.000	19.250	22.624	23.000	23.250	28.750	29.000	29.250
12.500	47998	47462	46791	40859	40430	40162	36426	36319	36216
13.000	54472	53779	52915	45410	44878	44545	39960	39830	39705
13.500				50425	49765	49353	43746	43590	43438
14.000				55965	55148	54640	47805	47617	47435
14.500				62106	61096	60469	52159	51933	51715
15.000				68938	67689	66917	56833	56563	56303
15.500				76572	75027	74075	61857	61535	61225
16.000					83233	82057	67263	66879	66511
16.500							73087	72632	72195
17.000							79372	78833	78315
17.500							86165	85527	84915
18.000							93522	92767	92044
18.500							101506	100614	99761
19.000							110192	109137	108130
19.500							119667	118420	117231
20.000							130036	128559	127153

Table 6.7.27 *Expansion Factor, Y_2, Pipe Taps; Static Pressure Taken from Downstream Taps [14]*

| h_o/P_n Ratio | $\beta = d/D$ | | | | | | | | | | |
	.1	.2	.3	.4	.45	.50	.52	.54	.56	.58	60
0.0	1.000	1.000	1.000	1.000	1.000	1.000	1.000	1.000	1.000	1.000	1.000
0.1	1.0008	1.0008	1.0006	1.0003	1.0002	1.0000	0.9999	0.9998	0.9997	0.9996	0.9995
0.2	1.0017	1.0015	1.0012	1.0007	1.0004	1.0000	0.9999	0.9997	0.9995	0.9993	0.9990
0.3	1.0025	1.0023	1.0018	1.0010	1.0006	1.0000	0.9998	0.9995	0.9992	0.9989	0.9986
0.4	1.0034	1.0030	1.0024	1.0014	1.0008	1.0001	0.9997	0.9994	0.9990	0.9986	0.9981
0.5	1.0042	1.0038	1.0030	1.0018	1.0010	1.0001	0.9997	0.9992	0.9988	0.9982	0.9976
0.6	1.0051	1.0045	1.0036	1.0021	1.0012	1.0001	0.9996	0.9991	0.9985	0.9979	0.9972
0.7	1.0059	1.0053	1.0041	1.0025	1.0014	1.0002	0.9996	0.9990	0.9983	0.9975	0.9967
0.8	1.0068	1.0060	1.0047	1.0028	1.0016	1.0002	0.9995	0.9988	0.9980	0.9972	0.9962
0.9	1.0076	1.0068	1.0053	1.0032	1.0018	1.0002	0.9995	0.9987	0.9978	0.9969	0.9958
1.0	1.0085	1.0075	1.0059	1.0036	1.0021	1.0003	0.9994	0.9986	0.9976	0.9965	0.9954
1.1	1.0093	1.0083	1.0065	1.0039	1.0023	1.0003	0.9994	0.9984	0.9974	0.9962	0.9949
1.2	1.0102	1.0091	1.0071	1.0043	1.0025	1.0004	0.9994	0.9983	0.9972	0.9959	0.9945
1.3	1.0110	1.0098	1.0077	1.0047	1.0027	1.0004	0.9994	0.9982	0.9970	0.9956	0.9941
1.4	1.0119	1.0106	1.0083	1.0051	1.0030	1.0004	0.9993	0.9981	0.9968	0.9953	0.9936
1.5	1.0127	1.0113	1.0089	1.0054	1.0032	1.0005	0.9993	0.9980	0.9966	0.9950	0.9932
1.6	1.0136	1.0121	1.0096	1.0058	1.0034	1.0006	0.9993	0.9979	0.9964	0.9947	0.9928
1.7	1.0144	1.0128	1.0102	1.0062	1.0036	1.0006	0.9992	0.9978	0.9962	0.9944	0.9924
1.8	1.0153	1.0136	1.0108	1.0066	1.0039	1.0007	0.9992	0.9977	0.9960	0.9941	0.9920
1.9	1.0161	1.0144	1.0114	1.0070	1.0041	1.0008	0.9992	0.9976	0.9958	0.9938	0.9916
2.0	1.0170	1.0151	1.0120	1.0073	1.0044	1.0008	0.9992	0.9975	0.9956	0.9935	0.9912
2.1	1.0178	1.0159	0.0126	1.0077	1.0046	1.0009	0.9992	0.9974	0.9954	0.9932	0.9908
2.2	1.0187	1.0167	1.0132	1.0081	1.0048	1.0010	0.9992	0.9973	0.9952	0.9929	0.9904
2.3	1.0195	1.0174	1.0138	1.0085	1.0051	1.0010	0.9992	0.9972	0.9950	0.9927	0.9900
2.4	1.0204	1.0182	1.0144	1.0089	1.0053	1.0011	0.9992	0.9971	0.9949	0.9924	0.9896
2.5	1.0212	1.0189	1.0150	1.0093	1.0056	1.0012	0.9992	0.9971	0.9947	0.9921	0.9893
2.6	1.0221	1.0197	1.0156	1.0097	1.0058	1.0013	0.9992	0.9970	0.9945	0.9919	0.9889
2.7	1.0229	1.0205	1.0162	1.0101	1.0061	1.0014	0.9992	0.9969	0.9944	0.9916	0.9885
2.8	1.0238	1.0212	1.0169	1.0104	1.0063	1.0014	0.9992	0.9968	0.9942	0.9914	0.9882
2.9	1.0246	1.0220	1.0175	1.0108	1.0066	1.0015	0.9992	0.9968	0.9941	0.9911	0.9878
3.0	1.0255	1.0228	1.0181	1.0112	1.0068	1.0016	0.9993	0.9967	0.9939	0.9908	0.9874

(continued)

Table 6.7.27 *Expansion Factor, Y_2, Pipe Taps; Static Pressure Taken from Downstream Taps [14] (Continued)*

h_o/P_n Ratio	\multicolumn{11}{c}{$\beta = d/D$}
	.1	.2	.3	.4	.45	.50	.52	.54	.56	.58	60
3.1	1.0264	1.0235	1.0187	1.0116	1.0071	1.0017	0.9993	0.9966	0.9938	0.9906	0.9871
3.2	1.0272	1.0243	1.0193	1.0120	1.0074	1.0018	0.9993	0.9966	0.9936	0.9904	0.9867
3.3	1.0280	1.0250	1.0199	1.0124	1.0076	1.0019	0.9993	0.9965	0.9935	0.9901	0.9864
3.4	1.0289	1.0258	1.0206	1.0128	1.0079	1.0020	0.9994	0.9965	0.9933	0.9899	0.9860
3.5	1.0298	1.0266	1.0212	1.0133	1.0082	1.0021	0.9994	0.9964	0.9932	0.9896	0.9857
3.6	1.0306	1.0273	1.0218	1.0137	1.0084	1.0022	0.9994	0.9964	0.9931	0.9894	0.9854
3.7	1.0314	1.0281	1.0224	1.0141	1.0087	1.0024	0.9994	0.9963	0.9929	0.9892	0.9850
3.8	1.0323	1.0289	1.0230	1.0145	1.0090	1.0025	0.9995	0.9963	0.9928	0.9890	0.9847
3.9	1.0332	1.0296	1.0237	1.0149	1.0093	1.0026	0.9995	0.9963	0.9927	0.9888	0.9844
4.0	1.0340	1.0304	1.0243	1.0153	1.0095	1.0027	0.9996	0.9962	0.9926	0.9885	0.9840

h_o/P_n Ratio	\multicolumn{10}{c}{$\beta = d/D$}
	.61	.62	.63	.64	.65	.66	.67	.68	.69	.70
0.0	1.000	1.000	1.000	1.000	1.000	1.000	1.000	1.000	1.000	1.000
0.1	0.9994	0.9994	0.9993	0.9992	0.9991	0.9990	0.9989	0.9988	0.9987	0.9986
0.2	0.9989	0.9988	0.9986	0.9985	0.9983	0.9981	0.9979	0.9977	0.9974	0.9972
0.3	0.9984	0.9982	0.9979	0.9977	0.9974	0.9972	0.9969	0.9965	0.9962	0.9958
0.4	0.9978	0.9976	0.9972	0.9969	0.9966	0.9962	0.9958	0.9954	0.9949	0.9944
0.5	0.9973	0.9970	0.9966	0.9962	0.9958	0.9953	0.9948	0.9942	0.9936	0.9930
0.6	0.9968	0.9964	0.9959	0.9954	0.9949	0.9944	0.9938	0.9931	0.9924	0.9916
0.7	0.9962	0.9958	0.9953	0.9947	0.9941	0.9935	0.9928	0.9920	0.9912	0.9902
0.8	0.9957	0.9952	0.9946	0.9940	0.9933	0.9926	0.9918	0.9909	0.9899	0.9889
0.9	0.9952	0.9946	0.9940	0.9932	0.9925	0.9917	0.9908	0.9898	0.9887	0.9875
1.0	0.9947	0.9940	0.9933	0.9925	0.9917	0.9908	0.9898	0.9887	0.9875	0.9862
1.1	0.9942	0.9935	0.9927	0.9918	0.9909	0.9899	0.9888	0.9876	0.9863	0.9848
1.2	0.9937	0.9929	0.9920	0.9911	0.9901	0.9890	0.9878	0.9865	0.9851	0.9835
1.3	0.9932	0.9924	0.9914	0.9904	0.9893	0.9881	0.9868	0.9854	0.9839	0.9822
1.4	0.9928	0.9918	0.9908	0.9897	0.9885	0.9872	0.9859	0.9844	0.9827	0.9809
1.5	0.9923	0.9912	0.9902	0.9890	0.9877	0.9864	0.9849	0.9833	0.9815	0.9796
1.6	0.9918	0.9907	0.9896	0.9883	0.9870	0.9855	0.9840	0.9822	0.9804	0.9783
1.7	0.9913	0.9902	0.9889	0.9876	0.9862	0.9847	0.9830	0.9812	0.9792	0.9770
1.8	0.9908	0.9896	0.9883	0.9870	0.9854	0.9838	0.9821	0.9801	0.9780	0.9757
1.9	0.9904	0.9891	0.9877	0.9863	0.9847	0.9830	0.9811	0.9791	0.9769	0.9744
2.0	0.9899	0.9886	0.9872	0.9856	0.9840	0.9822	0.9802	0.9781	0.9757	0.9732
2.1	0.9895	0.9881	0.9866	0.9849	0.9832	0.9813	0.9793	0.9770	0.9746	0.9719
2.2	0.9890	0.9876	0.9860	0.9843	0.9825	0.9805	0.9784	0.9760	0.9734	0.9606
2.3	0.9886	0.9870	0.9854	0.9836	0.9817	0.9797	0.9774	0.9750	0.9723	0.9694
2.4	0.9881	0.9865	0.9848	0.9830	0.9810	0.9789	0.9765	0.9740	0.9712	0.9681
2.5	0.9877	0.9860	0.9842	0.9823	0.9803	0.9780	0.9756	0.9730	0.9701	0.9669
2.6	0.9873	0.9855	0.9837	0.9817	0.9796	0.9772	0.9747	0.9720	0.9690	0.9657
2.7	0.9868	0.9850	0.9831	0.9811	0.9788	0.9764	0.9738	0.9710	0.9670	0.9644
2.8	0.9864	0.9846	0.9826	0.9804	0.9781	0.9757	0.9730	0.9700	0.9668	0.9632
2.9	0.9860	0.9841	0.9820	0.9798	0.9774	0.9749	0.9721	0.9690	0.9657	0.9620
3.0	0.9856	0.9836	0.9815	0.9792	0.9767	0.9741	0.9712	0.9681	0.9646	0.9608
3.1	0.9852	0.9831	0.9809	0.9786	0.9760	0.9733	0.9703	0.9671	0.9635	0.9596
3.2	0.9848	0.9826	0.9804	0.9780	0.9754	0.9725	0.9695	0.9661	0.9625	0.9584
3.3	0.9843	0.9822	0.9798	0.9774	0.9747	0.9718	0.9686	0.9652	0.9614	0.9572
3.4	0.9839	0.9817	0.9793	0.9768	0.9740	0.9710	0.9678	0.9642	0.9603	0.9561
3.5	0.9835	0.9812	0.9788	0.9762	0.9733	0.9702	0.9669	0.9633	0.9593	0.9549
3.6	0.9832	0.9808	0.9783	0.9756	0.9727	0.9695	0.9661	0.9623	0.9582	0.9537
3.7	0.9828	0.9803	0.9778	0.9750	0.9720	0.9688	0.9652	0.9614	0.9572	0.9526
3.8	0.9824	0.9799	0.9772	0.9744	0.9713	0.9680	0.9644	0.9605	0.9562	0.9514
3.9	0.9820	0.9794	0.9767	0.9738	0.9707	0.9673	0.9636	0.9596	0.9551	0.9503
4.0	0.9816	0.9790	0.9762	0.9732	0.9700	0.9665	0.9628	0.9586	0.9541	0.9491

Table 6.7.28 *Expansion Factors: Flange Taps, Y_m; Static Pressure, Mean of Upstream and Downstream*

$$F_d = \frac{d_o}{d_i}$$

$\dfrac{h_w}{p_{fm}}$ Ratio	0.1	0.2	0.3	0.4	0.45	0.50	0.52	0.54	0.56	0.58	0.60	0.61	0.62
0.0	1.0000	1.0000	1.0000	1.0000	1.0000	1.0000	1.0000	1.0000	1 0000	1.0000	1.0000	1.0000	1.0060
0.1	0.9998	0.9998	0.9998	0.9997	0.9997	0.9997	0.9997	0.9997	0.9997	0.9997	0.9996	0.9996	0.9996
0.2	0.9995	0.9995	0.9995	0.9995	0.9994	0.9994	0.9994	0.9994	0.9993	0.9993	0.9993	0.9993	0.9992
0.3	0.9993	0.9993	0.9993	0.9992	0.9992	0.9991	0.9991	0.9990	0.9990	0.9990	0.9989	0.9989	0.9989
0.4	0.9991	0.9991	0.9990	0.9990	0.9989	0.9988	0.9988	0.9987	0.9987	0.9986	0.9986	0.9985	0.9985
0.5	0.9988	0.9988	0.9988	0.9987	0.9986	0.9985	0.9985	0.9984	0.9983	0.9983	0.9982	0.9981	0.9981
0.6	0.9986	0.9986	0.9985	0.9984	0.9984	0.9982	0.9982	0.9981	0.9980	0.9979	0.9978	0.9978	0 9977
0.7	0.9984	0.9984	0.9983	0.9982	0.9981	0.9979	0.9979	0.9978	0.9977	0.9976	0.9975	0.9974	0.9973
0.8	0.9981	0.9981	0.9981	0.9979	0.9978	0.9976	0.9976	0.9975	0.9974	0.9972	0.9971	0.9970	00.970
0.9	0.9979	0.9979	0.9978	0.9977	0.9975	0.9973	0.9973	0.9972	0.9971	0.9969	0.9968	0.9967	0.9967
1.0	0.9977	0.9977	0.9976	0.9975	0.9973	0.9971	0.9970	0.9969	0.9968	0.9967	0.9965	0.9964	0.9963
1.1	0.9975	0.9974	0.9974	0.9972	0.9971	0.9968	0.9967	0.9966	0.9965	0.9963	0.9961	0.9960	0.9959
1.2	0.9973	0.9973	0.9972	0.9970	0.9968	0.9966	0.9964	0.9963	0.9961	0.9960	0.9958	0.9957	0.9955
1.3	0.9971	0.9970	0.9970	0.9967	0.9965	0.9963	0.9961	0.9960	0.9958	0.9956	0.9954	0.9953	0.9952
1.4	0.9968	0.9968	0.9967	0.9965	0.9963	0.9960	0.9958	0.9957	0.9955	0.9953	0.9951	0.9949	0.9948
1.5	0.9966	0.9966	0.9965	0.9962	0.9960	0.9957	0.9955	0.9954	0.9952	0,9950	0.9948	0.9946	0.9945
1.6	0.9964	0.9964	0.9962	0.9960	0.9957	0.9954	0.9952	0.9950	0.9948	0.9946	0.9944	0.9943	0.9941
1.7	0.9962	0.9962	0.9960	0.9958	0.9955	0.9952	0.9950	0.9948	0.9946	0.9943	0.9941	0.9939	0.9938
1.8	0.9959	0.9959	0.9958	0.9955	0.9953	0.9949	0.9947	0.9945	0.9943	0.9940	0.9937	0.9936	0.9934
1.9	0.9957	0.9957	0.9956	0.9953	0.9950	0.9948	0.9944	0.9942	0.9940	0.9937	0.9934	0.9933	0.9931
2.0	0.9955	0.9954	0.9953	0.9950	0.9946	0.9942	0.9940	0.9938	0.9935	0.9932	0.9929	0.9927	0.9925
2.1	0.9953	0.9953	0.9952	0.9948	0.9945	0.9940	0.9938	0.9936	0.9933	0.9930	0.9927	0.9925	0.9923
2.2	0.9951	0.9951	0.9950	0.9947	0.9943	0.9938	0.9936	0.9934	0.9931	0.9927	0.9924	0.9922	0.9920
2.3	0.9949	0.9949	0.9947	0.9944	0.9940	0.9936	0.9933	0.9931	0.9928	0.9924	0.9921	0.9919	0.9917
2.4	0.9947	0.9947	0.9945	0.9941	0.9938	0.9933	0.9930	0.9927	0.9924	0.9921	0.9917	0.9915	0.9913
2.5	0.9945	0.9944	0.9943	0.9938	0.9935	0.9930	0.9927	0.9924	0.9921	0.9918	0.9914	0.9912	0.9909
2.6	0.9943	0.9942	0.9941	0.9936	0.9933	0.9928	0.9925	0.9922	0.9919	0.9915	0.9911	0.9909	0.9907
2.7	0.9941	0.9941	0.9939	0.9934	0.9930	0.9925	0.9922	0.9919	0.9916	0.9912	0.9907	0.9905	0.9903
2.8	0.9939	0.9938	0.9937	0.9932	0.9928	0.9922	0.9919	0.9916	0.9913	0.9909	0.9904	0.9902	0.9900
2.9	0.9937	0.9936	0.9934	0.9929	0.9925	0.9919	0.9916	0.9913	0.9909	0.9905	0.9901	0.9899	0.9896
3.0	0.9935	0.9934	0.9932	0.9927	0.9923	0.9917	0.9914	0.9911	0.9907	0.9903	0.9898	0.9895	0.9892
3.1	0.9932	0.9932	0.9930	0.9925	0.9920	0.9914	0.9911	0.9908	0.9904	0.9900	0.9895	0.9892	0.9890
3.2	0.9931	0.9931	0.9929	0.9923	0.9919	0.9912	0.9909	0.9905	0.9901	0.9897	0.9892	0.9889	0.9886
3.3	0.9929	0.9928	0.9926	0.9921	0.9916	0.9909	0.9906	0.9902	0.9898	0.9894	0.9889	0.9886	0.9883
3.4	0.9927	0.9926	0.9924	0.9918	0.9913	0.9907	0.9903	0.9899	0.9895	0.9890	0.9885	0.9882	0.9879
3.5	0.9925	0.9924	0.9922	0.9916	0.9911	0.9905	0.9901	0.9897	0.9893	0.9888	0.9883	0.9880	0.9877
3.6	0.9923	0.9923	0.9920	0.9914	0.9909	0.9902	0.9898	0.9894	0.9890	0.9885	0.9878	0.9876	0.9873
3.7	0.9921	0.9920	0.9918	0.9912	0.9907	0.9899	0.9895	0.9891	0.9887	0.9882	0.9876	0.9873	0.9870
3.8	0.9919	0.9918	0.9916	0.9910	0.9905	0.9897	0.9893	0.9889	0.9884	0.9879	0.9873	0.9870	0.9866
3.9	0.9918	0.9917	0.9915	0.9908	0.9902	0.9894	0.9890	0.9886	0.9881	0.9876	0.9870	0.9867	0.9863
4.0	0.9916	0.9915	0.9913	0.9906	0.9900	0.9892	0.9888	0.9884	0.9879	0.9873	0.9867	0.9864	0.9861

(continued)

Table 6.7.28 *Expansion Factors: Flange Taps, Y_m; Static Pressure, Mean of Upstream and Downstream (Continued)*

$$F_d = \frac{d_o}{d_i}$$

$\dfrac{h_w}{p_{fm}}$ Ratio	0.63	0.64	0.65	0.66	0.67	0.68	0.69	0.70	0.71	0.72	0.73	0.74	0.75
0.0	1.0000	1.0000	1.0000	1.0000	1.0000	1.0000	1.0000	1.0000	1.0000	1.0000	1.0000	1.0000	1.0000
0.1	0.9996	0.9996	0.9996	0.9996	0.9996	0.9996	0.9995	0.9995	0.9995	0.9995	0.9995	0.9995	0.9995
0.2	0.9992	0.9992	0.9992	0.9992	0.9991	0.9991	0.9991	0.9991	0.9990	0.9990	0.9990	0.9989	0.9989
0.3	0.9988	0.9988	0.9988	0.9987	0.9987	0.9987	0.9986	0.9986	0.9986	0.9985	0.9985	0.9984	0.9984
0.4	0.9984	0.9984	0.9984	0.9983	0.9983	0.9982	0.9982	0.9981	0.9981	0.9980	0.9980	0.9979	0.9978
0.5	0.9981	0.9980	0.9980	0.9979	0.9978	0.9978	0.9977	0.9977	0.9976	0.9975	0.9974	0.9974	0.9973
0.6	0.9977	0.9976	0.9975	0.9975	0.9974	0.9973	0.9973	0.9972	0.9971	0.9970	0.9969	0.9968	0.9967
0.7	0.9973	0.9972	0.9971	0.9971	0.9970	0.9969	0.9968	0.9967	0.9966	0.9965	0.9964	0.9963	0.9962
0.8	0.9969	0.9968	0.9967	0.9967	0.9965	0.9965	0.9964	0.9963	0.9962	0.9961	0.9960	0.9959	0.9957
0.9	0.9966	0.9965	0.9964	0.9963	0.9962	0.9961	0.9960	0.9959	0.9957	0.9956	0.9955	0.9953	0.9952
1.0	0.9962	0.9961	0.9960	0.9959	0.9958	0.9957	0.9955	0.9954	0.9953	0.9951	0.9950	0.9948	0.9947
1.1	0.9958	0.9957	0.9956	0.9955	0.9953	0.9952	0.9951	0.9949	0.9948	0.9946	0.9945	0.9943	0.9941
1.2	0.9954	0.9953	0.9952	0.9951	0.9949	0.9948	0.9946	0.9945	0.9943	0.9941	0.9940	0.9938	0.9936
1.3	0.9951	0.9949	0.9948	0.9946	0.9945	0.9943	0.9942	0.9940	0.9938	0.9936	0.9934	0.9933	0.9931
1.4	0.9947	0.9945	0.9944	0.9942	0.9941	0.9939	0.9938	0.9936	0.9934	0.9932	0.9930	0.9928	0.9926
1.5	0.9944	0.9942	0.9941	0.9939	0.9937	0.9935	0.9933	0.9932	0.9929	0.9927	0.9925	0.9923	0.9920
1.6	0.9940	0.9938	0.9937	0.9935	0.9933	0.9931	0.9929	0.9927	0.9925	0.9922	0.9920	0.9918	0.9915
1.7	0.9936	0.9934	0.9933	0.9931	0.9929	0.9927	0.9925	0.9923	0.9921	0.9918	0.9916	0.9913	0.9910
1.8	0.9932	0.9930	0.9928	0.9927	0.9925	0.9923	0.9920	0.9918	0.9916	0.9913	0.9911	0.9908	0.9905
1.9	0.9929	0.9927	0.9925	0.9923	0.9921	0.9919	0.9916	0.9914	0.9911	0.9908	0.9906	0.9903	0.9900
2.0	0.9923	0.9921	0.9919	0.9917	0.9915	0.9912	0.9910	0.9907	0.9904	0.9901	0.9899	0.9896	0.9892
2.1	0.9921	0.9919	0.9917	0.9915	0.9913	0.9910	0.9908	0.9905	0.9902	0.9899	0.9896	0.9893	0.9890
2.2	0.9918	0.9916	0.9913	0.9911	0.9908	0.9906	0.9903	0.9900	0.9897	0.9894	0.9892	0.9888	0.9884
2.3	0.9915	0.9913	0.9910	0.9908	0.9905	0.9902	0.9900	0.9897	0.9894	0.9891	0.9887	0.9884	0.9880
2.4	0.9911	0.9909	0.9906	0.9904	0.9901	0.9898	0.9895	0.9892	0.9889	0.9886	0.9882	0.9878	0.9875
2.5	0.9907	0.9905	0.9902	0.9900	0.9897	0.9894	0.9891	0.9888	0.9885	0.9881	0.9878	0.9874	0.9870
2.6	0.9904	0.9902	0.9899	0.9896	0.9893	0.9890	0.9887	0.9884	0.9880	0.9876	0.9873	0.9869	0.9865
2.7	0.9900	0.9898	0.9895	0.9892	0.9889	0.9886	0.9883	0.9880	0.9876	0.9872	0.9869	0.9864	0.9860
2.8	0.9897	0.9895	0.9892	0.9889	0.9886	0.9882	0.9879	0.9875	0.9871	0.9867	0.9864	0.9859	0.9855
2.9	0.9893	0.9891	0.9888	0.9885	0.9882	0.9878	0.9874	0.9871	0.9867	0.9863	0.9859	0.9855	0.9850
3.0	0.9890	0.9887	0.9884	0.9881	0.9877	0.9874	0.9870	0.9867	0.9863	0.9859	0.9854	0.9850	0.9845
3.1	0.9887	0.9884	0.9881	0.9877	0.9874	0.9870	0.9867	0.9863	0.9859	0.9854	0.9850	0.9845	0.9840
3.2	0.9883	0.9880	0.9877	0.9873	0.9870	0.9866	0.9862	0.9858	0.9854	0.9850	0.9845	0.9840	0.9835
3.3	0.9880	0.9877	0.9873	0.9870	0.9866	0.9863	0.9859	0.9854	0.9850	0.9845	0.9841	0.9836	0.9831
3.4	0.9876	0.9873	0.9869	0.9866	0.9862	0.9858	0.9854	0.9850	0.9845	0.9841	0.9836	0.9831	0.9825
3.5	0.9873	0.9870	0.9866	0.9863	0.9859	0.9855	0.9850	0.9846	0.9841	0.9837	0.9832	0.9826	0.9821
3.6	0.9869	0.9866	0.9862	0.9859	0.9855	0.9851	0.9846	0.9842	0.9837	0.9833	0.9827	0.9822	0.9816
3.7	0.9866	0.9863	0.9859	0.9855	0.9851	0.9847	0.9842	0.9838	0.9833	0.9828	0.9822	0.9817	0.9811
3.8	0.9863	0.9859	0.9855	0.9851	0.9847	0.9843	0.9838	0.9834	0.9829	0.9824	0.9818	0.9813	0.9807
3.9	0.9860	0.9856	0.9852	0.9848	0.9844	0.9839	0.9835	0.9830	0.9825	0.9820	0.9814	0.9808	0.9802
4.0	0.9857	0.9853	0.9849	0.9845	0.9840	0.9836	0.9831	0.9826	0.9820	0.9815	0.9809	0.9803	0.9797

Table 6.7.29 *Gauge Location Factors [14]*

Degrees latitude	Gauge elevation above sea level—lineal feet					
	Sea level	2,000′	4,000′	6,000′	8,000′	10,000′
0 (Equator)	0.9987	0.9986	0.9985	0.9984	0.9983	0.9982
5	0.9987	0.9986	0.9985	0.9984	0.9983	0.9982
10	0.9988	0.9987	0.9986	0.9985	0.9984	0.9983
15	0.9989	0.9988	0.9987	0.9986	0.9985	0.9984
20	0.9990	0.9989	0.9988	0.9987	0.9986	0.9985
25	0.9991	0.9990	0.9989	0.9988	0.9987	0.9986
30	0.9993	0.9992	0.9991	0.9990	0.9989	0.9988
35	0.9995	0.9994	0.9993	0.9992	0.9991	0.9990
40	0.9998	0.9997	0.9996	0.9995	0.9994	0.9993
45	1.0000	0.9999	0.9998	0.9997	0.9996	0.9995
50	1.0002	1.0001	1.0000	0.9999	0.9998	0.9997
55	1.0004	1.0003	1.0002	1.0001	1.0000	0.9999
60	1.0007	1.0006	1.0005	1.0004	1.0003	1.0002
65	1.0008	1.0007	1.0006	1.0005	1.0004	1.0003
70	1.0010	1.0009	1.0008	1.0007	1.0006	1.0005
75	1.0011	1.0010	1.0009	1.0008	1.0007	1.0006
80	1.0012	1.0011	1.0010	1.0009	1.0008	1.0007
85	1.0013	1.0012	1.0011	1.0010	1.0009	1.0008
90 (Ploe)	1.0013	1.0012	1.0011	1.0010	1.0009	1.0008

Note: While F_t values are strictly manometer factors, to account for gauges being operated under gravitational forces that depart from standard location; it is suggested that it be combined with other flow constants. In which instance, F_t becomes a location factor constant and F_m, the manometer factor agreeable with standard gravity remains a variable factor, subject to change with specific gravity, ambient temperature, and static pressure.

Table 6.7.30 F_m *Manometer Factors [12]*

Specific Gravity, G	Flowing Pressure, psig						
	0	500	1000	1500	2000	2500	3000
	Ambient Temperature						
0.55	1.000	0.9989	0.9976	0.9960	0.9943	0.9930	0.9921
0.60	1.000	0.9988	0.9972	0.9952	0.9932	0.9919	0.9910
0.65	1.000	0.9987	0.9967	0.9941	0.9920	0.9908	0.9900
0.70	1.000	0.9985	0.9961	0.9927	0.9907	0.9896	0.9890
0.75	1.000						
	Ambient Temperature						
0.55	1.000	0.9990	0.9979	0.9967	0.9954	0.9942	0.9932
0.60	1.000	0.9989	0.9976	0.9962	0.9946	0.9933	0.9923
0.65	1.000	0.9988	0.9973	0.9955	0.9937	0.9923	0.9913
0.70	1.000	0.9987	0.9970	0.9947	0.9926	0.9912	0.9903
0.75	1.000	0.9986	0.9965	0.9937	0.9915	0.9902	0.9893
	Ambient Temperature						
0.55	1.000	0.9991	0.9981	0.9971	0.9960	0.9950	0.9941
0.60	1.000	0.9990	0.9979	0.9967	0.9955	0.9943	0.9933
0.65	1.000	0.9989	0.9977	0.9963	0.9948	0.9935	0.9925
0.70	1.000	0.9988	0.9974	0.9958	0.9940	0.9926	0.9915
0.75	1.000	0.9987	0.9971	0.9951	0.9931	0.9916	0.9906
	Ambient Temperature						
0.55	1.000	0.9992	0.9983	0.9974	0.9965	0.9956	0.9948
0.60	1.000	0.9991	0.9981	0.9971	0.9960	0.9950	0.9941
0.65	1.000	0.9990	0.9979	0.9967	0.9955	0.9944	0.9934
0.70	1.000	0.9989	0.9977	0.9963	0.9950	0.9937	0.9926
0.75	1.000	0.9988	0.9975	0.9959	0.9943	0.9929	0.9918

Note: Factors for intermediate values of pressure, temperature, and specific gravity should be interpolated.
Note: This table is for use with mercury manometer type recording gauges that have gas in contact with the mercury surface.

(text continued from page 6-308)

The flowing temperature factor, F_{tf} is required to change the assumed flowing temperature of 60°F to the actual flowing temperature T_f (°R)

$$F_{tf} = \frac{519.67}{T_f}^{0.5} \qquad [6.7.32]$$

Reynold's Number Factor, F_r — The Reynold's number factor takes into account the variations of the discharge coefficient of an orifice with Reynold's number. The discharge coefficient of the orifice decreases as the Reynold's number increases until the Reynold's number is infinite, when the discharge coefficient will have the least possible value for that particular installation. Discharge coefficients vary considerably with the Reynold's number in the measurement of viscous liquids; they tend to become more constant as the velocity of the liquid decreases, and as the Reynold's number correspondingly increases. Within the limits of commercial measurement of gases, the discharge coefficient is practically constant, varying only slightly with Reynold's number. The variation is sufficiently slight, and the viscosities of commercial gases are sufficiently constant to arrant arbitrarily using an average value for the gas viscosity in computing Reynold's number factors. For all practical purposes, this leaves the Reynold's number factor as a function of the orifice and meter tube dimensions, the location of the differential pressure taps, and the rate of flow — a function of the pressure extension, $h_w p_f$.

The value of pressure extension may be estimated from a knowledge of the differentials and the static pressures at the meter. The pressure extension obtained in this manner will probably be sufficiently close to the average operating condition of the meter for the purpose of selecting the proper value of the Reynold's number factor. The variations in the factor corresponding to the variations in the pressure extension above or below the selected value will be compensating over any appreciable length of time.

The factor may also be based upon an average value of the pressure extensions obtained from the meter records. This is probably the simplest method since the sums of the 24 pressure extensions are generally obtained from each chart in order to compute the daily delivery.

Table of "b" factors from which the Reynold's number factor can be readily obtained are shown on Table 6.7.22 for flange taps, and Table 6.7.23 for pipe taps.

The Reynold's number factor, F_r, can be calculated by

$$F_r = 1 + \frac{b}{(h_w p_f)^{0.5}} \qquad [6.7.33]$$

Expansion Factor, Y — When a gas flows through an orifice, the change in velocity and pressure is accompanied by a change in specific weight and a factor must be applied to the coefficient to allow for this change. The expansion factor depends upon the location of the differential taps, the location of the static pressure tap, and the ratio of the orifice diameter to the diameter of the meter tube. It is also dependent upon the ratio of the differential pressure to the static pressure and upon the specific heats for the flowing gas. The variation in factor is slight and the ratio of the specific heats for commercial gases is sufficiently constant to warrant using an assumed constant ratio of specific heats. An assumed ratio of specific heats k = 1.3 was used in calculating the expansion factors in *AGA Report No. 3*. This permits the tabulation of the factors according to the diameter ratio, β, and the ratio of the differential to static pressure, h_w/p_f. in the tables p_f is indicated as p_{f1}, p_{fm}, or p_{f2}, depending upon whether the static pressure is obtained from the upstream tap, a mean of the upstream and downstream pressures, or the downstream tap.

The expansion factor may be based upon a value of the differential ratio obtained from the meter records or estimated from a knowledge of the differentials and the static pressures at the meter. The differential ratio obtained in this manner will probably be sufficiently close to the average operating condition of the meter for the purpose of selecting the proper value of the factor. The variations in the factor corresponding to the variations in the differential ratio above or below the selected value will be compensating over any appreciable length of time.

Table of expansion factor, Y, are shown on Table 6.7.24 through Table 6.7.28.

Supercompressibility Factor, F_{pv} — The supercompressibility factor accounts for the deviation from the "perfect gas" laws. In the basic flow equations, gas volumes are assumed to vary with pressure and temperature in accordance with Boyle's and Charles' laws (the "perfect gas" laws). Actually the volume occupied by individual gases deviate, by a slight degree, from the volumes, which the "perfect gas" laws indicate. The amount of deviation is a function of the composition of the gas and varies primarily with static pressure and temperature. The actual deviation is obtained by a test conducted on a sample of the gas, carefully taken at line conditions of pressure and temperature.

Practical relationships have been established by which this deviation can be calculated and tabulated for natural gases containing normal mixtures of hydrocarbon components, considering the presence of small quantities of CO_2 and nitrogen and also relating the deviation to the heating value of the gas.

The values of supercompressibility factor, F_{pv}, can be determined from Fig. 6.1.1.

Manometer Factor, F_m — The manometer factor is used with the mercury-type differential gage to correct the slight error in measurement caused by the weight of the unbalanced column of dense gas above the mercury. This factor is NOT applicable where measurement is made with a bellow-type differential gage because there is no unbalanced column of dense gas in a manometer of that type. Manometer factors are shown on Table 6.7.30.

The manometer factor, F_m, can be calculated as follows:

$$F_m = \frac{Y_m - Y_g}{846.324}^{0.5} \qquad [6.7.34]$$

where

$Y_m = 846.324[1 - 0.000101(T_a - 520)]$
$Y_g = 2.699053 \, SG_g(F_{pv})^2/T_a$
T_a = ambient temperature in degrees Rankin (air or atmospheric temperature surrounding the orifice meter)
SG_g = specific gravity of flowing gas
F_{pv} = supercompressibility factor for the gas at ambient temperature, specific gravity and static gage location

The orifice thermal expansion factor, F_a, is introduced to correct for expansion or contraction of the orifice, when operating at temperatures appreciably different from the temperature at which the orifice was bored. It is calculated as:

$$F_a = 1 + [0.0000185(T_f - 68)]$$

(for 304 and 316 stainless steel)

$$F_a = 1 + [0.0000159(T_f - 68)] \text{ (for monel)}$$

where T_f = flowing temperature of gas at the orifice in °F

The gage location factor F_1 is required when mercury orifice differential instruments are used. It compensates for

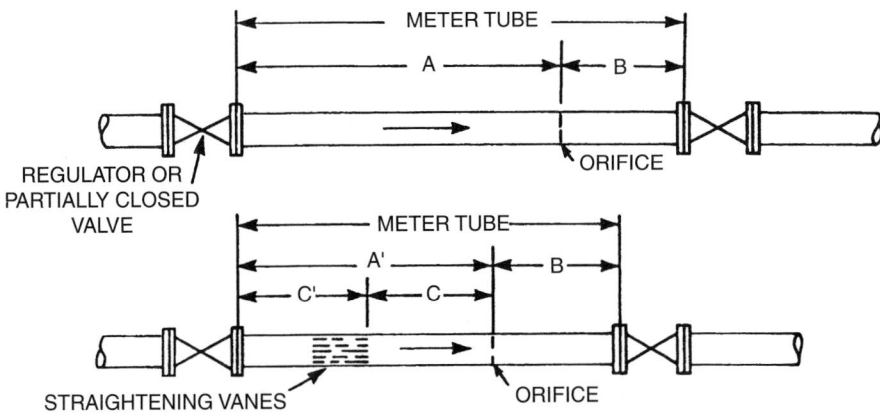

Figure 6.7.39 *Partly closed valve upstream of meter tube.*

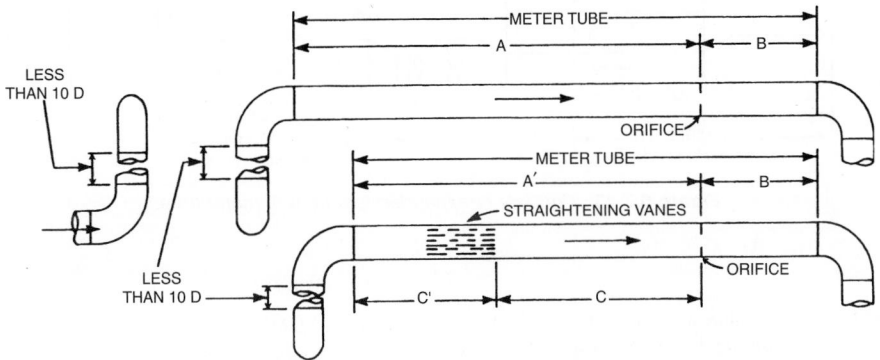

Figure 6.7.40 *Two ells not in the same plane upstream of meter tube.*

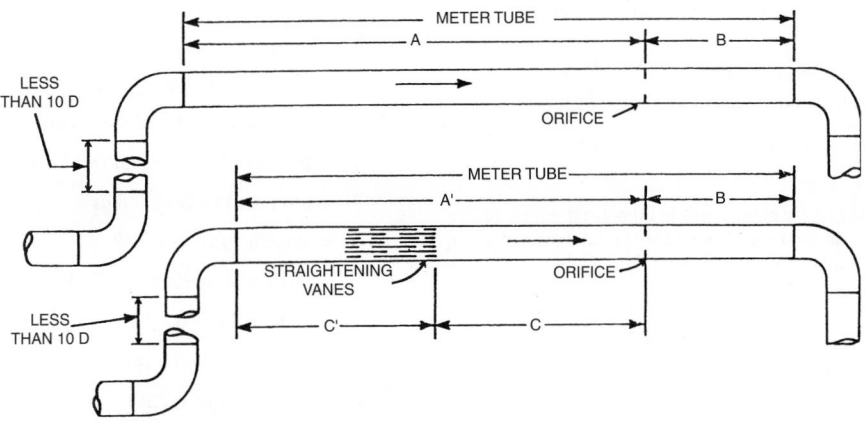

Figure 6.7.41 *Less than 10 pipe diameters (D) between two ells in the same plane upstream of meter tube.*

differences in the weight of the gas column above the mercury reservoir for meter locations other than the standard of 45° latitude at sea level. It is a constant for any given metering location and can be combined with other flow constants or stored in the computer master file for use in calculating

orifice flow. The equation for determining the gage location factor is:

$$F_1 = \left(\frac{g_1}{980.665}\right)^{0.5} \qquad [6.7.35]$$

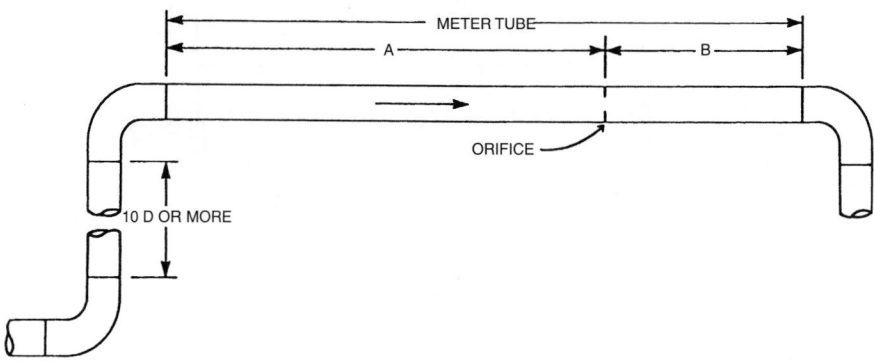

Figure 6.7.42   *Greater than 10 pipe diameters (D) between two ells in the same plane upstream of meter tube.*

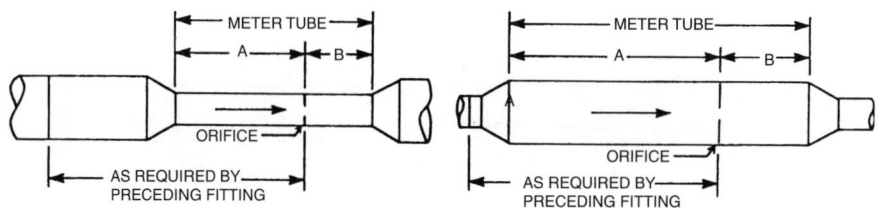

Figure 6.7.43   *Reducer or expander upstream of meter tube.*

where

g_1 = the ambient gravity value at the metering location can be obtained from the U.S. Coast and Geodetic Survey data in reference to aeronautical data, from the Smithsonian meteorological tables, or calculated for the midlatitudes, between 30° and 60°, as

$$g_1 = 980.665 + [0.087(°L - 45)] - 0.000094H$$

where °L = degrees latitude
H = elevation in lineal feet above sea level

This gage location factor value can be read in Table 6.7.29.

Example 6.7.17
A 0.7 specific gravity natural gas is measured in and orifice meter having flange taps. Determine the flow rate in scf/hr at 14.4 psia and 60°F if the following data apply:

$$h_w = 40 \text{ in. } H_2O$$

$$p_f = 143 \text{ psig (measured downstream)}$$

average flowing temp = 84°F

line size = 4.026 in.

orifice-plate opening = 1.50 in.

Solution
The respective factors are:

$F_b = 460.79$

$F_{pb} = 1.0229$

$F_{tb} = 1.0000$

$F_g = 1.1952$

$F_{tf} = 0.9777$

$F_r = 1.0004$

$Y = 1.0016$

$C' = 551.89$ (the product of the above numbers)

Then $q_g = 551.89[\{40\}\{143 + 14.7\}]^{0.5} = 43,810$ scf/hr.

Example 6.7.18
What is the flow rate in scf/hr for the following conditions?

d = diameter of orifice = 3.000 in.

D = internal diameter of meter tube = 6.065 in.

p_b = pressure base, 8 oz.above14.4 = 14.9 psia

T_b = temperature base = 60°F

G = specific gravity = 0.60

T_f = flowing temperature = 65°F

F_b = 1891.9 (Table 6.7.16)

F_{pb} = 0.9886 (Table 6.7.18)

F_{tb} = 1.0000 (Table 6.7.19)

F_g = 1.2910 (Table 6.7.20)

F_{tf} = 0.9952 (Table 6.7.21)

Static pressure obtained at downstream flange tap.
Average differential = 49.5 in.

Average static pressure = 80 psia (65.6 psig)

$\beta = 3.000 \div 6.065 = 0.49$

$(h_w p_f)^{0.5}$ = pressure extension (average) = 63

Using Table 6.7.22, for $\beta = 0.49$ and $D = 6.065$; $b = 0.0332$

$F_r = 1 + (0.0332 \div 63) = 1.0005$

Differential ratio, $h_w \div p_{f2} = 49.5 \div 80 = 0.62$

Using Table 6.7.26, for $\beta = 0.49$ and $h_w p_f = 0.62$;

$Y_2 = 1.0037$

Using Table 6.7.30, $F_{pv} = 1.0046$

Manometer factor F_m, using Table 6.7.31 $F_m = 0.9999$

Therefore, $C' = 1891.9 \times 0.9886 \times 1.000 \times 1.2910$

$\times 0.9952 \times 1.0005 \times 1.0037 \times 1.0046$

$\times 0.9999 = 2424.0$

For the average pressure extension,

$(h_w p_f)^{0.5} = (49.5 \times 80)^{0.5} = 62.929$

The flow rate is $Q_h = C'(h_w p_f)^{0.5} = 2424.0 \times 62.929$

$= 152,540 \text{ scf/hr}$

Example 6.7.19
For a meter equipped with pipe taps, what is the flow rate in scf/hr?

$d = $ diameter of orifice $= 3.000$ in.

$D = $ internal diameter of meter tube $= 7.981$ in.

$p_b = $ pressure base, 4 oz. above 14.4 $= 14.65$ psia

$T_b = $ temperature base $= 60°F$

$G = $ specific gravity $= 0.64$

$T_f = $ flowing temperature $= 60°F$

$F_b = 2012.7$ (Table 6.7.17)

$F_{pb} = 1.0055$ (Table 6.7.18)

$F_{tb} = 1.0000$ (Table 6.7.19)

$F_g = 1.2500$ (Table 6.7.20)

$F_{tf} = 1.0000$ (Table 6.7.21)

Static pressure obtained at upstream pipe tap

Average differential $= 64.5$ in.

Average static pressure $= 539.4$ (525 psig)

$\beta = 3.000 \div 7.981 = 0.38$

$(h_w p_f)^{0.5} = $ pressure extension (average) $= 187$

For $\beta = 0.38$ and $h_w \div p_{f1} = 0.12$

$F_r = 1 + (0.0214 \div 187) = 1.0001$

Differential ratio, $h_w \div p_{f2} = 0.12$

Using Table 6.7.25, $\beta = 0.38$ and $h_w \div p_{f1} = 0.12$; $Y_1 = 0.9983$

It is assumed in this example that the supercompressibility factor is not used, by terms of contract.

Using Table 6.7.31, the manometer factor, $F_m = 0.9989$

Therefore, $C' = 2012.7 \times 1.0055 \times 1.0000$

$\times 1.2500 \times 1.0000 \times 1.0001$

$\times 0.9983 \times 0.9989 = 2522.9$

For an average pressure extension of $(h_w p_f)^{0.5}$

$= (64.5 \times 539.4)^{0.5} = 186.52$

The flow rate is $Q_h = C'(h_w p_f)^{0.5} = 2522.9 \times 186.52$

$= 470,570 \text{ scf/hr}$

In orifice meter measurement of gases, the effect of compressibility equates to the relationship $(1/Z)^{0.5}$; this has been termed the "supercompressibility" of the gas. The supercompressibility factor, F_{pv} may be calculated from:

$$F_{pv} = \left(\frac{Z_b}{Z_{fl}}\right)^{0.5} \qquad [6.7.36]$$

The best way to obtain the Z calculation is to use the Hall-Yarborough equation [16], or figures of compressibility factors for natural gas Figure 6.7.44. Also the supercompressibility factor F_{pv} may be taken from the *AGA Report No. 3* [14] and/or empirical equations as follows [17]:

$$F_{pv} = 1 + \left(\frac{2.48 P_G \times 10^{5.0+2.02 SG_g}}{T_f^{3.825}}\right)^{0.5} \qquad [6.7.37a]$$

for $0.601 \, SG_g \leq 0.650$ and $P < 600$ psig

$$F_{pv} = 1 + \left(\frac{3.32 P_G \times 10^{5.0+1.81 SG_g}}{T_f^{3.825}}\right)^{0.5} \qquad [6.7.37b]$$

for $SG_g \leq 0.600$ and $P < 600$ psig

$$F_{pv} = 1 + \left(\frac{4.66 P_G \times 10^{5.0+1.6 SG_g}}{T_f^{3.825}}\right)^{0.5} \qquad [6.7.37c]$$

for $0.651 \leq SG_g \leq 0.750$ and $P < 600$ psig

$$F_{pv} = 1 + \left(\frac{7.91 P_G \times 10^{5.0+1.26 SG_g}}{T_f^{3.825}}\right)^{0.5} \qquad [6.7.37d]$$

for $0.0751 \leq SG_g \leq 0.900$ and $P < 600$ psig

where $P_G = $ gage pressure psig
$SG_g = $ specific gravity of gas
$T_f = $ flowing temperature in °R

Example 6.7.20
Compute the daily flowrate of a natural gas through an orifice meter for the following conditions:

Barometric pressure $= 14.5$ psia

Diameter of pipe $= 11.938$ in.

Orifice Diameter $= 4.000$ in.

Differential Pressure across meter $= 27.0$ in. of water

Static pressure on meter $= 600$ psig

Type of meter $= $ Flange taps

Temperature base $(T_b) = 60°F$

Flowing temperature $= 75°F$

Pressure base $(P_b) = 14.65$ psia

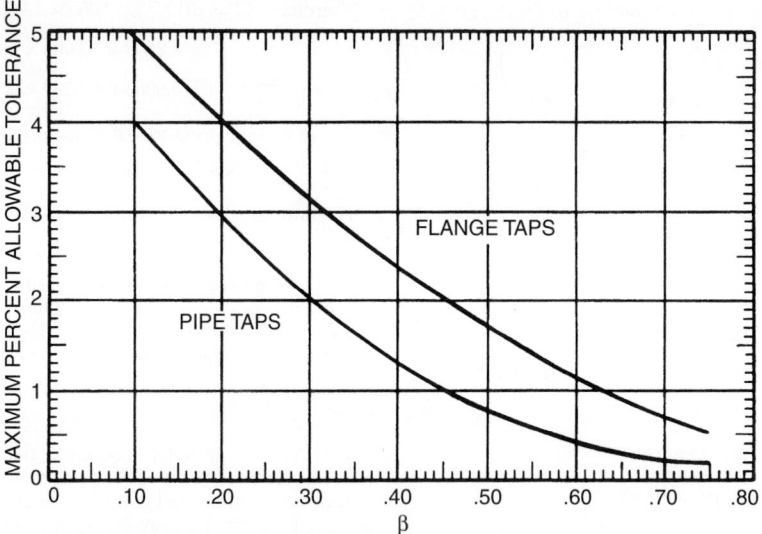

Note: Maximum percent allowable tolerance between measured upstream inside diameter andthe published inside diameter refer to the published inside diameter.

Figure 6.7.44 *Maximum percent allowable meter tube tolerance.*

Gas composition by volume = $C_1 - 85\%$, $C_2 - 5\%$, $C_3 - 3\%$,

$$i\text{-}C_4 - 1\%, N_2 - 6\%$$

Solution
1. Daily Flow Rate = $24 \times Q_N$
2. $(h_w p_f)^{0.5} = [27(600 + 14.5)]^{0.5} = 128.8$
3. Basic orifice factor for flange taps, F_b; see Table 6.7.16 for $D = 12.000$ in. (11.938) and $D = 4.000$, $F_b = 3,260$
4. Reynold's number factor, F_r:

$$F_r = 1 + \frac{b}{(h_w p_f)^{0.5}} = \frac{b}{128.8}$$

for b see Table 6.7.22 for given d and D

$$b = 0.0206$$

$$F_r = 1 + \frac{0.0206}{128.6} = 1.0001599 = 1.0002$$

(minimum four decimal places is required)
5. Expansion Factor Y, (function of β and h/P)

$$\beta = \frac{d}{D} = \frac{4.000}{11.938} = 0.33506$$

$$\frac{h}{P} = \frac{27}{692.5} = 2.8989 \times 10^{-2} = 0.038989 = 0.039$$

From Table 6.7.24 Y = 0.9996 by interpolation
6. Temperature base factor, F_{tb}:

$$F_{tb} = \frac{469.67 + 60}{519.67} = 1.0$$

7. Pressure base factor, F_{pb}:

$$F_{pb} = \frac{14.73}{14.65} = 1.0055$$

8. Flowing temperature factor, F_{tf}:

$$F_{tf} = \left(\frac{T_b}{T_f}\right)^{0.5} = \left(\frac{519.67}{459.67 + 75}\right)^{0.5} = 0.9859$$

9. Gas specific gravity factor, F_g or $F_{gr} = (1/SG_g)^{0.5}$

Comp.	Concentr.	$T_c(°R)$	$Z_i T_{ci}$	P_c (psia)	$Z_i P_{ci}$	M	$Z_i M_i$
C_1	0.85	344	292.4	673	572.1	16	13.6
C_2	0.05	550	27.5	712	35.6	30	1.5
C_3	0.03	666	20.0	617	18.5	44	1.3
iC_4	0.01	734	7.3	528	5.3	58	0.6
N_2	0.06	227	13.6	492	29.5	28	1.7
	Σ		360.9		661 psia		18.7

$$SG_g = \frac{M_{gas}}{M_{air}} = \frac{18.7}{28.9625} = 0.646$$

$$F_{gr} = \left(\frac{1}{SG_g}\right)^{0.5} = 1.2442$$

10. Supercompressibility factor, F_{pv}:

$$F_{pv} = \frac{1}{Z^{0.5}}$$

$$T_f = \frac{459.67 + 75}{360.9} = 1.481$$

$$P_r = \frac{692.5}{661} = 1.0477$$

from Figure 6.7.43, Z = 0.895

$$F_{pv} = \frac{1}{0.895^{0.5}} = 1.0564$$

or from Equation 6.7.

$$F_{pv} = 1 + \frac{3.32 \times 600 \times 10^{5.0+1.81 \times 0.646}}{534.67^{3.825}} = 1.0526$$

11. Daily flow rate:

$$24Q_v = 24 \times (F_b F_r Y F_{pb} F_{tb} F_{tf} F_{gr} F_{pv}(h_w p_f)^{0.5})$$

$$= 24 \times 3260 \times 1.0002 \times 0.9996 \times 1.0055 \times 1.0$$

$$\times 0.9859 \times 1.2442 \times 1.0526 \times 128.8$$

$$= 13.08 \text{ MMscfd}$$

Example 6.7.21
A 1.6 MMscm daily flow rate of 0.63 relative density natural gas is to be metered at 27°C and 7 MPa in an orifice meter using flange taps. At this condition the compressibility factor $Z = 0.895$. What size orifice plate and meter run should be used?

Assume $h_w = 52$ in. of water
$\quad\quad P_b = 14.5$ psia
$\quad\quad T_b = 59°F$

Ignore F_r & Y

$T = 27°C = 80.6°F$

$P_{\text{static on meter}} = 7 \text{ MPa} = 1,1015.26 \text{ psig}$

$Z = 0.895$

$P_b = 14.5 \text{ psia} = 100 \text{ kPa} = 1 \text{ bar}$

$T_b = 59°F = 15°C$

for P_{bar} assume it is the same as P_{base} if P_{bar} is unknown:

$$24 \, Q_v = (F_b F_r Y F_{pb} F_{tb} F_{tf} F_{gr} F_{pv}(h_w p_f)^{0.5})$$

1. Flow rate Q_v (scf/hr) at 14.73 psia and $T = 60°F$

$$24 \, Q_v = 1.6 \text{ MMscm}/24 \text{ hr}$$

$$Q_v = 1.6 \text{ MMscm/hr}$$

Standard conditions for SI system:

$$P = 100 \text{ kPa}$$

$$T = 15°C$$

or

$$P = 101.325 \text{ kP}$$

$$T = 0°C$$

Use the first set, where mole volume $= 23.96$ scm/kmol; then our T_b and P_b are exactly the same.

$$P_b = 14.5 \text{ psia} = 100 \text{ kPa}$$

$$T_b = 59°F = 15°C$$

Hence, $Q_v = 1.6 \times 10^6$ (scm/day) (1 kmol/23.96 scm) $= 66,777.96$ kmol/24 hr

$$= \frac{147,217.73}{24} \frac{\text{lb mol}}{\text{hr}}$$

Q_v is in the formula for orifice equations according the AGA Report No. 3 [12] and has standard conditions, $P_{sc} = 14.73$ psia and $T_{sc} = 60°F$. For those conditions the molal volume $= 378.4 \text{ ft.}^3/\text{lb mol}$.

Flowrate Q_v (scf/hr) at 14.73 psia and 60°F is equal to

$$Q_{v(AGA)} = \frac{147,217.73}{24} \frac{\text{lb mol}}{\text{h}} \times 378.4 \frac{\text{scf}}{\text{lb mol}}$$

$$= 2.321133 \text{ scf/hr}$$

2. Temperature base factor F_{tb}

$$F_{tb} = \frac{459.67 + 59}{519.67} = 0.9981$$

3. Pressure base factor F_{pb}

$$F_{pb} = \frac{14.73}{14.50} = 1.0159$$

4. Gas gravity factor F_{gr}

$$F_{gr} = \left(\frac{1}{SG_g}\right)^{0.5} = \frac{1}{0.63} = 1.2599$$

5. Supercompressibility factor F_{pv}

$$F_{pv} = \left(\frac{1}{Z_f}\right)^{0.5} = \frac{1}{0.895} = 1.057$$

6. Flowing temperature factor F_{tf}

$$F_{tf} = \left(\frac{T_b}{T_f}\right)^{0.5} = \left(\frac{459.67 + 59}{459.67 + 80.6}\right)^{0.5} = 0.9798$$

7. $(h_w p_f)^{0.5} = [52 \times (1015.26 + 14.5)]^{0.5} = 0.9798$
8. Solve for F_b

$$F_b = \frac{Q_v}{F_r Y F_{pb} F_{tb} F_{tf} F_{gr} F_{pv}(h_w p_f)^{0.5}}$$

$$= \frac{2,321,133}{(0.9981)(1.0159)(0.9798)(1.2599)(1.2599)(1.0570)(231.4)}$$

$$= 6,017.7$$

9. Selection of the meter run (see several combinations to orifice-to-pipe ratios) will fall within the recommended β ratio.

F_b	Pipe "D"		Orifice	
	Nominal	Size	"d"	β
7579.2	12	11.938	6.0	0.5026
7564.7	12	12.090	6.0	0.4963
7640.4	12	11.375	6.0	0.5274
777.8	8	8.071	5.75	0.7124

The larger the run, the lower β but the higher cost. Once a size has been chosen the values for F_r and Y_1 could be calculated, but the effect on the choice is trivial.

6.7.11 Uncertainty Limits and Field Problems
No two orifice meters can be built, as a rule, to give exactly the same reading when the same amount of gas is flowing. For this reason, uncertainties are necessary for the values of the constants given in this handbook. The commercial accuracy will be somewhat less than the accuracy indicated by the tolerance given for the orifice flow constants. Very exact duplication of orifice plates is not commercially possible. An example of the effect of uncertainties is provided below:

Example 6.7.22
Using the flow equation, find the overall measurement uncertainty of the flow, knowing percentage uncertainty of each variable. We use the equation:

$$Q_v = F_b F_r Y F_{pb} F_{tb} F_{tf} F_{gr} F_{pv}(h_w p_f)^{0.5}$$

Variable	% Uncertainty of Variable (+)	Effect Factor	
		Exponent	Square
F_b	0.5	1	0.25
F_r	0.1	1	0.01
Y	0.25	1	0.0625
F_{pb}	0.05	1	0.0025
F_{tb}	0.03	1	0.0009
F_{tf}	0.25	$\frac{1}{2}$	0.0156
F_{gr}	0.7	$\frac{1}{2}$	0.1225
F_{pv}	0.4	$\frac{1}{2}$	0.040
h_w	0.5	$\frac{1}{2}$	0.0625
p_f	0.5	$\frac{1}{2}$	0.0625
		Sum of squares	0.6265
		Square root of sum	0.7915%

As the table illustrates, the overall measurement uncertainty of the flow in this example is ±0.7915%. The above analysis assumes constant flowrate, which is a rather theoretical case.

System design must consider monthly variations in flow, daily variations in flow and even very short form variation in flow. One of the most difficulty metering conditions is when the producers has installed intermitters on the well production. Very often in these systems, the no-flow period (or low-flow period) is of relatively long duration when compared to the flow periods. The result is a chart interpretation that is much higher than actual flow.

Sometimes in vacuum gathering systems, backflow may occur. In such a case it is possible to pay for the gas twice. The gas flow through the metering system once as it is produced; it may flow backwards through the system into the reservoir when the compressor shuts down. When the compressor is restarted, it may be drawn through the system and again remetered. For that reason it is important to install check valves on the sublet of meter runs.

Electronic Measurement—The primary considerations for going to electronic measurement includes the fact that it is more accurate. Electronic metering units are more capable of following highly variable flows that a mechanical chart recorder. After installation of an electronic system it is possible to reduce the imbalance between the plant inlets vs. field volumes to less than 2% as compared with a balance that frequently runs 6% to 8% using mechanical recorders.

6.7.12 Gas Gathering

Gas gathering systems usually consist of the piping and the processing equipment between individual wells in a gas field and the compressor station at the inlet of the transmission or distribution lines. The smallest gathering system consists simply of two or more gas well interconnected by piping and tied directly into a distribution system. For large fields and for several interconnected fields involving hundreds of miles of piping, gathering systems may include such equipment as drips, separators, meters, heaters, dehydrators, gasoline plant, sulfur plant, cleaners, and compressors, as well as piping and valves.

Depending upon local conditions, there are a few types of gathering systems. The basic systems are axial, radial, loop and their combination (see Figure 6.7.45 and 6.7.46).

The choice between the gathering systems is usually economic, but technical feasibility is also important. Special attention is paid in the solution of complex transmission systems to express the various lengths and diameters of the pipe in the system as equivalent lengths of a common diameter or equivalent diameters of a common length. Equivalent, in this case, means that both lines will have the same capacity with the same total pressure drop.

6.7.13 Flow Equations for Steady State

Commonly used equations for gas flow are given below:

1. Basic equation

$$Q_{sc} = K \left(\frac{T_{sc}}{P_{sc}} \right)^{1.000} \left[\frac{(P_1^2 - P_2^2)D^5}{fSG_g LT_m Z_m} \right]^{0.5} \text{(E)} \qquad [6.7.38]$$

where K = 38.774 is in English units
= 5.62×10^5 in in SI units

2. Weymouth equation

$$Q_{sc} = K \left(\frac{T_{sc}}{P_{sc}} \right)^{1.000} \left[\frac{(P_1^2 - P_2^2)D^{5.333}}{SG_g LT_m Z_m} \right]^{0.5} \text{(E)} \qquad [6.7.39]$$

where K = 433.5, f = $0.008/D^{0.33}$ is in English units
= 1.162×10^7, f = $0.0109/D^{0.33}$ is in SI units

3. Panhandle A equation

$$Q_{sc} = K \left(\frac{T_{sc}}{P_{sc}} \right)^{1.0788} \left[\frac{(P_1^2 - P_2^2)D^{4.854}}{SG_g^{0.854} LT_m Z_m} \right] \text{(E)} \qquad [6.7.40]$$

where K = 435.9, f = $0.0192/(Q_{sc}SG_g/D)^{0.1461}$
is in English units
= 1.198×10^7, f = $0.0099/(Q_{sc}SG_g/D)^{0.1461}$
is in SI units

4. Panhandle B equation

$$Q_{sc} = K \left(\frac{T_{sc}}{P_{sc}} \right)^{1.020} \left[\frac{(P_1^2 - P_2^2)D^{4.961}}{SG_g^{0.961} LT_m Z_m} \right]^{0.51} \text{(E)} \qquad [6.7.41]$$

where K = 737, f = $0.00359/(Q_{sc}SG_g/D)^{0.03922}$
is in English units
= 1.264×10^7, f = $0.0030/(Q_{sc}SG_g/D)^{0.03922}$
is in SI units

5. The Clinedinst equation in English units

$$Q_{sc} = 288.7 \left(\frac{T_{sc}}{P_{sc}} \right) Z_{sc} P_{sc} \left[\frac{D^5}{SG_g T_m Lf} \right.$$

$$\times \left. \left(\int_0^{P_{r,1}} \frac{P_r}{Z} dP - \int_0^{P_{r,2}} \frac{P_r}{Z} dP_r \right) \right]^{0.5} \qquad [6.7.42]$$

Values of the integral functions $\int_0^{P_r} (P_r/Z) dP_r$ are tabulated in Table A-6 of Katz [144] or in Table 7-1 of Kumar [145].

The constants in the Equations 6.7.38 through 6.7.42 are as follows:

Q_{sc} = gas flow rate at T_{sc}, P_{sc}, in scf/d or scm/d
P, P_{sc} = absolute pressure at standard conditions in psia or kpa
T_m, T_{sc} = mean absolute temperature of line at standard conditions in R° or K
D = inside diameter of pipes in inches or meters
L = Pipe length in miles or meters
μ = viscosity in lb/(ft•s) or (Pa•s)
γ_g = relative gas density (air = 1.0)
Z_m = mean compressibility factor
E = pipeline efficiency
R_e = Reynold's number
P_r = pseudo critical pressure

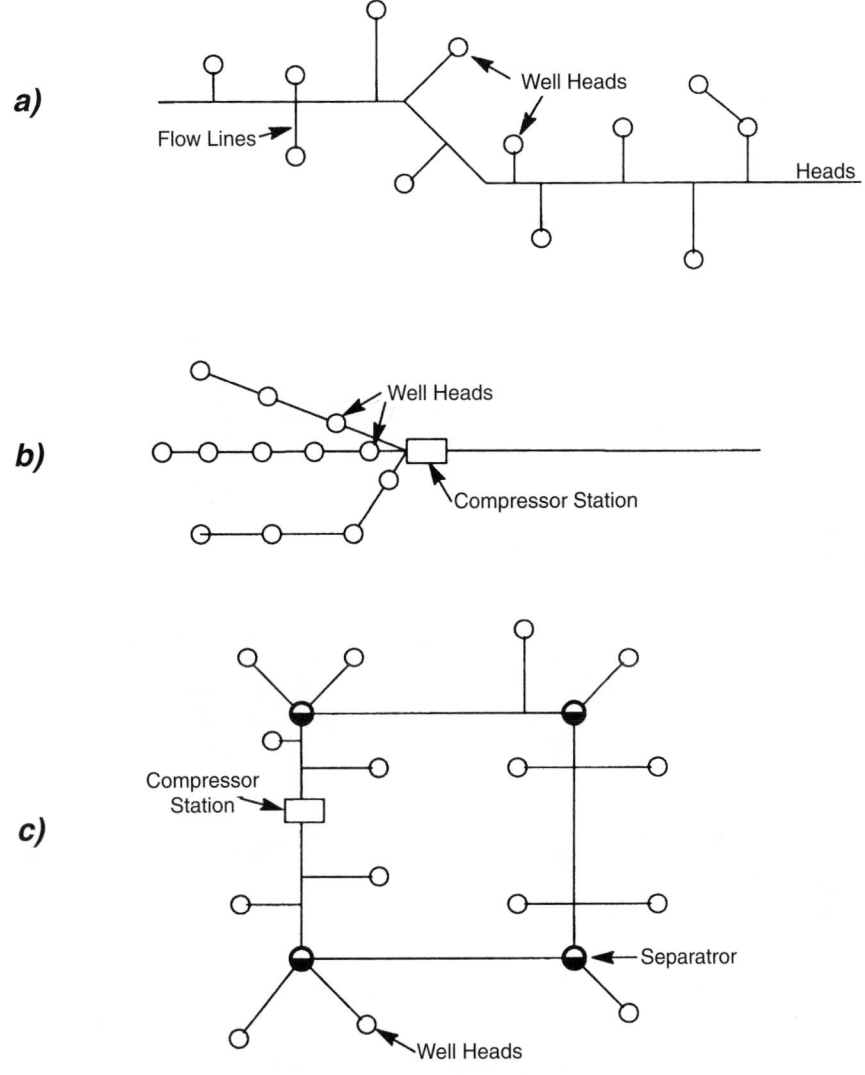

Figure 6.7.45 *Simple gathering systems: a) an axial, b) a radial, c) loop.*

Pipeline efficiency is a correction for small amounts of liquid, general debris, weld resistance, valve installations, line bends, and other factors that reduce the gas flow rate to a point below the basic equation of state. The design value of "E" in a new clean gas line usually is estimated at 0.92. Some companies arbitrarily use a graduated "E":

E = 1.0, new straight pipe without bends, very seldom used in design
= 0.95, excellent conditions (with frequent pigging)
= 0.92, average to good condition (normal design)
= 0.85, adverse, unpigged, old dirty line

Equations 6.7.38 through 6.7.42 are equations for steady-state flow of gas through a horizontal pipe. These equations assume:

- no mechanical work is done on the gas between the points at which the pressures are measured,
- negligible kinetic energy change

The Weymouth equation was devised for sizing gas lines operating at pressure from 35 to 100 psig if Z = 1.0. By including the compressibility factor, the Weymouth equation showed reasonable agreement with metered volumes, for filed gathering lines operating at 1,000 to 3,200 psig. The Weymouth equation is used most often for designing gas gathering and transmission systems because it generally maximizes pipe diameter requirements for a given flowrate and pressure drop.

Example 6.7.23
A pipe line is to be designed to transmit 300 MMsf/d measured at 60°F and 15.025 psia. Specific gravity of the gas is 0.71. Flowing temperature is expected to average 60°F while barometric pressure is to average 14.3 psia. It has already been decided that 24-in. pipe will be used (ID = 23.2 in.) and that discharge and suction pressures at the compressed stations will be 800 psig and 350 psig, respectively. How far

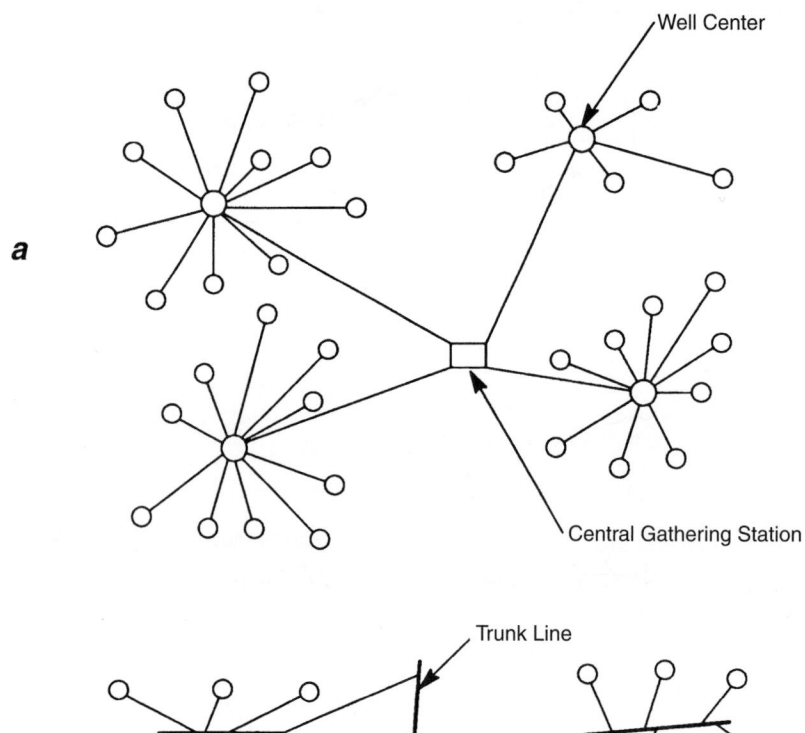

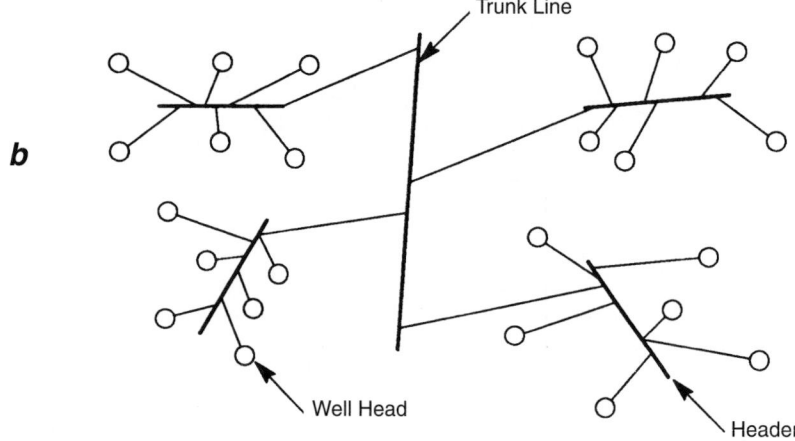

Figure 6.7.46 *A combination of gathering systems: a) well-center, b) trunk-line.*

apart will the compressor stations be? Use the Weymouth and Panhandle equations, where Z = 0.86, E = 0.92.

From Equation 6.3.39

$$Q_{sc} = \left(\frac{T_{sc}}{P_{sc}}\right)^{1.000} K \left[\frac{(P_1^2 - P_2^2)D^{5.333}}{SG_g LTZ}\right]^{0.5} \quad (E)$$

Substituting the given values into the equation gives

$$300 \times 10^6 = \left(\frac{519.67}{15.025}\right)^{1.000}$$

$$\times\, 433.5 \left[\frac{(814.3^2 - 364.3^2)(23.2)^{5.333}}{(0.71)L(519.67)(0.86)}\right]^{0.5} \quad (E)$$

$$2855.7 \times L = 228542\,E^2$$

$$= 80.03\,E^2$$

if E = 1.0, then L = 80.03 mi
if E = 0.82, then L = 67.74 mi

From Equation 6.7.40

$$Q_{sc} = K \left(\frac{T_{sc}}{P_{sc}}\right)^{1.0788} \left[\frac{(P_1^2 - P_2^2)D^{4.854}}{SG_g^{0.854} LTZ}\right]^{0.5394}$$

$$300 \times 10^6 = 435.87 \left(\frac{519.67}{15.025}\right)^{1.0788}$$

$$\times \left[\frac{530,370 \times 23.2^{4.854}}{(0.71)^{0.854}L(519.67)(0.86)}\right]^{0.5394}$$

$$\times \left[\frac{530,370 \times 4,246,964.97}{33.58L}\right]^{0.5394} \quad (E)$$

$$L^{0.5394} = 13.317\,E$$

If E = 1.0, then L = −121.5 mi
If E = 0.92, then 1 = 104.1 mi

From Equation 6.7.41

$$300 \times 10^6 = 737 \left(\frac{519.67}{15.025} \right)^{1.02}$$

$$\times \left[\frac{(814.3^2 - 364.3^2)23.2^{4.861}}{(0.71)L(519.67)(0.86)} \right]^{0.51} \quad \text{(E)}$$

$$L = 99.4$$

Panhandle equations "A" and "B" give good results for pressures higher than 100 psia and diameters larger than 8 in. and for systems operating at $R_e > 10^5$.

In our example, because of the large diameter of pipe being used, $L = 104.1$ mi is the correct answer:

Example 6.7.24
Suppose that the pipeline of example 18 has been built and that at some later date the capacity of the line is to be increased to 320 MMcf/d by lowering the suction pressure at the stations. If $E = 0.95$, what suction pressure would be required? Use the Panhandle "A" and "B" equations.

Using Panhandle "A" equation

$$P_2^2 = P_1^2 - \frac{Q_{sc}^{2/0.5394} p_{sc}^2 T_m Z_m \gamma_g^{0.854} L}{(KE)^{1/0.5394}(520)^2(23.2)^{4.854}}$$

$$= (84.3)^2 - \frac{(320)(10^6)^{1.8539}(15.025)(520)(0.86)(0.71)^{0.854}(104.1)}{(0.95)^{1.8539}(435.87)^{1.8539}(520)^2(23.2)^{4.854}}$$

$$= 100,183$$

$$P_2 = 316.5 \, \text{psia}$$

Using Panhandle "B" equation

$$P_2^2 = (814.3)^2$$

$$- \frac{(320)(10^6)^{1.96}(15.025)^2(520)(0.86)(0.71)^{0.961}(99.4)}{(0.95)^{1.96}(737)^{1.96}(520)^2(23.2)^{4.961}}$$

$$= 100,183$$

$$P_2 = 313.1 \, \text{psia}$$

6.7.14 Complex Gas Flow Systems
A situation often requires increasing the flowrate per unit pressure drop in a system. The common way to solve this problem is to place one or more lines parallel to the original, either partially or throughout the entire length of the system (see Figure 6.7.47a). If the lines are of equal length $(\Delta P/L)_A = (\Delta P/L)_B$ and if the line 1-2 is partially looped with B length, $L_A = L_B$

$$Q_A + Q_B = Q_C = Q_{total}$$

$$\Delta P_A = \Delta P_B$$

The pressure loss due to friction in the looped position plus that in the unlooped portion must equal the total friction loss:

$$\frac{\Delta P_f}{L_{nA}}(x) + \frac{\Delta P_f}{L_n}(L - x) = \text{total } \Delta P_f$$

The purpose of the second line B is to decrease the pressure drop between points 1 and 3, compensating for the increased pressure between points 3 and 2, the unlooped portion, with the required increased flow.

The basic Equation 6.7.38 may be used to anticipate rates and sizing of lines, which could be expressed as:

$$Q = K' \frac{D_5^{0.5}}{f_L} \quad \text{(E)}$$

The flow of gas in "n" number of horizontal parallel looped lines is shown as follows:

$$Q = K \left(\frac{T}{P} \right)_{sc} \left(\frac{P_1^2 - P_2^2}{SG_g TLZ} \right) \underbrace{\left(\frac{D_A^{2.5}}{f_A^{0.5}} + \frac{D_B^{2.5}}{f_B^{0.5}} + \cdots + \frac{D_n^{2.5}}{f_n^{0.5}} \right)}_{D_e} E$$

[6.7.43]

The equivalent diameter D_e of a single line having the same capacity as the foregoing set of parallel lines is shown:

$$D_e^{2.5} = \left(\frac{f_e}{f_A} \right)^{0.5} D_A^{2.5} + \left(\frac{f_e}{f_B} \right)^{0.5} D_B^{2.5} + \cdots + \left(\frac{f_e}{f_n} \right) D_n^{2.5}$$

[6.7.44]

Combining both an equivalent length, L_e, and equivalent diameter, D_e, terms:

$$\frac{D_e^{2.5}}{(f_e L_e)^{0.5}} = \frac{D_A^{2.5}}{(f_A L_A)^{0.5}} + \frac{D_B^{2.5}}{(f_B L_B)^{0.5}} + \cdots + \frac{D_n^{2.5}}{(f_n L_n)^{0.5}} \quad [6.7.45]$$

In Equation 6.7.45, one usually chooses a diameter D_e, equal in size to the unlooped portion, then calculates the equivalent length, L_e, (corresponding to the ΔP characteristic of the actual looped section). This, L_e, is added to the length of the unlooped portion. Table 6.7.31 summarizes other complex gas flows.

Example 6.7.25
A brand line is laid to move gas a distance of 60 mi from a junction on the main line to a city. The line is to transmit 23 MMcf/d of 0.73 gravity gas, measured at 60°F and 14.65 psia. Flowing temperature will be 65°F at 14.65 psia. Pressure at the junction is 620 psia and the gas is to be delivered to the city gate at 100 psia. The objective is to minimize the investment in pipe by laying a line consisting of two sizes in series. The following sizes are available:

Nominal Size (in.)	Internal diameter (in.)
6	6.065
8	8.071
10	10.192
12	12.090

Which two of the above pipe sizes should be used, and how much of each should be laid? Disregard the compressibility factor and pipeline efficiency in this problem.

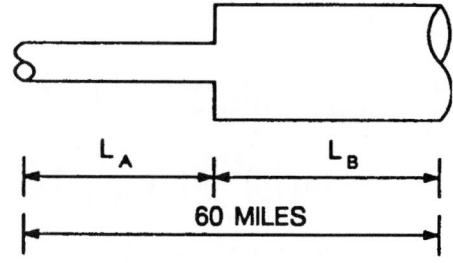

Find D_e from Weymouth Equation 6.7.39:

$$D_e = \left[\frac{SG_g T_m Z_m L}{(P_1^2 - P_2^2)} \left(\frac{Q_{sc} EP_{sc}}{KT_{sc}} \right)^2 \right]^{1/5.333}$$

$$Z_m = E = 1.0$$

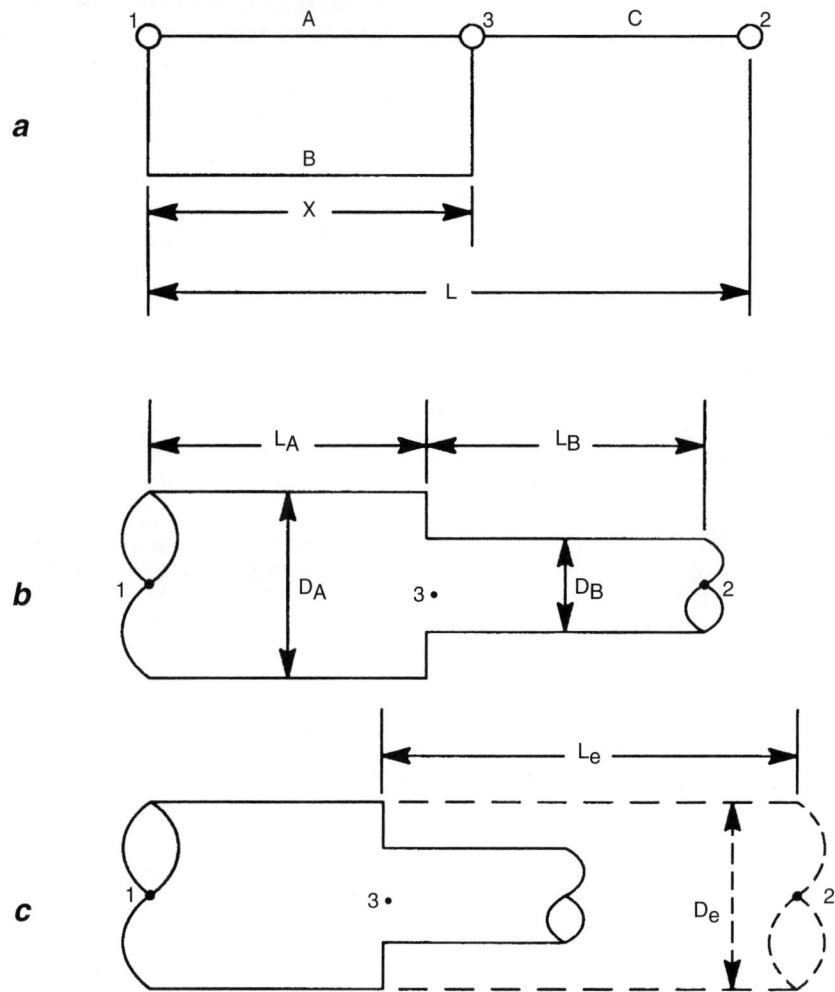

Figure 6.7.47 *Complex pipelines a) the looped pipeline system, b) the series pipeline system, c) the equivalent length and diameter system.*

$$D_e = \left[\frac{0.73 \times 525 \times 60}{620^2 - 100^2} \left(\frac{23 \times 10^6 \times 14.65}{433.5 \times 520} \right)^2 \right]^{1/5.333}$$

$$= 9.19023$$

For Case I, assume $D_A = 8.071$, $D_B = 10.192$, $L_A + L_B = 60$, $L_A = 60 - L_B$, $L_B = 60 - L_A$ or

$$D_e = D_B \left(\frac{L_A}{L_B} \right)^{3/16} = D_B \left(\frac{L_A}{60 - L_A} \right)^{3/16}$$

$$9.19023 = 10.192 \left(\frac{L_A}{60 - L_A} \right)^{3/16}$$

$$L_A = 21.92 \ (D_A = 8 \text{ in.})$$

and

$$L_B = 38.08 \ (D_B = 10 \text{ in.})$$

For case II, assume $D_A = 10.192$, $D_B = 8.071$

$$D_e = D_B \left(\frac{L_A}{L_B} \right)^{3/16}$$

$$9.19023 = 8.071 \left(\frac{L_A}{60 - L_A} \right)^{3/16}$$

$$1.9989 = \frac{L_A}{60 - L_A}$$

$L_A = 39.99 = 40$ mi $(D_A = 10$ in.$)$

$L_B = 20$ mi $(D_B = 8$ in.$)$

Both cases, more or less, in given conditions give similar results. As a solution, it is better to take $L_A = 21.92$ with $D_A = 8$ in. and $L_B = 38.08$ with $D_B = 10$ in.

Example 6.7.26

Two lines, AC (5-in. ID, 4.5 mi long) and BC (5 in. ID, 3.8 mi long), emanating from leases A and B, respectively, terminate at the gathering station at C. From C a single 11.5-mi, 8-in. ID pipeline leads into the regional trunk line at D, where the gas must be delivered at 520 psig. Lease A produces 5 MMscfd of 0.65 gravity gas, while lease B produces 7 MMscfd of 0.68 gravity gas. What pressure should the gas be compressed to at (a) lease A and (b) lease B to enable delivery of gas to the trunk line? $E = 0.92$.

Assume a flowing temperature of 85°F, and horizontal flow system. $P_{sc} = 14.73$ psia, $T_{sc} = 60$°F.

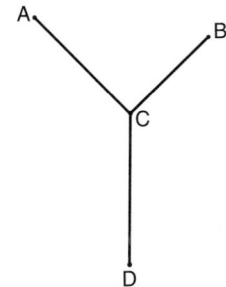

$AC = 4.5$ mi, ID $= 5$ in.,

$Q = 5$ MMscfd

$BC = 3.8$ mi, ID $= 5$ in.,

$Q = 7$ MMscfd

$CD = 11.5$ mi, ID $= 8$ in.

First, the pressure at point C is necessary to find:

$$SG_{gav} = \frac{5 \times 0.65 + 7 \times 0.68}{12}$$

$$= 0.6675$$

Apply the Weymouth equation for section C–D;

$SG_{gav} = 0.6675$, $T_{pc} = 378$°R, $P_{pc} = 668$ psia

$Z = 0.91$

$$12 \times 10^6 = 433.5 \left(\frac{520}{14.73}\right) \left[\frac{P_1^2 - (534.73)^2 8^{16/3}}{(0.6675)(11.5)(545)(0.91)}\right]^{0.5} 0.92$$

P_1 corresponds to P_c

$P_c = 572.8$

Section B–C

$SG_g = 0.68$, $T_{pc} = 384$°R, $P_{pc} = 668$ psia

$Z = 0.9$

$$7 \times 10^6 = 433.5 \left(\frac{60 + 460}{14.73}\right) \left[\frac{(P_B^2 - 572.8^2)5^{16/3}}{(0.68)(0.9)(545)(3.8)}\right]^{0.5} 0.92$$

$P_B = 621.8$ psia

Section A–C

$SG_g = 0.65$, $T_{pc} = 373$°R, $P_{pc} = 670$ psia, $Z = 0.9$

$$5 \times 10^6 = 433.5 \left(\frac{60 + 460}{14.73}\right) \left[\frac{(P_A^2 - 572.8^2)5^{16/3}}{(0.65)(0.9)(545)(4.5)}\right]^{0.5} 0.92$$

$P_A = 601.6$ psia

Example 6.7.27

A portion of a large gas-gathering system consists of a 6.067-in. ID line 9.4 mi long, handling 7.6 MMscfd of gas 0.64 gravity. The pressure at the upstream end of this section is 375 psig and the average delivery pressure is 300 psig. The average temperature is 73°F. Due to new well completions, it is desired to increase the capacity of this line by 20% by looping with additional 6.067-in. pipe. What length is required?

From Table 6.7.29, if the Weymouth equation is used, then

$$X_f = \frac{1 - \left(\dfrac{Q_1}{Q_2}\right)^2}{1 - \left(\dfrac{D_1^{8/3}}{D_1^{8/3} + D_2^{8/3}}\right)^2} = \frac{1 - \left(\dfrac{1}{1.2}\right)^2}{1 - \left(\dfrac{6.067}{6.067 + 6.067}\right)^2}$$

$$= 0.4074$$

The length of looping pipe required $= 0.4074 \times 9.4 = 3.83$ mi

6.7.15 Compressor Stations

A compressor station is one of the most important elements in a natural gas pipeline system. Compressor stations supply the energy to pump gas from production fields, overcome frictional losses in transmission pipeline and pump gas into storage reservoirs.

Depending on its purpose, a compressor station generally falls into one of three distinct classifications: production, storage and transmission. Compressors are used whenever it is necessary to flow gas from a lower pressure to a higher pressure system. The differential pressure is expressed in terms of overall compressor ratio, R_T, which is defined as follows:

$$R_T = \frac{P_d}{P_d} \qquad [6.7.46]$$

where R_T = overall compression ratio
$\qquad P_d$ = discharge pressure, psia
$\qquad P_d$ = suction pressure, psia

Production stations are used in the natural gas production field to pump gas from the well head to the pipeline system. They usually are designed for unattended operation with a life expectancy comparable to that of the production field. Unitized, skid-mounted, medium to high-speed, packages compressor units in the 50 to 1,000 hp (35 to 745 kW) range are often utilized for this type of service. Operating pressure in production systems vary widely. The suction pressure could be as high as 1,000 psig (6,900 kpa).

Storage stations are used primarily to pump gas from a transmission pipeline into an underground storage field during the injection period. In addition to injecting gas during the storage season, some are used to pump gas from the storage field to the pipeline during the withdrawal period. Most storage field compressors are designed to operate over a broad range of volumes and pressures. A typical storage station would contain several reciprocating compressor units ranging in size from 1,000 to 6,000 hp (745 to 4,475 kW) each.

Table 6.7.31 *Equations for Complex Gas Flow*

Equation	Basic	Weymouth	Panhandle A
	For Two Lines in Series		
Equivalent Diameter D_e	$D_B \left(\dfrac{L_A}{L_B}\right)^{1/5} \left(\dfrac{f_A}{f_B}\right)^{1/5}$	$D_B \left(\dfrac{L_A}{L_B}\right)^{3/16}$	$D_B \left(\dfrac{L_A}{L_B}\right)^{0.2060}$
Equivalent Diameter L_e	$L_B \left(\dfrac{D_A}{D_B}\right)^{5} \left(\dfrac{f_B}{f_A}\right)$	$L_B \left(\dfrac{D_A}{D_B}\right)^{16/3}$	$L_B \left(\dfrac{D_A}{D_B}\right)^{4.854}$
	For Two Lines in Series		
Equivalent Diameter or Length-Loops D_e or L_e	$\dfrac{D_e^{2.5}}{(f_e L_e)^{0.5}} = \dfrac{D_A^{2.5}}{(f_A L_A)^{0.5}} + \dfrac{D_B^{2.5}}{(f_B L_B)^{0.5}}$	$\dfrac{D_e^{8/3}}{L_e^{0.5}} = \dfrac{D_A^{8/3}}{L_A^{0.5}} + \dfrac{D_B^{8/3}}{L_B^{0.5}}$	$\dfrac{D_e^{2.616}}{L_e^{0.5394}} = \dfrac{D_A^{2.616}}{L_A^{0.5394}} + \dfrac{D_B^{2.616}}{L_B^{0.5394}}$
Loops-Diameters and Flows Vary X_f – Fraction Looped $D_R = \left(\dfrac{f_1}{f_2}\right)^{0.5} \left(\dfrac{D_2}{D_1}\right)^{5/2}$	$x_f = \dfrac{\left(\dfrac{Q_1}{Q_2}\right)^{0.5} - 1}{\left[\dfrac{1}{(1+D_R)^2} - 1\right]}$	$x_f = \dfrac{1 - \left(\dfrac{Q_1}{Q_2}\right)^2}{1 - \left[\dfrac{D_1^{8/3}}{D_1^{8/3}+D_2^{8/3}}\right]^2}$	$x_f = \dfrac{1 - \left(\dfrac{Q_1}{Q_2}\right)^{1.86}}{1 - \left[1 - \left(\dfrac{D_2}{D_1}\right)^{2.62}\right]^{1.86}}$
Entire Line Looped		$\dfrac{Q_2}{Q_1} = 1 + \left(\dfrac{D_2}{D_1}\right)^{8/3}$	$\dfrac{Q_2}{Q_1} = 1 + \left(\dfrac{D_2}{D_1}\right)^{2.618}$
Diameters of Original and Parallel Lines are the same X_f = Fraction Looped	$x_f = \dfrac{4}{3}\left[1 - \left(\dfrac{Q_1}{Q_2}\right)^2\right]$	$x_f = \dfrac{4}{3}\left[1 - \left(\dfrac{Q_1}{Q_2}\right)^2\right]$	$x_f = \dfrac{4}{3}\left[1 - \left(\dfrac{Q_1}{Q_2}\right)^{1.86}\right]$

Transmission stations are the most common type. They generally contain several large compressors units at one location, each having a power rating of 1,000 to 2,000 hp. Their purpose is to move gas in the transmission pipeline from one station to the next. Operating pressures seldom exceed 1,000 psig (6,900 kpa). Designed for long life, 50 years or more, and intended to operate year-round.

Independent of compressor stations, three types of compressors are used: reciprocating, centrifugal and axial. See Figure 6.7.48. Reciprocating compressors are considered for applications where the inlet gas rate is about 3,000 acfm or less. They are favored for low-flow, high pressure service, compression ratio 2:1 to 4:1. They are available in sizes up to 15,000 hp (11,190 kw).

Centrifugal compressors are used for an inlet gas rate 500 to 200,000 acfm, low compression ratio 1.1:2 and range of power 1,000 to 30,000 hp (746 to 22,380 kw).

For axial-flow compressors, with and inlet gas rate of 75,000 to 600,000 acfm, smaller axial compressors are available (down to 20,000 acfm), but centrifugal compressors are preferred at these capacities. Axial compressors are more efficient than centrifugal compressors.

6.7.15.1 Reciprocating Compressor Calculations

Figure 6.7.48 illustrates the ideal compression cycle on a pressure-volume cycle. [18]. The events of this cycle are as follows:

- *Compression (points 1 to 2)* — The piston has moved to the left, reducing the original volume of gas with an accompanying rise in pressure. Valves remain closed. The P–V diagram shows compression from point 1 to point 2 and the pressure inside the cylinder has reached that in the receiver.

Table 6.7.32 *Centrifugal Compressor Flow Range*

Nominal flow range (Inlet acfm)	Average polytropic efficiency	Average adiabatic (isentropic) efficiency	Speed to develop 10,000 ft head/wheel
100–500	0.63	0.60	20,500
500–7,500	0.74	0.70	10,500
7,500–20,000	0.77	0.73	8,200
20,000–33,000	0.77	0.73	6,500
33,000–55,000	0.77	0.73	4,900
55,000–80,000	0.77	0.73	4,300
80,000–115,000	0.77	0.73	3,600
115,000–145,000	0.77	0.73	2,800
145,000–200,000	0.77	0.73	2,500

- *Discharge (points 2 to 3)* — The discharge valves are opened just beyond point 2. Compressed gas in flowing out through the discharge valves to the receiver. After the piston reaches point 3, the discharge valves will close, leaving the clearance space filled with gas at discharge pressure.
- *Expansion (point 3 to 4)* — During the expansion stroke, both the inlet and discharge valves remain closed and gas trapped in the clearance space increases in volume, causing a reduction in pressure. This continues as the piston moves to the right, until the cylinder pressure drops below the inlet pressure at point 4.
- *Suction (point 4 to 1)* — At point 4 the suction valve opens and permits gas at suction pressure to enter the cylinder as the piston moves from point 4 to point 1.

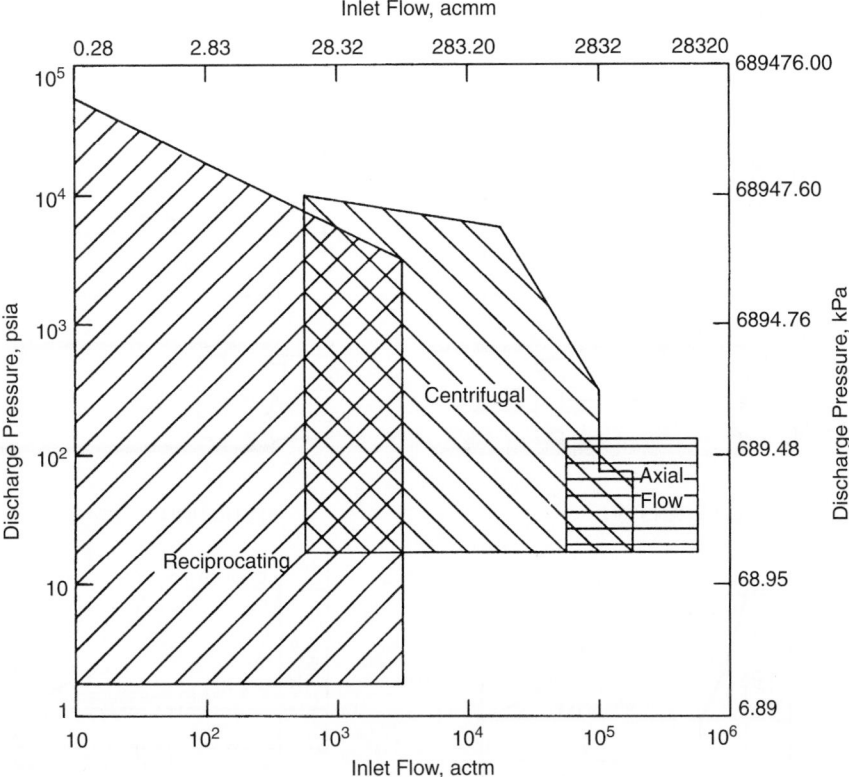

Figure 6.7.48 *Approximate ranges of application for reciprocating, centrifugal, and axial flow compressors.*

To determine the quantity of gas that a specific compressor can pump, the displacement volume of gas at suction conditions and suction volumetric efficiency must be known. The displacement of a compressor is expressed in cubic feet per minute (ft.³/min), cubic feet per second (ft.³/s) or in cubic meters per second (m³/s).

Figure 6.7.49 shows the cycle as related to a single-acting compressor (compression occurs only on one end of the piston). Most reciprocating compressors in the gas industry are double-acting. Gas is compressed alternately on each end of the piston.

For single acting compressors:

$$PD = \frac{A_H \times L \times M}{1,728} \quad (ft.^3/min) \qquad [6.7.47]$$

or

$$PD = \frac{A_H LN}{6 \times 10^9} \quad (m^3/s)$$

For double-acting compressors:

$$PD = \frac{(A_H + A_c) \times L \times N}{1,728} \quad (ft.^3/min) \qquad [6.7.48]$$

or

$$PD = \frac{(A_H + A_c) \times L \times N}{6 \times 10^9} \quad (m^3/s)$$

where PD = piston displacement in ft.³/min or m³/s

A_H = area of head end of the piston in in.² or mm²
L = length of stroke in in. or mm
N = compressor rotation speed in rpm
A_c = area of crank end of the piston in in.² or mm²

Most compressor manufacturers publish data listing displacement at full compressor speed for each size cylinder that is manufactured.

Volumetric efficiency E_v, is the ratio (%) of the actual delivered volume flow rate (at inlet conditions) to the piston displacement. The volumetric efficiency factor may be calculated as follows:

$$E_v = 100 - L - Cl \times \left[\frac{Z_s}{Z_d} \left(\frac{P_d}{P_s} \right)^{1/k} - 1 \right] \qquad [6.7.49]$$

where E_v = volumetric efficiency factor in %
Cl = cylinder clearance expressed as a percentage of piston displacement
P_d = pressure at the discharge flange in psia or kPa
P_s = pressure at the suction flange in psia or kPa
k = ratio of specific heat capacities (adiabatic exponent), C_p/C_v
L = volumetric efficiency correction factor in %, $L = P_d/P_s$
Z_s = compressibility factor at suction conditions
Z_d = compressibility factor at discharge conditions
P_d/P_s = r, the compression ratio

Compressor cylinder capacity Q may be found from the following:

$$Q = PD \times E_v \times P_s/P_a \times 14.4 \times 10^{-6} \ (MMscf/d) \qquad [6.7.50]$$

or

$$Q = PD \times E_v \times P_s/P_a \times 864 \ (m^3/d)$$

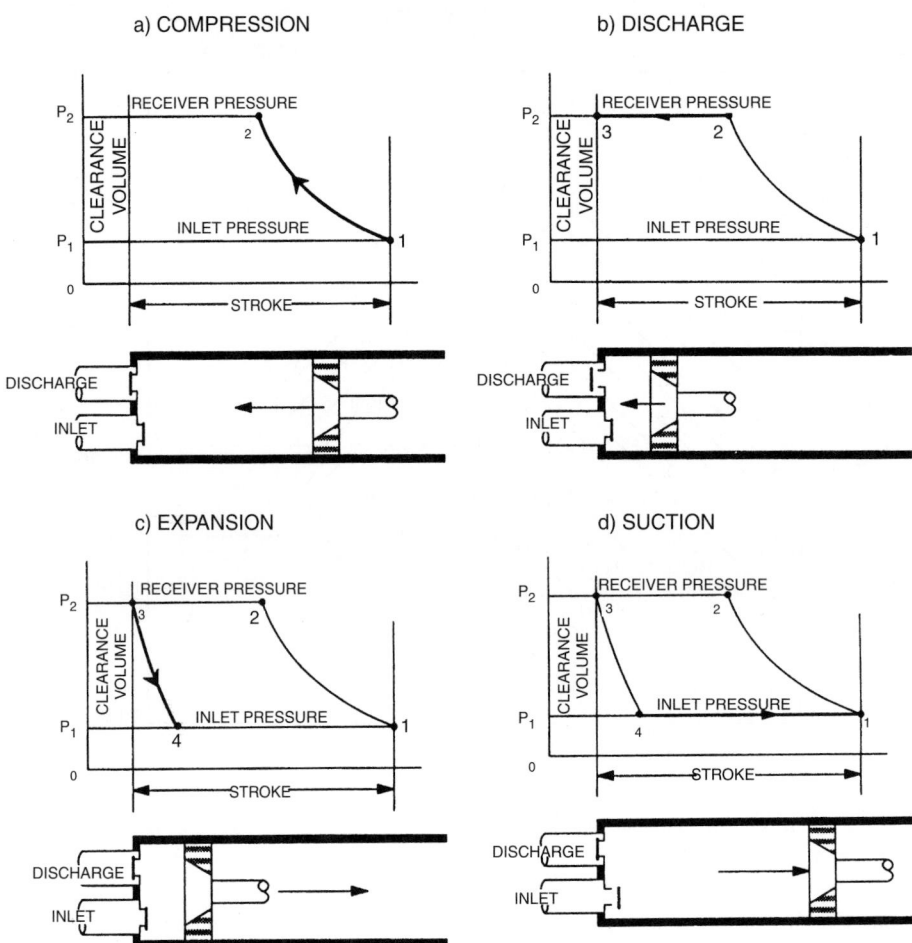

Figure 6.7.49 *P-N diagram illustrating ideal reciprocating compression of gases.*

where Q = cylinder capacity in desired units at the prevailing inlet temperature and the base suction pressure in MMscf/d or m^3/d

P_a = base reference pressure in psia or kpa

PD, E_v, P_s are as in Equation 6.7.49

If the base reference pressure is 14.4 psia (99 kpa), the cylinder capacity at the base reference pressure and prevailing temperature is:

$$Q = PD \times E_v \times P_s \times 10^{-6} \quad (MMscf/d) \qquad [6.7.51]$$

or

$$Q = PD \times E_v \times P_s \times 8.73 \quad (m^3/d)$$

Q represents the actual displaced quantity of gas in MMscf/d (m^3/d), corrected only for pressure to a base of 14.4 psia (99 kpa) at the prevailing suction temperature. According to ISO 5024, scm has only one set of reference conditions: P_a = 101.325 kpa and T = 15°C. in the United States, T = 15.5°C (60°F) and pressure differ in the various industries.

$$Q_{std} = \frac{Q \times P_a \times T_b}{Z_s \times P_b \times T_s} \qquad [6.7.52]$$

where

Q_{std} = rate of flow in MMscf/d or scm/d

Q = rate of flow in MMcf/d or m^3/d

P_a = reference base pressure in psia or kPa

P_b = pressure base or standard pressure in psia or kPa

T_b = temperature base or standard in °R (K)

T_s = suction temperature in °R (K)

Z_s = compressibility factor at inlet conditions

For power P, many methods are available to determine the power required to compress a given volume of gas between two levels. The theoretical adiabatic power requirement, taking volumetric efficiency into account, can be calculated as follows:

$$P_{ad} = 4.3636 \times 10^{-5} \times P_s \times PD \times k/k - 1 \ [r^{(k-1)/k} - 1]E_v \qquad [6.7.53a]$$

where P is in hp or, in the SI system

$$P_{ad} = 1.6672 \times 10^{-4} \times P_s \times PD \times k/k - 1 \ [r^{(k-1)/k} - 1]E_v \qquad [6.7.53b]$$

where P is in kw.

If compressor capacity is given in mass flow rate, then for case I,

$$V_1 = V_s \rightarrow Z_1 = Z_s = 1$$

$$P_{ad} = 0.2618 \times P_s \times V_s \times k/k - 1\,[r^{(k-1)/k} - 1] \quad [6.7.54a]$$

where P_{ad} power is in hp
P_s suction pressure is in psia
V_s flow rate at suction is in scf/s

And in the SI system;

$$P_{ad} = P_s \times V_s \times k/k - 1\,[r^{(k-1)/k} - 1] \quad [6.7.54b]$$

where P_{ad} = power in kw
P_s = suction pressure in kpa
V_s = (scm/s) = flow rate at suction (scm/s)
$V_s = V_s m(kg/s \times m^3/kg) \rightarrow (m^3/s)$

For case II

$$V_1 \neq V_s \rightarrow Z_1 \neq Z_s$$

$$P_{ad} = const\ V_s\ (P/T)_{sc} T_s Z_{avg}(k/k - 1)\ (r^{(k-1)/k} - 1) \quad [6.7.55]$$

where const = 0.2618 for the English system
= 1.0 for the SI system
V_s = is in scf/s or scm/s
P, T = standard conditions in psia, °R or kpa, K
T_s = suction temperature in °R or K
$Z_{avg} = (Z_s + Z_D)/2$
$T_D = T_s[(P_d/P_s)^{k-1/k}]$

The actual brake horsepower (ABHP), or the power to the shaft, is calculated as:

$$ABHP = \frac{P_{ad}}{\eta_{ad}\eta_m} = \frac{P_{ad}}{\eta_t} \quad [6.7.56]$$

The adiabatic efficiencies for commercial machines are as follows [19]:

- 0.83 to 0.94 for a cylinder of a reciprocating compressor
- 0.60 to 0.65 for Roots blower
- 0.80 to 0.85 for a screw-type rotary compressor

Compression ratio (Equation 6.7.46), $R_T = P_d/P_s$, when this value is higher than 3.5; a multiple-stage machine may have to be used. There are several reasons for considering multistage compression systems:

- to avoid high temperatures [180°C–200°C (350°F to 380°F)]
- to optimize the mechanical design of the compressor
- to minimize compression power

The total power is a minimum when the ratio in each stage is the same.

$$R_{opt} = (R_T)^{1/s} = (P_d/P_s)^{1/s}$$

where s = number of stages required
R_{opt} = optimum compression ratio per stage

If intercoolers are provided between the stages, reduce the theoretical intake pressure of each stage by about 3%, so

$$R_{opt} = (r_T/0.97)^{1/s} \quad [6.7.57]$$

Example 6.7.28

A compressor station is to handle 54 MMscfd of 0.65 relative density gas. Suction temperature will be 70°F and suction pressure 190 psia. Discharge pressure will be 1,500 psia. Calculate the brake horsepower required for two-stage compression with intercooling at 70°F. Other data observed: $\eta_{ad} = 0.85$, $\eta_m = 0.97$, $T_{sc} = 60°F$, $P_{sc} = 14.5$ psia.

Flow rate $V = 54 \times 10^6$ scf/day = 625 scf/s

$SG_g = 0.65$

$T_s = 70°F = 530°R$

$P_s = 190$ psia $P_d = 1,500$ psia

1. Single-stage compression, from Equation 6.7.55, is calculated by

$$P_{ad} = 0.2618V(P/T)_{sc}T_s Z_{avg}k/k - 1\,[(P_d/P_s)^{k-1/k} - 1]$$

$$Z_{avg} = (Z_s + Z_d)/Z$$

$$M = 28.96 \times 0.65 = 18.8$$

For the given molecular mass

$P_{pc} = 674$ psia, $T_{pc} = 324°R$

$p_r = 190/674 = 0.282$, $T_r = 530/324 = 1.64$, $Z_s = 0.97$

$T_d = 530(1500/190)^{(k-1)/k}$ $k = 1.26$ (from Figure 6.5.50)

if $T = 530°R$

$T_d = 811.8°R$

$P_r = 1500/674 = 2.23$, $T_r = 811.8/324 = 2.51$,

$Z_d = 0.99$

$$Z_{avg} = \frac{0.97 + 0.99}{2} = 0.98$$

$$T_{avg} = \frac{70 + 352}{2} = 211°F$$

$k = 1.23$ (from Figure 6.5.50)

$$P_{ad} = 0.2618 \times 625\left(\frac{14.5}{520}\right)530(0.98)\frac{1.23}{1.23 - 1}$$

$$\times \left[\left(\frac{1,500}{190}\right)^{0.23/1.23} - 1\right] = 5,977$$

$$ABHP = \frac{5,977}{\eta_{ad}\eta_m} = \frac{5,977}{(0.85)(0.97)} = 7,249$$

2. Two-stage compression is calculated by

$$r_t = \frac{1,500}{190} = 7.89 \text{ total compression ratio}$$

$$r_{opt} = (7.89/0.97)^{0.5}$$

$$= 2.85 \text{ comprerssion ratio per stage with}$$

intercoolers

The calculation for the first stage:

$$P_{d1} = rP_{s1} = 2.85 \times 190 = 541.9$$

For

$$T_s = 70°F \qquad k = 1.26$$

$$T_d = 530\left(\frac{541.9}{190}\right)^{0.26/1.26} = 658°R = 198°F$$

For

$$T_m = \frac{70 + 198}{2} = 134°F(294°R) \quad k \cong 1.26$$

$$Z_s = 0.97$$

$$P_{ad} = 0.2618 \times 6.25 \left(\frac{14.5}{520}\right) 530(0.97)$$

$$\times \frac{1.26}{0.26}\left[\left(\frac{541.9}{190}\right)^{1.26-1} - 1\right]$$

$$= 2,744.4 \text{ hp}$$

$$ABHP = \frac{P_{ad}}{\eta_{ad}\eta_m} = \frac{2,744.4}{(0.85)(0.97)} = 3,328.6 \text{ hp}$$

The calculation for the second stage:

Because of intercooling

$$T_s = 70°F = 530°R \qquad P_s = 541.9 \qquad k = 1.24$$

$$T_d = T_s \left(\frac{1,500}{541.9}\right)^{(1.24-1)/1.24} = 85.2°F$$

$$T_{avg} = \frac{85.2 + 70}{2} = 77.6 \qquad k = 1.27$$

$$P_r = \frac{(1,500 + 541.9)/2}{674} = 1.515$$

$$T_r = \frac{77.6 + 460}{324} = 1.66$$

$$Z_{avg} = 0.9$$

$$H_{ad} = 0.2618 \times 625 \left(\frac{14.5}{520}\right) 530(0.9)$$

$$\times \frac{1.27}{1.27-1}\left[\left(\frac{1,500}{541.9}\right)^{0.27/1.27} - 1\right] = 2,473.9 \text{ hp}$$

$$ABHP = \frac{2,473.9}{0.85 \times 0.97} = 3,000.54$$

Total ABHP = 3,000.54 + 3,328.6 = 6,329.1 hp

Example 6.7.29
A branch pipeline is to move 40.0 MMscf/d of 0.69 relative density gas from a junction on the main line to a city, a distance of 100 mi. The line will be a 10-in. line (internal diameter 9.562 in.). The gas will enter the branch line at a pressure of 750 psia and is to be delivered to the city gate at a pressure of 80 psia. The flowing temperature will be 50°F. Pressure on the pipeline is not to exceed 1,000 psia. One compressor station will be needed. If the compressor station should be located at the midpoint of the branch line, what would the required brake horsepower be? Where should the compressor station be located for the minimum horsepower requirements? Assume that $Z = 1.0$, $k = 1.30$, $E = 1.0$, $\eta_{ad} = 0.85$, $\eta_m = 0.97$, $T_{sc} = 60°F$, $P_{sc} = 14.696$ psia.

1. Solve the Weymouth equation if the compressor is in the midpoint:

$$P_s = ?$$

$$40 \times 10^6 = 433.5\left(\frac{519.67}{14.696}\right)\left[\frac{(9.562)^{16/3}(750^2 - P_s^2)}{0.69100(509.67)(1)}\right]^{0.5} \times 1.0$$

$$(1,188.16)^2 = 750^2 - P_s^2$$

It is impossible.

2. At what point downstream reaches 80 psia?

$$40 \times 10^6 = 433.5\left(\frac{519.67}{14.696}\right)\left[\frac{(9.562)^{16/3}(750^2 - 80^2)}{0.69(509.67)(X) \times 1}\right]^{0.5}$$

$$X = 39.4 \text{ mi}$$

3. What is the minimum discharge pressure for compressors?

$$40 \times 10^6 = 433.5\left(\frac{519.67}{14.696}\right)\left[\frac{(9.562)^{16/3}(P_d^2 - 80^2)}{0.69(509.67)(100 - 39.4)}\right]^{0.5} \quad (1)$$

$$P_d = 928.3 \text{ psia}$$

$$ABHP = \frac{0.2618(1.3)463 \times 14,696 \times 509.67}{(0.3)(0.85)(0.97)(519.67)}$$

$$\times \left[\left(\frac{928.3}{80}\right)^{0.3/1.3} - 1\right]$$

$$= 6,984.6 \text{ hp}$$

To minimize ABHP it is necessary to use stage-2 compressors because $r = 11.6$ and maximum r should not be higher than 6.

Alternative Solution
The equation for the upstream section is

$$40 \times 10^6 = \frac{433.5(519.67)}{14.696}\left[\frac{(9.562)^{16/3}(750^2 - P_s^2)}{(0.69)(509.67)(1)X}\right]^{0.5} \quad (1.0)$$

$$6,800,709X = 482.14(750^2 - P_s^2)$$

$$14,105.3X = 750^2 - P_s^2$$

The equations for the downstream section are

$$40 \times 10^6 = 14,338.5\left[482.14\frac{P_d^2 - 80^2}{100 - X}\right]^{0.5} \quad (1.0)$$

$$14,105.3(100 - X) = P_d^2 - 80^2$$

$$P_d^2 = 1,416,930 - 14,105.3X$$

Solving these two equations gives

$$14,105.3X = 1,416,930 - P_d^2$$

$$= 750^2 - P_s^2$$

If $\quad P_s = 80, \quad P_d = 927.8$

If $\quad P_d = 1,000, \quad P_s = 381$

$$ABHP = \frac{0.2619(1.3)(463)(14,696)(509.67)}{(0.3)(0.85)(0.97)(519.67)}$$

$$\times \left[\left(\frac{1,000}{381}\right)^{0.3/1.3} - 1\right]$$

$$= 2,790$$

The minimum hp requirement for suction pressure to compressor will be 381 psia. It corresponds to X equal to

$$14,105.3X = 750^2 - 381^2 = 417,339$$

$$X = 29.58 \text{ mi}$$

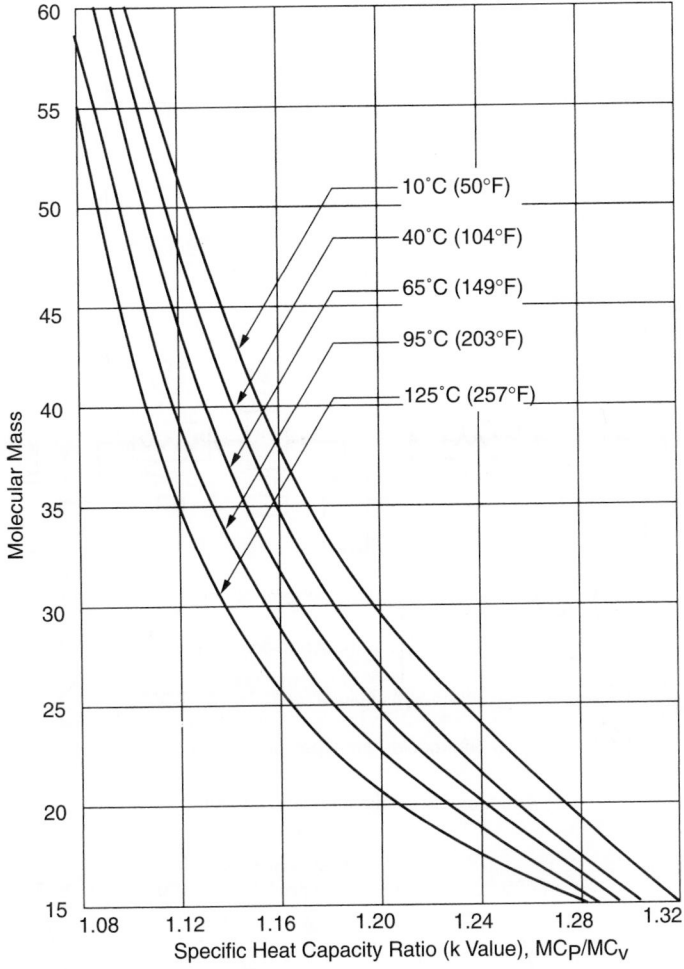

Figure 6.7.50 *Approximate specific heat capacity ratios of hydrocarbon gases.*

6.7.15.2 Using Mollier Charts for Compressor Calculations

Mollier charts are forms in which gas properties such as pressure, specific volume, temperature, entropy and enthalpy are presented. Gases are generally plotted as pressure against enthalpy or enthalpy-entropy. The enthalpy-entropy plot is a good technique for solving compression problems (isentropic compression). The key assumptions in this approach are that the process is adiabatic, $Q = 0$,

$$W = \Delta H = n(h_2 - h_1) \qquad [6.7.58]$$

where

$\qquad W = $ work done by compressor on the gas in Btu or kJ

$\quad \Delta H = $ change in enthalpy of the gas in Btu or kJ

$\qquad n = $ number of moles of gas being compressed in lb mole or kmol

$h_1, h_2 = $ enthalpies of the gas at the compressor inlet and discharge, respectively, in Btu/lb mole or kJ/kmol

Necessary data include inlet gas rate, pressure, temperature and composition plus outlet pressure. The corresponding outlet temperature is found from these data by assuming $S_1 = S_2$, using an iterative process.

Figure 6.7.51 shows a qualitive sketch of an enthalpy-entropy chart, illustrating two-stage compression with intercoolers and aftercoolers. Point 1 is the initial state of the gas as it enters the compressor. Path 1-2 shows the first stage of compression. The gas is then cooled in the intercoolers at constant pressure, path 2-3; the difference in enthalpy along this path is equal to the heat removed in the intercoolers. Path 3-4 shows the second stage of compression. Path 4-5 shows cooling at a constant pressure in the aftercoolers. The temperatures at points 2 and 4 are the temperatures of the gas at the end of the first and second stages of compression. The temperatures at points 3 and 5 are the temperatures to which the gas is cooled in the intercoolers and aftercoolers. The ideal compression power (or rate of work) required is given by:

$$P = \frac{W}{t} = \frac{n(h_2 - h_1)}{t} \qquad [6.7.59]$$

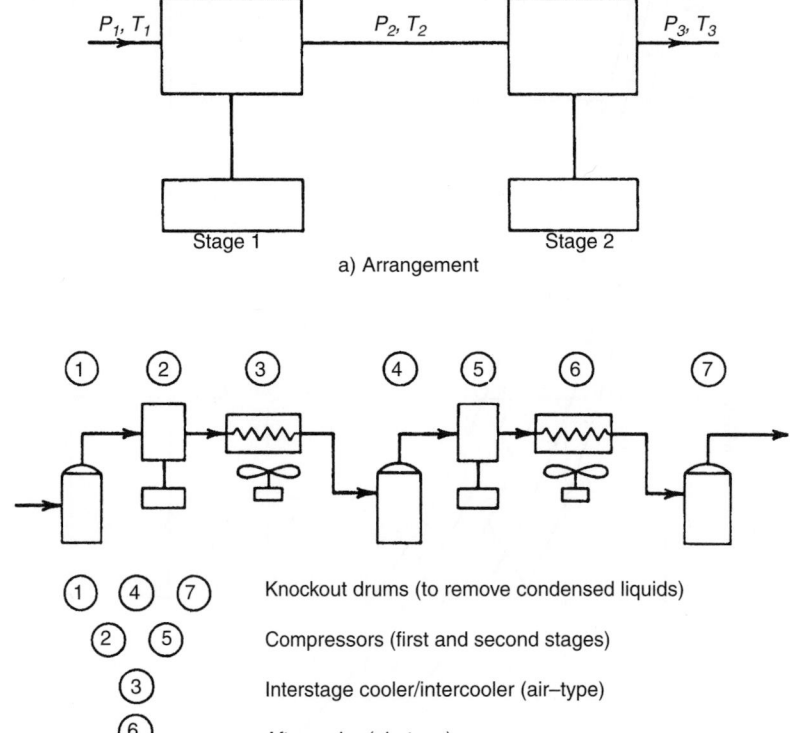

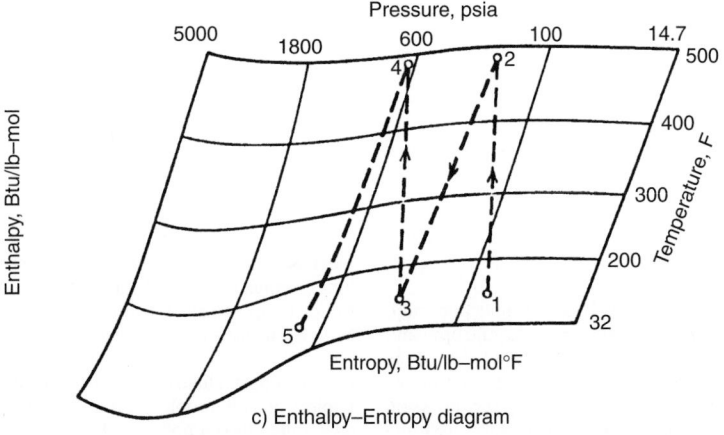

c) Enthalpy–Entropy diagram

Figure 6.7.51 *Two-stage compression.*

where

P = compression power required in Btu/day or kJ/day
t = time for compression in days

$$\text{Power Requirement} = \text{Btu/day} \rightarrow \text{HP}$$

$$HP = \frac{1\,\text{hp}}{33{,}000\,\text{ft.lb}_f/\text{min}} \cdot \frac{778.2\,\text{ft.lb}_f/\text{min}}{\text{Btu/min}} \cdot \frac{n(h_2-h_1)/t\,\text{Btu/d}}{1{,}440\,\text{min/d}}$$

$$= 1.6376 \times 10^{-5} \frac{n(h_2-h_1)}{t}$$

Example 6.7.30
A reciprocating compression system is to be designed to compress 10 MMscfd from 100 psia and 80°F to 2,200 psia of the gas as follows: $C_1 = 0.9233$, $C_2 = 0.0488$, $C_3 = 0.0185$, i-$C_4 = 0.0039$, n-$C_4 = 0.0055$. The gas is cooled with intercoolers and aftercoolers to 120°F.
    Find:

a. molecular mass of gas and compression ratio
b. brake hp using the analytical method
c. brake hp using Mollier diagram method

d. estimate the cooling requirements from the results of part (c)
e. From the results of part (c), determine whether the first stage can be handled by a compressor with a speed of 1,200 rpm, piston diameter of 12 in. and stroke length of 3 ft. Assume that $\eta_t = 0.82$, CL $= 0.08$, $Z_s = Z_D = 1.0$

(A) For given composition calculate molecular mass

$$M = \sum Y_i M_i = 17.64 \text{ lb}_m/\text{lb mole}$$

$$SG_g = \frac{17.64}{28.97} \cong 0.61$$

$$T_{pc} = 363°R \quad P_{pc} = 668$$

Using $P_s = 100$ psia, $T_s = 80°F$, We find T_r and P_r and use charts

$$Z_s = 0.985, k = 1.28 \text{ at } T = 80°F \text{ (from Figure 6-36)}$$

Compression ratio is calculated from

$$r = \frac{P_d}{P_s} = \frac{2,200}{100} = 22$$

Assuming two-stage compression

$$r_{opt} = (22)^{0.5} = 4.69$$

if intercooling is not provided or

$$r_{opt} = (22/0.97)^{0.5} = 4.76$$

if intercooling is provided.

(B) Brake horsepower using the analytical method, two-stage compression with intercooling, is calculated:

$$P_{ad} = 0.2618Q \left(\frac{P}{T}\right)_{sc} T_s Z_{avg} \left(\frac{k}{k-1}\right) (r^{(k-1)/k} - 1)$$

$$Q = 10 \text{ MMscfd} = 115.7 \text{ scf/s}$$

$$P = 14.7 \quad T = 520$$

$$T_d = T_s \left(\frac{P_d}{P_s}\right)^{(k-1)/k} = 540 \left(\frac{476}{100}\right)^{0.28/1.29} = 760$$

$$T_{avg} = \frac{540 + 760}{2} = 650°R = 190°F$$

for $T_{avg} = 190°F \quad k = 1.26$

$$P_{ad} = 0.2618(115.7) \left(\frac{147}{520}\right) 540(1) \left(\frac{1.26}{1.26 - 1}\right)$$

$$\times (4.76^{0.26/1.26} - 1)$$

$$= 851.1 \text{ hp}$$

Stage II

$$T_D = 580 \left(\frac{2,200}{476}\right)^{0.26/1.26} = 795°R$$

$$T_{avg} = \frac{580 + 795}{2} = 688°R = 228°F$$

$$k = 1.25 \quad \text{for} \quad T_{avg} = 228°F$$

$$P_{ad} = 0.2618(115.7) \left(\frac{14.7}{520}\right) (580)(1) \left(\frac{1.25}{0.25}\right)$$

$$\times (4.62^{0.26/1.26} - 1)$$

$$= 922.4 \text{ hp}$$

$$ABHP = \frac{851.1 + 922.4}{\eta_t} = \frac{1,773.5}{0.82} = 2,163 \text{ hp}$$

(C) Brake horsepower using the Mollier diagram (use any H–S chart for natural gas with $\gamma_g = 0.61$) is calculated. At the first stage:

$$P_1 = 100 \text{ psia at } 80°F \text{ and } P_2 = 476 \text{ psia}$$

when

$$\Delta h_{1-2} = h_2 - h_1 = 2,200 - 400$$

$$= 1,800 \text{ Btu/lb mol}$$

Cooling in intercoolers at constant pressure of 476 psia to T $= 580°R$ (120°F) is calculated as

$$\Delta h_{2-3} = h_3 - h_2 = 600 - 2,200 = -1,600 \text{ Btu/lb mol}$$

At the second stage:

$$P_s = 2,200/4.76 = 462 \text{ at } 120°F \quad P_D = 2,200 \text{ psia}$$

$$\Delta h_{3-4} = h_4 - h_3 = 2,900 - 600 = 2,300 \text{ Btu/lb mol.}$$

Cooling in aftercooler at constant pressure of 2,200 psia is

$$\Delta h_{5-4} = h_5 - h_4 = -200 - 2,900 = -3,100 \text{ Btu/lb mol}$$

Gas flow in number of moles is

$$\frac{h}{t} = \frac{10 \times 10^6 \text{ scf}}{\text{day}} \frac{\text{lb mol}}{379.3 \text{ scf}} = 26,364 \frac{\text{lb mol}}{\text{day}}$$

$$HP = 1.6376 \times 10^{-5} \left(26,364 \frac{\text{lb mol}}{\text{day}}\right)$$

$$\times \left(1,800 \frac{\text{Btu}}{\text{lb mol}} + 2,300 \frac{\text{Btu}}{\text{lb mol}}\right) = 1,770$$

$$ABHP = \frac{1,700}{0.82} = 2,159$$

(D) The cooling requirements are the following:
Heat load on intercooler is

$$\Delta h_{2-3}(n) = \left(-1,600 \frac{\text{Btu}}{\text{lb mol}}\right) \left(26,364 \frac{\text{lb mol}}{\text{day}}\right)$$

$$= -4.218 \times 10^7 \text{ Btu/day} = -691 \text{ hp}$$

Heat load aftercooler is

$$\Delta h_{4-5}(n) = \left(-3,100 \frac{\text{Btu}}{\text{lb mol}}\right) \left(26,364 \frac{\text{lb mol}}{\text{day}}\right)$$

$$= -8.173 \times 10^7 \text{ Btu/day} = -1,338 \text{ hp}$$

Total cooling requirements are

$$-691 - 1,338 = -2,029 \text{ hp}$$

(E) From the results of part (D), determine whether the first stage can be handled by a compressor with a speed of 1,200 rpm from Equation 6.7.49:

$$E_r = 100 - L - Cl \left[\frac{Z_s}{Z_d} \left(\frac{P_d}{P_s}\right)^{1/k} - 1\right]$$

$$L = \frac{P_d}{P_s} = 4.76$$

$$E_v = 100 - 4.76 - 8 \left[\left(\frac{476}{100}\right)^{1/1.26} - 1\right] = 75.6\%$$

From Equation 6-327a

$$HP = 4.3636 \times 10^{-5} \times 100 \times PD$$

$$\times \frac{1.26}{0.26} (4.76^{0.26/1.26} - 1)76\% = 1,726.2$$

$$PD = \frac{A_H \times L \times N}{1,728} = \frac{\pi r^2 \times 36 \times 1,200}{1,728}$$

$$= 2,827.4 \text{ ft}^3/\text{min}$$

Answer
Yes, it can easily handle the necessary rate. The first-stage theoretical power needed is 851.1 hp using the analytical method; it is 777.1 hp using Mollier charts.

6.7.15.3 Centrifugal Compressors
Centrifugal compressors have become very popular because the offer more power per unit weight and are essentially vibration free. This makes them particularly attractive for offshore locations or where air transportation to remote locations is necessary. The performance of centrifugal compressors usually is measured and described in terms of head, volume flow, efficiency and power.

Head is a measure of the energy input per unit mass into the gas stream by the compressor. It is produced by the velocity changes in the gas flow resulting from the action of the impeller. Since the compressor geometry remains constant (as contracted with a reciprocating or positive-displacement compressor), a variation in flow will cause changes in velocity. This, in turn, causes a varying head. A compressor operating on a given gas at a fixed speed will produce a head versus volume flow relationship that essentially remains constant; that is, for each value of flow there is a corresponding value of head. Since head is produced by velocity changes that are dependent upon volume rather than gas properties or conditions, the characteristic head versus volume characteristic can be considered constant, the pressure ratio and volume changes vary greatly from one gas to another. Theoretical head may be calculated as follows:

$$H_{ad} = Z_s R T_s \frac{k}{k-1} \left[\left(\frac{P_d}{P_s} \right)^{(k-1)/k} - 1 \right]$$

where

H_{ad} = adiabatic head is in desired units, ft × lb/lb or J/kg
Z_s = compressibility factor at compressor suction conditions
k = mean ratio of specific heat capacities, c_p/c_v, where $k = (k_s + k_d)/2$
R = specific gas constant for a particular gas: $1545.3/M$ [ft × lb/lb °R)] or $8.314/M$ [J/(kg × K)]
M = molecular mass of gas in lb/(lb mol) or kg/kmol

In the above equation, the value of the polytropic exponent "n" may be substituted for "k". This will yield a polytropic head, which more closely will represent the actual process. The polytropic exponent must be developed experimentally. It can be seen from the above equation that the head is dependent upon the ratio of specific heat capacities, molar mass of the gas, inlet temperature, and compression ratio. Conversely, the molar mass of the gas, inlet temperature, compressibility factor and k value determine the pressure ratio resulting from the head developed by a stage of compression.

6.7.15.4 Volume Flow Rate
Variations in volume flow rate affect the head developed by an impeller. The specific shape of the head-flow curve is a function of the stage geometry, that is, the inlet guide vane angle, impeller blade angle, and outlet diffuser construction. When describing centrifugal compressor performance, it is necessary to convert the mass flow in lb/h (kg/h), or volume flow in MMscf/d (scm/d) to actual volumetric rate at the inlet conditions. This is the volume of gas that actually is entering the eye of the impeller.

To convert the volumetric flow rate from reference conditions to suction conditions, the following is used:

$$Q_{act} = 694 Q_{std} \frac{P_b}{P_s} \frac{T_s}{T_b} \quad (ft.^3/min) \qquad [6.7.60a]$$

or

$$Q_{act} = 1.1574 \times 10^{-5} Q_{std} Z_s \frac{P_b}{P_s} \frac{T_s}{T_b} \quad (m^3/s) \qquad [6.7.60b]$$

where Q_{act} = actual inlet volumetric flowrate in desired units: ft.^{3}/min or m^3/s at T_s, P_s
Q_{act} = volumetric flowrate at base conditions in MMscf/d or scm/d at T_b, P_b

When the original data are reported as mass flow per hour, the actual volumetric flowrate can be determined by multiplying the mass flowrate by the specific volume value [ft.^{3}/lb (m^3/kg)] and dividing it by the appropriate factor to account for differences in time base [60 min/hr (3,600 s/hr)].

6.7.15.5 Power
The theoretical equation for gas hp or kw for a centrifugal compressor is as follows:

$$P_{ad} = 8.179 \times 10^{-5} H_{ad} Q_{act} \gamma_g \frac{P_s}{A_s T_s} \qquad [6.7.61a]$$

and in the SI unit system

$$P_{ad} = H_{ad} Q_{act} SG_g \frac{P_s}{287 Z_s T_s} \qquad [6.7.61b]$$

where P_{ad} = input power to the gas stream in desired units, hp or kw
SG_g = gas relative density

To account for the aerodynamic losses, it is necessary to replace H_{ad} with H'/η', where H'/η' is the adiabatic or polytropic head divided by the corresponding adiabatic or polytropic efficiency.

In addition to the aerodynamic losses, there are mechanical losses due to seal and bearing friction. For service design purposes, the mechanical losses for centrifugal and rotary compressors over 1,000 hp (746 kW) can be assumed to be 35 hp (26 kW) for bearings and 35 hp (26 kW) for oil-type shaft seals; below that size, losses will amount to 1% to 3%. In any case, these losses are included in the mechanical efficiency factor. Therefore, the required shaft power is:

$$P_{shaft} = P/\eta_m \qquad [6.7.62]$$

where P_{shaft} = compressor shaft input power in desired units, hp or kw
η_m = mechanical efficiency (0.97 to 0.99)

6.7.15.6 Comparison of Reciprocating and Centrifugal Compressors
The advantages of a centrifugal compressor over reciprocation machine are:

- lower installed first cost where pressure and volume conditions are favorable
- lower maintenance expense
- greater continuity of service and dependability
- less operating attention
- greater volume capacity per unit plot area
- adaptability to high-speed low-maintenance-cost drivers

The advantages of a reciprocating machine over a centrifugal machine are:

- greater flexibility in capacity and pressure range
- higher compression efficiency and lower power cost
- capability of delivering higher pressures
- capability of handling smaller volumes
- less sensitive to changes in gas composition and density

Example 6.7.31

Perform centrifugal compressor calculations using the following data: $M = 18.3$, $P_{inlet} = 156$ psia, $T_{inlet} = 105°F$, $Q_{std} = 12,000$ scfm, $P_{discharge} = 310$ psia. The base conditions are $P = 14.73$ psia, $T = 60°F$

- Calculation of the compressibility factor Z_s and k

$$T_s = 105°F = 565°R$$

$$M = 18.3 = SG_g = 0.63$$

$$T_{pc} = 370°R, \quad P_{pc} = 670 \text{ psia}$$

$$T_r = 565/370 = 1.53, \quad P_r = 156/670 = 0.233$$

$$Z_s = 0.98$$

$$k_s = 1.27 \text{ at } 565 \text{ (from Figure 6.7.48)}$$

$$T_d = T_s(P_d/P_s)^{(k-1)/k} = 565(1.99)^{0.27/1.27} = 654°R$$

$$= 194°F$$

$$k_d = 1.25 \text{ at } 194°F$$

$$k = (k_s + k_d)/2 = 1.26$$

- The actual inlet capacity, using Equation 6.7.60a, is

$$Q_{act} = 694 \times 17.28 \times \frac{14.73}{156} \frac{565}{520} = 1,230.3 \text{ acfm}$$

- The theoretical head, using Equation 6.7.61, is

$$H_{ad} = Z_s R T_s (k/k - 1)[r^{(k-1)/k} - 1]$$

$$R = \frac{1,545.3}{M} = \frac{1,545.3}{18.3} = 84.44$$

$$H_{ad} = 0.98 \times 84.44 \times 565 \times \frac{1.26}{1.26 - 1}(1.99^{0.26/1.26} - 1)$$

$$= 34,570$$

- The theoretical power, is

$$P_{ad} = 8.179 \times 10^{-5} H_{ad} Q_{act} \gamma \frac{P_s}{Z_s T_s}$$

$$= 8.179 \times 10^{-5} \times 34,570 \times 1,230.3$$

$$\times 0.63 \frac{156}{0.98 \times 565} = 617 \text{ hp}$$

- Adiabatic efficiency $\eta_{ad} = 0.7$ mechanical efficiency assuming 0.98.
- The required shift power is

$$P_{shaft} = \frac{P_{ad}}{\eta_{ad}\eta_m} = \frac{617}{0.7 \times 0.98} = 899.4 \text{ hp}$$

6.7.16 Engine and Compressor Foundation [18]

Machinery foundations are one of the most important aspects of compressor station engineering. Such factors as soil analysis, reinforced concrete design, and dynamics of foundation systems require particular attention. There are two basic types of foundations used for compressor units in the gas industry (Figures 6.7.52 and 6.7.53). The first is a simple concrete pad that is used for most small skid-mounted reciprocating units and for high-speed centrifugal compressors. The second incorporates a combination of reinforced block and mat concepts. The block and mat design is common for large, low or medium-speed reciprocating compressors. The block provides the required load distribution so that the allowable soil-bearing capacity is not exceeded. The following minimum data should be available to the foundation designer:

- total weight of the package
- base dimensions of the package

- operating speed of the machine
- radius of the crankshaft throws
- weight of the reciprocating parts
- length of the connecting rods
- weight of the unbalanced rotating parts

The base dimension of the concrete pad or block should be such that there is 6 to 12 inches of concrete beyond all sides of the base of the unit. Manufacturers have recommended from 0.091 to 0.141 yd^3 of concrete per engine hp (0.093 to 0.144 m^3/kw) for low speed machines. Manufacturers also recommend a foundation that is two to five times the weight of the machine.

The combined resonant frequency of the machine, foundation and soil system is next in importance.

$$f_n = \frac{1}{2\pi}\left(\frac{K \times g}{W_f + W_s}\right) \qquad [6.7.63]$$

where

f_n = natural frequency of the foundation and soil system in Hz
K = spring constant of soil mass in lb/ft or kN/m
W_f = weight of machine and foundation in lb/ft or kN
W_s = weight of soil system in lb or kN
g = acceleration of gravity in 32.174 ft/s or 9.81 m/s^2

The spring constant "k" is a function of Poisson's ratio, shear modulus and the radius of the loading area. In general, the natural frequency of the system should be at least twice the frequency of the applied forces for low-speed units, or less than one-half of the resonant frequency for machine that operate at a frequency greater than the natural frequency of the system. The natural frequency of the foundation system can be decreased by:

1. increasing the weight of the foundation
2. decreasing the base area of the foundation
3. reducing the shear modulus of the soil
4. placing the foundation deeper into the soil

6.7.16.1 The Gas Piping System

The gas piping system in a compressor station includes the valving, pulsation control equipment, overpressure protection devices, cathodic protection facilities and structural supports to route the gas through the compressor and gas conditioning facilities. Figure 6.7.53 shows a schematic diagram of such a system at a small compressor station. Figure 6.7.54 gives a general flow diagram for a storage/withdrawal compressor operation. In designing the piping system the following items must be taken into consideration:

- operating pressure and capacity
- pressure drop
- gas velocity
- mechanical strength and composition of pipe materials
- temperatures
- effect of corrosion and erosion
- safety
- efficiency of operations
- economy of installation

Pressure drops in compressor station piping result from energy losses caused by friction between the gas and pipe wall and should not exceed 1 to 2% of the total station power requirements.

Gas velocity can be determined as follows:

$$V = \frac{127.3 \times 10^3 Q P_b T_f Z}{D^2 P_m T_b} \qquad [6.7.64a]$$

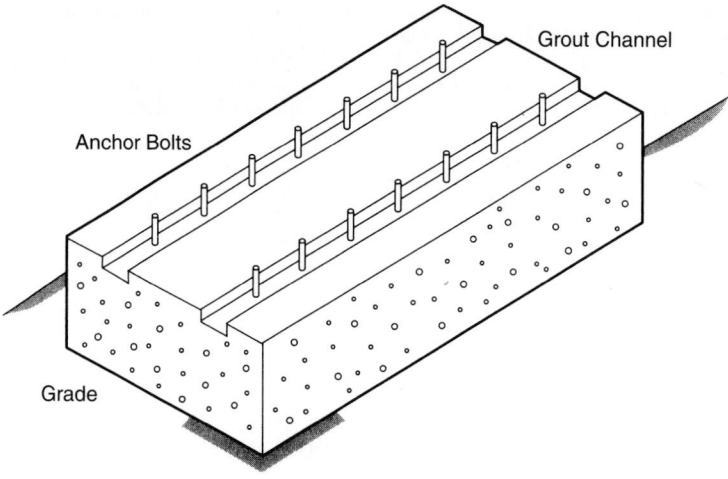

a) Continuous Pad Foundation

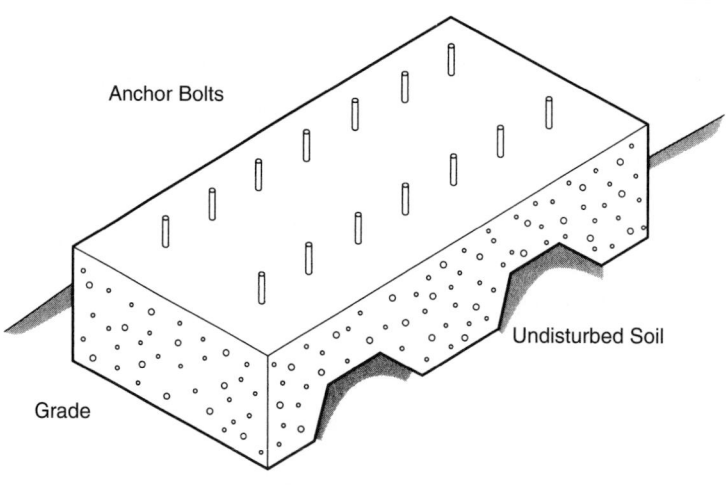

b) Waffle Type Pad Foundation

Figure 6.7.52 *Typical concrete pad compressor foundations.*

or in the SI system

$$V = \frac{14.73 Q P_b T_f Z}{D_2 P_f T_b} \qquad [6.7.64b]$$

where V = velocity in desired units in ft./min or m/s
Q = volume rate of flow in MMcf/d or m^3/d
P_b = base pressure in lb/in.^{2} or kPa absolute
T_f = flowing temperature in °R or K
Z = compressibility
D = diameter of pipe in in. or mm
P_f = flowing pressure in lb/in.^{2} or kPa absolute
T_b = base temperature in °R or K

Gas piping used in compressor stations must be manufactured in compliance with the standards (DOT) Code, Title 49, Part 192. (See Table 6.7.32).

6.7.16.2 Gas Cooling

Gas cooling equipment is necessary in compressor stations to remove heat from the gas stream between stages of compression (intercoolers) or after the final stage of compression (aftercoolers). Heat exchangers are used to remove heat from gas streams, reduce cooling water temperature, lower oil temperature or to cool compressed air. Several types of gas coolers are available, but the aerial cooler is one of the most common. Aerial coolers make use if air to cool hot gas. There are two kinds: forced draft and induced draft (see Figure 6.7.56), [20].

The advantages of forced draft are:

- slightly lower horsepower since the fan is in cold air; horsepower varies directly as the absolute temperature
- better accessibility of mechanical compound for maintenance

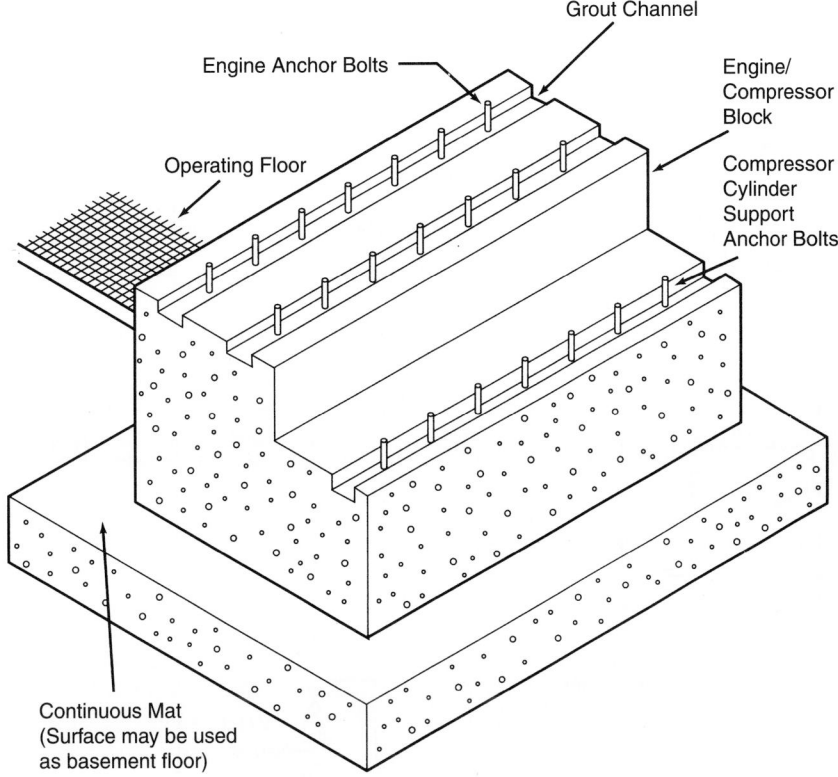

Figure 6.7.53 *Concrete block and mat compression.*

- easily adaptable for warm air recirculation for cold climates.

  The disadvantages of forced draft are:

- poor distribution of air over section
- greatly increased possibility of hot air recirculation, due to low discharge velocity from the sections and absence of stack
- low natural draft capability on fan failure due to small stack effect
- total exposure of tubes to sun, rain, and hail

  The advantages of induced draft are:

- better distribution of air across section
- less possibility of the hot effluent air recirculating around to the intake of sections; the hot air is discharged upward at approximately 2.5 times the velocity of the intake or about 1,500 ft./min.
- less effect of sun, rain and hail, because 60% of the face area of the sections is covered
- increased capacity in the event of fan failure, since the natural draft stack effect is much greater with induced draft

  The disadvantages of induced draft are:

- Higher horsepower since the fan is located in the hot air.
- Effluent air temperature should be limited 200°F to prevent potential damage to fan blades bearing, V belts, or other mechanical components in the hot air stream.

- The fan drive components are less accessible for maintenance, which may have to be done in the hot air generated by natural convection.
- For inlet process fluids above 350°F forced draft design should be used: otherwise, fan failure could subject the fan blades and bearings to excessive temperature.

  In sizing gas coolers the most important parameters are the actual heat load and design air temperature. The heat load may be found from the following:

$$Q_H = Q_{std}\rho C_p(t_1 - t_2) \quad \text{for SI units} \quad [6.7.65]$$

$$Q_H = Q_{std}\gamma C_p(t_1 - t_2) \quad \text{for English units} \quad [6.7.66]$$

where Q_H = heat load in desired units: Btu/n or W
  Q_{std} = gas flow rate in scf/n or scm/s
  C_p = specific heat capacity at average temperature $(T_s + T_d)$ /2 in Btu/(lb°F)or(kg°C)
  t_1, t_2 = inlet and outlet temperatures in °F or °C
  ρ = density (kg/scm)
  γ = specific weight (lb/scf)

  The design air temperature is usually average daily maximum temperature for the hottest month of the year at the gas cooler location plus 3 to 5°F(1.5 to 2.5°C) correction value for the location environmental conditions.

6.7.17 Cooling Towers [20]
Cooling towers allow water to be cooled by ambient air through evaporation. They have two types of air flow, crossflow and counterflow. In crossflow towers, the air moves

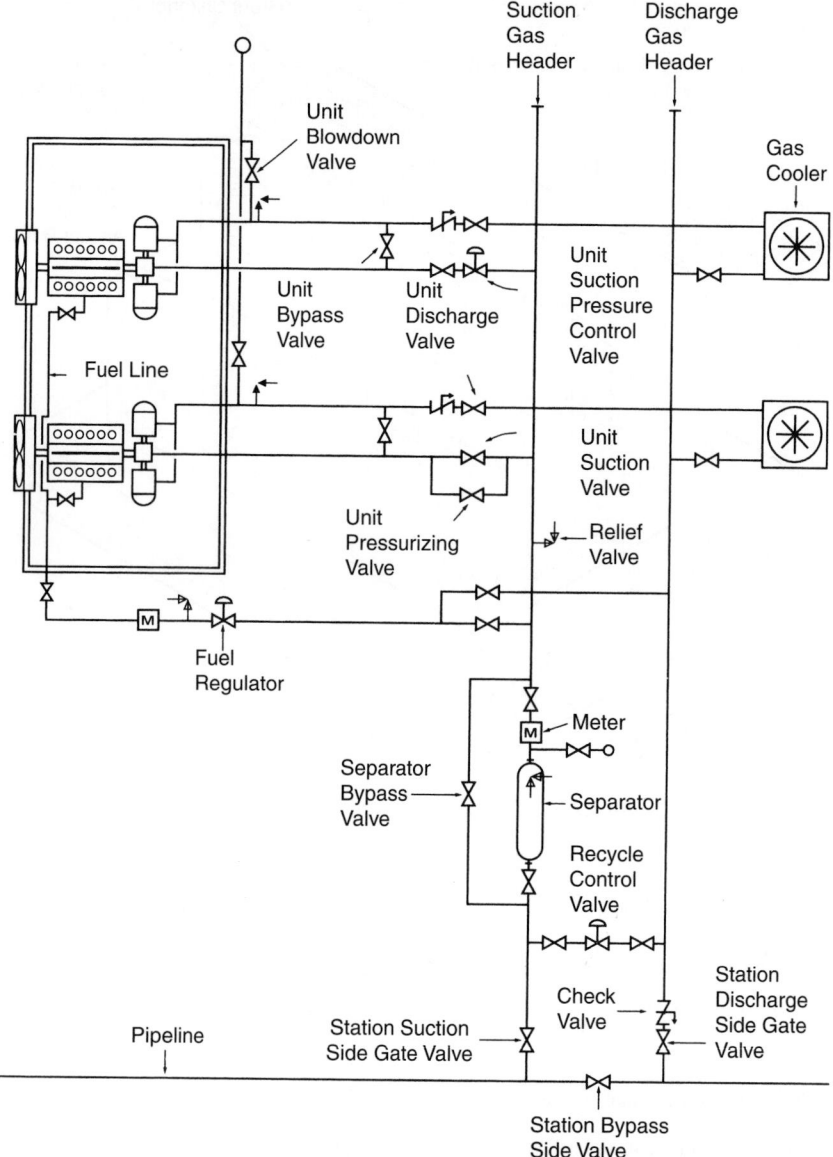

Figure 6.7.54 *Foundation typical reciprocating compressor station piping.*

horizontally across the downward flow of water. In counterflow towers, the air moves vertically upward against the downward fall of the water.

There are a few types and sizes of cooling towers explained as follows:

Mechanical draft towers (Figure 6.7.57) are characterized as follows:

- Forced draft towers, where the fan is located on the airstream entering the tower.
- Induced draft tower, where the fan is located on the airstream leaving the tower.
- Coil shed towers, where the atmospheric coils or sections are located in the basin of the cooling tower.

Natural draft towers (Figure 6.7.58) are characterized as follows:

- Atmospheric spray towers are dependent upon atmospheric conditions; no mechanical devices are used to move the air.
- Hyperbolic natural draft towers, where a chimney or stack is used to induce air movement through the tower.

The performance characteristics of various types of towers will vary with height, fill configuration and flow arrangement: crossflow and counterflow. When rough characteristics of a specific tower are required the performance characteristic nomograph (Figure 6.7.59) can be used.

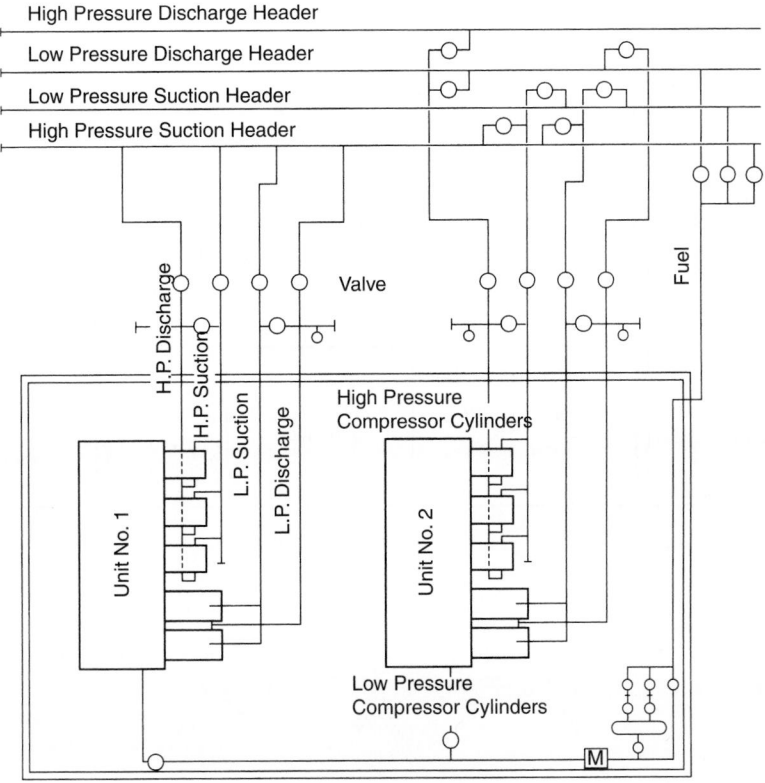

Figure 6.7.55 *Compressor unit gas flow diagram for storage injection and withdrawal service.*

Example 6.7.32

The use of the nomograph is illustrated by the following examples covering typical changes in operating conditions. Assume a cooling tower is operating at these known conditions: flow = 1,000 gal/min, hot water at 110°F, cold water at 86°F, wet bulb temperature = 70°F. This is commonly referred to as 110° − 86° − 70° or 24° range (110° − 86°) and 10° approach (86° − 76°).

1. What is the effect of varying wet bulb temperature (WB) on cold water temperature (CWT)? What is new CWT when WB changes from 76 to 65°F with gal/min and the range remains constant? Enter the nomograph at 86° CWT, go horizontally to 76°F WB, then vertically down to 65°F WB, then read new CWT of 79.5°.
2. What is the effect of varying the cooling range on the cold water temperature? What is the CWT now when the cooling range is changed from 24° to 30° (25% increase in heat load) with gal/min and WB held constant? Enter the nomograph at 86° CWT, go horizontally to 76° WB, vertically to 24°R, horizontally to 30°R, vertically downward to 76° WB, then read new CWT 87.4°.
3. What is the effect of varying water circulating rate and heat load on cold water temperature? What is the new CWT when water circulation is changed from 1,000 to 1,200 gal/min (50% change in heat load at constant range)? Varying the water rate, particularly over wide ranges, may require modifications to the distribution system. Enter the nomograph at 86°F CWT, 90 horizontally to 76°F WB, vertically to 24°R, horizontally to performance factor of 2.88. Obtain a new performance factor (PF), multiplying (2.88)(1,200/1,000) = 3.46, then enter

the nomograph at PF = 3.46, go horizontally to 24°R, vertically down to 76°F WB, then read new CWT 88.5°F.
4. What is the effect of varying the WB temperature, range and water circulating rate on the cold water temperature? What is the new CWT when the WB changes from 76° to 65°, R changes from 24° to 25°, gal/min changes from 1,000 to 1,500²? Enter the nomograph at 86° CWT, go horizontally to 76 WB, vertically to 24°R, horizontally read PF = 2.88, then multiply (2.88)(1,500/1,000) = 4.32 (new PF). Enter the nomograph at PF = 432, go horizontally to 25°R, vertically down to 65° WB, then read 86.5° as the new CWT.
5. What is the effect of varying the fan hp input on the cold water temperature? What is the new CWT if the motor is changed from 25 to 30 hp in example (d)? The air flowrate varies as the cube root of the horsepower and performance varies almost directly with the ratio of water rate to air rate; therefore, the change in air flowrate can be applied to the performance factor. Increasing the air flowrate (or hp) decreases the performance factor. PF correction factor = $(30/25)^{1/3} = 1.0627$.

   Divide PF by the PF correction factor to get the new PF. Applying this to example (d), we get 4.32/1.0627 = 4.07. Enter the nomograph at 4.07 PF (instead of 4.32), go horizontally to 25°R, vertically down to 65 WB, then read 85.5 CWT.
6. The correction factor shown in example (e) could also be used to increase the gal/min instead of decreasing the CWT, as was done in example (e). In example (d) we developed a new CWT of 86.5 when circulating 1,500 gal/min at 25°R and 65 WB. If the motor hp is increased from 25 to 30 under these conditions, with a

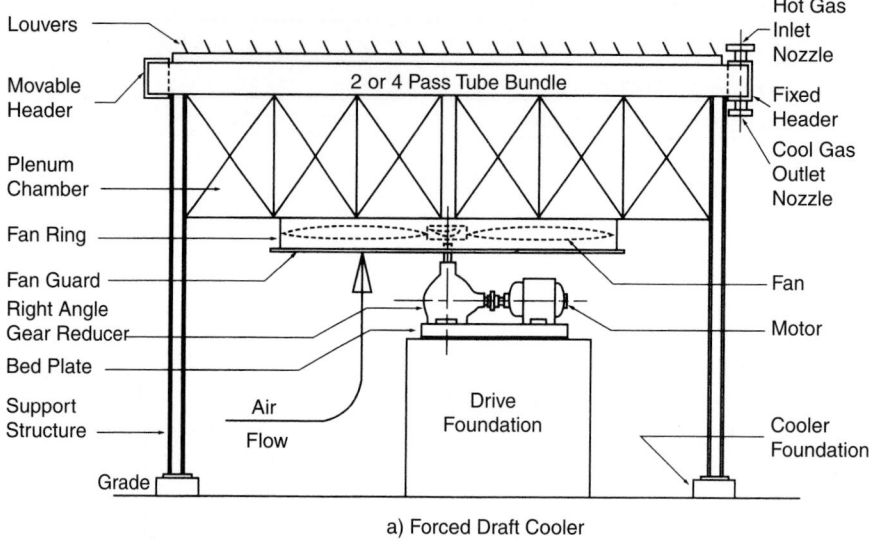

a) Forced Draft Cooler

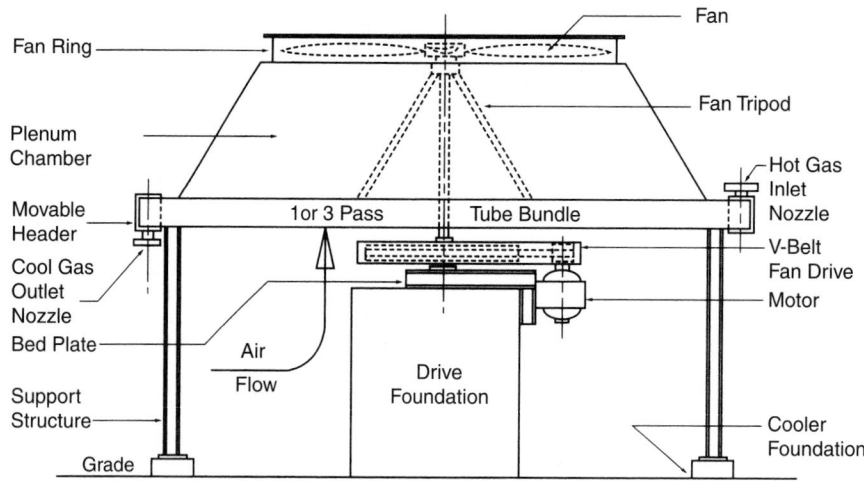

b) Induced Draft Cooler

Figure 6.7.56 *Types of aerial heat exchanges.*

PF correction factor of 1.0627 (as shown above), gal/min could be increased from 1,500 to $1,500 \times 1.0627 = 1,594$ gal/min.

7. Calculate the concentrations and blowdown rate for the following cooling tower.
   Circulation rate = 8,000 gal/min
   Water temperature drop through tower = 20°F
   Type of tower = mechanically induced draft
   Blowdown rate = 20 gal/min or 0.2% of circulation rate

Solution
Evaporation loss = 2% (1% for each 1° temperature drop) (All rates are based on a percent of circulation rate.) Windage loss = 0.3% (maximum for mechanical draft tower) Number of concentrations (cycles) is

$$N = (E + B)/B = \frac{2.0 + (0.2 + 0.3)}{(0.2 + 0.3)} = 5.0$$

If the resultant concentrations are excessive and a desired concentration of 4.0 is required, what must the blowdown rate be?

$$B = \frac{E}{\text{cycles} - 1} = \frac{2.0}{4.0 - 1} = 0.67\%$$

The windage component of B is 0.3%, therefore, the blowdown rate required would be $0.67 - 0.3 = 0.37\%$ or (8,000 gal/min) (0.0037) = 29.6 gal/min.

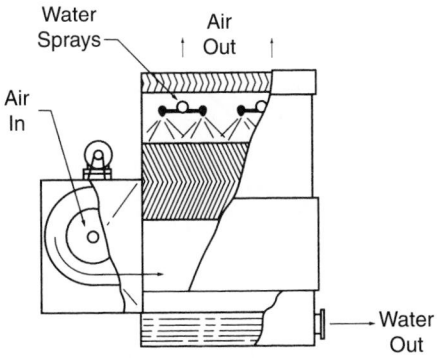

a) Forced draft, counterflow, blower fan tower.

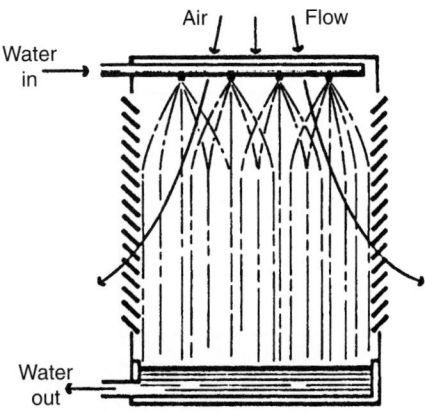

a) Atmospheric spray tower.

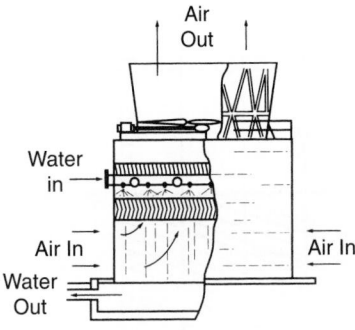

b) Induced draft counterflow tower.

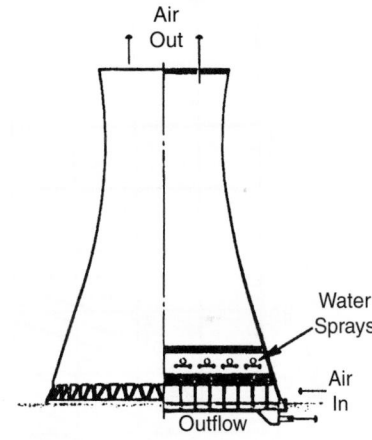

b) Counterflow natural draft tower.

Figure 6.7.58 *Natural draft towers.*

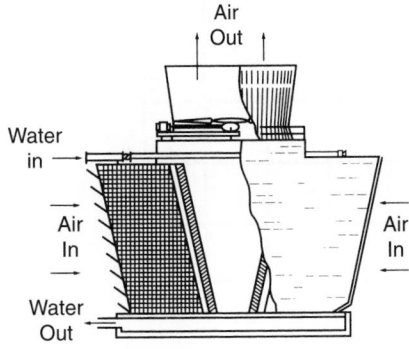

c) Double-flow, crossflow tower

Figure 6.7.57 *Mechanical draft towers.*

References

1. Campbell, J. M., *Gas Conditioning and Processing*, Vols. 1 and 2, Campbell Petroleum Series, Norman, Oklahoma, 1984.
2. *Engineering Data Book*, 11th Edition, Vols. 1 and 2, Gas Processors Suppliers Association, Tulsa, Oklahoma, 1998.
3. Robinson, J. M., et al. "Estimation of the Water Content of Sour Natural Gases," Trans. AIME, Vol. 263, p. 281, August 1977.
4. Arnold, K. and Stewart, M., *Surface Production Operations*, Vols. 1 and 2, Gulf Publishing Company, Houston, Texas, 1989.
5. Carr-Brion, K., *Moisture Sensors in Process Control*, Elsevier Applied Science Publishers, London, 1986.
6. Katz, D. L., et al. *Handbook of Natural Gas Engineering*, McGraw-Hill Book Co., New York, 1959.
7. *Chandler Engineering Handbook*, Chandler Engineering, Tulsa, Oklahoma, 2002
8. Makogon, Y. F., *Hydrates of Natural Gas*, Pennwell Books, Tulsa, Oklahoma, 1981.
9. Katz, D. L., "Prediction of Conditions for Hydrate Formation", Trans AIME Vol. 160, p. 140, 1945.
10. Carson, D. B., and Katz, D. L., Trans AIME Vol. 146, p. 150, 1942.
11. Sivalls, C. R., *Glycol Dehydrator Design Manual*, Sivalls Inc., Box 2792, Odessa, Texas, 2002.

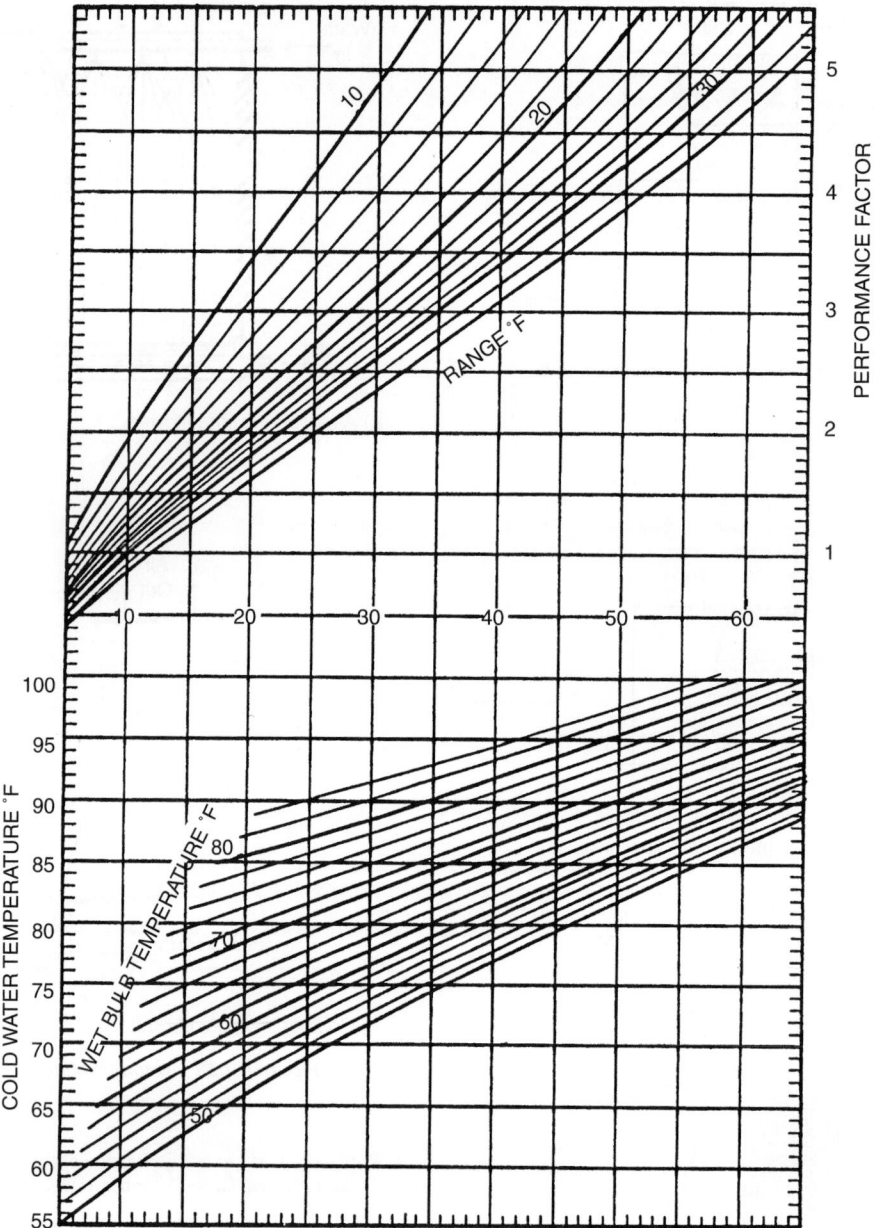

Figure 6.7.59 *Pressure characteristic nomograph.*

12. *Instrument Engineers handbook*, Liptak, B. G., and Venczel, K., Chilton Book Co., Radnor, Pennsylvania, 1982.
13. *Orifice Meter Constants*, Handbook E-2, American Meter Company, Inc., 1955.
14. American Gas Association, Gas Measurement Committee, Report No. 3.
15. Bradley, H. B., *Petroleum Engineering Handbook*, Ed., Society of Petroleum Engineers, Richardson, Texas, 1987.
16. Yarborough, L., and Hall, K. R., SPE Publication # 13, pp. 233–235, Richardson, Texas.
17. Miller, R. W., *Flow Measurement Engineering Handbook*, McGraw-Hill Book Co., New York, 1983.
18. *Compressor Operations Handbook*, The American Gas Association, GEOP Book T-@, Arlington, Virginia, 1985.
19. *Compressor Application Engineering*, Vols. 1 and 2, Gulf Publishing Co., Houston, Texas, 1985.
20. *Cooling Tower Fundamentals*, The Marley Cooling Tower Co., Kansas City, Missouri, 1982.

6.8 CORROSION IN PRODUCTION OPERATIONS

6.8.1 Introduction

The corrosion problems and mitigation techniques related to oil and gas production operations are discussed here.

The production environments that were discussed in the Chapter on "Drilling and Well Completion Corrosion" are still pertinent to corrosion in production operations that is the subject of this chapter. However, there are some very important differences. First, the exposure to the production fluids is not modified by controlled drilling fluids. Second, the down hole production environment and the equipment exposed to it are not cooled through the use of recirculated or injected fluids. Third, the equipment and their constitution is markedly different from drilling equipment. Fourth, generally the service life of production equipment is measured in years. Fifth, there are new types of environments that are introduced that are different from or modify the production environment such as completion brines, CO_2 injection and water flood operations.

A wide variety of engineered tools and structures are used in the production side of the oil and gas industry. Production equipment includes production tubing, production packers, safety valves, tubing hangers, chokes, flow control equipment, valves and wellheads, gas lift equipment, sucker rods and rod pumps, submersible pumps, compressors, flow lines, and pipelines. Each of these products has different engineering requirements that equates to different metals and conditions. These products may also be exposed to a different set of environmental conditions that pose different corrosion potentials. These metals and their condition exhibit varying susceptibilities to the various potential corrosion reactions that are prevalent in the oilfield production environments.

There has been a large volume of information developed over the years on the subject of corrosion problems encountered in production operations. The solutions to these corrosion problems that we have encountered have created a knowledge base from which we can draw to avoid the many pitfalls that potentially await us in operations.

This chapter is structured to provide an introduction to the causes of corrosion in production, the equipment used in production, and the common techniques to mitigate the effects of corrosion.

6.8.2 Causes of Corrosion in Production Operations

The causes of corrosion in production operations are dependent on the stage of production operation we are considering and the varied constituents and conditions that occur there. The segments of the production operation (or where in the production stream) are defined here as follows:

1. Downhole Segment — from the formation, the production fluids flow up production tubing; downhole accessories may assist with production or control it.
2. Wellhead — the production fluids are directed and controlled by valves and associated equipment as they exit the production tubing on the wellhead.
3. Flowlines/Pipelines — with the assistance of storage facilities and compressors, the production fluids are transported to treatment and refining facilities.

Many operational variables will change with location, and temperature is the one of the most significant. The temperature is highest at the formation and cools as it comes out of the earth through the wellhead and into pipelines and flowlines. An example of a bottom hole temperature of 175°C (350°F) may cool to about 85°C (185°F) at the wellhead and ultimately to the range of ambient temperatures in the pipelines. Another factor is phase stability

because temperature and pressure will effect what phases are present. A discussion of the primary corrosion causing environments is presented below with how the environment changes or is affected by where in the production stream it is.

6.8.2.1 Acid Gas Corrosion

In production operation, corrosion is usually associated with acidity in the bulk or in localized regions. The acidity in the bulk environment is caused by the presence of acid gases that naturally occur in the formation coexisting with petroleum products or as the result of CO_2 gas injection. Acid gases, hydrogen sulfide H_2S and CO_2, are the gases that form acids in aqueous environments. As the concentration of these gases increases, the amount of acid in the water phase increases. The more acid (lower pH) there is, the greater the corrosion rate. Listed below are some factors that affect the corrosion from these acid gases.

Factors that affect acid gas corrosion

- acidity or pH from the acid gas concentration
- water/gas/oil phases
- produced water chemistry
- temperature
- chlorides
- flow rates
- composition and condition of the metals

6.8.2.1.1 CO_2 Partial Pressure and Temperature

As presented in the corrosion discussion in drilling and completion operations, the pH or acidity is directly related to the concentration of these acid gases that is usually expressed in parts per million (ppm), partial pressure (bar or psi), or mole fraction (percent). Early researchers found that the corrosion in CO_2 containing environments was related to this concentration and temperature. DeWaard and Williams [1] developed an equation to predict corrosion rates, and this equation is presented below.

$$\text{Log } r = 8.78 - 2.320/(T + 273) - 5.55 \times 10^{-3} \times T$$
$$\times \log \text{Pco}_2 \qquad [6.8.1]$$

where r = corrosion rate in mils per year (mpy)
T = temperature, degrees C
Pco_2 is the partial pressure of CO_2 in psi

The nomograph that was the result from this equation is presented in Figure 6.8.1.

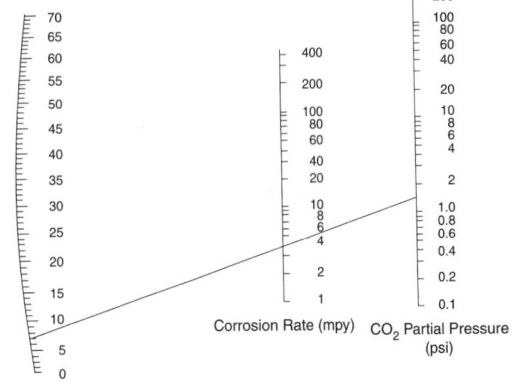

Figure 6.8.1 *Nomograph for calculation of corrosion rates as a function of carbon dioxide partial pressure and temperature [1].*

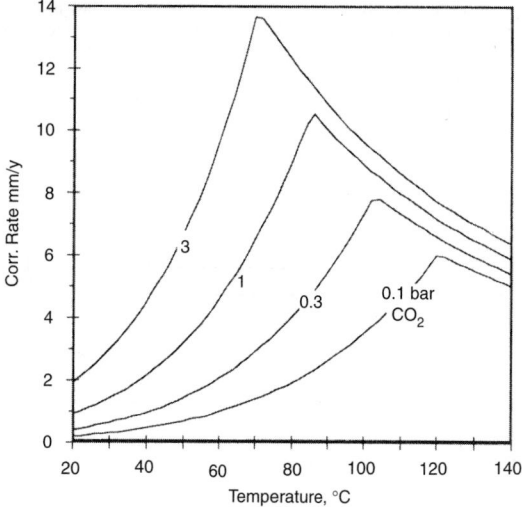

Figure 6.8.2 *Effect of temperature on CO_2 corrosion [2].*

Table 6.8.1 *Available Prediction Models for CO_2 Corrosion*

Model Name	Type (note 1)	Scale effect	Oil effect	Multiphase flow	H_2S effects
CASSANDRA	E+L	Low	—	—	—
NORSOK	E+L	High	—	—	—
CORMED	E+F	—	—	—	—
LIPUCOR	E+I+F	—	—	Yes	Yes
HYDROCOR	M	Moderate	Yes	Yes	Yes
KSC	M	High	—	—	—
TULSA	M	High	—	—	—
PREDICT	E	High	Yes	Yes	Yes
SWEETCOR	E	—	—	—	Yes
CORPOS	E	—	Yes	Yes	Yes
OHIO	M+E	High	Yes	—	—
ULL	M+E	—	Yes	—	—
DREAM	E	High	—	—	—
OLI	M	Yes	—	—	Yes
ECE	E	—	Yes	—	Yes

E, empirical; L, lab tests; F, field data; M, mechanistic.

Since that early work, much progress has been made refining these relationships taking into account the stability of protective corrosion product films. The corrosion rate of steels in CO_2 containing oilfield environments as a function of partial pressure of CO_2 and temperature is presented in Figure 6.8.2.

6.8.2.1.2 Prediction Models

The relationship between environmental variables and corrosion rates in CO_2 environments continues to be refined as Figure 6.8.2 is an example. Still considering CO_2 corrosion but now allowing for small amounts of H_2S, several prediction models for corrosion of oil and gas pipelines have been developed. These models are based on either mechanistic modeling of the different processes involved in CO_2 corrosion of carbon steel or empirical correlations with laboratory or field data. The models differ considerably in how they predict the effect of protective corrosion films and the effect of oil wetting on CO_2 corrosion, and these two factors account for the most pronounced differences between the various models [3].

A list of some of the CO_2 corrosion models is presented in Table 6.8.1 [adapted from 4].

6.8.2.1.3 Water Content

Water content (also known as watercut) is one of variables that is related to the probability of corrosion. The most severe corrosive environments have been found to contain above approximately 40% watercut [5,6]. When there are no appreciable amounts of acid gases present, corrosion is generally not a problem when water cuts are below about 25%. The experience found between watercut and the number of corrosive wells is presented in Figure 6.8.3.

6.8.2.1.4 Oil Phase and Flow Effects

The other phases present has a large effect as well. The presence of oil is generally a positive effect in that corrosion rates may be reduced or corrosion may be prevented. The corrosion reaction requires an electrolyte to conduct the flow of electrons, oil being nonpolar, is not a conductor. The heavier the oil (more viscous), the greater the protection provided by the oil.

In addition to the presence of a corrosion reaction that is contingent upon a contacting water phase, the characteristics of the resultant corrosion products have a role in corrosion kinetics. As discussed previously, the carbon and low alloy steels form iron carbonate films that provide corrosion resistance. In flowing conditions, these films could be removed or compromised by hydrodynamic erosion.

There are relationships between the relative amounts of the different phase, the flow regime and the potential for corrosion. The types and dispersion of phases in liquid plus gas phases in a vertical and horizontal pipe are presented in Figures 6.8.4 and 6.8.5, respectively.

The different types of flow that are predicted as a function of the relative gas and liquid velocities in Figures 6.8.4 and 6.8.5 can be used to predict corrosion severity. For example, the maximum corrosion rates under a specific set of environmental conditions predicted as a function of the type of flow is presented in Figure 6.8.6.

6.8.2.1.5 Corrosion and Oil Weight

The corrosion rates are a function of watercut, flow rate, and a given weight of oil. The predictive relationships for API specific Gravities 38 and 45 oils are presented in Figures 6.8.7 and 6.8.8.

What has not been discussed yet is the effect of pipe angles that are off the vertical on these flow regimes. Three modes of water wetting have been defined as they relate to the angle off vertical [8]. These modes are presented in Figure 6.8.9.

In Figure 6.8.9, mode I relates to completely emulsified liquid wetting, mode II to wetting by accumulated water volumes at locations of high deviation, and mode III to wetting by coalesced water droplets. Mode II was found to be very dependent on the angle of deviation and results in water contact with metal surfaces even at low watercuts [8].

6.8.2.1.6 Velocity and Steels

Investigating the effects of velocity with other variables provides additional clues that we can use. For example, J.R. Vera et al [9] examined the composition of the corrosion products as a function of flow velocity for a variety of low alloy products including API 5CT [10] Grades L80 and C90 tubulars. The primary difference between the L80 and the C90 pipe was the chromium content, which was 0.5% for the L80 and 1.0% for the C90, and molybdenum content, which not present in the L80 and was 0.5% in the C90. The L80 had corrosion

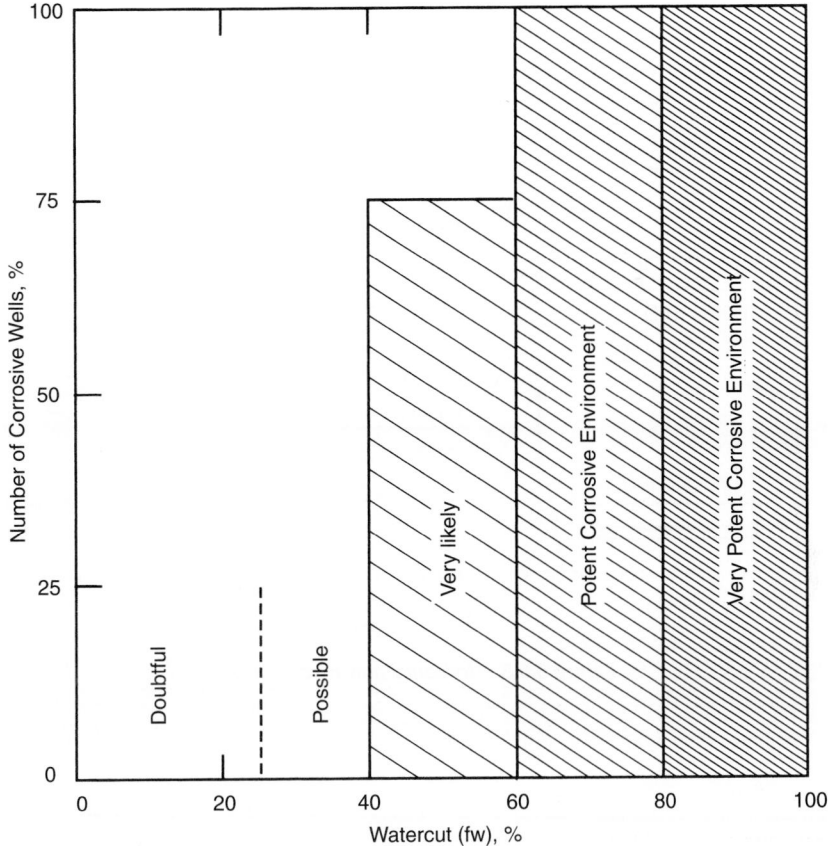

Figure 6.8.3 *Relationship between watercut, and number of corrosive wells [5,6].*

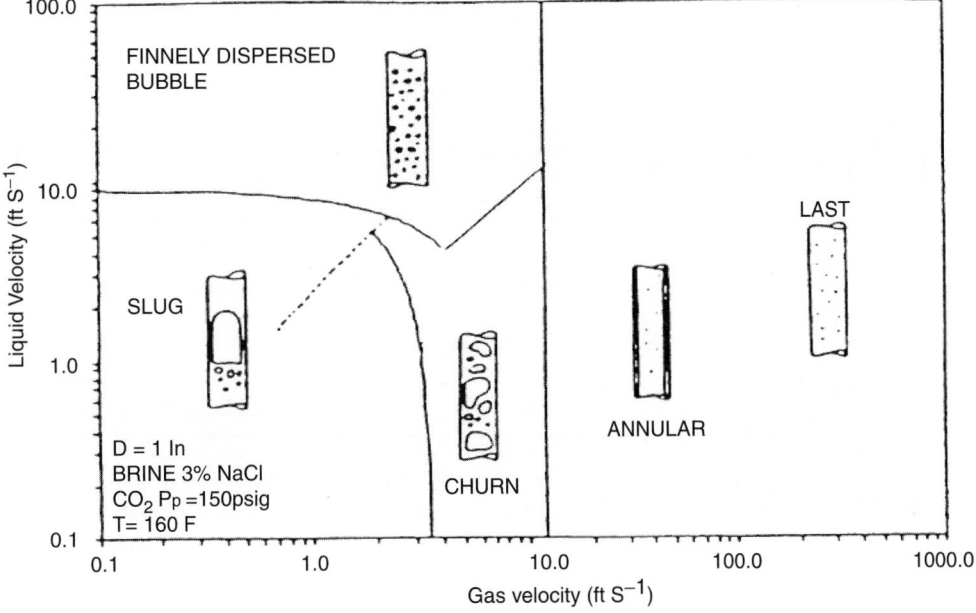

Figure 6.8.4 *Vertical flow map [after 7].*

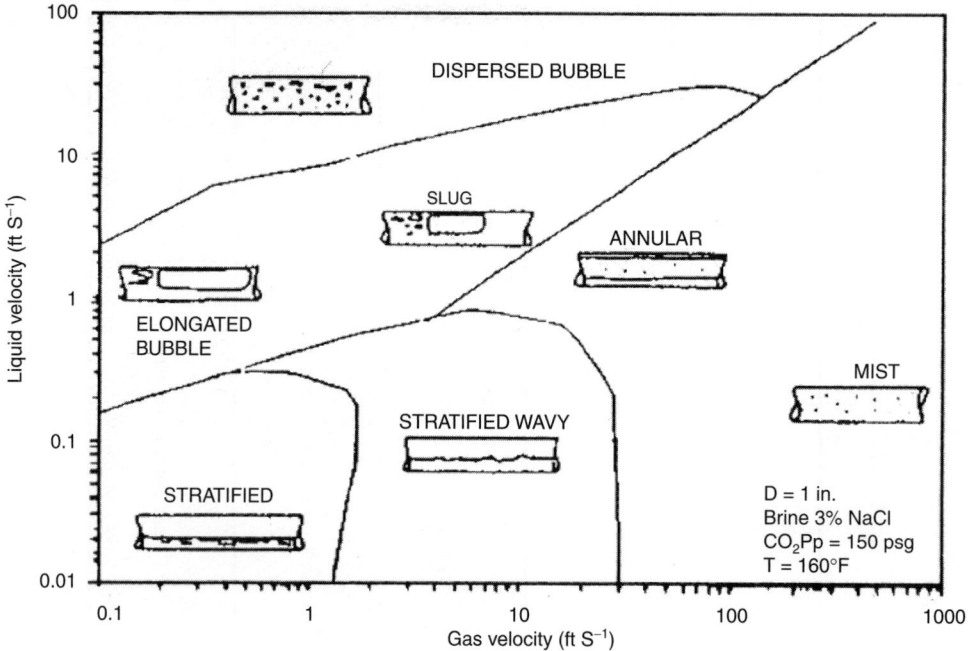

Figure 6.8.5   *Horizontal flow map [after 7].*

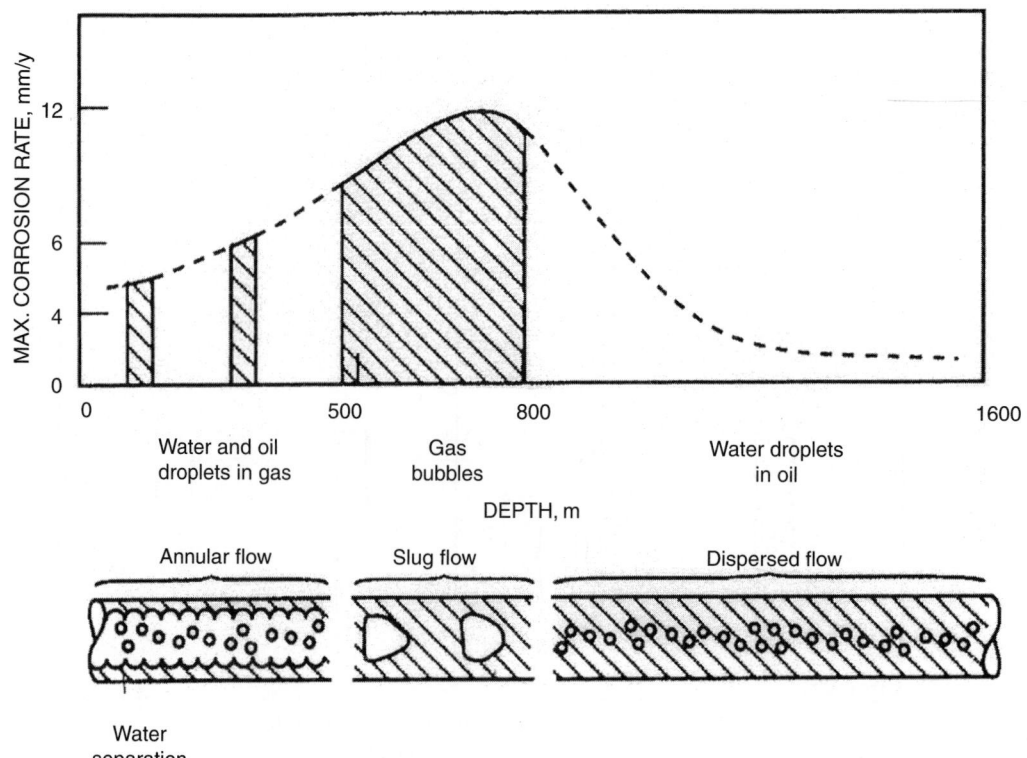

Figure 6.8.6   *Horizontal flow and corrosion predictions [after 7].*

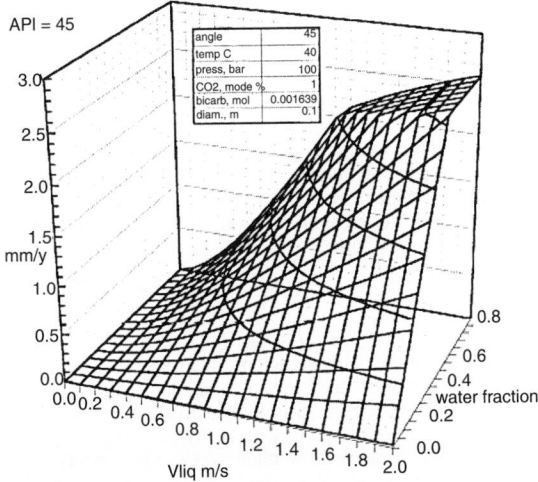

Figure 6.8.7 CO₂ corrosion rates as a function of velocity and watercut with API specific gravity 45 oil [after 8].

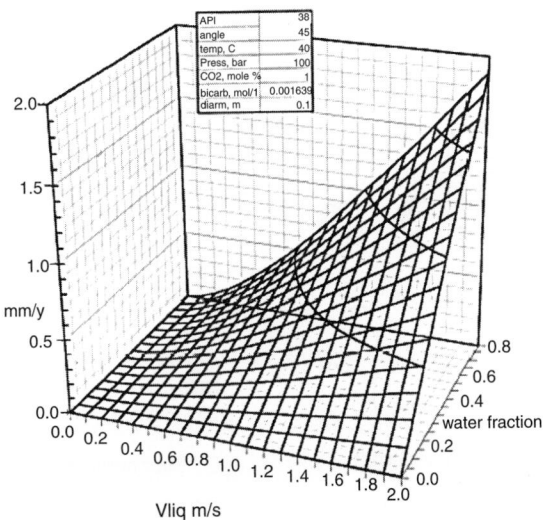

Figure 6.8.8 CO₂ corrosion rates as a function of velocity and watercut with API specific gravity 38 oil [after 8].

rates that increased with velocity up until about 4 m/s and then decreased with increasing velocity. It was found that the C90 had much lower corrosion rates at low velocities but exhibited pitting corrosion and an increasing corrosion rate at higher velocities.

These differences in corrosion rate were correlated to the proportion of corrosion product constituents that comprised the surface scale. These proportions as expressed as strong, medium, weak, and undetected are presented in Figure 6.8.10.

For steels, the combined effects of velocity and pH are presented in Figure 6.8.11.

6.8.2.1.7 Velocity and Corrosion-Resistant Alloys

The velocity effects of CO_2 environments on corrosion-resistant alloys (CRAs) have also been investigated. The velocity versus corrosion rates of 13% chromium, 13% chromium with nickel and molybdenum additions (S13Cr or Super 13Cr), and 22% chromium duplex stainless steels are presented in Figure 6.8.12.

6.8.2.1.8 Chloride Effects

There is a small or mild effect of chloride concentration on corrosion rates. The effect is related to how the chloride interferes or influences the film stability. The effect of chloride content on chromium containing low alloy steels and 13% chromium stainless steel is presented in Figure 6.8.13.

6.8.2.1.9 H_2S Effects on CO_2 Corrosion

When even small amounts of H_2S are present, generally the sulfide corrosion products dominate. The iron sulfide corrosion products are much less soluble and much more stable than the iron carbonates.

Some of the models previously noted that predict CO_2 corrosion take into account the presence of small amounts of H_2S. The potential problem is that, although the H_2S contributes to increasing surface passivity, thereby lowering general corrosion rates, the H_2S increases the tendency toward localized pitting corrosion. The presence of chlorides increases this tendency.

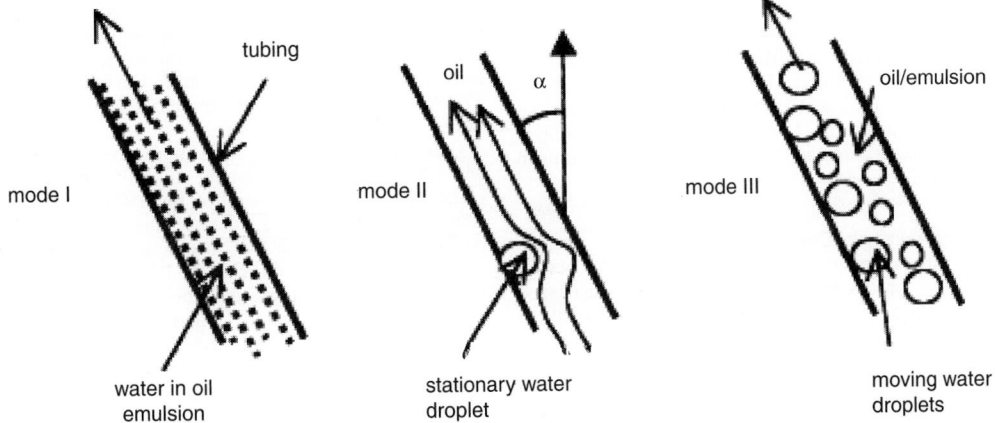

Figure 6.8.9 Modes of water-oil mixtures in production tubing at an angle of deviation, α [after 8].

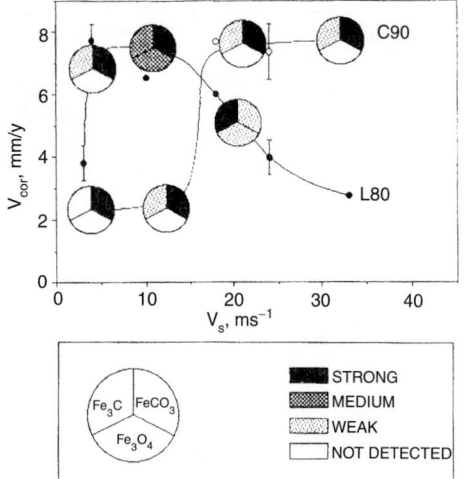

Figure 6.8.10 *Velocity effects on corrosion product composition [after 9].*

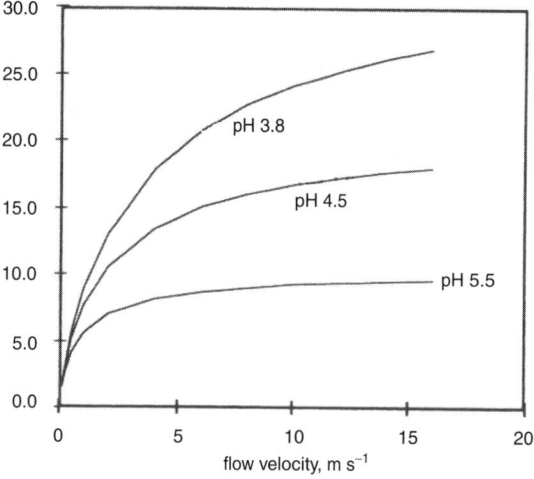

Figure 6.8.11 *Velocity and pH effects on steel corrosion in mm/y [after 2].*

Environmental cracking is the biggest concern with the presence of H_2S. This subject was discussed for carbon and low alloy steels in the chapter on "Corrosion in Drilling and Completion Environments." The environmental cracking in corrosion resistant alloys and hydrogen induced cracking in pipelines are discussed next.

6.8.2.2 Environmental Cracking
There are three types of environmental cracking that occurs in oil and gas production environments that is discussed here. First, there is cracking that is a result of hydrogen adsorption, resulting in embrittlement or cracking. Second, there is stress corrosion cracking where the cracking agent is chloride but the mechanism is assisted when H_2S is also present. Third, there is sulfide stress cracking (SSC), and this a cracking phenomenon that is related both to internal hydrogen effects and stress corrosion cracking.

The stress corrosion cracking and SSC mechanisms requires a susceptible material that is exposed to an environment that causes stress corrosion cracking of the specific material under the application of a tensile stress. The tensile stress may be imposed, tensile component of an imposed shear stress or residual stress. There is at least one set of environmental conditions in which most metals can exhibit stress corrosion cracking. The hydrogen embrittlement (result in hydrogen induced cracking) requires only a hydrogen producing corrosion reaction and a material that is susceptible to cracking.

6.8.2.2.1 H_2S—Sulfide Stress Cracking
Sulfide stress corrosion cracking, also known as SSC, is a stress corrosion cracking phenomena that occurs in the presence of H_2S. H_2S is known to cause or assist in the mechanism of cracking with many alloy types, including low carbon steels, low alloy steels, alloy steels, tool steels, martensitic stainless steels, ferritic stainless steels, austenitic stainless steels, duplex stainless steels, nickel–copper alloys, and nickel base alloys. For many of these materials, there coexists the potential for chloride stress corrosion cracking. The presence of hydrogen sulfide often increases the severity of the cracking or enlarges the range of environmental conditions where cracking occurs.

There are several competing mechanisms that result in SSC, including internal hydrogen stressing, hydrogen embrittlement, stress corrosion cracking, and combinations of these mechanisms. These competing mechanisms have complicated the identification of a universally accepted molecular mechanism that explains all the observed phenomena, that are involved. In this section, SSC will be discussed, and the cracking that is related or complicated by other cracking mechanisms will be presented in the next section, "Stress Corrosion Cracking."

The amount of H_2S that causes SSC is conditional on the other environmental variables but the historical limit that has been listed in NACE MR0175 is 0.05 psi partial pressure of H_2S [12,13]. The environmental factors that influence the existence and extent of SSC include the following:

1. H_2S concentration
2. acidity (pH)
3. time
4. temperature
5. water content (also gas/oil ratio and solids content)
6. chloride content

The effects of these variables are interrelated and also very much conditional on the specific alloy, its condition, and stress state. The other influencing variables, oxygen, and elemental sulfur are discussed in "Stress Corrosion Cracking." The effects listed above are complicated because they are interrelated.

Generally, cracking becomes more prevalent or severe when the acidity, water content, chloride content, and H_2S content increases. The severity of the SSC generally lessens as temperature increases; however, increasing temperature could activate a potential SCC mechanism. These competing mechanisms make it difficult to determine which cracking mechanism will dominate. The complexity also makes selection of materials and estimating the environmental conditions that render them most susceptible to cracking difficult.

For example, martensitic stainless steels such as 13% chromium are not very susceptible to chloride stress corrosion cracking but are susceptible to SSC. The 13% chromium stainless steels would be expected to exhibit a maximum

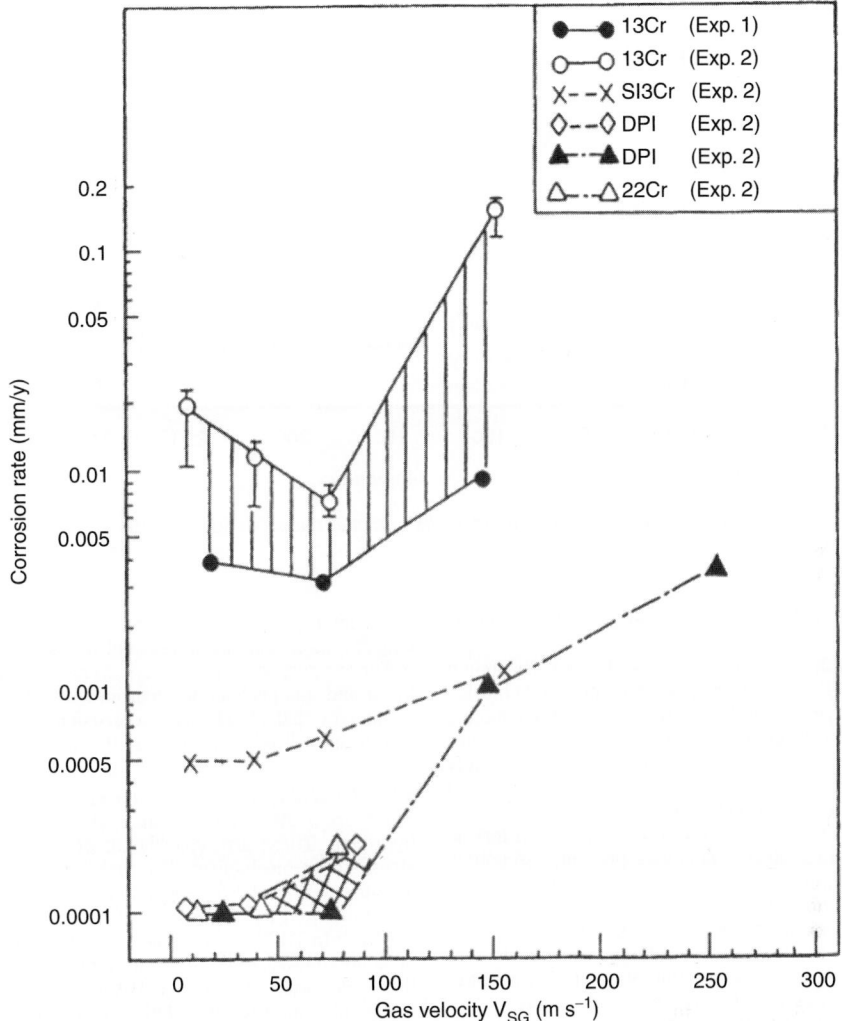

Figure 6.8.12 *Velocity effects on CRAs in CO_2 environments [after 11].*

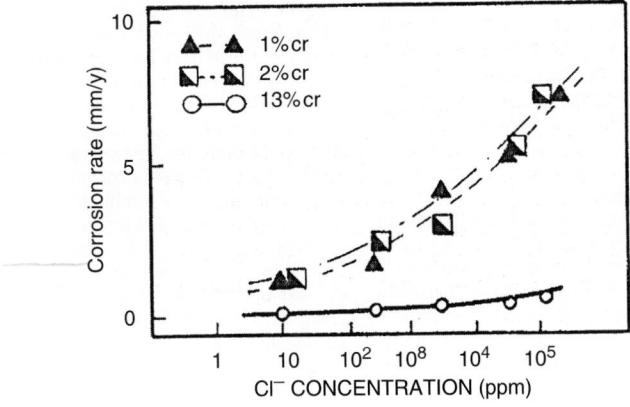

Figure 6.8.13 *Chloride concentration effects on chromium containing low alloy steels and 13% chromium stainless steel in CO_2 environments [after 11].*

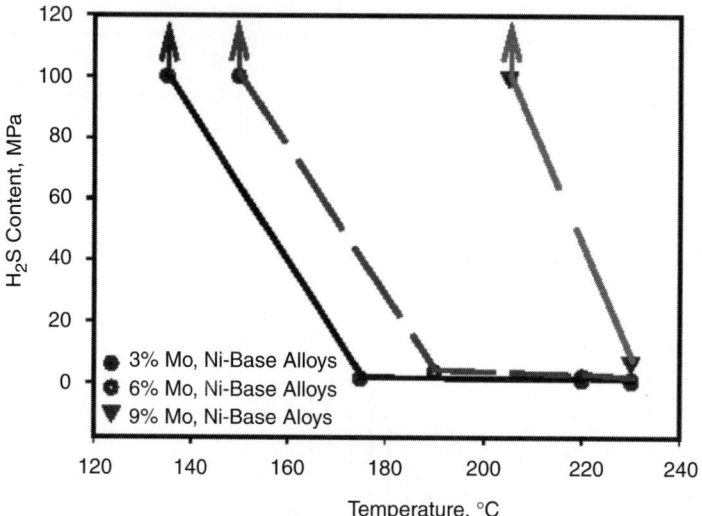

Figure 6.8.14   *Maximum use limits of nickel-based alloys [after 15].*

cracking susceptibility at temperatures closer to room temperature.

Austenitic stainless steels, such as the 18% chromium 8% nickel varieties, are primarily susceptible to chloride stress corrosion cracking. For these materials, the potential for cracking steadily increases with increasing temperature and chloride contents. The incidence rates of SCC greatly increase with increasing H_2S levels. The presence of oxygen has dramatic effects on cracking even at low levels.

In duplex stainless steels, where there are both ferritic and austenitic microstructural phases present, susceptibility to both chloride stress corrosion cracking and SSC exists. Because both susceptibilities exist, the temperature of maximum susceptibility would be expected to be some intermediate range. Barteri [14] conducted tests and demonstrated that the cracking susceptibility was greatest in the 70°C to 110°C range (160°F to 230°F range).

In nickel-based alloys, the potential cracking mechanism is controlled by both chloride and H_2S and is generally only present at elevated temperatures. The relative susceptibilities of nickel-based alloys categorized by molybdenum contents as a function of H_2S content and temperature is presented in Figure 6.8.14.

With respect to time, failures could occur in minutes in very susceptible materials in a very aggressive environment. In situations in which more corrosion cracking–resistant materials are used, cracking and failures could be delayed to many months with some material environmental combinations.

6.8.2.2.2 Stress Corrosion Cracking

As discussed in the preceding section, the identification of the mechanism is complicated because there are a variety of processes and conditions that can cause stress corrosion cracking. For example, we know that stress corrosion cracking can occur as result of either anodic or cathodic corrosion reactions; it is also possible to have both anodic and cathodic reactions promoting stress corrosion cracking simultaneously.

For each alloy and condition under a specific set of environmental conditions, there is a threshold stress intensity below which stress corrosion cracking will not occur. The

stress intensity is affected by the bulk-applied stress, stress concentration factors, and residual stresses.

For our purposes in considering environmental cracking in oil and gas production environments, the environmental factors that affect stress corrosion cracking are not significantly different from those that affect SSC.

6.8.2.2.3 Hydrogen-Induced Cracking

There are many forms of damage that occur as a result of hydrogen. These are typically categorized by the role of stress in the mechanism. The three types of damage that are observed are (1) hydrogen-induced cracking (HIC), (2) stress-oriented hydrogen-induced cracking (SOHIC), and (3) SSC. In HIC, the externally applied stress is low and does not interact with the cracking mechanism. In SOHIC, this is essentially the same as HIC except that the external stress influences the orientation or direction of the crack propagation. In SSC, the stress is an integral component of the cracking mechanism and does not occur without the application of applied stress.

In the oil and gas production environments, these cracking mechanisms have been experienced in pressure vessels, pipelines, and flowlines. In HIC, failures can occur at applied stress levels that are a fraction of the yield strength. For example, there have been many well-documented [16] failures due to HIC in low-strength pipeline steels in gas transmission lines containing H_2S.

Hydrogen is evolved during corrosion-reactions, and H_2S plays a pivotal role in hydrogen related damage mechanisms. The hydrogen evolved is mono-atomic or nascent hydrogen, and its recombination into molecular hydrogen is inhibited or poisoned by the hydrogen sulfide. The result of this inhibited reaction is greater quantities of the mono-atomic hydrogen that readily diffuses into the metal lattice; the size of the mono-atomic hydrogen is minute and moves easily through metal structures. The reactions of hydrogen sulfide and iron are presented below:

$$Fe + H_2S \rightarrow FeS + 2H$$

$$2H \rightarrow H^2$$

The second reaction above is inhibited or poisoned.

Table 6.8.2 *Comparison of Standard Test Methods for Cracking in H_2S*

Test Source	Method	Description	Test Type
NACE TM0177 & EFC 16	A	Uniaxial tensile	Sustained load
NACE TM0177	B	Three-point bent beam	Constant strain
EFC 16	B	Four-point bent beam	Constant strain
NACE TM0177 & EFC 16	C	C-ring	Constant strain
NACE TMO177 & EFC 16	D	Double cantilever beam	Fracture toughness
EFC 16	E	Slow strain rate	Strain to failure
NACE TM0198	—	Slow strain rate	Strain to failure

There are several theories that explain the various types of hydrogen damage [17]. Although numerous models and theories exist with respect to HIC, the phenomenon of trapping best quantifies the metal susceptibility to hydrogen damage [18,19]. The cracks occur when the atomic hydrogen migrates to these trap sites and then becomes "trapped."

We know that inclusions, especially elongated manganese sulfide inclusions, increase susceptibility to HIC damage. The other known major factor is alloys segregation in the form of banding. Banding is particularly a problem in plate and pipe formed from plate products but can be seen in seamless pipe as well.

To avoid HIC, clean steels with uniform microstructures have the best resistance to this form of cracking.

6.8.2.2.4 Testing

There are a variety of standard test methods exist to determine susceptibility to environmental cracking as a function of material and environmental details.

The tests that examine materials susceptibility to HIC is based on measuring the amount of diffusible hydrogen or quantifying the cracking that occurs in standard test environments. The most widely accepted standard test method for measuring susceptibility to HIC is NACE's TM-0284 [20]. This test method uses unstressed beams and internal cracks are assessed as a function of crack direction.

A related phenomenon of cracking that occurs that is oriented by applied stress is SOHIC. A test method to determine the resistance of plate steels for SOHIC resistance is NACE's TM-0103 [21].

The standard tests that measure resistance to room temperature and elevated temperature SSC are found in NACE TM0177 [22], EFC Publication 16 [23] and NACE TM0198 [24]. The tests referenced in these documents are summarized in Table 6.8.2.

The selection of tests is somewhat dependent on specimen geometry. For example, thin-walled tubulars can be tested using C-ring and either three-point or four-point bent beam type test specimens. Bars, forgings, castings and thick wall thickness tubulars can be tested by any of the tests described above.

Different values or results that indicate relative resistance to environmental cracking are determined from these tests. The uniaxial tension, three- and four-point bent beam, and C-ring tests results in a threshold stress that defines the maximum stress level at which cracking does not occur. These tests are usually run at either a minimum stress level where no failures occur in a standard period of time (30 days) or in several stress levels at which the threshold stress can be determined.

The double cantilever beam (DCB) test results in a fracture toughness constant (K_{1SSC}) that is relative to the environment in which the test is conducted. The slow strain rate tests (SSRTs) result in ductility ratios between tests conducted in air versus tests conducted in a test environment.

Jean Leyer et al. [25] compared several of the standard test methods to evaluate test selection on threshold stress. In tests using a steel that was susceptible to environmental cracking, API 5CT [10] type P110, they found the following ranking of tests in order of increasing severity:

Four-point bent beam < C-ring (ID tension)

< C-ring (OD tension) < uniaxial tension test

Therefore, the uniaxial tension test is the most severe of those evaluated.

In applying these tests for qualification purposes, the selection of tests and the test environment for steels is straight forward. The environments that are most often used are either the standard test conditions defined in NACE TM0177 [22] or a mock-up of the worst case actual production environment. The stress level required for qualification purposes by NACE/ISO 15156 [13] in the TM0177 type conditions would be at a minimum of 80% of the actual yield strength, whereas the mock-up environments would require a minimum of 90% of the actual yield strength.

Testing of corrosion-resistant alloys varies in that often more than one type test or test environment is required to determine resistance to more than one potential failure mode. For example, many stainless steels are tested for room temperature resistance to SSC and elevated temperature testing for resistance to stress corrosion cracking with H_2S.

One of the areas of concern with benchmarking CRAs has been the selection of a standard test environment. This concern has been partly alleviated by the standard test levels defined in NACE MR0175 [12] and reproduced in ISO 15156 [13]. These levels are defined in Table 6.8.3.

Most of the new materials and changes proposed to the standards that govern materials for use in H_2S containing environments will be based on tests that are conducted to one or more of the standard test conditions in Table 6.8.3.

6.8.2.3 Other Fluids in Production Operations

In addition to production fluids, production equipment is often exposed to other fluids. For example, during well completions, weighted salt (brine) environments are frequently used. These are often referred to as *packer fluids* or *packer brines*. During the life cycle of a producing well, the production from the formation will decrease and production can be stimulated through injection of acids into the formation. Often naturally occurring formation salts will deposit along metals walls, and these may clog or restrict production; these are removed through the use of scale inhibitors or removers.

In addition to production wells, there are also injection wells where CO_2 steam of fresh or salt water is injected into the formation to stimulate production. Often the same wells

Table 6.8.3 *Description of Standard Test Levels*

Test Level	Temperature	H$_2$S	CO$_2$	NaCl	pH	Other
I	25°C	Note 1	None	Note 1	Note 2	Note 1
II	25°C	15 psi	None	5%	2.7–3.6	None
III	25°C	15 psi	None	5%	2.7–3.6	Coupled
IV	90°C	0.4 psi	100 psi	150,000 mg/L	Note 2	Note 3
V	150°C	100 psi	200 psi	150,000 mg/L	Note 2	Note 3
VI	175°C	500 psi	500 psi	150,000 mg/L	Note 2	Note 3
VII	205°C	500 psi	500 psi	150,000 mg/L	Note 2	Note 3

Note 1: Level 1 was intended as a lower H$_2$S SSC test, conditions to be listed.
Note 2: The pH level and buffering to be described.
Note 3: Other environmental constituents such as elemental sulfur to be described.

that are used to stimulate are used to produce the hydrocarbons, these will, therefore, alternate between being injectors and producers. Another method used to enhance production it gas lift. In these applications, produced gas is reinjected into the production tubing.

6.8.2.3.1 Brines — Completion Fluids

A variety of completion brines are used. The selection of brine is based first on the desired weight of the fluid usually expressed in pounds per gallon (ppg) and second on compatibility with contacting equipment. The brines that have been used include the following: sodium chloride, calcium chloride, calcium bromide, zinc bromide, and cesium formate. The completion brines are often mixtures of more than one of these constituents.

There are several potential problems that stem from these completion brines. First, many of the brines are acidic and therefore are corrosive. The most acidic of the above brines is zinc bromide. An associated problem with zinc bromide is the effect on elastomers, nitrile based elastomers are degraded by contact with this brine. Second, the intrusion of production acid gases (CO$_2$ and H$_2$S) in high-pressure high-temperature (HPHT) gas wells into the brines will acidify it. When the brines become acidified, there is an increased risk of corrosion and stress corrosion cracking. There are many examples of these corrosion [26] and environmental cracking problems [27].

There continues to be incidences of failures in brines, and there is continuing efforts to use safer weighted brines. For example, there has been recent work reported on mixture of potassium and cesium formates as packer brines [28].

6.8.2.3.2 Injected Fluids — Acidizing and Scale Removers

To enhance production and remove wall deposits, a variety of fluids are injected into the formation, production tubing string, and flowlines. The primary potential problem with these fluids is accelerated general or localized corrosion attack. The selection of fluid with the proper additives (neutralizers or corrosion inhibitors) is important to prevent this corrosion attack.

During acidizing operations, even with proper fluid selection and inhibitor additions, there are two other potential problems that require evaluation. These are selection of inhibitor for all the metals being contacted and the problems with spent acid returns. The inhibitor selection is important because inhibitors often do not protect all types of metals and may lose effectiveness under certain downhole conditions. An example of a corrosion problem on a stainless steel component in an improperly applied inhibited acid treatment is presented in Figure 6.8.15.

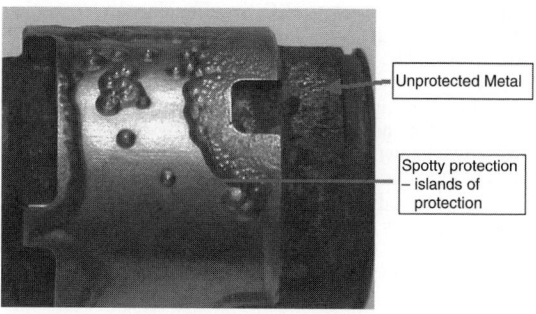

Figure 6.8.15 *Example of improperly applied corrosion inhibition during acidizing.*

6.8.2.3.3 Injected Gases — Gas Lift and CO$_2$ Flood

Produced gas is used in gas lift operations as a means to enhance production. The injected gas in injected at points downhole in the production string to "lift" the fluids. The produced gas may contain acid gases, so the same care is required in lift operations as in the production string.

In CO$_2$ flood operations, the formation is stimulated by injected of CO$_2$. Because CO$_2$ forms carbonic acid in water, this acidifies the formation. In the producing wells in fields where CO$_2$ flood operation are performed, the production tubing and facilities require attention to prevent corrosion.

6.8.2.4 Microbiologically Induced Corrosion

Microorganisms are present throughout our environment. These organisms can result in corrosion by producing a deposit on metal surfaces and/or by increasing the acidity/reducing the pH of the environment. This altering the environment can occur in the bulk of locally. This type of corrosion can be found in flowlines and pipelines where these microorganisms are inadvertently introduced during inspection and pigging operations.

There are two basic classes of microbes, anaerobic and aerobic microorganisms. Anaerobic microorganisms flourish in the absence of oxygen such as deep sea or downhole environments. Aerobic microorganisms require oxygen to survive and flourish and are found in all surface environments including salt and fresh waters.

One example of an anaerobic microorganism that is troublesome in the oilfield is the genus Desulovibrio. This is a sulfate-reducing bacteria (SRB), and the resultant corrosion mechanism involves naturally occurring sulfates to H$_2$S. The resultant acid from the H$_2$S can cause or accelerate corrosion or cause HIC or SSC as discussed previously. A simplistic

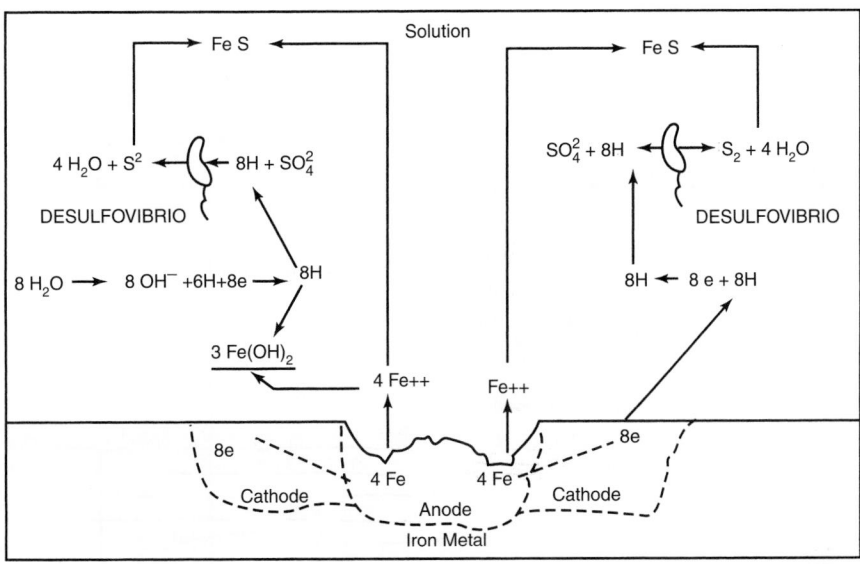

Figure 6.8.16 *Diagram of polarization of local cathode by a film of hydrogen gas bubbles (cathodic area to right of anode is polarized) [29].*

chemical mechanism illustrating corrosion attack on steel by SRB is presented in Figure 6.8.16.

There are many iron oxidizing types of bacteria such as Gallionalla and Crenothrix that can accelerate corrosion by releasing CO_2 as one of the corrosion products as illustrated by the following equation.

$$4FeCO_3 + O_2 + 6H_2O \rightarrow 4Fe(OH)_3 + 4CO_2$$

$Fe(OH)_3$ is not very soluble and will precipitate on the metal surface. These deposits can thus cause oxygen concentration cells resulting in corrosion beneath the deposits.

In addition to the bacteria examples above, there are also molds, yeasts, algae, and protozoa. These, like bacteria, can have either a waste product that enhances corrosion or can result in slimes or deposits beneath which oxygen concentration cells/localized corrosion attack can result.

6.8.3 Production Equipment

The operations of the various products used in the production of oil and gas are discussed elsewhere in this handbook. In this section, the different types of completions and the corresponding materials of construction are presented.

Open hole completions are not normally used today but can be used where there is only one production zone. There are many disadvantages to this type of completion including poor control over production and frequent cleanout operations being required due to formation intrusion. The general appearance of this type of completion is presented in Figure 6.8.17.

More commonly, open hole completions are assisted with the use of screen liners and slotted liners. These products keep the formation from collapsing and filter out formation sands. A schematic of this type of completion is presented in Figure 6.8.18.

Perforated casing is the common type of completion and utilize packers to isolate production from the annulus between the tubing and casing. This completion method can be used to produce single or multiple zones. Schematics of

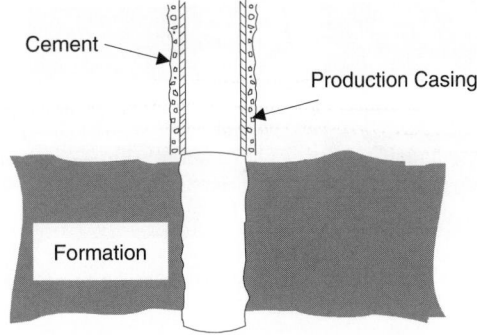

Figure 6.8.17 *Open hole completion.*

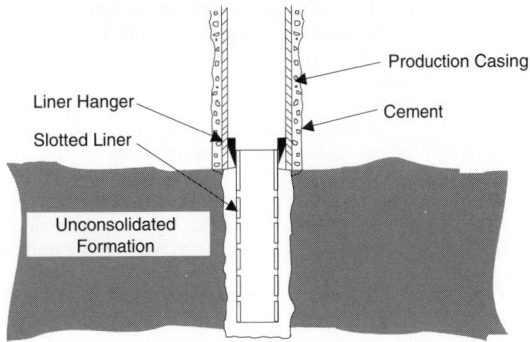

Figure 6.8.18 *Open hole completion with slotted liner.*

multiple-zone single-string and multiple-zone multiple-string completions are presented in Figures 6.8.19 and 6.8.20.

For small-diameter wellbores at relatively shallow depths, a tubing-less type of completion can be used. In these

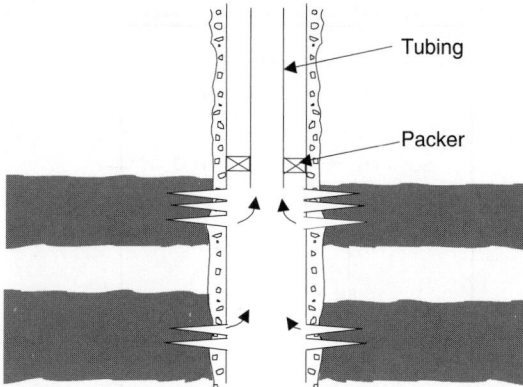

Figure 6.8.19 *Multiple-zone single-string completion.*

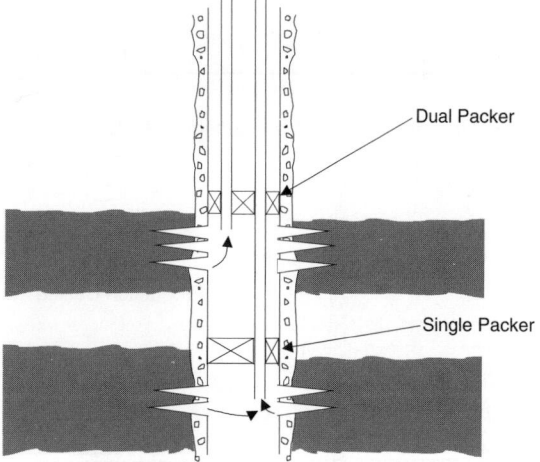

Figure 6.8.20 *Multiple-zone multiple-string completion.*

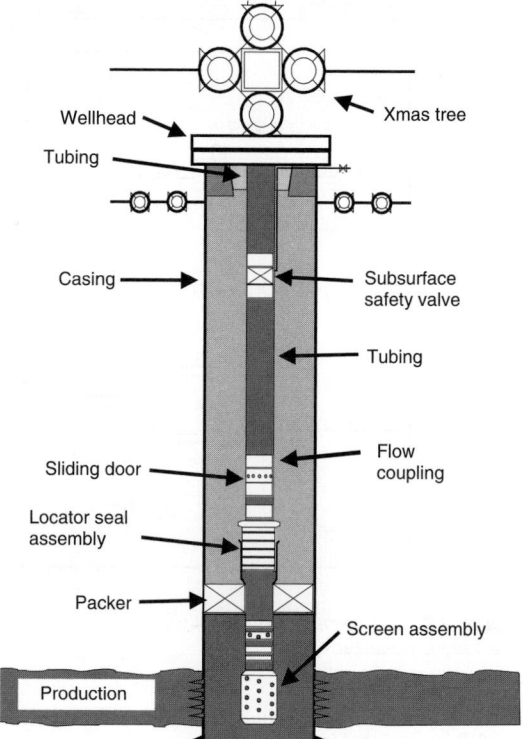

Figure 6.8.21 *Schematic of single-zone deep gas completion.*

Table 6.8.4 *Composition of Common API 5CT Tubular Grades*

Grade	Type		C	Mn	Mo	Cr	Ni	Cu
J55	—	Max.	—	—	—	—	—	—
K55	—	Max.	—	—	—	—	—	—
L80	1	Max.	0.43*	1.90	—	—	0.25	0.35
L80	9Cr	Min.	—	0.30	0.90	8.00	—	—
		Max.	0.15	0.60	1.10	10.0	0.5	0.25
L80	13Cr	Min.	0.15	0.25	—	12.0	—	—
		Max.	0.22	1.00	—	14.0	0.5	0.25
C90	1	Min.	—	—	$0.25^{\dagger}$	—	—	—
		Max.	0.35	1.00	0.75	1.20	0.99	—
T95	1	Min.	—	—	$0.25^{\dagger}$	—	—	—
		Max.	0.35	1.20	0.85	1.50	0.99	—
P110	—	Max.	—	—	—	—	—	—
Q125	1	Max.	0.35	1.00	0.75	1.20	0.99	—

*The carbon content may be increased to 0.50 when oil-quenched.
†The Mo content may be less for thin walled tubulars (< 17.78 mm).

instances, the small-diameter casing such as 2-7/8″ is cemented into place and conventionally perforated in the production zone. The casing then becomes the tubing string. The advantage of this type of completion is that it provides a low-cost completion technique.

Deep well completions require higher strength tubulars to be able to withstand the large tensile and differential pressure stresses that are encountered. The casing programs require multiple casing sizes that telescope from the surface down into the producing formations. These completions also require specialized equipment to manage tubing movement due to expansion and contraction during heating and cooling respectively. A schematic of a typical gas well completion is presented in Figure 6.8.21.

The various components and functions of the production equipment illustrated in the preceding schematics are discussed elsewhere in this handbook. The following sections describe the typical variety of metallic materials that are used in the construction so that an understanding of the potential corrosion problems can be obtained.

6.8.3.1 Production Tubing and Casing
In normal operations, the casing is protected from the production fluids. The corrosion risk for casing lies when the casing is used for production or when formation fluids migrate through the barriers that are used such as packers and cement.

The tubing is the conduit through which petroleum products are removed from the wells. The tubing is one of the largest investments made by operators during the completion phase of oilfield operations. The tubing is therefore the single largest investment that is subjected directly to corrosive production environments and requires evaluation of potential corrosion and cracking reactions.

Table 6.8.5 *Mechanical Properties of Common API 5CT Tubular Grades*

Grade	Types	Yield Strength, ksi (MPa) Min.	Yield Strength, ksi (MPa) Max.	Tensile Strength, Ksi (MPa), Min.	Hardness HRC, Max.
J55	—	55.0 (379)	80.0 (552)	75.0 (517)	—
K55	—	55.0 (379)	80.0 (552)	95.0 (655)	—
L80	1, 9Cr, 13Cr	80.0 (552)	95.0 (655)	95.0 (655)	23
C90	1, 2	90.0 (621)	105.0 (724)	100.0 (689)	25.4
T95	1, 2	95.0 (655)	110.0 (758)	105.0 (724)	25.4
P110	—	110.0 (758)	140.0 (965)	125.0 (862)	—
Q125	1, 2, 3, 4	125.0 (862)	150.0 (1034)	135.0 (931)	—

Table 6.8.6 *Corrosion-Resistant Alloy Tubular Products*

UNS No.	Common Name	Type
S41425	Super 13Cr	Martensitic stainless
S41426	Super 13Cr	Martensitic stainless
S41427	Super 13Cr	Martensitic stainless
S31254	254 SMO	Super austenitic stainless
N08028	Alloy 28	Super austenitic stainless
N08904	904L	Super austenitic stainless
N08367	AI-6XN	Super austenitic stainless
S31803	2205	Duplex stainless steel
S32750	2507	Duplex stainless steel
S32760	Zeron 100	Duplex stainless steel
N08032	NIC 32	Solid solution nickel base alloy
N08042	NIC 42M	Solid solution nickel base alloy
N08825	825	Solid solution nickel base alloy
N06255	2550	Solid solution nickel base alloy
N06985	G3	Solid solution nickel base alloy
N10276	C276	Solid solution nickel base alloy

Table 6.8.7 *Nominal Composition of Corrosion-Resistant Alloy Tubular Products*

UNS No.	Cr	Ni	Mo	Cu	Fe	N	Others
S41425	13.8	5.5	1.8	—	Bal.	0.1	—
S41426	12.5	5.5	2.3	—	Bal.	—	—
S41427	12.5	5.3	2	—	Bal.	—	—
S31254	20	18	6	0.8	Bal.	0.20	—
S08028	27	32	3.5	1.0	Bal.	—	—
N08904	21	25.5	4.5	1.5	Bal.	—	Co = 4.5
N08367	21	24.5	6.5	—	Bal.	0.22	—
S31803	23	5.2	3	—	Bal.	0.15	—
S32750	25	7	4	—	Bal.	0.28	0.2
S32760	25	7	3.5	0.8	Bal.	0.2	W = 0.8
N08032	21.5	32	4.5	—	Bal.	—	—
N08042	21.5	42	6	2.2	Bal.	—	Ti = 0.9
N08825	21.5	42	3	2.2	Bal.	—	Ti = 0.9
N06255	24.5	49.5	7	—	Bal.	—	—
N06985	22.3	Bal.	7	2.0	19.5	—	—
N10276	15.5	Bal.	16	—	5.5	—	W = 3.8

API 5CT [10] is the primary industry specification that governs casing and tubing materials. The compositions and mechanical properties of common API 5CT Grades are presented in Tables 6.8.4 and 6.8.5, respectively.

The L80 type 1, C90 and T95 grades above are generally known as the grades used when cracking resistance in H_2S containing environments is required. Grades J55, K55, L80 types 9Cr and 13Cr have some lesser resistance to H_2S cracking and the higher strength grades (P110 and Q125) can be used in H_2S-containing environments only at higher temperatures per NACE MR0175 [12] and ISO 15156 [13].

The L80 types 9Cr and 13Cr are used extensively in CO_2-containing anaerobic production environments.

The requirement for greater corrosion resistance and greater resistance to environmental has resulted in a number of corrosion-resistant alloys being used for tubing and, to a lesser extent, casing. A survey conducted by NACE in the mid-1990s indicated that the majority of CRA tubulars in use to that point in time were of the 13% chromium variety but that the use of corrosion resistant alloys was on the rise [30]. A list of some of the more popular alloys is presented in Table 6.8.6 and the nominal compositions are presented in Table 6.8.7.

With the exception of martensitic stainless steels listed in Tables 6.8.6 and 6.8.7, the CRAs above derive the mechanical properties through cold work. The typical minimum yield strength levels in the cold worked alloys range from about 758 MPa (110 ksi) to about 1034 MPa (150 ksi).

6.8.3.2 Packers, Safety Valves, and Other Downhole Accessories

The downhole accessories such as packers and safety valves are typically made from similar materials to the tubing string. Generally, the materials are selected such that the accessory corrosion resistance is equal to or slightly better than the tubing string. One major difference between the accessories and the tubulars is the form of the raw materials; most accessories are tubular in nature but they are typically machined from bar stock. This translates to a different set of metals being used because of the inability to achieve the required mechanical properties from cold worked tubular corrosion resistant alloys in heavy sections.

To achieve the required strength levels, a different hardening mechanism is required. The common materials used in these applications utilize either a martensitic transformation, such as that present in steels and other martensitic alloys, or a precipitation hardening mechanism, such as that present in precipitation hardening stainless steels and nickel base alloys. A list of some of the common precipitation hardening stainless steels and nickel base alloys is presented in Table 6.8.8 with the typical specified minimum yield strength(s). The nominal compositions of these alloys is presented in Table 6.8.9.

6.8.3.3 Wellhead/Valves

The wellhead valves and assorted hardware have different material constraints than the tubulars and downhole accessories previously discussed. The equipment is typically

Table 6.8.8 *Common Precipitation Hardening Corrosion-Resistant Alloys*

UNS No.	Common Name	Typical Minimum Yield Strength
S17400	17-4PH	724 Mpa (105 ksi)
S45000	Custom 450	655 Mpa (95 ksi)
S66286	A286	586 Mpa (85 ksi)
N05500	K500	621 Mpa (90 ksi)
N07718	718	827 Mpa (120 Ksi)
N07725	725	896 Mpa (130 ksi)
N07750	X-750	724 Mpa (105 ksi)
N09925	925 Alloy	758 Mpa (110 ksi)

Table 6.8.9 *Nominal Composition of Common Precipitation Hardening Corrosion Resistant Alloys*

UNS No.	Cr	Ni	Mo	Cu	Fe	Nb	Others
S17400	16.5	4	—	4	Bal.	0.3	—
S45000	15	6	0.8	1.5	Bal.	0.3	—
S66286	15	25.5	1.3	—	Bal.	—	Ti = 2.1, V = 0.3
N05500	—	66	—	Bal.	—	—	Ti = 0.6, Al = 2.7
N07718	19	52	3	—	Bal.	5.0	Ti = 0.9, Al = 0.5
N07725	20.7	57	8.3	—	Bal.	3.4	Ti = 1.4
N07750	15.5	Bal.	—	1	7	3.3	Ti = 2.5, Al = 0.7
N09925	21.5	42	3	2.3	Bal.	—	Ti = 2.2, Al = 0.3

larger because it does not have to fit through the casing. This permits new material options that encompass forgings and castings, and where corrosion resistance is required, linings and overlays can be used.

The larger equipment results in lower stress states; therefore, lower strength materials may be used. The wellhead equipment and material properties are defined by API Specification 6A [31]. The API 6A material strength requirements are presented in Table 6.8.10.

A Christmas tree is named because of its appearance as a tree with ornaments hanging off it. It is comprised of a number of valves, spools chokes, etc. An example of a Christmas tree is presented in Figure 6.8.22.

Wellhead equipment has been made from a wide variety of wrought and cast steel products. Some solid corrosion-resistant alloys have been used such as the 13Cr family of materials. A variety of overlays and hardsurfacings are used in corrosion resistant wear or flow path/controlling components. Generally the selection is based on functional requirements first and corrosion resistance second. A list of some of the common CRAs with typical strength levels is presented in Table 6.8.11.

API 6A also defines classes of service and the types of materials that are used for each class; these classes are presented in Table 6.8.12.

The relative corrosivity of production environments has been related by API 6A to the partial pressure of CO_2 present in the production fluids. This corrosivity as it is defined in API 6A with service type is presented in Table 6.8.13.

6.8.3.4 Artificial Lift Equipment

The purpose of lift equipment is to assist with "lifting" oil out of the ground. In most cases, the downhole pressure is insufficient for the liquid petroleum products to rise unassisted to the surface. There are many forms of lift equipment such as gas lift, plunger lift, rod pumps, hydraulic pumps, and electric submersible pumps (ESPs). Schematics of these types of lift equipment are presented in Figure 6.8.23 for gas lift, Figure 6.8.24 for plunger lift, Figure 6.8.25 for rod pumps, Figure 6.8.26 for hydraulic pumps, and Figure 6.8.27 for electric submersible pumps.

The gas lift equipment is similar in nature to downhole accessories in that most of the materials used in gas lift equipment can also be found in gas lift. Additionally, there are copper nickel alloys such as alloy 400 and alloys 405 that are found gas lift equipment.

Most rods for rod pumps are made from carbon and low alloy steels. The various types of pumps described above utilize a variety of materials. Some of the most common materials that are used in these pumps are presented in Table 6.8.14.

Many of the steel materials above utilize surface hardening techniques such as carburizing, nitriding, induction hardening, arc spray hardsurfacing, high-velocity oxyfuel (HVOF) hardsurfacing, electroless nickel plating, and chromium plating.

6.8.3.5 Flowlines and Pipelines

Flowlines and Pipelines are mostly carbon steels and carbon steels with small alloying additions (high-strength low-alloy or HSLA steels) such as those defined by ASTM A106 [32], and API 5L [33]. These specifications do not define the chemical composition except to limit elements that effect weldability and impact/fracture toughness. The flowline and pipline materials are generally field welded and, therefore, must be very weldable. After welding, most pipeline welds are left in the as-welded condition without the benefit of postweld heat treatment.

The mechanical property requirements of API 5L pipeline grades for PSL 2 (product specification level 2) are presented in Table 6.8.15.

For the transportation of corrosive production fluids in flowlines and pipelines, there have been numerous applications of clad pipe and corrosion resistant alloy pipe. The internal cladding has been a variety of austenitic and duplex stainless steels and nickel-based alloys.

Because of the welding requirements for the pipe products, most of the corrosion-resistant alloys used for pipe have been either low-carbon 13% chromium martensitic stainless steels or 22% to 25% chromium duplex stainless steels.

Table 6.8.10 *Standard API 6A Material Designations*

API Material Designation	Yield Strength Min Mpa (Ksi)	Tensile Strength Min Mpa (Ksi)	Elongation Min	Reduction of Area Min
36 K	248 (36.0)	483 (70.0)	21%	No requirement
45 K	310 (45.0)	483 (70.0)	19%	32%
60 K	414 (60.0)	586 (85.0)	18%	35%
75 K	517 (75.0)	655 (95.0)	17%	35%

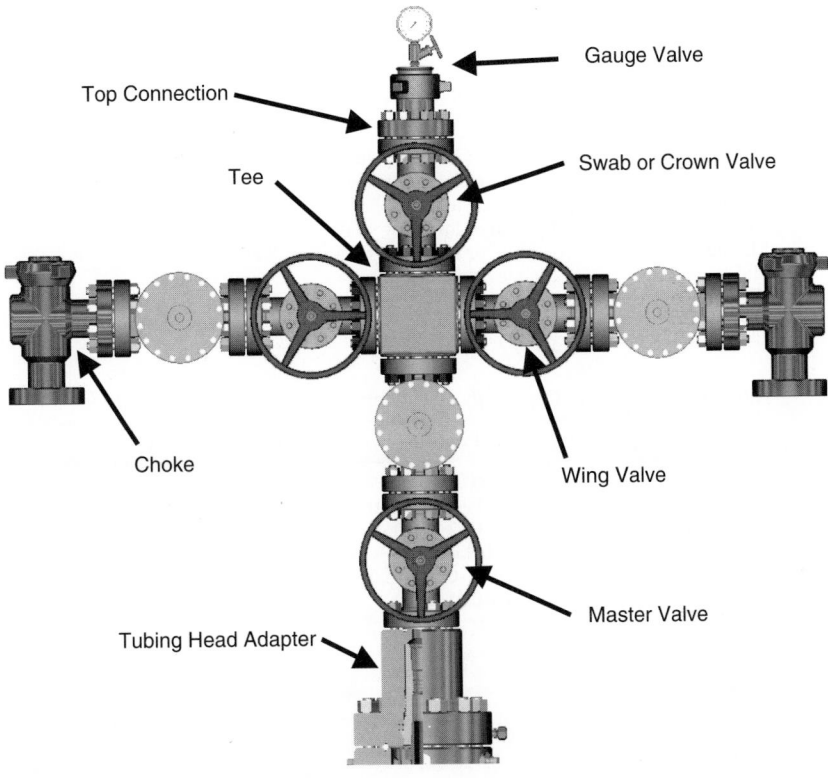

Gauge Valve

Top Connection

Swab or Crown Valve

Tee

Choke

Wing Valve

Master Valve

Tubing Head Adapter

Figure 6.8.22   *Example of a Christmas tree.*

Table 6.8.11   *Common Wellhead Corrosion-Resistant Alloys*

UNS No.	Common Name	Typical Minimum Yield Strength
S41000	410	517 Mpa (75 ksi)
S42000	420	517 Mpa (75 ksi)
J91150	CA15	517 Mpa (75 ksi)
J91540	CA6NM	517 Mpa (75 ksi)
S41500	F6NM	517 Mpa (75 ksi)
S17400	17-4PH	724 Mpa (105 ksi)
J92180	Cb7Cu-1	655 Mpa (95 ksi)
S30400	304	207 Mpa (30 ksi)
S31600	316	207 Mpa (30 ksi)
S31803	22Cr	414 Mpa (60 ksi)
S32750	25Cr	517 Mpa (75 Ksi)
N05500	K500	621 Mpa (90 Ksi)
N07718	718	827 Mpa (120 ksi)

6.8.4 Mitigation Techniques

To combat corrosion, an understanding of why it occurs and knowledge of the environment is necessary. The next step is to understand the life cycle requirements of the equipment and how the operations and environments may change over the life of the equipment. Once we know these variables, we proceed to examining the equipment on which we need to control corrosion to determine where corrosion needs to be prevented and the material requirements of the components. Finally, we assess the various techniques available to us so that we can prevent or control corrosion.

The most common methods to control corrosion in the petroleum industry are (1) materials selection for corrosion resistance; (2) chemical inhibition of the corrosion reactions; (3) coatings that are either sacrificial, corrosion resistant, or barriers to corrosion; (4) cathodic protection; and (5) designing with a corrosion a allowance. In this section the topics of material selection, coatings, and inhibition are presented.

6.8.4.1 Materials Selection

In addition to corrosion resistance, the factors that are to be considered in materials selection include strength level, product form, impact toughness, meeting code requirements, and cost. It is frequently the case that some of the other product requirements take precedure over the corrosion resistance. It may also be that permitting corrosion and adding a corrosion and adding a corrosion allowance to the design is the most cost effective materials design selection. The consequences of not applying mitigation techniques are component failures with all the risks that may be associated with those failures. An example of corroded production tubing in a CO_2 containing environment is presented in Figure 6.8.28.

The common materials used for typical production equipment such as production tubing and casing, downhole accessories, wellhead equipment, artificial lift equipment, and pipelines have been discussed previously under this topic of production corrosion. Within that range of materials, some of the basis for materials selection is presented herein.

Looking at alloying elements, different elements exhibit different roles in corrosion resistance. In iron-based metals such as low-alloys steels, alloy steels, and stainless steels,

Table 6.8.12 *Description of Material Classes from API 6A*

Material Class	Minimum Material Requirements	
	Bodies, Bonnets, End, and Outlet Connections	Stems and Pressure Controlling Parts
AA-General service	Carbon or low alloy steel	Carbon or low alloy steel
BB-General service	Carbon or low alloy steel	Stainless steel
CC-General service	Stainless steel	Stainless steel
DD-Sour service	Carbon or low alloy steel	Carbon or low alloy steel
EE-Sour service	Carbon or low alloy steel	Stainless steel
FF-Sour service	Stainless steel	Stainless steel
HH-Sour service	CRAs	CRAs

All materials for sour service shall be in compliance with NACE MR0175.

Table 6.8.13 *Relative Corrosivity as Indicated by CO_2 Content from API 6A*

Service Type	CO_2 Partial Pressure	Relative Corrosivity
General service	< 7 psi (< 0.05 Mpa)	Noncorrosive
General service	7–30 psi (0.05–0.21 Mpa)	Slightly corrosive
General service	> 30 psi (> 0.21 Mpa)	Moderate to highly corrosive
Sour service	< 7 psi (< 0.05 Mpa)	Noncorrosive
Sour service	7–30 psi (0.05–0.21 Mpa)	Slightly corrosive
Sour service	> 30 psi (> 0.21 Mpa)	Moderate to highly corrosive

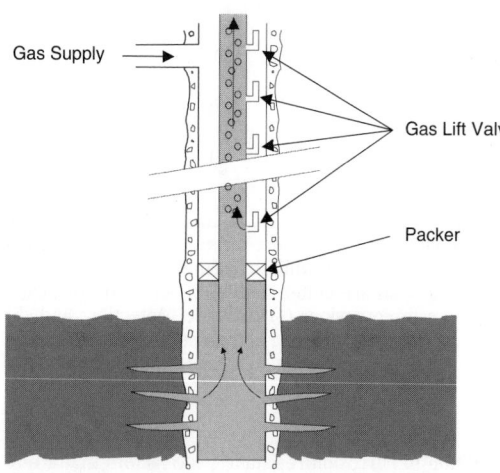

Figure 6.8.23 *Schematic of gas lift method of artificial lift.*

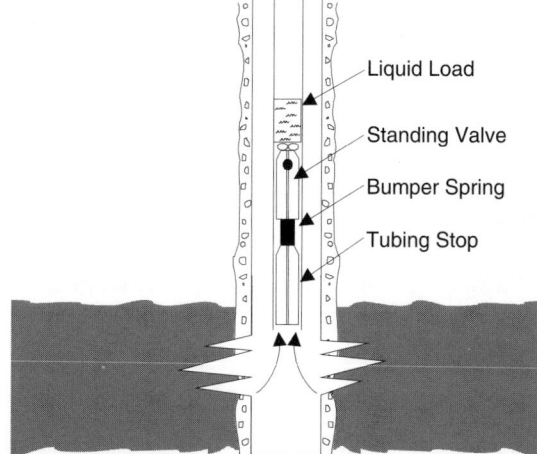

Figure 6.8.24 *Schematic of plunger lift method of artificial lift.*

the first alloying element considered is chromium. It is chromium that makes the stainless steels resistant to corrosion. The effect of chromium on corrosion rate in steel is presented in Figure 6.8.30.

The rationale for selection of 5% chromium, 9% chromium, and 12% chromium stainless steels is the corrosion resistance as illustrated in Figure 6.8.29. Other alloying elements in steels also increase corrosion resistance such as molybdenum and tungsten. There have been several relationships developed that relate the composition of iron-based CRAs to the relative corrosion and pitting resistance in the presence of dissolved chlorides and oxygen (salt solutions) such as salt water. The PREN relationship that exists in the ISO Standard 15156 [13] is presented below:

$$PREN = W_{Cr} + 3.3(W_{Mo} + 0.5W_W) + 16W_N$$

where

W_{Cr} is the chromium content (% of total composition)
W_{Mo} is the molybdenum content (% of total composition)
W_W is the tungsten content (% of total composition)
W_N is the nitrogen content (% of total composition)

These percentages are based on the mass fraction of chromium in the alloy expressed as a percentage of the total composition. An example of a PREN calculation for a duplex stainless steel such as UNS S39377 with 24.8% chromium, 7% nickel, 2.5% molybdenum, 1.1% tungsten and 0.28% nitrogen would yield the following result:

$$PREN = 24.8 + 3.3 \times (2.5 + 0.5 \times 1.1) + 16 \times 0.28$$
$$PREN = 24.8 + 10.1 + 4.5 = 39.4$$

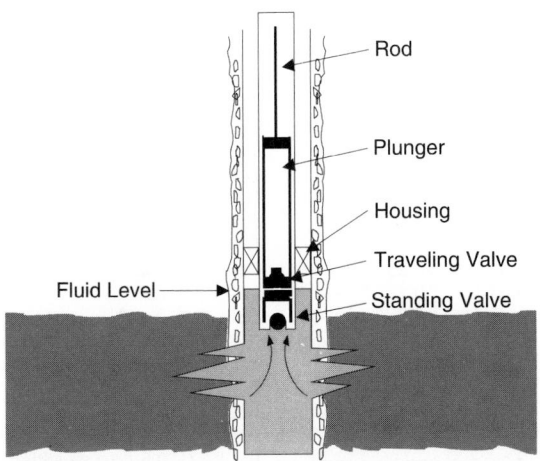

Figure 6.8.25 Schematic of rod pump lift method of artificial lift.

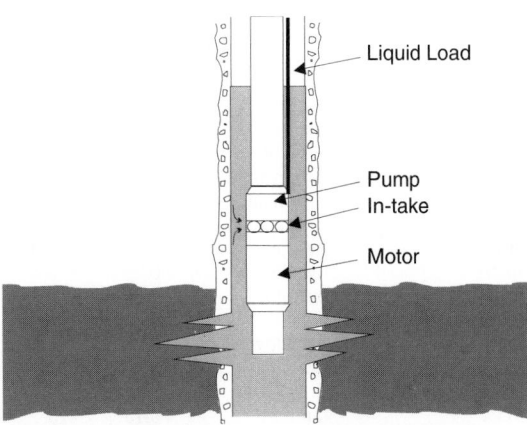

Figure 6.8.27 Schematic of electric submersible pump lift method of artificial lift.

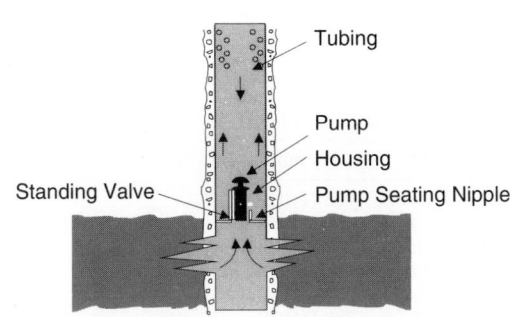

Figure 6.8.26 Schematic of hydraulic pump lift method of artificial lift.

This 39.4 value of the PREN can be used to match up with design criteria or compare with other alloys to rank relative corrosion and pitting resistance. A note of caution for the PREN relationship presented here is that the relationship was developed for stainless steels, and the relationship may not be a good predictor of relative ranking when the alloys being compared includes other types of alloys such as nickel-based alloys.

From the above relationship, the higher the PREN number, the higher the resistance to general corrosion and pitting corrosion. To mitigate the effects of corrosion, we can select alloys that exhibit a high PREN number. For most metals there is a critical temperature below which pitting does not occur. This temperature is dependent on the composition and the specific environment. For stainless steels, the critical pitting temperature is related to the PREN number as expressed above.

Although this evaluation of PREN permits an initial assessment and ranks alloys by alloy content, the PREN is often not sufficient by itself as a guide to select the most economical material for the application. For example, the resistance to environmental cracking and the other property requirements of the equipment are not assessed. The PREN can be used as a tool in conjunction with (1) company experience, (2) field and customer experience, (3) literature searches, (4) industry wide experience and consultants, (5) material testing, and (6) prototype equipment tests.

For example, with production tubing and casing, there is abundant literature on performance envelopes of a wide variety of corrosion resistant alloys. This literature is based on laboratory testing and/or field experience. Some of the data that permits the determination of a performance envelope for 13% chromium stainless steels is presented in Figures 6.8.30 and 6.8.31.

Several good sources exist for corrosion data such as NACE and ASM International. An example of corrosion rate data in a wide variety of environments and materials is *Handbook of Corrosion Data* [36]. An example of performance envelopes developed for the petroleum industry is *Selection Guidelines for Corrosion Resistant Alloys in the oil and Gas Industry* [37].

Some of the performance envelopes developed for production environments in the petroleum industry are presented here as an initial guide in selecting candidates for applications. The generally acceptable regions for 13Cr type martensitic stainless steels and 316 stainless steel are presented in Figures 6.8.32 and 6.8.33 respectively. The acceptable regions defined in these figures are for those downhole environments in the absence of oxygen and H_2S. For 13Cr stainless steels, these constituents can lead to severe corrosion pitting and, in the case of H_2S, could lead to environmental cracking. For 316 stainless steel, both oxygen and H_2S can lead to environmental cracking.

The generally acceptable regions for 22Cr type duplex stainless steel and 28Cr austenitic stainless steel are presented in Figures 6.8.34 and 6.8.35, respectively.

The generally acceptable regions for 825 and 625 nickel-base alloys are presented in Figures 6.8.36 and 6.8.37 respectively.

These alloys have been presented in order of increasing corrosion resistance and cost. For example, the 13Cr stainless steels have the least corrosion resistance and lowest cost, whereas the 625 nickel-based alloys has the greatest corrosion resistance and cost. The relative cost relationship between low-alloy steel and some common CRAs are presented in Table 6.8.16.

In Table 6.8.16, the data are standardized such that the relative cost of 4130 to 4140 low-alloy steels was set to 1.0. The reader is cautioned in that these relative costs often fluctuate with commodity pricing and will vary accordingly. The relative costs in Table 6.8.16 were valid in May 2004.

Table 6.8.14 *Nominal Composition of Materials Used in Artificial Lift Pumps*

Alloy	Alloy Type	C	Cr	Ni	Mo	Cu	Fe	Mn	Others
Gray iron	Cast iron	3.3	—	—	—	—	Bal.	0.7	Si = 2.5
White iron	Cast iron	3.2	—	—	—	—	Bal.	0.6	Si = 0.5
Aust. Gray iron	Cast iron	2.5	2	16	—	6.5	Bal.	1.0	Si = 1.9
1018–1026	Carbon steel	0.20	—	—	—	—	Bal.	0.6	—
1040–1045	Carbon steel	0.44	—	—	—	—	Bal.	0.9	—
1527	Carbon steel	0.27	—	—	—	—	Bal.	1.4	—
4130	Low alloy	0.30	1.0	—	—	—	Bal.	0.5	—
4620	Low alloy	0.20	—	1.8	0.25	—	Bal.	0.55	—
8620	Low alloy	0.20	0.5	0.65	0.2	—	Bal.	0.9	—
8640	Low alloy	0.40	0.5	0.65	0.2	—	Bal.	0.9	—
9310	Low alloy	0.1	1.2	3.3	0.1	—	Bal.	0.55	—
Nitriding alloy	Low alloy	0.40	1.6	—	0.35	—	Bal.	0.6	Al = 1.1
5 Cr steel	Alloy steel	0.1	5	—	0.5	—	Bal.	0.5	—
13 Cr	Stainless	0.15	13	—	—	—	Bal.	0.5	—
440A	Stainless	0.7	17	—	—	—	Bal.	0.5	—
303	Stainless	0.08	19	9	—	—	Bal.	0.5	S added
304	Stainless	0.05	19	9	—	—	Bal.	0.5	—
316	Stainless	0.05	17	12	2.5	—	Bal.	0.5	—
R400 & R405	Nickel copper	—	—	Bal.	—	31	1	0.5	—
260/360 brass	Brass	—	—	—	—	70	—	—	Zn = 30
464 brass	Brass	—	—	—	—	60	Bal.	—	Zn = 40

Table 6.8.15 *API 5L Tensile Requirements for PSL2*

Grade	Yield Strength, MPa (ksi)		Tensile Strength, MPa (ksi)	
	Minimum	Maximum	Minimum	Maximum
B	241 (35.0)	448 (65.0)	414 (60.0)	758 (110.0)
X42	290 (42.0)	496 (72.0)	414 (60.0)	758 (110.0)
X46	317 (46.0)	524 (76.0)	434 (63.0)	758 (110.0)
X52	359 (52.0)	531 (77.0)	455 (66.0)	758 (110.0)
X56	386 (56.0)	544 (79.0)	490 (71.0)	758 (110.0)
X60	414 (60.0)	565 (82.0)	517 (75.0)	758 (110.0)
X65	448 (65.0)	600 (87.0)	531 (77.0)	758 (110.0)
X70	483 (70.0)	621 (90.0)	565 (82.0)	758 (110.0)
X80	552 (80.0)	690 (100.0)	621 (90.0)	827 (120.0)

Figure 6.8.28 *Corroded production tubing.*

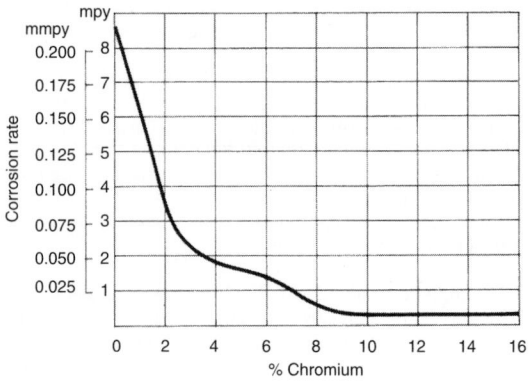

Figure 6.8.29 *Effect of chromium content on corrosion rate [after 34].*

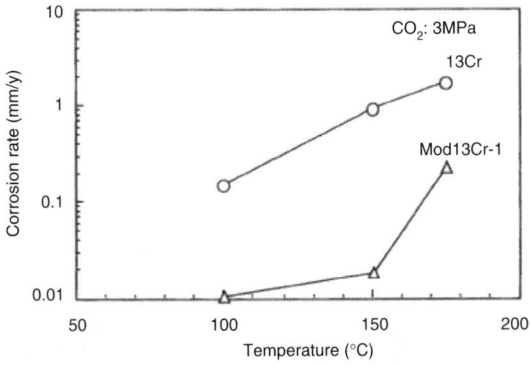

Figure 6.8.30 *Corrosion rates as a function of temperature for 13Cr and modified 13Cr stainless steel in a CO_2 containing production environment [after 35].*

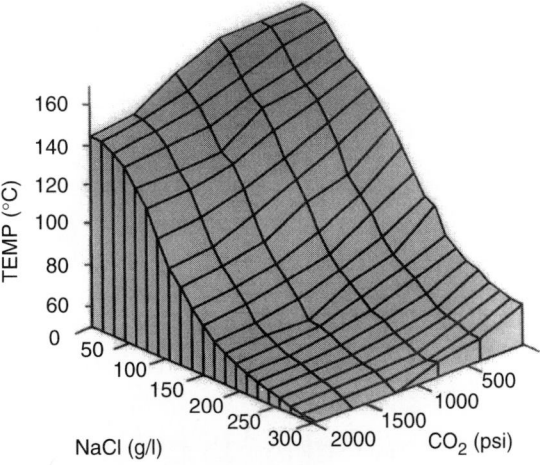

Figure 6.8.32 *Acceptable application range for 13Cr stainless steels [after 37].*

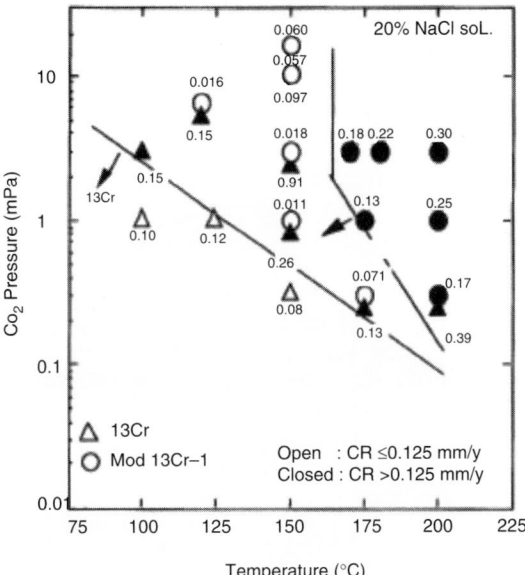

Figure 6.8.31 *Corrosion regime as a function of CO_2 and temperature on 13Cr and modified 13Cr stainless steel [after 35].*

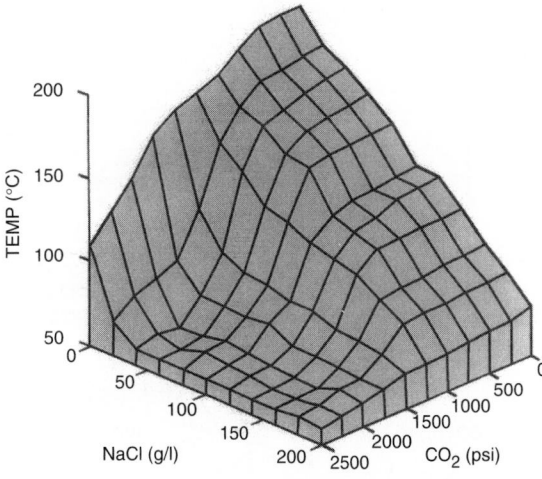

Figure 6.8.33 *Acceptable application range for 316 stainless steel [after 37].*

6.8.4.2 Coatings and Surfacings

There are several types of coatings that are used to prevent corrosion in production equipment. These can be broadly classed as metallic coatings, inorganic coatings and organic coatings. With all of these classes, the one feature that has a huge role in the final success rate is the surface preparation. The surface needs to be clean and free of contaminants and exhibit the proper surface profile for the specific coating being applied.

Metallic coatings include (1) hot-dip galavanizing, (2) electroplating of nickel, chromium, copper, tin and zinc, (3) electroless nickel plating, (4) high-velocity (oxy-fuel) spray coatings — also known as HVOF, (5) metallizing (arc or flame spraying), (6) plasma vapor deposition (PVD) and ion implantation, (7) chemical vapor deposition (CVD) coatings, (8) cladding, and (9) weld overlays. The major use of these metallic coatings is to prevent corrosion on carbon and low-alloy steels. There are applications of cooper plating and hard surfacing for reasons other than corrosion control.

The plating metals used most often in the petroleum industry are chromium and nickel. Both of these can be electroplated, and nickel is also available as an electroless plating. Chromium platings are generally very hard and exhibit numerous microcracks that precludes its use as a corrosion barrier. This problem is exasperated by the potential galvanic difference between the chromium and the substrate that can drive the initiation and propagation of localized corrosion attack, ultimately resulting in deep penetrating corrosion pits and disbondment of the chormium plating. There are very thin dense chromium platings with a coating thickness less than 0.25 mm (0.001 in.) that can be dense and crack-free. Chromium platings are often plated over thin electroplated nickel, where the nickel provides the corrosion barrier and the hard chromium provides the desired wear resistance.

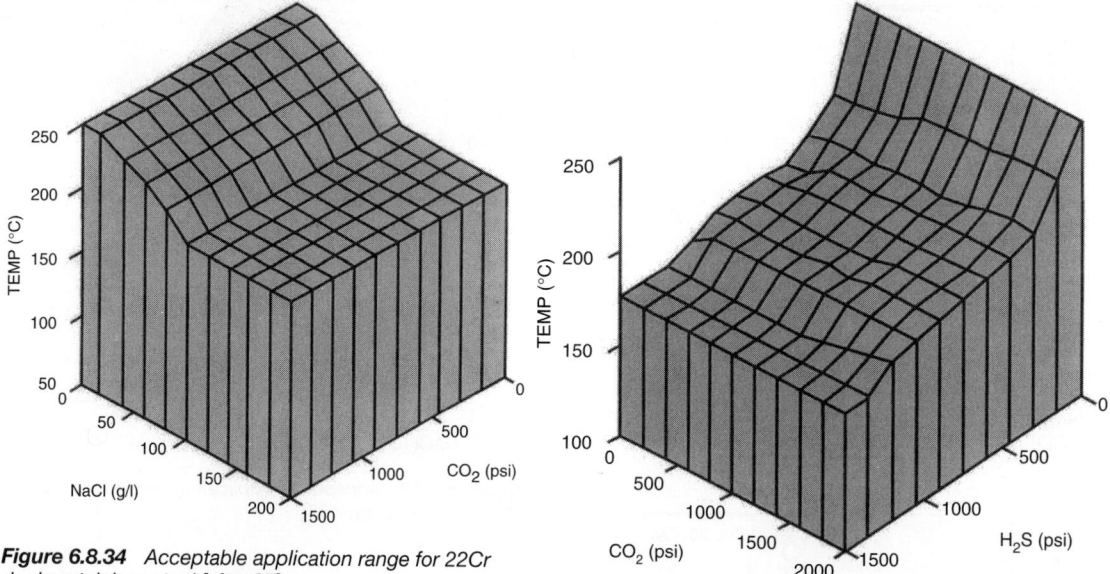

Figure 6.8.34 *Acceptable application range for 22Cr duplex stainless steel [after 37].*

Figure 6.8.36 *Acceptable application range for 825 nickel-based alloy [after 37].*

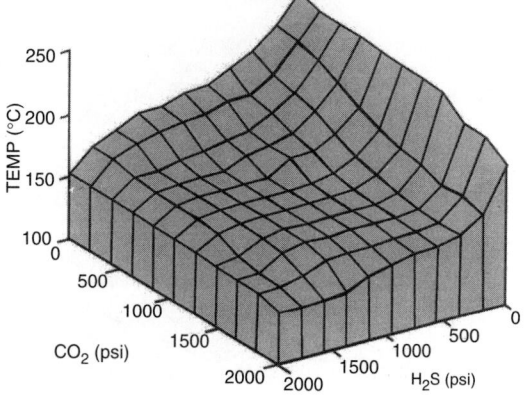

Figure 6.8.35 *Acceptable application range for 28Cr austenitic stainless [after 37].*

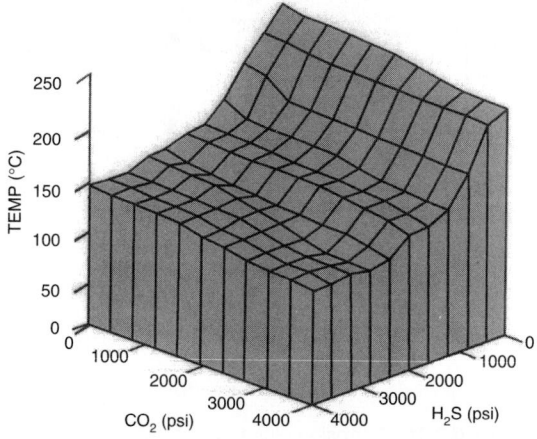

Figure 6.8.37 *Acceptable range of 625 nickel-based alloy [after 37].*

Another type of plating that is not driven by applied current is electroless nickel, in which an activated metal substrate plates autocatalytically. An advantage of the electroless nickel process is that, because current density is not an issue, there are no shape and distance effects with respect to anode (source of current to complete circuit) to cathode (substrate being plated) placement. This results in a uniform plating thickness. One other aspect of electroless nickel is that phosphorus is intentionally codeposited, and this forms a hardened plating that starts to approach hardness levels found in chromium platings.

Inorganic coatings includes include (1) surface conversions, (2) inorganic zinc silicate, (3) HVOF ceramic coatings, and (4) flame and arc spray ceramic coatings. The common surface conversions are anodizing on aluminum, phosphating and oxidizing processes on steels and chromate surface conversions on steel.

Organic coatings include (1) epoxy paints, (2) acrylic paints, (3) polyurethane paints, (4) organic zinc paints, (5) phenolic coatings, (6) epoxy-phenolic coatings, (7) fluorocarbon based coatings, (8) molybdenum disulfide coatings, and (9) silicone paints and coatings. Generally, where the word "paint" appears in the above list, the coating consists of at least two components, one a carrier and one the paint. In many cases, with two or more part products that are mixed, the paint undergoes a chemical reaction, causing or assisting in the curing and/or drying cycles. The coating listed above that do not list paint generally require a heat curing cycle that sets the coating onto the substrate.

Organic internal pipe and tubing coatings definitely provide a benefit by reducing power costs, increasing flow and eliminating corrosion [39]. Rehabilitating a pipeline with inplace internal coating versus replacement returns the line to active service. During research of flow improvement due to pipe cleaning, additional studies were made to determine the improvements in pipeline flow and corrosion protection

Table 6.8.16 *Relative Costs for Material*

Alloy	Alloy Type	Relative Cost
4130–4140	Low-alloy steel	1.0
8630	Low-alloy steel	1.25
4340	Low-alloy steel	1.35
9Cr-1Mo	Highly alloyed steel	2.0
13Cr	Stainless steel	2.2
316	Austenitic stainless steel	4.5
22Cr	Duplex stainless steel	5.5
25Cr	Super duplex stainless steel	6.5
28Cr	Super austenitic stainless steel	6.5
825	Nickel-based alloy	13
K500	Nickel copper alloy	15
718	Nickel-based alloy	17
625	Nickel-based alloy	20

with use of coatings [40]. The conclusion was that both flow improvement and corrosion protection were evident.

The coatings that best meet the needs of the pipeline industry are the polyamide epoxies and amine-adduct-type coatings [41]. A comparison of the two types is presented in Table 6.8.17.

McConkey has listed the following properties required for a pipeline coating [42]:

- ease of application
- good adhesion to pipe
- good resistance to impact
- flexibility
- resistance to soil stress
- resistance to flow
- water resistance
- electrical resistance
- chemical and physical stability
- resistance to soil bacteria
- resistance to cathodic disbandment

These properties relate either to the economic viability of the coating or to the ability of the coating to protect the pipeline from damage and possible failure due to corrosion.

6.8.4.3 Inhibition
An *inhibitor* is a substance that slows or stops a chemical reaction. When an inhibitor is present in sufficient enough

quantity to slow a corrosion reaction, the corrosion is prevented. The basic mechanisms by which inhibitors function are as follows [38]:

1. They absorb onto the corroding material as a thin film.
2. They induce formation of a thick corrosion product, which forms a passive layer.
3. They change the characteristics of the environment either by producing protective particulates or by removing or inactivating an aggressive constituent in the environments.

The corrosion inhibitors that are used in the petroleum industry are typically organic compounds, and they are usually employed at low concentration levels, generally on the order of 0.1%.

The decision to use inhibitors is determined by the following:

1. the predicted corrosion rate without inhibition
2. the predicted effectiveness of an inhibition program
3. the predicted cost of the inhibition program
4. the required equipment life and predicted life of the field
5. the cost of other mitigation techniques such as coatings and corrosion resistant alloys

The effectiveness or performance of an inhibitor is related to the temperature, flow rate (wall shear stress), metal surface condition, alloy type, oxygen, and the other specific environmental constituents. Generally inhibition is very effective in anaerobic systems with low pressures, low temperatures, and low flow rates. Conditions that includes high temperatures, high CO_2 partial pressures and/or high velocities decrease the effectiveness of the inhibition treatments. Successful inhibition under these conditions generally require greater concentrations of inhibitors, and the selection of specific inhibitor becomes increasingly important because the inhibitor needs to be effective while wetting the surface(s) to be protected.

6.8.4.3.1 Classification of Inhibitors
Corrosion inhibitors are generally classed by the type of inhibitor or by their solubility.

As a type of material, the inhibitor may be an organic or an inorganic compound. Inorganic inhibitors include chromates (CrO_4^{-2}), silicates (SiO_3^{-2}), and phosphates (PO_4^{-2}). The three basic types of inorganic inhibitors are anodic passivating inhibitors, cathodic inhibitors, and cathodic precipitators. There are also mixed inhibitors that suppress both anodic and cathodic reactions.

Table 6.8.17 *Coating Comparison [41]*

	Coating	
Property	Amine	Polyamide
Hardness	5H to 6H pencil hardness, faster initial cure	4H to 5H pencil hardness, slower initial cure
Tolerance to inadequately prepared surface	Good	Very good
Brittleness	Very brittle in a few months	Same resiliency remains after a few months
Sag resistance (5–7 m wet)	Good	Good
Adhesion	Good	Good
Application characteristics	Very good	Excellent
Flexibility	—	Best
Abrasion resistance	Equal	Equal
Water resistance	—	Best

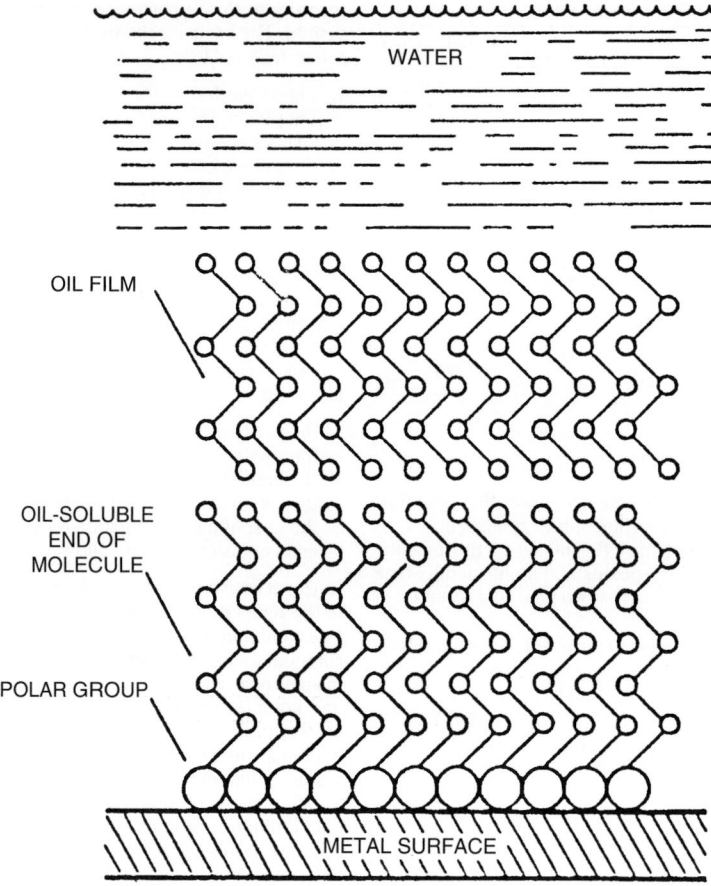

Figure 6.8.38 *Schematic of film-forming inhibitor [44].*

Anodic inhibitors suppress or prevent anodic reactions by assisting the natural passivation of metal surfaces or by forming films/deposits that isolate the metal from the environment. These anodic inhibitors are sometimes categorized by being either highly oxidizing or much less oxidizing. Examples of the highly oxidizing anions include chromates, nitrites, and nitrates.

The low-oxidizing anodic inhibitors include phosphates, molybdates, and tungstates. The low-oxidizing anions require an external source of oxygen for passivation. These low-oxidizing anions are sometimes referred to as dangerous inhibitors because they can actually cause pitting when present in low concentrations. Anodic inhibitors are also considered dangerous because cathodic reaction may not be inhibited on unprotected areas [43].

The cathodic inhibitors suppress the cathodic chemical reaction, and these are further subgrouped by the way in which the cathodic reaction is affected. The three basic types are cathodic poisons, cathodic precipitators, and oxygen scavengers.

The organic type of inhibitors affect the entire surface of the metal by forming films on the surface. The organic inhibitors are semipolar in that the make-up of the inhibitor includes a polar (conductive) and a nonpolar (nonconductive) component. The semipolar nature renders one end of the molecule hydrophobic (water repelling) and the other hydrophilic (water loving).

These semipolar molecules bond or attach to the metal surface on the polar (conductive) end of the chemical compound. A thin film only a few molecules thick is adsorbed according to the ionic charge of the inhibitor molecule and the charge of the metal surface. These organic inhibitors can preferentially attach to either the cathode or the anode of the system; these inhibitors are termed cationic inhibitors (positively charged) and anionic inhibitors (negatively charged). Amines are examples of organic cationic inhibitors and sulfonates are examples of organic anionic inhibitors. The bonding of these inhibitors is illustrated in Figure 6.8.38.

One advantage of the organic inhibitors is that they are typically self-healing when the inhibitor is present in sufficient concentration. If the film is damaged or removed through flowing conditions, the film reforms. Examples of organic inhibitors include primary amines, carboxylated amines, amides, imidazoline, and quaternary ammonium ion.

The other way to classify inhibitors is by their solubility. The classes based on primary solubility are presented below:

1. oil soluble
2. water soluble
3. oil or water dispersible (emulsions)
4. oil and water insoluble
5. gas phase

However, most inhibitors at least partly soluble or partly dispersible in both oil and water.

Water soluble inhibitors are used in systems with a significant water phase. Examples of water soluble inhibitors include quaternary amines, amine salts and salted imidazolines. Oil soluble inhibitors include long-chain primary amines, imidazolines, fatty acids, and phosphate esters. These inhibitors can be used in both batch and continuous inhibitor injection systems.

Batch treatments usually rely on oil-soluble or oil- and water-insoluble inhibitors. The reason is that there needs to some insolubility in the water phase if there is a persistent film when water wet. Gas-phase inhibitors are usually low-molecular-weight amines that have high vapor pressures such as some carboxylated amines.

The inhibition systems are classified by the method by which the inhibitors are introduced into the production fluids (continuous, batch, and squeeze).

6.8.4.3.2 Choosing Inhibition for Wellhead and Downhole Equipment

During production operations, failures of tubular goods and equipment are mainly caused by the presence of carbon dioxide, hydrogen sulfide, oxygen, and water. The following general procedures for a new design are recommended.

1. Evaluate the type of corrosion that might occur. In production operations, hydrogen sulfide, carbon dioxide, oxygen, and water are the main causes of corrosion.
2. When H_2S is present, select steels and alloys that are resistant to cracking in the environmental conditions present including acidity, phases present, temperature and chloride content in addition to the level of H_2S present. While designing the tubular goods and equipment most compatible to the environment encountered, consider the metallurgical factors and alloying elements for "sweet" and "sour" oil and gas environments.
3. Select the best and most economically feasible inhibitor by considering similar situations and past experience with conditions close to those encountered. In the case of a unique condition, applicable laboratory tests should be performed to confirm the effectiveness of the inhibitor.
4. Determine a method of inhibitor application by considering its effectiveness, economy, and frequency of application.
5. Monitor and inspect the equipment and keep track of the performance to evaluate and update the program treatment.

The control the corrosion in the tubing, casing, and surface equipment for a system already in operation where the material cannot be chosen since it is already in use. In this situation, corrosion can be controlled by a judicious selection of inhibitors and practical method of applying the inhibitor. A general procedure in an existing system is as follows:

1. Select the most effective inhibitor for the situation. If possible, investigate its performance for similar situations in field history and in the laboratory tests. For H_2S, filming amines and organic phosphates are reported to be effective corrosion inhibitors. Filming amines are also reported to be effective for carbonate dioxide.
2. Determine the most effective concentration and amount of the inhibitor.
3. Use a practical method to apply the inhibitor to the system. The continuous method, especially with a capillary system, is reported to be the most effective and economical.
4. Determine the frequency of application.

5. Monitor and evaluate the performance of the system, and adjust the treatment program according to the observation.

In water flooding not only is corrosion more severe than in primary production, but also there are scale problems. Because water is open to atmosphere, it dissolves more oxygen. Necessary water treatments include:

1. *Filtration*—to remove the suspended solids from water.
2. *Oxygen scavengers*—to reduce the oxygen concentration.
3. *Scale inhibitors*—either to prevent crystallization or prevent the adherence of crystals to the surface. It should be mentioned that improper use of scale causes pseudoscale; therefore, extensive field and laboratory tests should be performed to prevent this problem.
4. *Chemical treatment for bacteria*—to reduce or eliminate the bacteria population.
5. *Corrosion inhibition*—basically this is the same as primary production. It is more desirable to use inhibitors that reduce scale as well as effective corrosion controllers.

All three techniques, batch, continuous, and squeeze, have been used for inhibition systems for the protection of downhole equipment. Of the three, the continuous method is the most effective and preferable from a performance perspective. The continuous injection of inhibitor into the fluid stream has been recognized as the optimum treating method since the inception of corrosion control by chemical injection [45–47].

Continuous downhole injection is reported to increase the functional life of the tubing 10- to 20-fold, where no other treating methods could increase the functional life of the tubing more than three to fivefold [48]. The continuous treating method generally involves introducing the inhibitor on a continuous basis at sufficient concentrations to form a protective film on the tubing. It can be used to protect topside equipment such as wellheads, separation facilities and flowlines as well as downhole equipment.

The batch technique is only used to handle special problems such as the protection of well tubing when downhole injection is difficult. The squeeze technique is used for scale inhibitors but is not typical for corrosion inhibitors because of the potential damage to the formations.

The use of chemical injection tubing for the continuous or batch injection of inhibitors into the well bore is presented in Figure 6.8.39. Chemical injection through the annulus of a larger tube (concentric completions) is presented in Figure 6.8.40. The use of corrosion inhibition in a Y-block completion is presented in Figure 6.8.41. The use of gas lift valves to inject inhibitors with the lift gas is presented in Figure 6.8.42, and the use of a separate chemical injection line with a side pocket mandrel (gas lift) is presented in Figure 6.8.43.

Water floods are one of the common oilfield practices that results in additional corrosion considerations. This technique is used to increase the amount of oil recovered. The water injected is natural surface water sources or reinjected produced water. In water flooding not only is corrosion more severe than in primary production, but also there are scale problems. In many instances, the produced water has insufficient volume for injection purposes and additional water is required (naturally occurring surface water). Because water is open to atmosphere, it contains dissolved oxygen and this contributes greatly to the corrosion potential. Necessary water treatments include:

1. Filtration—to remove the suspended solids from water.
2. Oxygen scavengers—to reduce the oxygen concentration.

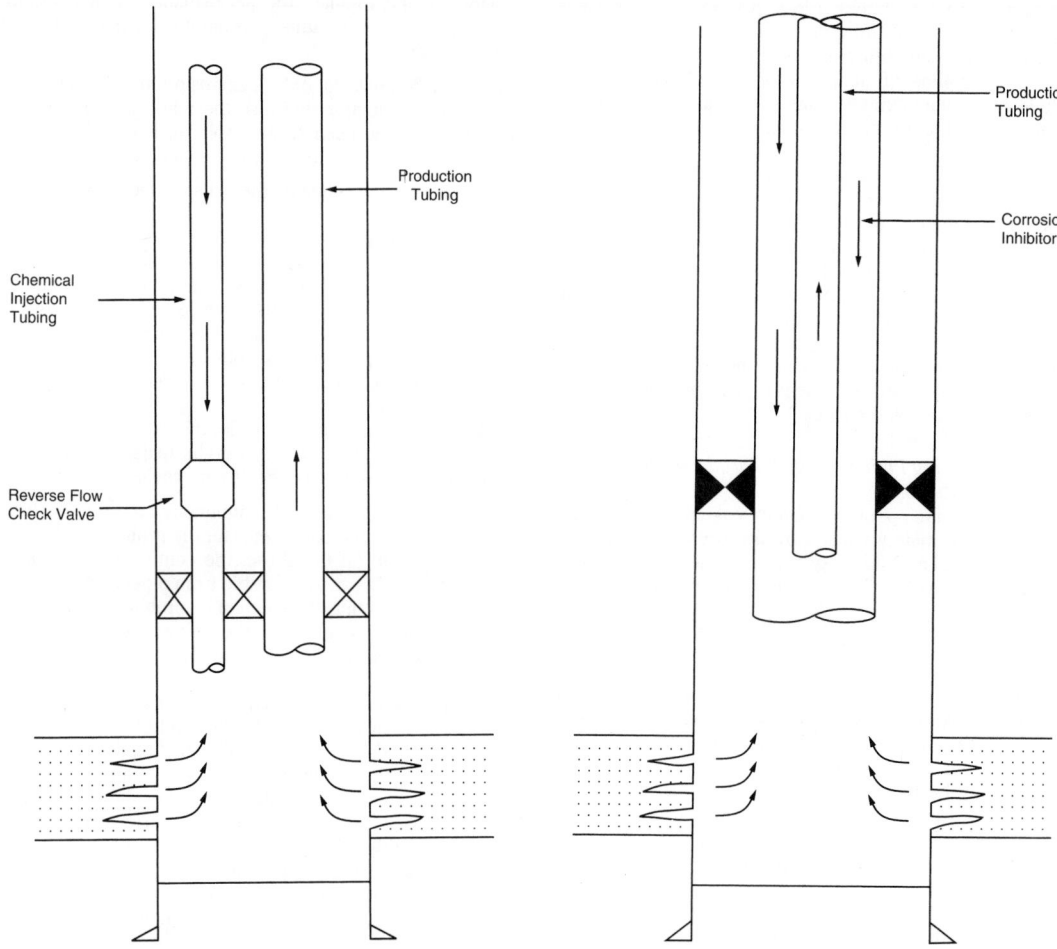

Figure 6.8.39   *Dual completion [48].*   **Figure 6.8.40**   *Concentric completion [48].*

3. Scale inhibitors—either to prevent crystallization or prevent the adherence of crystals to the surface. The improper use of scale inhibitors causes pseudoscale; therefore, field and laboratory tests should be performed to prevent this problem.
4. Chemical treatment for bacteria—to reduce or eliminate the bacteria population.
5. Corrosion inhibition—basically this is the same as primary production. It is more desirable to use inhibitors that reduce scale and are effective corrosion controllers.

An evaluation of the various corrosion inhibition methods is presented in Table 6.8.18 [45,46,48].

For water flood applications, the volume of water dictates that metal surfaces will be water wet. This results in the selection of inhibitors that will be soluble in water with good metal filming characteristics. The practical corrosion inhibitors in water flood applications are variations of the following compounds:

- primary monoamines
- polysubstituted monoamines
- diamines
- polyamines
- imidazolines
- quaternary ammonium compounds

6.8.4.3.3 Choosing Inhibition for Flowlines and Pipelines

Internal corrosion in product pipelines, in most cases, is caused by water and oxygen dissolved in the product. Dry refined products with normal additives are noncorrosive to steel pipelines. The products are corrosive because of associated water and air. A film of liquid water adheres to the pipeline surface, and oxygen is available from air dissolved in the product. The solubility of air in products varies, but refined products generally carry sufficient oxygen to support corrosion. Air is introduced into the products by tank mixers, turbulence, normal tank breathing, etc.

The outside of pipelines and flowlines are exposed to ground waters, saltwater, and industrial environments. The lines may be above ground, buried, or submerged beneath fresh or saltwater. Internal corrosion can be controlled by removing one of the active ingredients, water or air, by adding an inhibitor which will make the steel inactive; or using a barrier coating on the steel. A general procedure to

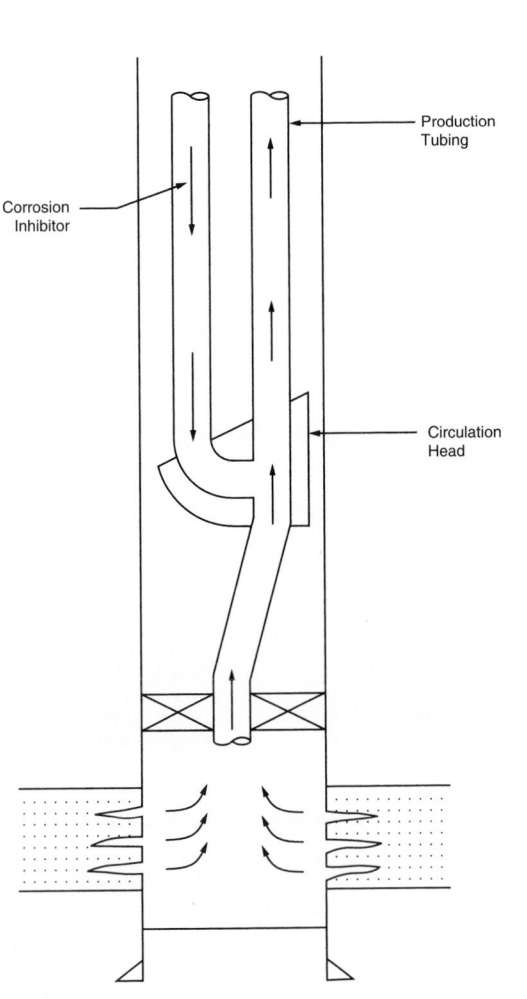

Figure 6.8.41 *Y-block completion [48].*

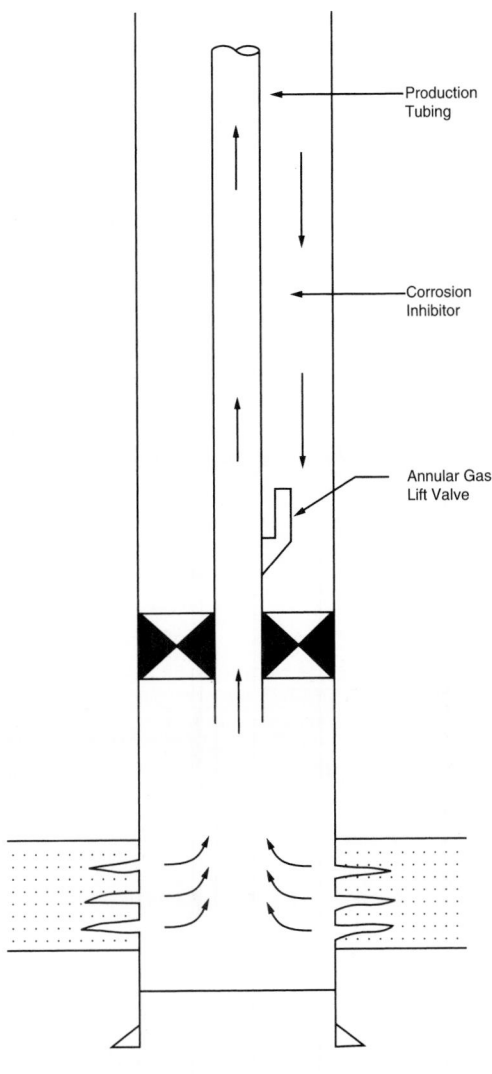

Figure 6.8.42 *Gas-lift well completion [48].*

mitigate corrosion includes the following:

1. Select coatings and wraps for external pipeline protection.
2. Investigate cathodic protection for buried and submerged flowlines and pipelines
3. Reduce sources of oxygen in the fluids transported.
4. Reduce water content and evaluate the use of dehydration.
5. Chemical treatment for bacteria—to reduce or eliminate the bacteria population.
6. Evaluate corrosion inhibition and candidate inhibitors.
7. Determine the most effective concentration and amount of the inhibitor.
8. Use a practical method to apply the inhibitor to the system. Because the continuous method, especially with a capillary system, is reported to be the most effective and economical, this method should be considered strongly if applicable.
9. Determine the frequency of application.

10. Monitor and evaluate the performance of the system and adjust the treatment program according to the observation. Basically this is the same as primary production. It is more desirable to use inhibitors that reduce scale and are effective corrosion controllers.

Even though the product is clear when it is placed in the pipeline, indicating absence of free water, temperature drops may occur during transit and cause water to separate. Table 6.8.19 shows increasing solubility of water in gasoline as temperature increases. Table 6.8.20 indicates the range of solubility of water in some pure hydrocarbons. The important factor is the change in solubility per increment of temperature change. From Table 6.8.20 a 8°C (17°F) increase in temperature doubles the solubility of water in n-pentane.

Dehydration can be used to control corrosion if the water is kept at low level. This method requires considerable equipment and manpower for maintenance, so the cost is high for control. Free water can be removed in the storage tank

Table 6.8.18 *Comparison of Corrosion Treatment Methods*

Gas	Production Gas cond.	Oil	Flow V	Efficiency %	Frequency (days)	Formation Damage	Shut-In Time
				Batch Treatment			
Yes	Yes	No	< Ve	60 to 80	< 30	Very unlikely	Yes
				Tubing Displacement			
Yes	Yes	Yes	< Ve	80 to 90	< 30	Unlikely	Yes
				Formation squeeze			
Yes	Yes	Yes	Ve	80 to 90	> 30	Possibly	Yes
				Continuous Treatment			
Yes	Yes	Yes	> Ve	95 to 100	—	None	No

(Ve = erosional velocity ft/sec)

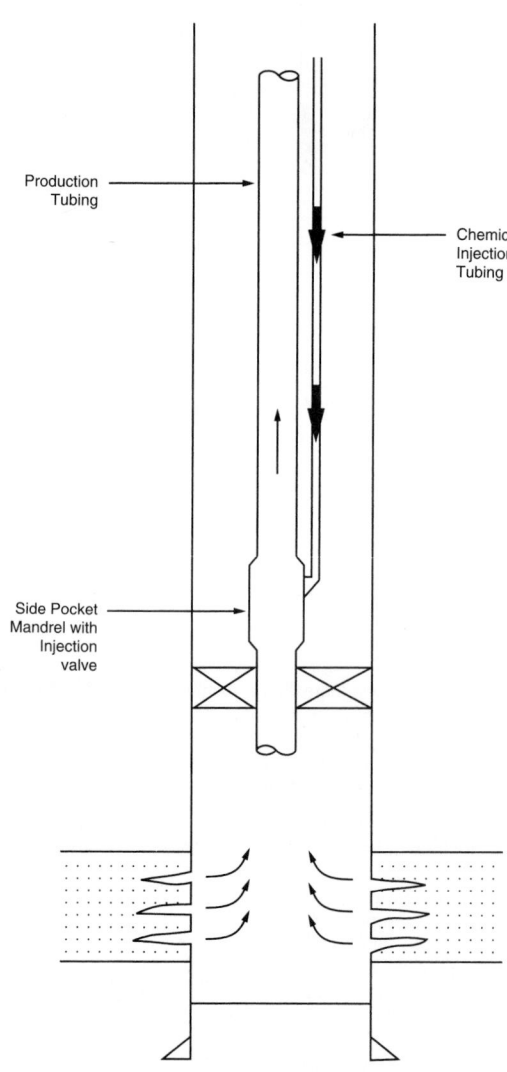

Figure 6.8.43 *Continuous downhole injection system (CDIS) [46].*

Production Tubing

Chemical Injection Tubing

Side Pocket Mandrel with Injection valve

Table 6.8.19 *Solubility of Water in Gasoline*

Temperature	Solubility
4°C (40°F)	1.8 gal/1,000 bbl
10°C (50°F)	1.8 gal/1,000 bbl
16°C (60°F)	1.8 gal/1,000 bbl
21°C (70°F)	1.8 gal/1,000 bbl
27°C (80°F)	1.8 gal/1,000 bbl
32°C (90°F)	1.8 gal/1,000 bbl
38°C (100°F)	1.8 gal/1,000 bbl
43°C (110°F)	1.8 gal/1,000 bbl

Table 6.8.20 *Solubility of Water in Specific Hydrocarbons*

Specific Hydrocarbon	Temperature °C(°F)	Solubility mg/100 g	Solubility gal/1000 bbl
n-butane	20 (68)	6.5	1.5
isobutene	19 (66.2)	6.5	1.7
n-pentane	15 (59)	6.1	1.6
	24.8 (76.6)	12.0	3.2
isopentane	20 (68)	9.4	2.4
n-hexane	20 (68)	11.1	3.1
cyclohexane	20 (68)	10.0	3.3
n-heptane	20 (68)	12.6	3.6
n-octane	20 (68)	14.2	4.2
benzene	20 (68)	43.5	16.1
heptene-1	20.5 (68.9)	104.7	30.8
butene-1	20 (68)	39.7	11.1

by allowing the water to settle out of the product. Free and entrained water can be removed by using water separators and coalescers.

Rust particles, dirt, etc., in the product will settle out in the low section of the pipeline if the flow velocity is not sufficient to keep the particles entrained. These particles, being of dissimilar electrochemical properties to the underlying metal, will form local corrosion cells in the presence of water. These particles will be filmed with a chemical inhibitor the same as the pipe, and this causes depletion of available inhibitor to protect the line. Most filters are designed to prevent free water from entering the system. Some of the common types of filters used include (1) hay tanks, (2) cartridge type, and

(3) centrifugal type. Selection of the filter will depend on the application and design of each pipeline.

For pipelines and flowlines, the most common method employed today in controlling internal corrosion is the use of organic or oil-soluble inhibitors. These inhibitors are generally hydrocarbons (with polar group attached) that tend to form a protective film on the pipeline's internal surface [49]. Water-soluble corrosion inhibitors are also used to solve internal pipeline corrosion problems but because of the relatively short film life of water-soluble corrosion inhibitors, they are continuously injected as opposed to the batch-treatment method.

Because water on the bottom of the pipe is a source of the corrosion problems, one might assume a water-soluble inhibitor to be the answer to treating a pipeline carrying crude oil and water. However, another problem must be considered. The portion of pipe occupied by the oil and water emulsion is also a corrosive environment. Under such conditions, a water-soluble inhibitor will probably not provide adequate protection. However, there is laboratory evidence that a portion is absorbed into the oil, and this can provide some protection in the oil phase (Figures 6.8.44 and 6.8.45) [50].

Oil-soluble corrosion inhibitors are widespread in systems handling both oil and water. This type of inhibitor may do a good job protecting small-diameter pipelines because high velocity or turbulent flow prompts mixing of the inhibitor and produced fluids. However, because of reduction in velocity or turbulence, it may not provide the desired protection in large flowlines or pipelines carrying these same fluids.

The oil-soluble inhibitors have the advantage that they may be injected at the refinery during the normal blending operations and will protect the refinery piping, the pipeline, the gasoline station tanks, etc. This advantage is fully realized when a company operates its own pipeline and handles its own products. Similarly, common-product pipelines can inject inhibitor at source points only, or can require shippers to supply inhibited products [51].

Inhibition can be accomplished by one of two general methods; these are batch (intermittent) treatment or by continuous injection. Batch treatment normally entails pumping a suitable-size slug or pure or high-concentration inhibitor solution through the line. Frequency of the treatment is governed by the effectiveness of the inhibitor remaining after a given time or after a specified amount of product has been moved through the line

Continuous injection consists of adding a constant volume of inhibitor to the product being transported through the pipeline. This method is probably the most desirable and widely used.

Injection facilities vary greatly in design and operation, and in general the installation consists of the following equipment units:

1. inhibitor storage vessel
2. injection pump

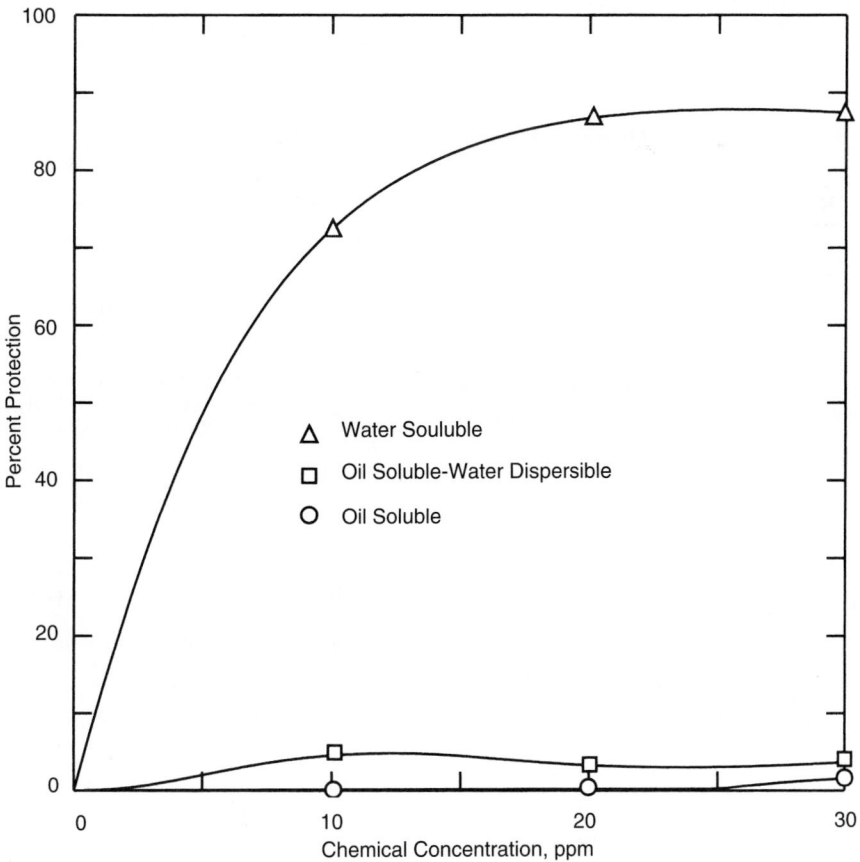

Figure 6.8.44 Sweet corrosion wheel test [50].

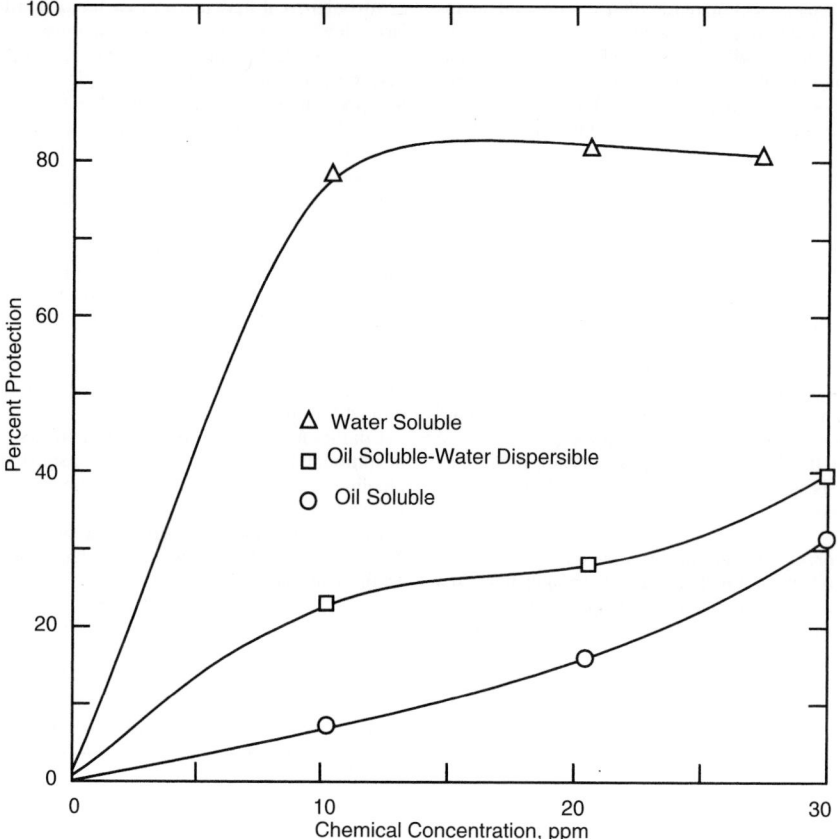

Figure 6.8.45 *Sour corrosion wheel test [50].*

3. measurement device (meter or calibrated sight glass)
4. connection to the pipeline

Positive-displacement chemical pumps, with adjustable capacity, are probably the most widely used on product pipelines [52].

Construction materials for the equipment should be suitable for continuous service when in contact with the inhibitor. Stainless steel should be considered for small-diameter piping or tubing in which even minor rusting could cause plugging and make pumping of more viscous liquids difficult. When handling nitrogen-based inhibitors (amines, amides, nitrites, etc.), the use of copper or copper-based alloys should be avoided, as stress corrosion cracking might result. Nonmetallic seals and packing material should be checked for compatibility with the inhibitor formulation [53].

Points of injection should be chosen to provide maximum benefit in the pipeline system. Injection on the suction side of pumps takes advantage of pump turbulence to promote mixing of the inhibitor with product. Injection through a tube into the center of the pipeline aids mixing.

Premixing or dilution of the inhibitor can improve handling and promote a more rapid phase-contract. Viscous inhibitors can be diluted with compatible hydrocarbon carriers to decrease viscosity, making pumping easier and metering more accurate, especially at low-dosage rates. Premixing water-phase inhibitors with water before injection greatly facilitates phase contact of inhibitor with entrained water.

6.8.4.3.4 Choosing Inhibition for Gas Transmission Lines
Gas transmission lines are very similar to fluid flowlines and pipelines. Here, oxygen, CO_2, and H_2S are the main corrosives in natural gas pipelines, and they are aggressive only when they are absorbed into water or condensed moisture in the lines. Most pipeline gas has been sweetened and dehydrated as well as treated otherwise before it is pumped into transmission lines. Consequently, most of the corrosion problems in gas systems occur in the various gathering lines and in the piping and equipment used for removing liquid hydrocarbons and sulfur.

Although inhibitors may be slugged, injected, or sprayed into a gas stream, they function only to the extent that they are absorbed onto or react with the steel to set up a barrier between it and the aggressive agents [54].

Natural gas pipelines can be classified as either wet or dehydrated. Most of the internal corrosion problems discussed thus far are characteristic or wet-gas pipelines.

Dehydrated systems are usually characterized by a water vapor content in the gas less than saturation and the sporadic movement of fluids through the system. Without a contacting water phase onto the pipe walls, there is a very low potential for corrosion reactions.

The major problem in treating these lines is the distribution of inhibitor to the corrodible areas. How do we get an effective, nonvolatile inhibitor to the top and sides of a pipeline? How do we get a nonvolatile inhibitor to carry through the entire length of a pipeline? Under proper conditions, these are all possible using the simplest of application methods. According to EnDean all we need to know are [55]:

1. diameter of the line
2. length of the line
3. volume of gas in MMcfd
4. pressure
5. pressure drop
6. temperature
7. volume of liquid water through the pipe
8. volume of liquid hydrocarbons

From these data, we can predict if there is annular flow in the pipeline. If there is annular flow, then all that is necessary is to add the inhibitor as a solution. Transport and distribution of the inhibitor along the pipeline is assumed. The eight variables listed define the flow conditions in a pipeline and thus determine the method and type of inhibitor treatment that is required [56].

To determine whether the concentration, based on gas volume, is sufficient, there are tests for chemical concentrations in water; these are called *residuals*. This information, combined with comprehensive monitoring, allows for maximum protection at minimum cost. It should be realized that residuals can fluctuate widely in systems where water volumes are subject to great fluctuations. These residuals are still very useful when monitored over a period of time and a statistical average determined.

6.8.4.3.5 Scale Removers

Scale can occur in downhole production tubing, flowlines, and pipelines. Scaling is the precipitation of adherent deposits on metal surfaces. Normally, precipitation of scale-forming salts occurs when solubilities are exceeded because of high concentrations or unfavorable temperatures. The problem of scale in water flooding occurs all the way from the water injection facilities to the producing well. Generally, there are six important regions where scaling can occur during and after injection operations [57–59]. These are:

1. in the injector wellbore
2. near the injection-well bottomhole
3. in the reservoir between the injector and the producer
4. at the skin of the producer well
5. in the producer wellbore
6. at the surface facilities

The formation of these scales is harmful in many ways in the petroleum industry. Scale build-up in tubulars decreases production capability and wireline and production logging operations can also be affected. Scales can also reduce the efficiency of heat exchanger surfaces in process and refining equipment downstream.

Wide variations in temperature, total pressure and changes in pressure during production make practical control of scale deposition difficult in water-flooding operations. $CaSO_4$ (calcium sulfate), $CaCO_3$ (calcium carbonate), $BaSO_4$ (barium sulfate), $SrSO_4$ (strontium sulfate), $FeCO_3$ (iron carbonate), and iron hydroxides are the most common scales in oilfield environments. There are a number of reasons for precipitating scale. Temperature, pressure, and water incompatibility are known to be the major causes of scale deposition in oilfield environments. Scales are usually formed in a supersaturation condition; that is, when the solubility product of a deposit-forming material is exceeded, it precipitates. A supersaturated solution is a solution that contains a higher concentration of a particular mineral than the solution can hold under the same set of conditions with its solute in equilibrium. Each scale is deposited under certain conditions that depend on temperature, pressure, and mineral content.

There are some scale deposits in oilfield environments that are called pseudoscale, that is, the deposit of a reaction product between two or more human-introduced chemicals, or between a naturally occurring fluid and one or more human-introduced chemicals. For example, most water-soluble corrosion inhibitors have a tendency to establish pseudoscale. Also, scale inhibitors such as phosphonates and polymers react with Ca^{2+} and/or Mg^{2+} ions in oilfield brines, thus forming pseudoscales that look and behave exactly like "real" scale [59,60].

There are two states of supersaturation: metastable and labile. Figure 6.8.46 represents a normal solubility curve and two different states of supersaturation.

A scale inhibitor tends to shift the dotted line, which is a function of fluid velocity, hydrocarbon content, agitation, temperature, pressure, and pH, toward the solid line that represents the normal solubility curve. Normally, crystallization of scale will not occur in the stable region (unstaurated). In the metastable region, spontaneous crystallization does not occur nor do scales form rapidly. In the upper portion, above the dotted line that represents the unstable, or libile region, spontaneous crystallization is probable [57,59].

$BaSO_4$ crystallization and adherence will increase with decreasing temperature. Other factors that have direct influence on the solubility of $BaSO_4$ are increasing pressure and increasing salt content of the brine. Figure 6.8.47 shows solubility of barium sulfate as a function of temperature.

Ethylenediaminetetraacetic acid (EDTA) and nitrilotriacetic acid (NTA) are known to be the common solvents for $BaSO_4$. These solvents and some other similar chelating compounds break down in $BaSO_4$ crystals by tying up the Ba^{2+} ions. The dissolved Ba^{2+} ions are masked or chelated. Therefore, they become incorporated into new chemical compounds that are soluble [59].

The mechanism of scale inhibitors in preventing scaling tendency is very difficult to predict under a given set of conditions. Some scale inhibitors are believed to prevent the crystallization of a scale, and others are believed to prevent the adherence of crystals to themselves or to metal surfaces in the field. Some of the common effective scale inhibitors include the derivatives of one of the following chemicals:

1. polymers (polymaleic acids, polyacrylates)
2. esters of phosphoric acid
3. phosphonates such as triethylenediaminetetra (methylene phosphonic acid)

The necessary concentration of scale inhibitor is a function of three variables: supersaturation, temperature, and the chemical composition of the scale. The higher the supersaturation and the higher the temperature, the higher will be the inhibitor concentration required to prevent the precipitation of a given amount of scale per unit volume of solution. Extensive laboratory and field experiments indicate the inhibitors listed in Table 6.8.20 are effective against the $BaSO_4$ scale [60,62].

There are a few disadvantages of some of the scale inhibitors that lead to formation of emulsions and pseudoscale. Basically, the inhibitor reactions with dissolved ions in oilfield brines are believed to be the main causes of pseudoscale. The pseudoscales are mostly caused by interactions between the dissolved ions such as Ba^{2+}, Ca^{2+}, and Mg^{2+} ions in the brine and the applied inhibitors. The pseudoscales act like a real scale; therefore, additional problems caused by deposition of pseudoscales can be encountered with the application of improper inhibitors. Oilfield operators have tried to solve emulsion problems by raising the temperature of the additions, but as temperature increases, more scaling is encountered. Therefore, the application of an effective

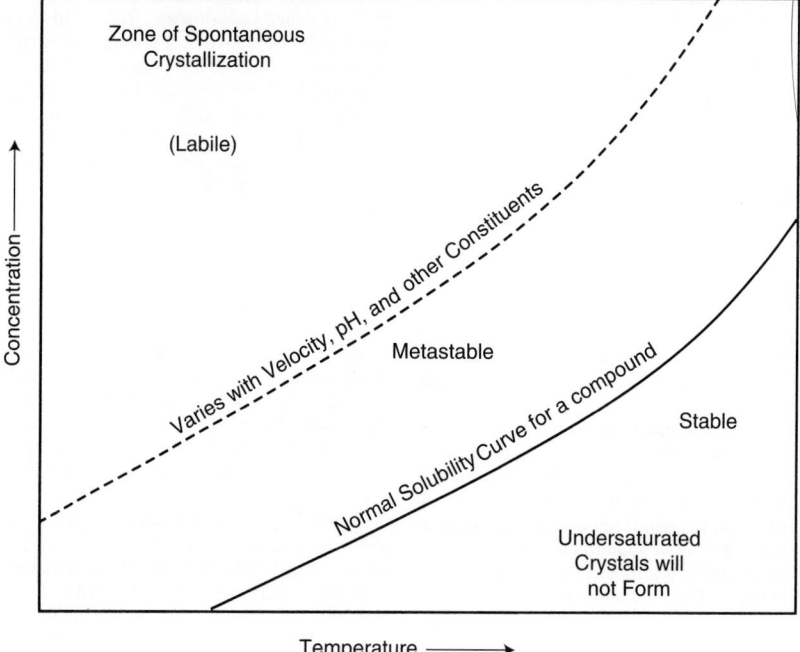

Figure 6.8.46 *Variation of solubility and supersaturation with temperature [57].*

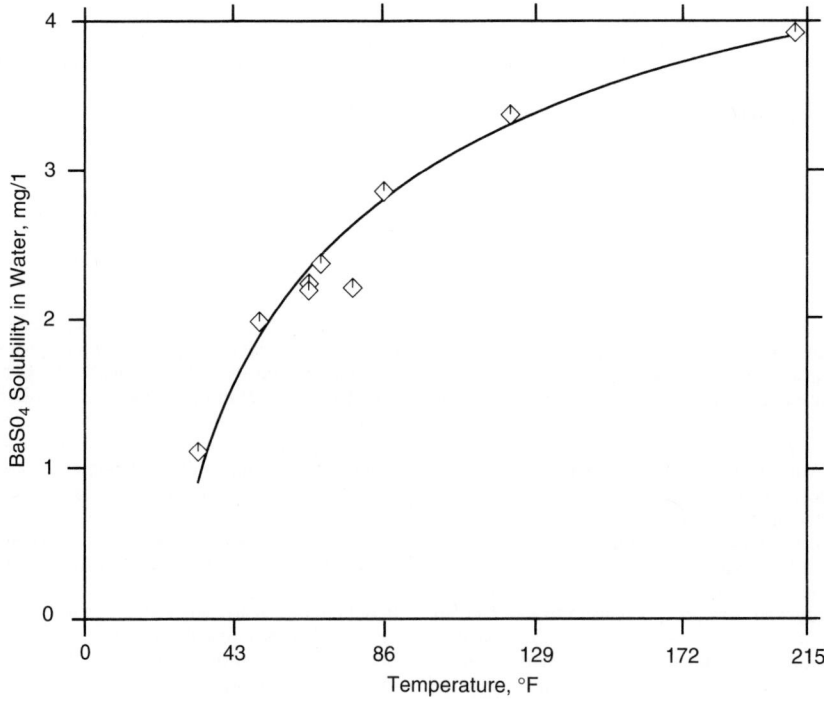

Figure 6.8.47 *BaSO₄ solubility in water [61].*

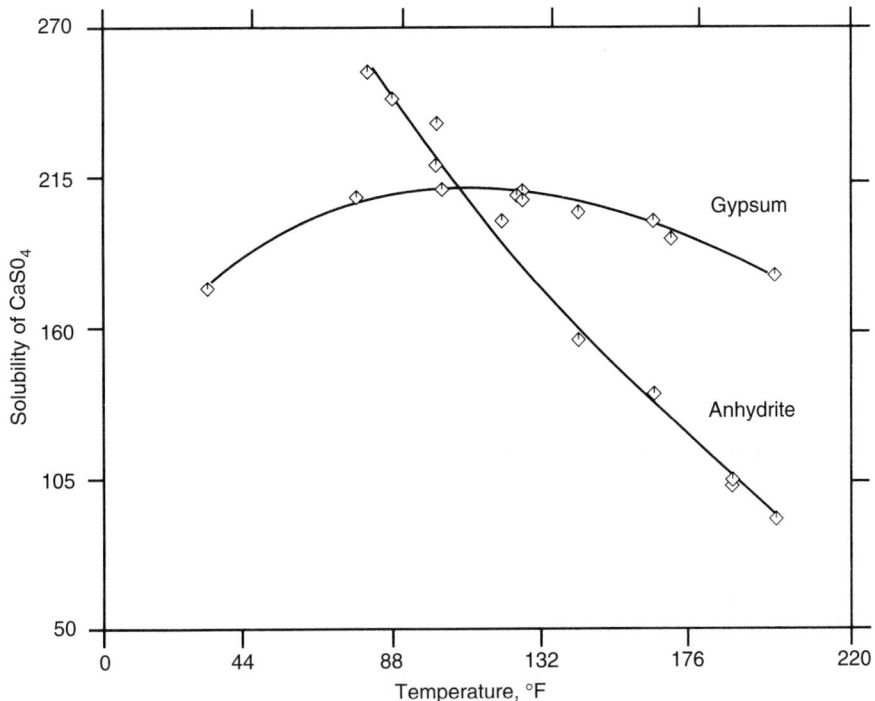

Figure 6.8.48 *CaSO₄ solubility in water [61].*

inhibitor under specific conditions requires laboratory studies and field tests. To predict the emulsification tendency of an inhibitor with crude oil, a dynamic test is usually required under specific given conditions employing kerosene and/or the specific crude oil [59,60].

$CaSO_4$ is also known as one of the most common scales in oilfield environments. The main reason for the scaling difficulties experienced with $CaSO_4$ is the solubility behavior of $CaSO_4$ in its three forms: dihydrate gypsum ($CaSO_4$-$2H_2O$), anhydrite ($CaSO_4$), and hemihydrate ($CaSO_4$-1/2 H_2O). Gypsum is less soluble at low temperatures, but anhydrite is formed at higher temperatures. Pressure drops play a major role in deposition of gypsum and anhydrite scales. Figure 6.8.48 shows solubility of calcium sulfate as a function of temperature. Some water-flooding operations in Texas and Wyoming have indicated significant production declines due to deposition of calcium sulfate. One of the major causes of $CaSO_4$ scale deposition in oil fields is known to be the presence of high quantities of native anhydrite in the reservoir formation as well as incompatibility of injection and reservoir waters [59,63].

Fracturing jobs could increase the scale problem unless scale deposition is effectively prevented. The squeeze treatment is known to be the most effective method for application of scale inhibitors to prevent calcium sulfate deposition. One of the most effective scale inhibitors to prevent $CaSO_4$ scale is a water solution of imidazolines and methylphosphonic acids [64].

This chemical composition as a scale inhibitor was tried in a well near Powell, Wyoming, in which $CaSO_4$ scale problems were known to be the only cause of the production decline of a well from 380 to 42 bpd. After physically breaking up the scale with explosives, the scale inhibitor at 130 ppm concentration was injected by a squeeze treatment. Table 6.8.21

is the production performance of the treated well with the inhibitor noted above [64].

The majority of laboratory and field experiments show that phosphonates and low-molecular-weight polyacrylate scale inhibitors are effective in preventing the formation of calcium sulfate scale at temperatures up to 150°C (300°F). Experiments using phosphonates and polyacrylates have show that metastable compounds such as gypsum scale were dehydrated to anhydrous $CaSO_4$ within the deposit. The additional influence of a phosphonate-type scale inhibitor at high temperature in a supersaturated calcium sulfate solution reduces the growth rate of rate of all three forms of calcium sulfate crystals but, at 130°C (265°F), the phosphonate-type scale inhibitors accelerate formation of anhydrite. To prevent the formation of anhydrite, the application of a low-molecular-weight polyacrylic acid is suggested [62,65–67].

$CaCO_3$ scale is normally encountered in primary oil production, water flooding, and CO_2 flooding. For a low CO_2 partial pressure, $CaCO_3$ shows lower solubility than it does at higher partial pressures. The most influential factors in precipitation of $CaCO_3$ include the temperature fluctuation, pH, injection water composition and produced water composition. Figure 6.8.49 shows the limiting pH level for $CaCO_3$ precipitation from normal seawater as a function of temperature. Other waters will have similar responses.

The solubility of $CaCO_3$ decreases with increasing temperature. The relationship between temperature and $CaCO_3$ solubility is presented in Figure 6.8.50. Normally, when CO_2 comes in contact with water, it dissolves and forms carbonic acid.

The effect of CO_2 pressure on $CaCO_3$ solubility is shown in Figure 6.8.51. Normally, as the water is produced, the CO_2 pressure decreases as the water reaches the surface. This phenomenon upsets the chemical equilibrium, and scale

Table 6.8.21 *Some Effective Scale Inhibitors for BaSO₄*

Scale Inhibitor		Maximum Temperature
Type	Concentration	
Esters	10–25 ppm	120°C(250°F)
Phosphonates	10–25 ppm	175°C(350°F)
Polyelectrolytes (polymer)	15–45 ppm	175°C(350°F)

Table 6.8.22 *Effect of Methyl Phosphonic Acid as a Calcium Sulfate Scale Inhibitor*

Days after Treatment	Inhibitor Concentration (in Production)	Production Rates, bbl/day	
		Oil	Water
Prior (0)	0	42	24
2	130 ppm	196	186
5	130 ppm	408	161
11	130 ppm	312	172
35	130 ppm	219	135
46	60 ppm	219	135
62	40 ppm	219	135
74	40 ppm	204	138
77	80 ppm	204	138
89	60 ppm	204	120
102	60 ppm	237	120
103	40 ppm	237	120
117	50 ppm	237	120
122	50 ppm	198	132

deposits. Because the $CaCO_3$ has a higher solubility at lower temperatures (Figure 6.8.50), $CaCO_3$ scaling tends to be lower at the surface facilities than at the production wellbore as predicted from temperature change. Table 6.8.22 shows solubility if $CaCO_3$ at different CO_2 partial pressures and temperatures [70].

The use of scale inhibitors is necessary to prevent ($CaCO_3$) crystallization and adherence in water-flooding and CO_2-flooding operations. Both organic and inorganic scale inhibitors have been found to be effective in preventing $CaCO_3$ scale formation. Most inorganic inhibitors are

Table 6.8.23 *Solubility of CaCO₃ as a Function of Pressure and Temperature*

CO₂ partial pressure	Solubility as a Function of Temperature, °C(°F)				
	38°C(100°F)	66°C(150°F)	93°C(200°F)	121°C(250°F)	149°C(300°F)
1 bar	0.261 g/L	0.094 g/L	0.040 g/L	0.015 g/L	0.006 g/L
4 bar	0.360 g/L	0.158 g/L	0.063 g/L	0.024 g/L	0.009 g/L
12 bar	0.555 g/L	0.221 g/L	0.091 g/L	0.036 g/L	0.012 g/L
62 bar	—	0.405 g/L	0.152 g/L	0.051 g/L	0.014 g/L

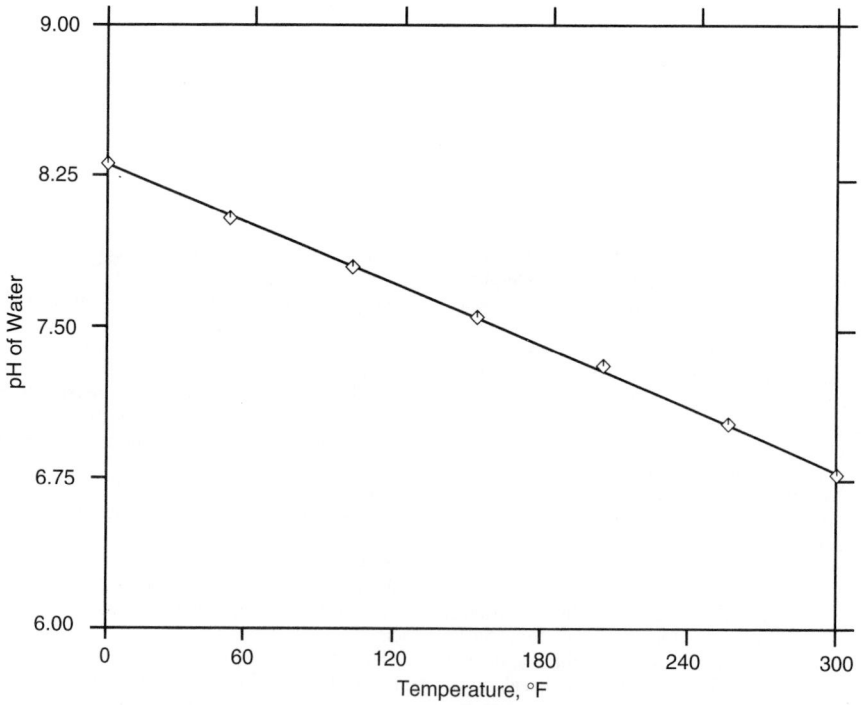

Figure 6.8.49 *CaCO₃ scaling threshold in normal seawater [68].*

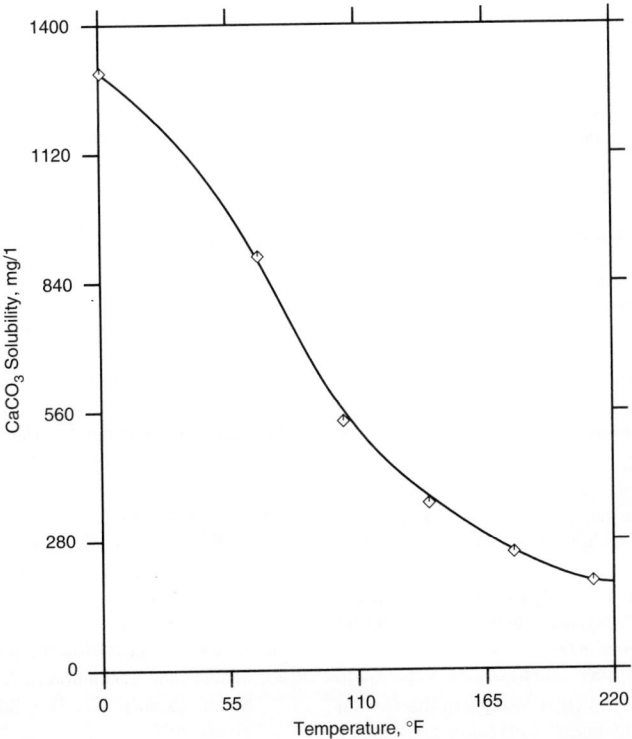

Figure 6.8.50 *Effect of temperature on the solubility of CaCO₃ [69].*

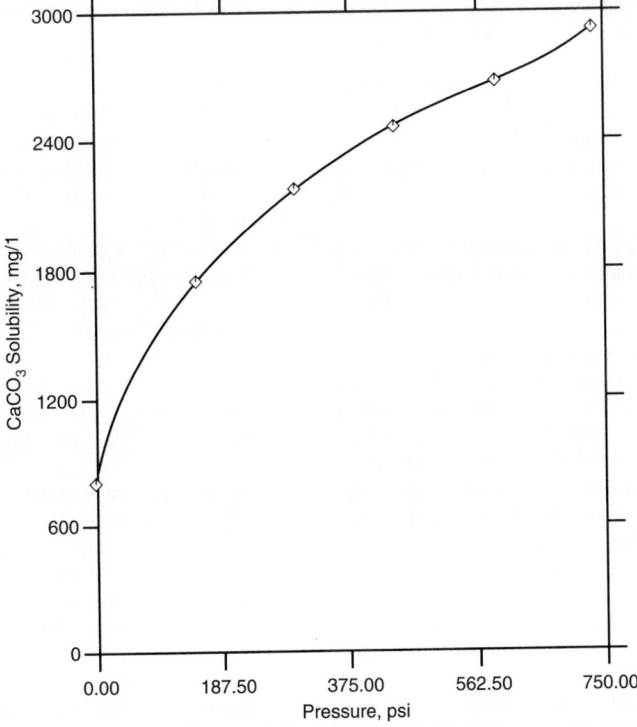

Figure 6.8.51 *Influence of CO₂ partial pressure on solubility of CaCO₃ [69].*

phosphates such as tetrasodium pyrophosphate, sodium triphosphate, sodium-calcium phosphate, and sodium-magnesium phosphate. These phosphate inhibitors are normally used in the 5 to 60 ppm range that depends upon watercut, treatment application, and type of scale [71].

The organic compounds have shown better results in inhibition of $CaCO_3$ and other scales formed at higher temperatures. Within these organic scale inhibitors, phosphonate materials are the most effective. The majority of experience in actual field environments indicates that the general types of organic materials used as scale inhibitors such as phosphonates, disphosphonates, phosphate esters and polyacrylates in concentration of 10 to 100 ppm, are effective in stopping scale deposition through squeeze treatments.

References

1. DeWaard, C., and Milliams, D.E., *Corrosion*, Vol. 31, no. 5, 1975.

2. DeWaard, C., and Lotz, U., "Prediction of Co_2 Corrosion of carbon steel," *Predicting CO_2 Corrosion in the Oil and Gas Industry*, European Federation of Corrosion publication 13, Insitute of Metals, London, 1994.

3. Nyborg, R., "Overview of CO_2 Corosion Models for Wells and Pipelines," Corrosion 2002 paper no. 02233 NACE, Houston, Texas, 2002.

4. Kapusta, S.D., Pots, B.F.M., and Rippon, I., "The Application of Corrosion Prediction Models to the Design and Operation of Pipelines," Corrosion 2004 paper no. 04633, New Orleans, Louisiana, 2004.

5. EnDean, E.J., "Corrosion Control in the Well Bore," *Petroleum Engineering International*, August 1976.

6. Weintritt, D.J., "Criteria for Scale and Corrosion," *Petroleum Engineering International*, August 1980.

7. Dawson, J.L., Shih, C.C., and Barlett, P.K.N., "Models and Predictions of CO_2 Corrosion and Erosion–Corrosion Under Flowing Condition," *Predicting CO_2 Corrosion in the Oil and Gas Industry,* European Federation of Corrosion publication 13, Institute of Metals, London, 1994.

8. DeWaard, C., Smith, L.M., and Craig, B.D., "The Influence of Crude Oils on Well Tubing Corrosion Rates," Corrosion 2003 paper no. 03629, NACE, San Diego, California, 2003.

9. Vera, J.R., Viloria, A., Castillo, M., et al., "Flow velocity Effect of CO_2 Corrosion of Carbon Steel Using a Dynamic Field Tester," *Predicting CO_2 Corrosion in the Oil Industry*, European Federation of Corrosion publication 13, Institute of Metals, London, 1994.

10. API 5CT, Specification for Tubing and casing, 7th Edition, American Petroleum Institute, Washington, DC, 2001.

11. Ikeda, A., and Ueda, M., "CO_2 Corrosion Behavior of Cr-containing Steels," *Predicting CO_2 Corrosion in the Oil and Gas Industry*, European Federation of Corrosion publication 13, Institute of Metals, London, 1994.

12. NACE Standard MR0175-2002, "Sulfide Stress Corrosion Cracking Resistant Materials for Oilfield Equipment," NACE, Houston, Texas, 2002.

13. NACE MR0175/ISO 15156-3: 2003, "Petroleum and Natural Gas Industries: Materials for Use in HS-containing Environments in Oil and Gas Production," NACE, Houston, Texas, 2003.

14. Barteri, M., Rindelli, G., Scoppio, L., and Tamba, A., "Cold working and Compositional Effect on the Performance of Duplex Stainless Steels for OCTG," Duplex Stainless Steels 91, Les Editions de Physique, 1991.

15. Hibner, E.L., and Patchell J.K., "Comparison of Corrosion Resistance of Nickel-Base Alloys for OCTGs and Mechanical Tubing in Severe Sour Service Conditions," Corrosion 2004 paper no. 04110, New Orleans, Louisiana, 2004.

16. Moore, E.M., and Warga, J.J., *Material Performance*, pp. 17–23, June 1976.

17. ASM International, *Metals Handbook*, Vol. 13: Corrosion, ASM International, Materials Park, Ohio, 1987.

18. *Hydrogen Effects in Metals*, I.M. Bernstein and A.W. Thompson, Eds. AIME, 1981.

19. Pressouyre, G.M., and Berntein, I.M., "A Quantitative Analysis of Hydrogen Trapping," *Metallurgical Transactions A*, Vol. 9A, 1978.

20. NACE Standard TM0284-2003, "Laboratory Test Procedures for Evaluation of SOHIC Resistance of Plate Steels Used in Wet H_2S Service," NACE, Houston, Texas, 2003.

21. NACE Standard TM0103-2003, "Laboratory Test Procedures for Evaluation of SOHIC Resistance of Plate Steels Used in Wet H_2S Service," NACE, Houston, Texas, 2003.

22. NACE TM0177, "Laboratory Testing of Metals for Resistance to Specific Forms of Environment Cracking in H_2s Environments," NACE, Houston, Texas, 1996.

23. Guidelines on Materials Requirements for Carbon and Low Alloy Steels for H_2S-Containing Environments in Oil and Gas production," European Federation of Corrosion Publication 16, 2nd Edition, Institute of Metals, London, 2002.

24. NACE TM0198, "Slow Strain Rate Method for Screening Corrosion-Resistant Alloys (CRA's) for Stress Corrosion Cracking in Sour Oilfield Service," NACE, Houston, Texas, 2004.

25. Leyer, J., Sutter, P., Linne, C.P., and Gunaltun, Y.M., "Influence of Test Method on the SSC Threshold Stress of OCTG and Line Pipe Steel Grades," Corrosion 2002 paper no. 0253, Houston, Texas, 2002.

26. Sutanto, H., and Semerad, V., "Annulus Corrosion in High Temperature Gas Well," SPE Conference paper 17678, New Orleans, Louisiana, 1990.

27. Rhodes, P.R., "Environment-Assisted Cracking of CRA's in Oil and Gas Production Environment: A Review," Corrosion, Vol. 57, no. 11, 2002.

28. Scoppio, L., Nice, P.I., Nodland, S., and Piccolo, E.L., "Corrosion and Environmental Cracking Testing of a High-Density Brine for HPHT Field Application," Corrosion 2004 paper no. 04113, New Orleans, Louisianna, 2004.

29. "The Role of Bacteria in the Corrosion of Oil Field Equipment," TPC3, NACE, Houston, Texas, 1976.

30. NACE Report 1F196, "Survey of CRA Tubular Usage" NACE, Houston, Texas, 1996.

31. API6A, Specification for Wellhead Valves and Christmas Tree Equipment, 17th Edition, American Petroleum Institute, Washington, D.C., 1999.

32. ASTM A106, "Specification for Seamless Carbon Steel Pipe for High-Temperature Service," ASTM, West Conshohocken, Pennsylvania, 1999.

33. API 5L, Specification for Line, 42nd edition, American Petroleum Institute, Washington, D.C., 2000.

34. "Design Guidelines for the Selection and Use of Stainless Steel," Committee of Stainless Steel Producers, AISI, Washington, D.C., 1977.

35. Kimura, M., Miyata, Y., Sakata, K., and Mochiduki, R., "Corrosion Resistance of Martensitic Stainless Steel OCTG in High Temperature and High CO_2 Environment," Corrosion 2004 paper no. 04118, New Orleans, Louisiana, 2004.

36. Handbook of Corrosion Data, 2nd Edition. B.D. Craig and D.B. Anderson, ASM International, Materials Park, Ohio, 1995.

37. Craig, B.D., "Selection Guidelines for Corrosion Resistance Alloys in the Oil and Gas Industry," 2nd Edition, Nickel Institute, Toronto, Canada, 2000.

38. Nathan, C.C., "Corrosion Inhibitors," NACE Houston, Texas, 1974.

39. Donnell, P.O., "Offshore Pipeline Interanlly Coated in Place," Oil & Gas Journal, October 21, 1975.

40. Senkovski, Jr. E., "Standard Laboratory test for Pipeline Coatings," Materials Performance, August 1979.

41. Kipin, P., "Internal Pipe Coatings Pay Off," Oil and Gas Journal, July 19, 1982.

42. McConkey, S.E., "Fusion-bonded Epoxy Pipe Coatings are Economical, Practical," Oil & Gas Journal, July 19, 1982.

43. Talbot, D., and Talbot, J., Corrosion Science and Technology, CRC Press, New York, 1998.

44. Betz Handbook of Industrial Water Conditioning, Betz Laboratories Inc., Trevose, Pennsylvania, 1980.

45. Mitchell, R.W., Grist, D.M., and Boyle, M.J., "Chemical Treatment Associated with North Sea Projects," Journal of Petroleum Technolgy, May 1980.

46. Cannon, J.H., EnDean, H.J., Todd, R.B., and Belanus, K., "Corrosin Protection by Downhole Continuous Inhibitor Transmission Via External Capillary," Materials Performance, February 1981.

47. LaFayetrte, C.R., Landrum, A.U., Atwood, J.E., and Mutti, D.H., "Corrosion Control for Gas-Lift Well Tubulars by Continous Injection Into the Gas-Lift Stream," Journal of Petroleum Technology, May 1980.

48. Karla S.K., and Bradburn, J.B., "Corrosion Mitigation-Critical Facet of Well-Completion Design," Journal of Petroleum Technology, September 1983.

49. Taylor, R.E., "How to Control Corrosion in product Pipelines," Pipeline & Gas Journal, February 1980.

50. Pettus, P.L., and Strickland, L.N., "Water Soluble Corrosion Inhibitors Help Sole Internal Corrosion Problems," Pipeline & Gas Journal, February 1979.

51. Cogram, G.E., "Organic Inhibitors Help Control Internal Pipeline Corrosion," Oil & Gas Journal, December 12, 1974.

52. Carradine, W.R., Hanna, G.J., and Grabois, R.N., "High Performance Flow Improves for Products Lines," Oil & Gas Journal, August 8, 1983.

53. Bauman, T.C., and Overstreet, L.T., "Corrosion and Piping Materials in the CPI," Chemical Engineering, August 3, 1978.

54. Nathan, C.C., Corrosion Inhibitors, NACE, Houston, Texas, 1974.

55. EnDean, H., "Design Corrosion Control Treatment," Champion Chemical Company Manual, Houston, Texas, 1982.

56. "Corrosion Inhibition of Gas Pipelines by Chemical Treatment," SPE 6596, 1977.

57. Weintritt, D.J., "Criteria for Scale and Corrosion," Petroleum Engineering International, August 1980

58. Hover, G.W., and Spriggs, D.M., "Field Performance of a Inhibitor Squeeze Program," Journal of Petroleum Technology, July 1972.

59. Vetter, O.J., "Oilfield Scale: Can we Handle it?" Journal of Petroleum Technology, December 1976.

60. "Corrosion Inhibitors in Secondary Recovery," A.K. Dunlop Corrosion Inhibitors, NACE, Houston, Texas, 1973.

61. Linke, W.F., "Seidell's Solubilities," The American Chemical Society, 4th edition, Vol. 2 pp. 1452–1455, Washington, DC 1965.

62. Kandarpa, U., and Vetter, O.J., "Scale Inhibitors for Injection of Incompatible Waters," SPE 10595, 6th SPE International Symposium on Oilfield and Geothermal Chemistry, Dallas, Texas, January 1982.

63. Mitchell, R.W., "The Forties Field Sea-Water Injection System," Journal of Petroleum Technology, June 1978.

64. Donham, J.E., U.S. Patent 3,699,118, October 17, 1972.

65. Ralston, P.H., "Scale Control with Aminomethylene-Phosphonates," Journal of Petroleum Technology, August 1969.

66. Gill, J.S., and Nancollas, G.H., "Formation and Dissolution of High-Temperature Forms of Calcium Sulfate Scales: The Influence of Inhibitors," SPE 7861, SPE-AIME; Houston, Texas, January 1979.

67. Sawada, K., Nancollas, G.H., "Formation of Scales of Calcium Carbonate Polymorphs: The Influence of Magnesium Ion and Inhibitors," Journal of Petroleum Technology, March 1982.

68. Place, M.C., "Corrosion Control: Deep Sour Gas Production," Presented at the 54th Annual Fall Conference & Exhibition, The Society of Petroleum Engineers of AIME, Las Vegas, Nevada, September 1979.

69. Miller, J.P., "A Portion of the System Calcium Carbonate, Carbon Dioxide, Water," American Journal of Solids, 1952.

70. Ellis, A.S., "The Solubility of Calcium Carbonate in Sodium Chloride Solutions at High Temperature," American Journal of Solids, 1963.

71. Ostroff, A.G., Introduction to Oilfield Water Technology, NACE Houston, Texas, 1979.

6.9 ENVIRONMENTAL CONSIDERATIONS IN OIL AND GAS OPERATIONS

6.9.1 Introduction

The key role of the engineer in daily operations is the design and maintenance of facilities. In designing a new facility or upgrading an existing one, the engineer must be aware of environmental concerns such as air emissions, water discharges, and hazardous waste generation. The engineer needs an in-depth understanding on the impact of the operation on the environment, and how considerations made before implementation of the project can circumvent or minimize these impacts. The engineer must also be aware of the regulations involved so that designs are legally feasible, and permits are acquired in a timely manner.

6.9.1.1 Oil and Gas Waste

The American Petroleum Institute (API) estimated that 149 million barrels of drilling waste, 17.9 billion barrels of produced water, and 20.6 million barrels of other associated wastes were generated in 1995 from exploration and production operations. Once generated, managing these wastes in a manner that protects human health and the environment is essential for limiting operators' legal and financial liabilities, as well as makes good business sense. Operators must also determine if the waste is subject to hazardous waste regulations. At times, this determination is misunderstood and can lead to improper waste management decisions. Prudent waste management decisions, even for nonhazardous wastes, should be based on the inherent nature of the waste. Not all waste options are appropriate for every waste. The preferred option for preventing pollution is to avoid generating wastes whenever possible (*source reduction*). Examples include process modifications to reduce waste volumes and material substitution to reduce toxicity. Understanding the procedures for determining the exempt and nonexempt status of a waste is a valuable tool, especially for operators who choose to develop voluntary waste management plans. When these procedures are used in conjunction with knowledge of the nature of the waste, the operator will be better prepared to develop site-specific waste management plans and to manage E&P wastes in a manner that protects human health and the environment [1].

6.9.1.2 Regulations and Permits

The average well and separation facility or battery requires minor human support on a daily basis. The pumper or lease operator may visit the site for a few minutes, confirm normal operations, and move on to the next location. Occasionally, a well may be worked over for maintenance, repair, or stimulation, and the activity around the wellsite or facility is increased for a short period of time. Even though the activity may be limited, the potential of environmental impacts associated with wellsites and separation facilities are not. Today pollution prevention is addressed everyday, as it requires continuous improvement in daily operating practices as well as in facility design. In fact there are plans and permits that must be prepared and obtained before to beginning development for new wells and facilities, as well as with modifications to existing ones. Modification in some cases includes an increase in production streams.

This section will attempt to address most of the regulations and permits that are typical for Permian Basin operators. Environmental health and safety issues have to be addressed, beginning with project conception and confinning with the company forever. Fines may be levied, individuals may be sentenced, pipeline connections may be severed if the paperwork is not filed on time and/or regulations not adhered to, and operations personnel may not go home if safety is an issue.

6.9.2 Rules and Regulations

Each state has a governing body or bodies that are involved in day-to-day oil and gas operations and environmental permits, or they may defer to federal rules and regulations (Regs). With regard to worker safety Occupational Safety and Health Administration (OSHA) regulations (Standards 29 CFR) apply. There are also additional considerations to be addressed on federal and tribal lands. Table 6.9.1 is a partial listing a federal and state Web sites for various agencies that are involved in oil and gas regulation. When more than one agency is involved there is typically a *memorandum of understanding* (MOU), which delineates specific authority for the agencies involved. This is a very dynamic side of the industry, with new rules and regs being approved and others modified each year. Key federal environmental law and rules are:

- *The Resource Conservation and Recovery Act (RCRA)* — enacted in 1976, required the Environmental Protection Agency (EPA) to establish procedures for indentifying solid wastes as either hazardous or nonhazardous and promulgate requirements for managing both. RCRA was amended in 1980, at that time the U.S. Congress recognized the special nature, high volume, and low toxicity of waste generated by oil and gas exploration and production operations (as well as by mining, geothermal operations, electric utilities, and cement kilns). As a result, the EPA was directed to study wastes and recommend appropriate regulatory action.
- *The Safe Drinking Water Act (SDWA)* — passed in 1974, it developed drinking water standards as well as passed regulation of underground injection wells under the Underground Injection Control program (UIC).
- *The Clean Water Act (CWA)* — enacted in 1972 primarily to control point source discharges into waters of the US.
- *The Clean Air Act (CAA)* — enacted in 1970 and most recently amended in 1990. Contains attainment provisions for air quality standards, permits, enforcement, hazardous air pollutants (HAP), acid deposition, stratospheric ozone protection, and motor vehicles and fuels.

The associated Code of Federal Regulations (CFR) and additional laws are rules listed in Table 6.9.2. Because of the dynamic nature of this topic, this is a partial listing. Be aware that interpretations of these vary, so legal counsel should be sought.

6.9.3 Waste Management Methods

The EPA has developed an hierarchy of waste management methods that is to be used to guide generators toward waste minimization, which is a major component of pollution prevention:

- *Source Reduction* — to reduce the amount of waste at the source through:

  - material elimination
  - inventory control and management
  - material substitution
  - process modification
  - improved housekeeping
  - return of unused material to supplier

- *Recycling/Reuse* — to recycle and reuse material for the original or some other purpose, such as materials recovery or energy production; this may to onsite or offsite.

Table 6.9.1 *Federal and State Regulatory Agencies (partial listing)*

Federal	Environmental Protection Agency (EPA)	http://www.epa.gov
	Storm Water Pollution Prevention Plan (SWPPP)	http://cfpub.epa.gov/npdes/stormwater/swphase2.cfm
Federal	Bureau of Land Management (BLM)	http://www.blm.gov/nhp/index.htm
Federal	Minerals Management Service (MMS)	http://www.mms.gov
Federal	Occupational Safety and Health Administration (OSHA)	http://www.osha.gov
Alabama	Alabama State Oil and Gas Board	http://www.ogb.state.al.us/
Alaska	Alaska Oil and Gas Conservation Commission	http://www.state.ak.us/local/akpages/ADMIN/ogc/homeogc.htm
California	California Department of Conservation — Division of Oil, Gas & Geothermal Resources	http://www.consrv.ca.gov/DOG/getting_started.htm
Colorado	Colorado Department of Natural Resources — Oil and Gas Conservation Commission	http://oil-gas.state.co.us/
Florida	Florida Department of Environmental Protection — Geological Survey	http://www.dep.state.fl.us/geology/rulesstatues/oilandgasrules.htm
Indiana	Indiana Department of Natural Resources — Division of Oil and Gas	http://www.in.gov/dnroil/
Kansas	Kansas Corporation Commission — Conservation Division	http://www.kcc.state.ks.us/conservation/conservation.htm
Louisiana	Louisiana Department of Natural Resources — Office of Conservation	http://www.dnr.state.la.us/CONS/Conserv.ssi
	Louisiana Department of Environmental Quality	http://www.deq.state.la.us/
Michigan	Department of Environmental Quality — Geological Survey Division	http://www.michigan.gov/deq/0,1607,7-135-3306_3334_3568—,00.html
New Mexico	New Mexico Energy, Minerals and Natural Resources Department — Oil Conservation Division	http://www.emnrd.state.nm.us/OCD/default.htm
New York	New York State Department of Environmental Conservation — Division of Mineral Resources	http://www.dec.state.ny.us/website/dmn
	Division of Environmental Permits — Sate Environmental Quality Review	http://www.dec.state.ny.us/website/dcs/seqr/index.html
North Dakota	North Dakota Industrial Commission, Oil and Gas Division	http://explorer.ndic.state.nd.us/
Ohio	Ohio Department of Natural Resources	http://www.ohiodnr.com/mineral/oil/index.html
Oklahoma	Oklahoma Corporation Commission — Oil and Gas Conservation Division	http://www.occ.state.ok.us/o&ginfo.htm
South Dakota	South Dakota Department of Environment and Natural Resources — Minerals and Mining Program — Oil & Gas Section	http://www.state.sd.us/denr/DES/Mining/Oil&Gas/O&Ghome.htm
Texas	Railroad Commission of Texas — Oil and Gas Division (RRC)	http://www.rrc.state.tx.us/
	Texas Commission on Environmental Quality (previously known as the Texas Natural Resource Conservation Commission (TNRCC))	http://www.tceq.state.tx.us/
Utah	The Utah Division of Oil, Gas and Mining	http://ogm.utah.gov/default.htm
	Administrative Rules for Title R649. Natural Resources; Oil, Gas and Mining; Oil and Gas.	http://www.rules.state.ut.us/publicat/code/r649/r649.htm
West Virginia	Division of Environmental Protection — Office of Oil and Gas	http://www.dep.state.wv.us/item.cfm?ssid=23
Wyoming	Wyoming Oil and Gas Conservation Commission	http://wogcc.state.wy.us/regchoice.cfm
	Wyoming Department of Environmental Quality	http://deq.state.wy.us/

Table 6.9.2 *Federal Environmental Legislations and Regulations (partial listing)*

	Code of Federal Regulations (CFR)
Resource Conservation and Recovery Act (RCRA)	• Definition of Solid Wastes — 40 CFR Part 261.2 • Listed Hazardous Waste — 40 CFR Part 261 Subpart D • Characteristically Hazardous Waste — 40 CFR Part 261 Subpart C • Mixture Rule — 40 CFR Part 261.3 • Quality Determination — 40 CFR Part 261.5 and 262
Safe Drinking Water Act (SDWA)	• UIC Program — 40 CFR Part 146 Subpart C • Drinking Water Standards — 40 CFR Part 41 Subpart B
Clean Water Act (CWA)	• NPEDS Point Sources — 40 CFR Part 435 • Storm Water Permits — 40 CFR Part 122.26 • Spill Prevention Control and Countermeasures Plan — 40 CFR Part 112.7 • Dredge and Fill — 33 CFR Parts 323 and 325
Clean Air Act (CAA)	
Toxic Substances Control Act (TSCA)	
Comprehensive Environmental Response, Compensation and Liability Act (CERCLA) and the Superfund Amendments and Reauthorization Act (SARA)	
Oil Pollution Act of 1990 (OPA-90)	
National Environmental Policy Act (NEPA)	
Federal Land Policy and Management Act (FLPMA)	
Endangered Species Act (ESA)	
Federal Oil and Gas Royalty Management Act (FOGRMA)	
Federal Insecticide, Fungicide and Rodenticide Act (FIFRA)	
Hazardous Material Transport Act	

- reuse
- reprocess
- underground injection for enhanced recovery
- use as fuel
- reclaim
- road spreading

- *Treatment* — destroy, detoxify, and neutralize waste into less harmful substances through:

  - filtration
  - chemical treatment
  - biological treatment
  - thermal treatment
  - extraction
  - chemical Stabilization
  - incineration
  - landfarming
  - landspreading

- *Disposal* — dispose of wastes through:

  - landfills
  - NPDES discharge
  - solidification
  - burial
  - underground injection for disposal

By incorporating waste minimization practices into a waste management program, the generators can protect public health and worker health and safety; protect the environment; meet company, state, and/or national waste minimization goals; save money by reducing waste treatment and disposal costs, as well as other operating costs, and reduce potential environmental liabilities.

To achieve pollution prevention and waste minimization goals, waste management needs to be viewed as an integrated system. A good waste management system should include the following key elements:

- A system for maintaing knowledge of pertinent laws and regulations
- A system for pollution prevention/waste minimization
- A health and safety program
- An incident response preparedness program
- A training program
- A system for proper waste identification
- A transportation program
- A proper waste storage and disposal program
- A system for waste tracking, inventories, and record keeping
- A waste management auditing program

Waste management plans need to be area specific, providing guidance for each waste generated in E&P operations. It needs to be written and distributed to field operations, and this is the basis for training, evaluation, and monitoring pollution prevention programs. The plan should be reviewed periodically and updated as technology and regulations dictate. Self-audits need to be conducted periodically to ensure

Table 6.9.3 *Activities, Pollutant Sources, and Pollutants*

Activity	Pollutant Source	Pollutant
Construction of: • access roads • drill pads • reserve pits • personnel quarters • surface impoundments	Soil/dirt, leaking equipment, and vehicles	Total suspended solids (TSS), total dissolved solids (TDS), oil, and grease
Well drilling	Drilling Fluid[1], lubricants, mud, cuttings, produced water	TSS, TDS, oil, and grease, chemical oxygen demand (COD), chlorides, barium, naphthalene, phenanthrene, benzene, lead, arsenic
Well completion/ stimulation	Fluids (used to control pressure in well), cement, residual oil, acids surfactants, solvents, produced water, sand	TSS, TDS, oil and grease, COD, pH, acetone, toluene, ethanol xylenes
Production	Produced water, oil waste sludge, tank bottoms, acids, oily debris, emulsions	Chlorides, TDS, oil and grease, TSS, pH, benzene, phenanthrene, barium, arsenic, lead, antimony
Equipment leaning and repairing	Cleaning solvents, lubricants, chemical additives	TSS, TDS, oil and grease, pH
Site closure	Residual muds, oils debris	TSS, TDS, oil and grease

[1]The potential contaminants to be found in drilling fluid varies from site to site, depending on the components of the fluid and any pollutants added due to use of the fluid. Storm water discharges that come into contact with used drilling fluids may include the following pollutants, among others: toluene, ethyl benzene, phenol, benzene, and phenanthrene. Used drilling fluids may also contain inorganic pollutants from additives or downhole exposure, such as arsenic, chromium, lead, aluminum, sulfur, and various sulfates.

the design of the plan and to identify any gaps. If gaps are found, they need to be addressed and additional training should be rolled out to all involved [2].

6.9.4 Federal Permits

One of the first permits that must be applied for is a construction general permit as the first step in the Storm Water Pollution Prevention Plan (SWPPP), which is part of the CWA–National Pollutant Discharge Elimination System (NPDES) Permitting Program. The SWPPP is in place to minimize the impact from pollution caused by water runoff associated with storms; a 1998 federal study of water quality showed that storm water is the major source of impairment. Sediment in storm water has shown to carry other pollutants attached to soil particles (metals, nutrients, etc.); siltation decreases water depth and impacts flood storage capacity; it directly impacts aquatic habitat and ecosystems; the increased turbidity decreases sunlight penetration; and finally costs of drinking water treatment are increased. The EPA has identified some storm water pollutants and sources typically associated with oil and gas facilities. These are shown in Table 6.9.3 [3].

The oil and gas production operation is brought into the SWPPP is because construction activity is disturbing between 1 and 5 acres of land, which requires a construction general permit. This includes all wellsites, access roads, facilities, right of ways, etc. What is exempt is "nonconstruction" industrial activity such as well drilling, well completion, well stimulation, production, equipment cleaning and repairing, and site closure activities, and is applicable when storm water is not contaminated or does not come into contact with raw material, overburden, etc. Any oil and gas facilities that discharges "contaminated" storm water is required to submit an industrial permit application under the storm water rule. For oil and gas facilities, this is anything with a "reportable quantity" (e.g., a sheen).

Section 404 of CWA authorizes civil and criminal penalties for violations, with civil enforcement and judicial referral if more than $137,500 or injunctive relief is required administrative if less than $137,500 and injunctive relief is not needed. The Administrative Penalty Order is broken down into two classes: class I, up to $27,000 but not more than $11,000 per violation; and class II, up to $137,500 but not more than $11,000 per violation. Both are filed with the regional hearing clerk, and the EPA files public notice in the newspaper. Both are resolved either by payment to the Central Accounting Finance Office and/or through a hearing before a regional judicial officer. [4]

Another EPA rule that addresses the oil and gas industry as part of the CWA, which prohibits the discharge of oil in quantities harmful to waters of the United States, is the Oil Pollution Prevention Regulation 40 CFR 112. This requires applicable facilities to implement Spill Prevention Control and Countermeasures Plans (SPCCs). In addition, the Oil Pollution Act of 1990 (OPA-90) requires facilities that because of their location could reasonably expect to cause substantial harm to the environment with an oil discharge to implement a Facility Response Plan (FRP). The reason for an SPCC plan is that oil discharges can have severe impact on human drinking water resources; in the Permian Basin this includes inland lakes, rivers, reservoirs, and groundwater. Per the regulation, permanent facilities that could be reasonably expected to discharge oil into navigable waters of the United States, and have:

• Aboveground capacity > 660 gallons (15.7 bbls) in a single container, or
• Total aboveground storage > 1320 gal (31.4 bbls), or
• Total underground storage > 42,000 gal (1000 bbls).

The main item to remember here is that navigable water of the United States are defined as oceans, interstate waters, and intrastate lakes, rivers, streams (including intermittent streams), wetlands, mudflats, sloughs, prairies potholes, wet meadows, and ponds for which the use, degradation, or destruction of which could affect interstate commerce, including any waters that could be used for recreation, or from which fish or shellfish could be taken and sold through interstate or foreign commerce or be used for interstate industrial commerce.

SPCC plan requirements for oil and gas facilities include the following:

- If one or more spills have occurred in the previous 12 months, description of spill, corrective actions taken, and plans for preventing recurrence are required.
- Where there is reasonable potential for spill (e.g., tank overflow), plan should include likely direction of flow and potential quantity of oil lost.
- Tank batteries and treating stations must have appropriate secondary containment to prevent discharge from reaching waterway (e.g., dikes, berms, spill diversion ponds). Drains should be closed at all times except to drain rainwater. Containment must be able to hold the entire contents of the largest tank.
- Tanks must be visually examined for maintenance needs on a scheduled basis.
- Tank batteries should be fail-safe engineered to prevent spills (e.g., adequate capacity to prevent overfills, equalization lines between tanks, high-level alarms).
- Flowlines should be examined on regular and scheduled basis (corrosion, flange connections, valves, drip pans, pipeline supports).
- Written procedures for inspections and inspection record keeping are needed.
- Personnel trained on operation and maintenance of equipment to prevent spills, and on pollution control laws, rules, and regulations.
- Have a designated person to oversee plan requirements and report to management.
- Plans must be certified by a professional engineer.
- Plan must be reviewed/updated every 3 years or amended if there are any changes in the facility design, construction, operation, or maintenance that may affect the facilities potential to discharge oil.

A FRP is also required for facilities that have the potential to cause substantial harm to the environment with a spill. Facilities that are applicable have:

- total oil storage > 1,000 bbls with over-water transfers
- total storage > 23,810 bbls with no secondary containment
- total storage > 23,810 bbls and near sensitive environments
- total storage > 23,810 bbls and near public drinking water intakes
- total storage > 23,810 bbls and spilled > 238 bbls in past 5 years

Key elements of FRP include facility description, diagrams, topography, evacuation and likely discharge paths; list of emergency equipment and personnel; identification of spill hazards and previous spills; identification of small, medium, and worst-case scenarios; a detailed implementation plan for response and disposal; records of training, exercises, and drills; security fencing; lighting; and emergency cut-off valves. If a FRP is not required, the SPCC plan must have a certification statement included in it stating that the facility does not require FRP [2].

6.9.5 State Permits
As stated previously, this section will attempt to address most of the regulations and permits that are typical for Permian

Basin operators that must be applied for and obtained from the appropriate agencies. In Texas, air quality, a PI-8, emissions screening must be completed on all collection facilities. If the emissions are too great, then a PI-7 must be submitted to an approved by the TCEQ. The PI-7 submission must include a plot plan of the facility, with all emission points noted with maximum outputs, and a current gas analysis. A copy of the calculations for those emissions, along with a process flow diagram and description is also required.

If the facility has a permanent lined emergency overflow pit, a pit permit is required from the RRC. Open pits are becoming a thing of the past, but those in existence must be maintained and netted so that wildlife does not get in. Inspection and maintenance records need to be maintained. All upset conditions, which require the use of the pit, must be reported and the pit drained when operations return to normal.

6.9.6 Hydrocarbon Contamination
As attested to by the federal requirements, spills and leaks are the most common releases encountered in oil and gas production. The spill of condensate, oil, or produced water onto the ground and the release of gas and vapors into the atmosphere exhibit the greatest liability to the environment or lease operator. Chemicals associated with operations may also leak and create an undesirable situation. Leaks and spills may be considered to be unpreventable due to equipment failure and "acts of God" (storms, natural disasters, etc.), but they are all considered containable mitigate able.

Containment of a spill is one preventable measure. Pumps and truck connections are regularly placed in containment pans, which are checked on a daily basis and vacuumed out on an as-needed basis in addition to a scheduled maintenance plan. The routine scheduled maintenance vacuuming is done to remove rainwater or blow dirt that can accumulate regularly depending upon the location. Historically, earthen berms have been placed around facilities to contain tank and piping failures. The berm design is based on the total production flowrate multiplied by the frequency of inspection. The berm design equation is

$$S_{vol} = 5.61 Q F_1 (SF) \qquad [6.9.1]$$

$$B_b = \frac{S_{vol}}{\left[B_1 B_w - n \frac{\pi}{n} D_t^2 \right]} \qquad [6.9.2]$$

where S_{vol} = spill volume in ft.^{3}
SF = safety factor (2.0 to 3.0)
Q = liquid flowrate to tank in bpd
F_i = frequency of inspection in days
B_h = height of berm in ft.
B_l = length of berm in ft.
B_w = width of berm in ft.
n = number of tanks in battery
D_t = tank diameter of n tanks in ft.

Soil containment by a spill or leak may be treated as an exempt waste provided the leaking substance is listed as an exempt waste [5]. Although all discharges must be reported to the appropriate government agency, this, in many cases, is to be verbally at the time the spill or leak is discovered and followed by the appropriate forms in a timely manner when information is compiled. In Texas, the Railroad Commission (TRRC) statewide rules that address this are as follows: Statewide Rule 8 provides that no person conducting activities subject to regulation by the commission may

Table 6.9.4 *Comparative Samples of Oil Waste Road Material and Asphalt*

Parameter	Oil Waste (ppm)	Asphalt (ppm)
Arsenic	0.04	0.10
Boron	0.10	0.38
Cadmium	0.001	0.01
Chromium	0.60	0.16
Lead	0.057	0.63
Zinc	0.14	0.99
Mercury	0.0001	0.0001
Selenium	0.0017	0.0002
Barium	0.55	2.80
Copper	0.06	0.36
Phenol	0.035	0.24

Table 6.9.5 *Phase Distribution of Certain Petroleum Products in Percentages in Air, Soil, and Water*

Compound	Adsorbed to Soil	Volatilization	Soluble in Groundwater
Aromatics			
Benzene	5	60	35
Toluene	5	75	20
Ethyl benzene	20	60	20
Xylenes	15	55	30
Aliphatics			
n-Pentane	<1	95	4
Iso-Pentane	<1	95	4
n-Hexane	<1	95	4
n-Heptane	<1	95	4
Cyclohexane	1	90	4
Polycyclic aromatic hydrocarbons (PAHs)			
Naphthalene	60	10	30
Benzo(a)pyrene	100	0	0
Anthracene	100	0	0
Phenanthrene	90	3	7
Alcohols			
Methanol	0	2	98
Ethanol	0	2	98
Phenol	9	>1	90

cause or allow pollution of surface or subsurface water in the state. Rule 20 provides that operators shall give immediate notice of a fire, leak, spill, or break, followed by a letter giving the full description of the event, including the volume of products lost. Rule 91 provides that cleanup requirements for hydrocarbon condensate spills and crude oil spills in sensitive areas will be determined on a case-by-case basis. [6] There are new products out on the market that form a liner that is blown onto the berm and ground within forming an impermeable surface, and are under evaluation by many companies.

The hydrocarbon/soil mixture may be left in place provided that all petroleum products stay on location and that it is de-watered. Operators have found that a mixture of crude and raw materials make an excellent road base and, if properly designed, create less of an environmental hazard than the traditional asphalt road base (Table 6.9.4 [7]). In Texas a minor permit from the TRRC and landowner permission is required before any spreading contaminated soil. A typical mixture of 13% heavy oil, 7.5% liquid (H_2O), and 79.5% solids make a road surface that can endure heavy road traffic [8]. This road application has been tested in northern climates where reduced volatilization of hydrocarbons is realized. In warmer climates the appearance of a sheen during runoff conditions may be apparent, thus increasing the likelihood of hydrocarbon contamination to the immediate area. It has been recognized that hydrocarbon contents above 20% in soil mixtures have a detrimental effect on plant life, with 1% showing no adverse effects [8]. In land farming, applications of 2% to 3% liquid hydrocarbons are common.

As a result of a spill, a hydrocarbon may pose a hazard to ground water, flora, fauna, and humans. Although many wastes produced at the wellsite are exempt, the operator is responsible for degradation of the surrounding due to imprudence, neglect, and accidental release. Hydrocarbon content in the soil that is in excess of the saturation limit may be transmitted to the surrounding environment through runoff and gravity. These hydrocarbons are loosely categorized as dense nonaqueous phase liquids (DNAPL) or light nonaqueous phase liquids (LNAPL), with DNAPLs being heavier than water. In the field both these species are present. The LNAPLs are more mobile. This mobility translates to greater relative conductivity in the soils and lesser adsorption onto the soil. However, the LNAPL possess a lower vapor pressure, allowing the liquid to more rapidly dissipate into the gaseous phase, leaving only traces of its product. Table 6.9.5 [9] shows how the hydrocarbon's partition among the phases Figure 6.9.1 shows the divisions graphically. An approximation for the volume of liquid hydrocarbon that may be retained by a volume of soil, the depth

of infiltration, as well as, the lateral spread of the oil onto the top of the water table is shown in the following equation [10]:

$$V_s = \frac{0.2V_{hc}}{\Phi S_{wr}} \qquad [6.9.3]$$

where

V_s = soil required to obtain residual saturation in yd^3
V_{hc} = volume spill hydrocarbon in bbl
Φ = soil porosity
S_{wr} = residual oil saturation capacity of the soil (Table 6.9.6 [11])

And knowing the area of infiltration of the soil, the depth of the spill may be determined by:

$$D = \frac{27V_S}{A} \qquad [6.9.4]$$

where

D = depth in ft.
A = observed area of infiltration in $ft.^2$

In the event the spill has been determined to reach the water table, the lateral spreading onto the top of the water table may be estimated as [12]:

$$S = \left(\frac{1,000}{F}\right)\left(V - \frac{Ad}{K}\right) \qquad [6.9.5]$$

where

S = maximum spread of pancake shaped layer in m^2
V = volume of infiltrating hydrocarbons in m^3
A = area of infiltration
d = depth to water table in m
K = constant (Table 6.9.7)

Oil, being immiscible in water, will either float on top of the water table in the case of the LNAPL or sink to the bottom. The amount of hydrocarbon that is dissolved in the water

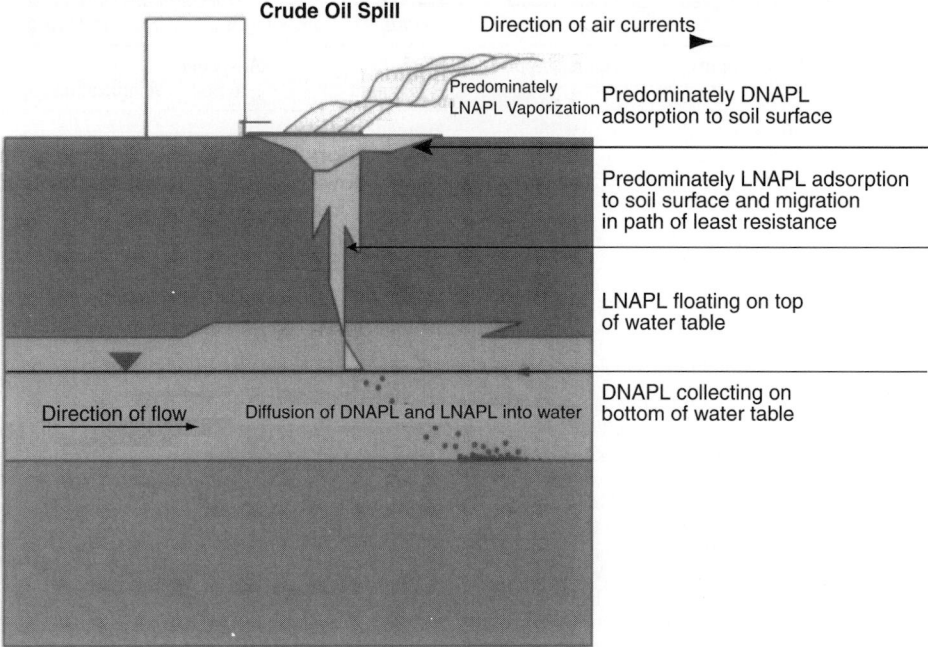

Figure 6.9.1   *Partitioning of hydrocarbons during a spill.*

Table 6.9.6   *Kerosene Retention Capacities in Unsaturated Soils*

Soil Type	Kerosene Retention Capacity	
	Gal/yd³	L/m³
Stone, course sand	1	5
Gravel, course sand	2	8
Course sand, medium sand	3	15
Medium sand, fine sand	5	25
Fine sand, silt	8	40

Table 6.9.7   *K-Values for Equation 6.9.5*

Soil Type	Gasoline	Kerosene	Light Gas Oil
Stone, course sand	400	200	100
Gravel, course sand	250	125	62
Course sand, medium sand	130	66	33
Medium sand, fine sand	80	40	20
Fine sand, silt	50	25	18

Table 6.9.8   *Aqueous Solubility Data for Selected Petroleum Products*

Product	Solubility (mg/L of H_2O)
Gasoline	50–100
1-Pentene	150
Benzene	1,761
Toluene	515
Ethyl benzene	75
Xylenes	150
n-Hexane	12
Cyclo-hexane	210
i-Octane	8 ppb
JP-4 jet fuel	<1
Kerosene	<1
Diesel	<1
Light fuel oil (#1 and #2)	<1
Heavy fuel oil (#4, #5, and #6)	<1
Lubricating oil	<1 ppb
Used oil	<1 ppb
Methanol	>100,000

depends primarily on the solubility and mass of the components. Only a few hydrocarbons are readily soluble in water, such as benzene, toluene, and xylene. Table 6.9.8 [13] shows the solubilities for several hydrocarbons.

When a land spill occurs, the liquid petroleum will simultaneously spread over the top of the site and seep into the soil. In the event the water table is shallow, the liquid will rest on top of or below the water table and then migrate at a rate proportional to the groundwater flow.

The evaporation rate of spilled hydrocarbons may be inferred from Fick's law:

$$N_A = D_A A \frac{dC}{dX} \qquad [6.9.6]$$

where

N_A = rate of mass transport in mol/s
D_A = mass diffusivity of hydrocarbon in air in ft.²/s
A = transport area in ft.²
C = concentration in M
X = length pf path traveled in ft.

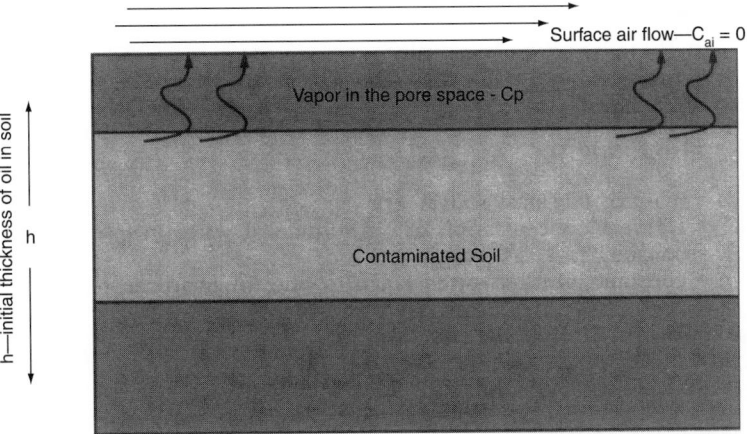

Figure 6.9.2 *Vapor migration through soil pore space.*

In a stagnant system, the length of the diffusing path is often unknown for a spill in a given field. In the event advection is also taking place in the presence of wind currents, the boundary layer may be inferred to be this length:

$$\delta = 5(x)(N_{Re})_x^{-0.5} \qquad [6.9.7]$$

For a finite flat plate in laminar flow, the thickness is [14]:

$$(N_{Re})_x = \frac{V \times \rho}{\mu} \qquad [6.9.8]$$

where

- δ = the thickness of the boundary layer in cm
- x = the length of the spill in cm
- N_{Re} = Reynolds Number
- V = free stream velocity of the wind in cm/s
- ρ = density of air in g/cm^3
- μ = viscosity of the air in g s^{-1}/cm

The concentration of the hydrocarbons is evaluated at the equilibrium concentrations of the vapor in the pores (C_p) and the concentration of the vapor at the soil-air interface (C_{ai}). In the presence of wind, C_{ai} may be assumed to be zero. The concentration of the vapor in the pores is estimated from Dalton's law

$$P_i = y_i P_T \qquad [6.9.9]$$

and Raoult's law

$$P_i = x_i P_i^0 \qquad [6.9.10]$$

where

- P_i = partial pressure of the hydrocarbon in atm
- y_i = mol fraction of the hydrocarbon in the gas phase
- x_i = mol fraction of the hydrocarbon in the liquid phase
- P_T = total pressure (1) atm in most cases)
- P_i^0 = vapor pressure of the hydrocarbon in atm

Knowing the mass of the hydrocarbon at the surface, the time for disappearance of the hydrocarbon from the surface may be determined by using Equation 6.9.11. In turn, the time for the spill to evaporate from the pore space may be estimated by Equation 6.9.12, using a correlation between the diffusivity in air and the hydrocarbon's diffusivity in the pore space [15]. Figure 6.9.2 illustrates this analogy.

$$D_p = D_a \frac{\phi}{\tau} \qquad [6.9.11]$$

where

- D_p = mass diffusivity of the hydrocarbon in the pore space in cm^2/hr
- Φ/τ = ratio of porosity to tortuosity (typically 1.5 to 3.0)

The time for the spill to evaporate from the pore space may be estimated by

$$T = \frac{hM}{ZAD_p(C_p - C_{ai})} \qquad [6.9.12]$$

where

- T = time in hrs
- M = mass of hydrocarbon in mol
- h = initial thickness of the contaminated soil in cm
- A = area perpendicular to flux in cm^2

To accurately infer the results of a spill, onsite and laboratory studies should be made for each representative site. The onsite investigation would take inventory on the effects of a worst-case scenario, such as total evacuation of a holding tank. The inventory would include the likely migration path of a spill in accordance to topography; potential threats to humans, livestock, and plants, and the relationship of the spill area to groundwater and surface water supplies. This study is required as part of the SPCC, refer back to "Federal Permits". A laboratory analysis would include hydrocarbon quantitative analysis for constituents creating potential health risks, soil analysis, and volatilization parameters.

The soil analysis determines the conductivity of the soil in regard to the liquid hydrocarbon and the absorption/desorption potential of the soil. Knowledge of this conductivity will help in determining the consequences of a spill and may in some cases divert an unwarranted emergency cleanup operation. Many times the heavier crudes will only penetrate the surface a few inches. The process of determining conductivity is analogous to determining relative permeabilities in a reservoir core sample. In the present analysis, a soil core is taken and left in the core barrel with the soil water content undisturbed. The site's liquid hydrocarbon is then introduced over the soil core at a constant representative head. The time to exit the barrel will infer the influx of the hydrocarbon at partial saturation. This time is critical due to fingering of hydrocarbons. The area affected

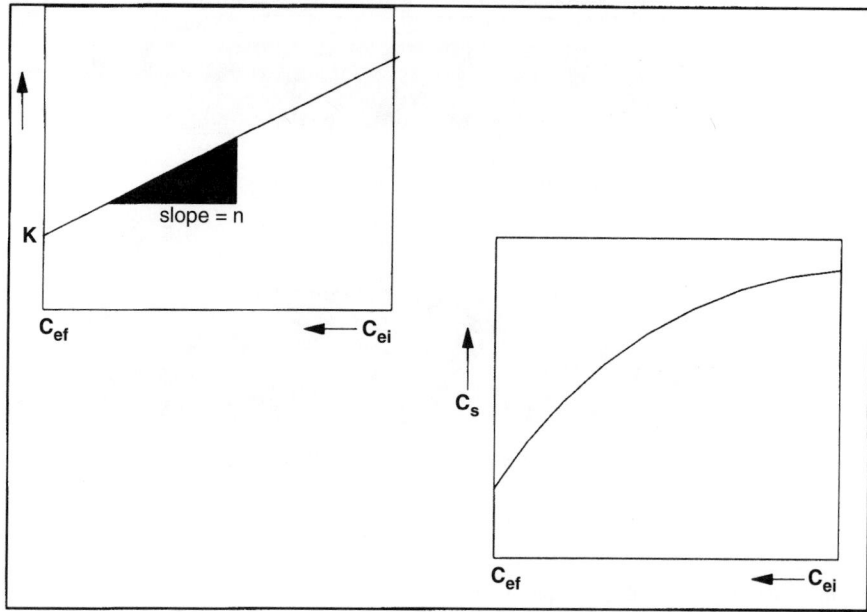

Figure 6.9.3 *Fruendlich type isotherms.*

initially may be much less than the total area of the barrel. The test is continued until the sample is saturated. The conductivity is then estimated using Darcy's Law (Equation 6.9.14). The mass flowrate may then be estimated from this conductivity relationship.

$$q = K \frac{dH}{dL} \qquad [6.9.13]$$

where

q = hydrocarbon flux in m/s
K = conductivity in m/s
dH = differential head on core in m
dL = length of core in m

After the conductivity is found, all residual oil is allowed to drain from the sample. The residual left in the core is the absorption capacity of the soil sample. The saturation level of a particular soil may also be found by plotting the data in the form of a sorption isotherm. After the residual oil is drained, the soil sample is washed at constant temperature with water until the hydrocarbon exists only in trace amounts in the product stream. The difference in the hydrocarbon mass introduced initially and the displaced hydrocarbon is the total residual saturation for that soil. Depending on the wellsite conditions, this amount of soil may be safely left in place without harm to the surrounding and will continue to decrease in quantity due to in situ microbial populations and through diffusion. The desorption isotherm may be constructed through the Fruendlich equation:

$$C_s = KC_e^h \qquad [6.9.14]$$

or

$$\ln C_s = \ln K + n \ln C_e \qquad [6.9.15]$$

where

C_s = mass of oil/mass of soil in g/g
C_e = effluent concentration of oil/volume of water in g/cm^3
K = constant in L/g
n = constant (unitless — often equal to one for hydrocarbons)

In the linearized form of the equation, K is recognized as the residual oil saturation for the soil sample. As shown in the Fruendlich-type isotherms.

The constant K is a measure of the soils' volumetric capacity of hydrocarbon saturation, whereas the constant 1/n is the measure of the mutual attraction between the constituents. Once the isotherm is defined, it may then be used to predict effluent concentrations of hydrocarbons that may be transported into the groundwater.

In general, because of the nonpolar nature of hydrocarbon, a soil comprised of a high percentage of organic matter will absorb more liquid hydrocarbons. Less-soluble, heavier hydrocarbons will absorb better than lighter soluble hydrocarbons and are considerably less mobile than LNAPLs.

Solutions of gases and crude oil follow Henry's law at atmospheric pressures. Henry's constant, more commonly referred to as an equilibrium constant in the petroleum industry, may be used to find the distribution of a hydrocarbon in the aqueous and the atmospheric phases:

$$K_H = \frac{y_i}{x_i} \qquad [6.9.16]$$

where K_H = Henry's law constant (unitless)
y_i = mole fraction of the hydrocarbons in air
x_i = mole fraction of the hydrocarbons in liquid

Henry's constant is most often referred to in the environmental industry in units of atm. In 1 L of water, there are 55.6 mol and 22.4 L of air per mol of air, such that

$$K_H = \frac{\text{partial pressure of gas in air}}{\text{mol hydrated gas/mol water}} \qquad [6.9.17]$$

Table 6.9.9 [16] gives some values of Henry's constant for gases in contact with air at 20°C.

6.9.7 Other Sources of Waste and Contamination

Glycol and other liquids used for dehydration may occasionally leak or be spilled on the ground. They also may

Table 6.9.9 *Henry's Constant for Gases at 1 Atm. 20°C*

Component	Henry's Constant (Water)
Methane	3.8×10^4
Benzene	2.4×10^2
H_2S	5.15×10^2
CO_2	1.51×10^2
Toluene	3.4×10^2 (25°C)

become contaminated, in which case the product must be replaced. Because of inherit physical properties, namely, viscosity, of dehydration liquids such as triethylene glycol (TEG), significant losses are not realized to soil seepage or evaporation. The desiccants on the whole are very toxic and completely miscible in water. Unless treated, ethylene glycol is colorless and odorless, and may be mistaken for water upon casual analysis. The greatest potential danger is then contamination of surface runoff. Any spillage should be retrieved and properly disposed of. Waste liquid desiccants should be recycled when possible. Class II injection is the most cost-effective disposal method. Solid desiccants, such as alumina, molecular sieves, and silica, become uneconomically regenerable with time. There is also loss due top structural failure. When exhausted, these materials may be sent to a disposal or buried on site. Any onsite burial must be with the landowners' approval.

In the sweetening processes, amine treatment is the most prevalent. The amines, such as monoethanoamine (MEA), DEA, and sulfinol combine with CO_2 and H_2S at moderate temperature and are released when the combined solution temperatures is slightly raised.

The wasted H_2S and CO_2 may then be flared. H_2S may be recycled into sulfur products, and the CO_2 may be collected and used in recovery projects. These gases may also be trapped in caustic solutions. In this case, the pH of the solution is lowered and a salt is formed. These salt solutions, as with most liquid wastes, may be disposed of in class II injection wells when recycling is not feasible (also refer to "Class II Injection Wells"). The pertinent chemical reactions involved here are:

$$CO_2(g) + NaOH \rightarrow NaHCO_3 \qquad [6.9.18]$$

$$H_2S + 2NaOH \rightarrow 2H_2O + Na_2S \qquad [6.9.19]$$

6.9.8 Production Site Remediation
In remediation, a site's contaminated material is reduced by either relocating the contaminated material to a disposal or on-site treatment. Physical treatment methods may include [17]

Gravity Separation	Dissolution
Sedimentation	Soil Washing/Flushing
Centrifugation	Chelation
Flocculation	Super Critical Solvent Extraction
Oil/Water Separation	
Dissolved Air Flotation	
Heavy Metal Separation	

Phase Changes	Size/Adsorptivity/Ionic Characteristics
Evaporation	Filtration
Air Stripping	Carbon Adsorption
Steam Stripping	Ion Exchange
Distillation	Reverse Osmosis

In addition to these physical methods, chemical alterations and biotransformation are also used in site remediation. Table 6.9.10 [18] shows a relative cost comparison between selected treatment techniques.

6.9.9 Landfarming
Land farming utilizes in situ and introduced microbes such as *Bacillus cereus, Bacillus polymixa, Arthrobacter globiformous*, and *Alcanigenes porasoxus* to degrade contaminants. Heterotrophic bacteria use organic compounds as energy and the carbon source for synthesis. The hertertrophs are classic oil spill degraders. Autotrophic bacteria use CO_2 as a carbon source and oxidize inorganic compounds for energy. They are most useful in drilling mud degradation.

In some instances biodegradation of contaminants may result in harmful intermediates, such as mercury. Mercury and other minerals, in their pure metal form, may remain immobile until the introduction of reducing microbes. The altering of conditions, pH and organic content, may change the redox state of the metal. Under such conditions, the metal may then be prone to migration and thus leach into the groundwater. Also, anaerobic microbes may reduce the metal to a sulfide, and solubility is enhanced.

In the case of oil spills, onsite bioremediation has proven successful. In situ microbial strains are active in most soils. If a previous spill has been documented, a culture from the older site may be introduced in addition to other cometabolites. The successive strains have shown to genetically adapt and become more successful in regards to rate and mass consumed. Composting of contaminated material has also proven effective[18]. Assuming equivalent environments, these successive stains of biomass maybe compared quantitatively by

$$dM/dT = \phi V_s K_b C_p \qquad [6.9.20]$$

where

K_b = biorate constant in hr^{-1}
C_p = concentration of petroleum in the pore space in mol/yd^3

In this method, it is assumed that the soil is homogeneously distributed with hydrocarbons, nutrients, and biomass and that the degradation occurs in a uniformly distributed manner. The biodegradation rate constant, assumed pseudo first-order, may then be used comparatively between sites. In the event that the addition of cometabolite results in a slower rate constant, it is assumed that the previous biomass was more efficient.

In landfarming, drilled soils, mud, and contaminated soils are distributed over an area in addition to microbial nutrients (fertilizer). The mixture may be turned and fertilized periodically to enhance degradation. The microbial growth can be quantified by a number of methods. The microcosm turbidity method can be used to estimate the viability of the indigenous species to degrade the hydrocarbons present. A sample of soil or cuttings is taken and allowed to grow. A turbidimeter is then used to quantify growth periodically, and mass is calculated as a function of time [19]. CO_2, generated by the microbes, collected in a caustic solution and the use of a respirometer are yet another growth measurement methods. Both of these use the mass of gas produced or consumed as a measure of cell growth. Figure 6.9.4 shows the characteristic growth phase of a pure culture of bacteria. Bacterial growth is typically substrate limited.

In the case of oil-based muds, several mixtures of hydrocarbon-to-fertilizer ratios have been tested for microbial response. The microbes have been found to react

Table 6.9.10 *Comparison of Treatment Processes*

Type of Treatment	Cost per Cubic Yard	Time Required Months	Additional Factors	Safety Issues
Incineration	250–800	6–9	Energy	Air pollution
Fixation	90–125	6–9	Transportation/ monitoring	Leaching
Landfill	150–250	6–9	Monitoring	Leaching
Biotreatment	40–100	18–60	Time	Intermediary metabolites and polymerization

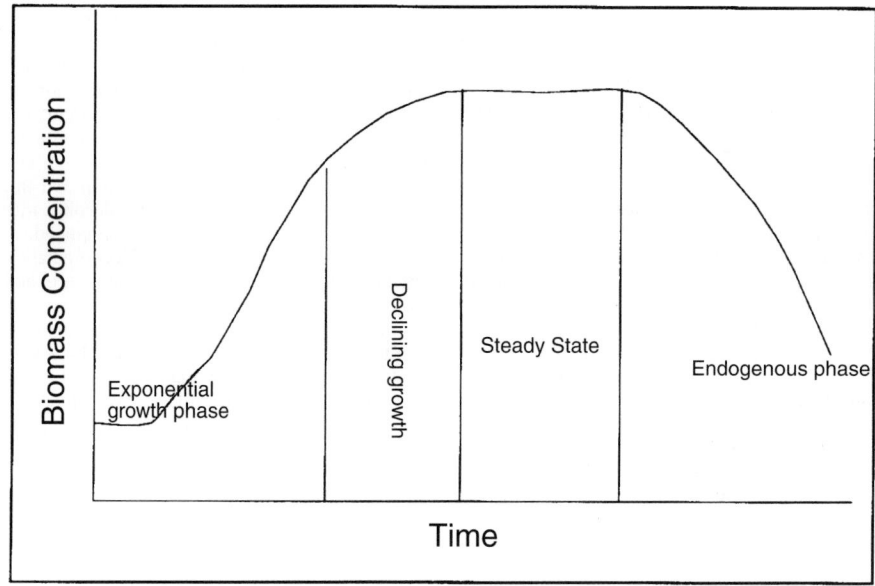

Figure 6.9.4 *Growth phases of bacteria.*

favorably to certain conditions. A hydrocarbon:ammonia-nitrogen:phosphate level ratio of 100:2:1 with the addition of trace minerals, improves moisture retention properties, and adjustment of pH to neutral levels was found to increase the half-life reduction of napthenic crude from more than 2 years to approximately 8 weeks[20]. In this same set of experiments, a hydrocarbon:amonia-nitrogen:phosphate level ratio of 100:1:0.2 had only marginal hydrocarbon biodegradation rates, and increasing this ratio to 100:2:0.15 while controlling pH and tilling did not significantly raise the rate. Another study concluded the most productive carbon-to-phosphorus ratio was 120:1 with a pH of 7.7 at 75°F to 95°F. The soil was treated with water spray loaded with common fertilizers. The initial hydrocarbon contents were 600 to 700 ppm and degraded to 415 ppm within 2 months $(k = 0.003 \text{ mg L}^{-1} \text{ min}^{-1})$ [21].

Slurry phase bioreactors maybe used to decrease the oil and grease levels in soil and drill cuttings. Here, drilling cuttings are introduced to a tank in the presence of nutrients, oxygen, and hydrocarbon degrading microbes. The vessels are operated at pH levels from 5 to 9 and nitrogen-to-phosphorus level of 10:20. Oil and grease psudeo first-order degradation rates of 0.60 mg L^{-1} min^{-1} (k) have been recorded with pH levels below 6.5 where phosphorous is more readily soluble [22]. The slurry reactor parameters

are typically designed as batch mixed at constant pressure using these design equations:

$$- r_a = k C_a \qquad [6.9.21]$$

$$T_R = \int_{N_{af}}^{N_{ao}} \frac{dN_a}{V_R(-r_a)} \qquad [6.9.22]$$

$$T_R = N_{ao} \int_0^{X_{af}} \frac{dX_a}{V_R(-r_a)} \qquad [6.9.23]$$

where $- r_a = - (dC_a/dt) =$ the degradation rate of species "a" (hydrocarbon)
$C_a =$ concentration of "a" in M
$K =$ first-order rate constant M$_s^{-1}$
$N_a =$ mol of "a" at time t
$N_{ao} =$ initial mol of "a"
$T_R =$ reactor residence time in s
$V_R =$ reactor volume in L
$X_a =$ fractional conversion of "a"

Here all the reactants are fed into the reactor initially and remain there for a mean residence time (T_R). Nutrients may be added occasionally until desired degradation is met. The volume attributed by the additional nutrients can usually be neglected. Equations 6.9.21 through 6.9.23 are valid for any batch-mix operation. Usually the kinetics of systems are established and then the reactor optimized to facilitate the operation [14].

Kinetics are controlled by oxygen level, moisture content, temperature, nutrients, pH, and concentration. Soils may have hydrocarbons up to 5% by weight without harmful effects, and half-lives of only 2 years may be realized for soils containing up to 14% hydrocarbons [23, 24].

6.9.10 Landspreading

Landspreading is differentiated from landfarming in that the area over which the contaminated soil or liquid is spread but not actively manipulated to increase degradation rates. One study showed that spreading oil-based cuttings over an area of soil at thickness not over 2 in. lead to natural degradation rates of approximately 80% in the first year and 95% of the hydrocarbons were eliminated after 3 years. Further analysis proved that no significant migration or leaching of the hydrocarbons had taken place [25]. Another method of landspreading includes the reliquidfying the solids and then spreading them evenly over an area. Here the loading factors take into consideration the soil and contaminant characteristics. A cumulative salt burden of 6,000 lb/acre (3000 ppm) is considered a safe level with a hydrocarbon loading of 40,000 lb/acre (2%). These numbers are based on a 2,000,000 lb/acre soil horizon [26].

The total salt burden of the soil is calculated from the total dissolved solids contained in the contaminant/soil mixture. Applications of freshwater gels may be added at 1,000 kg Cl/ha (KCL form) and 1,400 kg Cl/ha (NaCL from) and present no limitation to the soils and may actually enhance soil characteristics for plant growth. Chloride concentrations of 1,000 to 2,600 ppm have been safely used without loss of yield in grass growth. Gypsum has been added in when sodium activity ratio (SAR) levels have increased, thus relieving the problem.

Hydrocarbons and salts inherent in some production water and drilling muds may pose threats to the local ecosystems if not properly managed. The salt concentration may be estimated for a liquid from the electrical conductivity of the solution by

$$TDS = B \, (EC) \qquad [6.9.24]$$

$$I = 0.016 \, (EC) \qquad [6.9.25]$$

Electrical conductivity (EC) is measures in mmhos/cm, total dissolved solids (TDSs) as mg/L and ionic strength (I) as molarity. The values for B are held closely in agreement by several sources as 640 and 613. Although equations Equation 6.9.24 and 6.9.25 are only used as an estimate of ionic content in the soil or solution, they are accepted due to their general use in the soil sciences. EC values less than 4 mmhos/cm are generally satisfactory for crop yields. The addition of drilling mud with an EC of 1.3 to 5.3 mmhos/cm showed no adverse effects on Bermuda grass [27]. Generally an SAR<12 and ESP<15% is acceptable for land disposal. The ESP is another measure of sodium contained in the soil. This is

$$ESP = \frac{[NaX]}{CEC}(100) \qquad [6.9.26]$$

$$SAR = \frac{[Na]}{\sqrt{\dfrac{[CA] + [MG]}{2}}} \qquad [6.9.27]$$

where

NaX = the sodium in the solid phase
CEC = the cation exchange capacity of the soil phase in meq/100g
[ ] = molarity

Table 6.9.11 *Oil tolerance for Selected Grasses and Plants*

Grass/Plant Type	Oil and Grease for One Application
Lawn grasses	<0.5%
Rye grass, oats, wheat, corn	<1.5%
Red clover	<3.0%
Perennial grasses, trees	<3.0%

Table 6.9.11 shows the oil tolerance for selected grasses and plants [27]. API recommends a one-time application of <1%.

6.9.11 Air Stripping

In phase transformation, most often the liquid phase is let to a gaseous phase, through the addition of heat, a reduction in pressure or concentration, or may combination thereof. Evaporation may be used as applied through variations of Equation 6.9.33. The two film theory states the basis for most air stripping operations, and is mathematically defined by

$$dC/dt = -K_L a \, (c_s - c_t) \qquad [6.9.28]$$

where

C = concentration of the hydrocarbon in M
$K_L a$ = overall rate constant in hr^{-1}
  c_s = concentration of the hydrocarbon in the gas phase at time (t) in M
  c_t = concentration of the hydrocarbon in the liquid phase at time (t) in M

Here, a gas is passed through the contaminated liquid phase, and the resulting equilibrium concentration differences between the phases drives the vaporization of the liquid. The air stripping design may be simple bubbling of air through an open tank, as shown in Equation 6.9.28, or the more effective and costly packed tower that uses these design equation[16].

$$(HTU) \, (NTU) = \frac{L(C_i - C_e)}{K_L a} \frac{\ln \frac{DF_e}{DF_i}}{DF_e - DF_i} \qquad [6.9.29]$$

$$HTU = \frac{L}{K_L a} \qquad [6.9.30]$$

$$NTU = \frac{R}{R-1} \frac{\ln \frac{C_i}{C_e}(R-1)+1}{R} \qquad [6.9.31]$$

$$R = \frac{K_H G}{L} \qquad [6.9.32]$$

where

HTU = height of transfer unit in m
NTU = number of transfer unit
  L = liquid velocity through empty diameter of transfer unit in m/s
  G = gas velocity through empty diameter of transfer in m/s
  C_i = inlet concentration of contaminant in gas in M
  C_e = exit concentration of contaminant in gas in M
  Df_i = driving force at the inlet in M
  Df_e = driving force at the outlet in M
  R = stripping factor
  K_H = unitless Henry's constant for contaminate
$K_L a$ = (s^{-1})

Table 6.9.12 *Diffusion Coefficients for Some Petroleum-Related Contaminants*

Contaminant	Diffusion coefficient $\times 10^{10} \text{m}^2/\text{s}$ at 20°C
Methane	18.1
Trichloroethylene	8.37
Chloromethane	13.1
Benzene	8.91
Carbon dioxide	19.6 (at 25°C)
Oxygen	16.1 (at 25°C)
Hydrogen sulfide	20.3

Table 6.9.13 *Sherwood-Holloway Constants for Packing Materials as Based on the Desorption of Hydrogen, Oxygen, and Carbon Dioxide*

Packing	Size (mm)	m	n
Raschig rings	50	80	0.22
Raschig rings	38	90	0.22
Raschig rings	25	100	0.22
Berl saddles	28	160	0.28
Berl saddles	25	170	0.28

In the stripping tower design, determination of K_La is the most important. This value is particular to the type of packing used in the tower, the type of contaminant, and a number of other parameters. The Sherwood-Holloway equation may be used to determine the K_La for a given packing, liquid, and gas. This is

$$K_La = D_LM \left[\frac{L}{U_L} \right]^{1-n} \frac{u_L^{0.5}}{r_LD_L} \qquad [6.9.33]$$

where

$K_La = \text{hr}^{-1}$
$D_L = $ diffusion coefficient for gas of interest in water in ft.^{2}/hr
$L = $ liquid velocity in lb water/hr-ft.^{2}
$u_L = $ liquid viscosity in lb/hr-ft.
$r_L = $ liquid density in lb/ft.^{3}

Tables 6.9.12 and 6.9.13 [16] give values for the diffusion and the constants m and n, respectively.

6.9.12 Produced Water
Produced water varies widely in composition with dissolved ions being the major component in these water. The TDS content may be close to saturation at more than 300,000 mg/L TDS or be considered fresh at less than 500 mg/L TDS. The composition of the water depends on a number of factors. Some waters are the remains of the ancient sea held in formations thousands of feet below the surface. Some water may have been fresh originally but through geologic time has become saturated with the minerals contained in the surroundings rock. Still other water has changed in composition through migration through pore space and fractures, solubilizing minerals in the pathway. In a study characterizing production of produced waters from natural gas wells, it was found that less than 60% had any concentration of the heavier volatile constituents, but instead the bulk was composed of lighter hydrocarbons [28]. Table 6.9.14 shows the average water analysis for 19 natural gas wells. The components shown were found in 80% of the surveyed wells

By knowing the composition of the reservoir rock and assuming freshwater was in place at the time of deposition,

Table 6.9.14 *Produced Water from Natural Gas Sites*

Compound	Average Value
Total dissolved solids (TDS)	93000 mg/L
Total suspended solids (TSS)	132 mg/L
BOD$_5$	1486 mg/L
Chloride	55000 mg/L
COD	11200 mg/L
Ammonia as N	72.3 mg/L
Oil and grease (62.5%)	15.6 mg/L
PH	5.83 mg/L
TOC	2280 mg/L
Barium	87.0 mg/L
Calcium	6424 mg/L
Iron	100 mg/L
Magnesium	683 mg/L
Manganese	7.47 mg/L
Sodium	18800 mg/L
Lithium	91.3 mg/L
Potassium	539 mg/L
Silicon	9.86 mg/L
Strontium	10 mg/L
Sulfur	25.9 mg/L
Phenols (62.5%)	330 ppb
Naphthalene (62.5%)	39.8 ppb
Benzene	5980 ppb
Toluene	6440 ppb
Total xylenes (75%)	3420 ppb

water's TDS may be forecast with some accuracy for equilibrium equations. Produced water of course also carries a certain amount of hydrocarbons. These concentrations may also be calculated from equilibrium equations. For water in contact with the gas phase, Henry's law is used to predict the concentration of certain hydrocarbons. These gas-phase constituents, however, maintain equilibrium with the atmosphere when exposed. Thus, the organics may be stripped from the water by means of a packed tower, diffused aeration, or other means.

6.9.13 Solubility
Because of the dissolved mineral content in the water and the residual hydrocarbons, produced water must be treated for use or disposed. Safe disposal of this water is secured through injection into a secure geologic formation, evaporation in lined pits, or other approved means. The use of lined pits as the sole means for disposal has been eliminated in all but a few cases in which the evaporation rates are competitive with the well's production. Unlined pits are not considered a sound environmental approach in today's political climate. In the past, the percolation is such pits has allowed for a cost-effective means of disposal. In some cases, however, the practice has led to local groundwater contamination [29].

Guidelines have been made for the pit construction, including the following [30]:

- As much as practical, the pit shall be located on level ground and away from established drainage patterns, including intermittent/ephemeral drainage ways, and unstable ground or depressions in the area.
- The pit shall have adequate storage capacity for safe containment of all produced water, even in those periods when evaporation rates are at a minimum. The design shall provide for a minimum of 2 ft. of free-board.
- The pit shall be fenced or enclosed to prevent access by livestock, wildlife, and unauthorized personnel. If necessary, the pit shall be equipped to deter entry by birds. Fences shall not be constructed on the levees.

- The pit levees to be constructed so that the inside grade of the levee is no steeper than 1 (vertical): 2 (horizontal), and the outside grade no steeper than 1:3.
- The top of levees shall be level and least 18 in. wide.
- The pit location shall be reclaimed pursuant to the requirements and standards of the surface management agency. On a split estate (private surface, federal mineral) a surface owner's release statement or form is acceptable.

Lined pits shall be designed to meet following requirement and minimum standards in addition to those specified above:

- The material used in lining pits shall impervious. It shall be resistant to weather, sunlight, hydrocarbons, aqueous acids, alkalies, salt, fungi, or other substances likely to be contained in the produced water.
- If rigid materials are used, leak-proof expansion joints shall be provided, or the material shall be of sufficient thickness and length to withstand expansion without cracking, contraction, and settling movements in the underlying earth. Semirigid liners such as compacted bentonite or clay may be used provided that, considering the thickness of the lining material chosen and its degree pf permeability, the liner is impervious for the excepted period of use.
- If flexible membrane materials are used, they shall have adequate resistance to tears or punctures.
- Lined pits shall have an underlying gravel-filled sump and lateral system or other suitable devices for the detection of leaks.
- Method and schedules for removal of residual solids and saturated brine shall be established.

Pits used as a means of storage for treatment may be preferred over manufactured tanks due to volume considerations. Evaporation may be used in these wells and may in some instances be considerable in arid climates as calculated through the Meyers equation.

In the case of injection, steel tank storage is preferred to eliminate any contamination of the water supply by dust and other partials. In the case of class II injection, the water is often pumped into subsurface formations at less than the fracture pressure. Thus, the water must be free of particulate matter to avoid pore plugging. Pretreatment with biocides is for this same reason. The construction of a properly designed class II injection well often costs in excess of a million dollars. Thus, considerable planning is necessary to prevent early decline of injectivity.

In planning a class II well, the geology of the area must be considered as well as the number of viable injection zones presented. These zones must be separated from each other; at least one zone must be permeable and the other must impermeable from the underground source of drinking water (USDW) and be free of commercial quantities of oil or gas. They are must also be proven free of faults, cross bedding, and tonguing, and other permeability hindrances should be considered. Most states allow a radial extent of $\frac{1}{4}$ to $\frac{1}{2}$ mi. The are then available for storage is

$$V = C_t(V_f - V_i) D_{avg}Ah\phi \qquad [6.9.34]$$

where

V_f = final pressure gradient in less than frac pressure−psi/ft.
V_i = initial pressure gradient in psi/ft.
A = aerial extent in ft.
D_{avg} = average depth of injection zone
C_t = total compressibility in psi^{-1}
h = thickness of zone in ft.
ϕ = porosity

6.9.14 NORMS in Produced Water/Gas

Radioactive contamination of production equipment is an emerging concern in the petroleum industry. Over a 20-year period, the EPA estimates that the combined production of natural gas and crude oil in the United States resulted in accumulation of 13 million metric tons of Naturally Occurring Radioactive Material (NORM), as opposed to more than 21 billion metric tons associated with metal, uranium, and phosphate mining and processing [31]. NORM can accumulate as scale or sludge in well casing, production tubing, surface equipment, gas gathering pipelines, and by-product waste streams. The sources of most of the radioactivity are isotopes of uranium-238 and thorium-226, which are naturally present in the subsurface formations from which natural gas is produced. The primary radionuclides of concern are radium-226 in the uranium-238 decay series and radium-228 in the thorium-232 decay series. Other radionuclides of concern include those resulting from the decay of radon-226 and radon-228, such as radon-222. Pipe scale and sludge accumulations are dominated by radium-226 and radium-228, whereas deposits on the interior surfaces of gas plant equipment are predominantly lead-210 and polonium-210 [32]. Radiation is normally reported in curies (Ci), rads, rems, or bequerels (Bq). The commonest way of reporting is in picocuries pCi or 10^{-12} curies. A picocurie means the quantity of radioactive material producing 2.2 nuclear transforms per minute[33]. One curie is equal to $3.7 \times 1,010$ nuclear transformations per second or 1 radium. In contract, 1 uranium had 0.36×10^{-6} Ci of activity. A rad qualifies an absorbed dose, and a rem is a dose effect (0.1 mrem/yr = 10^{-6} excess lifetime cancer risk).

$$Xrems = Q (Yrads) \qquad [6.9.35]$$

The Q here is a quality factor based on the source of activity. Alpha particles inflect much more damage to human tissue than do beta emissions. The generation of these two particles caused the decay of the parent isotope and leads to production of the progeny, which are sometimes referred to as *daughters*. Isotopes with larger half-lives usually present lower radioactive risks. Gamma ray emission does not produce progenies Elements that have very short half-lives are not significant in that they change rapidly in transport. Radium-228 emits beta particles, where as uranium, radon, and radium-226 are alpha emitters. The average annual dose received by humans in the United States is 200 mrem, of which only 4 mrem are attributed to drinking water. Some public water supplies have show radioactivity in excess of 10,000 pCi/L, equivalent to a dose of 100 mrem/yr [33].

Contamination may be caused by natural sources, and as such termed NORM. Radioactivity is a concern due to the widespread occurrence of radium, uranium, and radon and inherent carcinogenicty attributed to them. National Primary Drinking Water Regulations (NPDWRs or primary standards) under the 1996 amendments to the SDWA are legally enforceable standards that apply to public water systems. Primary standards protect public health by limiting the levels of contaminants in drinking water (Table 6.9.15 [34]). In Canada, the maximum acceptable concentrations of radon-226 is 27 pCi/L (1 Bq/L) and for strontium-90 270 pCi/L (10 Bq/L). The World Health Organization sets a guideline value of 2.7 pCi/L (0.1 Bq/L) for the gross alpha activity and 27 pCi/L (1 Bq/L) for the gross beta activity.

Radon gas is common to natural gas production. It has long-lived decay products that contribute to the contamination of production equipment. Radon-222 is a highly mobile inert gas produced from radium-226. Other radon isotopes exist(thoron) but are rare and have extremely short half-lives, which cause decay even before the gas is produced at the wellhead. Because of radon's characteristics,

Table 6.9.15 *National Primary Drinking Water Regulations for Radionuclides*

Radionuclides	MCLG (mg/L)	MCL or TT (mg/L)	Sources of Contaminant in Drinking Water
Alpha particles	none* ——— zero	15 (pCi/L)	Erosion of natural deposits of certain minerals that are radioactive and may emit a form of radiation known as alpha radiation
Beta particles and photon emitters	none ——— zero	4 millirems per year	Decay of natural and manmade deposits of certain minerals that are radioactive and may emit forms of radiation known as photons and beta radiation
Radium-226 and radium-228 (combined)	none* ——— zero	5 pCi/L	Erosion of natural deposits
Uranium	zero	30 µg/L as of 12/08/03	Erosion of natural deposits

Maximum Contaminant Level (MCL) is the highest level of a contaminant that is allowed in drinking water. MCLs are set as close to MCLGs as feasible using the best available treatment technology and taking cost into consideration. MCLs are enforceable standards. **Maximum Contaminant Level Goal (MCLG)** is the level of a contaminant in drinking water below which there is no known or expected risk to health. MCLGs allow for a margin of safety and are non-enforceable public health goals.
Units are in milligrams per liter (mg/L) unless otherwise noted. Milligrams per liter are equivalent to parts per million.
*MCLGs were not established before the 1986 Amendments to the SDWA. Therefore, there is no MCLG for this contaminant.

Table 6.9.16 *Radon Concentrations Found at the Wellhead around the World.*

Location of Well	Radon Concentrations (pCi/L)
Borneo	1–3
Canada	
Alberta	10–205
British Columbia	390–540
Ontario	4–800
Germany	1–10
The Netherlands	1–45
Nigeria	1–3
North Sea	2–4
United States	
Colorado, New Mexico	1–160
Texas, Oklahoma, Kansas	1–1450
California	1–100

it freely travels in groundwater, oil, and natural gas without chemical reaction. Once produced at the wellhead, the radon-222 is compressed with the natural gas. In the refinery, radon may become concentrated in the production of propylene or ethane, with a boiling point being intermediate to these two. On the wellsite, concentrated radon gas is not expected. Table 6.9.16 [36] gives radon concentrations found worldwide at the wellhead.

Radon-222 has a short half-life at 3.8 days. Being inert, it does not accumulate in the body, 99% of the radon-222 decays into lead-210 within 25 days. Figure 6.9.5 [37] shows the complete decay series of uranium-238. In the wellsite facilities, the daughter products of radon accumulate inside the containing vessels. Because of the low concentrations, it may be difficult to measure the radioactivity from the outside of the vessels. Usually, laboratory analysis or specialized probes are needed to detect these quantities. An alpha/beta probe may register positive if held to the internal surface. Radon gas does have sufficient gamma energies to register on a gamma survey meter. These radioactive materials are not considered hazardous unless inhaled or ingested.

Radon-228 comes from the decay of thorium-232. Radon-228 has a beta intensity of 4.71MeV, and radon226 emits an alpha intensity of 0.024 Mev. Radon-226 has an initial registered gamma energy of 185.7 keB compared with 10.3 keV for radon-228. Table 6.9.17 shows Wyoming's produced water characteristics [38].

Tracer surveys are an important technology used in today's petroleum industry. The injection of low-level radioactive elements downhole allows us to follow fluid migration downhole through gamma ray logs. The federal government allows their use in confirmation for mechanical integrity in class II injection wells (40 CFR 146.8) [39]. The low-level radioisotopes are prepared in the laboratory and transported in approved containers to the site. Several different isotopes may be injected into any one well, depending on the number of fluids that needed to be traced. The isotopes of strontium, antimony, gold, and others are frequently used. Each isotope registers a different signature on the gamma ray log.

Water or gas contaminated by radioactive tracers is to be expected. In the case of an injection well, the contamination should be documented with no further considerations. In the case of a producing gas well, precautions must be taken to protect workers from inhalation of vented gas that may contain radioactive vapor. The completion or workover fluids retrieved from the tracer operation are nonexempt subtitle "C" hazardous wastes. As such, they must be properly contained, labeled, and transported to a specially permitted TSD facility authorized to dispose of low-level radioactive waste. Low amounts of waste may be stored onsite for extended periods of time.

Cation exchange is a very effective means of radium removal because it is preferred above all other common cations found in water. An even more efficient exchange mechanism is BaSO4 located in alumina or SAC resin. A resin impregnated with barium sulfate is used to decontaminate the spent regenerant solution from a conventional cation exchange process of radium. The process was tested on a small municipal system, the test was very successful and was still removing radium when the resin was

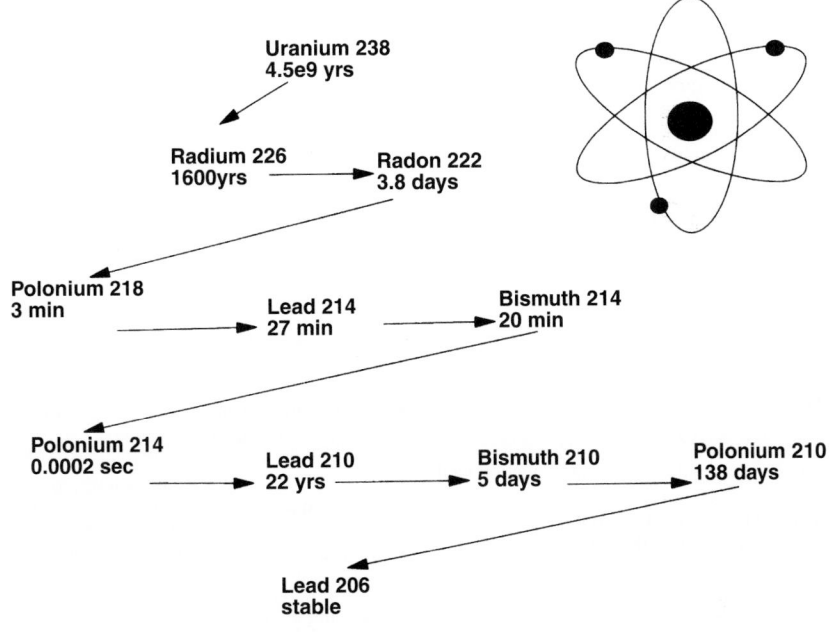

Figure 6.9.5 Uranium-235 decay series.

Table 6.9.17 *Concentrations of Radon-226 in Wyoming Produced Waters*

Parameter	Value
No. of wells tested	373
Max. recorded value of radon-226	2152 pCi/L
Max. recorded value of radon-226	0 pCi/L
Average value of radon-226	21.5 pCi/L
Median value of radon-226	3.7 pCi/L
No. of wells above 5 pCi/L	167

loaded at 2.7 million pCi/L. This level, however, posted a far greater disposal problem than the initial brine solution [40].

NORMs frequently enter the production system in the form of scale. This scale may be found in tubing, wellhead, and any surface equipment. The concentrations have been found up to 30,000 pCi/gm. In the production scale and 250 pCi. Gm in associated sediments such as BS&W. Soils surrounding locations may show radium levels up to 2000 pCi/gm [41].

The radioactive scale is often found in association with precipitates of barium, strontium, and calcium. Equilibrium calculations favor the formation of radium sulfate in excess of barium sulfate. The prevention of NORM contamination then may be prevented through scale inhibition. An equal mixture at 1 mg/L of animotrimethylene phosphonic acid and phosphinopoly-arboxylate has been found effective in eliminating radioactive scale.

The greatest health risk attributed to NORM-contained production equipment comes from production equipment comes from the cleaning and repair. The radioactive materials are not health hazards unless ingested or inhaled into the body. In the event of inhalation, dustborn particles containing NORM or a gas such as radon attach to the lung tissue where the emit alpha particles into the lung tissue, which may lead to cancer.

6.9.15 Class II Injection Wells

Class II injection wells afford an economic means of disposal of produced brine water. They are regulated through the UIC Program under the provisions of the SDWA. State and local governments may be granted primacy over UIC programs by the federal government. The underlying rules for the UIC program are found in 40 CFR 144 and 146 [42, 43].

6.9.16 UIC Criteria and Standards

USDW includes an aquifer and the portion that supplies a public water system, contains a sufficient amount of water to supply a public water system, contains less than 10,000 ppm TDS, and is not an exempted aquifer (40 CFR 146.3). An exempted aquifer or portion of such must meet the following criteria: does not currently serve as a source of drinking water, and cannot or will not serve as source of drinking water due to

1. commercial quantities of minerals, hydrocarbons, or geothermal energy
2. situated at a depth or allocation that makes recovery of water technically or economically impractical
3. so contaminated that recovery of water is technically or economically impractical for drinking water purposes
4. contains more than 3,000 ppm TDS but less than 10,000 ppm TDS and is not reasonably expected to supply drinking water
5. is located over a class II well mining area subject to collapse or subsidence (40 CFR 146.4)

A class II injection well inject fluids

1. that are brought to the surface in connection to conventional oil and natural gas production and may be commingled with wastewaters from integral gas plants,
2. for enhanced recovery of oil or natural gas
3. for storage of hydrocarbons that are liquid at standard temperature and pressure (40 CFR 146.5)

Class II injection wells are to be located in a formation separated from any USDW by a confining zone and be free of known open faults within the area of review. The area of review may be computed from a modified Theis equation.

$$r = \left[\frac{(2.25KH+)}{S10^x} \right]^{0.5} \quad [6.9.36]$$

$$x = \frac{4\pi KH \left(h_w - h_{bo} S_p G_b\right)}{23Q} \quad [6.9.37]$$

where

r = radius of endangering influence from injection well (length)
K = hydraulic conductivity (length/time)
H = thickness of injection zone (length)
t = time of injection
S = storage coefficient ft.
h_{bo} = observed original hydrostatic heat of injection zone as measured from the base of the lowermost SDW (length)
h_w = hydrostatic head height of USDW zone as measured from the base of the lowermost SDW (length)
$S_p G_b$ = specific gravity of the fluid in the injection zone

These equations assume homogeneous, unconfined, isotropic, and total penetration of the aquifer with $r_w \ll r_e$ (40 CFR 144.6).

Some states question the use of Equations 6.9.36 and 6.9.37 in determining the range of influence of an injection well and require a set radius. New Mexico and Texas require a set radius of influence of $\frac{1}{2}$ mi.

The requirements do not apply to those wells that were previously drilled under a state or federal regulatory system in compliance with these controls, provided well injection will not contaminate a USDW. Remedial work may be approved to bring newly drilled wells (dryholes) into compliance. Currently, amendments to 40 CFR 144 and 146 are proposed for modifying well construction and extending the scope of wells, necessitating documentation for "area of review" [44]. Under the new rulings, with some allowances, all new class II wells will be required to:

- Run surface casing through all USDWs containing ≤3,000 mg/1 TDS and cement same to surface
- Cement the long string casing to prevent the migration of fluid from the injection zone
- Have the tubing string landed with a packer.

6.9.17 UIC Permitting Process

To be granted a permit for a class II injection well, the operator must file an application to the ruling UIC program at the state, local, or EPA office. On federal leases, the BLM must also grant approval. The permit will address the following as stipulated in 40 CFR 146 [43].

1. depth of the well
2. depth to base of USDW
3. estimated maximum and average injection pressures
4. nature of the injection fluid
5. lithology of injection and confirming zones
6. external pressure, confining pressure, and axial loading
7. hole size
8. size and grade of all casing strings
9. class of cement

The permit is then reviewed and, once accepted, is let to the public for comment. After this comment period, the final permit decision is issued. Once completed the, well then may be used to dispose of:

- produced water
- waste fluids from drilling operations
- pigging fluids from within the field.
- used workover and stimulation fluids from injection, exploration, and recovery wells
- any gas used for enhanced recovery or pressure maintenance
- brine reject associated with enhanced recovery
- waste fluids from sweetening and dehydration of methane as long as the fluid is not hazardous at the point of injection
- waste oil and fluids from cleanup operations associated with primary production within the field
- waste fluids from cementing operations
- fresh water used as enhanced recovery base
- water containing chemicals used for enhanced recovery.

Monitoring of fluid composition, injection pressure, annulus pressure, flowrate, and volume is required. Reporting of regular activities is required annually, and a mechanical integrity test must be performed every 5 years, or as required by permitting authority, and each time tubing is moved. The mechanical integrity test requires that a static pressure test must be made on the well annulus and the pressure held steady for a minimum of 30 min.

References

1. EPA Exemption of Oil and Gas Exploration and Production Wastes from Federal Hazardous Waste Regulations, EPAA530-K-01-004, Washington, DC, January 2002, www.epa.gov/epaoswer/other/oil/oil-gas.pdf.
2. Federal Register, Vol. 67, no. 137, pp. 47042–47152, Washington, DC, July 17, 2002, http://www.epa.gov/oilspill/pdfs/40cfr112.
3. Federal Register, Vol. 60, no. 189, pp. 50913–51170, Washington, DC, September 29, 1995, www.dep.state.fl.us/water/stormwater/docs/msgp/ifp.pdf.
4. Federal Register, Vol. 65, no. 210, pp. 64746–64880, Washington, DC, October 30, 2000, http://www.epa.gov/earth1r6/6wq/npdes/sw/industry/msgp2000.pdf.
5. API E5 Environmental Guidance Document: Waste Management in Exploration and Production Operations, 2nd Edition, API, Washington, DC, February 1997.
6. Railroad Commission of Texas. Field Guide for the Assessment and Cleanup of Soil and Groundwater contaminated with Condensate From a Spill Incident (Statewide Rules 8, 20, and 91), Austin, TX, http://www.rrc.state.tx.us/divisions/og/key-programs/spillcleanup.html.
7. Kennedy, A. J., Esso Resources Canada LTD., "Oil Waste Road Application Practices at the Esso Resources LTD Cold Lake Project," Proceeding from the 1st International Symposium on Oil and Gas Exploration Waste Management Practices, pp. 689–701, 1990.
8. Deuel, L. E., "Evaluation of limiting constituents suggested for land disposal of exploration and production wastes," Proceeding from the 1st International Symposium on Oil and Gas Exploration Waste Management Practices, pp. 411–430, 1990.

9. Coile, G. M., *Assessment and remediation of Petroleum Contaminated Sites*, CRC Press, Boca Raton, Florida, 1994.
10. American Petroleum Institute, "The migration of petroleum in soil and groundwater, principles and countermeasures," no. 4149, API, Washington, DC, 1972.
11. CONCAWE, "Protection of Groundwater from Oil Pollution," NTIS PB82-174608, The Hauge, 1979.
12. CONCAWE Secretariate, "Inland Oil Spill Cleanup Manual," Report No. 4/74, The Hauge, 1974.
13. Riddick, J. A., Bunger, W. B., and Sakano, T. K. *Organic Solvents Physical Properties and Methods of Purification*, 4th Edition, John Wiley & Sons, New York, 1986.
14. Perry, R., and Green, D., *Perry's Chemical Engineering Handbook*, McGraw-Hill Inc., New York, 1984.
15. Nirmalakhandan, N., *Fate and Transport of Environmental Contaminants*, New Mexico State University Printing Press, Las Cruces, New Mexico, 1994.
16. American Water Works Association, *Water Quality and Treatment*, 4th Edition, McGraw-Hill, Inc., New York, 1990.
17. EPA "A Compendium of Technologies Used in the Treatment of Hazardous Wastes." Office of Research and Development, EPA/625/8-87/014, Washington, DC, 1987.
18. Levin, M. A., and Gealt, M. A., *Biotreatment of Industrial and Hazardous Waste*, McGraw-Hill Inc., New York, 1993.
19. Wolfram, J. H., et al. "Method Development Using Field Samples for Assessing Bioremediation Potential," SPE 25994, SPE/EPA Exploration and Production Environmental Conference, San Antonio, Texas, March 7–10, 1993.
20. McMillen, S. J., Kerr, J. M., and Gray, N. R., "Micro Studies of Factors that Influence Bioremediation of Crude Oils in Soils," SPE 25981, SPE/EPA Exploration and Production Environmental Conference, San Antonio, Texas, March 7–10, 1993.
21. Ratliff, M., "Construction and Operation of a Biological Treatment Cell for the Treatment of Hydrocarbon-Contaminated Soil in Alaska," SPE 25998, SPE/EPA Exploration and Production Environmental Conference, San Antonio, Texas, March 7–10, 1993.
22. Crews, B., and Malachosky, E., "The Effect of pH on Microbial Degradation of Oil Based Drilling Mud in a Slurry Phase Reactor," SPE 25992, SPE/EPA Exploration and Production Environmental Conference, San Antonio, Texas, March 7–10, 1993.
23. Whitfield, D. L., "Soil Farming of Oil Mud Drill Cuttings," SPE/IADC Conference, New Orleans, Louisiana, pp. 429–438, 1987.
24. Streebe, L. E., "Landtreatment of Petroleum Refinery Sludges" EPA Doc no. 600/2-84-193, 1985.
25. Ashworth, J., Scroggins, R. P., and McCoy. D., "Feasibility of Land Application as a Waste Management Practice for the Disposal of Residual Diesel Invert Base Muds and Cuttings in the Foothills of Alberta," Proceedings from the International Conference on Drilling Wastes, Calgary, 1988.
26. Shirazi, G. L., "Landfarming of Drilling Muds in Conjunction with the Pit Site Reclamation: A Case History," Proceedings from the 1st International Symposium on Oil and Gas Exploration Waste Management Practices, pp. 553–564, 1990.
27. Deuel, L. E., "Evaluation of limiting constituents suggested for land disposal of exploration and production wastes," Proceedings from the 1st International Symposium on Oil and Gas Exploration Waste Management Practices, pp. 411–430, 1990.
28. Wesolwski, A., et al. "Characterization of Produced Waters for Natural Gas Production Operations," Topical Report for the Gas Research Institute, Chicago, Illinois, December 1987.
29. Bureau of Land Management, "Unlined Surface Impoundment Remediation and Closure for Approximately 47,175-62,900 Unlined Surface Impoundments" (under the jurisdiction of the Farmington and Albuquerque Districts), Environmental Assessment NM-070-3004, December 1993.
30. Federal Register, Vol. 58, no. 172, Bureau of Land Management, Onshore Order no. 7, 43 CFR 3160. Anchorage, Alaska, September 8, 1993, http://www.anchorage.ak.blm.gov/oilngas/ord7.html#design.
31. EPA, "Radioactive Waste Disposal: Environmental Perspective" EPA 402-K-94-001 Washington, DC, August 1994.
32. EPA–Energy Information Administration, Natural Gas 1998: Issues and Trends, Chapter 2: Natural Gas and the Environment, http://www.eia.doe.gov/pub/oil_gas/natural_gas/analysis_publications/natural_gas_1998_issues_trends/pdf/chapter2.pdf.
33. EPA–40CFR Ch. 1 (7-1-00) Part 141: National Primary Drinking Water Regulations, http://www.epa.gov/safewater/regs/cfr141.pdf.
34. USGS, "Naturally Occurring Radionuclides in the Ground Water of Southeastern Pennsylvania," September 19, 2000, http://pa.water.usgs.gov/reports/fs012-00.html.
35. EPA–Ground Water and Drinking Water, Current Drinking Water Standards EPA816-F-02-013, July 2002, http://www.epa.gov/safewater/mcl.html#rads or http://www.epa.gov/safewater/mcl.html.
36. Gray, P. R., "Norm Contamination in the Petroleum Industry," *Journal of Petroleum Technology*, January 1993.
37. *Handbook of Chemistry and Physics*, CRC Press, Cleveland, Ohio, 1977.
38. Wagner, J. F., "Toxicity and Radium 226 in Produced Water: Wyoming's Regulatory Approach," Proceedings from the 1st International Symposium on Oil and Gas Exploration Waste Management Practices, pp. 987–994, 1990.
39. Osborne, P., "Program Overview: Underground Injection Control, Region VIII," 2nd Edition, EPA Region VIII, Denver, Colorado, 1991
40. Mangelson, K. "Radium Removal for Small Community Water Supply Systems," EPA/600/52-88/039, USEPA.
41. Miller, H. T., and Bruce, E.D., "Pathway Exposure Analysis and the Identification of Waste Disposal Options for the Petroleum Production Wastes Containing Naturally Occurring Radioactive Materials," Proceedings from the 1st International Symposium on Oil and Gas Exploration Waste Management Practices, pp. 731–744, 1990.
42. CFR Title 40 Protection of Environment Part 144—Underground Injection Control Program, http://www.access.gpo.gov/nara/cfr/cfrhtml_00/Title_40/40cfr144_00.html.
43. CFR Title 40 Protection of Environment Part 146—Underground Injection Control Program: Criteria and Standards, http://www.access.gpo.gov/nara/cfr/waisidx_99/40cfr146_99.html.
44. Smith, J. B., and Browning, L. A.,"Proposed Changes to EPA Class II Well Construction Standards and Area of Review Procedures," SPE 25961, SPE/EPA Exploration and Production Environmental Conference, San Antonio, Texas, March 7–10, 1993.

6.10 OFFSHORE OPERATIONS

Drilling and production technology in the oil and gas industry has advanced from producing oil and gas onshore to producing in distant offshore locations in water depths up to 6,000 ft.

Some 50 years ago, locations in the marshlands were built up above the water level with oyster shells. Lumber measuring 3 in. × 8 ft. and random lengths were laid as matting over the shell fill. A land rig was then moved in by tug and barge in a freshly dug canal to the location where the derrick and drilling equipment was unloaded and rigged up. Drilling and completion was accomplished as on land locations.

Later a barge rig was used. This was a derrick with drilling machinery mounted on a steel barge. This barge was sunk on location in water depth limited by the freeboard of the barge before sinking.

Because the barge rig was limited to very shallow depths, a posted barge was developed with the drilling deck separated from the barge by I-beams supporting the drilling deck. With this unit, the barge could be sunk completely below the surface of the water as long as the drilling deck was above the water. The barge rig and posted barge rig were used mostly in marshlands and shallow lakes. They are still used in such areas today.

Offshore drilling started in shallow water near the shore line on concrete structures. Concrete prestressed pilings were driven in the ocean floor, and a concrete platform was installed on the piling to support the land-type rig. After the well was completed, the rig was moved off the structure and production equipment moved in.

Steel structures were later fabricated onshore and moved to offshore locations by tug and crane barge. At location, the crane was used to lift the structure, swing it to one side, and lower it onto the ocean floor. Large-diameter pipe was hammered down through the larger steel-casing legs in order to anchor the structure to the ocean floor. After the platform was installed on the structure, the drilling rig was unloaded onto the platform and rigged up for drilling. After the well was drilled, cased, and completed, including the Christmas tree, the rig would move to another location that was ready to be drilled. Once the drilling rig left the completed well, flow lines were laid and all necessary equipment installed to transport the production either to a purchaser's line, to an onshore separation facility, or to a centrally located production platform before being transported to land.

Then came the mobile drilling rigs. These were the submersibles, Jack-ups, semisubmersibles, and floating drillships. Figure 6.10.1 shows the maximum water depth drilled in the Gulf of Mexico by year. Drilling has occurred in water depths exceeding 9,000 ft.

The major deepwater offshore provinces are the Gulf of Mexico, offshore West Africa, and Brazil. Other areas of offshore production include North Sea, Newfoundland, Southeast Asia, Caspian Sea, Mediterranean Sea, Lake Maracaibo (Venezuela), offshore Australia, and offshore India.

Production platforms in 600 to 700 ft. of water are anchored to the sea floor. The structures are massive to withstand all external forces, including currents, wind, tides, earthquakes, and floating material such as ice. The taller the structure, the greater the moment around the base. This greater moment is caused by the greater force (same force per unit length but acting over a longer distance), and this force is concentrated at a greater distance from the ocean floor. Smaller (marginal) field discoveries in shallow water demand less costly minimum-offshore fixed platform and decks [2].

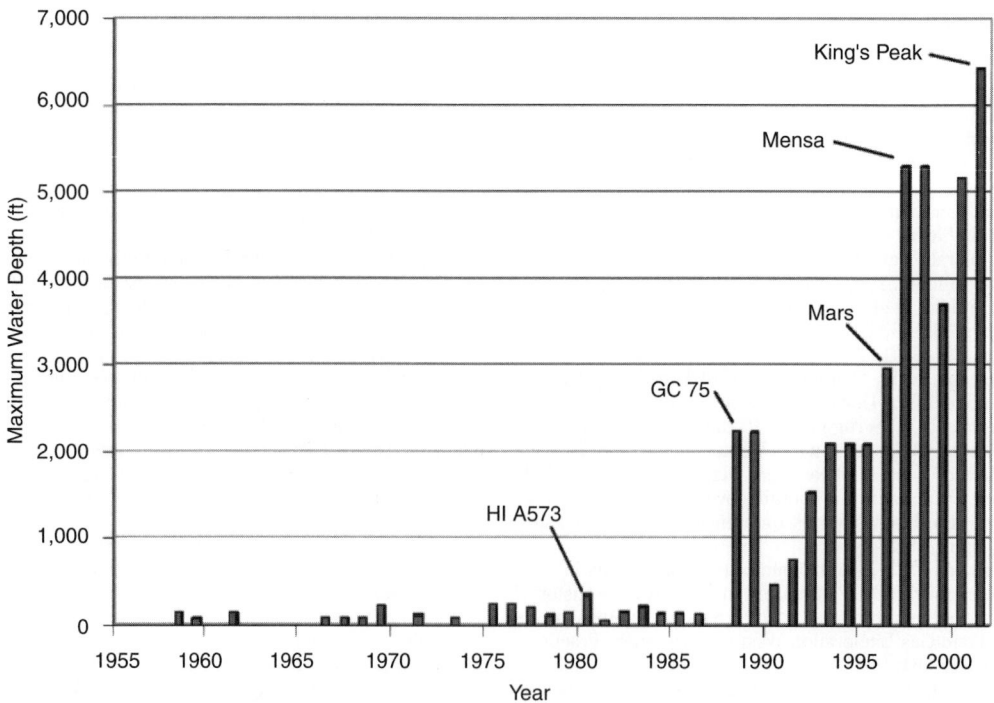

Figure 6.10.1   *Maximum water depth drilled in Gulf of Mexico [1].*

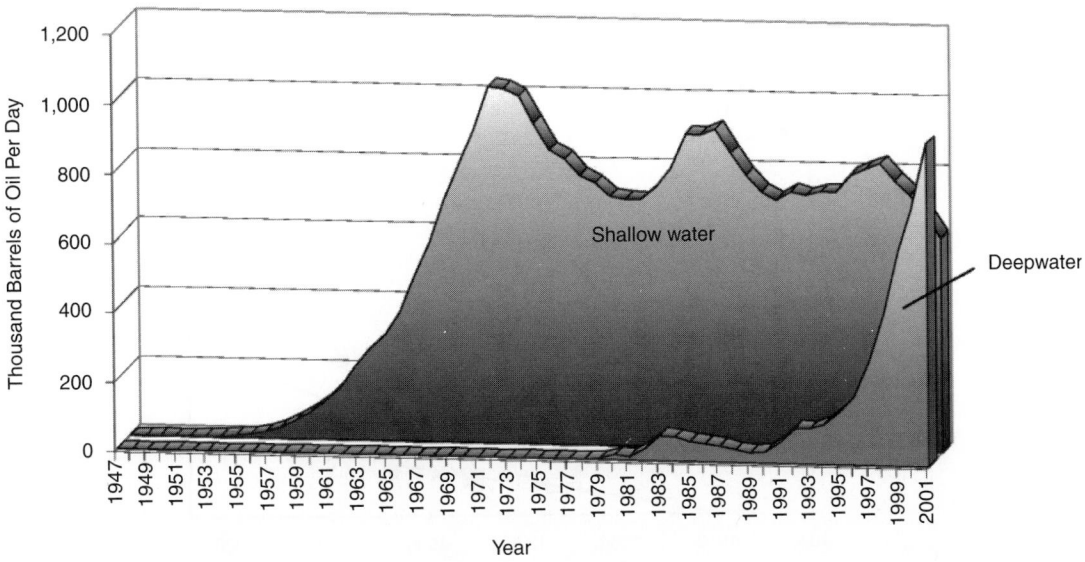

Figure 6.10.2 *Comparison of average annual shallow and deepwater oil production in the Gulf of Mexico [1].*

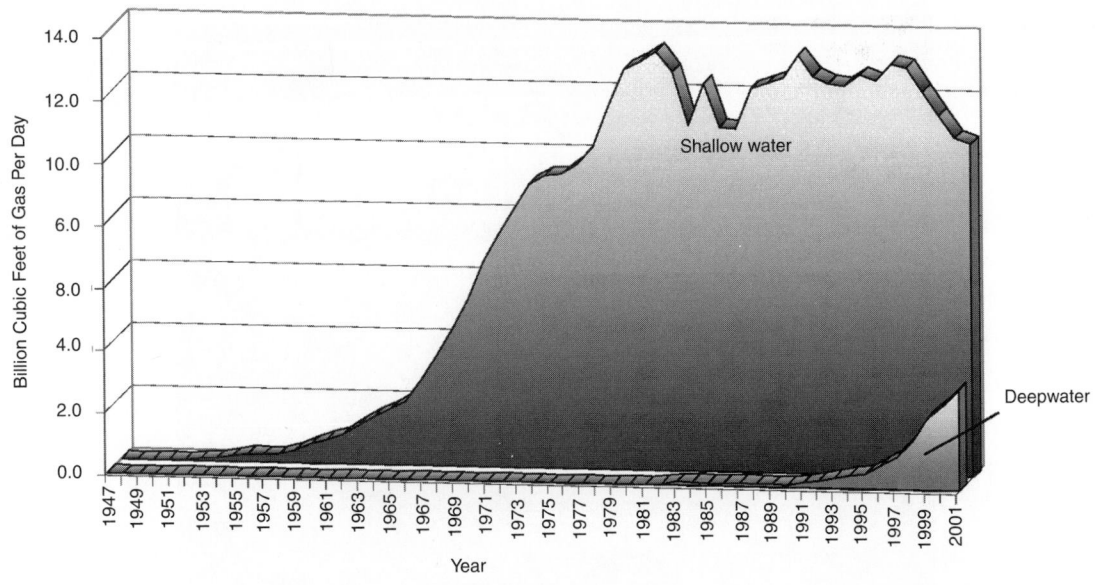

Figure 6.10.3 *Comparison of average annual shallow and deepwater gas production in the Gulf of Mexico [1].*

The U.S. Minerals Management Service defines shallow water as less than 1,000 ft., deepwater as 1,000 to 5,000 ft., and ultra-deepwater as more than 5,000 ft. Figures 6.10.2 and 6.10.3 show that more and more U.S. production is coming from the Gulf of Mexico deepwater.

Figure 6.10.4 shows the increasing water depth records over time for each of the dry tree concepts used to date.

The escalating weight and associated cost of ocean floor–anchored structures has led to the development of other types of platforms for deepwater applications. Systems in use today fall under three main categories: fixed platform, floating production system (FPS), and subsea system [1, 4].

Figure 6.10.5 shows bottom supported and vertically moored and FPS and subsea systems.

Table 6.10.1 identifies chronologically all systems used in productive deepwater Gulf of Mexico fields; deepwater systems worldwide are identified in *Offshore Magazine*'s Poster 36 [4].

6.10.1 Fixed Platforms

Fixed platforms can be either steel or concrete. Steel structures include traditional jackets with battered legs from top to bottom—the jacket's leg spacing increases to the

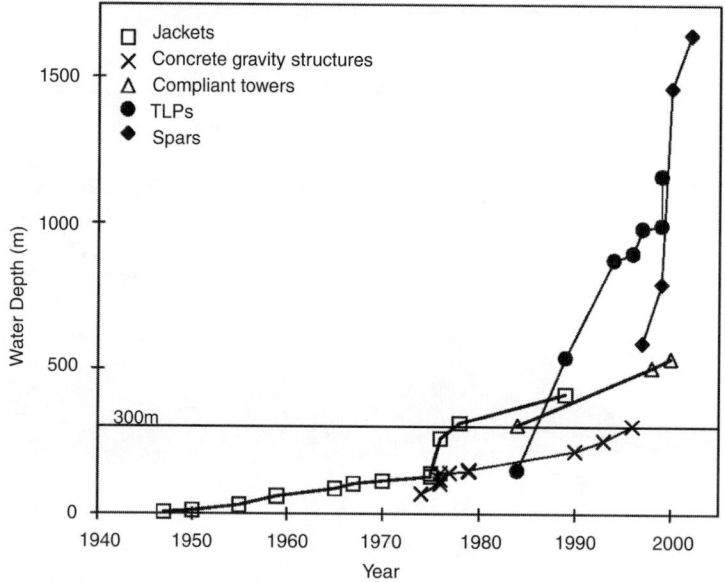

Figure 6.10.4   *Increasing water depth records by year [3].*

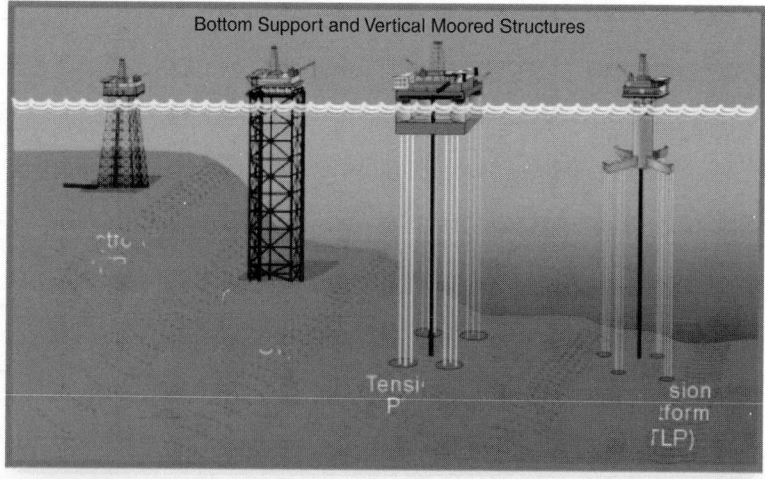

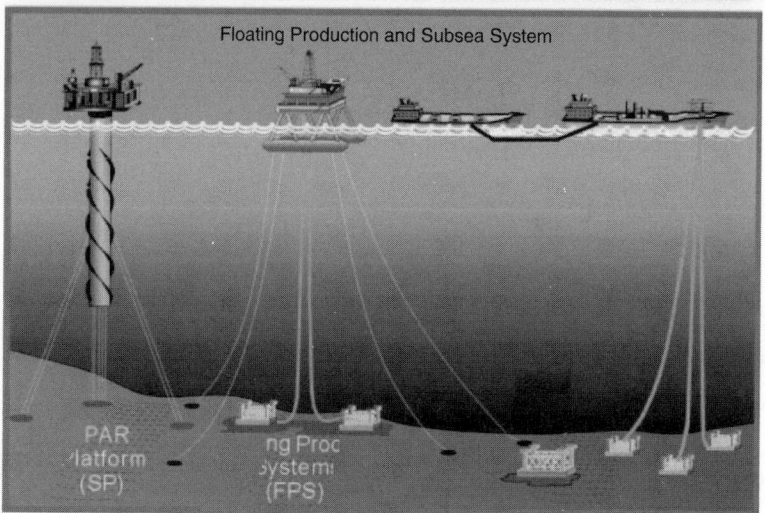

Figure 6.10.5   *Bottom supported, vertically moored FPS and subsea systems [1].*

Table 6.10.1 *Development of Productive Deepwater Gulf of Mexico Fields [1]*

Year of First Production	Field Nickname	System Type	Block	Water Depth (ft.)	Operator
1979	Cognac	Fixed Platform	MC 194	1,023	Shell
1984	Lena	Compliant Tower	MC 281	1,017	ExxonMobil
1988*	Unnamed	FPS	GC 75	2,172	Oryx
1988*	Unnamed	Semi-Submersible	GC 29	1,554	Placid
1989	Bullwinkle	Fixed Platform	GC 65	1,330	Shell
1989	Jolliet	TLP	GC 184	1,724	Conoco
1991	Amberjack	Fixed Platform	MC 109	1,050	BP
1993*	Diamond	Subsea	MC 445	2,095	Oryx
1993*	Seattle Slew	Fixed Platform/Subsea	EW 914	1,019	Tatham
1993	Zinc	Subsea	MC354	1,475	ExxonMobil
1994	Auger	TLP	GB 426	2,863	Shell
1994	Pompano/Pompano II	Fixed Platform/Subsea	VK 990	1,440	BP
1994	Tahoe/Tahoe II	Subsea	VK 783	1,391	Shell
1995*	Cooper	Semi-Submersible	GB 387	2,260	EEX
1995	Unnamed	Subsea	VK 862	1,043	Walter
1996	Mars	TLP/Subsea	MC 807	2,992	Shell
1996	Popeye	Subsea	GC 116	2,065	Shell
1996	Rocky	Subsea	GC 110	1,719	Shell
1997	Mensa	Subsea	MC 731	5,276	Shell
1997	Neptune/Thor	Spar/Subsea	VK 825	1,866	Kerr McGee
1997	Ram-Powell	TLP	VK 956	3,243	Shell
1997	Troika	Subsea	GC 244	2,679	BP
1998	Arnold	Subsea	EW 963	1,752	Marathon
1998	Baladpate	Compliant Tower	GB 260	1,604	Amerada Hess
1998	Morpeth/Klamath	TLP/Subsea	EW 921	1,747	Agip
1998	Salsa	Subsea	GB 171	1,121	Shell
1999	Allegheny	TLP/Subsea	GC 254	3,194	Agip
1999	Angus	Subsea	GC 112	1,901	Shell
1999	Diana	Subsea	EB 945	4,670	ExxonMobil
1999	Dulcimer	Subsea	GB 367	1,123	Mariner
1999	Genesis	Spar	GC 205	2,597	Chevron Texaco
1999	Gemini	Subsea	MC 292	3,488	Chevron Texaco
1999	Macaroni	Subsea	GB 602	3,691	Shell
1999	Pluto	Subsea	MC 718	2,748	Mariner
1999	Unnamed	Subsea	EW 1006	1,832	Walter
1999	Ursa	TLP	MC 810	3,877	Shell
1999	Virgo	Fixed Platform	VK 823	1,136	TotalFinaElf
2000	Europa	Subsea	MC 935	3,880	Shell
2000	Hoover	Spar	AC 25	4,806	ExxonMobil
2000	Marlin	TLP	VK 915	3,300	BP
2000	Northwestern	Subsea	GB 200	1,261	Amerada Hess
2000	Petronius	Compliant Tower	VK 786	1,753	Chevron Texaco
2001	Brutus	TLP	GC 158	2,952	Shell
2001	Einset	Subsea	VK 873	3,584	Shell
2001	Crosby	Subsea	MC 899	4,400	Shell
2001	Ladybug	Subsea	GB 409	1,357	ATP
2001	Madison	Subsea	AC 24	4,854	ExxonMobil
2001	Marshall	Subsea	EB 949	4,376	ExxonMobil
2001	Mica	Subsea	MC 211	4,337	ExxonMobil
2001	Nile	Subsea	VK 914	3,535	BP
2001	Oregano	Subsea	GB 559	3,400	Shell
2001	Prince	TLP	EW 958	1,493	Argo
2001	Serrano	Subsea	GB 516	3,359	Shell
2001	Typhoon	TLP/Subsea	GC 236	2,679	Chevron Texaco
2001	Unnamed	Subsea	EW 878	1,585	Walter
2001	Unnamed	Subsea	MC 68	1,214	Walter

*Fields that are no longer on production.

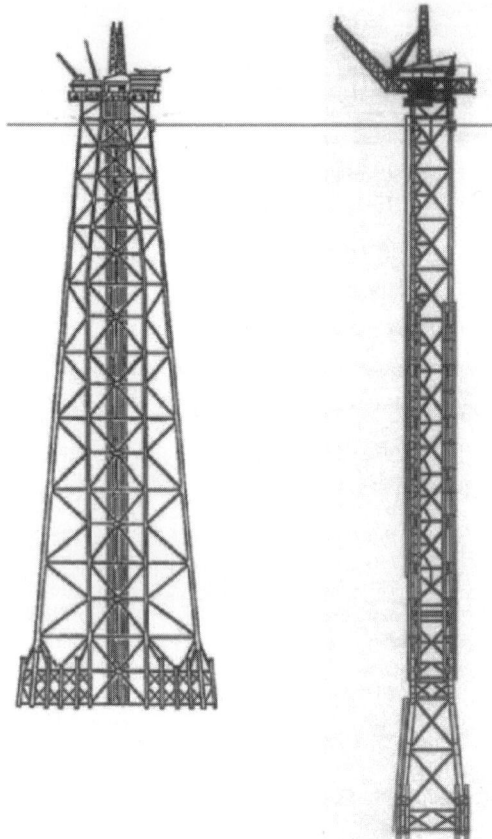

Figure 6.10.6 *Bullwinkle jacket (left) and Baldplate compliant tower [3].*

seafloor—and relatively new compliant towers where the leg spacing remains constant from seafloor to surface. Figure 6.10.6 shows a fixed platform and a compliant tower.

Fixed platforms use steel piles driven 200 to 400 ft. into the sea floor to anchor them in place. Thousands of traditional steel jackets have been installed in water to 600 ft. Seven deepwater platforms with jackets are installed with water depths ranging from 1,025 (Cognac field in 1978) to 1,353 ft. (Bullwinkle field in 1991; http://www.shellus.com/sepco/where/offshore/bullwinkle.htm). The Bullwinkle platform cost about $ 500 million (U.S.) to construct.

All three compliant towers installed to date have been in the Gulf of Mexico. The first was ExxonMobil's Lena in 1,000-ft. water in 1983, Amerada Hess' Baldplate in 1,650-ft. water in June 1998 (http://www.offshore-technology.com/projects/baldpate/index.html), and ChevronTexaco's Petronius in 1,754-ft. water in 2000 (http://www.offshore-technology.com/projects/petronius/index.html). Petronius' $70 million (U.S.), 3,600-ton south module was dropped to the sea floor as it was lifted for final installation when a lift cable broke. The module did not damage the compliant tower, but it took 12 months to rebuild the module and successfully se it in May 2000. The fourth compliant tower will be installed offshore Angola in ChevronTexaco's Benguela, Belize, Tomboco (BBT) field in 2004.

Concrete gravity structures (CGSs) can be more cost effective that steel jackets in relatively shallow water. A CGS rests on the sea floor, and gravity holds it in place.

Figure 6.10.7 shows a CGS used in the Gorgon field development offshore western Australia in 100-m water depth.

CGS have been used mainly in the North Sea. The deepest CGS installed to date is in Shell's Troll field offshore Norway in 994 ft. water in 1996. Shell's Malampaya field in the Philippines was installed with a CGS in 2001 (http://www.offshore-technology.com/projects/malampaya/index.html).

6.10.2 FPSs

FPSs generally fall into four categories: semi-submersibles, ship-shaped vessels, tension-leg platforms (TLP), and spars. The cost of these systems ranges from $100 million to $850 million (U.S.) and can take from 1 to 4 years to design and construct, depending on their processing capacity and the water depth installed. Production capacities can range from 40,000 bopd for spars and TLPs in the Gulf of Mexico to 250,000 bopd for floating production, storage, and offloading (FPSO) vessels offshore West Africa [4]. Other systems may include water injection up to 300,000 bwpd and gas re-injection up to 400 MMSCFD.

Although TLPs are anchored to the sea floor using tendons, other floating production systems are anchored using catenary mooring lines (e.g., a 12-point spread mooring for an FPSO or a nine-point spread mooring for a spar, effectively anchoring the system in place). The ship-shaped vessels, more often called FPSO vessels, can be anchored to the sea floor using a single-point mooring system (SPM) in harsh environments. The SPM can be external or internal to the vessel, allowing the vessel to weathervane around the mooring point to face prevailing weather. Figure 6.10.8 shows an example of an externally mounted SPM on the FPSO *Buffalo Venturer.*

6.10.3 Semi-submersibles

The first FPS was a converted semi-submersible installed on Hamilton Oil Co. Ltd.'s Argyll field offshore United Kingdom in June 1975. Thirty-five (35) semi-submersibles are installed worldwide [4]. The deepest unit installed to date was the Petrobras' *P-36* installed in 1998 in the Roncador field located offshore Brazil in 4,462-ft. water depth. In March 2001, it experienced two explosions and sank to the sea floor over the course of 5 days in March 2001. Figure 6.10.9 shows the *P-36* rig before it submerged.

This instance is the only time that a semi-submersible has been lost, and an extensive investigation was done to understand the cause [5]. The cause was determined to be a rupture of the emergency drain storage tank in the starboard aft column due to accidental entry of hydrocarbons, leading to explosion, damage, and loss of 11 lives. Nevertheless, the benefits of a properly designed and operated semi-submersible ensure that it will be used extensively in deepwater areas. Three new semi-submersibles are sanctioned to be installed in the Gulf of Mexico by 2005 in water depths of 6,000 to 7,000 ft.

6.10.4 FPSOs

The first FPSO was installed in 1984. Over 79 FPSOs are currently installed worldwide, and nine more have been sanctioned. Petrobras' *Seillean* is the deepest, installed in 6,080-ft. water in 1998. Detailed information such as operator, field, capacity and contractor(s) for installed FPSOs can be found in *Offshore Magazine* [6] and Oilfield Publications Limited (www.oilpubs.com) [7]. In January 2002, the U.S. Minerals Management Service (MMS) approved the general concept of using an FPSO in the Gulf of Mexico Outercontinental Shelf. No FPSO is planned at this time, but the first one could be installed as early as 2005. Figure 6.10.10 shows the concept approved by MMS for a Gulf of Mexico installation.

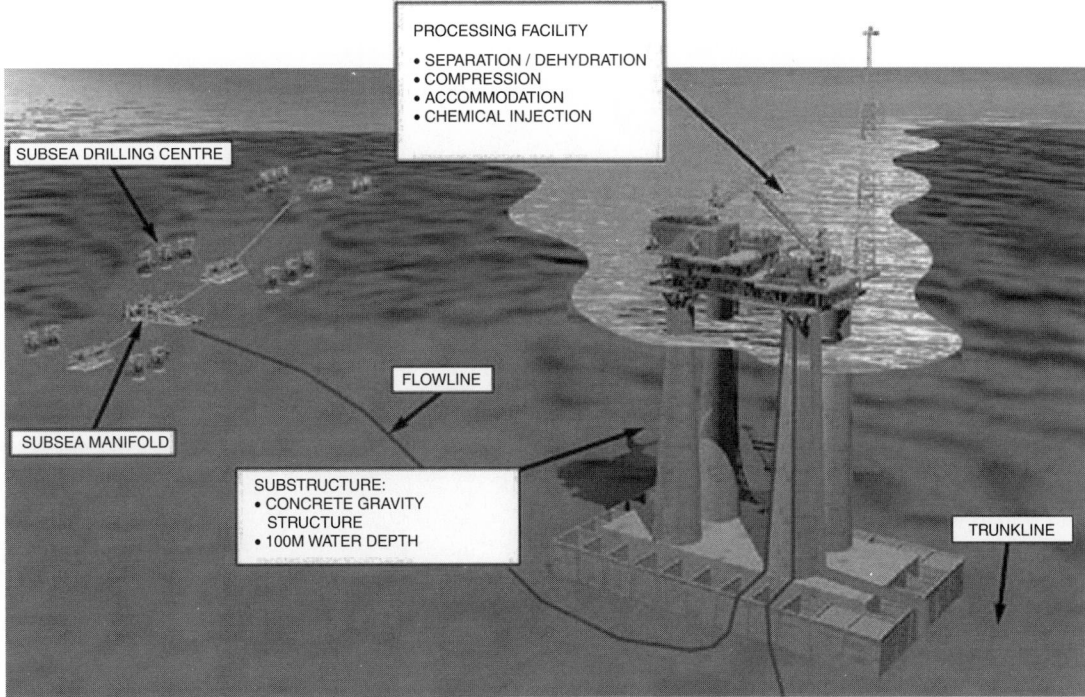

Figure 6.10.7 *CGS used in Gorgon field development offshore West Australia [http://www.offshore-technology.com/ projects/gorgon/].*

Figure 6.10.8 *Externally mounted SPM on FPSO Buffalo Venture [http://www.offshore-technology.com/project/ marlim/].*

6.10.5 TLPs
Conoco installed the first TLP on the Hutton field (now operated by Kerr McGee) located in the North Sea in 486 ft. water in 1984. Thirteen (13) TLPs are currently installed worldwide, and five have been sanctioned for installation in water depths ranging from 486 to 4,700 ft. Figure 6.10.11 shows examples of TLPs.

6.10.6 Spars
Kerr McGee installed the first spar in their Neptune field in 1,929-ft. water in 1997, and all five spars installed to date have been in the Gulf of Mexico. An additional seven spars have been sanctioned for installation by 2005. ExxonMobil's

Hoover-Diana is the deepest to date in water 4,800 ft. deep. Dominion's Devil's Tower spar to be installed in 2003 will be in water 5,610 ft. deep. Figure 6.10.12 shows examples of spars.

6.10.7 Subsea Systems
Subsea production systems are used in combination with new or existing FPSs or fixed platforms. A basic subsea system consists of a wellhead, manifold, flowlines, and electrical and hydraulic control umbilicals. Figures 6.10.13, 6.10.14, and 6.10.15 shows the layout, manifold skid, and subsea wellhead, respectively, for the Mensa field. More than 38 subsea are now producing throughout the Gulf of Mexico through over 200 wells.

Because of the water depth, remote-operated vehicles (ROVs) [8] are used to install and maintain subsea systems. ROVs typically have an attached cable to supply power and control from the surface while underwater. Industry is currently developing reliable autonomous underwater vehicles (AUVs) that do not require a cable.

Figure 6.10.16 shows the number of Gulf of Mexico subsea completions and Figure 6.10.17 indicates that Gulf of Mexico subsea wells are producing in water depths exceeding 6,000 ft. Subsea wells are tied back to new or existing facilities in shallower water for separation of oil, gas, and water. The record distance for an oil subsea tie-back is 29 miles (Mica field in 4,350-ft. water), and the record distance for a gas subsea tie-back is 68 miles (Mensa field in 5,300-ft. water) [4].

Large deepwater floating production systems are being installed with the intent that they will act like huds to accept the subsea tie-backs of relatively smaller fields. For example,

Text continued on Page 427

Figure 6.10.9   *P-36 semi-submersible shortly before it submerged offshore Brazil [http://www.offshore- technology.com/projects/roncador/].*

Figure 6.10.10   *FPSO concept approved by MMS in January 2002 for Gulf of Mexico installation [1].*

Figure 6.10.11 *Three different versions of TLPs (left to right); a SeaStar installed at ChevronTexaco's Typhoon field, a MOSES installed at El Paso's Prince field, and a conventional installed at Shell's Ursa field (photographs courtesy of ChevronTexaco, El Paso and Shell, respectively) [1].*

Figure 6.10.12   *Two competing versions of the production spar. A conventional spar shown in the artist's drawing at left, as installed on Kerr-McGee's Neptune field. The middle graphic shows the truss spar planned for BP's Holstein development. The photograph at the right is the world's first truss spar installed at Nansen field by Kerr-McGee (photographs courtesy of Oryx, BP, and Kerr-McGee, respectively) [1].*

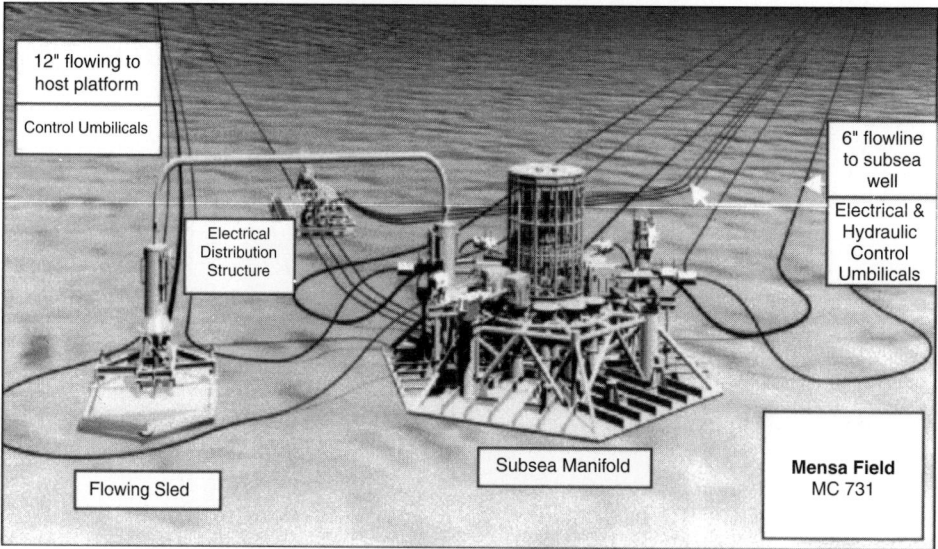

Figure 6.10.13   *Mensa Project subsea equipment layout [1].*

Figure 6.10.14 *Manifold skid for Mensa field [http://www.offshore-technology.com/projects/mensa/].*

Figure 6.10.15 *Subsea wellhead for Mensa field [http://www.offshore-technology.com/projects/mensa/].*

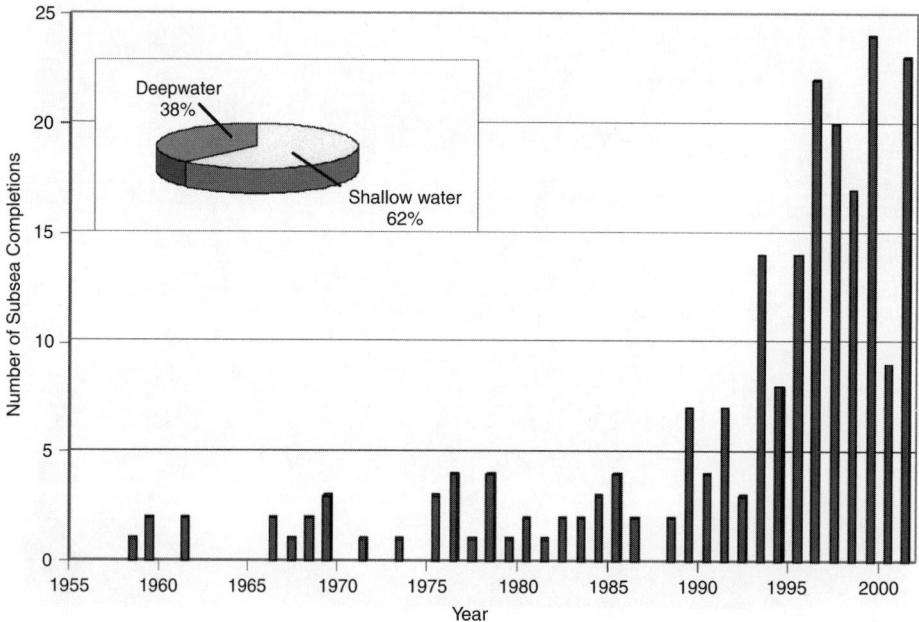

Figure 6.10.16   *Number of Gulf of Mexico subsea completions each year [1].*

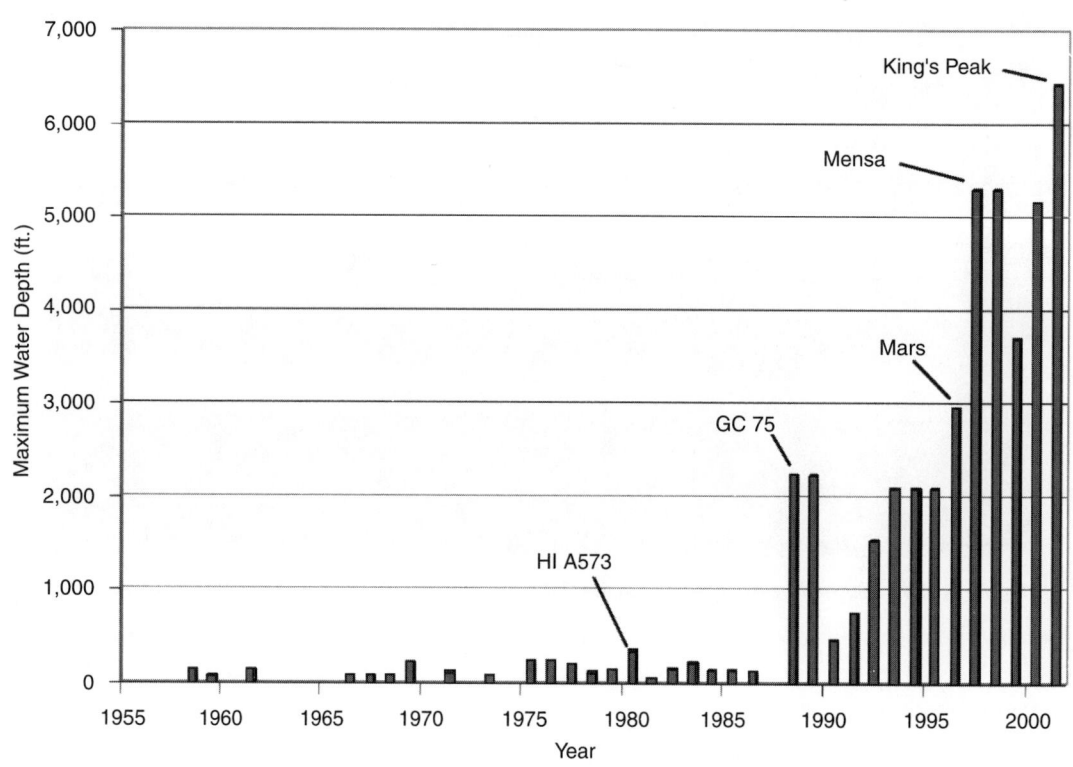

Figure 6.10.17   *Maximum water depth of Gulf of Mexico subsea completions [1].*

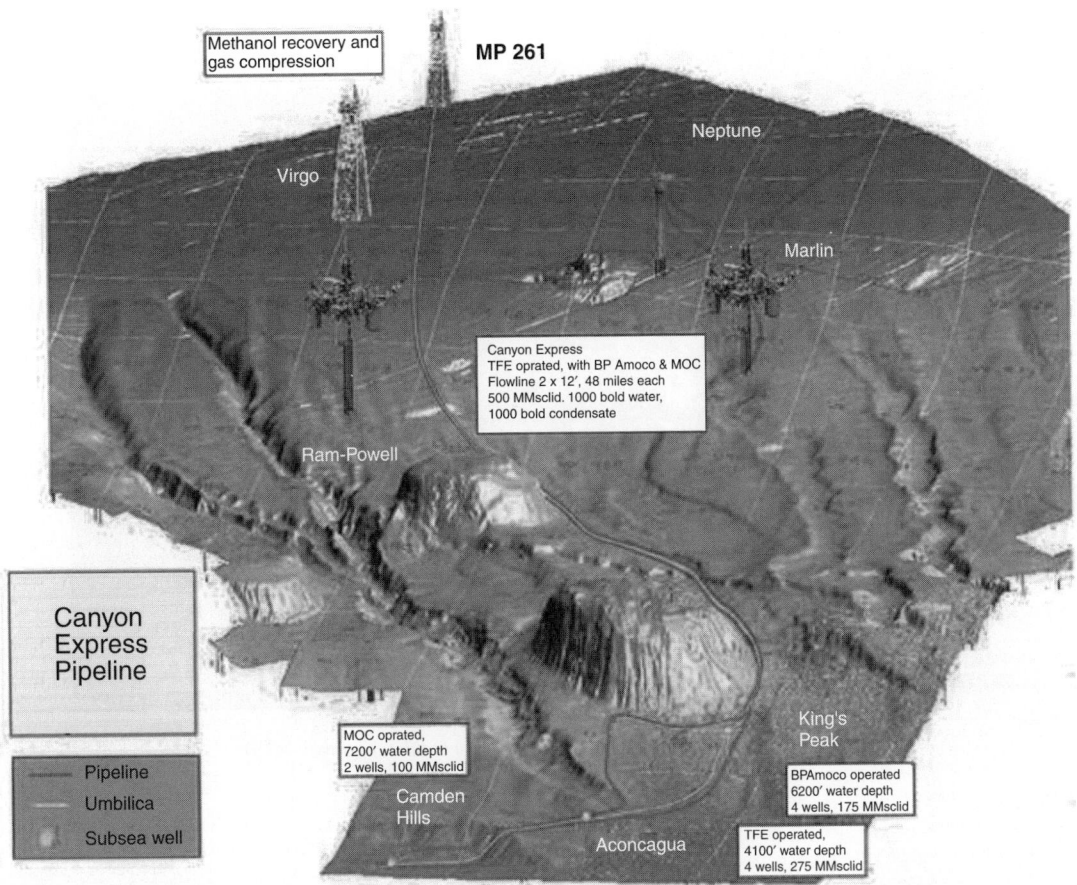

Figure 6.10.18 *Canyon Express Pipeline field layout [http://www.offshore-technology.com/projects/canyon/].*

the Auger TLP installed in 1994, now supports subsea tie-backs from Macaroni (1999), Serrano (2001), and Oregano (2001).

Another notable subsea development in the Gulf of Mexico is the Canyon Express development. Figure 6.10.18 shows the tie back of three separately owned fields—Marathon's Camden Hills, TotalFinaElf's Aconcagua, and BP's King's Peak—to a new fixed platform operated over 48 miles away in shallow water.

Current research is developing subsea processing equipment in an effort to move processing systems such as separation, multiphase booster pumps, and electrical power distribution from sea level to the sea floor, reducing the overall facility costs [9]. Prototypes include ABB's subsea separation and injection system (SUBSIS), shown in Figure 6.10.19. It has been in pilot operation in Norsk Hydro's Troll C oil field in 1,148-ft. water since June 2000. It removes bulk water from the wellstream and treats the water for re-injection at the seafloor.

6.10.8 Multiple Use of Structures
The surface structure for use offshore may be constructed as a drilling platform but may also be used as a production platform. In the case of dry trees used on a fixed platform or compliant tower, the structure acts as a stabilizer for the well casing above the sea floor to the Christmas tree located

above the surface of the water. Flowline risers, helicopter landing pads, and mooring facilities for crew boats and work boats are necessities that must be supported by the structure. In the case of dry trees on TLPs and spars, tensioning devices and buoyancy cans, respectively, are used to support the well casing. Flowline and pipeline risers and auxiliary equipment must still be supported by the structure.

FPSOs and semi-submersibles utilize wet trees with flow-lines leading from the subsea tree back to the surface. These flowline risers can impose tremendous forces on the production system and careful consideration must be paid to their design. Some risers are flexible while others are steel catenary riser (SCR).

6.10.9 Primary Considerations
The entire next section for a fixed platform is excerpted from API RP 2A-WSD [10]. Refer to API RP 2FPS [11] and API RP 2T [12] for guidelines specific to designing FPSs and TLPs, respectively. FPSs must be classified by an agency such as Det norske Veritas (DnV), Lloyds, or American Bureau of Shipping (ABS) [13]. Typically, "classed" floating facilities can be on station for 10 years before undergoing an extensive inspection, usually in a dry-dock.

One of the primary considerations in the design of a structure used in drilling and production is for the safety of personnel. Other considerations that must be built into the

Figure 6.10.19 *SUBSIS installed on Troll C pilot (courtesy ABB).*

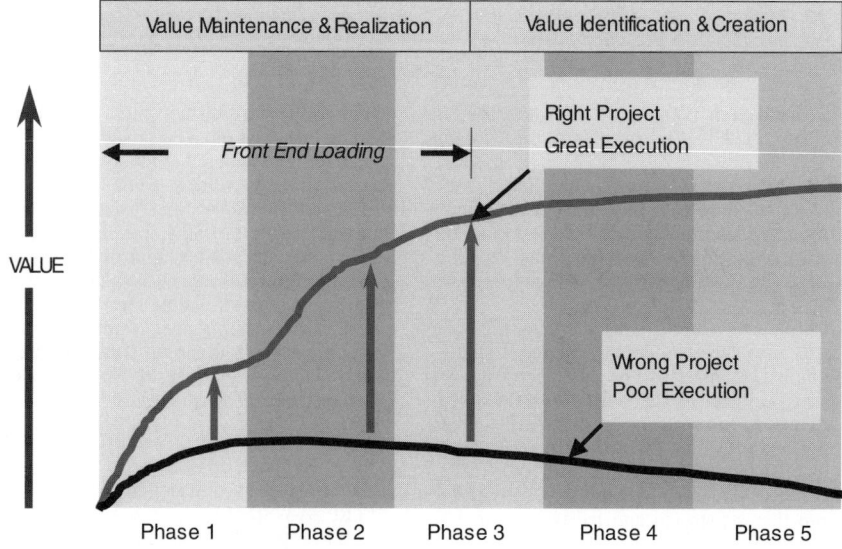

Figure 6.10.20 *Front end loading's impact on overall project value (courtesy Sapphire Offshore).*

Figure 6.10.21   *Value chain for offshore facilities development (courtesy Sapphire Offshore).*

structure with adequate safety factors are the loads applied on the structure that may be summarized as follows:

1. *Total weight of structure.* This is the total weight in air minus the buoyancy on the part of the structure below the water line.
2. *Drilling equipment weight.* This load shall be the weight of the drilling equipment placed on the platform, including such items as derrick, drawworks, mud pumps, and mud tanks.
3. *Production equipment weight.* Production equipment load shall be the weight of production equipment, which includes separators, compressors, tanks, and production manifolds.
4. *Drilling supply weights.* This load, which includes such things as mud weights, water, diesel fuel, and casing, will vary during production.
5. *Production supply weight.* Production supply weight will vary during production and include such things as fluid in tanks and separators.
6. *Drilling load.* The drilling load shall consist of any appropriate combination of derrick corner load, set-back, or rotary load.
7. *Dynamic loads.* Those loads, which act in addition to the equipment weight taken statistically, shall include due consideration of the following: (a) dynamic amplification of cyclic loads that excite the platform (or some component) or near natural frequency; and (b) impact from loads that are suddenly or dynamically applied.
8. *Installation loads.* Those are loads experienced during platform construction. They generally are forces that occur during loading out, launching or lifting operations.
9. *Environmental loads.* Those are loads imposed on the platform by the environment. Combinations and severity of environmental loads used for the design shall be consistent with the probability of occurrence in particular location. Loads to be considered are (a) Wave loads, (b) Current load, (c) Wind load, (d) Ice or debris loads, and (e) Earthquake load.
10. Natural simultaneous occurrence of those phenomena shall be recognized by proper superposition of items (a) through (d). Item (e), earthquake load, when applied, shall be in lieu of items (a)through (d).

6.10.10 Environmental Considerations
In establishing structure orientation, prevailing seas, swells, currents, and winds should be considered. Likewise, when planning for heliports, docking facilities, flare and relief systems, support cranes and hoists, and escape systems, oceanographic and meteorological influences should be introduced.

Weather conditions, such as temperature, precipitation, humidity, and winds, have a significant effect on the overall arrangement of the production facility structure. For example, in cold climates, enclosed structures are desirable. Enclosures in turn affect design considerations such as ventilation and communication systems.

General information on the various types of storms that might affect the platform site should be used to supplement other data developed for operational conditions. Statistics can be compiled, giving the expected occurrence of storms by the seasons, direction of approach, etc. Of special interest for construction planning is the duration. Also of major importance is the ability to forecast storms in the vicinity of a plotform.

The probability of personnel being quartered on the platform should be considered, and transportation made available to remove personnel from the platform on short notice.

6.10.11 Geographical Considerations
This section is excerpted from API RP 2A-WSD [10]. Refer to API RP 2FPS [11] and API RP 2T [12] for guidelines specific to designing FPSs and TLPs, respectively.

Knowledge of the soil conditions at the site of construction of any sizable structure is necessary for safe and economical design. Onsite soil investigations that outline the various soil strata and strength parameters should be made if sufficient knowledge has not been gained in a particular locality from previous soil investigations. Soundings should also be gathered during the onsite studies. These data should be combined with an understanding of the geology of the region to develop the required foundation design parameters. These studies should extend throughout the depth of the soil to be affected by installation of the foundation elements. The bearing capacity of mat and spread footing foundations and the lateral capacity of the pile foundations are largely determined by the strength of the soil close to the sea floor. Consequently, particular attention should be paid to developing complete information on these soils.

The distance between the platform and shoreside terminal will be a definite consideration when planning pipelines, shipping pumps, gas compressors, storage requirements, and waste water–handling facilities.

6.10.12 Operational and Design Considerations
Space is an important factor in promoting a safe operation. As the density of production facilities on a structure increases,

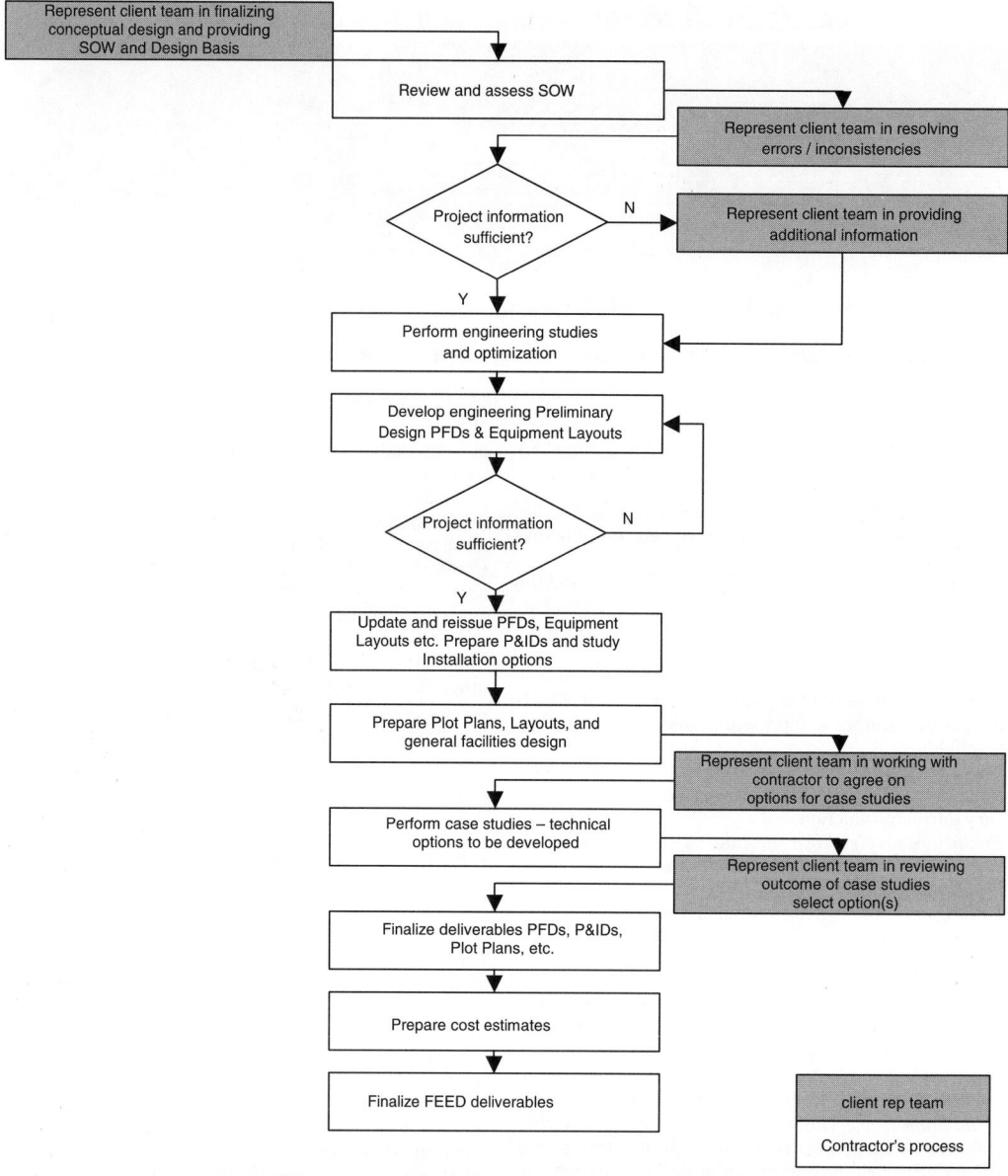

Figure 6.10.22 *Sample FEED process (courtesy Sapphire Offshore).*

operating and maintenance problems and the chance of failure also increase. The use of protective measures should be considered.

Adequate space should be provided around machinery, tanks, vessels, and pipe headers to permit easy access for maintenance. Craneways or lifting points should be provided for the ease of handling of equipment and supplies. Work areas should be well lighted and ventilated with adequate provisions for communication between personnel.

In determining spacing for production facilities on an offshore structure, many factors should be considered. Some of the major items to be considered are:

- space of operation and operating personnel
- space for maintenance access

- space to provide safety from inadvertent mechanical damage
- space to protect against sources of ignition
- space to provide access for control of fires
- space to limit exposure of important equipment and utilities to possible fire

Space limitations imposed by the nature of offshore structures will make compromises necessary. However, production facilities can be arranged to provide a safe, pollution-free operation.

The safety of operating personnel is the primary consideration in designing production facilities. Requirements for egress, or means of escape; personnel landings; guard rails; and lifesaving appliances are specified

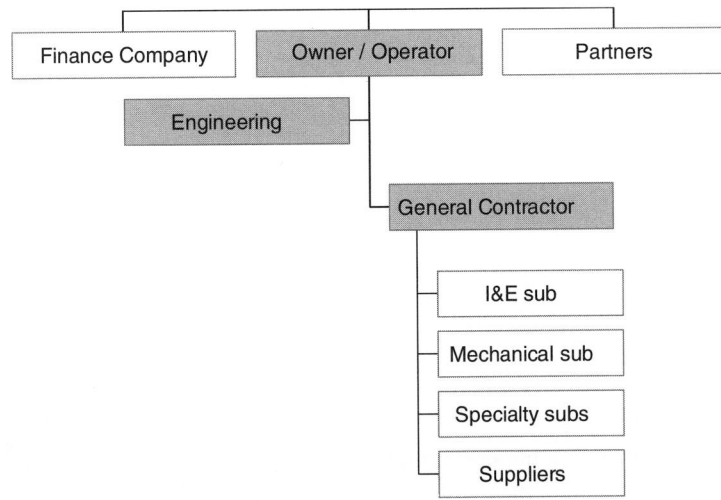

Figure 6.10.23 *Prime or general contract structure (courtesy Sapphire Offshore).*

29 CFR 1910: Occupational Safety and Health Standards and 29 CFR 1926: Safety and Health Regulations for Construction. Floating structures must also comply with Safety of Life at sea (SOLAS) published by the International Maritime Organization (IMO). The United Nations formed the IMO as an International body devoted exclusively to maritime matters such as improving maritime safety and preventing marine pollution.

All equipment should be designed in accordance with the latest standards and in compliance with correct government regulations. The Gas Processors Supplier Association's (GPSA) *Engineering Data Book* [14], API RP Series 12 [15], and API RP Series 11 [16] offer design considerations for equipment such as glycol units and separators. API RP Series 14 [17–20] offers design considerations for offshore safety and anti-pollution.

Equipment should be arranged to provide well-defined corridors of egress from all structural areas. Two exit routes in opposite directions from each area should be provided where possible. Enclosed areas containing a source of fuel should have at least two exits opening to a nonhazardous area.

Piping in all areas should be planned to minimize the number of bends, corrosion, and erosion and also provide easy access and egress from the functional parts of each piece of equipment.

There are many types of structures utilized in offshore operations. These vary from single-well structures to multi-well completely self-contained drilling and production handling structures. The utilities and quarters required vary with the type of structure and how it is utilized.

In planning the utility systems, consideration should be given to number and type of wells, oil and gas processing facilities, remoteness from shore, anticipated production volume, number of people to be housed on the structure, type of the fighting system, type of control system, and electric power source. For example, a single-well caisson structure typically used in 8- to 10-ft. waters offshore Louisiana does not require the installation of any utility system, whereas the self-contained manned structure further offshore may require all utilities listed.

6.10.13 Flowlines and Risers
Subsea flowlines are designed to ANSI B31.4 [21] for oil lines or ASNSI B31.8 [22] for gas lines. API RP 17A [23] and ABS guidelines [24] deal with external factors such as geotechnical conditions, environmental effects, strength criteria, and span rectification designing subsea pipelines and flowlines. Flow assurance is of the utmost importance as the sea floor temperature remains constant around 34°F, and both hydrates and paraffin deposition become major concerns. Additional information on hydrates may be found in the GPSA *Engineering Data Book* [14].

Design guidelines for risers are available in API RP 2RD [25] and API RP 1111 [26].

6.10.14 Safety Shut-Down Systems
A properly designed safety shut-down system will sense an abnormal operational or equipment condition and react to this condition by shutting in or isolating necessary system components or even the entire system. Other actions, such as sounding alarms, starting fire extinguishing systems, an depressuring all piping and pressure vessels, may also be initiated by the shut-down system. The actions to be taken will depend on the level of criticality of the abnormal conditions. The three primary purposes for installing shut-down systems are to:

- protect human life
- prevent ecological damage
- protect the investment

In planning and designing shut-down systems, it is first necessary to determine which events could endanger life, environment, or investment. Inspection, maintenance, and failure documentation are definite considerations in planning shut-down systems. Inspection procedures that call for in-place functional tests or component removal should be carefully planned. Production facilities arrangements should include locating shut-down system components for easy access for the inspection and test. The personnel who perform inspections should be educated and trained on a formal basis.

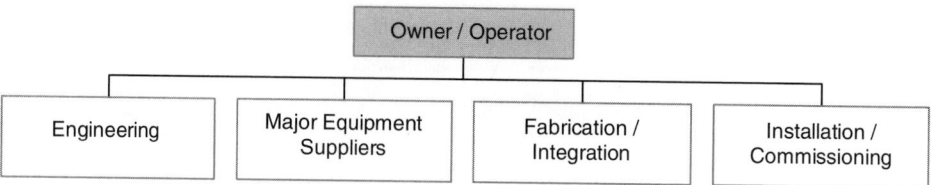

Figure 6.10.24 *Multiple prime contract structure (courtesy Sapphire Offshore).*

6.10.15 Flare and Emergency Relief Systems

Flare and emergency relief systems associated with process equipment should be designed and located with consideration of the amount of combustibles to be relieved, prevailing winds, location of other equipment (including rigs, personnel quarters, fresh air intake systems, and helicopter approaches), and other factors affecting the safe, normal flaring or emergency relieving of the process fluids and gases.

6.10.15.1 Relief/Blowdown System

The relief system is an emergency system for quickly discharging gas by manual or controlled means or by an automatic pressure relief valve from a pressured vessel or piping system to the atmosphere for the purpose or relieving pressures in excess of rated working pressures. It is also used to depressurize, or "blow down," the facility during unsafe conditions. The relief system normally includes a relief valve or rupture disc, the collection piping, a gas scrubber for liquid separation, and a gas flare or vent.

Major equipment items also have an automated blowdown valve to depressurize large gas volumes in the equipment in less than 1 to 2 min should an unsafe condition occur (e.g., fire). The tremendous pressure drops associated with relief valves and blowdown valves can create cryogenic conditions. So material selection for valves and downstream piping is a concern. Another consideration is the instantaneous gas flow rate in a relief or blowdown condition. This rate can exceed the design capacity of the scrubber and flare system if the system is not carefully designed. Additional design considerations for relief systems may be found in GPSA *Engineering Data Book* [14].

6.10.15.2 Flare System

Larger facilities may have a flare system for discharging gas through a control valve from a pressured system to the atmosphere during normal operations. The flare system normally includes a flare control valve, collection piping, the gas liquid scrubber, and gas flare. This discharge may be either continuous or intermittent. Continuous flaring is not allowed in the Gulf of Mexico and most countries. Produced gas must be processed for gas sales or re-injected. Continuous flaring still occurs in remote areas with no market for the gas. On some platforms, considerable quantities (e.g., upward of 300 MMSCFD on a single facility) are flared. However, these countries realize the lost resources are revenues can not be replaced, and they are moving to eliminate flaring.

Flare systems today normally have a dual flare tip (low pressure and high pressure) to enable quick, simultaneous depressurization of low- and high-pressure production systems. It is important that no air flow back into the flare system piping because it would create a potentially explosive mixture. Heat radiation from the flare tip during relief and blowdown is a major concern for not only operating personnel but also nearby equipment.

6.10.16 Ventilation

Enclosed structures require a thorough review to ensure adequate ventilation. Areas enclosed on all sides that contain equipment considered a source of ignition, such as control rooms, should be pressurized to prevent hydrocarbons entry. The air intake for the pressurizing system should be located to preclude entry of hydrocarbons into that system. Enclosed areas containing hydrocarbon fuel sources should be vented with an exhaust system to ensure removal of any escaping hydrocarbons. Also, enclosed areas where welding is to be conducted should be ventilated with an exhaust system to ensure removal of gas evolved during welding operations. Air intake for this system should be located to preclude entry of hydrocarbons.

Equipment areas located on open-type structures should be arranged to allow the natural ventilation caused by winds and convection currents. Care should be taken around fired process equipment to ensure that adequate draft for the equipment is provided. Also, the equipment should be arranged to take advantage of the prevailing winds to keep escaping hydrocarbons from being carried toward equipment considered to be a source of ignition. Care should be taken in the use of protective walls to ensure proper ventilation, and consideration should be given to ventilation of the wellhead areas. This area should be as open as possible, with a minimum of two sides of the structure open.

6.10.17 Transportation

In designing support facilities for the transportation of personnel and equipment on offshore structures, one must consider the prevailing meteorological and oceanographic conditions. The location of transportation facilities, vessels, and associated mooring lines relative to prevailing wind, waves, and currents may control the orientation and layout of the entire structure, including pipelines and flowlines.

On fixed platforms, boat landings and docks should be located on the lee side of the structure. Cranes in turn must be located over the boat landing for convenience in loading and offloading equipment. Storage areas for pipe and bulk materials should be located within or adjacent to the area covered by crane boom. On floating facilities, personnel that do not fly out by helicopter are usually transferred from boat to facility by a crane using a personnel carrier.

Helicopter pads should be located so as to give clear landing approaches for the helicopters. Stacks, guy wires, crane booms, antennas, etc., should be arranged so as not to intrude into the approach or departure paths of the helicopters. The lack of other structures in the area may dictate the need for landing space for two or more helicopters.

6.10.18 Pollution Prevention

Offshore production facilities must include methods for containment and proper disposal of any contaminants (liquids or solids containing liquid hydrocarbons, relatively high

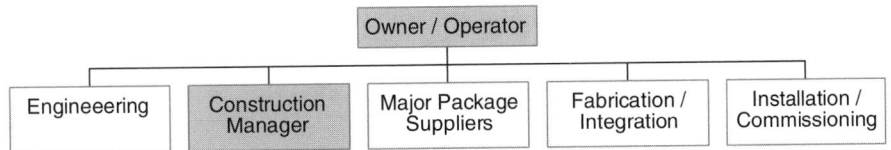

Figure 6.10.25 *Construction management contract structure (pure) (courtesy Sapphire Offshore).*

concentrations of caustic or acidic chemicals, raw sewage, trash, and inedible garbage). Guidelines are provided in API RP 75 [27].

All deck areas that have a source of oil leakage, spills, or drips (e.g., control valves around vessels and tanks) must be surrounded by a curbing or a continuous gutter. Normally, drip or skid pans are installed under equipment; liquids are routed to the open drain system. These drain pans serve as primary containment.

Secondary containment must also be provided for collecting spilled hydrocarbons and rainwater from all deck areas. For example, solid deck areas may be drained to a gutter and routed through an open drain system of gutters or piping to a tank or container where separation takes place due to a specific gravity difference. Liquid hydrocarbons may then be skimmed off and routed into the production system and the remaining water treated by further separation or filtration to meet the requirements of applicable government regulations before being pumped overboard.

In installations where toilets are installed and human waste is discharged into surrounding waters, the effluent must meet requirements of applicable government agencies.

Combustible solid wastes, such as paper of wood products, or other organic material, such as garbage may be disposed of by incineration (if permitted) in an approved dual burn chamber incinerator. Alternatively, the waste may be placed in containers and transported to shore for proper disposal.

6.10.19 Basic Surface Safety Systems

The primary reason to have a service safety system on an offshore production platform is for maximum protection of personnel, to minimize chances of fire, and to reduce chances of explosion and pollution as a result of equipment failure. All offshore Gulf of Mexico facilities must have surface safety systems per API RP 14C [17]. Many areas of the world also use this recommended practice, although installations in the North Sea have separate practices.

Safety in the offshore industry in general and the North Sea in particular was heavily influenced by the disaster on the Piper Alpha oil platform on July 6, 1988, in which 167 of 226 people lost their lives (http://www.ukooa.co.uk/issues/PiperAlpha/v0000864.htm). A public inquiry investigated the disaster and wrote The Cullen Report in November 1990, making several recommendations that have influenced the design and implementation of safety systems worldwide.

The design of each safety system must be engineered for the type of well or wells coming onto the platform. The design must be within guidelines established by the regulatory agencies. In the United States, the MMS enforces periodic testing of the various parts of the shut-down system, provided the platform in question is in federal waters. Platforms in state waters fall within the jurisdiction of the bordering state.

6.10.20 Regulatory Agencies Most Involved with Construction and Operations

6.10.20.1 United States

6.10.20.1.1 U.S. Army Corps of Engineers

In the United States, a permit is required from the U.S. Army Corps of Engineers to set a structure on or build on manmade islands in any federal or state waters. The geographic jurisdiction of the U.S. Army Corps of Engineers per the Rivers and Harbors Act of 1899 includes all *navigable waters* of the United States, which are defined (33 CFR Part 329) as, "those waters that are subject to the ebb and flow of the tide and/or are presently used, or have been used in the past, or may be susceptible to use to transport interstate or foreign commerce."

This jurisdiction extends seaward to include all ocean waters within a zone three nautical miles from the coast line (the "*territorial seas*"). Limited authorities extend across the outer continental shelf for artificial islands, installations and other devices (see 43 U.S.C. 333 [e]). Activities requiring permits include structures (e.g., piers, wharfs, breakwaters, bulkheads, jetties, weirs, transmission lines) and work such as dredging or disposal of dredged material, or excavation, filling, or other modifications to the navigable waters of the United States. The permitting process for the New Orleans district is available on the USACE's Web site at http://www.mvn.usace.army.mil/ops/regulatory/permitreg.asp.

6.10.20.1.2 U.S. Coast Guard

The U.S. Coast Guard (www.usgc.mil) must be notified well in advance of any movement of structures in state or federal waters beyond the coast line. This requirement enables the U.S. Coast Guard to inform all vessels navigating in the area of such movement and hazard. Refer to Code of Federal Regulations (CFR) 46 CFR: Shipping and 33 CFR: Navigation and Navigable Waters for further information.

In addition, the Oil Pollution Act of 1990 (OPA-90) [28] and the International MARPOL Treaty require owners/operators of certain vessels to prepare vessel response plans (VRPs) and/or shipboard oil pollution emergency plans (SOPEP) approved by the U.S. Coast Guard.

In February 2002, the U.S. Coast Guard issued the Proposed Final Draft for the National Preparedness for Response Program (PREP), addressing the exercise requirements for oil pollution response. PREP was developed to establish a workable exercise program that meets the intent of section 4202(a) of the OPA-90, amending section 311(j) of the Federal Water Pollution Control Act (FWPCA), by adding a new subsection (7) for spill response preparedness (33 U.S.C. 1321 [j][7]). The PREP is a unified federal effort and satisfies the exercise requirements of the U.S. Coast Guard, the Environmental Protection Agency (EPA), the Research and Special Program Administration (RSPA) Office of Pipeline Safety, and the MMS. Completion of the PREP exercises will satisfy all OPA-90 mandated federal oil pollution response exercise requirements.

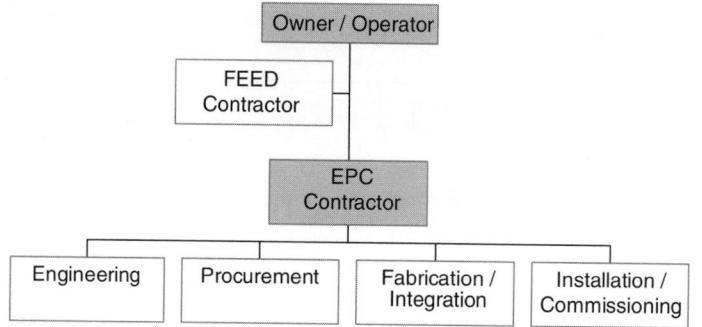

Figure 6.10.26   *EPC contract structure (maximum risk) (courtesy Sapphire Offshore).*

The PREP is a voluntary program. Plan holders are not required to follow the PREP guidelines and, if they choose not to, may develop their own exercise program that complies with the regulatory exercise requirements. All plan holders, whether participating in the PREP or following the exercise mandates of relevant agency regulations, will be subject to government-initiated unannounced exercises. Unannounced exercises are mandated by OPA-90.

6.10.20.1.3 U.S. MMS
Under the Outer Continental Shelf (OCS) Land Ace of 1953, the U. S. Department of the Interior (www.doi.gov) is charged with managing the exploration and development of mineral resources on the Federal OCS. The Secretary of the Interior vested this responsibility in the MMS (http://www.mms.gov/Assets/Banners/aboutMMS.htm). Offshore activity occurs in the Gulf of Mexico, Pacific, and Alaska OCS regions. OCS Report MMS 96-0039 [29] provides a quick reference to the various regulatory requirements of the MMS and other agencies having jurisdiction over hydrocarbon exploration, development, and production operations in the OCS. In managing OCS activity, MMS has two core responsibilities, which are:

- *Offshore Safety* (http://www.mms.gov/offshoresafety/) — ensure incident free minerals exploration and development on Federal Offshore Leases.
- *Environmental production*—(http://www.mms.gov/eppd/index.htm) — ensure that all activities on the OCS are conducted with appropriate environmental protection and impact mitigation.

The general procedure for filing with the MMS on a deepwater offshore field development follows [1]:

1. File an Exploration Plan (EP).
2. Drill exploratory wells.
3. File a Conceptual Deepwater Operations Plan (DWOP).
4. File a Development Operations Coordination Document (DOCD).
5. Drill development wells.
6. File a preliminary DWOP.
7. Begin production.

On April 1, 1988, revised regulations 30 CFR 250 [30] and 30 CFR 256 [31] that restructured and consolidated the rules governing OCS exploration, development, and production activities were published in the Federal Registrer and became effective May 31, 1988.

The former Gulf of Mexico OCS Orders, which dealt with the technical aspects of OCS activities, have been incorporated into the new rules (53 FR 10596-10777).

6.10.20.2 Worldwide
Each nation has governing bodies regulating their oil and gas resources and their development. Many countries have national oil companies that essentially set the rules for developing oil and gas fields in their country.

6.10.21 International Standards
The Geneva-based International Organization for Standardization (ISO) at (see Web site at www.iso.ch) has been issuing standards for the petroleum industry since 1999. ISO Technical Committee (ISO-TC67) is charged with preparing standards for a wide selection of materials, equipment, and offshore structures for the petroleum, petrochemical, and natural gas industries. A target list of standards they plan to create is available at www.tc67.net. These standards are based predominantly on American Petroleum Institute (API) standards, and in some cases, the resulting ISO standards have been adopted back by API as API standards [32]:

- ISO 10423 as API 6A for wellhead and Christmas tree equipment
- ISO 13535 as API RP 8B for hoisting equipment specifications

An example of a recent global standard is the one used for materials in services containing hydrogen sulfide (H_2S) gas. It is toxic to humans, and it can cause severe cracking corrosion of the pipes and process equipment used in the industry. Since 1975, The industry has mainly used the National Association of Corrosion Engineers (NACE) MR0175 to specify materials for use in "sour" service. The European Federation of Corrosion (EFC) also has Publications Nos. 16 and 17 to specify materials in "sour" service. Cooperation between NACE International, the EFC, and ISO has produced ISO 15156, using the latest versions of NACE and EFC publications to create a global standard.

References
1. "Deepwater Gulf of Mexico 2002: America's Expanding Frontier," OCS Report MMS 2002-021, U.S. Department of Interior, Minerals Management Service, Gulf of Mexico OCS Region, New Orleans, April 2002.
2. "2001 Worldwide Survey of Minimal Offshore Fixed Platforms & Decks for Marginal Fields," *Offshore Magazine* Poster, Houston, Texas, January 2001.
3. "Deepwater Production with Surface Trees: Trends in Facilities and Risers," SPE paper no. 68761, presented at SPE Asia Pacific Oil and Gas Conference and Exhibition held in Jakarta, Indonesia, April 17–19, 2001.

4. "2002 Offshore Oil & Gas Industry Deepwater Solutions for Concept Selection," *Offshore Magazine* Poster 36, Houston, Texas, May 2002.

5. "Final Report, Inquiry Commission P-36 Accident," Petrobras, Rio de Janeiro, Brazil, June 22, 2001.

6. "2001 Worldwide Survey of Floating Production, Storage and Offloading (FPSO) Units," *Offshore Magazine* Poster, Houston, Texas, August 2001.

7. Oilfield Publications Limited (OPL) publishes specialist reference maps, book, and vessel registers. Catalog available at http://www.oilpubs.com/v_catalog/welcome.asp.

8. "2000/2001 Worldwide Survey of Work Class ROV Operators," *Offshore Magazine* Poster, Houston, Texas, December 2000.

9. "2002 Survey of Separation Technology," *Offshore Magazine* Poster, Houston Texas.

10. API RP 2A-WSD, "Planning, Designing and Constructing Fixed Offshore Platforms: Working Stress Design," 21st ed, December 2000.

11. API RP 2FPS, "Recommended Practices for Planning, Designing and Constructing Floating Production Systems," 1st ed, March 2001.

12. API RP 2T, "Planning, Designing and Constructing Tension Leg Platforms," 2nd ed, August 1997.

13. "Guide for Building and Classing Facilities on Offshore Installations (June 2000)", American Bureau of Shipping, Houston, Texas, with Supplement, May 2001 and RCN 1, October 2001.

14. *Engineering Data Book*, 11th Edition, Gas Processors Association, Tulsa, Oklahoma, 1998.

15. Recommended Practices, Series 12: Lease Production Vessels, American Petroleum Institute (www.api.org).

16. Recommended Practices, Series 11: Production Equipment, American Petroleum Institute (www.api.org).

17. API RP 14C, "Analysis, Design, Installation and Testing of Basic Surface Safety Systems on Offshore Production Platforms," 7th Edition, March 2001.

18. API RP 14E, "Design and Installation of Offshore Production Platform Piping Systems," 5th Edition, October 1, 1991, Reaffirmed June 2000.

19. API RP 14F, "Design and Installation of Electrical Systems for Fixed and Floating Offshore Petroleum Facilities for Unclassified and Class I, Division 1 and Division 2 Locations," 4th Edition, June 1999.

20. API RP 14J, "Design and Hazards Analysis for Offshore Production Facilities," 2nd Edition, April 2001.

21. "Pipeline Transportation Systems for Liquid Hydrocarbons and Other Liquids," ASME B31.4, ASME International, ISBN# 0791825663, 1998.

22. "Gas Transmission and Distribution Piping systems," ASME B31.8, ASME International, ISBN#0791826082, 2000.

23. API RP 17A, "Design and Operation of Subsea Production Systems," 2nd Edition, December 1, 1996.

24. "Guide for Building and Classing Undersea Pipeline Systems and Risers," American Bureau of Shipping, Houston, Texas, March 2001.

25. API RP 2RD, "Design of Risers for Floating Production Systems (FPSs) and Tension-Leg Platforms (TLPs)," 1st Edition, June 1998.

26. API RP 1111, "Design, Construction, Operation and Maintenance of Offshore Hydrocarbon Pipeline and Risers," 3rd Edition, July 1999.

27. API RP 75, "Development of a Safety and Environmental Management Program for Outer Continental Shelf Operations and Facilities," 2nd Edition, July 1988.

28. Oil Pollution Act of 1990, 33 USC Sec. 2701-2761 (http://envirotext.eh.doe.gov/data/uscode/33/2701.shtml).

29. "Regulatory Compliance for OCS Oil and Gas Operations," OCS Report MMS 96-0039, (http://www.mms.gov/omm/pacific/offshore/ocsregs/96-0039.htm).

30. "Oil and Gas and Sulphur Operations in the Outer Continental Shelf," 30 CFR 250, Subchapter B–Offshore, United States Code of Federal Regulations.

31. "Leasing of Sulphur or Oil and Gas in the Outer Continental Shelf," 30 CFR 256, Subchapter B–Offshore, United States Code of Federal Regulations.

32. "Petroleum industry advances international standards," *Oil & Gas Journal*, pp. 41–46, July 2002.

6.11 INDUSTRY STANDARDS FOR PRODUCTION FACILITIES

The oil and gas industry uses codes and standards from many organizations. Industry standards used in each production facility are specified in the general and equipment specifications developed for the facility by the operator(s). Specified standards normally conform to the laws and legal codes of the host country or the laws of a country mutually agreed by both the operator and the governing agencies.

6.11.1 Commonly Referenced Industry Standards

Because the United States played a large part in the development of the oil and gas industry, many commonly used industry standards originated with organizations based in the United States. Some regions of the world with more severe environmental conditions, such as the North Sea, moved toward more localized standards. Other standards were developed for national reasons. For example, pipe thread standards include those of the National Standards Institute American (ANSI), British Standards Institute, and German Institute for Standardization. Standards from both the international maritime industry and the oil and gas industry are used in the development of floating production systems. Today, the oil and gas industry is moving toward international standards that are discussed in the following section.

The main standards body in the United States is the ANSI (www.ansi.org). It is a private, nonprofit organization that administers and coordinates the U.S. voluntary standardization and conformity assessment system. Founded in 1918, the Institute remains a private, nonprofit membership organization supported by private and public sector organization. ANSI does not develop standards but facilitates development by establishing consensus among qualified groups.

Many U.S. standards can be purchased at ANSI's Electronic Standards Store (http://webstore.ansi.org). Global standards can be found at NSSN: A National Resource for Global Standards at http://www.nssn.org. Tables 6.11.1 and 6.11.2 identify the organizations commonly referenced for the design and construction of production facilities along with some of their more common guidelines, standards, and codes.

6.11.2 International Standards Under Development

The scope of the Geneva-based International Organization for Standardization (ISO) (www.iso.org) is all standards other than those involving electrical, electronic, and related technologies that are the focus of the International Electrochemical Commission (IEC) (www.iec.org). The field of information technology is handled by a joint ISO/IEC committee known as JTC 1. IEC was founded in 1906 and has over 50 participating countries. ISO was established in 1947, and its membership comprises 130 countries. Both organizations are based in Switzerland.

Table 6.11.1 *Commonly Referenced Industry Organizations and Standards*

General Organizations and Standards

- API — American Petroleum Institute (www.api.org)

API RP Series 2	Offshore Structures
API RP Series 5	Tubular Goods
API RP Series 6	Valves and Wellhead Equipment
API RP Series 11	Production Equipment
API RP Series 12	Lease Production Vessels
API RP Series 14	Offshore Safety and Antipollution
API RP Series 17	Subsea Production Systems

  Drilling and Production Operations: Recommended Operating Practices
  Petroleum Measurement
  Pipeline Operations Publications

- ANSI — American National Standards Institute (www.ansi.org)

ANSI B16.5	Pipe Flanges and Flanged Fittings
ANSI B31.3	Chemical Plant and Petroleum Refinery Piping
ANSI B31.4	Pipeline Transportation Systems for Liquid Hydrocarbons and Other Liquids
ANSI B31.8	Gas Transmission and Distribution Piping Systems

- IMO — International Maritime Organization (www.imo.org)

SOLAS	International Convention for the Safety of Life at Sea
MARPOL	International Convention for the Prevention of Pollution from Ships

- ISO — International Organization for Standardization (www.iso.org)

ISO 8501	Preparation of steel substrates before application of paints and related products – Visual assessment of surface cleanliness. Parts 1, 2, and 3.
ISO 9001, 2000	Quality Management Systems – Requirements
ISO 14001, 1996	Environmental Management Systems

- IEC — International Electrotechnical Commission (www.ice.ch)
- ITU — International Telecommunication Union (www.itu.int)

Table 6.11.2 *Organizations and Standards Related to Specific Disciplines*

- NFPA — National Fire Protection Association (www.nfpa.org)

NFPA 70	National Electric Code

- NACE — National Association for Corrosion Engineers (www.nace.org)

NACE MR0175-2002	Standard Material Requirements – Sulfide Stress Cracking Resistant Metallic Materials for Oilfield Equipment
NACE RP0176-94	Recommended Practice for Control of Corrosion on Steel Fixed Offshore Platforms Associated with Petroleum Production
NACE RP0169-96	Control of External Corrosion on Underground or Submerged Metallic Piping Systems
NACE RP0274-98	High-Voltage Electrical Inspection of Pipeline Coatings Prior to Installation

- AISC — American Institute of Steel Construction (www.aisc.org)

AISC	Load and Resistance Factor Design (LRFD) Manual of Steel Construction

- ASME — American Society of Mechanical Engineers (www.asme.org)

ASME Section II	Boiler and Pressure Vessel Code: Materials
ASME Section VIII	Boiler and Pressure Vessel Code: Unfired Pressure Vessels
ASME Section IX	Boiler and Pressure Vessel Code: Welding and Brazing Qualifications

- ISA — Instrument Standards Associations (www.isa.org)

ISA 5.1	Instrumentation Symbols and Identification

- AGA — American Gas Association (www.aga.org)

	Gas Measurement Manual: Parts 1-15

(continued)

Table 6.11.2 *Organizations and Standards Related to Specific Disciplines (continued)*

- AWS — American Welding Society (www.aws.org)

AWS D1.1	Structural Welding Code

- ASTM — American Society for Testing Materials (www.astm.org)

ASTM A6	Standard Specification for General Requirements for Rolled Structural Steel Bars, Plates, Shapes, and Sheet Piling
ASTM A36	Standard Specification for Carbon Structural Steel
ASTM A105	Standard Specification for Carbon Steel Forgings for Piping Applications
ASTM A106	Standard Specification for Seamless Carbon Steel Pipe for High-Temperature Service
ASTM A193	Standard Specification for Alloy-Steel and Stainless Steel Bolting Materials for High-Temperature Service
ASTM A194	Standard Specification for Carbon and Alloy Steel Nuts for Bolts for High-Pressure or High-Temperature Service, or Both

- ASNT — American Society for Nondestructive Testing (www.asnt.org)

- SSPC — Steel Structures Painting Council (www.sspc.org)

  Surface Preparation
  Painting Systems and Coating Systems
  Pain Application

- TEMA — Tubular Exchanger Manufacturers Association (www.tema.org)

- NEMA — National Electrical Manufacturers Association (www.nema.org)

NEMA 250, 1997	Enclosures for Electrical Equipment (1000 Volts Maximum)

- EIA — Electronic Industries Alliance (www.eia.org)

- ACI — American Concrete Institute (www.aci-int.org)

- OCIMF — Oil Companies International Marine Forum (www.ocimf.com)

  Ship-to-Ship Transfer Guide (Petroleum)
  Single Point Mooring Maintenance and Operations Guide
  Hawser Guidelines
  Recommendations for Oil Tanker Manifold and Associated Equipment

- IALA — International Association of Lighthouse Authories (www.iala-aism.org)

  Recommendations for the Marking of Offshore Structures

- ABS — American Bureau of Shipping (www.eagle.org)

  Guide for Building and Classing Facilities on Offshore Installations
  Guide for Building and Classing Undersea Pipeline Systems and Risers
  Rules for Building and Classing Steel Vessels
  Rules for Building and Classing Single Point Moorings

- ASHRAE — American Society of Heating, Refrigerating, and Air-Conditioning Engineers (www.ashrae.org)

- AWWA — American Water Works Association (www.awwa.org)

- UL — Underwriters Laboratories (www.ul.com)

- ICBO — International Conference of Building Officials (www.icbo.org)

UBC	Uniform Building Code
UPC	Uniform Plumbing Code

- SIS — Swedish Standards Institute (www.sis.se/english)

SIS 055900	Pictorial Surface Preparation Standards for Painting Steel Surfaces

- BS — British Standards Institute (www.bsi-global.com)

- DIN — German Institute for Standardization e.V. (www.din.de)

- DNV — Det Norske Veritas (www.dnv.com)

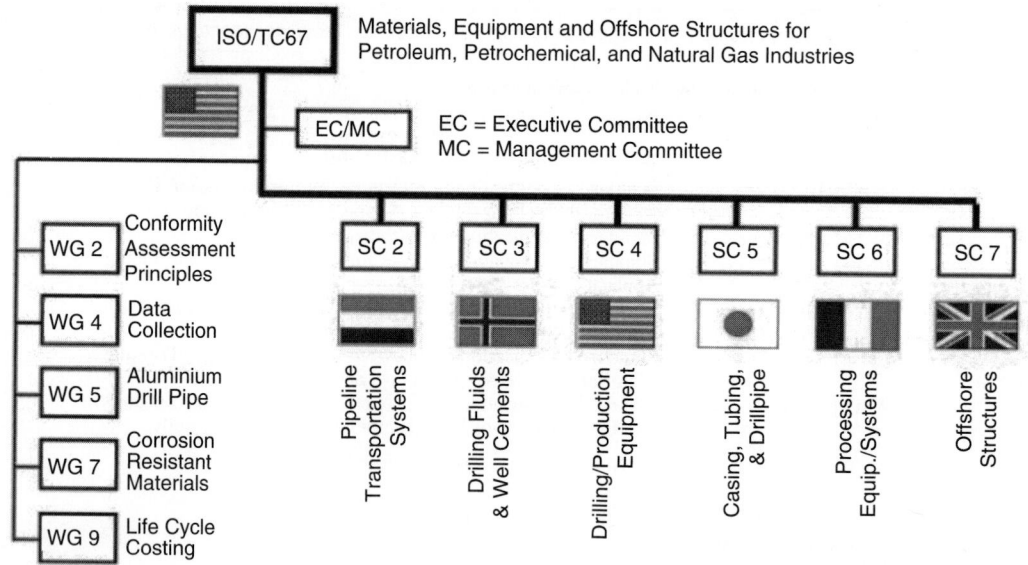

Figure 6.11.1   *Organization of ISO/TC67 (courtesy of International Organization for Standardization).*

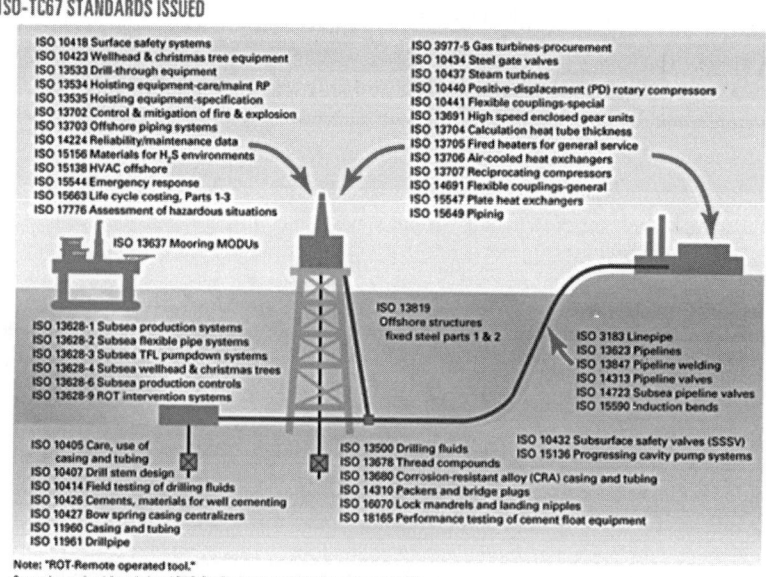

Figure 6.11.2   *ISO/TC67 Standards issued through 2001 [1].*

Telecommunications has traditionally been a government activity, and the International Telecommunications Union (ITU) (www.itu.int) is the global body in this field. Its membership is generally government telecommunications companies or regulators.

In the United States, ANSI is the official U.S. representative to the International Accreditation Forum (IAF), ISO, and, via the U.S. National Committee, the IEC. ANSI was a founding member of the ISO and plays an active role in its governance. ANSI is one of five permanent members of the governing ISO Council, and one of four permanent members of ISO's Technical Management Board.

In Europe, European Committee for Standardization (CEN) (www.cenorm.be) promotes voluntary technical harmonization in Europe in conjunction with worldwide bodies and its partners in Europe. In Europe, CEN works in partnership with the European Committee for Electrotechnical Standardization (CENELEC) (www.cenelec.org) and the European Telecommunications Standards Institute (ETSI) (www.etsi.org).

ISO has been issuing standards for the petroleum industry since 1999. ISO Technical Committee 67 (ISO/TC67) is charged with preparing standards for a wide selection of materials, equipment, and offshore structures for the petroleum, petrochemical, and natural gas industries. ISO/TC67 is composed of five work groups (WGs) and six subcommittees (SCs). Figure 6.11.1 shows the organization of ISO/TC67. More information, such as purpose, scope, and objectives, of these committees can be viewed under "ISO/TC67 Management Plan, 2001 Onwards" at www.tc67.net.

ISO/TC67 has issued several publications in 1999 and 2000 (Figure 6.11.2). Another 20 publications were issued in 2001. These standards are based predominantly on American Petroleum Institute (API) standards, and in some cases the resulting ISO standards have been adopted back by API as API standards:

- ISO 10423 as API 6A for wellhead and Christmas tree equipment

- ISO 13535 as API RP 8B for hoisting equipment specifications

An example of a recent global standard is the one used for materials in services containing hydrogen sulfide (H_2S) gas. It is toxic to humans, and it can cause severe cracking corrosion of the pipes and process equipment used in the industry. Since 1975, the industry has mainly used the National Association of Corrosion Engineers (NACE) MR0175 to specify materials for use in "sour" service. The European Federation of Corrosion (EFC) also has Publications nos. 16 and 17 to specify materials in "sour" service. Cooperation among NACE International, the EFC, and ISO has produced ISO 15156, using the latest versions of NACE and EFC publications to create a global standard.

In a parallel effort, CEN Technical Committee 12 (CEN/TC 12) is charged with standardization of the materials, equipment, and offshore structures used in drilling, production, refining, and the transport by pipelines of petroleum and natural gas, excluding on and supply systems used by the gas supply industry and those aspects of offshore structures covered by IMO requirements (ISO/TC 8).

To strive for truly international standards, ISO and CEN reached the ISO-CEN Vienna Agreement, 2001, providing for various forms of technical liaison and cooperation between ISO and CEN. Of primary interest to ISO/TC67 and to CEN/TC12 is that the work program adopted by CEN/TC12 can be transferred to ISO/TC67 in order to develop international standards with global application. ISO/TC67 then assumes the obligation to deliver standards that can be adopted by CEN. CEN/TC12 appoints European project leaders to the ISO Working Groups, whose role is to ensure that CEN interests are recognized. The Vienna Agreement reflects the priority of international standardization, establishing ISO as the lead on most standardization issues.

References

1. "Petroleum Industry Advances International Standards," *Oil & Gas Journal*, 2002, pp. 41–46.

7

Petroleum Economic Evaluation

Contributing Authors
Richard J. Miller

Contents

The purpose of petroleum engineering is to examine, define, and implement methods and procedures for developing and extracting oil, gas and associated products so as to (1) optimize profits, and (2) obtain a return-on-investment that is commensurate with the risk incurred in making the investment. The term *optimize* is used because there are factors other than financial considerations that influence both the decision to make an investment in an oil and gas project and the return to be obtained from the investment. The goal of this chapter is to outline the methods and processes used to evaluate the economic potential of an oil and/or gas property. In this discussion, the terms *property* and *project* are used interchangeably as the objects of evaluation. Also, to avoid repetition of the phrase *oil and/or gas*, the term *oil* is used as a reference to both oil and gas unless specific reference to one or the other is necessary.

Economic evaluation consists of two major objectives:

- Estimation of the amount of producible oil and/or gas attributable to a property or project and prediction of a schedule of recovery of the producible volume
- Estimation of the economic value of the predicted future production

These are not separate functions: some aspects of the first objective are necessary to accomplish the second objective, and in that respect, the two functions meld at the boundary so that one influences the other. Conversely, it is not necessary to complete the first objective in order to accomplish the second. The first part of this discussion will deal with the first objective of estimating the amount of producible oil and gas along with estimating the future production schedule. The second part will take up the examination of the methods for estimating the values of future production.

7.1 ESTIMATING PRODUCIBLE VOLUMES AND FUTURE OF PRODUCTION

7.1.1 The Concept of Resources and Reserves
One of the most prevalent, enduring, important and misunderstood concepts in the evaluation of oil and gas properties is that of *"resources"* and *"reserves"* and the relationship between the two. There is often confusion about the relation of reserves to resources and even greater confusion about the classification of hydrocarbon volumes within the two categories. Part of the confusion arises out of differing usage of the terms within the petroleum industry and among the financial/investment/regulatory communities associated with and directly or indirectly impacted by industry practice (Fig. 7.1.1).

7.1.1.1 What are Reserves?
Before delving into a comparison of various definitions and an extended discussion of the estimation of reserves, it may be useful to consider the concept of "reserves" and to examine the necessity for definition of the term. Irrespective of which of several definitions is used, reserves are expressed as a volume of oil gas, or other product that is a function of the *expected* future production of oil and of certain expected economic conditions. It is not a physical volume in the sense that it can be examined, measured and moved from place to place. Reserves do not connote a physical entity in the way that "building" connotes a three-dimensional structure that one can see and feel and walk into and out of if one so desires or the sense that "rock" connotes a hard object that has an obvious tactile presence, size, shape, color, and direct use as a tool. *Reserves* is an abstract concept that describes the total volume of future oil production that could be expected to be recovered, assuming that certain physical and economic conditions exists and continue to prevail for however long is required to obtain the production. A comparison to an office building is illustrative: an office building retains the same physical size and character whether the rents for the building go up or down. Reserves, on the other hand, may increase or decline with changes in oil and/or gas price.

The primary objective of oil production is the income to be derived from the sales of the produced oil, not the oil itself. However, potential total income is difficult to compare among properties in any useful manner. Therefore, "reserves" is used as an euphemism for the future income and provides a means of keeping track of that future income

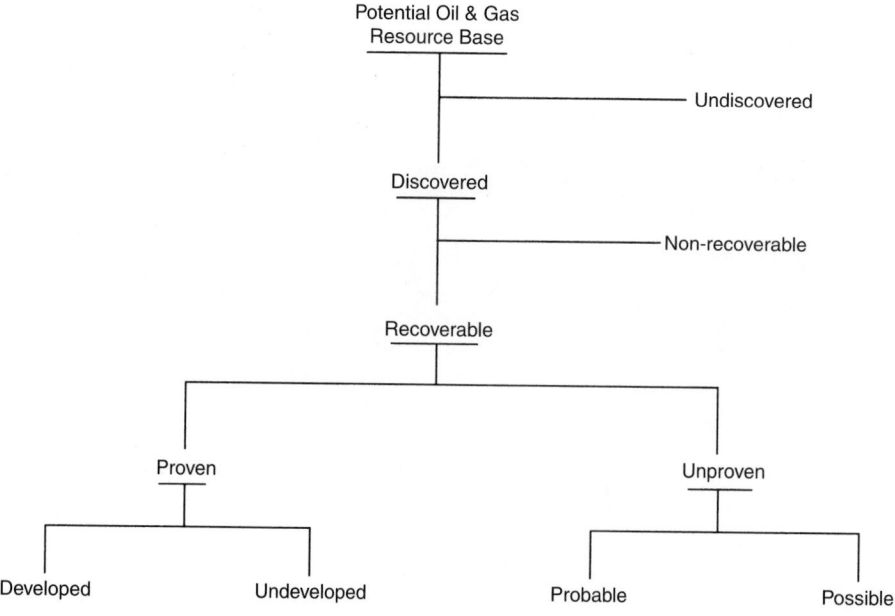

Figure 7.1.1 *Table of resources and reserves.*

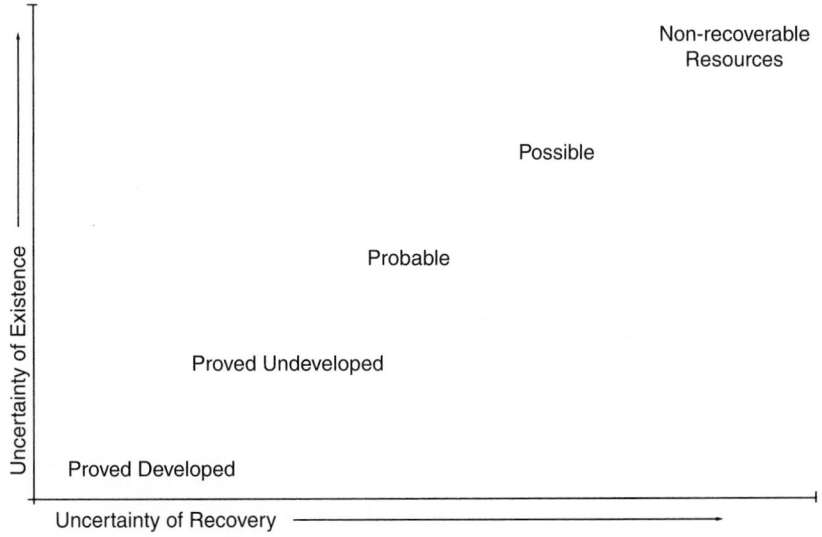

Figure 7.1.2   *Reserve class vs. uncertainty.*

and of comparing one property or company to another. Reserves are also used as a commonly accepted repository of the value of a property, even though the values is derived from the income resulting form production and sale of the oil that makes up the reserve. Unfortunately, both expected income and the related reserves are transitory, and neither is entirely satisfactory as a measure.

Because reserves depend on the expectation of future production and sales of oil and because different persons can have differing expectations for production and sales from a property, the reserves of that property can never be a fixed volume, but can and do change from observer to observer and from time to time, or both. For that reason, reserves as such cannot have value but can only represent value, and then only in an abstract sense. For example, if all conditions are identical, a property with greater reserves can be construed to have a greater value. But all conditions are never identical.

7.1.1.2 Reserves: Definitions and Conditions [1–5]
The distinction between *resources* and *reserves* is significant but can be described in relatively simple terms. Paraphrasing from the Department of Energy (DOE), "*The total resource base of oil and gas is the total volume formed and trapped-in-place within the earth before any production*" [1]. A large portion of this resource base is unrecoverable by current of foreseeable technology because either it is dispersed in low concentrations or it is simply cannot be extracted due to economics, intractable physical forces, or both. The remainder is the recoverable resource base that consists of discovered and undiscovered resources. Discovered resources include recoverable and nonrecoverable resources, where the recoverable resources include cumulative production, if any, to date and reserves. If there has been no production, then all the recoverable resource is classed as *reserves*. DOE describes discovered recoverable resources as "*. . . those economically recoverable quantities of oil and gas for which specific locations are known.*"[2] DOE further classifies resources as economically recoverable or economically unrecoverable.

On the other hand, U.S. Geological Survey (USGS) classifies the undiscovered resource base into accumulations that are of sufficient size and quality that they could be technically recoverable with existing technology, but without consideration of economic viability. This USGS concern is of abstract interest, but because the economic circumstances are not considered, the USGS concept falls outside the purpose of this discussion. Further, the DOE concept of resources and reserves, although of more direct interest, is important only from the standpoint of a definable framework within which to examine industry practice. DOE does make estimates of both resources and reserves as well as compiling reserves estimates provided by industry.

The petroleum industry and related interests have created an elaborate hierarchy of resources and reserves, coupled with a more restrictive terminology that includes a significant concern for economics. Although it is not possible in this space to explore all the definitions and terminology, the concept of reserves is vital to petroleum economics and must, of necessity, be thoroughly understood. The resources concept is not further discussed.

The American Petroleum Institute (API) is usually credited with publishing the first definition of proved reserves. However, the Society of Petroleum Engineers (SPE) has been the industry leader in establishing terms of reference for reserves and has over several decades proposed and recommended at lease four definitions for reserves of oil and gas starting in 1964–1965. The 1964 SPE definition considered only a portion of total reserves by defining proved reserves as:

"*The quantities of crude oil, natural gas and natural gas liquids which geological and engineering data demonstrate with reasonable certainty to be recoverable in the future from known oil and gas reservoirs under existing economic and operating conditions. They represent strictly technical judgments, and are not knowingly influenced by attitudes of conservatism or optimism.*"[3]

SPE adds some explanation regarding proved reserves:

"*When evaluating an individual property in an existing oil or gas field, the proved reserves within the framework of the above definition are those quantities indicated to be recoverable commercially from the subject property at current prices and costs, under existing regulatory practices, and with conventional methods and equipment."*[4]

Proved reserves are further categorized as proved developed producing (PDP), proved developed nonproducing (PDNP), and proved undeveloped (PUD), with the latter two categories deriving, in part, from the scale of expected investment necessary to obtain the production.

These early definitions, along with explanations, were relatively simple and succinct. However, as noted by the Society of Petroleum Evaluation Engineers (SPEE):

"The continuing process of updating and revising reserve definitions is a reflection of both the changing nature of the petroleum industry with time and the difficulty of writing a comprehensive definition of reserves and guidelines that will serve the broad needs of industry without excess complexity."[5]

The *reserves* concept becomes even more abstract when the number of conditions used to define reserves is expanded to include risk. In this case, the reserves can be subdivided into categories such as proved and unproved based on the relative risk of recovery of the reserves in each category. Here again, the risk is not in the reserves per se, but in the likelihood that the expectations for future production and economic conditions will be met. The risk is accounted for by assigning the anticipated future volume of oil production to a certain category of reserve based upon that risk (Fig. 7.1.2).

If *reserves* are a function of the interaction of a set of expectations for future production and economic conditions, and if those expected conditions are variable in both time and space, then *reserves* are relative. To avoid a situation in which *reserves* lose all meaning as a comparative measure, a definition for reserves is necessary in order to provide some broadly accepted constraints on the conditions that would result in the estimation of a particular volume of *reserves*. Without a definition as a guideline or reference, a volume labeled as *"reserves"* could be anything, but would have no meaning to anyone.

The events of the 1970s and resulting changes in the oil industry led to the issuance of definitions by the Security and Exchange Commission (SEC) in 1979, by SPE in 1981, and by the World Petroleum Congress (WPC) in 1983. SPE revised its definitions in 1987 and, in cooperation with WPC, again in 1997. Until 1983 all reserves definitions, at least those issued by recognized authorities such as API and SPE, were for proved reserves only. The WPC definitions of 1983 include d categories ranging from proved to "speculative" reserves. The 1987 SPE definitions were the first to formally define proved, probable. This does not mean that the terms were not in use or that probable and possible reserves had not been calculated and reported. Producing companies, consulting firms, banks, and government agencies developed their own criteria for unproved reserves and used them.

Regarding the 1987 SPE definition, SPEE wrote:

"The definitions for oil and gas reserves were designed to address many of these problems. Until now, there has been no industry standard for probable and possible reserves. The need to categorize such reserves has long been recognized. The many definitions that have been advanced for probable and possible reserves have contributed to the confusion of those attempting to use these higher-risk reserves estimates.

In addition, there has been a need to reexamine and expand definitions used for improved recovery reserves where technology has expanded rapidly. Supplemental recovery methods include both conventional methods (waterflooding and gas injection) and enhanced methods (steam injection, in situ combustion, polymer and miscible flooding)."[6]

The most recent definitions adopted by SPE and the WPC in 1997 are relatively complex and fill several pages. In the 1997 SPE/WPC definitions, *"reserves"* are:

". . . those quantities of petroleum which are anticipated to be commercially recovered from known accumulations from a given date forward. All reserve estimates involve some degree of uncertainty. The uncertainty depends chiefly on the amount of reliable geologic and engineering data available at the time of the estimate and the interpretation of these data. The relative degree of uncertainty may be conveyed by placing reserves into one of two principal classifications, either proved or unproved. Unproved reserves are less certain to be recovered than proved reserves and may be further subclassified as probable and possible reserves to denote progressively increasing uncertainty in their recoverability."[7]

Within the term *"reserves"* are two subcategories of proved and unproved where *proved reserves* are:

". . . those quantities of petroleum, which by analysis of geological and engineering data, can be estimated with reasonable certainty to be commercially recoverable, from a given data forward, from known reservoirs and under current economic conditions, operating methods and government regulations. Proved reserves can be categorized as developed or underdeveloped.

If deterministic methods are used, the term reasonable certainty is intended to express a high degree of confidence that the quantities will be recovered. If probabilistic methods are used, there should be at least an 90% probability that the quantities actually recovered will equal or exceed the estimate.

Establishment of current economic conditions should include relevant historical petroleum prices and associated costs and may involve an averaging period that is consistent with the purpose of the reserves estimate, appropriate contract obligations, corporate procedures, and government regulations involved in reporting these reserves."[8]

In addition to the SPE/WPC, other definitions of reserves, particularly Proved reserves, are offered for specific purposes. More often than not, these definitions incorporate the SPE definitions although they may change a word here or a phrase there. The DOE uses the SPE definition, as do many evaluation organizations and taxing authorities. The significant exception of this discussion is the SEC, which defines Proved reserves as:

"Proved oil and gas reserves are the estimated quantities of crude oil, natural gas, and natural gas liquids which geological and engineering data demonstrate with reasonable certainty to be recoverable in future years from known reservoirs under existing economic and operating conditions, i.e., prices and costs as of the date the estimate is made. Prices include consideration of changes in existing prices provided only by contractual arrangements, but not on escalations based upon future conditions."[9]

This is virtually identical to the then existing SPE (1965) definition except for the addition of the phrase, *"i.e., prices and costs as of the date the estimate is made"*[10] and also the final sentence.

The intent of Rule 4-10 was very simple. Having decided that reserves would be reported by public companies as notes to the financial statements, as standard and unambiguous definition was required. So-called SEC reserves serve only one purpose, which is to allow a comparison of public companies by providing some information regarding a company's reserves of oil and gas. This was to be done by estimating the volume of reserves on the last day of the company's fiscal year using the prices (and costs) in effect on that date with no escalation (increase or decrease) from those prices and costs. The company also reports a dollar value of these reserves using the fixed price/cost projection discounted at a standard 10% before income tax.

The purpose of including this information in the financial report of a company was ostensibly to provide investors with data with which to compare companies. In that context, the fixed price/cost and the 10% discount rate are appropriate to providing a relative comparison. It may not be accurate or emulate the real world, and it certainly does not represent market value, but is does allow a reasonably informed investor to compare Company A with Company B, particularly over time.

The SEC reserves definitions have very limited application and utility and will not be further discussed. However, it is essential to be aware of the differences between the SEC and the SPE definitions.

7.1.2 Classification of Petroleum Products [6–8]

The petroleum products that may be subject to evaluation can be grouped as:

- crude oil
- natural gas
- associated products

Crude oil is a liquid composite of many hydrocarbon compounds that, depending on the composition, has differing properties such as oil gravity, viscosity, and pour-point which, at least in part, define the quality of the oil for potential end use and also influence the methods that would be used to develop and produce the oil. Crude oil that has a high API gravity, and low viscosity is generally easier to produce than low-gravity, high-viscosity oil that may require stimulation to maintain production rates. The characteristics of the oil can, and often do, influence everything from well spacing to pump size to life of production — all of which has an impact on the economic value of the producing property.

Crude oil is a market commodity and is subject to considerable variation in its economic value. Oil prices vary based on location, gravity, sulfur content, and competition from other fuel sources. As an example, crude oil produced in California is consistently priced $3 to 4/bbl below similar oil in the midcontinent in part because of oil gravity but also because of the sulfur content and the lack of access to other markets. At times, such as under the windfall profit tax (WPT) crude oil has been economically stratified based on when it was put on production and the status of the producing company as defined by the regulating agency.

Natural gas is a fluid also composed of many hydrocarbon compounds, although generally not as complex as crude oil. The primary differences between gases are (1) the heating value (generally the methane content), (2) the nonhydrocarbon gas (N_2, O_2, H_2S, etc.) content, and (3) the amount of liquid of heavy ends that can be obtained as natural gas liquids (NGLs).

The economic value of natural gas is directly related to its composition. A gas with no nonhydrocarbons and that is entirely methane and ethane has a high heating value. If it contains propanes and other stripable ends, the value increases. Natural gas enjoys a relatively high economic value because it has a low cost of production. However, it is highly regulated through pipeline and utility controls on transportation and pricing so that it rarely achieves equivalent value with crude oil on a $/BTU basis. In addition, because of the ease of transporting gas, there is substantial competition among gas producing regions, which acts as a control on gas prices.

Associated products include (a) NGLs that can include propane, butane, natural gasoline, and virtually any other hydrocarbon that can be stripped from gas; (b) sulfur; (c) nonhydrocarbon gases. The NGLs can have substantial economic value but require investment in specialized striping plants. Pricing tends to be driven by the local market and can be volatile with demand. Sulfur is a common by-product in many Canadian fields and in some areas of the United States. The economic value tends to fluctuate considerably, and sulfur production may often be more of a nuisance than an economic benefit. Nonhydrocarbon gases such as nitrogen, carbon dioxide, and helium can be economic by-products, but in most cases where this occurs the nonhydrocarbon becomes the primary product and the natural gas is secondary.

The economics of crude oil, natural gas, and associated products can differ significantly depending on market conditions.

7.1.3 Methods For Estimating Reserves [9–11]

The true reserves of a property are known only after production has ceased and the property has been abandoned. Reserves determination is never more than an estimate. Some estimates are based on more knowledge and analysis than others, but a large part of the estimate is the evaluator's perception of future production and economic conditions. Information is critical and the proper usage of available data is vital.

A review of the reserves definitions that were discussed above suggest that the estimation of reserves could be broken down into a series of steps, each of which would narrow the final estimate. Those steps are built on the components of the reserves definitions.

> Step One: Estimate Oil-in-Place Producible Volume
> Step Two: Define Production Mechanism
> Step Three: Estimate Method and Rate of Production
> Step Four: Determine the Economically Producible Volume

The volume derived under Step Four could be the volume of reserves for most applications.

Steps One and Two can be accomplished separately, but as a practical matter the producible volume and the production mechanism are essentially inseparable. Step Three is interactive with Steps One and Two in the sense that the information gained from the analysis in those steps can lead to a decision about: whether to produce by flowing or pumped production, the development well density, whether to use horizontal or straight-hold wells, when to introduce a method of production stimulation, or when to begin enhanced oil recovery (EOR) operations. A decision on production mechanism may in turn influence the estimation of the producible volume and/or the rate(s) of production.

7.1.3.1 Primary Production in Oil Reservoirs

Primary production uses the natural reservoir energy of dissolved gas, encroaching water, gravity or other source as the recovery mechanism. There are a number of methods available for reserves estimation, which can be used alone or in concert. The use of the methods depends on the stage of life of the property and the amount and quality of data that are available. These methods can be generally grouped as (1) volumetric, (2) material balance, (3) production performance. Each method has its attributes and drawbacks.

Volumetric methods [12,13] are used early in the life of a property, before significant production has occurred. This is a subjective criteria; volumetric methods can be and often are used long after production has reached maturity. Volumetric methods attempt to determine the amount of oil and/or gas-in-place and reserves by calculating a volume from the physical properties of the reservoirs(s). The method requires a knowledge of the size of the reservoir, and the physical properties of the reservoir rock(s) and fluids(s). The volume of original oil in place (OOIP) in a segment of the reservoir is equal to:

Volume V = Amount of pore space

$$\times \text{ (amount of oil} - \text{amount of water)}$$

$$\text{OOIP} = V_0 = V(\phi) \times S_h \qquad [7.1.1]$$

where   V = specified volume of reservoir measured in acre feet

ϕ = porosity–% of void space in V

S_h = hydrocarbon saturation as a % of fluid content

= $1 - S_w$, where S_w is the water saturation as a % of fluid content

= $S_o + S_g$ if there is a free gas cap

In practical use, the equation is

$$V_0 = \frac{7758V\,(\phi)\,(1 - S_w)}{B_{oi}} \text{ bbl STO} \qquad [7.1.2]$$

in which 7,758 is the conversion from acre-feet to barrels and B_o is a factor to convert the fluid volume at reservoir pressure and temperature to stock tank barrels.

In free-gas reservoirs or in the gas cap of oil reservoirs, the volume of gas-in-place is

$$V_g = \frac{43,560V\,(\phi)\,(1 - S_w)}{B_{gi}} \text{ scf} \qquad [7.1.3]$$

where 43,560 is the conversion from acre-feet to cubic feet. The gas formation volume factor (B_g) may be estimated for various combinations of pressure, temperature and gas gravity from published tables.

The volume of dissolved gas in an oil reservoir is given by

$$V_g = \frac{7758V\,(\phi)\,(1 - S_w)\,R_s}{B_g} \qquad [7.1.4]$$

where R_s is the volume of dissolved gas per barrel of oil at reservoir conditions.

The volumetric method can be subject to considerable error because (1) it is often used to evaluate a property when little specific data may be available and (2) it requires the estimation of reservoir rock and fluid properties and reservoir volumes from spot measurements of the properties that are then applied to the entire reservoir. Porosity and water saturation are obtained from well logs and/or core samples that are measured from a small volume of the reservoir and which under the best circumstances, only approximate the conditions in the reservoir. The areal extent of the reservoir is rarely known until many wells are drilled, while volume is estimated using a zone thickness measured at one or more points in the reservoir. While techniques of core analysis and, especially, electric and other well log measurement and analysis have become very sophisticated, and have been advanced by three-dimensional and other seismic and geophysical methods, the volumetric method remains only a gross estimate of oil-in-place.

Conversion of a volumetric oil-in-place to ultimate recovery requires the use of a *recovery factor*, Rf, which can be either a unit recovery (bbl or Mcf/acre ft) or a percentage of OOIP.

Recovery factors can be determined from the performance of similar reservoirs, from laboratory analysis of cores, or computer simulation of anticipated performance.

Unit recovery can also be calculated assuming information if available from reservoir fluids and core analysis.

Ultimate recovery = V_0 × recovery factor

For solution gas drive or depletion drive reservoirs:

$$N = 7758\,(\phi)\left(\frac{1 - S_w}{B_{oi}} - \frac{1 - S_w - S_{gr}}{B_{oa}}\right) \text{ STB/acre ft}$$

$$[7.1.5]$$

where recovery is the difference between OOIP at initial reservoir conditions and OIP at abandonment conditions.

In a water drive reservoir:

$$N = 7758(\phi)\left(\frac{1 - S_{oi}}{B_{oi}} - \frac{S_{or}}{11B_{oa}}\right) \text{ STB/acre ft} \qquad [7.1.6]$$

The volumetric method has certain data requirements. In checklist form:

Reservoir volume:

- accurate mapping of gross and net sand
- determination of oil–water and gas–oil contacts
- calculation of reservoir volume (acre/ft)

Rock properties:

- determine porosity—from logs, cores or both
- determine water saturation—from logs, cores or both
- determine residual S_o and S_w—from core tests

Fluid properties:

- determine B_o at initial and abandonment conditions from PVT analysis.

The *material balance* [14–16] is a more complex method of estimating producible volume, but has the advantage of providing an estimate of recovery over time under certain conditions. The method has several forms and requires both an extensive pressure–volume–temperature (PVT) analysis of reservoir fluids and an accurate pressure history of the reservoir. The latter obviously requires that some production (5% to 10% of ultimate recovery) occur before the method can be used. The method is not a substitute for the volumetric method but can be used along with the volumetric method later in the life of the property and is often used to obtain a recovery factor for volumetric calculations. The material balance method could also be used if reliable pressure history data can be obtained for a reservoir with similar rock and fluid properties.

The material balance is a practical application of conservation of mass and energy principles used to balance the withdrawals from a reservoir with changes in volume of the original reservoir fluids and the influx of additional fluids. The method has been modified, adapted and simplified by many authors. The several forms attempt to model reservoir performance over time by equating the expansion of reservoir fluids, as pressure decline occurs during production, to the change in voidage of the reservoir caused by withdrawal of oil, gas, and water (less any water influx). The method requires an iterative solution of sequential pressure drops caused by production. The values for each successive step in the analysis are taken from PVT analysis of the reservoir fluids.

The general equation is:

$$N = \frac{N_p\left[B_t + 0.1781B_g\left(R_p - R_{si}\right)\right] - \left(W_e - W_p\right)}{B_{oi}\left[m\left(\dfrac{B_g}{B_{gi}}\right) + \dfrac{B_t}{B_{oi}} - (m+1)\left(1 - \dfrac{\Delta p\left(C_f + S_w C_w\right)}{1 - S_w}\right)\right]}$$

$$[7.1.7]$$

where N is the OOIP, which, by definition, must remain the same through each step of pressure decline. The general equation assumes that all components of the reservoir react to production. By using the pressure-production history of the reservoir to obtain a relation of N_p (production) to ΔP (change in reservoir pressure), a projection of $N_p/\Delta P$ can be used as the basis for a material balance estimate of future production to the point of pressure depletion. At that point, ultimate *primary* recovery is obtained and

$$R_f = N_p/N$$

The material balance may be thought of as:

Initial oil in place = oil remaining + oil produced at time (t)

However, it is more convenient to treat the balance as:

Initial gas in place = gas remaining + gas produced

In this form the general equation becomes:

$$N = \frac{N_p\left[B_o + B_g\left(R_p - R_s\right)\right] - B_w\left(W_e - W_p\right)}{mB_{oi}\left(\dfrac{B_g}{B_{gi}} - 1\right) + B_g\left(R_{si} - R_s\right) - \left(B_{oi} - B_o\right)} \quad [7.1.8]$$

where $N_p[B_o + B_g(R_p - R_s)]$ is the reservoir volume of produced oil and gas; $B_w(W_e - W_p)$ is the total amount of water influx retained in the reservoir; $mB_{oi}(B_g/B_{gi}-1)$ is the expansion of the gas cap; and $B_g(R_{si} - R_s)$ is the reduction in the amount of solution gas at reservoir conditions occurring with production of N_p barrels of oil [17,18].

If there is no gas cap and no water influx, then:

$$N = \frac{N_p\left[B_o + B_g\left(R_p - R_s\right)\right] + B_w W_p}{B_g\left(R_{si} - R_s\right) - \left(B_{oi} - B_o\right)} \quad [7.1.9]$$

If the reservoir is above the bubble point [19], then B_g remains constant; R_p, R_s and R_{si} are equal and:

$$N = \frac{N_p B_o + W_p B_w}{B_a - B_{oi}} \quad [7.1.10]$$

B_w can often be ignored if reservoir pressure is low; however, data on water compressibility are readily available. Where reservoir fluid property data indicate measurable change in the compressibility of oil, water, and rock with changes in pressure, then the equation(s) would be modified by using an effective oil compressibility term:

$$N = \frac{N_p B_o + W_p B_o}{C_{oe} B_{oi}\left(P_i - P\right)} \quad [7.1.11]$$

where:

$$C_{oe} = C_o + \left(\frac{S_w}{1 - S_w}\right) C_w + \left(\frac{1 - \phi}{\phi\left(1 - S_w\right)}\right) C_f \quad [7.1.12]$$

In water drive reservoirs [20,21]:

$$N = \frac{N_p\left[B_o + B_g\left(R_p - R_s\right)\right] + B_w W_p}{D_i} - \frac{W_e B_w}{D} = N_i - \frac{W_e B_w}{D_i} \quad [7.1.13]$$

where:

$$D_i = B_g\left(R_{si} - R_s\right) - \left(B_{oi} - B_o\right) \quad [7.1.13a]$$

Water influx (W_e) is rarely known or measurable; however, where water influx occurs, calculated values of N_i over time would increase. N could be estimated by plotting N_i versus time and extrapolating back to $t = 0$ where $N_i(t = 0) = N$. This approach includes a significant potential for error if the plotted points are not a straight line. Several authors have presented means of improving this approach (Figure 7.1.3).

In reservoirs with an active gas cap and no active oil/gas segregation with production:

$$N = \frac{N_p\left[B_o + B_q\left(R_p - R_s\right)\right] + B_w W_p}{mB_{ob}\left(\dfrac{B_g}{B_{gi}} - 1\right) + B_g\left(R_{sb} - R_s\right) - \left(B_{ob} - B_o\right)} \quad [7.1.14]$$

where subscript b refers to the bubble point (BP).

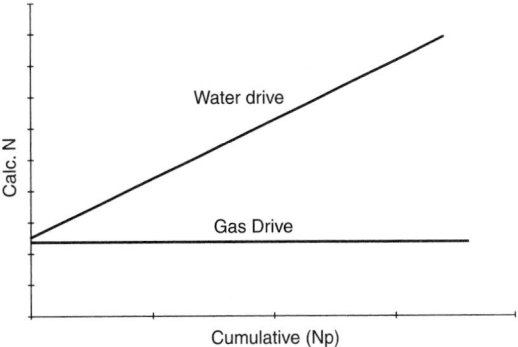

Figure 7.1.3 *Calculated N vs. cumulative N_p.*

In reservoirs with a combination of drives:

$$N = \frac{N_p\left[B_o + B_g\left(R_p - R_s\right)\right] - B_w\left(W_e - W_p\right)}{mB_{ob}\left(\dfrac{B_g}{B_{gi}} - 1\right) + B_g\left(R_{sb} - R_s\right) - \left(B_{ob} - B_o\right)} \quad [7.1.15]$$

or

$$N = \frac{N_p\left[B_o + B_g\left(R_c - R_s\right)\right] + B_w W_p}{D_i} - \frac{B_w W_e}{D_i} \quad [7.1.16]$$

where

$$D_i = mB_{ob}\left(\frac{B_g}{B_{gi}} - 1\right) + B_g\left(R_{sb} - R_s\right) - \left(B_{ob} - B_o\right) \quad [7.1.17]$$

Several authors have presented variations in the calculation procedure to obtain material balance recovery factors. Muskat's method [22–24] calculates the change in oil saturation with change in pressure as production occurs:

$$\frac{\Delta S_o}{\Delta p}$$

$$= \frac{S_o\left(\dfrac{B_g dR_s}{B_o dp}\right) + \left(1 - S_o - S_w\right)\left[B_g \dfrac{d\left(1/B_g\right)}{dp}\right] + S_o\left(\dfrac{\mu_o}{\mu_g} \dfrac{k_{rq}}{k_{ro}} \dfrac{dB_o}{B_o dp}\right)}{1 + \dfrac{\mu_o}{\mu_g} \dfrac{k_{rq}}{k_{ro}}} \quad [7.1.18]$$

At depletion,

$$S_{oa} = S_{oi} - \Delta S_o \quad [7.1.19]$$

$$R_f = \frac{S_{oi} - S_{oa}}{S_{oi}} \,(\%) \quad [7.1.20]$$

Muskat's method assumes uniform oil saturation (no gas segregation) and relatively low permeability. The method requires small pressure increments but is readily adapted to computer analysis, thereby reducing the tedium of the calculation. The results are converted into recovery per acre foot by:

$$N_p = 7758\,(\phi)\left(\frac{1 - S_w}{B_{oi}} - \frac{S_o}{B_o}\right) \text{ STB/acre ft.} \quad [7.1.21]$$

Cumulative recovery as a percentage of OOIP can be determined by:

$$\frac{N_p}{N} = 1 - \left(\frac{S_o}{1 - S_w}\right)\left(\frac{B_{oi}}{B_o}\right) \quad [7.1.22]$$

and gas/oil ratio performance by:

$$R = R_s + 5.615\left(\frac{B_o}{B_{oi}} \frac{\mu_o}{\mu_g} \frac{k_{rg}}{k_{ro}}\right) \text{ SCF/STB} \quad [7.1.23]$$

Relative production can be calculated from:

$$q_o = q_{oi} \left(\frac{k_o}{k_{oi}} \frac{\mu_{oi}}{\mu_o} \frac{p}{p_t} \right) \text{ STB/Day} \qquad [7.1.24]$$

Tarner's method [25,26] for use in solution gas reservoirs below the BP, requires a simultaneous solution of the material balance equation and the instantaneous gas/oil ratio equation. The procedure is to calculate the cumulative oil (N_p) and gas (G_p) for a pressure drop ($p_1 - p_2$) as follows:

1. $N_p = 0$ at BP.

2. $(G_p)_2 = (N_p)_2 (R_p)_2 = N \left[(R_{si} - R_s) - 5.615 \left(\dfrac{B_{oi} - B_o}{B_g} \right) \right]$

$$- (N_p)_2 \left(5.615 \frac{B_o}{B_g} - R_s \right) \qquad [7.1.25]$$

3. $(S_t)_2 = S_w + (1 - S_w) \dfrac{B_o}{B_{oi}} \left[1 - \dfrac{(N_p)_2}{N} \right] \qquad [7.1.26]$

4. Determine k_{rg}/k_{ro} at $(S_t)_2$:

$$R_2 = R_s + 5.615 \left(\frac{B_o}{B_g} \right) \left(\frac{\mu_o}{\mu_g} \right) \left(\frac{k_{ro}}{k_{rg}} \right) \quad \text{at} \quad p_2 \quad [7.1.27]$$

5. Compute

$$(G_p)_2 = (G_p)_1 + \frac{R_1 + R_2}{Z} \left[(N_p)_2 - (N_p)_1 \right] @p_2 \quad [7.1.28]$$

6. Make three good estimates of $(N_p)_2$ and the corresponding $(G_p)_2$ from steps 2 and 5. Plot $(N_p)_2$ versus $(G_p)_2$ for step 2 and for step 5 and take the intersection of the curves as satisfying both equations.

Published tables of recovery factors, calculated using the Muskat and Tarner methods, can be used where no detailed data regarding reservoir fluids or rock properties is available.

7.1.3.2 Non-Associated Gas Reservoirs [27,28]

The estimation of producible volumes for gas reservoirs is less complex than for oil reservoirs, if only because the fluid dynamics are simpler and similar field performance is more directly applicable. The basic principles of volumetric gas-in-place and recovery factor are the same as for crude oil.

$$\text{OGIP} = G = \frac{43,560 V_g (\phi) (1 - S_w)}{B_g} \quad \text{SCF} \quad [7.1.29]$$

$$\Delta G_p = 43,560 \, (\phi) (1 - S_w) \left(\frac{1}{B_{gi}} - \frac{1}{B_{ga}} \right) \text{ SCF/acre ft.}$$
$$[7.1.30]$$

The gas formation volume factor (B_g) at abandonment (B_{ga}) is calculated at abandonment pressure (P_a), which can be based on pipeline pressure or the minimum pressure to which the reservoir can be reduced for the conditions of reservoir depth, tubing size or other constraint(s).

7.1.4 Estimating Future Production
7.1.4.1 Production Performance—Oil Reservoirs

The volumetric and material balance methods of estimating reserves are valuable tools but are often limited by paucity of data; mathematical calculations that, in being simplified, leave out or assume certain reservoir conditions to be true; and/or the assumption of uniformity of conditions throughout the reservoir. In contrast, the *production performance* [29–33] *approach* implicitly includes all reservoir and production operating conditions that would effect performance. When production is not curtailed by regulatory or other artificial conditions, the volume of production from the well is a direct result of the interaction, however great or small or uniformly dispersed, of all reservoir rock and fluid properties

with the existing wellbore and operating conditions. Because oil reservoirs are finite in volume, production over time causes a reduction in pressure that, in turn, causes a decline in the rate of production per unit time. The combination of time, production rate and cumulative production can be used to determine both remaining reserves and productive life.

When there has been sufficient history to establish a production trend for a property, the three variables may be plotted as graphs, commonly known as *decline curves*, which can then be extrapolated to determine future production and reserves. The most common approach is the rate-time plot where time is plotted as the independent variable (X) and production rate is plotted as the dependent variable (Y). This curve, with sufficient definition, can be extrapolated into the future to estimate future production and reserves.

The only requirements for extrapolation are that the curve demonstrate uniformity of shape and that there be an end point. The uniformity of shape is required to ensure that performance is the result of interaction of reservoir and operating condition and is not being altered by changing operating or other artificial conditions. It is important to carefully analyze all data used to define production performance and to equate production during short production periods (February) and downtime (during pump changes or other well work) to production during "normal" periods. Changes in production caused by mechanical alterations, such as opening or closing flow valves, pump changes or pump speeds, must be equated and may require additional graphical analysis. The use of sales data must equate the time between sales to the production time. Often sales occur when a certain tank volume is accumulated, not on a strict time basis, so that, over time, the period between sales may increase while the volume appears to be constant. The production rate, however, may well be declining.

The decline curve may generally demonstrate one of three forms: *exponential*, essentially a straight line of constant slope; *hyperbolic*, a continuously flattening curve that can be described mathematically; and *harmonic*, a special case of the hyperbolic decline (Figure 7.1.4).

The decline can be described in two ways: the nominal decline rate is the negative slope of the curve of the natural log of the production ratio (q) at time (t), or

$$D = -\frac{d \ln q}{dt} = -\frac{dq/dt}{q} \qquad [7.1.31]$$

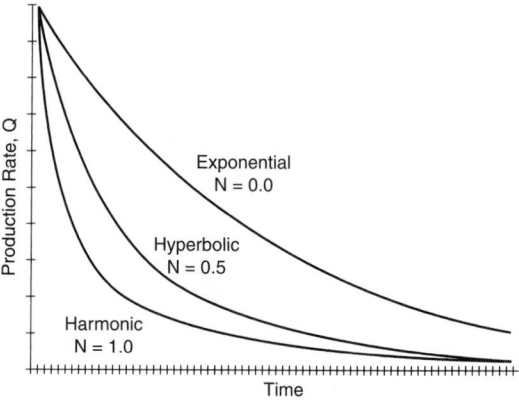

Figure 7.1.4 *Decline curve–rate/time (exponential, harmonic, hyperbolic).*

The effective decline rate is more common in actual practice and has the form of a loss rate:

$$D_e = \frac{q_i - q_1}{q_i} \qquad [7.1.32]$$

Decline rate is normally expressed as an annual rate for comparison among properties.

For exponential or constant-percentage decline, the nominal rate is

$$D = -\frac{dq/dt}{q} \qquad [7.1.33]$$

where, after integration,

$$q = q_i e^{-Dt} \qquad [7.1.34]$$

where t is the time period over which q is to be calculated and to which D must conform. Further integration yields

$$N_p = \frac{q_i - q}{D} \qquad [7.1.35]$$

where N_p is the cumulative production during the period between q_i and q. If q_a at abandonment is substituted for q, then N_p is the ultimate recovery at decline rate D.

Hyperbolic decline may only become evident later in the life of a property and may require careful analysis to be discerned. In fact, most primary production occurs as hyperbolic decline, but in practice, constant rate decline is often used to approximate future production. In the late life of production, hyperbolic decline approaches asymptotic conditions and can very well be approximated by an exponential decline. For the early life:

$$D = \frac{-dq/dt}{q} = bq^n \qquad [7.1.36]$$

where n is a fractional power of the production rate between 0 and 1 and b is a constant determined at initial conditions:

$$b = \frac{D_i}{q_i^n} \qquad [7.1.37]$$

$$q = q_i (1 + nD_i t)^{-1/n} \qquad [7.1.38]$$

$$N_p = \frac{q_i^n \left(q_i^{1-n} - q^{1-n} \right)}{(1 - n) D_i} \qquad [7.1.39]$$

For harmonic decline:

$$D = -\frac{dq/dt}{q} = bq \qquad [7.1.40]$$

where:

$$b = \frac{D_i}{q_i} \qquad [7.1.41]$$

and, after integrating,

$$q = \frac{q_i}{1 + D_i t} \qquad [7.1.42]$$

$$N_p = \frac{q_i}{D_i} \ln \frac{q_i}{q} = \frac{q_i}{D_i} \ln r \qquad [7.1.43]$$

Nominal and effective declines can be related:

Exponential: $D_e = 1 - e^{-D}$ [7.1.44]

$$D_n = -\ln(1 - D_e) \qquad [7.1.45]$$

Hyperbolic: $D_{ei} = 1 - (1 + nD_i)^{-1/n}$ [7.1.46]

$$D_{ni} = \frac{1}{n} \left[(1 - d_{ei})^{-n} - 1 \right] \qquad [7.1.47]$$

Harmonic: $D_{ei} = \frac{D_i}{1 + D_i}$ [7.1.48]

$$D_{ni} = \frac{D_{ei}}{1 - D_{ei}} \qquad [7.1.49]$$

Most decline curves are hyperbolic with values of n = 0.0 and 0.7, whereas the majority are between 0.0 and 0.4. Published tables relating time to loss rate and N_p for various decline rates may be useful.

7.1.4.1.1 Economic Limit

In addition to the requirement for a uniform trend, estimation of reserves from production performance requires an end point. This can be an imposed limit, such as the flowrate at a certain wellhead pressure for gas wells, but is most commonly an *economic limit*.

The economic limit is the production rate at which the revenue from sale of production equals the cost of production at the same time. Continued production at or below the economic limit rate creates no economic gain and would serve no economic purpose. Of course, there may be other reasons to continue production — as many operators did after the price declines of 1986–1990 — but the estimated amount of that production is not, by definition, reserves.

Economic limit = Production × product price
           minus royalty
           minus production and *ad valorem* tax
           minus operating costs
           = zero

or

$$\text{Economic limit} = \frac{\text{Costs of production/unit time}}{\text{Product price/bbl or Mcf}}$$

$$= \text{Production/unit time}$$

Other forms of production performance analysis may be useful as an adjunct to the time-rate curve.

7.1.4.2 Production Rate–Cumulative Production

A plot of exponential production rate versus cumulative production on cartesian scale often yields a straight line. Extrapolation of this line to the economic limit production rate yields the ultimate oil or gas recovery (under the assumed economic conditions).

Reserves = ultimate recovery − cumulative recovery

7.1.4.3 Water–Oil Ratio (WOR)–Cumulative Production

In water drive reservoirs or in most waterflood or streamflooding operations, ultimate recovery can be estimated by plotting the ratio of water to oil produced (WOR) against cumulative recovery (Figure 7.1.5). The maximum WOR that can be sustained under the assumed economic conditions can be used as an economic limit. The cumulative production at that point is ultimate recovery and reserves can be estimated as above. A variation on this method is to

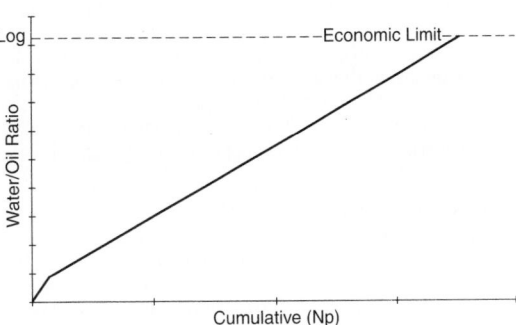

Figure 7.1.5 *Water/oil ratio and cumulative decline curve.*

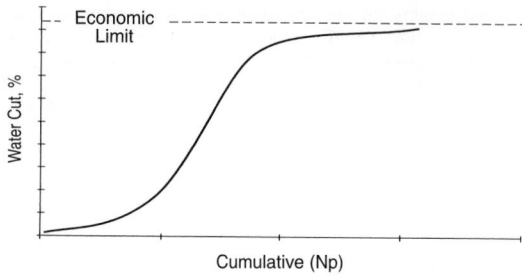

Figure 7.1.6 *Water cut and cumulative decline curve.*

use water cut (water as a % of total produced fluid) rather than WOR (Figure 7.1.6).

7.1.4.4 Gas–Oil Ratio (GOR)—Cumulative Oil Production

Use of a rate-cum or WOR-cum analysis can be used along with a material balance calculation to estimate periodic and ultimate GOR, which can then be converted to gas production. Conversely, if gas volume is a limitation on production (such as where gas cannot be sold) requiring that gas production be restricted, then GOR versus cumulative oil can be used to estimate recovery.

Analysis of production rate-cum and other variations using hyperbolic or harmonic decline trends may require semi-log or log-log treatment.

In the instance where the property being evaluated does not have sufficient history to be fully definitive, the *performance of similar properties* may be used to estimate producible volumes for the subject property. The similar property may be another well on the same lease if the subject property is a new well, or it may be a group of wells on an adjacent lease, or even a group of wells in a nearby field that produces from the same reservoir. "Similar" is a broad term and must be used with care. Similar does not necessarily mean identical — no two wells or fields are identical. The properties must, however, share the reservoir rock and fluid characteristics and operating conditions that would allow the evaluator to expect that they would perform much the same way over time. Wells in the same field (or portions thereof) and reservoir should perform in a similar manner, whereas wells in different reservoirs may perform very differently. In some cases, however, even wells in the same reservoir may perform differently if the conditions are changed. Analysis and comparison of the performance of a field or lease developed in the 1950s, for example, may not be relevant to a property developed with more modern methods in the late 1990s. The performance of infill wells may be very different from the original wells if the infill wells reduce the drainage areas of all wells.

Single-well comparisons require extreme care, but can be used by imposing the historical performance decline of one well on a new well. Multiwell or "family" curves combine the performance of several wells by overlaying the decline curve of each well and defining a composite curve through the set. The steps to this are:

1. Select the comparable wells or leases.
2. Plot the decline curves for each well or lease with initial production points set at (a) a common zero point of the wells that went on production at near the same time or (b) at date of first production. Some judgment is needed here.
3. Define a composite decline curve through the curve set.
4. Use the family curve to project the new well or lease.

Fetkovich [34,35] and others [36] have developed methods of decline curve (rate-time) analysis that are described as "type-curve" analysis. The methods are based on the idea that wells or groups of wells with similar reservoir characteristics will describe similar decline curves over time when compared on a dimensionless time-rate basis. The method requires production history sufficient to demonstrate some depletion but can then be used to determine (a) the form of the decline (including the hyperbolic constant), and (b) certain reservoir properties (assuming other data are available). Type-curve matching requires an overlay of actual rate-time data on a set of dimensionless time-rate curves. The dimensionless decline trend that best fits the actual rate-time curve can then be transferred directly to the actual curve to define the future decline trend. The method can employ a number of forms of dimensionless curves.

The match points can then be used to calculate certain reservoir properties, such as kh, using the dimensionless equations:

$$q_{Dd} = \frac{q(t)}{q_i} = q_D\left[\ln\left(\frac{r_e}{r_w}\right) - \frac{1}{2}\right] = \frac{\dfrac{q(t)}{kh(P_i - P_{wf})}}{141.3\mu B\left[\ln\left(\dfrac{r_e}{r_o}\right) - \dfrac{1}{2}\right]}$$

[7.1.50]

$$t_{Dd} = \frac{t_d}{\dfrac{1}{2}\left[\left(\dfrac{r_e}{r_w}\right)^2 - 1\right]\left[\ln\left(\dfrac{r_e}{r_w}\right) - \dfrac{1}{2}\right]}$$

$$= \frac{\dfrac{0.00634kt}{\phi\mu C_i r_w^2}}{\dfrac{1}{2}\left[\left(\dfrac{r_e}{r_w}\right)^2 - 1\right]\left[\ln\left(\dfrac{r_e}{r_w}\right) - \dfrac{1}{2}\right]}$$

[7.1.51]

Type-curve matching is similar-well analysis on a highly technical level. The method has a potential advantage in allowing an evaluator to define a decline form and quantify the hyperbolic constant earlier in the life of a property than would be the case using "eyeball" methods.

For primary reserves estimation, the production performance method is the most reliable method assuming the data are available and are properly analyzed. The characteristics of the various methods allow them to be used in progression from volumetric to performance with increasing accuracy and reliability. In practice, the methods can be shown to be complementary over time so that one method may be used as a check on another method, all other things being equal.

7.1.4.5 Production Performance in Gas Reservoirs

Estimation of future recovery from natural gas reservoirs uses the same rate-time and rate-cumulative relations as discussed for oil reservoirs, however, the fluid characteristics of gas gives greater effect to pressure. A useful relation for estimating gas reserves is the P/Z-cumulative relation (Figure 7.1.7) which plots reservoir pressure divided by the compressibility factor, Z, at time t against cumulative gas Ng. Extrapolation to the economic limit yields an estimate of ultimate recovery.

7.1.4.5.1 A Note on Production Data

Use of the production performance methods of evaluation, either alone or in combination with other methods, requires the collection and analysis of historical production data to formulate a trend that can be extrapolated into the future in the form of a production schedule. There are numerous

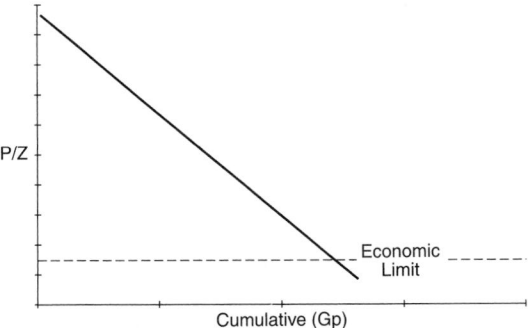

Figure 7.1.7 *Nonassociated gas—Pressure/cumulative decline.*

sources of data, including, but not limited to, company production records, sales records, well tests, reports to regulatory agencies and public data gathering and reporting firms. All data, regardless of the source, must be carefully reviewed for accuracy and to define any variations that may have occurred.

The most common variations in performance are those caused by (a) short producing periods, (b) mechanical changes, (c) regulatory restrictions, and (d) capacity restrictions.

Review and check production data for:
Duration of producing period:

- short months (February)
- wells produced less than full month

Mechanical changes:

- production reductions due to well downtime for repairs, etc.
- increase production due to change in flow valves; pump size, stroke length, or speed; workovers
- decline in production due to increasing fluid levels, etc.

Regulatory restrictions:

- proration (regulatory or contractual)
- limits on gas disposal
- limits on water disposal
- air emission requirements

Capacity restrictions:

- limited pipeline or other shipping facilities

7.1.4.6 Production Stimulation

A subset of the estimation of primary oil (and gas) reserves is the evaluation of properties that are subject to production stimulation techniques. These are methods such as cyclic steaming, fracturing, and acidizing, among others, that alter the reservoir and, possibly, operating conditions of a property but continue to rely on the natural reservoir energy to cause production to occur. These methods add no new energy and are therefore not secondary or enhanced recovery methods and, except in some arguable cases, do not add reserves; they simply act to speed the recovery of oil and/or gas that would be ultimately recovered at some time in the future. On the other hand, the stimulation of eventual production sooner rather than later *can* have important economic benefits and may result in additional reserves being credited to the property where *economic conditions* alone would suggest that the property should be shut-in or abandoned.

Cyclic steaming [37–40] is a common method in oil fields where the oil is high viscosity or has a physical composition that causes production to occur at low rates. Steam is injected into a producing well for periods of up to several days. The well is then returned to production after a period of time. The steam is used to transfer heat to the reservoir where the heat serves two primary purposes in varying degrees. Heat reduces the viscosity of the oil. It also "cleans up" the wellbore by steam cleaning the perforations and/or liner slots and the sand face of accumulated tar and sand thereby improving permeability near the wellbore.

The combination of effects may cause production to increase as much as 10 to 20 times the presteam rate. Wells that have been steamed generally increase to a peak immediately following the steaming and then decline in a definable manner over time. The production cycle may last from a few months to as much as 2 years. Wells can be steamed repeatedly, but the stimulation effect will noticeably diminish with repeated cycles as reservoir energy in the near wellbore region depletes. Eventually, the production declines to a point where the cost of steaming is equal to the incremental return and cycling steaming is stopped.

Fracturing [41–45] — the use of hydraulic energy to create fractures in the reservoir — and acidizing — the use of acid to improve near wellbore permeability — are production stimulation methods that alter the reservoir characteristics, permanently or temporarily, in such a way as to cause production to increase over a period of time. As with cyclic steaming, the methods generally increase the rate of production and, thereby, have an economic benefit but do not usually add reserves.

Estimation of reserves and future production from stimulated wells generally requires an overlay of production performance and volumetric methods. Some experience with performance using a stimulation method is necessary before a reliable projection method can be defined. Estimation of recovery from cyclic steaming must be based solely on experience and analysis of actual performance of the same or similar properties. The construction of "family" curves of wells, grouped according to the number of cycles common to the wells, can be used to estimate the change in q_p to q_s with each cycle and the point on the decline curve where the stimulated production will begin. Several authors have published methods of estimating production performance from fractured and acidized wells and these may be used as part of an estimation of reserves.

7.1.4.7 Secondary Recovery and EOR Methods

There are many differing definitions of secondary recovery and enhanced oil recovery (EOR). As used in this discussion, EOR will include recovery methods that might also be described as secondary or tertiary depending on the timing of the project and the circumstances of the reservoir. The essential difference between these methods and primary recovery is that EOR methods result in additional production by adding energy to the reservoir, whereas primary recovery uses only the natural energy of the reservoir. The purpose of EOR is to increase the reserves and production from reservoirs that can be, or have been, produced by primary means.

7.1.4.7.1 General Considerations

EOR methods can be put in three general groups:

1. Gas injection—the use of natural gas or other gas as an injection fluid
2. Water injection—use of unheated water as an injection fluid
3. Thermal methods including

   - steam and/or hot water injection
   - *in situ* combustion

These basic methods have many variants depending on reservoir conditions and the use of additives in the injection fluids. Indeed, each project design can be considered a unique method if the design is based on specific reservoir characteristics. It is not possible in this space to adequately discuss the many variations of the basic methods or even to discuss in detail the basic methods. The list of references at the end of this chapter is intended to provide a starting point for further discussion.

In virtually all EOR methods, a fluid is injected into the reservoir at one point with the intention of sweeping or flushing oil from that point to other points in the reservoir where the oil can be produced. For analysis purposes, this is most often pictured as a piston-type mechanism where the injected fluid occupies an increasingly larger part of the reservoir pushing a "bank" of oil ahead of it to producing wells.

The success or failure of an EOR project generally depends on:

1. The mobility ratio (M), where:

$$M = \frac{\text{Mobility of displacing fluid}}{\text{Mobility of displaced fluid}} = \frac{\lambda_1}{\lambda_2} \qquad [7.1.52]$$

and where

$$\lambda = K/\mu = \frac{\text{permeability}}{\text{viscosity}} \qquad [7.1.53]$$

If M = 1, the fluid mobilities are identical

Favorable $\langle M = 1 \rangle$ Unfavorable

2. *Areal and vertical sweep efficiency* — the relative proportion (%) of reservoir area and net thickness that is swept by the displacing fluid. This is largely a function of geology although mobility also plays a part.

Regardless of the EOR method used, the project must be designed to consider and account for both mobility and sweep efficiency. Most of the variations in the basic methods result from attempts to alter and improve mobility or sweep efficiency or both.

7.1.4.7.1.1 Gas Injection. The injection of natural gas either as a pressure maintenance or EOR method is by far the oldest form of injection method for increasing oil recovery. Before the time when natural gas became a major fuel source and could be transported, most produced gas was either burned off or injected into the same or another reservoir. Gas injection is virtually unknown in the United States today.

Gas can be a very efficient injection medium but is generally limited (by mobility ratio) to light oil reservoirs with thin sand sections. The method is most efficient when injection can occur in the top of a structure or into an existing gas cap where it then performs as a piston moving downward, pushing oil down or out to producing wells.

7.1.4.7.1.2 Water Injection. Water injection or waterflooding has been around for a long time. Waterflooding may be done by reinjecting produced water from the reservoir; injecting water from other reservoirs; or by mixing produced waters from various sources. Freshwater and seawater have also been used. In addition, there are many varieties of materials such as soap, carbon dioxide and other more exotic material that may be added to the water to improve mobility and/or sweep efficiency. Some of these agents act to reduce the viscosity of the oil or to increase the viscosity of the water while others act to improve the relative permeability to water which often results in stripping more oil off of such surfaces. Various mechanical methods are also used to improve sweep efficiency.

The basis for most waterflooding recovery methodologies is the Buckley-Leverett [46–48] or *frontal advance method*. The Buckley-Leverett method assumes a linear oil-bearing zone in a depleted state. The method requires a knowledge of the change in reservoir properties of k_o, k_w, μ_o, and μ_w with change in water saturation S_w. Assuming immiscible, steady-state conditions the flow through a linear block of reservoir can be modeled where

$$f_D = \frac{1 - \dfrac{1.127 k_o}{q_t \mu_o} \left[\dfrac{\partial Pc}{\partial \mu} + 0.434 \, (\Delta p) \sin \alpha \right]}{1 + \left(\dfrac{k_o}{k_w} \dfrac{\mu_w}{\mu_o} \right)} \qquad [7.1.54]$$

where f_D is the fraction of displacing fluid flowing at a given point in the system and q_i is the total flowrate per cross-sectional area. If the capillary pressure gradient $\partial Pc/\partial u$ is small and the zone is essentially flat, so that the gravitational function can be dropped, the equation can be simplified to

$$f_D = \frac{1}{1 + \left(\dfrac{k_o}{k_w} \dfrac{\mu_w}{\mu_o} \right)} \qquad [7.1.55]$$

The distance (u) that a plane of constant S_w has advanced at time (t), can be calculated as

$$u = \frac{Q_t}{A \phi} \left(\frac{\partial f_D}{\partial S_D} \right) t \qquad [7.1.56]$$

where Q_t is the injection rate which is assumed constant. The slope $\partial f_D / \partial S_D$ is obtained by plotting f_D for various values of S_D. The time to water breakthrough is then

$$t_B = \frac{AL\phi \left(\bar{S}_{wf} - S_{wi} \right)}{Q_t}$$

where S_{wf} is the mean water saturation in the system. S_{wf} is obtained by projecting a line from S_{wi} tangent to the $\partial f_D/\partial S_D$ curve to $f_D = 1$. If water is the displacing fluid and S_{wi} is the irreducible saturation, then oil production (q_o) until breakthrough is equal to water injected (q_{wi}).

If water is mobile, then $q_t = q_o + q_w = q_{wi}$ and the water/oil ratio is

$$WOR_{surf} = \frac{k_w \mu_o B_o}{k_o \mu_w B_w} \qquad [7.1.57]$$

and

$$q_o = \frac{(1 - f_w) Q_t}{B_o} = \frac{Q_t}{B_o + (WOR) B_w} \qquad \text{at time t} \qquad [7.1.58]$$

After breakthrough, oil production is determined by obtaining values of S_{wf} at various values of f_w up to an abandonment f_w; calculating t for each value of S_{wf}; determining WOR at each f_w; and calculating q_o for each value of WOR.

The Stiles [49] method takes a frontal advance approach but assumes the zone is composed of layers of constant thickness but each layer may have a different permeability. Stiles requires a constant q_i and predetermined S_o after flooding. The method extends the frontal advance approach to determine ultimate recoverable oil and then uses f_w to determine the recovered fraction as a function of time. The method allows a direct derivation of q_o, cumulative oil and WOR.

Suder and Calhoun presented another frontal advance method that allows calculation of injection rates and assumes radial flow up to a point. The method requires a predetermined value for S_{or}.

Dykstra and Parsons [50] also presented a method based on permeability stratification and Johnson [51] converted the method to a graphical approach. The method allows a consideration of permeability variation. The Johnson graphical approach is particularly useful for quick estimates before engaging in one of the more detailed calculations. The set of four graphs is defined at producing WOR of 1, 5, 25, and 100 and result in values of recovery factor R_f for each WOR. Assuming recoverable oil and injection rate can be estimated, the method allows calculation of production rates and cumulative recovery over time. The method works well as a first approximation and is simple enough to use that several analyses can be done for varying conditions

There have been many variations of these methods proposed and used. In addition, the inclusion of variations in reservoir fluid or rock conditions is only a matter of how much work the evaluator is willing to do. These basic methods underlie most if not all computer models used to simulate waterflooding.

7.1.4.7.1.3 Heat Injection Methods. [52] The primary purpose of heat injection using steam or hot water is to transfer heat from the injected fluid to the crude oil and reservoir rocks to reduce oil viscosity. Reduction of viscosity in the oil — at least at the water–oil contact — results in an improved mobility ratio, thereby allowing high viscosity oils to be recovered.

Steam is particularly useful for this purpose because steam can carry a much greater quantity of heat per unit volume than can hot water or heated gasses.

Although some attempts to use superheated steam have occurred, in practice steam at about 80% quality (80% vapor/20% liquid) is most commonly used. Heat injection is a costly process due to the requirement to burn fuel to generate steam or hot water. The major limitation on heat injection methods is heat loss in surface facilities, distribution lines, wellbores, and in the reservoir, to over and underlying rocks and to water in the reservoir. Because of the heat losses in the wellbore, steam injection is normally limited to 3,000 ft. or less in depth. Reservoir heat losses cause the injected fluid to continually cool as the steam front advances from the injection sand face. Continual steam injection is required to attempt to maintain a heated oil–water interface at the flow front. Many authors have described and quantified the heat losses in surface, wellbore and reservoir rocks and reference should be made to those sources.

Marx and Langenheim [53] have presented a series of equations designed to determine the radial distance at which the heat loss ratio in the reservoir equals the heat injection rate.

The inclusion of cost factors ($h and $0) allows the calculation to be thought of as an economic limit as well as a physical limit. The calculation assumes no breakthrough but otherwise could be used to define well spacing for a continuing flood.

$$\left[e^{x^2}\text{erfc}\,x\right]_1 = (5.618 \times 10^{-6})\left[\frac{\$_h M\,(\Delta T)}{\$_o\phi\,(S_o - S_{or})}\right] \quad [7.1.59]$$

where

$$x = \frac{2k_{ob}t^{1/2}}{Mh\sqrt{a_D}}, \quad \text{dimensionless time} \quad [7.1.60]$$

and

$$M = (1 - \phi)\,\rho_f C_f + S_w\phi\rho_w C_w + S_o\phi\rho_o C_o \quad [7.1.61]$$

The cumulative heated area (ft.^{2}) can be calculated at time (t):

$$A\,(t) = \frac{H_o Mh\alpha_D}{4k_{ob}^2\,(\Delta T)}\left(e^{x^2}\text{erfc}\,x + \frac{2x}{\sqrt{\pi}} - 1\right) \quad [7.1.62]$$

where H_o = constant injection rate in Btu/hr

Several approaches have been presented for estimating steam drive performance. In essence, steamflooding is analogous to cold waterflooding and the same principles apply with additional recovery due to:

1. thermal expansion of the oil

2. viscosity reduction, which improves mobility
3. steam distillation of some light oil components

Thermal expansion and viscosity reduction can be determined in laboratory analysis and the effects can be built into the calculation. Estimation of results from distillation are more difficult to assess and include.

One author has proposed methods of determining steam drive performance that envisions a series of displacements occurring. A cold water–oil displacement front; a hot water and condensate oil front; and a steam-condensate and hot water displacement which is partially miscible.

The radial position of the steam front is calculated as

$$R_{ST}^2 = \frac{14.6i_{st}H_{fg}}{k\,(T_{ST} - T_f)}\sqrt{\frac{\alpha}{\pi}}\left[\frac{\sqrt{t}}{2} - \frac{h}{8}\sqrt{\frac{\pi}{K}}\frac{(\rho C_p)_f}{(\rho C_p)_{OB}}\right.$$
$$\left. \times \ln\left(\frac{4}{h}\sqrt{\frac{\alpha}{\pi}}\frac{(\rho C_p)_{OB}}{(\rho C_p)_f}\sqrt{t} + 1\right)\right] \quad [7.1.63]$$

The steam injection rate must increase as the flood front expands in order to maintain temperature. The required steam rate is

$$i_{st} = \frac{\pi R_{ST}^2\,(T_{ST} - T_f)}{14.6H_{fg}t}\left[h\,(\rho C_p)_f + 4k\sqrt{\frac{t}{\pi\alpha}}\right] \quad [7.1.64]$$

In general, however, the Buckley-Leverett approach used for waterflood can be modified for steamflood. For a radial system,

$$\left[\frac{\partial\,(R^2)}{\partial t}\right]_{S=\text{const}} = \frac{i_w}{13.43}\left(\frac{\partial fw}{\partial sw}\right)_{S=\text{const}} \quad [7.1.65]$$

where i_w = effective injection rate in bbl/day of steam, and

$$fw = \frac{1}{1 + \dfrac{k_o\mu_w}{k_w\mu_o}} \quad [7.1.66]$$

Both i_w and f_w (as a function of S_w) are functions of temperature; therefore, the above equation is modified to

$$(R^2)_{S=\text{const}} = \frac{1}{13.43}\int_o^t i_w\left(\frac{\partial fw}{\partial Sw}\right)_{S=\text{const}} dt \quad [7.1.67]$$

which is amenable to graphical solution as outlined above for waterflood, assuming isothermal steps.

Given the complexity of these calculations and the many variations presented by authors, the evaluator is well advised to make maximum use of data from heat injection projects in similar fields to determine reservoir and production performance over time. Data on many EOR projects have been published in the 1970s and 1980s as the result of DOE projects and a high level of interest in industry.

In 1980, Gomaa [54] presented a method of estimating steamflood performance based on correlations to reservoir characteristics with steamflood injection rates, pattern sizes and shapes, and other factors from existing projects and laboratory studies. The method uses a series of graphical correlations to provide an estimate of reserves and recovery rate. The method is useful in quick evaluations of steamflood project potential.

Heat injection methods of EOR have numerous advantages and disadvantages.

Advantages:

- increase recovery from high viscosity oil reservoirs
- increase production rates from high viscosity reservoirs

Disadvantages:

- requires extensive reservoir, rock, and fluid analysis

- high capital investment of steam generators, new injection wells, and surface systems.
- high operating costs for steam or hot water generation, hot fluid and feedwater treating, and personnel
- subject to environmental and other regulatory constraints

References

1. Society of Petroleum Engineers, *Definitions of Oil and Gas Reserves*, 1997.
2. *Guidelines for Application of Petroleum Reserves Definitions: Monograph 1, Second Edition*, Society of Petroleum Evaluation Engineers, Houston Texas, October 1998.
3. *U.S. Crude Oil, Natural Gas, and Natural Gas Liquids Reserves: 2001 Annual Report*, Energy Information Agency, U.S. Department of Energy, Appendix G.
4. *Regulation S-X, Rule 4-10—Financial Accounting and Reporting of Oil and Gas Producing Activities*, U.S. Securities Exchange Commission. See also *Statement of Financial Accounting Standards No. 69: Disclosures about Oil and Gas Producing Activities*, Financial Accounting Standards Board, November 1982.
5. *Federal Securities Law Reports*, Commerce Cleaning House, 1981, Reg. 210.1-01, paragraphs 69, 101; Accounting Rules (Regulation S-X), Form and Content of and Requirements for Financial Statements, Securities Act of 1933.
6. Bradly, H. B., Ed. *Petroleum Engineering Handbook*, Society of Petroleum Engineers, Richardson, Texas, 1992, pp. 20-1–21-15.
7. Standing, M. B., *Volumetric and Phase Behavior Of Oil Field Hydrocarbon Systems*, Reinhold Publishing Co., 1952, pp. 1–9.
8. Gatlin, C., *Petroleum Engineering*, pp. 1–5, Prentice-Hall, Inc., Englewood Cliffs, New Jersey, 1964.
9. Bradley, H. B., (Ed.,) *Petroleum Engineering Handbook*, pp. 40-1–40-32, Society of Petroleum Engineers, Richardson, Texas, 1992.
10. Cronquist, C., *Estimating and Classification of Reserves of Crude Oil, Natural Gas, and Condensate*, Society of Petroleum Engineers, Richardson, Texas, 2001.
11. *Determination of Oil and Gas Reserves, Petroleum Society Monograph No. 1*, The Petroleum Society of Canada, Clgary, Alberta, 1994.
12. Bradley, H. B., (Ed.,) *Petroleum Engineering Handbook*, pp. 40–44, Society of Petroleum Engineers, Richardson, Texas, 1992.
13. Frick, T. C., *Petroleum Production Handbook*, pp. 37-10–37-20, McGraw-Hill Book Co., New York, 1962.
14. Bradley, H. B., (Ed.,) *Petroleum Engineering Handbook*, pp. 40-5–40-8, Society of Petroleum Engineers, Richardson, Texas, 1992.
15. Campbell, J. M., et al., *Mineral Property Economics*, pp. 23–44, Campbell Petroleum Series, 1978.
16. Amyx, Bass & Whiting, *Petroleum Reservoir Engineering*, pp. 561–598, McGraw-Hill Book Co., New York, 1960.
17. Wahl, W. L., Mullins, L. D., and Elfrink, E. B., "Estimating of Ultimate Recovery from Solution Gas Drive Reservoirs," *Journal of Petroleum Technology*, June 1958.
18. Higgins, R. V., "Calculating Oil Recoveries for Solution-Gas Drive Reservoirs," *RI 5226*, USBM, April 1956.
19. "Material Balances in Expansion Type Reservoirs Above Bubble Point," *Journal of Petroleum Technology*, October 1955.
20. Buckley, S. E., and Leverett, M. C., "Mechanism of Fluid Displacement in Sand," *Transactions of AIME*, 1942.
21. Welge, H. J., "A Simplified Method of Calculating Recovery by Gas or Water Drive," *Transactions of AIME*, 1952.
22. Woods, W. W., and Muskat, M., "An Analysis of Material Balance Calculations," *Transactions of AIME*, 1945.
23. Muskat, M., and Taylor, M. O., "Effect of Reservoir Fluid and Rock Characteristics on Production Histories of Gas Drive Reservoirs," *Transactions of AIME*, 1946.
24. Muskat, M., *Physical Principles of Oil Production*, p. 177, McGraw-Hill Book Co., New York, 1949.
25. Tarner, J., "How Different Size Gas Caps and Reserve Maintenance Programs Affect Amount of Recoverable Oil," *Oil Weekly*, June 12, 1944.
26. Bradley, H. B., (Ed.,) *Petroleum Engineering Handbook*, pp. 40-8–40-12, Society of Petroleum Engineers, Richardson, Texas, 1992.
27. Bradley, H. B., (Ed.,) *Petroleum Engineering Handbook*, pp. 40-21–40-26, Society of Petroleum Engineers, Richardson, Texas, 1992.
28. Craft, B. C., and Hawkins, M. F., *Applied Petroleum Reservoir Engineering*, pp. 14–48, Prentice-Hall Inc., Englewood Cliffs, New Jersey, 1959. See also *Applied Petroleum Reservoir Engineering*, 2nd Edition, revised by R. E. Terry, Prentice-Hall, 1991.
29. Bradley, H. B., (Ed.,) *Petroleum Engineering Handbook*, pp. 40-26–40-32, Society of Petroleum Engineers, Richardson, Texas, 1992.
30. Arps, J. J., "Estimation of Primary Oil Reserves," *Transactions of AIME*, 1956.
31. Arps, J. J., "Analysis of Decline Curves," *Transactions of AIME*, 1945.
32. Cronquist, C., *Estimating and Classification of Reserves of Crude Oil, Natural Gas, and Condensate*, pp. 120–142, Society of Petroleum Engineers, Richardson, Texas, 2001.
33. *Determination of Oil and Gas Reserves, Petroleum Society Monograph no. 1*, pp. 222–235, The Petroleum Society of Canada, Calgary, Alberta, 1994.
34. Fetkovich, M. J., "Decline Curve Analysis Using Type Curves," *Journal of Petroleum Technology*, June 1980.
35. Fetkovich, M. J., et al, "Decline Curve Analysis Using Type Curves: Case Histories," *SPE no. 13169*, 59th Annual Technical Conference, Houston, Texas, September 1984.
36. "The ability of rate-time decline curves to predict future production rates," MS thesis, University of Tulsa, Tulsa, OK, 1968.
37. "Review of Steam Soak Operations: California," *Journal of Petroleum Technology*, August 1972.
38. Hong, K.C., *Steamflood Reservoir Management: Thermal Enhanced Recovery*, pp. 19–22, Pennwell Books, Tulsa, Oklahoma, 1994.
39. "Steam Stimulation: Correlation of Performance," *Journal of Petroleum Technology*, November 1972.
40. Smith, C. R., *Mechanics of Secondary Oil Recovery*, pp. 471–476, Reinhold Publishing Co., 1966.
41. Van Poollen, H. K., et al., "Hydraulic Fracturing — Fracture Flow Capacity vs. Well Productivity," *Transactions of AIME*, 1958.
42. *Enhanced Oil Recovery*, National Petroleum Council., 1984.
43. *Determination of Oil and Gas Reserves, Petroleum Society Monograph no. 1*, pp. 187–189, The Petroleum Society of Canada, Calgary, Alberta, 1994.
44. Tinsley, J. M., et al., "Vertical Fracture Height — Its Effect on Steady State Production Increase," *Journal of Petroleum Technology*, May 1969.
45. McGuire, W. J., and Sikora, V. J., "The Effect of Vertical Fractures on Well Productivity," *Journal of Petroleum Technology*, October 1960.
46. Smith, C. R., *Mechanics of Secondary Oil Recovery*, Chapter 9, p. 282, Reinhold Publishing Co., 1966.

47. Smith, C. R., *Mechanics of Secondary Oil Recovery*, Chapter 5, p. 148, Reinhold Publishing Co., 1966.
48. Buckley, S. E., and Leverett, M. C., "Mechanism of Fluid Displacement in Sand," *Transactions of AIME*, 1942.
49. Stiles, W. E., "Use of Permeability Distribution in Water Flood Calculations," *Transactions of AIME*, 1942.
50. Dykstra, H., and Parsons, R. L., "The Prediction of Waterflood Performance with Variation in Permeability Profile," *Production Monthly*, 1950.
51. Johnson, J. P., "Predicting Waterflood Performance by Graphical Representation of Porosity and Permeability Distribution," *Journal of Petroleum Technology*, November 1965.
52. Marx, J. W., and Lagenhiem, R. H., "Reservoir Heating by Hot Fluid Injection," *Petroleum Transactions of AIME*, 1959.
53. Hong, K. C., *Steamflood Reservoir Management: Thermal Enhanced Recovery*, pp. 169–233; 313–469, Pennwell Books, Tulsa, Oklahoma, 1994.
54. Gomaa, E. E., "Correlations for Predicting Oil Recovery by Steamflood," *Journal of Petroleum Technology*, 1980.

7.2 ESTIMATING THE VALUE OF FUTURE PRODUCTION [1,4]

The earlier discussion of the various definitions of reserves included the reference to economics. Reserves are, by definition, not simply a physical volume but an economically recoverable volume. By going through the process of estimating (1) oil and/or gas-in-place, (2) the recoverable volume, and (3) the projected future production, the physical aspects of the reserves definitions are met. However, as defined, reserves are only the economically recoverable volume. There are two economic functions that are applicable to the final estimation of reserves. The more obvious is the economic limit that establishes a termination point of production. Less apparent, however, is that calculation of the economic limit and, to a large extent, the determination of the most efficient development and production methodology is dependent upon an economic evaluation of the property or project. That is, the economic limit cannot be determined without an analysis of future product prices and costs of production. The latter, in turn, is dependent upon the production mechanism and operating requirements of the property. This strongly suggest that the economic evaluation is a vital and interactive part of the process of estimating reserves in addition to the overriding role of estimating the value of the future production and of the property. No property evaluation for any purpose is complete without a determination of the economic value of future production.

7.2.1 The Market for Petroleum [5,6]

7.2.1.1 Supply and Demand

Crude oil, natural gas and the derivatives of both — are the primary sources of energy worldwide. According to studies published by the DOE, in 2001 crude oil was used to produce 156.5 quadrillion BTUs of energy, while natural gas was used to produce 93.1 quadrillion BTUs, amounting to 38.75% and 23.05% respectively of total energy consumption. The balance was made up of coal at 23.7%, along with nuclear and "other." The fastest growing energy source is natural gas, which is projected to expand by 2.8% per year through 2025, while crude oil usage is expected to increase by 1.8% per year. The industrialized countries of North America and Western Europe currently account for 45.5% of energy consumption, of which 64% is crude oil and natural gas. The fastest growing areas of energy consumption is the developing countries of Asia, including China where energy consumption is projected to grow at 3.0% per year through 2025, with the largest fuel growth being oil and gas.

These projections suggest that world crude oil demand will more than double from 2001 usage, while natural gas demand will triple. The same DOE analysis suggests that supply of crude oil and gas will be found to meet the projected demand. Petroleum is likely to retain this leading position primarily because there are very large volumes of oil and gas already discovered and, particularly in comparison to the cost of developing alternative sources of energy; they are available at relatively low prices. The primary sources of crude oil are the countries of the Middle East and Persian Gulf, which have combined proved reserves (USGS) of 685.64 billion barrels at 2000 or 56.5% of the estimated world total of proved reserves. The region also has the largest expected growth in reserves, so that the Middle East could represent the primary source of crude oil through 2025.

Weighed against this forecast are projections of world oil depletion and the increasing difficulty of exploration and development of new crude oil and natural gas sources in remote areas and harsher environments, coupled with increasing government and political resistance to resource development in the industrialized countries and environmentally sensitive regions.

The demand for crude oil, and to a lesser extent natural gas, is supported by the fact that known reserves are concentrated in the Middle East. Reserves can be produced at low cost when compared to production in the United States and other parts of the world. Persian Gulf reserves are large enough and production sufficiently controlled, even though somewhat erratically, by the Gulf countries, that the increase in usage of higher-cost energy sources such as nuclear or environmentally acceptable coal would be very difficult unless heavily subsidized.

7.2.1.2 The Shift Toward Natural Gas

One of the more important trends in the energy supply/demand equation is the shift, particularly in the heavily industrialized regions of North America and Western Europe, from crude oil, and to some extent coal, toward natural gas. Part of this shift is economic as large supplies of natural gas have become available, but the leading cause, particularly in the United States, is the pressure of regulations requiring "cleaner burning" fuels for power generation. A secondary source of demand for natural gas is the petrochemicals industry. The shift toward increased use of natural gas has lead to increased production of natural gas, which has reduced pressure on crude oil production.

Consumption of crude oil and natural gas are projected to continue at a combined growth rate of 2% per year through 2025, however, domestic production of crude oil has been in decline since 1970 with 1998 oil production at 65% of the peak 1970 rate, which did not include ANS production. Natural gas production has increased in the United States since the mid-1980s, but is still below the peak rates of the early 1970s, while proved reserves have declined sharply over the same period. As a result, the United States is a net importer of both crude oil and natural gas, with 30% of crude oil coming from Canada and Mexico and another 23% from South America, principally Venezuela.

In the United States and other areas where similar industry conditions exist, petroleum economics will be controlled, indirectly and directly, for the foreseeable future by the baseline oil price either set by or derived from major producing countries. Under market conditions, oil prices in the United States cannot rise much above the world market level but can readily fall below that level. Production in the United States is generally in decline so that the United States is a net importer of over 60% of crude oil demand, thereby tying the U.S. market more closely to world markets. In addition, U.S. production is very high cost relative to the Persian Gulf or anywhere else so that the difference between price (revenue) and cost of production can be, and often is, very narrow. As shown by the significant price declines of 1985–1986 and more recently in 1997–1998, a decrease in oil price is often enough to render a large volume of U.S. production uneconomic, causing wells and properties to be shut in and abandoned; resulting in the cancellation or deferment of new drilling, exploration and other capital investment projects; and bringing about the financial collapse of oil companies, service companies, and whole regions that depend on the oil industry.

The primary impact of the market for petroleum is, of course, on price. Prices for oil and natural gas have historically varied with demand and supply. In the United States, there have been periods of high production relative to demand causing prices to drop; as well as periods of high demand relative to production that resulted in price increases. For a long period, however, from the early 1930s until 1971, oil production was controlled by proration that limited production, particularly in Texas but also in other areas, to a certain "allowable" each month that was expected to fulfill, but not exceed, demand. The allowable production was set as a percentage of productive capacity. This regulation resulted in stable and relatively low oil prices for most of that period. The increase in allowables over time to 100% in the early 1970s was one of the major reasons that control of the world market passed from the United States to OPEC and from industry to governments in the mid-1970s.

OPEC control dominated the market for petroleum for most of the 1970s and early 1980s. The focus of the cartel was to maintain high prices by restricting supply. The first result was an increase in the wellhead price of crude oil from an average of $3.18/Bbl in 1970 to $31.77 in 1981. This rapid increase in oil price brought about an increase in the value of oil and gas properties, and allowed the development of new production in the North Sea, Africa, South America, offshore Gulf of Mexico and Alaska North Slope (ANS), making it possible to justify expensive EOR projects in the United States,

Canada and elsewhere. This activity peaked in the mid-1980s and had the effect of (1) increasing the supply of crude oil and natural gas, (2) diversifying the sources of supply away from OPEC countries, and (3) reducing demand in the industrialized countries through fuel conservation and a shift away from crude oil.

By 1986, OPEC could no longer control oil supply, and the price dropped so that by 1988 the average US wellhead price was $12.58. Since that time, OPEC has had influence at the margin, but government/cartel price controls have been largely ineffective. Oil prices have become more volatile since the late 1980s, as shown by wide swings in price during the Gulf War in 1991, the significant price decline of 1997–1998 and the recent escalation in price that has brought average wellhead prices to over $20 in 2003.

The U.S. economy is energy intensive. In this situation, changes in oil and gas prices take on national significance and can have immediate and serious impacts on the economy of regions and the country as a whole. The demonstration of this is in the relation between oil price and inflation. Since 1928, the first year that reliable inflation data in the form of the Consumer Price Index was kept, changes in oil price can be shown to be closely followed by changes in CPI or, more broadly, inflation. Major price increases such as 1971–1973 and 1979–1980 resulted in serious increases in inflation in those and following years. For most of the period from 1928 to 1974, however, oil production exceeded demand. Price was regulated through proration and in many years oil price declined. Inflation, however, being caused by many factors, continued even though at low rates so that, while the nominal or actual oil price may have increased slightly or remained essentially constant, the "real" price—the nominal price minus inflation—actually declined for many years and has, in fact, declined significantly since 1982 (Figures 7.2.1 and 7.2.2).

The important relation to be observed here is that the changes in oil and, to an extent, gas prices tend to cause changes in inflation, but changes in inflation do not necessary result in changes in oil or gas prices. As market commodities, neither oil nor gas is subject to inflation except in the cost of production. Therefore, although it is interesting

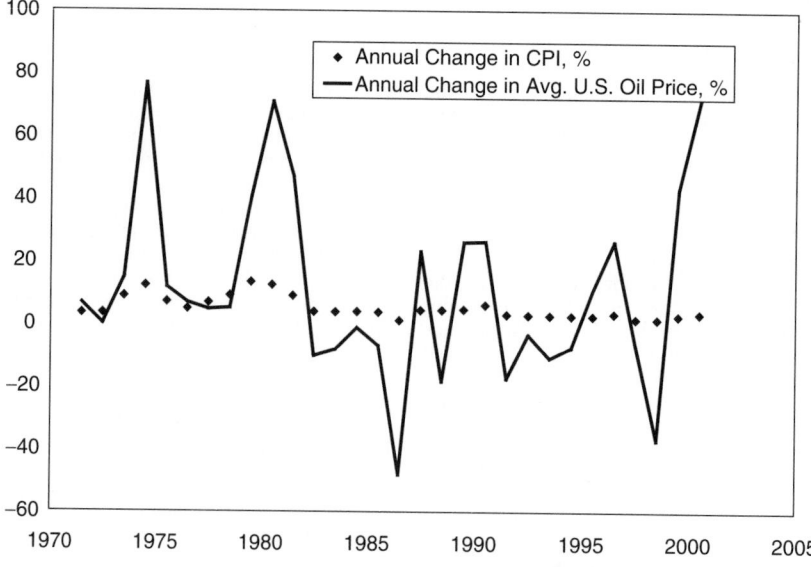

Figure 7.2.1 *Oil price history versus inflation.*

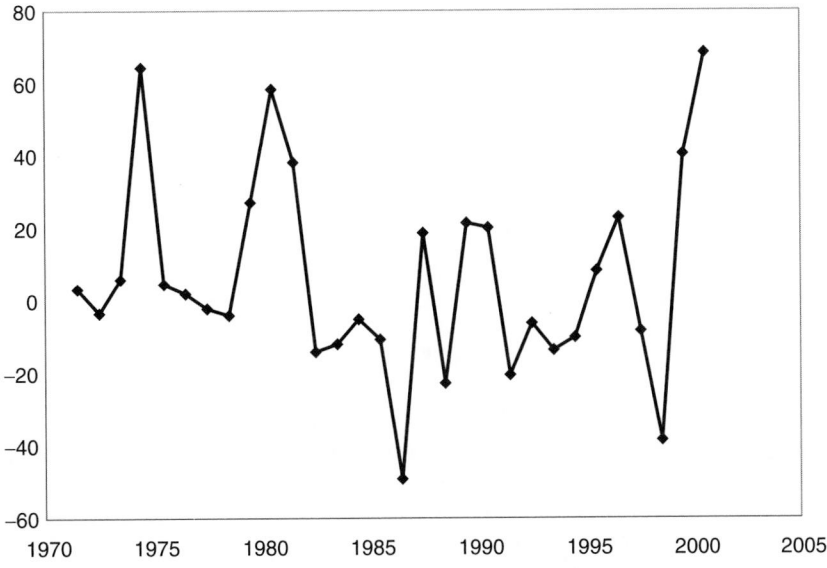

Figure 7.2.2 *Annual change in real oil price.*

to point out the decline in "real" oil prices and to speak in "real" terms, the commodity nature of crude oil and natural gas mean that only the nominal value or price should be of concern.

7.2.1.3 Economic Cycles and the Petroleum Industry [7]
During the period of the 1980s and 1990s, the oil business endured or, at times, benefited from several cycles that brought about extreme fluctuation in the prices for oil and gas, changes in interest rates from as low as 3% to over 20%, substantial declines in domestic U.S. production and continually increasing regulation of the physical and economic aspects of the industry worldwide. It is important to have an understanding of the effects of the cycles in prices, interest rates, regulation and other economic factors in the evaluation process and as a component of the valuation of oil properties for sale/acquisition, taxes, litigation, and other purposes.

Economic cycles come in all shapes, sizes and durations. Economic cycles may coincide with business cycles, or they may not. Economic cycles can be general and have an impact on all industries in the board economy, or they can be limited to a segment of the general economy. It is not unusual to have changes in the "economy" of the auto industry or the steel industry; there is currently both a general recession and a particularly strong downturn in the so-called 'tech' sector of computers, software and semi-conductors. Economic cycles are common in the oil and gas industry. In many cases, the cycle can be shown to have an impact on the industry as a whole, or only the effect may be limited to an industry segment, such as the natural gas or refining and not necessarily production.

Changes in the economic cycle measured by interest rates, etc. influence the general economy and have varying effects on the components of the economy — business, personal, and government. Such changes may be beneficial to one industry and adverse to another industry.

The petroleum industry is characterized by large capital investments in plant, equipment exploration, production development, and transportation facilities. A new refinery or production development project can require billions of dollars in capital investment, require years to complete and

then must depend upon many years of income from commodity sales to obtain a return of an on the initial investment. Capital intensive industries, such as oil and gas, are more susceptible to changing interest rates than service industries, but on the other hand, do not gain or lose directly as does the financial industry. Similarly, inflationary expectations, more so than the activity or inflation, can have an influence on industries with large investments to be recovered over long periods of time.

The economic cycles of the petroleum industry are relatively easy to identify and measure. Interest rates are not useful, because the same changes in prime rate and corporate bond rates that affect Ford, General Electric, and IBM also affect oil industry borrowers. The degree may be different, but there is no separate debt market for the oil industry. Inflation can have a disparate effect on the petroleum industry by causing an increase in (1) costs of operation, production, and (2) capital investment. However, since the overwhelming majority of oil industry income is dependent upon commodity prices for crude oil, gasoline, etc., and are not subject to inflation, periods of high inflation can significantly impact petroleum industry economics.

Equity returns can be used to measure economic cycles in the oil industry. Viewed from 1968 forward, AFIT return on book equity has gone through several very obvious cycles or changes from lows of 10% or less in 1973 and the 1986–1987 period, to peaks over 20% in 1979–1980 and 2000–2001. As shown in Figure 7.2.3, the changes in return-on-equity (ROE) have in some periods been very sharp, such as in 1973–1974, 1979–1980 and 1985–1987. When viewed as quarterly data, from 1988 forward the changes are a bit more moderate and do not appear as sharp until one looks at 1997–2002 quarter by quarter.

These equity returns contrast in some interesting ways with interest rates. Although oil industry ROE and corporate bond yields followed a similar path in the late 1960s and 1970s and both peaked in the 1980–1982 period, interest rates, while declining, take a shallower path than do oil industry equity returns, which drop precipitously after 1982 through 1988. Oil industry ROE fell below the BAA bond rate for a couple of years in the middle 1980s.

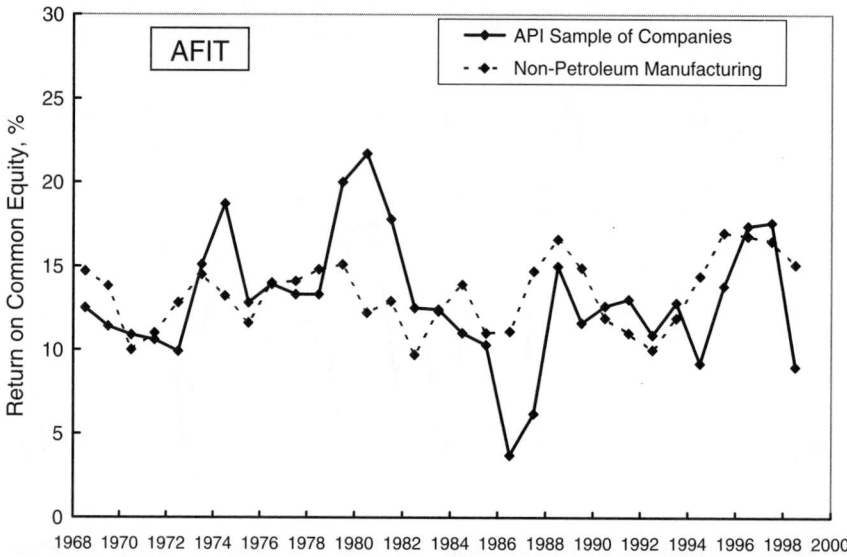

Figure 7.2.3 *Return on commodity equity comparison of petroleum industry to other industry.*

Oil industry equity industry returns have also varied to some degree from the equity returns from other industries, as shown in Figure 7.2.4. During the latter half of the 1990s, oil industry return on book equity lagged other industrials, and when it did catch up in the 1996–1997 period, other industrials began to decline.

7.2.1.3.1 Oil Price and Industry Economic Cycles
The fortunes of the oil and gas industry can be traced to the prices for two commodities — crude oil and natural gas. Of these, crude oil is arguably the more important and the more easily measured. The monthly average posted price for West Texas Intermediate (WTI) from 1946 through 1973 was essentially flat. After the 1973 oil embargo, as shown in

Figure 7.2.5, WTI price began to increase up to $40/bbl in 1980. This euphoric peak was followed by a steep decline to 1986, followed by an even sharper drop. Since 1987, WTI price has varied considerably in a broad range between $10 and $25/bbl.

7.2.1.3.2 Oil Price and Equity Returns
It should not be surprising that the oil industry ROE tracks relatively closely to the WTI posted price. Each marginal dollar change in oil price results in a more or less direct change in industry revenue, earnings and anticipated return-on-investment. A comparison of WTI posted price to industry ROE indicates a very close relationship, particularly since the spike in 1980. As might be expected, for most of the

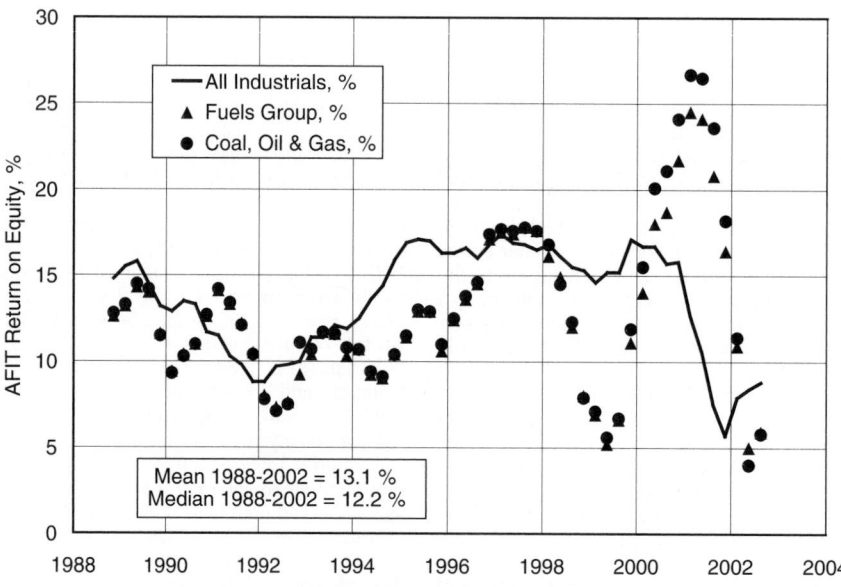

Figure 7.2.4 *Corporate after income tax return on book equity.*

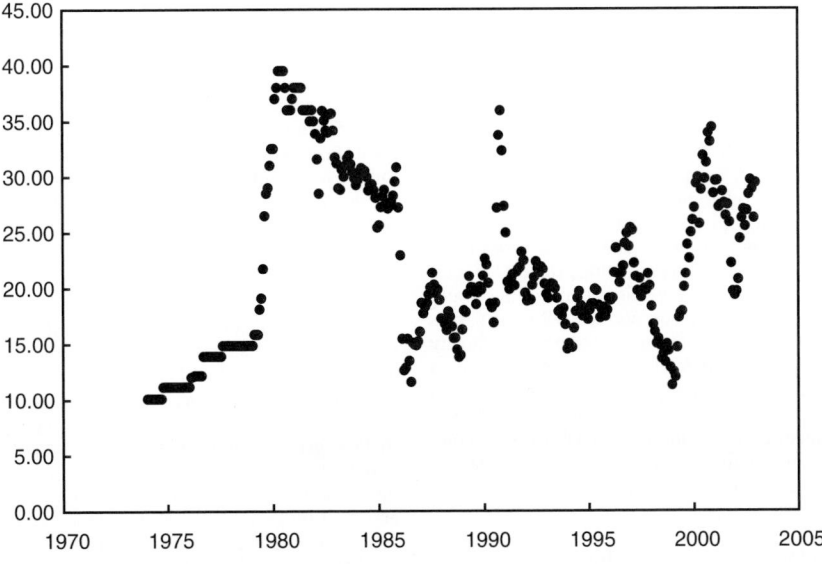

Figure 7.2.5 *West Texas intermediate posted price.*

period since 1980 changes in ROE lag changes in oil price. Of course, WTI is not the only crude oil, and oil company returns are made up of far more than revenues from crude oil, but it is hard to escape the relationship between the two. It might be instructive to compare returns to natural gas prices or to retail/wholesale gasoline prices. Another interesting view is shown in Figure 7.2.6, which adds in the returns from nonpetroleum manufacturing companies.

7.2.2 Economics and the Petroleum Engineer

The function of the petroleum engineer is to determine how to produce the most oil and/or gas from a property in the most efficient manner and at the least cost to optimize the economic return from that production. Producing the most

oil from a property in the most efficient manner explicitly includes production and reservoir management so as to optimize *both* production rate and ultimate recovery from the reservoir. The petroleum engineer has little or no control over product price. Therefore, he or she must attempt to minimize production costs and investment over the life of the property so that, whatever the price, the economic return is optimized. This means that virtually every decision made regarding the development and maintenance of production must consider the *economic costs and benefits* of the decision. This is true regardless of whether the project is drilling a new well, changing a pumping unit, considering a stimulation method, installing an EOR project or buying a property. The benefits may be increased production rate, increased reserves, or a reduction in costs, or some combination; while

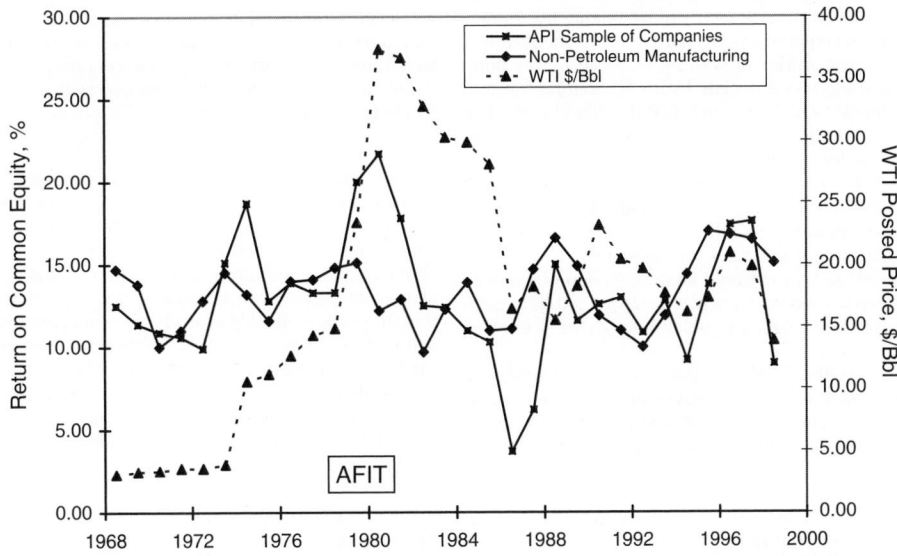

Figure 7.2.6 *Return on commodity equity comparison of petroleum industry to other industry.*

the costs may include new capital investment or increased operating costs.

The relation between benefits and costs can be measured in various ways. Among these are the undiscounted net profit, the net present value (NPV), the (internal) rate of return (ROR) and various other methods as will be discussed in detail later. Each of these differing views provides a somewhat different perspective on project evaluation: (1) measures a particular relationship between future income and capital investment, and (2) allow the engineer or other decision maker to not only evaluate a particular project, but allows rational comparisons when selecting projects for completion. As an example, if a new well is proposed that would result in a net profit of $1 million and a ROR of 19%, this project must be compared to other projects and to company criteria for net profit and ROR. If the company requires a minimum ROR of 15%, the project might be accepted. If, on the other hand, the net profit was $1 million but ROR was only 9%, the project ROR would not only fail to meet company guidelines, but might be less than could be obtained on investments where there is no risk.

7.2.3 Purposes for an Economic Evaluation

There are many reasons for estimating the economic value of an oil and gas property or project and a wide variety of methods used to make the estimate. However diverse the reasons, they can generally be placed in two groups: *investment analysis* and *regulatory compliance*. The latter are special valuation cases done for SEC filings, ad valorem tax valuation, and for other tax requirements such as estate tax.

In most cases, however, valuation is done for *investment analysis*. This could be anything from the evaluation of the potential economic benefit from a workover or equipment upgrade to drilling a new well to acquiring a producing property. Irrespective of the purpose, the practical aspects are identical. In investment analysis, the evaluator is seeking to determine the additional benefit in terms of future income, profit, and return-on-investment that could be obtained by making a particular investment. As an example, if it will cost $100,000 to workover a well, acidize a well, or open a new production interval, will the additional production increase revenues sufficiently to pay back the $100,000 in a reasonable time and provide a return on the investment that compensates for the risk incurred by making the investment? If payback requires 3 years, and there is a measurable possibility that the workover will not increase production, then there is some risk associated with (1) waiting 3 years to get the investment back, and (2) the likelihood of a lesser than expected income stream. The economic evaluation attempts to determine (1) if the future return is enough to compensate for the combined risks of the investment, and (2) whether the investment should be made. The same principles apply to spending $3 billion to acquire a producing property or to making any other investment.

For the most part, investment analysis is done by evaluators for internal corporate purposes and/or for financing support where the evaluation guidelines may be predetermined.

Regulatory evaluation also follows rules or guidelines, but they are established by a government agency for a specific purpose. The SEC requires valuations of proved reserves for all public oil and gas companies to provide equity investors with standardized information about those companies. The procedures incorporate standard petroleum engineering practices but require fixed prices and costs and the use of a 10% discount rate.

In Texas and California, ad valorem taxes are assessed against oil and gas properties based on the value of the future income expected from the property. Estimation of the tax generally requires a determination of the market value of the property. Estimation of value for the taxes to be paid on the transfer of an estate also requires estimation of the market value on a particular date. Depending upon the regulatory authority and existing laws, the evaluation for ad valorem or estate tax would make use of standard evaluation procedures for proved reserves but may impose certain conditions on the economic analysis, such as limits on pricing projections, operating costs, and/or discount rate.

7.2.3.1 Fair Market Value

Many investments, such as the acquisition or producing properties or valuations done for ad valorem and estate taxes, require a determination of the market value or fair market value of the property. Market valuation is a special case of investment analysis, where the evaluator is attempting not to estimate the value to himself or his company but to a hypothetical group of investors in oil and gas properties. This hypothetical group of investors is termed the *marketplace* for oil properties, and the value that would be determined and paid by this market is the *market value*. It is the consensus investment value of the parties that comprise the market. Fair market value is a legal concept that has generally been defined by the courts. One useful definition is the following:

> "... the highest price estimated in terms of money which the land would bring if exposed for sale in the open market, with reasonable time allowed in which to find a purchaser, buying with knowledge of all the uses and purposes to which it was adapted and for which it was capable of being used... the highest sum which the property is worth to persons generally..."
> [Sacramento Southern R.R Co. v. Heilbron 156 Cal. 408, 104 p. 979 (1909)]

There are many definitions of market value, but for the most part they are variations on the language above. The definition means that the true value of a property can only be established in the marketplace by the free interaction of buyers and sellers. The actual value of a property can only be known after the property has been sold under market conditions. Even then the value is usually the highest price paid, not necessarily the view of the market as a whole. However, the market value can be (and often must be) estimated through an evaluation. The difference is that market valuations require the use of price/cost projections and discount rates that reasonably approximate those used by the marketplace for the buying and selling of properties.

7.2.4 Preparation of an Economic Evaluation

An economic evaluation consists of five basic elements: (1) production schedule, (2) product prices, (3) ownership interests, (4) costs of production and (5) capital expenditures. The shorthand and commonly used terms for this construction are *income stream* or *cash flow*, which capture the concept that the purpose of oil/gas production is to generate revenue. Any of these segments of the cash flow may be expanded and made as simple or as complex as necessary. The cash flow can be done as a preincome tax or afterincome tax calculation. The cash flow is designed to model the production, price and cost expectations of the evaluator to an economic limit.

7.2.4.1 Production Schedule

An estimate must be made of the rates of future for oil and gas production plus any associated product, such as NGL or condensate, and is the starting point for any evaluation.

Previous discussion has described the methods of estimating future reserves and production. Extrapolation of existing or similar property production decline is the most direct method of estimating future production when either oil or nonassociated gas is the primary production stream. Associated gas may be projected as a function of oil production using a fixed or variable gas-oil ratio (GOR). Nonassociated gas may have associated condensate production that can be projected as a fixed or variable yield.

Where decline curve extrapolation is used, the form of the decline, whether exponential, hyperbolic or harmonic, must be defined and applied. The reserves obtained from production decline should be compared with reserves obtained from other methods such as rate-cum or WOR-cum curves or to volumetric and/or material balance calculations. When extrapolation of historical performance may not be appropriate, such as when projecting future production from a new field or reservoir or for an EOR project, production estimates can be obtained by converting the volume results of the material balance, frontal advance, or other method into annual or monthly production. This may require determining a limiting condition that can be identified, such as lifting capacity or injection rates and back-calculating a production rate. However it is done, the evaluator must compare the production schedule with other similar projects or fields and with good reservoir and operating practice. Also, the selection of a schedule may have economic impacts, such as the requirement for investment, increased operating costs, or in some circumstances royalties and/or taxes that may cause an alteration of the selected production schedule. Finally, as noted earlier, the proper analysis of the source and form of the production data is very important to a valid cash flow analysis.

7.2.4.2 Product Prices

Product prices are the market price of oil, gas, condensate, and/or NGL used in an economic evaluation may depend upon the purpose for the evaluation. Unless a particular price forecast is imposed by regulation or other authority, the estimate of future product price is a function of experience and foresight. A cash flow is usually done as of a point in time so the prices would be those in effect at that date. Two basic sources for initial product prices are actual sales and posted prices. The actual price for oil and/or gas being received on the property to be evaluated is the best source of a price for a cash flow. In using *actual prices*, determine if there are any shipping, pipeline, dehydration or other deductions that would reduce the price actually received for the oil or gas produced. These charges must be accounted for in the cash flow. Also, *gas sales* prices are often based on the heating or BTU value of the gas *not* on the McF or volume of gas. Since gas production is normally expressed in McF, a correction must be made if the gas price is in $/BTU. *Posted prices* (Figure 7.2.7) are the prices offered in the market by large purchasers of crude oil, such as major refiners. These posted prices are readily available for oil from various fields or for regions of the country or state. The posted price changes as economic conditions change. The posted price is listed as $/BBL for a certain gravity of oil, with a correction factor for gravity differentials and, sometimes, sulfur content.

The appropriate price for the oil in the property being evaluated can be estimated by obtaining one or more posted prices for oil of a similar gravity in or near the same field and making necessary adjustments for gravity and other factors. Gas prices can be estimated from standard prices offered by pipelines in the area. Gas sales are not always as straightforward as oil sales. During periods of high demand, the purchaser may be willing to take all the gas that can be produced and will build the connection lines to the property.

At other times, however, the purchaser may iimit sales to a percent of capacity and may not be willing to provide connection. These conditions could not only reduce revenue from sales but could require additional investment and may have the effect of reducing production of other products if the gas is associated gas. Whatever the conditions, they must be considered in the production schedule and cash flow.

Condensate is generally treated and priced as crude oil. NGL is generally treated as a by-product of gas sales where "wet" gas containing NGL is sold to a "gas plant" that strips out the liquids and sells the "dry" gas and liquids. The producer may receive either (1) revenue from the dry gas sales and a share of NGL sales or (2) receive a wet gas price. If the operator is also removing the NGL from the gas, the liquids may be mixed with the crude oil to raise the gravity of the oil prior to sale.

The *projection of product prices* into the future depends on the perspective of the evaluator regarding future economic conditions and, to some extent, the purpose of the cash flow. The simplest approach is to determine the appropriate price(s) as of the date of evaluation and hold these prices constant for the life of production. This was virtually the only way projections were done prior to the price increases of the 1970s and is still relatively common. Constant pricing is required for SEC evaluations. Since the 1970s it has become more common to attempt to estimate whether oil and/or gas prices would change over the expected life of production and to build those anticipated changes into the cash flow by escalating or deescalating oil prices at certain rates over time.

The question of whether or not to escalate prices in a cash flow depends on the information available to the evaluator and how well the evaluator can translate information into expectation. Major companies often have economics departments whose purpose is to estimate future oil prices that are often implemented in cash flows. Smaller companies and others are generally without such resources and must rely on other, published and unpublished sources and their own intuition. Since the early 1970s, oil and, to a lesser extent, gas have been open market commodities subject to a wide range of forces causing prices to rise and fall.

7.2.4.3 Use of Oil and Gas Futures Markets

In the absence of reliable guidance regarding future oil and gas prices, the evaluator can make use of the "futures" market for oil and gas as a source of data. The recognition of oil and gas as tradable commodities has led to the development of vehicles for estimating the price that would be paid for a future delivery of a certain volume of oil and/or gas. The New York Mercantile Exchange or NYMEX maintains markets in crude oil and natural gas for the trading of contracts for future delivery. These "futures markets" operate in the same manner as the stock exchanges, and the prices for future contracts are published daily. Oil contracts can be defined for 2 to 3 years while natural gas contracts extend to 5 to 6 years. The contract prices can give a sense of the direction of future prices (up to down) and the anticipated degree of change in prices.

7.2.4.4 Oil and Gas as Commodities

The prices used in cash flow projections are normally "nominal" prices that include expected inflation. This should not be construed to mean that oil or gas prices should be expected to increase at the rate of inflation. It is entirely possible for oil prices to decline in the face of strong inflation. Even in those periods when oil and/or gas prices were increasing, the rate of increase, or the escalation rate, has rarely exceeded the rate of inflation for more than 2 to 3 years (Figure 7.2.1) and that, after that time, the escalation rate may be equal to or less

Amoco Production Company
200 East Randolph Drive
P.O. Box 87689
Chicago, Illinois 60680-0689

Crude Oil Price Bulletin 90-50
(Supersedes Crude Oil Price Bulletin 90-49)

September 27, 1990

This bulletin shows prices and gravity adjustments posted by Amoco Production Company (Amoco) for purchases of crude oil and condensate. Prices are based on the use of 100% tank tables or mutually acceptable automatic measuring equipment with customary adjustment of volume and gravity for temperature and full deduction for basic sediment and water.

Subject to change without notice, Amoco will pay the prices shown for merchantable crude oil and condensate delivered for its account into the facilities of its authorized receiving agency. Merchantable crude and condensate is defined as virgin crude and/or condensate produced from wells which is free of injected or outside foreign contamination, added chemicals containing but not limited to halogenated organic compounds and oxygenated compounds and which is fit for normal refinery processing. Seller warrants that the crude oil and condensate delivered to Amoco shall be of merchantable quality as defined above and fit for normal refinery use. Prices are shown for 40 gravity and above, except as noted.

All prices may be subject to deductions for trucking and other charges where applicable. In the event government regulations require adjustment to Amoco's prices or effective dates, Amoco reserves the right to amend these prices and recover any excess payments by withholding payments from future settlements or by separate invoicing.

Effective September 27, 1990

Areas		Price: $/Barrel	For Gravity Adjustment See Column
Colorado	Western	39.00*	5
	Eastern	38.00*	7
Kansas	Sweet	37.75*	3
	Southern Louisiana Light	38.75*	3
	Hackberry, Charenton, Edgerly (30° Lowest Price)	38.10*	3
	Southern Louisiana Sour — Eugene Island	37.25*	3
Nebraska	West Panhandle Sweet	37.75*	1
North Dakota	Incl. Sheridan, Roosevelt, Richland Counties, Montana	37.65*	1
	Fryburg-Medora (Flat Price)	36.75*	None
Oklahoma	Sweet	38.25*	8
	Sour	37.50*	5
Texas	East — Asphaltic (32° Top Price)	36.00*	4
	—East Texas Field (Flat Price)	38.25*	None
	—Quitman Light	36.25*	5
	Gulf Coast—Sweet (Flat Price)	38.25*	None
	—Fairbanks Area, Gillock, High Island	38.25*	1
	North	38.25*	1
	West Central—Including King & Knox Counties	38.25*	1
	West Texas & New Mexico—Intermediate	38.25*	1
	—Sour (34° Top Price)	36.25*	6
Utah	Black Wax	38.75*	5
	Yellow Wax	38.00*	5
Wyoming	Southwestern Sweet (Carbon, Lincoln, Sublette, Sweetwater, Uinta Counties)	38.75*	5
	Sweet (Other)	38.00*	1
	Sour (Including Montana fields)	37.00*	2
	Sweet (Salt Creek Field)	38.50*	1

* indicates change

A recorded message containing information on Amoco Production Company's latest Crude Oil Price Bulletin is available by calling (312) 856-3114

Figure 7.2.7 *Example of posted prices.*

than the rate of inflation. On average the "real" price of oil (real = nominal minus inflation) has increased by an average of 1.1% over the 60 years from the late 1920s to the late 1980s, including the major increases of the 1970s (Figure 7.2.2). A projection of oil or gas price that exceeds inflation for an extended period should also assume that costs of production would increase at an even faster pace.

7.2.4.5 Ownership Interests
The percentages of the production revenue and cash flow from a property owned by various parties are the *ownership interests* in that property. A property can be entirely owned (100%) by one party, or the property ownership may be

highly fragmented among a number of owners. Ownership can be broadly divided into two classes: working and royalty interests. The basic difference is that the *working interest* (WI) pays all the operating costs and makes the investments and then receives a share of the revenue. The *royalty interest* (RI) receives a percentage of the revenue but pays no costs and makes no investment. The share of revenue received by the working interest is called the net revenue interest. If the WI is 100% or 8/8 and the royalty is 12.5% or one-eighth, then

Net Revenue Interest (NRI)
= Working Interest − Royalty Interest
= 100.0% − 12.5% = 87.5%

Table 7.2.1   *Example of Working Interest Reversion*

	Before Payout		After Payout	
	WI	NRI	WI	NRI
A	25%	21.875%	65%	56.875%
B	75%	65.625%	35%	30.625%

There are several types of royalty interest including landowner, overriding, and sliding scale. The sliding scale royalty is most common on federal or state government leases [8] but may occur on private leases and are very common in Canada [9] and other countries. The royalty interest is paid directly out of sales, often by the production purchaser, so that the WI pays all operating costs, etc., out of the NRI. In any property or project there can be, and often are, multiple WI owners and/or RI owners with varying shares of the WI and RI. If there is more than one working interest owner, the WI and NRI will generally be divided proportionately. Whatever the type of royalty, the effect on the cash flow is the same.

In some cases, the ownership interest may change during the expected life of production. These "reversionary" interests may occur when a certain cumulative volume of production is obtained but more commonly occur when a predetermined amount of net cash flow has been obtained by the WI. The situation often occurs in new drilling projects when a WI owner, generally an investor, receives a certain WI until his or her investment is recovered or "paid out" at which time the WI reverts to a reduced percentage. At that time the other WI owner(s) share would increase. For example, assume the operator (A) of a property wants to drill a new well and an investor (B) agrees to provide the $100,000 needed to drill the well and to accept a 75% WI reverting to 35% at payout. Assume a 12.5% royalty (Table 7.2.1).

Another form of ownership is a net profits interest (NPI), in which a third-party owns a share of any profit or positive cash flow that occurs to the property after operating costs and investment. This form is not common but occurs when (1) a royalty owner has negotiated a particularly good agreement, or (2) when a WI owner is not able to contribute his share of capital expenses. The NPI is termed *cumulative* if the capital investment and/or any negative cash flow must be recovered out of future profits. During that time the NPI owner receives no income.

A special case of oil property ownership involves the *unitization* of multiple properties and owners, both WI and RI, into a single production entity, or *Unit*. These combined properties are most often formed for the purpose of EOR operation where production and injection can be expected to cross lease lines, or where geology, geography or operational circumstances suggest that a combination of properties would improve efficiency. In the formation of a typical Unit, each property is assigned a proportional share of the Unit production, operating costs, investment and income based on agreed-upon parameters, such as oil-in-place or remaining reserves on each property as a percentage of the total. This percentage becomes the basis for the value of the properties forming the Unit and for the owner of an interest in that property. For example, if an estate tax value is required for the owner of a 12.5% royalty in Tract B of the Big Unit, then this value of the royalty is derived from 12.5% of the Tract B share of the Unit production and income.

In constructing a cash flow it is very important that the ownership interests be determined and accurately applied. The best source of ownership data are the "*division orders*" issued by product purchasers in most states.

7.2.4.6 Costs of Production

The costs of production, or operation costs, include all the costs, except royalty and investment, required to produce, treat and sell the oil, gas and any other products from the property being evaluated. There are a great many different types of costs and companies may account for them in various ways but they must be included in the cash flow. Costs of production can be broadly divided into (1) operating or lifting costs and (2) production taxes and charges.

Operating costs can be further subdivided into (a) fixed, (b) variable, (c) periodic and (d) overhead. *Fixed costs* include those costs that remain generally the same regardless of the volume of production, such as lease labor (pumper, treater, well pulling, maintenance, engineering, and/or supervisory staff); some fuel and power costs; lease maintenance and repair. While these costs may change over a long period of time, even as production and/or the number of wells changes, for cash flow purposes they may be considered fixed costs in dollars per month or per year. *Variable costs* are those that vary with the volume of gross (oil plus water) production or with the number of wells. Such costs are chemical treating, some fuel and power costs, water disposal and some labor costs. *Periodic costs* are those that do not occur constantly but recur with sufficient regularity that they can be scheduled in a cash flow. Such costs might be pump changes, hot oiling, dewaxing, acidizing, cyclic steaming or other stimulation or maintenance requirements. *Overhead* is often an indirect cost that is incurred off the lease but is then allocated to the property. Such costs might be district or head office supervisory and engineering staff. *Other costs* might include distributed costs of environmental, regulatory or other programs that are applied across several properties or that have no property-specific application.

The best source of operating cost data is the historical cost records for the property being evaluated or a similar property. It is important to review at least 12 to 24 months of previous costs to properly define the full range of costs and the variation of those costs with time and production changes, and to define any recurring costs. If certain costs remain essentially the same over time, they should be treated as fixed costs in the cash flow. On the other hand, costs that can be shown to vary with total fluid or oil production or the number of wells or some other criteria should be treated as variable costs in the cash flow and projected as a function of the gross fluid, number of wells, or other criteria. Periodic costs can be included in the cash flow as occurring at specific time intervals or can be spread over the cash flow as a monthly or annual charge.

Certain special cases require mention. If produced oil and/or gas is retained for use as *lease fuel*, the volume used for fuel can be included in the cash flow by (1) reducing the amount of production, or (2) by deducting the equivalent value of the fuel volume as an operating cost. The choice of treatment depends on such things as whether or not the lease requires payment of royalty on production used as lease fuel (most leases do not) or whether fuel usage can be deducted for production tax calculation (generally no). Deduction of fuel usage from production may distort the cash flow; therefore, it is probably best to treat fuel usage as an operating cost at the prevailing price for oil or gas used.

Production taxes and charges include, but are not limited to (1) severance taxes, (2) wellhead taxes, (3) *ad valorem* taxes, (4) regulatory impositions, and (5) certain charges based on income such as the *windfall profit tax*. These taxes not only occur individually but, in many areas, one or more of the taxes occur and are cumulative. It is important to identify the taxes applicable to the property being evaluated and to include them in the cash flow [10–12]. Most major producing states (Texas, Oklahoma, Louisiana) impose a *severance* tax on oil and gas production, which is a percentage of the sales price before any costs, including royalty or other taxes, are deducted. The tax may be a different percentage for oil and gas. They are normally paid to the state by the purchaser of the oil and gas. While some states have relatively high severance tax rates (7–8%), the tax is a fixed share of the price and fluctuates with the price. *Wellhead taxes*, on the other hand, are generally a fixed amount such as $/bbl or McF, which, while generally small, become a larger percentage of revenue if prices decline.

Ad valorem taxes are based on the value of the property as determined by the taxing authority such as a county tax assessor. The *ad valorem* tax is relatively minor in most states. The exception is California where there is no severance tax and the *ad valorem* tax can equate to as much as 3–5% of gross revenue. The difficulty with *ad valorem* taxes is that the assessed property values do not always follow changes in property value caused by changing economic conditions. For an evaluation, it may be unnecessary to review assessment records and/or prior tax statements to estimate future property taxes. For use in a cash flow, *ad valorem* taxes can be converted to a percentage of gross revenue or deducted as an annual amount. *Regulatory taxes* can consist of either percentage or fixed amount per unit taxes that are similar to severance or wellhead taxes but are directed toward specific uses, such as funding a state agency or environmental fund. These taxes vary greatly from one jurisdiction to another and may only occur for fixed time periods.

Some taxes are based on income but are not income taxes per se. The so-called *windfall profit tax* (WPT) is one example. The WPT, since expired, imposed a percentage tax, which varied depending on differing classes of sellers and types of oil production, on the adjusted difference between a so-called "*base price*" and the actual sales prices. At some points during the time it was effective, the WPT absorbed a large percentage of the price differential. The tax remained long after the "windfall" profits had evaporated.

7.2.4.7 Capital Expenditures

While the difference between periodically recurring costs and capital expenditures is not always clear and each company may have its own definition, capital expenditures can be broadly defined as relatively major infusions of capital, which may or may not be predictable, and which are generally expected to result in an increase in production or a decrease in costs or both as a direct result of the investment. Changing a downhole pump might be operating cost while changing a pumping unit could be a capital expenditure. Cyclic steaming might be a periodic cost but fracing might be an investment. There is a certain amount of subjectivity involved in the categorization of these costs. Any anticipated capital investment including eventual abandonment should be included in the cash flow and evaluated for appropriate return. Investment is deducted after all royalties, operating costs, production taxes.

7.2.4.8 Abandonment Costs

Eventually, all properties are either depleted or become uneconomic to future operations. At that point, most mineral leases, and a growing body of state, federal and local laws, require that the property be properly abandoned. This includes plugging the wellbore to prevent fluid migration; removing all downhole and surface equipment including flowlines and other associated installations; restoration of the surface to its original condition; and, if necessary, and as directed, conduct remediation of any soil and/or water contamination that may have occurred resulting from petroleum operations on the property.

As production in many areas of the US has matured, and as environmental issues have gained greater leverage, the consideration of abandonment, restoration, and remediation (AR&R) costs in oil property evaluation has received more attention, both because of the imminence of those costs and the potential economic impact of AR&R expenditures. In any property evaluation, it is necessary to estimate the future costs of AR&R and to include those costs as part of the evaluation. Estimation of remediation costs may require a third-party study and cost estimate.

AR&R is a liability against the property. In order to properly determine the value of a property, the future AR&R costs must be accumulated out of income from the property prior to the scheduled abandonment. There are several methods of doing this, such as deduction of (1) a percentage of annual income, (2) a fixed amount per year, or (3) the use of a sinking fund to accumulate the estimated amount. The reason for accumulating funds out of projected future income is that at the time of abandonment, by definition, there is no income to pay for AR&R.

7.2.4.8.1 Constructing the Cash Flow

The components discussed to this point are sufficient to develop a more or less detailed cash flow for any property which will allow the property to be evaluated for virtually any purpose for which a cash flow is used (Figure 7.2.8). In model form:

$$\text{Production} \times \text{ product price} = \text{ gross revenue}$$

$$\text{Net operating income} = \text{ gross revenue}$$
$$- \text{ royalty}$$
$$- \text{ production taxes}$$
$$- \text{ operating costs}$$
$$\text{Cash flow} = \text{net operating income}$$
$$- \text{ capital expenditures}$$
$$- \text{ AR\&R costs}$$

7.2.4.8.2 Economic Limit

When calculated on a monthly or annual basis to the *economic limit* of production, the cash flow provides a foundation for determining the value of the property and/or the profitability of capital expenditures; or, simply, the future flow of cash from the property. The economic limit is the minimum production rate which, at a given price, is required to exactly offset all the costs of production.

$$\frac{\text{Costs of production/unit time}}{\text{Price per Bbl or Mcf}}$$

$$= \frac{\text{Production taxes} + \text{operating costs}}{(\text{ Price per Bbl or Mcf})\,(\text{NRI})}$$

$$= \text{Economic limit (Bbls or Mcf/unit time)} \qquad [7.2.1]$$

The economic limit can be difficult to calculate, particularly if operating costs are complex and closely related to the volume of production. The economic limit must be calculated for the primary product, whether oil or gas, but the calculation should include the revenue from all products and all the costs associated with production and sale of all products.

7.2.4.8.3 Income Tax

Income taxes, both state and federal, exist and should be considered in any economic evaluation, particularly where large capital investments are involved. Even with the occasional tax reductions that have occurred, corporate and individual tax rates still exceed 30% to 35% of taxable income. That taxable income can be substantial depending on the deductions available at the time. The tax code is complex, is subject

Evaluation Texaco, 1

Run Date: 08 50 1990
Run Time: 13:45:55

As of Date: Jan 84

	NPV	10.0%	652.033 BFIT
Name: Gusher 1-1	NPV	15.0%	445.028 BFIT
Field: Oil Dorado Field	NPV	23.0%	158.646 BFIT
Location: Oz County	NPV	30.0%	56.478 BFIT
Formation: Yellow Brick Lime	NPV	35.0%	−27.553 BFIT
Operator: Oyl Bidness, Inc.	IRR		33% BFIT

Interests and Effective Date				Prices			Gross Reserves				
Cost	Revenue	Date		Beginning	Ending	Average	Cumulative	Remaining	Ultimate	%Remaining	
1.000000	0.850000	Jan 84	Oil	19.00	29.48	21.79	543.000	226.976	769.976	29.48	Oil
0.900000	0.800000	Jan 85	Gas	1.30	1.92	1.42	1629.000	680.927	2309.927	29.48	Gas
0.500000	0.400000	Jan 86	Cond	0.00	0.000	0.00	0.000	0.000	0.000	0.00	Cond
0.500000	0.300000	Jan 91									

Year	Gross Oil Production MBBLS	Gross Gas Production MMSCF	Net Oil Production MBBLS	Net Gas Production MMSCF	Average Oil Price $/B	Average Gas Price $/MSCF	Net Oil Sales M$	Net Gas Sales M$	Net Other Rev M$	Net Total Revenue M$
1984(12Mo)	34.168	102.505	29.043	87.129	19.000	1.300	551.813	113.268	0.000	665.081
1985	30.752	92.255	24.602	73.804	19.000	1.300	467.430	95.945	0.000	563.376
1986	27.676	83.029	11.070	33.212	19.475	1.300	215.596	43.175	0.000	258.771
1987	24.909	74.726	9.964	29.890	20.449	1.300	203.743	38.858	0.000	242.601
1988	22.418	67.254	8.967	26.902	21.471	1.365	192.536	36.721	0.000	229.257
1989	20.176	60.528	8.070	24.211	22.545	1.433	181.945	34.701	0.000	216.646
1990	18.158	54.475	7.263	21.790	23.672	1.505	171.934	32.792	0.000	204.726
1991	15.433	46.298	4.630	13.889	24.856	1.580	115.079	21.947	0.000	137.026
1992	12.346	37.039	3.704	11.112	26.098	1.659	96.663	18.436	0.000	115.099
1993	9.877	29.631	2.963	8.889	27.403	1.742	01.199	15.486	0.000	96.685
1994	7.902	23.705	2.371	7.112	28.773	1.829	68.210	13.009	0.000	81.219
1995(6Mo)	3.161	9.482	0.948	2.845	29.475	1.921	27.947	5.464	0.000	33.411
1996	0.000	0.000	0.000	0.000	0.000	0.000	0.000	0.000	0.000	0.000
1997	0.000	0.000	0.000	0.000	0.000	0.000	0.000	0.000	0.000	0.000
1998	0.000	0.000	0.000	0.000	0.000	0.000	0.000	0.000	0.000	0.000
Sub Total	226.976	680.927	113.595	340.785	21.786	1.420	2374.097	469.801	0.000	2843.897
Remaining	0.000	0.000	0.000	0.000	0.000	0.000	0.000	0.000	0.000	0.000
Tot 11.5Yr	226.976	680.927	113.595	340.785	21.786	1.420	2374.097	469.801	0.000	2843.897

Year	Net Total Prod Tax M$	Net Total Loe M$	Net Total Oper Exp M$	Net Oper Revenue M$	Net Leasehold M$	Net Total Investments M$	Net BFIT Cashflow M$	Cum BFIT Cashflow M$	BFIT CF Disc @ 20% M$	Cum BFIT CF Disc @ 20% M$
1984(12Mo)	19.952	24.000	43.952	621.129	0.000	1250.000	−628.871	−628.871	−667.658	−667.650
1985	16.901	22.680	39.581	523.794	0.000	0.000	523.794	−105.077	395.992	−271.666
1986	7.763	13.230	20.993	237.778	0.000	0.000	237.778	132.701	149.655	−122.010
1987	7.278	13.891	21.170	221.431	0.000	0.000	221.431	354.132	116.136	−5.873
1988	6.878	14.586	21.464	207.793	0.000	0.000	207.793	561.926	90.820	84.946
1989	6.499	15.315	21.815	194.831	0.000	0.000	194.831	756.757	70.961	155.908
1990	6.142	16.081	22.223	182.503	0.000	0.000	182.503	939.260	55.392	211.300
1991	4.111	16.885	20.996	116.030	0.000	0.000	116.030	1055.290	29.346	240.646
1992	3.453	17.729	21.182	93.917	0.000	0.000	93.917	1149.207	19.793	260.439
1993	2.901	18.616	21.516	75.168	0.000	0.000	75.168	1224.376	13.201	273.640
1994	2.437	19.547	21.983	59.236	0.000	0.000	59.236	1283.611	8.668	282.309
1995(6Mo)	1.002	10.262	11.264	22.146	0.000	0.000	22.146	1305.757	2.827	285.136
1996	0.000	0.000	0.000	0.000	0.000	0.000	0.000	0.000	0.000	0.000
1997	0.000	0.000	0.000	0.000	0.000	0.000	0.000	0.000	0.000	0.000
1998	0.000	0.000	0.000	0.000	0.000	0.000	0.000	0.000	0.000	0.000
Sub total	85.317	202.823	288.140	2555.757	0.000	1250.000	1305.757	1305.757	285.136	285.136
Remaining	0.000	0.000	0.000	0.000	0.000	0.000	0.000	0.000	0.000	0.000
Tot 11.5Yr	85.317	202.823	288.140	2555.757	0.000	1250.000	1305.757	1305.757	285.136	285.136

Figure 7.2.8 *Example of cash flow.*

to rapid change, and to individual case application so that a comprehensive treatment in a handbook is not appropriate.

7.2.5 Present Value of Future Income [13–15]
Oil property valuation can be thought of as having three components. The first component is the *income stream*, or *cash flow*, which comprises all the elements of the evaluation that are unique to the property being evaluated in the production projection: estimates of futures prices, operating costs, capital expenditures, and net income.

The second element of valuation is the present value factor, or *discount rate*, which reduces the future income to the third component, which is *value*. The prior discussion has covered the construction and composition of the income stream. This section covers the second and third components, discount rate and value.

7.2.5.1 Structure of the Discount Rate
A cash flow is a stream of income expected to be received over a period of time into the future. For example, if a cash

flow consisted of equal amounts of $5,000 over 5 years, the total income would be $25,000.

However, from the perspective of the date of valuation, the $5,000 to be earned in years 2, 3, 4, and 5 does not have the same value as the $5,000 to be earned in the first year. Because of inflation and other financial risks, the *present value* of income to be received in the future is reduced or discounted by an amount that varies with the evaluator's perceptions of the risks involved in waiting for the income. To equate future cash flow to present value, a discount rate or present value factor is applied to the cash flow. Present value calculation can be thought of as the inverse of compounding interest where, instead of increasing the value of a currently held amount over time, the present value factor discounts future amounts to a present value. For example, if a principal sum P is invested at an annual interest rate i for n years, the accumulated interest would increase P to a total future amount S as follows:

$$S = P(1+i)^N \qquad [7.2.2]$$

On the other hand, if S is the amount of income to be received in year N, and the anticipated return is i, then the present value of S would be

$$P = \frac{S}{(1+i)^N} \qquad [7.2.3]$$

A cash flow is a series of payments received over time. These successive payments are discounted to present value by setting each annual amount equal to S and setting N at the appropriate amount in years.

The selection of an appropriate N and i is very important. Cash flows for oil and gas production are rarely constant and generally decline over time. In addition, cash flows are often calculated on a monthly basis. The basic equation then must be modified to reflect the condition of the evaluation. A midyear discount is probably more common than year-end. In this case,

$$P = S\left(\frac{1}{\left(1+\frac{i}{2}\right)^{-0.5}}\right) \text{ in year 1} \qquad [7.2.4]$$

and

$$P = S\left(\frac{1}{(1+i)^{N-0.5}}\right) \text{ in later years} \qquad [7.2.5]$$

If discounting is done monthly, then

$$P = S\left(\frac{1}{\left(1+\frac{i}{12}\right)^N}\right) \qquad [7.2.6]$$

where S = monthly cash flow
       N = number of months from start of cash flow

This method is most commonly used in computer programs that calculate on a monthly basis. There are many convenient tables of present value factors published, and specialized tables are readily generated using the basic formulas.

7.2.5.2 The Discount Rate

The choice of the discount rate is one of the most important elections made in evaluating a property or project. This is easily illustrated by comparing the present value of a cash flow that provides $100,000 every year for 10 years. At increasing discount rate, the present value declines significantly as shown in Table 7.2.2. If, as in most oil and gas cash flows, the cash flow declines, the effect of the discount rate is even more pronounced as shown in Table 7.2.2.

The discount rate is a factor that is used to account for the perceived risk associated with the deferral of income to some

Table 7.2.2 *Effect of Discount Rate on Uniform and Declining Cash Flow*

Discount Rate, %	No Decline Present Value, $	Decline @ 10%/yr Present Value, $
0 (undisc)	1,000,000	651,322
5	772.183	523,961
10	614,457	432,785
15	501,877	365,525
20	419,247	314,562
25	357,050	275,017
30	309,154	243,677

time in the future or, more commonly, making an investment with the expectation of future returns. The discount rate is built out of four primary parts: (1) a safe nominal rate of return, (2) a return-of-the-original investment, (3) investment risk, and (4) specific property risk. The first three are financial factors. Any investor requires the return of the original investment (principal) plus a return on the investment (interest) that is commensurate with the length of the delay in receiving the income. In common industry practice, the return-of-investment is covered by requiring payback or payout in the first few years. All income received after Payout is return-on-investment.

Investment Risk, in the evaluation of oil properties, is the risk related to (1) the liquidity of the investment, and (2) the diversity of the income stream. Investments in oil properties are not as liquid and cannot be sold as readily as stocks and bonds, which are usually suggested as the comparable investment. Stocks and bonds can be sold quickly, often in minutes, but oil properties generally require weeks or months. This lack of liquidity imposes a price in the form of higher risk. Also, oil properties have only one stream of income from produced products, while a share of stock is supported by revenue from multiple sources. This lack of diversity of income is a risk that the expected income will not be received.

Specific Property Risk is the risk related to the property that is the revenue source. Risk elements related to a property include access to markets, maturity of production, class of reserves, production methods and ability to manage the operation of the property so as to achieve the expected production.

These four elements, when quantified and combined, constitute the rate-of-return required on the original investment in order to justify the investment. The rate-of-return is expressed as an annual percentage rate. The rate is unique to the property at the time of the investment — if conditions change, the rate could also change. The inverse of the rate-of-return is the Discount Rate. If the investor or evaluator can correctly assess and quantify the risks in the property to define the required rate-of-return, then that return can be used as the Discount Rate to estimate value from the income stream.

As noted above, the selection of a discount rate is pivotal in an evaluation. In many cases the discount rate may be provided by company guidelines, financial institution requirements or by government regulation. Absent these sources, the evaluator can rely on professional judgment and derive a discount rate from experience and/or personal knowledge. In either case, the foundation for discount rate selection rests on two information sources: (1) the marketplace for oil and gas properties, and (2) the cost-of-capital used to make investments in oil properties and projects.

Market information regarding discount rates used in oil property valuation is very limited. Oil property transactions are sparse and, aside from scanty information reported in press releases by public companies, are rarely made public. Obtaining marketplace discount rates from such sales is difficult, and the results may not be definitive. There have been a few professional papers that report analyses of sales data and provide some useful guidelines for discount rates. There are also a few studies sponsored by industry groups or published by government agencies that report discount rate data. However, unlike the marketplace for homes and surface acreage, there is no public database of oil property evaluation data. [16–23].

The *cost-of-capital* is a financial evaluation term and procedure that refers to the cost to the investor of the funds that would be used to make the investment. While space does not allow an extensive treatment of this topic, as there are many textbooks and articles on this subject, a background in the subject is important to understanding the basis for modern petroleum investment and evaluation practices. Investment capital consists of debt and equity. The cost-of-capital to an investor is the return that the investor must provide to the sources of any debt and/or equity funds that are used in an investment in a property or project. If the investment is financed with 50% debt from a lender and 50% equity from stockholders, then the *cost-of-debt* is the interest rate due to the lender and the *cost-of-equity* is appreciation in stock value expected by the equity holder. If the interest rate on debt is 10% and the expected equity return is 20%, the investor who makes use of these funds must provide each source of funds with the expected return. Equity expects a higher return because it is higher risk.

A commonly used financial guideline is the weighted average cost-of-capital (WACC) [24,25], which describes the return anticipated from a mix of debt and equity. In the above example:

$$\text{WACC} = (\% \text{ Debt} \times \text{Cost of Debt}) + (\% \text{ Equity}$$
$$\times \text{Cost of Equity})$$
$$= (0.5 \times 10\%) + (0.5 \times 20\%)$$
$$= 15\%$$

An investment made in a project using the above mix of funds would be expected to obtain a minimum 15% return. The cost-of-capital is the financial standard used in industry for investment analysis. The same cost-of-capital can serve as the foundation for an evaluation discount rate, subject to adjustments to account for the differences in risk between (1) stocks and bonds, and (2) petroleum properties and projects.

However the discount rate is chosen, it is applied to the cash flow to determine the present value of a future stream of income from a particular property or project, and possibly for a particular purpose. If the discount rate is properly chosen, the present value cash flow from that property or project is directly comparable with the present value cash flow from any other similar property or project.

7.2.6 Investment Analysis [25]

Profit is the amount of revenue remaining after payment of royalty, operating costs, investments and taxes, including income taxes. A distinction is often made between before-income tax (BFIT) profit and after-income tax (AFIT) profit for analysis purposes. If a firm has a large number of investments to evaluate and compare, use of pretax profit analysis may be useful if the investments have differing tax characteristics or benefits. However, comparison on only a BFIT basis may distort the ultimate profit and return because taxes do have to be paid.

7.2.7 Industry Methods of Measuring Profitability

As noted above, many evaluations are done to determine the economic value of an investment. To that end, there are many methods available to determine and compare the value of investments. These can be grouped as undiscounted methods and discounted methods. Undiscounted methods are based on an undiscounted cash flow that can be BFIT or AFIT or both. In 1985, Boyle and Schenck [26] published a study of investment evaluation methods used in the oil and gas industry. The study found that most companies (1) allocate capital to investments in a systematic manner and (2) tend to use certain basic methods of investment analysis. The study also defined the methods in most common use by companies. A similar study by Dougherty and Sarkar [27] in 1993 found comparable results. The most common methods found in both studies are:

7.2.7.1 Financial Analysis Methods
7.2.7.1.1 Net Present Value
This method compares the total present value cash flow discounted at a specific rate to the capital investment required to obtain the cash flow. When calculated prior to deduction of investment, the NPV can be compared to the amount of capital investment. If the NPV exceeds the investment, the project should be profitable.

7.2.7.1.2 Rate of Return
ROR is the discount rate that would reduce the cash flow to the amount of the investment or to zero if the investment is deducted in the cash flow. Many authors have suggested modifications on this method, and many variations are in use depending on individual or company preference. In general, the method requires a trial-and-error approach of discounting the cash flow at various DCRs (Table 7.2.2) until a match with the investment is achieved. The method can also be done graphically by calculating the PV of the cash flow at various DCRs defining a curve through the points, and finding the DCR at the point on the curve where PV cash flow equals the planned investment (Figure 7.2.9).

7.2.7.1.3 Payout or Payback
The time required to accumulate an amount equal to the investment from the incremental cash flow created by the investment is the payout or payback. A common rule of thumb is 2–4 years subject to the conditions of the cash flow. It is important to measure payout using only the *incremental cash flow* created by the investment, not cash flow that would have been obtained anyway.

Payout is the common term but, in a financial context, payout is the return-of-investment required by any investor.

7.2.7.2 Rules of Thumb
7.2.7.2.1 Profit/Investment
The ratio of AFIT cumulative cash flow derived from an investment to the amount of the investment is

$$\text{P/I} = \frac{\text{AFIT cash flow}}{\text{Investment}}$$

There are several variations on this ratio that include using BFIT cash flow, discounted investment, and other combinations. Whichever is chosen, it must be used consistently from one evaluation to another. There are a large number of special-purpose rules of thumb that have been developed over time, particularly for property acquisitions. These include, but are not limited to:

7.2.7.2.2 $/BOE
According to this formulation, a property has a value of X$ per bbl of recoverable reserves. At the time the rule enjoyed

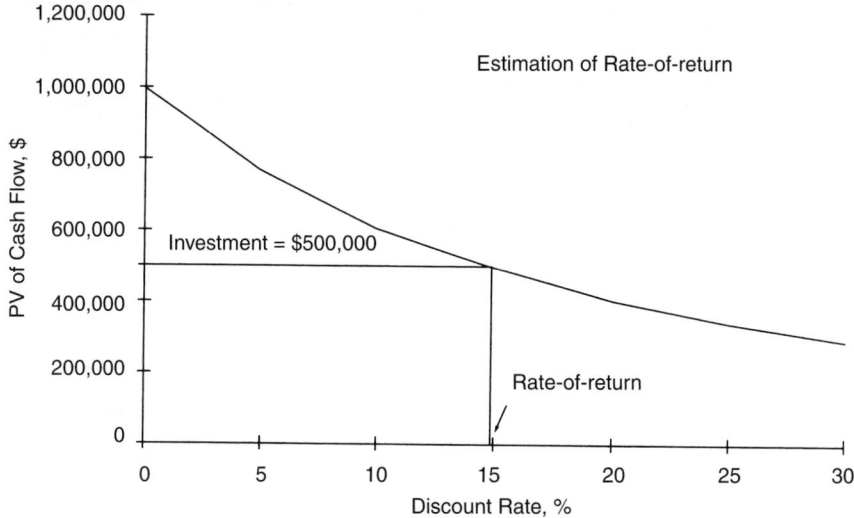

Figure 7.2.9 *Graphical ROR solution.*

general use (1940–1960s), the ratio ranged from 25 to 40% of the wellhead price of oil depending on the section of the country.

7.2.7.2.3 $/BOPD

Property value was X$ per bopd of production at the time of purchase. Values ranged from $5,000–15,000/bopd depending on location, etc.

Other rules of thumb are variations on the above. In practice, the development and use of such methods, including payout held some benefit simply because they were easy to use and did not necessarily require a cash flow. However, as production has changed from declining primary to variable secondary and enhanced recovery and as economics have become less predictable, the validity of most rules of thumb has diminished. In addition, the rapid growth in computers and software allows even the smallest operator to do relatively sophisticated economic evaluations so that reliance on generalizations are no longer necessary.

There are two basic discounted methods of defining project profitability, both of which have numerous variations. Either method can be used BFIT or AFIT.

7.2.8 Investment Decision Making [2]

Every company and individual has a unique approach to making investment decisions. As shown by Dougherty [27], many companies use two or more methods together. Companies often use a certain discount rate that may be known as a "hurdle" rate or *minimum required return* (MRR) to discount all cash flows. Assuming some consistency in other cash flow parameters, the various projects may then be compared to each other and the most favorable projects selected. As an example, a company has six projects in which it can invest funds but does not have sufficient funds for all six. The company has a cost of capital of 18% BFIT and an MRR of 20% BFIT. The company requires all cash flows to be discounted at MRR to determine NPV net of investment and also requires that IRR on each project meet or exceed MRR. The results are as shown in Table 7.2.3.

In this comparison, Project D has the highest NPV but the IRR of 19% is less than MRR and only 1% above the cost of funds and, therefore, has little margin for risk. Project B has

Table 7.2.3 *Comparison of Net Present Value and Internal Rate-of-Return*

	NPV	IRR
A	$300,000	22%
B	$150,000	26%
C	$100,000	24%
D	$500,000	19%
E	$200,000	21%
F	$ 20,000	25%

a considerably lower NPV but a high IRR. A combination of Projects A, B and C would have a total NPV greater than D and also a higher composite IRR. Depending on budget constraints, a ranking of A, B, C, E and F would be appropriate. There are circumstances when D would be accepted despite its low(er) IRR such as when intangible values come into play. In this example, all the projects were assumed to have the same level of risk. That may not always be the case and, if not, differing threshold levels of IRR may be used to account for the variance in risk.

7.2.9 Risk Analysis [28–33]

Risk is the possibility that events or conditions in the future may not occur as expected. When flipping a coin you may expect it to turn up heads, but there is a 50% possibility that it will be tails. In drilling a purely exploratory well, there is a very high possibility that the well will be dry. In projecting future oil prices, there is a very good chance that your projection will be wrong both as to direction and timing. Each example contains risk, but the ability to quantify that risk ranges from a fixed 50% to a statistical value for wildcat drilling to virtually unquantifiable for the price projection.

Risk has a significant impact on valuation of oil and gas properties. The oil business is a relatively high-risk endeavor to begin with and, within any property evaluation, there are numerous opportunities for risk to impact the value. The production projection and reserves determination can be done using all available data and the best methods, but the actual production is still subject to the vagaries of a natural system — the reservoir rocks and fluids — that cannot even be sampled and measured to any major extent, let alone be accurately modeled to eliminate the potential for variance from expectations. Reserves are classified as *proved, probable* and *possible* on the basis of the ability to estimate those reserves and likelihood of recovery. Added to that are various potential mechanical problems that impose uncertainty on any projection of production.

There is substantial risk in projecting future prices and operating costs because of the inherent uncertainties of trying to determine how economic conditions that affect oil and gas prices and operating costs over the often several year period of an oil property evaluation.

A property valuation should attempt to recognize and account for risk. There are several methods of analyzing risk and applying adjustments. The process has steps which can be generalized:

1. Define the source of the risk.
2. Determine if the risk is measurable.
3. Define a range of values for the risk.
4. Select a risk evaluation method(s).
5. Apply risk adjustments to the evaluation.

7.2.9.1 Define the Risk
Any evaluation of an oil and gas property has a wide range of risks associated with it, starting with the geologic risk that recoverable hydrocarbons exist to the economic risk posed by price/cost projections and selection of discount rate.

Geologic risk is very important in assessing drilling prospects but is much reduced in production development projects. There have been several treatises written on a drilling project risk that allows the risk to approach quantification.

Geologic risks:

- Does the zone exist?
- Field size?
- Sufficient recoverable hydrocarbons?

Performance risk:

- reservoir properties (Sw, ϕ, K, μ, etc.)
- zone thickness
- areal extent
- drive mechanism (effects R_f)
- decline rate
- depth
- production method needed
- well spacing required
- stimulation needed?

Economic risks:

- Price projection depends on gravity of oil and composition
- Operating costs projection depends on gravity and composition, depth, number of wells, etc.
- Royalty and production taxes.
- Required investment.
- Cost-of-capital.
- Income tax treatment.

7.2.9.2 Measurement of Risk
Each of these risk elements and many others can be measured to some degree although the degree of real quantification varies considerably. Geologic risk has been addressed by many authors to the point that such factors as field size distributions are quantified based on normal or log-normal analysis. Reservoir properties such as S_w and ϕ can generally

be estimated from similar properties and fields and can be analyzed statistically as data are available. Areal extent can be estimated from mapping depending on the quality of control data. Economic data can be well known, such as current price, within very narrow boundaries but future trends can be highly variable. Each factor must be reviewed to determine the error potential, degree of uncertainty and range of possible but realistic values.

7.2.9.3 Define Range of Values
Most of the risk factors that may occur in an evaluation and that can be measured can be quantified on a range of values about the most likely value. Porosity may be 20±5%, or areal extent might be expressed as drainage radius of 20 acres ±20%, whereas an overall field size of 100,000 barrels might have a 10±10% probability of occurring.

Each risk element must be assessed for the range of possible values that may occur or the likelihood that a certain value would occur. Selection of those values should be based on objective measurement where possible supplemented by analysis of similar field (property) data and tempered by subjective experience.

7.2.9.4 Select Risk Evaluation Method
The method used for evaluation of risk often depends on (a) the relation to the risk element to the overall evaluation and (b) the degree to which a probability can be assigned to the value or values in the range of values. As an example, such factors as porosity, Sw, zone thickness; areal extent; and starting prices and costs could have as much probability of being one value as another within the range of values, in which case sensitivity analysis or Monte Carlo analysis would provide sufficient consideration of the risk factors. On the other hand, where probabilities of occurrence can be assigned to such factors as field size or price/cost escalations an expected value approach might be more appropriate.

Of course, each risk element can be evaluated either on its own or in combination with other factors. Oil-in-place might be calculated using a range of equally likely values for ϕ, Sw, etc., which would result in a range of values for OOIP that could then be tested for field size distribution and assigned a probability. A set of production projections could be evaluated using a range of price escalations to determine the sensitivity of project value or return to various prices.

7.2.9.5 Apply Risk Evaluation Methods
7.2.9.5.1 Sensitivity Analysis
1. Select one risk element and measure the range of equally likely values.
2. Apply the range of values, with appropriately selected intervals, to the part of the evaluation which is sensitive to that element; i.e., how is OOIP affected by changing ϕ from 18% to 30% at 1% intervals.
3. Combine compatible risk elements to determine if the elements offset or enhance each other.
4. Determine a range of outcomes and analyze statistically to determine the most likely outcome and range probabilities.

7.2.9.6 Monte Carlo Analysis
This method essentially combines the sensitivity analysis approach with a system of randomly selecting the values to be used. The method is most effective in analyzing the interrelation of a large number of variable factors. The outcomes can be statistically analyzed for assignment of probabilities.

7.2.9.7 Expected Value Theory
Expected value is similar to sensitivity analysis with the major difference that the values used are considered to have

Table 7.2.4 *Probability Assignment*

Test #1	∅.%	Probability
1	20	40%
2	18	20%
3	22	20%
4	17	15%
5	24	10%
6	16	5%

a probability of occurrence and to be mutually exclusive. As an example, if ϕ were the element being evaluated, the range of values might be:

Assuming that all the other components of the OOIP calculation are fixed, an expected value of OOIP would be determined by calculating OOIP using each value of ϕ and then multiplying that OOIP by the probability. The sum of the adjusted OOIP is the expected value of OOIP.

In another example, probabilities could be assigned to a range of oil price escalation rates than an expected value of a property could be calculated by calculating a value using each escalation rate, multiplying the result by the probability and taking the sum as the expected value of the property.

These methods are relatively easy to apply once the risks (also read variables) have been analyzed and defined. They are particularly adaptable to computer analysis.

References

1. *Determination of Oil and Gas Reserves, Petroleum Society Monograph no. 1*, pp. 154–185, 200–203, The Petroleum Society of Canada, Calgary, Alberta, 1994.
2. Bradley, H. B., Ed., *Petroleum Engineering Handbook*, p. 41-1, Society of Petroleum Engineers, Richardson, Texas, 1992.
3. Wiggins, III, G. B., "Oil and Gas Property Evaluations," *The Business of Petroleum Exploration*, R. Steinmetz, Ed., pp. 125–162, American Association of Petroleum Geologists, Tulsa, Oklahoma, 1992.
4. *Determination of Oil and Gas Reserves, Petroleum Society Monograph no. 1*, pp. 253–305, The Petroleum Society of Canada, Calgary, Alberta, 1994.
5. "Annual Energy Outlook 2002 with Projection to 2020," Energy Information Agency, U.S. Department of Energy, Washington, D.C., December 2001.
6. *Twentieth Century Petroleum Statistics — 1999*, DeGloyer and MacNaughton, Dallas, Texas, December 1999.
7. Miller, R. J., "Economic Cycles and the Valuation of Oil and Gas Properties," *SPE 82007*, Society of Petroleum Engineers, Dallas, Texas, 2003.
8. Mercier, D., "Maximizing Field Value Using a Royalty Rate that Tacks Oil Prices," *SPE 37951*, Dallas, Texas, 1997.
9. *Determination of Oil and Gas Reserves, Petroleum Society Monograph no. 1*, pp. 254–257, The Petroleum Society of Canada, Calgary, Alberta, 1994.
10. *SPEE Oil & Gas Production Taxes, Section 29, and COPAS*, Society of Petroleum Evaluation Engineers, Houston, Texas, Updated annually.
11. *SPEE Ad Valorem Taxes — All States*, Society of Petroleum Evaluation Engineers, Houston, Texas.
12. Olds, D. R., "An Overview of Ad Valorem Taxes," *SPE 26390*, Society of Petroleum Engineers, Dallas, Texas, 1993.
13. Brealey, R. A. and Myers, S. C., *Principles of Corporate Finance*, pp. 11–49, 5th Edition, McGraw-Hill, New York, 1996.
14. Weston, J. F., and Copeland, T. E., *Managerial Finance*, Chapter 21, p. 579, The Dryden Press, 1986.
15. Bierman, Jr., H., and Smidt, S., *Financial Management for Decision Making*, Chapter 14, p. 420, MacMillan, New York, 1986.
16. Miller, R. J., and Vasquez, R., "Analysis of Oil and Gas Property Sales Transactions and Sales in California," *Journal of Petroleum Technology*, March 1988.
17. Miller, R. J., and Vasquez, R., "Discount Rates, Cost of Capital and Property Acquisition Evaluations," 63rd Annual Technical Conference, Houston, Texas, October 2–5, 1988.
18. Miller, R. J., "Fair Market Value Discount Rates: Analysis of Market Sales Data," SPE Hydrocarbon Economics and Evaluation Symposium, Dallas, Texas, March 26–28, 1995.
19. Miller, R. J., "The Cost-of-Capital and Fair Market Value Discount Rates," *SPE 52973*, Society of Petroleum Engineers, Dallas, Texas, 1999.
20. Miller, R. J., "Discount Rates from Market Sales vs Cost-of-Capital: From Whence Cometh the Difference," *SPE 71426*, Society of Petroleum Engineers, Dallas, Texas, 2001.
21. "Fair Market Value Transactions, Cost-of-Capital and Risk: California Oil and Gas Property Transactions 1983 through 2002," Richard J. Miller & Associates, Inc. for Western States Petroleum Association, Glendale, California, January 2003.
22. Manual for Discounting Oil and Gas Income," *Texas Property Tax*, Comptroller of Public Accounts, Austin, Texas, 1999.
23. *Survey of Economic Parameters Used in Property Evaluation*, Society of Petroleum Evaluation Engineers, Houston, Texas, May 2003.
24. *Stocks, Bonds, Bills and Inflation: Valuation Edition–2002 Yearbook*, Ibbotson Associates, Chicago, Illinois, 2002.
25. Weston, J. F., and Copeland, T. E., *Managerial Finance*, Chapter 6, p. 99, The Dryden Press, 1986.
26. Boyle, Jr., H. F., and Schenck, G. K., "Investment Analysis: U.S. Oil and Gas Producers Score High in University Survey," *Journal of Petroleum Technology*, April 1985.
27. Dougherty, E. K. and Sarkar, J., "Current Investment Practices and Procedures: Results of a Survey of U.S. Oil and Gas Producers and Petroleum Consultants," SPE Hydrocarbon Economics and Evaluation Symposium, Dallas, Texas, March 29–30, 1993.
28. Newendorp, P. D. and Schuyler, J. R., *Decision Analysis for Petroleum Exploration*, 2nd Edition, pg. 1, Planning Press, Aurora, Colorado, 2000.
29. Newendorp, P. D. *Decision Analysis for Petroleum Exploration*, Chapter 3, p. 58, Petroleum Publishing Co., 1976.
30. McCray, A. W., *Petroleum Evaluations and Economic Decisions*, Prentice-Hall Inc., Englewood Cliffs, New Jersey, 1975.
31. *Determination of Oil and Gas Reserves, Petroleum Society Monograph no. 1*, pp. 266–278, The Petroleum Society of Canada, Calgary, Alberta, 1994.
32. Newendorp, P. D. and Schuyler, J. R., *Decision Analysis for Petroleum Exploration*, 2nd Edition, pp. 327–387, Planning Press, Aurora, Colorado, 2000.
33. Rose, P. R., "Risk Analysis and Management of Petroleum Exploration Ventures," AAPG, Tulsa, Oklahoma, 2001.

Appendix: Units, Dimensions and Conversion Factors

The system of dimensions and units used in mechanics and related engineering disciplines are based on Newton's second law of motion, which is force equals mass multiplied by acceleration, or

$$F = m a \qquad \text{[A-1]}$$

The above equation can be used with any consistent systems of units.

Over the past four decades there has been a great deal of effort to try to make the English unit system have a methodology of practical use that works like the methodology of practical use of the International System of Units (or SI metric). This involved the creation of "new" artificial terms like lb_f and lb_m (and even N_f and N_m) together with g and g_c to adjust equations to one or the other system of units. The objective of this exercise was make the two unit systems "easier" to use. This did not happen. In fact, the exercise allowed scientists and engineers to avoid learning the two systems of units properly. In essence, the exercise that created these artificial units and their associated terms was unnecessary. To properly understand how to use the two systems of units it is only necessary to practice the two systems of units side by side with a simple dynamics problem. The two systems of units are used in exactly the same manner.

In the English unit system, scientists and engineers define a pound of force as the force required to accelerate one slug of mass at the rate of one foot per second per second. One slug of mass has a weight of approximately 32.2 lb when acted upon by the acceleration of gravity present at the surface of the earth. Thus, using Equation A-1 in the English units is

$$1 \text{ lb} = (1 \text{ slug})\left(1 \text{ ft./s}^2\right)$$

Using the above it can be found that

$$1 \text{ slug} = 1 \frac{\text{lb} - \text{s}^2}{\text{ft.}}$$

In SI metric, engineers define a Newton of force as the force required to accelerate on kilogram of mass at the rate of one meter per second per second. Thus, Equation A-1 in the SI metric is

$$1 \text{ N} = (1 \text{ kg})\left(1 \text{ m/s}^2\right)$$

Using the above, it can be found that

$$1 \text{ kg} = 1 \frac{\text{N} - \text{sec}^2}{\text{m}}$$

The mass units in both systems are made up of force, length, and time units. The mass unit of either system cannot be measured directly. On the surface of the earth it is necessary to measure the mutual attraction force between the mass and the mass of the earth (this force is called weight). Then knowing the acceleration of gravity on earth, that force can be converted to the mass unit. The mass units in both systems are not independent units. They are made up of pounds, feet, and seconds in the English unit system and N, m, and sec in the SI metric system.

The English system of units is called a *gravitational system* since it denotes any mass by its weight (i.e., lb). Weight is a specially named force that is the mutual attraction force between an object mass and the mass of the earth. The SI metric system is called an absolute system since it denotes any mass by its mass unit (i.e., kg). Except for this distinction between the weight of an object and the mass of an object, the two systems are used in exactly the same manner in making practical calculations. Both require the measurement of the weight before the mass unit can be determined. We list mass in the SI metric system to define an object and we list weight in the English unit system to define an object. In the

English unit system, we simply need to divide the weight by the acceleration of gravity to determine the object mass. To determine the weight of an object in the SI metric system we simply need to multiply the mass by the acceleration of gravity.

The two systems can be used in all science and engineering calculations. However, often one system is preferred over another in most problem types (e.g., the SI metric system is usually preferred for dynamics problems and the English system of units is usually preferred for strength of materials problems).

This handbook avoids the use of these artificial terms and uses these two systems of units in a consistent manner as described below.

Table A-1 *Basic Definitions of the English Unit System*

Force	1 lb = 1 slug-ft./s^2
Area	1 acre = 43,560 ft.^{2}
Energy	1 Btu = 778 ft.-lb
Flowrate	1 cfs = 448.83 gpm
Length	1 ft. = 12 in. 1 yd = 3 ft. 1 statute mile = 5280 ft. 1 nautical mile = 6000 ft.
Mass	1 slug = 1 lb-s^2/ft.
Power	1 hp = 550 ft.-lb/s = 0.708 Btu/sec
Velocity	1 mph = 1.467 ft./s 1 knot = 1.689 ft./s = 1.152 mph
Volume	1 ft.^{3} = 7.48 U.S. gal 1 U.S. gal = 231 in.^{3} = 0.1337 ft.^{3} (8.34 lb of water) 1 British imperial gal = 1.2 U. S. gal (10 lb of water)
Weight	1 U.S. (short) ton = 2000 lb 1 British (long) ton = 2240 lb

Table A-2 *Basic Definitions of the SI Metric System*

Force	1 N = 1 kg-m/s^2
Area	1 Hectare (ha) = 10^4 m^2 = 100 m square
Energy (or work)	1 Joule (J) = 1 N-m
Flowrate	m^3/s = 60,000 l pm
Length	1 cm = 10 mm 1 m = 100 cm 1 km = 1000 m
Mass	1 kg = 1000 g 1 metric ton = 1000 kg
Power	1 Watt = 1 N-m/s
Velocity	1 kph = 1000 m/hr = 0.277 m/s
Volume	1 m^3 = 1000 L 1 L = 10^3 cm

Table A-3   *Basic quantities in English and SI Metric*

	English Unit	SI Unit
Acceleration of gravity	32.2 ft./s^2	9.81 m/s^2
Density of water (at 39.4°F or 4°C)	1.94 slug/ft.^{3}	1000 kg/m^3
Specific weight of fresh water	62.4 lb/ft.^{3}	9810 N/m^3
Standard sea level atmospheric pressure		
API (at 60°F)	14.696 psia	
ASME (at 68°F)	14.7 pisa	
UK (at 60°F)	30.00 in. of Hg	
Continental Europe (at 15°C)	750 mm of Hg	

Table A-4   *Other important English and SI Metric Conversions*

Engineering gas constant **R**
  English $\mathbf{R} = 53.36$ ft.-lb/lb-°R
  Metric $\mathbf{R} = 29.28$ N-m/N-K

Energy
  English   1 ft.-lb = 1 lb-ft.
  Metric   1 N-m = 1 joule (J)

Heat
  English   1 Btu = 252 cal (heat required to raise 1.0 lb of water 1.0°R)
  Metric   1 cal = 4.187 J (heat required to raise 1.0 g of water 1.0 K)

Temperature
  English   °R = 459.67° + °F
  Metric   K = 273.15 + °C

Pressure
  English   1 lb/ft.^{2} = 144 psi
  Metric   1 Pascal (Pa) = N/m^2

Power
  English   1 horsepower = 550 ft.-lb/s
  Metric   1 watt (W) = 1 J/s (or 1 N-m/sec)

Absolute viscosity
  English   1 lb-s/ft.^{2} = 47.88 N-s/m^2
  Metric   1 Poise (P) = 10^{-1} N-s/N^2

Kinematic viscosity
  English   1 ft.^{2}/s = 0.0929 m^2/s
  Metric   1 Stroke (St) = 10^{-4} m^2/s

Table A-5   *Commonly Used Prefixes for SI Metric Units*

Factor by which unit is multiplied	Prefix	Symbol
10^{12}	tera	T
10^{9}	giga	G
10^{6}	mega	M
10^{3}	kilo	k
10^{2}	hecto	h
10	deka	da
10^{-1}	deci	d
10^{-2}	centi	c
10^{-3}	milli	m
10^{-6}	micro	μ
10^{-9}	nano	n

Table A-6 *Alphabetical List of Units (symbols of SI units given in parentheses)*

To Convert From	To	Multiply By	
abampere	ampere (A)	1.0	E + 01
abcoulomb	coulomb (C)	1.0	E + 01
abfarad	farad (F)	1.0	E + 09
abhenry	henry (H)	1.0	E − 09
abmho	siemens (S)	1.0	E + 09
abohm	ohm (Ω)	1.0	E − 09
abvolt	volt (V)	1.0	E − 08
acre-foot (U.S. survey)[1]	meter^3 (m^3)	1.233 489	E + 03
acre (U.S. survey)[1]	meter^2 (m^2)	4.046 873	E + 03
ampere hour	coulomb (C)	3.6	E + 03
are	meter^2 (m^2)	1.0	E + 02
angstrom	meter (m)	1.0	E − 10
astronomical unit	meter (m)	1.495 979	E + 11
atmosphere (standard)	pascal (Pa)	1.013 250	E + 05
atmosphere (technical = 1 kgf/cm^2)	pascal (Pa)	9.806 650	E + 04
bar	pascal (Pa)	1.0	E + 05
barn	meter^2 (m^2)	1.0	E − 28
barrel (for petroleum, 42 gal)	meter^3 (m^3)	1.589 873	E − 01
board foot	meter^3 (m^3)	2.359 737	E − 03
British thermal unit (International Table)[2]	joule (J)	1.055 056	E + 03
British thermal unit (mean)	joule (J)	1.055 87	E + 03
British thermal unit (thermochemical)	joule (J)	1.054 350	E + 03
British thermal unit (39°F)	joule (J)	1.059 67	E + 03
British thermal unit (59°F)	joule (J)	1.054 80	E + 03
British thermal unit (60°F)	joule (J)	1.054 68	E + 03
Btu (International Table)-ft./(hr-ft.^{2}-°F) (thermal conductivity)	watt per meter kelvin [W/(m·k)]	1.730 735	E + 00
Btu (thermochemical)-ft./(hr-ft.^{2}-°F) (thermal conductivity)	watt per meter kelvin [W/(m·k)]	1.729 577	E + 00
Btu (International Table)-in./(hr-ft.^{2}-°F) (thermal conductivity)	watt per meter kelvin [W/(m·k)]	1.442 279	E − 01
Btu (thermochemical)-in./(hr-ft.^{2}-°F) (thermal conductivity)	watt per meter kelvin [W/(m·k)]	1.441 314	E − 01
Btu (International Table)-in./(s-ft.^{2}-°F) (thermal conductivity)	watt per meter kelvin [W/(m.k)]	5.192 204	E + 02
Btu (thermochemical)-in./(s-ft.^{2}-°F) (thermal conductivity)	watt per meter kelvin [W/(m.k)]	55.188 732	E + 02
Btu (International Table)/hr	watt (W)	2.930 711	E − 01
Btu (thermochemical)/hr	watt (W)	2.928 751	E − 01
Btu (thermochemical)/min	watt (W)	1.757 250	E + 01
Btu (thermochemical)/s	watt (W)	1.054 350	E + 03
Btu (International Table)/ft.^{2}	joule per meter^2 (J/m^2)	1.135 653	E + 04
Btu (thermochemical)/ft.^{2}	joule per meter^2 (J/m^2)	1.134 893	E + 04
Btu (thermochemical)/(ft.^{2}-hr)	watt per meter^2 (W/m^2)	3.152 481	E + 00
Btu (thermochemical)/(ft.^{2}-min)	watt per meter^2 (W/m^2)	1.891 489	E + 02
Btu (thermochemical)/(ft.^{2}-s)	watt per meter^2 (W/m^2)	1.134 893	E + 04
Btu (thermochemical)/(in.^{2}-s)	watt per meter^2 (W/m^2)	1.634 246	E + 06
Btu (International Table)/(hr-ft.^{2}-°F) (thermal conductance)	watt per meter^2 kelvin [W/(m^2·k)]	5.678 263	E + 00
Btu (thermochemical)/(hr-ft.^{2}-°F) (thermal conductance)	watt per meter^2 kelvin [W/(m^2·k)]	5.674 466	E + 00
Btu (International Table)/(s-ft.^{2}-°F)	watt per meter^2 kelvin [W/(m^2·k)]	2.044 175	E + 04
Btu (thermochemical)/(s-ft.^{2}-°F)	watt per meter^2 kelvin [W/(m^2·k)]	2.042 808	E + 04
Btu (International Table)/lbm	joule per kilogram (J/kg)	2.326	E + 03
Btu (thermochemical)/lbm	joule per kilogram (J/kg)	2.324 444	E + 03
Btu (International Table)/(lbm-°F) (heat capacity)	joule per kilogram kelvin [J/(kg·K)]	4.186 8	E + 03
Btu (thermochemical)/(lbm-°F) (heat capacity)	joule per kilogram kelvin [J/(kg·K)]	4.184 000	E + 03

(continued)

Table A-6 Alphabetical List of Units (symbols of SI units given in parentheses) (continued)

To Convert From	To	Multiply By	
bushel (U.S.)	meter^3 (m^3)	3.523 907	E − 02
caliber (inch)	meter (m)	2.54	E − 02
calorie (International Table)	joule (J)	4.1868	E + 00
calorie (mean)	joule (J)	4.190 02	E + 00
calorie (thermochemical)	joule (J)	4.184	E + 00
calorie (15°C)	joule (J)	4.185 80	E + 00
calorie (20°C)	joule (J)	4.181 90	E + 00
calorie (kilogram, International Table)	joule (J)	4.1868	E + 03
calorie (kilogram, mean)	joule (J)	4.190 02	E + 03
calorie (kilogram, thermochemical)	joule (J)	4.184	E + 03
cal (thermochemical)/cm^2	joule per meter^2 (J/m^2)	4.184	E + 04
cal (International Table)/g	joule per kilogram (J/kg)	4.186	E + 03
cal (thermochemical)/g	joule per kilogram (J/kg)	4.184	E + 03
cal (International Table)/(g·°C)	joule per kilogram kelvin [J/(kg · K)]	4.1868	E + 03
cal (thermochemical)/(g·°C)	joule per kilogram kelvin [J/(kg · K)]	4.184	E + 03
cal (thermochemical)/min	watt (W)	6.973 333	E − 02
cal (thermochemical)/s	watt (W)	4.184	E + 00
cal (thermochemical)/(cm^2·min)	watt per meter^2 (W/m^2)	6.973 333	E + 02
cal (thermochemical)/(cm^2·s)	watt per meter^2 (W/m^2)	4.184	E + 04
cal (thermochemical)/(cm · s· °C)	watt per meter kelvin [W/(m·K)]	4.184	E + 02
capture unit (c.u. = 10^{-3} cm^{-1})	per meter (m^{-1})	1.0	E − 01
carat (metric)	(kg)	2.0	E − 04
centimeter of mercury (0°C)	pascal (Pa)	1.333 22	E + 03
centimeter of water (4°C)	pascal (Pa)	9.806 38	E + 01
centipoise	pascal second (Pa·s)	1.0	E − 03
centistokes	meter^2 per second (m^2/s)	1.0	E − 06
circular mil	meter^2 (m^2)	5.067 075	E − 10
clo	kelvin meter^2 per watt [(K·m^2)/W]	2.003 712	E − 01
cup	meter^3 (m^3)	2.365 882	E − 04
curie	becquerel (Bq)	3.7	E + 10
cycle per second	hertz (Hz)	1.0	E + 00
day (mean solar)	second (s)	8.640 000	E + 04
day (sidereal)	second (s)	8.616 409	E + 04
degree (angle)	radian (rad)	1.745 329	E − 02
degree Celsius	kelvin (K)	$T_K = T_{°C} + 273.15$	
degree centigrade (see degree Celsius)			
degree Fahrenheit	degree Celsius	$T_{°C} = (T_{°F} − 32)/1.8$	
degree Fahrenheit	kelvin (K)	$T_K = (T_{°F} + 459.67)/1.8$	
degree Rankine	kelvin (K)	$T_K = T_{°R}/1.8$	
°F·hr·ft.^{2}/Btu (International Table) (thermal resistance)	kelvin meter^2 per watt [(K·m^2)/W]	1.781 102	E − 01
°F·hr·ft.^{2}/Btu (thermochemical) (thermal resistance)	kelvin meter^2 per watt [(K·m^2)/W]	1.762 250	E − 01
denier	kilogram per meter (kg/m)	1.111 111	E − 07
dyne	newton (N)	1.0	E − 05
dyne-cm	newton meter (N·m)	1.0	E − 07
dyne/cm^2	pascal (Pa)	1.0	E − 01
electronvolt	joule (J)	1.602 19	E − 19
EMU of capacitance	farad (F)	1.0	E + 09
EMU of current	ampere (A)	1.0	E + 01
EMU of electric potential	volt (V)	1.0	E − 08
EMU of inductance	henry (H)	1.0	E − 09
EMU of resistance	ohm (Ω)	1.0	E − 09
ESU of capacitance	farad (F)	1.112 650	E − 12
ESU of current	ampere (A)	3.335 6	E − 10
ESU of electric potential	volt (V)	2.997 9	E + 02
ESU of inductance	henry (H)	8.987 554	E + 11
ESU of resistance	ohm (Ω)	8.987 554	E + 11

(continued)

Table A-6 *Alphabetical List of Units (symbols of SI units given in parentheses) (continued)*

To Convert From	To	Multiply By	
erg	joule (J)	1.0	E − 07
erg/cm^2·s	watt per meter^2 (W/m^2)	1.0	E − 03
erg/s	watt (W)	1.0	E − 07
faraday (based on carbon-12)	coulomb (C)	9.648 70	E + 04
faraday (chemical)	coulomb (C)	9.649 57	E + 04
faraday (physical)	coulomb (C)	9.652 19	E + 04
fathom	meter (m)	1.828 8	E + 00
fermi (femtometer)	meter (m)	1.0	E − 15
fluid ounce (U.S.)	meter^3 (m^3)	2.957 353	E − 05
foot	meter (m)	3.048	E − 01
foot (U.S. survey)[1]	meter (m)	3.048 006	E − 01
foot of water (39.2°F)	pascal (Pa)	2.988 98	E + 03
sq ft.	meter^2 (m^2)	9.290 304	E − 02
ft.^{2}/hr (thermal diffusivity)	meter^2 per second (m^2/s)	2.580 640	E − 05
ft.^{2}/s	meter^2 per second (m^2/s)	9.290 304	E − 02
cu ft. (volume; section modulus)	meter^3 (m^3)	2.831 685	E − 02
ft.^{3}/min	meter^3 per second (m^3/s)	4.719 474	E − 04
ft.^{3}/s	meter^3 per second (m^3/s)	2.831 685	E − 02
ft.^{4} (moment of section)[3]	meter^4 (m^4)	8.630 975	E − 03
ft./hr	meter per second (m/s)	8.466 667	E − 05
ft./min	meter per second (m/s)	5.080	E − 03
ft./s	meter per second (m/s)	3.048	E − 01
ft./s^2	meter per second^2 (m/s^2)	3.048	E − 01
footcandle	lux (lx)	1.076 391	E + 01
footlambert	candela per meter^2 (cd/m^2)	3.426 259	E + 00
ft.-lb	joule (J)	1.355 818	E + 00
ft.-lb/hr	watt (W)	3.766 161	E − 04
ft.-lb/min	watt (W)	2.259 697	E − 02
ft.-lb/s	watt (W)	1.355 818	E + 00
free fall, standard (g)	meter per second^2 (m/s^2)	9.806 650	E + 00
cm/s^2	meter per second^2 (m/s^2)	1.0	E − 02
gallon (Canadian liquid)	meter^3 (m^3)	4.546 090	E − 03
gallon (U.K. liquid)	meter^3 (m^3)	4.546 092	E − 03
gallon (U.S. dry)	meter^3 (m^3)	4.404 884	E − 03
gallon (U.S. liquid)	meter^3 (m^3)	3.785 412	E − 03
gal (U.S. liquid)/day	meter^3 per second (m^3/s)	4.381 264	E − 08
gal (U.S. liquid)/min	meter^3 per second (m^3/s)	6.309 020	E − 05
gal (U.S. liquid)/hp-hr (SFC, specific fuel consumption)	meter^3 per joule (m^3/J)	1.410 089	E − 09
gamma (magnetic field strength)	ampere per meter (A/m)	7.957 747	E − 04
gamma (magnetic flux density)	tesla (T)	1.0	E − 09
gauss	tesla (T)	1.0	E − 04
gilbert	ampere (A)	7.957 747	E − 01
gill (U.K.)	meter^3 (m^3)	1.420 654	E − 04
gill (U.S.)	meter^3 (m^3)	1.182 941	E − 04
grad	degree (angular)	9.0	E − 01
grad	radian (rad)	1.570 796	E − 02
grain (1/7000 lbm avoirdupois)	kilogram (kg)	6.479 891	E − 05
grain (lbm avoirduposis/7000)/gal (U.S. liquid)	kilogram per meter^3 (kg/m^3)	1.711 806	E − 02
gram	kilogram (kg)	1.0	E − 03
g/cm^3	kilogram per meter^3 (kg/m^3)	1.0	E + 03
gram-force/cm^2	pascal (Pa)	9.806 650	E + 01
hectare	meter^2 (m^2)	1.0	E + 04
horsepower (550 ft.-lbf/s)	watt (W)	7.456 999	E + 02
horsepower (boiler)	watt (W)	9.809 50	E + 03
horsepower (electric)	watt (W)	7.460	E + 02
horsepower (metric)	watt (W)	7.354 99	E + 02

(continued)

Table A-6 *Alphabetical List of Units (symbols of SI units given in parentheses) (continued)*

To Convert From	To	Multiply By	
horsepower (water)	watt (W)	7.460 43	E + 02
horsepower (U.K.)	watt (W)	7.457 0	E + 02
hour (mean solar)	second (s)	3.600 000	E + 03
hour (sidereal)	second (s)	3.590 170	E + 03
hundredweight (long)	kilogram (kg)	5.080 235	E + 01
hundredweight (short)	kilogram (kg)	4.535 924	E + 01
inch	meter (m)	2.54	E − 02
inch of mercury (32°F)	pascal (Pa)	3.386 38	E + 03
inch of mercury (60°F)	pascal (Pa)	3.376 85	E + 03
inch of water (39.2°F)	pascal (Pa)	2.490 82	E + 02
inch of water (60°F)	pascal (Pa)	2.488 4	E + 02
sq in.	meter^2 (m^2)	6.451 6	E − 04
cu in. (volume; section modulus)[4]	meter^3 (m^3)	1.638 706	E − 05
in.^{3}/min	meter^3 per second (m^3/s)	2.731 177	E − 07
in.^{4} (moment of section)[3]	meter^4 (m^4)	4.162 314	E − 07
in./s	meter per second (m/s)	2.54	E − 02
in./s^2	meter per second^2 (m/s^2)	2.54	E − 02
kayser	1 per meter (1/m)	1.0	E + 02
kelvin	degree Celsius	$T_{°C} = T_K − 273.15$	
kilocalorie (International Table)	joule (J)	4.186 8	E + 03
kilocalorie (mean)	joule (J)	4.190 02	E + 03
kilocalorie (thermochemical)	joule (J)	4.184	E + 03
kilocalorie (thermochemical)/min	watt (W)	6.973 333	E + 01
kilocalorie (thermochemical)/s	watt (W)	4.184	E + 03
km/h	meter per second (m/s)	2.777 778	E − 01
kilopond	newton (N)	9.806 65	E + 00
kilowatthour (kW-hr)	joule (J)	3.6	E + 06
kip (1000 lbf)	newton (N)	4.448 222	E + 03
kip/in.^{2} (ksi)	pascal (Pa)	6.894 757	E + 06
knot (international)	meter per second (m/s)	5.144 444	E − 01
lambert	candela per meter^2 (cd/m^2)	1/π	E + 04
lambert	candela per meter^2 (cd/m^2)	3.183 099	E + 03
langley	joule per meter^2 (J/m^2)	4.184	E + 04
league	meter (m)	(see Footnote 1)	
light year	meter (m)	9.460 55	E + 15
liter	meter^3 (m^3)	1.0	E − 03
maxwell	weber (Wb)	1.0	E − 08
mho	siemens (S)	1.0	E + 00
microinch	meter (m)	2.54	E − 08
microsecond/foot (μs/ft.)	microsecond/meter (μs/m)	3.280 840	E + 00
micron	meter (m)	1.0	E − 06
mil	meter (m)	2.54	E − 05
mile (international)	meter (m)	1.609 344	E + 03
mile (statute)	meter (m)	1.609 3	E + 03
mile (U.S. survey)[1]	meter (m)	1.609 347	E + 03
mile (international nautical)	meter (m)	1.852	E + 03
mile (U.K. nautical)	meter (m)	1.853 184	E + 03
mile (U.S. nautical)	meter (m)	1.852	E + 03
sq mile (international)	meter^2 (m^2)	2.589 988	E + 06
sq mile (U.S. survey)[1]	meter^2 (m^2)	2.589 998	E + 06
mile/hr (international)	meter per second (m/s)	4.470 4	E − 01
mile/hr (international)	kilometer per hour (km/h)	1.609 344	E + 00
mile/min (international)	meter per second (m/s)	2.682 24	E + 01
mile/s (international)	meter per second (m/s)	1.609 344	E + 03
millibar	pascal (Pa)	1.0	E + 02
millimeter of mercury (0°C)	pascal (Pa)	1.333 22	E + 02
minute (angle)	radian (rad)	2.908 882	E − 04
minute (mean solar)	second (s)	6.0	E + 01

(continued)

Table A-6 *Alphabetical List of Units (symbols of SI units given in parentheses) (continued)*

To Convert From	To	Multiply By	
minute (sidereal)	second (s)	5.983 617	E + 01
month (mean calendar)	second (s)	2.628 000	E + 06
oersted	ampere per meter (A/m)	7.957 747	E + 01
ohm centimeter	ohm meter (Ω·m)	1.0	E − 02
ohm circular-mil per ft.	ohm millimeter^2 per meter [(Ω·mm^2)m]	1.662 426	E − 03
ounce (avoirdupois)	kilogram (kg)	2.834 952	E − 02
ounce (troy or apothecary)	kilogram (kg)	3.110 348	E − 02
ounce (U.K. fluid)	meter^3 (m^3)	2.841 307	E − 05
ounce (U.S. fluid)	meter^3 (m^3)	2.957 353	E − 05
ounce-force	newton (N)	2.780 139	E − 01
ozf-in.	newton meter (N·m)	7.061 552	E − 03
oz (avoirdupois)/gal (U.K. liquid)	kiligram per meter^3 (kg/m^3)	6.236 021	E + 00
oz (avoirdupois)/gal (U.S. liquid)	kiligram per meter^3 (kg/m^3)	7.489 152	E + 00
oz (avoirdupois)/in.^{3}	kiligram per meter^3 (kg/m^3)	1.729 994	E + 03
oz (avoirdupois)/ft.^{2}	kiligram per meter^2 (kg/m^2)	3.051 517	E − 01
oz (avoirdupois)/yd^2	kiligram per meter^2 (kg/m^2)	3.390 575	E − 02
parsec	meter (m)	3.085 678	E + 16
peck (U.S.)	meter^3 (m^3)	8.809 768	E − 03
pennyweight	kilogram (kg)	1.555 174	E − 03
perm (°C)[5]	kilogram per pascal second meter^2 [kg/(Pa·s·m^2)]	5.721 35	E − 11
perm (23°C)[5]	kilogram per pascal second meter^2 [kg/(Pa·s·m^2)]	5.745 25	E − 11
perm-in. (0°C)[6]	kilogram per pascal second meter [kg/(Pa·s·m)]	1.453 22	E − 12
perm-in. (23°C)[6]	kilogram per pascal second meter [kg/(Pa·s·m)]	1.459 29	E − 12
phot	lumen per meter^2 (lm/m^2)	1.0	E + 04
pica (printer's)	meter (m)	4.217 518	E − 03
pint (U.S. dry)	meter^3 (m^3)	5.506 105	E − 04
pint (U.S. liquid)	meter^3 (m^3)	4.731 765	E − 04
point (printer's)	meter (m)	3.514 598	E − 04
poise (absolute viscosity)	pascal second (Pa·s)	1.0	E − 01
pound (lb)	newton (N)	4.448 222	E + 00
lb/in.^{2}	pascal (Pa)	6.895	E + 00
lb-ft.	newton meter (N·m)	1.355 818	E + 00
lb-ft./in.	newton meter per meter [(N·m)/m)]	5.337 866	E + 01
lb-in.	newton meter (N·m)	1.129 848	E − 01
lb-in./in.	newton meter per meter [(N·m)/m]	4.448 222	E + 00
lb-s/ft.^{2}	pascal second (Pa·s)	4.788 026	E + 01
lb-ft.	newton per meter (N/m)	1.459 390	E + 01
lb-ft.^{2}	pascal (Pa)	4.788 026	E + 01
lb-in.	newton per meter (N/m)	1.751 268	E + 02
lb-in.^{2} (psi)	pascal (Pa)	6.894 757	E + 03
quart (U.S. dry)	meter^3 (m^3)	1.101 221	E − 03
quart (U.S. liquid)	meter^3 (m^3)	9.463 529	E − 04
rad (radiation dose absorbed)	gray (Gy)	1.0	E − 02
rhe	1 per pascal second [1/(Pa·s)]	1.0	E + 01
rod	meter (m)	(see Footnote 1)	
roentgen	coulomb per kilogram (C/kg)	2.58	E − 04
second (angle)	radian (rad)	4.848 137	E − 06
second (sidereal)	second (s)	9.972 696	E − 01
section	meter^2 (m^2)	(see Footnote 1)	
shake	second (s)	1.000 000	E − 08
slug	kilogram (kg)	1.459 390	E + 01
slug/(ft.-s)	pascal second (Pa· s)	4.788 026	E + 01
slug/ft.^{3}	kilogram per meter^3 (kg/m^3)	5.153 788	E + 02
statampere	ampere (A)	3.335 640	E − 10
statcoulomb	coulomb (C)	3.335 640	E − 10

(continued)

Table A-6 *Alphabetical List of Units (symbols of SI units given in parentheses) (continued)*

To Convert From	To	Multiply By	
statfarad	farad (F)	1.112 650	E − 12
stathenry	henry (H)	8.987 554	E + 11
statmho	siemens (S)	1.112 650	E − 12
statohm	ohm (Ω)	8.987 554	E + 11
statvolt	volt (V)	2.997 925	E + 02
stere	meter^3 (m^3)	1.0	E + 00
stilb	candela per meter^2 (cd/m^2)	1.0	E + 04
stokes (kinematic viscosity)	meter^2 per second (m^2/s)	1.0	E − 04
tablespoon	meter^3 (m^3)	1.478 676	E − 05
teaspoon	meter^3 (m^3)	4.928 922	E − 06
tex	kilogram per meter (kg/m)	1.0	E − 06
therm	joule (J)	1.055 056	E + 08
ton (assay)	kilogram (kg)	2.916 667	E − 02
ton (long, 2240 lb)	kilogram (kg)	1.016 047	E + 03
ton (metric)	kilogram (kg)	1.0	E + 03
ton (nuclear equivalent of TNT)	joule (J)	4.184	E + 09[12]
ton (refrigeration)	watt (W)	3.516 800	E + 03
ton (register)	meter^3 (m^3)	2.831 685	E + 00
ton (short, 2000 lb)	kilogram (kg)	9.071 847	E + 02
ton (long)/yd^3	kilogram per meter^3 (kg/m^3)	1.328 939	E + 03
ton (short)/hr	kilogram per second (kg/s)	2.519 958	E − 01
ton-force (2000 lb)	newton (N)	8.896 444	E + 03
tonne	kilogram (kg)	1.0	E + 03
torr (mm Hg, 0°C)	pascal (Pa)	1.333 22	E + 02
township	meter^2 (m^2)	(see Footnote 1)	
unit pole	weber (Wb)	1.256 637	E − 07
watthour (W-hr)	joule (J)	3.60	E + 03
W·s	joule (J)	1.0	E + 00
W/cm^2	watt per meter^2 (W/m^2)	1.0	E + 04
W/in.^{2}	watt per meter^2 (W/m^2)	1.550 003	E + 03
yard	meter (m)	9.144	E − 01
yd^2	meter^2 (m^2)	8.361 274	E − 01
yd^3	meter^3 (m^3)	7.645 549	E − 01
yd^3/min	meter^3 per second (m^3/s)	1.274 258	E − 02
year (calender)	second (s)	3.153 600	E + 07
year (sidereal)	second (s)	3.155 815	E + 07
year (tropical)	second (s)	3.155 693	E + 07

(1) Since 1893 the U.S. basis of length measurement has been derived from metric standards. In 1959 a small refinement was made in the definition of the yard to resolve discrepancies both in this country and abroad, which changed its length from 3600/3937 m to 0.9144 m exactly. This resulted in the new value being shorter by two parts in a million. At the same time it was decided that any data in feet derived from and published as a result of geodetic surveys within the U.S. would remain with the old standard (1 ft. = 1200/3937 m) until further decision. This foot is named the U.S. survey foot. As a result, all U.S. land measurements in U.S. customary units will relate to the meter by the old standard. All the conversion factors in these tables for units referenced to this footnote are based on the U.S. survey foot, rather than the international foot. Conversion factors for the land measure given below may be determined from the following relationships:

$$1 \text{ league} = 3 \text{ miles (exactly)}$$
$$1 \text{ rod} = 16\,1/2 \text{ ft. (exactly)}$$
$$1 \text{ chain} = 66 \text{ ft. (exactly)}$$
$$1 \text{ section} = 1 \text{ sq mile}$$
$$1 \text{ township} = 36 \text{ sq miles}$$

(2) This value was adopted in 1956. Some of the older International Tables use the value 1.055 04 E + 03. The exact conversion factor is 1.055 055 852 62* E + 03.
(3) This sometimes is called the moment of inertia of a plane section about a specified axis.
(4) The exact conversion factor is 1.638 706 4*E − 05.
(5) Not the same as reservoir "perm."
(6) Not the same dimensions as "millidarcy-foot."

Table A-7 *Conversion Factors for the Vara**

Location	Value of Vara in Inches	Conversion Factor, Varas to Meters	
Argentina, Paraguay	34.12	8.666	E − 01
Cadiz, Chile, Peru	33.37	8.476	E − 01
California, except San Francisco	33.3720	8.476 49	E − 01
San Francisco	33.0	8.38	E − 01
Central America	33.87	8.603	E − 01
Colombia	31.5	8.00	E − 01
Honduras	33.0	8.38	E − 01
Mexico		8.380	E − 01
Portugal, Brazil	43.0	1.09	E + 00
Spain, Cuba, Venezuela, Philippine Islands	33.38**	8.479	E − 01
Texas			
Jan. 26, 1801, to Jan. 27, 1838	32.8748	8.350 20	E − 01
Jan. 27, 1838 to June 17, 1919, for surveys of state land made for Land Office	33-1/3	8.466 667	E − 01
Jan. 27, 1838 to June 17, 1919, on private surveys (unless changed to 33-1/3 in. by custom arising to dignity of law and overcoming former law)	32.8748	8.350 20	E − 01
June 17, 1919, to present	33-1/3	8.466 667	E − 01

*McElwee, P. G., *The Texas Vara*, available from commissioner, General Land Office, State of Texas, Austin, Texas (April 30, 1940). Courtesy of Society of Petroleum Engineers.

Table A-8 *"Memory Jogger" — Metric Units*

Customary Unit	"BallPark" Metric Values; (Do *Not* Use As Conversion Factors)	
acre	4000	square meters
	0.4	hectare
barrel	0.16	cubic meter
British thermal unit	1000	joules
British thermal unit per pound-mass	2300	joules per kilogram
	2.3	kilojoules per kilogram
calorie	4	joules
centipoise	1*	millipascal-second
centistokes	1*	square millimeter per second
darcy	1	square micrometer
degree Fahrenheit (temperature *difference*)	0.5	kelvin
dyne per centimeter	1*	millinewton per meter
foot	30	centimeters
	0.3	meter
cubic foot (cu ft)	0.03	cubic meter
cubic foot per pound-mass (ft.^{3}/lbm)	0.06	cubic meter per kilogram
square foot (sq ft)	0.1	square meter
foot per minute	0.3	meter per minute
	5	millimeters per second
foot-pound-force	1.4	joules
foot-pound-force per minute	0.02	watt
foot-pound-force per second	1.4	watts
horsepower	750	watts ($\frac{3}{4}$ kilowatt)
horsepower, boiler	10	kilowatts
inch	2.5	centimeters
kilowatthour	3.6*	megajoules
mile	1.6	kilometers
ounce (avoirdupois)	28	grams
ounce (fluid)	30	cubic centimeters
pound-force	4.5	newtons
pound-force per square inch (pressure, psi)	7	kilopascals
pound-mass	0.5	kilogram
pound-mass per cubic foot	16	kilograms per cubic meter
section	260	hectares
	2.6	million square meters
	2.6	square kilometers
ton, long (2240 pounds-mass)	1000	kilograms
ton, metric (tonne)	1000*	kilograms
ton, short	900	kilograms

*Exact equivalents
Courtesy of Society of Petroleum Engineers

Subject Index

Notes: Page numbers are styled as chapter number:page number. For example, 6:304 refers to chapter 6, page 304. Page numbers followed by "f" refer to figures; page numbers followed by "t" refer to tables.